ENCYCLOPEDIC DICTIONARY OF

GENETICS, GENOMICS, AND PROTEOMICS

Second Edition

ENCYCLOPEDIC DICTIONARY OF

GENETICS, GENOMICS, AND PROTEOMICS

Second Edition

GEORGE P. RÉDEI

WILEY-LISS

A JOHN WILEY & SONS, INC., PUBLICATION

9/07

Artwork by Author.

First published 1998

Cover photograph: Mitotic HeLa cells stained with DAPI, anti-tubulin (green) and human anti-centromere autoantibody (red), chromatin (blue). Courtesy of Kevin F. Sullivan, Department of Cell Biology, The Scripps Research Institute, San Diego, CA and Don W. Cleveland, Ludwig Institute for Cancer Research, University of California, San Diego, CA

Published by John Wiley & Sons, Inc., Hoboken, New Jersey.

Published simultaneously in Canada.

Library of Congress Cataloging-in-Publication Data:

Redei, G. P.
Encyclopedic dictionary of genetics, genomics, and proteomics / George P. Redei.—2nd ed.
p. ; cm.
Rev. ed. of: Genetics manual. c1998.
Includes bibliographical references and index.
ISBN 0-471-26821-6 (cloth : alk. paper)
1. Genetics—Encyclopedias. 2. Genomics—Encyclopedias. 3. Proteomics—Encyclopedias.
[DNLM: 1. Genetics—Encyclopedias. 2. Genomics—Encyclopedias. 3. Proteome—Encyclopedias. QH 427 R314e 2003] I. Redei, G. P. Genetics manual. II. Title.
QH427 .R43 2003
576.5′03—dc21

2002011131

Printed in the United States of America.

10 9 8 7 6 5 4 3 2

♥ TO PAIGE, GRACE, and ANNE ♥

PREFACE TO THE SECOND EDITION

The majority of the users of the first edition considered this book as an encyclopedia because the cross-references tied the short entries into comprehensive reviews of the topics. In contrast to big encyclopedias, in this work relatively few entries exceed a couple of thousand words, making it much faster to find the specific concept or term of interest. Unlike multiauthor works, this book is practically free of redundancy and it is compact in size but not in depth of information. One of the reviewers pointed out that many of the topics covered could not be found in any other single book, including encyclopedias, dictionaries, or glossaries. Another reviewer appreciated it as a broad resource of information that may take a lengthy search to uncover without it.

Since the publication of the first edition, I have steadily updated and improved on the topics. I have added many new concepts, illustrations, references, and database addresses. (The database addresses are in an unfortunate flux and some may be out existence by the time you wish to log in; therefore, I provided several to minimize the problem beyond my control.) This second edition contains about 50% more information and more than twice as many illustrations than the first edition. A new feature is the predominantly current, over 7,000 text references to journal articles. Their bibliographies may help to locate additional key and classical papers. The General References at the end includes about 2,000 books. For additional medical genetics references, I suggest the use of OMIM at the National Center for Biotechnology Information (<http://www.ncbi.nlm.nih.gov/>; see also Grivell, L. 2002, **EMBO Rep.** 3:200). I have greatly expanded the cross-references among the entries because the users found this feature especially useful. Color figures have been added in a separate section in the center of the book. At the end of the sections there are some historical vignettes.

Since the publication of the first edition, the need for such a book became even more evident. In the literature, unexplained concepts, terms, and acronyms are on the increase, and even a name, DAS (*dreaded abbreviation syndrome*), has been coined for the malaise (*Science*, [1999] 283:1118). The users of the first edition agreed with the Nobel-laureate geneticist, H. J. Muller, who posed and answered the still current problem: "Must we geneticists become bacteriologists, physiological chemists and physicists, simultaneously with being zoologists and botanists? Let us hope so" (*Am. Nat.* 1922, 56:32).

The vision of genetics today is not less than the complete understanding of how cells and organisms are built, how they function metabolically and developmentally, and how they have evolved. This requires the integration of previously separate disciplines based on diverse concepts and tongues. Whatever your specialization or interest, I hope you will find this single volume helpful and affordable.

Although I had the aim of comprehensiveness beyond all the available compendiums, there were hard decisions of what to include and what to pass. The same science may appear different depending on who and when looks at it, as Gerald H. Fisher's art at left (man or woman or both) illustrates the point.*

Thank you for the appreciation of the first edition. I will be much indebted for your comments and suggestions.

GPR
<redeia@mchsi.com>

---

*By permission of *Perception and Psychophysics*, 4:189, 1968.

PREFACE TO THE FIRST EDITION

The primary goal of this encyclopedic dictionary is the facilitation of communication and understanding across the wide range of biology that is now called genetics. The emphasis is on recent theoretical advances, new concepts, terms and their applications. The book includes about 18,000 concepts and over 650 illustrations (graphs, tables, and equations). Most of the computational procedures are illustrated by worked-out examples. A list of about 900, mainly recent books, is provided at the end of the volume, and additional references are located at many entries and illustrations. The most relevant databases are also listed. The cross-references following the entries connect to a network within the book, so this is not just a dictionary or glossary. By an alphabetical search, comprehensive, integrated information can be obtained as you prepare for exams, or lectures, or develop or update a course, or need to review a manuscript, or just wish to clarify some problems. In contrast to standard encyclopedias, I have used relatively short but a greater variety of entries in order to facilitate rapid access to specific topics. This book was designed for students, teachers, scientists, physicians, reviewers, environmentalists, lawyers, administrators, and to all educated persons who are interested in modern biology. Concise technical information is available here on a broad range of topics without a need for browsing an entire library. This volume can always be at your fingertips. Despite the brevity of the entries, the contents are clear even for the beginner. Herbert Macgregor made the remarkable statement that in 1992 about 7,000 articles related just to chromosomes were scattered among 627 journals. Since then, the situation has become even more complicated. Many publications — beyond a person's specialization — are almost unreadable because of the multitude of unfamiliar acronyms and undefined terms. Students and colleagues have encouraged me to undertake this effort to facilitate reading of scientific and popular articles and summarize briefly the current status of important topics. According to Robert Graves, "[a good poem] makes complete sense and says all that it has to say memorably and economically." I hope you will appreciate the sense and economy of this book. I will be much indebted for any comments, suggestions, and corrections.

GPR
3005 Woodbine Ct.
Columbia, MO 65203-0906
Telephone: (573) 442-7435
e-mail: <redeig@missouri.edu> or <redeia@mchsi.com>

"I almost forgot to say that genetics will disappear as a separate science because, in the 21st century, everything in biology will become gene-based, and every biologist will be a geneticist."

Sydney Brenner, 1993

ACKNOWLEDGMENTS

I thank my wife Magdi for her patience while writing and for critical reading of the text. My daughter Mari introduced me into word processing. My son-in-law, Kirk was very supportive. Granddaughters, Grace, Paige and Anne are most inspirational. I am grateful to countless numbers of colleagues on whose work this material is based and to whom I could not refer because of limitations of space. I am indebted to colleagues, especially to Dr. Csaba Koncz, Max-Planck-Institut, Cologne, Germany for many useful discussions. My students at the University of Missouri, Columbia, MO, and students and colleagues, particularly Dr. András Fodor, at the Eötvös Lóránd University of Basic Sciences, Budapest, provided purpose for undertaking this project. Some of the chemical formulas are based on the Merck Index, on the Aldrich Catalog and Fluka Catalog.

I appreciated the comments from the readers on the first edition. The first e-mail from Dr. S. L. C. said: "Thank you for assembling such concise explanations of all genetic concepts in a single volume". Similar letters came from many others that I cannot all quote here.

I am thankful to the public reviews (listed in chronological order) for the constructive comments and suggestions: Drs. R. Adler (*Choice* 1998, p. 544), Anna Wedell (*Acta Paediatrica* 87:1211), Michael Cummings (*Doody's Health Sciences Book Review Journal* 1998, *Annals of Internal Medicine* 130:168), Fredrick A. Bliss (*HortScience* 33:1274), Zoltán Király (*Acta Phytopathologica et Entomologica* 33:217 and *J. Phytopathology* 147:623), James N. Thompson (*Annals of the Entomological Society of America* 91:984), Jolán Miklós (*American Journal of Medical Genetics* 83:145–46), John G. Scandalios (*Quarterly Review of Biology* 74:74), John S. Wassom (1999, *Mutation Research Forum 4:2* [http://www.elsevier.com/homepage/sah/mutform/doc/bookrev.html]), J. Sutka (*Acta Agric. Hung. Acad. Sci.* 42:237).

I am much indebted to Senior Editor Luna Han for her encouragement, and enthusiasm. Thanks are due to Editorial Assistant Kristen Hauser for her efficiency. I owe special gratitude to Associate Managing Editor Kristin Cooke Fasano for the conscientious production of this book. I appreciated that she has graciously accepted and incorporated last minute updates.

The sample of comments of the following reviews on the first edition as well of some earlier books of mine mean a great deal to me:

* NATURE (Lond) *302*:169. "...sections dealing with molecular genetics...are remarkably clear and up to date." "...this one is one of the best textbooks of general genetics". "The book is extremely well illustrated..."
* THEORETICAL AND APPLIED GENETICS *66*:38. "...a balanced treatment of almost all genetics disciplines and their important findings."
* HEREDITY 33:123 "...unusually good value."
* QUARTERLY REVIEW OF BIOLOGY *74*:74. "The author clearly met his goal: to facilitate communication and understanding across all biological sciences". "...this book should prove very useful as a quick desk reference for students, professionals and nonprofessionals."
* AMERICAN JOURNAL OF MEDICAL GENETICS *83*:145. "He undauntedly covers the entire field of contemporary genetics including not only the classical areas of formal genetics and cytogenetics but also the latest global molecular advances and their applications." I found the contents very useful..."
* ACTA PÆDIATRICA *87:1211*. "The strength of the book lies in its brief explanations..." "...it is invaluable for anyone interested in this rapidly evolving area, including scientist, teachers, students, and laymen."
* ANNALS OF INTERNAL MEDICINE *130*:168. "This book is a valuable reference tool and will be of interest to a wide audience. Other books, including glossaries and dictionaries of genetics are available but are more limited in coverage."
* ANNALS OF THE ENTOMOLOGICAL SOCIETY OF AMERICA *91*:894. "The preparation of a book like this is a massive task, If it is done well, it can provide a valuable central resource to information that might otherwise take a lengthy search to uncover". "In this case...the overall effort has been very successful."
* HortScience 33:1274. "The books is attractive and user-friendly, beginning with the opening pages and continuing throughout." "This manual is by far the best I have used".
* MUTATION RESEARCH FORUM 1999 May. "...this book is a true information warehouse". "The words in this book came through as if someone were talking directly to me." "The investment...is worth it."
* JOURNAL OF PHYTOPATHOLOGY *147*:623. "This giant volume is recommended to those who are in contact with the genetic aspects of questions in their field of research or practice." ... "The most important feature of this book is that all concepts are explained not just defined" ... "The style is clear even for the beginner".
* ACTA AGRONOMICA HUNGARICA *47*(2):237. "...although a number of dictionaries or glossaries of genetics already exist, this...is quite different from the others in-as-much as it is much more comprehensive, and not only defines the concepts but explains them in concise plain language" ... "fills a void."
* CHOICE *1998*, 36–1563. "...outstanding compendium of genetics" ... "Highly recommended for biologists."

Cudweed by Konrad von Gesner (1597), the famous Swiss savant and zoological and botanical illustrator. Some of his work has been borrowed by many, among them the German Joachim Camerarius (1500–1576) of Tübingen, a great authority on classics, religion and science and whose descendant Rudolph Jacob Camerarius (1665–1721) became the first experimental geneticist and who discovered the love life of plants.

FIRST TO READ FOR THE SECOND EDITION

The organization of this expanded Second Edition is slightly different from that of the First Edition. The material is still in alphabetical order, of course, but there are some slight differences that conform to standard rules of alphabetization. Numbers involved with the entries do not affect the order. Entries beginning with Greek letters are sorted as if spelled out, e.g. "α average inbreeding coefficient" follows "Alpert syndrome." Words followed by a comma and another word precede entry words without a comma, e.g. "antibody, secondary" is followed by "antibody detection." Hyphenated entries are sorted as if they are single words. There is some variation in the spelling of certain technical terms in the literature (e.g., use of "e" versus "ae"). Here, the most common usage is favored. Although every effort has been made to be consistent, you will find certain instances of variation in this text (e.g., single word versus hyphenation and "c" versus "k"). Some entries are qualified by another word added after, and in others the qualifier comes first. An attempt has been made to guide the reader to the desired entry when necessary. In rare instances, you may need to search by synonym or related term to find the desired entry. Thank you for your patience.

This book contains a large number of Internet addresses under the heading of databases and after certain entries. It must be remembered that data are not knowledge. The data must be integrated into science. Every effort has been made to keep the web addresses current. Unfortunately, they are altered frequently and it is likely that some will cease to exist or will change by the time the book is in your hands.

The content of the entries is based on the best information available in the literature at the time of the completion of the writing. Moreover, because the sources of information, like most human products, are not perfect, the author cannot claim perfection or assume legal responsibility for these contents.

"Knowledge ... built on opinion only, will not stand."
Linnaeus (1735)

A

A *See* adenine.

2,5-A oligonucleotide Adenine oligonucleotide generated by 2,5-A synthethase from double-stranded RNA. These oligonucleotides activate RNase L, which attacks infecting viruses of vertebrates. If the two genes encoding these two enzymes are transformed into plants, they provide resistance against RNA viruses. *See* host—pathogen relationship; ribonuclease.

a Atto-, 10^{-18}, e.g., attomole/amole.

Å (ångström) Unit of length, 1/10 of 1 nm; 10^{-7} mm.

A6 *Agrobacterium tumefaciens* strain with a Ti plasmid coding for octopine production in the plant cell. *See* *Agrobacterium*; octopine; opines.

A20 Cytoplasmic Zn-finger protein that limits TNF-induced NF-κB responses. It reduces apoptosis. Its deficiency may increase inflammation and may result in death. *See* apoptosis; NF-κB; TNF. (Kumar-Sinha, C., et al. 2002. *J. Biol. Chem.* 277:575.)

AAA protein ATPases, enzymes cleaving off phosphates from ATP. They are equipped with Walker boxes. AAA domain proteins are molecular chaperones and have important roles in vesicular transport, organelle biogenesis, microtubule rearrangements, etc. Some of this group of enzymes is related to the prokaryotic RuvA proteins. The Mgs1 protein (maintenance of genome stability) of yeast has homologs in all prokaryotes and eukaryotes. Their defects contribute to increased mitotic recombination. *See* ATP; ATPase; chaperone; MDl1; RuvABC; Walker boxes. (Dalal, S. & Hanson, P. I. 2001. *Cell* 104:5; Gadal, O., et al. 2001. *EMBO J.* 20:3695; Hishida, T., et al. 2001. *Proc. Natl. Acad. Sci. USA* 98:8283.)

AAAS *See* Aladin.

AAF *See* alpha accessory factor.

a/α-factor *See* yeast, mating type.

AAR1 *See* TUP1.

Aarskog syndrome (Aarskog-Scott syndrome) Autosomal dominant, autosomal recessive, X-linked (Xq12) recessive short stature, hypertelorism (increased distance between organs or parts), scrotum (the testis bag) anomaly, pointed hairline (Widow's peak), broad upper lip, floppy ears, etc. The basic defect involves the RHO/RAC member of the RAS family of GTP-binding proteins. *See* faciogenital dysplasia; head/face/brain defects; RAS; stature in humans.

AATAAA Consensus of 10–30 bp upstream from a CA dinucleotide at the site where cleavage then polyadenylation of the mRNA commonly take place. This consensus may also be a signal for transcription termination, although normally RNA polymerase II continues to work after passing it. *See* mRNA tail; polyadenylation signal; transcription termination. (Curuk, M. A., et al. 2001. *Hemoglobin* 25:255.)

AATDB *Arabidopsis thaliana* database provides general information on all aspects of the plant, including genes, scanned images of mutants, nucleotide sequences, genetic and physical map data, cosmid and YAC clones, bibliographical information, etc. Access is available without password through <http://www.weeds.harvard.edu/index.html> or by e-mail <curator@frodo.mgh.harvard.edu>. *See* AIMS; *Arabidopsis thaliana*; database.

AAUAAA Consensus for polyadenylation of the mRNA. Apparently the poly-A RNA polymerase enzyme and associated protein attach to this sequence before cleavage of the transcript, and posttranscriptional polyadenylation takes place. Yeast does not have this consensus. *See* AATAAA.

ABA (abscisic acid, 3-methyl-5-[1′-hydroxy-4′-oxo-2′-cyclyhexen-1′-yl]-cis-2,4-pentadienoic acid) Terpenoid synthesized from mevalonate and xanthins, apparently through two pathways. It has multiple physiological functions in concert with other plant hormones, particularly with gibberellins and cytokinins, by regulating seed dormancy, germination, leaf abscission, stomatal opening, drought response, etc. Some *aba* genes have been cloned. In the ABA signal transduction, farnesyl transferase seems to be involved. Cyclic ADP-ribose, regulated by Ca^{2+}, seems to be another signaling molecule for ABA. Ca^{2+} ion channels are activated by H_2O_2 produced by the guard cells upon the induction of ABA and thus the stomatal opening/closing is controlled. Responses to ABA are regulated by ABRC (ABA response complex) in the genes that include an ACGT box and a variable coupling element. *See* abscisic acid; farnesyl pyrophosphate; ion channels; plant hormones; prenylation; stoma. (Lopez-Molina, L., et al. 2001. *Proc. Natl. Acad. Sci. USA* 98:4782.)

abasic site In the DNA where glycosylases (base exchange repair) have removed bases by cleaving the glycosylic bond. According to an estimate, approximately 100,000 abasic sites are generated per cell daily, and their number increases by senescence. These sites may be intermediates in chemical mutagenesis and repair. DNA polymerases ζ and η can insert nucleotides opposite 8-oxoguanine (C) and O⁶-methylguanine sites (C or T). The repair may not be highly efficient. *See* apurinic site; A rule; DNA polymerases; DNA repair; glycosylase; oxidative damages. (Haracska, L., et al. 2001. *J. Biol. Chem.* 276:6861.)

abaxial Not in the axis of body or of an organ.

ABCB *See* multidrug resistance.

ABCD model Environmental matrix for the study of the performance of species. *See* box.

	Environment 1	Environment 2
Genotype 1	A	B
Genotype 2	C	D

ABC excinuclease 260,000 M_r protein complex containing the subunits coded for by the *uvrA*, *uvrB*, and *uvrC* genes of *Escherichia coli*. UvrA is an adenosine triphosphatase and also brings into position UvrB, which after attaching to the DNA cuts it at the 3′ position, and that provides the opportunity for UvrC to incise at the 5′ position. UvrD, a helicase, releases the damaged oligomer along with UvrC. Following these events, DNA polymerase fills in the correct nucleotides. In yeast the RAD1, 2, 3, 4, 10, and 14 carry out the same tasks as the ABC excinucleases of bacteria. In humans, the XPA (a damage-recognition protein, comparable to UvrA) binds to the XPF-ERCC1 (excision repair cross-complementing protein) heterodimer and to the human single-strand-binding replication protein, HSSB. XPF (3′ cut) and XPG (5′ cut) are nucleases. The gap-filling polymerases are polδ and polε. XPB and XPD are helicase subunits of the TFIIH transcription factor. The excinuclease complex is released at the end of the process with the aid of the proliferating cell nuclear antigen (PCNA). This complex is capable of excision of cyclobutane pyrimidine dimers, 6 − 4 photoproducts (adjacent pyrimidines cross-linked through $C^6 - C^4$), and nucleotide adducts (molecules with added groups) formed by mutagenic agents. *See* adduct; base flipping; cyclobutane; DNA ligase; DNA polymerases; excision repair; helicase; PCNA; pyrimidine-pyrimidinone photoproduct; transcription factors. (Zou, Y., et al. 2001. *Biochemistry* 40:2923.)

ABC transporters ATP-binding cassette transporters (9q22-q31). Constitute a large family of proteins that hydrolyze ATP and mediate transfers through membranes. These are frequently called TAPs. The ABC transporters have two membrane-spanning (MSD) and the dimeric ATP-nucleotide-binding domains (NBD). The MSDs may show greater variations, depending on whether they operate as a pump or a conductance channel. The NBD subunits play the role of the engines of the transport and interact through their arm 1 with the two MSDs. The ABCA4/ABCR mutations may account for Stargardt disease (STGD1), fundus flavimaculatus (FFM), retinitis pigmentosa (RP), and cone-rod dystrophy (CRD), all recessive with somewhat overlapping retinal symptoms. By 2001, 48 ABC transporters belonging to 7 gene families had been identified. *See* Byler disease; cone dystrophy; high-density lipoprotein; multidrug resistance; multiple drug resistance; protein-conducting channel; pseudoxanthoma; retinitis pigmentosa, signal hypothesis; SRP; Stargardt disease; translocon; translocase; Tangier disease; TAP; TRAM. (Dean, M., et al. 2001. *Genome Res.* 11:1156; Neufeld, E. B., et al. 2001. *J. Biol. Chem.* 276:27,584; Chang, G. & Roth, C. B. 2001. *Science* 293:1793; Borst, P. & Elferink, R. O. 2002. *Annu. Rev. Biochem.* 71:537.)

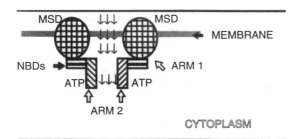

abdomen in *drosophila* The body segment between the thorax and telson. *See also* Drosophila.

Abelson murine leukemia virus oncogene (*abl*) Mammalian homolog of the avian Rous sarcoma virus. It codes for a plasma membrane, tyrosine kinase. This enzyme acquires constitutive catalytic activity in the presence of the Philadelphia chromosome and causes chronic myelogenous leukemia (CML) in humans. It is treated effectively by STI-571 tyrosine kinase inhibitor. *See* leukemia; oncogenes; Philadelphia chromosome; Rous sarcoma; tyrosine kinase. (Wang, J. Y., et al. 1984. *Cell* 36:349.)

aberrant genetic ratio Occurs when the chromosomes carrying the wild type or mutant allele of a gene, respectively, have reduced transmission through meiosis or the viability of the gametes is diminished. Depending on the chromosomal location of the defect, either the one (wild type) or the other (recessive) allele may appear in excess of expectation of normal phenotypic ratios.

aberration, chromosomal *See* chromosome breakage.

abetalipoproteinemia (microsomal triglyceride transfer protein deficiency, 4q22-q24) Involves very low levels of the very low density (VLDL), low density (LDL, apolipoprotein B, 2p24), and high density (HDL) of these lipoproteins (microsomal triglyceride transfer protein [MTP] defect). The rare recessive anomaly is accompanied by excretion of lipoproteins, malabsorption of fat, acanthocytosis (see below, thorny-type erythrocytes rather the normal doughnut-shaped), retinitis pigmentosa (sclerosis [hardening], pigmentation and atrophy [wasting away]) of the retina of the eye, and irregular coordination of the nerves (ataxia). *See* beta-lipoprotein; hyperlipoproteinemia; hypobetalipoproteinemia; neuromuscular disease.

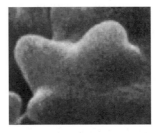

ABF-1 Nuclear transcriptional repressor belonging to the basic helix-loop-helix family. *See* helix-loop-helix. (Wong, J., et al. 2001. *DNA Cell Biol.* 20:465.)

Abf Autonomously replicating sequence binding factor is involved in silencing of yeast mating types. Also, it may bind to various promoters and thus may initiate replication or transcription *See* ARS; HML and HMR; mating type, yeast; ORC.

ABH antigens Human secretion in the saliva and other glycoprotein-containing mucus in the presence of the *Se* (dominant allele, human chromosome 19q13.13), and the gene codes for the α2L-fucosyltransferase enzyme. The secreted

Fucose
|
Galactose(β1-3,)N-acetyl-D-glucosamine-R

glycoproteins, A and B, are about 85% carbohydrate and about 15% protein. Approximately 75–80% of Caucasoids are secretors (homozygous or heterozygous for *Se*). The precursors of the antigens are galactose(β1-3)*N*-acetyl-D-glucosamine-R and Gal(β1-4)*N*-acetyl-glucosamine-R (where R stands for the extension of the carbohydrate chain). Antigen H has the critical structure of that shown above. Antigenic determinant A is formed by *N*-acetylgalactosamine, and the B antigen by galactose addition at the nonterminal position to the H antigen. Thus, the A, B, and H antigens are different from each other by these carbohydrates and in some variants by the number of fucose molecules. The A and B alleles are codominant. The recessive O blood group lacks a fucosidase activity that places a fucose, by an α1-2 linkage on a galactose. The Lewis blood group (Le [Les], 19q13.1-q13.11) is distinguished on the basis that its dominant allele Le places fucose in an α-1,4 linkage to the *N*-acetylglucosamine. Individuals that have no secretor activity but are Le belong to the Lewis blood group Le^a, whereas when both Se and Le are expressed, they represent the Le^b type. *See also* ABO blood group; Bombay blood type; fucosyl transferase; secretor. (Domino, S. E., et al. 2001. *J. Biol. Chem.* 276:23,748.)

ab initio   From the beginning in Latin, e.g., genes *ab initio* indicates the genes as first recognized by sequencing, but the exact exon/intron structure was not yet identified.

abiogenesis   Spontaneous generation of life, origin of living cells from organic material during the early history of the earth. *See* origin of life; spontaneous generation.

ABL   *See* abetalipoproteinemia.

ABL   Abelson murine leukemia virus oncogene is located to human chromosome 9q34.1 and mouse chromosome 2. When translocated to human chromosome 22, it may transcribe a fusion protein with an abnormal protein tyrosine kinase activity, and this is probably the cause of chronic myeloid leukemia. Acute lymphocytic leukemia is also associated with a similar translocation, the Philadelphia chromosome, but it appears that tyrosine kinase activation is different from that of the fusion protein. The ABL gene has about 300 kb intron downstream from the first exon. This intron appears to be the target of the translocations and causes acute lymphocytic leukemia. Insertion of DNA sequence into the *abl* gene of mouse results in several morphological alterations and death. Abl also controls differentiation, cell division, and stress responses. The SH3 domain negatively regulates Abl activity, and deletion of SH3 makes it an oncogenic protein. Mutations in the SH3, or in the catalytic domain or in the linker region between the SH3 and SH2 domains, are also oncogenic. ABL deficiency in mice also leads to osteoporosis. *See* ARG; leukemia; oncogenes; osteoporosis; Philadelphia chromosome; SH2; SH3. (Maru, Y. 2001. *Int. J. Hematol.* 73:308.)

abl   B-cell lymphoma (Abelson leukemia) oncogene encoding a nonreceptor protein tyrosine kinase. Ionizing radiation and alkylating agents activate this oncogene. In the Philadelphia

chromosome the contact between BCR and ABL most commonly leads to myelogenous leukemia. In the absence of ABL, the JNK/SAP kinases (Jun kinase) are not stimulated. *See* BCR; JNK; JUN; leukemia; lymphoma; Philadelphia chromosome; tyrosine kinase.

ablation   Mechanical removal of cells or tissues of stem cells or plant meristems to study the role of those cells in differentiation and development. The purpose also can be achieved by obtaining genetic deletions in these areas heterozygous for appropriate marker genes. The deletion of the dominant allele reveals the function of the recessives and permits tracing cell lineages on the basis of the visible sectors formed. Familial retina ablation may occur in animals as a hereditary abnormality. *See* cell lineages; deletion; gene fusion; intercellular immunization; pseudodominance.

ABM paper   *See* diazotized paper.

ABO blood group   Represented by three major type of alleles (human chromosome 9q34) displaying codominance. These blood types are extremely important because inappropriate mixing (in blood transfusion) results in agglutination, which prevents blood flow through the veins and oxygen transfer and is potentially lethal. These antigens are actually carbohydrates (attached to polypeptides), and the genes A and B specify α-D-*N*-acetylgalactosaminyltransferase and α-D-galactosyltransferase enzymes, respectively. Gene O is not active as an enzyme. The A and B enzymes (M_r about 100,000) are dimeric and structurally similar to each other. The A and B molecules are identified as A and B antigens. Occasionally maternal antibodies against the A and B antigens may enter through the placenta, the fetal bloodstream, and adversely affect the erythrocytes, causing anemia and hyperbilirubinemia. In such cases, medical treatment may be required. The ABO system also has a limited use in forensic medicine in paternity suits, in typing blood stains, semen, and saliva in criminal cases. Immunologically active forms may be recovered in old human remains and can also be used in archaeological research. This blood group provided some correlative information in cancer research, e.g., in O individuals afflicted with carcinomas, A antigen may be detected in 10–20% of the cases. The major clinical characteristics are shown below.

Blood Group (frequency in Caucasoids*)	Genotype	Antigens Formed	Antibodies Formed	Clumping With	Blood Type Acceptable for Transfusion
O (0.45)	$i^O i^O$	neither	anti-A anti-B	A, B AB	O
A (0.44)	$i^A i^A$ or $i^A i^O$	A	anti-B	B, AB	A, O
B (0.08)	$i^B i^B$ $i^B i^O$	B	anti-A	A, AB	B, O
AB (0.03)	$i^A i^B$	A, B	neither	neither	A, B, O

*The frequency of these alleles varies in different populations. For the calculation of frequencies, see gene frequencies. Actually, the A type exists in A_1 and A_2 forms; in about 1–2 % of the A_2 and 25% of the A_2B individuals, anti-A_1 antigens occur.

It appears that changes in glycosyltransferase activity are not uncommon in several types of tumors. The frequency of the various ABO alleles varies a great deal in the world's population. It has been shown that the O blood type provided some protection against the most severe form of syphilis (*Treponema pallidum*) but somewhat higher susceptibility to diarrhea caused by some viral and bacterial infections. The B blood group may have afforded some protection against smallpox, plague, and cholera. *See* ABH antigen; blood groups; forensic genetics; Lewis blood group; *Treponema pallidum*. (Race, E. E. & Sanger, R. 1975. *Blood Groups in Man*. Blackwell, Oxford; Chester, M. A. 2001. *Transfus. Med. Rev.* 15:177.)

aborigine   First group of inhabitants, humans, animals, or plants.

abortion, medical   Induced during the early period of pregnancy usually by antiprogestin (mifepristone) in combination with prostaglandins (in countries where approved) or by the less costly and not very effective misoprostol. Progestin is a synthetic progesterone. *See* contraceptives; genetic screening; mifepristone; progesterone; prostaglandins; selective abortion.

abortion, spontaneous   Frequently caused by disease, incompatibility genes, or chromosomal aberrations. Various types of chromosomal defects were cytologically detected in 30–50% of the aborted fetuses. About 15–20% of the verified human pregnancies are aborted spontaneously and an estimated 22% of the abortions occur before pregnancy is clinically detected. Different molecular mechanisms may account for abortion. Th2 lymphocytes and IL-10 and TGF-β may suppress incompatible paternal antigens in the fetus. Th1, cytokines, IL-2, INF-γ, and TNF-α may contribute to abortion. By catabolizing tryptophan, indolamine 2,3-dioxygenase (IDO) may help in suppressing allospecific maternal T cells in the lining of the uterus (decidua). When activated by α-galactosyl-ceramide or by glycosyl-phosphatidylinositols (the latter is present in blood parasites), the special Vα14 natural killer T cells (NKT) may cause abortion. Perforin, TNF-α, and INF-γ of the NKT cells may destroy the embryonic trophoblasts. *See* allospecific; ceramides; chromosomal rearrangement; chromosome breakage; TNF, Il-2; IFN; inosities; perforin; selective abortion; trisomy, T cells; trophoblast. (Hamerton, J. L. 1971. *Human Cytogenetics*. Academic Press, New York; Hallermann, F. B., et al. 2001. *Eur. J. Hum. Genet.* 9:539.)

abortive infection   Bacteria are infected with a phage capsule that carries bacterial rather than phage DNA and thus cannot result in the liberation of phage particles. Abortive response by infection of mammalian cells may be caused by any deficiency of the interacting system. (Hosel, M., et al. 2001. *Virus Res.* 81:1.)

abortive transduction   The transduced DNA is not incorporated into the bacterial genome, and in the absence of a replicational origin, it can be transmitted but cannot be propagated. Therefore, the transduced fragment is contained in a decreasing proportion of the multiplied bacteria. *See* transduction, abortive. (Stocker, B. A. D., et al. 1953. *J. Gen. Microbiol.* 9:410; Benson, N. R. & Roth, J. 1997. *Genetics* 145:17.)

A box   Internal control region of genes (5S ribosomal RNA and tRNA) transcribed by DNA-dependent RNA polymerase III; the consensus is 5′-TGGCNNAGTGG-3′. The tRNA genes also have an essential *intermediate segment* of about a dozen bases that has no consensus, yet its length is necessary for function. Also, there is another regulatory consensus nearby, the B box 5′-GGTTCGAANNC-3′. The matrix attachment region (MAR) is also an A box (with a consensus of AATTAAA/CAAA). *See* MAR. (Borovjagin, A. V. & Gerbi, S. A. 2001. *Mol. Cell Biol.* 21:6210.)

abrin   Agglutinin, a toxic lectin and hemagglutinin extracted from the seed of the tropical leguminous plant jequirity (*Abrus precarius*). Abrins A, B, C, and D are glycoproteins of two polypeptide chains. The small A chain is an inhibitor of aminoacyl-tRNA binding and has nothing to do with agglutination. Abrin is more toxic to a variety of cancer cells (ascites, sarcomas) than to normal cells. *See* aminoacyl-tRNA synthetase; hemagglutinin; lectins; RIP. (Wu, A. M., et al. 2001. *Life Sci.* 69:2027.)

abrine   N-methyl-L-tryptophan (α-methylamino-β-[3-indole] propionic acid) is an inflammatory drug, and is unrelated to abrin. (Richou, R., et al. 1966. *C.R. Acad. Sci. Hebd. D [Paris]* 263:308.)

abscisic acid   Plant hormone regulating a variety of physiological processes, including modification of the action of other plant hormones. Originally it was detected as a substance involved in the abscission of leaves. *See* ABA; plant hormones. (Hugouvieux, V., et al. 2001. *Cell* 106:477; Finkelstein, R. R. 2002. *Plant Cell* 14:S15.)

abscission zone   Thin-walled tissue layer (low in lignin and suberin) formed at the base of the plant organs before abscission (falling off) takes place. *See* abscisic acid.

absinthe   Green liqueur containing thujone. It is a GABA antagonist. *See* GABA.

absolute dating   Determines the age of archaeological objects by using radiometry (rate of decay of radioactive isotopes), electron spin resonance (measures age of crystals from a few thousand to 300,000 years), or thermoluminescence (heated objects release light and energy, and when they are heated again, you can estimate the time elapsed since they were last heated (Renne, P. R., et al. 2000. *Sci. Progr.* 83:107.)

absolute linkage   There is no recombination between (among) the genes in a chromosome. See linkage; recombination.

absolute weight   Mass of 1,000 seeds or kernels after appropriate cleaning.

absorption   Uptake of compounds through cell membranes or through the intestines into the bloodstream.

absorption spectrum   Characteristic absorption peaks of a compound at various wavelengths of light; e.g., guanine has maximal absorption at about 278 nm at pH 9, but its maximum at pH 6.8 is at ca. 245 nm ultraviolet light; chlorophyll-a has an absorption maximum in benzene at ca. 680 and 420 nm visible light, whereas chlorophyll-b maxima are at ca. 660 and 460 nm, respectively. These characteristics vary according to the pH and the solvents used and are determined by spectrophotometers.

abundance Average number of molecules in cells.

abundant mRNAs Small number of RNAs that occur with great numbers in the cells. *See* mRNA.

abzymes Monoclonal antibodies with enzyme-like properties. If these antibodies can recognize the transition state analogs of enzyme-substrate reactions, they might have enzymatic properties. These abzymes would have numerous chemical and pharmaceutical applications. *See* antibody; catalytic antibody; monoclonal antibody; transition state. (Takahashi, N., et al. 2001. *Nature Biotechnol.* 19:563.)

acanthocytosis *See* abetalipoproteinemia; elliptocytosis.

acanthosis nigricans Hyperkeratosis and hyperpigmentation of the skin that may accompany Crouzon syndrome and Berardinelli-Seip syndrome. *See* achondroplasia; Crouzon syndrome; Donahue syndrome; lipodystrophy.

AcAP Anticoagulant protein isolated from *Ancylostoma caninum* hookworm.

Hookworm.

ACAT *See* sterol.

acatalasemia (CAT, 11p13) Rare dominant/semidominant/recessive trait involving the deficiency of the enzyme catalase. This enzyme has a protective role in the tissues by removing the H_2O_2. Symptoms include small painful ulcers around the neck, gangrenes in the mouth and atrophy of the gum, and very low catalase activity in the blood and other tissues. The heterozygotes have intermediate levels of catalase activity. Acatalasemia may be classified into different groups according to the clinical symptoms, both in humans and in animals. The gene extends to 34 kb with 14 introns. It is closely linked to WAGR. *See* Wilms' tumor.

acatalasia Same as acatalasemia.

ACC (1-aminocyclopropane-1-carboxylic acid) Precursor of the plant hormone ethylene. *See* ethylene.

accelerator mass spectrometry (AMS) Quantifies isotopes such as C^{14}, H^3, Ca^{41}, Cl^{36}, Al^{26}, etc., in biological, archaeological, pharmacological, or other materials with attomole sensitivity and high precision. It can be used to study the tissue distribution, metabolism, pharmacokinetics, and radiological hazards of isotopes. It is also a potent tool for paleontological analysis and dating archaeological remains. *See* MALDI/TOF/MS.

acceptor splicing site Junction between the right end of one exon and the left end of the next exon. *See* introns; splicing.

acceptor stem Part of the tRNA, including the site (5'-CCA-3') where amino acids are attached. *See* aminoacyl-tRNA.

accessibility Genetically determined ability of the genome to provide access for the V(D)J recombinase to rearrange the immunoglobulin genes. The accessibility depends on the increased activity of the loci, i.e., status of transcription, demethylation, and increased DNase sensitivity. *See* CDR; RAG, immunoglobulins; RSS; V(J)D recombinase.

accession number In bioinformatics this number permanently identifies a particular molecular sequence submitted to a database (e.g., BankIt, Bioseq, gi, ASN.1). The accession number is also used by various biological collections for the identification of specimens such as plants in an herbarium, differently acquired strains of organisms, etc.

accessory cells (companion cells) Epidermal cells next to the guard cells around the plant stomata that appear different from the usual epidermal cells. In animals they promote adaptive immunity, although they are not directly involved in antigen recognition.

accessory chromosome *See* B chromosome.

accessory DNA Product of DNA amplification in the cell. *See* amplification.

accessory gland Relatively minor tissue aiding the function of a gland. *See* epidimys.

accessory pigments Complement chlorophylls in absorbing light (carotenoids, xanthophyll, phycobilins).

accessory proteins For example, transcription factors that bind to upstream DNA elements for controlling transcription and other binding proteins that take part (not necessarily the main part) in a particular function. Accessory host proteins are also involved in the orientation or directionality of transposons. *See* transcription factors; transposable elements; transposons.

accessory sexual characters Structures and organs of the genital tract including accessory glands and external genitalia, but not the gonads, which are the primary sexual characters. *See* gonad; secondary sexual character; sex, phenotypic; sex determination.

access time Time interval between calling in a piece of information from a storage source to the actual delivery of that information to the caller. *See* real time.

accuracy Percentage of correct identification of carcinogens and noncarcinogens on the basis of mutagenicity tests. The mutagenicity tests are much faster and much less expensive than direct carcinogenicity assays, but it is important to know how well these simpler tests reveal the carcinogenic (or noncarcinogenic) properties of the chemicals tested. *See* bioassay in genetic toxicology; predictivity; sensitivity; specificity, mutagen assay. (Rédei, G. P., et al. 1984. In *Mutation, Cancer and Malformation*, Chu, E. H. Y. & Generoso, W. M., eds., p. 689. Plenum, New York.)

accuracy of DNA replication *See* DNA replication error.

Ac-Ds (*Activator-Dissociator*) First transposable element system recognized on the basis of its genetic behavior in maize. It contains 4563 bp and is bordered by 11 bp imperfect, inverted repeats. The independently discovered *Mp* (*Modulator* of *p*1 [pericarp color]) is basically the same transposon. *Ac* is an autonomous element and can move by its own transposase function. The *Ac*/*Mp* element makes a 3.5 kb transcript, initiated at several sites upstream, and a 2,421 base mRNA. A defective (deleted) version of it, *Ds* (*Dissociator*), is nonautonomous and requires the presence of *Ac* for transposition. *Ds* was originally discovered on the basis of frequent chromosome breakage associated with it. The *Ds* elements are quite varied in size but practically identical at the terminal sections to *Ac*. These elements have been identified first on the basis of mutation at known loci (*a*, *Adh*, *sh*, *wx*, etc.) upon insertion and reversions when the inserted element

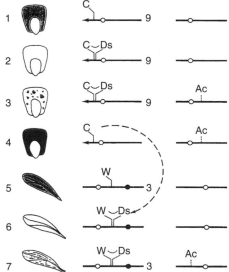

The possible phenotypic expression of genes in the presence of the *Ac-Ds* elements in maize. (1) The expression of the *C* allele in chromosome 9 in the absence of the transposable elements. (2) If *Ds* is introduced into the locus, the function of *C* is disrupted and the kernel becomes colorless (2). If *Ac* (transposase) is introduced into any other location of the genome, it may cause the movement of the transposable element, and colored spots appear (3). In case *Ds* is entirely dislodged from the germline, in the following generation full or partial function of the *C* gene is restored, depending on whether the original site was completely restored or some modifications took place, and only diluted color appears (4). The *W* allele in chromosome 3 controls the development of green leaf color (5). If *Ds* moves into the gene, it may disrupt its function, and albinism is observed (6). In case *Ac* is introduced by crossing, *Ds* may move as indicated by the green stripes (7). Remember, *Ds* lacks transposase function, although *Ac*, which carries the transposase, may move it.

is evicted. More recently, it has been shown that many of the insertions do not lead to observable change in the expression of the genes, or their effect is minimal, and only sequencing of the target loci may then reveal their presence. The *Ac* element is transposed by a nonreplicative manner, and after meiosis only one of the sister chromatids displays *Ac*/*Mp* at the original site (called *donor site*). In the other chromatid, the element may be at another location (*recipient site*), and the original location becomes "empty." The recipient sites are most commonly in the same chromosome and quite frequently within the vicinity of the donor site. The *Ds* element frequently initiates a series of events resulting in chromosomal breakage by the mechanism of *breakage-fusion-bridge cycles* and duplications between the original donor and recipient sites. The *Ds* element may move in an inverted manner to the vicinity of a locus and thus the revertants may still contain a *Ds* element. In the control of transposition, the 11 bp inverted terminal repeats and, in addition, sequences 0.05 to 0.18 kb, have importance. The *Ac-Ds* target sites display 8 bp duplication, which remains even after the removal of the element. The empty target sites may show internal deletions and rearrangements. A transposase enzyme that can mobilize the *Ac* element, which codes for it, mediates the transposition, but it may act on the *Ds* elements too (which are transposase-defective *Ac* elements). It appears that an increase in the number of some but not of other *Ac* elements results in proportionally smaller revertant sectors, and the genetic background, developmental specificities (e.g., somatic or germline tissues), and physiological factors may influence the timing and frequency of transposition. There is evidence in favor of methylation being one of the factor(s) affecting *Ac* expression. This family of transposable elements has additional members, such as *MITE* (miniature transposable element), which has the same termini but is very short. The *Ds1* element is similar to *Ds*, but it carries retrotransposons within its sequences. *Ac* has been successfully transferred to other species such as tobacco, *Arabidopsis*, and yeast, and it functions there similarly as in maize. *See* controlling elements; hybrid dysgenesis; insertional mutation; transposable elements; transposase. (Fedoroff, N. V. 1989. In *Mobile DNA*, Berg, D. E. & Howe, M. M., eds., pp. 377–411. *Amer. Soc. Microbiol. Washington*, DC; Ros, F. & Kunze, R. 2001. *Genetics* 157:1723.)

ACE (1) Angiotensin-converting enzyme. *See* angiotensin. (2) Affinity capillary electrophoresis is a procedure to test the binding strength of ligands.

ACEDB *Caenorhabditis elegans* (a nematode, useful for genetic analyses) database. *See Caenorhabditis elegans*.

acenaphthene Spindle fiber poison and thus polyploidization agent. It is also a fungicide and insecticide. *See* colchicine; polyploid; spindle poison.

Acenaphthene.

acentric fragment Broken-off piece of a chromosome that lacks centromere and therefore its distribution to the poles during nuclear divisions is random and commonly it is lost.

Acentric fragments are frequent consequences of irradiation of cells with X rays and other ionizing radiations. Chromosomal inversions may generate bridges (shown between the two poles) and three acentric chromosome fragments of substantial size that are drifting in the middle of the cell and are not distributed to the poles. *See* centromere; chromosome morphology.

(The photomicrogram is the courtesy of Dr. Arnold Sparrow.)

aceruloplasminemia (3q23-q24) Generally recessive deficiency of ceruloplasmin resulting in dementia, ataxia, diabetes, etc. Ceruloplasmin mediates the peroxidation of transferrin FeII to the FeIII form. *See* ceruloplasmin; iron metabolism; transferrin. (Hellman, N. E., et al. 2002. *J. Biol. Chem.* 277:1375.)

acervulus Disk-like, conidia-bearing, reproductive structure of fungi. *See* conidia.

ACESIMS Affinity capture-release electrospray mass spectrometry uses biotinylated tags similarly to ICAT to capture conjugates in complex biological mixtures and to target specific enzymes that have a role in metabolic defects such as disease. *See* ICAT; MALDI/TOF/MS; proteomics. (Turecek, F. 2002. *J. Mass Spectrom.* 37:1.)

Acetabularia Single-celled green alga that may reach the size of 2–3 cm and may be differentiated into rhizoids, stem and cap. It can survive enucleation for several months. The rhizoids, containing the nucleus, may regenerate into complete plants; $x \approx 10$. *See* enucleate.

acetocarmine *See* stains.

acetonitrile (methyl cyanide) Highly poisonous liquid with ether-like odor, flash point 12.8°C (beware of the vapors). Polar solvent used (among others) for the separation of oligonucleotides by reverse-phase chromatography on silica gels.

acetoorcein *See* stains.

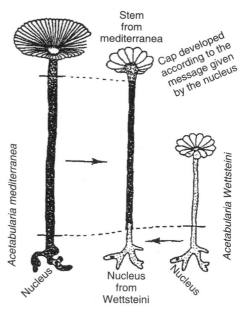

Acetabularia species are useful for developmental genetic studies and dramatically show the role of the cell nucleus. Grafting of the nucleus-containing section of the cell of *A. wettsteinii* to *A. mediterrania* caused *A. mediterranea* to develop a cap according to the instructions of the nucleus donor species. Experiment of J. Hämmerling during the 1940s. (Modified after Goldschmidt, R. B. 1958. *Theoretical Genetics*. Univ. California Press. Berkeley, CA.)

acetosyringone 4-acetyl-2,6-di-methoxyphenol and hydroxyacetosyringone are produced in plant cells (tobacco) and are one group of the compounds that induce the *vir* gene system of the *Agrobacterium* plasmid. *See Agrobacterium*; transformation of plants; virulence genes, *Agrobacterium*.

acetylation Histone acetylation opens the nucleosomal structure for transcription of the DNA. Acetylation of H3 and H4 histones may generate bromodomains for protein–protein interactions. Several nonhistone proteins involved in the regulation of transcription are also acetylated. HMG proteins, nuclear import proteins, and tubulins are acetylated primarily at selected lysine sites. In the DNA-binding transcription factors (p53, E2F1, EKLF, GATA), the sites near to the binding domain are acetylated, and this increases binding. Acetylation of some of the HMG-BOX proteins results in reduced binding to DNA. Acetylation of TCF may disrupt its binding to other proteins or acetylation may prevent binding together of some regulatory proteins. Acetylation may increase protein half-life (e.g., E2F1, α-tubulin) and may enhance protein targeting (e.g., p53). Signaling molecules may provide cues for acetylation. The roles of acetylation may bear similarity to that of kinases, although the number of acetyltransferases is much smaller than that of kinases. *See* GCN5; histone acetyltransferases. (Kouzarides, T. 2000. *EMBO J.* 19:1176.)

acetylcholine (M_r 149) Acetylcholine receptor provides the connection between synapsing neurons and it is thus a signal transmitter. When acetylcholine binds to a receptor, a Na^+/K^+ channel opens. The muscarinic acetylcholine receptors are activated by the fungal alkaloid, muscarine, whereas the nicotinic acetylcholine receptors are operating in the nerve and muscle cells. Acetylcholine receptors are diffusely distributed on the embryonic myotubes but become highly concentrated in a minute area in the postsynaptic membrane and they tether the synaptic cytoskeletal complex. *See* acetylcholine receptors; agrin; cytoskeleton; ion channels; memory; muscarinic acetylcholine receptors; myotube; neuregulin; neurotransmitters; rapsyn; synapse. (Smit, A. B., et al. 2001. *Nature* 411:261.)

acetylcholine receptors Acetylcholine-regulated cation (Na^+, K^+, and Ca^{2+}) channels between the motor neurons and the skeletal muscles. The receptor in the skeletal muscle contains five transmembrane polypeptides, encoded by four separate yet similar genes. When acetylcholine attaches to the receptor, a conformational change ensues, resulting in a brief opening of the channel. The peptides easily isolated from the electric organs of some fishes. *See* agrin; ion channels; muscarinic acetylcholine receptors; nicotinic acetylcholine receptors. (Brejc, K., et al. 2001. *Nature* 411:269.)

acetylcholinesterase (ACHE) Encoded in human chromosome 3q25.2 by codominant alleles. It hydrolyzes acetylcholine into acetate and choline and it restores the polarized state in the postsynaptic nerve membranes. ACHE inhibitors are insecticides and drugs. Nerve gases are also ACHE inhibitors. *See* acetylcholine; acetylcholine receptors; pseudocholinesterase deficiency.

acetyl-CoA (acetyl coenzyme A, ACoA) Heat-stable cofactor involved in the transfer of acetyl groups in many biological reactions (citric acid cycle, fatty acid metabolism, etc.). It has three major domains: the β-mercapto ethylamine unit, the panthothenate unit, and adenylic acid. *See* epinephrine.

acetyl-CoA carboxylase deficiency (ACAC) Recessive ACACA is in human chromosome 17q21. The cytosolic ACACA is expressed primarily in the liver and in adipose tissues. ACACB (12q24.1) is in the mitochondria and expressed mainly in the heart and muscles. ACAC causes multiple interference with gluconeogenesis, fatty acid, and the branched-chain amino acid metabolism. ACACB deficiency leads to continuous oxidation of fatty acids and reduced fat storage in mice. Acetyl-CoA carboxylase (also called ACCase) is a biotin-dependent enzyme in the pathway of long-chain fatty acids located in the cytosol and in the chloroplasts of plants. This enzyme is the target of oxyphenoxypropionate and cyclohexanedione herbicides. *See* branched-chain amino acids; herbicides.

Acetyl coenzyme A, Li salt.

acetyl coenzyme A *See* acetyl-CoA.

acetylglutamate synthetase deficiency Form of autosomal-recessive hyperammonemia. *See* hyperammonemia.

acetyl group Derived from acetic acid CH_3COOH; the R stands for different chemical groups. *See* acyl group.

$$(R-\overset{\overset{\textstyle O}{\|}}{C}-CH_3)$$

Acetyl group.

ACF (ATP-utilizing chromatin assembly and remodeling factor) See chromatin remodeling

achaete-scute complex In *Drosophila*, a complex X-chromosomal (1–0.0) locus regulating bristle formation and nerve differentiation. The posterior dorsocentral bristles are usually missing and the hairs are also sparse in that area. The *achaete* phenotype is generally due to some type of chromosomal rearrangement or loss. *See* complex locus.

Illustration from Bridges, C. & Brehme, K. 1944. *Carnegie Inst. Washington* 552:12.

achalasia-Addisonianism-alacrima syndrome (AAA, 12q13) Triple A syndrome. It is a complex glucocorticoid/adrenal deficiency causing failure of some muscles to relax, hypotension and weakness, failure in shedding tears normally, and various nervous anomalies. The basic defect may involve a WD-repeat protein. See WD-40. (Handschug, K., et al. 2001. *Hum. Mol. Genet.* 10:283.)

acheiropodia Recessive 7q36 developmental anomaly (incidence ~0.000004) involving bilateral amputation of the extremities. The corresponding mouse locus is *Lmbr1*.

achene Single-seed dry fruit.

achiasmate Nuclear division without the formation of chiasmata. *See* chiasma; distributive pairing; meiosis.

Achilles' heel technique Applicable to systems where there is abundant sequence information, and it permits the cleavage of only a small set of restriction sites. DNA sequences around the site or set of sites are synthesized and added to the genomic DNA along with RecA and a methylase. After deproteinization, a restriction enzyme is added. All the (methylated) restriction sites are protected from cleavage except those that were covered by the RecA-DNA complex. *See* DNA sequencing; methylase; methylation, DNA; Rec; restriction enzyme. (Szybalski, W. 1997. *Curr. Opin. Biotechnol.* 8:75.)

achondrogenesis Has been described in two or more autosomal-recessive forms involving deficiency in bone formation at the hip area and large head, short limbs, stillbirth or neonatal death. The phenotypes show variations, and clear-cut differentiation of the symptoms is difficult. *See* achondroplasia; collagen; hypochondroplasia; stature in humans.

achondroplasia (ACH) Rather common chromosome 4p16.3-dominant (homozygous perinatal lethal) type of human dwarfness observed, e.g., in Denmark, at a frequency of 1.1×10^{-4}. Its mutation frequency (predominantly of paternal origin) was estimated to be within the range 4.3 to 7×10^{-5}.

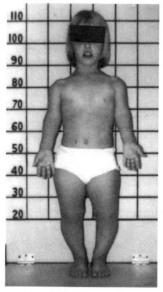

Autosomal-dominant-type achondroplasiac adolescent. (Courtesy of Dr. D. L. Rimoin, Harbor General Hospital, Los Angeles, CA, and Dr. Judith Miles.)

The proximal bones in the limbs are most reduced. Large head with disproportionally small midface; abnormal hip and hands are characteristic. The heterozygotes are generally plagued by heart, respiratory, and other problems. Hypochondroplasia appears to be allelic to achondroplasia. The so-called Swiss-type achondroplasia is recessive and the afflicted individuals show reduced amounts of leukocytes (lymphopenia) and agammaglobulinemia. Pseudoachondroplastic dysplasias (PSACH, spondyloepiphyseal dysplasia) are autosomal recessive (19p13.1), but some ambiguities were noted regarding the pattern of inheritance because of apparent gonadal mosaicism. PSACH is apparently due to a defect in the cartilage matrix. The different forms do not have clear phenotypic distinctions within the group and from the dominant achondroplasia. Some of the skeletal reductions and defects are aggravated by face

and eye defects, cleft palate, and muscle weakness. Achondroplasia is caused by defects in the fibroblast growth-factor receptor 3 (FGFR-3), located in human chromosome 4p16.3. A recurrent missense mutation in a CpG doublet of the transmembrane domain of FGFR-3 caused an arginine substitution for glycine. Achondroplasia with developmental delay and acanthosis nigricans and thanatophoric dysplasia are also defective in fibroblast growth-factor receptor 3. Achondroplasiacs usually display normal intelligence. *See* acanthosis nigricans; achondrogenesis; agammaglobulinemia; cleft palate; dwarfism; fibroblast growth factor; hypochondroplasia; pseudoachondroplasia; receptor tyrosine kinase; stature in humans; thanatophoric dysplasia.

achromatic Parts of the cell nucleus that are not stained by nuclear stains. A microscope lens that does not refract light into different colors.

achromatopsia Recessive inability to distinguish colors, low visual acuity, and involuntary eye movements (nystagmus) are a rod monochromatism due to defects in the α-subunit or β-subunit of the cone cyclic nucleotide-gated cation channel (8q21-q22). This normally generates the light-evoked electrical responses of the cone receptors. Another locus 2q11-q12 with a defect in the α-subunit of the cGMP-gated ion channel debilitates the cone photoreceptor, and a third locus (Xp11.4) with cone dystrophy causes achromatopsia.

***A* chromosome** Member of the regular chromosome set in contrast to a B or supernumerary chromosome. *See* accessory chromosome; B chromosome.

acid-base catalysis Acids and bases are common catalysts of organic reactions in proportion of the presence of H^+ or OH^- ions in the medium. Enzymes are particularly well-suited catalysts because they can carry out either acid or base or simultaneously both acid and base catalysis. Ribozymes are also potential acid/base catalysts. *See* transition state.

acid blob Sequence of amino acids (negatively charged) responsible for activation of a transcription factor. *See* transcription factors. (Almlof, T., et al. 1995. *J. Biol. Chem.* 270:17535.)

acid fuchsin Histological stain used to detect connective tissue and secretion granules (Mallory's acid fuchsin, orange G, and aniline blue, and in the Van Gieson's solution of trinitrophenol staining of connective tissue of mammals). *See* stains.

acidic dye Stains basic cellular residues.

acidic sugar *See* sialic acids.

acid maltase deficiency Type II glycogen storage disease involving defect(s) in α-1,4-glucosidase activity. The disease causes accumulation of glycogen in most tissues, including the heart. The first symptoms appear by 2 months after birth and by 5–6 months death results due to cardiorespiratory (heart and lung) failures. Although it is classified as an autosomal-recessive trait in humans (GAA, 17q25.2-q25.3), the heterozygotes may be distinguished clinically. *See* Gaucher disease; glucosidase; glycogen storage diseases.

acidosis Reduction of buffering capacity of the body resulting in lower pH of fluids.

acid phosphatase Cleaves phosphate linkages at low pH. Its levels are increased in most lysosomal storage diseases, particularly in Gaucher's diseases involving glucosyl ceramide lipidosis (defect in lipid metabolism involving cerebrosides, a complex of basic amino alcohols [sphingosine], fatty acids, and glucose). Other diseases may also cause an increase of acid phosphatase. In plants only acid phosphatases are found in appreciable quantities. Yeast has at least four genes with acid phosphatase function; one of them is constitutive and others are repressed by inorganic phosphate. ACP1 is in human chromosome 2p25, ACP2 in 11p12-p11. *See* alkaline phosphatase.

acid sensing Mediated by proton-gated ion channels in the sensory neurons. *See* ion channels.

acinar cells Exocrine cells such as the mammary gland cells that secrete milk, lacrimal cells that secrete tears, etc. Acinar cells resemble sacs. *See* exocrine.

ACINUS Apoptotic chromatin condensation inducer in the nucleus is apparently the substrate of caspase-3, and this cleavage activates pycnosis in the cell nucleus. *See* apoptosis; CAD; caspase; karyorrhexis; pycnosis. (Seewaldt, V. L., et al. 2001. *J. Cell Biol.* 155:471.)

AcMNPV (*Autographa californica* nuclear polyhedrosis virus) Can be used for the construction of insect and mammalian transformation vector. *See* baculovirus; polyhedrosis virus.

acne Inflammation of the sebaceous glands (that secrete oily stuff on the skin). It does not appear to be under strict genetic control but rather various environmental conditions cause it, including bacterial infections, mechanical irritation, cosmetics, etc. It usually appears in puberty and disappears afterward but may leave behind permanent scars. Occasionally it occurs in infants. *See* skin diseases.

aconitase Enzyme controlling the dehydration of citrate to cis-aconitate and the hydration of the latter to isocitrate. This enzyme also has an important role in the transport of iron. Iron-containing proteins regulate many processes in both prokaryotes and eukaryotes. In eukaryotic cells the level of the storage protein ferritin increases when the soluble iron level increases in the cytosol. The control of the process is mediated by a 30-nucleotide *iron-response element* to which aconitase binds and then blocks the downstream translation of RNA. Aconitase is an iron-binding protein, and the increasing level of iron within the cell dissociates it from the ferritin mRNA, resulting in about a two orders of magnitude increase of ferritin by releasing the translation suppressor from the ferritin mRNA. The increased level of iron also decreases the stability of several mRNAs encoding the receptor that binds the iron-transporting transferritin and thereby reduces the amount of the receptor. Aconitase also binds to the 3′ untranslated tract of the transferrin receptor mRNA and enhances the production of the receptor, probably by stabilizing the mRNA. The human ACO1 gene is in chromosome 9p22-p14 and the mitochondrially located ACO2 is encoded in 22q11-q13. *See* ferritin; rabbit reticulocyte *in vitro* translation; translation repressor proteins.

acquired Alteration occurred during the lifetime of an individual. *See* constitutional.

acquired characters, inheritance of Ancient idea supposing that the minor and major environmental effects may cause long-lasting heritable changes in the genetic material. This view was proven incorrect by the advances of biology in the 19th century. It had been revived, however, in the Soviet Union by the poorly trained ideologues of Marxism, the followers of Mitchurin and Lysenko. It seems to be resurrected periodically by modern biologists who claim the existence of environmentally inducible selective mutations. Most of these recent experiments also remain controversial because alternative explanations of the experimental data seem to be as good or even more satisfactory (*see* directed and local mutagenesis). Advantageous frameshift backmutations may take place, however, under selective conditions by recombination. The inheritance of acquired characters has also been attributed to a mechanism of canalization. The change in the environment permits selection of hidden variations in chaperones adapted to the environmental change. Even after the release of the stress, the selected new forms of the chaperones may persist and simulate inheritance of acquired characters. *See* backmutation; canalization; chaperones; evolution; frameshift; Lamarckism; Lysenkoism; Mitchurin; recombination; Soviet genetics; transformation. (Zirkle, C. 1946. *The Early History of the Idea of the Inheritance of Acquired Characters and Pangenesis*. Amer. Philos. Soc. Philadelphia, PA; Lindegren, C. C. 1966. *The Cold War in Biology*. Planarian Press, Ann Arbor, MI.)

acquired immunity (adaptive immunity) Consequence of natural infection or vaccination or direct transfer of antibodies or lymphocytes from an appropriate donor. The acquired immunity is based on potential variations in the immunoglobulins in response to invading antigens. This immunity system consists of CD4$^+$ and CD8$^+$ T cells. T cells recognize antigens after they have been processed by the antigen-presenting cells (dendritic cells, macrophages, and B cells), which express MHC class II molecules. After the recognition T-helper cells (T$_H$-1 and T$_H$-2), differentiation begins. T$_H$-1 cells characteristically produce gamma interferon (INF-γ), which attacks intracellular invader microbes. For T$_H$-2 cells, interleukin-4 (IL-4) is diagnostic. T$_H$-2 cells require MHC class I molecules while T$_H$-1 cells depend on MHC class II. Both helper T cells also utilize a variety of cytokines for the development of effector function, i.e., to be fully activated. *See* immune system; immunity; innate immunity. (Crowe, J. E., Jr., et al. 2001. *J. Immunol.* 167:3910.)

acquired immunodeficiency syndrome (AIDS) Apparently caused by the HIV-1 (HTLV-III) and HIV-2 (human immunodeficiency virus [lentivirus]) retroviruses. The general structure of the HIV-1 virus includes three major structural

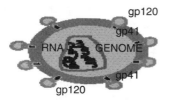

The schematic structure of the HIV-1 virus. The gp12-gp41 heterodimer associates in a trimer to form the spikes. This and the envelope determine antigenicity and immunogenicity (gp indicates envelope glycoproteins).

proteins: gag, pol, env; and several regulatory and accessory proteins: vif, vpr, vpu, vpt, tev/tnt.

The gag proteins serve as structural elements: 132 amino acid matrix (MA), 152–231 amino acid capsid (CA), 55 amino acid nucleocapsid (NC), and 51 amino acid p6 (vpr-binding protein). The pol is processed into the dimer of two 99 amino acid proteases (PR), reverse transcriptase (RT) is a heterodimer of 560 and 440 residues, and the tetramericand is the tetrameric integrase (IN) of 288-residue monomers. The reverse transcriptase generates the enzyme, which transcribes RNA into DNA, and this viral copy can be inserted into the human chromosome and survives there for a long time as a provirus. The protease processes the polyproteins into the various enzymes of the virus, and integrase facilitates the entry of the virion into the host cells. The env envelope protein includes a surface glycoprotein, gp120 (SU), and a transmembrane glycoprotein, gp41 (TM), which are processed from the gp160 molecule. The envelope protein vpr (14 kDa) accelerates replication and infection. Vpr facilitates the transport of the viral core into the nucleus, stimulates the expression of viral genes, and mediates cell cycle arrest at the G_2 stage (de Noronha, C. M. C., et al. 2001. *Science* 294:1105).

Rev (19 kDa) is transcribed from two exons, regulates viral replication, and its basic amino acid domain (nuclear export signal, NES) interacts with the Rev response element (RRE, within *env*) targeting the viral transcripts to the cell nucleus. Within the nucleus the exportin-1/CRM1 protein represents a receptor for NES. Tat (14 kDa, two exons) is the primary regulator of the virus. Vpu (15–20-kDa) membrane protein attacks CD4 with the assistance of the proteasome degradation pathway.

Nef (25–27 kDa) mediates the degradation of CD4 on the cell surface and promotes endocytosis through the clathrin-coated pits. Nef and Tat proteins may be produced before the integration of HIV into the chromosome. These two proteins activate quiescent T cells, a requisite for viral integration and replication. Active genes are preferential targets for integration (Schrödet, A. R. W., et al. 2002. *cell* 110:521). Nef also inhibits the cellular protein ASK1, an apoptosis signaling serine/threonine kinase. This protects the infected cells from apoptosis, although neighboring cells may die through bystander effect. Successful entry and productive infection also require the cooperation of the cellular protein cyclophilin A. In case cyclophilin is inhibited, HIV cannot infect neighboring cells even if HIV is within the originally infected cell. Similarly blocking the activity of MAPK, virulence of HIV is reduced.

The viral Vif protein (23 kDa) is also required for the assembly of the viral coat proteins after infection. Non-permissive host cells produce the CEM15 protein, which prevents viral infectivity of Vifdeficient HIV. CEM15 is absent from permissive cells and this permits infection by Vifdeficient virus (Pomerantz, J. R. 2002. *Nature* 418:594). For entry into the cell nucleus, the virion needs the nuclear localization signal (NLS) provided by the uncoated viral nucleoprotein preintegration complex (PIC). Viral protein Vpr interacts with PIC and thus assists nuclear localization of HIV. The virus is not transmitted through the germline. The *tat* gene functions only through the 5′ RNA hairpin TAR (transactivation response element, 59 nucleotides) present within the repeat region (R) of the 5′ LTR. The 5′ LTR also includes the basal core promoter, the core enhancer, and a modulatory region. The eukaryotic eIF2 elongation initiation element recognizes TAR. The 5′ LTR serves as the binding sites for a large number of host transcription factors (Pereira, L. A., et al. 2000. *Nucleic Acid Res.* 28:663).

The Tat (14 kDa) primarily regulates the elongation of the transcript, generated by host RNA polymerase II. Pol II starts working at the 5′ LTR. The Tak-associated kinase (TAK) complex phosphorylates the COOH end (CTD) of transcriptase pol II. The phosphorylation is the job of Cdk9 (formerly called (PITALRE). Cdk9 is bound to Tat by cyclin T (CycT) and enhances the specificity of Cdk9 to 5′ TAR. The TATA box is situated −24 to −28 positions from the GGT initiator codon. Further upstream in the enhancer region are the binding sites for the USF (upstream enhancer), Ets-1 (thymocyte-enriched protein), LEF (lymphocyte-specific high-mobility group protein), NK-κB (nuclear factor κ binding protein), and Sp1 (a mammalian transcription factor)-binding proteins within the region −166 to −45 in the 5′ LTR. Around the transcription initiation site are the overlapping SSR (initiator) and IST (initiator of short transcripts) sequences. The virus does not have a known genetic repair system and displays great antigenic variability; therefore it is difficult to develop an effective vaccine against it (Rossio, J. L., et al. 1998. *J. Virol.* 72:7992, Gaschen, B., et al. 2002. *Science* 296:2354).

The Nef protein protects the infected primary cells from cytotoxic T cells. The viral-coding RNA genome is about 9 kb. HIV1 and lentiviruses are suitable for the construction of transformation vectors that may integrate into nondividing cells. Two of the viral proteins interact with nuclear import and mediate the active transport of the HIV preintegration complex into the nucleus through the nuclear pores. The infection begins when the virus penetrates the cell membrane, its own lipid membrane fuses with the cell membrane, and the viral core is released into the cell. Inside the cell, the viral reverse transcriptase synthesizes DNA copies of its RNA genome, and this DNA provirus integrates into the host genome with the

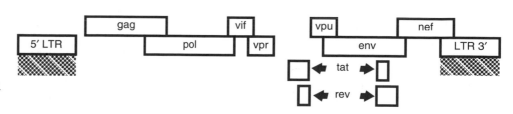

The genetic organization of the HIV1 retrovirus.

aid of its terminal repeats, also characteristic for all types of insertion elements, yet HIV is not transmitted through the germline. The HIV contains genes for proteins and their regulation. HIV does not have a lytic phase, so it does not kill the cells directly. Instead, it assembles its particles in the cytoplasm and then infects other cells.

Upon HIV infection, monocytes, macrophages, endothelial cells, and fibroblasts overproduce IL-1, IL-6, and TNFα. The antiinflammatory IL-1ra and IL-10 are also hyperproduced. The latter ones inhibit the synthesis of the inflammatory lymphokines and IL-12. Soluble tumor necrosis factor receptors (sTNFR) hinder the binding of TNF to the cell membrane receptors. *Staphylococcus*-stimulated monocytes produce an order of magnitude less TNF than IL-12.

After HIV infection, CD4$^+$ T cells lower the output of IL-2. Since IL-2 stimulates several players of the immune system, the immune response decreases. The dysregulation of cytokine balance results in a deficiency of cell-mediated immunity. The delayed-type hypersensitivity reaction (DTH) cannot then control the intracellular microorganisms. The main cause of the immunological failures is the defect in the CD4$^+$ T cells, in the antigen-presenting cells, and the destruction of the CD4$^+$ T cells, although several billion CD4$^+$ T cells are produced every day after the infection.

The primary targets are the helper T lymphocytes carrying the CD4 receptors. Impairing these cells, the immune system is debilitated, and that is the primary cause of the disease. In the endoplasmic reticulum of the infected cell, 845–870 amino acid protein precursors of the viral envelope are formed. After the addition of asparagine-linked mannose chains, the glycoprotein gp160 precursor is synthesized. The trimeric gp160 is carried to the Golgi apparatus, where through proteolysis the gp120 envelope protein and gp41 transmembrane proteins are formed.

Targeting the gp41 carboxy-terminus by a small protein, called 5-Helix, the entry of HIV-1 into the cell is inhibited. Binding of CD4 on the lymphocytes, monocytes, dendritic cells, and brain microglia by the gp120 viral surface protein results in a conformational change in gp120. These changes may make binding sites available for chemokine receptors (primarily CCR5 and ligand CXCR4) to secure the necessary second receptors for the viral entry into the cell. Polymorphism of these receptors and the stromal-derived factor (SDF-1) may either accelerate or retard the progression of the disease. The feline immunodeficiency virus directly uses the chemokine receptors, and the V3 loop of the variable region is the most important for binding of the chemokine receptors. The constant regions in between the V regions are folded into the core of the glycoprotein. A so-called bridging sheet that binds to CD4 connects the outer and inner domains of the core. Mutations in the core area may influence infectivity and may serve as targets for medical attack on the virus. The CD4-induced antibodies (CD4i and its 17b epitope) may block the binding of the gp120-CD4 complexes to the chemokine receptors.

For neutralizing HIV, probably the CD4BS epitopes, directed to the gp120 inner core, are most significant. The 2G12 antibody recognizes the outer domain of gp120. The antigenic surface of gp120 is largely shielded from humoral immune responses by the glycosylation and other barriers. Conformational changes in gp120 provide additional structural means for the evasion of the immune reactions. Generally the first sign of the disease is the susceptibility to *Pneumocystis*

carinii, an opportunistic fungal pathogen causing influenza-like symptoms. This happens because the AIDS patients have only 200 CD4 helper cells per mL of blood versus 800 in normals. The other, most critical, diagnostic feature of AIDS is the development of Kaposi's sarcoma, a disease causing bluish eruptions all over the body that become cancerous. In tissue culture the infected and uninfected cells fuse into syncytia, and this and immunological methods are being used as laboratory diagnostic procedures for the infections.

AIDS is one of the most dreaded diseases of our time. It is battled with the most advanced techniques of molecular genetics, yet no effective cure has been devised. Avoidance of infection through body fluids (blood, semen, saliva, etc.) is the only effective defense until immunization or a cure can be developed. Although the majority of the specialists in medical virology and molecular biology maintain the view that the causative agent of the disease is HIV, some reject this assumption and others take a look at the mechanism with some reservations. They find it likely or conceivable that AIDS is the result of the synergistic action of viral and other requisites such as the use of drugs (antibiotics, etc.), some types of autoimmune predisposition, etc., and thus has multifactorial origin. The AIDS disease has infected >21 million people worldwide, and their numbers are increasing daily by 8,500. There are now three main groups of HIV-1: (1) M type, which is the most widely spread; (2) the O group in Cameroon, Gabon, and equatorial Guinea; and (3) an N type found in 1998 in Cameroon, Africa. There are also 10 known subtypes of the virus. The three main types appear to have originated independently from the chimpanzee virus SIVcpz. The HIV-2 strains seem to have originated in West Africa from the simian virus strain SIVsm. Suggestions that the AIDS epidemics originated by SIV contamination of the early polio vaccines do not seem to have scientific basis.

The pharmaceutical industry is developing various drugs to combat the disease. None of the drugs so far provide a full cure or prevention, yet definite progress is being made. The first and best-known chemicals (AZT) attack the viral replication system; protease inhibitors are aimed at the assembly process of the viral coat protein in order to prevent multiplication of HIV. The virus depends on cutting and processing of host cellular protein and uses protease to this end. Unfortunately the protease inhibitors may cause very unpleasant side effects, and HIV may develop resistance against the drugs primarily by inhibiting the mitochondrial DNA polymerase γ and possibly other DNA polymerases (Feng, J. Y., et al. 2001. *J. Biol. Chem.* 276:23832). None of the current drugs actually kill and/or remove the virus; they only limit its functions. Halting the treatment, the virus may re-emerge from its hiding place in the lymph nodes.

Other potential drug targets may be the cells' entry sites and CXCRE4 and CCRS receptors. According to some estimates, HIV may produce 10 billion virions daily. Since its genome contains 10^4 nucleotides, the virus can readily test all possible mutational combinations. The estimated number of mutations/replication is 3×10^{-5}. Drug resistance is based primarily on new mutation rather than on transmission of resistant virus. In addition, recombination facilitates the production of new variants. The combination of two nucleoside analogs and protease inhibitor may reduce the level of the viral RNA copies from 20,000 and 1 million copies/mL plasma below detectability (i.e., below 200–400 copies/mL). These figures are concerned with viral levels in the blood. The newer techniques may detect even 20 virions/cell and reveal that usually the best medication available by 2002 cannot completely deplete the virus from the body. The lymphnodes and other sanctuaries may regenerate the virus after the discontinuation of therapy.

Additional problems may arise from the irreversibility of the tissue (thymus) damage. HIV-1 replication requires the REV oncogene cofactor and the eukaryotic peptide elongation factor EIF-5A. Some mutations in the elongation factor retained the ability to bind to the HIV-1 REV response element:REV complex and were expressed in human cells. When such T lymphocytes were infected with replication-competent virus, replication was inhibited, however. RNA decoys of the Tat and Rev genes may mimic the viral TAR and RRE RNAs but are nonfunctional, yet they sequester HIV-1 regulatory functions needed for the viral replication and gene expression. RNA polymerase III synthesizes these decoys and the transcript is a tRNA-TAR chimera. The decoys may, however, tie up some cellular molecules that could interfere with TAR and RRE.

Antisense/ribozyme RNAs against various (TAR, U5, *tat, rev, pol, vpu, gag, env*) transcripts have also been explored. The latent provirus, integrated into the cells, may possibly be eliminated from the body by induced apoptosis of the cells harboring it. CD4 protein, conjugated with ricin or *Pseudomonas* exotoxin, may home on the gp120 viral surface proteins and may destroy the infected cells. For the incapacitation of the HIV virus, self-inactivating E-vectors to remove the encapsidation signal (Ψ, psi) from the 5' LTR have been designed. Vectors containing the Cre/LoxP system are also capable of deleting the packaging signal (Ψ) and replacing it with a desired sequence.

Other current research attempts are focusing on the immune system to prevent infection. Although several interpretations are available, it is still uncertain why the period required for the development of full-scale AIDS varies after the initial infection. It had been assumed that the immune system is weakened by the ever-increasing viral diversity. Others believe that an immune dysregulation is responsible for the outbreak. Others suggest that the cellular immune system against AIDS should be directed to both conserved and variable epitopes. It is assumed that the cytotoxic T lymphocytes alone cannot completely eliminate the virus and there is a need to achieve a balance between the viral load and the CD4$^+$ T lymphocytes. After a period of time, the increasing variations in the HIV-1 population deplete and foul up the immune system. In the so-called nonprogressor individuals, AIDS may not develop for more than 10 or even 20 years after the infection (HIV-exposed persistently seronegatives [HEPS]). In the cells of such a person, there are high levels of CD8$^+$ CD38$^-$ cytotoxic lymphocytes, high peripheral blood CD8$^+$ major histocompatibility class I–restricted anti-HIV cytotoxic lymphocytes and those stay at an even level. Also, there is a strong CD8$^+$ non-MHC-restricted HIV suppressor activity and a high level of antibody against HIV. CD8$^+$ cytotoxic T lymphocytes should be stimulated by any vaccine developed against HIV.

Studies indicate that in the nonprogressors the gene coding for the chemokine receptors (of the nonsyncytium-inducing viral isolates [NSI]) CCR2 and CCR5 is mutated (contains deletion[s]), and thus it appears that CCR5 and 2 assist infection by HIV-1. These receptors are ligands for a group of CC chemokines (CC, CXC), MIPα and β, and RANTES produced by CD8$^+$ T lymphocytes. These and siRNA

are able to suppress HIV replication in vitro probably by competition for the CCR5 receptors (Quin, X.-F. et al. 2003 *PNSA* 100:183). Mutation in HIV may facilitate the use of other members of the chemokine receptor family, including CCR3 and CCR2. Mutations in CX$_3$CR1 reduce the binding to the chemokine fractalkine and enhance the progression toward the development of AIDS. In late-disease-stage cases, the chemokine CXCR4/SDF1 may be used for entry into the cell. The CCR5 receptors may be polymorphic. Homozygosity and heterozygosity for the mutant alleles of CCR5 of the cells also appear to convey reduced susceptibility to infection. The viral protein gp120 reduces the response to chemokines. The frequency of the mutant alleles in Caucasian populations is about 0.092 and thus the predictable frequency of homozygosis ($0.092^2 \approx 0.0085$) is about 1%. This type of mutation seems to be much less common in African and Japanese populations. It is also likely that some nonprogressors were infected with less aggressive HIV variants.

The residual genetic constitution of the infected individual may affect the course of the disease. Recent studies revealed that homozygosity for a mutation in the chemokine receptors CCR5 (synonym CKR5) and CCR2 protects against HIV infection, and heterozygosity may also be of some advantage. This is the receptor through which HIV infection takes place. A homozygous mutant form of the chemokine SDF-1 gene, which codes for the principal ligand of a coreceptor of CXCR4 of the CD4 T cells of the HIV-1 virus, substantially restricts AIDS pathogenesis. It seems to offer better protection than the CCR5 and CCR2 chemokine receptor variants. Heterozygosity for the HLA class I loci A, B, and C conveys longer survival after infection by HIV-1. But the presence of alleles B*35 and Cw*04 potentiates rapid progression of the disease.

Screening of the blood donations for possible HIV infection is based on the determination of the proportion of CD4/CD8 molecules. The normal ratio is about 2, and in infected blood it is below 1. It has been claimed that in infants the HIV-1 infection may be transient, but further analysis of these cases did not confirm the claims. None of the HIV vaccines tested so far provided protection. Attenuated live HIV with *nef* gene deletions first appeared successful, but because of the high viral mutation rate, infective virus is recovered by time. In macaques the *nef*-defective SIV vaccine was protective. Recombinant envelope protein subunit vaccines also failed to elicit envelope-specific CTL or antibody-specific immune responses that could effectively neutralize HIV-1 in humans.

Attempts to provide immunological protection against the V3 hypervariable loop of the viral envelope protein (essential for the viral gp120–CD4–chemokine interaction) are still being explored. Recombinant vaccinia virus carrying HIV protein fragments raised some hopes because similar constructs were effective in monkeys against SIV. Another possibility is to use engineered avian pox viral vectors, which have shown some promise (displaying some CTL activity), yet the immunogenicity generated may be too low. Unfortunately, in immunosuppressed humans, serious side effects were encountered and the vaccines became impractical. BCG and other bacteria have been considered as potential vaccine vectors. DNA vaccines provide CTL activation and immune response, but so far the levels are very low to be effective. In rhesus monkeys infected with SIV lacking N-linked glycosylation at the 4th, 5th, and 6th sites of the envelope protein reduced the immune evasion of the virus. Normally cytotoxic T lymphocytes (CTL) recognize the invading HIV by their surface Tat peptides. Unfortunately, through mutation this Tat peptide mutates very rapidly and becomes unrecognizable by CTLs. Immunization before infection by an appropriate Tat vaccine may provide a headway for CTL to gain control over the virus. Antisense RNA, complementary to the viral genome or to messenger RNA, may also curtail viral functions by blocking transcription, translation, or activation of RNase H.

Gene therapy using a suicide gene under the control of the HIV promoter may be activated by TAT, and all the infected lymphocytes may be eliminated before the virus replication gets out of control (Caruso, M., et al. 1995. *Virology* 206:495). RNA decoys that curtail replication of the virus have also been targeted at TAT. Transdominant Rev has been used to limit productive infection (Escaich, S., et al. 1995. *Hum. Gene Ther.* 6:625). Intrabodies were explored as a protective measure (Marasco, W. A., et al. 1999. *J. Immunol. Methods* 231:223). siRNA may also inhibit HIV-1 infection (Novina, C. D., et al. 2002. *Nature Med.* 8:681).

So far the most effective protection from AIDS is behavioral, i.e., the avoidance of exposure to the virus. Infection by the virus is the easiest through blood cells, plasma, or cerebrospinal fluids. Semen transmits 10 to 50 times more viruses than vaginal/cervical fluids. In the United States, the major route of infection is male homosexual contacts, whereas in Africa heterosexual copulation is the predominant means of spreading the disease. *See* antisense technology; apoptosis; AZT; *Bacillus Calmette-Guerin*; biolistic transformation; CCR; CD4; CD8; chemokine; clathrin; *Cre/LoxP*; CTL; CXR; cyclophilin; DC-SIGN; decoy RNA; DNA flap; endocytosis; enhancer; E vector; exotoxin; fusin; gene therapy; herpes; HIV-1; HLA; HMG; hominidae; immune system; immunization, genetic; Kaposi's sarcoma; kissing loop; liposome; MAPK; MIP-1α; Nevirapine; NF-κB; numt, *Pneumocystis carinii;* primates; protease sequences; proteasome; RANTES; retroviruses; reverse transcriptase; ribozyme; ricin; RNAi; SDF-1; seronegative; SIV; Sp1; TBP; T cells; telomere; therapeutic vaccine; thymus; TIBO; vaccinia virus. (*Science* 288:2129ff; Amara, R. R., et al. 2001. *Science* 292:69; *Nature* 410:963ff; Poignard, P., et al. 2001. *Annu. Rev. Immunol.* 19:253; Englert, Y., et al. 2001. *Hum. Reprod.* 16:1309; Wu, Y. & Marsh, J. W. 2001. *Science* 293:1503; <http://hivdb.stanford.edu>.)

acridine dyes   For example, proflavin, acriflavine, acridine orange. They are potential frameshift mutagens by intercalating between the nucleotides of DNA. Some acridines act by photosensitization of the DNA. Acridine dye has been used to cure bacteria from plasmids (by selective removal) and to induce respiration-deficient mitochondrial mutations in yeast. *See* curing, plasmid; fluorochromes; frameshift mutation; mtDNA.

Acriflavin.

acriflavine   *See* acridine dyes; see formula above.

acrocentric chromosome   Has a near terminal centromere and one arm is very short. Acrocentric chromosomes may fuse or become translocated and may generate biarmed chromosomes. *See* Robertsonian translocations.

acrocephalosyndactyly   *See* Apert syndrome; Pfeiffer syndrome.

acrodermatitis enteropathica (8q24.3)   Recessive blistering of the skin usually accompanied by lack of hairs on the head, eyebrows, and eyelashes, and partial pancreatic hyperplasia and thymus hypoplasia. The deficiency in zinc binding is characteristic and causes low levels of zinc and alkaline phosphatase (a zinc metalloenzyme) in the plasma. Treatment with zinc is very successful. *See* alkaline phosphatase; hemochromatosis; hyperzincemia; Menke disease; skin disease; Wilson disease; zinc fingers. (Wang, K., et al. 2002. *Am. J. Hum. Genet.* 71:66.)

acrodysostosis   Autosomal-dominant defect of bone development of paternal origin and increased occurrence by the age of the father.

acromegaly   Increased growth due to overproduction of the pituitary hormone.

acropetal   The youngest leaf on the stem is at the tip of the stem of the plant.

acrosomal process   Spike-like actin structure on the head of the sperms of several animals and at their base. The acrosome is a sac of hydrolytic enzymes destined to facilitate sperm penetration through the gelatinous coat of the egg. Before acquiring competence for fertilization, the spermatozoa must be activated by bicarbonate-mediated soluble cAMP. The process is enhanced by progesterone, probably by acting on a $GABA_A$ receptor. In the starfish, egg jelly *ARIS* (polysaccharide with repeating units of sulfated pentasaccharide), *Co-Aris* (steroid saponin), and *asterosap* (a variety of 34 amino acid peptides) are required for the acrosomal process. In sea urchins, FSP (sulfated fucose polymer) activates the acrosomal process. In mammalian egg, the three ZP (zona pellucida) proteins bind to the receptors on the sperm plasma membrane and stimulate the exocytosis of the acrosomal vesicle in the front part of the sperm. Activated sperm contains nitric oxide synthase. Nitric oxide is important for fertilization. Phospholipase Cδ4 is also required for the process. *See* fertilization; GABA; progesterone; sperm. (Tulsiani, D. R. & Abou-Haila, A. 2001. *Zygote* 9:51; Kang-Decker, N., et al. 2001. *Science* 294:1531.)

acrosome   See acrosomal process

acrosome reaction   *See* acrosomal process.

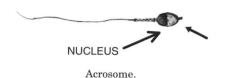

NUCLEUS

Acrosome.

acrosyndesis Spurious end-to-end pairing of the chromosomes during meiosis.

acrylaldehyde Toxic compound made from allylalcohol by the enzyme alcohol dehydrogenase. Cells defective in this enzyme permit the selective survival on allylalcohol as it is not converted to acrylaldehyde. *See* alcoholdehydrogenase; mutant isolation.

acrylamide In the presence of ammonium persulfate and TEMED (N,N,N′,N′-tetramethylenediamine) it is polymerized into chains with various lengths depending on the concentration used. In the presence of N,N′-methylenebisacrylamide it becomes cross-linked, and pores are formed, depending on the length of the chains and the degree of cross-linking. It can be used to separate nucleotides by electrophoresis from 2000 to 6 bp, depending on the pore size of the gels. Acrylamide is a potent neurotoxin and potential carcinogen. It can be absorbed through the skin. Although the polymerized form is considered nontoxic, it should be handled only with gloves because of the trace amounts of monomers. Acrylamide may be formed in small quantities in deep-fried starchy food. *See* electrophoresis; gel electrophoresis.

ACT Activator of CREM (in the testis) by binding through a LIM domain. *See* CREM; LIM.

ACTH Adrenocorticotropin controls adrenocortical growth and steroidogenesis. The hypothalamus controls the ACTH-releasing factor and in response the anterior pituitary releases this hormone. ACTH is encoded in human chromosome 2. *See* adrenocorticotropin; animal hormone; brain; cAMP; cortisol; glucocorticoids; hormone-response elements; melanocortin; pituitary gland; POMC; steroid hormones.

actin Protein of the cytoskeleton and the thin muscle fibers. Actin gene number varies in different organisms; yeast has only one, *Dictyostelium*, eight; *Drosophila*, six; *Caenorhabditis*, four; humans, about two dozen at dispersed locations. The cytoplasmic actins involved in cellular motility are similar in all eukaryotes. α-Actins are located in the smooth, skeletal, and cardiac muscles. The smooth muscles also have γ-actin. In the cytoplasm of mammals and birds, there are β- and γ-actins. The amino acid sequence and composition of the actins is rather well conserved and differences exist mainly at the amino terminals. Actin genes have different numbers of introns and pseudogenes, permitting evolutionary inferences partly because the flanking sequences are much more variable than the genes. Some proteins bind actin in monomeric or filamentous form, such as myosin (a contractile protein in muscles), α-actinin (involved in cross-linking actins), profilin (mediates the formation of actinin bundles), fimbrin (cross-linking parallel actin filaments), filamin (promotes the gel formation by actins), tropomyosin (strengthens actin filaments), spectrin (attaches filaments to plasma membranes), gelsolin (fragments

filaments), etc. *See* cardiomyopathy, dilated; CDC42: cofilin; cytoskeleton; filaments; glomerulosclerosis; microfilaments; myofibril; myosins; nemaline myopathy; pollen; Wiskott-Aldrich syndrome. (Geeves, M. A. & Holmes, K. C. 1999. *Annu. Rev. Biochem.* 68:687; Higgs, H. N. & Pollard, T. D. 2001. *Annu. Rev. Biochem.* 70:649.)

actin contractile ring Formed prior to the separation of the dividing chromosomes, and it contracts after anaphase. It may be involved in the formation of the septum between the two cells. *See* cell cycle.

actinin, α (M_r 120 K) Antiparallel peptides in muscle Z lines, focal adhesion, and intermediate junction structures. *See* actin; CAM, junction complex; glomerulosclerosis.

actinomorphic Structure (flower) in multiple symmetry patterns.

Actinomycetes Filamentous prokaryotes rather than fungi as they once were assumed to be. *See* actinomycin D; *Streptomyces*; streptomycin.

actinomycin D Antibiotic from *Streptomyces*. It is an inhibitor of transcription because it intercalates into the DNA between neighboring GC base pairs and hinders the movement of the transcriptase on the DNA template without interfering with replication. It is used in reverse transcriptase reactions to prevent self-primed second-strand synthesis. Actinomycin D is a teratogen and carcinogen. (There are several other actinomycin antibiotics.) *See Actinomycetes*; reverse transcription; *Streptomyces*; transcriptase. (Graves, D. E. 2001. *Methods Enzymol.* 340:377.)

action potential Rapid, transient, self-propagating electrical excitation in muscle or neuron membranes. It may mediate long-distance nerve signaling.

action spectrum Representation of a degree of response to certain types of treatment, e.g., photosynthesis in relation to wavelength of irradiation. *See* photomorphogenesis.

activating enzyme *See* aminoacyl-tRNA.

activation analysis Nuclear technique used for the very sensitive detection of radionuclides for various purposes including forensic analysis. *See* radionuclide.

activation energy Energy required for converting 1 gram molecular weight of a compound from the ground state to the transition state. It is required, from outside, by molecules and atoms to undergo chemical reaction(s). *See* transition state.

activation of genes *See* activator proteins; suppression.

activation of mutagens   Many mutagens (and carcinogens) require chemical alterations to become biologically active. The mutagenic and carcinogenic properties of many agents overlap and thus are active in mutagenesis and carcinogenesis. The activation generally requires enzymatic reactions. The most important enzymes are the mixed-function oxidases contained by the cytochrome P-450 cellular fraction. These reactions require NADPH and molecular oxygen and the general process is: **RH** (reduced reactant) $+$ NADPH $+$ $H^+ + O_2 \rightarrow$ **ROH** $+$ NADP$^+$ $+$ H_2O. These enzymes occur in multiple forms and can utilize a variety of substrates, hydrocarbons, amines and amids, hydrazines and triazines, nitroso compounds, etc. They occur in different tissues of animals, primarily in the endoplasmic reticulum of cells what is generally called microsomal fraction after isolation followed by grinding and centrifugal separation of the cellular fractions. These enzymes are subject to induction by phenobarbitals, methylcholanthrene, and a variety of substrates. Other related activating enzymes are flavoprotein N-oxygenases, hydrolases, and reductases. Other enzymes of activating ability include various transferases that add glucuronyl, sulfuryl, glutathion, acetyl, and other groups and either detoxify the compounds or further enhance their reactivity. The cellular and membrane transport, protein binding, and excretion affect these reactions. Genetic differences exist among the species and individuals. Differences by age, circadian rhythm, nutritional status, etc., are known. If the clearance of these compounds from the body is slow, the risk for the individuals increases. *See* Ames test; bioassays in genetic toxicology; cytochromes; environmental mutagens; mutagen assays; P450; promutagen; proximal mutagen; ultimate mutagen. (Baum, M., et al. 2001. *Chem. Res. Toxicol.* 14:686.)

activation tagging   Random insertions of transcriptional enhancers of the 35S cauliflower mosaic virus promoter with the aid of the *Agrobacterium* vector into the plant genome, resulting in misexpression and overexpression of many different genes. *See* cauliflower mosaic virus, T-DNA, Ti plasmid. (Weigel, D., et al. 2000. *Plant Physiol.* 122:1003.)

activator   *Ac*, the autonomous element of the *Ac-Ds* controlling element system of maize (*see Ac*). Also, any DNA-binding protein that enhances transcription, a positive modulator of an allosteric enzyme. More than one activator of transcription may operate at different activator-binding sites of a single promoter. Their action may be synergistic or each may have special affinity to a separate surface of the RNA polymerase II molecule, or one stabilizes the other activator. *See Ac-Ds*; allosteric control; modulation; transcriptional activator.

activator A   *See* RF-C.

activator I   *See* RF-C.

activator proteins   Stimulate transcription of genes by binding to TATA-box-binding protein (TBP) and the recruitment of TFIID complex to the promoter. Sometimes they require coactivator metabolites for function. Their primary role is probably the remodeling of the nucleosomal structure so that DNA-binding proteins can access their target. The DNA has multiple binding sites for activators in the promoter region. The potency of the activation domains of the activators may vary. An activator may turn into a repressor by binding a corepressor. *See* chromatin; chromatin remodeling; coactivator; corepressor; enhancer; GCN5; nucleosome; promoter; recruitment; regulation, gene activity; signal transduction; suppressor gene; TBP; transcriptional activators; transcription factors; VDR. (Evans, R., et al. 2001. *Genes Dev.* 15:2945.)

active immunity   *See* immunity.

active site   Special part of an enzyme where its substrate can bind and where the catalytic function is performed, the catalytic site. *See* enzymes; catalysis; substrate.

active telomeric expression site   Variable-surface glycoprotein (VSG) genes are responsible for the diversity of antigenic variants in *Trypanosomas*. They generate different antigenic properties of the parasite. There are about a thousand genes in this gene family and their activation is interpreted by their transposition to the vicinity of the telomere, the site of the silent copies. *See* mating type, yeast; silent sites; *Trypanosomas*. (Borst, P. & Ulbert, S. 2001. *Mol. Biochem. Parasitol.* 114:17.)

active transport   Passing solutes through membranes with the assistance of an energy donor. *See* passive transport.

activin   Soluble protein that may contribute to the formation of dorsal and mesodermal tissues in the developing animal embryo; its activity may be blocked by follistatin. Activins belong to the transforming growth factor-β superfamily of proteins. They are serine/threonine protein kinases. Activins respond to Smad signal transducers. Activin receptor-like kinase 1 (ACVRLK1) modulates TGF signaling in angiogenesis. Its absence or deficiency results in lack of response to TGF-β family growth factors, and mice afflicted die by midgestation due to fusion of major arteries and veins. *See* angiogenesis; follistatin; organizer; protein kinases; SMAD; TGF. (Luisi, S., et al. 2001. *Eur. J. Endocrinol.* 145:225.)

activity coefficient   Obtained by multiplying activity by the concentration of a solute to obtain its thermodynamic activity.

activity-based protein profiling   Monitors the expression dynamics of a family of proteins on the basis of cheminal tagging with common inhibitors. *See* Liu, Y. et al. 1999. *Proc. Natl. Acad. Sci. USA* 96:14694.

actomyosin   Complex of actin and myosin. *See* actin; myosin.

acuminate   Tapered.

Acuminate.

acute transforming retrovirus Contains a v-oncogene, which is an efficient oncogenic transformation agent. *See* oncogenes; retrovirus; v-oncogene.

acyclovir *See* ganciclovir.

acylation Attaching one or more acyl groups to a molecule. *See* acyl group.

acyl-CoA dehydrogenase deficiency (ACAD) Three different diseases have been described with short (SCAD, 12q22-qter), long (LCAD, 2q34-q35), and medium chain defects (MCAD, 1p31) involving β-oxidation of fatty acids. The clinical symptoms include hypoglycemia, dicarboxylaciduria, hyperammonemia, fatty liver, etc. Some of the symptoms overlap with those of Reye syndrome and maple syrup urine disease. *See* isoleucine-valine biosynthetic pathway; isovaleric acidemia; Reye syndrome.

acylcyclohexanedione Inhibitor of gibberellin biosynthesis. *See* plant hormones.

acyl group (boxed) Where R can be a number of different chemical groups. *See*
$$\boxed{\begin{array}{c} \text{O} \\ \| \\ \text{R--C--} \end{array}}$$
acetyl group.

Ad4BP *See* SF-1.

Ad5 E1B Adenovirus oncoprotein. *See* adenovirus; oncogenes.

ADA Zn-containing protein that transfers methyl groups to its own cysteine and thus repairs aberrantly methylated DNA (e.g., 6-O-methylguanine). By binding to specific DNA sequences it activates genes involved in conveying resistance to methylation. Some of the ADA proteins are involved in repression of transcription. *See* adenosine deaminase; histone acetyltransferase.

adactyly Absence of digits of the hand or foot. It may occur as part of several syndromes. *See* ectrodactyly; Holt-Oram syndrome; polydactyly.

Adactyly.

ADAM (a disintegrin and metalloprotease) Family of metalloprotease enzymes, such as KUZ (kuzbanian), responsible for partitioning of neural and nonneural cells during the development of the central and peripheral nervous system. ADAMs may have proteolytic, cell adhesion, signaling, and fusion functions in cell surface molecules. ADAM is apparently involved with the Notch receptor function. ADAM motifs are found in the aggrecanase enzyme eroding cartilage in arthritis. ADAM 10 (α-secretase) also cleaves the amyloid precursor protein (APP). *See* Alzheimer disease; arthritis; bone morphogenetic protein; CAM; cardiomyopathy, hypertrophic, familial; cyritestin; fertilin; metalloproteinase; neurogenesis; *Notch*; secretase. (Stone, A. L., et al. 1999. *J. Protein Chem.* 18:447.)

ADAM complex Acronym for amniotic deformity, adhesions, mutilations phenotype complexed with other anomalies caused by mechanical constriction of the amniotic sac, but there is also evidence for the role of autosomal-recessive inheritance. The phenotype may include bands on fingers and loss of finger bones or even parts of legs (amputations), etc. *See* limb defects, human. (Keller, H., et al. 1978. *Am. J. Med. Genet.* 2:81.)

Adams-Oliver syndrome Autosomal-dominant mutilations of the limbs, skin, and skull lesions yet apparently normal intelligence.

adaptation The organism has fitness in a special environment. Mutation provides the genetic variations from what the evolutionary process selects the genes that convey the best adapted. Adaptation may be acquired by major mutations, although most commonly it is based on mutations with small cumulative effects without deleterious pleiotropic or epistatic consequences. R. A. Fisher expressed adaptation in "geometric" terms as is shown algebraically below where r is the distance to which a mutation moves the population in a sphere (d) from the sphere of previous adaptation of A. If r is very small, the chances are equal for bringing improvement or becoming deleterious. However, when r moves beyond the sphere of A, Fisher considered no chance for improvement. "The chance of improvement thus decreases steadily from its limiting value 1/2 when r is zero, to zero when r equals d." Fisher concluded that the probability for adaptive change is rapidly diminished when the change ($d/\sqrt{n}$) has manifold effects (n). In physiology, it defines adjustment to specific stimuli. *See* fitness; shifting balance theory of evolution. (Travisano, M. 2001. *Curr. Biol.* 11:R440.)

$$\frac{1}{2}\left(1 - \frac{r}{d}\right)$$

adapter ligation PCR *See* capture PCR; polymerase chain reaction.

adaptins (AP-1, AP-2, AP-3) Major coat proteins in a multisubunit complex on vesicles. These proteins bind the clathrin coat to the membrane and assist in trapping transmembrane receptor proteins that mediate the capture of cargo molecules and deliver them inside the vesicles. *See* AP1; AP180; arrestin; cargo receptors; clathrin; endocytosis; epsin. (Robinson, M. S. & Bonifacione, J. S. 2001. *Curr. Opin. Cell Biol.* 13:444.)

adaptive amplification Somewhat similar concept to adaptive mutation. It has been argued that the two are different because adaptive amplification, unlike mutation, is a flexible and more readily reversible alteration. The argument has been that if the days required to reform colonies equals the number of days after selection, when, e.g., the original *Lac*+ colony arose, the original revertant was preexisting. If the days required to reform colonies is less than the number of days after selection when the revertant emerged, then the alteration permitting growth on lactose is attributed to an adaptive response to the selective condition. *See* adaptive mutation. (Hastings, P. J., et al. 2000. *Cell* 103:723.)

adaptive convergence   Similarity in morphology and function among unrelated species within a particular environment, e.g., fins on fishes and mammalian whales.

adaptive enzyme   Inducible enzyme. *See Lac* operon.

adaptive evolution   The theory claims that evolution is largely based on mutations that increase fitness of the individuals and species involved. In contrast, the neutral mutation theory postulates the significance of the role of random neutral mutations based on synonymous codon substitutions. The relative abundance of synonymous (D_s) and nonsynonymous (D_n) mutations can be estimated as (a = adaptive substitution):

$$a = D_n - D_s(P_n/P_s),$$

where P_n and P_s stand for the numbers of nonsynonymous and synonymous substitutions, respectively. Hence α (the amino acid substitutions brought about by positive selection) is $\alpha = 1 - (D_s P_n)/(D_n P_s)$. By using this method, 45% of the amino acid substitutions in some *Drosophila* species appeared to be adaptive. *See* fitness; mutation, beneficial; mutation, neutral synonymous codons. (Smith, N. G. C. & Eyre-Walker, A. 2002. *Nature* 415:1022.)

adaptive immunity   Develops in response to an antigen.

adaptive landscape   Represents the frequency distribution of alleles corresponding to fitness of the genotypes, e.g., *AAbb* and *aaBB* means the fixation (*peak*) of the allelic pairs. The *pits* of fitness may mean the fixation of *AABB* or *aabb*, and the *saddle* usually corresponds to the polymorphic condition *AaBb*. A two-dimensional model of the allelic topography may represent the allelic constitutions with the corresponding fitness in a third dimension showing "mountain ranges" and "valleys." The landscape may be subject to evolutionary change due to change in allelic frequencies and the environment. The landscape may become complex if several allelic pairs are considered. The (+) and (−) signs indicate high and low fitness, respectively. The adaptive landscape may also be represented in a three-dimensional plot. The term is the same as adaptive topography. See fitness.

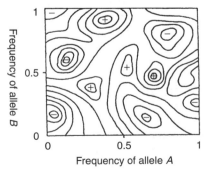

Adaptive landscape (Modified after Mohay, J. 1986. Genetika, Natura, Budapest).

adaptive mutation   Occurs at higher frequency in response to conditions of selection, although the mutation is not induced by the conditions of selection. *See* adaptive amplification;

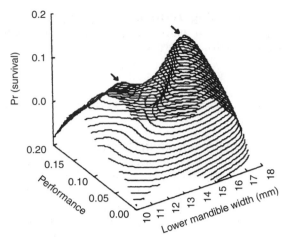

Two adaptive peaks →→ adaptive topography in 3-dimensional representation. (From Smith, T. B. & Girman, D. J. 2000 in Adaptive Genetic Variation in the Wild, p. 139. Mousseau, T. A., et al. eds © Oxford Univ. Press. New York.)

directed mutation. (Foster, P. L. 1999. *Annu. Rev. Genet.* 33:57; Hall, B. G. 2001. *Mol. Biol. Evol.* 18:1389.)

adaptive peak   Highest value(s) of fitness in an adaptive landscape. *See* adaptive landscape.

adaptive radiation   Phyletic lines spread over a variety of different ecological niches resulting in a rapid adaptation to these locales and appearing in strikingly different forms. *See* diversity; niche; phylogeny.

adaptive response   Induction of (bacterial) repair enzymes that activate glycosylases or O^6-methylguanine methyltransferase and thereby mutated DNA is repaired. The name comes from the property of adaptation to higher doses of mutagens after an initial shorter exposure; it is mediated by the *Ada* gene product (37 kDa) of *E. coli. See* alkylating agent; chemical mutagens; DNA repair; glycosylases; methylation, DNA.

adaptive value   *See* fitness.

adaptor   tRNA is called an adaptor by the older literature because it adapts the genetic information in DNA through mRNA to protein synthesis. *See* aminoacyl-tRNA synthetase; protein synthesis; tRNA. (Ibba, M., et al. 2000. *Trends Biochem. Sci.* 25:311.)

adaptor proteins   Play key roles in cellular signaling such as phosphorylation, dephosphorylation, signal transduction, organization of the cytoskeleton, cell adhesion, regulation of gene expression—all distinct yet interacting systems. Proteins equipped with the Rous sarcoma oncogen (Src) homology domains SH2 and SH3 mediate the interactions between the phosphotyrosine kinase receptors of mitogenic signals and the RAS-like G proteins. The SH2 domain selects the phospho-Tyr-Glu-Glu-Ile sequences. Phospholipase C (PLC-γ1) and protein tyrosine phosphatase (PTPase) recognize several hydrophobic residues following pTyr on the ligand-binding molecule. The SHC homology proteins and the insulin receptor substrate (IRS-1) recognize somewhat different sequences: Asn-Pro-X-pTyr. The SH3-binding sites involve about 10 proline-rich amino residues. The SH3-binding peptides can bind either in $NH_2 \rightarrow COOH$ or in the reverse orientation. The SOS (son of sevenless) adaptor protein binds to Grb2 (growth factor receptor–bound protein) and is attached in $C \rightarrow N$ orientation. The pleckstrin domains are widespread in occurrence (serine/threonine and tyrosine kinases, and their substrates, phospholipases, small GTPases, dynamin, cytoskeletal proteins, etc.). Pleckstrin domains occur in cytoplasmic and membrane signaling molecules. The LIM domains facilitate binding of signaling molecules, transcription factors, as well as the units of the cytoskeleton. These adaptor proteins may form partnerships with a variety of proteins and thus generate complex networks of signaling. (Hübener, C., et al. 2001. *Immunogenetics* 53:337; Beer, S., et al. 2001. *Biochim. Biophys. Acta* 1520:89.)

ADAR Adenosine deaminase acting on RNA, ADAR1, ADAR2, ADAR3. ADAR2 edits the pre-mRNA of the glutamate-sensitive ion channel receptor B subunit and adenine is converted to inosine, which behaves in coding as guanine. Thus, glutamine is replaced by arginine in the protein with over 99% efficiency. ADAR1 functions overlap with ADAR2 and ADAR3. ADAR3 is specific to the brain. The enzymes contain a double-strand-binding domain and a catalytic deaminase domain. Thus, alternative forms of the protein appear. At the NH_2 terminus it includes domain $Z\alpha$, responsible for the high-affinity binding to Z DNA. Point mutations in AMPA (α-amino-3-hydroxy-5-methyl-4-isoxazole propionate) receptor may compensate for lethality in ADAR2-deficient mice. *See* adenosine deaminase; DRADA; mRNA; RNA editing; Z DNA. (Wang, Q., et al. 2000. *Science* 290:1765.)

ADCC Antibody-dependent cell cytotoxicity. Mediates some of the immune responses. *See* antibody; immune system. (Hinterberger-Fischer, M. & Hinterberger, W. 2001. *Expert Opin. Biol. Ther.* 1:1029.)

addiction Complex phenotype relative to the abuse of drugs, alcohol, smoking, or habituation to other nonnatural behavior. It is generally controlled by multiple genes, deeply influenced by several social conditions, and commonly associated with antisocial behavior. It is assumed that the long-term abuse of these substances causes molecular changes in the neuronal signaling. The adaptation may modify the autonomic somatic functions, causing dependence, and when the agent is withdrawn, withdrawal anomalies result. The agent may alter the motivational control system, resulting in craving. Chronic use of morphine upregulates components of the cAMP signal transduction pathway. In mice, with a deletion of the CREBα element, the withdrawal symptoms were reduced, indicating that CREB-dependent gene transcription is a factor of opiate dependence. The major receptor for opiates (morphine, heroine) is the trimeric G protein–linked μ. Long-term opiate use decreases the μ-opiate receptor signaling without reducing the number of receptors and leading to tolerance and dependence. In nonaddicted individuals the opiate receptor opens an outward rectifying K^+ channel and reduces the phosphorylated state of an Na^+ channel. In an addicted individual the K^+ channel is shut off, however, and the G protein–adenylate cyclase activates a protein kinase (PKA), and the phosphorylated Na^+ channel moves sodium inward to the locus ceruleus, a pigmented structure at the floor of the brain. As a consequence, the cyclic AMP response element (CRE)–binding proteins (CREB) stimulate the transcription of RNA required for adaptation to the addictive drug. The psychoactive effects of cocaine can be superseded in rats by active immunization using a cocaine conjugate, GNC-KLH (a hapten and keyhole limpet hemocyanin). Addictive agents usually raise the dopamine level in the nucleus accumbens of the brain. Recently the glutamate receptors gained attention for their role in addiction and for their chemical blocking for a cure. The glutamate receptor seems to control the ΔFosB transcription factor and enhances the sensitivity to cocaine. The genetic component of various addictions (alcohol, opiates, cocaine, etc.) may exceed 50%. The genetic bases of addiction are generally polygenic and may be studied by the methods of quantitative genetics, although the specific genetic factors responsible are hard to identify and few genetic tools exist for the identification of the determination. Serotonin, nicotinic cholinergics, dopamine, cannabinoid-like receptors, and neuropeptides may affect responses to various drugs. *See* adenylate cyclase; alcoholism; behavioral genetics; bulimia; cAMP; CRE; CREB; dopamine; glutamate receptors; G protein; ion channels; keyhole limpet hemocyanin; neurotransmitters. (Nestler, E. F. & Landsman, D. 2001. *Nature* 409:834.)

addiction module Represents a prokaryotic system with resemblance to apoptosis in eukaryotes. The module includes the products of two genes: one is long-lasting and toxic; the other is short lived and protects against the toxic effect. The "addiction" is a dependence on the antagonist of the toxin. This system is usually controlled by plasmid elements. In the *hok-sok* module of the R1 plasmid, the *sok* gene product is an antisense RNA subject to degradation by a nuclease. Homologs of this plasmid system have also been found in the main bacterial chromosome. Encoded by the bacterial *rel* operon, the MazE antitoxin protein is subject to degradation by the clpPA serine protease. It protects from the toxic effects of MazF toxin protein. MazE-MazF is regulated by the level of ppGpp, which itself is toxic to the cells. MazE-MazF expression is also regulated by 3′,5′-bispyrophosphate, synthesized by the RelA protein under amino acid starvation. *See* apoptosis; partitioning. (Engelberg-Kulka, H. & Glaser, G. 1999. *Annu. Rev. Microbiol.* 53:43.)

Addison disease (adrenocortical hypofunction, Xq28) Auto-somal-dominant defect of the kidney cortical layer, resulting in excess potassium and sodium in the urine, decreased levels of cortisol, and hyperpigmentation of the skin. *See* adrenoleukodystrophy; cortisol; kidney diseases; pigmentation defects.

Addison-Schilder disease Xq28 chromosome–linked adrenocortical (kidney outer layer) atrophy and diffuse cerebral sclerosis (hardening). The cerebral lesions resemble the symptoms of multiple sclerosis. *See* multiple sclerosis.

addition line Carries an extra chromosome(s) coming from another genome. *See* alien addition; *Haynaldia villosa*.

additive effect In relation to genes, each allele contributes quantitatively to the phenotype of an individual that carries it — i.e., there is no dominance. *See* polygenic inheritance.

additive gene Each allele has a definite quantitative contribution to the phenotype without dominance within a locus and without epistasis between loci or overdominance between alleles. *See* additive variance; epistasis; heritability; overdominance; quantitative genetics.

additive variance Each allele contributes a special value (quantity) to the phenotype and there is no interallelic (overdominance) or interlocus (epistasis) effects on the variance. *See* genetic variance; heritability; QTL.

additivity of genetic maps Ideally it means that the distance between genes A–C is equal to the sum of the distance between A–B and B–C if the order of genes is A B C. Exceptions exist to this generally valid rule because of genetic interference. *See* coincidence; interference; mapping; genetic; recombination frequency.

adducin Membrane protein mediating the binding of spectrin to actin. *See* actin; spectrin.

adduct (1) To draw to the median plane or axial line. (2) The complex of two or more components such as the cyclobutane ring of pyrimidine dimers, benzo(a)pyrene-guanine, and other alkyl groups of mutagens added to nucleic acid bases. Lipid peroxidation generates various DNA adducts with mutagenic effects similar to those caused by exogenous carcinogens. This may be the cause by the "spontaneous" cases of carcinogenicity by high-fat diet. *See* ABC excinuclease; benzo(a)pyrene; ethylmethane sulfonate; excision repair; malondialdehyde; pyridine dimers; pyrimidopurinones.

adelphogamy Sib pollination of vegetatively propagated individual plants. *See* sibling.

adenine Purine base in either DNA or RNA. *See* purines.

adenine phosphoribosyltransferase (APRT, 16q24.3) Its recessive deficiency may lead to dihydroxyadenine accumulation in the urine and kidney disease. APRT may repair cyclobutane pyrimidine dimers. *See* cyclobutane dimer.

adeno-associated virus (AAV) Simple, icosahedral, non-enveloped, 22 nm, nonpathogenic, single-stranded DNA (5 kb) virus infecting a wide range of cell types in various species and integrating preferentially into human chromosomal site AAVS1 at 19q13.2-qter. Its loading capacity is ~4.7 kb. It has only two internally situated genes, *rep* (replication) and *cap* (capsid), encoding four replication (Rep78, Rep68, Rep52, Rep40) and three viral coat proteins (VP1, VP2, VP3) using different promoters and alternative splicing. AAV has two ~145 nucleotide inverted terminal repeats (ITR). Cis-acting functions required for replication (rrs) and binding to the human chromosome site (Sp1, 19q13.3-qter), helicase, site-specific and strand-specific endonuclease activity, packaging, integration, excision, and initiator (*inr*) of transcription site for RNA are encoded within the ITR. AAV has been considered as a vector for gene therapy. The AAV vectors usually cannot carry more than 4.5 kb, but because of the small size of the particles they can penetrate easier small targets (e.g., skeletal muscles, neurons). These vectors greatly benefit by the presence of adenovirus (AV) or herpes virus helper. AAV vectors transfect nondividing cells or dividing cells, and because of concatamer formation, they express transgenes for long periods and do not suffer immune rejection. The recombinant AAV molecule

with deletions within the viral *rep* gene is called rAAV. The *rep* guides the integration into the host chromosome and facilitates nonhomologous recombination. AAV vectors can target homologous mammalian chromosomal locations and alter ~1% of the cells without additional mutations. AAV may protect the cells against human melanoma or cervical carcinoma due to the product of the *rep* gene transcribed from the open reading frame beginning at the map position of promoter p5. The *rep*-minus strains do not integrate to preferential sites. For infection AAV requires a helper function provided by adenovirus or herpes virus. In the absence of a helper, AAV becomes latent but can be rescued. The AAV-based vectors are frequently used as a two-plasmid cotransfection system by relying on the complementary ITR promoter transgene and ITR *rep-cap* packaging system. Unlike the adenoviral vector, AAV vectors stay put in the cell. The host immune system usually does not react much to the AAV vectors beyond the initial stage of introduction. AAV vectors appear safe for the subjects and the environment, although in some instances unexpected tumors and chromosomal defects were observed. *See* adenoma; autonomous parvovirus; concatamer; gene therapy; herpes; parvoviruses; viral vectors. (Russel, D. W. & Kay, M. A. 1999. *Blood* 94:864; Miller, D. G., et al. 2002. *Nature Genet.* 30:147.)

adenocarcinoma Cancer of glandular tissues. *See* pancreatic adenocarcinoma.

adenoma Usually benign gland-shape epithelial tumor. The deficit of E-cadherin-mediated cell adhesion is one of the control steps in the change from adenoma to carcinoma. *See* cadherins; carcinoma; endocrine neoplasia. (Sieber, O. M., et al. 2002. *Proc. Natl. Acad. Sci. USA* 99:2954.)

adenomatosis, endocrine, multiple (MEN) In the autosomal-dominant MEN I (human chromosome 11q13), pancreatic adenonomas are prevalent; in MEN II, pheochromocytoma and thyroid carcinoma (10q11); and in MEN III, cancers of the nerve tissues are most common, although the latter two conditions appear allelic in the pericentric region of human chromosome 10q. *See* adenoma; cancer; pheochromocytoma; SHC.

adenomatous polyposis coli (APC) *See* FAP; Gardner syndrome.

adenosine Adenine with a ribose added. *See* adenine; nucleoside.

Adenosine.

adenosine 3′,5′ cyclic monophosphate (cAMP) Formed from ATP by adenylate cyclase enzyme. It has important regulatory functions as "second messenger" for microbial and animal cells. *See* adenylate cyclase; cAMP; G protein; second messenger, signal transduction.

adenosine deaminase (ADA) Enzyme that hydrolyzes adenosine monophosphate to inosine. Its deficiency causes severe immunodeficiency, and the patient's lymphocytes are disabled to fight infections successfully. ADA is synthesized in the cells for the purpose of detoxification of excessive amounts of adenosine or its analogs; it inactivates 9-β–D-xylofuranosyl adenine, a DNA-damaging chemical, and thus it can be used as a dominant selection agent in tissue culture. Its synthesis in the cells can be overproduced more than 10,000 times by a strong inhibitor of ADA, 2′-deoxycoformicin (dCF), and a transition-state analog for adenine nucleotide enzymes. It is encoded in human chromosome 20q13.11. RNA-specific adenosine deaminase (ADAR2, 21q22.3) is an editing enzyme. *See* ADAR; adenosine deaminase deficiency; immunodeficiency; mosaic; SCID; severe combined immunodeficiency; transition state.

adenosine deaminase deficiency (ADA deficiency) Also called severe combined immunodeficiency disease (SCID), and it may be treated with gene therapy or bone marrow transplantation or by enzyme replacement with polyethylene glycol adenosine deaminase (Peg-ADA). In case of the defect, deoxyadenosine (dAdo) or deoxyadenosine triphosphate reaches toxic levels. Into T lymphocytes isolated from the patients, retroviral vectors introduce the normally functional human ADA gene and the lymphocytes injected into the afflicted children. Upon periodically renewed treatment, symptoms of the disease (chronic infections, diarrhea, and muscle weakness) usually recede. Postexercise cramping of the muscles may be caused by inadequate levels of this enzyme. The ADA gene was located to human chromosome 20q13 area. *See* adenosine deaminase; deamination; gene therapy; gout; immunodeficiency; Lesch-Nyhan syndrome; lymphocytes; SCID; viral vectors.

adenosine diphosphate   *See* ADP.

adenosine monophosphate deaminase (AMPD1, 1p21-p13) Its recessive deficiency leads to cramps, myopathy, and weakness after exercising. Most commonly the enzyme in the skeletal muscles is affected and therefore it is called myoadenylate deaminase deficiency.

adenosine receptors Mediate the activities of diverse cell types (neurons, platelets, lymphocytes, muscle cells) in response to adenosine released by the degradation of ATP. The four types of receptors are A_1, A_{2a}, A_{2b}, and A_3. *See* aggressiveness.

adenovirus Large mammalian (1.8×10^8 Da), icosahedral (diameter about 80 nm), double-stranded DNA virus with ca. 36 kbp genetic material. The human adenovirus DNA has a 55 kDa protein covalently bonded to both 5′ ends. The initiation of replication depends on a viral 80 kDA protein reacting with the first deoxycytidylic residue, and its 3′-OH group serves as the starting point. The complementary DNA strand is a template. After replication, the 80 kDa protein is cleaved off, but the 55 kDa protein stays on. The replication does not require the synthesis of Okazaki fragments because one of the strands is first completed with the aid of the protein-dCTP primer, then the other strand is replicated. Replication can start at either end because the protein primer is used. Both strands of the DNA are transcribed into overlapping transcripts. The integrated viral DNA is generally smaller than the genome of adenoviruses. After lytic infection, a cell may release about 100,000 virus particles. Upon infection it may produce a flu-type ailment, and upon integration into the genome it may cause cancer.

In humans adenovirus is not carcinogenic. Because it stays episomal, it does not induce insertional mutation. The adenoviruses have a broad host range, which makes them suitable for veterinary vaccine production. The adenovirus oncoprotein E1A induces progression of the cycle by binding to a protein complex p300/CBP. A histone acetylase (P/CAF) competes in this with E1A and inhibits its mitogenic activity. The main function of E1A is to disrupt the association between p300/CBP and the histone acetylase. The viral E1B gene-encoded 55 kDa protein inactivates the p53 tumor suppressor gene and the cancerous proliferation begins. However, a mutant form of adenovirus (dl1520) that cannot express this 55 kDa protein still can replicate in cells that are defective in the p53 suppressor and consequently can lyse these defective cells. This finding offers a promise for the destruction of p53-deficient cancer cells by injection with dl1520 mutants. Normally wild-type p53 (apoptosis) is a requisite for the productive infection (destruction of the cells) by wild-type adenovirus. Adenoviruses have been used as vectors for genetic transformation after some regulatory sequences have been deleted and replaced.

The maximal carrying capacity is about 6–8 kbp. Since adenovirus preferentially infects the respiratory tract, it may be used for somatic gene therapy of, e.g., cystic fibrosis. It has also been used to transfer genes to skeletal muscles. The cell can take up adenovirus vector and its load DNA by a specific virus receptor and the $\alpha_V\beta_3$ or $\alpha_V\beta_5$ surface integrins. The adenoviral vector, taken up by nondividing or dividing cells, is not integrated into the human genome and thus does not lead to permanent genetic change, and it has to be reapplied periodically (in weeks or months). The adenovirus proteins evoke rapid immune responses, which may be the cause of the short duration of the transformation effects. Also, the current vectors may cause inflammation because of antivector cellular immunity. Immunosuppressive drugs (cyclosporin A, cyclophosphamide) may mitigate the immune response but may cause undesirable side effects.

The general design of an adenoviral vector. ITR = inverted terminal repeat, Y = Ψ, packaging signal, E1 is replaced by the transgene and E3 is deleted in the majority of vectors. The removal of E1 is important to prevent the infectivity of the viral DNA. For the lost E1 function, the host cells may provide complementation in trans. Recombination, however, may provide a chance for regeneration of replication competent virus.

The immunogenic property of the adenoviral vectors may eventually be exploited for immunological destruction of the targeted cancer cells. An advantage of this vector is that it can be used in very high titers (10^{11}–10^{13} particles/mL). Adenovirus is not known to induce human cancer. Adenoviral vector–mediated interleukin-12 gene therapy seems to protect mice with metastatic colon carcinoma. Construction of improved vectors is of major interest. Gutless or fully deleted adenoviral vectors are helper-dependent because most of the viral genome is removed to reduce the risk of adverse immune reaction and increase the duration of expression. *See* adeno-associated virus; antivector immunity; cancer; CAR1; cystic fibrosis; gene therapy; icosahedral; Okazaki fragment; p53; replication; titer; tumor vaccination; viral vectors. (Benihoud, K., et al. 1999. *Curr. Opin. Biotechnol.* 10:440, Frisch, S. M. & Mymryk, J. S. 2002. *Nature Rev. Mol. Cell Biol.* 3:441.)

adenylate   Salt of adenylic acid.

adenylate cyclase (adenylyl cyclase) Integral membrane enzyme with an active site facing the cytosol, generating cAMP (cyclic AMP) from ATP and releasing inorganic pyrophosphate (two phosphates). $G_{s\alpha}$ subunit–bound GTP activates this enzyme. The enzyme also has a weak GTPase activity that eventually breaks the $G_{s\alpha}$–GTP links and thus turns off the cyclase function. The activation of the cyclase function is initiated by the hormone epinephrine, which binds to a membrane receptor and activates the G_s proteins. cAMP itself is degraded by cyclic nucleotide phosphodiesterase. The cellular level of Ca^{2+} regulates oscillation of its level. The type I enzyme is stimulated by neurotransmitters, which elevate the level of Ca^{2+}. The type II enzyme requires stimulation by $G_{\alpha S}$ in the presence of the $G_{\beta\gamma}$ subunits of the G protein. *See* adenosine 3′,5′ cyclic monophosphate; animal hormones; anthrax; calcium ion channel; cyclic AMP-dependent protein kinase; epinephrine; G protein; G_S; GTPase; GTPase-activating protein; junction of cellular networks. (Jaiswal, B. S. & Conti, M. 2001. *J. Biol. Chem.* 276:31698; Onda, T., et al. 2001. *J. Biol. Chem.* 276:47785.)

adenylate kinase deficiency (AK1) Dominant in human chromosome 9q34.1; it causes hemolytic problems. Several AKI alleles were named. *See* hemolysis; hemolytic diseases.

adenylic acid   Phosphorylated adenosine.

adenylosuccinase deficiency (adenylosuccinate lyase, ADSL, 22q13.1) Normally the enzyme catalyzes the reaction: succinylaminoimidazole carboxamide ribotid → aminoimidazole carboxamide ribotide and also the removal of fumarate from adenylosuccinate to yield adenosinemonophosphate. Homozygosity for the recessive mutations may cause autism, mental and psychomotor retardations. *See* autism.

adenylyl cyclase   *See* adenylate cyclase.

adenylylation   Addition of adenine to an amino acid near the active site in a protein (by the enzyme adenylyl transferase); it may regulate the activity of the target.

ADEPT   Antibody-directed prodrug therapy. A prodrug is supplied to an organism afflicted by cancer. The prodrug itself is not toxic until it is enzymatically activated. Care should be taken so that the enzymes of the body do not convert the prodrug automatically into a toxin. An antibody, specific for the cancer, is conjugated with an activating enzyme. The conjugate seeks the cancer cells, and by generating a locally high concentration of the toxin, it is expected to kill the cancer cells. It would be desirable that the toxin not diffuse from the tumor into normal cells. Enzymes of potential use with prodrug → toxin:
Pseudomonas carboxypeptidase/glutamic acid derivatives → benzoic acid mustards, *E. coli* β-lactamase/cephalosporin derivatives → nitrogen mustards, yeast cytosine deaminase/5-fluorocytosine → 5-fluorouracil, almond β-glucosidase/amygdalin → hydrogen cyanide, etc. *See* magic bullet; vascular targeting. (Xu, G. & McLeod, H. L. 2001. *Clin. Cancer Res.* 7:3314.)

ADH   (1) Antidiuretic hormone. A short peptide (vasopressin). *See* antidiuretic hormone. (2) Alcohol dehydrogenase is an enzyme catalyzing the reversible reaction:

$$\text{acetaldehyde} + \text{NADH} + \text{H}^+ \rightleftharpoons \text{ethanol} + \text{NAD}^+$$

The ADH subunits (α, β, γ) are encoded in human chromosome 4q21. *See* acetaldehyde dehydrogenase; *adh*⁻; allyl alcohol; mutant isolation.

***adh*⁻**   Mutant with a defective ADH enzyme. *See* acrylaldehyde; mutation detection.

adhalin   *See* muscular dystrophy.

ADHD   *See* attention-deficit hyperactivity disorder.

adherence reaction   Binding of molecules to the complement receptors of the cell surface or agglutination of antibody and antigen complexes. *See* antibody; complement; complement fixation.

adherens junction   Cell surface where actin filaments attach. AJ is regulating cell adhesions mediated by Rap1 GTPase. *See* β-catenin; RAP1A. (Knox, A. L. & Brown, N. H. 2002. *Science* 295:1285.)

adherin   Chromosomal proteins that are similar in function to cohesin. *See* sister chromatid cohesion.

adhesion   Sticking together, e.g., water molecules clinging to various surfaces. *See* adherens junction; cadherins; integrins; plakoglobin; selectins; talin; vinculin.

adhesion belt   Adherens belt connects neighboring cells. *See* adherens junction.

adhesion plaque (focal contact) Spot where a cell is anchored to the extracellular matrix by transmembrane proteins.

adipocere (grave wax) Hydrolyzation product of body fats after death; its formation may help DNA preservation of the brain in some ancient animal/human remains. *See* ancient DNA.

adipocyte   Fat storage cell; it is used as a depository of excess caloric intake or reserve when expenditure exceeds intake of calories. The white adipose stores energy as triglycerides; the brown adipose tissue is involved in thermogenesis. Adipocyte differentiation is regulated by the CCAAT/enhancer-binding proteins, adipocyte differentiation determinant (ADD1)/sterol response element-binding protein (SREBP1), and peroxisome proliferator-activated receptors (PPAR). In addition, retinoic acid, vitamin D_3, and thyroid hormone receptors are involved. The retinoic receptors (RXR) are indispensable for the viability of mice. *See* animal hormones; brown fat; obesity; PPAR; resistin; retinoic acid; vitamin D. (Gregoire, F. M., et al. 1998. *Physiological Rev.* 78:783; Rosen, E. D. & Spiegelman, B. M. 2000. *Annu. Rev. Cell Dev. Biol.* 16:145.)

Fat

adiponectin Cytokine produced by adipocytes in response to metabolic or extracellular signals. It lowers blood glucose level and reduces the level of triglycerides in the muscles and the production of fat. It may be exploited for treatment of insulin-resistant diabetes. *See* cytokines; diabetes; triglyceride. (Yamauchi, T., et al. 2001. *Nature Med.* 7:887; Berg, A. H., et al. 2001. *Nature Med.* 7:947.)

adipose Related to fat, e.g., adipose tissue = fat tissue.

adjacent disjunction Neighboring members of translocation rings or chains move to the same pole; adjacent-1 when centromeres are nonhomologous, adjacent-2 when centromeres are homologous (nondisjunctional) at the poles. *See* translocation, chromosomal.

adjacent distribution *See* adjacent disjunction.

adjuvant, immunological If the immune response to an antigen is unsatisfactory because of the small amount present, the immune reaction may be enhanced by protecting the antigen from degradation and promoting slow release, and may increase its uptake by macrophages. For this purpose, mineral oils, alum (a hydrated aluminum oxide), charcoal, Freund adjuvant, specific nucleotide sequences, CD154, etc., can be used. *See* antigen; CD154; Freund adjuvant; immune response; immunization, genetic.

admixture in populations May take place when two potentially interbreeding populations share a habitat for a period of time. When the frequency of about 50 or more markers is analyzed, statistical information may be derived on the extent and time of the admixture. *See* introgression, linkage disequilibrium. (Chikhi, L., et al. 2001. *Genetics* 158:1347.)

aDNA Ancient DNA. *See* ancient DNA.

A DNA *See* DNA.

ADNFLE *See* epilepsy.

AdoMet S-adenosyl-L-methionine (current abbreviation is SAM) is a methyl donor for the enzyme guanosine 7-methyl transferase, the 2′-O-methyl transferase enzymes in the cap of pre-mRNAs, and other methylation reactions. *See* cap, homocystinuria; methionine adenosyl transferase; methylase; methylation, DNA; methylation, RNA; SAM. (LeGros, L., et al. 2001. *J. Biol. Chem.* 276:24918.)

adopted children Frequently used in human genetics to determine the relative effects of genes and the environment. These studies are sometimes hampered, however, because either the families do not have biological children or the biological parents of the adopted children are not available for examination. According to civil law, adopted children may lose any legal ties to and identity with the natural parents. This loss of identity may carry some genetic caveats because of chances of inbreeding by the lack of information about descent.

These problems may be similar to the ones encountered by artificial insemination using anonymous sperm donors. *See* artificial insemination; twinning.

adoptive cellular therapy Infusion of immune effector cells (NK cells, macrophages, $\gamma\delta$ T cells, $\alpha\beta$ T cells, B cells, etc.) for the treatment or prevention of disease (lymphoma, leukemia, myeloma). Although it may appear to be an attractive alternative to chemotherapy or radiation, the allogeneic cells may induce graft-versus-host disease (immune rejection). To avoid these complications, the herpes simplex virus thymidine kinase gene (HStk) may be targeted to the malignant cells by a Moloney murine leukemia retroviral vector. This vector selectively seeks dividing cells. Since cancer cells divide more frequently than normal cells, the vector does not usually hit normal cells. The cells are then infused by ganciclovir, which is incorporated into DNA and RNA when metabolized to ganciclovir triphosphate, resulting in selective termination of replication in the HStk$^+$ cells by this suicide technique. *See* cell therapy; ganciclovir; gene therapy; Moloney mouse leukemia oncogene; suicide vector; TK; viral vectors. (Link, C. J., et al. 2000. *Stem Cells* 18:220.)

ADP Adenosine 5′-diphosphate, a phosphate group acceptor in various cellular processes. It is produced by hydrolyis of ATP; it can also regenerate ATP by oxidative phosphorylation.

ADP-ribosylation factor *See* ARF.

Adr Positive regulator protein of transcription.

adrenal Adjacent or pertinent to the kidney.

adrenal hyperplasia, congenital (CAH) Occurs in both X-linked and autosomal forms and is apparently controlled by several loci. Cortisol deficiency is involved. The X-linked form is attributed to gonadotropin deficiency. The steroid hormone overproduction indicates a defect in steroid 21-hydroxylase, encoded within the boundary of the HLA complex in human chromosome 6p21.3. The affected female babies are masculinized. Masculinization is preventable by the administration of dexamethasone, but side effects may occur. It may be associated with Addison disease (hypotension, anorexia, weakness, and pigmentation). One form of the disease is accompanied by difficulties in salt retention in newborns. The prevalence is about 7×10^{-5}. The 17-α-hydroxylase deficiency is encoded at 10q24.3. It involves an excessive amount of corticosterones and hypertension and hypoalkemic alkalosis (increase of bases [e.g., K$^+$] yet lower pH). An autosomal form in chromosome 8q21 is deficient in 11-β-hydroxylase and/or corticosteroid methyl oxidase II (HSDB, 1p13.1) occurs at a frequency of 1×10^{-5}. Masculinization may be caused by several other genetic anomalies and by various medications administered to the mother or maternal androgen-producing tumors. Prenatal diagnosis may use allele-specific PCR on DNA from chorionic villi. *See* Addison disease; adrenal hypoplasia; chorionic villi; congenital adrenal hyperplasia; dexamethasone; genital anomaly syndromes; hermaphroditism; HLA; PCR; STAR.

adrenal hypoplasia, congenital (AHC) Characterized by abnormal underdevelopment (hypoplasia) of the genitalia and the gonads, insufficient function of the kidneys, hypoglycemia (reduced blood sugar), seizures, etc. Several forms of the disease (hypoadrenocorticism, polyglandular autoimmune syndrome, Addison disease) were reported with autosomal-recessive inheritance. The X-chromosome-linked DAX1/AHC (Xp21.3-p21.2) locus encodes a dominant negative regulator of transcription, a nuclear hormone receptor protein with a DNA-binding domain. The DAX-1 transcription is mediated by the retinoic acid receptor. AHC may modify male determination by SRY and it seems to also regulate the steroidogenic factor Sf-1, but it does not affect ovarian development. *See* Addison disease; adrenal hyperplasia; dominant negative; epilepsy; hypogonadism; Kallmann syndrome; RAR; retinoic acid; Sf-1; SRY; transcriptional activator.

adrenaline   *See* animal hormones; epinephrine.

adrenergic receptors   Come in the forms of α_1, α_2, β_1, β_2 distinguished on the basis of their responses to agonists and antagonists and tissue specificity. They all respond to the adrenal hormones, epinephrine, and norepinephrine. *See* agonist; antagonist; arrestin, hypotension; epinephrine; membrane proteins; receptors.

adrenocortical   Pertaining to the cortex (the outer layer) of the kidney.

adrenocorticotropin   *See* ACTH.

adrenodoxin   Electron carrier iron-sulfur protein in the mitochondria of the kidney cortex and assists cholesterol biosynthesis. *See* cerebral cholesterinosis.

adrenonoleukodystrophy (ALD, Addison disease, Xq28) The X-linked neonatal form is a defect in peroxisome assembly. The disease is associated with very long-chain fatty acid (VLCFA) acyl coenzyme A synthase defects. Neural degeneration and blindness are the consequences. Autosomal forms encoding different peroxins and peroxin receptors are at 2p15, 12p13.3, and 7q21-q22. *See* microbodies; peroxins; Refsum diseases; Zellweger syndrome.

adrenomedullin   22 amino acid vasodilator, a calcitonin-related peptide. In its absence, hydrops fetalis may develop. *See* calcitonin; hydrops fetalis.

Adriamycin   *See* doxorubicin.

adrogenital syndrome   Complex genetic disorder based on anomalies of steroid biosynthesis and adrenal hyperplasia. Gene frequencies vary a great deal in different populations from 0.026 of Alaskan Eskimos to 0.004 in Maryland. *See* adrenal hyperplasia; allelic frequency; steroid hormones.

adsorption   Tendency of molecules to adhere to a surface (different from absorption, which is uptake through a membrane).

adsorption chromatography   *See* column chromatography; thin-layer chromatography.

adsorptive endocytosis   *See* receptor-mediated endocytosis.

advantageous mutations   Favored by a particular environment and expected to propagate under steady-state conditions by a rate per generation: $v = \sigma\sqrt{2s}$, where σ is the standard deviation caused by diffusion (migration) and s is the selective advantage in the absence of dominance, e.g., if $\sigma = 10$ km and $s = 0.02$, then the advance per generation in kilometers will be $10\sqrt{2} \times 0.02 = 2$; then it would take 250 generations to advance 500 km. *See* migration; mutation; mutation, beneficial; selection coefficient. (Cavalli-Sforza, L. L. & Bodmer, W. F. 1971. *The Genetics of Human Populations.* Freeman, San Francisco, CA.)

adventitia   Outer coating of organs by lose connective tissues composed mainly of fibrillin and elastin.

adventive embryos   Developing from the diploid tissues of the plant nucellus (without fertilization); they occur commonly in citruses. *See* apomixia; nucellus.

aecidiospore   Dikaryon of plant rust fungi formed through a sexual process that did not involve nuclear fusion; the aecidiospores are products of the aecidium, a group of sporangia. *See* aecidium; dikaryon; fungal life cycle; sporangium.

aecidium   Or aecium, a fruiting structure of fungi (Basidiomycetes-Uredinales) such as *Puccinia graminis tritici*. Aecidia are formed only on the intermediate host, barberry, but the spores infect only wheat.

Aedes egypti   Mosquito that transmits yellow fever and dengue fever viruses. (Rai, K. S. & Black, W. C. 1999. *Adv. Genet.* 41:1.)

Aegilops caudata   Diploid representative of the *Triticum* genus (2n = 14) carrying the 7-chromosome C genome (current name *Triticum dichasians*). *Aegilops cylindrica*: an allotetraploid of the wheat genus containing the CD genomes (C from *T. dichasians* and D from *T. tauschii*). Current name *Triticum cylindricum*. *Aegilops squarrosa*: *Triticum tauschii* by current name, is a diploid species in the wheat genus

Glume of *T. peregrinum*. (Courtesy of Drs. Gordon Kimber and Moshe Feldman.)

with the D genome. *Aegilops umbellulata*: by current name *Triticum umbellulatum*, a diploid species of the wheat genus with the C^u genome. *Aegilops variabilis*: currently *Triticum peregrinum*, a species of the wheat genus; occurs in nature both as tetraploid (DM) and hexaploid (DDM) genomic constitution. *See Triticum*.

aegricorpus Genetic-physiological complex determined by the host–pathogen interaction; the phenotype of the disease in plants. *See Flor's model; host–pathogen relations*.

aequorin (GFP) Luminescent protein (green fluorescent protein of 238 amino acids) from jellyfish (*Aequorea victoria*); its activation is dependent on the level of the available Ca^{2+}, and on this basis minute quantities and differences in this ion can be measured by optical means within the range of 0.5–10 mM. This may be of major importance because calcium may play regulatory functions in all eukaryotic cells. The chromophore results from the cyclization and oxidation of the Ser^{65} (Thr^{65})Tyr^{66} Gly^{67} amino acid sequence in the central helix of the 11-stranded β barrel. Aequorin also has the advantage that it is a noninvasive and nondestructive label in various organisms. Its excitation peaks are at 395 and 475 nm, and the emission peak of the pure GFP is at 470 nm. Mutant proteins with higher absorption peaks and different colors (red, yellow) exist. By the use of fluorescence resonance energy transfer (FRET), linking GFP reversibly to peptide spacers, results in conformational changes and altered light emission (color). GFP is sensitive to pH, temperature, and prior illumination. Excitation by 488-nanometer light increases fluorescence 100 fold and remains stable for days (Patterson, G. H. & Lippincott-Schwartz, J. 2002. *Science* 297:1873). The gene has been cloned and sequenced and has been widely used in animals and plants, in various modified forms, as a reporter gene. GFP is a strong immunogen. *See barrel; BFP; calmodulin; drFP583; FERT; Renilla GFP*. (*Methods in Enzymology*, vol. 302. 1999. Tsien, R. Y. 1998. *Annu. Rev. Biochem.* 67:509; Labas, Y. A., et al. 2002. *Proc. Natl. Acad. Sci. USA* 99:4256; van Roessel, P. & Brand, A. H. 2002. *Nature Cell Biol.* 4:E15.)

AER Apical ectodermal ridge is a projection on the ectoderm. *See ectoderm*.

aerobe Organism that uses oxygen as the terminal electron acceptor in respiration. *See electron acceptor; respiration*.

aerobic Reaction (or organism) that requires oxygen or takes place in the presence of oxygen.

AF Activation function protein activates and recruits coactivators of gene expression. *See co-activator*.

affected individual Expresses a particular (disease) trait.

affected-sib-pair method Nonparametric method for linkage analysis of susceptibility genes. For this purpose the risk ratio (λ_s) is determined — that is, the risk of a sib of an affected proband compared to the average prevalence in the population, e.g., diabetes has a prevalence of 0.004 in the general population, but its incidence among sibs of affected individuals is 0.06, hence $\lambda_s = 0.06/0.004 = 15$. This λ_s is for all loci responsible for the phenotype. The larger λ_s, the higher is the genetic

contribution. It is also affected by the interaction (can be multiplicative or additive) of the various factors contributing to the phenotype. The strength of the proof for linkage depends on the so-called *maximal lod score* (MLS) symbolized by T and means the log odds in favor of linkage. Usually, the estimation is carried out in steps by selecting linkage at each step with markers increasing from $T > 1.0$. Statistically valid linkage is expected when the T score reaches or exceeds 3. The T value may also increase by the use of larger populations. Recombination decreases MLS and the use of multiple loci increases the estimate. Another advantage of this approach is that both recessive and dominant alleles can be studied. It is also applicable for quantitative traits. This procedure may not necessarily be applicable only to sibs; cousins or other close relatives (uncles, aunts) may also be included. *See allele sharing; lod score; nonparametric tests; recombination frequency; maximum likelihood method applied to recombination*. (Dupuis, J. & Van Eerdewegh, P. 2000. *Am. J. Hum. Genet.* 67:462.)

affective disorder Psychological illness, psychosis. *See attention deficit hyperactivity disorder; autism; bipolar mood; manic depression; neurodegenerative diseases; obsessive-compulsive behavior; paranoia; schizophrenia; Tourette's syndrome; unipolar depression*. (Evans, K. L., et al. 2001. *Trends Genet.* 17:35.)

afferent Conducting or transferring toward the middle.

affine gap cost Expresses the "penalty" for gaps in a sequence alignment according to the length of a gap. *See contig; genome projects*.

affinity Unlinked genes segregate to the same gamete more frequently (quasi-linkage) or less frequently (reverse linkage) than expected on the basis of randomness. In immunogenetics: the intensity of interaction between a particular antigen receptor and its epitope. *See epitope*. (Bailey, N. T. J. 1961. *Introduction to the Mathematical Theory of Genetic Linkage*. Oxford, Clarendon Press, England; Michie, D. & Wallace, M. E. 1953. *Nature* 171:26.)

affinity capture *See acesims*.

affinity chromatography Polyadenylated mRNA can be separated from other RNAs by adsorption on oligo T (thymine) cellulose or sepharose columns by virtue of the complementarity of the A and T bases. Similar procedures are used for the purification of antibodies on immobilized antigen media (antibody purification), and DNA-binding proteins can be isolated and enriched by a factor of 10^4 with the aid of affinity chromatography. *See cDNA library screening; gel retardation assay*.

affinity-directed mass spectrometry Detects interaction between proteins, receptors-ligands, proteins-nucleotides, etc. *See mass spectrum*.

affinity labeling Most commonly a photo-affinity hapten is used (i.e., one that is activated only upon illumination). The affinity label is bound to the antigen-binding site amino acids of the antibody and thus reveals the site on the antibody where the attachment is taking place. *See antibody; antigen; hapten*.

affinity maturation Selection of cells with high affinity for the antigen as clonal selection progresses. It takes place by accumulation of mutations in the germinal center of a lymphoid follicle in the paracortex of a lymphoid node and combinatorial assembly of the variable, joining, and diversity sequences of the immunoglobulin genes. These alterations take place in response to the antigens arriving there through small capillary veins on the surface of the antigen-presenting cells and helper T cells. Immunoglobulin G (IgG) usually responds well after the second immunization. Affinity maturation is also a process for the selection of memory cells. TRAF and CD40 regulate the affinity maturation. *See* antibody; antigen-presenting cell; clonal selection; hapten; immune response; immunoglobulins; lymphoid organs; memory, immunological; repertoire shift; TRAF, CD40. (Ahonen, C. L., et al. 2002. *Nature Immunol.* 3:451; Meffre, E., et al., 2001. *J. Exp. Med.* 194:375.)

affinity purification Required unless the antibody reacts with more than one antigen. If this is not the case, an affinity chromatography column is prepared by using pure antigen. Alternatively, monoclonal antibody must be used or the immunoglobulin library must be carefully analyzed for true or false positive immune reactions. *See* antibody; antigen; monoclonal antibody.

AFI Amaurotic familial idiocy, now called Tay-Sachs disease (TSD). *See* Tay-Sachs disease.

afibrinogenemia 4q28 recessive deficiency of fibrinogen (blood coagulation factor I). The afflicted individuals bleed very heavily after injury. Periodic blood accumulation under the skin (ecchymosis), nose bleeding (epistaxes), bloody tumors (hematomas), bloody cough (hemoptysis), or stomach-intestinal or genitourinary bleeding occur. Characteristically, for longer periods of time no symptoms appear. Therapy is intravenous injection of concentrated human fibrinogen. *See* antihemophilic factors; dysfibrinogenemia; fibrin-stabilizing factor; hemophilia.

aflatoxins Group of heterocyclic mycotoxins produced under appropriate conditions by the *Aspergillus flavus* and *Aspergillus parasiticus* fungi. The aflatoxins are extremely carcinogenic because they affect DNA synthesis. The LD_{50} of aflatoxins orally administered to monkeys may be as low as 1750 μg/kg. Aflatoxin may be a contaminant on grains, peanuts, dry chili pepper, and on many other material that humans and animals eat or are exposed to. Aflatoxins frequently cause mutations in *p53* tumor suppressant at codon 249 (AGG → AGT), resulting in hepatocarcinomas. The 8,9-oxide of aflatoxin B forms a mutagenic adduct in the DNA at N^7-guanine. See adduct; environmental mutagens; mycotoxins; p53; toxins. (Smela, M. E., et al. 2001. *Carcinogenesis* 22:535; Cary, J. W., et al. 2002. *Biochim. Biophys. Acta* 1576:316.)

AFLP Anonymous fragment length polymorphism, amplified fragment length polymorphism. DNA fingerprinting technique involving restriction enzyme digestion and amplification of special fragments by PCR. It can be used for mapping genes or for the estimation of nucleotide diversity in populations. See anonymous DNA segment; DNA fingerprinting; PCR; restriction enzymes; VNTR. (Vos, P. R., et al. 1995. *Nucleic Acids Res.* 23:4407; Saunders, J. A., et al. 2001. *Crop Sci.* 41:1596.)

African green monkey (*Cercopithecus aethiops*) Kidney cells are the best laboratory host for the propagation of SV40 (Simian virus 40). *See* Cercopithecidae; cos; SV40.

AFS see affected-sib-pair method

after morning pill *See* hormone receptors (RU-486).

agameon Species without sexual reproduction.

agamic Species reproducing asexually (without gametes).

agammaglobulinemia Occurs as an X-chromosomal (congenital) and autosomal defect in the synthesis of γ-globulin, a component of the heavy chain of antibodies. The X-chromosome-linked (Xq21.33) is frequently called Bruton's agammaglobulinemia (XLA). The protein responsible is tyrosine kinase of 659 amino acids, encoded by 19 exons. The manifestation of XLA may differ in different families, indicating the involvement of several genes. It is conceivable that the defect is caused by rearrangement of the involved genes. Some of the individuals have truncated V regions of the antibody. In the afflicted persons, the IgG and IgM content is generally no more than 1% of normal. The absence of plasma cells from the lymph nodes, spleen, intestine, and bone marrow is also a basic defect. The patients are very susceptible to pyogenic infectious bacteria (staphylococci, pneumococci, streptococci, and *Haemophilus influenzae*). Pus-forming inflammation of the sinuses, pneumonia, meningitis (inflammation of the brain), and furunculosis (boils) are common but can be prevented by the use of antibiotics or raising the γ-globulin levels by regular injections. Without treatment these infections may become fatal. The afflicted children are not more susceptible to viral, enterococcal, gram-negative bacteria, protozoan or the majority of fungal infections. Another X-chromosome-linked or autosomal agammaglobulinemia causes susceptibility to bacterial, fungal, and viral infections and leukemia. This is generally accompanied by lymphopenia (decrease of lymphocytes in the blood). This disease is generally detected after the discontinuation of breastfeeding of the babies or near the end of the first year of life. Agammaglobulinemia may also occur as an acquired disorder with onset at different ages, generally as a follow-up to other diseases. The prevalence is about 0.5 to 1×10^{-5}. *See* achondroplasia; antibody; BTK; cancer; gammaglobulin; hypogammaglobulinemia/common variable immunodeficiency; immune system; immunoglobulins; immunodeficiency.

agamospecies Reproduces by nonsexual means, e.g., parthenogenesis, apomixia. *See* apomixia; asexual reproduction; parthenogenesis; species.

agamospermy Seed production without fertilization, apomixis. *See* adventive embryos; apomixis; apospory; diplospory; parthenogenesis.

aganglionosis Congenital lack of intestinal ganglions. *See* Hirschsprung disease.

agar Gelling agent produced from marine algae with various degrees of purification (bacteriological agar, noble agar) and used for microbial and plant cell culture media. *See* agarose; gellan gum.

agarose Purified linear galactan hydrocolloid isolated from marine algae. In the crude form it is generally contaminated with salts and other substances, polysaccharides, and proteins. Some commercial products are highly purified. It is used for electrophoretic separation of oligonucleotides and polynucleotides from 0.1 to 60 kb range, depending on the concentration of this matrix. The higher concentration (2%) separates the smallest molecules, whereas the lowest concentration (0.3%) permits the separation of the largest fragments. The most commonly used concentration range separating 0.4 to 7 kb fragments is 0.9–1.2%. Contaminations of the agarose may interfere with further enzymatic handling of the eluted DNA. *See* electrophoresis; gel electrophoresis.

Agave (sisal) Basic chromosome number $x = 30$, and the various plant species may be diploid, triploid, or pentaploid. The plant has been used for medicinal purposes as a laxative and its juice may cause abortion.

AGE Advanced glycation end product is a sugar-derived carbonyl group added to a free amine that forms an adduct after rearrangement. AGE may cross-link amino groups in macromolecules and thus may promote aging, accelerate diabetes, and participate in other reactions. The cross-links may be broken by N-phenacetylthiazolium bromide and may have therapeutic application. *See* adduct; aging; Alzheimer disease; diabetes. (Pushkarsky, T., et al. 1997. *Mol. Med.* 3:740.)

age The time since the birth of an individual. Prenatal age is more difficult to determine. Ultrasonic measurements are frequently compared with tables obtained by empirical data.

age and mutation in human populations Expressed by the formula $\mu_t = \alpha t + \mu_0$, where mutation rate at a given time is μ_t, α is the mutation rate per cell divisions, and μ_0 is the initial frequency of mutation. It is expected that mutation rate increases as the number of cell divisions increases in the spermatogonia and oogonia. The available data indicate that chondrodystrophy (achondroplasia, a dominant dwarfness) and acrocephalosyndactyly (Apert's syndrome [pointed top of the head] and syndactyly [webbing in between or attachment of the fingers and toes]) increases at birth by about 2- to 4-fold with paternal age from 25 to 45 years. Other dominant mutations show similar tendencies but with much less clear differences. The human eggs may be different because new egg cells are not formed in the female babies after birth; the oogenesis is almost complete in the newborn. Nevertheless some age differences are still expected. Chromosomal aberrations (trisomy) in the eggs may increase, however, from 1/2300 at age 20 to 1/46 after age 45, probably because of the prolonged meiotic dictyotene stage (diakinesis). Some of the eggs complete meiosis before each ovulation, a period extending over 30 to 40 years. Trisomy in sperm is much less common, partly because it is the product of new divisions, partly because the disomic sperm

may be at a disadvantage in competition for fertilization. A normal human ejaculate may contain 25–40 million sperm cells. The increase of mitochondrial mutation rate by aging is equivocal. Accumulation of mutations by aging is organ-specific in mice. *See* Apert syndrome; gametogenesis; gonads; longevity; mutation rate; syndactyly; trisomy. (Dollé, M. E. T., et al. 2002. *Nucleic Acids Res.* 30:545.)

age correlation between mates It is much higher in consanguineous marriages than in unrelated mates. On the average, age correlation makes first-cousin marriages about twice, second-cousin marriages about 1.7 times, and third-cousin marriages about 1.4 times as frequent as if there were no correlation between the ages at marriage. Since some of the human hereditary diseases have late onset, older marrying ages may reduce the reproduction of genes with late manifestation. Also, the afflicted persons may not marry or may choose not to have children if they marry. *See* consanguinity.

agent orange ($Cl_3C_6H_2OCH_2COOH$) Herbicide containing mainly the synthetic auxin 2,4,5-trichlorophenoxy acetic acid (2,4,5-T). It had been used as a defoliating agent and brush killer. The LD_{50} of 2,4,5-T for mammals is 500 mg/kg; however, there are reports of much lower doses of high toxicity particularly at subcutaneous injection. It is frequently contaminated by dioxin, a carcinogen. The symptoms of agent orange exposure can be anorexia, hepatotoxicity, chloracne, gastric ulcers, porphyrinuria, porphyria, teratogenesis, leukemia, etc. 2,4,5-T may be degraded by genes in the plasmids derived from *Pseudomonas ceparia*. *See* acne; anorexia; chloracne; hepatotoxicity; LD_{50}; porphyria; teratogenesis; ulcer.

age of onset of disease The probability can be calculated: $(1 - \phi_1)(1 - \phi_2) \ldots (1 - \phi_{x-1})$, where $\phi_x =$ the probability of onset between ages x and $x + 1$; the probability of surviving to age x before onset is $l_x = (1 - q_1)(1 - q_2) \ldots (1 - q_{x-1})$, where $q_x =$ the probability of dying at age x before the onset of the condition. *See* aging.

age of parents and secondary sex ratio Slightly decreasing from 0.517–0.516 at parental age group 15–19 to 0.512–0.511 at parental age 45–49. On the basis of very large samples examined, the age gap between parents does not significantly affect the sex ratio. *See* sex ratio.

age-specific birth and death rates The probability that an individual of a certain age dies (or gives birth) within the following year is determined by population projection matrices. The numbers of birth and death rates in a time interval can be determined by $1 - B/N$ for birth per women extracted from available census figures. One study in 1966 found that in the age groups of women 15–30 and 30–45, the mean number of children born per woman was 1.37 and 0.465, respectively. Also, the study found that the average survival of age groups 0–15, 15–30, and 30–45 was 0.992, 0.988, and 0.964, respectively. Thus, in a sample of 30 women of age group 0–15, they will give birth to $30 \times 1.37 = 41.1$ children. In the age group 30–45 of 20 females, the prediction is $20 \times 0.465 = 9.3$, and so on. Similarly the survivors expected in the age group 0–15 is $40 \times 0.992 = 39.68$; in the age group

15–30, the expectation is $30 \times 0.988 = 29.64$; and so on. The natural logarithm of the annual growth rate is called the *intrinsic rate of natural increase of the population*, and it means that once a stable equilibrium is reached for the various age groups, it will increase by this intrinsic rate per year. Example: If the population growth at equilibrium at 15-year cycles is 1.307, then the annual $r = ln(1.307)^{1/15} \cong 0.0178$. The age-specific birth and death rates and r must be determined for each population because considerable variations may exist from time to time even in the same group, depending on cultural and economic conditions. A statistical survey indicates that women with later onset of menopause live longer. The estimated maximal human life span is now about 120 years. *See* human population growth; longevity; menopause.

AGGA box   Upstream transcriptional regulatory site. *See* promoter; transcription factors.

agglutination   Clumping. It occurs when two different blood types are mixed or when bacteria are exposed to specific antisera. The basis of this phenomenon is a component of the complement on the antibody ($C1_q$) protein that binds to the Fc region of the IgG heavy chain and is followed by a change in conformation of the antibody. The binding of the epitope to the antibody triggers this process. *See* antibody; complement; epitope; immunoglobulins.

agglutinin   Antibody that causes agglutination of cognate antigen. *See* abrin.

aggrecan   Chondroitin sulfate proteoglycan of the cartilage. (Schwartz, N. B., et al. 1999. *Progr. Nucleic Acid Res. Mol. Biol.* 62:177.)

aggrecanase   *See* arthritis.

aggregation   Of proteins/fragments of proteins may impair the ubiquitin-proteasome degradatory system. *See* ubiquitin.

aggregation, familial   Increased incidence of genetically determined traits among relatives, e.g., among natural children and parents compared with unrelated adoptive children. *See* heritability.

aggregation chimera   Produced in vitro by the assembly of genotypically different early embryonic cells. *See* allophenic; chimera; multiparental hybrid.

aggregulon   Protein complex involved in activation and repression of genes; the term *reglomerate* was used in the same sense.

aggresome   Aggregate sink of insoluble misfolded proteins in the endoplasmic reticulum or close to the microtubule organizing center containing chaperones, proteasomes and proteasome activator complexes. *See* chaperone; endoplasmic reticulum; proteasome. (Kopito, R. R. 2000. *Trends Cell Biol.* 10:524.)

aggression   Behavioral trait with great variance in animal and human populations, and it may be an expression of innate self-assertion, frustration, or a response to antisocial behavior. It may be the consequence of affective disorders and mental illness (paranoia). Evolutionists attribute aggression to the means of survival in the struggle for life, and as such it is observable among the majority of animals. Accordingly in subhuman beings it is instinctive and largely depends on the species concerned. Among humans it has an animal component, but it is also determined by the ethical and cultural factors of the individual and the standards of the population. Whereas animals are not credited with conscientious value judgments, in human societies, the moral, ethical, religious, and cultural principles may predominate. All human ethnic groups appear to have a condemning attitude toward violence. Yet mainly humans display violent aggression within species. It has been suggested that the human species lacks the ability of submission, a widely common ability among other mammals. In the male vole, antidiuretic hormone (vasopressin) may be responsible for aggression. The genetic basis of aggressive behavior is generally not understood, although it is known that a deficiency—e.g., in hypoxanthine-guanine phosphoribosyl transferase—may result in hostile and self-mutilating behavior. The major problem is concerned with the large nonbiological but cultural component of aggression. Unfortunately, human societies treat the cultural problem with double standards: killing and violent behavior is condemned yet even major religions approve patriotic or holy wars with the weapons of mass destruction. The questions remain unsettled whether capital punishment is appropriate for killers, whether induced abortion is an act of aggression, and whether euthanasia is a merciful act or just another form of taking life. To what extent are criminals predestined by their genetic endowment to aggression and what is the role of the social environment, and the free will? Obviously, some of the answers are beyond the scope of genetics. Mice deficient in α-calcium-calmodulin kinase II display reduced levels of serotonin and aggressive behavior. Mice with knocked-out adenosine receptor ($A_{2a}R$) display high blood pressure and aggressiveness. *See* adenosine receptors; antidiuretic hormone; behavior, in humans; behavior genetics; calmodulin; ethics; instincts; Lesch-Nyhan syndrome; morality; nitric oxide; paranoia; personality; serotonin; submission signal.

aggressiveness   Plant pathogen aggressiveness is measured by the evocation of a disease phenotype, depending on the genotype of the pathogen, that of the host, and environmental factors.

aging   Exponential increase in mortality as a function of time or cell divisions. Some type of irreversible alterations in the DNA may determine aging. In older cells the chromosomal telomeres are shortened; the frequency of nondisjunction dramatically increases by age, e.g., the incidence of Down syndrome may increase 200-fold in the offspring of just premenopausal mothers. The autosomal-recessive Werner syndrome (gene frequency 1 to 5×10^{-3}) involves premature aging (graying of hair, atrophy of skin, osteoporosis, decreased libido, and increased risk of cancer) and is also characterized by nonketotic hyperglycinemia. Progeria (Hutchinson-Gilford syndrome), another autosomal-recessive trait, causes very early senescence. Aging has been attributed to defects of the immune system and to diminished activity of superoxide

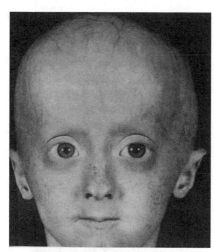

Progeria. (From Bergsma, D. ed. 1973 Birth Defects. Atlas and Compendium. By permission of the March of Dimes Foundation.)

dismutase, an enzyme normally destroying the highly reactive radicals that arise due to irradiation and aerobic metabolism.

It has been suggested that aging is the result of degenerative changes in the mitochondria: the formation of aberrant DNA circles. Recent information fails to support increased point mutation rate in the control region of cultured fibroblast mitochondrial DNA among normal individuals or persons with neurodegenerative diseases (Chinnery, P. F., et al. 2001. *Am. J. Hum. Genet.* 68:529). Aging mitochondria frequently show deletions (Bodyak, N., et al. 2001. *Hum. Mol. Biol.* 10:17). By-products of oxidative phosphorylation, hydrogen peroxide, and superoxide may accumulate during senescence. Aging primarily causes functional losses in the neurons rather than large-scale losses. Atrophy, decrease of receptors, accumulation of fluorescent pigments, and cytoskeletal abnormalities in the brain occur in aging mammals. Microarray hybridization of aging neocortex and cerebellar tissues of mouse involving 6347 genes indicated inflammatory responses, oxidative stress, and reduced neurotrophic support. Caloric limitations retarded some of the symptoms. In *Caenorhabditis*, superoxide dismutase and catalase reduced aging significantly. This causes delays in mitochondrial replication. The slow replication leaves the D loop of mtDNA unprotected, possibly increasing the chances for deletions and mutations. Misregulation of mitosis caused by gradual defects in cell cycle–control proteins, chromosomal movement, etc., may also be players in aging.

Hereditary premature aging is also known in animals. Voltage-activated Ca^{2+} influx into the brain neurons is accelerated during aging. The mouse autosomal-recessive gene *klotho* seems to be a regulator of several symptoms of aging. The product of this gene shares sequence similarities with β-glucosidase proteins. In *Caenorhabditis*, mutations are known that in combination may extend the life of the nematodes through two different pathways up to five-fold. Mutation in succinate dehydrogenase cytochrome b causes oxidative stress and premature aging in the nematodes. In *Drosophila*, the *mth* (*methuselah*) mutant line, apparently encoding a protein with homology to GTP-binding proteins with seven-transmembrane domain receptors, extends the life by about one-third. In yeast, activated GTPase (RAS), inactivation of the *LAG1* gene (encoding a membrane-spanning protein), and the *SIR* silencing complex extend life span.

Aging in mammals seems to be associated with aging of the lymphocytes and their function. The telomerase enzyme has also been implicated in aging. Since aging usually occurs after the reproductive period, it is no longer the object of natural selection (antagonistic pleiotropy). Population geneticists entertain two genetic mechanisms for aging: the accumulation of deleterious mutations and the increase in antagonistic pleiotropy among gene loci. In rodents, calorie-restricted (CR) diet has a substantial antiaging effect.

Aging has also been attributed to the gradual loss of the telomeric DNA repeats. It has been suggested that the secretion of inflammatory cytokines may be a contributing factor. The life expectancy in years in the United States changed from 47 to 76 from 1900 to 2000. In Sweden, the maximum life expectancy between 1861 and 1999 was 0.44 year per decade. The heritability of aging based on twins of humans is about or less than 0.35. The evidence for the role of mitochondria in aging has been questioned. Age-associated decline of cognitive abilities seems to be correlated to neuronal atrophy in the subcortical regions of the brain, and the process may be prevented by neurotrophin or neurotrophin gene therapy. *See* Alzheimer disease; apoptosis; Bloom syndrome; chromosome breakage; Cockayne syndrome; control region; cytokines; disposable soma; DNA repair; estradiol; gene therapy; heritability; ion channels; killer plasmids; longevity; lymphocytes; MARS model; mating type, yeast; mitochondrial mutations; mortality; neurotrophins; p53; pleiotropy; progeria; RAS; ROS; selection; senescence; silencer; superoxide dismutase; telomerase; telomere; Werner syndrome. (Cortopassi, G. A. & Wong, A. 1999. *Biochim. Biophys. Acta* 1410:183; Jazwinski, S. M. 2000. *Trends Genet.* 16:506; Kenyon, C. 2001. *Cell* 105:165; Finch, C. E. & Rivkun, G. 2001. *Annu. Rev. Genomics Hum. Genet.* 2:435; Pletcher, S. D., et al. 2002. *Current Biol.* 12:712.)

aglycon Protein or lipid linked to a polysaccharide.

agonadism, familial Absence of gonadal tissue; usually part of a syndrome. *See* azoospermia.

agonist Activates a receptor. *Inverse agonists* are antagonists of overexpressed receptors.

agonistic behavior Combative behavior. *See* aggression.

agouti Alternating light and dark bands on individual hairs of the fur in mammals such as mouse, rat, and rabbit. The genes *agouti* and *extension* determine the relative amounts of eumelanin (brown-black) and pheomelanin (yellow-red) pigments. *Extension* encodes the receptor of the melanocyte-stimulating hormone (MSH) and *agouti* is a signal sequence in the hair follicle, inhibiting eumelanin production and the melanocortin receptor, an MSH receptor. Agouti has been cloned and sequenced; it contains five exons, but two of them are not translated. The secreted protein products have 131 amino acid residues. The alleles that produce increased amounts of pheomelanin make the mice more prone to late-onset obesity and diabetes. An agouti-related protein (AGRP), a neuropeptide, may increase several-fold in obese mice. A^y and A^{vy} increase the liability to neoplasias, and others cause embryonic lethality. *See* pigmentation of animals, melanin, melanocyte-stimulating hormone, melanocortin, ghrelin, (Dinulescu, D. & Cone, R. D. 2000. *J. Biol. Chem.* 275:6695.)

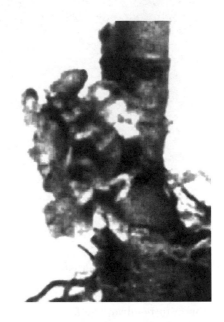

agretope Part of an antigen that interacts with a desetope (antigen-binding site) of an MHC (major histocompatibility) molecule. *See* antigen; desetope; epitope; histotop.

agricultural productivity Affected by genetic improvement of plants and animals and the improved husbanding and culture practices. Between 1951 and 1980, the overall plant productivity in the United States increased to 166% and that of animals increased to 144%. The yield of maize after the introduction of hybrids increased to about 500%. *See* heterosis; QTL.

agrin Natural glycoprotein (200 kDa) causing the aggregation of acetylcholine receptors on muscle cells in vitro and in vivo, used for the formation of neuromuscular junctions and T lymphocytes. The process also requires a muscle-specific protein kinase (MuSK). Agrin-deficient mutant mouse is inviable. Agrin may restore function in muscular dystrophies caused by mutation in laminin. *See* acetylcholine receptors; laminin; neuregulins. (Trautmann, A. & Vivier, E. 2001. *Science* 292:1667; Moll, J., et al. 2001. *Nature* 413:302.)

agrobacterial vector *See* binary vector; cointegrate vector; transcriptional gene fusion vector; translational gene fusion vectors.

agrobacterium mini-plasmid Carries T-DNA, including its borders, but it is free of other segments, including *vir* genes. *See Agrobacterium tumefaciens*; T-DNA.

Agrobacterium rhizogenes Bacterium closely related to *A. tumefaciens*. It induces hairy roots rather than crown gall on the host plants. The genes responsible for the formation of hairy roots reside in the *Ri* plasmid. The hairy root tissues, unlike crown gall, readily regenerate into plants. The *Ri* plasmid and the *Ti* plasmid have been used similarly to construct plant transformation vectors. In *Nicotiana glauca*, DNA sequences (*Ngrol*) homologous to the left segment of the T-DNA of the Ri plasmid have been detected. This observation indicates horizontal interspecific gene transfer. *See Agrobacterium tumefaciens*; crown gall; *Ri* plasmid. (Moriguchi, K., et al. 2001. *J. Mol. Biol.* 307:771.)

Agrobacterium tumefaciens Soil-borne plant pathogenic microorganism of the family of *Rhizbiaceae*. It is responsible for the crown gall disease (tumor) of the majority of wounded dicotyledonous plants and it also infects a few monocots (*Liliaceae, Amaryllidaceae*). Several of its characteristics are similar to *Rhizobium, Bradyrhizobium*, and *Phyllobacterium* species. The pathogenicity is coded in genes within the T-DNA of its Ti (tumor-inducing plasmid). T-DNA–containing plasmids are the most important transformation vectors of plants. The T-DNA is ~21 kb segment of the Ti plasmid with two direct repeat flanks bordering the oncogenes (responsible

for tumorigenesis in the wild-type plasmids) and some of the opine genes.

Molecular biologists most widely use the Agrobacterial strains A6 and C58, containing octopine- and nopaline-encoding Ti plasmids, respectively. Certain *Agrobacteria* strains have limited host range (LHR) caused by an altered *virA* gene in the Ti plasmid. The supervirulent strains, on the other hand, overproduce the VirG protein. The infection of some species of plants is limited. The transfer of the T-DNA to other cells, including plant cells, requires the formation of a conjugation tube (pilus) controlled by virulence genes *virA, virG, virB1* to *virB11*. *Agrobacterium tumefaciens* C58 has one ~2.1 Mb linear chromosome and three circular DNA plasmids (~2.8 Mb, ~0.54 Mb, ~0.21 Mb). Its total genome is ~5.67 Mb. The total number of assigned protein-coding genes is 1286, 1715, 333, and 141, respectively. The T-DNA is located in the 0.21 Mb plasmid. *Agrobacteria* can transfer DNA to yeast and some other fungi. The T-DNA can integrate from binary vectors into human HeLa cells. *See* BIBAC; host–pathogen relation; T-DNA; Ti plasmid; transformation, genetic; virulence genes, *Agrobacterium*. (Koncz, C., et al. 1992. *Methods in Arabidopsis Research*, C. Koncz, et al., eds., p. 284. World Scientific Publ. Co., Singapore; Tzfira, T., et al. 2000. *Annu. Rev. Microbiol.* 54:187; Kunik, T., et al. 2001. *Proc. Natl. Acad. Sci. USA* 98:1871; Wood, D. W., et al. 2001. *Science* 294:2317.)

agrocinopine Phosphorylated sugar, an opine, produced in octopine plasmids from mannopine by the enzyme agrocinopine synthase. *See Agrobacterium*; octopine; opines.

agroinfection Method of plant transformation. More than one genome of the double-stranded DNA of cauliflower mosaic virus is inserted in tandem within the T-DNA of *Agrobacterium tumefaciens*. Such a construction permits the escape of the viral DNA from the bacterial plasmid once it is introduced into plants. Geminiviruses can be introduced into plants in a similar way. See cauliflower mosaic virus; geminiviruses; transformation, genetic. (Grimsley, N., et al. 1989. *Mol. Gen. Genet.* 217:309.)

agropine Bicyclic phosphorylated sugar derivative of glutamic acid; it is synthesized by *Agrobacterium* strain Ach5. *See* opines.

Agropyron ($x = 7$) Genus of grasses; their chromosomes are homoeologous to those of several species within the genus of wheat and can be substituted to introduce agronomically useful genes (e.g., disease resistance). Some hybrids are known as perennial wheat, a forage crop. *See* alien transfer; chromosome substitution; homoeologous chromosomes.

AGRP Agouti-related protein. *See* agouti; obesity.

Ahonen blood group Rare type, distinct from ABO, MNS, P, Rh, Duffy, Kidd, and Dembrock. *See* blood groups.

AHR *See* arylhydrocarbon receptor.

AIB Steroid receptor implicated in breast cancer. See breast cancer; steroid hormones. (Anzick, S. L., et al. 1997. *Science* 277:965.)

Aicardi-Goutières syndrome (3p21) Progressive encephalopathy with calcification of the basal ganglia, excessive number of lymphocytes in the cerebrospinal fluid (lymphocytosis), brain atrophy, and early death. Elevated level of interferon-α in the brain fluids — in the absence of infection — is a marker for the disorder. (Crow, Y. J., et al. 2000. *Am. J. Hum. Genet.* 67:213.)

AICD *See* memory, immunological.

AID

(1) Artificial insemination by donor. *See* acquired immunodeficiency syndrome; AIH; ART; artificial insemination.

(2) Activation-induced deaminase. Its deficiency obliterates somatic hypermutation and class switching in the development of the secondary repertoire of antibody molecules, the last step in the generation of functional antibodies. *See* antibody; antibody gene switching; class switching; immune system; immunoglobulins. (Martin, A., et al. 2002. *Nature* 415:802.)

AIDS See acquired immunodeficiency syndrome.

AIF Apoptosis-inducing factor. A mammalian mitochondrial flavoprotein (M_r 57K) with homology to prokaryotic oxidoreductases. The encoding gene was located to human chromosome Xq25-q26. From purified mitochondria, AIF liberates — by increasing membrane permeability — cytochrome-c and caspase-9, and when injected into the nuclei it causes DNA breakage and thus promotes apoptosis. *See* ACINUS; APAF; apoptosis; CAD; L-DNase II; mitochondrial diseases in humans; mtPTP.

AIG Anchorage-independent growth. Normal mammalian cells grow in a monolayer anchored to a solid surface. Tumor cells grow independently of anchorage. *See* anchorage; cancer; CATR1; oncogenes; tumor.

AIH Artificial insemination by husband. *See* AID; ART; artificial insemination.

AIMS *Arabidopsis* information database at Michigan State University. *See Arabidopsis thaliana*.

AIR Mitotic kinase acting on histone.

AKAP A kinase anchoring proteins are cytoplasmic proteins binding to cyclic adenosine 3′, 5′ monophosphate (cAMP-dependent protein kinase PKA, calcineurin [phosphatase 2B], and protein kinase C [PKC]). They appear to have a regulatory role as a scaffold for the cellular signaling system. *See* condensin; signal transduction; T cell. (Colledge, M. & Scott, J. D. 1999. *Trends Cell Biol.* 9:216.)

AKI Adenylate kinase.

A-kinase cAMP-dependent protein phosphorylating enzyme; the phosphorylation is dependent on a sufficiently high level of cAMP. *See* cAMP.

AKR mice Long-inbred albino, specially selected strain of the animals containing the genes *Akv-1* and *Akv-2* that code for ecotropic retroviruses, causing thymic lymphosarcoma (leukemia). Ecotropic viruses replicate only in cells from which they have been isolated originally. AKR strains have a relatively short life span, are sensitive to ionizing radiation, and are highly susceptible to the carcinogenic effect, but resistant to the teratogenic effect of ethylnitrosourea. *See* ecotropic retrovirus; ENU; replicase.

Akt oncogene (PKB, serine/threonine protein kinase B) Isolated from thymomas (cancer of the thymus) of AKR mice transformed by an ecotropic virus. In the mouse genome it is located in chromosome 12. A homolog of it is found in human chromosome 14q32.3, and it is frequently associated with chromosomal breakage. The Akt protein is a threonine/serine protein kinase (protein kinase B/PKB) targeted by PI3-kinase-generated signals. Akt is involved in the regulation of cellular proliferation/apoptosis, glycogen synthase kinase (GSK3), endothelial nitric oxide synthase, and protein synthesis. Akt is also regulated by an insulin-like growth factor (IGF) and nerve growth factor (NGF). It is often called a cell survival kinase, activated by phosphoinosite kinase (PIK) and downregulated by the RAS oncoprotein or by a phosphatase. Akt reduces apoptosis by phosphorylating protein BAD and inhibiting Bcl and caspase-9 in human cells. Akt regulates NF-κB, which promotes the expression of antiapoptotic genes. When Akt phosphorylates Raf, the Raf-MEK-ERK signaling pathway is inhibited and cellular proliferation is initiated. The antiapoptotic PDGF also seems to be under Akt influence. TNF-α may or may not be involved in the regulation of NF-κB through IKK. Apoptotic FAS protein synthesis is apparently limited through blocking the FKK protein by Akt. *See* AKR; apoptosis; BAD; BCL; caspase; ecotropic retrovirus; FKK; glycogen; IKK; insulin-like growth factor; mouse; nerve growth factor; NFKB; nitric oxide; oncogenes; PDK; phosphoinositides; PKB; platelet-derived growth factor; signal transduction. (Datta, S. R., et al. 1999. *Genes & Development* 13:2905; Madrid, L. V., et al. 2001. *J. Biol. Chem.* 276:18934.)

ALA Aminolevulinic acid, a first compound in the synthesis of porphyrins from glycine and succinyl CoA. *See* heme.

Aladin (AAAS, triple-A syndrome) Neurological disorder resulting in alacrima (lack of tears), achalasia (failure of the smooth muscles to relax), and adrenal insufficiency because of a defect in a WD-repeat family of regulatory proteins encoded at 12q13. *See* WD-40.

Alagille syndrome Autosomal dominant (human chromosome 20p11.2), involving obstruction of the bile duct (cholestasis) and jaundice, lung anomalies (pulmonary stenosis), deformed vertebrae, arterial narrowness, deformed iris, altered eye pigmentation, facial anomalies, etc. The biochemical defect is in the Notch-ligand, Jagged-1. In some cases, translocations or deletions of the region accompany it. The multiplicity of the symptoms has been considered as a contiguous gene syndrome. The incidence is ~4 × 10⁻⁴. *See* BRIC, Fallot's tetralogy; Byler disease; cholestasis; contiguous gene syndrome; face/heart defects. (Spinner, N. B., et al. 2001. *Hum. Mutat.* 17:18.)

Aland Island eye disease *See* albinism, ocular.

alanine L-alanine ($CH_3CH[NH_2]COOH$) is a nonessential amino acid for mammals. The enantiomorph D-alanine may not be metabolized by some organisms and may even inhibit their growth. β-alanine ($CH_3CH_2CH_2CO_2COOH$) is synthesized by several microorganisms from aspartate, but it occurs only in trace amounts in animal tissues, possibly through the action of intestinal microorganisms; γ-butyric acid ($H_2NCH_2CH_2CH_2COOH$) is structurally related. Its dipeptides with histidine are carnosine, homocarnosine, and anserine. *See* alanine aminotransferase; alaninuria; carnosinemia.

alanine aminotransferase (glutamate-pyruvate transaminase, GPT) Autosomal-dominant gene (8q24.2-qter) encodes the enzyme that catalyzes the reversible transamination of pyruvate and α-ketoglutarate to alanine. This enzyme exists in cytosolic and mitochondrial forms. *See* alanine; alaninuria; amino acid metabolism; glutamate pyruvate transaminase.

alanine-scanning mutagenesis *See* homologue-scanning mutagenesis.

alaninuria (with microcephaly, dwarfism, enamel hypoplasia, and diabetes mellitus) The autosomal-recessive condition is accompanied by the clinically demonstrable excessive amounts of alanine, pyruvate, and lactate in the blood and urine. Both lactate and alanine are derived from pyruvate. *See* alanine; alanine aminotransferase; amino acid metabolism; diabetes; hypoplasia.

Albers-Schönberg disease *See* osteopetrosis type II.

albinism Pigment-free condition in plants and animals. The absence of skin and hair pigmentation in mammals is generally determined by homozygosity of recessive genes controlling melanin synthesis. Melanocytes are specialized cells for melanin synthesis. During embryonal development, melanoblasts, precursor cells of melanocytes, move to surface

Tyrosinase (*Tyr*, chromosome 7-44.0) alleles of mouse (left *c/c*, right *c/c⟨ch⟩ p/p*). Courtesy of Dr. Paul Szauter, <http://www.informatics.jax.org/mgihome/other/citation.shtml>.

Cuna Indian Albinos (Courtesy of Dr. C. Keeler).

areas. Melanin is synthesized in special cytoplasmic organelles called melanosomes. The precursor of melanin is the amino acid tyrosine, and the conversion is catalyzed by the aerobic oxidase, tyrosinase (polyphenol oxidase). In one type of albinism, tyrosinase activity in the hair follicle is still present, however. Albinism may involve the entire melanocyte system of the body or it may be limited to the eye (11q14-q21). In this case, a pigment-specific integral glycoprotein product of the (oculocutaneous) OCA1 gene is specially targeted to the intracellular melanosomes. OCA2 was assigned to 15q11.2-q12. OCA3 is at 6q13-q15. Albinism of the eye may occur in a sectorial manner in females heterozygous for the Xp22.3-22.2–linked recessive gene (OA1). Albinism of the eye may involve vision problems, involuntary eye movements (nystagmus), and head nodding. OA2 (Xp11.4-p11-23) males may be partially color blind (protonomalous). Albinism is controlled by numerous single genes. The prevalence of albinism varies from 1/14,000 to 1/60,000 depending on the gene and the ethnicity of the population; it is generally more frequent among negroids than caucasoids. Ocular albinism may occur very frequently among some Indian tribes (1/150). Albinism involves hypersensitivity to light and increased susceptibility to some forms of cancer. The absence of hair pigmentation may be locally corrected by gene therapy using the normal tyrosinase gene. Albinism may be a component of complex syndromes and may be associated with deafness, neuropathy, and bleeding disorders. Albina condition occurs with very high frequency in progenies of plants exposed to ionizing radiation. Over 100 mutated genes can cause albinism in plants. *See* Chédiak-Higashi syndrome; color blindness; hair color; Hermansky-Pudlak syndrome; Himalayan rabbit; light sensitivity diseases; motor proteins; oculocutaneous albinism; piebaldism; pigmentation, animal; tyrosinase; xeroderma pigmentosum. (Toyofuku, K., et al. 2001. *Biochem. J.* 355[pt2]:259.)

albino Animals defective in melanin synthesis. In plants, more commonly *albina* is the designation of leaf-pigment-free individuals (the Latin word *planta* is of feminine gender).

albizzin (L-2-amino-ureidopropionic acid) Glutamine analog. *See* asparagine synthetase.

albomaculata (or status albomaculatus) Green-yellow-white variegation caused by mutation in extranuclear genes in plants. *See* chloroplast genetics.

Albright hereditary osteodystrophy (pseudohypoparathyroidism, PHP1A, PHP1B, 20q13.2) Autosomal-recessive pseudoparathyroidism is based on a defect in a G-protein mutation. The locus actually encodes two G-protein subunits, Gs and XLα_s. Gs is expressed from both parents; XLα_s is expressed only from the paternal chromosome. There is also a third maternal transcript of the gene that encodes the neurosecretory NESP55 protein. The three transcripts share exons 2 to 11 but have different first exons. The NESP55 coding region is within exon 1 and the rest of the exons remain untranslated. An X-linked dominant form is based on a defect in the parathormone-adenylatecyclase-G$_s$-protein complex. *See* G-protein; hyperparathyroidism; imprinting; parathormone.

albumin Protein that is soluble in water and in dilute salt solutions, such as bovine serum albumin (BSA) used in chemical analyses. In the fetal serum of mammals the predominant protein is the albumin α-fetoprotein, transcribed from two genes in humans, and they have about 35% homology and are immunologically cross-reactive despite their substantial divergence. Both serum albumin and α-fetoprotein, products of the same gene family, are synthesized in the liver and gut. After birth the production of the latter drops dramatically, whereas the former is produced throughout life. Their tissue specificity of expression resides within 150 bp from the beginning of transcription. The α-fetoprotein gene carries three enhancer elements, 6.5, 5, and 2.5 kbp, upstream, which may increase the level of transcription up to 50-fold. The liver specificity for the serum albumin gene is controlled by the PE (proximal element), which is most important for promoter activity. It is located between the TATA and CCAAT boxes. The distal element (DE I [around -100 base from the initiation of translation]) is most important for liver specificity. There

are also DE II (around -116) and DE III (around -158). The PE element (5'-GTTAATGATCTAC-3') is quite similar to sequences in the promoters of other liver-expressed genes. The PE-binding protein, HNF-1 (88 kDa), is also shared by other liver genes, including the hepatitis virus promoter. The binding proteins associated with the DEs do not appear liver-specific inasmuch as they are used by a variety of other genes of ubiquitous expression, or the specificity is modified by unknown cofactors. *See* CCAAT box; enhancer; fetoprotein; HNF; promoter; serum; TATA box; transcription.

alcaptonuria *See* alkaptonuria.

alcohol Organic molecule formed from hydrocarbons by substituting $-OH$ for H. The simplest representative is ethanol (ethylalcohol, CH_3-CH_2-OH, MW 46.07, boiling point 78.3°C). Ethanol usually contains 5% water. Absolute ethanol is very hygroscopic; for disinfection the 60–80% solutions are most effective. Moderate alcohol consumption may protect against ischemic heart disease. This protection is attributed to modulation of blood lipoproteins and reduced activation of platelets and thrombosis. Protein kinase C(ε) signals may also be involved in the protection at physiological levels of blood alcohol (>10 mM). *See* ischemia; protein kinases; thrombosis.

alcohol dehydrogenase *See* ADH; mutation detection.

alcohol fermentation Conversion of sugar into alcohol in the absence of air by glycolysis. *See* glycolysis.

alcoholism Chronic and addictive use of the chemical is a behavioral trait with some hereditary component of the manifestation. Alcoholism may involve fatal or very serious consequences in certain diseases, in pregnancy, and when certain types of medicines or drugs are used. Fetal alcohol syndrome includes microcephaly (small head), folded skin at the side of the nose, defective eyelids, upturned nose, etc. In adults it may cause cirrhosis of the liver, leading to further (fatal) complications. Maternal alcohol blocks the NMDA glutamate receptors and activates the GABA$_A$ receptors in the fetus, resulting in a long-lasting process of neurodegeneration. Unfortunately, no association between alcoholism and any particular gene or chromosomal segment has been firmly established; it is apparently under polygenic control.

Alcohol abuse during pregnancy may expose the fetus and the newborn to serious developmental harm (fetal alcohol syndrome, FAS), including physical and mental retardation that may seriously affect the lifelong health and function of the individuals. FAS is an increasingly serious social problem along with other abuse of drugs. About 0.001–0.002% of children are suffering from it. In mice, the higher alcohol consumption appeared to be associated with defects in the 5-HT$_{1b}$ serotonin receptor and genetic variations (Lys487Glu) of the aldehyde dehydrogenase 1 locus. The (ADH1 and 2) aldehyde dehydrogenases are present in some Far East human populations and may increase the proclivity to alcohol consumption by a factor of 5 to 10. In some other populations low in ADH1, the alcohol consumption is moderate because of poor tolerance. ADH actually has a protective effect against alcoholism.

In mice, the *Alp1* locus may be responsible for 14% and the *Alcp2* for 18% of the total alcohol preference. Interestingly the former gene is acting only in males and the latter only in females. Other loci controlling alcohol withdrawal sensitivity (chromosome 1), alcohol-induced hypothermia and amphetamine-induced hyperthermia, and another hyperthermia loci seem to be in chromosome 9 of mice. These loci do not appear to be controlling general tendencies for substance abuse. Alcohol preference appears to be a quantitative trait with ~0.39 heritability and ~60% concordance between monozygotic twins. Sensitivity to the effects of alcohol may be affected by a GABA$_A$-type receptor and mediated by protein kinase Cε. *See* addiction; aldehyde dehydrogenase; behavior, in humans; Dubowitz syndrome; GABA; glutamate receptor; NMDA receptor; polygenic inheritance; protein kinases; QTL; serotonin; substance abuse; teratogenesis. (Almasy, L. 2001. *Am. J. Hum. Genet.* 68:128; Sillaber, I., et al. 2002. *Science* 296:931; Weiss, F. & Porrino, L. J. 2002. *J. Neurosci.* 22:3332; Yao, L., et al. 2002. *Cell* 109:733.)

aldehyde dehydrogenase (ALDH, acetaldehyde dehydrogenase)   Form ALDH1 is encoded in human chromosome 9q21. A low level of this form of the enzyme is responsible for poor alcohol tolerance. ALDH2 functions in the liver and it is encoded in 12q24.2, and ALDH3 is encoded in human chromosome 17. ALDH also oxidizes cyclophosphamide-derivative aldophosphamide to nontoxic carboxyphosphamide. *See* alcohol dehydrogenase; cyclophosphamide.

Aldehyde

$$\begin{array}{c} H \\ | \\ R-C=O \end{array}$$

aldolase-1 (fructose-1,6-bisphosphate aldolase)   The ALDOA isozyme has been mapped to human chromosome 16q22-q24, ALDOB (fructose intolerance) in human chromosome 9q22, ALDOC in human chromosome 17cen-q12. There is also a deoxyribose-5-phosphate aldolase, but its deficiency is apparently harmless. ALDOA deficiency may be involved in a form of hemolytic anemia (γ-glutamylcysteine synthetase deficiency). *See* anemia; fructose intolerance.

aldose   Sugar that ends with a carbonyl group (=C=O).

aldosterone (18-aldo-corticosterone)   Main electrolyte-regulating steroid hormone of the kidney cortex. *See* aldosteronism; steroid hormones.

aldosteronism (hyperaldosteronism, glucocorticoid-remediable aldosteronism [GRA])   Controlled by two autosomal-dominant genes. It is due to overactivity of aldosterone synthase (ADOS) and steroid 11β-hydroxylase (CYP11B2), coded in human chromosome 8q21. These two genes are quite similar in structure and frequently also form somatic recombinants. Increased aldosterone production and hypertension result. This hyperaldosteronism is suppressible by glucocorticoids and dexamethasone. The chimeric genes are under adrenocorticotropic hormone control. As a consequence, aldosterone is secreted and causes water and salt reabsorption and high blood pressure. *See* aldosterone; dexamethasone; glucocorticoid; hypertension; hypoaldosteronism; mineral corticoid syndrome; pseudohypoaldosteronism.

aleuron   Protein-rich outer layer of the endosperm of monocotyledonous kernels. In wheat and maize there is only one cell layer of aleurone, in barley there are three, and in rice the number is variable. There are about 250,000 cells in the maize aleurone and about 100,000 in barley. Aleurone color genes have proven to be very useful chromosomal markers (such as loci *A*, *C*, *R*, *Bz*, and *B* in maize and some of the functional homologs in other cereals). The dominant alleles can be identified in the seeds of the heterozygotes, and in the case of maize, they can be classified on the cob in immobilized condition and in large numbers. *See* maize.

Alexander's disease   Autosomal-recessive anomaly of lipid metabolism accompanied by megaencephaly (synonym macroencephaly), a pathological enlargement of the brain. Astrocyte fibers and small heat-shock proteins may be involved. Defects in NDUFV may be responsible for the symptoms. *See* NDUFV. (Brenner, M., et al. 2001. *Nature Genet.* 27:117.)

alfalfa (*Medicago sativa*)   Leguminous forage plant. Its closest wild relatives are *M. coerulea* ($2n = 16$) and the somewhat more distant *M. falcata* ($2n = 16$). *M. sativa* is autotetraploid ($2n = 4x = 32$). It is also called lucerne. See <http://www.tigr.org/tdb/tgi.shtml> for gene index.

alfalfa mosaic virus   Genetic material is four RNAs of 1.3, 1.0, 0.7, and 0.34×10^6 Da.

alga   Can be prokaryotic, such as blue-green algae, or eukaryotic, such as *Chlamydomonas reinhardtii*, *Ch. eugametos*, *Euglena gracilis*, seaweeds, etc.; they are photosynthetic microorganisms. *See* Chlamydomonas; Euglena.

algeny   Genetic alteration of an organism by nonnatural means such as genetic engineering, gene therapy, genetic surgery, or transformation. *See* genetic engineering; genetherapy; genetic surgery; transformation.

alginate   Polymer of mannuronic acid and guluronate in the cell wall of brown algae. *See* biofilm. (Wong, T. Y., et al. 2000. *Annu. Rev. Microbiol.* 54:289.)

ALGOL   Algorithmic-oriented language is a computer language set by international procedures. *See* algorithm.

algorithm   Set of rules and procedures for solving problems in a finite number of sets; usually the repetitive calculation is aimed in finding the greatest common divisor of two members. The computer programs include algorithms.

alien addition   Chromosome(s) of another species added to the genome of polyploids without seriously disturbing genic balance, in contrast to diploids, where even small duplications or deletions may become quite deleterious. The procedure of addition is crossing the higher chromosome number species as pistillate parent with the lower chromosome number pollen donor. The F$_1$ is generally sterile, but doubling their number (with colchicine) may result in a fertile amphiploid. Upon recurrent backcrossing the amphiploid with the recipient parent, monosomy results for the donor's chromosomes. After repeated backcrossing in large populations, one may obtain plants with single monosomes for all chromosomes of the donor. These are called single monosomic addition lines. Upon selfing such individuals, disomic additions are obtained. These carry one pair of extra chromosomes. The purpose of addition is that occasionally the added chromosome, containing agronomically useful genes, may be substituted for its homologue, then a substitution line results. *See* addition line; amphidiploid; disomic; homologue; monosome; pistillate; pollen; substitution line. *See also* diagram at right. (Sears, E. R. 1953. *Am. J. Bot.* 40:168.)

alien substitution   Chromosome(s) of another species replacing own chromosome(s) of a species. Alien substitutions may be obtained from alien addition lines. Most commonly, however, monosomic lines are used. In polyploids, monosomic lines can be maintained without too much difficulty because the genomes are better balanced. Monosomic plants can produce some nullisomic eggs. These recipients are then crossed with a donor species. In F₁, the chromosome absent in the nullisomic will appear as a monosome of the donor. These monosomic individuals are repeatedly backcrossed (6–8 times) with the recipient until in some individuals all the chromosomes of the donor are eliminated, except that particular monosome.

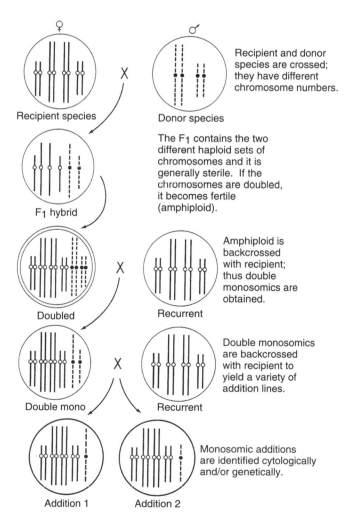

Recipient and donor species are crossed; they have different chromosome numbers.

The F₁ contains the two different haploid sets of chromosomes and it is generally sterile. If the chromosomes are doubled, it becomes fertile (amphiploid).

Amphiploid is backcrossed with recipient; thus double monosomics are obtained.

Double monosomics are backcrossed with recipient to yield a variety of addition lines.

Monosomic additions are identified cytologically and/or genetically.

Alien addition continued from preceding page. The general scheme for the generation alien addition lines in wheat. The same procedure is applicable to other polyploid plant species.

Upon selfing this monosomic substitution line, a disomic substitution results that has the same chromosome number as the euploid line from which the nullisomic arose, but one pair of the chromosomes will represent the donor. *See* alien addition; alien transfer lines; backcross; chromosome substitution; euploid; monosomic; nullisomic; selfing. (Sears, E. R. 1972. *Stadler Symp.* 4:23.)

alien transfer lines   Alien substitution lines are genetically interesting, but the plant breeder is rarely satisfied with the substitution of an entire chromosome because it may contain, besides the desirable gene(s), some undesirable ones. Homologous pairing and crossing over can borrow short segments or even single genes from the alien chromosome. Also, induced translocations may achieve this goal, and the result of these operations is called a transfer line. *See* alien substitution; translocation. (Sears, E. R. 1956. *Brookhaven Symp. Biol.* 9:1.)

AlignACE   Is a computer program to locate upstream regions of regulons. It enables the identification of genes in a functional pathway or genes homologous to known regulons or group of genes derived from conserved operons. (See regulon, McGuire, A. M., et al. 2000. *Genome Res.* 10:744.)

AlignMaker   Computer program that produces a file resembling MapSearch (<http://www.owlnet.rice.edu/~bios311/bios312day3.html>) output so that it can be read by the AlterMap program.

alignment   Finding the nucleotide or amino acid linear sequence matches in nucleic acids and polypeptides, respectively. High alignment scores indicate good similarities between sequences. Free aligning software may be obtained. *See* BLAST; homology.

ALK   *See* anaplastic lymphoma.

alkaline chromatography   Used for the rapid separation of less than 150 base-long DNA probes on sepharose CL-4B (a beaded [60–14 μm pore size], cross-linked agarose). *See* sepharose.

alkaline lysis   Procedure to extract plasmid DNA from bacterial cells by 0.2 N NaOH and 1% SDS. The plasmids may be further purified by CsCl-ethidium bromide gradient ultracentrifugation or by polyethylene glycol precipitation at 10,000 rpm. *See* SDS.

alkaline phosphatase   Cleaves phosphates at a pH optimum of about 9; it is present in microorganisms and animal cells but absent from plant tissues. The alkaline phosphatase of *E. coli* is 86 kDA and contains two subunits. Human intestinal alkaline phosphatase (ALP1) encoded in chromosome 2q37, the placental (ALPPP/PLAP) enzyme, is also in chromosome 2q37; the liver enzyme is coded by ALPL in chromosome 1p36.1-p34. The intestinal alkaline phosphatase is at 2q36.3-q37.1. Higher- or lower-than-normal levels of these enzymes may have deleterious consequences. Hypophosphatasia for ALP may have very serious or even lethal consequences for infants. *See* acid phosphatase; acrodermatitis enteropathica.

alkaloids   Diverse (more than 2500), mainly heterocyclic, organic compounds containing nitrogen, generally alkalic in nature, and secondary metabolites of plants; frequently of strong biological activity at higher concentrations (such as nicotine, caffeine, cocaine, morphine, strychnine, quinine, papaverine, atropine, hyosciamine, scopolamine codeine, capsaicine, lupinin, etc.). Of particular interest is colchicine (obtained from the lily, *Colchicum autumnale*), which blocks the microtubules and causes chromosome doubling in cells. Vincristine and vinblastine (from *Vinca rosea*) are antineoplastic drugs. The "animal alkaloid," ptomain, found in decomposing cadavers, is actually a microbial product. *See* caffeine; capsaicine; cocaine; colchicine; *Datura* alkaloids; lupine; morphine; nicotine; vinblastine. (Facchini, P. J. 2001. *Annu. Rev. Plant Physiol. Mol. Biol.* 52:29.)

alkalosis   Diminished buffering capacity of tissues, resulting in higher pH.

alkane   Aliphatic molecule joined by a single covalent bond, e.g., CH_3-CH_3. *See* alkene.

alkaptonuria (alcaptonuria, AKU)   Recessive metabolic disorder (prevalence about 1/40,000) with a defect in the enzyme homogentisate 1,2-deoxygenase; therefore, homogentisic acid is not metabolized into maleyl- and fumaryl-acetoacetic acids and eventually to acetoacetic acid and fumaric acid. The degradation of the aromatic amino acids follows the pathway:

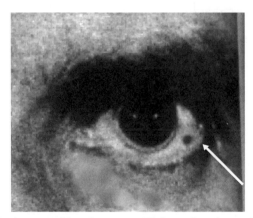

Phenylalanine → Tyrosine → pHydroxyphenylpyruvicacid → **Homogentisicacid** → ■ Maleylacetoaceticacid ⇒ Fumarylacetoacetic acid ⇒ Fumaricacid & Acetoaceticacid.

The accumulated homogentisic acid is excreted in the urine and it is readily oxidized into a dark compound (see on the cornea ↑), alkapton, darkly staining the diapers of affected newborns. Also, it involves dark pigmented spots in the connective tissues and bones, and later during development arthritis may follow. This human hereditary biochemical defect was first recognized in 1859, and Sir Archibald Garrod identified its genetic control in 1902. The gene (AKU) was located to human chromosome 3q21-q23. *See* amino acid metabolism; phenylketonuria; tyrosine; tyrosinemia.

alkene   Hydrocarbons with one or more double bonds, e.g., $H_2C=CH_2$. *See* alkane.

alkylating agent   Alkylates other molecules; many of the chemical mutagens and carcinogens are alkylating agents. The mutationally most effective alkylation site in the DNA is the O^6 site of guanine. The alkyl may be removed from the DNA in *E. coli* by an alkyltransferase enzyme. The acceptor may be a cysteine residue of that protein. *See* alkyltransferase; carcinogen; chemical mutagens; environmental mutagens; mutagenic potency; mutagens-carcinogens. *See* also diagram of alkylation by ethylmethane sulfonate below.

alkylation   Addition of a CH_3 group (or other member of the alkane series) to a molecule. Alkylation of DNA bases may lead to mutation through mispairing and base substitution. Also, alkylation may lead to disruption of the sugar-phosphate backbone of the DNA through depurination by AP nucleases. Thymine is alkylated at the O^4 position and adenine at the N3 position. *See* alkyltransferases; AP endonuclease; base pairing; chemical mutagens; hydrogen pairing; methyltransferase; tautomeric shift. (Bautz, E. & Freese, E. 1960. *Proc. Natl. Acad. Sci. USA* 46:1585.)

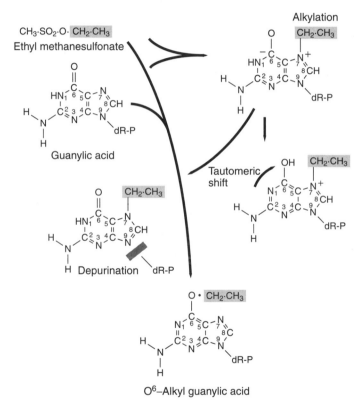

Most common reactions of the alkylating agent, ethylmethane sulfonate with guanine.

alklyltransferase   Protects the DNA against alkyl adducts by transferring the methyl or ethyl (alkyl) groups to cysteine and repairing the damage. The enzyme present in different organisms displays an active site consensus V(I)PCHRV(I). If the level of these enzymes is reduced either by inhibitors, e.g., O^6-benzylguanine, or by mutation, the efficiency of alkylating agents for treatment of cancer increases. *See* DNA repair. (Reese, J. S., et al. 2001. *Oncogene* 20:5258.)

allantois   Tubular part of the hindgut, later forming the umbilical cord of the fetus, and it fuses with the chorion. It participates in the formation of the placenta. *See* amnion; chorion.

Allegro Computer program for multipoint linkage, free at <allegro@decode.is>.

allele Alternative states of a gene (e.g., a^1 and a^2). Hybrids of a^1/a^2 are commonly of mutant phenotype, although they may show incomplete (allelic) complementation. Two alleles are identical if their base sequence is identical although different from the wild type. *Nonidentical alleles* are still in the same gene (and are noncomplementary), yet their expression may be distinguishable. *Homoalleles* are affected in the same codon, but at the same site a different nucleotide occurs, and therefore they cannot be separated by recombination in a heterozygote for the locus. *Heteroalleles* have their differences at nonidentical sites within the codon or in another codon; therefore, they can be separated by recombination. *Isoalleles* convey wild phenotype, yet under special circumstances they can be recognized by appearance. *Multiple alleles* are alleles of the same locus, but more than two alternatives exist. *Superalleles* are additional mutations in *cis* to an allele within a gene that reinforces their expression. In some organisms, alleles of the *a1* locus are symbolized as *a1-1*, *a1-2*, etc. Molecular geneticists involved in physical mapping of the DNA use this term for any DNA difference (e.g., restriction fragment) that displays Mendelian inheritance and occupies the same chromosomal site. *See* coalescent; gene symbols; Mendelian segregation; mutation, age of; RAPD; RFLP. (Slatkin, M. & Rannala, B. 2000. *Annu. Rev. Genomics. Hum. Genet.* 1:225.)

allele-sharing methods Used for the detection of linkage by examining pedigrees for whether a particular genetic locus (chromosomal fragment) is more common among individuals in a pedigree than expected by random segregation. It is basically a nonparametric method. The probability of allele sharing may be denoted by Y; the probability that R alleles are shared among

$$2N = \binom{2N}{R} Y R (1-Y)^{2N-R},$$

and for R shared alleles the maximum lod score $= R log_{10} R + (2N - R) log_{10}(2N - R) - 2N log_{10} N$. *See* lod score; nonparametric tests. (Nyholt, D. R. 2000. *Am. J. Hum. Genet.* 67:282.)

allele-specific probe for mutation (ASP) In principle, this would detect single base change mutations because oligonucleotide probes (ASO) under very high stringency of hybridization would hybridize only to an exactly matching sequence but not to another that has one base pair substitution. This would also identify heterozygotes because they would hybridize to both types of probes, mutant and normal. This procedure requires high skills but can be semiautomated. *See* ASO; hybridization; mutation detection; probe; SNIP. (Prince, J. A., et al. 2001. *Genome Res.* 11:152.)

allelic association *See* linkage disequilibrium.

allelic combination In gametes (at independent loci) can be predicted by $2n$ where n is the number of different allelic pairs, and it produces $4n$ gametic combinations, and the number of genotypes is $3n$. If the number of loci is n and each has a number of alleles, the number of zygotic genotypes can be calculated at 1 locus as $[a \times (a+1)]/2$ and for n loci: $\{[a \times (a+1)]/2\}^n$. Thus, e.g., for 100 loci, each with 3 alleles $[(3 \times 4)/2]^{100}$, 6.53×10^{77} zygotic genotypes are possible. *See* gametic array; Mendelian segregation; multiple alleles.

allelic complementation Partial or incomplete complementation among mutant alleles of a gene representing different cistrons. If the alleles are defective when homozygous, they do not contribute to the synthesis of functional proteins. Each of the two in a heterozygote has another nonoverlapping defective polypeptide product. In the cytoplasm the correct polypeptide chains may combine in the heterodimeric or heteropolymeric proteins, and due to the right assembly, the function of the protein may be restored. Since the available correct polypeptide chains are reduced in number relative to that in the wild type, only a reduced number of good protein molecules can be formed. Therefore, allelic complementation is incomplete. The beneficial effect of the nondefective peptide chains may also be brought about by conformation correction, i.e., the conformation of the defective chains is brought into line as an effect of the other polypeptide chain as long as there is no defect at the active site. The extent of allelic complementation can be best determined by in vitro enzyme assays when regulatory genes cannot modify the functions by higher intensity or prolonged transcription of the relevant cistrons. *See* allelism test; complementation mapping; conformation; nonallelic noncomplementation; step allelomorphism. (Li, S. L. & Rédei, G. P. 1969. *Genetics* 62:281.)

allelic dropout Occurs when during amplification (by PCR) a microsatellite locus is not replicated by the DNA polymerase. *See* PCR. (Miller, C. R., et al. 2002. *Genetics* 160:357.)

allelic exclusion Only one of the two alleles at a locus is expressed or one type of chain rearrangement is functional. Such conditions are found in immunoglobulins, in various receptors, interleukin-2, and imprinted genes. Protein kinase C (PKC) modulates both the differentiation and allelic exclusion during thymocyte differentiation. *See* immunoglobulins; imprinting; interleukin; monoallelic expression; PKC; thymocytes. (Michie, A. M., et al. 2001. *Proc. Natl. Acad. Sci. USA* 98:609; Borst, P. 2002. *Cell* 109:5)

allelic fixation In a random mating population allelic fixation takes place when one allele completely replaces another, and the process depends on the coefficient of selection and the size of the populations (see diagrams next page).

allelic frequency Can be determined on the basis of the Hardy-Weinberg theorem; i.e., the genotypic composition of a random mating population is $p^2 + 2pq + q^2$, where p^2 and q^2 are the frequencies of the homozygous dominants and recessives, respectively. Thus, if we consider a single allelic pair, A and a, and diploidy, the frequency of the A allele = double the number of homozygous dominants plus the number of heterozygotes. The frequency of the a allele = double the number of homozygotes plus the heterozygotes because the homozygotes have two copies of the same allele, whereas the heterozygotes have only one of each kind. The frequency of the recessive alleles in an equilibrium population is simply $1 - p (= q)$. In case of dominance, the heterozygotes may not

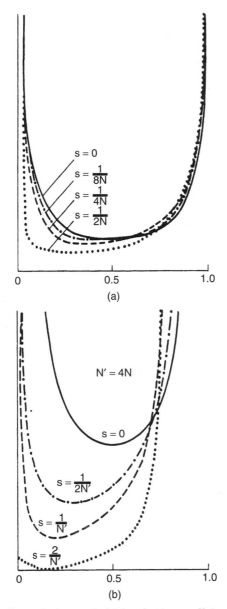

(a)

(b)

In very small populations only high selection coefficients can bring about changes in allelic distribution (**A**). When the size of the population increases four-fold, relatively small selection coefficients effectively modify the direction of fixation (**B**). The ordinate represents allelic densities; the abscissa shows allelic frequencies from loss, (0) to complete fixation (1.0); s = selection coefficient, N = population size. (From Wright, S. 1931. *Genetics* 16:97.)

be directly recognized; therefore, this procedure may not work. However, in case the population is at equilibrium and the mating is at random, the frequency of the recessive alleles is $q = \sqrt{q^2}$. If the size of the homozygous-recessive class is very small, the vast majority of the recessive alleles are in the heterozygotes. In case of sex linkage, the male carries one dose (XY), whereas the female is XX. Thus, the males more frequently display a recessive trait than the females, which express the trait only when homozygous. If the expression of an X-linked recessive allele is 0.10 in the males, in the females it is expected to show up at $(0.10)^2 = 0.01$. The frequency of alleles in a population may change by selection, mutation, random drift, and migration. At random mating the total

variance of allelic frequency with two alleles of $2n$ genes is computed from the Weir formula

$$V(p) = \frac{p(1-p)}{2n}[F_{ST}(2n-1)+1$$

and

$$F_{ST} = \frac{\sigma 2 - 1}{2n - 1}.$$

See Hardy-Weinberg theorem; mutation; random genetic drift; selection. (Hung, S.-P. & Weir, B. S. 2001. *Genetics* 159:1365; <http://info. med.yale.edu/genetics/kkidd>; <http://alfred.med.yale.edu/alfred/AboutALFRED.asp>.)

allelic interaction   *See* overdominance.

allelic recombination   Takes place between the same sites of the homologous chromosomes. *See* ectopic recombination; homologous recombination.

allelic rescue   Procedure for cloning a mutant allele. A vector carrying the wild-type allele, which has, however, an internal deletion overlapping the mutant site, transforms the mutant cell. When the cellular (gap) repair system fills in the deleted sequences using the mutant template, the plasmid vector carries a copy of the mutant gene, and it can then be isolated along with the vector. *See* DNA repair; marker rescue; transformation.

allelism test   Carried out by complementation test. If two recessive genes are allelic, they fail to complement each other in the F_1 hybrids (i.e., the hybrid is of mutant phenotype). In case the hybrid of two recessive individuals is of wild phenotype (i.e., they complement each other), the two genes are not allelic. Thus, the number of complementation groups reveals the number of different loci. In practice, when people talk about complementation groups, this is understood as different complementation groups. However, allelic genes at the same multicistronic locus may show partial or allelic complementation. *See* allelic complementation; nonallelic noncomplementation. (Demerec, M. & Ozeki, H. 1959. *Genetics* 44:269.)

allelomorph   Historical term for allele. *See* allele.

allelotyping   Determination of the spectrum and frequency of allelic variations in a population. Polymorphism may be determined by restriction fragment length, SNIPS, LOH, PCR, etc. *See* genotyping; LOH; polymerase chain reaction; RFLP; SNIPS. (Girard, L., et al. 2000. *Cancer Res.* 60:4894.)

allergen   Any substance causing allergy. *See* allergy.

allergy   Sensitivity to a particular antigen(s) showing an immunological reaction. Common forms are hay fever after exposure to pollen (ragweed), or drug, food, bacteria, cold, etc. The allergic reaction may be a hereditary property (atopy). In asthma, hay fever, and various other allergic reactions, the regulatory *Re* gene was implicated in the decrease of immunoglobulin E, and in *re/re* individuals it seemed to be higher. The frequency of the *re* gene was estimated to be about 0.49.

The IgE response in about 60% of the atopy cases was assigned to chromosome 11q13-q12, the site of the high-affinity IgE receptor (FcεRI-β) gene. The IgE response is apparently controlled by IgE receptor (FcεRI), and it is regulated by interleukin 4 (IL-4). It also appears that IgG Fc receptor, FcγRIII, affects FcεRI assembly. The ragweed sensitivity was assigned to the HLA complex in human chromosome 6. Elevated levels of immunoglobulin E, controlled by an autosomal-dominant gene with incomplete penetrance, were detected in the neutrophil chemotaxis defect, characterized by chronic eczema, repeated infections by staphylococci, and eosinophilia (cytological structures readily stained with eosin).

Asthma and other allergies are apparently under the control of multiple genetic loci. Allergies may be alleviated by desensitization that involves exposure to increasing amounts of the allergen, and this downregulates IgE production. DNA immunization may stimulate Th1 immunity by the production of IgG2a and IFN-γ or Th2 response and the production of IgE and IgG and the increase in interleukins (IL-4, -5, -10). Actually, Th1 cells antagonize the inflammatory reaction, whereas Th2 cells, with the aid of IL-3, IL-5, and GM-CSF, stimulate eosinophils through IL-3, IL-4, IL-6, and IL-9 regulate mast cells and inflammation. Dendritic cells negatively regulate Th2 cells with the aid of IL-12, and Th2 cells lower the response of dendritic cells to allergens by IL-10. Bacteria and viruses promote the production of IL-12 and thus stimulate Th1 cells. Th1 cells boost the defense against intracellular pathogens (bacteria and viruses) by increased production of IFN-γ and synthesis of IgG2a. The Th2 cells and IgE and IgG1 mediate defense against large extracellular pathogens. The allergens of fungi and other parasites boost the level of Th2 cells and elevate IgE level in the serum. IL-4 and IL-13 enhance IgE production by B lymphocytes and thus evoke inflammatory responses. The CD23 receptor of IgE may promote or hinder antibody presentation depending on the circumstances. In grass hay fever, CD4$^+$ CD30$^+$ Th2 cells react to the allergen. Glucocorticoids and IL-4 enhance Th2 activity, whereas dehydroepiandrosterol favors Th1 cells. IFNα, IL-12, and TGF-β expand Th1. IL-10, IL-6, and IL-4 skew the balance toward Th2. Allergen recognition through MHC class II peptides and organ localization and response to allergens have clear genetic components. However, environmental factors have very important roles as the overall incidence of allergy and asthma is on the rise. *See* anaphylaxis; asthma; atopy; CD4$^+$; CD23; CD30$^+$; eczema; $\gamma\delta$ T cell; glucocorticoid; histamine; HLA; hypersensitive reaction; IFN; immunization, genetic; immunoglobulins; IL-3; IL-4; IL-5; IL-6; IL-10; IL-12; IL-13; interferons; interleukin; ragweed; TGF. (*Nature* 402[6760] Suppl.)

alligator *Alligator missisipiensis*, $2n = 36$.

Allium (onion, garlic) Monocot genus, $2n = 16$ or 32; well suited for cytological analysis.

alloantibody (isoantibody) Produced by an individual of a species against alloantigens within the species. These may be due to preceding transfusions or pregnancies and may cause hyperacute rejection in case of transplantation in another individual of the same species. *See* alloantigen.

alloantigen Genetically determined antigen variant within the species. It may also be called neoantigen when the epitope appears the first time. It is recognized within the same species by the lymphocytes with different haplotypes. *See* alloantibody; antigen; epitope; haplotype; isoallogen; lymphocyte.

alloantisera Antibodies that can recognize a certain protein in a different individual.

allocatalasia Characterized by normal catalase activity and stability, yet the protein is a different variant. *See* catalase.

allocation Differential distribution of cellular resources to specific structures or organs in an individual organism.

allochronic species Do not occur at the same time level with others during evolution.

allocycly Chromosomal regions or chromosomes or genomes may show cyclic variation in coiling and heteropycnosis. *See* Barr body; heterochromatin; heteropycnosis; lyonization.

allodiploid Polyploid that has chromosome sets (genomes) derived from more than one ancestral organism, e.g., hexaploid bread wheat has A, B, and D genomes. *See* autotetraploid; *Triticum*.

allogamous The individuals are not pollinating the pistils of the same plant; they are cross-pollinating, and they are also called exogamous species.

allogamy Fertilization by gametes coming from a different individual(s). *See* autogamy.

allogeneic Antigenic difference exists between two cells (in a chimera). *See* allograft; antigen; autologous; isoallogen; xenogeneic.

allogeneic inhibition In mice, parental cells do no accept a graft from the F$_1$, but the reciprocal graft may be successful. *See* grafting. (Mathew, J. M., et al. 2000. *Transplantation* 70:1752.)

allogenic Same as allogeneic.

allograft Transplantation of tissues that carry cell surface antigens not present in the recipient. The result may be rejection and destruction of the graft and harm to the recipient. *See* complement; graft; grafting in medicine; heterograft; HLA; isograft; xenotransplantation.

allohaploid Haploid cell derived from an allodiploid. *See* allodiploid.

allolactose Inducer of the *Lac* operon; it is the intermediate product of lactose (a disaccharide) digestion by β-galactosidase, and it is then converted to galactose and glucose by the same enzyme. *See* galactosidase; *Lac* operon; lactose.

allometric development Different growth (development) rate of one part of the body relative to other parts.

allometry Study of growth of organs in different dimensions of space and time within an individual, populations, or during evolution.

allomixis Cross-fertilization, allogamy.

allopatric speciation Involved in geographic adaptation and sexual isolation of species living in nonidentical habitats. *See* postzygotic isolation; speciation.

allophenic The original meaning was that some genes may not be expressed in one cell type but can act as gene activators in other tissues. Also, the expression of genes in chimeric tissues of an embryo or adult that has been produced through in vitro fusion of two or more genetically different (chimeric) blastomeres. These blastomeres developed upon the fusion of the gametes of two parents, and several different blastomeres can be fused, resulting in (tri-, quadri-, hexa-parental, etc.) multiparental offspring developing from these allophenic individuals. The fused blastomeres are implanted into the uterus of pseudopregnant animals, which carry the developing mosaic embryos to terms. The opposite of allophenic chimeras is splitting up eight-cell embryos, in two steps, into separate blastomeres and insertion of four such cells into an empty zona pellucida. Subsequently these "quadruplets" can be transferred into the uterus of a rhesus monkey, which has produced a viable, normal offspring by this procedure (Chan, A. W. S., et al. 2000. *Science* 287:317). This type of cloning offers the means for producing progeny identical in both nuclear and cytoplasmic hereditary components. (See biparental; blastomere; chimera; multiparental, LoCascio, N. J., et al. 1987. *Dev. Biol.* 124:291; Petters, R. M. & Markert, C. L. 1980. *J. Hered.* 71:70.)

allophycocyanin A fluorochrome; it is excited at wavelength 610 and 640 nm and emits bright red light at 650 nm. It is used for flow cytometry. *See* flow cytometry; fluorochromes; phycobilins.

alloplasmic The cytoplasm and the nucleus are of different origin. *See* cell genetics; nuclear transplantation.

allopolyploid Containing two or more kinds of genomes from different species, e.g., *Triticum turgidum* (macaroni wheat) is an allotetraploid containing the AABB genomes and *Triticum aestivum* (bread wheat) is an allohexaploid with AABBDD genomes or *Triticum crassum* (a wild grass) hexaploid DDDDMM. *Nicotiana tabacum* is an allotetraploid ($2n = 48$) containing the genomes of *N. tomentosiformis* ($2n = 24$) and *N. sylvestris* ($2n = 24$). When *N. tabacum* is crossed with either of the progenitors in the F_1, there will be 12 bivalent (12″) and 12 univalent (12′) chromosomes. On the basis of chromosome pairing and chiasma frequency, the degree of homology between genomes can cytologically be determined in meiosis. Allopolyploids generally acquired genes during evolution that suppress multivalent pairing of the chromosomes; therefore the gene segregation pattern resembles that of diploids with more than one pair of alleles. A duplex autotetraploid may segregate in 35:1 to 19.3:1 (depending on the distance between gene and centromere), but an allotetraploid is expected to display a 15:1 ratio, and

an allohexaploid a 63:1 proportion if there are four and six copies of the genes, respectively. Some genes (which have only two alleles) in hexaploids may display a 3:1 segregation. *See* allopolyploid, segmental; autopolyploid; duplex, sesquidiploid.

allopolyploid, segmental Participating genomes have partial (segmental) homology yet are sufficiently different to cause some sterility. *See* allopolyploid.

alloproteins Contain nonnatural amino acids. (*See* Kiga, D., et al. 2002. *Proc. Natl. Acad. Sci. USA* 99:9715.)

allopurinol (hydroxypyrazole pyrimidine) Inhibitor of de novo pyrimidine synthesis and xanthine oxidase activity. It is a gout medicine. *See* gout; xanthine.

alloreactive (allorestrictive) T cell that recognizes foreign antigen and mobilizes cellular defense against it, e.g., graft rejection. *See* antigen; T cell.

all-or-none trait Either present or absent, and there are no intermediates.

allospecific Specificity is different from the standard normal.

allostatin Juvenile hormone inhibitor in insects with highly conserved six C-terminal amino acids. *See* juvenile hormone.

allosteric control Modification of the activity of an enzyme by alteration at a site different from the active site by another molecule affecting its conformation without a covalent attachment. *See* active site; conformation; intrasteric regulation.

allosteric effector Molecule involved in bringing about allosteric control. *See* allosteric control; allostery.

allostery Conformational change in a protein (ribozyme) through the effect of a ligand molecule; the process is often called allosteric shift. *See* allosteric control. (Monod, J., et al. 1963. *J. Mol. Biol.* 6:306.)

allosyndesis Synapsis between non–entirely homologous chromosomes in an allopolyploid. *See* chromosome pairing; homologous chromosomes.

allotetraploid *See* amphidiploid.

allotopic expression Gene that is not organellar (mitochondrial or plastidic by origin) is targeted and expressed in an organelle. *See* ectopic expression.

allotype Difference in antibody (or antigen) caused presumably by allelic substitution mutation in the same constant region genes. *See* antibody; immunoglobulins; isotype.

allozygote Individual that at one or more loci possesses alleles not derived from the same common ancestor, i.e., are not identical by descent. *See* autozygous; coancestry; inbreeding.

allozyme Different form of an enzyme due to allelic differences of the genes.

all-walking approach Program used in physical mapping of DNA in connection with YACs. The sequence-tagged sites (STS) are derived from the ends of YAC inserts. Its advantages are that the position of the STS is defined compared to when the STS is internal; it identifies chimeric YACs, and the end-STS YACs tend to be larger than the others. *See* STS; YAC.

allyl alcohol Eye-irritating liquid that permits positive selection of alcohol dehydrogenase mutations because the wild-type cells (adh^+) in a suicidal manner convert this compound to acrylaldehyde and thus only the adh^- cells can survive. *See* mutant isolation.

almond (*Prunus amygdalus*) Basic chromosome number $x = 7$, but $2x$ to $6x$ forms are known.

alopecia Hair loss; baldness probably caused by an autoimmune condition occurring in different forms. In some cases it is accompanied by psychomotor epilepsy (involuntary movements) or palm and sole keratosis (callosity), nail dysfunction, and lower mental capacity. In humans it appears as autosomal dominant. A recessive mutation (ACA, [Thr] → GCA, [Ala]) in human chromosome 8p12 causing total baldness (alopecia universalis) may be based on a defect in a zinc-finger transcription factor. One form is caused by mutation in the Hfh11^{nu} (forkhead/winged helix) transcription factor family. In mice, asebia (rudimentary sebacious glands) and alopecia may be caused by mutation in two genes: *Scd1* encoding stearoyl-CoA desaturase 1 and *Scd2* coding for stearoyl-CoA desaturase 2. Some cancer chemotherapies induce alopecia, which can be mitigated by topically applied inhibitors of CDK2 such as the analogs of 3-(benzylidene)indolin-2-ones, etoposides, and cyclophosphamide-doxorubicin. *See* autoimmune disease; baldness; connective tissue disorders; cyclophosphamide; doxorubicin; etoposide; hypotrichosis; keratosis; nude mouse; zinc finger.

Alpers progressive infantile poliodystrophy Involves degeneration of the gray matter of the brain and cirrhosis of the liver.

Alpert disease (AFP) Alpha-fetoprotein deficiency encoded at human chromosome 4q11-q13. *See* fetoprotein.

Alpert syndrome Acrocephalosyndactylia, pointed head top and fused fingers and toes, an autosomal-dominant or -recessive disease of humans involving the fibroblast growth factor receptor (FGFR2), also defective in Pfeiffer syndrome. The estimated mutation rate is $3–4 \times 10^{-6}$. *See* Pfeiffer syndrome.

α Average inbreeding coefficient, $\alpha = \Sigma p_i F_i$, where p_i is the relative frequency of inbred individuals with F_i coefficient of inbreeding. This value in most human populations is less than 0.001 while in isolated human groups it may exceed 0.02 or 0.04. *See* error types; inbreeding coefficient.

alpha accessory factor Enhances the affinity of pol α and primase for the DNA template. *See* Okazaki fragment; pol α; primase.

alpha amanitin Protein synthesis inhibitor fungal octapeptide. It blocks RNA pol II (0.1 µg/mL); RNA polymerase III is also blocked by it but at much higher concentrations (20 µg/mL), but pol I is insensitive to it even at 200 µg/mL. LD$_{50}$ in albino mice is 0.1 mg/kg. *See* LD$_{50}$; pol; RNA polymerase. (Begun, D. J. & Whitley, P. 2000. *Heredity* 85:184.)

αβ T cells Recognize the major histocompatibility complex-bound peptide antigens. *See* γδ T cell; MHC; T cell.

α-amylase Hydrolyzes α-1-4 glucosidic linkages of amylose, amylopectin, and other carbohydrates and yields maltose, α-dextrin, and maltotriose. β-amylase hydrolyzes starch into maltose. The human AMY genes are in chromosome 1p21.

alpha complex Translocation complex of chromosomes that transmits only through the females, whereas the beta complex is transmitted only through the males (in *Oenothera*). *See* complex heterozygotes; translocation complex. (Cleland, R. E. 1972. *Oenothera: Cytogenetics and Evolution*. Academic Press, New York.)

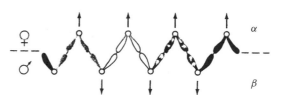

Alternate distribution of a four-translocation complex at anaphase I (α and β).

α-CPM α-connecting peptide domain connects the α and β chains of the αβ T-cell receptor, but it is absent from the γδ T-cell receptor. This domain is required for positive selection of T cells, although negative selection may take place in its absence. *See* lymphocyte positive selection; T-cell receptor; T cells.

α Fetoprotein (AFP) Expressed in the embryonic yolk and liver of mammals. Serum albumin and α-fetoprotein genes are linked at about 15 kb apart, and each encodes about 580 amino acids. The two proteins are immunologically cross-reactive and display about 35% homology. The rate of transcription of AFP drops four orders of magnitude after birth. Regulat-ory sequences are positioned within 150 bp 5′ and there are also enhancers 6.5, 5, and 2.5 kbp upstream from the transcription initiation site. The gene (human chromosome 4q11-q22) is a classical example for tissue-specific and developmental regulation. *See* MSAFP. (Mizejewski, G. J. 2001. *Exp. Biol. Med.* 226:377.)

a/α-factors *See* mating type determination in yeast.

α-1,4-glucosidase deficiency *See* acid maltase.

αGT Glucosyltransferase uridine 5′-diphosphate galactose: β-D galactosyl-1,4-N-acetyl-D-glucosaminide α(1–3)galacto-syltransferase, E. C.2.4.1.151 synthesizes the carbohydrate epitope Galα 1–3Galβ1–4GlcNAc-R, which reacts with natural antibodies, forming a barrier to xenotransplantation. Murine bone marrow cells transgenic for αGT may overcome the production of xenoreactive antibodies. This principle may enhance the development of techniques to facilitate organ transfer between animals and humans. See antibodies; epitope; transgenic; xenograft.

alpha helix Hydrogen-bonded secondary structure of polypeptides when the polypeptide backbone is tightly wound around the longitudinal axis of the peptide bonds and the side groups of the amino acids are protruding along the generally right-handed helical structure. Most commonly each turn takes 3.6 amino acids. Also frequently represented graphically as a cylinder. See pitch; protein structure.

Alpha helix.

alpha lactose Milk sugar is converted into allolactose by the β-galactosidase gene of the *Lac* operon of *E. coli* and the latter then becomes the inducer of the operon. See *Lac* operon.

alpha mating-type factor of yeast See mating-type, yeast.

alphameric Symbols using letter identifications.

alphanumeric Set of characters using letters and numerals and possibly other character symbols.

alpha parameter Provides a combined estimate of the frequency of quadrivalent association (q), meiotic exchange (e), and favorable anaphase distributions (a). From these it predicts the frequency of double reduction, i.e., the production of *aa* gametes when the parental constitution is *AAAa* (see diagram of chromosome mechanics below). The derivation of the α parameter is as follows. In a triplex the cytogenetic constitutions can be represented as shown. The letters W, X, Y, and Z stand for the centromeres, the chromatids are symbolized by the dashed lines, and the dominant and recessive alleles are numbered from *A1* to *A6* and *a1* to *a2*, respectively (see below).

(i) In the *absence of recombination* the association of chromatids, alleles, and centromeres is *A1-A2, A3-A4, A5-A6*, and *a1-a2*.

(ii) In case of *recombination* between gene and centromere, the following arrays are formed: *A1-A3, A1-A5, A3-A5* are the possible recombinant associations of dominant alleles, which were originally attached to different centromeres.

(iii) *A1-a1, A3-a1*, and *A5-a1* are the three possible dominant-recessive recombinant associations when only one chromatid originally attached to a centromere

is considered in the quadrivalent. The total frequency of the gametes is 1, the frequency of group (i) is designated as α, and the chance of each of the four types of associations within this group is α/4. The combined frequency recombinant group (ii) and (iii) associations is $1 - \alpha$. Since groups (ii) and (iii) have three representatives each, and six combined, the frequency of each of the recombinant associations is $(1 - \alpha)/6$.

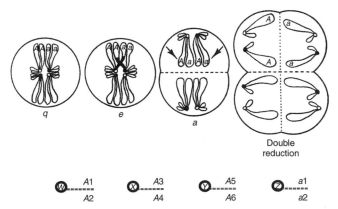

Double reduction

The chromosome mechanics (double reduction) are represented by the diagram (empty arms carry the A allele).

Group (i) has three double dominants (*A1-A2, A3-A4, A5-A6*) among the total of four, each has a frequency of α/4, and the combined frequency is $3 \times (\alpha/4)$. Group (ii) also has three double dominants (*A1-A3, A1-A5, A3-A5*) with individual frequencies $(1 - \alpha)/6$, and their combined frequency is $3 \times (1 - \alpha)/6$. Thus the total frequency of gametes with a dominant allele in both chromatids is $[3 \times (\alpha/4)] + [3 \times (1 - \alpha)/6] = [3\alpha/4] + [(3 - 3\alpha)/6]$. After dividing both numerator and denominator of the second term by 1.5, we obtain $[3\alpha/4] + [(2 - 2\alpha)/4] = (2 + \alpha)/4 =$ the frequency of *AA* gametes. The frequency of the *Aa* gametes is obtained similarly. Group (iii) has three *Aa* gametes with individual frequencies of $(1 - \alpha)/6$, and their combined frequency upon dividing numerator and denominator by 1.5 becomes $3 \times (1 - \alpha)/6 = (3 - 3\alpha)/6 = (2 - 2\alpha)/4$. The frequency of the double recessive gametes (*aa*) as shown above is α/4. Thus the total gametic output of the triplex is $[(2 + \alpha)/4\ AA]:[(2 - 2\alpha)/4\ Aa]:(\alpha/4\ aa)$. The practical meaning of the α parameter is best illustrated by an example. Let us assume that the cytological analysis indicates the value of $q = 0.7$, $e = 0.25$, $a = 0.333$, and thus $\alpha = 0.7 \times 0.25 \times 0.333 \times 0.5 = 0.02914$. This value can then be substituted into the formulas of the table next page on the appropriate line.

Accordingly, for the simplex (line 3 at tetrasomy) *AA* becomes $\alpha/4 = 0.02914/4 = 0.007285$, *Aa* is obtained in frequency $2(1 - \alpha)/4 = 0.48543$, and *aa* = $(2 + \alpha)/4 = 0.507285$. Thus the double dominant gametes are expected below 1%, the *Aa* and *aa* in near equal frequency around 50%.

The proportion of the double reduction (*aa* gametes) gives some indication of the relative distance of the gene from the centromere, albeit not in precise units, directly convertible to map distances but approximating. Theoretically, these calculations are elegant; unfortunately, the determination of the variable components of α requires great experimental

ZYGOTES	GAMETES				Simplex :	1 _A_
	TETRASOMY				Duplex:	2 _A_
	AA	**Aa**	**aa**	DIVISOR	Triplex:	3 _A_
Triplex	$2 + \alpha$	$2(1 - \alpha)$	α	4	Quadruplex:	4 _A_
Duplex	$1 + 2-\alpha$	$4(1 - \alpha)$	$1 + 2\alpha$	6	Pentaplex:	5 _A_
Simplex	α	$2(1 - \alpha)$	$2 + \alpha$	4	Nulliplex:	no _A_

	HEXASOMY				DIVISOR
	AAA	**Aaa**	**Aaa**	**aaa**	
Pentaplex	$3 + \alpha$	$3 - 2\alpha$	α	0	6
Quadruplex	$3 + 3\alpha$	$9 - 5\alpha$	$3 + 3\alpha$	α	15
Triplex	$1 + 3\alpha$	$9 - 3\alpha$	$9 - 3\alpha$	$1 + 3\alpha$	20
Duplex	α	$3 + \alpha$	$9 - 5\alpha$	$3 + 3\alpha$	15
Simplex	0	α	$3 - 2\alpha$	$3 + \alpha$	6

The _A_ value indicates the number of dominant alleles in the zygotes.

The expected gametic frequencies of polyploides. Each term on the corresponding line has to be divided by the _divisor_ shown at the right column. (After Fisher, R. A. & Mather, K. 1943. _Annals of Eugenics_ 12:1.)

skills and very favorable conditions. Therefore the analysis of segregation in polyploids is very difficult. See autopolyploids; centromere mapping. (Mather, K. 1935. _J. Genet._ 30:53; Mather, K. 1936. _J. Genet._ 32:287.)

alpha particles Helium nuclei (contain two protons and two neutrons) emitted by radioactive decay. They release their high energy in a very short track; even in air they move only for a few centimeters. In living material they have minimal penetration, yet because of their short track and high ionizing energy, they are very destructive (break chromosomes). Mean number of lethal lesions/per cell: α rays: 0.01, γ rays: 0.001, 10 MeVB neutrons: 0.005. See ionizing radiation; linear energy transfer.

α-repeat 171 bp abundant (up to 1 million) repeat in the human genome, localized primarily in the centromeric regions of the chromosomes. This and the Alu repeats constitute 5–10% of the human genome. See Alu family.

alpha satellite The centromeric DNA that is normally heterochromatic, but it may have an important role in controlling chromosome segregation and other centromere functions. The 171 bp tandemly repeated sequence has been found in all human centromeric areas. It is connected by 17 bp (missing in the human, mouse, and green monkey Y centromeres) with protein CENP-B, a common autoimmune antigen. All human centromeres and neocentromeres apparently also include CENP-A, a histone H3-like protein. Introduction of exogenous alphoid DNA into the cells may cause chromosomal instabilities. The CENP-B protein bears sequence similarity to the pogo transposases. See centromere; heterochromatin; human artificial chromosome; hybrid dysgenesis; meiosis; microchromosome; neocentromere; satellite DNA; segregation. (Buno, I., et al. 2001. _Genome_ 44:120.)

alpha-satellite DNA Centromeric repetitive DNA. See repetitious DNA; satellite DNA.

α-thiophosphate-dNTP Point mutagen when incorporated into gapped DNA by DNA polymerase I. The thiophosphates are not effectively removed by the $3' \rightarrow 5'$ editing function of the DNA pol I enzyme. See pol.

alpha tocopherol (vitamin E) Tocopherols are plant products, but they are required in mammals for the maintenance of fertility and for the prevention of muscle degeneration. They appear to be antioxidants for unsaturated lipids. Lipid peroxidation may result in cross-linking of proteins and may cause mutation, the appearance of age pigments (lipofuscin), etc. See fatty acids; unsaturated fatty acids.

alphavirus Single, negative-strand RNA virus with 240 molecule basic capsid proteins surrounded by a lipid bilayer of 240 glycoprotein heteromeric envelopes. It can infect a variety of cells, and its genomic RNA is translated into nonstructural proteins to begin, then the replication of the viral RNA. Although it is a cytocidal virus, it may be engineered into a vector for transient gene therapy or used for vaccination. This group of viruses includes the Sendai virus and the Simliki forest virus. See Sendai virus.

alphoid DNA See α satellite.

Alport's disease Exists in different forms; autosomal-dominant, -recessive, and X-linked types have been described. The phenotypes vary, but most commonly inflammation of the kidney(s) and deafness are present. Most probably the basic defect involves the basement membrane of the kidney glomerules and the Goodpasture antigen, leading to kidney failure and hypertension. The basement membrane defect is an Xq22-coded collagen α-chain anomaly. Basically Alport syndrome is the same as Epstein syndrome. The autosomal-dominant Alport syndrome with leukocyte inclusions and macrothrombocytopenia (also called Fechtner syndrome) was assigned to 22q11-q13. See basement membrane; collagen; Goodpasture syndrome; kidney diseases; May-Hegglin anomaly; thrombocytopenia.

ALPS-1, ALPS-2 Autoimmune lymphoproliferative syndrome. ALPS-1 (10q24) is caused by malfunction (mutation in Fas or FASL [1q23]); ALPS-2 is due to mutation in caspase-10 (2q33-q34). In either case, the apoptotic process is interfered with, causing neurodegeneration, tumorigenesis, and other disorders. The disease is rare and it occurs in childhood. T cells accumulate but carry no CD4 or CD8, and the immune system turns against the red blood cells or the platelets. *See* apoptosis; autoimmune diseases; CD4; CD8; FAS; FasL; T cells. (Chun, H. J., et al. 2002. *Nature* 419:395.)

ALS *See* amyotrophic lateral sclerosis.

Alström syndrome Autosomal-recessive human defect involving obesity, retinitis pigmentosa, deafness, and diabetes. Its frequency is elevated in some Louisiana and Nova Scotia populations of French origin. *See* Bardet-Biedl syndrome; obesity.

ALT Mechanism alternative to telomerase for the maintenance of telomere length integrity. ALT relies on recombination and depends on the Rad52 protein mediating homologous recombination. In cancer therapy—in order to prevent the restoration of telomere length—perhaps both telomerase and ALT need to be targeted. *See* telomerase. (Grobelny, J. V., et al. 2001. *Hum. Mol. Genet.* 10:1953.)

AlterMap Computer program replacing sections of the Kohara map of *E. coli* with the MapSearch alignments of the DNA fragments. *See* Kohara map.

alternate disjunction Occurs in a translocation heterozygote when each pole receives a complete set of the genetic material and consequently is genetically stable. *See* adjacent distribution; translocation, chromosomal; translocation complex.

alternation of generations Cycles of haploid and diploid generations such as the gametophytic and sporophytic generations of plants. Also refers to the cycles of sexual and asexual generations that coexist in some species. *See* apomixis; fission; gametophyte; life cycles; meiosis; mitosis; parthenogenesis; sporophyte.

alternative splicing mRNA alternative splicing generates different protein molecules from the same genes. There are splicing factors such as hnRNP and a serine-arginine protein to carry out these functions. hnRNP A1 normally favors a distal 5′ splice site, but under the influence of p38 protein kinase signals, splicing may be switched to a proximal 5′ splice site. Typically alternative splicing occurs in immunoglobulin synthesis (among other mechanisms) and T cell receptors to generate a greater repertoire of antibodies from a lesser number of genes (Wang, J., et al. 2002. *Science* 297:108). A survey of 528 human genes indicated 22% alternative splicing. Some other estimates indicated 2.5 to 3.2 as the number of different transcripts per gene. Thus alternative splicing requires fewer genes to carry out different functions. Alternative splicing is about the same in humans and other eukaryotes. Aberrant alternative splicing is a common cause of human genetic disease (Cáceres, J. F. & Kornblihtt, A. R. 2002. *Trends Genet.* 18:186). Tandemly duplicated exons (occurring in about 10% of eukaryotic genes) may be responsible for alternative splicing. *See* hnRNA; introns; p38; splicing. (Lopez, A. J. 1998. *Annu. Rev. Genet.* 32:279; Tollervey, D. & Caceres, J. F. 2000. *Cell* 103:703; Standiford, D. M., et al. 2001. *Genetics* 157:259; Modrek, B., et al. 2001. *Nucleic Acids Res.* 29:2850; Hu, G. K., et al. 2001. *Genome Res.* 11:1237; Gravely, B. R. 2001. *Trends Genet.* 17:100; Brett, D., et al. 2002. *Nature*

Genet. 30:29; Kadener, S., et al. 2002. *Proc. Natl. Acad. Sci. USA* 99:8185, Letunic, I., et al. 2002. *Hum. Mol. Genet.* 11:1561, Gravely, B. R. 2002. *Cell* 109:409, Maniatis, T. & Tasic, B. 2002. *Nature* 418:236, <http://cbcg.nersc.gov/asdb>; <http://isis.bit.uq.edu.au>; <http://www.bioinformatics; ucla. edu/HASDB>; putative alternative splicing [PALS]: <http:// palsdb.ym.edu.tw>.)

altruistic behavior Evolutionary feature in animals where members of the species protect other members, especially the young, even at their own peril. Through this behavior the survival of the population is promoted. Altruism may also be manifested in mating behavior. In a pride of animals, some males may refrain from reproduction and may allow more powerful kin to mate with the females available. Apoptosis involves some elements of altruism of single cells that are suicidal to assure differentiation of a tissue or defend the organism against mechanical or biological injuries or attacks. *See* aggression; apoptosis; behavior genetics; inclusive fitness; kin selection. (Agrawal, A. F. 2001. *Proc. Roy. Soc. Lond. B Biol. Sci.* 268:1099; Abbot, P., et al. 2001. *Proc. Natl. Acad. Sci. USA* 98:12068.)

Alu-equivalent Group of genomic sequences similar to the Alu family. *See* Alu family.

Alu family About 150–300-kb-long nucleotide sequence monomers associated head-to-tail, and repeated about 300,000–500,000 times in the primate genome. These nucleotide sequences are cut by the Alu I restriction enzyme (recognition site AG ↓ CT) and hence the name of this gene family. Its members are also considered transposable elements that depend on other elements for transposition. The Alu sequences are specific for the human genomes, but homologs appear in other mammals. The Alu sequences appear to have evolved from the 7SL RNA genes about 15–30 Myr ago. Alu insertional mutations were identified in the antihemophilic factor IX, neurofibromatosis, Apert syndrome, adenomatous polyposis cancer, X-linked immunodeficiency, and breast cancer genes, and it is most likely that many more will be identified in the completely sequenced human genome. Alu sequences as well as other repeats also increase the instability of the genome by recombination. *See* LINE; selfish DNA; SINE. (Stenger, J. E., et al. 2001. *Genome Res.* 11:12; Roy-Engel, A. M., et al. 2001. *Genetics* 159:279; Batzer, M. A. & Deininger, P. L. 2002. *Nature Rev. Genet.* 3:370.)

aluminum tolerance Can be bred into plants by expression of the transgene of *Pseudomonas aeruginosa* citrate synthase. The transgenic plants exude citrate by the roots and lower the pH. Aluminum in alkalic or neutral soil is toxic to many crop plants. (Ma, J. F., et al. 2001. *Trends Plant Sci.* 6:273.)

Alzheimer disease (AD, FAD) Presenile/senile dementia (loss of memory and ability of judgment as well as general physical impairment) involves the accumulation of amyloid protein plaques in the brain, resulting in degeneration of neurons and neurofibril tangles. Four major genes are responsible for AD. The amyloid-β peptide (Aβ, of 40–42 amino acids) comes from a larger amyloid precursor protein (βAPP) that is synthesized in normal brain, processed in a number of ways, and encoded in chromosome 21q21.3-q22.05 as rare early-onset dominant (AD1). The largest protein spans the cell membrane (AD3, 14q24.3). One of the extracellular domains is a protease inhibitor. In normal cells this domain may be released, but in the diseased cells the amyloid protein is processed incorrectly. Genes responsible for amyloid protein synthesis have been cloned and others mapped to human chromosomes 1q31-42 encoding STM2/AD4, a seven-transmembrane integral protein (presenilin 2); chromosome 14q24.3 encodes protein S182/AD3 (presenilin 1), which is 67% homologous to STM2. Human chromosome 19q13.2 encodes (AD2) apolipoprotein E (APOE), which controls the late onset of Alzheimer disease. Homozygosity for the APOE-4 allele (frequency ~16%) increases the chances of onset of the disease about 20 times, whereas a single copy of the APOE-2 (frequency ~7%) only doubles this chance. Chromosome 21q21.3-q22.05 (AD1) encodes βAPP (early onset, dominant), and it seems to be involved with the disease.

Aβ is the major component of the brain plaques, and its ligand is a ~50K relative molecular weight protein, identical to RAGE (receptor for advanced glycation end product) or AGE/βAPP and for amphoterin controlling neurite outgrowth (an inflammatory process). RAGE/AGER receptor (6p21.3) mediates the interaction of Aβ with endothelial cells and neurons, causing oxidative stress. Its interaction with microglia results in cytokine production, chemotaxis, and binding movements. A spurious, unconfirmed interaction among AD1/APP, AD2/APOE, A2M (α-macroglobulin, encoded at 12p13.3-p12.3), and a low-density lipoprotein-related protein (LRP, encoded at 12q13.1-q13.3) have been reported. The PAR-4/PRKC (prostate apoptosis response) mutations in the gene encoding 342 amino acids, including a leucine zipper and a death domain, mediate neuronal degeneration and mitochondrial dysfunction in case of defects in presenilin 1.

The sporadic Alzheimer disease may be associated with the very low-density lipoprotein (VLDL) receptor gene. Alzheimer disease is common among Down syndrome (due to chromosome 21 trisomy where βAPP is located) individuals. Some psychotropic drugs (affecting the nervous system) may alleviate some of the symptoms. The incidence of Alzheimer disease (AD) increases from 0.1% below age 70 to double after age 80. The risk for first-degree relatives varies from 24 to 50% by age 90. The concordance rate of monozygotic twins is 40–50% and among dizygotic twins 10–50%. Genetic screening for the disease is not considered appropriate. The sporadic (apparently nongenetic) cases of this disease may be caused by frameshift mutation in the RNA during transcription or after transcription in the β-amyloid precursor and/or in ubiquitin-B. The identification of AD is difficult without the detection of the brain plaques by biopsy or autopsy.

Using fluorescence correlation spectroscopy, the aggregation of Aβ can be detected if the polymerization is promoted by "seeding" with synthetic Aβ probe in femtoliter samples of the cerebrospinal fluid using Cy2 fluorophore. The difference between afflicted and healthy individuals is clear and the procedure may be of potential value for diagnosis. Mice transgenic for a mutant (Val^{717} → Phe^{717}) human APP when immunized with Aβ_{42} either prevented or reduced the neuropthological symptoms. The most likely time course of the development of AD is: mutations in the amyloid and presenilin genes → production of Aβ_{42} fragments → formation of plaques in the brain cortex → hyperphosphorylation of tau, oxidative stress, and the formation of tangled fibers → neuronal dysfunction and neuronal death → mental deterioration. The hyperphosphorylation of tau is mediated by the cyclin-dependent kinase Cdk5. When the p35 regulatory subunit of Cdk5 is cleaved into a truncated p25 fragment, phosphorylation is enhanced and tau binds less efficiently to microtubules. Also, Cdk5/p25 promotes apoptotic cell death of neurons.

Alzheimer disease involves inflammation of the brain. β-amyloid seems to stimulate CD40-CD40L interaction, causing the activation of microglia. Microglial cells are important players in AD pathogenesis because they promote the degeneration of neurons. Although AD is generally attributed to the accumulation of pathological levels of Aβ_{1-42}, it is possible that the other cause of the disease is the lack of clearance of the plaques by the neprilysin-like neutral endopeptidase (NEP) and related enzymes. Insulin-degrading enzyme (IDE, 10q23-q25) in neurons and microglia degrades Aβ. In a mouse model, Aβ immunization appeared to reduce the plaques and fibrils in the brain and improved the cognitive functions and memory. In late-onset Alzheimer disease (LOAD), the Aβ42 level is elevated by genetic factors in human chromosome 10q. (Myers, A., et al. 2002. *Am. J. Med. Genet.* 114:235). *See* AGE; BACE; behavior, human; β-amyloid; CD40; CDK; CD40 ligand; corticotropin-releasing factor; Creutzfeldt-Jakob disease; Down syndrome; encephalopathy; ERAB; fluorophore; frameshift; humanin; LDL; mental retardation; microglia; mitochondrial disease, human; neprilysin; NF-κB; p35; presenilins; prion; scrapie; secretase; statins; tau; ubiquitin; VLDL. (Chapman, P. F., et al. 2001. *Trends Genet.* 17:254; Wiltfang, J., et al. 2001. *Gerontology* 47:65, Selkoe, D. J. 2001. *Physiol. Rev.* 81:741; Baekelandt, V., et al. 2000. *Curr. Opin. Mol. Ther.* 2:540, mouse model: Wong, P. C., et al. 2002. *Nature Neurosci.* 5:633; Sekoe, D. F. & Podlishy, M. E. 2002. *Annu. Rev. Genomics Hum. Genet.* 3:67. <www.alzforum.org/members/index.html>.)

amacrine cell   Retinal neurons with short axons. *See* axon; neurogenesis; retina.

amanitin ($C_{39}H_{54}N_{10}O_{13}S$)   *See* α-amanitin.

amaranth   Subtropical or tropical American seed plants, $2n = 2x = 32$.

amastigote   *See Trypanosoma.*

amatoxin   Bicyclic octapeptides (e.g., α-amanitin) produced by the fungus *Amanita phalloides* inhibit the function of RNA polymerase II (occasionally pol III) of eukaryotes, but they do not affect the transcriptases of the prokaryotic type, e.g., those in the cytoplasmic organelles of eukaryotes. *See* pol III; RNA polymerase; RNA replication; transcription.

Amauris   African species of butterflies are models to mimic by another butterfly, *Papilio dardanus*. The purpose of the mimicking is that the models are distasteful to predators and thus the mimicking species improves its survival. *See* Batesian mimicry.

amaurosis congenita (Leber congenital amaurosis, LCA) Group of autosomal-recessive blindness or near-blindness diseases caused by a defect of the cornea (keratoconus). About 10% of the blinds are suffering from it. Mutations in the photoreceptor guanylate cyclase (RETGC) or retinal pigment epithelium (RPE) and genes responsible for phototransduction and photoreceptor maintenance also may cause LCA. An autosomal-dominant photoreceptor-specific homeodomain gene (CRX) is responsible for cone-rod dystrophy of the retina. Leber congenital amaurosis (LCA) has been mapped to human chromosome 17p13.1. Another mutation was mapped to a retinitis pigmentosa locus (RET3C11) at 1q321-q32.1. This locus is called Crumbs Homolog 1 (CRB1) of a *Drosophila* gene. *See* eye diseases. (Seeliger, M. W., et al. 2001. *Nature Genet.* 29:70, Cremers, F. P. M., et al. 2002. *Hum. Mol. Genet.* 11:1169.)

amaurotic familial idiocy (AFI)   Old name for Tay-Sachs disease. *See* Batten disease; Tay-Sachs disease.

amber   (1) Fossil tree resin up to millions of years old. It has hardened and become resistant to most environmental factors. Frequently it contains microbes, plants or animals, or residues of them in a usually well-preserved state and thus may provide very valuable information on old organismal specimens, including the genetic material. *See* ancient DNA. (2) Chain-terminator codon (UAG).

amber mutation   Generates a chain-termination polar effect (the name has nothing to do with function; rather it was named for Felix Bernstein, whose German family name translates into amber). *See* code, genetic; polar mutation. (Epstein, R. H., et al. 1963. *Cold Spring Harbor Symp. Quant. Biol.* 28:375.)

amber suppressor   Mutations in the anticodon triplet (3′-AUC-5′) of a tRNA, so the amber mutation (5′-UAG-3′) may be read as a tyrosine codon and thus the translation is not terminated. *See* supC; supD; supE; supF; supG; supU. (Kiga, D., et al. 2001. *Eur. J. Biochem.* 268:6207.)

ambidextrous   *See* handedness.

ambient signal   The position of a particular cell determines its response to certain environmental stimuli.

ambiguity in translation (mistranslation, miscoding)   May be caused by antibiotics, or modification of the tRNA or the ribosomes (16S subunit), and consequently an amino acid different from the correct one is incorporated into the nascent polypeptide. Apparently the cognate tRNAs have ca. four orders of magnitude higher recognition rate than the noncognate ones, measured on the basis of GTPase action rate in the EF-Tu-GDP ternary complex. The estimated error is 10^{-4} per amino acid under normal conditions. *See* aminoacylation error; protein synthesis; RAN. (Dong, H. & Kurland, C. G. 1995. *J. Mol. Biol.* 248:551; Ardell, D. H. & Sella, G. 2001. *J. Mol. Evol.* 53:269.)

ambiguity of restriction enzymes   They can cut more than a single sequence, although not with equal efficiency, e.g., Hind I: GTT ↓ GAC, GTT ↓ AAC, GTC ↓ GAC.

ambisense virus (e.g., some bunyaviruses, arenaviruses)   Transcribed into mRNA, but also the 5′-end of the RNA genome may function as mRNA.

A medium   For *Escherichia coli*, g/L: K_2HPO_4 10.5, KH_2PO_4 4.5, $(NH_4)_2 SO_4$ 1.0, Na-citrate.$2H_2O$, 0.5 plus glucose 0.4%, thiamin 1 mg/L, $MgSO_4$ 1 mM, and an appropriate antibiotic. *See* culture media. (For different bacterial culture media, see Winkler, U., et al. 1976. *Bacterial, Phage and Molecular Genetics. An Experimental Course.* Springer-Verlag, New York).

amelia   *See* limb defects in humans; phocomelia; thalidomide.

amelioration of genes   DNA sequences incorporated into a genome by horizontal transfer tend to adapt to the codon usage of the recipient organisms during evolution. *See* codon usage; transmission.

amelogenesis imperfecta (AI)   The autosomal-dominant forms (ameloblastin and enamelin encoded within 4q11-q21) involve softness of the tooth enamel caused by lack of calcium. Similar symptoms but with calcium deposits in the kidneys and other variations indicate autosomal-recessive inheritance. Two Xq24-q27.1-linked forms are distinguished. One of them is very similar in phenotype to the autosomal-dominant form. In one, the thickness of the enamel is about normal but soft; in the other, the enamel is hard but very thin. The combined prevalence in Sweden is $\sim 1.4 \times 10^{-3}$. *See* tooth. (Rajpar, M. H., et al. 2001. *Hum. Mol. Genet.* 10:1673.)

amelogenin test   Forensic and archaeological sex-typing tissue test. The X– and Y chromosome–derived amelogenin sizes are different and thus the test indicates sex of the specimen. In rare instances, the AMELY gene (Yp11, Xp22.3-p22.1) is deleted from the Y and the sample from the male is not distinguishable from that of the female unless other markers (e.g., the SRY gene) are involved in the test. Amelogenin is a dental enamel protein. See sex determination; SRY. (Buel, E., et al. 1995. *J. Forensic Sci.* 40:641.)

amenorrhea   Absence of menstruation may be caused by physiological factors (over- or underweight, pregnancy), hormones, age, disease, or may have genetic causes such as pseudohermaphroditism, Turner syndrome, absence of ovaries, uterus, or vagina, etc. In the absence of structural deficiencies, selective estrogen-receptor modulating (SERM) drugs may be beneficial. Secondary amenorrhea occurs when menstruation is suspended or ceased after a period of time. In such cases hormone replacement therapy may be indicated. *See* hermaphroditism; Turner syndrome.

amensalism   One organism is inhibited by another, which is unaffected by this relationship.

American-type culture collection (ATCC)   Maintains and catalogs microbial stocks, viruses, and cultured cells. (<http://www.atcc.org/>.)

Ames test   Bacterial assay based on backmutation of different histidine-requiring strains of *Salmonella typhimurium*. The reversions are capable of detecting various types of base substitutions and frameshift. A single plate generally detects mutations in 100,000 or more cells.

In some strains the *his⁻* genes are present in multicopy plasmids to further enhance their targets. The bacteria also carry mutations that interfere with genetic repair. The testing medium includes microsomal fractions of mammalian liver that can activate promutagens into ultimate mutagen. Thus the mutagenic effectiveness of the majority of chemicals may be increased by three orders of magnitude. The results of this assay are highly correlated with carcinogenicity of the compounds, yet it requires only 2 days compared with the rodent tests, taking possibly several months for complete evaluation. It is also inexpensive and permits the evaluation of a large number of compounds at low cost. *See* base substitution mutation; bioassays in genetic toxicology; frameshift mutation; microsomes; mutagen activation; reversion assays, *Salmonella*, and *E. coli* in genetic toxicology. (Kim, B. S. & Margolin, B. H. 1999. *Mutation Res.* 436:113; Maron, D. M. & Ames, B. N. 1983. *Mutation Res.* 113:173.)

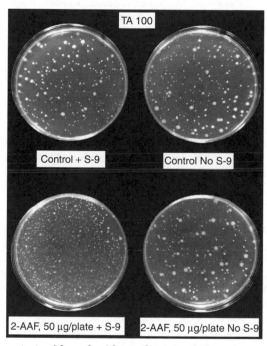

The Ames test with and without the use of the microsomal (S-9) fraction. It is clear that the microsomal enzymes do not affect the frequency of reversions without 2-AAF (2-acetylamino-fluorene). Also, 2-AAF without activation is not mutagenic. The S-9 fraction is usually prepared from rodent liver homogenate, and in its presence, the promutagens can also be assayed. The bacteria lack the activating enzyme component. (Courtesy of Dr. D. M. Zimmer.)

amethopterin Inhibitor of dihydrofolate reductase, an important enzyme in the de novo biosynthetic pathway of purine and pyrimidine nucleotides. Synonymous with methotrexate, used as an antitumor drug and a selective agent in genetic transformation. It has also been used to treat rheumatoid arthritis and psoriasis. It is extremely toxic; in concentrations of 10^{-8} to 10^{-9} it may shut down the biosynthesis of nucleotides. It may cause headaches, rashes, diarrhea, and cirrhosis of the liver. *See* aminopterin; methotrexate; psoriasis.

amide bond Carbonyl group linked to an amine. *See* peptide bond.

Amide bond.

amifostin ($C_5H_{15}N_2O_3P\ S$) Biological radioprotector. *See* radioprotectors.

amiloride ($C_6H_8ClN_7O$) Potassium-sparing diuretic regulating K^+ and Na^+ balance in the cells. *See* ion channels.

amino acid Relatively simple yet diverse chemical compound. All amino acids have at least one NH_2. Where **R** can be a *nonpolar aliphatic* group: glycine (Gly), alanine (Ala), valine (Val), leucine (Leu), isoleucine (Ile), proline (Pro); *aromatic*: phenyl alanine (Phe), tyrosine (Tyr), tryptophan (Trp); *polar uncharged*: serine (Ser), threonine (Thr), cysteine (Cys), methionine (Met), asparagine (Asn), glutamine (Gln); *negatively charged*: aspartic acid (Asp), glutamic (Glu); *positively charged*: lysine (Lys), arginine (Arg), histidine (His). About 20 natural amino acids are the building blocks of proteins. Archaea and eubacteria encode also pyrrolysine (UAG) and selenocysteine (UGA, the latter also in animals). Some amino acids are modified in certain types of proteins. Cysteine and methionine always contain sulfur. The α-amino acids have both the amino and carboxyl group(s) attached to the same C atom. The common natural amino acids in living organisms occur as the L enantiomorphs. In the brain, however, upon glutamate stimulation, the astrocytes enzymatically synthesize D-serine, which facilitates synapse between neurons by stimulation of the NMDA receptors. *See* amino acid symbols in protein sequences; astrocyte; enantiomorph; essential amino acids; genetic code; NMDA receptor; nonessential amino acids; pyrrolysine; selenocysteine.

amino acid activation *See* aminoacylation.

amino acid analyzer   Automated equipment similar to a high-pressure liquid chromatography apparatus to separate and quantify the amino acid composition of digested protein. *See* chromatography.

amino acid metabolism   Amino acids are derived from the compounds in the glycolytic, citric acid, and pentose phosphate pathways. The biosynthetic systems in different evolutionary categories may vary. Bacteria and plants are normally able to synthesize all 20 primary amino acids, whereas animals depend on diet for the essential amino acids. Genetics of microorganisms play an important role in elucidating the pathways. Single-gene mutations generate a special requirement for all amino acids that can be met by feeding the amino acid or the appropriate precursor. In higher plants, auxotrophy exists for very few amino acids, probably because amino acids may be synthesized by parallel pathways or functionally duplicated genes. In humans and other mammals, genetic defects are known that in some way affect all the natural amino acids and many of their derivatives, thus causing inborn errors of metabolism. *See* alanine aminotransferase; alaninuria; alkaptonuria; argininemia; asparagine synthetase; aspartate aminotransferase; aspartoacylase deficiency; carnosinemia; citrullinemia; cystathionuria; cystin-lysinuria; cystinosis; cystinuria; dibasicaminoaciduria; glutamate decarboxylase; glutamate dehydrogenase; glutamate forminotransferase deficiency; glutamate oxaloacetate transaminase; glutamate pyruvate transaminase; glutamate synthesis; glutaminase; glycine biosynthesis; glycinemia; histidase; histidinemia; histidine operon; homocystinuria; hydroxymethyl glutaricaciduria; 3-hydroxy-3-methylglutaryl CoA lyase deficiency; hyperlysinemia; hyperprotinemia; hypervalinemia; isoleucine-valine biosynthetic pathway; isovalericacidemia; leucine metabolism; lysine biosynthesis; methionine adenosyl transferase deficiency; methionine biosynthesis; methionine malabsorption; methylcrotonyl glycinemia; methylglutaconicaciduria; methylmalonicaciduria; ornithine aminotransferase; ornithine decarboxylase; ornithine transcarbamylase; phenylalanine; phenylketonuria; proline biosynthesis; sarcosinemia; serine; threonine; tryptophan; tyrosine; urea cycle; valine; vitamin B_{12} defects.

amino acid regulation   Translation of a specific amino acid mRNA or global protein synthesis may be regulated by amino acids and hormones (e.g., insulin) through an integrated pathway of signals.

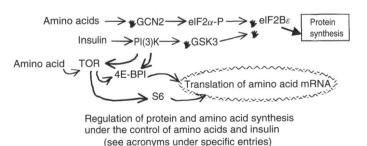

Regulation of protein and amino acid synthesis
under the control of amino acids and insulin
(see acronyms under specific entries)

amino acid replacement   Takes place by base substitution in the codons, e.g., a glutamic acid (GAA) residue may be replaced by glutamine (CAA), lysine (AAA), glycine (GGA), valine (GTA), alanine, (GCA), aspartic acid (GAT), and so on. The rate of amino acid substitution per site in protein was estimated to be on the average of 10^{-9}/year during evolution. This average may vary by 3–4 orders of magnitude among different proteins. *See* PAM. (<http://www.genome.ad.jp/aaindex/>.)

amino acid sequencing   Can be carried out in different ways. Today most frequently the putative amino acid sequence is deduced indirectly from the codon sequences in the DNA. Direct estimates can be obtained from polypeptides cleaved by proteolytic enzymes (trypsin, chymotrypsin, pepsin, and various other proteases, cyanogen bromide) to obtain manageable smaller fragments of proteins. These agents have preferences for certain cleavage points represented by particular amino acids. Also, chemical breakage of disulfide bonds is utilized. Edman degradation is then employed, which uses end labeling and removal of single amino acids. Eventually, the sequenced fragments must be ordered on the basis of overlapping ends. *See* amino acid analyzer; databases; DNA sequencing; Edman degradation; sequenator. (Rajagopal, I. & Ahern, K. 2001. *Science* 294:2571; key amino acid positions in structurally similar proteins: <http://ckaaps.sdsc.edu>.)

amino acid starvation   *See* stringent control; stringent response.

amino acid symbols in protein sequences   Alanine A, aspartic acid or asparagine B, cysteine C, aspartic acid D, glutamic acid E, phenylalanine F, glycine G, histidine H, isoleucine I, lysine K, leucine L, methionine M, asparagine N, proline P, glutamine Q, arginine R, serine S, threonine T, valine V, tryptophan W, unknown X, tyrosine Y, glutamic acid or glutamine Z. *See* amino acid.

aminoacidurias   Diverse groups of hereditary diseases are characterized by urinal excretion of cystine (cystinosis), tyrosine (tyrosinemia), all kinds of amino acids (fructose intolerance), very large quantities of primarily threonine, tyrosine, and histidine (Hartnup disease), and hypervalinemia. Many diseases of the kidneys show excessive amino acid excretion. Dicarboxylic aminoaciduria is a defect of glutamate/aspartate transport at 9p24. Dibasic aminoaciduria is a defect of cystinuria and the failure or normal transport off dibasic amino acids at 9q13.1. *See* blue diaper syndrome; Fanconi renotubular syndrome; Hartnup disease; homocystinuria; iminoglycinuria; neuromuscular diseases; Rowley-Rosenberg syndrome; tyrosinemia.

aminoacylation   To the acceptor arm (CCA-OH) of tRNA an amino acid is attached by its NH_2 end by an ATP-dependent enzymatic process. Rate of mischarging is 3×10^{-3} in prokaryotes and may be higher in yeast or higher eukaryotes. Although this reaction requires a protein enzyme, a ribozyme may also be adapted to carry out the aminoacylating function in a manner analogous to the ribozyme, peptidyl transferase. Certain aminoacyl-tRNA synthetases have a particular site where the misactivated amino acid tRNA complex is destroyed to maintain the correct protein structure. Aminoacylation takes place within the eukaryotic nucleus before the correct tRNA is released to the cytosol. *See* aminoacylation error; aminoacyl–tRNA synthetase; EF-TU·GTP; operational RNA code; ribozyme; tRNA. (Rodnina, M. V. & Wintermeyer, W. 2001. *Annu. Rev. Biochem.* 70:415; Hendrickson, T. L., et al. 2002. *Mol. Cell* 9:353.)

aminoacyl-tRNA Amino acid–charged tRNA at the 3′ end. *See* aminoacylation; aminoacyl-tRNA synthetase; protein synthesis; tRNA.

aminoacyl-tRNA synthetase Enzymes carrying out aminoacylation of tRNA. The first step is the attachment of the amino acid to the α-phosphate group of an ATP molecule that is accompanied by the removal of an inorganic pyrophosphate group. The aminoacyl adenylate is then bound to the active site of one of the two types of aminoacyl-tRNA synthase enzymes. Class I mainly monomeric (except*) enzymes handle Arg, Cys, Gln, Glu, Ile, Leu, Met, Trp*, Tyr*, and Val. Class II dimeric enzymes are involved with Ala, Asn, Asp, Gly, His, Lys, Phe, Pro, Ser, and Thr (for these abbreviations, see amino acids). Class I synthase first attaches the aminoacyl-A to the 2′-OH of the terminal A of the amino arm of tRNA and subsequently it is moved to the 3′-OH by transesterification. Class II enzymes bypass the 2′-OH transfer step. The enzymes recognize the appropriate acid among the 40–80 or more tRNAs, and this rather complex recognition process is directed by the so-called second genetic code. Several sites on the tRNA determine the recognition of the proper tRNAs, most importantly by the anticodon. In *Drosophila* there is a glutamic acid–proline tRNA synthetase (GluProRS). The amino-terminal domain is active for Glu and the C-terminal fragment is functional for Pro.

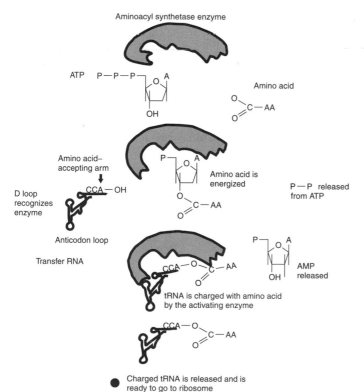

Aminoacyl synthetase enzyme

ATP   P—P—P   A

Amino acid

OH

D loop recognizes enzyme

Amino acid–accepting arm

CCA—OH

Anticodon loop

Transfer RNA

Amino acid is energized

P—P released from ATP

AMP released

tRNA is charged with amino acid by the activating enzyme

CCA—O

Charged tRNA is released and is ready to go to ribosome

In some Archaea, e.g., *Methanococcus janaschii*, a single aminoacyl-tRNA synthetase exists for proline and cysteine, ProCysRS. In these organisms a single subunit carries out the acylation of tRNAs for both amino acids, but it never makes ProtRNA^Cys or CystRNA^Pro (Stathopolous, C., et al. 2000. *Science* 287:479). In *E. coli* the anticodon is the most important for recognition for 17 of the 20 amino acids. For many of the isoaccepting tRNAs, the 73 position of the amino acid–accepting arm is also very important along with the anticodon. The enzyme also capable of correcting errors in recognition, e.g., isoleucyl-tRNA, cannot entirely prevent valine from attaching to its binding site and may form a valyl-adenylate. This activated valine cannot, however, attach to either tRNA^Val or tRNA^Ile; rather it is hydrolyzed by tRNA^Ile, so no erroneous valyl-tRNA^Ile is formed.

Another way to eliminate translational errors is to modify the amino acids attached to the wrong tRNA. Thus rarely can these misacylated tRNA be used for peptide elongation. Actually, misacylation may occur as an intermediate step, but the mentioned quality control prevents most of the ambiguities and errors. Misacylation of amino acids is subjected to correction by an editing complex of the tRNA (Bishop, A. C., et al. 2002. *Proc. Natl. Acad. Sci. USA* 99:585). tRNA-dependent amino acid modifications are the only means for the formation of formylmethionyl-tRNA and some others (e.g., Asp-tRNA^Asn, Glu-tRNA^Gln) in some bacteria, archaea, and organelles. The majority of the aminoacyl-tRNA synthetases either discriminate against the D-enantiomers at activation or use, e.g., D-Tyr-tRNA^Tyr deacylase, to prevent the D form from incorporation into protein.

The elongation factors (EF-Tu, EF-1α) also prefer the L enantiomorphs. The C-terminal domain of the tyrosyl-tRNA synthetase has ~49% homology with a cytokine (endothelial monocyte-activating polypeptide II [EMAPII]). This cytokine causes phagocytotic cells to express *tissue factor* and TNFα and migrates to the inflammation sites. The overall mistake in amino acid incorporation is about 1/10,000 residues. Nuclear genes encode the aminoacyl-tRNA synthetases of the organelles. The enzymes are organelle-specific, however. Some nuclear genes may encode both types of enzymes by differential transcription and processing. The reaction, catalyzed by the aminoacyl-tRNA synthetase, can also be catalyzed by a reactive RNA. The aminoacyl-tRNA synthetases of higher eukaryotes form multiprotein complexes. *See* arginyl tRNA synthetase; cytokines; EF-TU·GTP; EMAPII; glutamyl-tRNA synthetase; histidyl tRNA synthetase; leucine tRNA synthetase; methionyl tRNA synthetase; missing genes; operational RNA code; protein synthesis; ribosomes; ribozyme; threonyl tRNA synthetase; tmRNA; tRNA; tryptophanyl tRNA synthetase; valyl tRNA synthetase; wobble. (Jakubowski, H. & Goldman, E. 1992. *Microbiol. Rev.* 56:412; Carter, C. W. Jr. 1993. *Annu. Rev. Biochem.* 62:715; Ibba, M. & Söll, D. 2000. *Annu. Rev. Biochem.* 69:617; Ribas de Pouplana, L. & Schimmel, P. 2001. *Cell* 98:191; <http://rose.man.poznan.pl/aars/>.)

aminobenzyloxymethyl paper Diazotized (using 1-[(m-nitrobenzyloxy)-methyl] pyridinium chloride [NBPC]) Whatman 540 or other comparable paper used for Northern blotting. *See* Northern blotting.

amino end   The amino end of a protein is where the synthesis starts on the ribosome. It is commonly a methionine residue, although during processing of the protein the first amino acid(s) may be removed. The amino end of the polypeptide corresponds to the 5' end of the mRNA. *See* amino terminus; protein synthesis.

amino group   Derived from ammonia (NH_3) by replacing one of the hydrogens by another atom (H_2N-).

aminoglycoside phosphotransferase (NPTII, aph(3')II) Phosphorylates aminoglycoside antibiotics and causes resistance against these antibiotics. The genes for the two related enzymes were isolated from Tn5 and Tn60 bacterial transposons, respectively, and are used as dominant selectable markers (with appropriate promoters) in transformation of animal and plant cells. *See* geneticin resistance; kanamycin resistance; neo[R]; neomycinphosphotransferase. (Wright, G. D., et al. 1998. *Adv. Exp. Med. Biol.* 456:27; Boehr, D. D., et al. 2001. *J. Biol. Chem.* 276:23929.)

aminoglycosides   Group of antibiotics with a common characteristic of a cyclic alcohol in glycosidic linkage with amino-substituted sugars. They (streptomycin, kanamycin, neomycin, gentamicin, paromomycin, etc.) affect the A site (16S rRNA in the 30S ribosomal subunit) of the prokaryotic/organelle ribosomes where the codon–anticodon interacts, and thus interferes with initiation of translation, fidelity of decoding of the codon, peptidyl transfer, and peptide translocation. Inhibition of eukaryotic ribosome function requires about 20-fold higher concentration of the antibiotic. Widespread resistance may be accounted for by ability of the cells to expel the antibiotics, or enzymatic modification either of the antibiotic or the cellular target. *See* A site; gentamicin; kanamycin; neomycin; phenotypic reversion; protein synthesis; ribosome. (Ryu, H. & Rando, R. R. 2001. *Bioorg. Med. Chem.* 9:2601.)

aminolevulinic acid (ALA)   Precursor of porphyrin required for the production of hemoglobin and chlorophylls. The ALA dehydratase (ALAD) is coded in human chromosome 9q34. *See* chlorophyll; hemoglobin; porphyrin.

$$H_2NCH_2COCH_2CH_2CO_2H$$

Aminolevulinic acid

aminopterin   Inhibits the activity of dihydrofolate reductase at 10^{-8} to 10^{-9} M concentrations. This enzyme is required for the biosynthetic pathway of both pyrimidine and purines. This drug is also used in the HAT medium to shut down the de novo synthetic pathway of nucleotides when thymine kinase and hypoxanthine–guanine phosphorybosyl transferase mutations are screened for in mammalian cell cultures. *See* amethopterin; DHFR; HAT medium.

2-aminopurine (AP)   Adenine analog that may incorporate into DNA in place of adenine and can form normal hydrogen bonds with thymine, but it is prone to mispair with cytosine either with a single hydrogen bond in its normal state or after tautomeric shift with two hydrogen bonds. The mispairing may result in a replacement of an AT pair by a GC pair and thus results in mutation. AP may be highly mutagenic in some prokaryotes but not in eukaryotes. *See* base analogs; base substitution; hydrogen pairing.

2-Aminopurine.

aminoterminal   The only amino acid in a polypeptide chain with a free α-amino group, and it occurs at the end of the chain. *See* amino end.

aminotransferase   Transaminase enzymes that transfer α-amino groups from amino acids to α-keto acids.

3-amino-1,2,4-triazole   Carcinogenic standard (nonmutagenic in the Ames test); now a banned herbicide. *See* Ames test.

Amish   Mennonite religious group with strict and conservative principles and lifestyle. Their communities are relatively isolated from the surrounding populations. Actually the Amish population in the United States was in three approximately equal-size groups of ~14,000 people in the 1960s. Gene frequencies distinguish the three related groups. Endogamy and consanguineous marriages are higher and certain genetically determined conditions are relatively frequent in these populations. The recessive Ellis–Van Creveld syndrome, pyruvate kinase deficiency, cartilage-hair hypoplasia, limb-girdle muscular dystrophy, and Christmas disease are relatively common. Amish brittle hair syndrome (also recessive) involving short stature, somewhat lower intelligence, brittle hair, reduced fertility, and low sulfur content of the nails was first recognized in such a population. *See* cartilage-hair hypoplasia; Christmas disease; consanguinity; Ellis–Van Creveld syndrome; endogamy. (McKusick, V. A. 1980. *Endeavour* 4(2)52.)

amitochondriate   Lacks mitochondria, e.g., some microsporidian, eukaryotic parasites of mammals with genomes of less than 3 Mb.

amitosis   Nuclear divisions without the characteristic features of the mitotic apparatus and involving the small (21–1,500 kb) acentric chromosomes in the macronucleus of some *Protists*. No mitotic spindle is evident and the nuclear membrane seems intact during the entire division. Nevertheless, the distribution of the chromatin is not entirely random. *See* acentric; chromatin; fission; mitosis; *Paramecia*. (Prescott, D. M. 1994. *Microbiol. Rev.* 58:233.)

amixis   In fungal genetics, apomixis. *See* apomixis.

AML1   Acute myeloid leukemia oncogene, a DNA-binding protein encoded in human chromosome 21q22. *See* leukemias.

ammonification   Release of ammonia upon decomposition of compounds (e.g., amino acids).

ammunition   Gene tagging with nonautonomous P elements of *Drosophila* that stays put after the removal of the helper (complete) element. *See* hybrid dysgenesis; smart ammunition.

amniocentesis   Prenatal diagnosis of the genetic constitution of a fetus by withdrawing fluid or cells from the abdomen (amniotic sac) of a pregnant woman. Amniotic fluid is sufficient for this procedure after about 16 weeks of pregnancy. The test can be cytological, enzymological, immunological, or molecular and may involve cell cultures to amplify the material. Amniocentesis can also be used for genetic counseling. Normally it entails minimal risk; to the fetus and the mother, yet it should be used only in cases when it is warranted by other parts of the diagnosis. *See* PCR; polymerase chain reaction; prenatal diagnosis; risk; counseling, genetic. (Trent, R. J., ed. 1995. *Handbook of Prenatal Diagnosis*. Cambridge Univ. Press, New York.)

amnion   Strong membrane that envelops the mammalian embryo and fetus and contains the amniotic fluid that helps in protecting the embryo during the entire period of development until delivery. A similar membrane is found in other animals. The amnion is closest to the embryo, covered by the allantoic mesoderm, and the outer layer is the chorion. See allantois; chorion.

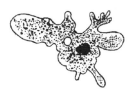

Amoeba proteus.

amoeba   Free-living or parasitic single-cell eukaryote. Some of the amoebae crawl by forming pseudopodia (leg-like extensions of the single cell). *Amoeba dubia* has a genome size (bp) of 6.7×10^{11} in n = several hundred chromosomes. *See* nuclear transplantation.

amorph allele   Inactive; it may also be a deletion. *See* allele.

AMOVA   *See* ANOVA.

amoxicillin   Inhibitor of cell wall–cross-linking transpeptidase and thus enhances the effect of β-lactam antibiotics. *See* clavulanate.

Amp (ampicillin, 6-[D(-)α aminophenylacetamid]-penicillinic acid)   Member of the penicillin family of antibiotics. The Amp^R genes are common in genetic vectors. *See* antibiotics; vectors, genetic.

Ampicillin.

AMP   Adenosine 5'-monophosphate (adenylic acid); when two additional phosphates are added, ATP is formed. *See* cAMP.

AMPA (α-amino-3-hydroxy-5-methyl-4-isoaxozolpropionate)   Member of the glutamate receptor family of proteins. It mediates the excitatory synaptic transmissions in the brain and the spinal cord. It also controls postsynaptic influx of Ca^{2+}, further regulating synapse. The AMPA receptors are built of four variable subunits having large extracellular amino ends, three transmembrane domains, and an intracellular COOH end. PDZ domains mediate cell targeting. The GRIP (glutamate receptor interacting protein) contains seven PDZ domains, interacts with the C end, and links AMPA to other proteins. *See* excitatory neurotransmitters; kainate; NMDA; PDZ domains; synapse. (Posser, R. A. 2001. *J. Neurosci.* 21:7815.)

Ampere (A)   1 A = 1 C/sec. 1 C (coulomb) = 1 As (amperesecond). *See* volt; watt.

amphibolic path   Of metabolism, involves both anabolic and catabolic reactions.

amphid   Chemoreceptor in nematodes, e.g., in *Caenorhabditis*. *See* Caenorhabditis.

amphidiploid   Contains two genomes from at least two different species; it is obtained by doubling the number of chromosomes of amphiploids. *See* amphiploid; chromosome doubling. (Kashkush, K., et al. 2002. *Genetics* 160:1651.)

amphigamy   In the usual type of fertilization the gametic nuclei fuse. *See* dikaryon.

amphihaploid   Haploid cell of an amphidiploid, an allohaploid. *See* allohaploid; amphidiploid; haploid.

amphimeric genome Inverted repeat separated by wide sequences. Amphimeric genomes may be generated in the mitochondrial DNA of yeast. Their origin appears to be due to illegitimate recombination between a pair of short inverted repeats. In amplified genomes they are relatively common and are presumably advantageous for DNA replication. (Royko, E. & Goursot, R. 1999. *Curr. Genet.* 35:14.)

amphimixis Sexual reproduction. *See* apomixis.

amphipathic Has a charged and a neutral face (e.g., some proteins forming an amphipathic helix); structures that have hydrophilic and (polar) hydrophobic (nonpolar) surfaces, e.g., lipids.

amphiphysin Nerve protein of the synaptic vesicle bound to synaptotagmin. It also participates in general endocytosis. *See* BIN1; endocytosis; synaptotagmin.

amphiploid Contains at least two genomes from more than one species. *See* allopolyploid; amphidiploid.

amphiprotic Can donate or accept protons and thus can behave weakly acid or alkalic, e.g., water or amino acids. *See* amino acids; proton.

amphiregulin Regulator with both (+) and (−) effects. It regulates the proliferation of keratocytes and some fibroblasts and inhibits the proliferation of various tumor cells. It competes for the epidermal growth factor (EGF) receptor. Amphiregulin is required for normal implantation of blastocytes and is regulated by progesterone. *See* EGFR; embryogenesis; keratosis; progesterone. (Akatsu, N., et al. 2001. *Biochem. Biophys. Res. Commun.* 281:1051.)

amphistomatous Leaves bearing stomata on both surfaces. *See* stoma.

amphithallism Homoheteromixis; both self-fertilization and outcrossing occur in fungi.

amphitropic molecule Carries out different functions at different sites.

ampholine Ampholyte used for polyacrylamide, agarose, and dextran gels, density gradient stabilizing in analytical and preparative electrofocusing. *See* isoelectric focusing.

ampholyte Amphoteric electrolyte. *See* amphoteric; electrolyte.

amphoteric Has dual, opposite characteristics such as behaving both as an acid and a base.

amphoterine *See* Alzheimer disease.

amphotropic retrovirus (polytropic retrovirus) Replicates both in the cells from where it was isolated and in other cell types. *See* ecotropic retroviruses; xenotropic retroviruses.

ampicillin Antibiotic that binds to bacterial cell membranes and inhibits the synthesis of the cell wall. The ampicillin resistance gene (*amp^r*) codes for a β-lactamase enzyme that detoxifies this antibiotic; the *Amp^r* gene is also used as a marker for insertional inactivation and the concomitant ampicillin susceptibility. *See* Amp; antibiotics; β-lactamase; insertional mutation; pBR322.

amplicon DNA fragment produced by polymerase chain reaction (PCR) amplification. Also the amount of DNA present in an amplified gene or chromosomal segment. A reduced-size viral construct used for genetic transformation. *See* PCR.

amplification Temporary synthesis of extra, functional copies of some genes in vivo or in vitro by the use of some forms of the polymerase chain reaction. Bacteriophage λ can be amplified by a series of nitrocellulose filter transfers after in situ hybridization. The addition of chloramphenicol (10−20 μg/mL) to pBR322 and pBR327 may amplify plasmid yield if the protein synthesis is not completely prevented. Cosmid libraries may be amplified by starting on solid plates followed by liquid cultures. Replica plating can amplify animal cell cultures. Approximately 5×10^4 colonies can be accommodated on a 138 mm filter, so about 30 filters are required to obtain a representative library of overlapping fragments.

DNA amplification can occur in a genetically programmed and predetermined manner in eukaryotes, e.g., in the ovarian follicle of *Drosophila* large quantities of an eggshell protein is needed during oogenesis. The need is met by a disproportionately favorable replication of the chorion gene clusters in the X chromosome and chromosome 3. DNA replication is initiated bidirectionally at a replicational origin and it generates multiple copies of the genes needed. A distance away, the replication tapers off and the flanking regions are amplified less and less in proportion to the distance from the origin. Similar programmed amplification takes place in the ribosomal genes of amphibia during intense periods of protein synthesis in embryogenesis. The approximately 500−600 genomic copies of rRNA genes may be increased by a factor of 1000. The replication of detached DNA sequences follows a rolling circle−type process, and the new DNAs (in about 100 rDNA repeats) are separately localized in micronuclei. The replicates of these nuclei are structurally similar to each other, indicating that they are the clonal products of a single replicating unit, but the new micronuclei generated in different cells may not be the same as judged by the length differences in the intergenic spacers.

Ribosomal DNA amplification takes place during the amitotic divisions of the protozoan, *Tetrahymena*. Here again the macronuclear rDNA copies may be selectively amplified in the 10^4 range, whereas in the micronuclear DNA there is only a single rDNA gene. A genetically nonprogrammed amplification takes place in several mutant cell lines to correct mutational defects. Producing multiple copies of the gene controlling low-efficiency enzymes may compensate for enzyme deficiencies. Transfection of ADA genes to mammalian cells may be amplified in the presence of dCF (*See* adenosine deaminase). Mammalian cells can be amplified if they are cotransfected with the *dhfr* (conveying methotrexate resistance) gene and other desired sequences. In the presence of methotrexate, the *dhfr* genes, as well the flanking DNA, may be amplified

1000-fold. The amplified DNA, in stable lines, is integrated into the chromosome in homogeneously stained regions (HSRs). In unstable cell lines, *dhfr* is in autonomously replicating elements called double-minute (DM) chromosomes that have no centromeres and can be maintained only in methotrexate-containing cultures. Amplification may generate fragile sites in the chromosomes by integration of DM chromosome sequences. Hypoxia may be an inducing factor of such integration.

Some general features of amplification are: (1) expansion of a particular locus and flanking regions or the generation of small supernumerary chromosomes called double minutes that contain the critical gene; (2) the amplified unit may undergo rearrangements; (3) the amplified sequences are not all identical and may change, but these changes are somewhat unusual because a larger number of copies may be altered simultaneously in an identical manner. In vivo amplification of genes during evolution may account for the presence of gene families. When the larger copy number is no longer advantageous, some of the amplified genes may acquire new functions without entirely losing their structural similarity to the ancestral sequences. Other members of the amplified group lose their function(s) through deletions and mutations and become pseudogenes. Carcinogenesis commonly involves amplification of some of the oncogenes and genes involved with the cell cycle (cyclins). Fragile sites in some chromosomes aid amplification. *See* ADA; adaptive amplification; bidirectional replication; chloramphenicol; chorion; cosmid library; DM chromosome; fragile sites; HSR; in situ hybridization; methotrexate; micronucleus; nitrocellulose filter; oogenesis; pBR322; PCR; pseudogene; rolling circle; unequal crossing over. (Romero, D. & Palacios, R. 1997. *Annu. Rev. Genet.* 31:91; Monni, O., et al. 2001. *Proc. Natl. Acad. Sci. USA* 98:5711; Dean, F. B., et al. 2002. *Proc. Natl. Acad. Sci. USA* 99:5261.)

amplification control elements Amplification of genes in chromosome 3 and the X chromosome of *Drosophila* is determined by less than 5 kbp DNA sequences that normally occur in the vicinity of the genes amplified under natural conditions of the genome (e.g., the chorion protein gene). If these control elements are isolated, inserted into genetic vectors (P-elements), and reintroduced at random sites into the *Drosophila* genome, they may amplify other sequences in their new neighborhood. *See* amplification; hybrid dysgenesis.

AMPLITAQ™ Taq DNA polymerase, a single polypeptide chain enzyme with minimal secondary structure. It is isolated from the bacterium *Thermus aquaticus*. Its temperature optimum is about 75°C but can withstand ≤ 95°C without great loss of activity. It lacks intrinsic nuclease function but has polymerization-dependent $5' \rightarrow 3'$ exonuclease activity. Taq DNA is a preferred enzyme for PCR. *See* DNA polymerase; exonuclease; PCR; Taq DNA polymerase.

Amplitype *See* DNA fingerprinting.

amputation *See* ADAM complex; limb defects.

Amsterdam criteria Established at a meeting in Amsterdam, NL for ascertaining the hereditary nature of nonpolyposis colorectal cancer: (1) at least three family members, of which two are first-degree relatives, are affected; (2) at least two generations are represented; (3) at least one family member is below age 50 at the time of onset. *See* hereditary nonpolyposis colorectal cancer, relatedness degree of.

amusia Deficit of music perception caused by a genetic or acquired brain anomaly. It may not affect any other brain function or intelligence. In some cases, it is associated with limitation of prosody, rhythm, and pitch of speech. *See* musical talent; prosody.

AMV oncogen (*v-amv*) *See* MYB.

amyloid angiopathy *See* amyloidosis (type VI).

amyloidosis Involves extracellular deposition of variable amounts of amyloids, a special fibrous glycoprotein of the connective tissues caused by protein misfolding. Some of the familial nephropathies (kidney diseases), heart diseases, and neoplasias involve amyloidosis. Genetically these are inhomogeneous groups of diseases mainly with a dominant pattern of inheritance. Amyloidosis occurring in some aging individuals develops similar symptoms to Alzheimer disease, and the dominant genes are mapped to the same region of chromosome 21 and to 20p12, the site of the prion gene. Swedish and Portuguese amyloidosis I is a dominant polyneuropathy encoded near the centromere in the long arm of human chromosome 18. Finnish amyloidosis type V is apparently due to an autosomal-dominant defect in gelsolin. Icelandic amyloidosis type VI involves a high incidence of hemorrhages due to accumulation of amyloids. The afflicted individuals (dominant) are low in cysteine proteinase inhibitor, cystatin C, encoded in the region of human chromosome 20q13. Ohio amyloidosis type VII involves ocular and mental affliction. German amyloidosis type VIII is a visceral and renal disease. Amyloidosis type IX is a skin disorder. Familial British dementia is a dominant, late-onset brain degeneration caused by the BRI gene in human chromosome 13q14. There are recessive amyloidoses affecting the gingiva (gum), eyelids, cornea (eyeball), and mental state. Gelatinous drop-like corneal dystrophy (GDLD, human chromosome 1p) is an amyloidosis caused by mutation in a gastrointestinal tumor-associated antigen. *See* Alzheimer disease; amyotrophic lateral sclerosis; β-amyloid; cold hypersensitivity; gelsolin; Mediterranean fever; prion; scrapie. (Pepys, M. B., et al. 2002. *Nature* 417:254.)

amyloid Fibrillar poorly soluble/insoluble protein-forming β sheet such as apolipoprotein. Proteoglycan–amyloid complexes are protected from proteolysis. Some starch-like substances are also called amyloids. *See* Alzheimer disease; amyloidosis; β sheet; protein structure; proteoglycan.

amylopectin Normally a minor variant of common starch. While starch (amylose) is an unbranched chain of D-glucose units of α-1-4 glycosidic linkages, amylopectin also contains branch points in α-1-6 linkages at every 24 to 30 residues. Most commonly cereal grains contain amylose as the principal

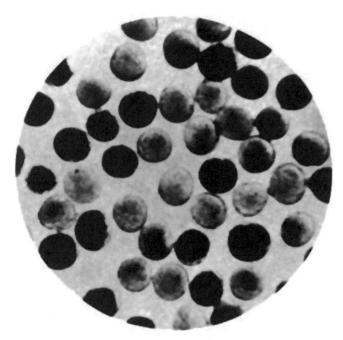

Stained sorghum pollen displays segregation for starch and amylopectin. (Courtesy of Dr. J. R. Quinby. See Karper, R. E. 1933. *J. Hered.* 24:257.)

storage polysaccharide, but recessive mutations may cause the predominance of amylopectin (dextrin). These two types of starches are easily distinguished in situ by a drop of iodine solution ($I_2$0.12 g + KI0.4 g in 100 mL H_2O); amylose stains blue-black and amylopectin appears red-brown. The amylose content of corn is desirable for the film and fiber manufacturing industry. Several genes (*ae, du*) may substantially increase the amylose content relative to that of amylopectin.

amyloplast   Plastid with a primary role of starch storage.

amylose   *See* amylopectin.

amyotrophic lateral sclerosis (ALS, Lou Gehrig disease) Hardening of the lateral columns of the spinal cord with concomitant muscular atrophy that may spread and may cause death in a few years after onset. According to a mouse model, it is probably caused by a defect in the enzyme Cu/zinc superoxide dismutase (SOD) in about 20% of familial cases. This enzyme breaks down superoxide radicals (highly reactive compounds) to less reactive products, although it may form other types of free radicals. SOD1 apparently causes neural death by acting on caspases that mediate apoptosis. The *Bcl-1* gene inhibiting apoptosis prolongs the life of mice affected by SOD. A subsequent study found, however, either the elimination or elevation of SOD activity in mice had no influence on the expression of ALS. Now it appears that zinc-deficient SOD plays a role in nitric oxide–dependent apoptosis of some motor neurons. The SOD transgene effect can be restrained by N-benzyloxycarbonyl-Val-Asp-fluoromethyl-ketone (zVAD-fmk), an inhibitor of caspases, and it delays the onset and mortality of ALS in mice.

The gene involved in the disease symptoms is a dominant gain-of-function mutation within the area 21q22.1-q22.2. The syndrome in different forms occurs at a frequency of about

1×10^{-5}. About 10% of the cases are hereditary and 90% are sporadic. It was named LGD after baseball infielder Henry Louis (Lou) Gehrig, who was elected to the U.S. National Hall of Fame in 1939 and suffered from this condition. ALS is sometimes associated with Parkinson's and Alzheimer disease–like phenotypes. This form may be caused or aggravated by nutritional factors (neurotoxins in the food, low calcium and magnesium uptake). A recessive autosomal type with an early onset from 3 to 20 years of age was assigned to human chromosome 2q33. The encoded protein, alsin, may directly affect motor neuron degeneration and may signal to GEFs. *See* Alzheimer disease; apoptosis; filament; gain-of-function; GEF; hypoxia; neuromuscular diseases; Parkinson disease; SOD. (Julien, J.-P. 2001. *Cell* 104:581; Yang, Y., et al. 2001. *Nature Genet.* 29:160; Giess, R., et al. 2002. *Am. J. Hum. Genet.* 70:1277.)

anabasine (neonicotine)   Alkaloid in chenopods and solanaceous plants; it is highly toxic (LDlo orally 5 mg/kg for humans). *See* LDlo.

anabolic steroid   Androgen that promotes protein synthesis, muscle, bone, and general growth. Some synthetic forms (methyltestosterone, oxymetholone, norethandrolone) have higher anabolic than testosterone activities and are used illegally by athletes to boost performance ("body builders"). Some of the androgenic and anabolic steroids are used as drugs for the treatment of impotence, anemias, and bone marrow aplasia. These compounds may cause liver adenomas that rarely may become cancerous. *See* adenoma; anemia; aplasia; impotence; steroid doping; steroid hormones.

anabolism   Energy-requiring synthetic processes of the cellular metabolism.

anaerobe   Organism that lives without atmospheric (free) oxygen.

anaerobic   Process in the absence of (air) oxygen.

anagenesis   Evolutionary change within a line of descent. *See* cladogenesis.

analbuminemia   Human chromosome 4 recessive absence or reduction of albumin from the blood serum that is not accompanied by very serious ailments, although fatigue, mild anemia, and mild diarrhea may be associated with it. *See* albumins.

analgesic   Medication alleviates pain without losing consciousness.

analogous gene   Has similar function without common evolutionary descent. *See* homologous genes.

analogue   Chemical compound similar to a natural one, but it may or may not function in metabolism or may even block the function of a normal metabolite or the enzyme involved.

analogy   Similarity is not based on common origin. See convergent evolution; homology.

analysis of variance Statistical method for detecting the components of variance. It is used for the evaluation of differences of experimental data involving different treatments. The square root of the quotient of the sum of squares of the variates and the mean square of the error variance is equal to t, and the corresponding probability, at each degree of freedom, can be read from a t-distribution table. The results are usually presented in a table form such as:

Variance Source	Degrees of Freedom	Sum of Squares (SS)	Mean Square (MS)	Mean Square Ratio (MSR)
Between Groups	k − 1	SSB	SSB/(k − 1)	SSB/(k − 1)
Within Groups	N − k	SSW	SSW/(N − k)	SSW/(N − k)
Total	N − 1			

The MSR also permits testing the significance of the data with the aid of an F table. Analysis of variance is also used in calculating heritability by intraclass correlation. *See* F distribution; t-distribution; variance intraclass correlation. (Sokal, R. R. & Rohlf, F. J. 1969. *Biometry*. Freeman, San Francisco.)

anaphase In *mitosis*, at anaphase the centromeres of the chromosomes split, which makes it possible for the spindle fibers to pull the two identical chromatids toward the opposite poles. This assures the genetic identity of the daughter cells. In *meiotic* anaphase I, the centromeres do not split and the chromatids are held together as they move toward the poles. Thus the chromosome number is reduced. Anaphase II of meiosis essentially resembles anaphase in mitosis. Microtubules and special motor proteins mediate the chromosome movements. The molecular mechanism of the process is partly known. In yeast the *MAD* (mitotic arrest deficient) and *BUB* (budding inhibited by benzimidazole) gene products seem to be the sensors of the kinetochores, which have not yet tackled the spindle fibers. Before the sister chromatids can separate, the anaphase-promoting complex (APC) degrades the inhibitors of the process (Pds1/Cut2). Proteins Cdc20/Sℓp1 and Hct1/Cdh1 digest other inhibitory proteins (Clb2 and Ase1). *See* APC; cell cycle; cohesin; meiosis; microtubules; mitosis; motor protein; sister chromatid cohesion; spindle.

anaphase-promoting complex *See* APC.

anaphylactic shock Immediate hypersensitivity to specific antigens or haptens resulting in dangerous loss of respiratory function. See anaphylatoxin; anaphylaxis.

anaphylatoxins Fragment released during activation of the serum complement C proteins of the antibodies. C3a, C4a, and C5a (each ~10 kDa) anaphylatoxins are proteolytically cleaved from the corresponding complement components. These activation peptides are called anaphylatoxins because they may elicit reactions similar to anaphylactic shock (violent reaction to antibodies and/or haptens that may be fatal). These fragments also cause contraction of the smooth muscles, release of histamine, other vasoactive amines, and lysosomal enzymes, and enhance vascular permeability. *See* antibody; complement; histamine; lysozymes. (Gerard, C. & Gerard, N. P. 1994. *Annu. Rev. Immunol.* 12:775.)

anaphylaxis Rapid serological (antigen-antibody) reaction of an organism to a foreign protein. Either the crystalline fragment of the antibody (Fc) or the complement is involved. Prior sensitization may make the reaction quite violent and may cause death. Anaphylaxis may be treated with adrenaline. *See* allergy; antibody; complement; immune system.

anaplasia Dedifferentiation.

anaplastic lymphoma (large-cell non-Hodgkin lymphoma) Lymphoma of children causing a 2p23:5q35 chromosomal translocation fusing a protein tyrosine kinase gene, ALK, to the nucleolar phosphoprotein genes (NPM). The resulting anomaly affects the small intestine, testis, and brain but not the lymphocytes. Alk is related to the insulin receptor kinases and may eventually cause malignancies. Translocations involving 1q21-q23, the site of the IgG Fc receptor (FcγRIIB), is also responsible for this malignant lymphoma. *See* antibody; Duncan syndrome; Hodgkin disease; immunoglobulins; leukemias; lymphoma. (Pulford, K., et al., 2001. *Curr. Opin. Hematol.* 8[4]:231.)

anaplerosis Biological repair or replacement.

anastomosis Formation of a reticulate arrangement, fusion between vessels.

anastral spindle Mitotic spindle without asters, such as in higher plants. *See* aster.

ancestral Inherited from a remote forebear or derived from a precursor molecule. Ancestral inheritance as a theory based on the false assumptions of a nonparticulate genetic material was developed by Francis Galton (1897) and lost meaning by Mendelism.

anchorage dependence Normal mammalian cells grow in culture in a monolayer attached to a solid surface; cancer cells are not contact inhibited and pile up on each other. It appears that the suppression of cyclin E-CDK2 activity is required for cell anchorage. In transformed fibroblasts the cyclin E-CDK2 complex is active regardless of anchorage. *See* AIG; anoikis; cancer cells; CATR1; cyclins.

anchor cell Gonadal cell of *Caenorhabditis* that induces the development of neighboring cells into the vulval opening. *See* *Caenorhabditis*; morphogenesis; organizer.

anchoring The DNA fragments obtained during the initial stages of physical mapping must be tied together by contigs. For the establishment of contigs, large-capacity YACs are used. These YACs must be correlated with molecular markers (anchors) along the length of the chromosome. Such anchors may be RFLPs, RAPDs, STSs, and even the recombination maps obtained by strictly genetic methods. The relative position of two YACs is revealed when they bridge two anchors. Anchoring may provide the means for correlating the strictly genetic linkage maps with the physical maps based on nucleotide sequencing. The principle of the procedure is diagrammed below. (Matallana, E., et al. 1992. *Methods in Arabidopsis Research*, Koncz, C., et al., eds., p. 144. World Scientific, Singapore.)

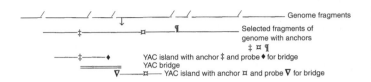

anchor locus Gene with well-known map position that can be used as a reference point for mapping new genes. *See* anchoring.

anchor residue Amino acids of the peptides attaching to MHC molecules. *See* MHC.

ancient DNA DNA of 50,000–100,000 years old or even older from ancient bones may still be analyzed. Samples preserved in amber may last longer. Mitochondrial DNA extracted from 80-million-year-old bones and amplified by PCR had sequences different from any other sequences known so far. The validity of these reports on very old DNA samples has been seriously questioned and contamination may not be ruled out. The condition of preservation is critical. It is very important that during PCR analysis the greatest caution is exercised to avoid contamination. It is advisable to test not just the sample but the immediate environment, the reagents themselves, verify that the sample conforms to that of the species and is suspect if the fragments are too long. In case protein is present, the high ratio between D and L aspartic acid indicates that most likely the DNA has been degraded. The purpose of the analysis of ancient DNA is to obtain information on individuals and groups or to assess evolutionary relations. The mtDNA (~17,000 bp) of two kinds of moa birds extinct for 400 years has been fully recovered. *See* ancient organisms; coproscopy; hominidae; ice man; mummies; out-of-Africa; PCR; Romanovs. (Hoftreiter, M., et al. 2001. *Nature Rev. Genet.* 2:353; Lambert, D. M., et al. 2002. *Science* 295:2270.)

ancient organism Now extinct species recognized as paleontological relics are difficult to study even by the most modern research techniques because the organic material has decayed. A 25–40-million-year-old bacterial spore discovered in the digestive tract of a now extinct bee species, preserved as an amber enclosure, was reported to be revived, and its 16S ribosomal RNA was quite similar to the living species of *Bacillus sphericus*. Actually, the calculated rate of nucleotide substitution in the 16S RNA encoding DNA segment appeared to be 1.8 to 2.4×10^{-9} per site per year. Although the isolation of the spore from the amber was carried out with extreme caution, some questions regarding possible contamination may be raised and newer studies have failed to confirm DNA in amber. More recently (2000), 250-million-year-old spore-forming bacilli have been revived from salt crystal. *See* amber; ancient DNA; ice man; mummies. (Hoftreiter, M., et al. 2001. *Nature Rev. Genet.* 2:353.)

ancient RNA Retrieved from extinct or very old specimens. *See* ancient organisms.

ANCOVA Analysis of covariance. *See* correlation.

Andalusian fowl Has been frequently used as an example for codominant segregation; when black and white fowl are crossed in the F_2, one black, two blue, and one white are found; the blue has black and white (white-splashed) feathers. *See* codominance.

Andalusian fowl.

Andersen disease (3p12) Autosomal-recessive deficiency of amylotransglucosidase(s) causing liver, heart, and muscle disease because of the defect in glycogen storage. *See* glycogen storage disease (type IV).

Andersen syndrome (KCJN2, 17q23) Periodic paralysis with heart arrhythmia and deformations. The basic defect is in an inwardly rectifying potassium channel. *See* ion channels. (Plaster, N. M., et al. 2001. *Cell* 105:511.)

Anderson disease Involves lipid transport defects of the intestines and the retention of chylomicrons. *See* chylomicron; lipids.

Anderson-Fabry disease Human X-chromosome-linked deficiency of α-galactosidase resulting in angiokeratoma (red or pink skin or mucous membrane lesions caused by dilation of veins). The gene is 12 kb, with 7 exons encoding a 427 amino acid protein. *See* angiokeratoma; galactosidase-β.

ANDi (inserted DNA [in reverse]) Name of the first transgenic (rhesus) monkey.

androdioecy Separate plants produce the male and hermaphrodite flowers. *See* dioecy; hermaphrodite. (Wolf, D. E., et al. 2001. *Genetics* 159:1243.)

androecium Male region of a flower (the stamens). *See* stamen.

androgen Hormone that promotes virility, but it is present at a lower level in females. Androgens are formed by hydroxylation of progesterone. The most important androsterone is testosterone. *See* animal hormones; aromatase; FGF; steroid hormones; testosterone.

androgenesis Development of the male gamete into a paternal haploid or diploid embryo under natural conditions; it can be obtained by in vitro culturing and regeneration of plants from microspores. In vitro androgenesis can be direct when the microspores develop directly into plantlets or indirect when the microspores first form a callus, and from that plantlets are regenerated in a second step. Androgenesis may also result when from a fertilized egg all the chromosomes of the female are lost and those of the male remain; again, paternal offspring results. Androgenesis occurs in the plant as well as in the animal kingdom. *See* anther culture; apomixis; embryo culture; gynogenesis; hemiclonal; hybridogenetic; hydatidiform mole; microspore culture. (Kermicle, J. L. 1969. *Science* 166:1422; Corley-Smith, G. E., et al. 1996. *Genetics* 142:1265.)

androgen insensitivity (Xq11-q12) Due to a defect in the dihydrotestosterone receptor. *See* Kennedy disease; Reifenstein syndrome; testicular feminization.

androgenital syndrome *See* pseudohermaphroditism.

androgenote Diploid embryo with only paternal sets of chromosomes. *See* androgenesis.

androgenous Pseudo- or true hermaphroditic stage in mammals or plants. *See* hermaphrodite.

androgen receptor Activated in the muscle cells by a 205 kDa actin-binding protein, supervillin. *See* gynecomastia; hormone response elements (HRE); Kennedy disease; testicular feminization. (Reid, K. J., et al. 2001. *J. Biol. Chem.* 276:2943; Ting, H.-J., et al. 2002. *Proc. Natl. Acad. Sci. USA* 99:661.)

andromerogony Development of an egg (or part of it) containing only the male pronucleus; the egg's own nucleus is removed prior to fusion with the male nucleus. *See* androgenesis; pronucleus.

andropause Period of decline of the free testosterone level in human males after its peak at age 30, yet at age 60 it is still comparable to that at age 20. Muscle strength may be increased by replacement therapy, but impotence is usually not cured. *See* menopause.

androsome Chromosome that normally occurs only in males. *See* sex chromosomes.

androstane Androstanol and androstenol steroids.

androstanol (5α-androstan-3α-ol) Mammalian pheromone, inhibitory to constitutive CAR-β. *See* androstane; androstenol; CAR-β.

androstenol (5α-androst-16-en-3α-ol) Mammalian pheromone, inhibitory to constitutive CAR-β. *See* androstane; androstanol; CAR-β.

anemia Reduction of the red blood cells and hemoglobin below the normal level. It occurs when the production of erythrocytes does not keep up with losses. Several human diseases involve anemia, including some hereditary ones such as the thalassemias, sickle cell anemia, glucose-6-phosphate dehydrogenase deficiency, etc. Some anemias appear under autosomal-dominant, autosomal-recessive, or X-linked control. *See* aceruloplasminemia; adenylate kinase deficiency; atransferrinemia; Cooley's anemia; diphosphoglycerate mutase deficiency; elliptocytosis; Fanconi's anemia; glutathione synthetase deficiency; hemochromatosis; hemolytic anemia; IRE; megaloblastic anemia; pyrimidine-5-nucleotidase deficiency; pyruvate kinase deficiency; sickle cell anemia; siderocyte anemia; thalassemia; transcobalamine deficiency.

anemophily Pollination by the wind.

anencephaly (spina bifida) Perinatal disorder of fetuses and newborns without brain (cerebrum and cerebellum). Many of the afflicted die before birth; 1/16 survive birth but rarely survive for a week. It may be due to a recessive mutation, but some of the cases are due to nongenetic causes. Its prevalence is less than 1/1000. A prenatal test may be carried out if family history indicates genetic causes. Microhydranencephaly maps to 16p13.3-p12.1. *See* Arnold-Chiari malformation; genetic screening; hydrocephalus; MSAPF; neural tube defects; prenatal diagnosis.

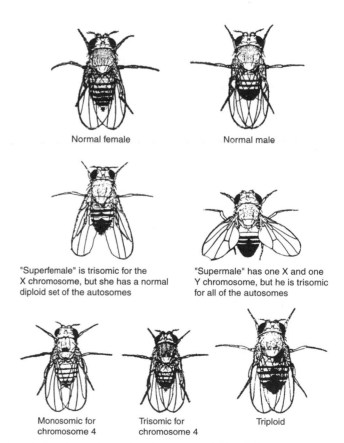

Normal female Normal male

"Superfemale" is trisomic for the X chromosome, but she has a normal diploid set of the autosomes

"Supermale" has one X and one Y chromosome, but he is trisomic for all of the autosomes

Monosomic for chromosome 4 Trisomic for chromosome 4 Triploid

Normal female, male and various aneuploids of *Drosophila*. (From Morgan, T. H., et al. 1925. *Bibl. Genet.* 2:3.); text next page.

anergy Unresponsiveness of the lymphocyte to an antigen because, e.g., a slightly modified peptide-MHC is attached to the T-cell receptor or some inducive factors are not functioning adequately. Anergized CD4$^+$ T cells are not completely idle but have some regulatory function. *See* HLA; MHC; T cell. (Jooss, K., et al. 2001. *Proc. Natl. Acad. Sci. USA* 98:8738.)

anesthetic Used to numb the nerve receptors; they are generally affecting the ligand-gated ion channels and cell membrane lipids and proteins. In mammals, stomatin and degenerin, and in *Caenorhabditis*, the product of the *UNC-1* gene, may affect the critical ion channels. *See* degenerin; ion channels; stomatin. (Humphrey, J. A., et al. 2002. *Hum. Mol. Genet.* 11:1241.)

aneugamy The chromosome number of the two gametes involved in the fertilization is different. *See* anisogamy; heterogametic; homogametic; isogamy.

aneuhaploid Haploid that has an incomplete set(s) of chromosomes. *See* aneuploidy; haploid.

aneuploidy Chromosome number is either more or less $2n \pm 1$ or ± 2 or ± 3, etc. Aneuploids are trisomics or monosomics, single or multiple. Aneuploidy is frequent in cultured cells and in cancer cells. After surveying 1291 spontaneous human abortions, Hamerton (1971) found 5% monosomics, 11.9% trisomics, 4.1% triploids, and 1.2% tetraploids (note that the triploids and tetraploids are polyploids but not aneuploids and their frequency is included only for comparison).

Among live human births chromosomal anomalies are close to 1%. Aneuploids are usually very deleterious, yet sex chromosomal aneuploidy, e.g., Turner syndrome XO, Klinefelter syndrome XXY, etc., is not generally lethal in humans or animals. In the 47 XXY individuals the anomaly in the majority of cases is due to nondisjunction of the XY bivalent. This high frequency may be attributed to the fact that usually even under normal conditions only a single chiasma occurs between the X and the Y (in the pseudoautosomal region), and if this chiasma fails to materialize, nondisjunction takes place. Monosomics $(2n - 1)$ have been very skillfully exploited for mapping genes to chromosomes in polyploid plants (wheat, oats, etc.). Microarray hybridization profiles reveal aneuploidy without cytological analysis because the expressions of large tracts of genes are detectable. Aneuploidy may lead to cancerous growth. One cause of aneuploidy is probably the abnormal organization of the centrosome of animals. *See* chiasma; centrosome; hyperploid; hypoploid; microarray hybridization; monosomic analysis; MSAFP; pentaploid; pseudoautosomal; triploid. (Jacobs, P. A. & Hassold, T. J. 1995. *Adv. Genet.* 33:101; Hassold, T. & Hunt, P. 2001. *Nature Rev. Genet.* 2:280; Yuan, L. et al. 2002. *Science* 296:1115; illustration on preceding page.)

aneurysm Formation of small sacs of blood caused by the dilation of the veins. It is under autosomal-dominant control in both abdominal (more common in females) and brain aneurysms. Aortic aneurism (15q21) is a heart disease involving fibrillin. Apparent linkage to 5q22-q31, 7q11, and 14q22 was also reported. *See* collagen; fibrillin. (Onda, H., et al. 2001. *Am. J. Hum. Genet.* 69:804.)

aneusomatic Somatic chromosome number varies among the body cells because of the presence of supernumerary chromosomes and their frequent somatic nondisjunction.

Aneusomy is one of the most common causes of cancer. *See* aneuploidy; nondisjunction; supernumerary chromosomes. (Fabarius, A., et al. 2002. *Proc. Natl. Acad. Sci. USA* 99:6778.)

aneusomy, segmental *See* contiguous gene syndrome.

Angelman syndrome (happy puppet syndrome) Apparently an autosomal-recessive human defect with somewhat irregular inheritance. Cytologically and molecularly detectable deletion in the 15q11-q13 region (similar to Prader-Willi syndrome) was observed. The unusual feature of this condition is that the syndrome is transmitted only through the mother, whereas in Prader-Willi syndrome the transmission is paternal. Imprinting has been suggested for the phenomenon. It has been proposed that in the female germline the so-named BD RNA transcripts induce methylation in the promoter of snRNP genes. The affected individuals have motor function defects, mental retardation, epilepsy, speech defect or absence, and frequent protruding tongue accompanied by excessive laughter (hence the HPS term). It appears that the syndrome is due to an abnormal ubiquitin-mediated degradation of a brain ligase (UBE3A) of the E6-AP class. *See* disomic; E3; head/face/brain defects; imprinting; imprinting box; mental retardation; Prader-Willi syndrome; snRNP; Ube3; ubiquitin. (Jiang, Y.-H., et al. 1999. *Am. J. Hum. Genet.* 65:1.)

angioedema Dilation of the subcutaneous capillary veins leading to skin, respiratory tract, and gastrointestinal fluid accumulations. The hereditary dominant form has been attributed to mutations in a serpin gene or to complement inhibitory factor deficiency (C1-INH). The condition may be haplo-insufficient. See complement; haplo-insufficient; serpin.

angiogenesis Formation of blood vessels and chronic inflammation. The vascular endothelial growth factor and its two receptors, Flt-1 and Flk-1/KDR, are required in rodents for angiogenesis. Vasculogenesis factor (VEGF), peptide hormones secreted by tumors, increase blood supply and ensure neoplasias and their further growth. There are two angiogenesis pathways. The fibroblast growth factor or tumor necrosis factor-α initiated path depends on integrin $\alpha_v\beta_3$, whereas angiogenesis initiated by vascular endothelial growth factor, transforming growth factor-α, or phorbol ester uses the $\alpha_v\beta_5$ path. Disruption of matrix metalloproteinase 2 binding to integrin inhibits angiogenesis and it may be relevant to tumor control. The tumor necrosis factor-α induced angiogenesis uses the B61 cytokine-inducible ligand for the Eck protein tyrosine kinase receptor (RPTK). Angiogenesis is also required for the proliferation of tumors, but the process can be restricted by the antibiotic minocycline, AGM and interferon $\alpha/\beta/\gamma$, angiostatin, endostatin, interferons, etc. Promoters of angiogenesis include cytokines (EGF, TGF, TNF), various carbohydrates, angiogenin, and several other molecules. The ligands of the Tie receptors, Angi1 and Angi 4, regulate angiogenesis positively, whereas Angi3 is a negative regulator. *See* angiopoietin; angiostatin; cancer; CXCR; EGF; endostatin; fibroblast growth factor; Flk; Flt; hemangioblast; Id proteins; integrin; leptin; maspin; metalloproteinases; neuropilin; phorbol esters; PTEN; TGF; TNF; tumor; tumor necrosis factor; VEGF. (Carmeliet, P. & Jain, R. K. 2000. *Nature* 407:249; Folkman, J. 2001. *Proc. Natl. Acad. Sci. USA* 98:398; Kuo, C. J., et al. 2001. *Proc.*

Natl. Acad. Sci. USA 98:4605; Jones, N., et al. 2001. *Nature Rev. Mol. Cell Biol.* 2:257; Isner, J. M. 2002. *Nature* 415:234; <http://angiodb.snu.ac.kr>.)

angiogenin RNase stimulating blood vessel formation. *See* ribonucleases.

angiokeratoma Recessive X-chromosome-linked disease involving dilation of the small veins, warty growth, thickening of the epidermis primarily on the fingers, toes, and scrotum. *See* Anderson-Fabry disease; fucosidosis; Kanzaki disease.

angioma Tumor of the blood or lymph vessels or a neoplasia that forms blood and lymph vessels. Many forms exist in humans; they are controlled by dominant genes at human chromosomes 7q11.2-q21 (CCM1), 7p15-p13 (CCM2), and 3q25.2-q27 (CCM3). The CCM1 locus encodes RAP1A-interacting KRIT1 protein. *See* hemangioma; RAP1.

angioneurotic edema, hereditary Dominant chromosome 11q11-q13.1 deficiency of complement C1 inhibitor causing edema of the air passageway. The reduced level of the inhibitor leads to excesses of the C4 and C2 kinin fragments. *See* angioedema; complement; kinin.

angiopoietin-1 Blood vessel differentiation factor that promotes tissue vascularization. Angiopoietin-2 is an antagonist of angiogenesis. *See* angiogenesis; VEGF.

angiosperm Plant that bears seeds within an ovary; the majority of higher plants belong to this taxonomic category. Fossil evidence points to their presence in the Jurassic period (137–190 Mya). *See* geological time periods; Mya.

angiostatin 38 kDa protein with some homology to plasminogen. It is an anticancer agent that deprives cancer cells of new blood vessel development. Resistance mutation against it (common for anticancer drugs) has not been observed. The therapeutic effectiveness in the cure of human cancer has not been completely accepted, although positive results were obtained, especially in combination with radiation treatment. Angiostatin may act by inhibiting endothelial ATP synthase, which may be required for supplying the energy for tumorigenesis. *See* angiogenesis; cancer; cancer therapy; endostatin; plasminogen. (Moser, T. L., et al. 2001. *Proc. Natl. Acad. Sci. USA* 98:6656.)

angiotensin Asp-Arg-Val-Tyr-Ile-His-Pro-Phe peptide stimulates the smooth muscles of the blood vessels, reduces the blood flow through the kidneys and decreases the excretion of fluid and salts, increases the secretion of aldosterone, and stimulates the reabsorption of sodium. It is involved in the hereditary disorders of adrenocortical steroid biogenesis. Angiotensin I receptor AT_1 mediates the pressor (higher blood pressure) and angiotensin II receptor AT_2 has the opposite (depressor) effect. Angiotensin II cell surface receptor is directly stimulated by the Jak/STAT signal transduction pathway. The angiotensin-converting enzyme (ACE, 17q23) is a dipeptidyl carboxypeptidase (kininase), and it catalyzes the conversion of angiotensin I to angiotensin II. *See* aldosterone;

angiostatin; BBB; DCP1; ecclampsia; hypertension; pseudoaldosteronism; signal transduction; tachykinin. (Zhu, X., et al. 2001. *Am. J. Hum. Genet.* 68:1139; Morimoto, S., et al. 2002. *Physiol. Genet.* 9:113.)

angstrom (Å) 1 Å = 1/10 nanometer (nm).

angular transformation (arcsine transformation) Used with percentages and proportions. In a binomial distribution the variance is a function of the mean. The arcsine transformation prevents $\theta = \text{arcsine}\sqrt{p}$, where p is a proportion that stands for an angle whose sine is the given quantity. The transformation stretches out both tails of a distribution of percentages and compresses the middle. It may be usefully applied to genetic data when the figures fall outside the 30% and 70% ranges. *See* arcsine; sine.

anhidrosis Reduction or lack of sweating. An X-linked hypo- or anhidrotic ectodermal dysplasia is caused by mutation in a transmembrane protein. Ectodermal dysplasia is part of about 150 syndromes. *See* ectodermal dysplasia; pain insensitivity.

anhidrotic ectodermal dysplasia *See* ectodermal dysplasia.

anhydride Result of a condensation reaction where water has been eliminated (between carboxyl and phosphate groups).

animalcule The pioneer microscopist Anthony Leuwenhoek (17th century) believed he saw small encapsulated animals in the sperm of various animals. This observation supported his view that inheritance is only through the sperm and the females serve only as incubators. His observations led to the notion of preformation rather than epigenesis. *See* epigenesis; preformation; sperm.

animal hormone First chemical messenger secreted by some tissues and carried by the bloodstream to the specific sites of action where regulatory functions are carried out. Animal hormones regulate either the synthesis or activity of enzymes or affect membrane transport in cooperation with the second messengers, cyclic adenosine monophosphate and cGMP. They are of three major types. *Peptide hormones*: Secreted by the hypophysis of the pituitary gland: somatotropin (general growth hormone, GH), corticotropin (adrenocorticotropin, ACTH in the kidneys), thyrotropin (thyroid-stimulating hormone, TSH), follitropin (FSH, in gonads), lutrotropin (luteinizing hormone, LH, in gonads), prolactin (in mammary glands). Secreted by the neurohypophysis: oxytocin (controls uterine contractions and milk production), vasopressin (antidiuretic hormone controls water reabsorption of the kidneys and blood pressure), secreted by the middle section of the hypophysis. Melanotropins (control melanin pigments). The pancreas secretes insulin (controls carbohydrate, fatty acid, cholesterol metabolism) and glucagon (stimulates glucose production by the liver). The ovary produces relaxin (controls pelvic ligaments, the uteral cervix, thereby labor); the thyroid gland is the source of parathyrin (involved in calcium and phosphorus metabolism); the kidneys release erythropoietin (a glucoprotein involved in erythrocyte production by the bone marrow) and renin (causes constriction of the blood vessels); the digestive tract secretes gastrin (promotes digestive enzymes), enterogastrone (controls the gastric secretion), cholecystokinin (regulates gall bladder), secretin (controls pancreatic fluids and bile production), and pancreozymin (of duodenal origin stimulates pancreatic functions). *Amino acid hormones*: thyroxin and triiodothyronin, secreted by the thyroid gland, affect many functions in the body. The kidney tissues secrete epinephrine (adrenaline) and norepinephrine (triiodothyronin), which regulate blood pressure and heart rate; the pineal gland (a cone-shaped epithelial body at the base of the brain) produces melatonin, which affects the pigment-producing melanophore cells. The nerve cells produce serotonin (5-hydroxytryptamine), affecting contraction of the blood vessels and nerve function. Serotonin controls the central nervous system, e.g., alertness, sleep, mood, aggressiveness, etc. *Steroid hormones*: produced in the testes (testosterone, regulates male reproductive capacities), in the ovaries (estrogen [estradiol-17β], involved in female reproductive functions), in the corpus luteum of the ovary, and in the Schwann cells of the peripheral nervous system progesterone is made. It functions during menstrual cycles and pregnancy and in myelin formation. In the kidney cortex, cortisol (corticosterone) is synthesized, affecting glucose utilization and glucose levels in the blood. Estrogen is typically a female hormone, yet extremely high concentrations occur in the testis fluids, and it is important for male fertility. Progesterone is necessary for the maintenance of pregnancy. It binds to the oxytocin receptor (OTR) and prevents uterine contractions. *Eicosanoid (hormone-like) substances* are prostaglandins (triggering smooth muscle contraction, control fever and inflammations), leukotrienes (secreted by the white blood cells and affecting hypersensitivity reactions and pulmonary functions), thromboxanes (produced by the blood platelets and other cells, and are involved in blood clotting, blood vessel constriction, etc.). There are a large number of other hormones with important functions. *See* hormones; hormone receptors; hormone response elements; opiocortin; oxytocin.

animal host cell   Used for genetic transformation. *Xenopus* oocytes are well suited for such studies because they can propagate foreign genes in appropriate vectors quite efficiently. Similarly, COS cells of mice and other somatic cells have been used effectively. More recently, techniques have become available for the transformation of animal zygotes and embryos, and thus genetic information can be added or replaced in the germline and transmitted to the sexual progeny. *See* COS; germline; transformation of animal cells; vectors, genetic.

animal model   Certain biological phenomena cannot be studied in humans because mutants are not available and cannot be produced or manipulated effectively. In such cases, animals such as *Caenorhabditis, Drosophila*, and mice are used for experimentation (in behavioral genetics, neurobiology, various diseases, etc.). Animal models may

WILD-TYPE PROTOONCOGENE

MUTATIONALLY ACTIVATED ▥▥▥ AND GENETICALLY ENGINEERED ONCOGENE CONTAINS THE SELECTABLE MARKER ▭ AND IT IS INSERTED INTO THE WILD-TYPE LOCUS

SUCH A CONSTRUCT MAY RECOMBINE WITH THE WILD-TYPE LOCUS AND PRODUCE A FUNCTIONAL TUMOR SUPPRESSOR:

OR AN ACTIVE ONCOGENE:

A genetic construct simulating the sporadic occurrence of oncogenic mutations. (See Johnson. L., et al. 2001. *Nature* 410:1111.)

have an important role in improving the techniques of gene therapy. The "shiverer" deletion of mice, resulting in convulsions because of the loss of a gene coding for a myelin protein, had been genetically cured by transfection of the wild-type allele into the gamete. Similarly, the size of mice could be genetically increased by transformation using the rat somatotropin (RGH, growth hormone) gene fused to and regulated by a metallothionein promoter. The following monogenic human genetic disorders have mouse models (abbreviations h. chr. = human chromosome, m. chr. mouse chromosome): *adenomatous polyposis* (protrusive growth in the mucous membranes, h. chr. 5q21-q22, mouse homolog Apc^{Min}, chr. 18); *androgen insensitivity* (sterility, h. chr. Xq11.2-q12, mouse gene AR^{Tfm}, m. chr. X), *X-linked agammaglobulinemia* (deficiency of γ globulin in blood, h. chr. Xq21.33-q22, mouse gene Btk^{Xid}, m. chr. X); *Duchenne muscular dystrophy* (an early muscular disability, h. chr. Xp21.3-p21.2, mouse Dmd^{mdx}, m. chr. X); *Greig cephalopolysyndactyly* (multiple fusion of digits, h. chr. 7p13, mouse gene $Gli3^{Xt}$ m. chr. 13); *mucopolysaccharidosis type VII* (a type of lysosomal storage disease, h. chr. 7q22, mouse gene Gus^{mps}, m. chr. 5); *α-thalassemia* (defect in the hemoglobin α chain, h. chr. 16p13.3, mouse gene Hba^{th}, m. chr. 11); *β-thalassemia* (defect in the β-chain of hemoglobin, h. chr. 11p15.5, mouse gene Hbb^{th}, m. chr. 7); *piebaldism* (color patches on the body, h. chr. 4p11-q22, mouse gene Kit^W, m. chr. 5); *ornithine transcarbamylase* (defect in the transfer of a carbamoyl group, $H_2N-C=O$, from ornithine to citrulline, h. chr. Xp21.1, mouse gene Otc^{Spf}, m. chr. X); *tyrosinase-positive type II* (oculocutaneous albinism, h. chr. 15q11-q12, mouse gene p^p, m. chr. 7); *phenylketonuria* (phenylalanine hydroxylase deficiency, h. chr. 12q22-q24.2, mouse gene Pah^{enu2}, m. chr. 10); *Waardenburg syndrome type 1* (h. chr. 2q35-q37, mouse gene $Pax3^{Sp}$, m. chr. 1); *aniridia* (absence of the iris, h. chr. 11p13, mouse gene $Pax6^{Sey}$ m. chr. 2); *pituitary hormone deficiency* (h. chr. 3q, mouse gene $Pit1^{dw}$, m. chr. 16); *Pelizaeus-Merzbacher disease* (central brain sclerosis, h. chr. Xq21.33-q22, mouse gene Plp^{jlp}, m. chr. X); *Charcot-Marie-Tooth disease type 1A* (a progressive neuropathic muscular atrophy, h. chr. 17p12-p11.2, mouse gene $Pmp22^{Tr}$, m. chr. 11); *retinitis pigmentosa* (sclerosis and pigmentation of the retina, h. chr. 6p21.2-cen, mouse gene $RD2^{Rd2}$, m. chr. 17); *gonadal dysgenesis* (underdeveloped germ cells in the testes, h. chr. Y11.2-pter, mouse gene Sry^{Sxr}, m. chr. Y); *tyrosinase negative oculocutaneous albinism* (see albinism, h. chr. 11q14-q21, mouse gene Tyr^c, m. chr. 7). By disruption of hexosaminidase α subunit, a model for Tay-Sachs disease has been generated in mice. Interestingly, these animals suffered no obvious behavioral or neurological deficit. Disrupting the hexoseaminidase β subunit (Sandhoff disease model) resulted in massive depletion of spinal cord axons and neuronal storage of ganglioside G_{M2}. These latter two examples indicate possible complications with animal models.

Mouse polygenic disorders with similarities to human conditions (human problem—mouse strain): alcoholism and opiate drug addictions, C57BL/6J; asthma, A/J; atherosclerosis, C57BL; audiogenic (sound-induced) seizures, DBA; cleft palate (fissure in the mouth), A; deafness, LP; dental disease, C57BL, BALB/c; diabetes, NOD; epilepsy, EL, SWXL-4; granulosa cell tumors in the ovary, SWR; germ cell tumors in the ovary, LT; germ cell tumors in the testes, 129; hemolytic anemia, NZB; hepatitis, BALB/c; Hodgkin disease (pre-B-cell lymphoma), SJL; hypertension, MA/My; kidney adenocarcinoma, BALB/c Cd; leprosy (*Mycobacterium leprae*), BALB/c; leukemia, AKR/J, C58/J, P/J; lung tumors, A, Ma/My; measles, BALB/c; osteoporosis, DBA; polygenic obesity, NZB, NZW; pulmonary tumors, A/J; rheumatoid arthritis, MRL/Mp; spina bifida (defect of the bones of the spinal cord), CT; systemic lupus erythematosus (a skin degeneration), NZB, NZW; whooping cough, BALB/c.

Some of the diseases (e.g., various types of cancer) occur sporadically because of mutations during animal/human development. The introduction of a functional oncogene into the germline or into the soma line of a person cannot appropriately represent the conditions emerging in sporadic cases. Usually the sporadic occurrence of mutation in cancer is predominant. Mutations in the soma are generally surrounded by normal cells, and through bystander effect these may modify the expression of the mutant cells and their clonal derivatives, unlike cases when the mutation has occurred in the male/female germline before fertilization. The former condition can be simulated in a genetic model if, e.g., through recombination between one of the wild-type RAS oncogenes and its mutant and potentially proliferative allele is activated. The construction of such a model may be represented graphically.

animal pole   Dorsal end of the animal egg opposite the lower end, the vegetal pole, and where the sperm entry is located. After the entry, the egg cortex rotates slightly, and in some species a gray crescent is formed at the side opposite the entry. *See* vegetal pole.

animal species hybrid   The most familiar examples are the hybrids of the mare (*Equus caballus*, $2n = 64$) and the jackass (*Equus asinus*, $2n = 62$), and the stallion and the she-ass. The hybrid males do not produce viable sperm, although they may show normal libido. The females may have estrus and ovulate, but there are no proven cases of fertility. Zebras ($2n = 44$) also may form hybrids with both donkeys and horses. Buffalo (*Bison bison*, $2n = 60$) may be crossed reciprocally with cattle (*Bos taurus*, $2n = 60$), but their offspring (cattalo) have reduced fertility. The domesticated pig (*Sus crofa*, $2n = 38$) forms fertile hybrids with several wild pigs with the same number of chromosomes. The sheep (*Ovis aries,* $2n = 54$) interbreeds with the wild mouflons, but the sheep × goat (*Capra hircus,* $2n = 60$) hybrid embryo only rarely can be kept alive. Some monkeys can be interbred, but primates are generally sexually isolated. There is no sexual barrier among the various human races, indicating close relationship, but no hybrids are known between humans and any other species. These general rules do not hold for somatic cell hybrids because human cells can be fused with rodent or plant cells, but they cannot be regenerated or even maintained successfully for indefinite periods of time. The hybridization barrier is not identical with other functional barriers. Somatic cell hybridization and transformation may yield hybrid cells of different species. *See* goat–sheep hybrids; somatic cell hybrids; transformation, genetic.

Hybrid of the male grant's zebra and the female black arabian ass, gloucester zoo. (From Gray, A. P. 1971. Mammalian Hybrids. Commonwealth Agric. Bureau. Farnham Roal, Slough, UK.)

animal transformation vector   Most commonly Simian virus 40 (SV40)– and Bovine papilloma virus (BPV)–based vectors are used. The BPV vectors can be used for the synthesis of large amounts of proteins specified by the gene(s) carried by the expression vectors. In addition, the BPV vectors can be maintained for long periods of time in cell cultures and may yield 10 mg specific protein(s) per liter of culture/24 hr. The SV40 vectors can also be used for gene amplification in COS cells. Both of these vectors can serve as shuttles between animal and prokaryotic cells. *See* adeno-associated virus; adenovirus; BPV and SV40 constructs; COS cells; gene therapy; lentivirus; retroviral vectors; vaccinia virus.

animal virus   Includes both invertebrate and vertebrate viruses. The Rhabdoviridae and the Bunyoviridae may also infect plants. The *double—stranded DNA* viruses may be *enveloped*: Baculoviridae, Poxviridae, Herpesviridae, Hepadnaviridae, Polydnaviridae; and double-stranded DNA viruses *without envelope*: Iridoviridae, Adenoviridae, Papovaviridae. The Parvoviridae have *single-stranded DNA* and they are *not enveloped*. The *single-stranded RNA* and *enveloped* group includes the Togaviridae, Bunyaviridae, Rhabdoviridae, Coronaviridae, Paramixoviridae, Toroviridae, Orthomyxoviridae, Arenaviridae, Flaviviridae, Retroviridae, and Filoviridae. The *single-stranded RNA* and *nonenveloped* viruses are Picornaviridae, Tetraviridae, Nodaviridae, and Caliciviridae. The *double-stranded RNA* and *nonenveloped* viruses are Reoviridae and Birnaviridae. Their genetic material varies in size from 5 kb in the Parvoviridae to 375 kbp in the Poxviridae. The Polydnaviridae may have several copies of double-stranded circular DNAs. The Papovaviridae have only single double-stranded DNA genetic material. The others may have two or more segments of linear nucleic acid genetic material. *See* viruses.

anion Negatively charged ion.

anion exchange resin Polymer with cationic groups; traps anionic groups and thus can be used in chromatographic separation.

aniridia Absence or reduction of the iris in the eye. It is frequently accompanied by cataract (opacity of the eye[s]), glaucoma (increased intraocular pressure causing deformation of the optic disk), nystagmus (involuntary movement of the eyeball), etc. The condition is caused by dominant defects in human chromosomes 2 and 11. In a Michigan population, the rate of mutation appeared to be 4×10^{-6}. It may involve Wilms' tumors and genital abnormalities due to a deletion in human chromosome 11p13. Aniridia may be haplo-insufficient. The *Drosophila* locus *eyeless* and the mouse *Sey/Pax-6* are the corresponding homologs. *See* deletion; eyeless; haplo-insufficient; WAGR; Wilms' tumor.

anisogamy Gametes are not identical, e.g., male and female (+ or −) are distinguishable. *See* isogamy.

Anisomycin Antibiotic isolated from *Streptomyces griseolus*. It inhibits peptidyl transferase during protein synthesis on the ribosomes. It also inhibits pathogenic fungi (e.g., mildew) in plants and was found to be useful against infection by various species of the parasitic flagellate, *Trichomonas*, causing inflammation of the gum in the mouth, diarrhea, and vaginal discharge and irritation in humans and animals (particularly poultry and pigeons). *See* antibiotics; protein synthesis.

anisotropic Material varies in different directions, responds differently to external effects depending on directions.

Anj1 Heat- and other stress-inducible membrane-associated chaperone of higher plants. The Cys-Ala-Gln-Gln C-terminus may be subject to farnesylation. *See* chaperones; DnaK; heatshock proteins; prenylation.

ankyloblepharon Fused eyelids. *See* Hay-Wells syndrome.

ankylosing spondylitis (AS) Autosomal-dominant rheumatism-type disease with reduced penetrance. The greatest susceptibility to AS is associated with MHC (HLA B27), but spurious or fair linkage was observed with 1p, 2q, 6p, 9q, 10q, 16q, and 19q. Onset is after age 20. *See* autoimmune disease; connective tissue disorders; HLA; immunodeficiency; penetrance. (Laval, S. H., et al. 2001. *Am. J. Hum. Genet.* 68:918.)

ankyrin Protein motifs capable of binding fibrous proteins (e.g., spectrin) of the cytoskeleton and thus may be involved in some polar transports within the cell. Several ankyrin and ankyrin-like proteins are encoded in different human chromosomes (8p11.2, 4q25-q27, 10q21, etc.). *See* cytoskeleton; elliptocytosis; poikilocytosis; IκB; spectrin; spherocytosis; tankyrase. (Hayashi, T. & Su, T.-S. 2001. *Proc. Natl. Acad. Sci. USA* 98:491.)

anlage Group of cells of the embryo initiating specific biological structures. *See* primordium.

annealing Formation of double-stranded nucleic acid when two complementary single-stranded chains meet (nucleic acid hybridization, attachment of a primer). The process is used to estimate DNA complexity, for identifying the presence of homologous sequences in the genome by radioactively labeled or fluorescent homologous and heterologous probes. *See* chromosome painting; c_0t curve; DNA hybridization; FISH; primer; probe.

annexin Protein composed of four or eight conserved 70 amino acid domains with variations mainly at the amino end. In mammals, there are at least 10 annexins, and others exist in lower eukaryotes. Annexins bind to negatively charged phospholipids in the membranes. Annexins V and VII form voltage-regulated ion channels for different cations, whereas VII is specific for Ca^{2+}. Annexin II may assist exo- and endocytosis. An annexin-like protein may be involved in mitigating H_2O_2 stress. Annexin 7 (ANX7, 10q21) is a tumor suppressor. *See* endocytosis; exocytosis; ion channels. (Bandorowicz-Pikula, J., et al. 2001. *Bioessays* 23:170.)

annotation *See* genome annotation.

annulus (a ring) For example, specialized cells in a sporangium involved in opening.

anoikis Loss of cell anchorage to a substrate may lead to apoptosis and may be the requisite for metastasis. Some cell lines resistant to anoikis display increased metastasis because the probability of apoptosis is reduced. Rac GTP-ase may protect against anoikis. *See* anchorage dependence; apoptosis; metastasis. (Coniglio, S., et al. 2001. *J. Biol. Chem.* 276:28113.)

anomalous genetic ratio Caused by many different mechanisms. Defective chromosomes or chromosomes carrying deleterious genes are transmitted at lower than normal frequencies and reduce the expression (transmission) of the genes residing in that chromosome (conversely the other allele may appear in excess). Monosomy and trisomy also modify segregation ratios. The genetic ratios may be altered by preferential segregation of certain chromosomes in meiosis. Similarly segregation distorter genes can cause dysfunction of the sperm that carry them. Meiotic drive in a population can work against the more fit alleles. *See* aneuploidy; certation; chromosomal breakage; deletion; drift, genetic; gametophyte factor; gene conversion; meiotic drive; Mendelian segregation; monosomic analysis; Muller's ratchet; penetrance; preferential segregation; segregation distorter; trisomic analysis.

anomalous killer cell (AK) T cell grown in the presence of IL-2; acquires natural killer cell (NK)–like properties. *See* killer cell; *Paramecium*.

anomers Stereoisomers of sugars differing only in the configuration of the carbonyl residue, e.g., α-D-(+)-glucose and β-D(+)-glucose.

anonymous DNA segment Mapped DNA fragment without known gene content.

anonymous gene Mapped gene without information about its molecular mechanisms but known to affect the expression of a quantitative response such as a behavioral trait. If it displays two allelic states, it can be used for (DNA) mapping. *See* behavior; behavior genetics.

anonymous probe DNA probe with no known gene(s) in it and its function is unknown. Nevertheless, it provides information on the presence of sequences homologous to it and thus may be useful for taxonomic or evolutionary studies. See microsatellites; physical mapping.

***Anopheles* mosquito** Host and vector of the protozoan *Plasmodium falciparum*, the cause of malaria. *Anopheles gambiae* carries one major and two minor genes that control the formation of melanin-rich capsules in the midgut, thus disarming the *Plasmodium*. Anopheles control may be a major objective of fighting malaria. *See* sickle cell anemia; thalassemia; *Wolbachia*. (Atkinson, P. W. & Michel, K. 2002. *Genesis* 32:42; *Science* 298, 4 Oct. 2002.)

anophthalmos Autosomal-recessive bilateral defect in the formation of the optic pit. It has also been reported as an Xq27-encoded fusion of the eyelids and other complications. *See* eye diseases; microphthalmos.

anorexia Lack of appetite, or anorexia nervosa, is a psychological disturbance of adolescents (primarily females) caused by an abnormal fear of gaining weight and therefore refusing to eat. Characterized by self-induced vomiting, unnecessary use of laxatives leading to emaciation, irregular or lack of ovulation, reduced interest in sex, and other anomalies. Medical treatment may be required. The melanocyte-stimulating hormone, α-MSH, and analogs may be responsible for anorexia and weight loss. Oleyethanolamide may be a regulator of feeding. Susceptibility loci appear to be in chromosomes 1, 2 and 13. *See* bulimia; leptin; melanocyte-stimulating hormone; obesity. (Rodríguez de Fonseca, F., et al. 2001. *Nature* 414:209; Adan, R. A. & Vink, T. 2001. *Eur. Neuropsychopharmacol.* 11[6]:483; Devlin, B., et al. 2002. *Hum. Mol. Genet.* 11:689.)

ANOVA Analysis of variance. *See* analysis of variance.

anoxia Absence or deficiency of oxygen; reduces chromosomal damage during irradiation. *See* radiation effects; ARE.

anserine (β-alanine-1-methylhistidine) Dipeptide occurring in birds and some mammals but not in humans. *See* carnosinemia.

ANT (*Formica sanguinea*) $2n = 48$.

antagonist Blocks biological receptor activation. *See* agonist.

anteater (*Tamandua tetradactyla*) $2n = 54$.

antecedent Precursor, forerunner.

antelope (*Antilocapra americana*) $2n = 58$.

antenatal diagnosis Determination of a particular condition before birth by amniocentesis or blood samplings or by other means. *See* amniocentesis; fetoscopy; prenatal diagnosis.

antenna Feeler organ on the head of insects. *See Drosophila.*

Antenna.

Antennapedia *Drosophila* gene (*Antp*; map location 3–47.5, salivary bands 84B1-2) with numerous alleles. The null alleles result in embryonic lethality. Initially the locus was recognized by mutations that transform the antennae into mesothoracic legs. Numerous other homeotic changes may accompany the mutations. The different alleles may involve various types of homeotic changes at the locus. The gene occupies about 100 kb and contains eight exons. These exons are transcribed from promoters P1 or P2 or both. The transcripts may undergo alternate splicing. The homeobox motif is in exon 8. Actually *Ant* promotes leg differentiation by suppressing antenna-determining genes *extradenticle* (*exd*, 1–54) and *homothorax* (*hth*, 3–48). *See* homeotic genes; morphogenesis; *Polycomb*.

antenna pigment In the chloroplasts it collects light energy that is transmitted to the reaction centers for photochemical use. *See* chlorophyll; chloroplasts; photosynthesis.

anterior Indicates a direction in front of something or toward the head.

anterior-posterior polarity Head to tail anatomical direction.

anterograde Ahead or forward moving. *See* retrograd.

anther Pollen-containing part of male flowers. *See* gametogenesis.

Anther

anther culture Used for the isolation of haploid plants. The culture may start with microspores that are directly regenerated into plantlets (without an intermediate callus stage), or haploid tissues are isolated from anthers and a callus is formed, then the calli are regenerated into plants. Both procedures are using tissue culture methods under aseptic conditions. The haploid cells may diploidize spontaneously or by induction, which results in perfect homozygosis of the plants. *See* androgenesis; *Asparagus*; embryo culture; gametogenesis; YY plants. (Jahne-Gartner, A. & Lörz, H. 1999. *Methods Mol. Biol.* 111:269.)

antheridium   Male sex organ (gametangium) of lower plants and fungi.

anthesis   Time of pollen shedding or receptivity of a flowering plant.

anthocyanin   Plant flower pigments (delphinidine, cyanidin, pelargonidine, peonidine, petunidine, malvidine, etc.) from phenylalanine via trans-cinnamic acid and cinnamoyl-CoA, chalcones, and flavonones. CH_3 and OH groups on the B-ring determine the color produced; glycosylation (hexose or pentose) at the 3 and 5 positions (or at both) increases stability of the pigments, and these glycosides are called anthocyanidins. Each enzymatic step is controlled by different genes, and these original discoveries, beginning in the early 20th century, prepared the way for biochemical genetics. The color is also affected by the pH of the vacuoles under genetic control. By the use of antisense constructs of the gene chalcone synthase (CHS), the activity of this enzyme and chalcone flavonone isomerase (CHI) could also be reduced, indicating that CHS also regulates the expression of CHI. (Markham, K. R., et al. 2000. *Phytochemistry* 55:327; Rasusher, M. D., et al. 1999. *Mol. Biol. Evol.* 16:266; van Houwelingen, A., et al. 1998. *Plant J.* 13:39.)

anthranylic acid   Synthesis begins with the condensation of erythrose-4-phosphate + phosphoenolpyruvate, and from this shikimate and then chorismate are formed. Chorismate through prephenate contributes to phenylalanine and tyrosine and through another path it is a precursor of the amino acid tryptophan (actually indole-3-glycerol phosphate → indole and serine are converted to this amino acid). *See* phenylalanine; tyrosine.

anthrax   Toxin produced by *Bacillus anthracis*. The toxin primarily affects herbivorous animals, but it may spread to carnivorous predators and also to humans through the skin, by ingestion or inhalation of dust contaminated by the bacteria. The toxin consists of three proteins: (1) protective antigen (PA) facilitates the formation of a membrane channel for the (2) edema factor (EF, an adenylate cyclase) and (3) lethal factor (LF, a metalloprotease and selective inhibitor of MAPK and MAPKK). Although LF targets mainly MAPKK, apparently it also hydrolyzes a number of peptide hormones: granuloliberin R, dynorphin A (a 17 amino acid neuropeptide), kinetensin, and angiotension-1 (brain peptides). The chemical

PD09859 is also a MAPKK inhibitor, but it acts differently from LF. Mutation in PA may prevent the uptake of EF and LF, and thus in a dominant negative manner may become a potential tool for preventing toxic effects. Another preventive approach is to block the formation of the heptameric cell-binding subunit of the toxin by a synthetic polyvalent inhibitor (Mourez, M., et al. 2001. *Nature Biotechnol.* 19:958.) *See* adenylate cyclase; bioterrorism; MAPKK; metalloproteases; toxins. (Sellman, B. R., et al. 2001. *Science* 292:695; Mock, M. & Fouet, A. 2001. *Annu. Rev. Microbiol.* 55:647; Bhatnagar, R. & Batra, S. 2001. *Crit. Rev. Microbiol.* 27[3]:167; Schuch, R., et al. 2002. *Nature* 418:884.)

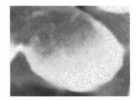

A spore of *Bacillus anthracis*.

anthropometric traits   Physical or physiological characters of humans (such as weight, head circumference, hair color, protein differences, behavior, etc.) that may be used for the characterization of human populations.

anti   Conformation of nucleotides, the CO and NH groups in the 2 and 3 positions of the pyrimidine ring (1, 2, 6 positions in the purine ring) are away from the glycosidic ring while in the SYN conformation they lie over the ring. The anti conformation is most common in nucleic acids and free nucleotides. (Kornberg, A. 1982. *DNA Replication.* Freeman, San Francisco, CA.)

antiauxin   Interferes with the action of auxins, e.g., 2,3,5-triiodobenzoic acid inhibits the growth-promoting action of 2,4-D (dichlorophenoxyacetic acid) or the indoleacetic acid (IAA) analog 5′-azido-indole-3-acetic acid interferes with enzymes involved with IAA. *See* plant hormones.

anti-4-1BB monoclonal antibody   Is a co-stimulatory receptor expressed on activated T cells and it may be effective in amplifying T-cell-mediated immunity in cancer therapy. When used for intra-tumoral adenoviral gene transfer it improved survival rate and reduced metastasis substantially. (See cancer gene therapy, Martinet, O., et al. 2002. *Gene Ther.* 9:786.)

antibiotic   Chemical produced by microorganisms and plants (also now by organic laboratory synthesis) that is toxic to

Pelargonidin displays an OH group at position 4′, cyanidine has two OH groups at 3′ and 4′, and delphinidine has three OH groups (3′, 4′ and 5′). Peonidine (not shown) has 3′ OCH_3 and 4′ OH. Petunidine: 3′ OCH_3, and 5′ OH. Malvidine: 3′ and 5′ OCH_3 and 4′OH. Further color variations may be brought about by glycosylation and acetylations of the A ring(s) (at left).

Delphinidin (purple)   Cyanidin (red)   Pelargonidin (salmon)

other organisms. The major types of antibiotics are *penicillins*, *ampicillin*, and *cephalosporins* (interfere with bacterial cell wall biosynthesis). *Chloramphenicol* binds to 50S ribosomal subunit and blocks the peptidyl transferase ribozyme function during protein synthesis of prokaryotes. *Tetracyclines* inhibit the entry of the charged tRNA to the A site of the ribosome in prokaryotes. *Streptomycin* blocks the process of prokaryotic peptide chain elongation and causes reading errors during translation. *Spectinomycin* inhibits the function of the 30S ribosomal subunit. *Kanamycin, geneticin (G418), neomycin, gentamycin*, and *hygromycin* bind to 30S and 50S ribosomal subunits and prevent protein synthesis or cause misreading. *Erythromycin* inhibits the translocation of the nascent peptide chain on the prokaryotic ribosomes. *Lincomycin* inhibits chain elongation on the prokaryotic ribosome by its effect on peptidyl transferase but does not have the same effect on eukaryotic ribosomes.

Rifampycin interacts with the β subunits of the prokaryotic RNA polymerase. *Fusidic acid* interferes with the binding of aminoacylated tRNAs to the ribosomal A site by inhibiting the release of prokaryotic elongation factor EF-G and also eukaryotic elongation factor eEF-2. *Kasugamycin* blocks the attachment of tRNA^{fMet} to the P site of the prokaryotic ribosome. *Kirromycin* actually promotes the binding of elongation factor EF-TU-GTP complex to the prokaryotic ribosome but then inhibits the release of the elongation factor. *Thiosrepton*, from *Streptomyces azureus*, blocks prokaryotic peptide elongation from both prokaryotic and eukaryotic ribosomes. *Cycloheximide* interferes with peptide translocation on the eukaryotic ribosome. *Anysomycin* blocks the peptidyl transferase on the eukaryotic ribosomes and is comparable in effect to chloramphenicol in prokaryotes.

Streptolydodigins do not block RNA initiation but interfere with the elongation of the RNA chain in prokaryotes. *Ciprofloxacin* interacts with DNA gyrase. *Actinomycin D* primarily inhibits RNA polymerase II and to a lesser extent the other RNA polymerases in prokaryotes and eukaryotes but not DNA polymerase. *α-amanitin* also inhibits eukaryotic RNA polymerase II and in very high concentration pol III but not pol I. *Pactamycin* blocks the eukaryotic initiator tRNA^{Met} to attach to the P site of the ribosome. *Showdowmycin* interferes with the formation of the eukaryotic eEF−tRNA^{Met} complex. *Sparsomycin* is a eukaryotic peptide chain translocation blocker. *Cefotaxime* (synonym *claforan*), *carbenicillin*, and *vancomycin* are more effective as antibacterial agents and are frequently used in plant tissue culture to prevent bacterial growth. Antibiotics, which interfere with protein synthesis on prokaryotic ribosomes, cause similar damage to the ribosomes of eukaryotic organelles (mitochondria, plastids).

The availability of antibiotics in the 1940s opened a new era in medicine and they became the most important selectable markers for the construction of vectors for genetic engineering in the 1970s. They are used for selective isolation of various genetic constructs in microbial, plant, and animal cell genetics. The need for new antibiotics is continuously increasing because microorganisms develop resistance to the old drugs. *Staphylococcus aureus* bacteria are now resistant to all antibiotics except vancomycin, and it will be only a matter of time before resistance mutations develop to it, as well. There are already *Enterococcus faecium* strains that are resistant to vancomycin. *See* antibiotic resistance; antimicrobial peptides; bleomycin; cell genetics; protein synthesis; selectable marker;

vectors. (Walsh, C. 2000. *Nature* 407:775; Palumbi, S. R. 2001. *Science* 293:1786.)

antibiotic resistance Brought about by enzymatic inactivation of the antibiotic, modification of the target, active efflux of the substance, or sequestration by binding to special proteins. Genes in bacterial plasmids and transposons generally determine it. The mechanisms of resistance vary: penicillins and cephalosporins (β-lactamase hydrolysis), chloramphenicol (detoxification by chloramphenicol transacetylase that acetylates the hydroxyl groups or interferes with uptake), tetracyclines (interference with uptake or maintenance of the molecules), aminoglycosides (streptomycin, kanamycin, etc., enzymatic modification of the drug [phosphorylation] interferes with uptake or action), erythromycin, lincomycin (methylation of the small ribosomal subunit). Tetracycline, pactamycin, and hygromycin B modify the 30S ribosomal subunit in special ways and affect the decoding of mRNAs. Antibiotic resistance acquired through conjugative transfer of the resistance factors or mutation poses serious problems to medicine, e.g., the recent resistance of *Mycobacterium tuberculosis* to all known antibiotics. Antibiotic resistance genes are generally used to assure the removal (by carbenicillin or claforan [cefotaxime]) of the carrier *Agrobacteria* after infection with plant transformation vectors. Also, the transformed bacterial, fungal, animal, and plant cells are selectively isolated on the basis of antibiotic resistance. Insertional mutagenesis in bacteria is monitored by the inactivation of the resistance genes upon integration. Various antibiotics are used all over the world in animal feed to increase animal productivity by 4–5%. Unfortunately, some of the antibiotic resistance genes may become incorporated into (facultative) human pathogens through animal products and waste and may pose a threat to human health. The advantages gained by antibiotics in the feed may be partially compensated for by improved animal hygiene. *See* aminoglycoside phosphotransferases; amoxicillin; antibiotics; clavulanate; decoding on ribosomes; lactamase; pBR322. (Witte, W. 1998. *Science* 279:996; Walsh, C. 2000. *Nature* 407:775; Walker, E. S. & Levy, F. 2001. *Evolution* 55:1110; Schlünzen, F., et al. 2001. *Nature* 413:814.)

antibody Specific immunoglobulin that reacts, as a cellular defense with foreign antigens. Antibodies contain two light chains, either κ or λ, and one of the five heavy chains ($\mu, \delta, \gamma, \varepsilon, \alpha$) and their variants. Both light and heavy chains contain variable and constant regions. The specificity resides in the variable regions. Antibodies have specificities to about a million different antigens. This specificity is achieved with the aid of a much smaller number of antibody genes by differential processing of the transcripts, mutation, recombination, gene conversion, and transposition within the families of immunoglobulin genes. Antibodies are made by the lymphocytes and may be attached to their membrane or may become humoral antibodies (secreted into the bloodstream by the B lymphocytes). One particular B cell synthesizes only one type of antibody molecule. Each B cell deposits the first 100,000 antibodies it makes in its plasma membrane and serves as an antigen receptor. When a particular antigen binds to the B cell, it stimulates its clonal division and the production of more antibodies. These series of the antibody are made at the amazing rate of about 2,000 molecules/second, then secreted

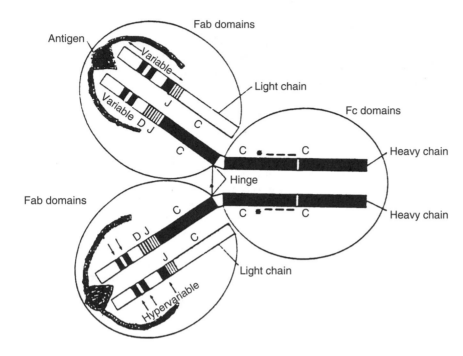

Antibody structure: Antibodies have three main domains: the two Fab domains (fragment antigen binding), including the light chains and parts of the heavy chains, and one Fc (fragment crystalline) domain. The light chains have a size of about 23 kDA; the heavy chains vary from 53 to 70 kDa. X-ray crystallography revealed the domains as 2×4 nm oval or cylindrical in shape and the polypeptide chain in each domain is folded in pleated β-sheets. The dimeric structure of the light and heavy chains is held together by disulfide bonds in variable numbers, depending on the particular molecules. The inter- and intrachain disulfide bonds are not shown, except at the proline-rich hinge area that provides the molecules with some flexibility. The IgM and IgE antibody monomers lack hinges but have an additional C-terminal heavy-chain domain. At the amino end of the light and heavy chains are the variable and hypervariable regions that determine the specificity of the antibody. This region includes approximately 100 to 115 amino acids. The specificity is determined by complementarity between antibody and antigen in the antigen-binding "pocket" or surface as outlined by the hatched arms of the antigen around the area. The various specificities are determined by combinations of the variable (V), diversity (D), and junction (J) genes that account for about 25% of the amino acid residues, and the remaining 75% are considered the framework. CDR1, CDR2, and CDR3 (shown by the dark bands) generally identify the complementarity-determining regions. The variable regions in the light and heavy chains are homologous. The constant regions (C) show very little variability within a species. There is a glycosylation site in the constant heavy-chain region within the Fc domain (*). Also in the constant heavy chains there are sites for binding of the activator of the complement (—). The complement consists of about 30 different proteins of catabolic functions that are activated in a cascading manner after the binding of the antigen to the antibody and carry out the destruction of the foreign antigen.

into the blood plasma. An individual can make about 10,000 different heavy-chain variants and about 1,000 different light-chain variants. Since these chains can combine freely, the total number of different antibodies can be $10^4 \times 10^3 = 10^7$. IgM-type antibodies (containing γ-immunoglobulin chains) occur at the largest concentration in the blood serum and their half-life is the longest. The general structure of the antibody molecules is diagrammed here. Each antibody molecule has two identical antigen-binding sites (see diagram). The majority of the antigens have, however, several to many antigenic determinants (epitopes). Some of these antigens may be built of repeating units and in these cases they are *multivalent* because they have multiple copies of the epitope. The binding between epitopes (*e*) and antibody (*a*) is a concentration-dependent, reversible process: $(a + e) \Leftrightarrow (ae)$. When the concentration of the epitope increases, the binding to the antibody increases and the intensity of the reaction is expressed by the *affinity constant*: $(k) = (ae)/(a)(e)$. When half of the (*a*) sites are filled, $k = 1/e$. The values of (*k*) range from 5×10^4 to 10^{12} moles.

The *avidity* of an antibody for an antigenic determinant also depends on how many binding sites are available. The affinity

increases with time after immunization (affinity maturation). Antibodies are involved in the destruction of invaders either through stimulating the macrophage cells to phagocytosis or by ions using the complement enzymes or activating the killer cells. It was discovered that antibodies can generate H_2O_2 by oxidation of water with the aid of singlet oxygen ($^1O_2^*$). This ability adds a chemical to their repertory of defense (Wentworth, P., et al. 2001. *Science* 293:1806). Usually their turnover is rapid; the half-life of antibodies is days to a few weeks. By chemical modifications, antibody/ligand complexes can be generated that do not dissociate and do not cross-react appreciably with other ligands (Chmura, A. J., et al. 2001. *Proc. Natl. Acad. Sci. USA* 98:8480). About 20% of the total plasma proteins represent a diverse set of antibodies. After the B lymphocytes respond to an antigen and differentiate into plasma cells, their rate of antibody production may reach 1,000 molecules/second after the immunization (affinity maturation). Receptors (FcRn) of the Fc domain (see diagram) contribute toward the phagocytotic functions, cytotoxicity, and neonate immunity.

In the maternal uterus, FcRn/IgG has been detected. The FcRn receptors transfer maternal humoral immunoglobulins to the newborn before the immune system of the progeny is activated. During nursing the FcRn class receptors mediate the transfer of the IgG/FcRn complex by the milk. Antibody genes can be expressed not just in lymphoid cells but also ectopically, e.g., in bacterial cells when introduced by transformation. In such a system, they may form inclusion bodies either in the cytoplasm or in the periplasmic space or may be present as soluble proteins secreted into the cytoplasm. In the periplasmic space, disulfide isomerase-like and proline cis-trans isomerase (rotamase) proteins may exist that mediate folding of the antibodies or fragments. The prokaryotic chaperones may also participate in the folding. In the *Camelidae* (camels and llamas) the antibodies contain only single heavy chain, and from the constant region the first domain is absent, although it is present in the genome but is not retained during mRNA processing. *See* antibody, polyclonal; antibody engineering; anti-DNA antibody; antigen; antigen-presenting cell; anti-idiotype antibody; B-lymphocyte receptor; catalytic antibody; chaperone; complement; HLA; hybridoma; immune system; immunization alloantibody; immunoglobulins; internal image immunoglobulin; killer cell; lymphocytes; MHC; monoclonal antibodies; natural antibody; neutralizing antibody; periplasm; recombinant antibody; rotamase; T cell; TCR. (Heyman, B. 2000. *Annu. Rev. Immunol.* 18:709; Ravetch, J. V. & Bolland, S. 2001. *Annu. Rev. Immunol.* 19:275.)

antibody, antigenization   Modification of the hypervariable region of an antibody by protein engineering in order to enhance the recognition of the new antibody to foreign epitopes by the B and T lymphocytes. *See* antibody; antigen; B lymphocyte; epitope; T cells.

antibody, bispecific   Has affinity to two different antigens. Such antibodies may not exist.

antibody, bivalent   Has two antigen-binding sites.

antibody, chimeric   Can be produced with the aid of genetic engineering by fusing the variable regions of one type to the constant region of another antibody. It can also be produced in vivo by homologous recombination in hybridoma cells or by using the *Cre-loxP* system. *See* Cre/loxP; HAMA; hybridoma; primatized antibody.

antibody, intracellular   By introducing specific antibody genes into a cell, and if the transgene is expressed, interaction between macromolecules, fixing enzymes in an active or inactive state, modifying (binding) ligands, targeting intracellular signals, etc., can be explored. Tissue-targeting vectors can be constructed for the introduction of genes to specific locations. In these systems antibodies are coupled to viral vectors, liposomes, or directly to passenger DNA. Antibodies can be targeted to T-cell receptors. Bispecific antibodies can be used to retarget effector cells to tumors. *See* immune system; KDEL; liposome; monoclonal antibody therapies; T-cell receptor; viral vectors.

antibody, monoclonal   *See* monoclonal antibody.

antibody, monovalent   Has only a single binding site for an antigen. Normally the antibody is divalent, i.e., it has antigen-binding sites at both light- and heavy-chain variability regions. By linking together multiple binding sites, the avidity of the antibody increases. *See* antibody; antibody, bispecific.

antibody, neutralizing   The loss of infectivity, which ensues when antibody molecule(s) bind to a virus particle, and usually occurs without the involvement of any other agency. As such this is unusual of antibody paralleled only by the inhibition of toxins and enzymes (Dimmock, N. J. 1995. *Rev. Med. Virol.* 5:165; Finbe, D., et al. 2003. *Proc. Natl. Acad. Sci. USA* 100:199).

antibody, polyclonal   Human polyclonal antibodies can be obtained by transferring into bovine embryonic cells and thus into calves both the heavy and the lambda-chains of immunoglobulin gamma genes on human artificial chromosome vector. *See* antigen; epitope; monoclonal antibody; recombinant antibody; human artificial chromosome, nuclear transplantation. (Kuroiwa, Y., et al. 2002. *Nature Biotechnal.* 20:889.)

antibody, secondary   A molecule, cell, or tissue may be labeled with the cognate antibody (primary antibody). Then to boost the level of recognition, the primary antibody is reacted with another antibody (secondary antibody) and labeled with an isotope (e.g., I^{125}) or a fluorochrome so that a stronger signal can be obtained. *See* antibody; fluorochromes.

antibody detection   Possible by several procedures: Antibodies bound to proteins expressed in *E. coli* are detected by I^{125} (isotope)–labeled antibodies that react to the species-specific determinants of the primary antibodies. Protein A labeled with I^{125} second antibody, conjugated to horseradish peroxidase (HRP) or HRP coupled to avidin, may be used to detect a second antibody coupled to biotin or a second antibody conjugated to alkaline phosphatase using radiolabeled ligands. Antibodies can also be detected by agglutination and complement fixation. In agglutination a precipitate is formed upon the reaction. One of the procedures is the *Ouchterlony assay*. By placing the antibody and the antigen in neighboring wells of agar plates, upon diffusion a visible precipitate is formed about midway between the two wells if the antigen (e) and antibody (a) recognize each other. The complement-fixing procedure has a unique feature inasmuch as the complement binds only to the antibody that is complexed with the antigen. Adding red blood cells and cognate antibody to the reaction mix, resulting in no hemolysis, is the proof for fixation of the complement, and the procedure can be quantitated by employing a series of dilutions. *See* antibodies; complement; immunostaining.

Precipitate between wells   No precipitate   ← Wells

Ouchterlony assay.

antibody domain   Segment of light- and heavy-chain polypeptides separable by chemical treatments. *See* antibody.

antibody effector function Carried out by activation of the complement system and by interactions of the antigen through the Fc domain receptors (e.g., FcγR) leading to ADCC. *See* ADCC; antibody; complement; FcγR.

antibody engineering Involves genetic modification of the immunoglobulin genes, particularly the complementarity-determining regions of the antibody. *See* antibody, antibody polymers; bispecific monoclonal antibody; CDR; Fv; gene fusion; humanized antibody; immunotoxin; monoclonal antibody; phage display; plantibody; transgenic. (Maynard, J. & Georgiou, G. 2000. *Annu. Rev. Biomed. Eng.* 2:339.)

antibody fusion Gene fusion most commonly involving the antibody heavy chain and enzyme-coding sequence (nuclease, glucuronidase, etc.) or toxin (e.g., angiogenin toxin, neurotoxin), cytokinins (interleukin 2, TNF, IGF), and labeling proteins (aequorin, avidin).

antibody gene switching Preceded by pairing constant heavy-chain gene families and loop formation between members of the antibody, which are then cut off at the stem. The deletion brings different heavy-chain elements in the vicinity of the J (junction) genes. The site-specific switch permits the gene expression in the vicinity of the J genes after the stem of the loop is cut off and the DNA strands are religated. The transcript is then further processed by removal of the introns. This is one of the mechanisms for generating greater diversity in the heavy-chain antibody proteins. The switching is stimulated by cytokines secreted by the T$_H$ lymphocytes. In mouse cells, IL-4 induces the switch from IgM to IgG1 or IgE. Interferon-γ causes switching from IgM to IgG2a and TGF-β mediates the switch from IgM to IgG2b or IgA. Activation-induced cytidine deaminase (AID) seems to be involved in the mediation of switching. Switching is different from the V(D)J recombination process. *See* antibodies; class switching; ectodermal dysplasia; germline transcript; hypermutation; immune system; immunoglobulins; somatic hypermutation; T$_H$; V(D)J. (Kataoka, T., et al. 1981. *Cell* 23:357; Revy, P., et al. 2000. *Cell* 102:565; Stavnezer, J. 2000. *Science* 288:984; Honjo, T., et al. 2002. *Annu. Rev. Immunol.* 20:165.)

antibody lattice When the cognate antibody is in excess of the antigen, an alternating antigen–antibody complex is formed between the Fc domain of the IgG and the antigen. *See* antibody; antigen.

antibody mimic Small synthetic polypeptide with specificity for a particular natural or synthetic epitope. *See* antibody; epitope.

antibody polymer Fusion of the immunoglobulin μ-chain tailpiece to the C end of the γ-chain may increase by two orders of magnitude the activity of the complement system. Also simple IgM, IgG tetramers are more effective than dimers. *See* antibody; complement; immunoglobulins; pIgR; tailpiece.

antibody preparation An animal is injected with a pure antigenic molecule. After 2 to 3 weeks, it develops antibodies against the epitope, then the animal is bled and from the serum the antibody is removed by precipitation with the cognate antigen and further purified. Hundreds of different antibody preparations are commercially available from biochemical supply companies.

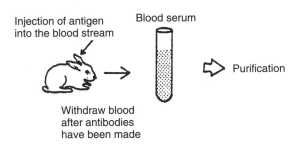

antibody purification The protein antigen may be coupled to a cyanogen-bromide–activated Sepharose. The epitope then retains the cognate antibodies while all other antibodies flow through. Breaking the complex (with potassium thiocyanate, low pH buffers, etc.) can then retrieve the antibody. The methods must be adapted to the different proteins.

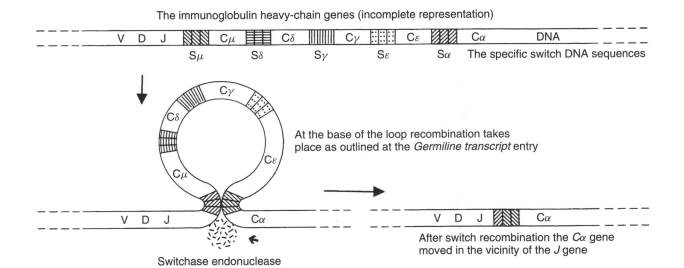

Another procedure is to adsorb antibodies to protein antigens immobilized on diazotized paper or nitrocellulose filters following electrophoresis by SDS-polyacrylamide gels. The antibodies are then eluted with a suitable buffer. Antibodies can be used for qualitative and quantitative assays of antigens, including immunoprecipitation, Western blotting, and solid-phase radioimmunoassays (RIA). *See* cyanogen bromide; diazotized paper; electrophoresis; epitope; immunoprecipitation; nitrocellulose filter; radioimmunoassay; SDS-polyacrylamide gels; Sepharose.

antibody valency Specifies the number of antigen-binding sites. *See* antibody, monovalent.

anticancer agent Includes alkylating agents, cytotoxic and cytostatic agents, antibiotics (bleomycin, chlorambucil), topoisomerase inhibitors (etoposide, podophyllotoxin), ionizing radiation, etc. *See* cancer gene therapy; cancer therapy; chemotherapy; ionizing radiation.

anticarcinogen *See* antimutagens.

antichaperone Protein factor promoting aggregation of other proteins. *See* chaperone.

antichromatin State of the chromatin not conducive for active transcription. *See* chromatin; pro-chromatin.

anticipation In successive generations it may appear as if the genetic trait (disease) would have occurred with an earlier onset in the more recent generations. However, frequently this is an artifact because when the investigator knows what is expected, the recognition becomes easier. There is also the possibility that individuals with early onset of the disease died early or failed to leave offspring. In the cases of diseases based on expansion of trinucleotide repeats, there is a possibility of increased severity and earlier onset if the patient leaves offspring. *See* ascertainment test; trinucleotide repeats. (Kovach, M. J., et al. 2002. *Am. J. Med. Genet.* 108:295.)

anticlinal selection The selection takes different directions in different environments compared to the *synclinal selection* in which the direction is the same. *See* cline.

anticoagulation Blood coagulation is positively regulated by antihemophilic factors. Negative regulation (shutting down the coagulation pathway) is mediated by thrombomodulin, which binds thrombin and activates protein C, which in turn binds protein S, and factors Va and VIIIa are degraded. Thrombomodulin (an epidermal growth factor–like molecule) works by binding to thrombin at an exosite where thrombin would otherwise bind to fibrinogen. Coumarin impairs the procoagulant thrombin, antihemophilic factors Xa, IXa, and VIIa, and anticoagulant proteins C and S. Heparin enhances the inhibition of thrombin and factor Xa by antithrombin III. *See* antihemophilic factors; antithrombin; blood clotting pathways; exosite; protein C; protein S; thrombin; vitamin K.

anticoding strand Transcribed strand of DNA. *See* antisense RNA; coding strand; plus strand; sense strand; template strand.

anticodon Part of the tRNA that recognizes an mRNA code word by complementarity, and it is one of the means of tRNA identity. In the mitochondria the "universal" genetic code does not entirely prevail, but different eukaryotic mitochondria (except higher plants) use a somewhat different codon dictionary. In these systems the anticodons are also different inasmuch as there are no separate tRNAs for each of the synonymous codons. Rather the mtDNA codons recognized in pairs or in four-member sets of codons and the anticodon–codon interaction is by G•U pairing, or the 5′-terminal U of the anticodon of the four-member set can pair with any of the four bases in the mRNA codon. Although there are 61 different sense codons in eukaryotes, there are only 54 anticodons in the universal code, and 46 species of tRNAs and anticodons are sufficient for protein synthesis on the ribosomes. *See* genetic code; tRNA; wobble. (Jukes, T. H. 1984. *Adv. Space Res.* 4[12]:177.)

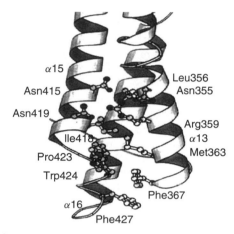

The anticodon-binding α-helix bundle of bacterium tRNA[Met] synthetase. The stick-and-ball structure shows the exposed side chains of the amino acids (Courtesy of Professor M. Konno. See also Sugiura, I., et al. 2000. *Structure* 8:197).

antideterminant Ribonuclease III, which processes about 20 bp double-stranded RNAs, may not cut at any position because some Watson-Crick pairs interfere with scission and serve as antideterminants. Such an antideterminant is, e.g., a 3 bp sequence from the selenocysteine-accepting tRNA (tRNA[Sec]) and is an antideterminant for EF-Tu binding to this tRNA. *See* EF-TU-GTP; ribonuclease III; selenocysteine. (Evguenieva-Hackenberg, E. & Klug, G. 2000. *J. Bacteriol.* 182:4719; Mohan, A., et al. 1999. *RNA* 5:245.)

antidiuretic hormone (vasopressin) Small peptide hormone (ADH, M_r 1040) that increases water reabsorption in the kidney and increases blood pressure; it affects a variety of functions, including learning and behavior (aggression). *Nephrogenic diabetes insipidus*, an X-chromosomal human disease with problems of maintaining water balance, fails to respond to ADH, which is very similar to oxytocin; only a difference of two amino acids exists between ADH and oxytocin. The structure of vasopressin is:

$$\text{Cys—Tyr—Phe—Gln—Asn—Cys—Pro—Arg—GlyNH}_2$$

with S—S bridge connecting the two Cys residues.

It binds to receptor molecules in the plasma membrane in the kidney and blood vessels and activates a specific membrane phospholipase. The phospholipase then breaks the bond between glycerol and phosphate in phosphatidylinositol-4,5-bisphosphate and releases inositol-1,4,5-triphosphate and diaglycerol. Vasopressin is encoded in the short arm of human chromosome 20 along with oxytocin. *See* diabetes insipidus; diaglycerol; inositol; nocturnal enuresis; oxytocin; phosphoinositides; phospholipase.

anti-DNA antibody DNA is a poor antigen, although antibodies bind to DNA in the autoimmune disease lupus erythematosus. Most DNA antibodies are not entirely specific because they bind to repetitive sequences. DNA tracts with stably bound proteins and can, however, be used as antigens to specific sequences. *See* antibody; antigen; autoimmune disease. (Stollar, B. D. 1986. *CRC Crit. Rev. Biochem.* 20:1; Cerutti, M. L., et al. 2001. *J. Biol. Chem.* 276:12769.)

antiestrogen Binds to the estrogen receptors and antagonizes the effects of the hormones. Some, however, may have various levels of agonist activity. *See* raloxifene; tamoxifen.

antifreeze protein Present in several species of fishes living in the northern regions. The glycoprotein binds, through free OH groups of amino acids, to the first ice crystals and thus prevents the expansion of the ice, so the fish are protected. In fish there are more than eight forms, encoded as different proteins, yet they all contain the tripeptide (Thr-Ala-Ala/Pro-Ala-) repeats. Mainly Leu/Phe-Ile/Asn-Phe spacers link the monomers into a large polyprotein. The AFGP (antifreeze glycoprotein) genes usually contain two exons (the small for a signal peptide and the large for the antifreeze) separated by a single intron. Somewhat similar proteins may play a role in other organisms. A very efficient antifreeze protein was isolated from the insect *Tenebrio monitor*. A 36 kDa glycoprotein isolated from cold-acclimated carrot taproots is similar in sequence to polygalacturonase inhibitor proteins. The antifreeze protein in perennial ryegrass (*Lolium perenne*) appears to control more ice crystal growth than prevent freezing per se. *See* cold hypersensitivity; hysteresis; mealworm; thermotolerance. (Miao, M., et al. 2000. *Eur. J. Biochem.* 267:7237; Tomczak, M. M., et al. 2001. *Biochim. Biophys. Acta* 1511:255; Haymet, A. D., et al. 2001. *FEBS Lett.* 491:285; Fairly, K., et al. 2002. *J. Biol. Chem.* 277:24073.)

antifungal response Insects defend themselves against fungi and microorganisms by the production of proteolytic enzymes, phagocytosis, and the production of antimicrobial peptides. In *Drosophila*, antifungal drosomycin and several antimicrobial/antibacterial peptides — cecropins, dyptericin, drosocin, attacin, and defensin — are produced. The *spätzle*, *Toll*, *cactus*, and *dorsal* dorsoventral regulatory genes (corresponding to the mammalian NF-κB cascade) and the immunodeficiency gene *imd* mediate these responses. *See* antimicrobial peptides; host–pathogen relationship; morphogenesis in *Drosophila*; NF-κB.

antigen Substance (usually a protein) that alone or in combination with a protein elicits antibody formation. The protein antigen may be a large molecule with more than a single specificity due to its different subunits. A particular specificity of the antigen is determined by the epitope or a hapten conjugated with the protein molecule to form an antigen that reacts with the paratope of the antibody. *See* antibody; epitope; paratope; superantigen; TI antigens. (Kurosaki, T. 1999. *Annu. Rev. Immunol.* 17:555; Zinkernagel, R. M. & Hengartner, H. 2001. *Science* 293:251.)

antigen, male specific *See* grafting in medicine; H-Y antigen.

antigenic determinant *See* antibody; epitope.

antigenic distance Indicates the degree of similarity between/among antigens.

antigenic drift The surface antigens of a pathogen may change by mutation. *See* antigenic variation; *Borrelia*; phase variation; *Trypanosoma*.

antigenic shift Rearrangement in the genetic material of a virus resulting in an escape of the normal immune reaction. *See* antigenic drift; antigenic variation.

antigenic sin Individuals who were previously exposed to one virus and later encountered another virus variant of the same subtype can make antibodies against the original viral hemagglutinin (HA) and also against the new one. This happens because the memory B cells or the T cells were activated in a specific way for the progenitor virus. In some instances, the variant virus may escape the immune defense of the host because the lymphocyte receptor is altered by mutation, although the histocompatibility class I molecules may bind normally. *See* antigen; hemagglutinin; HLA; immune system. (Good, M. F., et al. 1993. *Parasite Immunol.* 15:187.)

antigenic variation Property of prokaryotic and eukaryotic microorganisms to switch on the synthesis of different surface proteins to escape the immunological defense system of the host organisms. This goal is reached generally by transposition of genes relative to the promoter. The bacterium *Neisseria gonorrhoeae* (responsible for a venereal disease manifested primarily in males but transmitted through both sexes) relies on gene conversion for this purpose. *See Borrelia*; cassette model of yeast; gene conversion; phase variation; serotype; *Trypanosoma brucei*. (Barry, J. D. & McCulloch, R. 2001. *Adv. Parasitol.* 40:1; Brayton, K. A., et al. 2001. *Proc. Natl. Acad. Sci. USA* 98:4130.)

antigen mimic Short polypeptide used for screening for specific paratope sites. *See* antibody mimic; paratope.

antigenome In the replicative form of the viral genetic material it serves as a template for the synthesis of the genome. *See* RF.

antigen presenting cell (APC) Binds antigens, internalize, process and expresses them on their surface in conjunction with class II–type molecules (one of the two types of molecules coded for by the MHC genes). T cells recognize the presented antigen through their receptors. Helper T cells can be activated only in the presence of APC cells. Macrophages, dendritic (branched) cells, and B-lymphocyte cells express class II antigens and thus they can serve as APC in vitro; in vivo macrophages and dendritic cells are apparently the most important as APC. The activation of helper T cells requires that the T cells and the APC are derived from animals (mice) syngeneic in region *I* of the MHC, and requires the production of the lymphokine, interleukin-1 (IL-1) family member CD80. *See* affinity maturation; antigen; CD1; CD40; CD80; clonal selection; cross presentation; cytotoxic T cell; HLA; immune system; interleukins; lymphokines; MHC; proteasomes; syngeneic; T cell, T-cell receptor. (Jenkins, M. K., et al. 2001. *Annu. Rev. Immunol.* 19:23; Guermonprez, P., et al. 2002. *Annu. Rev. Immunol.* 20:621.)

antigen processing and presentation Antigen-presenting cells mediate the association of the native antigen with an MHC molecule and thereby the antigen is recognized by the T lymphocytes. The antigenic protein must be degraded to some extent by immunoproteasomes and processed for presentation to the MHC molecules. The processing takes place either within endosomal compartments of the cell or by the proteases secreted onto the surface of the immature dendritic cells. The MHC I–associated peptides are generally shorter (9 ± 1 amino acids) than those associated with MHC II molecules derived from excreted proteins or other external proteins. Usually the peptides enter the endoplasmic reticulum before their epitope is presented to the MHC molecules. If the proteins lack the signal peptide to be transferred to the endoplasmic reticulum, their epitope may still be presented to the MHC molecules. The MHC class II molecules are associated with an invariant I_i polypeptide that mediates the folding of the MHC II molecules in the endoplasmic reticulum and compartmentalizes the MHC II molecules for special peptide binding in the endosomes. The processing is mediated by cathepsins, but an asparagine-specific cysteine endopeptidase may also be involved in degrading microbial antigens. *See* antigen-presenting cell; cathepsins; CLIP; endosome; HLA; immunoproteasomes; lymphocytes; major histocompatibility complex; TAP; T cell. (York, I. A. & Rock, K. L. 1996. *Annu. Rev. Immunol.* 14:369; Watts, C. 1997. *Annu. Rev. Immunol.* 15:821.)

antigen receptor Molecule on lymphocytes that is responsible for recognition and binding of antigens and antigen-MHC. *See* HLA; lymphocytes; receptor editing.

antigene technology Used for triple helix formation. *See* triple helix formation.

antihemophilic factor Blood coagulation requires the formation of complexes between serine protease coagulation factors and membrane-bound cofactors. Tissue thromboplastin, an integral membrane glycoprotein (factor III, encoded at 1p22-p21), and proconvertin (VII, encoded at 13q34) are required to activate factors IX and X. Factor VIIa is a trypsin-like serine protease that also plays a key role in blood coagulation after binding to the γ-carboxyglutamic acid-containing domain. Plasma thromboplastin antecedent (XI, 4q35) activates factor IX. Blood coagulation factor VIII (Xq28), a 293 kDa plasma glycoprotein, acts in concert with factor IXa, a proteolytic enzyme, to activate factor X (Stuart factor, 13q34). The latter in turn activates prothrombin II (11p11-q12) to thrombin, which acts on fibrinogen (I, 4q28) to convert it to fibrin (responsible for loose clot); then the fibrin-stabilizing factors (XIII, α-chain 6p25-p24; β-chain 1q31-q32.1) generate the firm clots required for blood clotting. Hageman factor (XII, 5q33-qter) activates thromboplastin antecedent (XI, 4q35). Accelerin (V, 1q23) stimulates the activation of prothrombin (II, 11p11-q12). In classic recessive X-chromosomal hemophilia, factor VIII is defective. Factor IX (454 amino acids, Xq27.1-q27.2) deficiency, a partially dominant disorder of hemostasis (arrest of blood flow), is involved in Christmas disease. Blood clotting also requires calcium and thromboplastin (lipoprotein released into blood from injured tissues). A thromboplastin antecedent (XI, 4q35) deficiency is responsible for hemophilia C. The level of factor IX increases with age and may be responsible for the increase of cardiovascular and thrombotic disorders in the elderly. *See* afibrinogenemia; anticoagulation; APC; blood clotting pathways; coumarin-like drug resistance; dysfibrinogenemia; fibrin-stabilizing factor; Hageman trait; hemophilia; hemostasis; hypoproconvertinemia; LINE; parahemophilia; platelet abnormalities; prothrombin deficiency; PTA deficiency; Stuart disease; thrombopoietin; tissue factor; vitamin K–dependent clotting factors; Warfarin von Willebrand's disease. (Bajaj, S. P., et al. 2001. *J. Biol. Chem.* 276:16302; Hockin, M. F., et al. 2002. *J. Biol. Chem.* 277:18322; Tuddenham, E. 2002. *Nature* 419:23.)

antihormone Antagonist of hormones by altering the conformation of the hormones or by binding to the hormone receptor sites, preventing the attachment of hormones to the hormone-responsive elements (HRE) in the DNA and thus blocking the transcription of the hormone-responsive genes. *See* conformation; hormone-responsive element.

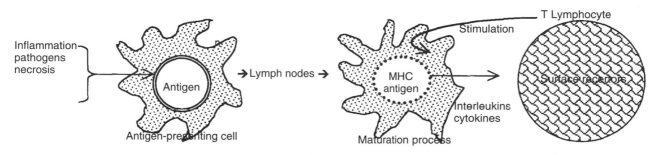

Antigen presentation.

anti-idiotype antibody Specific antibody that recognizes a particular paratope (idiotype) of an antibody and binds to it rather than to the epitope of the antigen. Homologous anti-idiotypic antibodies are produced within the species, whereas in different species heterologous anti-idiotypic antibodies are produced. The anti-idiotypic antibody may be generated in the laboratory by first exposing the cell to the epitope, an antigen. This specific antibody may give rise to another antibody, a mimic of the original. Other antibodies may arise in a similar manner that may respond to the original and to a mutant antigen. These antibodies may be capable of stimulation of B lymphocytes and T cells, and thus both humoral and cellular immunity can be generated. Anti-idiotype vaccine production may become feasible for particular cases, e.g., against the poorly antigenic bacterial polysaccharides or mutant p53 proteins that do not suppress tumor formation. *See* antibody; epitope; idiotype; internal image immunoglobulin; p53; paratope. (Birebent, B., et al. 2001. *Crit. Rev. Oncol. Hematol.* 39:117; Bhattacharya-Chatterjee, M., et al. 2001. *Curr. Opin. Mol. Ther.* 3:63.)

antilog Inverse logarithm obtained if the base is raised to the power of the logarithm. The antilogarithm for $\log_{10} x$ is 10^x and for $\ln(x)$ the antilogarithm is e^x. *See* logarithm.

antilope (blackbuck, *Antilope cervicapra*) The male is $2n = 31–33$; the female is $2n = 30–32$. *See* antelope.

antimetabolite Compound that binds to an enzyme but is not generally utilized as a substrate and thus interferes with normal metabolism. *See* metabolism; metabolite.

antimicrobial peptide Occurs on or in animals and plants as a defense system. Antimicrobial peptides can be linear molecules such as *cecropin* (moths, pig, *Drosophila*), making pores by lysis; *magainin* (frog skin)-forming pores; or *bactenein* (bovine neutrophiles), affecting membrane permeability. Disulfides: *defensins* (in several organisms, Hoover, D. M., et al. 2001. *J. Biol. Chem.* 276:39021), making pores; *tachyplesins* (in horseshoe crab), affecting potassium efflux; *protegrins* (pig leukocytes). Many of these peptide genes contain attachment sites for transcriptional activators related to NF-κB, *Rel/Dorsal* oncogene. *Serprocidins* are high-molecular-weight protease-like molecules: *protease 3* and *azurocidin* in mammals and *cathepsin G* in human neutrophils inhibit metabolism. Lipopolysaccharide-binding proteins and bactericidal/permeability-increasing proteins, collectins, are components of the mammalian defense system. A defensin-like peptide is expressed exclusively in the epididymis of rats. Most cells respond to invaders by the mediation of Toll receptors, which initiate and activate the production of antimicrobial peptides. *See* antibiotics; antifungal response; nisin; opsonins; Toll. (Hoffmann, J. A., et al. 1999. *Science* 284:1313; Khush, R. S. & Lemaitre, B. 2000. *Trends Genet.* 16:442; Zasloff, M. 2002. *Nature* 415:389.)

antimitotic agent Blocks or inhibits mitosis, e.g., ionizing radiation, radiomimetic chemicals, and inhibitors of the cell cycle. *See* cancer therapy; cytostatic; radiation effects.

antimongolism chromosome Chromosome 21 deletion in humans that compensates in some respects for the syndrome accompanied by trisomy of complete chromosome 21 (Down syndrome; by old name mongoloid idiocy). This deletion, GI, causes the formation of large ears, prominent nasal bridges, antimongoloid slant of the eyelids, long fingers and toes, micro- or dolicephaly, and hypo-γ-globulinemia (rather than an excess as in Down syndrome). *See* agammaglobulinemia; dolicephaly; Down's syndrome; microcephaly.

antimorph Dominant mutation that antagonizes the function of the wild-type allele (by competing for the substrate). *See* dominant negative; killer genes.

antimüllerian hormone (AMH/AMS) Produced by the Sertoli cells. Masculinizes XX rodents, whereas males deficient in AMH become pseudohermaphrodites. SF-1, SOX9, and WT1 regulate AMH. *See* gonads; pseudohermaphrodite; Sertoli cells; SF-1; SOX; Wilms' tumor.

antimutagen Protects against the mutagenic effect(s) of other agents. Generally hypoxia, reducing agents (such as dithiothreitol) lower the damage of ionizing radiation. Inhibitors of microsomal mutagen-activating enzymes (such as 9-hydroxyellipticine, gallic and tannic acids, carbon monoxide, selenium, etc.) may reduce the mutagenic effectiveness of chemicals. *See* antimutator; caffeic acid; methylguanine-O⁶-methyltransferase; mutagen; polyphenols. (Novick, A. 1955. *Brookhaven Symp. Biol.* 8:201.)

antimutator Lowers mutation rate. Increased level of nuclease activity (editing function) and all other genetic repair mechanisms may act this way. Compounds that inactivate microsomal enzymes involved in conversion of promutagens into mutagens are also antimutagens. *See* ABC excinucleases; AP nucleases; DNA repair; mismatch repair; mutator; proofreading. (Reha-Kranz, L. J. 1998. *Genetics* 148:1551.)

antioncogene Normal allele of some genes, which in the mutant state incites tumors, e.g., the cloned normal allele of the human retinoblastoma gene codes for a DNA-binding protein and the cancer cells transformed by this gene are suppressed in proliferation. *See* tumor suppressor genes.

antipain Protease inhibitor (1–2 μg/mL) effective against cathepsin A and B, papain, and trypsin protease enzymes.

antiparallel pairing At the same end of the double helix of a polynucleotide chain, there is one 5′ end and one 3′ end of the paired nucleotides:

antiphosholipid syndrome The endocytosis defect involves greater risk of thrombosis, thrombocytopenia, and recurrent spontaneous abortions. The antibodies may attack phospholipids, or the protein–phospholipid complex or proteins such as β2-glycoprotein. *See* endocytosis.

antipodal Haploid cells (nuclei) located in the plant embryo sac at the end opposite the egg and the micropyle. *See* embryo sac.

antiport Membrane transport of substances in opposite directions, or a substance is sequestered through the antiport within another compartment.

antipyretic Fever-reducing drug.

antirepression Transcription factors bound to the DNA upstream of the promoters may interfere with the binding of unspecific DNA-binding proteins, which would exert repression.

antirestriction mechanism Prevents DNA cleavage by different mechanisms, e.g., methylation of critical bases (e.g., phages T2, T4, SPβ), inhibition of the endonuclease (e.g., T3, T7 phages), enhancing host-encoded methylase (phage λ), carrying hydroxymethyl cytosine in place of cytosine (T-even phages), 5-hydroxymethyluracil substitution for thymine (SPO1, SP8, ϕ25), reducing certain vulnerable restriction sites in their DNA (ϕ29), etc. *See* restriction endonucleases; restriction methylation. (King, G. & Murray, N. E. 1995. *Mol. Microbiol.* 16:769.)

Antirrhinum majus (snapdragon) Higher plant of the *Scrophulariaceae* family, (2n = 16). A favorite and attractive autogamous flower for genetic and cytogenetic studies. A large collection of mutants and transposable elements are available. *See* peloric; snapdragon; TAM.

antisense DNA Library can be used for transformation and isolation of mutations or for other purposes of preventing gene expression. The 25-base oligodeoxyribonucleotide phosphorothioate (TCTTCCTCTCTCTACCCACGCTCTC; Hybridon. Inc., trade name GEM® 91) binds to the translation initiation site of the *gag* gene of the HIV-1 pathogen of acquired immunodeficiency and may inhibit the production of new infectious particles because of the defect in packaging. The human L1 (LINE) retrotransposon has two promoters, a sense promoter that directs the transcription of the full-length L1 tract and an antisense promoter that drives in the opposite direction and into adjacent sequences and thus generates chimeric transcripts. Antisense transcripts occur in tumors but also in normal cells. Both promoters are situated within the 5' nontranslated region of L1 (Nigumann, P., et al. 2002. *Genomics* 79:628). *See* acquired immunodeficiency; antisense RNA; antisense technologies; aptamer; *Bcl*; cancer gene therapy; G3139; HIV; OL(1)p53; peptide nucleic acid; phosphorothioate. (Zhang, Y. M., et al. 2001. *J. Nucl. Med.* 42:1660; Lehner, B., et al. 2002. *Trends Genet.* 18:63.)

antisense oligodeoxynucleotide (AS ODN) *See* antisense DNA; antisense RNA.

antisense RNA Transcript of a gene or transposon that may inhibit translation by pairing with the 5' end of the correct (sense) mRNA and thus prevent its ribosome binding and expression. In several bacterial plasmids, by inhibition of the synthesis of the replication initiator protein, the antisense RNA limits copy number. Some synthetic oligonucleotide analogs may block replication and transcription, interfere with splicing of exons, disrupt RNA structure, destabilize mRNA by interfering with 5' capping of mRNA, inhibit polyadenylation, and activate ribonuclease H. When coupled to alkylating agents, they can cross-link nucleic acids at the recognized sequences, can be used as vehicles for targeted DNA cleavage, may inhibit receptors, etc. The various functions require a large variety of specific antisense constructs. Usually the antisense oligonucleotides are 12–50 nucleotides long. According to calculations in the human genome, any 17-base sequence occurs only once, and in the mRNA populations, 13mer residues are unique. Shorter sequences do not have sufficient specificity. Long antisense sequences may have self-binding tracts that may cause lowered affinity for their target. Natural antisense RNA transcripts occur in all types of biological systems from viruses to higher eukaryotes. This fact indicates that in eukaryotes both strands of the DNA may be transcribed.

Antisense RNA (or DNA) was expected to become an important therapeutic tool for fighting infections and cancer. This technology is still under development for cure against cytomegaloviruses, HIV1, papillomavirus, autoimmune diseases (arthritis, etc.), leukemia (CMV), and for blocking the immune system in organ transplants. Antisense RNA may trigger an immune reaction because the CpG blocks are unmethylated and the animal immune system responds to them as to bacterial molecules. In bacteria these bases are largely unmethylated, in contrast to eukaryotes, where a substantial fraction of the DNA is methylated. The phosphorothioate oligodeoxynucleotides or oligoribonucleotides are taken up by a variety of cell types, including some prokaryotes (*Vibrio*), and bind either to DNA, RNA, or protein. Antisense RNAs may block embryonic development and can be used to inhibit gene expression at defined stages. Although antisense RNA is supposed to be very specific for the intended target, actually it may affect several genes that have short or long sequences homologous to the target. In addition, antisense RNA may bind and affect different proteins as an aptamer. Furthermore, the nucleic acid degradation products concomitant or following the administration of antisense RNA may result in unspecific inhibition in the cells. (See anthocyanin; anticoding strand; antisense technologies; aptamer; AS ODN; autoimmune disease; cap; coding strand; cosuppression; cytomegalovirus; fruit ripening; G quartet; host–pathogen relations; hybrid arrested translation; leukemia; methylphosphonates; papillomavirus; peptide nucleic acid; phosphorothioates; pseudoknot; RIP; RNA, double-stranded; RNA; sense strand; transplantation antigens; triple strand formation; triplex. (Helene, C. & Toulme, J. J. 1990. *Biochim. Biophys. Acta* 1049:99; Matveeva, O. V., et al. 2000. *Nucleic Acids Res.* 28:2862; Sohail, M., et al. 2001. *Nucleic Acids Res.* 29:2041.)

antisense strand of DNA   Template strand of DNA from which mRNA or other functional, natural RNAs are replicated as complementary copies. *See* antisense RNA.

antisense technology   Uses RNA and DNA targets for the suppression or modification of gene expression. The antisense molecule then blocks the synthesis of RNA and protein. Various forms of antisense molecules have been used (*see* antisense RNA); for antisense DNA technology, the nucleotides are ligated by, e.g., phosphorothioate linkage, not by the normal phosphodiesterase linkage, in order to protect the antisense strand from nuclease attack. For the production of antisense nucleic acids, the oligonucleotides are modified either in the base or the sugar or changes in the sugar phosphate background. The good antisense molecules are expected to allow for RNase H activity to remove the natural target (preventing its translation), then bind stably to DNA and block protein synthesis. The modification usually prevents enzymatic disposal of the antisense constructs.

The antisense sequences may have side effects. Guanine-rich antisense sequences may have undesirable effects on the telomerase enzyme, may form quadruplex structures and interfere with chromosome replication, and may bind to proteins and modify their function. In order to minimize the deleterious consequences, various alterations have been attempted. The number of phosphorothioates is reduced, or in a 5-base sequence the terminals ("wings") are modified, whereas in between an RNase H-competent 2'-deoxyoligodeoxynucleotide "window" is preserved. This approach is basically the generation of an artificial restriction endonuclease site. Another possibility is targeting a mutant-activated oncogene by single mismatch antisense RNA. The mismatch is expected to reduce the chance of cleavage at the heteroduplex site but increase the chance of cleavage by RNase H at the oncogenic mutation so that the perfectly matched mutant mRNA and malignancy suppression may be achieved. Besides these changes, good uptake should be secured, e.g., by the use of cationic lipids, then assuring membrane permeabilization with stability of the internalized oligonucleotides. The nerve cells apparently take up oligonucleotides more readily than other types of tissues if introduced by injection, but there is a blood/neuron barrier after intravenous or intraperitoneal applications. Antisense constructs readily target the liver and kidney, but the degradation and excretion are most rapid in and from these tissues. Endocytosis and pinocytosis can take up antisense oligonucleotides, but then they are usually locked up in the vesicles within the cells. When injection or electroporation introduces the antisense molecules, they may reach the nucleus. The half-life may be less than 5 minutes, and within 10 hours half the antisense oligonucleotides are lost. The antisense construct may have a variety of effects on the cells, and the observed consequences may be the results of a nonantisense type of action. The antisense DNA oligonucleotides are usually targeted to the AUG initiator codon, to the 5' cap, to the first splice acceptor, to the polyadenylation, or to the translocation breakage point site in cancer. The actively transcribed RNA is a superior target. At the proper dosage, AO has minimal or no effect on normal cells but may be quite effective. *See* antisense RNA; aptamer; BCL; cancer gene therapy; cationic lipid; fomivirsen; fruit ripening; G3139; methylphosphonates; mixed backbone oligonucleotides; peptide nucleic acid; phosphoramidate; phosphorothioates; PKA; ribonuclease H; TFD; triple helix formation, tricyclo-DNA. (Galderisi, U., et al. 1999. *J. Cell Physiol.* 181:251; Cotter, F. E., et al. 1999. *Biochim. Biophys. Acta* 1489:97; Kushner, D. M. & Silverman, R. H. 2000. *Curr. Oncol. Rep.* 21:23; Astriab Fisher, A., et al. 2002. *J. Biol. Chem.* 277:22980; Fu, C., et al. 2002. *Anal Biochem.* 306:135.)

antiserum   Blood serum that contains specific antibodies obtained from an animal after natural or artificial exposure to an antigen. Antisera are collected from the blood of fasted animals by centrifugation and allowed to clot at room temperature. The clot is then discarded and the straw-colored serum may be preserved either by lyophilization and stored at room temperature or at 4° with 0.02% sodium azide, or deep frozen at −20° to −70°C. The antisera generally contain polyclonal antibodies. *See* antibody, polyclonal; monoclonal antibody.

antiserum purification   Of polyclonal antibodies by affinity chromatography on protein A-Sepharose columns. Protein A binds the Fc domain of IgG of various sources but not with equal intensity. Further purification may be obtained by affinity chromatography with immobilized antigen of high purity. *See* antibody; antibody purification; immunoglobulins.

anti–Shine-Dalgarno sequence   CCUCC is complementary to the GGAGG Shine-Dalgarno consensus near the 3' end of the 16S rRNA molecule. *See* Shine-Dalgarno.

antisuppression   Inactivates suppressor genes. *See* suppressor gene; suppressor tRNA.

antitermination   Permits the RNA polymerase to ignore transcription termination instructions such as bacterial *rho* and thus it proceeds through the termination signal. In phage λ, after the transcription of two immediate early genes the RNA polymerase should stop. The switch to transcribe the next set of genes is controlled by gene *N*, transcribed from the left promoter (PL) and terminated by the rho-dependent tL1 terminator and *cro*, transcribed from the right promoter (PR) and terminated by the rho-dependent tR1 terminator. The product of the *N* gene is protein N (pN), an antiterminator that permits readthrough to the delayed early genes in both tL1 and tR1. Although pN has a half-life of about 5 min, transcription is maintained because *N* is part of the delayed early transcription. Gene *Q* is also part of the delayed early

Phosphodiester

Phosphorothioate (PS)

transcription and its product pQ is an antitermination protein that allows by readthrough the transcription at the late promoter *PR*. The recognition site for pN is upstream at the N utilization sites *NutL* and *NutR*; the former is near the promoter, but the latter is near the terminator. pN can act on both rho-dependent and rho-independent systems. Different phages have different *nut* sites, yet these all work in a similar manner. The *nut* elements include *boxA* and *boxB*; the former is required for binding the bacterial antitermination proteins, used by phages as well as by bacteria. The *boxB* is a phage-specific element.

Mutations in bacteria (*rpoB*) interact with pN. The *nus* loci (*A*, *B*, *G*) are involved with transcription termination; *nus E* codes for a protein in the 30S ribosomal subunit (p10). The product of *nusA* is a general transcription factor interacting with p10 and affecting termination by binding to *boxA*. Gene *nusG* organizes the various Nus proteins, which together control rho-dependent termination, whereas the *nusA* product combined with pN may interfere with termination where it normally is supposed to take place. In *E. coli*, antitermination also involves the ribosomal *rrn* genes. This operon has in its leader sequence a *boxA* where the NusB-S10 protein dimer binds to the RNA polymerase as it passes through. This binding enables pol to continue transcription through the rho-dependent terminators. Protein NusA does not bind to the bacterial RNA polymerase when it is associated with the σ-factor, but after pol attaches to the promoter, σ may be released, providing an opportunity for transcription and for the formation of the core polymerase-Nus complex. After termination of transcription, the pol complex is released from the DNA and the separation of Nus from pol takes place. Thus the polymerase core enzyme may be in two alternative states: one with σ for transcription and another with Nus with the potential for termination of transcription. Antitermination may then be mediated through pN after the polymerase binds Nus. Gene *Q* of phage λ also has a role in antitermination by permitting through its product the passage over the terminator signals. Transcription is modulated by preventing termination of transcription at T-rich sequences that occur randomly within the gene, but they dissociate RNA polymerase from the DNA when it arrives at the T-rich region at the end of the gene where termination of transcription is expected. Other antitermination (attenuation) proteins are acting in the amino acid operons of bacteria and allow the expression of the operon only after the protein that mediates attenuation of transcription is made. Thus attenuation is not always dependent on the presence of an excess of charged specific tRNAs, which slow down transcription when the supply of this particular amino acid is sufficient. In eukaryotes, the Pol II-transcribed U1-RNA is involved in proper processing of the 3′ ends of the RNA used for reinitiation of transcript elongation. *See* attenuation; half-life; lambda phage; rho; rho factors; RNA polymerases; rrn; σ; terminator; transcription; transcription termination in eukaryotes; transcription termination in prokaryotes; T box; tryptophan operon. (Mason, S. W. & Greenblatt. 1992. *J. Biol. Chem.* 267:19418; Yarnell, W. S. & Roberts, J. W. 1999. *Science* 284:611; Grundy, F. J., et al. 2002. *Proc. Natl. Acad. Sci. USA* 99:0121.)

antithrombin (AT-III, 1q23-q25) α-globulin neutralizing the blood-clotting contribution of thrombin. Antithrombin—especially when cleaved at the COOH-terminal loop—blocks angiogenesis and tumor development. *See* angiogenesis; anticoagulation; antihemophilic factors; blood clotting pathways; dysfibrinogenemia; protein C; protein S; thrombin.

antitoxin *See* immunization.

antitrypsin gene (AAT or PI) In human chromosome 14q32.1 it prevents the activity of the protease trypsin and elastase, and the α-antitrypsin gene is supposed to be involved in pulmonary emphysema (increase of lung size because of dilatation of the alveoli [the small sacs] of the lung) and liver disease. Different mutations may lead to one or the other or to both of these diseases. The so-called Z mutant group prevents the exit of the AAT protein from the liver, where it is synthesized, and as a consequence liver disease (cirrhosis) appears. Smoking may increase the chances of the development of cirrhosis in individuals of ZZ genotype by 3 orders of magnitude. The incidence of AAT deficiency is about 8×10^{-4} in the white population of the United States. The total length of the α-antitrypsin gene is 10.2 kb with coding sequences of 1,434 bp. Oral administration of 4-phenylbutyric acid facilitates the release of AAT from the endoplasmic reticulum and as a "chemical chaperone" may prevent the injuries resulting from AAT deficiency. The 14q32 chromosomal site includes the serpin gene encoding the corticosteroid-binding globulin (CBG) and a DNase-1 hypersensitive site. The AAT gene can be inserted into sheep eggs and under favorable condition the milk may contain the protein it encodes. *See* cirrhosis of the liver; corticosteroid; DNase hypersensitive site; emphysema; endoplasmic reticulum; liver cancer; serpin. (Crystal, R. G. 1989. *Trends Genet.* 5:411; Brigham, K. L., et al. 2000. *Hum. Gene Ther.* 11:1023.)

antivector cellular immunity *See* human gene transfer.

antiviral antibody Immunization against some viral diseases is not fully successful. Monoclonal (or enriched polyclonal) antibody therapies have been considered against human respiratory syncytial virus (RSV), rabies, hepatitis B and C, herpes simplex viruses, cytomegalovirus, and acquired human immunodeficiency (HIV). These preparations may be administered intramuscularly. *See* monoclonal antibody therapies.

antizyme Protein that binds to enzymes and directs their degradation by proteasomes without ubiquitin. Their synthesis is induced by the proximal or distal products of the enzymes they inhibit. Antizymes regulate polyamine enzymes such as ornithine decarboxylase. *See* polyamines; proteasome. (Coffino, P. 2000. *Proc. Natl. Acad. Sci. USA* 97:4421; Chattopadhyay, M. K., et al. 2001. *J. Biol. Chem.* 276:21235.)

Antley-Bixler syndrome (trapezoidocephaly-synostosis syndrome) Defect in bone formation, abnormality of the face, and other developmental anomalies due to mutation in the fibroblast growth factor receptor 2 (FGFR2) gene. *See* Apert or Apert-Crouzon syndrome; craniosynostosis syndromes; fibroblast growth factor.

anucleate A cell after the nucleus is removed. *See* cytochalasins; cytoplast; nuclear transplantation.

anus   End opening of the intestinal tract.

anxiety   *See* stress.

aorta   Main arterial vein (carrying blood away from the heart) originating in the left heart ventricle and passing through the chest and abdomen. *See* coarctation of the aorta.

aortic stenosis   *See* coarctation of the aorta.

aotus (owl monkey)   *See* cebidae.

AP   *See* amino purine; base analog mutagen.

AP1, AP2, AP3, AP4, AP5 (activated protein)   Group of transcription factors. AP1 is similar to the one coded for by the chicken virus oncogene *v-jun*. The human gene at chromosomal location 1p32-p31 shows 80% homology to the avian viral protein gene; binding is greatly enhanced by the *fos* oncogene. AP1 generally appears as a heterodimer of Jun and FOS; FraIn yeast AP1 has a homolog, GCN4, and the mammalian homolog is TFIID. The yeast and mammalian factors can substitute for each other. This family of genes encodes the AP transcription factors where the binding motif is well conserved, but other sequences may vary. These proteins bind to 5′-TGANTCA-3′ consensus in DNA. AP2 binds only to TC-II but not to TC-I of the two identical and adjacent TC motifs (5′-TCCCCAG-3′) upstream in the promoter of eukaryotic genes. AP2 binding affects enhancer activity. AP2 is an essential morphogenetic factor; in its deficiency, head development is impaired. AP2 seems to have negative control on the cell cycle, possibly by activation of p21 protein. AP3 binds to TC-II and to the adjacent GT-I motif (5′-G[C/G]TGTGGA[A/T]TGT-3′), and also to the so-called core enhancer sequence (5′-GTGG[A/T][A/t][A/T]G-3′) that is similar to parts of viral and prokaryotic enhancers but does not function by itself. AP4 binds to the 5′-CAGCTGTGG sequence that partially overlaps the GT-II motif (identical to GT-I except two bases). AP5 binds to GT-II and adjacent sequences (5′-CTGTGGAATGT-3′) and is present in some cell types but not in others. The mouse *jun* genes (chromosomes 4 and 8) are inducible by serum and the phorbol ester, 12-o-tetradecanoyl phorbol 13-acetate (TPA). The *AP* loci of *Arabidopsis* are completely different and mean *apetala*, a defective flower type. *See* adaptin; endocytosis; epsin; Fos; Fra; Jun; oncogenes; transcription factors. (Shaulian, E. & Karin, M. 2002. Nature Cell Biol. 4:E131.)

AP180 (assembly protein)   Mediates the assembly of clathrin for endocytosis. It is built of four adaptin proteins (100, 100, 50, 25 kDa, respectively). *See* endocytosis.

Apaf-1 (apoptotic protease activating factor/CED4)   Interacts with caspase-9 after being activated by cytochrome c and dATP. Then caspase-3 triggers the process of apoptosis. Somehow caspase-3 is linked to an endonuclease that cuts up chromosomal DNA in the cells destined for apoptosis. The *Apaf* gene (and some others) may be disabled by methylation, apoptosis is interfered with, and the road opens to carcinogenesis such as happens in chemotherapy-resistant metastatic melanoma. *See* AIF; apoptosis; caspase; Crohn disease; melanoma. (Bratton, S. B., et al. 2001. *EMBO J.* 20:998; Soengas, M. S., et al. 2001. *Nature* 409:207.)

apandry   Development of diploid fruiting body of fungi by fusion of two female nuclei without the involvement of any male gamete.

APC   **(1)** *See* antigen-presenting cell; ASE1; Gardner syndrome. **(2)** Anaphase-promoting complex (also called cyclosome) is a 12-subunit, 1500 kDa ubiquitin ligase protein complex containing CDC27, CDC16, CDC23, CDC26, Apc1p, Apc2p, Apc4p, Apc5p, APC9, APC10/DOC, Apc11p, Apc13, and bimE. APC is required for the progression from metaphase to anaphase. It is regulated by CDC20 and CDH1 in humans, *fzy* and *fzr* in *Drosophila*, and the APC complex mediates ubiquitination of the superfluous cyclins and anaphase-inhibitory proteins. This degradation is a requisite for exit from each phase of the cell cycle and entry into the next one. APC recognizes a 9-amino-acid destruction box at the N-terminus of cyclins and some other proteins such as Pds1p/Cut2p anaphase inhibitors of yeasts or the spindle protein Ase1. *See* bimE; CDC20; CDCs; CDH; CDH1; cell cycle; cohesion; cullin; D box; E2; mitotic exit; MPF; PDS; Rbx1; SCF; securin; separin; sister chromatid; tetratrico sequences; ubiquitin. (Page, A. M. & Hieter, P. 1999. *Annu. Rev. Biochem.* 68:583; Schwab, M., et al. 2001. *EMBO J.* 20:5165; Peters, J.-M. 2002. *Mol. Cell* 9:931.) **(3)** Activated protein C mediates cleavage and inactivation of antihemophilic factors Va and VIIIa with the cooperation of protein S. Its mutation-conveying resistance to blood coagulation may increase the risk of thrombosis 5–10-fold and is the most common genetic cause of thrombosis. *See* antihemophilic factors; protein C; protein S; thrombosis.

APE   Apurinic/apyrimidinic endonuclease recognizes these sites and cleaves the nucleic acid backbone as part of the repair function. *See* excision repair.

APECED   Autoimmune polyendocrinopathy, candidiasis, ectodermal dystrophy is a human autoimmunity syndrome involving a Zn-finger-like protein (a transcription factor) encoded in chromosome 21q22.3 by the AIRE gene (autoimmune regulator). *See* autoimmune disease; Zinc fingers.

AP endonuclease (APE)   Basically a repair enzyme in both prokaryotes (two enzymes) and eukaryotes (encoded in humans by HAPIm BAP1, APE/APEX) that cuts DNA 5′ or 3′ to modified (alkylated or otherwise mutated) DNA bases or at apurinic and apyrimidinic sites from where glycosylases already removed damaged purines or pyrimidines. Usually the first step is the recognition of the altered bases, and the DNA sequence is cut in the vicinity. Then the exonuclease activity removes the damaged section and creates a gap. A repair synthesis adds the correct bases to the 3′-OH ends using the undamaged strand of the double helix as a template. Ligation by covalent bonds restores the integrity of the DNA. The glycosylases have some specificity for deaminated cytosine residues, uracil-N-glycosylase removes uracil residues, and hypoxanthine-N-glycosylase removes hypoxanthines formed by deamination of adenine. These endonucleases have antimutator activities. The eukaryotic DNA uses pol β or pol δ and pol ε for filling the gap. *See* antimutator; AP site; DNA polymerases; DNA repair; glycosylases. (Sobol, R. W. & Wilson, S. H. 2001. *Progr. Nucleic Acid Res. Mol. Biol.* 68:57.)

aperiodic crystal Term used for chromosome by the physicist Erwin Schrödinger in 1944. (Stent, G. S. 1995. *Ann. NY Acad. Sci.* 758:25.)

Apert or Apert-Crouzon syndrome Involves acrocephaly (top of the head is pointed), syndactyly (fingers fused), and mental retardation, although some individuals have near normal intelligence. The symptoms vary. Many of the cases are sporadic; in others, autosomal-dominant inheritance is most likely; chromosomal rearrangement may also be present in some cases. This condition may also be caused by a defect in FGFR2 (fibroblast growth factor receptor), a protein tyrosine kinase, encoded at 10q25-q26. An insertion of an Alu element in the gene results in an alternately spliced keratinocyte growth factor receptor (KGFR). It is allelic to Crouzon and Pfeiffer syndromes. *See* craniosynostosis syndromes; Crouzon syndrome; Jackson-Weiss syndrome; mental retardation; Pfeiffer syndrome; syndactyly; tyrosine kinase receptor.

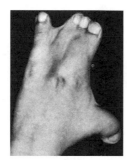

Syndactyly. (From Bergsma, D., ed. 1973. Birth Defects. Atlas and Compendium. By permission of the March of Dimes Foundation.)

ape The closest animal to a to human. The human nonrepetitive DNA sequences appear 98.7% identical to chimpanzees and 98.38% to gorillas. *See* primates. (Hacia, J. G. 2001. *Trends Genet.* 17:637.)

apex Top part of a cell, organ, or any structure. *See* apical; arrayed primer extension; meristem.

APH Aminoglycoside phosphotransferases are enzymes phosphorylating aminoglycoside antibiotics, resulting in resistance to the antibiotics when the enzyme is present (introduced by transformation). *See* APH[3′]II; aminoglycosides; antibiotics.

APH(3′)II Aminoglycoside phosphotransferase enzyme inactivates kanamycin, neomycin, and geneticin, commonly used antibiotic resistance markers for transformation in tissue culture; synonymous with NPTII. *See* aminoglycoside; antibiotics.

apheresis Separation of a certain component(s) of a patient's blood and reinfusion of the remainder.

aphid (*Aphididae*, homoptera) Small sucking insects, parasites of almost all plant species. At the site of the infestation the plants secrete honeydew, which may attract other types of insects. They reproduce sexually at the end of the growing season after males have differentiated. During the rest of the year, only females are found that reproduce parthenogenetically, and their ca. 20 generations produce three to seven nymphs daily. Thus the progeny of a single individual may run into billions during the year. Besides the direct damage by sucking, they spread plant viral diseases. They can be controlled by contact or systemic insecticides. Aphids harbor 60–80 large cells (bacteriocytes) that contain the symbiotic *Buchnera* bacteria with ~100 copies of a genome of 640,681 bp. The bacteria supply essential amino acids to the aphids and rely on the host for cell-surface molecules, regulator genes, and defense. *See* parthenogenesis. (Abbot, P., et al. 2001. *Proc. Natl. Acad. Sci. USA* 98:12068.)

Doralis faba nymph (right), female (left).

aphidicolin Tetracyclic diterpene of fungal (*Cephalosporium*) origin capable of blocking cell division and of antiviral activity; it is an inhibitor of DNA polymerase α. *See* pol; terpenes.

apical Indicates top position. *See* apex.

apical dominance Terminal bud of the main stem of a plant preventing or suppressing the formation of lateral buds or branches.

apical ectodermal ridge (AER) Group of cells at the tip of the limb bud involved in the differentiation of the limbs of animals. *See* morphogenesis; organizer; ZPA.

apicomplexan plastid *See* apicoplast.

apicoplast (apicomplexan plastid) Acquired ~35 kb DNA-containing plastid-type body (from endosymbiosis by green algae) in several parasites (e.g., *Toxoplasma*, *Plasmodium*). It has an as-yet undefined essential role in the survival of the parasite and can be targeted by antibiotics as a measure of defense. *See* toxoplasmosis. (Wilson, R. J. M. 2002. *J. Mol. Biol.* 319:257.)

apigenin Flavone plant pigment.

Apis mellifera (honeybee) Social insects with three types of individuals: diploid egg-laying queen ($2n = 32$), haploid drones, and sexually undifferentiated diploid workers. The drones hatch from unfertilized eggs. The difference between the queen and the workers is due to different nutrition of the larvae. *See* arrhenotoky; honey bee. (Robinson, G. E., et al. 1997. *Bioessays* 19:1099.)

aplasia Failure of the development of an organ or a type of tissue.

aplastic anemia Condition of several blood diseases when the bone marrow may not produce the cellular elements of the blood. *See* anemia; Duncan syndrome.

AP lyase Releases apurinic and apyrimidinic sites from the DNA. *See* apurinic site; apyrimidinic site; DNA repair.

Aplysia Sea mollusk, an invertebrate small animal, frequently used for behavioral studies.

APM Affected pedigree member. APM is used in determining identity by descent and as a nonparametric method to detect linkage. See IBD.

apnea (familial obstructive sleep, snoring) Breathing disorder of any age; responsible for sudden infant death. The genetic basis is unclear. The composer Johannes Brahms might have been afflicted by it. *See* narcolepsy.

APO1 *See* Fas.

Apo-2 *See* TRAIL.

apoaequorin *See* aequorin.

apocytochrome b gene (cob) Located in mitochondrial DNA of yeast; cytochromes are heme-containing proteins involved in electron transport. *See* mitochondrial genetics; mtDNA.

apoenzyme Enzyme protein without the cofactors required for activity.

apoferritin Protein part (M_r 460,000) of ferritin that contains ferric hydroxide clusters. About 20–24% of it is iron. Ferritin is the most readily available iron storage facility in the body. *See* ferritin.

apogamety (apogamy) Embryo formation without fertilization from a cell of the embryo sac other than the egg cell. *See* apomixis.

apoinducer DNA-binding protein that stimulates transcription. *See* transcription.

apolar Molecules are generally insoluble in water because they do not have symmetrical positive and negative charges.

apolipoprotein Lipid-binding protein in the blood that transports triaglycerols, phospholipids, cholesterol, and cholesteryl esters within the body. Apolipoproteins are the most important parts of high-density lipoproteins (HDL) because they reduce the risk of coronary heart disease. Different classes are distinguished: APOA1 (human chromosome 11q23.2-qter, mouse chromosome 9), APOA2 (1q21-q23), APOC2 (19q13.2), APOC3 and APOA4 (in the same region). Apolipoprotein B (human chromosome 2p24) exists in two lengths due to different editing of the transcript. APOC cluster is in human chromosome 19q13.2 and APOE

appears to be linked to it. APOE deficiency causes hyperlipidemia and atherosclerosis. APOE isoform E4 is involved in dementia associated with HIV infection and Alzheimer disease. Some other apolipoproteins are genetically less well defined. Apolipoprotein A-IV may protect against atherosclerosis without an increase in HDL levels. Apolipoprotein B deficiency reduces male fertility in knockout mice. *See* abetalipoproteinemia; AIDS; Alzheimer disease; arteriosclerosis; atherosclerosis; cholesterols; fatty acids; HDL; hyperbetalipoproteinemia; hyperlipidemia; hyperlipoproteinemia; hypo-α-lipoproteinemia; lipoprotein lipase; megalin; Tangier disease. (Mahley, R. W. & Rall, S. C., Jr. 2000. *Annu. Rev. Genomics Hum. Genet.* 1:507; Pennachio, L. A., et al. 2001. *Science* 294:169.)

apomeiosis Gamete develops without a meiotic process. *See* apoxis; meiosis.

apomict Reproduces by apomixis. *See* apomixis; *Rosa canina*.

apomixia Parthenogenesis, common in *Caenorhabditis elegans*, bees, wasps, aphids, in some crustacea, lizards, isopoda, lepidoptera, etc.; it does not occur in humans. *See* apomixis; parthenogenesis.

apomixis Embryo (zygote) development without fertilization in plants and fungi. It occurs regularly in certain species, e.g., in the polyploid *Festuca*, hawkweeds (*Hieracium*), etc. Some apomicts reproduce sexually after doubling the chromosome number. Apomicts may make the fixation of heterozygous conditions possible. Apomixis may be genetically very different from somatic embryogenesis. If apomixis is preceded by meiosis and the egg parent is heterozygous, segregation may occur among the apomictic progeny. In aposporous apomixis the megagametophyte develops from a somatic cell of the ovule. *See* agamospermy; androgenesis; apogamety; apomixia; parthenogenesis. (Koltunov, A. M. 1993. *Plant Cell* 5:1425; van Dijk, O. & van Damme, J. 2000. *Trends Plant Sci.* 5:81; Grimanelli, D., et al. 2001. *Trends Genet.* 17:597.)

apomorphic Species trait evolved from a more primitive state of the same. *See* plesiomorphic; synapomorphic; symplesiomorphic.

apopain (caspase 3, human chromosome 4q35) *See* apoptosis.

apoplast Intercellular material.

apoptosis (programmed cell death, PCD) The cells and the nuclei shrink and generally are absorbed after fragmentation. Apoptosis is an indispensable process for the majority of organisms. Unneeded cells are disposed of, room is made for differentiated cells, and it is a safeguard against cancerous growth. A generalized outline of the apoptotic cell death pathway is represented next page in this entry. Ceramide is one of the regulatory molecules of the process. Tumor necrosis factor (TNF) is an inducer of apoptosis. The metabolites of ceramides, sphingosine and sphingosine-1-phosphate, prevent the symptoms of apoptosis. These two molecules are supposedly

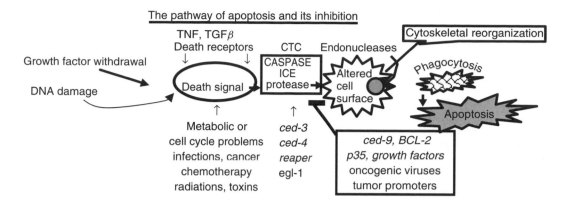

The pathway of apoptosis and its inhibition

second messengers for cell proliferation mediated by platelet-derived growth factor. Activation of protein kinase C brought about by sphingosine kinase and the increase of the level of sphingosine-1-phosphate inhibits ceramide-mediated apoptosis. The latter molecules also stimulate the ERK-controlled reactions and inhibit the stress-activated kinases SAPK/JNK.

In *Caenorhabditis*, more than a dozen *ced* (cell death) genes have been identified. Ced3 protein is an interleukin-1-converting (ICE) cysteine protease enzyme involved in ceramide production. Ced9 is a suppressor of cell death. The latter has 23% identity to the human oncogene BCL-2, controlling follicular lymphoma. If this human gene is transfected to the nematode, it suppresses apoptosis, indicating that the same function is controlled over a wide evolutionary range. The *reaper* gene (*rpr*) of *Drosophila* is an activator of apoptosis, and it is homologous to *ced-3* of *Caenorhabditis*. Another *Drosophila* gene, *hid* (*head-involution defective*), is linked to *reaper*. The 65-amino-acid RPR protein also has similarity to the "death domain" of the tumor necrosis factor receptor (TNFR) family. TNFR1 and Fas induce cell death when activated by ligand binding or when overexpressed. Other death receptors are DR3 (Ws1, Apo3, TRAMP, Lard) and Dr4. They may directly connect to the Fas-associated death domain (FADD/MORT1) or to the tumor necrosis factor–associated death domain receptor (TRADD). FADD recruits procaspase 8. TRAIL is another killer protein for which DR3 and Dr4 serve as receptors. Curiously, the latter receptors are present in nonapoptotic cells. The lack of killing effect in these cells is explained by other TRAIL receptors: TRID (TRAIL receptor without an intracellular domain) or DcR1 (decoy receptor 1). These receptors have glycophospholipid-anchored cell surface portions that trap TRAIL but do not allow the transfer of the death signal to FADD and actually function as a decoy, preventing the death signal transduction even for Dra3 and Dra4.

The dead cells are generally disposed of by the macrophages without producing inflammation in the tissue. The apoptotic cells are generally recognized by the altered sugar groups or phosphatidylserine on their surface. The macrophage secretes an extracellular protein, thrombospondin, which recognizes apoptotic cells. The *Alg-2* (apoptosis-linked gene) encodes a Ca^{2+}-binding protein that is required for T-cell receptor-, Fas-, and glucocorticoid-induced apoptosis. *Alg-3* is a homologue of the Alzheimer disease gene, which is basically a senescence gene. Apoptosis may be a very natural response for cell disposal when cells are no longer needed. In some cancers

the proliferation is out of hand because regulators of the process go awry. The baculovirus apoptosis inhibitor proteins (Cp-IAp and Op-IAP), as well as neuronal apoptosis inhibitor proteins (NAIP), located in human chromosome 5q13.1, are defective or deleted in spinal muscular atrophy.

BAX is a heterodimeric protein that works in the opposite direction as BCL2 (chronic lymphocytic leukemia, B cell). The gain-of-function BAX mutations—knockouts—were viable, but in the lymphocyte cell lineages apoptosis was induced. In other cell lineages hyperplasia was observed. Thus the BAX expression depends on cellular context. The wild-type BCL2 gene functions similarly to *Ced-9* of *Caenorhabditis*, i.e., it suppresses apoptosis. The enzyme apopain, cleaving poly(ADP-ribose) polymerase (PAR), is also necessary for apoptosis to proceed. Apopain is generated from the proenzyme CPP32, a protein related to ICE and CED-3. Lymphocyte apoptosis may be mediated by type 3 inositol 1,4,5-trisphosphate receptor in the plasma membrane by promoting the influx of calcium. Apoptosis of neurons is mediated by the activation of JNK (JUN [oncogene] NH2-terminal kinase) in a process opposing the effect of ERK (extracellular signal-activated kinase) in the absence of NGF (nerve growth factor). The p35 protein of the Baculovirus *Autographa californica* has similar antiapoptotic properties for insects and for mammals as the Ced-9 gene product of *Caenorhabditis*. Ced-9/Bcl-2 gene product inhibits both the apoptosis-promoting and -protecting effects of the Ced-4/Apaf-1 products. CED-4 interacts with CED-3 and they oligomerize. Their oligomers may associate with BCL, and this results in processing (in the presence of ATP) of CED-3, leading to cell death. In order to activate the death pathway, the EGL-1 protein stops the interaction of CED4-CED3 with BCL.

Ced-9 is homologous to the mammalian BCL-2. *Ced-3* and *Ced-4* are considered as promoters of apoptosis. Actually, *Ced-4* has two transcripts; the short transcript promotes apoptosis, but the long transcript is somewhat protective. Overexpression of *ced-4L* actually can prevent programmed cell death. It is interesting to note the structurally unrelated *BCL-x* and *Ich-1* genes are also involved in apoptosis; similar to *Ced-4*, all have two alternative transcripts. This indicates that RNA splicing may have an important role in programmed cell death.

Before caspase can be fully functional, ced-4 moves from the mitochondrion to a perinuclear location. The basic leucine zipper proteins (bZIP), PAR (proline and acid rich) and other members of the protein family, can also control apoptosis. Inappropriate activation of apoptosis may be involved in diseases such as AIDS, degeneration of the nervous system

(Alzheimer disease, amyotrophic lateral sclerosis, Parkinson's disease, retinitis pigmentosa), constriction or obstruction of the blood vessels (ischemic stroke), anemias, liver diseases, autoimmune diseases, etc. The diagram above is modified after Thompson, C. B. 1995. *Science* 267:1456.

The generic name CASPASE has been suggested (c for cysteine protease, aspase for cleaving at aspartate) for the ICE-ced protease enzyme system. The individual enzymes are also designated by numbers. After the initiation of the apoptotic pathway, cytochrome c is released from the mitochondria by the action of the Bax protein, which in turn leads to the activation of the caspases by Apaf-1. Actually, some say there are two pathways of apoptosis: one through the Fas and another through the Apaf/mitochondrial route. Protein BAR (member of the BCL family) may coordinate these two pathways. BCL-2/CED-9 protein blocks caspase activation. "BH-only" proteins of the Bcl family (Bad, Bil Blk, Hrk/Dp5, Bid, Bim, Noxa, EGL-1) share 9-6 amino acids, and they are initiators of apoptosis. The X-chromosome-linked IAP (inhibitor of apoptosis) directly inhibits the caspase 3 and 7 proteases.

Another protease inhibitor is known by the synonyms FLIP, Casper, Flame, Cash, and I-FLICE. The CED-3 enzymes are cysteine proteases, accompanied by nucleases that cut the DNA to approximately 180–200 bp fragments, the size of a nucleosomal unit (hence the name caspase-activated DNase [CAD]). There is also another closely associated protein, an inhibitor of CAD (ICAD or DFF45 in humans). ICAD releases CAD after caspase-3 is cut and it is a chaperone. CAD is produced as a complex with ICAD; the action of caspase-3 permits CAD to move into the nucleus from the cytoplasm. The site of cutting is between two nucleosomes by the 343 amino acid nuclease, a basic protein. Actually, the whole complex process involves a number of other proteins that interact and affect the outcome of cell death or proliferation.

Apoptosis is used for multiple purposes besides aging. Differentiation, homeostasis, and cellular defense mandate this process of suicide for damaged or unnecessary cells. The purpose of the apoptotic process is to free the system from unwanted cells and stop proliferation (e.g., prevent cancer and autoimmune disease) and maintain healthy conditions. Glucocorticoids are effective stimulators of apoptosis. After stroke or Alzheimer disease, overactive apoptosis may, however, damage the brain. A low level of apoptosis may be the cause of follicular B-cell lymphoma. Tumor formation may result from overexpression of Bcl-2 suppressor of apoptosis. Suppression of Bcl-2 may, however, promote apoptotic death of cancer cells. Suppression of survivin, an apoptosis inhibitor, may also cause death of cancer cells. TRAIL, caspases, and caspase inhibitors, respectively, have also been considered for cancer therapy. p53 tumor suppressor may channel damaged cells to an apoptotic path.

It has been estimated that 10 billion cells in a human body suffer apoptosis daily and about the same number of cells arise again by mitosis.

Programmed cell death also occurs in plants during the differentiation of the vascular system (xyleme), fruit ripening and senescence, and the hypersensitive defense reaction against pathogens. *See* acinus; acquired immunodeficiency; addiction module; aging; AIF; altruism; Alzheimer disease; amyotrophic lateral sclerosis; anoikis; Apaf; aplastic anemia; apopain; ARF; BAK; BCL; BID; ceramides; chaperone; CTC; cysteine proteases; DAP kinase; death signaling; DISC; Down syndrome; DR; endonuclease G; ERK; FADD/MORT-1; Fas; FLICE; fragmentin-2; glucocorticoid; granzymes; Hayflick's limit; hypersensitive reaction; IAP; ICE; IEX; IImtPTP; interleukins; L-DNase; leukemia; lymphoma; macrophage; mitochondrial diseases in humans; Myc; necrosis; nur77; p35; p53; perforin; phagocytosis; porin; retinitis pigmentosa; SAPK; signal transduction; smac; sphingosine; survival factor; survivin; T cell; T-cell receptors; TGF; TNF; TNFR; TRAIL; transmission; TRF2. (<http://www3.shinbiro.com/~virbio/index.html>; <http://vl.bwh.harvard.edu/apoptosis.shtml>; *Nature* 407:769ff; Huang, D. C. S. & Strasser, A. 2000. *Cell*: 103:839; Vousden, K. H. 2000. *Cell* 103:691; Fesik, S. W. 2000. *Cell* 103:273; Strasser, A., et al. 2000. *Annu. Rev. Biochem.* 69:217; Engelberg-Kulka, H. & Glaser, G. 1999. *Annu. Rev. Microbiol.* 53:43; Aravind, L., et al. 2001. *Science* 291:1279; Joza. N., et al. 2001. *Nature* 410:549; Wei, M. C., et al. 2001. *Science* 292:727; Hunot, S. & Flavell, R. A. 2001. *Science* 292:865, *Nature Cell Biol.* 2002. June issue for several papers, Igney, F. H. & Krammer, P. H. 2002. *Nature Rev. Cancer* 2:277.)

apoptosis inhibitor Viral (baculovirus) protein aimed at overcoming the host defense against viral infection by cell death. Similar proteins targeting primarily caspase 9 and apoptotic mitochondrial cytochrome c occur in insects, mammals, and humans.

apoptosome Complex of apoptosis proteins including caspases, Apaf, etc. *See* Apaf; apoptosis; caspase. (Acehan, D., et al. 2002. *Mol. Cell* 9:423.)

aporepressor Repressor protein that requires another molecule, the corepressor (frequently a late product of the metabolic pathway) to be active in controlling transcription. *See* transcription; tryptophan operon; tryptophan repressor.

aposematic coloration Warning display of animals and plants against invaders, e.g., bright color of poisonous snakes or poisonous mushrooms or plants. *See* Batesian mimicry; Müllerian mimicry. (Brodie, E. D. III & Agrawal, A. F. 2001. *Proc. Natl. Acad. Sci. USA* 98:7884.)

apospory Seed formation without fertilization from diploid cells of the nucellus or integumentum. *See* adventitious embryo; agamospermy; amixia; apomixis; diplospermy.

apostatic selection Predators often prefer the most abundant types of prey and thus maintain the polymorphism of the prey population. *See* selection types. (Bond, A. B. & Kamil, A. C. 2002. *Nature* 415:609.)

aposymbiotic Organism cured from the symbiotic partner. *See* symbionts.

apothecium Open fruiting body of fungi on which the asci develop; it is similar to perithecium, but the latter is a closed fruiting body. Among the genetically widely used organisms, *Ascobolus* develops apothecia. *See* perithecium.

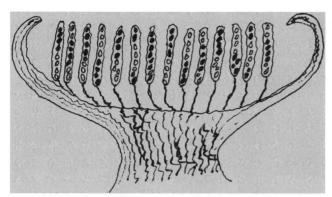

Apothecium (after Weier, T. E., et al. *Botany*. 1974. Wiley & Sons, New York).

apotransferrin Transport protein without its ligand. *See* ligand; transporters.

apo-VLDL *See* apolipoprotein.

APP Amyloid precursor protein. *See* Alzheimer disease; amyloids; secretase.

AP-PCR Arbitrarily primed PCR. *See* polymerase chain reaction.

apple (*Malus* spp.) About 25 species all with $x = 17$. Most of them are diploid, although tetraploid and triploid varieties also occur. Apples are frequently self-incompatible but cross-fertile with other apples and do not easily hybridize with pears (*Pyrus*), but they hybridize with *Sorbus* (mountain ash). *See* pears.

application programs (computer) Programs serving special purposes such as word processing, graphics, telecommunication, DNA sequencing, data management, etc.

appressor Cylindrical or globular fungal organ at the end of the hyphae with a rigid wall. It serves for infection by rupturing the cell wall of plants and invasion of the tissues with the aid of the penetration peg. The process may rely on cutinase, cellulase, and other enzymes, but it may utilize high-turgor mechanical pressure.

appressorium Enlarged fungal structure at the point of invasion of the host.

APRF Acute-phase response factor, 17q21 is a transcription factor related to the p91 subunit of the interferon-stimulated gene factor-3α (ISGF-3α). It is phosphorylated by the mediation of cytokines, and along with Jak1 kinase, it is associated with gp130. Interleukin-6, leukemia inhibitory factor (LIF), oncostatin M (OSM, 252 amino acids, encoded at human chromosome 22q12.1-q12.2 at the same general location as LIF), ciliary neurotrophic factor (CNTF, 11q12.2), cytokines, neurokines, and neuronal differentiation factors are involved. LIF and OSM may promote atherosclerosis by inhibiting the replacement of defective endothelial cells. *See* atherosclerosis; ciliary neurotrophic factor; cytokines; embryogenesis; gp130; leukemia inhibitory factor; signal transduction.

apricot (*Prunus armeniaca*) $x = 7$; $2x$ and $3x$ forms are known.

April A proliferation-inducing ligand is a member of the TNF ligand family with homology to CD95. The tumor necrosis factor ligand 13B (encoded at 13q32-q34) is also called TNFSF-13B and April. *See* BAFF; CD95; TNF. (Stein, J. V., et al. 2002. *J. Clin. Invest.* 109:1587.)

aprotinin Inhibitor (at concentrations 1–2 µg/mL) of proteases kallikrein, trypsin, chymotrypsin, and plasmin but not papain. *See* chymotrypsin; kallikrein; papain; plasmin; protease; trypsin.

APSES (present in ASM-1-Phd1-StuA-EFGTF1-Sok2 proteins [among others]) Helix-loop-helix-like structure regulating developmental processes. *See* DNA-binding protein domains.

AP site *See* apurinic site; apyrimidinic site.

aptamer Oligo-RNA, oligo-DNA, or a protein–oligo-RNA complex that can specifically bind a particular protein or other molecule, e.g., a thrombin-binding aptamer will inhibit the action of thrombin in blood clotting and thus prevent the formation of blood clots. Human neutrophil elastase, fibroblast growth factors, vascular endothelial growth factor, selectin, and antibodies have been successfully isolated by the SELEX procedure. Short RNA aptamers inserted into the 5′ untranslated region of mRNA may bind various ligands and may facilitate the control of translation behind it. *See* antibody; antisense RNA; elastase; FGF; mRNA display; selectin; SELEX. (Hermann, T. & Patel, D. J. 2000. *Science* 287:820.)

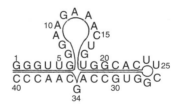

An RNA aptamer of ATP.

aptazyme Ribozyme with an aptamer. *See* aptamer; ribozyme.

apurinic endonuclease *See* AP endonucleases.

apurinic site (AP) Where a purine has been removed from the nucleic acid. It has been estimated that mammalian cells lose about 10,000 purines daily. The apurinic sites are removed by excision repair and deoxycytidyltransferase can include deoxycytidine across the apurinic site. DNA polymerase ζ makes further repair. *See* abasic sites; glycosylases. (Lindahl, T. & Nyberg, B. 1972. *Biochemistry* 11:3610; Haracska, L., et al. 2001. *J. Biol. Chem.* 276:6861.)

apyrase Acid tri- and diphosphatase enzyme that also degrades nucleotides.

apyrimidinic endonuclease   *See* AP endonucleases.

apyrimidinic site   Where a pyrimidine has been removed from a nucleic acid.

aquaporin (AQP2, 12q13, APQ4, 18q11.2-q12.1)   Six-transmembrane domain proteins (M_r 28K) and water channel in fluid-absorbing and fluid-secreting cells. The AQP1 monomer contains 269 amino acids, forming two tandem repeats of three membrane-spanning domains and amino and carboxy termini located on the cytoplasmic side of the membrane. The AQP family includes proteins with a wide distribution in diverse species across the plant, animal, and microbial world. cAMP-dependent mechanisms or PKA activate AQP. It is important for various types of cells, diabetes, kidney function, *Drosophila* neural development, nematodal infestation of plants, etc. Deletion of aquaporin-4 in mice reduces brain edema. *See* cAMP; cell membranes; CHIP; forskolin; ion channels; PKA. (Borgnia, M., et al. 1999. *Annu. Rev. Biochem.* 68:425; Murata, K., et al. 2000. *Nature* 407:599; Sui, H., et al. 2001. *Nature* 414:872).

aqueous   Prepared with water (e.g., a solution) or watery in appearance.

Aquifex aeolicus   Chemolithoautotrophic bacterium capable of growth at 95°C. Its completely sequenced DNA genome is 1,551,335 bp.

***Arabidopsis* mutagen assay**   In the mature embryo of the plants two diploid cells represent the inflorescence. Therefore, if the seeds are exposed at such a stage to a mutagen, in the progeny of the emerging plants the segregation is about 7:1 for recessive mutations. This also indicates that one of the two apical cells became heterozygous for the newly induced mutation. For mutagen assays it is sufficient to open up the immature fruits of the plants before the seed coat becomes opaque (about 10–14 days after fertilization) and albina or other color mutations in the cotyledons or embryo defects can be determined. Within a single fruit the segregation for recessives is 3:1 (see above). The fruits on the plants emerging from the treated seed already contain the F_2 generation. Generally, two opposite fruits next to each other are examined because the phyllotaxy index assures that these are sufficient for complete sampling. Such a test permits the identification of about 80% of the spectrum of visible mutations. Since the plants are diploid and the "germline" consists of two cells, the mutation rate on a genome basis is calculated by counting all independent mutational events and dividing that number by the total number of plants tested × 4. *Arabidopsis* can activate many types of promutagens and therefore provides an efficient means to assess genotoxic effects of a wide variety of agents in a single culture at low cost. *See Arabidopsis thaliana*; bioassays for genetic toxicology; phyllotaxy. (Rédei, G. P., & Koncz, C. 1992. *Methods in Arabidopsis Research*, Koncz, C., et al., eds., p. 16. World Scientific, Singapore, Hays, J. B. 2002. *DNA Repair* 1:579.)

Open *Arabidopsis* fruit.

Arabidopsis thaliana   Autogamous plant of the crucifer family, $2n = 10$; genome size has been estimated to be 9×10^7 to 1.5×10^8 bp (the current best estimate is ~120 Mbp). Its life cycle may be as short as 5–6 weeks; its seed output may exceed 50,0000 per plant. Its mitochondrial DNA is 366,924 bp, 10% duplicated, and apparently codes for 58 genes. Because of its small size, thousands of individuals may be screened even on a Petri plate. The plant's plastid DNA is approximately 154 kbp, encoding 79 proteins, and 17% is duplicated. It is the first higher plant with the genome completely mapped and sequenced in 2000. Transposons constitute 14% of the nuclear DNA and 4% of the mitochondria; the plastids appear free of moving elements. Chromosome 2 and 4 sequences became available by late 1999. (*Nature* 402:761 and 769.) The current data indicate higher gene number (25,498) than in *Drosophila* or *Caenorhabditis*.

All chromosomes display on the average ca. 60% duplication. Although many of the genes show homologies to those of other organisms, the identity with the human breast cancer gene (BRC2; 38.9%), Werner syndrome (37.4%), and Niemann-Pick disease (42.7%) is remarkable. Interestingly, many genes with apparently so far unidentified function and specific for plants were revealed by the sequences. Annotation of the full-length cDNA: Seki, M., et al. 2002. *Science* 296:141. Information is available through Arabidopsis Biological Resource Center, Ohio State University, 1735 Neil Ave., Columbus, OH 43210; Tel.: 614-292-1982 (Scholl, seeds), 614-292-2988 (Ware, DNA). E-mail: <arabidopsis+@osu.edu>. Orders: by fax 614-292-0603. E-mail <seedstock@arabidopsis.org>, <dnastock@arabidopsis.org>, NASC <http://nasc.nott.ac.uk>, TAIR: <http://arabidopsis.org>, <http://www.tigr.org/tdb. agi/>. Annotations: <http://baggage.stanford.edu/group/arab-protein>, <http://luggagefast.Stanford.EDU/group/arabpro-tein/>, <arabidopsis.org/home.html>, <http://mips.gsf.de/proj/thal/db>. Insertions: <http://genoplante-info.infobiogen. fr>, *Science* 282:6612. Physical maps: Marra, M., et al. 1999. *Nature Genet.* 22:265; Mozo, T., et al. 1999. *Nature Genet.* 22:271; *Nature* 408:796 [2000]; Bennetzen, J. L. 2001. *Nature Genet.* 27:3; Allen, K. D. 2002. *Proc. Natl. Acad. Sci. USA* 99:9568.)

arabinose operon   Consists of three juxtapositioned structural genes—*araB* (L-ribulokinase), *araA* (L-arabinose isomerase), and *araD* (L-ribulose-4-epimerase)—transcribed in this order into a polycistronic araBAD mRNA starting at the O_{BAD} operator and initiated by the P_{BAD} promoter. The repressor-activator site functions by positive or negative control, is transcribed in the opposite direction from the O_C

operator, and uses the P_C promoter. These genes are near the beginning of the *E. coli* map, while another gene, *araF*, is located at about map position 45. They form a common regulatory system: a regulon. Activation of the *ara* operon requires the presence of the substrate arabinose and that the catabolite-activating protein cyclic adenosine monophosphate complex be attached to the promoter. This operon is subject to catabolite suppression, and as long as glucose is present in the medium (even when arabinose is also available), its transcription cannot begin. *See* cAMP; catabolite-activating protein; negative control; operator; operon; polycistronic. (Schleif, R. 2000. *Trends Genet.* 16:559.)

arabinosuria Early name of pentosuria, but subsequently it turned out that L-xylulose was misidentified as arabinose; the current name of the recessive disorder is (essential) pentosuria. *See* pentosuria; xylulose.

arachidonic acid (arachidate) Unsaturated fatty acid, called arachidonate when there are four double bonds in the molecule (it is synonymous with eicosatetraenoate). It occurs in lipids and plays a role in mediating signal transduction. Cyclooxigenase mediates the formation of prostaglandins, prostacyclins, and thromboxanes, whereas lipoxygenase catalyzes the synthesis of leukotrienes from arachidonic acid. *See* atherosclerosis; cyclooxygenase; fatty acids; lipoxygenase signal transduction.

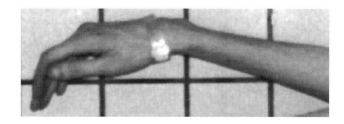

arachnodactyly (5q23-q31) Characteristic of Marfan syndrome, involving unusually long fingers and toes. Unlike Marfan syndrome (FBN1 15q21.1), here the mutation involves the fibrillin gene FBN2. *See* fibrillin; Marfan syndrome. (Belleh, S., et al. 2000. *Am. J. Med. Genet.* 92:7.)

ARAF oncogene Has been assigned to the mouse X-chromosome, whereas in humans ARAF 1 is in chromosome Xp11-p11-2 and ARAF 2 is localized to either 7p11.4-q21 or 7p12-q11.21. These oncogenes are homologous to the RAF1 oncogene and are supposed to encode a serine/threonine kinase. *See* RAF1.

arbovirus Parasite of blood-sucking insects and vertebrates. The genetic material of arboviruses is RNA.

ARC *See* DRIP.

archaea Third major group of living systems besides bacteria and eukarya. *Methanococcus jannaschii* DNA (1.66 megabase) was sequenced by 1996, and 1,738 predicted protein-coding genes have been identified. The organism has 58 kb and 16 kb extrachromosomal elements. Only 38% of its genes appear similar to genes (by nucleotide sequences) of other fully sequenced bacteria or budding yeast. The metabolic genes bear similarities to bacterial genes, whereas the genes involved in transcription, translation, and replication resemble eukaryotic genes. The complete nucleotide sequence is available at <http://www.tigr.org/tdb/mdb/mjdb.html>. *See* life form domains. (Whitman, W. B., et al. 1999. *Genetics* 152:1245; Podani, J., et al. 2001. *Nature Genet.* 29:54.)

archaebacteria Groups of prokaryotes that appear to have some similarities to eukaryotes, e.g., displaying nucleosome-like structures in their DNA, introns in their genes, unlinked 5S RNA genes, and their transcriptase enzyme is somewhat related antigenically to similar enzymes in lower eukaryotes. *See* archaea.

archaeogenetics Studies the descent of humans mainly on the basis of mitochondrial and Y-chromosomal population information.

archegonium Female sexual organ (gametangium) of lower plants where the eggs develop.

archeogenetics Application of molecular genetics techniques to ancient populations, their bone remains, or otherwise preserved biological samples and their evolving descendants. The studies are based on mitochondrial DNA, which can reveal the pattern(s) of human/animal evolution and migration of the females, and the analysis of Y-chromosomal makeup provides comparable information on the male lineages. *See* Eve, foremother of mtDNA; Y chromosome.

archeozoic Geological period 400 to 100 million years ago when protists (unicellular organisms) evolved.

archespore Ancestral, enlarged cell that develops into the megasporocyte (megaspore mother cell) in plants. *See* megagametophyte.

archezoa *See* microsporidia.

architectural editing Proteins from the endoplasmic reticulum are selectively transported. The new proteins are retained until properly folded and the misfolded chains are degraded. *See* endoplasmic reticulum.

architectural protein Modulates DNA structure in such a way that transcription factors gain better access to the promoter area. *See* high-mobility group proteins; UBF.

archival DNA Had been stored in museum, herbarium, or other preserved samples of long dead cells; can be amplified with PCR techniques for analysis to find information on old populations or extinct species. *See* ancient DNA; PCR.

archtype Hypothetical ancestral form in evolution.

arcsine Inverse of sine (sine^{-1}); stands for the angle of a given sine. *See* sine.

arcsine transformation *See* angular transformation.

ARE Anoxia response element. DNA sequences regulating responses to anaerobiosis. ARE responding genes represent detoxification and antioxidant defense. *See* anoxia. (Li, J., et al. 2002. *Physiol. Genomics* 9:137.)

ARF ADP-ribosylation factor, p14^{ARF} (human)/p19^{ARF} (mouse). GTP-binding protein of the monomeric Raf family of G proteins involved in the transport between the endoplasmic reticulum and the Golgi apparatus and within the Golgi complex. ARFs alter membrane lipid composition. ARF is required for the physiological effect of cholera and pertussis toxins. p^{ARF} regulates tumor suppressors p53 and RB through the E2F-1 transcription factor. ARF is turned on by Sec7 guanine nucleotide exchange factor domain proteins, which are inhibited by brefeldin. *See* ARNO; brefeldin; cholera toxin; E2F1; Golgi; G proteins; GTPase; guanine nucleotide exchange factor; MDM2; p16^{INK4}; p53; pertussis toxin; raf; retinoblastoma; SecA; SecB; signal transduction; translocase; translocon; tumor suppressor. (Sherr, C. J. 1998. *Genes & Development* 12:2984; Randle, D. H., et al. 2001. *Proc. Natl. Acad. Sci. USA* 98:9654.)

ARF1 **(1)** GTPase protein-activating phospholipase D. It is activated by PtdIns (phosphatidyl inositol) and participates in the recruitment of coatomer and trans-Golgi network (TGN) clathrin. *See* clathrin; coatomer; COP transport vehicles; GEF; phosphoinositols; Sec; trans-Golgi network; Ypt. **(2)** Auxin response factor modulates the action of auxin response elements (AuxRE, TGTCTC) in combination with transcription factors. *See* auxins; plant hormones.

arfaptin Adaptor mediating cross-talk between ARF and small G-proteins Rac, RHO, and RAS in signal transduction. *See* ARF; cross-talk; G-proteins; signal transduction. (Peters, P. J., et al. 2002. *Nature Cell Biol.* Feb. 18.)

ARG Oncogene related to ABL in human chromosome 1q24-q25 and in mouse chromosome 1. It encodes a tyrosine kinase, different from that of the ABL product. *See* ABL; oncogenes.

Arg Arginine.

arginase *See* argininemia.

arginine (2-amino-5-guanidinovaleric acid) Essential amino acid. *See* urea cycle.

argininemia (hyperargininemia) Accumulation of high levels or arginine in the blood and urine caused by autosomal-recessive arginase deficiency (ARG1 and ARG2 genes). ARG1 (6q23) coded enzyme represents 98% of the arginase activity in the liver and its deficiencies the common argininemia. Arginine accumulates in the blood because it is not degraded. Argininemia is a relatively rare disease. Treatment with benzoate and restriction of arginine intake may ameliorate the condition. Shope virus infection may restore arginase activity in the cells. *See* amino acid metabolism; citrullinemia; citrullinuria; urea cycle.

argininosuccinic aciduria Rare hereditary disorder (human chromosome 7cen-q11.2) involving mental retardation, seizures, hepatomegaly (enlargement of the liver that may become cancerous), intermittent ataxia, brittle and tufted hair, and accumulation of large quantities of arginosuccinic acid (an intermediate in the arginine-citrulline [urea] cycle) in the blood, urine, and cerebrospinal (brain and spinal cord) fluid. Early- and late-onset types have been distinguished. The basic defect is arginosuccinase or arginosuccinate lyase deficiency. *See* urea cycle.

arginyl tRNA synthetase Enzyme that charges the appropriate tRNA with arginine. The encoding gene is located at human chromosome 5. *See* aminoacyl tRNA synthetase.

ARGONAUTE Mediates the degradation of mRNA in respose to interfering dsRNA. This protein may also function downstream of Dicer or RNA-dependent RNA polymerase in gene silencing. *See* RNAi. (Williams, R. W. and Rubin, G. M. 2002. *Proc. Natl. Acad. Sci. USA* 99:6889; Martinez, J., et al. 2002. *Cell* 110:563.)

argos Secreted *Drosophila* protein containing a single EGF motif. It is a repressor of eye and wing determination and it acts against *Spitz*. *See* DER; EGF; *Spitz*.

argosome Epithelial membrane vesicle capable of moving cargo between cells. (Greco, V., et al. 2001. *Cell* 106:633.)

Arias syndrome Probably the same as Gilbert syndrome or Crigler-Najjar syndrome II.

arithmetic mean $\bar{x} = \Sigma x/N$ is the sum of all measurements (x) divided by the number of measurements (N). *See* mean.

arithmetic progression Series with elements increasing by the same quantity, e.g., 1, 3, 5, 7, 9. *See* geometric progression.

armadillo *Euphractus sexcinctus* $2n = 58$; *Dasypus novemcinctus* $2n = 64$; *Cabassous centralis* $2n = 62$; *Chaetophractus villosus* $2n = 60$.

Armadillo (*arm*, 1–1.2) Homozygosity of the recessive allele is lethal. The normal allele of *Drosophila* is involved in embryonic differentiation in connection with other genes. Its vertebrate homologue encodes β-catenin. It is positively regulated by *Wg* (*wingless*) and downregulated by axin. *See* axin; morphogenesis in *Drosophila*; *wnt*.

Armitage-Doll model Interprets carcinogenesis as a multistage process developing by series of subsequent mutations. *See* Knudson's two-mutation theory; Moolgavkar-Venzon model.

arm ratio Relative length of the two arms of a eukaryotic nuclear chromosome. *See* chromosome arm; chromosome morphology.

ARMS Amplification refractory mutation system. It is used with PCR and may detect the strand that contains a known mutation or identify polymorphism of a particular DNA stretch. Two sets of primers are used for amplification, and one of the primers has a difference at the site of the suspected mutation. The different nucleotide is inserted at the 3′ end of the primers and extension of the strands follows. The penultimate base frequently leads, however, to a mismatch in both mutant and wild-type primers and it may be difficult to find a primer suitable for sequence-specific amplification. *See* mutation detection; PCR; primer; primer extension. (Chiu, R. W., et al. 2001. *Clin. Chem.* 47:667; Carrera, P., et al. 2001. *Methods Mol. Biol.* 163:95.)

arms of bacteriophage λ When the stuffer segment is removed, a left and a right segment (arms) of the genome remains, and these are used for vector construction. *See* lambda phage; stuffer DNA.

aRNA Ancient RNA. *See* ancient DNA; ancient organisms.

ARNO ARF nucleotide-binding site opener is a 399-amino-acid human protein involved in the GDP⇌GTP exchange of ARF. This and similar proteins contain an amino-terminal coiled coil, a central secretory protein domain (Sec), and a C-terminal pleckstrin domain. It is a homologue of the yeast Gea1, and both are inhibited by brefeldin. *See* ARF; brefeldin; GTP; pleckstrin.

Arnold-Chiari malformation Multifactorial recessive brain anomaly. The brain stem is herniated into the foramen magnum (interconnecting the brain and the vertebral column), also resulting in hydrocephalus and anencephalus. *See* anencephaly; hydrocephalus.

ARNT Arylhydrocarbon-receptor nuclear translocator is a helix-loop-helix heterodimeric transcription factor. The AHR and other receptors mediate the metabolism of xenobiotics. *See* aryl hydrocarbon receptor; helix-loop-helix; xenobiotics.

aromatase (ARO) ~500 amino acid cytochrome 450 (CYP19) protein (estrogen synthetase) converting C19 androgen into C18 estrogen. It is encoded at human chromosome 15q21.1. It is present in the skin, muscle, fat, ovary, placental and nerve tissues. Its deficiency in females causes virility and pseudohermaphroditism and reduced fertility in male mice. Its excess in human males may cause gynecomastia. *See* estradiol; gynecomastia; pseudohermaphroditism; steroid hormones.

aromatic molecule Closed-ring molecule with C in the ring, linked by alternating single and double bonds; they are frequently conjugated with other compounds.

ARP Autonomously replicating pieces. *See* macronucleus

ARP2/3 complex Actin-related protein complex is an actin assembly complex involved in the movement of cells and mitochondria during cell division and budding of yeast. *See* actin. (Robinson, R. C., et al. 2001. *Science* 294:1679; Cooper, J. A., et al. 2001. *Cell* 107:703.)

ARPKD *See* polycystic kidney disease; renal-hepatic-pancreatic kidney disease.

arrayed library Cloned DNA sequences are arranged on two-dimensional microtiter plates where they can be readily identified by row and column specifications. *See* DNA library; microarray.

arrayed primer extension (APEX) Array of oligonucleotide primers immobilized by their 5′ end on a glass surface. The DNA is amplified by PCR, digested enzymatically, and annealed to the immobilized primers. Template-dependent DNA polymerase extends the sequence using fluorescent-labeled dideoxynucleotides. Mutation is revealed by change in the color code of the primer sites. The procedure is suited for analysis of DNA polymorphism. *See* dideoxynucleotide; PCR; primer. (Kurg, A., et al. 2000. *Genet. Test.* 4:1.)

array hybridization Designed to identify single (or a small number) of nucleotide changes in genomic DNA. The procedure requires a large array of oligonucleotide probes obtained by light-directed parallel chemical synthesis. In each oligonucleotide set, four probes are used that differ only in one of the four bases, A, T, G, C, and the flanking bases are kept identical. The complementary synthetic probes then query each sequence of the target:

```
Target:    5′ ...   TGAACTGTATCCGACAT...3′
Probes     3′          TGACATAGGCTGTA       match
                       TGACATCGGCTGTAG    ⎫
                       TGACATGGGCTGTAG    ⎬ mismatches
                       TGACATTGGCTGTAG    ⎭
```

The single-base hybridization probes are distinguished on the basis of the signals provided by the differences in the fluorescence labels and hybridization intensities. The procedure also permits the identification of more than one base and deletions. A confocal device can thus scan an entire genome, and its great merit is fastness. *See* DNA chips; light-directed parallel synthesis. (Chee, M., et al. 1996. *Science* 274:610.)

arrest, transcriptional Transcription is stopped because the supply of one or more kinds of nucleotides has run out, or protein factors may also be the cause. It usually can be restarted by the missing building block. T-rich sequences in the nontemplate DNA strand are frequently liable for the arrest. Genes such as *LexA*, the *lac* repressor, the CAAT-box-binding protein, and other binding proteins may block or impede transcription. In some instances, the RNA polymerase may either bypass or remove the binding proteins in its way. The nucleosomal structure may not interfere with transcription, although in some cases it may slow it down. The degree of interference may depend on the dissociation of the protein. Strong positive or negative supercoiling of the DNA may impede RNA elongation. Some RNA polymerases may transcribe through gaps of a few nucleotides, but the transcript will have deletions. *See* *lac* operon; *lexA*; nucleosome; pause, transcriptional; RNA polymerase; supercoiled DNA.

arrestin 45 kDa phosphoprotein that regulates the phototransduction and β_2-adrenergic pathways (by nonvisual arrestins) in animals. Arrestin is dephosphorylated when it interacts with the trimeric G-protein-coupled signal receptor.

It may serve as a deactivator of the G-protein-mediated signaling path by binding to the SH$_3$ domain of the cellular Src molecules. Arrestin may also recruit clathrin to the receptor complex, resulting in the internalization of the complex into clathrin-coated pits. In such a situation it may stimulate cross-talk with the MAP kinase pathway. *See* adaptin; adrenergic receptor; AP180; cargo receptors; clathrin; crosstalk; densensitization; endocytosis; MAP; Oguchi disease; PDZ; phototransduction; retinal dystrophy; signal transduction; Src. (Krupnick, J. G. & Benovic, J. L. 1998. *Annu. Rev. Pharmacol. Toxicol.* 38:289.)

arrhenotoky   Mechanism of sex determination. The males are haploid and the females are diploid for the sex genes (as in bees, wasps). The males develop from unfertilized eggs (a form of parthenogenesis) and display one or the other allele(s) for which the females (queens) are heterozygous. The homozygous diploid males are either sterile or lethal or destroyed by the workers in the colony. Actually about 20% of animals (mites, white flies, scale insects, thrips, rotifera, etc.) use this type of sex determination. *See* chromosomal sex determination; complementary sex determination; sex determination.

arrhythmia, cardiac   *See* LQT.

ARS   Autonomously replicating sequences are about 100-bp-long origins of replication of yeast chromosomal DNAs. The different ARS sequences share a consensus of 11 base pairs (5′-[A/T]TTTAT[A/G]TTT[A/G]-3′), and there are some additional elements around them that vary in the different chromosomes from which they were derived. ARS1 contains subdomains A, B1, which are recognized by ORC. Subdomain B2 unwinds DNA, and B3 is where ABF1 binding factor is attached. Artificial yeast plasmids must contain ARS sequences to be maintained, and they may remain stable as long as selective pressure exists for their maintenance, i.e., they carry essential genes for the survival of the yeast cell (missing from or inactive in the yeast nucleus). ARS elements also occur in organellar and other DNAs. *See* Abf; cell cycle; DUE; MCM; ORC; YAC; yeast vectors. (Marilley, M. 2000. *Mol. Gen. Genet.* 263:854.)

arsenic (As^{3+} or As^{5+})   Common contaminants of coal burning and glass manufacturing. Arsenic may be a serious environmental poison and carcinogen (although not for rodents). It may cause chromosomal deletions in rodents and humans by the generation of oxyradicals. (Basu, A., et al. 2001. *Mutation Res.* 488:171.)

ART   Assisted reproductive technology may be helpful to about 15% of couples who are infertile. Various techniques are available to help conception. *See* artificial semination; counselling, genetic; GIFT; ICSI; insemination by donor; intrafallopian transfer of gamete and zygote; intrauterine insemination; in vitro fertilization; IUGTE; micromanipulation of the oocyte; oocyte donation; preimplantation genetics (PGD); sex selection; sperm bank; surrogate mother. (Baritt, J. A., et al. 2001. *Human. Reprod.* 16:513; Trounson, A. & Gardner, D., eds., 2000. *Handbook of In Vitro Fertilization.* CRC Press, Boca Raton, FL.)

artemis   Is a single-strand-specific 5′ → 3′ exonuclease and upon activation by a protein kinase (DNA-PK$_{CS}$) it gains endonuclease function for 5′ and 3′ overhangs and hairpins. Its mutations render DNA hypersensitive to double-strand breaks and the loss of B and T lymphocytes results in severe combined immune deficiency. (See endonuclease, exonuclease, lymphocyte, severe combined immunodeficiency, Ma, Y., et al. 2002. *Cell* 108:781; Fulgesi, H., et al. 2002. *Proc. Natl. Acad. Sci. USA* 99:11501.)

arteriosclerosis   Thickening and hardening of the arterial vein walls, a common form of heart disease. *See* atherosclerosis.

arthritis   Inflammation and erosion of the joints is caused by several factors with incomplete penetrance and expressivity. Arthritis is common with familial gout, a hyperuricemia (excessive uric acid production). Rheumatoid arthritis is generally considered an autoimmune disease. It appears to be autosomal dominant, but the genetic control is not entirely clear. The erosion of aggrecan (a component of the cartilage) is mediated by aggrecanase (a metalloproteinase with thrombospondin, glycoprotein secreted by the endothelium). IL-6, IL-8, and GM-CSF promote inflammation, whereas IL-10, IL-1ra, and soluble TNF-R reduce inflammation in rheumatoid arthritis. Anti-TNF-α antibody treatment may offer some promise. In some forms the basic problem is that the lymphocytes target the cell's glucose-6-phosphate isomerase. The prevalence is about 1% in the general population. *See* ADAM; arthropathy; arthropathy-camptodactyly; *Borrelia*; connective tissue disorders; GM-CSF; IFN; IL; IL-1; IL-6; IL-8; IL-10; metalloproteinase; NF-κB; osteoarthritis; rheumatoid arthritis; TNF; TNF-R. (Ota, M., et al. 2001. *Genomics* 71:263; Feldmann, M. & Maini, R. N. 2001. *Annu. Rev. Immunol.* 19:163.)

arthrogryposis   Unclear (autosomal) genetic determination of this malformation of low recurrence affecting limb deformations, hip dislocation, scoliosis (crooked spine), frequently short stature, amyoplasia (poor muscle formation), etc. X-linked forms have also been described. *See* connective tissue disorders; limb defects.

arthroophthalmopathy   *See* Stickler syndrome.

arthropathy   Any disease that affects the joints.

arthropathy-camptodactyly (synovitis)   Based on autosomal-recessive inheritance. It involves inflammation of the joints (synovial membranes), resembling arthritis. It may have an early childhood onset. *See* connective tissue disorders.

arthropod   Invertebrate animal with segmented body, e.g., insects, spiders, crustaceans, etc.

Arthus reaction   Inflammatory immunological reaction to antigens introduced into sensitized animals. The lesion causes activation of the complement and the large number of infiltrating neutrophils release lysosomal enzymes, causing tissue destruction. *See* antibody; antigen; complement; immune response; lysosomes; neutrophil.

artichoke (*Cynara scolymus*)   Vegetable crop; $2n = 2x = 34$.

artifact   Something that was man-made or came about by human handling of the object, rather than due to entirely natural causes.

artificial chromosome   *See* BAC; human artificial chromosome; PAC; YAC.

artificial insemination   May be used to overcome the consequences of male infertility in humans or to obtain a larger number of offspring from male animals with economically desirable characters and high productivity. Generally the sperm is obtained from sperm banks, where the semen is preserved at very low temperature. The cryopreservation may protect against sexually transmitted disease. *See* AID; AIH; ART; bioethics; intrauterine insemination; sperm bank; surrogate mother.

artificial intelligence   Device (computer) with the ability to function similarly to human intelligence, i.e., capability of learning, reasoning, and self-improvement.

artificial seed (synthetic seed)   May usually be formed from somatic embryos that are enclosed by an Na-alginate (polymer of mixed mannuronic and glucuronic acids) capsule in the presence of a calcium salt ($CaNO_3$ or $CaCl_2$). Within the capsule an "artificial endosperm" of nutrients may be included. Partially dehydrated embryos may also be used as artificial seeds. Artificial seeds may be used for studying the physiology of such constructs and possibly for micropropagation of some plants. *See* micropropagation.

artificial selection   Artificial selection may alter the structure of the populations in a way similar to natural selection. When the selection is relaxed or reversed due to genetic homeostasis, selection may still be effective for various traits (e.g., bristle number in *Drosophila*, oil or protein content in plants, etc.) that are under polygenic control. *See* gain; homeostasis; selection; selection conditions; selection index; illustration above.

A rule   Adenylic acid is the preferred nucleotide for incorporation opposite to an abasic site of the DNA during repair. *See* abasic site; DNA repair. (Otterlei, M., et al. 2000. *EMBO J.* 19:5542.)

arylesterase (ESA, paraoxonase)   Encoded in human chromosome 7q22. The enzyme breaks down parathion and related insecticides. *See* cholinesterase; pseudocholinesterase.

aryl hydrocarbon receptor (ARH)   Mediates the carcinogenic and teratogenic, immunosuppressive, etc., responses to arylhydrocarbons present in many environmental toxins (dioxin, benzo(a)pyrene, cigarette smoke, polychlorinated and polybrominated biphenyls, etc.). ARH-regulated genes include cytochrome P450; uridine diphosphate-glucuronosyl transferase, growth factors, and proteins. *See* ARNT; benzo(a)pyrene; P450; polychlorinated biphcnyl; uridine diphosphate glucuronosyl transferase.

arylsulfate   Aromatic molecule with bound sulfate. Deficiency of arylsulfatases is involved in the lipidosis group of

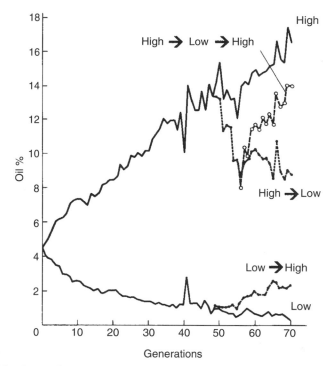

Selection and reversed selection of oil content in maize. (After Dudley, J. W. 1973. *Rep. 28th Annu. Corn Sorghum Res. Conf. Am. Seed Trade Asssoc.* Washington, DC, p. 126.)

diseases, collectively designated as metachromatic leukodystrophy. *See* Krabbe's leukodystrophy; lipidoses; metachromatic leukodystrophy.

AS   *See* asparagine synthetase.

ASAP   Cell adhesion molecule of the ARF family of proteins. ASAP1 is a regulator of protein sorting through membranes, activates GTPase, and regulates the cytoskeleton. *See* ARF; cytoskeleton; polycystic kidney disease; protein sorting.

asbestos   Mineral silicate fibers. They are carcinogenic supposedly by being phagocytized, then accumulate around the cell nucleus, where they may interfere with chromosome segregation. The mechanical irritation may contribute to mesothelial (lung) cancer. (Tweedale, G. 2002. *Nature Rev. Cancer* 2:311.)

Ascaris megalocephala (horse threadworm)   Has very unusual chromosome behavior. It has only one pair of large chromosomes in the germline, but during somatic cell divisions these large chromosomes are fragmented into a large number of small chromosomes. This organism made the discovery of reductional division in meiosis possible for Van Beneden in 1883, a cornerstone of the cytological basis of Mendelian segregation.

   A. megalocephala univalens has $2n = 2$, *A. megalocephala bivalens* $2n = 4$, and *A. lumbricoides* $2n = 43$ chromosomes. *See* chromosome breakage, programmed.

ascertainment test   Generally required in larger mammals with few offspring to determine the segregation ratios on

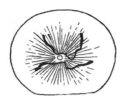

Ascaris megalopcephala univalens (After T. Boveri).

the basis of pooled data of several families. The problem involved in biased sampling (because those families where the parents are heterozygous for the recessive gene escape identification if no homozygotes are observed among the progeny) can be corrected for. The solution is mathematical. According to Mendelian expectation 3/4, of the single-child families have no affected children. Among the two-child families, $(3/4)^2 = 9/16$ is the probability that neither will be of recessive phenotype. Of the remaining 7/16 of the families, 6/7 will have 1 recessive and 1 dominant and 1/7 should have 2 recessives. Thus, the average expectation is $(1) \times (6/7) + (2) \times (1/7) = 8/7 = 1.143$. In the three-child families, 27/64 will have no affected offspring, 9/37 will have 2, and 1/37 are expected with 3 recessives. Therefore, the average expected is $(1) \times (27/37) + (2) \times (9/37) + (3) \times (1/37) = 1.294$. In the same manner, the average expectation of recessives for various sizes of families can be determined.

Number of Children →	1	2	3	4	5	6	7	8
Average Homozygotes	1.000	1.143	1.297	1.463	1.639	1.825	2.020	2.223

With this information the number of observed and expected data for affected and unaffected of different-size families can be analyzed with the chi-square procedure and the goodness of fit can be evaluated:

Number of Sibs/Family	Families	Number of Affected Sibs Observed	Number of Affected Sibs Expected	Number of Unaffected Sibs Expected	Number of Unaffected Sibs Observed
1	7	7	$7 \times 1 = 7.00$	0	0
2	10	8	$10 \times 1.143 = 11.43$	12	8.57
3	4	6	$4 \times 1.297 = 5.19$	6	6.81
5	2	4	$2 \times 1.639 = 3.28$	6	6.72
		Total 25	26.90	24	22.10

$$\chi^2 = \frac{(25 - 26.9)^2}{26.9} + \frac{(24 - 22.1)^2}{22.1} = 0.298;$$

the degree of freedom = 1, and the probability of fit is >0.5 (for χ^2 only, values below 0.05 would have some ground for doubting the fit). A simpler (and less reliable) procedure for determining the average number of recessives ($\hat{q}$):

$$\hat{q} = \frac{R - N}{T - N},$$

where R is the number of recessive segregants observed, N is the number of families showing recessives, and T is the total number of children of these families.

The *ascertainment bias* (the correction for truncated/incomplete selection of the families on the basis of probands) can also be estimated by the Bernstein formula: expected number of affected recessives $E_r = sn_s(p/1 - q^s)$, where s is the number of sibs per family, n_s is the number of families with s number of sibs, p is the segregation ratio, and $q = 1 - p$. The results of the ascertainment may not be valid for populations, which were not part of the samplings in complex cases. Using DNA sequence information, the ascertainment bias may be eliminated because penetrance or expressivity does not affect the correct molecular information. *See* chi square; expressivity; penetrance; proband; sib. (Burton, P. R., et al. 2000. *Am. J. Hum. Genet.* 67:1505; Lake, S. L., et al. 2000. *Am. J. Hum. Genet.* 67:1515; Haghighi, F. & Hodge, S. E. 2002. *Am. J. Hum. Genet.* 70:142; Epstein, M. P., et al. 2002. *Am. J. Hum. Genet.* 70:886.)

Aschheim-Zondek test (AZT)   Uses subcutaneous injection of the urine of human females into immature female mice to test for early pregnancy. Swelling, congestion, and hemorrhages of the ovaries and precocious maturation of the follicles in the mice are positive indicators of pregnancy of the tested person. Today, either hemagglutination or chorionic gonadotropin tests are used. Pregnancy immediately raises the level of this hormone dramatically. *See* gonadotropin; hemagglutinin.

ASCI   Plural of ascus. *See* ascus.

ascidian   Invertebrate (chordate) sea animal with sexual and asexual reproduction.

ascites   Condition in which abdominal fluid (may also contain cells) is excreted in response to cell proliferation in the abdominal cavity because of a neoplasia. The fluid is serum that contains polyclonal antibodies. Cirrhosis or hypoalbuminemia and experimental injections may also cause ascites. *See* albumin; cirrhosis of the liver.

ascobolus   Fungal genus; advantageously exploited for tetrad analysis. (See tetrad analysis.)

Ascobolus immersus   Ascomycete where the dissection of the ascospores is very simple; the spores spring off when touched and can be captured on microscope slides. This fungus has been extensively used for studies of recombination and gene conversion; $x = 12, 16, 18$.

ascocarp   Site in fungi where the perithecia and apothecia (fruiting bodies) develop.

ascogenous hyphae   Diploid or bikaryotic hyphae leading to the formation of fruiting bodies in fungi. *See* fruiting body; hypha.

ascogonium   Gametangium (oogonium), the female sexual organ of fungi (also called protoperithecium).

ascomycete   Large group of different fungi producing asexual conidiospores and/or ascospores within asci as a consequence of meiosis. *See* tetrad analysis.

ascorbic acid (vitamin C)   Antiscurvy (antiscorbutic) substance. It is required for proper hydroxylation of collagen, and its deficiency leads to skin lesions and damage to the blood vessels, which are symptoms of scurvy. It is a reducing compound, and upon oxidation it is converted into dehydroascorbic acid. Together with Fe(II) and O_2, ascorbic acid is a hydroxylating agent for aromatics. In the process, H_2O_2 is also formed. It has been claimed that in high daily doses it reduces the risks of the common cold and other ailments. Also, it has been found to be weakly mutagenic, probably because of its ability to generate free radicals. Primates and guinea pigs cannot synthesize this vitamin and depend on dietary supplies. The biosynthetic pathway in plants differs from that in animals, algae, and fungi. *See* vitamin C. (Lee, S. H., et al. 2001. *Science* 292:2083; Smirnoff, N., et al. 2001. *Annu. Rev. Plant Physiol. Plant Mol. Biol.* 52:437.)

ascospore   Haploid product of meiosis formed within an ascus. *See* ascus; tetrad analysis.

ASCT1   Zwitterionic amino acid transporter. *See* transporters; zwitterion.

ascus   Sac-like structure in the *Ascomycete* fungi, containing the four products of meiosis (spores). In many fungi, the number of ascospores may become eight due to a mitotic division following meiosis. The spores in the asci may be arranged in the same linear order as in the linear tetrad of meiosis (ordered tetrads such as *Neurospora*, *Ascobolus*, *Aspergillus*, etc.) or may be scrambled (unordered tetrad such as in yeast). Asci have been used very effectively to study the mechanics of recombination because the results of single meiotic events can be analyzed separately. *See* tetrad analysis.

ascus-dominant   Mutation or even a deletion affecting (preventing) the expression of the dominant allele within an ascospore. It has been attributed to reduced dosage, defects in internuclear communication, and transvection. *See* transvection.

ASE1   Anaphase spindle elongation is a gene encoding MAP required for elongation of the mitotic spindle and separation of the spindle poles. The anaphase-promoting complex (APC) degrades it. *See* cell cycle; centriole; MAP; spindle.

aseptic   Culture is free from contaminating microorganisms. *See* autoclaving; axenic; filter sterilization; pasteurization.

asexual reproduction   Does not involve fusion of gametes of opposite sex or mating type. *See* reproduction.

ASF   *See* SF2/ASF.

ASF1   Antisilencing function protein is a chaperone for newly synthesized histones, and it participates in nucleosome assembly, DNA replication, and repair. *See* chaperones; NHEJ; nucleosomes.

AS-FISH   Antisense fluorescent in situ hybridization. The sense strand of the DNA is labeled by the probe and thus it may make it possible to differentially label the transcribed and nontranscribed heterologous DNA introduced by transformation into the cell. *See* antisense strand; FISH.

ash   Mineral residue of tissues left after igniting the organic material.

ash tree   Forest and ornamental trees (*Fraxinus excelsior*, $2n = 46$; *F. americana*, $2n = 46, 92, 138$).

Ashkenazi (m)   Jewish population that lived during the Middle Ages in German lands, although migrated from there to Eastern Europe and other parts of the world. They preserved their ethnic identity and a special gene pool. Therefore, certain hereditary conditions such as Tay-Sachs disease, Gaucher disease, Niemann-Pick disease, Bloom syndrome, higher IQ, etc., occur at increased frequencies in the population compared to some other ethnic groups. *See* Jews and genetic diseases; Sephardic.

asialoglycoprotein receptor   Normally many soluble glycoproteins have sialic acid residues attached to their end. The sialic acid residues determine whether the glycoprotein is circulated in the bloodstream. If the sialic acid is lost, the glycoprotein may bind to the plasma membrane of the liver cells (hepatocytes) and become asialoglycoprotein receptors. Glycoproteins attached to these receptors are generally degraded by the lysosomes of the liver. *See* sialic acid.

Asilomar Conference   In 1975, at the beginning of the more widespread use of recombinant DNA, scientists convened at this California conference to work out voluntary guidelines for protection against the potential hazards of new technique applications. *See* containment. (Berg, P., et al. 1975. *Proc. Natl. Acad. Sci. USA* 72:1981.)

A site (decoding site)   Compartment on the ribosome. At the beginning of the translation process the first codon, Met or fMET, lands at the P site, and the next amino acid is delivered to the A site. Then the elongation of the peptide chain proceeds. The decoding site of the 16S ribosomal RNA has the universally conserved ⚲ A1492 and ⚲ A1493 nucleotides as the location is shown at below. *See* aminoacyl-tRNA synthetase; protein synthesis; ribosome. (Rodnina, M. V. & Wintermeyer, W. 2001. *Annu. Rev. Biochem.* 70:415.)

```
← UGAA      1498
GC    ⚓ GUG G
CG    ⚓ CAC C
→ UCAC      1412
```

ASK1 Apoptosis signal regulating kinase is a member of the mitogen-activated MAP protein family; it is activated by TNF-α. It induces apoptosis but may inhibit TNF-α-induced apoptosis. It stimulates JNK activation and interacts with the TRAF family, especially with TRAF2- induced JNK activation. *See* apoptosis; JNK; MAP; TNF; TRAF.

ASLV Avian sarcoma-leukosis virus. *See* retroviruses.

ASMD Anterior segment mesenchymal dysgenesis. Encoded by the dominant PTX3 gene (10q25), affecting the development of cataracts and later midbrain, tongue, incisors, chest bone (sternum), vertebrae, and limbs. *See* cataract; eye diseases; Rieger syndrome.

ASN.1 Abstract Syntax Notation describes the format in sequence databases to which all other files correspond; asn.all describes the formats of both literature and genetic sequence messages. *See* accession; Bioseq; gi.

Asn-Pro-X-Tyr Amino acid sequence responsible for the internalization of low-density lipoproteins (LDL) of the membranes. *See* LDL.

ASO Allele-specific oligonucleotide probe. Screening can be carried out by semiautomated procedures. *See* allele-specific probe.

AS ODN Antisense oligodeoxynucleotide may bind oncogene mRNA, may inhibit cancer growth, may regulate the formation of megakaryocytes, and may be used in gene therapy. *See* antisense technologies; cancer gene therapy; cytofectin; gene therapy; megakaryocytes.

ASP analysis Used to estimate linkage in cases where a particular trait is under polygenic control. The cosegregation of multiple markers is followed in individuals who express the particular trait and which of the markers are most consistently present in the individuals displaying the trait of primary interest is determined. The analysis still requires multiple segregating families. The obtained information is evaluated by the MAPMAKER/SIBS computer program. *See* interval analysis; MAPMAKER; QTL.

asparagine (α-aminosuccinamic acid) NH$_2$COCH$_2$CH (NH$_2$)COOH; its RNA codons are AAU, AAC.

asparagine synthetase (AS) Of bacteria, uses ammonia as an amide donor rather than glutamine as the mammalian enzyme. Cells expressing the bacterial AS will grow on asparagine-free medium if the glutamine analog, albizzin, is present. In AS-transfected mammalian cells, the gene can be amplified in the presence of β-aspartyl hydroxamate, an analog of aspartate, and thus AS can be used as a

dominant amplifiable marker in mammalian cell cultures. The mammalian genes are present in human chromosomes 7q21-q31, 8pter-q21, and 21pter-q22. The AS genes do not have TATA and CAAT boxes in the promoter. They are homologous to the hamster *ts11* gene required for passing the cell cycle through the G1 stage. *See* amino acid metabolism; CAAT box; cell cycle; housekeeping genes; TATA box.

asparaginyl tRNA synthetase (ASNRS) Charges the appropriate tRNA with asparagine In human cells it has been located in chromosome 18. *See* aminoacyl tRNA synthetase.

Asparagus officinalis (a dioecious monocot, $2n = 20$) Sex determination by XX pistillate and XY staminate plants. By anther culture, YY plants can be obtained that can be vegetatively propagated, or by pollination they produce exclusively male progeny. The male plants are of special economic value because their yield/area of the edible spears is substantially higher. Almost half of the human populations excrete methanethiol in their urine after eating this vegetable. The excretion trait appears to be autosomal dominant. The ability to smell this particular odor may also be under dominant control. *See* olfactory genetics; YY asparagus.

aspartame (Nutra-Sweet) N-L-α-aspartyl-L-phenylala-nine-1-methyl ester, an artificial low-calorie food and beverage sweetener; about 160 times as sweet as sucrose. Not recommended for phenylketonurics because it contains phenylalanine. *See* fructose; phenylketonuria; saccharine.

aspartate aminotransferase (glutamate oxaloacetate transaminase, GOT1, GOT2) One of the functional forms of this enzyme, GOT1, is encoded in human chromosome 10q24.1-q25.1 and is expressed in the cytosol. A homolog GOT2 is encoded in human chromosome 16q12-q21 and is expressed in the mitochondria. Pseudogenes of the latter were located at 12p13.2-p13.1, 1p33-p32, and 1q25-q31. In liver, the mitochondrial enzyme is largely present, whereas in the serum mainly the cytosolic enzyme is present. *See* amino acid metabolism; asparagine synthetase.

aspartate phosphatase *See* two-component regulatory systems.

aspartic acid (HOOCCH$_2$CH[NH$_2$]COOH) *See* amino acids; ancient DNA; aspartate aminotransferase.

aspartic acid racemization *See* ancient DNA.

aspartoacylase deficiency (aminoacylase-2 deficiency, Canavan disease, ACY2) This enzyme cleaves acylated amino L-acids into an acyl and amino acid group; amino-acylase-1 (ACY-1) similarly cleaves all acylated L-amino acids except L-aspartate. The autosomal-recessive disorder has an early or late onset, resulting in debilitating muscle and eye defects, mental retardation, and spongy degeneration of the white matter of the brain. There may be a 200-fold increase of N-acetylaspartic acid in the urine. Its incidence is increased among Jews of Ashkenazi extraction and in Saudi Arabian populations. Chromosomal location is 17pter-p13. *See* amino acid metabolism; eye diseases; Jews and genetic diseases; mental retardation; neuromuscular defects.

aspartylglucosaminuria (AGA) Chromosome 4−recessive defect of the enzyme aspartylglucoseaminidase (4q32-q33) may eliminate an important S−S bridge of the protein, resulting in neurological-mental and other defects. Its frequency is higher ($\sim 4 \times 10^{-5}$) in populations of Finnish descent. *See* amino acid metabolism; disulfide bridge; sialidosis. (Saarela, J., et al. 2001. *Hum. Mol. Genet.* 10:983.)

AS-PCR Allele-specific PCR. *See* polymerase chain reaction.

ASPD Artificially Selected Proteins/Peptides Database. *See* phage display.

Aspergillus Ascomycetes; *Aspergillus nidulans* ($n = 8$, 2×10^7 bp) is a favorite organism for studies of recombination. One meiotic map unit is about 5−10 kbp. It has been extensively used for mitotic recombination. Asexual reproduction is by conidiospores (3–3.5 µm). This is a homothallic fungus and thus does not have different mating types. In the cleistothecium there are up to 10,000 binucleate ascospores in eight-cell linearly ordered asci. Transformation systems are available. It yields about 5×10^3 transformants/µg DNA. *Aspergillus flavus* is responsible for the production of aflatoxin, an extremely poisonous toxin developing on infected plant residues, seeds, etc. *See* aflatoxins; cleistothecium; conidia; mitotic recombination; recombination; tetrad analysis. (<http://aspergillus-genomics.org/>.)

aspermia Lack of ejaculating ability of the male.

aspirin (salicylic acid acetate) Analgesic, antifever, antiinflammatory drug. It inhibits cyclooxygenases, IKK, and JNK. *See* cyclooxygenases; host-pathogen relations; IKK; JNK; salicylic acid. (Kurumbail, R. G., et al. 1996. *Nature* 384:644.)

Aspirin.

asplenia One form (Ivemark syndrome) is usually sporadic or autosomal recessive, and it is associated with absence or enlargement of the spleen or multiple accessory spleens and cardiac or other organ malformations. Another form of asplenia most conspicuously involves cystic liver, kidney, and pancreas. *See* spleen.

assay Test for mutagenic effectiveness or efficiency, the velocity of a chemical reaction catalyzed by enzymes, or the function of any biological process.

assembly initiation complex Minimal elements required for the completion of the assembly of the viral components. *See* bacteriophages.

assignment test *See* somatic cell hybrids.

assimilation Conversion of nutrients into the cell constituents. Also, blending of an initially different ethnic (cultural) group into the general population. *See* genetic assimilation.

association Joint occurrence of pathological symptoms that do not have an expected common functional basis. *See* syndrome.

association constant (K_a) Association between components of a complex. The larger the K_a, the stronger the association.

association mapping Identifies chromosomal regions containing disease susceptibility or other genes on the basis of their association (linkage) with other markers. The association may not necessarily indicate linkage because selective forces may bias the observations in small populations. Also, recent migration or other admixture may create bias. A transmission disequilibrium test may provide a remedy for spurious association. *See* linkage disequilibrium; transmission disequilibrium test. (Sham, P. C. 2000. *Am. J. Hum. Genet.* 66:1616.)

association phase Coupling phase in linkage, a term used in fungal genetics. *See* coupling phase; crossing over; linkage; repulsion.

association site Periodically distributed, microscopically detectable multiple interstitial association points are also called nodules. The distance between the paired chromosomes is about 0.4 µm. *See* meiosis; recombinational nodule; synaptonemal complex; zygotene stage.

association test Basically a 2×2 contingency chi-square test based on a panel: where a, b, c, d represent the number of observations $(++)$, $(-+)$, $(+-)$, and $(--)$, respectively; n = the total number of observations. If $b = c = 0$, there is no association. The significance of the association is tested:

$$\chi^2 = \frac{n(|ad - bc| - 0.5)^2}{(a+c)(a+b)(b+d)(c+d)},$$

and the probability of a greater chi square can be determined by a χ^2 table or χ^2 chart for 1 degree of freedom. The association test is most useful within a homogeneous population. A particular association may not be an indication of genetic linkage or a physiological or cause−effect relationship but may provide useful information on the relationship between two diseases or whether the reciprocal crosses would be identical. *See* chi square. (Lange, C. & Laird, N. M. 2002. *Am. J. Hum. Genet.* 71:575.)

		First variable	
		+	−
Second variable	+	a	b
	−	c	d

assortative mating Mates are chosen on the basis of preference or avoidance (positive or negative assortative mating) rather than at random, e.g., tall people frequently choose tall spouses; educated, higher economic or social status individuals most commonly marry within their group. Traits unknown to the majority, such as blood groups, usually do not come into consideration in mate selection. Assortative mating may contribute only slightly to the average coefficient of inbreeding (f) in human populations:

$$\bar{f} = \frac{r}{2n_e(1-r)+r},$$

where r = correlation coefficient, n_e = an equivalent number of genes

$$\left(n_e = \frac{\sum_{ij} \sigma_i \sigma_j}{\sum_i \sigma_i^2} \right).$$

Assortative mating may have some effect on the expression of a quantitative trait, and the heritability becomes $h^2 = \hat{h}^2(1 - [1 - \hat{h}^2]A)$, where A is the product of the average heritability and the phenotypic correlation, i.e., $r\hat{h}^2$. *See* controlled mating; correlation; inbreeding; mating system. (Rice, T. K. & Borecki, I. B. 2001. *Adv. Genet.* 42:35.)

astacin Zinc metalloprotease. *See* bone morphogenetic protein.

aster *See* centrioles.

asthenozoospermia Less than 25–50% of the spermatozoa show forward motility. It appears that reduced OXPHOS activity in the mitochondria affects motility. Defects in dynein axonemal heavy chain (DNAH1, 3p21.3) may also have an effect. *See* dynein; OXPHOS. (Ruiz-Pesini, E., et al. 2000. *Am. J. Hum. Genet.* 67:682; Neesen, J., et al. 2001. *Hum. Mol. Genet.* 10:1117.)

asthma Respiratory disease with multiple causes affecting ~155 million people worldwide. Nasal polyps or an elevated level of immunoglobulin A (IgA) or IgE may cause some autosomal-recessive forms. Key players in the development of asthma are interleukin-13 (IL-13) and IL-10 because they guide immature T cells into the development of T_H2 lymphocytes. IL-4 controls the development of B cells (which produce IgE), and IL-5, IL-3, and GM-CSF also play a role through eosinophils, which are required for allergic inflammation. Mast cells and basophils affect the production of histamines; cytokines and chemokines control acute symptoms of asthma. The mast cells respond to IgE and the allergens. The interleukin gene cluster is located in human chromosome 5q31-q33. Genes in human chromosomes 1p32, 2q, 5q31, 6p21, 7, 8p23, 11q21, 12q12, 13q, 14q, 15q13, and perhaps, other sites appear to be associated with the expression of asthma. In different populations, different loci may play a major role. The α-chain of IL-4 receptor binds IL-13 to T_H2. Susceptibility to asthma is controlled by a few genes, and the heritability has been estimated as ~75%. Asthma—as well as some other anomalies of the immune system—also has a maternal effect. The risk of maternal transmission seems to be four-fold. This may be caused either by allelic exclusion, imprinting, placental transfer, or breast feeding. Indeed the IgE receptor (FCεRI-β, IL-5) has been mapped to a chromosome (11q13) that commonly affects imprinting. The metalloprotease ADAM33 (20p13) seem to be an important regulator of the disease (Van Eerdewegh, P., et al. 2002. *Nature* 418:426). Glucocorticoids are most commonly used for medication. *See* ADAM; allelic exclusion; allergy; atopy; hypersensitive reaction; IL-3; IL-4; IL-5; IL-10; IL-13; immunoglobulins; imprinting; platelet-activating factor; polyp; protease inhibitor; T cells, γδ T cell. (Xu, J., et al. 2001. *Am. J. Hum. Genet.* 68:1437; Niimi, T., et al. 2002. *Am. J. Hum. Genet.* 70:718; Umatsu, D. T., et al. 2002. *Nature Immunol.* 3:715.) <http://cooke.gsf.de>)

astral Compendium to protein structures. *See* protein structure; SCOP; structural classification of proteins. (<http://astral.stanford.edu>.)

astrocyte Type of branching cell that supports the nervous system. *See* glial cells.

ASV Avian sarcoma virus of birds is an oncogenic RNA virus that can induce sarcoma in rodents. *See* sarcoma.

asymbiotic nitrogen fixation Proceeds by a microorganism without dependence on cohabitation with other organisms such as by the members of the soil bacterial species *Azotobacter* and *Clostridium*. *See* nitrogen fixation; symbiosis.

asymmetric carbon Atom has four different covalent attachments. *See* covalent bond.

asymmetric cell division Requisite for embryonal differentiation. These divisions specify the dorsoventral and anterior-posterior polarities of the body pattern. Several protein factors specify the process. The orientation of the spindle in *Drosophila* involves the localization of the Numb and Prospero proteins in the basal cells, and the polarity instructions may come from the product of the *inscruteable* (*insc*), *partner of inscruteable* (*pins*), and other loci. Yeast (Ash1p) and *Caenorhabditis* (SKN-1) also have controls similar to Numb and Prospero; *she* and *par* genes, respectively, are analogous to *inscruteable*. In *Drosophila* epithelium, the adherens junctions inhibit asymmetric divisions. Cdc2 appears to link the asymmetric division machinery and the cell cycle. *See* adherens junction; axis of asymmetry; Cdc2; morphogenesis in *Drosophila*; spindle. (Grill, S. W., et al. 2001. *Nature*

409:630; Knoblich, J. A. 2001. *Nature Rev. Mol. Cell Biol.* 2:11; Adler, P. N. & Taylor, J. 2001. *Curr. Biol.* 11:R233; Knust, E. 2001. *Cell* 107:125.)

asymmetric heteroduplex DNA   *See* Meselson-Radding model of recombination.

asymmetric hybrid   Lost some of the chromosomes of one or the other parent. *See* somatic hybrids.

asymmetric replication   At the replication fork, DNA synthesis on the leading and lagging strands proceed in opposite directions, relative to the base of the fork. *See* replication; replication fork.

asynapsis   Failure of chromosome pairing. *See* desynapsis; synapsis.

atabrine   Preparation of quinacrine, an antimalaria and antihelminthic (intestinal tapeworm) drug. The quinacrine mustard (ICR-100) is a radiomimetic mutagen. *See* quinacrines; radiomimetic.

at least one hypothesis   Every NK cell in an individual expresses at least one inhibitory receptor molecule specific for a self-MHC class I molecule. Accordingly, self-tolerance is increased, since many NK cells are capable to destroying any autologous cells that have downregulated MHC class I molecules. *See* killer cells; MHC; self-tolerance. (Valiante, N. M., et al. 1997. *Immunity* 7:739.)

ATase   *See* Utase.

atavism   Recurrence of expression of traits of ancestors beyond great-grandparents. It is based on recessive, complementary recessive, or recombination of genes or special environmental conditions. For a period of time in the 20th century, it was no longer used in the genetic literature. Atavism may, however, have real basis in the genetic material and may present in an altered form of ancient genetic sequences expressed in an "atavistic" manner if appropriately activated by a developmental program shift. Under such circumstances, occasionally hind limb bones may develop from the rudimentary limb buds of whales. Hypertrichosis in humans, encoded in chromosome Xq24-q27.1, also represents this atavistic reprogramming. Basically these atavistic changes may not be much different from expression of homeotic genes. (*Atavus* in Latin means great-great-grandfather.) *See* homeotic genes; hypertrichosis. (Verhulst, J. 1996. *Acta Biotheor.* 44:59.)

ataxia telangiectasia (AT)   One of about a dozen human ailments involves ataxias: poor coordination of the muscles because of dilations in the brain blood vessels, reduced immunity, elevated level of α-fetoprotein, etc. Its appearance in human diseases is attributed to instability and breakage of chromosomes 14, 7, 2, 11, and 12, although the major locus (150 kb genomic DNA and 66 exons transcribed into 13 kb RNA) appears to be at chromosome 11q22-q23. Leukemias and other malignancies are very common among the patients. Cultured cells of the affected individuals are highly sensitive to both X-ray and UV damage. Also,

standard radiation therapy for malignant tumors may be fatal to these individuals. The basic defect in AT is either in a DNA-dependent phosphatidylinositol protein kinase (M_r 350K) that controls progression of the cell cycle (p53) or in DNA repair and recombination (its homologues are MEC1, SAD3, ESR1). Alternatively, it has been found that inositol 1,4,5-trisphosphate receptor (IP_3R1)–deficient mouse mutants either die in utero or display severe ataxia when born and die very shortly afterward. It appears now that the mutant AT protein (ATM, 11q22.3) interacts with c-Abl oncogene, resulting in radiation sensitivity and the arrest of the cell cycle at the G1 phase. An SH3 domain of c-ABL interacts with a DPAPNPPHFP amino acid sequence in ATM. As a consequence of radiation, the tyrosine kinase activity of c-Abl is reduced in the ATM cells. Homozygosity for this recessive human gene has a frequency of about 5×10^{-5}, and the frequency of the carriers, prone to breast cancer and other malignancies, is about 1%. The *spinocerebellar ataxia* (SCA5) of human chromosome 11 is caused by an instability of the CAG trinucleotide repeats. SCA6 involves defects in the α-subunits of Ca^{2+} ion channel. SCA1 is in chromosome 6p23.5-p24.2 and has CAG instability, resulting in polyglutamine protein misfolding. SCA2 maps to 12q23-q24.1, SCA3 in 14q24.3-qter, SCA4 in 16q. The *autosomal-dominant cerebellar ataxia* type III (SCA11) maps to 15q14-21.3 region. SCA10 is in human chromosome 22. The autosomal-dominant cerebellar ataxia (ADCA type II) with pigmentary muscular dystrophy is coded in human chromosome 3p12-p21.1, and ADCA-like recessive gene is at 9q14. *Episodic ataxia* is associated with defects in the potassium ion channel or in the α-subunits of Ca^{2+} channel functions. *See* abl; AVED; breast cancer; cancer; carcinogenesis; cell cycle; DNA repair; DNA replication in eukaryotes; excision repair; Friedreich ataxia; β-galactosidase; gangliosidoses; Hartnup disease; ion channels; light sensitivity diseases; Mantle cell lymphoma; metachromatic leukodystrophy; Mre11; myotonia; neurofibromatosis; neuromuscular diseases; Niemann-Pick disease; Nijmegen breakage syndrome; olivopontocerebellar atrophies; p53; phosphoinositides; RAD3; Refsum diseases; SH3; telangiectasia; trinucleotide repeats; Usher syndrome.

ataxin   Protein responsible for SCA1 ataxia associates with a cerebellar leucine-rich acidic protein and alters the nuclear matrix. *See* ataxia; spinocerebellar ataxia.

ATCC   American Type Culture Collection maintains cell cultures of prokaryotes and lower and higher eukaryotes.

ateles (spider monkey)   *See* cebidae.

ATF2   Activating transcription factor. A family of proteins containing homologous basic/leucine zipper (bZIP)–binding domains; it is regulated by the JNK signal transduction pathway. Mutations in ATF2 interfere with the retinoblastoma and E1A oncogene's transcription-suppressing activities. *See* adenovirus (E1A); bZIP; JNK; retinoblastoma. (Bhoumik, A., et al. 2002. *J. Clin. Invest.* 110:643.)

athanogene   Generates antiapoptotic function. *See* apoptosis; BAG1.

atherosclerosis   Hardening, then degeneration of the walls of arteries by the deposition of fatty acid nodules on the inner wall and obstruction of blood circulation. In the first phase, lipid-filled foam cells (macrophages) appear. Next, fibrous plaques of lipids and necrotic cells are formed, covered by smooth muscle cells and collagen. The final-phase lesion involves platelet and fibrous clots (thrombus). This group of vascular diseases is one of the most common causes of heart disease. Coronary heart disease causes 490,000 deaths per year in the United States, and stroke causes 150,000 deaths. The underlying genetic mechanisms vary, and nongenetic factors are substantial. In human atherosclerotic lesions, CD40 and its ligand, CD40L, are expressed. By blocking these latter signaling molecules, atherosclerosis and some autoimmune symptoms may be reduced. Atherosclerosis develops with a high level of the enzyme ACAT (acylcoenzyme A : cholesterol acyltransferase). Apolipoprotein E (APOE) deficiency in mouse atherosclerosis, caused by oxidation of arachidonic acid, can be reduced by oral administration of vitamin E. Monocyte chemoattractant protein (MCP-1), a chemokine, low-density lipoprotein (LDL) deficiency, and substantially reduced lipid deposition in the arteries are characteristics. MCP-1 apparently recruits monocytes to the arterial epithelium during the earliest stages of the disease. *See* APRF; arachidonic acid; cardiovascular diseases; CD40; CETPl; HDL; heart disease; LDL; MCP; monocytes; sterol; vitamin E. (Lusis, A. J. 2000. *Nature* 407:233; Welch, C. L., et al. 2001. *Proc. Natl. Acad. Sci. USA* 98:7946; Glass, C. K. & Witztum, J. L. 2001. *Cell* 104:503.)

AT hook   DNA minor groove-binding peptide common in chromatin-associated proteins. The AT hooks generally contain a conserved GRP (glycine-arginine-proline) core surrounded by basic amino acids. They assist other proteins in binding to DNA.

at least one hypothesis   Every NK cell in an individual expresses at least one inhibitory receptor molecule specific for a self-MHC class I molecule. Accordingly, self-tolerance is increased, since many NK cells are capable to destroying any autologous cells that have downregulated MHC class I molecules. *See* killer cells; MHC; self-tolerance. (Valiante, N. M., et al. 1997. *Immunity* 7:739.)

ATLAS™ human cDNA   Contains commercially available arrays of cDNAs (Clontech, Palo Alto, CA) on membranes in several quadrants, each specific for 96 genes of different specificity of expression. The membranes can be used for hybridization probes for identification of genes with unknown function in different tissues or healthy and diseased conditions. (See microarray hybridization, Sehgal, A., et al. 1998. *J. Surgical Oncol.* 67:234.)

ATM   Ataxia telangiectasia mutated, 370 kDa. ATM involves an altered phosphatidylinositol kinase. ATM kinase may activate p53 in response to radiation stress, but if ATM is defective, p53 is not responding to, e.g., ionizing radiation, and in the absence of apoptosis the chance for cancer may increase. ATM or loss of AT increases the chance of oxidative damage to the cell. ATM is homologous to *MEC1* and *rad53* of yeast and *mei-41* of *Drosophila. See* apoptosis; ataxia

telangiectasia; breast cancer; Chk2; double-strand break; p53; PIK; telangiectasia; X-ray repair. (Pincheira, J., et al. 2001. *Mutagenesis* 16:419.)

atomic force microscope (AFM)   Instrument that can image the surfaces of conductor and nonconductor molecules even in aqueous media. It can reveal molecular structure of surfaces, adhesion forces between ligands and receptors, and other biological processes in real time. It can also be adopted for DNA sequencing. *See* nanotechnology; STM. (Ljubchenko, Y. L., et al. 1995. *Scanning Microsc.* 9:705; Lyubchenko, Y. L., et al. 2001. *Methods Mol. Biol.* 148:569; Müller, D. J., et al. 2002. *Progr. Biophys. Mol. Biol.* 79:11.)

atomic radiation   Killed 100,000 and injured 60,000 in Hiroshima and Nagasaki at the end of World War II; caused substantial increase (about four-fold or more at the epicenter) of cancer but showed no significant increase in human mutation. The incidence of cancer displayed variations by distance from the epicenter, age, sex (higher in females), by the type of cancers, and by some unexplained factors (such as geographic location, Hiroshima or Nagasaki).

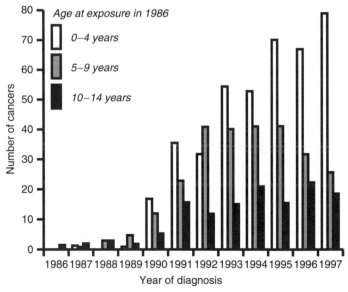

The incidence of thyroid cancer in Belarus following the Chernobyl accident in 1986. (From *Sources and Effects of Ionizing Radiation.* United Nations Scientific Committee Report 2000. Vol. II. New York.

The cause of the scarcity of mutations is not that these radiations were genetically ineffective; rather the human breeding system, avoiding marriage between relatives, did not favor homozygosity of the recessive mutations, i.e., lethality. Studies of the populations exposed to the radiation caused by the failure of the Chernobyl power plant (26 Apr. 1986) indicate not only an increase in cancer but also of mutation. Most likely, some of the mutations induced will be maintained in the exposed populations and may contribute to an increase of the genetic load. The total radiation from natural sources (cosmic radiation, disintegration of terrestrial isotopes [uranium, thorium, potassium], etc.) reaching the human gonads was estimated to be 100–125 millirads per year. The atomic bomb tests conducted during 1956 to 1965 contributed an average of about 76 millirads and were expected to have exerted their effect mainly up to the year 2000 through the short half-life radioactive elements (Cesium[137], Strontium[90]) and to have substantially decayed by then. The long half-life Carbon[14] will continue to pollute by an additional estimated 167 millirads even after year 2000. The meltdown of the atomic power plant in the Ukraine near the Byelorussia border in the spring of 1986 exposed nearby populations to up to 75 rem, whereas the exposure of the entire Byelorussia area was about 3.3 rem.

In 1986 in Byelorussia, the total number of thyroid cancer cases in children was 2, and by 1992 it reached about 60 cases; by 1999, more than 800 children who drank milk from exposed cows developed thyroid cancer, and the figures are still increasing. The thyroid cancer has been attributed to iodine[131] released during the fallout. In the human minisatellite DNA, the mutation rate doubled, and in the feral populations of voles (*Microtus*) the base-pair substitution frequency in the mitochondrial DNA was found to be in excess of 10^{-4} — an increase of more than two orders of magnitude above the appropriate control groups. Nevertheless, the rodent populations appeared in good condition and their fertility was also good. This 1996 report about the high mutation rate at Chernobyl was retracted by the authors in 1997. (*Nature* 390:100). In barn swallows an increase of mutations at two microsatellite loci was observed (*Nature* 389:593 [1997]). The Hanford Nuclear Reservation in the state of Washington exposed nearby populations in excess of 33 rads over a period of 3 years. The official estimates place 0.025 rads per year as safe for airborne pollution by nuclear weapon plants for the civilians living in the neighboring area and 5 rad for the entire body per year for the workers in those plants. According to some estimates based on irradiation of mice, 20–40 rad is the doubling dose of mutation for ionizing radiation. It was estimated that the radioactive fallout from weapon testing may have increased the genetic risks by 2% over the natural background effects and by 8% for leukemia. The effects of atomic radiation on mutation rates in the minisatellite DNA remain controversial because of the difficulties of finding appropriate (concurrent) controls. The mutation rate in these very sensitive DNA sequences is much affected by environmental factors (pollution), age, etc.

When considering the harmful effects of radiation potentially released by atomic power plants, one must consider the harmful pollution generated by the coal-fired industry and the carcinogenic hydrocarbons released by the combustion in wood fireplaces, etc. Also, the shortage of energy may directly or indirectly cause substantial sufferings and even death to the genetically more vulnerable part of the populations, especially to children.

Estimation of the risk is very complicated because of the many modifying factors involved (angle of the radiation, age, sex, length of exposure, genetic susceptibility to cancer, lifestyle [smoking, drug use, etc.]). One simple formula for assessing the excess relative risk (ERR) is $1 + \beta z$, where $\beta = $ ERR and $z = $ radiation dose. Some variations of the following formula based on least squares regression models have also been developed for the estimation of excess risk: $(\text{Cases.PYR})_d = \alpha + \beta d + \varepsilon$, where PYR is the dose-specific person years, α denotes the intercept of the regression term, β stands for the contribution of the doses of the radiation as an excess risk, and ε is the error (formula after D. A. Pierce). *See* control; correlation; cosmic radiation; doubling dose; isotopes; mutation detection; mutation in human populations; nuclear reactors; plutonium; public opinion; rad; radiation hazard assessment; rem. (Dubrova, Y. E., et al. 2002. *Science* 295:1037; Williams, D. 2002. *Nature Rev. Cancer.* 2:543; Dubrova, Y. E., et al. 2002. *Am. J. Hum. Genet.* 71:801.)

atom microscopy Being developed for imaging atomic structures. The equipment is using monoenergetic sodium atoms ejected into a vacuum chamber and carried by noble gases such as argon. The beam is broken up into subcomponents on a silicon nitride grid. The phase shift generated by two beams is then measured.

atopy Familial allergy, including asthma, hay fever, and eczema. The blood serum carries an increased level of immunoglobulin E (IgE). An IgE responsiveness locus was assigned to 11q12-q13. Human chromosome 5q31-q33 harbors an asthma susceptibility region. *See* allergy; asthma; immunoglobulins; Netherton syndrome. (Wheatley, A. P., et al. 2002. *Hum. Mol. Genet.* 11:2143.)

ATP Adenosine-5′-triphosphate is a universal carrier of metabolic energy. It transfers the terminal phosphate to various acceptors and results in ADP (adenosine diphosphate) that is recycled to ATP by either the chemical energy of oxidative phosphorylation or the solar energy of photosynthesis. Besides the thermodynamic role, ATP also has catalytic activity, e.g., in nitrogen fixation. ATP provides binding energy through noncovalent interactions with various molecules in order to lower activation energy. It provides energy for charging tRNA with amino acids, for DNA synthesis, and for bioluminescence mediated by the firefly luciferase. It is indispensable in carbohydrate metabolism; it serves as a precursor of cyclic AMP, which has a major role in signal transduction and protein phosphorylation, etc. The major catabolic pathways (glycolysis, citric acid cycle, fatty acid and amino acid oxidation, and oxidative phosphorylation) are coordinately regulated in the production of ATP. The relative abundance of ATP and ADP controls electron transfers in the cell. ATP is generated in the mitochondria and chloroplasts. ATP is the major link between anabolic and catabolic reactions mediated by enzymes. UTP (uridine triphosphate), GTP (guanosine triphosphate), and CTP (cytidine triphosphate) are also important in similar processes but have relatively minor roles compared to ATP. *See* ATPase; ATP synthase; cAMP. (Pfeiffer, T., et al. 2001. *Science* 292:504.)

ATPase Enzymes are required for active transport of chemicals and other functions in the cells. The *P-type* ATPases maintain low Na^+, low Ca^{2+}, and high K^+ levels inside the cells; generate low pH within cellular compartments; activate proteases and other hydrolytic enzymes of eukaryotes; and generate transmembrane electric potentials. The *V-type* (vacuolar) ATPases secure low pH inside lysosomes and vacuoles of eukaryotes. The *F-type* ATPases (energy coupling factors, F_1-F_0-ATPase), located in the plasma of prokaryotes and in the mitochondrial and thylakoid membranes of eukaryotes, are actually ATP synthase enzymes generating ATP from ADP and inorganic phosphate. The *DNA-dependent* ATPases are type I restriction endonucleases that depend on Mg^{2+}, ATP, and SAM for cutting of DNA strands. After cleavage they function only as ATPases. *See* ATP; SAM. (Palmgren, M. G. 2001. *Annu. Rev. Plant Physiol. Plant Mol. Biol.* 52:817.)

ATP synthase ~500 kDa multisubunit protein complex forming ATP from ADP and phosphate (oxidative phosphorylation) on plasma membranes (bacterial, mitochondrial, chloroplast). It is a motor protein. *See* ATP; ATPase. (*Annu. Rev. Biochem.* 66:717; Yoshida, M., et al. 2001. *Nature Rev. Mol. Cell Biol.* 2:669.)

A-tract Includes four or more AT base pairs in the DNA without a 5′-TA- 3′ step. Such elements cause curvature a the helix axis and influence nucleosome packaging and base pair opening due to the C^5 methyl of thymine. Such structures modulate sequence-specific ligand binding and gene expression. (See Wärmländer, S., et al. 2002. *J. Biol. Chem.* 277:28491.)

ATR ATM-related protein is a phosphatidylinositol kinase related to ATM and yeast gene product RAD3. It controls cell cycle checkpoints. *See* ATM; checkpoint; *RAD3*. (Cortez, D., et al. 2001. *Science* 294:1713.)

atransferrinemia (3q21) Defect in the synthesis of the iron regulatory protein transferrin, resulting in hypochromic anemia. *See* anemia; transferrin.

atrazine (Lasso) *See* herbicides; photogenes.

atresia (1) Closure of an organ (e.g., vagina, which can be surgically corrected to permit procreation), parts of the digestive tract (pyloric atresia), etc. *See* pyloric stenosis. (2) Mediates the elimination — by apoptosis — of oocytes with mutant mitochondria. Although the primordial germ cells produce millions of oocytes in humans, only a small fraction of them reach the ovulation stage. Thus, atresia serves as a genetic quality control. In the male germ cells (which do not transmit mitochondria), atresia was not observed. *See* apoptosis; mtDNA. (Krakauer, D. C. & Mira, A. 1999. *Nature* 400:125.)

atrial septal defect Autosomal-recessive-type developmental heart disease displays increased recurrence when transmitted through the males, although the prevalence is greater in the females. The dominant form encoding a transcription factor is in human chromosome 6. Dominant defects in the NXX2-5 gene (encoded at 5q35) affect cardiac septation and are responsible for congenital heart disease. Gene TBX5 (12q24) is responsible for ventricular septal defects. *See* heart diseases.

AT-rich DNA Common in the repetitive sequences, and it is generally not transcribed. Some of the petite colony mutants of yeast mitochondrial DNA contain mainly AT sequences. *See* mitochondrial genetics; mtDNA.

at-risk-motif (ARM) Increases instability of the genome such as inverted repeats, palindromes, or insertion elements either by illegitimate or homologous recombination or rearrangements. *See* Alu family; instability, genetic; palindrome RecA-independent recombination; repeat, inverted. (Gordenin, D. A. & Resnik, M. A. 1998. *Mutat. Res.* 400:45.)

atrium Entrance to an organ.

Atropa belladonna Plant of *the Solanaceae* family ($n = 50$, 72), a source of alkaloids. *See* henbane.

atrophine-1 Protein is encoded by the human gene DRPLA and affects other trinucleotide repeat genes. *See* dentatorubral-pallidoluysian atrophy; Huntington's chorea.

atrophy Under- or lack of nutrition, wasting away of cells and tissues. *See* dystrophy; Kennedy disease; Kugelberg-Welander syndrome; muscular dystrophy; neuromuscular diseases; spinal muscular atrophy.

atropine Highly toxic alkaloid. *See* henbane.

ATRX Helicase protein encoded at human chromosome Xq13 (>220 kb). It has similarity to the RAD54 and SWI/SNF proteins. It is implicated in psychomotor functions, DNA methylation, regulation of transcription, DNA repair, and chromosome segregation. Mutations have been found in cases of thalassemia/mental retardation and in Juberg-Marsidi syndrome. *See* helicase; SWI/SNF; thalassemia; Juberg-Marsidi syndrome.

attached X chromosomes Two X chromosomes fused at the centromere. They were exploited for cytogenetics. Among others, they were first used to carry out half-tetrad analysis in *Drosophila*. Females with attached X produce eggs, but half of them have only autosomes and no X chromosome. If the attached X chromosomes carry different alleles of a locus, a double dose of the same allele in the eggs can be achieved only if there is a recombination between that gene and the centromere because the first meiotic division is reductional and the second is equational. *See* compound X chromosomes; half-tetrad analysis. (Anderson, E. G. 1925. *Genetics* 10:403.)

Attached X chromosomes (↑) in the oogonium of an XXY *Drosophila*. (Drawing by Curt Stern in the 1920s.)

attachment point (*ap*) Mappable site in the chromosome of the chloroplast of *Chlamydomonas reinhardi* green alga, representing a hypothetical centromere-like element. It is called *ap* because it attaches to the chloroplast membrane and assists the disjunction of the ring DNA during division. In genetic recombination this is taken as the 0 coordinate of marker segregation. *See* chloroplast genetics; mapping of chloroplast genes.

attachment site *See att* site.

attention deficit hyperactivity disorder (ADHD) Observed in 2 to 5% of elementary school children; causes learning disabilities and emotional problems. Boys have about a five-fold higher chance of being affected than girls. It frequently goes into remission by progressive age, but some personality disorders (hyperactivity, antisocial behavior, alcoholism, hysteria) may persist even in adulthood. About 25 to 30% of the parents of affected children had some of the symptoms in childhood. The genetic basis is unclear. The dopamine receptor 4 encoded in human chromosome 11p15.5 may be responsible for the behavioral anomalies but not necessarily for the attention deficit. *See* affective disorders; behavior genetics; dyslexia. (Fisher, S. E., et al. 2002. *Am. J. Hum. Genet.* 70:1183; Wilens, T. E., et al. 2002. *Annu. Rev. Med.* 53:113.)

attenuate Tapered appearance.

attenuation Regulatory process in bacteria. *See* antitermination; attenuator region; host–pathogen relations; tryptophan operon.

attenuation, viral Reduction in virulence that is achieved by subculturing in a new cell population where after a period of time numerous adaptive mutations occur that although permit them to grow well in the original cells but with diminished virulence. These mutations generally occur in the 5′-nontranslated region and modify the translation of the viral RNA, although attenuating mutations may occur all over the viral RNA genome. *See* attenuator region.

attenuator region Where RNA polymerase may stop transcription when all the cognate tRNAs are charged, the mRNA assumes a special secondary structure that leads to a temporary reduction of transcription by a factor of 8 to 10. It is one of the regulatory mechanisms of bacterial amino acid operons. A type of attenuation also regulates the pyrimidine operon of *E. coli*. The operon is induced by a low concentration of uridine triphosphate. When the UTP level increases, slippage occurs at the promoter, incorporating long stretches of uridylic acid, and the RNA polymerase cannot escape the promoter. The cytosine deaminase/cytosine transport locus behaves similarly. The histidine operon does not even use the more common type of operator repressor/inducer system. Sucrose, β-glucoside, and β-glucan utilization enzymes in bacteria use RNA-binding proteins that inactivate transcription termination and thus promote transcription. The elongation of some lipid biosynthesis RNAs may be also negatively controlled. *See* antitermination; repressor; slippage; TRAP; tryptophan. (Yanofsky, C. 2000. *J. Bacteriol.* 182:1.) See Diagrams next page.

attractin Human serum glycoprotein-regulating cell-mediated immunity homologous to the *mg* locus of mouse. It is a low-affinity receptor for agouti protein. *See* agouti; obesity. (He, L., et al. 2001. *Nature Genet.* 27:40.)

att site Lysogenic bacteria and temperate phage have consensus sequences at the position where site-specific integration and excision takes place. The *att* sites are about 150 nucleotides long in λ and 25 bp in the bacterium, and 15 bp sequences are identical in both.

The underscored sequences are then reciprocally recombined and POB′ and BOP′ sequences are generated from the left (attL) and right (attR) sequences. The integration requires the phage-coded INT and the bacterial-coded HF proteins. The excision requires an additional protein XIS coded by the bacterial gene *xis* probably because it is not exactly the reverse type of process, since the original attP and attB elements were not identical except the 15 bp. *See* integrase; lambda-phage. (Williams, K. P. 2002. *Nucleic Acids Res.* 30:866.)

The phage (POP′) sequence is:
```
G C T T T T T A T A C T A A
C G A A A A A A T A T G A T T
```

The bacterial (BOB′) sequence is:
```
G C T T T T T T A T A C T A A
C G A A A A A A T A T G A T T
```

Auberger blood group *See* Lutheran blood group.

AUC Area under the curve. See diagram at below.

Auer bodies Clusters of granules or bundles of rods in the nuclei of acute promyelocytic leukemia cells. *See* leukemias.

AUG codon In mRNA the only codon that specifies methionine, yet there are two different tRNAs for methionine. In the majority of cases in prokaryotes, one of the methionine-tRNAs is formylated at the amino group by N^{10}-formyltetrahydrofolate, and this formylmethionine tRNA initiates translation, whereas the other methionine-tRNA carries methionine to all other sites in the polypeptide. In eukaryotes the *initiator methionyl-tRNA* is not formylated; the primary structure and conformation of the tRNA specify its initiator attribute. Thus, the overwhelming majority of the nascent proteins start at the NH_2 end with a methionine. In the mature protein this methionine may be absent because of processing. *See* genetic code.

Auger emission ^{125}I (iodine) or ^{195m}Pt (platinum) isotopes may emit electrons or Auger positrons (^{64}Cu) when excited by external radiation. These isotopes when incorporated may deliver high doses within the radius of a cell and can be used to damage tumor cells.

AU-rich element (ARE) May target the mRNAs of protooncogenes, cytokines, and lymphokines for rapid degradation in the 3′ untranslated region. However, heat shock, UV, hypoxia, stimulation, and oncogenic transformation stabilize these AU-rich mRNAs. The ELAV family of proteins, such as HuRs, may bind AREs. *See* ELAV; HuR. (Stoecklin, G., et al. 2001. *RNA* 7:1578.)

auricle Small projection at the upper part of the leaf sheath in cereals. Auricles are of important for taxonomic characterization.

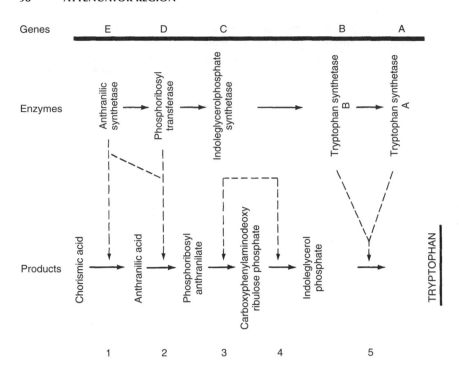

Genes

Enzymes

Products

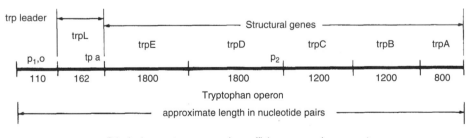

The biosynthesis of tryptophan from chorismic acid in *E. coli bacteria* requires five steps, mediated by five enzymes. The sequence of the encoding genes in the bacterial genetic map corresponds to the sequence of the metabolic steps. This was the first case of recognition of such a coordinate regulation in bacteria. The five genes are controlled primarily by a repressor, and attenuation provides an additional fine-tuning. Some of the enzymes are composed from more than a single functional unit. Indoleglycerolphosphate synthetase catalyzes two synthetic steps as shown at right.

p₁ : Principal promoter   p₂ : Low-efficiency secondary promoter
o : Operator   a : Attenuator
tp: Tryptophan pause

A genetic and molecular map of the tryptophan operon in *Escherichia coli* bacterium. Attenuation may dictate an early termination of transcription. The site of attenuation (a) is within the tryptophan leader sequence (*trpL*) and the site of the transcription pause (*tp*) precedes it. Transcription is primarily under the control of the promoter-operator region, and the process begins at the left end of the *trpL* site. The RNA polymerase pauses at the *tp* site before proceeding further. In case most of the tryptophanyl tRNAs (tRNA^{Trp}) are charged with tryptophan and therefore there is no need for additional molecules of this amino acid, transcription is momentarily terminated at the attenuator (a) site. If, however, the tRNA^{Trp} is largely uncharged because of a shortage of tryptophan and active protein synthesis, the transcriptase RNA polymerase passes through the a site without interference. This passage is made possible by alterations in the secondary structure of the RNA transcript of the operon. The initial segment of the leader sequence encodes a short tryptophan-rich peptide. In case there is a scarcity in tryptophan, translation on the ribosome is stalled at the tryptophan codons in the leader sequence. During the pause (*tp*), the mRNA transcript assumes a hairpin-like structure by base pairing between segments marked by (2) and (3) and thus the passage through the *attenuator* (a) site is facilitated. In case, however, most of the cognate tRNAs are charged, the transcript shows base pairing between segments (3) and (4), resulting in stoppage of transcription until the oversupply is exhausted by protein synthesis. The tryptophan operon also relies on suppressive transcriptional controls. See base sequences of the attenuator at the entry *tryptophan operon*. (Modified after Yanofsky, C. 1981. *Nature* 289:751.)

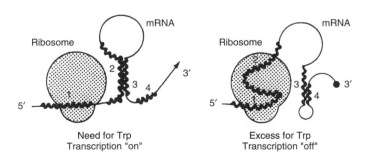

Need for Trp   Excess for Trp
Transcription "on"   Transcription "off"

aurone Plant flavonoid conveying yellow color to flowers. Aurones are synthesized from chalcones by aureusidine synthase (39 kDa copper glycoprotein) and a member of polyphenol oxidases.

Aurora Is a threonine/serine protein kinase regulating the mitotic spindle, chromosome segregation, cytokinesis, etc. (See passenger proteins, Taguchi, S., et al. 2002. *FEBS Lett.* 519:59.)

Austin disease *See* mucosulfatidosis.

Australopithecus Extinct fossil (1–5 million years old) of the bipedal Hominidae of the Old World. Its brain size is intermediate between modern humans and apes. Exact relation to existing species is unclear. (*Austral* means Southern.) *See* hominidae.

autism Human behavioral anomaly involving reticence, self-centered, subjective thoughts and actions, learning and communication difficulties. Its prevalence is 0.02 to 0.05% in the general population, and the recurrence risk in families may be 6 to 8%. The concordance between monozygotic twins appeared to be 36 to 96%; between dizygotic ones 0 to 24% was observed. Its incidence is higher in males than females. Generally it is associated with mental retardation and other psychological disorders. Although the incidence within affected families is 50 times that in the general population, no single gene could be identified as a causative agent; serotonin is suspected to play a role. Autism may be associated with genes in several chromosomes, but 7q31 and 1p appear to be the most likely locations of major factors. Other research has revealed several other putative linkage relations (Liu, J., et al. 2001. *Am. J. Hum. Genet.* 69:327). An aminophospholipid-transporting ATPase situated in the imprinted region 15q11-q13 near the ubiquitin ligase E3A and the Angelman syndrome genes is associated with a few percent of the autisms (Herzing, L. B. K., et al. 2001. *Am. J. Hum. Genet.* 68:1501; Folstein, S. E. & Rosen-Sheidley, B. 2001. *Nature Rev. Genet.* 2:943.)

Infantile autism becomes apparent during the first year of life. Apparently it is under polygenic control. In dominant autism with onset after an initial normalcy (Rett syndrome, RTT, prevalence 1×10^{-4}), the symptoms are shared, but progressive dementia, uncoordination, and deterioration of all mental functions follow. The latter type appears to be coded at Xq28 as MeCpG2. Rett syndrome affects primarily females. The short arm of the X chromosome has also been implicated. *See* affective disorders; MeCpG2; mental retardation. (Fombonne, E. 1999. *Psychol. Med.* 29:769; Folstein, S. E. & Mankoski, R. E. 2000. *Am. J. Hum. Genet.* 67:278; Geschwind, D. H., et al. 2001. *Am. J. Hum. Genet.* 69:463; Shao, Y., et al. 2002. *Am. J. Hum. Genet.* 70:1058; Yu, C.-E., et al. 2002. *Am. J. Hum. Genet.* 71:100.)

autoallopolyploid Polyploid in which the genome(s) is duplicated from one or more species, e.g., AAAABBBB or AAAABB. *See* allopolyploid; sesquidiploid.

autoantibody Antibody that is formed against the body's own antigens, such as in autoimmune disease. *See* autoimmune disease.

autoantigen (self-antigen) Normal cellular protein, yet it may be attacked by the cellular immune system similarly to what happens in autoimmune disease. *See* immune reaction; immune system.

autocatalytic function of DNA, is the process of replication. Also, any reaction that is promoted by its own product. Although self-replication is the most common property of nucleotide chains, peptides and other molecules may be involved in autocatalysis and cross-catalysis (i.e., the formation of other molecules). *See* heterocatalysis; replication.

autochthonous Located at its original site or a graft of an individual at another position within the same body.

autoclaving Heating under pressure (1 atmosphere above sea level) by steam, usually at 121°C for a minimum of 15 min, to kill non-spore-forming bacteria and other cells. *See* filter sterilization; sterilization.

autocorrelation, spatial Compares data (e.g., DNA sequences and haplogroup frequencies) within arbitrary areas in order to study diversity distribution. Measures of overall genetic similarity are evaluated in each distance class and the degree of genetic similarity at the different genetic distances is determined. A variable can be autocorrelated either (+) or (−) if its value at a given point in space is associated with its measures at other locations. (Simoni, L., et al. 2000. *Am. J. Hum. Genet.* 66:262.)

autocrine Signal production within a cell in response to external stimuli.

autocrine stimulation Cells infected by protooncogene-carrying virus secrete a growth factor that further stimulates the cell's proliferation. *See* paracrine stimulation; protooncogenes.

autoecious Parasite that completes its life cycle on the same host.

autogamy Self-fertilization common in hermaphroditic and monoecious plants; autogamy in the unicellular animals, *Paramecia*, is preceded by meiosis, and one the four haploid products survives. This cell then divides into two cells by mitosis. These two identical cells may fuse and a genetically homozygous diploid zygote may be formed. *See* allogamy.

autogenesis *See* Lamarckism.

autogenous control Gene product regulates the coding gene either in a positive or negative way. *See* negative control; positive control.

autogenous evolution Structures and organelles evolved through differentiation of the cell's own system. *See* exogenous evolution.

autogenous suppression The *Salmonella* RF2 translation termination protein occasionally fails to recognize or misreads the UGA stop codon, resulting in readthrough by suppressing termination. *See* readthrough; recoding; stop codon; translation termination.

autograft Tissue transplantation is within one individual. *See* homograft.

autoimmune disease The immune system fails to recognize the cell's own antigens and attacks them. In many instances, altered glycosylation is responsible for the pathogenesis. Normally the lymphocytes with defects in self-antigen recognition are eliminated by apoptosis. It has been shown that receptor tyrosine kinases (Tyro 3, Axl, Mer) play an essential regulatory role in the development of the immune response. Normally these receptors control the function of antigen-presenting cells by supplying growth-promoting and pro-survival molecules. They seem to also have negative control. Mutation of these receptors may disable the binding of gamma-interferon, and the inability of clearing dying cells results in overactivity of the macrophages, which then attack the body's own cells (Lu, Q. & Lemke, G. 2001. *Science* 293:306).

Lupus erythematosus cells (a variety of skin and possibly visceral inflammations) make antibodies against their own DNA and RNA. In insulin-dependent diabetes the insulin producer β-cells of the pancreas are attacked by the body's immune system, coded for by the major histocompatibility genes. Rasmussen's encephalitis, a rare form of epilepsy, and paraneoplastic neurodegenerative syndrome (PNS) are caused by autoantibodies against the glutamate receptors of the nervous system. Herpes simplex virus type 1 expresses a coat protein that recognizes autoreactive T cells targeting mouse corneal antigens and may cause stromal keratitis (inflammation of the fibrous coat of the eye). Autoimmune diseases include a series of different anomalies (p = prevalence, r = risk of siblings relative to risks in the general population): psoriasis (p: 2.8, r: 6), rheumatoid arthritis (p: 1, r: 8), goiter (p: 0.5, r: 15), insulin-dependent diabetes (p: 0.4, r: 1.6), ankylosing spondylitis (p: 0.13, r: 54), multiple sclerosis (p: 0.1, r: 20), lupus erythematosus (p: 0.1, r: 20), Crohn disease (p: 0.06, r: 20), narcolepsy (p: 0.06, r: 12), celiac disease (p: 0.05, r: 60), cirrhosis of the liver (p: 0.008, r: 100).

Autoimmune diseases have been attributed to increased V(J)D recombination in a class of B (B-1) lymphocytes as a result of increased RAG activity. Several autoimmune diseases (multiple sclerosis, rheumatoid arthritis) are more prevalent in females. The cause is apparently the difference in response to hormones of the T_H1 and T_H2 lymphocytes. Low-estrogen-level T_H1 cells secrete IL-2, INF-γ, and lymphotoxins, and multiple sclerosis and rheumatoid arthritis are aggravated. High-estrogen-level (increased progesterone, testosterone) T_H2 cells promote IL-4, IL-5, IL-6, and IL-10. As a consequence, during pregnancy the symptoms of multiple sclerosis and rheumatoid arthritis are mitigated, but lupus erythematosus may be aggravated. *See* ALPS; antigen-presenting cell; APECED; B cell; Borrelia; bullous pemphigoid autoimmune disease; caspase; complement; Goodpasture syndrome; hemolytic anemia; HLA; immunoglobulins; interferon; interleukins; lymphotoxins; monoclonal antibody therapies; NF-κB; RAG; signal transduction; Sjögren syndrome; TGF; T-helper cell; V(J)D. (Marrack, P., et al. 2001. *Nature Med.* 7:899; Leadbetter, E. A., et al. 2002, *Nature* 416:603.)

autoinduction Type of cell-to-cell interaction in bacteria and other organisms. The cells release small extracellular signaling molecules, which are taken up again by the cells. Autoinduction adjusts gene expression in the cells, responding to a level appropriate for the local density of the signaling cells. The signals may be acylated homoserine lactones, Tra proteins, amino acids, short peptides, and pheromones. *See* autoregulation; pheromones; quorum sensing; tra. (Tata, J. R. 2000. *Insect Biochem. Mol. Biol.* 30:645.)

autointerference Defective virions may interfere with the replication of intact ones.

autologous Its origin is within the cell or individual; it is a self-made molecule.

autologous transplantation Used in cancer therapy by implanting, e.g., genetically modified bone marrow cells of the same individual so that immune rejection may be avoided. *See* cancer gene therapy; gene therapy; immune system.

autolysis Decomposition of cells and cell content by the action of the natural enzymes of the cells. It generally takes place in injured cells.

automixis Self-fertilization.

automutagen Metabolite of the organism may become mutagenic, e.g., tryptophan.

autonomous controlling element Plant transposable element carries the transposase function and controls its own movement, e.g., Ac versus Ds in maize; the latter is a defective form of Ac, incapable of moving by its own power unless the autonomous (intact) Ac is present in the cell, *See Ac-Ds*; *Spm*; transposable elements.

autonomous developmental specification Maternal information or prelocalized morphogenetic information regulates the initiation of transcription of morphogenetic genes. *See* morphogen.

autonomously replicating pieces *See* macronucleus.

autonomously replicating sequences *See* ARS.

autonomous parvovirus Uses the host system for productive replication. Only strain B19 is pathogenic in humans. These viruses display antineoplastic properties in Ehrlich ascites tumors. *See* ascites; parvoviruses.

autonomy Cells transplanted into tissues of a different genotype or forming parts of genetically different sectors still maintain the expression encoded by their genotype; they are not—or barely—affected by the genetically different tissue environment.

autophagy Destruction of cytoplasmic particles within a cell by delivering dispensable structures or molecules (in autophagosome vehicles) to lysosomes or vacuoles. This process gets rid of and reutilizes the molecules during adverse conditions (e.g., cell starvation). Degradation within the lysosomes is also called *microautophagy*. In *macroautophagy* the subcellular membranes are altered and part of the cytoplasm is sequestered into double-membrane-surrounded autophagic vacuoles (autophagosomes). *See* beclin 1; lysosomes; pexophagy; ubiquitin. (Klionsky, D. J. & Emr, S. D. 2000. *Science* 290:1717; Subramani, S. 2001. *Developmental Cell* 1:6; Ohsumi, Y. 2001. *Nature Rev. Mol. Cell Biol.* 2:211; Khalfan, W. & Klionsky, D. J. 2002. *Current Opin. Cell Biol.* 14:468.)

autophene Genetically determined trait expressed independently of the position in case of transplantation. *See* allophenic.

autophosphorylation Upon binding a ligand to a receptor, there is rapid phosphorylation of the receptor by its own subunits, e.g., by members of a dimeric molecule generally at tyrosine sites. *See* receptor tyrosine kinase.

autoploid (autopolyploid) More than two complete sets of an identical genome per cell. If the number of chromosomes in the somatic cells is reduced to half, they become polyhaploids. Autopolyploids may be [auto]tetraploid ($2n = 4x$), hexaploid ($2n = 6x$), octaploid ($2n = 8x$), etc.

Autotetraploids in meiosis may pair as quadrivalents; however, at a particular point of the chromosomes, only two synapse. In autopolyploids, pairing may also be as two bivalents, one trivalent and one univalent, and may form four univalents. If all chromosomes pair as bivalents, it is called selective pairing, and the segregation of genes will resemble that of diploids with duplicate genes. Autotetraploids may carry a different allele in each of the four chromosomes; therefore, they can produce a larger variety of gametes than diploids. The maximal number of gametic combinations can be determined by the formula:

$$\begin{bmatrix} n \\ x \end{bmatrix}$$

where n = the total number of alleles, and x = the number of alleles in a gamete, thus in autotetraploids it becomes

$$\begin{bmatrix} 4 \\ 2 \end{bmatrix}$$

for octaploids it is

$$\begin{bmatrix} 8 \\ 4 \end{bmatrix}$$

and these can be rewritten as

$$\frac{4 \times 3}{2 \times 1} = 6$$

for autotetraploids and

$$\frac{8 \times 7 \times 6 \times 5}{4 \times 3 \times 2 \times 1} = 70$$

Gametic Output of Autotetraploids

Parent	Absolute Linkage*			Independence from Centromere[†]		
	AA	Aa	aa	AA	AA	aa
AAAa	1	1	0	13	10	1 (4.2%)
AAaa	1	4	1 (16.6%)	2	5	2 (22.2%)
Aaaa	0	1	1 (50.0%)	1	10	13 (54.2%)

Phenotypic segregation ratios in autotetraploids in case the dominance is complete in F_2

Mating	Absolute Linkage*		Independence from Centromere[†]	
	Dominant	Recessive	Dominant	Recessive
AAAa selfed	1	0	575	1
AAaa selfed	35	1	19.3	1
Aaaa selfed	3	1	2.4	1
AAAa x AAaa	1	0	107	1
AAAa x Aaaa	1	0	43.3	1
AAAa x aaaa	1	0	23	1
AAAa x Aaaa	11	1	7.3	1
AAAa x aaaa	5	1	3.5	1
Aaaa x aaaa	1	1	1	1.2

*No recombination between gene and centromere (chromosome segregation).
[†]The distance between gene and centromere is 50 map units or more, and therefore recombination occurs freely as if they (gene and centromere) would not be syntenic (chromatid segregation or maximum equational segregation).

for autooctaploids, and this means that the number of allelic combinations with 4 different alleles, 6 types of gametes are possible, and with 8 different alleles in an octaploid, the total number of gametic types is 70. In case all four alleles are dominant, AAAA, the individual is a quadruplex, AAAa = triplex, AAaa = duplex, Aaaa = simplex, and aaaa = nulliplex. The segregation ratios in F_2 depend on whether there is crossing over between the gene and the centromere, the type of pairing (as indicated above), and the type of disjunction at anaphase II (alpha-parameter). The phenotypic proportions in F_2 are determined by the gametic output of the parents or selfed individuals. The gametic output and F_2 segregation of autopolyploids is very difficult to generalize because the genes are rarely linked absolutely to the centromere, and the frequency of recombination may vary from 0 to 50%. There are additional variables that may be estimated by the alpha-parameter. Segregation ratios at a higher level of polyploidy can be predicted only theoretically. The actual results may be quite different, however. *See* alpha parameter; bivalent; maximum equational segregation; synteny; trivalent; univalent. (Haldane, J. B. S. 1930. *J. Genet.* 22:359; Rédei, G. P. 1982. *Genetics*. Macmillan, New York.)

autopodium Skeletal portion of the hand and foot.

autoprocessing Sequence of a protein (e.g., the C-terminal) is involved in its processing.

autorad Lab slang for autoradiogram. *See* autoradiography.

autoradiography Labeling technique by which a radioactive substance reveals its own position in a cell or on a chromatogram when brought into contact with photographic film. For cytological analyses, most commonly H^3-labeled thymidine is used because it gives the clearest resolution of chromosomal regions without serious DNA breakage, whereas in molecular genetics the much higher energy P^{32}-labeled compounds are usually employed. *See* immunoprobes; nonradioactive labels. (Taylor, J. H., et al. 1957. *Proc. Natl. Acad. Sci. USA* 43:122.)

autoreactive   The lymphocytes recognize the individual's own molecules and develop an immune reaction to them. *See* autoimmune disease.

autoreduplication   Self-duplication.

autoregulation   Compound (or system) controls the rate of its own synthesis, e.g., the bacterium *Klebsiella aerogenes* uses glutamine dehydrogenase to make glutamate from α-ketoglutarate and ammonia if the concentration of the latter exceeds 1 nM. If the concentration of ammonia is low, glutamate dehydrogenase cannot function to an appreciable extent. In this case, the ammonia + glutamate are converted into glutamine by glutamate synthetase. The active form of glutamine synthetase is nonadenylylated. In the presence of a high concentration of ammonia, the enzyme is adenylylated and thus the activity is reduced by this mechanism of autoregulation. In its nonadenylylated states it represses glutamate dehydrogenase. *See* diagram next page; nitrogen fixation; regulation of gene activity. (Magasanik, B., et al. 1974. *Curr. Top. Cell. Reg.* 8:119; Chandler, D. S., et al. 2001. *Nucleic Acids Res.* 29:3012.)

autosegregation   May take place in an apomictic or vegetatively multiplied organism due to chromosomal loss or somatic mutation. *See* apomixis; mutation.

autosexing   Sex is identified by genetic markers rather than by the genitalia. Silkworm breeders and poultry producers have exploited this procedure. Homozygosity for the *B* (barring) genes suppresses the appearance of colored spots on the head of the newly hatched chicks controlled by this sex-linked gene (remember that in birds the males are homogametic). The *B* gene is dominant, yet it shows clear dosage effect. In the females that are heterogametic, the spot is evident. Thus they can separate the hens from the roosters early when the recognition of gender is very difficult by anatomy. Because most of the roosters will be used for meat production and the hens will be used for egg production, they can be fed and managed accordingly. In the silkworm the male cocoons (chrysalis) produce 25 to 30% more silk than the females; therefore, autosexing may have economic advantage. An electronic device may sort the silkworm eggs

Left: Autosexing in the silkworm. The dominant gene in the Y chromosome permits the distinction between the eggs, which will hatch to become a male (pale color) and a female (dark). **Right**: The homozygous male (B/B) chicks are light colored while the hemizygous B/O females develop colored spots on the head. (In the Lepidoptera and birds the males are WW and the females are WZ.)

according to color (sex). *See* chromosomal sex determination; sexing.

autosomal-dominant mutation   Readily detected and identified in many instances because the novel type appears suddenly without precedence in the pedigree. Achondroplasia in humans is frequently cited as such an example. The homozygotes generally suffer perinatal death. Therefore, most of the achondroplasiac dwarfs are heterozygotes and new mutants. These dwarfs are of normal and frequently superior intelligence. Remember that over 70 gene loci are responsible for various types of dwarfing in humans. Autosomal-dominant mutation rates (per gamete/generation) in human populations for 10 diseases vary from 4 to 100×10^{-6}. *See* achondroplasia; mutation rate.

autosomal-recessive lethal assay   Tester stock used for the detection of recessive second chromosomal lethals in *Drosophila* is of the following genetic constitution: *Cy L/Pm*, where *Cy*, (*Curly*), *L* (*Lobe*), and *Pm*, (*Plum*) are heterozygous-viable but homozygous-lethal dominant genes. The *Cy* chromosome generally carries three inversions to prevent the recovery of crossovers. The heterozygotes of either sex are crossed with a mate that before the test carried no mutation in either of the two second chromosomes. Single F$_1$ males are then backcrossed with the *Cy L/Pm* female

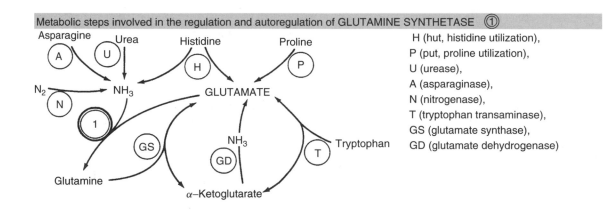

Metabolic steps involved in the regulation and autoregulation of GLUTAMINE SYNTHETASE ①

H (hut, histidine utilization),
P (put, proline utilization),
U (urease),
A (asparaginase),
N (nitrogenase),
T (tryptophan transaminase),
GS (glutamate synthase),
GD (glutamate dehydrogenase)

tester. From their offspring, *Cy L* individual sibs are mated. From this mating an F_2 is obtained. If all the survivors are *Cy L*, this indicates that a new lethal mutation occurred in the grandfathers' or grandmothers' second chromosome, and therefore *non-Curly* and *non-Lobe* homozygous individuals could not live. The diagram does not show the genotypes in the F_2. *See* sex-linked recessive lethal assays.

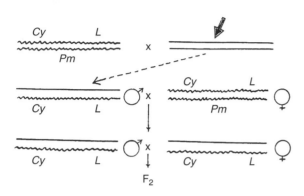

autosome   Chromosome that is not a sex chromosome. *See* chromosomal sex determination.

autosyndesis   May take place between homoeologous chromosomes in the absence of homoeologous pairing-suppressor genes. *See* chromosome 5B; homoeologous.

autotoxic enterogenous cyanosis   Obsolete name for human familial NADH-methemoglobin deficiency. *See* methemoglobin.

autotroph   Can synthesize cellular C- and N-containing molecules from carbon dioxide and ammonia.

autozygous   Genotype, which is not just homozygous, but the alleles at the locus are *identical by descent*. *See* allozygous; coancestry; homozygosity; inbreeding coefficient.

auxanography   Method of mutant selection. Minimal medium is overlayered with auxotrophic spore or cell suspension and small quantities of various substances that the cells may need for growth are added to different segments of the plate. Where growth occurs, the cells utilize the added compounds and their nutritional requirement is identified.

auxilin   100 kDa brain-specific chaperone with C-terminal homology to DnaJ. *See* clathrin; DnaJ.

auxin   Phytohormone (morphogen) produced by plant metabolism such as indole-3-acetic acid, or of synthetic origin such as α-naphthalene acetic acid or 2,4-dichlorophenoxyacetic acid. Auxins play an important role in cell elongation, signal transduction, and as required supplements for proliferation and regeneration in tissue culture. The transport of auxins in the plant tissues is regulated by chemosmosis aided by various transporter proteins. Genes (*iaaH, iaaM*) in the Ti plasmid of *Agrobacterium* have instructions for their production and regulation, and these play a role in crown gall formation (in cooperation with cytokinins). An auxin response element first identified in the octopine synthase (*ocs*) gene of *Agrobacterium tumefaciens* (AuxRe

[named *as-1* in cauliflower mosaic virus]) is an enhancer and is present in many genes. The *ocs/as-1* consensus consists of a more or less well-conserved 20 bp direct repeat with a 4-base spacer: TGACGTAAGCGCTGACGTAA. These elements respond to various auxins, salicylic acid, methyljasmonate, and many other diverse compounds. The binding transcription factors have basic leucine zipper (bZip) motifs. Indole-3-acetic acid is biosynthesized mainly from tryptophan (aminotransferase) through indole-3-pyruvate (decarboxylase). *See* ARF1; bZip; chemosmosis; crown gall; embryogenesis, somatic; plant hormones; SAUR; Ti plasmid. (Guilfoyle, T. J. & Hagen, G. 1999. *Inducible Gene Expression in Plants*, Reynolds, P. H. S., ed., p. 219. CABI, New York; Sabatini, S., et al. 1999. *Cell* 1999:463; Zhao, Y., et al. 2001. *Science* 291:306; Gray, W. M., et al. 2001. *Nature* 414:271; Leyser, O. 2002. *Annu. Rev. Plant Biol.* 53:377.)

auxonography   *See* auxanography.

auxotroph   Mutant that requires nutritive(s) not needed by the wild type (prototroph). Auxotrophic mutations have been extensively used for the study of biochemical pathways and for the identification of enzymes catalyzing particular metabolic steps. In genetic analysis, auxotrophs facilitate selective techniques in backmutation, recombination, transformation, etc. A pyridoxine deficiency, causing seizures in humans, has been identified. True auxotrophic animal mutations are exceptional and they are also rare in higher plants. In *Arabidopsis* over 200 mutations were obtained in the thiamine pathway without any obligate auxotrophy for other metabolites. The scarcity of auxotrophic mutations may be

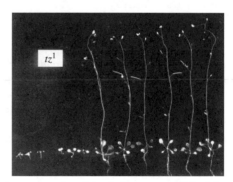

Left: Thiazole auxotrophs of *Arabidopsis* on basal medium; **Right**: on thiamine medium.

due to redundancy of the genomes, alternative metabolic pathways, and compensatory effects in large genetic networks. *See* autotroph; genetic network; pyridoxine; redundancy. (Rédei, G. P. 1982. Genetics, Macmillan, New York.)

AuxRE   Auxin-response element. *See* ARF1; auxin.

AVED   Ataxia with vitamin E deficiency is caused by mutation in the large subunit of microsomal triglyceride transfer protein (encoded in human chromosome 8q13). Although the intestinal absorption of α-tocopherol is normal, the hepatic secretion into the blood is defective. The condition is very similar to Friedreich ataxia. *See* Friedreich ataxia.

Avena   Genus of grasses (oats) with basic chromosome numbers $x = 7$; they form an allopolyploid series, 2n, 4n, 6n.

average Arithmetic mean, i.e., the sum of all measurements (x) divided by the number of measurements (N):

$$\bar{x} = \frac{\Sigma x}{N}.$$

See mean; median; mode.

average inbreeding coefficient *See* α.

avian Pertaining to the taxonomic class of *Aves* (sing. *Avis*, bird).

avian erythroblastosis (*erbB*) Viral oncogene (prevents maturation of the red blood cells in fowl) has its cellular homologue as a protooncogene in several eukaryotes. It is a protein kinase that primarily phosphorylates tyrosine residues. The normal allele specifies a plasma membrane receptor of epidermal growth factor (EGF). *See* erythroblastosis fetalis.

avian MC29 myelocytomatosis Viral oncogene (*myc*, causes carcinoma, sarcoma, and myelocytoma [a kind of leukemia]). It is present as a cellular protooncogene in vertebrates and its homologues are also present in plant cells. *See* carcinoma; leukemias; oncogene; protooncogene; sarcoma.

avian myeloblastosis *See* MYB oncogene.

avian sarcoma virus *See* ASV.

avidin Ca. 68,000 M_r protein of four subunits, each with a strong affinity to biotin. It binds strongly to any molecule complexed with biotin such as nucleic acids and to biotin-containing enzymes. It is widely used for nonisotopic labeling of nucleic acids. Originally avidin was isolated from raw egg white. Eating raw eggs may cause biotin deficiency (cooking inactivates it). It is also isolated from *Streptomyces avidinii* under the name streptavidin. *See* biotinylation; genomic subtraction.

avidity *See* antibody.

avirulence Lack of competence for causing pathological effects by an infectious agent.

Avogadro number Number of molecules (6.02×10^{23}) in a gram molecular weight, a constant for all molecules.

avoidance learning Classical test of animal behavior. In a two-compartment box, one compartment is electrically wired to provide an electric shock to the test animals after a light turns on. After a learning period, some of the animals immediately move to the safe compartment as they see the light signals and learn the coming shock in one compartment. The learning ability of inbred mice strains is genetically different. In some, about half of the individuals "learn"; in others, only 10% associate the light signal with the shock. *See* behavior genetics.

avuncular Ancestral relatedness such as existing between nephews/nieces and uncles/aunts.

axenic Pure culture of organisms or cells without any contamination by other (micro) organisms. *See* aseptic culture; tissue culture.

Axenfeld-Rieger anomaly (FKHL7/FOXC1, 6p25) Anterior eye segment defect and glaucoma caused by mutation in the human homologue of the *Drosophila* forkhead gene, FOXC1. Additional loci at 4q25 (PITX2, a bicoid-related protein), 13q14, and 16q22-q24 (FOXC2 forkhead-like). *See* bicoid; forkhead; glaucoma. (Priston, M., et al. 2001. *Hum. Mol. Genet.* 10:1631; Lines, M. A., et al. 2002. *Hum. Mol. Genet.* 11:1177.)

axial element Lateral element of the tripartite synaptonemal complex. *See* synaptonemal complex.

axillary Formed in the axil, the upper surface of the area between the leaf petiole and stem.

axin Protein controlling body axis formation. Mutant axin may interfere with the normal developmental pathway and cause cancer, especially when the mismatch repair is defective. It is homologous to conductin. *See* conductin; DNA repair; wnt.

axis of asymmetry Through which the objects or molecules form mirror images. In the body of the majority of organisms, three axes are recognized: anterior-posterior (front-hind), dorsal-ventral (back-abdominal), and left-right. Genes have been identified that control asymmetry of the body. It has been known for a long time that changing the placement of internal organs has multiple deleterious consequences (*See* situs inversus viscerum). It has now been shown in the *lefty* mouse mutant that the expression of the transforming growth factor (TGFβ) family plays the role of a morphogen in controlling asymmetry by expressing only in the left half of the gastrula. This asymmetry is transient and sets on before lateral asymmetry becomes visible. Similar genetically controlled mechanisms have been discovered in chickens and other organisms. *See* activin; asymmetric cell division; Kartagener syndrome; morphogen; situs inversus visceri; TGF. (Lall, S. & Patel, N. H. 2001. *Annu. Rev. Genet.* 35:407.)

Axl Receptor tyrosine kinase is human myeloid leukemia transforming protein. *See* leukemias.

axon Long nerve fiber that generally communicates impulses in a bundle surrounded by a myelin sheath between the central and the peripheral nervous system. Organelles and molecules can be transported along the nerve axons outward from the cell or back to the cell. *See* axotomy; comm; netrin;

neurogenesis; neuropilin; Robo; Slit. (Kamal. A., et al. 2000. *Neuron* 28:449.)

axoneme Cylindrical structure of microtubule doublets and about 250 attached polypeptides that is the major part of cilia, flagella, and sperm. There are two rows of the motor protein dynein situated along the microtubules. *See* dynein; microtubule.

axon guidance Axons grow and move through the embryonal body toward their targets and allow for synaptic connections of the neurons. Many proteins guide their advance. Some axons follow the same path bundled together by fasciculation. Brain wiring and axon guidance can be monitored with the aid of the *PLAP* (placental alkaline phosphatase) vector equipped with an IRES site 5′ to the *PLAP* gene. The vector includes another part carrying *β-galactosidase* and *neomycin phosphotransferase* (*G*418′). This portion is expressed by virtue of its fusion to neural body cell-specific promoters. Transformants can be selected on neomycin media. The *β-gal* gene marks the cell body by blue color on X-gal medium; the *PLAP* gene is expressed exclusively in the dendritic part of the neurons. Thus the wiring pattern of the brain can be monitored without laborious chemical purification. *See* axon; *β-galactosidase*; G418; IRES; neuron; X-gal. (Leighton, P. A., et al. 2001. *Nature* 410:175; Lin, M. Z. & Greenberg, M. E. 2000. *Cell* 101:239; Stein, E. & Tessier-Lavigne, M. 2001. *Science* 291:1928; Patel, B. N. & Van Vactor, D. L. 2002. *Current Opin. Cell Biol.* 14:221.)

axoplasm Cytoplasm of axons. *See* axon.

axotomy Lesion of axons; may affect expression of genes. *See* regulation of gene activity.

5-azacytidine Pyrimidine analog (and suspected carcinogen) that interferes with methylation of DNA bases. It may affect differentiation and development because hypomethylated genes are preferentially transcribed. It is noteworthy that some small eukaryotic genomes (yeast, *Drosophila*) do not contain methylcytosine, yet their genomes are regulated during development. *See* fragile X; housekeeping genes; methylation of DNA; trichostatin.

5-Azacytidine.

azaguanine Toxic analog of guanine readily incorporated into RNA or DNA. *See* HAT medium.

8-Azaguanine.

8-azaguanine resistance Commonly used marker in mammalian cell cultures; the resistance is based on a deficiency of the enzyme azaguanine-hypoxanthine phosphotransferase and therefore this toxic purine cannot be processed by metabolism. *See* HAT medium.

azaserine (O-diazoacetyl-L-serine) Alkylating, antitumor, antifungal, and mutagenic agent. The oral LD50 for rodents is 150–170 mg/kg. *See* LD50.

6-azauracil Antineoplastic pyrimidine analog; its nucleotide is inhibitory to orotidylic acid decarboxylase and may repress the synthesis of orotidylic acid pyrophosphorylase — key enzymes in the de novo pathway of nucleotide synthesis. *See* TIIFS.

azide Compound with NH_3; sodium azide, a respiratory inhibitor, is a strong mutagen for certain organisms at low pH but not for others. Nitrogenase enzymes may reduce azides to N_2 and NH_4. *See* nitrogenase.

azidothymidine *See* AZT.

azoospermia Human gene AZF (azoospermia factor) appears to be the expression of the DAZ (deleted in azoospermia) site and has been assigned to human chromosome Yq11. At the AZF site, three long palindromic sequences encoding 11 transcription units have been identified (Kuroda-Kawaguchi, T., et al. 2001. *Nature Genet.* 29:279). At Yq11.2, in the vicinity of AZF, is the DFFR (*Drosophila* fat-facet related), another spermatogenesis control gene. The frequency of DAZ causes sterility in about 1.25×10^{-4} men. This gene is substantially (42%) homologous to the *Drosophila* gene *boule* (*bol*) controlling meiotic G2-M transition. Mouse gene *Dazla* is 33% homologous to DAZ. Both the mouse and the *Drosophila* genes are, however, autosomal, yet they also involve male sterility. It has been shown that the human AZF gene was originally in the short arm of human chromosome 3 (where highly homologous sequences still exist) and it was transposed to the Y chromosome, amplified and pruned. The human Y chromosome encodes an RNA recognition motif, which is active particularly in the testes, and the deletion of this motif may cause azoospermia. *See* agonadism; *boule* [*bol*]; CBADV; holandric genes; infertility; oligospermia; *pelota*; RBM; *twine*. (Hackstein, J. H. P., et al. 2000. *Trends Genet.* 16:565, Xu, E. Y., et al. 2001. *Proc. Natl. Acad. Sci. USA* 98:7414.)

Azorhizobium *See* nitrogen fixation.

Azotobacter *See* nitrogen fixation.

AZT (azidothymidine, also called zidovudin) Thymidine analog with an azido (N_3) substitution of the 3'-OH group; it may slow down the reverse transcriptase activity of HIV virus by preferentially selecting this analog that has only a minor effect on DNA polymerase of the mammalian cells. Unfortunately, some bone marrow damage is associated with the drug, and this limits its usefulness in protecting against the full-scale development of AIDS. Mutations in the HIV reverse transcriptase may result in resistance against the drug by removing the AZTMP that blocks transcription. Eventually it may also debilitate the cells by the inhibition of DNA polymerase γ. The Aschheim-Zondek test for pregnancy is also abbreviated as AZT. *See* acquired immunodeficiency; AIDS; HIV; mtDNA. (Lim, S. E. & Copeland, W. C. 2001. *J. Biol. Chem.* 276:23616.)

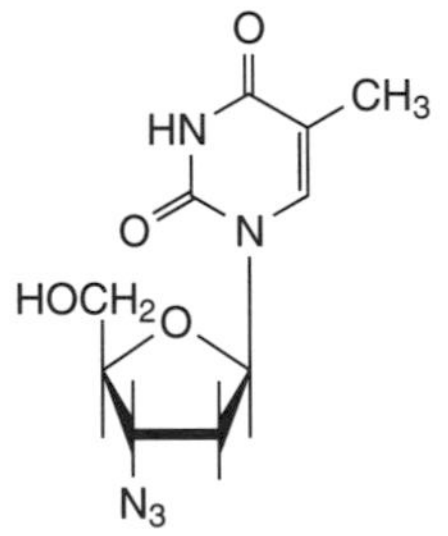

azurocidin *See* antimicrobial peptides.

The Cambridge (Massachusetts) City Council were not the first to disapprove of recombinant DNA. Joshuah Sylvester (1563–1618) answers the "New objection of Atheists, concerning the capacitie of the Ark":

"O profane mockers! if I but exclude
Out of this Vessell a vast multitude
Of since-born mongrels, that derive their birth
From monstrous medly of *Venerian* mirth:
Fantastick Mules, and spotted Leopards,
Of incest-heat ingendred afterwards:
So many sorts of Dogs, of Cocks, and Doves,
Since, dayly sprung from strange & mingled loves,
Wherein from time to time in various sort,
Dedalian Nature seems her to disport:
If plainer, yet I prove you space by space,
And foot by foot, that all this ample place,
By subtill judgement made and *Symmetrie*,
Might lodge so many creatures handsomely,
Sith every brace was *Geometricall*:
Nought resteth (*Momes*) for your reply at all;
If, who dispute with God, may be content
To take for current, Reason's argument."

— *The Complete Works of Joshuah Sylvester*, Vol. 1, ed. Rev. Alexander B. Grosart, printed for private circulation, 1880, p. 136.

A historical vignette of the 16th century.

B

B Backcross generation; the numbers of backcrosses are indicated by subscripts, e.g., B_1. *See* backcross.

7B2 (secretogranin V, chromogranin) 25–29 kDa pro-hormone processed to a 18–21 kDa neuroendocrine chaperone (distantly related to chaperonins 60/10) in the secretory pathway. It has a wide occurrence in animals, and it is encoded in human chromosome 15q13-q14 as SGNE-1. It is an inhibitor/activator of the pro-hormone convertase PC2 enzyme but not of other PCs. *See* chaperonins; Golgi apparatus. (Umemura, S., et al. 2001. *Pathol. Int.* 51:667.)

B104 *Drosophila* retroposon similar to copia, gypsy, and others. *See* copia.

baboon (*Papio*) *See* Cercopithecidae; xenotransplantation.

BAC Bacterial artificial chromosome. Bacterial cloning vector (derived from F plasmid) that can accommodate up to 350 kb (most commonly 120–150 kb) DNA sequences and has a much lower error rate than the larger-capacity yeast artificial chromosome (YAC). BACs usually exist in a single copy per cell. *Random BACs* are selected from a genomic library, then are shotgun sequenced. Most BAC vectors lack selectable markers suitable for mammalian cell selection but can be retrofitted by employing the Cre/loxP site-specific recombination system. *See* BIBAC; Cre/loxP; F plasmid; genome projects; PAC; selectable marker; shotgun sequencing; vectors; YAC. (Wang, Z., et al. 2001. *Genome Res.* 11:137; <http://www.nih.gov/science/models/bacsequencing/end_sequencing_project.html>.)

BACE Beta site APP-cleaving enzyme. Amyloid precursor protein (APP) cleavage enzyme — a β-secretase (membrane-bound aspartyl protease) — involved in the production of brain plaques in Alzheimer disease. The transmembrane BACE splits APP into soluble β-sAPP and a membrane-attached carboxy-terminal fragment, CTF-β. The latter is expected to be the substrate for γ-secretase. BACE1-deficient mice do not generate $A\beta$ peptide, which is responsible for the Alzheimer plaques, and appear to be normal. BACE is encoded at human chromosome 11q23.3. Its homologue, BACE2, is in chromosome 21, near the critical region in Down syndrome trisomy. Drug development for Alzheimer disease research has a likely target in BACE. *See* Alzheimer disease; Down syndrome; presenilin; secretase. (Roberds, S. L., et al. 2001. *Hum. Mol. Genet.* 10:1317.)

BACH BRCA1-associated carboxy-terminal helicase is a DEAH family protein. *See* breast cancer; DEAH box proteins.

Bach, Sebastian (1685–1750) One of the greatest geniuses of classical music had a family with over 50 more or less renowned organists, cantors, and musicians. His first marriage to his second cousin, Maria Barbara Bach, included 3 musicians among the 4 surviving children (inbreeding

coefficient 1/64 [3 offspring died in infancy]). From his second marriage to unrelated singer Anna Magdalena Wilcken (assortative mating) 13 babies were born, and among the 5 survivors, 3 were musically talented. This family tree indicates that musical ability may be controlled by relatively few genes, and the cultural environment may also have a major role. Studies have demonstrated that musical talent is correlated with stronger development of the left planum temporale, increased leftward asymmetry of the brain cortex. *See* dysmelodia. (Wolff, C. 2000. *Johann Sebastian Bach: The Learned Musician.* Norton, New York.)

Bacillus Rod-shape bacterium. *See Bacillus subtilis*; *Bacillus thüringiensis*.

Bacillus Calmette-Guerin (BCG) Attenuated form of *Mycobacterium bovis* bacillus used for vaccination against the human *Mycobacterium tuberculosis*; may serve as a suitable vector for *B. burgdorferi* and HIV virus. BCG differs from the virulent *M.t.* by the deletion of ~91 open reading frames and ~38 additions. BCG has been introduced into liver cancer cells, skin tumors, and other cancer cells with some beneficial effects on slowing down metastasis and/or delaying recurrence. *See* acquired immunodeficiency; *Borrelia*; metastasis; mycobacteria. (Sassetti, C. M., et al. 2001. *Proc. Natl. Acad. Sci. USA* 98:12712.)

Bacillus subtilis Gram-positive, rod-shaped soil bacterium; it lives on decayed organic material and is thus harmless. In starvation most of the cell content, particularly the DNA, moves to one end of the cell. This area, constituting about 10% of the cell, is walled off and becomes a spore. The spore is extremely resistant to various environmental effects that would kill vegetative cells. Under favorable conditions the spore regenerates the bacterium. The size of its cells is similar to those of *E. coli*, and its genome of 4,214,810 bp (4,100 ORF) was completely sequenced by 1997 (*Nature* 390:248). The DNA has a different base composition from that in *E. coli*, A + T/G + C ratio in the former 1.38, in the latter 0.91, indicating that *B. subtilis* has more A + T than *E. coli*. The genome includes many repeats in only half of the chromosome at both sides of the replicational origin. For transcription it uses 18 different σ-factors, although its major RNA polymerase is similar to that of *E. coli* ($\alpha\alpha\beta\beta'\sigma$). The 43 kDa ($\sigma^{43}$) recognizes some of the consensus sequences of *E. coli* promoters. Its best-known phage is SPO1, which is transcribed either by a phage RNA polymerase or by the host. It contains

4,100 protein-coding genes with an average length of 890 bp. Seventy-eight percent of them start with ATG. Seventy-five percent of the genes are transcribed in the direction of the replication. Fifty-three percent of the genes occur only once (singlets), whereas the putative ATP-binding transporter family paralogues appear to be 77 (14% of the genome). The DNA is not methylated. *See E. coli*; endospore; forespore; paralogous; sporulation. (Hecker, M. & Engelmann, S. 2000. *Int. J. Med. Microbiol.* 290:123; <http://genolist.pasteur.fr/>.)

Bacillus thüringiensis   Gram-positive bacterium that produces a BT toxin. The toxin (delta-endotoxin) is within the crystalline inclusion bodies produced during sporulation. In an alkaline environment (such as in the midgut of insects) the crystals dissolve and release proteins of M_r 65,000 to 160,000 that are cleaved by the proteolytic enzymes of the insects into highly toxic peptides. These toxins are most effective against *Lepidopteran* larvae (caterpillars), but some bacteria produce toxins against *Diptera* and *Coleoptera*. The extract of these bacteria can be used directly as a powder or suspension against the insects. The effectiveness is generally very good, but the cost is substantially higher than that of chemical insecticides. A great advantage is that these toxins are not harmful to mammals and many other insects. The most economical solution is to transform plants (tobacco, cotton, maize, etc., which can be transformed) with the *Bt2* gene that codes for the 1,115-amino-acid residue pro-toxin protein. Actually, a smaller polypeptide, M_r 60K, and even smaller fragments, are still fully active. The transgenic plants kill invading caterpillars within a couple of days and remain practically immune to any damage. The activity of the transgene has been further enhanced by the use of high-efficiency promoters in the T-DNA constructs. In some instances, transgenic cotton was still overpowered by bollworms. There are differences in the spectra of the bacterial toxins produced by different strains of the *Bacillus*, and this provides an opportunity to extend the range of toxicity to other insect species. For the production of corn rootworm resistant transgenic plants, the toxin gene of *B. thüringiensis tenebrionis* has been used. Mutation in insect aminopeptidase receptors, in a cadherin superfamily gene, and in β-1,3-galactosyltransferase may impart resistance to the BT toxin. Transfer of the BT toxin gene into commercial varieties may have some deleterious effect on the Monarch butterfly and other lepidopteran larvae; however, under field conditions these adverse effects may not be very serious. *See* cadherins; Cry9C; GMO; insect resistance in plant; pest eradication by genetic means; *Photorhabdus luminescens*; promoter; transformation of plants. (Gahan, L. J., et al. 2001. *Science* 293:857; Griffits, J. S., et al. 2001. *Science* 293:860; series of articles in *Proc. Natl. Acad. Sci. USA* 98:11908–11937 [2001].)

backcross   The F_1 is crossed (mated) by either of its two parents (*See* test cross). Each backcrossing reduces by 50% the genetic contribution of the nonrecurrent parent; thus, after (r) backcrosses it will be $(0.5)^r$. The percentage of individuals homozygous for the (n) loci of the recurrent parent in (r) number of backcrosses $= ([2^r - 1]/2^r)^n$. The chance of eliminating a gene linked to a selected allele is determined by the intensity of linkage (p) and the number of backcrosses (r) according to the formula $1 - (1 - p)^{r+1}$.

background, genetic   The (residual) genetic constitution not considering particular loci or genes under the special study.

Knowing the genetic background may be of substantial importance because different sets of modifier genes may influence the expression of particular genes. *See* modifier genes.

background radiation   Natural radiation coming from cosmic or terrestrial sources. *See* cosmic radiation; terrestrial radiation.

background selection   Recurrence of deleterious mutations reduces the effective size of the population. The selection is directed against the chromosomal background carrying the particular allele(s). The balance between hitchhiking and background selection determines the extent of genetic variation in a population. *See* effective population size; hitchhiking. (Charlesworth, D., et al. 1995. *Genetics* 141:1619.)

backmutation   Mutation (recessive) reverts to the wild-type allele. *See* reversion.

backreaction   Property of RNA polymerase to move backward and cleave the synthesized RNA if a nucleotide needed for the forward (synthetic) reaction is not available. *See* dead-end complex.

bactenein   *See* antimicrobial peptides.

bacteria   Broad taxonomic group of microscopically visible (prokaryotic) organisms with DNA as the genetic material (nucleoid) not enclosed by a distinct membrane within the cell. Bacteria may have various numbers of extrachromosomal elements—plasmids that constitute from about 2 to 20% of the DNA per cell. Their size varies enormously (0.2–750 μm). Bacteria are capable of protein synthesis and of independent metabolism even if they are parasitic or saprophytic. Their cell wall is mucopolysaccharide and protein. Their cells contain ribosomes (70S) but no mitochondria, plastids, endoplasmic reticulum, or other compartments. They divide by fission in an exponential manner as long as nutrients and air are not in limiting supply. The division rate of bacteria during exponential growth can be expressed as $N = 2^g \times N_0$, where N is the number of cells after g generation of growth and N_0 is the initial cell number. After the exponential phase, unless their increasing need is met, the growth either declines or may become stationary. The generation time of *E. coli* under standard conditions may be 20–25 minutes.

Bacteria can be classified into three main groups: *Archebacteriales*, *Eubacteriales*, and *Actinomycetales*. *Eubacteriales* includes *Pseudomonadaceae* (*Pseudomonas aeruginosa*), *Azotobacteriaceae* (*Azotobacter vinelandii*), *Rhizobiaceae* (*Rhizobia, Agrobactria*), *Micrococcaceae* (*Micrococcus pyogenes*), *Parvobacteriaceae* (*Haemophilus influenzae*), *Lactobacteriaceae* (*Diplococcus pneumoniae, Streptococcus faecalis*), *Enterobacteriaceae*, (*Aerobacter aerogenes, Escherichia coli, Salmonella typhimurium*), and *Bacilliaceae* (*Bacillus subtilis, B. thüringiensis*). *Actinomycetales* includes *Mycobacteriaceae* (*Mycobacterium phlei, Mycobacterium tuberculosis*) and *Streptomycetaceae* (*Streptomyces coelicolor, S. griseus*). There are many other types of classification systems. It is common to identify bacteria as gram-positive (indicating that they retain the deep red color of the gram stain [crystal-violet and iodine] after treatment with ethanol) or gram-negative (fail to retain deep red color and may appear colorless or just slightly pinkish). These properties depend on the composition and structure of the cell wall. The gram-positive bacteria are surrounded by peptidoglycan outside of the plasma membrane; the gram-negative cells have an outer membrane enveloping the peptidoglycan wall. The peptidoglycans are polymers of sugars and peptides and are cross-linked by pentaglycines that determine the shape of the cell wall and the bacterium. There are at least 10^{30} bacteria on the planet and at least as many bacteriophages. *See* bacteria counting; bacterial recombination frequency; bacterial transformation; conjugation; recombination molecular mechanisms in prokaryotes.

bacteria counting

Done either by counting the number of colonies formed or by determining cell density in a volume using a photometer. By the first procedure, an inoculum of a great dilution of a culture is seeded on a nutrient agar plate and incubated for a period of time (e.g., 2 days). Each colony formed represents the progeny of a single cell and the number of colonies indicate the number of *live* bacteria in the volume of the inoculum. By the second procedure, the optical density indicates the cell density, which becomes meaningful only if information is available on the correlation between light absorption and cell number, determined earlier by the plating technique described above. If the plate was seeded by 2 mL of the culture diluted 10^7 times and 100 colonies are observed, then the number of live cells is $(100/2) \times 10^7 = 5 \times 10^8$ cells/mL. *See* bacteria; lawn, bacterial.

bacterial artificial chromosome

See BAC.

bacterial recombination frequency

In case of conjugation transfer of genes, recombination frequency is determined by the time in minutes since the beginning of mating. This procedure is useful for genes that are more than 2 to 3 minutes apart. It takes about 90–100 minutes at 37°C to transfer the entire genome (more than 4 million nucleotides) from an Hfr donor bacterium to an F⁻ recipient cell. The efficiency of transfer also depends on the nature of the Hfr strain used. Thus, approximately $5-6 \times 10^4$ nucleotides are transferred per minute. Bacterial recombination does not permit the recovery of the reciprocal products of recombination, and all detected crossover products are double crossovers. If bacterial genes are closer than 2 to 3 minutes, then recombination mapping is used. For bacterial recombination, generally selectable (auxotrophic) markers are used so that the phenotypes can be easily recognized. The recipient strain carries genes, e.g., a and b^+, and the donor strain carries the alleles a^+ and b, defining the interval where recombination is studied. In order to be able to measure the number of successful matings, the donor strain also carries the prototrophic gene (c^+) and the recipient is marked by the auxotrophy allele (c) of the same locus. The c gene does not have to be very close to the interval studied:

$$\frac{a\ b^+}{a^+\ b} \quad \frac{c^+}{c}$$

The *frequency of recombination* (p) is then calculated:

$$p = \frac{\text{number of cells } a^+b^+ \text{ constitution}}{\text{number of } c^+ \text{ cells}}$$

The $a^+\ b^+$ recombinants are the result an exchange between a^+ and b^+ and also beyond the c^+ site as shown by the arrows: $a^+ \uparrow b^+\ c^+ \downarrow$

To determine *gene order by recombination*, at least three loci must be used in a reciprocal manner:

Would the gene order be <u>a b c</u>:				To obtain triple
Donor	a	$\downarrow b^+$	$c^+\downarrow$	prototrophs, **double**
Recipient	a^+	b	c	exchange is
Donor	$\downarrow a^+$	$\downarrow b$	c	sufficient in both of
Recipient	a	b^+	c^+	the reciprocal crosses.

Would the gene sequence be <u>b a c</u>:			In order to obtain triple
Donor	$\downarrow b^+$	$\downarrow a$	$c^+\downarrow$	prototroph recipients,
Recipient	b	a^+	$\uparrow c$	the number of exchanges
Donor	b	$\downarrow a^+$	$\downarrow c$	must be at least
Recipient	b^+	a	c^+	**quadruple**, as shown by

the arrows. In the reciprocal cross, only **double** recombination is required to produce $b^+a^+c^+$ prototrophs.

Thus, depending on whether the gene order is abc or bac, we can tell from the frequency of prototrophs in the reciprocal crosses. The higher number of exchange are less frequent.

Recombination frequency in bacteria within very short intervals, such as between alleles within a gene, can also be determined by transduction. If the constitution of the donor DNA is a^+b^+ and the recipient is $a\ b$, the *frequency of transduction* (recombination) is

$$\frac{[a^+\ b] + [a\ b^+]}{[a^+\ b] + [a\ b^+] + [a^+\ b^+]}$$

Gene order in bacteria can also be determined by a three-point transformation test as illustrated below in a hypothetical experiment when the donor DNA is $a^+\ b^+\ d^+$ and the recipient is $a^-\ b^-\ d^-$, and the reciprocal products of recombination are not recovered:

Genes	Genotypes of Transformants						
a	+	−	−	−	+	+	+
b	+	+	−	+	−	−	+
d	+	+	+	−	−	+	−
Number of cells	12,000	3,400	700	400	2,500	100	1,200

Recombination for a particular interval is calculated by the number of recombinants in the interval(s) divided by the total number of cells transformed in that interval. We have seven classes of cells. Recombination is calculated in three steps: in the ab, bc, and ad intervals:

In the **ab** interval,

$$\frac{[3400 + 400 + 2500 + 100]}{[1200 + 3400 + 400 + 2500 + 100 + 1200]} \approx 0.33$$

In the **bd** interval,

$$\frac{[700 + 400 + 100 + 1200]}{[12000 + 3400 + 700 + 400 + 100 + 1200]} \approx 0.14$$

In the **ad** region,

$$\frac{[3400 + 700 + 400 + 2500 + 100 + 1200]}{[12000 + 3400 + 700 + 400 + 2500 + 100 + 1200]} \approx 0.41$$

Although the frequency of recombination in the three-point transformation test is never exactly additive, it is clear that the $a-d$ distance is the longest, and thus the conclusion that the gene order is abd appears to be reasonable. Recombination may affect population structure and permits evolutionary conclusion. See bacterial transformation; conjugation; crossing over; generalized transduction; mapping, genetic; physical mapping; transduction. (Lederberg, J. 1987. *Annu. Rev. Genet.* 21:23; Feil, E. J. & Spratt, B. G. 2001. *Annu. Rev. Microbiol.* 55:561.)

bacterial transformation Genetic alteration brought about by the uptake and integration of exogenous DNA in the cell that is capable of expression. The exogenous DNA is generally supplied at a concentration of 5 to 10 µg/mL to transformation-competent cells. Competence is a physiological state in which the cells are ready to accept and integrate the exogenous DNA. Competence is maximal in the middle of the logarithmic growth phase. The donor DNA may synapse with the recipient bacterial chromosome, and naked DNA generally replaces a segment of the bacterial genetic material rather than adding to it. The entire length of the donor DNA may not be integrated into the host, and the superfluous material is degraded. The integrated DNA may form a permanent part of the bacterium's chromosomal genetic material. During integration, only one or both strands of the donor DNA may be integrated. Transformation may be regarded as one of the mechanisms of recombination and can be used for determining gene order in the bacterial chromosome. The frequency of transformation in prokaryotes is generally less than 1%, and most commonly it is within the range of 10^{-3} to 10^{-5}. Transformation of bacterial protoplasts (spheroplasts) may occur at much higher frequency. Transformation may also mean the transfer and expression of plasmid DNA in the cell. These plasmids may stay as autonomous elements within the bacteria. Transformation with the aid of plasmids is much more efficient. Also, the competence can greatly be enhanced by some divalent cations and by other means. See bacterial recombination frequency; competence of bacteria; transformation, genetic; vectors. (Hotchkiss, R. D. & Gabor, M. 1970. *Annu. Rev. Genet.* 4:193; Oishi, M. & Cosloy, S. D. 1972. *Biochem. Biophys. Res. Commun.* 49:1568.)

bacteriocin Natural bacterial products that may kill sensitive bacteria. See colicins; pesticin; pyocin. (Riley, M. A. 1998. *Annu. Rev. Genet.* 32:255.)

bacteriophage Viruses infecting bacteria. See development; filamentous phages; icosahedral; lambda-phage; MS2; mu-bacteriophage; phage; phage life cycle; phage morphogenesis; phage therapy; ϕX174; T4; T7; temperate phage; virulence.

(Knipe, D. M. & Howley, P. M., eds. 2001. *Fundamental Virology*. Lippincott Williams & Wilkins, Philadelphia; Brüssow, H. & Hendrix, R. W. 2002. *Cell* 108:13; <www.phage.org>.) The major types and characteristics of bacteriophages are shown below.

Phage	Type	Host	$Da \times 10^6$	Morphology
MS2, f2, R17	RNA, ss, virulent	E. coli	1	icosahedral
ϕ6	RNA, ds, virulent	Pseudomonas	3.3, 4.6, 7.5	icosahedral
ϕX174, G4, St-1	DNA, ss, virulent	E. coli	1.8	icosahedral-tail
M13, fd, f1	DNA, ss, virulent	E. coli	2.1	filamentous
P22	DNA, ds, temperate	Salmonella	26	icosahedral-tail
SPO1	DNA, ds, virulent	Bacillus subtilis	91	isosahedral-tail
T7	DNA, ds, virulent	E. coli	26	octahedral-tail
lambda	DNA, ds, temperate	E. coli	31	icosahedral-tail
P1, P7	DNA, ds, temperate	E. coli	59	head-tail
T5	DNA, ds, virulent	E. coli	75	octahedral-tail
T2, T4, T6	DNA, ds, virulent	E. coli	108	oblong head-tail

ss = single-stranded, ds = double-stranded.

bacteriorhodopsin Light receptor protein in the plasma membrane of some bacteria; it pumps protons upon illumination. See rhodopsin.

bacteriostasis Preventing the reproduction of bacteria, which may lead in the long run to their destruction but not immediate killing. Many antibiotics have such an effect. See antibiotics.

bacteroid Specialized, modified form of bacteria such as found in the root nodules, where bacteroids act in the fashion of intracellular "organelles" in nitrogen fixation. See nitrogen fixation. (Li, Y., et al. 2002. *Microbiology* 148:1959.)

bactigs Contigs of BACs. See BAC; contig.

bactotryptone Peptone rich in indole (tryptophan) used for bacterial cultures and classification of bacteria on the basis of activity.

bacto yeast extract Water-soluble fraction of autolyzed yeast containing vitamin B complex.

baculovirus Large (130 kbp) double-stranded DNA virus used for the construction of insect transformation vectors. The baculoviruses do not transform mammalian or plant cells. The baculovirus vectors accommodate large amounts of DNA and the foreign DNA replacing the polyhedrin gene is expressed under the powerful polyhedrin promoter. The majority of the proteins within the insect remain soluble. The extracellularly present virus particles appearing *late* in the infection are called nonoccluded virus. The occluded virus particles occur in the cell nuclei and appear *very late* in the infection phase. The polyhedra viral protein coating is responsible for the occlusion. *See* AcNBPV; insect viruses; polyhedrosis; transformation; viral vectors. (Grabherr, R., et al. 2001. *Trends Biotechnol.* 19[6]:231.)

baculum Bony structural element above the urethra in the penis of many species (e.g., rodents, carnivores, primates) but absent in humans.

BAD Apoptosis-promoting protein when phosphorylated by MAPK/RSK or Akt or a c-AMP-dependent protein kinase. When it is dephosphorylated (by calcineurin), it may interfere with the apoptosis suppression of Bcl proteins. Also, when RSK phosphorylates CREB, cell survival is facilitated. *See* Akt; apoptosis; BCL; CaM-KK; CREB; MAPK; RSK; survivin. (Konishi, Y., et al. 2002. *Mol. Cell* 9:1005.)

badger (*Taxidea taxus*) $2n = 32$.

badnavirus Double-stranded DNA virus of plants. *See* pararetrovirus.

BAF complex Similar to the SWI/SNF proteins, and it regulates chromatin remodeling. *See* chromatin remodeling; SWI/SNF. (Liu, R., et al. 2001. *Cell* 106:309.)

BAFF TNF receptor ligand, which among other proteins regulates B-cell proliferation and differentiation. *See* APRIL; Blys; NF-κB; TNFR. (Thompson, J. S., et al. 2001. *Science* 293:2108; Schiemann, B., et al. 2001. *Science* 293:2111.)

BAG1 (Bcl2-associated athanogene, 9p12) Part of an anti-apoptotic complex affecting cell division, cell migration, and differentiation. BAG family proteins recruit molecular chaperones and thus have roles in regulating protein conformation. *See* athanogene; BCL. (Takayama, S. & Reed, J. C. 2001. *Nature Cell Biol.* 3:E237.)

BAIT *See* two-hybrid system.

BAK Member of the Bcl protein family that promotes apoptosis by opening the permeability transition pore complex channel. *See* apoptosis; BCL; BID; porin. (Wei, M. C., et al. 2000. *Genes Dev.* 14:2060; Korsmeyer, S. J., et al. 2000. *Cell Death Differ.* 7:1166; Cheng, E. H., et al. 2001. *Mol. Cell* 8:705.)

Bal 31 Exonuclease simultaneously removes nucleotides from the 3' and 5' ends, and thus it can be used for mapping functional sites in DNA. The fragments can then be separated by electrophoresis and assayed after transformation. *See* deletions, unidirectional; exonuclease electrophoresis. (Wei, C.-F., et al. 1983. *J. Biol. Chem.* 258:13506.)

0 time	a b c d e	original DNA
after 1 time unit	b c d	digest
after 2 time units	c	digest

balanced alleles This population model assumes that at the majority of loci several different alleles are present that are kept in a dynamic equilibrium by the continuous but variable selective forces. *See* balanced polymorphism; fitness; Hardy—Weinberg theorem; selection.

balanced lethals Genetic stocks heterozygous for two or more nonallelic linked recessive lethal genes. Since both homozygotes die, only heterozygotes survive, which are the phenotypically wild type or in some cases exhibit mutant phenotype and keep on producing both types of the lethals. Such stocks can be maintained indefinitely as long as recombination between the linked loci can be prevented. For the balancing, generally spanning inversions are used that eliminate the crossover gametes because of the duplications and deficiencies generated by recombination within the inverted region. The first balanced lethal of *Drosophila* contained the gene *Bd1* (*Beaded*, incised wing in heterozygotes, lethal in homozygotes, at map location 3–91.9 [slightly different in some other alleles]) and *l(3)a*, a spontaneous lethal mutation (map position 3–81.6) within the inversion *In(3R)C*. In multiple translocations of plants of different *Oenothera* species, there are also gametic and zygotic recessive lethal genes prevented from becoming homozygous, thus helping to maintain these lethal genes by balanced heterozygosity. Besides the biological advantage of balanced lethals, they may be useful for various types of research. The *Bd* alleles have also been extensively studied at the molecular level, and the developmental functions could be revealed by the availability of heterozygotes for the mutations. *See* lethal equivalent; lethal factors; translocation ring. (Muller, H. J. 1918. *Genetics* 3:422.)

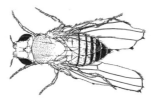

Beaded (From Bridges, C. B. & Morgan, T. H. 1923. *Carnegie Inst. Was.* 327:152).

balanced polymorphism When the fitness (reproductive success) of heterozygotes exceeds both homozygotes at a locus, a stable genetic equilibrium may be established and the heterozygotes may reproduce the homozygotes in equal frequencies. This type of heterozygote advantage may lead to balanced polymorphism, i.e., the population may maintain several genotypes in stable proportions even if some of the homozygotes have low adaptive value. *See* autosomal-recessive lethal assay; balanced lethals; balance of alleles; fitness; Hardy—Weinberg theorem; Muller's ratchet; selection; selection coefficient. (Rucknagel, D. L. & Neel, J. V. 1961. *Progr. Med. Genet.* 1:158.)

balanced translocation Reciprocal translocation where each of the interchanged chromosomes has a centromere. Unbalanced translocations have an acentric piece due to the interchange. *See* translocation, chromosomal.

balancer chromosome Structurally modified (by inversions, translocation), so the recombinants (because of duplications or deficiencies in the meiotic products) are not recovered in the progeny and facilitate the maintenance of certain chromosomal constitutions without recombination. Balancer chromosomes have been developed with the use of clastogenic agents. By inserting the *LoxP* gene in opposite orientations and bringing about recombination with the aid of the *Cre* recombinase, inversions can also be generated in, e.g., mouse cells at particular intervals. *See* autosomal-recessive; balanced lethals; *Basc*; *ClB* method; *Cre / Lox*; inversion; lethal assay; *Oenothera*; Renner complex; targeting genes; translocation, chromosomal.

balancing selection Includes heterozygote advantage (overdominance) or alleles differently selected by sex, season, niche in the habitat, or in a frequency-dependent manner. *See* frequency-dependent selection; overdominance; selection; sexual selection.

BALB/c mice Albino inbred laboratory strain used frequently in immunoglobulin (antibody) and cancer research. It is highly susceptible to *Salmonella*. *See* mouse.

Albino mouse (Courtesy of Dr. Paul Szauter, http://www.informatics. jax.org/mgihome/other/citation.shtml).

Balbiani ring Puff (bloated segment) of the polytenic chromosome indicating special activity (intense RNA transcription) at the site and generating loosening up of the multiple elements of the chromosome. *See* BR RNP; polytenic chromosomes; puff.

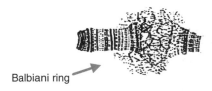

Balbiani ring

baldness Sex-influenced trait in humans being more common in males than in females (particularly with later onset), probably depending to some extent on the level of androgen. It generally displays a developmentally manifested pattern starting from the front hairline toward the top of the head. This condition has a strong hereditary component, but it may be caused by certain diseases (alopecia) and exposure to higher doses of ionizing radiation and certain carcinostatic drugs. Early baldness may be determined by an autosomal-dominant gene (10q24) encoding steroid 17α-hydroxylase with better penetrance in the males than in the females. *See* alopecia; androgen; hair; monilethrix; penetrance.

Baldwin effect Physiological homeostasis permits the survival of a species until mutation may genetically fix the adaptive trait in an originally inhospitable environment. *See* canalization; homeostasis.

BamH1 restriction enzyme with recognition sequence G↓GATCC.

banana (*Musa acuminata*, $x = 11$) Fruit plant. The diploid fruits are loaded with seeds and have minimal edible pulp. The majority of the edible fruits are harvested from seedless triploid plants. When the triploids are crossed with diploids, the progeny is partly tetraploid ($2n = 44$) and heptaploid ($2n = 77$) because of the high frequency of unreduced $3x$ and $6x$ gametes, and their fruits are also seedless. Some of the related species have chromosome numbers $2n = 14$, 18, and 20. *See* seedless fruits; sugar beet; triploidy. (Simmonds, N. W. 1966. *Bananas*. Longman, London.)

band Element of the cross-striped chromosome. The banding (perpendicular to the length of the salivary gland chromosomes and continuous across the giant chromosome) may be due to condensation of the juxtapositioned chrommeres or to specific staining of the chromatin (*See* bands of polytenic chromosomes; chromosome banding). The average DNA content in a single natural band of the *Drosophila melanogaster* salivary gland chromosome is 26.2 kb. The total number of salivary chromosome bands is 5072.

Electrophoretic separation of restriction enzyme–digested DNA or pulsed-field electrophoresis-separated small chromosomes, as well as various proteins subjected to separation in electric field, generate bands in the substrate (gel) when visualized either by staining or by special illumination. *See* chromosome banding; coefficient of crossing over; electrophoresis; FISH; pulsed-field electrophoresis.

band cloning Amplifying DNA bands extracted from electrophoretic gels in genetic cloning vectors for molecular analyses. *See* cloning vectors.

banding pattern Distribution of chromosome bands reflecting genetic differences or differences in expression of genes displaying more or less loose puffs. *See* chromosome banding; lampbrush chromosomes; polytenic chromosomes; puff.

band-morph mutation Distinguished by electrophoretic analysis of the proteins. Mutations resulting in amino acid replacement of different charges appear as mobility differences in the electric field. Although these studies were very popular during the 1960s and 1970s, they have very poor resolution because only one-fourth or less of the mutations are detectable. The advantage of band-morph mutation was that large populations could be screened for mutations that would not necessarily have other phenotypic effects. *See* band; electrophoresis. (Harada, K., et al. 1993. *Jpn. J. Genet.* 68:605.)

band III protein Transmembrane protein consisting of about 800 amino acids. *See* spectrin. (Low, P. S., et al. 2001. *Blood Cells Mol. Dis.* 27:81.)

band-sharing coefficient (S_{xy}) Indicates the proportion of shared DNA fragments separated by electrophoresis; $S_{xy} = (2n_{xy})/(n_x + n_y)$, where n_x and n_y are the number of bands in x and y samples and n_{xy} is the number of shared bands. This coefficient may be used to determine the genetic composition of populations on the basis of DNA. In multilocus forensic tests the British legal system used the formula $(0.26)^k$ for calculating the match probabilities of alleles of at least 4 kb in length. The 0.26 is an empirical constant and k is the average number of matching alleles. This latter formula lacks sufficient robustness. The single-locus probes (SLP) based on the profiles of 6–8 short tandem repeat loci (STR) are more popular today. This procedure is useful with DNA samples as low as 100 pg when amplified by PCR. *See* DNA fingerprinting. (Zhu, J., et al. 1996. *Poultry Sci.* 75:25.)

band shifting *See* gel retardation assay.

bands of polytenic chromosomes Deeply stained prominent crossbands on the chromosomes where the chromomeres of the elementary strands are appositioned. The salivary chromosomes of *Drosophila* display about 5,000 bands, and it was once assumed that each corresponded to a gene locus. Today, it is known that the number of genes is about 2.5 times the number of bands. In region 2B of the X chromosome of *Drosophila*, the bands may appear different, and rather than perpendicular to the axis, they may be roughly parallel to the axis. In situ hybridization with molecular probes suggests that this unusual structure is caused by inverted repeats in the DNA. *See* band; coefficient of crossing over for the tip of the X chromosome; polyteny; salivary gland chromosomes.

BankIt GenBank submission form for protein coding sequences. It generates a GenBank accession number. Its address is <http://www.ncbi.nlm.nih.gov/BankIt/>.

Bannayan-Riley-Ruvalcaba syndrome *See* Bannayan-Zonona syndrome.

Bannayan-Zonona syndrome Autosomal-dominant macrocephaly with multiple lipomas and hemangiomas. It also involves hamartomatous polyposis cancer susceptibility. Haploinsufficiency of PTEN may play a role in its manifestation. *See* hemangioma; lipomatosis; multiple hamartomas; PTEN.

BAP (1) 6-benzylaminopurine, a plant hormone. *See* plant hormones. (2) B-cell receptor associated proteins. Three (32, 37, 41 kDa) proteins associated only with the IgM membranes. The 32 and similar 37 kDa molecules form heterodimers and seem to be inhibitors of cell division. BAP29 (240 amino acids) and BAP31 (245 amino acids) are 43% homologous and bind mainly to IgD and somewhat to IgM. *See* B-lymphocyte receptor.

BAPG *See* bullous pemphigoid autoimmune disease.

BAR Regulator of the Fas- and Apaf-mediated pathways of apoptosis. *See* apoptosis.

bar code, genetic (molecular) Bar codes generally represent two or four different vertical widths that correspond to digits 0 and 1, and can in turn specify numbers 0 to 9. An optical laser can read the bar-coded information and through computer the scanner can identify various types of information, including properties of a gene, phenotypic expression, etc. Molecular bar codes can be generated by ~20 base oligonucleotides (UPTAG, DOWNTAG) introduced by transformation into special (deletion) cells. *See* DNA chips; targeting genes. (Gad, S., et al. 2001. *Genes Chromosomes Cancer* 31:75.)

BARD1 BRCA1-associated ring domain protein inhibits polyadenylation of mRNA in cooperation with Cstf-50 (cleavage stimulation factor). *See* breast cancer; cleavage stimulation factor. (Joukov, V., et al. 2001. *Proc. Natl. Acad. Sci. USA* 98:12078.)

Bardet-Biedl syndrome (BBS6, MKKS, 20p12) Heterogeneous recessive disease involving retinal dystrophy (retinitis pigmentosa), polydactyly, and other anomalies of the limbs, obesity, underdeveloped genitalia, kidney malfunction, and diabetes. Mental retardation is also common. Chromosomal locations 11q13 (BBS1), 2q31 (BBS5), 3p12-p13 (BBS3), 15q23 (BBS4), and 16q21 (BBS2) have also been reported. The major form of BBS shares chromosomal position with McKusick-Kaufman (MKKS, 20p12) syndrome. It seems that the basic problem involves a chaperonin that folds several proteins improperly. *See* chaperonin; eye diseases; kidney diseases; McKusick-Kaufman syndrome; triallelic inheritance. (Beales, P. L., et al. 2001. *Am. J. Hum. Genet.* 68:606; Mykytyn, K., et al. 2001. *Nature Genet.* 28:188; Katsanis, N., et al. 2001. *Hum. Mol. Genet.* 10:2293.)

Bare Lymphocyte Syndrome (BLS, 19p12 [RFXANK], 16p13 [MHC2TA], 1q21.1-q21.3 [RFX5], 13q14 [RFXAP]) Group of severe recessive immunodeficiency diseases caused by defects in the regulation of the major histocompatibility system by any of the four loci identified above. Some forms are due to a defect(s) in the HLA class I or class II genes involving lymphocyte differentiation. Some of the defects involve the RFX proteins, which bind to the X box of the MHC2TA promoter. MHC II deficiency may also result from mutation in MCC2TA transactivator. The MHCII molecules are heterodimeric transmembrane proteins. The current therapy is bone marrow transplantation. In the future, gene therapy may be feasible. *See* gene therapy; HLA; immunodeficiency; lymphocyte; MHC; RFX. (Reith, W. & Mach, B. 2001. *Annu. Rev. Immunol.* 19:331.)

barley (*Hordeum*) Cereal crop used for feed, food, and in the brewery industry. The cultivated *H. sativum* is diploid $2n = 14$. Some of the wild barleys are polyploids. *See* haploid (*H. bulbosum*); *Hordeum*; <http://www.tigr. org/tdb/tgi.shtml>.)

BAR mutation of *Drosophila* (*B*, map position 1–57.0), reduces the eye to a vertical bar with about 90 facets in the male and about 70 in the female compared with 740 in normal males and 780 in normal females; the heterozygous females have 360. The *B* mutation is actually a tandem duplication of salivary band 16A that arises by unequal (oblique) crossing over. Thus the "normal" allele has 16A, the *Bar* 16A-16A, and the *Ultrabar* 16A-16A-16A constitution. The phenotype is actually a position effect, not the cause of a dosage effect, as the genetic analyses demonstrated. The process of unequal crossing over may be repeated; as many as 9 copies of band 16A can accumulate in a single X chromosome. Also, the 16A band may be lost, resulting in reversion by the loss of the *roo* transposable element. *B* mutations may also be induced by the P hybrid dysgenesis element, whereas chemical mutagens never produce this mutation. These facts indicate that the *breakage points in the duplications cause the Bar phenotype*. The *Bar* phenotype may be the result of breakage in a regulatory element or within an intron, causing abnormal fusion of the exons. The *Bar* locus is very large; it spans at least 37 kb DNA. *See ClB*; duplication; exon; intron; position effect; unequal crossing over. (Bridges, Sturtevant, A. H. 1925. *Genetics* 10:117; Bridges, C. B. 1936. *Science* 83:210.)

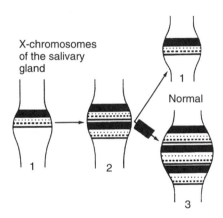

barnase   *Bacillus amyloliquefaciens* ribonuclease is a ribonuclease that may be associated with chaperones. *See* barstar; chaperones; ribonucleases.

Ribbon representation of barnase. (From Alm, E. & Baker, D. 1999. *Proc. Natl. Acad. Sci. USA* 96:11305.)

Barr body   Dark-stained (heteropycnotic) structure visible at the periphery of the interphase nuclei of cells that have more than one X chromosome. XY cells do not have a Barr body, whereas normal XX female cells have one. The number of Barr bodies (named after M. L. Barr) is always one less than the number of X chromosomes, indicating that the nonactive X chromosomes remain condensed (dosage compensation). Barr bodies are also present in XXY males. The Barr body is sometimes called sex chromatin. In the leukocytes the Barr body is enclosed in a special nuclear appendage, called a drumstick because of its shape. Methylation of CpG dinucleotides is the mechanism of inactivation. *See* dosage compensation; lyonization; methylation of DNA. (Heard, E., et al. 1997. *Annu. Rev. Genet.* 31:571; Hong, B., et al. 2001. *Proc. Natl. Acad. Sci. USA* 98:8703.)

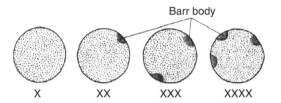

barrage   Sign of vegetative incompatibility in fungi. At the zone of contact between the two types of mycelia a distinguishable zone is formed as the result of antagonism between the two strains. (Rizet, G. 1952. *Rev. Cytol. Biol. Végét.* 13:51.)

barrel   Protein β-sheets closing the interior and α-chains on the exterior. *See* protein structure.

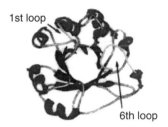

See color plate in Color figures section in center of book. α/β-Barrel folds of phosphoribosylanthranylate isomerase. The β-sheets in the center are darker. (From Gerlt, J. A. 2000. *Nature Struct. Biol.* 7:171.)

barren stalk   (*ba*) Maize genes (in chromosomes 3, 2, and 9) affecting the tassel or ear or both and causing partial/full sterility. The tassel is male inflorescence. *See* tassel seed.

barring gene (*B*)   *See* autosexing.

barstar   Inhibitor of barnase. *See* barnase. (Hartley, R. W. 1989. *Trends. Biochem. Sci.* 14:450.)

Barth syndrome   *See* endocardial fibroelastosis.

Bartter syndrome   Type 1 (15q15-q21.1) is a defect in NaKCl transporter. Type 2–dominant human chromosome 11q24–encoded disease is characterized by salt wasting and low blood pressure, accompanied by excessive amounts of calcium in the urine. The basic defect involves an inward rectifier potassium ion channel. Type 3 (1p36) involves chloride channel

B. *See* Gitelman syndrome; hypertension; hypoaldosteronism; hypokalemia; ion channels; Liddle syndrome; ROMK.

basal   At or near the base.

basal body   Group of microtubules and proteins at the base of cilia and flagella of eukaryotes. *See* cilia; flagellum; microtubule.

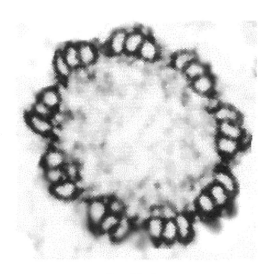

Basal body.

basal cell carcinoma   *See* nevoid basal cell carcinoma.

basal lamina   *See* basement membrane.

basal-level element (BLE)   Has enhancer-type functions in gene regulation and can occur at several positions. *See* enhancer.

basal promoter   Generally situated 100 bp upstream of the transcription initiation site; contains various regulatory elements of transcription. *See* core promoter; promoter; transcription factors.

basal transcription factor   *See* transcription factors.

BASC   One of several similar *Drosophila* genetic stocks; contains the dominant *Bar* (*B*), the recessive eye color allele, *apricot* (*w*^{a}), and several *scute* inversions. The *B* and *w*^{a} markers identify the untreated chromosomes of the untreated females and eliminate the crossovers with the treated X chromosomes of the males. The mutagenic effectiveness is determined on the basis of the reduced proportion of males in F$_2$ if a lethal or sublethal mutation is induced in the X chromosome by the treatment. This type of analysis is called Muller 5 technique after H. J. Muller, who designed the first stocks. The advantage of these stocks is that both males and homozygous females are completely fertile, whereas the XO males are poorly viable and no crossovers appear along the X chromosome.

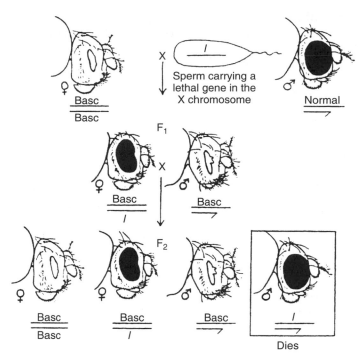

ℓ Indicates lethal mutation.

Variegation may occur in some unlinked genes. Rarely some exceptional females are also detected due to an unequal sister chromatid exchange in the inversion heterozygote females. *See* autosomal-dominant mutation; autosomal-recessive mutation; ClB. (Inoue, Y. 1992. *Genetica* 87:169; Forbes, C. 1981. *Mutation Res.* 90:255.)

Exceptional females.

BASC   BRCA1-associated genome surveillance complex includes tumor suppressors, DNA repair proteins, DNA replication factor C, etc. *See* breast cancer. (Wang, Y., et al. 2000. *Genes Dev.* 14:927.)

base   Lowest part of a structure of a compound or ion that can combine with protons to yield salt. *Nitrogenous bases* such as are the pyrimidines and purines of nucleic acids. *See* nucleic acid bases.

base analog   Nucleic acid base or nucleoside similar to the normal compounds, but it causes mutation when incorporated into DNA either by incorporating in the wrong place or by mispairing with the incorrect base. The most commonly used mutagenic base analogs are 5-bromouracil (thymine analog) and 2-aminopurine (adenine analog). *See* base substitution; 2-aminopurine; bromouracil; hydrogen bonding; point mutation; universal base. (Freeze, E. 1959. *J. Mol. Biol.* 1:87.)

base calling   Identifying the correct nucleotide in a sequence in the DNA. Identifying the correct base sequence on the basis of hybridization in microarrays compared to the actual direct sequencing information. Miscalls are the false identifications. *See* microarray hybridization; PHRAP; PHRED. (Walther, D., et al. 2001. *Genome Res.* 11:875.)

Basedow disease   *See* goiter.

base excision repair (BER)   *See* DNA repair.

base flipping   Some enzymes such as methyltransferases, glycosylases, T4 endonuclease V, *E. coli* phospholyase, and endonuclease III, etc., must access the bases inside the sugar-phosphate backbone of the B DNA double helix in order to be recognized by the active site of the protein. Some bases are swung out of the helix into an extrahelical position in order to meet the requirement. *See* ABC excinucleases; cyclobutane ring; DNA repair; endonuclease; glycosylases; methylation of DNA; photolyase. (Cheng, X. & Roberts, R. J. 2001. *Nucleic Acids Res.* 29:3784; Patel, P. H., et al. 2001. *J. Mol. Biol.* 308:823.)

basement membrane (basal lamina)   Less than 500-nanometer-thick laminated condensation of the extracellular matrix (including laminin, collagen IV, and other proteins) on the basal surfaces of epithelia and condensed mesenchyma. The basement membrane is an attachment platform and a barrier to cell mixing. Several human diseases involve anomalies of basement membranes and/or associated proteins. *See* Alport's disease; collagen; extracellular matrix; Goodpasture syndrome; laminin; proteoglycan.

base-modifying agent   Nitrous acid causes oxidative deamination; hydroxylamine converts cytosine into hydroxylaminocytosine (a thymine analog); alkylating agents place alkyl groups at several possible positions to purines and pyrimidines. *See* base substitution; chemical mutagens; hydrogen bonding; point mutation.

base pair (bp)   Hydrogen-bonded A=T and G≡C in DNA or A=U in double-stranded RNA. *See* hydrogen pairing; mismatch; mispairing; universal bases; Watson-Crick model)

base pair opening (base flipping)   Nucleoside unit swivels out of the DNA helix and inserts into the recognition pocket of a protein. Such nucleoside extrusion and extrahelical recognition may take place by processing the DNA by various glycosylases and endonuclease action. *See* base flipping.

base promoter   *See* core promoter.

base sequencing   *See* DNA sequencing.

base-specific reagents for DNA single strands   (1) For dimethylsulfate (DMS) + hydrazine, the methylated cytosine is cleaved at the 3′ position. (2) DMS alone methylates guanine. (3) Osmiumtetroxide or potassium permanganate oxidize the C5-C6 double bonds in thymidine.

(4) Diethylpyrocarbonate (O[CO₂C₂O₅]₂) preferentially modifies adenine at N-7, although it affects other purines, too. *See* DNA sequencing (Maxam & Gilbert method).

base stacking   The nucleotides in parts of a polynucleotide chain may lay in such a way that the faces of the rings are appositioned. The stacking is most likely to occur by noncovalent forces near the chain termini where the bases move somewhat. It conveys some rigidity to the strand(s). The stacking is detectable by physical methods such as circular dichroism and optical rotatory dispersion. Reagents, which weaken hydrophobic reactions, eliminate the stacking, and heating reduces stacking, resulting in hyperchromicity. Destruction of hydrogen bonding also reduces stacking in double-stranded DNA. Base stacking may occur in double-stranded molecules where the pairing is weakened by deletions or mismatches. *See* circular dichroism; hyperchromicity; optical rotatory dispersion. (Kool, E. T. 2001. *Annu. Rev. Biophys. Biomol. Struct.* 30:1.)

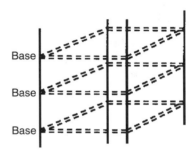

base substitution   When a pyrimidine in the DNA is replaced by another pyrimidine or a purine is replaced by another purine, the change is a *transition*. When a purine is replaced by a pyrimidine, or vice versa, a *transversion* takes place. These changes cause mutation in the DNA, and they may cause amino acid replacement in the protein if the base substitution involves a nonsynonymous codon. Many mutagens cause base substitutions, e.g., hydroxylamine (NH₂OH) targets cytosine (C) and hydroxyaminocytosine is formed, which is a thymine analog. As a result, a C≡G base pair is replaced by a T=A bp. Similarly 5-bromouracil may cause a T=A to be replaced by a C≡G through tautomeric shift, and 2-aminopurine, an adenine analog, causes a tautomeric shift that may cause the replacement of an A=T pair by a G≡C pair. Some other chemicals, e.g., nitrous acid, by deamination, changes cytosine to uracil and converts adenine into guanine. These base substitutions may also cause reversions. A hydroxylamine-induced mutation may be reverted by bromouracil, etc. Generally it is assumed that base substitutions occur independently and coincidental double substitutions should be rare. Some genetic repair mechanisms may, however, bring about more than single replacements. A large-scale evolutionary study indicates that double substitutions may occur in the serine codons of primates at the high frequency of 0.1/site/billion years. *See* base analogs; chemical mutagens; DNA; evolution and base substitutions; hydrogen bonding; incorporation error; mutation and DNA replication; point mutation; replication error; substitution mutations. (Freese, E. 1959. *Brookhaven Symp. Biol.* 12:63.)

BASH (B cell-restricted adaptor protein, BLNK/SLP-64) After ligation, Sly tyrosine kinase phosphorylates BASH and it binds various B-cell signaling proteins that control B-cell development. Its role is similar to that of SLP-76 for T cells. *See* B lymphocyte; ITIM; SLP-76. (Tsuji, S., et al. 2001. *J. Exp. Med.* 194:529.)

basic chromosome number Found in the gametes of diploid organisms; it is represented by x; in polyploids, the haploid (n) number may be $2x$, $3x$, and so on, depending on the number of genomes contained. The basic number is frequently called a genome. *See* chromosome numbers; genome; polyploid.

basic copy gene Silent copy of a *Trypanosoma* gene that is activated by transposition to an activation site in the telomeric region of the chromosome. *See* telomere; *Trypanosomas*.

basic dye Stains negatively charged molecules. *See* stains.

Basidiomycetes Taxonomic group of fungi bearing the meiotic products in basidia. *See* basidium; mushroom.

Basidium with four meiotic spores.

basidium Fungal reproductive structure generally of club shape where meiosis takes place, then the haploid basidiospores are released for infection of host plants. *See* stem rust.

basonuclin Cell type–specific Zn-finger protein with nuclear localization sequence and a serine stripe (serine-rich region). It is abundant in the human keratinocyte nuclei, but in the absence of phosphorylation it is in the cytosol. Basonuclin also occurs in the epidermal cells and the germ cells of the testis and ovary. It binds to the rRNA promoter and apparently regulates rRNA transcription. *See* nuclear localization signal; Zinc finger. (Tian, Q., et al. 2001. *Development* 128:407.)

basophil Type of white blood cell, well stainable with basic cytological dyes. Basophils contain conspicuous secretory granules and release histamine and serotonin in some immune reactions. Any other (acidic) structure or molecule with an affinity for positive charges. *See* blood; granulocytes; immune system.

BASTA Glufosinate ammonium herbicide, pesticide, and a selective agent in plant transformation. *See* herbicides; transformation, genetic. (Rathore, K. S., et al. 1993. *Plant Mol. Biol.* 21:871.)

bastard Hybrid (in German); an illegitimate or undesirable offspring (in English).

BAT *Carollia perpicillata* $2n = 21$ in males, 20 in females; *Glossophaga soricina* $2n = 32$; *Desmodus rotundus murinus* $2n = 28$; *Atropzous pallidus* $2n = 46$; *Eptesicus fuscus* $2n = 50$; *Myotis velifer incautus* $2n = 44$; *Nysticeius humeralis* $2n = 46$.

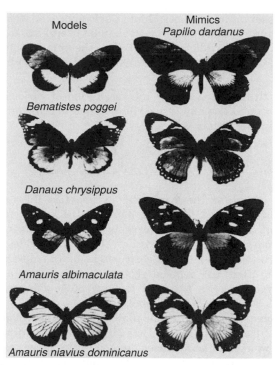

Examples of batesian mimicry of the butterfly *Papilio* is shown at right. (From Sheppard, P. M. 1959. *Cold Spring Harbor Symp. Quant. Biol.* 24:131.)

Bateman's principle Reproductive success of males shows greater variation than that of females because of greater competition among males and a larger number of male gametes.

Batesian mimicry Adaptive evolutionary device. Certain species develop phenotypic characteristics of sympatric species (models) in order to secure better survival. The models are repugnant (distasteful) to certain predators and avoid them, and so the mimickers when mistaken for the models also escape destruction. The mimicry is more common among females than males (butterflies) because females are more subject to predation than males. *See* adaptation; Müllerian mimicry; natural selection.

Batten disease Recessive human chromosome 16p12.1-p11.2 juvenile onset familial amaurotic idiocy caused by lipid accumulation in the nerve tissues and vacuolization of the lymphocytes. Its incidence is $\sim 5 \times 10^{-5}$. The disease seems to be associated with low vacuolar pH. In yeast *btn1* defects, chloroquine reverses the phenotype. *See* amaurotic familial idiocy; ceroid lipofuscinosis; chloroquine; PPT; Vogt-Spielmeyer disease. (Luiro, K., et al. 2001. *Hum. Mol. Genet.* 10:2123.)

Batten-Turner syndrome *See* myopathy, congenital.

bauplan Pattern of body organization.

BAX *See* porin.

Bayesian mapping Can be applied to QTL. Inferences are made on the basis of the joint posterior distribution of all unknown variables given the prior distribution of all unknowns

Bayes' theorem   Permits the estimation of various conditional probabilities and is used for decision-making processes. In the simplest general form:

$$P(A|B) = \frac{P(B|A)P(A)}{P(B|A)P(A) + P(B|A')P(A')}$$

Let us illustrate the use of it in genetics by assuming that we have 3 individuals: 2 homozygotes for a semilethal dominant factor and 1 heterozygote for the same semilethal and incompletely dominant gene. There are problems with visual classification at an early stage of the development assuming that the 2 homozygotes (A) have 60% and the heterozygote (A') is expected to have 80% viability. Thus, (P[B|A]) = 0.6, and P(B|A') = 0.8.

The chance of selecting one individual of either genotype is P(A) = 2/3 and P(A') = 1/3. The individual picked turns out to be very weak, and being uncertain about the choice, we would like to figure out — in view of the information available — what is the probability that we have chosen the heterozygote:

$$P(A|B) = \frac{(0.8)(0.33)}{(0.8)(0.33) + (0.6)(0.67)} \cong 0.4$$

Basically the Bayesian method considers the classical population parameters as random variables with a specific a priori probability of distribution. Then the *conditional probability* is estimated on the basis of the a priori distribution. The conditional probability is thus a property of a posteriori distribution because the accepted or supposed a priori distribution is used for the estimation of an existing situation in a population. *See* conditional probability; probability; risk. (Shoemaker, J. S., et al. 1999. *Trends Genet.* 15:354; Bernardo, J. M. & Smith, A. F. M. 1994. *Bayesian Theory.* Oxford Univ. Press, Oxford, UK.)

BB-1   Same as B7 or CD80.

BBB   Blood-brain barrier seriously limits transport to the central nervous system because of the tight junction of the endothelial cells of the brain capillaries. Molecules larger than 180 MW and viruses may be excluded. Lymphocytes may enter the central nervous system but are not retained unless foreign antigens are present. Neurons and some other cells may poorly express MHC proteins and escape the effects of cytotoxic T cells. In such a situation, different viruses (rubella, measles, polyoma JC, herpes simplex, rabies, mumps) may infect the brain. In the absence of such a barrier, serum may leak into the brain, causing edema. Angiotensin may be required to maintain BBB. *See* angiotensin; glucose transporters; MHC; multiple sclerosis; protein transduction. (Asahi, M., et al. 2001. *J. Neurosci.* 21:7724.)

B1 B cell   Fetal and early infant B cells, but they may exist in greater proportion than normal in leukemias and autoimmune diseases. *See* B lymphocyte; leukemias.

B box   Part of the internal control region of tRNA and some other genes. *See* A box; internal control region of pol III genes; pol III; tRNA.

B cell   *See* B lymphocyte; B lymphocyte receptor.

B1, B2 repeat   Highly dispersed SINE elements in the mouse genome. *See* SINE.

B-cell differentiation factor   *See* interferon β-2 (IFNB2).

B-cell growth factor   *See* IL-4; interleukins.

BCG   *See Bacillus Calmette-Guerin.*

BCGF   B-cell growth factor. 12 kDa cytokine produced by activated T cells. See B cell; cytokine; lymphocytes; T cell.

B chromosome   Accessory (supernumerary) chromosome. B chromosomes are generally heterochromatic and carry no major genes yet may be present in several copies in many species of plants. B chromosomes have no homology to the regular chromosomal set (A chromosomes) and are prone to nondisjunction because their centromeres appear to be defective. If A-B translocations are constructed, the

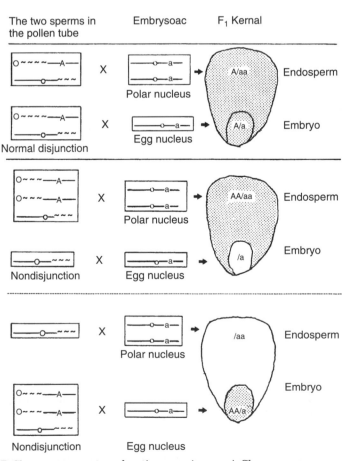

B-Chromosome translocation mapping. A-Chromosome segments ———— B-Chromosome segments ~~~~~~~~ Dominant allele *A* expression shaded, recessive allele *a* is indicated blank.

placement of genes to chromosomes, arms, or even shorter regions may be facilitated. The principle of the use of A-B chromosome translocations is outlined preceding page. The A chromosome or a translocated segment carries the dominant *A* allele and the B chromosome has no counterpart to it; therefore, a null phenotype (*a*) appears in its absence. In the diagram, the male has the translocation and the female is homozygous recessive for the *a* allele. In case there is no B-chromosomal nondisjunction — when the chromosomal constitution is as diagrammed — both endosperm and embryo express the dominant gene. In case of nondisjunction, the phenotypic effects depend on the constitution of the sperm, which fertilizes the diploid polar nucleus of the endosperm or the embryo, respectively (see middle and bottom parts of the diagram). This type of difference reveals the approximate physical and genetic position of the locus. Had the dominant allele been outside the translocated segment, the recessive allele would not have been unmasked. The example shows the most favorable case when the consequence of the translocation can be identified without tissue specificity.

B chromosomes have been reported in ∼1300 species of plants and in ∼500 species of invertebrate and vertebrate animals. The transmission of the B chromosomes varies in different species, and it may be preferential in the female (e.g., in grasshoppers) or in some wasps by the male. *See* centromere; mapping, genetic; translocation, genetic; trisomic analysis. (Beckett, J. B. 1982. *J. Hered.* 73:29; Page, B. T., et al. 2001. *Genetics* 159:291; Camacho, J. P. M., et al. 2000. *Philos. Trans. R. Soc. Lond. B* 355:163.)

BCIP 5-bromo-4-chloro-3-indolyl phosphate is used in combination with nitroblue tetrazolium (NBT; it reveals precipitated indoxyl groups) for the detection of antigen–antibody–antibody-AP (alkaline phosphatase) complexes. *See* antibody; antigen.

BCL1, BCL2, BCL3, BCL5, BCL6 (B-cell lymphoma) Leukemia oncogenes, but they are also upregulated in various other types of cancers. Bcλ-1 is cyclin D1. Bcℓ-2 (18q21.3) suppresses apoptosis (by phosphatase action when bound to calcineurin) as a defense against malignant tumorigenesis and suppresses signaling by NF-AT. BCL2 apparently shuts off the voltage-dependent anion channel on the mitochondrial membrane and prevents the leakage of apoptotic cytochrome c into the cytosol to guard against apoptosis. It also promotes regeneration of severed cells in the central nervous system. The Bcℓ-2 protein also functions as an ion channel and a docking protein. BCL-2 is located in the outer membrane of the mitochondria, nuclei, and endoplasmic reticulum. Bcℓ6 regulates STAT and cytokine signaling. The Bcℓ protein is usually upregulated in lymphomas, and gene therapy involving deoxyoligonucleotides (such as 5′-TCTCCCAGCGTGCGCC<u>CAT</u>-3′) targeted to the <u>AUG</u> initiator codon has been used. There are over a dozen members of the Bcℓ family. Bcℓ-2 is homologous to *Ced-9* of *Caenorhabditis*. Antimycin A, a complex of highly toxic antifungal substances, inhibitors of electron transport, binds Bcℓ and Bcℓ-X and favors apoptosis, thus possibly protecting against cancerous growth. Bcℓ-2 interferes with RAD51-controlled recombination that may mediate error-free repair and thus may promote cancer-prone conditions independently from its antiapoptosis effect. *See* apoptosis; BAD; BAK; BAX; BID; calcineurin; cyclin; cytokine; G3139; G3854; IA; leukemia; lymphoma; malignant growth; NF-AT; STAT. (Petros, A. M., et al. 2001. *Proc. Natl. Acad. Sci. USA* 98:3012; Saintigny, Y., et al. 2001. *EMBO J.* 20:2596; Vander Heiden, M. G., et al. 2001. *J. Biol. Chem.* 276:19414.)

BCMA One of the B-lymphocyte receptors. *See* BAFF; BCR; Blys.

BCR B-cell antigen receptor. *See* B lymphocyte.

BCR (bcr) Break point cluster region in the Philadelphia chromosome is an area where multiple chromosomal breakages have been observed, leading to cancerous transformation (leukemia, ABL). The BCR polypeptide contains 1,271 amino acids and its normal function is a protein serine/threonine kinase. *See* ABL; leukemia; Philadelphia chromosome.

B DNA Conformation of the DNA most common in hydrated living cells. *See* DNA types.

BDNF Brain-derived neurotrophic factor is an autocrine growth substance of neurons. *See* autocrine; neuron; Parkinson disease.

beacon, molecular Hairpin-shaped single-stranded oligonucleotide genetic probe that becomes fluorescent after hybridization to the homologous target. One end is covalently bound to a fluorophore and a nonfluorescent quencher is attached to the other end. If the probe does not find homology, there is no fluorescence because the quencher prevents it in the hairpin. If the probe locates a homologous sequence, unwinding removes the quencher from the vicinity of the fluorophore and the site lights up.* *See* spectral genotyping; (Tyagi, S. & Kramer, F. R. 1996. *Nature Biotechnol.* 14:303; Heyduk, T. & Heyduk, E. 2002. *Nature Biotechnol.* 20:171) diagram below.

beads-on-a-string Originally this meant the (light microscopically visible) chromosome structure in pachytene where the chromomeres could be seen. In the 1920s, chromomeres were equated with genes, units of function, mutation, and recombination. Today, beads on a string denote DNA strands wrapped around the eight histones (nucleosomes) as detected by the electron microscope. *See* chromomeres; complex locus; nucleosome; pachynema.

Beagle *See* copia.

bean (*Phaseolus* spp)   Pulse crop with $2n = 2x = 22$ chromosomes, including the most common *P. vulgaris* (French or navy bean) and *P. lunatus* (lima bean).

bear   *Ursus americanus* (black bear) $2n = 74$; *Tremarctos ornatus* (spectacled bear) $2n = 52$.

Beare-Stevenson syndrome   Autosomal-dominant disorder involving furrowed, corrugated skin (cutis gyrata), head bone fusions, facial anomalies, abnormal digits, umbilical and genital malformations, and early death. The basic defect is in the fibroblast growth factor receptor 2, encoded at 10q26. Some heterogeneity exists. *See* FGF.

beaver   *Castor canadensis*, $2n = 40$. A large (>80 long) brown rodent.

Becker muscular dystrophy (BMD)   *See* muscular dystrophy

Beckwith-Wiedemann syndrome (EMG syndrome)
Caused by a dominant gene in the human chromosome 11p15.5 area. The symptoms include enlarged tongue (detectable at birth) and generally umbilical anomalies (omphalocele = herniated intestines at the belly button area), hypoglycemia, enlargement of the internal organs (visceromegaly), frequent concomitant kidney and liver anomalies, tumorous striated muscles (rhabdomyosarcoma), etc. It may be associated with trisomy for chromosome 11 and it has been suggested that it is caused by paternal or maternal disomy when the normal (most commonly paternal) chromosome is lost from the trisomic cell lineage, or imprinting. A gene encoding a cyclin-dependent kinase inhibitor (p57^{KIP2}) that is imprinted and preferentially expressed by the maternal allele may be responsible for some of the cases. Insulin-like growth factor (IGF2) also has a regulatory effect. Deletion, duplication, and balanced translocation are also suggested for some cases. It has been shown recently that KVLQT1 gene, encoding a putative potassium channel and mapped to the 1p15.5 region, controls imprinting not just the Beckwith-Wiedemann gene but also Jervell and Lange-Nielsen syndrome and LQT heart arrhythmia. *See* cancer; cyclin; disomic; disomic, uniparental; Golabi-Behmel syndrome; imprinting, insulin-like growth factors; ion channels; Jervell and Lange-Nielssen syndrome; LQT; p57; rhabdomyosarcoma; Simpon-Wilms' tumor. (Alders, M., et al. 2000. *Am. J. Hum. Genet.* 66:1473; Bliek, J., et al. 2001. *Hum. Mol. Genet.* 10:467.)

beclin 1 (17q21)   Mammalian autophagy gene that is deleted in 40–75% of the sporadic ovarian and breast carcinoma cells. *See* autophagy.

becquerel   1 disintegration of radioactive material/second; 1 becquerel $= 2.7027 \times 10^{-11}$ curie ($\approx$27 picocuries). *See* curie.

bedwetting   *See* nocturnal enuresis.

beech (*Fagus*)   Hardwood tree, *F. sylvestris*, $2n = 2x = 24$.

Beethoven (Bth)   Mouse mutation with dominant progressive hearing loss at the locus homologous to the human gene TMC1, causing DFNA36. Beethoven, the great composer, was afflicted by the same mutation. *See* deafness.

begonia (*Begonia semperflorens*)   Ornamental plant, $2n = 34$.

behavior, in humans   Has long been suspected to be genetically determined, but with a few exceptions (Lesch-Nyhan syndrome, Tay-Sachs disease, Huntington's chorea, etc.) the genetic control is not completely understood. The majority of the behavioral traits are under the control of several genes. In such instances, the tools of quantitative inheritance are needed, such as heritability, comparison of monozygotic-dizygotic twins with the general population, QTL mapping, etc. The approximate ratios of monozygotic:dizygotic concordance for alcoholism for females (1.1) versus males (1.7), dyslexia (1.7), Alzheimer's disease (2.1), major affective disorders (2.4), schizophrenia (2.7), autism (6.7). Heritabilities determined by intraclass correlation were: for memory (0.22), mental processing speed (0.22), scholastic achievement in adolescence (0.38), spatial reasoning (0.40), adolescent vocational interest (0.42), neuroticism (0.46), verbal reasoning (0.50), general intelligence (0.52). (Data based on Plomin, Owen & McGuffin, 1994. *Science* 264:1733.)

Cognitive abilities are also studied as part of the developmental genetic pattern (longitudinal genetic analysis). Multivariate genetic analysis determines the covariance (see correlation) among multiple traits. Although some genetic effects are specific to certain abilities, the majority of the genetic components have overlapping effects. The studies must also consider assessing behavioral cognitive genetic traits that form a continuum, and the anomaly in a proband or several individuals may be just the extreme form of a normally existing behavioral pattern. Behavioral traits generally have about 50% or greater environmental components. These effects include family relationships and changes in that (e.g., divorce, death, accidents), social environment (economic status, schools, drug use, neighborhood), etc. The quantitative genetic approaches assume that behavioral traits are complex and are the end result of cooperative action of individual genes expressed as a phenotypic class rather than one gene–one disorder (OGOD). Some behavioral anomalies may show cosegregation with DNA markers such as those used in QTL analysis, although some of the mental anomalies, such as phenylketonuria (single recessive defect in phenylalanine hydroxylation), may account for about 1% of the affliction in mental asylums (*See* fragile-X syndrome). More than 100 single-gene-determined human diseases include mental retardation as part of the syndromes. Defects in the X-chromosomally encoded gene product, mitochondrial enzyme, monoamine oxidase A (MAOA), were attributed to violent behavior and schizophrenia. MAOA degrades serotonin, dopamine and norepinephrine. *See* aggression; autism;

behavior genetics; cocaine; cognitive abilities; determinism; differential psychology; dopamine; dyslexia; ethology; heritability; homosexuality; human intelligence; instinct; MAOA; morality; norepinephrine; personality; QTL; self-destructive behavior; serotonin.

behavior genetics   Analyzes the genetic determination and regulation of how organisms behave. The majority of these traits (courtship, bird and frog songs) is under multigenic control and they depend a great deal on the influence of the environment. In a few cases, large effects of single genes have also been shown. In the honeybee, a single gene controls the habit of uncapping the honeycombs containing dead larvae, but another gene is required for the removal of the dead brood. If both genes are present, the colony becomes resistant to the bacterial disease foulbrood because of the improved hygienic behavior.

Alcoholism, criminality, etc., in humans may be determined by several genes and by the social environment. Lesch-Nyhan syndrome is caused by a deficiency of the enzyme hypoxanthine-guanine phosphorybosyl transferase, and because of this, the salvage pathway of nucleic acid is inoperational. As a consequence, purines accumulate and uric acid is overproduced, leading to gout-like symptoms, but more importantly the nervous system is also affected, leading to antisocial behavior and self-mutilation. This gene has been isolated and cloned and may be transferred to the afflicted human body for gene therapy.

Since behavioral traits are determined by the nervous system, neurogenetics may provide the answer for many serious conditions such as Alzheimer disease (an amyloid-accumulating presenile dementia), neurofibromatosis (a soft tumor of the nervous system affecting the entire body and involving a protein resembling a GTPase activator), etc. Molecular analysis of memory and learning ability has made progress through studies of simple organisms such as the slug *Aplysia* and *Drosophila*. Several genes involved in the development of the nervous system of *Drosophila* have been cloned.

Studies show that mice without the *fos* gene fail to nurse their pups, presumably because of brain lesions. For genetic and developmental analysis of nerve functions, the nematode, *Caenorhabditis*, is particularly well suited because its entire nervous system consists of only 302 cells. In the tobacco hornworm (*Manduca sexta*) the feeding preference depends on an acquired recognition template. The naïve larvae can feed on different plant species, but once they have been exposed to a steroidal glycoside (indioside D) present in tobacco leaves, they become "addicted" to that host.

Although all traits and attributes have some biological basis, the influence of the environment has a very substantial role in the development of behavioral traits. Frequently, for unknown reasons, identical behavioral traits are expressed differently in isogenic strains maintained in different laboratories under apparently identical conditions. This fact tells us that genes cannot exonerate criminal behavior; neither should people be condemned on the basis of collective responsibility by supposed sharing of a common gene pool.

The molecular basis of a few mouse genes is now known. Deletion from the RORα gene causes the *staggerer* phenotype. The *vibrator* mouse carries a retroposon in an intron of the phosphatidylinositol transfer protein. The *weaver*

ataxia is a serine → glycine replacement in a G-protein-gated inwardly rectifying K-ion channel. The different types of estrogen receptors control mating behavior. Oxytocin knockout male mice are afflicted by an olfactory recognition, a deficit in the amygdala of the brain and can be corrected by injection of oxytocin. Galanin, a neuropeptide with inhibitory action on neurotransmission and memory, causes Alzheimer disease–like behavior in mice. Enkephalin (opioid peptide) knockout and/or the loss of its receptor involves increase in anxiety. Mutation in the dopamine receptor D2 results in Parkisonism-like phenotype. Loss of the serotonin receptors increases anxiety. NO synthase knockout mice show greater aggressiveness. Single genes encoding pheromone-binding protein(s) may regulate complex social behavior, such as recognition of conspecific individuals in social insects. (Keller, L. & Parker, J. D. 2002. *Current Genet.* 12[5]:R180). Mice deficient in the TRP2 gene lose their ability to distinguish between sexual types, and the males mate with both males and females. *See* addiction; affective disorders; aggression; alcoholism; altruistic behavior; Alzheimer disease; attention deficit hyperactivity disorder; autism; avoidance learning; behavior; behavior, in humans; cognitive abilities; courtship in *Drosophila*; cross-fostering; dyslexia; ethics; eugenics; fate mapping; FOS; homosexual; human intelligence; Huntington's chorea; instinct; mental retardation; morality; personality. (Pfaff, D. 2001. *Proc. Natl. Acad. Sci. USA* 98:5957; McGuffin, P., et al. 2001. *Science* 291:1232; Toye, A. A. & Cox, R. 2001. *Curr. Biol.* 11:R473; Krieger, M. J. B. & Ross, K. G. 2002. *Science* 295:328; Bucan, M. & Abel., T. 2002. *Nature Rev. Genet.* 3:114; Stowers, L., et al. 2002. *Science* 295:1493; Kucharski, R. & Maleszka, R. 2002. *Genome Biol.* 3[2]:res.0007.1; Rankin, C. H. 2002. *Nature Rev. Genet.* 3:622.)

Behcet syndrome (TAP)   Rare mouth and genital inflammation in humans; probably autosomal dominant.

Behr syndrome   Recessive infantile optical nerve atrophy. *See* optic atrophy.

BEL   *See copia.*

bell curve   *See* normal distribution.

Bellevalia   Subspecies of lilies ($2n = 8$ or 16) with large and well-stainable chromosomes.

bellophage   1 kb RNA phage, encoding a nucleocapsid protein, a replicase component, and an integrase. It has some retroviral-like properties. The small replicase binds to the host DNA polymerase and modifies it in such a way that the enzymes act as an RNA-directed DNA polymerase. With the aid of the integrase, the formed DNA is inserted into the host genome as a prophage. The nucleocapsid, because of its leucine zipper motif, can associate with its helper phage. The helper is originally the *Salmonella* phage Ω, but it may recruit adenovirus or influenza virus for this function if mutation alters the leucine zipper. In chickens and some apes, the provirus integrates into the mtDNA rather than into the nucleus. Thus, it opens the possibilities of inserting foreign DNAs into mitochondria and chloroplasts for genetic engineering. *See* integrase; leucine zipper; transformation of organelles; viral vectors.

BEM1   Protein with SH_3 domain involved in signal transduction. *See* SH_3; signal transduction.

Bematistes pongei   African butterfly mimicked by *Papilio dardanus*. *See* Batesian mimicry.

Bence-Jones protein   Some immunoglobulin heavy-chain diseases (HCD) such as the lymphoproliferative neoplasms may contain only the antibody heavy chains (IgM, IgG, IgA), and even those are truncated and are deficient in most parts of the variable region. The γ- and α-type HCD cells do not synthesize light chains, but the μ-HCD cells secrete an almost normal light chain that is detectable in the urine and is called Bence-Jones protein. Some of the bone marrow cancer (myeloma) patients also produce Bence-Jones protein in their urine. These light-chain immunoglobulins are generally homogeneous because they are the products of a clone of cancer cells and were very useful historically to gain information on antibody structure. *See* antibody; immunoglobulins; monoclonal antibody; myeloma. (Beetham, R. 2000. *Ann. Clin. Biochem.* 37(5)563.)

beneficial mutation   The majority of new mutations are less well adapted than the prevailing wild-type allele in a particular environment or at best they may be neutral. Beneficial mutations are rare because during the long history of evolution the possible mutations at a locus had been tried and the good ones preserved. Nevertheless, if a new mutation has 0.01 reproductive advantage, the odds against its survival in the first generation are $e^{-1.01} = 0.364$. Its chances to be eliminated by the 127th generation are reduced to 0.973 compared to a neutral mutation that would be eliminated by a chance of 0.985. Even mutations with an exceptionally high selective advantage may have a good chance of being lost ($e^{-2} \cong 0.1353$). Under normal conditions the selective advantage(s) is generally very small and the chance of ultimate survival is $(y) = 2s$ and the chance of extinction is $(l) = 1 - 2s$. For the mutation to have a better than 50% probability of survival, the requisites must be $(1 - 2s)^n < 0.5$ or $(1 - 2s) > 2$. Hence, $-nln((1 - 2s) > ln2$ or approximately $-n(-2s) > ln2$, and therefore $n > (ln2)/2s$ or $\cong (0.6931)/2s$. If $(s) = 0.01$ and $(n) =$ number of mutations, (n) must be larger than $0.6931/(2 \times 0.01) \cong 34.66$. That is, at least 35 mutational events must take place with at least 1% selective advantage of the mutants over the wild type for one to ultimately survive. If the rate of mutation is 10^{-6}, a population of about 35 million may provide such a mathematical chance. Under evolutionary conditions, neutral or even deleterious mutations may make it, however, by random drift or chance in small populations. *See* mutation, neutral; mutation, spontaneous; mutation in human populations; mutation rate. (Fisher, R. A. 1958. *The Genetical Theory of Natural Selection*. Dover; Dobzhansky, T. & Spassky, B. 1947. *Evolution* 1:191; Miura, T. & Sonigo, P. 2001. *J. Theor. Biol.* 209:497.)

benign hereditary chorea   Dominant childhood chorea encoded at 14q. *See* chorea.

Benton—Davis plaque hybridization   Involves selection of recombinant bacteriophages on the basis of DNA hybridization with ^{32}P probes on an appropriate (nitrocellulose or nylon) membrane. For the screening of a mammalian or other large library, hundreds of thousands of recombinants are necessary. In a 150 mm petri dish, 5×10^4 plaques may be used. *See* DNA hybridization; DNA library; Grunstein—Hogness screening; recombination molecular mechanisms; plaque; plaque-forming unit; prokaryotes. (Benton, W. D. & Davis, R. W. 1977. *Science* 196:180; Lewis, J. A., et al. 1983. *Mol. Cell Biol.* 3:1815.)

benzimidazole   Tubulin-binding chemical used as an herbicide (trifluralin, oryzalin) or fungicide (benomyl). Oral dose of LD_{50} for mice is $\sim$2910 mg/kg. *See* tubulin.

benzo(a)pyrene   Highly carcinogenic polycyclic hydrocarbon generated by combustion at relatively lower temperature by polymerization of organic material. It occurs in automobile exhausts, by burning of coal, in cigarette smoke, fried and grilled meat (in char-broiled T-bone steaks more than 50 µg/kg has been detected). It has been estimated that 13,000 tons are annually released into the world's atmosphere by these processes. A single 0.2 mg intragastric dose per mouse resulted in 14 tumors in 5 of the 11 animals treated. Skin exposure and inhalation of fumes are high and fast carcinogenicity routes. Benzo(a)pyrene is also a promutagen requiring metabolic activation in *E. coli*, yeast, *Drosophila*, various rodents, and the plant *Arabidopsis*. Benzo(a)pyrene forms adducts with guanine by binding to the N2 position, but it also forms adducts with deoxyadenosine. It causes sister-chromatid exchange and the formation of micronuclei. Exposure to benzo(a)pyrene results in the expression of cytochrome P450 (*cyp1a1*) in the skin and liver of mice if the aryl hydrocarbon receptor (AhR) is active. Cyp1a2 gene expression did not need AhR. For carcinogenesis by benzo(a)pyrene AhR is a requisite. *See* adduct; Ames test; arylhydrocarbon receptor; bioassays in genetic toxicology; carcinogens; cytochromes; environmental mutagens; micronucleus formation as a bioassay; sister chromatid exchange. (Chiapperino, D., et al. 2002. *J. Biol. Chem.* 277:11765.)

benzyladenine (6-benzylaminopurine)   *See* plant hormones.

BER (base excision repair)   *See* DNA repair; excision repair.

Berardinelli disease   *See* lipodystrophy, familial.

Bernard-Soulier syndrome   22q11.2– and 17pter-p12– recessive dysfunction of the platelets; thrombocytopenia is a potentially lethal bleeding disease. A platelet membrane receptor, glycoprotein Ib-IX-V, is absent, and the platelets do not agglutinate by interaction with the von Willebrand plasma factor. *See* platelet; thrombocytopenia; von Willebrand disease.

Bernoulli process   Independent experiments that can provide only two outcomes: yes or no. Success or failure is called a Bernoulli trial and the two-event classes and their probabilities are called the Bernoulli process, where $p =$ probability of success and $1 - p = q =$ probability of failure. If in a sequence of 10 trials there are 4 successes, the probability of that sequence is $p^4 q^6$; if

$$p = \frac{2}{3},$$

then the probability of the sequence becomes

$$\left(\frac{2}{3}\right)^4 \left(\frac{1}{3}\right)^6$$

The general formula becomes

$$P(r \text{ success}|N, p) = \binom{N}{r} p^r q^{N-r} \ldots\ldots \{1\}$$

where $p =$ the probability of success, $r =$ the exact number of successes, and $N =$ the number of independent trials. Let us assume that we observe a monogenic segregation where the penetrance of the mutant class is reduced from 25% to 20%, then the probability of finding a recessive mutant among 3 individuals will be according to {1}:

$$P(1 \text{ mutant among 3 individuals}) = \binom{3}{1}(0.20)^1(0.80)^2$$

$$= \left(\frac{3!}{2!(2-1)!}\right) 0.2 \times 0.64$$

$$= 0.384$$

If the population is increased to say 210, the chances increase to 70 for finding a mutant. *See* binomial probability.

BERT   Background equivalent radiation time. If a diagnostic X-ray exam uses 360 mrem, that corresponds to 1 BERT/year (approximate average in the United States). To the natural background contributing sources in mrem: radon (200), cosmic sources (100), medical treatment (39), consumer and industrial products (11), air travel (6) and nuclear industry (<1). The total may vary, however, from 100 to 600 mrem or even more at certain locations. The general public usually overestimates the risk of the nuclear industry and underestimates that from medical diagnosis and treatment. *See* cosmic radiation; radiation hazard assessment; rem; risk.

berylliosis (CBD)   Granulomatous (nodular inflammation) lung disease among people exposed to beryllium dust. Homozygosity of a rare major histocompatibility allele (MHC) predisposes certain individuals. *See* MHC.

Best disease   *See* macular dystrophy.

BeT (best hit)   Feature of orthologous sequences displaying homologies among individual genes (COGs) in different species. *See* COG.

β   *See* error types; power of a test.

β-amyloid   Exists as extracellular deposits of the brain plaques in Alzheimer disease. It is split off the amyloid precursor protein (APP) by secretase. In the neuronal tissue,

APP_{695} is prevalent. If Ile, Phe, or Gly replaces Val642, the substitutions lead to fragmentation of nucleosomal DNA in the neurons and presumably contribute to neurotoxicity. *See* Alzheimer's disease; amino acid symbols in protein sequences; amyloidosis; prion; scrapie; secretase.

The major amyloid fibers (Aβ1-42) in Alzheimer disease are truncated at the C terminus:

Aβ1-40: DAEFRHDSGYEVHHQLVFFAEDVGSNKGAIIGLMVGGVV

Aβ1-42:
DAEFRHDSGYEVHHQLVFFAEDVGSNKGAIIGLMVGGVVIA

Aggregation of the fibers may lead to plaque formation seen in amyloidosis. The aggregation may be initiated by "seeding" like a crystallization process

---

β-ARK   β-adrenergic receptor kinase. *See* adrenergic receptors barrel.

β-barrel   Polypeptide chain of a membrane protein forms a folded-up β-sheet arranged in the shape of a barrel. *See* membrane proteins; protein structure; barrel.

β-catenin   Component of the cadherin-catenin cell adhesion complex. *See* adherens junction.

β-conformation   Extended conformation of a peptide chain; a type of secondary structure. *See* protein structure.

beta complex   One of the alternately distributed translocation complexes of the plant *Oenothera*. *See* alpha complex alternative; complex heterozygote; multiple translocations.

beta distribution   Very similar to the binomial probability function, but the beta distribution is continuous and the binomial distribution is discrete.

$$f(x) = \frac{x^{\alpha-1}(1-x)^{\beta-1}}{B(\alpha, \beta)}.$$

See binomial distribution.

β-galactosidase   Enzyme (lactase) that splits the disaccharide lactose into galactose and glucose. It can act on some lactose analogs too, e.g., on ONPG (o-nitrophenyl-β-D-galactopyranoside). When exposed to the active enzymes (10^{10} molecules/mL), this substrate (10^{-3} M or less) yields a yellow product (that has an absorption maximum at 420 nm) and can be used to measure the activity of the enzyme. In cells grown in A medium with Z buffer, the activity of galactosidase is determined by the formula:

$$1000x \frac{OD_{420} - (1.75 x\, OD_{550})}{t x (0.1 x\, OD_{660})},$$

where OD is the optical density at the wavelength indicated, and t is the time of the reaction run in minutes. On petri plates the activity of β-galactosidase is detected on EMB agar (containing eosin yellow, methylene blue, and lactose), and in case the sugar is fermented, a dark red color develops. Bacterial galactosidase is an inducible enzyme; induction takes place by allolactose that is formed upon the action of the residual few galactosidase molecules in the noninduced cells. A gratuitous inducer (inducing the synthesis of the enzyme, although is not a substrate itself) is isopropyl β-D-thiogalactoside (IPTG). Constitutive mutants of the *E. coli z* gene can be identified on Xgal media containing 5-bromo-4-chloro-3-indolyl-β-D-galactoside dissolved generally in dimethylformamide (20 mg/mL). This compound is not an inducer of the enzyme, but it is cleaved by it and thus a blue indolyl derivative is released. (*See* also galactose utilization, Fabry's disease sphingolipidoses, galactosidase; gangliosidosis general, Krabbe's leukodystrophy, lactosyl ceramidosis, *Lac* operon, Xgal. (Pshezhetsky, A. V. & Ashmarina, M. 2001. *Progr. Nucleic Acid Res. Mol. Biol.* 69:81.)

ß-glucuronidase *See* GUS.

β lactamase
Enzyme (synonym: penicillinase) capable of cleaving the β-lactam ring of antibiotics of the penicillin family. Their activity is determined by the R group attached to the lactam ring. The majority of the synthetic penicillins are not susceptible to penicillinase action. The coding gene was originally detected in Tn*3*. The ampicillin resistance gene in the pBR322 plasmid codes for a 263-amino-acid-residue preprotein containing a 23-amino-acid leader sequence that directs the secretion of the protein into the periplasmic space of the bacterium. The transcription of the gene starts counterclockwise at pBR322 coordinate 4146 and ends at 3297. Its mRNA in vitro contains a 5′-pppGpA terminus. The tetracycline resistance gene in pBR322 is transcribed from another promoter clockwise, starting at coordinate 244 or 245. The Tc^R gene encodes a polypeptide of 396 residues. Penicillinases occur naturally only in bacteria with peptidoglycan cell wall. The lack of the enzyme, in the absence of antibiotics, is inconsequential for the bacteria. The β-lactamase genes are used extensively in vector construction (to convey antibiotic resistance) and for the detection of insertional events that inactivate the enzymes. The lactamase enzyme can be used for real-time monitoring gene transcription. A substrate (e.g., cephalosporin) complexed with a fluorochrome (7-hydroxy-coumarin) upon hydrolysis may generate a wavelength shift (from blue to green) in the emission of the substrate located in the plasma membrane. With the aid of a cell sorter, transcription can thus be monitored in single cells. β-lactamase is inhibited by clavulanic acid (an oxygen containing β-lactam). *See* antibiotics; cell sorter; coumarin; fluorochrome; PBP; penicillin; periplasma; Tn*3*; vectors. (Daiyasu, H., et al. 2001. *FEBS Lett.* 503:1.)

ApoB-48 fraction is made in the gut and the ApoB-1000 in the liver by differential processing of the transcript of the same locus. *See* abetalipoproteinemia; hyperbetalipoproteinemia; hypobetaliproteinemia.

β-oxidation
Degradation of fatty acids at the β-carbon into acetyl-coenzyme A. *See* acetyl CoA; fatty acids.

beta-particle
Electron emitted by radioactive isotope. The mass of a β-particle is 1/1837 that of a proton. The negatively charged form of it is an electron, whereas the positively charged is a proton. The beta-particles have no independent existence; they are created at the instance of emission. In the biological laboratory the most commonly used isotopes emitting β-radiation (with energy in MEV) are H^3 (0.018), C^{14} (0.155), P^{32} (1.718), S^{35} (0.167), and I^{131} (0.600 and 0.300 but also emits γ-radiations of various energy levels). The mean length of the path of H^3 is about 0.5 µm and that of P^{32} is about 2600 µm. *See* isotopes; linear energy transfer.

β-pleated sheet
Extended polypeptide chains in parallel or antiparallel arrangement linked by hydrogen bonds between the amino and carboxyl groups. *See* protein structure.

β-sheet
Secondary structure of proteins where relaxed polypeptide chains are running in close parallel or antiparallel arrangement. *See* protein structure.

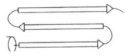

β-sheet breaker peptides (iAβ5)
Four amino acids (LVFF) in the 17–20 N-terminal domain of amyloid-β protein (Aβ) may be substituted (particularly by proline) in this region and may alter the conformation of β-sheets, as well as reducing formation of and disassembling the already formed amyloid fibrils characteristic for Alzheimer disease. A charged amino acid may be added to increase solubility (Leu, Pro, Phe, Phe, Asp). Shorter or different peptides may have opposite effects. The iAβ5 molecules may provide an approach for the treatment of Alzheimer disease. *See* Alzheimer disease. (Soto, C., et al. 2000. *Lancet* 355:192.)

Beta vulgaris
Beets (*Chenopodiaceae*), basic chromosome number 9. Sugar beets, fodder beets, mangold, and chards are all important food and feed crops. The sugar beets represent a glowing example of the success of selective plant breeding by increasing sugar content (about 2% in the mid-18th century) by over 10-fold in some modern varieties. The most productive current varieties display triploid heterosis and improved disease resistance, and the monogerm "seeds" facilitate mechanization of cultivation, etc. *See* heterosis; monogerm seed; triploid.

R
|
C=O
|
HN
| H   S   CH₃
H—C—C   C
| | CH₃
C — N   C
|| *β-Lactam ring* H COO⁻
O

beta-lipoprotein (apolipoprotein, 2p24)
Component of the low-density lipoprotein fraction (LDL) in the plasma. The

betel nut (*Arecia catechu*)
Seed palm tree; used as a stimulant; $2n = 4x = 32$.

bethedging Sexually mature individuals reduce their reproductive potentials due to environmental circumstances. (Menu, F., et al. 2000. *Am. Nat.* 155:724.)

Bethlem myopathy Dominant human disorder involving contractures of the joints, muscular weakness, and wasting. It is associated with mutations of collagen type VII genes in human chromosome 21q22.3 and 2q37. *See* collagen; laminin.

BEV Baculovirus expression vector is a potential tool to control insect populations by biological means. *See* baculovirus; biological control; viral vectors.

bFGF Basic fibroblast growth factor.

BFP Blue fluorescent protein is similar to GFP (green fluorescent protein). Excitation at 368 nm causes light emission at 445 nm, which excites the Ser65Cis mutant of GFP and causes light emission at 509 nm. *See* aequorin.

BGH *See* bovine growth hormone.

BHK Baby hamster kidney cell; cultured fibroblasts of the Syrian hamster. *See* hamster.

bHLH Basic helix-loop-helix protein. *See* helix-loop-helix.

b/HLH/Z motif Basic amino acid sequence at the N terminus, probably required for DNA binding, helix-loop-helix structure, and leucine zipper. This general structure is widespread among biologically active proteins involved in DNA binding. *See* binding proteins; DNA-binding protein domains; helix-loop-helix.

bialaphos Inhibitor is a glutamine synthetase normally produced by *Streptomyces hygroscopicus*. Upon splitting off two alanine residues, it is activated into phosphinotricin. *See* herbicides.

biallelic For gene expression in diploids both alleles must be present. *See* allele.

bi-armed chromosome Has two arms at the opposite sides of the centromere. *See* chromosome morphology; telochromosome.

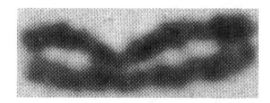

bias Average error of an estimate. False difference of an observation from the correct value.

BIBAC Binary bacterial artificial chromosome is a plant genetic expression vector that can be propagated in *Agrobacterium tumefaciens* and *E. coli*, and it can deliver large (160 kb) foreign DNA sequences to plant chromosomes. *See Agrobacterium*; BAC; transformation; vectors.

bicoid (*bcd*) Maternal effect mutation in *Drosophila*. The larvae lack a head, thorax, some abdominal segments, and duplicate telsons. In the wild type the *bicoid* mRNA is localized in the anterior part of the egg. The mammalian homologs are Pitx1 and Pitx2. *See Drosophila*; left-right asymmetry; maternal effect; morphogenesis in *Drosophila*; telson. (Cha, B.-J., et al. 2001. *Cell* 106:36; Houchmandzadeh, B., et al. 2002. *Nature* 415:798.)

BID Pro-apoptotic protein that links proximal signals to the apoptotic pathways (FAS and TNFR). The inactive cytosolic form of BID (p22) is split into the 15 kDa fragment, tBID (truncated BID), which then moves to the mitochondria. Its BH3 domain then oligomerizes with BAK, causing mitochondrial dysfunction and the release of cytochrome c. Postproteolytic N-myristoylation switches on the BID-induced apoptosis. *See* apoptosis; BAK; Bc𝓁; myristic acid. (Wei, M. C., et al. 2000. *Genes Dev.* 14:2060; Zha, J., et al. 2000. *Science* 290:1761.)

bidirectional gene organization The genes in large genomes are not dispersed uniformly but clusters are formed. Among the 144 and 319, known genes of human chromosome 21 and 22, respectively 22% and 18%, respectively, are divergently arranged within ~1 kb from each other whereas the average spacing distance for all genes is ~85 kb. The spacing islands of the bidirectional genes in most cases carried a CpG island that generally overlapped partially or entirely the first exon or rarely more than one. This organization apparently has regulatory advantage. (See operon, clustering, Adachi, N. & Lieber, M. R. 2002. *Cell* 109:807.)

bidirectional replication From the replicational origin the replication moves in opposite directions in the DNA. *See* replication, bidirectional; replication fork.

BIDS *See* hair—brain syndrome.

biennial Plants with life span extended into 2 years, yet the total length is less than 2 years; e.g., they germinate in the fall and mature and die before the end of the following year.

bifunctional antibody   Carries an antibody variable-region fragment that secures its ability to recognize certain molecules (antigen). In addition, a fused heterologous component conveys either enzyme activity or carries a toxin or a pro-drug, etc. Such a complex can home on the cognate molecules and destroy or alter them according to the specificity of the heterologous portion. *See* bispecific antibody.

bifunctional enzyme   Has apparently evolved with the potential to adapt the amino and carboxy terminal tracts to fulfill the needs of metabolic requirements of different tissues. *See* one gene–one enzyme theorem. (Kitzing, K., et al. 2001. *J. Biol. Chem.* 276:42658.)

bifunctional mutagen   *See* functionality of mutagens.

Big Blue®   Commercially available mouse strain carrying the bacterial *LacI* gene (in about 40 copies) as a stably integrated lambda vector, making it very useful for the laboratory detection of in vivo mutagenic/carcinogenic effects. The transgene is extracted from mice, and with the aid of the λ-shuttle vector, it is introduced into *E. coli* bacteria. In the presence of Xgal substrate, mutations in the *LacI* gene give rise to blue plaques because of the loss of inhibition of the galactosidase gene, *LacZ*. Thus, mutations in the prokaryotic gene within the animals, activated by animal metabolism, are screened in bacteria because of the convenience of detection in the prokaryotic system. This system permits the detection of mutation in different animal tissues, e.g., liver, brain, etc., at different developmental stages and under different conditions. *See* β-galactosidase; Xgal, host-mediated assay; Lac operon; Muta™ Mouse. (Gossen, J. A., et al. 1989 *Proc. Natl. Acad. Sci. USA* 86:7971; Nohmi, T., et al. 2000. *Mutation Res.* 455:191.)

BigSeq   Computer program with similar purpose as Mask. Contains information on millions of contigs. *See* contig; physical map.

bilateral symmetry   *See* zygomorphic.

Bilateral symmetry.

bilayer   A membrane bilayer consists of amphipathic lipids (and proteins), and the nonpolar phase faces inward. The majority of cellular membranes are double membranes. *See* amphipathic.

bile salt   Detergent-type steroid derivative involved in digestion and absorption of lipids.

bilineality   More than a single locus determines a particular trait, and they may segregate independently, making chromosomal localization by genetic techniques very difficult.

bilirubin   Bile pigment formed by the degradation of hemoglobin and other heme-containing molecules such as cytochromes. It is circulated in the blood as a complex with albumin. When deposited in the liver, it forms bilirubin diglucoronide. It may arise from biliverdin, a breakdown product of heme, through reduction. Bilirubin is a strong anti-oxidant and may cause brain damage in neonatal jaundice. *See* Alagille syndrome; Byler disease; cerebral xanthomatosis; cholestasis; Crigler-Nijjar syndrome; Dubin-Johnson syndrome; Gilbert syndrome; hyperbilirubinemia; jaundice; steroid dehydrogenase; steroid reductase. (Tomaro, M. L. & del C. Battle, A. M. 2002. *Int. J. Biochem. & Cell Biol.* 34:216.)

BimC   Family of motor proteins of the kinesin group involved in the separation of the mitotic chromosomes by the spindle. *See* mitosis; monastrol; motor proteins; spindle.

BimD   Negative regulator of the cell cycle progression in *Aspergillus*; mediates recombination and chromosome morphology.

BimE   *Aspergillus* protein subunit of the APC complex, homologous to APC1. *See* APC; cell cycle.

bimodal distribution   When the population, represented graphically, displays two peaks or, in general, when the data are clustered in two modes, in two classes.

BIN   Group of markers (microsatellite DNA) mapped to the same location.

BIN1   Box-dependent Myc-interacting protein-1 is a tumor-suppressor protein (human chromosome 2q14) interacting with the Myc oncoprotein. It is related to amphiphysin, which serves a similar purpose in breast cancer, and to the RVS167 cell cycle control gene of yeast. *See* MYC; tumor suppressor. (DuHadaway, J. B., et al. 2001. *Cancer Res.* 61:3151.)

Bin2 (Cct3, TriC)   Synonymous with the CCT γ-chaperonin subunit. *See* chaperonins.

Bin3 (Cct2)   Synonymous with CCT β-subunit of chaperonins. *See* chaperonins.

binary   Any condition, choice, or selection with two possibilities or a numeration system with a radix of 2. *See* founder cells; radix.

binary fission   Splitting into two parts. Bacteria, chloroplasts, and mitochondria that do not have a mitotic mechanism reproduce this way after DNA replication has been completed.

binary targeting   Specific recombinase gene (*Cre, Flp*) is carried by one of the mating pairs and the site-specific recombination site (*loxP*, *FRT*) is present in the other partner. *See* Cre/LoxP; Flp/FRT; targeting genes.

binary vector   of *Agrobacterium* — carrying two plasmids, one has the T-DNA borders and other sequences (special genes, selectable markers) that will integrate into the transformed cell's chromosomes; the other is a helper plasmid carrying the Ti plasmid virulence genes required for transfer, but no part of the latter plasmid is integrated into the host genome during transformation. *See* agrobacterial vectors; cointegrate vector; T-DNA; transformation, genetic; virulence genes of *Agrobacterium*. (Bevan, M. 1984. *Nucleic Acids Res.* 12:8711.)

binase   12 kDa dimeric ($\alpha\beta$) endonucleolytic ribonuclease binding to N-1 of 3'-guanine monophosphate. It has 82% amino acid identity with barnase. *See* barnase. (Wang, L., et al. 2001. *Proc. Natl. Acad. Sci. USA* 98:7684.)

bindin   Acrosomal protein mediating the species-specific binding between gametes during fertilization. *See* acrosomal process; acrosome; fertilization; sperm. (Glaser, R. W., et al. 1999. *Biochemistry* 38:2560.)

binding energy   Derived from noncovalent interaction between ligand and receptor, enzyme and substrate. *See* ligand; receptor.

binding proteins   Great variety; they control gene expression generally at the level of transcription (transcription factors, hormones, heat-shock proteins, etc.). The majority binds to upstream consensus sequences. The cap-binding proteins regulate the stability of the mRNA. Some of them are transcription termination factors, such as rho in bacteria or the Sal I box-binding proteins in mice. Their position and function at the DNA level is studied by footprinting. Some proteins bind to the cellular membranes and control imports and exports; others mediate signal transduction. These proteins may have a combinatorial hierarchy and thus are capable of influencing a multitude of processes in the cell, far in excess of their individual numbers. *See* affinity-directed mass spectrometry; DNA-binding protein domains; footprinting; single-strand-binding protein; signal transduction; transcription factors; UAS. (Ren, B., et al. 2000. *Science* 290:2306.)

Binet test   *See* human intelligence.

binomial coefficient   *See* binomial probability.

binomial distribution   Useful in genetics for the direct estimation of segregation ratios in case of dominance by expansion of $(3 + 1)^n$, where n is the number of heterozygous loci (note: the $3 + 1$ must not be added). By expansion, the binomial becomes

$$1 \times 3^n + [n!]/1!(n-1)!] \times 3^{n-1} + [n!/2!(n-2)!] \times$$
$$3^{n-2} + \cdots + [n!/(n-1)!] \times 3^{n-(n-1)} + 1 \times 3^{n-n}$$

The *exponent* of a base gives the number of loci with the dominant phenotype, the *power* identifies the frequency of that phenotype, and the *coefficients* show how many times quadruple,

triple, etc., dominant phenotypic classes will be expected theoretically. The solution for four heterozygous pairs of alleles is

$$(1 \times 3^4) + \left(\frac{4 \times 3 \times 2 \times 1}{1 \times 3 \times 2 \times 1} \times 3^{4-1}\right) + \left(\frac{4 \times 3 \times 2 \times 1}{2 \times 1 \times 2 \times 1} \times 3^{4-2}\right)$$
$$+ \left(\frac{4 \times 3 \times 2 \times 1}{3 \times 2 \times 1 \times 1} \times 3^{4-3}\right) + (1 \times 3^{4-4})$$
$$= (1 \times 3^4) + \left(\frac{24}{6} \times 3^3\right) + \left(\frac{24}{4} \times 3^2\right) + \left(\frac{24}{6} \times 3^1\right) + (1 \times 3^0)$$
$$= (1 \times 81) + (4 \times 27) + (6 \times 9) + (4 \times 3) + (1 \times 1)$$

Translated into genetic language in case of an Aa Bb Cc Dd heterozygote's F_2 progeny, the phenotypic classes are **81** ABCD, (**27** ABCd, **27** ABcD, **27** AbCD, **27**, aBCD), (**9** ABcd, **9** AbCd, **9** AbcD, **9** aBCd, **9** aBcD, **9** abCD), (**3** Abcd, **3** aBcd, **3** abCd, **3** abcD), (**1** abcd), or *81*:*108*(4 × 27):*54*(6 × 9):*12*(4 × 3):*1*. For the calculation of genotypic classes among the segregants, *See* trinomials and multinomials. *See* Mendelian segregation; Pascal triangle.

binomial nomenclature   *See* taxonomy.

binomial probability P is the complete binomial probability function, whereas the $n!/(x!(n-x)!$ is the binomial coefficient, an integer that shows how many ways one can have x combinations of n, and $p = 0.75$ and q = 0.25 (because of the 3:1 segregation). In genetic experiments this shows — if we have n = independently segregating gene loci, and the inheritance is dominant — how many ways can have x combinations of n; e.g., if we deal with $n = 5$ loci and we wish to know the chance that at 3 ($= x$) loci the dominant phenotype would appear, then

$$[5!/(3!2!)] = (0.75^3)(0.25^2) \cong 0.263672$$

$$\boxed{\begin{array}{l} P = \binom{n}{x} p^x q^{(n-x)} \\[2mm] \binom{n}{x} = \dfrac{n!}{x!\,(n-x)!} \end{array}}$$

The binomial distribution is obtained from the expansion of the binomial terms $(p + q)^n$; its standard deviation

$$\sigma = \sqrt{\frac{pq}{n}} \text{ and } p + q = 1;$$

n is the exponent. *See* Bernoulli process; binomial distribution; Pascal triangle; transmission disequilibrium; trinomial distribution.

binuclear zinc cluster   Domain of a DNA-binding transcriptional activator containing 2 zinc ions about 3.5 Å apart and regulated by 6 cysteines. *See* zinc finger.

bioassay (biological assay)   Used for determining the biological effect(s) of chemicals, drugs, or any other factor on live animals, plants, microorganisms, or cells.

bioassays in genetic toxicology Designed to assess mutagenic (and indirectly carcinogenic) properties of factors to which human, animal, plant, and microbial populations may be exposed. Their range varies from testing chromosome breakage and point mutations in a wide variety of organisms using different end points. All the different procedures cannot be discussed or even enumerated here, but the major types of tests include excision repair, reversion studies in *Salmonella* and *E. coli,* sister chromatid exchange, mitotic recombination, host-mediated assays, specific locus mutation assays, micronuclei formation, chromosome breakage, sex-linked lethal assays, unscheduled DNA synthesis, sperm morphology studies, cell transformation assays, dominant mutation, somatic mutation detection, *Arabidopsis* mutagen assays, human mutagenic assays, mitotic recombination as a bioassay in genetic toxicology, *Tradescantia* stamen hair somatic mutation assay (popular for monitoring environmental pollution), zebrafish assay for mutagens in aquatic media, and mouse lymphoma test. *See* Big Blue®; genotoxic chemicals; hemiclonal; Muta™ Mouse; mutation detection; transgene mutation assay.

biocatalyst Enzyme that mediates metabolic processes. Biocatalysts have numerous industrial applications, and with the aid of molecular biotechnology, means exist for improvement of their efficiency. (Burton, S. G., et al. 2002. *Nature Biotechnol.* 20:37.)

biocenosis Different organisms living together within the same environment; some of them may be dependent on others for survival or may interact in various ways.

biochemical genetics Studies the genetic mechanisms involved in the determination and control of metabolic pathways. *See* inborn errors of metabolism.

biochemical mutant The chemical basis of the mutant function is identified. *See* auxotroph.

biochemical pathway Chemical steps involved in a biological function are represented in a sequence. Enzymes encoded by separate genes usually mediate the individual steps. *See* one gene–one enzyme theorem, genetic networks, transcriptome. (Rison, S. C. & Thornton, J. M. 2002. *Current Opin. Struct. Biol.* 12:374.)

biochips *See* DNA chips.

biocrystallization DNA may be protected in prokaryotes by co-crystallization with the stress-induced protein Dps. Dps dodecamers protect against oxidative damage (and nucleases) in a manner similar to ferritins. *See* ferritin.

biodegradation Decomposition; destruction of substances by bacterial or other organisms. *See* bioremediation; oil spills; *Pseudomonas.* (<http://umbbd.ahc.umn.edu/>.)

biodiversity *See* species, extant. (<http://www.sp2000. org/>; <www.eti.uva.nl/Database/WBD.html>).

bioengineering Replacement of body parts by man-made devices created either by mechanical or biological means. (*Science* [2002] 295:998–1031.)

bioethics Application of ethical principles to biotechnology, medical, genetic, and related fields. *See* biotechnology; embryo research; ethics; human subjects; informed consent; nuclear transplantation. (Merz, J. F., et al. 2002. *Am. J. Hum. Genet.* 70:965; <http://guweb. georgetown.edu/nrcbl>; <http://www.bioethics.net>; <http://209.143.140.244/napbc/ tissue.htm>; <http://bioethics.gov/cgi-bin/bioeth_counter.pl>; <http://www.ornl.gov/hgmis/elsi/elsi.html>; <http://www. nhgri.nih.gov/ELSI/>.)

biofilm Community of single-cell organisms established as a surface layer with chemical communication among the components. Bacterial aggregates of a single or dozens of different species (also including fungi) surrounded by a foamy substance and resistant to many types of disinfectants, antibiotics, or antibodies. Cells within the biofilm differentiate differently from free-living bacteria. The polysaccharide (alginate) film may corrode pipes and medical equipment and may be responsible for dental plaques, lung, kidney, prostate, etc. infections, and inflammation. The development of the biofilm usually requires quorum-sensing signals. Some fungi (*Candida albicans*) can also form biofilm. *See* algin; cystic fibrosis; quorum sensing. (O'Toole, G., et al. 2000. *Annu. Rev. Microbiol.* 54:49; Whitely, M., et al. 2001. *Nature* 413:860.)

Bio-Gel Commercial ion exchange chromatography medium suitable for the separation of RNA from DNA, purification of oligonucleotides and linkers, etc.

biogenesis Cells arise only from cells rather than from nonliving organic material. *See* spontaneous generation.

biohazard Working with pathogenic organisms or transgenic material containing potentially dangerous genes (coding for toxins). Containment (P1 to P4, the latter the most stringent) is necessary and the appropriate safety regulations must be complied with. Information for particular cases can be obtained from the local biohazard committees or from the National Institute of Health, Building 31, Room A452, Bethesda, MD 20205. *See* biological containment; laboratory safety; recombinant DNA and biohazards. (<ftp://potency.berkeley.edu/pub/tables/hybrid.other.tab>.)

bioinformatics Use of computers for developing algorithms, information gathering, storage and analysis of molecular biological data. *See* databases; GenBank; GSDB; image analyzer; NCBI. (Basset, D. E., Jr., et al. *Nature Genet.* Suppl. Vol. 21; Searls, D. B. 2000. *Annu. Rev. Genomics Hum. Genet.* 1:251; Jenssen, T. K., et al. 2001. *Nature Genet.* 28:21; Letovsky, S., ed. 1999. *Bioinformatics. Databases and Systems.* Kluwer, Boston; Davidson, D. & Baldock, R. 2001. *Nature Rev. Genet.* 2:409; Yardell, M. D. & Majoros, W. H. 2002. *Nature Rev. Genet.* 3:601; <http://www.abrf.org/index.cfm/news.annDetails/3htm>; <http://www.ncbi.nlm.nih.gov/Tools/index.html>.)

biolistic transformation (biological-ballistic) Introduces genes into the nuclei of cells (of the germline) by shooting DNA-coated particles into the target cells, propelled by high-power air or helium guns. It is a most useful procedure when other methods of transformation are not sufficiently successful. It can also accomplish transformation in terminally differentiated cells and chloroplasts and mitochondria. Gene guns are also used for cancer gene therapy; *See* cancer gene therapy; chloroplast genetics; mitochondrial genetics; transformation. (Klein, T. M., et al. 1987. *Nature* 327:70; Maenpaa, P., et al. 1999. *Mol. Biotechnol.* 13:67.)

biological clock Frequently called circadian rhythm; measures daily periods and responses to alternation of light and dark cycles in various organisms. The endogenous rhythms also influence gene activity and developmental patterns. *See* circadian. (<http://www.cbt.virginia.edu>.)

biological containment Preventive measures to avoid the spread of potentially hazardous organisms outside the laboratory. Recombinant DNA-containing organisms with unknown biological impact in the environment may be prevented from accidental spreading using transformation vectors that lack the *bom* and *nic* sites that facilitate plasmid mobilization. Also, the cloning bacteria may have an absolute requirement for diaminopimelic acid (lysine precursor), deficient in excision repair (*uvrB* deletion), auxotrophic for thymidine and *rec⁻* (recombination and repair deficient). Thus, even after accidental escape, assuming a mutation rate of 10^{-6} for each of the 5 loci, they would require $(10^{-6})^5 = 10^{-30}$ chance to succeed in the environment. *E. coli* bacterial cell number 10^{30} has a mass of about 10^{11} metric tons. The mass of the Earth has been estimated to be 10^{20} tons. *See* biohazards.

biological control Pathogens or parasites are contained by propagation of their natural enemies: other pathogens or parasites (e.g., *Aphelinus mali*) or genetically engineered organisms. *Colletotrichum truncatum* fungus is used as a weed-killer bioherbicide; *Sesbania exaltata* in various crops such as soybeans, rice, and cotton. Attack of plants by armyworm (*Spodoptera exigua*) is concomitant with the oral secretion of N-(17-hydroxylinolenoyl)-L-glutamine (volicitin), triggering the emission by plant chemical signals that attract parasitic wasps, predators of armyworm larvae. Although biological control is frequently considered the safest method of protection, some biologists worry about the general environmental impact. Some of the control organisms may adversely affect useful native species. *See* antisense RNA; *Bacillus thüringiensis*; BEV; Dengue virus; genetic sterilization. (Howarth, F. G. 1991. *Annu. Rev. Entomol.* 36:485; Pemberton, R. W. & Strong. D. R. 2000. *Science* 290:1896.)

biological mutagen A large number of natural products present in different organisms may be mutagenic for others. Spontaneous mutation may be increased by endogenous factors such as defective DNA polymerase or defects in the genetic repair system. *See* mutator genes; transposable elements; transposons.

biological systems *See* systems biology.

biological weapon May contain highly pathogenic microorganisms such as *Bacillus anthracis* (anthrax), *Corynebacterium diphtheriae* (diphtheria), *Pasteurella pestis / Yersinia pestis* (bubonic plague), *Francisella tularensis* (tularemia), or various viruses, etc. The *Clostridium botulinum* toxin or castor bean toxin, ricin, may be very dangerous. (The Botox toxin has been used recently for the very questionable cosmetic purpose of reducing facial wrinkles by immobilizing muscles through blocking acetylcholine at the ends of the motor nerves.) In case the genetic signature of all organisms potentially usable for biological warfare would be available, rapid identification and effective protection may be facilitated. *See* anthrax; diphtheria toxin; pox virus ricin; signature of molecules; Yersinia. (Stone, R. 2001. *Science* 293:414; Kortepeter, M. G., et al. 2001. *J. Env. Health* 63[6]:21; Hawley, R. J. & Eitzen, E. M., Jr. 2001. *Annu. Rev. Microbiol.* 55:235; <http://www.virology.net/garryfavwebbw.html>.)

biologics The material is of a biological nature, the processing is biological, and the quality of the product is determined by biological methods.

bioluminescence *See* aequorin; luciferase.

biomarker Any product of the body (e.g., a metabolite) that may respond to adverse environmental effects (carcinogens, mutagens, etc.).

biometric Electronic code of human physical features (fingerprints, eye iris scans); can be used for digital personal identification.

biometry Mathematical statistical principles applicable to the study of genetic and nongenetic variation in biology. *See* population genetics; quantitative genetics.

biomining Certain bacteria obtain energy by oxidizing inorganic materials. This process may release acid, which in turn can wash out metals from ores. *Thiobacillus ferrooxidans* can release copper and gold; *Pseudomonas cepacia* may assist phosphate mining. Eventually this biotechnology may become economical, especially for low-grade ores. Some plants, e.g., *Brassica* spp. and *Impatiens* spp., may accumulate gold from the soil. *See* bioremediation. (Mergeay, M. 1991. *Trends Biotechnol.* 9:17; Guiliani, N. & Jerez, C. A. 2000. *Appl. Environ. Microbiol.* 66:2318.)

biomonitoring Surveying potential mutagens, carcinogens, or other health hazards using biological means such as organismal bioassays for mutagens, blood cells, human buccal cells, nasal mucosal cells, scalp hair follicles, sputum, detached colon cells, cervical epithelia, exfoliated bladder cells, spermatozoa, etc. The tests may involve cytological or molecular methods.

bionics Construction of mechanical devices with the technology of engineering and biology.

biophore Hypothesized hereditary unit of the premendelian era. *See* pangenesis.

biophore Compound or structural element of a biophore may exert potential biological (carcinogen) activity. *See* CASE; MULTICASE; SAR.

biophysics Theory and practice of application of physical methods for the study of biological structures (e.g., nucleic acids, proteins) and mechanisms of function (energy conversions, thermodynamics). (<www.biophysics.org/biophys/society/btol>.)

biopoesis Evolution of living cells from chemical substances rather than from other cells. *See* evolution, prebiotic; origin of life.

bioprospecting Searching for natural products (genes) potentially useful for pharmaceutical or agricultural applications. *See* biotechnology.

biopterin Pterin-derived cofactor of enzymes functioning in oxidation-reduction processes.

Biopterin.

bioreactor Large-scale (industrial) culture of cells for the purpose of production and extraction of pharmaceuticals, enzymes, polypeptides, biodegradable plastics, etc. The use of transgenic organisms extended the range of utility of these procedures. Some constructions simulate conditions for growth in outer space. *See* cell culture; tissue culture; transgenic. (Baoudreault, R. & Armstrong, D. W. 1988. *Trends Biotechnol.* 6:91.)

bioremediation Procedure of adding organisms to an environment for the purpose of promoting degradation of harmful or undesirable properties of that environment. Some observations indicate that $44 \pm 18\%$ of polycyclic hydrocarbons of the atmosphere are captured by the vegetation and eventually incorporated into the soil. Many polycyclic hydrocarbons are carcinogenic and mutagenic and pose serious health hazards to people and animals. Their removal from the atmosphere is desirable; however, the consequences of eating plants that have absorbed these semivolatile compounds is not clear. Several organic compounds can be degraded by sequential exposure to anaerobic and aerobic bacteria. Bacterial mercuric ion reductase gene in a reengineered form was introduced into *Arabidopsis* plants by transformation and the transgenic plants became resistant to $HgCl_2$ and Au^{3+}. The transgenic plants evolved substantial amounts of Hg^0 (vapors). Plants can extract toxic substances from the soil (phytoextraction) and from water (rhizofiltration) and thus facilitate the cleaning up of the environment. By a genetic engineering technique, cytochrome P450 monooxygenase genes can be combined with toluene dioxygenase genes in, e.g., *Pseudomonas*. Such bacteria can then degrade polyhalogenated compounds such as 1,1,1,2-tetrachloroethane (a powerful narcotic and liver poison) to 1,1-dichloroethylene and eventually to formic and glyoxylic acids, which are still irritants but occur in natural products of ants and fruits, respectively, but do not pose a serious threat at low concentrations. *See* biodegradation; biomining; environmental mutagens. (Lovley, D. R. 2001. *Science* 293:1444; Kramer, U. & Chardonnens, A. N. 2001. *Appl. Microbiol. Biotechnol.* 55:661.)

bi-orientation The sister kinetochores are attached to spindle fibers that connect them to the opposite spindle poles. *See* spindle fibers, microtubule, kinetochore. (Tanaka, T. U. 2002. *Current Opin. Cell Biol.* 14:365.)

biosensors Analyzes macromolecular interactions in real time in intact cells. Among the different systems, ligand-receptor binding and signal transduction pathways may be the most sensitive, especially when coupled to fluorescent stains. *See* aequorin; fluorochromes; ligand; signal transduction; surface plasmon resonance. (Aravanis, A. M., et al. 2001. *Biosens. Bioelectron.* 16:571.)

Bioseq Contains relevant information about a biological sequence beyond what is included in the ASN.1. *See* accession number; ASN.1; gi.

biosphere Range of habitat of organisms living in and on the soil, in bodies of water and the atmosphere.

biostratigraphy Relative dating of the succession of different evolutionary forms of organisms on the basis of paleontological relics.

biosynthesis Synthesis of molecules by living cells.

biota Community of all living organisms in an environment.

biotechnology Purposeful application of biological principles to industrial, medical, and agricultural production, such as the molecular alteration of enzymes, cloning recombinant DNA and its translated products (e.g., human insulin produced by transgenic bacteria), replacement of defective genes by site-specific recombination, gene medicine (introducing genes capable of producing the medication required transiently into cells), transfer of desirable genes into domestic animals and plants by genetic transformation to improve their economic value, cleanup of environmental pollutants by modified microorganisms capable of digesting crude oil, etc. *See* bioethics; bioprospecting; genetic engineering; genomics; GMO; agricultural biotechnology: <http://www.cid.harvard.edu/cidbiotech/homepage.htm>. (Daar, A. S., et al. 2002. *Nature Genet.* 22:229.)

bioterrorism Produces fear and harm among selected individuals or in the general population, or harms plants, animals, or the environment with the use of agents such as bacteria, viruses, fungi, or toxins derived from biological agents. *See* biological weapons. (Henderson, D. A. 1999. *Science* 283:1279; Atlas, R. M. 2002. *Annu. Rev. Microbiol.* 56:167; <http://www.hopkins-biodefense.org/>, <bob.nap.edu/shelves/first>.)

biotic   Related to living organisms.

biotidinase deficiency (same as multiple carboxylase deficiency, 21q22.1, 3p25)   Autosomal-recessive disease, yet the heterozygotes may also be identified but by much less obvious symptoms. The biochemical basis is a deficiency of an enzyme (multiple carboxylase) that splits biocytin (biotin–ε-lysine) and thus generates free biotin from protein linkages. The symptoms that may have late onset or appear in neonates are hypotonia (reduced tension of muscles), ataxia (reduced coordination of muscles), neurological deficiencies (hearing, vision), alopecia (baldness), skin rash, susceptibility to infections, etc. Generally, administration of biotin alleviates the symptoms and may restore normality. The prevalence varies within the 10^{-5} range. A simple procedure is available for testing the blood by color on filter paper without purification. *See* biotin; genetic screening.

biotin   Vitamin, a mobile carrier of activated CO_2, its major biological role involves pyruvate carboxylase. It combines with avidin and thus is used for nonradioactive labeling. *See* biotinylation; fluorochromes; nonradioactive labeling. (Mardach, R., et al. 2002. *J. Clin. Invest.* 109:1617.)

Biotin.

biotinylation   Very sensitive nonradioactive labeling generated by incorporation into the DNA, with the aid of nick translation, biotinylated deoxyuridylic or deoxyadenylic acid. Biotin in the DNA has great affinity for streptavidin carrying a dye marker, and the labeled DNA can thus be identified in light either cytologically or on membrane filters. *See* biotin; FISH; fluorochromes; labeling. (Demidov, V. V., et al. 2000. *Curr. Issues Mol. Biol.* 2:31.)

biotrophic   Parasite living on a live host. *See* saprophytic.

biotype   Physiologically distinct race within a species.

BiP   Soluble heat-shock protein 70, a chaperone. An immunoglobulin-binding protein. *See* chaperone; heat-shock proteins; Sps70.

biparental inheritance   Nuclear genes are usually transmitted by the female and male parents, in contrast to cytoplasmic organelles (and their genetic material), which are most commonly inherited only through the egg, and therefore the inheritance is uniparental (through the female). *See* allophenic; chloroplast genetics; meiotic drive; mitochondrial genetics; mtDNA.

Bride & groom, Ceramics by Margit Kovács.

bipedal   Animals walking on two feet as an evolutionarily developed characteristic.

bipolarity   Both strands of the DNA are transcribed in opposite $\overrightarrow{\leftarrow}$ directions.

bipolar mood disorder   Complex human disorder involving manic depression fluctuating with euphoria. Putative genetic determinants have been found in human chromosomes 1p33-p36, 2q21-q33, 3p14, 3p21, 3q26-q27, 4p15.3-p16.1, 5p15, 6q21-q22, 8q24, 8p21, 10q25-q26, 7, 13q11, 13q31-q34, 14q12-q13, 15, 16, 17, 18, and 21q22. Apparently a major factor is associated with human chromosome 22q. *See* affective disorders; lithium; manic depression; QTL; unipolar depression. (Kelsoe, J. R., et al. 2001. *Proc. Natl. Acad. Sci. USA* 98:585; Cichon, S., et al. 2001. *Hum. Mol. Genet.* 10:2933; Mitchell, P. B. & Malhi, G. S. 2002. *Annu. Rev. Med.* 53:173.)

BIR   (1) Chromosome break-induced replication is a mechanism to repair a broken single strand of a chromosome by new DNA synthesis. (Kraus, E., et al. 2001. *Proc. Natl. Acad. Sci. USA* 98:8255.) (2) Baculoviral IAP repeats. N-terminal motifs in the IAP proteins, in one or several copies. The Bir1p protein is an inhibitor of apoptosis, and in cooperation with the kinetochore proteins Ndc10p, Cep3p, Ctf13p, and Skp1p, it controls chromosome segregation. *See* IAP; kinetochore.

birch (*Betula*)   Silver birch, hardwood tree *B. pubescens* is $2n = 28$, and the *B. verrucosa* is $2n = 56$; $x = 14$.

Betula alba leaf.

birth control   *See* hormone receptors; menstruation; sex hormones.

birth defect   Perinatal anomaly of either hereditary or extraneous cause.

Birt-Hogg-Dubé syndrome Genodermatosis (genetic skin disease) involving tumorous hair follicles, renal neoplasia, lung cysts, and pneumothorax (air accumulation in the serous membrane of the chest). It is linked to the pericentromeric region of human chromosome 17p. (Schmidt, L. S., et al. 2001. *Am. J. Hum. Genet.* 69:867.)

birth rate *See* age-specific birth and death rates.

birth weight May be affected by a number of intrauterine factors such as maternal nutrition, disease, smoking, and also genetic causes. Some of the initial relative differences may or may not be eliminated during subsequent development. *See* glucokinase.

bisexuality May be a case of hermaphroditism or just a behavioral anomaly. In the fruit fly, some losses in the brain olfactory centers or receptors lead to a defect of the interpretation of pheromones, causing anatomically male flies to court females as well as males. *See* hermaphrodite; homosexual; olfactogenetics; olfactory; pheromones; sex determination.

bison American buffalo (*Bison bison*), $2n = 60$.

bispecific monoclonal antibody (diabody) One of the two arms of the antibody has the recognition site for the surface antigens of a tumor cell, the other for the antigens of a killer lymphocyte. The bispecific antibody is thus expected to bring these two cells together and destroy the tumor cells. *See* antibodies; antibody engineering; diabody; monoclonal antibodies; quadroma; triabody.

bistable systems Can toggle between two alternative steady-states but cannot rest in intermediate states. Such systems have importance for signal transduction, feedback and differentiation. (See signal transduction, feedback, Ferrell, J. E., Jr. 2002. *Current Opin. Cell Biol.* 14:140.)

Biston betularia (peppered moth) Frequently used example for adaptive natural selection. The moth had predominantly overall grayish tones until the industrial revolution in the vicinity of Birmingham, England, deposited black soot on the tree barks and favored the propagation of the dark-colored (carbonaria) form of the moth, which could hide better from predators. In unpolluted areas, the light peppered form remained. (Cook, L. M. & Grant, B. S. 2000. *Heredity* 85:580.)

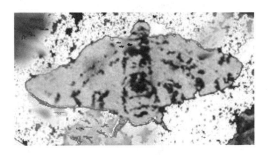

Biston betularia light.

bit (1) Binary digit with a two-way choice such as a value of 1 or 0, on or off, etc. (2) The smallest unit of information a computer recognizes.

Bithorax (*bx*, 3–58.8) *See* morphogenesis in *Drosophila*; Polycomb. (Duncan, I. 1987. *Annu. Rev. Genet.* 21:285.)

From Bridges & Morgan 1923, *bx*.

bitmap Bits representing a graphic image in the memory of a computer.

bitnote Message communicated through the computer (e-mail) using the Bitnet system, an IBM mainframe connection to the Internet. *See* Internet.

bitransgenic regulation The *A* transgene to be expressed must have its regulator transgene *R* present in the cell. For normal function, the appropriate ligand(s) must also be available within the cell. *See* ligand; transgenic. (Yao, T. P., et al. 1992. *Cell* 71:63.)

BITSCORE Raw quantitated sequence alignment score that indicates the statistical property of the scoring system. *See* BLAST.

bivalent Two homologous chromosomes consisting of a total of four chromatids paired in meiotic prophase. *See* chromatid; heteromorphic bivalent; interlocking bivalent; synaptonemal complex.

bivariate distribution Joint distribution of two random variables.

bivariate flow cytometry Sorting chromosomes tagged with two fluorochromes (Hoechst 33258) specific for A=T, and chloromycin A3 specific for G≡C, and excited by laser. *See* flow cytometry; laser. (Nunez, R. 2001. *Curr. Issues Mol. Biol.* 3:67.)

BKM sequence Tetranucleotide GATA and GACA repeats in the W chromosomes (comparable to the Y chromosome) of birds and reptiles, occasionally in other eukaryotic chromosomes. *See* satellite DNA; tetranucleotide repeats; Y chromosome.

BK virus Has 80% homology to Simian virus 40 with a somewhat different host range (human, monkey, hamster, and other rodent cells). It may occur as an episomal element in two dozen to hundreds of copies. Its autonomous replicon may be useful for propagating DNA and genes in human cell cultures.

black box Slang expression for a piece of equipment that is too complicated inside to be generally understood. Figuratively, a living cell was considered to be a black box because some of its functions were observed, yet all the mechanisms that drove these functions were not fully understood. Geneticists knew segregation of genes and chromosomes, but the molecular mechanisms underlying these processes were largely shut inside the "black box" until the discoveries of DNA replication, transcription, translation, gene regulation, cell cycle, etc.

black locust (*Robinia pseudoacacia*) Leguminous tree with fragrant flowers; $2n = 20$.

black pepper (*Piper nigrum*) Southeast Asian spice. Basic chromosome number probably 12, 13, or 16, and $2n = 46, 52, 104$, and 128 have been reported.

blank allele Not expressed.

BLAST Basic local alignment search tool is used for comparison of nucleotide sequences in apparently related (homologous) DNA (Altschul, S. F., et al. 1990. *J. Mol. Biol.* 215:403). Email: <blast@ncbi.nlm.nih.gov>. *See* BLOSUM; databases; DNA sequencing; evolutionary tree; FASTA; homology. (*Trends Supplement* [Elsevier Science] 1998; Wolfsberg, T. G. & Madden, T. L. 1999. *Short Protocols in Molecular Biology*, Ausubel, F. M., et al., eds. Wiley, New York; <http:www.ncbi.nlm.nih.gov/BLAST/>.)

blast cell Cell that may give rise to a progeny cell(s) different from itself.

blastid Site in the fertilized egg where cellular organization takes place.

blastocoel *See* blastula.

blastocyst Early embryonal stage when the blastocoel is enveloped by a trophoblast cell layer, a preimplantation stage of the animal zygote when the zona pellucida (the envelop of the egg) is still visible and the blastula begins to develop its inner cell mass. *See* blastula; trophoblast.

blastocyte Undifferentiated cell of an early zygote.

blastoderm Single layer of cells at the embryonic stage of insects surrounding a fluid-containing cavity (blastocoel) at the blastula stage of cell division.

blastoma Cell in the early stage of differentiation or a neoplastic tissue containing embryonic cells.

blastomere The large fertilized egg through cleavage divisions produces smaller cells, the blastomeres. These divisions are extremely fast, and during the short process, RNA synthesis ceases and protein synthesis depends on reserve mRNAs. *See* blastoderm; cleavage; founder cells.

blastopore Near the site of the gray center of the animal pole where invagination of the blastula begins and eventually encompasses the vegetal pole. The *dorsal lip* of the blastopore organizes gastrulation. *See* animal pole; blastula; gastrulation; organizer; vegetal pole.

blastula Product of the cleavage of the early zygote when it becomes a spherical structure in which the blastoderm envelops the blastocoel cavity. *See* blastula.

BLASTX Computer program for gene searches. (*Nature Genet.* 3:266 [1993].)

BLATTNER NUMBER Refers to the position of genes in the sequenced genome of *E. coli* bacterium. *See* E. coli. (Blattner, F. R., et al. 1997. *Science* 277:1453; Riley, M. & Serres, M. H. 2000. *Annu. Rev. Microbiol.* 54:341; Liang, P., et al. 2002. *Physiol. Genomics* 9:15; <http://genprotec.mbl.edu/>.)

Blau disease Form of ulcerative colitis (intestinal inflammation) under the control of more than a single gene. Its symptoms overlap with those of Crohn disease. *See* Crohn disease. (Miceli-Richard, C., et al. 2001. *Nature Genet.* 29:19.)

bleeder disease *See* hemophilia.

blending inheritance Considered erroneously to account for some of the variations observed in nature up to the time of Mendel. A contemporary of Darwin, Fleeming Jenkins, an engineer, pointed out that if this were true, unique traits of single organisms would disappear in panmictic populations like a drop of ink in the sea. *See* particulate inheritance.

bleomycin (BLM) Antitumor (malignant lymphomas, squamous cell carcinoma) antibiotic. It degrades DNA (and also RNA), especially when supported by redox-active metal ions (Fe, Cu) with specificity to 5′-GT-3′ and 5′-GC-3′ sequences in double or single strands. However, Fe(II)BLM does not cleave all RNAs and shows preferences to some tRNAs, although it may cleave a variety of RNAs (mRNA, 5S rRNA, DNA-RNA heteroduplexes, etc). Mg^{2+} is inhibitory. BLM also affects lipid peroxidation. *See* cancer therapy; ribonuclease. (Hoehn, S. T., et al. 2001. *Nucleic Acids Res.* 29:3413.)

blepharophimosis Defect of the eyelids and nose, also associated with ovarian atrophy and small uterus. The dominant disorder was assigned to human chromosome 3q21-q23; another locus is at the 7p21-p13 region. *See* polled. (De Baere, E., et al. 2001. *Hum. Mol. Genet.* 10:1591.)

blepharoplast *See* basal body.

Blimp-1 Repressor of c-Myc oncoprotein expression that blocks plasmacytoma formation and terminal differentiation of B lymphocytes. *See* B lymphocyte; MYC.

blindness *See* eye diseases.

blister Vesicle-like skin abnormality. Autosomal-dominant, -recessive and X-linked types exist. Autosomal-dominant class: *Bullous erythroderma* involves a hyperkeratosis with anomalies of the keratin tonofibrils. In one form, the blisters are limited to hands and feet and appear only in warm weather after heavy exercise. Another form (*epidermolyis bullosa dystrophica*) may appear already at birth, and the blisters may appear on the ears and buttocks as well as on the extremities. The latter type accumulates and secretes sulfated glycosaminoglycans. In an early and transient form, the blisters disappear generally by the end of the first year and do not return. In *herpetiform epidermolysis bullosa*, the larger vesicles appear in clusters on the palms, soles, neck, and around the mouth apparently due to a mutation in the keratin 14 gene. Another epidermolysis bullosa appears to be due to deficiency of *galactosylhydroxylysyl glucosyltransferase*. In one form, a human chromosome 12-coded gelatin-specific *metalloprotease* deficiency may be involved. A *mottled* type (pigmented spots) displays recurrent blistering beginning at birth and premature aging. The *epidermolysis bullosa with absence of skin and deformity of nails* has a perfect penetrance. Autosomal-recessive types: *Epidermolysis bullosa dystrophica* (human chromosome 11q11-q13) is caused by excessive collagenase activity affecting primarily the hands, feet, elbows, and knees at birth or infancy but may also affect other organs. *Epidermolysis bullosa letalis* may kill infants within about 3 weeks from birth, but occasionally some survive to the first decade of life. In some forms, the distal opening of the stomach (pylorus) may be constricted and atrophied; in other forms, congenital deafness and muscular dystrophy may appear. X-linked epidermolysis with multiple complications (baldness, hyperpigmentation, dwarfism, microcephaly [small head], mental retardation, finger and nail malformation) and death before adult age is also known. *See* ichthyosis; keratosis; skin diseases.

BLK   *See* SRC oncogene family.

BLM   *See* Bloom syndrome.

Bloch-Sulzberger syndrome   *See* incontinentia pigmenti.

blocked reading frame   Translation is interrupted by nonsense codons. *See* nonsense codons.

blocking buffer   3% BSA (bovine serum albumin) in phosphate-buffered saline also containing 0.02% sodium azide. BSA blocks the binding sites on nitrocellulose filter that are not occupied by proteins transferred from (SDS polyacrylamide) gels. *See* gel electrophoresis.

block mutation   Affects more than a single nucleotide in the cell, e.g., deletion; such mutations may not yield wild-type recombinants if the defects overlap.

BLOCKS   Internet tool for the search of functional DNA motifs: <blocks@howard.fhcrc.org> or <http://blocks.fhcrc.org>.

blood   Fluid that carries nutrients and oxygen by circulating through the blood vessels in the animal body. It is composed of red (nonnucleated mature erythrocytes) and white (nucleated leukocytes) cells. The white cells include granulocytes, neutrophils, eosinophils and basophils, B and T lymphocytes, and natural killer cells. The differentiation of these various types of cells from the multipotential hematopoietic stem cells is determined by a combination of growth factors, such as interleukins, stem cell factors, colony-stimulating factors, etc. In the lymphoid developmental path, the *Pax5* gene and its product, BSAP (B-cell-specific activator protein), play a key role. The blood also contains platelets (thrombocytes) and plasma, the noncorpusculate yellowish fraction. In *Drosophila* there are crystal cells (contain defense enzymes), plasmatocytes (the main phagocytotic defense cells), and lamellocytes (develop from plasmatocytes with similar functions). *See* antihemophilic factors; blood groups; B lymphocyte; dendritic cells; hematopoiesis; hemolytic disease; immune system; macrophages; Pax; T cell.

blood brain barrier   *See* BBB.

blood clotting pathways   *Intrinsic pathway* involving the successive participation of the Hagemann factor (XII), plasma thromboplastin antecedent (PTA XI), Christmas factor (IX), antihemophilic factor (VIII), Stuart factor (X), phospholipid, and proaccelerin; and *extrinsic clotting pathway* requiring proconvertin (VII), Stuart factor (X), proaccelerin, and calcium ions. With the aid of lentiviral vectors directed to the mouse liver, the human factor IX gene can be expressed and stably maintained for months. *See* antihemophilic factors; tissue factor; vitamin K. (Tsui, L. V., et al. 2002. *Nature Biotechnol.* 20:53.)

blood coagulation   *See* antihemophilic factors; blood clotting pathway.

blood formation (hematopoeisis)   During early embryonic development the yolk and the aorta-gonad-mesonephros (AGM) region is involved; later the function in the embryo is switched to the liver and after birth the bone marrow is involved.

blood groups   Incomplete list of the types found in this volume: ABO, ABH, Ahonen, Colton, Diego, Dembrock, Duch, Duffy, En, Gerbich, I system, Kell-Cellano, Kidd, Lewis, Lutheran, LW, MN, Newfoundland, OK, P blood group, Radin, Rhesus, Scianna, Ss, Webb, Wright, Yt, and Xg. These are distinguished mainly by the epitopes on the erythrocytes. *See* epitope; erythrocyte.

blood pressure   Pressure of the blood on the blood vessels (arteries). *See* hypertension.

***blood* transposable element**   *See* copia.

blood typing   Identification of a person's blood group. *See* blood groups.

Bloom syndrome (BS, BLM)   Recessive human dwarfism; increases the frequency of chromosomal aberrations (particularly sister chromatid exchanges) and various forms of cancer (leukemia); sensitivity to sunlight (red blotches over face) and usually shorter than normal life expectancy. It was attributed to a DNA ligase I deficiency, but the cloning and sequencing of the gene indicates that this is not the primary defect; rather, a DNA helicase-like protein (RecQ family, homologue of budding yeast genes *SGS1* and *SRS2*), encoded in human chromosome 15q26.1, is involved. SGS1 mutations greatly enhance gross chromosomal aberrations and the rate of recombination. The wild-type SGS1 represses chromosomal aberrations. The *Drosophila* homologue is *Dmblm*. In BS, sister chromatid exchange and mitotic recombination are elevated, but genetic repair seems to be normal. In *Dmblm*, an extra copy of Ku70 compensates for sterility. The BS helicase physically interacts with the Werner syndrome helicase. *See* carcinogenesis; Cockayne syndrome; cosuppression; DNA repair; Ku; light sensitivity diseases; Rothmund-Thomson syndrome; Werner syndrome; xeroderma pigmentosum. (Luo, G., et al. 2000. *Nature Genet.* 26:424; Myung, K., et al. 2001. *Nature Genet.* 27:113; von Kobbe, C., et al. 2002. *J. Biol. Chem.* 277:22035.)

BLOSUMs   Amino acid substitution matrices used to determine evolutionary changes in proteins. *See* BLAST; FASTA. (Henikoff, S. & Henikoff, J. G. 1992. *Proc. Natl. Acad. Sci. USA* 89:10915.)

blotting   Macromolecules separated by electrophoresis in agarose or polyacrylamide are transferred to a cellulose or nylon membrane and immobilized there for further study. *See* colony hybridization; immunoprobe; Northern blot; Southern blot; Western blot.

BLOTTO   Bovine Lacto Transfer Technique Optimizer is a 5% solution of nonfat evaporated milk in 0.02% sodium azide (NaN$_3$). (It may contain RNase activity). In 25-fold dilution it may be used for blocking background annealing in Grunstein-Hogness hybridization, Benton-Davis hybridization, dot blots, and nonsingle-copy Southern hybridization. *See* Benton-Davis plaque hybridization; Denhardt reagent; dot blot; Grunstein-Hogness screening; heparin.

blueberry (*Vaccinium* spp)   Fruit shrub with $x = 12$; *V. corymbosum* (high-bush blueberry) is tetraploid; *V. angustifolium* (low-bush blueberry) is diploid.

blue diaper syndrome   Intestinal failure to transport tryptophan. *Pseudomonas aeruginosa* bacteria convert the amino acid into indole, which upon oxidation stains bluish. *See* amino acidurias.

blue grass (*Poa pratensis*)   Lawn and pasture plant; $2n = 36$-123 in the polyploid series.

blue light response   Photomorphogenetic reaction (of plants) to illumination in the range of 400–500 nm wavelength. *See* cryptochromes; photomorphogenesis.

bluescript M13   2.96 kb genetic vector containing the bacteriophage M13 replication origin and a polycloning insertion site flanked by T7 and T3 phage promoters in opposite orientation and useful for generation of single-stranded DNA or RNA complementary to the double-stranded DNA insert. Several variations exist (e.g., λZAP, bluescript SK). The name comes from the bacterial *Lac* fragment that upon expression

of β-galactosidase in Xgal medium forms an easily detectable blue color. (Xgal, Short, J. M., et al. 1988. *Nucleic Acids Res.* 16:7583; Snead, M., et al. *Methods Mol. Biol.* 81:255.)

blunt end   Of double-stranded DNA, is generated by nonstaggered cut and terminates at the same base pair across both strands of the double helix. Bacterial DNA polymerase I (Klenow fragment) or phage T4 DNA polymerase can also generate 3′ blunt ends of DNA by 5′ → 3′ exonucleolytic activity. *See* blunt end ligation.

BLUNT ENDS OF DNA →   ═══════

blunt-end ligation   T4 phage DNA ligase joins nonstaggered DNA ends or adds chemically synthesized duplexes to double-stranded blunt ends. *See* DNA ligases; linker.

BLUP   Best linear unbiased prediction is a statistical procedure based on covariance analysis of gametic genetic disequilibrium of QTL and other types of markers in multibreed populations. *See* multibreed; QTL. (Wang, T., et al. 1998. *Genetics* 148:507.)

blym   Chicken bursal lymphoma oncogene, located to human chromosome 1p32. It is homologous to transferrin, a glycoprotein with an important role in the synthesis of ribonucleotide reductase, and thereby in DNA replication and mitosis. *See* lymphoma; oncogenes; transferrin.

B lymphocyte (B cell)   Responsible for humoral antibody synthesis and secretion. There are two types of B cells: B1 and B2. T cells activate the common B2 cells. B-cell differentiation from hematopoietic cells (plasma cells) depends on transcription factors (XBP-1) of the bone marrow, cytokines, and antigens, T$_H$ cells, and a series of nonreceptor and receptor tyrosine kinases and phosphatases. Proteins mediating the pathway shown in italics in the diagram below outline the developmental pathway of B lymphocytes. The B1 cells also employ an RNA editing system for the diversification and amplification of their antigen receptors in contrast to T cells, which rely on the V(D)J recombination system and appear independent of T-cell activation. *See* BASH;

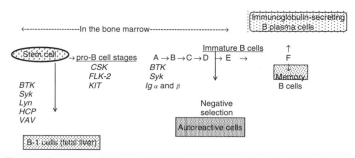

During the pro-B cell stages in the bone marrow, first the IgH and then the IgL immunoglobulin chains are rearranged through the signaling by the pre-BCR (B-cell receptor). After expression of the BCR, some cells leave the bone marrow and mature into IgM and IgD cells that move between the peripheral lymphoid organs.

Blimp-1; blood; bone marrow; BTK; CD40; CpG motifs; EBF; germinal center; immune reactions; immune system; immunoglobulins; Pax; RNA editing; surrogate chains; TAPA-1; T cells; thymus; V(D)J; XBP. (Fagarasan, S. & Honjo, T. 2000. *Science* 290:89; Hardy, R. R, & Hayakawa, K. 2001. *Annu. Rev. Immunol.* 19:595; Reimold, A. M., et al. 2001. *Nature* 412:300; Berland, R. & Wortis, H. H. 2002. *Annu. Rev. Immunol.* 20:253, chicken B cells: <http:genetics.hpi.uni-hamburg.de/dt40.html/>.)

B-lymphocyte receptor (B-cell receptor, BCR) Constructed from the membrane-bound immunoglobulin molecules IgM and IgD as receptors, and after the attachment of the antigen B-lymphocyte receptors can also use IgG, IgA, and IgE. The intracellularly linked IgA and IgB heterodimer that is the signaling portion of the receptor and the complex transmits the immunoglobulins recognized by BCR. All the immunoglobulins attached to the receptors associate with the IgA and IgB heterodimer through their heavy-chain terminal amino acids. This tail consists of 3 amino acids in IgM and IgD, but it has 28 amino acids in IgG and IgE. Both parts of the

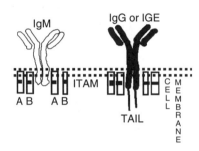

B lymphocyte receptor.

A-B heterodimer include an ITAM (immunoreceptor tyrosine-based activation motif) that is instrumental in activating the Sky and Lyn protein tyrosine kinases, which mediate the switching to the IgG, IgA, and IgE molecules when the appropriate antigen is presented. These tails are required for the endosomal targeting of the immunoglobulins. The B-cell linker protein (BLNK) is also required for the normal development of B lymphocytes. *See* agammaglobulinemia; BAP; BASH; B lymphocyte; endocytosis; immune system; ITAM; ITIM. Meffre, E., et al. 2001. *J. Clin. Invest.* 108:879; Vilches, C. & Parham, P. 2002. *Annu. Rev. Immunol.* 20:217.)

BLYS B-lymphocyte stimulator is a human chromosome 13q34-encoded protein of the tumor necrosis factor family involved in B-cell proliferation and immunoglobulin secretion. *See* B lymphocyte; TNF.

BMI *See* body mass index.

BMK1 Big MAP kinase. *See* ERK.

BMP *See* bone morphogenetic protein.

BMT Transformed monkey cell line expressing the T antigen of SV40, driven by a mouse metallothionein promoter. *See* metallothionein; SV40.

BMYC Oncogene isolated from rat has extensive homology to the MYC oncogene, but it maps to a different location than the other members of the MYC family: LMYC, NMYC, PMYC, and RMYC. *See* MYC; oncogenes.

Bni Member of the formin proteins involved in polar morphogenesis and cytokinesis of eukaryotic cells. Bni1 protein is associated with CDC42 protein, with actin, profilin, and Bud6. *See* actin; Bud; CDC42; profilin.

Bob *See* OBF.

BOB' *See* att sites.

bodipy *See* fluorochrome.

body map Human and mouse gene expression database. (Kawamoto, S., et al. 2000. *Genome Res.* 10:1807; http://body-map.ims.u-tokyo.ac.jp>.)

body mass index, in humans (BMI) Determined as a measure of obesity by the formula: weight in kg/height in meter^2. In morbid obesity, BMI >40 kg/m^2. Obesity (BMI > 30) is detrimental to health, although for centuries it was associated with fertility. Correlation of BMI between monozygotic twins is about 0.74 versus 0.32 for dizygotic twins, indicating very high heritability. The correlation among parents and biological offspring was found to be 0.19 versus adopted children 0.06, indicating the major role of hereditary factors. The pro-opiomelanocortin locus in human chromosome 2p21 is a significant contributor to body mass according to some studies. (Perola, M., et al. 2001. *Am. J. Hum. Genet.* 69:117) failed to detect any linkage to BMI by QTL in Finnish populations. *See* leptin; melanocortin; morbidity; obesity. (Barsh, G. S., et al. 2000. *Nature* 404:644.)

Fertility goddess from Willendorf, Austria ∼15,000 B.C. (From Gowans, C. S. 1974. *Stadler Symp.* 6:113.)

bol *See* boule.

bom Bacterial gene (basis of mobilization) required for the transfer of plasmids. *See* Hfr; mob; plasmid mobilization.

Bombardia lunata Ascomycete, $n = 7$.

Bombay blood type Relatively rare blood type discovered in India and subsequently on the Reunion Island in the Indian Ocean. This blood type has two main forms determined by the recessive allele h (for H-type red cell antigen). In the h/h se/se individuals, the enzyme fucosyltransferase 1 is also inactive; in the h/h Se/se individuals, a weak expression of the H antigen may be observed; Se is apparently coding for fucosyltransferase 2. *See* ABH antigen; fucose; Lewis blood group; secretor.

bombesin (protein, $C_{71}H_{110}N_{24}O_{18}S$) Modulates smooth muscle contraction, hormone traffic, metabolism, hyperglycemia, hypertension, and eating behavior. *See* obesity.

Bombyx mori *See* silk worm.

bond energy Required to break a chemical bond.

bone disease *See* achondroplasia; adactyly; brachydactyly; campomelic dysplasia; collagen; diastrophic dysplasia; dwarfism; dyssegmental dysplasia; Ellis–van Creveld syndrome; exostosis; head/face/brain defects; Larsen syndrome; osteochondromatosis; osteogenesis imperfecta; osteolysis; osteopetrosis; osteoporosis; osteosarcoma; Paget disease; PAPS; polydactyly; pseudoachodroplasia; pycnodysostosis; SED; trichorhinophalangeal syndrome.

bone marrow Red spongy tissue inside the bones gives rise to lymphocyte stem cells and erythrocytes; the yellow bone marrow is mainly fat cells. *See* thymus.

bone morphogenetic protein (BMP) Maternally expressed factor in *Xenopus* embryos. In addition to bone differentiation, it is involved in dorsoventral organization of the embryo. Osteoblast differentiation and proliferation is controlled by BMP and Smad. BMP-1 is a procollagen protease (PCP) that assembles collagen within the extracellular matrix. The other BMPs belong to the transforming growth factor (TGF-β) family. BMP-4 regulates apoptosis in neural crest cells, affecting skeletal bone and muscle formation. BMP-3 is a negative regulator of bone density. The growth/differentiation factors of mouse (GDF) belong to this family, and their mutation shortens the limb bones (brachypodism) without affecting the axial skeleton. The CBFA-1 gene seems to be a major factor in ossification. BMP is the vertebrate homologue of *decapentaplegic* (*dpp*) in *Drosophila*. Sog (short gastrulation) and the Chd (chordin) proteins in vertebrates and invertebrates negatively regulate the BMP/Dpp system, respectively. The metalloprotease *Xld* (Xolloid)/*Tld* (tolloid) releases the Bpm/Dpp from inactive complexes. Thus, a balance between Sog/Chd and Xld/Tld determines a morphogenetic gradient. Noggin and follistatin inhibit BMP and dorsalize the embryo. The process is, however, more complex, since the other serine protease(s) may also be involved (astacin, furin). The Kuz metalloprotease (a reprolysin) regulates the Notch cell surface receptor by proteolytic cleavage. BMP in coordination with other signal proteins regulates the specification of teeth development: $Msx\text{-}1^+\text{-}Barx\text{-}1^-$ activity state leads to the development of incisors, whereas $Msx\text{-}1^-\text{-}Barx\text{-}1^+$ state promotes molar formation in the oral mesenchyme in mouse. *See* collagen; decapentaplegic; fibrodysplasia ossificans progressiva; furin; GLI; noggin; Notch; organizer; osteoblast; osteoclast; pulmonary hypertension; Smad. (Olsen, B. R., et al. 2000. *Annu. Rev. Cell Dev. Biol.* 16:191; Ray, R. P. & Wharton, K. A. 2001. *Cell* 104:801.)

Bonferroni correction Guards against type I error at some α-values. In a small number of tests, it is satisfactory and simple. *See* error types; significance level. (Altman, D. G. 1991. *Practical Statistics for Medical Research*. Chapman & Hall, London.)

bookmark Indicates an address on the Internet or other computer-related items where you wish to return.

Book syndrome Autosomal-dominant defect of tooth development, high degree of sweating, and premature loss of hair color. *See* hair color.

Boolean algebra Developed by George Boole (1815–1864) for the use of formal logic. He supposed that in binary forms thinkable objects could be defined. Thus, if $x =$ horned and $y =$ sheep, then by selecting x and y the class of horned sheep is defined. Also $1 - x$ would define all things of the universe that are not horned, and $(1 - x)(1 - y)$ would identify all things that are neither horned nor sheep. This approach defines sets and subsets in discrete forms without intermediates, yet is capable of defining mutual relationships. Using simple symbols, syllogisms could be developed in mathematical forms. Learning of concepts by humans appears to be proportional to their Boolean complexity, i.e., to the length of the shortest logically equivalent proposition. The switch gear of telephone systems and modern digital computers were developed on the basis of Boolean binary logic.

bootstrap Statistical device that was introduced for computer operations (versus the classical-type computations). The standard error by the classical method is computed as

$$se(\overline{x}) = \left\{ \sum_{i=1}^{n} (x_i - \overline{x})^2 / (n[n-1]) \right\}^{1/2}$$

in comparison with the bootstrap procedure

$$se(t[x]) = \left\{ \sum_{b=1}^{B} (t[x^{*b}] - \overline{t})^2 / (B-1) \right\}^{1/2}$$

where $se(t[x])$ is the standard error of the bootstrap statistic, $t(x)$, $B =$ bootstrap samples of size n from the data, $\overline{t}$ is the average of the B bootstrap replications $(t[x^{*b}]$.

The bootstrap algorithm can be applied to the majority of statistical problems and is widely used for estimating the confidence level in evolutionary trees. The data points x_i need not be single numbers; they can be vectors, matrices, or more general quantities, such as maps and graphs. The statistic $t(x)$ can be anything as long $t(x^*)$ can be computed for every bootstrap data set x^*. Data set x does not have to be a random sample from a single distribution. Regression models, time series, or stratified samples can be accommodated by appropriate changes. For details and specific references, see B. Efron & R. J. Tibshirani. 1993. *An Introduction to the Bootsrap*. Chapman & Hall, New York. *See* jackknifing. (Kerr, M. K. & Churchill, G. A. 2001. *Proc. Natl. Acad. Sci. USA* 98:8961;

Davison, A. C. & Hinkley, D. V. 1997. *Bootstrap Methods and Their Application*. Cambridge Univ. Press, Cambridge, UK.)

Bora Bora   220 kb centromeric sequences in *Drosophila*. *See* centromere.

border sequence   *See* T-DNA.

Borjeson syndrome (Borjeson-Forssman-Lehman syndrome) Face, nervous system, endocrine defects, hypogonadism, assigned to human chromosome Xq26-q27. *See* head/face/brain defects; RBM.

leukocyte function–associated antigen-1 (hlFA-1). Thus, it seems that the apparently antibiotic-resistant individuals have an autoimmune reaction to this major histocompatibility class peptide encoded by the dominant DRB*0401 allele. *See* antigen; arthritis; autoimmune disease; BCG; HLA; *Ixodoidea*; mucosal immunity; serotype; serum. (Ohnishi, J., et al. 2001. *Proc. Natl. Acad. Sci. USA* 98:670; Kumaran, D., et al. 2001. *EMBO J.* 20:971; Revel, A. T., et al. 2002. *Proc. Natl. Acad. Sci. USA* 99:1562.)

Borromean ring   Interlocked rings.

The simplest form of Borromean rings : DNA may be arranged in this or more complex ways.

Borrelia   Spirochete bacteria; about 28 species (*B. burgdorferi*, *B. hermsii*, etc.) are responsible for relapsing fever or Lyme disease and other human ailments all over the world. The *Borrelia burgdorferi* genome B31 contains 910,725 bp linear DNA and 17 linear and circular plasmids. This bacterium, like *Mycoplasma genitalium*, has no genes for cellular biosynthetic functions, but there are 853 genes for transcription, translation, transport, and energy metabolism. The filamentous bacteria are 8- to 16-μm-long flagellate cells infectious for birds and mammals. Their generation time is about 6 hours, and in about 5 days within a single animal their population may exceed 10^6 cells, which coincides with the major symptoms (erythema migrans [enlarged red spots]) of the infection. Within weeks or months, the bacteria may invade all major organs of the body, primarily the joints, causing arthritic symptoms. If untreated, Lyme disease may be fatal.

Intravenous injection of rocephin or other antibiotics (also orally administered doxycycline) may be the cure, although some of the effects may persist for years. The vectors of the bacteria are the *Ixodes* arthropods (ticks) that live on grasses and low-growing bushes in wildlife (deer, mice, birds)–frequented rural and suburban areas. Identification of the disease is difficult because of the complexity of the symptoms. Serological detection encounters problems because the outer membrane of the bacteria displays variable serotypes. *Borrelias* harbor several copies of approximately 23–50 kb linear plasmids with genes for Vmps/Vsps (variable major proteins). Transposition within and recombination between the plasmids assures great antigenic variation in these organisms. New serotypes appear at an estimated frequency of 10^{-4} to 10^{-3} per cell per generation. This fact accounts for the difficulties in developing effective immunsera. An attenuated strain of *Mycobacterium bovis*, the bacillus Calmette-Guerin (BCG), may serve as a suitable vector for the *B. burgdorferi* surface protein antigen A and may secure more than year-long protection by mucosal delivery. About 10% of Lyme disease patients appear resistant to antibiotic treatment and display arthritis symptoms long after spirochetal DNA in joint fluids is no longer detectable. The arthritis is an immune response to the outer surface protein A (OspA) of the bacterium. Actually, OspA-reactive type 1 T-helper lymphocytes are found for many years after the cured infection in the joints. OspA has homology to human

boss   Transmembrane protein product on the R8 photoreceptor in the eyes of *Drosophila*, encoded by gene *boss* (*bride of sevenless*, 3.90.5). It is the ligand and activator for the receptor tyrosine kinase, encoded by the *sev* (*sevenless*, 1–33.38) gene. *See* daughter of sevenless; receptor tyrosine kinase; rhodopsin; sevenless; son-of-sevenless.

Bos taurus (cattle) $2n = 60$.

botany   Basic scientific field concerned with plants. (Herbaria and botanists: <www.nybg.org/bsci/ih>; <plants.usda.gov/plants/index.html>; Plant Name Index: <www. ipni.org>.)

bottleneck effect   If the size of the population is periodically reduced substantially, genetic drift may alter gene frequencies. Usually bottlenecks reduce variation. Occasionally after bottlenecks an increase of variation has been reported, and it is attributed to dominance and epistasis. Bottleneck effect is quite common in the transmission of mtDNA because only a small portion of it is passed through the germline, and the heteroplasmy may be altered. *See* genetic drift; heteroplasmy; mtDNA. (Galtier, N., et al. 2000. *Genetics* 155:981.)

bottom-up analysis   *See* top-down analysis.

bottom-up map   Relying on STS-based information. Bottom-up maps are useful for relatively short chromosomal distances. Two STSs are "singly linked" if they share at least one YAC and "doubly linked" if they share at least two YACs. Single linkage is generally not useful because of the high degree of chimerism among the YACs. In the first step, STSs are assembled into doubly linked contigs. Then the doubly linked contigs are ordered either on the basis of radiation hybrids or traditional genetic recombination information. Finally, single linkage can be used to join contigs to the same short genetic region. *See* bottom-down mapping; contig; mapping, genetic; radiation hybrid; STS; YAC. (Carrano, A. V., et al. 1989. *Genome* 31:1059.)

BOTULIN   Highly toxic product of Clostridium bacteria and frequent cause of potentially lethal food poisining. It is approved for treatment of strabismus and blepharospasm (a nervous eyelid problem) and it is also a cosmetics for treatment of 'crow's feet' facial signs of aging. *See* strabismus. (Moore, A. 2002. *EMBO Reports* 3:714.)

boule (*bol*) Autosomal gene in *Drosophila* encoding a cell cycle protein regulating G2-M transition. It is homologous to the human Y-chromosomal gene DAZ responsible for azoospermia. Its suspected function is translation and localization of mRNA. *See* azoospermia; cell cycle; *Dazla*; fertility; infertility; *pelota*; *twine*.

boundary element (barrier) Limits the function of cis-regulatory elements or the spread of heterochromatinization. *See* CTCF; heterochromatin; insulator; RAP.

bouquet (polarization) In leptotene the chromosomes are attached by their ends to a small area of the nuclear membrane while the rest of the chromosome length is looped across the nucleus. *See* horsetail stage; leptotene stage; meiosis.

Polarization.

bovine growth hormone (BGH) Somatotropin; it has been commercially produced by genetic engineering and it significantly boosts milk production. *See* somatotropin.

bovine papilloma viral vector By genetic manipulations, with pBR322 bacterial plasmid sequences added, it can be converted into a shuttle vector carrying genes between mice and *E. coli*. *See* shuttle vector; transformation, genetic; vectors; viral vectors.

bovine papilloma virus (BPV) Papova virus (about 7.9 kbp DNA) responsible for warts in animals. The BPV$_{69T}$ segment of its genome (5.5 kbp) has been used as a large-capacity vector that can multiply into 10 to 200 copies. It can stay as an episome or can be integrated into the chromosomes of mammals.

bovine spongiform encephalopathy (BSE) *See* Alzheimer disease; Creutzfeldt-Jakob disease; encephalopathies; prions; scrapie.

bowel disease (chronic inflammatory bowel disease) *See* CARD15; CIBD; Crohn disease.

Bowman-Birk inhibitors Prepared from plant tissues can bind and inhibit simultaneously or independently trypsin and chymotrypsin. They may decrease cancer development in animals exposed to alkylating agents or displaying sporadic tumorous growth. (*See* Witchi, H. & Espiritu, I. 2002. *Cancer. Lett.* 183[2]:141.)

box Generally used for a consensus sequence in the DNA, such as a homeobox; domains of the internal control regions (box A, box B, box C) that are the sequences where transcription factors bind. Sometimes protein boxes are also distinguished.

box gene Clustered mutations in exons or introns (in mosaic genes of mtDNA). *See* mtDNA.

bp Base pair.

B7 protein Required for the activation of B cells. It is recognized by CD28 on the surface of the antigen-presenting cells. The other required signal for B-cell activation is a foreign peptide antigen associated with class II MHC molecules on the surface of an antigen-presenting cell. The same as BB-1 or CD80. *See* antigen-presenting cells; B cell; CD40; costimulator; ICOS; MHC. (Yoshinaga, S. K., et al. 2000. *Int. Immunol.* 12:1439.)

BPV *See* bovine papilloma virus.

Brachmann-De Lange syndrome *See* De Lange syndrome.

brachydactyly Abnormally short fingers and toes controlled by autosomal-dominant genes. In type E the metacarpus and metatarsus (the bones between the wrist and the fingers of the hand and the corresponding bones in the foot) are shortened. In still other types, nervous defects, hypertension, and shortened bones of the arm accompany the hand and foot problems. The expression may vary. Most commonly, the middle bones (phalanx/phalanges) are affected (type A); in some cases, not all the fingers express the gene. In type B (9q22, receptor tyrosine kinase ROR2), in addition to the middle phalanges, the terminal ones are also short or absent. In type C, more than three phalanges may appear. Type D involves short and flat terminal phalanges of the big toe and the thumb. In type E, the metacarpus and metatarsus are shortened. In still other types, nervous defects, hypertension, and shortened bones of the arm accompany the hand and foot bone problems. In an autosomal-recessive form, the brachydactyly involves a small head (microcephaly). In another recessive form, primarily the great toe is affected, but the proximal (near the wrist) joints do not move. The dominant brachydactyly with severe hypertension gene has been assigned to human chromosome 12p. Brachydactyly type A-1 is due to mutation in the Indian hedgehog gene. *See* cartilage; hedgehog; polydactyly; Robinow syndrome; syndactyly. (Schwabe, G. C., et al. 2000. *Am. J. Hum. Genet.* 67:822; Gao, B., et al. 2001. *Nature Genet.* 28:386.)

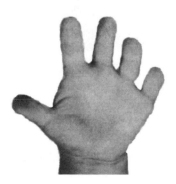

brachymeiosis Second meiotic division is missing. *See* meiosis.

brachyury Homozygous (*TT*) dominant lethal (after 10 days of conception) gene in mice. The *Tt* heterozygotes are viable and have reduced tail (tailless); the homozygous *tt* also dies in 5 days. The different alleles of the complex locus have different developmental effects. The anomaly (chromosome 17) involves a genetic defect in notochord development. The somites undergo differentiation but are resorbed before birth. There are also defects in the posterior parts (limbs, allantois, umbilical vessels). The *t* alleles may display meiotic drive; from the male *t/+* heterozygotes more than 90% of the progeny may receive the *t* allele. The distortion of the transmission (TRD) is controlled by at least 6 loci that do not normally recombine because of the presence of inversions. The rare recombinants are called partial *t* haplotypes. In addition, there are at least 16 lethality loci within the *t* haplotype, but they are not the primary causes of the distorted segregation. The distorted segregation is due to the cis-acting so-called T-complex responder (*Tcr*). The distorted ratio is due to the Smok (sperm motility kinase) located at the C-terminus of the ribosomal Rsk3 kinase. Smok apparently phosphorylates the axonemaldynein of the microtubules. The T-complex distorters (*Tcd*) are trans-acting factors that increase transmission of the *Tcr*-bearing chromosome. Despite the preferential transmission of the *t* haplotype by the heterozygous males, the populations carry only 10–25% *t* haplotypes. The brachyury transcription factor is embedded by its carboxy-terminal into the minor groove of the DNA contacting a guanine residue, but it does not bend the DNA. It is an important transcription factor for mesodermal specification. Brachyury-like anomalies also occur among cats, dogs, sheep, cattle, and pigs. *See* axoneme; dynein; haplotype; Holt-Oran syndrome; killer spore; Manx in cat; meiotic drive; notochord; somite; TCP-1. (Schimenti, J. 2000. *Trends Genet.* 16:240.)

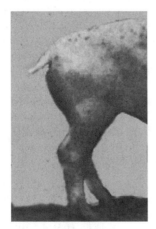

Short-tail pig.

bracken fern (*Pteridium aquilinum*) Carcinogenic plant used as food in Japan.

bract Small modified leaf from which flowers may develop, or a leaf on the floral axis subtending the flower. BRACT

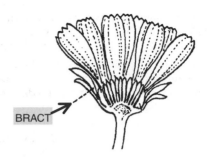

BRACT

Bradford method (*Anal. Biochem.* 72:248) For 1 to 100 μg protein. Prepare a standard solution (0.5 mg/mL) of bovine serum albumin (BSA), make a dilution series 5 to 20 μL, and dilute to 100 μL with 0.15 M NaCl. Also prepare 0.15 M NaCl blanks. Make a series of dilutions from the unknown quantity of the protein to be tested. Add 1 mL Coomassie brilliant blue to all and mix thoroughly. After 2 min, determine absorption at 1 cm path length at 595 nm wavelength in a spectrophotometer and extrapolate the concentration of the sample from the standard series. *See* Kjeldahl method; Lowry test.

bradykinin *See* kininogen.

bradytelic evolution Has a very slow pace and involves species whose adaptive environment extends over very long (geological) periods. In contrast, tachytelic evolution progresses at a fast pace. Horotelic evolution appears to show an average rate. *See* evolution.

brain, human Very complex structure, and here only a few major landmarks are outlined as reference to several entries dealing with the central nervous system. The seven-layered hippocampus, consisting of gray matter, is not shown, although this is the most important area at the basal-temporal region involved in memory and learning. The functional areas of the brain can be identified by the

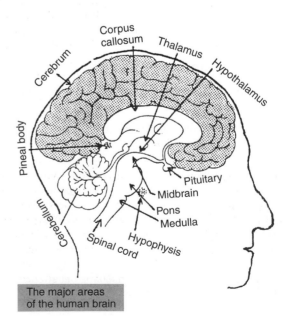

The major areas of the human brain

increased blood flow upon stimulation by using positron emission tomography or functional magnetic resonance imaging. Microelectrodes applied to individual nerve cells reveal electrical activity at a single cell level. The brain size of hominids approximately tripled in less than 3 million years of evolution. Brain volume and cognitive abilities seem to be positively correlated ($r = +0.4$), yet the correlation between brain size and cognitive performance is very low within families. The positive correlation may be mainly nongenetic and influenced by socioeconomic status, cultural influences, etc. *See* cerebellum; human intelligence; memory; nerve function; nuclear magnetic resonance spectrography; tomography. (Nichols, M. J. & Newsome, W. T. 1999. *Nature. Suppl.* 402:C35; <www.loni.ucla.edu>; <braininfo.rprc.washington.edu>; mouse brain: <www.mbl.org>.)

brain disease *See* Addison-Schilder syndrome; affective disorders; craniofacial synostosis syndromes; epiloia; mental retardation; prions.

brain stem Medulla + pons + midbrain. *See* diagram of brain.

branched-chain amino acids *See* isoleucine-valine biosynthetic steps.

branched RNA Intermediate of RNA splicing. *See* introns.

branchio-otorenal syndrome (BOR) Human chromosome 8q13.3–dominant syndrome with incomplete penetrance and expressivity. It involves defects in the appearance of the ears and underdevelopment of the middle ear structures (malleus, incus, stapes) and the inner ear (cochlea), resulting in mild to severe hearing loss. This is accompanied by underdevelopment of the kidney and the urinary tract. A variant form without the kidney symptoms maps to 1q31. The gene bears homology to the *Drosophila* gene *eyes absent* (*eya*). (Kumar, S., et al. 2000. *Am. J. Hum. Genet.* 66:1715.)

branch migration During the process of molecular recombination, the exchange point between two fixed sites of the DNA single strands can move left or right. When the two single strands are separated, they can simultaneously reassociate in an exchanged manner in both double helices. This strand invasion brings about heteroduplexes. In *E. coli*, the RuvA (a specificity factor) and RuvB (an ATPase) proteins—induced by ultraviolet radiation damage to the DNA—bind to the Holliday junctions and increase the length of the heteroduplex. Mismatches in the synaptic strands may interfere with branch migration. *See* heteroduplex; Holliday juncture; Holliday model; mismatch repair; recombination, molecular mechanisms; Ruv ABC; Walker box. (Putnam, C. D., et al. 2001. *J. Mol. Biol.* 311:297; Constatinou, A., et al. 2001. *Cell* 104:259; Fabisiewicz, A. & Worth, L., Jr. 2001. *J. Biol. Chem.* 276:9413.)

branch point sequence Short RNA tract (YNCUR⌐A⌐Y, [Y stands for pyrimidine, R for purine, and N can be either]) in the primary transcript near (18–38 base upstream) the 3′ end of an intron (**AG**) of mammals. After exon1-intron boundary is severed and the intron end is released with a 5′-**GU** pair at the end, it forms a loop as it folds back by **G**, making a 2′ → 5′ bond with the boxed A shown above. Subsequently a cut is made at the 3′ end of the intron, the intron is released, and exon1 is attached to exon 2. *See* introns. (Peled-Zehavi, H., et al. 2001. *Mol. Cell Biol.* 21:5232.)

Brassica oleracea (cabbage, kale) Vegetable crops. Basic chromosome number is controversial 5 or 6, although cabbage is $2n = 18$, and there are some indications of being an amphidiploid. *See* mustards; radish; rapes; swedes; turnip; watercress. (Howell, E. C., et al. 2002. *Genetics* 161:1225.)

brassinolide *See* brassinosteroid.

brassinosteroid Synthesized through the pathway campesterol → campestanol → cathasterone → teasterone → 3-dehydroteasterone → typhasterol → castasterone → brassinolide. The latter compound has been shown to remedy deetiolation, derepression of light-induced genes, miniaturizing, male sterility, and other symptoms of stress-regulated genes. These brassinosteroids bear close similarity to ecdysones, the animal molting hormones. The phytoecdysones have been known for two decades in plants. All these plant hormones interact with each other in various ways and regulate signal transduction and gene activities. Unlike most animal steroid hormones, plant hormones are of small molecular size (except brassinosteroids), generally in the range of 28–350 Da. Brassinolide has a MW of about 480. A putative brassinosteroid receptor kinase shows similarities to the *ERECTA* and *CLAVATA1* gene products of *Arabidopsis* and shares leucine-rich repeat with disease-resistance genes. The *BAS1* locus of *Arabidopsis* regulates the level of brassinosteroids and the response to light. *See* deetiolation; hormones; photomorphogenesis; plant hormones; steroid hormones. (Szekeres, M., et al. 1996. *Cell* 85:171; Clouse, S. D. & Sasse, J. M. 1998. *Annu. Rev. Plant. Physiol. Mol. Biol.* 49:427; Neff, M. M., et al. 1999. *Proc. Natl. Acad. Sci. USA* 96:15316; Kang, J.-G., et al. 2001. *Cell* 105:625; Li, J. & Nam, K. H. 2002. *Science* 295:1299; Yin, Y., et al. 2002. *Cell* 109:181; Bishop, G. J. & Koncz, C. 2002. *Plant Cell* 14:S97.)

BRCA1 Breast cancer antigen is an exclusively nuclearly located protein. *See* breast cancer.

BrdU Bromodeoxyuridine. *See* bromouracil; chemical mutagens; hydrogen pairing.

breakage and reunion The chromatids or DNA single strands are broken at the position of chiasmata and reunited in an exchanged manner during genetic recombination. This process is a physical event not requiring (normally) DNA replication as it was one time hypothesized with the copy choice idea. It is now known that recombination also takes place by replication. *See* Holliday model; recombination by replication; recombination molecular models. (Creighton, H. B. & McClintock, B. 1931. *Proc. Natl. Acad. Sci. USA* 17:492; Stern, C. 1931. *Biol. Zbl.* 51:547; Meselson, M. 1964. *J. Mol. Biol.* 9:734.)

breakage-fusion-bridge cycle   May cause variegation in the tissues because some of the genes may not be present in one of the daughter cells, whereas the other cell receives two copies. If this dominant gene determines color, its presence is immediately recognized in the cell lineages. In telomerase-deficient and p53 mutant mice, epithelial cancer development is promoted by breakage-fusion-bridge cycles. In *Saccharomyces cerevisiae*, dysfunctional telomeres may increase mutation rate 10- to 100-fold. *See Ac-Ds*; cancer; p53; telomerase. (McClintock, B. 1941. *Genetics* 26:234; Hackett, J. A., et al. 2001. *Cell* 106:275.)

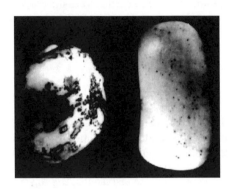

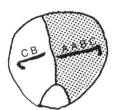

The left side of the kernel is colorless

The right side is colored because of the *AA* genes

CBA

End of chromosome is broken ===

CBA
CBA

After replication broken ends fuse   →

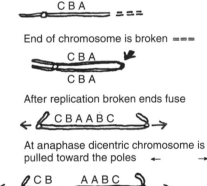

CBAABC

At anaphase dicentric chromosome is pulled toward the poles   ←   →

CB   AABC

Bridge

CB   AABC

Chromosome breaks apart at bridge

No *A* gene at left   Two *A* genes at
                                       right

Breakage-fusion-bridge cycles may occur in the endosperm of maize plants if the end of the chromosomes is broken. The genetic and cytological consequences of such events are diagramed here. The relative size of the sectors (detectable when appropriate color markers are used) indicates the developmental time of the cycle. Early events involve large sectors, late events are indicated by small sectors. If the event occurs repeatedly, several sectors are observed. In case the two centromeres move toward the same pole, no bridge is formed at anaphase and an intact dicentric chromatid is recovered. In case a double bridge is formed and both chromatids break, two monocentric (most commonly defective chromosomes) go to the poles. The breakage-fusion-bridge cycle may not continue in the tissues of the growing plants because of apparent healing of the broken ends. The healing is attributed to the acquisition of new telomeres. (Photographs courtesy of Barbara McClintock.)

breakpoint mapping   *See* inversions.

breast cancer (BRCA)   Breast cancer is one of the most common diseases of women. There is a greater than 10% chance that women who live to 90 or over will develop breast cancer and ∼5–10% of the cases are caused by mutation either in the BRCA1 or BRCA2 genes. Deficiency of BRCA2 leads to impaired homologous recombination but maintains normal nonhomologous end joining (Xia, F., et al. 2001. *Proc. Natl. Acad. Sci. USA* 98:8644). Even in nonhereditary (sporadic) cases of breast cancer, the BRCA1 gene is frequently lost, inactive, or rearranged. BRCA1 is actually a tumor suppressor (in human chromosome 17q21) and it is responsible for 25% of the cases diagnosed before age 30. BRCA1 along with CREB is part of the pol II holoenzyme. It activates transcription when it is associated by its C-terminus with RNA helicase A and pol II. In BRCA1, of the 12 RNA polymerase subunits the activation of hRPB2 and hRPB10α is most critical. Predisposition to breast cancer is inherited as a dominant trait, but the somatic expression (manifestation of cancer) requires that in these heterozygotes the normal allele is lost or inactivated during the lifetime of the individual who inherited one BRCA susceptibility allele. The BRCA1 reading frame encodes 1,863 amino acids (22 exons) with a Zn-finger domain at the NH_2 end. Amino acids 1,528 to 1,863, i.e., the C-terminal domain, appear to be transcriptional activators. The primary single transcript is 7.8 kb and is expressed primarily in the testis and the thymus but also in the breast and the ovary. The transcript displays alternative splicing.

Gene Id4 is an important negative transcriptional regulator of BRC1 expression. BRCA1 is an important component of the 18-protein complex (SWI/SNF) involved in chromatin remodeling. BRCA1 strongly binds DNA, protects the DNA from nucleolytic attack without sequence specificity, and is involved in double-strand DNA repair. BRCA1 binds to BRG1. As a transcription factor, it enhances the expression of several genes including p53. The BRCA1 sequences are well conserved in mammals but absent from chicken. The defects in the gene vary in the different kindreds from 11 bp deletion to frame shifts, nonsense, missense mutations, or other alterations causing instability. The phenotype of the patients varies between the kindreds, indicating that the specific mutations at the locus may affect its expression. It is noteworthy that there have been female carriers of the mutation(s) who by age 80 failed to develop breast or ovarian cancer. Apparently, the expression is affected to some extent by extraneous genetic and environmental factors.

The BRCA1 protein may also be aberrantly localized in the cytoplasm and complicates the expression pattern. BRCA2 in human chromosome 13q12-q13 encodes 3,418 amino acids within a 6 cM region; it is also a dominant early-onset disease responsible for about 45% of all *hereditary* breast cancers. Its highly conserved third exon is homologous to the c-Jun oncogene where the JNK protein binds. The 18–60 amino acid residues are potential activation sites. On both sides of exon 3 there are inhibitory regions (IR1 & 2). BRCA2 does not create a substantial risk for ovarian and other cancers, but the chance for breast cancer in males may be slightly elevated in contrast to BRCA1. Deletion 185delAG in BRCA1 and deletion 617delT in BRCA2 occur with carrier frequencies of 1.09% and 1.52%, respectively, in Ashkenazy Jewish populations. The BRCA2 protein is cytoplasmically located. Its nuclear localization factor resides within the C-terminal 156 amino acids and deletion in that region (e.g. 617delT) prevents the translocation of the protein to the nucleus; consequently, it loses its ability to suppress tumorigenesis. In 36% of the BRC families, chromosomal rearrangements (deficiencies, duplications) may be present.

The normal allele of BRCA1 transactivates the cyclin-dependent protein kinase inhibitor $p21^{WAF1/CiP1}$ without the cooperation of p53 and thus the entry into the S-phase of the cell cycle is prevented. This process depends, however, on the normal allele of p21. BRCA1 appears to be involved in transcription-coupled repair of oxidative DNA damage, and the defective BRCA1 conveys hypersensitivity to ionizing radiation and hydrogen peroxide. Human Cds1 phosphorylates the BRCA1 protein at serine 988 after DNA damage, thus assisting cell survival. BRCA2 appears to be a cofactor of the radiation hypersensitivity gene Rad51 mediating double-strand DNA repair by homologous recombination or transcription-coupled repair upon phosphorylation by ATM or ATR (an ataxia telangiectasia-mutated and RAD3-related protein). BRCA1 and BRCA2 proteins apparently interact in the cell. The existence of a BRCA3 gene in chromosome 13 is controversial; however, there are several low-penetrance genes with potential importance. Homozygosity of mutant (truncated) BRCA2 (unexpectedly) also blocks cell proliferation and causes chromosomal breakage, but mutations in spindle assembly checkpoint genes (p53, Bub1, Mad3L) relieve this growth arrest and restart (neoplastic) proliferation (Lee, H., et al. 1999. *Mol. Cell* 4:1).

One of the remaining types of breast cancer is supposed to be due to a mutation in the *KRAS2* (Kirsten sarcoma) gene in human chromosome 6 when at codon 13 a G → A transition takes place, resulting in Gly → Asp substitution. Ductal breast cancer was attributed to the loss of genes in human chromosome 1q21ter, but chromosomes 2, 14, and 20 were also implicated. In families with a high incidence of breast cancer, a small secreted protein gene expressed only in human breast cancer was mapped to chromosome 21q22.3. A dominant

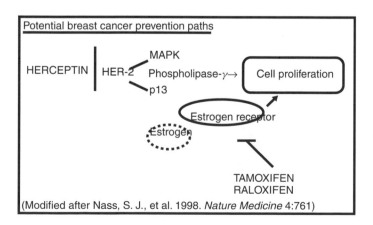

Potential breast cancer prevention paths

(Modified after Nass, S. J., et al. 1998. *Nature Medicine* 4:761)

gene product serologically reacts with murine monoclonal antibody DF3. In the region of chromosome 17p13.3, the *TP53* regulator of tumor protein p53 was found, but it could not be ruled out that a regulator exists 20 megabases telomeric to *TP53*. AIB1 steroid receptor (20q, member of the SRC-1 family oncogenes) is either amplified or overexpressed in the tumors. AIB1 seems to be a coactivator. Some forms of breast cancer also affect males, and the incidence of breast cancer in Klinefelter (XXY) men is almost as high as that in women.

Some types of breast cancers are associated with cancers in other organs. The risk of recurrence of breast cancer among first-degree relatives may be as high as 50%. The risk of breast cancer increases with age, unmarried status, obesity, radiation exposure, etc. Several environmental chemicals (heterocyclic amines, o-toluidine, dibromoethane, glycidol, etc.) may cause breast cancer. In North America, the lifetime risk of breast cancer in females is about 10%. Of all breast cancer incidences, about 5–10% are due to inherited causes. The incidence of breast cancer rises steeply between age's 25 and 50, then levels off apparently because of the decreasing level of estrogen. The development of breast cancer for BRCA1 and BRCA2 carrier females by age 70 is 28–87% and for ovarian cancer ~25–30%. The male carriers also have an increased risk for breast, prostate, colon, pancreas, gall bladder, bile duct, and stomach cancers and melanoma (Liede, A., et al. 2000. *Am. J. Hum. Genet.* 67:1494).

The contralateral occurrence of breast cancer is much higher than the prevalence of breast cancer in the general population. BRCA1 gene is partly responsible for ovarian and prostatic cancers. DNA repair defect seems to be involved. BRCA1 forms a complex with proteins hRAD50-hMRE11-p95. BIC (breast cancer information core) address: <http://www.nchgri.nih.gov/Intramural_research/Lab_transfer/bic/>. The genes responsible for Peutz-Jeger syndrome, Cowden syndrome, androgen receptor, p53, ataxia telangiectasia, AM Muir-Torre syndrome, Li-Fraumeni syndrome, and Nijmegen breakage syndrome may also increase breast cancer risk. For environmental factors in breast cancer: <http://www.cfe.cornell.edu/bcrf>. *See* ataxia telangiectasia; AIB; ATM; ATR; BRG1; cancer; Cdc1; contralateral; cyclin D; estrogen receptor; genetic screening; granin; HER-2; herceptin; homologous recombination; hormone receptors; JNK; Jun; Klinefelter syndrome; Li Fraumeni syndrome; mouse mammary tumor virus; multiple hamartomas; NHEJ; Nijmegen breakage syndrome; oncogenes; ovarian cancer p13; p21; p53; p95; phospholipases; RADSO; raloxifen; RNA helicase; RNA polymerase; tamoxifen; tumor suppressor; ZAG. (Welcsh, P. L., et al. 2000. *Trends Genet.* 16:69; Cui, J., et al. 2001. *Am. J. Hum. Genet.* 68:420; Risch, H. A., et al. 2001. *Am. J. Hum. Genet.* 68:700; Nathanson, K., et al. 2001. *Nature Med.* 7:552; Welcsh, P. L. & King. M.-C. 2001. *Hum. Mol. Genet.* 10:705; Nathanson, K. L. & Weber, B. L. 2001. *Hum. Mol. Genet.* 10:715; Narod, S. A. 2002. *Nature Rev. Cancer* 2:113; Bhatia, S. & Sklar, C. 2002. *Nature Rev. Cancer* 2:124; Chlebowski, R. T. 2002. *Annu. Rev. Med.* 53:519.)

BREATHING OF DNA A reversible, short-range strand-separation below the melting temperature. *See* melting temperature.

breeding system Mating within a population may be random (each individual has an equal chance to mate with any member of the opposite sex), or it may be self-fertilization (in monoecious species) or inbreeding (in dioecious species). Also, random mating and inbreeding both take place within the group. The term *breeding system* denotes these alternatives. The predominant breeding system of a few plant species is tabulated (A: apomictic, D: dioecious, I: self-incompatible, M: monoecious, O: outbreeding, S: selfing)

Alder M	Datura S	Potato O-S-I
Alfalfa O-S	Elm M	Ramie O
Almond O-I	Eggplant O	Radish O-I
Antirrhinum O-S	Fescue O-S	Rape seed O-I
Apple O-I	Flax S	Rice S
Apricot O	Grape O	Rose O
Arabidopsis S	Hemp D-M	Rubber O
Ash M-D	Hop D	Rubus O
Asparagus D	Lentil S	Rye O-I
Barley S	Lespedeza O	Rye grass O-I
Basswood O	Lettuce S	Sorghum O-S
Beach M	Lupine O-S	Soybean S
Bean S	Maize M	Spinach D
Belladonna O	Maple O	Spruce M
Beet O-I	Meadow foxtail O	Squash M
Birch M-I	Millet O-S	Stock S
Blue grass O-A-S	Mulberry D	Strawberry D-S
Broad bean S	Mustard O-I	Sugar beet O
Brome grass O	Oak M	Sugarcane O-I
Buckwheat O-I	Oat S	Sunflower O-I
Cabbage O-I	Oenothera O-S	Sweetclover O-S
Carnation S	Onion O	Sweetpea S
Carrot O	Orchard grass O-I	Sycamore M
Castorbean M	Osage orange D	Tea O
Celery O	Parsley O	Teosinte M
Cherry O-I	Pea S	Timothy O
Chestnut O	Peach O	Tobacco S
Citrus O-I	Peanut S	Tomato S
Clover O-I	Pear O	Tripsacum M
Coffee O-I	Petunia O-S	Triticale S
Collinsia S	Pine M	Walnut M
Cotton O-S	Pineapple O-I	Wheat S
Cucumber M	Plum O-I	
Date palm D	Poplar D	

breeding value   Quantitative value of a genotype judged on the basis of the mean performance of the offspring. Actually, it is twice the mean deviation of the offspring from the mean of the parental population. The doubling is used here because each parent contributes a haploid gamete to the offspring and thus half of its genes. Breeders frequently call it the additive effect. The observed performance of individuals is called the *phenotypic value* (P), which is measured as the mean value of the population. The average value of two homozygotes is called the *midpoint*. It is equal to zero (only in cases where the frequency of the two alleles is equal, 0.5) because the two parents deviate from it by a quantity of $(+a)$ and $(-a)$, by definition, and their sums cancel out each other. The *genotypic value* of the heterozygotes is designated by (d). In the absence of dominance $d = 0$, with complete dominance $d = a$, with overdominance $d > a$:

aa	0		Aa	AA
$\leftarrow$	$-a$	$\rightarrow \leftarrow$	$+a$	$\rightarrow$
		$\leftarrow$ d	$\rightarrow$	

The mean value (x) is usually calculated as the weighted mean, i.e., multiplied by the genotypic frequencies in the population. If the population is in equilibrium, the mean is $\bar{x} = p^2(a) + 2pq(d) + q^2(-a) = (p^2 - q^2)(a) + 2pq(d) = (p + q)(p - q)(a) + 2pq(d)$ and because $p + q = 1$, $\bar{x} = (a)(p - q) + 2pq(d)$: and if several loci are involved, $\bar{x} = \Sigma[(a)(p - q) + 2pq(d)]$. *See* gain; Hardy-Weinberg equilibrium; heritability; merit. (Falconer, D. S. and Mackay, T. E. C. 1996. *Introduction to Quantitative Genetics*. Longman/Addison Wesley, White Plains, NY.)

brefeldin A (γ,4-dihydroxy-2-[6-hydroxy-1-heptenyl]-4-cyclopentanecrotonic acid λ-lactone)   Inhibitor of passing peptides from the endoplasmic reticulum and Golgi complex. In the absence of brefeldin, yeast cells suffer chromosome instabilities. *See* ARF; ARNO; translocase. (Wigge, P. A., et al. 1998. *J. Cell. Biol.* 141:967; Lang, B. D., et al. 2001. *Nucleic Acids Res.* 292567.)

bremsstrahlung (brake radiation, [from German])   Electromagnetic radiation resulting from retardation or acceleration of a high-energy particle.

BRENDA   Database of at least 40,000 enzymes or 6,900 organisms. (<http://www.brenda.uni-koeln.de.)

BRF   Binding factor in RNA synthesis initiation.

BRG1   1,613-amino-acid DNA-dependent human ATPase active in SWI/SNF, RSC-mediating chromatin remodeling. *See* breast cancer; chromatin remodeling; SWI/SNF. (Strobeck, M. W., et al. 2001. *J. Biol. Chem. Online*, Nov. 21; Barker, N., et al. 2001. *EMBO J.* 20:4935.)

BRIC   *See* Byler disease.

bridge   Anaphase tie between separating centromeres in dicentric chromosomes. *See* breakage-fusion-bridge; inversion.

Meiotic anaphase single and double bridges and 1 or 2 chromatid fragments resulting from 2- and 4-strand recombination in a paracentric inversion heterozygote. (Courtesy of Dr. Arnold Sparrow.)

BRIDGE β-lactamase reporter for imaging downstream gene expression. *See* β-lactamase.

bridge protein   Facilitates the interaction between viral particles and cell surface receptors.

bridging cross If two genetically distant sexually incompatible species (A and B) are to be selected for gene transfer by sexual means. The problems can possibly be overcome by first mating one of the species with an intermediate compatible form (C), then crossing the hybrid (A × C) to the other (B). C serves as the bridge.

bright-field microscopy Ordinary light microscopy. *See* dark-field microscopy; fluorescent microscopy; Nomarski; phase contrast microscopy; stereomicroscopy.

bright paramecia *See* symbionts, hereditary.

bristle *See* chaetae.

Britten & Davidson model Suggested as a working hypothesis in the 1960s for interpreting the processes involved in the regulation of eukaryotic gene functions. The external stimuli were supposed to be directed to *sensor* genes that activated *integrator* genes, which in turn transmitted the signals to *receptor* genes, which affected *structural* genes, coding for protein. These systems might have operated in a series of interacting batteries. (Britten, R. J. & Davidson, E. H. 1969. *Science* 165:349; Britten, R. J. 1998. *Proc. Natl. Acad. Sci. USA* 95:9372.)

BRM Animal chromatin remodeling ATPases of ~1,600 amino acids. *See* chromatin remodeling.

Brn Eukaryotic transcription factors with POU domain controlling terminal differentiation of sensorineural cells. They are homologous to unc-86 in *Caenorhabditis*. *See* POU; *unc-86*.

B-RNA *See* cowpea mosaic virus.

broad bean (*Vicia faba*) $2n = 2x = 12$ and large chromosomes have been used extensively for cytological research. It is an important crop in cool climates. *See* favism.

broad-betaliproteinemia Hyperlipoproteinemia. *See* apolipoproteins; hyperlipoproteinemia.

broad sense heritability *See* heritability.

Brody disease (ATP2A1) Rare recessive disease encoded in human chromosome 16p12.1-6p12.2; it involves impairment of muscle relaxation, stiffness and cramps in the muscles. The basic defect is associated with the muscle sarcoplasmic reticulum calcium ATPase (SERCA1). *See* Darier-White disease; neuromuscular diseases; sarcoplasmic reticulum.

broken tulip Variegation (sectors) in the flowers caused by viral infection. These sectorial tulips have commercial value in floriculture. In the 17th–18th centuries, the bulbs were so highly valued that they fetched gold of equal weight. *See* infectious heredity; symbionts, hereditary.

Photo courtesy of Stichtung Laboratorium Bloembollenonderzoek, Lisse, The Netherlands.

bromodomain More or less cylindrical shape association of four helices of about 100 amino acids that form the docking sites in the chromatin for a large number of proteins. The lysine-acetylated H3 and H4 histones may be fitting into the bromodomains and are the conditions for gene transcription in eukaryotes. Bromodomains appear to anchor histone acetylase to the chromatin. *See* chromatin remodeling; histone acetyltransferases; histones; p300; PCAF; SAGA; SNF; TAF; TAF$_{II}$250. (Ornaghi, P., et al. 1999. *J. Mol. Biol.* 287:1; Dhalluin, C., et al. 1999. *Nature* 399:491.)

bromophenol blue *See* tracking dyes.

bromouracil (BU) Pyrimidine base analog that is mutagenic in prokaryotes because it may lead to base substitution after tautomeric shift. When incorporated into eukaryotic chromosomes, it may cause breakage upon exposure to light. On this basis, it has been used successfully as a selective agent in animal cell cultures. The nongrowing mutant cells failed to incorporate it and survived while the growing (wild-type) cells were killed upon illumination. Eosinophil peroxidase may produce 5-bromodeoxycytidine, and the latter is incorporated into DNA as 5-bromodeoxyuridine, which by mispairing with guanine may become mutagenic. *See* base substitution; chemical mutagens; eosinophil; hydrogen pairing; tautomeric shift. (Benzer, S. & Freeze, E. 1958. *Proc. Natl. Acad. Sci. USA* 44:112.)

Bronze Age About 5,000 years ago marked the development of crafts and urbanization.

brood Offspring of a single mating or the cluster of eggs (clutch) laid by a bird or reptile.

brown fat Adipose tissue that dissipates mainly heat. Humans with pheochromocytoma have large deposits of brown fat. *See* adipocyte; pheochromocytoma.

Brownian ratchet Hypothesis explaining transport of molecules into the mitochondria by Brownian movement (thermal agitation of molecules). Mitochondrial Hsp70-bound ATP complex associates with Tim proteins, and nuclear-encoded preproteins destined to mitochondria slide into import channels. The Mge1 protein initiates ADP exchange, thus preventing backward movement, and the protein is imported into the organelle. *See* ADP; ATP; Mge1; mitochondrial import. (Brokaw, C. J. 2001. *Biophys. J.* 81:1333.)

Brownian-Zsigmondy movement   Colloidal particles in solution may be in a continuously agitated motion due to collision with the medium. It can microscopically be observed in the cytosol of living cells.

BR RNP (Balbiani ring ribonucleoprotein)   Transcripts of the Balbiani ring associated with about 500 protein molecules of a total molecular size of 10^6 daltons. *See* Balbiani ring.

Bruce effect   Termination of pregnancy in mice by olfactory influence on a pregnant female of a male that is genetically different from the inseminator. *See* olfactogenetics; pheromones.

Brugada syndrome (SCN5A, 3p21)   Dominant LQT-type heart disease (idiopathic ventricular fibrillation) due to defect in exon 28 of a sodium channel (SCN5A) gene. SUNDS (sudden unexplained nocturnal death syndrome), relatively common in Southeast Asia, is controlled by an allelic gene. *See* heart diseases; ion channels; LQT. (Vatta, M., et al. 2002. *Hum. Mol. Genet.* 11:337.)

brush border   Dense lawn of microvilli on the intestinal and kidney epithelium that facilitates absorption by increasing the surface. *See* microvilli.

Bruton's tyrosine kinase   *See* agammaglobulinemia; BTK.

bryophyte   Moss, liverwort, and hornwort. Bryophytes are green plants similar to algae, but the organization of their body is more complex. Their gametangia is either unicellular or multicellular and show some cell differentiation. They usually have haploid and diploid life forms. The majority of them are terrestrial. *See* alga.

BS   *See* Bloom syndrome.

BSAP   *See* pax.

BSE   Bovine spongiform encephalopathy. *See* Creutzfeldt-Jakob disease; encephalopathies.

BSL   Biological safety level is specified by governmental regulations. BSL-1 is the minimal and BSL-4 is the most stringent, depending on the hazards involved.

bt   Prefix for *Bos taurus* (bovine) DNA or protein.

BTF2   Same as TFIIH. *See* transcription factors.

BTG   Antiproliferative protein encoded in human chromosome 1q32. Its synthesis is regulated by p53. *See* p53.

BTK   Bruton's tyrosine kinase belongs to a family of nonreceptor tyrosine kinases. Its deficiency results in immunodeficiency. Syk and Lyn activate BTK with the mediation of BLNK. *See* agammaglobulinemia; B-lymphocyte receptor; Lyn; Sky. (Liu, W., et al. 2001. *Nature Immunol.* 2:897.)

BUB1, BUB2   Checkpoint proteins arresting mitosis when the spindle attachment assembly is defective or there are problems with sister chromatid separation. *See* cell cycle; MAD. (Krishnan, R., et al. 2000. *Genetics* 156:489; Geymonat, M. et al. 2002. *J. Biol. Chem.* 277:29439.)

bubonic plague   *See Yersinia.*

buccal smear   Sample of the epithelial cells from the inner surface of the cheek is spread onto microscope slides and used for rapid determination of the number of Barr bodies with 4–5% error rate. This procedure (mucosal swab) is noninvasive and the sampling is painless. The sampled cells may also be used for DNA analysis by PCR after rapid (ca. 1 hr) extraction. *See* Barr body; PCR.

buckwheat (*Fagopyrum*)   Feed and food plant ($2n = 2x = 16$). Eaten by animals or by humans; may increase sensitivity to light and skin rash. *See* favism.

Buckwheat.

bud (budding protein of yeast)   Cytoskeleton assembly-mediating protein family. Bud proteins associated with GTPase-activating protein (GAP) are required for axial and bipolar budding patterns. *See* Bni; cytoskeleton; polarity, embryonic; profilin.

budding   Asexual reproduction by what the cell's cytoplasm does not divide into two equal halves, yet the bud eventually reproduces all cytoplasmic elements and grows to normal size after receiving a mitotically divided nucleus. Budding of enveloped viruses takes place by acquiring their membrane of lipid bilayer and proteins. They direct their surface glycoproteins into one or another type of cell membrane. They are pinching off either from the cell surface or into the cell lumen. The lipids of the bilayer come from the cell, whereas the virus DNA specifies the proteins. *See Saccharomyces cerevisiae.*

budding yeast   *See Saccharomyces cerevisiae.*

BUdR (5-bromodeoxyuridine)   Animal cells deficient in thymidine kinase enzyme are resistant to this analog. It is a base analog mutagen substituting for thymidine and may cause mutation. Its mutagenic efficiency is low, especially in eukaryotes, although if cells with BUdR-substituted chromosomes are exposed to visible light, chromosome breakage is induced. When chromosomes with BUdR substitutions in at least one of the strands are exposed to low LET ionizing radiation, double-strand breaks occur in proportion to the amount of substitutions. The presence of reducing agents (e.g., $e^-_{aq\bullet}$) favors breakage. See base substitution mutation; bromouracil; hydrogen pairing; sister chromatid exchange.

bud scar Chitin ring formed at the junction of the mother and daughter yeast cells that persists even after separation of the two cells. The number of bud scars may indicate the number of cell divisions (age) as well as polyploidy, which is characterized by a different pattern of bud scars.

bud sport Genetically different sector (due to somatic mutation) in an individual plant (chimera).

Buerger disease Autosomal-recessive predisposition to thromboangiitis (inflammation of the blood vessels); its frequency is relatively high in some oriental ethnic groups.

buffalo Asiatic swamp buffalo (*Bubalus bubalis*) $2n = 48$, Murrah buffalo (*Bubalus bubalis*) $2n = 50$, African buffalo (*Syncerus caffer caffer*) $2n = 52$, and *Syncerus caffer nanus* $2n = 54$.

buffer (1) Chemical solution capable of maintaining a level of pH within a particular range depending on the components of an acid-base system. (2) Special storage area in the memory of a computer from where the information can be utilized at different rates by different programs; e.g., the printer can store information faster than it can print it.

buffering, genetic Homeostatic mechanism that is supposed to maintain the function of genes at a certain level. Redundancy (duplications), feedback, epistasis, temperature sensitivity, modifier genes, signal transduction, chaperones, apoptosis, etc., may mediate it. (Kitani, T. & Nadean, J. H. 2002. *Nature Genet.* 32:191.)

buffer sequence To avoid the loss of indispensable 5' and 3' tracts from linearized transforming DNA, protective sequences may be added to the constructs. These buffers should not have cryptic splice or regulatory sites. Introns may be used.

Bufo vulgaris (toad, $2n = 36$) Primarily terrestrial small frog species living in water environment during the mating season. See frog; *Rana*; toad; *Xenopus*.

bulge Unpaired stretches in the DNA. They are involved in binding of regulatory protein domains, enzymatic repairs, and slipped mispairing in the replication of microsatellite DNA; they are intermediates in frame shift mutations and essential elements for naturally occurring antisense RNA. See antisense RNA; binding proteins; DNA repair; frame shift mutation; microsatellite; mispairing.

bulimia Psychological disorder involving excessive eating based on serotoenergic abnormality. See addiction; anorexia; obesity; serotonin.

bulked segregant analysis Used in mapping recombinant inbred lines. The individuals in the population are identical at a particular locus, but unlinked regions in the chromosomes are represented at random. See RAPD.

bulk-flow model Postulates that export of molecules from the endoplasmic reticulum is not regulated by targeting signals. Experimental evidence indicates, however, that even the constitutively excreted proteins carry some target specificities. See endoplasmic reticulum.

bulking Plant breeding procedure in which selection of segregants after a cross is delayed to later generations when the majority of individuals become homozygous by continued inbreeding. The number of heterozygotes by F_n is expected to be $0.5^{(n-1)}$ for a particular locus, where n stands for the number of generations selfed. Note that the the F_1 is produced by crossing; therefore, we use $n - 1$. See inbreeding, progress of; inbreeding rate.

Buller phenomenon Nuclei from dikaryotic mycelia may move into monokaryotic ones.

bullous pemphigoid autoimmune disease Human chromosome 6p12-p11−dominant autoimmune disease manifested as vesicles on the skin. It is based on a defect involving the ca. 230 kDa glycoprotein (antigen BPAG1e). This antigen also affects the nerve fibers. See autoimmune disease; filament.

bundle Protein α-helices running along the same axis. See protein structure.

bundle sheath Cells wrapped around the phloem bundles. Their chloroplasts do not show grana and synthesize carbohydrates through the C_3 pathway, although the other chloroplasts in the same plant operate by the C4 system. See chloroplasts; C3 plants; C4 plants.

bungarotoxin See toxins.

buoyant density A molecule (e.g., DNA) suspended in a salt density gradient (such as a CsCl solution, spun for 24 hr at 40,000 rpm in an ultracentrifuge tube) comes to rest in the salt gradient at the position where the medium (CsCl) density is identical to its own. The buoyant density of, e.g., *Chlamydomonas* nuclear and chloroplast DNA is 1.724 and 1.695, respectively. Higher buoyant density reflects higher $G + C$ content in the DNA. The refractive index determined by a refractometer capable of 5-digit resolution can be used to determine the density (ρ) of a cesium chloride solution in any sample withdrawn from the centrifuge tube. The relevant relationships at 25°C are the following: The density of *E. coli* DNA is about 1.710 and that of *Mycobacterium phlei* is 1.732, and a deoxy A-T polymer has a (ρ) value of 1.679. Since none of these values are directly readable from the tabulation, interpolation is required, which can be represented graphically. See density gradient centrifugation.

CsCl % Weight	Density g/mL	Refractive Index	Molarity
50	1.5825	1.3885	4.700
55	1.6778	1.3973	5.481
56	1.699	1.3992	5.651
57	1.7200	1.4012	5.823
58	1.7410	1.4032	5.998

buphthalmos   *See* glaucoma.

Burbank, Luther (1849–1926)   American breeder credited with the production of over 800 new varieties and strains of plants, mainly in Santa Rosa, California. In addition, his success had a stimulating effect on the development of plant breeding. Regrettably his mystic Lamarckian ideas also hindered scientific plant breeding. (Crow, J. 2001. *Genetics* 158:1391.)

burdo   Graft hybrid between tomato and nightshade (*Belladonna*), forming a periclinal chimera. The graft hybrids were assumed to be the result of fusion between the cells of different species combined at the grafting site. From this site somatic hybrid cells regenerated into plants. *See* periclinal; somatic cell hybrids. (Winkler, H. 1907. *Ber. Dtsch. Bot. Ges.* 25:568.)

Burkitt's lymphoma   Human cancer caused by the Epstein-Barr virus is most common in central Africa but occurs in other parts of the world. It is associated with nasopharyngeal (nose and throat) carcinoma and neoplasias of the jaws and abdomen. It frequently involves a translocation between human chromosomes 8 to 14. The receptor of Burkitt's lymphoma is activated by a chemokine that is targeted to B cells in the lymphoid follicles. *See* B cell; chemokine; Epstein-Barr virus; lymphoma.

bursa   Generally a sac-like pouch; the bursae Fabricius are located in the intestinal tract of birds and produce B lymphocytes. *See* lymphocytes.

burst size   Average number of phage particles released by the lysis of bacteria.

bus   In a computer it is the circuit system that transmits information within the hardware or the cables that link various computer devices together.

bushmen   Nomadic people who live in the wilderness (bush), such as the Australian aborigines or people in the Kalahari Desert. The anthropological characteristics are shared within the group, but there is no known evolutionary relationship among the bushmen inhabiting different geographical regions.

BvgS   *Bordatella pertussis* (bacterial) kinase affecting virulence regulatory protein BvgA. *See* pertussis toxin.

Byler disease (PFIC1, 18q21)   Progressive intrahepatic cholestasis. It is apparently allelic to the benign recurrent intrahepatic cholestasis (BRIC). Bile acid secretion is defective due to a defect(s) in the ATP-binding cassette (ABC) transporter. *See* ABC transporter; Alagille syndrome; cholestasis.

bypass replication   Capable of bypassing a DNA defect and continues the process beyond it. *See* DNA repair.

Byr 2   Serine-threonine kinase of *Schizosaccharomyces pombe*. An analog of RAF. *See* raf; signal transduction; serine/threonine kinase.

bystander activation   Hypothesis for the origin of autoimmune reaction. Viruses may induce inflammation, resulting in cytokine production. The cytokines may reactivate dormant T cells with a low activation threshold; these T cells may then attack self-antigens that normally escape their attention. *See* autoimmune disease; bystander effect; immune response; immune tolerance; self-antigen. (Fournie, G. J., et al. 2001. *J. Autoimmun.* 16:319.)

bystander effect   Genetic vectors targeted to a particular cell type do not spread to neighboring cells, yet their synthesized transgene product (e.g., a toxin) may diffuse and also kill the surrounding cells. The bystander effect is a complex result of intercellular communication through gap junctions, apoptotic cell death, release of cytokines, blocking of angiogenesis, etc. *See* angiogenesis; apoptosis; cancer gene therapy; cytokines; gap junctions; gene therapy; MoMulLV; transgene; vectors. (Zheng, X., et al. 2001. *Mol. Pharmacol.* 60:262.)

byte   Computer unit of information consisting of a number of adjacent bits. Most frequently, 1 byte is 8 bits, which represent a letter or other characters that the computer uses. *See* bit.

bZIP   Basic leucine zipper. *See* DNA-binding protein domains; leucine zipper.

B-ZIP protein   DNA-binding protein containing a basic amino acid zipper domain. *See* binding proteins.

William Bateson, the most ardent Mendelian, among many of his original contributions, discovered that the interactions among gene products modify the Mendelian ratios without compromising the validity of the basic principles. In 1926 T. H. Morgan eulogized Bateson with these words: "His intellectual rectitude was beyond all praise and recognized by friend and foe alike" (*Science* **63**:531).

Bateson enjoyed popularity in America and was personally familiar with most of his American colleagues. In 1922, at the University of Pennsylvania, he concluded a memorial lecture with the following warning:

"I think we shall do genetical science no disservice if we postpone acceptance of the chromosome theory in its many extensions and implications. Let us distinguish fact from hypothesis. It has been proved that, especially in animals, certain transferable characters have a direct association with particular chromosomes. Though made in a restricted field this is a very extraordinary and most encouraging advance. Nevertheless the hope that it may be safely extended into a comprehensive theory of heredity seems to me ill-founded, and I can scarcely suppose that on a wide survey of genetical facts, especially those so commonly witnessed among plants, such an expectation would be entertained. For phenomena to which the simple chromosome theory is inapplicable, save by the invocation of a train of subordinate hypotheses, have been there met with continually, as even our brief experience of some fifteen years has abundantly demonstrated" (*J. Genet.* **16**:201 [1926]).

W. Bateson
1922

A historical vignette.

C

C Cytosine. *See* cytosine.

C6 **(1)** Rat glioma cell line (tumor of tissues supporting the nerves). *See* glioma. **(2)** Fungal zinc-binding protein cluster.

c25 Prolactin/cycloheximide-responsive transcription factor similar to IRF-1 (interferon regulator factor); controls proliferation and antiproliferation responses in cells. *See* interferon; IRF; transcription factors.

CAAT BOX (CCAAT) Consensus sequence in the untranslated promoter region of eukaryotic genes, recognized by transcription factors. Housekeeping genes may not have this box or a TATA box. *See* AP; asparagine synthetase; C/EBP; G box; housekeeping genes; promoter; TATA box.

CaaX box Membrane-binding protein motif.

CAB *See* chlorophyll-binding protein.

cabbage *See Brassica.*

CaBP Calcium-binding protein. An endoplasmic reticulum chaperone protein (49 kDa) with two thioredoxin-kind domains. *See* chaperones; PDI.

CAC Chromatin assembly complex includes histones H3 and H4 and the chromatin assembly factor. CAC subunits are p160, p60, and p48. *See* CAF; chromatin; histones; nucleosome.

cacajo *See* Cebidae.

cacao (*Theobroma cacao*) Tropical plant; a source of chocolate and cocoa that is derived from the dried oily cotyledons of the seed. The economically useful species are $2n = 2x = 20$.

cache Memory in the computer that increases the speed and efficiency of the machine.

cachectin Hormone-like protein product of macrophages releasing fat and lowering the concentration of fat synthetic and storage enzymes, encoded by a gene situated within the HLA cluster in human chromosome 6p21.3. *See* HLA; macrophage. (Jue, D. M., et al. 1990. *Biochemistry* 29:8371.)

CACHET Condensation of amplification circles after hybridization of encoding tags. *See* padlock probes.

cachexia Condition of emaciation, wasting away of muscles. It may be caused by a 24 kDa proteoglycan as the consequence of cancer or other debilitating conditions. Tumor necrosis factor α may have an important role in it. The proteasome activity is aided by glucocorticoids, but NF-κB aids the process by downregulating MyoD, which replenishes muscle fibers. *See* MyoD; NF-κB; obesity; proteasome; proteoglycan; TNF.

CACN1A4 α-subunit of a brain-specific voltage-gated neuronal calcium ion channel encoded at human chromosome 19p13.1. *See* ion channels; migraine; spinocerebellar ataxia.

caco Colon adenocarcinoma, a malignant adenoma. *See* adenoma; carcinoma.

cactus Protein product of the *cact* gene of *Drosophila* (2–52) controls dorsoventral differentiation by maternal effect in the embryo. Its action is similar to Dorsal. Protein Toll dissociates Cactus from Dorsal and subsequently Dorsal moves into the nucleus. Pelle (serine/threonine kinase) and Tube mediate the signals from Toll to the Dorsal-Cactus complex. *See* morphogenesis in *Drosophila*.

CAD **(1)** Caspase-activated DNase degrades chromatin by cleavage between the histone-DNA complex of the nucleosomes during apoptosis. *See* acinus; apoptosis; caspase; chromatin; DNase; endonuclease G; ICAD; nucleosome. (Enari, M., et al. 1998. *Nature* 391:43.) **(2)** Coronary artery disease. *See* coronary heart disease. **(3)** Three-enzyme complex trifunctional protein catalyzing the first three steps in the de novo pyrimidine pathways (carbamoyl phosphate synthetase II, aspartate transcarbamylase, and dihydroorotase). All the three identical polypeptide subunits (Mr 230,000 each) have active sites for the three reactions. In Syrian hamster the gene coding for it is 25 kbp, has 37 introns, and the mRNA transcript is 7.9 kb. When amplified, this complex is 500 kbp. The synthesis of carbamoyl phosphate synthetase II is stimulated by epidermal growth factor (EGF) and ERK MAP kinase system. *See* EGF; pyrimidine; signal tranduction. (Chen, S., et al. 2001. *Proc. Natl. Acad. Sci. USA* 98:13802.)

CADASIL Cerebral autosomal-dominant arteriopathy with subcortical infarcts and leukoencephalopathy is a complex syndrome encoded in a 800 kb region of human chromosome 19p13.2-13.1 (Notch 3). It involves diffuse white matter in the brain, defects in the brain blood vessels, stroke, progressive mental illness, paralysis of the face, headaches, severe depression, etc. The basic defect is in a glycosylated transmembrane receptor protein, homologous to Notch of *Drosophila*. *See* Alzheimer disease; brain, human; leukoencephalopathy; migraine; morphogenesis in *Drosophila*; Notch; presenilins; stroke; transmembrane proteins.

cadherin Cell adhesion molecule (glycoprotein) dependent on the presence of Ca^{2+}. The quantity and quality of cadherins determine how cells of the same type stay together and how different types segregate during embryonic development, i.e., the nurse cells, follicle cells, and the embryo segregate to different poles in the egg chamber of *Drosophila* (see diagram at maternal effect genes). Cadherins are encoded in human chromosome 16q22.1. The human protocadherin genes PCDH α, β, γ are in chromosome 5q31, but PCDH7 is at 4p15 and PCDH22 is at Yp11.2. Procadherin genes are also found at 13q21.1 and 10q21-q22. The protocadherins are classified as nonclassic cadherins because — unlike the classic cadherins—they do not interact with catenins. They have a basic role in normal development, and a reduction in their level increases the invasiveness of many types of cancerous growth. They have membrane-spanning and cytoplasmic domains. The former assures cell-to-cell contacts; the latter attaches to the cytoskeleton. Dominant negative cadherin mutants develop Crohn disease–like symptoms, spina bifida, and adenomas. Dysadherin, a 178-amino-acid cell membrane glycoprotein, downregulates E-cadherin and promotes metastasis. *See* adenoma; catenin; Crohn disease; deafness; gastric cancer; integrin; snail; spina bifida; Usher syndrome. (Nollet, F., et al. 1999. *Mol. Cell Biol. Res. Comm.* 2:77; Poser, I., et al. 2001. *J. Biol. Chem.* 276:24661; Alagramam, K. N., et al. 2001. *Hum. Mol. Genet.* 10:1709; Ino, Y., et al. 2002. *Proc. Natl. Acad. Sci. USA* 99:365.)

cADP-ribose   *See* cyclic ADP-ribose.

Caenorhabditis elegans   Small nematode feeding on bacteria. It completes its life cycle in about $3\frac{1}{2}$ days. Approximately 99.8% of the animals have the chromosomal constitution of five pairs of autosomes and two X chromosomes and are hermaphrodites; 0.2% of the populations are XO males generated by nondisjunction. Its genome is about ~97 million bp and has been sequenced. It includes a little more than 19,000 genes, and 40% of them appear homologous to those of other organisms.

The nervous system includes only about 302 cells. The neurons represent 118 structural classes and the number of positions of identifiable chemical synapses appears to be 7,600. More than 250 genes were identified by mutational analysis to be involved in behavior. All of the genes have been cloned and sequenced. *C. elegans* is one of the best organisms for molecular developmental studies. The simplest developmental pathway of animals is seen in this nematode, with approximately 958 somatic cells, including its nervous system, in a thin 1.2-mm-long body of the adult. The egg usually develops hermaphroditically or by fertilization by the rare males into a 550-cell embryo in the eggshell. After hatching and passing, further divisions take place by moltings through four larval stages. The entire process may be completed in about 3 days.

Through the transparent body the migration of the embryonal cells and the formation of the organs can be traced under the microscope using Nomarski differential interference contrast optics. The pattern of differentiation and development displays very little variation. The somatic tissues generally have multicellular origin, whereas the intestinal cells and the germline cells are monoclonal. For differentiated functions, cytokinesis is not absolutely required. Differentiation is generally not controlled by the cellular milieu but by the identity of the particular cell (demonstrated by ablation experiments). Nevertheless, signal transduction among cells may also be required for the formation of the egg-discharging mechanisms. Through involvement of a single *anchor cell*, the *vulva* is formed, and the *uterus* passes the egg through the anchor and vulva on the culture medium contained in a petri plate where development is completed. A somatic "distal tip cell" controls the mitotic activity of the "germ cells," which undergo meiosis and produce the gametes. The hermaphroditic XX females are self-fertilized, but by nondisjunction they also produce male gametes at a frequency of 0.1%. When mated with an XX female, the XO males then produce males and females in equal proportion. Of 44 human disease genes analyzed, 32 had significant similarities to the genes *of C. elegans. See* Nomarski differential interference contrast microscopy; unc-86. (Hodgkin, J. & Herman, R. K. 1998. *Trends Genet.* 14:352; sequence biology 1998. *Science* 282:2011; <http://www.wormbase.org>, <elegans.swmed.edu>.)

CAF   Chromatin assembly factor facilitates the structural organization of the chromosomal elements in association with acetylated histones. Its three subunits are Cac1, Cac2, and Cac3. The small CAF-1 subunit p48 is identical to the retinoblastoma-associated protein 48. Deletion of any one single Cac increases UV sensitivity. CAF defect in *Arabidopsis* results in fasciation. *See* ACF; CAC; chromatin; chromatin remodeling; fasciation; nucleosomes; retinoblastoma. (Tyler, J. K., et al. 2001. *Mol. Cell Biol.* 21:6574.)

café-au-lait SPOT   Light brown skin macule; diagnostic signs of neurofibromatosis. *See* neurofibromatosis.

caffeic acid (3,4-dihydroxycinnamic acid phenethyl ester, CAPE)   Related to flavonoids present in honeybees' glue. It

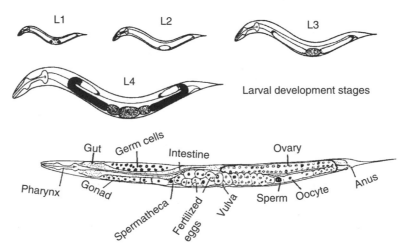

Caenorhabditis elegans adult hermaphrodite

has an antiviral, antiinflammatory, immunomodulatory effect and inhibits tumor growth, lipid peroxidation, lypoxygenase, ornithine decarboxylase, protein tyrosine kinase, and the activation of NF-κB. *See* antimutagen.

caffeine Modified purine molecule present in coffee, tea, cola nuts, and other plants. It is a stimulant and diuretic. Caffeine itself is not a clastogen, but it interferes with the repair of spontaneously or otherwise caused chromosomal damage, and thus may enhance the mutagenic potential of irradiation and chemicals by overriding checkpoint controls in the cell cycle. A ubiquitin-conjugating protein (encoded by fission yeast gene *rhp6*) required for entry into mitosis may be affected by caffeine. Interference with Cdc25 phosphatase and Cdc2 opens the path to mitosis and does not leave enough time for DNA damage. Oral LDLo is 192 mg/kg for humans. LD_{50} orally for male mice is 127 mg/kg, for females 137 mg/kg. *See* alkaloids; Cdc2; Cdc25; checkpoint; LD_{50}; LDLo; purine; theobromine. (Schlegel, R. & Pardee, A. B. 1986. *Science* 232:1264.)

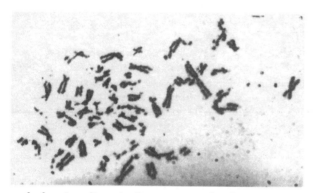

Human leukocyte culture in the presence of caffeine with multiple chromosome breakage. (From Ostertag, W. & Greif, B. J. 1967. *Human Genet.* 3:282.)

caged compound Rendered inactive by combining it with a photosensitive molecule and then introduced into the cell by microinjection, electroporesis, or protoplast fusion and subsequently irradiated by UV light of a specific wavelength or by laser beam, leading to liberation of the molecules (e.g., Ca^{++}, cAMP, GTP) from the sensitizer (e.g., nitrobenzyl side chain). This procedure permits the study of localized effects within the cytoplasm. Fluorochromes can also be caged for monitoring the behavior of subcellular structures, e.g., the function of tubulins. *See* electroporation; fluorochromes; laser; microinjection; protoplast fusion; tubulin; UV.

CAHMR Very rare autosomal-recessive cataract, hypertrichosis, and mental retardation syndrome. *See* cataract; cerebro-oculo-facio-skeletal syndrome; mental retardation;

hypertrichosis. (Temtamy, S. A. & Sinbawy, A. H. H. 1991. *Am. J. Med. Genet.* 41:432.)

Cairns structure DNA molecule undergoing bidirectional θ-replication. *See* bidirectional replication; θ-replication. (Cairns, J. 1963. *Cold Spring Harb. Symp. Quant. Biol.* 28:43.)

Cajal body (coiled body) Nuclear organelles, usually at the nucleolar periphery. They contain p80, coilin, fibrillarin, snRNP, etc., and are involved in processing mRNA, rRNA, and snRNA; they interact with histones and neuronal functions, etc. *See* coiled body; coilin; p80; snurposome. (Gall, J. G. 2000. *Annu. Rev. Cell Dev. Biol.* 16:273.)

Cak (cyclin-dependent-activating kinase) Generally a component of the TFIIH transcription factor, but in *Saccharomyces* it is not part of TFIIH. It activates CDC28. *See CDC28*; Cdk; cyclin; MO15; transcription factors.

CAK β *See* CAM.

calcineurin (serine/threonine protein phosphatase-IIB, PP-2B) Its action is dependent on calcium and calmodulin. Calcineurin is the target of cyslosporin and FK506. By binding to NF-AT, calcineurin incites the immune system and activates neuronal and muscle development but not muscle growth. If this binding could be selectively blocked without affecting other functions, immunosuppression could be controlled, which would be a great benefit for transplant patients. Calcineurin and Mpk1 (Map kinase) and Zds1 (regulator also of Cdc42) proteins regulate the action of gene *SWE1* (a member of the *WEE1* family) at transcription and posttranscription and the calcium-induced delay in the G2 phase of the cell cycle. The product of the DSCR1 (Down syndrome candidate region) gene is an inhibitor of calcineurin and increases senile Alzheimer plaques and neurofibrillary tangles two to three times. *See* Alzheimer disease; BAD; Bcl; calmodulin; cardiomyopathy, hypertrophic familial; Cdc42; cyslosporin; NF-AT; MAP, FK506; immunosuppressant; serine/threonine phosphoprotein phosphatases; T cells; Wee. (Crabtree, G. R. 2001. *J. Biol. Chem.* 276:2313; Ermak, G., et al. 2001. *J. Biol. Chem.* 276:38787; Luan, S., et al. 2002. *Plant Cell* 14:S389.)

calcitonin (CT, 11p15.2-p15.1) Oligopetide hormone of the thyroid gland controlling calcium and phosphate levels and an antagonist of parathyroid hormone. Calcitonin gene-related peptide and substance P induce inflammation by dilation of the blood vessels. Calcitonin receptors are encoded at 2q31-q32 and 7q21.3. *See* adrenomedullin; animal hormones; immune privilege.

calcium ion channel The voltage-regulated channel is built of several subunits, each encoded at different human chromosomal sites: CACNL1A2 (neuroendocrine/brain) at 3p14.3. The CACNL1A4 isoform is at 19p13; CACNA1S (skeletal muscle) at 1q32; CACNA1C (cardiac muscle) at 12p13.3. CACNA2D1 in several tissues modulates the activity of the channel at 7q21-q22; the β-subunit, CACB1, is at the same location. The γ-subunit, CACNG1, is encoded nearby at 17q24. *See* calmodulin; dihydropyridine receptor; ion channels; neurotransmitters; ω-agatoxin; ω-conotoxin;

second messengers. (Catterall, W. A. 2000. *Annu. Rev. Cell Dev. Biol.* 16:521; Toyoshima, C. & Nomura, H. 2002. *Nature* 418:605.)

calcium-phosphate precipitation   May significantly enhance the chances of DNA uptake by (*E. coli* or mammalian) cells to be transformed (transfected). *See* bacterial transformation; cotransfection.

calcium signaling   Calcium plays the role of a second messenger and activates many enzymes in the cell. It regulates synaptic activity of the neurons, cell adhesion, motility, proliferation, etc. When phospholipase C (PLC) is activated, the cells release intracellular Ca^{2+} through the calcium-release activated calcium channels (CRACs), receptor-operated calcium channels (ROCs), and store-operated calcium channels (SOCs). The incoming calcium then replenishes the Ca^{2+} store in the cell; it is also called CCE (capacitative Ca^{2+} entry). Hormones may activate CCE by immune reactions or neural receptors. Light signal transduction in *Drosophila* requires PLC activation, synthesis of IP3, and the release of Ca^{2+}. For normal vision the flies must be able to maintain Ca^{2+} homeostasis. Mutants are known (and cloned) that represent gene loci involved in calcium regulation. Nuclear gene expression may be regulated through the CRE element, whereas the cytoplasmic signaling may be mediated by SRE. Large transient rise of Ca^{2+} activates transcriptional regulators such as NF-κB and JNK. NFAT is activated by low Ca^{2+} levels. Intracellular Ca^{2+} triggers the activation of T lymphocytes by antigens. *See* calcineurin; calcitonin; calcium ion channels; calmodulin; CRE; Darier-White disease; homeostasis; IP3; JNK; NFAT; NF-κB; PLC; SRE. (Wallingford, J. B., et al. 2001. *Curr. Biol.* 11:652; Lewis, R. S. 2001. *Annu. Rev. Immunol.* 19:497; Carafoli, E. 2002. *Proc. Natl. Acad. Sci. USA* 99:1115; Sanders, D., et al. 2002. *Plant Cell* 14:S401.)

calcofluor white (Tinopal)   Fluorescent brightener that can be used to monitor cell wall formation in protoplast suspensions. *See* fluorochromes.

caldesmon   83 kDa actin- and calmodulin-binding protein. During mitosis it dissociates from the microfilaments as a consequence of mitosis-specific phosphorylation of actomyosin ATPase. *See* ATPase; calmodulin; myosin. (Krauze, K., et al. 1998. *Biochem. Biophys. Res. Commun.* 247:576.)

calico cat   Heterozygous *female* animals with black-yellow-white fur patches. The alternation of black and yellow fur is due to inactivation of the genes residing in one of the two X chromosomes. In males, this pattern occurs only in connection with the Klinefelter (XXY) constitution. The white fur is an autosomal trait controlled by the *S* (*spotted*) gene. *See* lyonization; tortoiseshell fur; X-chromosome inactivation.

Callicebus   *See* Cebidae.

Callithricidae   New world monkeys (marmosets and tamarins). *Callithrix argentata* $2n = 44$; *Callithrix humeralifer* $2n = 44$; *Callthrix jaccus* $2n = 46$; *Callimico goldi* $2n = 48$ some males 47; *Cebuella pygmaea* $2n = 44$; *Leontocebus rosalia* $2n = 46$; *Sanguinis fuscicollis illigeri* $2n = 46$; *Sanguinus oedipus*

$2n = 46$; *Tamarinus mystax* $2n = 46$; *Tamarinus nigricollis* $2n = 46$. *See* primates.

callose   Carbohydrate (glucan) formed on injured plant tissue.

callus   Solid mass of plant cells (generally) on synthetic media, thickening of animal skin, unorganized bone growth.

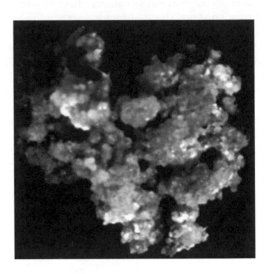

calmegin Ca^{2+} protein with 54% homology to calnexin, but its expression s limited the stage of pachytene to spermatid development. It is a chaperone. *See* calnexin; chaperone.

calmodulin (CaM)   17,000 M_r. acidic protein with 4 Ca^{2+}-binding sites affecting (among others) membrane transport, chromosome movement, and processes in fertilization; it functions as a regulatory unit of several enzymes, although calmodulin itself is not an enzyme. Its binding results in conformational changes of the target proteins. The presence of Ca^{2+}/CaM is required for phosphorylation of several proteins (myosin light-chain kinase, phosphorylase kinase). CaM-kinase II mediates the secretion of neurotransmitters. It activates tyrosine hydroxylase required in catecholamine biosynthesis. It is involved in the control of such brain functions as memory and learning. CaM-kinase II is capable of autophosphorylation even in the absence of Ca^{++}. CaM regulates adenylate cyclase (cAMP), and cAMP regulates CaM. A-kinase, regulated by CaM, phosphorylates the IP_3 receptor, and cAMP and CaM-kinases control CREB. The delta subunit of phosphorylase kinase is CaM. CaM activates cyclic nucleotide phosphodiesterase, an enzyme degrading cAMP. CaM is ubiquitous among eukaryotes and it is one of the most conserved proteins. *See* aequorin; autophosphorylation; calcium signaling; cAMP; CDC31; CREB; IP_3; neurotransmitters; phosphorylases; signal transduction. (Fujisawa, H. 2001. *J. Biochem.* 129:193; Soderling, T. R., et al. 2001. *J. Biol. Chem.* 276:3719; Corcoran, E. E. & Means, A. R. 2001. *J. Biol. Chem.* 276:2975; Cyert, M. S. 2001. *Annu. Rev. Genet.* 35:647; Hoeflich, K. P. & Ikura, M. 2002. *Cell* 108:739; Luan, S., et al. 2002. *Plant Cell* 14:S389.)

calnexin   Membrane-bound chaperone with preference for glycoproteins; it may also associate with an antigen before entering the endoplasmic reticulum. Its soluble homologue is calreticulin. *See* BiP; calreticulin; chaperones; Cne; endoplasmic reticulum. (Danilczyk, U. G. & Williams, D. B. 2001. *J. Biol. Chem.* 276:25532; Schrag, J. D., et al. 2001. *Mol. Cell* 8:633.)

calorie   Amount of heat required to elevate the temperature of 1 gram of water from 14.5 to 15.5°C (4.186 international joules). *See* joule.

calpain   Calcium-dependent neutral cysteine protease of the papain family. Calpain may cleave neural growth activator protein 35 (p35), resulting in p25. Accumulation of p25 causes the mislocalization of CDK5, then CDK5 and p25 hyperphoshorylate tau, a microtubule-associated protein; consequently, the cytoskeleton disruption leads to death of neurons and to the apoptosis seen in Alzheimer disease. CALPAIN10 (2q37.3) seems to be linked to diabetes (NIDDM1) and may be identical to it. Two other proteases, caboxypeptidase E/H in human chromosome 4 and pro-hormone convertase-1 (5q15-q21), also seem to be associated with diabetes. *See* Alzheimer disease; CDK; cytoskeleton; diabetes; microtubule; papain; protease; tau. (McGrath, M. E. 1999. *Annu. Rev. Biophys. Biomol. Struct.* 28:181; Horikawa, Y., et al. 2000. *Nature Genet.* 26:163; Sorimachi, H. & Suzuki, K. 2001. *Biochem. J.* 129:653; Perrin, B. J. & Huttenlocher, A. 2002. *Int. J. Biochem. & Cell Biol.* 34:722.)

calphostin C   Inhibitor of diaglycerol and C^{2+}-dependent phosphokinase CD (PKC). *See* diaglycerol; PKC; T cells.

calreticulin   Calcium storage protein within the endoplasmic reticulum and also present in the nucleus. It may prevent the glucocorticoid receptor from binding to its response element, and it is thus a transcription factor and a chaperone. Calreticulin is also required for calcium signaling and cell adhesion by integrins. *See* calnexin; endoplasmic reticulum; glucocorticoid; hormone response element; integrin; tapasin. (Nakamura, K., et al. 2001. *J. Clin. Invest.* 107:1245; Fadel, M. P., et al. 2001. *J. Biol. Chem.* 276:27083.)

calsenilin Ca^{2+}-binding protein that mediates apoptosis and Alzheimer disease plaque protein, $A\beta$. *See* Alzheimer disease; apoptosis; presenilin.

Calvin cycle   Pathway of fixation of CO_2 (1C) into 3-phosphoglycerate (3C), 1,3-bisphosphoglycerate (3C), and glyceraldehyde-3-phosphate (3C). In this reaction 3 molecules of ATP and 2 molecules of NADPH are used for each CO_2 converted into carbohydrate. *See* Krebs-Szentgyörgyi cycle; photosynthesis.

calypso   *Drosophila* transposable element (7.2 kb) generally present in 10 to 20 copies per cell. *See* hybrid dysgenesis; transposable elements.

calyx   Collective name of sepals, the basal whorl of the flowers. *See* flower differentiation.

cam   Chloramphenicol transacetylase conveys resistance to chloramphenicol, which would otherwise block peptidyl transferase on the 70S ribosomes in prokaryotes and in eukaryotic organelles. *See* antibiotics; chloramphenicol; protein synthesis.

CaM   *See* calmodulin.

CAM   (1) Cell adhesion molecule regulates monolayer formation in cultured mammalian cells and mediates neuronal connections with the aid of fibroblast growth factor (FGF). Cell-to-cell adhesion is mediated by cadherins, immunoglobulins, selectins, and integrins. Integrins and transmembrane proteoglycan mediate cell matrix adhesion. Focal cell adhesion, integrin-mediated contact between cells and the extracellular matrix, is linked to the activation of pp125^{FAK} protein kinase. This enzyme is a member of the family of FakB, PYK2/ CAKβ, and RAFTK protein tyrosine kinases. PYK2 regulates calcium ion channels and the MAPK signaling pathway. The carboxy terminal of pp125^{FAK} is expressed as a nonkinase pp41/43^{FRNK}. This latter protein is an inhibitor of pp125^{FAK} and the phosphorylation of tensin and paxillin adhesion proteins. PYK2 activity is also coupled with the JNK signaling pathway. NCAM is the neural cell adhesion molecule. *See* ADAM; cadherins; extracellular matrix; fibronectin; ICAM; immunoglobulins; integrins; ion channels; JNK; L1; MAPK; protein kinases; protein zero; proteoglycan; selectins; tenascin; Usher syndrome. (Chen, L., et al. 2001. *J. Cell. Biol.* 154:841; Voura, E. B., et al. 2001. *Mol. Biol. Cell* 12:2699.) (2) *See Crassulacean* acid metabolism.

camalexin   Phytoalexin that may be a player in the defense of plants against fungal diseases. *See* host-pathogen relationships; phytoalexins. (Pedras, M. S. & Khan, A. Q. 2000. *Phytochemistry* 53:59.)

cambium   Meristemic tissue layer around the stem of plants providing for growth in diameter. *See* meristem.

Cambrian   Geological era 500–600 million years ago. The fossil plant relics are poorly preserved. All major animal groups except the vertebrates were already present. *See* evolution; geological time periods.

camel   (*Camelus bactrianus*) $2n = 74$, and its American relative, the vicuna (*Vicugna vicugna*), is also $2n = 74$. Their antibodies lack light chains and are highly soluble. *See* antibody. (Hamers-Casterman, et al. 1993. *Nature* 363:446.)

Camfak syndrome   Apparently autosomal-dominant cataract, microcephaly, failure to thrive, kyphoscoliosis (curved spinal column), arthropgryposis (flexure and/or contracture of the joints), and mental retardation. It has also been termed CAMAK. It bears similarities to Cockayne syndrome, cerebro-oculo-facio-skeletal syndrome, and other developmental anomalies. *See* cerebro-oculo-facio-skeletal syndrome; scoliosis.

CaMK   (calcium-calmodulin-dependent protein kinase) CaMKII is necessary for several physiological processes including learning and memory. NMDA receptors control its translocation. Camk4 serine/threonine kinase is important for mice spermiogenesis. *See* calmodulin; kinases, NMDA; titin. (Yang, Y., et al. 2001. *J. Biol. Chem.* 276:41064.)

CaM-KK   CaMK kinase that activates PKB (protein kinase B), which phosphorylates the apoptotic protein BAD.

BAD then interacts with protein 14-3-3 and cell survival is facilitated. *See* apoptosis; BAD; protein 14-3-3; protein kinases. (Tomitsu, H., et al. 2001. *J. Biol. Chem.* 275:20090.)

CaMO   *See* cauliflower mosaic virus.

C amount of DNA   Content of DNA in the gametes is 1C; it is 2C in the zygotic cells before S phase, 4C after S phase. The 1C amount actually means that the chromosomes have only one chromatid containing a single DNA double helix. After replication each chromosome becomes double stranded, i.e., composed of two chromatids held together at the centromere. The 4C stage usually is an indication of a diploid or zygotic cell. *See* cell cycle; C-value paradox.

cAMP   adenosine 3′:5′ monophosphate (cyclic AMP, second messenger); has a crucial role in signal transduction and general gene regulation in bacteria and animals. For many years, there was no convincing evidence for cAMP in plants. More recently, several facts indicated the role of cAMP in plant development, and in 1997, adenylyl cyclase was identified in tobacco tissues. Some molecules can increase the level of cAMP (adenylate cyclase) by binding to certain transmembrane receptors, whereas other cellular signal molecules are inhibitory. The stimulatory G_s protein activates adenylate cyclase. Actually, the α_s subunit dissociates from the

The inhibitory trimeric G_i has a special α_i subunit, and it is activated by other types of cellular signals. In G_i, the $\beta\gamma$ subunits are dissociated from the α_i subunit; these subunits then directly and indirectly interfere with adenylate cyclase. More importantly, G_i opens K^+ ion channels in the plasma membrane. The pertussis toxin (due to *Bordetella* bacterial whooping cough)—in contrast to the cholera toxin—mediates the adenylation of an a_i subunit that interferes with responding to the receptors, GDP remains bound to the G-protein, adenylate cyclase activity is not blocked, and potassium channels are not opened.

One of the most important functions of cAMP is the activation of the four-subunit *protein kinase A*. This phosphorylase enzyme then adds phosphate groups from ATP to serine and threonine residues in certain proteins. Protein kinase A is activated by cAMP through forming a complex with two of its regulatory subunits while the two other separated catalytic subunits are turned on. The enzyme exists in two forms: one is cytosolic, whereas the other is bound to membranes and microtubules.

cAMP controls the transcription of several other genes in a positive or negative manner. CREB (cAMP response elements)—located in the upstream region of genes—promotes transcription upon phosphorylation of a specific serine residue in the binding proteins, whereas CREM α and β are repres

Adenosine 3′–phosphoric acid (3′–adenylic acid) — Adenine

Adenosine 3′, 5′–phosphoric acid (cyclic adenylic acid)

cAMP

Adenosine 5′–phosphoric acid (adenylic acid; 5′–adenylic acid) — Adenine

two other chains, binds to and hydrolyzes GTP, then binds to adenylate cyclase and boosts the production of cAMP. Upon binding to adenylate cyclase, GTPase activity increases and an inactive G_s is formed again by recombining α_s with $\beta\gamma$. Cholera toxin (produced upon infection by the bacterium *Vibrio cholera*) mediates the transfer of adenylate to α_s, which prevents the hydrolysis of its GTP and therefore the adenylate cyclase function stays on, resulting in increased levels of cAMP. This condition then opens a very active sodium and water efflux through the intestinal walls, causing debilitating diarrhea and dehydration of the entire body.

sors. cAMP also controls the secretion of cortisol by the kidney cortex, the thyroid hormone, the secretion of the luteinizing hormone (progesterone), adrenaline and glucagon production, glycogen and triglyceride breakdown, heart muscle and other muscle functions, etc. *See* adenosine-3′,5′ monophosphate; adenylate cyclase; cAMP-dependent protein kinase; CBP; CREB; cyclic ADP-ribose; epinephrine; forskolin; G-protein; progesterone; signal transduction. (Montminy, M. 1997. *Annu. Rev. Biochem.* 66:807.)

Campbell model   Of recombination, suggested the mechanism of integration of the (pro)phage into bacteria by reciprocal

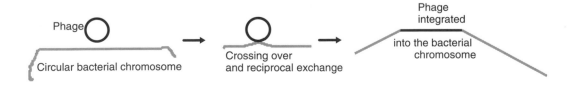

recombination between the circular bacterial and temperate phage DNA molecules (see model below). *See* bacteriophages; *E. coli.* (Campbell, A. M. 1962. *Advances Genet.* 11:101.)

cAMP-dependent protein kinase *See* phosphorylase *b* kinase; protein kinase A.

campomelic dysplasia (CD, CMD1)   Most likely a recessive/haploinsufficiency (rather than dominant) bone formation and gonad development defect due to mutation in the SOX9 gene in human chromosome 17q24.1-q25.1. The congenital bowing of the skeletal bones and malformation of the head bones usually causes death in early infancy; however, some individuals survive to early adulthood. This defect is often associated with chromosomal rearrangements and sex reversal. *See* gonad; sex reversal; SOX; SRY. (Preiss, S., et al. 2001. *J. Biol. Chem.* 276:27864.)

cAMP receptor protein (CRP)   Assists the binding of *E. coli* RNA polymerase to the operator of carbohydrate operons (positive regulation), but it may also act as an activator of the negative regulator CytR and as such becomes a corepressor. The DNA-binding site is 22 bp arranged in a rotational symmetry. The binding results in DNA bending. *See* corepressor; DNA bending; Lac operon; negative control; positive control; promoter; repressor; RNA polymerase.

camptodactyly   Autosomal-dominant bent fingers.

Campodactyly.

camptothecin ($C_{20}H_{16}N_2O_4$)   Plant alkaloid is an antibiotic and anticancer agent targeting DNA topoisomerase I–DNA complexes. *See* topoisomerases. (Arimondo, P. B., et al. 2002. *J. Biol. Chem.* 277:3132.)

Campylobacter jejuni   Gram-negative, spiral, flagellate bacterium, pathogenic to the stomach and an inducer of Guillain-Barré syndrome. Its circular DNA chromosome contains 1,641,481 bp encoding 1,654 proteins and 54 stable RNAs. Its genome is virtually free of insertion or phage-associated sequences and has rare repeats but contains short hypervariable sequences. *See* Guillain-Barré syndrome. (Parkhill, J., et al. 2000. *Nature* 403:665.)

Camurati-Engelmann disease (CED, diaphyseal dysplasia, 19q13.1-q13.3)   Progressive bone formation on the shaft of the long bones and the skull. Muscular weakness and pain, facial paralysis, hearing and vision problems, etc., may be caused by a defect in the transforming growth factor–β1 subunit. *See* TGF.

CaMV   *See* cauliflower mosaic virus.

Canale-Smith syndrome (ALPS)   Autoproliferative disease caused by deficiency of the Fas protein and a defect in apoptosis. *See* apoptosis; Fas; lymphoproliferative diseases.

canalization   Genetic buffering mechanism that reduces the visible variations beyond what is expected on the basis of genetic diversity. It is the developmental path modulated by environmental inputs, the epigenetic landscape. It permits the maintenance of hidden genetic variations and thus facilitates the conservation of a "normal" phenotype. Also, it eliminates from the populations those genotypes that cannot adjust to environmental fluctuations. Mutants generally have reduced buffering capacity compared to the wild type (that is best canalized for survival) and can be readily eliminated. *See* epigenesis; genetic assimilation; genetic homeostasis; homeostasis; reaction norm. (Waddington, C. H. 1940. *J. Genet.* 41:75; Newman, S. A. & Muller, G. B. 2000. *J. Exp. Zool.* 288:304.)

Canavan disease   *See* aspartoacylase deficiency.

canavanine   Competitive inhibitor of arginine (natural plant product). *See* competitive inhibitor.

cancer   Cells continue to divide when cell divisions are not expected; it is an uncontrolled growth, a malignant growth. Cells may spread through the bloodstream to other locations in the body and initiate secondary foci of malignant growth (metastasis). Cancer may occur in a variety of forms such as leukemia, adenoma, lymphoma, or sarcoma, but it is not exactly known how these different types are specified. It appears that there are more than 200 types of cancers. In each cancer, several (3–6) different mutations may be found. (Hahn, W. C. & Weinberg, R. A. 2002. *Nature Rev. Cancer* 2:331). The anomaly in malignant growth is that cells do not remain in a quiescent stage (G_0) and proceed either through terminal differentiation or death (apoptosis), but from G1 phase, and independently from normal cellular regulation, continue indefinitely to S phase, mitosis, and cell division. Normal cells pause at the G1 *restriction point* and respond to the instructions coming through cyclins. The cyclins (CLNs), coupled with the labile cyclin-dependent kinases (CDK4, CDK6), are receptive to mitogens, but they are kept in check by several INK (inhibitors of kinase) proteins.

When tumor suppressors such as the retinoblastoma (RB) and other proteins (E2F family) normally become phosphorylated, the cells exit from G1 and embark on DNA synthesis. The phosphorylated tumor suppressors transactivate CDKs, and unless cell divisions are blocked at various checkpoints, divisions continue and proceed in an accelerated and uncontrolled manner. The G1-phase-specific cyclin E-CDK2 complex stimulates more phosphorylation of the tumor suppressors RB and E2F, and the dependence on mitogens is diminished. CLN-A–CLN-B–dependent kinase (CDK2) reinforces the phosphorylation process, and dephosphorylation does not occur until the completion of mitosis. CLN-E– and CLN-A–associated CDK2 assist the DNA replication machinery. CLN-dependent kinases are also suppressed by CDK inhibitors, p21^{CIP1}, p27^{KIP1}, and p57^{KIP2}, but if the latter ones are deleted or mutated, both cell numbers and size increase.

Methylation of the promoter of tumor suppressor genes may lead to their silencing and neoplasia. Hypomethylation of oncogenes or retroelements or latent viral sequences in the genome may increase cancerous transformation. The patterns of hyper- and hypomethylation may be quite variable at different restriction enzyme recognition sites. Hypomethylation of some DNA sites may facilitate chromosomal instability, a common cause of neoplasia. Comparative genomic data are summarized in Struski, S., et al. 2002. *Cancer Genet. Cytogenet.* 135:63.

Germline mutations are responsible for only about 1% of all cancer cases, and about 10–15% of all cancer has substantial hereditary components. Many cancer genes (oncogenes) have now been cloned, but several genes acting in the same biochemical pathway may be responsible for one type of cancer. Single genes may be involved in different types of cancer. It has been observed that the range of expression of germline mutations of the same cancer gene is different from those of somatic mutational events. Allelic differences within a gene may cause different cancers. The majority of cancer cases are the result of somatic mutations arising during the lifetime of the individuals, and some are due to unknown environmental factors, thus representing phenocopies.

Cancerous growth may be initiated by mutations in structural genes, suppressor genes, methylation of tumor suppressors, transmembrane receptors, transcription factors, regulator genes, cell cycle genes, aneuploidy (hypo- and hyperploidy) repair functions, chromosomal rearrangements (translocations, inversions, transpositions, duplications, deletions), and viral insertions in the chromosomes. Although a single mutation may suffice for uncontrolled cellular proliferation, several additional factors may be involved for full-scale development. The initial mutation may take place in protooncogenes. Mutations or loss of the CLN-D1 locus (human chromosome 11q13) and the CDK4 gene (human chromosome 12q13), INK4a (CDK^{INK4a}, chromosome 9p21), p16, and p53 are common in many types of cancers. p53 also regulates p21, p27, and p57 proteins. The gene encoding p53 is in human chromosome 17p13.15-p12, and it is altered in a large number of different cancers. The normal function of p53 is required for the development of the centrosome and proper segregation of the chromosomes.

cellular functions (some of the protooncogenes, e.g., *Ras*, encode G-proteins and play an important role in healthy signal transduction), but mutation may alter their normal role and they become (cellular) c-oncogenes. The v-oncogenes are viral counterparts of the c-oncogenes. When c-oncogenes are inserted into chromosomes of animals, they may provide efficient promoters for cellular genes involved in growth.

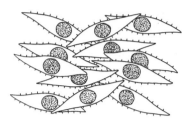

Cultured cancer cells lose their contact inhibition and proliferate in an overlapping fashion. (After Dulbecco, R. 1967. *Sci. Amer.* 216:[4] 28.)

In the development of cancer, the products of more than a dozen tumor suppressor genes (e.g., p53, p16) play a major role. Because of mutation, they no longer are capable of controlling genes of the cell cycle. Normal mammalian cells grow in the cultures in an anchorage-dependent fashion, in monolayer. Cancerous cells differ in their growth habit by growing in disarray, in layers (see drawing above). Since the human body contains about a billion cells per gram of solid tissue, an average adult human body may have about 10^{14} to 10^{15} diploid cells according to the various estimates. Average mutation rates are within the range of 10^{-5} to 10^{-8} per genome, per generation; therefore, all human individuals must have suffered numerous mutations with neoplastic potentials, yet only a fraction of the human population is afflicted by this group of diseases. The causes of this discrepancy are diploidy, which masks recessive mutations, and the immunological surveillance system recognizes mutant surface antigens on cancer cells and destroys them before they get out of control. When the immune system weakens by advancing age, by certain diseases (e.g., AIDS), or by taking immunosuppressive

Some protooncogenes and their role in carcinogenesis

Chromosome	5q		12p12	18q23	17q	
Alteration	del.		activat.	del	del	
Gene product	APC	DDM-1 demethylation	KRAS2 oncoprot.	DCC netrin	p53 suppr.	Various metabolic regulators

Normal epithelium →proliferation→adenoma→adenoma →adenoma→Carcinoma→ Metastasis
(Abnormal) (Early) (Intermediate) (Late)

del: deletion, activat = activation, oncoprot = oncoprotein, netrin = guidance proteins, suppr = tumor suppressor
(Modified after Fearon, E.R. & Vogelstein, B. 1990. *Cell* 61:757.)

About 100 protooncogenes have now been discovered, and their possible malfunction due to mutation, chromosomal rearrangement, or loss is many-fold (see diagram above). These mutations, similarly to other mutations, are generally (not always) recessive and not expressed in diploid cells. When through another event(s) they become homozygous or hemizygous, cancerous growth may follow. This may be why a long period is required after the initial genetic lesion for the development of cancer. These genes have normal

drugs, the incidence of cancer increases. The cancerous growth itself is not necessarily lethal, but it generally deprives the body of its normal metabolism and the patients succumb to secondary, opportunistic diseases.

The general assumption is that cancer develops monoclonally, i.e., in the tumor each cell has a single common ancestor. This assumption does not preclude, however, that additional mutations occurred in the same cell lineage during the process of multiplication, although experimental evidence does not

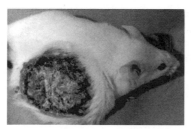

Benzo(a)pyrene-induced cancer in the laboratory by painting the carcinogen on the skin of a rodent. (Courtesy of Drs. Nesnow, S. and Slaga, T.)

show appreciable frequency of mutation during the progression of cancer. In very rare cases, it has been shown that a particular cancer in an individual may not be monoclonal but more than a single founder cell contributed to its formation. Also, in some cancer cells, multiple, different genetic mutations, primarily chromosomal alterations, were observed. Some of these multiple alterations may be due to rearrangement of mini- or microsatellite sequences; some of these are just the consequences of defects in the genetic repair system.

With the technique of intersimple sequence repeat PCR (INTER-SS PCR), many chromosomal alterations (insertions, deletions, translocations) are detectable. The PCR product is produced by using a single primer homologous to dinucleotide repeats and attached at 3′ by two nonrepetitive sequences as diagrammed below (R = purine, Y = pyrimidine): The number of alterations (genetic instability) is determined by the formula:

basic cause of cancerous transformation (Li, R., et al. 2000. *Proc. Natl. Acad. Sci. USA* 97:3236). The aneuploid cells grow slow and are less competitive than normal cells, which would explain the delay of onset of cancer. Furthermore, aneuploids are unstable, and they may affect the expression of many genes over the course of time.

Genetic tests for cancer risk have serious technical limitations, as is obvious from this discussion. In a few instances, such as intestinal polyposis, the detection of the cancer gene may warrant earlier and continued surveillance by colonoscopy. In the presence of defects (MEN2) in RET alleles, prophylactic thyroidectomy may be considered. Identification of individuals with the BRC1 or BRC2 breast cancer genes may warn of the increased possibilities of breast or ovarian cancer; however, preventive surgery may be too high a price for the uncertain manifestation of the neoplasia. Metastatic conditions may be detected by blood DNA analysis for specific oncogenic markers. RT-PCR and PSA levels may indicate (not just prostate cancer) micrometastatic conditions. For these types of analyses, molecular methods are available. These may detect LOH, microsatellite heterogeneity, telomerase deficiency, or eventually SAGE and DNA chip analyses may become practical.

After two decades of epidemiological research, no convincing evidence has been found for the putative leukemogenic effect of low-frequency electromagnetic fields for children. About 85% of human cancers involve solid tumors most commonly associated with the loss of several genes regulating the cell cycle, growth hormones, cell surface receptors, cell adhesion molecules, etc. Mitotic crossing over may also generate homozygosity

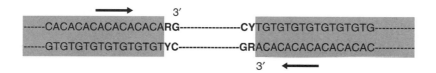

$$N = [(\text{No. of altered bands/PCR}) \times (\text{Total genome size}) \times$$
(No. uniquely altered bands)]/[Σ size of PCR fragments] (after Boland, C. R. & Ricciardiello, L. 1999. *Proc. Natl. Acad. Sci. USA* 96:1465). In colorectal cancer cells and premalignant polyps, Stoler, D. L., et al. (*Proc. Natl. Acad. Sci. USA* 96:15121) found approximately 11,000 alterations per cell with the above procedure. They concluded: "... genomic instability being a cause rather than an effect of malignancy."

The point-mutational origin of cancer has been questioned because (1) there are carcinogens that do not induce gene mutation or are very ineffective mutagens, e.g., asbestos, nickel (Ni^{2+}), butter yellow, urethan, etc. (should be noted that authors list arsenic as a nonmutagenic carcinogen, but in barley plants arsenates are strong mutagens); (2) the majority of mutations in the various types of cancers are not present in all types of cancers; (3) no genes isolated from cancers induce mutations in animal or human cells, and the loss of the oncogenes, e.g., RAS, does not revert cancer cells to normal cells; (4) the long period of latency does not seem to be consistent with the mutational origin; (5) the cancer cells are unstable, whereas mutations are most commonly stable; and (6) virtually all solid tumors are aneuploid. Therefore, Peter Duesberg and co-workers assume that aneuploidy is the

for oncogenes. Aneuploidy is common in solid tumors. In tumors of the hematopoietic or lymphatic system, chromosome translocations are frequent. Some of the multiple chromosomal rearrangements in cancer follow a preferred path. In human cells, chemical and physical agents (carcinogens) can induce cancerous growth in culture. The transformation of the coding sequence for the large catalytic subunit of the telomerase (hTERT), the T oncogene of Simian virus 40, and the RAS oncogene into epithelial and fibroblast cell cultures leads to cancerous proliferation and maintenance of this state.

Cytokines regulate tumor development at several levels. *Drosophila* has many genes that are homologous to mammalian oncogenes, tumor suppressors, and other genes responsible for abnormal cellular proliferation. Cancer is initiated and maintained by many different mechanisms, yet all of them share the properties of independence of external growth signals, limitations to respond to growth-inhibitory and apoptotic signals, and steady proliferative potentials supported by angiogenesis and metastasis. The incidence of cancer increases dramatically after age 40 in humans, but a plateau is reached after age 80. Epithelial carcinomas (breast, prostate, lung, colon, etc.) represent about 9% of cancer in children, but by adulthood the share grows to above 80%. *See* adenomatosis; adenovirus; agammaglobulinemia; aneuploid; angiogenesis; apoptosis; ataxia telangiectasia; Beckwith-Wiedemann syndrome; Bloom syndrome; breast cancer; cancer gene therapy; cancer prevention; cancer susceptibility; carcinogens; cell cycle; chromosome breakage; colorectal cancer; cytokines; DNA chips; DNA repair; Down syndrome; environmental mutagens; Epstein-Barr virus; exostosis; Fanconi anemia; focal dermal hypoplasia; Gardner syndrome; gatekeeper gene; genetic tumors; gonadal dysgenesis; hamartoma; immunological surveillance; imprinting; keratosis; Klinefelter syndrome; Knudsen's two-mutation theory; leiomyoma; lentigens; leukemias; Li-Fraumeni syndrome; lipomatosis; liver cancer; LOH; melanoma; MEN; metastasis; methylation of DNA; microsatellite; multiple hamartomas; neuroblastoma; neurofibromatosis; nevoid basal cell carcinoma; oncogenes; *PEG-3*; phorbol esters; polyposis; prostate cancer; protein tyrosine kinases; proteomics; PSA; retinoblastoma; ROS; Rothmund-Thompson syndrome; RT-PCR; SAGE; telomerase; tissue microarray; tuberous sclerosis; tumor antigen; tumor suppressor; von Hippel–Lindau syndrome; Werner syndrome; Wilms' tumor; Wiskott-Aldrich syndrome; xeroderma pigmentosum. (for cancer characterization by microarray: *Nature Medicine* 4:844; Hoeijmakers, J. H. J. 2001. *Nature* 411:366; Sánchez-García, I, 1997. *Annu. Rev. Genet.* 31:429; Knudson, A. G. 2000. *Annu. Rev. Genet.* 34:1; Bertram, J. J. 2000. *Mol. Aspects. Med.* 21:167; Balmain, A. 2001. *Nature Rev. Cancer* 1:77; Hahn, W. C. & Weinberg, R. A. 2002. *Nature Rev. Cancer* 2:331, mouse models: Jonkers, J. & Berns, A. 2002. *Nature Rev. Cancer* 2:251, molecular markers: Sidransky, D. 2002. *Nature Rev. Cancer* 2:210, <www.cancergenetics.org >; Cancer Gene Anatomy Project [CGAP]: <http://www.ncbi.nlm.nih.gov/dbEST/index.html >; <http://www.ncbi.nlm.nih.gov/ncicgap>; <http://oncolink. upenn.edu/edu/disease/>; cancer gene expression databases: <www.ncbi.nlm.nih.gov/SAGE>; <www.ncbi.nlm.nih.gov/ geo>; <www.ebi.ac.uk/arrayexpress>. Cytogenetics: <http:// www.infobiogen.fr/services/chromcancer>; cancer resources: <resresources.nci.nih.gov>, <www.cancerhandbook.net>.)

cancer classification Many of the human genetic diseases involve syndromes, and symptoms of different syndromes frequently overlap. Therefore, clear separation of a disease may not be straightforward. Similar and even worth is the situation in various types of cancers. Cancers with similar phenotypes may respond differently to therapies. Also, early detection and classification of cancers may increase the chances of successful treatment. The expression of cancer may depend on a good number of genes. Thus, identifying the expression patterns of the genes may help classification of cancers of similar phenotypes. The cancer investigator faces two main problems: discovery of the main type and possible subtypes of the neoplasia and predicting which

class(es) the cancer investigator/therapist deals with. Using microarray hybridization of thousands of genes expressed in the tumors, some smaller groups of genes can be selected for detailed analysis. *See* cancer; cluster analysis; microarray hybridization; predictor gene; recursive partitioning; RLGS. (Golub, T. R., et al. 1999. *Science* 286:531; Wooster, R. 2000. *Trends Genet.* 16:327; Zhang, H., et al. 2001. *Proc. Natl. Acad. Sci. USA* 98:6730.)

cancer death Usually initiated by several somatic mutations; only about 5% of the cases are attributable to germline mutations. In the United States, 1.6–2.5 million new cancer cases occur annually, and the death rate is about 560,000/year. The worldwide cancer death is about 6.6 million/year. In men, the most common types of cancer are prostate (43%) and lung and bronchus cancer (13%). In women, breast cancer is most common (30%); lung/bronchus (13%) and colon/rectum cancer (11%) follow. Lung cancer claims more deaths (32% and 25%, respectively) in both sexes than any other types. Although the rate of cancer deaths is declining somewhat, the incidence of cancer is not. About 70% of cancer victims may show some remission, but 50% of them relapse and become unresponsive to currently available treatments. *See* cancer; cancer gene therapy; cancer therapy.

cancer drug trials (<www.rhic.bnl.gov>).

cancer family syndrome *See* Li-Fraumeni syndrome; Lynch cancer family syndrome.

cancer gene therapy Transforming by (1) cytokine genes; (2) introducing genes encoding foreign antigens; (3) using antisense RNA or DNA constructs; (4) functional tumor suppressor alleles; (5) introducing wild-type dominant oncogenes; (6) ribozyme technology; and (7) transferring the *Herpes simplex* virus thymidine kinase (HSVTK) gene, making the tumor sensitive to ganciclovir (GCV), may be exploited. Phosphorylation of GCV (GCV-TP) inhibits DNA polymerase, which stops cancer cell proliferation; (8) using the multiple-drug resistance gene, MDR and (9) magic bullets have been attempted; (10) some approaches of tumor vaccinations are now in clinical trials. Cancer vaccines are based on whole cells, cell extracts and specific antigens, including carbohydrate-protein conjugates, GM2, GD3 and fucosyl GM1 gangliosides, globo H, Lewis blood group antigens, and mucin core structures. Immunological defense is a problem because cancer cells are self-originating and the immunological response to them is not good. Rejection of the malignant cells is very effective if tumor vaccines, e.g., after viral infection or transfection, elicit the expression of a foreign gene by a viral gene. Some individuals are probably immunosuppressed because some suppressor T

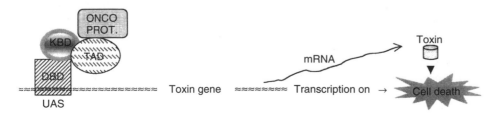

The toxin gene is transactivated and turned on, killing the cancer cell after it has been exposed to the transactivator by connecting the circuit through the oncoprotein (onco prot) to the upstream activator sequence (UAS). (Diagram modified after Da Costa, et al. 1996. *Proc. Natl. Acad. Sci. USA* 93:4192.)

cells downregulate the effector T cells. In some cases, the signaling to the T cell receptors seems to be disabled or diminished.

Some studies indicate an activation of the immune system by introduction of cytokine genes into tumors. This might be the result of better signaling or the recruitment of granulocyte-macrophage–colony-stimulating factor (GM-CSF). The tumor cells may display unusual epitopes (mutated β-catenin, caspase) on their MHC molecules and thus be recognized by the immune system, causing the rejection of the parent-type tumor cells. These epitopes may actually occur as differentiation antigens also present in reduced amounts on normal cells; the treatment must avoid hurting the normal cells. There may be substantial variations among the different tumors; thus, the immunotherapy faces complications. Some of the human cancers are of viral origin, and in these cases proteins essential for the maintenance of the cancerous state can be targeted. The therapeutic strategy must also consider whether cell-mediated (T-cell) or antibody-mediated defense (Th1 helper cells) may be the most effective route.

If patients were exposed to radiation or chemotherapy, they may be immunosuppressed or tolerant; the immunotherapy reduces the chance for success. In some instances, "generic" immunotherapy may help, but "custom" therapies using the mutated self-antigens of the patients may have a higher probability for success. Tumor cell–based vaccines, immunization with tumor peptides, DNA- or RNA-based vaccination, stimulation of the lymphocytes, or reintroduction into the patients of their own in vitro–stimulated lymphocytes or antigens are some of the therapeutic possibilities.

When mice experimental tumors were rendered transgenic for TNF (tumor necrosis factor), IL-2, IL-4 (interleukin), IFN-γ (interferon), and GM-CSF (granulocyte–macrophage colony-stimulating factor), the tumors went into remission or rejection. Actually, IL-2 itself does not directly curb the proliferation of cancer cells; rather, it stimulates the expansion of lymphocytes, which have antitumor activity. CD8$^+$ T lymphocytes recognize 8–10 amino acid peptides generated by in vitro sensitization of T cells, then cytoplasmic proteins by proteasomes, and displayed on the cancer cell surface by class I HLA molecules. CD4$^+$ T cells — in contrast — recognize extracellular proteins engulfed and digested in the endosomes and displayed on the cell surface by class II HLA molecules. Based on these properties, cancer antigens can be purified and characterized. Cancer antigens can be tested on intact cancer cells by this *reverse immunology*.

Another newer technology is represented by SEREX. Thousands of genomic instabilities may occur in cancer cells, generating a great variety of new types of cancer antigens. Some success was achieved in the rejection of melanoma and lymphoproliferative cancers by immunization via the sensitized lymphocytes (adoptive therapy). Administration of IL-2 cytokines substantially enhances the effectiveness of these immunotherapies (Rosenberg, S. A. 2001. *Nature* 411:380).

New technologies are continuously being developed. One such cancer therapy design would convert the cancer genes (oncoproteins) into cancer killer genes by activating a toxin gene according to a construct outlined on p. 158. The oncoprotein has a DNA-binding domain (DBD), transactivator domain (TAD), and killer-binding domain (KBD). Another possibility is to fuse the *Pseudomonas* exotoxin (PEA) to the extracellular domain of the HER2 transmembrane receptor

protein. Transformed lymphocytes can then introduce the toxin that inactivates elongation factor-2 (FE-2) into the cancer cells. Transfection of antisense RNA–producing constructs has been proven successful in animal models. Human tissue factor (TF), an important receptor of the blood coagulation cascade, can be targeted to endothelial blood vessels of the tumor by an appropriate antibody. TF complexed with antihemophilic factor VII activates the serine protease zymogen factors IX and X, resulting in the formation of thrombin and blood clotting.

Specifically and locally blocking the blood supply (antiangiogenesis) to the tumor may cause its substantial regression. Small peptides that are identified by the technique of phage display and have a special affinity for the tumor blood vessels can be linked to potent antibiotics (such as doxorubicin) and appear to be a promising alternative to chemotherapy. Unlike the dangers of drug resistance in chemotherapy, antiangiogenesis in endothelial cells does not lead to resistance. The cancer gene therapy vectors must be designed to recognize and destroy the cancer cells without inflicting damage to the normal cells even if they are interspersed with tumor cells. By employing cancer cell–specific promoters (such as the melanocyte-specific tyrosinase promoter; or the α-fetoprotein promoter specific for liver carcinomas; or the c-erb-B2 specific for breast cancer, ovarian cancer, and neuroblastoma), the expression of the genes controlling cytotoxic substances can be localized. Radiation-responsive and hormonal receptor regulatory systems may permit the combination of radiation treatment and hormone therapy along with the cytotoxicity transgene. Cancer gene therapy is an innovative research area and more work is required before it can become a general clinical practice. *See* adeno-associated virus; adenovirus; angiostatin; antihemophilic factors; antisense technologies; bystander effect; camptothecin; cancer death; cancer prevention; cancer therapy; cell therapy; cytokines; cytosine deaminase; dendritic cell vaccine; EF-2; endostatin; epitope; ERBB2; exotoxin; ganciclovir; gangliosides; gene therapy; globo H; HIV; immune system; immunotherapy; immunotherapy, adoptive; informed consent; Lewis blood group; liposome; lymphokines; magic bullet; multidrug resistance; MDR; MHC; mini-organ therapy; MoMuLV; OL(1); oncogene antagonism; oncolytic viruses; p16^{INK}; p21; p53; phage display; Philadelphia chromosome; plasmovirus; retinoblastoma; retroviral vectors; ribozymes; SEREX; suicide vectors; T cell; tissue factor; topoisomerases; transfection; tTA; tumor suppressor genes; tumor vaccination; vaccinia virus. (Nettelback, D. M., et al. 2000. *Trends Genet.* 16:174; Kudryashov, V., et al. 2001. *Proc. Natl. Acad. Sci. USA* 98:3264; McCormick, F. 2001. *Nature Rev. Cancer* 1:130, Wadhwa, P. D., et al. 2002. *Annu. Rev. Med.* 53:437, Ye, Z., et al. 2002. *Nature Med.* 8:343.)

Cancer Genome Anatomy Project (CGAP) Collaborative network of cancer researchers to study the origin, formation, and progression of cancer and for the dissemination of information. (Riggins, G. J. & Strausberg, R. L. 2001. *Hum. Mol. Genet.* 10:663; <http://cgap.nci.nih.gov/>.)

cancer prevention Different from cancer chemotherapy inasmuch as it is not expected to fully kill cancerous cells; rather it is aimed at blocking either the primary or the secondary malignancy routes. The chemical treatment may reverse neoplastic development or block preneoplastic conditions without substantial toxicity. Chemicals that stop metabolic activation of procarcinogens are also sought. Cyclooxygenase (COX-2) enzyme appears to increase during the development of intestinal tumors. Blocking the activity of this enzyme by MF-tricyclics without side effects is one approach. Breast cancer may be blocked by tissue-specific estrogen inhibitors (raloxifen, LY353381). Prostate cancer prevention is feasible by interference with estrogen receptors (flutamid). Lung cancer development may be arrested by restoration of the expression of a retinoic acid receptor-β by 13-cis-retinoic acid. Approaches to vaccination against cancer are being worked on. Glycopeptides resembling cell surface tumor antigens may be produced by laboratory techniques. These may then trigger immune responses. *See* breast cancer; cyclooxygenase; immunological surveillance; peptide vaccination; raloxifen; RAR; retinoic acid; tamoxifen. (Levi, M. S., et al. 2001. *Curr. Med. Chem.* 8:1439.)

cancer promoter   *See* phorbol esters.

cancer susceptibility   May be affected by environmental factors and genotypic differences. The most effective environmental carcinogens are polycyclic aromatic hydrocarbons and aromatic amines. The genetic predisposition depends on the individual differences in the activation and detoxification systems. The major risk factors are mutation(s) in tumor suppressor genes. Somewhat less important are the mutations in activating enzymes, but they may become important under conditions of longer or repeated exposure to carcinogens. The detoxification process itself may involve activation of the procarcinogens into carcinogens. The metabolic factors include the P450 cytochrome system, glutathione-S-transferase (GST), epoxide hydrolase, N-acetyltransferase (NAT), and defects in apoptosis, cell cycle, genetic repair, etc. Individuals with mutations in the enzymes modulating responses to carcinogens may have an increased risk. Generally the contributing factors include interactions among several of the modulatory enzymes. Therefore, the risk on the basis of analyzing a single enzyme may not be sufficient. Ethnic, sex, and age differences exist in susceptibility. Some of the differences may be caused by cultural and environmental factors (diet, smoking, etc.). So far no general method is available for cancer screening. *See* activation of mutagens; cancer; carcinogens; environmental mutagens; epoxide; glutathione-S-transferase; P450. (Taningher, M., et al. 1999. *Mutation Res.* 436:227.)

cancer therapy   *See* adenovirus; alkyltransferase; angiostatin; antimitotic agents, cytostatic; antisense technologies; BCL; bleomycin; cancer gene therapy; cancer prevention; chemotherapy; clonogenic test; cytokines; cytostatic; electroporation; endostatin; immunological surveillance; immunotoxin; magic bullet; multiple drug resistance; NF-κB; plasmovirus; radiation effects; tamoxifen; telomerase; telomeres; tumor-infiltrating lymphocyte; Yttrium. (Hurley, L. H. 2002. *Nature Rev. Cancer* 2:188.)

cancer vaccine   *See* cancer gene therapy; vaccines.

candela   Unit of luminous intensity. A standard radiator produces 60 candela/cm^2 at the freezing temperature of platinum ($-2,046$ K); 1/60 candela = 1 candle (new unit). 1 foot candle = 10.76 lux; 1 lux = 1 lumen/m^2; 1 lumen = the total visible energy emitted from 1 candle point luminous intensity.

"candidate gene"   Already mapped in a chromosomal region and possibly included in the DNA fragment to be mapped, but it is not known whether it is involved in a particular function or phenotype associated with a mutation. The functional identity of the isolated DNA requires rigorous biochemical proof. In case multiple alleles of the gene are available, the verification of the identity is greatly facilitated. In case the wild-type allele of the isolated gene is returned by transformation in a mutant stock and the normal phenotype is restored, there can be no doubt about identity. See gene isolation; transformation, genetic. (Ernst, J. F. 2000. *Microbiol.* 146:1763.)

candidiasis, familial chronic mucocutaneous (FCMC) Apparently autosomal-recessive (2p) immunodeficiency involving T lymphocytes resulting in infections of the mucous membranes, skin, and nails by *Candida* fungi. The diploid *C. albicans* ($n = 7$, 16×10^6 bp) is the most prevalent pathogen in this type of disease and it is responsible for the very common and frequently lethal nosocomial infections. *See* immunodeficiency; nosocomial. (De Backer, M. D., et al. 2000. *Annu. Rev. Microbiol.* 54:463; Chibana, H., et al. 2000. *Genome Res.* 10:1865.)

candle   *See* candela.

cannabinoids   About 60 related compounds synthesized by the hemp plants (*Cannabis*) and present in the psychoactive drug marijuana. Besides the psychoactive effects, they have various immunosuppressive properties and may control pain initiation and transmission through modulation of the rostral ventromedial medulla of the brain and tremor in multiple sclerosis. Some "endocannabinoids" are present in cocoa, milk, and other human food but do not appear to be responsible for chocolate cravings or other psychoactive effects. Δ^9-Tetrahydrocannabinoid may reduce sexual activity and fertility by affecting steroid metabolism. In cooperation with leptin, endocannabinoids regulate food intake and body weight; leptin restricts food craving, whereas endocannabinoids increase appetite. The endocannabinoid anandamide in low concentration promotes receptivity of the mouse uterus to embryo transplantation, whereas at higher levels the receptivity is reduced. So far there is no evidence for the normal control of fertility, although obese mice may be sterile until leptin is supplied. Endocannabinoids also activate the pleasure sensations, e.g., in eating tasty food, which may explain addiction. *See Cannabis sativa*; immunosuppressants; leptin. (Iversen, L. L. 2000. *The Science of Marijuana*. Oxford Univ. Press, New York; Di Marzo, V., et al. 2001. *Nature* 410:822; Carlson, G., et al. 2002. *Nature Neurosci.* 5:723.)

Cannabis sativa (hemp) Generally a dioecious species (*Moraceae*) with a diploid chromosome number of 20, including either XX female or XY male flowers. The XXXY tetrasomics are also gynoecious. Hermaphroditic forms are known that produce more seed and only a somewhat reduced amount of fiber. Low-cannabinoid genetic stocks are known. *See* cannabinoids; hop.

cannibalism Animals eating members of their own species. In heterocannibalism the eaten individuals are not the offspring of the cannibal. Filial cannibals devour their own offspring. The latter class is generally less frequent because it is usually counterproductive for fitness. Filial cannibals apparently lack the ability to identify their own offspring. Males are more likely to commit filial cannibalism than females because they have much less investment in the progeny. *See* fitness. (DeWoody, J. A., et al. 2001. *Proc. Natl. Acad. Sci. USA* 98:5090.)

canola Erucic acid-free rape, *Brassica napus* ($2n = 38$, AACC genomes) oil crop. *See* erucic acid.

canonical sequence Typical set of nucleotides in several genes (e.g., in a Pribnow box or Hogness box or other conserved elements).

canyon Deep structural site on the viral surface where receptors bind.

cap Methylated guanylic residue linked at transcription to the 5′ end of the eukaryotic mRNA (7ME G5′ppp-mRNA) or 2,2,7-trimethyl guanosine in U RNA. The prokaryotic mRNA has only three phosphates at the 5′ nucleotide end. The cap of the (long-life) eukaryotic mRN stabilizes it (makes it less sensitive to nucleases), assures the transport of the mRNA to the cytosol, and facilitates its binding to the ribosome and initiation of transcription by lending itself as an anchor to eIF4F eukaryotic initiation factors. The cap also regulates splicing of the first intron. A capping complex (CBC) composed of cap-binding nuclear proteins CBP80 and CBP20 mediates the effect of Cap on pre-mRNA splicing. CBC then mediates the export of RNA. The U3 snRNA also has a trimethylated cap that is added in the nucleus and the RNA remains in the nucleus. Binding between the cap and the poly(A) tail promotes translation. Picornavirus mRNAs lack the 5′ caps and thus translation initiation is cap-independent; it begins at an IRES site. *See* capping enzyme; eIF4; FLAG; IRES; poly(A); regulation of gene activity; ribosome scanning. (Rottman, F., et al. 1974. *Cell* 3:197; Efimov, V. A., et al. 2001. *Nucleic Acids Res.* 29:4751.)

CAP (1) Catabolite activator protein is a homodimer with two subunits of M_r 22,000. It has a binding site for cAMP and DNA with a helix-turn-helix motif. As long as glucose is available in the nutrient medium, the *Lac* operon of *E. coli* is inactive because glucose, a catabolite of lactose, represses the operon (catabolite repression). In order to turn on the operon, the CAP protein must be attached to the CAP site in a process mediated by cAMP. The adenyl cyclase enzyme forms the latter from ATP, and this process is inhibited by glucose. When glucose is used up, cAMP is formed and the latter binds to CAP; the complex binds to its palindromic site in the DNA (GTGAGTTAGCTCAC) near the promoter of the operon and transcription by the RNA polymerase is activated. Without CAP the promoter is very weak. In order to pursue normal transcription, the grip of the repressor protein must also be lifted, and this is mediated by lactose (allolactose). The cAMP-CAP complex also regulates the arabinose and galactose operons. *See* arabinose operon; cAMP; galactose operon; *lac* operon; negative control; palindrome; positive control; regulation of gene activity; transcriptional activation. (Johnson, C. M. & Schleif, R. F. 2000. *J. Bacteriol.* 182:1995.) (2) Ceramide-activated protein. CAP kinase mediates tumor necrosis factor and interleukin-1β functions. It phosphorylates Raf1 on Thr 269 and increases Raf affinity for ERK kinases. *See* ceramides; ERK; interleukin; RAF; TNF. (Chalfant, C. E., et al. 2000. *Methods Enzymol.* 312:420.)

CAP3 N-terminal domain of FLICE. *See* FLICE.

Cap-binding protein *See* eIF; translation initiation. (Mazza, C., et al. 2001. *Mol. Cell* 8:383.)

Cap-binding protein complex (CBC) Binds to the 5′-end of the mature mRNA and facilitates U snRNA export. *See* eIF; export adaptors; nuclear pore; protein synthesis; translation initiation; U RNA.

capillary Structure resembling hair; usually bearing a small bore through which liquids can move such as in the capillary veins or capillary tubes, or soil capillary spaces.

capillary electrophoresis Uses gel-filled glass/quartz tubes of 50−100 μm diameter and 20−50 cm long to facilitate fast separation of substances in the electric field by applying high voltage (10−30 kV) without excessively heating the system because of the better dissipation of the heat. UV or fluorescence detects labeling. *See* DNA sequencing; electrophoresis; ultrathin-layer chromatography. (Guttman, A. & Ulfelder, K. J. 1998. *Adv. Chromatogr.* 38:301.)

capillary transfer Used to draw a buffer by wicks from a reservoir to an electrophoretic gel containing separated DNA fragments. The gel is in contact with the absorbent papers. The moving stream elutes the DNA from the gel and deposits it onto a nitrocellulose or nylon filter in immediate contact with the gel, in between the gel and the stack of papers topped by a glass plate and weighted down. *See* Southern blotting.

cap′n′collar gene (CNC) Basic leucine zipper transcription factor regulating homeotic genes. *See* DNA-binding protein domains; homeotic genes.

capping enzyme Generates the mRNA cap by using GTP, and the diphosphate splits off from the first nucleotide triphosphate of the first residue in the pre-mRNA. The 5′

Methyl guanine
in the cap

At the 3' **P** the chain
continues

The cap of the eukaryotic mRNA.

terminal triphosphate is replaced by the guanyl group of GTP, and it loses the γ and β PO_4 groups.

No capping takes place if there is a monophosphate at the terminal position. The capped G indicates the beginning of the transcript. The G is methylated by guanosine-7-methyl transferase at the 7 position. The 2'-OH is subsequently methylated by 2'-O-methyl transferase using SAM as a methyl donor. The capping reaction is associated with pol II and the triphosphate termini of U6 RNA; 5S RNA and the pre-tRNAs transcribed by pol III are not capped. The caps stabilize the mRNA and facilitate the ribosomal attachment. Initiation factor eIF-4E recognizes the cap and mediates its binding to the 40S ribosomal subunit. Picornaviruses do not need the cap. They inactivate the cap-binding proteins of the host and thus turn off the synthesis of host proteins. *See* cap; class II genes; eIF; FLAG; picornaviruses; pol II eukaryotic; transcription factors. (Cho, E. J., et al. 1997. *Genes. Dev.* 11:3319; Shuman, S. 2000. *Progr. Nucleic Acid Res. Mol. Biol.* 66:1; Changela, A., et al. 2001. *EMBO J.* 20:2575; Pei, Y., et al. 2001. *J. Biol. Chem.* 276:28075.)

CAPS Cleaved amplified polymorphic sequences are produced by digesting PCR products by restriction enzymes to find polymorphism in the DNA. *See* AFLP; PCR; restriction enzyme.

capsaicin Pungent substance of *Capsicum* peppers. It may be anticarcinogenic due to its antioxidative function. The capsaicin (vanilloid) receptor is a cation channel required for heat and pain perception. Bradykinin and nerve growth factor activate G-protein-coupled and tyrosine kinase receptors, and phospholipase C signals to the primary afferent neurons. The potentiation requires the VR1 heat-sensitized ion channel on sensory neurons. Capsaicin-containing fruits are generally avoided by wild mammals but not by birds (Jordt, S.-E. & Julius, D. 2002. *Cell* 108:421). It is apparently a self-defense substance in the plants because birds are efficient dispensers of the seed, but mammals are not because the seed passing through their alimentary channel looses germination. *See* alkaloids; kininogen; nerve growth factor; nociceptor; phenolics; phospholipase; signal transduction. (Chuang, H.-H., et al. 2001. *Nature* 411:957.)

Capsaicin.

capsid Protein shell of the viral genetic material; in complex viruses nucleic acids may also be found in the shell. *See* viruses.

cap snatching Viral RNA replication secures primers for its initiation by cleaving off about a dozen nucleotide long pieces from the 5'-end of nuclear RNA polymerase II transcripts. The cleavage is mediated by a virus-encoded endonuclease. The priming does not involve hydrogen bonding between the primers and the 3'-end of the viral RNA template. *See* primer. (Duijsings, D., et al. 2001. *EMBO J.* 20:2545.)

capsomer Protein subunit of the viral capsid.

capsule Polysaccharide coat of bacterial cells or in general structure with content or a fungal sporangium, or a seed capsule of plants formed by fusion of two or more carpels. *See* carpel.

capture PCR Facilitates the isolation of DNA sequences next to known nucleotide segments. Linkers of two base-paired oligonucleotides are ligated to restriction enzyme-digested DNA ends. By using a biotinylated primer of known sequence, the construct is extended. This permits the capture of the extended products on a streptavidin-coated medium. These products are further amplified by PCR with the aid of another specific oligonucleotide hybridized to the 3'-end of the biotinylated sequence. The simultaneous isolation of a large number of fragments is greatly facilitated by the use of a manifold connected to each individual well of a microtiter plate. *See* biotinylation; microarray hybridization; PCR; restriction enzyme; streptavidin. (Lagerstrom, M., et al. 1991. *PCR Methods Appl.* 1:111.)

CaR Extracellular calcium receptor and thus regulator of diverse metabolic functions.

CAR Cyclic AMP receptor. *See* cAMP.

CAR1 Coxsackie and adenovirus receptor is a 368-amino-acid receptor that recognizes the avian leukosis-sarcoma virus envelope and induces apoptosis upon infection. It is homologous to the TNF/NGF mammalian family (TRAIL) receptors. CAR is also a cellular attachment receptor for adenovirus, a useful vector in gene therapy. CAR-transgenic mice are a promising model for human gene therapy. *See* adenovirus; apoptosis; coxackie viruses; NGF; TNF; TRAIL. (Tallone, T., et al. 2001. *Proc. Natl. Acad. Sci. USA* 98:7910.)

CAR-β   Constitutive androstane receptor is negatively regulated by androstanes. *See* androstane.

C3a-R   Receptor of C3a complement component. Its mass varies (83 to 104 kDa). *See* complement.

C5a-R   Receptor of the C5a complement component. The size of this protein encoded in human chromosome 19q13.3 varies greatly (8–52 kDa) depending on the cells on which it is found. *See* complement.

carbachol (carbamylcholine chloride)   Cholinergic agonist resistant to cholinesterase; carbachol may increase phosphorylation of RAS. *See* cholinesterase; RAS.

carbamoyl phosphate synthetase deficiency (CPSI, 2q35)   Hypermonemia is caused by defects in several enzymes in the mitochondria; others are encoded by several autosomes and are cytosolic. Carbamoylphosphate synthetase/aspartate transcarbamoylase/dehydroorotase (2p21) is a three-enzyme locus involved in pyrimidine synthesis. Ornithine transcarbamylase deficiency (OTC) is encoded at Xp2.11. *See* CAD; channeling; urea cycle.

carbenicillin   Semisynthetic antibiotic of the penicillin family; effective against gram-negative and some gram-positive bacteria. *See* antibiotics.

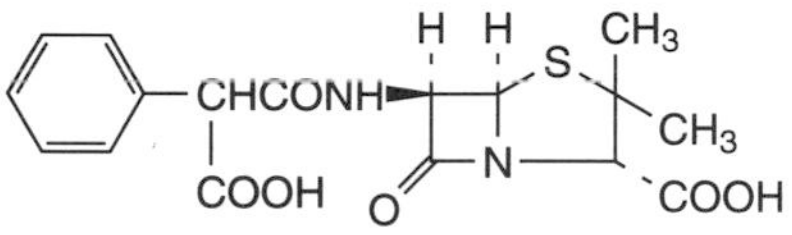

Carbenicillin.

carbocation   *See* carbonium ion.

carbohydrate   sugars and their polymers. (<http:www.dkfz.de/spec2/sweetdb>.)

carbon dating   *See* evolutionary clock.

carbon fixation   Photosynthetic organisms form sugars from atmospheric CO_2. See C3 plants; C4 plants.

carbonic anhydrase (CA)   Catalyzes the reaction $H_2CO_3 \rightleftarrows H_2O + CO_2$, i.e., it provides an equilibrium between carbonic acid and carbon dioxide, and the additional reaction of $CO_2 + H_2O \rightarrow H^+$ plus the formation of a carbonium ion from carbon dioxide. A Zn enzyme of about M_r 30,000 is common in various eukaryotic tissues (1–2 g per L of mammalian blood). It is an extremely active enzyme, but a much lower rate also catalyzes the hydration of acetaldehyde. Seven isozymic forms have been identified in humans. CA 1, 2, and 3 were located to chromosome 8q13-q22. CA2 is about 20 kb apart from CA3, and it is transcribed in the same direction, while CA1 and CA3 are separated by about 80 kb and their direction of transcription is opposite. CA1, 2, and 3 are common in the muscles. CA2 (8q22) deficiency is involved in osteopetrosis with renal tubular acidosis. CA4 (17q23) and CA9 (17q21.2) are closely linked. CA5 (16q24.3), an apparently mitochondrial protein, does not appear to be of specific significance; it may participate in gluconeogenesis. CA6 gene is present in chromosome 1p37.33-p36.22 and the gene is expressed in the saliva. CA7 gene is in another chromosome (16q21-q23) and is specific for the kidney, lung, and liver mitochondria. CA10 (chromosome 7) codes for α-carbonic anhydrase. CA11 (19q13.2-q13.3) is primarily a brain enzyme, but it is also expressed in the spinal cord and other tissues. CA12 (15q22) shows slightly elevated expression in some tumors. CA14 is expressed primarily in the adult brain and several visceral organs but only in the fetal heart (1q21). *See* gluconeogenesis; mitochondrial genetics; mtDNA; osteopetrosis. (Comprehensive series of reviews in *EXS* [2000], 90.)

carbonium ion   Group of atoms containing only six electrons (rather than the normal octet); it is considered highly reactive, and it is supposed to be involved in the reactions following alkylative processes in chemical mutagenesis and carcinogenesis. *See* alkylating agents; carcinogen; mutagen specificity.

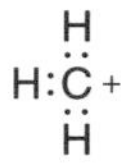

Methyl
carbonium ion

carbonyl group   Such as occurs in aldehydes and ketones.

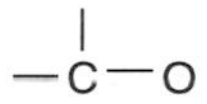

carboxyl group

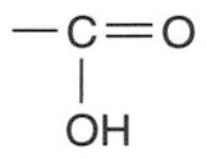

carboxyl terminus   *See* C-terminus.

carboxypeptidase   Zinc-containing proteolytic enzyme cleaving the polypeptide chain at the carboxyl end after substrate binding caused an induced fit, i.e., a conformational change in the enzyme protein. The 'electronic strain' due to the presence of Zn accelerates catalysis.

carcinoembryonic antigen (CEA)   It may be expressed in normal epithelium of the colon, but it is frequently found in colorectal, gastric, pancreatic, and some breast and non-small-cell lung carcinomas. It may be targeted by genetic immunization. *See* immunization, genetic. (Jacobsen, G. K., et al. 1981. *Oncodev. Biol. Med.* 2:399.)

carcinogen Agent that is capable of causing cancerous transformation of cells. Carcinogens include chemical, physical, and viral agents. De novo carcinogenicity may have several steps: initiation (DNA adduct formation, mutation), promotion (DNA methylation, clonal expansion of the altered cells), progression (increased methyltransferase activity, additional mutational events and/or other chemicals modulating the process), and invasive events (metastasis). Some of the carcinogens do not act, however, by mutation. The chemical compounds may be genotoxic and act directly on the DNA (e.g., ethyleneimine, various epoxides, lactones, sulfate esters, mustard gas, 2-naphthylamine, nitroamides and nitrosoureas, etc.). They can be procarcinogens that require enzymatic activation (e.g., polycyclic or heterocyclic hydrocarbons such as benzo[a]pyrene, benzanthracene, etc.) or inorganic compounds or elements that interfere with the fidelity of DNA replication. Some carcinogens act by some sort of physical means (e.g., various polymers, asbestos).

Some hormones may also promote carcinogenesis in an indirect manner. Phorbol esters, n-dodecane, etc., may not be the primary cause of cancer but are considered to be promoters of cancerous growth. The latter group of chemicals not acting directly on DNA are frequently called *epigenetic carcinogens*. Carcinogens include most of the mutagens, many industrial and laboratory chemicals, pesticides, insecticides, fungicides, drugs, cigarette smoke, benzo(a)pyrene, medicines (cyclosporin, estrogens, tamoxifen, etc.), diethylstilbestrol, cadmium and nickel compounds, cosmetics, food preservatives, food additives, flame retardants, cross-linking agents, plastics, solvents, paints, adhesives, exhaust fumes, other products of combustion, tar, soot, benzenes, naphthalenes, carbamates, cyanates, metals, nitroso compounds, alkylating agents, terpenes, some fibers (asbestos), arsenics, etc. Other possible carcinogens are natural plant products (e.g., pyrrolyzidine alkaloids), safrole, mycotoxins (e.g., aflatoxins), antibiotics (e.g., streptozotocin), viruses such as the Epstein-Barr virus, Simian virus 40, HIV, hepatitis B and C virus, some papilloma viruses, adenoviruses, etc. Many food products such as oxidized fats, overcooked meats, etc., are also potential carcinogens. The direct assays of carcinogens involve testing the induction of skin or lung tumors in rodents, breast tumors in young female Wistar or Sprague-Dawley strains of rats, examination of rodent livers for carcinogenic response, etc.

The determination of carcinogenicity at relatively low potency is extremely difficult by direct animal assays because tumorigenesis may occur only after a long delay (months or years) following exposure and because the required population size may be practically prohibitive. All experiments must include an equal-size concurrent control to obtain reliable information. The application of the carcinogen to the test animals may be by painting of the skin, subcutaneous or intravenous injection, feeding in the diet or drinking water, inhalation, etc. Since many of the carcinogens are also mutagens, the preliminary tests are generally conducted by mutagenic assays that permit the evaluation in large populations within a short time and at a low cost. The mutagenic assays (*See* Ames test) usually try to substitute for the animal activation system by human or animal liver (microsomal) fractions.

In human and animal populations, the high caloric diet may be an important factor in tumor formation. The response of humans to carcinogens is not entirely identical to that of animals; e.g., prostate, pancreas, colon, and cervix/uterus cancers are low in lab rodents but high in humans. In contrast, liver, kidney, forestomach, and thyroid gland cancers are frequent in animals but relatively rare in humans. Arsenics are human carcinogens but not for laboratory rodents. Humans cannot be subjected to direct cancer tests and most reliable cancer information could come from epidemiological studies. Humans defective in epoxide hydrolase and glutathione S-transferase M1 have increased susceptibility to carcinogens. Unfortunately complex factors, and low frequencies frequently bias carcinogenesis data. *See* adduct; bioassays in genetic toxicology; cancer; cigarette smoke; cocarcinogens; Epstein-Barr virus; genetic tumors; hepatitis B virus; hepatitis C virus; *IARC Monographs*; ionizing radiation; k_e test; MTD; mutagen activation; mutagen assays; neoplasia; oncogenes; peroxisome; radiation effects; radiation hazard assessment. (Kitchin, K. T., ed. 1999. *Carcinogenicity*. Marcel Dekker, New York; <http://www.iarc.fr/monoeval/allmonos.htm>.)

carcinogenesis Process of cancer induction and progression. *See* cancer; carcinogen. (Ponder, B. A. J. 2001. *Nature* 411:336.)

carcinoma Malignant cancer tissue of epithelial origin.

carcinostasis Tumor growth inhibition.

CARD Caspase recruitment domain is required for the recruitment and activation of caspase-9 by Apaf-1 in apoptosis. CARD15 contains a nucleotide-binding domain and 10 COOH-terminal leucin-rich domains. Mutations in CARD 15 are involved in inflammatory bowel diseases such as Crohn disease and ulcerative colitis. *See* Apaf; apoptosis; caspase; colorectal cancer; Crohn disease. (Bouchier-Hayes, L., et al. 2001. *J. Biol. Chem.* 276:44069; Lesage, S., et al. 2002. *Am. J. Hum. Genet.* 70:845.)

cardiac arrhythmia *See* cardiomyopathy, arrhythmogenic ventricular. LQT.

cardiac conduction defect (PCCD, Lenegre disease, Lev disease) Common degenerative heart disease frequently requiring the insertion of a pacemaker. The defects are associated with sodium or potassium ion channels and are encoded in different chromosomes (19q13.3, 3p21).

cardio-auditory syndrome *See* Lange-Nielssen syndrome.

cardiomyopathies Group of noninflammatory heart diseases affecting about 25,000 people annually in the United States. *See* acyl-CoA dehydrogenase deficiencies; Barth syndrome; Becker muscular dystrophy; cardiac arrhythmia; cardiomyopathy, arrhythmogenic ventricular; cardiomyopathy, dilated; cardiomyopathy, hypertrophic; cardiovascular diseases; Duchenne muscular dystrophy; histocytoid cardiomyopathy; mitochondrial diseases in humans; myopathy; superoxide dismutase. (Seidman, J. G. & Seidman, C. 2001. *Cell* 104:557.)

cardiomyopathy, arrhythmogenic ventricular (ARDV) Autosomal-dominant (14q24.3, 14q12-q22, 1q32, 10p12-p14) and possibly -recessive condition involves fibrous/fatty replacement of heart muscles, particularly in the right ventricle. It causes unusual palpitation, fainting (syncope), heart failure, and possibly sudden death. *See* cardiomyopathies.

cardiomyopathy, dilated (CMD) Involves thinner than normal ventricular heart wall, reduced contractility, and heart failure. Prevalence in the United States is about 4×10^{-4} and about 25% of these cases are hereditary and may benefit from heart transplantation. Defects may involve recessive β-oxidation of fatty acids, including acyl-CoA dehydrogenase, carnitin palmitoyl transferase II, and impaired mitochondrial oxidative phosphorylation. Missense mutation in cardiac actin gene (ACTC, human chromosome 15q11-qter) seems to be involved. Several genes for idiopathic dilated cardiac myopathy (IDC) have been located in human chromosomes Xp21, Xq28, 1p1-1q21, 1q32, 2q31, 2q14-q22, 3p22-25, 5q33-q34, 6p24, 6q12-q16, 9q13-q12, 10q21-q23, 14q11, 15q14, 15q22, and 17q21. Defects in dystrophin and myosin may also be involved. Cardiomyopathy CMD1A involves a defect in lamins (1q21.2-q21.3). Autoimmune dilated cardiomyopathy may be due to deficiency of the PD-1 receptor. Homoplasmic point mutations in mitochondrial tRNA^{His} may predispose for CMD. *See* acetyl coenzyme A; actin; cardiomyopathies; cardiovascular diseases; carnitin; dystrophin; fatty acids; lamins; mitochondrial diseases in humans; PD-1. (Chien, K. R. 1999. *Cell* 98:555; Schönberger, J. & Seidman, C. E. 2001. *Am. J. Hum. Genet.* 69:249.)

cardiomyopathy, hypertrophic, familial (FHC) Heterogeneous autosomal-dominant symptoms (thickening of the heart's ventricular walls, shortness of breath, arrhythmia, and sudden death.) Four genes are known to be involved in controlling contractile heart proteins such as β-myosin heavy chain (human chromosome 14q11-12), α-tropomyosin (15q2), troponin T (1q3), and cardiac myosin-binding protein C (11p11.2). Additional gene loci have been implicated in human chromosomes 7q3, 18, and 16. Cardiac hypertrophy may be blocked by inhibitors of ADAM12 protein, which sheds heparin-binding epidermal growth factor (HB-EGF), responsible for hypertrophy. *See* ADAM; cardiomyopathies; cardiomyopathy, dilated; cardiovascular diseases; myosin; troponin. (Asakura, M., et al. 2002. *Nature Med.* 8:35.)

cardiomyopathy, restrictive Affects the expansion (diastole) of the heart without much effect on the contraction (systole). Only about 10% of the cases are clearly hereditary and X-chromosomal or autosomal dominant. *See* cardiomyopathies.

cardiotrophin (CT-1, ~22 kDa, encoded at 16p11.1-p11.2) Cytokine regulating cardiac muscle cells. It shares a receptor with LIF and other cytokines. *See* cytokine; LIF.

cardiovascular disease Affects the heart and the vein system. *See* cardiomyopathies; coarctation of the aorta; coronary heart disease; Ehlers-Danlos syndrome; familial hypercholesterolemia; homocystinuria; hypertension; hyperthyroidism; hypo-α-proteinemia; hypobetalipoproteinemia; idiopathic ventricular fibrillation; lipidoses; lipoproteins; LQT; lysosomal storage disease; Marfan syndrome; mucopolysaccharidosis;

Norum disease; sickle cell disease; supravalvular stenosis; Tangier disease; telangiectasia, hereditary hemorrhagic; Williams syndrome.

CaRE cis-acting element responsible for induction of the c-fos protooncogene by calcium. *See* cis-acting element; FOS.

caretaker gene Repairs genetic defects that may lead to instability or cancer. *See* gatekeeper; tumor suppressor genes.

CArG Conserved promoter element [CC(A/T)$_6$GG] closely related to SRF. *See* promoter; SRF.

CArG box *See* SRE.

cargo receptors Special molecules in or on the plasma membrane that are recognized by adaptins associated with clathrin-coated vesicles, thus trapping various molecules in the process of endocytosis. *See* adaptins; clathrin; endocytosis; receptors. (Moroianu, J. 1999. *J. Cell Biochem. Suppl.* 32–33:76.)

CARL *See* comparative anchor reference loci.

CARM1 Coactivator associated with arginine methyltransferase-1, related to AdoMet. It may mediate transcription when recruited to the steroid receptor coactivator (SRC-1) and cofactors p300 and PCAF. *See* AdoMet; p300; PCAF; SRC-1. (Chen, S. L., et al. 2002. *J. Biol. Chem.* 277:4324.)

Carney complex Multiple and usually pigmented neoplasias of soft tissues and the heart (myxoma). One form is CNC2, located at 2p16; the other is CNC1 at 17q23-q24. The latter is presumably a dominant mutation in a tumor suppressor regulatory alpha unit of a cAMP-dependent protein kinase (PRKAR1A).

carnitine (γ-trimethylamino-β-hydroxybutyrate) Facilitates the entry of fatty acids into mitochondria. It occurs in all organisms and it is most abundant in the muscles (0.1% of dry matter). Systemic carnitin deficiency involves progressive cardiomyopathy, skeletal myopathy, hypoglycemia, hyperammonemia, and sudden infant death syndrome. The basic defect is in a gene encoding sodium-dependent carnitin transporter. Carnitin defects may be coded in human chromosomes 5q33.1, 3p21, 1p32, 11q13, and 9q34. *See* TAP.

carnitine palmitoyl transferase deficiency (CPT I, 11q13) CPT I regulates the carnitin transfer through mitochondrial membranes, and its deficiency may be lethal, but it can be effectively treated by medium-chain triglycerides.

carnosinemia Carnosine is a neurotransmitter dipeptide of β-alanine and histidine. Enzymatically it may be split by carnosinase into two components, or by the action of a methyltransferase it may be converted into anserine. An autosomal-recessive defect in carnosinase may lead to the excretion of carnosine, anserine, and homocarnosine and neurological disorders. *See* anserine; neuromuscular diseases.

Caroli disease   *See* polycystic kidney disease.

carotene, β   Widely considered as an anticarcinogen, but recent studies indicate that it enhances the activity of several cytochrome (CYP)-activating enzymes. Large-scale epidemiological studies indicate increase in cancer caused by a variety of carcinogens. *See* activation of mutagens; carotenoids; cytochromes. (Paolini, M., et al. 1999. *Nature* 398:760.)

carotenoids   Accessory light-absorbing pigments of yellow, red, or purple, including carotene and xanthophylls. (See Park, H., et al. 2002. *Plant Cell* 14:321, Isaacson, T., et al. 2002. *Plant Cell* 14:333.)

carp (*Cyprinus carpio*) $2n = 100–104$. $2n = 98$, genome size $bp/n = 1.7 \times 10^9$.

carpel   Floral leaf forming the (enclosure) site for the ovules. *See* flower differentiation; gynoecium; ovule.

Carpels surrounding iris seeds.

carrier   Human heterozygote for a recessive gene that is not expressed in that individual but may be transmitted to the progeny. *Obligate carrier* is identified on the basis of family history; the natural parents of a homozygous-recessive individual must be carriers unless a rare mutational event happened in the heterozygous embryo after fertilization or the natural father is not identical with the legal one.

The posterior probability that chromosome A is the carrier of the mutation in question is

$$R = \frac{P(Ma|A = D)}{P(Ma|A \neq D)} x \frac{P(Mb|B = D)}{P(Mb|B \neq D)}$$

$R/(R + 1)$ (see R at left), and Ma and Mb represent the marker information for chromosomes A and B, respectively. D stands for the mutation-bearing chromosome, so $P(Ma|A = D)$ would indicate the probability that chromosome A would be carrying the mutation and $P(Mb|B = D)$ would be the same for chromosome B. Identification of carriers of human diseases using molecular, enzymological, cytological, or other techniques may be highly desirable for early treatment of various human diseases. Genetic counseling may take advantage of the information of the carrier status of the prospective parent(s). The knowledge of carrier frequencies may allow predictions about the occurrence of genetic diseases

and may affect governmental health care as well as insurance policies. *See* Bayes theorem; counseling, genetic; heterozygote; microarray analysis. (Watts, Y. H., et al. 2002. *Ann. J. Hum. Genet.* 71:791.)

carrier DNA   Nonspecific DNA that may be mixed with specific DNA to facilitate transformation or other manipulations. *See* transformation.

carrier protein   Transports solutes through membranes while its conformation is altered. Some carrier proteins transport a single type of molecule (uniporter); others carry more in the same (symporter) or opposite directions (antiporter).

carrot (*Daucus carota*)   All cultivated forms are $2n = 2x = 18$.

carrying capacity   In a particular environment only a certain number of species or individuals of a species can survive. Their survival depends on their genetic adaptation. The carrying capacity of a vector is the size of DNA that it can accommodate. *See* ecogenetics; vectors.

CART   Cocaine- and amphetamine-regulated transcript is a leptin-dependent molecule that suppresses craving for food and antagonizes the feeding stimulatory neuropeptide Y. *See* leptin; neuropeptide Y.

Carter-Falconer mapping function   Based on the assumption that there is substantial positive interference along the length of the chromosome: map distance = $0.25\{0.5[\ln(1 + 2r) − \ln(1 − 2r)] + \tan^{-1}(2r)\}$, where r = the observed recombination fraction, ln = natural logarithm, tan = tangent. *See* Haldane's mapping function; Kosambi's mapping function; mapping function. (Carter, T. C. & Falconer, D. S. 1950. *J. Genet.* 50:307.)

cartilage   Fibrous connective tissue; also it may be converted into bone tissue during postembryonic development. It is rich in collagen and chondroitin sulfate. Ror2 receptor-like tyrosine kinase deficiency leads to skeletal abnormalities and brachydactyly. *See* brachydactyly; chondrocyte; collagen.

cartilage-hair dysplasia   *See* chondrodysplasia, McKusick type.

cartilage-hair hypoplasia (CHH)   Dwarfism, short fingernails, cartilage deficiency, and weak immune system due to a chromosome 9 defect. The condition is common among the Lancaster County Amish and some Finnish populations. *See* Amish.

caruncle   Small outgrowth on animal and plant tissues.

caryonide   Clonal derivative of a cell, which after conjugation retains the original macronucleus in ciliates, and all macronuclei in the subclones are derived from a single macronucleus. *See Paramecium.*

caryopsis "Seed" (kernel) of some monocots containing the single embryo and endosperm and also tissues derived from the fruit (pericarp).

Caryopsis.

CAS (1) Chemical Abstract Service Registry that identifies chemicals by specific numbers. (2) (Crcas). ≈130 kDA human microtubule-associated protein involved in the control of chromosome segregation, cell adhesion, cell migration, growth factor stimulation, cytokine receptor engagement, bacterial infection, actin stress fiber formation, and Src oncogene-induced transformation. *See* microtubule; spindle fibers. (3) (Cse 1p). Transport factor, a nuclear export receptor of α-importin. *See* importin; karyopherin; nuclear pore.

casamino acids Hydrochloric acid hydrolysate of casein, containing amino acids (except tryptophan, which is destroyed by the process). Total nitrogen content 8 to 10%; NaCl, 14 to 38%.

cascade Sequence of events depending on specific consecutive steps such as feedback control, signal transduction, and differentiation.

cascade hybridization Procedure for enriching a certain fraction of the DNA transcribed at a particular developmental stage. Total cDNA is hybridized in a cascade of events with 20, 50, and 100 excess amounts of mRNAs synthesized at stage 2. The hybrid is then adsorbed to hydroxyapatite column. Unbound molecules are passed through. The unbound emanate is then hybridized with 100 excess of stage 1 mRNAs. Thus, cDNA transcribed only in stage 1 is much enriched. *See* cDNA; hydroxyapatite; mRNA; subtractive cloning. (Timberlake, W. E. 1980. *Dev. Biol.* 78:497.)

cascade testing Basically a genetic screen for carriers in a population beginning with the natural parents of a proband, then extended to relatives. Such a test has a much higher chance to locate carriers than a random one, and it is more economical. *See* carrier; genetic screening; proband. (Krawczak, M., et al. 2001. *Am. J. Hum. Genet.* 69:361.)

CASE Computer-automated structure evaluation relates the correlation between chemical structure and biological activity of chemicals. *See* biophore; MULTICASE; SAR.

case-control design Often used in human genetics when the genetic determination of polygenic disease condition is compared to unaffected controls. Such studies on gene frequencies are frequently loaded by errors because of differences in ethnicity. Inferences based on sibs provide better information. *See* genomic control; sib. (Wilson, J. F. & Goldstein, D. B. 2000. *Am. J. Hum. Genet.* 67:926.)

CASH *See* apoptosis.

caspase Class of 14 cysteine-dependent aspartate-directed heterotetramer proteases that regulate apoptosis and oogenesis. Caspases are activated by different mechanisms. When a virus infects an organism, the natural killer cells (NK) secrete to the surface of the infected cell perforin, which allows (among other proteins) the entrance of granzymes into the infected cells. Granzyme B then processes procaspases into active caspases, and these mediate the suicidal process of apoptosis. The binding of the Fas ligand (FASL) to FAS receptors and the recruitment of procaspases can activate caspases and subsequently lead to apoptotic suicide. Caspase-8/Mach (2q33) initiates apoptotic death signals downstream of the death receptors located on the plasma membrane through effector caspases 6 (4q25), 3 (11q22), and 7 (10q25).

Cellular/mitochondrial damage may lead to the release of cytochrome c into the cytosol, resulting in the activation of procaspase-9 (1p34) and eventually apoptosis. The effector caspases then target the CAD–Inhibitor-CAD (I^{CAD}) complex and release CAD (a DNase) from the cytoplasm into the nucleus. Mutation in caspase-10/Flice (2q33) may lead to

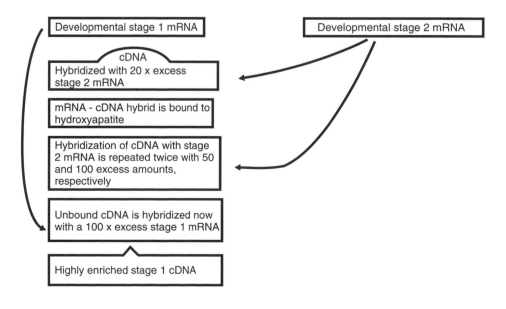

Cascade hybridization.

autoimmune lymphoproliferative syndrome, ALPS. Caspase-12 is activated by stress of the endoplasmic reticulum and may contribute to neurotoxicity by amyloid-β. Caspase 2 (12q21.33-q23.1) has homology to ICH1. Activation of caspase may have a therapeutic effect by slowing down or preventing the accumulation of the huntingtin protein, and it is conceivable that the production of Aβ in Alzheimer disease is proportional to the cleavage activity of caspase(s). There is no agreement among research workers whether the mitochondria would have the leading role in the initiation of apoptosis or if the mitochondrial events are only part of the apoptosis process already underway. Caspase-mediated cleavage of inhibitor proteins may lead to gain in function. Cysteine proteases may be required for infection of plants by potyviruses, and their inhibitors may confer resistance to tobacco etch virus and potato virus Y. *See* AKT; ALPS; Alzheimer disease; Apaf; apopain; apoptosis; Bcl; BIR; CAD; CARD; cytochromes; DIABLO; Egl; FAS; granzyme; Huntington's chorea; IAP; ICAD; ICE; ICH; killer cells; oogenesis; perforin; RGD; scaffold-mediated activation; Smac. (Earnshaw, W. C., et al. 1999. *Annu. Rev. Biochem.* 68:383; Goodsell, D. S. 2000. *Stem Cells* 18:457; Goyal, L. 2001. *Cell* 104:805.)

Casper (Flip)   *See* apoptosis; FLIP.

cassava (*Manihot esculenta*)   Belongs to the Euphorbiaceae. Originally an American perennial shrub and an important source of carbohydrates. The 98 species all have $2n = 36$ chromosomes.

cassette   *See* vector cassette.

cassette mutagenesis   Synthetic DNA fragment is used to replace a short sequence of DNA and thus alters the genetic information at a site. *See* localized mutagenesis; mutation induction; TAB mutagenesis; targeting genes. (Cho, S. W., et al. 2001. *Eur. J. Biochem* 268:3205.)

cassettes model of sex expression   Explains switches of the mating type in homothallic (*HO*) yeast by the transposition of either one or the other (*a* or *α*) silent, distant elements (cassettes) to the sex locus *MAT* where they can be expressed: *See* mating type determination in yeast.

caste   Specialized group within an insect society, e.g., workers among bees. The social and ethnic complexities of the castes of the Indian subcontinent are very difficult to define. Although intermarriage among the castes was limited, some ethnic mixing diluted the cultural isolation. During the 1500–1200 B.C. period, several Aryan groups invaded the subcontinent and some of the descendants of the invaders, the Brahmins, retained their elite, priestly status. The Brahmins were less bound by the strict rules of marriage than the rest of the castes, i.e., they could take wives from outside their caste. For their origin and evolution, mainly legal and legendary evidence existed. Recent DNA studies based on mitochondrial and Y-chromosomal evidence indicate that the only egg-transmitted DNA displays the same variations as the average of the populations. The Y-chromosomal DNA is apparently more similar to the European human races. These data confirm the earlier anthropological views about their origin and also prove that women had chances for upward mobility, whereas the male lineages were more strictly controlled. *See* Eve, foremother of mitochondrial DNA; Y chromosome. (Majumder, P. P. 2001. *Genome Res.* 11:931; Bamshad, M., et al. 2001. *Genome Res.* 11:994.)

castor bean (*Ricinus communis*)   Oil crop and ornamental plant adapted to a wide range of climates. The viscosity of its oil is rather constant at various temperatures and thus a good lubricant for high-speed engines. It is a very potent irritant and laxative and may be used for medical purposes. The plant contains the lectin ricin, which is one of the most toxic compounds known; 2–5 mg/kg is lethal for humans; in mice the intraperitonial minimal lethal dose is 0.001 µg ricin nitrogen/g body weight. All species are $2n = 20$.

castration   Surgical removal of the gonads or chemical prevention of testosterone production to prevent reproduction and reduce sex drive. A male deprived of testes is called a eunoch. The surgical removal of the female gonads is spaying (oophorectomy). *See* androgen; estrogen; eunoch; ovariectomy. (Kolvenbag, G. J., et al. 2001. *Urology* 58 (2 Suppl.): 16; Muss, H. B. 2001. *Semin. Oncol.* 28:313.)

CAT   (1) Chloramphenicol acetyltransferase conveys resistance to the antibiotic chloramphenicol. (2) *Felis catus*: the domesticated cat, $2n = 38$; the majority of wild cats have $2n = 38$ or 36. Cats were domesticated about 10,000 years ago. Among more than 30 distinct breeds, about 200 genetic diseases are known, and many of them are homologous to human hereditary diseases. (Gold, L. 1996. *Cats Are Not Peas: A Calico History of Genetics*. Copernicus.)

catabolism   Degradative metabolism of chemical substances in the cells, mediated by enzymes, most commonly for energy utilization.

catabolite activator protein (CAP)   *See* CAP.

catabolite repression   Carbohydrate (glucose) represses the synthesis of enzymes involved in carbohydrate metabolism or other metabolic steps through decreasing the level of cyclic AMP. In some instances, the glucose repression does not depend on cAMP because cAMP does not relieve its repression (e.g., pyrroline dehydrogenase, putrescine aminotransferase). Some bacteria fail to synthesize sufficient amounts of cAMP (*Bacillus megaterium*) yet display intense glucose effect. Catabolite repression may be strongly modulated by the catabolite modulation protein factor (CMF). The catabolite repressor–cAMP system may also function as a gene activator. Catabolite repression may switch to activation in case the gene carries overlapping promoters or when divergent dual promoters exist. Some organisms (e.g., yeast) can discriminate among different nitrogen sources (and turn off the utilization of the less desirable ones) by a mechanism called nitrogen catabolite repression. *See* cAMP; cytosine repressor (CytR); DNA bending; feedback controls; FNR, glucose effect; *Lac* operon; promoter. (Gancedo, J. M. 1998. *Microbiol. Mol. Biol. Rev.* 62:1092; Stülke, J. & Hillen, W. 2000. *Annu. Rev. Microbiol.* 54:849.)

catalase (CAT)   Enzyme that mediates the reaction $2H_2O_2$ (hydrogen peroxide) $\rightarrow 2H_2O + O_2$. CAT is encoded in human chromosome 11p13. In plants the catalase enzymes are essential under conditions of photorespiration and for lipid metabolism. *See* peroxide.

catalysis   Mediates chemical reactions without being used up in the process. Enzymes are biocatalysts. *See* enzymes.

catalytic antibody   Contains catalytic and selective binding sites in one molecule and performs enzymatic reactions as enzyme mimics. Their catalytic efficiency is generally suboptimal; however, the range of the specificity may be broader than the regular enzymes. *See* abzyme. (Hilvert, D. 2000. *Annu. Rev. Biochem.* 69:751, Wentworth, P., Jr. 2002. *Science* 296:2247.)

catalytic RNA   *See* ribozymes.

catalytic site   *See* active site.

cataplexy   Muscular defect evoked by emotional effects. It is usually associated with narcolepsy. *See* narcolepsy.

cataract   Disease of the eyes causing opacity of the lens. Cataracts are under X-chromosomal (Xp), autosomal-dominant (connexin, 1q21-q25; crystallin, 2q33-q35; developmental regulator *PITX3*, 10q25, 16q22.1, 17q24, 13q11-q12), and autosomal-recessive (galactosemia [17q24], chondrodysplasia punctata, Zellweger syndrome [7q11, 12q11-q13]) control. The normal eye displays no opacity, whereas the cataract appears in various shapes and distribution of opacity on the lens. In a juvenile-onset hereditary cataract, a single amino acid substitution (Arg-14→Cys) in the γ-crystallin protein was responsible for the progressive opacity. The dominant disorder MIP at 12q14 represents a membrane-bound aquaporin protein. An autosomal-dominant cataract has been assigned to 15q21-q22. A recessive locus at 9q13-q22 is responsible for a progressive adult cataract. *See* aquaporin; ASMD; chondrodysplasia; connexin; eye diseases;

Lowe disease; microbody; Wilms' tumor. (Héon, E., et al. 2001. *Am. J. Hum. Genet.* 68:772.)

Normal   Cataract

catastroph, mitotic   Disruption of nuclear division by chemical, physical, or other agents.

catatonia, periodic   Form of schizophrenia with psychomotor disturbances controlled by a major factor at 15q15. *See* schizophrenia. (Stöber, G., et al. 2000. *Am. J. Hum. Genet.* 67:1201.)

CATCH   Cardiac defect, abnormal face, thymic hypoplasia, cleft palate, and hypocalcemia are symptoms associated with a deletion in human chromosome 22q11.2. *See* cleft palate; DiGeorge syndrome; hypocalcemia; velocardiofacial syndrome.

cat-cry syndrome   *See* cri du chat.

catecholamine   Neurotransmitter. *See* neurotransmitters.

catenanes   Interlocked DNA circles.

Catenanes.

catenated   Attached like two links of a chain.

catenin   Intracellular attachment protein connecting the inward-reaching carboxyl end of cadherins to actin filaments within the cell; β-catenin is involved in the cell proliferation pathway and the development of adenomatous polyposis of the colon, breast cancer, and other tumors; its degradation predisposes to Alzheimer disease. β-catenin activates the transcription of cyclin D1 and the promoters, which display TCF/LEF-binding sites. p21[ras] also activates cyclin D1 at promoters with Ets and CREB sites. In adenomatous polyposis cells, the degradation of β-catenin is reduced. A dominant negative TCF is inhibitory to the expression of cyclin D1. Overproduction of β-catenin increases hair growth in mice. *See* actin; *Armadillo*; cadherin; conductin; CREB; cyclin D; ETS; Gardner syndrome; LEF; Tcf, p21[ras]; polyposis adenomatous, intestinal; presenilin; *wingless*. (Huber, A. H. & Weis, W. I. 2001. *Cell* 105:391.)

cat-eye syndrome   The pupil appears vertical because a deformity of the iris, heart anomaly, imperforate anus, and various degrees of mental retardation occur. The anomaly is apparently caused by the presence of an extra copy of a very short metacentric chromosome 22 (partial trisomy or tetrasomy). The symptoms vary depending on the structure of this extra chromosome. *See* coloboma; duplication; eye disease; mental retardation; tetrasomy; trisomy.

catfish (*Ictalurus punctatus*) One of the economically most important freshwater fish species. (Waldbieser, G. C., et al. 2001. *Genetics* 158:727.)

cathepsin (CTS) Intracellular cystein proteases generally relegated to the lysosomes. In humans, CTSK and CTSO are encoded at 1q21; in mouse four cathepsins are in as many different chromosomes. In mice, deficiency in cathepsin C may not affect health, but their cytotoxic lymphocytes (CTL) are inactive because granzymes A and B are not processed. Cathepsins may be involved in processing of keratins. Secreted cathepsin L may generate endostatin from collagen XVIII. Mutation in cathepsin K may underlay osteopetrosis or osteosclerosis. It is substantially expressed in some breast cancers and suspected to favor metastasis. *See* antimicrobial peptides; collagen; endostatin; granzyme; keratin; lysosomes; periodontitis; pycnodysostosis; Toulouse-Lautrec. (McGrawth, M. E. 1999. *Annu. Rev. Biophys. Biomol. Struct.* 28:181.)

cathode ray Electromagnetic radiation emitted by the cathode toward an anode in a vacuum tube. After exposure of a metal target to this radiation, X-rays are generated. *See* electromagnetic radiation; X-rays.

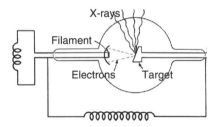

cation Positively charged atom or radical. *See* electrolyte; ion.

cation exchange Replacement of one positive ion by another on a negatively charged surface.

cation-π interaction Plays a role in determining protein structure. It is more likely that there would be an interaction when a cationic side chain of lysine or arginine is close to an aromatic amino acid (phenylalanine, tyrosine, or tryptophan) than with a neutral amine. Arg and Trp are most likely to be involved. *See* protein structure.

cationic amino acid *See* Cat transporters.

cationic lipid *See* cytofectin GS2888; lipid, cationic; lipofection; liposome.

cationic liposome *See* liposome.

catkin Male inflorescence of some plants.

Oak catkin.

CATR1 Genetic element encoding 79 amino acids and a long untranslated region, expressed in malignant tumors. This element does not have homology with other oncogenes or tumor suppressor genes. It has been localized to human chromosome 7q31-q32. *See* AIG; cancer; tumor oncogenes; tumor suppressor. Li, D., et al. 1995. *Proc. Natl. Acad. Sci. USA* 92:6409.)

CATS Comparative anchor tagged sequences. *See* anchoring; comparative maps.

CAT scan *See* tomography.

cattalo Hybrid of buffalo (2*n* = 60) and cattle (2*n* = 60) with reduced male fertility.

Cattanach translocation Involves the X chromosome and autosome 7 of mouse. It includes several X-chromosomal fur color genes that may be subject to lyonization. *See* lyonization; translocation. (Cattanach, B. M. 1975. *Annu. Rev. Genet.* 9:1.)

cattle (*Bos taurus*) 2*n* = 60. Genetic evidence indicates that the European breeds originated from Near East herds, rather than the now extinct European wild ox (*Bos primigenius*). The domesticated African cattle seem different from other domesticated breeds on the basis of mitochondrial DNA. (Troy, C. S., et al. 2001. *Nature* 410:1088; Hanotte, O., et al. 2002 Science 296:336, <http://www.tigr.org/tdb/tgi.shtml>.)

CAT transporter Transports through membranes of cationic amino acids (arginine, lysine, ornithine, histidine). *See* ABC transporters; translocon; transporters. (Vékony, N., et al. 2001. *Biochemistry* 40:12387.)

Caucasoid (Caucasian) Commonly used misnomer for ethnic groups of European, North African, Near Eastern, and Indian descent. When the term was coined erroneously, anthropologists believed that the origin was in the Caucasus area. It is sometimes used as a collective term for white-skinned people. In fact, the group includes many brown-skinned populations.

caudal Tail oriented.

caudate structure Tail-like appendage.

caulescent Leaves are born separated by visible internodes on a stem. *See* rosette.

Caulescent.

cauliflower mosaic virus (CaMV) ~8 kbp double-stranded DNA virus with limited potential use for genetic engineering. It infects cruciferous plants such as turnip, broccoli, *Arabidopsis*, etc. The DNA of the virus shows three discontinuities ("gaps"), one in one of the strands and two in the other. The promoter of the 35S transcript of the virus drives high-level constitutive expression of genes spliced to it and it is widely used; the promoter of the 19S peptide gene has not been proven nearly as useful for biotechnology. *See* activation tagging; agroinfection; retroid virus. (Balázs, E., et al. 1985. *Gene* 40:343; Hirth, L. 1986. *Microbiol. Sci.* 3:260; Hohn, T. & Futterer, J. 1992. *Curr. Opin. Genet. Dev.* 2:90; al-Kaff, N. & Covey, S. N. 1994. *J. Gen. Virol.* 75 [Pt. 11]:3137.)

Caulobacter Dimorphic gram-negative group of prokaryotes. The cells may have either a long flagellum or a thicker stalk of about two-thirds of the 2 μm length of the cell. The formation of these appendages has been extensively studied and the hierarchy of genes regulates them. *Caulobacter crescentus* genome of 4,016,942 bp encodes ~3,767 genes. *See* gram negative/gram positive. (Nierman, W. C., et al. 2001. *Proc. Natl. Acad. Sci. USA* 98:4136.)

caulonema In the gametophyte of some mosses a special type of cells are formed subapically and separated by oblique walls.

caveolae (plasmalemmal vesicles) ~70-nm-diameter flask- or Ω-shaped detergent-insoluble glycolipid vehicles/rafts moved by actin motors on the surface of T cells and other endothelial cells. The caveolae are loaded with kinases, protectin/CD59, decay-accelerating factor (DAF), alkaline phosphatase, Thy-1 glycoprotein, and signaling adapter molecules. After DC28 engages the T-cell receptor, the rafts are attracted to the area where the antibody contacts the TCR. The caveolae concentrate lipids, proteins, prenylated proteins, glycosylphospatidyl inositols, various membrane receptors, actin, myosin, ezerin, NSF, signal transducing proteins, etc. Caveolae have a role in the transmission of the Pr^Sc prions, the transfer into the cells pathogens (SV40, *Campylobacter*), and protein toxins of pathogens (cholera, *Plasmodium*, *Trypanosoma*, *Leishmania* toxins). Caveolae may play a role in cardiovascular diseases because they transport cholesterol and lipoproteins and affect blood clotting. *See* antigen-presenting cell; CD48; DAF; DC28; flotillin; potocytosis; prion; protectin; raft; signal transduction; TCR; Thy-1. (Shin, J.-S. & Abraham, S. N. 2001. *Science* 293:1447; Galbiati, F., et al. 2001. *Cell* 106:403.)

caveolin (CAV) CAV-3 (300–350 kDa), encoded at human chromosome 3p25 as homooligomers consisting of 12–14 monomers, associated with the sarcolemma, and colocalized with dystrophin. The same RNA encodes CAV-1 (21–24 kDa) and CAV-2, but it is processed differently. Caveolins are integral membrane proteins and bind G proteins. Caveolin-1 is a negative regulator of caveolae-mediated endocytosis to the endoplasmic reticulum. *See* dystrophin; muscular dystrophy; prostate cancer; sarcolemma. (Engelman, J. A., et al. 1998. *Am. J. Hum. Genet.* 63:1578; Anderson, R. G. W. 1998. *Annu. Rev. Biochem.* 67:199, Pol, A., et al. 2001. *J. Cell Biol.* 152:1057; Le, P. U., et al. 2002. *J. Biol. Chem.* 277:3371.)

cavitation After the formation of the 32-cell stage of the mammalian embryo, a fluid secretion occurs during the blastocyte stage that initially accumulates between the cells, then is collected in the blastocoele. *See* blastocoele; blastocyte.

C banding Type of banding pattern near the centromere (and some other limited areas, such as telomeres) obtained after staining chromosomes with Giemsa stain (a mixture of basic dyes), particularly when the chromosomes were exposed to the protease trypsin prior to staining. *See* chromosome banding; G banding; R banding. (Chen, T. R. & Ruddle, F. H. 1971. *Chromosoma* 34:51.)

CBAVD *See* congenital bilateral aplasia of the vas deferens.

CBC *See* Cap-binding protein complex.

CBF/NF-Y Trimeric CCAAT-binding (factor) protein, which with other proteins facilitates transcription. *See* CAAT. (Maity, S. N. & de Crombrugghe, B. 1998. *Trends Biochem. Sci.* 23:174.)

C57BL Strains (B6 and B10) of black inbred mouse commonly used for genetic studies.

CBL2 (oncogene) Three cellular homologues of this viral oncogene are expressed in mammalian hematopoietic (blood cell–forming) systems. In humans, CBL2 was located to chromosome 11q23.3. Its translocations to chromosome 4 are associated with acute leukemia and B-cell lymphoma. The oncogene is present in the ecotropic Cas-Br-M virus; the form present in lymphomas is a recombinant between the virus and the cellular oncogene. The 100 kDa transforming fusion protein has sequence homology to the yeast transcription factor GCN4 and to *sli*-1 regulator of vulval development in *Caenorhabditis*. Apparently, this family of genes modifies receptor tyrosine kinase–mediated signal transduction. Cbl-b controls the dependence of T-cell activation by CD28. *See* CD28; Jacobsen syndrome; oncogenes; signal transduction; SLAP; Src; Syk. (Thien, C. B. F. & Langdon, W. Y. 2001. *Nature Rev. Mol. Cell Biol.* 2:294.)

CBP (CREB-binding protein, p300) Associated with CREB and mediates the induction of some promoters by cAMP. It interacts with p300 (or may be the same), the nuclear hormone receptors, and the basic transcription machinery. In addition, it interacts with a range of transcriptional activators and coprecipitates with RNA polymerase holoenzyme. CBP has a signal-regulated transcriptional activation domain, regulated by Ca^{2+} and calmodulin-dependent protein kinase and cAMP. With histone acetyltransferase activity, CBP may function as a coactivator of p53. The loss of CBP may be the basis of Rubinstein-Taybi syndrome. *See* calmodulin; cAMP; cap; CREB; E1A; histone acetyltransferase; p53; p300; Rubinstein syndrome; transcriptional activators; transcription factors. (Mayr, B. M., et al. 2001. *Proc. Natl. Acad. Sci. USA* 98:10936.)

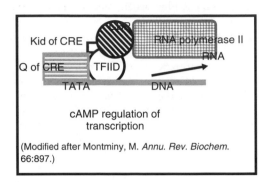

Kid of CRE

Q of CRE   TFIID   RNA polymerase II   RNA

TATA   DNA

cAMP regulation of
transcription

(Modified after Montminy, M. *Annu. Rev. Biochem.*
66:897.)

CBP2   Cellular-binding protein required for splicing class I mitochondrial introns. The catalytic RNA domains must be in a folded form before CBP2 binds to them. Subsequently, other proteins are attached to the 5′ domain. In this ribozyme the catalytic component remains the RNA, but the protein associated with it increases the splicing rate by three orders of magnitude. *See* intron; mtDNA; ribozyme; spliceosome. (Shaw, L. C. & Lewin, A. S. 1997. *Nucleic Acids Res.* 25:1597.)

CBP-II   Similar to eIF-4F translation factor. *See* eIF-4F.

CbpA   Protein of *E. coli* similar in function and structure to the DnaJ co-chaperone, but its transcription is initiated with stationary phase σ^{38} subunit of the RNA polymerase. *See* co-chaperone; DnaJ; Hsp70. (Ueguchi, C., et al. 1995. *J. Bacteriol.* 177:3894.)

Cbs   Chromosome breakage sequence is a 15 bp nucleotide sequence at 50 to 200 sites in the micronuclear DNA of ciliates near the breakage sites where the genome rearrangement occurs in the macronuclear chromosomes. *See Tetrahymena*. (Fan, Q. & Yao, M. C. 2000. *Nucleic Acids Res.* 28:895.)

CBS (conserved sequence elements)   Three short sequences in the mtDNA where the transition from DNA to RNA synthesis takes place. *See* DNA replication; mitochondria; mtDNA.

CC57   Very popular black inbred mouse line; low spontaneous mammary tumors but highly prone to lung tumors. A large number of variants exist.

CCAAT   *See* CAAT box.

CCCH zinc finger proteins   Are mRNA inhibitors and antiviral agents. They contain CCCH amino acids (3 cysteine, 1 histidine) in the tandem Zinc-finger domain. *See* DNA binding protein domains. (Lai, W. S., et al. 2002. *J. Biol. Chem.* 277:9606.)

CC CKR5   Chemokine receptor. *See* CCR; chemokines; CXCR; RANTES. (Combadiere, C., et al. 1996. *J. Leukoc. Biol.* 60:147.)

CCE   Cell cycle element has 11 bp that bind the histone nuclear factor (HiNF-M), required for the activation of histone 4. *See* HiNF; IFN; immune surveillance; IRF-2.

CCG-1   Hamster gene responsible for G1 phase cell cycle arrest.

CCL39   Hamster lung fibroblast cell line.

CCR   Chemokine receptors; same as CKR. CCR2 is the receptor for MCP-1 and related chemokines MCP-3, MCP-4, MCP-5. $CCR7^+$ cells express lymph node homing receptors without having immediate effector function. $CCR7^-$ possesses effector function, and they differentiate from $CCR7^+$ cells upon activation by cytokines and antigens. *See* acquired immunodeficiency syndrome; chemokine; CKR5; CXCR; effector; MCP-1; metastasis. (Luther, S. A. & Cyster, J. G. 2001. *Nature Immunol.* 2:102.)

CCT   Cytosolic chaperonins. *See* Bin2p; Bin3p; chaperonins; TCP20; TRiC.

CCW   Counterclockwise.

Counterclockwise

CD1   Antigen-presenting molecules — distantly related or unrelated to MHC proteins — for the T lymphocytes. CD1 can deliver endosomal lipoglycan antigens to T cells and can deliver other nonpeptide ligands to T cells, which are then stimulated to produce γ-interferon and interleukin-4. CD1 is encoded in human chromosome 1q22-q23. *See* antigen-presenting cell; cytokines; HLA; interferon; major histocompatibility antigen; T cell. (Park, S-H. & Bendelac, A. 2000. *Nature* 407:788.)

CD2   T-cell-surface glycoprotein, mediating cell adhesion (to antigen-presenting cells) and the transduction of signals; it is encoded in human chromosome 1q22-q23. *See* CD48.

CD3   Immunoglobulins of γ (25 K), δ (20 K), ε (20 K), and ζ (16 K) molecular (weight) chains expressed on all T cells and assisting in transducing signals when the major histocompatibility-antigen complex binds to the surface. CD3δ-negative lymphocytes fail to induce ERK kinase and are unable to undergo positive selection. They are encoded in human chromosome 11q23. *See* ERK; HLA; immunoglobulin; LAT; signal transduction; T-cell receptor; tumor vaccination.

CD4   Cell surface protein (55 kDa) with domain binding major histocompatibility (MHC II) molecules on helper T cells. $CD4^+$ T cells induce clonal proliferation of B cells along two paths. Foreign antigens switch on through the B-cell antigen receptors. CD40L (CD40 ligand) and interleukin-4 (IL-4) promote the proliferation of B cells. CD95 (Fas) and Fas ligand (FasL) limit the access of mitogenic signals and mediate apoptosis in response to autoantigens. T cells are costimulated by CD80 (B7.1) or CD86 (B7.2) by binding to CD28 on the antigen-presenting cells (APC). For normal function, both CD40 and Fas functions must be maintained. CD4 is encoded in human chromosome 11q13. *See* APC; apoptosis; autoantigen; B cells; CD8; CD28; CD40; CD80; CD86; FAS; HLA; interleukins; LCK oncogene; MHC; T cells. (Groux. 2001. *Microbes Infect.* 3:883.)

CD5 Transmembrane protein on the surface of T cells and some B cells. It may be a negative regulator of the T-cell receptor (TCR)–mediated signal transduction. Encoded at human chromosome 11q13. *See* T-cell receptor. (Weston, K. M., et al. 2001. *J. Mol. Recogn.* 14:245.)

CD8 Homo- or heterodimer proteins (M_r 70 K) on cytotoxic T cells (CTL), binding class I major histocompatibility molecules. CD8 lymphocytes recognize the endogenously generated antigens, associated with MHC class I molecules, such as those formed in cancer cells. CD8A and CD8B are encoded in human chromosome 2p12. CD8 is an active player in the T-cell recognition complex. *See* CD4; HLA; LCK oncogene; MHC; T cell. (Fujii, S., et al. 2001. *Blood* 98:2143.)

CD9 Transmembrane protein of the tetraspanin family. It is a molecular facilitator. The eight exons of murine CD9 extends to 20 kb. Homozygous knockouts are physically normal in both males and females; however, only 50–60% of the females produced offspring. Some reduction in fertility was also observed in the homozygous males. The pup mortality was also higher. CD9-associated integrin $\alpha6\beta1$ is apparently required for the egg-sperm fusion. *See* CD81; CD82; fertilin; fertilization; integrin; knockout; tetraspanin. (Charrin, S., et al. 2001. *J. Biol. Chem.* 276:14329.)

CD11a/18 Binds CD54 (ICAM) and mediates cell adhesion. CD11 is α-integrin (human chromosome 16p12-p11); CD18 is β-integrin (human chromosome 21q22.3). CD11b/CD18 are the CR3 complement receptors binding C3b complement component. Salmonella infection is spread from the gastrointestinal tract to the bloodstream by CD18-expressing phagocytes. *See* complement; ICAM; integrin.

CD14 Glycosylphosphatidylinositol-anchored membrane protein sensing lipopolysaccharides of invading microbes as part of the defense mechanism. It enhances B-lymphocyte and monocyte differentiation and it is encoded at the human chromosome 5q23-q31 region. *See* innate immunity; *Toll*. (Glück, T., et al. 2001. *Eur. J. Med. Res.* 6:351.)

CD15 (α-fucosyltransferase, 11q21) Important for the function of leukocyte adhesion molecule LAD2. *See* leukocyte adhesion. (Nakayama, F., et al. 2001. *J. Biol. Chem.* 276:16100.)

CD16 (FcγRIII, immunoglobulin G [IgG] Fc receptor III) Neutrophil-specific antigen encoded at 1q23. It may be involved in cytokine signaling to lymphocytes. *See* antibody. (Paolini, R., et al. 2001. *Proc. Natl. Acad. USA* 98:9611.)

CD18 *See* CD11a/CD18.

CD19 (16p11.2) Important for the differentiation of B cells, it is a tyrosine kinase receptor. On mature B cells it is associated with CD81 and CD21. *See* B cell; BCP; genistein;

leukemia. (Uckun, F. M., et al. 1993. *J. Biol. Chem.* 268:21172; Otero, D. C., et al. 2001. *J. Biol. Chem.* 276:1474.)

CD21 C3dg/C3d complement receptor 2 (CR2), 150 kDa. CD21 provides signals for B-lymphocyte survival in the germinal center. *See* complement; germinal center; TAPA-1; TCL. (Cherekuri, A., et al. 2001. *J. Immunol.* 167:163.)

CD22 Associated with immunoglobulin of B-cell membranes. Tyrosine phosphorylated CD22 activates SHP protein tyrosine phosphatase, and this downregulates signaling through the immunoglobulin (Igμ). If CD22 is prevented from binding to μIg, the B cell may become 100-fold more receptive. CD22 is encoded at human chromosome 19q13.1. *See* B cell; lymphocytes; signal transduction. (van Rossenberg, S. M., et al. 2001. *J. Biol. Chem.* 276:12967.)

CD23 Low-affinity receptor of immunoglobulin E (IgE) playing a role in allergy. *See* allergy. (Kilmon, M. A., et al. 2001. *J. Immunol.* 167:3139.)

CD25 Interleukin-2 receptor α-chain encoded at human chromosome 10p15-p14. *See* interleukins. (Sutmuller, R. P., et al. 2001. *J. Exp. Med.* 194:823.)

CD26 (dipeptidyl-peptidase IV, DPP IV) Pluripotent exopeptidase expressed on the membrane of memory T cells, endothelial cells, and epithelial cells. It interferes with immune reactions. It processes chemokines. *See* chemokines; CXCR. (Collebaut, C., et al. 1993. *Science* 262:2045; Herrera, C., et al. 2001. *J. Biol. Chem.* 276:19532.)

CD27 TNF receptor glycoprotein encoded at 12p13; it regulates immune reactions. The CD27-ligand, CD70, is encoded at 19p13. *See* TNF; TRAF. (Jacquot, S., et al. 2001. *Int. Immunol.* 13:871.)

CD28 Homodimeric immunoglobulin (M_r 80 K), encoded in human chromosome 2q33-q34, present on the surface of helper T cells. The CD28/B7 and LFA-1 provide costimulation to T-cell activation. CD28 is also a specific activator of JNK or NF-κB in the presence of T-cell receptor (TCR). Costimulation initiates the active transport of protein and lipid domains to the area of the cell-to-cell contact, depending on myosin. *See* anergy; B cell; caveolae; Cbl; CD80; CD86; CTLA; ICOS; JNK; motor proteins; NF-κB; T cell; TCR. (Salomon, B. & Bluestone, J. A. 2001. *Annu. Rev. Immunol.* 19:225.)

CD30 Member of the tumor necrosis factor/nerve growth factor receptor family, also including TNF-R1, TNF-R2, CD40, CD27, Fas, etc. Its extracellular domain has cysteine-rich repeats. It interacts with TRAFs and indirectly with NF-κB. Deficiency for CD30 disarms CD8-positive T cells very aggressively, destroys pancreatic islets, and contributes to the autoimmune reaction that leads to diabetes in cooperation with other genes. CD30 is encoded at human chromosome 1p36. *See* allergy. (Hombach, A., et al. 2001. *Gene Ther.* 8:891.)

CD31 Inhibitory receptor of myeloid, platelet, endothelial, and some T cells, encoded at 17q23. (Balduini, C. L., et al. 2001. *Br. J. Haematol.* 114:951.)

CD33 (human chromosome 19q13.3) Inhibitory receptor of myeloid cells with sialic acid ligand. (Mingari, M. C., et al. 2001. *Immunol. Rev.* 181:260.)

CD34 Sialomucin-like adhesion protein expressed on 1–3% of the (CD34$^+$) bone marrow cells, and it is associated with hematopoietic function. It is encoded in human chromosome 1q32. *See* hematopoiesis; SDF; selectin. (Pratt, G., et al. 2001. *Br. J. Haematol.* 114:937.)

CD35 Receptor CR1 (190 kDa) of the C3b complement component encoded at human chromosome 1q32. *See* complement. (Klein, M. A., et al. 2001. *Nature Med.* 7:488.)

CD36 (Fat) Gene is in rat chromosome 4 and is responsible for the quantitative effects of diabetes Type 2, obesity, hyperlipidemia, essential hypertension, and platelet formation. The human gene was mapped to 7q11.2. The gene encodes fatty acid translocase. CD36 deficiency (common in some African and Asian human populations) may increase susceptibility to malaria. CD36 and CD51 regulate dendritic cells and thereby the immune response. *See* diabetes; hyperlipidemia; hypertension; immune response; malaria. (Urban, B. C., et al. 2001. *Proc. Natl. Acad. Sci. USA* 98:8750.)

CD38 (Okt10/p45) 45 kDa antigen, encoded in human chromosome 4p15, of acute lymphoblastic leukemia cells with activity similar to adenylyl ribosyl cyclase. *See* cyclic ADP-ribose; leukemia. (Cakir-Kiefer, C., et al. 2001. *Biochem. J.* 358 [pt2]:399.)

CD40 (TNFR/SF5) 48–50 kDa transmembrane glycoprotein (encoded at 20q12-q13.2) expressed on the surface of B cells; their interaction with CD40 ligands is a requisite for their activation by helper T cells. This system appears critical for the development of humoral immunity. CD40 signaling may prevent Fas-induced apoptosis of B cells by crosslinking the immunoglobulin M complex. It induces B-cell differentiation and Ig isotype switching and the expression of CD80. CD40 cytoplasmic tail interacts with CRAF1, a tumor necrosis receptor–associated protein. CD40 is also expressed in dendritic cells, activated macrophages, epithelial cells, and several tumors. In the absence of CD40, immunodeficiency arises and initiation of germinal centers suffers. CD40 is upregulated by IFN, IL-1, TNF, and possibly IL-6. CD40 signaling to lymphocytes is mediated by NF-κB through upregulation of the antiapoptotic proteins Bcℓ-x and Bfℓ-1. *See* adjuvant, immunological; Alzheimer disease; apoptosis; B7; BCL, B lymphocyte; CD40; CD80; CD154; CRAF; dendritic cell; Fas; germinal center; hyper-IgM syndrome; IFN; Igα; IL-1; IL-6; immunodeficiency; immunoglobulins; ligand; macrophage; NF-κB, T cells; TNF. (Tone, M., et al. 2001. *Proc. Natl. Acad. Sci. USA* 98:1751.)

CD40 ligand (gp39/CD40L/TBAM; Xq26-q27.2) Membrane-bound signaling molecule associated with CD40 transmembrane protein found on the lymphoid follicles of CD4$^+$ T lymphocytes. It may regulate adhesion and movement. Its defect is responsible for X-linked hyper-IgM immunodeficiency. *See* CD40; hyper-IgM syndrome; immunodeficiency; T cell. (Vidalain, P. O., et al. 2001. *J. Immunol.* 167:3765.)

CD43 Cell surface sialoglycoprotein of blood cells, encoded in human chromosome 16p11.2; it is deficient in Wiskott-Aldrich immunodeficiency. *See* Wiskott-Aldrich syndrome. (Bagriacik, E. U., et al. 2001. *Immunol. Cell Biol.* 79:303.)

CD44 Family of cell surface receptors involved in adhesion and movement. The cytokine osteopontin (Eta-1) is one of its ligands, activating chemotaxis but not cell aggregation. Hyaluronate (a carbohydrate ligand), on the other hand, affects growth. Thus, metastasis of cancer cells may be controlled by the state of CD44, encoded in human chromosome 11pter-p13. Sulfated CD44 stimulated by TNF-α results in binding of B leukocytes to inflammatory sites. *See* diabetes mellitus; lymphocytes; osteopontin; TNF. (Ponta, H., et al. 1998. *Int. J. Biochem. Cell Biol.* 30:299; Teder, P., et al. 2002. *Science* 296:155.)

CD45 (leukocyte common antigen) Protein tyrosine phosphatase (JAK phosphatase), a transmembrane glycoprotein, activated by antigens on the surface of red blood cells; regulates T and B lymphocytes. Antibodies that react with this leukocyte antigen may prevent rejection of allografts. It is encoded in human chromosome 1q31. In some families with multiple sclerosis, CD45 is altered. *See* allograft; antibody; lymphocytes; multiple sclerosis. (Virts, E. L. & Raschke, W. C. 2001. *J. Biol. Chem.* 276:19913.)

CD46 Complement regulatory protein encoded at human chromosome 1q32. *See* antibody; complement; MCP. (Evlashev, A., et al. 2001. *J. Gen. Virol.* 82 [pt9]:2125.) Marie, J. C., et al. 2002. *Nature Immunol.* 3:659.)

CD47 Integrin-associated protein serving as a marker of self-identity on red blood cells. *See* integrin. (Blazar, B. R., et al. 2001. *J. Exp. Med.* 194:541.)

CD48 Membrane (glycosylphosphatidylinositol-associated) protein (1q21.3-q22) cooperating with CD2 in the interaction of the antigen-presenting cell with the T-cell receptor. Uptake of bacteria by mast cells may depend on CD48 caveolae. *See* antigen-presenting cell; caveolae; CD2; T-cell receptor. (Veréb, G., et al. 2000. *Proc. Natl. Acad. Sci. USA* 97:6013.)

CD54 (19p13.3-p13.2) *See* ICAM.

CD55 *See* decay-accelerating factor.

CD59 *See* protectin.

CD63 Transmembrane protein of the tetraspanin family involved in suppression of metastasis. *See* tetraspanin. (Ryu, F., et al. 2000. *Cell Struct. Func.* 25:317.)

CD64 (FcγRI, 1q21.2-q21.3) Receptor 1 of the antibody G Fc domain. *See* antibody. (Shen, L., et al. 1987. *J. Immunol.* 139:534.)

CD66 (19q13.2) Inhibitory receptors on granulocytes and some lymphocytes (T, B, NK). (Nair, K. S. & Zingde, S. M. 2001. *Cell. Immunol.* 208:96.)

CD70 *See* CD27.

CD80 Member of the interleukin family of proteins encoded at 3q21; same as B7.1 and BB-1. *See* interleukins, anergy; antigen-presenting cells; B7; CD4; CD28; CD40; costimulator. (Hattori, H., et al. 2001. *Clin. Exp. Allergy* 31:1242.)

CD81 (target of antiproliferative antibody/TAPA) Tetraspanin cooperating with CD9. *See* CD9; TAPA; tetraspanin. (Charrin, S., et al. 2001. *J. Biol. Chem.* 276:14329.)

CD82 Transmembrane protein of the tetraspanin family involved in suppression of metastasis and viral infection. *See* CD9; tetraspanin. (Pique, C., et al. 2000. *Virology* 276:455.)

CD85 *See* ILT.

CD86 (B7.2) T-lymphocyte costimulatory molecule (encoded at 3q21), binding to the CD28 receptor of antigen-presenting cells. *See* anergy; antigen-presenting cell; CD28. (Flo, J., et al. 2001. *Cell. Immunol.* 24:156.)

CD95 *See* Fas.

CD98 Regulator protein of integrin-mediated cell adhesion and transport (encoded at 11q13); it indicates T-cell activation. *See* integrin; lymphocytes. (Suga, K., et al. 2001. *FEBS Lett.* 489:249.)

CD117 Receptor for stem cell factors in thymic lymphocyte precursors. *See* steel factor; stem cell factor; T cells. (Shimizu, M., et al. 2001. *Exp. Cell Res.* 266:311.)

CD120 (12p3) TNF family ligand recognizing the TWEAK receptors. *See* TNF; TWEAK.

CD137 It stimulates the response of CD8$^+$ T cells to viruses. *See* T cell. (Halstead, E. S., et al. 2002. *Nature Immunol.* 3:536.)

CD146 (S-Endo 1 Ag, Mel-Cam, MUC18) Primarily endothelial integral membrane protein, but it also occurs in nonmalignant and malignant (melanoma) cells. (Anfosso, F., et al. 2001. *J. Biol. Chem.* 276:1564.)

CD152 *See* CTLA.

CD154 (M$_r$ ∼ 39 K) CD40 ligand; it is an immunological adjuvant. A member of the TNF receptors encoded at 20q12-q13.2. Antibodies blocking CD154 interfere with the rejection of tissue transplants in monkeys. *See* adjuvant, immunological; CD40. (Pierson, R. N., et al. 2001. *Immunol. Res.* 23:253.)

CD200 (OX2) Membrane glycoprotein negatively regulating macrophages. *See* macrophage. (Clark, D. A., et al. 2001. *Semin. Immunol.* 13:255.)

CD antigen (cluster of differentiation antigens) Large number of antigens of the leukocytes that can be classified and identified by monoclonal antibodies. *See* CD proteins. (Mason, D., et al. 2001. *J. Leukoc. Biol.* 70:685.)

C/D box *See* snoRNA.

CDC (cell division cycle) *cdc* genes encode cyclin-dependent kinase. More than 100 CDC genes have been discovered, and in 2001, Leland Hartwell, Paul Nurse, and Tim Hunt were awarded the Nobel prize for the original discoveries. *See* CDK; cell cycle.

CDC1 Regulatory subunit of DNA polymerase δ. CDC1 seems to regulate the cellular Mn^{2+} level. *See* CDC27; pol δ. (Reynolds, N. & MacNeill, S. A. 1999. *Gene* 230:15.)

CDC2 When it is associated with mitotic cyclin proteins, CDC2 becomes the MPF (maturation promoting factor) of the oocytes, a serine/threonine protein kinase. CDC2 controls the transition from the G$_1$ to the S phase and prevents the reinitiation of the cell cycle at G$_2$. CDC2 also interacts with ORC. Phosphorylation of CDC2 on threonine-14 (by Xenopus protein Myt1) and tyrosine-15 (by fission yeast protein Wee1) inhibits the activity of CDC2. For the activity of CDC2, CAK (cyclin-dependent kinase) phosphorylates threonine-161. At the G$_2$ → M transition, CDC25 protein dephosphorylates Thr^{14} and Tyr^{15}; consequently, CDC2 suppresses MPF during interphase. High CDC2 activity inhibits anaphase but not the degradation of securin. *See* CDK; cell cycle; mitosis; MPF; ORC; Plx1; securin; separin. Wee1, Mik1, CDC25, LATS. (Vas, A., et al. 2001. *Mol. Cell Biol.* 21:5767; Karaiskou, A., et al. 2001. *J. Biol. Chem.* 276:36028; Stemmann, O., et al. 2001. *Cell* 107:715.)

CDC4 (Fbw7/Ago) Protease required for the transition from the G1 to the S phase during the cell cycle in cooperation with CDC34, CDC53 (cullin), and SKP1. CDC4 contains an F box, which interacts with SKP1 and eight WD-40 repeats common for proteins involved in protein-protein interactions. *See* APC; CDC34; cell cycle; SKP1; WD-40. (Desautels, M., et al. 2001. *J. Biol. Chem.* 276:5943.)

CDC5 Fission yeast checkpoint control gene. The CDC5 proteins are related to Myb and conserved through evolution. CDC5 is required for splicing pre-mRNA. *See* checkpoint; Myb oncogene; splicing. (Lee, S. E., et al. 2001. *Curr. Biol.* 11:784.)

CDC6 Protein DNA polymerase δ that appears in G1 and early S phase and late mitosis. Apparently it takes part in the formation of the prereplicative complexes and ORC. It is encoded at 17q21.3. *See* cdc18; cdt1; cell cycle; DNA polymerases; geminin; MCM; ORC. (Yanow, S. K., et al. 2001. *EMBO J.* 20:4648; Gowen Cook, J., et al. 2002. *Proc. Natl. Acad. Sci. USA* 99:1347.)

CDC7 Cell division cycle serine/threonine kinase with histone (H1) specificity. *See* cycle; DBF4; histones. (Guo, B. & Lee, H. 1999. *Somat. Cell Mol. Genet.* 25:159.)

CDC10 Transcription factor for ribonucleotide reductase (CDC22). CDC10 also mediates budding in yeast. *See* ribonucleotide reductase. (Jeong, J. W., et al. 2001. *Mol. Cells.* 12:77.)

CDC13 Protein is a regulator of gene *cdc2* in cooperation with genes *wee1* and *cdc25*; it is required for the maintenance of telomeres in yeast. *See cdc25*; cell cycle; Stn1; telomere; *wee1*. (Pennock, E., et al. 2001. *Cell* 104:387; Chandra, A., et al. 2001. *Genes & Dev.* 15:404.)

CDC14 Monocyte differentiation factor (5q31.1), a phosphatase; inactivates cyclinB/Cdk1 and activates Swi5, which facilitates the transcription of Sic1. Sic1 is normally phosphorylated by cyclinB/Cdk1. Cdc14 dephosphorylates Cdh1, resulting in the activation of APC. Cdc14 is initially detectable in the nucleolus, where it is sequestered by another protein, Cfi1 (Cdc14 factor inhibitor) but eventually spreads into the nucleus and even into the cytoplasm. Cdc14 regulates mitotic exit in cooperation with other proteins. *See* APC; Cdh1; Cdk1; cell cycle; cyclin B; E2F; FEAR; mitotic exit; Net1; Sic; Swi. (Guertin, D. A. & McCollum, D. 2001. *J. Biol. Chem.* 276:28185.)

CDC15 Inhibitor in MEN, controlling exit from mitosis along with TEM1. *See* MEN; TEM1. Mah, A. S., et al. 2001. *Proc. Natl. Acad. Sci. USA* 98:7325.)

CDC16 (APC6) *See* APC; CDC27.

CDC17 *Saccharomyces cerevisiae* gene for DNA polymerase α and *cdc17*+ is a DNA ligase I gene of *Schizosaccharomyces pombe*. *See* DNA polymerases; DNA replication. (Adams, M. A., et al. 2000. *Mol. Cell. Biol.* 20:786.)

CDC18 Rate-limiting activator (initiator) of replication; it interacts with ORC2. In human chromosome it is encoded at 17q21.3. In budding yeast it is CDC6. *See* CDC6; cdt1; cell cycle; geminin; ORC2. (Yanow, S. K., et al. 2001. *EMBO J.* 20:4648.)

CDC19 (Nda1) Component of the mini-chromosome maintenance complex (MCM). The human homologue is encoded at 3q21. *See* MCM. (Liang, D. T., et al. 1999. *J. Cell. Sci.* 112 [pt4]:559.)

CDC20 DNA polymerase ε and anaphase spindle checkpoint (repair) protein. It is encoded at human chromosome 9q12-q22. *See* anaphase; APC; CDH1; cell cycle; checkpoint; spindle. (Pfleger, C. M., et al. 2001. *Genes Dev.* 15:2396.)

CDC20-50 Potentiate passing beyond M phase to anaphase in the absence of spindle formation. *See* CDC20; CDC50; cell cycle. (Schott, E. J. & Hoyt, M. A. 1998. *Genetics* 148:599.)

CDC21 Component of the mini-chromosome maintenance complex (MCM) encoded at human chromosome 8q11.2 *See* MCM. (Satoh, T., et al. 1997. *Genomics* 46:525.)

CDC22 Ribonucleotide reductase acting on purine and pyrimidine nucleoside di- and triphosphates and catalyzing the formation of DNA precursors. *See* cdt1; ribonucleotide reductase. (Fernandez Sarabia, M. J., et al. 1993. *Mol. Gen. Genet.* 238:241.)

CDC23 (APC8) Member of the APC protein complex, encoded at human chromosome 5q31.1. *See* APC. (Goh, P. Y., et al. 2000. *Eur. J. Biochem.* 267:434.)

CDC24 Guanine nucleotide exchange factor (GEF) in signal transduction and a DNA replication factor responsible for chromosome integrity. *See* CDC42; GEF. (Bose, I., et al. 2001. *J. Biol. Chem.* 276:7176.)

CDC25 Cyclin-dependent CDC25 (A, B, and C at 5q31) phosphatases remove inhibitory phosphates from tyrosines and threonines to faciliate the transition from G_2 to mitosis during the cell cycle. In human and mouse cells, they represent a multigene family. In about a third of breast cancers CDC25B is overexpressed. In *Saccharomyces cerevisiae*, *CDC25* genes regulate RAS/cAMP pathway and their mutation causes defects in G_1 phase of the cell cycle. *Cdc25* is thus a protooncogene, and when growth factors are exhausted, it can induce apoptosis. The *Drosophila* homologue is *stg* (*string*). *See* budding yeast; Cdc2; CDF; cdk; cell cycle; checkpoint; Chk1; GEF; MYC; p53; parvulins; Plx1; protein 14-3-3; ras. SOS recruiting system. (Forrest, A. & Gabrielli, B. 2001. *Oncogene* 20:4393.)

CDC27 (APC3) Member of the tumor necrosis factor receptor family; restricts DNA replication to one round per cell cycle in cooperation with CDC16. It is also a component of the anaphase-promoting complex encoded at 17q12-q13.2. Cdc27/p66 is a component of DNA polymerase δ. *See* anaphase-promoting complex; TNF. (Shikata, K., et al. 2001. *J. Biochem.* [Tokyo] 129:699; Bermudez, V. P., et al. 2002. *J. Biol. Chem.* 277:36853.)

CDC28 (in *Saccharomyces cerevisiae*), **Cdc2** (in *Schizosaccharomyces pombe*) Genes are responsible for the start of mitosis in the cell division cycle. Cdc28p/Cdk1p protein is a cyclin-dependent kinase (CDK). Cyclin-3 binding activates Cdc28. The human homologue encoded at 2q33 is a Ser/Arg-rich protein required for splicing of pre-mRNA. *See* Cdc2; CDC34; Cdc45; CDK; cell cycle; mRNA; Sic. (Russo, G. L., et al. 2000. *Biochem. J.* 351[pt 1]:143.)

CDC30 Member of the tumor necrosis factor receptor family. It is involved also in the control of sporulation of yeast. *See* TNF. (Dickinson, J. R., et al. 1988. *J. Gen. Microbiol.* 134 [pt 9]:2475.)

CDC31 (centrin) Ca^{2+}-binding protein in the microtubule organizing center. *See* calmodulin; centrin; MTOC. (Ivanovska, I. & Rose, M. D. 2001. *Genetics* 157:503.)

CDC33 Encodes the eIF-4E translation initiation protein. *See* eIF-4E. (Brenner, C., et al. 1988. *Mol. Cell Biol.* 8:3556.)

CDC34 Ubiquitin-activating enzyme required for the G1 → S transition in the cell cycle. The human homologue is encoded at 19p13.3. It facilitates the destruction of cyclin 2 (CLN2) and cyclin 3 (CLN3) and degrades CDK inhibitor, SIC1. CLN2 and CLN3 are phosphorylated by CDC28 before CDC34-dependent ubiquitination. *See* cell cycle; cyclins; glucose induction; SIC1; ubiquitin; Wee1. (Ptak, C., et al. 2001. *Mol. Cell Biol.* 21:6537.)

CDC39 Gene with a glutamine-rich repressor product affecting G1/S-phase transition. *See* cell cycle. (Collart, M. A. & Struhl, K. 1994. *Genes Dev.* 8:525.)

CDC40 Member of the tumor necrosis factor receptor family. It is a pre-mRNA splicing factor. *See* TNF; splicing. (Russel, C. S., et al. 2000. *RNA* 6:1565.)

CDC42 RHO (Rac) family GTPase protein (encoded at human chromosome 1p36.1) involved in signal transduction pathways. It affects the mating pheromone signaling in yeast and binds several proteins. Cdc42 is an activator of the JNK and ERK pathways, including the one responsible for Wiskott-Aldrich syndrome. It causes depolarization of actin and regulates the exit of proteins from the trans-Golgi network. Its GDP-GTP exchange factor is Cdc24. It is a substrate for caspases and cooperates with the Fas apoptotic pathway. Integrin-mediated activation of CDC42 is involved in controlling astrocyte polarity with the cooperation of PKC and dynein. *See* actin; apoptosis; astrocyte; Bin; Cdc24; cell migration; dynein; ERK; faciogenital dysplasia; Fas; Golgi; GTPase; integrin; JNK; mating type determination in yeast; PAK; PKC; Rac; RHO; signal transduction; Wiskott-Aldrich syndrome. (Tu, S. & Cerione, R. A. 2001. *J. Biol. Chem.* 276:19656; Etienne-Manneville, S. & Hall, A. 2001. *Cell* 106:489.)

The structure of Cdc42 (Courtesy of Laue, E. D. From Morreale, A., et al. 2000. *Nature Struct. Biol.* 7:364.)

CDC44 *Saccharomyces cerevisae* gene encoding the large subunit of RFC. *See* Okazaki fragment; RFC. (McAlear, M. A., et al. 1996. *Genetics* 142:65.)

CDC45/CDC46Mcm5 Proteins (receptor-like transmembrane protein tyrosine phosphatases) are required for the initiation of chromosomal replication in association with CDC28 and other cyclin kinases. The human homologue of CDC45 is encoded at 22q11.2; the CDC46MCM is at 22q13.1. Inactivation of CDC45 leads to lymphoproliferation and autoimmunity. *See* autoimmune disease; CDC28; cell cycle; DiGeorge syndrome; MCM. (Ehrenhofer-Murray, A. E., et al. 1999. *Genetics* 153:1171.)

Cdc48 Is an AAA family protein with roles in the cell cycle, membrane fusion, endoplasmic reticulum assembly, nuclear fusion, Golgi re-assembly, degradation of ubiquitin fusion proteins, transcription factor processing, degradation of endoplasmic reticulum associated proteins, apoptosis and chaperone-like activity. *See* terms at separate entries. (Thoms 2002. *FEBS Lett.* 520:107.)

CDC50 Controls START in the G1 phase of the cell cycle. *See* CDC20; START. (Radji, M., et al. 2001. *Yeast* 18:195.)

CDC53 (cullin, CUL) CUL2 was mapped to human chromosome 10p11.2-p11.1. CUL3 is in human chromosome 2. *See* CDC4; cullin; glucose induction. (Botuyan, M. V., et al. 2001. *J. Mol. Biol.* 312:177.)

CDC55 Mammalian homologue of the yeast Cdc20, a mitotic protein. *See* Cdc20; spindle. (Wang, Y. & Burke, D. J. 1997. *Mol. Cell Biol.* 17:620.)

CDC68 (Spt16) Transcription elongation factor on active chromatin. *See* chromatin remodeling; FACT; transcription factors. (Formosa, T., et al. 2001. *EMBO J.* 20:3506.)

CDE element *See* CDF; centromere.

CDF Facultative repressor of the cell division cycle. It regulates the expression of cdc25, cyclin A, and MYB. CDF-1 acts on repressor sites CDE (GGCGG) and CHR (ATTTGAA). In late S phase, CDF binding to promoters leads to derepression of transcription. Binding E2F1 to the promoter upregulates transcription in late G1 phase. Binding both E2F and CDF results in intermediate kinetics. *See* cdc25; CDE; cell cycle; CHR; cyclins; E2F1; MYB. (Nettlebeck, D. M., et al. 1999. *Gene Ther.* 6:1276.)

CD fraction Constant dosage of the DNA in a chromosome including the functionally known genes. They are expected to be balanced with each other within a chromosome. These genes may, however, be amplified without serious detriment, but changing the dosage of the syntenic genes (aneuploidy) may have very undesirable consequences. *See* aneuploidy.

CDH1 (Hct1) APC regulatory WD protein with functions similar to CDC20. *See* Apc; Cdc14; CDC20; mitotic exit; WD-40. (Pfleger, C. M., et al. 2001. *Genes Dev.* 15:1759.)

CDK Cyclin-dependent kinases involved in cell division (replication) or apoptosis. These kinases do not operate without cyclin. Cyclin D–CDK4 and CDK6 are involved in G1 of the cell cycle; cyclin E–CDK2 drive G1-S phase and DNA replication; cyclin A–CDK is active in S phase; cell division requires cyclin A–CDK1 and cyclin B in G2 and M phases. Full activity is achieved by phosphorylation by CAK. Some cells (myocytes) may be protected from apoptosis by p21[CIP1] and p16[INK4a] inhibitors of Cdk. p16[INK4a] prevents the binding of cyclin D to CDK4/CDK6. Among the CDK inhibitors, p15, p16, and p18 specifically inhibit CDK4 and CDK6, whereas p21, p27, p28, and p57 are inhibitors of a wide range of CDK cyclin complexes. CDK4 is tyrosine phosphorylated in G1, and dephosphorylation is required for the progression into S phase. UV irradiation may prevent dephosphorylation, and cells are arrested in G1. If the CDK4 is not phosphorylated in G1, chromosomal breakage increases and cell death may result.

Cdk-activating kinase is a component of the Cak complex and part of the carboxyterminal of transcription factor TFIIH. The CDK proteins are similar in size (35–40 K) and display >40% identity in the different organisms where the somewhat different enzymes are denoted differently. Cdc2 is the typical enzyme for fission yeast. In budding yeast, the cyclin box comparable protein is CDC28, whereas in human cells it is CDK1/CDK. The 300-amino-residue catalytic subunit is inactive as a monomer or in the unphosphorylated form. In the inactive state the substrate-binding site is blocked and the ATP-binding sites are not readily available for the phosphorylation required for activity. The binding of CDK to the ≈100-amino-acid cyclin box is indispensable for function. CDK5/p35 is involved in neural development. When p35 is cleaved into a p25 fragment, Cdk5 excessively phosphorylates tau and Alzheimer neurofibrillar tangles appear in the brain.

Human CDK7 is homologous to the yeast Kin28 and has subunits of the general transcription factor TFIIH. It phosphorylates the carboxy-terminal domain of RNA polymerase II after the formation of the preinitiation complex. CDK7/Kin28 promotes transcription. CDK8 is homologous to the yeast Srb10 and is a subunit of the Pol II enzyme. It is also capable of phosphorylation of the C-terminal domain of Pol II before the formation of the preinitiation complex. CDK8/Srb10 actually inhibits transcription. The difference between Kin28 and Srb10/11 is in the temporal sequence of action. The effect of Srb10 does not generally apply to all genes, however, but it affects genes determining cell types, meiosis, and sugar utilization. Srb2, 4, 5, and 6 are positive regulators of transcription. Cdk9 (encoded in human chromosome 12) is a cofactor of lentiviral transcription.

CDC2 may associate with a few different cyclins, whereas CDC28 may be attached to nine different cyclins during the course of the cell cycle. The level and form of cyclins may vary during the cell cycle and their destruction is mediated by ubiquitins. In yeast, activation of CDC28 by the G1 cyclins stimulates cyclins CLN1 and CLN2 to degrade mitotic cyclins (CLB). After the G1 phase, CLB levels may be elevated, leading to the repression of G1 cyclins. When CDC28 is activated, CLB decay begins. CDK-cyclin complex may be inhibited by phosphorylation near the amino end of CDC2 and CDK2 (at Thr 14 and Tyr 15). Phosphorylation at these two residues is followed by a rise of mitotic cyclins (CLB). At the end of the G2 phase, Thr 14 and Tyr 15 are dephosphorylated by CDC25 phosphatase and CDC2 is activated. CDK may be inactivated by protein CKI (a family of inhibitory proteins to the cyclin-CDK complex) by attaching to the complex. The inhibitory subunits include p21 and p27 and other proteins. Eventually, the CKIs also decay and the cyclic events continue. Selective inhibitors of these kinases may be potential therapeutics. Cdk1 is also called Cdc28. Human CDK8 and cyclin C carry out the same function as Srb10 and Srb11 negative regulators of transcription in yeast. *See* Alzheimer disease; apoptosis; CAK; CDC28; cell cycle; centrosome; CIP; CKI; CLB; cyclin; cyclin T; KIN28; kinase; KIP; lentiviruses; p16; p21; p27; p53; p57; p16^{INK}; PHO85; PITALRE; preinitiation complex; Srb; tau; ubiquitin; UV. (Andrews, B. & Measday, V. 1998. *Trends Genet.* 14 (2):68; for CDK inhibitors: Knockaert, M., et al. 2002. *J. Biol. Chem.* 277:25493.)

CDKN2A (CDKN2/ TP16/p16^{INK4}, 9p21) Cyclin-dependent kinase inhibitor and general tumor suppressor. *See* CDK; p16. (Palmieri, G., et al. 2000. *Br. J. Cancer* 83:1707.)

cDNA DNA complementary to mRNA (made through reverse transcription) that does not normally contain introns. *See* mRNA; reverse transcriptases; transcription. (<http://www.kazusa.or.jp/huge>.)

cDNA library Collection of DNA sequences complementary to mRNA. *See* mRNA; processed genes.

cDNA library screening Can be carried out by probing the DNA sequences either with special binding proteins for the purpose of high degree purification or they can be probed with any DNA or RNA in order to identify genes or gene products isolated from different organisms. *See* cloning; gel retardation assay; probe.

CDP (1) Mammalian displacement protein at the CCAAT sequence of DNA competing for binding of CP1, a CCAAT box–binding protein. It is a transcriptional repressor. *See* CAAT box; CP1. (Ellis, T., et al. 2001. *Genes Dev.* 15:2307.) (2) Cytidine diphosphate. Nucleotide involved in the biosynthesis of phospholipid.

CDPK Ca^{2+}-dependent protein kinases regulate signaling pathways. *See* signal transduction. (Allwood, E. G., et al. 2001. *FEBS Lett.* 499:97; Zhang, X. S. & Choi, J. H. 2001. *J. Mol. Evol.* 53:214.)

CD protein (clusters of differentiation) Accessory protein on the surface of T cells. *See* ICAM; integrins; LFA; T cells.

CDR Complementarity determining region of the antibody's hypervariable region that binds the antigen; it is the antigen-binding site (idiotype). *See* antibody; idiotype; paratope. (Furukawa, K., et al. 2001. *J. Biol. Chem.* 276:27622.)

CDS Protein-coding sequences in the genome. *See* Raddatz, G., et al. 2001. *Bioinformatics* 17:98.)

Cds1 Fission yeast kinase, an inhibitor of Cdc2. Human Cds1 phosphorylates the breast cancer protein (BRCA1) at serine 988 after DNA damage and assists cell survival. *See* breast cancer; Cdc2. (Boddy, M. N., et al. 2000. *Mol. Cell Biol.* 20:8758.)

Cdt1 Fission yeast protein (with similarities to deoxyribonuclease I) required for assembly of the DNA prereplication complex in concert with Cdc22, Cdc18/Cdc6, and MCM. Its homologues are present in vertebrates. CDTs (cytolethal distending toxins) have the special feature of affecting DNA rather than cellular proteins. *See* CDC6; CDC18; CDC22; DNA replication; geminin; MCM; replication fork; replication licensing factor. (Yanow, S. K., et al. 2001. *EMBO J.* 20:4648.)

CD-tagging method Uses a DNA cassette for insertion mutagenesis; when inserted into an intron, transcribed, and spliced, the mRNA contains a special tag (guest tag).

Upon translation, the polypeptide also carries a tag (guest peptide). The latter can be identified by monoclonal antibody specially prepared for this epitope. The mRNA and the DNA sequences can be identified by PCR. Thus, this method simultaneously labels DNA, RNA, and the peptide; hence its name central dogma tagging. *See* epitope; insertional mutagenesis; monoclonal antibody; PCR. (Jarvik, J. W., et al. 1996. *Biotechniques* 20:896.)

CDw32 (FcγRII) Antibody Fc domain receptor. *See* antibody. (McKenzie, S. E. & Schreiber, A. D. 1994. *Curr. Opin. Hematol.* 1:45.)

ce Prefix for *Caenorhabditis elegans* DNA, RNA, or protein.

Cebidae Families of New World monkeys. *Aotus trivirgatus trivirgatus* $2n = 54$; *Aotus trivirgatus griseimembra* $2n = 52$, 53, 54; *Ateles geoffroy* $2n = 34$; *Callicebus moloch* $2n = 46$; *Callicebus torquatus* $2n = 20$; *Cacajo* $2n = 46$; *Cebus albifrons* $2n = 54$; *Lagotrix ubericolor* $2n = 62$; *Pithecia pithecia* $2n = 48$; *Scaimii sciureus* $2n = 44$. *See* primates.

C/EBP (CEBP) CAAT/enhancer-binding protein is transcription factor AP1, product of JUN and FOS oncogenes. The protein is essential for the differentiation of granulocytes. Its mutation is common in acute myelogenic leukemia (Pabst, T., et al. 2001. *Nature Genet.* 27:263). These proteins regulate different cellular functions, including adipogenesis. *See* AP; enhancer; FOS; granulocyte; JUN; leukemia. (Lekstrom-Himes, J. & Hanthopolus, K. G. 1998. *J. Biol. Chem.* 273:28545; McKnight, S. L. 2001. *Cell* 107:259.)

Cebus (capuchin monkey) *See* Cebidae.

cecropin *See* antimicrobial peptides.

CED *See* apoptosis.

cefotaxim (Claforan) Cephalosporin-type general medical antibiotic with relatively mild toxicity to plant cells. It is widely used to free plant tissue from *Agrobacterium* infected for the purpose of genetic transformation. *See* Agrobacterium; genetic transformation. (Husson, M. O., et al. 2000. *Pathol. Biol [Paris]* 48:933.)

ceiling principle Statistical procedure for the conservative estimation of the odds for the likelihood that DNA fingerprints would match. The odds against chance match is usually determined on the basis of the frequency of a particular genetic marker in a certain population (such as Caucasians, blacks, Hispanics, Orientals, etc.). The markers used for DNA fingerprint analysis are supposed to be of low frequency, below 10% or 5%, but for the majority of subpopulations, such information is not yet available. In such cases, they take into account, say, 0.1 as a maximal frequency (a ceiling). The chance that a particular person would have the same DNA marker as another individual in his or her group would be $0.1 \times 0.1 = 0.01$. The probability that 8 markers would be identical by chance would be $(0.1)^8 = 1/100,000,000$, and if 0.05 is chosen as a ceiling, it would be approximately 2.6×10^{-10}. The world's population of 6 billion is about 23% of 2.6×10^{10}. Some population geneticists disagree with the use of the rather arbitrary ceilings and advocate the theoretically more valid use of mean frequencies with the pertinent confidence intervals. Today, more information is available on gene frequencies; therefore, direct frequencies can be used for the majority of the genes involved. Abandoning the ceiling principle increases the accuracy of establishing individual liabilities and does not allow unwarranted advantage for criminals. The recommended genetic markers for forensic comparison (in lieu of the ceiling principle) are variable number tandem repeats, VNTR (D2S44, 75 alleles, and D1S80, 30 alleles), short tandem repeats, STRs (HUMTHO1, 8 alleles), simple sequence variations, SSV (DQA, 8 alleles, polymarker [5 loci, 972 combinations]), and mtDNA D-loop with >95% diversity. *See* allelic frequency; confidence intervals; DNA fingerprinting; Frye test. (Slimoewitz, J. R. & Cohen, J. E. 1993. *Am. J. Hum. Genet.* 53:314.)

ceinsulin Insulin-like growth hormone in *Caenorhabditis elegans*. *See* Caenorhabditis; insulin.

celery (*Apium graveolens*) The stalks are used as food or the celeriac is a root vegetable; $2n = 2x = 22$.

celiac disease (coeliac disease) In certain individuals the intestinal enzymes do not digest some water-insoluble proteins, such as the gliadin in wheat (also in other cereals). This protein then causes inflammation of the intestinal lining and bloating of the abdomen. Its incidence in the general population is about 4/1,000, and the genetic recurrence risk in the brothers and sisters of afflicted sibs is about 2 to 3%. The genes responsible for A2-gliadin synthesis are located to the long arm of chromosomes 6A, 6B, and 6D of hexaploid wheat. Some gliadin genes are also in chromosome 1. There are substantial quantitative differences among the different chromosomes concerning the production of this protein. In the intestinal endomysium (reticular sheath of muscle fibers), a tissue-specific transglutaminase exists. Deamidation of gliadin opens up epitopes, which then bind HLA-DQ2 on antigen-presenting cells and facilitate the recognition by intestinal T lymphocytes. It appears particularly important for the inflammation response when glutamine 148 is converted into acidic glutamate. The CD4$^+$ T cells stimulate helper T cells (T$_H$) to secrete cytokines such as TNF-α, which in turn induces the release of metalloproteinases, which degrade fibrillar collagen, proteoglycans, and matrix glycoproteins. The T$_H$ cells facilitate immunoglobulin A (IgA) production by B lymphocytes. IgA turns against gliadin and gliadin complexes.

Some anthropologists suggested that the ancient Egyptians consumed high gliadin wheat varieties or some of the pharaohs were more susceptible to the disease (these families practiced a high degree of inbreeding) because of the extended bellies observed on several royal mummies. This disease is under the control of the HLA-DQ2 genes in humans, but one 6p locus 30 cM from the telomere (thus outside HLA) has been identified for the predisposition. Low-lod score putative linkage was observed with several chromosomal sites. *See* glutenin; HLA; immune system; immunoglobulins; metalloproteinases; T cells; T$_H$; *Triticum*. (King, A. L., et al. 2000. *Ann. Hum. Genet.* 64:479; Sollid, L. M. 2000. *Annu. Rev. Immunol.* 18:53. Kumar, R., et al. 2002. *J. Mol. Biol.* 319:593; Fleckenstein, B., et al. 2002. *J. Biol. Chem.* 277:34109.)

cell *See* cell comparisons; cell structure; single-cell analytical methods.

Cell comparisons.

CRITERIA	PROKARYOTES	PLANTS	ANIMALS
Cell wall	Present	Present	Absent
Nucleus	Nonenveloped nucleoid	Enveloped	Enveloped
Plastids	Absent	Present	Absent
Mitochondria	Absent	Present	Absent
Ribosomes	70S	80S	80S
(organellar)	Not applicable	70S	70S
Endoplasmic reticulum	Absent	Present	Present
Centrioles	Absent	Absent	Present
Spindle fibers	Absent	Present	Present
Microtubules	Absent	Present	Present
DNA location	Nucleoid	Nucleus	Nucleus
	Plasmids	Mitochondria, plastids	Mitochondria
Chromosomal composition	DNA	DNA	DNA
	Minimal protein	Protein	Protein
	or RNA	RNA	RNA
Division of the genetic material	Replication and partition	Replication Mitosis, meiosis	Replication Mitosis, meiosis

cell adhesion molecule *See* CAM.

cell autonomous The product of the gene is limited to the cell expressing it; it does not diffuse to other cells.

cell body Main part of the nerve cell containing the nucleus and excluding the axons and dendrites. *See* neurogenesis.

cell comparison Cells of various organisms have common features, yet differences exist that can be compared by the tabulation above.

cell cortex On the inner surface of the animal plasma membrane there is an actin-rich layer of the cytoplasm that mediates movement of the cell surface.

cell culture Generally the culture of isolated cells of higher eukaryotes is meant, although growing bacteria or yeast is also cell culturing. *See* tissue culture.

cell cycle The phases of cell reproduction and growth are G1 → S → G2 → M and cytokinesis, the generation of two daughter cells from one. The duration of the cell cycle varies among different organisms. It is influenced by several factors (temperature, nutrition, age, stage, etc.). In *Drosophila* embryos, it may be completed within 8 minutes, and in other early embryos, the cycle may be completed within 30 minutes. In other cells, the approximate numbers of hours required is in the table.

The S and M phases are present in all dividing tissues; the G phases can be clearly distinguished only in cells where the divisions are slower because of differentiation.

Before the cell enters the cell cycle, the prereplication complex (pre-CR) is assembled. This complex consists of the origin recognition complex (ORC), the cell division cycle 6 protein (Cdc6p), and MCM (mini-chromosome maintenance proteins).

The cell cycle has been studied most frequently in vitro by the dividing egg of mammals manipulated by microinjection of various cellular components or by fusion of cells of different developmental stages. In plants, it is analyzed in somatic cells where division is triggered by the application of hormones and in yeast by accumulating conditional (temperature-sensitive) mutations, then by the introduction of the wild-type allele through transformation. Throughout the cell cycle (except mitosis), RNA and protein synthesis takes place. The G1 phase (Gap 1, named not very felicitously) is actually a phase of cell growth when a commitment is made for DNA replication. This point of commitment is called START by yeast cell biologists, and by animal cell biologists RESTRICTION POINT.

For START, the cell requires the activity of a cyclin-dependent protein kinase composed of the catalytic subunit of protein Cdc28 and one of the three cyclins (C*l*n1, −2 or −3). After START, cyclin 5B and cyclin 6 kinases are required. These kinases are inhibited by protein Sic1, and the latter must be inactivated by proteolysis carried out by the ubiquitin-conjugated Cdc34. Some cells may stay for a long time at this preparatory stage; then it is called G_0 stage. In S phase, DNA

CELL TYPES	G1	S	G2	M	TOTAL	SOURCE
Onion roots	1.5	6.5	2.4	2.3	12.7	Van't Hoff
Mouse fibroblasts	9.1	9.9	2.2	0.7	22.0	John Lewis
Xenopus early gastrula	3.5	4.5	8.0	0.5	16.5	John Lewis
Xenopus late gastrula	2.0	2.0	3.5	0.5	8.0	John Lewis
Saccharomyces	0.45	0.45	0.45	0.15	1.5	Fante
Schizosaccharomyces	0.26	0.24	1.85	0.16	2.5	Fante

Proteolytic controls during the progression of mitosis

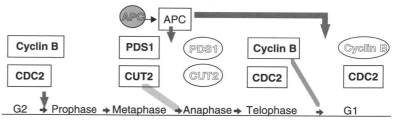

The heavy lines indicate activation; the gray lines stand for blocking the transition. The symbols in thin characters indicate lack of activity, and the gray circle stands for degradation. CDC2 = cell division cycle protein 2; PDS and CUT are noncyclin proteins; APC = anaphase-promoting complex.

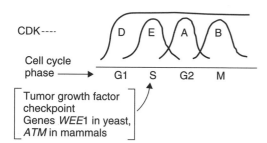

In the various phases of the cell, cyclin-dependent kinases (CDKD, CDKE, CDKA, and CDKB) reach their peak activity while other proteins (promoters and inhibitors) are also changing and balancing each other's effect on the progression of the cell cycle. The diagram shows only the cyclin activities in relation to the mitotic phases and tumor formation checkpoint.

synthesis is completed. The cells cannot enter G2 (Gap 2, another cell growth phase) before the completion of S phase.

Several genes were found to cooperate in *checkpoints* that provide clearance before the next phase can be entered. CLB-Cdc28 prevents reinitiation of the S phase by phosphorylating Cdc6 and removing of MCM2-7. Thus, the cell avoids polyteny or polyploidy. G2 is committed to the preparations of mitosis, a task that follows if during interphase all the nutritional requisites for carrying out mitosis build up. The two G phases and S phase combined are frequently named *interphase*, i.e., the phase between nuclear divisions (mitosis). In very rapidly dividing embryonic tissue, G1 and G2, involved in cell growth, may be extremely brief or even absent; therefore, the daughter cells become only half the size of the mother cells by each division. This is possible because the egg cell is generally very large at the time of fertilization. Some mature animal eggs may be thousands or tens of thousands of times larger than an average body cell and they are loaded with nutritious material. In such cases, the cell cycle may include only S (DNA synthetic) and M (mitosis, nuclear division) phases. Feeding labeled nucleotides, which are then incorporated into the DNA, can identify the S phase. The fraction of the cells doing so, the *labeling index*, can be determined. From the fraction of cells undergoing mitosis in a tissue, the *mitotic index* is derived.

Cell division is an extremely complex process involving the cooperation of a very large fraction of all the genes and affecting the expression of many others. Cell division requires the presence of several *protein kinases* (phosphorylases), *phosphatases*, and other activating proteins such as *cyclins*. The cyclins were named so because they are synthesized during the cycles of mitoses, but not much is made during the intervening (interphase) periods. In fission yeast, gene *cdc2 (cell division cycle)* is involved in the control through its 34 kDa phosphoprotein product (p34^{cdc2}), which is a

serine/threonine kinase activated by cyclins, and the complex becomes a cyclin-dependent protein kinase (Cdk). The cyclin-dependent protein kinase associated with cyclin B is also called MPF (*maturation protein factor*). In phosphorylation, a cyclic AMP–dependent protein kinase (cAPK) also has a role. MPF-dependent activation of cAMP–protein kinase A (PKA) and cyclin degradation are required for the passage from mitosis into interphase. There are a large number (over 70) of *cd* genes. The *cdc2* homologue in budding yeast is *cdc28*. Most of these genes have been well preserved during eukaryotic evolution. Homologues are present in yeasts, animals, and higher plants.

There are several different cyclins. The G1 cyclins are cyclin C; a number of different cyclin D's and cyclin A control the onset of the S phase when bound to the Cdk protein(s). The mitotic cyclin (cyclin B, encoded by *cdc13* in fission yeast) binds to Cdk before the onset of mitosis. MYC in cooperation with RAS also mediates the progression of the cell cycle from G1 to the S phase through induction of the accumulation of active cyclin-dependent kinase and transcription factor E2F. Actually, similar genes and functions occur in all eukaryotes, but they are named differently. CLN and numbers denote the cyclin homologue genes in budding yeast. Remember that in fission yeast usually the genes are symbolized with lowercase letters and plus or minus superscripts depending on whether wild-type (+) or mutant genes ( ) are represented. In budding yeast, the wild-type allele is capitalized and the mutant is lowercase. In both yeasts, the genes are italicized, whereas the protein symbols are not.

Upon the binding of Cdc2 and cyclin, the conformation of the former is altered, allowing the phosphorylation of the threonine at position 161 of Cdc2; the complex becomes a fully active promoter of mitosis. The Thr 161 phosphorylation is mediated by Cdk (cyclin-dependent kinase), also called Cak (Cdk-activating kinase). cAPK is autophosphorylated at a Thr197 residue. Cyclin binding is also followed by dephosphorylation of the tyrosine 15 residue by phosphatases. Before the cell can enter the M phase in the majority of organisms, phosphorylation of this protein decreases. In fission yeast, other proteins (encoded by genes *wee1, nim1, mik1*) exert negative control over the passage into the M phase. Eventually, Cdc2 protein is dephosphorylated (gene *cdc25* regulates a phosphatase activity) and cyclin is degraded, making the Cdc2 monomers available for another round of association with newly synthesized cyclins. If the DNA is damaged, Chk1 kinase is activated, which prevents the exit from the G2 phase into the M phase.

The MAPKK mitogen-activated serine/threonine kinase is also required for the G2 → M transition. The degradation of the cyclins is mediated by ubiquitin-dependent proteolytic cleavage pathways controlled by MPF. Genes that block entry into the M phase regulate exit from the M phase. The product of gene *suc1*, p13^{Suc1}, may be required (among other proteins) for

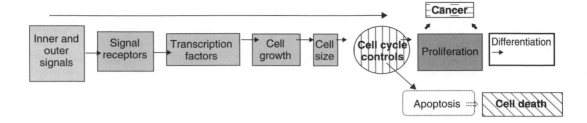

the termination of mitosis. If S phase takes place but mitosis is not completed, endopolyploidy may result, i.e., the chromosome number is multiplied (polyploidy). By this time, the organellar material (plastids, mitochondria) is also readied for fission.

There are some differences in the cell cycles of yeasts and other fungi from higher eukaryotes. In the former group, the nuclear envelope is present throughout the cell cycle, whereas in the latter it disappears from view from metaphase through telophase and is reformed after late telophase. In the fission yeast, the cell division resembles that of higher eukaryotes by forming a cell plate in between the two daughter nuclei. In budding yeast, one of the daughter nuclei moves into an extrusion of the cell, a bud, and eventually grows into a normal-size cell. In fungi, the spindle apparatus is located inside the nucleus rather than in the cytoplasm as in higher eukaryotes. These processes also require other regulatory mechanisms. It must be assured that DNA replication (S phase) produces complete sets of all essential genes and preferably *not* in multiple copies unless such amplification is required. In the regulation of the cell cycle, ubiquitin appears to have an important role. Three yeast proteins, CD16, CD23, and CD27 (or their homologues in other organisms), mediate the attachment of ubiquitin to cyclin, resulting in its degradation at the end of mitosis. The initiation of a new cycle is hampered until an inhibitor (p27) is removed from the cyclin-CDK complex. This inhibitor is degraded by ubiquitin placed on the proteins of the spindle by CD16 before a new S cycle is entered.

As anaphase is initiated, the cell cycle becomes irreversible, and as the cohesion between sister chromatids breaks down, the spindle fibers pull the new chromosomes (they were called chromatids until now) toward the poles. This process is mediated by the anaphase-promoting protein complex (APC/NR/TSG24), also called *cyclosome*. APC is a ubiquitin-protein ligase apparently managing for the chromosome to be properly arraigned in the metaphase plane and for other proteins to be functional. CDC20 and CDH1/HCT1 activate APC. Anaphase entry and mitosis exit are promoted by APC-dependent proteolysis of cell cycle proteins. CDC20 promotes the degradation of the early-acting proteins and HCT1 assists the degradation of Clb2 type cyclins.

The M-phase cyclins are now lysed by the 26S proteosome after the telophase is completed. Other protein factors that are no longer needed are also ubiquitinated. APC operates through the proteolytic pathway shown above. Another proteolytic pathway during the cell cycle is mediated through protein CDC34 (see CDC34).

The p34^{cdc2} and homologous proteins in cooperation with other factors (MPF) mediate the condensation of chromosomes through activation of H1 histone and control lamins to mediate the breakdown of the nuclear envelope (except in yeast). MPF controls tubulins and actins for the function of the mitotic spindle, etc. The cell cycle is intimately associated with signal transduction pathways, DNA topoisomerases, DNA polymerase, DNA ligase, RNA polymerases, transcription factors, etc. After all these events are successfully passed, the cell divides into two daughter cells and the cycle may be resumed depending on environmental conditions. Aphidicolin blocks DNA synthesis, hydroxyurea interferes with the formation of DNA nucleotides, and therefore DNA synthesis is halted. In case caffeine is added along with hydroxyurea, an abortive DNA replication and mitosis result in cell death.

Some of the mutations involving defective DNA repair inhibit the cell cycle and cell divisions. The various growth factors are directly or indirectly involved with the cell cycle. The p53, p16, and p21 proteins are regulators of the cell cycle; some of their mutations no longer control the pace of orderly cell divisions and are thus instrumental in tumorigenesis. Breakdown of some cell cycle signals causes failures in attachment of the spindle fibers to the kinetochore resulting in nondisjunction and aneuploidy. An overview of events leading to overall fate of cells through the cell cycle can be represented below.

Meiosis has similar controls, but the modulations are different. The primary oocyte is at an arrested G2 stage until a hormonal stimulation pushes the process to the first meiotic phase (reduction division), resulting in the formation of the first polar body in animals. In plants, this is the stage of the megaspore dyad. In mammalian oocytes, the diplotene stage (called dictyotene) may then last from early embryonic development (about the third month in humans) to puberty. This is followed by the formation of the second polar body and the egg. In plants, from the four products of the "female" meiosis, only one megaspore (most commonly the basal one) remains functional, and unlike in animals, it undergoes three more divisions to eventually form the egg.

Upon fertilization of the interphase egg, diploidy is restored, and cleavage divisions follow. The cell cycle has central importance for various processes of differentiation. The p21 protein regulates CDKs and, combined with PCNA controls, DNA replication. p21 also blocks keratinocyte differentiation. The Cak subunit of TFIIH transcription factor is involved with RNA polymerase II preinitiation complex. Cyclins regulate tumor suppressor gene RB (retinoblastoma) and MyoD muscle differentiation factor. RB binds to the E2F family of cell cycle transcription factors and the RB-E2F complex blocks transcriptional activation by recruiting histone deacetylase (HDAC). The HDAC-SWI/SNF nucleosome-remodeling complex inhibits the cyclin E and A genes and arrests the cell cycle at the G1 phase, preventing the exit from the G1 phase. The RB-SWI/SNF complex regulates the exit from the S phase. BRG and BRM may assist RB in incapacitating E2F. In the human cell cycle, oligonucleotide array analysis detected the involvment of about 700 genes. *See* amplification; APC;

apoptosis; Ase1; asparagine synthetase; ataxia telangiectasia; BRG; cancer; CDC; Cdc14; Cdc25; CDC27; Cdc28; CDF; CDK; Cds; cell division; Chk; CKI; cullin; cyclin; differentiation; E2F1; endomitosis; gametogenesis; HiNF; histone deacetylase; IFR; keratin; licensing factor; MCH; MCM; meiosis; mitogen-activated protein kinase; mitosis; MYC; MyoD; ORC; p21; p15$^{\text{INK4B}}$; p16$^{\text{INK4B}}$; PCNA; Pds; PIK; PIN1; polyploidy growth factors; polyteny; proteasomes; RAS; RB; regulation of gene activity; replication; retinoblastoma; SCF; senescence; signal transduction; SKP1; SWI/SNF; transcription factors; tumor suppressor gene; ubiquitin. (Cho, R. J., et al. 2001. *Nature Genet.* 27:48; Israels, E. D. & Israels, L. G. 2001. *Stem Cells* 19:88; Simon, I., et al. 2001. *Cell* 106:697; Groisman, I., et al. 2002. *Cell* 109:473; Vandepoele, K., et al. 2002. *Plant Cell* 14:903, <www.nature.com/celldivision>; <http://genome-www.stanford.edu/cellcycle/>; <http://cellcycle-www.stanford.edu/>.)

cell division   In eukaryotes, involves two steps: nuclear division (karyokinesis), followed by division of the rest of the cell (cytokinesis). A summary of the end results of the cell cycle is doubling of the chromatids as a result of the DNA replication during the S phase, and changes in the C values of the nuclei. The number of cell divisions in the germline of human males before puberty is ~30, whereas in the females it is ~22. *See* cell cycle; cytokinesis; diagram; mitosis; partitioning. (For bacterial cell divison: Rothfield, L., et al. 1999. *Annu. Rev. Genet.* 33:423; Nanninga, N. 2001. *Microbiol. Mol. Biol. Rev.* 65:319.)

cell fate   Program that determines the morphology and function of the undifferentiated cells in the embryo. *See* cue; fate map; morphogenesis.

cell fractionation   Separation of the different subcellular organelles generally by differential centrifugation in variable-density media. *See* centrifuge; density gradient centrifugation.

cell-free extract   Prepared by grinding cells (tissues) in a buffer or other solutions and removing insoluble particulate material by filtration or centrifugation. Such extracts may be used for enzyme assays or for the purification of soluble cellular constituents.

cell-free protein synthesis   In vitro protein synthesis in the presence of ribosomes, mRNA, tRNA, aminoacylating enzymes, amino acids, and all the complex of translation factors and energy donor nucleotides. *See* rabbit reticulocyte assay; wheat germ assay.

cell-free translation   *See* cell-free protein synthesis.

cell fusion   Means to generate somatic cell hybrids. In contrast to hybridization by gametic fusion when the two nuclei fuse but usually only the maternal cytoplasm is preserved, in the fusion of somatic cells the entire content of the cells is combined in the somatic hybrid. For the fusion to take place, polyethylene glycol, a high concentration of calcium, or a higher pH medium have been used. (Inactivated Sendai virus also promotes the fusion of animal cells.) Carbon fiber ultra-microelectrodes may make the fusion of selected cells or cells and liposomes possible. For the selective isolation of somatic cell hybrids, both of the two types of cells generally carry recessive mutations that interfere with the survival of the cells on basic media, e.g., thymidine kinase–deficient animal cells die because they cannot synthesize thymidylic acid (DNA); hypoxanthine-guanine phosphoribosyl transferase–deficient cells cannot make purine nucleotides (DNA). The fused cells are heterozygous at nonallelic loci and are functional, however, and they can be selectively isolated in large cell populations. Somatic cell fusion has had important contributions to genetics involving human and other animals because it has made allelism tests possible, facilitated the assignment of genes to chromosomes, and identified the functional significance of chromosomal regions (in case of deletions). Somatic cells of very distantly related or entirely unrelated organisms, e.g., chicken and yeast, tobacco and human, human and rodent cells, all can be fused, although their further division usually may not be possible. *See* liposomes; microfusion; protoplast fusion; radiation hybrids; somatic cell hybrids. (Cocking, E. C. 1972. *Annu. Rev. Plant Physiol.* 23:29; Ephrusi, B. 1972. *Hybridization of Somatic Cells.* Princeton Univ. Press, Princeton, NJ; Hotchkiss, R. D. & Gabor, M. H. 1980. *Proc. Natl. Acad. Sci. USA* 77:3553.) McKay, R. 2002. *Nature Biotechnol.* 20:426.)

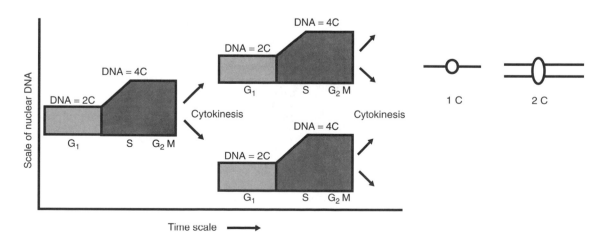

The relation between the nuclear cycles and cell divisions. During S phase the amount of DNA doubles and each 1 C chromosome will have 2C amounts (2 chromatids), and a diploid cell will have a total of a 4 C amounts of DNA before mitosis takes place and then cell division (cytokinesis) ensues.

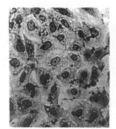

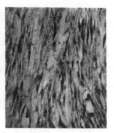

Hamster   Hamster + Man   Man
(Courtesy of Drs. R. Wang & L. Dardick)

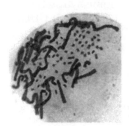

SOYBEAN (2n = 40) + VETCH (2n = 12)
(Courtesy of Dr. L. O. Gamborg. from constabel, G.*et al.*
1977. C.R. Hebd, Séances, Acad, Sci. Paris Ser D. 285:319.)

cell genetics Actually, nearly all genetics is cell genetics because geneticists generally think at the cellular level; in the narrow sense, this term is applied to the genetic manipulations with isolated cells of multicellular organisms. *See* cell fusion; FL-REX; fusion of somatic cells; mitotic crossing over; nuclear transplantation; somatic cell hybrids; somatic cells; transformation. (Ruddle, F. H. & Creagan, R. P. 1975. *Annu. Rev. Genet.* 9:407; Puck. 1974. *Stadler Symp.* 6:47; Dudits, D., et al., eds. 1976. *Cell Genetics of Higher Plants.* Akad. Kiadó, Budapest.)

cell growth In any particular time, $N = 2^g N_0$, where N is the final cell number, N_0 is the initial number of cells, and g is the time required for a complete cell cycle. This equation is valid as long as there is no limitation on multiplication by nutrients, air, differentiation pattern, etc. In the absence of any limitation, cell growth indicates the cell-doubling process. Actually, growth is frequently used in place of cell proliferation, growth is however an increase in volume or size, not in cell number. *See* growth retardation; plating efficiency; proliferation; turbidity.

cell hybridization Fusion of somatic cells. *See* cell fusion; cell genetics.

cell interaction Influence of cells on each other during differentiation and development. *See* contact inhibition; maternal effect genes; morphogenesis; nurse cells.

cell junction Area involved in the connection and communication between and among cells and extracellular matrices. *See* contact inhibition; extracellular matrix; morphogenesis; transmembrane proteins. (Tepass, U., et al. 2001. *Annu. Rev. Genet.* 35:747.)

cell lethal Mutations may not be isolated or ascertained because the cells involved cannot live.

cell line (Homogeneous) population of cells (of eukaryotes) that can be maintained in live (growing) conditions. *See* clonal analysis; clone.

cell lineages Traces of the path of growth (multiplication) of the cells through several cell divisions. The descent of the germline cells or the signs of visible mutations in the somatic tissues are shown by the pattern of the sectors formed in chimeric organisms. *See* clonal analysis; fate maps; founder cells; nondisjuinction; phyllotaxis diagram below. (Stern, C. D. & Fraser, S. E. 2001. *Nature Cell Biol.* 3:E216; Liu, Y. J. 2001. *Cell* 106:259.)

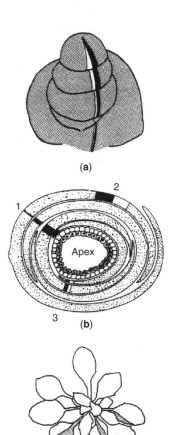

(a)

(b)

(c)

Mutation, deletion, and nondisjunction during embryogenesis may potentially be identified during and after embryogenesis if the organism is heterozygous for a distinguishable somatic marker(s). (**a**) Organization of the apical dome of a wheat plant showing diagrammatically the consequence of nondisjunction. (**b**) In the monocot apex the leaf initials wrap around the central axis and overlap each other; the humps at the leaves indicate the midribs. The oldest leaf initials are outside. Note that the outermost leaf and the one below it have their midrib at opposite sides. Sector 1 is very narrow at the surface (old) leaf and it becomes wider in the (younger) ones below. Also, the oldest sector is left from the midrib of the first leaf, but it is at the right side of the one just below and again at the left side in the third. Sector (2) representing nondisjunction and twin sectors (black and white) occurred only in one leaf because of a tangential event in a region of the embryonal apex. Sector (3) is a late-occurring nondisjunction indicated by the narrow twin sectors. (**c**) Somatic mutation in a single cell of the mature embryo of the dicot *Arabidopsis*. Three leaves displayed white sectors (shaded) because they differentiated from the same cell line of the apex. Nonsectorial leaves appeared in between the mutant sectors because of phyllotaxis.

cell-mediated immunity *See* immune system; T lymphocytes.

cell-mediated mutagenesis Chemical mutagen activation is provided by addition of suitable activated (liver) cells as feeders to the culture. These feeder cells may be genetically modified to express high levels of the activating enzymes. *See* activation of mutagens; host-mediated assay. (Rudo, K., et al. 1987. *Cancer Res.* 47:5861; Langenbach, R. & Nesnow, S. 1983. *Basic Life Sci.* 24:377.)

cell membrane All cells are surrounded by membranes; inside the cells there are membrane-enclosed bodies (nucleus, mitochondria, plastids, vacuoles, Golgi bodies, dictyosomes, lysosomes, peroxisomes). The endoplasmic reticulum, mitochondrial crests, and thylakoids are all membraneous structures. Cellular imports and exports pass through the membranes by active and passive mechanisms. The bulk of the plasma membranes consists of proteins and lipids (phospholipids, cholesterol, other sterols and glycolipids, triaglycerols, steryl esters, etc.). The composition varies in the different organisms and according to the particular membranes. The ultrastructure of the various membranes has common features and specificities. The basic structural element is the lipid bilayer of about 5 to 8 nm in thickness. In the double structure, the polar head of the lipid faces the aqueous environment and the inward tails are hydrophobic. See diagram below.

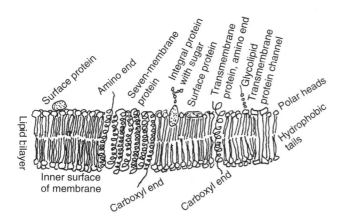

Generalized diagram of membrane structure.

Unsaturated lipids are concentrated in the inner layer of the structure. The outer surface of the membrane is also different from the inner surface that envelops organelles or vesicles. The inner side carries charged groups on the surface; the outward surface may have a variety of peripheral proteins (glycoproteins) that determine the surface antigenicity of the cells. Some other proteins are integral parts of the membrane sunken in the fluid lipid bilayer. The fluidity is somewhat stabilized by the presence of sterols. The so-called seven-membrane proteins traverse the lipid bilayer and form within it a cluster of seven folds, the amino end at the outside and the carboxyl end inward. Other transbilayer polypeptides have hydrophilic domains both outside and inside, and outward are ports for communication (ion channels) with special proteins and lipids (transporters, permeases). Some of the peripheral proteins are attached to the membrane by electrostatic forces and H bonds.

The proteins are regulators of membrane-bound enzymes (e.g., phospholipase C) and mediate signal transduction. Membranes have a flexible structure to curl up into vesicles that ferry within the cell lipids and proteins in a protected manner. The membranes have the ability to fuse with another membrane at the delivery target. Exocytosis and endocytosis are the means of traffic. Fusion of the egg with the sperm, fusion between protoplasts of plant cells, somatic cell hybridization, protein synthesis within the endosplasmic reticulum, etc., are mediated by membrane functions. Membranes can be targeted through labeled myristoylated and palmitoylated proteins. Such modifications of the membranes may affect the membrane-attached G proteins involved in signal transduction. Ca^{2+}-regulated exocytosis repairs damaged plasma membranes. *See* cell structure; fatty acids; myristic acids; prenylation. (Reddy, A., et al. 2001. *Cell* 106:157. Maxfield, F. R. 2002. *Current Opin. Cell Biol.* 14:483.)

cell memory Property of differentiated cells that they reproduce similar specialized cells through numerous cell divisions to which they have been committed. *See* differentiation; memory, immunological; morphogenesis. (Anfossi, N., et al. 2001. *Immunol. Rev.* 181:269.)

cell migration Common during animal development. Precursors of the blood cells, the germ cells, neurons, cells of the somites, metastasis, etc., move through the embryo and are guided by cell surface receptor proteins aided by the extracellular matrix of fibronectin, whereas chondroitin sulfate proteoglycan interferes with the movement. The Kit protein in the membrane of the migrating cells and the ligand Steel factor produced by the cells, which are contacted by the migrant, also control the movement. During oogenesis in *Drosophila*, EGF receptor signals through a TGF-α-like ligand as a guidance cue. Gastrulation requires cell movement of mesodermal cells through the fibronectin-rich matrix in the blastocoele. Contraction of actin and myosin is aided by changes in focal adhesion. Cdc42 stimulates actin polymerization in the forward direction of filopodia and Rac aids the formation of lamillopodia. For cell migration in vertebrates, lysosphingolipids have importance. The Dock2 proteins regulate lymphocyte migration. *See* actin; Cdc42; chondroitin sulfate; EGF; FAK; fibronectin; filopodium; integrin; KIT; lamellipodium; metastasis; microtubules; oncogene; proteoglycan; Rac; sphingolipids; Steel factor; tenascin; TGF. (Lauffenburger, D., et al. 1996. *Cell* 84:359; Fukui, Y., et al. 2001. *Nature* 4121:826.)

cell number In a small *Arabidopsis* plant, about 5×10^6; in an *Arabidopsis* seed, about 6,000 to 7,000; in a wheat embryo 10 days after fertilization, about 40,000, and 150,000 at maturity (including the scutellum); at the surface of a maize endosperm, about 1,400; in the human body, about 6×10^{15} per 60 kg weight (ca. 1 billion per gram tissue) (Robert DeMars, personal comm.). The number of cells in a tissue depends on cell division, cell death, and possibly on migration (in animals). The number of cells per particular structure is also affected by cell size. The local number of cells may be controlled by the tissue environment or extrinsic factors. The number of cells depends primarily on the cell division cycles controlled by a large number of hormones and other proteins. Protein p27 in mouse appears to be a potent inhibitor of the growth of cell number and cell size by controlling cyclin and cyclin-dependent protein kinases. In diploids the body size apparently depends on cell number rather than size (Trumpp, A., et al. 2001. *Nature* 414:768), but in polyploids the size of the cells is larger. *See* cell cycle.

cell-penetrating peptide   11 to 34 amino acid residues long; extremely effective in entering cells. This ability also includes transport of various cargoes into the cell, some instances even traversing the blood-brain barrier. *See* BBB; CRM1; nuclear pore. (Zenklusen, D. & Stutz, F. 2001. *FEBS Lett.* 498:150; Galouzi, I.-E. & Steitz, J. A. 2001. *Science* 294:1895.)

cell plate   Precursor of new cell wall in dividing plant cells. *See* cytokinesis. (Verma, D. P. S. 2001. *Annu. Rev. Plant Physiol. Plant Mol. Biol.* 52:751.)

cell receptor   *See* receptors; signal transduction; transmembrane proteins.

cell sap   Fluid, nonparticulate cell content.

cell size   Varies a great deal depending on organisms and function. An *E. coli* cell is about 800 × 2,000 nm. Plant and animal cells generally have a diameter of 20 to 60 μm and their length is much more variable. Some fibrous cells may be 20 cm long. An important regulator of cell size is the S6 kinase. In budding yeast, the gene *WHI3* increases cell size, whereas increasing the number of copies of *CLN3* decreases the cell volume. *See* cell number; S6 kinase. (Conlon, I. J., et al. 2001. *Nature Cell Biol.* 3:918.)

cell sorter   Cells can be labeled by cognate antibodies coupled with a fluorochrome or by other incorporated material. In a mixture where only one in a few thousands cells carries this distinctive label, the latter ones can be separated using an electronic device. When a file of cells passes in front of a laser beam, the fluorescent cells receive a different electric charge from the unstained ones. The high-intensity electric field down in the path then separates the positively charged fluorescing cells from the negatively charged (unstained) ones. *See* antibody; cell cycle; fluorochromes; labeling index; laser; segregation distorter; sex selection. (Asai, J., et al. 1999. *Clin. Neurol. Neurosurg.* 101:229.)

cell strain   Animal cell culture obtained directly or recently from an organism. It usually has a limited, less than 50, generation life span. These cultures are not immortalized. *See* HeLa; cell culture; immortalization.

cell structure   *See* animal and plant cells diagram on page 188.

cell theory   Proposed in the 19th century. It states that the cells are the elementary units of life and they can be produced only from preexisting cells by mitosis (or meiosis). Abiogenetic reproduction of cells, assumed to exist in the 17th century, has not been shown to exist in the present geological period. *See* origin of life; spontaneous generation.

cell therapy   Transferring specific cells to an organ in the body with the purpose of letting them propagate there, restoring a defective function. The transplanted cells may cure diseases such as Duchenne muscular dystrophy; replace degenerated retinal macula or dopaminergic neurons in Parkinson's disease; or replace bone marrow cells to restore the hematopoietic system or Langerhans islets to fight diabetes, etc. Cell therapy may be part of a cancer treatment. Before radiation or chemotherapy, hematopoietic stem cells may be withdrawn, multiplied, and after the anticancer therapy, they can be reintroduced into the body to restore the immune system. *See* adoptive cellular therapy; cancer gene therapy; diabetes; gene therapy; hematopoiesis; muscular dystrophy; Parkinson's disease; retinal dystrophy; stem cells; transplantation of organelles; xenotransplantation. (Strom, T., et al. 2002. *Curr. Opin. Immunol.* 14:601.)

cell transformation assays in genetic toxicology   Generally hamster embryo cells or mouse prostate cells are exposed to chemicals and tumorigenicity is tested after introduction of the cells into live animals (rodents). In vitro the transformed cells do not grow in monolayer as do normal cells but form a dense mass or colony on top of the monolayer. Also, the activation of c-oncogenes (AKR, adenovirus) by chemicals is investigated. *See* bioassays in genetic toxicology.

cellular immunity   Mediated by the T cells. *See* T cells.

cellulase   Enzyme digesting cellulose. Generally a collection of enzymes is used for removal of the cell wall and gaining plant protoplasts, such as *Onozuka R-10*. *See* cellulose; macerozyme; protoplast.

cellulosome   Macromolecular complex that degrades cellulose and associated polysaccharides. (Shoma, Y., et al. 1999. *Trends Microbiol.* 7:275.)

cellulose   Polysaccharide consisting of glucose subunits. It strengthens the plant cell wall and forms the plant vascular system. *See* cellulase. (Delmer, D. P. 1999. *Annu. Rev. Plant Physiol. Plant Mol. Biol.* 50:245.)

cell wall   Exists in bacteria, fungi, and plants; animal cells are surrounded only by membrane. In bacteria the wall is made of polysaccharides, protein, and lipids. In plants it is mainly cellulose (polysaccharide), but lignin (a hard phenylanine and tyrosine polymer), suberin (a corky wax), and cutin (a fatty acid polymer) occur. In fungi it may contain chitin, a linear polysaccharide differing from cellulose by a replacement at the C-2 OH group of an acetylated amino group.

CEN   Symbol of centromere DNA sequences. *See* centromere.

cenancestor   Most recent common ancestor of two taxa. *See* taxon.

Cenozoic   Geological period dating back to 75 million years ago when mammals and humans appeared. *See* geological time periods.

CENP   Protein that is diffusely located in the cytoplasm during G2 and prophase. During prometaphase, it associates with the kinetochore until metaphase. At anaphase, it is located in the midzone of the spindle and degraded after cytokinesis. *See* cell cycle; CENP-A; CENP-B; CENP-C; centromere; kinetochore; mitosis; proteins; spindle. (Fukagawa, T., et al. 2001. *Nucleic Acids Res.* 29:3796.)

centimorgan Unit of eukaryotic recombination; 1% meiotic recombination is one map unit (m.u.) = 1 centimorgan (cM); ≈10 kb DNA in humans. *See* CentiRay; mapping, genetic; mapping function; recombination frequency.

centiray (cR) Chromosomal span within which a break can be induced with 1% probability by a specified dose of X-radiation. 1 cR ≈ 3 × 10⁴ bp DNA. *See* radiation hybrid.

centisome Quantitative unit of genomic sites.

central body (centrosome) *See* centrosome.

central core disease (19q13.1) Nonprogressive muscle weakness; the core of the muscle fibers is generally absent. It is a complex disorder usually with ryanodine receptor defects. *See* ryanodine.

central dogma Concept that the flow of genetic information follows the path DNA → RNA → protein; it had to be slightly modified by the discovery of reverse transcriptases and ribozymes. *See* reverse transcriptases; ribozyme. (Crick, F. H. C. 1958. *Symp. Soc. Exp. Biol.* 12:138.)

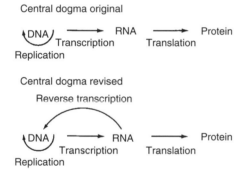

central limit theorem Variable large sample representing the sum of many components is expected to approach the normal distribution. This basic statistical principle has important implications for genetics when population samples and segregation data are evaluated. *See* normal distribution. (Klein, E. K., et al. 1999. *Theor. Popul. Biol.* 55[3]:235.)

central nervous system Brain and spinal cord. *See* brain, human.

central tendency Statistical index—such as the median, mean, and mode—of the typical or average distribution of some characteristics of a population. *See* mean; median; mode.

centric fusion Fusion of two telocentric chromosomes into a biarmed single chromosome. *See* acrocentric; misdivision; Robertsonian translocation; telocentric chromosome.

centric shift Changing the position of the centromere and thus the relative arm length of a chromosome by pericentric inversion or transposition. *See* inversion; shift; transposition.

centrifuge Instrument for sedimenting or separation of material by centrifugal force according to density. Low-speed centrifuges generally spin the suspended material at less than 5,000–6,000 rpm (revolution per minute); tabletop centrifuges usually reach a maximum speed of 10,000–12,000 rpm; high-speed centrifuges may reach about 20,000 rpm and usually are refrigerated so that biological material suffers only minimal degradation. Ultracentrifuges, using refrigerated vacuum chambers, may reach much higher speeds, may exceed 300,000 × *g* force, and can even separate molecules. The conversion of revolution per minute into *g* force is generally done on the basis of tables provided by the manufacturers. The actual centrifugal force

$$F = \frac{\pi^2 S^2 M R}{900},$$

where $\pi = 3.14159$, S = revolutions per minute, M = the mass in grams, and R = the radius in centimeters. It is more convenient to express it as ×*g* force, where *g* = 980.665 cm/sec/sec. Thus, e.g., when the maximal rpm is 20,000, the relative centrifugal force (RCF) may be 41,320 × *g*, but this actually varies from the maximum at the bottom of the centrifuge tube to a lower value at the top, etc. *See* buoyant density centrifugation; density gradient centrifugation; ultracentrifuge.

centrin Cellular motor of polar bodies, centrioles. *See* CDC31; mitosis in unicellular protists; spasmoneme; spindle pole body. (Middendorp, S., et al. 1997. *Proc. Natl. Acad. Sci. USA* 94:9141; Salisbury, J. L., et al. 2002. *Current Biol.* 12:1297.)

centriole Hollow cylinders formed by nine microtubule triplets surrounded by a dense area in the centrosome. The two centrioles in each centrosome serve as the attachment point for the spindle fibers during nuclear divisions and along with the radial array of microtubules form the two *asters* in animal cells. RanGTP mediates this organization in association with the nucleotide exchange factor RCC1, other proteins such as γTurRC (γ-tubulin ring complex), and microtubule-associated protein. The nuclear mitotic apparatus protein (NuMA) is involved in the organization of asters. NuMA also interacts with importin-β, making a link between spindle assembly and nuclear import. *See* centromere; centrosome; Ran; RCC1; spindle fibers. (Wiese, C., et al. 2001. *Science* 291:653; Marshall, W. F. 2001. *Curr. Biol.* 11:487.)

centromere Region of the attachment of chromatids after chromosome replication and of spindle fiber attachment at the kinetochore (localized within the centromere) during nuclear divisions. The centromere used to be called the primary constriction of the chromosomes because by light microscopic techniques it frequently appears as a short, slender region. (Secondary constrictions mark the juncture of the chromosomal satellites.) In some organisms (*Juncaceae, Parascaris, Spirogyra, Scenedesmus*, etc.), the centromeres are diffuse, i.e., their position spreads over the length of the chromosome (holocentric, polycentric chromosome). Under some conditions, "neocentromeres" are visible, i.e., the spindle fibers may associate at more than one location within the chromosomes.

Idealized PLANT CELLS as viewed through

Light microscope   *Electron microscope*

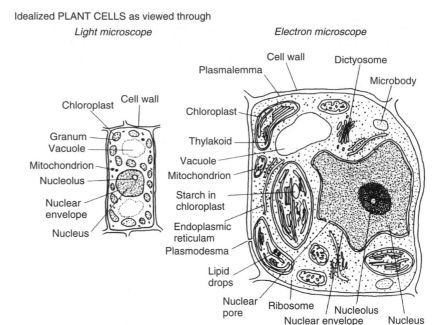

Idealized ANIMAL CELLS as viewed through

Light microscope   *Electron microscope*

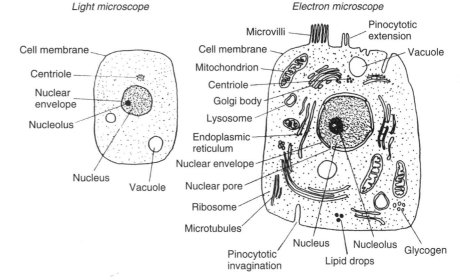

Generalized structure of plant and animal cells by light and electron microscopy. The diagrams do not show the cytoskeleton and various microtubules, filaments, or transport vesicles.

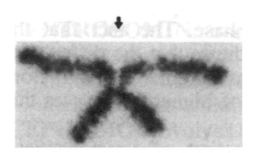

The cloned centromeric region of yeast extends to a minimum of 150 bp. Although the base sequences in various centromeres of budding yeast are not identical, there is substantial homology. All contain a core (element II) of 83–84 base pairs that are 93 to 95% A = T. In addition, there are TCAC and TG identical stretches in flanking element I (11 bp), and on the other flank (element III, ∼25 bp) there are GT and TG and T and CCGAA and TAAAA identical sequences separated by one to four different bases. The general structure of the yeast centromeres can be represented as shown next page.

The regions of the centromere elements (CDEI [8 bp], II [78–86 bp], and III [25 bp]) are less susceptible to nuclease attack than the rest of the chromosomes and seem to indicate that they are associated with different types of proteins. CDEIII is the site of the centromere-binding factor 3 (CBF3) that is essential for segregation of the chromosomes. CDEI contains a nucleotide octamer, similar to the *octa* sequences present in the promoters of several genes where it binds a 39 kDa transcriptional activator protein (Cpf1p/CP1/CBF1). The arrangement of these three elements is shown below. Centromere-specific DNAs are not preserved among evolutionarily distant species.

A number of known proteins, with not entirely known functions, assure the formation of the centromere-kinetochore complex and the attachment of the spindle fiber. The localized centromeres—in contrast to the holocentric ones—may be either *point* or *regional centromeres*. The point centromere contains about 250 bp tightly packaged into a nuclease-resistant structure and it binds only a single microtubule. The point centromere may be arranged around a special nucleosome. The H2B and H4 histones may be altered, and another highly variable protein, related to H-3 (Cse4), is present. Cse4 protein apparently interacts with centromere elements I and II but not III. (Cse4 is an exclusively centromeric protein in *Saccharomyces* like the Cid [centromere identifier] in *Drosophila* or HCP-3 in *Caenorhabditis*.) H-3 like histones are also present in the nucleosomes at the holocentric chromosomes of *Caenorhabditis*. The latter assists chromosome segregation. Point centromeres are found in yeasts (*Saccharomyces cerevisiae*, *Schizosaccharomyces uvarum*, *Kluyveromyces lactii*).

The regional centromeres may consist of several kilobases and may be quite polymorphic. Actually, the *Schizosaccharomyces pombe* centromere is more similar to the mammalian centromeres than to those of budding yeast. The *S. pombe* centromeres vary from 40 to 100 kb, and the core of 4 to 7 kb may be surrounded by direct and inverted repeats. There is a larger array of proteins associated with the regional centromeres. In the majority of higher organisms, the centromere is surrounded by heterochromatin and apparently not transcribed. The budding yeast centromeric DNA (CEN) in chromosome 3 (CEN3) appears to contain an open reading frame capable of coding for a peptide of 52 amino acids. There is no evidence for transcription to take place, however, in the centromere. Actually, this heterochromatic region suppresses the expression of open reading frames even if they are transposed or inserted in this region, indicating that "silencing" may be essential for the proper function of the centromere in chromosome disjunction.

The centromeric Swi6p (yeast) and HP1 (mammalian) proteins are repressors, and *Drosophila* protein Pc (*polycomb*) is a negative regulator of *Bithorax* (*BXC*) and *Antennapedia* (*ANTX*) complexes. In the *Drosophila* centromeric region, the 220 bp Bora Bora complex sequences, flanked by either 5′ or 3′ by ~200 bp "simple" sequence, have been identified. The former is believed to contribute to the kinetochore formation; the latter may control sister chromatid association. The centromeres of mammals display considerable variations and a larger number of proteins including centromere-binding proteins (CENP-A,-B,-C,-D, the kinesin-related MCAK and CENP-E, dynein, INCENPs [move to the microtubules in mitosis], etc). In the centromeric region of a wide range of organisms, a special histone-3-like protein is present in the nucleosomes. This protein is presumed to mark the centromere in higher organisms because this appears to be a common feature in yeast to mammals. In higher organisms, there is no DNA consensus for the centromeric region, unlike in budding yeast, where a unique ~125 bp sequence specifies the centromere. Heterochromatin—usually surrounding the centromere—does not appear to convey determination because the neocentromeres are not heterochromatic.

The centromeres may also have functions during interphase. In human cells, the centromeres seem to be associated with the nucleolus during interphase. The fact that the centromeres are different yet functional when interchanged seems to indicate that the differences are not all functional specificities. The centromeric DNAs in fission yeast are quite different from those of budding yeasts and show more similarity to the centromeres of higher eukaryotes, which are many times larger. The centromeres of higher eukaryotes are composed of repeated sequences, and some losses may not necessarily affect the transmission of the chromosomes. The telocentric chromosomes generated by misdivision usually have impaired

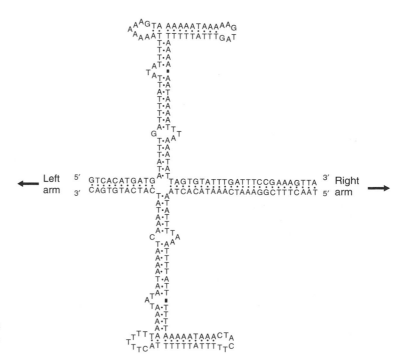

Nucleotide sequences of chromosome 3 centromeric region (627 bp) of yeast. (From Clark, L., et al. 1981. *Stadler Genet. Symp.* 13:9.)

transmission, and the same is true for the B chromosomes prone to nondisjunction.

Recombination is usually much reduced in the centromeric regions. The centromeres contain relatively few genes and pseudogenes, although in *Arabidopsis*, 5 and 12 genes per 100 kb were suggested in two centromeric regions in contrast to 25/100 kb in the normal arms. These genes seem to encode functions of mobile elements, tRNA, regulatory activity, and metabolic steps. Although genes within heterochromatic regions are suppressed, there is evidence for the transcription of some of the genes. The CEN regions include 180 kb repeats.

Although prokaryotes do not have structures exactly homologous to the eukaryotic centromere, the partitioning site of the plasmid DNA (involved in the distribution of the replicated DNA ring) has been called a centromere. This partitioning site, *parS*, forms a module, including protein ParA (an adenosine triphosphatase) and ParB (a *ParS*-binding protein). Similarly to the eukaryotic centromere, prokaryotic genes on the flanks of *parS* may be silenced.

The availability of cloned centromeric DNAs permitted the construction of yeast artificial chromosomes (YACs). When properly constructed, these YACs can be subjected to tetrad analysis. YACs are very useful tools for the physical mapping of larger DNAs. Centromeric proteins determine chromosome segregation and regulate the difference between anaphase I and anaphase II disjunction of chromosomes and chromatids, respectively. *See* α-satellite; aster; B chromosome; centromere A protein; centromere mapping; dynein; holocentric chromosome; kinesin; kinetochore; meiosis; microtubule; misdivision; mitosis; neocentromere; octa; Roberts syndrome; sister chromatid cohesion; spindle fibers; telochromosome; tetrad analysis YAC; yeast centromeric

centromere B protein (CENP-B) 80 kDa. It binds to the 17 bp CEN-B box present in human α-satellite and the mouse minor satellite DNA. CENP-B is not life-essential for mice, yet it leads to lower body weight and testis size. *See* α-satellite DNA; CENP; CENP-A; CENP-C; centromere. (Chen, C., et al. 1999. *Mamm. Genome* 10:13.)

centromere C protein (CENP-C) 140 kDa. It apparently binds DNA and is essential for embryo survival in mice beyond $3\frac{1}{2}$ days after conception. *See* CENP; CENP-A; CENP-B; centromere. (Pluta, A. F., et al. 1998. *J. Cell Sci.* 111 [pt14]:292; Fukagawa, T., et al. 2001. *Nucleic Acids Res.* 29:3796.)

centromere index Length of the short arm divided by length of the entire chromosome × 100. *See* chromosome arm.

centromere mapping in fungal tetrads *See* tetrad analysis.

centromere mapping in higher eukaryotes In heterozygous autotetraploid (triplex: AAAa) or trisomic (duplex AAa) progenies, the greater the proportion of recessive individuals for a particular marker, the further the locus is from the centromere, because only crossing over between gene and centromere (maximal equational segregation) can produce double recessive gametes. Thus, in very large populations the relative distances can be estimated. More precise estimates of gene centromere distance can be obtained in allopolyploids by using telochromosomes that are usually not transmitted through the pollen. The experimental design may be as follows (only one pair of chromosomes is shown):

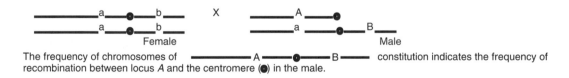

The frequency of chromosomes of ——A—●—B—— constitution indicates the frequency of recombination between locus *A* and the centromere (●) in the male.

factor. (Copenhaver, G. P., et al. 1999. *Science* 286:2468; Dobie, K. W., et al. 1999. *Curr. Opin. Genet. Dev.* 9:206; Gindullis, F., et al. 2001. *Genome Res.* 11:253; Sullivan, B. A., et al. 2001. *Nature Rev. Genet.* 2:584; Choo, A. K. H. 2001. *Developmental Cell* 1:165; Hennikoff, S., et al. 2001. *Science* 293:1098; Schueler, M. G., et al. 2001. *Science* 294:109; Hudakova, S., et al. 2001. *Nucleic Acids Res.* 29:5029; Smirnova, J. B. & McFarlane, R. J. 2002. *J. Biol. Chem.* 277:19817; Smith, M. M. 2002. *Current Opin. Cell Biol.* 14:279; Blower, M. D., et al. 2002. Developmental cell 2:319.)

centromere activation Transposition of the centromere to a new position within the chromosome. *See* centromere; holocentric; neocentromere.

centromere A protein (CENP-A) 17 kDa histone-3-like protein and part of the centromeric nucleosomes. Its disruption or loss severely affects mitosis and causes fragmentation of the chromatin. Its normal role is marking centromere organization in the chromosome. *See* CENP; centromere; centromere B protein; centromere C protein. (Murakami, Y., et al. 1996. *Proc. Natl. Acad. Sci. USA* 93:502; Sugimoto, K., et al. 2000. *Cell Struct. Funct.* 25:253.)

The centromeres of the chromosomes of rice were mapped using RFLP markers in telo- and isotrisomics. The distance was calculated on the basis of linkage intensities of markers on the opposite sides of the centromere, inferred by gene dosage. In mice, Robertsonian translocations can be exploited for centromere mapping. In most eukaryotes, the centromeric region has repeated sequences; these can also be used for centromere mapping of RFLPs. *See* allopolyploid; half-tetrad analysis; maximal equational segregation; RFLP; Robertsonian translocation; telochromosome; tetrad analysis; trisomic analysis. (Sears, E. R. 1966. *Hereditas*, Suppl. 2:370.)

centromeric fission *See* misdivision of the centromere; telochromosome.

centromeric fusion Joining two telocentric chromosomes into a single biarmed one.

centromeric vector *See* YAC.

centrosome Center (~1 μm) where spindle fibers (microtubules) originate (a microtubule organizing center) and develop from a pair of centrioles toward the centromeres during nuclear divisions in animals (and few lower plants). A pericentriolar material made of γ-tubulin and Asp (asymmetric spindle) protein of 220 kDa surrounds the centrioles. Asp has phosphorylation sites for p34 and other mitogen-activated kinases as well as binding domains for actin and calmodulin. The fast-growing (plus) ends of the microtubules project into the cytoplasm, whereas the slow-growing (minus) ends are embedded into the γ-globulin ring. The division of the centrosome (also called centrosome cycle) is essential for the completion of the cell cycle in animals and it may be blocked by mutation in gene *cdc31*.

For the initiation of the division of the centrosome in *Xenopus* egg calcium, calmodulin and calcium/calmodulin-dependent protein kinase II are required (Matsumoto, Y. & Maller, J. L. 2002. *Science* 295:499). The aurora-2/STK15 serine/threonine kinase (encoded in human chromosome 20q13.2) is associated with the centrosome, and its amplification interferes with the normal function of the centromere, resulting in aneuploidy in common cancer cells. Cdk2–cyclin E may cause abnormally high proliferation of the centrosomes as it phosphorylates nucleophosmin, a centrosomal protein. Hsp90 is also a component of the centrosomal core, along with several other proteins. The deficiency of the p53 tumor suppressor protein results in multiple centrosomes and unequal distribution of chromosomes. Similar are the consequences of defects in PLK1 and an ataxia telangiectasia (rad3)–related (ATR) mutation. Plants do not have such distinct structures. Mutations affecting the centrosomes have been isolated. The centrosome is paternally derived during fertilization in the majority of the animal species. Depending on the sperm donor, the size of the aster may vary. The centrosome is also required for DNA synthesis during the cell cycle. Centrosome defects may lead to aneuploidy and cancer. *See* actin; aneuploidy; aster; calmodulin; Cdk; centrioles; centromere; centrosome; centrosomin; dynein; Hsp90; kinesis; microtubule; mitosis; multipolar spindle; p34; p53; spindle; spindle pole body. (Brinkley, B. R. 2001. *Trends Cell Biol.* 11:18; Stearns, T. 2001. *Cell* 105:417; Bornens, M. & Piel, M. 2001. *Curr. Biol.* 12:R71; Bornens, M. 2002. *Current Opin. Cell Biol.* 14:25; Meraldi, P. & Nigg, E. A. 2002. *FEBS Lett.* 521:9.)

centrosomin One of the essential protein components of the centrosome. *See* centrosome. (Vaizel-Ohayon, D. & Schejter, E. D. 1999. *Curr. Biol.* 9:889.)

CEPH Centre d'Étude du Polymorphism Humain, Paris, France, research institute where human cells lines were collected and are maintained from four grandparents, two parents, and their multiple children in order to map their genes and study their transmission. The institute is involved in the study of the human genome. (<http://www.cephb.fr>.)

cephalic Involves the head or indicates the direction toward the head.

cephalohepatorenal syndrome *See* Zellweger syndrome.

cephalosporin-type antibiotics Derived name from *Cephalosporium acrimonium*; include a number of natural and semisynthetic antibiotics. The latter may be resistant to the enzyme penicillinase. Their action involves interference with the cross-linking of peptidoglycans of the bacterial cell wall. *See* antibiotics; β-lactamase; lactam; penicillin. *Clin. Microbiol. Infect.* 2000 Suppl. 3:1.)

ceramide Structural unit of sphingolipids, a fatty acid attached by $-NH_2$ linkage to a sphingosine molecule. Ceramides mediate stress responses, apoptosis, cell cycle arrest, and senescence. *See* CAP; Farber's disease; sphingolipids. (Hannun, Y. A. & Luberto, C. 2000. *Trends Cell Biol.* 10:73.)

Cerberus Factor expressed in the organizer of *Xenopus* embryos, causing the development of ectopic heads, duplicated hearts and livers. (It was named after the mythological three-headed monster guarding the gate of the underworld.) *See* ectopic expression; organizer.

Cercopithecidae (Old World monkeys) *Allenopithecus nigroviridis* $2n = 48$; *Cercocebus torquatus* $2n = 42$; *Cercopithecus aethiops sabaceus* $2n = 60$; *Cercopithecus ascanius* $2n = 66$; *Cercopithecus cephus* $2n = 66$; *Erythrocebus patas* $2n = 54$; *Macaca fascicularis* $2n = 42$; *Macaca mulatta* $2n = 42$; *Miopithecus talapoin* $2n = 54$; *Papio* spp. $2n = 42$; *Presbytis melalophus* $2n = 44$; *Presbytis senex* $2n = 44$. *See* primates.

cerebellum Hind part of the brain; supposed to be involved in the coordination of movements. Recent information indicates that the cerebellum acquires and discriminates among sensory information rather than directly controlling movements. The Purkinje cells in the cortex are involved in the information output. Each Purkinje cell is innervated by the *mossy fiber system*, up to 200,000 parallel fibers originating from the deeper layers of the cerebellum, and by a single *climbing fiber* originating from the oliva (a mass of cells) below the surface of the cerebellar cortex. *See* brain, human; cerebrum; Purkinje cells.

cerebral cholesterinosis (CTX) Human chromosome 2q33-qter recessive deficiency of sterol-27 hydroxylase, mitochondrial P-450, and other mitochondrial proteins, as well as adrenodoxin reductase, causes lipid (cholesterol) accumulation in the tendons, brain, lung, and other tissues. *See* adrenodoxin; cholesterol; mitochondria. (Rosen, H., et al. 1998. *J. Biol. Chem.* 273:14805.)

cerebral gigantism (Sotos syndrome) Rare autosomal-dominant (3p21 or 6p21) condition involving excessive bone growth and usually mental retardation. Male-to-male transmission is predominant. *See* mental retardation.

cerebral palsy *See* palsy.

cerebro-oculo-facio-skeletal syndrome (COFS, 10q11) Recessive, progressive brain (microcephaly, atrophy), eye, and joint anomaly with resemblance to Cockayne syndrome and CAMFAK syndrome. Defective nucleotide exchange repair may be involved. *See* CAHMR; CAMFAK; Cockayne syndrome; Martsolf syndrome. (Graham, J. M., et al. 2001. *Am. J. Hum. Genet.* 69:291.)

cerebrosides Sphingolipids, sugars linked to a ceramide. In the membranes of neural cells, the sugar is generally galactose; in other cell membranes, it is generally glucose. *See* sphingolipids.

cerebrotendinous xanthomatosis Cerebral cholesterinosis.

cerebrum   Major part of the brain in two lobes filling the upper part of the cranium. *See* brain, human; cerebellum.

Cerenkov radiation   When charged particles pass through an optically transparent material at a speed exceeding that of light, they cause emission of visible light. Cerenkov radiation is used in high-energy nuclear physics and molecular biology for the detection of charged particles and the measurement of their velocity.

ceroid lipofuscinosis (NCL)   Apparently autosomal recessive (assigned to several chromosomes). Brown ceroid (waxlike) deposits in several internal organs, including the nervous system, cause spasms and mental retardation. The infantile subtype (CNL1) was located to chromosome 1p32, and it involves rapidly progressing mental deterioration due to a deficiency of palmitoylprotein thioesterase. Its prevalence is about 1/12,500. CNL3 (16q12.1) or Batten disease/Vogt-Spielmeyer disease involves neuronal degeneration, loss of brain material, and retinal atrophy. This affects either a lysosome-associated membrane protein or neuronal synaptophysin. Its prevalence at live birth is ~ 4–5×10^{-6} to 5×10^{-5}. CLN2 (11p15.5, Jansky-Bielschowsky disease) is a late juvenile type. The late infantile neuronal ceroid lipofuscinosis (CLN5, 13q22) was attributed to a pepstatin-insensitive lysosomal peptidase or lysosomal transmembrane protein. CLN6 (15q21-q23) is another late infantile form. Some other variants with granular osmiophilic deposits and others have been described. *See* Batten disease; epilepsy; mental retardation; pepstatin; prevalence; synaptophysin. (Lehtovirta, M., et al. 2001. *Hum. Mol. Genet.* 10:69; Gao, H., et al. 2002. *Am. J. Hum. Genet.* 70:324; <http://www.ucl.ac.uk/ncl/>.)

certation   Competition among elongating pollen tubes for fertilization of the egg. Genetically impaired pollen tubes are at a disadvantage, which may cause a distortion of the phenotypic ratios because certain phenotypic classes may not appear or appear at a reduced frequency. *See* cytoplasmic male sterility; gametophyte; gametophyte factor; last-male sperm precedence; male sterility; meiotic drive; segregation distortion; selection conditions. (Nilsson, H. 1915. *Lunds Univ. Aarsskr. N.F. Adf.2* (12):1; Konishi, T., et al. 1990. *Jap. J. Genet.* 65:411.)

ceruloplasmin   Blue copper-transporting glycoprotein in the vertebrate blood. It is located in human chromosome 3q. *See* aceruloplasminemia; Wilson disease.

cervical cancer   Appears to be associated with infection by the human papilloma virus; however, there seems to be a genetic predisposition to susceptibility. *See* papilloma virus.

cesium   Alkali metal element; its salts CsCl (MW 168.4) and Cs_2SO_4 (MW 361.9) are used as density gradient solutions for preparative and analytical ultracentrifugation, respectively. *See* buoyant density; ultracentrifugation.

CETP   Cholesterylester transfer protein mediates the catabolism of HDL and the transfer of cholesterol to the liver and may thus be antiatherogenic. See atherosclerosis; HDL.

cetyl pyridinium bromide (CPB)   Cationic detergent used for the precipitation of (radiolabeled) oligonucleotides.

cetyl trimethyl ammonium bromide (CTAB)   Detergent suitable for the precipitation of DNA. Generally used in a stock solution in 0.7 M NaCl. *See* formula at CTAB.

CFTR   Cystic fibrosis transmembrane conductance regulator. *See* cystic fibrosis.

CFU   Colony-forming unit. Number of cells/mL capable of propagation in in vitro culture. *See* pfu.

CG   Dinucleotide; where the cytosine is most commonly methylated in vertebrates. The so-called maintenance methylase enzyme acts on it when paired in a complementary manner in the DNA. The methylation may be transmitted through DNA replication. *See* methylation of DNA.

CGAP   Cancer Gene Anatomy Project attempts to identify the function(s) of all human genes one time estimated to be 60,000–150,000, but now the number appears to be $\sim$30,000–40,000. (<http://www.ncbi.nlm.nih.gov/UniGene/genediscovery.html>.)

C gene   Gene coding for the constant region of the antibody molecule (such as $C\mu$, $C\delta$, $C\gamma$, $C\varepsilon$, $C\alpha$. *See* antibody; immunoglobulins.

CGH   *See* comparative genomic hybridization.

CGIAR   International food and agricultural policy institute.

cGMP   Cyclic guanosylmonophosphate, a second messenger. *See* cAMP; second messenger.

C2H2   Ubiquitous zinc-finger regulatory protein domain.

CH$_{50}$ In vitro assay for the activity of the complement. *See* complement.

chaetae   Bristles of insects, sensory organs of the peripheral nervous system. The large ones, macrochaetae, are mechanical sensory organs, and the smaller ones of different types are microchaetae (a fraction of them are chemoreceptors). *See* microchaetae; tormogen; trichome.

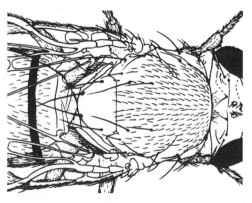

Chaetae.

Chagas disease   Nonhereditary, potentially fatal disease caused by infection by *Trypanosoma cruzi*. *See* paratransgenic; *Trypanosoma*. (Cohen, J. E. & Gürtler, C. E. 2001. *Science* 293:694.)

chain-sense paradox   Expresses the problem of conformational switch from B to Z DNA. *See* DNA types.

chain shuffling of antibodies *See* phage display.

chain termination *See* initiator codon; nonsense codons; transcription termination. DNA sequencing [Sanger method].

chalaza (1) Site on the plant seed where the funiculus unites with the ovule. (2) Points in the bird eggs where the yolk is connected to the egg shell.

chalcone Phenylalanine is converted to *trans*-cinnamic acid and cinnamoyl-CoA, which through a condensation reaction yields chalcone from which a series of other plant pigments (flavonones, flavones, flavonols, anthocyanidins, etc.) are derived through single-gene-controlled biochemical steps.

chalone Water-soluble glycoprotein that can inhibit mitosis.

Chambon's rule Splice points at the ends of intervening sequences are generally GT · · · AG (except in tRNA genes and other minor classes of genes). *See* introns; splicing.

chameleon protein Contains a short amino acid sequence that may fold either as an α-helix or a β-sheet depending on its position. *See* protein structure.

Chanarin-Dorfman disease (ichthyotic neutral lipid storage disease, CDC, human chromosome 3p21) Triglyceride storage disease with defective long-chain fatty acid oxidation. The basic defect involves an esterase/lipase/thioesterase protein. Within the cells, triacylglycerol droplets are found, and liver, muscle, and eye anomalies may accompany ichthyosis. *See* ichthyosis. (Lefèvre, C., et al. 2001. *Am. J. Hum. Genet.* 69:1002.)

chance Statistical probability or uncertainty. *See* likelihood; probability.

change of state Different levels of methylation of a genetic sequence. *See* Spm.

channel Path through which signals (molecules) can be transmitted.

channeling (1) Tunneling. Transfer of a common metabolite between two enzymes in a sequential and parallel function; e.g., a mutation may shut down ✋ the carbamyl phosphate pool leading to arginine synthesis, but an overflow through a tunnel from the carbamyl phosphate precursor in the pyrimidine pathway may substitute for the defect and eliminate the dependence on exogenous arginine because of the channeling ⮕ of the accumulated carbamyl-P_{Pyr} into the arginine pathway when another mutation blocks ✋ the pyrimidine path. Tryptophan synthase, glutamine phosphoribosylphosphate amidotransferase, asparagine synthetase, and other multifunctional enzymes also display this phenomenon. Channeling may go both ways between two metabolite pools. *See* regulation of gene activity. (Ovádi, J. & Srere, P. A. 2000. *Int. Rev. Cytol.* 192:255; Huang, X., et al. 2001. *Annu. Rev. Biochem* 70:149.) (2) Topologically constrained intramolecular recombination of transposons. Such an event may lead to the formation of a new element rather than destruction of it.

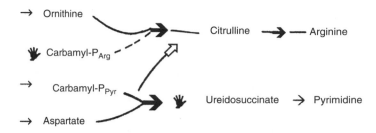

chaos System extremely sensitive to minute perturbations; also called the butterfly effect. In common usage, chaotic conditions are meant to be uncontrollable and unpredictable because many degrees of freedom should be dealt with simultaneously. There are, however, high-dimensional systems where the attractor (the dynamics) is low-dimensional. Among the infinite varieties in the system, one can be selected and stabilized by minute changes in one parameter. The task is to locate the critical points in a multitude of "noise." Various mathematical algorithms have been worked out to deal with the problems. Systems of chaos are characterized by nonlinear dynamics. The control of chaotic systems has great relevance to biology and genetics, where a large number of genetic (epistatic), epigenetic, and environmental factors interact, such as the chaos in the heart, neuronal information processing, epileptic seizures, development, differentiation, etc. Chaos control (anticontrol) may then be applicable to harness the system and manipulate controls over it even when all the details are not understood. *See* algorithm; fractals. (Williams, G. P. 1997. *Chaos Theory Tamed.* Taylor & Francis, Bristol, PA; Bar-Yam, Y. 1987. *Dynamics of Complex Systems.* Addison-Wesley, Reading, MA.)

chaotrope Interaction between ion-protein complexes are more attractive than uncomplexed proteins. This is called also salting-out. When interaction between protein-ion complexes are more repulsive (kosmotrope) than in uncomplexed proteins, salting-in is observed. (See Curtis, C. A., et al. 2002. *Biotechnol. Bioeng.* 79:367.)

chaperone Protein mediating the conformational change or assembly of polypeptides (usually) without becoming a permanent part of the final product (e.g., the heat-shock protein families Hsp70 and Hsp 60, rubisco, etc.). Their major function is prevention of inappropriate conformation, aggregation, and interaction with incorrect ligands. Chaperones may restore some aggregates to native conformation in case the denaturation is not irreversible. The chaperones come in greatly different molecular sizes that are shown in their designation (in kDa). snRNAs may mediate the folding of rRNAs (Weeks, K. M. 1997. *Curr. Opin. Struct. Biol.* 7:336). An RNA-dependent ATPase may chaperone RNAs (Mohr, S., et al. 2002 Cell 109:769). The heat-shock proteins may also play a role in signal transduction by modifying the folding of steroid hormone receptors. The chaperones may be classified into the Hsp70 (heat-shock protein 70) and the chaperonin families such as the Hsp60s. The chaperones recognize short extended polypeptides rich in hydrophobic residues that are released upon ATP hydrolysis. Chaperones may facilitate protein degradation within the cells. The chaperonins are large oligomeric ring complexes. The mitochondrial proteases

(Lon, Afg3p, Rca1p) can carry out chaperone functions and assemble mitochondrial proteins. The small *intramolecular chaperones* (IMC) are different from the *molecular chaperones* inasmuch as they do not require ATP for folding, are very highly specific and not reusable, and can change structure of mature proteins. Misfolding of proteins may lead to the development of Alzheimer disease, prions, etc. *See* Alzheimer disease; antichaperone; chaperonins; cue; DnaJ; flexer; GroEL; heat-shock proteins; HSE; HSP; Hsp70; PDI; PPI; prion; protein folding; trigger factor. (Ellis, R. J. & van der Vies, S. M. 1991. *Annu. Rev. Biochem.* 60:321; Frydman, J. 2001. *Annu. Rev. Biochem.* 70:603; Dobson, C. M. 1999. *Trends Biochem. Sci.* 24:329.)

chaperonin (cpn)   Ring proteins (chaperonin 60 and 10) mediate the assembly of 12 identical phage (λ, T4, T5)-encoded polypeptides; these serve as a template for lining up the phage head precursors. Chaperonin 10 releases the phage proteins from chaperonin 60. Homologous proteins occur in bacteria and mitochondria and chloroplasts of eukaryotes. Their most essential function is (generally ATP- and K$^+$-dependent) folding of proteins. The best-known chaperonin proteins belong to two families: the GroEL/GroES (include Hsp60 and rubisco-binding proteins) and the TRiC. These form a porous cylinder of 14 subunits. The folding of proteins has numerous implications for normal function of proteins as well as for pathological conditions. The thermosome (TF55/TCP1) chaperonins occur in the thermophilic Archaea bacteria. The latter are related to other cytosolic chaperonins of eukaryotes (animals and plants). The *cytosolic chaperonins* (CCT, ca. $2–3 \times 10^5$ complexes per mammalian cell) function similarly to the organellar chaperonins (ATPase activity, folding nonnative proteins), although structurally they are different. All chaperonins have the molecular mass of 800–1,000 kDa, built of ca. 60 kDa subunits, $(\alpha, \beta, \gamma, \varepsilon, \zeta, \eta, \delta, \theta)$ encoded by *Cct* genes. The GroES heptamer has 10 kDa subunits. The folding takes place within a GroEL chamber after it has entered through the small GroES capping proteins. *See* 7B2; chaperones; Cpn10; Cpn21; Cpn60; GreEL; heat-shock proteins; rubisco; TCP-1; TRiC. (Ang, D., et al. 2000. *Annu. Rev. Genet.* 34:439; Gottesman, M. E. & Hendrickson, W. A. 2000. *Curr. Opin. Microbiol.* 3:197; Thirumalai, D. & Lorimer, G. H. 2001. *Annu. Rev. Biophys. Biomol. Struct.* 30:245; Hartl, F. U. 2001. *Cell* 107:223.)

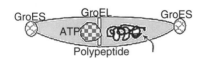

character   (1) Trait in genetics that may or may not be expressed through inheritance. (2) Any symbol used for conveying information, e.g., letters, numerals, punctuation marks, etc.

character displacement   Occurs when two species occupying similar but not identical habitats share a common area, and in this shared zone each differs more from the other regarding a particular trait(s) than in the nonshared area.

character matrix   Device of classification of different groups regarding a trait as have it (1) or not (0). On this basis, a

similarity index can be obtained; the distinguished groups are called *operational taxonomic units* (OTU). If the differences are counted or measured, a *distance matrix* is obtained. On this basis of the similarities/differences, branching *phenograms* or *dendrograms* can be constructed. *See* species. (Ward, B. B. 2002. *Proc. Natl. Acad. Sci. USA* 99:10234.)

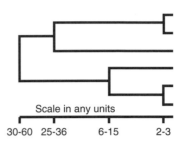

character process model   Analyzes the changes in a trait during a process and the correlation of stages of the process according to the expanded formula: $G(s, t) = v_g(s)v_g(t)\rho_g(|s - t|)$, where v_g = genetic variance, s and t = stage/time variations, and $\rho_g(|s - t|)$ = genetic correlation between stages/time.

Charcot-Marie-Tooth disease (CMT, hereditary motor and sensory neuropathy)   Known in multiple forms with autosomal-recessive, -dominant, or even Xq13-linked types. Its frequency is about 0.0004. The disease affects the nervous system and hearing; it is debilitating but not lethal. The average age of onset is in the early teens but is more clearly manifested by the late 20s. The defect type 1B involves human chromosomes 1p21-23 or HNPP at 17p12-p11.2 encoding myelin 22. Its presence can be detected by genetic screening in the affected families. In the basic defect, connexin 32 encoded at Xq13.1 has also been implicated. Type 2 disease was found to be associated with human chromosomes 1q21.2-q21.3, 3q13-q22, and 7p14. The recessive demyelinating dual-specifity phosphatase (myotubularin-related protein 2) is encoded at 11q22. This disease may also be caused by a duplication due to unequal crossing over in the peripheral myelin gene (chromosome 17) induced by a transposable element resembling *mariner* in *Drosophila*. An autosomal-recessive form is determined by chromosome 8q21.1 (GDAP1), and in 1p36 another locus has been implicated. An axonal form of the disease maps to 19q13.3 encoding the two PDZ domain proteins, periaxins. In the mouse, the *trembler* mutation and some others are involved in hypomyelination. *See* connexins; genetic screening; HNPP; hypomyelinopathies; *mariner*; MITE; MLE; myelin; neuropathy; Pelizaeus-Merzbacher disease. (Leal, A., et al. 2001. *Am. J. Hum. Genet.* 68:269; Sherman, D. L., et al. 2001. *Neuron* 30:677; Zhao, C., et al. 2001. *Cell* 105:587; Cuesta, A., et al. 2002. *Nature Genetics* 30:22.)

Chargaff's rule   In double-stranded DNA, nucleotides are paired as A = T and G ≡ C (*see* hydrogen pairing, DNA). Therefore, the amount of adenine equals that of thymine, and the quantity of guanine is about the same as that of cytosine in double-stranded DNA. In the majority of organisms AT is not equal in amount to GC. The recognition of this fact contributed significantly to the construction of the Watson-Crick model of DNA. *See* Watson and Crick model. (Forsdyke, D. R. & Mortimer, J. R. 2000. *Gene* 261:127.)

CHARGE Coloboma, heart anomalies, choanal atresia (obstruction of the nasal passageway by bony malformation), retardation, genital, and ear anomalies is a complex human disease association syndrome; about 8% familial. Chromosomes 14q22-q24.3 and 22q11 have been implicated but unequivocally not demonstrated. A CHARGE-like X-chromosomal syndrome has also been reported. *See* coloboma, familial.

Charge-coupled devices (CCD)   Are used for highly sensitive imaging of fluorescence, bioluminescence and tomography. The devices are based on semiconductors arranged in a manner that the output of one serves as input for the next.

charged tRNA   Transfer RNA carries an amino acid. *See* aminoacyl tRNA synthetase; protein synthesis; tRNA.

charge-to-alanine scanning mutagenesis *See* homologue scanning mutagenesis.

charomid   Specially constructed phage lambda-derived vector. The constructs must be of a minimum of 38 kb; otherwise, they cannot be packaged into infectious particles. A conventional cosmid vector is 5 kb; therefore, the minimal size of an insert would be 33 kb. When the fragment to be cloned is much smaller, charomids are used that carry repeating units of about 2 kb fragments of plasmid pBR322 in head-to-tail arrangement as space fillers. Thus, depending on the nature of charomids, they can be used for cloning DNAs from 2 to 45 kb length. In *recA⁻* bacteria (ED8767), these vectors are quite stable and can be used just like cosmids. *See* cosmids; *Rec*; vectors. (Saito, I & Stark, G. R. 1986. *Proc. Natl. Acad. Sci. USA* 83:8664.)

Charon vector   Modified phage λ-plasmids (pronounce kharon; named after the ferryman of mythology who carried the dead souls through the infernal Styx River). The Charon vectors are primarily replacement vectors, i.e., they can place DNA into deleted parts of the stuffer region (between λ-genes J and N, *see* diagram at lambda phage). There are a large number of Charon vectors with somewhat different features. *Charon 4* was used mainly to generate eukaryotic genomic libraries. In the replacement (stuffer) region, there are E.coRI of 6.9 kb and 7.8 kb containing the β-galactosidase (*Lac Z*) and the biotin (*bio*) genes, respectively. Successful replacement is recognized by the inability of *lac⁻* bacteria carrying this vector to develop blue color on Xgal medium because the vector can no longer provide the galactosidase function. Replacement of the *bio* gene results in biotin dependence in a similar way. The vector is also *Spi⁻* (wild-type lambda-phages are unable to grow in bacteria containing prophage P2 and have the designation *Spi⁺* [sensitive to P2 interference]). The Spi selection is inoperable with Charon 4. Charon vectors that lose the *red* and *gam* functions (required for recombination) can grow in P2 lysogens and are called *Spi⁻*. The bacteria have the *rec⁺* gene (mediating recombination) and the phage has its own *chi* element (a substrate for the *recBC* system of recombination), or the inserted foreign gene contains one

(mammalian DNA has numerous *chi* elements). Also, *supE* and *SupF* (amber suppressors) must be present in the host to allow for selection of recombinant libraries by in vivo recombination because vector *A* and *B* genes carry amber mutations (chain termination codon UAG). Besides the EcoRI sites, there is an XbaI site in the stuffer region that accepts up to 6 kb insertion. *Charon 32* vector permits cloning DNA fragments at EcoRI (substitution up to 19 kb), HindIII (substitution up to 18 kb), and SacI sites (insertion up to 10 kb). In these cases, *recA⁻* hosts suffice because the vector is *gam⁻*. When Charon 32 is used as a substitution vector for EcoRI-SalI, SalI-XhoI, or EcoRI-BamHI fragments, the host must be *recA⁺* because *gam* is lost. *Charon 34* and *Charon 35* are suitable for cloning fragments up to 21 kb at polycloning sites, using BamHi, EcoRI, HindIII, SacI, XbaI, and SalI, as well as their combinations. These vectors retain *gam* functions. *Charon 40* is useful for cloning fragments from 9.2 to 24.2 kb. In the stuffer region, there are 16 restriction sites in opposite orientation near the ends. The poly-stuffer can be broken down into small fragments by the use of restriction enzyme NaeI (GCC ↓ GGC) and can be collected by precipitation with polyethylene glycol. The vector retains *gam*. *See Lac* operon; lambda phage; restriction enzyme; *Spi+*; *supC*; *supD*; *SupE*; *supF*; *supG*; *supU*; vectors; Xgal. (Chauthaiwale, V. M., et al. 1992. *Microbiol. Rev.* 56:577.)

Char syndrome   *See* patent ductus arteriosus.

Chase method   *See* haploids.

chasmogamy   Pollination takes place after the flower opens. *See* cleistogamy.

CheA, CheB, CheR, CheW, CheY, CheZ   Bacterial cytoplasmic proteins mediating the transduction signals of chemoeffectors through transducers to switch molecules. CheA is an autophosphorylating (histidine) kinase that also phosphorylates CheY and CheB. CheY autophosphorylates spontaneously in a few seconds and the process is accelerated by CheZ. CheA is central to information processing by the four transducers in cooperation with CheW. CheR and CheB

Structure of the CheY protein. (From Alm, E. & Baker, D. 1999. *Proc. Natl. Acad. Sci. USA* 96:11305.)

are cytoplasmic bacterial proteins mediating the return of the chemotaxis excited cells to the normal state (adaptation). CheR is a methyltransferase, CheB is a methylesterase. *See* autophosphorylation; chemotaxis; effector; esterases; excitation; switch, genetic; transducer proteins; transferases enzymes. (Sourjik, V. & Berg, H. C. 2000. *Mol. Microbiol.* 37:740; Bray, D. 2002. *Proc. Natl. Acad. Sci. USA* 99:7.)

checkerboard Representation of the genotypic array of snapdragon flowers in the style of a checkerboard where the top line and the left-most column of symbols display the gametic combinations. Note the 1:2:1 segregation in each of the four boxes. The left top individual is homozygous for both dominant alleles; the right bottom is homozygous for both recessives. At the diagonal from bottom left to top right, all genotypes are heterozygous for the two genes. The diagonal from top left to bottom right represents homozygotes at both genes. Along the other left-to-right diagonals within the four boxes, the immediate neighbors are of identical constitution. This representation is also called Punnett square. *See* Mendelian laws; Mendelian segregation; modified checkerboard; Punnett square. Diagram below.

	IncRad	Incrad	incRad	incrad
IncRad	Red, Normal	Red, Normal	Pink, Normal	Pink, Normal
Incrad	Red, Normal	Red, Peloric	Pink, Normal	Pink, Peloric
incRad	Pink, Normal	Pink, Normal	White, Normal	White, Normal
incrad	Pink, Normal	Pink, Peloric	White, Normal	White, Peloric

Checkerboard.

checkpoint Critical phases in the progress of cell division where the cycle can be stopped and kept, yet when conditions become appropriate, progress may be resumed. The main role of the checkpoint is to prevent progression to G_2 phase until the S phase is completed. There may be another checkpoint at the G_2 phase, preventing the completion of the cell cycle. The checkpoint indicates a step initiating a new direction in the progression; before the cell embarks on a new path, the preceding steps must be completed. The purpose of the checkpoints is to prevent mitosis of defective cells. Several proteins have now been identified in different organisms to mediate checkpoints. The checking is generally mediated by different cyclin-dependent kinases (CDKs).

In yeasts there are fewer checkpoints, whereas in animal cells they are more specialized. The CDKs may be activated/deactivated by phosphatase/kinase action; in addition, binding proteins, cyclin kinase inhibitors (CKI), and other modifiers may participate. These proteins are also subject to proteolytic destruction. In multicellular higher organisms, apoptosis is involved, which does not stop the cell cycle at a stage but eliminates the cell with a defect.

In yeasts a network of checkpoints have been identified including *RAD9, RAD17, RAD24, RAD53, DDC1, PDS1, POL2, PRI1, RFC5, MEC3*, and *MEC1* genes. Protein 14-3-3σ is also a component of the G_2 checkpoint; in the absence of its function, the cell cycle fails to stop and proceeds to death. Protein 14-3-3 normally binds Cdc25 phosphorylated at Ser[216]; the cell cycle proceeds, but phosphorylated Cdc25 is ferried out of the nucleus after DNA damage and the G2 checkpoint is abrogated. Chk1 phosphorylates Cdc25 at 216. Protein 14-3-3σ is transcribed upon the action of p53, and it sequesters cyclin B1 and Cdc2 in the cytoplasm, preventing their entry into the nucleus; thus, the cell cycle is not completed when the DNA is damaged. *See* APC; ATR; cdc2; CDC5; cdc25; Cdc28; cell cycle; chk1; cyclin B; FK506; FKH; p21; p53; p56[chk1]; PDS; PIK; protein 14-3-3; RAD; replication, S phage in eukaryotes; restriction point; SCF; sister chromatid cohesion; wee. (Skibbgens, R. V. & Hieter, P. 1998. *Annu. Rev. Genet.* 32:307; Nigg, E. A. 2001. *Nature Rev. Mol. Cell Biol.* 2:21; Melo, J. & Toczyski, D. 2002. *Current Opin. Cell Biol.* 14:237.)

Chédiak-Higashi syndrome (CHS) Autosomal-recessive defect of the cytotoxic T cells in human chromosome 1q42.1-q42.2. The afflicted individual displays reduced pigmentation of the hair and eyes, avoidance of light, reduction in the number of neutrophilic lymphocytes (neutropenia), high susceptibility to infections, and lymphoma. In the heterozygotes, the lymphocytes appear abnormally granular. Similar diseases occur in many mammals. Molecular investigations reveal a defect in a protein with carboxy-terminal prenylation and multiple potential phosphorylation sites (LYST). It may be involved as a relay integrating cellular signal response coupling. Its mouse homologue is the *beige* locus. *See* albinism; Hermansky-Pudlak syndrome; lymphocytes; lymphoma; neutrophil.

CHEF Contour-clamped homogeneous electric field that alternates between two orientations. Multiple electrodes along a polygonal contour generate the electric field. The method

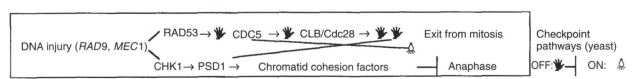

(Diagram modified after Sanchez, Y. *et al.* 1999 Science 286:1166.)

applies the principles of electrostatics (statical electricity) to gel electrophoresis of very large molecules such as the entire DNA of small chromosomes. *See* pulsed-gel electrophoresis.

cheirology (dactylology)   Study of hands, fingerprints. *See* fingerprinting.

chelation   Holding of a hydrogen or metal atom between two atoms of a single molecule. Hemin and chlorophyll are chelated. Chelation may improve solubility of metals and is also used for relieving metal poisoning.

chemical genetics   Develops small molecules that can modify proteins or nucleic acids and thus alter gene function. The results may be applicable to the development of drugs and reveal developmental mechanisms. *See* combinatorial chemistry. (Bishop, A. C., et al. 2000. *Nature* 407:395. Tan, D. S. 2002. *Nature Biotechnol.* 20:561.)

chemical mutagen   Includes an extremely large number and wide variety of different chemical groups. Their effectiveness varies within a great range. About 80% of the chemical mutagens are also carcinogens. Some of these agents pose health hazards as industrial, agricultural, and medical chemicals. A much smaller fraction is used for the experimental induction of mutation for research purposes. A very general classification follows: (1) DNA base and nucleoside analogs 5-bromouracil or 5-bromo-deoxyuridine (BuDR), 2-aminopurine (2-AP). These compounds are of relatively very low efficiency and can be utilized only under highly selective conditions in microbial populations. They act primarily through incorporation at the wrong place in the DNA or by replacing their normal counterparts followed by mispairing (tautomeric shift), in both cases causing base substitution. (2) Chemical modifiers of nucleic acid bases such as nitrous acid (HNO_2), which oxidatively deaminates cytosine into uracil, adenine into hypoxanthine (resulting actually in A → G transitions), and guanine into xanthine (causing lethal effects because of interference with replication). Nitrous acid is not an effective mutagen for higher eukaryotic cells because of the high protein content of the chromosomes. Nitrous acid has destructive effects on proteins, but it has been used very advantageously for tobacco mosaic virus and other viruses. Nitrous acid may

OH
|
N═C-$_5$C—Br
| ||
HO—C-$_1$-C
    N |
           H

5-Bromouracil

H
|
C-C-N
| || ‖
N-$_1$ N-CH
| || |
H_2N—C-$_2$ C-N
    N |
            H

2-Aminopurine

nitrosate, however, several plant products (black pepper, beer [if the barley is improperly dried], soybean and fava bean

products, and cruciferous vegetables [cabbage, etc.]). Hydroxylamine targets primarily cytosine and thus changes a G ≡ C pair into A = T. (3) Alkylating agents of a diverse group, such as sulfur and nitrogen mustards, epoxides, ethyleneimins, unsaturated lactones (aflatoxins), alkyl, and alkene sulfonates, are powerful mutagens for different organisms. The nitrogen mustards primarily induce chromosome breakage like X-rays and are frequently referred to as radiomimetic agents. The mustards are no longer used as laboratory mutagens because of the relatively low efficiency compared to their lethal effects and because of the risks involved in routine handling. Among these, the most commonly used alkylating mutagens are methyl and ethyl methanesulfonate ($CH_3SO_2OCH_2CH_3$), which are highly effective in practically all organisms. They alkylate the guanine at the 7 position, but the major mutagenic action results from the alkylation of the O^6 of guanine. To a lesser extent, they also mutagenically alkylate other bases. Alkylation may break the linkage between the ribose and the N^9 of the purines, resulting in depurination and breakage of the DNA strand. (4) N-nitroso compounds such as nitrosamines, nitrosoureas, methyl-nitro-nitrosoguanidine (MNNG, $C_2H_5N_5O_3$), etc., are very potent mutagens. The latter one predominantly induces point mutations at pH 6, but it must be handled in the dark because it rapidly decays in light. (5) A variety of compounds that produce free radicals such as hydrazines and hydrazides, hydrogen peroxide or organic peroxides, aldehydes, and phenols are mutagenic but usually not employed for the induction of mutation, except perhaps maleic hydrazide, which induces high-frequency chromosomal breakage. (6) Acridine dyes (proflavin, acriflavin, acridine orange, etc.) are intercalating agents and induce frameshift mutations at neutral pH. *See* alkylation; Ames test; biological mutagenic agents; chemicals hazardous; chromosome breakage; dupurination; environmental mutagens; frameshift; laboratory safety; physical mutagens; radiomimetic; tautomcric shift. (Hollaender, A. ed. 1971–80. *Chemical Mutagens. Principles and Methods for Their Detection.* Plenum, New York.)

chemical mutation   Alters the active site of an enzyme by modifying the (tertiary) structure of the protein. *See* protein structure.

chemicals, hazardous   The total number of chemicals identified exceeds 6 million. The various industries use over 60,000 chemical compounds, and it is estimated that 700 new chemicals are introduced annually for various uses. Since the biological effects (mutagenicity, carcinogenicity) of only a relatively small fraction is known with certainty, the majority of chemicals should be regarded with appropriate caution. *See* carcinogens; chemical mutagens; environmental mutagens; laboratory safety. (<http://toxnet.nlm.nih.gov>.)

chemiosmotic coupling   Uses a pH gradient through a membrane to drive energy-requiring processes. *See* chemosmosis.

chemoattractant   *See* chemotaxis.

chemoautotroph   The organism obtains energy from inorganic chemical reactions.

chemoeffector   Elicits bacterial response by bacterial signal transducers in chemotaxis.

chemogenomics Seeks out potential new drug targets by proteomic technology. Genes identified by genomic analysis are expressed as proteins. Potential drug-like libraries are tested whether the compounds bind the protein targets. The outcome of these tests facilitates either the selection or synthesis of additional, improved structurally similar compounds. Selected molecules are then tested in biological systems for therapeutic effectiveness. *See* combinatorial chemistry. (Agrafiotis, D. K., et al. 2002. *Nature Rev. Drug Discovery* 1:337.)

chemoheterotroph The organism obtains energy from the breakdown of organic molecules.

chemokine (chemotactic cytokine) Chemical (moving protein) involved in defense systems or much activation by chemical agents. Chemokines are classified according to their cysteine motifs into CXC, CC, C, and CX3C groups. Chemokines are participants in specific inflammatory responses. Some of them are identical to specific lymphokines. Chemokines also lure the lymphocytes to the sites of infection and inflammation. These small proteins have four conserved cysteines of which two are either adjacent or one other amino acid is in between the two. Chemokine-α is represented by IL-8 and others encoded in close vicinity to each other in human chromosome 4q12-q21. On the surface of human neutrophils, 20,000 high-affinity receptors have been found. They are encoded in human chromosome 2q34-q35. The β-chemokine family includes MIP; the others are homologous to 28–73% and are encoded in close linkage in human chromosome 17q11-q21. The MIP/RANTES receptor is encoded in 3q21. Chemokines cause a rise of free calcium and release of microbicidal oxygen radicals and bioactive lipids, storage granules containing proteases from neutrophils and monocytes, histamine from basophils, and cytotoxic proteins from eosinophils. Chemokines interact with seven-transmembrane G-protein-coupled receptors, CCR, CXCR, and CX3CR, of about 40 kDa. CXC or α-chemokines act on neutrophils and T cells; CC or β-chemokines, such as monocyte chemoattractant protein (MCP-1), act on monocytes, basophils, eosinophils, T cells (NK), and dendritic cells but usually not on neutrophils. Chemokines trigger the activation of phosphoinositide kinase (PIK). They mediate cell polarization in differentiation, movement of HIV-1 infection, and immune response. Homo- or heterodimerization of the chemokine receptors activates specific signaling pathways. Eotaxins select only eosinophilic and basophilic granulocytes. Lymphotactin and fractalkine select monocytes and neutrophils. *See* acquired immunodeficiency syndrome; antimicrobial peptides; blood; CCR; CXR; cytokines; fusin; G proteins; histamine; lymphoid organs; lymphokines; MIP; PIK; radical; RANTES; SDF; signal transduction. (Cyster, J. G. 1999. *Science* 286:2098; Mellado, M., et al. 2001. *Annu. Rev. Immunol.* 19:397; Mellado, M., et al. 2001. *EMBO J.* 20:2497; Schwarz, M. K., et al. 2002. *Nature Rev. Drug Discovery* 1:347.)

chemolithoautotroph For biosynthesis the organism relies on inorganic material and inorganic chemical energy. *See* chemoheterotroph.

chemoprevention of cancer Is generally based on regulation of cell division, ligands of nuclear receptors, selective modulation of the estrogen receptors, vitamin D analogs, inhibitors of cyclooxygenase, NF-κB, anti-inflammatory agents (curcumin [anti-cancer phenolic], resveratrol [antioxydant],

caffeic acid [phenolic antioxydant]), chromatin modifiers (histone deacetylase inhibitors, methyl transferases), agents that control signal transduction (herceptin), enhancers of the SMAD complex pathway boosts TGF-β signaling that is usually down-regulated in some cancers. (See cancer therapy, nuclear receptors, cyclooxygenase, SMAD, TGF, herceptin, NF-κB. (Sporn, M. B. & Suh, N. 2002. *Nature Rev. Cancer.* 2:537.)

chemoprevention *See* cancer therapy.

chemoreceptor Gene product activated by a chemical signal(s). In *Caenorhabditis*, ~1000 seven-transmembrane receptor genes have been identified that may be chemoreceptors, although only half of them seem to be functional. This is the largest gene family in this nematode. *See* olfactogenetics; taste. (Gestwiczki, J. E. & Kiessling, L. L. 2002. *Nature* 415:81.)

chemosensitivity The administered drugs affect many genes at the same time, and the response of individual genes in different patients may vary. By exposing cells to various compounds at different concentrations, 50% growth inhibition (GI_{50}) can be scored. The most sensitive and the most resistant cell lines can then be evaluated with the aid of microarray hybridization. The information may be of help for predictions of drug response, e.g., by cancer patients. *See* microarray hybridization; multidrug resistance. (Staunton, J. E., et al. 2001. *Proc. Natl. Acad. Sci. USA* 98:10787.)

chemosmosis Chemical reaction across membranes.

chemostat Apparatus for culturing bacteria at a steady level of nutrients, aeration, temperature, etc., so that cell divisions are continuously maintained.

chemotaxis Movement toward chemical attractants and away from chemical repellents. Bacteria have very high sensitivity receptors for chemical signals. Upon binding the chemical ligand, the receptors propagate in clusters on the bacterial surface, but single receptors also exist and thus they can adapt (respond) to a very wide concentration range (five orders of magnitude) of the chemicals. The receptor then transmits the signal through PLC and PIK to the cellular interior for further processing. Macrophages, neutrophils, eosinophils, and lymphocytes are attracted to a wide variety of substances, causing inflammation. *See* blood cells; CheA; PIK; PLC. (Mori, I. 1999. *Annu. Rev. Genet.* 33:399.)

chemotaxonomy Study of evolutionary relatedness on the basis of chemical compounds.

chemotherapy Curing a disease by chemical medication or fighting cancerous proliferation by antimitotic (cytostatic) chemicals. *See* cancer; cancer therapy; cytostatic; multiple drug resistance.

Cheney syndrome (acroosteolysis) Autosomal-dominant bone disease with similarities to pycnodysostosis, but rather than hardening of some of the bones (osteosclerosis), bone loss (osteoporosis) is accompanied with early loss of teeth, laxity of the joints, etc. In this syndrome the stature of the patients is not necessarily short. *See* pycnodysostosis.

Chernobyl   *See* atomic radiation.

cherry (*Prunus cerasus*) $x = 7$, but a wide variety of diploid and polyploid forms are known.

cherubism   Dominant (4p16.3) proliferation of the lower (mandibula) or upper (maxilla) jawbone usually beginning at age 2–5 years and receding after puberty. The distortion of the face pulls down the lower eyelids and the eyes seem to be gazing upward. The distorted teeth and visual anomalies may remain. Fibroblast growth factor (FGF3) mapping to the same general area may be involved. Recent evidence indicates mutations in the SH3-binding protein SH3BP2. *See* FGF. (Mangion, J., et al. 1998. *Am. J. Hum. Genet.* 65:151; Tiziani, V., et al. Ibid., p.158; Ueki, Y., et al. 2001. *Nature Genet.* 28:125.)

chestnut (*Castanea* spp)   Monoecious tree, $2n = 2x = 24$.

chestnut color   In horses, is due to homozygosity for the *d* allele; actually this color is expressed when the genetic constitution of the animals is *AAbbCCdd*.

chetah   *Acionyx jubatus*, $2n = 38$.

chiasma (plural chiasmata)   Chromatid overlap in prophase (appearing through the light microscope like the Greek letter chi [χ]), resulting in genetic exchange between bivalents. If the number of chiasmata truly represents the points of genetic exchange of chromosomes, from the cytological observation of chiasma frequency the length of the genetic map could be inferred because each crossing over corresponds to 50% recombination. Thus, the number of chiasmata multiplied by 50 should be equal to the sum of map distances. For example, C. D. Darlington observed 3 chiasmata in chromosome 3 of maize; thus $3 \times 50 = 150$. Actually, according to the modern map, the length of this chromosome is about ~167 m.u. According to B. Lewin, the number of chiasmata per meiocyte in *Drosophila melanogaster* was 6.6, and $6.6 \times 50 = 300$, and the new cytogenetic map appears to be of ~297 m.u.

Multiple chiasmata in a male grasshopper chromosome. (Courtesy of Dr. B. John.) Below: Schematic representation of one chiasma in each arm.

Analyses of plant recombination and chiasma data find that the frequency of recombination is higher than the chiasma frequency. This revelation may or may not affect the general validity of the correspondence suggested earlier because of the cytological difficulties in obtaining very precise estimates on chiasmata. Also, the recombination data may have some inherent errors even if mapping functions are used. It is well known that interference varies along the length of the chromosome and according to species, etc., and no general mapping function can fully appreciate that fact. Chiasmata take place as the bivalents represent 4 strands (chromatids).

In about 6–10% of the X chromosomes, there are no chiasmata, whereas 60–65% of the bivalents display one, and 30–35% show two, but a higher number of chiasmata are very rare. There is evidence that the occurrence of chiasma facilitates orderly segregation of the chromosomes and reduces nondisjunction. In the majority of organisms, chiasmata are rare at the tip and at the centromeric regions. Special genes regulate chiasmata. In *Saccharomyces*, the Rad50 protein (an ATP-dependent DNA-binding protein, localized at interstitial sites) seems to be involved in the development of the axial structure of chromosomes, pairing, and recombination. *See* achiasmate; association point; chromatid interference; crossing over; count-location models; desynapsis; interference; isochores; mapping genetic; mapping function; meiosis; nondisjunction; recombination frequency; recombination nodule; separin; sister chromatid cohesion; stationary renewal. (Tease, C. 1998. *Chromosoma* 107:549.)

chiasma interference   *See* chromatid interference; interference.

chiasmata (plural of chiasma)   *See* chiasma.

chiasma terminalization   *See* terminalization of chromosomes.

Chicago classification   Of human chromosomes, based on banding and morphology; it has been refined since 1966. *See* Denver classification; human chromosomes; Paris classification.

chicken   *Gallus domesticus* $2n = $ ca. 78. Genome size bp/n $= 1.2 \times 10^9$. (<http://poultry.mph.msu.edu/>; <http://www.tigr.org/tdb/tgi.shtml>.)

chi elements (χ)   Crossing-over hot spots inciting DNA sequences in both prokaryotes and eukaryotes. The prokaryotic chi-elements have the consensus 5'-GCTGGTGG-3' and nicking by the RecBCD complex takes place 4 to 6 bases downstream during recombination. In phage DNA, chi-sequences are required for the function of the bacterial genes *RecBC* coding for exonuclease V. The product of phage gene *gam* inhibits this enzyme. In *E. coli*, there are about 500 to 600 chi-elements. Chi promotes recombination in a region within 10 kb from its location. It may be activated by double-strand breaks within a range several kb downstream. Chi probably facilitates the production and availability of the free 3' end for recombination. The hepatitis B virus encapsidation signal carrying a 61 bp sequence (called 15AB) is a hot spot for recombination, and a cellular protein binding to this sequence appears to be a recombinogenic protein. The same nucleotide sequence (5'-CCAAG**CTGTG**CCTTGGGTGGC-3') has been identified with approximately 80% homologies in the rat, mouse, and human genomes. Note the pentanucleotides in bold; they are also present in the prokaryotic chi-elements as shown above. It appears that this element is responsible for some of the chromosome rearrangements observed in hepatocarcinomas. *See* crossing over; hepatoma; molecular mechanisms of recombination; *rec*; recombination. (Anderson, D. G., et al. 1999. *J. Biol. Chem.* 274:27139; Lao, P. J. & Forsdyke, D. R. 2000. *Gene* 243:47.)

chi form   Recombinational intermediate representing consummated chiasmata. *See* chiasma.

chick pea (*Cicer arietinum*)   Grain legumes representing the pulses. The chromosome number is generally $2n = 2x = 16$.

CHILD syndrome   Congenital hemidysplasia with ichthyosiform erythroderma and limb defects is caused by deficiency of 3-β-hydroxysteroid-δ-8, δ-7 isomerase at Xp11.23-p11.22 (EBP) or by the NAD(P)H steroid dehydrogenase-like protein at Xq28 (NSDHL, ConradiHunermann syndrome).

chimaerin   Family of Rac-GTPase-activating proteins. *See* Rac.

chimera   (1) Mytological monster that had a serpent's tail, goat's body, and lion's head and vomited flames through her mouth. (2) Mixture of genetically different tissues within an individual or other structures of two or more different fused elements; frequently it displays visible sectoring. *See* mericlinal chimera; multiparental hybrids; periclinal chimera.

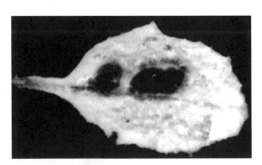

Chimeric leaf.

chimeraplasty   Designed to repair (with the aid of Rec A, MutS) single-nucleotide defects in mammalian (human) cells and plants. The perfected procedure is expected to create specific mutations. The complementary DNA is hybridized with 2'O-methyl-RNA to protect the construct from nuclease attacks. The delivery is by injection in liposomes, and it is supposed to be taken up by a sialoglycoprotein receptor. The targeting in this system is supposedly excellent. The initial experiments involved mismatch repair in the Criggler-Najjar cells and hepatocytes. *See* Criggler-Najjar disease; liposome; mismatch repair; MutS; RecA; sialic acid. (Stephenson, J. J. 1999. *Am. Med. Assoc.* 281:119.)

chimeric clones   Two segments of DNA derived from noncontiguous regions of the chromosomes are joined together. A high level of homologous recombination may cause this. These joined areas are cloned together (co-cloning). Co-cloning is disadvantageous for the construction of physical maps. *See* chimeric DNA; physical map.

chimeric DNA   *See* recombinant DNA.

chimeric plasmid   Contains genetic sequences from other genomes along with their own DNA. *See* plasmid; vector.

chimeric protein   May be used to gain additional function(s) by the same molecule, which can be constructed by adding new domains. The desired domain may be provided with "sticky ends" through PCR procedures. The sticky ends are supposed to pair with homologous portions of a target in a single-strand vector. After replication, one of the new strands of the double-stranded DNA will contain the sequences coding for a polypeptide corresponding to the donor molecule. Chimeric proteins may also be obtained by the use of translational fusion vectors. *See* primer extension; translational gene fusion vector. (Louis, J. M., et al. 2001. *Biochemistry* 40:11184.)

chimeric YAC   Produced when more than one piece of DNA is ligated to the same vector arm. It is generally undesirable for chromosome walking toward a particular locus because it may direct toward a different direction(s) than the region of interest. *See* chromosome walking; YAC.

chimpanzee (*Pan*, $2n = 48$)   Closest to humans among the great apes by DNA sequences. The mean pairwise sequence difference (MPSD) among chimpanzees is about four times higher (0.13%) than among humans (0.037%). Similarly, the mtDNA is more variable among different chimpanzee subspecies. This information also indicates the longer evolutionary history of this ape. *See* Pongidae; primates. (Wildman, D. E. 2002. *BioEssays* 24:490; Muchmore, E. A. 2001. *Immunol. Rev.* 183:86.)

Chinese restaurant syndrome   Adverse reaction (headache, stiffness of the neck and back, nausea, etc.) to the flavor enhancer monosodium glutamate in certain foods such as soy sauce, hot dog, etc. It may be controlled by a recessive gene. *See* monosodium glutamate. (Walker, R. & Lupien, J. R. 2000. *J. Nutr.* 130 [4S Suppl.]:1049S.)

ChIP (chromatin immunoprecipitation)   Assay can be used to identify definite regulatory sequences in the DNA by the use of antibodies against specific transcription factors. (Zeller, K. I., et al. 2001. *J. Biol. Chem.* 276:48285.)

CHIP   Channel-forming integral protein. Member of water transporters to various types of cells and tissues in cellular organisms. CHIP interacts with several homeodomain proteins involved with the differentiation and development of eukaryotes. In humans, the aquaporin-1 gene in chromosome 7p14 encodes it. *See* aquaporin.

chip   Electronic circuit within a single piece of semiconducting material, e.g., silicon. *See* semiconductor.

chipmunk *Eutamias amoenus* $2n = 38$; *Eutamias minimus* $2n = 38$; *Funambulus palmarum* $2n = 46$; *Funambulus pennanti* $2n = 54$; *Glaucomys volans* $2n = 48$.

chiral compound *See* enantiomorph.

chirality Dissymmetry of a molecule, i.e., its plane mirror image cannot be brought to coincide with itself. *See* enantiomorph.

Chironomus species ($2n = 8$) Favorable organisms (dipteran flies) for cytology because of the very conspicuous differences in the banding of the salivary gland chromosome among species and within species, reflecting the activity of genes by the pattern of puff formation. *See* giant chromosomes; puff; *Rhynchosciara*; *Sciara*. (Phillips, A. M., et al. 2000. *Methods Mol. Biol.* 123:83.)

chi square (χ^2) Statistical device for testing the goodness of fit to a particular (null) hypothesis.

$$\chi^2 = \sum \frac{[\text{observed number} - \text{expected number}]^2}{\text{expected number}}$$

where Σ stands for sum. When the degree of freedom is 1, the use of the Yates correction may be justified, and the formula becomes

$$\chi^2 = \sum \frac{[|\text{observed} - \text{expected}| - 0.5]^2}{\text{expected}}$$

The Yates correction may be applied for other degrees of freedom in case the size of any particular class is 5 or less. It may be a better practice, however, to avoid using this correction factor, keeping in mind that our chi-square figure is conservative. Although χ^2 is very useful, it must be remembered that it has a general weakness because it was derived from the principles of normal distribution but is actually applied for discrete classes. An alternative to the above χ^2 statistics is the likelihood ratio criterion, χ^2_L, which is easier to compute in some cases and may give a more realistic estimate:

$$\chi^2_L = 2 \sum \text{observed} \times \ln\left(\frac{\text{observed}}{\text{expected}}\right)$$

The degree of freedom is calculated the same way as in other cases.

A slightly different type of formula is used in tetrad analysis to determine whether the frequency of parental ditype (PD) tetrads really exceeds that of nonparental ditypes (NPD).

$$\chi^2 = \frac{(PD - NPD)^2}{PD + NPD}$$

In linkage, PD should exceed that of NPD. When the population size exceeds 30, use $\sqrt{2\chi^2} - \sqrt{2n-1}$ as a normal deviate. *See* association test; chi-square table; degrees of freedom; homogeneity test; lod score; tetrad analysis.

chi-square table Displays the probability of a greater χ^2. Determine the value of χ^2 by using the chi-square formulas. Locate the nearest higher value in the body of the table on the appropriate line of degrees of freedom (df). On the top line at the intersecting column, you find the corresponding level of probability (**P**). For a particular degree of freedom, the smaller χ^2 value indicates a better fit to the theoretical expectation (null hypothesis). By statistical convention when $P > 0.05$, the fit is not questionable; when P is between 0.05 and 0.01, the fit is more or less in doubt. When P is below 0.01, the null hypothesis is no longer tenable. (In the majority of statistical books, more extensive chi-square tables can be found.)

P→	0.99	0.90	0.75	0.50	0.25	0.10	0.05	0.01	0.005
df↓									
1	0.00	0.02	0.10	0.45	1.32	2.71	3.84	6.64	7.90
2	0.02	0.21	0.58	1.39	2.77	4.60	5.99	9.92	10.59
3	0.11	0.58	1.21	2.37	4.11	6.25	7.82	11.32	12.82
4	0.30	1.06	1.92	3.36	5.39	7.78	9.49	13.28	14.82
5	0.55	1.61	2.67	4.35	6.63	9.24	11.07	15.09	16.76
6	0.87	2.20	3.45	5.35	7.84	10.65	12.60	16.81	18.55
7	1.24	2.83	4.25	6.35	9.04	12.02	14.07	18.47	20.27
8	1.64	3.49	5.07	7.34	10.22	13.36	15.51	20.08	21.97

chitin Poly-N-acetylglucosamine is part of the exoskeleton of insects, of other lower animals, and of the cell wall of fungi. The chitinase enzyme plays a role in the protection of plants against fungal and insect damage. In chitinases, the nonsynonymous mutations in the DNA exceed the synonymous ones, indicating the tendency of adaptive evolution in the plant defenses. *See* exoskeleton; host–pathogen relations; insect resistance. (Bishop, J. G., et al. 2000. *Proc. Natl. Acad. Sci. USA* 97:5322.)

Chk1 476-amino-acid G2-cell-cycle-phase checkpoint serine/threonine kinase, which prevents the cell to enter the M phase in case the DNA is damaged. Cdc25 protein phosphatase in yeasts and human cells dephosphorylates Cdc2 at tyrosine 15, thus activating it and turning on the cell cycle. Cdc25 phosphorylated at serine-216 by Chk1 binds to proteins 14–3-3, such as Rad24, and the complex brings about the cell cycle arrest while the DNA damage is corrected. Cdc25 is basically a cytoplasmic protein, but it also enters the nucleus to activate Cdc2 under normal conditions. After the cell is damaged, e.g., by irradiation, Rad24 facilitates the nuclear export of Cdc25 and the checkpoint arrest is created. *See* Cdc2; Cdc25; cell cycle; checkpoint; p53; protein 14-3-3. (Chen, P., et al. 2000. *Cell* 100:681.)

CHK2 Checkpoint kinase, a mammalian homologue of Rad53 in *Saccharomyces cerevisiae* and Cds1 in *Schizosaccharomyces pombe*. This protein is phosphorylated in response to replication arrest or DNA injury. In vitro CHK2, like CHK1, phosphorylates Cdc25C and thus prevents entry into mitosis; it seems to be involved with ATM. *See* ataxia; ATM; CDC25; FHA; p53; RAD53. (Ward, I. M., et al. 2001. *J. Biol. Chem.* 276:47755.)

Chlamydia Pathogenic bacteria. *C. pneumoniae* (1.23 Mb) respiratory and atherosclerosis pathogen, *C. trachomatis* (1.05/1.07 Mb), causes blindness. Several other chronic diseases may be affected by *Chlamydia* and other microbes. (Zimmer, Z. 2001. *Science* 293:1977.)

Chlamydomonas eugametos ($n = 7$) Unicellular green alga.

Chlamydomonas reinhardtii Unicellular green alga showing both haploid and diploid stages. The fusion of ($+$) and ($−$) mating-type gametes (not identifiable by morphology) gives rise to diploid cells that immediately undergo meiosis and release haploid zoospores. This progeny can be subjected to unordered tetrad analysis. On solid media they produce only rudimentary flagellae, but in liquid culture they are flagellated. The zoospores may readily divide by mitosis, but they sexually differentiate under nitrogen starvation and gametic fusion follows. Their basic chromosome number is $n = 8$ (1×10^6 bp). They contain one large chloroplast (with about 196 kbp DNA). Chloroplast genes normally show uniparental inheritance in 99% of the progeny. When the male (*mt-*) cells are irradiated with UV before mating, biparental plastids are formed. The heterozygotes for plastid genes (cytohets) display recombinations of plastid genes in their single-plastid progeny. Recombination is either reciprocal or nonreciprocal among the plastid genes. On this basis, strictly genetic maps could be constructed for the circular plastid DNA. Physical maps are also available and are used mainly for determining gene positions. Molecular analysis revealed that one plastid gene, *psa* (a photosystem protein), contains three exons; exon 1 is 50 kb away from exon 2, and exon 3 is 90 bp away from exon 2. In between, there are several other transcribed genes. Further complication is that exon 1 is in opposite orientation to the other two. It is supposed that for the expression of this gene transsplicing is used, i.e. (in contrast to the regular, common mechanism of splicing neighboring exons), here distant transcripts are brought together in the mRNA. The mitochondrial DNA is about 15.8 kb. Some alga mutants (*minutes*) are apparently mitochondrial and resemble *petites* in yeast. Since insertional mutagenesis became feasible, genetic and molecular analysis of the photosynthetic apparatus is greatly facilitated. Gene targeting, homologous recombination, site-directed mutagenesis, etc., techniques are available. (The name of this alga is often spelled *Chlamydomonas reinhardi*.) See chloroplast DNA; chloroplast genetics; eyespot; mitochondria; petite colony mutants. (Rochaix, J.-D. 1995. *Annu. Rev. Genet.*

29:209; Rochaix, J.-D., et al., eds. 1998. *The Molecular Biology of Chloroplasts and Mitochondria in Chlamydomonas*. Kluwer, Dordrecht; Harris, E. H. 2001. *Annu. Rev. Plant Physiol. Mol. Biol.* 52:363; <http://www.biology.duke.edu/chlamy>.)

chlamydospore Thick-walled persistent asexual spore.

chloracne Eruption on the skin caused by exposure to chlorine and related compounds.

chlorambucil (p-[di-2-chloroethylamino]phenylbutyric acid) Radiomimetic nitrogen mustard derivative primarily causing deletions and translocations. It had been used as an antineoplastic drug and it is a carcinogen. *See* carcinogen; radiomimetic.

chloramphenicol Antibiotic affecting peptidyl transferase in bacterial and mitochondrial protein synthesis. In human populations, about 5×10^{-5} of the individuals may be very sensitive to the drug and develop anemia. *See* cycloheximide; mitochondrial human disease.

Chloramphenicol.

chloramphenicol acetyltransferase (CAT) Gene has been extensively used as a reporter for transformation in cell culture by becoming resistant to the antibiotic chloramphenicol and thus selectable. *See* antibiotics; transformation.

chlorate Has been used for the isolation of nitrate reductase–deficient mutations that are not poisoned by chlorate. Some mutations, however, are hypersensitive to chlorate. Chlorate-sensitivity is apparently also based on uptake problems. *See* nitrate reductase.

chlorenchyma Tissue with green plastids (chloroplasts).

chloride diarrhea, congenital Recessive defect involving ion transport, encoded in human chromosome 7q31. *See* chlorate.

chloronema Gametophytic cell row in mosses. *See* gametophyte.

chlorophyll Magnesium-porphyrin complexes (chlorophyll-a, -b, protochlorophyll, bacteriochlorophyll) in green plants and bacteria, receptors of light energy for carbon fixation and photosynthetic phosphorylation. *See* Calvin cycle; chlorophyll-binding proteins. (Suzuki, J. Y., et al. 1997. *Annu. Rev. Genet.* 31:61.)

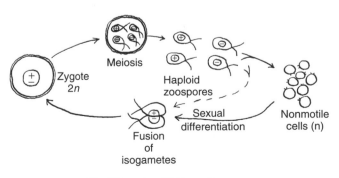

The life cycle of *Chlamydomonas*.

chlorophyll-binding protein (CAB, chlorophyll A- and B-binding proteins, light-harvesting chlorophyll protein complex, LHCP) Situated in the membrane of the thylakoids. They modify the plane of orientation of chlorophylls. Due to this modification, the chlorophyll does not fluoresce when excited by visible light (as it would do without CAB). Rather, the light energy absorbed by (an antenna) chlorophyll is transferred to a neighboring chlorophyll molecule, then excites this second chlorophyll while the first one returns to the ground state. This *resonance energy transfer* is continued to further neighbors until the *photochemical reaction center* is reached. In this molecule an electron is raised to higher-energy orbital, and this electron is transferred to the *electron transfer chain* of the chloroplast, resulting in an electron hole (empty orbital). The electron acceptor thus gains a negative charge and the lost electron by the reaction center is replaced by another electron coming from a neighbor molecule, which therefore becomes positively charged. As a consequence, the light sets into motion an oxidation-reduction chain and the generation of ATP and NADPH. About 16 genes encode the CAB complex, and a LHCPII system has also been identified. The red algae and cyanobacteria, which have only chlorophyll a, use phycobilisomes for light harvesting. Several types of accessory pigments (carotenoids, pteridins, phycoerythrobilin, etc.) may be associated with the light-harvesting complex. *See* Calvin cycle; photophosphorylation; photosynthesis; photosystems; phycobilins; Z scheme. (Grosman, A. R., et al. 1995. *Annu. Rev. Genet.* 29:231.)

chloroplast Chlorophyll-containing organelle (2–20 μm) in green plants with double-stranded circular DNA (cpDNA of 120–180 kbp in 10–100 copies/plastid) genetic material and a capacity of transcription and translation. The cpDNA codes for 16S, 23S, 5S, and 4.5S ribosomal RNAs and has ribosomes of about 70S size, resembling those of prokaryotes. This genome codes for about 100 polypeptides and 35 different RNAs (rRNA, tRNA). The genes are frequently transcribed into polycistronic RNA, resembling bacterial gene clusters. Their rRNA genes (10–30 kbp) are generally inversely repeated in land plants. Eighteen genera of the legumes *Fabaceae* do not have such inverted repeats. The unique sequences have a small (15–25 kbp) and a large (80–100 kbp) tract. *Pelargonium hortorum* (geranium) displays a 76 kbp repeat, and this includes some usually "unique" sequences (genes) here, however, duplicated. The cpDNA usually contains other inverted repeats. The cpDNA of *Chlamydomonas* algae is larger (195–294 kbp) but has similar inverted repeats, and the unique sequences consist of two about equal tracts, with gene order differing from land plants. They have a recombination system acting among the different repeats. The *Euglena* cpDNA is similar in size (130–152 kbp) to that of land plants, yet it lacks the two inverted repeats of rRNA, but it has triple tandem repeats and two-fold tandem 16S rRNA genes. The tRNAs are clustered 2 to 5 each. Variations are found in several miscellaneous genes, too. Other algae show additional variations. The colorless algae and plants (*Epifagus*) have smaller genomes (about 70 kbp) and lack about 95% of the genes encoding the photosynthetic apparatus, yet they have genes required for protein synthesis (rRNA, 17 tRNAs, 80% of the ribosomal proteins, etc.).

The major function of chloroplasts is photosynthesis. Actually, the chloroplasts reduce CO_2 and split water and release O_2. Enzymes of the chloroplast stroma convert CO_2 into carbohydrates. Some of the chloroplast genes have introns and some sequences in the cpDNA have homologies, with nuclear, mitochondrial, and *E. coli* DNA. In bleached mutants or in antibiotic-sensitive plants, callus growth can be maintained on carbohydrate-supplied culture media. The genetic code in the plastid DNA is the universal one, unlike in mitochondria. Nuclear genes encode several plastid functions, and the proteins or RNAs may be imported into the plastids from the cytosol with the assistance of transit peptides. An assembly of nuclear and plastid genes encode some of the plastid protein subunits. The chloroplast genome has apparently evolved through endosymbiosis from an ancestral prokaryote(s). The size of the chloroplasts varies from 1 to 3 μm in diameter and about two to three times as much in length. The double membrane enclosing its content has no pores. The internal flattened membrane vesicles called *thylakoids* are stacked into *grana* and they harbor the photosynthetic apparatus. The grana are connected by the *stroma lamellae*. The stroma is the fluid phase of the chloroplasts. The photosynthetic apparatus of green bacteria and some lower algae is structurally simpler. *See* chloroplast endoplasmic reticulum; chloroplast envelope; chloroplast genetics; chlorosome; chromatophore; compatibility, organelles and nuclear genome; cpDNA; differentiation, plastid nucleoids; Dinoflagellates; endosymbiont theory; evolution by base substitutions; introns; nucleoid; nucleomorph; organelle sequence transfer; photorespiration; photosystems; plastid male transmission; *Plasmodium*; plastid number; ribosomal RNA; spacers. (Osteryoung, K. W. & McAndrew, R. S. 2001. *Annu. Rev. Plant Physiol. Plant Mol. Biol.* 52:315.)

Stacked-up thylakoids form a granum.

chloroplast endoplasmic reticulum *See* endoplasmic reticulum; nucleomorph.

chloroplast envelope Double membrane surrounding this organelle; it controls the uptake of metabolites and the transport of proteins encoded by nuclear genes. Furthermore, it participates in the biosynthesis of many plastid molecules. A few of the plastid envelope components are coded, however, by ctDNA. *See* chloroplasts.

chloroplast genetics Most successfully studied in the *Chlamydomonas* alga. *C. reinhardtii* has only one chloroplast with about 80 cpDNA molecules of 196 kbp each. The transmission of the cpDNA genome is largely uniparental, i.e., inherited most commonly through the mt+ (comparable to egg) cytoplasm, although in 1–10% of the cases biparental transmission may take place. The chloroplast nucleoid (DNA) transmitted by the mt− mate is completely digested within 10 minutes after zygote formation, whereas the mitochondrial nucleoid still remains intact. Exceptionally (e.g., conifers), uniparental male transmission may also exist. The uniparental mt$^+$ transfer may sometimes be spurious in cases where the coding of the subunits of a particular protein is under the control of nuclear and organelle genes, respectively. In diploid vegetative zygotes of *Chlamydomonas*, the biparental transmission of the extranuclear genes is most likely. Incubation in the dark or postponing the meiosis by nitrogen starvation, however, favors the uniparental transmission.

In higher plants, the transmission of the plastid is usually through the female, but biparental or only male transmission also occurs. The plastid nucleoids may be eliminated from or degraded in the sperm during the first or second nuclear division in the pollen, or they are left behind when the generative nucleus enters the egg.

The most common algal mutations (at different loci) require acetate as a carbon source because they cannot fix CO_2. Antibiotic resistance (or antibiotic-dependent) mutations involve the rRNA or ribosomal proteins. Fluorodeoxyuridine is a specific mutagen for cpDNA. Arsenate and metronidazole are selective for nonphotosynthetic mutations. Photosynthesis-defective mutations (nuclear or cpDNA) can be screened under long-wavelength UV. Also, in higher plants, streptomycin, spectinomycin (16S rRNA), lincomycin (23S rRNA), etc., resistance mutations can be induced and isolated.

The partial inactivation of the mt$^+$ cells by UV permits the transmission of cpDNA genes by the mt$^-$ cells, and this makes recombinational studies; possible however, some progeny cells are heteroplasmic even after three divisions of the zygote. In cpDNA, only closely situated genes display linkage with about 1 kb/map unit. Mapping is practical by physical methods: either by deletional analysis or by positional cloning or in interspecific crosses when RFLP exists by cosegregation with restriction endonuclease fragments (*C. eugametos* x *C. moewusii*). The nature of the recombination map was hotly debated until it turned out that genes within the inverted repeats map as if the area were linear, whereas recombination of genes outside this region reflects the circular nature of the cpDNA.

A single recombination between two circular molecules may result in a cointegrate. Recombination of cpDNA genes in higher plants occurs, albeit apparently quite rarely. The chloroplasts usually do not "synapse." Within cells of interspecific (intergeneric) somatic fusion, recombination of antibiotic resistance markers has been demonstrated in *Solanaceae*. Cosegregation of mitochondrial traits (cytoplasmic male sterility and chloroplast antibiotic markers) was also shown. Transforming of appropriate genes (Atrazine resistance) into the cell nucleus using T-DNA vectors may alter chloroplast functions. Transformation—using

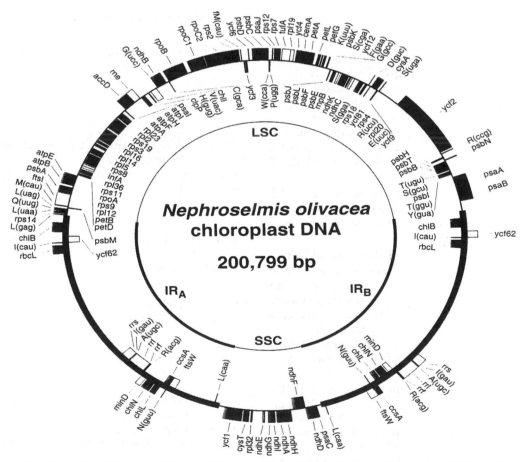

The chloroplast DNA map of the alga *Nephroselmis*. LSC large single-copy sequences, SSC: small single-copy sequences. Genes outside the large circle are transcribed clockwise and the inner set is transcribed counterclockwise. The thicker lines on the inner circle represent the two copies of the rRNA genes (IR). The gene sequences differ from those of some other algae and land plants. (From Turmel, M., et al. 1999. *Proc. Natl. Acad. Sci. USA* 96:10248.)

biolistic procedures—of cpDNA can be accomplished at high frequencies (up to 1×10^{-4}) in *Chlamydomonas* and higher plants. The insertion takes place around the passenger and vector junctions, and the insert replaces the resident copy in the nucleoid. The insertion takes place by site-specific recombination and thus it is nonrandom. In tobacco, the integration may be the outcome of multiple recombinational events.

Cotransformation of antibiotic resistance genes situated in the inverted repeats and photosynthetic genes (situated in the unique regions) can be accomplished by the simultaneous use of two vectors. Alternatively, the bacterial *aadA* gene (which detoxifies antibiotics) is used. The gene is equipped by cpDNA transcription and translation elements and surrounded by the appropriate target sequences. Whether the transformation involves correction of a resident gene or insertion of a foreign gene (e.g., *uidA* [glucuronidase]), the integration requires homologous recombination within the flanking sequences. Transformation has also been achieved by direct transfer and integration of the cloned gene into protoplasts in the presence of polyethylene glycol or by biolistic methods. Under selective conditions, sorting out of the transgene takes place rapidly. The segregation and sorting out of chloroplast genes were subjected to analysis by the methods of population genetics and computer simulation. The biological observations do not seem to support the stochastic models of sorting out.

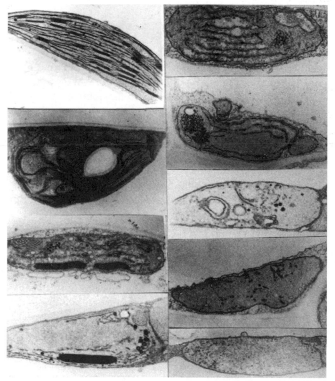

Mutant plastids in *Arabidopsis* induced by a nuclear mutator gene. At top left, a normal chloroplast is shown. Because of the presence of the mutator, within single cells morphologically different plastids are visible. The mutator shows biparental recessive inheritance. The plastid mutations are transmitted only through the egg. The mutator effect in *Arabidopsis* has been attributed to defects in mitochondrial DNA. A similar mutator gene in *Oenothera* appears to be responsible for template slippage during DNA replication; however, the mutations are transmitted through the chloroplast DNA. (From Rédei, G. P. 1982. *Genetics*. Macmillan, New York.)

Mobile genetic elements such as group I introns, encoding the I-*Cre*I (or I-*Ceu*I) endonuclease and locating in the large ribosomal subunit gene or in the cytochrome-b gene (I-*Csm*I endonuclease) in *mt*$^+$ (mating-type) plastid nucleoids, have been found in different *Chlamydomonas* chloroplasts. The I-*Cre*I endonuclease has a recognition site of 24 bp. These mobile elements resemble those of I-Sce in the mitochondria of yeast.

The promoter regions of the cpDNA genome are similar to those of prokaryotes. There are generally 10 nucleotides between the TATA box (TATAAT or longer) and the first translated codon, and again 17–19 bases separate the TATA box from the 5'-TTGACA promoter consensus at about −35. Internal and further upstream promoters have also been identified. Many genes are transcribed into polycistronic RNAs. In spinach, 18 major RNAs were made from a single polycistronic transcript. The 3' termini of the transcripts generally contain inverted repeats that have probably only some processing and/or stabilizing functions along with some 5'-untranslated sequences. Higher plants appear to have a second DNA-dependent RNA polymerase, which is encoded by the nucleus. This polymerase transcribes a different set of chloroplast genes. Binding proteins (3') seem to be involved in RNA processing. Some observations indicate that polycistronic transcripts may bind to the ribosomes and be translated without processing, perhaps with reduced efficiency. Translation of the chloroplast mRNA appears to be light regulated. Apparently, an activator protein binds to the upstream untranslated region of mRNA, and the regulation is mediated through the redox state of this protein.

Endonucleolytic processing of the transcripts may provide alternative leader sequences and binding sites for transcription factors. Ribosomal proteins may exert activation by induction and modulation of translation. The chloroplast mRNAs are not capped and the initiation of transcription is regulated in a manner similar to that in prokaryotes. In the untranslated upstream regions (UTR), there are binding sites for nuclearly encoded proteins that regulate transcription. Other proteins may bind to the 3' downstream sequences. Nuclear proteins mediate translational control. Chloroplast DNA most commonly encodes ~30 tRNAs. In the plastids of nongreen plants, tRNA gene number is reduced to about half. *See* β-glucuronidase; binding protein; biolistic transformation; chloroplast mapping; chloroplasts; cointegration; ctDNA; deletion mapping; endosymbiont theory; heteroplastidic; introns; mating type; maturase; metronidazole; mitochondrial genetics; mutation in cellular organelles; nucleoid; nucleomorph; physical mapping; polycistronic; processing; promoter; redox reaction; RFLP; RNA editing; σ; sorting out; transcription; transformation of organelles; translation; twintrons. (Sugiura, M., et al. 1998. *Annu. Rev. Genet.* 32:437; Jarvis, P. 2001. *Curr. Biol.* 11:R307; Hagemann, R. 2000. *J. Hered.* 91:435; Rodermel, S. 2001. *Trends Plant Sci.* 6:471; Ogihara, Y., et al. 2002. *Mol. Genet. Genomics* 266:740; <http://www.ncbi.nlm.nih.gov/ PMGifs/Genomes/organelles. html>.)

chloroplast import   The chloroplasts do not have structures for through-membrane traffic comparable to the nuclear pores. Four proteins are involved in import through the outer membrane of the chloroplast envelope and two with import through the inner membrane (IAP = import intermediate-associated proteins). Proteins enter the chloroplasts through the Toc (transport outer chloroplast membrane) complex (M_r 159K). The 159K complex may contain Toc132 or Toc120. Toc86 has proteolytic function, Toc34 appears to be a GTP-regulated import receptor, and Toc75 forms a transport channel within the membrane. Toc75 is immunologically related to heat-shock protein 70. Two others (also called IAP34 and IAP85) are guanosine triphosphate–binding proteins. Protein transport within the chloroplast and into the thylakoid lumen requires proteins SecA and SecY, homologous to translocation proteins present in bacteria. In addition, plastocyanin and the 23 kDa and 17 kDa subunits of the oxygen-evolving complex (OEC) are involved in thylakoid transport. The Clp chaperone may also be involved in folding the imported proteins in the chloroplast stroma. *See* Clp; mitochondrial import; plastocyanin; Sec. (Bauer, J., et al. 2001. *Cell. Mol. Life Sci.* 58:420.)

chloroplast mapping   In some algae the chloroplasts can be mapped by genetic means, which is not a completely natural process in most lower or higher plants. In addition, in higher plants the number of chloroplasts per cell may be quite high, and recombination is a process involving simultaneously multiple events. Physical maps based on molecular techniques are much more practical and provide more detailed information on the organization of the plastid genome. In the majority of higher plants (about 150 kbp) and *Chlamydomonas* algae (about 195 kbp), the 16, 23, 4.5, and 5 S ribosomal RNA genes occur in two repeats separated by a long and a shorter sequence, coding for about 35 tRNAs and about 100 proteins. There are also different organizations. In *Euglena* alga with about 145 kbp genome, the repeated rRNA clusters are adjacent. In *Pisum* (pea), the genome is only about 120 kbp and the 16S and 23S rRNA genes are located in a single cluster. For the first completely sequenced higher plant chloroplast genome of tobacco (155844 bp), the inverted rRNA repeats include 25,339 bp and the two single-copy sequences contain 86,684 and 18,482 bps. *See* chloroplast genetics; chloroplasts. (Sager, R. 1972. *Cytoplasmic Genes and Organelles*. Academic Press, New York; Shinozaki, K., et al. 1986. *EMBO J.* 5:2043–2049; Palmer, J. D. 1991. in *Molecular Biology of Plastids*, Bogorad, L. & Vasil, I .K., eds., pp. 5–53. Academic Press, San Diego, CA.)

chloroplast stroma   Fluid phase of the chloroplast content, the site of CO_2 fixation, RUBISCO, the Calvin cycle, chlorophyll synthesis, metabolism of amino acids, fatty acid synthesis, etc. Proteins can be exchanged among plastids. *See* Calvin cycle; RUBISCO.

chloroquine   Antimalaria drug and an immunosuppressant compound that may be used for posttreatment of animal cells transfected by the calcium phosphate precitation method. It may increase the expression of the introduced DNA. *See* malaria; transformation, calcium phosphate precipitation.

chlororespiration   Interaction between photosynthetic and respiratory electron transports. *See* photorespiration; photosystems. (Nixon, P. J. 2000. *Philos. Trans. R. Roy. Soc. London B Biol. Sci.* 355:1541; Bennoun, P. 2001. *Biochim. Biophys Acta* 1506:133.)

chlorosis   Condition of plants when the chlorophyll content is reduced either by a nutritional deficiency (iron) or infection by viruses or other parasites. *See* chlorophylls.

chlorosome   Nonmembraneous light-harvesting structures in green bacteria.

chlorosulfuron   (2-chloro-N-[(4-methoxy-6-methyl-1-3,5-triazin-2-yl-amino carbonyl)   Herbicide to which resistant mutations have been isolated at the rate of about 1.2×10^{-7}. *See* herbicides. (Haughn, G. S. W. & Somerville, C. R. 1990. *Plant Physiol.* 92:1081.)

Ch-No38   Chicken protein involved in shuttle functions between nucleus and cytoplasm and assembly of ribosomes. *See* ribosomes.

CHO   (Chinese hamster ovary cells)   Frequently used for mutation studies in cell culture because the hemizygosity permits the identification of recessive mutations in these diploid cells ($2n = \pm44$). *See* radiation hybrid. (Puck, T. T. 1974. *Stadler Symp.* 6:47.)

cholecystokinin   Peptide (neuropeptide) hormone controlling appetite and several other physiological processes related to steroid hormones. *See* obesity. (Malendowicz, L. K., et al. 2001. *Endocrinology* 142:4251.)

cholera toxin   Causes the severely debilitating intestinal efflux of water and Na^+ as a consequence of infection by the *Vibrio cholerae* bacterium. This enterotoxin enzyme mediates the transfer of adenylate from NAD^+ to the G_s subunit of the G-proteins. As a consequence, $G\alpha_s$ stops acting as a GTPase. Therefore, adenylate cyclase continuously synthesizes cAMP, resulting in the disturbance of the water and salt balance. Until recently, the etiology of the disease posed some hard problems because very often the *Vibrio* appeared entirely harmless. It has been shown that the toxin production depends on the acquisition of the *ctx* gene complex from the filamentous phage CTX. The ctx complex can then be transferred from one bacterium to others. The information transfer requires that the bacterium have active pili. These toxin-coregulated pili (TCP) are encoded within the pathogenicity islands (VPI) of the *Vibrio* bacteria. The pathogenicity island itself is also a single-stranded DNA phage. The new information explains the difficulties of immunization against this infection affecting hundreds of thousands in some years, particularly in Southeast Asia. *See* ARF; G-proteins; G_s; pathogenicity island; pilus; *Vibrio cholerae*. (Tsai, B., et al. 2001. *Cell* 104:937.)

cholestasis   Recessive benign recurrent intrahepatic (liver) cholestasis (BRIC or Summerskill syndrome) and recessive familial intrahepatic cholestasis (PFIC1 or Byler disease), a more serious form of impaired bile flow. Both are located at human chromosome 18q21-q22 and should be named as FIC1 (familial intrahepatic cholestasis). The normal alleles encode a P-type ATPase most likely involved in the transport of aminophospholipids. The similar PFIC2 is at 2q24 and controls a bile salt export pump. PFIC3 is a mutation of the multidrug resistance glycoprotein gene at 7q21. Cholestasis-lymphedema is a severe childhood disease of the bile. *See* ABC transporters; ATPase; bile salts; cirrhosis of the liver; phospholipid; steroid 5-beta reductase. (Bull, L. N. 2002. *Current Op. Genet. Dev.* 12:336.)

cholesterol   Amphipatic (lipid) molecules of the membranes with a polar head and a nonpolar hydrocarbon tail (*see* cell membranes). Cholesterols are the principal sterols in animals; they also occur in plants (stigmasterols, sitosterol) and in fungi (ergosterol). Bacteria generally lack sterols. Cholesterols are synthesized from acetate through mevalonate, isoprenes, and squalenes to a four-ring steroid nucleus. The principal regulatory enzyme is 3-hydroxy-3-methyl-glutaryl–coenzyme A (HMG-CoA) reductase. Most of the synthetic activity takes place in the liver and is used as bile acids (cholesteryl esters). The excess cholesterol is generally degraded into bile acids and catalyzed by cytochrome P450 cholesterol 7α-hydroxylase (CYP7A1). The farnesoid X receptor (FXR) and four other orphan nuclear receptors are involved in the transcriptional control. The level of bile acids also regulates the activity of CYP7A1. A small heterodimer partner (SHP) and the liver receptor homologue 1 (LRH-1) modulate the process. Cholesterols are parts of membranes, steroid hormones, and precursors of vitamin D. Cholesterols are indispensable for numerous functions and are generally synthesized in sufficient amounts by the human body, although dietary cholesterol may increase their level. Accumulation of excessive amounts of cholesterol bound to low-density lipoproteins (LDL) may lead to occlusion of the blood vessels (atherosclerotic plaques), leading to coronary heart disease. HMG-CoA and Δ^7-reductase deficiency result in embryonic malformation in rodents. The Δ^7-reductase deficiency in humans results in Smith-Lemli-Opitz syndrome. Δ^7-reductase and Δ^{24}-reductase control the step preceding cholesterol from 7-dehydrocholesterol and desmosterol, respectively. Recessive desmosterolosis (1p31.1-p33) is due to a defect of 3β-hydroxysterol Δ^{24}-reductase (Waterham, H. R., et al. 2001. *Am. J. Hum. Genet.* 69:885). Defects in cholesterol transport and uptake are also deleterious to embryonal (brain) development in mice. The receptors of the LDL and VDL lipoproteins are apparently of minor importance for the embryo, but deficiency of the scavenger receptor (SR-BI) expressed on the surface of the mouse yolk endodermal cells and within the placenta may cause embryo lethality.

Animal cholesterol (left), plant sitosterol (right).

Similarly, deficiency of LDL receptor-related protein (LRP) involves lethality. Defects in another lipoprotein receptor, megalin, cause death after birth by holoprosencephaly (a neural tube defect) and other anomalies. Cholesterols modify the hedgehog protein involved in developmental signaling.

Cholesterol biosynthesis is regulated through LXR receptors. LXRα forms a dimer with receptor RXR, and they bind to a DNA-response element after being activated by two oxysterols. LXRα-deficient mice are unable to efficiently convert cholesterols to bile acids. *See* apolipoprotein; cerebral cholinesterosis; CETP; CHILD syndrome; farnesoid X receptor; hedgehog; high-density lipoprotein; holoprosencephaly; hypercholesterolemia, familial; hypertension; lipids; lovastatin; Niemann-Pick disease; prenylation; raft; Smith-Lemli-Opitz syndrome; sphingolipids; Tangier disease; Wolman disease. (Krieger, M. 1999. *Annu. Rev. Biochem.* 68:523; Repa, J. J. & Mangelsdorf, D. J. 2000. *Annu. Rev. Cell Dev. Biol.* 16:459; Goldstein, J. L. & Brown, M. S. 2001. *Science* 292:1310; Nwokoro, N. A., et al. 2001. *Mol. Genet. Metabol.* 74:105; Haas, D., et al. 2001. *Neuropediatrics* 32:113.)

cholesterol efflux regulatory protein (CERP)   *See* ABC transporters; Tangier disease.

cholesteryl ester storage disease   *See* Wolman disease.

cholic acid ($C_{24}H_{40}O_5$)   Bile acid produced from cholesterol by 7α-hydroxylase and 12α-hydroxylase. Gene CYP7A1 (8q11.13) and CYP7B1 (8q21.3) defects encoding these enzymes may result in cholestasis (blockage of bile) and cirrhosis of the liver (fibrous alteration of the parenchymal tissue). *See* cholesterol; cirrhosis of the liver; farnesoid receptors.

choline   In the form of acetylcholine it is a neurotransmitter and affects the development of the brain. Choline is also a methyl donor for methionine, and it contributes to the biosynthesis of membrane and signaling phospholipids, sphingomyelin, phosphatidylcholine, etc. It is recommended that pregnant and lactating mothers include 450–550 mg/day, respectively, in their diet. Lysphosphatidylcholine is a ligand of the immunoregulatory G2A receptor and a defect in the system may lead to autoimmune disease and atherosclerosis. *See* acetylcholine; atherosclerosis; autoimmune disease; sphingolipids. (Kabarowski, J. H. S., et al. 2001. *Science* 293:702.)

cholinesterase   Hydrolase that splits acyl groups from choline; it is required for normal function of the nervous system. Organophosphorous insecticides include parathion, chlorpyrifos (Dursban®), diazinon, and the nerve gas, sarin poison cholinesterase. Paraoxonase (PON1), a high-density lipoprotein–associated enzyme, may detoxify organophosphate compounds, but substantial human polymorphism exists in PON1. *See* pseudocholinesterase. (Stewart, R. 2001. *Lancet* 358:73; Walker, A. W. & Keasling, J. D. 2002 Biotechnol. Bioeng. 78:715.)

chondriolite   Nucleoid of the mitochondrion. *See* mitochondria; nucleoid.

chondriome   Genome in the mitochondria.

chondriosome   Old name of the mitochondrion. *See* mitochondria.

chondrocyte   Cartilage cell of mesenchymal origin that can secrete collagen and glucosaminoglycans (polysaccharides with alternating sequence of more than one type of sugars). *See* cartilage; nitric oxide; SOX.

chondrodysplasia (CD)   In the autosomal-*dominant* form (Conradi-Hünermann disease), extra bone formations take place at the epiphysis (bone ends), and other abnormal bone formations but relatively rare (less than 30%) cataracts and skin anomalies are also found. The dominant CPDX2 Conradi-Hünermann syndrome gene at Xp11.22-p11.23 encodes a 3β-hydroxysteroid-Δ^8, Δ^7-isomerase involved in cholesterol biosynthesis, converting cholest-8(9)-en-3β-ol into cholest-7-en-3β-ol (lathosterol). Excessive calcium deposits accompany the autosomal-recessive forms, and hence the name chondrodysplasia punctata, because the cartilage appears stippled, similar to Zellweger syndrome. About two-thirds of the cases have cataracts. The symptoms may be phenocopied by maternal exposure to the warfarin pesticide that depletes vitamin K–dependent blood coagulation, although autosomal CD punctata is accompanied by hereditary blood coagulation factor deficiencies that may be cured by vitamin K. Another *dominant X-chromosomal* dwarfism gene (supposedly at Xq28) also shows bald or scar-like spots (punctata) apparently caused by lower peroxisomal functions. The rare autosomal-dominant Murk-Jansen-type CD is characterized by short legs, extreme disorganization of the limb (metaphysis) and foot bones, and defects of the spine, pelvis, and fingers involving accumulation of calcium but reduced phosphate levels in the blood. The basic defect is in the gene encoding the parathyroid hormone–parathyroid hormone-related peptide involving replacement of histidine[223] by arginine. The autosomal-recessive Hunter-Thompson acromesomelic chondrodysplasia (the shortening of the limbs is most pronounced in the distal bones) is caused by defects in the cartilage-derived morphogenetic protein (CDMP1), a member of the TGF-β superfamily of growth factors. The *metaphyseal chondrodysplasia* (McKusick type) is a recessive 9p1 short-limb dwarfness accompanied by sparse blonde hair; it is caused by mutation RNase MRP. *Rhizomelic chondrodysplasia punctata* (RCDP1, 6q22-q24) is caused by mutation of the peroxin 7 gene (DAPAT), but similar symptoms occur in other diseases involving peroxisomes. The name *punctata* comes from spotted calcification of the cartilage. The Schmid-type metaphyseal chondrodysplasia (6q21q22.3) is a defect in the α-chain of collagen X. *See* antihemophilic factors; CHILD syndrome; collagen; dwarfness; GDF; microbodies; MRP; peroxins; peroxisome; RNase; Schwartz-Jampel syndrome; Smith-Lemli-Opitz syndrome; TGF; vitamin K; vitamin K–dependent blood-clotting factors; warfarin; Zellweger syndrome.

chondroitin sulfate   Heteropolysaccharide composed of alternating units of glucuronic acid and acetyl-glucoseamine; with related compounds, chondroitin sulfate forms the ground substance, an intracellular cement, of the connective tissue. *See* mucopolysaccharidosis; spondyloepiphyseal dysplasia.

chondrome   Complete set of the mitochondrial genes. *See* mtDNA.

CHOP   *See* GADD153.

chopase   Recombination enzyme making double-strand breaks versus the nickase, which cuts only one DNA strand. *See* recombination, molecular mechanisms.

chordin   *See* bone morphogenetic protein; organizer.

chordoma (7q33)   Apparently dominant, rare, malignant tumor of the notochordal remnants that may metastasize. *See* metastasis; notochord. (Kelley, M. J., et al. 2001. *Am. J. Hum. Genet.* 69:454.)

chorea   Complex, involuntary, jerky movements. *See* benign hereditary; Huntington's chorea.

chorea acanthocytosis (CHAC, 9q21)   Neurodegeneration (jerky movements) and erythrocyte malformation with an onset generally between ages 25 and 45. The basic defect is due to 3,174-amino-acid protein, chorein. *See* acathocytosis. (Ueno, S.-I., et al. 2001. *Nature Genet.* 28:121.)

choreoathetosis, kinesigenic paroxysmal (PKC)   Recurrent involuntary movements caused by a neurological dominant defect at 16p11.2-q12.1. Males are affected three to four times more frequently than females.

chorion   Outermost envelope of the mammalian embryo (fetus), the noncellular membrane around the egg of arthropods, fishes, etc. *See* allantois; amnion; chorionic villi.

chorionic villi   Thread-like protrusions, tufts on the surface of the chorion, an embryonic tissue. They are used for prenatal genetic examination by amniocentesis. *See* amniocentesis.

chorismate   Precursor of the biosynthesis tryptophan, phenylalanine and tyrosine, and their various derivatives. In the following pathway, in between the metabolites, the enzymes involved are shown in parentheses: Phosphoenolpyruvate + Erythrose-4-phosphate $\rightarrow$ (*2-keto-3-deoxy-D-arabinoheptulosonate-7-phosphate synthase*) $\rightarrow$ 2-Keto-3-deoxy-D-arabinoheptulosonate-7-phosphate $\rightarrow$ (*dehydroquinate synthase*) $\rightarrow$ 3-Dehydroquinate $\rightarrow$ (*3-dehydroquinate dehydratase*) $\rightarrow$ 3-Dehydroshikimate $\rightarrow$ (*shikimate dehydrogenase*) $\rightarrow$ Shikimate $\rightarrow$ (*shikimate kinase*) $\rightarrow$ Shikimate-5-phosphate $\rightarrow$ (*enolpyruvylshikimate-5-phosphate synthase*) $\rightarrow$ 3-Enol-pyruvyl-shikimate-5-phosphate $\rightarrow$ (*chorismate synthase*) $\rightarrow$ Chorismate. *See* alkaptonuria; phenylalanine; phenylketonuria; tryptophan; tyrosine. (Dosselaere, F. & Vanderleyden, J. 2001. *Crit. Rev. Microbiol.* 27:75.)

choroid   Inner layer of the eyeball supplying blood to nerves. *See* retinal dystrophy.

choroidal osteoma   Autosomal-dominant eye neoplasias with bony-like cells. The choroidal sclerosis and choroid plexus calcification are autosomal recessive. *See* choroid.

choroideremia   X-linked recessive atrophy of the choroid and the retina. *See* choroid; eye diseases; retina.

choroidoretinal degeneration Type of X-linked retinitis pigmentosa distinguished by a brilliant patch at the macula of the eye. *See* foveal dystrophy; macula; macular degeneration; macular dystrophy; retinitis pigmentosa.

choroid plexus It forms in the blood-brain barrier and produces cerebrospinal fluid. *See* BBB.

Chotzen syndrome (Saethre–Chotzen syndrome, acrocephalosyndactyly, ACS) Dominant inheritance of syndactyly of fingers and toes, asymmetric and narrow head, etc., encoded in human chromosome 7p21-p22 by a gene appearing homologous to *Twist* of *Drosophila*, coding for a 490-amino-acid protein containing a basic helix-loop-helix motif. ACS may involve haploinsufficiency. *See* Apert syndrome; craniosynostosis syndromes; helix-loop-helix; limb defects; Robinow syndrome; syndactily; *Twist*. (Yousfi, M., et al. 2002. *Hum. Mol. Genet.* 11:359.)

CHR *See* CDF; cluster homology region; corticotropin releasing factor.

CHRAC Chromatin accessibility complex. *See* nucleosome.

Christmas disease *See* antihemophilic factors.

chromaffin cell Specifically receptive to staining by chromium salts.

chromatid Chromosomal strand containing one DNA double helix.

After replication, each chromosome usually contains two chromatids held together at the left and right sides of the centromere. *See* centromere; chromosome.

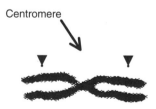

Centromere

Chromatids.

chromatid bridge Whenever dicentric chromosomes are formed (paracentric inversion heterozygotes, breakage-bridge-fusion cycles, ring chromosomes), the spindle fibers are pulling the chromosomes to the opposite poles and at one region the chromosomal material is stretched (not unlike a pulled rubber band). This thin connecting tie is called a chromosome bridge. The bridge eventually will break, which may lead to an unequal distribution of genes to the poles, resulting in duplications and deficiencies in the daughter cells. *See* breakage-bridge-fusion cycles; bridge; inversion; ring chromosome.

Pole ←   Bridge   → Pole

chromatid conversion *See* tetrad analysis.

chromatid interference Crossing over within a chromatid reduces the chance for another to occur. Chromatid interference can genetically be determined by tetrad analysis, where in the absence of chromatid interference, 2-strand, 3-strand, and 4-strand double crossing over should occur in the proportion 1:2:1; any deviation from this proportion is chromatid interference. The evidence for positive chromatid interference is rare. *See* coincidence; double crossing overs; interference. (Zhao, H., et al. 1995. *Genetics* 139:1057; Copenhaver, G. P., et al. 2002. *Genetics* 160:1631.)

chromatid segregation Takes place after recombination, and in case of polysomy it has a detectable effect on the segregation ratios. *See* autopolyploidy; trisomy.

chromatin Material of the eukaryotic chromosome (DNA, RNA, histones, and nonhistone proteins). The DNA stretches in the chromatin of a chromosome from telomere to telomere and attracts various acidic proteins and RNA in the matrix. The DNA has short (80–100 bp) supercoiled stretches. An "average" eukaryotic chromosome at metaphase may be 10 to 15 μm long, but the DNA in it may be 100,000 times longer when fully extended. The problem of accommodation is comparable to fitting a 2.5-km-long thread into a 2.5-cm-long skein 3 mm in diameter. The folding must be extremely orderly and stable to assure perfect synapsis preceding recombination and still flexible enough to make error-free replication possible within a few hours. The elementary DNA double helices form nucleosomal structures with histone protein. When the majority of the protein is digested away from the DNA, the remaining structure appears in the form of a protein scaffold with DNA loops attached. The region where the DNA is attached to the nuclear matrix is called MAR (matrix attachment region). MAR sequences are generally rich in AT. They seem to contain transcriptional activators and recognition sites for topoisomerase II. Condensed chromosomes by various nuclear stains (Giemsa stain, a mixture of basic dyes) display characteristic C (centromeric region) and G bands that help to identify several regions along the chromosomes. The nature of the specificity of the G staining is not known. That *rotational positioning* of the nucleosome in which the histone octamer is facing away from the minor groove of the DNA permits DNase I to cleave the chromatin into 10 bp sequences. *Translational positioning* of the nucleosomes in the chromatin defines the position of the nucleosomes relative to the site of transcription initiation of a gene about 300 to 150 bp away. In yeast, the MFα2 repressor regulates positioning. When a histone octamer is at the TATA box, transcription is hindered. Histone 1 is a repressor of all three eukaryotic RNA polymerases. During active transcription the nucleosomes are apparently not removed but only reconfigured. Genes in the chromatin may be attached to the nuclear scaffold at areas of their separation and thus are insulated from each other by these boundary elements. The locus control element (LCR) in the chromatin seems to be regulating the activity of groups of genes. Genes that are co-expressed are frequently clustered in the chromatin. These *open chromatin* regions are thus distinguished from the *closed chromatin* regions lacking active genes (Roy, P. J., et al. 2002. *Nature* 418:975). *See* antichromatin; chromatin code;

chromosomal proteins; chromosome banding; chromosome painting; euchromatin; FISH; heterochromatin; high-mobility group of proteins; hypersensitive sights; LCR; nuclease; nuclease-sensitive sights; nucleosome; prochromatin; stains. (Widom, J. 1998. *Annu. Rev. Biophys. Biomol. Struct.* 27:285; Cremer, T. & Cremer, C. 2001. *Nature Rev. Genet.* 2:292; Gasser, S. M. 2002. *Science* 296:1412; Ishii, K., et al. 2002. *Cell* 109:551; Kadam, S. & Emerson, B. M. 2002. *Current Opin. Cell Biol.* 14:262, Cavalli, G. 2002. *Current Opin. Cell Biol.* 14:269; Hansen, J. C. 2002. *Ann. Rev. Biophys. Biomol. Struct.* 31:361; <http://www.cstone.net/~jrb7q/chrom.html>.)

chromatin assembly factor (CAF) During genetic repair the chromatin and the nucleosomal organization has to be destabilized; after repair it must be reorganized. The nucleosomal structure, replication complex, various transcription factors, etc., must be restored after excision of the DNA defects. This process requires a series of sequentially interacting proteins. *See* chromatin; Rad53. (Emili, A., et al. 2001. *Molecular Cell* 7:13.)

chromatin code Hypothesized system regulating the folding of chromatin fibers of eukaryotes. *See* chromatin.

chromatin diminution Occurs when pieces of chromosomes are excised or entire genomes are fragmented to generate mini-chromosomes. Such a phenomenon is normal during the 2nd to 8th cleavage divisions of *Ascaris* and related nematodes and during the formation of the macronuclei of ciliated protozoa. The fragments may be reintegrated again into much larger chromosome(s) in the generative cell nuclei or the germline may develop from a single cell that has not undergone chromosome diminution. The fragmentation is followed by the addition of 2–4 kb telomeric repeats (TTAGGC). Such repeats may also be added at other chromosomal breakage region (CBR) sites. AT-rich sequences near the CBR (approximately one-fourth of the germline DNA, including the two types of Tas retrotransposons) are eliminated during diminution. *See* *Ascaris megalocephala; Paramecium*; telomere. (Müller, F. & Tobler, H. 2000. *Int. J. Parasitol.* 30:391; Redi, C. A., et al. 2001. *Chromosoma* 110:136; Goday, C. & Esteban, M. R. 2001. *Bioessays* 23:242.)

chromatin filament 30 nm in diameter nucleosomal DNA fiber appearing as beads-on-string structure by electron microscopy; the beads are the nucleosomes. The folding of the fiber may be the consequence of H1 histone-induced contraction of the internucleosomal angle as the salt concentration approaches the physiological level. Decondensation required for transcription may be the consequence of depletion of the linker histone or acetylation of the core histone tail domain. *See* chromatin; histones; lamins; nucleosome. (Moir, R. D., et al. 2000. *J. Struct. Biol.* 129:324.)

chromatin negative Cells do not display heterochromatic Barr bodies. *See* Barr body; sex chromatin.

chromatin remodeling Change in the nucleosomal structure by establishing nuclease hypersensitive sites in front of active genes. Several protein factors have been identified with this type of function. The first is the ATP-dependent yeast SWI/SNF (switch/*Saccharomyces* nuclear factor, 11 subunits in yeast, 8 subunits in humans) recognized by transcriptional activation. The Snf2-like subunits display helicase-like function. In *Drosophila*, the NURF (nucleosome remodeling factor) and an ATP-dependent 4-subunit protein was found. Its critical subunit, ISWI (imitation switch, 140 kDa; human homologue of hSNF2h), is shared by yeast. Isw2 complex represses some genes upon its recruitment by Ume6p to specific loci. Ume6 is a sequence-specific binding protein. ACF has an ISWI subunit of ATP, and also topoisomerase II activity. The RSC (remodeling *Saccharomyces* chromatin, 15 subunits) protein is working on the chromatin around the centromere and specialized probably for the regulation of the kinetochore function. The bromodomain of Gcn5/p300 coordinates the remodeling of nucleosomes. RSF (remodeling and spacing factor) and FACT (facilitator of chromatin transcription) mediate the initiation of transcription on the chromatin template depending on ATP and the other protein factors. They also facilitate elongation of the transcript through the nucleosomes. A mutant form of the p53 protein assumed a gain-of-function switch, regulating the cell cycle and causing polyploidy in Li-Fraumeni syndrome. SWI/SNF and RSF can also facilitate the expression of transcriptional repressors. The NUR protein has both chromatin remodeling and histone acetylation capabilities. Some of these factors participate in the assembly of nucleosomes and remodeling. Besides histone acetyltransferases and phosphorylases, methylation of lysine residue 9 near the N-end of histone 3 regulates the activity of chromatin. Methylation of Lys^9 interferes with phosphorylation of $serine^{10}$ and may involve aberrant mitotic divisions. Phosphorylation of histone 3 (by ERK, RSK, MSK, p38) at serine 10 is linked to transcriptional activation of DNA. *See* ACF; BRG; BRM; bromodomain; CAF; CHRAC; chromatin; Coffin-Lowry syndrome; DiGeorge syndrome; DNA methylation; DNase hypersensitive site; ERK; GCN5; histone acetyltransferase; histone deacetylase; histones; kinetochore; Li-Fraumeni syndrome; MSK; nuclear receptors; nucleosome; p38; p53; p300; PCAF; PIC; *Polycomb*; promoter; RAG1; RSK; SAGA; SANT; Sin3; SNF; SSRP; SWI; TAF_{II}; $TAF_{II}230-250$; topoisomerase; transcription factors; Williams syndrome. (Tyler, J. K., et al. 1999. *Cell* 99:443; Knoepfler, P. S. & Eisenman, R. N. 1999. *Cell* 99:447; Gebuhr, T. C., et al. 2000. *Genesis* 26:189; Cheung, P., et al. 2000. *Cell* 103:263; Cosma, M. P., et al. 2000. *Cell* 97:299; Nakayama, T. & Takami, Y. 2001. *J. Biochem.* 129:491; Fry, C. J. & Peterson, C. L. 2001. *Curr. Biol.* 11:R185; Sassone-Corsi, P. 2002. *Science* 296:2176. Tsukiyama, T. 2002 *Nature Rev. Mol. Cell Biol.* 3:422; Olave, I. A., et al. 2002 *Annu. Rev. Biochem.* 71:755.)

chromatin state fixation Hypothetical mechanism that stabilizes the function of certain genes and keeps others dormant (by heterochromatinization). *See* heterochromatin.

chromatography Partitions molecular mixtures between a stationary base (sugar, cellulose, silica gel, sepharose, hydroxyapatite) and a water base or organic liquid phase.

chromatophore Pigmented cell or the pigment-rich invagination of the cell membrane.

chromatosome Part of the nucleosome obtained as an intermediate during digestion with micrococcal nuclease. It contains a core particle of a histone octamer wrapped around by about 1 3/4 turns of ~146 bp plus 20 on each side, held together at the entry and exit points by H1 histone. *See* nucleosome. (Zlatanova, J., et al. 1999. *Crit. Rev. Eukaryot. Gene Expr.* 9:245.)

chromocenter Heterochromatin aggregate such as the common attachment point of the polytenic chromosomes in the salivary gland nuclei. *See* salivary gland chromosomes. (Clark, D. V., et al. 1998. *Chromosoma* 107:96.)

chromodomain Conserved domain in the family of *Polycomb* genes, and also in heterochromatin protein HP1. Chromodomains are supposed to be involved in the maintenance of chromatin structure by interaction between proteins and RNA, and they repress transcription. Chromodomains may be regulators of dosage compensation in males. *See* chromatin; dosage compensation; heterochromatin; *Polycomb*; SET; Swi6. (Jones, D. O., et al. 2000. *BioEssays* 22:124.)

chromogranins (CGA, CGB) are inositol 1,4,5-triphosphate-sensitive Ca^{2+} storage proteins of the secretory granules of neuroendocrine cells. They occur in the cytoplasm and in the nucleus. CGB appears to control (+ or −) the transcription of several genes including those encoding transcription factors. *See* $InsP_3$; transcription factors. (Yoo, S. H., et al. 2002. *J. Biol. Chem.* 277:16011.)

chromokinesin Protein that holds the chromosomes on the mitotic/meiotic spindle. *See* NOD; spindle; Xklp1. (Levesque, A. A. & Compton, D. A. 2001. *J. Cell Biol.* 154:1135.)

chromomere Densely stained bead-like structures along the chromosomes at early prophase. The chromomeres represent increased coiling of the chromatin fibers. In the lampbrush chromosomes, the loops seem to emanate from the chromomeres. In the salivary chromosome bands, there are appositioned chromomeres of a large number of chromatids. Chromomeric structures were recognized and Balbiani observed their constant number in 1876. In the 1920s, John Belling counted their numbers in the genome and assumed that the chromomeres are physically identical with the genes, and the 2,193 chromomeres in *Lilium pardalinum* would be the number of genes in

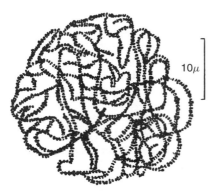

Chromomeres as illustrated by J. Belling [1928] (University of California Publ. Bot 14:307).

lilies. The chromomeric structure of the chromosomes was useful in identifying chromosomal aberrations by light microscopy. In the human autosomes, the chromomere number varied between 386 and 499. *See* chromosome morphology; gene number; lampbrush chromosomes; pachytene analysis; salivary gland chromosomes. (Judd, B. H. 1998. *Genetics* 150:1; Jagiello, G. M. & Fang, J. S. 1982. *Am. J. Hum. Genet.* 34:112.)

chromomethylase Enzymes methylate CpXpG chromosomal sites, especially within transposons. *See* methylation of DNA; transposon. (Tompa, R., et al. 2002. *Curr. Biol.* 12:65.)

chromomycin A3 Used as a stain, specific for GC nucleotides in the DNA.

chromonema Term used for the chromosomes of prokaryotes and for the smallest light-microscopically visible chromosome thread, chromatin filament, or genophore. *See* chromatin filament; chromosome structure; genophore; Mosolov model. (Nicolini, C., et al. 1997. *Mol. Biol. Rep.* 24[4]:235.)

chromophore (dye) Chemical substance that gives color to a structure upon binding to other compounds. *See* fluorochromes; stains.

chromoplast Plastids in which red and yellow pigments (rather than chlorophylls) predominate, e.g., in the fruits of mature tomatoes and other plants, e.g., *Capsicum*. *See* plastids.

chromoshadow domain *See* heterochromatin. (Lechner, M. S., et al. 2000. *Mol. Cell Biol.* 20:6449.)

chromosomal aberration *See* chromosomal rearrangements; chromosome breakage; X-ray breakage of chromosomes. *Mutation Res.* 504, issues 1–2 (2002).

chromosomal DNA *See* chromatin.

chromosomal inheritance Somewhat outdated term for inheritance of nuclear genes, since frequently organellar DNA molecules are also called chromosomes.

chromosomal instability May occur spontaneously in cultured cells of animals and plants; frequency depends on the age and composition of the nutrient media. Generally on liquid media the frequency of the anomalies is lower. Mitotic anomalies (nondisjunction), as well as polyploidy, aneuploidy, and rearrangements, occur. One of the major contributing factors may be the lack of coordination between nuclear divisions and cell divisions. Chromosomal instabilities may be found in intact organisms caused primarily by natural insertion and transposable elements, and also by insertions introduced by transformation. *See* hybrid dysgenesis; isochores; RIP; RIZ; Roberts syndrome; transposable elements. (Kolodner, R. D., et al. 2002. *Science* 297:552.)

chromosomal interchange, reciprocal *See* translocation.

chromosomal mosaic Not all the cells in the body have the same chromosomal constitution, i.e., patches of different chromosomal morphology or number coexist. *See* chimera.

chromosomal mutation Mutations that involve defects detectable by the light microscope, and in general mutations involving structural and numerical alterations of the chromosomes and mutations other than base substitutions. *See* chromosomal rearrangements; chromosome breakage; deficiency; duplication; inversion; translocation.

chromosomal polymorphism In the population more than one type of chromosomal morphology or arrangement is present.

chromosomal protein Includes histones and a variety of nonhistone proteins involved in the determination of the structure, replication, transcription, and regulation of these processes in eukaryotes. In sperm, protamines are found in place of histones. In the majority of fungi, histones are absent, but eukaryotic viruses, e.g., SV40 chromosomes, have nucleosomal structures. Prokaryotic chromosomes contain a minimal variety and quantity of proteins associated with the chromosomes in comparison with eukaryotes. *See* chromatin; histones; nonhistone proteins.

chromosomal rearrangement Includes internal deletions, terminal deficiencies, duplications, transpositions, inversions, and translocations. The majority of the chromosomal rearrangements are deleterious because they involve loss or altered regulations of functions (chromosomal aberrations are concomitant with several types of cancers), yet duplications and inversions play a role in evolution. Besides neoplasias, several hereditary human diseases involve chromosomal alterations. Chromosomal rearrangements play an important role in the etiology of cancer. Deletions may eliminate genes that control important checkpoints in the cell cycle (tumor suppressants). Inversions and translocations may result in gene fusions, and proliferative processes are activated (oncogenes, transcription factors, transcriptional activators, etc.). If the c-MYC oncogene is inserted into the immunoglobulin heavy chain or into an immunoglobulin κ or λ gene or T-cell receptor gene, the oncogene may be activated. When the ABL oncogene (9q34.1) is translocated to the breakpoint cluster (BCR) in the Philadelphia chromosome (22q), myelogenous and acute leukemia may develop as a consequence of elevated protein tyrosine kinase activity.

Inversion of human chromosome 14q (11;q32) involving the T-cell receptor (14q11) and the immunoglobulin heavy chain (14q32.33) results in the fusion of the variable region of the immunoglobulin and the T-cell receptor (TCRα) and may cause lymphoma. Translocations may facilitate protein dimerization and changes in transcription. Site-directed chromosome rearrangements can be induced by the techniques of molecular biology. The *loxP* prokaryote gene can be inserted site-specifically by recombination into two selected chromosomal positions. Then the *Cre* prokaryotic recombinase is introduced into the same embryonic mouse stem cells by a transient expression transformation procedure. In such a manner, translocation involving the MYC oncogene (chromosome 15) and an immunoglobulin (IgH) gene (chromosome 12) were reciprocally translocated. The selectable markers neomycin phosphotransferase (*neo*) and hypoxanthinephosphoribosyl transferase (*Hprt*) facilitated the selective isolation of the translocations. The frequency of this type of exchange is in the 10^{-6} to 10^{-8} range.

In budding yeast, mutations in genes RFA1, RAD27, MRE11, XRS2, and RAD50 may increase gross chromosomal aberrations 600–5,000–fold. During the evolution of humans and mice, apparently one chromosomal rearrangment was fixed per Myr of evolution. Among cats, cows, sheep, and pigs, significantly fewer (0.2/Myr) rearrangements took place during evolution. Among plants, the rearrangement rates varied substantially from 0.15 to 0.41 per Myr to 1.1 to 1.3 per Myr. Chromosomal aberrations of larger size can be detected by light microscopic analysis of meiotic cells or sometimes in mitosis employing old-fashioned stains (carmine or orceine) or the various banding procedures. The use of variations of FISH technology greatly facilitates microscopic identifications. Very small chromosomes or chromosomal fragments can be studied by pulsed-field gel electrophoresis. *See* aceto-carmine; aceto-orcein; Alagille syndrome; ataxia; Bloom's syndrome; cancer; Charcot-Marie-Tooth disease; chromosomal mutations; chromosome banding; chromosome breakage; Cockayne syndrome; Cre/loxP; cri du chat; DiGeorge syndrome; Fanconi anemia; FISH; gene fusion; helicase sterility; homing endonuclease; HPRT; immunoglobulins; Langer-Gideon; Lynch syndromes; Miller-Dieker syndrome; MYC; Myr; neo; position effect; Pelizaeus-Merzbacher disease; Prader-Willi syndrome; pulsed field gel electrophoresis; RecQ; Robetsonian translocations; Rubinstein-Taybi syndrome; salivary gland chromosomes; Smith-Magenis syndrome; targeting genes; WAGR; Werner syndrome; Williams syndrome; Wiskott-Aldrich syndrome; Wolf-Hirschhorn; xeroderma pigmentosum. (Shaffer, L. G. & Lupski, J. R. 2000. *Annu. Rev. Genet.* 34:297; Yu, Y. & Bradley, A. 2001. *Nature Rev. Genet.* 2:780; Stankiewicz, P. & Lupski, J. R. 2002. *Trend Genet* 18:74; Inoue, K. & Lupski, J. R. 2002. *Annu. Rev. Genomics Hum. Genet.* 3:199.)

chromosomal sex determination In the majority of eukaryotes, females have two X chromosomes (XX) and males have one X and one Y chromosome (XY); birds and butterflies have heterogametic females (WZ) and homogametic males (ZZ). In the nematode *Caenorhabditis*, males have only one X chromosome (XO) and hermaphrodites have two (XX); a similar mechanism exists in several fishes. In wasps and bees, females and workers hatch from fertilized eggs, whereas males are the products of unfertilized eggs or hatch from eggs that lost the paternal set of chromosomes after fertilization, although in the body cells males may double their DNA content during development. In mammals, the chromosomal constitution determines the gonads/sex, but the phenotype is controlled by hormones. *See* arrhenotoky; deuterotoky; hormonal effect on sex expression; mealy bug; mountjack; sex determination; thelyotoky; X-chromosome counting. (Charlesworth, B. 1996. *Curr. Biol.* 6:149.)

chromosomal virulence loci (*chv*) The majority of the virulence loci (*vir,* controlling transfer of the T-DNA of the Ti plasmid) of *Agrobacterium* are situated in the plasmid, but the *chv* genes are in the main DNA (chromosome) of the bacteria. *See* Agrobacteria; Ti plasmid; virulence genes of *Agrobacterium*. (Suzuki, K., et al. 2001. *DNA Res.* 8:141.)

chromosome Nuclear structure containing DNA, embedded in a protein and RNA matrix of eukaryotes. Also, the DNA strings of prokaryotic nucleoids and mitochondria and chloroplasts (sometimes called genophores because the latter ones are associated with only small amounts of proteins in comparison to the eukaryotic nuclear chromosomes). The morphology (length, arm ratio, appendages [satellites], and banding pattern [natural or upon staining by special dyes]) and chromosome number are characteristic for the species in eukaryotes. The nuclear chromosomes of eukaryotes are generally linear but the organellar DNAs are circular chromosomes. Bacterial and viral chromosomes are generally circular. *See* chromosome morphology; chromosome structure; satellite.

Chromosome, supernumerary *see* B chromosome; cat-eye syndrome.

chromosome abnormality database e-mail address: <simon@bioch.ox.ac.uk>.

chromosome addition *See* alien addition.

chromosome arm Portion of a chromosome on either side of the centromere. *See* arm ratio; chromosome morphology; isobrachial, heterobrachial.

chromosome assignment test *See* somatic cell hybridization.

chromosome 5B A gene (*Ph*) in the B genome of wheat suppresses homoeologous pairing in this allohexaploid species. In case of loss of this chromosome or the pairing-controlling region of it, synapsis may take place among all homoeologous chromosomes, and multivalents are formed. Similar regulator genes occur in other chromosomes of wheat and other allopolyploid species. *See* allopolyploid; chromosome pairing; multivalent; synapsis; *Triticum*. (Dworak, J. & Lukaszewski, A. J. 2000. *Chromosoma* 109:410.)

chromosome banding *See* C-banding; G-banding; Q-banding; R bands; T bands.

chromosome breakage Chromosomes may break spontaneously or by the effects of chemical and physical agents (ionizing radiation). The breakage may involve only one of the chromatids or both (isochromatid breaks). Single breaks may lead to terminal losses of the chromosomes (chromosome deficiency), or double and multiple breaks may cause internal deletions and various rearrangements such as transposition, inversion, and translocation. Deletions require only single breaks to occur at first-order kinetics; chromosomal rearrangements generally follow second- or multiple-order kinetics. At first-order kinetics, the breakage occurs in a linear proportion to the dose of the agent causing it; at second-order kinetics, the number of breaks is proportional to the square of the dose (exponential response), i.e., at low doses the rise of the number of breaks is slow and at higher doses their numbers rise more (*see* kinetics). The electrons may directly hit the DNA or may generate reactive OH• radicals. Chromosomal double breaks may also be caused by the localized hits of two or more OH• radicals. The radical may affect only one strand at a site and a second hit may take place 10 bp away (*hybrid attack*) on the other strand of the double helix. A very low-energy electron transfer may be sufficient to break the second strand, thus resulting in a double-strand break that may follow single-electron absorption. When several hits are delivered at close proximity, clusters of complex damage may take place (Boudaïffa, B., et al. 2000. *Science* 287:1658). The final outcome is modified by the efficiency of repair mechanisms. Cancerous growth is frequently associated with chromosome breakage, although it is unknown how many were caused by chromosome breakage and how often this occurred only during the process of abnormal cell proliferation. Infection by adenovirus, cytomegalovirus, herpes, Epstein-Barr virus, etc., may cause human chromosomal breakage. Several human syndromes are accompanied by an increased frequency of chromosome breakage and/or deficiency of genetic repair: Fanconi's anemia, Bloom's syndrome, ataxia telangiectasia, xeroderma pigmentosum, Cockayne syndrome, and leukemia. Leukemias and solid tumors frequently involve chromosome breakage and concomitant gene fusion. Tumorigenesis is commonly initiated by a recombination-like event involving the immunoglobulin genes or the sequences in the T-cell receptor (TCR) genes. Chromosomal translocations may bring transcriptional activators, enhancers, DNA-binding proteins, protein ligands, or transcription factors in the vicinity of genes, resulting in the production of growth and cell cycle factors. *See* binding proteins; cancer; chromosomal rearrangements; chromosome breakage, programmed; deletion; DNA repair; enhancer; environmental mutagens; immunoglobulins; inversion; isochores; mutator genes; position effect; Roberts syndrome; T-cell receptor; transcriptional activators; transcription factors; translocation; transposition; X-ray-caused chromosome breakage. (Sánchez-García, I. 1997. *Annu. Rev. Genet.* 31:429; Anderson, R. M., et al. 2002. *Proc. Natl. Acad. Sci. USA* 99:12157.)

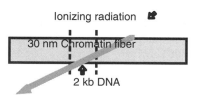

The ionizing radiation traversing an elementary chromosome fiber may simultaneously inflict damage at multiple sites within 2,000 base pairs hit. (After Rydberg, B., et al. 1995. *Radiation Damage to DNA*, p. 56; Fuciarelli, A. F. & Zimbrick, J. D., eds., Batelle, Columbus, OH.)

chromosome breakage, as a bioassay Many mutagenic and carcinogenic agents cause cytologically (by the light microscope) detectable chromosome breakage involving single-chromatid lesions, double (isochromatid) breaks, deletions, transpositions, translocations, inversions, chromosome fusions, dicentric and acentric fragment formation, etc. The frequency of these alterations can be quantitated during mitotic and meiotic nuclear divisions of suitable plant (root tips, flower buds) and animal systems (cultured lymphocytes, fibroblasts, cells withdrawn from amniotic fluids during gestation, bone marrow cells, and spermatocytes). The cytological assays can also reveal aneuploidy; polyploidy, which does not involve chromosome breakage but nondisjunction, may be the result of damage either to the centromere or to the spindle fibers. *See* bioassays in genetic toxicology; heritable translocation assays. (Chu, E. H. Y. & Generoso, W. M., eds. 1984. *Mutation, Cancer and Malformation,* Plenum, New York.)

chromosome breakage, programmed   Occurs during the development of ciliated protozoa (*Tetrahymena*) and ascarid nematodes (*Ascaris megalopcephala*) and converts the larger chromosomes of the germline into many smaller chromosomes of the macronucleus. In Tetrahymena, the 5 basic chromosomes contain 50 to 200 specific breakage sites and generate fragments of about 800 kb that persist during the vegetative life. The different breakage sites in *Tetrahymena thermophila* share a 15 bp conserved tract (5'-TAAACCAACCTCTTT-3'). After the breakage, the broken ends 5–25 bp away form new telomeres and lose about 25–65 bp around the breakage point. *See Ascaris megalocephala*; *Tetrahymena*. (Fan, Q. & Yao, M-C. 2000. *Nucleic Acid Res.* 28:895.)

chromosome breakage syndrome   *See* ataxia; Bloom syndrome; Fanconi anemia; fragile X syndrome; trinucleotide repeats.

chromosome bridge   *See* bridge; chromatid bridge.

chromosome coiling   Status of condensation of chromosomes. During interphase, the chromosomes are almost entirely stretched out, but as the cell cycle proceeds, the coiling increases, reaching a maximum at metaphase. The two chromatids may be twisted around each other during prophase (relational coiling). This type of coiling does not permit the coiled strands to separate entirely unless the separation begins at one end and is completed at the other end (plectonemic coiling). In case the coiling resembles pushing two spirals together, they can be separated in a single movement because they are not entangled (paranemic coiling). The coiling is not usually detectable in all cytological preparations unless special treatment (e.g., ammonia vapor) is employed. *See* concatenate; Mosolov model; SMC; supercoiling. (Dietzel, S. & Belmont, A. S. 2001. *Nature Cell Biol.* 3:767.)

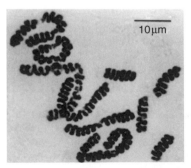

The *Tradescantia virginiana* photo is courtesy of Vosa, C. G; from Clowes, F. A. & Juniper, B. E. 1968. *Plant Cells.* Blackwell, Oxford, UK.

chromosome compaction   Folding and packaging of the elementary chromosome fibers into the chromosomes, visible through light microscopy. *See* chromatin.

chromosome complement   Haploid chromosome set. *See* haploid.

chromosome condensation   Increasingly tight winding of the chromosomal coils from interphase to metaphase. *See* chromosome coiling; condensin; DNA packaging; meiosis; mitosis; packing ratio. (Uhlmann, F. 2001. *Curr. Biol.* 11:R384.)

chromosome configuration (meiotic configuration)   Manner of pairing or assembly of chromosomes during meiosis. *See* inversion; meiosis; translocation.

chromosome conformation   Relative spatial disposition of the chromatin fiber. The conformation affects gene expression, interaction, and genetic repair processes. (Dekker, J., et al. 2002. *Science* 295:1306.)

chromosome contamination   *See* hybrid dysgenesis.

chromosome core   Central axial part of the chromosome; well visible in lampbrush chromosomes. *See* lampbrush chromosome; synaptonemal complex.

chromosome crawling   Inverse PCR. *See* inverse PCR.

chromosome crisis   Abnormalities in the cell lead to telomere dysfunction and to other chromosome anomalies leading eventually to malignant transformation, cancer. (*See* telomere, cancer. (Maser, R. S. & DePinho, R. A. 2002. *Science* 397:565.)

chromosome dimer   Is formed by recombination between two DNA rings or two ring chromosomes.

chromosome diminution   Fragmentation of the large (polycentric) meiotic chromosomes in the soma line of *Ascaris megalocephala univalens* with 1 pair of meiotic chromosomes, and in *A. bivalens*, 2 meiotic pairs, into 52 to 72 and 62 to 144 small chromosomes in the soma, respectively. *See* internally eliminated sequences; macronucleus; *Paramecia*. (Niedermaier, J. & Moritz, K. B. 2000. *Chromosoma* 109:439.)

chromosome doubling   Can be brought about by chemical or physical agents that block the function of the spindle fibers during meiosis or mitosis. Most commonly the alkaloid colchicine is used, but others, e.g., acenaphtene (a petroleum product of industrial, pesticide, and plastic manufacturing), as well as other agents, have also been employed. The purpose of the chromosome doubling is induction of polyploidy and in species hybrids to restore fertility of the hybrids, which would be sterile without doubling the chromosome number, because the distantly related chromosomes would not have homologues to pair with. *See* amphidiploid; colchicine; polyploid. (Otto, S. P. & Whitton, J. 2000. *Annu. Rev. Genet.* 34:401.)

chromosome drive   *See* meiotic drive.

chromosome elimination   During cleavage, divisions of dipteran and hemipteral insect somatic chromosomes may be lost as a natural process, but the germline cells retain the entire intact genome. In the *Ascaridea* and some other species, certain chromosomal segments may be lost as part of chromosomal differentiation during mitosis. In *Ascaris megalocephala univalens*, there is only one pair of chromosomes during meiosis, and that is fragmented into several smaller ones during somatic cell divisions. The macronuclei in *Paramecia* have only metabolic function and disintegrate after the exconjugants are formed following fertilization. Only the micronulei are retained. The macronuclei are reformed after mitoses of the diploid zygotes. Nondisjunction may also result in elimination because both of the nondisjoined chromosomes pass to one pole. The gene *polymitotic* (*pol,* map location 6S-4) of maize may eliminate several or even all of the chromosomes after meiosis during successive divisions because nuclear divisions do not keep up with the rapid succession of cell divisions. In the pentaploid *Rosa canina* ($2n = 35$), 7 bivalents are formed both at male and female meiosis, but in the male generally all the univalents are lost; in the female, one set of the 7 chromosomes derived from the bivalents and the univalents are retained in the basal megaspore and in the egg. Thus, upon fertilization, the 35 chromosome number is restored in the zygotes. Chromosome elimination occurs regularly in species with supernumerary chromosomes that have unknown function and a defective centromere. Because of the nature of the centromere, the supernumerary chromosomes commonly display nondisjunction and additional losses after fragmentation. Chromosome elimination occurs when *Hordeum bulbosum* is crossed either with *H. vulgare* or hexaploid wheat. In somatic cell hybrids of human and mouse cells, the human chromosomes are gradually eliminated unless they carry genes essential for the survival of the cybrids. *See Ascaris megalocephala*; assignment test; B chromosomes; bivalent; chromosome diminution; cybrid; haploids; *Hordeum bulbosum*; *Paramecium*; *Rosa canina*; univalent. (Ruddle, F. H. & Kucherlapati, R. S. 1974. *Sci. Am.* 231[1]:36; Goday, C. & Esteban, M. R. 2001. *Bioessays* 23:242.)

chromosome engineering   Generates rearrangements in the genomes, making alien additions, translocations; facilitating homoeologous pairing and recombination, alien transfers, alien substitutions, monosomics, chromosomal rearrangement, targeting genes, etc., with a primary goal to improve the species for agronomic purposes. *See* Cre/loxP. (Sears, E. R. 1972. *Stadler Symp.* 4:23; Higgins, A. W., et al. 1999. *Chromosoma* 108:256; Choo, K. H. 2001. *Trends Mol. Med.* 7:235; Mills, A. A. & Bradley, A. 2001. *Trends Genet.* 17:331.)

chromosome hopping   *See* chromosome jumping.

chromosome inheritance   Determined by the mitotic and meiotic apparatus. Disturbance of the normal transmission of the chromosomes may cause genetic anomalies, disease, and cancer. The several proteins control the complex process. (Dobie, K. W., et al. 2001. *Genetics* 157:1623.)

chromosome interference   *See* interference.

chromosome jumping   Special type of chromosome walking, taking advantage of the breakpoints of chromosomal rearrangements as guideposts; it permits the cloning of the two ends of a DNA sequence without the middle section. The procedure may take advantage of existing chromosomal rearrangements or the genomic DNA is partially digested with any restriction endonuclease or with enzymes, which cut very rarely. The DNA fragments are circularized with the aid of DNA ligase and cloned in such a way that the cloning vector contains a known *E. coli* sequence between the ligation sites. The cloned product is then digested by a restriction enzyme that does not cut within the special *E. coli* sequence. Thus, the generated fragments contain the *E. coli* sequence flanked by the cloned target DNA sequence that was originally far away (100–150 kb) in the chromosome. The *E. coli* sequence containing fragments are then recloned in a phage vector and the DNA is probed to a DNA library to identify the clones that contain sequences far away in the eukaryotic genome. This procedure thus facilitates the rapid movement toward the genetic cloning target. It may also be combined with chromosome walking to approach the desired gene. *See* chromosome walking; jumping library. (Bender, W., et al. 1983. *J. Mol. Biol.* 168:18.)

chromosome knob   Dark-stained structure in the chromosomes best recognized during pachytene stage, representing local condensation of the chromatin. It is a characteristic feature of certain genomes within a species. The presence of knobs may affect recombination frequencies in their vicinity and may be involved in preferential segregation. *See* chromosome morphology; karyotype; knob; pachytene analysis; preferential segregation. (Ananiev, E. V., et al. 1998. *Genetics* 149:2025.)

Pachynema of maize chromosome 9 TIP, homozygous for large knob K^L. (Courtesy of Dr. Gary Kikudome.)

chromosome landing   Approach to gene isolation from large eukaryotic genomes by first identifying linkage to close physical markers. It is a substitute for chromosome walking, which is frequently impractical in these organisms. *See* chromosome walking. (Tanksley, S. D., et al. 1995. *Trends Genet.* 11:63.)

chromosome library   Collection of individual chromosomes isolated by flow cytometric separation or pulsed-field gel electrophoresis. Such a library may facilitate the manipulation of large eukaryotic genomes. *See* flow cytometry; pulsed-field electrophoresis. (Zeng, C., et al. 2001. *Genomics* 77:27.)

chromosome maintenance region 1 (CRM1, also called exportin 1)   Karyopherin-like protein also involved in nuclear import along with NES (nuclear export signal), a leucine-rich protein. CRM1 may affect chromosome segregation. *See* exportin; importin; karyopherin; nuclear pore; RNA export. (Lindsay, M. E., et al. 2001. *J. Cell. Biol.* 153:1391.)

chromosome map   *See* mapping, genetic; physical mapping; radiation mapping.

chromosome marker   Morphological (e.g., knob, satellite, band) or molecular (restriction enzyme recognition site) signpost on a chromosome. *See* banding; chromosome knob; satellite.

chromosome mobilization   *See* conjugation; *mob*.

chromosome morphology   Generally identified at metaphase and accordingly meta-, submeta-, acro-, and telocentric chromosomes are distinguished. Furthermore, secondary constrictions and appendages (satellites) are distinguished. The various banding techniques permit the analysis of individual chromosomes on the basis of differential staining. By the application of probes with fluorochromes (chromosome painting), different details (translocations, transpositions) can be identified. In interphase and prophase (pachytene), some chromosomes display chromomeres, crossbands (salivary gland chromosomes, giant chromosomes), or natural knobs. The nonnuclear, prokaryotic and viral chromosomes are usually circular. *See* centromere; chromosome banding; chromosome knobs; chromosome painting; FISH; fluorochromes; karyotype; nucleolar organizer; organelles; pachytene analysis; PRINS; secondary constriction.

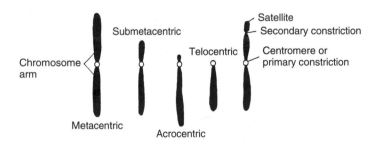

chromosome mutation   Any change (beyond the size of a nucleotide or codon) involving the structure or number of the chromosomes. *See* chromosomal aberration; chromosomal rearrangement; chromosome doubling; codon; point mutation.

chromosome number   Variable among the different species and may vary between species by polyploidy, although it is an important, stable taxonomic feature. The basic number is represented by x, the gametic number by n, and the somatic number by $2n$. Thus, in a diploid species like *Arabidopsis*, $x = 5 = n$. In hexaploid wheat $6x = 2n$. The chromosome number may vary between males and females as a mechanism of sex determination. Also, centromeric fusion may generate one biarmed chromosome from two telocentrics (acrocentric) or the opposite may take place. In cultured cells of mice, the chromosomes may become acrocentric and

double in number. The majority of bacteria have a single circular chromosome. Some other species, e.g. *Rhodobacter sphaeroides*, *Agrobacterium tumefaciens*, *Brucella melitensis*, *Vibrio cholerae*, also have multiple chromosomes. Numbers of chromosomes of genetically often-studied organisms are shown here and additional numbers are found under the English (or scientific) name of the different organisms. *See* acrocentric; animal genomes; chromosome arm; genome; haploid; polyploid; Robertsonian translocation; sex chromosomes; telocentric. (<mesquite.biosci.arizona.edu/mesquite/mesquite.html>.)

MICROORGANISMS	x
Aspergillus nidulans	8
Chlamydomonas reinhardi	8
Dictyostelium discoides	7
Neurospora crassa	7
Saccharmoyces cerevisiae	16
Saccharomyces pombe	3
PLANTS	
Arabidopsis thaliana	5
Barley (*Hordeum vulgare*)	7
Broad bean (*Vicia faba*)	6
Datura sp.	12
Epilobium sp.	18
Haplopappus gracilis	2
Lily (*Lilium sp.*)	12
Maize (*Zea mays*)	10
Oenothera lamarckiana	7
Petunia hybrida	7
Potato (*Solanum tuberosum*, 2n = 48 = 4x)	12 + 12
Tobacco (*Nicotiana tabacum*, 2n = 48 = 4x)	12 + 12
	12
(*Nicotiana plumbaginifolia*, 2n = 24)	
Tomato (*Lycopersicum sp.*)	12
Wheat (*Triticum aestivum*, 2n = 42 = 6x)	7 + 7 + 7
	7 + 7
(*Triticum turgidum*, 2n = 24 = 4x)	
	7
(*Triticum monococcum*, 2n = 14)	
ANIMALS	
Cattle (*Bos taurus*)	30
Caenorhabditis elegans (female 2n = 12, male 2n = 11)	6
Chimpanzee (*Pan troglodytes*)	24
Cricket (*Gryllus campestris*, female 2n = 30, male 2n = 29)	15
Drosophila melanogaster	4
Hamster (*Mesocricetus auratus*)	22
Honeybee (*Apis mellifera*, female 2n = 32, male 1n = 16)	16
Housefly (*Musca domestica*)	6
Homo sapiens	23
Mouse (*Mus musculus*)	20
Rat (*Rattus norvegicus*)	21
Sea urchin (*Strongylocetrotus purpuratus*)	18
Silkworm (*Bombyx mori*)	28
Swine (*Sus scrofa*)	19
Tetrahymena pyriformis	5
Toad (*Xenopus laevis*)	18
Wasp (*Habrobracon sp.*, female 2n = 20, male 1n = 10)	10

chromosome painting   Identification of chromosomes by in situ hybridization using fluorochrome-labeled probes.

With recent refinements of these techniques, each human chromosome can be distinctly identified by color and various rearrangements can be detected in an unprecedented manner. *See* chromosome morphology; FISH; fluorescence microscopy; fluorochromes; in situ hybridization; spectral karyotyping; telomeric probes; USP; WCPP. (Fauth, C. & Speicher, M. R. 2001. *Cytogenet. Cell Genet.* 93:1, COLOR PLATES between pages 692 and 693.)

chromosome pairing (synapsis) Ability of homologous (or under some circumstances homoeologous chromosomes) eukaryotic chromosomes to associate intimately during the prophase of meiosis and form bivalents. The bivalents represent the essentially identical homologous paternal and maternal chromosomes. In some organisms chromosomes may also pair during mitoses, but this association is generally not considered equally intimate, although in the salivary glands of *Drosophila* (and other dipterans) homologous chromosomes are tightly associated (somatic pairing). Synapsis most commonly begins at the termini of chromosomes at the zygotene stage and proceeds toward the centromere. By pachytene the pairing is complete, and if it is not complete, it is not going to completion later and some areas will remain unpaired. The pairing is genetically determined, and single, specific genes may prevent pairing such as *as1* (*asynaptic*, chromosomal location 1–56 in maize). Curiously, in *as1* maize, crossing over may increase. In hexaploid wheat, the *Ph* gene suppresses homoeologous association of chromosomes, but when it mutates or is deleted (monosomics and nullisomics for chromosome 5B), a high degree of homoeologous pairing occurs even in hybrids of related species. Some *desynaptic* genes terminate pairing precociously. In polyploids (polysomics) the homologous chromosomes may display multivalent association, but at any particular point the synapsis is only between two chromosomes. In the salivary glands of trisomic flies, the three chromosomes may be paired all along their length. In diptera, where the X and Y chromosomes share homologous euchromatic termini, a short-duration delayed pairing occurs (touch-and-go pairing). Synapsis is facilitated by the formation, beginning in leptotene, of the synaptonemal complex, a tripartite protein structure, formed between the paired chromosomes. Synapsis provides the opportunity for the homologous chromosomes to experience crossing over and recombination. When chiasma and recombination takes place, the distribution of the chromosomes at anaphase is orderly (*exchange pairing*), whereas in the absence of chiasmata

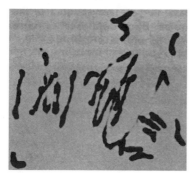

Multivalent pairing in F1 hybrid of *Triticum aestivum* (AABBDD) x *Aegilops variabilis* (C^uC^uS^1S^1) in the absence of chromosome 5B (pairing inhibitor). (Courtesy of Dr. Gordon Kimber.)

(*distributive pairing*) the chance for nondisjunction increases. There are various types of pairings quite distinct from synapsis; these nonspecific associations at the chromocenter of salivary gland chromosomes, association of telomeric heterochromatin in monosomes, or self-pairing in certain univalents may be observed if they possess more or less homologous sequences. *See* crossing over; distributive pairing; illegitimate pairing; meiosis; pachytene analysis; parasexual mechanisms; recombination; somatic pairing. (Sybenga, J. 1999. *Chromosoma* 108:209.)

chromosome partioning In prokaryotes, after replication one of the two chromosomes, each, is delivered to the daughter cells. *See* DNA replication, prokaryotes; mitosis (for comparison). (Lemon, K. P., et al. 2001. *Proc. Natl. Acad. Sci. USA* 98:212.)

chromosome positioning In prokaryotes, the old and new (replicated) chromosomes tend to move to opposite poles of the cell. A defect in positioning may involve a condition analogous to nondisjunction in eukaryotes, i.e., the 0–2 distribution. Chromosome positions in the interphase eukaryotic nucleus are not random. The centromeres tend to interact with nuclear lamina at the pores. *See* anaphase. (McEwen, B. F., et al. 2001. *Mol. Biol. Cell* 12:2776; Marshall, W. F. 2002. *Current Biol.* 12[5]:R186.)

chromosome puffing Takes place in the polytenic chromosomes (of animals and plants) when genes are activated and begin synthesizing large amounts of RNA. When transcription is terminated, the puffs recede. Puffing may be stimulated by the administration of hormones (e.g., ecdyson in insects). *See* ecdyson; puff. (Thummel, C. S. 1990. *Bioessays* 12:561.)

chromosome region maintenance *See* CRM.

chromosome replication numbers in humans prior to a sperm produced at age A can be determined by the formula $N_A = 30 + 23(A - 15) + 5$, and thus in males of age 20, 30, 40, and 50, N_A is 150, 380, 610, and 840, respectively. Therefore, it is not unexpected that the offspring of older fathers may be loaded by more new mutations. In the human females, the number of replications is much smaller as a function of age. Indeed, the mutation of paternal genes is significantly higher in many hereditary diseases. There is an almost complete absence of mutant males in 13 X-linked traits that are lethal or sterilizing in females. This is due to the generally high male mutation rate. Affected males would have heterozygous mothers. If the rate of mutation in the females is low, the observations make sense. In the diseases Duchenne muscular dystrophy and neurofibromatosis, based on very large genes, the mutations are generally not of paternal origin. In these latter cases, the mutations are not caused by replicational errors but by deletions and duplications. In hemophilia, there is a higher male rate for point mutations and a higher female rate for deletions. *See* dictyotene stage; hemophilias; muscular dystrophy; mutation rate; neurofibromatosis; replication. (Crow, J. F. 1999. *Genetics* 152:821.)

chromosome rosette At the prometaphase stage (lasting about 5–10 min in human cells), the chromosomes are arranged like a wheel, the centromeres oriented toward the hub and the arms assuming an arrangement like the spokes. If the chromosomes are painted by fluorochrome labels, the homologues appear at opposite positions of the rosette. *See* FISH; metaphase; mitosis. (Munkel, C., et al. 1999. *J. Mol. Biol.* 285:1053.)

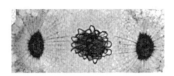

Monocystis magna rosette.

chromosome scaffold   Structurally preserved form of the chromosome freed from histone proteins. *See* chromatin; chromosome structure. (Stack, S. M. & Anderson, L. K. 2001. *Chromosome Res.* 9:175.)

chromosome segregation   Basis of Mendelian inheritance (*See* mitosis and meiosis). In autopolyploids, the chromosomes may segregate reductionally, e.g., at anaphase I in an autotetraploid, chromosomes with A, A, A, A may move toward one pole and chromosomes a, a, a, a toward the other. This is called reductional segregation (R). Alternatively, the distribution may be $A\,a$, $a\,A$ and $a\,A$, $A\,a$, or $a\,A$, $a\,A$ and $A\,a$, $A\,a$, respectively; i.e., equational segregation (E) occurs. These two types of separations follow the proportion of 1R:2E. The term *chromosome segregation* is used for cases in polyploids when genes are closely linked to the centromere and crossing over does not take place between them, in contrast to *maximal equational segregation*, where one crossing over takes place between the gene and the centromere in each meiocyte. *See* autotetraploid; meiosis.

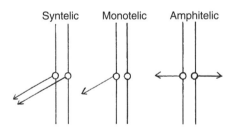

The segregation of the chromatids (vertical lines) is controlled by anaphase promoting complex (APC) regulated by the Aurora B kinase. If the conditions are normal, the centromeres segregate equationally in mitosis (amphitelic segregation) and the two centromeres (circles) are pulled to the opposite poles by the spindle fibers (arrows). This facilitated by destroying cyclin B and securin proteins. When syntelic or monotelic movement are sensed Aurora inhibits APC at a checkpoint. Such a control assures that the daughter cells will be normal. *See* meiosis, autotetraploid, Aurora, centromere, spindle fibers, cyklin B, securin, APC. (Nasmyth, K. 2002. *Science* 297:559.)

chromosome set   Group of chromosomes representing all the chromosomes of the haploid set; the genome. It is represented by x. *See* genome; haploid.

chromosome sorting   *See* flow cytometry.

chromosome stickiness   Apparent adhesion of chromosomes that tend to stay together.

chromosome structure   Electron microscopic image of a chromosome reveals much more details than a light-microscopic image. The light microscope has, however, great advantage for the study of chromosomal behavior in mitosis and meiosis, such as chiasmata, nondisjunction, misdivision, etc. *See* chromatin; Mosolov model; nucleosomes. (Woodcock, C. L. & Dimitrov, S. 2001. *Curr. Opin. Genet. Dev.* 11:130.)

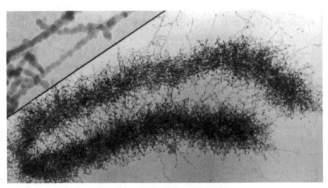

The image represents a loosely packaged acrocentric human chromosome isolated from a Burkitt's lymphoma cell. The individual fibers are shown on the inset at approximately 85,000x magnification. (Photomicrograph from Lampert, F., Bahr, G. F & DuPraw, E. J. 1969. *Cancer* 24:367.)

chromosome substitution   Can be alien substitution or intervarietal substitution when one chromosome of a variety is replaced by the corresponding chromosome of another variety in polyploids where monosomic or nullisomic lines can be propagated Thus, some desirable genes can be transferred without altering the genetic background. Although E. R. Sears developed chromosome substitution primarily for hexaploid wheat plants since the 1930s, it is applicable to any other system for the localization of QTLs or other genes. *See* alien substitution; genetic engineering; linkage in breeding; QTL.

chromosome symbols   *See Drosophila*; gene symbol.

chromosome ten tumor suppressor   *See* PTEN.

chromosome territory   *See* perichromatin fiber.

chromosome theory   Developed at the beginning of the 20th century, stating that the genetic material is contained in the chromosomes and the Mendelian laws are based on the mechanisms of meiosis. *See* Mendelian laws; Mendelian segregation. (Sutton, W. S. 1903. *Bull. Biol.* 4:231.)

chromosome transfer (chromosome-mediated gene transfer)   *See* chromosome substitution; chromosome uptake.

chromosome territories   Are the radial position of the chromosomes within the interphase nucleus. Gene-rich, early replicating chromosomal regions are generally clustered in the internal areas whereas gene-poor, late-replicating regions seem to be located at the periphery (Tanabe, H., et al. 2002. *Mutation Res.* 504:37.)

chromosome uptake   Outlined on page 222.

chromosome walking   Mapping the position of a DNA site or a gene by using overlapping restriction fragments. The principle is somewhat similar to classical cytogenetic mapping with overlapping deletions. It is also used for map-based isolation, then cloning of specific genes. The success of isolation of genes by this method is greatly affected by the size of the genome and even more importantly by the distance that must be "walked" from a known genomic position toward the desired gene. Some means must also be found for determining the function of the gene so that its identity can be verified. In large eukaryotic genomes, the procedure is facilitated if physical maps are already available. *See* chromosome jumping; chromosome landing; cosmid vectors; map-based cloning; position effect; YAC vectors. (Bender, W., et al. 1983. *J. Mol. Biol.* 168:17; Kneidinger, B., et al. 2001. *Biotechniques* 30:248, diagram on page 221.)

chromosomin   Nonhistone protein of the chromatin. *See* chromatin.)

chronic granulomatous disease (CGD)   Complex disease based on the inability of the phagocytizing neutrophils to destroy infectious microbes because of defects in delivering high enough levels of oxygen to the neutrophil membranes. Laboratory diagnosis is generally based on the failure of the phagocytes to reduce nitroblue tetrazolium. Prenatal diagnosis exists for males and carriers. Those afflicted are liable to infections. The human gene was localized to Xp21, and it apparently has a defect in the cytochrome b system, probably most commonly in the β subunit (CYBB), whereas the autosomal form (human chromosome 16q24) is defective in the α subunit (CYBA). In some variants other functions may also be involved. *See* cytochromes; immunodeficiency; neutrophil; phagocytosis. (McBride, O. W. & Peterson, J. L. 1980. *Annu. Rev. Genet.* 14:321.)

chronic lymphocytic leukemia (CLL)   *See* leukemia.

chronic radiation   Radiation dose(s) is delivered continuously without interruptions. *See* fractionated dose.

chronic wasting disease (CWD)   Prion disease-like encephalopathy of ~4% wild deer and ~1% elk in the American Northwest. Its transmission and symptoms are similar to other encephalopathies, although the epidemiological information is still limited. It may pose a danger to cattle, to eaters of venison, and to taxidermists. *See* encephalopathies.

chronology of genetics   *See* genetics, chronology of.

CHUK   IκB-α kinase. *See* IκB.

Chrysanthemum (*C. indicum*)   Herbaceous perennial ornamental; $2n = 20, 45-30$.

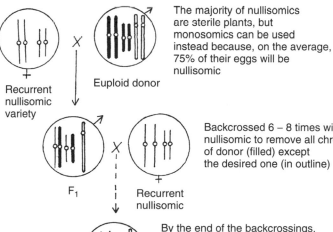

The majority of nullisomics are sterile plants, but monosomics can be used instead because, on the average, 75% of their eggs will be nullisomic

Recurrent nullisomic variety

Euploid donor

Backcrossed 6 – 8 times with nullisomic to remove all chromosomes of donor (filled) except the desired one (in outline)

F_1

Recurrent nullisomic

By the end of the backcrossings, all the chromosomes should be identical to the recurrent line. Except the desired one (shown in outline)

Monosomic substitution line is selfed to obtain

Disomic substitution line

Chromosome substitution.

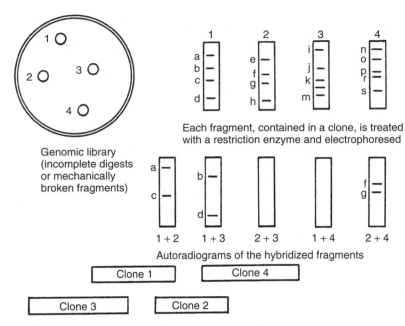

Genomic library
(incomplete digests
or mechanically
broken fragments)

Each fragment, contained in a clone, is treated
with a restriction enzyme and electrophoresed

1 + 2   1 + 3   2 + 3   1 + 4   2 + 4

Autoradiograms of the hybridized fragments

Clone 1   Clone 4

Clone 3   Clone 2

On the basis of hybridization, the overlapping sequences within the
set of fragments can be determined, and the pattern of overlaps reveals
their sequence within the chromosome

Chromosome walking.

Churg-Strauss vasciculitis Deficiency of the C1-INH complement component resulting in allergic inflammation of the vessels system. *See* complement.

chv Chromosomal virulence loci in *Agrobacterium*. *See* virulence genes of *Agrobacterium*.

chylomicrin Transporter of lipoproteins ingested or synthesized in the small intestines. *See* Anderson disease; hyperlipoproteinemia.

chymase Similar to chymotrypsin; hydrolyzes peptide bonds near the carboxyl end of hydrophobic amino acids. *See* chymotrypsin.

chymotrypsin Protease (M_r 25 K) cleaving near aromatic amino acids, nonpolar groups, and ester bonds. Chymotrypsin is targeted by nerve gas (DFP).

Ci (Curie) Measure of radioactivity; 1 Ci = 3.7×10^{10} disintegrations/sec. *See* isotopes; radioactive label; radioactivity.

cI Phage-λ repressor. *See* lambda phage.

CIBD Chronic inflammatory bowel disease. Its prevalence in the Western world is 2×10^{-3}. *See* Crohn disease.

CIC Chloride ion channel. It controls excitability of skeletal muscles, blood pressure; acidifies endosomal compartments; regulates GABA responses. *See* endosome; GABA; ion channels.

CID Collision-induced dissociation is a fragmentation of molecules, e.g., peptides at particular bonds, which sheds information on the peptide sequence analyzed by mass spectrometry in proteomics. *See* MALDI/TOF/MS; mass spectrometer; neutral-loss scan.

CIGAR CGGAAR (R = purine) enhancer motif of herpes simplex virus. *See* herpes; TAT-GARAT.

cigarette smoke Contains dozens of various combustion products including the most potent carcinogens and mutagens, e.g., benzo(a)pyrene about 20–40 ng/cigarette. It is responsible for the majority of cases of lung cancer cells, which carry mutations in the p53 tumor suppressor gene, mainly G → T transversions, at codons 157, 248, and 273. The codon 157 hot spot is absent from other types of cancers. These hot spots are the sites of adduct formation by benzo(a)pyrene-diol-epoxide and guanine-N^2. *See* benzo(a)pyrene; cancer; hot spot; p53; transversion. (Izzotti, A., et al. 2001. *Mutation Res.* 494:97.)

CIITA (class II transactivator) Apparently non-DNA-binding modulator of the synthesis of class II MHC molecules. When it binds GTP, it moves to the cell nucleus, where it interacts with the complex RFX. RFX is bound to MHC II gene promoters. CIITA transcription uses four promoters. Promoter 1 specifically controls its expression in dendritic cells, promoter 3 is relied on in B and T lymphocytes, and promoter 4 is activated by interferon-γ. CIITA is also a global coactivator of the human leukocyte antigen-D (HLA-D) genes, and it regulates import through the nuclear pore. *See* dendritic cell; HLA; interferon; MHC; nuclear pores; RFX; transactivator. (Wiszniewski, W., et al. 2001. *J. Immunol.* 167:1787.)

cilia (singular cilium) Hair-like structures formed from microtubules; they are extensions of the basal bodies. They are used for locomotion (swimming) in watery media or on viscous films by a vibratory or lashing movement. *See* microtubule.

- The transferred segment may be 17 to 1000 kbp and microscopically invisible (microtransgenome)
- Microscopically visible segments (macrotransgenomes) constitute about 15% of the transformants
- Integration is generally not by homology, rather by random events like translocations
- Transferred genes are expressed, frequently at higher intensity because of larger copy number
- Both partially degraded and more or less intact fragments become stable after integration

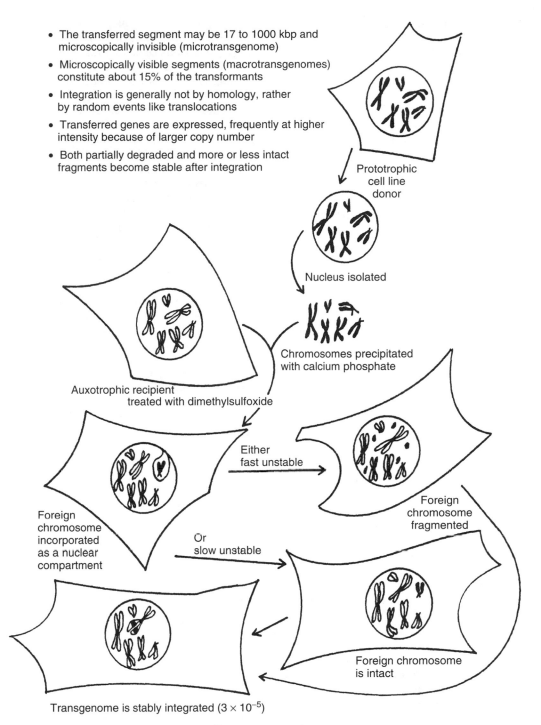

Prototrophic cell line donor

Nucleus isolated

Chromosomes precipitated with calcium phosphate

Auxotrophic recipient treated with dimethylsulfoxide

Either fast unstable

Foreign chromosome fragmented

Foreign chromosome incorporated as a nuclear compartment

Or slow unstable

Foreign chromosome is intact

Transgenome is stably integrated (3×10^{-5})

Chromosome uptake.

ciliary dyskinesia, primary The basic defect is in the axonemal heavy chain dynein at 7p21. Half of the Kartagener patients have the same mutation. *See* Kartagener syndrome.

ciliary neurotrophic factor (CNTF, human chromosome 11q12.2) Released by glial cells to repair damaged neurons. CNTF is involved in the general differentiation of nerve cells (gliogenesis) from multipotential precursor cells. Its 41 kDa receptor, CNTFR, is encoded at 9p13. CNTF specifically stimulates the JAK-STAT signal transduction pathway leading to cell fate determination. CNTF deficiency does not involve any known human disease, although CNTF may have some beneficial effect for amyotrophic lateral sclerosis. It has some effects like leptin. *See* amyotrophic; APRF; lateral sclerosis; leptin; neurogenesis; neuron; nerve growth factor; obesity; signal transduction. (Linker, R. A., et al. 2002. *Nature Med.* 8:620.)

CIMP CpG island methylator phenotype activates neoplasias by interfering with the expression of tumor suppressor genes. *See* CpG island; methylation of DNA; tumor suppressor genes.

CIN Chromosome instability factors. *See* chromosomal instability.

Cin Invertase. *See* Cin4; invertases.

Cin4 *See* hybrid dysgenesis I-R; nonviral retroposons.

C1 inhibitor Proteinase inhibitor of the serpin family; it is a regulator of blood clotting. *See* serpin.

CIP Calf intestinal alkaline phosphatase is used for the removal of 5′ phosphate from nucleic acids and nucleotides. It is a general inhibitor of CDKs. *See* CDK; KIPp21; p/CIP.

ciprofloxacin Fluoroquinolone antibiotic; it inhibits bacterial DNA gyrase and topoisomerase IV cleavage, resealing, and thus DNA replication. It is used effectively against anthrax. *See* antibiotics; gyrase; topoisomerase.

Ciprofloxacin.

CIR Cardiac inwardly rectifying ion channel. *See* ion channels, I_{KAch}.

circadian rhythm A daily (24 h) periodicity; from the Latin *circa diem* (around the day). Such periodicity (rhythm) may occur in plants in opening and closing flowers, translocation of metabolites, or may affect the daily behavioral and metabolic pattern of animals (such as sleep, social activities, etc.). Usually the circadian rhythm is more or less temperature independent. In *Arabidopsis*, nearly 500 genes are affected by the daily rhythm. A clock-controlled promoter sequence (AAAATATCT) appears to be repeated many times in most of the cycling genes. In *Drosophila*, the *per* (period) mutation has, however a dimerization domain, PAS, which is subject to temperature effects. The *per* genes encode Thr-Gly repeats in 14, 17, 20, and 23 dipeptide lengths in the various ecological races of different *Drosophila* species. The 17 and 20 alleles are predominant in the Mediterranean and northern regions of Europe, respectively. In animals, the control spot for the circadian periodicity resides in the suprachiasmatic nucleus of the hypothalamus in the brain. Glutamate, methyl-D-asparate, or nitric oxide injection into the brain produces the same effect as light in controlling the periodicity. Intracellular Ca^{2+}-ryanodine channels have a regulatory role.

The circadian rhythm is involved in olfactory responses. Some prokaryotes also display this rhythm, but *Saccharomyces* and *Schizosaccharomyces* eukaryotes apparently lack it. The molecular mechanism of the circadian oscillations seems complex, but it appears that it is dependent on activation of the E-box and transcription of genes *Per*, *Tim*e, by the gene *Clock-MOP3* complex. After the mRNAS are translated, the protein may eventually feedback inhibits *Clock* and the daily oscillations come to a full circle. The expression of *Vrille* is a requisite for *Clock*. In *Drosophila*, the gene *Doubletime* product (DBT) has been identified, which destabilizes Per in the evening until a sufficient quantity of Tim accumulates and protects it from DBT. The *Doubltetime* homologue of Syrian hamster (*TAU*) encodes a casein kinase 1 epsilon (CK1ε) protein. Then Per and Tim pass to the nucleus and cause their feedback inhibition. By the time, in the morning, these two gene products fade away, and Clock and other proteins again turn on Per and Tim. *Drosophila* controls the circadian functions by five major genes. Similar systems operate in other animals based on homologous genes and proteins. In plants, the control is slightly different. The circadian light receptor in mammals resides in the eyes, and the pineal gland and the signals are chemically transmitted to the organs (deep in the body) not accessible to light. In the transparent body of *Drosophila* or zebrafish, internal organs also have light oscillators. A common circadian rhythm disorder is *jet leg*, based on maladjustment to longer than 5 hours of time zone differences, especially by traveling from west to east. Evidence is accumulating for the role of PDF (pigment-dispersing factor, a neuropeptide) in controlling genetic clock genes. An analog of the Clock gene (NPA2) expressed in the forebrain of mammals is regulated by the redox state of NAD cofactors. *See* brain, human; clock genes; cryptochrome; E box; endogenous rhythm; entrainment; hypothalamus; melatonin; NAD; nitric oxide; oscillator; oxidation-reduction; *per* mutation in *Drosophila*; pineal gland; rhodopsin; ryanodine; *zeitgeber*; *zeitnehmer*. (Young, M. W. 1998. *Annu. Rev. Biochem.* 67:135; Dunlap, J. C. 1999. *Cell* 96:271; Scully, A. L. & Kay, S. A. 2000. *Cell* 100:297; Lowrey, P. L. & Takahashi, J. S. 2000. *Annu. Rev. Genet.* 34:533; Shearman, L. P., et al. 2000. *Science* 288:1013; Clayton, J. D., et al. 2001. *Nature* 409:829; McClung, C. R. 2001. *Annu. Rev. Plant Physiol. Mol. Biol.* 52:29; Rutter, J., et al. 2001. *Science* 293:510; Ueda, H. R., et al. 2002. *J. Biol. Chem.* 277:14048, <http://www.cbt.virginia.edu>.)

circular dichroism Difference between the molar absorptivities for left-handed and right-handed polarized light; it is observed in chiral molecules (enantiomorphs). Circular dichroism detects Z DNA structures and the interaction of drugs and carcinogens/mutagens with the DNA. *See* chirality; enantiomorph; Z DNA.

circular DNA Covalently closed ring-shaped DNA such as the genetic material of bacteria, eukaryotic organelles, and the majority of plasmids.

circularization Forming a circle, like plasmids; unwanted circularization of DNA can be prevented by directional cloning or treatment with bacterial phosphatase to remove terminal phosphates needed for joining DNA ends. *See* directional cloning.

cirrhosis of the liver Mainly autosomal-recessive diseases characterized by fibrous structure of the liver. It may be precipitated by various environmental factors such as copper toxicity, alcoholism, antitrypsin deficiency, and hepatitis virus infection. It may occur as a component of other syndromes. Cirrhotic liver loses its ability to regenerate normal cells, apparently because of diminished telomerase activity. Childhood cirrhosis characterized by cholestasis was localized to 16q22. Telomerase gene, carried by adenoviral vector, improved cirrhosis in telomerase-deficient mice. *See* adenovirus; alcoholism; antitrypsin; autoimmune diseases; cholestasis; cholic acid; galactosemia; gene therapy; transaldolase deficiency; Wilson's disease. (Friedman, S. L. 2000. *J. Biol. Chem.* 275:2247.)

CIS (CISH, cytokin-inducible-SH2-containing protein, 3p21.3) Contains 222 amino acids. It binds to tyrosine-phosphorylated IL-3 and erythropoietin receptors. It may be part of the system turning off cytokine signaling. Its deletions are frequent in lung and kidney tumors. *See* cytokines; erythropoietin; IL-3; JAB; SH2 domain; SOCS-box; SSI-1. (Uchida, K., et al. 1997. *Cytogenet. Cell Genet.* 78:209.)

cis arrangement Two genes (or two different mutant sites) of a locus are within the same chromosome strand, e.g., a b. *See* cis-trans test; trans arrangement.

cis-acting element Must be in the same DNA strand as its target to act on in transcription. *See* cis-regulatory modules; LCR; trans-acting element.

cis-dominant Dominance affecting only alleles in *cis* position but not those in *trans*. *See* cis arrangement.

cis-Golgi Side of the Golgi apparatus where molecules enter the complex. *See* Golgi apparatus; trans-Golgi network.

cis-immunity Property of transposons to prevent integration of another element within the boundary of the insertion element. *See* self-immunity; transposon.

cisplatin (cis-diamminedichloroplatinum, cis-DDP) DNA cross-linking agent and an effective anticancer drug with some specificity for testicular tumors. High-mobility-group proteins are attracted to the DNA distorted by cis-DDP and may mediate antitumor activity. Cisplatin may cause kidney damage. *See* cancer therapy; high-mobility group of proteins. (Lippert, B., ed. 1999. *Cisplatin: Chemistry and Biochemistry of a Leading Anticancer Drug.* Helvetica Chimica Acta Vlg., Zürich. Jung, Y., et al. 2001. *J. Biol. Chem.* 276:43589.)

cis preference Protein products of the L1 retrotransposons usually (not necessarily) bind to the encoding RNA. *See* LINE.

cis-regulatory module (CRM) Modular unit of a few hundred base pairs in the DNA; mediates the multiple binding of transcription factors. Cis-regulatory modules play important roles in development and tissue-specific expression of genes. *See* tissue specificity. (Berman, B. P., et al. 2002. *Proc. Natl. Acad. Sci. USA* 99:757.)

CISS hybridization Chromosome in situ suppression. Before DNA probes are applied to chromosomes for in situ hybridization, the chromosomes are treated with DNA to block the nontarget sequences so that they will not interfere with annealing of the specific labeled probes. *See* in situ hybridization; probe. (Sadder, M. T., et al. 2000. *Genome* 43:1081.)

cis-syn dimer Mutagenic cyclobutane UV photoproduct. *See* cyclobutane; DNA polymerases (Polη); translesion; UV photoproducts. (McCullough, A. K., et al. 1998. *J. Biol. Chem.* 273:13136.)

Cis-syn dimer

cisterna Membrane-enclosed space frequently containing fluid.

cis-trans test Procedure for determining allelism. If two independent *recessive* mutations are made heterozygous in a diploid (or by using an F′ plasmid in bacteria) are in the opposite strands (in trans-position $\dfrac{m1+}{+m2}$ and fail to complement each other (i.e., the phenotype is mutant), then the two mutations are allelic (occupy the same cistron). When, however, the two recessive mutations are in the same DNA strand (in cis $\dfrac{++}{m1m2}$ position) and are made heterozygous (or merozygous in prokaryotes), they are expected to be complementary, i.e., nonmutant in phenotype, because the strand containing the two nonmutant sites permits the transcription of a wild-type (uninterrupted) mRNA in the heterozygote. Molecular geneticists involved in physical DNA mapping use the term *allele* for any physical variation inherited by a Mendelian fashion and occupying the same chromosomal locus. *See* allele; allelism test; cistron. (Lewis. E. B. 1951. *Cold Spring Harbor Symp. Quant. Biol.* 16:159.)

cistron Segment of the DNA coding for one polypeptide chain or determining the base sequence in one tRNA or in one rRNA subunit. Mutant sites within a cistron generally do not fully complement, i.e., the heterozygotes are not wild type, although in rare cases weak allelic complementation may be observed. *See* allelic complementation; cis-trans test. (Benzer, S. 1957. *The Chemical Basis of Heredity*, McElroy, W. D. & Glass, B., eds., p. 70. Johns Hopkins Univ. Press, Baltimore.)

cis-vection Position effect, operational only in case the genetic elements are syntenic, e.g., the promoter, some DNA-binding protein elements (UAS), and structural genes. *See* cis-acting element; operon; position effect; pseudoalleles; synteny; UAS. (Lewis. E. B. 1950. *Advances Genet.* 3:73.)

Citation Index Publication of the Institute of Scientific Information (ISI, <http://www.isinet.com/isi/>) listing the number of times a particular journal paper has been cited in the scientific literature by first author. *See* impact factor.

citric acid cycle *See* Krebs-Szentgyörgyi cycle.

citron RHO-regulated protein kinase mediating myosin-based contractility of the spindle fibers during cytokinesis. Its overproduction may lead to multinucleate cells. *See* cytokinesis; myosin; PDZ; RHO; spindle fibers. (Madaule, P., et al. 2000. *Microsc. Res. Tech.* 49:123.)

citrullinemia (ASS)   Chromosome 9q34 (CTLN1) and 7q21.3 (CTLN2) recessive defects in the enzyme argininosuccinase. Normally citrulline is converted via aspartate into argininosuccinate. If the latter cannot be cleaved, citrulline and ammonia accumulate; as a consequence, incontinence, insomnia, sweating, vomiting, diarrhea, convulsions, psychotic anomalies, and even periods of coma may result. The disease has an early onset and may proceed progressively into adulthood; rarely the onset is during adult life. Craving for high-arginine food (legumes) and avoidance of low-arginine food and sweets is noticeable. The ASS genes may be present in 10 copies per human genome scattered over several chromosomes according to hybridization by a DNA probe. The multiple copies are presumably pseudogenes. *See* amino acid metabolism; arginine; argininemia; pseudogene; urea cycle.

citrullinuria Citrullinemia.

citrus (*Citrus* spp)   The taxonomy is unclear, but several species are known, $x = 9$, and diploid as well as tetraploid forms exist among lemons, oranges, mandarin, lime, grapefruit, etc.

civilization *See* humanized antibody.

CJD *See* Creutzfeldt-Jakob disease.

CJM Cell junction molecule. *See* gap junction.

CKB (casein kinase genes)   Their product is required for the completion of anaphase. *See* cell cycle. (McKay, R. M., et al. 2001. *Dev. Biol.* 235:378.)

CKI Inhibitor of the CDK-cyclin complex. *See* CDK; FAR.

C kinase Ca^{2+}-dependent protein kinase attached to the plasma membrane. Diaglycerols and/or phosphatidylserine may activate it. It may be indirectly involved in $+/-$ regulation of genes. *See* RACK. (Hartness, M. E., et al. 2001. *Eur. J. Neurosci.* 13:925.)

CKR Chemokine receptor. *See* acquired immunodeficiency; CCR; chemokines.

CKS1 Subunit of cyclin-dependent kinases (CDKs) is an essential cofactor in the ubiquitination of p27 CDK inhibitor by SCF. *See* CDK; SCF; ubiquitination. (Harper, J. W. 2001. *Curr. Biol.* 11:R431.)

clade Group of distinct families/subfamilies descended from a common ancestral taxonomic entity by an evolutionary split.

cladistic Representation of descent in the manner of a dendrogram, i.e., the divergence of taxonomic groups is shown by links and nodes. *See* character index; dendrogram; evolutionary tree; homoplasy; parsimony; stratocladistics.

cladogenesis Evolutionary change involving branching of lineage of descent. *See* anagenesis.

cladogram *See* character index; evolutionary tree.

Claforan *See* cefotaxim.

clamp loader Conformation of some proteins capable of assisting ring-shaped DNA polymerase processivity factors to be placed on the DNA. Its is generally a component of the eukaryotic DNA polymerase holoenzyme. It is essential for DNA replication and repair recombination in prokaryotes and eukaryotes. In its absence mutation rate may increase probably by an impaired mismatch repair system. *See* DNA polymerase; DNA repair; processivity; replication; replication machine; sliding clamp. (Ellison, V. & Stillman, B. 2001. *Cell* 106:655.)

class I genes (eukaryotic) Transcribed by RNA polymerase I; these include 5.8S, 18S, and 28S ribosomal RNAs. *See* pol I, eukaryotic.

class II genes Transcribed by eukaryotic RNA polymerase II; these include mRNA and snRNA (except U6 RNA). They carry in the mRNA transcript a 7-methyl guanine cap (with the exception of the picornaviruses) and a 2,2,7-trimethyl guanine in the U RNAs. In lower eukaryotes, 75 to 125, in vertebrates 200 to 300 residues-long poly-A tail is added posttranscriptionally. Histone mRNAs and U RNAs have no poly-A tails. Some mRNAs include N_6-methylated adenine, U RNAs have modified uracils. They are regulated by cis- and trans-acting elements. The cap is associated with cap-binding protein, the capping enzyme. *See* capping enzyme; cis-acting; pol II, eukaryotic; polyA mRNA; trans-acting; transcription; transcription factors; transcription termination in eukaryotes; transcription unit; U RNA.

class III genes of eukaryotes, transcribed by RNA polymerase III; they include 5S ribosomal RNA and some small cytoplasmic RNAs. *See* pol III, eukaryotic.

classical genetics Studies functions based on phenotype and genotype of the genetic material; serves as the primary guidance to the understanding of the mechanisms involved, in contrast to reversed genetics, where the analysis begins with molecules. *See* reversed genetics.

classical hemophilia *See* hemophilia.

classification Sorting out phenotypes (or genotypes) by groups. It may be difficult in case of continuous variation or when the penetrance or expressivity is low. *See* expressivity; penetrance.

class switching Change in expression of immunoglobulin (antibody) heavy-chain genes during cellular differentiation of an antibody-producing lymphocyte by changing the production from one immunoglobulin heavy chain to another, e.g., the IgM constant region is replaced by the constant region of another class of immunoglobulins such as IgG, IgA, or IgE. *See* AID; antibody gene switching; hyper-IgM syndrome; immunoglobulins. (Stavnezer, J. 2000. *Curr. Top. Microbiol. Immunol.* 245:127; Kinoshita, K. & Honjo, T. 2001. *Nature Rev. Mol. Cell Biol.* 2:493; Petersen, S., et al. 2001. *Nature* 414:660.)

clastogen Any agent that can cause chromosomal breakage directly or indirectly by affecting DNA replication. Clastogenic agents may be ionizing radiation, bleomycin, hydroxyurea, maleic hydrazide, etc. *See* chromosome breakage.

clathrin 192 kDa triskelion proteins that in cooperation with smaller (~35 kDa) polypeptides form the polyhedral coat on the surface of coated vesicles involved in intracellular transport between cellular organelles. Before fusing with the target, the vesicle coats are stripped with the assistance of chaperones (hsp70) and another cofactor, auxilin. *See* adaptin; cargo receptors; chaperone; coatomer; endocytosis; lysosomes; triskelion. (Kirchausen, T. 2000. *Annu. Rev. Biochem.* 69:699; Ford, M. G. J., et al. 2002. *Nature* 419:361.)

claudin-11 (3q26.2-q26.3) Oligodendrocyte transmembrane protein controlling a paracellular barrier of tight junctions required for normal spermatogenesis and nerve conduction. *See* tight junction. (Gow, A., et al. 1999. *Cell* 99:649.)

clavulanate Product of streptomycetes is suicide substrate for β-lactamase; it enhances the effect of amoxicillin and is an effective weapon for circumventing some antibiotic resistance. *See* amoxicillin; antibiotics; β-lactamase.

claw-foot (Roussy-Levy hereditary areflexic dysplasia) Autosomal-dominant anomaly usually involving paternal transmission. It bears resemblance to the Charcot-Marie-Tooth disease but is accompanied by hand tremors. *See* Charcot-Marie-Tooth disease.

claw-like fingers and toes (curved nail of fourth toe) Rare, apparently autosomal-recessive nail deformity of the fourth (and fifth) toes and fingers.

CLB Mitotic cyclin protein. CLB6 activates only the early origins of replication of the chromosome, whereas CLB5 activates both early and late replicational origins. A complex of CLB-Cdc28 prevents an additional cycle of replication when one is completed to prevent polyteny or polyploidy. *See* CDK; cyclin; DNA replication, eukaryotes; DNA replication during the cell cycle.

ClB method Detects new sex-linked lethal mutations among the grandsons of *Drosophila* males on the basis of altered male:female ratios (*C* is an inversion, eliminating cross-over chromosomes, *l* is a lethal gene, and *B* stands for *Bar* eye [narrow]). If a new recessive lethal mutation (*m*) occurred in the X chromosome of the grandfather, the grandson receiving this or the *ClB* chromosome would die. In F_2, the males carrying either the *ClB* or the mutant X chromosome would die; without the new mutation (*m*), the female:male ratio is 2:1. In case of a new lethal mutation, no male grandsons may survive, or in case the expressivity of the new lethal gene is reduced, some males may survive. *See* autosomal dominant; autosomal recessive mutation; *Basc*. (Muller, H. J. 1928. *Genetics* 13:279.)

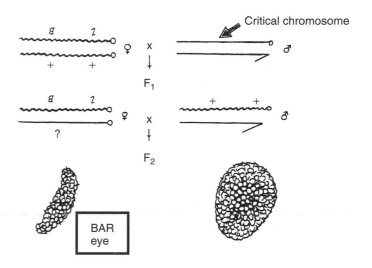

clean-room Laboratory or manufacturing facility where the biological or chemical material is relatively safe from contamination and aseptic conditions can be maintained during use and in between use. Generally four grade levels are distinguished by government regulations on the basis of the maximum floating particle size (0.5 μm to 5 μm) and number per m³, flow-hood circulation, etc. The institutional safety officer should be able to provide up-to-date information.

cleavage furrow Early embryonal division (that gives rise to the blastomeres) by which the larger fertilized egg breaks up into several smaller cells without growth—in general, splitting cells, organelles, and macromolecules into two. The cleavage furrow is formed by an actomyosine structure of the *contractile ring*. *See* actomyosin. (Glotzer, M. 2001. *Annu. Rev. Cell Dev. Biol.* 17:351.)

cleavage nucleus Nucleus of the dividing egg.

cleavage stimulation factor Mediates the polyadenylation of the majority of mRNAs. CstF is a heterotrimeric protein of 77, 64, and 50 kDa subunits and recognizes the G + U-rich element downstream of the RNA transcript. CstF-64 binds to the RNA. CstF-77 (a homologue of suppressor of forked in *Drosophila*) bridges the 64 and 50 subunits and interacts with the AAUAAA polyadenylation signal-binding factor CPSF (cleavage-polyadenylation specificity factor). CstF-50 contains WD-40 repeats and interacts with the largest subunit carboxy-terminal domain of RNA polymerase II. *See* BARD-1; forked; polyadenylation signal; WD-40. (Gross, S. & Moore, C. 2001. *Proc. Natl. Acad. Sci. USA* 98:6080.)

cleft palate Oral fissure frequently associated with harelip (lip fissure). Its incidence in the general population is about 4 to 20/10,000. Its recurrence risk in sibs of an affected individual is about 2% or more; in monozygotic twins, 50%. Actually, these developmental anomalies may be parts of autosomal or X-linked-dominant or autosomal-recessive syndromes and some trisomies. Human chromosomal regions 2q32, 4p16-p13, and 4q31-q35 may be associated with the disorder. In mice, deficiency of the β3 subunit of the A-type γ-aminobutyric acid receptor results in cleft palate. In mice, mutation in the LIM domain of a homeobox gene *Lhx8* appears to be critical for the determination. In humans, the condition is associated with variation in the epidermal growth factor level. The CLPED1 (cleft lip/ectodermal dysplasia) gene encoding nectin-1 (PRR1) was assigned to 11q23. Nectin is a cell adhesion protein and the principal receptor for α-herpesvirus. Its mutation may convey resistance to these viruses. Cleft palate with ankyloglossia (adherent tongue) has been located to human chromosome Xq21 and is due to a mutation in the T-box transcription factor TBX22. *See* CATCH; EGF; GABA; harelip; herpes; homeobox; LIM; recurrence risk; sib; trisomy; Van der Woude syndrome. (Braybrook, C., et al. 2001. *Nature Genet.* 29:179.)

cleidocranial dysostosis (cleidocranial dysplasia, 6p21, CCD) Autosomal-dominant and -recessive mutations may lead to deficiency of closure of the skull sutures, supernumerary teeth, chest, shoulder, hip and finger anomalies, short stature, etc. Autosomal-dominant forms may also show reduced jaw size, absence of thumbs or toes, and loss of distal digital bones. A mutation in mouse chromosome 6 may involve the CBFA-1 (core-binding factor) gene (apparently homologous with CCD) encoding or regulating osteocalcin, a protein that controls osteoblast formation. This molecule may function as a transcription factor for bone-forming genes. The CCD symbol is used now for the human central core disease of the muscles in humans due to mutation in the ryanodine receptor (RYR1, 19q13.1). *See* central core disease; dysostosis; osteogenesis imperfecta; pycnodysostosis; stature in humans.

cleistogamy Shedding pollen before the flowers open, thus resulting in self-fertilization in plants. *See* autogamy; protandry; protogyny.

cleistothecium Closed, spherical fruiting body of ascomycetes such as in *Aspergillus* and powdery mildew fungus. The asci containing the spores are released after the rupture of the wall of the fruiting body. *See* ascogonium; gymnothecium; perithecium.

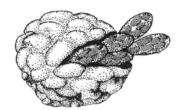

Cleistothecium.

CLIA Clinical Laboratory Improvement Amendments. A U.S. government organization oversees the laboratory procedures for clinical testing of genetically determined diseases.

cline Gradual change in the distribution of genotypes or phenotypes within a large population caused by the environment, population density, or other factors. *See* aclinal.

clinical genetics Deals with practical genetic problems encountered during patient care. *See* bioethics; counseling, genetic; empirical risk; genetic counseling; human genetics; informed consent; medical genetics.

clinical tests for heterozygosity Many genes fail to reveal their anomalous alleles by simple visual observation. Some of these are detectable by enzyme or serological assays in biopsies, by amniocentesis, and cell cultures. These tests may not be able to identify with certainty the genetic constitution because the same function may be affected by different gene loci. Some mutant alleles can be identified by RFLP or PCR analysis. *See* amniocentesis; carrier; PCR; RFLP.

clinical trial phases Conducted in three steps: (1) The first trial is with a few humans and is expected to provide information on safety/maximal dose, timing, and body reaction. (2) Larger numbers of individuals are tested for safety as well as efficacy. (3) Large-scale experimental treatment is expected to reveal efficacy compared with alternative methods of treatments and side effects. *See* informed consent.

CliP *See* sister chromatid cohesion.

CLIP Class II–associated invariant-chain complex protein. *See* major histocompatibility complex.

clipboard Of a computer, storage area of the memory system from which information can be transferred to different documents created in the computer.

clitoris Oval body (homologous to the male penis) in the inner side of the cleft between the opening of the urethra and the vagina of mammalian females. Its expansion results in relaxation following sexual excitement (climax, orgasm) without ejaculation.

CLN Cyclin genes regulating cell division. *See* CDK; cyclin.

clock gene Affects the biological clock such as the diurnal rhythm and endogenous rhythm or aging. *See* aging; circadian rhythm; endogenous rhythm. (Lakin-Thomas, P. L. 2000. *Trends Genet.* 16:135.)

clonal analysis Study of pattern formation (genetic mosaics) as a consequence of mutation, deletion, recombination, and nondisjunction. Such an analysis permits tracing the event to its origin, estimation of the number of cell divisions that have taken place since the event, the number of cells in the primordium involved, etc. Usually a requisite of such an analysis is that appropriate genetic markers would be available. Cell-autonomous genes are particularly useful for clonal analysis because genes of neighboring cells do not affect their expression. Heterozygotes for recessive color markers are frequently used because the loss or mutation of the dominant allele will result in the formation of recessive sector(s), pseudodominance. An unstable (ring) X chromosome in *Drosophila* may lead to the formation of gynandromorphs. Nondisjunction or somatic recombination may lead to the formation of twin spots (sectors). In *Drosophila*, developmental compartments of polyclonal origin may be revealed on the basis of what structures and organs are affected simultaneously or consequently. Mutations formed in the last three divisions of the wing imaginal disk involve changes in cuticular elements, e.g., hairs and wing veins. The presence of sectors also reveals that a gene is cell autonomous and whether its expression is modified by the product of other genes. Extracts of wild-type lymph, mRNA, or protein injected into mutants may correct genetic defects in a localized form, according to a gradient or in a particular pattern, depending on the nature of function of the corresponding mutant allele, being a specific transcription factor, a transmembrane protein, a signal receptor, etc. DNA labeled by fluorochromes hybridized in the tissues may identify the regions of mRNA distribution of a particular gene. The function of particular genes in regulation of others can be detected by immunostaining of particular loci with specific antibodies. *See* cell lineage; fate map; fluorochromes; GUS; immunostaining; luciferase; morphogenesis; probe. (Steffensen, D. M. 1968. *Am. J. Bot.* 55:354; Hotta, Y. & Benzer, S. 1972. *Nature* 240:527; Duchmann, R., et al. 2001. *Clin. Exp. Immunol.* 123:315.)

clonal interference Competition between beneficial mutations in asexual lineages within populations.

clonal restriction Proliferating cells are propagated only within the preordained pattern of differentiation. See cell lineages; clonal analysis.

clonal selection Specificity for specific antigens exists in the lymphocytes before exposure to these antigens. A few antibody-producing lymphocytes committed to the production of a specific antibody are selected for clonal proliferation and stimulated for the synthesis of that specific antibody when infected by the pertinent antigen. This way immunity may build up. The B lymphocytes with lesser selective value are eliminated by apoptosis. *See* antibody; apoptosis; B cell; immune system; immunoglobulins; lymphocyte; memory, immunological; receptor editing. (Meffre, E., et al. 2000. *Nature Immunol.* 1:379; Silverstein, A. M. 2002. *Nature Immunol.* 3:793.)

clone Progeny of asexual reproduction or molecular cloning, yielding identical products. *See* asexual reproduction; cloning; cloning vectors; molecular cloning; nuclear transplantation. (McLaren, A. 2000. *Science* 288:1775.)

clone-based map The genomic DNA is incompletely digested by restriction endonuclease to about 150 kbp pieces and propagated by BAC vectors. Subsequently, the clones are completely digested by restriction enzymes to generate a collection of smaller fragments. The fragments are aligned by chromosome walking and finally sequenced repeatedly to obtain an accurate order of the nucleotides. This is one of the basic principles used for sequencing larger genomes. *See* BAC; chromosome walking; restriction enzyme.

CloneConvert Computer program that converts a file into the miniset.dat format to individual MapSearch Probes.

clone validation In large-scale genome sequencing terminology, indicates that the clones accurately represent the genome. *See* DNA sequencing; genome projects.

clonidine α-2-adrenoreceptor agonist. At low concentration it decreases presynaptic firing of noradrenergic cells. Its antagonist is idazoxan. *See* adrenergic receptor; agonist; synaps.

cloning Asexual reproduction in eukaryotes or replication of DNA (genes) with the aid of plasmid vectors in appropriate host cells (*molecular cloning*). *Reproductive cloning*, i.e., generating an embryo from human stem cells, is prohibited in the majority of countries for ethical reasons. *Therapeutic cloning*, i.e., generating human tissues for medical purposes, seems to be favored by biologists and medical researchers. *See* clone; cloning vectors; DNA; DNA library; expression cloning; nuclear transplantation; plasmids; stem cells; therapeutic cloning; transplantation of organelles.)

cloning, animals and humans Technically feasible by the use of the procedures of nuclear transplantation, although — because of technical problems — cloning of higher animals cannot be used as a routine procedure for propagation. Currently, 15% or more of the cloned mammalian embryos suffer from chromosomal or metabolic defects. There is rather widespread opposition to this type of research (reproductive cloning) on the basis of less than well-informed ethical considerations. Naturally, this and almost any technology can potentially be abused, and most of the condemning arguments are weak on biological justification. Before the practical application of this and any other medical technology, sensible guidelines are required. Politically based banning of the research may deprive the society of expanding the repertory for remedying human disease. In 1997 (*Science* 278:2130), by nuclear transplantation the gene for antihemophilic factor IX was introduced into sheep, opening a new avenue to cure hemophilia B (Christmas disease). Many dread the ethical consequences of cloning humans. Others fear the biological consequences. Unfortunately, the ethical criteria are rather subjectively ill defined. The biological consequences of poor technologies are real. It must be considered that by using somatic donor nuclei the risks of undesirable combination of homozygous deleterious genes can be avoided if the nuclei are taken from adults who have already withstood the test of being disease- or malformation-free. The potential future risk of inbreeding should be and can be managed. In human cloning, distinction must be made between cloning cells for therapeutic purposes and cloning as a means of reproducing intact organisms. *See* bioethics; embryo research; hemophilia; nuclear transplantation; stem cells; therapeutic cloning. (Solter, D. 2000. *Nature Rev. Genet.* 1:199; Jaenisch, R. & Wilmut, J. 2001. *Science* 291:2552; Lanza, R. P., et al. 2002. *Science* 294:1893; O'Mathuana, D. P. 2002. *EMBO Rep.* 3:502.)

cloning bias   Deviation from randomness in the representation of fragments in a DNA library. The bias may be caused by rearrangement of direct repeats in *Rec+* bacterial strains (but can be avoided by *recA* hosts). Palindromic sequences may become unstable in some λ-phage and plasmid vectors (can be avoided by the use of *recB* and/or *recC* and *sbcB* strains of *E. coli*, base modification of the DNA, host restriction enzymes, etc.). *See* cloning vectors; DNA library; restriction enzyme.

cloning distortion   Different DNA fragments, because of their nature, length, etc., may be replicated at different rates in the vectors and may bias the representation of the sequences in the library. *See* library; replication; vectors.

cloning site   Recognition sequence for restriction enzymes within genetic vectors or other recipients where passenger DNA can be inserted. *See* passenger DNA; restriction enzymes; vectors.

cloning strategy   Plan that will permit the identification of the cloned copy either by a suitable probe (DNA, RNA, or antibody) or a positional cloning or PCR-based procedure. *See* gene isolation; PCR; positional cloning; probe.

cloning vector   Generally a plasmid, phage, or eukaryotic virus-derived linear or circular DNA capable of reproduction (most commonly in bacteria or yeast) and producing (large number) molecular clones of the DNA inserted into it. Cloning vectors must have replicator mechanisms (replication drive unit) for self-propagation, multiple cloning sites (single or few recognition sites for several restriction enzymes), selectable markers (for verifying the success of molecular recombination and uncontaminated maintenance), regulatory elements for their copy number in the host (generally smaller plasmids can be present in a larger number of copies), mechanisms for equal partition among the daughter cells, and genetic stability to prevent rearrangement by host enzymes. It is frequently desirable to propagate in more than one host cell (shuttle vectors). Some of the cloning vectors are used only for propagation of DNA; others permit expression of the genes carried; and others may be useful to isolate functional elements of the hosts (promoters, enhancers) by virtue of in vivo gene fusion. *See* agrobacterial vectors; BAC; ColE1; cosmids; lambda vectors; PAC; phagemids; plasmids; retroviral vectors; shuttle vector; SV40; YAC.

cloning vehicle   Cells suitable to propagate the cloning vectors, e.g., *E. coli*, yeast, *Agrobacterium*, etc. *See* cloning vector.

clonogenic test   Isolated cancer cells seeded onto culture plates are exposed to radiation or other anticancer treatment and incubated for about 2 weeks. Solid tumor cells in this assay die only when they divide, e.g., due to chromosome breakage. This fact indicates that the mechanical injury to the chromosomes rather than apoptosis causes the demise of the cells as evidenced by failure to form clones. Such an assay may reveal some information regarding the prospects of treatment of a particular cancer by different agents. *See* cancer; cancer therapy.

closed-loop model of translation   mRNA, which is supposed to be translated, is visualized as a molecule circularized by the 5′ — 3′ ends. Support comes from the observation that the poly(A) tail promotes translation and electron microscopy also reveals RNA circles The circularization is mediated by trans-acting protein factors such as eIF4G, eIF4F, and polyA-binding protein. *See* eIF4F; eIF4G; polyA tail. (Jacobson, A. 1996. *Translational Control*, Hershey, J. W. B., et al., eds., p. 85. Cold Spring Harbor Lab. Press, Cold Spring Harbor, NY; Sachs, A. 2000. *Translational Control of Gene Expression*, Sonnenberg, N., et al., eds., p. 447. Cold Spring Harbor Lab. Press, Cold Spring Harbor, NY.)

closed promoter complex   Transcriptase attached to the target promoter cannot start transcription because the DNA strands are not separated. *See* open promoter complex; pol prokaryotic RNA polymerase.

closed reading frame   Chain termination (nonsense) codons block its translation.

closure of mapping   When approaching completion and two genome-size DNAs have been mapped, the use of random clones is very inefficient; therefore, nonrandom clones are employed. *See* physical map.

clotting factor   *See* antihemophilic factors.

clover (*Trifolium* spp)   Includes about 250 species with the most prominent representatives: white clover (*T. repens*, $2n = 32$), fragrant alsike (*T. hybridum*, $2n = 16$), strawberry (*T. fragiferum*, $2n = 16$), red clover (*T. pratense*, $2n = 14$), crimson clover (*T. incarnatuum*, $2n = 14$), and subterranean clover (*T. subterraneum*, $2n = 16$).

cloverleaf   Representation of the tRNA displaying the single-strand stem (amino acid arm) and the D (dihydrouracil), AC (anticodon), and T (thymine) loops reminding to stem and the three leaflets of a leaf of a clover plant. *See* tRNA.

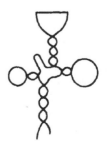

tRNA cloverleaf.

Cℓp (CℓpX) Different prokaryotic proteins involved in chaperone or protease activities belonging to the Hsp100 family, although the size of the different molecules may vary substantially. They include inducible and noninducible forms and are distributed widely among eukaryotes and prokaryotes. Cℓp proteins also occur in the plant chloroplasts and in the plastids of *Plasmodia*. The budding yeast Hsp104 Cℓp protein and the *E. coli* CℓpB are intramolecular chaperones and as such do not depend on ATP, but CℓpA and CℓpX are chaperones and ATP-dependent proteases. Cℓp-dependent proteolysis may protect against degradation of unmodified bacterial DNA by type I restriction endonucleases *See* AAA proteins; chaperone; chloroplasts; Hsp; *Plasmodium*; protease; proteasome; protein repair. (Neuwald, A. F., et al. 1999. *Genome Res.* 9:27.)

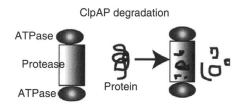

ClpAP degradation

clubfoot (talipes) Hereditary malformation of the foot with a prevalence of below 0.1% and with a recurrence risk among the sibs of an afflicted child of about 4%. Thus, it appears to be under the control of more than one gene and may have substantial environmental influences because even monozygotic twins may not both be afflicted. Various forms have been classified and only one is shown below. *See* limb defects.

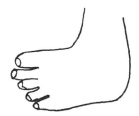

CLUSTAL W Improving the sensitivity of progressive multiple sequence alignment through sequence weighting, position-specific gap penalties, and weight matrix choice. A freely available computer program published is one of the ten most cited papers during the year 2000. (Thompson, J. D., Higgins, D. G. & Gibson, T. J. 1994. *Nucleic Acids Res.* 22:4673.)

cluster analysis Monitors simultaneously the expression patterns of thousands of genes — based on DNA arrays — at various stages of development or in response to any particular environmental influence. The method is utilizing mathematical cluster analysis suitable for classification of multidimensional complex data. The GENECLUSTER computer program can perform the calculations and assists in the interpretation of the biological meaning of the information collected. Alternatively, the differential expression of particular genes involved in similar function can also be tested statistically. *See* DNA chips; microarray hybridization; support vector machine. (Tamayo, P., et al. 1999. *Proc. Natl. Aced. Sci. USA* 96:2907; Eisen, M. B., et al. 1998. *Proc. Natl. Acad. Sci. USA* 95:14863; Miki, R., et al. 2001. *Proc. Natl. Acad. Sci. USA* 98:2199; Thomas, J. G., et al. 2001. *Genome Res.* 11:1227; Harris, R. A., et al. 2002. *Proteomics* 2:212; Ramoni, M. F., et al.. 2002. *Proc. Natl. Acad. Sci. USA* 99:9121.)

cluster homology region (CHR) Homologous DNA sequences located in different chromosomes, probably the result of gene duplication. In some cases, the coded protein is specialized to tissue- or organelle-specific functions or its function has changed. *See* clustering of genes; duplication; evolution and duplication. (Nagai, K. 2001. *Gene* 270:161.)

clusterin (complement lysis inhibitor, CLI) Evolutionarily highly conserved 75–80 kDA heterodimeric glycoprotein encoded in human chromosome 8p21. Along with vitronectin, it prevents the attack of the cell membrane by the C5b-9 complex. It also regulates lipoprotein metabolism, neuroendocrine functions, and germ cell differentiation and is involved in the development of inflammatory diseases such as Alzheimer disease and Niemann-Pick disease. Its level may be elevated in some neurological disorders. *See* Alzheimer disease; complement; Niemann-Pick disease; vitronectin. (Bailey, R. W., et al. 2001. *Biochemistry* 40:11828; Jones, S. E. & Jomary, C. 2002. *Int. J. Biochem. & Cell Biol.* 34:427.)

clustering of genes Bacteriophage genes are in a linear order of morphogenesis. Several bacterial genes are clustered in the exact order of the biosynthetic pathway (tryptophan operon); others are only within a group but not in strict biosynthetic order (histidine operon) and they are under coordinated regulation. In the lower eukaryotes (fungi), some of the histidine genes and chorismic acid genes are in groups, although they are not transcribed into a polycistronic RNA. The vertebrate homeotic genes of the HOX families are in functional groups and may be regulated by shared global enhancers. The ribosomal and tRNA genes are in a linear array in prokaryotes and eukaryotes and are processed after transcription into individual molecules. The histone genes in *Drosophila* and sea urchin are in the same region but separated by spacers. Some of the antibody genes in mammals are clustered in gene families and are repeated many times. Some of the highly expressed mammalian housekeeping genes are clustered (Lercher, M. J., et al.. 2002. *Nature Genet.* 31:180). Some genes of the nematode *Caenorhabditis* are even polycistronic. Some (e.g., ribosomal) of the choroplast genes are clustered according to the pattern of their prokaryotic ancestors. In mice, genes involved in spermatogenesis are not distributed at random in the genome, e.g., 10/25 spermatogonia-specific genes are in the X chromosome and three are in the Y chromosome. Of the 36 genes found to be involved in sperm production, only 23 are scattered among the 18 autosomes. *See* attenuator region; coordinate regulation; chorismate; *His* operon; histones; homeotic genes; operon; polycistronic mRNA; rRNA; spacer DNA; synexpression; tRNA. (Wicker, N., et al. 2002. *Nucleic Acids Res.* 30:3992.)

clustering of phenotypes May be caused by exposure of similar environmental effects. Epidemiological factors such as carcinogens in the environment and viral infections may precipitate the expression of phenotypes in an unusual distribution.

clustering of recombinants   May be found in the progeny or in the gametes if recombination in the germline (mitotic recombination) preceded meiosis. The reality of clustering may need statistical verification by using the formula

$$V_W = \frac{1}{N} \sum_{i}^{k} \left[\frac{u_i{}^2}{n_i} - \frac{U^2}{N} \right]$$

(Tanaka, M. M., et al. 1997. *Genetics* 147:1769), where u_i = numbers in category u of brood i, k = total number of broods, $U = \Sigma u_i$ = total counts in category u, n_i = total number of progeny in brood i, and $N = \Sigma n_i$ = total number of progeny in the set of data. *See* brood; mitotic crossing over.

clusters of differentiation   Antigens associated with distinct processes of differentiation are immunologically detectable. *See* CD proteins.

clutch   Cluster of eggs laid by a bird.

cm or cM   centi-Morgan = 1% recombination; 1 map unit. *See* mapping, genetic; recombination.

C-meiosis   Meiosis is arrested because colchicine is poisoning the spindle fibers. *See* colchicine; meiosis.

cMG1   Protein related to TIS11 (a 67-amino-acid region is 72% identical), responds to epidermal growth factor plus cycloheximide.

CMI   Cell-mediated immunity. *See* immunity.

C-mitosis   The poisonous effects of colchicine block mitotic anaphase; consequently, the cell and its progeny may become polyploid. Endoreplication may take place and the two-chromatid chromosomes may sometimes be detected in juxtaposition. *See* C-meiosis; colchicine; partial karyotype below.)

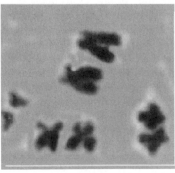

C-mitosis. (After W. V. Brown, 1972. *Textbook of Cytogenetics*. Mosby, St. Louis.)

CMRF35H (17q24)   Inhibitory immune receptor on leukocytes.

cms   Cytoplasmically determined male sterility. *See* cytoplasmic male sterility.

CMT   Monkey cell line expressing SV40 T antigen. *See* SV40.

CMV   Cytomegalovirus; hCMV: human cytomegalovirus. *See* cytomegalovirus.

Cne1, CneX   Yeast homologues of calnexin. *See* calnexin.

CNF   Cytotoxic necrotizing factor inhibits GTPase activity and thus contributes to the activation of RHO (p21). CNF is a virulence factor of *E. coli* bacteria. *See* p21; RHO. (Thomas, W., et al. 2001. *Infect. Immun.* 69:6839.)

CNK   Connector enhancer of KSR is a multidomain protein involved in RAS signaling. CNK^{N-term} enhances RAS signals, whereas CNK^{C-term} interferes with signaling when overexpressed. *See* KSR; RAS. (Anselmo, A. N., et al. 2002. *J. Biol. Chem.* 277:5940.)

cNMP cyclase   Cyclic nucleotide monophosphate cyclase is involved in the biosynthesis of cAMP and cGMP. *See* cAMP; cGMP. (McCue, L. A., et al. 2000. *Genome Res.* 10:204.)

CNS   (1) Central nervous system. *See* brain, human. (2) Conserved noncoding sequences. *See* noncoding sequences.

CNTF   Ciliary neurotrophic factor. *See* APRF; ciliary neurotrophic factor; neurotrophins.

Co^{60}   *See* isotopes.

CoA   *See* acetyl coenzyme A.

coacervate   Colloidal aggregate of organic compounds. They probably have played a role in organic evolution. *See* prebiotic evolution. (Jensen, S. A., et al. 2000. *J. Biol. Chem.* 275:29449.)

coactivator (AF, activator function)   Molecules required to activation of gene transcription in addition to TBP, TAF, and the general transcription factors. Coactivators seem to acetylate histones, whereas histone deacetylases appear to be transcriptional corepressors. *See* chromatin remodeling; high-mobility group of proteins (HMG); nuclear receptor; TAF; TATA box; TBP; transactivator; transcription factors. (Näär, A. M., et al. 2001. *Annu. Rev. Biochem.* 70:475.)

coadapted gene   Represents genotypes capable of expression of a satisfactory (fit) phenotype. *See* fitness; outbreeding depression. (Dobzhansky, T. & Pavlovsky, O. 1958. *Proc. Natl. Acad. Sci. USA* 44:622.)

coagulation factors   *See* antihemophilic factors.

coalescence   Point or node of an evolutionary tree where two lineages merge (diverge) at a time or at any other scale. *See* evolutionary tree; MRCA.

coalescent   Statistical parameter of the genealogical information of genetic data. It is an approximation from a random sample of genes in a population with constant size over many generations without selection and recombination within the chromosomal sequences considered. The analysis should reveal the number of generations that the chosen entities have undergone since they were separated from the common ancestor. The analysis thus reveals the *most recent common ancestor* (MRCA) on which the genetic samples coalesce. The coalescence represents the dynamic (demographic) history of the populations. *See* F_{ST}, $(\delta\mu)^2$, mutation age of. (Donnelly, P. & Tavare, S. 1995. *Annu. Rev. Genet.* 29:401; Rosenberg, N. A. & Nordborg, M. 2002. *Nature Rev. Genet.* 3:380.)

coancestral   Gene(s) identical by descent in two individuals, e.g., uncle and niece, first cousins, etc. *See* coefficient of coancestry; consanguinity; inbreeding coefficient.

coarctation of the aorta   Apparently autosomal polygenic narrowing of the blood vessels, leading to congenital heart failure. *See* cardiovascular disease; heart disease; supravalvar aortic stenosis.

coassortment   *See* macronucleus.

coat color   *See* pigmentation of animals, fur color.

(Courtesy of Dr. Paul Szauter, http://www.informatics.jax.org/mgi home/other/citation.shtml.)

coated pit   Generated on the surface of coated vesicles by invagination and pinching off, thereby facilitating transport by losing the coat and fusing with other intracellular vesicles (lysosomes). *See* dynamin; lysosomes.

coated vesicle   Clathrin-coated vesicle. *See* clathrin.

coatomer (coat protomer)   Large protein complex on the surface of vesicles (Golgi) mediating nonselective transport within cells. Their assembly requires ATP. After the transfer of the cargo, the coatomer is still retained and docks with another membrane. *See* clathrin; COP transport vesicles. (Sullivan, B. M., et al. 2000. *Mol. Biol. Cell* 11:3155.)

coat protein   Protein(s) of the viral capsid. *See* capsid; capsomer.

cob   Apocytochrome *b* gene in the mitochondrium (yeast) may exist without introns and with introns (called boxes). Its exons code for the apocytochrome 1 protein, whereas the box 3 intron codes for a maturase protein that excises the introns from the long-form gene and a box 7–coded protein splices the exons of the adjacent cytochrome oxidase (*oxi3*) gene. For the stability of the cob mRNA, the nuclear gene product Cbp1 must interact with a CCG sequence in the 5′ untranslated region. *See* cytochromes; mtDNA.

cobalamin (cyanocobalamin)   Vitamin B_{12}, a coenzyme for methylmalonyl CoA mutase; it has therapeutic use in anemia and acidosis. *See* transcobalamin; transcobalamin deficiency.

cob-box gene   Encodes mitochondrial cytochrome oxidase with introns. The intron boxes may have independent functions in processing the apocytochrome transcripts. *See* mtDNA.

COBRA-FISH (combined binary ratio-FISH)   *See* combinatorial labeling; FISH.

cobratoxin   Major protein toxin in the cobra venom. It irreversibly blocks nicotinic receptors and cholinergic transmission at the neuromuscular junctions. *See* toxins.

coca (*Erythroxylon coca*)   Source of cocaine; $2n = 2x = 24$. *See* cocaine.

cocaine ([3-benzoyloxy]-8methyl-8-azabicyclo[3.2.1] octane-2-carboxylic acid methyl ester)   Topical anesthetic and euphoriant. It is an addictive drug obtained either from *Erythroxylon* plants or produced by chemical synthesis. Cocaine dependence and intense craving may return even after prolonged abstinence. The priming of the relapse seems to be mediated by the D_2-like receptor agonists of the dopamine system, whereas the D_1-like receptor agonists prevent cocaine-seeking behavior. The effect of cocaine seems to be the blocking of the dopamine transporter protein. In cocaine addiction, one specific serotonin receptor may be involved. Continued use of cocaine increases cAMP activity and PKA in the brain. Overexpression of CREB in the rat brain nucleus accumbens (site in the brain responding by rewarding actions to opiates) decreases the craving for cocaine and actually promotes aversion to it. Overexpression of a dominant negative CREB mutation intensifies the cocaine reward. Blocking the opioid receptors by dynorphin may antagonize the CREB effect. Cocaine immunoconjugates may substantially block the effects of the drug. *See* agonist; alkaloids; cAMP; CREB; dopamine; dynorphin; opiate; PKA; serotonin. (Yarmolaieva, O., et al. 2001. *J. Neurosci.* 21:7474.)

co-carcinogen   May not be carcinogenic alone but may act as a tumor promoter such as phorbol ester. *See* carcinogens; phorbol esters.

coccus   Small ($\sim 1 > \mu$m) bacterium of sphere shape.

co-chaperone   Assists the function of a major chaperone, e.g., DnaJ for Hsp70 in *E. coli*. *See* chaperone; CbpA; DnaJ; heat-shock proteins; Hsp70.

co-chaperonin   GroES in bacteria assists protein folding by GroEL chaperonin. The function of the bacteriophage-encoded

Gp31 protein is similar. It can assist GroES or can substitute for it. *See* chaperonin.

Cochliomya hominivorax (screwworm) Tropical and subtropical fly, an obligatory parasite of warm-blooded animals. Its infestation is causing myasis (weight loss and sometimes death). This fly punctures the skin (hide) and causes multimillion-dollar annual losses to animal breeders; it is also menacing to people. For its biological control, the mass release of genetically sterile (irradiated) males has proven effective. *See* genetic sterilization; myasis.

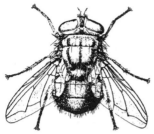

Cochliomya, from *insects*, USDA Yearbook, 1952, Stefferud, A., ed., by permission.

Cockayne syndrome (CS) Characterized by very short stature, precocious aging, deafness, eye degeneration, mental retardation, and sunlight sensitivity. It seems to be associated with a DNA repair defect, but unlike other DNA repair problems (ataxia telangiectasia, xeroderma pigmentosum, Bloom syndrome, Fanconi anemia), it does not involve increased proclivity to cancer. Some of the mutations (TAM) seem to be associated with the transcription-coupled nucleotide exchange repair of oxidative damage. It may be controlled by nonallelic autosomal-recessive loci. CSA was assigned to human chromosome 5, CSB has been located to human chromosome 10q11 (and it functions as transcript elongation protein), CSD is in chromosome 19, and CSG is in chromosome 13. The yeast has homologous genes *RAD28* and *RAD26*. All CS loci are involved in nucleotide exchange repair or base-excision repair or transcription-coupled repair. *See* cerebro-oculo-facio-skeletal syndrome; chromosome breakage; DNA repair; light sensitivity diseases; transcript elongation; transcription factors (TFIIH); trichothiodystrophy; ultraviolet sensitivity syndrome; xeroderma pigmentosum. (Le Page, F., et al. 2000. *Cell* 101:159; Lee, S.-K., et al. 2002. *Cell* 109:823.)

cockroach *Blatta germanica*, $2n = 23$ male, 24 female.

Female cockroach.

co-cloning *See* chimeric clones.

coconut (*Cocos nucifera*) Source of copra (endosperm); used for food, and the oils are processed for margarine, etc.; $2n = 2x = 32$. The coconut milk had been used for plant tissue culture before synthetic plant hormones became commercially available.

coconversion Two neighboring sites are simultaneously involved in gene conversion. *See* gene conversion.

cocultivation Permitting plant cell proliferation on agar plates in the presence of *Agrobacterial* suspension (equipped with a vector plasmid) for about 1 1/2 to 2 days. During this period, the T-DNA is transferred to the plant cells. Then the bacteria are stopped or killed by an appropriate antibiotic (e.g., cefotaxim or carbenicillin) and the transformed cells are selectively grown further on a medium to which they are expected to be resistant. *See* antibiotics; carbenicillin; cefotaxim; transformation, genetic; vectors.

code, comma-free In the 1950s before the genetic code was experimentally identified, various speculations suggested that the code needed some marks to avoid ambiguity. Eventually it turned out that punctuation marks were unnecessary because the triplet codons can be read flawlessly. Actually, frameshift mutations may alter the meaning of the text, yet there are no signals between triplets. The nucleotide sequence needs only initiator codons and stop codons; some larger genes have additional demarcations by exons and introns. All nucleotide coding sequences are commaless, *See* exon; frame shift; insulator; intron. (Crick, F. H. C., et al. 1957. *Proc. Natl. Acad. Sci. USA* 43:416.)

code, genetic Specifies in adjacent nucleotide triplets (commaless code) each amino acid in the polypeptide chain. The triplets are generally called *codons* and are represented in mRNA nucleotides. The sequence of the amino acids is determined by the codon sequences in the mRNA recognized by the anticodons in the tRNAs. The genetic code is redundant because some amino acids have up to 6 codons. The redundancy is also called *degeneracy* because several codons are degenerated (reduced) to the meaning of one amino acid. The code is *almost universal* from prokaryotes to eukaryotes, i.e., identical codons are used for the same amino acids across taxonomic boundaries. Exceptions to these rules exist in some mycobacteria, ciliates, and mitochondrial DNAs. The 4 regular nucleotides in all possible combinations of 4 (4^3) generate 64 triplets from which 61 are *sense codons*, i.e., specify amino acids, and 3 are *nonsense* codons (stop codons), because at their position in the mRNA the translation into protein stops and the polypeptide chain is terminated. In the customary codon table, the triplets are arranged into 16 families (*See* table of genetic code) where the first 2 nucleotides may be sufficient for specifications, although all 3 must be present. In higher organisms, usually only one of the DNA strands is transcribed into mRNA codons; in prokaryotes both strands may be coding but in opposite orientation (always 5′ to 3′). Although individual codons are *not overlapping*, the genes may be read in overlapping registers (*See* overlapping genes). This facilitates better use of the relatively very small amount of DNA for a greater variety of functions in some bacteriophages. The *usage* of the redundant codons may vary from gene to gene

(*See* codon usage). Some of the suppressor mutations in DNA may alter the anticodons of the tRNAs and the meaning of the stop or other codons may be altered, although the mRNA codons remain the same. Changes in the codons require mutation in the DNA. Synthetic nontriplet codons may also be translated. *See* amino acid symbols in protein sequences; codon; genetic code; RNA editing.

codeletion analysis Genes situated very closely to each other are jointly lost more frequently than those separated by a larger distance, and this fact permits the construction of deletion maps. *See* deletion mapping.

CODEN Abbreviation of literature references as given by the Periodical Tables of the Chemical Abstracts Service.

coding capacity Of an organism or organelle, is determined by the number and length of open reading frames. *See* open reading frame.

coding dictionary *See* genetic code.

coding joint Juncture of the VDJ segments of the immunoglobulin and T-cell receptor genes. *See* immunoglobulins; TCR.

coding sequence Transcribed into a functional RNA. *See* noncoding sequence; RNA polymerase; transcript.

coding strand The DNA strand has the same base sequence as the functional RNA within the cell, except that in DNA, thymine occurs at the place of uracil. The terminology has some ambiguity because in some cases both strands of the DNA may be transcribed into RNA. Also, in some older papers, coding strand is defined as the DNA strand, which is transcribed. Larger-scale sequencing of *Drosophila* genome does not indicate much difference in the transcription of the two strands of the DNA (121:108), and in the areas crowded by genes the direction of transcription seems to alternate, although in general it appears to be random. There is a possibility that transcripts of both strands of a DNA tract are combined for the translation of a particular protein. *See* anticoding strand; antisense RNA; sense strand; template. (Labrador, M., et al. 2001. *Nature* 409:1000.)

codominance Both alleles at a locus are simultaneously expressed in the heterozygote. At the phenotypic level, observed without in-depth analysis, codominance is not very common. At the protein level (using electrophoresis or serological tests), the majority of the heterozygotes (H) display the products of both alleles (that of P_1 and P_2) at a locus. *See* dominance.

codon (a nucleotide triplet) 61 triplets specify the 20 natural amino acids, with 1 to 6 codons for each, and 3 codons (nonsense codons) signal the termination of the polypeptide chain. Synthetic 4-base codons can be recognized by tRNAs, with 4-base anticodons and unnatural amino acids incorporated into proteins. Similarly, 5-base codons can be translated with tRNA with synthetic 5-base anticodons. Actually, in an *E. coli* mutant, a GUGUG codon was translated into valine by a 3-base anticodon tRNA^{Val}. These unusual constructs are useful for studying the mechanisms of frameshift mutations and the effect of nonnatural amino acids on protein function. *See* anticodon; genetic code; protein synthesis. (Jukes, T. H. 1978. *Adv. Enzymol.* 47:375; Hohsaka, T., et al. 2001. *Nucleic Acids Res.* 29:3646.)

Codon Adaptation Index (CAI) Measures the relative usage of each codon by particular genes and the codon usage by organisms. *See* codon usage. (Sharp, P. M. & Li, W. H. 1987. *Nucleic Acids Res.* 15:1281.)

codon bias *See* codon usage.

codon choice *See* codon usage.

codon family (synonymous codons) Group of codons encoding the same amino acid, e.g., CGU, CGC, CGA, CGG, AGA, and AGG all code for arginine. *See* genetic code.

codon preference *See* codon usage.

codon recognition *See* aminoacyl tRNA; anticodon; protein synthesis.

codon usage Synonymous codons may not be selected at random, but their usage varies from gene to gene, e.g., in the MS2 RNA phage the UAC codon for tyrosine is used 3 times as frequently as the UAU codon. In the δ-chain of human hemoglobin (146 amino acid residues), the CUU codon for leucine and the GUA codon for valine were not used at all, but the CUG and GUG codons were relied on 12 and 13 times, respectively. In plants, the CUG codon for leucine was used on the average in 9%, whereas the CUU codons were employed in 27–28%, and the AAG lysine codon was used almost twice as frequently as the AAA codon. Similar variations occur in the use of other codons in the majority of organisms and genes. Nucleotides following the codons (N_1 context) generally affect codon usage in the sequenced eukaryotes (Fedorov, A., et al. 2002. *Nucleic Acids Res.* 30:1192). The most highly expressed genes tend to use codons corresponding to the major tRNAs, whereas the less frequently expressed genes rely mainly on rarer tRNAs. The highly expressed genes containing A and T chose C for 3rd position, and if the first two bases are G and C, the third is preferentially T. TAA stops the translation of highly expressed genes, and TAG and TGA nonsense signals block those of lower expression. The rate and accuracy of the translation is lowered by substitution of rare codons into the mRNA, which may also result in frameshifts, skipping, or termination. The codon usage in eukaryotes also vary in the different isochores. Gene length and higher recombination frequency may increase codon bias. Highly expressed genes may show higher codon bias and lower synonymous substitutions than genes with a low level of expression. The optimally chosen codons are recognized by the most common tRNAs. There is, however, a bias in tRNA usage. In humans, the highly used CUG^{Leu} codon has only 6 cognate tRNAs genes, whereas for the relatively rare UGU and UGC cysteine codons 30 tRNA genes are available. The nature of the second codon in the open reading frame may affect the efficiency of translation. *See* amino acids; antisense DNA; coding strand; codon adaptation index; gene amelioration; genetic code; Grantham's rule; isochore; sense DNA. (McVean, G. A. T. & Vieira, J. 2001. *Genetics* 157:245; Dunn, K. A., et al. 2001. *Genetics* 157:295; Sato, T., et al. 2001. *J. Biochem.* 129:851; <http://www.kazusa.or.jp/codon/>; <http://pbil.univ-lyon1.fr/pbil.html>.)

Co-eIF-2A   Similar to eIF-2C.

coefficient of coancestry (kinship, consanguinity)   Indicates the probability that one allele derived from the same common ancestor is identical by descent in two individuals. All diploid individuals have two alleles (paternal and maternal) at a locus, and each parent has a 50% chance for transmitting one or the other of these alleles to the offspring. Thus, an allele of a grandmother has a 0.5 chance to be transmitted to her daughter or son, and that individual again has a 0.5 chance for transmitting it to the grandchildren. The probability that two first cousins would have the same allele of the grandmother is 0.5^4. The degree of consanguinity varies a great deal in different cultures and within ethnic groups. Overall in Europe, by the middle of the 20th century, it remained below 1% (with rare exceptions). In North, Central, and South America, the frequency of consanguineous marriages was generally higher. In Asia, it was significantly higher, especially in southern India and Pakistan, where in some areas the majority of the marriages were consanguineous. Similar situations existed in many African regions. Increased urbanization and expanding education result in reduction of inbreeding in human populations. *See* genetic load; inbreeding; inbreeding coefficient; lethal equivalent, incest; relatedness degree.

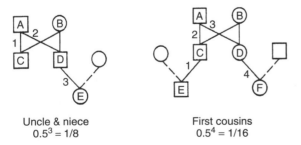

Uncle & niece
$0.5^3 = 1/8$

First cousins
$0.5^4 = 1/16$

The coefficient of consanguinity can be derived from the pedigrees. The solid-line paths through which a particular allele is transmitted are numbered.

coefficient of coincidence   Designed by H. J. Muller in 1916 for estimating interference on the basis of dividing the *number* of double crossovers observed by the *number* of double crossovers expected by the probability of single crossover frequencies. In case the two crossing-over events are independent, the coefficient of coincidence is 1 and there is no interference. Genes far apart generally have higher coincidence than those that are closely linked. Example for computation: The size of the test-cross progeny is 3,000 individuals. Frequencies of single-recombinant individuals are 0.0687 and 0.2973. Thus, the expected frequency of double recombinants is $0.0687 \times 0.2973 = 0.0204$. Hence the expected number of double recombinants is $0.0204 \times 3000 = 61.2$. If the number of double recombinants observed is 50, the coefficient of coincidence is $50/61.2 = 0.817$. (Note that coincidence must be computed from integers and not from fractions.) The *positive interference* is in the example shown here: $1 - 0.817 = +0.183$. In case of *negative interference*, the actually observed double recombinants exceed the expectations based on single events. Thus, e.g., if the observed number of double recombinants is 300 and the expected number is 50, the coincidence is 6, and hence the interference is $1 - 6 = -5$. In some cases, *high negative interference* has been observed, which may be as high as -100 or more. Recent investigations indicate that coincidence is not related to the physical distance of genes (base pair or micrometers) but rather to the genetic distance expressed in map units. Accordingly, the coefficient of coincidence as a function of map distances for separated intervals can be calculated by the Foss equation (*Genetics* 133:681):

$$S_4 = (m + 1)e^{-y} \sum_{i=0}^{\infty} \frac{y^{m(m+1)i}}{[m + (m + 1)i]!}$$

where m = fixed number of recombinational events resolved without crossing over between neighboring crossover events resolved with crossover, y = mean number of events ($y = 2[m + 1]x$) per tetrad that can result in gene conversion disregarding accompanying crossovers in a test interval, and S_4 = coefficient of coincidence for separated intervals; x = map distance in Morgan units (mean number of events resolved per tetrad in a given interval). *See* gene conversion; interference; map unit; tetrad analysis. (Zhao, H., et al. 1995. *Genetics* 139:1045; Chase, M. & Doermann, A. H. 1958. *Genetics* 43:332.)

coefficient of consanguinity   *See* coefficient of coancestry.

coefficient of crossing over   Term of Calvin Bridges (1937) stating the relation of the physical length relative to map distance. He found in *Drosophila* that 1 map unit corresponded to 4.2 μm length of the salivary gland chromosomes. Today, the genetic length of chromosomes can be expressed in nucleotide numbers. The estimates among the organisms may vary greatly depending on the amount of the DNA and resolution of the genetic markers. In *Arabidopsis*, 1 map unit is about 140 kb, whereas in maize it is about 240,000 kb. These ratios become important for efforts of map-based isolation and cloning of genes. Until a genome is completely sequenced, these estimates may not be accurate. Another problem with the estimation of the coefficient of crossing over is the variation in genetic recombination frequencies among different lines of the same organisms. *See* bands of polytenic chromosomes; gene number; map-based cloning; mapping,

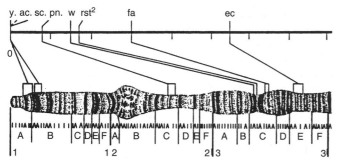

y. ac. sc. pn. w rst²   fa   ec

The distal end of the X chromosome of *Drosophila*. (After Bridges, C. B. 1935. *J. Heredity* 26:69.)

genetic; mapping function; physical mapping; recombination hot spots.

coefficient of inbreeding   *See* inbreeding.

coefficient of kinship   *See* coefficient of coancestry.

coefficient of selection   *See* selection coefficient and fitness.

coelom   Inner cavity of the embryo. In mammals there are two such cavities: the chest and the abdomen.

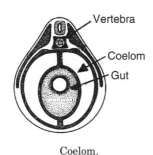

Coelom.

coelomocyte   Scavenger cell, which continuously and non-specifically endocytoses fluids of the body cavity of *Caenorhabditis*. *See* endocytosis. (Fares, H. & Greenwald, I. 2001. *Genetics* 159:123.)

coenobium   Colony of unicellular organisms enclosed by a single membrane.

coenocyte   Multinucleate protoplast or cell aggregate without separation by walls.

coenzyme   Cofactor required for normal activity of enzymes; many coenzymes are vitamins. Coenzyme A (acetyl coenzyme A) transfers acyl groups within cells.

coevolution   Evolution of two populations is concomitant because they have some mutual relationship, e.g., host and parasites, predators and preys, or two gene loci evolve as a unit. *See* frequency-dependent selection; high-dose/refuge

strategy; selection types. (Bergelson, J., et al. 2001. *Annu Rev. Genet.* 35:469.)

cofactor   Inorganic or organic substance required for enzyme activity. *See* enzymes.

coffee (*Coffea* spp)   About 90 species with $x = 11$, and the plants are either diploid or tetraploid.

Coffin-Lowry syndrome (CLS)   Xp22.3-p22.1. Dominant mental retardation, face anomalies, tapered fingers, and lysosomal storage defects. The MAPK/RSK signaling pathway is responsible for the expression of the Rsk-2 kinase affected in this anomaly. Rsk-2 is required for histone H3 phosphorylation activated by the epidermal growth factor (EGF). *See* bromodomain; chromatin remodeling; EGF; head/face/brain defects; lysosomal storage diseases; MAPK; mental retardation; RSK.

Coffin-Siris syndrome (fifth-digit syndrome)   Does not have clear criteria, but generally it involves mental retardation, broad nose, tapered fingers, sometimes missing distal digits, poor nail development, hypertrichosis on the body but hypotrichosis on the scalp, etc. It is probably autosomal dominant with low penetrance, although autosomal recessivity has also been claimed. *See* Coffin-Lowry syndrome; hypertrichosis; patella aplasia-hypoplasia.

cofilin (actin depolymerizing factor, ADF)   Small (M_r 15–20 K) actin- and phosphatidylinositol-binding protein mediating the rapid turnover of actin in the cytoskeleton. Its phosphorylation by LIM kinase is activated by ROCK. Cofilin and Arp2/3 mediate barbed-end formation involved in cell motility. *See* actin; cytoskeleton; LIM kinase; phosphoinositides; ROCK; Williams syndrome. (Bamburg, J. R., et al. 1999. *Trends Cell Biol.* 9:364; Pfannstiel, J., et al. 2001. *J. Biol. Chem.* 276:49476; Ichetovkin, I., et al. 2002. *Curr. Biol.* 12:79.)

COG   Cluster of orthologous groups are genes that have common ancestry and related function. They may also be members of paralogous groups of genes. *See* orthologous; paralogous.

cognate   Has the ability of recognizing a particular molecule, like ligand and receptor, enzyme and substrate. *See* ligand; receptor; substrate.

cognate tRNA   Recognized by an aminoacyl synthetase. *See* aminoacylation.

cognitive ability   The major components are verbal and spatial abilities, memory, speed of perception, and reasoning. These components may have subcategories such as verbal comprehension, verbal fluency, vocabulary, etc. *See* behavior genetics; human intelligence.

Cohen & Boyer patents   US #4.237.224 and 4.468.464 were obtained by Stanford University on the construction of cloning plasmid chimeras (pSC101), and for the products commercially produced with the aid of them. Users must thus pay reasonable royalties for these key procedures. The fact of patenting became the subject of controversy because of its unusual nature, but it called attention to institutions of learning benefiting directly from the commercial exploitation of the results of scientific discoveries. *See* patent.

Cohen syndrome Highly complex autosomal-recessive anomaly (prominent incisors, high nasal bridge, eye defects, mental retardation, hypothyroidism, etc.) with suggested chromosomal locations 15q11, 5q33, 7p, 8q22. *See* eye diseases; mental retardation.

cohesin At least four-protein (Smc1, Smc3, Scc1, Scc3) complex regulating in combination with another four proteins—including APC—the cohesion of sister chromatids primarily during interphase after the replication of DNA. The complex of Scc2 and Scc4 mediates the association of cohesin with the chromosomes. Cohesin (Scc1/Rad21) is removed by separase before the onset of anaphase and cytokinesis. Rad21-cohesin requires a special heterochromatin protein, Swi6, for the association at the centromeres but not for that along the chromosome arms. *See* adherin; APC; cytokinesis; heterochromatin; Rec8; Scc1; sister chromatid cohesion; securin. (Tanaka, T., et al. 1999. *Cell* 98:847; Nasmyth, K., et al. 2000. *Science* 288:1379; Bernard, P., et al. 2001. *Science* 294:2539.)

cohesion-tension theory Interprets the movements of sap through the vessel system of plants. *See* guttation; transpiration. (Steudle, E. 2001. *Annu. Rev. Plant Physiol. Plant Mol. Biol.* 52:847.)

cohesive ends Two DNA molecules have base complementarity ends that can anneal.

These fit together

cohort In population studies it is used to designate a particular group of individuals with common characteristics. (age, treatment, ethnicity, education, taxonomic group, etc.).

coiled body (CB, Cajal body) Intranuclear elements (1–10 per nucleus) attached to the surface of the nucleolus. The CB forms 0.15 to 1.5 μm tangled threads of snRNA and proteins during G_1 phase of the cell cycle, but disassembled during mitosis. *See* Cajal body; coilin; gemini of coiled bodies; nucleolus; snRNA.

coiled coil Two α-helices of polypeptides wound around each other. *See* α-helix.

coilin 80 kDa proteins in the nuclear coiled body (Cajal body) involved in processing of nucleolar and extranucleolar RNA. *See* Cajal body. (Raska, I., et al. 1991. *Exp. Cell Res.* 195:27.)

coincidence *See* coefficient of coincidence.

coinducer Chemical substance required, in addition to the inducer substrate, to activate the gene, e.g., in the arabinose operon, besides the P protein, cAMP is required for turning on the genes. *See Arabinose* operon; coactivator; *Lac* operon.

cointegrate vector Of *Agrobacterium*, is a plant transformation vector containing both the T-DNA and the virulence genes in a single circular molecule. These vectors are usually free of the oncogenes and carry both bacterially and plant

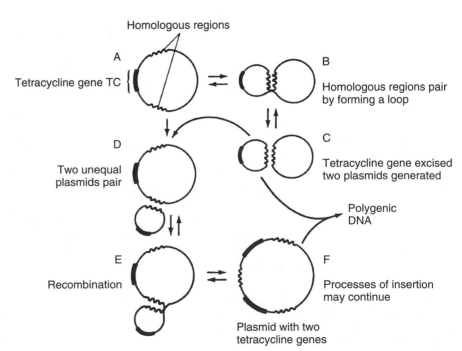

Polygenic plasmids by cointegration. (After Clewell, D. B., Yagi, Y. & Bauer, B. 1975. *Proc. Natl. Acad. Sci. USA* 72:1720.)

selectable markers besides some engineered genes. *See* binary vector; selectable marker; Ti plasmid.

cointegration   Two circular plasmids combine without loss into a double-size plasmid. Also, a circular donor plasmid carrying a transposable element fuses with a recipient circular plasmid and at each point of juncture there will be a copy of the transposable element. This process requires the breakage of the phosphodiester bond and duplicating the target site and the insertion element. This is replicative transposition.

The cointegrate has the two plasmids and two transposable elements. Upon resolution of the cointegrate, two plasmids are produced, each carrying a transposable element. *See* plasmid; transposable element.

coisogenic   Has identical genes with another strain, except at a single locus. *See* congenic; inbred; isogenic; subline; substrain.

coitus   Sexual intercourse.

Col   *See* colicin

Colcemid (demecolcine, N-deacetyl-N-methylcolchicine) Synthetic colchicine. *See* colchicine.

colchicine   Alkaloid produced by *Colchicum autumnale* and other liliaceous plants. Its tropolone ring specifically and strongly binds and disassembles microtubules of the mitotic spindle and thus the nuclear divisions are arrested in mitosis; cells skip at least one division, resulting in doubling (or multiplying) of the chromosome number (polyploidization). It is particularly useful for doubling the chromosome number of interspecific and intergeneric hybrids, thereby making the otherwise sterile individuals fertile by securing a pair for all chromosomes. Colchicine alleviates some of the symptoms of gout, a mammalian disease caused by uric acid overproduction. Colchicine is synthesized from phenylalanine and tyrosine. The LDLo oral dose for humans is 5 mg/kg; therefore, it is a dangerous substance, taken up also through the skin. *See* alkaloids; allotetraploid; autotetraploid; chromosome doubling; c mitosis; colcemid; gout; LDLo; microtubule. (Eigsti, O. J. & Dustin, P., Jr. 1955. *Colchicine*, Iowa State College Press, Ames.)

Colchicine.

cold hypersensitivity   Autosomal-dominant genes may be responsible for the development of urticaria (allergy-like rash), joint discomfort, and fever after exposure to cold. It may be associated with amyloidosis (accumulation of fibrillar proteins) in various tissues. Amyloid nephropathy may be fatal. Familial cold urticaria (FCU) and Muckle-Wells syndrome encode a protein with a pyrin domain and leucine-rich repeat at human chromosome 1q44. The latter syndrome has similar symptoms (rash, conjuctivitis, pain in the joints, and fever), but the symptoms are not elicited by cold. *See* amyloidosis; asthma; cold-regulated genes; hyperthermia; Muckle-Wells syndrome; temperature-sensitive mutation; urticaria familial cold. (Hoffman, H. M., et al. 2001. *Nature Genet.* 29:301.)

cold-regulated gene (COR)   Required for acclimation to low temperature. Some plant mutants may become very sensitive to cool temperature. An *Arabidopsis* mutant was killed at 18°C after exposure for a few days. The chilling may cause electrolyte leakage, and changes in the synthesis of steryl esters. Cold acclimation is frequently based on microsomal stearoyl coenzyme A desaturase activity. The unsaturated phosphoglycerides restore fluidity of cold-rigidified cell membranes. COR genes are regulated by binding of the CBF transcriptional activator to the CRT/DRE (C-repeat/drought-responsive) element. *See* antifreeze proteins; CBF; cold hypersensitivity; fatty acids; glycerophospholipid; thermotolerance. (Lee, H., et al. 2001. *Genes. Dev.* 15:912; Karlson, D., et al. 2002. *J. Biol. Chem.* 277:35248.)

cold sensitive   Mutant fails to grow normally at low temperature, although it may be entirely normal at higher temperature. The TRP (transient receptor potential) family ion channels mediate thermosensation. *See* cold hypersensitivity; temperature-sensitive mutants. (McKemy, D. D., et al. 2002. *Nature* 416:52.)

cold-shock protein   Produced in response to abrupt change to low temperature. In mammals, uncoupling proteins (UCP1, UCP2) pass a proton electrochemical gradient across the inner membrane of the mitochondria, resulting in generation of heat rather than ATP. A plant protein homologue (StUCP) appears to have a similar function. *See* heat-shock proteins; Hsc66; uncoupling agent. (Phadtare, S., et al. 1999. *Curr. Opin. Microbiol.* 2:175; Somerville, J. 1999. *Bioessays* 21:319; Manival, X., et al. 2001. *Nucleic Acids Res.* 29:2223; <http://www.chemie.uni-marburg.de/~csdbase/>.)

cold spot   Chromosomal area where mutation, recombination, or insertion is rare. *See* hot spot.

ColE1 replicon   Similar in nature to pMB1. The replicon produces 15–20 copies per cell and does not require any plasmid-encoded function for replication. It uses DNA polymerase I and III, DNA-dependent RNA polymerase, and the products of bacterial genes *dnaB*, *dnaC*, *dnaD*, and *dnaZ*, which have a very long life. In the presence of protein synthesis inhibitors (chloramphenicol, spectinomycin), when cell replication ceases this replicon can produce 2,000 to 3,000 plasmid copies per cell. The majority of modern bacterial vectors utilize this replicon. The name is derived from natural plasmids coding for colicine production. The 4.2 megadalton plasmid is nonconjugative. *See* cloning vectors; colicins; phagemids; plasmids; replicon. (Mruk, I., et al. 2001. *Plasmid* 46:128.)

coleoptile Membrane-like first leaf in monocotyledons enclosing succeeding leaves at the stage of germination. *See* embryogenesis in plants.

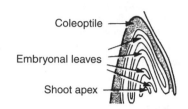

Coleoptile; embryonal leaves; shoot apex.

coleorhiza Envelope of the root tip of germinating of grasses.

Coleus blumei Leafy, shade-tolerant ornamental plant; $2n = 24$.

colicin Genetically (plasmid) controlled bacterial toxins that kill sensitive bacteria even at very low concentration of this substance, produced by killer *E. coli* and *Shigella sonnei* bacteria. Colicins require binding to the extracellular receptor (BtuB, vitamin B_{12} receptor) of the target cells. Then Ton or Tol proteins mediate the translocation to the periplasm, followed by voltage-gated depolarization of the inner cell membrane and/or nuclease action on ribosomes or DNA. Colicin E1 and colicin K inhibit active transport; colicin E2 may contribute to the degradation of DNA; colicin E3, E4, E6 interfere with protein synthesis of the sensitive bacteria by attacking the 16S rRNAs at the 49th base from the 3′ end and removing a small fragment from their 3′ terminus. E5 colicin splits tRNAs of Tyr, His, Asp, and Asn, which contain the Q (queuine) at the wobble site of the anticodon. Colicin E5 bacteria that harbor Col plasmids are immune to each of these lethal effects of colicins, and this property has been used for selection of Col-transformed bacterial cells. Each cell may normally carry about 20 copies of these plasmids. Colicin resistance mutations occur at appreciable frequencies in bacterial populations. The ColE1 plasmid has a molecular size of about 4.2 MDa and is nonconjugative. The ColE1 plasmid replicon has been used for the construction of genetic vectors and has been engineered into cosmids and other plasmids. *See* ColE1; ion channels; killer strains; queunine. (Riley, M. A. 1998. *Annu. Rev. Genet.* 32:255; Stroud, R. M., et al. 1998. *Curr. Opin. Struct. Biol.* 8:525; Lazdunski, C. J., et al. 1998. *J. Bacteriol.* 180:4993; Smajs, D. & Weinstock, G. M. 2001. *J. Bacteriol.* 183:3949.)

colicinogenic bacteria Killer strains producing colicins (bacterial toxins).

coliform Enteric, gram-negative bacteria, related to *E. coli*. *See E. coli.*

colinear *See* collinearity.

coliphage Bacteriophage, infectious for *E. coli* bacterium.

collagen Fibrous protein built mainly from hydrophobic amino acids (35% Gly, 11% Ala, 21% Pro and hydroxyproline). It forms a left-handed helix with 3 residues per turn and is made up of repeated units with glycine having every third position. The collagens form triple helixes from 3 polypeptide chains, and they have structural roles in the cell. The 38 kbp chicken collagen gene (similar also in humans or mice) has 52 introns and short exons (54 to 108 bp) built of tandem repeats of 9 bases. There are about 9 types of collagens encoded by about 17 gene loci in humans. This protein is the major component of the cuticle, tendons, and cartilage. Some of the collagen diseases can be detected by prenatal analysis. In *Caenorhabditis*, about 150 collagen genes determine cuticle organization. Several mammalian diseases are based on defects in collagen. *See* achondrogenesis; Alport syndrome; aneurism, aortic; chondrodysplasia; dermatoparaxis, cattle; Ehlers-Danlos syndrome; epidermolysis; epiphyseal dysplasia; hypochondrogenesis; Kniest dysplasia; Marfan syndrome; metastasis; muscular dystrophy; osteoarthritis; osteogenesis imperfecta; spondyloepiphyseal dysplasia; Stickler disease; Weissenbacher-Zweymuller syndrome. (Nimni, M. E., ed. 1988. *Collagen*, CRC Press, Boca Raton, FL.)

collagenase Cell surface metalloproteinase involved in remodeling the cellular matrix and thus shaping cells and facilitating cell migration. *See* metalloproteinases.

collapsin Member of the protein family of semaphorin; it seems to be inhibitory to axon outgrowth. *See* axon; CRMP-62; neurogenesis; semaphorin. (Liu, B. P. & Strittmatter, S. M. 2001. *Curr. Opin. Cell Biol.* 13:619.)

collateral Accessory.

collateral relative Animal in a breeding program that has one or more common ancestors but is not a direct descendant of these ancestors.

collectin Broad-spectrum, complement-like antimacrobial proteins in mammals. *See* complement. (Ohtani, K., et al. 2001. *J. Biol. Chem.* 276: 44222.)

collenchyma Parenchyma cells in the stem of plants that fit together closely; they have thickened walls especially at the corners.

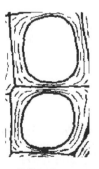

Collenchyma

Colletotrichum circinans Fungal parasite producing smudge on sensitive onions. Generally, white onions are susceptible because they lack the gene for the synthesis of parachatechuic acid, which inhibits the parasite. Red onions have the *W* allele, which conveys resistance. Some varieties of onions have an *I* (inhibitor) allele, which prevents the expression of *W*; thus, in the F_2 of the double heterozygotes, there are 13 white susceptible and 3 red and resistant segregants.

collie eye Eye anomaly of very high frequency among collie dogs; detectable only by ophthalmoscope. It frequently impairs the dog's vision.

collinearity Amino acid sequences in the polypeptide correspond to the codon sequences in nucleic acids; the 5′ end of the mRNA matches with the NH_2 end of the polypeptide chain. Some of the *Drosophila* genes, e.g., within the *ANTC* (*Antennapeadia* complex, chromosome 3-47.5), appear in the same sequence in the map as the morphogenetic function they control (*lab, Pb, Dfd, Scr, Antp*). Similar collinearity has been shown in the *BXC* (*Bithorax* complex) *Ultrabithorax* segment. In the homeotic complexes, generally there is another position effect; the products of the more posterior-acting genes appear to be more abundant than those of the anterior ones. *See* homeotic genes; *Lac* operon; lambda phage; morphogenesis in *Drosophila*. (Yanofsky, C. & Horn, V. 1972. *Biol. Chem.* 247:4494.)

Collinsia (Scrophulariaceae) Approximately 20 plant species, $2n = 2x = 14$; used for cytogenetic studies.

collochore Short heterochromatic sequence in the chromosomes; supposed to be involved in chromosome association, especially in the absence of chiasmata. *See* heterochromatin. (Virkki, N. 1989. *Hereditas* 110:101.)

collodion Alcohol and ether solution of pyroxylin (mainly nitrocellulose), used in microtechnical preparations and as a veterinary skin protector.

colloid Particles in the range of about 0.1 to 0.001 μm in diameter that can exist in fine suspensions (in gas, liquid, or solids) or emulsions (in water).

Colobidae (Old World primates, langurs) *Nasalis larvatus* $2n = 48$; *Presbytis crystatus* $2n = 44$; *Presbytus entellis* $2n = 44$; *Presbytis obscurus* $2n = 44$; *Pygathrix nemaeus* $2n = 44$. *See* primates.

coloboma Appears as a missing or defective sector involving the iris, retina, or optic nerve. It may be associated with brachydactyly, abnormal movements, and retardation. It may be controlled by autosomal-dominant or -recessive genes, but X-linkage has also been suggested for some forms. *See* brachydactyly; cat eye syndrome; eye diseases.

colon cancer *See* colorectal cancer.

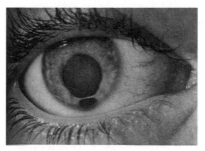

Coloboma of the IRIS. (From Bergsma, D., ed. 1973. *Birth Defects. Atlas and Compendium.* By permission of the National March of Dimes Foundation.)

colonization Evolutionary factor of settling outside of the original habitat. *See* gene flow.

colony Group of microbial cells grown at the same spot. They may have originated from a single cell or from several. The shape of the colony for a certain bacterial strain may vary according to the nutrient content and diffusion, the movement and reproduction of the bacteria, and local cell communication. Colony characteristics are used as classification criteria.

colony hybridization Isolated DNA is cut into fragments with appropriate restriction endonucleases. A library of the fragments is established by cloning them with the aid of a cloning vector (e.g., cosmid, YAC, etc.). The bacteria, presumably each carrying the DNA fragments in a chimeric plasmid, are seeded at low density on agar plates so that each colony would be separate. After the colonies are formed, a replica plate is established from the master plate by pressing a membrane filter over it. After denaturation of the DNA on the nitrocellulose filter, it is hybridized with a labeled DNA or RNA probe. After washing off the unbound probe, the filter is autoradiographed and the colonies containing the desired molecules of DNA are identified. Since the position of the colonies corresponding to the black dots on the photographic film (dot blot) can be identified, the bacteria containing the vector with that specific DNA can be further propagated. In the reverse dot blot, the labeled DNA or RNA (the probe) is immobilized on the hybridization filter and the cloning vector is annealed to it. *See* probe, autoradiography; cloning vectors; cosmids; denaturation; DNA library; plaque lift; replica plating; YAC. (Grunstein, M. & Hogness, D. S. 1975. *Proc. Natl. Acad. Sci. USA* 72:3961.)

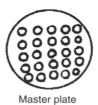

Master plate   Replica plate after autoradiography

colony stimulating factor (CSF-1) Cytokine protein activating RAS p21 protein by increasing the proportion of GTP-bound molecules. CSF2 is encoded within the IL-3 gene cluster.

See CSFR; IL-3; M-CSF; RAS. (McMahon, K. A., et al. 2001. *Biochem. J.* 258[pt2]:431.)

color blindness Exists in different forms. Complete or nearly complete light sensing (monochromatism or achromatopsia) has a prevalence in the 10^{-5} range and it may not involve any alteration in the retina (achromatopsia 1: human chromosome 14, achromatopsia 2: 2q11, achromatopsia 3: 8q21-q22). Adjacent chromosomal areas frequently modulate the X-linked recessive gene. One form of achromatopsia (autosomal recessive) involves light sensitivity. The affected individuals are bothered by light but have better-than-average vision under dim conditions (day blindness). Complete achromatopsia involves defects in the retinal cones and in the α-subunit of the cone photoreceptor of a cyclic guanine monophosphate-gated cation channel. This gene (CNGA3 is at human chromosome 2q11. The X-linked incomplete *congenital stationary night blindness* gene CSNB2 encodes at Xp11.23, a retina-specific Ca^{2+} channel α_1-subunit of 48 exons with 1,966 amino residues. The complete form of this anomaly was mapped to Xp11.4. Partial color blindness (green color blindness, *deuteranopia*) is also Xq28-linked recessive, and it may affect 8% of males of Western European descent. A cluster of genes that may recombine and undergo gene conversion, explaining the variations and the high frequency of this condition, encodes the green color vision. The red color blindness (*protanopia*) appears to be determined by two X-chromosomal-recessive loci (Xq28) with a frequency of 0.08 in males. Another type of color blindness involves loss of blue and yellow sensors but retains those for red and green (*tritanopia*), may exist in X-linked-recessive or autosomal-dominant forms (7q31.3-q32), and may occur at a very high frequency of 0.02 in some populations, whereas in others it may be an order of magnitude or even less frequent. The blue cone pigment gene displays high homology to that of rhodopsin and substantial homology with the red and green pigments. In the latter form, the rhodopsin receptor may be defective. The great chemist John Dalton suffered from "daltonism" or deuteranopia. *See* color vision; deuteranomaly; eye diseases; hemeralopia; ion channels; nyctalopia; rhodopsin. (Neitz, J., et al. 1999. *Nature Neurosci.* 2:884; Crognale, M. A., et al. 1999. *Vision Res.* 39:707.)

colorectal cancer May be controlled by a large number of genes involved in the cancer family syndrome. In many cases, the mismatch repair genes have mutated, compared to Gardner syndrome, where the APC tumor suppressor is mutant. Genetic instabilities causing this type of cancer can be dominant, involving changes in chromosome number, or if recessive, concerned with microsatellite instabilities. The most common types of cancers are adenocarcinoma of the colon and endometrial cancer (inner mucous membrane) of the uterus, but other types, such as breast, ovarian, and brain tumors, as well as leukemia, may be under similar controls (Lynch type I). About 75% of the carcinomas show deletions of the short arm of human chromosome 17. This chromosomal segment may be responsible for the transition from the benign to the malignant state of the carcinomas. Nonpolyposis colorectal carcinoma causes 3.8 to 5.5% of the colorectal cases, whereas 0.2% are the contributions of adenomatous and 0.6% by ulcerative colitis. About one-third of the alterations involve the RAS oncogene (KIS = Kirsten murine sarcoma oncogene, 12p12.1). Deletions involve human chromosomes 22, 5, 6, 12q, 15, 17, 18. The human chromosome 18q21 locus encodes a TGFβ-regulated serine/threonine kinase receptor, MADR2. DCC (deleted in colorectal cancer) is based on a defect in a netrin receptor. Netrin is a human homologue of the *Caenorhabditis* gene products UNC-6 and UNC-40 involved in the guidance of neuronal axons (Forcet, C., et al. 2002. *Nature* 417:443.) These findings indicate that besides the activation of the major oncogene, it is necessary to inactivate several tumor suppressors. UV light–induced cyclobutane pyrimidine dimers are more readily removed from the transcribed strand of the DNA than from the other. This is called transcription-coupled repair. Such a repair system seems to be defective in several types of colorectal cancers. About 50% of the Western populations develop this type of cancer by age 70 and 10% of the cases become malignant. An estimated 15% of the cases have a strong dominant hereditary component. *See* adenomatous polyposis; cancer; cyclobutane ring; DNA repair; evolutionary clock; FAP; Gardner syndrome; hereditary nonpolyposis colorectal cancer; Lynch cancer families; mismatch repair; mitochondrial disease, human; mitochondrial genetics; neuron; p16; p53; PMS; polyposis; PRL-3; RAS; RIZ. (Abdel-Rahman, W. M., et al. 2001. *Proc. Natl. Acad. Sci. USA* 98:2538.)

colorless testa Mutations occur in different plant species. These recessive mutations display delayed inheritance because the seed coat is maternal tissue, but in F_3 usually about one-fourth of the individuals uniformly have the recessive seed-coat color. *See* delayed inheritance; testa.

color vision In the human X chromosome, there is an array of middle- (2–7) to long-wavelength (2–4)–sensitive visual pigments. Humans with normal color vision typically have a single long-wavelength gene and two or three middle-wavelength sensitivity genes. The multiple copies probably arose by unequal recombination, and those with more copies may be able to see the differences better in hues. Unlike humans, pigeons perceive ultraviolet light. *See* color blindness; opsin; rhodopsin; unequal crossing over. (Yokoyama, S. & Radwimmer, F. B. 2001. *Genetics* 158:1697.)

Coltan (Co) Relatively rare blood type apparently encoded in human chromosome 7. *See* blood groups.

column chromatography (adsorption chromatography) Separation of (organic) mixtures on sugar, resin, sephadex, silica gel, or other columns established in glass tubes and eluting the components stepwise by different solvents or solvent mixtures. The eluates collected by fraction collectors can be monitored by spectrophotometry in the samples. *See* sephadex; chromatography; spectrophotometry.

comb Comb-like device to make wells in the electrophoresis gel.

GEL BED COMB.

combed DNA color bar coding Technique for the rapid detection of deletions in DNA. An appropriate genetic sample is stretched out on a treated glass surface and analyzed with fluorescent probes to detect structural changes. (Gad, S., et al. 2001. *Genes Chromosomes Cancer* 31:75.)

combination Generating from **n** number of individuals all possible sets containing only *x* numbers in each set. Mathematically: $\binom{n}{x} = \frac{n!}{x!(n-x)!}$. *See* binomial distribution; permutation; variation.

combinatorial chemistry Method aiming at generating permutations of small molecular building blocks with the goal of finding the most effective pharmaceuticals. This process is then combined with bioassays to find agonists and antagonists of the biochemical process, respectively, to discover new, effective drugs. The completion of the genome sequencing projects and the developing information on molecular structure facilitate the drug discovery process. *See* agonist; antagonist; chemical genetics; chemogenomics; pharmaceuticals. (Lehn, J.-M. & Eliseev, A. V. 2001. *Science* 291:2331; Bhattacharyya, S. 2001. *Curr. Med. Chem.* 8:1383.)

combinatorial diversification The large number of various immunoglobulin genes may enter into different combinations and generate an enormous array of antibody molecules. *See* affinity maturation; antibody; immunoglobulins; junctional diversification; somatic hypermutation.

combinatorial gene control The transcription of genes is regulated by the cooperative action of general and specific transcription proteins. It is not known how many such proteins exist, but their number can be much smaller than the number of genes, yet they can assure a high degree of specificity. If we assume that there is a total number (*n*) of 20 different inducible transcription factors (certainly an underestimated figure) and each gene requires 5 (*x*), the total number of specificities is $\binom{n}{x} = \frac{n!}{x!(n-x)!} = \frac{20!}{5!(15!)} = 15,504$, or if there are 27 inducible transcription factors and 5 are used by each gene, the number of specificities $\binom{27}{5} = 80,730$ is enough to regulate all of the estimated human genes (ca. 35,000 to 40,000). *See* gene number; regulation of gene activity. (Darimont, B. D., et al. 1998. *Genes Dev.* 12:3343.)

combinatorial labeling Cytogenetic method of chromosome analysis simultaneously using more than one fluorochrome-conjugated nucleotide. The number of useful combinations is $2^N - 1$. *See* chromosome painting; FISH; fluorochromes; ratio labeling.

combinatorial library Antibody heavy- and light-chain cDNAs are amplified separately by PCR, then ligated and cloned in vectors. Thus, a random combinatorial array of constructs is generated. *E. coli* cells infected with the vectors produce both antibody chains. But only the heavy chain contains the variable region and the first constant domain and the Fab region (*See* antibody diagram). This protein binds to the antigen but lacks the effector domain. The library can be screened by radioactively labeled antigen, and after washing off the unbound radioactivity, the sought antigen-antibody can be spotted on the plate by the fixed radioactivity. This method permits a very efficient selection among millions of types of antibodies. This selection appears a thousand-fold more efficient than the monoclonal method. A further improvement is provided by using filamentous phages (M13) that display antibodies on their surface. Screening can be done in liquid media and subjected to adsorption chromatographic purification. This procedure is called also epitope screening. *See* antibody; antigen; chromatography; epitope; filamentous phage; monoclonal antibody; phage display. (Pelletier, J. & Sidhu, S. 2001. *Curr. Opin. Biotechnol.* 12:340; Pinilla, C., et al. 2001. *Cancer Res.* 61:5153.)

combing *See* molecular combing.

combining ability Term used in quantitative genetics and animal and plant breeding. *General combining* indicates that a particular stock has better-than-average performance in any hybrid combination. *Specific combining ability* indicates a better-than-average performance only in certain hybrids. From (*n*) lines $\frac{n(n-1)}{2}$, single crosses and $\frac{(n-1)(n-2)(n-3)}{8}$ double crosses are possible. *See* double cross; heterosis; hybrid vigor. (Henderson, C. R. 1952. in *Heterosis*, Gowen, J., ed. Iowa State College Press, Ames.)

comb traits Of poultry, are determined by two allele pairs, and the interaction of their gene products specifies four comb types —*RRPP* and *RrPp*: walnut, *RRpp*: rose, *rrPP*: pea, and *rrpp*: single comb in the proportion of 9:3:3:1. *See* epistasis; Mendelian segregation; walnut comb.

comet assay Detects chromosomal lesions by exposing alkaline-treated (pH 13) cells to electrophoresis and microscopically examines the comet shape of the destabilized nucleus that suffered chromosome breakage by the mutagenic agent. The tail (length of the comet) and the head (diameter) ratios are evaluated. *See* mutation detection. (Hartmann, A., et al. 2001. *Mutation Res.* 497:199.)

← Comet tail

Comet assay.

Comm (*commisureless*) *Drosophila* gene counteracts the effects of *Robo*. *See* axon; *Robo*.

commensalism Species sharing the same natural resources without necessarily benefiting or suffering from the relationship. *See* pathogenenic; symbiont.

commingling test Analyzes whether in large populations the phenotypic distribution is based on the contribution of a single group or of an admixture of several groups. Such a study may be important for the analysis of quantitative traits and in genetic epidemiology. (Khoury, M. J., et al. 1993.

Fundamentals of Genetic Epidemiology. Oxford Univ. Press, New York.)

commisure Joining of corresponding parts of organs, e.g., of the lips of the mouth or vagina or between the axons crossing the midline of the body.

commitment *See* determination.

commitment point Determination (start) point of the cell division cycle. *See* cell cycle; restriction point.

comorbid is a condition of a co-occurring, additional disease with the primary one. It may alter the diagnosis and may aggravate the condition.

ComP *Bacillus subtilis* kinase affecting competence through regulator ComA.

COMP Cartilage oligomeric matrix protein is a pentameric member of the thrombospondin family. *See* pseudoachondroplasia; thrombospondin. (Oldberg, A., et al. 1992. *J. Biol. Chem.* 267:22346.)

compaction Tight binding of cells to each other such as occurs when the blastomeres of the embryo form the morula. *See* morula.

companion cell Associated with sieve tubes in plant conductive tissues. *See* sieve tube.

comparative anchor reference loci (CARL) Span the genomes and facilitate comparative genome mapping of different species of mammals. (Lyons, L. A., et al. 1997. *Nature Genet.* 15:47.)

comparative expressed sequence hybridization (CESH) Relatively rapid method that gives a genome-wide view of chromosomal location of differently expressed genes within tissues. mRNA or cDNA prepared from two different tissues, e.g., healthy or cancerous, is differentially labeled with fluorochromes before hybridization. No prior knowledge of genes or cloning is necessary, and minimal amounts of tissue can be used. Expression profiles are achieved in a manner similar to the identification of chromosomal imbalances by comparative genomic hybridization analysis. The method reveals chromosomal regions where genes are overexpressed as a consequence of drug-resistance, cancer or other functional differences. *See* comparative genomic hybridization. (Lu, Y.-J., et al. 2001. *Proc. Natl. Acad. Sci. USA* 98:9197.)

comparative genomic hybridization (CGH) Method of cytological localization mutant DNA sequence. The normal sequence and the mutant sequence of DNA are labeled by different fluorochromes, and in situ hybridization is carried out by the mixture. The relative hybridization signals of the bands are monitored by fluorescence microscopy. This method permits the analysis of DNA sequence dosage (loss or gain) in cancer tissues compared to normal cells. It is also suitable for scanning genomes for evolutionary

and cancer studies. *See* FISH; fluorochromes; in situ hybridization; microarray hybridization. (Pinkel, D., et al. 1998. *Nature Genet.* 20:207; Lomax, B., et al. 2000. *Am. J. Hum. Genet.* 66:1516; Lin, J. Y., et al. 2002. *Genome Biol.* 3:research 0026.1)

comparative genomics (sequenced genomes) Prospects for comparable DNA sequences among various prokaryotic and eukaryotic genomes across phylogenetic ranges. (Cliften, P. F., et al. 2001. *Genome Res.* 11:1175; Wassaarman, K. M., et al. 2001. *Genes Dev.* 15:1637; <http://www.ncbi.nlm.nih.gov/COG>; <http://www.ncbi.nlm.nih.g ov/XREFdb/>; <http://gib.genes.nig.ac.jp>.)

comparative map Reveals the evolutionary conservation of genes and nucleotide sequences across phylogenetic boundaries. For this purpose, comparative anchor-tagged sequences (CATS) are needed in the species where substantial amount of nucleotide sequence information is available. The information available in various species suggests that 1 to 10 chromosomal rearrangements might have occurred per million years. *See* comparative anchor reference loci; evolution; unified genetic maps; mouse. (Smith, E. J., et al. 2001. *Poultry Sci.* 80:1263.)

compartmentalization Certain groups of cells give rise to clones, which are different by morphology and/or function from their surrounding tissues. Adjacent compartments do not intermingle. The compartment may further differentiate during development. Compartmentalization also takes place within the cells by assignment of special functions in a polar fashion or to special subcellular organs. *See* differentiation; morphogenesis; mRNA targeting. (Helweg-Larsen, J., et al. 2001. *J. Clin. Microbiol.* 39:3789.)

compatibility Crossing or mating results in (normal) offspring. Also, tissue transplantation does not involve adverse immunological reaction or when any other simultaneous treatments are without undesirable consequence(s). *See* incompatibility.

compatibility, organelle and nuclear genome Can be determined in species where biparental transmission occurs and organelles between species can be transferred. Also, on the basis of cybrids, it appears that compatibility differences are real and thus various cytoplasms can be classified. The reciprocal crosses are usually informative. *See* ctDNA; cybrids; mtDNA; paternal leakage; plasmid male transmission.

compatibility group Plasmids that may or may not coexist in the same bacteria.

compensasome RNA and protein complex that mediates dosage compensation. *See* dosage compensation.

compensatory mutation Mitigates the genetic disadvantage of another mutation. The streptomycin resistance gene in *Salmonella* or *E. coli* endows the cells with great selective advantage on streptomycin media, but under conditions where the antibiotic is not present the primary mutants are disadvantaged because the modified ribosomes (conveying the resistance) are less efficient in translation. With several passages through streptomycin-free media, the fitness of the bacteria improves without reversion to the wild-type gene. The improvement is generally due to mutations that compensate for the shortcoming of the ribosome. *See* fitness; reversion; second site reversion. (Guan, Y., et al. 2001. *J. Virology* 75:11920.)

competence, embryonal   State of being receptive to stimuli for differentiation. It may depend on specific transcription factors, presence of the right signaling molecules, absence of interfering signals, presence of the appropriate receptors and ligands, modifying factors such as kinases and phosphatases, availability of the proper environmental cues, etc., and their complex interactions. (Duranthon, V. & Renard, J. P. 2001. *Theriogenology* 55:1277.)

competence of bacteria   Physiological state of the (bacterial) cell when transformation (uptake and integration of DNA) is successful; it generally coincides with the second half of the generation time, or its peak is near the end of the exponential growth phase. Divalent cations and a combination of them can induce competence. For more than four decades, *E. coli* was refractory to transformation, until in 1970 it was discovered that cold $CaCl_2$ makes the uptake of phage DNA possible. One milligram of supercoiled plasmid DNA yields 10^5 to 10^6 transformations. This frequency can further be increased by 2 to 3 orders of magnitude by the use of improved protocols also involving DMSO (dimethylsulfoxide, a wide-range solvent and penetrant). The treated bacteria keep their transformation competence if stored at $-70°C$. Competence in *Bacillus subtilis* is regulated by the secreted competence stimulating factor (CSF, 520–720 Da peptide) and the pheromone ComX (the $\approx$10 amino acid C-terminal section of the 55 amino acid peptide). The process of competence development requires transcription factors and specific nutritional conditions. In *Neisseria* bacteria, competence is expressed constitutively. In *Haemophilus influenzae*, arrest of cell division results in competence. *See* exponential growth; supercoiling; transformation, genetic. (Grossman, A. D. 1995. *Annu. Rev. Genet.* 29:477; Peterson, S., et al. 2000. *J. Bacteriol.* 182:6192.)

competition   Rivalry between or among free-living organisms for the available resources. It may be a passive *exploitative* process when one or more competitors use up the resources and the facilities or supplies become limited for others. By *interference* one organism actively attacks or eliminates another by the production of deleterious substances or mechanical obstacles. An *apparent* competition results when an organism of a certain genotype evokes a host defense mechanism that prevents others from infecting the same host. *See* frequency-dependent selection; pollen competition; selection; superinfection.

competitive exclusion   Two similar species generally do not coexist in the same niche indefinitely. They will either coalesce (be interbreeding) or one will be extinguished.

competitive inhibition   The inhibitor competes with the substrate for the active site of the enzyme and its blocking effect can be relieved by an increase in the concentration of the substrate. *See* regulation of enzyme activity. (Takahashi, Y. & Kamataki, T. 2001. *Drug Metab. Rev.* 33:37.)

competitive regulation   Besides the minimal enhancer, accessory cis elements are needed to ensure proper gene expression during the different developmental stages.

competitive release   Species or races are selected for general performance rather than for high adaptive specialization under conditions where there are limited or no competitive species.

complement   When antigen and antibody form a complex, in the Fc domain of the heavy chain of IgG and IgM the activation of the complement takes place (*See* antibody). The complement has numerous functions. One of the major functions is mediated through the antibody-related lytic pathway and the other is through the antibody-independent pathway, outlined below. Besides this, it may protect the immune complexes from precipitation, it facilitates the solubilization of the immune complex, and makes possible its transportation by the circulatory system of the body. The complement (except C5b-9) may interact with surface receptors of the hematopoietic cells of the immune system to stimulate inflammation and mount an immune response.

The complement may promote oposonization by macrophages. Graft rejection may be based on the activation of the complement system. Nonbiological, foreign material used during surgery or other medical interventions (various drugs, radiographic contrast media) may activate the complement system. Medical treatments may apply complement deactivating or neutralizing drugs (e.g., aspirin, cobra venom, specific antibodies, protease inhibitors) in autoimmune or other inflammatory disease. The complement consists of about 30 different proteins that have a role in the destruction of the foreign cells by lysis and activate the leukocytes to engulf the invaders by phagocytosis. Immunoglobulin M (IgM, 950 kDa) and the subclasses of immunoglobulin G (IgG, 150 kDa) bind complement C1q (459 kDa) components of C1 protein, encoded in human chromosome 1p34-p36.3. The binding causes the sequential activation of C1r (83 kDa, serine protease, encoded at 12p13), C1s (83 kDa, also a serine protease component, encoded at 12p13), C4 (95, 75, 33 kDa, encoded at location of the MHC III in chromosome 6, activator of C2, C3, C5), C2 (90–102 kDa, encoded at the MHC locus), and the cleavage of C3 (185 kDa, encoded at the end of 19q). C2 binds to C4b to form C4b2, which forms the C4b2a complex after cleavage by C1s, then activates C3.

There are a number other proteins that modulate these reactions: C1-inhibitor (C1-INH is a serpin [serine-protease inhibitor], 53–71 kDa, encoded at 11p11.2-q13); C4b-binding plasma glycoprotein (570 kDa, encoded at 1q) is an inhibitor of C3 convertase (cleaving protein); IgA, IgE, polysaccharides, and endotoxins can activate the cleavage of complement component C3; proteins B (93 kDa, encoded in the MHC complex), D (24 kDa), P (properdin, 220 kDa, encoded at chromosome Xp), factor I (dimeric 50 and 38 kDa, serine protease), and modulator H (155 kDa encoded in chromosome 1) are also involved in the cleavage of C3. The largest fragment of C3b then interacts with another protein group, properdin, and participates in a positive feedback system to stimulate the cleavage of C3. Finally, C3b initiates the cleavage of complement components C5 (192 Da, human chromosome 9), C6 (104–128 kDa), C7 (92.4 to 121 kDa), C8 (152 kDa, 9q), and C9 (71 kDa). Proteins 6,7, and 9 are encoded in chromosome 5p. The latter proteins form the so-called MAC (membrane attack complex). In the so-called *classical pathway*, proteins C1q, C1r, C1s, C4, C2, and C3 are involved.

The alternative means of activation follows the lectin or *mannose-binding* protein (MBP) path involving C4, C2, and C3 and the mannose-binding-associated serine proteases (MASP). This alternative MBP pathway—in contrast to the classical pathway—attacks the foreign proteins without prior antibody synthesis. The majority of gram-negative bacteria activate the MBP reactions, but many gram-negatives do not. Activating the binding of the H protein to C3b may prevent the activation. A variety of cell wall components of the pathogens also prevent the initiation of the MBP pathway. The complement components react with cell-specific receptors and some receptors react with more than one complement component. This reaction results in enhanced phagocytosis, antibody-dependent cellular cytotoxicity (ADCC), lysis, B-lymphocyte proliferation, etc.

Components C1, C4, and C2 attack viral invaders, and their interaction increases permeability. C3a and C5a have anaphylatoxic effects, including the release of cellular histamine and regulate humoral immune reactions. The C1q molecules may bind to specific receptors (C1q-R) such as the collectin receptor (60 kDa cC1q-R), or the 100 kDa cC1q-R, or the 30 kDa gC1q-R present on a variety of cells. C3b and C4b bind to surface receptors of several mammalian cells and to bacteria and cause their destruction. According to the "doughnut" hypothesis, C5b-C9 components of the complement form transmembrane channels (maximally 10 nm in diameter) through which the lytic components can penetrate the attacked cells. The target cells may be protected by elimination of the C5b-9 channels. The presence of Ca^{2+} may assist in the (partial) elimination of the channels, but Ca^{2+} may also enhance cell death. C5b-9 binding initiates hydrolysis of membrane lipids, and small amounts of C5b-9 may lead to the synthesis of prostaglandins, leukotrienes, growth factors, and a number of cellular proteins. The complement has a particular defense role before antibody formation has been completed. HIV and human T-cell leukemia virus are, however, resistant to the human complement. The Gal(α1–3)Gal terminal carbohydrate antigens are present on the endothelium lining in the majority of mammals, except humans, because human cells do not have this type of functional galactosyltransferase (although several galactosyltransferases are encoded by the human genome). If the porcine enzyme is transfected into human cells, the retroviruses become sensitive to the human serum.

Immunochemical and hemolytic properties measure the amount of the complement. CH_{50} is a measure of the complement, indicating that 50% of the antibody-sensitized erythrocytes released their hemoglobin. Genetically determined deficiencies are known for the complement proteins. Homozygotes may entirely miss a particular protein, whereas heterozygotes may display only reduced amounts. *Lupus erythematosus*, an autoimmune disease, is caused by C2 deficiency. Other deficiencies in the terminal components of the complement pathway may contribute to the symptoms of rheumatoid

immune diseases. Susceptibility to various types of infections is also related to deficiencies in the components of the complement proteins. The complement is basically an innate component of the immune system and it is complementary to the function of the macrophages, mast cells, T cells, and B cells. The complement can be activated not only by binding of the antibody and antigen but by surface structures of microbes, and in an antibody-independent manner by serine proteases associated with mannan-binding lectin. Hereditary deficiencies have been identified for most of the complement components. Many of these deficiencies are not lethal per se but make the individuals more liable to infections. *See* allograft; anaphylatoxins; angioedema; angioneurotic edema; antibody; C5a-R; CD11/CD18; CD21; CD22; CD35; Churg-Strauss vasculitis; convertase; decay accelerating factor; endotoxins; Galα1–3Gal; glomerulonephritis; hemolysis; histamine; HLA; humoral antibody; hyperacute reaction; immune system; immunodeficiency; immunoglobulin; lectins; Leiner's disease; leukotrienes; lupus erythematosus; lymphocytes; macrophages; mannan; mast cells; MCP; membrane attack complex; Niemann-Pick disease; opsonins; paroxysmal nocturnal hemoglobinuria; prostaglandins; Reynaud's disease; rheumatoid arthritis; TAPA-1; TCC; T cells; vitronectin; xenograft. (Morley, B. J. & Walport, M. J., eds. 2000. *The Complement Facts Book*. Academic Press, San Diego.)

complementarity   Of nucleic acid bases; adenine pairs with thymine and guanine with cytosine by two and three hydrogen bonds, respectively, and two complementary polynucleotide chains pair in an antiparallel manner. *See* Chargaff's rule; DNA structure.

complementarity determining region   *See* CDR.

complementary alleles   Belong to different gene loci, although partial complementation may occur between alleles belonging to different cistrons of the same gene locus. *See* allele; allelism test; cistron; complementation mapping; complementation test in vitro.

complementary base (nucleotide) sequence   The nucleotides can form hydrogen bonds according to the base-pairing rules in double-stranded DNA or double-stranded RNA. *See* Chargaff's rule; hydrogen bonding.

complementary DNA   *See* cDNA.

complementary genes   Homozygosity of recessive alleles at different loci but controlling functions in the same biosynthetic pathway may be expressed by identical (or very similar) phenotype, and in the F_2 of a double heterozygote, expected to display the phenotypic proportions 9 wild type and 7 mutants (homozygous for one [3] + for the other [3] + for both [1] of the recessive alleles = 7). This is a modification of the 9:3:3:1 (9/16, 3/16, 3/16, 1/16) ratio. *See* modified Mendelian ratios.

complementary segregation   *See* complementary genes.

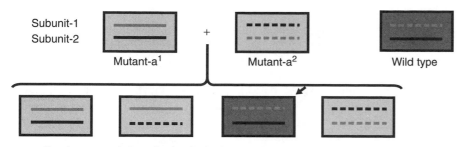

Subunit-1
Subunit-2

Mutant-a¹ + Mutant-a² Wild type

Random association of subunits in the cytoplasm of the compound

Allelic complementation may be based on random cytoplasmic association of protein subunits of the multimeric enzymes. The mutant subunits may not be functional, yet if they assemble favorably, some enzyme molecules can show activity.

complementary sex determination May occur in animals with arrhenotoky. The females are heterozygous for complementary alleles at the sex locus and diploid. The normal males are haploid, whereas diploidy in the males is deleterious or lethal. *See* arrhenotoky.

complementation, donor strand Formation of a stable complex by the interaction of two proteins. One donor may be a chaperone molecule. *See* chaperones.

complementation, extracellular By diffusion, the product of one gene may compensate for the defect in a nearby cell. *See* complementation.

complementation, tetraploid *See* tetraploid complementation.

complementation groups Recessive mutations that complement in trans arrangement belong to different complementation groups, i.e., they represent different gene loci, whereas the noncomplementary alleles belong to the same complementation group. *See* allelic complementation; allelism test; cis; complementation test; epistasis group; trans.

complementation mapping Used for determining the pattern of the genetic differences among complementary alleles and the extent of the genetic lesions involved. *See* diagram below.

complementation test Test of allelism. Recessive alleles of the same cistron generally fail to complement in trans arrangement in heterozygotes, e.g., the $\frac{Ab}{aB}$ heterozygote is nonmutant if a and b alleles belong to different loci. If two recessive mutations in the F_1 display mutant phenotype, the two mutants are allelic. *See* allelism.

complementation test, in vitro Carried out by cell extracts (enzyme assays). This is the most reliable test of complementation because in vivo the genes are regulated and that may boost complementation when the complementation may actually be quite week, e.g., 5% activity of an enzyme may appear as if it would be of wild phenotype in vivo, but in vitro, in the absence of regulation, it can be readily distinguished from the 100% activity. *See* allelism. (Loper, J. C. 1961. *Proc. Natl. Acad. Sci. USA* 47:1440.)

complementation unit Cistron. *See* cistron; complementation test.

complement fixation Serological measure of the degree of antigen-antibody reaction. *See* antibody detection; CH_{50}; complement.

complete digestion Reaction with a restriction endonuclease is continued until all potential cutting sites in the DNA are cleaved by the enzyme. *See* restriction enzymes.

complete dominance In the presence of a completely dominant allele, the phenotype controlled by the recessive allele is not detectable under the conditions of study. *See* codominance; Mendelian laws; Mendelian segregation; modified Mendelian ratios; semidominance.

complete flower Has sepals, petals, stamens, and carpels. *See* flower differentiation.

complete linkage Genes fail to recombine because of their extreme closeness in the chromosome or because of genetic factors interfering with crossing over (chromosomal aberrations), inviability of the recombinants (deficiency or duplication gametes in inversion heterozygotes), or absence of recombination in the heterogametic sex of male *Drosophila* or female silk worm. *See* linkage; recombination. (Sturtevant, A. H. & Beadle, G. W. 1939. *An Introduction to Genetics*, Dover, New York.)

complete medium Contains all the nutrients that potentially auxotrophic cells may require for growth. *See* auxotroph; minimal medium.

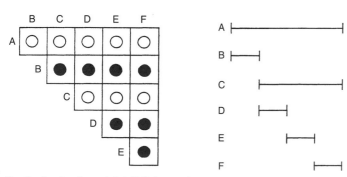

On the basis of partial (allelic) complementation of various alleles of a locus controlling a multimeric enzyme, genetic mapping is possible. This is not necessarily a physical map, rather a functional one, and therefore it may not always be linear. Open circles indicate lack of complementation; solid circles stand for complementation of recessive alleles identified here by A, B, C, D, E, and F. (From Catcheside, D. G. 1960. in *Microbial Genetics*. Hayes, W. & Clowes, R. C., eds., Cambridge Univ. Press, Cambridge, UK.)

completion Production of high-quality nucleotide sequence map of both unique and repetitive DNA of a genome. *See* finishing; genome projects; high-quality sequence.

complex Functional aggregate of molecules without covalent bonds among them.

complex disease Multiple genes and diverse environmental factors determine manifestation and susceptibility. Neither statistical nor molecular analyses facilitated, to an entirely satisfactory extent, the identification of the factors and mechanisms involved, although several approaches are available. (Altmüller, J., et al. 2001. *Am. J. Hum. Genet.* 69:936.)

complex heterozygote Viable heterozygotes for multiple reciprocal translocations when translocation homozygotes are lethal. It is actually a balanced lethal system, and heterozygosity is permanent for the genes within the translocation complexes. If there are two multiple translocation complexes, it appears as if there would be only two groups of linked genes even when the basic number of chromosomes is larger. In meiosis, the distribution of the chromosomes is alternate, and paternal and maternal complexes always go to opposite poles. Recombination between the complexes is rare, but if it occurs, it may change the complexes. The size of the complexes may vary depending on the number of chromosomes involved in the reciprocal translocation complexes. *Oenothera lamarckiana* carries the seven-chromosome *gaudens* complex ("happy" in Latin, i.e., the chromosomes do not contain dominant or semidominant deleterious genes) and the seven-chromosome *velans* translocation complex ("concealing" in Latin, since it contains semidominant genes that are responsible for paler color and narrower leaves). *Oenothera hookeri* also has 14 chromosomes, but they are not involved in reciprocal translocations. When these two *Oenotheras* are crossed, the F₁ is not uniform as in normal Mendelian crosses, but they form *twin hybrids*. The maternally transmitted *gaudens* complex makes the hybrids a normal shape and color, whereas the maternally transmitted *velans* complex–containing hybrids are pale and have narrow leaves. Thus, the F₁ reminds us of a test cross involving one pair of heterozygous alleles because of the two complexes. In *Oenothera*, these complexes behave differently. Some kill the male; others kill the female gametes; others cause zygotic lethality in the homozygotes, but only the complex heterozygotes are found in the sporophytic plants. The multiple translocations are identified by the light microscope as translocation rings. Such systems are common among

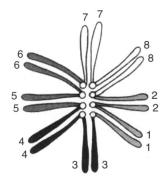

A diagram of multiple reciprocal translocations.

Oenothera plants but also occur in other species. *See* beta complex; gaudens; *Oenothera*; translocations; zygotic lethal. (Cleland, 1972. *Oenothera: Cytogenetics and Evolution*, Academic Press, New York.)

complexin Proteins of the nerve termini, binding syntaxin. They regulate Ca²⁺-dependent release of neurotransmitters. *See* neurotransmitter; syntaxin. (Reim, K., et al. 2001. *Cell* 104:71.)

complex inheritance Found when multiple loci and interactions are involved in the expression of the gene products. (Morton, N. E. 1996. *Proc. Natl. Acad. Sci. USA* 93:3471.) Morton published parametric and nonparametric statistics for estimating linkage, considering various pedigrees. *See* complex trait; multifactorial trait; polygenic inheritance; QTL. (Zwick, M. E., et al. 2000. *Annu. Rev. Genomics Hum. Genet.* 1:387.)

complexity of DNA Indicates the size of the DNA molecule as determined from the c₀t curve. *See* c₀t₁/₂; c₀t curve; kinetic complexity; kinetics.

complexity of function Complexity can be defined as the number of metabolic steps required for biological function(s). Before different genomes were sequenced, the assumption appeared valid that the number of genes determines biological complexity. In broad terms, this is still valid because the MS2 phage requires only 4 genes, whereas the bacterium *E. coli* contains 4,288 protein-coding genes. The simple worm, *Caenorhabditis elegans*, has 18,424 genes, whereas the more complex fruit fly, *Drosophila melanogaster*, has only 13,601, but the small plant *Arabidopsis* plant has about 25,498. Therefore, there must be a better way to define biological complexity. Since genes may be spliced in alternative manners, this feature may boost complexity. Also, the expression of not being housekeeping genes is highly regulated by the recruitment of specific transcription factors and effector proteins. There is apparently a great variety of ways that the different proteins can enter into complexes, and single polypeptide chains may participate in several functionally different aggregates. Therefore, it appears that the cooperating (enhancing and inhibitory) networks have a major role in determining biological complexity. *See* microarray hybridization; proteome. (Szathmáry, E., et al. 2001. *Science* 292:1315.)

complex locus Generally large cluster of functionally related but not entirely similar cistrons, and alleles within the locus may show partial complementation (*scute* in *Drosophila*, *t* in mouse, HLA and immunoglobulins in mammals, *R* in maize). Some of the complex loci appear to be more mutable than other genes because unequal crossing over between the repeated sequences generates new phenotypes. *See* allelic complementation; Bar; pseudoallelism; step allelomorphism; operon. (Carlson, E. A. 1959. *Quart. Rev. Biol.* 34:33; Demerec, M., et al. 1955. *Proc. Natl. Acad. Sci. USA* 41:359.)

complex promoter Besides the usual promoter elements, the complex promoter contains an insertion element (e.g., Ty) with its own promoter situated within the long terminal repeat. As a consequence, transcription may initiate from the Ty promoter (δ) located within the interval between the TATA and the UAS sequence, and the transcript of the yeast gene involved may not be functional. The δ-promoter also contains its own binding regulatory elements. *See* promoter; Ty. (Pilpel, Y., et al. 2001. *Nature Genet.* 29:153.)

complex trait Its inheritance is not based on single dominant or recessive alleles, but multiple factors may be involved, and various environmental effects contribute to their expression. *See* complex inheritance; QTL. (Moore, K. J. & Nagle, D. L. 2000. *Annu. Rev. Genet.* 34:653.)

complex transcription unit The transcript of the gene may be processed in more than one way, and the translated products vary according to cell- or tissue-specific functions.

compliant mutation Readily identifiable by genetic testing; reveals the existence of a certain risk for disease. *See* genetic testing.

complicon Is a complex chromosomal translocation and commonly co-occurs with oncogenic transformation. *See* cancer. (Zhu, C., et al. 2002. *Cell* 109:811.)

composite cross Individuals of various genetic constitution are hybridized in a mass for the purpose of studying the effect of natural selection or to obtain improved varieties.

composite promoter *See* complex promoter.

composite transposon Carries genes (e.g., antibiotic resistance) beyond those required for transposition. *See* insertion element; transposon.

compound chromosome Results from the fusion of telocentric chromosomes into biarmed monocentric, or by Robertsonian translocation between acrocentric chromosomes, or by translocations, or by intrachromosomal transposition. *See* compound X chromosome; Robertsonian translocation; telocentric; translocation; transposition.

compound, genetic Heterozygote for two mutant alleles of the same gene and may display a phenotype that is intermediate between that of the two recessive homozygotes. The loci may be in cis or trans position. *See* cis.

compound eye In Arthropods (insects), the eye is composed of several, each structurally complete elements (ommatidium, about 800) in each eye of *Drosophila*). *See* CℓB; *Drosophila*; ommatidium.

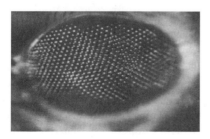

compound leaf Composed of several leaflets.

compound X chromosome Two X chromosomes are fused to the same centromere. E. Novitski distinguished the following six types: reversed metacentric or attached X (RM), reversed acrocenric or double X (RA), reversed ring (RR), tandem metacentric or tandem attached X (TM), tandem acrocentric (TA), tandem ring (TR). (Novitski, E. 1954. *Genetics* 39:127.)

Compound X chromosomes of *Drosophila*. The heavy lines represent heterochromatic regions.

compressions in gels In DNA sequencing gels abnormal intra-strand structures may form and cause anomalous pattern of migration, especially in DNAs with high $G + C$ content. The compression may be avoided by using another DNA polymerase or by 2′-deoxyinosine-5′-triphosphate or 7-deaza-2′-deoxyguanosine-5′-triphosphate. *See* base-calling; deaza-nucleotides; DNA sequencing.

Compton effect As the energy of electromagnetic radiation increases above 0.5 MeV, the radiation of electrons may scatter and recoil depending on the surface hit and the angle of the incidence of the radiation. This may affect the effective dose absorbed by the object and may create hazards if the source is not effectively protected in all directions. *See* cathode rays; X-ray. (van't Veld, A. A., et al. 2001. *Med. Phys.* 28:738.)

computerization of DNA and protein sequence data The sequence information must be entered in a computer file. For proper handling, there may be a need for a *sequence manipulator*, which is included with the software. Some programs have an audio feature (e.g., SeqSpeak) that spells out the data as entered. The entered data should be saved in text files or ASCII (American standard code for information exchange). The files are supposed to contain the relevant laboratory identification. The various software packages use different formats. Digitizing pads permit reading the sequence information directly from the autoradiogram if the quality of the sequencing gel is good. Automated gel readers are available commercially. When a multiplex method is used, a known sequence can be incorporated as an internal standard along with the gel. Automated DNA sequencers use special fluorochrome-tagged primers, and no manual work is required for the entire operation. The automated scanners may come with editing features so that if a mistake is made it can be manually corrected. To minimize sequencing errors, generally both strands are sequenced and generally more than once. Also, the reading may need repeating. Homology searching can automatically align the data for comparisons. The sequencing programs usually include the means to predict the presumable collinear amino acid sequence of the open reading frames. The programs may detect fragment sequence overlaps in order to generate contigs. The contigs can then be compared with restriction maps if available. Some of the programs may generate graphic restriction maps. For the identification of functional regions, the detection of repeats and stem-loop structures in DNA or RNA may be used. The GC/AT content may reveal Z DNA or bent DNA structures.

Some programs assist in the identification of likely secondary structures (stem-loop) and folding of the RNA, including base stacking. Programs exist for the design of PCR primers. Gene families with related functions can be revealed by the use of degenerate oligonucleotide probes. In order to identify protein-coding tracts, each strand is translated in three reading frames (six all together) to obtain a contiguous, uninterrupted open reading frame (ORF). It may be difficult to reveal the possible splice sites, although various programs have this goal. A scoring matrix that quantitates the changes per 100 amino acids detects evolutionary changes in protein sequences. The Basic Local Alignment Search Tool (BLAST) permits homology searches even through the Internet. FASTA serves a similar purpose through e-mail. Helpful publications and software sources are listed in Ausubel, F. M., et al., eds. 1999. *Short Protocols in Molecular Biology*, Wiley, New York. *See* proteomics.

conalbumin   Egg-white iron-binding protein, encoded by a gene with 17 introns and regulated by estrogen and progesterone.

c-onc (cellular oncogene)   Normal gene that may lose its ability to limit cell divisions, then initiates cancerous growth. *See* v-oncogene.

concanavalin   Agglutinin protein that preferentially agglutinates cancer cells, used also as a probe for cell surface membrane dynamics. Concanavalins are mitogenic. *See* cell adhesion; lectins; mitogen. (Wallach, D. F. & Schmidt-Ullrich, R. 1976. *J. Cell Physiol.* 89:771.)

concatamer   Repeated phage genomes associated in a linear array of DNA molecules in a head-to-tail fashion, formed during normal replication, and must be cut to head capacity size for packaging into the capsid by a terminase gene product (endonuclease). *See* catenane; headful rule; lambda phage; nonpermuted redundancy; permuted redundancy. (O'Donnell, R., et al. 2001. *Nucleic Acids Res.* 29:716.)

concatenane   Interlocked DNA rings or chains *See* knotted DNA.

conceptacle   Cavity in the fern leaves (fronds) bearing gametangia. *See* gametangia.

conception   *See* fertility; fertilization; hormone receptors; sex hormones.

conceptus   Fertilized egg during the entire development including embryo and fetus stages and the extra-embryonic membranes until birth.

concerted evolution   Copies of redundant DNA sequences are rather well conserved in the genomes, although one would have expected divergence by repeated mutations. *Intragenic concerted evolution* may exert more mutational and recombinational pressure on internally repetitive exons of a gene than on the introns. Exons are crucial for the function of the protein product of the gene. As a consequence, extensive homology is observed within species and little homology between related species (such as found in intergenic spacer nucleotides and LINEs). *See* LINE; molecular drive. (Liao, D. 2000. *J. Mol. Evol.* 51:305.)

concordance   Identity of traits within twins or groups of individuals. Among monozygotic twins, the concordance is expected to be ~100%, whereas among dizygotic twins it is between 25% and 50%. Lower concordance is an indication of (substantial) environmental contribution to the phenotype. *See* adopted children; discordance; dizygotic twins; expressivity; monozygotic twins; penetrance; twinning; zygosis.

concurrent control   In an experiment involving a certain type of treatment it is required that an adequate group of untreated individuals are studied *simultaneously* to permit a reliable comparison and assessment of the effect of the treatment. *See* control; historical control.

condensation reaction   During the formation of a covalent bond, water is released as a by-product.

condensin   Complex of five proteins involved in chromosome condensation in the presence of topoisomerase I during mitosis in association with XCAP-C and XCAP-E. Condensin is activated by phosphorylation using Cdc2 kinase and ATP as a phosphate donor. *See* ATP; Cdc2; mitosis; Mosolov model; sister chromatids; SMC; topoisomerase; XCAP. (Hirano, T. 2000. *Annu. Rev. Biochem.* 69:115; Kimura, K., et al. 2001. *J. Biol. Chem.* 276:5417.)

conditional dominance   *See* dominance reversal.

conditional lethal   Dies only under certain conditions (e.g., high nonpermissive [restrictive] temperature) but is viable under others (e.g., low permissive temperature). Many auxotrophs are conditional lethal because they survive only when the required nutrient is provided. *See* auxotroph; conditional mutation; temperature-sensitive mutation.

conditionally dispensable chromosome B chromosome.

conditional mutation Expressed only under the condition(s) required. Such mutations may be extremely useful for the study of conditional lethal genes. Conditional expression may also be regulated by agents that affect transcription and/or translation or conformation of the proteins (heat shock, heavy metals, hormones, repressors, DNA-binding proteins, dimerization of transactivator, signal transduction, etc). *See* auxotrophs; conditional lethal mutation; temperature-sensitive mutation. (Lewandoski, M. 2001. *Nature Rev. Genet.* 2:743; Genesis vol. 32:49–191 [2002].)

conditional probability Not based on absolute frequencies but, e.g., one group is fixed in a matrix. When we have three genotypes, AA, Aa, and aa, their frequencies add up to 1. If one of the groups is fixed, then another must be of one of the two genotypes. In other words, probability depends on what events have taken place previously. *See* Bayes' theorem; risk.

conditional targeting Gene-targeting procedure aimed either at specific tissues or developmental stages or both. *See* knockdown; knock-in; knock-out; targeting genes.

conductance A nonconjugative plasmid can be transmitted to a recipient cell by cointegration into a mobile, conjugative plasmid. *See* cointegration; conjugation; plasmid.

conductin 840-amino-acid protein equipped with a β-catenin-, an RGS- (regulator of G-protein signaling), and a glycogen synthase kinase-3β-binding domain. This complex interacts with various fragments of APC (adenomatous polyposis) tumor suppressor protein. Conductin contributes to the degradation of β-catenin, whereas APC interferes with this degradation. Conductin has homology to axin and is capable of similar functions. *See* adenomatous polyposis; axin; catenin; GSK; RGS. (Siderovski, D. P., et al. 1999. *Crit. Rev. Biochem. Mol. Biol.* 34:215.)

cone Fruiting structure bearing sporangia, e.g., a pine cone (*see* sporangium). Also, the retinal cone in the eye is a visual cell. *See* ommatidium; retinitis pigmentosa; retinoblastoma.

cone dystrophy Group of dominant diseases involving loss of color vision, photophobia, reduced central vision acuity occurring at a frequency of $\sim 1 \times 10^{-4}$. Responsible genes have been assigned to 17p12-p13, 6q25-q26, 6p21.1, and Xp21.1-p11.3. The 6p21.1 mutations involve guanylate cyclase-activating protein. *See* ABC transporter. (Wilkie, S. E., et al. 2001. *Am. J. Hum. Genet.* 69:471.)

cone pigment Present in the retina and mediates color vision. *See* cone.

cone-rod dystrophy *See* ABC transporters.

confidence interval Population parameters can be estimated by *point estimates* and by *interval estimates*. The former specifies the parameter itself; the latter defines the range of values within which the parameter is expected at a certain level of confidence (probability). If we obtain an average value in a population, we would like to know the probability within what range the real, true average might fluctuate by 95% or another confidence of choice. Confidence intervals (CI) for reasonably large populations can be determined with the formula: $\mathrm{CI} = p \pm z\sqrt{pq/n}$, where p and q are the proportions observed

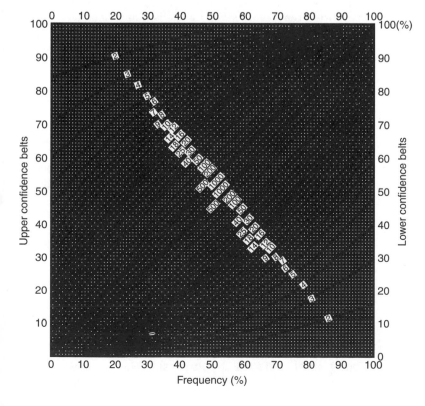

Use Of The Confidence Chart: Calculate from the experimental data the frequencies p and q. At the intersections of the vertical line corresponding to "frequency" and the two belts (curves) numbered and indicating the size of the sample (population), one can read the percentage of upper and lower confidence. This chart also facilitates an estimate of the (1) repeatability of the fractions obtained, (2) the statistical range of the fractions, and (3) the size of the population where a chosen fraction will be represented by a number of individuals (counts) within the specified range, e.g., the frequency observed is 0.20. The confidence limits at probability 0.95 for a population of 100 will be 29 and 12. Or if the expected frequency is 0.25 and one is willing to accept a latitude of 10% (i.e., a frequency between 0.20 and 0.30), the minimum population size required is 200 at $P = 0.95$ because this latitude is bracketed by the 200 belts. (From Koller, S. 1956. *Biochem. Taschenbuch*, Rauen, H. M, ed., Springer-Verlag, Berlin.)

$(p + q = 1)$, z is the critical value for the normal distribution at a given level of confidence; $z = 1.96$ (for 95%) 2.58 (for 99%), 3.29 (for 99.9%), and it is not very useful to go for even higher z values because then the range will be so wide that it will become almost meaningless; n = the numbers in the population counted or measured. Example: A population of 140 consists of two groups represented by 60 (p = 60/140 ≈ 0.43) and 80 (q = 80/140 ≈ 0.57). CI for $p = 0.43 \pm 1.96\sqrt{[0.43][0.57]/140} = 0.43 \pm 1.96 \times 0.04184 = 0.43 \pm 0.08$. Therefore, the frequencies and 0.51 and 0.35 are the 95% confidence limits of the experimentally observed p. Alternatively, the 95% confidence belts can be used.

confidentiality   The code of ethics and many state laws in the United States prohibit physicians from disclosing any medical information to a third party even in court. Some courts permit disclosure if the third party is in imminent danger by genetic or infectious disease. *See* bioethics; counseling, genetic; genetic privacy; informed consent; paternity test; wrongful life. (Cohen, P. E. & Wolpert, C. 1998. *Approaches to Gene Mapping in Complex Human Diseases*, p. 131. Haines, J. L. & Pericak-Vance, M. A., editors, Wiley, New York.)

conflict, evolutionary   Spreading of an allele that lowers the fitness of the individual or its progeny. In some cases, one allele may be advantageous in one sex but disadvantageous in the other. To neutralize its effect, suppressor genes may evolve. *See* meiotic drive.

confluent   The cell culture extends over the entire surface of the medium.

confocal microscopy   Permits the formation of a three-dimensional image by focusing laser or other light through a pinhole and with a dichroic mirror to a distinct small area of the object. The fluorescent light emitted from that small focal point is then reflected through another (confocal) pinhole (i.e., this second pinhole is exactly in the focus of that area of the object) and thus a sharp image is received on a detector. The images of the areas not in focus are thus obliterated. Subsequently, the light is focused at several other points and the image of each point is registered on the video screen, building up in a three-dimensional integrated picture. If the pinholes are moved in a synchronous manner, a scanning real-time image can be obtained. *See* microscopy.

conformation   Arrangement of a molecule in space that does not require the severance of any bond because of its freedom of rotation. Point mutations and sequence rearrangements do not substantially alter the original conformation. *See* chaperone; folding. (Sinha, N. & Nussinov, R. 2001. *Proc. Natl. Acad. Sci. USA* 98:3139.)

conformation correction   Multimeric proteins may be inactivated by mutation in a subunit that affects the conformation of that polypeptide chain. Another mutation may alter the conformation of another subunit in such a way that the first defect in conformation is corrected and the activity of the enzyme is at least partially restored. It was assumed that some observed allelic complementation is based on such

a correction. *See* allelic complementation; dominant negative. (Soto, C. 2001. *FEBS Lett.* 498:204.)

confounding   Omitting a particular experimental comparison in order to reduce the error variation. In a factorial design, some of the interactive elements are not distinguishable alone but only in combination. *See* factorial experiment.

congeneic   Congenic; it is frequently spelled this way in immunogenetics.

congener   Related to something by origin and/or function, biologically or chemically.

congenic resistant lines of mice (CR)   Contain a new histocompatibility gene locus introduced from another inbred (or any other) line. Earlier these were called IR (isogenic resistant) lines. Mating inbred line #1 with known histoincompatible line #2 generates such CR lines. The hybrid accepts grafts of #1. In the progeny of hybrids mated *inter se*, some segregants do not reject transplants of #1. These will be further backcrossed with #1. After 7 backcrosses, the hybrid will have the same chromosomes as parent #1 in over 99% (1-0.5^7). Upon continued brother-sister matings (usually for 20 generations), an individual may show up that rejects grafts from #1. This is further backcrossed to #1 and the progeny will be selected to resistance against transplants to #1. Such a line is congenic resistant, i.e., almost identical (with the exception of the histocompatibility gene) with #1. Some very closely linked genes to the histoincompatibility locus of #2 may not be eliminated, however, from the CR line. Today the selection may be facilitated by serological assays rather than by expensive transplantation experiments. Also, the desired genes can be transferred by transfection without the need for repeated backcrosses. The development of a large number of CR lines permitted the identification of allelism and complementation groups of histocompatibility genes. The serological assays permitted the identification of strong and weaker responses and on this basis major and minor histocompatibility genes could be classified. *See* HLA; MHC. (Fortin, A., et al. 2001. *Proc. Natl. Acad. Sci. USA* 98:10793.)

congenic strains   They are identical, except at one locus or a very limited region of a chromosome; they are obtained by repeated (10–20) backcrosses. These lines can be used to determine the effect of a particular gene on a selected genetic background. Using eggs obtained by hormone-induced superovulation of prepubertal (3-week-old) mice and in vitro fertilization followed by embryo transfer, the generation time can be shortened to 6 weeks, and within a year the desired results may be obtained. *See* ART; coisogenic; congenic resistant lines of mouse; inbreeding; isogenic; subline; substrain.

congenital   Born with a condition regardless of whether it is due to direct genetic or developmental causes. *See* familial; hereditary.

congenital adrenal hyperplasia   Has a prevalence of about 0.0002–0.0001, but in some populations it may be orders more frequent. It is based on the deficiency of an enzyme in cortisol (steroid) biosynthesis. The accumulation of the precursor results in virilization of female babies. The condition can be successfully treated by glucocorticoids given to the mother. The autosomal-recessive gene(s) is closely linked to the HLA loci in human chromosome 6. Prenatal identification is feasible by linkage with appropriate DNA probes. There are, however, other types of steroid hydroxylase deficiencies encoded in other chromosomes. *See* adrenal hypoplasia; adrenal hyperplasia, congenital; cortisol; genetic screening; glucocorticoid; HLA.

congenital biplasic aplasia of the vas deferens (CBAVD) Autosomal-recessive absence of the excretory channel(s) of the testes resulting in azoospermia. *See* azoospermia; cystic fibrosis; P2X; vas deferens.

congenital disorders of glycosylation (CDG) Type Ib (15q22-ter) is a defect in mannose phosphate isomerase, and type Ia (16p13.3-p13.2) is a defect of phosphomannomutase-2. The basic problem is the unsatisfactory glycosylation of glycoproteins. The clinical consequences are neurological defects, blood coagulation problems, eye malformations, heart and kidney disease, abnormal transferrin, etc. The symptoms in type Ib disease can be alleviated by mannose. Type Ic (1p22.3) is an abnormality of the transfer of glucose to lipid-linked oligosaccharides. Several other variants have also been distinguished.

congenital hypothyroidism Defective development of the thyroid gland causing goiter, mental retardation, deafness, etc., because of deficiency of thyroid hormone. *See* goiter.

congenital trait Evident at birth and is due to either hereditary or other causes. *See* familial trait.

congression Assembly of chromosomes in the metaphase plane. *See* meiosis; mitosis.

Congression (Courtesy of A. Sparrow).

congruence analysis Tests the appropriateness of conclusions reached by different methods in the study of evolutionary trees. *See* homology; xenology.

congruent gene Shares sequences in the chromosome.

conidia (conidiospores) Asexual, uninucleate fungal spores that appear externally on a hypha by abstriction. Such a hypha is a conidiophore. *See* fungal life cycle; macroconidia; microconidia. (Kellner, E. M. & Adams, T. H. 2002. *Genetics* 160:159.)

conidiophore *See* conidia.

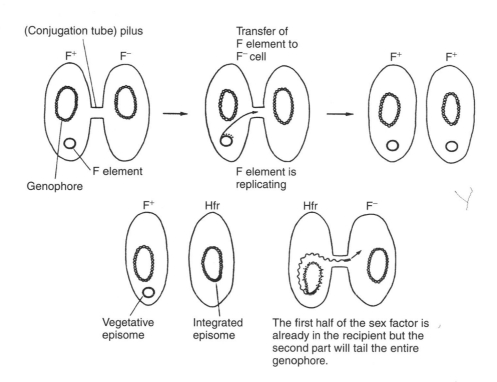

(Conjugation tube) pilus
F⁺   F⁻
Genophore
F element

Transfer of F element to F⁻ cell
F element is replicating

F⁺   F⁺

F⁺   Hfr
Vegetative episome   Integrated episome

Hfr   F⁻
The first half of the sex factor is already in the recipient but the second part will tail the entire genophore.

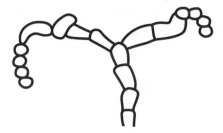

Conidiophore, round conidia at tip.

conjugated protein Contains prosthetic group(s); e.g., an iron or magnesium heme in hemoglobin and chlorophyll, respectively.

conjugate redox pair Electron donor and the corresponding electron acceptor, e.g., NADH and NAD$^+$.

conjugation, bacterial Generally means mating. In bacteria, the physical contact between an F$^+$ donor and an F$^-$ recipient cell and the unidirectional transfer of the (Hfr) chromosome by a rolling circle replication procedure through the conjugation tube. Approximately 12 μm DNA is transferred per minute. The standard genetic map of bacteria is based on the time in minutes required for the conjugational transfer of genes from donor to recipient.

The conjugational transfer may be clockwise or counterclockwise depending on the orientation of the (Hfr) element in the bacterial chromosome. The transmission of plasmids from one cell to another requires a conjugational mechanism. The conjugation requires cellular contact controlled by the mating pair formation system (Mpf). In *Agrobacteria*, besides the T-DNA, multisubunit protein complexes are transmitted to the eukaryotic cell. The Tra and Mpf proteins organize the conjugative pilus. The transforming plasmids generally lack the *mob* gene required for mobilization of a chromosome (genophore) or plasmid, so recombinant DNA can be contained. For some of the plasmids, the mobilization factor

can be provided in trans by a helper plasmid (ColK). Col plasmids code for a protein that opens up the circular DNA at the *nic* site close to *bom* (bacterial origin of mobilization). Some plasmids lack the *nic/bom* system and cannot be transferred through conjugation (nonconjugative plasmids). The latter type is favored for containment of recombinants. In the majority of bacteria, conjugation is not a standard mode of reproduction, unlike in true sexual organisms. Recent reexamination of the data indicates linkage disequilibrium in bacterial populations. Using the RK2 plasmid system, bacterial DNA can be transferred to yeast cells as well as to Chinese hamster ovary cells (Waters, V. L. 2001. *Nature Genet.* 29:375). *See* bacterial recombination frequency; CHO; conjugation, *Paramecia*; conjugational mapping; F plasmid; Hfr; linkage disequilibrium; nonplasmid conjugation; *ori*$_T$; pilus; relaxosome; RK2 plasmid; *tra* genes. (Ippen-Ihler, K. A. & Minkley, E. G., Jr. 1986. *Annu. Rev. Genet.* 20:593; Frost, L. S., et al. 1994. *Microbiol. Rev.* 58:162; Lanka, E. & Wilkins, B. M. 1995. *Annu. Rev. Biochem.* 64:141; Matson, S. W., et al. 2001. *J. Biol. Chem.* 276:2372.)

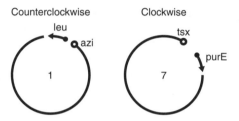

conjugation, *Paramecia* Sexual process of these unicellular protozoa that most commonly reproduce asexually by fission. During conjugation, two cells of opposite mating types appose and meiosis proceeds in the micronuclei. From the four meiotic products, only one survives in both cells, and that divides twice by mitosis, generating for haploid nuclei in the two cells. One of the gametes is then passed over to the other cell through a conjugation bridge in a reciprocal

Markers:	*thr*	*leu*	*azi*	T1	*lac*	T6	*gal*	λ	21	424
Hfr transfer (%)			>90	70	40	35	25	15	10	3
minutes	8	8½	9	11	18	20	24	26	35	72

(*On the basis of F. Jacob, and E. L. Wollman*, 1961. Sexuality and genetics of bacteria, *Academic Press, New York.*)

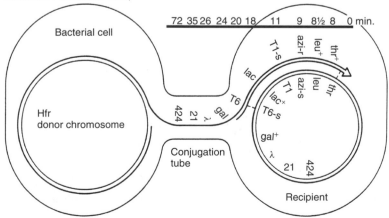

manner, resulting in mutual fertilization and in two diploid conjugants. Subsequently, the pair separates and the *exconjugants* are formed. The macronucleus has only a metabolic but no known genetic role; it disintegrates in both cells. The diploid micronucleus undergoes two mitotic divisions, and two of the four mitotic products fuse to regenerate the macronucleus. The remaining two nuclei form diploid micronuclei. *Paramecia* may also reproduce by *autogamy* in the absence of conjugation. Meiosis may take place, and four haploid nuclei arise from which again only one survives. The survivor divides again into two, which generate a diploid homozygote after fusion. If the conjugation lasts for a longer period, cytoplasmic elements are also transferred along with the gametes. *See Paramecium.* (Preer, J. R. 1971. *Annu. Rev. Genet.* 5:361.)

conjugation mapping   In bacteria the transfer and integration of genes of the Hfr chromosome to the recipient is measured by interrupted mating and the map is constructed on the basis of minutes required for the linear transfer of a particular marker(s). For the complete transfer, ca. 90–100 minutes are required (40–45 kb/2–3 minute intervals). By conjugational recombination the complementary products of the exchange are not recovered, and the surviving products are the results of double recombination. The transferred single strand becomes double-stranded after entering the recipient (F^-) cell. *See* bacterial recombination frequencies; conjugation, bacterial; F plasmid; Hfr. (Jacob, F. & Wollman, E. L. 1961. *Sexuality and the Genetics of Bacteria.* Academic Press, New York.)

conjugation tube   *See* conjugation, bacterial; mating, bacterial; pilus.

conjugative plasmid   *See* conjugation of bacteria; plasmid.

connectin   Titin. *See* titin.

connective tissue disorder   Involves cells with substantial extracellular matrix, which provides structural support, such as bone, cartilage, etc. In the majority of disorders, collagen synthesis is affected. *See* arthritis; autoimmune disease; collagen; Ehlers-Danlos syndrome; HLA; Kniest dysplasia; lupus erythematosus; Marfan syndrome; osteogenesis imperfecta; pseudoxanthoma elasticum; Reiter syndrome; skin diseases; Stickler syndrome.

connexin   Provides elements of gap junctions. Connexins are built of six polypetides. All connexins have four membrane-spanning domains, two extracellular loops, a cytoplasmic loop, and cytoplasmic amino and carboxyl termini. Connexins may form connexons, the pore of gap junctions. Connexins may have a role in heart diseases, infertility, cataracts, deafness, heterotaxy, etc. They are encoded in human chromosomes 1p35, 6q21-q23.2, 13q11, Xq13.1. *See* cataracts; Charcot-Marie-Tooth disease; deafness; ectodermal dysplasia; erythrokeratoderma variabilis; gap junctions; innexins. (Kelsell, D. P. 2001. *Trends Cell Biol.* 11:2.)

connexon   *See* connexin.

conplastic   Has the mitochondrial genome of one strain, but the nuclear genome is derived from another. Conplastic strains are obtained by at least 10 backcrossings of a female by a nuclear genome donor male. The probability that the nuclear genome is of the donor type is $1 - 0.5^n$, n = number of backcrosses. *See* chloroplast genetics; mitochondrial genetics; mtDNA.

Conradi-Hünermann disease   *See* CHILD syndrome; chondrodysplasia.

consanguinity   *See* coefficient of coancestry; inbreeding coefficient.

consciousness   Awareness, sensory discrimination of events in the outside world and the body. Unfortunately, neurophysiology does not yet have adequate tools to tackle most of the problems involved.

CONSED   One of the frequently used DNA sequence alignment/editing programs. *See* PHRAP; PHRED; PolyPhrap. (Rieder, M. J., et al. 1998. *Nucleic Acid. Res.* 26:967.)

consensus   Basically common (although generally not entirely identical) nucleotide sequence at certain positions among some DNAs or amino acids in proteins. Consensus is indicative of a common functionally important role such as the TATA box, transcription termination signals, etc. On the basis of the organization of consensus sequences, consensus maps may be constructed indicating phylogenetic relationships. *See* core sequences.

conservation genetics   Concerned with the population genetics principles involved in the maintenance of feral species. The maintenance of the populations depends on inbreeding, outbreeding, effective population size, deleterious mutation, reproductive success, adaptation to captivity, reintroduction, vagility, outbreeding depression, extinction, etc. *See* species, extant. (Frankham, R. 1995. *Annu. Rev. Genet.* 29:305.)

conservative replication   An early idea of DNA replication suggesting that after replication each double-stranded molecule would contain either two old or two new single strands. *See* DNA replication; replication; replication fork; semiconservative replication.

conservative substitution   In a polypeptide chain an amino acid is replaced by another with similar properties. *See* radical amino acid substitution.

conservative transposition   *See* transposition.

conserved sequence   *See* consensus.

consilience   Search for scientific/philosophical principles that are very widely applicable.

consomic   After 10 or more backcrosses of a male to an inbred recipient female strain, all of the chromosomes (0.5^{10}) will belong to the recipient strain—at very high probability—except the Y chromosome. *See* conplastic.

conspecific Strains or varieties belonging to the same species.

constant genome paradigm Visualized the genetic material in a stable form because insertion or transposable elements was not yet known. The fluid genome idea was developed after the discovery of the mobile genetic elements, which are capable of restructuring the genome. *See* transposable elements; transposons.

constitutional It was present in the individual at birth. *See* acquired.

constitutional translocation Between 11q13 and 22q11 in AT-rich regions; presumably the only recurrent translocation besides the Robertsonian translocation. It leads to different types of human hereditary diseases.

constitutive enzyme Functions at a rather constant level in all cells all of the time. *See* housekeeping genes; inducible enzymes.

constitutive gene The rate of its transcription is not subject to the effect of regulator gene(s). *See* housekeeping genes.

constitutive heterochromatin Heterochromatic at all stages and in all cells, indicating that these sequences are never transcribed. *See* euchromatin; heterochromatin.

constitutive mutation Lost its regulatory element(s) and is in "on" position. *See* constitutive gene.

constitutive splicing Of primary RNA transcripts, occurs when the exons are spliced together in a single pattern consistent with their order in the gene. *See* alternative splicing; introns.

constitutive triple response (CTR) Reaction of plants to ethylene, namely retention of the plumular hook, prevention of the geotropic response, and reduction of stem elongation. *See* plant hormones. (Guzman, P. & Ecker, J. R. 1990. *Plant Cell* 2:513.)

constriction, chromosomal The primary constriction (C) is the centromere and the secondary constriction may tie an appendage to the end of the chromosome by a relatively thin stalk. These secondary constrictions are frequently called *satellites* (S) and they are associated with the nucleolus (nucleolar organizer region). The nucleolus contains RNA, and that region of the chromosome was believed not to contain DNA; hence SAT for *sine acido thymonucleico* (without thymonucleic acid); at that time (in the early 1930s), DNA was called thymonucleic acid. *See* centromere; satellite.

Satellites (arrows).

construct Most commonly used for the designation of a specially built plasmid or engineered chromosome.

contact guidance The movement of axons may be guided by the physical environment in the tissue. *See* axon.

contact inhibition Normal animal cells are anchorage dependent in culture and grow in monolayer because of inhibition by neighbor cells. Cancer cells lose the dependence on anchorage and constraints in growth by neighbors. This is also associated with changes in cell morphology. The oncogenic transformant cells thus can pile up in an apparently disorganized manner into tumors. *See* cancer; metastasis; saturation density. (Baba, M., et al. 2001. *Oncogene* 20:2727.)

containment Safe place from which hazardous material, including certain types of biological vectors, cannot presumably escape, and thus can be worked with safely. *See* biohazards; laboratory safety.

context, genetic Contribution or effect of genes outside the locus of primary concern.

contig Set of (partially overlapping) DNA fragments that includes a complete region of the chromosome without gaps. For the determination of contigs, large-capacity vectors are used, e.g., YACs (up to 1–2 megabase), BAC (~150 kb), P1 plasmid (100 kb), and cosmids (40 kb). *See* anchoring; BAC; cosmid; genome project; tiling; YAC. (Hall, D., et al. 2001. *Genetics* 157:1045.)

contiguous gene syndrome Deletions resulting in phenotypes consistent with overlapping functional sequences of more than one gene. Synonymous with segmental aneusomy. *See* Alagille syndrome; Angelman syndrome; Beckwith-Wiedemann syndrome; deletion; DiGeorge syndrome; glycerol kinase deficiency; granulomatous gene syndrome; Lange-Giedion syndrome; McLeod syndrome; Miller-Dieker syndrome; muscular dystrophy; Norrie disease; polyposis adenomatous; Prader-Willi syndrome; retinoblastoma; Wilms' tumor.

contingency table *See* association test.

continuity, genetic Assured by the mitotic nuclear divisions. Equal halves of each chromosome (derived by equational division) are shared between the two daughter chromosomes that are identical to those of the maternal cells (barring mutation, chromosomal breakage, or accidents). *See* mitosis.

continuous trait Displays a range of expression (such as weight, height, etc.) rather than an all-or-none appearance (such as white or red). Continuous traits are usually under polygenic control and subject to substantial environmental influence in expression. *See* continuous variation; polygenes; QTL.

continuous variation A trait shows a range of expression, forming a quantitative series, without sharply separate classes. The majority of such quantitative traits are based on the collective effects of numerous genes (polygenic systems) that may each have only a small effect, but cooperatively they may bring about large variations. Traits subject to continuous variation are characterized by their fitting to the normal distribution or some classes of it. These variations are also affected by environmental factors, and the outcome is a continuous or almost continuous series from relatively small to relatively large phenotypic effects. Continuous variations cannot be classified into discrete categories such as black or white, but they have a continuous spectrum. The characterization is made by counting or measuring or by systems of grading (such as learning scores, disease susceptibility, etc.). The study of continuous variation requires statistical tools such as mean, variance, standard deviation, tests of significance, correlation, regression, heritability, etc. *See* QTL.

contraceptive *See* hormone receptors; immunocontraceptive; menstruation; mifepristone; RU486; sex hormones.

contractile ring Made of actin filaments and positioned in the cell equator; it is instrumental in cutting apart the two daughter cells after the completion of mitosis. *See* cell cycle; cytokinesis; Fts Z ring.

contractile vacuole In ciliated algae near the base of the flagella. *See* flagellum.

contralateral Affects both sides (e.g., both breasts) versus ipsilateral. *See* ipsilateral.

contraselection By selecting for one desirable trait, e.g. high lactation, the high producers may become susceptible to disease or otherwise less vigorous and thus the outcome is negative for the breeder. The selection may affect linked genes or physiological correlation. (See counterselection.)

control (check) Standard to which the experimental data are compared. The standard must be identical and treated identically to the experimental material except for the special condition (genotype, developmental stage, time, chemicals, etc.) being studied. The *negative* control includes all elements except the genetic (or other) condition under investigation. The *positive* control is another similar sample, e.g., another DNA (female DNA extract if Y-chromosomal DNA is to be studied), just to see that the experimental system works. *Blind* control is used to test for possible contamination of the reagents, e.g., all of the extraction and purification steps are taken without the actual sample material. *No-template* control uses all of the DNA amplification reagents except DNA. Use of appropriate controls is indispensable for the objective evaluation of scientific data. *See* concurrent control; historical control; standard.

controlled mating Graphs below show the percentage of homozygosis in successive generations of controlled mating systems and the limit of eventual homozygosity after an infinite number of generations. Controlled indicates that the mating is not random but controlled by some plan. The different mating systems have important consequences for the genotypic composition of the population.

Controlled mating is generally used by plant and animal breeders and in all types of genetic experiments. Controlled mating of humans has been advocated by the discredited negative eugenics movement. The assortative matings based on ethnic, religious, cultural, or other bases within smaller human groups may also be controlled matings with the same biological consequences. Civil and religious laws impose some controlled mating rules involving close relatives. Some state laws prohibit marriage among mentally defective individuals and even include sterilization. Most of the controlled mating laws are no longer enforced in enlightened societies, except for the marriage between close relatives. Controlled mating is very important in animal and plant breeding. Insemination by donor sperm may possibly have some risk, since the paternity is generally not known to the offspring and incest may not be prevented. (Chart below was redrawn after Wright, S. 1921. *Genetics* 6:167.) *See* breeding system; mating system.

controlling elements Historical term for plant transposable elements that can occupy different positions in the plant

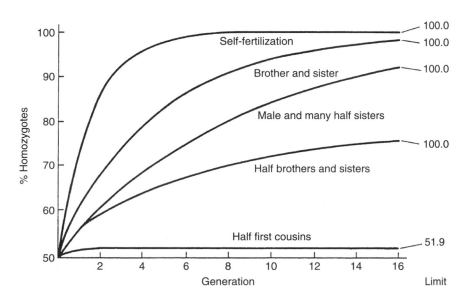

Controlled mating systems.

chromosome and regulate the expression of various genes, causing insertional mutations. *See* Ac-Ds; Dt; insertional mutation; insertion elements; Mu; Spm; TAM; transposable elements.

control region Mitochondrial DNA sequences (~1 kb in vertebrates) devoid of structural genes but containing replicational and transcriptional signals. Aging has been attributed to increased mutation rate in this region of the mtDNA. *See* aging; mitochondria; mitochondrial diseases in humans (for map position); mtDNA. (Howell, N. & Smejkal, C. B. 2000. *Am. J. Hum. Genet.* 66:1589.)

convection Alternative term for nononcogenic transformation of animal cells by exogenous DNA. *See* transfection; transformation.

convergence *See* convergent evolution.

convergent evolution (homoplasy)   Similarity is based not on common ancestry but on adaptive values; i.e., species from different lines of descent assume the forms, structure, and function most valuable for their survival, e.g., sea mammals (whales, dolphins) are more similar in some traits to fishes than to terrestrial mammals. Convergence at single sites may be achieved by mutation or gene conversion. *See* divergence; gene conversion. (Nevo, E. 2001. *Proc. Natl. Acad. Sci. USA* 98:6233.)

conversion **(1)** Hypothetical early step after the first mutational event in carcinogenesis. During this process, the cells become receptive to tumor promotion (propagation of the mutant cell line), and conversion is eventually followed by progression (involving additional stimulatory mutations) and malignancy. *See* cancer; carcinogen; malignant growth; phorbol esters; progression. (Boukamp, P., et al. 1999. *Oncogene* 18:5638.) **(2)** In human–rodent fusion hybrid cell lines, only one human chromosome may be retained; in this monosomic condition, the recessive allele can be expressed. The human chromosome is thus converted to a haploid state.

conversion asci   *See* gene conversion.

Segregation 5:3 rather than 4:4.

convertant   *See* gene conversion.

convertase   Large enzyme complex that activates components of the complement. Several enzymes cleave different components. *See* complement; properdin.

convulsion, benign familial infantile (BFIC1, 19q)   Dominant epileptic-type seizures beginning at age 1 to 3 years. Two additional loci BFIC2 (16p12-q12) and BFIC3 (2q24) have also been identified.

Coolie's anemia   *See* thalassemia.

Coomassie brilliant blue (acid blue, anazolene sodium)   Stain for proteins and a reagent for quantitative protein determination. Intravenal LD_{50} for mice is 450 mg/kg. *See* LD_{50}.

Coombs test   Detects the presence of erythrocyte antigens in blood typing and also autoimmune and hemolytic diseases. It is called antiglobulin test (AGT). The polymorphic serum gammaglobulin complex (IgG) in about 60% of some populations may interfere with the reaction. *See* agammaglobulinemia; gammaglobulin; immunoglobulins.

cooperativity   Requisite for the formation of molecular complexes by one ligand promoting the binding of the following ones. *See* ligand.

coordinate regulation   A group of genes are controlled by common regulatory elements such as is the case in operons and regulons. The regulation may be induction or repression.

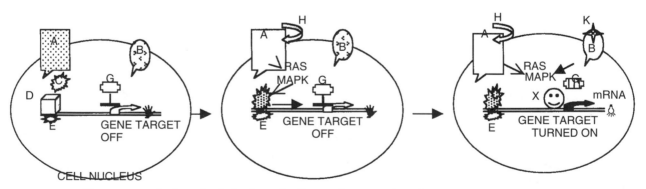

Three stages of a hypothetical scheme leading to the expression of a target gene. The lettered symbols stand for proteins in or entering the nucleus. The double line represents the chromosomal DNA. In the first step, D (cube) and G prevent transcription. In the second step, D is replaced by a modified C, yet that is insufficient for transcription of the target gene. In the last step, through the joint action of proteins, A + H and B + K, a signal transduction path is activated. When complex X, an aggregate of several proteins (smiley face), reaches the chromatin, it dislodges inhibitor G and transcription can proceed ⚥. Sign ✋ indicates blocks.

In the past, coordinated regulation was assumed only for prokaryotes, but correlated expression of adjacent genes has also been detected in yeast, a eukaryote. Developmental processes are based on the cooperation of many genes by both positive and negative control. *See Arabinose* operon; correlated expression; development; homeotic genes; junction of cellular networks; *Lac* operon; morphogenesis; regulon; SL1; SL2. (Finnegan, E. J. 2001. *Trends Genet.* 17:361.)

coorientation   By the end of metaphase, the bivalent chromosomes or homologous chromatids tend to assume positions, so the centromeres face the opposite poles in anticipation of anaphase. *See* meiosis; mitosis; Rabl orientation.

Coorientation (after M. M. Roades).

COP   12-subunit protein complex involved in the regulation of photomorphogenesis of plants. *See* neddylation; photomorphogenesis.

Cope's rule   Hypothesis that evolution usually involves an increase in body size. The idea does not have general validity.

copia   RETROTRANSPOSON elements occur in all types of eukaryotes; in *Drosophila* they come generally in 5–7 kb sizes with about 300 bp direct and much shorter inverted terminal repeats. They frequently occur in 30–40 families and constitute 5–10% of the genome. The copia elements along with about 30 other insertion and transposable elements are involved in the mutability of the genomes by their movement. The frequency of their rearrangement is within the range of 10^{-3} to 10^{-4}. The long terminal repeats (LTR) encode a viral transposase-like protein. Retrotransposition cycles are initiated from an RNA copy of a transposable element. The transcription begins at the 5' region of the long terminal repeat (LTR) and this transcript is copied into a double-stranded DNA by a reverse transcriptase coded within the transposon. The synthesis of the first DNA strand is primed by the 3'-OH group of a host tRNA that immediately anneals to the LTR (this is called tRNA PBS [primer binding site]).

Integration requires short inverted repeats (4–12 bp) at the end of the LTRs. Integration then involves a few bp duplications at the insertional target sites. These duplications are the consequences of the staggered cuts at the target that are filled in by complementary bases after integration. The length of the target-site duplications reflects on the specificity of the transposon-encoded integrase. *Drosophila* retrotransposons have a number of features corresponding to retroviruses. They contain direct terminal repeats that have the sequences required for the initiation of transcription and polyadenylation. The majority of the *Drosophila* retrotransposons carry short inverted repeats at the end of LTRs. The LTR at one terminus contains TG and at the other CA bases. The majority of the retroposons have a purine-rich sequence immediately upstream to the 3'-LTR. These sequences are the priming sites for the second-strand DNA (SSP [second-strand primer]).

Most of the *Drosophila* retrotransposons have sequences starting with TGG immediately downstream to the 5'-LTRs that is complementary to the 3'-end of the tRNAs (the amino acid acceptor arm). The *Drosophila* retrotransposons have reverse transcriptase (*pol*), group-specific antigen (*gag*) polyproteins just like the retroviruses of vertebrates. Actually some plant retrotransposon-like elements are also organized in a similar manner, but the majority of them are no longer able to transpose because some of their genes have been reduced to pseudogenic forms. During retroviral life cycles, extrachromosomal linear and circular DNA elements are formed with one or two LTRs. These features are also retained in some of the copia-like elements. These retroposons produce viral-like elements in the eukaryotic cells quite similar to the real retroviruses. Also, some of the copia-like elements generate "strong-stop" DNA elements in the cells like retroviruses. These strong-stop DNAs and DNA-RNA heteroduplexes are leftovers of the first- and second-strand transcription by the reverse transcriptase enzyme. These copia-like elements transpose quite vigorously in both somatic and germline cells and induce mutations in the germline and the soma.

Recombination between the multiple copies of the retroposons may cause all types of chromosomal aberrations in the host such as deletions, inversions, and translocations. Perhaps the movement (insertion) of these elements causes the majority of the spontaneous mutations in the species that harbor transposable elements. The insertions may cause inactivation of exons, but more frequently they insert into AT-rich sequences. The major representatives in *Drosophila* are *17.6* (7.4 kb) occurring in about 40 copies per genome and generating a 4-base duplication at the insertion target site; the LTR is 512 bp. *297* (7 kb) has approximately 30 copies. Its LTR is 415 bp and shows 1.7 kb homology between the right-hand ends with *17.6*. Its transposition-replication is primed by a tRNA^{Ser}. *412* (7.6 kb), copy number 40, LTR is 481 bp, target site duplication 4 bases. tRNA^{Arg} primes its transposition. *1731* (4.6 kb) is present in about 10 copies and generates a target site replication of 5 bp. Its reverse transcription is probably primed by a fragment of the initiator tRNA^{Met}. *3S18* (6.5 kb) has a target site duplication of 5 bp and occurs in about 15 copies. *BEL* (7.3 kb) is present in about 25 copies and the termini are very similar. *blood* (6 kb) occurs in 9 to 15 copies, LTR 400 bp, and generates a target site duplication of 4. The primer may be tRNA^{Arg} similarly to elements *412* and *mdg1*. *copia* (5 kb) is present in 60 copies, LTR is 276 bp, with 5 bp target site duplication. Its name came from copious amounts of its polyadenylated transcripts in the cells. The initiator tRNA^{Met} primes for the reverse transcription of the copia RNA. Left end of the element: TGTTGGAATA TACTTATTCAA CCTACAAAG TAACGTTAAA; right end: TATTAAAGAAA GGAAATATAA ACAACA. *gypsy* (synonymous with *mdg4*), 7.3 kb, with 10 copy number; LTR is 479 bp and generates 4 bp target-site duplication.

These elements are associated with many mutations suppressed by *su(Hw)* (*suppressor of Hairy wing*). The product of this gene binds to an enhancer-like sequence within *gypsy* and affects the expression of adjacent genes including *gypsy*. The phenotype of some of the mutations caused by *gypsy* insertions are affected by *su(f)* (*suppressor of forked*). The gypsy element is considered to be a retrovirus and its movement is controlled primarily by the X-chromosomal mutation *flam* (flamingo). *H.M.S.Beagle* (7.3 kb) has 50 copies, LTR is 266 bp and shows 4-base target-site duplication. *mdg1* (7.3 kb), may be present in 25 copies, its LTR is 442 bp, and it creates 4 bp target-site duplication; 14/18 of its primer binding sites are identical to those of *412*, as well as the 27 bp adjacent to the left LTRs. tRNA^{Arg} is the most likely primer for their reverse transcription. *mdg3* (5.4 kb) has LTR of 267 kb; the target-site duplications are 4 bp. *micropia* (5.5 kb) hybridizes with *copia* in the Y chromosome of *D. hydei*. *NEB* (5.5 kb), *opus* (8 kb). *roo* (8.7 kb, formerly called *B104*), LTR 429 bp, and generates 5 bp target-site duplication. *springer* (8.8 kb), LTR 405 bp, and generates 6 bp target-site duplication. *See* Cin4; *Drosophila*; hybrid dysgenesis; polyprotein; retroposon; retrotransposon; retroviruses; reverse transcription; suppressor; transposable elements; transposase; tRNA. (Bowen, N. J. & McDonald, J. F. 2001. *Genome Res.* 11:1527; Mejlumian, L., et al. 2002. *Genetics* 160:201.)

copolymer   Molecule built from more than one type of components, e.g., a nucleic acid made of adenine and thymine units. If the sequence and quantity of the components do not follow a particular system, it may be a random copolymer. If the units display periodic repetitions, it is called a repeating copolymer.

copper homeostasis disease   *See* Menke's disease; Wilson disease.

copper-inducible system   *See* metallothionein.

copper malabsorption   *See* Menke's disease.

coprocessor   Auxiliary processor to assist the main processor in special heavy tasks. Generally it also speeds up the processing of computers. *See* processor.

coprolite (fossilized dung)   Reveals the diet and food digestion of extinct species or ancestral populations of modern animals. (<http://www.scirpus.ca/dung.shtml>.)

coprophagy   The nymphs of some insects feed on the feces of adults and utilize the nutrients left in there while ingesting their abundant bacterial symbionts. *See* paratransgenic.

coproporphyria (CPO, 3q11.2)   Dominant deficiency of coproporphinogen oxidase resulting in excessive excretion of coproporphyrin III, an intermediate in porphyrin synthesis. The protein is mitochondrially localized. *See* harderoporphyria; porphyria; porphyrin. (Lamoril, J., et al. 2001. *Am. J. Hum. Genet.* 68:1130.)

coproscopy   Molecular analysis of coprolites. After N-phenacylthiazolium bromide cleavage of protein DNA cross-links, various short DNA sequences can be extracted, amplified by PCR, and sequenced. Information may be obtained on animals that are thousands of years old and their plant food. *See* coprolite; PCR.

COP transport vesicle   COPI is built of ARF, coatomer subunits, and a GTPase and mediates the transport within the Golgi apparatus and between the Golgi and the endoplasmic reticulum. The sorting signals are interpreted by GTPase switch (GTP$\rightleftharpoons$GDP) on the COPI vehicles. The COPII vesicles are made of SEC and Sar proteins. *See* ARF; clathrin; endocytosis; endoplasmic reticulum; Golgi apparatus; p24; SAR; SEC. (Barlowe, C., et al. 1994. *Cell* 77:895; Lederkremer, G. Z., et al. 2001. *Proc. Natl. Acad. USA* 98:10704.)

copulation   Sexual intercourse between female and male. Evolutionary selection of characteristics of the process would indicate that leaving more offspring even at the expense of decreased survival might be favored. Actually, this has been shown by the redback spider (*Latrodectus hasselty*), where hungry females frequently cannibalize their mates that copulate longer and thus transfer more sperm, thereby contributing more of their genes to the population.

copy choice   Hypothesis so named in the 1950s. It assumes that in recombination there may be no physical breakage and union between exchanged strands rather a replication and choice. This hypothesis bears similarity to Bateson's (1906) reduplication theory and gene conversion (Lindegren, 1949). Evidence has now been presented that prokaryotic DNA polymerase III holoenzyme may slip frequently during replicating direct repeats. Recombination between RNA viruses may be brought about by template switching during the elongative RNA synthesis by RNA-dependent RNA polymerases. *See* breakage and reunion; gene conversion; recombination in RNA viruses; reduplication hypothesis. (Kim, M. J. & Kao, C. 2001. *Proc. Natl. Acad. Sci. USA* 98:4972; Moumen, A., et al. 2001. *Nucleic Acids Res.* 29:3814.)

copy number paradox   Genetically functional copies of organellar DNA appear much smaller than the physical copy number estimates. *See* chloroplasts; C value paradox; mtDNA. (Jenuth, J. P., et al. 1996. *Nature Genet.* 14:123.)

copyright   Right to reproduce, publish, or sell an intellectual property. (<http://www.loc.gov/copyright>.)

copy-up mutation   Initiates runaway replication of the plasmid, i.e., the copy number is increasing beyond that in the wild type. *See* runaway plasmids. (Blasina, A., et al. 1996. *Proc. Natl. Acad. Sci. USA* 93:3559; Toukdarian, A. E. & Helinski, D. R. 1998. *Gene* 223:205.)

CoR-box   Amino-terminal ligand-binding domain of the thyroid and retinoic acid receptors. *See* ligand; N-CoR; retinoic acid; thyroid.

cordate   Heart-shaped.

cordycepin ($C_{10}H_{13}N_5O_3$)   Nucleoside analog (3′-deoxy adenosine); inhibits transcription and polyadenylation. *See* polyadenylation signal; polyA mRNA; polyA tail.

core, protein Region common to the majority of structures in a superfamily or in a common fold.

core binding factor (CBF) Heterodimeric transcription factor. Core binding factors are essential for normal development and health. *See* runt; transcription factors.

core DNA Wraps around the histone octamer in the nucleosome core and does not include H1. *See* histones; nucleosome.

DNA (black ribbon) winds around the 8 histones.

core enzyme Prokaryotic transcriptase has two identical α- and two $\beta(\beta\beta')$-subunits, and to this core a fifth, σ-subunit may be attached that is not essential for transcription, but it seeks out the position of proper promoters. *See* RNA polymerase.

core particle *See* core DNA; histones; nucleosome.

core polymerase Composed of only 3/10 subunits of prokaryotic DNA polymerase III (pol III). These subunits are α (polymerase), ε (proofreading $3' \rightarrow 5'$ exonuclease), and θ (enhances the activity of ε). *See* DNA polymerases.

core promoter Most essential sequence within a promoter to carry out transcription. The core promoter of the eukaryotic RNA polymerase II usually includes the TATA box and may also have an *initiator site* (Inr), enhancer, and the TATA box–associated general transcription factors, such as the TFIIB recognition element (BRE) and the downstream promoter element (DPE). When present, DPE is situated 30 nucleotides downstream of the transcription initiation site. Generally the DPE element has Inr but no TATA box. The TAF subunits of TFIID bind DPE and Inr. The core promoter is usually situated ± 40 nucleotides from the transcription initiation site (TATA box). The TATA box is bound by TBP, which is a subunit of transcription factor TFIID. At a distance of -50 to -200 bp of the TATA box, there are binding sites for NF-1, CBF, and NF-Y. In prokaryotes, the core promoter complex includes only the RNA polymerase holoenzyme and the σ-subunit. In eukaryotes, instead of the σ, an assembly of the general transcription factors is found. In addition, the TATA box–binding proteins (TBP), TATA box–associated proteins (TBA), initiator-binding protein (IBP), and enhancer-binding proteins (EBP) are usually present. The modulation and regulation of transcription requires series activators, coactivators, and different specific transcription factors. *See* base promoter; CBF; DSTF; NF-1; NF-X; null promoter; open promoter complex; PWM; TAF; TBP; transcription factors; transcription unit. (Hamada, M., et al. 2001. *Mol. Cell Biol.* 21:6870; Morimoto, M., et al. 2001. *Arterioscler. Thromb. Vasc. Biol.* 21:771.)

core proteome Number of distinct families of proteins within a genome. In *Haemophilus influenzae*, of the 1,709 protein-coding genes, 1,247 ($\sim 73\%$) do not have sequence relatives within the genome. In yeast, there are ~ 6241 protein-coding genes; from these, 4,383 ($\sim 70\%$) represent the core proteome. The core proteomes in *Drosophila* and *Caenorhabditis* are 8,065 ($\sim 62\%$) and 9,453 ($\sim 51\%$), respectively. *See* duplications; gene number; genome; proteome.

core sequence Usually invariable short tract within more or less well-preserved consensus sequence of the DNA. *See* consensus.

coreceptor CD^8 and CD^4 T-cell surface proteins along with the T-cell receptor recognize MHC I and MHC II molecules. The CD cytoplasmic domains associate with a SRC protein tyrosine kinase. *See* CD^4; CD^8; MHC; protein tyrosine kinase; SRC; T-cell receptor.

coremium *See* hypha.

cORF Composite open reading frame occurs by tandem duplication of genes. *See* ORF.

corepressor Metabolite or protein which in combination with a repressor protein interferes with transcription and thereby with enzyme synthesis. The corepressor may not bind to DNA directly. A corepressor when recruited to an activator may turn the latter into a repressor. *See* aporepressor; cAMP receptor; chromatin remodeling; cosuppression; Groucho; N-CoR; nuclear receptor; regulation of gene activity; repression; repressor; Tup1. (Yoh, S. M. & Privalsky, M. L. 2001. *J. Biol. Chem.* 276:16857; Zhang, Q., et al. 2002. *Science* 295:1895.)

coretention analysis Had been used in genetic mapping of mitochondrial genes. More or less large deletions frequently take place in the mitochondrial DNA, resulting in simultaneous loss and retention of antibiotic resistance markers. The codeletion and coretention frequencies provide converse estimates on linkage. Both codeleted and coretained markers must be present in groups, indicating their physical position relative to each other. *See* deletion mapping; linkage; mitochondrial genetics; physical mapping. (Heyting, C. & Menke, H. H. 1979. *Mol. Gen. Genet.* 168:279.)

coriaceous Leathery.

corm Underground shoots, modified for storage.

Corm.

corn *See* maize.

corneal dystrophy Has been described in various autosomal-dominant and -recessive forms involving defects in the keratin filaments in the eye. *See* filaments; keratin.

cornea plana Characteristics are extreme farsightedness (hyperopia) and opacity of the cornea, especially at the margins. It exists in autosomal-dominant and -recessive forms; most common in Finland. *See* eye diseases.

Cornelia de Lange syndrome *See* De Lange syndrome.

corolla Collective term for petals. *See* flower differentiation.

coronary heart disease Generally involves deposition of lipid plaques within the arteries surrounding the heart (atherosclerosis). Low-density lipoprotein in the blood (above 200 mg/100 mL) increases the risk proportionally. On the other hand, high-density lipoprotein is favorable for avoidance. These account for the majority of all heart diseases and afflict 6–7% of the populations (predominantly males) in the Western industrialized countries. The underlying organic defects vary and many nonhereditary factors (diet, smoking, drug and alcohol consumption, age, etc.) and independent or concomitant diseases and conditions (blood pressure, diabetes, temperament, etc.) aggravate the liabilities. In Finnish populations, two likely susceptibility loci were found at 2q21-q22 (LOD score 3.2) and Xq23-q26. *See* cardiovascular disease; cholesterol; cholesterol acetyltransferase deficiency; diabetes; familial hypercholesterolemia; familial hyperlipidemia; homocysteinuria; hyperlipoproteinemia; hypertension; lecithin; lipoprotein; Marfan syndrome; mucopolysaccharidosis; pseudoxanthoma elasticum; Tangier disease. (Breslow, J. L. 2000. *Annu. Rev. Genet.* 34:233.)

corpuscular radiation Emitted by unstable radioisotopes (β particles [electrons] by ^{3}H, ^{14}C, ^{32}P, etc., and α-particles [helium nuclei]) by uranium or fast and thermal neutrons during nuclear fission. *See* electromagnetic radiation; ionizing radiation; isotopes; physical mutagens; radiation effects; radiation hazard assessment; radiation measurement.

corpus luteum (yellow body) Formed by luteinization of an ovarian follicle after the discharge of an ovum. In case of fertilization of the egg, the corpus luteum increases in size and persists for several months. If there is no fertilization, the CL disintegrates. The CL secretes progesterone. *See* animal hormones; corticotropin; Graafian follicle; luteinization; luteinizing hormone-release factor; ovary; ovum; progesterone; relaxin.

correction *See* DNA repair.

correlated expression In the eukaryote yeast, adjacent genes, irrespective of their orientation in the chromosome, are more likely to be expressed at the same time than nonadjacent ones. Only $r > 0.7$ was accepted as valid indication of correlated expression. The significance of correlated adjacent pairs was determined statistically on the basis of the cumulative binomial distribution by using the formula shown below.

$$P(n \geq n_0) = \sum_{n=n_0}^{N} p^n (1-p)^{N-n} \left[\frac{N!}{n!(N-n)!} \right]$$

In some cases, only one of the gene pairs carries an upstream activating sequence. Some genes utilize the same UAS even when they are not immediately downstream from it. On this basis, chromosome correlation maps can be constructed. *See* coordinate regulation; operon; regulon; UAS. (Cohen, B. A., et al. 2000. *Nature Genet.* 26:183.)

correlation Interdependence of two variates. This relation may be statistical or physiological. The statistical correlation does not necessarily reveal any cause and effect link. The correlation may be positive (when the change of the variables follows the same direction) or negative (when the increase of one variable involves the decrease of the other). Thus, the value of the correlation coefficient may vary between +1 and −1.

The correlation coefficient is independent from the scale of quantitations used, e.g., one can measure the correlation between intelligence and wearing a necktie. For the calculation of the correlation coefficient, *covariance* has to be determined, i.e., the average product of the deviations of two variables from their respective means. It is estimated by dividing the sum of the products of the deviations from their means by the appropriate degrees of freedom. Thus, **covariance** is: $(\mathbf{w}) = \{\Sigma[(x_i - \overline{x})(y_i - \overline{y})]\}/(n-1)$. For actual calculation, the mathematically equivalent but computationally more convenient equation is used: $(\mathbf{w}) = \{\Sigma(X_i Y_i) - [\Sigma(X_i)\Sigma(Y_i)]/n\}/(n-1)$.

The analysis of covariance has many uses in biology and particularly in genetics. It may be used to separate the genotypic effects from treatment effects. It may reveal the relations among various types of variables. This type of analysis is useful to study the relationships among multiple

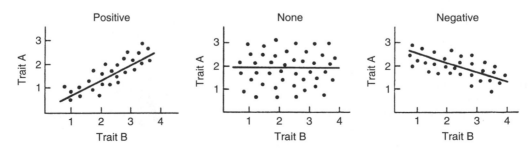

Graphical representation of correlation. the dots correspond to the points of the measured values of two quantitative traits.

classifications such as may occur if the experiments involve organisms of different genotypes, in different age groups and developmental stages and environments. In this book it is not possible to work out examples for all these different applications; only the basic procedure is illustrated with step-by-step simple calculations.

The actual use of the formulas can best be shown by a hypothetical example. Let us assume that we measured (i) number of variates (in the following case i is 1 to 10) in two groups. In the columns x_i and y_i, these measurements are listed from 1 to 10. To make the calculations easier (without changing the outcome) for each measurement, a quantity (the same quantity, close to the means) is subtracted, and we name these X_i and Y_i, respectively, as shown at the top of columns 3 and 4. Column 5 is the product of the lines in columns 3 and 4: X_iY_i. Columns 6 and 7 display the power of values in columns 3 and 4, respectively: $(X_i)^2$ and $(Y_i)^2$.

Computation scheme for covariance

(1) x_i	(2) y_i	(3) $X_i = x_i - 150$	(4) $Y_i = y_i - 150$	(5) X_iY_i	(6) $(X_i)^2$	(7) $(Y_i)^2$
148	149	−2	−1	2	4	1
158	152	+8	+2	16	64	4
150	155	0	+5	0	0	25
143	142	−7	−8	56	49	64
162	160	+8	+10	80	64	100
150	160	0	+10	0	0	100
156	153	+6	+3	18	36	9
160	159	+10	+9	90	100	81
153	158	+3	+8	24	9	64
150	152	0	+2	0	0	4

$\bar{x} = 153$ $\bar{y} = 154$ $\Sigma X_i = +26$ $\Sigma Y_i = +40$ $\Sigma X_iY_i = 286$ $\Sigma(X_i)^2 = 326$
$\Sigma(Y_i)^2 = 452$

Substituting the values into the covariance formula:
$\mathbf{W} = \{\Sigma(X_iY_i) - [\Sigma(X_i)(\Sigma(X_i)]/n\}/(n-1) = \{286 - [(26) \times (40)]/10\}/9 = \mathbf{20.22}$

Variances:
$\mathbf{V_x} = \{\Sigma(X_i)^2 - [(\Sigma X_i)^2]/n\}/(n-1) = \{(326) - [(26)^2]/10\}/9 = \mathbf{+28.71}$

$\mathbf{V_y} = \{\Sigma(Y_i)^2 - [(\Sigma Y_i)^2]/n\}/(n-1) = \{(452) - [(40)^2]/10\}/9 = \mathbf{+32.44}$
The **coefficient of correlation**: $\mathbf{r} = W/\sqrt{V_xV_y} = 20.22/\sqrt{28.71 \times 32.44} = \mathbf{+0.663}$

Indicating that an increase in the values of X involved a positive in the values of Y as shown at the top of columns 3 and 4. Column 5 is the product of the lines in columns 3 and 4, X_iY_i. Columns 6 and 7 display the power of values in Columns 3 and 4, respectively: $(X_i)^2$ and $(Y_i)^2$.

In genetic analysis the coefficient of regression is often used. It measures in quantitative units how much the dependent variable (Y) is changing as a function of the independent variable (X), e.g., we can determine how much the offspring's weight or height regresses to the weight (in kg) or height (in cm) of the parents or to the mother or father. This is in contrast to correlation that could state only plus or minus and strong or weak correspondence but not in actual quantitative units. The **coefficient of regression**: $\mathbf{b} = W/V_x =$ and in the example of the table $\mathbf{b} = 20.22/28.71 = 0.704$. For a predictive value, we use the linear regression equation: $Y = a + bx$ from which $a = Y - bx$, where (a) is the intercept of the straight line on the (Y) coordinate; (b) is the slope indicating how much (y) changes by changes in (x). After substituting the data of our calculations into the equation:

$a = Y - bx$, since $(\bar{y} = Y = 154)$, and $\bar{x} = X = 153$, and $b = 0.704$; thus, $a = 154 - (0.704 \times 153) = 46.29$. Therefore, if the ($x$) independent variable is 158 kg, the dependent variable (Y) is expected to be $46.29 + (0.704 \times 158) = 157.52$ kg. When the independent variable is 150 kg, the dependent variable (Y) is expected to be $46.29 + (0.704 \times 150) = 151.89$ kg. The example also testifies for the name of regression. Originally it was observed that large and small parents' children both tend to follow more the population's mean, i.e., they regress toward the mean. The offspring–parent regression is actually a measure of heritability. The linear regression is frequently represented graphically as shown above. *See* genetic correlation; heritability; intraclass correlation.

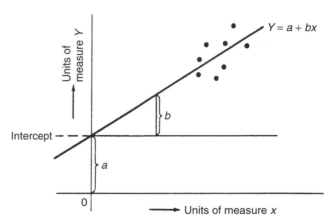

The linear regression line is specified by the equation: $Y = a + bx$. In this function, a is the value of the intercept, b defines the slope of the line, equal to the increase in Y per units of x. The solid circles correspond to the intersections of the values of the two variables. According to the table shown, we can consider the x axis as the units of measures of one of the variates and Y as the axis for the measurements of the other variates.

correspondence analysis Computational method for the study of associations between variables. It displays low-dimensional projection of the data into a plane for both variables simultaneously to reveal their association. It is applicable to microarray data of different complexities. *See* microarray hybridization. (Fellenberg, K., et al. 2001. *Proc. Natl. Acad. Sci. USA* 98:10781.)

cortex Outer layer of various tissues (egg, brain, kidney, tree bark, etc.).

cortical granule Small secretory vesicle under the egg membrane. By releasing some of its contents, the egg is protected from being fertilized by multiple sperms. *See* fertilization; polyspermy.

cortical inheritance Cytoplasmic organization may affect the expression of a trait without alteration in the genetic material. In *Paramecia*, the movement of the cortically located cilia may be oriented in the same direction. If a piece of the cortical cytoplasm is grafted in the reverse orientation, the graft will beat in the opposite direction, and this ciliary movement pattern may be transmitted through generations. Similarly grafted vestibules may appear as an additional ingestatory apparatus in the progeny. These cortical layers contain no DNA or RNA. The pattern of development seems to be fixed within the organizational structure. *See* epigenesis; non-Mendelian inheritance; illustration on p. 268. (Beisson, J. & Sonneborn, T. M. 1965. *Proc. Natl. Acd. Sci. USA* 53:275.)

Gullet and vestibule (→) of *Paramecium aurelia* that can be surgically transferred to another site on the animal and it is then cortically inherited. (Courtesy of Dr. Tracy Sonneborn.)

cortical reaction   *See* cortical granules; fertilization.

corticobasal degeneration   Fronto-temporal dementia generally with parkinsonism caused by mutation in the tau protein. *See* parkinsonism; tau.

corticosteroid   21-carbon steroids synthesized in the outer firm yellowish layer of the adrenal (kidney) gland. The glucocorticoids regulate carbohydrate and protein metabolism, whereas the mineralocorticoids regulate salt and water traffic. *See* animal hormones; antitrypsin; glucocorticoids; transcortin deficiency.

corticotropin (adrenocorticotropin, ACTH) 39-amino-acid peptide hormone of the anterior pituitary; regulates corticosteroid synthesis in the adrenal cortex. Its release — in response to stress, e.g. — is controlled by the releasing factors of the hypothalamus such as thyrotropic-releasing factor (TRF), luteinizing hormone–releasing factor (LRF), and corticotropin-releasing factor. *See* corticotropin releasing factor; opiocortin; stress. (Lin, X., et al. 2001. *Mol. Endocrinol.* 15:1264.)

corticotropin releasing factor (CRF/CRH)   It is supposed to be associated with the brain cognitive response. In Alzheimer's disease, CRF is very low, although CRF receptors accumulate. CRF mediates endocrine, autonomous, behavioral, and immune responses to stress (alcohol and drug withdrawal, etc.). *See* Alzheimer disease; maternal tolerance; stress; urocortin. (Eckart, K., et al. 2001. *Proc. Natl. Acad. Sci. USA* 98:11142.)

cortisol   Derived from progesterone (along with aldosterone) in the kidney cortex. It is chemically very similar to *cortisone*. The production of cortisone is controlled by the pituitary hormone corticotropin and cAMP. The glucocorticoids promote gluconeogenesis and the deposition of glycogen in the liver, inhibit protein synthesis in the muscles, mediate fat and fatty acid breakdown in the adipose tissue, and control inflammatory responses. *See* glucocorticoids. (Johannson, A., et al. 2001. *J. Clin. Endocrinol. Metab.* 86:4276.)

cortisone   *See* cortisol.

Corynebacterium   *See* biological weapons; diphtheria toxin.

cos   Cohesive ends (12 bp) of phage λ where linearization and circularization of the DNA take place:

pGGGCGGCGACCT−−−−−

−−−−−CCCGCCGCTGGAp

See lambda phage. (Wieczorek, D. J. & Feiss, M. 2001. *Genetics* 158:495.)

COS cell (cell, origin, Simian)   African green monkey cells with a chromosomally inserted simian virus (SV40) DNA, which is defective in the origin of replication but contains the intact T antigen. COS cells thus replicate multiple copies (small circles) of the harmless viral vector and the inserted passenger DNA independently from the monkey cell chromosome. The foreign DNA in the SV40 vector is transcribed and translated into the appropriate protein with the assistance of the metabolic machinery of the cell. *See* African green monkey; SV40; diagram below. (Gerard, R. D. & Guzman, Y. 1985. *Mol. Cell Biol.* 5:3231.)

cosexual   *See* hermaphrodite.

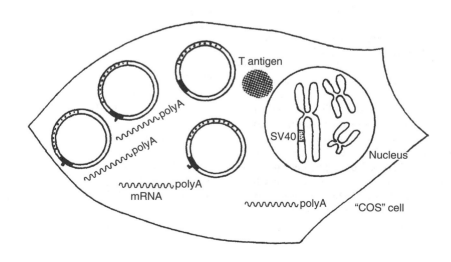

cosegregation Analysis for genetic linkage between two different genetic markers, e.g., a visible mutant phenotype and antibiotic resistance.

CO₂ sensitivity Of some strains of *Drosophila*, is manifested as paralysis and death after anesthesia with the gas. The condition is caused by infection with rhabdovirus sigma. This RNA virus resembles the vesicular stomatitis virus (VSV) of horses (an acute febrile [fever-causing] infection on the tongues, mouth membranes, and lips) and some fish viruses (PFR, SVC), which can also elicit carbon dioxide sensitivity in *Drosophila*, but the fly virus cannot infect vertebrates. In the nonstabilized state, only the *Drosophila* females transmit the virus. In the stabilized state, the transmission by eggs is 100%; it is also transmitted by the sperm, although stabilized infection does not ensue. The *ref* mutants (in chromosomes X, 2, 3) are refractory to this type of infection. *See Drosophila*; *Rhabdoviridae*. (L'Héritier, P. 1948. *Heredity* 2:325.)

cosmic radiation Strikes the earth from interstellar space, partly directly from the radiating bodies, partly (primary) emanating from the collision of nuclei with cosmic (secondary) radiation in the upper atmosphere. The amount of cosmic radiation varies according to altitude; it is very high at the height of the space vehicles. The amount on the surface of the earth is about 0.028 to 0.045 rad per year and increases substantially at higher altitudes. At 1,000 ft, it is ~2, at 5,000 ft >20, and at 9,000 ft ~70 mrem. From food and water, on the average in the United States, one may ingest ~40 mrems. These figures should be corrected by a factor of about 0.8 because of protection by housing. For comparison, an average medical diagnostic chest X-ray delivers about 0.2 rad (but the gonads are protected). *See* atomic radiation; isotopes; rad; radiation, natural; radiation hazard assessment; rem. (Curzio, G., et al. 2001. *Rad. Prot. Dosimetry* 93[2]:125.)

cosmid Approximately 5 kb cloning vectors are derived from phage-λ DNA, containing one or two cohesive (*cos*) sites in the same orientation, an origin of replication (*ColE1*), and a selectable marker (*amp^R*). The cloning capacity of the plasmid is about 35–47 kbp because nothing substantially smaller or larger than the λ genome (≈49 kbp) can be packaged into the capsid (*see* in vitro packaging). For cloning smaller DNAs, *charomids* must be used. Into the linearized cosmid the foreign DNA is ligated at appropriate cloning sites flanked by cosmid sequences. With the assistance of the phage terminase protein (an endonuclease with a specificity of cutting near *cos* sites), the *concatemer* is cut to head size to fill the capsid. After infection of *E. coli*, the cosmid recircularizes within the bacterial cell, and due to the presence of the antibiotic resistance marker(s), the bacteria carrying the recombinant cosmid can be isolated.

Cosmids can recombine with other plasmids (having homologous sequences) within *recA⁺* bacterial strains, and the cointegrates can be isolated if both carry independent selectable markers. There may be a problem, however, with rearrangements because of the active *recA*. A number of specially designed cosmids are available and they have been used for particular cloning needs. Cosmid **pJBS** has been used

successfully for chromosome walking. After the recombinant cosmid is isolated from a library, the cloned foreign DNA segment can be cut up with restriction enzymes without destroying the vector, which can be transformed into bacteria, yielding small fragments that can be used as probes for overlapping fragments in the original library. Cosmid **c2RB** carries two *cos* sites and ampicillin and kanamycin resistance. Concatenation can be suppressed after insertion of the foreign DNA by phosphatase treatment (prevents the formation of phosphodiester bonds) or by directional cloning. There are two EcoRI sites in the vector flanking a single BamHI site, making it easier to walk from one recombinant site to the next because the two small generated fragments can be used as probes to rescreen the original library for overlapping clones. Cosmid pcos1EMBL carries tetracycline and kanamycin resistance, and the origin of replication of plasmid R6K (a replicator unrelated to Col1). With this plasmid, the screening for recombinants can be carried out in vivo.

Special cosmid vectors have been developed for the transformation of animal cells. These carry antibiotic markers selectable in eukaryotic cells such as neomycin (*neo*), dihyrofolate (*dhfr*), hygromycin (*hph*), and other eukaryotic genes such as hypoxanthine-guanine phosphoribosyl transferase (*hgprt*), thymine kinase (*tk-*), etc. The **pWE** series was used advantageously because of the ease of walking from one cosmid clone to another. They carry the phage T3 and T7 promoters on either side of NotI, an eight-base recognition site (GC↓ GGCCGC) restriction endonuclease. The foreign DNA may be cloned into a BamHI site located between the phage promoters. After cloning, the cosmid is cut by restriction enzymes that do not affect the phage promoters. The cleavage products contain small fragments adjacent to the promoters. The fragments downstream of the promoters are then transcribed into labeled RNA. These (radioactive) probes are used to rescreen the library for overlapping fragments. Since NotI sites are rare in mammalian DNA, on the average about 1500 bp fragments are generated; frequently, the NotI digests include the entire cloned fragment. These large fragments facilitate the construction of physical maps and may be useful in transformation because they may include the gene that is to be expressed. Some other pWE vectors may have polycloning sites rather than only the BamHI site. This may make the cloning easier but may create problems in the hybridization probes because of background hybridization of the labeled RNAs. Cosmids are also used with plant DNA. *See* charomids; cosmid library; directional cloning; lambda phage; library; promoter, downstream; Rec; vectors. (Sambrook, K. J. & Russel, D. 2001. *Molecular Cloning*. Cold Spring Harbor Lab. Press.)

cosmid library Collection of DNA fragments of a genome cloned in cosmid vectors. *See* cosmids; library.

cosmid mapping Physical mapping of cosmid-contained DNA sequences. *See* cosmid; physical map.

cosmid walking Applies the chromosome walking procedure to DNA sequences in a cosmid library. *See* genome projects.

cospeciation Indicates joint evolution of two organisms, e.g., host and parasite. The measure of cospeciation is that the two phylogenies (on the basis of protein or DNA primary structure) are more similar than expected by chance. *See* phylogeny; speciation.

Costello syndrome Autosomal-recessive/dominant short stature, skin, hair anomalies, and reduced mental abilities. A deficiency of 67 kDa elastin-binding protein may be one of the basic problems. *See* Marfan syndrome. (Hinek, A., et al. 2000. *Am. J. Hum. Genet.* 66:859.)

costimulator Cell surface proteins that encourage the interaction between T-cell receptors and the antigen-presenting cell. Genes encoding such proteins may be transformed into cancer cells and interleukins, and other cytokines or cytokine-like proteins may reinforce their function. Painting on the tumor cells' costimulatory proteins is simpler than transfection yet effective in vaccination. The effectiveness may be enhanced by the use of GPI anchors. *See* CD4; CD28; CD80; CD86; CTLA-4; GPI anchor; transfection; tumor vaccination. (Schwartz, J.-C. D., et al. 2002. *Nature Immunol.* 3:427.)

cost of evolution Evolution proceeds by replacing old alleles with new ones; as a consequence, some individuals are sacrificed; these pay the cost of evolution according to Haldane. The proportion of eliminated zygotes $= sq^2$ and the proportion of survivors $= \overline{w} = 1 - sq^2$, or in general terms, the *cost of natural selection*, $C = \int_0^t (sq^2/\overline{w})dt$, where $s = $ selection coefficient $[\ell n(p_t) - (\ell n(p_0)]$, $q = $ allelic proportion, $\overline{w} = $ fitness, $t = $ time, $dt = $ integral differentiation. *See* allelic frequencies; evolution; fitness; natural selection; selection coefficient.

cosuppression When a gene is introduced by transformation into a cell, neither the resident nor the transgene copy of the same gene is expressed (repeat-induced gene silencing); or increasing the gene copy number reduces the degree of expression. Interestingly, in higher plants nontransgenic duplication does not result in cosuppression. The cosuppression may be complete or incomplete, and it may be reversible during development. It may result in sector formation. Cosuppression was discovered in plants but may occur in other organisms, too. The cause of the phenomenon may be methylation of the genes, posttranscriptional degradation of the mRNA, special allelic interaction, or antisense RNA. The mechanism might have developed to suppress viral propagation or suppress transposable elements. In *Neurospora*, the *qde-3*–encoded RecQ family DNA helicase is a posttranscriptional silencer. *See* antisense technology; Bloom syndrome; corepressor; epigene suppression; host-pathogen relationship; imprinting; methylation; MSUD; paramutation; posttranscriptional gene silencing; quelling; QTL; repetitive DNA; RIP; RNAi; silencer; transgene; transvection; Werner syndrome. (De Buck, S., et al. 2001. *Plant Mol. Biol.* 46:433.)

cotransduction More than a single gene is transferred by transducing phage if the genes are closely linked. *See* bacterial recombination; transduction.

cotransfection Transformation (genetic) by two or more genetically linked genes simultaneously. In the calcium phosphate precipitated donor DNA granules, nonlinked DNA molecules may also be included and integrated into the animal cell simultaneously. *See* calcium phosphate precipitation; transformation, genetic.

cotransformation Donor DNA molecules containing closely linked genes may be taken up and integrated simultaneously into the bacterial chromosome or into the chromosomes of other organisms. *See* cotransfection; transformation mapping.

cotranslational transport Proteins synthesized on the ribosomes associated with the endoplasmic reticulum (rough endoplasmic reticulum) pass into the lumen of the endoplasmic reticulum during the process of translation. *See* endoplasmic reticulum; protein synthesis; signal sequence. (Wilkinson, B. M., et al. 2000. *J. Biol. Chem.* 275:521.)

cotransport Two solutes passed through a membrane simultaneously by a transporter.

cotton *See Gossypium.*

co-twin Pair of twins. *See* twinning; zygosis.

cotyledon Seed leaf (an actually imprecise term because the seed before emergence may already have the initials of several leaves). At emergence, the dicotyledonous plants have two cotyledons, which have a nutritive storage role before and for a while after germination. The monocots have only a single very small cotyledon that frequently has a digestive role only. In dicots, occasionally three or more cotyledons are formed as a developmental anomaly, but tricotyledony is only exceptionally inherited. In zoology, the cotyledons mean the tufts or subdivisions on the placental surface of the uterus.

coulomb (C) Unit of radiation measurement; $1 \text{ C/kg} = 1 \text{ R} \times (2.58 \times 10^{-4})$. *See* R.

coumarin Plant metabolic product arising from 4-hydroxycinnamate through oxidation. It gives rise to dicoumarol, a vitamin K antagonist. These compounds are abundant in some varieties of sweet clover (*Melilotus*), and if the coumarin content is high, the forage value is much reduced because the animals may not eat it due to the bitter taste. After fermentation (silage), dicoumarol is formed, which is very toxic due to its anti–vitamin K and hemorrhagic effects. Plant breeding efforts have successfully reduced the coumarin content of some *Melilotus albus* varieties. *See* coumarin-like drug resistance; prothrombin deficiency; vitamin K.

coumarin-like drug resistance Due to an autosomal-dominant gene, some individuals are resistant to coumarin-like compounds such as warfarin, bishydroxycoumarin, and phenindione (anticoagulants). These compounds are therapeutic agents used in some surgeries and in treatment for thromboembolic (blood-clot-forming diseases blocking arteries) diseases. *See* antihemophilic factors; blood clotting; warfarin.

counseling, genetic Human geneticists can make predictions about the probability of recurrence risks in families affected by hereditary conditions and diseases such as birth defects, metabolic disorders, developmental problems, neurological or behavioral abnormalities, fertility problems, exposure to hazardous environment, consanguinity, problems associated with pregnancy beyond age 35, etc. These predictions are based on family histories, cytological, biochemical, and molecular analyses.

The completion of the physical mapping of the human genome will greatly improve the potentials of accurate genetic counseling by the availability of appropriate probes and nucleotide sequence information. Genetic counseling should be sought before marriage or procreation in families where risk is indicated. Participation is voluntary. The analyses may involve the prospective parents or the fetus may also be examined by amniocentesis. This procedure surgically withdraws with a syringe amniotic fluids 6–20 weeks following conception and analyzes the sloughed-off, floating fetal cells by various methods. Under some circumstances of very high risks, termination of pregnancy may be opted for where there are no moral and/or legal objections. Only the family within the limits of the law can make the decision. The physicians or geneticists provide all the relevant, confirmed facts but do not recommend the action(s) to be taken. Sometimes the counselor may, however, face the dilemma of whether to reveal nonpaternity if this information does not entail health-related problems or the diagnosis might jeopardize the emotional status of the counseled.

From the viewpoint of genetics, selective abortion raises problems. If the families compensate for the aborted fetuses with new pregnancies, the frequency of the deleterious genes will rise in the population because the carriers may transmit the defective alleles, so the problems are only postponed. If the carriers of serious hereditary defects refrain from reproduction, the frequency of these genes is supposed to decline eventually. The genetic counselor is a physician with thorough training in medicine and genetics involving cytogenetic, molecular, and statistical aspects of the field, and is expected to be familiar with all the relevant techniques involved. Very often, because of small family size, penetrance, expressivity problems, variability of the syndromes, and difficulties in obtaining candid information of family histories, especially about hereditary diseases with social stigmas, even the best qualified counselor may encounter difficulties. Besides being a good geneticist and an experienced physician, he/she must have sufficient background in psychology and ethics.

The genetic and medical facts must be explained in terms easily understandable by the families involved whose level of education may be quite variable from case to case. The counseling may be of limited effectiveness because of language problems or inadequate biological/genetical education of the counselee. Counseling is not supposed to follow any eugenic goals and is expected to be nondirective. The clients must be aware that genetic counseling may prevent family traumas and may offer relief from nagging anxieties in cases when the recurrence risks are low, e.g., in Down syndrome and some other aneuploidies. Also, some anomalies may be phenocopies and the genetics risks are practically nil. Since 1981, the American Board of Medical Genetics or the American Board of Genetic Counseling certifies genetic counselors. *See* amniocentesis; ART; bioethics; carrier; confidentiality; empirical risk; gene therapy; genetic counseling; genetic privacy; genetic risk; genetic screening; informed consent; OMIM; prenatal diagnosis; psychotherapy; recurrence risk; risk; selective abortion; supportive counseling; utility index, genetic counseling; wrongful life. (Mahowald, M. B., et al. 1998. *Annu. Rev. Genet.* 32:547; Weil J. 2002. *EMBO Rep.* 3:590.) (<http://www.genetests.org/>, <http://www.geneclinics.org/>.)

countercurrent distribution   Partitioning relatively small molecules between two liquids that are polar to a different extent. Both liquids are moved in a special apparatus in an opposite direction between steps of equilibration.

counter-ion condensation   Association of ions with the polyelectrolyte DNA; in B DNA, there is about 1 anionic charge per 1.7 Å distance. In case the line charge spacing changes, the DNA may become unstable. *See* DNA types. (Fahey, R. C., et al. 1991. *Int. J. Radiat. Biol.* 59:885.)

count-location model   Considers that chiasmata are distributed uniformly and independently along the length of the chromosome. *See* chiasma; mapping functions; stationary renewal process. (Browning, S. 2000. *Genetics* 155:1955.)

counterselection   In a hybrid population the two parental strains (each resistant e.g. for a different antibiotic) may not survive in a medium containing both of the antibiotics but the recombinants endowed with both of the resistance genes will survive. Thus there is a counterselection for the parental cells or strains. (See contraselection.)

count per minute   *See* cpm.

coupled reaction   The energy released by one reaction is utilized by the next reaction.

coupling phase   Two or more recessive (or dominant) alleles are in the same member of a bivalent chromosome (e.g.AB or $\underline{ab}$). Some geneticists call this arrangement *cis*. *See* bivalent; recombination; repulsion.

courtship in *Drosophila*   Extensively studied behavioral trait. Many genes are apparently involved, but all appear to be pleiotropic, and the mutations recovered affect more than one unrelated function. Some are affected in sex determination; others are visual or olfactory mutants or affect female receptivity or male fertility, or circadian rhythm; others are involved in the courtship song of the flies (generated by the vibration of the wings). A better understanding of neuronal function and the availability of selection techniques will promote progress in this field. Courtship, as well as other behavioral traits, can now be subjected to genetic and neurobiological analyses. *See* behavior genetics; pheromone. (Greenspan, R. J. & Ferveur, J.-F. 2000. *Annu. Rev. Genet.* 34:205.)

cousin   The child of the siblings of an individual's father or mother is a first cousin. The child of a first cousin is a second cousin or first cousin once removed. The cousins can be twice . . . tenth, and so on removed, depending on the steps in the relationship. *See* cousin, german.

cousin-german   First cousin. *See* german.

cousin marriage May increase the chances of defective offspring depending on the coefficient of inbreeding. Infant death rate during the first year of life among children of first cousins approximately doubles relative to that in the general population. (*See* coefficient of inbreeding; controlled mating). Between 1959 and 1960, 0.08% of Roman Catholic marriages were inter-first cousins, and among American Mormons from 1920 to 1949, it was 0.61%. In India, the first-cousin marriages in some societies was as high as 30% in the recent past. *See* coefficient of coancestry; genetic load; inbreeding coefficient; incest. (Fraser, F. C. & Biddle, C. J. 1976. *Am. J. Hum. Genet.* 28:522.)

covalent bond Chemical linkage through shared electron pairs, e.g.:

$$H\cdot + \cdot H \longrightarrow H{:}H \quad 3H\cdot + \cdot\overset{\cdot\cdot}{N}{:} \longrightarrow H{:}\overset{H}{\underset{H}{\overset{\cdot\cdot}{N}}}{:} \quad 4H\cdot + \cdot\overset{\cdot}{C}\cdot \longrightarrow H{:}\overset{H}{\underset{H}{C}}{:}H$$

covalently closed circle Circular macromolecule (plasmid) in which all the building blocks are covalently linked and thus there are no open ends.

covariance *See* correlation.

covarion theory Interprets evolution on the basis of varying codons in a lineage-specific manner among the sites. *See* codon; evolution. (Penny, D., et al. 2001. *J. Mol. Evol.* 53:711.)

coverslip Usually 0.13–0.25-mm-thick glass (plastic) covers on microscope slides for flattening of specimens and/or protecting during examination by temporary or permanent seal.

Cowden syndrome Multiple hamartomas. *See* multiple hamartoma syndrome.

cowpea (*Vigna unguiculata*) Food and fodder crop primarily of the tropics and subtropics. All the 170 species have $2n = 2x = 22$ chromosomes.

cowpea mosaic virus Its genome contains B-RNA and M-RNA. The former codes for Vpg in a polyprotein complex at the carboxy terminus. *See* Vpg.

COX Symbol for cytochrome oxidase and cyclooxygenase proteins and genes. Mitochondrial DNA encodes 3 subunits and the nuclear DNA codes for 10. *See* cyclooxygenase; cytochromes.

Coxsackie viruses (Picornaviridae) These viruses have ~7,500-base single-stranded RNA genetic material of 22–30–diameter particles. Humans, monkeys, and suckling mice are susceptible to them and they may cause damage to the nervous system, respiratory and alimentary tracts, muscles, internal organs, and may even induce autoimmune disease. Coxsackie B4 viral infection is associated with insulin-dependent diabetes and shares similarity with Langerhans islet autoantigen glutamic acid decarboxylase. *See* CAR1; diabetes mellitus; Langerhans islet; picornaviruses. (Wan, Y. Y., et al. 2000. *Proc. Natl. Acad. Sci. USA* 97:13784.)

coyote *Canis latrans* ($2n = 78$) North American canine; can interbreed with domestic dogs ($2n = 78$) and the wolf ($2n = 78$).

C1p (C*l*p) Member of the Hsp100 ATPase family. *See* HSP; ssrA tag.

CP complex Nuclear dimer of Cdc68 and Pob3 proteins; regulates gene expression (activation and repression) by interacting with the promoter of budding yeast and modifying chromatin structure, respectively.

CP1 (1) Mammalian transcription factor binding to the CCAAT box. *See* CAAT box; CDP. (2) (Cbf1/Cpf1). Centromere-binding ([A/G]TCAC[A/G]TG) yeast protein; it is part of the kinetochore. The protein may affect chromosome segregation and methionine synthesis. *See* kinetochore.

C11p11 DNA probe used for the identification of certain cancer sequences.

CPB (1) *See* cetylpiridinium bromide. (2) mRNA cap-binding protein. *See* eIF; translation initiation.

CPD Cross-linked adjacent pyrimidine dimers, photoproducts. *See* pyrimidine dimers.

cpDNA (chloroplast DNA) Its size varies generally between 8 and 13×10^7 Da; it is a circular double-stranded molecule with 20 to 40 copies per chloroplast of higher plants and 80 copies in the *Chlamydomonas reinhardtii* alga. *See* *Chlamydomonas*; chloroplast genetics; chloroplasts.

C period Full cycle of the replication of DNA measured in time units.

CpG motifs (cytosine-guanine) of bacteria; in an un-methylated state, these motifs induce murine B cells to proliferate and secrete antibodies. In the mammalian genomes, these doublets are most commonly methylated. The methylated cytosine may be transmitted epigenetically through the germline. *See* B cell; epigenesis; methylation of DNA.

CpG islands Regions of 500 bp or less, or more in 5′ upstream of genes with these doublets. Their function is regulatory. The doublets in these islands (especially in housekeeping genes) are protected from methylation and are instrumental in the transcription of about half of the mammalian genes. The CpG islands of imprinted and lyonizing genes may be methylated and silenced. Methylation may increase by aging and in vitro culture. The 5′-CpG role in silencing is mediated by the methyl-CpG–binding protein (MeCP) that interacts with a histone deacetylase complex. In cancer cells, the tumor suppressor genes appear to be methylated. Some cancer cells contain hypermethylated

CpG islands. The hypermethylation may silence the tumor suppressor genes. *See* histone deacetylase; housekeeping genes; imprinting; isochores; lyonization; methylation of DNA; trichostatin; trinucleotide repeats; tumor suppressor. (Jones, P. A. & Taki, D. 2001. *Science* 293:1068.)

CpG suppression Actually not a suppression mechanism. Some regions of the genomes are relatively low in CpG sequences and they may be methylated during most of the life of eukaryotes. *See* CpG islands.

cPLA$_2$ Catalyzes the hydrolysis of glycerophospholipid, yielding arachidonic acid (*see* fatty acids) and lysophospholipid (membrane lipid). It is activated by MAP kinase in the presence of Ca^{2+}. *See* integrin; MAP. (Blaine, S. A., et al. 2001. *J. Biol. Chem.* 276:42737.)

C$_3$ plant Produces three-carbon molecules (phosphoglycerate) as the first step in photosynthesis. By genetic engineering some of the C$_4$ enzymes can be expressed in C$_3$ plants. *See* Calvin cycle; photosynthesis. (Matsuoka, M., et al. 2001. *Annu. Rev. Plant Physiol. Mol. Biol.* 52:297.)

cpm Count per minute is a measure of radioactivity; 1 μCurie $\cong$ 1,000,000 cpm. *See* dpm.

cpn *See* chaperonins.

Cpn10 Mammalian homologue of the bacterial GroEs chaperonin in the mitochondria. *See* GroES.

Cpn21 Chloroplast chaperonin (homologue of GroES) encoded in the nucleus; it interacts with Cpn60 (homologue of GroEL). *See* GroEL.

Cpn60 Mammalian mitochondrial and plant chloroplast analogs of the bacterial GroEL chaperonin. *See* GroEL.

CPP32 Proenzyme of apopain. *See* apopain; apoptosis; cysteine proteases; Yama.

Cpr *See* cyclophilins.

CPR (cell cycle progression restoration) Cyclin-dependent human genes involved in relief of cell cycle arrest by pheromones; function as molecular chaperones and transcription factors; control morphogenetic pathways (carcinogenesis), etc.

CPSF (cleavage-polyadenylation protein factor) Mediates the formation of the 3'-end of mRNA. *See* polyadenylation signal; transcription factors.

cR *See* CentiRay.

CR line of mice *See* congenic resistant.

CRAb (chelating recombinant antibody) Antibody specific for two adjacent and nonoverlapping epitopes of a single

antigen molecule, generated by chelating, which increases specificity. *See* antibody; antigen; chelation; epitope.

CRAF-1 Protein factor interacting with the CD40 cytoplasmic tail by a region similar to the tumor necrosis factor receptor (TNF-α)–associated factors (TRAF). CRAF is required for CD40 binding and dimerization. CRAF has five Zn-fingers and a Zn-ring finger. It participates in signal transduction. *See* CD40; ring finger; signal transduction; TNF; TRAF; zinc fingers. (Luttrell, L. M., et al. 2001. *Proc. Natl. Acad. Sci. USA* 98:2449.)

craniometaphyseal dysplasia (CMDJ, 5p15,2-p14.1) Dominant malformation (outgrowth and sclerosis) of the head bones. Mutation is in the ankylosis gene that encodes a 492-amino-acid transmembrane protein controlling the pyrophosphate level; it is expressed in the joints and in other tissues. A recessive form (CMDR) is encoded at 6q21-q22. *See* head/face/brain defects. (Reichenberger, E., et al. 2001. *Am. J. Hum. Genet.* 68:1321.)

crarniorodigital syndrome (otopalatodigital syndrome type II) X-chromosomal head/face/brain defect that has overlapping symptoms with otopalatodigital syndrome; it is encoded in the same area of human chromosome Xq28. *See* head/face/brain defects; otopalatodigital syndrome.

craniosynostosis syndrome Occurs in great variety and involves premature closure of the sutures of the skull, resulting in facial malformations. The defect is in one of the fibroblast growth factor receptors (FGFR). Prevalence is in the 4×10^{-4} range. Most of these syndromes are under autosomal-recessive control. *See* Antley-Bixler syndrome; Apert syndrome; Chotzen syndrome; Crouzon syndrome; Gorlin-Chaudhry-Moss syndrome; Marfanoid; MSX; Muenke syndrome; Pfeiffer syndrome; Shpritzen-Goldberg syndrome. (Cohen, M. M., Jr. & MacLean, R. E., eds. 2000. *Craniosynostosis*, Oxford Univ. Press, New York.)

cranium Skeleton of the head, excluding the bones of the face; the container of the brain. *See* brain, human.

crassulacean acid metabolism Some succulent plants can store large amounts of acids (malate), which are formed during the night, and their level drops during the day. This system of photosynthesis fixes carbon with the aid of phosphoenolpyruvate carboxylase and generates the C$_4$ oxaloacetate, like other C$_4$ plants, rather than through ribulose-1,5-bisphosphate carboxylase/oxygenase, like the C$_3$ plants. *See* C$_3$ plants; C$_4$ plants. (Luttge, U. 2000. *Planta* 211:761.)

CRE Cyclic AMP-response elements (TGACGTCA); DNA sequences ~100 nt (nucleotide) upstream of the TATA box of the transcription unit of genes responding to cAMP. Gene expression by cAMP also requires protein kinase A action on the 43 kDa accessory protein of CREB. *See* cAMP; CREB; Cre/LoxP; protein kinase; TATA box.

creatine ([N-aminoiminomethyl]-N-methylglycine) Form of phosphocreatine is the major source of energy generation (kcal/mol = 10.3). Creatine kinase catalyzes the reaction: phosphocreatine + ADP ⇌ ATP + creatine.

creatine deficiency Guanidoacetate methyltransferase deficiency (GAMT, 19p13.3). Amidinotransferase catalyzes the conversion of glycine into guanidinoacetate, and from that GAMT generates creatine with S-adenosylmethionine methyl donor. The clinical symptoms involve postural and locomotive defects due to brain and muscular anomalies and very low creatine excretion. A creatine transport defect (CT1/SLC6A8, Xq28) also generates creatine deficiency. (Salamons, G. S., et al. 2001. *Am. J. Hum. Genet.* 68:1497.)

creationism Doctrine about how the universe came into existence based on oracle; suggests that organisms are as we see them at present because they have been so created and ordained. Another important aspect is "that the intellectual soul is created by God at the end of human generation, and this soul is the same time sensitive and nutritive" (Thomas Aquinas, 13th century). According to the *Tertullian* (2nd–3rd centuries), *transducianism*, both soul and body are conceived and formed at exactly the same time. The fossil record is a serious obstacle to the logic of creationism and has difficulty in explaining its divine purpose. *See* evolution; intelligent design. (Numbers, R. L. 1982. *Science* 218:538; Evans, E. M. 2001. *Cognit. Psychol.* 42:217.)

CREB CRE (cyclic AMP-response element)–binding protein requires phosphorylation at a serine residue at position 119 or 133 for transcriptional activity. Transactivation of CREB by PK-A requires the glutamine-rich constitutive activator domain (Q2) and the kinase-inducible domain (KID, 58-amino-acid modulatory sequence), which endow it with independent functions but may work synergistically. KID recognizes PK-A and the transcription complex and is important for transactivation. There are about 50,000 CREB molecules/cell, mainly bound to chromatin. The CREB family also includes CREM and ATF-1, all subject to phosphorylation by protein kinase A. Active calcium channels also stimulate CREB activity. CREB apparently activates T lymphocytes involving phosphorylation of CRE, which induces transcription factor AP1, leading to the production of interleukin-2 and to the progression of the cell cycle. The CREB kinase is apparently identical with RSK2, a member of the RAS family. Mutations in CREB have been implicated in Rubinstein-Taybi syndrome, in fusions with the MOZ (monocytic leukemia zinc-finger protein) in case of acute amyloid leukemia (AML), and histone acetyl transferase displacement by AP1, leading to transformation. The CREB-binding protein, CBP and p300, by histone acetyltransferase activity, assist the assembly of the transcriptional complex at the promoter. *See* AP1; ATF; calmodulin; cAMP; CBP; cell cycle; CRE; CREM; ICER; immune system; leukemia; MEK; p300; PKA; RAF; RAS; RHA; RSK; Rubinstein-Taybi syndrome; signal transduction; TAF; TAX; T cell transactivator; transcription factors; trinucleotide repeats. (Shaywitz, A. J. & Greenberg, M. E. 1999. *Annu. Rev. Biochem.* 68:821; Vo, N. & Goodman, R. H. 2001. *J. Biol. Chem.* 276:13505.)

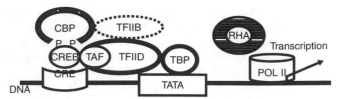

One of the functions of the CREB protein. CRE (cyclic AMP response element) is bound by CREB phosphorylated by CBP (CREB-binding protein). TBP (TATA box–binding protein) and the general transcription factors TFIID, TFIIB, as well as TAF130 (transcription-associated protein) congregate at the TATA box. RHA (an RNA helicase) and POLII (RNA polymerase II) are required for transcription. the system may require activation by hormones through G proteins, cAMP, protein kinase a (PKA), and the RAS, RAF, MEK, RSK signal transduction pathways. (Modified after Shaywith, A. J. & Greenberg, M. E. 1999. *Annu. Rev. Biochem.* 68:821.)

Cre/loxP P1 phage recombinase system affecting specific target sites; it can also be used in various eukaryotes for mediating site-specific recombination or chromosomal breakage. The Cre/loxP system has been successfully exploited for site-specific recombination and the generation of knockouts. The Cre protein (38-kDa) is a recombinase with specific recognition for the 34 bp *locus of crossing over of P1 (loxP)*, a pair of palindromic sequences, and recombination takes place within the 8 base central core (underscored at the sites delineated by ~). Note the inverted repeats left and right from the ~ signs. When Cre cuts the DNA, a phosphotyrosine intermediate is formed. Eliminating the phosphodiester bonds at ~ allows recombination to take place. The Cre recombinase can be targeted by choosing tissue-specific promoter (Zhou, L., et al.. 2002. *FEBS Lett.* 523:68)

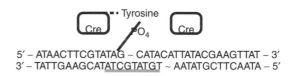

5′ – ATAACTTCGTATAG ~ CATACATTATACGAAGTTAT – 3′
3′ – TATTGAAGCATATCGTATGT ~ AATATGCTTCAATA – 5′

In animal cell cultures, the Cre recombinase may induce single- and double-strand breaks in the DNA and may slow down cell proliferation. At low concentration, the disadvantageous features can be avoided. *See* chromosomal rearrangements; Flp/FRT; resolvase, homing endonuclease; integrase; knockout; ligand-activated site-specific recombination; site-specific recombination; TAMERE; targeting genes. (*Genesis.* 2000. 26:99–165; Sauer, B. 1998. *Methods* 14:381; Zheng, B., et al. 2000. *Mol. Cell Biol.* 20:648; Van Duyne, G. D. 2001. *Annu. Rev. Biophys. Biomol. Struct.* 30:87; Loonstra, A., et al. 2001. *Proc. Natl. Acad. Sci. USA* 98:9209; Pfeifer, A., et al. 2001. *Proc. Natl. Acad. Sci. USA* 98:11450; Seligman, L. M., et al. 2002. *Nucleic Acids Res.* 30:3870.)

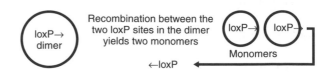

Recombination between the two loxP sites in the dimer yields two monomers

CREM a and b (modulator of CRE)   CREB-related transcriptional repressors, but CREB is a transcriptional activator. *See* ACT; CRE; CREB.

cremello   Color of horses of *AAbbCCDD* genetic constitution.

Cremello.

crenate   Has a tooth-like round protrusion.

Crepis   Composite flowers with chromosomes favorable for cytological studies. *C. capillaris* $2n = 6$; *C. tectorum* $2n = 8$; *C. rubra* $2n = 10$; *C. flexuosa* $2n = 14$; *C. biennis* $2n = 40$.

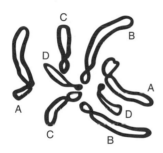

Crepis parviflora karyotype.

Cretaceous period   137 to 63 million years ago when the first human ancestors appeared. *See* geological time periods.

cretinism   Hereditary or congenital deficiency of thyroid hormone, causing mental and physical retardation. *See* goiter.

Creutzfeldt-Jakob disease   Degenerative nerve disorder that begins with forgetfulness and nervousness — most commonly at middle age, sometimes earlier or later. After a year or two, it progresses into jerky movements of the hands, insecure walk, and expressionless face. These symptoms overlap with Gerstmann-Straussler disease, and the two are probably basically identical, although different types of manifestations have been observed in both diseases. The diseases are not limited to humans; similar nerve degenerations have been described in sheep (scrapie), goat, and rodents. In the brain of the afflicted individuals, amyloid protein deposits are found.

The biochemical basis of the disorder was attributed to defects in the prion protein (PRP), a 27 to 30 kDa glycolipoprotein. The structural gene of PRP was assigned to mouse chromosome 2 and another gene in chromosome 17 was held responsible for the length of the incubation period of the disease. In humans, the PRP gene is in chromosome 20p12-pter. It appears that PRP is a normal protein of the nervous system, but proteolytic cleavage, amino acid replacements, insertions of 144 to 150 base pairs and insertions of 5 to 9 or more octapeptide repeats in between the amino acids encoded by codons 51 and 91 may trigger the disease. The most commonly observed Pro → Leu replacement at codon 102 was attributed to the ataxia symptoms, but changes at codons 117 (Ala → Val), 200 (Gln → Lys), and others were found to be associated with the degenerative phase of the PRP.

The injection of brain material into chimpanzees and other animals reproduces the disease. Since the infectious material does not contain any detectable amount of nucleic acid, scrapie, Creutzfeldt-Jakob, and Gerstmann-Straussler diseases are considered to be the first infectious protein diseases. They occur in a familial manner. It is not exactly known when and how the genetic determination of the degeneration occurs. It has been detected among all ethnic groups; in some its frequency is much higher than in others. Among Jews of Lybian origin, the incidence was reported to be 4×10^{-5}, nearly 50 times higher than in the general population. In addition, 41 to 47% of the observed cases were familial, whereas in some other populations only 4 to 8% appeared familial. The nv (new variant) CJD is supposed to be the result of human infection by animal encephalopathy–induced prions. *See* bovine spongiform encephalopathy; encephalopathies; fatal familial insomnia; Gerstmann-Straussler disease; kuru; prion.

cricket (*Gryllus campestris*)   Orthoptera, $2n$ male = 29, female $2n = 30$.

cri du chat (cat cry)   Deletion in the short arm of human chromosome 5 involves a mewing-like voice, mental and growth retardation, and other disorders.

Crigler-Najjar syndrome (bilirubin/uridine glucuronosyl transferase deficiency, UDPGT)   Recessive human chromosome 1q21-q23 defect; causes a nonhemolytic jaundice; in case of total deficiency of UDPGT, early infant death results. Partial deficiency of the enzyme (in type II form) is tolerated. *See* Dubin-Johnson syndrome; Gilbert syndrome; uridine diphosphate glucuronyl transferase.

criminal behavior   Has some genetic component, but the overwhelming motivation is provided by the family, social and economic conditions. Drug use, alcoholism, broken family ties, poverty, unemployment, etc., are the major factors. *See* behavior, human; behavior genetics.

CRINKLY-4   Plant TNF/EGF transmembrane receptor. *See* TNF.

CRIP   Group of proteins with LIM and an additional domain(s). *See* CRP; LIM; LMO; PINCH.

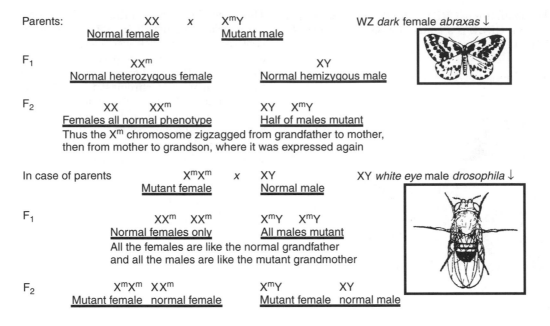

Parents: XX   x   X^mY   WZ *dark* female *abraxas* ↓
                     Normal female   Mutant male

F_1   XX^m   XY
              Normal heterozygous female   Normal hemizygous male

F_2   XX   XX^m   XY   X^mY
        Females all normal phenotype   Half of males mutant
        Thus the X^m chromosome zigzagged from grandfather to mother,
        then from mother to grandson, where it was expressed again

In case of parents   X^mX^m   x   XY   XY *white eye* male *drosophila* ↓
                        Mutant female   Normal male

F_1   XX^m   XX^m   X^mY   X^mY
              Normal females only   All males mutant
        All the females are like the normal grandfather
        and all the males are like the mutant grandmother

F_2   X^mX^m   XX^m   X^mY   XY
        Mutant female   normal female   Mutant female   normal male

crisis in animal cell culture   *See* Hayflick's limit.

criss-cross inheritance   Characteristic for X-chromosome-linked recessive mutations. The recessive genes follow the pattern of inheritance of the X chromosome; they are expressed in the hemizygous male but only in the homozygous female. In case the female is the heterogametic sex and the male is homogametic, e.g., in the *Lepidoptera* or birds, the inheritance follows the mirror image of what is shown on the diagram. *See* sex linkage diagram below. Criss-cross inheritance can also be observed in humans. All the offspring of a red-green color-blind father are expected to be normal, but all the male children of a color-blind mother are expected to be affected. Some of the male grandchildren of color-blind grandfathers will be color-blind, although their sons (the fathers of the grandsons) have normal vision.

cristae   Invaginations of the inner membranes of the mitochondria. *See* mitochondria.

critical population size   Although in a monogenic Mendelian F2 generation one-fourth of the population is expected to be homozygous recessive, it rarely happens that every fourth individual meets this expectation. Therefore, it may be important to know how many individuals are needed in F2 to find at least one of this desired phenotype. The statistical solution is a device that rules out the case when all the individuals would be the undesired type (3/4), e.g., $(3/4)^n = 1 - P$, where the (3/4) is the nonrecessive class, (n) is number of individuals required in the population, and P is the probability. Thus, $(3/4)^n = 1 - 0.95$ must be solved for (n), $n(\log 3 - \log 4) = \log 0.05$; hence $n = (\log 0.05)/(\log 3 - \log 4) = -1.30/(0.477 - 0.602) = 10.4$, i.e., 11 (because fractions of individuals do not exist and the 0.95 probability is valid for 10.4 or more). Therefore, at 0.95 probability only 11 individuals give us an assurance of finding at least one double recessive. The procedure is similar when we wish to determine the critical population size with a segregation ratio of 15:1 at 0.99 P: $n = \log 0.01/(\log 15 - \log 16) \simeq 72$. Similar calculations are useful for finding the minimum population size required for the recovery of a mutant individual after mutagenic treatment if we know (or guess) the induced mutation rate. *See* genetically effective cell number; mutation rate. (Mather, K. 1957. *The Measurement of Linkage in Heredity*, Wiley, New York.)

criticism on genetics   Essentially shares the same elements as those of other scientific fields. In most cases in which genetics is criticized, the blame is attributable largely to perverse political systems, but the participating biologists also deserve condemnation. Examples are attempts to pursue negative eugenics, racism, experimentation with biological warfare, inappropriate use of atomic energy, careless use of industrial, agricultural, and medical chemicals, distortion of population genetic principles applied to environmental problems, cloning of animals and possibly humans, genetic discrimination on the basis of genotyping information, modifying genes of plants, animals, and humans by genetic engineering or selecting special traits on the basis of preimplantation screening, etc. Some of the critiques argue that science cannot be left to scientists; the general public must be vigilant and reserve decision-making rights. Some scientists also support this view. The problem is how to make decisions without being fully familiar with a particular field of science. A solution appears to be increasing education on the progress of science. The motif behind the accusations is frequently sensationalism or attempts to gain political and financial advantage. *See* ART; atomic radiation; bioethics; biological containment; chemicals, hazardous; environmental mutagens; ethics; eugenics; genetic engineering; GMO; informed consent; mutation; nuclear transplantation; public opinion; radiation effects; selection conditions. (Reilly, P. R. 2000. *Annu. Rev. Genomics Hum. Genet.* 1:485; American Society of Human Genetics, Commentary 2002. *Am. J. Hum. Genet.* 70:1.)

Crk   Adaptor protein in a signal transduction pathway, containing SH2-SH3-SH3 domains; it requires phosphorylation between the two SH3 domains (by the Abl tyrosine kinase). *See* abl; SH2; SH3; signal transduction. (Feller, S. M. 2001. *Oncogene* 20:6348.)

CRM Cross-reacting material is a serologically identifiable protein (generally the product of a gene that fails to display enzyme activity). CRM⁻ designates a phenotype without an immunologically detectable protein. *See* immune response.

CRM1/XPO1 (exportin) Nuclear protein that mediates the export of polypeptides with Leu-rich sequences of the nuclear export signals. It works in cooperation with RanGTP. *See* nuclear export sequences; nuclear localization complex; nuclear pore; RAN; snRNA. (Fornerod, M., et al. 1997. *Cell* 90:967.)

CRMP-62 Collapsin response mediator protein is a M_r 62 K protein required for axon extension in chickens and *Xenopus*. *See* collapsin; UNC-33. (Ricard, D., et al. 2001. *J. Neurosci.* 21:7203.)

cRNA (complementary RNA) DNA transcript. *See* cDNA.

Crohn disease (CD, regional enteritis, 14q11-q12) Autosomal-recessive inflammation of the bowel is a familial condition; 10% of the affected individuals have relatives with the same affliction, yet the genetic control is unclear. It is likely that more than a single genetic factor is involved in Crohn disease. A recessive gene for inflammatory bowel disease (IBD1) is at 16p12-q13. Within this site, mutations in the NOD2/Apaf gene have been identified. The wild-type allele of this gene activates NF-κB, which senses the presence of bacterial lipopolysaccharides. Some NOD2 variants may thus alter the susceptibility to bacteria. TNF-α has been implicated in the symptoms. Genetic factors for CIBD and ulcerative colitis (UC) have also been located to human chromosomes 3, 7, and 21. Two new bowel diseases have been mapped to 5q31-q33 (Rioux, J. D., et al. 2001. *Nature Genet.* 29:223). and 19p13 (Rioux, J. D., et al. 2000. *Am. J. Hum. Genet.* 66:1863). Some studies indicate the presence of *Mycobacterium paratuberculosis* RNA sequences in the clinical samples obtained from patients. The drinking water may be the source of infection. *See* Apaf1; Blau disease; cadherin; CARD; CIBD; TNF. (Parkes, M., et al. 2000. *Am. J. Hum. Genet.* 67:1605; Ogura, Y., et al. 2001. *Nature* 411:603; Lawrance, I. C., et al. 2001. *Hum. Mol. Genet.* 10:445; <http://www.niddk.nih.gov/health/digest/pubs/crohns/crohns.htm>.)

Cro-Magnon *See* Neanderthal people.

crop plants (<http://synteny.nott.ac.uk>.)

***Cro* repressor** Of λ-phage, cooperatively with *cI*, regulates lysogeny. *See* lambda phage.

cross Mating between individuals of not-identical genetic constitution.

cross-breeding *See* cross-fertilization.

cross-feeding *See* channeling, syntrophic.

cross-fertilization Takes place when the sperm and the egg are produced by two different individuals of different genotypes (allogamy).

cross-fostering Procedure to test how much of a behavioral trait is hereditary and how much is the influence of the postnatal environment. Pups are separated from the natural mother and are given to foster mothers belonging to another inbred strain, then the behavioral differences are compared with individuals reared with the birth mother.

crossing Mating of two parental types of different genetic constitution.

crossing barrier *See* incompatibility; incompatibility alleles.

crossing over Process of reciprocal exchange between chromatids. Genes within a pair of homologous chromosomes may be linked in two fashions, either by coupling or by repulsion. Independent segregation observed by Mendel has the limitation of linkage, and association of genes within a chromosome is not absolute, either, because crossing over and recombination may separate genes depending on their genetic (physical) proximity. Chiasmata during meiotic prophase is the physical basis of crossing over (the exchange between chromatids), which is then detected in the genetic segregation as recombination. Crossing over takes place at the 4-strand stage, i.e., when each of the bivalents is composed of two chromatids associated at the centromere. This is the tetrad stage of the meiocyte. A single crossing over between two genes within a tetrad creates two reciprocally recombinant chromatids (for exceptions, *see* gene conversion), whereas the other two chromatids remain unaltered (parental). Since 2/4 chromatids are crossovers, the frequency of recombination caused by a single crossing over event is 50% for that particular tetrad. Each individual heterozygous for linked genes has numerous meiocytes, and crossing over does not take place in all of them at the same time; therefore, in a population of meiocytes, the frequency of crossing over may vary from 0 to 50% depending on the genetic distance between the genes. The maximal frequency of recombination in meiosis is thus 50%. Occasionally higher than 50% recombination has also been observed in apparent contradiction to the principle described. This value violates the principle because it is the result of selection during or after meiosis and gametogenesis or fertilization. One percent recombination is considered, by convention, 1 genetic map unit (m.u.) or 1 cM (centimorgan). Within a single meiocyte, more than a single crossing over may simultaneously take place. A single crossing over, however, always produces 50% recombination. If within the same genetic interval a second crossing over occurs, it may prevent the genetic detection of the first crossing-over event, because the second crossing over may restore the noncrossover-type arrangement of the genes. The third crossing over within the same interval again restores the recombinant arrangement of the genes demarcating that interval:
Thus, each odd-numbered crossover generates detectable recombinants, and the even-numbered ones restore the original linkage phase of the alleles. Since multiple crossing overs

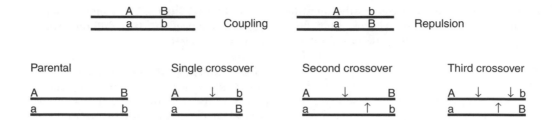

are expected to occur at the product of the frequencies, double crossing within a meiocytes does not usually affect the other meiocytes.

Therefore, if the frequency of a single crossing over is 0.30, double crossing over occurs by 0.30^2, and triple crossing over by the frequency of 0.30^3. Thus, if after the first crossing over the frequency of Ab + aB gametes is 0.30, after the second event it may be $0.30 - 0.30^2 = 0.21$, and after the third event $0.30 - 0.30^2 + 0.30^3 = 0.237$. The incidence of crossing overs may not occur as predicted by probability, but the first crossing over may hinder the occurrence of a second one (positive interference) or it may stimulate it (negative interference). Double crossing over may involve two, three, or four strands within a meiocyte (see illustration below). In yeast, each bivalent has at least one crossing over and usually not more than two. The probability of non−crossing over is commonly less than 0.1%. Usually crossing overs are not near to each other. Recombinational interactions are large, but only a few of them progress to develop into crossovers (minimization). Noncrossover interactions exceed actual crossovers by a factor of 2 in *Neurospora*, 4 in *Drosophila*, and based on the number of early recombination nodules in onions, the excess may be 30−40. It appears that the decision about crossing over is made before the formation of the synaptonemal complex, but the resolution of the Holliday junctions is delayed by or after pachytene.

Crossing over is generally limited to meiosis when the homologous pairs of chromosomes pair. The chromosomes of some organisms, or under certain circumstances, may also pair during mitoses, which may result in somatic recombination. Mitotic crossing over resembles the meiotic event, but the mechanism of exchange may not be identical. *See* association point; chiasma; chromosome pairing; coefficient of crossing over; coincidence; crossover; cytological evidence for crossing over; Holliday juncture; Holliday model; interference; mapping; mapping function; meiosis; mitotic crossing over; mitotic recombination; oblique crossing over; pachytene; recombinational nodule; recombination frequency; recombination mechanism, eukaryotes; recombination models; sperm typing; synapsis; tetrad analysis; time of crossing over.

crossing over, cytological evidence for   First in 1931 in maize by using a line heterozygous for chromosome 9 bearing a large terminal knob and the *C/c* and *Wx/wx* genes. McClintock and Creighton found that the cytologically detected knob followed the syntenic genetic markers and corresponded to the expectation in the recombinants. At the same time, Curt Stern used a fragmented X chromosome in *Drosophila* marked with *carnation* eye (*cr*) and *Bar* eye (*B*) and another X chromosome with the wild-type alleles; a fragment of the Y chromosome was attached to it. Again, the genetically observed recombination was associated with the physical exchange of the cytologically marked chromosomes. *See* crossing over; diagram below.

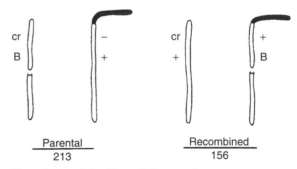

An outline of one of the *Drosophila* experiments on correspondence between genetic and physical exchange of chromosomes. (After Stern, C. 1931. *Biol. Zbl.* 51:547.)

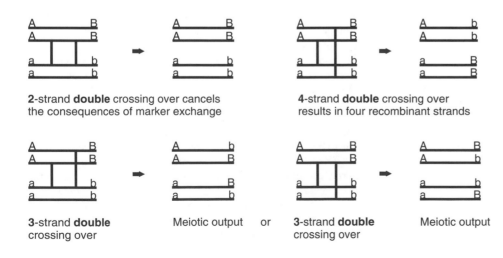

2-strand **double** crossing over cancels the consequences of marker exchange

4-strand **double** crossing over results in four recombinant strands

3-strand **double** crossing over   Meiotic output   or   **3**-strand **double** crossing over   Meiotic output

Two-strand, three-strand, and four-strand crossing overs occur normally in the proportion of 1:2:1 (in the absence of chromatid interference). The frequency of recombination by all four types of double crossing-over events combined is 50%.

crossing over, male   *See* male recombination.

crossing over, oblique   If the duplicated chromosomal region is obliquely paired, recombination may take place in between the segments in the following way:

$$\frac{AAA}{aaa} \rightarrow \text{Aaa and aAA or} \rightarrow \text{AAa and aaA}$$

Such an event may also lead to an increase or decrease of the chromatin and may cause position effect. *See Bar*; position effect; unequal crossing over.

crossing-over modifier   *See Rec* alleles.

cross-link   Various carcinogens and mutagens (nitrogen mustards, mitomycin C, psoralen dyes, cisplatin, etc.) cause DNA inter- and intrastrand cross-links. These lesions are repaired by ABC excinucleases, which generate dual incisions and repair DNA synthesis. *See* ABC excinucleases; cisplatin; mitomycin C; nitrogen mustards; psoralen dyes.

cross-linking   Establishing bonds between two molecules or linking together by covalent bond nucleotides in the same DNA or RNA. Chemical cross-linking agents are, e.g., bis-(2-chloroethyl)-methylamine and other mutagens and carcinogens (particularly the alkylating agents). The resulting quaternary structure leads to disruption of DNA functions. UV light can also produce cross-linking within and between nucleic acids and proteins. *See* alkylating agents; DNA repair; pyrimidine dimers. (Kuraoka, I., et al. 2000. *J. Biol. Chem.* 275: 26632; De Silva, I. U., et al. 2000. *Mol. Cell. Biol.* 20:7980; Dronkert, M. L. & Kanaar, R. 2001. *Mutation Res.* 486:1053.)

crossover   Recombinant chromatid, chromosome or individual, originated by genetic exchange between homologous chromosomes through the process of crossing over. The crossovers can be single or multiple. *See* coincidence; crossing over; crossover fixation; mapping, genetic; mapping function; non-crossover recombinant; recombination.

crossover connection, protein   Structural cores are connected by opposite ends of the cores across the surface of the domain.

crossover fixation   Hypothesis to explain the relative homogeneity in repeat units of satellite DNA. According to this model, the repeats can undergo frequent unequal crossing over during short evolutionary periods, then the same unit may be either propagated or eliminated after the recombinational event. If maintained, it can account for the homology of the sequences because there was not yet enough time to accumulate mutations even in the sequences that are not coding and are exempt from selection pressure. *See* crossing over; crossover; satellite DNA; unequal crossing over. (Fletcher, H. L. & Rafferty, J. A. 1993. *J. Theor. Biol.* 164:507.)

crossover interference   *See* chromosome interference.

crossover suppressor   during the early days of *Drosophila* genetics, parental inversions were erroneously considered as crossover suppressors in the progeny. Actually, in most of the cases the duplication or deficiency strands (generated by recombination within the inversion strands) of the inversion heterozygotes were not transmitted or caused lethality rather than suppressing crossing over *per se*. In the close vicinity of the inversion breakpoints chiasmata is usually reduced, however, but this is a different phenomenon. (See inversion)

cross-pathway regulation   Biosynthesis of several different amino acids is coordinately regulated in fungi. *See* channeling; tryptophan operon.

cross-presentation   Transfer of antigen from antigen-bearing cell to antigen-presenting cell. The process may involve MHC class I– but also class II–restricted antigens. The result may be either cross-priming of the T cell or T-cell tolerance (cross-tolerance). Cross-presentation may involve many types of antigens, e.g., tumors, grafts, viruses, and even self-tissues or enzymes and other proteins delivered by cell adsorption. A high dose of the antigen is favored. Cross-presentation occurs with a few days delay of the host compared what occurs with professional antigen-presenting cells. Cross-presentation is different from the mechanism used by professional antigen-presenting cells. *See* antigen-presenting cell; MHC; T cell. (Heath, W. R. & Carbone, F. R. 2001. *Annu. Rev. Immunol.* 19:47.)

cross-priming   Antigens are obtained from *donor* cells by bone marrow–derived antigen-presenting cells and are delivered to cytotoxic T lymphocytes (CTL) on MHC class I molecules. Most commonly, CTLs recognize antigens that are localized in the cytoplasm of the target cells. Also, GM-CSF transfected cancer cells may recruit a number of tumor antigen-presenting cells to T cells. *See* antigen-presenting cell; cytotoxic T cell; HLA; tumor vaccination.

cross-protection   *See* host-parasite relations.

cross-reacting material   *See* crm; cross-reaction.

cross-reaction   Binding an antibody to an antigen; usually ooccurs when the formation of the antibody was not stimulated by the antigen but by a very similar one. *See* antibody; antigen; crm; nonspecific binding.

cross-reactivation   *See* marker rescue; reactivation.

cross-sterility   *See* incompatibility alleles.

cross-talk   Transmission of signals from receiver to sensor molecules between signal transduction pathways. *See* epistasis; microarray hybridization; morphogen; signal transduction.

Crouzon syndrome (craniofacial dysostosis) Autosomal-dominant phenotype of ossified cartilages and various anomalies of the face, particularly protruding eyeballs. It is allelic to Jackson-Weiss syndrome and Pfeiffer syndrome. The basic defect appears to be due to the fibroblast growth factor receptor 2 in human chromosome 10q25-q26. *See* eye disease; fibroblast growth factor; receptor tyrosine kinase.

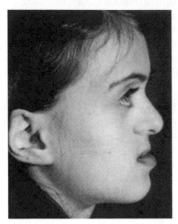

Crouzon syndrome with moderate exophthalmos (protrusion of the eyeballs) and malocclusion (improper position of the lower jaw). (From Tessier, P. 2000 in Craniosynostosis p. 228, Cohen, M. M. & MacLean, R. E., eds., © Oxford Univ. Press. By permission.)

crown gall Tumorous disease (mainly) of dicotyledonous plants caused by the soil-borne *Agrobacterium tumefaciens*. The tumor development is initiated by the large (200 kb) Ti (tumor-inducing) plasmid containing oncogenes responsible for the production of auxins and cytokinin plant hormones. Removal of the bacteria about 2 days after infection does not stop the disease because by that time the genes in the T-DNA part of the plasmids have integrated into the plant genome and are expressed in the plant cells. *See* Agrobacterium; T-DNA. (Burr, T. J. & Otten, L. 1999. *Annu. Rev. Phytopath.* 37:53.)

Crown gall (photo courtesy of Dr. A. C. Brown).

crozier Ascogenous, dikaryotic hyphae of ascomycetes form a three-cell hook-type structure; in the cell at the tip, the two nuclei fuse (karyogamy) to become diploid, then an ascus is formed where meiosis takes place. *See* ascus; fungal life cycle; *Neurospora*.

Crozier

CRP (1) cAMP receptor protein. *See* cAMP. (2) proteins with LIM, LMO domains. *See* LIM; LMO. (3) Catabolite response protein. *See* catabolite repression.

CRSP Transcriptional cofactor complex required for activation of Sp1-TFIID ($M_r \sim 700$ K). It has 9 subunits with MWs of 30 K to 200 K. *See* Sp1; transcription factors.

cruciform DNA May be formed as a recombination intermediate between single strands of double-stranded DNA (*see* Holliday model); in the case of inverted repeats, palindromes occur in both strands in a double-stranded DNA, and within each strand these palindromes fold back and pair. *See* Holliday model; palindrome; repeat, inverted; transposition sites.

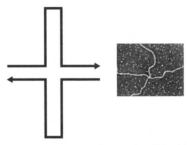

Cruciform recombination intermediate of DNA. (Courtesy of Drs. Potter, H. & Dressler, D.)

Cruzigard Technology is based on a similar principle as the paratransgenic method. Insect vectors are exposed to different bacterial strains transformed by antitrypanosomal genes as bait. The insect nymphs feeding on them release the antibiotic within the insect gut and debilitate the protozoon. *See* paratransgenic; *Trypanosoma*.

crwydryn Centromere-specific retrotransposon family in grasses. *See* retrotransposon.

CRX Transcription factor regulating photoreceptor outer segment proteins. *See* photoreceptor.

Cry9C Bacterial crystal protein with insecticidal property used in genetically modified maize (corn). The protein belongs to the family of proteins (Bt) present in *Bacillus thüringiensis*, but it is apparently more effective. Other insecticidal proteins of this family are Cry5A, Cry5B, and Cry6A. According to some claims, it is an allergen for humans. *See Bacillus thüringiensis*; GMO. (Crickmore, N., et al. 1998. *Microbiol. Mol. Biol. Rev.* 62:807.)

CRYO-EM (cryoelectron microscopy) Permits viewing of three-dimensional structures below a resolution of 10 Å. It may reach beyond nuclear magnetic resonance or X-ray crystallography. *See* nuclear magnetic resonance spectroscopy; X-ray diffraction analysis.

cryopreservation   Maintaining viability of biological tissues, enzymes, etc., at very low temperature, below −80°C, or in liquid nitrogen, −195.8°C. Even frozen mouse embryos and sperm, and ovaries can now be stored, distributed, reconstituted, and used for reproduction. *See* artificial insemination; sperm bank; vitrification. (Ludwig, M., et al. 1999. *Hum. Reprod.* 14(Suppl):162; Rall, W. F. 1992. *Animal Reprod. Sci.* 28:237; Watson, P. F. & Holt, W. V., eds. 2001. *Cryobanking the Genetic Resource. Wildlife Conservation for the Future?* Taylor & Francis, London, New York.)

cryptic dominance   Failure of recessive alleles of different loci to display complementation because of interaction of the gene products. *See* epistasis.

cryptic element   Such as a plasmid or transposon, which does not express a particular phenotype.

cryptic plasmid   Does not have a known phenotype; cryptic plasmids contain genes only for their maintenance and transmission. *See* plasmid. (Burian, J., et al. 1999. *J. Mol. Biol.* 294:49.)

cryptic prophage   Can no longer exit from the chromosome of the bacterium and develop into a virion, although some its genes are still transcribed. *See* prophage.

cryptic satellite   Satellite DNA is not displayed by ultracentrifugation as a separate (band) peak, but it is masked within the main band of the DNA. *See* buoyant density; satellite DNA; ultracentrifuge.

cryptic simplicity   Originally found in microsatellite regions where many point mutations occur, and the repetitiveness is broken down or a few intermixed states are found. These are most common in noncoding regions but may occur in coding sequences. *See* microsatellite.

cryptic splice site   Unusual juncture where splicing of exons may take place in case the usual site is changed by mutation. *See* introns; splicing.

cryptobiosis   Suspended life or reversible death.

cryptochrome   Linked to plant developmental (photoperiodism, circadian clock) processes under high-intensity blue light. Cryptochromes are the blue light photoreceptors of the animal eye and are modifiers of the circadian rhythm. They display similarities to photolyases. The mammalian cryptochrome genes (CRY) may have a light-independent regulatory role. In plant morphogenic responses, cryptochromes cooperate with phytochromes through the Ca^{2+}-binding protein SUB1. Cryptochromes are relevant to various human health problems such as seasonal depression, sleep disorders, jet lag, and even breast cancer. *See* circadian rhythm; photolyase; photoperiodism; phytochromes. (Yang, H-Q., et al. 2000. *Cell* 103:815;

Sancar, A. 2000. *Annu. Rev. Biochem.* 69:31; Lin, C. 2002. *Plant Cell* 14:S207.)

cryptogamic plant   Does not develop flowers or seeds and multiplies by spores such as ferns, mosses, and algae. *See* spermatophytes.

cryptogene   Mitochondrial DNA gene with primary transcript subject to pan editing that almost hides the original RNA sequences. *See* kinetoplast; pan editing; RNA editing; *Trypanosoma*.

cryptopolyploidy   Increase in nuclear DNA content among related families with the same chromosome number. *See* polyploid. (Sparrow, A. H. & Nauman, A. F. 1976. *Science* 192:524.)

cryptorchidism   Failure of the testes to descend from the abdominal cavity into the scrotum. This anomaly affects 2% of human births. Mutation in the insulin-like hormone (Insl3) may regulate the growth and differentiation of the gubernaculum (ligament of transient existence connecting the testis and epididymis to the scrotum) and testis descent. In females it may result in sterility. *See* epididyis; scrotum.

crystal cell   Hematoidin (hemoglobin-derived bilirubin-like) body in the blood.

crystal lattice   *See* semiconductor.

crystallin, α-, β−, γ-   Small heat-shock protein occurring in large (700–800 kDa) heteroploymers where it fends off protein denaturation. These crystallins are implicated in neuro- and muscle degenerations and cataracts (opacity of the eye lens). *See* desmin; eye diseases; sHsp. (Wang, K. 2001. *Biochem. Biophys. Res. Commun.* 287:642.)

crystallization   Formation of crystals; a procedure preparatory to X-ray crystallographic analysis of macromolecular structures. *See* X-ray diffraction analysis.

Cs (crossovers)   Recombinational events.

^{137}Cs   Cesium isotope emitting β- and γ-radiation; has a half-life of 33 years. *See* isotopes.

CSA   Compartmentalized shotgun assembly. *See* shotgun sequencing.

CSAID   Cytokine-suppressive antiinflammatory drug is an inhibitor of cytokine biosynthesis. *See* CSBP; cytokines.

CSBP   CSAID-binding proteins are mitogen-activated protein kinases. *See* CSAID; p38.

CSE   Negative regulator element of chromosome segregation in yeast that belongs to the Mediator family. *See* Mediator.

CSF   *See* cytostatic stability factor.

CSF (CSF-1 at human chromosome 1p21-p13, CSF-2 at 5q31.1, CSF-3 at 17q11.2-q12)   Colony-stimulating factor, involved in autophosphorylation. The CSF receptor (product of the FMS oncogene) is homologous to the product of the KIT oncogene, a protein tyrosine kinase. *See* colony-stimulating factor; FMS oncogene; KIT oncogene; protein tyrosine kinases.

CSF1R   Colony-stimulating factor receptor, product of the FMS oncogene. *See* colony-stimulating factor; CSF; FMS oncogene; KIT oncogene.

CSGE   Conformation-sensitive gel electrophoresis uses gel electrophoresis of PCR fragments for the detection of variations in (mitochondrial) DNA in order to provide data for evolutionary analyses. *See* gel electrophoresis; PCR. (Finnilä, S., et al. 2000. *Am. J. Hum. Genet.* 66:1017.)

Csk   Cytoplasmic protein tyrosine kinase; it has two amino-terminal protein-protein interaction domains (Src homology 2 and 3) and a carboxy-terminal catalytic domain. It is necessary for development of live mouse. In csk$^-$ cells, Src, Fyn, and Lyn phosphotyrosine kinase activity is decreased. Csk mediates the selection of T cells and regulates antigen receptors. *See* B lymphocytes; Fyn; Lck; protein-tyrosine kinase; Pyk; SFK; Src. (Wang, D. & Cole, P. A. 2001. *J. Am. Chem. Soc.* 123:8883.)

CSL (C-promoter binding factor-1/suppressor of hairless/LAG-1)   Family of transcription factors that mediate transcriptional activation through the Notch receptor. Activation of Hes1 and Hes5 may inhibit neurite extension. The Numb, Numb-like, and Deltex proteins inhibit Notch signaling and facilitate neurite extension. *See Notch.*

CstF   Cleavage-polyadenylation protein factor; may be associated with RNA polymerase II CTD. *See* CPSF; CTD; poll II; transcription factors. (Calvo, O. & Manley, J. L. 2001. *Mol. Cell* 7:1013.)

c$_0$t$_{1/2}$ Index of the reassociation of nucleic acid molecules when the reaction is half completed. It is proportional to the unique sequences in the reannealing molecules. *See* c$_0$t curve.

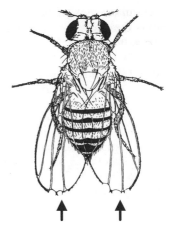

Notch mutant of *Drosophila.*

c$_0$t curve (pronounce cot)   When single-stranded DNA molecules are mixed, they may reassociate, depending on the homology between the two types of strands. The facility of reassociation is determined by the degree of homology of the reactants, their concentration, and the time allowed for the annealing. These three parameters are expressed by the c$_0$t curve. The reaction greatly depends on the amount of redundant sequences in the DNA and the complexity of the DNAs. Redundant sequences and palindromic DNAs can anneal rapidly because of the similarities. For unique sequences, it takes more time for the complementary sequences to collide. For the characterization of DNAs of different types, generally the half c$_0$t values are used, i.e., the time when the reassociation is half-completed (*see* diagram). The reassociation of *E. coli* DNA generally takes place between c$_0$t values of about 0.1 and 10 (ca. 4.7×10^6 bp, mainly unique DNA), whereas in a large and complex genome of rye containing about 80% redundancy in the 7.9×10^9 bp DNA, it is within 5 orders of magnitude of c$_0$t (between 0,001 and 100) in contrast to the 2 orders of magnitude in the bacterium. If single-stranded homologous molecules are allowed to reanneal, the process can be represented as

$$\frac{dc}{dt} = -kc^2,$$

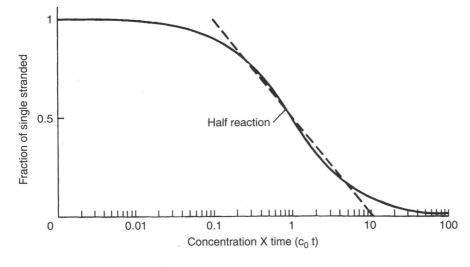

The time course of an ideal second-order reaction illustrates the features of the c$_0$t curve of nucleic acid reassociation. The ordinate corresponds to the fraction of single-stranded molecules. The abscissa denotes the mole value per liter of the material multiplied by time (usually in seconds) on a logarithmic scale. (Modified after Britten, R. J. & Kohne, D. E. 1968. *Science* 161:529.)

where $c =$ the molecular concentration, $t =$ time of the reaction, $k =$ constant that depends only on the length of the nucleic acid molecules, and d means the differential integral. After integration, we get:

$$\int_t^0 \frac{dc}{c^2} = -kdt,$$

and hence

$$\frac{c_0}{c} \bullet \left(-\frac{1}{c}\right) = -\mathrm{k}_t^0(\mathrm{t})$$

and

$$\frac{1}{c} - \frac{1}{c_0} = kt;$$

if $t_{1/2} =$ the time when $c = c_0/2$ and

$$\frac{2}{c_0} - \frac{1}{c_0} = \frac{1}{c_0} = kt_{1/2},$$

and if the concentration is expressed as mass (g/L), then: $c_0 = C_0/M_r$, where the molecular weight (M_r) is in daltons and the complexity becomes

$$C_{0t_{1/2}} = \left[\frac{1}{k}\right] M_r.$$

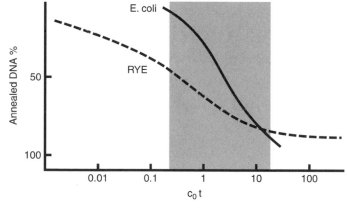

The reassociation kinetics of two genomes (7.9×10^9 bp of rye and 4.7×10^6 E. coli) of different complexities. The stippled area extends to 2 orders of c_0t values and covers most of the range of E. coli DNA. In contrast, 11% of the rye DNA reassociated at 0.001 to 100 c_0t. The half-reassociation value of the nonrepeated sequences of rye may be estimated by dividing 7.9 billion by 4.7 million, and it is ca. 1,681; after multiplying this figure by 2.5 (the c_0t values at half-reaction of the E. coli DNA), we obtain $1,681 \times 2.5 \approx 4,202 =$ the half-reassociation value of the nonrepeated rye. (Diagram and calculations were modified after Smith, D. B. & Flavell, R. B. 1977. Biochim. Biophys. Acta 474:82.)

C_0t value Index of the rate of reassociation of single-stranded DNA molecules. See c_0t curve; kinetics of reassociation.

CTAB ($[CH_3(CH_2)_{15}N^+(CH_3)]_3Br^-$) See cetyl trimethylammonium bromide.

CtBP Translational corepressor by interacting with E1A. Ctbp 1 is encoded at 4p16 and Ctbp2 at 21q21.3. See corepressor; E1A. (Chinnadurai, G. 2002. Mol. Cell 9:213.)

CTC See cytotoxic T cell.

CTCF (NeP1) Histone deacetylase and repressor of transcription. A CTCF-dependent enhancer-binding element acts as a chromatin insulator, and it may regulate imprinting. See boundary element; imprinting; insulator. (Hark, A. T., et al. 2000. Nature 405:486; Filippova, G. N., et al. 2001. Nature Genet. 28:335; Chao, W., et al. 2002. Science 295:345.)

CTD Carboxy-terminal domain, e.g., the repeat unit of the largest subunit of RNA polymerases (RNAP); can substitute for TATA box in genes that lack TATA sequences. This CTD of the α-subunits is the contact site for transcription activator proteins and the upstream promoter element (UP) in E. coli. The CTDK kinase, TFIIH, and FCP (yeast) phosphorylate the CTD of the transcriptase. The CTD domains of various proteins have structural and functional specificities. See RNA polymerase; TATA box; transcription factors.

ctDNA cpDNA or chloroplast DNA, photo p. 382.

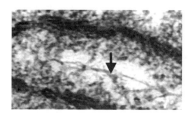

See ctDNA threads at arrow.

C terminus Carboxyl end of a polypeptide chain. See amino end; collinearity.

CTF CCAA motif-binding proteins. See binding proteins.

CTL Cytotoxic lymphocyte. See T cell.

CTLA-4 Cytotoxic T lymphocyte antigen (2q33) is a T-cell regulatory molecule along with antigens bound to the major histocompatibility complex (MHC) on antigen-presenting cells. CTLA-4 inhibits TCR signaling by binding to its ζ chain and interferes with tyrosine phosphorylation after T-cell activation. CTLA-4 has signals opposite from the stimulatory proteins CD28, CD80, CD86, and ICOS. See antigen-presenting cell; CD28; CD80; CD86; ICOS; ITAM; ITIM; MHC; T cell; TCR. (Chambers, C. A., et al. 2001. Annu. Rev. Immunol. 19:565; Egen, J. G., et al. 2002. Nature Immunol. 3:611.)

CTR See constitutive triple response.

Cubitus interruptus *(ci, 4.0)*   *Drosophila* gene locus with many different alleles. The cubital vein is interrupted at various lengths in the mutants. (Aza-Blanc, P. & Kornberg, T. B. 1999. *Trends Genet.* 15:458.)

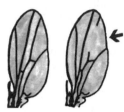

The *cubitus interruptus* phenotype (wild type, left; mutant at right). (Redrawn after Stern, C. & Kodani, M. 1955 *Genetics* 40:343.)

Cucurbits *(Cucurbitaceae)* *Cucumis sativus* (cucumber) $2n = 14$; *Cucumis melo* (muskmelon) $2n = 24$; *Citrullus lanatus* (watermelon) $2n = 22$; *Cucurbita pepo* (summer squash) and other squashes $2n = 40$.

Cue   Signal or stimulus that initiates or specifies a developmental fate. The *extrinsic signals* may be diffusible molecules (exocrine or paracrine hormones), cell membrane–attached proteins, extracellular matrix factors, morphogens, pheromones, or other environmental signals such as light or temperature. The *intrinsic cues* can be transcription factors, activators, transactivators, signal-transducing adaptor molecules, G proteins, etc. The intrinsic cues are considered to be cell autonomous if they are able to affect cell fates irrespective of the surrounding tissue. *See* activators; chaperone; exocrine; G proteins; morphogenesis; morphogens; paracrine; pheromones; signal transduction; transactivation; transcription factors.

Cullin (CUL1)   Family of proteins involved in ubiquitination of G_1-phase cyclins and cyclin-dependent kinase inhibitors. *See* CDC53; cell cycle; cyclin; ubiquitin. (Schwechheimer, C. & Deng, X. 2001. *Trends Cell Biol.* 11:420.)

cultigen   Form of cultivated plants. *See* cultivar; variety.

cultivar   Genetically distinct variety of crop plants generally adapted to a region; has some agronomic value for the grower and the consumer. *See* cultigens; variety.

culture collection   Maintains cell lines of microbial, viral, and higher organims and makes specimens available for a fee. (<www.ukncc.co.uk>.)

culture media   For bacteria, see Sambrook, J., et al. 1989. *Molecular Cloning*, Cold Spring Harbor Lab. Press; for plants. Gamborg medium; Murashige & Skoog medium. (animal cells: Butler, M. 1996. *Animal Cell Culture and Technology*. IRL/Oxford Univ. Press, New York.)

cumulus ovaricus   Group of diploid follicular cells surrounding the mammalian egg. Such cells have been successfully used for nuclear transplantation after removal of the resident nucleus, resulting in cloning. *See* nuclear transplantation.

curare   *See* toxins.

curated   Database generated/maintained by human supervision, not only by mechanical/electronic means.

curie   Basic unit of radioactivity contained in 1 g radium, i.e., 3.7×10^{10} disintegrations per second (dps). Most commonly, 1/1,000, millicurie (mCi) or 1/1,000,000 microcurie (μCi, 2.2×10^6 desintegrations per minute [dpm]) are used in laboratory work. In most equipment, only about half of the disintegrations are detectable; thus, 1μC corresponds to 1 million counts per minute (cpm). Since the Ci unit defines the rate of disintegrations/time unit and the half-life of the different isotopes may vary greatly depending on the species, the shorter half-life isotopes lose their isotopic atoms faster. 1 becquerel $= 2.7027 \times 10^{-11}$ curie (≈ 27 picocuries). *See* isotopes.

curing   Process of chemical hardening of physical agents.

curing, plasmid   Removal of a plasmid or prophage from a bacterium, e.g., by the use of chemicals (acridine dye, Novobiocin, Coumeromycin) or irradiation. The $[PSI^+]$ form of the yeast prion can be "cured," i.e., converted into $[psi^-]$ form by treatment with guanidine hydrochloride or methanol, in a reversible process. *See* acridine dye; prion.

currants *(Ribes* spp)   All are $2n = 2x = 16$.

Currarino syndrome (HAS, 7q36)   Dominant hereditary sacral agenesis (defect in the formation of the bone just above the hip). The responsible homeobox gene HLXB9 encodes a 403-amino-acid protein. (Ross, A. J., et al. 1998. *Nature Genet.* 20:358.)

curvans   *See rigens.*

curvature of DNA   Involves deflection(s) of the local helical axis. It causes retardation of movement in the electrophoretic field. (Hardwidge, P. R. & Maher, L. J. 2001. *Nucleic Acids Res.* 29:2619.)

Cushing syndrome (mixoma)   Also called NAME (nevi, atrial mixoma, mixoid neurofibromata, endocrine overactivity). Critical features are adrenal, hypothalmic, pituitary, and lung tumors and hippocampal atrophy as a consequence of overproduction of glucocorticoids (hydrocortisone). *See* adrenal; atrophy; glucocorticoids; hippocampus; neurofibromatosis; nevus; pituitary. (Yano, H., et al. 1998. *Mol. Endocrinol.* 12:1708.)

cut-and-paste   Mechanism of conservative transposition (see diagram below) transposition. Transposase (e.g., Tn5) and host proteins bind the transposable element and target, double-strand DNA breaks at the junction in a staggered manner, insertion follows, and the gaps are filled with complementary nucleotides (e.g., 9 bp [depending on the nature of the target cuts]). Duplications are generated at the end of the "simple insert." *See* transposable elements; transposons.

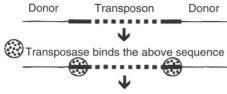

Donor   Transposon   Donor

↓

Transposase binds the above sequence

↓

Dimerization of transposase is followed by excision

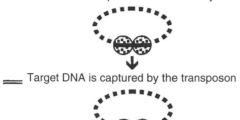

↓

Target DNA is captured by the transposon

↓

The catalytically active transposon with the aid of an activated water molecule hydrolyzes one strand of the target and liberates A 3′-OH group at the end of the transposon. The 3′-OH end makes a nucleophilic attack on the target and manages transfer of the transposon into a new location.

Model is modified after Davis, D. R., et al.. 2000. *Science* 289:77.

cuticle Layer of substance (collagen) over the surface of epidermal or epithelial cells. *See* collagen.

cutin Waxy material on the surface of plant cells that is slightly permeable to water or gaseous substances.

cutis gyrata *See* Beare-Stevenson syndrome.

cutis laxa Includes autosomal-dominant, recessive, or X-linked forms. The hereditary forms appear early in life and show loose skin and joints and folding skin. The recessive I form involves lung, heart, and digestive tract anomalies. Type II has bone malformation (dystrophy) and retarded development. Some other forms have the signs of early aging of the skin or emphysema and anemia. The X-linked (Xq12-q13) form causes the formation of bony horns on the foramen magnum (the opening of the cranium to the vertebral canal), deficiency of lysyl oxidase, and disturbance in copper metabolism. *See* collagen; Ehlers-Danlos syndrome; Menkes syndrome; pseudoxanthoma elasticum; skin diseases; Williams syndrome.

C value Amount of DNA in a single chromatid or chromosome before the DNA has replicated. Thus, the C value in the gametes is 1, in the diploid zygote it is 2, and after S phase before cell division takes place it may be 4.

C-value paradox Lack of relationship between evolutionary status (complexity of an organism) and its genome size, e.g., the size of the genome of the plant *Fritillaria* is $\sim 3 \times 10^8$ kbp, whereas that of *Homo* is $\sim 3 \times 10^6$ kbp. *See* copy number paradox; genome; H-value paradox; junk DNA; N-value paradox; pseudogenes; redundancy; repetitious DNA. (Petrov, D. A. 2001. *Trends Genet.* 17:23; Gregory, T. R. 2001. *Biol. Rev.* 76:65.)

CW Clockwise.

CXCR Chemokine coreceptor of the CD4 T cells, a member of the serpine receptor family. CXCR4 and its ligand SDF-1 (a member of a chemokine family) are involved in hematopoiesis, lymphocyte migration and activation, differentiation of the vascular system, development of neuronal patterning, and acquired immunodeficiency symptoms. CXCR is a seven-trans-membrane-spanning G-protein-coupled receptor. *See* acquired immunodeficiency; angiogenesis; CCR; CD34; chemokine; G protein; metastasis; neuron; PBSF; SDF; serpines; seven membrane proteins; ZAP-70. (Blanco, J., et al. 2000. *FEBS Lett.* 477:123.)

CXR Chemokine receptor. *See* chemokines.

Cxs Gene conversions accompanied by crossing over. *See* gene conversion; crossing over.

cyanide (HCN) Very powerful inhibitor of cellular oxidation by complexing with cytochrome oxidase; thus, it is an extremely dangerous poison. Sodium thiosulfate and sodium nitrate are antidotes. Cyanides may be present in various foods (almond) and feed (Sudan grass, white clover).

cyanidine *See* anthocyanin.

cyanobacteria (blue-green algae) Photosynthetic prokaryotes; they contain phycobiliproteins and chlorophyll a. Their genetic material is DNA and can be manipulated in a way similar to other prokaryotes. *See* chlorophyll; phycobilins; phycocyanins.

cyanogen bromide (BrCN) Very toxic, lacrimant gas. It has been used to cleave polypeptides and fusion proteins (at the C-terminal side of methionine) and also as a ligand for activated chromatography matrices for Western blotting. *See* ligand; Western blotting.

cyanogenic glucoside *See* insect resistance in plants; lotoaustralin.

cyanosis Bluish discoloration in body areas caused by the accumulation of reduced hemoglobin and methemoglobin resulting from mutation or other factors. *See* methemoglobin.

cybernetics Theory of control and communication such as exists in the nervous system and also in mechanical devices such as the computer. *See* cyborg.

CYBORG (cybernetic organism) Futuristic human body modified by substitution of artificial structures for natural ones, e.g., electronics for the nervous system. Such a system may permit better understanding of the working of an organic system. Also, it may correct natural defects in hearing, vision, or muscle control, etc., by the use of microchips (silicon or germanium pieces used in computer circuits. (Sequiera, S. 2001. *Trends Neurosci.* 24:834.)

cybrid   Contains two different cytoplasms in fused cells. *See* cell fusion; cell genetics.

cycle cloning   Mammalian cell cultures are subjected to multiple rounds of phenotypic selection in order to isolate cells of certain biological properties. *See* MaRX.

cyclic ADP-ribose   Apparently synthesized from NAD^+ by ADP-cyclase or by CD38; serves as a (abscisic acid) signaling molecule when activated by Ca^{2+} . The ryanodine receptor, regulating calcium ion channels, may be its receptor. It also regulates calcium signaling in T lymphocytes. *See* ABA; c-AMP; CD38; NAD; ryanodine; T cell.

cyclic AMP   *See* cAMP.

cAMP

cyclic electron flow   Electrons emanating from *photosystem I* in light returning to their origin. *See* photosynthesis; Z scheme.

cyclic GMP   *See* cGMP.

cyclic nucleotide-gated ion channel   *See* ion channels.

cyclic permutation   *See* permuted redundancy.

cyclic photophosphorylation   Cyclic electron flow drives ATP synthesis.

cyclin (CLN)   Proteins synthesized during cell divisional cycles are responsible players in the process by activating protein kinases (CDKs) and controlling the cyclic sequence of divisional steps. In *Saccharomyces* yeast cells, there are 22 cylins associated with one of the five CDKs. *See* CDK; cell cycle; CLB. (Stein, G. S., et al., eds. 1999. *The Molecular Basis of Cell Cycle and Growth Control*. Wiley, New York.)

cyclin A   Similar to cyclin B but appears earlier in the cell cycle. Cyclin A1 is expressed primarily in the germline, in the testes, and in myeloid leukemia cells. Cyclin A1−deficient

males are sterile. Cyclin A2 deficiency results in embryo lethality. *See* CDF; CDK; cell cycle; LATS. (Coverly, D., et al. 2002. *Nature Cell Biol.* 4:523.)

cyclin B   Has no known enzymatic activity; it is part of MPF and has a role accessory to the protein kinase (*cdc2*) subunit. It seems to determine the substrate specificity of MPF. *See* CAK; cdc2; CDK; cell cycle; MPF. (Groisman, I., et al. 2002. *Cell* 109:473.)

cyclin C   Is involved in mRNA processing (Barette, C., et al. 2001. *Oncogene* 20:551.)

cyclin D (CCND1, 11q13; CCND2, 12p13; CCND3, 6p21)   Proteins involved in the G1 phase of the cell cycle. They have over 50% homology. Cyclin D proteins are also oncogenic. CCND1-deficient mice are somewhat resistant to breast cancer mediated by the NEU/ERBB2 and RAS oncogene loci. CCND1 deficiency, however, does not affect the pathway to breast cancer mediated by c-Myc or Wnt-1 oncogenes. *See* breast cancer; CAK; catenins; CDK; cell cycle; c-Myc; ERBB1; RAS; Wnt. (Yu, Q., et al. 2001. *Nature* 411:1017.)

cyclin-dependent protein kinase (CDK)   In association with cyclin, CDK phosphorylates proteins and thus promotes cell divisional events. In yeast cells, there are five types of CDK enzymes: Cdc28/Cdk1, Pho85, Kin28, Srb10, and Ctk1. The inhibitors of kinase activity may inhibit cell divisions and can be used for therapeutic purposes. *See* CDK; CDK2; cell cycle. (Francis, D., Dudits, D. & Inzé, D., eds. 1998. *Plant Cell Division*, Portland Press, London, UK.)

cyclin E   Subunit of CDK2. Under normal conditions it accumulates between G_1 and S phase and degrades after S phase. In cancer cells its level is maintained or increased and may lead to chromosomal instability. *See* CDK2. (Moberg, K. H., et al. 2001. *Nature* 413:311; Strohmaier, H., et al. 2001. *Nature* 413:316; Payton, M. & Coats, S. 2002. *Int. J. Biochem. & Cell Biol.* 34:315.)

cyclin F   Displays the F-box motif in various protein-protein combinations, which recognizes proteins for proteolytic degradation and it promotes mitosis. *See* F-box; SCF. Kong, M., et al. 2000. *EMBO J.* 19:1378.)

cyclin G   Forms quaternsary complex with the B′ subunit of phosphatase PP2A and also binds Mdm2 resulting in the dephosphorylation of the latter. (*See* PP2A; MDM2. (Okamoto, K., et al. 2002. *Mol. Cell* 9:761.)

cyclin K   Phosphorylates with cdk9 the carboxyterminal of RNA polymerase II (Edwards, M. C., et al. 1998. *Mol. Cell Biol.* 184291.)

cyclin L   And an associated kinase are involved in processing of pre-mRNA (Dickinson, L. A., et al. 2002. *J. Biol. Chem.* 277:25465.)

cyclin T   Part of the Cdk9 complex; assists in binding Cdk9 to the Tat lentiviral RNA polymerase cofactor at the 5′-TAR site. *See* acquired immunodeficiency; CDK; PITALRE; TEFb. (Yang, Z., et al. 2001. *Nature* 414:317.)

5′,8-cycloadenosine
And other other cyclopurine deoxynucleosides may be generated in the DNA when they are exposed to reactive oxygen under hypoxic conditions. 5′,8-cycloadenosine may be formed in stereoisomeric forms and blocks the 3′ to 5′ exonuclease function of the DNase III repair enzyme. DNA polymerase η may use 5′,8-cycloadenosines as templates for translesion synthesis, although this enzyme prefers the normal deoxypurine. *See* DNA repair; translesion. (Kuraoka, I., et al. 2001. *J. Biol. Chem.* 276:49283.)

Cycloadenosine.

cyclobutane dimer
Cyclobutanes have 4 C atoms in a ring. Such a structure may also be formed by cross-linking adjacent pyrimidines in the DNA upon exposure to UV light. The most common genetically effective alteration in ultraviolet light–exposed DNA is the formation of thymine dimers (as shown below). The formation of cyclobutane dimers is dependent on the wavelength of the radiation. At 280 nm, it is induced almost five times as efficiently as at 240 nm. The dimer physically distorts the DNA and interferes with normal DNA replication and incorporation of nucleotides at wrong sites; the substitutions may lead to mutation. Visible light-activated enzymes may split the dimers by light repair. Adenine phosphoribosyltransferase may also repair the damage at both strands of the DNA. *See* adenine phosphoribosyltransferase; cis-syn dimer; DNA repair; glycosylases; photolyase; 5′,8-purine cyclodeoxynucleosides; pyrimidine dimer; pyrimidine-pyrimidinone photoproduct. (Zheng, Y., et al. 2001. *J. Biol. Chem.* 276:15786; You, Y.-H., et al. 2001. *J. Biol. Chem.* 276:44688.)

Cyclobutane pyrimidine dimer.

cyclodeoxynucleoside
5′,8-purine cyclodeoxynucleoside.

cycloheximide
Antibiotic that blocks the peptidyl transferase on the 80S ribosomes but does not affect protein synthesis on the 70S ribosomes. *See* antibiotics; peptidyl transferase; protein synthesis; ribosome; signaling to translation.

Cycloheximide.

cycloidea allele
Determines the radial symmetry of flowers.

Cycloidea snapdragon mutant.

cyclooxygenase
(COX) Enzyme involved in the synthesis of the eicosanoids prostaglandins, prostacyclins, and thromboxanes from arachidonic acid. This enzyme thus plays an important role in inflammation, pain, fever, and embryo implantation. It suppresses immunosurveillance and stimulates tumorous growth. Resveratrol, a product of grapes and the legume *Cassia quinqueangulata*, acts as a chemopreventive of the tumorigenic effect of cyclooxygenases. The common drug aspirin acetylates both COX-1 (prostaglandin-endoperoxide synthase 1; 9q32-q33.3) and COX-2 (prostaglandin-endoperoxide synthase 2; 1q25.2-q25.3) and irreversibly inactivates COX-2. Acetylated COX-1 has antithrombotic (protection against heart diseases) but also inflammatory effects (causing heartburn, etc.), whereas shutting off COX-2 is beneficial for immunological and neoplastic diseases. *See* arachidonic acid; aspirin; IKK; lipoxygenase; polyposis adenomatous intestinal; prostaglandins; thrombosis; thromboxanes. (Smith, W. L., et al. 2000. *Annu. Rev. Biochem* 69:145; Turini, M. E. & DuBois, R. N. 2002. *Annu. Rev. Med.* 53:35.)

cyclopea
Developmental anomaly resulting in the formation of only a single eye. The buildup of cholesterol and defective signaling through the Sonic hedgehog protein appear to be causative factors. *See* holoprosencephaly; sonic hedgehog.

cyclophilin
(Cpr, Cyp, rotamase) Evolutionarily highly conserved peptidyl-prolyl-isomerase and catalyzed cis-trans isomerization of X-Pro peptide bonds. Peptidyl-prolyl-isomerases are encoded at several human chromosomal locations. The cytoplasmic cyclophilin A (18 kDa, 7p13) is bound to and may be inhibited by the immunosuppressive antibiotic, cyclosporine. Cyclosporine bound to cyclophilin inactivates calcineurin. Cyclophilin B and cyclophilin C (each 23 kDa) are found in

the secretory pathway. Cyclophilin D (20 kDa, 4q31.3) is mitochondrial. Cyclophilin 40 (40-kDa) is cytoplasmic and usually heat-inducible and seems to participate in heat-shock protein-90 and heat-shock protein-70–dependent signal transduction. *See* calcineurin; cyclosporine; heat-shock proteins; immunophilins; immunosuppressant; mitochondrial import; nina; peptidyl-prolyl isomerases; PPI; SCC; signal transduction. (Schiene-Fischer, C. & Yu, C. 2001. *FEBS Lett.* 495:1.)

cyclophosphamide (cytoxan, endoxan, $C_7H_{15}Cl_2N_2O_2P$) Alkylating carcinogen, clastogen, antineoplastic agent, and immunosuppressive drug. *See* aldehyde dehydrogenase.

Cyclophosphamide.

cyclosome APC. *See* cell cycle.

cyclosporine Peptide antibiotic. It inhibits the activation of T lymphocytes by blocking the synthesis of IL-2, and it is thus an immunosuppressant. Cyclosporine may also become carcinogenic not only by its immunosuppressive effects but also by boosting the synthesis of TGF-β. *See* antibiotics; calcineurin; cyclophilin; cyclophosphamide; FK506; immunological surveillance; immunophilins; immunosuppressant; TGF.

Cyclosporine.

cyclosporine-binding protein *See* cyclophilins.

cyclotide Naturally synthesized circular plant peptides of ~30 or fewer amino acids occurring in *Rubiacea*, *Viola odorata*, and some other plants. The molecules form a distorted triple-stranded β-sheet and a cystine-knot motif. They are antibiotics and have protective effects against insects. *See* cystine knot. (Craik, D. J., et al. 1999. *J. Mol. Biol.* 294:1327; Jennings, C., et al. 2001. *Proc. Natl. Acad. Sci. USA* 98:10614.)

cyclotron Accelerator of electrically charged particles and atomic nuclei. It is used to cause transmutations in atomic nuclei, e.g., a normal magnesium into radioactive sodium.

cylindromatosis (CYLD, turban tumor) Skin tumor (primarily on hairy areas) caused by rare inactivation of a dominant gene in human chromosome 16q12-q13. The gene encodes a cytoskeleton-associated protein.

cyme Inflorescence where the growth of the apex ceases early to the benefit of the branches.

Cyme.

CYNOMOLGUS Most commonly means *Macaca irus* laboratory monkey. *See* Cercopithecidae; Rhesus.

Cyp *See* cyclophilins.

CYP *See* cytochrome P450.

cyritestin ADAM family protein regulating the binding and fusion of the membranes of the sperm and egg. *See* ADAM; fertilization. (Grzmil, P., et al. 2001. *Biochem. J.* 357[pt 2]:551.)

Cys₄ receptor Contains 4 cysteine-zinc domains in eukaryotic transcriptional regulators of α-helix-loop motif. The N-termini of the α-helix contacts the nucleotide base, the N-terminal loop binds to the phosphate backbone, and the C-end is the dimerization interface. *See* binding proteins; zinc finger.

cyst In general, bridges (ring canal) connect a closed cavity in the body or the cluster of germline-derived cells that have undergone only a limited cytokinesis. Spermatogenesis in most animals (except nematodes) takes place in cysts. The oocyte develops within ovarian cysts. *See* fusome; gametogenesis; maternal effect genes; morphogenesis in *Drosophila*. (de Cuevas, M. A., et al. 1997. *Annu. Rev. Genet.* 31:405.)

Cyst.

cystathionine β-synthetase (21q22.3) Its overexpression seems to be linked to decreased atherosclerosis and higher survival rates if stricken by acute myeloblastic leukemia (AML) in Down syndrome. Its anomalies may cause homocystinuria.

See AML; homocystinuria. (Ge, Y., et al. 2001. *J. Biol. Chem.* 276:43570.)

cystathioninuria Recessive (human chromosome 16) deficiency of γ-cystathionase. Consequently, cystathionine cannot be cleaved into cysteine and homoserine, resulting in benign defects. *See* amino acid metabolism; aminoacidurias; cystinelysinuria; cystinosis; cystinuria; homocystinuria.

cystatine (cysteineprotease inhibitor) Loss of the gene (*Cstb*) causes myoclonous epilepsy of the Unverricht-Lundborg type and an increase of cathepsin S, C1qB chain of the complement, microglobulin β_2, glial fibrillary acidic protein, apolipoprotein D, fibronectin 1, and metallothionein II. (Lieuallen, K., et al. 2001. *Hum. Mol. Genet.* 10:1867.)

cysteamine Decarboxylation product of cysteine. It is a radioprotective compound.

cysteine ($HSCH_2CH[NH_2]COOH$) Upon hydrolyis of proteins in air, it is converted to cystine. It is a radioprotective molecule.

cysteine-histidine finger *See* DNA-binding protein domains; Zinc finger.

cysteine knot cysteine-rich inhibitory motif of proteins, held together by disulfide bonds.

cysteine protease ICE, ICH1, NEDD2, ICH2/TX, CPP32/YAMA/apopain, and MHC2 are implicated in apoptosis. *See* apoptosis; caspase.

cysteine-scanning mutagenesis Single amino acid residues in a protein domain are replaced by cysteine and modified by sulfhydryl-specific iodoacetanide derivatives of 7-nitrobenz-2-oxa-1,3-diazolyl dye. The activity of the mutated protein is then determined. *See* site-directed mutagenesis. (Kunkel, T. A. 1985. *Proc. Natl. Acad. Sci. USA* 82:488.)

cysteine string protein (CSP) Contains palmitoylated cysteines on the peripheral membranes involved with the nerve synaptic vessels. *See* fatty acids; synapse.

cystic fibrosis Apparently one of the most common serious recessive hereditary defects (human chromosome 7q31-q32), involving fibrous degeneration of the pancreas, bile ducts, respiratory system, intestinal glands, sweat glands, male genital system, etc. The primary defect is apparently in chloride transport due to alteration of the gene regulating transmembrane conductance caused by membrane lipid imbalance. The electrolyte transport is mediated by syntaxin, and the Munc18 protein blocks the latter. It afflicts ∼1/2,000 white newborns (the frequency of the responsible alleles may exceed 0.02). In American blacks, the prevalence is more than an order of magnitude less, and it is even much less common among Orientals. The large CF gene (27 exons) has been cloned (250 kbp). Hundreds of different mutations, including several deletions, have been identified. Prenatal diagnosis is 98% effective on the basis of determining the high level of immunoreactive trypsin in the serum, characteristic for CF. Sweat test and DNA test can also be used for diagnosis. Inhaling DNase solutions and avoiding respiratory infections may provide pulmonary relief by the use of antibiotics. Replacement and proper diet may alleviate pancreatic symptoms. The intestinal mucilage may be removed by surgery.

Cystic fibrosis can be detected on the basis of the pattern of heat inactivation of the enzymes acid phosphatase and α-mannosidase. Homozygotes display practically no activity, heterozygotes 40–60%, and the absence of the defective allele is indicated by 80–100% activity. The testing of γ-glutamyltranspeptidase, aminopeptidase M, and alkaline phosphatase from the second-trimester amniotic fluid permits prenatal diagnosis. Genetic screening can be carried out in newborns on the basis of immunoreactive trypsin. The normal allele of the cystic fibrosis gene is involved in the regulation of sodium and chloride absorption; therefore, it has been called cystic fibrosis transmembrane conductance regulator (CFTR, 27 exons), protein kinase A, and ATP-regulated Cl⁻ ion channel. CFTR also affects HCO_3^- transport. A large deletion in the CFTR gene (ΔF508) conditions deficiency of internalization of *Pseudomonas aeruginosa* bacteria—to which cystic fibrosis patients are hypersusceptible—and because of this, the epithelial cells are unable to clear the mucosa from the lungs by desquamation. *P. aeruginosa* infection causes apoptosis by activation of the CD95/CD95-ligand system. The *P. aeruginosa* strains in CF patients generally carry mutator genes that facilitate the bacterial adaptation to antibiotic treatment.

Salmonella typhi (the human pathogen) enters the intestinal mucosa through wild-type CFTR. The human CFTR⁺ gene, equipped with the rat intestinal fatty acid–binding protein promoter, corrected a lethal intestinal defect in mouse when transfected. Mice heterozygous for the cystic fibrosis gene secreted 50% of the normal fluid and chloride ions in response to cholera toxin; thus, CF may convey heterozygote advantage in natural selection. It has been suggested that the relatively high prevalence of cytic fibrosis in humans is due to a selective advantage in cases of *Vibrio cholerae* and *Escherichia coli* infection causing diarrhea by their toxin, which stimulates intestinal chloride secretion (Högenauer, C., et al. 2000. *Am. J. Hum. Genet.* 67:1422).

Many different mutations exist in the CF genes, ranging from nil to only reduced synthesis of CFTR and various types of defects in regulation, processing, and altered conductance. CFTR is a member of the multidrug resistance protein family. Defects in CF may be associated with defects in the excretory channel of the testis (congenital bilateral aplasia of the vas deferens) and azoospermia, asthma (breathing difficulties), nasal polyposis, and hypertyrosinemia. Experiments indicate that gene therapy for cystic fibrosis is possible using primarily adenoviral, adeno-associated, and retroviral vectors, but the efficiency of local transformation is low, < 0.1%. For effective remedy, 10% efficiency may be required. Liposomal transfers and receptor-mediated gene transfer have also been attempted. The target should be the columnar cells lining the airways. Unfortunately, these cells are rather refractory to gene transfer. It has been shown that DNase I treatment has a beneficial effect on the removal of mucosa from the respiratory channel. Unfortunately, DNase is inhibited by F-actin, secreted by the leukocytes that infiltrate the airways in response to infections. Actin-resistant DNase has been engineered to alleviate this problem. *See* ABC transporters; adeno-associated viruses; adenovirus; azoospermia; biofilm; CBAVD; CD95; cholera toxin; cystic fibrosis antigens; endoplasmic reticulum; gene therapy; genetic screening; infertility; liposomes; multidrug resistance; munc; polyp; receptor-mediated gene transfer; retroviral vectors; syntaxin; tyrosinemia; (Zielinsky, J. & Tsui, L.-G. 1995. *Annu. Rev. Genet.* 29:777; Sheppard, D. N. & Welsh, M. J. 1999. *Physiol. Rev.* 79:S[1] 23; Mateu, E., et al. 2001. *Am. J. Hum. Genet.* 68:103; Bobadilla, J. L., et al. 2002. *Hum. Mut.* 19:575.)

cystic fibrosis antigen (calgranulin A and B)   Determined by dominant genes in human chromosome 1q12-q22. Homozygosity for the cystic fibrosis gene (7q31-q32) is accompanied by absence of these proteins, whereas the symptomless carriers have an intermediate level. It appears that these independent proteins track the basic defect in cystic fibrosis. Antigen A (also called calgranulin A because it is most abundant in the granulocytes) has Mr 11,000 and antigen B (calgranulin B) has Mr 14,000. These two proteins are virtually identical with calcium-binding proteins in other sources. *See* cystic fibrosis.

cystine knot   Two disulfide bridges link adjacent antiparallel strands of a peptide chain and form a ring that is penetrated by the third one. Such cystine knots are found in NGF, TGF-β, and PDGF-BB growth factors. *See* platelet-derived growth factor; TGF-β. (Hymowitz, S. G., et al. 2001. *EMBO J.* 20:5332.)

cystine-lysinuria (diaminopentanuria)   Autosomal-recessive increase of diamines (cadaverin) in the urine, causing ataxia and mental degeneration. *See* amino acid metabolism; cystinosis; cystinuria.

cystinosis (CTNS)   Semirecessive autosomal hereditary disorder under the control of more than one locus involving up to 100-fold amounts of cystine in the cells (lysosomes). Although it may not involve phenotypically obvious symptoms, it may eventually cause kidney failure, eye defects, growth arrest, and rickets. The heterozygotes can be identified by an increase in cystin in their cells. A cystinosis gene has been assigned to the short arm of human chromosome 17p13. It is actually a lysosomal membrane. The onset may be early or late. It is actually a lysosomal membrane protein (cystinosin) disease. The symptoms may be relieved by cysteamine. Renal transplantation may only provide a transient cure because the metabolic root is not localized to the kidney; kidneys only accumulate cystine. *See* amino acid metabolism; cystine lysinuria; cystinuria; Fanconi renotubular syndrome; homocystinuria. (Touchman, J. W., et al. 2000. *Genome Res.* 10:165.)

cystinuria   Recessive disease in several allelic forms causing variable degrees of cystin deposits in the kidney cysts and the bladder. Besides the increase of cysteine in the urine, larger-than-normal-amounts of other amino acids (lysine, arginine, and ornithine) may also be excreted. High fluid intake may prevent or alleviate the amino acid deposits. Type I is encoded at human chromosome 2p16.3 at the same site as the rBAT transporter protein. Another cystinuria gene was revealed at 19q12-q13.1. *See* amino acid metabolism; cystine lysinuria; cystinosis; homocystinuria; rBAT/4F2hc. (Font, M., et al. 2001. *Hum. Mol. Genet.* 10:305.)

(Photograph is the courtesy of Dr. D. L. Rimoin, Harbor General Hospital, Los Angeles, CA.)

cystoblast   In the *Drosophila* germarium, it divides four times and produces 16 cells (cystocytes), of which one becomes the oocyte; the remaining 15 become nurse cells. The oocyte develops from one of the two, which are connected to four others. *See* germarium; maternal effect genes.

cystocarp   Spore-bearing structure formed in red algae after fertilization.

cystocyte   See cystoblast.

cytidine   Pyrimidine base, cytosine, is associated with ribose or deoxyribose.

Cytidine (cytosine riboside).

cytidine 5′-triphosphate synthetase (CTPS, 1p34.1) Mediates the conversion of uridine triphosphate into cytidine triphosphate. Its deficiency may result in mutator effects and multidrug resistance. *See* multidrug resistance.

cytidylate   Salt of cytidylic acid.

cytidylic acid   Cytidine plus phosphate (a DNA or RNA nucleotide).

cytoblast   Mitotic cell of the germarium in insects. *See* germarium.

cytochalasin   Toxin that breaks cellular actin microfilaments, inhibits glucose transport, thyroid secretion, growth hormone release, phagocytosis, platelet aggregation, and is used to evict nuclei from animal cells to produce cytoplasts and

Cytochalasin.

karyoplasts. *See* actin; cytoplast; karyoplast; nuclear transplantation; phagocytosis; toxins.

cytochemistry Chemical analysis of isolated subcellular components, employing histochemical techniques for in situ tracing of their action.

cytochrome (CYP) Electron carrier heme protein with important roles in respiration, photosynthesis, and other oxidation-reduction processes including the activation of promutagens and procarcinogens. The human mitochondria encode 3 subunits of cytochrome oxidase (COX), and 10 are coded by the nucleus. The nuclear SCO genes (22q13, 17p13.1) assist the assembly. In the mitochondria, cytochromes c_1 and c oxidize quinols and reduce a terminal oxidase; in the chloroplasts and photosynthetic bacteria, cytochromes f and c_6 oxidize quinols and reduce a photooxidized reaction center. The cytochrome-b–cytochrome-c_1 complex, including heme c_1, Rieske [2Fe-2S] protein, and ubiquinol/ubiquinon sites within the mitochondrial inner membrane, plays an essential role in electron transfer along with cytochrome oxidase and NADH dehydrogenase. Cytochrome c nitrite reductase plays an essential role in the nitrogen cycle. Many other roles of cytochromes are also known. When the permeability of the mitochondrial membrane is altered, cytochrome c may be released into the cytoplasm and apoptosis may follow. *See* carotene; dehydrogenase; Leigh's encephalopathy; mutagen activation; NADH; P450; porin; procarcinogen; promutagens; SXR. (Berry, E. A., et al. 2000. *Annu. Rev. Biochem.* 69:1005.)

cytodifferentiation *See* differentiation; morphogenesis.

cytoduction Dominant cytoplasmic transmission of a particular hereditary state, introducing, e.g., mitochondria into a cell; heterokaryosis in yeast. *See* heterokaryon.

cytofectin GS2888 (dioleophosphosphatidylethanolamine) Transfecting agent for mammalian cells. It is coupled with a fusogenic compound and a cationic lipid (GS2888). It carries plasmids, AS ODNs, etc., into cells efficiently, and its toxicity is low. *See* AS ODN; fusigenic liposome; liposome; transfection. (Axel, D. I., et al. 2000. *J. Vasc. Res.* 37:224.)

cytogene Located in the cellular organelles (plastids, mitochondria), not in the nucleus. *See* chloroplast genetics; mitochondrial genetics.

cytogenetics Area of genetics involving the study of chromosomal structure and behavior in connection with inheritance; the study of chromosomal anomalies and accompanied pathological conditions. It involves cytological analysis of the evolution of chromosomes. Integrated cytogenetic and physical maps are available. (<http://www.ncbi.nlm.nih.gov/genemap>; <http://bioinformatics.weizmann.ac.il/udb>.) [Trask, B. J. 2002. *Nature Rev. Genet.* 3:769.]

cytohet Heterozygosity in cytopl asmic genes (in plastids and mitochondria) when the zygotes receive cytoplasmic material biparentally. *See* chloroplast genetics; mtDNA.

cytokine Peptide secreted in response to mitogenic stimulation. Cytokines participate in intercellular communication and cellular activation. Most of the cytokine receptors invoke tyrosine phosphorylation of cellular proteins. The various cytokine receptors permit the action of different protein tyrosine kinases (PTK) in different cell types. They are instrumental in the induction and regulation of the immune system, cellular differentiation, blood cell formation, apoptosis, tumor inhibition, cell migration, DNA synthesis, etc. The cytokine receptors belong to several superfamilies with some common features within the groups: (1) hematopoietin receptors, (2) interferon receptors, (3) TNF receptors, (4) interleukin-1 receptors, (5) TGF-β receptors, (6) immunoglobulin receptors (M-CSF-R, EGF-R, PDGF-R, etc.), and (7) chemokine receptors. Generally (one of) the α-chains provides the cytokine specificity, whereas a β-chain plays a boosting effect. Different cytokine receptors may share functions and more than one kind may be involved in the same process. The cytokines may have pleiotropic effects in the signal transduction paths. The different cytokines may recruit various docking proteins and utilize different Janus (JAK) kinases and Tyk2 to increase specificity. *See* aminoacyl-tRNA synthetase; chemokines; CIS; colony-stimulating factor; CSAID; EGF; interferons; interleukins; lymphokines; M-CSF; PDGF; PTK; signal transduction; SOCS-box; TGF. (Thompson, A. W., ed. 1998. *The Cytokine Handbook.* Academic Press, San Diego, CA; <www.copewithcytokines.de>.)

cytokinesis Division of the cytoplasm after nuclear division, leading to the formation of two cells from one. *See* cell cycle; cell plate; cleavage furrow; midbody; mitosis. (Nanninga, N. 2001. *Microbiol. Mol. Biol. Rev.* 65:319; Zeitlin, S. G. & Sullivan, K. F. 2001. *Curr. Biol.* 11:R514; Heese, M., et al. 2001. *J. Cell Biol.* 155:239; Glotzer, M. 2001. *Annu. Rev. Cell Dev. Biol.* 17:351; Guertin, D. A., et al. 2002. *Microbiol. Mol. Biol. Reviews* 66:155.)

cytokinin Cytokinins and auxins are the most important plant hormones regulating plant development. *See* plant hormones. (Werner, T., et al. 2001. *Proc. Natl. Acad. Sci. USA* 98:10487; Hwang, I. & Sheen, J. 2001. *Nature* 413:383; Mok, D. W. & Mok, M. C. 2001. *Annu. Rev. Plant Physiol. Plant Mol. Biol.* 52:89; Hutchison, C. E. & Kieber, J. J. 2002. *Plant Cell* 14:S47.)

cytological map Shows genetic sites in relation to microscopically visible structures such as chromosome bands, knobs, centromeres and satellites. *See* genetic map; physical map; RAPD; RFLP.

cytological marker   Unique feature of the chromosome (e.g., knob, satellite, etc.), number of nucleoli in the nucleus, defective plastids, etc., visible by cytological analysis. *See* knob; pachytene analysis; salivary gland chromosomes; satellite.

cytology   Study of the structure and related functions of the cell and subcellular elements. (<www.cellnucleus.org>.)

cytolysin   Secreted by cells to dissolve other cells, e.g., during the immune reaction, pore formation by bacteria. (Haas, W., et al. 2002. *Nature* 415:84.)

cytolysis   Dissolving the cells into their chemical components.

cytomegalovirus (CMV)   Includes the herpes viruses and Epstein-Barr virus. These potentially tumorigenic viruses have DNA genetic material and great and diverse mammalian host specificity. The name (*megalo*) comes from the observation that infection enlarges the host cells. The human cytomegalovirus can produce about 200 potentially antigenic polypeptides; however, the CMV early-expressed gene (gpUL40) may block antigen presentation by class I MHC molecules (HLA-E) and thus the defense function of CD8$^+$ cytotoxic T lymphocytes. Phosphorylation of the immediate early (IE) proteins may prevent the interference with antigen presentation and the CTL defense system. The human CMV may cause serious disease to congenitally infected infants and adults with a defective immune system. Homosexual males and female prostitutes may frequently be infected with CMV (up to 90%), depending on the number of sexual partners. The opportunistic CMV disease (most commonly expressed as inflammation and necrosis of the retina) may surface after latent infections in the case of AIDS. Treatment may employ the antisense oligodeoxynucleotide fomivirsen. The human cytomegalovirus may cause site-specific (human chromosomes 1q42 and 1q21) breakage due to viral adsorption/penetration but not by expression of CMV genes. It is a common cause of virus-induced birth defects. *See* AIDS; antigen-presenting cell; CTL; Epstein-Barr virus; fomivirsen; herpes; MHC; T cells.

cytoneme   Projection toward the signaling center of the imaginal disk. Cytonemes may transmit signals between the organizing centers and outlying cells. *See* imaginal disks; organizer; signal transduction.

cytonuclear disequilibria analysis   Method of evolutionary (population genetics) study of the association between nuclear markers to detect nonrandom mating, gene flow, population subdivision, mutation, and genetic drift. (Latta, R. G., et al. 2001. *Genetics* 158:843.)

cytopathic   Causes pathological changes to cells.

cytopenia   Reduction in the number of blood cells. It may occur as a consequence of chemotherapy or disease. *See* IVIG.

cytophotometry   Spectrophotometric chemical study of the content(s) of single cells. *See* microscopy; spectrophotometry.

cytoplasm   All material enclosed by the cell membrane, including cellular organelles, but the nucleus is excluded. *See* cell.

cytoplasmic drive   Promotes maternal inheritance by cytoplasmic factors.

cytoplasmic incompatibility   Involves the disruption of fertilization or embryogenesis because of bacterial infection in insect species. *See* incompatibility; infectious heredity; *Wolbachia*. (Charlat, S., et al. 2001. *Symbiosis: Mechanisms and Model Systems*, Seckbach, J., ed. Kluwer Academic, Dordrecht.)

cytoplasmic inheritance   Determined by nonnuclear genetic factors. Some the organelles (plastids, mitochondria) are endowed with independent genetic material. The endoplasmic reticulum and the Golgi apparatus do not have an independent genetic material. Their transmission to daughter cells requires attachment to the cytoskeleton or actin filaments or they may be also *de novo* synthesized. Their transmission, unlike that mediated by the mitotic spindle, may be asymmetric. *See* chloroplast genetics; mitochondrial genetics; mtDNA. Barr, F. A. 2002. *Current Opin. Cell Biol.* N14:496)

cytoplasmic male sterility (*cms*)   In maize, five different mitochondrial genomes have been identified; three of them (*T, Texas; S*, USDA; and *C*, Charrua) control cytoplasmically inherited male sterility. The *Rf* restorer genes restore fertility. Fertility restoration requires specific nuclear genes or the loss of the *T-urf13* site caused by an intramolecular mtDNA recombination between two 4.6 kb repeats followed by an intermolecular recombination of a 127 bp repeat, resulting in a 0.4 kb deletion involving *T-urf13*.

The joint action of *Rf1* and *Rf2* restorer genes is sporophytic, i.e., approximately all (95%) of the pollen produced is fertile in the plants heterozygous for these dominant nuclear genes. In the *S* cytoplasm, *Rf3* conveys gametophytic restoration, i.e., in the heterozygotes approximately half of the pollen is fully functional and half is aborted. *Rf4* also acts sporophytically in S cytoplasm. Although *cms* has been identified in over 100 maize stocks, all of them fall into the three groups.

The various circular mtDNAs have been physically mapped and their sizes are different (T is 540 kb). The male sterility in the *Tcms* stock was attributed to a mitochondrial locus, *T-urf13*, near a 4.7 kb repeat absent from normal mtDNA. Maize mitochondria contain several smaller DNAs besides the main genome. The loss of mitochondrial linear double-stranded DNA plasmids *S-1* (6,397-bp) and *S-2* (5,453-bp) from the S cytoplasm may or may not be required for the restoration of fertility, depending on the nuclear genome. In the cms-S mitochondrial DNA, there are 2 kb recombinational repeats (R) at the junctions $\sigma-\sigma'$ and $\Psi-\Psi'$. These repeats may become the sites of recombination between the integrated and the episomal R sequences, resulting in rearrangements of the mtDNA genomes. The R elements carry 210 bp sequences, which are highly homologous (~90%) with the terminal inverted repeats (TIR) of the S plasmids. These TIR

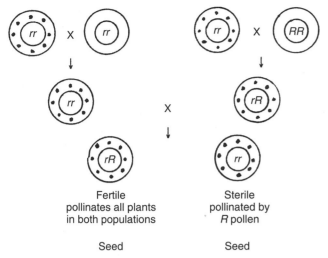

Fertile
pollinates all plants
in both populations

Sterile
pollinated by
R pollen

Seed   Seed

An outline of the procedure for obtaining double-cross hybrid seed by the use of cytoplasmic male sterile lines of maize. *R* stands for a dominant fertility restorer gene; *r* does not affect cytoplasmic male sterility. The inner circles represent the cell nucleus and a band around it symbolizes the cytoplasm. Dots in the cytoplasm stand for determinants of mitochondrial male sterility. Plants with dotted cytoplasm can produce seed only when fertilized by *R* pollen.

sequences are sites of frequent recombinations, resulting in complex rearrangements of the mtDNA. The size of the R elements varies in different maize stocks. Their homologues in some teosinte lines are named M1 and M2. Some of the nuclear fertility restorer genes (called *Rf* or *Fr* in different species) control the transcription and translation of the cms mtDNA. The "reversion" usually represents loss of parts of the R sequences and thus it is a permanent, irreversible change. The cytoplasmic male sterility of the *T* type is associated with susceptibility to *Helminthosporium maydis* (*Cochliobolus heterostrophus* Drechsler) blight and to other factors. The T cytoplasm also conveys susceptibility to the systemic insecticide methomyl (acetylcholinesterase inhibitor) and the toxin of the fungus *Phyllosticta maydis* (yellow corn leaf blight).

The URF13 protein contains 115 amino acids residing in the inner membrane of the mitochondria. URF13 is a receptor of the toxin of the blight pathogens and methomyl. When URF13 binds the toxin, it forms a mitochondrial pore and uncouples oxidative phosphorylation. *Rf1* restorer gene alters the transcript of the *T-urf13* gene, but *Rf2* does not do this, although it too inhibits male sterility. The RF2 protein is very similar to mammalian mitochondrial aldehyde dehydrogenase. Restoration of fertility does not abolish the susceptibility to the fungal toxins. The elements of these systems are also influenced by genetic background. In cultured cells, *Helminthosporium-r*esistant mitochondrial mutations occur that are also male fertile. Cytoplasmic male sterility systems were observed in several other plant species. The 3.7 kb *pvs* mtDNA sequence in common bean plants is associated with cytoplasmic male sterility. This sequence includes two open reading frames, *orf239* and *orf98*. The 27 kDa protein product of *orf239* localizes in the callose layer and the primary cell wall of the pollen. The mitochondrial *pcf* gene of *Petunia* encodes a 25 kDa protein in the CMS plants, but it is unknown how it affects sterility. Transforming *orf239* into tobacco (*N. plumbaginifolia*) without targeting it into

the mitochondria causes pollen disruption. Cytoplasmic male sterility is not based on single nucleotide change in the mtDNA, although nuclear encoded male sterility may be due to point mutations. The majority of the cytoplasmic male sterility cases are not the direct consequence of deletions. *See* cholinesterase; fertility restorer genes; heterosis; hybrid vigor; male sterility; mitochondrial abnormalities, plants; mitochondrial genetics; mitochondrial plasmids; mtDNA; RU maize; symbionts, hereditary. (Duvick, D. N. 1965. *Advances Genet.* 13:1; Williams, M. E. & Levings, C. S. III. 1992. *Plant Breeding Rev.* 10:23; Budar, F. & Pelletier, G. 2001. *C.R. Acad. Sci. III* 324:543; Bentolia, S., et al. 2002. *Proc. Natl. Acad. Sci. USA* 99:10887.)

cytoplasmic transfer *See* conjugation, *Paramecia*.

cytoplast Enucleated cell (cell that lost its nucleus). *See* karyoplast; transplantation of organelles.

cytosine (C) Pyrimidine base in RNA or DNA. *See* pyrimidines.

cytosine deaminase (CD) Bacterial (gene 1.3 kb) and fungal enzymes transform cytidine into uracil or deoxycytidine into deoxyuridine. Mammalian cells may lack this function. CD also deaminates cytosine arabinoside, an antileukemic drug, into inactive uracil derivatives. By suicidal action, CD converts 5-fluorocytosine (5-FC) into 5-fluorouracil (5-FU), a more effective anticancer drug. Transformation with the CD gene itself does not affect mammalian cells, but it may increase the sensitivity to 5-FC by 10^2- to 10^4-fold depending on the type of cancer. The effectiveness of CD gene therapy is determined by the effectiveness of delivering the vector to all cells of the tumor. The cancer cells, which do not incorporate the vector, may reinitiate tumor growth after an initial regression. The targeting of the vectors is also a problem even when it is equipped with tumor-specific monoclonal antibody. On the other hand, CD therapy may have a bystander effect. *See* bystander effect; cancer gene therapy; fluorouracil; suicide vector. (Ueda, K., et al. 2001. *Cancer Res.* 61:6158; Yoshikawa, K., et al. 2002. *Science* 296:2033.)

cytosine repressor (CytR) Regulates the transcription of at least eight genes of *E. coli* involved in nucleoside uptake and catabolism. All CytR-repressed promoters have a catabolite repressor (CRP)–binding site, and repression is the outcome of the cooperation between the CytR and the CRP-cAMP complex. *See* cAMP; catabolite repression.

cytoskeleton Bracing and scaffolding fibers (actin, microtubule and other filaments, diseases, cirrhosis of the liver) in the cytoplasm. The cytoskeleton may affect different cellular functions, progression of the cell cycle, differentiation, highways for intracellular vehicles moved by motor proteins, etc. *See* cirrhosis of the liver; filaments; intermediate filaments; motor proteins; pollen. (Fuchs, E. 1996. *Annu. Rev. Genet.* 30:197, Gachet, Y., et al. 2001. *Nature* 412:352; Kost, B. & Chua, N.-H. 2002. *Cell* 108:9.)

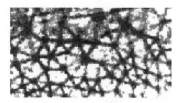

cytosol Soluble, nonparticulate material of the cell that suspends all cellular structures within the plasma membrane.

cytostatic Hinders cell division.

cytostatic factor *See* CSF.

cytostatic stability factor (CSF) Probably the product of c-mos protooncogene, it stabilizes MPF and thus prevents the exiting of the cell from the M phase of mitosis. RSK mediates the activity of CSF. *See* cell cycle; mitosis; MOS; MPF; protooncogenes; RSK. (Shibuya, E. K. & Masui, K. 1988. *Dev. Biol.* 129:253.)

cytotactin Extracellular matrix protein affecting cell movement.

cytotaxonomy Evolutionary studies based on the analysis of the cells and chromosomes.

cytotoxic Poisons or kills cells, primarily by inhibition of mitosis.

cytotoxic ribonuclease *See* colicins.

cytotoxic T cell (CTL, CTC) Type of T lymphocytes that destroys infecting cells or foreign antigens. Cytotoxic T cells recognize their targets by the antigenic peptides presented on dendritic cells. CTLs require activation mediated by CD4$^+$ T-helper cells (T$_H$), which have CD40 or the CD40L (ligand) proteins on their surface. CD40 and CD40L alone may be sufficient for activation of CTLs. Their presence on antigen-presenting cells (dendritic cells, macrophages, and B lymphocytes) enhances CTL activation. The natural killer cells (NK) lyse various types of cells without the cooperation of the MHC system. CTL and NK cells, however, may cooperate and induce the formation of cytolytic granules and lytic pores. In this process, perforin expression on CD8$^+$ T cells and sometimes on CD4$^+$ cells has an important role. The cytolytic granules may contain granulolysin, a saposin-like glycoprotein. Saposins are generated by cleavage of the 511-amino-acid presaposin, and the four types of saposins stimulate the hydrolysis of ceramides by stimulating the particular enzymes. Various leukodystrophies and Gaucher disease are attributed to saposin deficiencies. Another suggested mechanism of action relies on Fas-dependent apoptosis. Actually, these two routes may be used simultaneously. CD8$^+$ T cells recognize cytopathic and noncytopathic viruses by presentation of their peptide antigens on MHC molecules. The killing may also involve the secretion of antiviral lymphokines. Bacterial infections are handled by activated macrophages, NK cells, T-cell receptors, granulocytes, CD4$^+$ and CD8$^+$ T cells. IFNγ and perforin are the fighting molecules. In the autoimmunity disease, diabetes mellitus, perforin-dependent destruction of insulin-producing cells seems to be involved. Graft rejection is caused by the infiltration into the added tissue of NK cells, macrophages, CD4$^+$ and CD8$^+$ T cells. The latter two recognize either MHC II or MHC I differences, respectively. *See* antigen presenting cell; apoptosis; CD4; CD8; CD40; cytomegaloviruses; dendritic cell; diabetes; Fas; Gaucher disease; graft rejection; granzyme; killer cells; leukodystrophy; lymphocytes; macrophages; MHC; perforin; T cell; T-cell receptor. (Kägi, D., et al. 1996. *Annu. Rev. Immunol.* 14:297; Raulet, D. H., et al. 2001. *Annu. Rev. Immunol.* 19:291; Moretta, A., et al. 2001. *Annu. Rev. Immunol.* 19:197; Russell, J. H. & Ley, T. J. 2002. *Annu. Rev. Immunol.* 20:323.)

cytotoxicity Ability of any agent to harm, destroy, or poison cells.

CytR *See* cAMP receptor protein.

C$_6$ zinc cluster protein Group of transcriptional activators. Their DNA-binding sites have the conserved CGG...CCG triplets and in between a number of other bases in the different members of the family. Furthermore, there is a 19-amino-acid carboxy-terminal region at the zinc cluster side containing the linker and the beginning of the dimerization element that directs the protein to its preferred site. *See* transcriptional activator; transcription factors; zinc finger. (Akache, B., et al. 2001. *Nucleic Acids Res.* 29:2181.)

A.H. Sturtevant 1967 Reminiscences of T.H. Morgan. Genetics 159:1 [2001];

"...if you mix eggs and sperm from the same individual [*Ciona*, an ascidian]. Normally nothing happens. But sometimes self-fertilization does occur. And one of the questions was, Why? What brings this about? How does this happen? And Morgan had a nice hypothesis: maybe the acidity of the water is responsible. And let's see what pH changes will do. But being Morgan he didn't set up measured amounts or concentrations. What he did was to take a dish in which eggs and sperm were present and squeeze a lemon over it. And it worked. Then he studied it in more detail after that. This was one of the most successful experiments in the field."

A historical vignette.

D

d (dalton) *See* Dalton.

D Number of restriction enzyme recognition sites per DNA length. *See* restriction enzymes.

2,4-D *See* dichlorophenoxyacetic acid.

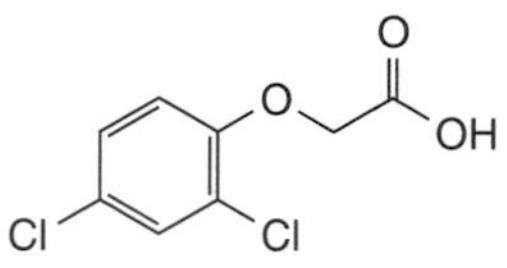

D-J *See* immunoglobulins.

Da See dalton.

DA TFIIA transcription factor associated with the TFIID–DNA complex. *See* transcription factors.

DAB Transcription factor TFIIA associated with TFIIB–TFIID–DNA. *See* transcription factors. (Maldonado, E., et al. 1990. *Mol. Cell. Biol.* 10:6335.)

dactylogram Fingerprint obtained by dactyloscopy (fingerprinting). *See* dermatoglyphics; fingerprints.

dactylology (cheirology) Study of hands, fingerprints. *See* fingerprinting.

daf (1) DNA amplification fingerprinting. *See* amplification; DNA fingerprinting. (2) *See* decay accelerating factor. (3) Dauer larva formation. *See* dauer larva.

DAG *See* diaglycerol.

DAI Interferon-induced protein; a double-strand RNA activated inhibitor involved in kinase function, regulating translation. When 25–30% of the factor is phosphorylated, protein synthesis is severely inhibited. *See* interferon; kinase; translation.

DALA δ-aminolevulinic acid.

dalton (Da) Measurement unit of molecular mass (M_r), generally used to estimate the size of macromolecules, 1 Da = 1.661×10^{-24} g (1/12 of the MW of the C^{12} isotope). 1 megadalton is 10^6 Da. 1 pg of DNA is about 0.60205×10^{12} Da; the M_r of a nucleotide pair is about 650 Da; 1 kbp DNA (Na salt) is about 6.5×10^5 Da. The average molecular weight of an amino acid residue in a protein is about 110–120 Da.

DALY Disability-adjusted life year is the sum of life years lost because of premature mortality and years of life with disabilities adjusted for the severity of the disability. On this basis, congenital anomalies occupy the 10th rank among 17 leading disabilities in the world. *See* genetic diseases.

dam Female mammal. *See* sire.

dam Deoxyadenine methylation factor in a GATC sequence. *See* methylation of DNA.

dam methylase Several main groups of methyl transferases (Hha I C5-cytosine methyltransferase, Taq I N6-adenine methyltransferase, and catechol *O*-methyl transferase). The majority of these enzymes use S-adenosyl-L-methionine (SAM) as a methyl donor. Their catalytic domain is rather well conserved and the various methylases can transfer methyl groups to DNA, RNA, proteins, and other molecules. *See* methylation, DNA; methylguanine-o^6-methyltransferase; methyltransferase. (Cheng, X. 1995. *Annu. Rev. Biophys. Biomol. Struct.* 24:293; Malygin, E. G., et al. 2001. *Nucleic Acids Res.* 29:2361.)

DAMD Directed amplification of minisatellite DNA uses PCR to produce probes for the determination of homologous variations among different species or genetic stocks for DNA fingerprinting, generally by RFLP. *See* DNA fingerprinting; minisatellite; PCR; RFLP. (Heath, D. D., et al. 1993. *Nucleic Acids Res.* 21:5782.)

D-amino acids Enantiomorphs of the natural L-amino acids. The D and L, respectively, are related to the optical rotation of the molecule; however, D or L molecules may have either (+) or (−) optical rotation. According to the original Fischer's model, D- and L-amino acids can be represented as shown below, but other representations are also used. The D forms frequently have inhibitory or antimicrobial effects. The genetic codons of both forms are the same and they are the products by posttranslational enzymatic reaction. *See* amino acids; enantiomorph.

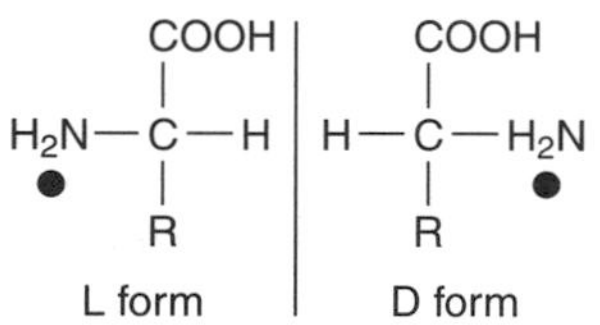

Dandy-Walker syndrome Form of autosomal-recessive hydrocephalus with considerable variations. *See* hydrocephalus.

Danon disease (Xq24) Caused by deficiency of the lysosomal-associated protein, LAMP-2. *See* LAMP; lysosomal storage diseases.

DAP *See* diaminopimelate.

DAPI 4′,6′-diamidino-2-phenylindole, a fluorochrome stain for chromosomes. When excited by UV light, blue light is emitted. *See* chromosome morphology; chromosome painting; FISH; fluorochromes.

DAP kinase (DAPK) Ca^{2+}/calmodulin-dependent serine/threonine phosphorylases of the cytoskeleton. They mediate apoptosis, metastasis, and lymphocyte function. *See* apoptosis; ITIM; killer cells; metastasis; methylation, DNA. (Cohen, O. & Kimchi, A. 2001. *Cell Death Diff.* 8:6.)

Darier-White disease (keratosis follicularis) Autosomal (12q23-q24.1)–dominant keratosis, prevalent in areas where sebaceous glands (excreting fatty substances and cellular debris) are located, e.g., on the scalp, face, chest, back, armpit, and groin. The basic defect is attributed to SERCA2-encoded Ca^{2+}-ATPase, regulating calcium signaling in cell-to-cell adhesion and epidermal differentiation. *See* Brody diseases; keratosis; SERCA1. (Zhao, X. S., et al. 2001. *EMBO J.* 20:2680.)

dark-field microscopy Receives only the scattered light arriving from a sideways illuminated object that appears light on a dark background. *See* fluorescent microscopy; microscopy; Nomarski; phase-contrast microscopy.

dark reaction Light-independent enzymatic reactions, following the light reactions of photosynthesis, leading to the formation of monosaccharides. *See* photosynthesis.

dark repair Repair of DNA not requiring light (excision repair). *See* DNA repair.

D arm *See* transfer RNA.

DARPP Dopamine-adenosine 3′,5′-monophosphate-regulated phosphoprotein is a 32 kDa molecule that becomes a protein phosphatase (PP-1) inhibitor in response to dopamine and thus regulates neurotransmission. It may actually display both serine/threonine phosphatase and kinase inhibitor activity depending on which of the amino acids is phosphorylated by Cdk5. DARPP-32 is required for progesterone-mediated sexual receptivity in female rodents. *See* Cdk5; dopamine; neurotransmitter; PP-1; progesterone. (Centonze, D., et al. 2001. *Eur. J. Neurosci.* 13:1071.)

darwin Measure of the rate of evolution $= (\ln[x_2] - \ln[x_1]/n$, where x_1 and x_2 indicate population sizes as they change during n million years (Myr), ln is the natural logarithm.

Darwinian fitness *See* beneficial mutation; fitness; neutral mutation.

DARWINISM Interpreting evolution as the outcome of natural selection, survival of the fittest. On a molecular basis, nonsynonymous amino acid replacement versus synonymous replacement may be an indication of selective or random events. *See* creationism; intelligent design; natural selection; neo-Darwinian evolution; nonsynonymous codons; social Darwinism; synonymous codons. (Gould, S. J. & Lloyd, E. A. 1999. *Proc. Natl. Acad. Sci. USA* 96:11904.)

DAS Distributed annotation system is a program facilitating the comparison of the annotations of different sources. *See* annotation. (<http://www.biodas.org/>, helpdesk@ensembl.org>.)

DASH Dynamic allele-specific hybridization in a well a short single-strand DNA sequence is hybridized with a known DNA probe. Upon pairing, the added fluorochrome lights up, indicating hybridization; upon denaturation, the fluorescent paint fades, and from the rate of fading, the strength of the hybridization can be estimated. Thus, it may reveal single-nucleotide differences between the two partners. *See* SNIPS. (Prince, J .A., et al. 2001. *Genome Res.* 11:152; Pitarque, M., et al. 2001. *Biochem. Biophys. Res. Commun.* 284:455.)

databases Provide information different subjects by electronic means. Addresses frequently change and many have overlapping information. Patience and experience are often required, but they can be very helpful. Internet addresses of a large number of databases are listed in *Trends Guide to Bioinformatics*, Supplement 1998, Elsevier Science, and *Nucleic Acid Research*, 29:1–349 (2001), describes hundreds of databases. (Baxevanis, A. D. 2002. *The Molecular Biology Database Collection: 2002 Update. Nucleic Acids Res.* 30:1; Wheler, D. L., et al. *Database Resources of the National Center for Biotechnology Information: 2002 Update. Nucleic Acids Res.* 30:13.) A catalog of more than 500 biological databases is found at <http://www.infobiogen.fr/services/dbcat>. Links to databases: <http://www/molbiol.ox.ac.uk/Links/database_links.htm>. Some information can be checked under separate entries in the alphabetical lists.

General Directories (Jump Stations)
National Center for Biotechnology:
  <http://www.ncbi.nlm.nih.gov/Entrez/>
SciCentral, Stanford Univ. general and specific information:
  <http://www.scicentral.com>
  <www.ebi.ac.uk>
  <Searchlight.cdlib.org/cgi-bin/searchlight>
  Yahoo (www subject directory):
  <http://www.yahoo.com>
Infoseek Yellow Pages: <http://www.infoseek.com/>
AltaVista: <http://www.altavista.com>
Google: <http://www.google.com>
Sherlockhound: <http://www.sherlockhound.com>
Netscape Net Search:
  <http://home.netscape.com/home/internet-search.html>
FTP (**f**ile **t**ransfer **p**rotocol): for several databases and free software packages in molecular biology:
  <ftp://ftp.ebi.ac.uk> or <ftp://ncbi.nlm.nih.gov/> European
Bioinformatics Institute: <http://www.embl-ebi.ac.uk/>
BioCatalog "yellow pages":
  <http://www.ebi.ac.uk/biocat/biocat.html> or

GenBank NCBI (National Center for Biotechnology Information): <http://www.ncbi.nlm.nih.gov/>, <http://www3.ncbi.nlm.nih.gov/Entrez>

GENOMIC AND GENETIC RESOURCES: <http://www.nhgri.nih.gov/Data> <http://www.ncbi.nlm.nih.gov/Sitemap/index. html> <http://www.ensembl.org/genome/central/>

Agriculturally important crop and animal genomes: <http://ars-genome.cornell.edu>

GenBank — general nucleotide sequence inquiries: <genbank%life@lanl.gov>

  sequence submission: <gb-sub%life@lanl.gov>

GenBank daily updated: <ftp://ncbi.nlm.nih.gov/genbank/gbrel.txt>

DDBJ — general nucleotide sequence inquiries: <ddjb@niguts.nig.junet>

  — submission: <ddjbsub@niguts.nig.junet>, <http://www.ddbj.nig.ac.jp>

EMBL — general nucleotide sequence inquiries: <datalib@embl.earn>

  — submission: <datasubs@embl.earn>. *See* European Bioinformatics, EMBL under separate entry.

EST Sequence Information: <http://www.ncbi.nlm.nih.gov/dbEST/index.html> <ftp://ncbi.nlm.nih.gov/repository/dbEST/>

STS, GenBank: <http://www.ncbi.nlm.nih.gov/dbSTS/index.html>

Genome sequences: <http://www.tigr.org>, <http://nature.com/genomics/> <http://www.ncbi.nlm.nih.gov/Entrez/Genome/org.html> <igwcb.integratedgenomics.com/GOLD>

Eukaryotic genes: <iubio.bio.indiana.edu/eugenes>

  TIGRE Gene Indices: <http://www.tigr.org/tdb/tgi.shtml>

Gene ontology: <http://www.geneontology.org>

Sequence similarity search: <http://www.ncbi.nlm.nih.gov/BLAST> <http://www.ebi.ac.uk/fasta3/>

GENOME RESEARCH TOOLS: <http://www.sanger.ac.uk./Software/EMBOSS/>

Genome Information Broker, Microbial: <http://gib.genes.nig.ac.jp/>

  DNA databank Japan: <http://www.ddbj.nig.ac.jp/>

  KEGG (Kyoto Encyclopedia of Genes and Genomes): <http://www.genome.ad.jp/kegg/>

DNA-based patents <http://geneticmedicine.org> or <http://208. 201.146. 119/>

Agricultural Patent Information: <www.cambiaip.org>

Image Library of Biological Macromolecules: <http://www.imb-jena.de/IMAGE.html>

SWISS-PROT (protein information): <http://www.expasy.ch/prosite/>

Protein sequence: <http://www.blocks.fhcrc.org> <http://protein.toulouse.inra.fr/> <http://www.sanger.ac.uk/Pfam>

Protein structure: <http://scop.mrc-lmb.cam.ac.uk/scop/> <http://www.ncbi.nlm.nih.gov/entrez/query.fcgi?db = Structure> <www.rcsb.org>,

Protein domains & functional sites: <www.ebi.ac.uk/interpro>

Structural classification of proteins: <http://scop.mrc-lmb.cam.ac.uk/scop/>

Protein complexes: <yeast@cellzome.com>

Interacting proteins: <http://www.doe-mbi.ucla.edu>

PepTool (for identifying function from primary structure): <www.biotools.com>

Protein structure images: <http://molbio.info.nih.gov/cgi-bin/pdb> <www.rcsb.org/pdb>

Protein function: <http://wit.mcs.anl.gov/WIT2> <http://www.genome.ad.jp/kegg/>

Protein modeling: <www.tc.cornell.edu/reports/NIH/resource/CompBiologyTools/>

Protein reviews (PROW): <http://www.ncbi.nih.gov/prow/>

Proteome (proteins): <http://www.expasy.ch>

Proteome (yeast, *Caenorhabditis*): <http://www.proteome.com>

Lipids: <http://www.lipidat.chemistry.ohio-state.edu>

Homology search: <info@ncbi.nlm.nih.gov>

  human-mouse homology: <http://www3.ncbi.nlm.nih.gov/Homology>

Meta-MEME (motif-based hidden Markov modeling): <metameme.sdsc.edu>

Cross-referencing model organism genes with human disease and other mammalian phenotypes: <http://www.ncbi.nlm.nih.gov/XREFdb/>

Linkage analysis and software: <http://linkage.rockefeller.edu/>

ANIMAL GENETICS:
  Poultry, Pig, Sheep, and Cattle genome: <http://www.ri.bbsrc.ac.uk/>

  Caenorhabditis: <http://genome.wustl.edu/gsc/C_elegans/>,

  Drosophila: <http://flybase.bio.indiana. edu/> or <http://edgp.ebi.ac.uk> or <http://fruitfly.bdgp.berkeley.edu/>

  Mouse Informatics Database: <http://www.informatics.jax.org>

    Encyclopedia of the Mouse Genome: <http://www.informatics.jax.org>

    Mouse, Rat, Human genome database <http://www-genome.wi.mit.edu>

    Recombinant inbred strain panels (mouse): Phone: 1-800-422 MICE or 207-288-3371. Fax: 207-288-3398

    Jackson Laboratory backcross DNA panel map service (mouse):<lbr@aretha. jax. org> or <meb@aretha.jax.org>

    TBASE (transgenic mice, knockouts): <http://www.jax.org/tbase/>

Zebrafish: <http://depts.washington.edu/~fishscop/>, <http://zebra.sc.edu>

Human gene mutation: <http://www.uwcm.ac.uk/uwcm/mg/hgmd0.html>

Human genetic variation: <http://www.ncbi.nlm.nih.gov/SNP/>

Locus link: <http://www.ncbi.nlm.nih.gov/LocusLink/>

Mendelian inheritance in man, online (OMIM): <http://www.ncbi.nlm.nih.gov/> <http://gdbwww.gdb.org>

Human Cytogenetic Network: <http://mcndb.imbg.ku.dk>

Human karyotypes: <http://www.pathology.washington.edu:80/Cyto-gallery/> or

Human physical map: <http://www.ncbi.nlm.nih.gov/genemap99/>

Human/mouse gene expression: <http://bodymap.ims.u-tokyo.ac.jp/>

Human mapping laboratories: <http://www.genethon.fr/> (information in French)

<http://www.sanger.ac.uk/> (includes also *Caenorhabditis* and other databases)

<http://www-shgc.stanford.edu/> (radiation hybrid maps primarily)

<http://www.well.ox.ac.uk/> (specific proteins, proteome and other information)

<http://www.incyte.com/>

<http://www.genome.wi.mit.edu/> (human and mouse primarily)

Genome Database: <http://gdbwww.gdb.org>)

Human Genome Project primer: <http://www.ornl.gov/hgmis>

Human cancer anatomy: <http://www.ncbi.nlm.nih.gov/ncicgap/>

Human anatomy: <http://www.le.ac.uk/pathology/teach/va2/titlpag1.html>

Human medical information: <http://www.medscape.com>

Disease control and prevention: <http://www.cdc.gov/health/diseases.htm>

GeneReviews: genetic tests <www.genetests.org>, <www.geneclinics.org>)

Genetic counseling: <http://www.geneclinics.org/>

PLANTS:

USDA, National Agricultural Library: <http://www.nal.usda.gov/pgdic/>

Mendel biotechnology: <http://www.mendelbio.com/>

Agricultural Genome Information System: <http://ars-genome.cornell.edu>

Forest tree genetics: <dendrome.ucdavis.edu>

Botany: <www.rrz.uni-hamburg.de/biologie/b_online/e00/contents.htm>

Herbarium species: <www.nybg.org/bsci/hcol/vasc>

Plants database: <http://plantsdatabase.com>

Generic model organisms: <http://www.gmod.org/>

PLANT PATHOLOGY: <http://www.ifgb.uni-hannover.de/extern/ppigb/ppigb.htm> <http://www.plantpath.wisc.edu/library>

ENTOMOLOGY: <http://ent.iastate.edu/list/>

EUKARYOTIC MICROBES:

Dictyostelium: <http://worms.cmsbio.nwu.edu/dicty.html>

Saccharomyces: <http://genome-www.stanford.edu/Saccharomyces/>

Saccharomyces Genome Database (SGD): <http://genome-www.stanford.edu> (includes also *Arabidopsis*, human data)

Yeast, Caenorhabditis, proteome: <http://www.proteome.com>

Yeast, European Network Informatics (MIPS): <http://speedy.mips. biochem. mpg.de/mips/yeast>

Neurospora: <http://biology.unm.edu/biology/ngp/home.html>

Microbes/prokaryotes for classrooms: <www.microbelibrary.org>

PROKARYOTES:

Microbiology directory: <http://www.microbiology-direct.com/>

Genome information broker: <http://gib.genes.nig.ac.jp/>

Haemophilus influenzae, Methanococcus, Mycoplasma, etc. <http://www.tigr.org>

E. coli: <http://cgsc.biology.yale.edu> or <ftp://ftp.pasteur.fr/pub/>

Microbial Strain Data Network (MSDN): <http://www.bdt.org.br/bdt/msdn/>

TIGR microbial resources: <http://www.tigr.org/tdb/tdb/index.shtml>

TIGR microbial database: <http://www.tigr.org/tdb/mdb/mdbinprogress. html>

Microbial genomes: <http://www.tigr.org/tdb/mdb/mdb.html>

Viruses, structure, biology: <medicine.wustl.edu/~virology/index.htm>

MICROSCOPIC GALLERY: <http://www.pbrc.hawaii.edu/~kunkel/>

Microscopy, histology: <www.itg.uiuc.edu/projects/atlas/>

GENATLAS (maps, functions, motifs, etc.): <http://www.infobiogen.fr/>

CARCINOGENS: <ftp://potency.berkeley.edu/pub/tables/hybrid.other. tab>

<cpdb@potency. berkeley.edu>

<http://www.iarc.fr/>)

GENETIC TOXICOLOGY (TEHIP): <http://sis.nlm.nih.gov>

MUTAGENS & TOXIC CHEMICALS: <http://toxnet.nlm.nih.gov>

Biosafety Information: <http://www.cdc.gov/od/ohs>

Biochemistry textbook:
<web.indstate.edu/thcme/mwking/home.html>

Cell biology and tools:
<www.vlib.org/Science/Cell_Biology/index.shtml>

VIRTUAL CELL (physiological modeling):
<http://www.nrcam.uchc.edu/>

Molecular structures:
<www.iumsc.indiana.edu/index.html>

Chemical terminology:
<http://www.chemswoc.org/chembytes/goldbook/>

Chemistry software and information resources:
<http://www.csir.org/>

MOLECULAR BIOLOGY ANALYTICAL TOOLS
by Internet:
<http://www.sdsc.edu/ResTools/cmshp.html>
<www.MolecularCloning.com>
<www.oup.co.uk/academic/science/biochemistry/
pas/online/>

Office of Biotechnology Activities:
<http://www4.od.nih.gov/oba>

DEVELOPMENTAL BIOLOGY
<www.luc.edu/depts/biology/dev.htm>

Teratology: <teratology.org/jfs/teratologyindex.html>

Taxonomy: <http://www.ncbi.nlm.nih.gov/Taxonomy/
taxonomyhome.html>

Biodiversity: <www.ecoport.org>

Parasites: <www.biosci.ohio-state.edu/
~parasite/home.html>

Insect pathogens:
<insectweb.inhs.uiuc.edu/Pathogens/EDWIP/
index.html>

Cognitive sciences: <http://cognet.mit.edu>

GRAPHICS: molecules, viruses, etc.
<http://www.csc.fi/lul/chem/graphics.html>

Biology textbook:
<www.ultranet.com/~jkimball/BiologyPages>

LITERATURE SEARCH:
<www.ncbi.nlm.nih.gov/entrez/query.fcgi>
<http://www.biomednet.com/db/medline/new>

National Library of Medicine:
<http://www.ncbi.nlm.nih.gov/>
<www.cshl.org/medline>
<www.researchindex.org>

EndNote: <www.endnote.com>

MedLine: <www4.ncbi.nlm.nih.gov/>

PubMed, GenBank alert service:
<http://www.pubcrawler.ie>

Full-text papers: <http://www.pubmedcentral.nih.gov/>
<http://highwire.stanford.edu>
<www.thescientificworld.com>
<http://www.freemedicaljournals.com/> free
full-length great biological and medical journals,
immediately or after a variable-length
waiting period

Biosis (journal article indexes):
<http://www.biosis.org/essentials/index.html>

Journal name abbreviations:
<http://www.uh.edu/~rmaddock/IRGO/journal-
titles.html> >

NATURE Journals: <http://www.nature.com/ng/>

Electronic Journals: <ejournal.coalliance.org>

Most cited papers in chemistry/biochemistry:
<www.cas.org/spotlight/index.html>

Hot papers in research, monthly:
<http://www.isinet.com/isi/hot/research/index.
html>

Highly cited papers: <http://isihighlycited.com/>

Biomedical engineering news:
<www.bmenet.org/BMEnet/>

Acronyms: <medstract.org/acro1.0/main3.htm>

WORDNET (word synonym sets):
<http://www.cogsci.princeton.edu/~wn/>

Statistical assistance:
<members.aol.com/johnp71/javastat.html>,
<faculty.vassar.edu/~lowry/VassarStats.html>

Computing dictionary: <www.foldoc.org>

Biodegradation: <www.labmed.umn.edu/umbbd>

Epidemics: <www.who.int/emc/index.html>

Clinical trials in progress:
<http://clinicaltrials.gov/ct/gui/c/r>

Radiocarbon dating:
<http://c14.sci.waikato.ac.nz/webinfo/index.html>

Genetics societies: <http://www.faseb.org/genetics/>

U.S. government research and development reports:
<www.osti.gov/fedrnd>

Pathways of discovery: <www.britannica.com>

Nobel laureates: <http://www.nobel.se>

Policy issues: <http://www.genesage.com/professionals/
geneletter/index.epl>

Research integrity: <http://ori.dhhs.gov>

PATENTS: <http://www.uspto.gov/>

JOBS: search the home pages of journals or
<http://www.careerpath.com> or
<http://www.ajb.dni.us/>

Cautionary note: Many symbols have multiple synonyms or single symbols standing for different genes or spelling and capitalization may vary for the same word and various errors may be encountered. Some databases cannot be entered without permission; the conditions for access can usually be obtained from the URL addresses. Addresses frequently change or fold (www-class.unl.edu/biochem/url/broken_links.html). Despite these problems, the databases provide invaluable information on details that could not be included in this book. The information must be critically read just as well as any publication coming from other sources. The addresses given above have been updated shortly before the completion of this book. (See, however, *Nature* 402:722 [1999].)

database management system (DBMS) Interprets the structure of data (schema) for a generation of databases in a specified form, e.g., Oracle™ and Sybase™. *See* Sybase.

data dredging Finding tendencies within the set of data unforeseen at the outset. *See* data mining.

data mining Searching for the biological meaning of nucleotide acid or amino acid sequences. *See* dredging data. (Baxter, S. M. & Fetrow, J. S. 2001. *Curr. Opin. Drug. Discov. Devel.* 4(3)291; Smyth, P. 2000. *Stat. Methods Med.* 9[4]:309; Perez-Iratxeta, C., et al. 2002. *Nature Genet.* 31:316; <http://cds.celera.com>.)

data model Set of bioinformatics constructs (sets, relations, objects) suitable to build databases. *See* bioinformatics, databases.

date palm (*Phoenix dactylifera*) One of the 12 species; dioecious, $2n = 2x = 36$.

dATP Deoxyadenosine triphosphate.

Datura Members of the *Solanaceae* family ($2n = 24$) have been used for genetic studies primarily involving cytological (trisomics) and cell culture methods. These species are the sources of the alkaloids atropin, hyoscine, hyoscyamine, and scopolamine. *See Atropa*; henbane.)

Datura stramonium.

Daubert rule 1993 modification of the principle of the Frye test; it allows the judge in the court to decide whether a scientific method is acceptable and recognized as an effective means to draw valid conclusions. The test in question must have been peer reviewed and must have a demonstrable accuracy. *See* Frye test. (Klee, C. H. & Friedman, H. J. 2001. *NeuroRehabilitation* 16[2]:79.)

dauer larva Represents an alternative form of larval development. At an early stage (e.g., after the second molt), the larva of *Caenorhabditis* becomes semidormant if feeding is inadequate or the culture is overcrowded. It is a safety option for survival. Such a larva does not feed, responds only to touching, and may stay alive for 30 to 70 days. Upon the appearance of a new food supply, the normal life cycle may be resumed. Dauer larvae do not age because when they resume the normal life the dauer stage does not affect the postdauer longevity. Dauer larva formation is under the control of bone morphogenetic (BMP)-like protein, a transforming growth factor (TGF-β) analog. *See* bone morphogenetic protein; *Caenorhabditis*; longevity; TGF. (Inoue, T. & Thomas, J. H. 2000. *Genetics* 156:1035; Houthoofd, K., et al. 2002. *Exp. Geront.* 37:1015.)

dauer modification Induced modification of the phenotype that may be transmitted to the progeny but persists only for a few generations (therefore it is not a mutation).

daughter cell Formed after division from the parental cell. *See* cell division.

daughter chromosome A replicated chromosome has two chromatids. When the chromatids are separated at the centromere during mitotic anaphase, they become two single-stranded daughter chromosomes. *See* centromere; chromatid; mitosis.

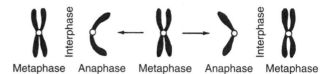

Metaphase Anaphase Metaphase Anaphase Metaphase

Formation of daughter chromosomes.

daughter of sevenless (DOS) Sevenless (SEV receptor tyrosine kinase) is a protein in the eye developmental pathway. Corcscrew (CSW) is a phosphotyrosine phosphatase in the signaling path and its substrate is DOS, a pleckstrin homology domain protein transmitting light signals for the eye between sevenless and RAS1 in *Drosophila*. In this pathway, Grb2 is an adaptor molecule with the son of sevenless (SOS) guanine exchange factor. The receptor sevenless tyrosine kinase (RTK) triggers neuronal differentiation in the single R7 cells of the ommatidia in response to the BOSS (bride of sevenless) ligand on the neighboring R8 photoreceptor cell. *See* boss; ommatidia; pleckstrin domain; rhodopsin; RTK; sevenless; son of sevenless. (Bausenwein, B. S., et al. 2000. *Mech. Dev.* 90[2]:205.)

DAX *See* adrenal hypoplasia.

day blindness (hemeralopia) Autosomal recessive; defective vision in bright light and total color-blindness. It is the result of defective cone-like bodies of the retina. *See* colorblindness; night blindness.

day neutral Does not respond to photoperiodic treatments. *See* photoperiodism.

days postcoitum *See* dpc.

Dazla *See* azoospermia.

DBA Old inbred, gray mouse.

DBF4 Cdc7-binding and -activating protein; may be required for the initiation of DNA replication. *See* Cdc7. (Ogino, K., et al. 2001. *J. Biol. Chem.* 276:31376; Jares, P., et al. 2000. *EMBO Rep.* 1[4]:319.)

DBL oncogene MCF2.

DBM paper *See* diazotized paper.

D box Recognition sequence for the ubiquitin ligase, APC (anaphase-promoting complex): RXXLXXXXN. *See* APC; ubiquitin.

DBMS *See* database management system.

DBP DNA-binding proteins are histones, suppressors, activators, silencers, DNA and RNA polymerases, and transcription factors. *See* binding proteins.

DCC Deleted colon carcinoma gene involved in cancerous growth. DCC also drives an apoptotic pathway by interacting with caspases 3 and 9. *See* apoptosis; caspases; colorectal cancer; p16; pancreatic adenocarcinoma; tumor suppressor genes. (Graziano, F., et al. 2001. *BMC Cancer* 1:9.)

dCF (deoxycoformicin) *See* ADA. (Jehn, U. & Heinemann, V. 1991. *Anticancer Res.* 11:705.)

dcm *See* methylation, DNA.

DCMU (3[3,4-dichlorophenyl]-1,1-dimethylurea) In-hibitor of photosystem II. *See* photosynthesis; Z scheme.

DCP1 Dipeptidyl carboxy peptidase is an angiotensin-converting enzyme. When it is active, it may protect against Alzheimer disease. *See* Alzheimer disease; angiotensin.

Dcp1p mRNA decapping nuclease. *See* cap.

DcR Death decoy receptor. *See* FAS.

DC-SIGN Dendritic cell-specific ICAM-3 grabbing nonintegrin is a protein on the surface of dendritic cells and communicates to the T lymphocytes. It also transfers HIV-1 virus from the mucosa of the cervix or rectum to the lymph nodes, where HIV infects the CD4$^+$ T cells and eventually causes AIDS by destroying the immune system. *See* acquired immunodeficiency; dendritic cell; ICAM; integrin. (Steinman, R. M. 2000. *Cell* 100:491.)

DCT1 Divalent cation transporter. Transports metal ions into the cells. *See* ion channels.

DctB Bacterial kinase, acting by phosphorylating protein DctD.

dCTP Deoxycytidine triphosphate (di-Na salt, MW 511.13).

dCTP.

DDB (aspartic, aspartic, glutamic acids) Appears to be the core motif in various transposases. *See* amino acid symbols in protein sequences.

DDB Damaged DNA-binding protein is a heterodimer of p127 and p48. Mutations at the xeroderma pigmentosum gene E result in loss of p48 activity and deficiency in global cyclobutane pyrimidine repair. *See* cyclobutane dimer; DNA repair; xeroderma pigmentosum. (Wakasugi, M., et al. 2002. *J. Biol. Chem.* 277:1637.)

DDBJ (http://www.ddbj.nig.ac.jp) Data Submissions, Laboratory of Genetic Information Analysis, Center for Genetic Information Research, National Institute of Genetics, 111 Yata, Mishiama Shizuoka 411, Japan, General inquiries about nucleotide sequence database, email: ddjb@niguts.nig.junet; submission forms: ddjbsub@niguts.nig.junet; telephone: 559 75 0771; <http://gib.genes.nig.ac.jp>.

DDT *See* dichlorodiphenyltrichloroethane.

***D* element** *See* nonviral retrotransposable elements.

dead-box proteins Family of ATP-dependent helicases, present in prokaryotes and eukaryotes; they can stabilize mRNA and facilitate translation with the involvement of the 43S complex containing eIF4A, eIF4B, and eIF4F. The 4F's have three subunits: eIF4A, eIF4E, and eIF4E. 4B and 4F form a helicase that binds the 5′-end of the untranslated RNA through the 4E subunit. The name DEAD comes from the single-letter amino acid symbols of proteins: Asp (D)–Glu (E)–Ala (A)–Asp (D), identifying a sequence present in eIF4A. *See* amino acid symbols in protein sequences; DEAH box proteins; degradosome; eIF-4A; helicases; RNA surveillance; translation initiation. (de la Cruz, J., et al. 1999. *Trends Biochem. Sci.* 24:192; <http://www.columbia.edu/~ej67/dbhome.htm>.)

dead-end complex If the required nucleotides are not available for transcription, the action of the polymerase is halted and may not be resumed after supplying the needed building blocks, presumably because of changes in the proper configuration at the 3′-end. Transcription may be resumed, however, if RNA hydrolytic proteins are supplied, that leads to redirection of the transcription complex. (Erie, D. A., et al. 1993. *Science* 262:867.)

deadenylation pathway Degradation of mRNA by removal of the poly(A) tail. *See* poly(A) tail. (Caponigro, G. & Parker, R. 1996. *Microbiol. Rev.* 60:233, Tucker, M., et al. 2001. *Cell* 104:377.)

DEAE cellulose As a membrane, it is used for trapping DNA from agarose gels. *See* gel electrophoresis.

DEAE dextran Polycationic diethylaminoethyl ether of dextran (a polysaccharide) that stimulates the uptake of proteins and polynucleotides into cells, promotes the infection of cells by viral RNA and DNA, may inhibit tumors in animals, and stimulates reactions to antibodies. *See* genetic animal cells; dextran; transformation.

DEAE-Sephacel DEAE-Sephadex, DEAE Sepharose

Used for gel filtration as ion exchangers and chromatographic media. *See* gel filtration; ion-exchange resins; Sephadex; Sepharose.

deaf mutism

Hereditary loss of hearing and speech with a prevalence of about 0.03 to 0.04% and with a recurrence risk among afflicted sibs of about 12%. There are an estimated 35 loci capable of causing this anomaly. The spontaneous rate of mutation is estimated to be about 4.5×10^{-4}. The incidence of deafness has a rather large environmental component. *See* deafness; Usher syndrome.

deafness

Hearing deficit within a broad range from slight hearing difficulties to complete loss; this may be a progressive phenomenon. By age 90, about half of the human population experiences some degree of hearing loss. About 68 loci are involved with nonsyndromic hearing deficit. The sensation of balance and hearing is mediated by the stereocilia (epithelial appendages) of the inner ear hair cells via electrical signals to the brain. About 75% of hearing problems have complete or partial genetic determination, and the rest may be environmentally induced. About 10-5% of the population develops hearing problems by advancing age and about 0.001 fraction of newborns are deaf or develop some kind of hearing loss by school age.

The first indication of an infant's hearing loss is the lack of articulate, understandable talking. About 87% of congenital deaf mutism is caused by recessive factors. Dominant inheritance determines deafness to some low-, middle-, and high-low-, middle-, and high-tone sounds and at other frequencies the hearing may be normal. Michel syndrome is responsible for a complete lack of internal ear formation. The hearing problems may have a wide range of organic bases but usually are classified as *conductive* (transmission) *hearing deficit*, which is caused by defects in the hearing canal or the middle ear. *Sensorineural* defects involve the inner ear and the associated nervous system. This *nonsyndromic* type (single phenotype is deafness only) is based on a mutation of the connexin gene (DFNA3) in chromosome 13q11-q12. DFNA2 (1p34) is due to mutations at a complex locus encoding the potassium ion channel KCNQ4 and possibly connexin 31. DFNA5 (7q15) is caused by deletion in intron 7 and premature termination of transcription. Deafness (DFNA15) in human chromosome 5q31 involves mutation in transcription factor POU4F3. Nonsyndromic DFNA17 (22q12.2-q13.3) is a mutation in the nonmuscle mysin gene, MYH9. DNFA23 (14q21) and DNFA25 (12q21-q24) encode nonsyndromic deafness. DFNB1 encodes connexin-26. Myo7, a mutation (DFNB2), is a nonsyndromic myosin VIIa defect. DFNB3 in human chromosome 17p11.2 (homologous with mutation shaker-2 in mouse) involves a defect in an unusual myosin molecule (Myo15), resulting in short stereocilia ([singular stereocilium] protoplasmic filaments on ca. 100 hair cells of the inner ear). DFNB4 encodes pendrin. DFNB9 is a mutation in otoferin and DFNB21 involves α-tectorin. DFNB29 gene (21q22.1) encodes the tight junction protein, Claudin-14. Nonsyndromic DFNA13 (human chromosome 6) is due to mutation in a collagen gene, COL11A2. DFNA1 appears to affect the cytoskeleton through connexin 32. DFNA16 is a dominant nonsyndromic deafness encoded at 2q23-q24.3. Mutation in human chromosome 11q22-q24 (DFNA8,

DFNA12) involves α-tectorin, an extracellular matrix protein over the inner ear hair cells. The latter mutation is synonymous with *shaker-1* of mouse chromosome 9. DFNB26 was assigned to 4q31. DFNM1 at 1q24 is a recessive deafness suppressor.

The classifications may not be absolute because the types may overlap and be further complicated in a number of syndromes. Conductive hearing problems occur in otopalatodigital syndrome, Treacher-Collins syndrome, osteogenesis imperfecta, Crouzon syndrome, and Turner syndrome. Sensorineural hearing defects are found in Alport syndrome, Jervell and Lange-Nielsen syndrome, Pendred syndrome, Usher syndrome, Stickler syndrome, Refsum syndrome, Wildervanck syndrome, Norrie's disease, albinism (coutaneous), Waardenburg syndrome, LEOPARD syndrome, and Ménière disease. Both conductive and sensorineural defects may be present in Klippel-Feil syndrome and some other cases.

Deletions or mutations in the Xq21 region may also cause deafness (DFN1, Charcot-Marie-Tooth disease, Mohr-Tranebjaerg syndrome, DFN3) in case of (surgical) injuries to the stapes (the stirrup-like bones in the ear), resulting in leakage of the fluid (perilymph) of the inner ear (gusher deafness). The molecular basis of this hearing impairment is in gene brain protein 4 (BRN-4/RHS2/POU3F4), encoding a transcription factor with a pou domain. Approximately 6% of the hereditary deafnesses are X-linked and are brought about by different mechanisms such as defects in the iris, cornea, or by ocular albinism or other types of albinisms. Thyroid hormone receptor β is essential for the normal development of the auditory function. Mitochondrially determined high sensitivity to aminoglycosides (streptomycin) may result in hearing loss. A nonsyndromic recessive deafness was located to human chromosomes 2p22-p23 and 21q22. Hereditary deafness occurs among children at a frequency of 5×10^{-4}. Some deaf children may be helped by stimulation of the auditory cortex through implants. Ames waltzer mouse's deafness is caused by mutation in a protocadherin gene. Genetic testing is feasible for connexin-based defects in hearing deficits. By 2001, 77 human gene loci were identified for nonsyndromic deafness (40 autosomal dominant, 30 autosomal recessive, and 7 X-linked). Additional genes are known to affect hearing (15 autosomal dominant, 9 autosomal recessive, 2 X-linked, 5 mitochondrial) and more than 32 genes control syndromic hearing deficits. *See* cadherin; Charcot-Marie-Tooth disease; connexin; Lange-Nielssen syndrome; mitochondrial diseases in humans; mucopolysaccharidosis; otosclerosis; pou; tight junction; Usher syndrome; Wolfram syndrome. (<http://www.uia.ac.be/dnalab/hhh/>; Steel, K. P. 1995. *Annu. Rev. Genet.* 29:675; Scott, H. S., et al. 2001. *Nature Genet.* 27:59; Kelsell, D. P., et al. 2001. *Am. J. Hum. Genet.* 68:559; Petit, C., et al. 2001. *Annu. Rev. Genet.* 35:589; Resendes, B. L., et al. 2001. *Am. J. Hum. Genet.* 69:923; Call, L. M. & Morton, C. C. 2002. *Current Op. Genet. Dev.* 12:343.)

DEAH box protein

Involved in the processing of precursor RNA and may be responsible for silencing of transgenes and transposons. *See* DEAD-box. (Wu-Scharf, D., et al. 2000. *Science* 290:1159.)

deamination

Removal of amino group(s) from a molecule. Cytosine is deaminated to uracil at the rate 3 to 7×10^{-13} sec^{-1} in double-stranded DNA, i.e., about 40 to 100 deaminations of this type occur daily in the human genome. Thus C≡G transitions to T=A are apparently of major significance for mutation and evolution. The deamination of 5-methyl-cytosine is 2 to 4 times higher than that of cytosine. In single-strand DNA, the rate of deamination is about 140 times higher than in double strands. Mismatched Cs are deaminated 8 to 26 times the rate of normally paired ones. Transcribed strands are about 4 times more likely to show deamination than nontemplate strands. Some of the deaminated nucleotides are, however, removed by uracil-DNA glycosylases. *See* DNA repair; nitrous acid; transition.

death   Irreversible stop of vital functions, especially that of the brain and the genetic material. *Genetic death* is a population genetics term for lack of reproduction. Lack of reproduction may be due to the presence of deleterious or lethal genes. The probability of survival depends on the number of recessive lethal factors (n), according to e^{-n}, e.g., in case of $n = 2$, $e^{-2} \cong 0.135$ = survival.

death domain   60–80 amino acids in the cytoplasmic regions of cytokine receptors that engage the apoptosis path. *See* apoptosis; cytokine; death receptor; scaffold-mediated activation; TNFR.

death rates   *See* age-specific birth and death rates; apoptosis, Hayflick's limit.

death receptors   Transduces the apoptotic signals with the aid of cysteine-rich extracellular domains to the intracellular death domains. *See* death domain; DED; FADD/Mort; FAS; FASL; mannose-6-phosphate; PLAD. (Ashkenazi, A. & Dixit, V. M. 1998. *Science* 281:1305.)

death signaling   (1) Ligands bind death receptors; (2) trimerization of receptor and initiation of DISC; (3) recruitment of FADD to DISC; (4) recruitment of procaspase 8; (5) formation of active heterotetrameric caspase 8; (6) activation of caspase cascades; (7) truncation of Bcl2 molecules; (8) release of *cytochrome c* and Smac/DIABLO; (9) cleavage of death substrates, leading to apoptosis. *See* apoptosis; Bcl; caspases; cytochromes; death receptors; DIABLO; DISC; FADD; Smac.

deazanucleotide   Analog of nucleotides; antiviral agent used for compression of sequencing gels. *See* compression in gels; DNA sequencing.

7-Deazaadenosine (tubercidine).

DEB   Diepoxybutane is an alkylating mutagen and carcinogen. *See* mutagens.

debranching enzyme   Converts a nucleic acid loop (lariat) into a linear molecule. *See* lariat RNA.

debrisoquin   Adrenergic-blocking drug used for treatment of hypertension. The response to the drug (human chromosome 22, dominant) depends on the cytochrome P450IID family of proteins. About 1 to 30% of the population, depending on ethnicity, may be poor hydroxylators of this and other similar drugs and may suffer serious side effects upon treatment. *See* cytochromes.

DEC-205   Integral membrane protein, homologous to the macrophage mannose receptor. It appears to have an important role in antigen presentation and processing in antigen-capturing (dendritic) T cells. *See* antigen-presenting cell. (Mahnke, K., et al. 2000. *J. Cell Biol.* 151:673.)

Decapentaplegic (*dpp*)   Complex gene/protein involved in dorsoventral and anterior/posterior (wing) differentiation in insects. Its vertebrate homologue is the bone morphogenetic protein (Bmp). An active dpp/Bmp favors ventral differentiation. DPP interacts with a large number of proteins. *See* bone morphogenetic protein; morphogenesis of *Drosophila*; *Mothers against decapentaplegic*; organizer.

decapping   Process of removal of the mRNA cap (encoded by *DCP1* in yeast) and degradation by enzymatic decay in the $5' \rightarrow 3'$ direction (by exonuclease Xrn1p). It is usually triggered by shortening of the poly(A) tail ($\leftarrow 3'$), but it may be brought about by other means. Mutation in the 7 *LSM* genes of yeast inhibits decapping and the decapping activator proteins Pat1/Mrt1. *See* cap; mRNA; polyadenylation signal; polyA mRNA. (Tucker, M. & Parker, R. 2000. *Annu. Rev. Biochem.* 69:571; Dunckley, T., et al. 2001. *Genetics* 157:27.)

decarboxylation   Removal of COOH group(s) from a molecule.

decatenation   Disentangling of the catenated sister chromatids formed during DNA replication with the aid of topoisomerase II. This is a requisite for chromatid condensation during the ensuing phases. *See* catenane; sister chromatid. (Downes, C. S., et al. 1994. *Nature* 372:467; Deming, P. B., et al. 2001. *Proc. Natl. Acad. Sci. USA* 98:12044.)

decay accelerating factor (DAF, CD55, 1q32) Erythrocyte membrane protein along with MCP and other members of the group; controls convertase activity. DAF may regulate translation without affecting mRNA in transgenic animals. *See* complement; convertase; erythrocyte; MCP. (Miyagawa, S., et al. 2001. *J. Biochem.* 129:795.)

decidua Membrane lining the uterus during pregnancy that is shed around delivery.

decoding Although the genetic code intrigued biologists before the nature of the genetic material was firmly determined, the complete genetic code was deciphered between 1961 and 1966. From frameshift mutations, Crick's laboratory concluded that the code is written most likely in triplets of nucleotides. Since 1961, random copolymers of RNA were used. e.g., in a 5A:1C copolymer the AAA triplets are expected to be more common than the CCC triplets. By chance alone, the AAA sequence is expected to have a frequency of $(5/6)^3 = 0.579$ (125/216). ACA is expected to have a frequency of $5/6 \times 1/6 \times 5/6 = 25/216$ (0.116). CCC is expected to have a frequency of $(1/6)^3 = 0.0046$ (1/216). Thus, the proportions of the three triplets were expected to be 125:25:1. In an in vitro protein synthesis assay, this copolymer most abundantly promoted the incorporation of lysine. Therefore, the codon of AAA was expected for lysine. The 2A:1C codons could be AAC, ACA, and CAA; therefore, additional copolymers were needed for determining their meanings. Repetitive ordered copolymers UUCUUCUUC permitted the incorporation of phenylalanine (UUC), serine (UCU), and leucine (CUU), depending on what register the sequence was read. The sequences GAUAGAUAG directed the incorporation of Asp (GAU) and Arg (AGA), then the translation was stopped. Thus, UAG was identified as the amber stop codon. The most precise method used ribosome binding. A collection of charged tRNAs was allowed to bind single known sequence triplets. Then tRNAs were charged with radioactively labeled amino acids. Cognate anticodons of specific charged tRNAs bound only one triplet and thus the codons were identified. The validity of the code was confirmed by recombination. At amino acid site 211 in the wild-type tryptophan synthetase glycine was identified. One mutation at this site resulted in a replacement by arginine and in another mutant by valine. A transduction experiment between the mutants restored glycine at site 211. This could be achieved if in the wild type there was CCT, and in the mutants GCT and CAT, respectively. Recombination between the first bases $\frac{GCT}{CGA}$ could produce CCT (glycine), thus verifying the codons for glycine, arginine, and valine. *See* genetic code; ribosome-binding assay; synthetic polynucleotides. (Yčas, M. 1969. *The Biological Code*, North-Holland, Amsterdam, The Netherlands.)

decoding, ribosomal Recognition of the proper mRNA triplet by the charged aminoacyl tRNA on the 30S ribosomal subunit.

decoration Minor protein fold(s) upon the basic structure that allows differences in function.

decoy receptor Binds various activator ligands and detracts them from their normal receptors; may suppress, e.g., inflammation. Some members of TNF family receptors such as lymphotoxin-β, osteoprotegrin, and TNFR2 may alleviate bowel inflammation, arthritis, autoimmune disease, etc. *See* arthritis; autoimmune disease; dumbbell oligonucleotides; receptor; TNF. (Mantovani, A., et al. 2001. *Trends Immunol.* 22[6]:328.)

decoy RNA May be used to decrease gene expression by sequestering viral RNA-binding regulatory proteins. *See* sequester; transdominant molecules. (Jayan, G. C., et al. 2001. *Gene Ther.* 8:1033.)

DED Death-effector domain of apoptosis proteins such as FLICE. *See* apoptosis; FLICE.

dedifferentiation Loss of cellular differentiation frequently followed by new cell divisions. It has been generally assumed that terminally differentiated mammalian cells (in contrast to those of lower animals and plants) are incapable of dedifferentiation. Cancerous transformation is, however, a type of dedifferentiation. Mammalian *msx1* gene can stimulate myotubes to dedifferentiate (in the presence of growth factors) and undergo transdetermination. *See* myotube; redifferentiation; transdetermination.

deer Chromosomally a very diverse group of Cervidae, yet the American white-tail deer (*Odocoilus virginianus*) is $2n = 70$ and the reindeer (*Rangifer tarandus*) is also $2n = 70$. *See* Slate, J., et al. 2002. *Genetics* 160:1587.

deetiolation Dark-grown plants usually elongate and fail to synthesize leaf pigments and thus show etiolation. Some mutations, however, show short hypocotyls and green pigment in the dark, i.e., they are deetiolated. *See* brassinosteroids.

default Preset instruction; it is followed until new instruction is given.

defective interfering particle (DI) Subgenome-size mutant due to deletion(s) that requires homologous virus for replication. These particles may have an advantage in replication over the helper virus and thus secure their maintenance. *See* helper virus.

defensin *See* antimicrobial peptides; Paneth's cell.

deficiency, chromosomal (Terminal) loss of a piece of the chromosomes (*See* deletion). Terminal losses of chromosomes can be readily induced by ionizing radiation at first-order kinetics (see kinetics) and they frequently behave as null alleles. If in a heterozygote the wild-type allele is destroyed or removed, the remaining recessive may be expressed (pseudodominance). An interstitial loss — if it is of substantial size — may be detected cytologically at meiotic pachytene because the wild-type strand bulges out across the lost tract as shown by the diagram. Nutritional or metabolic deficiency may be due to a mutant gene. *See* deletion; deletion mapping; duplication; duplication deficiency.

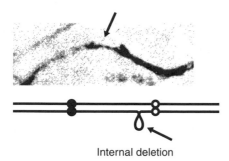

Internal deletion

defined medium Contains chemically identified and characterized nutrients.

DeFinetti diagram The genotype frequencies are represented as perpendicular lines from a point within an equilateral triangle in such a way that the length of the lines or the area correspond to the frequencies. (Li, C. C. 1976. *First Course in Population Genetics*, Boxwood Press, Pacific Grove, CA.)

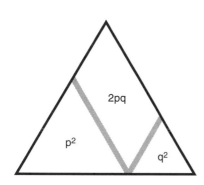

degenerate code The same amino acid is coded for by more than one type of nucleic acid triplet (e.g., the RNA code word for phenylalanine can be either UUU or UUC); other amino acids may have a single or up to six codons (i.e., six codons degenerate [go down] into one amino acid). *See* genetic code. (Crick, F. H. C. 1959. *Brookhaven Symp. Biol.* 12:35.)

degenerate oligonucleotide-directed mutagenesis Degenerate oligonucleotide (with several A residues is self-annealed) at the 3′-end with an eight-nucleotide palindrome. The 3′-sequence includes an EcoRI (G ↓ AATTC) restriction enzyme recognition site. The 5′-end should also encompass another restriction enzyme site (on the diagram below Dde I [C ↓ TNAD]). The central region is mutagenized (as shown by larger letters and a dot above the new base). The construct is treated with Klenow fragment (a fragment of bacterial DNA polymerase I that lacks 5′ → 3′ exonuclease activity) in the presence of the four deoxyribonucleotides to extend the nucleotide chains. The system produces two double-stranded mutant oligonucleotides after cleavage at the restriction endonuclease recognition sites with different base-pair substitutions. (Airaksinen, A. & Hovi, T. 1998. *Nucleic Acids Res.* 26:576.)

degenerin Member of a family of epithelial sodium channel proteins (interacting with collagen) in *Caenorhabditis* muscle contraction. Uncoordinated mutants (unc-105) have such a defect. Proteins MEC4 and MEC10, defective in mechanical signal transduction in the touch reception system, encode homologous proteins. Other homologues are the epithelial sodium channel (ENaC) genes of mammals. The MDEG1 protein, localized in the taste buds, appears to be a receptor for sour taste. *See* anesthetics; collagen; ion channels; Liddle syndrome; touch sensitivity. (Askwith, C. C., et al. 2001. *Proc. Natl. Acad. Sci. USA* 98:6459; Hamill, O. P. & Martinac., B. 2001. *Physiol. Rev.* 81:685.)

degradative plasmid *Pseudomonad* plasmids that have the ability of degrading salicylate, camphor, octane, chlorobenzoate, 2,4,5-trichlorophenoxyacetic acid, etc. (Cho, J. C. & Kim, S. J. 2001. *J. Mol. Microbiol. Biotechnol.* 3[4]:503.)

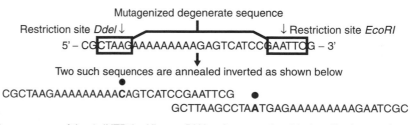

Degenerate oligonucleotide-directed mutagenesis. (Modified after Hill, D. E., et al. 1987. *Methods Enzymol.* 155:558.)

degradosome Multienzyme complex of polynucleotide phosphorylase (exoribonuclease), RNase E (endoribonuclease), RhlB (helicase), DnaK (heat-shock protein), and enolase (glycolytic enzyme) involved in the degradation of mRNA. The process is ATP-dependent. *See* glycolysis; helicase; mRNA; ribonuclease; RNA surveillance. (Capousis, A. J., et al. 1994. *Cell* 76:889; Blum, E., et al. 1999. *J. Biol. Chem.* 274:4009; Feng, Y., et al. 2001. *J. Biol. Chem.* 276:31651.)

degree of freedom Number of independent comparisons within numerical data; e.g., a 3:1 segregation has 1 degree of freedom because if one of the classes is specified within, say 4, the other can be either 3 or 1, i.e., there is only one choice. In cases where multiple comparisons can be made, e.g., in a segregation of 9:3:4 (recessive epistasis), the degree of freedom is 2, because if one class is specified as 4, the other two classes can be anything within 12/16; thus, there is another choice left. In case of 9:3:3:1 segregation, degrees of freedom are 3, because if one class is chosen, there are still three more from which to choose. In some experiments with multiple comparisons, e.g., in a variance analysis, careful consideration must be given to the number of independent comparisons before the correct degrees of freedom can be determined. *See* analysis of variance.

degree of relatedness *See* relatedness, degree of.

degron Signal element in the proteins liable to degradation by ubiquitin-mediated processes. *See* N-degron; PEST; ubiquitin. (Gardner, R. G. & Hampton, R. Y. 1999. *EMBO J.* 18:5994.)

DegS *Bacillus subtilis* kinase regulating degradative enzymes through protein DegU. (Kobayashi, K., et al. 2001. *J. Bacteriol.* 183:7365.)

dehydroepiandrosterone (DHEA) Multifunctional animal hormone enhancing immune response, with a possible antidiabetic, antiobesity, anticancer, neurotropic, memory-enhancing, and male antiaging properties. (Yen, S. S. C. 2001. *Proc. Natl. Acad. Sci. USA* 98:8167; Mazat, L., et al. 2001. *Proc. Natl. Acad. Sci. USA* 98:8145.)

dehydrogenase Enzymes mediating removal of hydrogen from molecules.

dehydrotestosterone Key hormone in maleness determination in mammals. It is made by steroid 5α-reductase enzyme from testosterone. Its deficiency causes defects in the development of the external male genitalia and the prostate, but the epididymis, seminal vesicles, and vas deferens are normal. The affected individuals are less prone to acne and baldness. The enzyme is encoded by two genes (SRD5A1, chromosome 5p15 and SRD5A2, 2p23). Dihydrotestosterone receptor deficiency leads to testicular feminization and Kennedy disease; both are located in human chromosome Xq11.1-q12. *See* animal hormones; Kennedy disease; testicular feminization.

Deinococcus radiodurans Gram-positive bacteria with 200-fold higher resistance to ionizing radiation and 20-fold higher resistance to UV than *E. coli*. The genome of two chromosomes of 2,648,638 and 412,348 bp, respectively, a megaplasmid of 177,466 bp and another plasmid of 45,704 bp, has been completely sequenced. *See* radiation sensitivity; Taq DNA polymerase. (White, O., et al. 1999. *Science* 286:1571; Makarova, K. S., et al. 2001. *Microbiol. Mol. Biol. Rev.* 65:44.)

Dejerine-Sottas syndrome (DSN, 17p11.2) Hypertrophic neuropathy, slow nerve conduction, abundant Schwann cell and basal lamina onion bulb formation, hypomyelination. Recessive mutations at the PRX locus (19q13.13-q13.2) encoding periaxin L and S are also responsible for DSN. The wild-type alleles are required for the maintenance of myelin of the peripheral nerves. DSN is also caused by mutation of the periaxin proteins encoded at 19q13.1-q132. *See* Charcot-Marie-Tooth disease; HMSN; HNPP; myelin; Schwann cell. (Boerkoel, C. F., et al. 2000. *Am. J. Hum. Genet.* 68:325.)

De Lange syndrome (Cornelia de Lange, Brachmann-de Lange syndrome) Most likely caused by new autosomal-dominant mutations. The afflicted individuals do not reproduce. Its frequency by a Danish study appeared to be 6×10^{-6}. The gene is situated in the area 3q21-qter. About 30% of cases are associated with various chromosomal anomalies, including duplication (and possibly deficiency) of the long arm of human chromosome 3. Some of the chromosomal anomalies may be unrelated to the syndrome characterized by the two eyebrows growing across the nose, hairy forehead and neck, long eyelashes, depressed nose bridge and uptilted nose tip, wide spacing of teeth, flat fingers and hands, altered palm print, mental retardation, etc. The physical anomalies are evident by the end of the second trimester. *See* head/face/brain defects; hypertrichosis; limb defects in humans; mental retardation.)

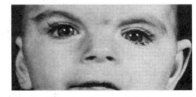

Note deformed ear, long eyelashes and flat nasal bridge. (Modified from Bergsma, D., ed. 1973 Birth Defects. Atlas and Compendium. By permission of the National Foundation of the March of Dimes.)

delayed early gene (DE) Turned on following the immediate early genes, about 2 minutes after phage infection. They use the early and new middle promoters. Their expression depends on protein synthesis. *See* immediate early genes; late genes.

delayed inheritance The expression of some traits depends on the genotype of the diploid oocyte rather then the genetic constitution of the zygote. In such cases, the reciprocal F_1 generation may be of two types (maternal or paternal), the F_2 may be uniform (because the genetic constitution of the F_1 is identical), and segregation is delayed to F_3. Similar phenomena are observed when the phenotype of the male gametophyte (pollen) is determined by the diploid microsporocyte rather than by the haploid nucleus of the microspore or pollen. *See* *Lathyrus odoratus*; *Limnaea*; testa.

delayed-response gene   Activated by a growth factor after a lag period (about an hour). *See* early gene; early-response gene; late gene.

deleterious mutation   Unfavorable for fitness. The rate of deleterious mutations is not easy to quantitate because small effects are difficult to identify. The statistical estimate for the mutation rate U is shown below where ΔM is the rate of change in the population and V_m is the variance of the mutation. The average homozygous mutation effect is $s = 2V_m/\Delta M$. A more accurate estimate can be obtained according to Keightley, P. D. & Bataillon, T. M. 2000. *Genetics* 154:1193, by using maximum likelihood.

$$\hat{U} = \Delta M^2/(2V_m)$$

deletion   Loss of a (internal) chromosomal segment; it is generally symbolized with −, or *d* or Δ or *del*. Small deletions may appear as recessive null mutations or larger deletions may have a dominant phenotype. Deletions are distinguished from mutations by failing to revert to the normal allele; they may also affect the frequency of recombination. Specifically

Chromatid deletion (Courtesy of B. R. Brinkley).

directed deletions can be obtained in mice by taking advantage of the *loxP* and *Cre* factors of bacteriophage P1 (*See* targeting genes). Deletions may be responsible for various human hereditary anomalies. Deletions can be generated in isolated DNAs by cutting the double-stranded molecules with the aid of restriction endonucleases leaving behind complementary single-strand overhangs. These ends can then be sliced back by Bal 31 exonuclease or S1 nuclease before ligation of the free ends. Deletions may be detected with the aid of a light microscope if the chromosomes have clear landmarks such as knobs or when they display banding patterns such as the salivary gland chromosomes; or deletions are revealed by special staining techniques or when the deleted chromosome is paired either during mitosis or meiosis with the normal chromosome. DNA deletions may be detected by electron microscopy if a deletion and a normal strand are hybridized in vitro. At the site of the deletion, the normal chromosome or the normal DNA strand buckles because it has no partner segment to pair with. Complete nucleotide sequences of parasitic prokaryotes indicate extensive losses of genes that were apparently not needed because the host could supply the compounds such as amino acids, fatty acids, nucleotides, enzyme cofactors, and enzymes of the Krebs cycle. Therefore, *Mycoplasma genitalium* has only 479 ORF, and *Mycoplasma pneumoniae* has only 679. *See* aging; Angelman syndrome; Bal 31; Beckwith-Wiedemann syndrome; contiguous gene syndrome; cri du chat; deficiency chromosomal; deletion mapping; DiGeorge syndrome; duplication; evolution; gene number; knockout; Langer-Giedion syndrome; Miller-Dieker syndrome; nested genes; overlapping genes; Prader-Willi

syndrome; pseudodominance; retinoblastoma; S1; Smith-Magenis syndrome; Wolf-Hirschhorn syndrome.

Deletions may be useful for identifying functional domains in the chromosome.

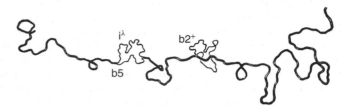

Deletion of phage-λ DNA (b2) (Westmoreland, B., et al. 1969. *Science* 163:1343).

Deletion in the salivary gland X chromosome of *Drosophila* (Painter, T. S. 1934. *J. Hered.* 25:465).

deletion analysis   Involves a number of diverse procedures, including pseudodominant expression of a recessive allele in a heterozygote when the dominant allele is deleted. Deletion mapping determines the extent of deleted segments on the basis of pseudodominance of linked genes; by genomic subtraction the normal DNA sequence extending over the gap can be isolated by molecular procedures; deletion of components of a gene (e.g., upstream regulatory elements) and their role can be identified, etc. *See* deletion mapping; genomic subtraction; linker scanning; pseudodominance.

deletion generator   The *deletion-generator* technology may produce losses in defined areas within 60 kb of the insertion sites. For this purpose in *Drosophila* a double transposable element, P{wHy} is available. It contains within a *P* element a *hobo* transposon with flanking *W* (white eye) and *Y* (yellow body) markers. The *hobo* element within the *generator* may recombine with any *hobo* insertion within the genome. The recombination then deletes any gene(s) between the *hobo* insertion and the *generator* insertion points including one of the flanking markers, *W* or *Y*, depending the orientation of the generator construct in the chromosome. The deletion of the *W* wild type allele produces white eye in the males and the loss of *Y* results in yellow body color. Therefore the deletions can be readily screened. Also a series of nested deletions can be obtained and the function of the unknown deleted genes can be annotated by the phenotype of the deletion animals. See also deficiency chromosomal, cri du chat, Wolf-Hirschhorn syndrome, retinoblastoma, Prader-Willi syndrome, Angelman syndrome, Smith-Magenis syndrome, Beckwith-Wiedemann syndrome, Langer-Giedion syndrome, DiGeorge syndrome, Miller-Dieker syndrome, Wolf-Hirschhorn syndrome, contiguous gene syndrome, aging, deletion mapping, pseudodominance, duplication, nested genes, overlapping genes, knockout, Bal 31, S1, gene number, evolution, gene number, hybrid dysgenesis, annotation. (Huet, F., et al. 2002. *Proc. Natl. Acad. Sci. USA* 99:9948.)

deletion mapping   Has been used in *Drosophila* and plants on the basis of pseudodominance, i.e., deletions of the normal sequences carrying the wild-type alleles of syntenic genes identify the length of the deletions in the heterozygotes and determine their relative position.

The fine structure of the *rII* gene of phage T4, about 300 sites, was mapped by about 2,000 mutants by the principle shown by the diagram below.

a   b   c   d   e   ← Recessive alleles in heterozygotes
                                ← Deletion of the entire area unmasks the 5 genes
                                ← Deletion unmasks *d* and *e* only
                                ← Deletion permits the expression of *a, b, c*
                                ← Overlapping deletions define the sequence *c, d,* and so on

Wild type was restored by recombination only between nonoverlapping deletions and the crosses with the long deletion indicated that the mutation and the short deletion were both within the range of the long deletion. *See* deficiency, chromosomal; deletion; pseudodominance. (Slizynska, H. 1938. *Genetics* 23:291; Benzer, S. 1961. *Proc. Natl. Acad. Sci. USA* 47:403; Khush, G. S. & Rick, C. M. 1968. *Chromosoma* 23:452; Glaever, G., et al. 2002. *Nature* 418:387.)

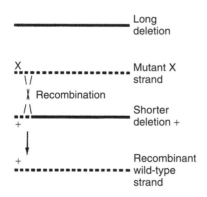

Recombination of deletion.

deletion mutation   Lacks a portion of the genetic material that may vary in extent from a single to millions of nucleotide pairs. Deletions may be generated by molecular biology techniques using restriction endonucleases and extension of the gaps by exonucleases. Since the restriction enzymes nick at specific locations, they can be used for site-directed deletions. *See* deletion, unidirectional; genomic subtraction; ionizing radiation; localized mutagenesis; radiation effects.

deletion, unidirectional   DNA molecules blocked at the 3′-end by sulfur-containing nucleotides (thionucleotides) are protected from 3′ → 5′ exonuclease action of T4 DNA polymerase and are sliced back in the opposite direction in a time-dependent extent. After removing the single strands

left behind, nested sets of deletions of various lengths are obtained. *See* Bal 31.

Delitto perfetto   RAD52-requiring technique is based on transformation of yeast by unpurified oligonucleotides leading to site-specific mutation in vivo within a window up to 200 bp. *See* site-specific mutation. (Storici, F., et al. 2001. *Nature Biotechnol.* 19:773.)

delphinidine   *See* anthocyanin.

Δ   Universal symbol for deletions.

δ   Yeast transposable element. *See* Ty.

delta 88   7 kb *Drosophila* transposable element. *See* transposable elements.

δ-deleting element (also called ΨJ$_\alpha$)   Mediates the deletion of the δ-gene from the α-gene in the T-cell receptor (TCR) site when there is no α-gene rearrangement. *See* T-cell receptor.

delta endotoxin   *See Bacillus thüringiensis*.

Δ*i*-mRNA   Fully spliced nuclear mRNA. *See* splicing.

Δ(lac⁻ proAB)   Deletion for the bacterial *lac* (lactose; map position 8) and *prolineA* and *prolineB* (map position 6 for both; blocks before glutamate semialdehyde) genes. *See Lac* operon.

$(\delta\mu)^2$ Measure of the genetic distance based on the square of the difference (δ) between average repeat size (μ) in two populations. At reproductive isolation, mutation and genetic drift in equilibrium within the ancestral and derivative populations indicates the time elapsed since separation. This parameter is highly variable and may be of limited use. *See* coalescent; evolutionary distance; F$_{ST}$.

ΔμH⁺   Energy donor, promotes translocation through cellular membranes.

demand theory   Gene expression is determined by negative control when there is a low demand for the product of the gene. In case of high demand, positive control is operating. The environment and evolutionary forces determine these cycles. *See* negative control; positive control.

deme   Interbreeding population (a Mendelian population) without reproductive isolation; it is also used for denoting a natural breeding group with one male and several females.

dementia   Deterioration of mental abilities, dullness of the mind due to innate defects of the brain or developmentally programmed disease (paranoia, schizophrenia, Alzheimer's disease, etc.) or caused by poisonous substances or drugs affecting the nervous system. Familial British (FBD) and familial Danish dementia (FDD) are caused by mutation in the ITM2B gene encoding the integral membrane protein 2B (ABRI). The mutation of FBD converts a stop codon to arginine at residue 267, thus extending the protein chain to 277 amino acids. In FDD there is a 10 bp duplication between residues 265 and 266. In both, some variations exist. The clinical symptoms resemble those of early-onset Alzheimer disease, but the amyloid deposits are systemic. *See* furin; neurodegenerative diseases. (Ghiso, J. A., et al. 2001. *J. Biol. Chem.* 276:43909; Kim, S.-H., et al. 2002. *J. Biol. Chem.* 277:1872.)

demethylation   May take place by demethylase enzymes or by base excision repair. Demethylation has significance for developmental controls in which methylation may turn off genes and demethylation may restore activity. The MeCP2 transcriptional repressor recruits histone deacetylases to the chromatin. In contrast, MBD2, which binds methylated DNA, was suggested to remove the methylation and to activate transcription. MBD2 is a member of the MeCP2 complex and—according to Ng, H.-H., et al. 1999. *Nature Genet.* 23:58—does not demethylate DNA; it silences transcription in an alternate manner. *See* histone deacetylase; methylation, DNA; trichostatin. (Cervoni, N. & Szyf, M. 2001. *J. Biol. Chem.* 276:40778; Trewick, S. C., et al. 2002. *Nature* 419:174.)

demography   Study of changes in the human population by migration, birth, mortality, marriages, health, occupations, education, etc. Genetics became the most important tool of the study of human populations and their origin by the use of data obtained from the analysis of mtDNA, the Y chromosome, and single-nucleotide polymorphism. *See* Eve foremother; mtDNA; SNIPs; Y chromosome. (Wakely, J., et al. 2001. *Am. J. Hum. Genet.* 69:1332.)

DEN   Diethylnitrosamine, an alkylating mutagen and carcinogen. *See* carcinogen; mutagen.

denaturation   Loss of native configuration of DNA (frequently separation of the complementary strands) or of proteins by damaging the noncovalent bonds at elevated temperatures or by chemicals, such as alkali, detergents, and others. The denaturation of the DNA is often reversible (renaturation). The forces between complementary DNA strands can be measured by atomic force microscopy. Adhesive forces between complementary strands of 20, 16, and 12 base pairs were found to be 1.52, 1.11, and 0.83 nanonewtons, respectively (1 newton[N] = 1 kg/m/sec^{-2} = 10^5 din). *See* atomic force microscope; c_0t; DNA thermal stability; renatura-

tion. (Ginoza, W. & Zimm, B. H. 1961. *Proc. Natl. Acad. Sci. USA* 47:639.)

denaturation mapping   Electron microscopic localization of strand mismatches in DNA. *See* deficiency, chromosomal; deletion; mismatch.

denaturing gradient gel electrophoresis (DGGE)   Permits the separation of DNA fragments of identical size on the basis of different susceptibility to denaturation because of mutation even in a single nucleotide. The DNA fragments are electrophoresed in polyacrylamide gels in which there is an increasing gradient of a denaturing agent such as urea or formamide or both. The increased temperature also promotes separation of the strands of the DNA double helix. The partially denatured fragments migrate at a lower speed in the gel. *See* electrophoresis; mutation detection. (Myers, R. M., et al. 1987. *Methods Enzymol.* 155:501.)

dendrimer   Branched molecular complex composed of monomers such as a DNA tract with a double-stranded central core and four single-stranded, noncomplementary tracts at the two ends. The complex of monomers has the potential to hybridize with complementary oligonucleotides at the open single strands. These structures can then be used as devices for microarray hybridization. Dendrimers have been used as synthetic gene delivery vehicles for in vivo gene therapy. *See* microarray hybridization. (Stears, R., et al. 2000. *Physiological Genomics* 3:93; Rudolph, C., et al. 2000. *J. Gene Med.* 2:269; Benters, R., et al. 2002. *Nucleic Acids Res.* 30[2]:210; <http://www. genisphere.com>.)

dendrite   Relatively short branch of neurons that receives information from other nerve cells. *See* nerve cells. (Häuser, M., et al. 2000. *Science* 290:739.)

dendritic cell   Heterogeneous leukocytes with the primary role of capturing MHC antigens and presenting them to T cells. The TNF receptor RANK ligand promotes their function. The granulocyte macrophage—colony-stimulating factor (GM-CSF) promotes their differentiation. TNF-α and IL-1β promote their maturation. They travel from the peripheral tissues to the lymphoid organs and promote immunity. Myeloid-type dendritic cells may stimulate T lymphocytes, whereas lymphoid dendritic cells may have the opposite effect. Fusion of human dendritic cells with breast carcinoma cells activates cytotoxic T cells against the tumors. The different CD8α dendritic cells develop from common myeloid progenitors. *See*

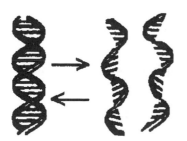

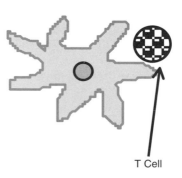

T Cell

Dendritic cell.

antigen-presenting cell; DC-SIGN; GM-CSF; IL-1; T cell; TNF. (Banchereau, J. & Steinman, R. M. 1998. *Nature* 392:245; Henri, S., et al. 2001. *J. Immunol.* 167:741; Pulendra, B., et al. 2001. *Science* 293:253; Mellman, I. & Steinman, R. M. 2001. *Cell* 106:255 and following reviews; Huang, Q., et al. 2001. *Science* 294:870.)

dendritic cell vaccine   Antigen-presenting cells primed with tumor antigens to incite CTLs against tumors. The dendritic cell may be transformed with the antigen gene using viral vectors. *See* antigen-presenting cell; cancer gene therapy; CTL; tumor vaccination.

dendrocyte   Wraps the nerve axon with the myelin sheath. Oligodendrocytes cause clustering of Na channels.

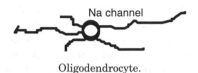

Oligodendrocyte.

dendrogram   Chart showing relationship among entities in a form resembling branches of a tree. *See* character matrix; Euclidean distance; evolutionary tree.

dengue virus   Positive-strand RNA virus translated into a single 350 kDa polyprotein, causing either dengue fever or hemorrhagic dengue. Both are debilitating diseases primarily in Southeast Asia. Mosquitoes transmit the virus. A way of protection was worked out involving the transformation of the mosquito (*Aedes egypti*) by a *Sindbis* virus vector carrying a 567-base antisense RNA of the premembrane coding region of dengue-2 virus. The transduced mosquitoes cannot support the replication and thus do not transmit the virus. *See* antisense RNA; biological control. (Diamond, M. S. & Harris, E. 2001. *Virology* 289:297.)

Denhardt reagent   Suppresses background hybridization in Northern hybridization, RNA probing, single-copy Southern hybridization, annealing DNA immobilized on nylon membrane. It is made of Ficoll-400 (a nonionic synthetic sucrose polymer), polyvinylpyrrolidone (an insoluble material removing phenolic impurities), and bovine serum albumin at a concentration of 2.5% (or less) each in water. *See* BSA; DNA hybridization; Northern blot.

denoising   Removal of noise and improvement in the quality of capillary electrophoresis. *See* capillary electrophoresis. (Perrin. C., et al. 2001. *Anal. Chem.* 73:4903.)

de novo   Starting anew (from "scratch").

de novo synthesis   Formation of molecules through synthesis from (simple) precursors rather than by cannibalization (recycling) of more complex processes of salvage.

density gradient centrifugation   Separation of subcellular organelles or macromolecules on the basis of their density.

The density of DNA (ρ) increases with increased GC content ($\rho = 1.660 + [0.098\{G + C\}]$). During buoyant density centrifugation, the DNA is positioned at the density of CsCl that corresponds to DNA density. The density of the CsCl solution is determined by refractometry. The nuclear DNA of most eukaryotes has an AT content of about 40%, corresponding to a density of about 1.7 g/mL. The DNA of cellular organelles may have a density different from that of the nucleus and thus forms a satellite band(s) in the ultracentrifuge. Single-stranded nucleic acids sediment faster and circular DNA molecules sediment slower. Among cellular organelles, nuclei sediment fastest, then chloroplasts and mitochondria occupy the highest position in the centrifuge tube. For the separation of organelles, either sucrose or percoll (polyvinylpyrrolidone-coated silica) are used most commonly. *See* buoyant density centrifugation; centrifuge; density labeling; ultracentrifugation. (Meselson, M., et al. 1957. *Proc. Natl. Acad. Sci. USA* 43:581.)

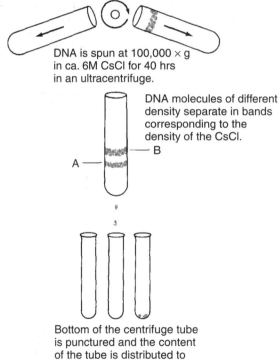

DNA is spun at 100,000 × g in ca. 6M CsCl for 40 hrs in an ultracentrifuge.

DNA molecules of different density separate in bands corresponding to the density of the CsCl.

Bottom of the centrifuge tube is punctured and the content of the tube is distributed to test tubes, 2 drops in each.

density labeling   Growing cells first in dense (heavy-isotope) medium, e.g., $^{15}NH_4Cl$, then transferring to normal $^{14}NH_4Cl$ medium; the heavy and light DNA or recombinant strands can be separated by density gradient equilibrium centrifugation. *See* density gradient centrifugation.

densovirus   *See* parvoviruses.

dental noneruption   Autosomal-dominant failure to cut teeth.

dentatorubral-pallidoluysan   atrophy (DRPLA, myoclonous epilepsy with choreoatheosis, Haw River disease)

Involves involuntary jerky muscle movements, epilepsy, feeble-mindedness, ataxia, brain degeneration due to a chromosome 12p13.31–dominant condition. The mutation causes an abnormal number (49–75) of CAG repeats in the atrophin-1 protein compared to 7 to 23 under normal conditions. *See* fragile sites; muscular atrophy; trinucleotide repeats.

DENT DISEASE (Xp11.22)　Recessive defect in the CLCN5 chloride ion channel responsible for hypophosphatemic rickets and kidney stones (nephrolithiasis). *See* nephrolithiasis; rickets.

denticle　Tooth-like extrusion (insects) or pulp stone in the mammalian tooth.

dentin dysplasia　The pulp is absent or poorly developed, the root canal frequently empty and/or enlarged, and the bluish teeth spontaneously lost. Several forms of autosomal-dominant expression have been observed.

dentinogenesis imperfecta　The dominant condition is encoded in human chromosome 4q21.3 causing blue-gray or brownish teeth due to dentin defect. The Brandywine type appears to be nonidentical. *See* dentin dysplasia; tooth. (Xiao, S., et al. 2001. *Nature Genet.* 2001. 27:201.)

Denver classification　Of human chromosomes in 1960, arranged the chromosomes into seven groups (A to G) on the basis of decreasing length and arm ratio. *See* Chicago classification; human chromosomes; Paris classification.

Denys-Drash syndrome　Very severe (dominant negative) mutation in the WT gene (Wilms, tumor), primarily affecting the female gonads and genitalia, internal male genitalia, and kidneys. *See* dominant negative; Frasier syndrome; Wilms tumor.

2-deoxyribonolactone　Oxidative damage to DNA may produce various modified products of deoxyribose such as 2-deoxypentos-4-ulose and deoxyribonolactone (dL). The latter creates problems in repairing the damage because the function of polymerase β is hampered at the AP endonuclease incised dL sites. *See* DNA repair. (DeMott, M. S., et al. 2002. *J. Biol. Chem.* 277:7637.)

2-Deoxyribonolactone.

deoxyribonuclease　Enzyme capable of breaking phosphodiester bonds in single- or double-stranded DNA. *See* DNase-free RNase; DNase hypersensitive sites; endonuclease; exonuclease; restriction enzymes.

deoxyribonucleic acid　*See* DNA.

deoxyribonucleotide　Contains only an H rather than an OH at the 2 position of the ribose, one of the nitrogenous bases (A, T, G, C), and phosphate. Deoxyribonucleotides are building blocks of DNA. *See* nucleotide chain growth; ribonucleotide.

deoxyribose　*See* ribose.

deoxyribozyme　Hammerhead ribozymes contain deoxyribonucleotides, yet they can cut RNA. Single-stranded DNAs may become Zn^{2+} /Cu^{2+} or Pb^{2+} metalloenzymes. The Zn^{2+} /Cu^{2+} enzymes may function as DNA ligase. *See* DNA ligase; ribozyme. (Carmi, N. & Breaker, R. R. 2001. *Bioorg. Med. Chem.* 9:2589; Wang, D. Y. & Sen, D. 2001. *J. Mol. Biol.* 310:723; Khachigian, L. M. 2000. *J. Clin. Invest.* 106:1189.)

depression　Pschychological state of sadness, despair, and low self-esteem generally accompanied by lack of appetite and sleeplessness. It is frequently associated with glucocorticoid overproduction, which may lead to hippocampal atrophy. The therapy may involve serotonin reuptake inhibitor drugs such as Prozac (fluoxetine). Mood is determined in the forebrain. Low activity of serotoninergic neurotransmission results in depression (dysphoria), whereas high activity causes euphoria. *See* affective disorders; glucocorticoid; hippocampus; serotonin. (Schafer, W. R. 1999. *Cell* 98:551; Zubenko, G. S., et al. 2002. *Amer. J. Med. Genet.* 114:413.)

depurination　Loss of purine from nucleic acids.

DER　Transmembrane hormone receptor tyrosine-kinase protein of the epidermal growth factor of *Drosophila*. It affects many phases of development including photoreceptor determination, wing vein formation, etc. The activator ligand is the Spitz protein (a homologue of TGF-α); Argos is an inhibitor of DER. *See* Argos; EGF; Spitz.

derepression　Removal of repression (so that protein synthesis can go on). *See* induction.

derivative　Mathematically is a function f' of a function f whose value at any point x_1 in the domain of f is as shown by the formula below if such limit exists.

$$\underset{\Delta x \to 0}{f(x_1)} = \lim \frac{(x_1 + \Delta x) - f(x_1)}{\Delta x}$$

derivative chromosome　Has been modified by chromosomal rearrangement(s).

dermatan sulfate　Glucosaminoglucan; repeating units are disaccharides, generally acetylgalactosamines linked to iuduronic acid. *See* mucopolysaccharidosis.

dermatitis, atopic (AD, 3q21)　Skin inflammation generally associated with itching, frequently evoked by allergic reactions to cosmetics, plants, animals, light, etc. The locus may be involved in paternal imprinting. AD genes are linked to psoriasis loci may be associated with 1q21, 3p24-p22, 13q14, 15q14-q15, 17q25-q21, and 20p. *See* imprinting; psoriasis; skin diseases. (Cookson, W. O. C. M., et al. 2001. *Nature Genet.* 27:372; Bradley, M., et al. 2002. *Hum. Mol. Genet.* 11:1539.)

dermatoglyphics Examination of dermal ridges and creases on fingers, toes, palms, and soles for the purpose of identification, diagnosis, and forensic investigations. *See* fingerprints. (Sodhi, G. S. & Kaur, J. 2001. *Forensic Sci. Int.* 120[3]:172.)

dermatome Cell layer generates the mesenschymal connective layer of skin.

dermatomyositis *See* polymyositis.

dermatosparaxis Hereditary disease in cattle. The procollagen peptidase that cleaves a peptide from the N-terminus of the chains is defective, causing disorganized, poor fiber formation, resulting in extreme brittleness of the hide. *See* collagen.

dermomyotome Primordium of the vertebrate skeletal muscles.

DES (diethylsulfate, $[C_2H_5]_2SO_2$) Potent alkylating mutagen and carcinogen.

De Sanctis-Caccione syndrome (xerodermic idiocy) Neurological defect frequently associated with xeroderma pimentosum. It is due to a deficiency of repair caused by mutation in ERCC6. *See* excision repair; xeroderma pigmentosum.

desaturase Enzymes on the endoplasmic reticulum of plants that introduce double bonds into the hydrocarbon portion (cis) of fatty acids in the presence of NADPH, light-generated ferredoxin, and activated O_2. Desaturation of some fatty acids may be desirable for some oil crops and can be accomplished by mutation techniques. Fatty acid desaturase may activate the plant defense system against pathogens. *See* fatty acids; host-pathogen relations. (Shanklin, J. & Cahoon, E. B. 1998. *Annu. Rev. Plant Physiol. Mol. Biol.* 49:611; Kacharoo, P., et al. 2001. *Proc. Natl. Acad. Sci. USA* 98:9448.)

desensitization Signal-response systems after prolonged stimulation display reduced responsiveness to stimulation by the same agent. The process is mediated through G protein–coupled receptors and arrestin. *See* arrestin; phosphoinositides; pleckstrin domain; signal transduction.

desert Chromosomal region with a paucity of genes or low frequency of recombination. *See* jungles.

desetope Antigen-binding site of the MHC molecule. *See* agretope; antigen; MHC.

desktop Of a computer, has the various menu bars and is where the actual work is performed.

desmin Muscle filament protein. Desmin-related myopathies (DRM) are neuromuscular diseases affecting skeletal, cardiac, and eye lens muscles. Missense mutations in the molecular chaperone αB-crystallin, encoded at human chromosome 11q21-q23, are responsible for one form. *See* crystallins; myopathy.

desmocollin Desmosome attached protein. *See* desmosome.

desmoid disease, hereditary (5q21-q22) Infiltrative fibromatosis (tumor-like fibrous nodules) due to mutation in the APC gene distal to the β-catenin domain. *See* Gardner syndrome.

Desmoplakin Protein involved in cell junctions. *See* desmosome; filament.

desmosome Protein plaque of cell junctions (between epithelial cells) into which desmin and keratin filaments of cells are tied. Besides mediating cell adhesion, desmosomal cadherins are involved in epithelial morphogenesis. A defect of desmosomes is responsible for various skin diseases. *See* cadherin; filaments; intermediate filaments; junction complex; pemphigus; plakoglobin; skin diseases. (Runswick, S. K., et al. 2001. *Nature Cell Biol.* 3:823.)

desmosterolosis Involves diverse lethal malformations due to deficiency of 3-β-hydroxysterol-δ-24-reductase. *See* chondrodysplasia. (Waterham, H. R. 2001. *Am. J. Hum. Genet.* 69:685; Kelley, R. I. & Herman, G. E. 2001. *Annu. Rev. Genomics Hum. Genet.* 2:299.)

desmoyokin Cell junction protein. *See* Nie, Z., et al. 2000. *J. Invest. Dermatol.* 114:1044.

desmutagen Chemical agents that can interact and directly inactivate mutagenic molecules.

destruction box Amino acid sequences at the N-terminal of cyclins and other proteins, making them targets for ubiquitinylation: **Arg** (Ala/Thr) (Ala) **Leu** (Gly) x (Ile/Val) (Gly/Thr) Asn. Those in bold are conserved; the others may vary. *See* amino acids; degron; N-degron; N-end rule; ubiquitins. (Burton, J. L. & Solomon, M. J. 2001. *Genes Dev.* 15:2381.)

desynapsis Loss of synapsis of the homologous chromosomes (after the completion of the recombination process). The desynaptic gene of maize reduced chiasma frequency in the majority of chromosomes, but it increased chiasma frequency in the short arm of chromosome 6 where the nucleolar organizer is situated. *See* asynapsis; chiasma; nucleolar organizer; sister chromatid cohesion; synapsis. (Dix, D. J., et al. 1997. *Development* 124:4595.)

detasseling Removal of the male inflorescence of maize. *See* heterosis.

Tassel.

detector protein Sense environmental signals (in bacteria) in the periplasmic region. *See* periplasma.

detergent The various types may have different chemical structures, but they all have a large nonpolar hydrocarbon end that is oil-soluble and a water-soluble polar end. The so-called soft detergents are biodegradable. In the laboratory, most commonly sodium lauryl sulfate (SDS, sodium dodecyl sulfate, n-$C_{11}H_{23}CH_2OSO_3^-$ Na^+) is used in gel electrophoresis. Detergents are also employed for the solubilization of membrane proteins and lipid components. *See* nonidet; polyacrylamide gels; SDS; Tween 20.

determinate inflorescence The stem terminates in a flower rather than in an apical meristem. *See* indeterminate inflorescence; meristem.

determination Establishment of a specific *commitment* to differentiation. It is a biochemical change within cells or tissues whereby they lose options for differentiation in all but one particular way. In plant cells, the determination may be reversible, however. In the processes controlled by homeotic genes, transdetermination may overrule the regular pattern. *See* heritability; homeotic genes; transdetermination; transplantation, nuclear. (Glotzer, M., et al. 2001. *Annu. Rev. Cell Dev. Biol.* 17:351.)

determinism, genetic Tenuous social theory that crime, immorality, disease, poverty, and all social ills are predetermined in families. Behavior genetics shows that inheritance plays a certain variable role in behavior and by logical extension in the judgment of human values. Also, the question of the degree of responsibility is opened. If behavior is genetic, what is the role of free will? Ethicists and legal scholars have difficulties defining values. If values are measured, can they be valued? The philosopher David Hume remarked: "Good sense and genius beget esteem: Wit and humor excite love." The problems certainly exceed the scope of this text, although the moral dilemmas are inescapable for the geneticist once Pandora's box has been opened. *See* ethics; human behavior. (Alper, J. S. 1998. *Soc. Sci. Med.* 46:1599.)

deterministic gene In its presence a certain condition or disease is most likely to occur.

deterministic model In a population the changes are based on fixed parameters such as the selection coefficient, mutation pressure, migration, etc. *See* stochastic model.

detoxification enzyme *See* glutathione-S-transferase, paraoxonase, superoxide dismutase.

detrimental mutation The mutant has low fitness, but its rate of survival is not below 10%. *See* beneficial mutation; fitness; mutation.

detritus Particulate decay product.

deubiquitinating enzyme Cleaves the ubiquitin-protein link and thus prevents degradation by protease. *See* ubiquitin. (Richert, K., et al. 2002. *Mol. Genet. Genomics* 276:88.)

deuteranomaly Deficiency of the photopigment, sensitive to middle wavelength, yet retains some trichromatic (three-color) vision, depending on how many short and long wavelength receptors there are. The mildest cases have differences from normal in exon 5 of the X-chromosomal gene. The next mildest anomaly involves exon 2 and either exon 3 or 4 but not both. In the most severe cases, the anomaly was commonly in exon 2 but usually not in 3 or 4. Deuteranomaly affects about 5% of U.S. males but only 0.25% of females because of X-linkage. *See* color-blindness; night blindness.

deuteranopia *See* color blindness.

deuterium ^2H, heavy hydrogen (atomic weight: 2.014725), a stable, nonradioactive isotope.

deuterostome Taxonomic group characterized by blastopores behind the mouth, such as echinoderms, hemichordates, and chordates. *See* blastopore; protostome.

deuterotoky Unfertilized eggs develop into male or female. *See* arrhenotoky; chromosomal sex determination; sex determination; thelytoky.

development Sequence of events beginning with determination and leading through differentiation to the various stages of life of the organism. In multicellular organisms, development begins with the fertilization of the egg and goes through epigenesis, terminating in death. The zygote grows, and at an early stage, polarity and segmentation become obvious. In animal embryos, the cells move in to organizing centers (gastrulation) and the development becomes rather rigidly determinate. The germline is laid down and is separated from somatic differentiation. Plant cells maintain a high degree of totipotency through development, and movement of cells is restricted by the rigid cell walls. In plants, *meristems* containing totipotent cells are organized and serve as the bases of growth differentiation and development. In animals, the *stem cells* assume the functions comparable to those of the meristem of plants. The development is based on a highly regulated cooperation of genes that are turned on, changed in the level of activity, and turned off. Although development is potentiated primarily according to the genetic blueprint, the realization of the plan

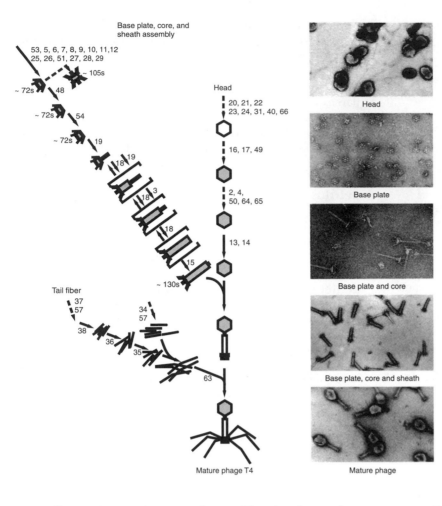

Base plate, core, and sheath assembly

The morphogenetic and developmental pathway of phage T4. The numbers refer to genes. In case of mutation, incomplete phage parts may develop within the host bacteria. Many of these structures are true phage precursors and may form viable phage particles in vitro from appropriate mixtures of mutants blocked at key developmental steps. (After Wood, W. B. 1980. *Quarterly Rev. Biol.* 55:353; and King, J. 1971. *J. Mol. Biol.* 58:693.)

generally requires environmental cues. The signal transduction pathways provide the motivation. Developmental events are the outcome of combinatorial action of a relatively limited number of signals and the expression of regulatory genes. The sensors of these signals may reside within the enhancers of the major genes concerned with the guidance of cellular differentiation. Development is genetically controlled in all organisms. Many genes of the bacteriophages have been identified that control individual steps of differentiation of the structures and their assembly into a mature phage particle.

The phage body is constructed according to the instructions of the viral genes, but the cellular metabolism is used for execution of the project. Thus, the cooperation of both host and the viral genomes is responsible for the product. The simplest developmental pathway of animals is seen in the nematode, *Caenorhabditis elegans*. About 90% of the genes of *Caenorhabditis* involved in the control of development are related to developmental genes of *Drosophila* and vertebrates, whereas only ~50% of the other protein coding genes are homologous to other taxonomic groups. Developmental mechanisms are of focal interest of research using diverse organisms. The complex interaction of many genes can be monitored by microarray hybridization. *See Caenorhabditis*; coordinate regulation; GeneEMAC; heterochronic RNA; junction of cellular networks; microarray hybridization; morphogenesis; morphogenetic pathways; segregation asymmetric; signal transduction. (Boys, D. C., et al. 2001. *Plant Cell* 13:1499; Rougvie, A. E. 2001. *Nature Rev. Genet.* 2:690; <http://www.ucalgary.ca/UofC/eduweb/virtualembryo/>.)

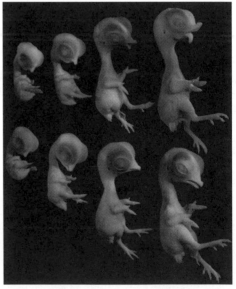

A clear example of the genetically programmed development of simple traits of higher animals. Bottom: normal. Top: "Donald Duck" chicken embryos after 8 to 11 days of incubation. The beak anomaly is the result of homozygosity of a recessive gene. The mutants cannot be identified until the 8th day of incubation, and within 2 days after the onset, both lower and upper beaks display the developmental defect. By later stages, the condition becomes more pronounced and the afflicted chicks cannot survive. (Courtesy of Abbott, U. K. & Lantz, F. H. 1967. *J. Heredity* 58:240.)

development, autonomous The process is independent of the (cellular, tissue) environment.

developmental clock The developmental fate of an organism or structure is determined as a function of time.

developmental cycle Processes during ontogenetic development in the alternating generations. *See* alternation of generations; ontogeny.

developmental field Group of cooperating cells destined to a differentiational fate.

developmental noise Variation that cannot be attributed to any verified cause.

developmental therapeutics program (DPT) Involves information on drug discovery: <http://epnws1.ncifcrf.gov:2345/dis3d/DTP.html>.

developmental regulator effect variegation Shows some similarities to the variegation-type position effect, but the variegation is induced when the *Polycomb* gene and its response element are inserted at euchromatic sites of *Drosophila*. This variegation is also sensitive to the products of the *trithorax* gene. *See* bithorax; *Polycomb*; position effect; *trithorax*; variegation.

deviation Difference from the usual type. *See* standard deviation; variance.

Dewar photoproduct UV-induced valence isomer of pyrimidine-pyrimidone 6-4 adduct. It is less mutagenic than the 6-4 adduct, which inserts preferentially a G opposite to a 3′ T and results in a highly mutagenic T → C transition in ~85% of the replications of *E. coli* cells. The Dewar product leads to a lower frequency but broader range of mutation by SOS repair. It leads to insertions of A in 72% across T, and in 21% of the cases a G occurs; in 13% of the cases, translesion errors occur. The Dewar product causes less distortion of the DNA structure than the 6-4 photoproduct and therefore causes less hardship for repair DNA polymerase. *See* DNA repair. (Lee, J.-H., et al. 2000. *Proc. Natl. Acad. Sci. USA* 97:4591.)

(6-4) adduct   Dewar product

dexamethasone ($C_{22}H_{29}FO_5$) Long-active synthetic glucocorticoid with antiinflammatory and immunosuppressive action. It is usually an inducer of glucocorticoid receptors in animals but not necessarily in plants. *See* glucocorticoids; GVG.

DexH/D (unwindase) Double-stranded RNA helicase that disrupts RNA-protein interaction and reorganizes ribonucleoproteins. *See* NPH; unwindase.

dextran α-1,6-linked poly-D-glucose, sometimes with α-1,4 or α-1,3–linked branches. It is present in cross-linked form in the gel filtration and anion exchanger agent Sephadex. It is synthesized by bacterial enzymes and is an important component (along with inorganic salts and lipids) of dental plaques responsible for tooth decay (caries) and gum disease. *See* DEAE-dextran; Sephadex.

dextrin *See* amylopectin.

dextrinosis *See* glycogen storage diseases (type III).

dextrorotatory Rotates the plane of polarized light clockwise.

dextrose Glucose.

df or DF *See* degree of freedom.

DFFR *See* azoospermia.

DFI Differential fluorescence induction of the green fluorescent protein gene during FACS. *See* aequorin; FACS.

DGGE *See* denaturing gradient gel electrophoresis.

dGTP Deoxyguanosine triphosphate.

Deoxyguanosine triphosphate.

DHAC 1 (RPD3) Histone deacetylase. *See* histone acetyltransferase; histone deacetylation.

DHFR Dihydrofolate reductase (22-kDa) controls the reaction: dihydrofolate + NADPH + H^+ → tetrahydrofolate + $NADP^+$; *dhfr*-deficient cells are resistant to methotrexate (amethopterin). Transformation to methotrexate resistance may be beneficial for cancer chemotherapy. The conversion of deoxyuridylic monophosphate to deoxythymidylic monophosphate requires tetrahydrofolate. *See* amethopterin; DM chromosome; folic acid; methotrexate; multidrug resistance; $NADP^+$. (Wallace, L. A. & Robert Matthews, C. 2002. *J. Mol. Biol.* 315:193.)

DHR domain PDZ domain.

diabetes insipidus Autosomal-dominant type. Like all diabetes, it involves an imbalance in electrolyte control, resulting in excessive urination and extreme thirst. In the case of the neurohypophysin type, the defect resides within the arginine V2 vasopressin receptor gene and the controlling antidiuretic neurohypophyseal hormone. The kidneys cannot reabsorb large amounts of water. A 10–15–fold higher-than-normal water intake (up to 20L/day) may be required. In the autosomal-dominant nephrogenic type, upon administration of antidiuretic hormone, the cyclic adenosine monophosphate level remarkably increases in the urine. There is also an early-onset juvenile insulin-dependent form. This may be recessive or polygenic and is hard to resolve because of the high frequency of the gene and the low penetrance. In another syndrome, diabetes mellitus, diabetes insipidus, and deafness occur together, probably under autosomal-recessive control. Some of the nephrogenic diabetes insipidus cases are either autosomal dominant or recessive and both are encoded at 12q13 or may be encoded at Xq28. The neurohypophyseal type has been assigned to 20p13. Other somewhat less-well-characterized complex forms are also known. *See* aquaporin; cAMP; diabetes mellitus; insulin; kidney disease; MODY; neurophysin; vasopressin; Wolfram syndrome. (Willcutts, M. D., et al. 1999. *Hum. Mol. Genet.* 8:1303; Pasel, K., et al. 2000. *J. Clin. Endocrin. Metab.* 85:1703.)

diabetes mellitus Recessive hereditary disease under the control of more than a single gene, causing hyperglycemia due to insulin deficiency and other factors, and by defects in glucose transport from the blood to the cells. Diabetes of the mother may seriously affect the fetus. The prevalence of diabetes in most human populations exceeds 3%. The recurrence risk among sibs varies depending on the type of genetic control or primarily a nongenetic type of diabetes. The penetrance and onset are variable, and dietary and other environmental factors affect it. Treatment involves dietary restrictions and insulin administration in insulin-dependent diabetes. In common usage, without qualification, diabetes means the mellitus form, although several other types of the condition have been identified.

Diabetes is characterized by excessive amounts of sugar in the blood and excretion of large amount of urine, which may or may not contain excessive amounts of sugar. Diabetes can be medically characterized into insulin-dependent or immune-dependent (IDDM, type I) and insulin-independent (NIDDM, type II) forms. Recent evidence seems to point to the cause of type II disease as the imbalance between insulin action and insulin secretion. The insulin receptor apparently acts on two proteins, the *insulin receptor substrates* (IRSs), and the type II disease IRS-2 may function in an aberrant manner. The IRSs apparently recruit a number of other proteins that are also involved in the pathways of insulin control.

In nonobese (NOD) mice, it appears that glutamic acid decarboxylase expressed in the β-islet cells of the pancreas is required for the activation of T cells, responsible for the development of the autoimmune reaction of type I diabetes (Greely, S. A. W., et al. 2002, *Nature Medicine* 8:399). In IDDM, the loss of glucose may cause hunger, great thirst, and frequent urination. The glucose loss may then involve increased catabolism of proteins and fat. As a consequence, weight loss may follow. When the mobilization of fats increases

and the oxidation of fatty acids becomes incomplete, ketone bodies and acetone may accumulate. The ketones appear in the urine and the acetone may be exhaled, giving the untreated patients a special odor.

IDDM is treated with insulin and a properly controlled diet. NIDDM may be caused by a defect in the glucagon receptor in human chromosome 17q25, near-telomeric region. A susceptibility locus was assigned for type II at 3q27-qter and 1q21-q24. NIDDM1 locus was assigned to chromosome 2 and NIDDM2 to chromosome 12. More substantial evidence is in favor of association of three protease locations (calpain10, [2q37.3]), carboxypeptidase E/H [chr. 4], and prohormone convertase-1 [5q15-q21] with diabetes NIDDM1. The calpain linkage was detected in Mexican populations and confirmed in German and Finnish individuals but rarely in other European lineages. In NIDDM, obesity is a frequent sign, but besides the increased blood-sugar content, other symptoms may vary, depending on the cause. The latter-type patients do not respond to insulin. The obesity in NIDDM seems to be regulated by TNF-α. The first clinical test for diabetes is the glucose tolerance test. About 100 g glucose is given in water orally to fasted individuals. In healthy individuals, the blood-sugar level returns to normal after 2–3 hours but not in the diabetic ones.

Various animals are also afflicted with this condition. Several forms of insulin-dependent diabetes are associated with HLA antigen DR3/4 or 3 or 4. Diabetic and glucose intolerance symptoms are associated with a large number of syndromes. A gene involved in the control of IDDM7 has been located in human chromosome 2q34. An early-onset insulin gene, IDDM2 (human chromosome 11p15.5), seems to be associated with tandemly repetitive DNA sequences that regulate insulin transcription. The principal gene for early onset (IDDM1) appears to be in human chromosome 6p21.3 and 1p, although 6 indicated reasonable linkage (lod scores ~3), yet lod scores of 1.0 to 1.8 with 6 other loci were found. Another 1998 study in England indicated only weak linkage (lod scores) with chromosomes 1p21 (1.8), 4p15 (1.9), 8p24-p21 (1.5), 12p13-pter (1.8), 16p11-q12 (2.2), and 21q11 (2.5). The glucokinase gene (GCK), located to 7p15, is also involved in IDDM. The X-chromosomal (Xp13-p11) IDDM locus is in fragment DXS1068. It was well established in people carrying the HLA-3DR antigen. Generally the male:female ratio is close to 1 at a high incidence of IDDM1, but it may be low in low-incidence countries of Asia and Africa. IDDM3 was located to chromosome 15q. IDDM4 appears to be in 11q13, IDDM5 in 6q25, and chromosome 18q harbors IDDM6; IDDM9 is in 3q21-q25; IDDM10 is near the centromere of chromosome 10. The literature reported IDDM locations at 15 different chromosomal regions, but a 1998 analysis could verify linkage with highly significant lod scores (>3) only at 6.p21.3 and 1p, although 6q indicated reasonable linkage, and lod scores of 1.0 to 1.8 with 6 other loci were found.

This is basically a polygenic autoimmune disease brought about by the destruction of insulin-producing β-cells in the Langerhans islets of the pancreas by the infiltrating of T lymphocytes, B lymphocytes, macrophages, and dendritic cells. The variable number of tandem repeats (VNTR) in the insulin genes may contribute to juvenile obesity. CTL may be activated by perforin-granzyme B release or FasL. TNF, lymphotoxin, and IFN-γ may also damage the pancreatic β-cells. Protein CD44 and its hyaluronic acid ligand have an adverse role in diabetes. T_H1 cells seem to play the primary role in the process, whereas $T_H 2$ cells may not have either a promoting or a protecting effect. There are some indications that retroviral infection may contribute to the expression of diabetes. The concordance between monozygotic twins is only 50%. The MHC class II factors are most important in determining diabetes susceptibility.

Diabetes may be responsible for blindness, kidney disease, heart attacks, strokes, and for the amputational needs of the lower extremities. Genetically engineered nonpancreatic cells carrying the insulin gene under the appropriate promoter may eventually cure diabetes. Such constructs would release insulin in a modulated manner like the normal pancreatic β-cells and would evade immune destruction. Mitochondrial dysfunction accounts for 0.5 to 1% of the cases of diabetes mellitus caused by loss or defect in pancreatic β-cells and diminished secretion of insulin. *See* adiponectin; autoimmune disease; blood cells; calpain; CD30; cytotoxic T cells; diabetes insipidus; FasL; glucagon; glutamate decarboxylase; granzyme; Hirschprung disease; HLA; HNF; IFN; insulin; insulin receptor substrates; ion channels; Langerhans islet; lipodystrophy; lod score; lymphotoxin; MODY; neurod; obesity; perforin; polygenic inheritance; PPAR; superantigen; T cells; thiazolidinediones; TNF; *VENTURE*; VNTR. (Watanabe, R. M., et al. 2000. *Am. J. Hum. Genet.* 67:1186; Bach, J.-F. & Chatenoud, L. 2001. *Annu. Rev. Immunol.* 19:131; *Nature* 414:782–827; *Amer J. Med. Genet.* 115, issue 1 [2002].)

diabetes type, I and II   *See* diabetes mellitus.

DIABLO (direct IAP-binding protein with low pI)   *See* caspase; IAP; pI; Smac. (Adrain, C., et al. 2001. *EMBO J.* 20:6627.)

diabody   Dimeric association of shorter-than-normal variable heavy and light chains of the antibody (ScFv). Such dimers may be monospecific, i.e., they are formed from two variable A immunoglobulin chains or bispecific when A and B chains are joined by a linker. *See* bispecific antibody; humanized antibody; recombinant antibody; triabody. (Kortt, A. A., et al. 2001. *Biomol. Eng.* 18[3]:95.)

diacylglycerol (DAG)   Lipid in the cell membrane that activates protein kinase C (PKC). DAG regulates growth and differentiation by its modulated level within the cell nucleus. The level of DAG in the nucleus is regulated by DAG-kinase-ζ (DGK-ζ). PKC phosphorylates DGK-ζ in the MARKCS (myristoylated alanine-rich C-kinase substrate) domain. *See* protein kinase.

diagenesis   Physical and chemical alterations that take place in the geological sediments after deposition of organisms.

diakinesis   Nuclear division phase characteristic only for meiosis. It very closely resembles the diplotene stage, but the condensation of the chromosomes is further increased. The chiasmata tend to move toward the termini of the chromatids (terminalization) and the paired chromosomes (bivalents) begin their separation from each other as approaching metaphase. *See* meiosis.

diallele analysis   Used in quantitative genetics and (plant) breeding to assess the breeding value of genetic stocks by crossing them in all possible combinations. The data reveal both nuclear and extranuclear contributions. The great amount of work involved limits, however, the number of stocks that can be studied. From n lines $n(n-1)/2$ single crosses and $[n(n-1)(n-2)(n-3)]/8$ double crosses are possible. Thus, from 10 lines, 45 single crosses and 630 double crosses are possible. Similarly from 40 inbred lines, 780 single crosses and 274,170 double crosses can be made. Therefore, some geneticists prefer heritability tests involving intraclass correlation. *See* double cross; heritability; intraclass correlation; single cross. (Hinkelmann, K. 1977. in *Proc. Int. Conf. Quant. Genet.*, Pollak, E., et al., eds., p. 719 Iowa State Univ. Press, Ames.)

dialog box   Requests further information from the user of the computer or gives some warnings accompanied by a beep sound.

dialysis   Removal of small molecules through a semipermeable membrane into water or into lower-concentration solutes.

diaminopimelate (DAP)   (Polylysine) peptidoglycan component of the cell wall of many bacteria. *E. coli* strain EK2 ($\chi 1776$) is blocked in its synthesis and prevents the organism from surviving in the mammalian gut. It was used as a host for genetic vectors for recombinant DNA to assure containment. *See* biohazards, biological containment.

2,6-diaminopurine   Mutagenic analogue of adenine.

Diamond-Blackfan anemia (DBA)   Human chromosome 19q13 erythroblastopenia (deficiency of red blood cells) caused by defects (breakage point) in the gene encoding ribosomal protein S19. Mostly neonates and infants are afflicted. The decrease of erythroid precursors is frequently associated with upper limb, face, and other malformations. Allografted bone marrow restores blood cell formation. *See* allograft; ribosomal proteins.

diandry   Fertilization of a single egg by two spermatozoa. *See* methylation, DNA.

diapause   Relatively inactive period in the life cycle of an organism such as during pupation, hibernation, seed stage, etc.

diapedesis   Exit of corpuscular elements (virus) through intact blood vessels. *See* extravasation.

diaphorase   *See* methemoglobin; NAD.

diaphyseal aclasis   *See* exostoses.

diarrhea   May have diverse origins. A rare autosomal-dominant type responded to steroid hormone treatment. Autosomal-recessive forms include a defect in chloride absorption and occur at a frequency of about 7.6×10^{-5}. Another different recessive form was affected in the Na^+/H^+ exchange, and a rare X-linked form was found to be very susceptible to infections and other complex problems, e.g., eczema, thyroid autoimmunity, diabetes, hyperimmunoglobulinemia D, etc.

diastase   Enzyme complex that can hydrolyze starch into sugar.

diastematomyelia   Autosomal-recessive disease causing a split in the spinal cord by either fibrous or bony material; each half is wrapped. It is often associated with spina bifida, atrophy of the legs, and other defects. *See* atrophy; neuromuscular diseases; spina bifida.

diastereomer   Stereoisomers that are not enantiomorphs, i.e., are not mirror images of each other, e.g., galactose/glucose. *See* enantiomorphs; galactose utilization.

diastole   Rhythmic expansions of the heart cavities and filling it with blood. Diastolic dysfunction is a common cause of heart failure. *See* hypertension; systole.

diastrophic dysplasia (DD)   Autosomal-recessive (human chromosome 5q31-q33 or q34) anomaly with curved spine (scoliosis), clubbed foot, backward-bending abnormal thumb, abnormal earlobes, premature calcification of rib cartilage, short stature, respiratory and cardiac insufficiencies, etc. Similar symptoms are shared by various hereditary bone diseases. *See* dwarfism; epiphyseal dysplasia; limb defects in humans; stature in humans.

diauxy   *See* glucose effect.

diazotized paper   Modified filter paper (Whatman 540, Schleicher & Schuell 589, or equivalent) is used for nucleic acid transfers and Western blots. The paper is first treated with nitrobenzyloxymethyl pyridinium (NBPC), which leads to the formation of NBM paper. This paper is then reduced with sodium bisulfite ($Na_2S_2O_4$) to aminobenzyloxymethyl (ABM) paper; through a reaction with nitrous acid (HNO_2), the amino group is converted into the diazo group of the DBM (diazotized) paper. NBM and ABM papers are commercially available. The stability of these modified papers depends on the conditions of storage. *See* Western blot.

dibasic amino aciduria   Autosomal-recessive defect involving accumulation of lysine, arginine, and ornithine when protein-rich food is consumed without any increase in cystine. The anomaly may lead to liver and bone defects. *See* amino acid metabolism; argininemia; citrullinemia; cystine-lysinemia; hyperlysinemia.

dicentric bridge   Formed when a dicentric chromosome is stretched because at anaphase the two centromeres are pulled

in opposite direction. *See* breakage-fusion-bridge cycles; bridge; dicentric chromosomes; paracentric inversions.

dicentric chromosome   Has two centromeres and is usually an unstable structure because at anaphase the two centromeres are pulled toward opposite poles, generating a bridge that may be ruptured at any point and leading to unequal distribution of the genes of that chromosome.

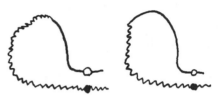

Complementary dicentric chromatids result from four-strand double-crossing over within a paracentric inversion.

The dicentric chromosomes usually do not transmit through meiosis by being trapped in between the two poles. In mammals, dicentric chromosomes with a short interval (4–12 Mb) between the two centromeres may function as if they were monocentric because one of the centromeres is inactivated. *See* breakage-fusion-bridge; inversion, paracentric; Robertsonian translocation; Turner syndrome. (Thrower, D. A. & Bloom, K. 2001. *Mol. Biol. Cell.* 12:2800.)

dicentric ring chromosome   Formed when there is a sister chromatid exchange in a ring chromosome. At mitotic anaphase, the sister centromeres are pulled toward the opposite poles, causing the chromatids to break at potential points of stress and leading to unequal distribution of the genes (duplications and deficiencies in the cells). The broken ends may fuse and the process may continue. The resulting somatic instability is revealed by sectors if appropriate genetic

Sectorial maize leaf homozygous for a deficiency and carring a ring chromosome.

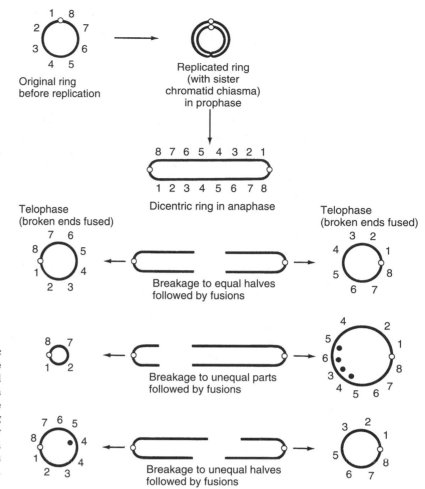

Dicentric ring chromosomes may be a source of genetic instability. If sister chromatid exchange takes place in the following anaphase, a double chromatid bridge is formed that may break equally, and in this case, the distribution of the genes to the poles is normal. (Numbers represent the genes.) The chromatids may also break unequally, resulting in unequal distribution of the chromatin (genes) that may be detected if appropriate markers are present. The broken ends usually heal and new dicentric ring chromosomes of different sizes are formed. (After B. McClintock. 1941. *Genetics* 26:542.)

markers are present. *See* ring, bivalent; ring chromosomes; translocation, chromosomal.

dicer *See* RNAi.

dichlorodiphenyltrichloroethane (DDT) Insecticide now almost entirely avoided in the industrialized countries of the Northern Hemisphere because it weakens the egg shells of birds preying on insects and it accumulates in mammalian fat tissues, etc.

Some insects develop a resistance to it by eliminating HCl from the molecule with the aid of increased levels of DDT dechlorinase enzyme. It is toxic to humans: oral LDLo 6 mg/kg. Although it was suspected to be carcinogenic, the final tests did not confirm this. It was discovered that the main metabolite of DDT, p,p′-DDE, is a potent antiandrogen that binds to the androgen receptor. It has not been used since the 1980s, yet it still persists in the environment (its half-life being ~100 years). It is blamed for reduced human sperm counts, increased testicular cancer, and other anomalies of the reproductive system. DDT has become an anathema, although it is the most effective weapon against mosquitoes responsible for malaria. In December 2000, the World Health Organization (WHO) allowed the temporary use of DDT in 25 countries with a very high incidence of malaria (Kapp, C. 2000. *Lancet* 356:2076). *See* LDLo; malaria; sperm.

2,4-dichlorophenoxyacetic acid (2,4-D) Synthetic plant growth hormone. *See* auxins; 2,4-D.

dichotomous trait Qualitative trait; however, its determination may be only partly genetic, and environmental factors have a substantial role in its etiology. *See* quantitative trait.

dichroic mirror Transmits light in one color, and when reflected, passes it in another color. *See* circular dichroism.

dicistronic translation Transcripts carrying internal ribosomal entry sites (IRES); may be cotranslated (expressed) in appropriate gene fusion vectors in the same sequence. *See* IRES; ribosome scanning.

dicot (dicotyledonous) Taxonomic category of plants with embryos having two cotyledons.

dictyosome *See* Golgi apparatus.

Dictyostelium discoideum Haploid ($x = 6$), unicellular amoeba with a genome size of ~50,000 kbp. They also contain diverse sets of high-copy-number nuclear plasmids. Diploid cells have been observed. The cells form colonies that upon starvation differentiate into structures resemble multicellular organisms. Actually, they form two types of cells: spores

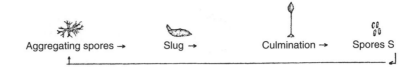

Aggregating spores →   Slug →   Culmination →   Spores S

and stalk cells. At a stage when they are about 70% prespore, and 30% prestalk, they form aggregates that are called slugs, appearing somewhat similar to gastropod animals. They make about 3,000 to 5,000 different mRNAs during the growth phase and during differentiation between 800 and 2,000, i.e., within the same order as lower animals. During the growth phase, *Pre-Starvation Protein Factor* (PSF) accumulates, and when its level reaches the critical point, it serves as a signal for development by activating adenylate cyclase, cAMP receptor, and a Gα protein required for signal transduction, although development may take place in the absence of cAMP. The steroid *Differentiation Inducing Factor* (DIF) is made by the end of aggregation. The interaction of these two factors leads to culmination and formation of the fruiting body. All of the functions required for growth and development involve the regulated expression of about 2,000 genes in 7 linkage groups. *See* adenylate cyclase; cAMP; Gα protein; signal transduction. (Wilkins, A. & Insall, R. H. 2001. *Trends Genet.* 17:41; <http://dicty.cmb.nwu.edu/dicty/dicty.html>; <http://www.tigr.org/tdb/tgi.shtml>; <http://dictybase.org>; <dictyworkbench.sdsc.edu>.)

dictyotene stage   Late prophase stage of meiosis in human females. Meiosis in the oocyte begins by about the fourth month of the fetus, but it is completed only before each ovulation. Thus, some cells remain in dictyotene for 30–40 years (until the end of ovulation). It has been hypothesized that this prolonged meiotic stage may be responsible for some of the chromosomal anomalies observed as a function of increasing age. In the human males, cell divisions in the germline continue to advanced age and point mutations increase in proportion to advanced age. *See* atresia; chromosome replication; Down's syndrome; meiosis.

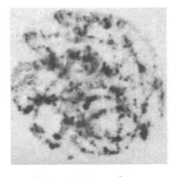

Late meiotic prophase.

dic(X)   Dicentric X chromosome. *See* dicentric chromosome.

dicytoplasmic system   The plastids are uniparentally transmitted (e.g., conifers) by the male, whereas the mitochondria are uniparentally inherited through the female.

dideoxy fingerprinting   Variation of the single-stranded conformation polymorphism (SSCP) technique that detects single-base or other subtle alterations in the DNA. One lane of the Sanger dideoxy termination reaction is electrophoresed in a nondenaturing gel. Mutations may show up as the presence of an extra segment or the absence of one, or by altered mobility. DNA sequencing Sanger; *See* single-strand conformation polymorphism.

dideoxynucleotide   *See* DNA sequencing.

dideoxyribonucleotide   Lacks an O atom at both the 2′ and 3′ positions of the ribose (see structural formulas see diagram above), and because of the latter feature, it is incapable of supporting DNA chain elongation required for the DNA polymerase to add nucleotides to the chain. Therefore, these molecules can be used for selective chain terminators in the Sanger method of DNA sequencing. *See* DNA sequencing.

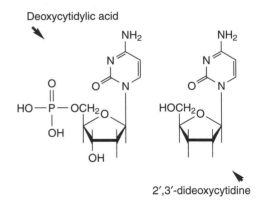

Deoxycytidylic acid

2′,3′-dideoxycytidine

Diego blood group system   Polymorphic in Mongolian, Chinese, and Native American populations. *See* blood groups.

diepoxybutane   *See* DEB.

diethylnitrosamine   *see* DEN.

diethylstilbestrol   Nonsterol yet sterol-like action chemical. It had been used for geese to increase the size of the liver. Until 1971, it was a prescription drug to prevent miscarriage in women. In utero exposed female offspring had a high incidence of malformations in their reproductive organs and adenocarcinoma. *See* estrogen.

DIF   Differentiation-inducing factor is basically a TNF involved in mitogen-promoting blood cell differentiation. *See* TNF. (Kanno, T., et al. 2001. *Dev. Growth Differ.* 43:709.)

DiFerrante syndrome   *see* mucopolysaccharidosis.

differential centrifugation   Partition of particles (cellular organelles) by their different rates of sedimentation in aqueous or organic liquid medium (sucrose, percoll, etc.) during centrifugation. *See* centrifuge.

differential display (DD) Revealing the profile of protein(s) or RNA differently expressed, e.g., in healthy versus diseased states. *See* RNA fingerprinting. (Liang, P. & Pardee, A. B. 1992. *Science* 257:967; Pei, L. & Melmed, S. 1997. *Mol. Endocrinol.* 11:433.)

differential psychology Study of behavior differences among individuals and groups, thus seeking out genetic differences in behavior rather than average characteristics; the object of experimental psychology. *See* IQ.

differential segment Region of the X and Y chromosomes that lacks homology, and those chromosomal areas — in general — fail to pair. Also, a particular borrowed chromosomal segment in a congenic strain. *See* congenic; holandric genes; sex linkage.

differentiation Morphological and/or functional specialization of originally totipotent cells to meet particular needs of the organisms. Differentiation is determined by morphogen gradients, asymmetric distribution, cell position, etc. The pattern of differentiation is regulated in anterior/posterior, dorsal/ventral, and by left/right gradients of the morphogens under genetic control. *See* compartmentalization; determination; development; intercellular immunization; intracellular immunization; morphogenesis; morphogenesis in *Drosophila*; mRNA targeting; polarity, embryonic; primitive streak; segregation, asymmetric; selector genes.

differentiation, plastid nucleoid Scattered nucleoids (SN) in the small plastids are distributed between the thylakoids and grana (land plants and many algae). The central nucleoids (CN) are located at the center of the plastids (red alga, undifferentiated proplastids of higher plants). Circular lamellar (CL) types occur as ring-shaped structures within the girdle lamella (brown algae). The peripherally scattered (PS) types lie near the inner plastid envelope and the SP type spreads around the pyrenoid. The proplastids may be of 0.2 μm diameter with 1–2 cpDNA genomes or 2–3 μm diameter with several cpDNA genomes. The smaller proplastid precursors may divide into structures like themselves or form the larger proplastids and change into colorless etioplasts in darkness; upon illumination, the latter differentiate into green chloroplasts. In the *Chlamydomonas* alga, the 5–6 nucleoids each may have 13–16 cpDNA, while in *Euglena*, the number of nucleoids may be 20–34 with 3–15 DNA rings each; in the chloroplasts of higher plants, the 12–25 nucleoids each may contain 2–5 cpDNAs. *See* chloroplasts; DNA replication in mitochondria; DNA replication in plastids.

diffuse large B-cell lymphoma (DLBCL) Highly morbid disease with less than 50% chance for cure. Microarrays of 6,817 genes in diagnostic tumors treated with cyclophosphamide, adriamycin, vincristine, and prednisone (a type of cortisone) classified by supervised learning revealed two potential survival categories (70% and 12%). Thus, there is a predictable target for successful therapy. (Shipp, M. A., et al. 2002. *Nature Med.* 8:68.)

diffusion Movement of molecules from higher to lower concentration in a solution. Diffusion of proteins within a cell from its origin along an axis can be determined as $S = (4Dt/\pi)^{1/2}$, where D is the diffusion coefficient and t = time. Examples: D of 10 μm$^2/S$ diffuses on average 4 μm in a 1 S period. A membrane-bound protein diffuses <0.8 μm and a transmembrane protein <0.25 μm. Cell signals are transduced either by protein-protein binding interactions or by second messenger-mediated interactions between signaling τ = the lifetime of the activated state. If τ = the time by 63% of the proteins to be inactivated, then the two-dimensional $r = (2D\tau)^{1/2}$, and for the three-dimensional diffusion $r = (3D\tau)^{1/.2}$. *See* signal transduction. (Teruel, M. N. & Meyer, T. 2000. *Cell* 103:181.)

diffusion, genetic Distribution in a wave-like manner of gene(s) in function of a variable such as a geographic region, infestation by a parasite, etc. The rate of advance per generation under steady-state conditions is $v = \sigma\sqrt{2s}$, where σ = standard deviation and s = selection coefficient. The length of the wave advance: $\sigma([1/2]s)^{1/2}$. The rate of advance is obtained by $\sigma(2s)^{1/2}$, and if we arbitrarily substitute $\sigma = 20$ and $s = 0.01$ the km, the spread per generation would be $20(2 \times 0.01)^{1/2} \approx 2.83$ km. *See* cline; legit.

DiGeorge syndrome Autosomal-dominant (human chromosome 22q11.2) defect of thymus and parathyroid formation and a variety of other anomalies (malformed ears, broad nasal bridge, far-apart eyes, abnormal U-shaped mouth, small jaws, heart disease, etc.), including immunodeficiency. Apparently an autosomal-recessive form has also been observed. This anomaly is frequently associated with deletions (1.5 to 3 megabase long) in chromosome 22, and some of the defects in this rather heterogeneous syndrome are due to a deletion in chromosome 10p13. This latter deletion also accounts for hypoparathyroidism associated with velocardiofacial syndrome (VCFS), sensorineural deafness, and renal anomaly (HDR). The 200 kb region involved in HDR contains the GATA3 zinc-finger transcription factor controlling vertebrate development. In some instances, transplantation of fetal thymus alleviated some of the symptoms. The 22q11 locus has been cloned and includes several components, e.g., WD40 repeats similar to the yeast transcriptional corepressors Hir1p and Hir2p most likely associated with chromosome remodeling by controlling histone transcription. Their mammalian homologue, HIRA, interacts with the Pax3 transcription factor. DiGeorge syndrome may be a defect in the CDC45 ligand. T-box haploinsufficiency may also elicit parts of the syndrome. *See* CATCH; CDC45; chromatin remodeling; deletions; face/heart defects; GATA; haploinsufficient; immunodeficiency; Pax; T box; velocardiofacial syndrome. (Jerome, L. A. & Papaioannou, V. E. 2001. *Nature Genet.* 27:286; Guris, D. L., et al. Ibid., p. 293; Baldini, A. 2002. *Hum. Mol. Genet.* 11:2363.)

digestion Enzymatic hydrolysis of molecules in vitro or in the digestive tract in vivo.

digital organism Simulation of self-replication, genome expression, interaction, and environmental responses by computer programs in virtual reality. (Lenski, R. E., et al. *Nature* 400:618.)

diglyceride *See* diacylglycerol.

digoxigenin Aglycone of the steroid nucleus digoxin. Digoxigenins are used in nonradioactive forms to label amino acids, DNA, and glycoconjugates. *See* nonisotopic labeling. (Broketa, M., et al. 2001. *J. Clin. Virol.* 23:17.)

Digoxigenin.

digynic Has two sets of maternal chromosomes in a normally diploid or triploid cell. *See* methylation, DNA.

dihaploid Individual having half the somatic chromosome number of that of a tetraploid from which it descended. Dihaploid plants are obtained generally by crossing tetraploids by diploids; among the offspring, diploids are selected that carry some advantageous genes of the tetraploid that were not present in the parental diploid. *See* haploid, autotetraploid.

dihybrid Heterozygous for two pairs of alleles, e.g., *A/a* and *B/b*. *See* checkerboard; Mendelian segregation; modified Mendelian ratios.

dihydrofolate reductase (DHFR) Enzyme synthesizes tetrahydrofolate, an essential precursor in the biosynthesis of thymine, purines, and glycine. *See* amethopterine; DHFR; methotrexate.

dihydropyridine receptor (DHPR) Voltage sensor in the L-type Ca^{2+} ion channels; ryanodine enhances the DHPR function. *See* ion channels; ryanodine.

dihydropyrimidine dehydrogenase (DPYD, 1p22) Mediates the catabolism of uracil and thymine. Its recessive deficiency leads to excessive amounts of pyrimidines in the urine. DPYD may catalyze the degradation of the anticancer drug 5-fluorouracil. DPYD deficiency may result in severe complex reaction to 5-fluorouracil. *See* 5-fluorouracil.

dihydrouracil May be formed by oxidizing effects and ionizing radiation of nucleic acids.

Dihydrouracil.

dihydrouridine (5,6-dihydro-2,4-dihydroxyuracil nucleoside) Posttranscriptionally modified uridine in the tRNA. *See* tRNA. (Bishop, A. C., et al. 2002. *J. Biol. Chem.* 277:25090.)

3,4-dihydroxyphenylalanine *See* DOPA.

dihydrozeatin Natural cytokinin. *See* cytokinins.

diisosomic In wheat, $20'' + i''$, $2n = 42$, ($'' = $ bivalent, $i = $ isosomic).

diisotrisomic In wheat $20'' + (i'')1'''$, $2n = 43$, ($i = $ isochromosome, $'' = $ disomic, $1''' = $ trisomic).

dikaryon A single cell has two nuclei; the nuclei may be genetically identical (homokaryon) or genetically different (heterokaryon). These regularly occur in fungi after fusion of somatic cells and in other eukaryotes if the nuclei do not fuse. *See* fungal life cycles.

dikaryote Cells with two unfused nuclei. *See* dikaryon.

dimer Association of two units (e.g., two polypeptides). Dimerization may regulate the function of certain enzymes. (Mellado, M., et al. 2001. *Cell Mol. Biol.* 47:575; Lodmell, J. S., et al. 2001. *J. Mol. Biol.* 311:475; Gomes, I., et al. 2001. *J. Mol. Med.* 79:226.)

dimer 14-3-3 Protein factor that assists dimerization of Raf, an oncogene and a key player in the mitogen signal transduction pathways. *See* protein 14-3-3; raf; signal transduction.

dimethylglycine dehydrogenase (DMGDH) Enzyme of the mitochondrial matrix; it is involved in the biosynthesis of choline. It causes fish-like body odor, and chronic muscle fatigue. The levels of the muscle form of creatine kinase and DMGDH in the serum are elevated. (Binzak, B. A., et al. 2001. *Am. J. Hum. Genet.* 68:839.)

dimethylsulfate *See* DMS.

dimethylsulfoxide (DMSO) Very effective solvent of a wide range of organic chemicals with moderate toxicity. It also protects cells against low temperatures. DMSO may facilitate cellular DNA uptake and thus transformation of eukaryotic and prokaryotic cells.

di-mon cross *See* Buller phenomenon.

dimorphic Displays two forms whether the two are chromosomes, cells, or organisms.

2,4-dinitrophenol Indicator of pH, wood preservative, and insecticide. It is highly toxic to animals by being an uncoupler of electron transport and oxidative phosphorylation. It blocks ATP formation by respiration. Uncoupling oxidative phosphorylation generates heat in hibernating animals, in those adapted to cold, or in newborns.

dinoflagellate Member of the marine plankton; may appear as red tide, and by vast numbers depletes the oxygen in its habitat and causes the death of sympatric animals. Dinoflagellates may secrete a neurotoxin that may be ingested by shellfish and thus may poison other creatures consuming shellfish. Their yellow, brown, or red color is due to the pigments of their primitive chloroplasts. These chloroplasts contain a very small number of genes and curiously each of them (~9) appears to be in a separate circular DNA plasmid. The majority of the normally plastid genes are within their nucleus. *See* chloroplasts. (Saldarriaga, J. F., et al. 2001. *J. Mol. Evol.* 53:204.)

dinucleotide abundance *See* dinucleotide odds ratio.

dinucleotide odds ratio Represents the frequency with which CpG or TpA sequences occur. Statistical survey of sequenced DNAs in different species indicates that the dinucleotide abundance within a species for different classes of DNAs (coding, intron, intergenic) tends to be similar but different. (Gentles, A. J. & Karlin, S. 2001. *Genome Res.* 11:540.)

dinucleotide repeat For example, GpT/ApC and ApG/CpT may cause genetic instabilities by slipped-strand mispairing and generating deletions or additions. *See* trinucleotide repeats. (Stalling, R. L., et al. 1991. *Genomics* 10:807; Renwick, A., et al. 2001. *Genetics* 159:737.)

dioecious The two sexes are represented by separate individuals like the majority of animals, fungi, and some plants (<4%) such as spinach, asparagus, date palm, poplar, osage orange, etc. *See* breeding system.

dioxin Family of carcinogens and (frameshift) mutagens. It may occur as a contaminant in various industrial chemicals. LD_{50} orally for mice is 114 µg/kg. *See* environmental mutagens, LD_{50}.

2,3,7,8-tetrahydrodibenzo-p-dioxin.

DIP Database of interacting proteins: <http://dip.doe-mbi.ucla.edu>.

diphosphoglycerate mutase deficiency (7q31-q34)
The enzyme controls oxygen affinity in red blood cells by binding to deoxyhemoglobin. Its deficiency leads to a type of hemolytic anemia. *See* anemia.

diphtheria toxin Single-chain (535-residue) very potent toxin produced by *Corynebacterium diphtheria* carrying a lysogenic phage. It has applied significance in cancer research because it blocks eukaryotic peptide chain elongation (eEF-2). Rat and mouse are resistant to this toxin because they lack its membrane surface receptor. Humans, guinea pigs, and rabbits are, however, sensitive to it. The human sensitivity is encoded in chromosome 5q23. The virulence of the bacterium is a response to the expression of the toxin gene (*tox*) regulated by the toxin repressor (DtxR) activated by transition metal ions. The metal ion (iron, Ni_{II}) triggers the two different DtxR subunits to embrace — at opposite sides — 33 bp of the DNA of the *tox* operator. Transformation by the diphtheria toxin gene has been employed for ablation of cells and to inhibit protein synthesis in targeted cells. *See* biological weapons; toxins. (Saito, M., et al. 2001. *Nature Biotechnol.* 19:746.)

diploblast Animal with only ecto- and endodermic germ layers. *See* ectoderm; endoderm; triploblast.

diplochromosome Displays four, rather than two chromatids in each arm because of replication without splitting of the centromere. *See* endoreduplication; salivary gland chromosomes.

Diplococcus pneumoniae (old name *Pneumococcus*)
Member of the genus in the tribe *Streptococceae* and the family *Lactobacillaceae*. This is the most common cause of pneumonia. The lanceolate cells occur in doublets. The virulent forms are encapsulated and form shiny (smooth) colonies; the avirulent forms lack the protective coat (form rough colonies) and are destroyed by the enzymes of the host cells. Genetic transformation was discovered with this bacterium in 1928. *See* transformation.

diploid Has two complete basic sets of chromosomes characteristic for the zygotes and for most of the body cells of the majority of animals and plants and for the premeiotic phase of several other eukaryotes (fungi, algae, etc.). Some diploids may have only a single sex chromosome in the somatic cells as a normal condition.

diplonema Structure of the chromosome thread at the diplotene stage of meiosis. *See* diplotene; meiosis.

diplont Diploid individual (cell).

diplontic selection Competition among diploid cells within multicellular organisms. This can take place only when dominant mutation or other dominant-acting changes occur.

diplophase Diploid phase of an organism that exists also as a haploid.

diplospory (apogamety) Nonreduced embryo sac develops without meiosis or by restitution. *See* restitution nucleus.

diplotene stage Of meiosis, is preceded by pachytene when the chromosomes begin to appear as clearly bipartite, doubled threads (diplonema) through the light microscope. The pairing (synapsis) of the bivalents appears somewhat relaxed, but the four chromatids are held together by the chi-shaped (χ) structures of the overlapping chromatids, the chiasmata. The synaptonemal complex is generally the most conspicuous (by

Diplotene chromosome of the grasshopper. (Courtesy of Dr. B. John.)

electron microscopy) when chiasmata are visible. *See* chiasma; chromatid; meiosis; synaptonemal complex.

dipole   Molecules have two equal and opposite charges separated in space. A simple dipole molecule is H_2O.

diquat   Can serve as an electron acceptor in photosystem I of photosynthesis. It is also a light-dependent herbicide. *See* herbicides; photosynthesis.

direct DNA transfer   Incorporation of DNA into plant protoplasts without bacteria using only plasmids or perhaps other naked DNA. *See* naked DNA; transformation.

directed mutation   Controversial (Lamarckian) notion that mutations of advantageous phenotype are preferentially induced in an adaptive environment, i.e., mutations would not be random and preferentially selected as required by the ideas of neo-Darwinian theory. Although supportive evidence has been put forth, none of these theories can be completely defended against the alternative, conservative, classical interpretations (see acquired characters). In-depth analysis of adaptive mutation (directed mutation) in *E coli* revealed that these adaptive reversions of the *Lac* alleles required F' plasmid transfer replication and homologous recombination involving F' plasmid elements that may contain revertant alleles and may thus account for the apparent directed mutations. In bacteria, the so-called *adaptive mutations* took place only in the presence of a functional RecBCD recombinational pathway. Others have pointed out that *Lac* locus may be amplified to cope with the reduced function due to mutation and the increased number of copies appears to be responsible for the rise in overall reversions without an increase of mutability in the individual copies. The notion of the existence of true adaptive mutations is thus controversial. Directed mutations can be obtained, however, by manipulating the DNA in vitro using the techniques of molecular biology and transformation. *See* acquired characters; cysteine-scanning mutagenesis; homologue-scanning mutagenesis; hypermutation; Kunkel mutagenesis; lamarckism; linker scanning; localized mutagenesis; lysenkoism; neo-Darwinian evolution; PCR-based mutagenesis; Soviet genetics; TAB mutagenesis. (Storici, F., et al. 2001. *Nature Biotechnol.* 19:773, staggered extension, RNA-peptide fusion, Xia, G., et al. 2002. *Proc. Natl. Acad. Sci. USA* 99:6597.)

directional cloning   Vector DNA termini are ligated with different linkers that produce different cohesive sequences at the two ends after restriction endonuclease cleavage. Thus the passenger DNA (insert) can be ligated in a chosen orientation only. *See* vectors. (Bubler, U. & Hoffman, B. J. 1983. *Gene* 25:263; Ohara, O. & Temple, G. 2001. *Nucleic Acids Res.* 29[4]:e22.)

directional evolution   Biased distribution, e.g., of the different length of microsatellites. *See* microsatellite. (Hutter, C. M., et al. 1998. *Mol. Biol. Evol.* 15:1620; Demetrius, L. 1997. *Proc. Natl. Acad. Sci. USA* 94:3491.)

directionality   Phage λ *chi* sites may enhance recombination more on the left than on the right side, or the transposase may be more functional in one orientation. *See* lambda phage; orientation selectivity.

directional selection   Alters the population mean in either a plus or minus direction. *See* selection types.

directory, computer   Reveals the contents of the folders (the documents the user has generated).

direct repeat   Adjacent or nonadjacent repeat of identical or similar nucleotide sequences of the same order such as ATG...ATG. These are common at the termini of insertion elements. *See* insertion elements; transposable elements; transposons.

direct suppressor   Modifies the final product of the gene so that the mutation is not (fully) expressed. *See* suppressor mutations.

Dirichlet distribution   Beta distribution of a multivariate type

$$f(x_1, x_2, \ldots x_q) = \frac{\Gamma(\nu_1 + \cdots + \nu_q)}{\Gamma(\nu_1) \ldots \Gamma(\nu_q)} x_1^{\nu_1 - 1} \ldots x_q^{\nu_q - 1}.$$

See beta distribution; multivariate normal distribution.

DIRVISH   Direct visual hybridization is a mapping procedure for DNA sequences using fluorochrome-labeled samples hybridized to highly stretched DNA strands. Their location is visualized with the aid of fluorescence microscopy. *See* FISH; fluorochromes. (Buckle, V. J. & Kearney, L. 1993. *Nature Genet.* 5:4; Aerssens, J., et al. 1995. *Cytogenet. Cell Genet.* 71:268.)

disaccharide   Two monosaccharides, covalently bound e.g., sucrose (glucose: fructose).

disaccharide intolerance   Collective name for the inability of proper metabolism of sugars. The sucrose intolerance is caused by sucrase-isomaltase malfunction. The enzyme is generally present but does not function normally, resulting in diarrhea when sucrose is consumed. Sucrose, maltose, and starch are well tolerated in lactase deficiency, but lactose cannot be reabsorbed and causes bloating, diarrhea, and general discomfort because of the high gas production by intestinal microbes. This defect is caused by a recessive mutation in human chromosome 2q21. Generally two types of lactase-phlorizin hydrolase activity are distinguished. One type appears generally as a developmental defect after age 5. Its frequency may be quite variable in different populations. Overall, its frequency is low where dairy farming is prevalent and high where milk is absent from the diet. The lactase persistence (LCT*P, 2q21) to adulthood is controlled by a genetic factor closely associated with the lactase gene locus. Its worldwide distribution follows the availability of dairy products in the diet. Several haplotypes for lactose tolerance/intolerance can be distinguished based on mutation and recombination of the different alleles. Lactose intolerance can be alleviated by taking tablets of microbial β-galactosidase before eating or with dairy products. A new possibility is gene therapy by oral administration of adeno-associated viral vectors with the good β-galactosidase gene. Rat experiments with such treatments produced somatic cure for up to 6 months because the vector was readily incorporated across epithelial barriers. The sucrase-isomaltase deficiency gene is at human chromosome 3q25-q26. *See* adeno-associated virus; fructose intolerance; gene therapy; glycosuria; lactose intolerance; phlorizin. (Hollox, E. J., et al. 2001. *Am. J. Hum. Genet.* 68:160; Enattah, N. S., et al. 2002. *Nature Genet.* 30:233.)

disarmed vector Agrobacterial genetic vector from which the oncogenes (encoding phytohormones) have been deleted in the T-DNA (and usually replaced by foreign, desirable genes). *See* Agrobacterium; plant hormones; Ti plasmid; transformation, genetic.

disassociation Separation of the two strands of DNA. *See* c_0t curve; Watson and Crick model.

disassortative mating The mating partners are less similar in phenotype than expected by random choice. *See* assortative mating; breeding systems; sexual selection. (Reusch, T. B. H., et al. 2001. *Nature* 414:300.)

DISC *See* disk.

DISC Death-inducing signaling complex of FAS, FADD, and caspase 8. *See* apoptosis; caspase; FADD; FAS/CD95.

discontinuous gene The translated exons are separated by introns. *See* exons; introns.

discontinuous replication *See* Okazaki fragments; replication fork.

discontinuous variation Caused by qualitative genes. The expression can be classified into discrete groups with relatively low environmental effects. *See* continuous traits.

discordance Dissimilarity of a trait between individuals because of genetic differences. The term is also used for gene loci carrying two different alleles in a diploid. *See* concordance; dizygotic twins; expressivity; monozygotic twins; penetrance; twinning; zygosis.

discriminant function Statistically estimates the overlap between two populations and it has uses for classification and diagnosis, for the study of relations between populations, and as a multivariate extension of the *t*-test. Example: A single

variate X is distributed in two populations with means μ_1 and μ_2; the standard deviations (σ) are assumed to be the same. We wish to know to which of the two populations the specimen X belongs. We classify X to population 1 if $X < (\mu_1 + \mu_2)/2$ and to population 2 if $X > (\mu_1 + \mu_2)/2$. If X is from population 1, our decision is wrong if $X > (\mu_1 + \mu_2)/2$ or when δ (the distance between the two means) $= (\mu_2 - \mu_1)$. $(X - \mu_1)/\sigma$ follows the normal distribution and misclassification is probable in the tail from $\delta/2\sigma$ and ∞ in both cases. The δ/σ must exceed 3.0 to consider the classification accurate. The δ/σ is also called the distance between two populations. Analysis of variance of the discriminant function can be used as follows: where D is the difference between $\overline{X}_1$ and $\overline{X}_2$. The significance of the difference can be determined by z. *See* analysis of variance; multivariate analysis; recursive partitioning; z. (Mather, K. 1965. *Statistical Analysis in Biology.* Methuen, London; Venables, W. N. & Ripley, B. D. 1994. *Modern Statistics,* Springer, New York.)

	Sum of Squares	Df*
Between Groups	$(n/2)D^2$	2
Within Group	D	$2n - 3$
Total	$D(1 + [n/2]D)$	$2n - 1$

*Df: degrees of freedom.

discriminator region In the -10 to $+1$ region of prokaryotic promoters, where 5'-diphosphate 3'-diphosphate (ppGpp) binds in amino-acid-starving cells. ppGpp is synthesized under starvation conditions and it represses rRNA protein genes and activates amino acid operons. The ppGpp-repressed promoters are GC rich; the ppGpp activated promoters are richer in AT in the discriminator region. *See* stringent control. (Pemberton, I. K., et al. 2000. *J. Mol. Biol.* 299:859.)

disease, human *See* genetic diseases.

disequilibrium Lack of equilibrium. *See* Hardy-Weinberg theorem; linkage disequilibrium and linkage equilibrium.

disheveled protein (Dsh) Mediates dorsal-ventral positioning and polarity in the embryo and the formation of a secondary axis.

disintegration Reversal of the integration process mediated by integrase. *See* dpm; integrase.

disintegrin Metalloproteinase controlling cell migration. *See* metalloproteinases.

disjunction Separation of bivalents (chromosomes) during meiosis I or of chromatids in mitosis. *See* meiosis; mitosis; nondisjunction.

disk (disc) Magnetic surface on which information can be recorded and later retrieved by the appropriate command to the computer. The capacity of amount of information stored is expressed in kilobytes (K) or megabytes (MB). The floppy disks (about 9×9 cm) can be lowdensity (400 K single-sided, or 800 K double-sided) or high-density disks (1.4 MB), the zip disk can store information from 100 to 250 Mb. CDs accommodate 700 to 800 Mb. The hard disks hold from 20 MB to several GB (gigabyte) or more information. 1 K = 1,024 characters, about 170 words in English.

dislocated hip (hereditary) Its prevalence in the human population is about 0.075% and its recurrence risk among sibs is about 5%. It is autosomal dominant and usually more common among females than males. It occurs sometimes as part of the recessive Marfan and Ehlers-Danlos syndromes. *See* Ehlers-Danlos syndrome; Marfan syndrome.

disome Two homologous chromosomes.

disomic Individual or cell with one specific chromosome or any of the chromosomes represented twice. *Maternal or paternal disomic* individuals arise from hybrids in case of nondisjunction for a particular chromosome in translocation heterozygotes (Robertsonian translocations). *Uniparental disomy* may be associated with trisomy when one of the three chromosomes derived from one of the parents is lost and the result is homozygosity (disomy) for the other parent's chromosomes. The frequency of human gametic disomy, estimated on the basis of trisomic conceptuses, varies according to chromosomes: X: sperm/ova 0.04/0.04; 16:0/1;13, 18:0/0.17; 21:0.03/0.40. *See* Angelmann syndrome; Beckwith-Weidemann syndrome; Prader-Willi syndrome; Robertsonian translocation; trisomy.

dispermic fertilization Two sperms enter into one egg. *See* double fertilization. (Palermo, G. D., et al. 1995. *Hum. Reprod.* 10 [Suppl. 1]:120.)

dispersed gene Members of (multi) gene families. *See* gene family.

dispersed repeat *See* LINE; SINE.

dispersion index Indicates the extent by which a set of data deviate from their mean. For the Poisson distribution, e.g., it is determined by the formula shown below. Also $R(t) = Var(S_t)/E(S_t)$, where S_t = number of amino acid substitutions at any site. *Var* and *E* stand for variance and expectation, respectively. In case of neutral substitutions, the variance-to-mean ratio, $R(t) \sim 1$, except when the mutations occur premeiotically (i.e., in clusters). In case the variances of the evolutionary rates exceed the expectations on the basis of Poisson distribution, the phenomenon is called overdispersion of the molecular clock. *See* evolutionary clock; Poisson distribution; variance.

$$D = \sum_{i=1}^{n} \frac{(x_i - \overline{x})^2}{\overline{x}}$$

dispersive replication The polynucleotide strands are a mosaic of old (straight) and newly synthesized (wavy) sequences. This old assumption has not been confirmed experimentally. *See* semiconservative replication. (Delbruck, M. &

Stent, G. S. 1957. in *The Chemical Basis of Heredity*, McElroy, W. D. & Glass, B. eds., p. 699. Johns Hopkins University Press, Baltimore, MD.)

displacement *(t)* In population genetics, t = the number of standard deviations difference between the mean values of homozygotes *AA* and *aa* (arbitrarily assuming that the variance within genotype is the same for each genotype). *See* genetic variances.

displacement loop *See* D loop.

display technologies allow the expression of large pool of modularly coded biomolecules and analyze and select diverse and specific proteins at very large scale. (See phage display, ribosome display, Ma, D. & Li, M. 2001. *J. Cellular Biochem. Suppl.* 37:34.)

disposable soma Theory that the evolution of aging requires the preservation of the body that limits the resources available for reproduction. In its simplest form, it says longevity is favored by producing fewer offspring. *See* aging. (Lycett, J. E., et al. 2000. *Proc. Roy. Soc. Lond. B Biol. Sci.* 267:31.)

disruptive selection As a consequence of the process, the population breaks up into two (contiguous) groups with different mean values. *See* selection types.

dissection, genetic Analyzes the mechanism of genetic determination of a process.

disseminated sclerosis *See* multiple sclerosis.

dissociation constant K_d is the equilibrium constant for the dissociation of a complex of molecules (AB) into components (A) and (B); K_a is the dissociation constant of an acid into its conjugate base and a proton. The smaller the K_d value

$$\frac{(A)(B)}{(AB)},$$

the tighter is the binding between the components.

distalization rule In differentiation, when new cells are generated (regenerated), each intercalary cell division produces progressively more distal cells until the circumferential filling of missing values is completed. *See* polar coordinate model. (French, V. 1981. *Philos. Trans. Roy. Soc. Lond. B. Biol. Sci.* 295:601.)

distal marker Situated in a direction away from the centromere or another gene, or in bacterial conjugation it is transferred after a particular site. *See* centromere mapping; conjugational mapping.

distal mutagen Formed when a promutagen through chemical modification is first converted into an intermediate (proximal mutagen), which finally becomes the ultimate or distal mutagen. *See* carcinogens; mutagens.

distance matrix *See* character matrix.

distichiasis Double eyelashes, lymphedema.

distorted segregation The transmission of one of the alleles is not the same as that of the other. The cause of this anomaly can be chromosomal defects, lethal or semilethal mutations, or genes like the *Segregation distorter* (*Sd*, map location 2–54) in *Drosophila*. The distortion may reduce either the dominant or the recessive class depending on the linkage phase and map distance of the marker to the factor that disturbs normal phenotypic or genotypic proportions. *See* certation; gametophyte factor; gene conversion; megaspore competition; meiotic drive; preferential segregation; segregation distorter. (Schimenti, J. 2000. *Trends Genet.* 16:240; Pardo-Manuel de Villena, F. & Sapienza, C. 2001. *Mamm. Genome* 12:331.)

distribution *See* Bernoulli process; binomial distribution; exponential distribution; gamma distribution; hypergeometric distribution; multinomial distribution; negative binomial; normal distribution; Poisson distribution; trinomial distribution.

distributive circuit in signal transduction *See* signal transduction.

distributive pairing The chromosomes may recombine during meiosis and are then distributed to the poles normally (*exchange pairing*). Chromosomes that are not involved in exchange display *distributive pairing* and may suffer nondisjunction. When homozygous, a *Drosophila* mutation *nod* (*no distributive disjunction*; map location 1–36) may display a high (800-fold) frequency of chromosome loss and nondisjunction of chromosome 4 during meiosis I. Information indicates that in the distribution of achiasmate chromosomes the heterochromatin adjacent to the centromere plays an important role. More than 70% of the nondisjunctions are the result of achiasmate meiosis. *See* achiasmate; chromokinesin; chromosome pairing; heterochromatin. (Hawley, R. S. & Theurkauf, W. E. 1993. *Trends Genet.* 9:310; Grell, R. F. 1985. *Basic Life Sci.* 36:317.)

disulfide bridge Covalent bond between two sulfur atoms (−S−S−) of two cysteine residues of a polypeptide chain(s) affecting conformation. DsbA (disulfide bond) is a small periplasmic protein. In the active form, it has two cysteines joined by a disulfide bond. Disulfide bonds are formed with an exchange between the oxidized DsbA and the reduced cysteine residues of the substrate. This results in inactive, reduced DsbA. DsbB reoxidizes the molecule and restores the active form. *See* protein structure.

ditelo-monotelosomic In wheat, $20'' + t'' + t'$, $2n = 43$, ($'' = $ disomic, $' = $ monosomic, $t = $ telosomic).

ditelosomic In wheat, the chromosome constitution is $20'' + t''$ ($2n = 42$) ($''$ indicates disomy, $t = $ telosomic).

ditelotrisomic In wheat, $20'' + (t'')1'''$, $2n = 43$, ($'' = $ disomic, $''' = $ trisomic, $t = $ telosomic).

dithioerythritol (DTE) 2,3-dihydroxybutane, Cleland reagent; it protects SH groups.

dithiothreitol (DTT) DL-threo-1,4-dimercapto-2,3-butanediol; protects SH groups.

ditype Tetrad with two kinds of spores. *See* tetrad analysis.

diurnal Response or behavior displaying daily cycles.

diuron Herbicide interfering with photosynthesis.

divergence Evolutionary differences in morphology, cytology, and/or in the primary structure of nucleic acids and proteins that are believed to have descended from common ancestry. The divergence can be quantitatively estimated on the basis of average chiasma frequencies of the genomes if the species can be crossed. The average frequencies of amino acid substitutions in the proteins or base substitutions in nucleic acids can be used to estimate genetic distance and the time required to achieve it. Some caution may be required in interpreting evolutionary divergence because similarity may also be based on evolutionary convergence. Mutation, chromosomal rearrangements, and recombination may bring about divergence. The recombinational force in *E. coli* appears to be 50 times higher than that of mutation in clonal divergence. *See* chiasma; convergent evolution; genetic distance.

divergent dual promoter Juxtapositioned promoters may carry out transcription in opposite directions. *See* catabolite repression; divergent transcription; promoter.

divergent transcription Proceeds from two promoters in opposite orientation. *See* promoter; transcription.

diversification See combinatorial diversification; junctional diversification.

diversity (Shannon-Weaver index) $H = (n(logN − \Sigma nlogn_i)/N$ where $H = $ diversity, $N = $ total number of individuals, and $n_i = $ number of individuals of different phenotypes. Molecular genetic diversity (π) can be assessed at the nucleotide level. Based on thousands of human genes, π values in the nontranslated 5′ and 3′ regions averaged 0.0003 ± 0.0001; nonsynonymous and synonymous values of π were found to be 0.0001 ± 0.0001 and 0.0005 ± 0.0002, respectively. Diversity at the protein level is inversely proportional to the number of subunits because of the greater functional constraints. Mutation rate, however, is increasing in proportion to the size of the genes because of the larger mutational target, although the mutations may be subject to selective screening for survival or even elimination. There are substantial variations among genes. Only a relatively small fraction of the diversity is limited to specific ethnic groups, although in general the most ancient populations, e.g., Africans, display greater diversity than the newer ones. *See* adaptive radiation; microarray hybridization; microsatellite; minisatellite; mutation rate; neutral mutation;

polymorphism; SNIP. (Chakravarti, A. 1999. *Nature Genet.* 21, Suppl. 56; Yu, N., et al. 2002. *Genetics* 161:269.)

dizygotic twins   Develop from two separate eggs fertilized by separate sperms. Dizygotic twinning seems to be linked to the peroxisome proliferator-activated receptor PPARG at 3p25. *See* monozygotic twins; pedigree analysis; PPAR; twinning; zygosis.

D loop   Formed when in replicating small circular DNA (mtDNA) one of the strands is displaced while the other strand is copied, or when in genetic recombination a single-strand DNA invades the RecA protein complex and displaces one strand of a duplex. *See* Meselson-Radding model; mitochondrial control regions; mtDNA; prokaryotes; recombination molecular mechanism; replication; strand displacement. (Clayton, D. A. 2000. *Hum. Reprod.* 15 [Suppl. 2]:11.)

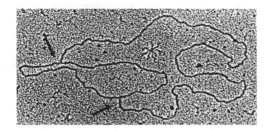

D loop of mtDNA replication. (Courtesy of Dr. K. Wolstenholme.)

dm   In front of a gene or protein, symbol indicates *Drosophila melanogaster* homology.

DM   HLA DMA and DMB gene products that are required to exchange CLIP from Class II major histocompatibility proteins so that the T cell can be loaded with the antigen. The DM α- and β-subunits are similar to the α- and β-chains of the Class II molecules. *See* CLIP; HLA; MHC.

DM (double minute) chromosome (DM)   Has no centromere and can be maintained only under selective pressure for the gene(s) it carries and is absent from the rest of the genome. In several types of cancer, oncogenes are amplified either in double minutes or in *homogeneously stained regions* of the chromosomes or both. Sublethal doses of hydroxyurea (HU) may cause a loss of DMs and thus the oncogenes. The loss of amplified epidermal growth factor receptor (EGFR) reduces tumor growth. Amplified dihydrofolate reductase genes (DHFR) in DMs in methotrexate-resistant mouse cells were also selectively reduced in the presence of MTX (methotrexate) and HU, but after a week, resistance to HU developed and DHFR was reamplified. *See* amplification; DHFR; homogeneously stained regions; hydroxyurea; methotrexate; YAC. (Hahn, P. J. 1993. *Bioessays* 15:477; Nevaldine, B. H., et al. 1999. *Mutation Res. Genomics* 406:55.)

DMBA (9,10-dimethyl-1,2-benzanthracene)   A carcinogen standard, a mutagen; forms covalent DNA adducts. *See* adduct; carcinogen.

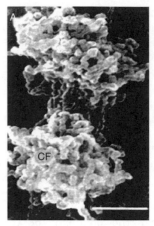

Scanning electron micrograph of a double-minute chromosome with ~30 looping interconnecting 17 nm fibers. The chromatin fibers (CFD) are ~32 nm in diameter. (From P. V. Schoenlein, et al. 1999. *Chromosoma* 108:121.)

DMC1 (disrupted meiotic cDNA gene)   Protein product in yeast controlling meiotic recombination. It is homologous to bacterial *RecA*. *See* RAD51; RecA. (Gupta, R. C., et al. 2001. *Proc. Natl. Acad. Sci. USA* 98:8433.)

DMS   Dimethylsulfate, an alkylating agent, mutagenic and carcinogenic. *See* alkylation.

DMSO   *See* dimethyl sulfoxide.

dna   Bacterial mutations involved in DNA replication.

DNA   Deoxyribonucleic acid, the genetic material of all eukaryotes and bacteria and many viruses. The most direct proof for DNA being the genetic material was provided by genetic transformation. DNA is measured either in base pairs (bp) or spectrophotometrically: 1 unit of optical density (OD) of double-stranded DNA (at 260 nm wavelength) is about 50 µg/mL. One OD of single-stranded DNA is about 40 µg/mL; 1 µg/mL DNA contains about 3.08 µM phosphate (*see* DNA types); 3000 nucleotides are about 1 µm in length; 1 pg DNA is about 0.60205×10^{12} Da $[10^{-12}/(1.6661 \times 10^{-24}$ g)]. *See* hydrogen pairing; spectrophotometer; transformation; Watson and Crick model.

Electron micrograph and interpretative drawing of double- and single-stranded DNA. (Courtesy of Professor Y. Aloni.)

DNA, blunted   Lacks ▬▬▬▬ (single-stranded) cohesive termini such as ▬▬▬▬ and cannot circularize without modification of ends. *See* cohesive ends.

DNA, breathing of   Reversible, short-range strand separation below the melting temperature. *See* melting temperature.

DNA, circular   The majority of plasmids, prokaryotic, and organellar (mitochondrial, plastid) DNAs form a covalently closed circle and a circular genetic map. *See* genetic circularity.

DNA, denatured   Hydrogen bonds are disrupted between the two strands; thus, the DNA is single stranded. *See* DNA denaturation.

DNA, genomic   Natural and complete nucleotide sequences of genes, introns included.

DNA, heavy chain   The chains of a DNA duplex can be separated and annealed with polyinosine-guanine (PI-G). The chains rich in cytosine bind more of PI-G, form a heavier band, and sediment faster upon ultracentrifugation in CsCl. Therefore, they are called heavy or C chains, whereas the complementary chains are called light chains (W). The density of DNA can be increased by substituting bromodeoxyuridine for thymidine or for (N^{15} or C^{13}) heavy isotopes. *See* bromouracil; ultracentrifuge.

DNA, light chain   *See* DNA; heavy chain.

DNA, native   The hydrogen bonds between the two strands are intact. *See* Watson and Crick model.

DNA, nonpermuted   Terminally redundant, e.g., **123456** 789012, and all molecules have the same sequence. *See* DNA, permuted.

DNA, permuted   Phage DNA may be repetitious as the concatameric molecules cut off at different positions but at the same length, so a collection of these molecules may appear (with numbers substituted for nucleotides). *See* DNA, nonpermuted; redundant.

<u>123456789012</u>   <u>345678901234</u>   <u>567890123456</u>

DNA, repetitive   *See* repetitious DNA; trinucleotide repeats.

DNA replication, eukaryote   The mechanism of replication in prokaryotes is better understood because more simple in vitro assay systems are available for the smaller DNAs. In yeast, the replicational origins are called ARS (*autonomously replicating sequences*). Although the genome of the eukaryote, *Saccharomyces cerevisiae*, is only about four times larger than that of the prokaryote *E. coli*, the former has 400 replicational origins, each including about 300 bp. The yeast ARS are not identical with each other, yet they have a short core sequence that is about 11 repeating A = T units. The single replicon of *E. coli* is over 4,000 kb, but the yeast replicons are only about 40 kb. The speed of replication in yeast is about 3.6 kb/min, but in *E. coli* it is almost 14 times as fast. In the large amphibian genome, it proceeds at a speed of about one-seventh that of yeast. If the replication of the toad DNA proceeds at only 500 bp/min, it would take more than 20 years to complete an S phase, when actually at the gastrula stage it takes about 4 to 5 hours because the large genome relies on more than 15,000 replicons. Replication in eukaryotes, just like in *E. coli*, proceeds bidirectionally, and electron microscopic examination

detects a large number of replication "bubbles" or replication "eyes" that are actually the replicational origins. As the replication within a bubble nears completion, the neighboring bubbles coalesce while replication is completed along the length of the chromosome. The molecular structure and function of the eukaryotic replication fork is best understood in the eukaryotic SV40 virus that has a 5,243 bp chromosome with nucleosomal organization.

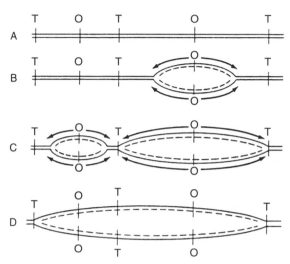

Multiple replicons are required for the large eukaryotic genomes. The horizontal lines represent the old DNA strands. The dashed lines show the new, bidirectionally synthesized DNA within the "eyes." O = origin, T = terminus. Eventually, the multiple replication bubbles coalesce. (Redrawn after Huberman, J. A. & Riggs, A. D. 1968. *J. Mol. Biol.* 32:327.)

Human and other mammalian cellular fractions combine with the large T antigen of SV40 and the viral genome in a plasmid replicate, indicating the similarities of the interchangeable elements of the systems. In the presence of ATP, a double hexamer of the viral T antigen (which is an initiator and a DNA helicase) binds to the viral origin and modifies it. This protein-DNA complex then binds the *cellular replication protein A* (RPA, subunits 70, 34, and 11 kDa). RPA/RFA/HSSB is a single-strand binding protein (SSB) similar in function to the helix-destabilizing proteins and in cooperation with the helicases assists in unwinding of the DNA duplex. The T antigen-RPA complex then binds eukaryotic *DNA polymeraseα* (in humans with subunits p180 [polymerase], p70 [assembles the primosome], p58 [stabilizer], and p48 [primase]) with RNA and DNA polymerase functions. The sizes of these subunits may be different in other eukaryotes. (Pol-α-primase functions may be inhibited by kinase cyclin A–CDK2. In human ataxia telangiectasia, the phosphorylation may be hindered.) This enzyme (pol α) then generates an RNA-DNA primer at the replicational origin.

Initiation of replication in mammalian (and presumably other eukaryotic cells) cells requires specific DNA sequences. The *cellular replication factor C* (RFC, subunits 140, 40, 38, 37, 36 kDa) binds to the 3′-end of the new DNA and brings to the growing point the *proliferating cell nuclear antigen* (PCNA, 29–36 kDa) and DNA-polymerase δ (pol δ, catalytic subunit 125 kDa, plus another 48 kDa subunit), then pol α is replaced by pol δ (RFC is also a DNA-dependent ATPase.) The

RCF/PCNA/pol-δ aggregate proceeds in elongating the leading strand. On the lagging strand, pol-α/primase complex remains active at the initiation of the Okazaki fragments, but after a switch the RCF/PCN/pol-δ complex also elongates this strand. The elongation of the two strands is a coordinated process. After the Okazaki fragments are completed, *RNase H* and 5′ → 3′ *exonuclease MF1* (Fen1 and Dna2 helicase) remove the RNA primer and DNA ligase binds the ends of the Okazaki fragments into a continuous strand. Eukaryotic replication also utilizes topoisomerases in a somewhat similar fashion to *E. coli*. When the replication fork encounters transcription initiation complexes, both in yeast (eukaryote) and prokaryotes (T4 page, *E. coli* bacterium) the progress of replication transiently slows down. Actually, DNA replication depends on the cooperation of a multitude of proteins forming higher orders of domains (factories). An experimental study detected about 22 sets of domains of DNA replication in mouse cells. The replication domains are distinct from the 16 higher-order transcriptional domains. The late-firing origins of replication are regulated during S phase by the Mec1 and Rad53 proteins of yeast. The majority of replication models assume that the replication factories move along the DNA. Alternative models suggest that the polymerases are in a fixed position within the factories and rather the DNA moves along and rotates during the process of replication. (*See* ARS; ataxia telangiectasia; cell cycle; CLB; DNA-PK; DNA polymerase; DNA replication, mitochondria; DNA replication, prokaryotes; MCM; Mec1; Okazaki fragment; PCNA; processivity; Rad27/Fen1; Rad53; replication, cell cycle; replication fork; replication machine; replisome; reverse transcription; ribonuclease H; SBD; steric exclusion model. (Kelly, T. J. & Brown, G. W. 2000. *Annu. Rev. Biochem.* 69:829; Kunkel, T. A. & Bebenek, K. 2000. *Annu. Rev. Biochem.* 69:497; Bell, S. P. & Dutta, A. 2002. *Annu. Rev. Biochem.* 71:333; Gerbi, S. A. & Bielinsky, A.-K. 2002. *Current Opin. Genet. Dev.* 12:243.)

DNA replication, mitochondria

The mammalian mitochondrial DNA is a very short duplex circle (16.5 kbp). In mouse it has two origins of replication, *ori-H* and *ori-L*. Before replication at the *ori-H* site, a displacement loop is formed of the H strand, apparently to make room for the unidirectional replication. When the H-strand loop is about two-thirds complete, replication of the L strand begins. The completion of the replication requires 2 hours as it proceeds 270 base/min, i.e., more than two orders of magnitude slower than the process in the *E. coli* chromosome. Priming is initiated by an RNA transcribed on the L strand. Replication and transcription appear to be primed by the same process. The transition to L (light)-strand synthesis in human mtDNA is at the template GGCCG. In the larger mitochondrial DNAs (e.g., in maize the T mtDNA is 540 kbp and N mtDNA is 700 kbp), more *ori* sites are used. The 80 kbp yeast mtDNA has at least seven replicational origins. At the latter, there are 3 shorter G ≡ C−rich regions interrupted by 2 longer A = T−rich sequences. Possible leftovers from the RNA primer are found in the DNA ribonucleotides. The principal replicase enzyme, DNA polymerase γ (180–300 kDa), has a very low rate of error (10^{-6}), which accounts for its high fidelity of replication. Nevertheless, mitochondria accumulate mutations at a higher rate than the nucleus, probably because of the high ROS activity. (Note that this is not identical with the γ subunit [52 kDa] of prokaryotic DNA polymerase III coded for by *E. coli* gene *dnaX*.)

The *Drosophila* mtDNA does not display D loops. The mtDNA may be divided into the daughter nucleoids by pinching off into two approximately equal pieces during nucleoid division (*Physarum*), or by dividing into two equal parts at the beginning of the splitting of the organelle (*Paramecium*), or by dividing prior to the beginning of the division of the organelle (*Nitella*). In another method of division, approximately 30 small mitochondria fuse into a long mitochondrion during the G₁ phase of the cell cycle following mtDNA replication. Subsequently, the large mitochondrion divides into two nucleoids, then fragments further into spherical nucleoids (*Saccharomyces*). Within each nucleoid, the number of mtDNA molecules may vary: 3–9 in yeast, 32 in the *Physarum* plasmodium, and 2–6 in mouse nucleoids. The number of mtDNA copies/cell may show substantial variations (1,000–8,000). In some fungi (*Candida*), the catenated mtDNA molecules may contain 7 units in a linear array. Most commonly, the mtDNA appears as covalently closed circles, yet in some cases (e.g., malignancy) two monomers may be catenated in a double circle or may be double-length chains. The 50 kbp mtDNA of *Tetrahymena* is linear, whereas in the *Paramecia* kinetoplasts mini- and maxicircles occur. *See* catenane; CSB; D loop; DNA polymerases, γ; DNA replication; kinetoplast; mitochondrial control regions; mitochondrial genetics; MRP; mtDNA; nucleoid; polymerase; plasmodium; replication error; R loop; ROS. (Moraes, C. T. 2001. *Trends Genet.* 17:199; Clayton, D. A. 2000. *Hum. Reprod.* 15[Suppl. 2:]11; Lecrenier, N. & Foury, F. 2000. *Gene* 246:37; Maier, D., et al. 2001. *Mol. Biol. Cell* 12:824; Gensler, S., et al. 2001. *Nucleic Acids Res.* 29:3657; Shadel, G. S. & Clayton, D. A. 1997. *Annu. Rev. Biochem.* 66:409.)

DNA replication, prokaryote

The genetic material must be faithfully replicated to assure heredity. Unlike in eukaryotes, pyrimidine deoxyribonucleotides are synthesized from ribonucleotide diphosphates rather than triphosphates. The replication takes place in a semiconservative manner (*see* semiconservative replication). Although the basic system of the process is very similar to viruses in eukaryotes, variations exist in the mode in which the final product is made. Viral DNA replication: Some viruses (φX174, G4, M13, fd) contain single-stranded viral DNA (V DNA) and thus the genetic material must be replicated via a complementary strand (C DNA) in the intermediate double-stranded step, the replicative form

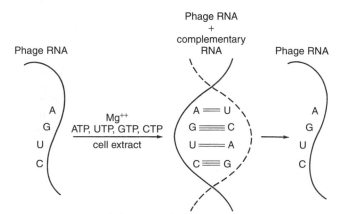

Broad outline of replication of a single-stranded nucleic acid molecule via a replicational intermediate double-stranded structure. (From August, J. T., et al. 1968. *Cold Spring Harbor Symp. Quant. Biol.* 33:73.)

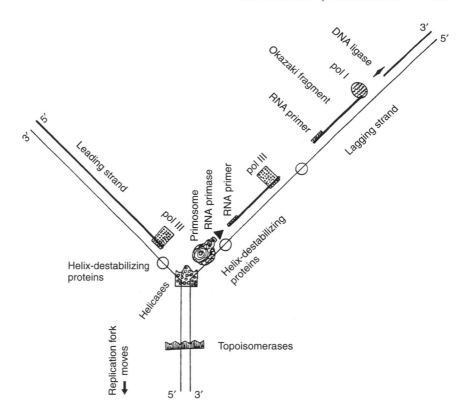

Overview of the DNA replication fork of prokaryotes.

(RF). All DNA types require a replication *origin* (*ori*) to begin the process. A common feature of the origins is that they are rich in $A = T$ to facilitate strand separation required for the semiconservative replication. The initiation of replication in the filamentous phage fdV DNA proceeds in the opposite direction of the C DNA in the RF from close-by points. In the G4 phage, the replicational origins of the V and C strands are widely separated, and in its close relative, ϕX174, there is one origin for the V strand but multiple origins for the C strand. The replication of the about 15-times-larger duplex DNA of phage T7 proceeds bidirectionally $\leftrightarrow$ from the origin and so does the *E. coli* duplex DNA that is about 800 times larger. Since the DNA single strands run antiparallel, simultaneous semiconservative replication has some special requirements. This problem is met in a simple way by the 36 kb DNA of adenoviruses: first the replication starts at one of the 3′-ends. When it is completed, the other strand is copied using the other OH-end as a template. Here each 5′-end cytidylic residue of the DNA is bound to a 55 kDa terminal protein. Then a 80 kDa viral protein displaces the end protein and uses the terminal deoxycytidylic acid as the beginning nucleotide template to lay down the first guanosinephosphate.

The 80 kDA protein binds to nucleotides 9 to 18 at the end, and host nuclear protein factor I is attached to the next 30 nucleotides. Since the 80 kDA protein has already formed a complex with the polymerase, replication can proceed.

The size of the replicational unit in *E coli*, the *replicon*, is thus the whole genome (4.7×10^6 bp), whereas in mouse the average length of approximately 2,500 replicons is about 150 kb. Despite the large number of replicons in higher eukaryotes, the pace of replication is much slower: in *E. coli*, about 50 kb/min, but in mice, about 2.2 bp/min.

The single origin in *E. coli* is in the *oriC* locus and the two forks must meet about halfway in the circular DNA. About

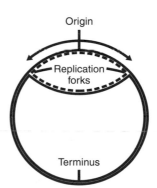

Replication in *E. coli*.

100 kb from this meeting point, there are *ter* (termination) regions of about 23 bp length; *terD* and *terA* signal termination on one of the branches of the fork and *terC* and *terB* on the other. The protein coded by the *tus* gene recognizes the terminators and halts replication after the replication forks pass the *ter* sites. DNA synthesis can be started only if appropriate hydroxyl ends are available. In *E. coli*, the hexameric *helicase* proteins (330 kDa) are required to unwind the two strands and generate the **Y** form. This process uses energy liberated by hydrolysis of ATP to ADP. Single-strand binding (SSB) helix-destabilizing proteins that make the bases free to be copied on both old strands assist the unwinding. The unwinding removes the negative supercoiling and may even reverse the original coiling into positive supercoiling. This process has limitations and the tension must be relieved, making single-strand cuts in the still unwound duplex. This nicking is carried out by topoisomerases (swivelases).

DNA *topoisomerase type I* enzymes make a single cut in one strand of the helix, then permit the nicked strand to rotate once around the intact strand, resealing the free ends and relieving negative supercoiling. *Topoisomerases type II* can cut and rejoin both strands of the duplex ahead of the fork and thus remove both negative and positive supercoiling. The latter group of enzymes include *gyrases*. They can make double cuts and convert positively supercoiled sections to negatively supercoiled ones. Type II topoisomerases can also convert catenated molecules ∞ into decatenated ones by permitting one duplex ring to slip out from the other ⇒○ ○. At the bifurcation, one of the templates runs $5' \rightarrow 3'$ and the other $3' \leftarrow 5'$. It is simple to proceed with replication of the old strand that faces the fork with its 5'-end because its complementary new strand can be elongated by adding bases to the 3'-OH terminus. This is called the *leading strand* (left branch of the fork on the diagram p. 325). While DNA synthesis (the new strand is shown by the heavier line) is going on, the strands at the fork must be kept separated by the *helix-destabilizing proteins* (also called SSB, tetrameric *single-strand binding* proteins of 74 kDa, shown as small circles). The principal enzyme of replication is *DNA polymerase III* holoenzyme (pol III, 10 to 20 copies per cell). The α-subunit of the enzyme (encoded by gene *polC*) carries out polymerization (chain extension) at the 3'-end of both *leading* and *lagging strands* (right branch of diagram). The ε-subunit (gene *dnaQ*) is endowed with a 3' to 5' exonuclease function that reduces errors from 7×10^{-6} to 5×10^{-9}. The two (α and ε), together with θ, represent the *pol III core*. The β-subunit (gene *dnaN*) apparently keeps the holoenzyme attached to the DNA. Additional polypeptide chains ($\gamma, \delta, \delta', \psi, \tau$, and χ) are also part of the huge holoenzyme (≈ 1 MDa).

The holoenzyme may be assembled in different ways under the control of gene *dnaX* products γ and τ. The lagging strand is synthesized in pieces (Okazaki fragments of 1 to 2 kb) because of the lack of a 3'-OH terminus to elongate a continuous strand starting from the base of the fork. Actually, even the leading strand is replicated in a discontinuous manner, but it has fewer initiation sites than the lagging strand. The lagging strand requires a primer to be initiated. For a primer, a short RNA sequence is used. The *primosome* is a protein complex of helicases (*dnaB, dnaC* gene products), prepriming protein (66 kDa, *dnaT* gene product), *priA* gene product (82 kDa monomer), which recognizes the primer assembly site and displaces the helix-destabilizing protein, and *primase* (60 kDa monomer product of gene *dnaG*). The primosome wraps around the DNA strands. Although this complex is shown only at the base of the lagging strand on the diagram, it probably moves along to other locations as its role requires. Since the lagging strand is synthesized in pieces (Okazaki fragments) that carry RNA at the initiation region (5'), there is a need for other functions.

DNA polymerase I (109 kDa monomer product of gene *polA*) serves triple roles. The larger C-terminal domain (68 kDa Klenow fragment, when cleaved by subtilisin) possesses a 5' to 3' polymerase and a 3' to 5' exonuclease function. It can elongate the 3'-OH DNA or RNA termini by adding 5' nucleotidyl phosphates and can slice off nucleotides in the 3' to 5' direction. Its N-terminal domain (35 kDa) is a 5' to 3' exonuclease, an activity blocked during the polymerization reaction. Thus, pol I can simultaneously extend the 3'-OH end of the Okazaki fragments and remove the primer RNA at the 5'-end, and it is capable of a replacement replication (nick translation) that is just needed for making the lagging

strand continuous. The 3' to 5' exonucleolytic activity also has the important function of "editing." If it finds "spelling errors" during synthesis, it removes the mismatched base and replaces it with the correct one. Its N-terminal domain reduces replicational errors from 10^{-5} to 10^{-7}, which is about the average spontaneous mutation rate in bacteria. After the Okazaki fragments reach full length, they are joined by *DNA ligase*, an *E. coli* enzyme that attaches adjacent 5'-phosphoryl and 3'-OH ends of nucleotide chains in the presence of NAD^+ (nicotinamide adenine dinucleotide). Most of the details of the information on DNA replication were obtained in in vitro systems where the single components can be added or withdrawn and the role can be established without the complications of the in vivo analyses. In *E. coli*, the transcriptionally most active genes are operating in the leading strand. *See* DNA chemical synthesis; Okazaki fragments; processivity; replication fork; replication machine; replication speed; RNA replication, viruses; rolling circle replication; steric exclusion model; θ replication. See color plate in Color figures section in center of book. (Wawrzynów, A. & Zylicz, M. 1997. in *Guidebook to Molecular Chaperones and Protein Folding Catalysts*. Gething, M.-J., ed., p. 481. Oxford Univ. Press, New York; Nossal, N. G., et al. 2001. *Molecular Cell* 7:31; Postow, L., et al. 2001. *Proc. Natl. Acad. Sci. USA* 98:8219.)

DNA, 7S Found in the mitochondria, where it facilitates the synthesis of the heavy strand of the mtDNA. In mammals, mtDNA replication proceeds from two promoters: the light-strand and the heavy-strand promoters, respectively. In this region an arrested H-strand DNA, the 7S DNA, is hybridized to the parental circular mtDNA and forms a triplex with the displaced replication loop. This 7S DNA is primed by a 7S RNA derived from the light-strand DNA. *See* mtDNA; RNA 7S. (Lee, D. Y. & Clayton, D. A. 1998. *J. Biol. Chem.* 273:30614; Gensler, S., et al. 2001. *Nucleic Acids Res.* 29:3657.)

DNA, selfish Has little or no known use for the organism, yet it commonly occurs, e.g., repetitive sequences. *See* introns; LINE; microsatellites; minisatellites; redundant DNA; repetitious DNA; SINE; transposable elements. (Crick, F. H. C. 1979. *Science* 204:264.)

DNA, ultraviolet absorption The maximal absorption is at about 260 nm, but the maximum may be influenced by the solvent and pH. The maximal absorption of polycytidylic acid is between 270 and 280 nm, whereas that of polyguanylic acid may vary between 250 and 290 nm, depending on the pH. Single-stranded molecules have increased absorption relative to double-stranded DNA. DNA isolated from plant or animal tissues generally have $280/260 \approx 2$ absorption ratios. The OD_{260} of 1 corresponds to about 50 µg/mL double-stranded and 40 µg/mL single-stranded DNA at about pH 8 in TE buffer. *See* buffer; cyclobutane dimer; OD; physical mutagens; TE buffer; UV.

DNA, unique DNA sequences that occur only in single copies in the genetic material.

DnaA DNA replication initiation protein in *E. coli*. It binds to a 9mer consensus the DnaA-box (5'-TT(A\T)TNCACA), and if it complexes with ATP, it binds to the ATP-Dna-box (5'-AGatct). *See* DnaB; *oriC*.

DNA alignment   *See* indel.

DNA amplification   *See* amplification.

DNA annealing   Reassociation of two complementary single strands into a double-stranded molecule. *See* c_0t curve; DNA hybridization.

DNA base composition   Varies among different organisms (approximate values):

Species	A	T	G	C*
Humans	30.7	31.2	19.3	18.8
Wheat (*Triticum aestivum*)	25.6	26.0	23.8	24.6
Budding yeast (*Saccharomyces cerevisiae*)	31.3	32.9	18.7	17.1
Escherichia coli	26.0	25.2	24.9	23.9
Mycobacterium phlei	16.5	16.0	34.2	33.2
Bacteriophage T4	32.4	32.4	18.3	17.0
Bacteriophage φ X174(single-stranded)	24.3	32.3	24.5	18.2

*Methyl cytosine and hydroxymethylcytosine are combined.

by over 100-fold and may also block gene expression. *See* DNA kinking; flexer; high-mobility-group proteins; lambda phage; looping of DNA; TBP. (Hagerman, P. J. 1990. *Annu. Rev. Biochem.* 59:755, Wu, J., et al. 2001. *J. Biol. Chem.* 276:14614; ibid., 14623.)

DNA binding   N-methyl imidazole and N-methylpyrrole amino acids can recognize specific base pairs or short base sequences in the DNA and bind to them. There are also many DNA-binding proteins in the cells. *See* DNA-binding protein domains.

DNA-binding protein domain   There are a variety of DNA-binding proteins and they have some common structural motifs belonging to four major groups: zinc fingers, helix-turn-helix, leucine zipper, and helix-loop-helix. *See* figures and more under the name of individual motifs. *See* binding proteins; DBP; DNA bending; hormone receptors; regulation of gene activity; RNA-binding proteins; single-strand binding proteins. (Karmirantzou, M. & Hamodrakas, S. J. 2001. *Protein Eng.* 14:465, Garvie, C. W. & Wolberger, C. 2001. *Mol. Cell* 8:937.)

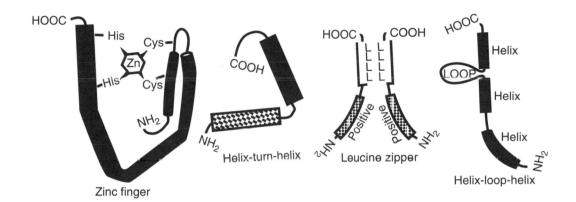

Zinc finger   Helix-turn-helix   Leucine zipper   Helix-loop-helix

The differences between the amounts of A and T, and G and C, respectively, were caused by analytical errors.

DNAase   *See* DNase.

DnaB   Hexameric DNA helicase involved in bacterial replication by polymerase III. Besides being a helicase, it regulates the synthesis of the primer and a factor of processivity. *See* DnaA; DNA replication in prokaryotes; *oriC*; O-some; processivity. (Carr, K. M. & Kaguni, J. M. 2001. *J. Biol. Chem.* 276:44919.)

DNA bending   Two distant DNA sites are brought closer together because of a nonstraight run of the double helix caused by phosphate neutralization due to protein binding. The DNA may also bend by attaching a sequence-specific ligand to two sites separated by, say, 10 nucleotides. The lambda-phage *Cro* repressor binds to three specific operator sites and causes DNA bending. Bending of chromatin may increase the chances of affinity of the TATA box-binding protein

DnaC   Prokaryotic DNA helicase. *See* DNA replication in prokaryotes.

DNA chemical synthesis   The 5'-end of the first nucleotide is protected by dimethoxytrityl (DMT) while the OH end is attached by a linker to silica. The reactive groups of all nucleotides are chemically protected. Afterward, DMT is removed by washing and the next nucleotide is activated and attached to the 3'-OH group. Using iodine, the 5' to 3' linkage is oxidized to generate a phosphotriester bond (one of the O of the phosphate group is methylated). The reaction is continued until the desired chain length is reached. About 70–80 residue polymers can be made this way. The process has also automated versions and routine synthetic services are commercially available. (Linkletter, B. A., et al. 2001. *Nucleic Acids Res.* 29:2370.)

DNA chimera   DNA molecule ligated together from originally different molecules.

DNA chips Combinatorial array of oligonucleotides synthesized on a solid support (modified glass or polypropylene) in situ at specific addresses in a checkerboard-like arrangement. The synthesis requires the use of masks to protect areas of the array. They may use the 4 nucleotides (A, C, G, T) in any sequence length (4^s) up to many thousands. These procedures are expected to be exploited for mass, automated sequence analyses, mutation detection, etc., in a way analogous to the electronic microchip technology. This type of technology can be used to synthesize a vast array of nucleotide sequences at a staggering speed. The procedure is a modification of the photolithography used by commercial printing for several decades.

A mercury light is directed to a photolithographic mask and through it to a solid surface. There it activates at specific areas of nucleotides for chemical coupling. The 5′ ends are protected. Subsequent exposure removes the protection and potentiates the reaction of the end with another nucleotide. The cycle is then continued. The specificity of the nucleotides to be chosen is controlled by the mask. Chips of 1/6 cm² can accommodate hundreds of thousands of sequences of 20 μm. The space occupied by an oligonucleotide sequence of millions of molecules is called *feature*. Each wafer contains a grid of 49–400 chips. The generation of the chips and the synthesis is mechanically controlled with minimal human labor, and the whole process requires about half an hour.

When the synthesis is complete, the chips are separated and hybridized with an appropriate fluorochrome-labeled nucleic acid (DNA or RNA) probe. After removal of the unhybridized probe sequences, the fluorescence analysis by a confocal microscope requires about half an hour. The resolution is going to be improved to 20 μm in the future and each chip may permit the resolution of 400,000 probes/chip. The ~3 billion–nucleotide mammalian genome may require only 10 chips. Detection of single-base-pair mutations is a much more elaborate process than monitoring the expression of genes. The cost of the technology is comparable to an automatic sequencer.

The DNA chip technology can be extended to mapping gene functions to chromosomal locations and obtaining information on all of the genes contributing to a particular function. The total genomic DNA is arranged in fragments over the wafers with a minimum of 20 oligonucleotide probes of 25 base length for each open reading frame of the yeast genome. Probes from the coding sequences are placed in order of their position in the genetic map (chromosome). Mismatches in the hybridization to the same array of the two genetically different strains are detected with the aid of a confocal laser scanner on the basis of the fluorochrome used for staining. The location of the chromatid exchanges can be physically identified by color and the information using known genetic markers can be verified by classical tetrad analysis. Since the total nucleotide sequence (representing all of the genes) of a large number of organisms is already known, these new technologies may detect simultaneously the turning on and off of thousands of genes and may distinguish the differences between a healthy and diseased body. The information may lend great help in designing more efficient drugs, and it should greatly further the knowledge of how a cell or billions of cells work together in the body (Southern, E. M. 1996. *Trends Genet.* 12:110; *Nature Genetics* 14, No 4; Fodor, S. P. A. 1997. *Science* 277:393; Winzeler, E. A., et al. 1998. *Science* 281:1194.) *See* genomics; microarray hybridization; nanotechnology; photolithography; SHOM.

Fluorescent image of an array containing more than 280,000 different 25-mer oligonucleotide probes in an area of 1.28 × 1.28 cm. The array contains probe sets for more than 6,800 human genes, and the image was obtained after overnight hybridization of an amplified and labeled human mRNA sample. (From Fields, S., et al. 1999. *Proc. Natl. Acad. Sci. USA* 96:8825.)

(<http://www.sciencemag.org/dmail.cgi?53241>; <www.gene-chips.com>; for optical lithography, Ito, T. & Okazaki, S. 2000. *Nature* 407:1027; Lockhart, D. J. & Winzeler, E. A. 2000. *Nature* 405:827.)

DNA cloning   *See* cloning.

DNA complexity   *See* c_0t curve.

DNA computer   Based on the four bases in the DNA. Theoretically it is possible to simultaneously conduct several different operations. At present 2003, the technical problems do not allow their practical construction, although in theory it is promising. *See* Hamiltonian path. (Chen, J. & Wood, D. H. 2000. *Proc. Natl. Acad. Sci. USA* 97:1328; Sakamoto, K., et al. 2000. *Science* 288:1223; Joachim, C., et al. 2000. *Nature* 408:541; Cox, J. C. & Ellington, A. D. 2001. *Curr. Biol.* 11:R336.)

DNA conformation   *See* DNA types.

DNA consensus   *see* consensus.

DNA crystal   May be used to explore molecular structures at the finest scale. Antiparallel-recombinant DNA molecules are tiled in order to generate lattices at periodic patterns. The crystal is an arrangement of molecular units in two- or three-dimensional periodicity. An understanding of the forces involved may assist in the design of special catalysts, enzymes, or industrial biochips, etc. *See* tiling, chip. (Packer, M. J. & Hunter, C. A. 2001. *J. Am. Chem. Soc.* 123:7399.)

DNA database   *See* databases.

DNA dating   *See* racemate.

DNA denaturation   Separation of the two strands of DNA duplexes by breaking the hydrogen bonds and the hydrophobic interactions among bases; it also results in reduced viscosity. This is accomplished by raising the pH or the temperature above 70° or 80°C. The melting temperature depends on the base composition because G≡C binds by 3 and A=T by 2 hydrogen bonds. By lowering the temperature below 60°C, renaturation may begin. *See* c_0t; denaturation; DNA; DNA thermal stability; hyperchromicity; hypochromicity.

DNA density   *See* density gradient centrifugation; ultracentrifugation.

DNAs of different densities separate in distinct bands in CsCl in the tube of the ultra-centrifuge and can be visualized by UV photography or by ethidiumbromide staining and the bands can be collected for further analysis.

DNA density shift   May occur if heavy isotopes (^{13}C or ^{15}N) or 5-bromodeoxyuridine is incorporated into the DNA during replication, replacing the normal atoms or nucleotides. *See* diagram above.

DNA-dependent protein kinase   *See* DNA-PK.

DNA-dependent RNA polymerase   Uses DNA template for the synthesis of RNA. *See* transcription.

DNA diagnostics   *See* genetic screening; prenatal analysis.

DNA-driven hybridization   In the molecular hybridization medium, the DNA is in excess relative to RNA. *See* DNA hybridization.

DnaE   Encodes the catalytic subunit of bacterial DNA polymerase III. *See* DNA polymerases. (Dervyn, E., et al. 2001. *Science* 294:1716.)

DNA extraction   There is a large number of procedures applicable to the major taxonomic groups. Only some basic outlines can be provided here. BACTERIA: *1.* Grow bacteria in culture medium of choice to a high density (1 to several days). *2.* Harvest cells by centrifugation. *3* Lyse cell pellet in TE buffer (Tris-EDTA) containing SDS (sodium dodecyl sulfate) and proteinase K (free of DNase) at 37°C for 1 hour. *4* Make it 0.5 M for NaCl and add CTAB (cetyl trimethylammonium chloride) to precipitate cell wall, polysaccharides, proteins, etc., but keep DNA in solution.

5 Add equal volume of chloroform:isoamyl alcohol and centrifuge to remove CTAB and polysaccharides. *6* Save supernatant. *7* Remove protein by phenol:chloroform:isoamyl alcohol and centrifuge. *8* From supernatant, precipitate DNA by isopropanol. *9* Wash the precipitated DNA with 70% ethanol. *10* Take up DNA in TE buffer and store it in refrigerator. PLANTS: *1* Use 10–50 g clean young tissue (from plants that have been kept in dark for 2 days to reduce starch). *2* Grind it to powder in liquid nitrogen. *3* Extract DNA in pH 8 Tris-EDTA buffer containing a detergent (SDS or Sarkosyl) for about 1 to 2 h at 55°C. *4* Pellet debris by centrifugation at 4°C (6,000 rpm, 10 min) and collect supernatant DNA. *5* Precipitate DNA from supernatant by 0.6 volume cold isopropanol (−20°C) by centrifugation (8,000–10,000 rpm, 4°C, 15 min). *6* Pellet is taken up in TE buffer. MAMMALIAN TISSUES: *1* Tissue or cell pellet (0.2 to 1 g) washed clean and powdered in frozen liquid nitrogen. *2* Lysis of cells (in NaCl, Tris buffer, EDTA, SDS, pH 8, proteinase K). *3* Extraction of DNA in phenol:chloroform:isoamyl alcohol and centrifuge (10 min, 10,000 rpm). *4* To supernatant, add 0.5 vol 7.5 M ammonium acetate and 2 vol cold ethanol to precipitate DNA by centrifugation (2 min, 5,000 rpm). *5* Wash DNA with 70% ethanol. *6* Suspend DNA in TE buffer. All of the above procedures are basically very similar. Further purification may be necessary by ultracentrifugation in CsCl. Quantity may be determined on the basis of absorption of ultraviolet light of 260 nm (quartz cuvettes) in a spectrophotometer. In a reasonably pure DNA preparation, the ratio of OD260: OD280 is about 2:1. One OD is about 50μg/mL for double-stranded DNA and 40μg/mL for single-stranded DNA; 1 pg DNA is about 6.5×10^{11} Da. *See* centrifuge; CTAB; EDTA; proteinaseK; Sarkosyl; SDS; spectrophotometer; TE buffer; Tris; ultracentrifuge.

DNA fingerprinting   Has been developed since the 1980s as a molecular tool for the genetic specification of an individual, a taxonomic groups, or other related groups. DNA fingerprinting is an important tool for the study of evolution and for establishing the degree of relatedness of genetic stocks. Its application is relevant to civil and criminal court decisions requiring identifications with high precision. The perfect definition of identity would come from the complete sequencing of the genomes, but that is not practical for routine purposes. The best approximation is to use Southern hybridization with some repetitive sequences as probes isolated from minisatellite DNAs of common occurrence in the eukaryotic genomes (*see* SINE). The first such probe has been obtained from the four 33 base-pair repeats in intron I of the human myoglobin gene. These repeats have a core sequence very similar to the *chi* elements responsible for meiotic recombination hot spots. These elements generate unequal crossing over at high frequency but are silent in the somatic cells. Because of this nature, they generate a high degree of specific genetic diversity. Actually, it appears that each and every individual is different from all others (except monozygotic twins or clones). These minisatellite DNA sequences differ from each other but share the highly conserved core.

The minisatellites are distributed widely in the human (and other eukaryotic genomes) within the chromosomes. In addition to the minisatellite probes, microsatellites can be used. These are generally mono- to tetranucleotide repeats occurring very frequently and dispersed in the eukaryotic genomes. When isolated genomic DNA is digested with one or more restriction endonucleases separated by agarose gel electrophoresis, Southern blotted, and hybridized to minisatellite probes that are radioactively labeled, the autoradiograms appear similar to the diagram shown below. The total DNA without probes cannot be used because its quantity and complexity are too large and the individual bands cannot be distinguished due to the fact that many of them share the same or almost the same position in the gel. Such a gel would not appear cross-banded, but it would look like a smear. Generally, even the bottom of probed gels contains a larger number of small fragments that are barely or not separated (not shown on the diagram). For the analyses, fresh DNA samples are preferred because in older specimens degradation may take place and the larger fragments (closer to the start point [top]) become very weak or vanish. Some fingerprinting may be feasible, however, in frozen flesh or in mummies more than 2,000 years old. In the same gel, appropriate full-range molecular-size markers are also run at both sides to facilitate identification of fragment length with appropriate accuracy. The fragment length estimated on the basis of mobility in the gel compared to that of a corresponding size marker might have an error of 3–4 %.

If one assumes that apparently comigrating bands in the gels of A and B individuals represent identical alleles of the same minisatellite locus, the probability x that an allele in A is also present in B is proportional to the frequency of that allele according to $x = 2q - q^2$. If the frequency of the allele in the homozygotes is low, then the mean probability $\hat{x} \approx 2\hat{q}$. Also, if it is assumed that there is little variance among allelic frequencies, then the number of alleles $n \approx 1/\hat{q}$ and the mean homozygosity is approximately

$$\sum_{1}^{n} q_i^2 \approx n\hat{q}^2 = \hat{q}.$$

A certain proportion of the comigrating bands in A and B belong to different minisatellite loci, and the estimates of mean allele frequency and homozygosity are maximal; the true estimates will depend on the accuracy of the resolution and estimation of the size of the fragments in the gels. The mean probability that all fragments detected by probe 33.15 (see table) are also present in individual B is $0.08^{2.9} \times 0.20^{5.1} \times 0.27^{6.7} \cong 2.78 \times 10^{-11}$. (Note that the probabilities are raised to the power of the corresponding mean fragment number). Because the current population of the earth is about 6×10^9, a > 5o-fold increase in the current population would be needed to perhaps find two individuals with identical genetic constitution. The precision of the analysis is further improved by the use of several probes. In the American forensic laboratory practice, four or five probes are used that detect only two allelic variations each, making the precise reading of the bands easier because only two are shown by each run. This type of analysis still provides 99.9% accuracy for determining the identity of a person. The power of this procedure is also indicated by the fact that 0.5 to 5 μg DNA (present in a drop of blood, in a few thousands of cells) may be enough for an RFLP test. The procedure described above is the DNA fingerprinting or RFLP (restriction fragment-length polymorphism). RFLP has superior analytical value to PCR because the former is based on the entire DNA, whereas in the latter only fragments of the DNA are amplified.

If RFLP is coupled with, say, 200,000-fold amplification of a sample using PCR, its efficiency may be further increased. The DNA sample prior to amplification may be as low as 500 pg (0.5 ng); however, somewhat larger quantities (2 ng) may give better results. The most efficient techniques may permit DNA analysis from palm swabs, briefcase handles or telephone handles, gloves, coffee mugs, etc. A single PCR amplification does not deal, however, with an entire genome-size DNA but only with an extremely small fragment, defined by the two 5′-terminal primers. Taq DNA polymerase is used, starting with pg quantities of DNA of 2 kb in length which can be increased up to 1 μg in 30–35 cycles, and after dilution it can be subjected to even further amplification. For RFLP tests, not just larger quantities of the DNA are required, but it must be fresh or nondegraded. PCR analysis can be carried out with more minute samples, but some degradation may not prevent its suitability for the test. For forensic PCR tests, generally the DNA of the second exon of the human leukocyte antigen (HLA)-DQ-α is used most commonly. Actually, one test kit, *Amplitype® HLA-DQ-alpha Forensic Kit*, marketed by the Perkin-Elmer Company, is commercially available. This procedure has an accuracy of 98–99%, but a newer PCR procedure, the *Poly-Marker Test* (PM), is supposed to have up to 99.7% precision. The PCR methods use dot blot hybridization and fluorochrome labels that eliminate the waiting period involved in exposing the Southern blot to X-ray film. The Amplitype relies on differences in six allelic types (n) that determine 21 different genotypes according to the general formula $[n(n+1)]/2$. The PCR method is less accurate than

Core G G A G G T G G G C A G G A $_G^A$

33.6	{(A G G G C T G G A G G)$_3$}$_{18}$	Trimeric section repeated 18 times
33.15	{(A G A G G T G G G C A G G T G G}$_{29}$	repeated 29 times in human genome

SCHEMATIC REPRESENTATION OF DNA FINGERPRINTS.
The purified genomic DNA is digested with one or more restriction endonucleases and then separated by agarose gel electrophoresis. The run is Southern blotted onto a nitrocellulose or other substrate, denatured, and hybridized with a radioactively labeled probe prepared from a minisatellite. After removal of the unbound label, it is autoradiographed at very low temperature for a few days. The film then reveals the restriction fragments that are homologous to the probe(s). The three code-bar-like patterns (at left) could be the DNA fingerprints of three different individuals or those of a single person probed with three different minisatellites. The largest fragments are at the top. The wide bands may indicate that several fragments of similar size did not separate well from each other.

In an experiment with 20 Englishmen, the following information was obtained. (Data and methods of calculations after A. J. Jeffreys, et al. 1985. *Nature* 316:76):

DNA Fragment Size, kb	No. of Fragments M± Standard Deviation Per Individual		Probability x that Fragment in A Individual is Present in B Individual		Max. Mean Allelic Frequency Per Heterozygosity	
Probes—	33.6	33.15	33.6	33.15	33.6	33.15
10–20	2.8 ± 1.0	2.9 ± 1.0	0.11	0.08	0.06	0.04
6–10	5.1 ± 1.3	5.1 ± 1.1	0.18	0.20	0.09	0.10
4–6	5.9 ± 1.6	6.7 ± 1.2	0.28	0.27	0.14	0.14

the RFLP test, but its value can approach that of the latter by the use of a larger number of probes. In both types of assays, the proper statistical procedure is very important. The power of the statistics depends on databases revealing the frequencies of the allelic variations in the general or ethnic populations so that the genotype of the individual to be tested can be compared to other similar or identical types. That is, the statistics will tell the probability of how many other individuals may have the same DNA fingerprint as the suspect. In court, along with the DNA fingerprints, other evidence is considered that may further improve identification (such as dermatoglyphics, alibi, time frame, etc.).

For DNA fingerprinting, a variable section of mitochondrial DNA can be successfully used. Since the size of this DNA is more limited than that of the nuclear DNA fragments, the potential information to be gained is less. DNA fingerprinting can identify members of a biological family with great certainty. If the total number of identifiable bands of putative mother (M) and father (F) is T, and of that number n is shared by offspring (O), then the probability that (O) would share all the bands in common between (M) and (F) is Y^T, where $Y = 1 - (1 - x)^2$ [for the definition of x, see above]. The probability that (O) would share the specific maternal or paternal fragments is x^{Mf} and x^{Ff}, where Mf and Ff are the number of mother- or father-specific fragments.

The use of DNA evidence in the criminal courts has been questioned on statistical grounds and because some of the crime or biological laboratory contractors may have based unwarranted conclusions on poorly performed and/or incompetently evaluated DNA tests. The statistical arguments brought forward are as follows. If the only evidence is the DNA profile and the probability of a match for a particular individual is 1/1,000,000, but there may be 500,000 individuals who could be responsible for the case if the DNA evidence is not considered, there are two possibilities: either the suspect is the criminal and he displays a perfect match to the DNA sample collected at the crime scene or another individual among the other 500,000 is the criminal and the DNA analyses match only by chance. The probability for the first case is $1/(500,001) \times 1 = 0.000001999$. The probability for the second case is $500,000 \times (1/500,001) \times (1/1,000,000) = 0.000,000,999$. The ratio of these two fractions is about 0.5, and $1.5/0.5 = 3$. Thus, the statistical chance that the suspect is innocent is $1/3 \approx 0.33$.

These methods are particularly useful to rule out identity, i.e., rule out that the blood, semen, or tissue sample left at the crime scene did not come from a particular suspect or that a particular person may not be a relative to an individual. Positive identification is also likely at an extremely high probability. The technology and the experience of the technicians have improved since the 1980s. Also, the removal of contaminants from the specimens (sulfur, dyes of the clothing, etc.) may be facilitated by new procedures of DNA purification. Several of the commercial testing laboratories follow the guidelines recommended by the *Technical Working Group on DNA Analysis Methods* (TWIGDAM). The greatest care is still required in the use of these powerful techniques, because in murder trials life or death of the suspects may depend on the correct identification. There are still some questions regarding the best procedures and the acceptance of DNA fingerprints and the statistical evaluation as evidence in different U.S. courts. The best statistical procedures for the criminal justice system should be to determine the ratio of the conditional probabilities (Pr) of the claims of the prosecution (C) and the defense (C') based on the evidence (E) as $Pr(E|C)/Pr(E|C')$. Currently—because of the great facility of accurate identifications—DNA fingerprinting is the most effective identification system in forensic repertory. DNA fingerprinting has wide applicability outside of the judicial system for identification of modern or ancient biological samples.

The mammalian mitochondrial DNA D loop is rich in variable base substitutions. In addition, the copy number of this DNA is very large, making its analysis rewarding even from samples highly degraded such as in exhumed corpses or ancient DNA. In such instances, DNA amplification is usually required. Techniques are being developed that permit DNA typing from single cells. Essentially the same total DNA procedures have been used for the analysis of feral populations of animals, anthropological studies, and ancient biological samples to check the authenticity of cell cultures or plant varieties, etc. DNA fingerprinting has been used for the generation of sequencing-ready large-insert BAC or PAC clones (Marra, M. A., et al. 1997. *Genome Res.* 7:1072). The inserts are labeled by ESTs and connected by chromosome walking. *See* allelic frequencies; autoradiography; Bayes' theorem; ceiling principle; chi elements; conditional probability; dot blot; electrophoresis; feral; fingerprinting; fluorochrome; forensic genetics; Frye test; gel electrophoresis; heteroplasmy; HLA; intron; labeling; microsatellite; minisatellite; molecular marker; m+DNA; MVR; myoglobin; nick translation; PCR; phylogenetic analysis; probe; RFLP; Romanovs; SINE; Southern blotting; utility index for genetic counseling; VNTR. (*The Evaluation of Forensic DNA Evidence.* Natl. Acad. Press, 1996; Krenke, B. E., et al. 2002. *J. Forensic. Sci.* 47:773.)

DNA flap Stable 99-nucleotide overlap in the center of the HIV-1 reverse transcription-generated DNA. HIV-1 synthesizes the DNA as two discrete half-genomic segments. A central polypurine sequence initiates the synthesis of the downstream part of the (+) strand. The upstream tract of the (+) strand is initiated at the 3'-end of the polypurine, proceeds to the center of the genome, and terminates after a discrete strand displacement. This event of the reverse transcription is mediated by the central termination sequence that removes the HIV reverse transcriptase at this stage. Thus, the final product of the (+) strand replication of the linear DNA has an overlap or flap. The integrity of the polypurine tract and the central termination sequence are essential for the retroviral life; the flap is also involved in the nuclear import of the HIV-1 genome. See acquired immunodeficiency; lentiviruses. (Whitwam, T., et al. 2001. *J. Virol.* 75:9407.)

DNA form Form I is superhelical and circular, II nicked circular, III linear. These conformational states affect electrophoretic mobility in gels. See electrophoresis; nick.

DNA glycosylase See glycosylase.

DNA groove A B DNA double helix displays a 3.4 nm pitch including a wider (major) and a narrower (minor) groove (furrow) along its helical structure. Molecules that specifically recognize base pairs in the major or minor groove provide a means of regulation of gene activity. The various binding proteins and transcription factors have been and still are thoroughly investigated, but so far no general recognition system between the nucleotide base (pairs) and amino acids is known. The four base pairs (A•T, T•A, G•C, C•G) may be distinguished in the major groove on the basis of hydrogen bonding properties. Oligonucleotides may recognize the double helix and form triple helix structures but do not provide sufficient single base (pair) specificity. Some antibiotics (such as distamycin [$C_{22}H_{27}N_9O_4$]), modified by replacing the pyrrole (Py) rings with an imidazole (Im) ring, provide limited ability to recognize G•C pairs. Two distamycins effectively recognize side-by-side A•T sequences in the minor groove. In 1998, new recognition codes were reported that are capable of distinguishing all four base pairs in the minor groove: Py/Im → C•G, Im/Py → G•C, hydroxypyrrole (Hp)/Py → T•A, and Py/Hp → A•T. This development does not provide discrimination among human genes because 17 base pairs would be needed to do so. New approaches may be open not just for gene regulation but also for drug design. See DNA types; pitch; Watson and Crick model. (Hélène, C. 1998. *Nature* 391:436; White, S., et al. Ibid. 468.)

Two major and one minor groove are shown in this space-filling model of DNA.

DNA gyrase DNA-binding protein, controlling coiling, site-specific nicking, and functional in replication, transcription, recombination, and DNA repair. These enzymes are members of the DNA topoisomerase family. See nick; topoisomerase.

DNA heteroduplex See heteroduplex, models of recombination.

DNA hybridization Annealing single-stranded DNA with complementary single-stranded DNA or RNA either for the purpose of measuring the degree of homology or for labeling it before selective isolation. Hybridization can be carried out in a solution or with DNA immobilized on a membrane filter (Southern blots) or in situ, in denatured chromosomes or in microbial colonies (dot blots). The material intended for hybridization is labeled either radioactively (^{3}H, ^{14}C, ^{32}P) or with biotin and fluorescent dyes. See Benton–Davis plaque hybridization; chromosome painting; FISH; Grunstein–Hogness screening; in situ hybridization; Northern blot; Southern blotting. (Marmur, D. & Lane, J. 1960. *Proc. Natl. Acad. Sci. USA* 46:453; Doty, P., et al. 1960. *Proc. Natl. Acad. Sci. USA* 46:461; Hall, B. D. & Spiegelman, S. 1961. *Proc. Natl. Acad. Sci. USA* 47:137.)

DNA immunization See immunization, genetic. (<http://www.genweb.com/Dnavax/dnavax.html>.)

DNA index Reveals the amount of DNA in a cell relative to the pre-S-phase of the diploid cell (=1). See cell cycle.

DNA isolation See DNA extraction.

DnaJ Family of chaperone proteins of *E. coli* (375 amino acids), belonging to the Hsp40 family of chaperones/cochaperones. They also regulate the ATPase and peptide-binding activity of Hsp70. They are present in prokaryotes and eukaryotes, animals, plants, and fungi. DnaJ stabilizes the DnaK-protein complexes with the cooperation of GrpE. It assists replication of DNA by activating RepA helicase binding to the origin of replication in *E. coli*. After the transcription of Hsp70, the complex may separate the heat-shock-specific σ^{32} subunit from the RNA polymerase and reactivate the heat-aggregated and thus inactivated regular σ^{70} subunit. See CbpA; chaperones; DnaK; GRP; HSP; Hsp70; RcsG; Rep; σ^{70}. (Rüdiger, S., et al. 2001. *EMBO J.* 20:1042.)

DNA joining see blunt-end ligation; cohesive ends; ligase.

DnaK *E. coli* chaperone protein of the Hsp70 family with high homology to eukaryotic Hsps. In cooperation with DnaJ and GroE, it carries out the main functions of controlling heat responses, protein folding, transport and degradation in concert with proteases, initiation of bacterial, plasmid, and phage replication, flagella synthesis, etc. DnaK synthesis is regulated by a specific σ^{32} subunit from the bacterial RNA polymerase. See Anj1; auxilin; chaperones; flagellum; GroEL; Hdj2; Hip; HsJ1; Hsp70; Ldj; Mtj1; RNA polymerase; σ; Sis1; Tid50; translocon; trigger factor; Ydj1; Zuotin. (Kedzierska, S. & Matuszewska, E. 2001. *FEMS Microbial. Lett.* 204:355.)

DNA kinking Regulation of transcription depends on protein-induced localized modification of the DNA structure (bending). This modification depends on the nature of the proteins recruited for the tasks and the base sequences in the DNA. The bases stacked in the major and minor grooves (the rungs) of the helix may cause four major categories of architectural alterations relative to each other: (1) twist (slight rotation of the rung); (2) roll (slightly lifting the broader side of a rung); (3) tilt (uplifting the rung at one side); and (4) rise (slipping the rungs away from each other). The elastic properties of the DNA determine how it is wrapped around histones in the nucleosomal structure, the packing of phage DNA into capsids, supercoiling, DNA looping, etc. These changes are reversible and depend on the physical environment of the DNA. *See* DNA bending; DNA looping; nucleosome; supercoiling; triple helix formation. (Beylot, B. & Spassky, A. 2001. *J. Biol. Chem.* 276:25243.)

DNA knots Catenated (interlocking ring) DNA. *See* catenated; catenene; knotted circle. (Podtelezhnikov, A. A., et al. 1999. *Proc. Natl. Acad. Sci. USA* 96:12974.)

DNA library Collection of restriction endonuclease-generated fragments, each containing different segments of the genome. *See* fragment recovery probability; restriction enzyme.

DNA ligases Enzymes joining DNA termini in the cell nucleus. DNA ligase I (125 kDA) is the most important ligase in animal cells; DNA ligase II (72 kDA) and DNA ligase III (100 kDa) have a minor role in mammalian cells. DNA ligase IV mediates joining of the ends of DNA in eukaryotic nonhomologous recombination. Other organisms have different DNA ligases. Bloom syndrome and acute lymphoblastic leukemia are caused by DNA ligase deficiency. *See* Bloom syndrome; deoxyribozyme; double-strand break; Ku; lymphoblastic leukemia; nonhomologous recombination; vectors. (Bentley, D., et al. 1996. *Nature Genet.* 13:489; Pierce, A. J. & Jasin, M. 2001. *Molecular Cell* 8:1160.)

DNA likelihood method Type of maximum likelihood procedure for the calculation of branch length in evolutionary trees. (*See* evolutionary distance; evolutionary tree; four-cluster analysis; least square methods; minimum evolution methods; neighbor joining method; unrooted evolutionary trees.

DNA looping *See* looping of DNA; LCR.

DNA marker Special isolated or identified DNA sequences such as restriction fragments, RAPDS, microsatellites, minisatellites, EST, and other sequences that can be used as probes or followed by molecular or genetic/molecular analysis.

DNA methylation *See* methylation, DNA.

DNA methyltransferase (DNMT) *See* methylation, DNA; methyltransferase DNA.

DNA microarray *See* DNA chips; microarray hybridization.

DNA migration in gels Affected by molecular size, configuration of the macromolecule, concentration of the support medium (agarose, polyacrylamide), voltage, changing direction in the electric field, base composition, presence of intercalating dyes, buffer, etc.

DNA mobility shift *See* gel retardation assay.

DNA modification *See* 5-azacytidine; methylation, DNA; restriction enzymes.

DNA mutation *See* mutation, spontaneous.

DNA nicking and closing *See* topoisomerases.

DNA overhang In a double-stranded DNA, one of the chains protrudes at the end as shown. Such a structure is frequently generated by restriction endonuclease cuts. *See* restriction enzyme.

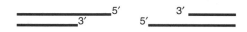

DNA packaging Insertion of a piece of (concatameric) DNA into a phage capsid. *See* condensation of chromosomes; packaging of DNA; packing ratio.

DNA pairing Formation of hydrogen bonds between complementary bases of two single-stranded DNA. *See* DNA; hydrogen pairing; Watson and Crick model.

DNA passenger *See* passenger DNA.

DNA photolyase Splits pyrimidine dimers during direct repair. *See* cryptochromes; cyclobutane dimer; DNA repair; pyrimidine dimer; ultraviolet light.

DNA-PK (DNA-dependent protein kinase, 8q11) Phosphatidyl inositol kinase family (PI[3]K) protein with a role in transcription, replication, immunoglobulin gene switching, cancer, and DNA repair. The sequence-specific enzyme complex works only when it is in cis-position to the site of phosphorylation. The Ku autoantigen, a transcription factor, attracts DNA-PK to specific DNA sequences. Ku binds to a specific negative regulatory element (NRE1) in the long terminal repeat and keeps expression of the glucocorticoid-induced transcription of mouse mammary tumor virus (MMTV) in check. Ku can also bind to other DNA sites, although it displays preference to DNA ends and to the MMTV glucocorticoid receptor and octamer transcription factor 1 (Oct-1). DNA-PK phosphorylates in vitro glutathione S transferase-Oct-1 fusion protein and the glucocorticoid receptor. Phosphorylation by DNA-PK is contingent on the presence of the glucocorticoid receptor and Oct-1-binding sites. DNA-PK is required for immunoglobulin (V[J]D) recombination and activation of innate immunity. DNA-PK mediates

Eukaryotes

pol α	pol β	pol γ	pol δ	pol ϵ
3 subunits	1 subunit	2 subunit	1 subunit	1 subunit
180–300 kDa	40 kDa	~180–300 kDa	~170–230 kDa	250 kDa
nuclear lagging	repair	polymerase	nuclear leading	$3' \rightarrow 5'$ leading
str. primase &	nuclear,	and	strand	nuclear
$3' \rightarrow 5'$	no exonuclease	exonuclease	replicase,	polymerase &
polymerase.		in mitochondria	$3' \rightarrow 5'$	$3' \rightarrow 5'$
no exonucl.			exonuclease	exonuclease

the binding of p53 protein to DNA and thus may play a role in the mammalian DNA damage control pathway. *See* autoantigen; double-strand break; immunoglobulins; innate immunity; Ku; ligase DNA; mouse mammary tumor virus; nonhomologous end joining; Oct-1; p53; p350; RAG; XRCC4. (Smith, G. C. M. & Jackson, S. P. 1999. *Genes Dev.* 23:916; Bryntesson, F., et al. 2001. *Radiat. Res.* 156:167; Chechlacz, M., et al. 2001. *J. Neurochem.* 78:141; Goiytisolo, F. A., et al. 2001. *Mol. Cell Biol.* 21:3642; Soubeyrand, S., et al. 2001. *Proc. Natl. Acad. Sci. USA* 98:9605; Mårtensson, S. & Hammersten, O. 2002. *J. Biol. Chem.* 277:3020.)

DNA polarity The first base in a DNA chain has a 5′ phosphate (triphosphate) at the beginning and the subsequent bases are added to it at the 3′-OH position. From the nucleotide triphosphate, two phosphates are removed, and the third one (α-group) forms a phosphodiester linkage with the first one. Thus, the chain growth is $3' \leftarrow 5'$. The complementary DNA chain also has polarity and the two run antiparallel.

$\rightleftharpoons$

DNA polymerase Prokaryotes and eukaryotes have several DNA polymerase enzymes with polymerase (and exonuclease) functions:

Prokaryotes

Pol I	Pol II	Pol III
109 kDa	90 kDa	1,000 kDa
$5' \rightarrow 3'$ polymerase	$5' \rightarrow 3'$ polymerase	$5' \rightarrow 3'$ polymerase
$3' \rightarrow 5'$ exonuclease	$3' \rightarrow 5'$ exonuclease	$3' \rightarrow 5'$ exonuclease
$5' \rightarrow 3'$ exonuclease	No $5' \rightarrow 3'$ exonuclease	None

DNA pol III holoenzyme forms a particle of 10 subunits, and it is responsible for the replication of the prokaryotic chromosome. The T7 phage DNA polymerase, encoded by gene 5, is 80 kDa. It associates with the bacterial-coded thioredoxin and polymerizes thousands of nucleotides. It also requires the hexameric protein of T7 primase-helicase and a T7 single-stranded binding protein that controls the synthesis of the leading and lagging strands at the replication fork. Prokaryotic DNA polymerases IV (DinB) and V (UmuD'$_2$ complex) enzymes have mutator functions in the SOS repair pathway. The majority of the pol IV mutations are small deletions. (Kobayashi, S., et al. 2002. *J. Biol. Chem.* 277:34087.)

In yeast, **pol ζ** was discovered; it does not have $3' \rightarrow 5'$ exonuclease function, although it is a repair enzyme that can more efficiently bypass thymine dimers than the other

polymerases. Pol ζ is error-prone, yet it has an essential role during the development of mouse embryos. DNA **pol η** (encoded by gene RAD30) of yeast has a function similar to pol ζ, but it is distinct from it. Pol η bypasses cis-syn thymine dimers efficiently and accurately. **Pol ι** (iota) has a high rate of misincorporation of deoxynucleotides opposite DNA lesions, but pol ζ immediately follows and extends the misrepair, although it alone does not insert mispaired bases. Pol ι also has base excision repair activity. DNA polymerase **pol θ** cannot bypass thymine dimers or 6-4 photoproducts or abasic sites. Its error rate with deoxyribonucleotides is 10^{-3} to 10^{-4}. It incorporates two adenines opposite the thymine dimer.

Telomerase and terminal nucleotide transferase are special DNA polymerases. **Polymerase κ** is essential for DNA replication of yeast and humans, along with pol α, δ, and ε, and it mediates sister chromatid cohesion. Enzyme pol κ, with some homology to pol β, in cooperation with other proteins, also mediates DNA repair. **Polymerase λ** is involved in base excision repair. **Polymerase μ** mediates nonhomologous end joining. **Polymerase σ** has a function in sister chromatid cohesion. **REV1** polymerase mediates bypass synthesis and **TdT** is involved in antigen receptor diversity. Pol5 (**DNA polymerase ϕ**) of yeast is localized in the nucleolus and it is involved in the synthesis of rRNA (Shimizu, K., et al. 2002. *Proc. Natl. Acad. Sci USA* 99:9133). The human genome encodes at least 15 DNA polymerases. The catalytic domains of the DNA polymerases are well conserved, although the other structural domains are variable. *See* ABC excinucleases; cis-syn dimer; DNA repair; DNA replication, eukaryote; DNA replication, prokaryote; mutator genes; PCNA; pol enzymes; recombination; replication fork; replisome; reverse transcriptase; reverse transcription; sister chromatid cohesion; sliding clamp; telomerase; terminal nucleotidyl transferase; thymine dimer. (Patel, P. H. & Loeb, L. A. 2000. *Proc. Natl. Acad. Sci. USA* 97:5095; Zhang, Y., et al. 2000. *Nucleic Acids Res.* 28:4147; Zhang, Y., et al. 2001. *Nucleic Acid Res.* 29:928;

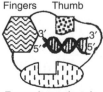
Synthetic mode
of the DNA polymerase

Fingers   Thumb

Exonuclease domain

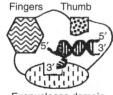
Editing mode
of the DNA polymerase

Fingers   Thumb

Exonuclease domain

Abstract, generalized representation of the polymerase and the exonuclease (editing) functions of DNA polymerases. (Modified after T. A. Steitz, 1999. *J. Biol. Chem.* 274 17395.)

Guo, D., et al. 2001. *Nucleic Acid Res.* 29:2875; Glover, B. P. & McHenry, C. S. 2001. *Cell* 105:925; Sutton, M. D. & Walker, G. C. 2001. *Proc. Natl. Acad. Sci. USA* 98:8342; Burgers, P. M. J., et al. 2001. *J. Biol. Chem.* 276:43487; Hübscher, U., et al. 2002. *Annu. Rev. Biochem.* 71:133.)

DNA polymorphism   See band-sharing coefficient; DNA fingerprinting; microsatellites; RFLP; SNIPs. (Satta, Y. 2001. *Genes Genet. Syst.* 76[3]159.)

DNA primase   See primase.

DNA probe   Radioactively or nonradioactively (fluorochrome) labeled single-stranded DNA sequence that anneals specifically to its homologous, complementary DNA or RNA tract and thus by virtue of the label identifies the homologous strand. *See* dot blots; FISH; fluorochrome; heterologous probe; insertional mutagenesis; labeling; nick translation; Northern hybridization; Southern hybridization.

DNA proofreading   See DNA repair; proofreading.

DnaQ   Bacterial gene encoding the proofreading ε-subunit of DNA polymerase III. *See* DNA polymerases; proofreading.

DNA rearrangement   See chromosome rearrangements; immunoglobulins; insertion elements; mating type determination in yeast; phase variation; recombination; transposons.

DNA reassociation kinetics   See c_0t curve.

DNA renaturation   Reannealing of denatured strands. *See* annealing; denaturation; renaturation.

DNA repair   May be brought about by different mechanisms. The frequency of single-strand breaks/Gy is about 25 times higher than that of double strands of DNA. The bulk of information is available from prokaryotic systems. The direct repair and excision repair systems are error-free because in the former no new DNA synthesis is required and in the latter the undamaged DNA strand provides a normal template, so there is no increase in errors beyond the normal level for the polymerase involved. The *SOS repair* is an error-prone repair mechanism operating when the path of the regular DNA polymerase is blocked by structural distortions and the enzyme is stalled; in distress (lacking appropriate template), this mechanism incorporates nucleotides in an unorthodox manner, thus leading to mutation.

I. DIRECT REPAIR systems remove pyrimidine dimers (cyclobutane dimers) by activating splitting enzymes, photolyases, upon exposure to visible light (light repair). These enzymes contain chromophore cofactors, such as reduced flavin adenine dinucleotide ($FADH_2$) and folate, to absorb light. The *Bacillus subtilis* spore photoproduct lyase breaks between C−C bonds of 5-thyminyl-5,6-dihydrothymine rather than cyclobutane dimers. Another direct repair enzyme is O^6-methylguanine methyltransferase, which accepts at a cysteine site an O^6 alkyl group from mutated guanylic acid or other methylated bases and thus restores the normal purine at the expense of its own inactivation. Direct repair does not involve any unscheduled DNA synthesis. Prokaryotes have two repair methyltransferases; mammals have only one.

II. EXCISION REPAIR SYSTEMS involve various and limited amounts of unscheduled DNA synthesis. (1) *Mismatch repair*: MutS protein binds to various mismatched bases. MutH protein binds to a GATC tract in the DNA and MutL protein links the other proteins into a repair complex. The dam methylase (*dam*) enzyme recognizes the old DNA strand at replication and methylates it at a C site within the 5′-GATC tract. At that time, the nascent DNA strand is still unmethylated. Mismatches are recognized (G—T, well; C—C, poorly) and repaired within a 1 kb range of the GATC sequence as long as the new strand is not yet methylated. The MutH protein is an endonuclease that cuts 5′ to GATC. The nicked strand then unwinds with assistance of helicase II (*uvrD* gene product). A single-strand binding protein (SSB, ssb) braces the site, and the defective single strand is sliced off 3′ to 5′ by exonuclease I (*sbcB*) through the defect. On the 3′ side of GATC, the mismatch is removed either by exonuclease VII or *recJ* 5′ to 3′ exonuclease. DNA polymerases (*pol I, pol II*, or *pol III*) repair the gap and the new ends are ligated (*dnaL*). (2) *Base excision repair* (BER): Cutting the N-glycosyl bonds by DNA glycosylases and leaving behind apurinic and apyrimidinic (AP) sites lifts defective bases out. The glycosylases remove deaminated cytosine (= uracil), deaminated adenine (= hypoxanthine), bases alkylated by alkylating mutagens, etc. Genes *ada* and *alkA* of prokaryotes are involved. After the base is removed, *AP endonucleases* cut out the deoxyribose phosphates and possibly more nucleotides. In humans, several glycosylases have been mapped (*see* glycosylases). The gap is filled in by pol I or pol β and ligation completes the repair process. Pol β is a highly error-prone polymerase. In mammalian cells, apurinic/apyrimidinic nuclease (APE1) may correct the process by exonucleolytically removing the incorrect bases from the ends (Chou, K. M. & Cheng, Y.-C. 2002. *Nature* 415:655). The prokaryotic pol I has an intrinsic proofreading exonuclease activity. (3) *Nucleotide excision repair (NER)*: Damages to the DNA involving longer tracts are cut at two sites by the *ABC excinuclease* system and the intervening sequences are removed. This complex is encoded by genes *uvrA* (coding for the ATPase subunit of endonucleases), *uvrB*, and *uvrC* (coding for the endonuclease subunits) of the excinuclease of *E. coli*. UvrD is a helicase that releases the excised oligomer. DNA pol I carries out repair synthesis and the M_r 75 ligase ties the ends. Excision repair may involve just a few and up to a few thousand nucleotides. The gaps are filled in by any of the four DNA polymerases of prokaryotes. The human excinuclease complex contains single or combined activities of 16 polypeptides. In humans, BER is the only mechanism to escape UV-caused genetic damage. (4) *Transcription-coupled repair*: A type of excision repair in which the pyrimidine dimers are more readily removed from the transcribed strand than from the nontranscribed strand. This type of repair may be coupled with the more common excision repair system (Svejstrup, J. Q. 2002. *Nature Rev. Mol. Cell Biol.* 3:21). Such a mechanism was first detected in *E. coli*, but it turns out that the failure of this mechanism is the most common cause of base mismatches in colorectal human cancer. The vertebrate enzyme DNA polymerase β is involved in excision repair and in the release of 5′-terminal deoxyribose phosphates from incised apurinic-apyrimidinic sites. The repair is confined to the transcribed strand and in mammals to genes transcribed by pol II. In humans, there is a requirement for the transcription repair coupling protein (TRC). Cockayne syndrome is a defect of two TRC factors. *E. coli* also requires a TRCF encoded by *mfd* gene. The human diseases xeroderma pigmentosum F and trichothiodystrophy are excision repair defects. Human DNA excision repair also requires XPA, a zinc-finger protein, and HSSB, a human single-strand binding protein, RPA, another damage-recognition protein, the transcription factor TFIIH, with multiple subunits of helicase, zinc finger, kinase, and cyclin functions. Furthermore, XPC, a ubiquitin, XPF (5′ cut) and XPG (3′ cut) nucleases, RFC, an ATPase, PCNA, an auxiliary protein to DNA polymerase δ, repair polymerases ε/δ, and DNA ligase are used (Sancar, A. 1996. *Annu. Rev. Biochem.* 65:43; Wood, R. D. 1996. *Annu. Rev. Biochem.* 65:135).

III. SOS REPAIR is required when there is extensive damage to the DNA (presence of cyclobutane rings and adducts formed by UV light, cross-linking caused by chemical mutagens, etc.) that prevents the movement of the DNA polymerase III enzyme along the strands so that the template cannot lend itself for accurate copying. The repair system is in distress and becomes error-prone. Shortly after the damage has taken place, the activity of the RecA protein increases up to 50 times. Next, the LexA repressor is cleaved, initiating the activity of a series of genes including the excinuclease system, *UmuC and UmuD* (repair functions), and *himA* (with multiple functions in replication, recombination, and regulation). The DinB prokaryotic and eukaryotic homologues make replication by polymerase IV error-prone. Rad30A operates the error-free polymerase η that bypasses the UV lesions. Rev1 deoxycytidyltransferase also assists in the bypass but it is error-prone. *LexA* repression targets an SOS box (containing a consensus CTG-N_{10}-CAG) in the vicinity of the promoter of its objects. Most commonly, dAMP is inserted opposite the abasic site. Eukaryotic DNA polymerase ζ and η can bypass pyrimidine dimers, but pol ζ may be hypermutagenic. SOS repair usually involves extensive DNA repair synthesis. SOS repair is considered to follow either the pathway of *damage avoidance* (DA), i.e., the complementary strand is used to rescue the blocked replication fork, or relies on *translesion synthesis* (TLS), i.e., the repair enzyme reads through the replication block. The block generated by the DNA-damaging agent determines the route of the SOS repair. In bacteria, DNA polymerase V—in the presence of protein UmuD′, RecA, and a single-strand DNA-binding protein—produces point mutations at a frequency of 2.1×10^{-4} per nucleotide, an increase of ~40-fold over that of the effect of DNA polymerase III holoenzyme. About 53% of the mutations are transversions. *See* translesion. (Johnson, R. E., et al. 1999. *Proc. Natl. Acad. Sci. USA* 96:12224; Sutton, M. D., et al. 2000. *Annu. Rev. Genet.* 34:479.)

Recombination repair.

IV. RECOMBINATIONAL REPAIR makes corrections by exchanging the defects with correct sequences through recombination (gene conversion). In double-stranded human DNA, the observed breakage rate, 5.8×10^{-3}/Mbp/Gy, was found for both 80 and 160 Gy, and about 75% rejoined correctly within 2 h (Löbrich, M., et al. 1995. *PNAS* 92:12050). RecA/Rad51 and Rad52 (in yeast) have a pivotal function in homologous recombination repair. Rad52 promotes the end-to-end joining of DNA. Ku70 and Ku80 proteins protect the ends from nuclease degradation. DNA ligase IV, XRCC, Mre11, and Rad50 mediate repair of DNA double-stranded breaks. Nonhomologous DNA end-joining (NHEJ) requires the presence of DNA-dependent protein kinase, XRCC4, and ligase IV. NHEJ is generally necessary for V(D)J recombination in generating the appropriate immunoglobulins. In *Drosophila*, chromosome breakage, induced by the P-transposable element, is repaired by recombination, and the efficiency of this repair is up to five times higher when the homologous sequence is syntenic within, at any position of the X chromosome, rather than somewhere in transposition in an autosome. In double-strand repair, the DNA remains in a fixed position within the nucleus; 30 min after the breakage, the Mre repair enzyme moves to the site of the damage. In bacteriophage T4, DNA replication and double-strand break repair are closely coupled. *See* DNA ligase; DNA polymerases; molecular mechanisms in prokaryotes; Mre11; P element; RAD; recombination; V(D)J; XRCC. (Cox, M. M., et al. 2000. *Nature* 404:37; Cox, M. M. 2001. *Proc. Natl. Acad. Sci. USA* 98:8173; George, J. W., et al. 2001. *Proc. Natl. Acad. Sci. USA* 98:8290; Bleuit, J. S., et al. 2001. *Proc. Natl. Acad. Sci. USA* 98:8298; Kuzminov, A. 2001. *Proc. Natl. Acad. Sci. USA* 98:8461; Fukushima, T., et al. 2001. *J. Biol. Chem.* 276:44413.)

Retrotransposons or parts of them may be inserted into double-strand breaks to heal the defects. The frequency of these events is not known, but most eukaryotic genomes have a large number of retrotransposon elements and this repair function may justify their evolutionary maintenance. Specialized retroposons such as HeT-A and TART of *Drosophila* or the Ty5 yeast element may use their reverse transcriptase and RNA template to heal telomeres in case the telomeres are shortened by aging or damaged by breaks. The repair mechanisms of eukaryotes are much less well understood than prokaryotes. The numerous *RAD* genes of yeast control radiation responses and are involved in repair. Short-patch DNA repairs (such as damages caused by alkylating agents) require the function of DNA polymerase β, a ubiquitous housekeeping enzyme. Repair of UV lesions involves DNA polymerase ε. For example, in xeroderma pigmentosum F, a human hereditary condition controlled by several dominant and recessive genes, the defect in excision repair of UV lesions may be connected with polymerase ε. It has been estimated that each cell under normal conditions loses more than 10,000 bases from the spontaneous breakdown of DNA, and the damaged spots have to be repaired. The efficiency of repair in eukaryotes also depends on the time available before the onset of the S phase of the cell cycle. Proteins p53, p21, PCNA, and other cellular proteins may have cell cycle–stalling functions. The efficiency of repair varies at different sites within a gene.

Generally areas near active genes are more efficiently repaired than areas in inactive regions. The most rapid repair is found in the transcribed strands of active genes. The DNA repair systems also play a role in the cell cycle and tumorigenesis. In 1994, *Science* magazine declared DNA repair enzymes the molecules of the year. DNA repair in eukaryotes is complicated by the nucleosomal organization of the chromatin. Progress in DNA sequencing identified ~130 genes involved in human DNA repair. (Wood, R. D., et al. 2001. *Science* 291:1284; *Biological Responses to DNA Damage.* 2001. *Cold Spring Harbor Symp. Quant. Biol.*, Vol. 65.)

Detection of DNA damage is important in mutation analysis, carcinogenesis, cancer therapy, aging, etc. The nature of the lesions can vary from base modifications and strand breaks to DNA protein and DNA chemical cross-links. Gas chromatography, mass spectrometry, high-performance liquid chromatography, immunoassays, etc., have been used for the identification of modified bases. A newer method using immunochemical recognition, capillary electrophoresis, and laser-detected fluorescence detection may identify lesions with a limit of 3×10^{-21} moles. (Le, X. C., et al. 1998. *Science* 280:1066). *See* aging; alkyltransferases; ataxia telangiectasia; Bloom syndrome; capillary electrophoresis; chromatin assembly factor; chromatography; Cockayne syndrome; colorectal cancer; DNA ligases; DNA polymerases; excision repair; Fanconi anemia; Gy; HEI; glycosylases; Ku; laser; male recombination; mating type determination in yeast; Mer⁻ phenotype; methylguanine-O⁶-methyltransferase; Mex; mismatch repair; mt DNA; NA-PK; NHEJ; p21p; p53; PARP; PCNA; photoreactivation; postmeiotic segregation; preferential repair; proofreading; RAD28; RAD50; radiation sensitivity; RecA; retroposons; retrotransposons; SNIPS; SSA; SSB; translesion; TRCF; Ty; UMU; xeroderma pigmentosum; X-ray repair; XRCC. (Aravind, L., et al. 1999. *Nucleic Acids Res.* 27:1223; Kunkel, T. A. & Bebenek, K. 2000. *Annu. Rev. Biochem.* 69:497; Sutton, M. D. & Walker, G. C. 2001. *Proc. Natl. Acad. Sci. USA* 98:8342; Wood, R. D., et al. 2001. *Science* 291:1284; Friedberg, E. C., et al. 2002. *Science* 296:1627.)

DNA repair-associated human disorders   Ataxia telangiectasia, Bloom syndrome, Breast cancer, Cockayne syndrome, De Sanctis–Caccione syndrome, Fanconi's anemia, Lynch cancer families, Rothmund-Thompson syndrome, trichothiodystrophy, Werner syndrome, xeroderma pigmentosum.

DNA repair synthesis   *See* unscheduled DNA synthesis.

DNA replication error   It has been estimated that in bacteria the individual nucleotide replacement error is about 10^{-9} to 10^{-10} per generation. *See* diversity; error-prone repair; proofreading; replication error. (Kunkel, T. A. & Bebenek, K. 2000. *Annu. Rev. Biochem.* 69:497.)

DNA replication types   Distinguished as type I and type II:

DNA-RNA hybrid   DNA is annealed with RNA in a double helix. *See* annealing.

DNase   Enzymes degrading DNA. *See* DNA polymerases; endonuclease; exonuclease; restriction enzymes.

DNase-free RNase   May be prepared by different procedures such as affinity chromatography on agarose 5'(4-aminophenylphosphoryl)-uridine-2'(3')-phosphate, by adsorption on Macaloid, or by heating in the presence of iodoacetate. Commercially available preparations may not always be trustworthy. *See* DNase; Macaloid; RNase; RNase-free DNase.

DNase hypersensitive site (HSS)   In the DNA, is very easily attacked by nucleases probably because they are unprotected by proteins (histones); their presence correlates with expression of adjacent genes. The generation of hypersensitive sites alters the nucleosomal structure of the chromatin. Hypersensitive sites are located in the vicinity of promoters or a promoter and enhancer. *See* antitrypsin; chromatin remodeling; enhancer; integrase; INI1; nucleosome; promoter.

DNA sequence alignment   *See* DNA sequence information; intel.

DNA sequence information   *See* databases.

DNA sequencing   Determines the order of nucleotides in DNA fragments. Two procedures have been most widely used: the protocol of Maxam and Gilbert (M&G) and a modification of the method of Sanger and co-workers. The M&G method can sequence both single- and double-stranded DNA. The P³² end-labeled DNA sample is divided into five aliquots, and by different, limited chemical breakage at one or two of the four bases, unique sets of labeled fragments are produced (G, G + A, A > C, C, C + T). From the five batches, fragments are separated according to length in five lanes in polyacrylamide-urea gels by electrophoresis. The gel is exposed to highly sensitive X-ray film to visualize the position at the labeled bands.

Since one end of each fragment in each batch terminates by the same nucleotide(s) according to the breakage mechanism indicated below and in each batch in which these ends are specific, the sequence of the bases in the different-length fragments can be read directly from the film of the gel (see illustration).

By the *Sanger method*, single-stranded DNA is sequenced that has been cloned in M13 (or related) phage vectors. A chosen DNA fragment is then replicated in the presence of the appropriate primer, α-P³²-dATP, a limited supply of the

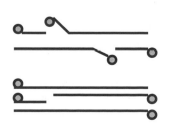

Type I is a linear duplex with one or more branches from the same DNA end

Type II molecules are partially single- and partially double-stranded as shown on the left

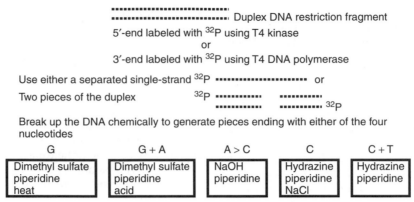

Duplex DNA restriction fragment

5'-end labeled with ^{32}P using T4 kinase

or

3'-end labeled with ^{32}P using T4 DNA polymerase

Use either a separated single-strand ^{32}P ·········· or

Two pieces of the duplex ^{32}P ··········   ··········
                                          ··········   ··········^{32}P

Break up the DNA chemically to generate pieces ending with either of the four nucleotides

G	G + A	A > C	C	C + T
Dimethyl sulfate piperidine heat	Dimethyl sulfate piperidine acid	NaOH piperidine	Hydrazine piperidine NaCl	Hydrazine piperidine

In the five batches the limited chemical breakage with yield fragments of different length that are all labeled at one of the ends by ^{32}P and the other end has either one of the two bases according to the specificity of the chemical breakage.

The reaction products are then separated in polyacrylamide-urea gels by electrophoresis and the individual fragments will move according to their length and their position is visualized after exposure to x-ray film because one of their ends carries the radioactive label. Thus the five gels will appear as shown above when run in the direction from bottom up.

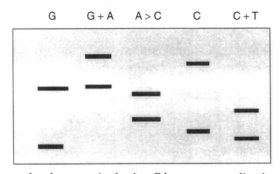

Because no band appears in the **A > C** lane corresponding in position in the **G + A** lane, or in the **C** lane corresponding to the **C + T** lane, the sequence must be **GTCATAGCA** (read from bottom to top). The Maxam Gilbert method can sequence both single- and double-strand DNAs, and the sequence can be read directly as illustrated below.

four normal deoxyribonucleotide triphosphates, and the large subunit of DNA polymerase I (Klenow fragment) or mainly Taq or phage T7 polymerases. In addition, the dideoxy analogs of the four nucleotides (dideoxyATP, dideoxyTTP, dideoxyGTP, dideoxyCTP) are added. When incorporated into the growing nucleotide chain, these analogs stop further elongation because they do not have OH at the 3' position where the new nucleotides normally attach during replication. As a result, the chain growth stops and the position of this nucleotide in the chain is marked. The fragments are separated by electrophoresis as described in the M&G procedure. From the results of the sequencing gel, the base sequences can be read directly.

Another method called sequencing by synthesis uses an iterative addition to the template of the four deoxynucleotide triphosphates, and as the DNA polymerase incorporates the complementary nucleotide monophosphate, the liberated pyrophosphate is determined. Actually, after addition, the nucleotides are enzymatically degraded in the sequence of incorporation. The detection is made possible by a sulfurylase reaction coupled with luciferase. Great technical advances have been made in DNA sequencing using laboratory equipment and computer technology. Megabase-size DNA

stretches can be sequenced in much shorter times. A large number of viral genomes, several prokaryotic genomes, and a score of eukaryotic (*Saccharomyces, Caenorhabditis, Drosophila*) genomes have been entirely sequenced and several others are nearing completion. The average estimated error rate is $1-3 \times 10^{-3}$ base. The sequencing of DNAs facilitates the understanding of how biological systems work. It opens new approaches for studying interacting proteins, facilitates more efficient design of drugs, deciphers the mechanisms of cancer and other diseases, provides new tools for forensic analyses, and even may identify biological warfare agents. *See* array hybridization; BLAST; BLASTX; BLOSUM; clone validation; computerization, DNA and protein sequence data; databases; DNA chips; DNA sequencing, automated; evolutionary distance; FASTA; luciferin; nucleic acid chain growth; pyrosequencing; sulfurylase. (*Methods in Enzymology* 59 & 65; Marziali, A. & Akeson, M. 2001. *Annu. Rev. Biomed. Eng.* 3:195; Green, E. D. 2001. *Nature Rev. Genet.* 2:573.) An outline of the most widely used dideoxynucleotide method of Sanger et al. is presented on the next page.

AN OUTLINE OF THE DIDEOXYNUCLEOTIDE SEQUENCING METHOD OF SANGER et al. (1977. *Proc. Natl. Acad. Sci. USA* 74:5463).

The genome projects use one of two automated sequencing procedures for large-scale sequence determinations. The hierarchical shotgun sequencing (HS) breaks up the (human) genome into overlapping BAC clones, sequences them, then reassembles them as merging clones. This procedure was used by the Human Genome Project. Whole-genome shotgun sequencing (WGS) uses the entire genome, then reassembles the entire collection into contigs and scaffolds. This procedure was designed and used by the Celera biotechnology firm. The two procedures are discussed by Waterson, R. H., et al. 2002. *Proc. Natl. Acad. Sci. USA* 99:3712. *See* computerization of DNA and protein sequence data; contig; DNA chips; genome projects; pyrosequencing; scaffold; sequatron; shotgun sequencing. New procedures' survey: Marziali, A. & Akeson, M. 2001. *Annu. Rev. Biomed. Eng.* 3:195. (<www.sequenceanalysis.com>.)

Insert into a cloning site (containing the *Lac* gene) of the single-stranded M13 phage. The insertion inactivates *Lac* in the phage and infect *Lac⁻* Bacteria. The transformants do not produce colored plaques on Xgal medium.

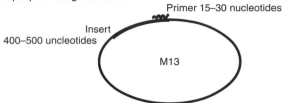

Primer 15–30 nucleotides

Insert
400–500 uncleotides

M13

To four tubes, each, add the sequencing vector (M13), ^{32}P-deoxyriboadenosine triphosphate, the normal deoxynucleotide triphosphates and a DNA polymerase (either the Klenow fragment or sequenase) and in each of the four tubes add also one of each of the four types of dideoxyribonucleotides:

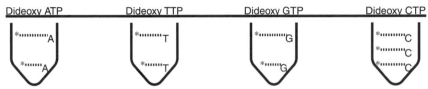

| Dideoxy ATP | Dideoxy TTP | Dideoxy GTP | Dideoxy CTP |

The dideoxynucleotides are incorporated in place of the normal nucleotides but lacking the 3'-OH end, the chain growth stops. Therefore, the synthesized fragments will terminate either with an **A** or **T** or **G** or **C**, respectively. Since normal nucleotides are also present, fragments of different length will be generated on the insert template according to the chance whether normal or blocking nucleotides will be incorporated by the polymerase. The fragments are then separated by length by polyacrylamide gel electrophoresis. The radioactive dATP marks one end of the fragments and they can be detected in the gel by autoradiogphay. The results may be as diagrammed:

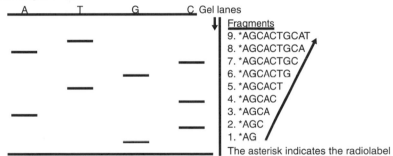

A   T   G   C Gel lanes

Fragments
9. *AGCACTGCAT
8. *AGCACTGCA
7. *AGCACTGC
6. *AGCACTG
5. *AGCACT
4. *AGCAC
3. *AGCA
2. *AGC
1. *AG

The asterisk indicates the radiolabel

Thus the **nucleotide sequence** can be read from the last base of fragment number 1 to fragment number 9: **G C A C T G C A T**

The start of the run is at the top and the longer is a fragment the shorter distance it moves in the electric field of the gel

DNA sequencing, automated

The nucleotide sequences on the X-ray film obtained by autoradiography can be read directly or can be scanned with the aid of computer programs. The computer can also be used to search for particular sequences, transcription signals, homeoboxes, termination signals, and any other type of conserved elements.

In addition, through computer connections to DNA databases, homologies to previously determined sequences can be identified By the most efficient methods (year 2000), more than one Mb DNA can be sequenced daily at a cost of less than $0.50 per nucleotide in a finished sequence. The genome projects use either a whole-genome shotgun method (Celera Genomics) or 150 kb clones in bacterial artificial chromosomes (International Human Genome Sequencing Consortium). *See* BAC; capillary electrophoresis; DNA sequencing; genome projects; MALDI/TOF/MS; nanopore technology; shotgun sequencing. (Meldrum, D. R. 2001. *Science* 292:515.)

DNA shuffling

In vitro homologous recombination in pools of randomly fragmented genes and reassembly by the

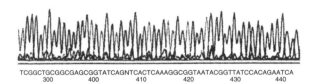

TCGGCTGCGGCGAGCGGTATCAGNTCACTCAAAGGCGGTAATACGGTTATCCACAGAATCA
300   400   410   420   430   440

Computer scan of a sequencing gel
each base is labeled by a different
fluorochrome

Section of a sequencing gel
each lane identifies another base

The peaks (in four colors) correspond to the pattern of the Sanger sequencing gel. The laser scanner feeds to the computer the information according to color that corresponds to specific bases of the DNA. (Scan after a Promega advertisement; gel after a Pharmacia catalog.)

polymerase chain reaction in order to test the consequences of the new sequences as evolutionary changes. Such a process may greatly increase the efficiency of genes if the DNA sequences are derived from different bacterial or viral species. Shuffling within DNA sequences from one species may also result in several-fold increases in enzyme efficiency. *See* localized mutagenesis; protein engineering; RCR-based mutagenesis. (Moore, G. L., et al. 2001. *Proc. Natl. Acad. Sci. USA* 98:326; Zhang, Y.-X., et al. 2002. *Nature* 415:644.)

DNA silencing   *See* silencer.

DNASIS   Computer program for DNA base sequence analysis.

DNA splicing   *See* splicing; splicing juncture.

DNA strand invasion   *See* branch migration; Holliday model.

DNA structure   *See* Watson and Crick model.

DNA subtraction   *See* genomic subtraction.

DNA supercoiling   *See* supercoiling of DNA.

DNA synthesis   *See* DNA chemical synthesis; DNA replication.

DNA synthesis, meiosis   During meiosis, there is no DNA synthesis, except what may be needed for genetic repair. Before meiosis, synthesis brings the level of DNA in the meiocytes to 4 C level. After the reductional division, each cell has 2 C amount of DNA, and at the end of meiosis, there is 1 C amount in each of the four gametes. *See* C amount DNA.

DNA thermal stability   Heat may disrupt the hydrogen bonds between the two strands of DNA. The stability depends on the base composition, since G≡C is linked by 3 hydrogen bonds, whereas A=T is linked by only 2. Generally, at temperatures exceeding 60°C, denaturation may begin, and upon prolonged exposure to 100°C, it is completed. The resulting single strands have higher UV absorption at 260 nm wavelength than the duplexes (hyperchromicity). *See* denaturation; DNA denaturation; hyperchromicity.

DNA topoisomerase   Capable of nicking and reclosing single strands of type I or causing scission in both strands and rejoining the cuts (type II). These enzymes are ubiquitous from bacteria to eukaryotes and have roles in relaxing superhelical configuration, replication, transcription, recombination, and repair. (Huang, W. M. 1996. *Annu. Rev. Genet.* 30:79; Champoux, J. J. 2001. *Annu. Rev. Biochem.* 70:369.)

DNA tracking   Following the path of the DNA, such as the movement of a polymerase along the template after the initiation of transcription or replication.

DNA transport   Concerned with segregation of the chromosomes, bacterial conjugation and chromosome transfer, phage infection and assembly into phage particles, transformation, etc. In bacteria, the SpoIII-like protein family is the most important instrument of the processes. *See* conjugation; infection; phage life cycle; transformation. (Errington, J., et al. 2001. *Nature Rev. Mol. Cell Biol.* 2:538.)

DNA transposition   *See* transposition.

DNA tumor virus   May be integrated into the animal host DNA. The Papovavirus family consists of a variety of types and may be responsible for warts and carcinomas. The Hepadnaviruses (hepatitis B) are responsible for liver carcinomas. The Herpes viruses (Epstein-Barr virus) may cause cancerous transformation of the lymphocytes (Burkitt's lymphoma and nasopharyngeal cancer). Several RNA viruses belong to the retrovirus groups, may be transcribed into DNA, and may activate cancerous growth. *See* isometric phage.

Adenovirus; an icosahedral; facultative DNA tumor virus.

DNA types   *See* table below. *See* H-DNA; Z DNA.

Helix	bp/turn	Degree of rotation per base pairs	Rise/bp angström	Diameter angström	Conditions of existence
A	11	+32.7	2.9	23	75% rel. hum.
B	10.4	+36.0	3.38	19	92% rel. hum., low ions
C	9.7	+38.6	3.34	19	66% rel. hum., Li ions
D	8	occurs only in guanine-free DNA			
E	7.5	occurs only in guanine-free DNA			
Z	12	−30.0	3.71	18	left-turned, alternating purines and pyrimidines, high salt the phosphate backbone is inside; 75% longer then B DNA
P	2.62				

(bp = base pair, rel. hum. = relative humidity)

DNA typing Determination of the individual specificity of a DNA sample. *See* DNA fingerprinting; DNA sequencing; microarray; multilocus sequence typing. (Burns, M. A., et al. 1998. *Science* 282:484.)

DNA uptake *See* oligodeoxyribonucleotide gated channel.

DNA vaccine *See* immunization, genetic; vaccines.

DnaX One of 10 different subunits of the prokaryotic DNA polymerase holoenzyme. *See* DNA replication in prokaryotes. (Song, M.-S. & McHenry, C. S. 2001. *J. Biol. Chem.* 276:48709.)

DNA-zymes (DNA enzymes, deoxyribozymes) Catalytic single-strand DNA molecules have been isolated in the laboratory by in vitro selection in random oligonucleotide pools. They are thus not "natural" enzymes in contrast to ribozymes. They can cleave and make phosphoester bonds and also catalyze other reactions (porphyrin metallation). They do not have the 2′-hydroxyl group of RNA yet can carry out the reactions very effectively. The so-called 10–23 DNA-zyme digests a specific motif of nucleotide sequences. The 15-nucleotide catalytic core of this DNA-zyme is flanked by two substrate recognition sequences that bind to complementary RNA by hydrogen bonds. It cleaves the phosphodiester bonds between unpaired purines and pyrimidines and generates 2′,3′-cyclic phosphate and 5′-hydroxyl ends. *See* ribozyme. Breaker, R. R. 2000. *Science* 290:2095.)

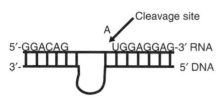

DNA-zyme (After Nowakowski, J., et al. 1999. *Nature Struct. Biol.* 6:151.)

DNP Deoxyribonucleoprotein; also dinitrophenol.

dNTP Deoxyribonucleotide triphosphate of any of the nucleic acid bases.

DOC (APC10) Cell cycle protein. *See* APC.

***Doc* element** *See* nonviral retrotransposable elements.

docking protein Signal-mediating protein at the cell membrane or inside the cell facilitating the cooperation of other proteins. *See* signal transduction. (Hadari, Y. R., et al. 2001. *Proc. Natl. Acad. Sci. USA* 98:8578; Wei, B. Q., et al. 2002. *J. Mol. Biol.* 312:339.)

dodecamer 12 units, composed of two hexamers.

dodecyl sulfate, sodium salt (SDS) Anionic detergent $(CH_3[CH_2]_{11}OSO_3Na)$. It is used to separate membrane proteins from the lipid layers because the detergent replaces the lipids at the hydrophobic tract of the membrane protein. It is also used for SDS-acrylamide-gel electrophoresis where the unfolded proteins move according to their size rather than by charge. Synonym: lauryl sulfate Na salt. *See* electrophoresis; membrane proteins.

DODO *See* parvulins.

dog (*Canis familiaris*) $2n = 78$; molecular data indicate its origin from wolves. Because of the high diversity in the mtDNA in both species, dogs apparently originated by more than a single event of divergence, although the difference between dog and gray wolf DNA is only 0.2%. Dogs and wolves are not isolated sexually. Dogs have been domesticated about 15,000 years ago and over 400 different breeds are known. About 370 mainly recessive genetic diseases have been identified and many of them (~58%) have homologues in humans. (Ostrander, E. A., et al. 2000. *Trends Genet.* 16:117; Breen, M., et al. 2001. *Genome Res.* 11:1784; <http://www.genetics.org/supplemental.)

dog rose *See Rosa canina*.

DOGS (dioctadecylamidoglycylspermine) Lipopolyamine used for delivering exogenous DNA into cells. *See* liposomes.

dolphin *Lissedelphis borealis* $2n = 44$; *Orcinus orca* and other species are $2n = 44$.

dolycephaly The head is abnormally long. It is a characteristic of several hereditary diseases.

domain Defined segments of proteins (with a characteristic tertiary structure that folds independently and usually represents specific functional properties) or cellular structures (chromosomes), e.g., DNA-binding domain of transcription factors. DNA sequence units or exons are also considered domains. Shared domains of macromolecules reveal evolutionary histories among organisms. *See* antibody domain; exon; protein domains; SMART.

domain swapping Generates from monomers of protein dimeric structures.

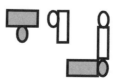

Dombrock system blood (DO) About 64% of northern Europeans are Do(a^+) blood type; it is coded in the short arm of human chromosome 1. *See* blood groups.

domestication Taming and breeding animals under human supervision and for human use began in the Neolithic (stone) age, about 10,000 years ago. Cultivation of plants started at about the same time. (Cowan, C. W. & Watson, J. 1992. *The Origins of Agriculture*. Smithonian Inst., Washington, DC; Salamini, F., et al. 2002. *Nature Rev. Genet.* 3:429.)

dominance Property of an allele to be expressed in a heterozygote (merozygote) in the presence of any other allele at a gene locus. In the majority of species, dominant mutations are much less common than recessive ones. In humans, however, the known dominant mutations seem to exceed the recessives, but many traits cannot be classified into either group. The term *dominance* was conceived before biochemical characterization of gene expression became feasible. Usually, at a finer level of analysis, both the dominant and the recessive alleles are expressed in the heterozygotes. In human or larger mammal pedigrees, because of the few offspring, it may be difficult to distinguish between dominant and recessive patterns of inheritance. The probabilities can, however, be analyzed statistically. Let us assume that 8 affected males have 6 affected female offspring and 5 unaffected male progeny. For nondominant inheritance, we would expect the probability to be $(0.5)^6 + (0.5)^5 = (0.5)^{11} \approx 0.00049$ (1/2048). The problems of penetrance and environmental factors must also be considered. *See* codominance; dominance reversal; penetrance; semidominance.

dominance, evolution of R. A. Fisher suggested in the 1930s that dominance evolves through the acquisition of modifier mutations that gradually convert the original recessive mutations into dominant ones. The basis of this idea was that under feral conditions the majority of alleles are dominant whereas the majority of new mutations are recessive. Sewall Wright (1934), J. B. S. Haldane (1930), and H. J. Muller (1932) hypothesized that dominance occurs by mutation and the rare advantageous dominant alleles have the selective advantage of masking the deleterious recessive alleles in the population. According to the molecular evidence that has come to light since then, the majority of mutations (except the loss of genes) show codominance when their function can be determined with greater precision. The development of dominance has been attributed to the control of processes regulating metabolic pathways and to the changes in activities of controlled enzymes. *See* codominance; dominance. (Bourguet, D. 1999. *Heredity* 83:1; Hurst, L. D. & Randerson, J. P. 2000. *J. Theor. Biol.* 205:641.)

dominance hypothesis Attributes the superior value of hybrids (hybrid vigor, heterosis) to the accumulation of favorable dominant genes. *See* hybrid vigor; overdominance; superdominance. (Rédei, G. P. 1982. *Cereal Res. Commun.* 10:5.)

dominance reversal Change of environment or developmental stage may alter and reverse the dominance-recessivity relationship of alleles. *See* conditional lethal genes; dominance; epistasis; temperature-sensitive mutation. (Shifriss, O. 1947. *Amer. Soc. Hort. Sci.* 50:330.)

dominance theory *See* dominance hypothesis.

dominance variance (in quantitative genetics) Is due to the expression of dominant alleles. *See* additive genes; genetic variances.

dominant allele Expressed fully in a diploid (polyploid) even in the presence of recessive alleles. *See* dominance; semidominant.

dominant individual In an animal society it has a higher rank and has better access to food and mates than the rest of the individuals.

dominant lethal assays in genetic toxicology Generally rodents are treated with mutagens during various stages of spermatogenesis. Subsequently, they are mated with untreated females. After about 2 weeks of pregnancy, the females are sacrificed and the number of lethal implantations is classified and counted. *See* autosomal-dominant mutation; bioassays in genetic toxicology; implantation.

dominant negative Inhibition of the activity of a normal (wild-type) protein when its subunits are mixed with mutant polypeptide subunits, altering the conformation and thus rendering it inactive. *See* allelic complementation; conformation correction; Denys-Drash syndrome; gain-of-function mutation; Marfan syndrome; osteogenesis imperfecta; RET oncogene; tetralogue.

donkey *Equus asinus* $2n = 62$. *See* hinny.

Donohue syndrome (leprechaunism, 19p13.2) Defect in the insulin receptor. Growth retardation; face anomalies; protuberant ears; abdomen, genitalia, hand, and feet enlargement; excessive hair on skin; acanthosis nigricans; hepatocyte vacuolization. *See* dwarfism; insulin; insulin receptor protein.

donor Provides genetic material (F^+) to the recipient (F^-) by bacterial conjugation or any other means of genetic transfer (including eukaryotes). Blood type O, which is acceptable to any individual with other alleles of the ABO blood group or other antigenic substances compatible to other serotypes, is also called a (universal) donor. *See* ABO blood group; conjugation; serotype.

donor site Original position in the DNA (map) of a nonreplicative transposable element from which it may move to another position within the genome, to the *recipient site*. *See* transposable elements.

donor splicing site *See* introns.

DOPA 3,4-dihydroxyphenylalanine, an intermediate in melanin synthesis; the derivative dopamine is a neurotransmitter. *See* neurotransmitter; phenylalanine.

dopamine Catecholamine formed by decarboxylation of dopa. It is a precursor of epinephrine, norepinephrine, and melanine. Dopamine hydroxylase converts dopamine into epinephrine. Dopaminergic means an activation or site of effect by dopamine. *See* animal hormones; DARPP; dopa; memory; neurotransmitter.

DOPE *See* lipids, cationic.

doping nucleotide A short nucleotide string is synthesized where some codons in the first strand have any of the four natural nucleotides and at the third position they have either G or C. In the second strand at the positions corresponding to the above named codons, there are inosines. The rest of the codons correspond to the usual amino acids. When such an insert is added to a vector, a variety of random mutations may occur in the protein domain coded by the sequence and yield a library of proteins with different amino acids at critical regions. *See* directed mutation; hypoxanthine; protein engineering. (Balint, R. F. & Larrick, J. W. 1993. *Gene* 137:109; Hutchison, C. A., et al. 1986. *Proc. Natl. Acad. Sci. USA* 83:710.)

Doppelgänger *See* prion.

dormancy State of low metabolic activity. In plants, seed dormancy requires the hormone abscisic acid, whereas gibberellic acid relieves dormancy. *See* plant hormones.

dorsal Relating to the back position of a body or upper surface of a structure or organ. Dorsal-ventral cell fate specification in the *Drosophila* embryo is partly under the control of an autoproteolysis cascade determined by genes *GD, Snk, and Ea* and the nerve growth factor ligand Spätzle (Dissing, M., et al. 2001. *EMBO J.* 20:2387). Dorsal represses the Zen (*zerknüllt*) homeobox by a 600 bp *ventral repression element* (VRE) containing 4 Dorsal-binding sites. There are 15 regions in the *Drosophila* genome that contain three or more binding sites within up to 400 bp sequences. (Markstein, M., et al. 2002. *Proc. Natl. Acad. Sci. USA* 99:763.)

dorsal closure After gastrulation of the embryo establishes the dorsal ectoderm by stretching—without proliferation—the lateral ectoderm over a transient epithelial structure (amnioserosa). This is a complex process mediated by the products of several known genes in *Drosophila*.

dorsalization The formation of dorsal elements is preferentially enhanced at the expense of ventral development during morphogenesis and embryo development. *See* morphogenesis.

dorsal lip *See* blastopore.

dorso-ventral Back-abdomen directional arrangement of anatomical structures.

dosage compensation A single dose of a gene has the same phenotypic effect as two or more. Examples abound in X-chromosome-linked genes where the phenotype for such genes is practically identical in the XY males and XX females or in the WZ females and ZZ males or the XO males and XX females. Such dosage compensation may occur with various doses of the X chromosome and to some extent it may occur in various aneuploids. MSL and MLE proteins appear to be involved in dosage compensation in *Drosophila* by hypertranscription in the single male X chromosome. MOF protein is also involved as a putative histone acetyl transferase. MLE seems to attach two noncoding RNA (roX1, roX2) to the male X chromosome, which may upregulate the transcription of genes in the X.

In mammals, however, the noncoding Xist RNA mediates X-chromosomal inactivation in the female. *Sex lethal* (*sxl*, 1–19.2, with many different alleles) controls both sexual dimorphism and dosage compensation through its 39 kDa protein product and promotes alternative splicing of *msl-2* RNA transcripts. Females with X:A (sex chromosome:autosome) ratio of 1 synthesize this protein and permit the transcription of both X chromosomes. Males with X:A of 0.5 do not make the SXL protein and increase the expression of the majority of the X-linked genes.

SXL permits the normal transcription of *Sxl* and *tra* in females, but in males it inserts stop codons into these two transcripts. There are binding sites for SXL in the 3′ untranslated region (UTR) in the *msl-1* transcripts, and these may cause downregulation in females. There are both 5′ and 3′ UTR-binding sites in the *msl-2* transcripts in females and thus other potentials for regulation. Msl-2 seems to be the main target and instrument of X-regulation. The SXL protein is suspected to have another function: downregulating some X-chromosomal genes in female flies. In the hermaphroditic XX females of *Caenorhabditis*, X-chromosomal genes are expressed at about the same level as in normal XO males (hypotranscription). In *Drosophila*, the single X chromosome is hypertranscribed to achieve dosage compensation. In mammals, only one of the two or more X chromosomes remains active.

Some of the (*sdc*) genes affect both sex determination and dosage compensation. In males, XOL1 prevents the assembly of the dosage compensation complex in contrast to XX females (hermaphrodites) where it activates the assembly. Other genes (dumpy series [*dpy*]) affect only dosage compensation. The DPY-27 protein is associated with the X chromosomes in hermaphrodites of *Caenorhabditis*. DPY-27–like proteins are also present in other organisms and control chromosome condensation and segregation. The DPY-26 locus affects Dpy-27 and Dpy-30 proteins for dosage compensation and Sdc2 and Sdc3 for the coordination of dosage compensation and sex determination. SDC-1 is a 139 kDa protein with Zn fingers. This family of proteins has some structural features characteristic for motor proteins. SDC-3 regulates sex determination by its ATP-binding domain and a Zn-finger-like domain seems to affect dosage compensation. Actually, both series are cooperating in reducing the level of X-chromosomal gene expression in hermaphrodites. A more recently discovered gene, *Pof* (painting of the 4th chromosome) involves heterochromatinization of the X chromosome in *Drosophila busckii* when translocated from *D. melanogaster*, where it is located in the 4th chromosome. In birds, observations on dosage compensation are not entirely unequivocal, although some of the genes seem to follow the rules in mammals. *See* epigenesis; hypertranscription; introns; JIL-1; lyonization; *Mle*; motor protein; Msl; MSL; neo-X-chromosome; numerator; sex determination; splicing; *Tsix*; *Xic*; *Xist*; *Xol*; Z inactivation; Zn finger. (Stuckenholz, C., et al. 1999. *Trends Genet.* 15:454; Meller, V. H. 2000. *Trends Cell Biol.* 10:54; Avner, P. & Heard, E. 2001. *Nature Rev. Genet.* 2:59; Park, Y. & Kuroda, M. I. 2001. *Science* 293:1083; McQueen, H. A., et al. 2001. *Curr. Biol.* 11:253.)

dosage effect The number of alleles of a certain type determines the degree of expression. *See* gene titration; mapping by dosage effect; quantitative gene number; quantitative trait.

dosage quotient (DQ) Used to determine whether a particular exon was deleted or duplicated in a multiplexed PCR reaction.

$$DQ = \frac{Area\ Locus\ exon\ 4(U)/Area\ locus\ exon\ 17(U)}{Area\ Locus\ exon\ 4(C)/Area\ Locus\ exon\ 17(C)},$$

where U stands for the unknown and C for control. Exons 4 and 17 are chosen arbitrarily. *See* exon; PCR multiplex. (Formula after Dabora, S. L., et al. 2001. *Am. J. Hum. Genet.* 68:64.)

dose, permissible   *See* radiation hazard assessment.

dose fractionation   The irradiation is delivered not in a single burst but with intervals between doses. This may permit intermittent repair. *See* radiation effects.

dose rate effects   Usually at low dosage the correlation between the frequency of mutations and mutagens (ionizing radiation) is linear, but at higher doses the response curve follows second- or higher-order kinetics because of the multiplicity of effects (hits) on the genetic material. *See* kinetics; radiation effects.

dosimeter, pocket   Contains an ion chamber and detects ionizations due to X- or γ-radiation. The direct reading types are held against light to make a reading against a precharged value. The useful range is generally 0 to 200 mR. False readings may sometimes be displayed. *See* radiation measurement.

dosimeter film (badge)   Contains photographic emulsion that can detect β-, γ-, and X-radiation by blackening when developed. The films are used within the range of 20 keV for X-rays and 200 keV for β-radiation. *See* radiation measurements.

DOT   Department of Transportation (USA) regulates shipment of certain substances. *See* carcinogens; environmental mutagens; mutagens.

dot blot   *See* colony hybridization.

dot matrix   Method for aligning complex sequences. The two sequences are placed along the vertical and horizontal axes of a square. The matching residues form a diagonal, which may be

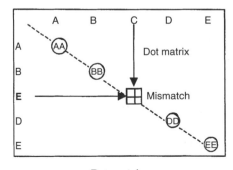

Dot matrix.

discontinuous if mutations, deletions, or insertions distinguish the two sequences. Such an arrangement is actually similar to a checker box or Punnet square of classical genetics. *See* Punnet square. (Sonnhammer, E. L. & Durbin, R. 1995. *Gene* 167[1–2]:GC1.)

DOTAP   *See* lipids, cationic; liposome.

DOTMA   *See* lipids, cationic.

DOTTED   *See Dt* gene.

double bridge   Visible at anaphase I if four-strand recombination occurs in paracentric inversion heterozygotes or in the case of sister chromatid exchange in ring chromosomes. *See* bridge; dicentric ring chromosome; inversion.

double-copy vector   A transgene, including its promoter, is inserted into the 3'-LTR. After replication (since the 3'-LTR is the template for the 5'-LTR), the transgene + promoter + regular 5'-LTR promoter are present in the new construct expected to display high activity of the transgene because of the duplications. The transgene may be inserted in forward or reverse orientation. In the former case, it may slow down the expression from the downstream copy. In reverse orientation, this may not happen, but the transgene may not find its own polyadenylation signal. Nevertheless, its stability may not suffer because the signal may be provided by adjacent cellular genes. *See* retroviral vectors; SIN vectors. (Wiznerowicz, M., et al. 1997. *Gene Ther.* 4:1061.)

double cortex   Human Xq21.3-q24 brain defect of the thinner cortex and disorganized neurons, resulting in mental retardation and seizures. *See* mental retardation; seizures.

double-cross hybrid   Were extensively used for the commercial production of hybrid corn. The seed companies produced the seed for the farmers according to the scheme:

single cross (1)   [a × B] × [C × D]   single cross (2)

DOUBLE CROSS SEED ↵

used by the farmers to produce corn for food, feed, or industrial raw material. Currently, most of the corn hybrids are single crosses. *See* cytoplasmic male sterility; heterosis; hybrid vigor.

double crossing over   A single recombination changes the synteny of linked markers: The frequency of double crossing over is expected to equal the product of the frequency of the single crossing overs. Two-strand double crossing over within an inverted segment generally does not have a harmful effect on inversion heterozygotes, and two normal and two inverted chromosomes are recovered in the gametes. Three-strand double crossing over within the inversion produces one acentric, one dicentric, one functional recombinant chromosome with a piece of the inverted segment and one functional inverted or noninverted chromosome. Four-strand double crossing within a paracentric inversion

From   A_____B   To   A_____b ▲
        a___X___b   a_____B ▼

Whereas double recombination involving the same strand restores the original synteny of the two genes: A_____B   To   A_____B
         a___X___X___b   a_____b

heterozygote leads to the formation of double bridges (two dicentric chromosomes), two acentric fragments, and most likely no viable gametes. In pericentric inversions, bridges do not occur, but the gametes receiving the exchange strands (except the two-strand double crossovers) will be duplicated or deficient and most likely nonviable. Double crossing over in translocation heterozygotes, depending on their site, may damage most of the gametes and reduces the fertility below 50%. *See* crossing over; inversion.

double exchange   *See* double crossing over.

double fertilization   Occurs in plants when one of the generative nuclei fuses with the haploid egg and gives rise to the zygote, whereas the other fuses with the two polar nuclei and initiates the development of the triploid endosperm. *See* adventive embryony; apomixis; embryogenesis in plants; gametophyte; heterofertilization; metaxenia; parthenocarpy; polyembryony; semigamy; xenia.

double flower   *See* flower differentiation; *Matthiola*; petal.

double helix   Two deoxyribonucleotide chains formed by phosphodiester linkages are joined in a helical arrangement through hydrogen bonding between bases as determined essentially by the Watson and Crick model. The two DNA strands form plectonemic coils (wound around each other) that can be partially and locally released during replication by the helicase and topoisomerase enzymes. *See* DNA types; helicase; topoisomerase; Watson and Crick model.

Double helix.

double heterozygote   Heterozygous at two gene loci.

double homozygote   Homozygous at two gene loci.

double infection   A bacterial cell may be infected with two different genotypes of the same or compatible bacteriophages, which may provide an opportunity for phage recombination. *See* multiplicity of infection.

double lysogenic   Bacterium carrying a normal lambda phage side-by-side with a λdg. This lambda defective-galactose phage lost a piece of its genetic material but acquired the bacterial *gal*⁺ gene and thus is capable of specialized transduction if another phage compensates for its defect. When such a phage is induced (to switch from prophage to vegetative

phage), it produces a *high-frequency lysate* and high *specialized transducing* ability. *See* lysogen; transduction.

double minute (DM)   Extrachromosomally amplified chromatin containing a particular acentric chromosomal segment (gene). The homogeneously stained sites in the chromosomes indicate their incorporation. They occur frequently in cancer cells. *See* amplification; cancer; DM chromosome; fragile sites; homogeneously stained sites; YAC. (Zimonjic, D. B., et al. 2001. *Cytogenet. Cell Genet.* 93:114).

double muscling   *See* myostatin.

double negative cell   Lymphocyte in the thymus without CD4 or CD8 proteins.

double PCR and digestion   Procedure intended to enrich mutant DNA sequences in the amplified mtDNA. During the first PCR step, both mutant and wild-type DNA are amplified. Then a restriction endonuclease is applied that degrades the wild-type but not the mutant DNA. A subsequent PCR step amplifies the full-length mutant sequence. *See* mtDNA; polymerase chain reaction.

double positive cell   Expresses both CD4 and CD8 proteins but may die within the thymus. *See* lymphocytes; T cell.

double reciprocal plot   *See* Lineweaver-Burk plot.

double recombination   Two recombinations in between two stands (chromatids, DNA) within a determined interval (*see* double crossing over). It is genetically detectable only if there are multiple markers in the chromosome.

double reduction   Chromatid segregation that may occur when recombination takes place in a trisomic or tetrasomic individual between a gene and the corresponding centromere. Consequently, in a duplex, double recessive gametes are produced and, e.g., an *AAa* individual produces *aa* gametes. The chromosome mechanism leading to double reduction in case of tetrasomy is represented at the entry *alpha parameter*. *See* alpha parameter; autopolyploid; centromere mapping in higher eukaryotes; trisomic analysis.

double replacement targeting   Can be used for the insertion of very small changes within a gene locus. The procedure is suitable for the HPRT marker where both positive and negative selections are practical. The first replacement is followed by a second step with another vector homologous to the same target and carrying the negative selectable gene,

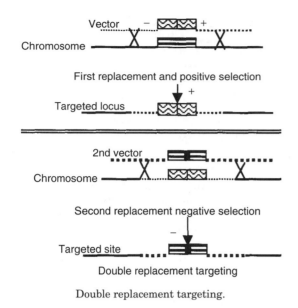

Double replacement targeting.

which is then screened. *See* diagram above. (Moore, R. C., et al. *Biotechnology* 1995 13:999.)

double restriction   The DNA is cleaved by two restriction enzymes. *See* restriction enzymes.

doubled haploid   Usually a diploid line arising by doubling the chromosome number of a haploid. Therefore, double haploids are expected to be homozygous at all gene loci until mutation occurs. *See* haploid.

double isotrisomic   In Wheat, $20'' + (i + i)1'''$, $2n = 43$ ($''$ = disomic, $'''$ = trisomic, i = isochromosome). *See* disomic; isochromosome; trisomic.

double monoisosomic   In Wheat, $20'' + i' + i'$, $(2n = 42)$ ($''$ = disomic, $'$ = monosomic, i = isosomic). *See* disomic; isosomic; monosomic.

double monotelosomic   In Wheat, $20'' + t' + t'$, $2n = 42$ ($''$ = disomic, $'$ = monosomic, t = telochromosome). *See* disomic; monosmic; telosome.

double-strand break   Chromosome double-strand breaking frequently occurs as a consequence of ionizing radiation of the nuclei, exposure to various chemicals (mutagens, carcinogens), free oxygen radicals, cellular enzymes, etc. Palindromic 'at-risk sequence motifs' (ARM), trinucleotide repeats, arrest of the cell cycle, meiotic recombination, mating-type switch, and intron homing may all cause such breaks. In the budding yeast, mitochondrial DNA sequences may be incorporated into chromosomal DNA in the vicinity of the LTRs of retrotransposons. Single-strand breaks occur 15–20 times more frequently. For repair of the breaks, Ku70, Ku80, DNA-dependent protein kinase, XRCC4, and DNA ligase IV are required. Double-strand breaks may be repaired by gene conversion. *E. coli* mutants in base excision repair because of defects in glycosylase activity are less susceptible to double-strand breaks, but overproduction of glycosylases results in more DNA lesions. The DNA breaks occur when the glycosylase attempts to remove the damaged DNA. In higher eukaryotic cells, more than eight double-strand breaks may occur daily. *See* ATM; chromosomal aberrations; chromosomal instability; DNA ligases; DNA-PK; DNA repair; end joining; gene conversion; glycosylases; intron homing; Ku; LTR; mismatch repair; nonhomologous end-joining; palindrome; PARP; RAD50; RAD51; retrotransposon; ROS; Szostak model; terminal deoxynucleotidyl transferase; trinucleotide repeats; V(D)J; X-ray caused chromosome breakage for illustration; X-ray repair; XRCC. (Haber, J. E. 2000. *Trends Genet.* 16:259; Paques, F. & Haber, J. E. 1999. *Microbiol. Mol. Biol. Rev.* 63:349; Blaisdell, J. O. & Wallace, S. S. 2001. *Proc. Natl. Acad. Sci. USA* 98:7426; Hopfner, K.-P., et al. 2002. *Current Opin. Struct. Biol.* 12:115.)

double-strandbreak repair model of recombination   *See* Szostak model.

double-strand RNA   *See* RNA, double-stranded.

double stranding   Sequencing both of the complementary strands of particular DNA stretches in order to minimize errors. *See* base calling; compression in gels.

doublet   (1) Double band in the salivary gland chromosome or a (2) double band in an electrophoretic gel. *See* gel electrophoresis; salivary gland chromosomes. (3) *Paramecia* that share a common endoplasm and a single macronucleus. *See* Paramecium.

Salivary gland chromosome.

double telotrisomic   In wheat, $20'' + (t + t) + 1'''$, $2n = 43$ ($''$ = disomic, t = telochromosome, $'''$ = trisomic). *See* disomic; telosome; trisomic.

doubling dose   Amount of a mutagen that doubles the spontaneous rate of mutation. Since mutation rate depends on the organism, its genotype, developmental stage, and environmental factors, the estimates arrived at by different investigators may not be identical. From mouse experiments, it has been estimated that 1 R chronic ionizing radiation produces a mutation rate of 2.5×10^{-8} per locus per generation, and the average spontaneous rate is considered to be 1×10^{-6}; hence $(1 \times 10^{-6})/2.5 \times 10^{-8} = 40$ R is assumed to be the doubling dose of ionizing radiation for humans, too. Other estimates consider the doubling dose for recessive mutations to be 32 R and for dominant ones 20 R. More recently, the doubling dose for radiation has been estimated as 100 cGray (1 Rad).

The majority of the data are systematically biased, depending on the genes used, because the range of inducible mutations in mice may vary within about an order of magnitude; therefore, the samples may not be true average representatives. The doubling dose may have important meaning for the estimation of genetic radiation risks.

Assuming that the average human dominant mutation rate is 10^{-5} to 10^{-6}, the chances for the occurrence of a human dominant mutation at 100 R dose (1 R = ca. 93 erg/wet tissue) exposure may be 32×10^{-5} to 32×10^{-6}, i.e., 1/3,125 to 1/31,250. Actually, these estimates may not be very accurate, but direct data cannot be readily obtained in human populations. Another problem is that the mutant gametes may not compete successfully and the absolute genetic damage may be much larger than these estimates. An additional problem is that in human populations the mating is random (or almost random); therefore, homozygous recessives may not appear most of the time unless the marriages are consanguineous.

Atomic radiations in Hiroshima and Nagasaki substantially increased the incidence of cancer and teratogenesis in the exposed generation, but an increase in human mutations in the following generation was not clearly detectable except by altered sex ratio. On the basis of the Hiroshima and Nagasaki human data of 31,500 children of parents exposed to the bombs within a 2 km range of the epicenter, James Neel estimated the doubling dose for low-level chronic radiation to be 3.4 to 4.4 Sv equivalent. On the basis of various mouse irradiation data, he estimated the average doubling dose in mice as 1.35 Gy. This latter figure was derived by pooling of recessive lethals (1.77), recessive visibles (3.89), dominant visibles (0.16), deletion or mutation of the dominant allele in male mice at 7 loci (0.44), protein loci (0.11), etc., studied in different mouse strains. However, the Chernobyl atomic power plant accident in 1986 has shown genetic effects and an increase in cancer caused by atomic radiation.

A medical X-ray machine may deliver 0.04 to 1.0 rem (100 rem = 1 Sievert [Sv]) to the organ or structure routinely examined. Although the gonads are generally shielded during these examinations, some general risk remains, particularly for carcinogenic effects. Molecular techniques developed to detect mutations in DNA base and amino acid sequences have also been used to ascertain doubling doses. *See* atomic radiation; BERT; coefficient of inbreeding; mutation in human populations; mutation detection; radiation effects; Rad; radiation hazard assessment; RBE; relative mutation risk; Rem; Sievert; specific locus mutations assay; teratogenesis. (Sankaranaraynan, K. & Chakrobarty, R. 2000. *Muation Res.* 453:183.)

doubling time   Time required for the completion of a cell division, measured in minutes, hours, or days for specific cell types under defined conditions. *See* cell cycle.

doubly uniparental inheritance   Displayed by the mitochondria of some marine mollusks (Mytilidae). They have two types of mitochondria: one (F) is transmitted by the females to both female and male offspring, and the other (M) is transmitted by the males only to the male offspring. *See* mitochondria; paternal leakage; uniparental inheritance. (Southerland, B., et al. 1998. *Genetics* 148:341.)

Douglas fir (*Pseudotsuga menziesii*)   Primarily a forest timber tree with $2n = 2x = 26$.

DOVAM-S (detection of virtually all mutations–single-strand conformation polymorphism)   Rapid method with very high efficiency. *See* single-strand conformation polymorphism. (Liu, Q., et al. 1999. *Biotechniques* 26:932, 936, 940.)

DOWEX Resins®   Ion exchangers; also used for gel filtration and chromatography. *See* chromatography; gel filtration; ion exchange resins.

down promoter   Slows down the rate of transcription.

downregulation   Decreasing activity by regulation. *See* regulation of gene activity; upregulation.

Down syndrome   Caused by primary trisomy or translocation of human chromosome 21 or 21q21-22.3. The incidence of this condition varies from less than 1/1,000 at maternal age around 20 years to close to 100/1,000 live birth when the mother approaches menopause. Trisomy 21 may affect only part of the body in a mosaic pattern. The transmission of disomic gametes is about 91% through the females and 9% through the males. The fertility of the afflicted person is reduced (especially males); fertility in afflicted females is less than 50%. The most important phenotypic signs in humans involves a relaxation of the muscles (hypotonia) controlling the eyefold, protruding tongue, flat face, shorter than expected height; two-thirds display the simian fold in the palm (see fig. below); generally (but not always) reduced mental abilities (IQ 25 to 50), although with proper training they can learn to read and write, enjoy life, and usually are quite sociable and affectionate. This trisomy is associated with a series of other anomalies such as heart disease and susceptibility to leukemia. Older literature calls this condition by the unfortunate term of mongoloid idiocy. Prenatal cytological examination of fetal cells in the amniotic fluid and sampling chorionic villi may reveal if the fetus has this condition. Unfortunately, amniotic cells may display anomalies that do not occur in the normal cells of the fetus.

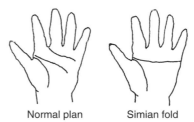

Normal plan   Simian fold

In a normal individual, the palm creases do not go all the way from one side to the other (left). In Down syndrome and some other congenital disorders, the typical simian fold is present. In addition, the hands may be shorter and the little finger crooked.

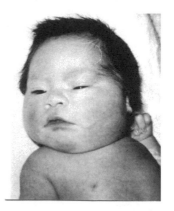

Down syndrome infant. (Courtesy of Dr. Judith Miles.)

There is a correlation between the low level of maternal α-fetoprotein (MSAFP) and fetal Down syndrome. The interpretation of the clinical analysis is further sharpened if unconjugated estriol (α-hydroxylation product of estradiol) and gonadotropin levels are also measured. Ultrasonic analysis may also indicate Down syndrome prenatally if heavy skin fold on the neck, excessive fluid accumulation, narrow small intestines, or short bones are indicated. The recurrence risk of trisomy is very low unless the family has a history of nondisjunctions and is heterozygous for translocations involving chromosome 21. In the expression of the bone characteristics of this trisomy, the ETS oncogene and transcription factor (human chromosome 21) may play the major role.

The brain of DS individuals has fewer neurons and abnormal neuron differentiation. The neurons become apoptotic earlier and predispose DS people to Alzheimer syndrome symptoms. These defects have been attributed to elevated levels of lipid peroxidation. The defects in the nervous system are also attributed to DYRK, a tyrosine phosphate–regulated kinase. Folate deficiency and hypomethylation due to defects in methylene tetrahydrofolate reductase and methylene synthase reductase may favor the nondisjunction involved. *See* Alzheimer disease; amniocentesis; amnion; apoptosis; estradiol; fetoprotein; homocystinuria; IQ; mongolism; MSAFP; nondisjunction; prenatal diagnosis; translocation; triple test; trisomic analysis; trisomy. (Hassold, P. A. & Jacobs, P. A. 1984. *Annu. Rev. Genet.* 18:69; Hunt, P. A. & LeMaire-Adkins, R. 1998. *Curr. Top. Dev. Biol.* 37:359; Kahlem, P. & Yaspo, M.-L. 2000. *Gene Funct. Dis.* 1:175; Epstein, C. J. 2001. *The Metabolic and Molecular Bases of Inherited Disease*, Scriver, J. R., et al., eds., p. 1123. McGraw-Hill, New York, Gardiner, K., et al. 2002. *Genomics* 79:833.)

downstream In the direction of the 3′ terminus of the polynucleotide; the nucleotide chain grows downstream because the nucleotides are added to the 3′ OH end of the preceding one.

downstream box Downstream enhancer of prokaryotic translation of mRNA. It appears that it is not acting by binding to the 16S rRNA (antidownstream box) as previously suggested.

downtag *See* bar code, genetic.

doxorubicin (adriamycin) Antineoplastic agent, immunosuppressant, inhibitor of reverse transcriptase, and suspected carcinogen.

Doxorubicin hydrochloride.

doxycycline Analog of tetracycline. *See* tetracycline.

Doxycycline.

Doyne honeycomb retinal dystrophy *See* macular degeneration.

DP1 Transcription factor activated by MDM2. *See* E2F1; MDM2.

DPA (DNA pairing activity) Protein (120 kDa) that promotes the formation of heteroduplexes in yeast. *See* eukaryotes; mechanisms; recombination.

dpc Days post coitum is the time after mating has taken place (in embryonic/fetal development).

DPC4 (deleted in pancreatic cancer at human chromosome 18q21.1) Gene missing in about 50% of the pancreatic cancer cells. It is homologous to the *Drosophila mad* (many abnormal [imaginal] discs, 3–78.6) gene, a member of the dpp (decapentaplegic) family, encoding a transforming signal related to TGF. *See* imaginal disc; morphogenesis in *Drosophila*; SMAD; TGF; tumor suppressor genes.

DPE Downstream promoter element. *See* core promoter; DSTF.

dpm Disintegration per minute is the number of disintegrations per 1 g radioactive radium = 1 μCi = 2.2×10^6 dpm, but usually about half as many counts per minute (cpm) are shown in the equipment that generally works at about 50% efficiency. *See* curie.

DPN (diphospho pyridine nucleotide) Current name NAD (nicotinamide adenine dinucleotide).

dpp (*decapentaplegic*) *See* morphogenesis in *Drosophila*.

D2 protein Histone-like protein in the nucleosomes of *Drosophila*. *See* nucleosomes.

DPT *See* developmental therapeutics program.

DQ antigen Encoded by the DQ segment of the HLA complex. *See* HLA.

DR (death receptor) *See* apoptosis; death domain; TRAIL.

DR antigen Encoded by the DR alleles of HLA. *See* HLA.

DRADA ds-RNA-dependent adenosine deaminase; an RNA editing enzyme in mammals; the same as dsRAD. *See* dsRNA; RNA editing.

draft genome sequence Assembly of scaffolds into a nucleotide sequence map, which still has some gaps and other deficiencies. *See* scaffolds in genome sequencing. (Katsanis, N., et al. 2001. *Nature Genet.* 29:88.)

DRAP *See* DSTF.

Dras *Drosophila* homologue of *ras. see* RAS oncogene.

DRB (5,6-dichloro-1-*β*-D-ribofuranosylbenzimidazole) Potent inhibitor of eukaryotic mRNA synthesis in crude in vitro transcription system but not with purified polymerase II. *See* P-TEFb; transcript elongation.

DREAM Downstream-regulatory element antagonist modulator is a transcriptional repressor. When its calcium-binding sites "EF-hands" are occupied by Ca^{2+}, it comes off from the DRE element of the gene and transcription may be facilitated. *See* calmodulin; CREB. (Ledo, F., et al. 2000. *Mol. Cell Biol.* 20:9120.)

DRES *Drosophila*-related expressed sequences.

DrFP583 Fluorescent protein that may change color from green to red after some elapse of time and may be used to monitor of gene expression in function of time. *See* aequorin.

drift, genetic *See* genetic drift.

DRIP (ARC [activator-required cofactor]) Transactivation complex of ~16 subunits in the chromatin. *See* chromatin; transactivator.

DriP (defective ribosomal products) About 30% of the proteins produced suffer translational error and are chopped up by proteasomes, the quality-control system of the cell. The amino acids are then recycled. Similar fate is waiting for some of the viral proteins forced upon the cells by the viral genetic material after infection. Aging cells may also be destroyed by apoptosis when their mission is completed. Viral proteins are disposed of mainly with the assistance of the major histocompatibility system class I (MHC I) that displays the peptide fragment of the surface of the cell. The MHC class II molecules similarly handle extracellular bacterial protein pieces. The MHC molecules reside inside the cells and the peptides are transported into the endoplasmic reticulum by TAP to meet the MHC system. The undesirable peptides are then loaded there onto the MHC molecules, which ferry them to the cell surface. The circulating T cells intercept the MHC–undesirable peptide complex and trigger an immune reaction. *See* apoptosis; immune system; MHC; proteasome; TAP; T cell.

driver excess hybridization *See* subtractive cloning.

DRK Downstream receptor kinase of the *Drosophila* light signal transduction pathway. Its vertebrate homologue is GRB2. It is equipped with SH2 and SH3 domains and it is frequently called as a mediator protein in signal transduction. *See* GRB2; SH2; SH3; signal transduction.

Droe1 24 kDa *Drosophila* mitochondrial homologue of GrpE. *See* Grp.

Drosophila Species are dipteran flies. Genetically most thoroughly studied is the cosmolitan *D. melanogaster* ($n = 4$). The species are reproductively isolated, i.e., they may intermate if the opposite sex from the same species is not available, but the F_1 offspring is sterile. The sterile eggs may be recognized by the lack of filaments (see diagrammatic life cycle). The size of the eggs is about 0.5 mm. Within less than a day, they hatch into the first instar stage larvae. (The instar is a larval growth stage in between moltings.) The larvae shed their cuticle by the process of molting. Three instar stages are distinguished. The larvae are very voracious feeders and reach a size of about 4.5 mm. The cuticle of the third instar darkens and hardens and becomes the puparium 4 days after hatching; in another 4 days, the metamorphosis is completed and the imagos (2-mm) emerge from the larvae. (The imago is a fully differentiated adult form.) This adult type emerges by the process of eclosion through the anterior (fore) end of the pupa. Within a day, its color darkens, the wings expand, and the abdomen becomes rotund.

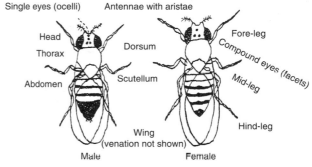

The major areas of the body of *Drosophila melanogaster*. Surface appendages (bristles, hairs) and wing venation are not shown here. (See more at figures in morphogenesis in *Drosophila*.)

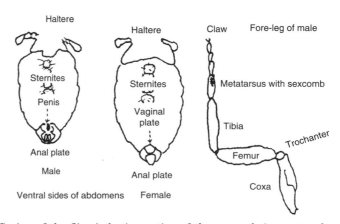

Sexing of the flies is by inspection of the sexcomb (present only on the metatarsus of the fore-leg of the male and the genitalia, visible on the posterior ventral part of the abdomen. For examination, the flies are anesthetized for 5–10 min with ether or other suitable fumes and placed under the dissecting microscope.

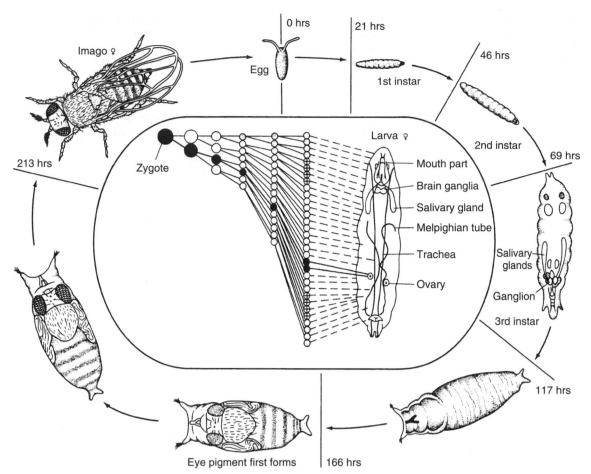

The life cycle of *Drosophila*. The female deposits more than a hundred filamentous eggs (0.5 mm). The hatching maggots (instars) burrow into the food, where they undergo two molts. The emerging larvae develop into pupae. The adults (imago) are about 2 mm long when they eclose from the pupal case. The average lifetime of the females is about 4 weeks; the males may live a little longer. The organs of the adults develop from the imaginal discs, present at the 1st instar stage. In the center (insert) the development of the germline is shown.

There is no further growth after emergence. Within 2 days, the imagos may be copulating after a mating courtship and may start laying eggs. The adult males live for about 33 days and the females die on the average after 26 days. See diagram of the life cycle and germline of *D. melanogaster*. Sexing (determining the gender) of *Drosophila* can be made by looking at the metatarsal segment of the fore leg that carries a structure called *sex comb* in the male only (see diagram p. 349). Sexing is also possible on the basis of the genitalia viewed from the ventral side of the abdomen. The abdominal segmentation is different in the two sexes. After copulation, the female stores the sperm in the spermatheca (located above the uterus) and in the uterus. Once the female has been mated, it may keep the sperm for a long time; therefore, for each genetically controlled mating, generally virgin females are used. The eggs are fertilized just before laying.

Drosophila melanogaster has XX (female) or XY, XO (male) sex chromosomes and three pairs of autosomes. The X chromosome (chromosome 1) is acrocentric (has one arm) and has a genetic length of about 73 map units; chromosomes 2, 3, and 4 have genetic lengths of about 110, 111, and 3 map

units, respectively. The Y chromosome has the *KL* male fertility complex on the long arm and the *KS* male fertility complex on the short arm and no other "visible" genes. The X chromosome contains the nucleolar organizer region in a tandem array of repeats at the *bb* (*bobbed*) locus, coding for the 5.8S, 18S, and 28S ribosomal RNAs. The polytenic X chromosome is divided into 20 regions: the 1st at the tip and the 20th at the centromere. It displays about 1,000 bands. The Y chromosome does not display polyteny. The second chromosome contains 20 sections in the left arm (Section 21 at the tip and Section 40 at the centromere), including about 869 to 927 bands, and the right arm starts at the centromere with section 41 and ends with section 60 at the other telomere, including 1,009 to 1,152 bands, depending on the classification of doublets and singlets. Chromosome 3 appears to be the longest. The left arm begins with section 61 at the tip and section 80 at the centromere, including 884 to 1,032 bands. The right arm starts with section 81 at the centromere through section 100 at the other end, containing about 1,147 to 1,233 bands, depending on the classification. Chromosome 4 is extremely short, does not display very clear polyteny, and does not contain more than about 40 to 50 bands. It is divided into sections 101 and 102, proximally and distally to the centromere, respectively.

Each section is further subdivided into A, B, C, D, E, F subsections by vertical lines positioned left to conspicuous bands. Within lettered subdivisions, individual bands are numbered. The size of the nuclear DNA is ca. 180 Mb and the circular mitochondrial genome contains 19,500 bp. The total euchromatin is ~121 Mb. The origin of replication of the mtDNA is within an AT-rich tract. The replication of the first strand is completed before the replication of the second strand begins. It codes for ribosomal RNA and 22 tRNAs. It also encodes cytochrome b, cytochrome c oxidase, two subunits of ATPase, and 7 subunits of the NADH reductase complex. These genes are crowded next to each other with very few nucleotides keeping them apart. They contain no introns. Translation is initiated with ATA, ATT, and ATG codons. TAA is the most common termination codon. Codon TGA spells Try rather than a stop like in the universal codon dictionary. ATA means Met rather than Ile, and AGA is a Ser codon rather than Arg. Codon usage is influenced by the predominance of AT pairs in this genome. Infectious heredity is also known in *Drosophilas*, caused by the (RNA) picornaviruses (DAV, DPV, DCV), the vesicular stomatitis virus-like agent (sigma virus), and the sex-ratio (*SR*) condition (females do not produce male offspring) caused by spirochaete-like protozoa. In *Drosophila melanogaster*, more than 4,000 genes and 9,000 chromosomal rearrangements are more or less well characterized. The genome has been almost completely sequenced and ~13,600 genes have been found (2000), which is substantially less than expected.

Alleles at a locus are generally distinguished by superscripts. Recessive alleles in a predominantly dominant allelic series are designated by capital letters, but a superscript *r* is added; alternatively, in a recessive series of alleles the dominant allele may be identified with lowercase letters, but with a *D* superscript their dominant expression is identified. The wild-type alleles may also be designated by a (+) superscript, and an *rv* superscript and a number indicates revertants of recessive alleles.

Revertants of dominant mutations are considered deficiencies and are symbolized as *Df* with a distinguishing number, e.g., *Df(1)1-D1* is a deficiency of chromosome 1 (X chromosome) and the latter part of the symbol is explained by words. A number symbolizes alleles specifying the absence of a protein with an *n* superscript and possibly a number superscript. *Proteins* are frequently designated by three Roman capital letters. Loci moved to new locations by transposable elements are enclosed in brackets followed by the new site in parentheses, e.g., [*ry*⁺] (*sd*) or [*Cp16*] (*52D*) indicates that the chorion protein gene was inserted by transformation into cytological site *52D*. *Mimics* (different loci but similar phenotype) are designated as, e.g., *tu-1a, tu1b, tu-2*, or by arbitrary numbers, *Sgs3, Sgs7*, or by polytenic location, *Act5C, Act42A*, or by molecular weight, *Hsp68, Hsp70*, added to the letter symbols. *Modifier genes* such as suppressors or enhancers are symbolized with *e* or *E* and *Su* or *su* followed by the symbol of the gene acted upon in parentheses, e.g., *su(lz³⁴)*, suppressor of a *lozenge* allele. *Translocations* are symbolized with *T*, e.g., *T*(1; *Y*; 3) indicates an X- and Y-chromosome interchange that may be followed by a number of other symbols. Chromosome rings are indicated as *R(1)2*, where *R* stands for ring, (1) for the X chromosome, and 2

for ring #2. *Inversions* are identified by *In* and additional specifications, e.g., *In(2L)Cy* indicates an inversion in the left arm of chromosome 2 involving the dominant wing mutation *Curly* or *In(3RL)* stands for a pericentric inversion of chromosome 3. Additional specifications may also be applied.

Transpositions are defined as three-break rearrangements (two are needed for excision and one at the target site). *Tp(3;1)ry* designates the movement of a third chromosomal *rosy* allele to the X-chromosome.

Duplications are identified by *Dp* and additional information, e.g., *Dp(1;1)y^{bl}* designates a duplication in the X chromosome containing the *yellow bristle* marker. When the duplication is a free fragment, the designation is *Dp(1;f)*, and it may be further specified by gene or band symbols. *Dp(1;1;1)* indicates triplication. *Deficiencies* are defined by *Df*, irrespective of whether a chromosomal segment or an entire chromosome (hypodiploid, $2n - 1$) is involved, e.g., *Df(2R)vg* is a deficiency of the *vestigial* gene in the right arm of chromosome 2. Complex chromosomal rearrangements can also be symbolized, but generally they require detailed descriptions. Deviations from the euploid normal chromosomes change the phenotypic sex because the balance of the sex chromosomes and autosomes affects sex in *Drosophila*.

The metafemales have the chromosomal constitution X/X/X;2/2;3/3;4/4; the triploid metafemales are X/X/X/X;2/2/2; 3/3/3;4/4/4 (or 4/4 only). Metamales are X/Y;2/2/2;3/3/3;4/4/4. Intersexes may be X/X;2/2/2;3/3/3;4/4/4 with a Y chromosome either present or absent, and the numbers of chromosome 4 may vary. Tetraploid females have been observed. Triploids are X/X/X;2/2/2;3/3/3;4/4/4 (or the sex chromosomes may be either X/X/X/Y or attached XX/X or XX/X/Y. Haploids of X;2;3;4 are known. Aneuploids nullo-X (Y/Y;2/2,3/3;4/4), nullo-X nullo-Y (0/0;2/2;3/3;4/4), tetra-4 (X/X;2/2;3/3;4/4/4/4), triplo-4 (X/X;2/2;3/3;4/4/4), haplo-4 (X/X;2/2;3/3;4), X0 male (X;2/2;3/3;4/4), XXY female (X/X/Y;2/2;3/3;4/4), XXYY female (X/X/Y/Y;2/2;3/3;4/4), XYY male (X/Y/Y;2/2;3/3;4/4), and XYYY male (either X/Y/Y/Y or attached XY/Y/Y plus normal autosomes) are also known. About 20% of the *Drosophila* genome consists of repeated sequences and displays four satellite bands (1.672, 1.686, 1.688, 1.705) upon CsCl density gradient centrifugation (not amplified in the salivary glands). Other repeated sequences, SINE (0.5 kbp) and LINE (5–7 kbp) and the over 30 different transposable elements are also parts of the genome. Detailed information on 4,000 gene loci and 9,000 chromosomal aberrations (including references) are described in *The Genome of Drosophila* by D. L. Lindsley and G. G. Zimm, 1992, Academic Press, San Diego, CA. *See* allele; aneuploid; courtship; deficiency; duplication; eclosion; fruit fly; hybrid dysgenesis; imaginal disk; infectious heredity; instar; inversion; LINE; metamorphosis; molting; morphogenesis; nucleolar organizer; P element; polytenic chromosomes; puparium; reproductive isolation; salivary gland chromosomes; SINE; suppressor; tetraploid; translocation; transposable elements; triploid; virgin. (<http://fruitfly.bdgp.berkely.edu/> or <http://flybase.bio.indiana.edu> or <http://edgp.ebi.ac.uk/>; Adams, M. D., et al. 2000. *Science* 287:2185; Rubin, G. M. & Lewis, E. B. 2000. *Science* 287:2216.)

drosopterin Bright red eye pigment in insects (*Drosophila*) and related pigments are common in other animals. The *se* (*sepia*) mutants of *Drosophila* (chromosome [3–26.0]) accumulate a yellow pigment (a dihydropteridine) but are unable to synthesize drosopterin or isodrosopterin because of the defect in PDA (2-amino-4-oxo-6-acetyl-7,8-didydro-3H,9H-pyrimido[4,5,6]-[1,4] diazepin) synthetase. *See* pigmentation of animals. (Wiederrecht, G. J., et al. 1981. *J. Biol. Chem.* 256:10399.)

DRPLA   *See* dentatorubral-pallidoluysan atrophy.

drug development   The *Drosophila* hybrid dysgenesis factor, the P element, is being explored to study transgene responses to drugs. The proteome analysis and combinatorial chemistry open new avenues to designing more effective medicine. *See* combinatorial chemistry; developmental therapeutics program; DNA grooves; hybrid dysgenesis; microarray hybridization; PCR-mediated gene replacement; phage display; pharmaceuticals; proteome; SAR by NMR. (*Nature Rev. Drug Discovery* [2002], 1[1].)

drug resistance   *See* antimetabolite; multiple drug resistance; resistance transfer factors. (Gottesman, M. M. 2002. *Annu. Rev. Med.* 53:615.)

drum stick   *See* Barr body.

Ds (*Dissociator*)   Defective transposable element of maize that can move only when *Ac* provides the transposase function. Its name came from the observation that it was frequently associated with chromosome breakage. *See Ac*; controlling elements; transposable elements.

DSB   Double-strand break.

DsbA   21.1 kDa monomeric *E. coli* periplasmic protein with strong oxidizing ability. DsbB reoxidizes it. DsbC has a similar function. DsbD and DsbE are thiodisulfide reductases. *See* glutaredoxin; PDI; thioredoxin.

dsDNA   Double-stranded DNA.

Double-stranded DNA.

DSE   Distal sequence element. *See* Hogness box.

DSIF   Regulatory dimeric (~14 and ~160 kDA subunits) protein of transcript elongation. It seems to be a target of DRB. The large subunit is homologous to the bacterial NusG. *See* DRB; NELF; nusA; TEFb; transcript elongation. (Renner, D. B., et al. 2001. *J. Biol. Chem.* 276:42601.)

DSP (dual-specificity protein phosphatase)   One of its active sites dephosporylates serine, threonine, and tyrosine in a protein, whereas the deeper active site is specific only for tyrosine.

DsrA   Short (87 base) RNA with three stem-loop motifs modulating transcriptional regulators H-NS (inhibitory) and RpoS (stimulatory). One of the stem loops binds and stabilizes RpoS mRNA and the second sequesters H-NS mRNA. (Lease, R. A. & Belfort, M. 2000. *Mol. Microbiol.* 38:667.)

dsRAD   *See* DRAD.

dsRNA   Double-stranded RNA. *See* RNA interference.

DSS (dosage-sensitive sex reversal)   Approximately 160 kb duplication in the short arm of the X chromosome upsets normal male gonad formation in XY individuals. Deletion of the same region, however, does not affect gonadal differentiation, although the hypogonadism results in infertility. A defect in the *DAX-1* gene enhanced the expression of the aromatase enzyme encoded by the *Cyp19* gene of mice. This enzyme converts testosterone to estradiol in the Leydig cells. *See* estradiol; sex determination; sex reversal; testosterone; Wolffian duct. (Wang, Z. J., et al. 2001. *Proc. Natl. Acad. Sci. USA* 98:7988.)

DST1   Yeast gene encodes the 38 kDa strand transfer protein (STPα). *See* eukaryotes; recombination mechanism.

DST2   Yeast gene encodes the STPβ protein (identical to Sep 1). *See* eukaryotes; recombination mechanism.

DSTF DPE-specific transcription factor   Facilitates the transcription by DPE promoters and represses the TATA box-driven promoters by binding to the TBP of the transcription factor TFIID. DSTF activity is associated with 43 kDa (NC) and 22 kDa (Drap) proteins. *See* core promoter; DPE; Drap; NC; TBP; TFIID. (Kutach, A. K. & Kadanoga, J. T. 2000. *Mol. Cell Biol.* 20:4754.)

dsx (*double sex*)   Abnormal location 3–48.1; regulates sexual differentiation in somatic cells of *Drosophila melanogaster*. The null allele converts males and females into intersexes. *See Drosophila*; sex determination. (Erickson, J. W. 2001. *Developmental Cell* 1:156.)

DT40   Chicken cell line with as high as 10 to 100% homologous recombination in nonisogenic DNA without high selection pressure. Thus, it provides an opportunity for high-efficiency gene transfer in vertebrate, including human cells. (Thompson, L. H. & Schild, D. 2001. *Mutation Res.* 477:131; <http:genetics.hpi.uni-hamburg.de/dt40.html>.)

dTAF$_{II}$ *See* transcription factors.

***Dt* gene**   Actually a transposable element that may be situated in several locations in the maize genome. It causes the colorless *a1-m* allele to produce anthocyanin-containing dots in the triploid aleurone tissue, depending on the dosage of the *Dt* elements containing a transposase. Also, it may cause reversions to the *A* allele in the germline. *Dt*, like other transposable elements, induces chromosome breakage and can newly arise through chromosomal breakage. The first *Dt* allele, *Dt1*, was located to the initial position of the short arm of chromosome 9. *Dt2* (chromosome 6L-44), *Dt3* (7L), *Dt4*

(4), *Dt5* (in the vicinity of *Dt1*), and *Dt6* (in the short arm of chromosome 4), not far from the centromere, were identified subsequently. *See* transposable elements. (Brown, J. J. 1989. *Mol. Gen. Genet.* 215:239.)

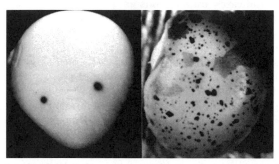

Maize kernels *aaa* (left) and *aaaᵐ Dt Dt Dt* (right) constitution. (Courtesy of M. G. Neuffer.)

DTH   Delayed-type hypersensitivity may cause acute rejection of allografts and autoimmune reactions due to nonspecific T-cell responses. *See* allograft; autoimmune; hypersensitivity; T cell. (Jager, D., et al. 2001. *J. Clin. Path.* 54 [9]:669.)

DTT (dithiotreitol, Cleland's reagent)   Protective agent for SH groups. *See* mercaptoethanol.

dual-gene operons   Synthetic constructs carrying two selected structural genes driven by a single promoter. Within the operon and within the untranslated 5′ and 3′ regions various regulatory elements (restriction enzyme sites) can be placed. Hairpin structures at the 3′-end may protect against exonuclease attack and transcription termination. Such systems permit the analysis of the differential expression of both genes under various controlled conditions. (*See* operon, Smolke, C. D. & Keasling, J. D. 2002. *Biotechnol. Bioeng.* 78:412.)

dual-specificity phosphatase   Recognizes more than a single phosphorylated amino acid in a protein and has an important role in signal transduction pathways, differentiation/development, and epigenesis. *See* epigenesis; serine/threonine kinase.

Duane retraction syndrome   Defect of moving the eyeball (globus), encoded in the region 2q31. About 5% of the strabismus cases show similar defect. *See* eye diseases; strabismus.

Dubinin effect   Change of dominance in alleles (variegation) because of transposition (of heterochromatin) in *Drosophila*. *See* dominance reversal; heterochromatin; position effect. (Dubinin, N. P. & Sidorov, B. N. 1934. *Am. Nat.* 68:377; Locke, J. & Tartof, K. D. 1994. *Mol. Gen. Genet.* 243:234.)

Dubin-Johnson syndrome (hyperbilirubinemia II, DJS, 10q24)   Recessive disease with a relatively high frequency (8×10^{-4}) among Iranian Jews. It involves jaundice, hyperbilirubinemia, and melanin-like deposits in the liver. It is a defect of the detoxification mechanism of the liver. Detoxification is mediated by conjugation with glutathione, glucuronides, or sulfates, resulting in negatively charged amphiphilic compounds that can effectively be secreted into the urine. The excretion is carried out with the assistance of canalicular (tubular) multispecific organic anion transporter (cMOAT). Hyperbilirubinemia is also caused by Gilbert syndrome (synonymous with Arias-type hyperbilirubinemia and Crigler-Najjar syndrome type II defects in UDP-glucuronosyl transferase) in human chromosome 2 and deficiency of factor VII in human chromosome 13q34. Rotor syndrome is apparently the same as DJS. *See* antihemophilic factors; Crigler-Najjar syndrome; glutathione-S-transferases; hyperbilirubinemia.

Dubowitz syndrome   Fetal-newborn physical and mental retardation complex under autosomal-recessive control. The symptoms may resemble those in fetal alcohol syndrome (nongenetic) and autosomal-recessive Fanconi anemia. *See* alcoholism; Fanconi's anemia.

Duch (DH blood group)   Extremely rare; it has been identified in Aarhus, Denmark. *See* blood groups.

Duchenne muscular dystrophy   *See* muscular dystrophy.

duck (*Anas platyrhynchos*) $2n = 78-80$.

DUE   DNA unwinding element is a part of the autonomously replicating sequence (ARS). *See* ARS.

Duffy blood group (*Fy*), alleles   In human chromosome 1q12-q21, they control a red blood cell antigen that is frequently used as a genetic marker in population studies; 99% of the Chinese and 65% of the Caucasians are positive, whereas 95–99% of the black Africans are of the negative type (*Fy⁻*/*Fy⁻*). Individuals lacking these antigens (Fy [a⁻ b⁻]) are protected against the African malaria-causing *Plasmodium vivax* protozoon. *See* blood groups. (Hamblin, M. T., et al. 2002. *Am. J. Hum. Genet.* 70:369.)

Dulbecco's medium (PBS)   Minimal animal tissue culture solution containing buffered sodium pyruvate, streptomycin, glucose, calcium chloride, and penicillin; various nutrients may be added. (Lawson, M. A. & Purslow, P. P. 2000. *Cells Tissues Organs* 167[2–3]:130; Baust, J. M., et al. 2000. *In Vitro Cell Dev. Biol. Anim.* 36[4]:262.)

dumbbell oligonucleotide   Short double-stranded DNA or DNA-RNA with closed loops at both ends. Ring-shape "dumbbells" are created by the fusion of two hairpins. They are very resistant to nucleases, can bind to very specific sequences, and block transcription factor access. They may carry antisense RNA or antisense DNA or may include sequences within the stem (in between the loops) where specific transcription factors can bind and exert interference with transcription. Dumbbell technology may have therapeutic value against viral infection. *See* antisense technologies; decoy receptor. (Clausel, C., et al. 1993. *Nucleic Acids Res.* 21:3405; Park, W. S., et al. 2000. *Biochem. Biophys. Res. Commun.* 270:953.)

dumping   Transfer of the contents of nurse cells to the oocyte via cytoplasmic bridges. *See* nurse cell; oocyte.

dumposome Protein Pdd1 of *Tetrahymena* that may be targeted to DNA degradation within cells of organisms. *See* apoptosis.

Duncan syndrome (X-linked lymphoproliferative disease) Xq25 immunodeficiency with immediate hypersensitivity to various allergens caused by higher-than-normal levels of IgE. The affected individuals are particularly sensitive to Epstein-Barr virus infections and usually die of lymphoma or aplastic anemia. *See* anaplastic lymphoma; aplastic anemia; Epstein-Barr virus; immunodeficiency; immunoglobulins; lymphoma; lymphoproliferative diseases, X-linked.

duplex In a polyploid (e.g., AAaa), triploid, or trisomic (e.g., AAa), a duplex carries two dominant alleles at a locus, and the other allele(s) at the same locus is recessive. *See* autopolyploid.

duplex DNA Double-stranded. *See* Watson-Crick model.

duplicate genes Convey the same (or very similar phenotype) but segregate independently; therefore, in F_2 the phenotypic proportions are 15 dominant:1 recessive. *See* modified Mendelian ratios.

duplicate ratio *See* duplicate genes.

duplication Eukaryotic chromosomal segment is repeated side-by-side (tandem duplication) or may be repeated at another location within the chromosome or in another chromosome. The duplication is cytologically detectable in duplication heterozygotes if it is of sufficient length. In prophase, when paired with the nonduplicated strand, it bulges out because the normal counterpart cannot pair with the extra chromosomal tract:

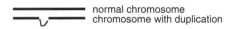

normal chromosome
chromosome with duplication

The duplication may be direct →→ or inverted →←. After insertion of transposable elements, duplications may occur at the target sites. Transposable elements generally also carry terminal direct or inverted repeats. Duplications and deficiencies are generated by crossing over within the inverted segments of para- and pericentric inversions, breakage-fusion-bridge cycles, crossing over between sister chromatids of ring chromosomes, unequal crossing over, adjacent I and adjacent II distribution of chromosomes in the meiosis of translocation heterozygotes, etc. Duplications may have evolutionary significance because the extra piece of DNA may be modified to carry out a new function(s). Molecular evolutionary studies have provided many examples of duplications followed by differentiation and divergence.

It has been estimated that after a duplication the α- and β-chains of the human hemoglobin separated about 500 million years ago. Subsequently, from the β-chain, the γ- and δ-chains evolved. (However, the rhesus monkey does not have the δ-chain.) The protein superfamily of trypsin, chymotrypsin, elastase, thrombin, kallikrein, plasmin, and bacterial trypsin all have structural similarities, indicating common ancestry. Complete nucleotide sequence data of prokaryotic genomes has revealed unexpected extensive duplications. In *Bacillus subtilis*, more than one-forth of the genome indicates duplications of recent origin. In *E. coli*, there are about 80 transport proteins of common origin. In *Haemophilus influenzae*, 284/1,709 (~17%), in *Saccharomyces cerevisiae*, 1,858/6,241 (~30%), in *Caenorhabditis elegans*, 8,971/18,424 (~49%), and in *Drosophila melanogaster*, 5,536/13,601 (~41%) of the protein-coding genes appear to have arisen by duplication.

The progress of sequencing of the human genome also revealed several duplications, mainly in the pericentromeric and telomeric heterochromatin of the same or different chromosomes. In *Drosophila*, the duplicated genes appear dispersed, although a cluster of 17 paralogous genes of unknown function has been identified. In *Caenorhabditis*, there is some clustering of 7 transmembrane domain paralogous genes. Almost 30% of the *Drosophila* proteins are orthologous with those of *Caenorhabditis* and ~20% of the fly and worm genes share common origins with yeast. About half of fly and about 36% of worm genes/proteins may have some similarity to those of humans, depending on the stringency of the comparisons; 744 families or domains of proteins were judged common to all three of these lower eukaryotic organisms. The flies, worms, and humans seem to have 300, 500, and more than 600 protein kinases and 85, 185, and more than 300 protein phosphatases, respectively. There are 450 and 260 peptidases in the fly and worm, respectively. The numbers of multidomain proteins in the fly, worm, and yeast are 2,130, 2,261, and 672, respectively. The number of G protein–coupled receptors (GPCR) is very high in the worm (~1,100), whereas in flies (160) and humans (~700) fewer were identified. The number of olfactory receptors in the mouse is ~1,000, whereas in the fishes there are only ~100. Of the 289 human genes where the molecular/cytological basis of a disease is known, 177 had orthologs in *Drosophila*, and 68% of the oncogenes had apparently orthologous fly counterparts by year 2000. The flies do not have breast cancer gene or leptin orthologues. Many cell cycle, cytoskeleton, cell adhesion, cell signaling, apoptosis, neuronal signaling, and cell defense (immunity) genes have counterparts in the lower animals. *See* breast cancer; cancer; chromosomal aberrations; core proteome; deficiency; deletion; duplication deficiency; evolution and duplications; gene number; heterochromatin; immune system; leptin; olfactogenetics; oncogenes; orthologous loci; paralogous loci; polyploidy; protein domains; redundancy. (Rubin, G. M., et al. 2000. *Science* 287:2204; Gu, X., et al. 2002. *Nature Genet.* 31:205; Conant, G. C. & Wagner, A. 2002. *Nucleic Acids Res.* 30:3378; <www.ebi.ac.uk/interpro>.)

duplication-deficiency Cells are produced by adjacent-1 and adjacent-2 distribution of chromosomes in translocation heterozygotes, in para- and pericentric inversion heterozygotes when crossing over takes place within the inverted segment. They may occur if there is a sister chromatid exchange in ring chromosomes and in all other cases when dicentric chromosomes are produced (breakage fusion-bridge cycles), in cases of unequal crossing over, etc. *See* dicentric ring chromosome; inversion; translocation.

duplicons Chromosome-specific low-copy repeats at multiple areas of the human genome. Homologous recombination between duplicons may lead to chromosomal aberrations such as duplications, deletions, and inversions. Such alterations may occur at a frequency of about 10^{-3} and cause genetic diseases. *See* chromosomal rearrangements. (Ji, Y., et al. 2000. *Genome Res.* 10:597.)

durum wheat Member of the allotetraploid series with AB genomes. *See Triticum*.

DUST Computer program designed to identify low-complexity regions in DNA sequences.

duty ratio (duty cycle) Indicates the frequency of the time when a motor protein is attached to its filament and is supposed to move. *See* motor protein.

dwarfism Caused by several *dominant* genes in apparently different human autosomes. The distinguishing features include (1) narrow vertebrae; (2) thickening of the tubular bone associated with far-sightedness and a lower-than-normal level of blood calcium; (3) low birthweight, stiff joints, and eye defects; (4) Myhre syndrome involving pre- and postnatal growth retardation, anomalies of face morphology, poor mobility of joints, defects in sex organs, heart, and hearing. Some *autosomal-recessive* dwarfism, such as Dubowitz syndrome, involves intrauterine growth retardation, mental and hearing defects, sex organ anomalies, unusually high or low voice, no response to growth hormone, etc. Various other types are accompanied by mental retardation, heart anomalies, hip dislocation, very short arms and digits, and the absence of fibula (the smaller bone of the leg). *See* achondroplasia; animal hormones; chondrodysplasia; cleidocranial dysplasia; corticotropin; diastrophic dysplasia; Donohue syndrome; dyschondrosteosis; dyssegmental dysplasia; GHRH, GHRHR; growth retardation; hypochondroplasia; Kniest dysplasia; Langer-Nielssen mesomelic dwarfness; Laron-type dwarfness; limb defects; macroimelic dysplasia; Moore-Federman syndrome; Mulbrey nanism; osteogenesis imperfecta; pituitary dwarfism; pycnodysostosis; Pygmy; Seckel dwarfism; SHORT syndrome; somatotropin; spondyloepiphyseal dysplasia; stature in humans; thanatophoric dysplasia; Williams syndrome. Dwarfism also occurs in plants. This condition in plants may be remedied in some cases by the supply of gibberellic acid or ethylene. Some of the gibberellin-insensitive dwarf mutants of plants are defective in G proteins that have a key role in signal transduction. *See* brassinosteroids; deetiolation; plant hormones.

dyad Chromosome with two chromatids; also, the two cells (dyads) formed by the first meiotic division. *See* chromatid.

dyad, spaced *See* spaced dyads.

dye primer and dye terminator chemistry During DNA sequencing, it may correct errors due to gel compression or base calling. *See* base calling; compression in gels; double stranding.

dynactin Dynein-regulating protein with the role of moving organelles; it may interact with the kinetochore and anchor dynein. *See* dynein; spindle. (Muresan, V., et al. 2001. *Mol. Cell* 7:173.)

dynamic allele-specific hybridization (DASH) Has been designed to detect SNIPS. Two DNA sequences are hybridized initially at low stringency. A fluorescent marker that binds only double-stranded DNA and emits the signal monitors hybrid DNA. When the temperature in the reaction well is raised, the SNIP areas lose the signal because they no longer form the hybrid and the mismatch is identified. *See* allele-specific probe; SNIP. (Prince, J. A., et al. 2001. *Genome Res.* 11:152.)

dynamic instability May occur when microtubules oscillate between growth and shortening during mitosis along with treadmilling. *See* treadmilling.

dynamic molecular combing Physical method for mapping distances between fluorochrome-labeled DNA molecules stretched out on the surface of microscopic cover slides. *See* physical mapping. (Michalet, X., et al. 1997. *Science* 277:1518.)

dynamic state Biochemical concept stemming from the realization that the constituents of the living body, irrespective of whether they are metabolic or structural, are in a steady state of lux.

dynamin ~100 kDa protein with guanosine triphosphatase activity. Dynamins are associated with other proteins and play an essential role in coated vesicle formation. They are localized at the plasma membrane around the neck of emerging coated pits. Several forms of dynamins are known and they display tissue specificity. They are involved in vesicle recycling, nerve terminal depolarization, etc. *See* clathrin; coated pits; endocytosis; GTPase; synaptic vesicles; synaptogamin; synaptojanin. (Hinshaw, J. E. 2000. *Annu. Rev. Cell Dev. Biol.* 16:483.)

Dynamin ring around vesicle.

dynein Multisubunit protein (1.2 MDa) associated with bundles of microtubules (axoneme) in cilia and flagella, assisting their movement in an ATP-dependent process. Dyneins are supposed to move the microtubules toward the centrosome. Dynein is also involved in the development of left-right asymmetry along the axis of the body and it mediates the breakdown of the nuclear membrane. *See* ankyrin; asthenozoospermia; axoneme; cilia; dynactin; flagellum; Kartagener syndrome;

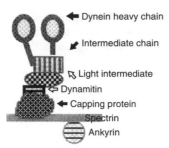

The dynein complex (Modified after Hirokawa, N. 1998. *Science* 279:519.)

kinesin; left-right asymmetry; microtubules; situs inversus viscerum; spectrin centrosome; tubulin. (Lee, I. H., et al., 2001. *Mol. Biol. Cell* 12:2195.)

dyne unit of force   This force is needed so that a body of 1 gram can be accelerated 1 centimeter per second.

Dynia factor   Blood factor controlling the antihemophilic factor. *See* antihemophilic factors; blood clotting pathways; hemostasis.

dynodes   Set of auxiliary electrodes that amplify electrons as they are hit by an electron beam. Dynodes are used in various types of photomultipliers and mass spectrometers in proteome research. *See* proteomics.

dynorphin   Opiate receptor ligand. *See* opiate.

dysautonomia (FD, 9q31)   Malfunction of the central nervous system resulting in hypertension, emotional instability, cold hands and feet, excessive sweating, and red skin spots. Generally the prospect of survival to age 40 is poor. It is usually caused by a defect in splicing of the RNA transcript of IKBAP (inhibitor of κ-light-chain polypeptide). This protein is part of the IKK complex that includes NF-κB-inducing kinase, IKK-β (inhibitor of κ-B kinase β), IKK-α (inhibitor of κ-B kinase α), NF-κB/RelA, and protein IKAP (IKK complex–associated protein). Its prevalence is high (~2–3×10^{-4} in Ashkenazi Jewish populations). *See* neuropathy; Riley-Day syndrome. (Anderson, S. L. 2001. *Am. J. Hum. Genet.* 68:753; Slaugenhaupt, S. A. & Gusella, J. F. 2002. *Current Op. Genet. Dev.* 12:307.)

dysbetaliporoteinemia   Defect in apolipoprotein E. *See* apolipoproteins; hyperlipoproteinemia.

dyscalculia   Brain defect manifested in difficulties with handling numerical problems.

dyschondrosteosis (Leri-Weill syndrome)   Dominant dwarfism of the forearms with about four times as high occurrence in females than in males. It may be caused by p22-q12 X-Y translocations or a large deletion at the Xp22.3 region involving the SHOX gene. The same gene is involved in Langer mesomelic dwarfism. *See* achondroplasia; dwarfness; hypochondroplasia; pseudoautosomal; short syndrome; stature in humans.

dyscrasia   Pathological condition of different manifestations in the blood or plasma, plasma proteins.

dysequilibrium syndrome   Autosomal-recessive cerebral palsy, mental retardation, muscular insufficiency, poor motor control, etc., based on malfunction of dopamine-β-hydroxylase activity.

dyserythropoietc anemia (HEMPAS for hereditary erythroblastic multinuclearity with positive acidified serum test) Endopolyploidy apparently limited to the bone marrow cells. As a consequence, anemia appears. The basic defect seems to be a deficiency in the enzyme N-acetyl-glucosaminyl transferase II affecting the biosynthesis of glycoproteins. GATA1 mutations result in this disease and thrombocytopenia. *See* GATA; thrombocytopenia.

dysferlin   Protein defective in both limb-girdle muscular dystrophy 2B and Myoshi myopathy. *See* muscular dystrophy. (Liu, J., et al. 1998. *Nature Genet.* 20:31.)

dysfibrinogenemia   In contrast to fibrinogenemia, where the synthesis of fibrinogen is reduced. These individuals fail to assemble the fibrinogen monomers into normally functional molecules. This human chromosome-4 dominant condition most commonly is not associated with heavy bleeding but may cause periodic clot formation in the blood vessels (thrombosis). *See* afibrinogenemia; antihemophilic factors; fibrin-stabilizing factor.

dysgenesis   Mechanism or process producing dysgenic individuals. *See* dysgenic, hybrid dysgenesis.

dysgenic   Deleterious, undesirable genetic trait. *See* hybrid dysgenesis.

dyskeratosis (DKC, Xq28; DKBI 4q35)   Autosomal-dominant forms involve hyperpigmentation of the skin, precancerous skin lesions, poor bone and nail development, and no dermal ridges (fingerprints). The benign form does not greatly affect the individual's well-being. The autosomal-recessive form is also similar. The Xq28-linked form afflicts primarily males who may be affected by testicle atrophy, anemia, cancer, and lacrimation because of defects in the lacrimal ducts. The DKC1 gene product, dyskerin (514 amino acids), is a nucleolar protein suspected of having a role in ribosomal function. Dyskerin is associated with telomerase function. By shortening the telomeres, it may hinder the proliferation of human somatic cells in the epithelium and blood. Dominant intraepithelial benign dyskeratosis (HBID) is caused by duplication at 4q35 and is responsible for red eye and other superficial lesions. A large 821 bp deletion of the 3′-end of a dyskerin gene, encoded at 3q, involves the RNA component of the telomerase enzyme. *See* fingerprints; keratoma; keratosis; skin diseases; telomeres. (Marciniak, R. A., et al. 2000. *Trends Genet.* 16:193; Vullamy, T., et al. 2001. *Nature* 413:432.)

dyskerin   *See* dyskeratosis.

dyslexia   Highly variable difference in the central nervous system causing difficulties in reading and understanding or tiredness of reading. In some forms it appears to be associated with left-handedness and speech defects. Dyslexia may not necessarily be associated with deficiencies of general intelligence (IQ) and can be compensated for by tutoring.

It has been suggested that a dominant gene in human chromosome 6p21.3 determines some specific forms of it with preferential penetrance in males. Chromosome 15 also harbors a gene for dyslexia. Several QTLs determine dyslexia. A major QTL has been assigned to 18p11.2 (Fisher, S. E., et al. 2001. *Nature Genet.* 30:86). Complicated orthography aggravates the condition, whereas it is less serious in near-phonetic languages. *See* IQ; mental retardation; QTL. (Wysman, E. M., et al. 2000. *Am. J. Hum. Genet.* 67:631; Paulesu, E., et al. 2001. *Science* 291:2165.)

dysmelodia Apparently autosomal-dominant gene with low penetrance causes reduced musical ability. There is also another autosomal-dominant gene (perfect pitch) enabling the individual to remember and play a tune. *See* Bach.

dysmorphology Abnormal morphological change. (Baraitser, M. & Winter, R. M. 2001. London Dysmorphology Database, Oxford Univ. Press, New York.)

dysostosis Defect in ossification (bone formation).

dysplasia Abnormal organization of the cells within the tissue.

dysploidy The basic chromosome number varies within a population either because of the presence of B chromosomes or because of Robertsonian translocations or misdivision at the centromere. The change in number does not involve an increase or decrease of an integer of the basic chromosome set. *See* aneuploidy; B chromosomes; misdivision; polyploidy; Robertsonian translocation. (Baldwin, B. G. & Wessa, B. L. 2000. *Am. J. Bot.* 87:1890.)

dyspnea Difficulty in breathing due to various genetic or other causes.

dyspraxia, developmental verbal (SPCH1, 7q31) Dominant expressive and receptive speech; grammar and language defect without substantial sensory or neurological impairment. It is caused by a defect in a putative transcription factor including the gene FOXP2, which encodes a polyglutamine tract and a DNA-binding fork-head domain. *See* FKH; speech and grammar disorder. (Lai, C. S. L., et al. 2001. *Nature* 413:519.)

dysreproductive genes Impair fertility of the individual by lowering or adversely affecting the reproductive system either by structural and developmental defects or as subvitals, semilethals, and lethals.

dyssegmental dwarfism (dyssegmental dysplasia) Autosomal-recessive phenotype involving abnormal development of the vertebrae. It is accompanied by various other symptoms such as dwarfism, cleft palate, hydrocephalus, etc. A defect in the normal formation of collagen is suspected. The DDSH gene (Silverman-Handmakers disease, 1p36.1) involves a defect in the synthesis of perlecan (heparan sulfate proteoglycan), a component of the basement membranes. *See* basement membrane; cleft palate; collagen; dwarfism; hydrocephalus. (Arikawa-Hirasawa, E., et al. 2001. *Nature Genet.* 27:431.)

dystasia (areflexic dystasia) *See* claw foot.

dystonia Lack of muscle coordination caused by a dominant factor at human chromosome 9q32-q34 in the region of the dopamine-β-hydroxylase locus. The prevalence of the disease (torsion dystonia) in Ashkenazi Jewish populations is high, $2-5 \times 10^{-4}$, and the penetrance is about 30%. Some dystonias respond favorably to DOPA. Autosomal-recessive expression (or possibly mitochondrial origin) may be accompanied by visual defects. In some X-linked forms, the symptoms include deafness. The hereditary progressive DOPA-responsive dystonia in human chromosome 14 is caused by a deficiency of GTP cyclohydrolase I. This enzyme controls biopterin biosynthesis required for dopamine. *See* biopterin; dopa; idiopathic torsion dystonia; myoclonous; neuromuscular diseases; sarcoglycan.

dystroglycan Part of the dystrophin-associated protein complex. It participates in the formation of the extracellular matrix in cells contacting basement membranes. It promotes the assembly of the acetylcholine receptors and the neuromuscular junctions. It is the receptor of *Mycobacterium leprae* and some arena viruses. *See* acetylcholine receptor; animal viruses; basement membrane; muscular dystrophy; Mycobacteria; sarcoglycan. (Michele, D. E., et al. 2002. *Nature* 418:417.)

dystrophin (DRP) Muscle protein that anchors muscle membranes to actin filaments in the myofibrils. The X-chromosomal gene contains 79 exons spanning 2,300 kb and the transcription requires about 16 hours. Utrophin is also a dystrophin-type protein (DRP1, 6q24), and DRP2, another dystrophin gene, has been mapped to human chromosome Xq22. *See* actin; caveolin; exon; gene size; muscular dystrophy; myofibril; sarcolemma.

dystrophy Inadequate nutrition (of the muscles). *See* atrophy; muscular dystrophy.

dyszoospermia Anomaly involving the formation of spermatozoa. *See* azoospermia; gametogenesis; spermatid; spermatogonia.

DZ Dizygotic twin. *See* twinning.

Henry Borsook noted in the Journal of Comparative Cellular Physiology. Suppl. 1:283 (1956) his recollections from the late 1920s.

Edwin Cohn, the physical chemist asked T.H. Morgan, the first Nobel-laureate geneticist what his plans are. And Morgan's answer was "I am not doing any genetics. I am bored with genetics. But I am going out to Cal Tech where I hope it will be possible to bring physics and chemistry to bear on biology."

Shortly after Morgan arrived to Cal Tech Einstein visited the place and posed about the same question. The answer was about the same as before. Einstein shook his head and said, "No, this trick won't work. The same trick does not work twice. How on earth are you ever going to explain in terms of chemistry and physics so important a biological phenomenon as first love?"

A historical vignette

E

E1 (Ubc1)   Ubiquitin-activating protein. *See* SCF; ubiquitin.

E2 (Ubc2)   Ubiquitin carrier, a ubiquitin-conjugating enzyme. Ubc2/Rad6 in yeast has a dual function of proteolysis and DNA repair. E2-C is involved in the degradation of cyclin B in cooperation with the APC/cyclosome. E2-C has been detected in many eukaryotes except *Saccharomyces cerevisiae*. *See* APC; cell cycle; SCF; UBC; UbcD1; ubiquitin.

E3 (Ubc3)   Family of ubiquitin ligases includes four types of proteins: (1) acting on the N-end rule proteins (~200-kDa), (2) E6-AP (papilloma virus E6 oncoprotein-associated protein, ~100 kDa); it also includes *hect* proteins (homologous to E6-AP); (3) the cyclosome/APC-associated enzyme acting on mitotic cyclins, anaphase inhibitors, and spindle apparatus proteins; (4) enzymes acting on the Bim mitotic motor proteins. *See* Angelman syndrome; APC; Bim; cell cycle; E2; N-end rule; SCF; ubiquitin.

E1A   Adenovirus oncoprotein; it can reverse the growth-inhibitory effect of transforming growth factor (TGF-β). E1A moves cells into the S phase by interacting with the retinoblastoma (Rb) and the p300 proteins. Rb recruits histone deacetylase and p300 attracts histone acetyltransferase. Both Rb and p300 are activated by phosphorylation at the G1/S checkpoint. *See* adenovirus; cell cycle; chromatin remodeling; CtBP; histone acetyltransferase; histone deacetylase; p300; retinoblastoma; T cells; TGF. (Fuchs, M., et al. 2001. *Cell* 106:297.)

E2A   Immunoglobulin enhancer-binding factor, a basic helix-loop-helix protein, encoded in human chromosome 19p13. *See* DNA-binding domains; immunoglobulins.

EAA   Excitatory amino acids. Glutamate and aspartate particularly can activate neurotransmitters. *See* neuro-transmitter.

Eadie-Hofstee plot   In a coordinate system, $v/(S)$ is plotted against v, where v is the enzymatic reaction velocity and (S) is the substrate concentration. *See* Linweaver-Burk plot; Michaelis-Menten equation.

EAR (extra annual risk)   Extra annual increase of incidence of cancer compared to (radiation) unexposed population. *See* Armitage-Doll model; ERR; risk.

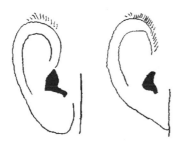

Ear lobes.

earlobe, attached (at right, dominant is at left).   Generally a recessive human trait.

early genes   Transcribed early during development; they are involved in the infection process of the virus before replication begins. *See* delayed-response gene; early-response gene; late gene.

early-response gene   Activated by growth factors within a few minutes without a prerequisite for protein synthesis. *See* delayed-response gene; early gene; late gene.

E3B1   Signal transducer between RAS and Rac. *See* EPS8; Rac; RAS; signal transduction.

EBF (early B-cell factor)   Transcription factor specific for B cells and expressed at antigen-independent stages. It regulates the immunoglobulin α-chain. *See* B lymphocyte; T cell.

ebgA⁰   Gene of *E. coli* (map position 67) which through two mutations permits galactose utilization, although this newly evolved galactosidase is immunologically distinct from the *LacZ*-encoded β-galactosidase enzyme. *See* lactose operon, (Hall, B. G. 1999. *J. Bacteriol.* 180:2862; Hall, B. G. 2001. *Mol. Biol. Evol.* 18:1389.)

EBNA-2 and -5   Epstein-Barr virus antigens and trans-activating oncoproteins. *See* Epstein-Barr virus; oncogenes; transactivation response element; transactivator.

Ebola virus (EBOV)   Member of the filoviridae, a negative-stranded, enveloped, nonsegmented RNA virus of 18,958 nucleotides, encoding 7 structural and 1 nonstructural proteins. It is a very dangerous hemorrhagic pathogen of humans and other primates. *See* animal viruses; negative-strand viruses. (Burton, D. R. & Parren, P. W. 2000. *Nature* 408:527.)

E box (Ebox)   Genes controlling endogenous rhythm (circadian oscillators) or other regulators are equipped by promoters containing the conserved core 5′-CACGTG- 3′ or 5′-CACATG-3′ to which the transcriptional activators of clock (and of other) genes bind. Myc, Max, Mad, Mxi, and related basic helix-loop-helix/leucine zipper proteins (TFE-3, TFE-B) preferentially bind to E boxes flanked by 5′ C and 3′ G but do not bind if the flanks are 5′ T or 3′ A. E-box motifs may reside at the exon-intron 1 junction and regulate transcription and splicing. *See* bHLH; circadian rhythm; clock genes; DNA-binding protein domains; endogenous rhythm; helix-loop-helix; HLH; LCR; MAD/MAX; MXI/MAX; Myc; TFE. (Comijn, J., et al. 2001. *Mol. Cell* 7:1267; Kim, C.-H., et al. 2001. *J. Biol. Chem.* 276:24797.)

EBP   Enhancer-binding protein. *See* Ct/EBP; enhancer.

4E-BP1 (PHAS-1) Binding protein of eIF-4E. After it is phosphorylated, it stimulates the activity of eIF-4F cap-binding protein, required for the beginning of translation. It appears that phosphorylation of 4E-BP1 releases eIF-4E from the inhibition by 4E-BP1. *See* eIF-4E; eIF-4F; translation initiation. (Heesom, K. J., et al. 2001. *Curr. Biol.* 11:1374.)

EBV *See* Epstein-Barr virus.

EC (1) Embryonal carcinoma cells. *See* carcinoma. (2) Enzyme classification number. *See* enzyme.

EC₅₀ Endpoint concentration 50. Concentration of a chemical that causes half of the individuals to reach an endpoint, e.g., anesthesia.

ECAF Endothelial attachment factor.

ecdysone Steroid molting hormone of insects produced by the prothoracic gland. It apparently stimulates transcription and activates puffing at different locations in the salivary gland chromosomes of *Sciara, Drosophila,* and other dipteran flies. This hormone and variants apparently regulate development in crayfish, arthropods, schisostomes, and nematodes. Similar compounds (β-ecdysones) occur in plants. Ecdysone does not play a natural role in mammals, yet the ecdysone receptor transgene controls ecdysone response. In *Drosophila,* hydroxyecdysone is produced after 6 hours before entering the prepupal stage and again 10 hours after pupariation. The production of the waves of these steroids is accompanied by distinct morphogenetic developments. The ecdysone, EcR (ecdysone receptor), and USP (*ultraspiracle* 1-[05]) complex (homologous to human retinoid receptor) then turns on various genes. Some of the feedback of the gene products (E75, DHR3) inhibits the complex and at the same time regulates (through mid-prepupal protein, βFTZ1) the onset of the second, smaller wave of ecdysone synthesis by these proteins as well as another set. The complete ecdysone response element (EcRE) is palindromic; some other response elements are homologous, although display variations of the basic sequence. Muristerone

```
AGGTCANTGACCT
TCCAGTNACTGGA
```

is an agonist of ecdysone; it occurs in animal and some plant tissues. *See* animal hormones; brassinosteroids; puffs; retinoic acid; salivary gland chromosomes; steroid hormones. (Segraves, W. *Hormones and Growth Factors in Development and Neoplasia,* Dickson, R. B. & Solomon, D. S., eds. p. 45. 1998. Wiley-Liss, New York; Koolman, J., ed. 1989. *Ecdysone.* Georg Thieme, Stuttgart, Germany; Arbeitman, M. N. & Hogness, D. 2000. *Cell* 101:67; Takeuchi, H., et al. 2001. *J. Biol. Chem.* 276:26819.)

ECE Extrachromosomal element, such as plasmid. *See* plasmid.

eceriferum loci In various plant species, they determine the cuticular waxes or lack of them. In barley, a large number of loci in different chromosomes have been analyzed genetically and biochemically. There is an amazing specific wax pattern associated with the mutations correlated with the biosynthetic relations among β-diketones, hydroxy-β-diketones, alkan-1-ol and alkan-2-ol esters. *See* fatty acids. (Wettstein-Knowles, V. P. 1975. *Molec. Gene Genet.* 144:43.)

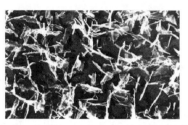

The scanning electron micrograph is by courtesy of Dr. Penny von Wettstein-Knowles.

ECGF Endothelial growth factor.

echocardiography Ultrasonic examination of the structure of the fetal heart as part of a repertory of prenatal diagnosis of congenital defects in anatomy. *See* prenatal diagnosis.

eclampsia/preeclampsia Hypertension, edema, proteinuria, kidney failure in the latter part of pregnancy, with a prevalence of 2–5% caused by etiological and genetic factors presumably at chromosome 2p. The heterodimer of angiotensin-1 receptor for angiotensin II and the B₂ receptor for bradykinin increases the susceptibility to preeclampsia 4–5-fold. *See* angiotensin; kininogen. (Moses, E. K., et al. 2000. *Am. J. Hum. Genet.* 67:1581; AbdAlla, S., et al. 2001. *Nature Med.* 7:1003.)

eclipse The (latent) period between viral infection of a bacterium and the burst of the new phage particles even if burst is induced. *See* burst; induction of a lysogenic bacterium; one-step growth. (Anderson, T. F. & Doerman, A. H. 1952. *J. Gen. Physiol.* 35:657; Abedon, S. T., et al. 2001. *Appl. Environ. Microbiol.* 67:4233.)

eclosion Hatching of the adult form (imago) of insects from the puparium. *See* pupa.

ECM *See* extracellular matrix.

EcoCyc *See* E. coli.

ecodeme Population adapted to a particular ecological condition.

ecogenetics (ecological genetics) Studies the genetically determined responses of organisms to the environment(s). *See* adaptation; adaptive radiation. (Hoffmann, A. A., et al. 1995. *Annu. Rev. Genet.* 29:349.)

E. coli (*Escherichia coli*, Colon bacillus; Enterobacteriaceae) Most predominant bacterium in the intestinal flora of mammals. In year 2000, ~28% of beef cattle in the United States were infected by *E. coli* (Elder, R. O., et al. 2000. *Proc. Natl. Acad. Sci. USA* 97:2999). Normally it is not pathogenic, but some strains (0157) may cause Winckel's disease (a possibly fatal jaundice of newborns), diarrhea, intestinal bleeding, and urinary infections. The size of single cells is about 1000 × 2400 nm, and their mass is about 2 pg. Their genome is 4,639,221 bp, containing 4,288 protein-coding genes, but about one-third have no known function so far; 2,357 of the genes are on the replicational leading strand and 1,929 are on the lagging strand. Transport and binding-protein genes (281) represent the largest group (6.55%).

For DNA replication, recombination, modification, and repair, 115 genes (2.68%) are allocated. Transcription, RNA synthesis, and modification require 55 genes, whereas translation and posttranslational modification involve 182 genes (4.38%). Phage, transposons, and plasmids occupy 2.03% of the genome. There are various groups of repeated sequences of different lengths. They can harbor a variety of plasmids, and can be lysogens. At 37°C, their generation time is about 20 min. *E. coli* is a genetically and biochemically thoroughly studied organism. By 1997, the genome of the K12 nonpathogenic laboratory strain had been completely sequenced almost independently in the United States and Japan. The enterohemorrhagic strain 0157:H7 had also been sequenced by 2001. The sequence revealed the existence of 1,397 new genes of different functions organized in specific clusters. *See* bacterial recombination frequency: conjugation; conjugation mapping; databases; recombination molecular mechanisms, prokaryotes. (<http://cgsc.biology.yale.edu>; <ftp://ftp.pasteur.fr/pub/GenomeDB>; <http://ecocyc.org/>; <http://gib.genes.nig.ac.jp/>; Blattner, F. R., et al. 1997. *Science* 277:1453; Riley, M. & Serres, M. H. 2000. *Annu. Rev. Microbiol.* 54:341; Perna, N. T. 2001. *Nature* 409:529; Schaechter, M., et al. 2001. *Microbiol. Mol. Biol. Rev.* 65:119; Fumoto, M., et al. 2002. *Nucleic Acids Res.* 30:66, see color plate in Color figures section in center of book.)

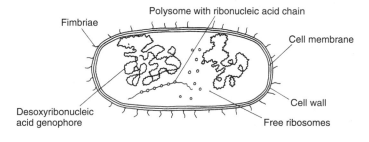

ecological race Distinctly adapted group of an organism without sexual isolation from the ancestral form. *See* sexual isolation.

ecology Study of the relation of living systems and the environment. Example: the increase of CO_2 concentration in the atmosphere and geosphere due to an increase of burning fossil fuels may increase the number of mitochondria in plants and alter the thylakoid structure in chloroplasts. The temperature change that has been claimed may alter the habitats of all organisms and affect biodiversity. *See* biodiversity; ecogenetics. (<www.natureserveexplorer. org>.)

ecomorph Species or group particularly adapted to a microhabitat (microenvironment.)

ecores Orthologous coding sequences. *See* orthologous.

EcoRI Type II restriction endonuclease with primary recognition site

$$5' \text{ pG} \downarrow \text{pÅpApTpC}$$

$$3'_{OH} \text{CpTpTpÅpAp} \uparrow_{OH} \text{G}$$

Å indicates potential base to be methylated and arrows indicate sites of cut; the staggered cuts have a receding OH and a protruding p end $----$ OH

$$---------- \text{ p}$$

Although the above 6-nucleotide sequence is the primary recognition site, alternate recognition sites with different preferences are also known. These secondary activities (named EcoRI*) depend on the composition of the incubation mixture. *See* restriction enzymes.

EcoSeq Database of a collection of DNA sequences of *E. coli* obtained from various sources.

ecosystem Relation of living organisms to each other and to all environmental factors. *See* species, extant.

ecotropic retrovirus Replicates only in cells of the species from which it was originally isolated. This specificity is determined by envelope glycoproteins that require specific receptors in the host. *See* amphotropic.

ecotype Population adapted to a particular ecological condition. Besides the main genes determining the adaptive trait(s), the population may not be genetically homogeneous; usually it is not an isogenic line. *See* adaptation; isogenic stocks.

ectoderm Surface layer of the embryo that develops into the epidermis, skin, nerves, hair, nails, ears, eyes, enamel of teeth, internal mouth, and anal tissues. *See* gastrulation.

ectodermal dysplasia (EDA) Several *autosomal-dominant forms* involve complex skin alterations and may eliminate the dermal ridges and alter palmar prints (*See* fingerprints). In ectodermal anhidrotic dysplasia, the patients do not sweat and have cleft lip and palate; in the hypohidrotic form, the sweating ability is only reduced, whereas in the hidrotic form there is normal sweating. The trichoodontochial form is accompanied by deficiency of tooth, hair, breast, and nipple formation. *EEC* (ectrodactyly [missing fingers]) *ectodermal dysplasia* has most of the symptoms mentioned above plus abnormal tear ducts. *Autosomal-recessive* ectodermal dysplasias also include complex features such as sweat gland tumors (accrine tumors and ectodermal dysplasia), anhidrotic type (due to mutation in connexin-30), dysplasia (HED or Coulson syndrome, 13q12) with neurosensory deafness, dysplasia with cleft lip and palate, mental retardation, syndactyly, hypohidrosis-hypothyroidism, and hypohidrosis-hypothyroidism-lung disease. The human chromosome 2q11-q13 locus has both dominant and recessive mutations with hypohydrotic ectodermal dysplasia. *X-linked recessive* (Xq11-q21.1) ectodermal dysplasia is anhidrotic, shows reduction in hair and tooth development, hyperpigmentation around the eyes, short stature, etc. The Xq13.1 locus involves hypohidrotic ectodermal dysplasia (HED) and incontinentia pigmenti, and the mutation is due to IKK-gamma (NEMO). Hypohydrotic ectodermal dysplasia with immunodeficiency is defective in switching from IgM to other classes of immunoglobulins. The latter gene is normally required for the activation of NF-κB, and its defect results in immunodeficiency. The X-linked ectodysplasin-A2 receptor (XEDAR) is different from the EDA-A1 receptor by the insertion of two amino acids. These receptors belong to the tumor necrosis receptor (TNFR) superfamily. *See* anhidrosis; antibody gene switching; cleft palate; connexin; cyst; dermatoglyphics; ectoderm; EEC syndrome; fingerprints; IKK; incontinentia pigmenti; NEMO; NF-κB; polydactyly; skin diseases; TNFR. (Döffinger, R., et al. 2001. *Nature Genet.* 27:277; Jain, A., et al. 2001. *Nature Immun.* 2:223.)

ectopic antibody Antibody molecule introduced exogenously into the cell. It may be genetically modified and specifically targeted. It may function as an effector or suppressor.

ectopic expression The organ-specific expression of a gene is altered by fusing the structural gene to a promoter with different or no organ specificity of expression. In more general meaning, the displaced condition of an organ or function. *See* transformation, allotopic expression; orthotopic; promoter.

ectopic integration Insertion of transforming DNA takes place at a target site different from that of the original position of the transforming DNA. *See* orthotopic; transformation.

ectopic pairing Association between sites in nonhomologous chromosomes due to intercalary heterochromatin. *See* chromosome pairing; heterochromatin. (Aragon-Alcaide, L. & Strunnikov, A. V. 2000. *Nature Cell Biol.* 2:812.)

ectopic pregnancy Abnormality of ~1.2–1.4% incidence when the fertilized ovum develops outside the uterus. Risk factors are pelvic inflammation, *Chlamydia* infection, smoking, tubal surgery, endometrosis, etc., but a role for chromosomal aberrations could not be verified. (Coste, J., et al. 2000. *Fertility & Sterility* 74:1259.)

ectopic recombination Recombination between nonhomologous chromosomal sequences or between a normal chromosomal site and an insert at a nonhomologous site. Recombination is between dispersed repetitive DNA such as ribosomal RNA genes, multicopy transposable elements (e.g., Alu), multigene families, transgenes, centromeric and telomeric DNA elements. Ectopic recombination may result in chromosomal rearrangements, aneuploidy, and the formation of nucleotide repeats causing various human diseases. *See* Alu;

homologous recombination; illegitimate recombination; intrachromosomal recombination; transgene; trinucleotide repeats. (Goldman, A. S. & Lichten, M. 2000. *Proc. Natl. Acad. Sci. USA* 97:9537; Goebel, P., et al. 2001. *J. Exp. Med.* 194:645.)

ectoplast Cytoplasmic membrane surrounding the protoplast.

ectrodactyly Split foot and hand, absence of fingers/polydactyly. It is most likely under autosomal-recessive control. Cleft lip and palate, skin, teeth, hair, nail defects, and closure (atresia) of the lacrimal ducts may accompany it. *See* adactyly; polydactyly; split hand/split foot malformation.

eczema Itching inflammation of the skin; it may be scaly or oozing and it may be caused by various factors even as simple as drying of the skin in winter, food allergy, infections, or Kaposi sarcoma, etc. *See* acquired immunodeficiency syndrome; allergy; dermatitis; Kaposi sarcoma; skin diseases; Wiskott-Aldrich syndrome.

EDG Endothelial differentiation gene. When overexpressed, it enhances cell aggregation and increases the expression of cadherins, depending on the RHO protein and activation by sphingosine-1-phosphate (SSP). *See* cadherins; sphingosine. (Marletta, M. A. 2001. *Trends Biochem. Sci.* 26:519; Robert, P., et al. 2001. *J. Mol. Cell Cardiol.* 33:1589.)

edible vaccine *See* plant vaccines.

editing Some DNA polymerases, aminoacyl tRNA synthetases, and aptamers also have nuclease functions and eliminate replicational mistakes or prevent translational errors. *See* antimutator; aptamer; DNA polymerases; editosome; error in aminoacylation; mtDNA; RNA editing; string edit distance.

editosome Complex determining the selection of RNA bases for editing. The determination may involve a protein or RNA. In the mitochondria of *Arabidopsis*, 8% of codons are edited (5,285 codons) and 16 have been edited twice. The specificity of selection of the correct cytosine residue may require only 2 to 8 nucleotides. *See* RNA editing. (Giegé, P. & Brennicke, A. 1999. *Proc. Natl. Acad. Sci. USA* 96:15324.)

Edman degradation Procedure for protein sequencing. The reagent phenylisothiocyanate causes the formation of phenylthiohydantoin and cleavage of the terminal residue. This amino acid can then be identified. The same process can be repeated many times and the amino acid sequences of the entire macromolecule can be determined step by step. The commercially available automatic protein sequenators use the same principle in an efficient way. (See Edman, P. & Begg, G. 1967. *Eur. J. Biochem.* 1:80.)

EDRF Endothelium-derived relaxing factor is nitric oxide. *See* nitric oxide.

EDTA Ethylenediaminetetraaceticacid is a chelating agent and as such an inhibitor of DNase; it is an anticoagulant and is also used for plant nutrient media for improving the solubility of iron, etc. *See* DNA extraction; DNA fingerprinting; embryogenesis, somatic; versenes.

Edward King of England, one of the three sons of Queen Victoria who did not inherit the hemophilia gene. *See* hemophilias.

Edward's syndrome Trisomy for human chromosome 18 with serious debilitating consequences; or prenatal or postnatal death within a few months or may survive up to a few years. Generally the head is elongated, ears are set low, eyelids are abnormal, clenched fingers, hypoplasia of nails, and almost all organs are affected. Its incidence is about 1 in 7,500 to 10,000 births, with a predominance in females. The disomic chromosome is maternally contributed in 95% of the cases, and 95% of the incidences fail to survive until birth. Increased maternal age and meiosis II errors are factors in the incidence. *See* hypoplasia; trisomic analysis; trisomy.

EEA-1 Early endosomal autoantigen 1 binds phosphoinositide-3-phosphate (PtdIns[3]-P) and Rab5 to the endosomes and mediates endosome fusion with the cooperation of phosphatidylinositol-3-OH kinase. *See* endocytosis; phosphoinositides; Rab.

EEC syndrome ectrodactyly, ectodermal dysplasia, cleft lip/palate) Autosomal (3q27)-dominant abnormality. The mutation involves the gene encoding p63, a homologue of p53, and it maps to a location identical with that of the Limb mammary syndrome. Components of EEC syndrome are parts of the defects attributed to mutations at split-hand/foot malformation syndromes 7q21.3-q22.1 (SHFM1), Xq26 (SHFM2), and 10q24-q25 (SHFM3). These human anomalies occur at an approximate frequency of 1.8×10^{-5}. *See* cleft lip; cleft palate; ectodermal dysplasia; ectrodactyly; limb defects in humans; Pallister-Hall syndrome; polysyndactyly; syndactyly.

eEF Eukaryotic elongation factors of peptide chain. Some of them recruit the aminoacyl-tRNA to the ribosome; others control the translocation of the ribosome on the mRNA. *See* Spt. (Negrutskii, B. S. & El'skaya, A. V. 1998. *Progr. Nucleic Acid Res. Mol. Biol.* 60:47.)

eEF-1A (eEF-1α) Eukaryotic translation factor, binds amino acid-tRNA, also a GTPase, and regulates cytoskeletal (microtubule) rearrangements. It occurs in two isoforms encoded by two genes with tissue-specific expression. It is similar to prokaryotic EF1A. *See* EF1A; protein synthesis. (Hotokezaka, Y., et al. 2002. *J. Biol. Chem.* 277:18545.)

eEF-1β (eEF1B or eEF-1$\beta\gamma\delta$) Mediates GTP-GDP exchange on eEF-1α in eukaryotic translation. It is similar to prokaryotic EF1B. *See* EF1B; protein synthesis.

eEF-1γ Mediates GTP exchange with the aid of eEF-1β in translation. *See* protein synthesis.

eEF-2 Eukaryotic translation factor; stimulates peptide chain translocation on ribosomes. When phosphorylated, it may slow the elongation rate. eEF-2 is also a GTPase. It is similar to prokaryotic EF-2. *See* EF-2; GTPase; protein synthesis; translation.

eEF3 Unique fungal translation factor without exact homologues in other taxonomic categories.

EEG (electroencephalogram) Record of the electric current developed in the brain, measured after electrodes were applied to the scalp, to the surface of the brain, or into the brain material. It reveals the functional state of the central nervous system. The EEG pattern depends on a number of factors but also has a hereditary component indicating psychological responses of the family.

EF-2 (EF-G) ~77 kDa translation elongation factor in prokaryotes; it catalyzes translocation on ribosomes. *See* protein synthesis.

E2F1 Family of transcription factors activated by RB and MDM2 oncogenes. They are involved in the regulation of the cell cycle, apoptosis, neoplasia, etc. E2F interacts with cyclinA/Cdk2 and is phosphorylated by the latter, thereby downregulating its DNA binding and transcription activation. E2F also regulates p53. E2F1 also regulates p73 protein and apoptosis by T-cell receptor activation–induced cell death even in the absence of p53. *See* ARF; cancer; CDC14; CDF; cell cycle; DP1; EMA; histone deacetylase; histone methyltransferases; MDM2; p53; p73; pocket; retinoblastoma. (Takahashi, Y., et al. 2000. *Genes & Development* 14:804; Wu, L., et al. 2001. *Nature* 414:457; Ogawa, H., et al. 2002. *Science* 296:1132.)

EF1A (EF-Tu) 44 kDa prokaryotic GTP-binding peptide chain elongation factor. It mediates the aminoacylated tRNA binding to the ribosomes. (LaRiviere, F. J., et al. 2001. *Science* 294:165.)

EF-1α Translation initiation factor in eukaryotes. *See* protein synthesis.

EF1B (Ef-Ts) Prokaryotic translation factor.

e1F-2B Translation factor involved in GTP-GDP exchange. *See* GTP; protein synthesis.

effective dose of radiation (E) Where W_T = the weighting factor, which varies according to the absorbing tissue, and H_T = equivalent dose to tissue. *See* radiation hazard; stochastic detriment.

$$E = \sum_T W_t H_T$$

effective mutagen Causes all types of mutations (including chromosomal changes) at high frequency. *See* efficient mutagen.

effective number of alleles Number of alleles maintained in the population. *See* effective population size.

effective number of loci Number of loci involved in a quantitative trait, also called segregation index. *See* gene number in quantitative traits.

effective population size (N_e) Number of individuals that leave offspring in a population. Although from several viewpoints (economic, agricultural, ecological, demographic, insurance, welfare) the total number of individuals may be the most important, geneticists are concerned primarily with that fraction of the population that passes on genes to future generations. The *genetically effective size of the population* is represented as (N_e). If the effective population size is small (even if the mating is random), the allelic and genotypic frequencies may be biased. In cases where unequal numbers represent the two sexes, the effective population size will be lower than the sum of the males and females. Each individual has 0.5 chance to contribute a particular allele to the offspring (through the egg and sperm) and the chance of contributing two of the same alleles is $0.5 \times 0.5 = 0.25$. The probability that the same female contributes two alleles is $(1/N_f)(1/4)$, and for a male it is $(1/N_m)(1/4)$, where N_f and N_m are the number of females and males, respectively. Therefore, the probability that any two alleles of the population come from the same individual is

$$\frac{1}{4N_m} + \frac{1}{4N_f} = \frac{1}{N_e};$$

hence,

$$N_e = \frac{4N_m N_f}{N_m + N_f}.$$

Fluctuations in population size from generation to generation as well as the nonrandom distribution of family size may affect the allelic sampling.

The sampling variance of the alleles is equal to

$$\sigma^2 = \frac{q(1-q)}{2N}$$

and hence

$$\sigma = \sqrt{\frac{q(1-q)}{2N}},$$

where q is the frequency of one of the alleles and N is the population size. If the frequency of the a allele is 0.5 and $N = 25$, the standard deviation of the frequency of the a allele becomes

$$\sigma = \sqrt{\frac{0.5 \times 0.5}{2 \times 25}} = \sqrt{0.005} \approx 0.071.$$

This indicates that chance alone can modify the frequency of the a allele in a small population. And 31.74% of the loci (because according to the normal distribution 68.26% of the population is supposed to be within $M \pm 1\sigma$ may carry the allele with frequencies more extreme than 0.429 to 0.571 rather than 0.5 ($0.50 \pm$ standard error). Thus, if the population is broken up into smaller breeding units, 31.74% may have gene frequencies in the range of 0.429 to 0.571. In cases of polygeny, N_e is reduced as determined by the formula: $4 \text{ MF}/(M + F)$, where M and F are the number of males and females, respectively. N_e for autosomal and Y populations is dependent on the ratio of mating men and mating women (R), and $N_e \approx (2 + 2R)/R$.

The effective number of breeding individuals can also be estimated from the number of heterozygotes, in excess of expectation, on the basis of Hardy-Weinberg equilibrium, because when the number of breeders is small, by chance, the allele frequencies will differ in the males and females. This procedure is applicable only to polygamous or polygynous populations with few breeders and many loci with multiple alleles (Luikart, G. & Cornuet, J.-M. 1999. *Genetics* 151:1211).

Any shift in the gene frequencies brought about by the random fluctuation in gametic sampling is called *random genetic drift*. The drift is completely accidental. Such events may take place in the population in any breeding season if the number of breeding individuals is reduced. Once such a sampling bias of gametes (in mating) has taken place, the process may continue in either direction, but there is a definite chance that the allelic composition of the small population will permanently change. A special case of random drift is the *founder effect*. A small number of individuals (immigrants) introduced into new habitats may not accurately represent the genetic constitution of the population of their origin. Their descendants then may form the basis of a divergent trend from the norm of their ancestors. Such a phenomenon is not uncommon if animal or plant species migrate to new parts of the world or to regions that are spatially isolated from their old homeland. Organelle DNAs (mitochondrial and chloroplast) have smaller genetically effective population sizes than the nuclear ones because they are inherited uniparentally and are effectively haploid (Birky, C. W., Jr., et al. 1989. *Genetics* 121:613). Also, their genetic diversity is more limited and evolves faster. *See* Amish; F_{ST}; isolation, genetic; normal distribution; polygamy; polygeny; speciation; standard error. (Anderson, E. C., et al. 2000. *Genetics* 156:2109.)

effector Small molecule that assists either in activating or deactivating a molecular event. Interaction of the RAS protein with effectors uses 32–40 amino acids, the effector region, varying in conformation in GTP- and GDP-bound RAS. *See* GTP; RAS; signal transduction.

effector cell Activated B cell or T_H cell that mediates the adaptive immune system. The activation results in an increase in antigen-specific killer cells. After an episodic infection, the effector CD8 T cells may migrate to nonlymphoid tissues and become long-lived memory cells. *See* B cell; CCR; CD34; immune system; killer cell; memory, immunological. (Gazitt, Y. 2000. *Stem Cells* 18:390.)

effector domain Of antibody, contains the region that specifically recognizes the cognate antigen. *See* antibody; antigen.

efficient mutagen Produces primarily point mutations and a relatively low amount of chromosomal alterations. *See* effective mutagen; point mutation.

efflux systems Carries out energy-requiring active pumping of harmful agents from the cells.

EF-G Translation factor and a motor protein that is also a GTPase. It mediates GTP-dependent transfer of peptidyl-tRNA-mRNA complex from the ribosomal A site to P site. Its domain IV has some structural similarity to the EF-Tu elongation factor. *See* aminoacylation; EF-Tu; protein synthesis; ribosome.

EF-Ts Prokaryotic translation factor protein involved in GTP-GDP exchange. *See* protein synthesis.

EF-TU·GTP Active prokaryotic elongation factor of protein synthesis in which EF-Tu interacts with aminoacyl-tRNA and promotes the translocation of peptidyl-tRNA from the ribosomal A to B site and the release from the ribosome of deacylated tRNA. *See* aminoacylation; protein synthesis; tRNA. (Frederick, J., et al. 2001. *Science* 294:165.)

E.GDP Bound (inactive) form of guanosine diphosphate in signal transduction. *See* E region of GTP-binding proteins.

EGF Epidermal growth factor; binds to receptors such as the protein coded for by protooncogene ERBB, triggering growth-promoting signals. The *v-erbB* viral oncogene encodes a truncated EGF that continuously binds to a ligand and provides a constitutive supply of growth signals. EGF controls a wide range of cellular processes. *See* growth factors; oncogenes; TGF. (Daub, H., et al. 1997. *EMBO J.* 16:7032; Pierce, K. L., et al. 2001. *J. Biol. Chem.* 276:23155.)

EGFR Epidermal growth factor receptor has been localized in the nucleus of different cells. It has been found in association with the promoter of cyclin D1, indicating that it is a transcription factor. *See* ERBB1; LDL receptor. (Bogdan, S. & Klämbt, C. 2001. *Curr. Biol.* 11:R292; Lin, S.-Y., et al. 2001. *Nature Cell Biol.* 3:802.)

EGG (ovum) Final haploid product of female meiosis. The eggs are huge cells (ca. 100 μm in diameter in humans) compared to the other ones in the body (about 20 μm). The egg cytoplasm contains yolk, a very condensed store of nutrients, especially in organisms that lay outside the eggs. In mammals, the yolk is comparatively minimal. (A generalized composite drawing is below.) In mammals the zona pellucida membrane surrounds the egg. In lower vertebrates there is a *vitelline layer* and other layers. In birds there is the *egg white*, and before laying the egg, the *shell* is added in the oviduct. The vitelline layer of insects is covered by the so-called *chorion*, secreted by the *follicle cells* of the *ovary*. In the layer under the plasma membrane (*cortex*) are the *cortical granules*. Their content protects the egg from fertilization by more than one sperm. During fertilization the sperm is attracted to the egg by chemotactic peptides, causing changes in the voltage of the membrane and altering the concentration of cAMP, cGMP, Ca^{2+}, and K^+ ion channels. *See* fertilization; gametogenesis; menopause; oocyte; ovary; polyspermic fertilization; RPTK; sperm.

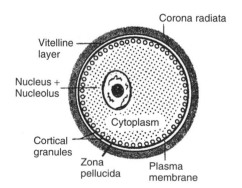

egg cylinder Very early mammalian embryonal structure including the cells that will form the fetus and the embryonal sacs (amnion, allantois), some extra embryonic tissue that forms the outer embryo sac (chorion), and the trophoblastic tissues of the placenta.

egg number in humans Each of the two human ovaries contains more than 200,000 primary oocytes yet only about 400 eggs develop to maturity during the period from puberty to menopause. The size of a human egg is comparable to a period in this print. *See* egg; menopause; oogenesis; oocyte; puberty.

eggplant (*Solanum melongena*) Tropical-subtropical vegetable with $2n = 2x = 24$ chromosomes. (Daganlar, S., et al. 2002. *Genetics* 161:1697.)

Egl-1 (Eglin) Proteinase inhibitor, it controls proapoptotic caspase-9 and other proteins *See* caspases. (Komiyama, T. & Fuller, R. S. 2000. *Biochemistry* 39:15156.)

Egr-1 Mitogen-induced transcription factor, a serum-inducible nuclear phosphoprotein with zinc finger. Egr2 is a Zn-finger protein regulating cell proliferation. Egr-3 (NGFI-C) is essential for muscle spindle development. *See* Krox; NGFI-A; platelet-derived growth factor; TIS8; Zinc finger. (Calogero, A., et al. 2001. *Clin. Cancer Res.* 7:2788.)

E.GTP Bound (active) form of guanosine triphosphate in signal transduction. *See* E region of GTP-binding proteins.

Ehlers-Danlos syndrome (EDS) Extremely complex disorder, usually dominant, involving loose joints that stretch excessively, excessive stretchability of the skin that is bruised easily, and frequently pseudotumors that appear after trauma. The anomalies are caused either by a deficiency of procollagen peptidase (recessive EDS type VII) or the lack of a collagen type. EDS I is deficient in COL5A2 and 5A1 encoded at 2q31 and 9q34.2, respectively.

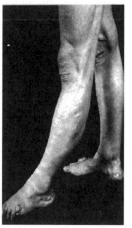

Ehlers-Danlos symptoms: Genu recurvatum (hyperextension of the knee or back knee), deformed toes, tumors under the ankle and on the front part of the foot. (From Bergsma, D., ed. 1973. *Birth Defects*. Atlas and Compendium. By permission of the March of Dimes Foundation.)

EDS II, the most common form of the disease, is also deficient in COL5A1 and COL5A2 encoded at 9q34.2-q34.3. COL1A1 is encoded at 17q21.3-q22. Low hydroxylysine content of the connective tissue prevents effective cross-linking. EDS type VII (dermatosparaxis), recessive/dominant at 5q23, is a deficiency of procollagen protease and lysyl oxidase (LOX).

Type VIA (encoded at 1p36.3–36.2) is deficient in lysyl hydroxylase, whereas in type VIB the hydroxylysine level is normal. Lysyl oxidase level is reduced in the individuals afflicted by EDS types VI, VII, and IX. Type IV (2q31, dominant) suffers from mutation in type III procollagen and increases the risk of aneurysm. Type V disease has a defect in the α-chain of collagen mapped to human chromosome 9q34, but type EDS V is X-linked and is caused by lysyl oxidase deficiency. EDS VIII is involved in periodontal disease.

EDS has many different variations from mild to severe and its incidence may be estimated to be in the $1-2 \times 10^{-4}$ range. Type X (2q34) involves platelet abnormality and is due to a deficiency of fibronectin. *See* aneurysm; cardiovascular disease; collagen, skin diseases; connective tissue disorders; fibronectin; Menkes syndrome; periodontal disease. (Schwarze, U., et al. 2000. *Am. J. Hum. Genet.* 66:1757; Wenstrup, R. J., et al. 2000. *Am. J. Hum. Genet.* 66:1766.)

Ehrlich ascites   *See* ascites.

EI   Ethyleneimine (C_2H_5N), a powerful alkylating mutagen and carcinogen. *See* alkylating agent.

eicosahedron   20-faceted body. *See* DNA tumor viruses.

Eicosahedron.

eicosanoid   Mammalian autocrine signaling molecule affecting muscle contraction, platelet aggregation, pain, and inflammation. Eicosanoids are produced at the expense of phospholipase-degraded phospholipids. Eicosanoid is the systematic name of arachidonic acid. *See* autocrine; fatty acids; leukotrienes; phospholipase; platelet; prostaglandins; signal transduction. (McMahon, B., et al. 2001. *Trends Pharmacol. Sci.* 22[8]:391.)

eIF   Eukaryotic initiation factor of protein synthesis. *See* IF. (Karim, M. M., et al. 2001. *J. Biol. Chem.* 276:20750.)

eIF-1   12.7 kDa (human) and 12.3 kDa (yeast) translation factor stimulating the 40S ribosomal subunit preinitiation complex. It interacts primarily with eIF-3. *See* protein synthesis; ribosome scanning.

eIF-1A (eiF-4C)   Ribosome dissociation factor (17–22 kDa) and 40S preinitiation complex-stimulating protein in translation. It shows ~20% homology with prokaryotic IF1. (Petroulakis, E. & Wang, E. 2002. *J. Biol. Chem.* 277:18718.)

eIF-2   Heterotrimeric (α,β,γ) translation elongation factor mediating GTP-dependent Met-tRNA binding to the 40S ribosomal subunit. It may repress protein synthesis if its α-subunit is phosphorylated. *See* aminoacylation; GTP; protein synthesis; ribosome. (Kimball, S. R. 1999. *Int. J. Biochem. Cell Biol.* 31:25.)

eIF-2A (eIF2α)   Heterotrimeric translation elongation factor controlling AUG-dependent Met-tRNA•GTP binding to the 40S ribosomal subunit. It is inactivated by phosphorylation through the hemin-regulated inhibitor kinase (HRI, PKR, PEK/PERK). GCN2 is activated by amino acid starvation. GCN4 activates the transcription of more than 30 amino-acid-synthesizing enzymes. *See* aminoacyl-tRNA synthetase; GCN4; HRI; PEK; protein synthesis.

eIF-2B   Heteropentameric guanine nucleotide exchange factor (GEF) involved in translation by controlling phosphorylation of eIF2. Its defect may lead to leukoencephalopathy with vanishing white matter. (Leegwater, P. A. J., et al. 2001. *Nature Genet.* 29:383.)

eIF-2C (Co-eIF-2A)   Translation initiation factor stabilizing ternary complexes (eIF2•GTP•tRNA) of translation. *See* protein synthesis; ternary.

eIF-3   ~600 kDa, 11-subunit elongation initiation factor activated by phosphorylation in translation. It stimulates the formation of the 40S ribosomal preinitiation complex. *See* protein synthesis; ribosome.

eIF-3A   Translation factor affecting ribosome dissociation by binding to the 60S ribosomal unit.

eIF-4   Eukaryotic translation (mRNA cap-binding proteins, eIF4E) factor (46 kDa), RNA-dependent ATP-ase, helicase (eIF4A), stimulates mRNA binding to ribosome. Several isoforms exist. Phosphorylation reduces its affinity for capped mRNA. *See* ATPase; DEAD-box proteins; FLAG; helicase; isoform; protein synthesis. (Gingras, A.-C., et al. 1999. *Annu. Rev. Biochem.* 68:913; Scheper, G. C., et al. 2002. *J. Biol. Chem.* 277:3303.)

eIF-4B   Eukaryotic translation initiation factor (mRNA cap-binding protein, 69 kDa in humans); elongation initiation of translation upon phosphorylation, a helicase, stimulates mRNA binding to ribosomes. *See* 4E-BP; helicase; mRNA; protein synthesis, ribosome.

eIF-4C   Function is similar to eIF-1A and it appears to be synonymous with it.

eIF-4D   Probably the same as eIF-5A.

eIF-4E   Translation factor involved in binding of the mRNA cap to the 40S ribosomal subunit. It may exist in different forms. *See* cap, 4E-BP; protein synthesis; ribosome.

eIF-4F   Heterotrimeric (including eIF4E, eIF4G, and eIF4A) key element of translation initiation; recognizes mRNA cap. It is a helicase elongation initiation factor, and when phosphorylated, it may lead to overexpression and cancerous growth. Viral genes using IRES may not absolutely depend on eIF-4F-cap complex. *See* cap; cancer; *CDC 33*; helicase; IRES; Pab1p; protein synthesis; ribosome; translation initiation.

eIF-4G (eIF-4γ)   Translation initiation factor (I, 171 kDa, and II, 176 kDa) for the 40S eukaryotic ribosomes. It functions as an adaptor between the cap-binding proteins eIF-4A, eIF-4E, eIF-3, and may bind to the PolyA tail-binding protein. The latter has been shown to involve translation stimulation. *See* cap; FLAG; IRES; Pab1p; protein synthesis. (Gallie, D. R. & Browning, K. S. 2001. *J. Biol. Chem.* 276:36951; Niedzwiecka, A., et al. 2002. *J. Mol. Biol.* 319:615.)

eIF-4H (25 kDa, monomeric)   Its stimulatory function resembles that of eIF-4B.

eIF-5   48.9 kDa (human) and 45.2 kDa (yeast) translation factors promoting joining of the small and large ribosomal subunits into the 80S ribosome. It is a GTPase activator. *See* protein synthesis; ribosome.

eIF-5A   Assists in the formation of the first peptide bond during translation. *See* peptide bond; protein synthesis.

eIF5B   Homologue of prokaryotic IF2. It interacts with eIF-2, promotes the hydrolysis of GTP, and generates the 80S initiation complex.

eIF-6   Similar to eIF-3A. The 25 kDA protein dissociates the 80S ribosome; facilitates protein synthesis by maintenance of the 60S subunit.

eIF-D   Similar in function to eIF-5A.

eigenvalue   Literally, this hybrid (German-English) word means proper value, and it is usually used (in physics) in the sense of a characteristic value.

Eincorn   *See Triticum* A genome.

Eincorn (*Triticum monococcum*). Courtesy of Drs. G. Kimber & M. Feldman.

ejaculate   *See* sperm.

EK2   Laboratory strain of *E. coli* defective in the synthesis of thymine and diaminopimelate. *See* diaminopimelate; *E. coli*; thymine.

EKLF   Erythroid Krüppel-like factor is a Zn-finger homologue of the *Drosophila Krüppel* gene involved in blood β-globin and blood nonglobin gene transcription regulation. *See* globins; Krüppel; LCR; morphogenesis in *Drosophila*; zinc fingers. (Huber, T. L., et al. 2001. *Curr. Biol.* 11:1456.)

elastase   Proteinase released by the lysosomes of blood granulocytes upon inflammation. It breaks down collagen if not inhibited by protease inhibitors. *See* lysosome; neutropenia.

elastic fiber disease   *See* ABC transporters; Costello syndrome; gangliosidoses; Hurler syndrome; Marfan syndrome; Menkes syndrome; pseudoxanthoma elasticum; supravalvular

aortic stenosis; Williams syndrome. (Urbán, Z. & Boyd, C. D. 2000. *Am. J. Hum. Genet.* 67:4.)

elastin ($M_r \cong 70$ K)   Rubber-like, cross-linked, glycine- and proline-rich, fibrous protein present primarily in the blood vessels near the heart but also in the ligaments; modest amounts in the skin and tendons. It controls arterial development and proliferation of smooth muscles. Its defect may cause aortic stenosis. *See* coarctation of the aorta; supravalvular aortic stenosis; Williams syndrome.

ELAV   Embryonic lethal abnormal vision system is a member of the evolutionarily conserved RNA-binding protein family. *See* AU-rich elements.

ELC   Expression linked copy. *See Trypanosoma*.

electroblotting   Transfer of macromolecules in an electric field from a (polyacrylamide) gel onto a membrane.

electrocardiography   Records the changes of the variations in the electric potentials of the heart muscles. The first deflections, denoted by P, are due to the excitation of the atria. The QRS indicates the depolarization phase of the excitation of the ventricles. The T wave indicates the complete repolarization of the ventricles. These intervals are measured in fractions of a second. The long Q-T (LQT) interval (bottom graph) may indicate abnormal function of either the potassium or sodium ion channel(s) or some other anomaly. The upper graph represents the normal QT. *See* HERG; Jervell and Lange-Nielsen syndrome; Ward-Romano syndrome.

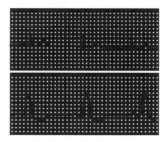

electrochemical gradient   Moves ions through biomembranes depending on their concentration and charge at the two sides.

electroelution, DNA   Separate DNA by electrophoresis in agarose gel containing 0.5 μg/mL ethidium bromide; locate the band by long-wavelength UV light, cut out the band, and place it into a dialysis tube filled with 1 × TAE buffer. Let the slice sink to the bottom of the tube, seal, and turn on the current in 1 TAE buffer (4 to 5 V/cm) for 2 to 3 hours. Remove the tube from the apparatus and collect the buffer content of the tube containing the eluted DNA. *See* electrophoresis; electrophoresis buffers.

electroencephalogram   *See* EEG.

electrolyte   Substance capable of dissociating into ions, then becoming a conductor of electricity. *See* ion.

electromagnetic radiation   The genetically effective range includes UV light, X-rays, and gamma rays. Under conditions of sensitization by various chromophores, it may extend into the visible range. Some of the radiations may have contaminating components with potential genetic effects. Extremely low-frequency electromagnetic fields (EMF) such as those generated by power lines and household appliances have also been suspected of stimulating gene activity (MYC) and increasing cancer risk. The experimental data are, however, insufficient to definitely rule in or out such an effect. *See* illumination; light intensities; MYC; radiation hazard assessment; radiation protection; diagram below.

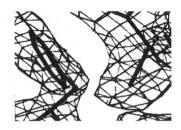

electronic PCR   STSs can be amplified and defined by a pair of PCR primers and used with the aid of an appropriate computer program to locate the STS to existing genetic or physical markers in a map without performing any experiment. *See* EST; PCR; STS. (Schuler, G. D. 1997. *Genome Res.* 7:541.)

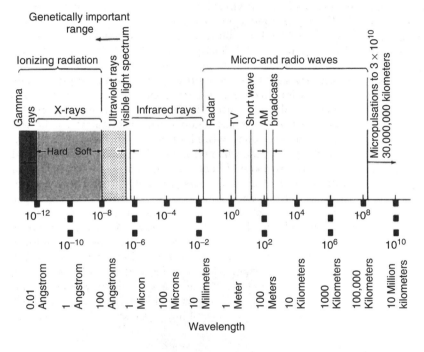

Spectrum of electromagnetic radiations. (After Lerner & Libby, 1976 Heredity, Evolution and Society, Freeman, San Francisco.)

electromorph   Genetically determined variation that can be revealed by electrophoresis. *See* electrophoretic polymorphism.

electron   Small negatively ([4.80294 ± 0.00008] × 10^{-10} absolute electrostatic units) charged particle; its mass is 1/1837 of the H nucleus and its diameter is 10^{-12} cm. All atoms contain one nucleus and one or more electrons. Cathode rays and beta rays are electrons. *See* beta particles; cathode rays; excitation.

electron acceptor   Takes up electrons and therefore becomes reduced, whereas the *electron donors* provide electrons to other molecules and become oxidized. *See* electrons.

electron density map of proteins   Determined by X-ray analysis; the results are interpreted by the Fourier method. The image represents the electron scattering by critical atoms in the molecule and reveals important structural properties. *See* FT-IR; protein structure; X-ray diffraction analysis. (Vagin, A. A. & Isupov, M. N. 2001. *Acta Crystallogr. D. Biol. Crystallogr.* 57 [pt 10]:1451.)

electron microscopy   In the light microscope the resolution of the objects is limited by the wavelength (within the 350 to 750 nm range) of the light used for the illumination. The wavelength of electrons accelerated at high voltage (100 KV) may become only 0.004 nm. Because of aberrations of the lenses available, this minimum cannot be attained, yet in the best equipment a resolution of about 0.1–2 nm, i.e., at about 100 to 1,000 higher than with the light microscope, may be possible. In the *transmission electron microscope*, instead of visible light the extremely thin (50–200 nm) specimen, specially fixed and stained, is placed on a grid and exposed to electron beams accelerated in a vacuum tube. In the areas where the specimens reduce the focused electron flux, an image appears that can be viewed or photographed and enlarged by procedures similar to light microscopy. Electron microscopic specimens are usually prefixed in buffered 3% glutaraldehyde and postfixed in 1% buffered osmium tetroxide, then dehydrated and embedded in a propylene oxide-resin mixture that requires polymerization before sectioning by glass or diamond microtomes. The sections are placed on a circular copper grid of 3 mm diameter covered by a carbon or plastic film. Either before or after sectioning, the biological specimen is usually treated with uranium or lead to assure contrast.

Electron microscopic specimens may be exposed to substrates of enzymes in order to localize an electron-dense precipitate if the enzyme is active. The specimens may be coupled with the specific fluorescent dyes or with colloidal gold, which may be recognized by an antibody, and thus its site is revealed. The material may also be *shadowed* — by spraying in an angle with other thin layers of metals (platinum, chromium) — for obtaining an apparent three-dimensional image. Macromolecules may be studied by *negative contrast*. The solution in 1% phosphotungstate (or uranyl acetate) is thinly spread on a carbon film and when dried an electron-dense layer is formed. At the position of the protein molecule, there is no tungstate. Thus, a negative image is generated. A special technique in electron microscopy is *freeze fracture*, which can provide images of the internal organization of delicate structures such as biological double membranes. This freezing technique does not require fixation, which usually denatures the material. The specimen is frozen in liquid nitrogen in the presence of a cryoprotectant (glycerol, dimethylsulfoxide) to prevent the formation of ice crystals. The frozen material is then fractured with a knife to crack the double membrane between the two layers. The generated surface is shadowed with a metal, and after the organic material is dissolved and removed, the remaining metal replica is examined. This replica is a negative image of the biological structure. A similar procedure is *freeze etching*. Again, the frozen sample is cracked, the moisture is removed by lyophilization, and the inner and outer exposed structures are shadowed. *See* freeze drying; microscopy; scanning electron microscopy. (Maunsbach, A. B. & Afzelius, B. A. 1998. *Biomedical Electron microscopy: Illustrated Methods and Interpretations*. Academic Press, San Diego, CA.)

electron paramagnetic resonance (EPR) Can be used to measure the radiation exposure in human or mammalian populations. When paramagnetic substances are placed in a stationary magnetic field and exposed to electromagnetic radiation, they register the field of strength and the frequency of the radiation. Paramagnetic substances contain molecules or atoms whose electrons move and produce weak magnetic fields. This method can then be used for the determination of the concentration of radiation-induced radicals in, e.g., hydroxyapatite, present in teeth and bones. The tooth enamel is 97% hydroxyapatite and does not metabolize ^{90}Sr, a common product of atomic fallout. Dentine, which is made up of only 70% hydroxyapatite, metabolizes this isotope. Therefore, ^{90}Sr will be found mainly in the dentine. The isotope emits short-range β-rays (0.6–1 mm path). One study showed that in humans with chronic radiation disease, the enamel received 1 Gy (gray unit), whereas the dentine showed 5.5 Gy γ-radiation. Thus, it is possible to determine whether the radiation came internally or externally. *See* fallout; radiation hazard assessment; radiation measurement. (Swenberg, C. E., et al. 1994. *Adv. Space Res.* 14[10]:181.)

electron transport Movement of electrons toward the lower level of energy by electron carriers.

electron volt (eV) 1 eV = 1.60207 ± 0.00007 × 10^{-12} erg. *See* erg.

electrophile Eager to accept electrons and can covalently bind nucleophiles. *See* electron; nucleophile; nucleophilic attack.

electrophoresis Separating charged molecules (such as polypeptides or polynucleotides) between the two poles of an electric field. The material is generally contained in a support medium (cellulose, starch, agarose polyacrylamide), bathed in an appropriate buffer, and may contain specific stains. The basis of the separation of the components of a mixture is the relative charge, shape, or molecular size of the components. Larger nucleic acid molecules are generally separated in agarose gels of various concentrations, depending on the size of the molecules. Smaller nucleic acid fragments are separated in polyacrylamide gels. SDS-PAGE electrophoresis treats the proteins with the detergent sodium dodecyl sulfate (SDS) and separates them by polyacrylamide gel electrophoresis (PAGE). The detergent breaks the noncovalent bonds of the proteins, and when β-mercaptoethanol is added, the disulfide bonds are eliminated. The treated proteins migrate in the gel according to their molecular weight rather than their charge. Nucleic acids have a uniform charge; thus, their separation is based on their molecular size. Ultrathin gel electrophoresis permits the use of higher voltage without deleterious heat effects because of increased heat transfer. Also, it increases the speed of separation by an order of magnitude, although it may decrease resolution and read length, and it is under experimental study. Another innovative approach is the matrix-assisted laser desorption and ionization method (MALDI), which permits single-charged ions from proteins as large as 300 kDa to be produced and analyzed. Mass spectrophotometry MALDI (MALDI-MS) is capable of discriminating length difference in short oligonucleotides due to a single base. *See* capillary electrophoresis; dodecyl sulfate sodium salt; gel electrophoresis; isoelectric focusing; pulsed-field electrophoresis; two-dimensional gel electrophoresis. (Rickwood, D. & Hames, B. D. 1990. *Gel Electrophoresis of Nucleic Acids: A Practical Approach*. IRL Press, New York; Hames, B. D. & Rickwood, D. 1990. *Gel Electrophoresis of Proteins: A Practical Approach*. IRL Press, New York.)

electrophoresis buffer TAE (Tris-acetate EDTA), TPE (Tris-phosphate EDTA), TBE (Tris-borate EDTA), and alkaline 50 mM NaOH-1 mM EDTA are commonly used.

electrophoretic karyotyping Small chromosomes are separated by pulsed-field gel electrophoresis according to size. *See* pulsed field. (Geiser, D. M., et al. 1996. *Current Genet.* 29:293; Shin, J. H., et al. 2001. *J. Clin. Microbiol.* 39:1258.)

electrophoretic polymorphism Protein or DNA variations in a population affecting charge or size. In proteins, it may be detectable on the bases of amino acid (glutamic and aspartic acid, lysine, histidine, arginine) substitutions by altered electrophoretic mobility or by size of polypeptides. Restriction enzyme digests of DNA provide such information after gel electrophoresis or by PCR analysis. *See* electrophoresis; isozyme; PCR; RFLP; SDS-polyacrylamide gels. (van der Bank, H., et al. 2001. *Biochem. Syst. Ecol.* 29(5)469; Zhu, X., et al. 2001. *Electrophoresis* 22:1930.)

electroporation Transport of DNA across cellular membrane with the aid of electric current pulses; the genes may display transient expression in the target cells. It is also a means of genetic transformation. *See* microfusion; transformation. (Neumann, E., et al. 1982. *EMBO J.* 1:841; Fromm, M. E., et al. 1986. *Nature* 319:791; Tsong, T. Y. 1991. *Biophys. J.* 60:297; Golzio, M., et al. 2002. *Proc. Natl. Acad. Sci. USA* 99:1292.)

electrospray MS (tandem mass spectrometry) Procedure for determining the molecular weight of unknown proteins and for the study of noncovalent interactions between proteins and other large molecules, thus facilitating their location and identification with the aid of databases. The nanoelectrospray is a low-flow device equipped with a needle of ~1 μm internal diameter, delivering a mixture of molecules, e.g., peptides, into a mass spectrometer. The peptides are separated in the first step and subsequently are fragmented (in this tandem process). The fragmentation in the mass spectrometer is achieved by collision with gas molecules. The fragments obtained from the NH_2 end of the protein are *b* ions and those from the COOH terminus are *y* ions. Another procedure involves the initial separation of the peptides by liquid chromatography, then sequencing as they are passed into the electrospray ion source. The fragments are sorted by the molecular mass of one amino acid. The location of the amino acid is then revealed in the peptide. Two amino acids with a known location (called a *peptide sequence tag*) suffice to fish out the peptide in large sequence databases. The extreme sensitivity of the procedures for 50 to 100 ng or even smaller amounts of proteins (femtomole range) may be analyzed. The method determines the mass-to-charge (m/z) of the polar molecules and the signal intensity is displayed by the algorithm used in discrete peaks. Some of the procedures are automated. *See* chromatography; genome projects; laser; laser desorption mass spectrum; MALDI; mass spectrum; proteome; STM. (Null, A. P., et al. 2001. *Anal. Chem.* 73:4518.)

element, chromosomal Conserved sequences (genes) within chromosomes.

elephant *Alphas maxis* 2n = 56; *Laxodonta africana* 2n = 56. (Roca, A. L., et al. 2001. *Science* 293:1473.)

elephant man Person with abnormally enlarged body parts. The cause of Joseph Merrick's anomaly has been attributed to neurofibromatosis and to Proteus syndrome, but the real cause is unclear. It is different from elephantiasis, a tropical infection of the lymphatic nodes, limbs, genitalia, etc., causing inflammation and hypertrophy as the result of the obstruction of the lymphatics by several species of nematodes. *See* neurofibromatosis; Proteus syndrome.

elicitor *See* host-pathogen relation.

ELISA Enzyme-linked immunosorbent assay. Capable of detecting *v*g quantities of protein per gram of tissue by the use of an antibody attached to a particular enzyme such as alkaline phosphatase (or peroxidase, urease). The enzyme + antibody complex binds to the protein antigen present in the reaction vessel and the cleavage of a chromogenic substrate (e.g., p-nitrophenyl phosphate) by the enzyme results in the development of color (absorbance detected by spectrophotometer) proportional to the amount of the antigen protein. *See* antibody; immune reaction; immunoglobulins; immunolabeling; immunoprecipitation; IRMA. (Davis, L. G., et al. 1986. *Basic Methods in Molecular Biology*. Elsevier, New York.)

ELK (oncogene) ELK1 (human chromosome Xp11.2) and ELK2 (14q32.3) are expressed in human lungs and testes. These oncogenes show homology with the ETS oncogenes. The Elk proteins bind to the SRE element of the chromosomes. After phosphorylation, they may activate transcription. *See* ETS; oncogenes; SAP-1; signal transduction; SRE.

ELK-1 Protein structure (top) complexed with DNA (below). (Courtesy of Marmorstein, R. From Mo, Y., et al. 2000. *Nature Struct. Biol.* 7:282.)

ELL Human RNA polymerase II elongation factors in chromosome 19–13.1. ELL1 is ~620 and ELL2 is ~640 amino acid size. They increase the rate of transcription after initiation has taken place because they suppress pausing of the polymerase along the DNA. Functionally they are similar to ELONGIN, another transcription elongation factor controlled by the von Hippel–Lindau tumor suppressor gene and the MLL gene products of leukemias (encoded at 11q23) by binding to the A subunit of ELL. The ELL proteins may also impede initiation by hindering the interaction of polymerase II with TBP. The ELLs have a conserved region homologous to the zonula occludens protein sequences that bind MAGUK. *See* elongation factors; elongin; leukemias; MAGUK; MLL; PITSLRE; TBP; TEFb; transcript elongation; von Hippel-Lindau syndrome; zonula occludens. (Luo, R. T., et al. 2001. *Mol. Cell Biol.* 21:5678; Eissenberg, J. C., et al. 2002. *Proc. Natl. Acad. Sci. USA* 99:9894.)

elliptocytosis (ovalocytosis) The shape of the erythrocytes (red blood cells) is not round as normal but elliptic, indicating an often fatal autosomal-dominant hemolytic anemia. In the *Camelidae*, elliptocytosis is a normal condition. Mutations in spectrin (SPTA1, 1q21) frequently lead to defects in the formation of the $\alpha\beta \sim \alpha\beta$ tetrameric spectrin molecules. This may be the basic cause, and a variety of this condition has been described. Some forms may be correlated with protection against malaria. The α-spectrin gene was assigned to human chromosome 1q21 (Rhesus-unlinked-type elliptocytosis). The Rhesus-linked type was assigned to 1p34-p33. An atypical elliptosis is apparently recessive. *See* anemia; ankyrin; poikilocytosis; spectrin; spherocytosis. (Wandersee, N. J., et al. 2001. *Blood* 97:543.)

Elliptic red blood cells.

Ellis-Van Creveld syndrome Human chromosome 4p16 recessive. It involves dwarfism, polydactyly, short extremities, heart malformation, dystrophy of fingernails, partial hare lip, teething by birth, heart defects, etc. It is quite common in some Amish populations. *See* Amish; dystrophy; face/heart defects; hare lip; McKusick-Kaufman syndrome; polydactyly; Weyers acrodental dysostosis.

ELMS (expected maximum lod score) Computed by weighting the sum of lod scores by their probability. *See* lod score.

ELOD *See* interval mapping; lod score.

elongation factor (EF) Proteins assisting translation on the ribosomes; *see* individual factors (eIF) separately listed. Factors eIF2, eIF3, eIF4A, eIF4B, eIF4F, and ATP are required for formation of *elongation initiation complex I*, which is unstable. The addition of eIF1 and eIF1A to complex I results in the formation of stable *elongation initiation complex II*. The biosynthesis of many proteins in response to intracellular attenuation environmental factors (e.g., heat shock) is regulated at the level of transcription. The prokaryotic proteins carry out similar functions to the eukaryotic ones, but they are structurally different. *See* attenuation; eIFs; ELL; promoter clearance; protein synthesis; rrn; TIIFS; transcript elongation; transcription factors; transcription shortening; transcription termination; translation termination. (Sonenberg, N., et al. eds. 2000. *Translational Control of Gene Expression*, Cold Spring Harbor Lab. Press, Cold Spring Harbor, NY.)

elongator (ELP) Protein associated with the hyperphosphorylated carboxyl domain of RNA polymerase II. Elongator activity is correlated with histone acetyltransferases. The complex replaces the mediator complex, which is associated with the unphosphorylated pol II during the initiation phase of transcription. *See* histone acetyltransferases; mediator complex; transcript elongation. (Li, Y., et al. 2001. *J. Biol. Chem.* 276:29628; Hawkes, N. A., et al. 2002. *J. Biol. Chem.* 277:3047.)

elongin Three-subunit (A, B, C; ~770, 118, and 112 amino acids, respectively) protein stimulatory to in vitro RNA transcript elongation. The A subunit has the major role, and B and C reinforce A. The elongin–von Hippel-Lindau–cullin complex may play a ubiquitin ligase role and target genes for degradation by the proteasome. *See* cullin; ELL; elongation factors; proteasome; transcript elongation; von Hippel–Lindau syndrome. (Luo, R. T., et al. 2001. *Mol. Cell. Biol.* 21:5678.)

ELSI (ethical, legal, social implications of genetics) *See* bioethics; copyright; ethics; genetic discrimination; genetic screening; patent; risk.

eluate Flows off a chromatographic column.

elutriation Sedimentation or separation of particles from suspensions. *Centrifugal elutriation* employs a special centrifuge in which a flow is generated in the direction opposite to the rotation. The latter can handle large volumes fast. *See* centrifuge.

EMA (E2F-binding site modulating activity) 2.4 kb DNA translated into a 272-amino-acid protein. EMA binds to transcription factors DP-1 and DP-2. It better recognizes the TCCCGCC rather than the TCGCGCC core sequences of the E2F-binding sites. EMA is a transcriptional suppressor. *See* E2F1. (Koziczak, M., et al. 2000. *Mol. Cell Biol.* 20:2014.)

EM algorithm Principle that can be applied to the calculation of recombination frequencies in special cases.

EMAPII Endothelial monocyte-activating polypeptide stimulates leukocytes, monocytes, chemotaxis, myeloperoxidase, tissue factor, and TNFα synthesis. EMPAII domains are evolutionarily conserved and their homologues occur at the C-terminal domain of aminoacyl-tRNA synthetases for one or more amino acids (eubacteria and *Caenorhabditis*: Met; yeast: Met, Gln; mammals: Tyr, Glu, Asp, Met, Pro, Glu, Arg, Lys); in yeast and mammals, other homologies may also be found. Tyr-tRNA synthetase released from the mitochondria, along with cytochrome c, prepares the cell for the apoptotic process. Secreted Tyr-tRNA synthetase in mammals undergoes proteolysis and is converted into an N-terminal catalytic fragment with IL-8 properties and into C-terminal cytokine-like fragments. EMAPII also migrates to the sites of inflammation. *See* aminoacyl-tRNA synthetase; cytokine; IL-8. (Shalak, V., et al. 2001. *J. Biol. Chem.* 276:23796.)

emasculation *See* castration.

EMB agar *See* β-galactosidase.

Embden-Meyerhof pathway Anaerobic process of glycolysis. The six-carbon sugars are converted into glucose-6-phosphate and through a number of steps to glyceraldehyde-3-phosphate and eventually to pyruvate. An oxidation step generates NADH, and pyruvate is further oxidized to form the acetyl units of acetyl coenzyme A, which may be oxidized in the citric acid cycle, becoming the core reactions of both carbohydrate and fat metabolism.

embedded gene Gene within the boundary of another. *See* φX174.

embedding Microscopic specimens requiring thin sections for examination may be surrounded (infiltrated) by paraffin wax, resins, or other suitable material before sectioning by microtomes. *See* microtomes; sectioning.

EMBL (http://www.ebi.ac.uk/embl/index.html) Stores data bases for macromolecular sequences; European Molecular Biology Data Library, P.O.B. 10.2209, D-6900 Heidelberg, Germany, telephone: 011-49-6221-387-258, EMBL Wellcome Trust Genome Campus, Hinxton, Cambridge CB10 1SD, UK, telephone: +44 1223 494444, email for general inquiries: <datalib@ebi.ac.uk>, for submissions and forms: <datasubs@ebi.ac.uk>, <http://www.ebi.ac.uk/embl/>, updates: <http://bio.ifom-firc.it/docs/EMBL/update.txt>, Users Manual: <http://www.ebi.ac.uk/embl/Documentation/>.

EMBL3 Lambda DNA vector carrying polylinkers at both sides of the *red* and *gam* sites; if equipped with adjacent promoters, genes cloned into the "stuffer" region can be directly transcribed by RNA polymerase without subcloning. *See* GeneScribe; λDASH; λFIX; lambda phage, stuffer region.

EMBRYO Differentiated zygote, in mammals up to about a one-fifth of the normal time of gestation. In plants, the embryo stage is considered to last up to the maturing and germination of the seed. *See* embryo culture; embryogenesis; fetus. (<http://visembryo.ucsf.edu>.)

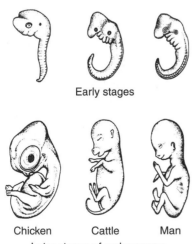

Early stages

Chicken   Cattle   Man
Later stages of embryogeny

Embryogenesis in vertebrates. (This widely accepted model of Ernst Haeckel has been seriously challenged.)

embryo culture When developing plant embryos are lifted from the ovaries about 10–14 days after fertilization, they can survive and mature in aseptic cultures. This procedure may be used for the rescue of hybrid embryos that may not survive in their natural environment because of the collapse of the endosperm, the nurturing tissue, or the overgrowth of the nucellus. Although this technique itself does not overcome hybrid sterility, viable seedlings can be produced that may become fertile after doubling the chromosome number (with colchicine). The embryos (ear, fruit) require disinfection by calcium hypochlorite (Ca[OCl]$_2$ 5%, for 8 to 10 min and several rinses in sterile H$_2$O). E2 culture medium mg/L H$_2$O, final concentration: NH$_4$NO$_3$ 400, KNO$_3$ 200, MgSO$_4$.7H$_2$O 100, CaH$_4$(PO$_4$)$_2$ 100, KH$_2$PO$_4$ 100, K$_2$HPO$_4$ 50. Before use, add 2.5 mL/L chelated iron (FeSO$_4$.7H$_2$O 556 mg or diethylenetriamine pentaacetic acid 786.4 mg, H$_2$O 100 mL). Supplement it with 3% sucrose, add agar to solidify without making it too hard (about 0.6%), and autoclave. Embryonic stem cells of mammals may also be cultured successfully, although it is more difficult to maintain totipotency of these cells and viable embryos. *See* embryo; in vitro; nuclear transplantation; ovary; tetraploid embryo complementation. (Rappaport, J. 1954. *Bot. Rev.* 20:201; Nagy, A., et al. 1993. *Proc. Natl. Acad. Sci. USA* 90:8424.)

embryogenesis, plant Takes place after the egg has been fertilized by the sperm within the embryo sac. The embryo sac is located within the ovule. The pattern of embryogenesis of higher plants is shown next page. The steps are genetically controlled and numerous mutations have been identified that are responsible for the different morphogenic steps.

There is less information regarding the molecular mechanisms involved in plants than those connected with animals. The seed coat is entirely maternal tissue; therefore, the inheritance of seed-coat traits is delayed. The endosperm develops from the fertilized fused polar cells and is generally triploid. In the majority of dicots, by the time the seed matures, the endosperm tissue is generally insignificant in amount or entirely absorbed. In the majority of monocots, the endosperm constitutes the bulk of the seed. In common language, the monocot fruit is also called a seed, but its proper designation is kernel. The outermost layer of the kernel is the maternal pericarp. Just under the pericarp is the membrane-like seed coat. The outer layer of the endosperm is called the aleurone. The scutellum of the monocots corresponds to the cotyledons of dicots.

Whereas the cotyledons emerge from the seed during germination, the scutellum does not. The germline of plants is not set aside during development as it is animals, yet the cells of the plant meristems form cell lineages that can be traced if visible genetic differences occur during ontogenesis. The size of the formed sectors permits a time estimate for when the genetic alteration has taken place. Large sectors indicate an early event; small sectors show late mutations. *See* development; *Drosophila*; embryo culture; endosperm; gametogenesis; gametophyte; megagametophyte; microgametophyte; morphogenesis; phyllotaxy. (Chaudhury, A. M., et al. 2001. *Annu. Rev. Cell Dev. Biol.* 17:677.)

embryogenesis, somatic Plant somatic embryos can be obtained by the techniques of tissue culture. The illustration shows embryogenesis in suspension culture and compares it with *in planta* embryogenesis. In most species it is not necessary to go to protoplasts, and embryos may develop from other tissue cells on regeneration media. Several nutrient media have been developed in various laboratories (*see* Gamborg 5; Murashige and Skoog). For tobacco and *Arabidopsis*, the following protocol is quite effective. *R4 Medium:* **I**. Major mineral salts, mg/500 mL H$_2$O: NH$_4$NO$_3$ 1,800, KNO$_3$ 800, MgSO$_4$.7H$_2$O 100, CaH$_4$(PO$_4$)$_2$ 100, KH$_2$PO$_4$ 100, K$_2$HPO$_4$ 50. **II**. CaCl$_2$ 15g/100 mL. **III**. KI 75 mg/100 mL.

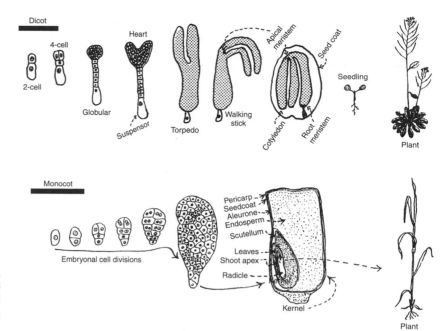

The pattern of embryogenesis in plants is different in dicots from that in monocots. Additional variations exist within these two groups. The dicot pattern is exemplified by the crucifer *Arabidopsis* and the monocot pattern is generalized on the basis of maize and wheat.

IV. Microelements mg/500 mL: H_3BO_3 6,200, $MnSO_4 \cdot H_2O$ 16,900, $ZnSO_4 \cdot 7H_2O$ 8,600, $NaMoO_4 \cdot 2 H_2O$, $CuSO_4 \cdot 5 H_2O$ 25, $CoSO_4 \cdot 7H_2O$ 29.54. **V**. Chelated iron: in 100 mL H_2O $FeSO_4 \cdot 7H_2O$ 556 mg, diethylenetriamine pentaacetic acid 786.4 mg. **VI**. Vitamins, mg/50 mL: myo-inositol 5,000, thiamin 500, nicotinic acid amide 50, pyridoxine.HCl 50. Final solution: **I**. 50 mL; **II**. 290 µL; **III**. 100 µL; **IV**. 50 µL; **V**. 500 µL; **VI**. 100 µL, pH 5.6, sucrose 3 g, fill up to 100 mL, agar about 600 mg (varies according to batch), or Gellan gum 180 mg (must be separately dissolved in distilled water on hot plate with magnetic stirrer). The culture generally requires five stages with hormones added in µg/mL: R4-1 (callus initiation) 9RiP 1.5, 2,4-D 0.025–0.050, R4-2 (callus growth and regeneration) 9Rip 2.0, NAA 0.1, R4-3 (leafy callus) 9RiP

1.5, NAA 0.1, R4-4 (shoots appear) BAP 1, NAA 0.1, R4-5 (rooting) BAP 0.0005, NAA 0.05. After R4-4, *Arabidopsis* can be transferred to E2 medium in vitro (*see* embryo culture), where they produce seeds even in the absence of roots. Alternatively, after R4-5, the seedlings can be transferred to soil or commercial soil substitute. (RiP: isopentenyl adenosine; 2,4-D: dichlorophenoxy acetic acid; NAA: α-naphthalene acetic acid; BAP: benzylamino purine). Heat-stable components are autoclaved; vitamins and hormones are filtered. *See* ART; cell genetics; embryo culture; tissue culture.

embryoid Develops through somatic embryogenesis. *See* embryogenesis, somatic.

embryoid body For example, endoderm, ectoderm, neurons, yolk, cartilage, and muscle cells formed from embryonic stem cells. *See* ectoderm; endoderm; stem cell.

embryology Study of embryogenesis from fertilization until birth (hatching) of an animal or the maturation of seeds of plants by genetic, morphological, and molecular techniques *See* cell lineages; embryogenesis; fate map.

embryonic induction Interaction between and among tissues leading to tissue differentiation in the embryo. Vg-1, activin, bone morphogenetic factor, fibroblast growth factor, etc., have been isolated from various sources and found to be effective in regulating embryonic differentiation in animals by binding to receptors, being restricted in diffusion through the extracellular matrix, affecting signal transduction pathways, etc. *See* embryogenesis; morphogenesis.

embryonic polarity Directional determination of the cells in the embryo along longitudinal, vertical, or other directions. The anterior/posterior axis of development is influenced by the sperm-derived centrosomes and the subsequent development of the microtubules.

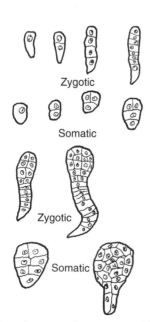

Zygotic and somatic embryogenesis at comparable stages in *Daucus carota*. (Redrawn after Street, H. E. & Withers, L. A. 1974. *Tissue Culture and Plant Science*. Academic Press, London, UK.)

embryonic stem cell (ES) Cells of the early embryo that are capable of continuous growth and differentiation of animals. They are comparable to the meristems of plants. They represent cells that have not undergone differentiation into somatic cell lineages. These cells can be maintained in vitro in this stage for prolonged periods. They can be reintroduced into preimplantation embryos and thus chimeric animals may be generated. They are suitable for transfection by transgenes, can be used to generate in vitro mutations, and are amenable to a variety of modifications such as gene trapping. *See* gene trap vectors; meristem; stem cells; trapping promoters.

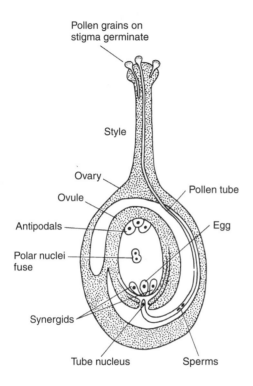

Pollen grains on stigma germinate

Style

Ovary

Ovule

Antipodals

Polar nuclei fuse

Synergids

Pollen tube

Egg

Tube nucleus

Sperms

embryo rescue *See* embryo culture.

embryo research (human) NIH recommendations: embryos (14) 18 days or older are not acceptable. Unacceptable procedures: transfer human embryos to animals for gestation, transfer research embryos or parthenotes (unfertilized eggs) to humans, separation of blastomeres for generating twins, cloning by nuclear transplantation, creation of any type of chimera (human-human, human-animal), creation of embryos in the lab from stem cells, cross-fertilization by human gametes with the exception of clinical laboratory sperm penetration tests with animal eggs, embryo transfer to cavities other than the uterus, sex selection with the exception to prevent sex-chromosome-linked disease, use of sperm or eggs without consent, use of sperm or egg from donors who were paid more than reasonably expected. The embryo research policies vary in different countries, and may/may not permit the use of embryos/tissues for studying infertility, contraception, genetic screening, gene therapy, cloning, construction of chimeras, interspecific implantation, sex selection, etc. *See* stem cells; diagram below.

embryo sac Of plants, develops after meiosis from one of the (generally basal) megaspores. The role of the three antipodals (at the top) is unclear. The diploid polar cell is fertilized with one of the sperms and gives rise to the triploid zygote and embryo. The synergids (flanking the egg) probably have an early role in nurturing the egg and zygote. The photo below shows clearly only four nuclei (which were in the same plane) in the embryo sac. *See* embryogenesis; gametogenesis; gametophyte; diagram at left.

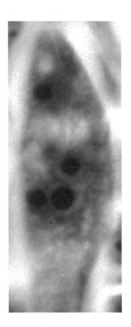

Plant embryo sac.

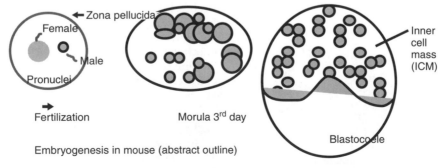

Zona pellucida

Female

Male

Pronuclei

Fertilization

Morula 3rd day

Inner cell mass (ICM)

Blastocoele

Embryogenesis in mouse (abstract outline)

Blastocyst at implantation 4-5th day

EMC (1) *See* enzyme mismatch cleavage; mutation detection. (2) Encephalomyocarditis virus with RNA genetic material.

emerin 254-amino-acid serine-rich protein encoded by the STA gene, responsible for Emery-Dreifuss muscular dystrophy. (Wolff, N., et al. 2001. *FEBS Lett.* 501:171.)

Emery-Dreifuss muscular dystrophy *See* emerin; muscular dystrophy.

EMF (extremely low-frequency electromagnetic fields) *See* electromagnetic radiation.

EMG syndrome *See* Beckwith-Wiedemann syndrome.

EMI1 (early mitotic inhibitor of mitosis) Zinc-binding F-box protein that controls the anaphase-promoting complex (APC) by binding to Cdc20. *See* APC; CDC20; F box; (Reimann, J. D. R., et al. 2001. *Cell* 105:645.)

EMLS Expected maximum lod score is the sum of lod scores weighted by their probability. EMLS for sharing values of Y (= probability of allele sharing) varies according to the size of the population studied. *See* allele sharing. (Risch, N. J. 2000. *Nature* 405:847.)

EMMA European Mouse Mutant Archive, Monterotondo, Italy, maintains mutant mouse/frozen embryos for research. *See* mouse.

Emmer wheat *See* Triticum.

emotion *See* stress.

emphysema, pulmonary Characterized by pathological inflation of the lungs and possible atrophy of the tissues. It appears to be due to autosomal-dominant genes or to environmental causes (smoking). Emphysema may show up in infancy as a complication of other ailments. Emphysema develops in case of deficiency of α-antitrypsin. It can be medicated by a regular supply of antitrypsin. The heme oxygenase (HMXO1, 22q12, inducible; HMXO2, 16p13.3, constitutive) genes may regulate susceptibility to chronic pulmonary emphysema (CPE) with lung antioxidant influence. *See* antitrypsin; serpin.

empirical risk Occurrence or recurrence of certain genetic defects cannot be predicted on the basis of theoretical principles or mathematical formulas because the genetic mechanisms and their control are not known. However, some observations in larger populations indicate a certain tendency of recurrence in families with an afflicted individual in the pedigree. The calculation may take into consideration the empirical experience of onset of a disease with both genetic and age components. Example: 100 families have 40 afflicted children older than 50 years and 170 unaffected who are still in the age group 20 to 50. Experience indicates that usually the age of onset begins at 50 and about half of the offspring of afflicted parents eventually express the condition. The empirical risk for the still healthy individuals then is $(170 \times 0.5)/210 = 0.41$. *See* clinical genetics; genetic risk; inbreeding coefficient; recurrence risk; risk.

EMS Very potent alkylating mutagen and carcinogen. *See* ethylmethane sulfonate.

EMSA Electrophoretic mobility shift assay detects in vitro the formation of protein complexes.

ENA-78 (epithelial cell-derived neutrophil attractant) 78-residue chemokine of epithelial cells, monocytes, neutrophils, fibroblasts, platelets, and some cancer cells. *See* chemokines.

ENaC (epithelial sodium channel) *See* hypoaldosteronism; ion channels.

enantiomorphs Stereoisomers; molecules with structures that are mirror images of each other. In some instances, only one of the enantiomorphs supports normal metabolism. Interestingly, in nature L amino acids and D sugars predominate. *See* chirality; D-amino acids.

En blood group Apparently limited mainly to individuals of Finnish descent; it is very rare. *See* blood groups.

encephalitis Brain inflammation. *See* encephalopathies; Rasmussen encephalitis.

encephalomyelitis Brain and spinal cord inflammation. *See* encephalopathies.

encephalopathy Usually recessive degenerative brain diseases. The childhood recurrent (autosomal-dominant) form involves impaired muscular coordination and speech problems. Gerstmann-Sträussler disease is also an autosomal-dominant encephalopathy with late onset (around age 50) and involves feeble-mindedness. The central nervous system shows myeloid plaques similar to those in Creutzfeldt-Jakob disease. Bovine spongiform encephalopathy (BSE) is responsible for the fatal mad cow disease, which bears similarities to scrapie in sheep. It was believed to be induced by a virus. Current views do not support the viral origin; rather it is attributed to an infectious protein (PrPSc).

In mice that have a null allele for the PrPC gene, i.e., they are PrnP0/0, even large doses of prions do not cause brain damage. There may be an undetermined chance that these encephalopathies are transmitted to humans from animals by meat consumption. Most of the encephalopathies usually have late onset. The incubation period for BSE may be several years. There is now apparently evidence that humans are infected by BSE in the form of a new variant of Creutzfeldt-Jakob disease (Scott, M. R., et al. 1999. *Proc. Natl. Acad. Sci. USA* 96:15137). It is also known that mice can be infected with scrapie. The major outbreak of BSE is attributed to the feeding of infected sheep offal to cattle. Infectivity can be abolished only by heating the animal products at high temperature and by some solvents. There are no sufficient data to confirm whether milk or eggs or poultry meat would transmit any source of infective material. After decisive control measures, BSE cases are on decline since the 1992 peak. An estimated 1 million cattle were infected by

BSE and about 160,000 died of the disease in England. BSE risk to humans is reduced when cattle over 30 months of age are excluded from consumption.

For complete eradication of the encephalopathies threatening humans, prion-resistant animals may have to be produced. It is known that sheep homozygous for Arg at position 171 in the PrPC protein escape scrapie. Kuru, BSE, and some forms of Creutzfeldt-Jakob disease, scrapie-transmissible mink encephalopathy, and feline spongiform encephalopathy are transmitted horizontally, whereas Gerstmann-Sträusler disease and some forms of Creutzfeldt-Jakob are transmitted vertically. Hepatic encephalopathy is an apparently nongenetic neuropsychiatric disorder caused by exposure to ammonia. Certain cyclic tetrapyrroles (deuteroporphyrin IX 2,4-bis[ethylene glycol] iron III, phthalocyanine tetrasulfonate) may inhibit the formation of protease-resistant prions. Prions are not inactivated by nucleases, UV irradiation, psoralen dyes that cross-link DNA, divalent cations, metal chelators with pH between 3 and 7, hydroxylamine, formalin, boiling, and proteases. Human prions are not contagious like microbial or viral infections, but they are transmitted by prion-containing animal residues or by prion transgenes.

Brain tissues (particularly the outermost membranes, the dura mater) are richest in prions, but infection by human blood is controversial, although rodent blood seems to transmit it. Peripheral nerves, the lung, and the liver seem to have low infectivity, and skeletal muscles, milk, blood, bone, hair, and urine may not transmit prions. There is also some risk that humans may develop a variant of Creutzfeldt-Jakob disease (vCJD) by ingesting industrial (e.g., gelatin, polysorbate) or medical (e.g., heparin, insulin) products made of or infected with animal material. There is a risk of transmission by inadequately sterilized surgical instruments and particularly with brain tissue specimens, even by aerosols generated by autopsy or through wounds penetrating the gloves during dissection. Fixation of tissue samples in formalin does not abolish infectivity. Autoclaving for 4.5 h at 132°C, denaturing in 50% phenol, guanidine isothiocyanate, or hydrochloride at >4 M or exposure to 1 N NaOH for 24 h or 2 N NaOH for 2 h inactivates prions. Animal assays for animal prions are generally very slow. Transgenic mice expressing the bovine prion (PrP) may respond faster. The incubation period after infection by the BSE (bovine spongiform encephalitis) agent is much affected by genetic factors of the recipient mouse. Immunoassays or Western blots for PrP^{Sc} (the scrapie protein) are effective in detection of the condition, but their sensitivity is low. Conformation-dependent immunoassay based on the difference between the α-helical PrP^C and the β-sheet component of PrP^{Sc} may be quite efficient. Ovine encephalopathy protein can now be propagated in vitro (Villette, D., et al. 2001. *Proc. Natl. Acad. Sci. USA* 98:4055). *See* chronic wasting disease; Creutzfeldt-Jakob disease; fatal familial insomnia; Gerstmann-Sträussler disease; kuru; neurogastrointestinal encephalomyopathy; porphyrin; prion; scrapie; Seitelberger disease.

enchondromatosis   Cartilage tumor(s) regulated by a parathyroid hormone-related protein and Indian hedgehog signals. *See* hedgehog; parathormone. (Hopyan, S., et al. 2002. *Nature Genet.* 30:306.)

enciphering   Information regarding the nature of a protein, such as a prion, is contained in its tertiary structure rather than in a nucleic acid. *See* prion.

Encyclopedia of the Mouse Genome   Contains genetic and cytogenetic information on mice, electronically obtainable through file transfer protocol (FTP), Gopher, or World Wide Web software (WWW) in Sun (UNIX) or Macintosh versions. *See* databases; mouse.

end   Bacterial genes coding for DNA-specific endonucleases.

endangered species   *See* species, extant.

endemic   Indigenous to a population rather then introduced; confined to a population or area.

endergonic reaction   Consumes energy.

end-joining   Restoration of broken chromosome ends, or fusion of broken chromosomes, e.g., in translocation or another type of breakage. *See* double-strand break; nonhomologous end-joining; translocation, chromosomal. (Smith, J., et al. 2001. *Nucleic Acids Res.* 29:4783.)

end labeling   The Klenow fragment of bacterial DNA polymerase I can add α-P^{32} dNTP to the 3′-OH end of a nucleotide (polynucleotide). Or transfer the γ-phosphate of ATP to the 5′-OH end of DNA or RNA (forward reaction). Or exchange the 5′-P of a DNA (in the presence of excess ADP) with the γ-P of radiolabeled ATP using T4 bacteriophage polynucleotide kinase. This 5′-labeling is used for the Maxam & Gilbert method of DNA sequencing or in any other procedure when 5′ labeling is required. *See* DNA sequencing; Klenow fragment.

endocannabinoid   *See* cannabinoids.

endocardial   fibroelastosis (Barth   syndrome) Human Xq28-linked recessive thickening of the heart wall muscles due to collagen proliferation (type I). In type II methylglutaconic aciduria, besides the defects in the heart muscles, neutropenia (decrease in the number of neutrophilic leukocytes) and morphological anomalies of the mitochondria are detectable by electron microscopy. *See* collagen; heart disease; methylglutaconic aciduria; neutropenia; neutrophil. (Bione, S., et al. 1996. *Nature Genet.* 12:385.)

endocarp   Inner layer of the fruit wall of plants such as the stony pit of peaches or cherries.

endocrine   Hormone-producing glands in response to peptide activators within an organism secrete their product into the bloodstream without relying on special ducts.

endocrine   neoplasia,   multiple (MEN) Autosomal-dominant MEN 1 (11q13) involves endocrine adenomas in several tissues (stomach, lung, parathyroid, pituitary, colon, pancreas, etc.). MEN1 gene has 10 exons and it is translated into a 610-amino-acid protein, *menin*. Dominant MEN2 and MEN2A, also involving pheochromocytoma and amyloid-producing medullary thyroid carcinoma, are apparently in chromosome 10q11.2. The latter is the location of RET protooncogene. MEN2B is similar to MEN2A but frequently shows neural hyperplasias of the mouth area and in the colon. All of these cases involve receptor-like tyrosine kinases. MEN3, also dominant, is located in the vicinity of MEN2A and shares the same symptoms; it also displays neural tumors. The prevalence is $1-10 \times 10^{-5}$. *See* adenoma; menin; pheochromocytoma; protein-tyrosine kinase; RET oncogene.

endocrinology
Study of hormones, other biological secretions, and their physiological roles. *See* animal hormones; endocrine; reverse endocrinology.

endocytosis
Uptake mechanism of cells involving invagination of the cell membrane, then cutting off the (clathrin-coated) vesicle (endosome) formed inside the cell. The cargo receptor transmembrane proteins recognize various molecules by some sort of specificity. An *adaptin* molecule (assembly protein) is attached to one terminus of the cargo proteins, which in turn recognizes clathrins. Then clathrin-coated vesicles are formed. When the clathrin and adaptin molecules separate from the membrane, a membrane-bound vesicle is formed that encloses the cargo molecules. The adaptins are multisubunit adaptor molecules that can recognize the (Tyr, X, Arg, Phe) peptide signals near the carboxyl end of the receptor reaching into the cytosol. The carboxyl end of phosphorylated M6P proteins are also recognized by adaptin molecules in the Golgi apparatus. M6P proteins are named for the mannose-6-phosphate groups linked to the amino ends of lysosomal enzymes.

M6P transmembrane proteins bind lysosomal hydrolases and assist in packaging them into transport vehicles that fuse with endosomes transporting molecules into lysosomes. The endosomes are membrane-enclosed transport vesicles. An endocytotic compartment of the B cells may accumulate class II type MHC molecules and antigens. The unique lipids in the internal membranes of the endosomes sort insulin-like growth factor 2, mannose-6-phosphate ligands, and lysosomal enzymes. These lipids are specific antigens for the antibodies of antiphospholipid syndrome. The endocytotic pathway also has a signaling function. *See* adaptin; antigen processing; antiphospholipid syndrome; ARF; arrestin; clathrin; coatomer; COP; endoplasmic reticulum; Golgi; insulin-like growth factors; lysosome; MHC; receptor-mediated gene transfer; SAR; Sec; triskelion. (Ungewickell, E. 1999. *Proc. Natl. Acd. Sci. USA* 96:8809; Itoh, T., et al. 2001. *Science* 291:1047; D'Hondt, K., et al. 2000. *Annu. Rev. Genet.* 34:255; Katzmann, D. J., et al. 2001. *Cell* 106:145; Di Fiore, P. P. & De Camilli, P. 2001. *Cell* 106:1; Brodsky, F. M., et al. 2001. *Annu. Rev. Cell Dev. Biol.* 17:517; Pelham, H. R. 2002. *Current Opin. Cell Biol.* 14:454.)

endoderm
Internal cell layer of the embryo from which the lung, digestive tract, bladder, and urethra are formed. *See* ectoderm.

endodermis
In plants, it is composed of heavy-walled cells without space among them; they are found around the vascular system, particularly in roots. *See* root.

endoduplication
Basically similar to endomitosis, but the term was first used to denote the diploidization of androgenetic or gynogenetic embryos. *See* androgenesis; c mitosis; endoreduplication; gynogenesis.

3′-end of nucleic acids
OH at the 3′ C atom on the ribose or deoxyribose. *See* nucleic acid chain growth.

5′-end of nucleic acids
The first nucleotide in the chain retains the three phosphates, whereas the subsequent ones form phosphodiester bonds with one phosphate between the 5′ end and the 3′ −OH position of another. *See* nucleic acid chain growth; phosphodiester bond.

endogamy
Mating within a group; a sort of inbreeding. *See* inbreeding.

endogenote
Tract of the recipient bacterial genome that is homologous to the donor DNA paired with it. In this merozygous condition, homo- and heterogenotes can be distinguished, depending on whether the donor sequences are identical or genetically different from those of the recipient. *See* conjugation, bacterial; exogenote; merozygote.

Endocytosis

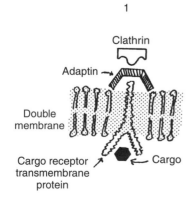
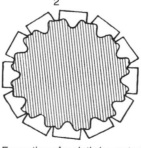
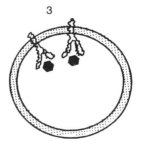

Clathrin

Adaptin →

Double membrane

Cargo receptor transmembrane protein

← Cargo

1

2
Formation of a clathrin-coated vesicle from several individual complexes as shown at 1. The contents are not shown but include the loaded cargo receptors and attached adaptins

3
After clathrins and adaptins are separated and removed, a membrane-coated vesicle is formed containing several cargo receptors and cargos

endogenous rhythm Periodic (oscillatory) changes in cells that occur without external influences. *See* E box.

endogenous virus Integrated into the host chromosome. Endogenous retroviruses may affect the transcription of genes where they are located. The development of superantigens responsible for autoimmune diseases has been attributed to HERV (human endogenous retrovirus) within the HLA gene cluster. The HERV elements retain the long terminal repeats but lack the functional viral envelope gene and usually are not expressed, although they are capable of expression. HERVs may constitute ~1–2% of the human genome. They have been used as phylogenetic markers by comparing the integration site polymorphism and for orthologous comparison of the changes in their nucleotide sequences. *See* HERVd; HLA; LINE; orthologous; provirus; retroposon; retrotransposon; retrovirus; SINE; superantigen; temperate phage. (<http://herv.img.cas.cz>.)

endoglin Homodimeric receptor glycoprotein on the vascular endothelial cells for transforming growth factor β, encoded in human chromosome 9q34.1. *See* telangiectasia, familial hemorrhagic; vascular targeting; VEGF. (Paquet, M. E., et al. 2001. *Hum. Mol. Genet.* 10:1347.)

endolysin Bacteriophage muralytic enzymes for degrading bacterial cell walls. *See* holin.

endometriosis Relatively common (10–15%) cause of female infertility due to a poorly understood physiological anomaly of the ovaries. Comparative genomic hybridization reveals losses in several human chromosomes, particularly in 1, 7, and 22. *See* comparative genomic hybridization; ovary. (Gogusev, J., et al. 1999. *Hum. Genet.* 105:444.)

endomitosis Chromosome replication is not followed by cell division and the result is polyploidy. In *Schizosaccharomyces*, it was found that gene *rum1+* overexpression permits repeated starts of the cell cycle without mitoses, and deletion of the gene allows successive mitoses without an S phase. It appears as if *rum1+* would regulate *Cdc2*, which apparently has two functional forms. Overexpression of *rum1+* would lock *Cdc2* in the form when it promotes the synthetic (S) phase and deletion of *rum1+* would switch *Cdc2* into the mitosis-promoting form. *See* CDC27; cell cycle; c mitosis; endoduplication; endoreduplication; licensing factor; polyploid.

endonuclease Enzyme that cuts DNA or RNA at internal positions. *See* homing endonuclease; exonuclease; restriction enzymes.

endonuclease III Removes oxidized pyrimidines from the DNA such as 5,6-dihydrouracil, 5,6-dihydrothymine, and thymine glycol. *See* glycosylases. (Katcher, H. L. & Wallace, S. S. 1983. *Biochemistry* 22:4071.)

endonuclease V T4 endonuclease V is called T4-pdg, T4-pyrimidine dimer glycosylase/AP lyase. *See* glycosylases; pyrimidine dimer.

endonuclease VIII Bacterial enzyme that removes oxidative-damaged bases from the DNA by glycosylase/lyase function. *See* DNA repair. (Burgess, S., et al. 2002. *J. Biol. Chem.* 277:2938.)

endonuclease G Mitochondrial enzyme that upon apoptotic stimuli moves to the cell nucleus and degrades DNA independent from the caspase pathway. *See* apoptosis; CAD; CED-3. (Parish, J., et al. 2001. *Nature* 412:90; Li, L. Y., et al., *ibid*, p. 95.)

endopeptidase Proteolytic enzymes cleaving internal bonds at cysteine, serine, aspartic, metallic, or other residues.

endophilin Lysophosphatidic acid acyl transferase. It is one effector of dynamin and is probably facilitating the invagination of membranes and the formation of synaptic-like microvesicles used in neurotransmission. *See* dynamin; lysophosphatidic acid; neurotransmitter; synaps. (Ramjun, A. R., et al. 2001. *J. Biol. Chem.* 276:28913.)

A portion of plant endoplasmic reticulum with ribosomes.

endoplasmic reticulum (ER) Internal membrane system within the eukaryotic cell with secretory channels. Within these membrane-bound compartments, proteins and lipids are synthesized; glycosylation and sulfation of secretory and membrane-bound proteins, proteoglycans, and lipids take place. The rough endoplasmic reticulum has attached ribosomes, causing a nonsmooth appearance (*rough endoplasmic reticulum*). Many nascent peptide chains are transferred to the ER, where the synthesis of the proteins is completed and the proteins are folded. If the folding slows down, the unfolded protein response (UPR) stimulates the transcription of ER-resident proteins. The Hac1 basic leucine-zipper protein is a transcription factor that binds to the UPR element in the promoter of the UPR genes. Splicing that bypasses the spliceosome is mediated by a tRNA ligase protein. After the completion of the synthesis, the proteins are translocated with the aid of the Hsp70 chaperones. The energy is provided by associated ATP, UDP, GDP, and CMP nucleotides. Within the endoplasmic reticulum, quality control takes place by selecting the ones that have the proper conformation. The folding is mediated by BiP, lectins, calnexin, calreticulin, and protein disulfide isomerase, GRP, and other ER chaperone proteins. The enzymes UDP-5′-diphosphate–glucose:glycoprotein glucosyl-transferase and glucosidases also play a role in substrate binding and release of ER proteins. Free cysteines participate in forming interchain disulfide bonds and aggregates and facilitate retention. These ER factors prevent the release of the proteins until their synthesis and folding is completed. The improperly folded polypeptides are degraded and only the correct ones are secreted to the cytosol. The structural surveillance in the ER prevents the release, then degrades some of the structurally mutant α_1-antitrypsin molecules, although they would be capable of function. In cystic fibrosis, the misfolded transmembrane conductance regulator protein may be degraded before it can be transported to the plasma membrane, although under conditions conducive to proper folding some function may be recovered. In other instances, the folding defect may cause precocious release of, e.g., β-lactoglobulin polypeptide. The released proteins may be transported with the aid of the cargo receptor to the Golgi apparatus where further processing may take place. *See* antitrypsin gene; architectural editing; BiP; calnexin; cell structure; chloroplast endoplasmic reticulum; COP transport vesicles; cystic fibrosis; endocytosis; folding; Golgi apparatus; GRP; Hac1; Hsp 70; inclusion body; lectins; protein folding; protein synthesis; Sec61 complex; signal peptide; signal sequence recognition particle; SsA; SsB; TAP; unfolded protein response; Ydj1. (Parodi, A. J. 2000. *Annu. Rev. Biochem.* 69:69; Yamamoto, K., et al. 2001. *EMBO J.* 20:3082; Lehrman, M. A. 2001. *J. Biol. Chem.* 276:8623; Glick, B. S. 2001. *Curr. Biol.* 11:R361; Nicchita, C. 2002. *Current Opin. Cell Biol.* 14:412; Hampton, R. Y. 2002. *Current Opin. Cell Biol.* 14:476.)

endopolyploid Cell with increased chromosome number because of endomitosis, i.e., the chromosomes have replicated, but cell division was skipped. Such phenomena commonly occur repeatedly in cultured plant and animal cells. *See* endoduplication; endomitosis.

endoproteases Family of enzymes involved in the processing of proteins, e.g., the large peptide hormone precursors, large secreted proteins, pheromones, etc.

endoreduplication Chromosome replication is not followed by centromere and cell divisions and therefore multistranded (polytenic) chromosomes are formed. The mechanism of this phenomenon requires prolongation of the S phase and suppression of the M phase. Polytenic chromosomes are present in the salivary gland cells of some flies (*Drosophila, Sciara*, etc.), but they also occur in some plant tissues such as the endosperm of cereals, in some cells (chalaza, antipodal) of *Allium ursinum, Aconitum ranunculifolium,* etc. *See* antipodal cells; chalaza; endoduplication; polytenic chromosomes; replication during the cell cycle. (Edgar, B. A. & Orr-Weaver, T. L. 2001. *Cell* 105:297.)

endorphins (endogenous morphine) Neuropeptide ligands of opiate receptors in the brain pituitary gland and peripheral tissues of vertebrates. Their physiological effects mimic morphines. They have potential therapeutic use in nervous disorders, pain perception, etc. Introducing the β-endorphin gene into the cerebrospinal fluid may mitigate pain for longer periods than drugs. *See* brain human; enkephalins; morphine; opiocortin; pain; pituitary; POMC. (Coventry, T. L., et al. 2001. *J. Endocrinol.* 169:185.)

endosome *See* endocytosis.

endosomolysis Disruption of the endocytotic vesicles and liberation of their contents. *See* endocytosis.

endosperm Nutritive tissue within the seed developed from the fertilized polar cells in the embryo sac. In some plants, the endosperm develops to a substantial mass and persists as the bulk of the seed (the majority of the monocots). Alternatively, it may be gradually consumed, and only traces are visible by the time the embryo fully develops, because that function is relegated to the growing cotyledons (majority of dicots). In monocots, the surface layer of the endosperm is called *aleurone*. Normally the embryo is diploid and the endosperm is triploid. Some mutations may permit the development of the endosperm without fertilization of the fused polar nuclei. In species crosses, the 2:3 proportion may be altered with deleterious consequences for the embryo. Crossing a diploid female with a tetraploid male, the embryo will be triploid and the endosperm will be tetraploid (3:4), but crossing a tetraploid female with a diploid male, the number of genomes in the embryo remains the same ($3n$), but the endosperm will be pentaploid (3:5). Thus, the gene dosage in the endosperm is usually different from that in the embryo. Some species of plants have different developmental patterns beginning with the embryo sac. *See* embryogenesis in plants; embryo sac. (Olsen, O.-A. 2001. *Annu. Rev. Plant Physiol. Mol. Biol.* 52:233.)

endosperm balance number Suggested term for the unusual observation that within some species the tetraploids do not effectively fertilize other tetraploids ($4x$) but may fertilize diploids ($2x$). When their chromosome number is doubled ($8x$), they may be crossed with tetraploids, except the progenitor tetraploid. Thus, a curious chromosome balance is observed. *See* endosperm.

endosperm mother cell Fused polar cells of the endosperm ($2n$) that give rise to the triploid endosperm tissue in the seeds of plants when fertilized by a sperm. *See* embryo sac; gametogenesis.

endospore Dormant bacterial cell resistant to most treatments that normally kill active cells. The process of sporulation has been extensively studied in *Bacillus subtilis. Clostridium tetani* and *Bacillus anthracis,* etc., also produce endospores. A fungal spore within a cell is also called an endospore *See* forespore.

endostatin Collagen XVIII C-terminal, ca. 20 kDa domain, which along with angiostatin can regulate angiogenesis and metastasis and may block cancerous growth by denying the blood supply to tumors. Endostatin-integrin interaction is relevant to angiogenesis. More recent evidence questions the antitumor effectiveness of endostatin. *See* angiogenesis; angiostatin; cancer; collagen; integrin; metastasis. (Sim, B. K., et al. 2000. *Cancer Metastasis Rev.* 19:181; Rehn, M., et al. 2001. *Proc Natl. Acad. Sci. USA* 98:1024; Kuo, C. J., et al. 2001. *Proc. Natl. Acad. Sci. USA* 98:4605.)

endosymbiont Organism that lives within the cell of another. *See* infectious heredity.

endosymbiont theory Evolutionary idea suggesting that plastids and mitochondria were originally free-living microorganisms and were later captured by nucleated cells (archaebacterium), then they became cellular organelles in eukaryotes (see Margulis, 1993, in general references [Evolution]). This old hypothesis is supported by DNA sequence analyses primarily of organellar and eubacterial and archaebacterial rRNA genes. It appears that the plastid rRNA core sequences resemble those of gram-positive cyanobacteria, and the mitochondrial rRNAs indicate descents from eubacteria. The dispute among evolutionists is still extant regarding monophyletic or polyphyletic endosymbiotic origin of these organellar genomes.

Secondary endosymbiosis is believed to occur when a eukaryote ingests and retains other eukaryotes with chloroplasts. In these cases, four-layer membranes may enclose the organelles. The old endosymbiotic theory fails to interpret the fact that there are eukaryotes without mitochondria or hydrogenosomes. (In hydrogenosomes, pyruvate is metabolized by pyruvate: ferredoxin oxidoreductase rather than by a pyruvate-dehydrogenase complex.) Archaebacteria do not have evidence for structures homologous to those of eukaryotes. Also, mitochondrion-free eukaryotes seem to have genes, which resemble mitochondrial DNA sequences. The *hydrogen hypothesis* has been advanced (Martin, W. & Müller, M. 1998. *Nature* 392:37). According to this theory, an autotrophic archaebacterium host enters a symbiotic relationship with a eubacterium capable of respiration. This symbiont produces molecular hydrogen as a waste product of its heterotrophic metabolism, which serves the host well and thus forges a stable relationship. In this system, hydrogenosomes and eventually mitochondria evolve from the symbiont. The symbiont transfers genes to the host for organic metabolism, enzyme generation, and ATP generation. Thus, heterotrophy evolves from autotrophy and the complex organism can utilize organic molecules. Eventually more complex cellular structures evolve, giving origin to eukaryotic cells. The obligate intracellular parasites of proteobacteria, *Rickettsia, Anaplasma*, and *Ehrlichia* genomes, appear to be closest to the mtDNA of animals. *See* chloroplasts; hydrogenosome; mitochondria; nucleomorph; Rickettsia; rRNA. (Lang, B. F., et al. 1999. *Annu. Rev. Genet.* 33:351.)

endosymbiosis   *See* endosymbiont.

endothelin   The veins and inner tissues of the heart secrete this peptide hormone. It controls heart muscle functions, hypertension, including myocardial infarction, protein kinase A (PKA) and chloride, potassium and calcium ion channel functions. Mutation in the endothelin may lead to hereditary hypoventilation. Endothelin mutations may account for Hirschprung disease and Shah-Waardenburg syndrome. The endothelin receptor (EDNRB) is required for melanoblast migration and development of the neural crest. Red wine reduces the synthesis of endothelin-1 and thus protects against coronary heart disease (Corder, R., et al. 2001. *Nature* 414:863). *See* Hirschprung disease; ion channels; melanocyte; neural crest; PKA; Shah-Waardenburg syndrome. (Hunley, T. E. & Kon, V. 2001. *Pediatr. Nephrol.* 16:752.)

endothelium   Layer of epithelial cells of mesodermal origin, and lining organ cavities.

endothermic   Chemical reaction that takes up heat.

endotoxin   Bacterial lipopolysaccharide toxin attached to the outer membrane, secreted only inward to the cell, and released only when the cell disintegrates. Endotoxins release large amounts of tumor necrosis factor (TNF) and interleukin-1 (IL-1). They may present a serious hazard to humans (diarrhea, bleeding, inflammation, increase of white blood cells, etc.). *See* IL-1; TNF; toxins.

endpoint   In chemistry it indicates the highest dilution during titration that still gives a detectable reaction with another substance; in genetic toxicology, the method of identification of the lowest effective dose of a mutagen or carcinogen that causes point mutations, chromosome breakage, unscheduled DNA synthesis, etc. *See* bioassays of genetic toxicology; unscheduled DNA synthesis.

end-product inhibition   *See* feedback control.

Enediynes   Anticancer antibiotics. They cleave the DNA by generating benzenoid diradicals when activated. They cleave the DNA backbone by extracting hydrogen atoms from it. *See* neocarzinostatin. (Lode, H. N., et al. 1998. *Cancer Res.* 58:2925; Liu, W., et al. 2002. *Science* 297:1170.)

energy charge   [(ATP) +1/2(ADP)]/[(ATP) + (ADP) + (AMP)]   It measures the phosphorylating capacity of the adenylate system. The energy charge is 0 if only AMP is available and it is 1 if all AMP is converted to ATP. *See* AMP; ATP.

energy coupling   Transfer of energy from one reaction path to another.

engrafting   Adding another tissue by surgical means, and if it is established, it is engrafted. *See* grafting.

engrailed *(en, Drosophila* gene, map location 2–62)   Among a variety of phenotypic consequences: half of the larval body segments are deleted and the remaining ones are duplicated. *See* groucho; morphogenesis in *Drosophila*.

enhanceosome   Protein complex of special gene activators and HMG proteins involved in enhancing the activity (transcription) of interferon and other genes in response to various signals and cues. The enhanceosome includes histone acetyltransferases, CREB-binding proteins, and associated activators (GCN5). HMGs are structural proteins facilitating their assembly after acetylation. Acetylation by CBP at lysine-65 destabilizes the complex, but acetylation by GCN5 at lysine-71 stabilizes the complex, facilitates transcription, and fends off acetylation by CBP. *See* activator proteins; CBP; CREB; enhancer; GCN5; HMG; interferon. (Carey, M. 1998. *Cell* 92:5; Munshi, N., et al. 2001. *Science* 293:1133.)

DNA L:OOP (continuous grey) is wrapped around by the enhanceosome.

enhancer Cis-acting positive regulatory elements positioned (frequently near Z DNA) either upstream or downstream (by several kb) of the initiation of transcription. The enhancer may also be located within the transcription unit. Supercoiling of the DNA may facilitate enhancer activation of genes at a distance over 2,500 bp. It alters chromatin structure to facilitate transcription by binding special proteins. The enhancers may activate a gene for transcription or may increase the basal rate of transcription by two orders of magnitude. Enhancers (one or more) are present in all eukaryotic genes. The size of the enhancers varies from 50 bp to 1.5 kbp. Prokaryotes usually do not use enhancers; however, some σ^{54} subunits of RNA polymerase make this enzyme enhancer-responsive. Some of the clustered genes (e.g., HOX) may share enhancers. The prokaryotic enhancer element may be situated upstream or downstream of the gene. For more about enhancers, *see* Simian virus 40. *See* activator; core promoter; G box; homeotic genes; octa; regulation of gene activity; silencer; TAF. (Bellen, H. J. 1999. *Plant Cell* 11:2271; Xu, Z., et al. 2001. *Gene* 272:149; Liu, Y., et al. 2001. *Proc. Natl. Acad. Sci. USA* 98:14883.)

enhancer competition The imprinted genes *H19* and *Igf2* share the same enhancer, but *H19* is expressed in the maternal chromosome, whereas *Igf2* is favored in the paternal chromosome. The two chromosomes are supposed to be in competition for the same enhancer. Newer evidence indicates that the differential control resides in ~2 kb region upstream from *H19*, the so-called *ICR* site (imprinting control region). *ICR* is apparently an allele-specific methylation-sensitive insulator, a boundary element. The conserved zinc-finger protein (CTCF) specifically binds to CG dinucleotide repeats, which occur in that region. *See* boundary element; imprinting; insulator. (Cai, H. N., et al. 2001. *Development* 128:4339.)

enhancer trapping *See* gene fusion.

enhancesome *See* enhanceosome.

En-I (enhancer-inhibitor) System of transposable elements of maize. *See* Spm.

enkephalin Endogenous opioid peptide regulating pain, stress response, aggression, and dominance behavior. The preproenkephalin gene, encoding the precursor of Met-enkephalin and introduced by herpes simplex viral vector into the mouse afferent nerves, may mitigate pain sensation for a longer term than analgesics. Enkephalin knockout mice display increased fear. Enkephalins regulate many neuronal and immunological functions. *See* endorphin; knockout; opiates; pain.

ENL Serine- and proline-rich protein (encoded in human chromosome 19p13); in chromosomal translocations it may be responsible for acute lymphocytic leukemia. *See* leukemias.

enolase (phosphopyruvate hydrolase, PPH, ENO) Catalyzes the conversion of 2-phosphoglycerate into phosphoenolpyruvate; ENO1 is encoded in human chromosome 1pter-p36.13. The neuron-specific enzyme (ENO2) is encoded in human chromosome 12p13, and the muscle-specific ENO3 is encoded in chromosome 17pter-p12.

enol form Of a molecule, contains an OH group, whereas the keto form has C=O. These may undergo enol-keto tautomerism; the enolic form may be subject to proton shift. *See* hydrogen pairing; tautomeric shift.

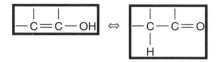

ENSEMBL (http://www.ensembl.org/) Automated annotation program for eukaryotic genes.

entelechy The vitalists postulated the inner force of organisms that is responsible for life and growth. In modern morphogenetic theory, it is the basis of the internal program of development.

enterobacteria Large group of gram-negative bacteria with a very wide distribution in insects, higher animals, and plants; represented by *Escherichia coli, Salmonella, Shigella, Serratia, Klebsiella, Proteus, Erwinia, Yersinia,* etc. Many of them are pathogenic, some are saprophytes, and others are facultative pathogens.

enterokinase deficiency Autosomal-recessive defect in the activation of pancreatic proteolytic proenzymes (such as chymotrypsinogen, procarboxypeptidase, proelastase) by this intestinal enteropeptidase; consequently, hypoproteinemia and general weakness results. *See* trypsinogen deficiency.

enteropeptidase Enterokinase.

enterotoxin Produced by enteric (intestinal) bacteria. *See* toxins.

enthalpy (H) Heat content or energy of a system.

enthalpy change (ΔH) Difference between the energy required for disrupting a chemical bond and gained by forming a new one(s).

Entner-Doudoroff pathway In bacteria, an aldolase generates pyruvate and triose phosphate from 6-phosphogluconate → 2-keto-3-deoxy-6-P-gluconate.

entoderm Endoderm. *See* endoderm.

entomology Study of insects. (<http://ent.iastate.edu/list/>.)

entomophagous Feeding on insects.

entrainment Exact match between the shift of an endogenous oscillator and the periods evoking the circadian rhythm. *See* circadian rhythm; oscillator.

entrapment Vector *See* gene trapping vector.

Entrez (http://www.ncbi.nlm.nih.gov/Entrez/) Source of nucleotide and protein sequence information. Batch Entrez allows the retrieval of several sequences at once from a database; after downloading, they can be analyzed on the local computer. Retrieval computer software and databases are distributed on CD-ROM. The software is public and available in Macintosh or IBM PC (Windows) format. General questions through INTERNET: <info@ncbi.nlm.nih.gov> or <net-info@ncbi.nlm.nih.gov>. *See* databases.

entropic trap array Nanofluidic device for the separation of 5,000- to 160,000-bp-long DNA molecules. *See* pulsed-field electrophoresis.

entropy Measure of energy unavailable for use within a system. The entropy increases by natural processes of aging.

ENU (N-ethyl-N-nitrosourea) Ethylating agent; one of the most potent point mutagens in mice (mutation rate/locus $6.6-15 \times 10^{-4}$); in *Arabidopsis*, within the concentration range of 0.25 to 1.25 mM (18 h exposure of seeds), caused embryo lethals from 2.8 to 85.1% of the plants. (Balling, R. 2001. *Annu. Rev. Genomics Hum. Genet.* 2:463; Herron, B. J., et al. 2002. *Nature Genet.* 30:185.)

$$O:C \cdot NH_2$$
$$|$$
$$N \cdot N:O$$
$$|$$
$$CH_2 - CH_3$$

ENU.

enucleate Cell without a nucleus. *See* cytoplast; enucleation; nuclear transplantation.

enucleation Removal of the cell nucleus. *See* cytochalasin; nuclear transplantation.

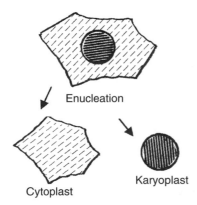

Enucleation

Cytoplast

Karyoplast

envelope Generally double-layer membranes (surrounding organelles) in the cell.

environmental deviation Quantitative characters are manifested in a phenotypic value, called P, which can be partitioned into a genotypic value, G, and an environmental deviation, E, i.e., $P = G + E$, where P is measured as the mean value of a population. The value of G can be determined by heritability estimates. *See* heritability.

environmental effects They are of significant importance for the expression of genes. The genetic constitution provides the blueprint, but the realization frequently has a variable environmental component. The genotype has been identified as providing a programmed *reaction norm*, but the temporal sequence and intensity of gene expression is determined by the internal regulatory DNA elements and the environment. Among the endless good examples, the expression of inducible enzymes, the immune reactions, the switching from fetal hemoglobin $\zeta\zeta\varepsilon\varepsilon$ in the early embryo to $\alpha\alpha G\gamma A\gamma$ by the 8th week, then to $\beta\beta\delta\delta$ shortly before birth, and by 6 months postnatal, it switches to $\alpha\alpha\beta\beta$ with about 2% $\alpha\alpha\delta\delta$. In plants, the onset of the flowering in many species requires vernalization and/or appropriate photoperiodism. Sometimes environmental effects persist through meiotic events. (*See* dauermodification; delayed inheritance; epigenesis; maternal effects. The environment exercises the greatest effect on the expression of quantitative genes. Many environmental factors are mutagenic, these may include food and feed additives, cosmetics, drugs, medicines, industrial and agricultural chemicals, natural food products, and radioactive fallout. *See* environmental mutagens; heritability; quantitative traits; cold-regulated genes salt tolerance. (Xiong, L., et al. 2002. *Plant Cell*) 14:S165.)

environmental load Some specific environmental conditions may favor and maintain alleles that convey inferior fitness to the individuals under usual, average conditions. These alleles represent the environmental load of the population. *See* genetic load.

environmental mutagens Carcinogens and teratogens occur in the laboratory, industry, and agricultural environment. According to the estimates of the U.S. National Research Council, approximately 70,000 chemicals are in use, and their number is increasing steadily. Their biological effects are incompletely known. A survey of about 80 products of engineered protein did not appear toxic (with the exception of interferon) and one humanized antibody + linker might have had some genotoxicity (*Mutation Res.* 436:137). Many potentially mutagenic agents are present in every household, in industry, and in research laboratories. The response to these potentially hazardous agents is determined or influenced by the genetic constitution of the individuals (susceptibility and resistance genes) and a number of factors in the environment besides the few agents named here. A very incomplete list follows:

*Acetaldehyde: an intermediate of organic solvents, preservative of certain food products; it may be present in small quantities in the sweet smell of ripening fruits.

*Acridine dyes: laboratory chemicals; may be present in some antimalaria drugs.

*Acrylamides: used for gel electrophoresis and are absorbed through the skin.

*Actinomycin D: antibiotic used to block transcription. In the laboratory it is a carcinogen.

*Aflatoxins: about a dozen different natural products of *Aspergillus* fungi detectable on peanuts, grains, and other feed and food. They cause mutation as well as liver cancer.

*Allylisothiocyanate: present in cruciferous vegetables, mustard condiments, horseradish.

*3-Aminotriazole: herbicide that is largely undetectable by mutagen assays, yet it is a powerful carcinogen.

*Aramite: insecticide used some time ago for fumigation of greenhouses.

*Asbestos: heat insulator and filtering agent, present in many buildings.

*Atabrine: antimalaria drug.

*Atrazine: herbicide that may be activated into a mutagen by the metabolism of plants.

*5-Azacytidine: interferes with DNA methylation.

*Benzimidazole: may be present in pharmaceuticals, preservatives, and insecticides.

*Benzidine: laboratory reagent.

*Benzo(a)pyrene: regularly formed during combustion of many organic materials (in fireplaces, automobile exhaust fumes, grilled-charcoal-broiled meat, tobacco smoke), refined mineral oils, and commercial wax products.

*β-Propiolactone: used as a plasticizer, in wood processing, tobacco processing, additive to leaded gasoline, insecticides.

*Bisphenol A: used for manufacturing resins and plastic coating of various containers. Bisphenols are estrogenic endocrine-disrupting chemicals (EEDC) mimicking the action of estradiols and affecting (sexual) development and responses to diseases.

*Bracken fern extracts.

*Caffeine: not mutagenic or carcinogenic itself, but interferes with genetic repair mechanisms.

*Captan: fungicide.

*Carbontetrachloride: solvent, seed fumigant.

*Chloral hydrate: microtechnical reagent, sedative.

*2-Chloroethanol: polymerizing agent, insecticide, herbicide. It is present in some sedatives.

*Chloroform: laboratory solvent, in some office supplies, may be formed in chlorinated drinking water.

*Chloroprene: in elastomers and adhesives.

*Chloropromazine: in some tranquilizers.

*Coal tar: present in soot, roofing tar, asphalt (bitumin).

*Colchicine: used for polyploidization. It is very toxic.

*Cycasin: present in cycade plants.

*Cyclosporin: immunosuppressant and antibiotic drug.

*Diamines: contained in some hair dyes.

*Diazomethane: laboratory reagent.

*Dichlorvos: insecticide.

*Diepoxybutane: industrial polymerization agent, may be present in some pharmaceuticals.

*Diethylsulfate: supermutagen.

*Dimethylsulfate: industrial methylating agent of cellulose, polymerizing compound, insecticide, stabilizer.

*Dioxin: contaminant in some herbicides, deodorant soaps (now banned).

*Dithranol: in some antidermatosis and ringworm drugs.

*EDTA (ethylenediamine tetraacetic acid): chelator and antioxidant in laboratory and industry.

*Epichlorohydrine: present in some epoxy resins, gum, paint, varnish, nail polishes, manufacturing of crease-resistant fabrics, paper processing, waterproofing.

*Epoxides: precursors of ethylene glycol, dioxane, carbowax, monoethanolamine, acetonitrile, plastics, gaseous sterilants. Epoxides may be trapped in the products or any material in contact with them may be contaminated.

*Estrogen (β-estradiol): fertility hormone used in ART is indirectly carcinogenic. Estrogen-like substances in the environment may adversely affect the reproductive system. *See* ART.

*Ethidium bromide: used as nucleic acid stain and as a mutagen for mitochondrial DNA.

*Ethylenedibromide: fumigant for grain storage.

*Ethyleneimine: used for the manufacture of flame-retardant clothing, crease-resistant and shrinkage controlling of fabrics, in insecticides, soil conditioners, synthetic fuels. It is a supermutagen.

*Ethylmercuric chloride: fungicide.

*Ethylmethane sulfonate: one of the most widely used and most potent laboratory mutagens. Some methane sulfonates are now banned ingredients of older prescription drugs.

*Fats (after oxidation in rancid food): may become mutagenic and carcinogenic.

*Formaldehyde: disinfectant, preservative of museum specimens of animals and organs, fixative, in adhesives, crease-, crush- and flame resistance aid, in automobile exhaust.

*Fumes: released by outdoor burning, defective woodburning fireplaces, coal furnaces.

*Glycidol: may be present in glycerol; used for manufacturing water-repellent fabrics, food preservative. It is a very powerful mutagen.

*Hair dyes: certain types are mutagenic/carcinogenic.

*Hydrazines: in photographic materials, rocket fuels, preservatives, solvents, gasoline additives.

*Hydrogen peroxide: used for bleaching flour, starch, paper, tobacco, cosmetics (hair lightener), stabilizer, plasticizer.

*Hydroxylamine: used in nylon manufacturing, photographic materials, adhesives, paints. It specifically converts cytosine into a thymine analog.

*Hydroxyurea: breaks heterochromatin in the chromosomes and is present in some antileukemic, dermatological drugs.

*Isotopic tracers for emitting β- and γ-radiation.

*Lindane: fungicide.

*Nicotine: addictive component of tobacco; smoking is a major cause of human cancer.

*Nitrofurans (AF-2): one-time widely used preservative in Japan for fish and soybean products.

*Nitrogenmustards: used as anticancer drugs, powerful radiomimetic agents.

*Nitrosoamines: extremely powerful mutagens but present in small quantities in nitrite-treated meat products; formed from nitrites by the action of stomach acids and intestinal microbes.

*Nitrosoguanidines: extremely potent laboratory mutagens.

*Nitrous acid: widely used preservatives for meat products are direct mutagens and can be converted by mammalian metabolism into nitrosoamines.

*Organomercurials: were widely used in fungicides.

*PCB (polychlorobiphenyls): present in electric transformers.

*Peroxides: bleaching agents, disinfectant; may be formed from tryptophan under UV light.

*Phenol: used in DNA extraction; extremely toxic, causes vesicles on skin and possibly mutagenic.

*Phenylmethylsulfonyl fluoride (PMSF): used in pulsed-field gel electrophoresis, is very toxic, and absorbed through skin.

*Polychlorinated biphenyls: once used for various industrial processes and in pesticides.

*Polycyclic aromatic hydrocarbons: volatile, semivolatile, and particulate materials.

*Propane sultone: may be present in detergents and dyes.

*Pyrrolizidine alkaloids: occur in several species of plants like *Senecio, Crotalaria,* etc.

*Quinacrine: antimalarial drug.

*Safrole: food additive, coloring agent; it is naturally present in root beer.

*Sodium azide: laboratory reagent and powerful mutagen in some organisms.

*Sodium bisulfite: preservative in fruit juices, wine, and dried fruits.

*Sodium nitrite: common preservative of coldcuts, fish, and cheese and can be converted in the body into nitrosoamines.

*Soot: contains hydrocarbons such as benzo(a)pyrene; 7,12-dimethylbenz(a)anthracene.

*Streptozotocin: prescription drug used against athlete's foot fungus, an alkylating agent.

*Styrene: in reinforced plastics, resins, and insulators. It is a clastogen.

*Thiotepa: used in flame-retardant, crease-resistant, and waterproof fabrics, in manufacturing dyes, adhesives, and drugs.

*Trichloroethylene: weak mutagen that may be present in dry-cleaning agents, degreasing fluids, paint, varnish, and food-processing solvents.

*Tri(2,-3-dibromopropyl) phosphate: flame retardant.

*Triethylenemelamine: cross-linking agent, used for finishing rayon fabrics, waterproofing of cellophane, insect chemosterilant and anticancer drug.

*Trimethylphosphate: gasoline additive, insecticide, flame-retardant, polymerizing agent, insect chemosterilant.

*Ultraviolet radiation: may cause skin cancer (melanoma) and mutation in tissues close to the surface; may also be hazardous through the peroxides generated. Wear protective goggles.

*Urethanes: used by the plastic industry; for manufacturing resins, fibers, textile finishes, herbicides, insecticides, and drugs.

*Vinyl chloride: monomer of polyvinyl plastics. May be liberated from the polymers by heat (in a locked-up car) and may contaminate the surface of plastic sheets, bags, water hoses. The monomer is a powerful carcinogen and mutagen requiring metabolic activation. It forms the adduct N^2-3-ethenoguanine.

Many of the mutagens and carcinogens require metabolic activation for effectiveness. There is an interindividual variability of susceptibility, particularly at low degrees of exposure. *See* activation of mutagens; adduct; Ames test; bioassays for environmental mutagens; carcinogens; choline esterase; mutagen essays; mutagenic potency; pseudocholinesterase deficiency. (Waters, M., et al. 1999. *Mutation Res.* 437:21; Albertini, R. J., et al. 2000. *Mutation Res.* 463:111; Johnson, F. M. 2002. *Env. Mol. Mutagen.* 39:69, <http://toxnet.nlm.nih.gov>.)

environmental variance *See* environmental deviation; genetic variance.

Env Z Phosphorylating enzyme (kinase) on the bacterial membrane envelope.

enzootic Infection may be present among animals all the time, but it shows up only sporadically or flares up only under certain conditions.

enzyme Generally protein molecules. Some ribonucleic acids and DNAs also have enzymatic functions (ribozymes, DNA-zymes). The protein part of the enzyme is the apoprotein or apoenzyme. Their catalytic function may require cofactors, such as metals (Cu^{2+}, Fe^{2+}, K^+, Mg^{2+}, Mn^{2+}, Mo^{3+}, Zn^{2+}, etc.), or organic compounds (vitamins, nucleotides, etc.) called coenzymes, also called the prosthetic group of the enzymes. The complete enzyme (apoenzyme + prosthetic group) is the holoenzyme. Enzymes are organic catalysts; they mediate biochemical reactions without becoming parts of the reaction products, and they enhance the reaction rates by ~10^{10}–10^{15}–fold. The site(s) of the enzyme molecule interacting with the substrate (the molecule to be acted on) is the active site of the enzymes.

Enzymes carry out gene functions. RNA is transcribed on the DNA and the RNA is translated into protein. Enzymes are named by adding: ase to either the name of the substrate or to the name of the reaction, e.g., DNase, phosphorylase. The International Union of Biochemistry (*Enzyme Nomenclature*, 1972, Elsevier, Amsterdam) classified enzymes into six major groups and assigned code names to them: (1) *oxidoreductases* (transfer electrons); (2) *transferases* (catalyze molecular group transfers); (3) *hydrolases* (cleave covalent bonds and transfer H and OH, respectively, to the products); (4) *lyases* (form or remove double bonds); (5) *isomerases* (rearrange molecules internally); (6) *ligases* (mediate condensations by forming C–C, C–N, C–O, and C–S bonds while cleaving ATP). In technical descriptions, enzymes are identified by EC (enzyme classification) numbers, where the first digit refers to the number of one of the above groups and the following digits more closely specify the nature of the mediated reaction mediated; thus, DNA polymerase I of *E. coli* has the EC number 2.7.7.7, whereas restriction endonuclease Eco RI is 3.1.21.4., bovine RNase is 3.1.27.5, and glucose-6-phosphate dehydrogenase is 1.1.1.49. Some enzymes are modular, i.e., they may have interchangeable components with other enzymes or may be components in different catalytic complexes. *See* allosteric control; competitive inhibition; DNA-zyme; Eadie-Hofstee plot; effector; enzymes, multifunctional; induction; inhibition; Lineweaver-Burk equation; Michaelis-Menten equation; protein synthesis; recombination mechanisms; regulation of enzyme activity; repression; restriction endonucleases; ribozymes; subunits. (*Nature* 409[2001]:225ff on biocatalysis, BRENDA; nomenclature database: <http://www.expasy.ch/enzyme/>.)

enzyme, multifunctional During the 1940s, the revolutionary one gene–one enzyme theory was proposed. Since that time, it has been recognized that attachment of different prosthetic groups and altering active sites, using different promoter sites, alternate processing, posttranslational modifications, etc., may contribute to the translation of different proteins from the same DNA sequence. (Perham, R. N. 2000. *Annu. Rev. Biochem.* 69:961.)

enzyme induction Requires the presence of the substrate or substrate analog. *See Arabinos*e operon; *Lac* operon.

enzyme-linked immunosorbent assay See ELISA.

enzyme mimic Catalytic antibody.

enzyme mismatch cleavage (EMC) Uses the resolvase enzyme of bacteriophages to recognize and cut at mismatches in double-stranded DNA that has been amplified by PCR. The PCR-amplified DNA is expected to have matching and nonmatching strands if there was mismatch in the original double-stranded DNA. This way, heterozygosity or mutation may be detected after gel electrophoresis. *See* mismatch; mutation detection; PCR; resolvase.

eomesodermin Transcription factor attaching to a T box (*see* MAR) and regulating trophoblast and mesoderm development in mice. *See* mesoderm; trophoblast.

eosinophil White blood cells (readily stainable by eosin dye) generally display bilobal large nuclei. They modulate allergic and inflammatory reactions of the animal body and mediate the destruction of parasites. *See* blood; granulocytes.

Eosinophil.

eotaxin C–C-type chemokine. *See* chemokines.

EP (1) Effector proteins stimulate the conversion of E.GTP into E.GDP. *See* GDP; GTP.

(2) Early pressure. In zebrafish, gynogenetic embryos are produced by exposing early embryos fertilized by UV-inactivated sperm to high hydrostatic pressure immediately after fertilization to suppress the second meiotic anaphase. A heat shock 15 minutes after fertilization suppresses the first mitotic division and leads to gynogenesis. *See* gynogenesis; UV; zebrafish.

EPA *See* RAP1.

ependymoma Brain or spinal chord neoplasia.

EPH Family of receptor tyrosine kinases involved in axon guidance and synapse regulation of integrin, cell migration, embryo differentiation, and indirect signals to the cytoskeleton. EPH proteins guard against mixing of different cell types (rhombomeres) during early embryonal development of animals. They form two subclasses: EphA and EphB. The EphA receptors are connected by ephrin-A ligands (glycosyl-phosphatidylinositol) to the cell membrane. The EphB receptors use ephrin-B ligands. EphB receptors regulate excitatory synapses by interacting with NMDA receptors. Alternative splice forms of a tyrosine kinase receptor control adhesion or repulsion of cells during embryogenesis. These ligands have different names in different organisms and tissues. *See* integrin; netrin; neurogenesis; NMDA receptor; receptor tyrosine kinases; rhombomeres; synapse; tyrosine protein kinase. (Drescher, U. 2000. *Cell* 103:1005; Wilkinson, D. G. 2001. *Nature Rev. Neurosci.* 2:155; Cowan, C. A. & Henkemeyer, M. 2001. *Nature* 413:174; Takasu, M. A., et al. 2002. *Science* 295:491; Kullander, K. & Klein, R. 2002. *Nature Rev. Mol. Cell Biol.* 4:475.)

ephelides *see* freckles.

ephrin *see* EPH.

epiallele Developmentally changed allele usually without DNA alteration; it may also arise as the result of mitotic gene conversion or recombination. *See* epigenesis; epimutation. (Rakyan, V. K., et al. 2002. *Trends Genet.* 18:348.)

epiblast Precursor of the ectoderm and other embryonal tissues. *See* ectoderm; organizer.

epiboly First embryonic cell movement whereby the upper portion of the yolk bulges in the direction of the animal pole and results in the formation of a dome. The blastoderm cells begin spreading dorsally over the yolk until the cell sheet meets at the ventral midline. *See* morphogenesis in *Drosophila*. (Marsden, M. & DeSimone, D. W. 2001. *Development* 128:3635.)

epicanthus Vertical skin fold near the eyelid; it is a normal characteristic for some Oriental human races and is also affected by some syndromes. *See* Down syndrome; ptosis.

epichromosomal Outside the chromosomes; an epichromosomal vector does not integrate into the chromosome and thus does not cause insertional mutation but may not be able to replicate continuously. *See* insertional mutation; vectors; (Rajcan-Separovic, E., et al. 1995. *Hum. Genet.* 96:39.)

epicotyl Stem section immediately above the cotyledons of plant embryos and seedlings.

epidemiology Study of the distribution, cause, and modifying factors of diseases and genetic defects in human and other populations. *See* founder effect; genetic drift; genetic toxicology; inbreeding; mutation detection; occupational hazard; population genetics.

epidemiology, genetic Studies the distribution of genetic variations of hereditary diseases, primarily using molecular tools such as blood groups, RFLP, and SNIP. *See* blood groups; linkage disequilibrium; RFLP; SNIP. (Bonassi, S. & William, W. A. 2002. *Rev. Mut. Res.* 511:73.)

epidermal growth factor *See* EGF.

epidermis Nonvascular outer cell layer derived from the embryonal ectoderm of animals; the surface cell layer of plants.

epidermoblast Gives rise to the epidermal cells.

epidermolysis Dissolution of the skin in spots, resulting in blisters (bullae), such as ichthyosis, keratosis, pemphigus, acrodermatitis, porphyria, etc. It is a characteristic symptom of many hereditary skin diseases. The epidermolysis bullosa simplex is encoded in human chromosome 8q24.13-qter and it is caused by mutation in plectin whereas a similar disease of mouse is associated with a defect in integrin. Junctional epidermolysis bullosa may be due to mutation either in collagen (Col17A1) or laminin protein subunits or in both. Epidermolysis bullosa is a very complex disorder due to defects at 1q32, 1q25-q31, 3p21.3, 10q24, 12q13, 17q11-qter, 17q12-q21. *See* collagen; integrin; intermediate filaments; laminin; plectin; skin diseases. (Fine, J. D., et al. 2000. *J. Am. Acad. Dermatol.* 42:1051.)

Epidermolysis bullosa. (From Bergsma, D., ed. 1973 Birth Defects. Atlas and Compendium. By permission of the March of Dimes Foundation.)

epididymis Narrow pouch-like structure attached to the testis; it serves for storage, maturation, and forwarding of the spermatozoa into the vas deferens during ejaculation. *See* cryptorchidism.

epigene conversion If the gene is introduced into the genome in an additional copy (by transformation of plants) that is hypermethylated, the other copy is silenced by transcription because the methylated promoter is presumably imposed on the unlinked promoter. *See* cosuppression; silencer.

epigenesis Process by which differentiation takes place from undifferentiated cells as programmed by the genome without any mutational event. It is sometimes defined as the mechanism of different types of gene expression from the same nucleotide sequences due to structural changes in the chromatin and the cooperative effect of transcription factors, other regulatory elements, and methylation of the gene(s). Movement of transposable elements is often considered an epigenetic event despite the fact that it may reorganize and alter the genome. It is expected that during meiosis the epigenetic alterations are erased and subsequently reversion takes place at the embryonic stage. In fission yeast, the mating-type conversion genes (*mat*) are, however, transmitted not just mitotically but meiotically. Similar examples in higher organisms are the paramutant state, imprinting, and position-type variegation. Meiotic transmission of the epigenetic state has been observed in fission yeast and *Drosophila* that do not methylate much DNA. Hyperacetylation of histone 4 (H4) may be involved in the transmission of the epigenetically activated state. Epigenetic alterations have decisive roles in carcinogenesis. *See* cosuppression; CpG motif; heterochromatin; histones; imprinting; mating type determination; methylation of DNA; MIP; morphosis; paramutation; peloric; phenocopy; position effect; preformation; prion; reaction norm; RDRP; reprogramming; RIGS; RIP; RLGS; RNAi. (Lindroth, A. M., et al. 2001. *Science* 292:2077; Jones, P. A. & Takai, D. 2001. *Science* 293:1068; Martienssen, R. A. & Colot, V. 2001. *Science* 293:1070.)

epigenetic Phenotypic change that may not be inherited because it is not based on changes in the genome. It is within the range of gene expression. Much of the epigenetic phenomena are based on differential DNA methylation during the life of the cell or organism or alteration in chromatin structure. *See* cortical inheritance; epimutation; nucleolar dominance; reaction norm; RLGS. (Surani, M. A. 2001. *Nature* 414:122; Petronis, A. 2001. *Trends Genet.* 17:142.)

epigenetics Study of epigenesis, epigenetic phenomena, and epimutation.

epigenomics *see* genomics.

epigyny The plant ovary is embedded in the flower receptacle and other flower organs are above it. *See* flower differentiation; receptacle.

epilepsy Suddenly recurring impairment of consciousness, involuntary movements, nervous disturbances occurring simultaneously or separately, caused by acquired or genetic factors. Epilepsy is a very complex paroxysmal disorder affecting 1–2% of the world's population. Some seizures may occur in early infancy; others are triggered by light (photogenic epilepsy) or by reading. According to one study among 27 monozygotic twins, 10 pairs were affected (both twins; concordant) and 17 pairs were not, and among 100 dizygotic twins, only 10 pairs had epilepsy (both twins). Thus, some epilepsies appear to have a strong genetic component (heritability, $h^2 > 0.40$). *Grand mal* epilepsy is a severe form of seizures occurring in 4–10% of the epileptics, and it generally develops fully by adult age. *Petit mal* (*absence*) epilepsy is a milder form, and it is less frequent. In mice, a similar mutation in the γ-subunit of a Ca^{2+} ion channel (VS) is responsible for the *stargazer* epileptic phenotype.

Autosomal-dominant nocturnal frontal lobe epilepsy (ADNFLE, human chromosome 20q13.2-q13.3) is caused by mutation in the neural nicotinic acetylcholine receptor $\beta2$ subunit. *Idiopathic epilepsy* is not associated with detectable brain lesions, whereas *symptomatic epilepsy* (acquired epilepsy) may be the consequence of a definable brain injury or disease. Idiopathic epilepsy is caused by mutation in the $GABA_A$ receptor $\gamma2$-subunit (Baulac, S., et al. 2001. *Nature Genet.* 28:46). The same molecular defect is responsible for childhood absence epilepsy (CAE) and febrile seizures (Wallace, R. H., et al., ibid., p. 49). (Some seizures or spasms are associated with a number of genetic maladies (phenylketonuria, epiloia, Zellweger syndrome, Menke syndrome, ceroid lipofuscinosis, adrenal hypoplasia, multiple sclerosis, glycogen storage diseases, lysosomal storage diseases, myoclonic epilepsy, porphyria, glutamate decarboxylase deficiency, rickets, galactosemias, Friedreich ataxia, West syndrome, Lesch-Nyhan syndrome). A gene for partial epilepsy was located to human chromosome 10q. *Familial partial epilepsy with variable foci* (FPEVF) was also assigned to 22q11-q12. The rare recessive *pyridoxine-dependent epilepsy* has been mapped to 5q13; it responds favorably to pyridoxine. The underproduction of γ-aminobutyrate (GABA, 5q31), a molecule with an important role in neurotransmission, may also lead to epileptic seizures and encephalitis. A rare but severe form of childhood epilepsy is caused by autoantibodies that turn against the brain's glutamate receptors. A homologue of this gene is found at 8q24. A voltage-gated sodium ion channel $\beta1$ subunit gene mutation in human chromosome 19q13.1 is one of the causes of *generalized epilepsy with febrile seizures plus* (GEFS$^+$). A second (GEFS$^+$) locus has been found at 2q24-q33. Febrile seizures are also associated with chromosome 8q13 region. Myoclonus epilepsy with ragged red fibers (MERRF) is determined by mitochondrial defects. The myoclonic epilepsy (EPM1) in human chromosome 21 is based on repeat insertions. *Idiopathic epilepsy* is characterized by seizures, which apparently show no metabolic structural cause, yet have a relatively high heritability. *Benign familial neonatal convulsion* (BFNC) diseases are attributed to defects in human chromosome 20q13.3, encoding a potassium channel (KCNQ2/KCNQ3). *Benign adult familial myoclonic epilepsy* is at 8q23.3-q24.1. *Familial recessive idiopathic myoclonic epilepsy of infancy* maps to 16p13. About 40 different types of epileptic syndromes have been described. Epilepsy occurs also in animals; in chickens it is controlled by a single recessive gene with reduced penetrance. The *unc* (*uncoordinated*) mutants of the nematode *Caenorhabditis* are also affected in GABA-mediated functions. *See* autoimmune diseases; GABA; glutamate decarboxylase deficiency disease; ion channel; mitochondrial disease in humans; myoclonic epilepsy; paroxysmal; seizures. (Meisler, M. H., et al. 2001. *Annu. Rev. Genet.* 35:567.)

Epilobium   Herbaceous plant species ($2n = 36$) in the family of *Onagraceae* distributed widely in Europe, Asia and Africa. It became a favorite organism for the study of nonnuclear inheritance since the early decades of this century. A variety of cytoplasmic mutations were discovered in the different species that were transmitted maternally. The identity of the cytoplasmic genetic material (plasmon) was preserved even when *Epilobium luteum x E. hirsutum* was backcrossed with *E. hirsutum* for 25 generations and less than 3×10^{-8} chance remained for the presence of *E. luteum* genes in the nucleus. Some of the information indicated, however, an interaction between nuclear genes and cytoplasmic genetic elements (plasmon-sensitive genes). Some of the cytoplasmic elements were assigned to the plastids and the others were presumably present in the mitochondria. These nonnuclear genes affected pigment variegation, male and female fertility and a number of morphological changes of the leaf, plant height, etc. From the developmental patterns of the cell lineages (sector formation and sector size), the number of plasmon and plastome (chloroplast-coded elements) were estimated. It was a tragic fact that after the retirement and death of the primary research worker, Peter Michaelis, the majority of the *Epilobium* mutants were lost because of the lack of an organized system for their maintenance. Epilobium species attracted also pharmacological interest. *See Oenothera*. (Michaelis, P. 1965. *Nucleus* 8:83; Battinelli, L., et al. 2001. *Farmaco.* 56[5–7]:341; <http://plantsdatabase.com/genus/Epilobium/>.)

epiloia (tuberous sclerosis, TSC) Caused by more than one dominant gene; human chromosomes 9q34.3 (TSC1), 3p26 and 12q23, and 16p13.3 (TSC2) have been implicated. Its prevalence is about 2×10^{-4}. Characteristic symptoms are seizures, mental retardation, papules on the face, low pigmented spots on the trunk and limbs, etc. Inactivation of either TSC1 or TSC2 results in increased cell size. In the mutant cell lines, the level of cyclin E and cyclin A is higher, which disturbs the cell cycle (Tapon, N., et al. 2001. *Cell* 105:345). The symptoms may vary substantially, yet mental retardation is apparent in 50% of the afflictions. Brain nodules may be revealed by tomography for diagnosis even in individuals who are otherwise nonsymptomatic. The majority of cases are due to new mutations; thus, sibs may have no risk of recurrence. When the anomaly is present in the pedigree, the recurrence risk may approach 50%, although the penetrance may be incomplete. The TSC1-encoded protein, hamartin, is about 130 kDa. The two proteins apparently affect the insulin signaling pathway. *See* mental retardation; tomography. (Dabora; S. L., et al. 2001. *Am. J. Hum. Genet.* 68:64; Inoki, K., et al. 2002. *Nature Cell Biol.* 4:648; [photo by courtesy of Dr. C. Stern, after Dr. V. McKusick].)

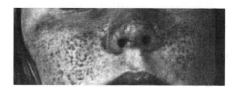

epimastigote   *See Trypanosoma.*

epimers   Stereoisomeric compound that differs at one asymmetric configuration, e.g., glucose and mannose or ribulose 5′-phosphate and xylulose 5′-phosphate; they are reversibly interconverted by epimerase enzymes. *See* enantiomorph.

epimorphosis   Cell-division-requiring regeneration of surface structures of animal cells. *See* morphallaxis; regeneration in animals.

epimutagenic   Physiological factor that causes epigenetic alteration. *See* epigenesis.

epimutation   Epigenetic change during development without change in the DNA sequence. Usually the epimutation, based on methylation of certain segments of the DNA, is erased by passing through the germline. *See* epiallele; epigenesis; imprinting; methylation of DNA; peloric; transgenerational effect. (Frevel, M. A., et al. 1999. *J. Biol. Chem.* 274:29331.)

epinasty   Downward bending of a plant organ because of the more rapid division of the upper cell layer(s). Hyponasty is the opposite phenomenon.

epinephrine (adrenaline)   Animal hormone produced by the kidney (adrenal medulla). When it binds to specific transmembrane receptors of the G_s proteins, they may be phosphorylated to the $G_{s\alpha}$-GTP form, facilitating the activation of adenylate cyclase enzyme, which produces 3′, 5′ cyclic AMP. cAMP mediates the function of protein kinase A and sets into motion a cascade of protein phosphorylations, resulting in the activation of glycogen synthase, phosphorylase b kinase (causing glycogen breakdown), acetyl coenzyme A, carboxylase (required for the synthesis of fatty acids), pyruvate dehydrogenase (oxidation of pyruvate to acetyl-CoA), triacylglycerol lipase (resulting in mobilization of fatty acids in the mitochondria and plant peroxisomes), phosphofructokinase, and fructose-2,6-bisphosphatase (involved in glycolysis and gluconeogenesis. *See* adenylate cyclase; β-adrenergic receptor; G_s; phosphorylase a.

epiphyseal dysplasia, multiple (MED)   Human chromosome 20q13.3–dominant defect in the α-chain of type IX collagen involves early-onset osteoarthritis, walking defects and short stature; it is a sulfate transporter. A recessive form is caused by mutation in the diastrophic dysplasia (5q32-q33.1) sulfate transporter. There is apparently a dominant MED gene

at 2p24-p23, encoding matrilin 3, an oligomeric protein of the cartilage matrix containing a von Willebrand factor A-type protein domain. Several different mutations have been identified at multiple loci for collagen and thus cartilage defects. *See* arthritis; collagen; diastrophic dysplasia; von Willebrand disease; (Chapman, K. L., et al. 2001. *Nature Genet.* 28:393; Czarny-Ratajczak, M., et al. 2001. *Am. J. Hum. Genet.* 69:969.)

episomal vector   *See* transformation, genetic; yeast episomal vector.

episome (plasmid)   Dispensable genetic element in bacteria that can exist in a free state within the cell or integrated into the main genetic material, comprising a couple of thousand to a few hundred thousand nucleotides. A typical episome is the bacterial sex plasmid (F); the temperate phages also behave like episomes. Originally, the term was coined for a hypothetical chromatin attached to the *Drosophila* chromosome. Plasmids of prokaryotes and eukaryotes are also considered episomes. Some plasmids may be diluted out during division of the cells, but the self-replication-competent plasmids may reach high levels in the cells. *See* F element; mitochondrial plasmids; plasmids. (Jacob, F. & Wollman, E. L. 1958. *C. R. Hebd. Séances Acad. Sci. Paris* 247:154.)

epistacy   Quantitative interaction among the products of more than one allelic pair. The term, originally suggested by R. A. Fisher in 1918, has been replaced by *epistasis*.

epistasis   Interaction between products of nonallelic genes, resulting in modification or masking of the phenotype expected without epistasis. The segregation ratios in F_2 may be altered depending on the type of epistasis involved. At *recessive epistasis* (a recessive allele of a locus is epistatic to another recessive allele of another independently segregating locus), the phenotypic classes in F_2 are 9:4:3; at *dominant epistasis* 12:3:1, rather than the common 9:3:3:1. Epistasis may be brought about by modification of gene function due to alterations in the signal-transducing pathway. *Synergistic epistasis* reinforces the consequences of other epistatic factors, e.g., deleterious mutations may enhance each other's detrimental effects. *Indirect epistasis*, in contrast, is not an intracellular phenomenon. A pregnant mother with some metabolic defect may exert a deleterious effect on the genetically normal (heterozygous) developing fetus by placental transfer of harmful metabolites. Similarly, breastfeeding by such mothers may elicit the symptoms of the disease in the nursing babies. Such a situation may exist in cases of phenylketonuric and myasthenic mothers. Epistasis is of interest for quantitative genetic analysis, and it may require sophisticated statistics to separate additive effects from interactions. Thus, epistasis has significance for animal and plant breeding and also for human genetics to determine the role of genes in the development of cancer and other traits under multiple controls. Actually, the majority of genes do not operate independently from each other. *See* crosstalk; epistacy; genetic network; interaction deviation; modified Mendelian ratios; morphogen; myasthenia; phenylketonuria; recombinant congenic; signal transduction; suppressor gene; synergism; two-hybrid method. (Bateson, W. 1907. *Science* 26:649; Mcmullen, M. D., et al. 2001. *Genome* 44:667.)

epistasis group Nonallelic genes with protein products, which may affect each other's function(s). *See* allelism complementation group.

epistatic selection Consecutively occurring mutations may decrease fitness synergistically. *See* fitness; selection.

epistemology Study of the nature and limitations of science.

epithelial cell Cell of the surface layer of organs or bodies.

epithelioma Autosomal-dominant benign or cancerous human skin lesions, encoded in the long arm of human chromosome 9. *See* skin diseases.

epitope Binding site on the antigen for the paratope of the antibody, the antigenic determinant. The *polymorphic* (private) *epitopes* are specific for one MHC, whereas the *monomorphic* (public) epitopes may be shared by more than a single MHC allele. The *cryptic epitopes* are usually expressed at a very low level, but they may be induced to overexpression *Immunodominant epitopes* commonly bind to antibodies. The α-galactosyl epitope (Galα1-3Galβ1-4GlcNAc-R), expressed on the surface of most mammalian cells (except humans and Old World primates), is the most common factor of tissue rejection. This epitope is present on several potentially pathogenic viruses, bacteria, and protozoa. Retroviral vectors also carry this epitope and appear to be the major cause for triggering the complement cascade against them. α-1,2-fucosyl transferase (H-transferase) may reduce α-galactosyl residues on the viral and cellular surface and thus reduce the complement-mediated immune reaction. The elimination of the α-galactosyl epitope either by gene knockout or by increasing H-transferase expression may facilitate xenotransplantation. *See* antibody; antigen; complement; epitope tagging; gene therapy; knockout; MHC; paratope; xenograft.

epitope screening If an antibody is available for a protein, it can be used to screen expression libraries of the antigen with the antibody. The antigen-containing cells are plated, then transferred to nitrocellulose filter replicas and subsequently submerged in a solution of the antibody. The epitope-containing colonies tightly bind the antibody; incubating the filter reveals the position of the complex by a second radiolabeled antibody that binds to the first one. This procedure is of medical significance for the development of immunological defense systems. *See* antibody; antigen; epitope; phage display; radioactive label.

epitope tagging Immunochemical/immunocytochemical method that employs a short polynucleotide encoding a 6- to 10-amino-acid epitope inserted into the gene, which is then transfected into a cell. This epitope already has a known antibody, and when it is expressed in a transgenic cell, it facilitates the tracking of the tagged transgene. It is superior as far as specificity is concerned to natural immunogens. Inserting tandem epitopes improves sensitivity. Generally, either the N or the C end of the proteins is labeled. For identification, Western blots, immunoprecipitation, immunofluorescence, electron microscopy, and other methods are used. Epitope tagging has wide applicability in basic proteomics and

eventually in clinical problems. Some epitope-antibody combinations are commercially available. (Jarvik, J. W. & Telner, C. A. 1998. *Annu. Rev. Genet.* 32:601; Heintz, N. 2000. *Hum. Mol. Genet.* 9:937; Ferrando. A., et al. 2000. *Plant. J.* 22:553.)

epizootic Disease of insects.

EPO Erythroid progenitor cell regulator primarily affecting magakaryocyte differentiation. *See* erythrocyte; megakaryocyte; TPO.

eponym Designation of a phenomenon or principle by the name of the person(s) who discovered it or who was associated with it as a proband, e.g., Punnett square, centi-Morgan, Hogness box, Abraham Lincoln hemoglobin, Lepore hemoglobin. (<www.whonamedit.com>.)

epothilone Polyketide produced by myxobacteria (*Sorangium cellulosum*) in low quantities. It is a spindle fiber poison and is a potential anticancer drug, similar to taxol. The epothilone gene has been cloned and can be produced more efficiently by other transgenic bacteria, such as *Streptomyces coelicolor*. *See* myxobacteria; polyketide; taxol. (Chen, H., et al. 2001. *Chem. Biol.* 8:899.)

epoxide Contains a three-member (oxirane) ring, e.g., ethylene oxide. It is highly reactive.

Epoxide ring

Eppendorf® Although this company merchandises other types of laboratory tools, generally the conical polypropylene microcentrifuge tubes of 1.5 mL capacity are meant. Several companies market 0.5 to 1.5 mL tubes of equal quality.

EPS Exopolysaccharide.

EPS8 Transduces signals between Rac and RAS. *See* E3B1; Rac; RAS; signal transduction.

ε-element At the 5′-end of viral RNA, the minus strand facilitates synthesis and encapsidation of the genome into the viral core. *See* retroid virus.

EPSIN (EPS) Clathrin-associated protein binding the adaptin subunit of AP2. It seems to transmit signals from the endocytotic pathway to the nucleus. *See* adaptin; AP1; clathrin; endocytosis. (Drake, M. T., et al. 2000. *J. Biol. Chem.* 275:6479.)

Epstein-Barr virus (EBV) 172 kbp linear DNA herpes virus with multiple direct repeats of 0.5 kb at both ends. Within the cells the EBV may make up to 100 circular, nonintegrated copies or it may integrate into the chromosomes. Some of its RNA genes are transcribed by RNA polymerase III and use upstream elements for regulation. Herpes viruses cause infectious mononucleosis, Burkitt's lymphoma, nasopharyngeal (nose-throat) carcinoma, Marek's disease (an avian tumor), etc. About 90% of the human populations harbor EBV. EBV infects and immortalizes B lymphocytes with the aid of the EBNA1 and EBNA2 (Epstein-Barr virus nuclear antigen) genes and may be used for the (sometimes inefficient) production of monoclonal antibodies. Replication of EBV is apparently preceded by nucleosomal destabilization brought about by EBNA1. EBNA2 is also a transactivator of other genes without binding to DNA. Responsive promoters are targeted through the CBF1 DNA-binding transcriptional repressor protein. CBF1 binds to a consensus of GTGGGAA at a considerable distance from the EBNA2-responsive cellular promoters. EBNA2 counters transcriptional repression by CBF1.

EBV is the most effective tumorigenic virus of humans. It targets B lymphocytes and transforms them into active immunoglobulin-producing immunoblasts. More than 90% of the infections lay dormant because of the immunological surveillance of the T cells. In Hodgkins disease and nasopharyngeal cancer cells EBV is detectable. In case of a mutation in the cells, EBV displays the X-linked lymphoproliferative syndrome (XLP, Xq24-q25) upon infection. The syndrome includes infectious mononucleosis, acquired hypogammaglobulinemia, and malignant lymphoma. The XLP/SH2D1A gene encodes a 128-amino-acid protein with a single SH2 domain. The XLP/SH21A gene product appears to be the same as SAP (SLAM-associated protein). SLAM (signaling lymphocyte activation molecule, alias CDw150) is a glycosylated transmembrane protein involved in the activation of B and T cells and interaction between B and T cells. The EBV antigen stimulates the proliferation of CD8$^+$ and CD4$^+$ T lymphocytes. The CTLs keep the virus under control until its proliferation is evoked by certain conditions.

EBV can be used as a lymphoid tissue–specific genetic vector. This specificity depends on the presence of the CD21 complement receptor type II protein. The CD21 ligand is a gp350/220 glycoprotein. The vectors form only 2 to 4 copies in mammalian cells. They can carry 35 kb inserts and can be used as a shuttle vector. EBV establishes latency in the target cells, making it a desirable genetic vector. The engineered EBV minireplicons need nontransforming (EBNA2-defective) helper cell–line packaging. The safety of using EBV-based vectors for gene therapy may not be completely assured. *See* Burkitt's lymphoma; cancer; carcinoma; CBF; complement; CTL; Duncan syndrome; herpes; heterohybridoma; Hodgkin disease; hypogammaglobulinemia; lymphocytes; lymphoma; lymphoproliferative diseases; mononucleosis; oncogene; SAP; shuttle vector; SLAM. (Khanna, R. & Burrows, S. R. 2000. *Annu. Rev. Microbiol.* 54:19; Avolio-Hunter, T. M., et al. 2001. *Nucleic Acids Res.* 29:3520; Grinstein, S., et al. 2002. *Cancer Res.* 62:4876.)

Epstein syndrome Macrothrombocytopathy. *See* Alport syndrome.

equational division *See* mitosis.

equational separation, chromosome Takes place in mitosis when the sister chromatids go to opposite poles during anaphase. Similarly, the separation of the chromosomes at anaphase I of meiosis may be equational in case of

recombination or in the absence of mutation. In heterozygous autopolyploids when crossing over takes place and the *Aa* and *aA* chromatids go to the same pole, the separation is equational. When the distance between a gene and its centromere is at least 50 map units (or more), maximal equational segregation occurs. Such a separation may permit the formation of *aa* gametes in an *AAAa* (triplex) and increase the frequency of *aa* gametes in duplex (*AAaa*) and simplex (*Aaaa*) individuals. *See* autopolyploidy; meiosis; mitosis; reductional separation.

equatorial plane Middle region of a dividing cell nucleus where the chromosomes congregate before anaphase begins and where the nucleus will divide into two. Note, sometimes it is called plate, but there is no physical plate at that stage. *See* mitosis.

equatorial plate *See* equatorial plane.

equilibrium State without a net change.

equilibrium, heterozygote Both homozygotes in a random mating population are equally disadvantaged compared to the heterozygotes, and heterozygote equilibrium takes place relative to homozygotes. *See* genetic equilibrium; Hardy-Weinberg theorem.

equilibrium, mutation The occurrence of a number of mutations in a random mating population has the same consequence as immigration, i.e., the frequency of the allele may increase. Unless the new mutation has a substantial selective advantage, its survival (fixation) or death (extinction) is equally likely. If the mutation is frequent and random, elimination (drift) is insignificant; maintenance of the mutant allele may be assured, however, even when it does not have a selective advantage. If a mutation from *A* to *a* is a regular event, the frequency of *a* may increase at the expense of *A*:

$$q_{n+1} = q_n + \mu(1 - q_n) \qquad \{1\}$$

where q = frequency of the recessive allele and $(1 - q)$ = frequency of the dominant allele, n = the number of generations, and μ = the mutation rate.

After a number of generations (n), the initial frequency of the recessive allele (q_0) may increase to q_n by the acquisition of the same mutant alleles ($A \rightarrow a$) as represented:

$$e^{-n\mu} = (1 - q_n)/(1 - q_0) \qquad \{2\}$$

and hence

$$-n\mu = ln[(1 - q_n)/(1 - q_0)] \qquad \{3\}$$

If, e.g., $q_0 = 0.05$ (as we hypothesize for an example), the number of generations required to double its frequency to $q_n = 0.10$ can be computed if the value of μ (mutation rate) is known, e.g., $\mu = 10^{-5}$. According to $\{3\}$, $ln[(1 - 0.10)/(1 - 0.05)] = ln[0.90/0.95] = ln\, 0.94737 = 0.05407 = -n\mu$. When 0.05407 is divided by 0.00001 (the mutation rate given above, 10^{-5}), we get 5,407, meaning that under the conditions specified by this hypothetical example, 5,407 generations are required to double the frequency of the recessive allele. Since this change depends not on the number of years but on the number of generations, species with many generations annually may change more rapidly than the ones with long times to sexual maturity and gestation. Therefore, it is easier to test these mathematical models with bacteria, *Drosophila*, or mice than with humans or elephants.

The rate of change of allelic frequencies in a random mating population can be expressed as:

$$\Delta q_n = \mu(1 - q_{\hat{n}}) \qquad \{4\}$$

where Δ indicates change in the frequency of the recessive allele (q). Thus, the rate of change $\{4\}$ for the hypothetical experiment is $10^{-5}(0.90) = 0.000009$. The larger the number of alleles, which can mutate, the larger is the chance for the change. One must also take into account the fact that mutations may revert. Accordingly,

$$\Delta q = \mu p - rq \qquad \{5\}$$

where μp represents the mutation $A \rightarrow a$, and rq stands for backmutation $A \leftarrow a$ (r = reversion). At mutational equilibrium,

$$\hat{q} = \mu/(\mu + r) \qquad \{6\}$$

and

$$\hat{p} = r/(\mu + r) \qquad \{7\}$$

where $\hat{q}$ and $\hat{p}$ represent the equilibrium frequencies of the recessive and dominant alleles, respectively.

It is evident if p is larger than q, a larger r is required to keep equilibrium with a smaller μ, e.g., if $p = 0.8$ and $q = 0.2$ ($p + q$ is always 1) and $\mu = 0.00001$, r must be 0.00004 to make $0.8 \times 0.00001 = 0.2 \times 0.00004$. Usually in nature the frequency μ (forward mutation) is larger because $A \rightarrow a$ change may occur by more mechanisms than base substitution alone (e.g., deletion, frameshift, etc.), whereas reversion (r) is expected to take place mainly by nucleotide replacement. Therefore, in the maintenance of allelic equilibrium, selection usually has a major role. *See* allelic frequencies.

equilibrium centrifugation   The centrifugation is continued in a density gradient until each macromolecule (subcellular organelles) reaches a position corresponding to its density. *See* density gradient centrifugation; ultracentrifugation.

equilibrium constant (K$_{eq}$)   Concentrations of all reactants and products at equilibrium under specified conditions. *See* dissociation constant.

equilibrium dialysis   Known quantities of antigen and antibody are placed in a dialysis bag. Only small antigens that cannot react with the antibody can pass the bag membrane. This way, the number of specific binding sites can be determined on the basis of antibody inside and outside the bag. *See* antibody; antigen; dialysis; immune reaction; valence.

equilibrium dissociation constant (Kd)

$$K_d = \frac{(A)(B)}{(AB)}$$

$$K_a = \frac{(AB)}{(A)(B)}$$

Expresses the strength of interaction between two molecules where (AB) is the concentration of the complex and (A) and (B) are the concentrations of the separate molecules (K_a). Its reciprocal (K_d) is the equilibrium association constant.

equivalence group   Cells with equal differentiational potential at a particular time.

ERAB   Endoplasmic reticulum–associated binding protein, a putative hydroxysteroid dehydrogenase, binds to the amyloid-β peptide and appears to cause neuronal dysfunction in Alzheimer disease. *See* Alzheimer disease. (He, X. Y., et al. 2001. *Eur. J. Biochem.* 268:4899.)

ERBA   Genes present in multiple copies in both the human (chromosome 17q22-q23 or 17q11-q12) and the mouse genomes. ERBA was held responsible for the potentiation of ERBB1; the protein is a receptor for thyroid hormones and functions in the nucleus as a transcription factor. It has the highest expression in the nervous system, and, unlike other thyroid receptors, not in the liver. It is related to the avian erythroblastic leukemia virus oncogene and is involved in leukemia and other cancers in humans including some translocations. *See* erbB; ERBB1; hormone response elements; hormones; oncogenes; regulation of gene activity; TRE. (Andersson, M. L. & Vennstrom, B. 2000. *Oncogene* 19:3563.)

erbA   Retroviral oncogene produces a transcription factor (member of the steroid receptor family).

erbB   Avian erythroleukemia oncogene (receptor for EGF and other cellular and viral proteins). Its overexpression may confer resistance to anticancer chemotherapy and radiation treatment. Herceptin and taxol may block erbB. Erb-specific tyrosine kinase inhibitors may reduce its effect on chemoresistance. *See* avian erythroblastoma; ERBB1; herceptin; taxol. (Olayioye, M. A., et al. 2000. *EMBO J.* 19:159.)

ERBB1   Oncogene (human chromosome 7p12-p22, mouse chromosome 7) encodes the glycoprotein, epidermal growth factor receptor (EGFR), a transmembrane protein tyrosine kinase. The protein has two subunits: one contains phosphotyrosine and phosphothreonine, and the other contains, in addition, phosphoserine. ERBB2 has been assigned to human chromosome 8p22-p11. The synonyms of ERBB2 are NEU and HER2. Probably ERBB2 is a normal growth factor, but when its expression is enhanced, it becomes a protooncogene. The amplification of its expression has been detected in adenocarcinomas and gastric cancer. ERBB2 product is present in neuroblastomas, breast and ovarian cancers. ERBB3 is a related (mammary) oncogene in human chromosome 12q13. *See* breast cancer; cyclin D; EGF; glycoprotein; Heregulin; neuroblastoma; oncogenes; protein kinase; protooncogene.

ERBBA   Two human genes in chromosome 19 (locus EAR2, with preferential expression in the liver) and chromosome 5 (locus EAR3) encode steroid and thyroid hormone receptors. The homologous retroviral gene is *erb*. *See* animal hormones; retrovirus.

ERC (1) Extrachromosomal rDNA circle. Repeated sequences in ribosomal DNA (rDNA) may recombine and yield extrachromosomal circles. It has been suggested that accumulation of these circles may result in aging. (2) Entorhinal cortex. Part of the central nervous system where neoritic degeneration occurs in Alzheimer disease. (Du, A. T., et al. 2001. *J. Neurol. Neurosurg. Psychiatry* 71:441.)

ERCC *See* excision repair; RAD25.

ERDA1 Expanded repeat domain CAG/CTG with 50 to 90 repeats at 17q21.3. *See* trinucleotide repeats. (Bowen, T., et al. 2001. *Psychiatr. Genet.* 10:33.)

E region of GTP-binding proteins Interacts with effectors (amino acids 32–42 in RAS). Mutation in this region may abolish oncogenic transforming ability without affecting GTP binding. *See* GTP; oncogenes; RAS.

eRF Recognizes stop codons in mRNA and terminates translation; stimulates peptidyl-tRNA cleavage and release. *See* stop codon; tRNA. (Ohta, M., et al. 2001. *Plant Cell* 13:1959.)

erg Unit of energy or work in dyne acting through a distance of 1 cm. *See* dyne unit of force.

ERG DNA-binding protein, oncogene, encoded in human chromosome 21q22.

ergodic rule Rare events are more likely to be found when the population size is large.

ergosterol Most common sterol in some fungi. It is a D_2 provitamin and is used as an antirachitic drug (for the prevention of certain bone deficiency problems). *See* cholesterol.

ergot Dry sclerotia (compact dry mycelial mass) of *Claviceps purpurea* fungus on rye ears (and some other grasses) containing a large number of alkaloids and peptide alkaloids, some of which promote the contraction of the uterus (oxytocic) and are highly toxic. Lysergic acid is one of the main alkaloids of ergot. It affects the nervous system, causes confusion, and has been used as a psychomimetic. It was supplied to the defenders of the Soviet political "conception" trials to admit crimes they did not commit. *See* alkaloids; psychomimetic. (Mukherjee, J. & Menge, M. 2000. *Adv. Biochem. Eng. Biotechnol.* 68:1.)

ERK Family of extracellular signal-regulated protein kinases. ERK5/BMK1 is required for epidermal growth factor to induce cell proliferation. ERK kinases are important for histone (H3) phosphorylation and activation of transcription. *See* EGF; MAP kinase; SAPK; signal transduction; (Chen, R. H., et al. 1992. *Mol. Cell. Biol.* 12:915; Adachi, M., et al. 2000. *J. Cell Biol.* 148:849; Howe, A. K., et al. 2002. *Current Opin. Genet. Dev.* 12:30.)

ERM Ezrin-radixin-moesin is a protein cofactor complex in its active state of GTPase reactions. It affects the cytoskeletal system and anchors filaments to the cell surface. *See* cytoskeleton; GTPase; merlin; neurofibromatosis. (Bretscher, A. 1999. *Curr. Opin. Cell Biol.* 11:109.)

ERP Transcription factor of the ETS family. *See* ETS; transcription factors.

Erp61 Endoplasmic reticulum stress protein with functions similar to PDI. *See* Erp72; PDI.

Erp72 Endoplasmic reticulum protein with thioredoxin- and PDI-like functions. *See* PDI; thioredoxin.

ERR Extra radiation risk expresses the increased cancer or other risk compared to the unexposed population. *See* EAR; risk.

error-free repair *See* DNA repair.

error in aminoacylation Approximately 3×10^{-4} is the rate of charging a tRNA with the wrong amino acid in prokaryotes and eukaryotes, 10^{-5} in yeast, 10^{-4} and 10^{-3} in higher forms. Several aminoacyl-tRNA synthetases carry out editing functions by hydrolyzing the misactivated aminoacyl adenylates and aminoacyl-tRNAs. DNA aptamers are also involved in editing functions. Amino acids larger than the cognate amino acid are eliminated in a single step. Amino acids smaller than the critical ones are selected in a second screening step. This is called a double-sieve model. *See* ambiguity in translation; aminoacyl tRNA synthetase; aptamer; editing; error in replication. (Parker, J. & Holtz, B. 1984. *Biochem. Biophys. Res. Commun.* 121:482; Mori, N., et al. 1985. *Biochemistry* 24:1231; Jakubowski, H. & Goldman, E. 1992. *Microbiol. Rev.* 56:412; Nordin, B. E. & Schimmel, P. 1999. *J. Biol. Chem.* 274:6835.)

error in replication Leads to base replacement and thus mutation. The rate of replicational errors of the different DNA polymerases varies, but it is partly compensated for by the editing function of the $3' \to 5'$ exonuclease activity of the polymerases. The error rate of the α-subunit of DNA polymerase III of *E. coli* is about 10^{-5}, but it is reduced by about two orders of magnitude by exonuclease subunit ε. The base substitution error of the *E. coli* DNA polymerase III holoenzyme is within the range 5×10^{-6} to 4×10^{-7}. The repair polymerase, pol I of *E. coli*, also has an error rate of about 1×10^{-5}, but the $3' \to 5'$ exonuclease activity again reduces the errors by two orders of magnitude. The T7 DNA polymerase has an error rate of 10^{-3} to 10^{-4}, but the repair system lowers it to 10^{-8} to 10^{-10}. The RNA polymerases of RNA viruses do not have proofreading and editing functions and their error rate may vary within the 10^{-3} to 10^{-4} range per nucleotide. The mitochondrial DNA polymerase

γ has a base substitution error of $\sim 3.8 \times 10^{-6}$ to 2.0×10^{-6}. This appears to be one or two orders of magnitude higher than in the nucleus of mammals. The proofreading rate of the exonuclease depends a great deal on the availability of replacement nucleotides and therefore the difference in the final fidelity may not be as much. *See* DNA replication mitochondria; editing; mutation rate; proofreading; reverse transcriptases. (Kunkel, T. A. & Bebenek, K. 2000. *Annu. Rev. Genet.* 69:497; Johnson, A. A. & Johnson, K. A. 2001. *J. Biol. Chem.* 276:38090; Johnson, A. A. & Johnson, K. A. 2001. *J. Biol. Chem.* 276:38097; Pesole, G., et al. 1999. *J. Mol. Evol.* 48:427.)

error in transcription Is higher (10^{-3} to 10^{-5}) than in replication because the DNA-dependent RNA polymerases do not have editing functions. The consequence of error in RNA is not serious relative to mutation in DNA. The yeast protein SII and the *E. coli* proteins GreA and GreB used to be credited for stimulating the cleavage of RNA with misincorporated bases. Experimental data indicate that neither of these proteins is indispensable and their proofreading value appears low. *See* ambiguity in translation; error in replication; mutation rate; RNA polymerase. (Paolini-Giacobino, A., et al. 2001. *Hum. Genet.* 109:40; Shaw, R. J., et al. 2002. *J. Biol. Chem.* 277:24420.)

error in translation *See* ambiguity in translation.

error-prone repair *See* DNA repair; mutasome; SOS repair; Y-family DNA polymerases. (Goodman, M. F. 2002. *Annu. Rev. Biochem.* 71:17.)

error types Type I (α) rejecting a true hypothesis, type II (β) accepting a wrong hypothesis on the basis of statistical analysis. *See* power of a test; significance level.

error variance Variance arising from agents, conditions beyond the ability to control an experiment and with which the apparent effect of any controlled factor must be compared to obtain meaningful evaluation. *See* analysis of variance.

erucic acid (13-eicosenoic acid) Monoethenoid acid in the seeds of *Cruciferae* (rapeseed, mustard, horseradish, etc.) and *Tropaeolaceae* plants. It may constitute 50 to 80% of the fatty acids in these plants. It has poor digestibility, but even worse, it accumulates in the muscles and livers of animals (humans) and causes serious and irreversible pathological changes. In the newly developed varieties of rapeseed (canola, *Brassica napus*), the erucic acid content has been reduced to 0.3% and thus this plant is a valuable oil-seed crop, particularly in cooler climates where soybeans do not fare well. The mutation blocks the conversion of oleic acid (18:1) to eicosenoic acid (20:1) and erucic acid (22:1). *See* canola; fatty acids. (Hans, J., et al. 2001. *Plant Mol. Biol.* 46:229.)

ERV *See* endogenous retrovirus.

Erwinia Group of gram-negative enterobacteria. It is a pathogen for a variety of plant species. *Drosophila* appears to be a vector.

erythroblast Nucleus-containing erythrocyte, or in a looser sense, immature red blood cells.

erythroblastoma Tumor-type mass of nucleated red blood cells.

erythroblastosis The circulating blood contains immature red blood cells.

eryhtroblastosis fetalis (neonatarum) When red blood cells of the developing fetus enter the maternal blood circulation of immunogenic mothers, an immunization reaction may be generated. It may be particularly intense when the mother is Rh-negative and the fetus has Rh-positive blood. The antibodies formed by the mother may then enter the fetal blood supply and cause increased bilirubin production (causing nerve damage) and agglutination and lysis of the fetal blood, resulting in intrauteral death or severe anemia unless the newborn's blood is replaced immediately after delivery. Monitoring for bilirubin or Rh-antibody accumulation, especially in the later stages of the pregnancy, may prevent the development of erythroblastosis. If the tests are positive and before Rh antibodies accumulate, the mother may be given anti-Rh γ-globulin that can be transferred through the placenta to the fetus and affords protection against erythroblastosis. Erythroblastosis may also be caused by infection by the avian retrovirus, AEV, and ABO blood group incompatibilities. The latter is quite common, but its effects are usually mild. *See* Rh blood type; SU antigen in pigs. (Stockman, J. A., III & de Alarcon, P. A. 2001. *J. Pediatr. Hematol. Oncol.* 23[6]:385.)

erythrocyte (red blood cell) Disk-shaped, biconcave cell. The red color of erthyrocytes is due to the red oxygen carrier molecules known as hemoglobin. The mature erythrocytes no longer contain nuclei or nuclear DNA. Primarily DNaseII of the macrophages mediates the enucleation during development. *See* blood; hemolytic disease; sickle cell anemia. (*Ann. Rev. Genet.* 31:33.)

erythroderma, ichtyosiformis, non-bullous (ALOX, 17p13.1) Due to mutations in lipoxygenase. *See* ichthyosis; lipoxygenase. (Jobard, F., et al. 2002. *Hum. Mol. Genet.* 11:107.)

erythrokeratodermia variabilis (1p34-p35) Dominant hyperkeratosis with transient red patches, encoded by gene GJB3, responsible for the connexin 31/30.3 component of gap junctions. *See* Charcot-Marie-Tooth disease; connexins; deafness; gap junctions; skin diseases.

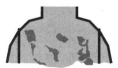

Sharp red keratotic spots on the human back.

erythroleukemia Based on an autosomal-dominant gene causing neoplastic growth of the immature and mature red blood cells, increase in the size of the liver and spleen, and acute anemia. It is considered a malignant disease. *See* leukemia.

erythermalgia Autosomal-dominant disease with intense burning pain in the feet and hands, redness, heat sensation, and swelling. A susceptibility locus was found at 2q31-q32. *See* pigmentation defects. (Drenth, J. P. H., et al. 2001. *Am. J. Hum. Genet.* 68:1277.)

erythromycin Group of three antibiotics produced by *Streptomyces erythreus*. They inhibit amino acid chain elongation on the ribosomes (*See* antibiotics). The *ery* (erythromycin resistance) gene is the most distal marker to *ap* (attachment point) in the chloroplast DNA of *Chlamydomonas reinhardi* according to recombination tests in cytohets (heterozygous for ctDNA). *See* chloroplast genetics; chloroplast mapping.

erythropoietin (EPO) Highly specific erythrocyte-stimulating glycoprotein factor (M_r 51,000), produced mainly in the kidneys and acting in the bone marrow. It was isolated from anemic sheep, but it is now also produced by recombinant DNA technology. It is used as a medication to avoid the need for blood transfusions in patients prone to anemia. The EPO receptor (EPOR) is a protein of the cytokine receptor family. It is formed as a dimer with an extracellular ligand-binding, a short, single-pass transmembrane segment, and a cytoplasmic domain without kinase activity. Its signaling relies on the Jak/STAT pathway. EPO/EPOR regulates caspases, which in turn control the fate of GATA-1 and apoptosis with the aid of FAS. EPO may induce tumor regression and antitumor immune response in murine myeloma. Cross-talk between Janus kinase and NF-κB signaling cascades results in neuroprotection by erythropoietin. *See* apoptosis; CIS; FAS; GATA; growth factors; hematopoiesis; Jak kinases; megakaryocyte; NF-κB; polycythemia; signal transduction. (Spivak, J. L., et al. 1991. *Blood* 77:1228; Digicaylioglu, M. & Lipton, S. A. 2001. *Nature* 412:641.)

ES Embryonic stem cell; cells that can proliferate in an undifferentiated state but can give rise to differentiated cells as well and can be returned to an embryo to become part of it. These pluripotent cells are taken from blastocyst-stage embryos. They are functionally similar to the meristem cells of plants. *See* pluripotency; stem cell; totipotency.

ESAG Expression site–associated genes. *See* Trypanosoma.

escape commitment After initiation of transcription by RNA polymerase II and synthesis of four nucleotides, the polymerase proceeds to elongation with an *escape commitment* and leaves the promoter. *See* transcription factors.

Escherichia coli See E. coli

ESE Exon splicing enhancer. About six to eight nucleotide sequences within exons promote efficient splicing. They generally bind SR proteins with an arginine/serine-rich tract. *See* exon; NAS; splicing; SR motif. (Woerfel, G. & Bindereif, A. 2001. *Nucleic Acids Res.* 29:3204; Tian, H. & Kole, R. 2001. *J. Biol. Chem.* 276:33833.)

ESI Electrospray ionization. A source of ions in mass spectrometers for the analysis, e.g., of proteins. *See* electrospray; MALDI/TOF/M MS/MS; mass spectrometer. (Han, X. 2002. *Anal. Biochem.* 302:199.)

E site (exit site) On the ribosome to which the deacylated tRNA moves from the P site after the peptidyl-tRNA–mRNA complex moves from the A to the P site. The process may be reversible. Elongation factor EF-T mediates the regeneration of the active EF-Tu-GTP from EF-TU-GDP. Thus, besides the initially identified A and P sites, this third site, E, is now known. In contrast to the A and P sites, the E site may not be a permanent structure; rather it is just a transient intermediate stage of the peptide elongation process. *See* amino acid activation; EF-TU-GTP; protein synthesis; ribosome.

ESMP Expressed sequence marker polymorphism is actually a SNIP marker. *See* SNIP.

ESS Exon splicing silencer. *See* ESE; NAS; NMD. (Tange, T. O. & Kjems, J. 2001. *J. Mol. Biol.* 312:649.)

Ess1 *See* parvulins

essential amino acids Cannot be synthesized de novo by vertebrates and must be provided in the diet: arginine (in young), histidine, isoleucine, lysine, methionine, phenylalanine, threonine, tryptophan, valine. *See* high-lysine corn; kwashiorkor.

essential athrombia *See* thrombopathia.

essential fatty acids Required by mammals, although they cannot synthesize them (linoleate [18:2 cis-Δ^9, Δ^{12}] and α-linolenate [18:3 cis-Δ^9, Δ^{12}, Δ^{15}] because they are unable to introduce additional double bonds beyond C-9. *See* fatty acids.

essential gene Cannot be dispensed of without lethal consequences. In *Saccharomyces cerevisiae* yeast, more than 80% of the genes are nonessential. (Tong, A. H. Y. Y., et al. 2001. *Science* 294:2364.)

EST *See* expressed-sequence tag.

established cell line Of animals, consists of fused cells (with one component being myeloma cancer cells) and is capable of indefinite growth without senescence. *See* HeLa; immortalization.

ester Formed when the OH end of an alcohol is combined with the COOH end of an acid, leading to the removal of H_2 such as R—O—CO—R. Nucleotides may form ester linkages during chain elongation. *See* nucleotide chain growth; phosphodiester bond.

esterases Enzymes involved in hydrolysis of ester bonds. Some proteolytic enzymes catalyze the reaction shown below. In biochemical systems, the esters may be phosphate-, glycerol-esters and many other molecules. Esterase A-4 (ESA4) is encoded in human chromosome 11q13-q22; ESD has been located to 13q14.11; other human variants exist.

$$R_1-\overset{\overset{\textstyle O}{\|}}{C}-O-R_2 + H_2O \rightleftharpoons R_1-C\underset{O^-}{\overset{O}{<}} + HO-R_2 + H^+$$

Ester   Acid   Alcohol

estradiol (estrogen) Steroid hormone produced by the animal (human) ovaries and placenta. It regulates secondary sexual characters and female estrus and implantation. 17β-estradion and raloxifene indirectly activate the estrogen-response element. The β-estradiols are carcinogenic through their metabolites (4-hydroxyesterone, 16α-hydroxyesterone, semiquinones) by contributing to reactive oxygen and lipid compounds. Estrogen-like substances in the environment pose reproductive hazards to humans and wildlife. Estradiol (17β) susceptibility in mice may vary by a factor of 16 or more among different strains. Although estrogen does not appear to be mutagenic in most of the standard bioassays, it generates DNA-estrogen adducts and damages DNA by free radicals. It produces 8-hydroxyguanine and may induce chromosomal aberrations. Estrogens may act as cancer promoters in cases where cancer cells are already present. Some estrogens or estrogen-like compounds are not steroids, such as diethylstilbestrol, hexestrol, and the flavone and isoflavone phytoestrogens (coumestrol, genistein), mycoestrogens, and some alkylphenols, arylphenols, and the nonaromatic pesticides, e.g., dieldrin, endosulfan, chlordane, some carbamates, and the herbicide atrazine. Some pesticides such as DDT and methoxychlor are estrogenic polluters. Similarly, the industrially used polychlorinated biphenyls (PCB), automobile exhaust, and Agent Orange, etc., may be considered endocrine disrupters and pose environmental hazards to humans and animals.

Estradiol.

Estrogens act through the estrogen receptors and various ligands and other cooperative molecules and may regulate + and − the transcription of many different genes (breast cancer, prolactin, dopamine, plasminogen activator, peroxidase, renin, etc.). Estrogens may have posttranscriptional and posttranslational regulatory influences. They affect the cholinergic activity in the brain, and when this function diminishes with advanced age, learning and memory are impaired. Estrogens are involved in many metabolic processes. 17β-hydroxysteroid dehydrogenase 2 and cytochrome P_{450} monooxygenases metabolize estrogens. The latter enzymes may produce carcinogenic catechol estrogen. The antiestrogen, tamoxifen, also appears carcinogenic to the liver and with prolonged use to the endometrium of the uterus. Some of the estrogens appeared mutagenic in various types of assays. Several diseases (breast, endometrial, cervical, and prostate cancers, thyroid diseases) increase hydroxylation of estradiol, whereas osteoporosis and obesity may decrease hydroxylation. In pregnancy the level of estrogens rises rapidly and the maintenance of an appropriate level is required for fetal development. It has been suggested that estrogens regulate the level of 5-hydroxy-tryptamine/serotonin in the brain and a lower supply of estrogens may thus be a contributing factor to menopausal and postpartum depression. Estrogens may control glucose transport to the brain and may protect against neurodegeneration such as associated with aging (Alzheimer disease). Gonadal hormones show protective effects against heart disease by contributing to the integrity of the vascular wall. Estrogens and antiestrogens modulate the immune system in various ways. A low level of estrogen

is present normally in mammalian males. Its deficiency due to mutation in the aromatase gene lowers the fertility of male mice, but its role in the human male is less obvious, although an increase in estrogen may cause gynecomastia. Human males with aromatase deficiency show increased levels of estradiol. Under normal conditions, 17β-estradiol level in males is higher than in postmenopausal women. *See* Alzheimer disease; androgen; animal hormones; antiestrogens; aromatase; choline; depression; estrogen response element; gynecomastia; nitric oxide; phytoestrogen; raloxifene; steroid hormones; tamoxifen. (Petterson, K. & Gustafsson, J. Å. 2001. *Annu. Rev. Physiol.* 63:165.)

estriol 16α-hydroxylation product of estradiol. *See* estradiol.

estrogen *See* animal hormones; estradiol.

estrogen receptor (ER) Dimeric (ERα + ERβ), and to become functional, ERα is phosphorylated at Ser[118] by mitogen-activated protein kinase (MAPK) in the presence of epidermal growth factor (EGF) and insulin-like growth factor (IGF). Dimeric ER associated with the proper ligands can bind to the estrogen receptor element in the chromosome and activate transcription of the estrogen-receptive genes. ER can bind also to AP1 transcriptional activator protein. ERα bound through the Jun and Fos oncoproteins to an AP1 site is activated by estradiol-17β. In contrast, ERβ-bound estradiol-17β inhibits transcription. The antiestrogens, tamoxifen and raloxifene, are transcriptional activators of ERβ at the AP1 site. These pieces of information indicate that the two ERs differ in their signaling properties depending on the response element and the ligand. Practical significance of the finding is that this may provide a clue of why tamoxifen is an inhibitor of breast cancer but at the same time carcinogenic to the uterus, whereas raloxifene may be beneficial for the chemotherapy of breast cancer without a risk to the uterus. There are some observations that after a few (2–5) years raloxifene may alter the cellular machinery and promote tumorigenesis. *See* breast cancer; EGF; estrogen; IGF; MAPK; MAPKK; phytoestrogen; raloxifene; tamoxifen. (Shang, Y., et al. 2000. *Cell* 103:843; Klinge, C. M. 2001. *Nucleic Acids Res.* 29:2905.)

estrogen response elements (ERE) Generally located about 200–300 bp upstream of the transcription initiation site. Despite their diversity, they generally share a low homology consensus, e.g., the *Xenopus* vitellogenin and the chicken ovalbumin gene carry the dyad GGTCANNNTG^A_TCC *See* animal hormones; hormone response elements; raloxifene; regulation of gene activity; tamoxifen. (Driscoll, M. D., et al. 1998. *J.Biol. Chem.* 273:29321; Shang, Y. & Brown, M. 2002. *Science* 295:2465.)

estrus Recurrent sexually receptive periods of female mammals (except humans) accompanied by a sexual urge (heat); it is induced by cyclic ovarian hormonal activity (oestrus).

Eta *See* osteopontin.

ET cloning Transformation of *E. coli* by using homologous recombination with the aid of a plasmid expressing the products of gene *RecE*, exonuclease VIII, and RecT. The gene to be transferred is flanked by about 40 nucleotides. *See* RecE; RecT.

ethenobases Metabolites of vinyl chloride through the action of the cytochrome P450 system produces chloroethylene oxide, which leads to the formation of several carcinogenic adducts. Among these are etheno bases. Ethenodeoxyguanine pairs with adenine and despite the relatively small amounts formed it is primarily responsive for the mutagenic/carcinogenic effects in the liver and brain. *See* environmental mutagens. (Morinello, E. J., et al. 2002. *Cancer Res.* 62:5183.)

1,N⁶-ethenodeoxyadenine   3,N⁴-ethenodeoxycytosine

N²,3-ethenodeoxyguanine

ethanol Common laboratory sterilizing agent used in 70–80% aqueous dilutions. *See* alcoholism.

Ethernet Wiring and software required to share data through the Internet and local talk.

ethics Rules of animal and human behavior based on genetic and instinctive or social and moral traditions. The *Gene Letter* URL (<http://www.geneletter.org>) is a source for current problems and information on genetics. *See* behavior genetics; bioethics; embryo research; ethology; genetic privacy; misconduct, scientific; nuclear transplantation; publication ethics. (<http://www.ornl.gov/hgmis/elsi/elsi.html>; <http://www.nhgri.nih.gov/ELSI/>.)

ethidium bromide (EB) Polycyclic fluorescent dye. Its fluorescence increases about 50-fold when it intercalates between DNA bases, and the sensitivity of the staining is so high that 2.5 ng DNA is detectable. It is used for revealing DNA in agarose or polyacrylamide gels and buffers (0.5 mg/mL). The UV light absorbed by DNA at 254 nm is transferred to EB, which itself has two absorption maxima at 303 and 366 nm and emits red-orange fluorescence at 590 nm maximum. DNA more than 250 ng/mL can quantitatively be measured spectrophotometrically. The detection of single-strand DNA or RNA is much less effective by EB. Intercalation into closed circular DNA is much less efficient than into linear DNA. Therefore, EB-stained plasmid or organellar DNA can be separated by buoyant density centrifugation from linear DNA by becoming of lesser density. EB-stained gels can be destained with water or by 1 mM $MgSO_4$ for 20 min. EB can be removed from the DNA by washing several times with an equal volume of water-saturated 1-butanol or isoamylalcohol. EB is a powerful mutagen that must be handled carefully with gloves; the used solutions after dilution to no more than 0.5 mg/L require decontamination by 1 vol. 0.5 M $KMnO_4$ and careful mixing of 1 vol 2.5 N HCl for several hours, followed by neutralization with 1 vol. 2.5 N NaOH. This procedure reduces mutagenicity to 3×10^{-3} of the untreated solutions.

ethnicity Human population related by biological features identifiable by anthropometric, cultural, linguistic, biochemical, serological, and molecular characteristics and gene frequencies. The identification may not use all of these methods, depending on the nature of the genetic distances in the comparisons. The various ethnic groups and isolated communities provide valuable information for population genetics research. Protection of privacy of individuals (*Nature Genet.* 23:275) deserves serious consideration, however. HGDP (Human Genome Diversity Project) has the goal of surveying the world population for genetic diversity at the DNA level, disease susceptibility/resistance genes, evolution, population genetics, etc. Mapping by admixture linkage disequilibrium (MALD) is a potential analytical statistical method. *See* genetic distance; Jews and genetic diseases; MALD. (Guglielmino, C. R., et al. 2000. *Ann. Hum. Genet.* 64:145; Pastinen, T., et al. 2001. *Hum. Mol. Genet.* 10:2961; Arab genetic diseases: Teebi, A. S., et al. 2002. *Hum. Mut.* 19:615; Finnish diseases: Sopilä, K. & Aula, P. 2002. *Hum. Mut.* 19:16; <http://gdbwww.gdb.org>.)

ethological isolation Type of sexual isolation. Males or females refuse to mate for some "psychological" reason(s). *See* sexual isolation.

ethology Study of behavior. *See* aggression; behavior genetics; ethics; instinct; morality.

ethyl ($-CH_2CH_3$) Alcohol radical.

ethylene (C_2H_4) Simplest plant hormone, yet it controls a wide variety of morphogenetic and developmental processes including stem and root growth, seed germination, fruit ripening, senescence, and sex determination. Ethylene is synthesized as a side branch of the Yang cycle. The key enzymes in ethylene synthesis are ACC synthase and ACC oxidase. A multigene family encodes ACC synthase and is regulated by hormonal, physical, and environmental signals. In *Arabidopsis*, a series of mutants are available that determine constitutive ethylene response and are overproducers. Ethylene-insensitive mutations include a histidine kinase (*ein1, etr1*); *ain* is ACC-insensitive; *ctr1* is a putative serine/threonine kinase similar to the members of MAPK. *EIN-2* has a central role in the pathway, encoding an integral membrane protein, and it is a transducer of ethylene and stress responses. Gene *ERS* (ethylene response sensor) has some structural similarity to *ETR*, but its role is upstream in the pathway. The order of gene action has been determined for most of the numerous mutations. Ethylene plays a key role in plant signaling and, from practical points of view, in fruit ripening and disease resistance. Several of the disease- and stress-resistance genes and other genes controlling differentiation contain GCC box repeats that are also ethylene-response elements. The ethylene-insensitive *ETR1* gene of *Arabidopsis* conveys ethylene binding in this fungus when transferred to yeast. The ethylene nonresponding mutation *hookless1* seems to be a regulator of auxins. *See* fruit ripening; SAR, hormone-response elements; hypersensitive reaction; MAPK; MAPKK; plant hormones; Yang cycle. (Bleecker, A. B. & Kende, H. 2000. *Annu. Rev. Cell. Dev. Biol.* 16:1; Johnson, P. R. & Ecker, J. R. 1998. *Annu. Rev. Genet.* 32:227; Wang, K. L.-C., et al. 2002. *Plant Cell* 14:S131.)

ethyleneimine *See* EI.

ethylene oxide (C_2H_4O) Colorless, explosive gas used as a sterilizing agent for instruments, textiles, some food, soil, etc. It is hazardous to handle, irritant to mucous membranes and lungs, and may cause pulmonary edema at higher concentrations. *See* sterilization.

ethyl methane sulfonate (EMS, $CH_3SO_2OCH_2CH_3$) Alkylating mutagen and carcinogen. Its prime target for mutagenesis is the alkylation of the O^6 position of guanine, although it preferentially alkylates the 7 position of guanine, but this rarely leads to base substitution. It alkylates all other nucleic acid bases, too, and may cause strand scission by depurination. It is one of the most useful chemical mutagens for a wide range of organisms. *See* alkylation; carcinogens; depurination; hydrogen pairing; mutagens.

etiolated Plants grown in the absence of light display elongation and lack of the typical leaf pigments (chlorophyll and carotenoids). The chloroplasts do not develop and the plastids arrive only to the etioplast stage (thylakoid membrane system incomplete). The leaves are not fully expanded. *See* deetiolation; photomorphogenesis; phytochrome.

etiology Study of cause(s) of disease and its development. Chemical etiology investigates the origin and cause(s) of the development of molecules.

etioplast Lacks the typical chloroplast pigments and displays a prolamellar body. *See* chloroplasts; etiolated; prolamellar body; proplastid.

ETL Economic trait loci determine phenotypes that are important for practical agriculture.

etoposide Synthetic cytostatic compound; it inhibits topoisomerase II, cancer, nucleoside transport, and CDK2. It may promote the function of nucleotidyl transferase and gene conversion in somatic mammalian cells. Etoposide may induce chromosomal aberrations and aneuploidy. When rubbed on the skin of rats treated by chemotherapy, hair loss was prevented in 50% or more of the animals. *See* aneuploid; CDK; cytostatic; terminal deoxynucleotidyl transferase; topoisomerases. (Jacob, S., et al. 2001. *Cancer Res.* 61:6555.)

ETS External transcribed spacers are located between pretRNA gene clusters and at other genes controlling growth and development:

ETS − 5′ − 18S − ITS − 5.85 − ITS − 28S) − ETS

The ETS family members usually bind to different protein monomers or dimers. *See* GABP; ITS; PU.1; SAP; spacer DNA. (Borovjagin, A. V. & Gerbi, S. A. 2001. *Mol. Cell Biol.* 21:6210.)

ETS oncogene ETS1 is located to human chromosome 11q23-q24 and is involved in acute monocytic leukemia (AMol) when interferon β gene (IFB-1, chromosome 9p22) is translocated to it. ETS-1 and ETS-2 genes are physically contiguous in birds. These homologous genes are coordinately expressed, and the proteins are both in the cytoplasm and nucleus. In humans, ETS-2 genes are in chromosomes 21q22 and 1-q22.3 and their expression is not coordinate; ETS-1 protein is cytoplasmic, whereas ETS-2 is nuclear. The translocation t(8;21)(q22;q22) is common in patients with acute myeloid leukemia (AML-M2). The breakage point in the latter cases is not within the ETS-2 sequences. The ETS-2 gene lacks TATA box and CAAT box, and alternative elements are substituted for them. ELK oncogenes display homology to ETS oncogenes. This family of transcription factors has about 35 known members and regulates various aspects of growth, development, and lymphocyte pool maintenance. PEA3, a member of the ETS family of proteins, binds to the 5′-AGGAAG-3′ DNA motif and suppresses the expression of HER-2 oncogene. Ets opposes p16^{INK4a} in cooperation with Id1 during senescence. *See* ELK; ERP; HER; interferon β-1; leukemia; lymphocytes; oncogenes; SAP. (Blair, D. G. & Athanasiou, M. 2000. *Oncogene* 19:6472.)

Eubacteria The majority of bacteria belong to this subkingdom of prokaryotes; the other subkingdom is *Archaebacteria*.

eucaryote *See* eukaryote; prokaryote.

euchromatic Chromosomal regions containing euchromatin. *See* euchromatin.

euchromatin Does not absorb the common nuclear stains during interphase and is normally transcribed into mRNA. In the human genome, the euchromatic fraction varies a great deal among the chromosomes. In chromosome 2, an almost 160 Mb sequence is less than 60% euchromatic, whereas a more than 30 Mb sequence in chromosome 21 is almost entirely euchromatic. *See* heterochromatin. (Yu, A., et al. 2001. *Nature* 409:951; <http://www.euchromatin.net/>.)

EUCIB European Collaborative Interspecies Backcross is a cooperative work group for plants.

Euclidean distance Calculations are used to construct dendrograms and evolutionary trees. In the formula below, x and y represent observations in a set of multivariate data. *See* dendrogram; evolutionary tree; multivariate analysis.

$$d_{xy} = \sqrt{\sum_{i=1}^{q} (x_i - y_i)^2}$$

Eug 65 kDa yeast protein in the endoplasmic reticulum with disulfide isomerase activity. *See* Mpd; PDI.

eugenic reform, need for According to R. A. Fisher (*The Genetical Theory of Natural Selection*): "The various theories which have thought to discover in wealth a cause of infertility, have missed the point that infertility is an important cause of wealth." *See* eugenics.

eugenics Application of genetic principles to the breeding of the human race with the purpose of improvement. *Positive eugenics* wishes to enrich the frequency of "favorable" genes, whereas *negative eugenics* wants to eliminate the "undesirable" genes by selective breeding. The word *eugenics* was coined by Sir Francis Galton in 1883, unaware of Mendel's discoveries, but his statistical data indicated that about 25% of the sons of eminent fathers were also eminent. He concluded that heredity has a role in talent, intellect, and behavior. These fields are highly controversial because of the lack of scientific bases for the objective measurement of "desirable" and "undesirable" and the potentials of exploitation for political and unethical goals. The eugenics movement generally emerged during economic downturns in an effort to find scapegoats for the ills of society, and the support generally came from people without any understanding or training in genetics. By 1917, in 16 U.S. states laws were approved for compulsory sterilization of the feebleminded, insane, rapists, some criminals, and other "hereditary unfit" individuals. Actually, by 1935, about 30,000 sterilizations were carried out on the basis of these state laws.

With the rise of scientific population genetics, it became obvious that this type of negative selection was ineffective. According to the Hardy-Weinberg theorem, if the frequency of the homozygous recessive "undesirable" individuals is $1/20,000$ (5×10^{-5}), then at genetic equilibrium the frequency of that gene is $\sqrt{0.00005} \cong 0.0071 = q$ and the frequency of the carriers (heterozygotes) is $2 \times p \times q = 2 \times 0.9929 \times 0.0071 \cong 0.014099$, i.e., about 1/71. Thus, about 99.3% of this "undesirable" allele will be present in the heterozygotes of the population and about 1/71 individuals will have a 50% chance of passing it on to their children. If all the "undesirable" individuals are prevented from reproduction by sterilization, the frequency of their "bad" gene (q_n) will change in 100 generations to

$$q_n = \frac{q_0}{1 + 100 q_0} = \frac{0.0071}{1 + (100 \times 0.0071)} \cong 0.004152$$

Thus, the initial gene frequency of 0.0071 will be reduced to about 0.004152 and the number of carriers by about 41% to 1/121 in about 3,000 years. Also, behavioral traits are under the control of multiple genes scattered in the genome, and each of them contributes partly to the phenotype. Furthermore, these characteristics are under polygenic control and are greatly affected by environmental influences. Thus, the negative eugenic measures are biologically ineffective and ethically unacceptable in enlightened societies. Nevertheless, in the name of eugenics, Hitler's regime exterminated 6 million Jews and millions of others of different ethnic groups as well as sterilized more than 250,000 people. Negative eugenics, although it is quite ineffective as shown, may still have some justification in some forms chosen by the individuals at high genetic risk of disease. The simplest humane and intellectually and ethically correct solution is refraining from reproduction. Therapeutic abortion is counterproductive because it may increase the number of carriers, although the "defective" individuals (homozygotes) are eliminated.

Positive eugenics has some biological and ethical problems because human values cannot be adequately assessed. Certain measures may, however, be acceptable and are practiced in societies without naming them as eugenics, e.g., scholarship to college students may facilitate family support and presumably intellectually better individuals may not be prevented from procreation because of economic hardship. Also, reproduction at a younger age may reduce chromosomal defects (*see* Down syndrome). The Nobel-laureate geneticist H. J. Muller advocated positive eugenics through his entire career. He considered it a necessity to fight genetic load through *germinal choice*, meaning that spousal love should be separated from the procreational role in marriage. He suggested relying on gene banks for artificial insemination of women and recommended systematically screening the sperm donors on the basis of health, intellect, and social consciousness. The sperm of the selected individuals was supposed to be stored frozen and used only some years after their death to make an objective and reliable assessment of value. Although such a program may appear reasonable, problems in value judgment remain. Muller believed that in a true Socialist system these problems could be overcome. His own disenchantment with the Marxist society of the USSR proved, however, otherwise.

Some general fears of intervention in the human system of reproduction are still not completely dissolved. Will the controlled insemination reduce the gene pool? Is it conceivable that the selection may foster the increase of some as-yet unforeseen genetic defects either by lack of recognition or by hitchhiking (linkage)? Is there a risk that political systems impose their selfish will upon the biologically desirable systems of reproduction? The problems of positive eugenics may not be ignored, however. The progress in medical technologies prevents the selection against formerly inferior traits. The use of medication, prosthetics, somatic gene therapy, etc., are devices of contraselection. Perhaps the replacement of genes of the germline involved in clinically proven defects may offer a solution. The methodology appears clear in principle, but the consequences are still untested. Thus, the "brave new world" is not yet at hand, primarily because our values cannot be defined in a simplistic manner. Eugenics must be separated from the racist views that discredit the field. Multiracial and multicultural societies offer unique advantages in the diversities for the betterment of life. Biologists must find the facts, but the application of the principles requires the democratic decisions of the ethicists and society. Will ethicists know enough biology? *See* eutelegenesis; Galtonian inheritance; gene therapy; genetic counseling; Hardy-Weinberg theorem; sterilization in humans. (Kempthorne. O. 1997. *Genetica* 99:109; Li, C. C. 2000. *Hum. Hered.* 50:22; Micklos, D. & Carlson, E. 2000. *Nature Rev. Genet.* 1:153; Gillham, N. W. 2001. *Annu. Rev. Genet.* 35:83; historical display on origin and flaws: <http://vector.cshl.org/>.)

Euglena   Green (has chloroplast) flagellate protozoon, $n = 45$. (Sheveleva, E. V., et al. 2002. *Nucleic Acids Res.* 30:1247.)

eukaryote   Organism with enveloped cell nucleus, mitosis, and meiosis such as fungi, plants, and animals. (euGenes information system: <http://iubio.bio.indiana.edu/eugenes/>.)

Eulerian path   Mathematical solution for DNA fragment assembly in a novel way named after the mathematician L. Euler (1707–1783), who solved basic problems of sets like the Venn diagrams. *See* Venn diagram. (Pevzener, P. A., et al. 2001. *Proc. Natl. Acad. Sci. USA* 98:9748.)

eunuch   Person whose testes have been removed by castration, preventing sexual function. If castration takes place early in childhood, it reduces expression of secondary sexual characteristics. Eunuchs were used in the Middle East and North Africa as guardians of the harems and chamberlains. The Chinese emperors employed hundreds of them in the forbidden city as counselors. In Italy, until 1871, the castrati dominated the operatic stages and church choirs because these male sopranos and contraltos had high-pitched voices with a great range. The removal of the testes contributed to the increased development of the vocal folds and, combined with greater capacity of the chest and lungs, facilitated finer expression than expected from women who were banned from such roles on "moral" grounds. Males are called eunuchs in the medical field if their sexual function is diminished by genitalia underdevelopment or mutation in sex hormone genes. *See* castration.

euphenics   Corrective measures for genetically determined defects with the aid of nongenetic means (e.g., spectacles, prosthetics, insulin, etc.). *See* eugenics.

euploid   Cells or organisms with one or more complete genomes. *See* aneuploid.

eusocial   Altruistic behavior of the sterile worker class in social insects supporting the fitness of the colony by tending to the needs of the reproductive casts. *See* altruistic behavior; fitness; social insects.

eutelegenesis   Form of positive eugenics where selected sperm sources are used for voluntary artificial insemination. The expected impact on improving IQ has been defined as

$$\overline{x}_1 = p\frac{\overline{x} - M}{2}h^2 + M$$

where $\overline{x}$ = the mean IQ of the sperm donors, p = the percentage of females involved, M = the mean, h^2 = heritability. *See* eugenics; heritability; human intelligence.

eutely   In some species of nematodes (*Caenorhabditis* and relatives), all individuals have the same cell number and all organs in all individuals show identical cell numbers. There are some exceptions to this rule, however. *See Caenorhabditis elegans*. (Azevedo, R. B. R. & Leroi, A. M. 2001. *Proc. Natl. Acad. Sci. USA* 98:5699.)

euthanasia   Method of causing "painless" death practiced on handicapped or seriously incapacitated or fatally ill persons. Infanticide and late-term abortion may be the most cruel forms of it. These are morally unacceptable procedures in the majority of socially enlightened societies. The morality of euthanasia has been questioned even when it is a form of suicide. Euthanasia has no eugenic consequence because the objects are not expected to reproduce due to age or illness.

eutherian   True placental mammal but not the marsupial or monotrene. *See* marsupials; monotrene.

eV   *See* electron volt; volt.

E value   Value in an alignment match (by BLAST) that is equivalent or better than expected by chance alone. The lower the E value, the better the score. *See* alignment; BLAST.

Evan's blue (direct blue)   Strong oxidant and suspected carcinogen. It can be used for the isolation of intact protoplasts, which do not stain by it, whereas the damaged ones appear blue-green. *See* protoplast.

Eve, foremother of molecular mtDNA   Evolutionary hypothesis attempting to trace the origin of the human race on the basis of mitochondrial DNA. Mitochondrial DNA in humans is transmitted only through the females. Because of the uniparental transmission, recombination cannot reshuffle the base composition of mtDNA. Due to technical and computational difficulties, the assumption that a single woman's descendants populate the earth could not unequivocally be placed in Africa, and the investigators remained divided concerning the notion that all humans descended from a single female or even from a single group of females. Most likely, Eve represented a human population of 10,000 to 100,000 living about 150 to 200 kya (kiloyears ago). Some recent studies (based on the decline of linkage disequilibrium) indicate the possibility that paternal mitochondria may be transmitted during fertilization and recombination may alter the maternal mtDNA sequences in larger blocks. *See* archeogenetics; effective population size; F_{ST}; genetic drift; mitochondrial genetics; mtDNA; mutation rate; out-of-Africa; paternal leakage; Y chromosome. (Owens, K. & King, M.-C. 1999. *Science* 286:451; Richards, M. & Macaulay, V. 2001. *Am. J. Hum. Genet.* 68:1315; Malhi, R. S., et al. 2002. *Am. J. Hum. Genet.* 70:905.)

E vector (self-inactivating vector, Ψ-vector)   Construct of large direct flanking repeats around the encapsidation signal (E/Ψ) may bring about a very high frequency of template switching of the reverse transcriptase. As a consequence, the viral (spleen necrosis virus, mouse leukemia virus) encapsidation protein may be deleted in 99% or even higher frequency and the virus will thus be inactivated. Also, suicide genes or other DNAs can be reconstituted by the same mechanism. Such a vector, when it includes selectable markers (e.g., neo), can be efficiently selected for in the presence of G418 antibiotic in gene therapy by retroviral vectors. *See* G418; neo; SIN vector. (Julias, J. G., et al. 1995. *J. Virol.* 69:6839; Delviks, K. A. & Pathak V. K. 1999. *J. Virol.* 73:8837.)

eversporting   Organisms that carry unstable gene(s) and display somatic or germinal variegation. Many of the eversporting conditions are caused by the presence of transposable elements. *See* retroposons; retrotransposons; transposable elements; transposons.

evicting plasmid   Incompatible with another type of plasmid. If selection is favoring the evicting plasmid (e.g., carries a gene for antibiotic resistance), the other plasmid can be eliminated. These operations can be used for the construction of particular genetic vectors. *See* vectors.

EVI oncogene   Integration site for retroviral insertions in human chromosome 3q24-q28, translocated to chromosome 5q34, resulting in nonlymphocytic leukemia. The Evi-1 mouse gene may cause murine leukemias. Its protein product has numerous zinc-finger domains and is presumably a transcription factor. The EVI2A gene is in the proximal part of the long arm of human chromosome 17. EVI2B is in the same chromosome about 15 kb apart. Its product is apparently a transmembrane protein with surface receptor function. A homologue of these genes is associated with murine myeloid leukemia. *See* leukemia; oncogenes; SMAD.

evocation   Induction of differentiation. *See* differentiation.

evo-devo   New term for evolutionary developmental biology. It studies the effect of development on the direction of evolution, the relation between micro- and macroevolution, developmental genes across phylogenetic boundaries, etc. (Arthur, W. 2002. *Nature* 415:757.)

evolution Process and theory of the biological and physical changes that bring about the variety of the living world and its environment at the global and cosmic ranges. Biological evolution is based on changes in gene frequencies or function brought about by mutation, horizontal gene transfer, rearrangement by transposable elements, selection, migration, adaptation, and fixed by reproductive isolation. The general evolutionary theory integrates all ideas about the nature, origin, and future of the universe. Organic evolution is concerned with the origin, development, and relationship of past and present living organisms. According to the most widely accepted current views, evolution came about in three major pathways: (1) bacteria (including among others proteobacteria and cyanobacteria); (2) eukarya (animals, fungi, and plants); and (3) archaea (euryarchaeota, crenarchaeota). The proteobacteria became symbionts in eukarya and evolved into mitochondria. The mechanism of replication, transcription, and translation of the archaea group is more similar to that of eukarya than to bacteria. Cyanobacterial symbionts in plants developed into chloroplasts. Evolutionary studies are greatly facilitated by genome-wide sequence information. The majority of protein-coding genes have homologues across widely different taxa. Safe information on function cannot always be obtained from nucleotide sequence data. In *Drosophila*, some genes with substantial homology to chitinase genes actually encode imaginal disk growth factors. Genes with clear phenotypes or high levels of expression are more likely conserved across phylogenetic boundaries.

Evolution is frequently contrasted with the views of creation as described in the Bible and other holy books of religions. The science of evolution is concerned with facts that can be experimentally studied with the available technological tools and has no room for faith, whereas the primary criterion of religion is the faith in its teaching. Thus, it is improper to compare these principles because they are not of the same nature and they are not required to support or exclude each other. The philosopher K. R. Popper defined the scientific method as "the criterion of potential satisfactoriness is thus testability, or improbability: only a highly testable or improbable theory is worth testing and is actually (and not merely potentially) satisfactory if it withstands severe tests—especially those tests to which we could point as crucial for the theory before they were ever undertaken." He also says, "I refuse to accept the view that there are statements in science which we have, resignedly, to accept as true merely because it does not seem possible, for logical reasons, to test them." Thus the testability and refutability are the most basic cornerstones of evolution. Evolution is using the methods of cytogenetics, population genetics, molecular biology, and the geological fossil records to establish the relationship, origin, and variation in the living world. Thoughts about evolution that are not supported by the methods referred to above are just ideas, not science. With the progress of sequencing DNAs, RNAs, and proteins, evolutionary studies are shedding light on mechanisms common to a wide range of organisms and are assisting in the solution of problems in health-related areas as well as improving the productivity of economically important organisms. Evolution is generally believed to be a *stochastic* (random) process in finite (small) populations and drift plays out the differences. In case the population is infinitely large, evolution may be *deterministic* (predictable on the basis of selective values), and selection has major importance. *See* acquired characters, inheritance of; covarion; creationism; duplication; Eve foremother; evolutionary distance; evolutionary trees; fossil records; genome projects; genomics; Hardy–Weinberg theorem; horizontal gene transfer; hovergen; human evolution; macroevolution; microevolution; microsporidia; migration; molecular evolution; mutation; orthologous loci; paralogous loci; PAUP; phylogeny; population genetics; reductive evolution; selection; speciation; unified genetic map; Y chromosome. (Rouzine, I. M., et al. 2001. *Microbiol. Mol. Biol. Rev.* 65:151; Cronk, Q. C. B. 2001. *Nature Rev. Genet.* 2:607; Medina, M., et al. 2001. *Proc. Natl. Acad. Sci. USA* 98:9707; Joyce, E. A., et al. 2002. *Nature Rev. Genet.* 3:462; <www4.nas.edu/opus/evolve.nsf>; <mesquite.biosci.arizona.edu/mesquite/mesquite.html>.)

evolution, cost of Evolution proceeds by replacing old alleles with new ones. The replacement may sacrifice some individuals. *See* genetic load.

evolution, directed Involves *in vitro* synthesis of proteins or polynucleotides with new functions. *See* mutation directed. (Xia, G., et al. 2002. *Proc. Natl. Acad. Sci. USA* 99:6597.)

evolution, in vitro Mutations in RNA molecules can occur in vitro during cell-free replication and may improve their catalytic and amplification rate. *See* evolution, prebiotic.

evolution, non-Darwinian Supposed to take place by the fixation of neutral mutations, in contrast to Darwinian evolution, which postulates evolution by survival of the fittest. Because of the degeneracy of the genetic code, many mutations will leave the phenotype or function unaltered. There is no reason why these neutral mutations would not be fixed, although it must be remembered that the codon usage in different genes and/or organisms is not identical and the codon selection may have an adaptive nature. Also, there is no reason to doubt that certain amino acid substitutions have no affect on protein structure and function and thus appear without selective effect. Rapid evolution is favored by the fixation of neutral changes. If neutral mutations are fixed, there is no need for concomitant elimination of the old genotypes. Random genetic drift may also lead to changes in gene frequencies and possibly to evolution. *See* beneficial mutation; drift, genetic; neutral mutation. (Matsuda, H. & Ishii, K. 2001. *Genes Genet. Syst.* 76[3]:149; King, J. L. & Jukes, T. H. 1969. *Science* 164:788.)

evolution, prebiotic Before living cells appeared on Earth, organic molecules must have been formed. In 1953, Stanley Miller and Harold Urey showed that when mixtures of ammonia, methane, hydrogen, and water were exposed to electric sparks (simulating stormy conditions of the early atmosphere), carbon monoxide, carbon dioxide, amino acids, aldehydes, and hydrogen cyanide were generated. More extensive experiments later revealed hundreds of organic molecules under such simulated abiotic conditions. It was of particular significance that more than 10 amino acids, various carboxylic acids, fatty acids, adenine, sugars, etc., could be obtained in abiotic experiments with increased sophistication. HCN and heavy-metal ions present in the early environment and in laboratory simulation could promote polymerization. Thus, the conclusion appeared logical that primitive macromolecules could be formed de novo and the macromolecules required for life could eventually evolve.

Amino acids under anaerobic aqueous conditions at 100°C and pH 7 to 10, in the presence of NiS, FeS, and CO, and H_2S or CH_3SH as catalysts—coprecipitate and form peptides. *See* abiogenesis; evolution in vitro; evolution of proteins; evolution of the genetic code; origin of life; spontaneous generation, unique and repeated. (Brack, A. 1999. *Adv. Space. Res.* 24:417.)

evolution, retrograde *See* retrograde evolution.

evolution and base substitution

Generally estimated on the basis of number of changes per site per 10^9 years. Accordingly, the rates in various genomes are: mammalian nuclear 2–8, angiosperm nuclear 5.4, mammalian mitochondrial 20–50, angiosperm mitochondrial 0.5, angiosperm cpDNA single-copy regions 1.5, angiosperm cpDNA inverted repeats 0.3. The high mutation rate of mammalian mtDNA may be the result of the deficiency in excision, photoreactivation, and recombinational repair, as well as the low level of selection and the ability of tRNAs of mtDNA to recognize all four synonymous codons within a family of amino acids. Base substitution rates in fungi and plant mtDNAs are much smaller than those in mammalian mtDNAs. Molecular evolution is generally interpreted on the basis of common descent of the nucleotide sequences. The orthogonal theories must accommodate the apparent facts of horizontal transfer by way of symbiosis and infections in both prokaryotes and eukaryotes. *See* base substitutions; diversity; evolution; evolutionary distance; Jukes-Cantor estimate; orthologous loci; parallel substitution; plasmids; retroposons; retrotransposons; retroviruses. (Salse, J., et al. 2002. *Nucleic Acids Res.* 30:2316.)

evolution and duplication

Molecular evidence indicates that duplications have played an important role in evolution. Duplication may arise by unequal crossing over between similar genes and may occur repeatedly. The example below shows the molecular consequences of an unequal crossing over in the DNA at the level of the protein products. Human haptoglobins have two α- and two β-chains, and when an unequal crossing over takes place, duplication is also detectable in the protein product. *See* cluster homology region; deletion; duplication; unequal crossing over. (Ohno, S. 1970. *Evolution by Gene Duplication*. Springer, New York; Ku, H.-M., et al. 2000. *Proc. Natl. Acad. Sci. USA* 97:9121; ibid., p. 4168; Vallente Samonte, R. & Eichler, E. E. 2002. *Nature Rev. Genet.* 3:65) diagram below.

evolutionary clock (molecular clock)

Measures the time required in 1% amino acid replacement in a protein in a million years (MY), based on the assumption that mutations are neutral and occur at random, and the amino acid changes reflect the time elapsed since a particular event. The molecular clocks may be biased because of selection. A better procedure is to determine the rate of nucleotide substitution in homologous genes because mutation in synonymous codons does not lead to amino acid replacement. The rate of protein evolution is frequently expressed as the accepted number of point mutations per 100 residues in the protein. This parameter is generally symbolized with the acronym PAM. The rate of amino acid substitution per site per year was estimated to be on average 10^{-9}, which is frequently referred to as the *pauling* unit of molecular evolution (named after the Nobel-laureate chemist, Linus Pauling).

Proteins	Amino Acid Substitutions Per Residues Per Million Years
Fibrinopeptides	90
Pancreatic Ribonuclease	33
Hemoglobins	14
Cytochrome C	3
Histone 4	0.06

Individual proteins evolve at very different rates (see table below). Perhaps better comparisons can be made if enzymes rather than other proteins are used. There are differences in substitution rates among different organisms. Mitochondrial base substitutions involve similar differences. The data may be biased if orthologous and paralogous loci cannot be safely distinguished. From the comparison of enzymes it appears that eukaryotes and eubacteria shared common ancestors about 2 billion years ago. The divergence of plants and animals took place about 1 billion years ago. The similarity of fungi to animals is somewhat better than to plants. The molecular evolutionary clocks cannot adequately estimate evolution because of the highly different rates of amino acid or base substitutions in individual proteins and genes, respectively. Multiprotein sequences have been used with promise for improved estimates. There is no comforting proof that the molecular clocks ticked uniformly through the ages. Another problem is the not-infrequent discrepancy between the molecular and the paleontological estimates. Unfortunately, there are not many choices for the evolutionists. In a somewhat similar manner, the progression of a cancerous tumor can be reconstructed. Colorectal tumors may be caused by the loss of genes controlling mismatch repair (MMR). The loss of MMR also involves expansion of microsatellite sequences. Microsatellites are expanded during cell divisions and thus the expansion, relative to the noncancerous or nonexpanded

these α1 chains have 84 amino acid residues and Hp^{IF} a lysine (K) has at position 54, whereas Hp^{15} has glutamic acid (E) at the same site

the total length of Hp^2 (α2) chain originating by an unequal crossing over (χ) is 153 amino acids long, containing both the K and the E sites state, indicates the number of cell divisions since the loss of MMR. A mass spectrometric method of deamidation of glumaminyl and asparaginyl residues in peptides has been developed as a molecular timer. *See* colorectal cancer; Eve foremother; microsatellite; orthologous; paralogous; racemate; Y chromosome. (Robinson, N. E. & Robinson, A. B. 2001. *Proc. Natl. Acad. Sci. USA* 98:944; Nei, M., et al. 2001. *Proc. Natl. Acad. Sci. USA* 98:2497.)

evolutionary distance   Can be numerically estimated on the basis of allelic differences (amino acid differences in proteins, RFLPs, nucleic acid base sequences, etc.). It must be remembered that evolution may proceed either by divergence or convergence of whatever criteria are used. When a larger number of loci is used, the greater is the accuracy of the estimate. There are several procedures in the literature. Here the method of Nei (*Molecular Population Genetics*, Elsevier/North Holland, 1975) is illustrated. The normalized identity of alleles in populations is determined by dividing the arithmetic means of the products of allelic frequencies by the geometric means of the sum of squares of the homozygote frequencies in each population to be compared:

$$I = \frac{\{[p_1 \times p_2] + [q_1 \times q_2]\}/L}{\sqrt{\{[p_1^2 + q_1^2] \times [p_2^2 + q_2^2]\}/L^2}}$$

where I is the index of identity, p and q are allelic frequencies in the two populations, and L is the number of loci studied (including even those with single-allele representation). The evolutionary (genetic) distance is then calculated from the natural logarithm of I, i.e., $D = -\log_e I$. This type of calculation is meaningful and reliable if a minimum of 25 loci are compared. It can be used with any type of alleles, including proteins of different electrophoretic mobilities. An example for the procedure is in the table below. By using the outlined procedure, the human racial distance and the number of years of divergence have been estimated (after M. Nei & A. K. Roychoudhury, 1974. *Am. J. Hum. Genet.* 26:421) and it turned out to be minimal:

Human Races	Genetic Distance	Years of Divergence
Caucasoid–African Negroid	0.023	115,000
Caucasoid–Oriental Mongoloid	0.011	55,000
African Negroid–Oriental Mongoloid	0.024	120,000

The evolutionary distance can be estimated more precisely on the basis of nucleotide sequence and replacements (M. Kimura. 1980. *J. Mol. Evol.* 16:111): the pyrimidine $\rightleftarrows$ pyrimidine and the purine $\rightleftarrows$ purine substitutions are designated as P and the pyrimidine $\rightleftarrows$ purine or purine $\rightleftarrows$ pyrimidine transversions are represented by Q The evolutionary distance K is

$$K = -(0.5)\log_e[(1 - 2P - Q)\sqrt{1 - 2Q}]$$

The standard error of K is

$$s_k = \frac{1}{\sqrt{n}}\left\{\sqrt{[a^2P + b^2Q]} - [aP + bQ]\right\}$$

where

$$a = \frac{1}{[1 - 2P - Q]}$$

and

$$b = 0.5\left[\frac{1}{1 - 2P - Q} + \frac{1}{1 - 2Q}\right]$$

In a sequencing study of the 438 nucleotide β-globin genes of chicken and rabbit, the P value was $58/438 = 0.132$ and $Q = 63/438 = 0.144$, and after the appropriate substitutions the evolutionary distance thus appeared to be 0.347 ± 0.0329.

The time of divergence (in millions of years) varies according to the specific gene and taxonomic category considered, e.g., in Echinodermata–Chordata, Annelida–Chordata for ATPase: 786 and 1059, for cytochrome: 883 and 1078, for cytochrome oxidase I: 1160 and 1465, for cytochrome oxidase II: 608 and 773, for 18S RNA: 1,288 and 1,214, respectively. *See* amino acid sequencing; array hybridization; $(\delta\mu)^2$; DNA likelihood method; DNA sequencing; evolutionary tree; Fitch-Margoliash test; four-cluster analysis; F_{ST}; genetic distance; least square methods; minimum evolution method; neighbor joining method; nucleotide diversity; protein likelihood method; transformed distance. (Tamura, K. & Nei, M. 1993. *Mol. Biol. Evol.* 10:512.)

evolutionary substitution rate   It estimates the number of mutations leading to an evolutionary divergence in millions of years:

$$\sum \frac{(d_{ij} - 2d_{ik})}{n_i n_j} \times \frac{1}{t_j}$$

where d_{ij} and d_{ik} are the number of the DNA base substitutions per number of sites in genotypes i and j, and between genotype i and evolutionary tree node k, respectively; $n_i n_j$ are the number of pairwise comparisons, t is the time in millions of years when the i genotype was deposited in the rock stratum, and k is the node furthest from the common root. *See* evolutionary distance; evolutionary tree; PAUP.

evolutionary tree   Displays the descent of organisms from one another. Initially trees were constructed on the basis of morphological traits and later on the basis of chromosome numbers and pairing affinities in hybrids and chiasma frequencies. The former criteria are not well quantifiable; the latter features are applicable to only relatively closely related forms because of the sexual isolation among distant species. The numbers of substitutions at a large or all comparable sites of macromolecules provides the most accurate information. Evolutionary trees based on nucleotide sequences are more reliable than those based on proteins because silent genes are not included in the proteins. Since DNA sequences are available in several organisms, nucleotide sequence alignment and percentage of homologies are frequently used. Caution may be necessary to interpret evolutionary trends because the similarities or differences can be brought about either by convergence or divergence. From the nucleotide sequences in the DNA, the amino acid sequence in the protein can be relatively easily inferred with some caution on the basis of the genetic code. Before nucleotide sequencing became practical

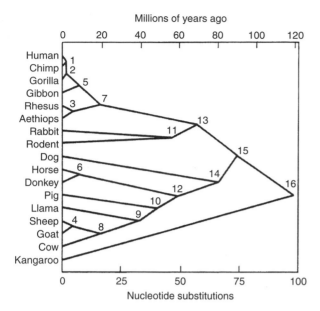

Millions of years ago

An evolutionary tree of mammals was constructed on the basis of molecular clocks of seven proteins. (After Fitch, W. M. & Langley, C. H. 1976. *Fed. Proc.* 35:2092.)

GTA	GTG	VALINE
↓	↓	
TTA	TTG	LEUCINE
↓	↓	
TCA	TCG	SERINE

in the late 1970s, amino acid sequencing was available, and many of the evolutionary trees were based on protein primary structure.

An evolutionary tree of mammals was constructed on the basis of molecular clocks of seven proteins. The right end of the tree was fixed by paleontological information on marsupial divergence from the placental mammals an estimated 120 million years ago. The nodes of divergence are based on the putative nucleotide substitutions in the seven coding genes. (After Fitch, W. M. & Langley, C. H. 1976. *Fed. Proc.* 35:2092.)

Since amino acids have up to six synonymous codons, there is no simple way to determine the nucleotide sequence in the gene from the amino acid positions in the protein. Furthermore, a present nucleotide or amino acid sequence may have evolved in different ways. Molecular evolutionists rely on the principle of *maximal parsimony*, i.e., they suppose that the actual evolution took place in the simplest possible way (which may not be true). Let us assume that the nucleotide sequence A C C A is the result of evolution. Through two mutational events (indicated by →) the original sequence may have changed in the following ways:

A	-	A	-	A	-	A		A	→	C	-	C	→	A		C	-	C	-	C	→	A
		↓		↓												↓						
A		C		C		A		A		C		C		A		A		C		C		A

Let us assume that in three different taxons (or orthologous loci) at a particular site, valine, leucine, and serine are found. Valine has four code words (GTT, GTC, GTA, GTG), leucine has six (TTA, TTG, CTT, CTC, CTA, CTG), and serine has six (TCT, TCC, TCA, TCG, AGT, AGC). A single mutation may replace the first G of the valine codon by T and a second mutation may lead to a T → C change at the second nucleotide; thus, in place of a valine residue, serine may occur. From the four valine triplets, only two may result in serine codons via two mutational events.

There are many ways to construct evolutionary trees. Each has advantages and disadvantages under a particular condition. Evolutionary trees of genes may not be the same as phylogenetic trees because individual genes evolve at different rates. The evolutionary trees constructed by different methods appear different, depending on type and number of markers used. *See* amino acid sequencing; bootstrap; DNA likelihood method; DNA sequencing; Euclidean distance; evolutionary distance; exhaustive search; Fitch-Margoliash test; four-cluster analysis; gene tree; heuristic search; homoplasy; indel; least square methods; METREE; microarray hybridization; neighbor joining method; patristic distance; PAUP; population tree; protein likelihood method; rooted evolutionary tree; transformed distance; unrooted evolutionary tree; UPGMA. (Miyamoto, M. M. & Ctacraft, J. 1992. *Phylogenetic Analysis of DNA Sequences.* Oxford Univ. Press for detailed discussions; software: <http://evolution.genetics.washington.edu/phylip/software. html>.)

evolution of DNA amounts   There is a tendency toward increased amounts of genetic material in the more highly evolved organisms. Viruses generally have less DNA than bacteria, and microorganisms have smaller genomes than higher eukaryotes (*see* genome). It is not entirely clear, however, why the total size of the genome varies by orders of magnitude among higher eukaryotes and, e.g., amphibians, and several plants have much more DNA than humans. This is frequently called the C-value paradox. *See* C amount of DNA; C-value paradox; deletion; duplication; genome.

evolution of chemicals   *See* evolution, prebiotic.

evolution of the genetic code   The genetic code is practically universal with the exception of a few mitochondrial codons. Furthermore, the structures of tRNAs reveal much greater similarities than expected on the basis of random assembly of the nucleotides. The most plausible interpretation of these facts is that these nucleotide sequences developed by an evolutionary process from common ancestral sequences, also indicating a relation of descent of eukaryotes from prokaryotes. It seems that the most archaic genetic code existed in RNA. The original sequences were probably relatively short, although they probably contained meaningless tracts mixed with useful ones. For the development of well-organized nucleotide sequences specifying an ancestral gene, mechanisms were required for the recognition and correction of errors. This ability probably evolved when the first DNA sequences appeared.

Modern DNA polymerases are generally endowed with synthetic, proofreading and editing (endonuclease) functions, which has been indispensable for the development of larger and conservative genetic molecules. The first codon probably specified those amino acids, which were most abundant among the molecules formed under abiotic conditions. It was suggested that the earliest codons were of the GNC type (where N is any base). This idea is supported by the fact that

G≡C base pairs having three hydrogen bonds are more stable than A=T pairs, and under simulated prebiotic conditions, glycine (GGC), alanine (GCC), aspartic acid (GAC), and valine (GUC) are produced in relative abundance. It is conceivable that these four amino acids formed the first protein involved in replication. The next three amino acids might have been glutamic acid, serine, and phenylalanine. These seven early amino acids could later have served as precursors for others, first through abiotic pathways, and later the synthesis might have been facilitated by enzymes. The earliest codons might have been somewhat ambiguous aggregates of nucleotides. Later, each amino acid probably shared its codon with its daughter amino acid(s).

The successive subdivision of codon domains is still reflected in the similarities of the codons among structurally related amino acids. Apparently, the expansion of the amino acid repertory took place in parallel with the evolution of the code. Eventually, the assignment of all 64 codons was completed. In the primordial dictionary, probably only the first two bases were required for the specification of an amino acid. Today, in most of the degenerate codon domains, the first two positions are still identical. With the acquisition of new amino acids and new codons, the stability, specificity, and efficiency of the primordial proteins adaptively changed. The archaic, ambiguous codons with lesser precision must have then been eliminated, culminating in the development of a codon dictionary common to prokaryotes and eukaryotes as we know it today. Through functionally improved enzymes, faster and more reliable replicational processes emerged. As the efficient and precise replication, transcription, and translation began, the prebiotic synthesis was no longer needed in this era, about 3 billion years ago. Some investigators have doubted the idea that high G + C content was required for origin of the code under high temperature. Also, on the basis of evolutionary calculations on ribosomal RNA, it was suggested that life may not necessarily have arisen in a hot environment, but the high G + C content of thermophilic microbes might have been due to adaptive evolution. *See* genetic code; operational RNA code; origin of life; prebiotic; spontaneous generation; translation; transcription. (Rodin, S. N. & Ohno, S. 1997. *Proc. Natl. Acad. Sci. USA* 94:5183; Wakasugi, K., et al. 1998. *EMBO J.* 17:297; Davis, B. K. 1999. *Progr. Biophys. Mol. Biol.* 72:157; Szathmáry, E. 1999. *Trends Genet.* 15:223; Knight, R. D., et al. 2001. *Nature Rev. Genet.* 2:49; Sella, G. & Ardell, D. H. 2002. *J. Mol. Evol.* 54:638.)

evolution of the karyotypes Chromosome number is a rather good characteristic of related species. Cytotaxonomists define homologies among taxonomic groups on the basis of the karyotype and the morphology of the chromosomes at (generally) metaphase. Some important landmarks of the chromosomes (chromomeres, band, knobs) can be better visualized at prophase (pachytene). Some chromosomal aberrations (inversions) are also identified at anaphase. Salivary gland (polytenic) chromosomes reveal a great deal of morphological information. With the introduction of banding techniques (Giemsa, C-banding), various taxonomic groups can be rather well defined. Various chromosomes or chromosomal segments can be identified by in situ hybridization with DNA probes "painted" with fluorochromes of various colors (such as fluorescein isothiocyanate [FITC, green], Spectrum Orange [red], 4′,6′-diamidino-2-phenyl indole

[DAPI, blue fluorescence in UV], biotin derivatives, etc.), and the genomes in amphiploids can be distinguished. However, the number of chromosomes may vary even in closely related taxonomic groups because of centromeric fusion or fission of telochromosomes and biarmed inversions:

By analysis of the natural banding pattern of salivary gland chromosomes, inversions can be traced among related species. Acrocentric chromosomes may undergo Robertsonian translocation changes in chromosome numbers by making one biarmed chromosome from two acrocentrics (quite common in mice). If, e.g., chromosome arms A and B, as well as A and D, and similarly D and C, as well as C and B, are fused in a diploid cell at meiotic metaphase I, the configuration shown below is displayed. Such a situation may lead to sterility,

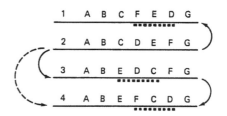

Method of tracing the origin of chromosomal inversions.

reproductive isolation, and eventually speciation. Although inversion heterozygosity may lead to sterility depending on the frequency of crossing over in the inverted segment, many wild populations contain various types of inversions. Paracentric inversions do not affect the fertility of females (either in plants or animals) because the inversion bridge prevents the incorporation of the crossover chromosomes into the eggs (*see* inversions, paracentric). Chromosomal inversions can frequently be traced in the polytenic chromosomes. Inversions may occur repeatedly and may partially involve the same chromosomal segment each time. The serial origin of such inversions can be detected. Molecular analysis of *Drosophila* chromosomes confirms the cytological evidence that extensive reshuffling has taken place during evolution. *See* acrocentric; chromosome painting; deletion; duplication; inversions; karyotype; karyotype evolution; polyteny; reproductive isolation; Robertsonian translocation; telochromosome. (Ranz, J. M., et al. 2001. *Genome Res.* 11:230; Nurminsky, D., et al. 2001. *Science* 291:128; Gonzalez, J., et al. 2002. *Genetics* 161:1137.)

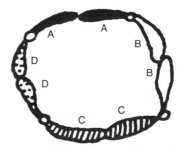

Robertsonian translocation ring.

evolution of organelles The cytoplasmic organelles of eukaryotes, mitochondria, and plastids contain DNA, but the base composition of this DNA displays very little similarity to that of the nucleus. The organellar DNA is usually circular versus the nuclear, which is linear. The ribosomes in the organelles bear more similarities to prokaryotic ribosomes, being also 70S, in contrast to the ribosomes in the cytosol, which are 80S. Several mitochondrial codons do not conform to the universal code words (*see* mtDNA) and no known organisms would have the same nuclear codon dictionary. Plastids use the universal codons. The mitochondrial genome probably has a polyphyletic origin with major similarity to purple nonsulfur photosynthetic bacteria. The plastid genome may have derived from noncyanobacterial oxygenic bacteria, and it may be closest to *Prochloron*. The plastid genome most likely descended from more than a single species. It is assumed that these organelles were captured by means of endocytosis of some lower organisms, perhaps repeatedly during evolution.

The double membranes wrapping the cytoplasmic organelles may have their origin partly from the ancestral donor and partly from the plasmalemma of the recipient cells. In the Euglenoids, the plastids have triple membranes, suggesting that this is the relic of double endocytotic events. The size of the mitochondrial genome is very variable among the various organisms. In mammals, it is about 16 kbp, but in the muskmelon it may be up to 2,400 kbp. In the majority of other plants, it generally varies between 200 and 500 kbp, thus exceeding the plastids, which are rather conservatively between 120 and 270 kbp. The transcriptional machinery in plastids and mitochondria resembles that of prokaryotes. Some of the chloroplast genes have upstream elements similar to the promoter sequences of bacteria. The plastid RNA polymerase is not inhibited by rifampicin, whereas this antibiotic blocks transcription in prokaryotes. Both plastids and mitochondria have genes for rRNAs and tRNAs; these also resemble the prokaryotic ones more than those of eukaryotes in the cytosol. More than a dozen of the plastid genes and about three of the mitochondrial genes have introns, an uncommon feature of prokaryotes. Less than one-third of the ribosomal proteins in the plastids are coded within this organelle; the rest are of cytosolic origin. This and other facts indicate that the acquisition of the organelles was followed by loss of some functions that could be taken care of by genes in the nucleus. There is also a regular import of proteins into the organelles from the cytosol, aided by the transit peptides of the nuclear-coded proteins. Nuclear genes encode the photosynthetic rubisco protein small subunits, whereas the large subunits are transcribed and translated by plastid DNA. Also, organellar gene functions, including mutability, are regulated by nuclear genes. It appears that some DNA sequences have homologues among the nucleus, plastids, and mitochondria, indicating the availability of some DNA transfer mechanism(s). *See* chloroplast; chloroplast genetics; mitochondria; mtDNA; organelle sequence transfers; ribulose1,6-bisphosphate carboxylase-oxydase. (Margulis, L. & Bermudes, D. 1985. *Symbiosis* 1:101; Delihas, N. & Fox, G. E. 1987. *Ann. NY Acad. Sci.* 503:92; Takemura, M. 2001. *J. Mol. Evol.* 52:419; Sato, N. 2001. *Trends Plant Sci.* 6:151; Selosse, M., et al. 2001. *Trends Ecol. Evol.* 16:135.)

evolution of proteins The amino acid sequence in the proteins reflects the coding sequences in the DNA or RNA genetic material. Although the genetic code has synonymous codons, the primary, secondary, tertiary, and quaternary structure reveals important information on how proteins function. Adaptive evolution relies primarily on the expression of genes rather than on their structure, although the latter is important for tracing the path of this evolution. Certain proteins such as cytochrome c (coded by the nucleus but present in the mitochondria) reveal similarities and differences in the amino acid sequence in the widest range of eukaryotes and display similarities even to bacterial cytochrome c2. Such analyses carried out decades ago have shown that similar functions require similar structures. Furthermore, the similarities within apparently related groups are greater than those among more distant groups by any type of classification. These studies brought the recognition that some gene loci are *orthologous*, i.e., they seem to be directly connected by descent across phylogenetic groups, whereas other genes, arising by duplication (*paralogous loci*), may show greater differences in the primary structure because the ancestral copies of the genes could continue providing the needed function while the duplications were more free for (adaptive) evolutionary experimentation. New catalytic properties can now be created on existing protein scaffolds in the laboratory using biochemical techniques. Proteins encoded by linked genes appear to evolve at similar rates apparently due to the force of stabilizing selection. It is not entirely certain that organic evolution proceeds through the same path, yet the new enzymes are functional and the technology provides means for the generation of enzymes not found in nature that may have an advantage for biotechnology. *See* homology; PRINTS; protein families; proteome; ribozymes. (Altamirano, M. M., et al. 2000. *Nature* 403:617; Williams, E. J. & Hurst, L. D. 2000. *Nature* 407:900; Gerlt, J. A. & Babbit, P. C. 2001. *Annu. Rev. Biochem.* 70:209; Aravind, L., et al. 2002. *Current Opin. Struct. Biol.* 12:392; Kinch, L. N. & Grishin, N. V. 2002. *Curr. Opin. Struct. Biol.* 12:400; Davis, B. K. 2002. *Progr. Biophys. Mol. Biol.* 79:77.)

evolution of recombination Influenced by physiological needs of the organisms such as improvement of genetic repair, chromosome distribution in mitosis and meiosis, etc. Alternatively, the breakage of linkage to disadvantageous genes may have adaptive value. The evolution of sexual reproduction aids both of these mechanisms. Under conditions of high (deleterious) mutations, the increase in recombination frequency may be beneficial and the reduction of linkage disequilibrium is expected. *See* linkage disequilibrium.

evolvability Ability to evolve. Evolvability in viruses is generally considered an adaptive response to natural selection. In RNA virus ϕ6, the experimental data indicate that advantageous genes favor the accumulation of advantageous mutations in their vicinity. *See* hitchhiking.

Ewing sarcoma *See* EWS; neuroepithelioma; sarcoma.

EWS Ewing sarcoma oncogene; a binding protein, encoded at human chromosome 22q12. In Ewing sarcoma translocations, t(11;22)(q24;q12) are common, but several other types also occur. *See* FLI.

ex vivo Operation carried out outside the body, but the manipulated organ or gene is returned to the body. *See* in vitro; in vivo.

exaggeration Recessive mutation in one chromosome and a deficiency for the same site in the homologous chromosome results in a more extreme mutant phenotype than homozygosity for the same recessive mutation. (Mohr, O. L. 1923. *Indukt. Abstamm-VererbLehre* 30:279.)

exaptation An adapted function is modified to become useful for a different biological function. An originally nonadaptive feature may become adaptive during evolution.

exchange pairing Chromosomes that have paired and recombined in meiosis are distributed normally to the pole. *See* chromosome pairing; distributive pairing.

exchange promoting protein (EP) Increases the separation of GDP from the bound form (E.GDP) in signal transduction. *See* E.GDP; GDP.

excimer *See* pyrene.

excinuclease *See* DNA repair.

excision Release of a phage, insertion element, episome, or any other element or DNA sequence from a nucleic acid chain. In *precise excision*, for the affected or involved sequence the wild-type DNA is copied from the homologous DNA strands and thus substituted for the stretch. Any interruption of this repair copying may result in retaining some parts of the inserted piece and thus in *imprecise excision*. *See* episome; insertion elements; T-DNA; transposable elements; transposons.

excision repair (dark repair, NER) Removal of damaged DNA segments followed by localized replacement (unscheduled DNA synthesis). In excision repair, a DNA glycosylase excises the damaged base, then the abasic deoxyribose is removed by AP endonucleases. In nucleotide excision repair, an enzyme system hydrolyzes the phosphodiester bonds on both sides of the damaged single-strand tract and the sequence including the defect (12–13 nucleotides in prokaryotes, 27–29 nucleotides in eukaryotes) is removed by excinucleases. Subsequently, the gap is refilled by repair polymerase and the ends are religated. The gap-filling process by DNA polymerases δ and ε is highly accurate, involving $\sim 10^{-5}$ errors. These enzymes correct incorporation errors by $3' \rightarrow 5'$-exonuclease function. The first 15 nucleotides are added by pol α and β, but they lack exonuclease function and start replication much less faithfully. By repairing single-nucleotide gaps, DNA pol β makes errors in the range of $3-5 \times 10^{-3}$. Fortunately, DNA ligase III refuses to join mismatched ends and a mammalian exonuclease (homologue of the bacterial pol III subunit encoded by DnaQ) may then correct the wrong base. Similarly, in prokaryotes, pol III is endowed by both the $3' \rightarrow 5'$ and $5' \rightarrow 3'$ editing function.

Another solution to avoid DNA damage without excision is provided by the yeast DNA polymerase ζ (and its homologue in humans), which can bypass pyrimidine dimers and other adducts and may incorporate the correct nucleotides in their place. The Rev1 (human hREV1) protein encodes a template-dependent deoxycytidylic acid transferase that can mediate translesion and insertion of a base opposite to a gap. The DNA polymerase η can insert two adenylic residues opposite a thymine dimer, thus saving the otherwise damaging situation. Excision repair genes have been identified in many organisms including mammals. XPA (human single-strand-biding protein [HSSB]) and its homologues (Rpa in yeast) are essential binding proteins for nucleotide exchange repair. The damage-binding proteins (DDB or UV-DDB) generally recognize defective areas caused by photoadducts. These proteins may recruit others such as XPCC (human) and some similar proteins of yeast (Rad4 and Rad23). The transcription complex TFIIH is composed of several subunits required for repair (XPB and XPD in humans) and by the homologs of yeast (Ss12 and Rad3). TFIIH also contains kinases (Cdk2). The human excision repair genes were identified by genetic transformation of UV-sensitive Chinese hamster cells with human DNA (excision repair cross-complementing genes, ERCC). Gene ERCC3 (human chromosome 2q21) appears to be at the locus responsible for xeroderma pigmentosum type II (B) and it controls an early step of excision repair involving a DNA helicase function; it is also associated with the basal transcription factor TFIIH. ERCC1/XPF (*RAD10*) endonuclease appears to be in human chromosome 3 and it is involved with xeroderma pigmentosum F. ERCC2 is in human chromosome 19, ERCC4 in chromosome 16, and ERCC5 (*RAD2*) in chromosome 13q22. Excision repair gene UV-135 (ERCM2) appears to be in human chromosome 13 and UV-24 in chromosome 2. Transcription factor TFIIH is also involved in nucleotide excision repair in transcriptionally nonactive DNA in cooperation with Rad 2 and 4. Pol δ and pol ε in cooperation with PCNA (proliferating cell nuclear antigen) and RFC (replication factor) auxiliary proteins may carry out repair synthesis. *See* ABC excinucleases; adducts; AP endonucleases; Bloom syndrome; Cockayne syndrome; cyclobutane dimer; DNA ligase; DNA polymerases; DNA repair; excision repair bioassays; glycosylases; light sensitivity diseases; mismatch repair; phosphodiester bond; photolyase; pyrimidine dimer; transcription factors; translesion; trichothiodystrophy; ultraviolet light; UVRBC; xeroderma pigmentosum; X-ray repair. (Sancar, A. 1996. *Annu. Rev. Biochem.* 65:43; de Laat, W. L., et al. 1999. *Genes & Development* 13:768; Mol, C. D., et al. 1999. *Annu. Rev. Biophys. Biomol. Struct.* 28:101; Moldave, K., ed. 2001. *Progr. Nucleic Acid Res. Mol. Biol.* 68.)

excision repair bioassay Essentially a biochemical procedure. The mutagen may cause the formation of cyclobutane rings, alkylation of bases, formation of covalent adducts between bases and the reactive chemical, depurination of DNA, interstrand cross-links in DNA, cross-links between DNA and protein, breakage of phosphodiester bonds, intercalation of the mutagen between DNA bases, etc. The defects are detected in the extracted DNA cleaved by acids (formic acid or trifluoroacetic acid). Acid hydrolysis does not give results as reliable as DNase or phosphodiesterase enzymes. Free bases and pyrimidine dimers are analyzed by one-dimensional or two-dimensional paper, thin-layer or column, or Sephadex or Dowex chromatography. By similar procedures, alkylated bases can be identified. Adducts identification has been attempted by using radiolabeled mutagens and detecting the sites of the adduct formation. This procedure, however, is not completely reliable. The use of single-strand-specific nucleases may detect the consequences of strand breakage resulting in strand separation, then liable to digestion by these nucleases. DNA breakage has also been evaluated by centrifugation of extracted DNA in 5–20% alkaline sucrose gradients where the intact DNA peaks closer to the bottom of the centrifuge tube while the damaged

DNA floats at the lower sucrose concentration areas. A different approach is to provide radioactively labeled bases after the mutagenic treatment and monitor the extent of repair replication that is supposed to be increased if the mutagen damages the DNA. The extent of repair is reflected by the increase in radioactivity in the DNA. Adding bromodeoxyuridine (BrDU) to the mutagen-treated DNA can aid in assessing extensive repair replication. Incorporation of BrDU increases the buoyant density of DNA, which is detectable upon separation by ultracentrifugation. Another variation is the exposure of BrDU containing DNA tracts to irradiation by 313 nm UV-B. The patches containing the analog are cleaved by this and reveal the original sites of damage by photolysis. *See* bioassays in genetic toxicology; buoyant density; mutation detection; site-specific mutation; ultracentrifuge; ultraviolet light; unscheduled DNA synthesis. (Rédei, G. P., et al. 1984. *Mutation Cancer and Malformation*, Chu, E. H. Y. & Generoso, W. M., eds., p. 689. Plenum, New York.)

excision vector   Carries the prokaryotic *Cre* gene. If a mouse is transformed by *Cre* and its mate carries a gene closely flanked by loxP, then their F_1 offspring (or any heterozygote) may evict the targeted gene as shown in the diagram below. *See Cre*; homologous recombination; targeting genes.

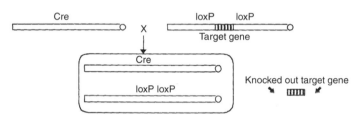

Excision vector.

excitation   Illumination with ultraviolet light may transiently raise the orbital electrons of an atom to a higher level of energy. Such excitation may lead to electron loss and may be sufficient to cause mutation, although the level of energy may drop back to ground level. *See* electron; ultraviolet light.

excitatory neurotransmitter   Opens cation channels and facilitates the influx of Na^+ to depolarize postsynaptic membranes. *See* acetylcholine; glutamate; neuro-transmitter; serotonin.

excited state   Energy-rich atom or molecule after the absorption of light energy.

excitotoxicity   Natural and synthetic compounds with glutamic acid resemblance may excite the central nervous system and cause excessive Ca^{2+} influx across the postsynaptic membrane, resulting in neural death by mitochondrial permeability transition.

exconjugant   Ciliates (*Paramecia*) may pair (conjugate). The paired cells mutually fertilize each other, Separation of the two, now diploid cells i.e., the exconjugants follows. *See Paramecium*.

exencephaly   Overgrowth of midbrain neural tissue (the brain may be outside the cranium) caused by a defect in protein p53. *See* Gadd45; neural tube defects; p53.

exergonic reaction   Releases free energy.

exflagellation   Transformation of the nonmotile *Plasmodium* male gametocyte into a very motile male gamete in the mosquito, triggered by xanthurenic acid, a derivative of tryptophan. *See* malaria; *Plasmodium*.

exhaustive search   Procedure of constructing an evolutionary tree by analyzing several individuals to calculate the length of the branches of evolutionary trees. *See* evolutionary tree; heuristic search; PAUP.

exine   Outer layer (of pollen grains).

exobiology   Study of the possibilities of extraterrestrial life forms. So far no firm evidence exists in spite of several claims. Biogenic hexa- and octahedral magnetite crystals in Martial meteorites is one recent suggestive indication. *See* extraterrestrial. (Thomas-Keprta, K. L., et al. 2001. *Proc. Natl. Acad. Sci. USA* 98:2164; Friedmann, E. I., et al. 2001. *Ibid.*, 2176; Cohen, J. & Stewart, I. 2001. *Nature* 409:1119.)

exocarp   Outer layer of the fruit wall, the peel of citrus fruits, the skin of peaches.

exocrine   Secretory glands that release substances (enzymes, hormones, etc.) through ducts in an outward direction. *See* endocrine.

exocyst   Protein complex that marks the docking position of transporting vesicles on the surface of the plasma membrane.

exocytosis   Secretion of molecules from the cell into the medium or transfer of molecules into membrane-enclosed storage compartments in the cell. Exocytosis is one of the mechanisms by which cytotoxic T lymphocytes (CTL) release, in a Ca^{2+}-dependent process, perforin and granzymes to lyse the target cells. *See* apoptosis; exocytotic vesicles; granzymes; perforin; SNARE; syntaxin. (Lin, R. C. & Scheller, R. H. 2000. *Annu. Rev. Cell Dev. Biol.* 16:19.)

exocytotic vesicle   Serves as a vehicle for transport by exocytosis. *See* exocytosis.

exogamy   Cross-fertilization.

exogenic heredity   Based not on biological inheritance but on cultural transmission of traditions, scientific information, laws, ethics, values, etc.

exogenote   When the bacterial genetic material recombines, the recipient genome is called an endogenote and the corresponding segment of the donor genome is an exogenote. *See* conjugation, bacterial.

exogenous   Influences coming from outside.

exogenous evolution   Evolution of cells of higher organisms by inclusion of prokaryotic cells such as plastids and mitochondria. *See* autogenous evolution.

exon   Segments of eukaryotic mosaic genes that are represented in the mature mRNA and translated into protein. The exon intron boundaries are usually conserved, e.g.:

5'-ACTGCAgtaagg...tttcctctctctagTGGGCG-3' DNA
  exon   intron   exon

All coding genes have exons. Introns are exceptional in prokaryotes, yet they also occur in the plastid and mitochondrial genes. The parts of RNA genes that are retained in the mature transcripts are also called exons. The length of individual exons is usually short, 10 to 400 nucleotides. The total average exon size per genes in mice is about 2,300 and in humans about 3,400, although large differences exist among different genes. Generally the human genes display many short exons (average 50 codons) and are separated by long introns up to 10 kbp larger. The splice sites of exons may vary. In humans, more than one-third of the exons may not be translated. The largest exon number, 234, was found in the human titin mRNA, whereas the average is about 7–9. Exons are approximately 1% of the human genome. *See* branch point sequence; gene; introns; mosaic genes; spliceosome; titin. (<http://intron.bic.nus.edu.sg/exint/exint.html>.)

exon connection   Method for isolation of genes by following these steps: (1) Isolate RNA from a cell line. (2) Obtain cDNA. (3) Use PCR with primers of suspected adjacent exons. (4) Clone the product of PCR. (5) Sequence is tested for connection of exons. (6) Southern blot is performed with the generated probe. (7) Identify and isolate the whole gene. *See* cDNA; DNA sequencing; exon; gene isolation; introns; PCR. (Bardos, J., et al. 1997. *Int. J. Cancer* 73:137.)

exon parsing   Determining the exact boundaries of exons or genes.

exon scrambling   In eukaryotes the exons may be spliced correctly but in different order from those in the genomic DNA. This scrambling is believed to be mediated by loops in the pre-mRNA and makes possible the alternative processing of the transcripts. These exons are not polyadenylated. *See* exons; introns. (Caldas, C., et al. 1998. *Gene* 208[2]:167.)

exon shuffling   Exons of the same gene may be processed and thus expressed in more than one pattern and may be recruited for the synthesis of more than one protein. It has been suggested that the role of introns in the earliest cells during evolution was to facilitate the assembly of new genes by exon shuffling. Exon shuffling is most common in vertebrates, but it has been detected in the cytosolic glyceraldehyde-3-phosphate dehydrogenase gene of plants. The ubiquitous retrotransposons of eukaryotic cells may move exons and promoters into other genes, thus generating new composite genes with different functions. L1 retrotransposons may be copied into the genes and they can mobilize the L1 flanking 3' sequences into the genes. The combination of different exons by recombination or genetic engineering may produce enzymes with potential advantages. *See* exon; intron; LINE; retrotransposon. (Petthy, L. 1996. *Matrix Biol.* 15:301; Kolkman, J. A. & Stemmer, W. P. C. 2001. *Nature Biotechnol.* 19:423.)

exon skipping   Outsplicing of an exon, i.e., skipping the inclusion into the mRNA. *See* splicing.

exon theory   Suggested that split genes arose by aggregation of exons during prebiotic or early evolution. Convincing experimental evidence is still lacking.

exon trapping   In physical mapping identifies candidate genes in transcribed sequences. By the use of an appropriate (commercially available) vector and a transfection technique, an exon is inserted into an adjacent intron. The transcript then appears longer and can be identified by Northern blot. *See* candidate gene; exon; Northern blotting; physical mapping; trapping promoters. (Wapenaar, M. C. & Den Dunnen, J. T. 2001. *Methods Mol. Biol.* 175:201.)

exonuclease   Enzyme that digests a polynucleotide chain beginning at either the 5' or the 3' terminus. *See* Bal1; endonuclease; mismatch repair; recombination mechanisms; restriction enzymes. (Grunberg-Manago, M. 1999. *Annu. Rev. Genet.* 33:193; Shevelev, I. V. & Hübscher, U. 2002. *Nature Rev. Mol. Cell Biol.* 3:364.)

exonuclease I   Degrades single-stranded DNA 3' → 5'.

exonuclease III   Multifunctional enzyme; in *E. coli* it works as a DNA repair endonuclease, exonuclease 3' → 5', phosphomonoesterase, and ribonuclease. *See* endonuclease; exonuclease; phosphomonoesterase; ribonuclease.

exonuclease V   Digests double-stranded DNA and displays DNA-dependent ATPase activity; it is encoded by the *recBCD* complex of *E coli*.

exonuclease VII   Heterodimer, cuts single-stranded DNA from both 3' and 5' ends without requirement for Mg^{2+}.

exonuclease λ   Encoded by phage λ, binds to the free end of a DNA duplex and degrades one strand in the 5' → 3' direction, releasing about 3,000 5'-mononucleotides at a rate of about 12/ sec. The 3' overhangs participate in genetic recombination with the assistance of the bacterial RecA protein. Alternatively, two homologous single strands may anneal without Rec and form double-stranded recombinants, *See* lambda phage; recombination mechanisms.

exoribonuclease   Enzyme that digests RNA from the terminus. Exoribonucleases belong to six superfamilies and occur in all organisms. In *E. coli*, 3' → 5' exoribonucleases are polynucleotide phosphorylase, RNase II, RNase D, RNase BN, RNase T, RNase PH, RNase R, and oligoribonuclease. Eukaryotes also have similar enzymes. All eukaryotes have 5' → 3' exoribonucleases, which are apparently absent from *E. coli*. *See* exosome. (Deutscher, M. P. & Li, Z. 2000. *Progr. Nucleic Acids Res. Mol. Biol.* 66:67; Zuo, Y. & Deutscher, M. P. 2001. *Nucleic Acids Res.* 29:1017; Gagliardi, D., et al. 2001. *J. Biol. Chem.* 276:43541.)

exosite   Anion-binding site on the surface of a protein molecule. *See* anticoagulation.

exoskeleton   The hard shell of the body (insects, crustaceans, etc.), including the vertebrate nails, hoofs, hair, and other epidermal structures, is considered as exoskeleton. *See* chitin.

exosome (1) Putative segment of DNA associated with, but not integrated into, the chromosome, yet it can express its genetic information. Exosomes may have a role in the quality control of mRNA. Either the hypopolyadenylated or hyperpolyadenylated molecules are permitted to exit from the nucleus and may be translated in case the nucleolytic complex of the exosome is defective. See episome; RNA surveillance. (Fox, A. S., et al. 1971. *Proc. Natl. Acad. Sci. USA* 68:342; Hilleren, P., et al. 2001. *Nature* 413:538.) (2) Complex of exoribonucleases, RNA-binding proteins, and helicases in prokaryotes and eukaryotes degrading RNA in the $3' \rightarrow 5'$ (or $5' \rightarrow 3'$) direction. In yeast, each of the 11 components is essential for viability. The complex present in the cytoplasm and in the nucleus trims rRNA primary transcripts and mRNA and degrades some other RNAs. Other eukaryotes may have somewhat different types of exosomes. See exoribonuclease; mRNA degradation; RNA processing. (Suzuki, N., et al. 2001. *Genetics* 158:613; Estavez, A. M., et al. 2001. *EMBO J.* 20:3831; Brouwer, R., et al. 2001. *Arthritis Res.* 3:102; Koonin, E. V., et al. 2001. *Genome Res.* 11:240.) (3) Dendritic cells constitutively secrete 50–90-nm-diameter vesicles that present antigens. They may stimulate antitumor responses by ca. 24 proteins. (Thery, C., et al. 2001. *J. Immunol.* 166:7309; Denzer, K., et al. 2000. *J. Cell Sci.* 113:3365.)

exostosis (EXT, diaphyseal aclasis) Autosomal-dominant (EXT1, human chromosome 8q24.1-p13, EXT2 in 11p11.1, EXT3 in 19p) phenotypes involving growth of extra cartilage or bony projections at the end of bones (mainly on the hands and fingers but rarely on the head, except the ear). EXT1 in chromosome 8 encodes a transmembrane glycoprotein and affects the cell surface heparan sulfate glycosaminoglycans. Exostosis is generally accompanied by short stature. EXT2 encodes α-1,4-N-acetyl hexosaminyltransferase. The incidence of such defects in Western populations is about $1-2 \times 10^{-5}$ and the estimated rate of mutations is $6-9 \times 10^{-6}$. The chance for bone cancer is increased to 0.5–2% of the cases. A deletion at 11p11.2 causes oval defects of a nerve/vein opening in the parietal bones (of the head) called Potocki-Shaffer syndrome. The highly homologous EXTL (EXT-like, EXTL1, 1p36.1, EXTL2, 1p12-p11) genes encode α1,4-N-acetyl glucoseaminyl transferase involved in the biosynthesis of heparan sulfate and heparin. See cancer; Ehlers-Danlos syndrome; fibrodysplasia; glycosaminoglycan; heparan sulfate; Langer-Giedion syndrome; limb defects in humans; metachondromatosis; stature in humans. (Cheung, P. K., et al. 2001. *Am. J. Hum. Genet.* 69:55.)

exothermic reaction Releases heat.

exotoxin Toxin secreted outside the cells (body) such as the protein toxins of several bacteria, e.g., *Clostridium botulinum* nerve toxin, etc. See toxins.

expansin Plant cell wall protein (ca. 26 K molecular mass) that facilitates the elastic growth of cell walls, loosening the stigmatic surface to aid penetration of the grass pollen tube and fruit ripening. Expansins may have a role as allergens. (Shie, M. W. & Cosgrove, D. J. 1998. *J. Plant Res.* 111:149; Pien, S., et al. 2001. *Proc. Natl. Acad. Sci. USA* 98:11812.)

expansion card Circuit board that can be inserted into some computers, permitting the user to perform special functions.

ExPaSy (Expert Protein Analysis System) Links to the Glaxo Institute of Molecular Biology, Geneva, two-dimensional electrophoretic database to the SWISS-PROT protein sequence database. See databases; electrospray MS; laser desorption MS; MELANIE II; SWISS-PROT.

expected phenotypic superiority $i_p \sigma p_1$, where i is the selection intensity when fraction p is selected in a breeding program. See selection intensity.

experiment Conducted to test a hypothesis (deductive method) or to generalize from empirical data (inductive method). In either case, the work has clearly defined objectives. Genetics is basically an experimental science. See genetics; science.

explant Cut-out piece of (plant) tissue used for in vitro culture. See tissue culture.

exponential distribution May represent the amount of time until the first event of a series occurs and the distribution of time between occurrences of the subsequent events in case they occur according to a Poisson process of independent events. See distributions; Poisson distribution.

$$P \ (r \text{ occurrences in } T \text{ units} | \lambda = \frac{e^{-\lambda T}(\lambda T)^r}{r!} \text{ where } \lambda = \text{the intensity per time units } t \text{ and } T = \text{the number of units of } t; \lambda T = \text{intensity/T units}$$

exponential growth Takes place when nutrients and other factors required for growth are optimal. Cell multiplication is determined by the exponent of 2; e.g., after 10 divisions of a cell, the number of cells becomes $2^{10} = 1,024$, and if the number of initial cells was 8, then after 10 divisions the number of cells becomes $2^{10} \times 8 = 8,192$. This type of growth is also called logarithmic growth because $\log_2[1,024] = 10$. The conversion to $\log_2$ from $\log_{10}$ is carried out as follows: $\log_2[1,024] = \log_{10}[1,024](1/\log_{10}[2]) = 3.010299957 \times \log_{10}(1/0.301029995) = 3.010299957 \times 3.321928095 \approx 10$. See growth curve.

export adaptor Proteins that attach to proteins of RNAs and assist their export or import, respectively, between nucleus and cytoplasm. Leucine-rich nuclear export signals (NES) move., e.g., the 5S rRNA to the cytoplasm through the nuclear pore complex. Members of the hnRNP family (~20 proteins) are involved in the transport of mRNA. See CBC; Gle1; Mex67; nuclear localization sequence; nuclear pore; Ran; RNA export; RNA transport; RRE; snRNA; transportin. (Segref, A., et al. 2001. *RNA* 7[3]:351.)

exportin See chromosome maintenance region, nuclear pore.

export signal Can be RNA-binding proteins specific for RNA classes (mRNA, tRNA, rRNA, etc.) and mediate the interaction with export receptor molecules. The recognition of U snRNAs depends on their monomethylguanosine caps, the cap and tail structures, HIV-1 Rev-responsive element, and 5S RNA-binding sites for TFIIIA; or they may bind to proteins with leucine-rich nuclear localization signals. *See* acquired immune deficiency syndrome; Cap; export adapter; hnRNA; nuclear localization sequences; ribosomes; transcription factors; U RNA. (Rashevsky-Finkel, A. & et al. 2001. *J. Biol. Chem.* 276:44963.)

expressed-sequence tag (EST)   Probe of short nucleotide sequences for genes that are expressed in a particular tissue, although no information may be provided regarding their function or role. The use of ESTs greatly facilitated the analysis of the functional fraction of eukaryotic genomes. It was important for the progress of the genome projects as usually single-pass tracts, although they may involve various artifacts. Patents for them were sought; the ethical justification for their patenting has been criticized. *See* gene discovery; ORESTES; patent; physical mapping; UniGene. (Adams, M. D., et al. 1991. *Science* 252:1651; Gemünd, C., et al. 2001. *Nucleic Acids Res.* 29:1272; <http://www.ncbi.nlm.nih.gov/dbEST/index.html>; <http://www.ncbi.nlm.nih.gov/ncicgap>; <http://www.tigr.org/tdb/tdb/index.shtml>.)

expression   Phenotype, which may be any morphological trait or a (protein) product of a gene. *See* genotype; phenotype.

expression cassette   Contains all the sequences required for the expression of a gene and can be inserted into various expression vectors for transcription; also, the structural gene may be replaced by another structural gene. *See* expression vector; structural gene.

expression cloning (reverse biochemistry)   Cloned cDNA is expressed in a cellular or cell-free translation system in order to study the pure product of a gene. *See* cDNA; FL-REX; molecular cloning; rabbit reticulocyte in vitro translation system; wheat germ in vitro translation system.

expression library   Includes cDNAs in expression vectors, permitting their ready use for transformation and expression in several hosts. *See* expression vector.

expression-linked copy   In the *Trypanosomes*, gene transposition is required for the expression of a different type of antigen. *See* Trypanosoma brucei.

expression matrix   May be used to study genetic data obtained by microarray hybridization. These are tables where rows designate genes, columns represent the various samples of tissues or conditions of expression, and within each cell of the grid the level of gene expression is shown. *See* microarray hybridization. (Brazma, A. & Vilo, J. 2000. *FEBS Lett.* 480:17.)

expression profile   Collection of genes transcribed (expressed) from genomic DNA, thus determining the cellular phenotype. It reveals developmental and extended interacting metabolic pathways. *See* DNA chips; microarray hybridization; transcriptome. (Miki, R., et al. 2001. *Proc. Natl. Acad. Sci. USA* 98:2199.)

expression vector   Carries the structural gene and all the regulatory signals required for expression to the target cell by transformation. *See* vectors.

expression-verified gene (EVG)   Coregulated exon as detected by microarray hybridization. *See* exon; microarray hybridization.

expressivity   Degree of gene expression. *See* penetrance. (Oleksiak, M. F., et al. 2002. *Nature Genet.* 32:261.)

EXProt   Database of proteins with experimentally verified function. *See* metabolism. (<http://www.cmbi.kun.nl/EXProt>.)

extein   Protein sequences remaining after the removal of inteins. *See* intein.

extensible markup language   Structural format for documents on the WEB. (<http://www.w3.org/XML>.)

extensins   Plant cell wall glycoprotein. (Yoshiba, Y., et al. 2001. *DNA Res.* 8:115.)

external guide sequences (EGS)   RNA oligonucleotide that can bind to specific RNA sites in an RNA molecule and guide ribonuclease P, which cleaves at specific sites, to this location. *See* antisense technology; ribonuclease P. (McKinney, J., et al. 2001. *Proc. Natl. Acad. Sci. USA* 98:6605.)

external transcribed spacer   See ETS.

extinction   Loss of a genotype (phenotype) from a population; disappearance of a conditioned response; suppression of cell-type-specific function in fused somatic cells. In spectrophotometry, it means the intensity of the absorption (ε_{max}) at the absorption peak (λ_{max}). *See* equilibrium of mutations; OD; species extant.

extirpation   Complete removal of an organ, a type of cell(s), or tissues. *See* ablation.

extracellular DNA   Cell-free, naked DNA that may be a waste of laboratory operations and may persist in the environment as a polymer and a potential hazard. In soil or other aqueous environments, the polymer rapidly decays. Plasmid or other supercoiled DNA may be less prone to degradation. Antibiotic-resistant DNA may not pose serious hazards unless selective conditions exist. (Doblehoff-Dier, O., et al. 2000. *Trends Biotech.* 18:141.)

extracellular matrix (ECM)   Complex of polysaccharides and proteins secreted by cells. Its role in the tissues is structural and physiological. ECM includes collagens, a variety of fibrous proteins of triple helical coiled structure, glycoproteins (laminin, fibronectin), proteoglycans, etc. ECM is linked to the intracellular cytoskeleton by integrins. Without anchorage, most cells succumb to apoptosis. *See* basement membrane; collagen; elastin; fibronectin; ICAM; integrin; laminin; proteoglycans.

extrachromosomal inheritance Determined by nonnuclear elements. *See* chloroplast genetics; mitochondrial genetics; mtDNA; plasmids.

extragenic The genetic factor is outside the boundary of the gene that it affects, e.g., suppressor tRNA. *See* suppressor gene; suppressor tRNA.

extranuclear gene In the cellular organelles (mitochondria, plastids) except the cell nucleus.

extranuclear inheritance Determined by genes in cellular organelles other than by those in the nucleus. *See* ctDNA; mtDNA; symbionts, hereditary.

extrapolation Calculation of values beyond an interval on the basis of the knowledge of values known within the interval. This is usually done with the assistance of the linear regression equation, $Y = a + bx$. Extrapolation may involve some risks of error if the value of x is increasing beyond the known range and even more seriously if the linear regression equation (Y) values do not represent the values properly, but in effect may be a curved line beyond the interval. *See* correlation; interpolation, linear.

extraterrestrial life Living creatures outside the earth have been claimed on the basis of various observations, but so far no unchallenged evidence has been available. (Wilson, T. L. 2001. *Nature* 409:1110.)

extravasation Passing cells from the blood into tissues. Such events occur during metastasis. *See* diapedesis; intravasation; metastasis.

extreme-value distribution theory The probability of an alignment score is caused by chance in physical mapping of DNA. *See* physical mapping. (Pagni, M. & Jongeneel, C. V. 2001. *Brief Bioinform.* 2[1]:51.)

extremities Body appendages such as the limbs.

exudative vitreoretinopathy, familial (11q13-q23) Dominant/recessive retinal detachment and vitreous hemorrhages. The symptoms are similar to vitelliform macular dystrophy. *See* macular dystrophy; retinal dystrophy.

exules Successive generations of insects (aphids).

eye color In humans, a polygenically controlled trait. It is frequently assumed that one gene locus (BEY, 15q11-q15) is responsible for brown/blue and another is responsible for green/blue (GEY, 19p13.1-q13.11) color. Brown (dark) tends to be dominant over blue, and green over blue. It is not too uncommon that blue-eyed parents have both blue-eyed and brown-eyed offspring. This is no basis for suspecting illegitimacy. Brown sectors in blue eyes (sometimes almost invisibly small) are attributed to somatic mutation. Such individuals are more likely to have brown-eyed children even if their spouses have blue eyes. Pigmentation of the retinal layer of the iris and the nature of the semiopaque layers in front of the iris determine the eye color. A blue layer may be present in all individuals, but it may be masked by another layer of melanin in the front part of the iris. Therefore, blue-eyed babies may develop brown eyes when melanin is accumulating in the front part of the iris at later stages. Green and hazel eyes indicate that melanin partially masks the reflections from the deeper layers of the iris. Gray eyes are variants of the blue color. Black eyes indicate a very deep brown melanin layer. When all pigments are absent in albinos, the eyes appear pink or very pale blue because of the reflections of the blood vessels. Some eye diseases involving coloboma may change eye color. Phenylketonuriacs may display light eye color because of the defect in pigment synthesis. It is not too uncommon that the color of the two eyes of a person does not match (heterochromia iridis) due to a developmental anomaly. In insects, the bright red color is derived from pteridines and the brown from ommochromes. The intermediate shades develop from the presence of both pteridines and ommochromes. *See* coloboma; eye diseases; phenylketonuria; pigmentation in animals.

eye disease (ophthalmologic disease) Concerned with anatomical and pathological conditions related to vision. More than 0.5% of the U.S. population is afflicted with some serious eye problems and almost 0.02% are legally blind. In other parts of the world, eye-related diseases may be even more prevalent. The genetic component of these ailments is variable and it is usually part of complex syndromes affecting anatomical and physiological disorders. *Clouding of the cornea* occurs in Hurler, Marquio, and Maroteaux-Lamy syndromes and Fabry's disease. *Cataracts* may be present in Wilms' tumors, Lowe syndrome, galactosemia, Werner syndrome, and myotonic dystrophy. *Cherry-red spot* may appear on the retina in sphingolipidoses, gangliosidoses, and mucolipidoses. *Reduced vision* occurs in neuroaminidase deficiency. Retinitis pigmentosa (most commonly clumped pigments and atrophy of the retina, contraction of the field of vision), Usher, Laurence-Moon, and Bardet-Biedl syndromes, and Refsum disease may involve eye pigment disorders. *Amaurosis congenita* and *retinoschisis* are malformations or deficiency of the retina. *Loss of eye pigments* and reduced vision in albinism and *blue sclera* (the normally white outer surface of the eye) in osteogenesis imperfecta are characteristic. *Variegation of eye color* (iris) is found in Waardenburg syndrome. *Retinal neoplasia* is the most critical feature of Norrie disease. *Hypoplasia of the iris* occurs in Rieger syndrome and *reduction in eye size* is caused by microphthalmos. *Myopia* is common in Stickler syndrome, Marfan syndrome, and Kniest dysplasia. *Tumors of the eye tissues* develop in retinoblastoma and von Hippel–Lindau syndrome. *Tumorous eyes* are also found in Crouzon syndrome. In human trisomy 21 (*Down syndrome*), slanted eyelids and white spots around the iris appear. *Autoimmune eye disease* may involve ocular cicatricial (scar-like) pemphigoid. Approximately 100 mapped genes are responsible for blindness. *See* anophthalmos; aspartoacyclase deficiency; cataracts; cat-eye syndrome; choroidal osteoma; choroideremia; choroidoretinal degeneration; Cohen syndrome; coloboma; color blindness; cornea plana; Duane retraction syndrome; ferritin; focal dermal hypoplasia; foveal dystrophy; galactosemia; glaucoma; GDLD; Hirschprung disease; Leber optical atrophy; macular degeneration; macular dystrophy; Michel syndrome; microphthalmos; mitochondrial diseases, human; myopia night blindness; nystagmus; oculodentodigital dysplasia; Oguchi disease; ophthalmoplegia; pemphigus; retinal dystrophy; retinitis pigmentosa;

Rieger syndrome; Rothmund-Thompson syndrome; S-cone syndrome; Stargardt disease; strabismus; Zellweger syndrome. (<http://mutview.dmb.med.keio.ac.jp/>.)

eyelashes, long (trichomegaly) May be an autosomal-dominant anomaly associated with cataracts, spherocytosis, and other ailments. An autosomal-recessive form was also studied and involved mental retardation, retinal degeneration, and a number of other symptoms with unclear relevance to the formation of the eyelashes. Multiple rows of eyelashes were observed to apparently follow either recessive or dominant inheritance or as only a nongenetic anomaly. *See* emphedema distichiasis; spherocytosis.

eyeless *Drosophila* mutation controlling eye differentiation but not the ocelli. The *Small eye* (*Sey*/*Pax-6*) of mice and the human Aniridia genes are homologous. Several other genes also affect eye development in *Drosophila*. *See* Aniridia; microphthalmos.

eyepiece Microscope lens through which the eye directly views the object through the tube. The eyepiece only enlarges the image but does not afford better resolution, unlike the lens of the objective. *See* microscopy; objective.

eyespot Complex light-perception structure in algae involved in the movement of the flagella (singular flagellum). By the light microscope in *Chlamydomonas reinhardtii* it appears as an orange spot in the chloroplast stroma. It is also involved with a Ca^{2+} channel. The eye spot is controlled by several genes. *See* Chlamydomonas.

Eyk Receptor tyrosine kinase encoded by a chicken protooncogene. *See* protooncogene; tyrosine kinase.

Ezrin Cytoskeletal protein. *See* caveolae; ERM; ICAM; T-cell receptor. (Ng, T., et al. 2001. *EMBO J.* 20:2723.)

Bacteriophages were discovered by F. W. Twort (1915, *Lancet* 11:1241). In 1925 F. D'Herelle (then Directeur du Service Bacteriologique du Conseil Sanitaire, Maritime et Quarantenaire d'Egypte) wrote a remarkable book of 629 pages about this agent, which he named bacteriophage because it could eat bacteria. The preface of the book emphasizes the practical significance of bacteriophages:

"Although up to the present time but few authors have undertaken the study of the behavior of the bacteriophage within the organism there are a relatively large number who have carried out investigations on what might be termed the other side, that is to say, on the application of the bacteriophage to the therapy of various infectious diseases. Of particular interest are the contributions of Hauduroy, who has applied this mode of treatment in colon bacillus infections, of Gratia, who has conclusively demonstrated a therapeutic effect in staphylococcus infections, and of da Costa Cruz who, after having confirmed my first studies on the specific treatment of the bacillary dysenteries, has undertaken a large scale demonstration of this mode of therapy with such results that all other modes of treatment have today been abandoned in Brazil."

Today bacteriophages are not used as therapeutic agents, but their application as a tool for biological research led to the development of molecular biology, which in turn is making a great impact on medicine.

At the 16th Cold Spring Harbor Symposium (p. 436) J. Lederberg et al., discussing bacterial recombination, say: "The exposure of sensitive cells to suspensions of the free phage, which we named "λ," by analogy to a killer factor in *Paramecium*, results in the lysis of a variable proportion of the cells.... The speculation that λ might be involved in genetic recombination needs no further mention."

A historical vignette.

F

F Coefficient of inbreeding expresses the probability for homozygosity of alleles, identical by descent, at a locus. It can simply be determined from the pedigree of an individual in a family. Each gamete has 0.5 chance to transmit a particular allele through the paths available. Thus, in the progeny of a half-brother and half-sister mating, being three relevant ancestors involved, the coefficient of inbreeding, $F = 0.5^3$. Similarly, the F value of the offspring of first cousins is 1/16 because through two loops, represented by E-C-B-D-F and F-D-A-C-E, from the great-grand-parental gametes the same allele can be transmitted to the I offspring, $0.5^5 + 0.5^5 = 0.03225 + 0.03125 = 0.0625 = 1/16$. Note that the looping is done backward from the I offspring. F_{IS} is the coefficient of inbreeding of an individual relative to the subpopulation in which it lives. Its value is determined by analysis of its descent (pedigree). F_{ST} is the index of fixation. F_{IT} is the coefficient of inbreeding of an individual (I) relative to the total population (T). Employing F_{IS}, F_{ST} and F_{IT} is called F statistics. *See* coefficient of coancestry; fixation index; metapopulation; relatedness. (Wang. J. 1997. *Genetics* 146:1453, 1465.)

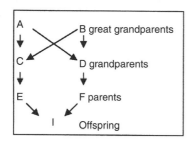

F′ (F prime) Fertility factor that regularly carries bacterial chromosomal genes. (Clark, A. J. & Adelberg, E. A. 1962. *Annu. Rev. Microbiol.* 16:289.)

F⁺ Bacterium carrying a sex element (F, fertility factor). *See* F plasmid.

$F_1, F_2, \ldots F_n$ Subsequent filial generations; the parents of the F_1 are homozygous for different alleles of the corresponding gene(s).

ϕ value Biophysical characteristic of a macromolecule. It is obtained by dividing the change in transition-state free energy accompanying a side-chain truncation by the change in native-state free energy. *See* transition state.

FAA Histolological fixative containing formalin, acetic acid, and ethanol. *See* fixatives.

Fab domain Of the antibody, includes the whole light chain and part of the heavy chain (can be split off by digestion with papain) including the paratope (the antigen recognition site). *See* antibody; epitope.

Fabry disease Xq22 chromosome disorder caused by a deficiency of α-glycosidase (α-galactosidase A) enzyme. The hemizygous males have skin lesions, opacity of the eye, periodic fevers, burning sensation in the extremities, and edema (fluid accumulation) due to kidney malfunction. The heterozygous females have much milder symptoms, yet they also develop opacity of the eyes. The afflicted individuals survive to adulthood when death is caused by heart, kidney, and brain problems. The heterozygotes can be identified and the condition is detectable by amniocentesis; thus, genetic counseling is effective. Galactosylgalactosylglucosyl ceramide accumulates in the endothelium (the tissue lining the heart, blood, and lymph vessels and other cavities), nerve system, epithelium (surface cell layers) of the kidney, and cornea (transparent outer coat in front of the eyes). Attempts have been made to correct the condition by retroviral vectors with α-glycosidase A (α-galactosidase A) expression. *See* galactosidases; gene therapy; sphingolipidoses; sphingolipids; viral vectors. (Ioannou, Y., et al. 2001. *Am. J. Hum. Genet.* 68:14.)

FAC *See* Fanconi's anemia.

face/heart defects *See* Alagille syndrome; DiGeorge syndrome; Noonan syndrome, Ellis-van Creveld syndrome; velo-cardiofacial syndrome; Williams syndrome.

facet Small flat surface on a structure or component of the insect eye. *See* compound eye; ommatidium.

facilitation By repeated impulses to the nerve cells, the amount of neurotransmitter is increased, but eventually this process may lead to exhaustion of the neurotransmitter supply.

faciogenital dysplasia (FGD1, Xp11.21) Hypertelorism (long distance between parts of the face), cryptorchidism (hidden testes), bone deformation, cup-shape earlobes, etc., caused by defects in a guanine nucleotide exchange factor that activates Cdc42. *See* Aarskog syndrome; CDC42; GEF; guanine nucleotide exchange factor. (Estrada, L., et al. 2001. *Hum. Mol. Genet.* 10:485.)

FACS Fluorescence-activated cell sorting. *See* cell sorter; flow cytometry.

FACT (complex of p140 and p80) Facilitates transcription through nucleosomes by chromatin remodeling. *See* chromatin remodeling. (Orphanides, G., et al. 1998. *Cell* 92:105.)

factor D Synonymous with TIF-1B.

factorial $1 \times 2 \times 3 \cdots \times (n-1) \times n = n!$, the products of integers from 1 to n, the factorial of n numbers. Note: $0! = 1$.

factorial experiment Simultaneously analyzes the effects of several factors, e.g., a_1b_1, a_2b_1, a_1b_2, a_2b_2. The comparison $a_2b_2 - a_1b_2$ assesses the main effect of A when B remains constant and $a_2b_1 - a_2b_2$ determines the main effect of B when A is constant. Such experiments are very economical for simultaneously testing the consequences of several factors and their interactions (AB) and are frequently used to determine dose responses by using different levels of the individual elements such as a_1 and a_2. *See* Graeco-Latin square. (Fisher, R. A. 1937. *The Design of Experiments*. Oliver & Boyd, Edinburgh, UK.)

facultative   No absolute single determination; it may show alternative forms or functions.

facultative heterochromatin   Behaves as heterochromatin only under certain conditions, such as in the mammalian X chromosome when present in more than a single copy. *See* heterochromatin.

FAD   (1) Flavin adenine dinucleotide, a riboflavin-containing coenzyme in some oxidative-reductive processes. (2) Familial Alzheimer disease. *See* Alzheimer disease.

FADD/MORT-1 (FAS-associated death domain, 26 kDa) The FLIP proteins interact with FADD in the presence of FLICE and inhibit apoptosis. FADD interacting with Fas permits the recruitment of caspase-8 (FLICE/MACH), which activates the protease cascade of apoptosis. FADD also signals to the TNFR-associated death domain (TRADD) and to developmental pathways. FADD regulates the pre-T-cell receptor and may act as a tumor suppressor. *See* apoptosis; death domain; FAS; FLICE; TNFR. (Zhang, J. & Winoto, A. 1996. *Mol. Cell Biol.* 16:2756; Gómez-Angelats, M. & Cidlowski, J. A. 2001. *J. Biol. Chem.* 276:44944.)

Fallopian tube (uterine tube, oviduct)   Connects the mammalian ovary to the uterus. Upon maturation of the egg, it is released into the fallopian tube and fertilization may take place after the sperm injected into the vagina travels to the egg through the cervical canal. *See* fertilization; gonads.

Fallot's tetralogy (TOF)   Pulmonary stenosis, atrial septal defect, and right ventricular hypertrophy (constituting the *trilogy*); the *tetralogy* = trilogy + the right shifting of the aorta overriding the interventricular septum, thus receiving both arterial and venous blood; the *pentalogy* = the tetralogy + open foramen ovale or atrial septal defect. Mutations at JAG1 (20p12) ligand of NOTCH may cause TOF. Its prevalence is ~1/3,000 live birth. *See* Alagille syndrome; heart disease. (Eldadah, Z. A., et al. 2001. *Hum. Mol. Genet.* 10:163.)

fallout   Radiation emitted by isotopes released into the atmosphere by applying or testing nuclear weapons, by nuclear explosions, and by atomic power plants. *See* atomic radiations.

false allelism   A series of overlapping deficiencies may mimic an allelic series at a gene locus. For an illustration, see the consequences of overlapping deficiencies observed at the tip of the short arm of chromosome 9 of maize. Note particularly that the *yg / pyd* heterozygotes are normal green, demonstrating that in the *pyd* deficiency the wild-type allele is present despite the phenotype of the homozygotes, and the allelism is only apparent. *See* allelism; position effect.

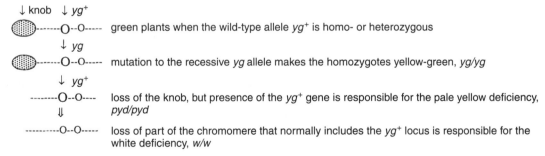

↓ knob   ↓ *yg$^+$*

━━━O--O----- green plants when the wild-type allele *yg$^+$* is homo- or heterozygous

↓ *yg*

━━━O--O----- mutation to the recessive *yg* allele makes the homozygotes yellow-green, *yg/yg*

↓ *yg$^+$*

--------O--O---- loss of the knob, but presence of the *yg$^+$* gene is responsible for the pale yellow deficiency, ⇓   *pyd/pyd*

----------O--O---- loss of part of the chromomere that normally includes the *yg$^+$* locus is responsible for the white deficiency, *w/w*

In the F$_1$, all *yg$^+$* homo- and heterozygotes as well as the *yg/pyd* plants are normal green, but the *yg/w* plants are yellow-green and the *pyd/w* individuals are pale yellow. (Modified after B. McClintock, 1944. *Genetics* 29:478.)

Fahr disease (14q)   Idiopathic basal ganglia calcification, a neurobehavioral and cognitive disorder.

FAK (focal adhesion tyrosine kinase)   Protein tyrosine kinase occurring at a high level in SRC transformed cells. Its role is in mediating cell adhesion and migration. FAK is localized at the cell membrane integrin receptors. Association of integrin with fibronectin results in binding of the Gerb2 adaptor to SRC and FAK protein tyrosine kinases and in the activation of MAPK (mitogen-activated protein kinase) in the signal transduction pathway. FAK interacts with phospholipase C-γ-1. *See* CAM; cell migration; FLT; Grb; integrin; MAPK; phospholipase; PTEN; signal transduction; SRC. (Xie, B., et al. 2001. *J. Biol. Chem.* 276:19512.)

false negative   The test performed indicates the absence of a condition, but in fact it is an incorrect observation. *See* false positive.

false positive   The test results indicate the presence of a particular condition, but the observation is incorrect. The determination of the rate depends on the prevalence of the condition. *Example*: On the basis of existing evidence, the incidence of a disease in a particular population is 3/10,00, but the clinical test has an error rate of 6%. The probability that a person diagnosed positive has actually the condition is (3/10,000)/0.06 = 0.005, i.e., 0.5%, and that he/she represents a false positive case is 1 − 0.005 = 0.995, or 99.5%. Hoffrage, et al. (*Science* 290:2261 [2000]) use the following illuminating example: Assume that 0.003 fraction (0.3%) of the population may be afflicted by colorectal cancer. The clinical laboratory test (blood in the feces) has a sensitivity of only 50% and the false positive rate is 3%. What is the probability that a positive test is actually correct? According to the rate of incidence stated above (0.3%), out of 10,000 individuals, 30 (10,000 × 0.003) have colorectal cancer, but because of the efficiency of the test (50% sensitivity) only 15 will be diagnosed as positive. Actually, of 9970 individuals (10,000 − 30), 300 individuals will test positive (3% of 10,000 = 300). Thus, the frequency of those actually afflicted among the positives is 15/(15 + 300) = 0.047. The correct estimation of the rate of false positives/negatives may be a question of life or death in medical or forensic cases. *See* false negative; sensitivity.

familial adenomatous polyposis   *See* FAP; Gardner syndrome.

familial hypercholesterolemia (FHC, 19p13.2)   Involves an increase in low-density lipoprotein (LDL) bound cholesterol, and as a consequence, at birth the homozygotes develop surface tumors filled with lipids (xanthomas) between the first two digits. Later on, these plasias appear at the tendons; the cornea of the eye may also display gray lipid-containing little sacks, and by adulthood coronary heart disease is expected. The basic defect affects the cell membrane or its receptor of LDL. Under normal conditions, the LDL passes into the lysosomes, where upon degradation the liberated cholesterol feedback-represses hydroxy-methylglutaryl coenzyme A reductase (HMGR), an enzyme responsible for cholesterol biosynthesis. The fungal products lovastatin and compactin are inhibitors of this enzyme and may be considered for treatment of FHC. The LDL receptor with 18 exons codes for 13 polypeptides with homologies to different proteins of a large superfamily. The dominant gene has been located to human chromosome 19p13.2-p13.11 and has various base substitutions, frameshift and deletion mutations. Some of the deletions are apparently caused by unequal recombination in the Alu sequences of introns 4 and 5 or others and missplicing. The heterozygotes may show the same symptoms but less evidently. The serum cholesterol and low-density lipoprotein levels in the homozygotes may exceed twice the normal levels (75−175 mg/100 mL) or more. About 5% of the myocardial infarction patients are heterozygous for this condition. The prevalence of FHC homozygotes is in the 10^{-6} range and that of the heterozygotes is higher than 0.002. A cholesterol-lowering gene has been assigned to 13q. Animals transplanted with autologous but genetically engineered hepatocytes developed a persisting 30–50% decrease of serum cholesterol levels. FHC may be identified at birth. This condition may not result in obesity, high blood pressure, or some other common indicators of susceptibility to heart ailments. Using retroviral vectors carrying the LDL receptor gene, gene therapy is promising, although improvement of the efficiency of the transformation technology is still desirable. FHC may involve haploinsufficiency. Recessive mutations at 15q25-q26, 1p34-p32, and 2p24 are also factors in FHC. Other genetic factors may contribute to the disease. *See* Alu; cholesterols; coronary heart disease; gene therapy; genetic screening; haploinsufficient; HDL; hypertension; LDL; lipids; lipoprotein; lovastatin; missplicing; sitosterolemia; sphingolipidoses; statins; unequal crossing over; VLDL. (Eden, E. R., et al. 2001. *Am. J. Hum. Genet.* 68:653.)

familial hypertriglyceridemia   Autosomal-dominant hyperlipoproteinemia leading to coronary heart disease. *See* coronary heart disease; hyperlipoproteinemia.

familial medullary thyroid carcinoma (FMTC)   *See* RET oncogene.

familial trait   Appears in certain families; its cause may be hereditary or, e.g., deafness may result from about two dozen genetic conditions, but ear infection, diabetes of the mother, birth injury, or senescence may be responsible for it. *See* congenital trait.

family   Offspring descended from a common mother (and father). In taxonomy it means a group of genera. *See* gene family; nuclear family; protein families.

family box   Group of RNA nucleotide triplets with identical nucleotides near the 5′-end but different at the 3′-end, yet coding for the same amino acid. *See* code, genetic.

family history   Essential source for the genetic counselors to determine recurrence risks. *See* empirical risk; genetic counseling; genetic risk; recurrence risk; risk.

family of genes   *See* gene family.

FAN   WD-repeat protein signaling to neutral sphingomyelinase activation. *See* sphingolipids; WD-40.

Fanconi anemia (FA, aplastic anemia, Fanconi's syndrome, Fanconi's disease, Fanconi pancytopenia)   Recessive genetic disorder assigned to eight complementation groups. The A complementation group (FANC-A/H) was localized to human chromosome 16q24.3, the FANC-C group to 9q22.3, the FANC-D to 3p25.3, FANCE (6p21-p22), FANCF (11p15), the FANC-G to 9p13, and it is the same as XRCC9. The earlier assignment to chromosome 20q could not be confirmed by newer data. The disease shows a wide range of symptoms: leukopenia (less than 5,000 leukocytes/mL blood), thrombocytopenia (reduced platelet count), pigmentation of the skin, and various malformations at different degrees of expression. The homozygous cells suffer high-frequency chromosome defects. Both red and white blood cells appear normal. It is generally fatal by the teen years. The genetic basis is chromosomal instability and associated leukemia and other malignancies. Genitourinary anomalies, cystinuria, heart disease, dwarfism, skeletal problems, microcephaly (small head), deafness, etc, accompany it. The genetic repair system appears to be defective in the complementation group C (FANC-C) protein (163-kDa), which seems to possess a nuclear localization signal. The A complementation group–coded protein is cytoplasmic and its precise function is unclear; however, a general regulatory role is suspected. Because of apparent reverse mutations, there is evidence for somatic mosaicism. Gene therapy for the C protein appears promising, although it does not remedy all of the problems of the disease. The FANC-F (11p15) encodes a 347-amino-acid protein, which is homologous to the prokaryotic RNA-binding protein, ROM. The FA cells are very sensitive to bifunctional alkylating agents such as mitomycin C and diepoxybutane. FA involves about a 3-fold increase of telomere defects and a substantial increase of extratelomeric and intrachromosomic

TTAGGG signals characteristic for telomeres. *See* Bloom syndrome; cancer; DEB; DNA repair; Dubowitz syndrome; gene therapy; hemostasis; light sensitivity diseases; limb anomalies; mitomycin C; mutagen functionality; PGD; platelet anomalies; ROM; stature in humans; XRCC. (de Winter, J. P., et al. 2000. *Am. J. Hum. Genet.* 67:1306; Medhurst, A. L., et al. 2001. *Hum. Mol. Genet.* 10:423; Joenje, H. & Patel, K. J. 2001. *Nature Rev. Genet.* 2:446; Callén, E., et al. 2002. *Hum. Mol. Genet.* 11:439.)

Fanconi-Bickel syndrome *See* glucose transporter (GLUT2).

Fanconi renotubular syndrome (15q15.3) Autosomal-dominant, late-onset kidney disease resulting in aminoaciduria and glycosuria; sometimes it is associated with bone problems. There are types with and without cystinosis. *See* aminoacidurias; cystinosis; glycosuria.

FANCY (functional analysis by coresponses in yeast) Analytical method for the expression of genes with minimal effect based on metabolite concentrations rather than metabolic flux. *See* Raamsdonk, L., et al. 2001. *Nature Biotech.* 19:45.

FANTOM (functional annotation of microarrays) *See* annotation; gene expression; interaction of gene products; microarray hybridization. (<http://fantom.gsc.riken.go.jp>.)

FAO Food and Agricultural Organization of the United Nations, Rome, Italy.

FAP Familial adenomatous polyposis. Genetically controlled polyps in the colon that may develop into colorectal cancer after additional somatic mutations. *See* Gardner syndrome; polyp.

FAP-1 Protein tyrosine phosphatase regulating the activity of the Fas cell surface receptor. *See* Fas.

FAPY (formamidopyrimidine) DNA base analog. It is removed from the DNA by Fpg (a corresponding glycosylase). The lesion is repairable. *See* glycosylases. (Ide, H. 2001. *Progr. Nucleic Acid Res. Mol. Biol.* 68:207.)

FAR Protein required for pheromone-induced cell cycle arrest at G1. *See* cell cycle; CKI.

Farber's disease (8p22-p21.3) Concerned with a defect of the enzyme ceramidase. It is manifested in early infancy as irritability, swelling of the joints, motor abnormalities, mental retardation, and early death. *See* ceramide; sphingolipids.

farnesoid X receptor (FXR) Involved in the regulation of bile acid metabolism. Chenodeoxycholic acid binds to FXR, then cholesterol 7α-hydroxylase is downregulated and bile acids are transported from the gut to the liver. Chenodeoxycholic acid indirectly regulates the proportion of bile acid and itself. Other proteins playing a key role in the synthesis of cholic acids from cholesterol are PPAR and LXRα. Oxysterol is the ligand for the latter in the liver. FXR and RXR (retinoid X receptor) jointly regulate cholesterol transport and bile acid synthesis. *See* cholesterol; cholic acid; PPAR; sterol. (Kast, H. R., et al. 2001. *Mol. Endocrinol.* 15:1720.)

farnesylation *See* prenylation.

farnesyl pyrophosphate Intermediate in the synthesis of cholesterol; has a role in mediating binding of proteins to membranes. *See* prenylation.

Far-Western hybridization Method for isolating genes, which encode interacting proteins. *See* two-hybrid method; Western blot. (Blanar, M. A. & Rutter, W. J. 1992. *Science* 256:1014.)

Fas (fibroblast-associated surface antigen, CD95, APO-1, 10q24.1) Member of the tumor necrosis factor and nerve growth factor receptor protein family including Cd40, CD27, CD30, OX40, and SFV-2. FasL: Fas ligand, encoded in human chromosome 1q23. FasL expression is induced by PMA. Fas plays a key role in apoptosis and autoimmune disease by binding the FasL, a TNF protein. PMA and protein tyrosine kinase inhibitors (herbimycin, genistein) can rapidly induce FasL and cyclosporin inhibits the induction of FasL. The antibody to FasL is immunoglobulin M (IgM), whereas the antibody to APO-1 is IgG3. Human Fas is a transmembrane protein containing 325 amino acids. Fas is upregulated by interferon-γ and tumor necrosis factor-α in human B and T cells. Fas signaling leads to apoptosis; it is triggered by cross-linking of Fas with Fas antibodies and by cells expressing FasL. Fas-induced apoptosis is faster than that induced by TNF-receptor. Fas and TNF induce apoptosis in the cells of the immune system. Therefore, in cancer gene immunotherapy, Fas-L may be considered an appropriate target. Fas mediates two apoptotic pathways: One involves the FADD adaptor protein and pro-caspase-8. The other path is mediated by Daxx (death-associated protein, human chromosome 6q21.3) and the activation of JNK-MAPKKK through ASK1 (apoptosis signal-regulating kinase 1). FAS-FasL expression, however, does not lead to apoptosis under all circumstances. DcR3 (decoy receptor, human chromosome 20q) may associate with FasL and inactivate the apoptotic pathway in lung and colon cancer cells. Apo-3 is a CD120A-homologous TNF/NGF receptor encoded at 1p36.3. *See* apoptosis; autoimmune disease; Canal-Smith syndrome; caspase; CD120; cyclosporin; FADD; FAP-1; immune system; JNK; Lyell syndrome; NGF; PMA; signal transduction; survival factors; TACE; TNF. (Nagata, S. 1999. *Annu. Rev. Genet.* 33:29.)

FAS Fatty acid synthase. *See* fatty acids.

fasciation Abnormal development resulting in flat stem, club-shaped fruits with four rows of seeds rather than two in *Arabidopsis*. In zoology it designates sheets of tissues. *See* CAF. (Kaya, H., et al. 2001. *Cell* 104:131.)

Fasciation (at right, normal at left).

fascicle (fasciculus)   Bundle of (nerve, muscle, etc.) fibers.

fasciclin   Cell adhesion molecule of the immunoglobulin superfamily expressed in motor neurons. It acts with semaphorin and connectin/titin. *See* semaphorin; titin. (Cheng, Y., et al. 2001. *Cell* 105:757; Wright, J. W. & Copenhaver, P. F. 2001. *Dev. Biol.* 234:24.)

FasL   *See* FAS.

FASTA   Program used for sequence comparisons in DNA, RNA, and proteins. *See* BLAST; BLOSUM. (Pearson, W. R. 1991. *Genomics* 11:635.) Email address <fasta@ebi.ac.uk>.

fast blue   Fluorescent (neuronal) tracer dye.

fast component   Of the nucleic acid reassociation reaction, represents the repetitive sequences. *See* c_0t curve; c_0t value.

fast green   Histochemical stain for basic proteins; it is generally used in combination with other stains, e.g., safranin or pyronin.

FASTLINK   Computer program for genetic linkage analysis.

fast neutrons   Particulate, ionizing radiations released at atomic nuclear fission. They are highly effective mutagens. *See* physical mutagens; radiation. (Hacker-Klom, U. B., et al. 2000. *Radiat. Res.* 154:667.)

fat   Fatty acid–glycerol ester. Its caloric value (9.3 kcal/g) is approximately 2.3 times of that of carbohydrates and proteins. Oxidation of fats (rancidity) may result in mutagenic compounds in natural products. *See* atherosclerosis; cholesterol; fatty acids; triaglycerols.

fatal familial insomnia (insomnia-dysautonomia thalamic syndrome, FFI)   Autosomal-dominant degenerative disease of the thalamic nuclei (the basal; part of the brain involved in transmission of sensory impulses). A progressive insomnia, a defect in the autonomous nervous system, a speech defect, tremor, and seizures may eventually lead to death. The pathological symptoms may resemble those of Creutzfeldt-Jakob disease (CJD), with the difference that the spongy transformation of the cells is limited to the thalamus. In CJD, the mutation at amino acid position 129 results in a valine replacement, whereas in FFI at 129 there is a methionine. In mice devoid of prion protein (PrP^C) because of mutation, the circadian activity rhythm, including the sleeping pattern, is altered. Because of somatic mutation, sporadic fatal insomnia occurs. *See* Creutzfeldt-Jakob syndrome; encephalopathies; prion. (Collins, S., et al. 2001. *J.Clin. Neurosci.* 8[5]:387.)

fat body   Insect organ comparable to the liver of higher animals. It secretes antimicrobial peptides and its role is similar to that of the immune system of higher animals. (Khush, R. S. & Lemaitre, B. 2000. *Trends Genet.* 16:442.)

fate map, morphogenetic   Indicates the positions in the blastoderm from which adult structures (legs, eyes, nerves, etc.) develop. In basic principles it has some resemblance to the genetic mapping of genes to chromosomes (both ideas were developed by A. Sturtevant). The procedure is generally as follows: The investigator constructs a diploid individual with an unstable chromosome with the wild-type allele, and the homologous chromosome carries the recessive gene, whose expression is to be traced to a developmental origin or control center. When the unstable chromosome is lost, the critical recessive allele can display pseudodominance in the sector, which no longer carries the wild-type allele. Most commonly, the defective chromosome in *Drosophila* is an X chromosome. Thus, its loss generates a gynandromorphic sector. The association between the gynander sector and the critical gene expression can be classified and the developmental origin of the function determined. The developmental distance calculated from fate mapping is expressed in *sturt* units; 1 sturt is the fate-mapping quotient multiplied by 100. That is, 1 sturt means that two cell clusters are different in 1% of the mosaics.

The results of a fate-mapping experiment using an unstable ring X chromosome and the *drop-dead* mutation of *Drosophila* are shown below. The homo- and hemizygous individuals walk in an uncoordinated manner and suddenly die about 10 days after eclosion. (Abridged from Y. Hotta & S. Benzer, *Nature* 240:527.)

Behavior ↓	Number of individuals with the constitution indicated in the head cuticle and in the abdominal cuticle			
	Head		Abdomen	
	XX	XO	XX	XO
Normal	91	8	54	28
drop-dead	6	72	23	51
Total	97	80	77	79

The frequency of *drop-dead* gynander and non-gynander (normal) heads and abdomens can be calculated similarly to recombination frequencies. Thus,

$$\text{HEAD:} \quad \frac{6+8}{97+80} = 0.079 \qquad \text{ABDOMEN:} \quad \frac{23+28}{77+79} = 0.327$$

The decimals above indicate that the *drop-dead* phenotype is more closely associated with the head cuticle gynandromorphy than with that of the abdomen, showing that this behavior is determined by the development of the head. Anatomical studies have shown perforations in the brain of flies, confirming the conclusions of the fate-mapping experiments. Fate maps can also be constructed on the basis of the distribution of molecules in the developing embryo. According to the suggestion of Sydney Brenner, the fate of individual cells may follow the "European plan," i.e., the ancestry of a cell determines its developmental fate; or the "American plan," when interaction with the neighboring cells is the most critical for their fate. The distinction between these alternatives may not always be possible. In more complex organisms such as mice, fate mapping may be possible by injecting cells equipped with a site-directed recombination or mutation system. *See* cell lineages; deletion mapping; gynandromorphs; mapping; physical mapping; site-directed mutation; site-directed recombination. (Inoue, T., et al. 2000. *Dev. Biol.* 219:373.)

fatty acids Aliphatic carboxylic acids of long-chain structure; parts of the membrane phospho- and glycolipids, cholesterols, also in fats and oils. Saturated fatty acids are formic (1:0), acetic (2:0), propionic (3:0), butyric (4:0), lauric (12:0), myristic (14:0), palmitic (16:0), stearic (18:0), arachidic (20:0), behenic (22:0), and lignoseric (24:0) acids. (Within the parentheses, the number of carbon atoms in the chain, and after the colon, the number of double bonds, are indicated.) Unsaturated fatty acids are crotonic (4:12), plamitoleic (16:1), oleic (18:1), vaccenic (18:1), linoleic (18:2), linolenic (13:3), and arachidonic (20:4) acids. Triacylglycerides (glycerol esters with three fatty acids) are one of the major sources of energy in mammals, particularly during hibernation (when nearly all of the energy comes from these compounds).

Linoleate and linolenate are essential fatty acids for mammals. Lipids are complexes of fatty acids with phosphates, sterols, or sugars and have indispensable functions in cellular membranes. Lipids associated with proteins have the role of transporting fatty acids. When fatty acid nodules are deposited on the inner walls of the blood vessels, atherosclerosis results, a major cause of heart disease. Fatty acid biosynthesis is important not just from the viewpoint of human disease, but knowledge may help in developing plant varieties for better diet and for the production of special raw material for industrial purposes. FAS (fatty acid synthase) is a primary enzyme in fatty acid synthesis. Inhibition of FAS has antitumor effects. Fatty acids positively and negatively regulate the transcription of some genes. *See* apolipoproteins; atherosclerosis; fat; lipidoses; lipids; sphingolipidoses; sphingolipids; triaglycerols. (Duplus, E., et al. 2000. *J. Biol. Chem.* 275:30749.)

favism Hemolytic anemia caused by eating even extremely small quantities of *Vicia faba* (broad bean) or inhaling its pollen. The initial reaction (headache, dizziness, nausea, chills, etc.) may occur within seconds, and it may be followed within a day by jaundice and blood in the urine. People who are deficient in glucose-6-phosphate dehydrogenase are susceptible to this condition. Light sensitivity of the skin after eating the seeds of *Fagopyrum vulgare* (buckwheat) is different; a photodynamic substance causes this. *See* broad bean; buckwheat; glucose-6-phosphate dehydrogenase deficiency; vicine. (Burbano, C., et al. 1995. *Plant Foods Hum. Nutr.* 47[3]:265.)

FB *See* hybrid dysgenesis.

FBI site (fold-back inhibition) The transcript of Tn*10* transposase folds back a region (UGGUC) complementary to the modified Shine-Dalgarno sequence (AUCAG) and prevents ribosome binding, thus reducing transposition. *See* Tn*10*

F-box Cyclin F interacting protein domain recruiting various proteins for proteolytic degradation (ubiquitination). F-box proteins are involved with the cell cycle, developmental processes, immune reactions, etc. *See* cyclin F; glucose induction; IκB; proteosome; SCF; ubiquitin. (Strohmaier, H., et al. 2001. *Nature* 413:316.)

FBS Fetal bovine serum.

Fc Crystallizable fragment of immunoglobin containing the C end of heavy chains. Fc receptors are expressed in monocytes and macrophages and some other cells of the immune system. *See* antibody; FcRn; immune system.

FcεRI Immunoglobulin IgE (Igε) crystallizable fragment high-affinity receptor on lymphocytes. It controls IgE-mediated antigen presentation. It participates in several signaling pathways of the allergic responses. *See* allergy; antibody; antigen presentation; asthma; immunoglobulins.

FcγR Receptors (I, II, III) of immunoglobulin G antibody crystalline fragments. *See* antibody; ITIM.

FCP *See* transcription complex.

FcRn Crystallizable Fragment receptor of the neonatal antibody. It transfers antibodies from the placenta to the fetus and secures immunity of newborns right after birth. *See* antibody; Fc.

FCS Fluorescence correlation spectroscopy, facilitates measuring diffusion constants of various proteins, the ratios of free and bound proteins, and protein-protein connections within the cell by microscopic examinations. *See* FRAP; FRET.

F-distribution Used for testing the significance of the difference between variances. The *F-test* is based on $F = (s_1)^2/(s_2)^2$ (s = standard deviation) computed from two different populations or samples. In this procedure, the null hypothesis is that the two samples are identical. If, however, the variance ratio exceeds the values, at the appropriate degrees of freedom in the body of the table the differences are significant. (The name *F* is in honor of R. A. Fisher; see table on page 419.)

fDNA (fossil DNA) DNA extracted from ancient bones or other relics. *See* ancient DNA.

F-duction A gene is transferred to the F$^-$ bacterial cell by an F plasmid, called F′; it is the same process as that of sexduction. *See* sexduction. (Hanson, R. L. & Rose, C. 1979. *J. Bacteriol.* 138:783.)

FEAR (Cdc fourteen, early anaphase release) Regulates the mitotic exit network (MEN) with the joint action of the polo kinase Cdc5, the separase Esp1, kinetochore-associated proteins Slk19 and Spo. *See* Cdc14; kinetochore; MEN; polo; separins; *SPO*. (Stegmeier, F., et al. 2002. *Cell* 108:207.)

feature *See* DNA chips.

Fechtner syndrome *See* Alport disease; May-Hegglin anomaly.

fecundity Reproductive ability determined by the quantity of gametes produced per time units. *See* fertility.

F-DISTRIBUTION at 5% and 1% probability levels. At significance the F value computed must exceed the numbers in the table at the intersection of the degrees of freedom (df₁ and df₂). (Condensed by permission from S. Koller, *Biochemisches Taschenbuch*, H. M. Rauen, ed. Springer-Verlag, Berlin.)

Df₂ for Denominator	Degrees of Freedom for Numberator Df₁												
	1	2	3	4	5	6	8	10	15	20	30	50	∞
						5%							
1	161	200	218	225	230	234	239	242	246	248	250	252	254
2	19	18	19	19	19	18	19	19	9	9	20	20	20
3	10.1	9.6	9.3	9.1	9.0	8.9	8.8	8.6	8.7	8.7	8.5	8.6	8.5
4	7.7	6.9	6.6	6.4	6.3	8.2	6.0	6.0	5.9	5.8	5.8	5.7	5.6
5	6.6	5.8	5.4	5.2	5.1	5.0	4.8	4.7	4.6	4.6	4.5	4.4	4.4
6	6.0	5.1	4.8	4.5	4.4	4.3	4.2	4.1	3.9	3.9	3.8	3.8	3.7
7	5.6	4.7	4.4	4.1	4.0	3.9	3.7	3.6	3.5	3.4	3.4	3.3	3.2
8	5.3	4.5	4.1	3.8	3.7	3.6	3.4	3.4	3.2	3.2	3.1	3.0	2.9
9	5.1	4.3	3.9	3.8	3.5	3.4	3.2	3.1	3.0	2.9	2.9	2.8	2.7
10	5.0	4.1	3.7	3.5	3.3	3.2	3.1	3.0	2.9	2.8	2.7	2.6	2.5
12	4.3	3.9	3.5	3.3	3.1	3.0	2.9	2.8	2.6	2.5	2.5	2.4	2.3
14	4.6	3.7	3.3	3.1	3.0	2.9	2.7	2.6	2.5	2.4	2.3	2.2	2.1
16	4.5	3.6	3.2	3.0	2.9	2.7	2.6	2.5	2.4	2.3	2.2	2.1	2.0
18	4.4	3.6	3.2	2.9	2.8	2.7	2.5	2.4	2.3	2.2	2.1	2.0	1.9
20	4.4	3.5	3.1	2.9	2.7	2.6	2.5	2.4	2.2	2.1	2.0	2.0	1.8
25	4.2	3.4	3.0	2.8	2.6	2.5	2.3	2.2	2.1	2.0	1.9	1.8	1.7
30	4.2	3.3	2.9	2.7	2.5	2.4	2.3	2.2	2.0	1.9	1.8	1.8	1.6
40	4.1	3.2	2.8	2.6	2.5	2.3	2.2	2.1	1.9	1.8	1.7	1.7	1.5
50	4.0	3.2	2.8	2.6	2.4	2.3	2.1	2.0	1.9	1.8	1.7	1.6	1.4
60	4.0	3.2	2.8	2.5	2.4	2.3	2.1	2.0	1.8	1.8	1.7	1.6	1.4
80	4.0	3.1	2.7	2.5	2.3	2.2	2.1	2.0	1.8	1.7	1.6	1.5	1.3
100	3.9	3.1	2.7	2.5	2.3	2.2	2.0	1.9	1.8	1.7	1.6	1.5	1.3
∞	3.84	3.00	2.60	2.37	2.21	2.10	1.94	1.83	1.57	1.57	1.46	1.35	1.00
						1%							
1	4100	5000	5400	5600	5800	5900	6000	6000	6200	6200	6200	6300	6400
2	98	99	99	99	99	99	99	99	99	99	99	100	100
3	34	31	29	28	28	28	27	27	27	27	27	26	26
4	21	18	16	16	16	15	15	15	14	14	14	14	13
5	16	13	11	11	11	11	10	10	9.7	9.6	9.4	9.2	9.0
6	14	11	8.8	9.2	8.8	8.5	8.1	7.8	7.6	7.4	7.2	7.1	6.9
7	12	9.6	8.5	7.8	7.5	7.2	8.8	6.6	6.3	6.2	6.0	5.9	5.7
8	11	8.7	7.6	7.0	6.6	6.4	6.0	5.8	5.5	5.4	5.2	5.1	4.9
9	11	8.0	7.0	6.4	6.1	5.8	5.5	5.3	5.0	4.8	4.7	4.5	4.3
10	10	7.6	6.6	6.0	5.6	5.4	5.1	4.9	4.7	4.4	4.3	4.1	3.9
12	8.3	6.9	6.0	5.4	5.1	4.8	4.5	4.3	4.0	3.9	3.7	3.6	3.4
14	8.9	6.5	5.6	5.0	4.7	4.5	4.1	3.9	3.7	3.5	3.4	3.2	3.0
16	8.5	6.2	5.3	4.8	4.4	4.2	3.9	3.7	3.4	3.3	3.1	3.0	2.8
18	8.3	6.0	5.1	4.6	4.3	4.0	3.7	3.5	3.2	3.1	2.9	2.8	2.6
20	8.1	5.9	4.9	4.4	4.1	3.9	3.6	3.4	3.1	2.9	2.8	2.6	2.4
25	7.8	5.6	4.7	4.2	3.9	3.6	3.3	3.1	2.9	2.7	2.5	2.4	2.2
30	7.8	5.4	4.8	4.0	3.7	3.5	3.2	3.0	2.7	2.6	2.4	2.3	2.0
40	7.3	5.2	4.3	3.8	3.5	3.3	3.0	2.8	2.5	2.4	2.2	2.1	1.8
50	7.2	5.1	4.2	3.7	3.4	3.2	2.9	2.7	2.4	2.3	2.1	2.0	1.7
60	7.1	5.0	4.1	3.7	3.3	3.1	2.8	2.6	2.4	2.2	2.0	1.9	1.6
80	7.0	4.9	4.0	3.6	3.3	3.0	2.7	2.6	2.3	2.1	1.9	1.8	1.5
100	6.9	4.8	4.0	3.5	3.2	3.0	2.7	2.5	2.2	2.1	1.9	1.7	1.4
∞	6.64	4.60	3.78	3.32	3.02	2.80	2.51	2.32	2.04	1.88	1.70	1.52	1.00

feedback control A late metabolite of a synthetic pathway regulates synthesis at earlier step(s). It can be negative or positive. Two specific sites on an early enzyme in the biosynthetic pathway mediate the feedback control. One site serves for recognition of the substrate; the other site recognizes a later product in the biosynthetic path. When this late product (end product) accumulates because it is not utilized in proportion to its synthesis, it may combine with the early enzyme's feedback recognition site, resulting in a reversible conformational change and cessation or lowering of enzyme activity. The feedback is an economy device of the cell; the production of a metabolite is slowed down in the absence of a need. Feedback systems may operate in a number of self-explanatory ways (see below); for the compensating feedback a note is needed. E inhibits the path between C and D, but product F may alleviate the inhibition by reducing the activity of enzyme A; thus, E never accumulates excessively. If feedback sensitivity is eliminated through mutation, the system may be overproducing the end product, which may be beneficial from an economic viewpoint in industrial or agronomic organisms.

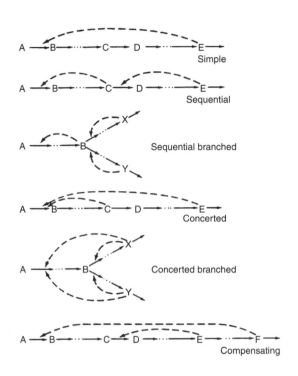

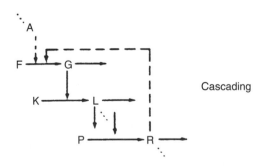

The major types of feedback controls. The broken lines indicate the feedback loops. The solid lines show the biochemical synthetic pathway. (Redrawn after Savageau, M. A. 1972. *Curr. Top. Cell Reg.* 6:63.)

These are the most common regulatory mechanisms in eukaryotes. *See* attenuation; inhibition; negative control; positive control; regulation of enzyme activity; repression. (Savageau, M. A. 1972. *Curr. Top. Cell. Reg.* 6:63.)

F element   Dot-size (smallest) chromosomes, common to all *Drosophila* species.

***F* element**   *See* nonviral retrotransposable elements. *See* FB; F factor; hybrid dysgenesis I-R.

feline   Pertaining to cats.

female   Individual producing the larger, usually nonmotile, gamete (egg). Its symbol is the Venus mirror ♀ or a circle in pedigree charts. This is the most ancient symbol of genetics, used 5,000 years ago on stone tablets in Asia Minor. *See* Mars shield; pedigree analysis.

Probably the earliest (~5000 years old) pedigree chart excavated in east Ur of mesopotamia employs the female symbol in upside-down form of the presently used symbol (encircled by GPR). (From Amschler, W. 1935. *J. Hered.* 26:223.)

female gametophyte   *See* gametophyte.

femto- 10^{-15}, e.g., femtomole.

FEN   *See* Rad27.

Fenton reaction   In the presence of iron (Fe^{2+}) and copper (Cu^{2+}) catalyzes the formation of $OH^{\bullet} + H_2O$ from H_2O_2. *See* $OH^{\bullet}$; ROS. (Dikalova, A. E., et al. 2001. *Proc. Natl. Acad. Sci. USA* 98:13549.)

FEO   Familial expansile osteolysis. *See* Paget disease.

feral population   Inhabits nature rather than living under laboratory or domesticated conditions (e.g., feral mice, feral *Drosophila*, feral pigs, etc.).

fermentation   Energy-producing anaerobic degradation of carbohydrates to lactate or ethanol.

fermentors   Precisely controlled mass culture vessels that may be used for industrial manufacturing of biological products (such as alcohol, antibiotics, proteins, etc.) by bacterial, fungal, animal, or plant cells.

fern   Lower plant of the *Pterophyta* taxonomic group. Ferns exist in two generations: sporophyte and gametophyte. The sporophytes develop roots, rhizomes, and leaves. These diploid plants generally form one type of sporangium on the lower surface of the leaves in aggregates, called sori. Meiosis takes place within the sporangia, and the released haploid products develop into generally heart-shaped gametophytes that (unlike in higher plants) form an independent organism on soil (thallus). When mature, on the lower surface of the gametophyte, eggs develop within the female archegonia and sperm in the male antheridia (6–10 cells). After a swimming spore fertilizes an egg, the diploid zygote grows on the gametophyte until the young sporophyte becomes ready (rooted) for independent existence. The gametophytes then die and through the sporophytes the life cycles are repeated. Ferns have a wide range of chromosome numbers of relatively large size, and many species are well suited for cytological analyses because of the long haploid phase.

ferredoxin Iron-containing protein involved in electron transport.

ferritin Ubiquitous, iron-storing cellular protein, thus protecting the cells from the toxic effects of this heavy metal. Also, it transports iron to the sites of need. Ferritin consists of 24 subunits around an iron core encoded in human chromosomes at 11q13 and 19q13.1. An intronless 242 precursor of a ferritin-like protein encoded at 5q23.1 is targeted to the mitochondria. An iron-regulatory protein (IRP) and an iron-responsive element (IRE) CAGUGU determine the synthesis. Mutation in the latter results in the dominant hyperferritinemia that may cause cataracts. Neuroferritinopathy (19q13.3) is a dominant disease of the basal ganglia caused by a gene mutation encoding the light polypeptide of ferritin. In Alzheimer disease and Parkinson disease, ferritin increases as iron accumulates in the brain. See aconitase; Alzheimer disease; apoferritin; eye diseases; ganglion; Hallervorden-Spatz syndrome; hyperferritinemia; MYC; Parkinson disease. (Levi, S., et al. 2001. *J. Biol. Chem.* 276:24437; Curtis, A. R. J., et al. 2001. *Nature Genet.* 28:350.)

fertilin Sperm cell surface protein mediating membrane adhesion and sperm-egg fusion, migration from the uterus to the oviduct, and binding to the zona pellucida. Fertilins belong to the ADAM family of proteins. The disintegrin domain binds to an integrin receptor of the egg plasma membrane during fertilization. See acrosomal process; ADAM; CD9; fertilization; integrin. (Evans, J. P. 2001. *Bioessays* 23:628.)

fertility Production of viable offspring (gamete) per individual during the reproductive period. *Effective fertility* denotes the reproductive rate of individuals afflicted by a disease compared to that of normal, healthy individuals. The ability to form fertile hybrids has been generally used to define species. Different species are not expected to produce fertile hybrid progeny. This distinction has some problems because, e.g., the female hybrids of *Drosophila melanogaster* female and *D. simulans* male are viable and fertile, whereas the males of the same cross die at the larval/pupal stage of development. In the reciprocal crosses, *D.simulans* female × *D.melanogaster* male, most of the females also die as embryos; only a few escape death. Human susceptibility to conception is about 6 days prior to and including the day of ovulation. The raise in hormone level, particularly by the raise of luteinizing hormone (LH) level, can clinically monitor ovulation. See animal hormones; fecundity; ovulation; reproductive isolation, species, speciation, infertility; spermiogenesis, transmission.

fertility factor See F factor; F plasmid.

fertility inhibition The conjugal function of the bacterial cell is prevented by an inhibitory plasmid. See bacterial conjugation.

fertility restorer genes May assist to overcome the cytoplasmically determined male sterility. Their ability to function also depends on the nature of the cytoplasm. In the S cytoplasm of maize, the restoration of fertility by *Rf 3* is gametophytic, i.e., only 50% of the pollen is fertile in the heterozygous plants. In the T cytoplasm, the restoration of fertility is under sporophytic control and the heterozygotes for the two complementary fertility genes *Rf1* and *Rf2* produce nearly 100% viable pollen. The *Rf4* gene is a sporophytically expressed restorer in the C cytoplasm. Fertility restoration may involve some additional minor factors. The cytoplasmic male sterility in these cytoplasms is controlled by mitochondrial plasmid genes. The fertility restorer genes have applied significance in the commercial production of hybrids. T (Texas) cytoplasmic male sterility is associated with susceptibility to *Helminthosporium* blight. The fertility restorer genes do not alleviate, however, the symptoms of the disease, indicating that the latter is controlled by additional means other than fertility. Cytoplasmic male sterility occurs in several plant species. See cytoplasmic male sterility; mtDNA; pollen sterility. (Duvick, D. N. 1965. *Advances Genet.* 13:1; Liu, F., et al. 2001. *Plant Cell* 13:1063; When, L. & Chase, C. D. 1999. *Curr. Genet.* 35:521.)

fertilization Involves the fusion of gametes of opposite sexuality. It may take place in different forms in hermaphroditic organisms such as the majority of plants and some of the animals: self-fertilization (autogamy), cross-fertilization (allogamy), or a mixture of the two. In the dioecious species of the majority of animals and a few plants, only cross-fertilization can happen. Fertilization results in the formation of zygotes and the restoration of the chromosome number to the 2*n* level. In many plant species, double fertilization takes place: the generative sperm fertilizes the egg and the vegetative sperm unites with the fused polar nuclei. As a result, the triploid endosperm mother cell is generated. In some plant species, the sperm egg nuclei fail to fuse and each contributes independently to the development of the embryo: *semigamy*. In some plants, the generative sperm and the vegetative sperm may come from a different pollen grain and thus *heterofertilization* results. If the genetic constitution of these two sperms is not identical, there is discordance (noncorrespondence) between the genetic constitution of the embryo and the endosperm.

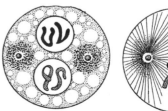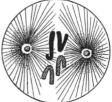

The paternal and maternal pronuclei within an *Ascaris* egg (left), the pronuclei have fused into a zygote nucleus, ready for mitosis (right).

Embryo development without fertilization is called *apomixis* in plants and *parthenogenesis* in animals. In the majority of plants and animals, only the nucleus of the sperm enters the embryo sac and egg, respectively, and the cytoplasmic genes are transmitted only through the female in these cases. Some plants transmit cytoplasmic elements by the sperm during fertilization, resulting in complete biparental inheritance. Only a limited number of pollen tubes in plants grow down the pistil. More than a thousand sperms in animals may attach to the surface of the egg, but only one succeeds in fertilization, although normal ejaculates may contain 20 million sperm. In autogamous plants the pollen count is much more limited than in the allogamous species. A large quantity of sperm is essential for normal fertilization of plants and animals. Human males with sperm counts per ejaculate below 20,000 are generally sterile. The number of eggs per ovule in plants or number of ova released during ovulation in animals may also be only one or a few. After a single injection by the males, *Drosophila* females can store about 700 sperm for a prolonged period of time, and fertilization may take place gradually. After a second mating within 2 weeks, the leftover sperm may be destroyed.

Usually the fertilization takes place within the female body; in others the fertilization is an external process or the eggs may be deposited in a pouch of the female, where fertilization and further development follow. Even in cases of internal fertilization, a variable amount of time may elapse between pollination or copulation and the actual penetration and fusion of the sperm and egg nuclei. In mammalian fertilization, the first step after the sperm reaches the follicular cells is binding to the egg membrane. The sperm is "capacitated" to this task by changes in the sperm plasma membrane through the secreted products of the female (Wu, C., et al. 2001. *J. Biol. Chem.* 276:26962). The sperm then binds to the zona pellucida membrane. This membrane is made up of a cross-linked network of glycoproteins ZP1, ZP2, and ZP3. The latter ones are *sperm receptors*. In mouse, the sp56 spermatid-specific protein recognizes ZP3. After this step, the protease and hyaluronidase content of the acrosome is released (*acrosomal reaction*), facilitating the penetration and passing through the zona pellucida.

After one sperm has fused with the egg plasma membrane, the so-called cortical reaction blocks the entry of other sperm as the zona hardens. The *cortical reaction* is mediated by an increase in cytosolic calcium through an inositol-phospholipid signaling pathway. On the surface of the egg coat, microvilli develop in such a manner that they firmly hold the sperm. After fertilization, the egg becomes a zygote and the process is completed when the male and female pronuclei fuse into a *zygote nucleus*. The sperm donates to the egg not just the male pronucleus containing a complete haploid set of chromosomes but also the *centriole* that is not available in the egg. The centrioles are then associated with the female *centrosome* of animals. This makes possible the mitotic divisions and embryogenesis. The mammalian zygote is formed through the fusion of the highly compacted nucleoprotamine-containing sperm chromatin and the metaphase II—arrested chromosomes of the oocyte. Following fertilization, remodeling of the chromatin of the two gametic chromosome sets into a diploid genome takes place. The sperm genome is much more highly methylated than that of the egg. Within 7–8 hours after fertilization, the paternal genome is significantly demethylated. Demethylation of the maternal genome follows after the second division of the zygote. The global demethylation exempts, however,

the imprinted genes. *See* acrosomal process; androgenesis; apomixis; artificial semination; bindin; cd9; centrosome; certation; cyritestin; embryogenesis; fertilin; fertility; gametogenesis; hormone receptors; imprinting; IVF; megaspore competition; methylation, DNA; pollen tubes; polyspermic fertilization; RPTK; selective fertilization; sex hormones; sperm; synergids. (Primakoff, P. & Myles, D. G. 2002. *Science* 295:2183.)

FES Feline fibrosarcoma viral oncogene (*fes*). It is in the long arm of human chromosome 15, and in most cancer cells it is translocated to chromosome 17, causing leukemia. Its protein product is a protein tyrosine kinase. *See* ABL; ERBA; PTK.

FET Genes involved in iron uptake and metabolism.

fetal alcohol syndrome *See* alcoholism.

fetal tissue research *See* embryo research; stem cells.

α Fetoprotein (AFP)   Expressed in the embryonic yolk and liver of mammals. Serum albumin and α-fetoprotein genes are linked at about 15 kb apart, and each encodes about 580 amino acids. The two proteins are immunologically cross-reactive and display about 35% ho-mology. The rate of transcription of AFP drops four orders of magnitude after birth. Regulatory sequences are positioned within 150 bp 5′ and there are also enhancers 6.5, 5, and 2.5 kbp upstream from the transcription initiation site. The gene (human chromosome 4q11-q22) is a classical example for tissue-specific and developmental regulation. *See* MSAFP. (Mizejewski, G. J. 2001. *Exp. Biol. Med.* 226:377.)

fetoscopy   Viewing (or possibly sampling of tissues) of the fetus within the womb to detect probable developmental or biochemical anomalies in case there is a good indication for them. Fetoscopy may involve up to 10% risk to the fetus, and if other less invasive (e.g., sonography) methods are available, it should be avoided. It is also used for the detection of fetal heartbeat. *See* amniocentesis; echocardiography; prenatal diagnosis; sonography.

fetus   Unborn child of viviparous animals at the stages following the embryonal state after substantial differentiation has been completed. In humans the fetal period is from 9 weeks after conception to birth. *See* embryo; embryo research; vivipary.

Feulgen   Microtechnical staining method detecting deoxyribose and thus DNA in warm acid-hydrolyzed tissue followed by staining with leucobasic fuchsin. *See* stains.

FeV (feline leukemia virus)   Retrovirus causing immunodeficiency in cats.

fever, periodic   Observed in different diseases. Familial Mediterranean fever appears recessive (16p13) and may be associated with amyloidosis. The dominant form (12p13.2) is based on a defect in tumor necrosis factor receptor. The recessive Dutch type (12q24) is a mevalonate kinase deficiency. *See* amyloidosis; Fabry disease; fibroblast growth factor; hyperimmunoglobulinemia D. (Aksentijevich, I., et al. 2001. *Am. J. Hum. Genet.* 69:301.)

FFA Focus-forming activity. A 170 kDa protein required for the association of replication protein A (RPA) with the focus-forming units of replication (the 100 to 1,000 DNA loop [replication bubbles] centers of replication along the euchromatic chromosome). *See* replication bubble; replication protein A; Werner syndrome.

F factor Bacterial fertility element (F plasmid, sex plasmid). *See* episome; F plasmid; Hfr. (Bernstein, H. L. 1958. *Symp. Soc. Exp. Biol.* 12:93.)

Ffh GTPase of the signal recognition particle of bacteria. *See* FtsY; SRP.

FFU Focus-forming unit is the measure of focus formation in transformed (cancer) cells.

FGF Fibroblast growth factor. It is also a FGF-related oncogene. FGF-B (the bovine form, FGF2) was assigned to human chromosome 4q25. FGF-6 and FGF-4 (human chromosome 2q13) bear similarity to the NT2 oncogene. FGF may have a critical role in organ differentiation. Fibroblast growth factor receptor (FGFR) has a cytoplasmic tyrosine kinase domain. Mutations in FGFR1 and FGFR2 have been associated with Crouzon, Jackson-Weiss, Pfeiffer, and Beare-Stevenson syndromes. FGFR3 is involved in hypochondroplasia. Defects in FGFR1c may lead to diabetes type II. FGF1 (in human chromosome 5q31) is implicated in angiogenesis. FGF3 (11q13) is homologous to a mouse mammary carcinoma gene, FGF5 (4q21). FGF7 is a keratocyte growth factor. FGF8 is an androgen-induced growth factor. FGF9 is the glia-activating factor. FGF10 is essential for limb and lung development in mice. The activation of FGF requires appropriate transmembrane receptors that are activated by oligomerization in the presence of heparin-like molecules. The human genome contains 30 and *Drosophila* and *Caenorhabditis* each show 2 FGFs. *See* achondroplasia; angiogenesis; Apert syndrome; Beare-Stevenson syndrome; craniosynostosis; Crouzon syndrome; diabetes; embryogenesis; glial cell; growth factors; heparin; HST oncogene; hypochondroplasia; INT oncogene; Jackson-Weiss syndrome; keratosis; organizer; Pfeiffer syndrome; sex reversal; signal transduction; thanatophoric dysplasia; tyrosine kinase. (Dubrulle, J., et al. 2001. *Cell* 106:219; Vasili-auskas, D. & Stern, C. D. 2001. *Cell* 106:133; Ahmad, S. M. & Baker, B. S. 2002. *Cell* 109:651; Sun., X., et al. 2002. *Nature* 418:801.)

FHA (forkhead-associated domain) Interacts with phosphorylated molecules. It has a role in DNA damage signaling and CHK2-dependent tumor suppression. *See* CHK2. (Durocher, D., et al. 2000. *Mol. Cell* 6:1169.)

FHIT *See* renal cell carcinoma.

fialouridine FIAU, [1-(2-deoxy-2-fluoro-D-arabinofuranosyl)-5-iodouracil]) Inhibitor of DNA polymerase γ, responsible for the synthesis of mtDNA. *See* DNA polymerases; mtDNA.

Fibonacci series The denominator doubles in each subsequent term while the numerator becomes the sum of its two preceding numerators, e.g., 1, 1/2, 2/4, 3/8, 5/16, 8/32, 13/64, and so on. This has applications in predicting the proportion of heterozygotes in an interbreeding population at each locus that was initially heterozygous ($H = 0.5$). Thus, six generations of sib mating leaves 13/64, i.e., ~20.31% heterozygous for each originally heterozygous locus, and for its complement ($100 - 20.31$), 79.69% will be expected to become homozygous.

fibrillarin Fibrillar protein present in many snRNPs within the nucleolus; it is also an autoantigen. *See* autoantigen; snoRNA; snRNP. (Snaar, S., et al. 2000. *J. Cell Biol.* 151:653.)

fibrillin Large cystein-rich glycoprotein molecule involved in the formation of microfibrils along with elastin. The fibrillin microfibrils are encoded by two genes in human chromosomes 15q-21.1 (FBN1) and 5q23-q31 (FBN2), respectively. Mutation of *Fbn-2* in mice may cause syndactyly. *See* aneurysm, aortic; arachnodactyly; Marfan syndrome; Marfanoid syndrome; syndactyly. (Chaudhry, S. S., et al. 2001. *Hum. Mol. Genet.* 10:835.)

fibrin Insoluble protein formed during blood clotting from fibrinogen (4q28) by the action of thrombin. *See* fibrinogen; fibrin-stabilizing factor deficiency; thrombin.

fibrinogen *See* fibrin.

fibrinolysin Serine endopeptidase. *See* plasmin.

fibrin-stabilizing factor deficiency Controlled by X-chromosomal-recessive factors causing a deficiency of blood coagulation factor XIII that normally stabilizes clot formation by cross-linking fibrin. The wound heals slowly, the umbilicus bleeds days after birth, bleeding may occur inside the joints

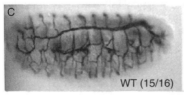
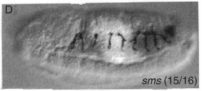

In *Drosophila* the *stumps* gene is required for the FGF-dependent migration of the tracheal and mesodermal migration in the embryo. The tracheal phenotype of the wild type (Left) and the *sms* mutant (Right) is shown at stage 15/16. (Courtesy M. A. Krasnow, see *Genetics* 152:307.)

FGFR Fibroblast growth factor receptor. *See* FGF.

FGR oncogene Located to human chromosome 1p36.2-p36.1; encodes an actin-like and tyrosine kinase sequence homologous to the viral gene *fgr*. Mouse with reduced FGR1c receptor.

(hemarthrosis), in the genitourinary tract (hematuria), and abortion with bleeding and intracranial bleeding (generally lethal) may occur. Apparently, the condition is caused not by the lack of factor XIII but by the formation of defective molecules. *See* afibrinogenemia; antihemophilic factors; defibrinogenemia; hemophilia; von Willebrand disease.

fibroblast Elongated connective tissue cell (gives rise to tendons and other supportive structures) that may produce collagen; it is relatively easy to culture in vitro. *See* collagen.

A fibroblast cell. They may occur in many shapes.

fibroblast growth factor See FGF.

fibrochondrogenesis Rare autosomal lethal recessive malformation of the hip cartilage (rhizomelic chondrodysplasia).

fibrodysplasia ossificans progressiva (FOP) Rare autosomal (4q27-q31)-dominant progressive ectopic (malplaced) bone formation. The basic defect appears to be the inappropriate production of bone morphogenetic protein (BMP-4) by the lymphocytes. Sometimes short fingers, anomalous vertebrae, deafness, baldness, and mental retardation accompany it. *See* bone morphogenetic protein; exostosis; metachondromatosis.

fibroin Silk protein. *See* silk fibroin; silk worm. (Hayashi, C. Y. & Lewis, R. V. 2001. *Bioessays* 23:750.)

fibronectin Dimeric (~220-kDa monomers, 2q24) extracellular matrix glycoprotein involved in development, wound healing, and antitumorigenesis (antimetastasis); suppresses angiogenesis. For the metastasis of melanoma, fibronectin is required, however. Fibronectin binds to the Fc fragment of immunoglobulins, particularly to IgG. It may play a role in autoimmune diseases, pulmonary fibrosis (formation of fibrous tissues), and glomerulonephritis. Fibronectin deficiency is characteristic for type X Ehlers-Danlos syndrome. *See* angiogenesis; autoimmune disease; CAM; Ehlers-Danlos syndrome; glomerulonephritis; immunoglobulins; integrin; L1; melanoma; metastasis; RGD. (Yi, M. & Ruoslahti, E. 2001. *Proc. Natl. Acad. Sci. USA* 98:620.)

fibulin-5 (DANCE) Integrin ligand required for tissue elasticity. *See* integrin. (Nakamura, T., et al. 2002. *Nature* 415:171.)

FICOLL® Synthetic polymer of sucrose. It is used in Denhardt reagent and in gel-loading buffers to increase the density of the sample applied to the wells in agarose gel. *See* Denhardt reagent; gel electrophoresis.

fidelity, gene conversion The converted allele is identical to the converter. *See* gene conversion.

fidelity, replication *See* error in replication.

fidelity, transcription May be mediated by the very low level of $3' \to 5'$ ribonuclease activity of eukaryotic RNA polymerase II. *See* error in transcription; RNA polymerase.

fidelity, translation Errors occur approximately in the 10^{-4} range. *See* ambiguity in translation.

fidelity paradox *E. coli* DNA polymerase II makes 3–5–fold more incorporation errors at the AT-rich sequences than at the GC-rich sequences. The polymerization activity of the same enzyme is 2–6–fold higher at the AT sequences. Normally one would expect higher proofreading by the 3'-exonuclease function in the less stable AT-rich tracts. It seems that the increased polymerization activity prevents proofreading efficiency. *See* proofreading. (Wang, Z., et al. 2002. *J. Biol. Chem.* 277:4446.)

fiducial limit Synonymous with confidence limits, confidence intervals. *See* confidence intervals.

field-flow fractionation (FFF) Size-based fractionation of large molecules through thin channels; from the retention time, hydrodynamic properties are estimated.

field-inversion gel electrophoresis *See* FIGE.

fig (*Ficus carica*) The genus includes about 2,000 species with $x = 13$, and the majority of them are diploid ($2n = 26$), although triploid and tetraploid forms are also known. It is a freeze-sensitive fruit tree with about three times higher Ca^{2+} in the fruits than in other plant species.

FIGE Field-inversion (10–0.02 Hertz) gel electrophoresis; it has been used to separate DNAs in the range of 15–700 kb, similar in size to the genetic material in the chromosomes of lower eukaryotes. Its advantage over pulsed-field gels is that the runs are straight. *See* Hertz; pulsed-field electrophoresis.

figure 8 Configuration of cointegration of two-ring DNAs before completion of the process. ⬤⬤ *See* cointegration.

filaments Myosin and actin fibers in animal tissue. The microfilaments (6 nanometers in diameter), the intermediate filaments (10 nm), and the microtubules (23 nm) form the structural mesh of the cytoplasm. The intermediate filaments (IF) carry out a variety of different functions depending on the ca. 50 genes that encode them in the human cells. Defects or deficiency of IF may lead to genetic lesions of the epidermal keratin and thus blistering. Defects in the neurofilaments may account for some cases of amyotrophic lateral sclerosis, muscular atrophy, and sensory neuropathy. *See* amyotrophic lateral sclerosis; corneal dystrophy; cytoskeleton; desmosome; epidermolysis; intermediate filaments; keratosis; monilethrix; neuropathy; pachyonychia; plakin; stamen.

filamentation Bacteria may grow as long filaments during SOS repair. *See* SOS repair.

filamentous growth *See* pseudohypha.

filamentous phages Frequently used as a cloning vector in genetic engineering: M13 (6408 b), f1 and fd (6408 b) single-stranded DNA phages. They have many advantages: They are easy to purify, have little constraint on packaging, and accept relatively large inserts (upto 5 times their original DNA). M13 and its derivatives are generally used in DNA sequencing by the Sanger method. They contain multiple cloning sites within a truncated *E. coli lacZ* gene. The vectors containing a successful insertion produce white plaques, whereas the plaques of the phage without the insertion are blue on X-gal medium because *LacZ* is not interrupted and thus expressed. The f1 phage—unlike the lytic ones—does not destroy the host bacterium in order to exit. The phage encodes and excretes protein pIV that is used to create a tightly gated channel in the bacterial cell membrane. Through this gate the single-stranded phage genome emerges dressed in the coat protein and fully formed without disruption of the bacterial host. *See* bacteriophages; cloning site; cloning vector; DNA sequencing; lysis; phage display; phagemids; pUC vectors; Sanger; Xgal. (Horiuchi, K. 1997. *Genes Cells* 2[7]425; Marvin, D. A. 1998. *Curr. Opin. Struct. Biol.* 8[2]:150; Cabilly, S. 1998. *Methods Mol. Biol.* 87:129; Larocca, D., et al. 1999. *FASEB J.* 13[6]:727.)

filamins Diverse, extended dimeric proteins that cross-link actins involved in membrane receptors and signaling molecules. Filamin-A is a 280 kDa protein encoded at Xq28 and its defect is responsible for heterotopia. Filamin-B (encoded at 3p14.3) is similar in size to filamin-A and it is bound to it. Filamin-C is encoded at 7q32-q35. Filamin defects may be the cause of several human maladies such as Alzheimer disease, Graves' disease, platelet abnormalities, von Willebrand disease, epilepsy, etc. (Stossel, T. P., et al. 2001. *Nature Rev. Mol. Cell Biol.* 2:138.)

file Unit of related records in the computer.

file server Permits additional users to share files and application programs on a computer.

filial Offspring (generation). *See* F₁.

filler DNA Nucleotides inserted at the junctions and excision sites of transposable elements and genetic vector-carried sequences or at sites of nonhomologous end joining. *See* insertion elements; NHEJ; transposable elements.

filopodium Slender amoeba-like cell involved in movement and/or growth; the spikes on the growth cones. *See* growth cone. (Wood, W. & Martin, P. 2002. *Int. J. Biochem. & Cell Biol.* 34:726.)

filterable agent Former name of tobacco mosaic virus and other viruses before their nature was revealed. The name comes from the fact that they passed through bacterial filters and retained their biological activity. *See* TMV.

filter enrichment *See* filtration enrichment.

filter hybridization One of the nucleic acid components is immobilized on a (nitrocellulose) filter and the other labeled liquid component is allowed to anneal with it. *See* DNA hybridization.

filtering nucleotide and protein sequences Removal of low-complexity (repetitive) sequences that might inflate the homology between/among macromolecules and introduce errors in conclusions concerning common ancestry. Some of the BLAST search programs automatically filter the data. *See* BLAST.

filter mating The bacterial cells at high density conjugate on the surface of a filter.

filter sterilization Removal of microbes from liquids without heating by the use of 0.45 or 0.2 μm pore-size membranes. *See* autoclaving; sterilization.

filtration enrichment Method of selective isolation of all types of nutritional mutants used primarily in fungi. Wild-type mycelia will grow on minimal media, but cells that have any type of special nutritional requirement most likely will not. A filter will retain the wild-type mycelia while the nongrowing spores pass through and thus are separated from the bulk. This mutant-enriched filtrate can be analyzed for the specific nutritional requirement. *See* mutant isolation; mutation detection. (Fries, N. 1947. *Nature* 159:199; Woodward, V. W., et al. 1954. *Proc. Natl. Acad. Sci. USA* 55:872.)

fimbriae Flagellum-like appendages in the surface of bacteria help, e.g., *Salmonella* in adhering to and infecting intestinal epithelia. *See E. coli*; pilus. (Shembri, M. A. & Klemm, P. 2001. *EMBO J.* 20:3074.)

fimbrin Actin filament bundling protein encoded by gene *SAC6* in yeast. Its overproduction may be lethal. *See* actin; ankyrin; spectrin.

fi plasmid Inhibits the fertility factor of bacteria. *See* F element; F plasmid.

fine structure mapping Genetic mapping of the position of mutations (alleles) within a gene locus. *See* allele; locus.

fine structure of genes Initially this meant recombinational analysis at high resolution in large populations frequently based on selective identification of the recombinants that permitted mapping of the sites of different alleles within the locus. The ultimate fine-structure analysis uses DNA (RNA) sequencing of cloned genes. *See* deletion mapping; mapping, genetic; physical mapping.

finger *See* Ring finger; zinc finger.

fingerprint Analyzed by forensic and police investigations to determine personal identity through dactyloscopy, a branch of dermatoglyphics. Ink is applied to the fingertips and impressions are made on paper. There may be a slight difference in the fingerprint patterns of the corresponding left- and right-hand fingers, yet the overall pattern is characteristic

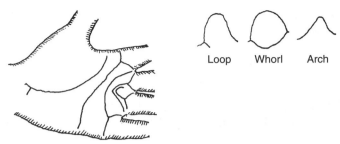

Fingerprints and palm-print patterns. The main types of the fingerprints are "loop", "whorl" and "arch" as shown but the additional dermal ridges within and outside are omitted here. The average number of ridges for women is about 127 and for men about 144. The far-left palmprint illustrates the main angles of the palmar triradial ridges. These are separate from the palmar creases affected in certain hereditary anomalies. See Down syndrome.

for an individual. Fingerprints are determined by a relatively small number of genes, yet each person has a unique fingerprint pattern. There is about 0.43 correlation between paternal and offspring fingerprints. The pattern is determined by the end of the third month of the fetus. The forensic value of fingerprints depends on the quality of the prints collected.

Fingerprints are influenced very little by environmental effects. Even if the epidermis is destroyed, regrowth reestablishes the original pattern. Toe and palm prints may provide additional confirmatory evidence for identity. There are three major types of fingerprints: *loops*, *whorls*, and *arches*; 65, 30, and 5% of the fingerprints respectively fall into these broad categories (see illustrations above). Autosomal-dominant genes either eliminate the dermal ridges or alter their pattern. There are elaborate systems for their classification into up to a million subgroups within which still individual differences are detectable. The uniqueness of fingerprints became known by the second half of the 16th century at about the same time as the same anatomist, Nehemiah Grew, identified sexuality of plants. Francis Galton, the founder of quantitative genetics, used it for forensic purposes in the 19th century. The forensic value of fingerprints may not always be perfect because the analysis may be subject to personal errors. Although lip prints have not been used for the same purpose as fingerprints, they may permit individual identifications. See DNA fingerprinting; dyskeratosis; ectodermal dysplasia; iris pattern. (Alter, M. 1966. *Medicine* 46:35; Williams, G., et al. 2001. *J. Forensic Sci.* 46:1085.)

Whorl-type fingerprint.

fingerprint clone contig   Clones arranged in contigs on the basis of fingerprints generated by restriction enzyme digests. (Barillot, E., et al. 1991. *Proc. Natl. Acad. Sci. USA* 88:3917.)

fingerprinting, of macromolecule   Two-dimensional separation (by chromatography and/or electrophoresis) of digests of proteins or electrophoretic separation pattern of restriction enzyme digests of DNA for the purpose of characterization. See chromatography; DNA fingerprinting; electrophoresis; motifs; RNA fingerprinting. (Ingram, V. M. 1956. *Nature* 178:792; Baglioni, C. 1961. *Biochim. Biophys. Acta* 48:392; Tao, Q., et al. 2001. *Genetics* 158:1711.)

finishing   DNA sequencing step that eliminates most of the errors and gaps in the genomic sequence and leads to a first-draft sequence. See first-draft sequence; gaps. (Deloukas, P., et al. 2001. *Nature* 414:865.)

fir (*Abies* spp.)   Timber trees, $2n = 24$.

fire   Intragenic (negative) regulatory element in the c-fos protooncogene; it must be relieved before Fos can be expressed. See FOS.

first division segregation   See tetrad analysis.

first-draft sequence   Before entirely completing a genome project, a nearly contiguous sequence may still have an error rate of 10^{-3}. See genome projects; high-quality sequence.

FirstEF   Computational program based on discriminant function that detects promoters and first exons with over 80% accuracy. See discriminant function. (Davuluri, R. V., et al. 2001. *Nature Genet.* 29:412.)

first-male sperm precedence   Method of sperm competition in case of multiple mating during a period of receptivity of the same female. In the newt, the process is controlled by the female in contrast to other animals and last-male precedence may occur. See certation. (Jones, A. G., et al. 2002. *Proc. Natl. Acad. Sci. USA* 99:2078.)

first messenger   Peptide hormones. See animal hormones.

first-strand DNA   Immediate product of reverse transcription. See cDNA; reverse transcription.

FIS   Factor for inversion stimulation is a protein involved in the movement of inversion and transposable elements. This element (11.2 kDa) may activate rRNA transcription (7x) after binding upstream promoter sites at -71, -102, and -143. See insertion elements; invertases; promoter; rrn; transposable elements.)

FISH   Fluorescence in situ hybridization detects eukaryotic chromosomal sites by the use of nonradioactive, fluorescence-labeled probes. In human chromosomes it permits the resolution of about 0.05–10 million base pairs as a band. The multiplex FISH technique using several fluorochromes simultaneously permits the identification of quite complex chromosomal rearrangements in amazing colors. Using DNA fibers, cloned probes can be mapped with a resolution down to 1 kb. The effectiveness of the FISH approach has greatly advanced by the use of epifluorescent filters and computer

Estimation of fitness and equilibrium frequencies of the alleles at heterozygote advantage in a hypothetical population

Observed (N = 2000):	$AA = 900$,	$Aa = 1000$,	$aa = 100$
Representation of alleles ($\Sigma = 4000$):	$A: (2 \times 900) + 1000 = 2800$,	$a: (2 \times 100) + 1000 = 1200$	
Allelic frequencies:	$A: 2800/4000 = 0.7 = p$,	$a: 1200/4000 = 0.3 = q$	
Expected no. of genotypes N ($p^2 + 2pq + q^2$)	$AA: 0.49 \times 2000$, $AA: 980$	$Aa: 0.42 \times 2000$, $Aa: 840$	$aa: 0.9 \times 2000$ $aa: 180$
Fitness $\dfrac{\text{observed number of genotypes}}{\text{expected number of genotypes}}$	$AA: \dfrac{900}{980} = 0.92$	$AA: \dfrac{1000}{840} = 1.19$	$aa: \dfrac{1000}{180} = 0.56$
Standardized fitness (rel. to w_2)	$AA: \dfrac{0.92}{1.19} = 0.77$	$Aa: \dfrac{1.19}{1.19} = 1.00$	$aa: \dfrac{0.56}{1.19} = 0.47$
Selection coefficients	$s = 1 - 0.77 = 0.23$		$t = 1 - 0.47 = 0.53$
Allelic frequencies at equilibrium	$A: \dfrac{t}{s+t} = \dfrac{0.53}{0.23 + 0.53}$ 0.7,	$a: \dfrac{s}{s+t} = \dfrac{0.23}{0.23 + 0.53} = 0.3$	

evaluation, permitting the simultaneous identification of 27 different DNA probes. This *multiplex FISH* (M-FISH) has been described by Speicher, M. R., Gwyn, S. & Ward, D. C. 1996. *Nature Genet.* 12:368; Henegariu, O., et al. 1999. *Nature Genet.* 23:263. *See* chromosome painting; combinatorial labeling; DIRVISH; fluorochromes; FRET; GISH; IRS-PCR; in situ hybridization; nonradioactive labels; probe; ratio labeling; spectral karyotyping; stretching chromosomes; telomeric probes; WCPP. (Azofeifa, J., et al. 2000. *Am. J. Hum. Genet.* 66:1684; Lee, C., et al. 2001. *Am. J. Hum. Genet.* 68:1043, see color plate in Color figure section in center of book.)

Fisher-Muller hypothesis   The advantage of sex is that beneficial mutations occurring in two different individuals can be combined in one.

Fisher's exact test   Exact probability test for independence (see in detail Sokal, R. R. & Rohlf, F. J. 1969. *Biometry.* Freeman) that can be used similarly to goodness-of-fit and contingency tests. The probabilities are determined with the aid of hypergeometric distribution tables; if the populations are large, the binomial distribution may satisfactorily approximate the hypergeometric distribution. *See* association test; chi square; hypergeometric distribution.

fish-eye disease   *See* lecithin:cholesterol acyltransferase deficiency.

fish orthologous gene   (<http:www.evolutionsbiologie. uni-konstanz.de/Wanda>.)

fission   Mode of asexual reproduction involving the division of a single cell or organelle by cleaving (generally) equal progeny (daughter) cells or organelles into two. *See* alternation of generations; life cycles.

fission yeast   See *Schizosaccharomyces*

FITC   Fluorochrome, a conjugate of fluorescein with isothiocyanate, used for cellular, chromosomal labeling. *See* chromosome morphology; chromosome painting; FISH; fluorochromes.

Fitch-Margoliash method for TD   Using n number of species, an evolutionary distance is computed by minimization of the branches of the evolutionary tree:

$$S_{\text{Fitch-Margoliash}} = \left[\frac{2\Sigma i < j\{(d_{ij} - e_{ij})/d_{ij}\}^2}{n(n-1)} \right]^{1/2} \times 100,$$

where d_{ij} is the observed distance between species i and j and e_{ij} is the patristic distance. *See* evolutionary distance; evolutionary tree; patristic distance; transformed distance.

fitness   Reproductive value in a population under specified conditions, determined by a genotype. For the method of calculation, see table above. Fitness in natural populations when hybrids have superior fitness can be estimated by the following equation: $w_{F1} = [(1 - p_1)p_3]/[p_1(1 - p_3)]$, where p_1 and p_3 are the frequencies of hybrids at the beginning and at the end of the study (Ebert, D., et al. 2002. *Science* 295:485).

See hybrid vigor; inclusive fitness; mutation rate in fitness; selection coefficient. (Loeske, E. B., et al. 2000. *Proc. Natl. Acad. Sci. USA* 97:698; Wloch, D. M., et al. 2001. *Genetics* 159:441. (Rozen, D. E., et al. 2002. *Current Biol.* 12:1040.)

fitness, standardized   See selection coefficient.

fix   *See* nitrogen fixation.

fixation   The allele becomes homozygous in every individual of the population.

fixation index   Average coefficient of inbreeding in a population. In case of random mating, the probability that an offspring would have exactly the same two ancestral alleles at a locus is $(1/2)N$, where N is the number of diploid individuals in the population. The probability of having two different alleles

Fission

at the same locus is $1 - (1/2)N$. The coefficient of inbreeding of the first generation of this population is also $(1/2)N$ by definition of inbreeding. In each succeeding generation, the noninbred part of the population will have a chance to produce offspring with an allele pair identical by descent. Therefore, the coefficient of inbreeding in the next generations will be $(1/2)N + [(1 - (1/2)N] \times F$, where F is the inbreeding coefficient of the preceding generation. After the gth generation, the coefficient of inbreeding of this population will be

$$F_g = (1/2)N + [1 - (1/2)N]F_{g-1}$$

and this is called the *index of fixation*. Its complement is the *panmictic index* (P_g), which represents the average noninbred fraction of the population:

$$P_g = 1 - F_g$$

The probability for the offspring to have two identical A or a alleles is Fp_{AA} and Fq_{aa}, respectively. Also, the probability of two alleles of a locus being nonidentical by descent is $1 - F$, and the proportions of AA, Aa, and aa are p^2, $2pq$, and q^2 (according to the Hardy-Weinberg theorem). Because the population will have both inbred and non-inbred components, its genetic structure will be

$$F(p_{AA} + q_{aa})$$

and

$$(1 - F)(p^2_{AA} + 2pq_{Aa} + q^2_{aa})$$

When a population is completely inbred, only homozygotes are found, which is a change in genotypes but may not be a change in allelic frequencies if both alleles have equal fitness. The change may actually be from $AA + 2Aa + aa \rightarrow AA + AA + aa + aa$, which is mathematically the same. The ultimate probability of fixation

$$P_f = \frac{1 - e^{-N_e sp}}{1 - e^{-N_e s}}$$

may also be estimated on the basis of the *initial frequency of the gene* $(= p)$, the *selection advantage* $(= s)$, and the *effective population size* (N_e). (The base of the natural logarithm $= e \approx 2.71828$). *See* Hardy-Weinberg theorem; hybrid vigor; inbreeding; inbreeding rate; mutation, beneficial; mutation, neutral; panmixis.

fixative Agent(s) required to treat biological materials before stain is applied for microscopic examination. The fixative rapidly kills the cells, immobilizes the structures, and assures better staining. Many different types of fixatives have been developed since the introduction of microscopic techniques. Some of the most widely used (especially for cytology) with their composition in parts (volume) are as follows:

Fixatives

Designation	Ethanol	Propionic Acid	Acetic Acid	Chloroform	
Farmer's	3	0	1	0	
Farmer's modif.	3	1	0	0	
Carnoy A*	6	0	1	3	*fresh
Carnoy B*	6	0	3	1	*fresh

Newer fixatives may contain 5% formalin, 5% glacial acetic acid, 90% or 70% ethanol, glutaraldehyde (4% in 0.025 M phosphate buffer pH 6.8), etc., and detergents (Tween 20) may be added for facilitating penetration, or aspiration is applied for a few minutes. Fixation time is a day or two. The fixed material can be stored for months in 70% ethanol in the refrigerator. Different fixatives are required for electron microscopy. The fixatives used in photographic processing are completely different. *See* electron microscopy; light microscopy; stains.

FixL *Rhizobium* kinase regulating N_2 fixation by FixJ. *See* nitrogen fixation.

FK506 (Tacrolimus) An immunosuppressive protein in combination with rapamycin binds to cellular proteins FKB12/FKBP13. FK506 intercepts the signal of the T lymphocyte receptor while rapamycin interferes with the signal of cytokines and growth factors. The FKBP12-FK506 complex inhibits the serine-threonine phosphatase, calcineurin. The rapamycin-FKBP12 complex binds to FRAP and regulates p70 ribosomal protein S6 kinase required for the progression from the G_1 phase of the cell cycle. Also called TOR, RAFT, FRAP. Members of this protein family are common in animals (FKBP25, FKBP12, FKBP51, etc.) and occur in fungi (Npi46) and plants (FKBP73). *See* ataxia telangiectasia; calcineurin; cardiomyopathy, hypertrophic familial; cell cycle; check point; cyclosporine; FkpA; immunophilins; immunosuppression; NF-AT; peptidyl-prolyl isomerases; PPI; signaling to translation. (Glynne, R., et al. 2000. *Nature* 403:672; Aghdasi, B., et al. 2001. *Proc. Natl. Acad. Sci. USA* 98:2425.)

FK1012 Lipid-soluble ligand, a dimeric form of FK506, and it directs the interaction between proteins linked to the FKBP12 receptor. *See* FK506.

FKB12/FKBP13 Enzyme complex involved in the catalysis of peptidyl-prolyl cis-trans isomerization of proteins. The FKBP12 molecules may be complexed with hormones or other proteins and facilitate the export of molecules from the endoplasmic reticulum. *See* COP; FK506; Ire. (Marx, S. O., et al. 2000. *Cell* 101:365; Gaburjakova, M., et al. 2001. *J. Biol. Chem.* 276:16931.)

FKH (forkhead) Embryonic lethal *Drosophila* gene at 3–95. A human homologue (13q14) is a DNA-binding protein; the human gene (FKHL7) at 6p25 encodes a transcription factor responsible for a dominant glaucoma. Translocations to 2q35 (PAX) may result in rhabdosarcoma (epithelial tumor). The forkhead homologue of *Caenorhabditis* (DAF-16) controls longevity of the worms and in humans it indirectly mediates apoptosis and the cell cycle. The protein Chfr (checkpoint with FHA and ring finger) seems to control the entry into metaphase in the cell cycle. The human homologue, FOXC1 at 6p25, is a glaucoma gene. *See* Akt; Alvarez; checkpoint; FHA; glaucoma; PAX. (Dyspraxia, B., et al. 2001. *Nature* 413:744; Tran, H., et al. 2002. *Science* 296:530.)

FkpA Periplasmic FK506-binding protein. *See* FK506; periplasma.

FLAG Eukaryotic translation initiation factor (eIF-4G2, encoded at 11p15); provides support for the capping protein eIF-4F by binding to eIF-4A. FLAGGED proteins are usually repressed. *See* cap; capping enzymes; eIF-4A; eIF-4F; eIF-4G.

flagellar antigen Of *Salmonella*, encoded by the *H1* and *H2* genes. At a particular time, either one or the other flagellin is expressed. The expression is controlled by phase variation, i.e., a transposition (inversion) of the 970-bp-long DNA segment. In one of the positions the *rh1* regulatory element represses *H*1; after the inversion, *H*2 is expressed. *See* flagellin; mating type determination in yeast; phase variation; *Trypanosoma*. (Masten, B. J. & Joys, T. M. 1993. *J. Bacteriol.* 175:5359.)

flagellin Protein material of the (bacterial) flagellum. *See* flagellum; phase variation. (Samatey, F. A., et al. 2001. *Nature* 410:331.)

flagellum (plural flagella) Cell appendage used for back-and-forth movement of microbial cells; in bacteria it is controlled by about 50 genes. *See* flagellin.

Flagellum.

FLAME *See* apoptosis; FLICE.

flanking DNA (flanking gene) Nucleotide sequences adjacent to a gene or adjacent genes.

flap nuclease Has both endo- and exonuclease activity. *See* RAD27/FEN1. (Debrauwere, H., et al. 2001. *Proc. Natl. Acad. Sci. USA* 98:8263; Dervan, J. J., et al. 2002. *ibid.* 99:8542.)

FLASH (FLICE-associated huge protein) Caspase regulatory protein in apoptosis. It interacts with FAS/CD95. *See* apoptosis; caspase; DISC; Fas; FLICE. (Choi, Y.-H., et al. 2001. *J. Biol. Chem.* 276:25073.)

flatworm (*Planaria torva*) $2n = 16$.

flavin nucleotide Riboflavin-containing coenzyme. *See* FAD; FMN.

flavone Along with, flavonones, flavonols are plant pigments derived from chalcones. *See* anthocyanidins; chalcone.

flavonoid Pigment with trimeric heterocyclic nucleus, and frequently glycosylated. The synthetic enzymes may assemble as macromolecular complexes. *See* flavone.

flavoprotein Tightly bound with a flavin nucleotide prosthetic group. *See* flavin nucleotides.

flax (*Linum usitatissimum* ($2n = 30, 32$) Includes the fiber crop and the seed crop, linseed, altogether with about 200 species. In the highly variable *Linum* genus, the basic chromosome numbers vary: $x = 8, 9, 12, 14, 15, 16$, but the most common $2n$ numbers are 18 and 30. Some strains show nonnuclear inheritance.

flea *See copia.*

Fletcher factor Causing a usually asymptomatic hereditary anomaly involved in the early regulation of the intrinsic blood clotting pathway. *See* antihemophilic factors; blood clotting pathways; hemostasis; kallikrein. (Weiss, A. S., et al. 1974. *J. Clin. Invest.* 53:622.)

flexer Protein that determines DNA folding and bending similarly to the chaperones of proteins. *See* chaperone; DNA bending; folding. (Lavoie, B. D., et al. 1996. *Cell* 85:761.)

FLI One of the consistent translocation breakpoints (22q12) in Ewing sarcoma.

FLICE (I-FLICE/caspase 10, human chromosome 2q33) Inhibitory participant in apoptosis; it is a member of the ICE family proteases, apparently the same as MACH (mort-associated Ced-3 homologue). Other apparent synonyms are FLIP, Casper, FLAME, and CASH. *See* apoptosis; caspases; FLASH; ICE. (Poulaki, V., et al. 2001. *Cancer Res.* 61:4864.)

FliG, FliM, FliN Bacterial switch proteins controlling flagella rotation. *See* phase variation. (Lux, R., et al. 2000. *J. Mol. Biol.* 298:577.)

F₂ linkage estimation In some species where controlled mating or crossing is difficult, linkage intensities (recombination frequencies) can be estimated in the selfed progeny of F₁ or in the offspring of brother-sister matings. In test crosses the recombination frequencies can be determined in a straightforward manner by dividing the number of recombinants by the sums of the parental plus recombinant individuals in the progeny. In F₂, the computations rely on indirect means using statistical tables (see table on next page). The product ratio tables were constructed on the basis of the formula

$$\frac{ad}{bc} = \frac{T(2+T)}{(1-T)^2}$$

where the meaning of a, b, c, d is given below, $T = (1-p)^2$ for coupling, and $T = (p)^2$ for repulsion data; p is the recombination frequency. The procedure can be best described by an example. Let us consider a hypothetical F₂ population with four phenotypic classes (a, b, c, d):

AB (**a**)	Ab (**b**)	aB (**c**)	ab (**d**)	Total
660	38	40	183	921

Linkage is obvious because the frequency of the double homozygous recessive class (*ab* or briefly **d**) exceeds that of the single heterozygotes (*Ab* and *aB*, classes **b** and **c**); the linkage phase is *coupling* in the two parents (*AB* and *ab*).

We can use the product ratio method (formula R for repulsion) and formula C in coupling:

Recombination fractions determined by the product ratio method using F₂ data. Modified after F. R. Immer, *Genetics* 15:81. Recombination fraction = y, product ratio in repulsion: *R*, product ratio coupling: *C* factor to be divided by √N, to obtain standard error of y, for repulsion *SeR* and for coupling *SeC*. N = total number of individuals in F₂. When the value of the product ratio does not correspond to a close-enough value of y, interpolation is required. More precise estimates can be obtained by the table of W. L. Stevens (*J. Genetics* 39:171), using 5 decimals, rather than 3, as in this table. since recombination fractions are generally variable, for most purposes this accuracy may be entirely satisfactory. For the method of use of this table, see text on preceding page and next page.

y	R	C	SeR	SeC	y	R	C	SeR	SeC
0.005	0.00005000	0.00003361	1.0000	0.0707	0.255	0.1536	0.1396	0.9243	0.5191
0.010	0.00020005	0.0001356	0.9999	0.1001	0.260	0.1608	0.1467	0.9214	0.5244
0.015	0.0004503	0.0003076	0.9997	0.1226	0.265	0.1682	0.1540	0.9186	0.5297
0.020	0.0008008	0.000516	0.9996	0.1417	0.270	0.1758	0.1616	0.9158	0.5351
0.025	0.001252	0.0008692	0.9993	0.1585	0.275	0.1837	0.1695	0.9128	0.5404
0.030	0.001804	0.001262	0.9988	0.1736	0.280	0.1919	0.1717	0.9099	0.5456
0.035	0.002458	0.002733	0.9985	0.1877	0.285	0.2003	0.1861	0.9069	0.5509
0.040	0.003213	0.002283	0.9979	0.2007	0.290	0.2089	0.1948	0.9039	0.5560
0.045	0.004070	0.002914	0.9975	0.2129	0.295	0.2179	0.2038	0.9008	0.5612
0.050	0.005031	0.003629	0.9969	0.2246	0.300	0.2271	0.2132	0.8977	0.5663
0.055	0.006096	0.04429	0.9962	0.2357	0.305	0.2367	0.2228	0.8946	0.5714
0.060	0.007265	0.005318	0.9956	0.2463	0.310	0.2465	0.2328	0.8913	0.5764
0.065	0.008540	0.006296	0.9947	0.2565	0.315	0.2567	0.2432	0.8882	0.5815
0.070	0.009921	0.007366	0.9939	0.2663	0.320	0.2672	0.2538	0.8850	0.5864
0.075	0.01141	0.008531	0.9930	0.2758	0.325	0.2780	0.2649	0.8817	0.5914
0.080	0.01301	0.009793	0.9920	0.2850	0.330	0.2892	0.2763	0.8784	0.5963
0.085	0.01471	0.01116	0.9910	0.2339	0.335	0.3008	0.2881	0.8750	0.6012
0.090	0.01653	0.01262	0.9899	0.3025	0.340	0.3127	0.3003	0.8716	0.6061
0.095	0.01846	0.01419	0.9889	0.3109	0.345	0.3250	0.3128	0.8684	0.6110
0.100	0.02051	0.01586	0.9977	0.1392	0.350	0.3377	0.3259	0.8648	0.6157
0.105	0.02267	0.01765	0.9864	0.3272	0.355	0.3508	0.3393	0.8614	0.6205
0.110	0.02495	0.01954	0.9850	0.3351	0.360	0.3643	0.3532	0.8580	0.6254
0.115	0.02734	0.02156	0.9837	0.3428	0.365	0.3783	0.3675	0.8544	0.6301
0.120	0.02986	0.02369	0.9822	0.3503	0.370	0.3927	0.3823	0.8509	0.6347
0.125	0.03250	0.02594	0.9809	0.3578	0.375	0.4076	0.3977	0.8473	0.6395
0.130	0.03527	0.02832	0.9793	0.3652	0.380	0.4230	0.4135	0.8437	0.6442
0.135	0.03816	0.03083	0.9776	0.3723	0.385	0.4389	0.4298	0.8400	0.6488
0.140	0.04118	0.03347	0.9760	0.3792	0.390	0.4553	0.4467	0.8363	0.6534
0.145	0.04434	0.03624	0.9744	0.3862	0.395	0.4723	0.4641	0.8328	0.6580
0.150	0.04763	0.03915	0.9726	0.3930	0.400	0.4898	0.4821	0.8291	0.6626
0.155	0.05105	0.04220	0.9708	0.3999	0.405	0.5079	0.5007	0.8252	0.6672
0.160	0.05462	0.04540	0.9689	0.4064	0.410	0.5266	0.5199	0.8215	0.6718
0.165	0.05832	0.04875	0.9670	0.4129	0.415	0.5460	0.5398	0.8178	0.6762
0.170	0.06218	0.05225	0.9650	0.4194	0.420	0.5660	0.5603	0.8139	0.6808
0.175	0.06618	0.05591	0.9629	0.4258	0.425	0.5867	0.5815	0.8101	0.6853
0.180	0.07043	0.05973	0.9610	0.4320	0.430	0.6081	0.6034	0.8062	0.6897
0.185	0.07464	0.06371	0.9588	0.4383	0.435	0.6302	0.6260	0.8024	0.6941
0.190	0.07911	0.06787	0.9567	0.4445	0.440	0.6531	0.6494	0.7985	0.6986
0.195	0.08374	0.07220	0.9545	0.4506	0.445	0.6768	0.6735	0.7945	0.7029
0.200	0.08854	0.07671	0.9521	0.4565	0.450	0.7013	0.6985	0.7907	0.7073
0.205	0.09351	0.08140	0.9499	0.4624	0.455	0.7766	0.7243	0.7867	0.7116
0.210	0.09865	0.08628	0.9475	0.4683	0.460	0.7529	0.7510	0.7827	0.7159
0.215	0.1040	0.09136	0.9452	0.4741	0.465	0.7801	0.7786	0.7787	0.7204
0.220	0.1095	0.09663	0.9426	0.4799	0.470	0.8082	0.8071	0.7747	0.7247
0.225	0.1152	0.1021	0.9401	0.4857	0.475	0.8374	0.8366	0.7705	0.7288
0.230	0.1211	0.1078	0.9376	0.4913	0.480	0.8676	0.8671	0.7665	0.7331
0.235	0.1272	0.1137	0.9351	0.4970	0.485	0.8990	0.8986	0.7623	0.7374
0.240	0.2334	0.1198	0.9324	0.5026	0.490	0.9314	0.9313	0.7583	0.7416
0.245	0.1400	0.1262	0.9297	0.5081	0.495	0.9651	0.9651	0.7542	0.7459
0.250	0.1467	0.1328	0.9271	0.5136	0.500	1.0000	1.0000	0.7500	0.7500

$$\frac{a \times d}{b \times c} = R$$

and

$$\frac{b \times c}{a \times d} = C$$

$(b \times c)/(a \times d) = (38 \times 40)/(660 \times 183) \cong 0.012584865 = C$. In the table on the next page, this exact fraction is not found; therefore, interpolation is necessary to find the right value. Interpolation is expected to provide a more accurate estimate than values in the product ratio table where the recombination fraction is shown in increments of 0.005. Corresponding to $y = 0.085$ and 0.090 in the C column, the table shows 0.01116 and 0.01262. The procedure of **linear interpolation** is illustrated below.

	1	2	3
C values	0.01116	0.012584865	0.01262
y values	0.085	?	0.090

Calculation:
$$\frac{C2 - C1}{C3 - C1} = \frac{0.012584865 - 0.01116}{0.01262 - 0.01116}$$
$$= \frac{0.001424865}{0.00146} = 0.975934931 = \alpha$$

and $1 - \alpha = 0.024065068$; **y2** (the recombination fraction sought) $= (y1)(1 - \alpha) + (y3)(\alpha) = 0.085(0.024065068) + 0.090(0.975934931) = 0.087834143$ or $\approx$**0.08783**.

The interpolated SeC value is obtained:
$0.2339(0.024065068) + 0.3025(0.975934931) = 0.300849136$.

The standard error of the y2 is calculated by

$$\frac{0.300849136}{\sqrt{921}} = 0.009913316.$$

Thus, the interpolated recombination fraction, (y2) = **0.08783 ± 0.00991**.

The efficiency of the product ratio method in the F_2 coupling phase equals that of the test crosses when the recombination frequency is low (below 8%), but approaching independent segregation, the efficiency is decreasing, and maximally in F_2, populations 2.5 times larger may be required to obtain equally dependable results. In repulsion, especially at close linkage (below 8% recombination), the F_2 and product ratio procedures are very inefficient. The product ratio method is very useful when both recessive alleles reduce gametic or zygotic viability but not when only one of the alleles causes differential mortality. Generally, the product ratio method is equal in efficiency to the maximum likelihood procedure. The dependability of the results is determined primarily by the data collected, not by the statistical procedures. *See* coupling; mapping, genetic; mapping functions; repulsion, maximum likelihood method applied to recombination frequencies; recombination. (Stevens, W. L. 1939. *J. Genet.* 39:171.)

F_3 linkage estimation

May be used in both repulsion and with coupling phases. In repulsion, the information is not very trustworthy unless the recombination frequency is about 0.1 or less. Recombination can be estimated by combining F_2 and F_3 data as follows:

$$x = \frac{a/a\ B/b}{n} = \frac{\text{Number of segregating lines}}{\text{Number of } a/a \text{ lines}}$$

then

$$p \text{ (recombination fraction)} = \frac{x}{2 - x}$$

and the standard error of p,

$$s_p = \frac{2\sqrt{x[1 - x]}}{[2 - x]^2 \sqrt{n}}$$

FLIP (1) Fluorescence loss in photobleaching. Occurs in living cells repeatedly spot-bleached by a laser beam. The loss of fluorescence indicates the dissociation of, e.g., green fluorescent protein from a particular compartment and reveals the protein traffic within the cell nucleus. *See* aequorin. (2) FLICE inhibitory protein, Casper *See* FLICE; FLP/FRT. (Pkada, Y., et al. 2001. *Cytokine* 15[2]:66.)

flip-flop recombination Occurs between two inverted repeats (in organelle genomes) and produces equal mixtures of two isomeric forms. *See* chloroplast genetics; mtDNA. (Hudspeth, M. E., et al. 1983. *Proc. Natl. Acad. Sci. USA* 80:142.)

flip-flop transcription A single locus-control region may move back and forth by looping in between members of a large complex gene cluster and thus permitting the transcription of different members of the complex. *See* gene cluster; locus control region. (Kano, M., et al. 1998. *Biochem. Biophys. Res. Commun.* 248:806.)

Flk-1 Receptor tyrosine kinase with supposed role in endothelial and B-lymphocyte differentiation, angiogenesis, and formation of solid tumors. Its ligand is VEGF (vascular endothelial growth Factor). *See* FLT; Flt-1; KIT; Tie-1; Tie-2; vascular endothelial growth factor. (Matsumoto, K., et al. 2001. *Science* 294:559.)

flood factor Blood protein resembling blood Factor VII, it shortens the slightly long prothrombin action time in asymptomatic individuals. *See* antihemophilic factors; blood clotting pathways; hemostasis.

floor plate Organizer center in the ventral midline of the neural tube directing the development of the central nervous system of vertebrates. *See* neural tube; organizer; roof plate.

floppy disk *See* disk.

flora Description of higher plant communities growing in a particular area or the vegetation present in an area.

floral dip Simple method of transformation of *Arabidopsis* plants by agrobacterial vectors. The inflorescence is dipped into a bacterial suspension containing 5 g sucrose and 500 µL Silwet L-77 surfactant. No in vitro tissue culture is required. *See* transformation, genetic/plants. (Clough, S. J. & Bent, A. F. 1998. *Plant J.* 16:735.)

floral evocation Process of commitment to flower differentiation of plants. *See* commitment; determination. (Nelson, D. C., et al. 2000. *Cell* 101:331.)

floral induction   Internal and external factors bringing about floral evocation. Induction of flowering is a complex process and may require low-temperature (vernalization), long- or short-day photoperiodic regimes, respectively. Recessive mutations of single genes of *Arabidopsis* at several unlinked loci may dramatically affect the flowering response. *See* flower differentiation. (Suárez-Lopez, P., et al. 2001. *Nature* 410:1116; Simpson, G. G. & Dean, C. 2002. *Science* 296:285; Mouradov, A., et al. 2002. *Plant Cell* 14:S111.)

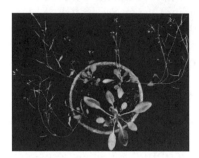

floret   Individual small flower being part of an inflorescence.

Flor's model   Developed by the plant pathologist, H. H. Flor, in the 1950s, claiming that for each virulence gene of the pathogen a gene exists in the host and thus corresponding gene pairs (GCP) exist. These can be expressed in four different categories. Actually, the models were further elaborated and only pathogenicity and host reaction alleles were recognized. The terms *virulence* and *resistance* were deemed unnecessary. The expression of the disease symptoms (aegricorpus), depending also on environmental factors, thus required an interorganismal genetic system. This rather complicated, interacting system was considered a great idea by many plant pathologists to explain the genetic control of host-pathogen relationships. Many geneticists dismissed it as commonplace. They argued that all functions in all organisms are genetically controlled and the interaction of gene products determines the phenotypes within organisms and between organisms, and disease is no exception. Thus, the model — they argued — in the absence of biochemical or molecular facts is not helpful. Since several plant genes have been cloned, there will be new opportunities to study host-pathogen relationships in physicochemical terms. The resistance of maize against *Cochliobolus carbonum* (*Helminthosporium*) seems to support the latter argument inasmuch as a plant enzyme degrades the fungal toxin. The discoveries that some resistance genes encode elicitors of the signal transduction pathways in one way or another still do not invalidate the simplest interpretation. Recent experiments using the yeast two-hybrid system with the *Lac* reporter indicate that the *AvrPto* bacterial virulence gene interacts with the tomato resistance gene *Pto*. This reaction was evident only in this genic combination. Analysis of the Pto protein also revealed that a 95-amino-acid stretch (129 to 224) of the protein was alone responsible for the specific interaction between the pathogen and the host. *See* aegricorpus; host-pathogen relations; immune response; quadratic check; two-hybrid system. (Flor, H. H. 1971. *Annu. Rev. Phytopath.* 2:131; Ellis, J., et al. 2000. *Curr. Opin. Plant Biol.* 3:278; Bergelson, J., et al. 2001. *Science* 292:2281.)

flotillin   Protein of the caveolae. *See* caveolae. (Garin, J., et al. 2001. *J. Cell Biol.* 152:165.)

floury (*fl-2*)   Gene of maize improves the nutritional value of the kernels by reduction of the contents of prolamine and zein, resulting in an increase in lysine, tryptophan, and methionine. *See* kwashiorkor; opaque.

flowchart   Graphical display of procedures of analyses for the solution of a particular problem or subset of that problem or an experimental protocol.

flow cytogenetics (flow karyotyping)   Analysis of chromosomes by flow cytometry. (Davies, D. C., et al. 2000. *Flow Cytometry*. Oxford Univ. Press, Oxford, UK.)

flow cytometry   In the flow cytometer, suspended particles (cells, chromosomes, etc.) are stained with fluorochromes and the dyes are excited by a laser beam. The particles are then sorted with the aid of a computer according to their special properties. The procedure directly estimates cell (rather than population) characteristics such as nucleic acid content, enzyme activity, membrane potential, calcium flux, cell surface receptors, cytoplasmic constituents, etc. It requires sophisticated instrumentation. *See* bivariate flow cytometry. (Ormerod, M. G., ed. 2000. *Flow Cytometry*. Oxford Univ. Press, Oxford, UK; Gygi, M. P., et al. 2002. *Nucleic Acids Res.* 30:2790.)

flower differentiation   Under very strict genetic control. Numerous genes have been identified which change the basic pattern of the morphogenesis. An idealized wild type dicot flower is shown at left, and the most common three types of homeotic conversions are indicated on the diagram. The homeotic transformations may not be complete and thus e.g., carpelloid sepals, stamenoid petals or petaloid stamina, etc. were identified. Usually, an entire whorl of the flowers ([1] sepals, [2] petals, [3] stamina, [4] and carpels) are affected. The conversions generally involve adjacent whorls and simultaneously more then two whorls may be affected. On the basis of mutational evidence it has been suggested that the four whorls belong to A (1 + 2), B (2 + 3) and C (3 + 4) identity groups in the floral meristem.

The actual phenotype of the mutants is determined, which identity groups are affected by the mutation(s), e.g., a type II change is the result of interaction between domains A (*Apetala 1* [*AP1*] and 2 [*AP2*] genes) and B (*Apetala 3* [*AP3*] and *Pistillata* [*PI*] genes). Mutations in the C and B domains result in the *agamous* (*AG*) in *Arabidopsis* and in either the *plena*, *pleniforma*, or *petaloida* alterations in *Anthirrinum*. AP1, AP3, PI, and AG proteins all contain a MADS box, but AP2 represents another DNA-binding protein. The plant MADS box proteins contain another conserved element: the K box, which may form amphipathic α-helices. AG also binds to another consensus ([CC(A/T)$_6$GG), the CArG box. These proteins interact with each other and may form AP1/AP1, AP3/PI, and AG/AG dimers, including the truncated AG/PI heterodimer. The association of these proteins is mediated to a large degree by the so-called L (linker) region of these proteins,

Genetic determination of flower organs (shown at A). Homeotic transformation affects neighboring whorls (B). The Meyerowitz-Coen-Yanofsky model of flower organ identity (C). This is also called the abc model represented by the bold letters in the three boxes (at top) and including the symbols of the genes (in italics) responsible for the control of differentiation of these organs. box **A** alone specifies sepals, **C** alone controls carpels, and **A** and **B** are responsible for petals, whereas **B** and **C** for stamens. **A** counteracts the activity of **C** and **C** is antagonistic to **A**. If **C** is not active, an indeterminate number of floral whorls appear. The lower two boxes illustrate the floral development in double mutants when both **B** and **C** genes mutate (b^-c^-) or when mutation occurs at the *Sepallata* (formerly denoted as *agamous-like AGL*) sites. When all three *(sep1, sep2, sep3)* are mutant, only **A** activity is displayed, i.e., in all three whorls sepals appear.

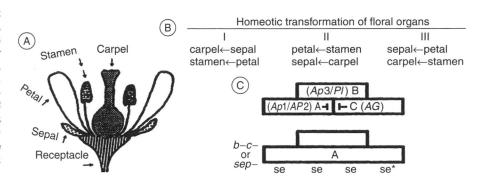

which involves amino acids 31–35 situated between the MADS box and the K (keratin homology) box:

N-terminus < MADS DOMAIN > < L REGION >

< K BOX > C terminus

The MADS box appears to represent a large family of genes with critical functions in plant development.

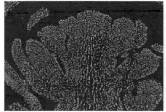 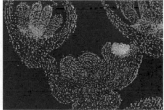

Apical meristem of *Arabidopsis*. At left, the *lfy (leafy)* mutant; at right, the wild type transgenic for the GUS reporter under the control of the *AG* regulatory sequences. *LEAFY* encodes a transcription factor, which determines whether the apical meristem will produce flowers instead of leaves or shoots. The original picture was taken by dark-field microscopy in color. The arrow points to the expression of the transgene in orange color that cannot be represented faithfully and in full beauty in black and white. See color plate in Color figure section in center of book (Courtesy of Dr. Detlef Weigel.)

Flower differentiation seems to be controlled at the level of transcription, and the specific RNA transcripts appear in regions of the flower primordia, which are affected by the specific genes. These genes are studied under the control of different promoters introduced by transformation. Studies indicate that at least some of the genes are also regulated posttranscriptionally.

Cell fate signaling in the apex of *Arabidopsis* is mediated by a receptor encoded by the *CLAVATA1* transmembrane protein kinase gene and it is activated by the ligand encoded by *CLAVATA3* regulating the balance between growth and differentiation. The *WUSCHEL* gene interferes with the activity of *CLAVATA* and maintains the undifferentiated status of the meristem. In Arabidopsis, the *LEAFY* and

SPLAYED genes appear to be activators of the floral homeotic genes such as *APETALA1* and *AGAMOUS*. The quartet theory of floral differentiation has been proposed by Honma, T. & Goto, K. 2001. (*Nature* 409:525). According to this model, four MADS box proteins in four different combinations primarily determine the identity of floral parts in *Arabidopsis*. Petals are under the control of genes *AP1, AP3, PI,* and *SEP* for carpels (*AG, SEP*)$_2$; for stamens, *PI, SEP, AP3, AG;* and for sepals, *AP1, AP1,* and two as-yet unidentified proteins. Normally *Arabidopsis* develops flowers following the development of whorl of leaves in the vegetative rosette. The *EMBRYONIC FLOWER* (*EMF*) eliminates the rosette stage and promotes the development of a flower at the apical meristem of each shoot. *See* homeotic genes; MADS box; morphogenesis; peloric; petals. (Ma, H. 1998. *Trends Genet.* 14:26; Fletcher, J. C., et al. 1999. *Science* 283:1911; Busch, M. A., et al. 1999. *Science* 285:585; Ma, H. & dePamphilis, C. 2000. *Cell* 101:5; Pelaz, S., et al. 2000. *Nature* 405:200; Lohmann, J. U., et al. 2001. *Cell* 105:793; Aubert, D., et al. 2001. *Plant Cell* 13:1865; Wagner, D. & Meyerowitz, E. M. 2002. *Curr. Biol.* 12:85; Lohmann, J. U. & Weigel, D. 2002. *Developmental Cell* 2:135.)

flower evocation Determination of the flowering process. *See* floral induction.

flower pigment *See* anthocyanin; chalcones; flavone.

floxing *See* targeting genes.

FLP/FRT *Saccharomyces* recombinase system that can also be expressed in *Drosophila*. FLP is a recombinase and FRT defines target sites of transposable elements. The system is very similar to the Cre-loxP of bacteriophage P1. The 43 kDa "flip" (FLP) is encoded by the 2 μ circular yeast plasmid. Recombination (X) takes place within the core sequence and between FLP-binding sites. This, as well as the *Cre/loxP* system, can be used for site-specific integration of transgenes and also for engineering chromosomal rearrangements in higher eukaryotes. *See* chromosomal rearrangements; *Cre/loxP*; homing endonuclease; recombinase; targeting genes; transposable elements. (Buchholz, F., et al. 1998. *Nat. Biotechnol.* 16:857.)

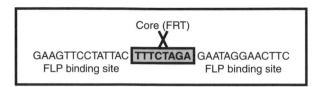

Core (FRT)

GAAGTTCCTATTAC **TTTCTAGA** GAATAGGAACTTC
FLP binding site   FLP binding site

Flpter   Fractional length from the hybridization signal to the terminus of the short arm of the chromosome. It is used for the localization of FISH labels on the chromosome. *See* FISH.

FL-REX (fluorescence localization–based retrovirus-mediated expression cloning)   mRNA is isolated and reverse-transcribed into cDNA. Then the cDNA (about 1000 bp) is inserted into a multiple cloning site of a green fluorescent protein (GFP) gene within a retroviral vector. With the aid of the vector, packaging cells are transfected and the isolated retroviruses are used to infect cultured mammalian cells. The GFP fusion permits the screening of the infected cells and subcloning individual cell lineages for a variety of different proteins with subcellular expression. Also, the different DNAs can be amplified by PCR and sequenced. In a similar manner, GFP fusion proteins can be localized within the cytoplasm of plants with the aid of agrobacterial vectors. *See* aequorin; DNA sequencing; expression cloning; lipofection; packaging cell lines; polymerase chain reaction; ψ; reverse transcription; subcellular; subcloning. (Misawa, K., et al. 2000. *Proc. Natl. Acad. Sci. USA* 97:3062.)

FLRTED   Transmissible variations generated by the Flp/FRT system of site-specific recombination. *See* Flp/FRT. See above at right.

FLT oncogene   In human chromosome 13q12; it encodes a protein tyrosine kinase. It shows homology to FMS and ROS. Flt-1 is essential for animal embryonal vasculature but not for endothelial differentiation. The FLT3/FLK2 receptor tyrosine kinase is closely related to KIT and FMS. *See* FAK; Flk-1; FMS; KIT; oncogenes; receptor tyrosine kinase; ROS; Tie-1; Tie-2; vascular endothelial growth factor. (Maru, Y., et al. 2001. *Biochim. Biophys. Acta* 1540:147.)

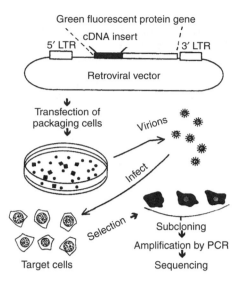

Green fluorescent protein gene
cDNA insert
5′ LTR   3′ LTR
Retroviral vector
Transfection of packaging cells
Virions
Infect
Selection   Subcloning
Amplification by PCR
Target cells   Sequencing

fluctuation test   Originally designed to determine whether the observed mutations were induced by a particular treatment or the treatment merely was a means of screening for that particular class of mutations that preexisted in the cultures. The design of the fluctuation test by Luria and Delbrück in 1943 initiated experimental bacterial genetics. It can be used in other microorganisms, in animal and plant cell cultures, and the principle can be adapted to most types of mutation experiments. The principle is as follows: From the same original culture, two series are generated. In series I, the culture is continued until a particular cell density is reached. In series II, small inocula are placed in, say, 10 (or more) vessels and the culture is continued until the same cell density is reached in all vessels of series I and II. Then plating is made in, say, 10 petri plates from the single-vessel culture I and one petri plate from each of the 10 test tubes of series II. After incubation, the number of mutations are scored separately on each, the total of 20 petri plates. The abridged data of an experiment of M. Demerec (*J. Bacteriol.* 56:63) are tabulated on the next page. The averages ($\bar{x}$) and the variances ($(\Sigma x^2 - n\bar{x}^2)/n - 1$) are determined. If it appears that the average number of mutational events in both series I and II are practically the same but the variance in series II

	SERIES I			SERIES II	
Culture No.	Resistant Cells (X)	X^2	Culture No.	Resistant Cells (X)	X^2
1	146	21,316	1	67	4,489
2	141	19,881	2	159	25,281
3	137	18,t69	3	135	18,225
4	128	18,384	4	291	84,681
5	121	14,641	5	75	5,625
6	110	12,100	6	117	13,689
7	125	15,625	7	73	5,329
8	135	18,225	8	129	16,641
9	121	14,641	9	86	7,396
10	112	12,544	10	101	10,201
SUMS	1,279	164,126	SUMS	1,233	191,557
Average $\bar{X}$	127.6		Average $\bar{X}$	123.3	
Variance	154		Variance	4,392.0	

significantly exceeds that of series I, the conclusion is that the treatment was not the cause of the mutations. A fluctuation test of bacterial mutation for streptomycin resistance series I showed an average of 127.6 mutations, and the variance was 145.4. In series II, the average was almost the same, 123.3, but the variance was 4,392.0. Thus, streptomycin only revealed the presence of preexisting mutations for streptomycin resistance. The logic of this argument is that in the single vessel I the preexisting mutants were distributed uniformly and therefore upon plating the variance was low. In the 10 vessels of series II, mutations occurred or did not before plating. If mutations occurred early, many mutants appeared on that petri plate; if mutation occurred late in the test tube, few mutations were detectable after plating; if no mutations occurred in a particular test tube, the petri plate failed to display any. The propagation of the preexisting mutations caused the large number of mutant colonies on some plates, and when there were no mutations prior to the treatment, none appeared after plating in series II. In series I in the single batch, the preexisting mutations were uniformly dispersed without fluctuation in the variation. Thus, the different fluctuations on the identical series of petri plates were the critical argument for the conclusion that the treatment was ineffective. *See* mean, variance. (Luria, S. E. & Delbrück, M. 1943. *Genetics* 28:491.)

fluid genome paradigm   *See* constant genome paradigm.

fluidity   Lipids can move within membranes. *See* cell membranes.

fluorescein   Fluorochrome with excitation at 490 nm wavelength of light and emission at 525 nm. Generally used in conjugates with avidin, isothiocyanate, or antibodies.

Fluorescein; R = radix of different types.

fluorescein diacetate   Vital stain for protoplasts.

fluorescence   Property of chemical compounds to emit radiation (light) upon absorption of radiation from another source. The fluorescent radiation generally has a longer wavelength than the absorbed one, e.g., nucleotides irradiated by 260 nm UV light display a visible purple color. The fluorescence lasts only as long as the exposure to the irradiation. *See* UV.

fluorescence microscopy   Uses a microscope with an illuminator equipped with a light filter that assures the stage receives only the narrow spectral band of light required for the excitation of the fluorochrome used for staining the specimen. In front of the ocular, there is another filter that transmits only

the fluorescence emitted but shuts out the exciting wavelength light. By staining with more than one dye and changing filters, in the same specimen different structures (molecules) may be distinguished by different bright colors. Some of the fluorochromes permit also viewing living cells. *See* chromosome painting; FISH; fluorochromes; microscopy.

fluorescent dye   *See* fluorochrome.

fluorescent focus assay   Determines titers of nonkiller viruses. Permeabilized (by acetone, methanol) cells are incubated with antibody against the virus. Then another antibody is added that recognizes the first one and is conjugated with a fluorescent dye. The fluorescing particles are counted by UV light microscopy. (Yang, D. P., et al. 1998. *Clin. Diagn. Lab. Imunol.* 5:790.)

fluorescent *in situ* hybridization   *See* FISH.

fluorochromes   Nonradioactive labels such as DAPI, fluorescein, rhodamine B, FITC, Texas Red, R-phycoerythrin, RED613, RED670, allophycocyanin, isothiocyanate, aequorin, etc. The dipyrrometheneboron difluoride dyes may have special advantages for automatic sequencing of DNA. *See* aequorin; biotinylation; FISH; fluorescent microscopy; FRET. (Han, M., et al. 2001. *Nature. Biotechnol.* 19:631.)

fluorography   Enhances the sensitivity of autoradiographic detection by adding to the sample scintillants such as 2,5-diphenyloxazole (PPO) or sodium salicylate. *See* autoradiography; scintillation counters.

fluorophore   Pigment that upon activation brings forth color. *See* fluorochrome.

fluoroscope   Medical apparatus for X-ray examination of deep-seated tissues. The image is seen on a fluorescent screen and is coated with calcium tungstate or zinc cadmium sulfide and other materials. It was favored in the past because it could detect motions. It has become of lesser significance with the introduction of television cameras and because it uses about 10-fold higher doses of radiation than diagnostic X-ray machines and poses radiation hazards to both operator and patient. *See* sonography; tomography; X-rays.

5'fluorouracil (FU)   Uracil analog incorporated into RNA. It is metabolized into deoxyfluorouridine monophosphate and triphosphate and can incorporate into DNA. It is a potent antineoplastic agent, although resistance can develop. *See* neoplasia.

FU.

flush-crash cycle When the size of a natural population fluctuates greatly.

flush end *See* blunt end (of DNA).

flux Spread of a label from a metabolic precursor to other molecules in the cell.

flux, genetic Alterations in the cell in response to internal and external stimuli.

fMet Formylmethionine is generally the beginning (modified) amino acid in the translation on the ribosome of prokaryotes mediated by tRNA^{fMet-i} (i for initiator). At internal peptide sites, even at prokaryotes, the tRNA^{Met-e} (e for elongation) is functional. The formylation is the duty of methionyl-tRNA^{Met} formyltransferase. This enzyme recognizes the tRNA as the initiator by a CA mismatch at position 1–72. The eukaryotic (and Archaea) initiator carries the A1-T72 base pair at the comparable site. In the mitochondria, there may be only a single type of tRNA^{Met}, which resembles the initiator type, or it may have the C1-A72 mismatch. Trypanosomal mitochondria lost their tRNA genes and imports all the tRNAs from the cytosol, formylating some of the imported elongation tRNAs with the aid of an enzyme similar to formyltransferases but with twice the mass of that of the prokaryotic enzyme. *See* formylmethionine; protein synthesis. (Tan, T. H. P., et al. 2002. *Proc. Natl. Acad. Sci. USA* 99:1152.)

FMN Flavin mononucleotide is composed of flavin (riboflavin) phosphate and dimethylisoalloxazine base (yellow enzyme); it is a coenzyme for oxidation-reduction-mediating proteins.

FMR1 *See* Fragile sites; KH module; trinucleotide repeats.

FMRFamides Neuropeptide. (Parker, S. L. & Parker, M. S. 2000. *Can. J. Physiol. Pharmacol.* 78[2]:150.)

FMRI Functional magnetic resonance imaging. *See* MRI.

FMR1 mutation Involves expansion of the CGG trinucleotides, which are frequently methylated on C and thus silenced. Hypermethylation of FMR1 is called *full mutation*, whereas the *premutation* is an intermediate range of expansion (50–200) of these repeats. *See* fragile sites; human intelligence; KH module; trinucleotide repeats. (Crawford, D. C., et al. 2001. *Genet. Med.* 3[5]:359.)

FMS Feline sarcoma virus protooncogene is a receptor of colony-stimulating factor-1 (CSF) and macrophage colony-stimulating factor. This gene is located in human chromosome 5q33.2-q33.3. It is involved in different types of leukemias. Its sequence is homologous to oncogene FLT, a tyrosine kinase. *See* colony stimulating factor; FGR; FLT; leukemias; protooncogene.

FNR Ferrous/ferric-binding bacterial aerobic/anaerobic regulator protein with some similarity to the catabolite repressor protein. *See* catabolite repression. (Moore, L. J. & Kiley, P. J. 2001. *J. Biol. Chem.* 276:45744.)

foamy viral vector Derived from Spuma retrovirus, which may not be pathogenic for humans, although these viruses infect many cell types in several hosts. Their carrying capacity is ~12 kb, somewhat larger than the murine leukemia virus (MLV), which accepts 9–10 kb. In addition, foamy viral vectors transduce nondividing cells somewhat more efficiently than MLV vectors. *See* retroviral vectors. (Vassilopoulos, G., et al. 2001. *Blood* 98:604.)

focal contact Adhesion plaque on the surface of a cell that is attached to the extracellular matrix by transmembrane proteins (integrin).

focal dermal hypoplasia (Goltz syndrome, FDOF) X-linked dominant lethal in males involving atrophy, skin pigmentation in a linear pattern, papillomas (epithelial neoplasms), polydactyly, underdevelopment of teeth, small defective eyes, mental retardation, etc. *See* cancer; eye diseases; mental retardation; pigmentation defects; polydactyly; skin diseases.

focus formation Neoplastic (cancerous) cells grow up in dense clusters. *See* cancer.

focus forming unit *See* FFU.

fodrin (spectrin) α-fodrin is nonerythroid α-spectrin (a cytoskeletal protein) with the critical amino terminus of RQK-LEDSY RFQFFQRDAEEL. βII-spectrins (2p21, 11q13, 15q21) are schwannomin- and other membrane-binding proteins. *See* neurofibromatosis; schwannoma; Sjögren syndrome; spectrin. (Takahashi, K., et al. 2001. *Eur. J. Pediatr.* 160:520.)

foetus *See* fetus.

FokI Endonuclease recognizes the 5'-GGATG-3' sequence and cleaves DNA 9 and 13 bases away.

folate Salt of folic acid. *See* folic acid.

folate malabsorption Defect in the transport of folate through the blood-brain barrier, resulting in anemia, movement problems, mental defects, seizures, etc.

fold-back DNA Palindromic inverted repeat sequences in a single strand of polynucleotides pair within the same strand

DNA Foldback.

fold-back inhibition site *See* FBI site.

fold-back transposon (FB) Transpose through a DNA-intermediate carrying imperfect inverted repeats ranging from 300 bp to several kbp. They may or may not contain open reading frames. *See* open reading frame; repeat, inverted; transposon.

folded leaf   Protein α-helices wrapped around a hydrophobic core. *See* α-helix; hydrophobic.

folding   Thermodynamically reversible process involving macromolecules (proteins), resulting in a functional conformational state. The folding may follow different pathways from simple two-state to sequential complex ones. Thiol (SH)$\rightleftharpoons$disulfide (-S-S-) conversions are generally involved. *See* chaperone; conformation; protein folding. (Thomas, P. J., et al. 1995. *Trends Biochem. Sci.* 20:456; Baker, D. & Agard, D. A. 1994. *Biochemistry* 33:7505; Frydman, J. 2001. *Annu. Rev. Biochem.* 70:603; Mirny, L. & Shakhanovich. E. 2001. *Annu. Rev. Biophys. Biomol. Struct.* 30:361.)

foldon   Protein domain involved in folding.

folic acid (pteroyl glutamic acid)   Water-soluble vitamin; it is required for de novo nucleotide synthesis and amino acid conversions.

As a derivative of pteridines, it is involved in many oxidation reactions, mediating animal coloring, and various light reactions. Tetrahydrofolic acid is an important coenzyme. Folic acid deficiency during pregnancy may lead to congenital malformations. Experimental data indicate that the folic acid–binding FOLR1 receptor may be insufficient. Folic acid deficiency may increase single-strand and double-strand chromosome breakage ($\sim$4 million/cell) because of uracil incorporation into the DNA in place of thymine. Leukemia (ALL but not AML) and colon cancer risk may also increase, especially when C is replaced by T at site 677 of methylenetetrahydrofolate reductase. The same genetic alteration may increase the risks of Down syndrome, and various types of cancer. Diets high in fruits and vegetables (and thus folic acid) and micronutrients such as vitamins B_{12}, B_6, niacin, C, E, and iron and zinc may be beneficial. *See* aminopterin; formiminotransferase deficiency; formyltetrahydrofolate synthetase; fragile sites FMR1; homocystinuria; methotrexate; 5,10-methylenetetrahydrofolate dehydrogenase; nondisjunction; phosphoribosylglycinamide formyltransferase. (Ames, B. N. 1999. *Proc. Natl. Acad. Sci. USA* 96:12216; Hobbs, C. A., et al. 2000. *Am. J. Hum. Genet.* 67:623.)

follicle   Small secretory sac or gland. The cell layer covering the ovary; in botany, a simple dry fruit, dehiscing along one suture and formed of a single carpel. *See* carpel; ovary.

follicle stimulating hormone   *See* FSH.

follistatin   Maternally expressed protein that by binding activin may interfere with activin-induced mesoderm formation. Follistatin-deficient mice die within hours of birth because of multiple defects, indicating its requirement for several proteins of the transforming growth factor family. *See* activin; bone morphogenetic protein; mesoderm; organizer. (Michel, U., et al. 1993. *Mol. Cell. Endocrinol.* 91:1; DePaolo, L. V. 1997. *Proc. Soc. Exp. Biol. Med.* 214:328.)

Fomivirsen (Isis 2922)   Antisense DNA phosphorothioate (5′-dGCGTTGCTCTTCTTCTTGCG-3′) drug against the 55 kDa protein of the cytomegalovirus, causing RNase H degradation of mRNA and inhibition of human CMV replication. In addition, it may interfere with viral adsorption. *See* antisense technologies; cytomegalovirus. (Perry, C. M. & Balfour, J. A. 1999. *Drugs* 57[3]:375.)

fonts   Various styles of characters and scripts that the computer can use.

footprint, transposable element   After a mobile genetic element (insertion or transposable element) moves, it usually leaves behind some nucleotides (as a footprint) at the original target site. *See* insertional mutation; transposon footprint; transposon mutagenesis. (van Houwelingen, A., et al. 1999. *Plant Cell* 11:1319.)

footprinting   Fragments of 5′-labeled double-stranded DNA are partially degraded by a DNase in the presence and in the absence of a protein expected to bind to certain sequences in the DNA. Subsequently, both samples (with and without the binding protein) are sequenced (Maxam and Gilbert method) and a comparison reveals the nucleotide sequences protected from DNase by the binding protein. Thus, from the path (footprints) of the DNase, the position of the binding protein is revealed. These sites may have importance for transcription initiation. *In vitro footprinting* is basically very similar to the methylation interference technique, but it is performed in living cells and followed by DNA extraction and electrophoresis. *See* methylation interference assay; regulation of gene activity. (Fox, K. R. & Waring, M. J. 2001. *Methods Enzymol.* 340:412; Metzger, W. & Heumann, H. 2001. *Methods Mol. Biol.* 148:39.)

footprinting, genetic   The role of sequenced genes without known function can be determined by inserting transposable elements (Ty) in a sequential manner, then determining the genetic consequences of the disruption. This process can screen yeast cell populations in 10^{11} range and detects a wide range of mutations of variable severity or fitness. *See* mutation detection; transposon footprint; Ty.

foramen   Natural, anatomical opening or passageway.

Forbes disease   Defect or deficiency of amylo-1,6-glucosidase and/or oligo-1,4-1,4-glucantransferase.

forelock, white   Autosomal-dominant human trait (a lock of white hairs in the front part of the scalp). *See* Waardenburg syndrome.

White forelock.

forensic genetics   Genetic studies used for legal purposes or in the judiciary courts relying on fingerprints, blood groups, other antigens, isozymes, and VNTR (variable number tandem repeats of DNA) by employing RFLP (restriction fragment length polymorphism) or other heritable criteria to identify biological relationships, paternity, criminals, etc. Most of the chemical analyses require small samples of blood (60 μL), semen (5 μL), or hair roots. For some of the tests, dried blood or semen spots are useful even if they are weeks, months, or years old.

On the basis of polymorphic proteins, identity can be defined to higher than 99% probability; for exclusion of an individual, much less effort is required. Generally, the spectrum of the protein components (separated by electrophoresis) in the sample obtained from the person to be identified is compared with the spectrum of the same protein components within the same (ethnic) population. The product of the frequencies expected by chance is compared with the actually observed data. The first test is generally for the ABO blood group, but this alone rarely suffices because of the limited variations and the failure to identify all heterozygotes. Other proteins assayed for polymorphism are adenylate kinase (AK), adenylate deaminase (ADA, must be examined within 6 months), carbonic anhydrase (CA-II, within a week), erythrocyte acid phosphatase (EAP, within 6 months), esterase (EsD, within 1 month), glyoxylase (GLO), hemoglobin (Hb), peptidase A (pepA), phosphoglucomutase (PGM, within 6 months), gammaglobulin (Gm, displays about a dozen antigens determined by very closely linked genes and the clusters have different frequencies in different ethnic groups), Lewis antigens (Le^a, Le^b), and rhesus antigens (47 Rh determinants, within 6 months). In semen samples, generally ABO, GLO, Pep A, PGM, and Le are used. If the semen sample is obtained by vaginal swabs in a rape case, it may be contaminated by vaginal fluids (may obscure semen fluids), proteolytic enzymes (that may degrade the proteins), or bacteria, thus possibly interfering with blood typing. Actually female cells may be separated from sperm by digestion with sodium dodecylsulfate/proteinase K, which does not destroy sperm. Later, the sperm can be digested in the same reaction mixture with (dithiothreitol) DTT added. The greatest specificity of identification can be obtained by analysis of the DNA in body fluids or skin or other tissue samples. By June 1998, the Forensic Science Service in the United Kingdom collected 320,000 DNA samples and removed 51,000 of them from the data bank after the suspects were exonerated. The establishment and maintenance of such national databases—although very useful—may be quite demanding because about 30% of the male population under age 30 may have at least one felony conviction. See ADA; blood groups; DNA fingerprinting; fingerprinting; forensic index; Frye test; gammaglobulin; hemoglobin; PGM; phylogenetic analysis; RFLP, sodium dodecil sulfate; VNTR. (National Research Council Technology in Forensic Science by Natl. Acad. Sci USA, Washington, DC, Masters, J. R., et al. 2001. *Proc. Natl. Acad. Sci. USA* 98:8012; Benecke, M. 2002. *EMBO Rep.* 3:498.)

forensic index   Provides statistical information regarding the probability that the evidence (*E*) collected at a crime scene or from other potentially incriminating objects would belong to the perpetrator (*P*) or to a suspect (*S*). Let us assume that the perpetrator and the suspect have DNA (VNTR or STR) or protein profile (blood group or enzymes) *A* and we must find what is the probability for the suspect being liable for that event or fact (*C*), or the suspect and the perpetrator are different individuals (event *C'*). These conditional probabilities are called *L*:

$$\text{(forensic index)} \rightarrow L = \frac{\Pr(E|C)}{\Pr(E|C')} = \frac{\Pr(S = A | P = A, C)}{\Pr(S = A | P = A, C')}$$

S = *A* indicates that the profile of *S* is *A*, i.e., the match is *L* times more probable if *S* and *P* are the same persons. Actually, the decisions are more complex because in the match the relatedness within the population (population structure) must be considered. See Bayes theorem; conditional probability; DNA fingerprinting; fixation index; inbreeding and population size; inbreeding coefficient; paternity index; STR; VNTR.

forestomach   Found in mice, rats, and hamsters between the esophagus (the channel of food from the throat to the stomach) and the glandular stomach. Humans do not have it. The forestomach is a common target of chemical carcinogens.

Fore tribe   Of New Guinea, is most affected by kuru disease because of a behavioral tradition: cannibalizing dead relatives. See kuru.

FORKED (*f*, 1–56.7)   *Drosophila* gene affecting the ends of microchaete, macrochaete, and trichomes. This phenotype may be suppressed by *suppressor of forked, su(f)* (1–65.9), *and suppressor of Hairy wing, su(Hw)* (3–54.8), in an allele-specific manner. See chaeta; cleavage stimulation factor; trichome.

forked tongue   Enables snakes to assess different signals (pheromones) simultaneously.

forkhead   Family of transcription factors. See FKH.

formal genetics   *See* classical genetics.

forma specialis   Genetically distinguishable race of a pathogen that can primarily infect a particular host or a host of a defined genotype/phenotype.

formiminotransferase deficiency   Autosomal-recessive physical retardation without mental retardation, anemia, etc., caused by oversupply of folate. See folic acid.

formylmethionine   Translation initiation amino acid in prokaryotes and cytoplasmic organelles (plastids and mitochondria) of eukaryotes, but it is not used in the cytosol of eukaryotes. It is carried to the 70S ribosomes by a formyl-methionine tRNA (tRNA$_i^{Met}$ or tRNA^{fMet}) distinct from the regular tRNA^{Met}. *See* protein synthesis. (Takeuchi, N., et al. 2001. *J. Biol. Chem.* 276:20064.)

10-formyltetrahydrofolate synthetase Key enzyme of folic acid metabolism. *See* folic acid.

forskolin Diterpene isolated from the plant *Coleus forskohlii*. It is an activator of adenylate cyclase and some other mechanisms that depend on cAMP. *See* adenylate cyclase; cAMP; signal transduction.

FORTRAN (formula translating system) Computer languages with specific problem orientations.

forward mutation Mutation from wild-type to mutant allele. *See* reversion.

fos Murine osteosarcoma (chondrosarcoma) protooncogene, general transcription factor (AP1). The fos-jun heterodimers bind to the 5'-TGAGTCAA- 3' sequence. Fos also controls complex behavioral traits such as nurturing and the development of the dorsal closure during embryogenesis in cooperation with *jun* in *Drosophila*. *See* AP1; apoptosis; FOS oncogene; jun; JUN; protooncogene; sarcoma; transcription factors.

fosmid Very low, single-copy-number *E. coli* F-replicon-based stable vectors that are particularly useful for cloning large eukaryotic genes. *See* F plasmid; vectors.

FOS oncogene In human chromosome 14q21-31; homologous to the v-oncogene *fos*. The c-FOS protein is a transcription factor of the Jun family. The human FOS has a normal expression in fetal membranes almost as high as that detectable in osteosarcomas of mice. Products of FOS and Jun contribute to the formation of the AP1 transcription factor and participate in multiple ways in tissue differentiation. The c-FOS oncogene is regulated by Ca^{2+}/CRE, SRE, and SIE elements in upstream regions of the gene. These elements are under the control of neurotransmitters, neurotrophins, and cytokines, respectively. Studies indicate that Fos knockouts in mice fail to nurse their pups. *See* AP1; apoptosis; behavior genetics; CRE; cytokines; JUN; knockout; neurotransmitters; neurotrophins; oncogenes; SIE; signal transduction; SRE. (Takeuchi, K., et al. 2001. *J. Biol. Chem.* 276:26077.)

fossil Petrified remains or impression of an organism of past geological ages preserved in the earth or rocky layers. *See* fossil record.

fossil, genomic Inactive retroviral elements that are relics of past infection during evolution.

fossil record Used to reveal the pattern of macroevolution (evolution of taxonomic categories above the species level) at the geological scale. Macroevolution was traditionally inferred from the paleontological data and the appearance of petrified taxa in the successive geological strata. The age of the remains is estimated by *radioisotope dating*. If the fossils are less than 40,000 years old, their age is inferred from their amount of carbon-14 (^{14}C). This isotope is produced at a relatively constant rate from nitrogen-14 (^{14}N) under the bombardment of cosmic radiation through the ages. This ^{14}C is utilized by the organisms in the same way as the more common ^{12}C. The former is unstable, however, and it decays to half in each 5,730-year cycle. The amount of ^{14}C in the organic material serves as a clock with an accuracy of ± 1 to 2%. The age of fossils over 40,000 years old is inferred from the age of the sedimentary rocks where the organism died (if that is the site of the fossil and it was not moved by geological changes in the strata). The age of the rocks is estimated by the decay of other isotopes. Uranium-238, e.g., decays into lead-206 with a half-life of 4.51×10^9 years. Therefore, the proportion of these two elements in the rocks indicates their geological age. The evolutionary relation of the fossils can be better defined if protein and DNA analyses are also feasible. *See* half-life; isotope.

foulbrood Disease of the honeybee caused by *Bacillus alvei*. Resistance is based on homozygosity of two nonallelic recessive genes determining behavior. One gene is responsible for uncapping the honeycombs when the larvae die; the other gene is responsible for the removal of the dead. *See* behavior genetics; honey bee.

foundation stock In rodents, consists of 10–20 monogamous brother × sister pairs. This breeding/maintenance regime is necessary so that the new, spontaneous, recessive mutations can be readily detected before they spread through the multiplication stocks. In mice, breeding tests, skin grafts (immunological reaction), and biochemical tests (electrophoresis) can monitor authenticity/identity of the strain designation. In plant breeding, the elite seed from the multiplied authentic registered varieties is used. *See* multiplication stock.

founder cell Early embryonic cell that contributes to the different cell lineages during differentiation and development. During first cleavages of the embryo in *Caenorhabditis elegans* 5 founder cells are generated. The first of these cells (AB) gives rise to 389 of the total 558 nuclei present at hatching. The AB cell lineage is specified by at least five inductions before gastrulation. The inductions do not actually specify the final tissues, but 8 blastomeres contribute to the final body plan. The specification usually requires binary (0 or 1) switches. In the mature seed of *Arabidopsis*, there are 12–16 founder cells. *See* blastomere; cell lineage; fate map; gastrula; morphogenesis.

founder effect Same as founder principle.

founder mouse Chimeric animal obtained after transformation that may or may not involve the germline. *See* chimera; germline; microinjection.

founder principle A new population descends from a limited number of immigrants (because of sampling error), resulting in genetic drift. It is also the called founder effect. *See* drift, genetic; effective population size; porphyria variegata.

four-cluster analysis Procedure for determining the evolutionary relationships among four large groups of organisms such as animals, plants, fungi, and protists without consideration to the variation within each of these groups. If we designate the four monophyletic groups as A, B, C, and D, three unrooted evolutionary trees can be generated. They are $T_1 = [(AB)(CD)]$, $T_2 = [(AC)(BD)]$, and $T_3 = [(AD)(BC)]$, from which one is expected to be correct on the basis that the correct construct would have the shortest sum of tree branch length. The three sums of branch lengths may be designated as S1, S2, and S3. The differences S1 − S2, S1 − S3, and S2 − S3 are determined by an appropriate algorithm. *See* evolutionary distance; evolutionary tree; least square methods; neighbor joining method; unrooted evolutionary trees. (Rzhetsky, A., et al. 1995. *Mol. Biol. Evol.* 12:163.)

four-hybrid system   Construct suitable for the activation of specific genes such as *HIS* or *lacZ*. The cyclin, CDK, and MAT1 stable system is part of the general transcription factor TFIIH, or the cell cycle and the interaction turn on genes in the presence of the target protein **X** and the VP16 activator. *See* LexA; MAT1; transcription factors; two-hybrid system; VP16. Diagram below. (Sandrock, B. & Egly, J.-M. 2001. *J. Biol. Chem.* 276:35328.)

Fourier method   *See* FT-IR.

four-o'clock   *See Mirabilis jalapa.*

foveal dystrophy   Autosomal-dominant lesion of the macula in the eye fundus with aminoaciduria. *See* eye diseases; macula.

fox   These canid species are quite variable genetically and by chromosome number. *Vulpes velox* (kit fox) $2n = 50$; *Vulpes vulpes* (red fox) $2n = 36$; *Vulpes fulva* (American red fox) $2n = 34$; *Urocyon cinereoargentus* (gray fox) $2n = 66$; *Otocyon megalotis* (bat-eared fox) $2n = 72$; *Lyalopex vetulus* (hoary fox) $2n = 74$; *Cerdocyon thous* (crab-eating fox) $2n = 74$. *See* wolf. FKH

FPC   Finger-printed clone.

F⁻ phenocopy   Bacterial cell that has no F pilus, although it carries an F element, yet it is not in a conjugative state. *See* bacterial conjugation; conjugation mapping.

F-pili (or F-pilus or sex pilus)   Bacterial cell appendage that forms the conjugation tube through which the F element, conjugative plasmids, and the Hfr bacterial chromosome are mobilized into the F⁻ cells. *See* conjugation; F factor; Hfr; sex factor. (Anthony, K. G., et al. 1999. *J. Bacteriol.* 181:5149.)

F plasmid   Also called F factor or fertility plasmid of *E. coli*; it is about 100 kb and has four major regions. The (1) *inc* and *rep* tracts control its vegetative replication. When only this is retained, it is called a *miniplasmid*. The (2) *IS* sequences are insertion elements and Tn1000 is a transposon (also called $\gamma\delta$) similar to Tn3. In the (3) *silent region*, few functions are known. The (4) *tra* sites control transmission of the plasmid, which originates at the *oriT* site. The *tra* region includes more than two dozen genes. The *tra M* is regulated by the *traJ* gene, which in turn is negatively regulated by *fin* (fertility inhibition) gene products. The *tra* operon also includes genes for the formation of the sex pilus through which plasmid and/or chromosomal DNA is transferred to the recipient cells. The total number of genes in this plasmid is about 30. The plasmid may be present in one or two copies per F⁺ bacterial cells. It is an episome and can integrate clockwise or counterclockwise at various sites into the bacterial chromosome. When excised, it may become an F' plasmid. *See* conjugation; episome; F' plasmid; Hfr; mapping; pilus; transposon. (Cavalli, L. L., et al. 1953. *J. Gen. Microbiol.* 8:89.)

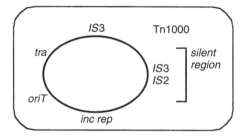

FPR (FKB)   *See* FK506; peptidyl-prolyl isomerase.

fps   Chicken sarcoma oncogene. *See* sarcoma.

FPS   Fixed pairing segment is a hypothesis that recombination is not entirely random in eukaryote but occurs in tracts that are either fixed at both ends or at the middle or at one end. During a single meiosis, only a fraction of these segments pair and there is positive interference in their vicinities. *See* interference.

FRA   Forms active heterodimer with Jun in the AP activator protein. *See* AP1.

fractal   Fractal geometry has been considered to describe chaotic systems such as the complexities of many biological phenomena. Fractals emerge from interaction of self-similar entities. The two-dimensional squares can be resolved into many squares and the three-dimensional cubes into many

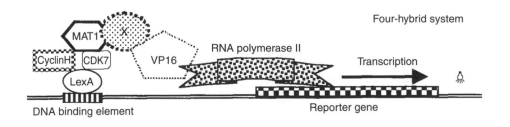

Four-hybrid system

smaller identical-looking cubes. Irregular-shaped objects also have self-similar fractions, but they cannot be adequately described like the square or cube in two or three dimensions; their characterization requires fractals, such as 2.2, rather than integers. *See* chaos.

fractalkine   CX3C-type chemokine. *See* acquired immunodeficiency; chemokines.

fractional mutation   Displaying mosaicism (variegation) in the tissues of the body. If the mutation was induced in the germ cells, it will indicate that the mutagenic agent was associated with only one strand of the DNA; therefore, mutant and nonmutant sectors arose and DNA repair occurred during the postfertilization stage. Fractionals may also be due to unstable genes, mitotic recombination, nondisjunction, transposable elements, etc. *See* gene conversion; mitotic recombination; nondisjunction, transposable elements; unstable genes. (Altenburg, E. & Browning, L. S. 1961. *Genetics* 46:203.)

fractionated dose   Irradiation is provided not in a chronic manner but with interruptions between each exposure, although the doses are summed up. *See* chronic radiation.

fraction 1 protein   Old name of ribulose bisphosphate carboxylase/oxygenase enzyme, which is the largest single protein encoded by the plastid and forms about 50% of the proteins in the chloroplasts. *See* rubisco.

fragile site (FRA)   Occurs in several human chromosomes. The overall frequency of autosomal fragile sites is about 2×10^{-3}. Generally three types are distinguished: (1) folate sensitive (shows up if the cell culture medium is deficient in folate), (2) elevated pH triggers their appearance; and (3) 5-bromodeoxyuridine (BdUR) is required for expression. The best studied is fragile X syndrome. Fragile sites were also identified at 2q11, 3p14.2, 6p23, 9p21, 9q32, 10q23 (folic acid sensitive), 10q 25 (BdUR sensitive), 11q23, 12q13, 16p12, 16q22 (appeared only in the presence of Epstein-Barr virus [EBV] antigen), 17p12, and 20p11. The fragile sites are generally dominant and involve overlapping syndromes, frequently mental retardation, cancer susceptibility, and other symptoms. In fragile X syndrome, the number of CGG repeats may run into hundreds, whereas under normal conditions only about 30 repeats are found. In Friedreich ataxia, GAA repeats are found in the introns of the frataxin gene. Five neurological disorders — spinal and bulbar muscular atrophy (Kennedy disease), spinocerebellar ataxia (olivopontocerebellar atrophy) type 1, Huntington's chorea, dentatorubral-pallidoluysian atrophy, and Machado-Joseph disease — display poly-CAG sequences within their genes and encode polyglutamine. Myotonic dystrophy is accompanied by CTG repeats in the untranslated last exon of a protein kinase gene. Recessive myoclonous epilepsy (human chromosome 21) contains multiple repeats at the 5′ and 3′ areas of the promoter of the cystatin B gene. Fragile sites may occur upon incorporation of double minutes into the chromosome. Hypoxia favors the formation of fragile sites, which seem to promote gene amplification and remodeling of the genome. In cancer, the most commonly amplified sites involve oncogenes ERBB, RAS, KRAS, MYC, and genes controlling the cell cycle. Fragile sites in humans and mice show rather high homology and conservation. *See* anticipation, ataxia; dentatorubral-pallidolyusian atrophy, double minutes; FMR1 mutation; glutamine-repeat diseases; human intelligence; Huntington's chorea; hypoxia; Kennedy disease; Machado-Joseph disease; mental retardation; myoclonous epilepsy, premutation; smoking; trinucleotide repeats; X chromosome. (Jin, P. & Warren, S. T. 2000. *Hum. Mol. Genet.* 9:901; Richards, R. I. 2001. *Trends Genet.* 17:339.)

Fragile site.

fragile X chromosome (FRAXA)   Displays poorly stainable sites under the light microscope that are liable to breakage and may result in mental retardation and cancer. The affected individuals also have macrocephaly (large head), prominent jaws, macroorchidims (enlarged testes), and a high-pitched voice. The condition is caused by folate deficiency leading to low levels of thymidylate. It involves amplification (up to >200) of a CGG repeat in the FMR-1 (fragile site mental retardation) gene (at Xq27.3, Martin-Bell syndrome) that occurs in human populations by a frequency of ~1/1,250–4,000 in males and 1/2,500–6,000 of females. Under normal conditions, there are 6–50 of the repeats; in fully expressed FRAXA, the number of repeats exceeds 200. When the number of repeats is between 50 and 200 (premutation), there is ~50% risk for full expression in the progeny of females but not in the immediate offspring of males. The premutational condition is not revealed by the phenotype but can be detected by Southern blotting or PCR. Methylation and deacetylation of histones H3 and H4 and gene silencing follow the amplification. Treatment of the fragile X cells with 5-aza-2′deoxycytidine restores transcription. Fragile X syndrome results in the absence of the RNA-binding FMR protein. The FRAXA protein (FMRP) defects appear to interfere with the translation template and thus with translation. FMRP binds intramolecular G quartets, which this may result in dysregulation of mRNA. Trichostatin restores acetylation of H4 but only minimally for H3 and does not result in transcription. The males are predominantly affected (80%); about one-third of the carrier females are also mentally retarded. Usually, 20% of the males with this X chromosome are phenotypically normal and their daughters are also normal, but their grandsons display the chromosome and the phenotype. Prenatal diagnosis is feasible. On folate-deficient cell culture media the critical X chromosome displays a constriction at the Xq27-p28 site. Other fragile sites may account for some of the less common forms of the disease. *See* azacytidine; fragile sites; G quartet; head/face/brain defects; histone deacetylase; histones; Jacobsen syndrome; mental retardation; PCR; Southern blot; trichostatin; trinucleotide repeats. (Tassone, F., et al. 2000. *Am. J. Hum. Genet.* 66:6; Toledano-Alhadef, H., et al. 2001. *Am. J. Hum. Genet.* 69:351; Li. Z., et al. 2001. *Nucleic Acids Res.* 29:2276; Brown, V., et al. 2001. *Cell* 107:477; Darnell, J. C., et al. 2001. *Cell* 107:489; Dombrowski, C., et al. 2002. *Hum. Mol. Genet.* 11:371.)

fragile X syndrome *See* fragile X chromosome.

fragment Gap-free, contiguous tract of the genome without any alien insert

fragmentin-1 Cytotoxic serine protease; it can trigger apoptosis in combination with perforin. *See* apoptosis; granzyme; ICE; perforin; RNKP-1. (Jans, D. A., et al. 1998. *J. Cell. Sci.* 111:2645.)

fragmentin-2 (granzyme B) Cytotoxic T cells and natural killer lymphocytes destroy their targets with the cooperation of perforin and fragmentins. *See* fragmentin-1; T cells. (Jans, D. A., et al. 1996. *J. Biol. Chem.* 271:30781.)

fragment recovery probability Indicates the number of genomic fragments to be screened in order to recover a desirable one with a chosen probability. Probability = $1 - (1 - f)^n$, where f = the size of an average fragment divided by the size of the genome and n = the required number of fragments to be cloned. Example: $P = 0.95$, average fragment size 1×10^6 Da, and the size of the genome is 2.6×10^9 Da. Then

$$n = \ln(1 - P)/\ln(1 - f)$$
$$= \ln(0.05)/\ln[1 - (1000000/2600000000)] \approx 7787$$

i.e., about 7,787 clones will include the wanted one by a probability of 95%. *See* DNA library; restriction enzymes.

fragrance Occurs in all types of organisms and serves various adaptive purposes. Among animals, the pheromones are a means of communication and are used both as attractants and repellents. In plants, the fragrances may be the means of aiding pollination or dispersal but are also insect repellents. Chemically the fragrances are diverse. Many plant fragrances are monoterpenes such as citral, thymole (in thyme), linaleol, and 1,8-cineol (in levander). In *Menthas*, carvon, piperitones, menthone, and menthole terpene rings represent the scents. In *Eucalyptus*, species geraniols and cineols occur. In the majority of the species of plants, the fragrances represent chemical complexes. Their inheritance is usually complex, and frequently in the hybrids the parental fragrances are hard to recover. By inserting the linaeol synthase gene into snapdragon, bergamot scent was expressed. *See* olfactogenetics; pheromones. (Vainstein, A., et al. 2001. *Plant Physiol.* 127:1383.)

frameshift, translational *See* overlapping genes.

frameshift mutation Insertion or deletion of bases changing the reading frame of the code words, leading to new amino acid sequences from the site toward the carboxyl end of the polypeptide. If one or two bases are either lost or gained, the genetic message from that site on is generally garbled, whereas if the loss or gain involves triplets, there is a possibility to continue reading in a normal manner. Frameshift mutations are frequently caused by acridine dyes and cross-linking mutagens. The discovery of frameshift mutagens contributes to the recognition that the genetic code relies on nucleotide triplets. Frameshift mutation can be represented by the following:

$$\downarrow$$

JOE AND BOB ATE THE BIG HOT DOG AND DID NOT SIP ICE TEA
JOE AND BOB ATE THE BIG HOT DOG AN**D IDN OTS IPI CET** EA

$$\Downarrow$$

JOE AND BOB ATE THE BIG HOT DOG AND **IDN OTS** SIP ICE TEA

Normal text: deletion ↓ and shift an addition ⇓ restores the meaning behind it.

See acridine dye; base substitution; codon; cross-linking; deletion mutation; duplication; genetic code; insertional mutation; mutation; reading frame; SNIPS. (Crick, F. H. C. & Brenner, S. J. 1967. *J. Mol. Biol.* 26:361; Farabaugh, P. J. 1996. *Annu. Rev. Genet.* 30:507; Hoffmann, G. R., et al. 2001. *Mutation Res.* 493:127.)

frameshift suppressor Generally insertion(s) or deletion(s) of nucleotides capable of restoring the normal reading frame within the gene. *See* frameshift mutation; reading frame; suppressor mutation.

framework amino acids Of antibodies, secure the scaffolding of the hypervariable region but not involving the CDR sequence. *See* antibody; CDR; immunoglobulins. (Holmes, M. A., et al. 2001. *J. Immunol.* 167:296; Jung, S., et al. 2001. *J. Mol. Biol.* 309:701.)

framework map Includes several (or only one) collections of genes or DNA sequence groups (RFLP, microsatellites) used to position loci or sequences (STS) relative to these panels. *See* microsatellite; radiation mapping; RFLP; skeletal map; STS.

FRAP (1) TOR, RAFT1. *See* FK506. (2) Fluorescence recovery after photo bleaching. Cells are stained by fluorochromes and bleached locally by laser beams to study the organization of nuclei, diffusion of various proteins, and viscosity within the cell. *See* FCS; FRET. (Lippincott-Schwartz, J., et al. 2001. *Nature Rev. Mol. Cell Biol.* 2:444.)

Frasier syndrome Recessive malformations caused by mutation at the WT gene (11p13) involving facial anomalies (hypertelorism), underdeveloped kidneys, fusion of the labia pudendi (the fleshy borders at the mons pubis of the external female genitalia), enlargement of the clitoris (the female erectile body [homologous to the penis of males]), defective fallopian tubes (connecting the ovaries with the uterus) and ovaries, etc. *See* Denys-Drash syndrome; genital anomaly syndromes; hypertelorism; kidney disease; Wilms' tumor.

frataxin *See* Friedreich ataxia.

fraternal Involving brothers; it is also used for describing dizygotic twins as fraternal twins even when girls are involved (in the latter case, the biologically correct usage should be *sororal twins*, but it is not used). *See* twinning; twins.

FRAX   Fragile X chromosome. *See* fragile X syndrome.

FRAXA   *See* fragile X syndrome.

FRAXE   Fragile X syndrome associated with the long arm of the human X chromosome–based expansion of CCG repeat tracts and hypermethylation of a CpG island, resulting in a rare form of mental retardation. *See* trinucleotide repeats.

Frazzled (*fra*, chromosome 2 of *Drosophila*)   Netrin receptor expressed all over the embryo and distributing netrin in a pattern different from the place of expression of netrin. *See* netrin.

freckles   Either ephelides, which are small, pigmented spots usually occurring on young, fair-skinned redheads or blonds; they usually show up less conspicuously with aging. Solar lentigines are due to spotty photodamage and an increase in frequency with age (age spots). Stimulation of the melanocortin-1 receptor (MC1R, 16q24.3) by α-melanocyte-stimulating hormone and proopiomelanocortin peptides may lead to increased synthesis of black eumelanin instead of red pheomelanin. The latter may promote the synthesis of free radicals in response to ultraviolet exposure and increase the risk of melanoma. There is substantial variation among MC1R alleles and susceptibility to melanoma. Freckles may also be due to several other genes. *See* lentigine; melanocortin; melanocyte; melanoma; opiocortin; xeroderma pigmentosum. (Bastiaens, M., et al. 2001. *Hum. Mol. Genet.* 10:1701.)

free energy (G)   Energy that can be obtained from a system for other purposes. *See* entropy.

Freeman-Sheldon syndrome (arthrogryposis)   Characterized by deformities of the limbs and other structures. One form has been mapped to human chromosome 5.5-pter-15p1.

freemartin   Somewhat masculinized sterile bovine (cattle, sheep, goat, pig, etc.) female born as a twin with a male. The sterility is attributed to the circulation of blood containing male-specific antigens and hormones. Freemartins do not occur in humans, although women treated with male hormones to prevent miscarriage have been reported to deliver female babies that after puberty may have shown some secondary virile characteristics. The exact origin of the term is unclear. In old English, a spayed heifer (neutered bovine female) was called martin. Also, St. Martin has been regarded as a protector of rogues (off-type creatures). In Scottish, ferrycow means a cow (temporarily) barren. *See* hormones in sex determination; puberty; spaying. (Ennis, S., et al. 1999. *Res. Vet. Sci.* 67:111; Kobayashi, J., et al. 1998. *Mol. Reprod. Dev.* 51[4]:390; Vigier, B., et al. 1988. *Reprod. Nutr. Dev.* 28[4B]:1113.)

free radical   Atom or group of atoms with an unpaired electron; therefore, it is extremely reactive. Free radicals may be produced by exposure of wet tissues to ionizing radiation, thus leading to physiological and genetic damage of the cells.

freeze drying (lyophilization)   Procedure for the preservation of biological specimens, bacteria, and enzymes frozen at about −50° and dehydrating under high vacuum in specially constructed equipment. The preserved samples are usually sealed in glass under vacuum for further storage. Generally, the activity of the enzymes is well maintained and the bacterial cells can be revived even after years of storage. Animal sperm may be preserved this way, stored at room temperature, and used for fertilization by intracytoplasmic injection into eggs.

freeze etching   Different from freeze fracture inasmuch as it allows the electron microscopic study of membrane surfaces rather than internal structures. The specimens are frozen in liquid nitrogen, the material is cracked, and the water is removed by sublimation in a freeze dryer. The etched parts are shadowed and viewed in the electron microscope. An improved version of this procedure involves *rapid freezing* with a copper block (−269°C, liquid helium) after the specimen is slammed against it and then lyophilized. This way the internal cell parts and filaments can be well visualized. *See* electron microscopy; freeze drying; freeze etching; membranes.

freeze fracture   Technique for preparation of membrane-containing specimens for electron microscopic examinations. The specimen is frozen in liquid nitrogen under the protection of antifreeze (cryoprotectant) to prevent ice crystal formation and concomitant distortion. After cracking the frozen blocks, some surfaces of the broken pieces expose the interior of cellular membrane bilayers. The membrane faces are then shadowed with platinum; after the organic material is removed, it can be viewed by electron microscopy. *See* electron microscopy; membranes; shadowing.

French pressure cell   Used for breaking up cell suspensions in combination with a hydraulic press at ~20,000 lb/inch^2 pressure.

frequency-dependent selection   *See* apostatic selection; selection types.

frequency distribution   Representation of a population in classes according to the frequency of individuals in each class. *See* negative binomial; normal distribution; Poisson distribution.

FRET   Fluorescent resonance energy transfer. A fluorophore donor molecule, which has an absorption maximum at a shorter wavelength, can be excited and then can transfer the energy of an adsorbed photon nonradioactively to an acceptor molecule that has an excitation maximum at a longer wavelength. The distance over which FRET can be measured is about 40 to 100 Å, and in general it depends on one-sixth power of the distance, but it is modified by several factors. This technology enables visual monitoring of protein interactions. *See* FCS; fluorochromes; FRAP; two-hybrid system. (Periasamy, A. & Day, R. N. 1999. *Methods Cell Biol.* 58:293.)

Freund adjuvant Water-light mineral oil emulsion containing an emulsifier and an antigen; sometimes, dry, dead *Mycobacterium butyricum* is also added. This preparation boosts the immune reaction in case of weak or small amounts of the antigen. *See* antigen.

Friedreich ataxia (FRDA) With optic nerve atrophy and deafness, it is an autosomal-dominant disease at 6p23. An autosomal-recessive form (9q13) is a rare ($\sim 2 \times 10^{-5}$) brain–spinal chord degenerative malfunction characterized by hypoactive knee and ankle jerks, poor coordination of the limbs, spasms, etc. The heterozygosity for the recessive form is as high as $\sim 10^{-2}$. Most of the cases are point mutations in the gene encoding the 210-amino-acid frataxin protein, but this condition frequently involves GAA repeats in the first exons/introns. The normal range of the FRDA repeats is 7–22, but with disease it increases to 200–900 or more. The increase of the trinucleotide repeats interferes with transcription. The DNA forms a triplex structure, which leads to increased mutability. The defect is concerned with a phosphatidylinositol-4-phosphate kinase and mitochondrial iron homeostasis. The critical protein has been located to the mitochondria, where it increases the concentration of iron and decreases respiration. The accumulated iron reacts with H_2O_2, resulting in lesions to proteins, lipids, and mtDNA. *See* ataxia telangiectasia; AVED; epilepsy; fragile sites; phosphatidylinositol; trinucleotide repeats. (Bradley, J. L., et al. 2000. *Hum. Mol. Genet.* 9:275; Patel, P. I. & Isaya, G. 2001. *Am. J. Hum. Genet.* 69:15; Salkamoto, N., et al. 2001. *J. Biol. Chem.* 276:27171; Puccio, H. & Koenig, M. 2002. *Current Op. Genet. Dev.* 12:272.)

Friend murine leukemia Caused by the replication-competent helper virus, F-MuLV (Friend murine leukemia RNA virus), and the replication-defective spleen focus-forming virus (SFFV). Pathogenicity depends on the chimeric SFFV envelope protein, gp55, in the presence of the erythropoietin receptor. Susceptibility is controlled by several Fv (Friend virus) loci. *See* FMS oncogene; leukemias.

frizzled *See Wingless.* (Strutt, D. I. 2000. *Mol. Cell* 7:367.)

frog *Rana pipiens* $2n = 26$, *Rana temporaria* $2n = 26$. *R. esculenta* and *R. catesbeiana* are considered delicacy foods, but some of the toads secrete poisonous or irritating substances as a defense. *See* toad. (<vize222.zo.utexas.edu/frog.html>.)

frond Leaf-like thallus of lichens or leaves of ferns.

frontotemporal dementia and parkinsonism *See* Pick disease.

FRP Human phosphatidylinositol kinase. *See* PIK.

FRS Fos-regulating kinase. *See* FOS.

fructose Monoketo-hexose; present in the disaccharide saccharose. It plays a key role in metabolism through phosphorylated derivatives (fructose-1-phosphate, fructose-6-phosphate, fructose-1,6-bisphosphate, etc.). Fructose utilization is not impaired in diabetes. Fructose is about twice as sweet-tasting as glucose, and its use may reduce caloric intake. The corn sweeteners contain fructose industrially produced from starch. Fructose and fructose-containing food and beverages, especially at acid pH and heating or just by long storage, may liberate furans, furaldehyde, and levulinic acid, which may be toxic. The photo displays the breakdown products of fructose autoclaved for 20 minutes: HMF = hydroxymethyl furfural, HAF = hydroxyacetyl furan. Furfural is actually an insecticide. Levulinic acid LD_{50} intraperitonially is 450 mg/kg for mice. *See* aspartame; fructose intolerance; fructosuria; saccharin. (Rédei, G. P. 1974. *Annals Bot.* 38:287.)

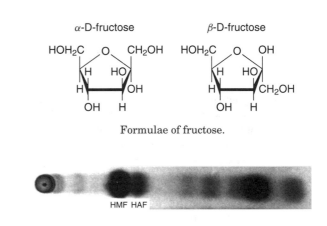

Formulae of fructose.

fructose-2,6-bisphosphatase Breaks down fructose-2,6-bisphosphate.

fructose intolerance (hereditary fructose intolerance) Human chromosome 9q22.3 recessive disorder caused by a deficiency of the enzyme fructose-1-phosphate aldolase. The patients begin sweating, trembling, feeling dizzy and nauseous 20 minutes after ingesting fructose. The immediate clinical findings are fructosuria, hypophosphatemia (abnormally low amounts of phosphate in the blood), aminoaciduria (amino acids in urine), fructosuria (moderate amounts of fructose in urine), hyperbilirubinemia (excess of bilirubin [red bile pigment] in blood), etc. The chronic symptoms include jaundice, enlargement of the liver (hepatomegaly), vomiting, dehydration, edema (excessive fluid in the tissues), ascites (fluids in the abdominal cavity), seizures, fructose accumulation in the urine (fructosuria) and in the blood (fructosemia), cirrhosis (destruction of cells and increase of connective tissues) in the liver, etc. On a diet low in fruits, honey, and fructose-containing sweeteners, patients may be quite normal, but consuming fructose may make them sick. Infants may even die if their formula contains fructose. The breakdown products (mainly furfural) of autoclaving cause the apparent toxic effect of fructose in plant cell cultures. Bacteria deficient in the phosphoenolpyruvate/glycose phosphotransferase system are unable to utilize fructose. *See* aldolase; fructosuria. (Santamaria, R., et al. 2001. *Biochem. J.* 359[pt3]:823.)

fructosuria (essential fructosuria, 2p23.3-p23.2) Rare autosomal disorder (prevalence is about 8×10^{-6}). The biochemical basis of this nondebilitating anomaly is a deficiency of fructokinase.

fruit Mature ovary of plants (may also include other parts of the flower); with the exception of the seed within, it is genetically maternal tissue. *See* flower differentiation; gametogenesis in plants.

fruit fly *Drosophila melanogaster* $2n = 8$; *D. ananssae* $2n = 10$; *D. melanica* $2n = 10$; *D. obscura* $2n = 10$; *D. pseudoobscura* $2n = 10$; *D. virilis* $2n = 10$; *D. willistonii* $2n = 6$. *See* Drosophila; karyotype evolution.

fruiting body Collective name of fungal organs (perithecium, cleistothecium, apothecium, locule) containing the haploid reproductive spores.

fruit ripening Consequence of changes in the composition and softening of the cell walls. The increase in respiration involves an increase of the production of the plant hormone ethylene, which affects the expression of a number of genes, notably polygalacturonase (PG), 1-aminocyclopropane 1-carboxylic acid synthase (ACCS), and 1-aminocyclopropane 1-carboxylic acid oxidase (ACCO; tomato gene pTOM13). By transforming tomatoes with a single PG antisense construct, PG activity could be reduced to 1%, but for a more efficient control of ripening, under polygenic control, the amount of the other enzymes is also reduced. For commercially effective storage without softening ripe-harvested tomato fruits, the use of antisense RNA of the LE-ACS2 and LE-ACS4 loci is required. The berries stored at 20°C remain firm for months but could be fully mature when exposed to ethylene (C_2H_4) or its analog (C_3H_6). The fruits handled this way are practically indistinguishable by color, scent, and consistency from vine-ripened fresh fruits. *See* antisense technology; ethylene; plant hormones. (Ferrándiz, C., et al. 1999. *Annu. Rev. Biochem.* 68:321; Giovannoni, J. 2001. *Annu. Rev. Plant Physiol. Plant Mol. Biol.* 52:725.)

frustration Interference by the phagocytotic activity of opsonins. *See* opsonins.

Frye test Legal ruling concerning admissible scientific evidence to the court. According to *Frye* v. *United States* (D. C. Cir. 1923, 293 Fed. 1013), new scientific methods must be generally accepted by the scientific community before evidence from such methods is admissible to the courts. This became a highly controversial issue because "general acceptance" is difficult to define. Criminal defense lawyers have frequently argued that electrophoretic pattern of proteins and DNA fingerprints and their statistical evaluation should not be presented to the jury because some scientists may dissent about some aspects of the data. *See* ceiling principle; Daubert rule; DNA fingerprinting; forensic genetics.

FrzE *Myxococcus xanthus* kinase affects FrzE and FrzG proteins regulating bacterial motility and development. (McBride, M. J. 2001. *Annu. Rev. Microbiol.* 55:49.)

F₂ segregation *See* Mendelian segregation.

FSH (follicle-stimulating hormone or follitropin, 2p21-p16) Controls ovarian follicles, estrogen secretion, menstrual cycles, spermatogenesis, and D2 cyclin. Ovarian and testicular tumors have high levels of cyclin D mRNA. FSH-deficient males have small testes yet are fertile to a variable degree. The 54 kb gene has 10 exons and encodes the 695-amino-acid protein. *See* animal hormone; cell cycle; differentiation; follicle; gametogenesis; gonadal dysgenesis; menstruation; NGFI-A; ovary. (Driancourt, M. A. 2001. *Theriogenology* 55:1211.)

F$_{St}$ Index for population diversity. For human autosomal loci, it is ~15–20%; for mtDNA ~19%; and for the Y chromosome ~65%. The complement of these percents approximately indicates the variations shared by any random population and the rest of the human race. F_{ST} for diploids $= 1/(1 + 4N\nu)$ and for Y and mitochondrial (haploid) systems $F_{ST} = 1/(1 + N\nu)$, where N is the effective population size (N_e), $\nu =$ the sum of migration and mutation (or more precisely, $m + \mu - m\mu = \nu$). *See* coalescent; diversity; effective population size; Eve foremother; G$_{ST}$; migration; mutation rate; R$_{ST}$; Y chromosome.

F-statistics *See* F.

F-table *See* F distribution.

FTDP-17 *See* Pick disease; tau.

F-test *See* F distribution.

FTICR Fourier-transformed ion cyclotron resonance mass spectrometer is used for the determination of the molecular masses of hundreds of proteins in one run. *See* MALDI.

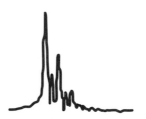

Mass spectrometric profile of proteins.

FT-IR Fourier transform infrared spectroscopy is a very sensitive method for the detection of intramolecular changes as function of time is converted into function of angular frequencies. *See* electron density map; Raman spectroscopy; tumorigenesis.

FTMS Fourier transform mass spectrometry is suitable for the identification of the mass of intact proteins. This type of analysis is particularly useful because it can obviate bias in mass due to alterations during the purification, e.g., oxidation of methionine residues. *See* proteomics. (Jensen, P. K., et al. 1999. *Anal. Chem.* 71:2076.)

FtsK Mediates coupling bacterial chromosome segregation with cell division.

FtsY Prokaryotic transport chaperone protein, related to mammalian SRP54. The related bacterial Ffh is ribonucleoprotein binding to FtsY in a GTP-dependent manner. *See* GTP; SRP. (Millman, J. S., et al. 2001. *J. Biol. Chem.* 276:25982.)

FtsZ *See* tubulins.

fuchsia The ornamental plant what Gregor Mendel was examining while photographed in 1862 with his monasterial colleagues. This plant has $2n$ chromosome numbers 22, 56, 66, and 77, and it was one of Mendel's fortunes that he did not experiment further with this erratic material, which would not have permitted the type of studies he conducted with the stable peas.

fuchsin (triaminotrimethylmethane) Red cytological stain.

fucose 6-deoxy sugar. It is present in several antigenic glycoproteins. It may be associated with several immunoglobulins and is present in plant cell walls.

fucosidase (FUCA2, 6q25-qter) Regulator of α-fucose in plasma and fibroblasts. It hydrolyzes fucose linkages in glycoproteins and glycosphingolipids. *See* sphingolipids.

fucosidosis (FUCA1) Recessive human chromosome 1p34 (and a pseudogene at 2q31) defect of α-fucosidase, causing neurodegeneration and angiokeratoma, although some forms are less detrimental. It is a lysosomal storage disease. The accumulation of fucose is detectable in amniotic fluid. *See* amniocentesis; angiokeratoma; lysosomal storage diseases.

fucosyltransferase (FUT2, FUT1, 19q13.3) In its deficiency the H blood antigen is absent. There are at least seven other FUT genes in the human genome scattered among the chromosomes. *See* ABH antigen; Bombay blood group; CD15; Secretor.

full sib Brothers/sisters with identical mother and father. *See* sib.

fumagillin ($C_{26}H_{34}O_7$) Fungal antibiotic and inhibitor of angiogenesis. It inhibits methionine aminopeptidase-2, a metalloenzyme cleaving methionine from the N end of proteins. *See* angiogenesis. (Owa, T., et al. 2001. *Curr. Med. Chem.* 8:1487.)

fumarate hydratase (FH) Encoded in human chromosome 1q42.1, but both cytoplasmic and mitochondrial forms of the enzyme exist, probably due to alternative processing of the transcript. Its deficiency results in mental and physical impairment. *See* mitochondrial diseases in humans.

Funaria hygrometrica (Bryophyte) Chromosome numbers vary: 14, 28, 56.

fun gene (functions unknown gene) *See* orphan genes; orphan receptor.

functional cloning In the first step, the protein encoded by the gene is isolated, its amino acid sequence is determined, and synthetic probes are generated on that basis. With the aid of the probe, the gene is fished out from a DNA library. *See* gene isolation; synthetic probe.

functional genomics Studies the function of all open reading frames revealed by DNA genome-wide sequencing. For this purpose, transposable elements, chemical and physical mutagenesis, and knockouts are being used. All of the techniques are currently not applicable to all organisms. In *Drosophila*, insertional mutations (using P elements) were accomplished for >25% of the vital genes by 1999 and ~85% is expected. *See* genome annotation; genomics; hybrid dysgenesis; insertional mutation; knockout. (Spradling, A. C., et al. 1999. *Genetics* 153:135.)

functionality of mutagens Number of chemical groups reacting in mutagenesis. *See* nitrogen mustard; sulphur mustard.

functional network Displays the functional relations among several (many) proteins based on *rosetta stone sequences*, *phylogenetic profiles*, and if applicable *gene neighbor methods*. Eventually the oversimplified relations of protein φ to other numbered proteins can be represented as shown by the chart below or a web.

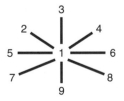

Functional networks.

functional redundancy According to this hypothesis, functions may be carried out by gene products with highly similar activity, but the genes as members of separate regulatory circuits would make the system more flexible than a combinatorial control of development. *See* combinatorial gene control.

function-valued trait Can best be described by a function of some variable(s), e.g., by the phenotypic changes during aging.

fundamentalism Religious beliefs that regard as the only absolute right what is written in the holy books (Bible, Koran) or in the works of the basic idealogues (e.g., Marx, Engels,

Lenin). Accordingly, only creationism properly explains the genesis of life and provides specific guidelines to human ethics and ideology or political or economical theory, respectively. *See* creationism; lysenkoism. (Bouchard, T. J., et al. 1999. *Twin Res.* 2[2]:88.)

fundamental theorem of natural selection *See* natural selection.

fundus flavimaculatus *See* ABC transporters.

fungal disease, human Can be due to primary infectious agents or to opportunistic fungi. Infection and pathogenicity are affected by a number of factors of the pathogen and the host. (van Burik, J.-A. H. & Magee, P. T. 2001. *Annu. Rev. Microbiol.* 55:743.)

fungal incompatibility Based on interaction of the products of nonallelic genes, thus preventing self-fertilization, similar to the outcome of S alleles of plants. In *Ustilago maydis* (a pathogen of maize), stable dikaryons can be formed only between different mating-type alleles of a multiallelic *b* locus recognized by pheromones. The same *b* locus is responsible for plant pathogenicity. The *bE* and *bW* alleles encode different homeodomain proteins. The E and W variants differ primarily in the N-terminal amino acids. Activity requires that the E and W allele products dimerize, which can happen only in appropriate allelic combinations; the majority of over 300 combinations are active. In other fungi, several multiallelic gene pairs coding for interacting homeodomains have been discovered; in yeast, only two mating types, *a* and *α*, exist. *Vegetative incompatibility* prevents the fusion of hyphae in case of *het* alleles, or nonallelic *het* genes are carried in their nuclei. *See* dikaryon; fungal life cycle; heterokaryon incompatibility; incompatibility, vegetative. (Glass, N. L., et al. 2000. *Annu. Rev. Genet.* 34:165.)

fungal life cycle Displays an enormous variety of specializations in the various taxonomic groups and cannot be

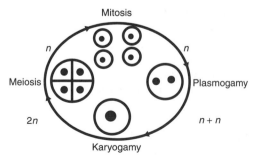

The approximately 2,000 genera of fungi have a variety of modes of reproductions that share some basic similarities. Between 9 o'clock and 3 o'clock are the haploid phases (*n*). Haploid cells fuse at 3 o'clock (plasmogamy), but the nuclei are still separate and a dikaryon (*n + n*) is formed. In a following step, at 6 o'clock, nuclear fusion takes place (karyogamy) and the cell becomes diploid (*2n*). This is followed by meiosis (9 o'clock) yielding four haploid sexual spores that may divide again by mitosis (12 o'clock) before the spores are released. These sexual spores may differentiate and the cycle will be reinitiated.

represented here. The general scheme is, however, relatively simple and shared by all fungi. *See* hypha.

fungi imperfecti Fungi without a known sexual reproduction mechanism. (Taylor, J. W., et al. 1999. *Annu. Rev. Phytopath.* 37:197.)

fungus (plural fungi) Eukaryotic thallophytes yet separate subkingdom from plants and bacteria. Includes many saprophytic, parasitic, and pathogenic species of enormous variety in structure and function. Fungal genetics provided and is providing an understanding for the basic genetic phenomena such as recombination, biochemical pathways, cell cycle, etc. The yeasts and other ascomycetes are among the most important tools of modern genetic research.

funiculus Vascular stalk of the plant ovule, a cord-like structure, and in animals it includes the umbilical cord, etc.

fur color Of animals, is determined by the melanin pigments formed in the melanocytes and the migration of the melanoblasts. The various genes involved then modify, reduce, or intensify pigmentation. In addition, the actual visible color depends on the dorsoventral distribution of the pheomelanin and eumelanin pigments. Superimposed on these are species- or genotype-specific striping, spotting, lyonization, temperature sensitivity, etc. *See* agouti; hair color; melanin; pigmentation of animals.

furin Golgi-associated proteinase; it may activate stromelysin and virulence factors of pathogens. Furin has an important role in antigen processing for presentation to the cytotoxic T lymphocytes. Furin processes the BRI-L and BRI-D fibrillogenic precursors of dementia peptides. *See* antigen processing and presentation; dementia; Golgi; stromelysin; T cell. (Gil-Torregosa, B. C., et al. 1998. *J. Exp. Med.* 188:1105; Bassi, D. E., et al. 2001. *Proc. Natl. Acad. Sci. USA* 98:10326; Thomas G. 2002. *Nature Rev. Mol. Cell Biol.* 3:753.)

FUS3 Protein kinase of the MAPK family. Indirectly it arrests yeast cells in G1 prior to mating and it downregulates Ty transposition. *Fus*$^{-/-}$ mutants of mice develop B-cell defects, chromosomal instability, and perinatal death. *See* Kss; MAPK; signal transduction; STE; Ty. (Cherkasova, V. & Elion, E. A. 2001. *Curr. Genet.* 40:13.)

fushi tarazu *(ftz)* *Drosophila* mutation of the pair-rule class; every other body segment is missing at the blastoderm stage. Photo displays the seven wild-type segments encoded by the gene. *See* morphogenesis in *Drosophila*; pair rule genes; SF-1.

Courtesy of Dr. Y. Nishida, et al. 1999. (*Genetics* 153:763.)

fusicoccin Toxin of the fungus *Fusicoccus amygdali*; an activator of plasma membrane H$^+$ ATP-ases, causing H$^+$ secretion and K$^+$ influx into guard cells, thus the opening of plant stomata. (Meinhard, M. & Schnabl, H. 2001. *Plant Sci.* 160:635.)

fusidic acid (C$_{31}$ H$_{48}$ O$_6$) Antibiotic (ramycin) isolated from *Fusidium coccineum*. This compound mimics the effect of *rel$^-$* (relaxed control) mutations and permits the synthesis of guanosine polyphosphates (ppGpp, pppGpp), ribosomal RNA, and ribosomal protein. Tetracycline has a similar effect. *See* antibiotics; relaxed control; stringent control. (Duvold, T., et al. 2001. *J. Med. Chem.* 44:3125.)

fusigenic liposome Similar to the liposome vectors except that fusigenic liposomes are engineered to carry hemagglutinating neuroaminidase (HNA) and a fusion protein on their surface. The hemagglutinating Japanese virus (HVJ) and the Sendai virus produce these proteins and make them capable of fusing with the cell membrane at neutral pH. HNA is required for binding to cell receptors containing sialoglycoproteins or sialolipids. The fusion protein is in an inactive form until it is hydrolyzed to two polypeptides, F1 and F2; F1 interacts with cholesterol to facilitate fusion, then the DNA carried by the liposome is delivered into the cell. *See* cholesterol; cytofectin; hemagglutinin; liposome; Sendai virus; sialic acid. (Kono, K., et al. 2001. *Gene Ther.* 8:5.)

fusin (LESTR, HUMSTR, CXCR4, stromal-derived factor [SDF1]) Coreceptor of the CD4 antigens required for fusion with the membrane and entry of a virus (HIV) into a cell. It is a heterotrimeric GTP-binding protein. *See* acquired immunodeficiency syndrome; CD4; CD8; HIV; MIP; RANTES. (Dragic, T., 2001. *J. Gen. Virol.* 82[p8]:1807.)

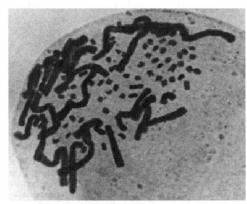

Soybean (short) and vetch (long) chromosomes in fused somatic cells. Courtesy of Dr. O. L Gamborg, et al.

fusion of somatic cells *See* cell fusion.

fusion protein Synthesized when neighboring genes are transcribed and translated together. It contains full or incomplete parts of two normal proteins. Also, a group of proteins mediating membrane fusion in cells. Fusion of ligands (e.g., transferrin) for lymphocyte cell surface receptors with particular immunoglobulins (IgG3) may potentiate the delivery of an antibody beyond certain (e.g., brain) barriers to the blood. Antibody-cytokine fusion may be useful in targeting tumors for destruction. *See* cancer; chromosome breakage; ferritin; gene fusion; immunoglobulins; leukemias; read-through proteins; transcriptional gene fusion vectors; translational gene fusion vectors.

fusome Structure that is formed in the germline cell of insects during mitotic divisions. It anchors the mitotic divisions, leading to the formation of the nurse cells and the oocyte. After the divisions are completed, it fades away. Antibodies to spectrin, a filamentous membrane protein, can detect its existence. *See* morphogenesis; nurse cell; spectrin. (Grieder, N. C., et al. 2000. *Development* 127:4253.)

futile cycle Apparently useless chemical reaction in the cell, e.g., fructose-6-phosphate is phosphorylated to fructose diphosphate and simultaneously it is hydrolyzed back to fructose-6-phosphate, resulting in cleavage of ATP into ADP and Pi (unnecessary ATPase activity). *See* substrate cycle.

fuzzy inheritance Statistical term used for cases when linkage information is computed by allele sets and set recoding. (O'Connell, J. R. & Weeks, D. E. 1995. *Nat. Genet.* 11:402[1995].)

fuzzy logic Either one or another different event can occur, e.g., frameshift or in-frame decoding of mRNA, or the translation is either terminated or read through. *See* pseudoknot; recoding. (Zimmermann, H. J. 1996 Fuzzy Set Theory and its Applications, Kluwer, Boston, MA.)

Fv Fragment variable is a functional antibody molecule composed of light- and heavy-chain antigen-binding sites of one region of the antibody separated from the rest by proteolysis. The single-chain (scFvs) of the light (V$_L$) and heavy chains (V$_H$) can be engineered to bind together by 8–15 amino acid linkages and can be cloned by various vectors. However, these engineered antibodies have somewhat reduced immunogenicity. These molecules may have applicability for radiolabeling and imaging of tumors and for the generation of immunotoxins. *See* antibody; antibody engineering.

Fv protein Homodimeric 175 kDa sialoprotein binding to the variable domains of immunoglobulin heavy chains. *See* immunoglobulins; sialic acid.

φX174 Single-stranded DNA bacteriophage with overlapping genes. Its genome was completely sequenced by 1975. Because of the simplicity of its genome, it has been used for many purposes in genetics. Mice cell lines trans-

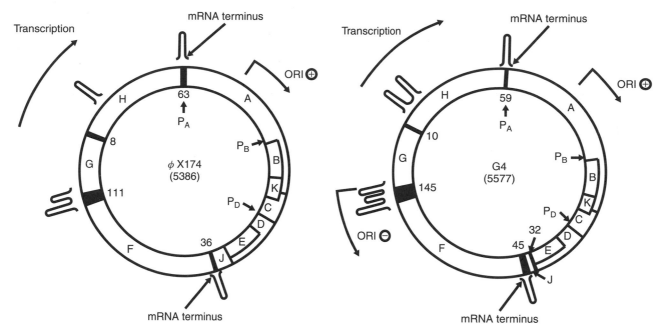

The physical map of bacteriophage φX174 with 10 genes A to H. Although some of the genes overlap each other (or included), there are nongenic (intergenic) sequences (shown in solid black). The related G4 phage is similar. (Courtesy of Godson, N. G. 1980. *Stadler Symp.* 12:143.)

genic for φ174 have been promising to study chemically induced forward mutation. (*See* bacteriophages; diagram above). (Valentine, C. R., et al. 2002. *Environ. Mol. Mutagen.* 39:55.)

Fyb   *See* SLAP. (Griffith, E. K., et al. 2001. *Science* 293:2260.)

FYN Oncogene is a member of the Src nonreceptor protein tyrosine kinase gene family. The protein controls memory and learning. Low levels of Fyn result in greater sensitivity to alcohol effects. *See* Csk; protein tyrosine kinase; Src; (Arold, S. T., et al. 2001. *J. Biol. Chem.* 276:17199.)

Johann Dzierzon, a Silesian priest and apiculturist, wrote the following comments in a regional technical magazine 11 years before the oral presentation of Mendel's "Experiments on Plant Hybridization" at a meeting of the Brünn Society of Naturalists:

"It must be determined that the queen belongs to a pure breed, because if she originated from a hybrid brood, she could not produce pure drones at all but only half Italian and half German ones, although strangely not by type but in numbers; apparently it is difficult for nature to blend the two species into an intermediate form" (*Der Bienenfreund aus Schlesien*, 1854).

A historical vignette.

G

g General intelligence. *See* human intelligence; intelligence quotient.

G G is used to denote generations after mutagenic treatment of mice: G_0, G_1, G_2, etc. This designation is somewhat confusing with the preempted cell cycle symbols. *See* cell cycle; guanine.

g^2 *See* genetic determination.

G418 ($C_{20}H_{40}N_4O_{10}.2H_2SO_4$) Aminoglycoside antibiotic. *See* geneticin.

G3139 18-mer full-phosphorothioate deoxyoligonucleotide (5'-TCTCCCAGCGTGCGCCAT-3' *Genta*, San Diego, CA) with sequence antisense to the first six codons of the open reading of gene BCL-2; it is used for therapy of lymphomas. *See* antisense technologies; Bcl-1; lymphoma; phosphorothioate.

G3854 20-mer full-phosphorothioate deoxyoligonucleotide with sequence antisense to open reading frame of gene BCL-2; it is used for therapy of lymphomas. It is similar to G3139 but 2 nucleotides longer. *See* antisense technologies; Bcl-1; lymphoma; phosphorothioate.

GA *See* gibberellic acid; plant hormones.

GABA (γ-aminobutyric acid) Plays an important role in neurotransmission of vertebrates and invertebrates. In the nematode *Caenorhabditis*, a series of *unc* (uncoordinated movement) genes respond to GABAergic neuronal effects. $GABA_A$ receptors mediate synaptic inhibition, but upon intense activation they may excite rather than inhibit neurons. GABA controls Cl^- ion channels by efflux (depolarization and excitation of the nerve cell) and by influx (hyperpolarization and reduced excitability). By an increase of the level of $GABA_A$ receptor, $\alpha 4$ premenstrual anxiety and susceptibility to seizures decrease. Progesterone also acts as a sedative by enhancing GABA function. Heteromeric GABA receptors ($GABA_B$R1a/b–$GABA_B$R2), in cooperation with G proteins, regulate potassium and calcium ion channels. The 19th-century alcoholic beverage, absinthe, containing wormwood oil (thujone), is antagonistic to the $GABA_A$ receptor channels, which explains its convulsant and other neurological effects. GABA transaminase (16p13.3) is responsible for the catabolism of GABA, and its deficiency leads to neuronal disorders. There are at least 13 GABA receptors encoded in different human chromosomes. *See Caenorhabditis*; cleft palate; epilepsy; glutamate decarboxylase deficiency disease; ion channels; neuron; neurotransmitter. (Ganguly, K., et al. 2001. *Cell* 105:521.)

GABA transaminase (16p13.3) *See* GABA aminobutyrate transaminase.

GABP GAA sequence (or their extension) binding heterotetrameric DNA-binding proteins (GABPα/β), members of

ETS domain protein families (about 40) regulating gene transcription in a combinatorial manner with other proteins. The α- subunit actually binds DNA, whereas the ankyrin repeats in β recruit other protein domains. *See* ankyrin; combinatorial gene control; ETS oncogenes; transcription factors.

GADD45 p53-inducible protein. It also binds PCNA. Deficiency of Gadd45a in mice leads to chromosomal instability, increased radiation sensitivity (cancer), and exencephaly. *See* exencephaly; p53; PCNA. (Takahashi, S., et al. 2001. *Cancer Res.* 61:1187; Kovalsky, O., et al. 2001. *J. Biol. Chem.* 276:39330.)

GADD153 (Growth arrest and DNA damage) Is a cellular enhancer-binding protein mediating stress of growth and differentiation. Under stress it may be activated by phosphorylation of Ser[78] and Ser[81] residues and consequently enhanced transcription and inhibited adipose cell differentiation. It is the same as CHOP. Gadd is activated under various stress conditions. It may act as an oncoprotein by suppressing differentiation, especially in various gene fusions brought about by chromosomal translocations. *See* cancer; chromosome breakage; DNA repair; enhancer. (O'Reilly, M. A., et al. 2000. *Am. J. Physiol. Lung Cell Mol. Physiol.* 278:L552; Jousse, C., et al. 2001. *Nucleic Acids Res.* 29:4341.)

GADS CD3 signaling adaptor that links SLP-76 to LAT. *See* CD3; LAT; SLP-76.

GAF DNA satellite-binding regulatory protein.

gag Group-specific antigen, a viral coat protein. *See* retroviruses.

GAG *See* glycosaminoglycan.

GAGA Multipurpose transcriptional activator binding to the GA/CT sites in the promoter. Its major function may be to rearrange the chromatin to facilitate transcription. GAGA activates chaperones and binds to the promoter of the *Ultrabithorax* and other *Drosophila* genes. *See* heat-shock proteins; heterochromatin; morphogenesis in *Drosophila*; position effect. (Basturia, A., et al. 2001. *Development* 128:2163.)

GAIA theory Organisms contribute to a self-regulating feedback that keeps the environment stable and suitable for life. The global environment (living and nonliving) determines the outcome of natural selection through interactive feedback processes. (Downing, K. & Zvirinsky, P. 1999. *Artif. Life* 5[4]:291.)

gain Practical measure of heritability frequently used by animal breeders. By this criterion, heritability, $h^2 =$ (gain)/(selection differential). See graphical representation at next page.

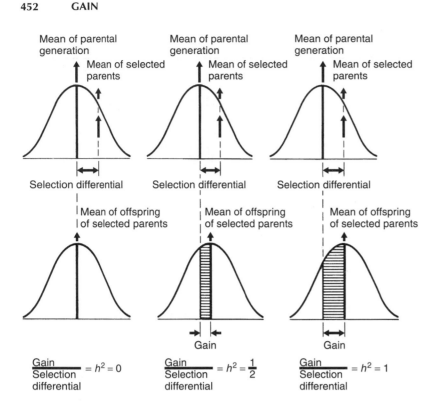

Gain / Selection differential $= h^2 = 0$

Gain / Selection differential $= h^2 = \frac{1}{2}$

Gain / Selection differential $= h^2 = 1$

(After Lerner, I. M. & Libby, W. J. 1976. *Heredity, Evolution and Society*. Freeman, San Francisco.)

The selection differential is the difference between the mean of the parental population and the mean of a portion of the parents selected for further reproduction to improve the herd. The gain/selection differential is frequently called *realized heritability*. The breeder may improve the gain either by increased heritability or by enhanced intensity of selection. Heritability estimates improve if environmental variation is kept at a low level by proper feeding and health care of the animals or appropriate tillage, fertilization, weed and pest control in plants. The intensity of selection is increased if the proportion of the individuals selected for parents is reduced. Although this may appear to be an easy approach to improve selection gains, the small populations may increase inbreeding and become counterproductive. In large mammals, the males generally have more offspring than the females. By the use of artificial insemination, the breeding value of the males can be determined even more precisely than that of the dams. Generally the estimates improve with the age of the animals because a larger number of offspring is available for evaluation. In practice, the selection is aimed simultaneously at several traits. Often these traits are negatively correlated because high performance may make the animals (plants) more susceptible to disease. Thus, the gain in one trait may mean a loss in the others. Therefore, breeders frequently use a *selection index* that weighs each trait by a score, and the total of the scores becomes the basis of the selection value. There are statistical methods for predicting the quantitative performance in a selective breeding program:

$$YO = \overline{Y} + Hn(Yp - \overline{Y}),$$

where YO is the predicted average performance of the progeny, Y_p is the average of the two parental families selected, $\overline{Y} =$ the average of the original population, and Hn is heritability in the narrow sense. Example: The average number of eggs laid per year in a flock of chickens is 250, the heritability is 0.25, the average of the selected family of parents is 274, then the expectation for the offspring $YO = 250 + 0.25(274 - 250) = 256$. The genetic gain from mass selection is computed from the covariance:

$$(XY) = w = \tfrac{1}{2}2\sigma A.$$

For determining the covariance, *see* correlation; $\sigma_A^2 =$ additive variance (*see* genetic variances); the plot-to-plot environmental variance is σ_e^2 and the plant-to-plant environmental variance $= \sigma_{we}^2 = \sigma_{wf}^2 + \sigma_{me}^2$; the genotype environmental variance, $\sigma_{G\times E}^2 = \sigma_{A\times E}^2 + \sigma_{D\times E}^2$.

If we consider the within-family variance $= 0$, then the gain for mass selection,

$$\Delta G_{\mathrm{m}} = \frac{\tfrac{1}{2}i\sigma^2 A}{\sqrt{\sigma_A^2 + \sigma_D^2 + \sigma_{A\times E}^2 + \sigma_{D\times E}^2 + \sigma_e^2 + \sigma_{me}^2}}$$

and $i =$ selection intensity $\sigma_D^2 =$ dominance, $\sigma_{A\times E}^2 =$ additive $\times$ environment, and $\sigma_{D\times E}^2 =$ dominance $\times$ environment variances. This procedure is applicable to large populations. In case of phenotypic recurrent selection, the equation for gain in mass selection for cycle needs to be multiplied by 2 because the selection is applied to both parents. Additional formulas for other types of selection can be found in Moreno-González, J. & Cubero, *J. Plant Breeding*, pp. 281–313, Hayward, M. D., et al., eds. 1993. Chapman & Hall, London, New York. *See* breeding value; correlation; heritability; intraclass correlation; polygenes; quantitative genes; recurrent selection; selection index.

gain-of-function mutation Generally mutations lead to loss of structures, e.g., hairs or bristles or certain function, e.g., auxotrophy. Some of the homeotic mutants, however, gain additional structures such as extra petals or stamens in the flowers or legs on the head in *Drosophila*. These gains are the result of homeotic transdetermination regulated by altered transcription and/or transcript processing. *See* dominant negative; flower differentiation; homeotic genes; Huntington chorea; loss-of-function mutation; muscular dystrophy; processing; transcription; transdetermination.

GAL   *See* galactose utilization.

GAL4   Positive regulatory protein of the yeast galactose genes; it binds to a specific upstream regulatory DNA sequence. *GAL4* is activated by the interaction of Gal80p and Gal3p in the cytoplasm. In various constructs introduced by transformation into other organisms, its activator domain is frequently utilized to boost the expression of selected reporter genes. *See* activator proteins; galactose utilization; Gene-Switch; p65; reporter gene; transcriptional activator. (Hartley, K. O., et al. 2002. *Proc. Natl. Acad. Sci. USA* 99:1377; Peng, G. & Hopper, J. E. 2002. *Proc. Natl. Acad. Sci. USA* 99:8548.)

galactan   Polymer of galactose. *See* galactose; hyperacute reaction.

galactokinase deficiency   May be due to autosomal-recessive defects at GALK1 (human chromosome 17q24) or GALK2 (chr. 15). Cataracts at infancy and hypergalactosemia occur. Galactokinase kinase converts galactose into galactose-1-phoshate. *See* galactose; galactosemias.

galactose   One of the most common six-carbon monosaccharides differing from glucose only sterically at carbon-4 chiral centers (an epimer of glucose). It can be converted to glucose by an epimerase enzyme (UDP-Gal → UDP-Glico). Galactosyl groups are present in some anthocyanins, collagens, and immunoglobulins. Lactose (the milk sugar) is a disaccharide of galactose + glucose split by the enzyme lactase. *See* chirality; epilepsy; epimers; eye diseases; galactose; galactosemias; galactose utilization; galactosidase; genetic screening.

galactose operon   *See* galactose utilization.

galactosemia   Autosomal hereditary disease in humans caused by the deficiency of either the enzyme galactokinase (GALK, human chromosome 17q24) or more commonly galactose-1-phosphate uridyltransferase (GALT, 9p13). As a consequence, galactose cannot be transformed into glucose. Since the milk sugar is a disaccharide of galactose and glucose, galactose accumulates in the blood and is excreted in the urine. The accumulating galactose causes severe intestinal problems and the accumulating galactose-1-phosphate may damage the liver, brain, eye lens (cataracts), and other organs. Unless this anomaly is detected right after birth, infant death may result. By a diet free of any source of galactose, damage may be prevented. This condition is quite common, about 4×10^{-4}. A human galactokinase gene GALK is also responsible for cataracts. Deficiency of the enzyme that converts UDP-galactose Δ UDP-glucose, galactose epimerase (GALE, chromosome 1p36-p35), also leads to galactosemia. GALT deficiency may cause neurological dysfunctions because of the reduction

of galactose available for galactosyl ceramides and glycosphingolipids and the accumulation of their precursors such as glucosyl ceramides. Deficiency of UDP-galactose:ceramide galactosyltransferase (CGT. 4q26) results in thinner myelin sheets and mild ataxias, low IQ, memory deficit, reduced visuomotor coordination, etc. Galactosemia (GALT) may cause ovary dysfunction because of the higher-than-normal levels of follicle-stimulating (FSH) and luteinizing hormones. *See* animal hormones; ataxia; ceramides; human intelligence; lipids; myelin; sphingolipidoses; sphingolipids. (Riehman, K., et al. 2001. *J. Biol. Chem.* 276:10634.)

galactose utilization   Coordinately regulated in prokaryotes and eukaryotes. The galactose genes of *E. coli* are either clustered (*galE* [UDP-galactose-4-epimerase], *galEo* [operator], *galElp* [promoter], *galE2p* [promoter of galEK], *galK* [galactokinase], *galT* [galactose-1-phosphate uridyltransferase]) all at map position 17 or galR (galactose regulator) at map position 61, *galP* (galactose permease) at map position 63, and *galU* (glucose-1-phosphate uridyltransferase) at map position 27. In yeast, the uptake of galactose is mediated by galactose permease (gene *GAL2*). In the presence of ATP, galactose is phosphorylated (Gal-1-P) by galactokinase (gene *GAL*). Galactose-1-phosphate (Gal-1-P) + uridine-diphosphoglucose (UDP-glucose) generate UDP-galactose + glucose-1-phosphate by the action of galactose-1-phosphate uridyltransferase (gene *GAL7*) while UDP-galactose-4-epimerase (gene *GAL10*) mediates the formation of UDP-glucose from UDP-galactose. Genes *GAL1*, *GAL7*, and *GAL10* form a cluster in yeast chromosome 2 and *GAL2* is in chromosome 12. These yeast genes are coordinately inducible up to 1,000-fold by the presence of galactose, although they are transcribed from separate promoters. Gene GAL4 (linkage group 16) and gene GAL80 (linkage group 13) regulate the GAL enzymes. *GAL4* is apparently a positive regulator of genes *1, 2, 7,* and *10,* whereas some *GAL80* mutations abolish the need for induction of the same genes and convert them either to constitutive forms or make them noninducible. It is assumed that the normal role of the product of gene *GAL80* is to prevent the transcriptional activation by *GAL4*, but the combination of the GAL1 protein with GAL80 protein inactivates the latter, then GAL1 activates GAL4, the activator of the system. This GAL1 protein is an enzyme as well as a regulator of transcription. The activation by *GAL4* depends on *upstream activating sequences* (UAS) located 200 to 400 base pair upstream of genes *1, 2, 7, 10,* and *GAL80*. The presence of two UAS was sufficient for full expression. The consensus within the 17 bp palindromic (↔) UAS is:

$$5'\text{-C G G A}^G_C \text{ G A } \textbf{CA} \text{ G T C}^G_C \text{ T C C G-}3'$$

←——————————   ——————————→

The protein product of gene *GAL4* is about 100 kDA and it contains three essential domains. Amino acids from 1 to 65 are involved in DNA binding; residues 65 to 94 are concerned with dimerization. Amino acids 148 to 196 and 768 to 881 mediate activation of transcription (activation domain) and at the C-terminus the sequence 851–881 binds the *GAL80* gene. At the N-terminus, amino acid residues 10–32 display a zinc-finger motif, common to binding proteins. At the C-terminus, there is a high density of acidic amino acids, a characteristic of regulatory proteins. The presence of inactivation by insertion elements in bacterial genomes was first recognized by a study of the *gal* operon in *E. coli*. *See* binding proteins; coordinated regulation; galactose; galactosemia; IS elements; operon; palindrome; regulation of gene activity; two-hybrid method; UAS; zinc fingers. (Weickert, M. J. & Adhya, S. 1993. *Mol. Microbiol.* 10:245; Frey, P. A. 1996. *FASEB J.* 10:461.)

Galactose   Glucose

galactosyl ceramide lipidosis *See* Krabbe's leukodystrophy.

Galago *See* Lorisidae.

Galα1–3Gal Terminal antigens present on the endothelial lining of blood vessels of the majority of mammals, except humans and most primates, because the latter higher animals do not have a functional 1,3-galactosyl transferase (GT). These antigens have the major role in organ graft rejection. In order to reduce rejection, antisense technology may be used to block the synthesis of GT mRNA. Alternatively, an inhibitory ligand (aptamer) or an enzyme (H transferase) is used to add fucose (rather than galactose) to the molecule to compete in the reaction. The use of α-galactosidase may destroy the antigenic galactose terminals. *See* grafting in medicine; immunity; xenotransplantation.

galanin Bioreactive peptide (in humans, Gly-Trp-Thr-Leu-Asn-Ser-Ala-Gly-Tyr-Leu-Leu-Gly-Pro-His-Ala-Val-Gly-Asn-His-Arg-Ser-Asp-Lys-Asn-Gly-Leu-Thr-Ser). Its composition is somewhat different in pigs or rats. In humans, it inhibits acetylcholine and glutamic acid release. Reduces excitability of spinal neurons and blocks voltage-activated Ca^{2+}-channels. It may be involved in behavioral and cognitive deficits in Alzheimer disease. Its effects in other mammals are similar. *See* acetylcholine; Alzheimer disease; glutamic acid; ion channels. (Steiner, R. A., et al. 2001. *Proc. Natl. Acad. Sci. USA* 98:4184.)

galectin β-galactoside-binding protein regulating growth and immunological responses. It may induce apoptosis in activated human T cells. Galectins-1 and -3 are part of the spliceosome, where they interact with the Gemin4 protein. *See* apoptosis; T cell. (Park, J. W., et al. 2001. *Nucleic Acids Res.* 29:3595.)

gall Generally undifferentiated tissue growth in plants caused by infection. *See* crown gall.

Gα G protein involved in hormonal stimulation of adenylate cyclase and may regulate ion channels or phospholipase C. The human $G_s1α$ gene contains 13 exons and 12 introns in a total size of 20 kbp. Gα types: $G_iα$, $G_oα$, $G_xα$, $G_tα$. $Gα_{12}$ and $Gα_{13}$ are subunits, which are stimulated by p115 RhoGEF. The latter bound to p115RhoGEF catalyzes G nucleotide exchange on RHO, but this may be inhibited by activated by $Gα_{13}$. These are very highly conserved proteins across phylogenetic ranges. In $G_sα$, only 1/394 amino acid difference was found between humans and rats, and the protein is entirely identical between humans and bovines. *See* G proteins; p115; RGS.

Gαβγ Heterotrimeric G proteins; the three subunits are of 39–52, 35–36, and 7–10 kDa size, respectively. In mammalian cells, several genes are known for each subunit and their cDNAs may generate additional variations by alternative splicing. *See* G proteins; splicing.

GALT Gut-associated lymphoid tissue. *See* Peyer's patches.

Galtonian inheritance The rules were formulated in different ways by Francis Galton (1822–1911), father of quantitative genetics. He recognized that his ideas on ancestral heredity were partly inconsistent with the observed facts. His application of the concept of regression to the inheritance of multifactorial traits remained essentially correct and contributed to the modern concepts of heritability. By rejecting Darwin's pangenesis and suggesting the concept of "stirpes," he essentially laid the path to the concepts of hard heredity and particulate inheritance. His idea still may be applicable to the inheritance of organelle-coded functions: "We appear, to be severally built up out of a host of minute particles, of whose nature we know nothing, any one of which may be derived from any one progenitor, but which usually transmitted in aggregates, considerable groups being derived from the same progenitor. It would seem that while the embryo is developing itself, the particles, more or less qualified for each post, as it were in competition to obtain it. Also that the particle that succeeds must owe its success partly to accident of position and partly of being better qualified than any equally well-placed competitor to gain lodgment." (Galton, F. 1889 Natural Inheritance, New York). *See* correlation; eugenics; hard heredity; heritability; pangenesis; polygenic inheritance; sorting out. (Roberts, H. F. 1965. *Plant Hybridization Before Mendel*. Hafner, New York; Kevles, D. J. 1995. *In the Name of Eugenics: Genetics and the Uses of Human Heredity*. Harvard Univ. Press, Cambridge, MA.)

gam *See* Charon vectors; lambda phage.

Gamborg medium (B5) For plant tissue culture, it is suitable for growing callus and different plant organs. Composition mg/L: KNO_3 2500, $CaCl_2.2H_2O$ 150, $MgSO_4.7H_2O$ 250, $(NH_4)_2SO_4$ 134, $NaH_2PO_4.H_2O$ 150, KI 0.75, H_3BO_3 3.0, $MnSO_4.H_2O$ 10, $ZnSO_4.7H_2O$ 2.0, $Na_2MoO_4.2H_2O$ 0.25, $CuSO_4.5H_2O$ 0.025, $CoCl_2.6H_2O$ 0.025, ferric-EDTA 43, sucrose 2%, pH 5.5, inositol 100, nicotinic acid 1.0, pyridoxine.HCl 1.0, thiamine.HCl 10, kinetin 0.1, 2,4-D 0.1–1.0. The microelements, vitamins, and hormones may be prepared in a stock solution and added before use. For kinetin, other cytokinins may be substituted such as 6-benzylamino purine (BA) or isopentenyl adenine (or its nucleoside); for 2,4-D (dichlorophenoxy acetic acid), naphthalene acetic acid (NAA) or indole acetic acid (IAA) may be substituted or a combination of the hormones may be used in concentrations that are best suited for the plant and the purpose of the culture. For solid media, use agar or gellan gum. Heat-labile components are sterilized by filtering through 0.45 μm syringe filters. This medium may be purchased from commercial suppliers in a dry mix ready to dissolve. *See* agar; cell culture; cell fusion; embryo culture; gellan gum; Murashige & Skoog medium; plant hormones. (Wetter, L. R. & Constabel, F., eds. 1982. *Plant Tissue Culture Methods. Pairie Res. Lab. Saskatoon, Saskatchewan, Canada.*)

gamergate   Worker of social insects that mates and can lay eggs. *See* social insects.

gametangia   Sex organs of fungi; oogonium in the female and antheridium in the male.

gamete   Haploid male or female generative cell (egg, sperm). Gametic fusion (formation of the zygote) takes place during sexual reproduction. The zygote ($2n$) has twice the number of chromosomes of the haploid (n) gametes. *See* fertilization; gametogenesis.

gamete competition   If multiple gametes are available, their success in fertilization may be determined by genetically controlled viability or vigor. It occurs commonly among sperm of animals and plants, pollen tubes (certation), and among eggs in multiparous animals or in plants where more than one megaspore of the tetrad may produce the egg. *See* certation; meiotic drive; preferential segregation; segregation distorter; selective fertilization.

game theory   Before a decision is made, the probabilities of a set of actions, e.g., $p(1)$ and $p(2)$ must be assessed, generally in a subjective manner. An essential feature is a strategy that assures the maximal rewards for the good decisions. Such a procedure is most widely used in the business world (marketing) under competitive conditions. It may be applied to natural sciences and evolution, where exact statistical methods are not practical due to the variability and uncertainty of the conditions. *See* prisoner's dilemma. (Binmore, K. & Samuelson, L. 2001. *J. Theor. Biol.* 210[1]:1; Sigmund, K. 2001. *Theor. Popul. Biol.* 59:3; Stearns, S. C. 2000. *Naturwiss.* 87[11]:476; Demetrius, L. & Grundlach, V. M. 2000. *Math. Biosci.* 168[1]9.)

gametic array   Of diploids, in case of independent segregation, can be determined by different procedures:

in a dihybrid: $(A + a) \times (B + b)$

$= AB, Ab, aB, ab;$

in a trihybrid: $(A + a) \times (B + b) \times (C + c)$

$= ABC, ABc, AbC, Abc, aBC, aBc, abC, abc;$

using any type of gene symbols such as I/i, R/r, or A/a, the combinations can be read from left to right by following the paths of the arrows, and at right we obtain the gametic arrays. In general, in diploids, in case of independent segregation, the gametic output can be determined by 2^n, where n corresponds to the number of allelic pairs, e.g., in a trihybrid cross $2^n = 2^3 = 8$,

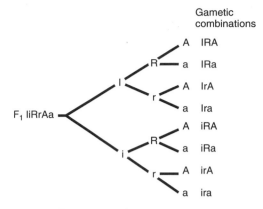

Gametic combinations

Derivation of gametic arrays.

as derived above. For gametic array in autopolyploids and trisomics, *see* autopolyploidy and trisomy, respectively. *See* Mendelian segregation.

gametic lethal   Death at the egg or sperm stage. *See* zygotic lethal.

gametocide   Any chemical that causes male sterility. They may be used in plant breeding to sparse the efforts of emasculation in large-scale hybridization or as birth-control agents when applied shortly before copulation. Some gametocidal genes induce chromosome breakage during interphase. *See* male sterility.

gametocyte   Cell that produces gametes. *See Plasmodium*.

gametogenesis   The animal egg is formed by differentiation without further cell division from a haploid product of meiosis (see figure next page) and so do the spermatozoa from the spermatids. The development of the female and male gametes of higher animals is represented on next page. Basically, gametogenesis in animals and plants shows substantial similarities because in both cases it is based on meiosis. The processes of gametogenesis are regulated in a complex manner by various hormones. *See* animal hormones; atresia; azoospermia; gametophytes; GDNF; hedgehog; Müllerian ducts; spermiogenesis; synergid; Wolffian ducts.

gametophyte   Cell resulting from the meiosis of plants that has half the chromosome number of the zygotes. The gametophytes (megaspores, microspores) form the gametes (egg and sperm). Selection at the gametophyte level is much more effective than in sporophytic generation when the intended target of the selection is expressed at this developmental stage. The effectiveness of selection is particularly needed when the frequency of the gene selected for is low. Selection at the haploid level is apparently successful for tolerance to herbicides, toxins secreted by pathogens, alcohol dehydrogenase mutations, and possibly against certain stress effects. *See* certation; cytoplasmic male sterility; gametophyte development; gametophyte factor; incompatibility alleles; male sterility; pollen competition; sporophyte; synergid. Diagrams on pages 456 and 458. (Kiesselbach, T. A. 1949. *Nebraska Agric. Exp. Sta. Bull. 161*; McCormick, S. 1993. *Plant Cell* 5:1265; Yang, W.-C. & Sundaresan, V. 2000. *Curr. Opin Plant Biol.* 3:53.)

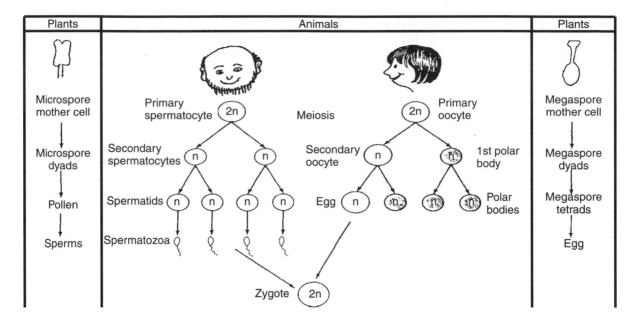

Plants	Animals		Plants

A comparative view of animal and plant gametogenesis

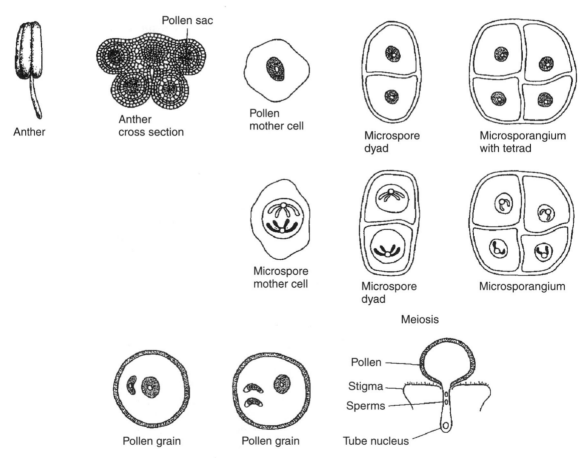

Development of the **male gametophyte** of higher plants. The meiotic stages (showing only one bivalent).

gametophyte factor Affects the haploid gametophyte and may be responsible for reduced transmission of the chromosome (gamete) that carries it in a heterozygote. Gametophyte factors generally have a more detrimental effect on the male, but in rare cases the female is also influenced to various degrees. *See* certation; gametophyte; meiotic drive; preferential segregation; selective fertilization; zygotic lethal.

gamma (γ) Risk for a disease or other attribute associated with a particular genotype. *See* risk.

gamma distribution Distribution of the amount of time until the Xth occurrence of an event by the Poisson distribution. It has been used to fit the evolutionary rate variation among protein sites. The gamma distribution and similar statistical concepts serve as the theoretical foundations for the t-distribution, F-test, and chi square frequently used in genetic analyses. *See* F-test; Poisson distribution; t-test.

$$f(x) = \frac{1}{\Gamma(n)} e^{-x} x^{n-1} \text{ the } \Gamma(n) = \int_0^\infty e^{-x} x^{n-1} \, dx$$

when n is an integer $\Gamma(n) = (n-1)!$

gamma field Area or space where chronic exposure is usually provided from a source of electromagnetic radiation (e.g., ^{60}Co). Such a field may help in studies assessing the effects of long-term exposures on various biological materials (mutation, chromosome breakage, physiological changes) in case of nuclear accidents. *See* electromagnetic radiation; gamma ray; radiation effects; radiation hazard assessment; radiation protection.

gammaglobulin Immunoglobulin (IgG) consisting of either light chain κ or λ; the heavy chains have one of the $C_{\gamma 3}$, $C_{\gamma 1}$, $C_{\gamma 2}$, $C_{\gamma 4}$ coded constant regions. *See* agammaglobulinemia; antibody; immunodeficiency; immunoglobulins.

γ-glutamyl carboxylase (GGC) Enzyme required for posttranslational modification of vitamin K–dependent proteins used for blood clotting and bone proteins. *See* vitamin K–dependent blood clotting factors.

gamma interferon activation site (GAS) TTNCNNNAAA. *See* interferon; ISRE; Jak-STAT; signal transduction; STAT.

gamma ray Ionizing radiation (photons, electromagnetic radiation) emitted by isotopes (such as ^{137}Cs, ^{60}Co, and others). They are similar to X-rays but have much higher energy and have an ability to traverse even several centimeters of lead. Gamma-rays from ^{60}Co (1.2–1.3 MeV) have a linear energy transfer 0.3 LET compared to hard X-rays (250 keV). (LET measures ionizing radiation in keV/nm path.) *See* electromagnetic radiations; eV; physical mutagens; Volt.

$\gamma\delta$ element (Tn*1000*) Insertion element (IE) of the bacterial F plasmid that may produce various Hfr bacterial strains either by cointegration or recombination. The pDUAL/pDelta vector series of the $\gamma\delta$ family vectors has been successfully used for generating (nested) deletions in both strands of a cloned insertion sequence. The plasmid replication origin and some selectable markers are located in both strands in such a way that none of the essential information would be outside the transposon. *See* cointegration; F factor; F plasmid; Hfr; nested; Tn3 family. (Broom, J. E., et al. 1995. *DNA Seq.* 5[3]:185.)

$\gamma\delta$ T cell Expresses any of the $V\gamma$ and $V\delta$ immunoglobulin genes, recognizes nonpeptidic antigens, and the antigen does not require processing by MHC class I or class II molecules and their ligands in order to be recognized by them. The $\gamma\delta$ T cells are very different from the most prevalent $\alpha\beta$ T cells, and they can be stimulated by nonpeptide antigens such as phosphocarbohydrates, X-uridine, X-thymidine-5′-triphosphates (TUBBag3 and TUBBag4, respectively), and isopentenyl pyrophosphate. The molecules may be the product of nucleic acid salvage pathways and intermediates of lipid metabolism. The $\gamma\delta$ T cells primarily mount an innate immune response, but because they stimulate chemokines and secrete cytokines, they also promote the acquired immune system ($\alpha\beta$ T cells). In the absence of $\gamma\delta$ T cells IgE and IgG1, IL-5 and eosinophils are reduced. Mice in this condition do not show allergic asthma of the airways in response to peptidic allergens. *See* allergen; $\alpha\beta$ T cells; asthma; eosinophil; HLA; IL-5; immune system; immunoglobulins; salvage pathway; T cells. (Allison, T. J., et al. 2001. *Nature* 411:820.)

γ satellite Repetitive heterochromatin in the pericentromeric area.

gammopathy Condition of defective immunoglobulin (gammaglobulin) synthesis.

gamodeme Same as deme. *See* deme.

ganciclovir (GCV) Guanine analog (9-[1,3-dihydroxy]-2-propoxymethylguanine) and a derivative of ganciclovir (2-amino-1,9-dihydro-9-[{2-hydroxy-ethoxy} methyl]-6-H-purine-6-one, DCV). Both are antiviral (herpes) and anticancer drugs when incorporated into DNA because the analogs prevent further replication of the genetic material. Actually, GCV requires activation, usually by herpes virus thymidine kinase (HSV-TK), which converts it to a monophosphate form and subsequently cellular kinases mediate the production of the toxic triphosphate. *See* adoptive cell therapy; suicide vector.

ganglion Group of nerve cells outside the central nervous system.

ganglioneuromatosis *See* MEN; tyrosine receptor kinase.

ganglioside Sphingolipid containing several units of acidic sugars attached to the fatty acid chain; gangliosides are common in nerve tissues. Their synthetic pathway is Uridine-diphosphate[UDP]-glucose + ceramide → glucosyl ceramide + UDP-galactose → galactosyl-glucosyl ceramide. Galactosyl-glucosyl ceramide + cytidine monophosphate-N-acetyl-neuraminic acid (CMP-NANA) → ganglioside G_{M3}. Ganglioside G_{M3} + UDP-N acetyl-galactoseamine → ganglioside G_{M2} + UDP. Ganglioside G_{M2} + UDP galactose → ganglioside G_{M1} + UDP. Ganglioside G_{M1} + nCMP-NANA → higher gangliosides. If the sialic acid group (acetyl neuraminic acid, glucosyl neuraminic acid) is removed, asialogangliosides are generated. Several diseases, sphingolipidoses, are involved in their accumulation and breakdown. *See* cancer gene therapy; gangliosidoses; sphingolipids; sphingolipidoses; Tay-Sachs disease. (Kolter, T., et al. 2002. *J. Biol. Chem.* 277:25859.)

gangliosidosis Includes a variety of forms. The general gangliosidosis type I is β-galactosidase deficiency (3p21.33), leading to severe, progressive degeneration of the brain and death by age 2. The overall symptoms resemble

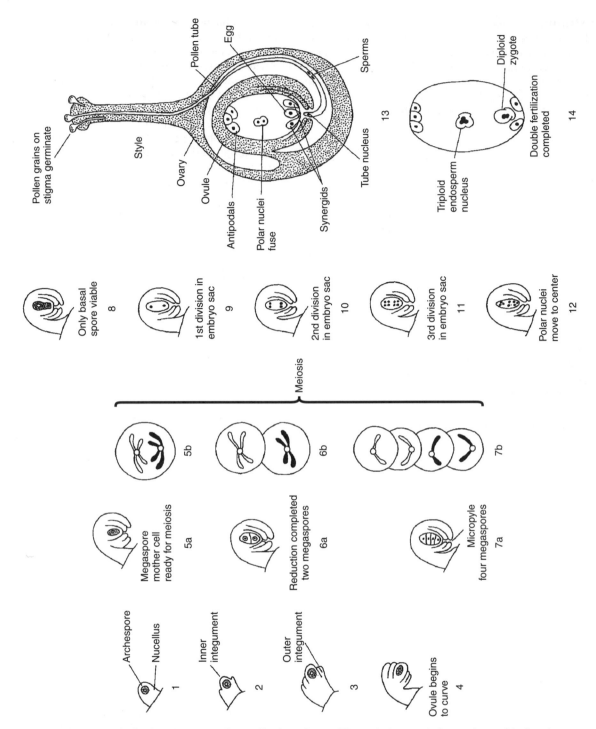

Development of the female gametophyte of higher plants. (The meiotic stages show only one bivalent.)

those of Tay-Sachs disease caused by hexoseaminidase A deficiency and Niemann-Pick disease are brought about by sphingomyelinase deficiency. Newborns show abnormally low activity accompanied by facial and other edemas (fluid accumulation). The distance between the upper lip and nose is enlarged, the ears are set low, there is light hairiness on the front and neck, the spinal column is deformed, the fingers are short, and poor appetite, lethargy, and general weakness are present. The liver and spleen become enlarged. Type II juvenile gangliosidosis has a later onset, and death is delayed to age 4 or 5. This form also has very low β-galactosidase

levels, yet another enzyme seems to be involved. In contrast to type I disease, liver and spleen enlargement, as well as bone deformities, are absent in type II. The heterozygotes can be identified by β-galactosidase assay and the recurrence may be avoided by genetic counseling. Type III is an adult form controlled by a locus different from that of type I. Besides autosomal type 3, there is an X-linked GM3-gangliosidosis that affects young children. The classification of gangliosidoses is quite complicated. *See* galactosidase; GM gangliosidoses; sphingolipidoses; Sandhoff's disease; sphingolipids; spleen; Tay-Sachs disease.

GAP (1) Unknown sequence in between contigs, gapped genome. Before a first-draft sequence is produced, filling must close the gaps in the sequences that may exist between contigs. *See* contig; first-draft sequence; genome projects. (Frohme, M., et al. 2001. *Genome Res.* 11:901.) (2) GTPase-activating proteins are encoded in the long arm of human chromosome 5. Tyrosine-phosphorylated GAP is in the cell membrane, whereas unphosphorylated GAP is mainly in the cytosol. RHO- and RAS-related GTPases are abundant in the cells, and they regulate signal transduction and the cytoskeleton. *See* GED; GTP; RanGTPase; RAS; RHO; signal transduction; von Recklinghousen disease. (Ross, E. M. & Wilkie, T. M. 2000. *Annu. Rev. Biochem.* 69:795.)

GAP gene In *Drosophila*, missing some segments or have fused segments. The body pattern along the longitudinal axis (posterior → anterior) of the wild-type *Drosophila* embryo. Some of the gap mutations delete and/or modify the segments. *See* knirps; Krüppel; morphogenesis; morphogenesis in *Drosophila*.

GAP junction Connecting channel (made of connexins) between apposed cells, permitting the transfer of molecules and electric signals between cells. The same task in plants is assigned to the plasmodesmata. Carcinogenic agents may block gap junctions. The function of gap junctions can be monitored through transfer of fluorescent dyes or by complementation of nutrients. *See* CAM; Charcot-Marie Tooth disease; connexin; innexins; plasmodesma. (Kumar, N. M. & Gilula, N. B. 1996. *Cell* 84:381.)

GAPO syndrome Autosomal-recessive defect characterized by retarded growth, reduced hair development (alopecia), toothlessness (pseudoanodontia), and progressive wasting of the optical nerves. It bears similarity to progeria. *See* Gombo syndrome; growth retardation; progeria.

GAP penalty When similarities are sought in nucleotide alignments for the sake of construction of evolutionary trees (or determine relatedness) and gaps are encountered, these are subtracted from the matches to avoid unwarranted conclusions regarding homologies.

Garden of Eden Evolutionary theory supposes that the human population originated from a single group, and about 100,000 years ago, after passing through several bottlenecks, they differentiated into several subpopulations and dispersed from Africa. *See* multiregional origin; out-of-Africa hypothesis. (Ambrose, S. H. 1998. *J. Hum. Evol.* 34:623.)

Gardner syndrome (APC, adenomatous polyposis of the colon, mutation in FAP) Autosomal-dominant (human chromosome 5q21-q22, 8535 bp) familial polyposis of the colon (FPC). Mutation or deletion occurring at a frequency of about

$2–3 \times 10^{-5}$ (in Ashkenazim populations the carrier frequency may exceed 7%), and it may cause adenomatous intestinal polyposis (a cancer) and breast cancer. Penetrance is very high and the prevalence in the general population is ~1/8,000. The syndrome includes several symptoms, especially if the genetic lesion extends to a larger segment in the region of several genes nearby. Congenital hypertrophy of the retinal pigment epithelium (CHRPE) may be an early diagnostic sign. In some cases, the polyps are limited only to the colon; in other cases, other parts of the intestinal tract, the stomach, the forehead, soft bony tissues, and epidermal cysts may also become tumorous. Some forms were associated with increased ornithine decarboxylase activity. Presymptomatic diagnosis may detect deletions by the use of appropriate DNA markers. Gene targeting experiments revealed the deletion of exon 14 of APC leads to rapid development of adenomas in mice. In the absence of WNT morphogenetic signal, APC interacts with glycogen synthase kinase (GSK) and β-catenin. In this case, the Tcf (T-cell transcription factor) and Lef (lymphoid enhancer factor) are blocked and the complex leads to tumor suppression. If APC is inactivated, monomeric β-catenin appears in the cytoplasm and tumors are formed. Catenin seems to be a transcriptional co-activator of Tcf and Lef. The APC tumor suppressor apparently has a nuclear export function. The c-MYC oncogene is repressed by wild-type APC protein but activated by β-catenin through Tcf-4-binding sites in the MYC promoter. APC appears to link the microtubules (spindle fibers) to the kinetochore, more specifically the Bub protein, a mitotic checkpoint control kinase. This large gene locus includes 15 exons with phenotypic difference among the different allelic mutations/deletions, and this causes some ambiguity in the nomenclature. APC is generally associated with chromosomal instabilities because mutation in the APC protein no longer controls the regular function of the kinetochore and its association with the spindle fibers. *See* adenoma; cancer; catenins; colorectal cancer; conductin; Cre/Loxp; desmoid disease, hereditary; exon; FAP; GSK; hereditary nonpolyposis colorectal cancer; kinetochore; mismatch repair; Muir-Torre syndrome; MYC; polyposis; skin diseases; spindle fibers; targeting gene; Turcot syndrome; WNT. (Bienz, M. & Clevers, H. 2000. *Cell* 103:311; Su, L.-K., et al. 2000. *Am. J. Hum. Genet.* 67:582; Livingston, D. M. 2001. *Nature* 410:536; Fearnhead, N. S., et al. 2001. *Hum. Mol. Genet.* 10:721.)

gargoylism Defect in L-iduronidase such as in Hurler and Hunter syndromes. *See* mucupolysaccharidosis.

garlic (*Allium sativum*) Spice, $2n = 16$. Its alliin (allicin) component inhibits cysteine proteinases and thereby may exert antibiotic effects on some parasitic microorganisms. Its ajoene (sulfur-rich) extract may be antimitotic, microtubule-regulatory and anticarcinogenic. *See* onion. (Li, M., et al. 2002. *Carcinogenesis* 23:523.)

GAS *See* gamma interferon activation site; interferons; signal transduction.

gas chromatography *See* chromatography; gas-liquid chromatography.

gas-liquid chromatography (GLC) Suited for the separation of volatile compounds according to their ability to dissolve in the material of the column bed. An inert gas (helium) driven through the column carries the volatile compounds, and they are sequentially eluted and collected. Some material must be converted to more volatile derivatives before applied to the columns. This method has been extensively used to separate and isolate fatty acids and other compounds. *See* chromatography.

gastric cancer (hereditary diffuse gastric cancer, HDGC, 16q22.1, 5q31.1)   As the chromosomal location indicates, two forms are known. One of them involves a methylation of 6–18 CpG nucleotides in the cadherin1 (CDH1, 16q22) promoter, leading to inactivation. Predisposing genes: nonpolyposis colon cancer, familial adenomatous polyposis, Peutz-Jeghers syndrome, and Cowden disease. *See* cadherins; Gardner syndrome; hereditary nonpolyposis colon cancer; multiple hamartoma syndrome; polyposis hamartomatous.

gastrin   Hormones of 14 to 34 amino acid residues released in the stomach; they regulate stomach acid secretion, other enzymes, and esophageal and gallbladder contraction.

gastroenteritis   Inflammation of the lining of the stomach and intestines.

gastrointestinal stromal tumor, familial   *See* KIT oncogene.

gastrula   Early stage of embryonic development following the blastula. Gastrulation patterns vary in different animal taxa. The general pattern is an invagination of the epithelial layer into the blastocoele (at the so-called vegetal pole) forming the endoderm, which gives rise to the gut. The outer layer, the epithelium, becomes the ectoderm, which will form the epidermis and the nervous system. The cells in between these two layers develop into a mesoderm, which differentiates into the notochord (a vertebral column or its substitute), into the connective tissues, bones, cartilages, fibers, muscles, urogenital system, and vascular system, including the heart and blood vessels. Gastrulation of the human embryo takes place during the third week of embryonal development (in mice 6–7 days postcoitum). In arthropods, gastrulation is followed by anterior-posterior segmentation and dorsal-ventral, medial-lateral identification of embryonal regions. *See* blastula; coitus; morphogenesis; organizer.

GATA   Mammalian transcription factors mediating the formation of erythrocytes. GATA-1 (Xp11.23) recognizes the $^{T\ G\ A\ T\ A\ G}_{A\ C\ T\ A\ T\ C}$ or very similar upstream DNA sequences. Several other GATA factors have been identified in other vertebrates. GATA-3 (10p15) is a hematopoietic factor responsible for the differentiation of T cells of the immune system. Loss of GATA3 results in noradrenaline deficiency and embryonic lethality in mice. GATA factors may include 4 zinc-finger domains. Mutation in GATA1 results in dyserythropoietic anemia and thrombocytopenia. GATA-2 and GATA-3 regulate adipocyte differentiation, and their deficiency may lead to obesity. GATA-4, -5, and -6 are expressed in the developing heart. *See* DiGeorge syndrome; DNA-binding protein domains; dyserythropoietic anemia; erythropoietin; hematopoiesis; noradrenaline; obesity; T cell; thrombocytopenia; transcription factors. (Charron, F. &

Nemer, M. 1999. *Semin. Cell Dev. Biol.* 10:85; Molkentin, J. D. 2000. *J. Biol. Chem.* 275:38949; Patient, R. K. & McGhee, J. D. 2002. *Current Op. Genet. Dev.* 12:416.)

Gatekeeper gene   Acts in the pathway of carcinogenesis by representing a certain threshold that must be passed before mutation of the tumor suppressor or activator gene(s) can mediate the development of the recognizable oncogenic transformation. Genes preventing other cellular processes are also called gatekeepers. *See* cancer; caretaker gene; oncogenes, oncogenic transformation; phorbol esters; progression. (Kinzler, K. W. & Vogelstein, B. 1997. *Nature* 386:761; Gomis-Ruth, F. X. & Coll, M. 2001. *Int. J. Biochem. Cell Biol.* 33[9]:939.)

Gating in cytometry   is used for typing different cells (e.g. CD4+, CD8+ lymphocytes) in cytometeres. The gates permit the selective identification and counting of specific types on the basis of fluorochrome or antibody, etc. labels. The procedure may facilitate the clinical evaluation of the status of e.g. AIDS patients. *See* acquired immunodeficiency. (Bergeron, M., et al. 2002. *Cytometry* 50:53.)

Gaucher disease   Chromosome 1q21–recessive complex of glucosyl ceramide lipidoses. (There is also a pseudogene 16 kb downstream.) Glucosyl ceramide sphingolipids accumulate in the reticuloendothelial Gaucher cells because of deficiency of a β-glucosidase (glucosylceramidase/glucocerebrosidase). These Gaucher cells occur in the lymphoid tissues, spleen, bone marrow, inside the veins, lung alveoli, and other tissues. Type I disease occurs in various age groups. The most characteristic symptoms are enlargement of the spleen and bone anomalies. The neuronopathic or malignant type II form appears before age 6 months and results in death by age 2. The cranial nerves and the brain stem are attacked, although there is not much lipid accumulation in these tissues. The less severe juvenile form (type III) may permit survival to age 30. Gaucher disease is of relatively common occurrence. Cure can be provided by enzyme replacement therapy. Prenatal diagnosis is feasible by the use of RFLP and enzyme assays. *See* glucosidase; Jews and genetic diseases; lysosomal storage disease; prenatal diagnosis; RFLP; sphingolipid; sphingolipidoses. (Koprivica, V., et al. 2000. *Am. J. Hum. Genet.* 66:1777.)

Gaudens   *Oenothera lamarckiana* contains a ring of 12 translocation chromosomes and one bivalent.

Gaudens   Velans

Oenothera pictures from H. deVries 1913 Gruppenweise Artbildung, Borntraeger, Berlin.

The translocations contain two complexes: *gaudens* (happy) conveys green color and *velans* (concealing) determines narrow leaves, pale color, and disease susceptibility. The (complex) heterozygotes appear normal. Because of the translocations and the recessive lethal genes they carry, half of the progeny is inviable (*gaudens/gaudens* and *velans/velans* homozygotes) and the other half (the balanced lethal translocation heterozygotes) breeds true and is of normal phenotype. *See* complex heterozygote; multiple translocations; *Oenothera*.

Gaussian distribution   *See* normal distribution.

gazella   In Dorcas gazella (*Gazella dorcas*) and Grant's gazella (*Gazella granti*), the male has 31 chromosomes and the female has 30. In *Gazella leptoceros*, the males are $2n = 33$, the females $2n = 32$. Thomson's gazella (*Gazella thomsoni*) is $2n = 58$.

G banding   Chromosome staining methods using Giemsa stain (a complex basic dye containing azures, eosin, glycerol, and methanol) after pretreatment with the proteolytic enzyme, trypsin. It permits the identification of dark cross-bands that vary among the individual eukaryotic chromosomes and usually facilitates their identification even when their length and arm ratio is similar. The darkly stained bands represent (AT-rich) heterochromatin and the lowest concentration of genes. *See* chromosome banding; rye; stains. (Dutrillaux, B. & Lejeune, J. 1974. *Adv. Hum. Genet.* 5:119.)

G banded human chromosome.

G-base   Genomic database of mice. For access, *See* Mouse Genome Database, Encyclopedia of the Mouse Genome. *See* databases; mouse.

G$_\beta$ *See* G$_{\alpha\beta\gamma}$

G box   Guanine-rich upstream cis element regulating transcription. Commonly it has the structure GACAACGTG GC. The G-box activator protein binds to the core sequence. It is commonly found in environmentally sensitive genes. *See* CAAT box; phytochromes; regulation of gene activity; Simian virus 40.

G-box element   Upstream binding site (GACAACGTGGC) in plants of which the critical part is the CAACGTG core sequence that binds the G-box protein, a transcriptional activator.

GBP   GSK-binding protein. *See* GSK.

Gbuilder   Nucleotide sequence visualization tool based on Java application of DNA clusters in EST data. *See* EST; Java. (Muilu, J., et al. 2001. *Genome Res.* 11:179.)

GC box   In eukaryotic promoters, the GC box generally contains the 5′-GGGCGG-3′ motif, a binding site for transcriptional regulator proteins.

GC (guanine-cytosine) content   Of DNA, is contributing to the higher buoyant density of the molecules. In the DNA of the majority of eukaryotes, the G-C content is about 40%. Higher organisms seem to display higher GC content yet genome size among eukaryotes does not involve higher GC content. *See* buoyant density; density gradient centrifugation; DNA base composition; ultracentrifugation.

GCN1/GCN20   Ribosome-binding complex bound by the N domain of GCN2. It is required for activation of the latter in starved yeast cells. (Kubota, H., et al. 2001. *J. Biol. Chem.* 276:17591.)

GCN2   It is dimeric and is required for the activation of GCN4. It has ribosome-binding, tRNA-binding, protein kinase, and GCN2/GCN20-binding domains. Heat-shock protein 90 assists in its maturation folding. Its sequence is conserved across fungi, insects, and mammals. (Marbach, I., et al. 2001. *J. Biol. Chem.* 276:16944.)

GCN4   Yeast transcription factor of a leucine zipper structure controlling the transcription of several genes. Its transcription is triggered by amino acid starvation when eukaryotic peptide elongation factor GCN2•eIF-2a becomes phosphorylated. The DNA-binding site consensus for GCN4 is $\frac{\text{ATGACTCAT}}{\text{TACTGAGTA}}$. *See* eIF-2a; HRI; leucine zipper; PEK. (Yu, L., et al. 2001. *J. Biol. Chem.* 276:33257; Natarajan, K., et al. 2001. *Mol. Cell Biol.* 21:4347.)

GCN5   Yeast transcriptional coactivator with a histone acetyl transferase domain including amino acids 99–262 of the 439-amino-acid protein. It consists of four antiparallel β-strands, an α-helix, and a fifth β-strand. This domain is shared by other histone acetyl transferases as well as by other acetyl transferases such as an aminoglycoside 3-N-acetyltransferase and serotonin N-acetyl transferase belonging to the GNAT superfamily. These enzymes transfer an acetyl group from acetyl-CoA to a primary but different amino group. *See* acetyl CoA; amino-glycosides; enhanceosome; histone acetyltransferase; p300; TAF$_{II}$; transcriptional activators. Diagram next page. (Kuo, M. H., et al. 2000. *Mol. Cell* 6:1309.)

GCR   G-protein-coupled receptor. *See* G proteins.

G-C skew $(G - C)/(G + C)$ Indicating the not entirely random distribution of G and C on the two DNA strands. In *E. coli*, there is 26.22% G in the leading strand, whereas in the lagging strand it is 24.58%. *See* DNA replication, prokaryotes.

G-CSF (granulocyte-colony stimulating growth factor)   Lymphokine modulating the effect of growth factors, fetal and postnatal development. *See* GM-SCF; M-CSF.

GD   *See* hybrid dysgenesis.

GDB   Genome database, the official depository of information of the human genome project. It can be accessed

Ribbon diagram of **GCN5** showing the conserved core in four acetyltransferases. (From Sternglanz, R. & Shindelin, H. 1999. *Proc. Natl. Acad. Sci. USA* 96:8807.)

by <http://gdbwww.gdb.org/>; <http://www.ncbi.nlm.nih.gov/Entrez/>.

GDF   Growth and differentiation factor is a member of the bone morphogenetic protein family. GDF5 mutations cause chondrodysplasia. *See* BMP; chondrodysplasia.

GDEPT   Gene-delivered enzyme-prodrug therapy. *See* prodrug.

GDI (guanine nucleotide dissociation inhibitor)   Inhibits the dissociation of GDP from certain G proteins. *See* GEF; G protein; signal transduction.

GDLD (gelatinous drop-like corneal dystrophy)   Rare hereditary amyloidosis. *See* amyloidosis.

gDNA   *See* genomic DNA.

GDNF   Glial cell line–derived neurotrophic factor assists the maintenance of central dopaminergic, noradrenergic, and motor neurons and peripheral and sympathetic neurons. It is a family with several functional members: neurturin, persefin, and artemin.. This protein is structurally related to the transforming growth factor (TGF-β) family and GDNF is a receptor tyrosine kinase. GDNF function requires a glycosylphosphatidyinositol-linked protein (GDNFR-α) and RET. It has been considered as a potential drug for Parkinson's disease, amyotrophic lateral sclerosis, and Alzheimer disease. GDNF also regulates spermatogenesis. *See* Alzheimer disease; amyotrophic lateral sclerosis; dopamine; kinase; MEN; neuron; Parkinson's disease; receptor tyrosine kinase; RET oncogene; TGF. (Hashino, E., et al. 2001. *Development* 128:3773; Bahuau, M., et al. 2001. *J. Med. Genet.* 38:638.)

GDP   Guanosine 5′-diphosphate.

GDRDA   *See* genetically directed representational difference analysis.

GECN   Genetically effective cell number. *See* genetically effective cells.

GED   GTPase effector domain located at the C-end of dynamin. *See* dynamin; GAP; GTPase.

GEF   (**1**) Translation factor similar in function to eIF-2B. *See* eI factors; TU. (**2**) Guanine nucleotide exchange (release) factor facilitates the dissociation of GDP from G proteins. It is similar to Cdc25. *See* ARF; Cdc25; faciogenital dysplasia; GAP; GDI; G protein; signal transduction; SOS. (Vetter, I. R. & Writtinghofer, A. 2001. *Science* 294:1299; Brugnera, E., et al. 2002. *Nature Cell Biol.* 4:574.)

Geiger counter   Geiger-Müller counter. Registers the rate of disintegration of radioactive isotopes. They are necessary in all isotope laboratories, also for monitoring contamination and spillage. The counter detects also environmental pollution of radioactivity in case of fallout. It measures β-radiation with an efficiency of 30–45% and for γ-radiation (shield closed) 5,000 counts per minute per milliroentgen (mR). A typical full-scale reading is 0.2 to 20 mR/hr. It is very useful for surveying because of the good sensitivity and response in seconds. Its shortcomings are energy dependence, saturation at high rates, and interference by ultraviolet and microwave radiations. A special adaptation of the Geiger counter is the strip counter, which detects radiation in chromatograms, membrane filters, blots, etc. *See* ionization chambers; radiation hazard assessment; scintillation counters.

geitonogamy   Pollination by neighboring plants of basically the same genetic constitution.

gelatinase   *See* metalloproteinases.

gel electrophoresis   Nucleic acid fragments are electrophoresed in agarose and polyacrylamide gels, depending on the size of the fragment. In agarose, larger fragments can be separated; in polyacrylamide, smaller fragments can be separated; e.g., in 0.3% agarose, 5–60 kb, in 0.7%, 0.8–10 kb fragments can be analyzed; in 5% polyacrylamide, 0.5–0.8 kb, in 2%, 0.04–0.1 kb fragments can be resolved (*see* DNA fingerprinting). Proteins can be electrophoresed in various media (paper, starch, polyacrylamide) by charge or by size in polyacrylamide-sodium dodecylsulfate (SDS) gels. Two-dimensional gel electrophoresis (2D gel) is an important tool for proteomics. *See* electrophoresis; two-dimensional gel electrophoresis.

gelding   Castrated male horse. *See* castration.

gelephyysic dysplasia   Mucopolysaccharidosis with happy face and dysostosis and heart problems. *See* mucopolysaccharidosis.

***G* element**   *See* nonviral retrotransposable element; retroposons; retrotransposons.

gel filtration   Porous polymers such as Sephadex and Bio-Gel (commercially available in various pore sizes) can be used to separate high-molecular-weight DNA or proteins from smaller molecules (unincorporated dNTPs, linkers, etc.). The large molecule is excluded while the smaller fragments are retained on the gel during chromatography. For rapid purification, it can be used in syringes. *See* linker; Sephadex.

gellan gum   Synthetic polysaccharide used for solidifying plant tissue and bacterial culture media (instead of agar). *See* agar; embryo culture.

gel mobility assay   *See* gel retardation assay.

gel retardation assay   Compared to DNA alone, DNA-bound protein retards the movement of the complex in the electrophoretic field (band shifting), and in this way, DNA-binding proteins can be isolated and analyzed. To the protein bound to DNA, other protein(s) may also bind, making the complex increasingly slow in moving from the start site. This process is called *supershift*. A more specific test uses DNA affinity chromatography. *See* affinity chromatography; DNA-binding domains; DNA-binding proteins; electrophoresis.

gelsolin   Actin-binding protein that regulates the cytoskeleton. *See* actin; amyloidosis; cytoskeleton; fodrin.

Gem   GTP-binding protein induced by mitogens; it is related to RAS. *See* GTP; mitogen; RAS.

GEM91   25-mer antisense phosphorothioate. *See* antisense technologies; phosphorothioates.

geminin   ~25-kDa protein preventing aberrant replication of the chromosomes after the S phase of mitosis. It keeps Cdc6/18 and Cdt1 in check. It accumulates during mitosis, but it is degraded at the transition from metaphase to anaphase. After it is degraded, Cdc6/18 and Cdt1 rebuild and associate with the chromatin in preparation for the S phase. *See* Cdc6; Cdc18; Cdt; cell cycle; MCM; mitosis; ORC. (Wohlschlegel, J. A., et al. 2000. *Science* 290:2309.)

gemini of coiled bodies   Consist of heterogeneous nuclear ribosomal proteins (hnRNP), coiled bodies, and "survival-of-motor-neuron" proteins (SMN). SMN forms are tightly associated with protein SIP1 (**S**MN **i**nteracting **p**rotein). SMN1 is located in human chromosome 5q13, and SMN2 is almost identical to SMN1 but lacks an exon-7 domain. SMN1 is frequently replaced by SMN2 in spinal muscular atrophy. *See* coiled bodies; hnRNP; spinal muscular atrophy. (Matera, A. G. & Frey, M. R. 1998. *Am. J. Hum. Genet.* 63:317.)

geminiviruses   Contain single-stranded small DNA genomes (~2.7 kb). Some can infect monocots; others infect dicots. Their capsules may be geminate (doubled) or their DNA may exist in two partially identical (200 bases common) rings. They may be used for plant vector construction. *See* agroinfection. (Lazarowitz, S. G. 1992. *Crit. Rev. Plant Sci.* 11:327.)

Geminivirus particles.

gemmules   Ancient term for the hereditary units.

GenBank   Data bank for information on nucleotide and protein sequences, Los Alamos Natl. Laboratory, Group T-10, Mail Stop K710, Los Alamos, NM 87545; Tel: (505) 665–2177; email: general inquiries genbank%life@lanl.gov, sequence submission and forms: gbsub%life@lanl.gov. *See* databases; DDBJ; EMBL; sperm bank. (Benson, D. A., et al. 2000. *Nucleic Acids Res.* 28:15; <www.ncbi.nlm.nih.gov/>; complete genomes: <http://www.ncbi.nlm.nih.gov/Genomes/index.html>; updating: <update@ncbi.nlm.nih.gov>.)

gender   Sexual type, e.g., female or male in societal or lexicographic context; in physiology or genetics, the word *sex* is more appropriate.

gender dimorphism   The two sexes are represented by separate individuals.

gender preselection   Separation of X- and Y- bearing sperms before fertilization. Success in such a procedure may prevent the transmission of X-chromosome-linked genetic defects. It may become an alternative to ethically or morally objectionable procedures of negative eugenenics. (It should be called sex preselection.) *See* abortion; eugenics. (*Fertility & Sterility* 75:861–864 [2001].)

gene   Specific functional unit of DNA (or RNA) potentially transcribed into RNA or coding for protein. A group of cotranscribed exons, but due to alternative splicing, exon shuffling, overlapping, using more than one promoter or termination signal, the same DNA sequence may encode more than a single protein. A common structural organization of protein-encoding genes in eukaryotes is

enhancer−promoter−leader−exons−introns

−termination signal−polyadenylation signal

−downstream regulators

The vast majority of human genes are mosaics containing 7–9 exons of 120–150 bps each. In some genes, the exon number and size may be much larger. In between exons, there are 1,000–3,500-bp-long introns. The size of the introns may also be several times larger. The number of the coding nucleotides generally varies between 1,100 and 1,300 bp, but the larger genes may have much longer coding sequences. In the exons + introns + 5′ and 3′ untranslated sequences combined, the genomic genes in general extend to 14–27 kb DNA. A large fraction of the human genes are alternatively spliced and thus the same "gene" may be translated into three or more kinds of proteins. The organization of other eukaryotic genes may vary quantitatively. Genic sequences are richer in GC nucleotides than the noncoding tracts. In prokaryotes, introns are rare and the size of the genes is smaller. Computational procedures might assist the identification of promoters and first exons in the human genome. *See* dystrophin; gene number; one gene–one enzyme; reaction norm; splicing; titin. (Davuluri, R. V., et al. 2001. *Nature Genet.* 29:412.)

gene, split   Contains introns; the vast majority of eukaryotic genes are therefore split into segments of exons. *See* exon; introns.

gene, synthetic   Have been produced since the 1970s. In 1976, Khorana synthesized the tyrosine suppressor tRNA genes of *E. coli* by classical methods or organic chemistry. In vitro synthesis of 26 oligonucleotide tracts were ligated into a 207 bp DNA containing the 86 nucleotide sequence of the gene plus leader, promoter, and terminator sequence. This gene turned out to be biologically active and suppressed an amber mutation when transformed into bacterial cells. *See* suppressor tRNA; synthetic genes; tRNA.

gene action   Type and mechanism of expression of genes.

gene activation   Turning genes on; initiates expression of genes.

gene activator proteins   *See* transcriptional activators.

gene alignment   Arranging the nucleotide sequences of functionally or evolutionarily related genes in such a manner that the homologous and nonhomologous stretches of nucleotides can be assessed. *See* dot matrix; homology.

genealogy   List and description of successive ancestors in a family. *See* coalescent; evolutionary tree; pedigree.

gene amplification   *See* amplification.

gene and protein names   *See* (<http://www.ba.cnr.it/keynet.html>).

gene assignment   Locating genes to chromosomes.

gene bank   *See* databases; DDBJ; EMBL; GenBank; sperm bank.

gene block   Group of syntenic genes. Gene blocks can be preserved in their original linkage phase if they are within inversions because the single recombinants are generally inviable and the double recombinants are very rare within short regions. Paracentric inversion testers have been used to locate advantageous gene blocks for utilization in plant breeding projects. Inversion homozygotes are backcrossed with inbred stocks, and the F_1 is backcrossed with the inversion-homozygote tester. This progeny is half homozygous for the inversion and half heterozygous. The two groups can be easily distinguished by genetic markers or by semisterility of the heterozygotes (if any crossing over takes place). If the heterozygotes surpass the parental forms in quantitative traits, the good performance is attributed to the inverted segment tested. Favorable gene blocks (quantitative gene loci, QTL) may also be identified by linkage to RFLP markers. *See* gene cluster; inversions; operon; QTL; RFLP.

gene cassette   Is a special type mobile genetic element, which carries most commonly only a single gene (e.g. antibiotic resistance) and a recombination sequence (59-be). For mobility it depends on another element, the integron. *See* integron. (Recchia, G. D. & Hall, R. M. 1997. *Trends Microbiol.* 5:389.)

gene center   Geographical area where the greatest genetic diversity within a species is found; therefore, it is considered as the evolutionary cradle of that species. *See* evolution.

gene-centromere distance   *See* alpha parameter; centromere mapping; tetrad analysis.

gene chips   *See* DNA chips; microarray hybridization.

gene cloning   Propagation of a piece of DNA, in identical copies, in a bacterial, viral, or yeast (or other) vector in order to increase its quantity. It is the same as molecular cloning.

gene cluster   Juxtapositioned genes sometimes with related functions. *See* immunoglobulin genes; operon; regulon; transcription.

gene conversion   Biological event that results in the change of one allele to another present in the homologous chromosome. It is a specific type of nonreciprocal recombination. As a consequence, the meiotic output is changing from 2:2 to 3:1, or if the conversion takes place in the opposite direction, to 1:3. In case octads are formed by an additional mitotic division following meiosis, other types of conversion asci can be identified as shown in the illustration below (the left octad is normal; the other five indicate gene conversions). Gene conversion within a locus proceeds in a polarized fashion following the direction of DNA replication. Gene conversion may involve *map expansion* because within very short distances the neighboring sites may be co-converted, thus reducing the chance of their separation. This is in contrast to classical recombination when the presence of multiple markers permits the detection of a higher number of recombinational events. The conversion is characterized by *fidelity*, i.e., the converter and the converted alleles are identical. The 3:1 and 1:3 spore ratios in the tetrads generally occur with equal frequency, which is called *parity*. Mitotic gene conversion may produce somatic sectors and may contribute variation (Chamnanpunt, J., et al. 2001. *Proc. Natl. Acad. Sci. USA* 98:14530). Although gene conversion is not a classical recombinational event itself, it is accompanied by exchange of markers at the flanking regions in about half of the cases. It has been assumed that if gene conversion is not associated with recombination of flanking markers, the donor of genetic information may be a cDNA. In yeast genes *RAD51*, *RAD55*, and *RAD57*, activity is involved in mitotic recombination and gene conversion between two DNA molecules. In RNA-mediated gene conversion, these are dispensable, but gene *RAD1* (encoding an endonuclease) is required.

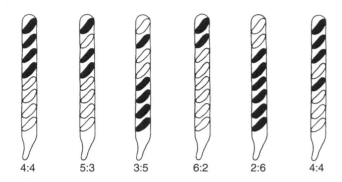

| 4:4 | 5:3 | 3:5 | 6:2 | 2:6 | 4:4 |

Originally gene conversion was discovered in ascomycetes; now it is believed to also occur in other eukaryotes, but the identification of gene conversion is still the easiest on the basis of tetrad data (or half tetrads in *Drosophila* with attached X chromosomes). The frequency of gene conversion may be highly variable from gene to gene, ranging from less than 0.5% to a few percent in yeast. The average size of the converted segment may extend to 1,000 bp. Not all aberrant spore output can be attributed to gene conversion. Polysomy, polyploidy, nondisjunction, premeiotic mitotic recombination, suppressor mutations, etc., may all produce aberrant asci similar to the results of gene conversion. The different types of gene conversions are best identified in the asci where meiosis is followed by an additional mitosis. The detection of gene conversion is relatively simple in organisms where the (pollen) tetrads are preserved (*Salpiglossis*, *Arabidopsis*). The use of PCR technology may permit the detection of converted sequences within any eukaryotic gene even in the absence of tetrad analysis. In some instances in human cells, the frequency of gene conversion much exceeded that of homologous reciprocal recombination, verified by molecular analysis. The mismatch repair of double-strand breaks may involve gene conversion. *See* conversion asci; half conversion; mismatch repair; PCR; recombination mechanisms; recombination models; sex circle model. (Fogel, S., et al. 1971. *Stadler Symp.* 1–2:89; Quintana, P. J. E., et al. 2001. *Genetics* 158:757.)

gene conversion, ectopic Interpretation for the variations in the nonrecombining regions of the X and Y chromosomes in feline species. Since chromosome pairing is a requisite for classical gene conversion, it is hypothesized that accidental pairing may occur. (Slattery, J. P., et al. 2000. *Proc. Natl. Acad. Sci. USA* 97:5307.)

gene copy number In lower organisms the majority of the genes occur in single copies per genome, yet even in bacteria, ribosomal genes may be present in seven copies. In the amphibian oocytes, during the great need for protein synthesis ribosomal genes may be amplified 1,000- to 1,500-fold and form more than 1,000 nucleoli. After meiosis, this excessive amount of rRNA genes is discarded. In maize plants, there may be 10,000 to 20,000 copies of ribosomal genes per diploid cell. *Xenopus* may have 24,000 copies of rRNA genes and 200 copies of each of the tRNA genes. *Drosophila* has about 500 copies of 5S RNA genes in the right arm of chromosome 2. Some particular sequences (SINE and LINE) occur in all eukaryotes, in high numbers. In many higher eukaryotes, repetitive sequences may exceed 80% of the genome. *See* amplification; gene number; LINE; SINE. (Romero, D. & Palacios, R. 1997. *Annu. Rev. Genet.* 31:91.)

gene delivery System of introduction of foreign genes into a cell. *See* biolistic transformation; electroporation; microinjection; receptor-mediated gene transfer; transfection; transformation, genetic; vectors.

gene density In *Drosophila*, 1/13.4 kb was reported, but it is variable (1/5.6 to 1/78 kb). It appears that in the human nucleus, the chromosomes with large gene density (e.g., 19) are more centrally located, whereas the chromosomes with lower gene density (e.g., 18) are situated more at the periphery. The size of the chromosomes does not affect their position. The marine chordate, *Oikopleura dioica*, has a minimum genome size of 51 Mb, about 15,000 genes, and a very short life cycle (2 to 4 days depending on the temperature). Its gene density appears to be only 1/5 kb, the lowest among chordates. (Boyle, S., et al. 2001. *Hum. Mol. Genet.* 10:211; Seo, H.-C., et al. 2001. *Science* 294:2506.)

gene discovery Completion of the genome sequencing still does not reveal the function of the open reading frames. Functional genomics seeks to identify what the individual or clusters of genes do within the cell. *See* EST; normalization; subtraction; UniGene.

gene disruption *See* insertional mutation; insertion elements; targeting genes; transposons.

gene diversity Estimated on the basis of allelic frequencies in a population. H = gene diversity at a locus, n = number of individuals, m = number of alleles at the locus, x_i = frequency of the ith allele. For self-fertilizing species, $n/(n-1)$ replaces $2n/2n - 1$. *See* diversity; evolutionary distance. (Nei, M. & Roychoudhury, A. K. 1974. *Genetics* 76:379; Nei, M. 1987. *Molecular Evolutionary Genetics.* Columbia Univ. Press, New York.)

$$H = \frac{2n}{(2n-1)}\left(1 - \sum_{i=1}^{m} x_i^2\right)$$

gene dosage Number of identical and repeated genes in the genome. *See* polyploidy.

gene duplication See duplication.

gene editing Parts of natural genes are replaced or completed by synthetic DNA chains or a natural repair process eliminates gaps or mismatches in the DNA (also called proofreading). *See* DNA polymerases; editing; mtDNA; RNA editing.

GeneEMAC Computerized method for the monitoring of gene expression during development by external marker-based automatic congruencing (EMAC). (Streicher, J., et al. 2000. *Nature Genet.* 25:147.)

gene evolution Process by which once similar genes diverged or different genes assumed similar structure and function. *See* convergence; divergence.

gene expression Realization of the genetic blueprint encoded in the nucleic acids. Gene expression may be modified, enhanced, silenced, and timed by the regulatory mechanism of the cell, responding to internal and external factors. Usually it is the transcription of DNA into RNA. *See* FANTOM; microarray hybridization; protein synthesis; regulation of gene activity expression; SAGE. Whitfield, M. L., et al. 2002. *Mol. Biol. Cell* 13:1977; Levsky, J. M., et al. 2002. *Science* 297:836. (Gene Resource Locator, GEO [Gene Expression Omnibus]: <http://www.ncbi.nlm.nih.gov/geo>, <www.HugeIndex.org>.)

gene expression map Combines information on expression, coexpression, and correlation of gene expression with all possible intrinsic (e.g., developmental) and extrinsic (e.g., heat shock) factors. These maps allow predictions on biological processes, functions, and phenotypes. (Kim. S. K., et al. 2001. *Science* 293:2087; Ihmel, J., et al. 2002. *Nature Genet.* 31:370.)

gene family Number of genes (paralogous loci) closely related by structure and generally also by function. They probably originated through duplications (and some divergence) during evolution. The members of these gene families are frequently (closely) linked but may also be dispersed in the genome. The complete genome sequences can reveal the relationships that were impossible to detect by earlier methods. *See* deletion; duplication; evolution, protein; exon shuffling; homoeologous alleles; homoeologous chromosomes; immunoglobulins; lateral transmission; orthologous loci; paralogous loci; paranome; protein families; protein isomorphs. (Thornton, J. W. & DeSalle, R. 2000. *Annu. Rev. Hum. Genet.* 1:41.)

gene farming cloning, transformation, and propagation of genes in another species.

genefinder Computer program for finding genes within DNA sequences on the basis of identifying likely splicing sites, translation starts, coding potential, intron sizes, etc., by statistical criteria based on log likelihood ratios. *See* gene identification. (Rogic, S., et al. 2001. *Genome Res.* 11:817.)

gene fission May occur during evolution by splitting one gene into two parts. This process has taken place most frequently in thermophilic archaea. *See* gene fusion.

gene flow Spread of genes in a population by migration of individuals. Gene flow, depending on its intensity, may rapidly alter gene frequencies in a population. Gene flow may be hindered or prevented by geographic isolation, physiological factors (differences in sexual maturity and breeding seasons), and genetical factors (by chromosomal rearrangements causing hybrid sterility, incompatibility alleles, and differences in chromosome number [polyploidy]). In neighboring populations at the overlapping borders, repeated backcrosses may occur, resulting in *introgressive hybridization* and permanent inclusion of new alleles into the gene pool in one or more populations. The availability of transformation techniques may overcome the natural gene flow and transfect genes among taxonomic groups that were previously unable to exchange genetic information because of complete sexual isolation. The wave of advance of an advantageous gene was calculated by R. A. Fisher: $r = \sqrt{2gm}$ where g = initial growth rate and m = migration rate per time and space. Molecular markers greatly facilitate the tracing of the path of genes in human and other populations. Analyses of mtDNA, X and Y chromosomes are the most useful tools for the migration of animals. *See* Eve mitochondrial foremother; introgressive hybridization; migration; transformation; Wahlund's principle; X chromosome; Y chromosome. (Wells, R. S., et al. 2001. *Proc. Natl. Acad. Sci. USA* 98:10244; Oota, H., et al. 2001. *Nature Genet.* 29:20; Weale, M. E., et al. 2002. *Mol. Biol. Evol.* 19:1008; Chikhi, L. 2002. *Annu. Rev. Genomics Hum. Genet.* 3:129.)

gene-for-gene Relationship between host and pathogen. *See* Flor's model; host-pathogen relation.

gene frequency Frequency of a certain allele relative to all alleles at a locus within a particular population *See* allelic frequencies; ceiling principle; DNA fingerprinting; drift; forensic genetics; genetic equilibrium; Hardy-Weinberg theorem; selection.

gene function Typical action of the product of the gene.

gene fusion Attaching a selected promoter or other element(s) to a structural gene by in vitro genetic engineering; or a promoterless structural gene is transformed into a host cell and it is expressed only if it can trap in vivo an appropriate host promoter (enhancer). The procedure permits a study of the nature of the fused heterologous genetic element. If gene fusion occurs between coding regions of two genes, the translation product becomes a *fusion protein* that contains amino acid sequences from two structural genes. This process may modify the function of the fusion protein. Fusing ablation factors, such as ricin or diphtheria toxin to site- or tissue-specific promoters, may facilitate the study of differentiation and development because certain cell types can be eliminated during critical periods. Gene fusion may also occur during evolution. Fused genes frequently are scattered in the evolutionary ranks indicating that they did not evolve by vertical common descent (Yanai, I., et al. 2002. *Genome Biol.* 3[5]:res. 0024.1). Metabolic enzymes of *E. coli* fuse three times more than other proteins. *See* ablation; diphtheria toxin; fusion protein; gene fission; read-through proteins; transcriptional gene fusion vectors; translational gene fusion vectors; trapping promoters. (Casadaban, M. J. 1976. *J. Mol. Biol.* 104:541; Silhavy, T. J., et al. 1984. *Experiments with Gene Fusions.* Cold Spring Harbor Lab., Cold Spring Harbor, NY; Lavasani, L. S. & Hiasa, H. 2001. *Biochemistry* 40:8438.)

gene gun   *See* biolistic transformation.

gene identification   May be required after a particular DNA tract has been sequenced but its function is unknown. The simplest approach is checking the DNA databases for homologous sequences among genes with known function. Extensive amounts of redundant sequences may make the comparisons difficult, but computer programs are available to identify repeats in human genes (<pythia@anl.gov> or <ftp://ncbi.nlm.nih.gov>) or BASTX for other organisms (<http://www.cshl.org/genomere/supplement/harris.htm>). BLAST and FASTA can search databases. *See* Blast; BLOCKS; databases; Fasta; genome annotation. (Fickett, J. W. 1996. *Trends Genet.* 12:316.)

gene indexing   Organizing information on groups of genes according to sequences/functions/EST using Unigene, STACK, and HGI. *See* HGI; STACK; Unigene. (Haas, S. A., et al. 2000. *Trends Genet.* 16:521; <http://www.dkfz.de/tbi/services/GeneNest/index>.)

gene interaction   Common misnomer; actually in most cases the products of the genes interact — with a few exceptions such as gene insertion, gene fusion, etc. *See* epistasis; gene product interaction; modified Mendelian ratios; morphogenesis in *Drosophila*.

gene isolation   The first gene isolation was reported in 1969. The *Lac* gene of *E. coli* was inserted in reverse orientation in bacteriophages λ and φ80 by a modification of specialized transduction. The DNA of these phages was extracted, denatured, and the heavy chain of λ was combined with the heavy chain of φ80. Since the base sequences of the two phage strands were not complementary, only the *Lac* sequences annealed, and the phage DNA sequences remained single-stranded. S_1 nuclease degraded the single strands, but the double-stranded *Lac* gene was preserved in pure form. Somewhat similarly, genes from F' plasmids could also be isolated. These ingenious methods did not have general applicability. A more general procedure was developed by isolation of cDNA from mRNA. The problem was that a eukaryotic cell may contain over 40,000 mRNA molecules at a time. To be reasonably certain (say at 99% probability) that the desired molecule was included, a very large number of molecules had to be isolated:

$$[1 - (1/40,000)]^n = 1 - P = 1 - 0.99 = 0.01$$

Hence,

$$n = \frac{\log 0.01}{\log[1 - (1/40,000)]} \cong 184,213$$

where n is the number of molecules to be screened to find at least 1 at $P(= 0.99)$ probability.

The desired mRNA may be enriched by cascade hybridization. The mRNAs can be extracted from cells at different developmental stages. Also, substrate induction, heat shock, drug-, hormone-, or pathogen-caused induction can be used for enrichment of mRNA. If a *DNA library* is available and the gene can be probed by colony hybridization, the fragments containing the gene or parts of it can be identified by the use of DNA probes. The simplest method of isolation of genes uses *heterologous probes*. Such a probe contains a homologous sequence of the gene to be isolated. The probe is labeled by *nick translation* using either radioactive nucleotides or biotinylation or any other nonradioactive fluorochromes or immunoprobes. The probe permits the selective isolation of the annealed DNA fragment. If the amino acid sequence of at least part of the gene product is known, synthetic probes can also be generated.

Genes can also be labeled by transposon mutagenesis or insertional inactivation using transformation (transfection). In case close genetic or physical mapping information is available, the gene may be isolated by the use of overlapping YAC clones and chromosome walking (*map-based gene isolation*). Linker scanning can identify essential regulatory elements of the gene. The identity of the gene generally requires confirmation by in vitro translation and testing the function of the protein so obtained. *See* biotinylation; candidate gene; chromosome walking; cloning; colony hybridization; cosmids; DNA library; DNA probes; EST; fluorochromes; functional cloning; heterologous probes; immunoprobes; insertional mutation; linker scanning; nick translation; plasmid rescue; positional cloning; radioactive labeling; synthetic probes; transfection; transformation; transposon mutagenesis; YAC vectors. (Nieuwlandt, D. 2000. *Curr. Issues Mol. Biol.* 2:9.)

gene knockout   *See* Cre/loxP; excision vector; gene disruption; knockout; targeting genes.

gene library   Collection of cloned genes. *See* cloning.

gene locus   *See* locus.

gene manipulation   *See* genetic engineering.

gene mapping   *See* mapping, genetic; mapping function; physical mapping.

gene marking   Insertion of a stable retroviral vector into some stem cells, blood cells, or other tissues detects its functional state or contamination with neoplastic cells or immunological reaction, etc. Besides diagnostic purposes, gene marking may serve therapeutic goals for hereditary disorders and viral infections. It can be employed for the introduction of drug-resistance genes and can test the therapeutic index (maximal, optimal, or toxic dose of pharmaceuticals). *See* retroviral vectors.

gene mutation   Molecular alteration within a gene (base substitution or frameshift, point mutation, substitution mutation). *See* mutation.

gene neighbor method Infers functional linkage of genes from the information of genetic linkage. It is primarily applicable to prokaryotes where operons are relatively common.

gene nomenclature assistance For human genes: <http://www.gene.ucl.ac.uk/nomenclature/> or <nome@galton.ucl.ac.uk>. *See* databases; gene symbols.

gene number The number of genes per genome of an organism can be estimated on the basis of mRNA complexity or by total sequencing of the genome. The estimates based on mRNA can be best determined when the entire genome is sequenced by this method. The single-stranded RNA phage, MS2, was found to have 4 genes. The gene number has also been estimated from mutation frequencies. If the overall induced mutation rate is 0.5 and the average mutation rate at selected loci is 1×10^{-5}, then the number of genes is $0.5/(1 \times 10^{-5}) = 50,000$. Although this method is loaded with some errors, the obtained estimates appear reasonable.

On the basis of mutation frequency in *Arabidopsis*, the total number of genes was estimated to be about 28,000 (Rédei, G. P., et al. 1984. *Mutation, Cancer, and Malformation*, p. 306, Chu, E. H. Y. & Generoso W. M., eds., Plenum). The protein-coding gene number of *Arabidopsis* is 25,498 after sequencing the entire genome. In *Drosophila*, ~17,000 genes were claimed on the basis of mRNA complexity. On the basis of the sequenced genome, the estimate is now ~13,600.

During the 1930s, C.B. Bridges counted ~5,000 bands in the *Drosophila* salivary chromosomes and for many years it was assumed that each band represented a gene. Nucleotide sequencing of 69 salivary bands in the long arm of chromosome 2 of *Drosophila* pointed to the presence of 218 protein-coding genes, 11 tRNAs, and 17 transposable element sequences within that ~2.9 Mb region. The shotgun sequencing of the *Drosophila* genome identified ~13,600 genes encoding 14,113 transcripts because of alternate splicing. In humans, 75,000–100,000 genes were expected on the basis of ESTs; of these, about 4,000 may involve hereditary illness or cancer. The human gene number estimates in 2001 still varied from ~27,000 to ~150,000. After the completion of the sequence, the number was estimated to be 28,000 to 35,000, yet the number of transcriptional units in humans appears to be 65,000 to 75,000 (Wright, F. A., et al. 2001. *Genome Biol.* 2[7]res. 0025).

The ways of alternative splicing and the use of more than a single promoter (initiation codon) by the same DNA tract complicate the difficulties in estimation of the number of functional units. In *Saccharomyces*, in the 5,885 open reading frames, 140 genes encode rRNA, 40 snRNA, and 270 tRNA. About 11% of the total protein produced by the yeast cells (proteome) has a metabolic function; 3% is involved in DNA replication; 3% in energy production; 7% is dedicated to transcription; 6% to translation; and 3% (200) are different transcription factors. About 7% are concerned with transporting molecules. About 4% are structural proteins. Many proteins are involved with membranes. In *Caenorhabditis*, 19,099 protein-coding genes are predicted on the basis of sequencing of the genome.

The minimal essential gene number has also been estimated by comparing presumably identical genes in the smallest free-living cells of *Mycoplasma genitalium* (470) and *Haemophilus influenzae* (1749), both completely sequenced. Insertional inactivation mutagenesis indicated the minimal number to be ~265–300. Gene knockouts indicate that some of the apparently minimally required genes of *Mycoplasma* are dispensable. Furthermore, only about 200 of the *Mycoplasma* genes are represented by orthologous genes in other organisms. In *Caenorhabditis elegans*, about 20 times more genes are indispensable for survival. In higher organisms, the number of open reading frames may be larger than the number of essential genes. By the late 1920s, John Belling counted 2,193 chromomeres in the pachytene chromosomes of *Lilium pardalinum* and assumed that this number corresponded to the gene number. The gene number may not accurately reflect the functional complexity of a genome or organism because the combinatorial arrangement of proteins may generate great diversity and specificity. When free-living organisms follow a parasitic lifestyle, a major fraction of their DNA is lost. *See* duplications; gene number in quantitative traits; knockout; proteome; transcriptome. (*Cell* 86:521 [1996]; *Science* 276:1962 [1997]; Adams, M. D., et al. 2000. *Science* 287:2185; Rubin, G. M., et al. 2000. *Science* 287:2204; Aparicio, S. A. J. R. 2000. *Nat. Genet.* 25:129; Koonin, E. V. 2000. *Annu. Rev. Genomics. Hum. Genet.* 1:99; Akerley, B. J., et al. 2002. *Proc. Natl. Acad. Sci. USA* 99:966; Moran, N. A. 2002. *Cell* 108:583.)

gene number in quantitative traits Has been estimated by various complex statistical procedures (Mather & Jinks. 1971. *Biometrical Genetics*. Chapman & Hall, London, UK), but none of the estimates are entirely reliable because the number of genes with a minor contribution or greatly influenced by environmental effects, genetic linkage, etc., confound the picture. Perhaps the number of polygenes controlling one quantitative trait may be only five or six rather than hundreds. Sewall Wright provided a very simple formula in 1913:

$$n = \frac{R^2}{8(s_1^2 - s_2^2)}$$

gene number (n) = where R is the difference between parental means, $[s_1]^2$ is the variance of the F_1 and $[s_2]^2$ is the variance of the F_2 generations. An improved model of Zeng considered linkage and variation in their effect, where $\bar{c}$ is the average recombination rate between loci and C is the coefficient of variation for the distribution.

$$\hat{n} = \frac{2\bar{c}\hat{n} + C^2(\hat{n} - 1)}{1 - \hat{n}(1 - 2\bar{c})}$$

See gene number; polygenes; QTL. (Jones, C. D. 2001. *J. Hered.* 92:274.)

gene ontology (GO) Set of gene classification rules regarding their molecular function, biological role, and cellular location. The same criteria are employed for the genomes of *Saccharomyces cerevisiae*, mice, *Drosophila melanogaster*, *Arabidopsis*, etc. GOs include categories (and additional groups within the main entry) such as nucleic acid binding proteins, cell cycle regulators, chaperones, motor proteins, actin binding, defense proteins, enzymes, enzyme activators, enzyme inhibitors, apoptosis proteins, signal transducers, storage proteins, structural proteins, transporters, ligands, ubiquitin, tumor suppressors, metabolism, organelle control, developmental regulators, sensory perception, behavior, etc. *See* genome projects. (Anonymous. 2001. *Genome Res.* 11:1425; WGS; <http://www. geneontology.org>.)

gene order in the chromosome Can be determined by three-point or multipoint test crosses in eukaryotes or by similar principles in prokaryotes. The conservation of the order permits evolutionary inferences among related species. *See* bacterial recombination; chromosome walking; mapping, genetic; physical map.

gene pool Sum of alleles that can be shared by members of an interbreeding population. *See* population genetics.

gene prediction *See* OTTO.

gene product Transcript(s) of a gene and by extension the processed transcripts and even the translated polypetides or RNAs. *See* polypeptide; processing; RNA; transcript.

gene product interaction Responsible for epistasis, additive, complementary, and suppressor types of modifications of Mendelian segregation ratios. These are frequently called gene interactions, but actually the gene products interact. Interaction among gene products is quite extensive in yeast; 250 sequence-specific regulators were found to affect the expression of ~6,000 genes. It appears that genes are coregulated at the level of transcription. *See* genetic network; microarray hybridization; modified Mendelian ratios; morphogenesis in *Drosophila*; phase display; protein-protein interaction; proteomics; two hybrid method. (Adamkewicz, J. I., et al. 2001. *J. Biol. Chem.* 276:11883; Ito, T., et al. 2001. *Proc. Natl. Acad. Sci. USA* 98:4569; Minton, A. P. 2001. *J. Biol. Chem.* 276:10577; von Mering, C., et al. 2002. *Nature* 417:399; <http://dip.doe-mbi.ucla.edu/>; <http://www.genome.ad.jp/ brite/>.)

gene rearrangement *See* chromosomal rearrangements; gene replacement; immunoglobulins; phase variation; sex determination in yeast; transposons; targeting genes.

gene regulation *See* regulation of gene activity.

gene relic Usually a member of a multigene family that does not have all of the elements necessary for function; it is actually a pseudogene. The existence of gene relics is explained by losses during evolution. *See* processed pseudogene; pseudogene.

gene replacement Can be accomplished by the aid of genetic vectors that carry a different allele and the flanking sequences of a chromosomal locus. This constitution permits intimate homologous pairing in the area. If double crossing over or gene conversion takes place, the allele in the vector may replace the one in the chromosome. Because the frequency of such an event is very low, selectable markers (*URA3* in the diagram in this integrating vector) must be used to screen

out the replacement in a large population. For the selection, one may use an antibiotic resistance gene with a defect in the upstream area and in the vector the same gene but with a defect downstream may restore antibiotic resistance; it can be selectively isolated on media containing the antibiotic. *See* Cre/LoxP; FLP/FRT; homologous recombination, knockout; localized mutagenesis; site-specific mutation; site-specific recombination; transformation; targeting genes. (Richardson, P. D., et al. 2001. *Curr. Opin. Mol. Ther.* 3[4]:327.)

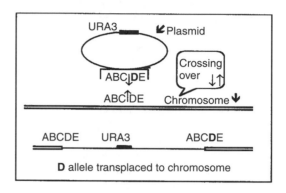

D allele transplaced to chromosome

gene rescue *See* plasmid rescue, marker rescue.

gene resource locator (GRL) Database on gene expression pattern, regulation, and alternatively spliced transcripts. *See* gene expression; regulation of gene activity. (<http://grl.gi.k.u-tokyo.ac.jp>.)

gene scanning *See* linker scanning.

GeneScape Computer program that seeks out miniset clones, DNA sequences, gene alignments to restriction maps, and allows for zooming from a display to the entire map of *E. coli*.

Gene-Scribe® Commercial transcriptional kit with a T7 phage RNA polymerase. It can be used to transcribe cloned genes without subcloning. *See* EMBL3; λDASH; λFIX.

gene silencing *See* silencer.

genesis Inception, origin, the processes of differentiation and development. Also in composite words such as embryogenesis, morphogenesis, neurogenesis, etc.

gene size Can be measured in different ways. If only the translated number of codons is considered, the smallest genes appear to be the 21 bp *mccA* coding for the antibiotic heptapeptide, microcin C7 (MW 1,177 Da) of *Enterobacteria*. Another is the pentapeptide encoded within the 23S ribosomal subunit by only 15 nucleotide pairs. The largest mammalian genes, including introns and upstream and downstream regulatory sequences, may be in the range of hundreds of kbp. The human dystrophin gene with 2.34×10^6 bp includes 79 exons and it is probably the longest gene known. The "average" gene may have 400 codons and thus encode 46 to 48 kDa proteins. The human genome of about 3×10^9 bp contains an estimated 30,000–40,000 genes. *Haemophilus influenzae* bacterium has 1,749 sequenced genes, whereas budding yeast in its 1.8×10^7 genome encodes 5,885 genes by 12,068 kb DNA; thus, its "average" gene is about 2,050 nucleotides long. By sequencing the genome of *Caenorhabditis*, 1 gene was found per ~5 kb; the average intron number was found to be 5. Upon completion of the sequencing of the entire genome in *Drosophila*, the average transcript size appeared to be ~3,058 bp with an average of ~4 exons. The intron sizes varied between 40 and >70 bp. The largest *Drosophila* protein, the cytoskeletal linker, Kakapo, contains 5,201 amino acids, and the smallest is the 21-amino-acid L38 ribosomal protein. The smallest gene numbers, four, are found in some viruses. *See* dystrophin; *Enterobacteria*; exons; genomic DNA; *Haemophilus influenzae*; introns; Mbp; ribosomal RNA; *Saccharomyces cerevisiae*.

gene-specific repair assay Detects the presence of functional repair enzymes acting at a specific gene. The procedure: (1) Expose cells (DNA) to mutagen/carcinogen. (2) Withdraw DNA after specific time periods. (3) Separate replicated and unreplicated DNAs on cesium chloride gradient ultracentrifugation. (4) Restriction enzyme digestion. (5) Employ repair enzymes (uvrABC, T4 endonuclease, Fpg, endonuclease III). (6) Electrophorese DNA and on Southern blots probe the gene of interest. (7) On the autoradiographed gel compare band intensities after allowing for repair, as in (5). In the absence of repair, the intensity of the critical band in the gel is independent from the time allowed for repair, whereas if repair is working, the intensity of the band increases by the length of the period of incubation. *See* autoradiography; density gradient centrifugation; DNA repair; endonuclease; Fpy; uvrABC. (Anson, R. M. & Bohr. V. A. 1999. *Methods Mol. Biol.* 113:257.)

gene substitution Replacing an old allele with a new one in a population; chromosome substitution.

gene switching Uses various ligands (tetracycline, rapamycin, estrogen analogs) that may downregulate gene expression without turning it off. *See* gene-switch cassette; tetracycline.

gene-switch cassette Construct facilitating the turning of genes on and off in a controllable manner. It encodes the DNA-binding domain of GAL4, the human progesterone receptor-ligand-binding domain, and the activation domain of the human p65 protein. In the presence of the antiprogestin mifepristone (RU486), the chimeric molecule binds to the upstream activating domain (UAS) and in a ligand-dependent manner transactivates the appropriate downstream genes. The construct can be inserted into *Drosophila* by adenoviral or P vectors. *See* GAL4; hybrid dysgenesis; p65; progesterone; RU486; tetracycline; transactivator; UAS. (Roman, G., et al. 2001. *Proc. Natl. Acad. Sci. USA* 98:12602; Osterwalder, T., et al. 2001. *Proc. Natl. Acad. Sci. USA* 98:12596.)

gene symbol The abbreviated representation of the function of the genes, or it designates it in a unique manner using one or more letters. Very frequently, the allele that fails to carry out the normal function provides the name for the locus, e.g., the white eye locus in *Drosophila* is symbolized as *w*, although the normal color of the eye is red. The symbols vary in different organisms. Generally the symbols begin with the first letter of the name, which is usually italicized. The recessive alleles in eukaryotes are symbolized with lowercase letters, whereas the wild-type alleles either begin with a capital letter or all of the letters are capitalized. Symbols of genes in the same chromosome strand are usually separated by a space. Genes in the homologous strands customarily have a slash in between them (a/b). A semicolon and one space separate genes in nonhomologous chromosomes (a; d). Multiple alleles in *Drosophila* are designated by the same letter(s) representing the locus and further identified by superscripts, e.g., w^a, w^{a2}, w^{aM}, w^{a79i}, or other additional distinctive signs. Recessive or dominant alleles in a series of mutant alleles are frequently symbolized as a^R or a^D, respectively. The common dominant allele may also be designated as a^+ or A^+. Absence of a gene or lack of its function may be symbolized by a lowercase letter such as a^-.

Isoenzyme-determining alleles may be designated as Adh^F Adh^S, the superscript indicating a fast or slow run in the electrophoretic field. A^n means null allele; a^l may be used for a lethal allele, if necessary with additional specifications. Nonallelic loci with similar phenotypes may be symbolized with the same letter(s) and subscripts: a_1 and a_2. Also, nonallelic loci with similar phenotypes (mimics) may be symbolized as *tu-1a*, *tu-1b*, *tu-2*. Different loci encoding similar proteins may be designated with the same letters but different numbers or an abbreviated form of the molecular weight of the protein (*Hsp68*, *Hsp83*).

Transpositions are symbolized with the designation of the original symbol with the new location in parentheses, $[ry^+](sd)$, indicating that the *rosy* gene was moved from chromosome 3-52 location to the *scalloped* locus in chromosome 1-51.5. The designation of transformants follows that of transpositions. Modifier genes, such as suppressors, may be designated as the symbol of the modifier, followed in parentheses by the modified gene: $su(lz^{34})$. Some symbols may carry the name of the discoverer or the location of discovery of the mutation or the mutagenic agent used. Reversion may be indicated by rv in superscript. Capital letters and additional specifications designate chromosomal aberrations. Translocations (reciprocal interchanges between/among nonhomologous chromosomes) are represented as $T(1; Y; 3)$, indicating that chromosomes 1 (X chromosome), Y, and 3 are involved. Each chromosome may be further specified using a capital letter superscript, indicating the approximate position of the break point as P (proximal to the centromere), D (distal), or M (median).

An X-chromosomal ring (of *Drosophila*) may be symbolized as $R(1)1$. Paracentric inversions are represented as $In(2L)$ or $In(2R)$, depending on whether the left or right arm of chromosome 2 is involved. $In(2L,R)$ indicates pericentric inversion of chromosome 2. Genes closest to the break points may be attached to this symbol. For transposition (nonreciprocal transfer of chromosomal segments), the symbol is Tn; the donor is in parentheses followed by the recipient chromosome, e.g., $Tn(2;3)$. Again, the gene(s) involved may be included in the symbols.

Deficiencies are symbolized by Df followed by the indication of the chromosome (arm) and locus involved: $Def(2R)vg$. Duplications are symbolized with Dp, such as $Dp(3;1)$, indicating that the duplicated segment of chromosome 3 is located in the X chromosome. When the duplicated segment has a centromere and it is a free element, it is symbolized with a letter *f*, e.g., *Dp(1;f)*. In case there are multiple repeats, *Dp(1;1;1)*. When a combination of multiple chromosomal rearrangements occurs, they are indicated with a plus sign between them. The location of break points may be designated by the euchromatic (1 to 102) or heterochromatic (h1 to h61) segment numbers. The older symbols in plants generally followed the customs in *Drosophila*. Today, largely for convenience of typing or printing, the subscripts are substituted with a number written with the gene symbol; the allelic number is attached and hyphenated: *a*2-5, the second *a* locus and allele 5 (rather than superscript 5). Mouse geneticists identify loci with three or four italicized letters, and the first is capitalized.

Human geneticists also use three or four (commonly not italicized) all-capital-letter symbols with additional numbers. In yeast and *Arabidopsis*, the new gene symbols use three italicized capital letters for the wild type and three italicized lowercase letters for the recessive alleles. In the majority of fungi, the wild-type alleles are designated with a superscript plus. Allelic designation frequently follows the locus designation in parentheses: *ilv(STL6)* or *pyr-3 (KS43)*. Suppressor mutation symbols may include the gene they modify: *su(met-7)-1*. The symbol *ssp* means supersuppressors. Mitochondrial mutations are designated as *mi* and additional numbers. RFLP fragments are identified with an italicized three-letter symbol of the laboratory and a serial number. Transposable elements are symbolized similarly to the genes. In human genetics, only capital-letter symbols (no more than 4 or 5 letters) are used without sub- or superscripts. Hyphens or punctuations in the symbols are exceptional. Different loci by the same symbol are numbered, e.g., BPAG1, BPAG2. Alleles may be indicated by an asterisk after the symbol and followed by other designations, e.g., ACY1*2. A slash between two symbols stands for the diploid genotype, hetero- or homozygous. Lack of synteny is indicated by semicolons between the symbols. If linkage is unknown, a comma is used. Gene order is usually started from the short arm down.

Bacterial geneticists designate the loci with three italicized lowercase letters followed by a capital letter, *lacI, lacZo, lacZ*, indicating the lactose utilization operon regulatory (inhibitor), the operator, and the β-galactosidase genes, respectively. The letters *p*, *o*, and *a* stand for promoter, operator, and attenuator, respectively.

Protein products of the genes are generally symbolized with the abbreviations of the genes, but they are in all capitals, or in yeast the first letter is capitalized and the rest are lowercase and Roman. Gene symbols have been periodically revised in some organisms, which may make reading the literature difficult. Creating new symbols is a cheap attempt to gain citations. If new symbolism is warranted, it should not be used retroactively. It is quite unfortunate that many genes have multiple synonyms and symbols. *See* databases (plants — Mendel); *Drosophila*; gene nomenclature assistance, for *Caenorhabditis*. (Horvitz, H. R., et al. 1979. *Mol. Gen. Genet.* 175:129; mouse: Maltais, L. J., et al. 2002. *Genomics* 79:471; humans: Wain, H. M., et al. 2002. *Genomics* 79:464; <www.gene.ucl.ac.uk/nomenclature>; <www.flynome.com>.)

gene synthesis Generating nucleotide sequences by the methods of organic chemistry. These sequences and their variations can then be tested for function after transformation into suitable host cells. The first entirely synthetic genes coded

for tRNAs. Gene synthesis may be carried out in a much simpler way. Sometimes an investigator wishes to remove or add a restriction enzyme recognition site or alter the coding properties, so a different amino acid is inserted into the protein. The desired sequence can be synthesized using pairs of 10–15mer oligonucleotides and annealing them at the 3′-ends of long oligonucleotides as templates and primers. At the same time, several sequences can be generated, each up to 400 nucleotides, then ligated before transforming them into a vector. The simplest diagrammatic representation is below: *See* genes synthetic. (Uhlmann, E. 1988. *Gene* 71:29.)

Annealed, then use T7 DNA polymerase and proceed with synthesis

→ Transformation

genet Genetically identical ramets that are clonal progeny of a single individual. *See* ramet.

gene tagging Places an insertion or transposable element or any other DNA sequence into a gene with the aid of genetic transformation (transfection). When the inserted sequence is known and can be probed by molecular hybridization and/or genetical inactivation or by altering the function (insertional mutation), it can identify the target gene. Some insertions going into intergenic or untranslated (intron) regions may not affect the expression of the gene involved. *See* biolistic transformation; insertional mutation; labeling; probe; targeting genes; transformation, genetic; transposons. (Johnson, G. C., et al. 2001. *Nature Genet.* 29:233.)

gene targeting Method of transformation using a cell-specific promoter attached to the prospective transgene in the vector. The goal is that the transgene expression should be localized to only one type of cells. *See* Cre/loxP; excision vector; *FLP/FRT*; gene replacement; knockout; localized mutagenesis; promoter; targeted gene transfers; targeting genes; transformation, genetic. (Reynolds, P. N., et al. 2001. *Nature Biotechnol.* 19:838.)

gene therapy Insertion of a functional gene into an organism for the purpose of correcting or compensating for its genetic defect. In contrast to biochemical compensation for a genetic defect (e.g., use of insulin), gene therapy may provide a dynamic supply of the missing or deficient metabolite rather than in discrete shots. The most important requisite of gene therapy is the correct identification of the genotype responsible for the phenotype determined by clinical means. It can be carried out either in somatic cells or in the germline (gametes, zygotes). Germline gene therapy may be risky because various chromosomal rearrangements may be caused in the transgenic cells. There is, however, a possibility of introducing genetically engineered DNA into the gametes or zygotes to achieve some of the goals. In vitro fertilization followed by screening of the 8-cell-stage embryos for some of the defects present in the heterozygous families may permit the transfer into the uterus of only those embryos free from the genetic defect. This procedure avoids most of the risks of directly manipulating the genome and it is the technology used by natural selection during evolution. The methods potentially available are microinjection of (foreign) DNA or transformation with retroviral or adenovirus vectors and liposomes (see human gene transfer; transformation of animals), and gene replacement by homologous recombination. Another possibility is knockout when the function of a deleterious gene is eliminated by insertional inactivation or deletion. The technology is available for transformation or in vitro mutation that can be followed by injecting embryonic stem cells into blastocytes introduced into the uterus of females to develop genetically modified embryos and eventually viable offspring. Introduction of hepatic cells that could proliferate in the defective liver of mice with induced tyrosinemia was successful in the laboratory with a model organism. The transformation technology needs refinements before it can be widely applied to humans.

The current gene therapy protocols (more than 300 available) involve altering the somatic cells. The techniques of embryo implantation are widely used to overcome female inability to conceive without surgical assistance. Before implantation, these fertilized embryos may be tested for expression of transferred remedial genes. These procedures may become potentially useful for preventing the expression of genetic diseases under the control of single genes such as Lesch-Nyhan syndrome, Tay-Sachs disease, cystic fibrosis, muscular dystrophy, Gaucher's disease, β-thalassemia, ADA, melanoma, neuroblastoma, multiple myeloma, lymphoma, breast cancer, colorectal cancer, and several others. One ADA patient treated by gene therapy appeared relatively well and still survives after more than 10 years of the use of retroviral transformation, although her immune system is below normal. Autologous CD34$^+$ and functional ADA gene transplantation by umbilical cord blood resulted in low frequency (1–10%) of ADA expressing T lymphocytes, which was too low to be effective for a cure. Polyethylene glycol–conjugated ADA enzyme treatment was much more effective yet not without undesirable effects. Liposomal vectors carrying human leukocyte antigen (HLA)-B7 and β_2 microglobulin cDNA to tumors can express these genes.

SCID-X1 patients were transfused by hematopoietic stem cells expressing the CD34 surface marker and transfected by a defective Moloney retroviral vector carrying the γc cDNA of the SCID-X1 human gene (Cavazzana-Calvo. M., et al. 2000. Science 299:669). The CD34 cells are capable of differentiating into all types of blood cells. The SCID-X1 gene is a cytokine receptor. The patients displayed normal lymphocytes and immune reactions 10 months after treatment. Actually, in 3 months they were able to leave the complete isolation of the hospital. The apparent success of this treatment—compared to the earlier attempts described above—was due to improvements in the cell culture techniques (using Flt3 protein). Flt3 is a natural embryo cell growth factor. Also, the new method avoided the administration of polyethylene glycol–conjugated ADA enzyme, which exerted an earlier toxic effect by deoxyadenosine. Another important technical improvement was the use of transfected stem cells rather than mature transgenic T cells.

Using adenovirus vector with the human cystic fibrosis transmembrane conductance regulator (CFTR) resulted in expression of the gene for 9 days in the nasal or bronchial membranes. With retroviral vector, the low-density lipoprotein (LDL) receptor that is important for familial hypercholesterolemia has been successfully transformed and functioned. Promising initiatives have been made by treatment of patients with retroviral vectors carrying interleukin-4 (IL-4) to fibroblasts, resulting in infiltration of the tumor with CD3$^+$ and CD4$^+$ T cells and cell adhesion molecules. The treatment has resulted in some trials in the increase of CD8$^+$ tumor-specific cytotoxic T lymphocytes (CTL) and eosinophils as well as CD16$^+$ killer cells. Another approach uses intrabodies. As a treatment of the HIV-1 virus infection, intrabodies are directed to the lumen of the endoplasmic reticulum of the cells, where they prevent the secretion of gp160 glucoprotein precursor (a viral envelope protein) and its transport to the cell surface. Using anti-gp120 intrabodies, the gp120-gp160 envelope proteins of the virus may be neutralized. Anti-tat antibody fragments introduced into the cells may prevent the activation of transcription by the viral TAT and the cellular NF-κB proteins. Intrabodies against the Rev splicing element may also reduce the replication of the virus. Intrabodies against fusin may hinder HIV entry into the cells. The somatic cell genetic therapy may target cancer cells with interleukins, tumor necrosis factor, and granulocyte macrophage colony-stimulating factor to reinforce the immune system or use monoclonal antibodies against cancer cells equipped with toxins or sources of radiation see lymphocytes; magic bullet). Some of the genes suitable for germline modification may be targeted to specific organs for alleviating or overcoming the symptoms of the disease. In some cases, e.g., neurological disorders, ex vivo methods have been sought for the restoration of the normal function (in Alzheimer disease) of nerve growth factor (NGF) or transplantation of dopamine-producing tissue (in Parkinson disease). Bone marrow transplantation may alleviate or reverse the course of lysosomal storage diseases. Targeting of the medication to specific cells, tissues, or organs may gain increased significance because it will permit greater effectiveness and higher dosage without side effects.

One problem of gene therapy is that the cell defense mechanism may inactivate the introduced gene by methylating its promoters or immunologically neutralizing foreign proteins. Integration of the vector into tumor suppressor genes very rarely results in cancerous transformation. Other problems arise by unexpected restitution of the retroviral pathogenicity through recombination of the vector and endogenous human viruses. These latter problems are reduced by the use of DNA (rather than retroviral) vectors. Some of the adenoviral vectors contain a much-truncated DNA to prevent viral replication and reduced immunological response against the vector, but other adenoviral proteins are actually immunosuppressors and their deletion may cause the elimination of the vector by the host cells. The adeno-associated and lentiviral vectors may be most promising.

Somatic gene therapy apparently involves smaller risks. The possible harmful consequences of germline alterations are much more difficult to assess. Some gene therapy protocols combine the procedure with chemical treatment. Introducing the herpes simplex virus thymidine kinase (HSTK) gene into a cancerous brain by a viral vector, working only with dividing cells, it is assured that the vector lands only in the tumor cells because the normal cells do not divide. After the establishment of the transgene, the patients are treated with ganciclovir. After phosphorylation by HSTK, this drug can be incorporated into the DNA, but it prevents DNA replication, resulting in the selective death of cancer cells. Recombinant proteins may be used to remedy diverse metabolic defects. Ideally the protein supply should be the same as under natural conditions, i.e., the amount would be variable in response to the need. For treatment success, it is usually important that the protein be rapidly secreted in response to orally administered drugs and the secretion be rapidly stopped after discontinuation of the drug. Also, the protein must not incite an immunologically adverse reaction.

Recently, in utero treatment of fetuses afflicted by α-thalassemia or severe combined immunodeficiency (SCID), which may harm the developing embryo before birth or immediately after birth, have been considered. Such treatments may still have unknown side effects both on the fetus and the mother. Some people oppose gene therapy on biological and/or ethical grounds. The arguments stem from the fears of unforeseeable damage to the human gene pool and the possibility of using these procedures for "genetic enhancement." Genetic enhancement would have similar goals as eugenics and eventually may be exploited to create "supersoldiers" or other antisocial individuals with "uniform" genetic makeup. These fears are frequently fanned by political agenda or by unfounded speculations. The argument in favor of gene therapy is that it provides a means to prevent the perpetuation of "disease genes" by specifically targeting the single defects. It may result not only in elimination of suffering but may reduce health maintenance costs in the long run.

In case of somatic gene therapy, unintended insertions into the germline may happen rarely. The U.S. Food and Drug Administration proposed that such insertion should be limited to less than 50 per μg of DNA employed; genetically this may mean less than 1 insertion/6,000 sperm. It is true that all the possible consequences of gene therapy have not been seen in evolutionary history. The same criticism may also apply to several drugs that are part of current medical practice. Many medicines have physiological and genetic side effects (e.g., diagnostic and therapeutic X-rays, some antibiotics, anticancer drugs, etc.), yet the benefits are supposed to outweigh their risks. In human gene transfer, there are some potential risks of new construct development by recombination with the viral vector. In some cases, the decision is very difficult, e.g., human dwarfism can be cured by the application of growth hormones or by functional growth hormone genes. Dwarfism is not an acute life-threatening anomaly, yet it interferes in many ways with the normal fulfillment of life. The question arises of how far social philosophy should be permitted to affect the life of an individual. Animal models can be successfully applied for the testing of the physiological and biochemical consequences of gene therapy but may not detect all the consequences for human behavior and mental abilities. Thus, gene therapy still has to face biological, technical problems as well as ethical ones.

Obviously, the public mistrusts new technologies, especially when the application suffers initial mishaps. The freedom of scientific inquiry and the innate human striving for knowledge should not be prevented, however. But the same caution may be necessary as was applied to the techniques of recombinant DNA. Some challenges include the development of more effective vectors and successful extrapolation from animal models to humans. Sporadic tragic misfortunes (Teichler

Zallen, D. 2000. *Trends Genet.* 16:272) with the application of this technology cannot be a rational cause for opposing these innovative and promising medical research efforts. As of July 2002 there are no U.S. Government approved gene therapy products on the market (Cimons, M. 2002. *Nature Med.* 8:646). The approximate share of the various genetic vectors in gene therapy experiments is as follows: retroviruses 40%, adenovirus 26%, liposomes 14%, plasmids 9%, vaccinia virus 5%, adeno-associated virus 2%, fowlpox virus 2%, canarypox virus 1%, RNA 1%, herpes simplex virus 0.3%. Gene therapy has potential applicability not only to hereditary diseases but to a wide variety of acquired illnesses. *See* adeno-associated virus; adenosine deaminase deficiency; adenovirus; adoptive cell therapy; antisense technologies; ART; cancer gene therapy; cell; disaccharide intolerance; epitope; ex vivo; ganciclovir; hemophilia; herpes; HIV; human gene transfer; immune system; immunization, genetic; immunostimulatory DNA; informed consent; intrabody; IUGT; lentivirus; lentivirus vectors; liposome; mitochondrial gene therapy; molecular breeding; MoMuLV; mosaic; NF-κB; receptor-mediated gene transfer; retroviral vectors; Rev; ribozyme; SCID; sickle cell anemia; targeting genes; T cells; thalassemia; transfection; transformation, genetic; viral vectors. (Anderson, W. F. 2000. *Science* 288:627; *Am. J. Hum. Genet.* [2000], 67:272; Romano, G., et al. 2000. *Stem Cells* 18:19; Hanazano, Y., et al. 2001. *Stem Cells* 19:12; Factor, P. H., ed. 2001. *Gene Therapy for Acute and Acquired Diseases.* Kluwer, Boston; Pfeifer, A. & Verma, I. M. 2001. *Annu. Rev. Genomics Hum. Genet.* 2:177; Opalinska, J. B. & Gewirtz, A. M. 2002. *Nature Rev. Drug Discovery* 1:503; OBA, <http://keats.admin.virginia.edu/gene/ Gene_Therapy_Regulatory_ Requirements.pdf>.)

gene therapy for infectious disease Gene therapy is most commonly directed against hereditary diseases or other acquired conditions for which drugs are not sufficiently effective. Gene therapy may also be designed against pathogenic microorganisms. The available or potentially working approaches to be considered are transformation by resistance genes, DNA vaccination, suicide genes, use of lytic phages, antibody genes, increase in the production of metabolites that interfere with the development of the disease, use of antisense technology, RNAi, transgenesis for antimicrobial peptides, ribozyme-mediated cleavage of RNA, activation of antimicrobial defense genes, increase in the dosage of antimicrobial genes to inactivate receptors required for infection; recruitment of antagonists of the parasites, etc. A great variety of strategies may be developed.

genethics *See* ethics.

genetically directed representational difference analysis (GDRDA) Targets and identifies traits that differ between congenic lines without prior knowledge concerning their biochemical function. It determines linkage to known genes or to polymorphic DNA markers. *See* congenic; DNA markers; RDA. (Higo, K., et al. 2000. *Exp. Anim.* 49[3]189.)

genetically effective cells Cells of the germline that actually contribute to the formation of the gametes and thus to the offspring. The number of genetically effective cells can be determined in autogamous species on the basis of the segregation ratios after mutation. In case the genetically effective cell number (GECN) is 1, the segregation is either 3:1 or 1:2:1. If the GECN is 2, the segregation of dominant:recessives is 7:1; in case GECN = 4, the expected ratio is 15:1 because only one of the cells of the germline segregates while the other cells provide only nonmutant offspring. Thus, the pooled phenotypic numbers yield the 7:1 $(4 + 3)$:1, and 15:1 $(4 + 4 + 4 + 3)$:1 proportions. These ratios may be (slightly) altered if the transmission of the gametes carrying the recessive alleles is impaired or if the viability of the recessive homozygotes is reduced. For the determination of GECN, the aberrant progenies should not be considered. *See* critical population size; mutation rate; planning of mutation experiments. (Rédei, G. P. & Koncz, C. 1992. *Methods in Arabidopsis Research*, Koncz, C., et al., eds., p. 16. World Scientific, Singapore.)

genetically effective population size *See* effective population size.

genetically modified organisms *See* GMO; pest eradication.

genetic assimilation Adaptive mechanism in a population to fix genes as permanent parts of the genome by selection. An initially epigenetic modification becomes fixed by heredity. *See* adaptation; canalization; epigenetic; fixation; fixation index.

genetic association Correlation between the presence of a genetic marker and a certain type of multifactorial disease. For the trustworthiness of the conclusions, large populations and high statistical probability are required. (Dahlman, I., et al. 2002. *Nature Genet.* 30:149; Xiong. M., et al. 2002. *Am. J. Hum. Genet.* 70:1257.)

genetic background All residual genes besides the one(s) of special interest.

genetic balance *See* balanced lethals; balanced polymorphism; balance of alleles.

genetic bar code *See* bar code; DNA chip.

genetic block Mutation in a gene may prevent or slow down the flow of a metabolic pathway. *See* leaky mutant; null allele.

genetic burden Same as genetic load.

genetic cascade Genes of a developmental pathway are activated in successive waves; the expression of "earlier" genes activates the next ones.

genetic circularity Consequence of circular DNA genetic material, i.e., the genetic map has no ends, although one point is generally designated as the origin. *See* DNA, circular.

genetic code Consists of 64 contiguous nucleotide triplets from which 61 specify >20 amino acids and 3 serve as signals for termination of translation on the ribosomes (see table on next page). In animals the 21th encoded (UGA) amino acid is selenocysteine and in Archaea and Eubacteria UAG may encode the 22nd amino acid, pyrrolysine. In addition,

programmable ribozymes may attach non-natural amino acids to tRNAs and incorporate them into engineered proteins (Bessho, Y., et al. 2002. *Nature Biotechnol.* 20:723). The number of triplet codons for a particular amino acid varies from 1 to 6. *See* it is of theoretical and practical importance to amplify the genetic code beyond the "magic number." Chemically modified, unnatural amino acids may alter protein function if incorporated in vivo in place of the natural ones. Incorporation may be achieved by modified special tRNAs and mutant aminoacyl-tRNA synthetases. Mutations at the editing site of *E. coli* tRNA^{Val} tRNA synthetase incorporated in higher than 20% aminobutyrate, a steric analog of cysteine, into the site of cysteine. For additional special differences and exceptions in the coding dictionary, see <http://www.ncbi.nlm.nih.gov/Taxonomy/taxonomyhome.html>. *See* amino acid symbols in protein sequences; aminoacyl-tRNA synthetase; evolution of the genetic code; codon; magic number; tRNA. (Knight, R. D., et al. 2001. *Nature Rev. Genet.* 2:49; Wang, L., et al. 2001. *Science* 292:498; Döring, V., et al. 2001. *Science* 292:501.)

genetic code, second *See* aminoacyl-tRNA synthetase.

genetic colonization Infection of plants by agrobacteria results in the expression of bacterial genes in plant cells. The gene products, such as opines, are utilized only by the bacteria. In population genetics colonization means also the establishment of breeding populations in new habitats. *See* *Agrobacterium*; founder principle; opines; transformation.

genetic complementation *See* allelic complementation; complementary alleles; complementation groups; complementation maps.

genetic conservation Preservation of species and subspecific genetic variations in protected areas, national parks, game reserves, botanical gardens, zoos, seed depositories, culture collections, sperm banks, etc.

genetic correlation Linked genes are expected to segregate together depending on the frequency of recombination. Members of the same family display correlation, and even assortative mating shows correlations, although the latter may not be genetic. From the correlation between certain phenotypes, the chromosomal location of genes can be predicted. The term *genetic correlation* in animal breeding is defined as a measure of the ratios of additive variances: $\text{Cov}(X, Y)/\sqrt{\text{Var}(X) x \text{Var}(Y)}$. *See* correlation.

genetic counseling Provides information, medical diagnosis on hereditary bases, recurrence risk, family history, and possible management of genetic anomalies for the benefit of the family. It has no eugenic purpose and it does not make recommendations for decisions; it merely informs the concerned individual(s). *See* counseling, genetic; recurrence risk; risk; utility index for genetic counseling. (Mahowald, M. B., et al. 1998. *Annu. Rev. Genet.* 32:547; Thornburn, D. R. & Dahl, H. H. 2001. *Am. J. Med. Genet.* 106:102; GENETests: <http://www.genetests.org/>.)

genetic death Occurs if an organism leaves no offspring or a gene is not transmitted to subsequent generations. *See* fitness;

The genetic code in RNA triplets

5′ Nucleotide	Second Nucleotide				3′ Nucleotide
	U	**C**	**A**	**G**	
U	Phe	Ser	Tyr	Cys	U
	Phe	Ser	Tyr	Cys	C
	Leu	Ser	ochre	opal	A
	Leu	Ser	amber	Trp	G
C	Leu	Pro	His	Arg	U
	Leu	Pro	His	Arg	C
	Leu	Pro	Gln	Arg	A
	Leu	Pro	Gln	Arg	G
A	Ile	Thr	Asn	Ser	U
	Ile	Thr	Asn	Ser	C
	Ile	Thr	Lys	Arg	A
	Met	Thr	Lys	Arg	G
G	Val	Ala	Asp	Gly	U
	Val	Ala	Asp	Gly	C
	Val	Ala	Glu	Gly	A
	Val	Ala	Glu	Gly	G

Ala = alanine (4)
Arg = arginine (6)
Asp = aspartic acid (2)
Asn = asparagine (2)
Cys = cysteine (2)
Glu = glutamic acid (2)
Gln = glutamine (2)
Gly = glycine (4)
His = histidine (2)
Ile = isoleucine (3)
Leu = leucine (6)
Lys = lysine (2)
Met = methionine (1)
Phe = phenylalanine (2)
Pro = proline (4)
Ser = serine (6)
Thr = threonine (4)
Trp = tryptophan (1)
Tyr = tyrosine (2)
Val = valine (4)

The universal genetic code for amino acids in RNA codons. The table shows underlined the three nonsense codons (chain-termination codon) and 61 sense codons coding or amino acids. The numbers after the amino acids (right-most column) indicates the number of synonymous codons for each

methionine and tryptophan each have only .... 1 amino acids.
asparagine, aspartic acid, cysteine, glutamic acid,
glutamine, histidine, lysine, phenylalanine, tyrosine, each has 2
isoleucine has 3
alanine, proline, threonine, and valine have 4
arginine, leucine, serine all have .... 6 codons.
The codon usage is not random, it varies among organisms and genes.

Exceptional codon meanings:

Mycoplasma capricolum	Tetrahymena thermophila	Euplotes octacarinatus	Mitochondria			
			Mammal	Drosophila	Yeast	Neurospora
UGA: Trp	UAA: Gln	UGA: Cys	AUA: Met	AUA: Met	AUA: Met	CUN: Thr
	CAG: Gln	UAA: stop	AUU: Met	AUU: Met	CUA: Thr	
	UAG: Gln	UAG: absent	AUG: Met	AUG: Met	CUC: Met	
			AUC: Met	....	CUU: Met	
			UGA: Trp	UGA: Trp	CUG: Met	
			AGA: stop	AGA: Ser		
			AGG: stop			

The UGA "universal" stop codon means Trp in the mitochondria of vertebrates, insects, molluscs, echinoderms, nematodes, platyhelminthes, fungi and ciliates. Selenocysteine is also coded by UGA in *E. coli* and mammals.

mutation, beneficial; mutation, neutral; selection coefficient; selection conditions; transmission.

genetic determination (g^2) Basically similar to heritability (measured by intraclass correlation).

$$\frac{MS_b - MS_w}{MS_b + (2n - 1)MS_w} = g^2,$$

where MS_b and MS_w stand for between- and within-strain mean squares, respectively. The $2n - 1$ in the denominator is used in testing inbred strains to compensate for the increase of additive genetic variances during inbreeding. *See* heritability; intraclass correlation; variance.

genetic discrimination Prejudicial treatment on the basis of phenotypic or genotypic constitution by employers, insurance companies, or any other person or institution. *See* bioethics; genetics and privacy. (Nowlan, W. 2002. *Science* 297:195; Rothenberg, K. H. & Terry, S. F. 2002. *Science* 297:196.)

genetic disease An estimated 4,000 human genes are directly or indirectly involved in the determination of human malformations and physical and mental disabilities. According to some estimates, 15 to 20% of newborns are afflicted by some hereditary problems and a large fraction of miscarriages are caused by chromosomal anomalies and/or recessive or dominant lethal genes. Approximately 25% of hospitalizations are due to maladies with a genetic component. Very often genes are not the absolute cause of the disease because many diseases can be prevented by proper lifestyle and preventive medication if disposition exists. The occurrence of genetic disease sometimes can be avoided or the risks reduced by proper education and genetic counseling. During gestation, transcription factors and enzyme defects are the most prevalent fraction of diseases of the fetus. After birth, defective enzymes have the largest share in the disorders. Anomalous genetic regulation accounts for the majority of developmental defects. Amino acid replacement mutations (based on six human diseases) indicate that evolutionarily conserved sites are most common in human disease. Polymorphic replacement mutations and silent mutations, however, appear to be randomly distributed. *See* consanguinity; DALY; eugenics; gene therapy; genetics and privacy; genetic screening; inbreeding; OMIM; prevalence; recurrence risk; risk; selection coefficient. (Miller, M. P. & Kumar, S. 2001. *Hum. Mol. Genet.* 10:2319; Perez-Iratxeta, C., et al. 2002. *Nature Genet.* 31:316; Dean, M., et al. 2002. *Annu. Rev. Genomics Hum. Genet.* 3:263; HOMOPHILA.) <archive.uwcm.ac.uk/uwcm/mg/fidd/index.html>, <http://www.bork.embl-heidelberg.de/g2d/>.)

genetic dissection   Analyzes the mechanism(s) of genetic determination and control of biological traits, morphogenesis, and/or other functions by the techniques of mutation, recombination, and pattern of inheritance in pedigrees or populations. *See* metabolic pathways; morphogenesis in *Drosophila*; one gene–one enzyme theorem.

genetic distance   Genetic distance (d) can be measured by different procedures. One simple solution based on a geometric model is $d^2 = 1 - \sqrt{p_1 p_2} - \sqrt{q_1 q_2}$, where p and q represent the frequencies of the two alleles of a locus in populations 1 and 2, respectively. For actual determination of the distance between two populations, more than one allelic pair must be considered. Genetic distance, F_{ST}, is also calculated as $V_p/\overline{p}(1 - p)$, where V_p is the variance between gene frequencies in a set of n populations and $\overline{p} =$ their average gene frequencies. *See* evolutionary distance; evolutionary tree. (Nei, M. 1972. *Am. Nat.* 106:283.)

genetic divergence   *See* divergence.

genetic diversity   Variations in the gene pool of a population or the genetic variations in the population. *See* diversity; gene pool; genetic conservation; genetic variation.

genetic drift   Change in gene frequencies by sampling error(s) of the gametic array, so the genes are not maintained on the basis of their fitness or the selective advantage they may convey, but the selection is the outcome of chance. In case of two alleles, selection by chance alone is determined by the frequency of the alleles, binomial distribution, and population size. Thus, if the frequency of allele A is p and that of a is q, the frequency of alleles by chance alone will follow the binomial distribution of $(p + q)^n$, where $n =$ the number of individuals that leave offspring surviving to the reproductive age; e.g., in case the allelic frequencies are equal and four individuals survive, the probability that all four will be homozygous recessives is 0.0625. *See* binomial distribution; effective population size; Eve foremother; founder

principle; hitchhiking; mutation, neutral; Pascal triangle. (Cavalli-Sforza, L. L. & Bodmer, W. F. 1971. *The Genetics of Human Populations.* Freeman, San Francisco; Wright, S. 1921. *Genetics* 6:111; Pritchard, J. K. 2001. *Am. J. Hum. Genet.* 69:124; Frost, S. D., et al. 2001. *Proc. Natl. Acad. Sci. USA* 98:6975.)

genetic endpoint   Classification of the types of genetic lesions such as mutation, chromosomal aberration, unscheduled DNA synthesis, etc., which are detected in mutagen testing. *See* bioassays in genetic toxicology.

genetic engineering   Construction of special chromosomes by cytogenetic manipulations, somatic cell fusions, or introduction of organelles into cells by mechanical means (genetic microsurgery). Isolation and propagation of DNA molecules in suitable hosts, molecular modification of genes and regulatory elements for special purposes, and transfer of genes among diverse organisms by bypassing the constraints of sexual reproduction and manipulating them for medical, industrial, and agricultural use are all part of genetic engineering. *See* alien addition; alien substitution; alien transfer lines; biotechnology; cancer gene therapy; chromosome substitution; cloning vectors; gene therapy; genomics; GMO; input trait; intercellular immunization; intracellular immunization; metabolite engineering; monoclonal antibody; monoclonal antibody therapies; pathogen identification; protein engineering; scaffold; stem cells; terminator technology; tissue engineering; transformation; transgenic.

genetic enhancement   *See* animal breeding; eugenenics; gene therapy; plant breeding.

genetic equilibrium   Exists when gene frequencies are stable for generations *see* mutations and genetic equilibrium. In a panmictic diploid equilibrium population, the frequency of heterozygotes is twice the square root of the product of the frequencies of the two homozygous classes: $2 = H/\sqrt{D \times R}$, where H, D, and R stand for heterozygotes, homozygous dominants, and homozygous recessives, respectively. This is derived from the middle term of the Hardy-Weinberg formula, $2pq = H = 2\sqrt{p^2 q^2} = 2\sqrt{D \times R}$ and hence $2 = H/\sqrt{D \times R}$. This principle can graphically be represented in the figure below. In an equilibrium population, the frequency of

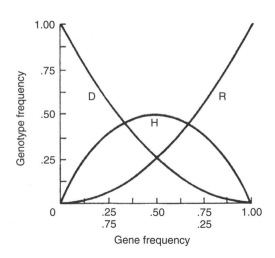

The four populations represented in the body of the table below all have identical gene frequencies, $p = 0.8$, $q = 0.2$ yet the genotypic proportions are quite different. According to the definition in the text only population 4 is in equilibrium

Populations	Genotypic AA	Frequencies	
		Aa	aa
1	0.80	0.00	0.20
2	0.70	0.20	0.10
3	0.60	0.40	0.00
4	0.64	0.32	0.04

heterozygotes is represented by a parabola as the proportion of the alleles vary from 0 to 1 to 0 as long as the three genotypes have equal fitness. With respect to an individual locus, equilibrium is attained within one generation of random mating.

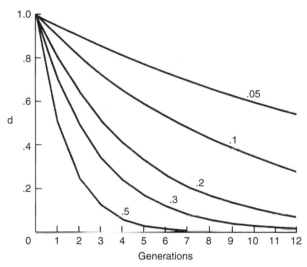

Progress toward genetic equilibrium in case two genes are linked in repulsion at zero generation time. At equilibrium, the repulsion and the coupling phases are equal. Four of the curves (0.05 to 0.3) represent the courses to equilibria at various intensities of recombination; 0.5 indicates independent segregation. (Modified after Falconer, D. S. *Introduction to Quantitative Genetics.* Longman, London.)

As long as random mating prevails and there is no selection, gene and genotypic frequencies do not change and the Hardy-Weinberg principle prevails. Multiple loci require more generations to attain equilibrium. Also, equilibrium depends on the intensity of linkage among the loci. Progress toward equilibrium is delayed if the genes are sex-linked. If in the original mating the homogametic sex is homozygous for a recessive allele ($X^a X^a$) and the heterogametic sex carries the other allele ($X^A Y$), the allelic frequencies in the two sexes will follow an oscillatory path during the generations because of the zigzag pattern of inheritance of the X chromosome. In equilibrium the allelic proportions in the two sexes will be represented by the proportions of the X chromosomes. A somewhat similar situation exists in hermaphrodites carrying self-sterility alleles (S). The mating of plant self-incompatibility alleles (S) produce the offspring shown in the table below. Half of the progeny is the same as the male, whereas the other half has a different constitution. The genetic constitution of the females does not reappear in the immediate progeny because of self-sterility. Therefore, if the frequencies of the alleles are not identical, the genotypes most common among the parents will be the least frequent among their offspring, although they reappear in advanced generations. Equilibrium is reached, however, if the frequencies of the alleles are equal.

Female	Male	Offspring
$S_1 S_2$	Either $S_1 S_3$ or $S_2 S_3$	$\frac{1}{2}S_1 S_3 + \frac{1}{2}S_2 S_3$
$S_1 S_3$	Either $S_1 S_2$ or $S_2 S_3$	$\frac{1}{2}S_1 S_2 + \frac{1}{2}S_2 S_3$
$S_2 S_3$	Either $S_1 S_2$ or $S_1 S_3$	$\frac{1}{2}S_1 S_2 + \frac{1}{2}S_1 S_3$

In polyploids, the progress toward equilibrium is quite complicated and can be determined according to C. C. Li (*First Course in Population Genetics*, Boxwood Press, CA). If the gametic output of an autotetraploid population is $G_0 \equiv x(AA) + 2y(Aa) + z(aa) = 1$, the frequency ($p$) of $A = x + y$ and the frequency (q) of $a = y + z$. The gametic proportions in the course of generations (n) is expressed as $d = (y^2 - xz) = y^2 - (p - y)(q - y) = y - pq$, and d is the index of divergence from the equilibrium condition. This index is reduced by two-thirds during each generation of random mating. The gametic proportions and gene frequencies can thus be obtained as

$$yn = pq + dn = pq + \left(\frac{1}{3}\right)nd \rightarrow pq,$$

$$xn = p - yn = p2 - \left(\frac{1}{3}\right)nd \rightarrow p2,$$

$$zn = q - yn = q2 - \left(\frac{1}{3}\right)nd \rightarrow q2$$

See autopolyploidy; Hardy-Weinberg; linkage disequilibrium; self-sterility; sex linkage; Wahlund's principle.

genetic essentialism Criticism of modern genetics for equating human (and other) beings with a molecular entity (DNA), including social, historical, and moral complexities and responsibilities. *See* vitalism.

genetic fine structure Analysis involves recombination within the boundaries of individual genes.

genetic fingerprinting *See* DNA fingerprinting.

genetic hazard *See* empirical risk; environmental mutagens; genetic risk, genotypic risk ratio; GMO; Kaplan-Meier estimator of survival; λ_S; radiation effects; radiation hazard assessment; recurrence risk; risk.

genetic homeostasis Property of a population to maintain its genetic composition and resist changes in gene frequencies by phenotypic regulation under variable environmental conditions. *See* artificial selection; canalization; homeostasis.

genetic homology Degree of similarity in the base sequences of DNA and RNA or the amino acid sequences in the proteins. *See* amino acid sequencing; databases; DNA sequencing; homology; protein structure; RNA sequencing.

geneticin (G418) Aminoglycoside antibiotic. *See* aminoglycoside phosphotransferase; antibiotics; kanamycin; neomycin.

Geneticin.

genetic information Instructions in the nucleic acids for the cellular machinery.

genetic instability *See* instability, genetic.

genetic isolation Lack of ability to interbreed (incompatibility) and/or hybrid inviability or sterility between/among different taxonomic groups. *See* isolation, genetic; speciation.

genetic load Sum of deleterious genes in the genome. Recessive alleles cannot generally be detected in the heterozygotes. These heterozygotes may continuously contribute homozygotes to the population. If the recessives are deleterious, they may adversely affect the fitness of the population, thus constituting a genetic load. The amount of hidden genetic variation is revealed by the coefficient of breeding. In F_1, 100% of the population is heterozygous. In successive generations of selfing, the heterozygosity decreases by $(0.5)^n$, where n = the number of selfed generations (e.g., by F_5 there are 4 selfings). Thus, the sum of the heterozygotes = $1 - (0.5)^n$. The coefficient of inbreeding, F, in the offspring of first cousins is 0.0625, whereas among unrelated individuals it is presumed to be 0. Thus, if the mortality range in a certain age group is say 11% in the general population and 16% among the children of first cousins, the difference is 5%. Therefore, $16 \times 0.05 = 0.80$; 80% would be the average mortality if the coefficient of inbreeding would reach 100%. Recessive zygotic lethality requires homozygosity at the same locus (present in both parental gametes). According to the Hardy-Weinberg theorem, the frequency of the double recessive genotypes is expected to be q^2, and the frequency of at least one lethal equivalent gene is then $\sqrt{0.80} \approx 0.89$, indicating that almost 90% of the gametes carry a lethal gene or a combination of genes that cause lethality at homozygosity. On this basis, the genetic load of this population is close to 1 lethal equivalent factor per gamete.

Some other investigations estimate the genetic load to be twice as high in some populations. The amount of the genetic load may vary in different populations. It is affected by exposure to environmental mutagens, drugs, exposure to chemicals in the food chain (natural toxins or insecticides, pesticides) or in industrial pollutants or occupational hazards, presence of mutator genes (transposable elements), natural or other types of radiations (X-rays, UV, etc.). Completely dominant lethal mutations do not contribute to the genetic load because they may eliminate the carriers of the genetic defect and thus no load is passed on to successive generations. Of course, some of the mutations may show intermediate types or conditional expression and may or may not contribute to the load. Some deleterious genes are closely linked to advantageous genes and thus transmitted beyond their merit by this "hitchhiking" effect. In such a situation, a recombinational load may exist. Environmental load is generated in highly variable environments where under certain conditions genes are selected that normally convey inferior fitness. Incompatibility load arises in cases of deleterious maternal-fetal interactions, such as those that may arise if the mother is Rh negative for this blood antigen and the fetus is positive, or if the mother expresses phenylketonuria but the fetus is heterozygous (maternal epistasis). *See* allelic frequencies; coefficient of breeding; consanguinity; death; epistasis; fitness; genetic risk; Hardy-Weinberg theorem; incest; incompatibility; lethal equivalent; Muller's ratchet; mutation beneficial; mutation neutral; mutation in human populations; selection coefficient; truncation. (Cavalli-Sforza, L. L. & Bodmer, W. F. 1971. *The Genetics of Human Populations*. Freeman, San Francisco; Muller, H. J. 1950. *Am. J. Hum. Genet.* 2:111; Drake, J. W., et al. 1998. *Genetics* 148:1667; Szafraniec, K., et al. 2001. *Proc. Natl. Acad. Sci. USA* 98:1107.)

The frequency of deleterious alleles is proportional to the mutation rate and selection coefficient: $\hat{q}^2 = u/s$. By rewriting the formula, the mutational load of recessive alleles becomes $u = s\hat{q}^2$, $\hat{q} = \sqrt{u/s}$.

The mutational load of dominant genes is $2u$. In the absence of dominance in a random mating population, the mutational load is $L = 2u/(1+u)$. The mutational load in the most common cases is proportional to the rate of mutation and not to the severity of the affliction.

---

genetic lottery (journalistic) Chance of individuals to inherit certain genes in a population.

genetic manipulation Application of genetic, cytological, or molecular techniques for constructing altered organisms. *See* chromosome engineering; genetic engineering.

genetic map Relative position of genes or other chromosomal markers represented in a linear manner on the basis of recombination frequencies. *See* deletion maps; linkage group; mapping, genetic; mapping function; physical map.

genetic marker Helps to identify nuclear chromosomes, cytoplasmic organelles, and isolated cells on the basis of their inherited behavior and facilitates the identification of the genetic mechanisms involved in special phenomena, such as recombination, gene conversion, mutation, chromosomal rearrangements, genetic transformation, cell fusion, selection, etc.

genetic material Either DNA (in eukaryotes and the majority of prokaryotes) or RNA (in some viruses). These nucleic acids can occur in either double- or single-stranded form. *See* ctDNA; mtDNA; prion; RNA; Watson and Crick model.

genetic milieu *See* genetic background.

genetic mosaic Individual with cell patches of different genetic constitution. It may come about by somatic mutation, movement of insertion or transposable elements, somatic recombination, nondisjunction, deletion, etc. *See* chimera; codominance.

genetic network Connections between DNA, RNA, protein, cis- and transacting regulators, operons, epistasis, signals and signal transducing systems, feedback; involves a large number of genetic and environmental inputs. Genes involved in common processes tend to be expressed in detectable hierarchical waves. Exposure of yeast cells to Cd^{2+} induced 54 proteins and the majority of the sulfur assimilated by the cells is utilized for the formation of glutathion and reduced the production of other sulfur-rich proteins. The regulation takes place at the mRNA level. Glutathione is required for detoxification (Fauchon, M., et al. 2002. *Mol. Cell* 9:713). On the basis of experimental data available as the result of molecular techniques, mathematics-aided models can be developed that may be applied to medical and biotechnological problems. (*See* epistasis; interlogs; microarray; protein complexes; proteome; regulon; synthetic genetic networks; transcriptome; überoperon. (Kalir, S., et al. 2001. *Science* 292:2080; Becskei, A., et al. 2001. *EMBO J.* 20:2528; Hasty, J., et al. 2001. *Nature Rev. Genet.* 2:268; Davidson, E. H., et al. 2002. *Science* 295:1669; Gavin, A.-C., et al. 2002. *Nature* 415:141; Saito, R., et al. 2002. *Nucleic Acid Res.* 30:1163; Guet, C. C., et al. 2002. *Science* 296:1466; Shen-Orr, S. S., et al. 2002. *Nature Genet.* 31:64; Wyrick, J. J. & Young, R. A. 2002. *Current Opin. Genet. Dev.* 12:130; Dietmann, S., et al. 2002. *Current Opin. Struct. Biol.* 12:362; Valencia, A. & Pazos, F. 2002. *Current Opin. Struct. Biol.* 12:368; Rison, S. C. & Thornton, J. M. 2002. *Current Opin. Struct. Biol.* 12:374; Gilman, A. & Arkin, A. P. 2002. *Annu. Rev. Genomics Hum. Genet.* 3:341; <http://dip.doe-mbi.ucla.edu>; <http://predictome.bu.edu>; <http://wwmgs.bionet.nsc.ru/mgs/systems/genenet>.) See color plates in Color figures section in center of book.)

genetic nomenclature *See* databases; gene symbols; Genew. (<http://www.ba.cnr.it/keynet.html>.)

genetic polymorphism Gene loci (or chromosomal arrangements, organelles) in a population are represented by more than one (allelic) form. *See* allele.

genetic privacy Right of an individual to keep his/her genetic record closed to the public. There are two aspects of this right: (1) protection from discrimination by employers, insurance companies, etc., and (2) it may hinder research on genetic disorders and development of new drugs. U.S. law recognizes the protection of medical information. *See* ethics;

genetic testing; wrongful life. (Hall, M. A. & Rich, S. S. 2000. *Am. J. Hum. Genet.* 66:293; Skene, L. 2002. *Trends Mol. Med.* 8:48.)

genetic profile Electrophoretic pattern of microsatellites, restriction fragments, PCR products, etc. *See* electrophoresis; PCR; RFLP.

genetic recombination *See* bacterial recombination frequency; chloroplast genetics; crossing over; gene conversion; homologous recombination; illegitimate recombination; intragenic recombination; mapping, genetic; mitochondrial genetics; molecular mechanism, prokaryote; mtDNA; recombination; recombination, variations of; recombination frequency; recombination mechanism, eukaryote; recombination models; sister chromatid exchange; site-specific recombination; targeting genes; unequal crossing over.

genetic repair *See* DNA repair.

genetic risk Chance that an offspring will be affected by a hereditary defect. The risk can be inferred from the heritability of a particular gene or gene complex in a population. In case of simple Mendelian inheritance such as cystic fibrosis, in some Caucasian populations in genetic equilibrium, the frequency of this anomaly is $\approx 1/2,000 = 0.0005$. Thus, if the frequency of the recessive allele is $\sqrt{0.0005} \approx 0.022 = q$. At genetic equilibrium the frequency of carriers (heterozygotes) is $H = 2pq = 2 \times (1 - q) \times q \cong 0.043 \cong 1/23$. If a person is heterozygous for such a deleterious gene ($q = 0.5$), and marries a spouse by random choice ($q = 0.022$), the chance that they will have an afflicted offspring is $0.5 \times 0.022 = 0.011$, i.e., approximately 1/91. If the same heterozygous person marries a first cousin who may have a 0.25 chance of carrying the same allele, the probability that they will have an afflicted child may be as high as $0.5 \times 0.25 = 0.125$, i.e., 1/8. If, however, an average Caucasian has an offspring with an average Japanese spouse ($q = 0.004$), the probability that their child will be afflicted by cystic fibrosis is only $0.022 \times 0.004 = 0.000088$ or 1/11,363.

The genetic risks will slowly rise with the application of medical care that compensates for the hereditary defects by medicine, e.g., administration of insulin to diabetics or using gene therapy without replacing the defective gene(s). The remedial treatments will not greatly affect the incidence of rare diseases in the short term. If the incidence of a dominant human anomaly is presently 1×10^{-5}, it may take 3,000 years (100 generations) to increase its prevalence to 1×10^{-3}. The incidence of recessive anomalies will rise much slower because the alleles are already sheltered from selection in the heterozygotes. The genetic risk can now be estimated with good precision if molecular information is available on the nucleotid sequences of a gene, e.g., in familial hypercholesterolemia in the gene encoding cardiac β-myosin, a substitution of Glu for Gly at position 256 involves only 0.56 chance for the penetrance of the disease, whereas a Gln → Arg change at position 403 predicts a 100% penetrance and thus sudden death. *See* allelic frequencies; amniocentesis; clinical tests for heterozygosity; empirical risk; genetic counseling; genetic load; genetic screening; genotypic risk ratio; Hardy-Weinberg theorem; mutation rate; prenatal tests; risk; transgenic. (Falconer, D. S. 1965. *Ann. Hum. Genet.* 29:51; Coulson, A. S., et al. 2001. *Methods Inf. Med.* 40[4]:315.)

genetics Study of inheritance, variation, and the physical nature and function of the genetic material. William Bateson suggested the term in 1906 for the entirely new discipline. Genetics may be pursued as a basic science where only the discovery of new principles and their integration into the store of knowledge are the goals. Alternatively, applied branches of genetics rely on the established genetic principles and are used for agricultural (plant and animal breeding) and industrial (biotechnology) purposes or for the improvement of human health (medical genetics). Applications of genetics are expanding into paleontology, archaeology, and forensics. The tools of genetics are integrated into all biological disciplines from taxonomy, evolution, cytology, development, behavior, and physiology, to biochemistry, biophysics, and molecular studies. Thus, genetics has escaped from its classical boundaries of heredity and cytology and has become the core and unifying element of biology. *See* clinical genetics; genetics criticism on; experiments; genetic engineering; GMO; heredity; human genetics; inheritance; medical genetics; population genetics; quantitative genetics; reversed genetics; science.

genetics, chronology of Very broad overview includes only the most important milestones of basic genetics compiled somewhat subjectively. Paraphrasing G. B. Shaw, who would dare to say who is greater than Shakespeare? To keep the length minimal, applied aspects of genetics are not included. (Rédei, G. P. 1974. *Biol. Zbl.* 93:385.)

200–300 B. C. Greek philosophers discuss heredity.

1694 Camerarius recognizes sex in plants.

1761 Kölreuter reports thousands of attempted and some successful plant hybridizations.

1839 Schleiden (plants) and Schwann (animals) discover cellular organization.

1865 Mendel recognizes the basic principles of inheritance.

1866 Haeckel points out the role of the nucleus in heredity.

1869 Galton lays down the foundations of statistics-based inheritance.

1871 Miescher reports about nuclein.

1873–on Mitosis, meiosis, chromosome numbers, supremacy, and continuity of chromosomes are recognized.

1900 Mendel's work is rediscovered.

1902 Sutton proposes the chromosomal theory of inheritance.

1902 Benda recognizes mitochondria.

1902 Garrod reports on alkaptonuria as an inherited biochemical trait.

1906 Bateson suggests the term *genetics*.

1909 Johannsen coins the terms *gene, genotype*, and *phenotype* and explains pure lines.

1909 Correns and Baur discover non-Mendelian inheritance of chloroplasts.

1910–11 Morgan discovers sex linkage and crossing over.

1910 von Dungern and Hirschfeld show that blood groups are inherited.

1913 Sturtevant constructs the first linear map of 6 genes of the *Drosophila* X chromosome.

1913–1925 Bridges and Sturtevant discover deficiency, nondisjunction, duplication, inversion, and translocation.

1926 Chetverikoff and Helena Timoféeff-Ressovskaya found experimental population genetics.

1926 D'Hérelle describes bacteriophages.

1927 Landsteiner and Levine lay the foundations of immunogenetics.

1927 Muller and then Stadler induce mutations by X-rays.

1928 Griffiths observes bacterial transformation.

1930–on Fisher, Wright, and Haldane, working independently, lay down the foundations of theoretical population genetics.

1939–on Delbrück and Luria initiate phage genetics.

1940 Beadle and Tatum conduct experiments leading to biochemical genetics and to the gene-polypeptide theory.

1944 Auerbach and Robson discover chemical. mutagenesis.

1944 Avery, MacLeod, and McCarty demonstrate that the transforming principle is DNA.

1946 Lederberg and Tatum show bacterial recombination.

1949 Chargaff discovers the variable-base composition and A=T, G≡C relations in different DNAs.

1951 McClintock discovers transposable elements.

1952 Lederberg reports transduction.

1953 Watson and Crick construct a valid DNA model.

1955 Fraenkel-Conrat and Williams prove that RNA can also be a genetic material.

1956 Kornberg shows in vitro replication of DNA.

1957 Taylor, and in 1958 Meselsohn and Stahl, show that DNA replication is semiconservative.

1957–on Beginning of the understanding of the machinery of protein synthesis.

1960 Marmur and Lane hybridize nucleic acids.

1960 Barski makes somatic cell hybrids.

1961 Brenner and coworkers explain the nature of mRNA.

1961 Nirenberg and Ochoa laboratories independently demonstrate the nature of the genetic code.

1961 Jacob and Monod propose the operon concept.

1965 Southerland discovers cAMP and opens inquiries on signal transduction and transcription factors.

1969 Shapiro et al. isolate the *lac* operon.

1970 Temin and Baltimore discover reverse transcription.

1970 Khorana synthesizes a tRNA gene in vitro.

1972 Transformation by recombinant DNA begins in Cohen, Berg and Lobban laboratories using plasmid vectors.

1977 Development of efficient DNA sequencing by Gilbert's and Sanger's laboratories.

1978 Shortle and Nathans make localized mutagenesis.

1980 Capecchi et al. and Ruddle et al. transform mice.

1980s Based on earlier studies by E. B. Lewis, Christiane Nüsslein-Volhard and E. Wieschaus establish a new approach to developmental genetics.

1981 Schell et al. transform plants by *Agrobacterium*.

1981 Cech discovers ribozymes.

1983 Varmus and others identify c-oncogenes.

1985 Mullis et al. develop the PCR procedure.

1989 Saiki, Walsh, and Erlich initiate microarray-type analysis of amplified DNA with immobilized sequence-specific probes.

1995–on Sequencing of complete DNA genomes of prokaryotes and the eukaryote yeast.

1996 Beginning of the mass identification of the function of the sequenced genes.

1999 The almost complete sequence of the 33.4-megabase human chromosome 22 was published by 217 authors. Craig Venter' Celera group, the Berkeley, Canadian, and the European Genome Projects publish the first "complete" sequence of the *Drosophila* genome. The sequencing of the *Arabidopsis* genome (2000) and the human genome (2001) followed this. Genetics progresses at breathtaking speed. Yet it is hard to give credit to the major current developments because there are so many and they are so intertwined. During the preceding decades, geneticists tried to reveal the function of single *good* genes or of genetically controlled pathways. By the turn of the millennium, the field is moving toward synthesis and integration. The goal of future research is not less than understanding the function of entire organisms (proteome), including their descent and cooperation. In the coming years, we can expect major progress in the understanding of developmental control, the organization and function of the nervous system, evolution, application of gene and cancer gene therapy; in developing more productive and safer agricultural plants and livestock; in moving from databases to complex information systems. Although genetics is again in a golden age, the excitement may last indefinitely. Yet one must keep in mind the words of the immunologist Peter Medawar: wise people may have expectations, but only the fools make predictions. (Lander, E. S. & Weinberg, R. A. 2000. *Science* 287:1777.)

genetics and privacy Information that is rapidly accumulating on risks based on various screening techniques and DNA sequencing may result in discrimination by insurance companies, potential employers, and possibly society in general. Therefore, there is considerable concern that such information should not be divulged without the consent of the individual and the privacy should be legally protected. *See* bioethics; genetic privacy. (Annas, G. J. 2001. *N. Engl. J. Med.* 345[5]:385.)

genetic scanning   *See* genotyping.

genetic screening   Applied to an asymptomatic population as (1) *prenatal tests* during pregnancy, such as for mucopolysaccharidosis, muscular dystrophy, cytological tests for Down syndrome and fragile X conditions, ultrasound test for developmental anomalies, tests for blood groups (Rh) and infections (syphilis, toxoplasmosis, cytomegalovirus) are mandated or voluntary. Screening of (2) *newborns* aims to reveal whether they are afflicted by autosomal-recessive disorders that require immediate medical attention to prevent severe later consequences. Most frequently, the tests include biotidinase deficiency, congenital hypothyroidism, galactosemia, hereditary tyrosinemia, homocystinuria, maple syrup urine disease, phenylketonuria, and sickle cell anemia. U.S. law mandates these tests. Congenital adrenal hyperplasia, cystic fibrosis, hyperphenylaninemia, arginosuccinase deficiency, galactokinase deficiency, phosphoglucomutase deficiency, homocytinuria, glucose-6-phosphate dehydrogenase deficiency, and others may also be involved.

The frequency of genetic defects may vary in different ethnic groups. Some of the tests are limited to families where history provides clues to potential risk. The tests may be performed on blood drawn from the neonates by specialized laboratories using standard and reliable procedures such as ELISA, enzyme assays, immunoassays, Guthrie test, DNA tests, etc. (3) *Carrier testing* detects heterozygotes for recessive disorders in order to facilitate informed decisions by prospective parents regarding risks, especially in populations where the frequency of the deleterious genes is expected to be high (Tay-Sachs disease among Ashkenazi Jews [0.02]). Thalassemia in people of Mediterranean ethnicity may occur at frequencies exceeding 0.1 in high malaria areas. Cystic fibrosis occurs with variable (generally about 0.02) frequency but is much higher in ethnic populations with a high degree of consanguinity. About 70% of those afflicted by cystic fibrosis have a CTT (Phe) deletion of codon 508 in exon 10 (ΔF508). This assay is not yet used widely. (4) *Presymptomatic* and susceptibility screening may be applied to younger individuals with liability to late-onset genetic anomalies such as autosomal polycystic kidney disease, Charcot-Marie-Tooth disease, Huntington chorea, familial hypercholesterolemia, and retinitis pigmentosa. Some tests provide predictions regarding susceptibility to diabetes mellitus, coronary heart disease, breast cancer, etc.

In some countries predictive premarital testing is mandated for some diseases. Testing for predispositions must require confidentiality because of the obvious relevance to finding jobs and health insurance. Genetic screening of individuals without previous indication of a disorder in the family may involve psychosocial and ethical issues. It is important that screening be conducted only for essential diseases or predisposition, since some conditions are medically treatable and informed choices are available in case the test proves positive. Appropriate and safe procedures should be available and the tests should not be objectionable to the population in general and should be acceptable by the subjects. Genetic screening raises several ethical questions regarding the right or advisability of withholding information, disclosure of information to members of the family, and storage, safe-keeping, and release of the information. *See* abortion, medical; ART; cascade testing; eugenics; false negative; false positive; genetic counseling; GMS; polymerase chain reaction; preimplantation genetics; prenatal diagnosis; RDA; selective abortion; sperm typing; tandem mass spectrometry. (Levy, H. L. & Albers, S. 2000. *Annu. Rev. Genomics Hum. Genet.* 1:139; Pastinen, T., et al.

2000. *Genome Res.* 10:1031; Chace, D. H. 2001. *Chem. Rev.* 101:445.)

genetic segregation   *See* meiosis; Mendelian segregation; modified segregation ratios; preferential segregation; somatic segregation.

genetic sexing line   Mechanical separation of insects by sex is often very difficult or nearly impossible at a larger scale, although this may be required for control by genetic sterilization. By genetic engineering, strains can be developed where under defined conditions either the female or the male individuals can be selectively eliminated upon induction. *See* autosexing; genetic sterilization; sexing. (Robinson, A. S. & Franz, G. 2000. *Insect Transgenesis: Methods and Applications,* Handler, A. M. & James, A. J. eds., p. 307. CRC Press, Boca Raton, FL.)

genetic similarity index   Expresses the similarities between different strains on the basis of the number of shared restriction fragments identified by probes such as DNA minisatellite sequences, etc. *See* genetic distance; microsatellite; minisatellite; probe.

genetics of behavior   *See* behavior genetics.

genetics of cancer   *See* cancer; cancer gene therapy; genetic tumors.

genetic stability   Good if the gene and chromosomal mutabilities are relatively rare, transposable elements are absent, and the population is in genetic equilibrium. *See* genetic equilibrium; genetic homeostasis; mutability; transposable elements. (Li, S. L. & Rédei, G. P. 1969. *Theor. Appl. Genet.* 39:68.)

genetic sterilization (sterile insect technique, SIT)   Heavy doses of ionizing radiation (X-rays) break the chromosomes but do not necessarily kill the irradiated animals and remain capable of mating. In their progeny, because of the chromosomal rearrangements that follow, sterility or lethality occur; or, although the irradiated males copulate, they cannot fertilize the eggs of the females and they leave no offspring. This basic genetic knowledge was successfully applied to insect eradication. The screwworm (*Cochliomya hominivorax*), a tropical and subtropical parasite of warm-blooded animals, produces larvae that hatch in the wounds of livestock and cause great damage to the hide, making it inferior for the leather industry. Additional damage results to agriculture by weight loss in cattle and sheep and to game animals, but the fly may also pose hazards to people. The chemical control of this insect is difficult on live animals and not without danger of pollution and health effects. Therefore, pupae were reared in a large laboratory and treated by about 7,500 R X-radiation, and every weeks 2 million irradiated males were released in the areas with heavy infestation. The monogamous females so mated either failed to produce offspring, or when more sophisticated chromosomal constructs were used, "genetic time bombs" were generated that kept on killing the offspring due to the chromosomal or genic defects (temperature-sensitive alleles). In some areas and in some years this pest control was so effective that the screwworm population was reduced to 1% of what it was before initiation of the program. A similar procedure was successfully applied to mosquito control. Particularly good results were observed in the control of lepidopteran insects with holocentric chromosomes where the delayed and sustained lethal effects could be best exploited.

Constructing a conditional lethal dominant genetic system may cause death without irradiation. The insect becomes lethal when specific conditions are met. In one construct designed in a *Drosophila* model, a tetracycline-repressible transactivator (tTa) protein was placed under the control of the Yp3 fat-body gene promoter. In the absence of tetracycline, any gene controlled by a tetracycline-responsive element (tRe) is normally expressed in the females. On a culture medium containing as low as $0.1\,\mu g/mL$ tetracycline, the females produced no progeny because the tTa prevented the synthesis of fat body (a yolk protein) that is required for nutrient storage and for the insect immune system. The progeny of the males was not affected because they do not produce eggs and do not need fat body for fertility. In a similar manner, a mutant allele of the *male-specific lethal 2* gene (*msl-2^{NOPU}*) selectively killed the females by activating the dosage compensation mechanism in both males and females. By the same principles, insect-mediated (insect vector) human viral (dengue fever, West Nile fever, yellow fever), bacterial (plague, typhus, Lyme disease), protozoan (malaria, Leishmaniasis, sleeping sickness, Chagas disease), and worm-inflicted diseases (river blindness, filariasis) may be controlled. With the increased knowledge of genetic transformation and the availability of various transposable elements, new approaches may open up in insect control. *See* GSM; high-dose/refuge strategy; holocentric chromosomes; *msl*; radiation effects; refractory genes; rtTA; sex determination; SIT; TetR; tetracycline; translocations; tTA. (Thomas, D. D., et al. 2000. *Science* 287:2474; Robinson, A. S. 2002. *Rev. Mut. Res.* 511:113.)

genetic surgery   Replacing a single or a few (defective) genes of an organism with the aid of (plasmid) vectors or introducing foreign genetic material (organelles, chromosomes) into cells with microsyringes or microcapillaries controlled by micromanipulators under microscopes. *See* gene replacement; gene therapy; genetic engineering; gene transfer by microinjection; targeting genes; transformation, genetic.

genetic switch   Mechanisms based on interaction between specific DNA and protein sequences to turn genes on and off. *See Borrelia*; DNA-binding protein domains; DNA-binding proteins; immunoglobulins; mating type determination in yeast; phase variation; regulation of gene activity; serotype; transposition; *Trypanosomas*.

genetic system   Prevalent mode of reproduction (selfing, inbreeding, random mating, assortative mating, etc.). Generally the mechanisms affecting variability (recombination, mutational mechanisms, etc.) are also included in this term. *See* mating systems.

genetic testing May reveal the liabilities of an individual to certain diseases and genetic anomalies. DNA sequencing identifies alterations within genes, although heterozygotes may not necessarily bear a direct burden or risk. Microarray hybridization may also reveal genetic defects, although the tests are more expensive and the results may be ambiguous in case of heterozygosity of the diploid cells. Single-strand conformation polymorphism and denaturing gradient gel electrophoresis are applicable techniques. An individual may benefit from genetic testing because glucose-6-phosphate dehydrogenase deficiency may make a person more susceptible to environmental oxidants (ozone, nitrogen dioxide). Thalassemias may increase the dangers of exposure to lead and benzene; porphyrias to chloroquine and barbiturates; pseudocholinesterase deficiency to organophosphate and carbamate insecticides, etc. Molecular tests may reveal nonsymptomatic heterozygosity for genetic diseases and may predict the risk for having various disorders in the offspring. On the other hand, employers and health insurance companies may discriminate against individuals on the basis of genetic records. Genetic testing may not be applicable for the identification to certain anomalies or diseases and the results of the tests for some conditions may not accurately predict the onset of a disease. With the approval of the Genetics and Insurance Committee of the United Kingdom, the private health insurance companies in the U.K. already use seven tests for mutant alleles for early-onset Alzheimer disease, breast cancer, familial adenomatous polyposis coli, Huntington disease, and three other monogenic diseases. Those who are positive are obligated to pay higher premiums to obtain life insurance over £100,000 and have mortgage insurance. In the U.K., extensive public welfare and health care systems are available. Nevertheless, the ethical aspects of such insurance policy have been questioned. *See* compliant mutation; conversion; DNA sequencing; genetic privacy; genetic screening; microarray hybridization; refractory mutation. (Yan, H., et al. 2000. *Science* 289:1890, Cutler, D. J., et al. 2001. *Genome Res.* 11:1913; <http://www4.od.nih.gov/oba/>, <www.genetests.org>, <www.geneclinics.org>.)

genetic toxicology *See* gene-tox.

genetic transfer May be mediated by the gametes during sexual reproduction, by cytoplasmic organelles, plasmids, episomes, infectious heredity, bacteriophages (transduction) or plasmids, fusion of somatic cells, transfer of isolated organelles, transformation, vectors, viruses, retroviruses, prions, microinjection, electroporesis, targeting genes.

genetic transformation *See* transfection; transformation, genetic; transformation, oncogenic.

genetic transfusion Transfer of organelles and cellular inclusions by protoplast fusion.

genetic translation *See* protein synthesis; regulation of gene activity.

genetic tumor More than two dozen tumor genes have been assigned to *Drosophila melanogaster* chromosomes 1, 2, and 3. The majority of these are not malignant and occur freely or attached to internal organs in the thorax and abdomen. They are distinguished at the third instar larva stage and persist through the life of the individuals. The majority of the tumors become melanotic. The melanotic tumors determined by genes *mbn* and *Tum* have malignant characteristics. In several inbred mice strains, ovary tumors, testis tumors, B-cell lymphoma, kidney adenocarcinoma, leukemia, and pulmonary tumors are under polygenic control. *Bilateral retinoblastoma* (tumor of the eyes) in humans is controlled by a dominant gene. Deficiencies involving the long arm of chromosome 13 may also induce retinoblastoma. Genes involved in the skin disease *xeroderma pigmentosum* are based on a deficiency in the genetic repair mechanism. Initially the disease involves excessive freckle formation and may become tumorous. Exposure to ultraviolet light (sunlight) enhances the formation of skin tumors, particularly in fair-skinned and albino individuals.

Genetic tumors of interspecific tobacco hybrids. (Courtesy of Dr. H. H. Smith.)

About 5–10% of human cancers (hereditary or sporadic) show definite genetic components. Cancer cells commonly display hypermethylation of promoter CpG islands and demethylation of the rest of the genome. The incidence of leukemia may increase in cases of trisomy or partial deficiency for chromosome 21. Both DNA (SV40, adenovirus, bovine papilloma virus, etc.) and RNA viruses (Epstein-Barr virus, retroviruses) can cause tumorigenesis in mammals. The loss or mutation in a gene controlling protein p53 may lead to tumorigenesis presumably due to lack of function of this suppressor gene. Genetic hybrids between the species of the platyfishes (*Xiphophorus*) are prone to develop melanoma. Approximately 30 species crosses of tobaccos may produce tumorous offspring that form callus in vitro cultures without a requirement for phytohormones. In *Nicotiana glauca* ($2n = 24$) × *N. langsdorffii* ($2n = 18$) hybrids, more than one locus is involved in tumor development. In the hybrids of *N. longiflora* ($2n = 20$) × *N. tabacum* ($2n = 48$), one chromosomal segment appears to be responsible for tumorigenesis. *N. saunderae* may inhibit the expression of tumors. In the majority of dicotyledonous and some monocotyledonous plants, agrobacterial infection and the insertion of T-DNA of the Ti plasmid may lead to tumor formation by genetic transformation. Certain viral infections also result in tumorous growth in plants. Several insects stimulate the formation of gall tumors in plant tissues through their metabolic products. For the in vitro development of plant tumors, the additions of phytohormones (primarily natural or synthetic auxins) are required. Some cultures, however, become "habituated" after a course of culture and the exogenous auxin supply may no longer be required. The plant tumors are nonmalignant. *See* adenoviruses; *Agrobacterium*; cancer; carcinogens; Knudson's two-mutation theory of cancer; retinoblastoma; retroviruses; reverse transcription; SV40; tumor; tumor suppressor. (Purello, M., et al. 2001. *Oncogene* 20:4877; Suhardja, A., et al. 2001. *J. Neurooncol.* 52:195; Esteller, M., et al. 2001. *Hum. Mol. Genet.* 10:3001; Smith, H. H. 1973. *Brookhaven Symp.* 25:309.)

genetic vaccine   *See* immunization, genetic.

genetic variability   Ability or proneness (proclivity) to hereditary change. *See* genetic homeostasis; homeostasis; mutation; mutator genes; transposable elements.

genetic variance   Caused by the various effects of the genotype (V_g). The variance observed is usually the phenotypic variance (V_p), which is the outcome of the mutual action of the genotype and the environmental variance (V_e). The genetic variance itself has three components: $V_g = V_a + V_d + V_i$, where V_a is the additive genetic variance or breeding value, V_d is the dominance variance, and V_i is the interaction. The interactions can be epistatic effects among the individual quantitative traits and the effect of the environment on gene expression. The additive genetic variance may also be expressed as $V_a = 2pqt^2(p[1-d]+qd)^2$ where t stands for displacement. The dominance variance $V_d = p^2q^2t^2(d-0.5)^2$. *See* breeding value; displacement; gain; genotypic risk ratio; heritability; midpoint value; polygenes; variance.

genetic variation   Hereditary differences within or between populations.

gene titration   Determining the quantitive effect of genes as a function of dosage or conditional expression. *See* dosage effect; titration. (Yinduo, J., et al. 2001. *Science* 293:2266.)

gene transfer   *See* gene transfer, lateral; human gene transfer; transformation.

gene transfer, lateral   Transmission of genes and genetic elements by infection, plasmids, transposable elements, and the acquisition of mitochondria and chloroplasts during evolution. *See* evolution; infectious heredity; organelle sequence transfer; plasmids.

gene transfer by microinjection   Principal means of transformation of animals in the 1980s. Today, gene targeting and other procedures are preferred. *See* diagram below gene replacement; targeting genes; transformation, genetic [animals].

gene trap vector (entrapment vector)   Equipped with a reporter gene that can insert at a splice acceptor site. The resulting gene fusion may facilitate the transcription of the reporter gene using a donor promoter. It may be used with (mouse) embryonic stem cells to detect genes expressed during early development. Actually, this procedure can tag any gene even if it is not expressed. *See* ES; gene fusion; insertional mutation; OMNIBANK; reporter gene. (Medico. E., et al. 2001. *Nature Biotechnol.* 19:579; Stanford, W. L., et al. 2001. *Nature Rev. Genet.* 2:756; Lai, Z., et al. 2002. *Proc. Natl. Acad. Sci. USA* 99:3651.)

Gene tree   Reveals when a population is divided into two subgroups on the basis that one of the subgroups has a particular mutation(s) and the other does not. Such an analysis can be continued for any number of genes. *See* evolutionary tree; population tree.

general acid-base catalysis   Proton transfer from and to a molecule with the exception of water.

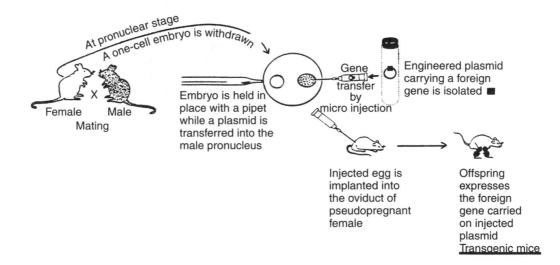

generalized transduction Can be mediated by either temperate or virulent bacteriophages (diagram below). The phage infects a donor bacterium (step 1) carrying the wild-type allele (a⁺) and then lyses it (step 2). Some phage shells randomly scoop up almost **any** bacterial DNA fragment rather than phage DNA. When these unusual phages infect a recipient cell, they can transfer the donor bacterial gene into the recipient (step 3).

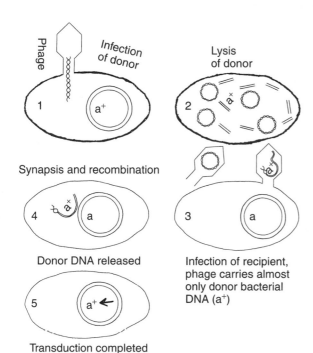

GENE-TOX Genetic toxicology; study of factors (physical and chemical agents) responsible for mutation and cancer or both. *See* carcinogen; environmental mutagens; toxicogenomics. (<http://medlars.nlm.nih.gov>, <http://toxnet.nlm.nih.gov>.)

Genew Human Gene Nomenclature Database. *See* gene symbols; genetic nomenclature; (<http://www.gene.ucl.ac.uk/cgi-bin/nomenclature/searchgenes.pl>.)

genic balance In some organisms (e.g., *Drosophila*) the proportion of the sex chromosomes and autosomes has a crucial role in sex determination. In Drosophila (1 X) and (2 sets of autosomes) = male, whereas (2X) and (2 sets of autosomes) = female (1:1). In general, all individuals with chromosomal ratios above 1 are females and those between 0.5 and <1 are intersexes. In humans and mice, the XO is female, whereas in *Drosophila* XO is male. In trisomics and nullisomics, the nature of the individual chromosome(s), present or absent, makes a great difference in the phenotype. *See* nullisomic; nullisomic compensation; sex determination; trisomy.

geniculate body (geniculate nucleus) Knee-like structure in the brain where optic and auditory fibers are received.

GENIE Gene predictor program. *See* GENSCAN. (Kulp, D., et al. 1996. *Proc. Int. Conf. ISMB* 4:134.)

genistein (4,5,7-trihydroxyflavone) Phytoestrogen, an inhibitor of protein tyrosine kinases frequently used for probing signal transduction pathways. *See* BCP; leukemia.

The transduced DNA and the indigent DNA can synapse (step 4) if they are homologous. By a double exchange, they replace the recipient's gene with that of the donor. This step completes the generalized transduction. In case the donor DNA carries alleles a^+b^+ and the constitution of the recipient is ab, recombination frequencies can be calculated as shown by the formula:

$$\frac{(a^+b)(ab^+)}{(a^+b) + (ab^+) + (a^+b^+)}$$

With generalized transduction, recombination can be estimated only within very short intervals, e.g., within genes. *See* abortive transduction; bacterial recombination frequency; marker effect; *pac* sites; specialized transduction. (Lederberg, J., et al. 1952. *Cold Spring Harbor Symp. Quant. Biol.* 16:413; Burke, J., et al. 2001. *Proc. Natl. Acad. Sci USA* 98:6289.)

general recombination Recombination between homologous sequences. *See* gene conversion; illegitimate recombination; recombination, genetic.

general transcription factor *See* transcription factors.

GeneReviews *See* genetic testing.

genital anomaly syndrome In the genetical males, these syndromes may involve hypospadias or cryptorchidism or micropenis. In hypospadias, the urethra may open at the lower side of the penis or between the anus and the scrotum. Cryptorchidism indicates that the testes do not descend from the abdominal cavity into the scrotum (the testicular bag). In females, most commonly either the ovaries, the uterus, the fallopian tubes (connecting the ovaries with the uterus), or the vagina fails to develop normally and clitoromegaly or fusion of the labia occurs. *See* adrenal hyperplasia; Bardet-Biedel syndrome; Fraser syndrome; gonadal dysgenesis; hermaphroditism; Opitz syndrome; pseudohermaphroditism; Robinow syndrome; Smith-Lemli-Opitz syndrome; testicular feminization; trisomy; Wilms' tumor; Wolf-Hirschhorn syndrome.

genius According to the Roman meaning of the word, it is a guarding spirit influencing a person for better or worse. C. D. Darlington (1964) defined genius as one who "changes the environment of others for his own and even for succeeding generations, for his own species and even for the whole living world." In his book, *Hereditary Genius* (1869), Francis Galton came to the conclusion that eminence is biologically inherited. Darlington had a less asserting view: "the sons of great men are given the best chances with the worst results".

Human intelligence cannot be exempted from the general biological laws (offspring-parent regression), although environment has a great influence on the development of hereditary qualities. A good example is Marie Curie, who became the first woman who received the Nobel Prize for physics in 1903, with her husband, Pierre, and she received it a second time in 1911 for chemistry. In 1935, their daughter, Irene, and her husband, Frédérick Joliot, were awarded the Nobel Prize for chemistry. *See* Bach Sebastian; human intelligence.

GENMAP   Computer program for mapping genetic data based on least squares. *See* least squares.

GENOMATRON   Gene-mapping machine.

genome   complete single set of genes of an organism (taxonomic unit) or organelle; also, the basic haploid chromosome set. The size of genomes in rounded nucleotide numbers varies in the different taxonomic categories:

Human mitochondrion	1.7×10^3	bp
MS2 (single-stranded RNA bacteriophage)	3.5×10^3	bases
φX174 (single-stranded DNA bacteriophage)	5.4×10^3	bases
SV40 (double-stranded animal DNA virus)	5.2×10^3	bp
Tobacco mosaic virus (single-stranded RNA)	6.4×10^3	bases
Influenza virus (single-stranded animal RNA virus)	1.4×10^4	bases
λ (double-stranded DNA bacteriophage)	4.9×10^4	bp
Vaccinia virus (double-stranded animal DNA virus)	1.9×10^5	bp
T2, T4 (double-stranded DNA phages)	1.7×10^5	bp
Chlamydia (bacteria)	6.0×10^5	bp
Escherichia coli bacterium	4.7×10^6	bp
Calotrix (bacteria)	1.3×10^7	bp
Saccharomyces cerevisiae (fungus, eukaryote)	1.2×10^7	bp
Drosophila melanogaster (insect)	9.0×10^7	bp
Caenorhabditis elegans (nematode)	1.0×10^8	bp
Arabidopsis thaliana (higher plant)	$\sim 1.2 \times 10^8$	bp
Rice	$\sim 4.0 \times 10^8$	bp
Mouse	$< 3.5 \times 10^9$	bp
Homo sapiens	$\sim 3.2 \times 10^9$	bp
Toad *(Bufo bufo)*	6.0×10^9	bp
Maize (higher plant)	2.5×10^9	bp
Hexaploid wheat ($n = 3x$)	1.7×10^{10}	bp
Trillium luteum (higher plant)	6.5×10^{10}	bp
Fritillaria davisii (higher plant)	1.5×10^{11}	bp

The average genome size of birds is about one-third of that of mammals, mainly because the avian introns are shorter. *See*

chloroplasts; mtDNA; plants. (Bennett, M. D. & Leitch, I. J. 1995. *Ann Bot.* 76:113, bacterial genomes: Casjens, S. 1998. *Annu. Rev. Genet.* 32:307; Genetica Vol. 115: issue 1 (2002), minimal genome size, C-value paradox, and animal genomes: (<mesquite.biosci.arizona.edu/mesquite/mesquite.html>; <http://www.cbs.dtu.dk/databases/DOGS/>.)

genome analysis   Initially it meant determining the origin of the component genomes in allopolyploid species on the basis of chromosome pairing, univalent(s) and multivalent associations, chiasma frequencies, chromosome substitution, chromosome morphology, chromosome banding, and hemizygous ineffective alleles. Today, it is used more generally for studying DNA base sequences, microsatellites, etc.

genome annotation   Identification of the function of open reading frames and other elements. In microbial genomes, when an annotated gene is linked to an unclassified one either by genetic linkage in a related species or microarray profile or protein–protein interaction, there is a high probability that they are members of the same functional category. The expansion of annotations will be the task of the proteome projects. *See* DAS; proteome. (Marcotte, E. M., et al. 1999. *Science* 285:751; Mount, S. M. 2000. *Am. J. Hum. Genet.* 67:788; Devos, D. & Valencia, A. 2001. *Trends Genet.* 17:429; Karlin, S., et al. 2001. *Nature* 411:259; Auburg, S. & Rouze, P. 2001. *Plant Physiol. Biochem.* 39:181; Stein, L. 2001. *Nature Rev. Genet.* 2:493; Yanai, I., et al. 2001. *Proc. Natl. Acad. Sci. USA* 98:7940; <http://gen100.imb-jena.de/~baumgart/rummage/register.html>; <http://www.fruitfly.org/GASP/index.html>; annotations for 77 genomes available by July 31, 2002: <http://cbcsrv.watson.ibm.com/Annotations/home.html>.)

genome database   (<http://gdbwww.gdb.org>.)

genome equivalent   The mass of the DNA/RNA is the same as that of a genome.

genome hitchhiking   *See* überoperon.

Genome Information Broker (GIG)   Database for genomics of prokaryotes, fungi, and *Arabidopsis*. *See* genomics. (<http://gib.genes.nig.ac.jp>.)

genome-linked viral protein   *See* VPg.

genome mutation   Affects chromosome numbers. *See* aneuploid; polyploid.

genome organization   *See* chromatin; euchromatin; genome; heterochromatin; repetitive DNA.

genome project   Focused on the physical mapping and sequencing of entire genomes of humans and other higher and lower eukaryotes as well as of prokaryotes. Upon completion of these projects, a detailed inventory of all genes will become available. This in turn will facilitate a new generalization of organization and function of the cells and will permit the application of the new principles and the new technologies to human economic fields as well as for preventing and curing diseases. The complete nucleotide sequence of the 4 genes of the MS2 RNA virus was determined by 1976, and by 1995, the 1,749 genes of *Haemophilus influenzae* bacterium had been sequenced. The genome of *Saccharomyces cerevisiae* yeast has also been completely sequenced.

The large eukaryotic genomes, such as those of humans, containing over 3 billion bps, are ordered first into sequential stretches by the use of overlapping fragments. The first step is breaking up the human chromosomal DNAs (average of 250 Mb) into 100–2,000 kb fragments and cloning them in YACs. The YACs are cleaved into an average of 40 kb fragments and cloned by cosmids. The contents of the cosmids are then cloned in 5–10 kb capacity double-stranded DNA plasmid vectors or into the single-stranded filamentous phage M13 vector of 1 kb load. The fragments at each step can be tied into contigs by chromosome walking. The nucleotide sequences of the smaller clones can be analyzed. The entire human genome requires a minimum of about 3,000 YAC,

20,000 BAC, 75,000 cosmid, 600,000 plasmid, or 3 million M13 phage clones.

An alternative approach, complete sequencing, is to proceed from sequence-tagged connectors (STC). The human chromosomes would be cloned in BAC vectors and sequence 300–500 nucleotides at the ends. The 600,000 BAC end sequences represent 10% of the genome and are scattered every 5 kb across the genome. They are called *sequence-tagged connectors* because they allow each BAC clone to be connected to about 30 others (150 kb insert/5 kb ≅ 30; Mahairas, G. G., et al. 1999. *Proc. Natl. Acad. Sci. USA* 96:9739). The BAC inserts are digested by a restriction enzyme to determine their size. The sequencing templates have pUC18-based plasmids with ~2 kbp templates. A "seed" BAC is sequenced and checked against the data of sequence-tagged connectors to identify the overlapping clones. In a following step, two BACs that show internal homology by the restriction enzyme digests and minimal overlap at their end are completely sequenced. By such a procedure, the entire human genome could be sequenced in 20,000 clones. The advantage of this proposal (Venter, J. C., Smit, H. O. & Hood, L. 1996. *Nature* 381:364) is that some of the low-resolution mapping (YAC and cosmid steps) could be eliminated and automatic sequencing procedures could be applied, reducing cost and labor. Many groups worldwide could

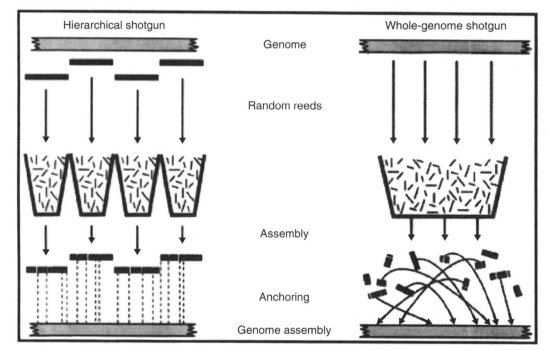

The human genome sequencing projects reported in February 15, 2001, *Science* 295, and *Nature*, February 16, 2001, used two somewhat different sequencing strategies. The *hierarchical shotgun* procedure cut up the genome into BAC clones and arranged them in a somewhat overlapping tiling path. After shotgun sequencing, each BAC clone was reassembled, then the sequences of adjacent clones were merged. This had the advantage that all sequence contigs and scaffolds derived from a BAC belong to a single compartment with respect to anchoring to the genome. The *whole genome shotgun* strategy shotgun sequenced the entire genome, then reassembled the entire collection. With this method, each contig and scaffold is an independent component that has to be anchored to the genome. In general, many scaffolds might have been difficult to anchor to the genome. (From Robert H. Waterston, Eric S. Lander, and John E. Sulston. 2002. *Proceedings of the Natl. Acad. Sci. USA* 99:3712–3716. Copyright 2002 National Academy of Sciences USA.)

do the BAC clone sequencing. The already known sequence-tagged sites (STS) and expressed sequence tags (EST) could be readily located and additional genes could be placed more easily. The suggested procedure would greatly facilitate the sequencing of other smaller genomes of interest.

The Perkin-Elmer Corporation and Craig Venter use very high-efficiency DNA sequencing apparatuses (230ABI PRISM 3700) based on capillary electrophoresis (~1,000 samples/day) and robotization to expedite the process and substantially reduce the cost of sequencing. The completed sequencing information of entire genomes reveals that the majority, but not all, of cellular functions are directed by homologous genes from *Saccharomyces* to *Canorhabditis, Drosophila, Homo* and *Arabidopsis.* Yet the regulation of the functions shows differences to account for the differences among these organisms. The number of genes used by the different organisms varies. Apparently and unexpectedly, *Arabidopsis* needs about twice as many genes as *Drosophila,* and *Caenorhabditis* relies on nearly 45% more genes than *Drosophila. Drosophila* has about 700 transcription factors versus 500 in *Caenorhabditis.* Among the 24 fully sequenced microbial genomes, on the average, about 25% of the open reading frames are unique to each organism. Dunham (2000, *Trends Genet.* 16:458) tabulates a chronology of the innovations facilitating the realization of the genome projects. *See* BAC; capillary; clone validation; contigs; cosmid; cosmid vectors; databases; DNA chips; DNA sequencing; DNA sequencing, automated; electrophoresis; EST; finishing; gap; gene ontology; parking; physical mapping; restriction enzyme; scaffolds, genome sequencing; seeding; sequence-tagged connectors; shotgun sequencing; STS; tiling; WGS; YAC. (Venter, J. C., et al. 1998. *Science* 280:1540; Mullikin, J. C. & McMurray, A. A. 1999. *Science* 283:1867; Adams, M. D., et al. 2000. *Science* 287:2185; Waterston, R. H., et al. 2002. *Proc. Natl. Acad. Sci. USA* 99:3712; Myers, E. W., et al. 2002. *Proc. Natl. Acad. Sci. USA* 99:3712; Internet guides to the majority of genome-related databases: *Nature Genet.* 32 suppl. 1–79 [2002], Birney, E., et al. 2002. *Annu. Rev. Genomics Hum. Genet.* 3:293; <http://www.ncbi.nlm.nih.gov/genome/guide>; <http://www.ncbi.nlm.nih.gov/Genomes/index.html>; <compbio.ornl.gov/channel>; <www. ensembl.org/genome/central>.)

genome scanning Cutting up the genome first by 8 bp–recognizing restriction endonucleases into large fragments, followed by using more frequent cutter enzymes to generate physical information on the entire genome. These fragments can then be used to establish a physical map. *See* gene finding; physical map; restriction enzyme. (Rouillard, J. M., et al. 2001. *Genome Res.* 11:1453; Beekman, M., et al. 2001. *Genet Res.* 77:129.)

genome sequence database (GSDB) (<http://www.ncgr. org>; <http://www.mips.biochem.mpg.de>).

genome sequence sampling (GSS) Chromosomal DNA digested with several restriction enzymes is cloned into cosmids. Hybridization with YAC clones of the same chromosomal DNA identifies all of the cosmids that contain sequences present within the YAC. The cosmids are then broken down into contigs and their ends are identified by hybridization to pure cosmid DNA. The 300–500 bp of the ends are sequenced

and aligned in sequence, permitting the generation of a rather high-density physical map.

genome size *See* genome.

genome defense model Generally, multiple, different transposable elements occur in all organisms, and their movements from one chromosomal location to another may bring about rearrangements in the genome. The cell keeps these transpositions in check by methylation of the transposase and thus restricts deleterious alterations in the genome. (Miura, A., et al. 2001. *Nature* 411:212.)

GenomeScan Gene identification algorithm. (Yef, R.-F., et al. 2001. *Genome Res.* 11:803.)

genometrics Biometric analysis of chromosomes. *See* biometry; genomics. (Roten, C.-A. H., et al. 2002. *Nucleic Acids Res.* 30:142; <http://www.unil.ch/comparativegenometrics>.)

genomic clone Prepared from chromosomal DNA rather than from cDNA. *See* cDNA; genomic DNA.

genomic control Statistical method for the estimation of population structure in a manner somewhat similar to case-control design or transmission disequilibrium tests. *See* case-control design; transmission disequilibrium test. (Bacanu, S. A., et al. 2000. *Am. J. Hum. Genet.* 66:1933.)

genomic DNA (gDNA) Native DNA including exons, introns, and spacer sequences (versus the processed genes transcribed from mRNA to DNA by reverse transcription and having only the coding sequences). *See* processed genes; reverse transcription.

genomic exclusion Takes place in the ciliate *Tetrahymena pyriformis* in case one of the two mates has a defective genome (micronucleus) that is therefore not included in the meiotic progeny. The first progeny becomes heterokaryotic, having only the normal diploid micronucleus and an old macronucleus that is genetically not concordant with the micronucleus. After a subsequent mating, the normal micronucleus forms a macronucleus concordant with its own genetic constitution. As a result, the strain is purged from the defect. *See* conjugation; *Paramecia.* (Cole, E. S., et al. 2001. *J. Eukaryot. Microbiol.* 48[3]:266.)

genomic formulas n = haploid, $2n$ = diploid, $3n$ = triploid, etc., and $n - 1$ or $2n - 2$ = nullisomic, $2n - 1$ = monosomic, $2n + 1$ = trisomic, $2n + 2$ = tetrasomic, etc., where n = haploid chromosome number. The basic chromosome number is, however, x, and the diploids may be $2x = 2n$. *See* aneuploids; polyploids.

genomic hybridization *See* comparative genomic hybridization.

genomic library Set of cloned genomic DNAs; it is expected that a good library includes at least one copy of all of the genes of a particular genome. *See* cloning; fragment recovery probability; genome.

genomic mismatch scanning   *See* GMS.

genomic prospecting
Searching of diverse species (e.g., different mammalian genomes) for DNA sequences that could alleviate human disease with the aid of gene therapy. *See* gene therapy. (O'Brien, S. J. 1995. *Nature Med.* 1:742.)

genomics
Study of the molecular organization of genomic DNA and physical mapping. *Structural genomics* studies the folds of macromolecules, the three-dimensional shape of biological molecules with the aid of physical instruments (X-ray crystallography, etc.), and bioinformatics and classifies these molecules into functional families. *Functional genomics* deals with genome-wide functional analyses, integration of structure of the DNA, and the molecular function and interaction of genes and gene products (Wu, L. F., et al. 2002. *Nature Genet.* 31:255). After completion of sequencing of the organisms, interest has turned to the determination of the function(s) of genetics. Such studies can now be conducted with high-thoroughput procedures (see diagram modified after Ross-Macdonald, P., et al. 1999. *Nature* 402:413). These types of methods are capable of identification of the function of thousands of ORFs in combination with macroarray analysis. This area of study integrates genetics, molecular biology, biochemistry, pharmacology (*pharmacogenomics*: designing drugs that best fit the genetic constitution of an individual), agriculture, medicine, and other disciplines. In 1996 alone, half a million patents were proposed from the field. *Epigenomics* studies the interaction between proteomes and genomes, global patterns of methylation, methylation signals, and surveys this type of information in different species. *Comparative genomics/phylogenomics* seeks to determine (1) the number of distinct protein families encoded by different genomes, (2) the distribution of the coding genes within the genomes, and (3) how many of the genes are shared by the different genomes. *Orthogenomics* deals with the genomes of orthologous descent, whereas *paragenomics* studies paralogous genomes. The study includes the composition and organization of protein domains in different organisms. *Toxicogenomics* seeks understanding in the complexities in the biological system responding to toxic, mutagenic, and carcinogenic factors. Special consideration is given to homologies of genes involved in controlling disease in the human genome, sharing fundamental functions such as cell cycle and structure, cell adhesion, signaling, apoptosis, neuronal controls, and the defense system (immune reactions). The finished genomic sequence is contiguous and errors do not exceed 1/10,000 bases. *See* biotechnology; *Cre/Lox*; DNA chips; DNA sequencing; duplications; gene numbers; genetic engineering; genome projects; genomic DNA; knockin; macroarray analysis; maldi/tof/ms; mass spectrometer; microarray hybridization; ORF; orthology; paralogy; physical mapping; proteome; SAGE; X-ray diffraction analysis. (Craig, A. G. & Hoheisel, J. D. 1999. *Automation: Genome and Functional Analyses.* Academic Press, San Diego, CA; *Trends Guide to Bioinformatics*, Sup., Elsevier, 1998; Rubin, G. M., et al. 2000. *Science* 287:2204; Koonin, E. V. 2001. *Curr. Biol.* 11:R155; Reboul, J., et al. 2001. *Nature Genet.* 27:227; Gopal, S. 2001. *Nature Genet.* 27:337;

genometrics: Meyerowitz, E. M. 2002. *Science* 295:1482; Aardema, M. J. & MacGregor, J. T. 2002. *Mutation Res.* 499:13; <www.nature.com/genomics/>; <http://gib.genes.nig.ac.jp>.)

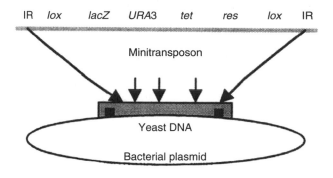

☆ Prepare and amplify yeast DNA from plasmid
☆ Cut at restriction sites (■) of yeast DNA
☆ Knock in the transposon into haploid yeast strains and identify the presence of the insertion of the markers within a large number of known yeast ORFs
☆ By the disruption the function of the genes is identified

genomic screening
Used for the localization of genetics markers (genes). For *random genomic screening*, usually anonymous polymorphic markers are employed. For *directed genomic screening*, specific polymorphic markers are suitable that have already been located in the vicinity of a targeted gene(s). The best markers are easy to recognize, are highly heterozygous, and have established chromosomal location. In human genetics the Généthon (http://www.genethon.fr/genethon_en.html) map containing more than 5,000 dinucleotide markers covering the entire genome by ~2 cM average spacing are used. Alternatively, the Cooperative Human Linkage Project (CHLC, http://www.chlc.org) covers the genome by 3,600 tri- and tetranucleotide markers with an average spacing of 1 cM; the Utah Human Genetics Institute (http://www.genetics.utah.edu/home.html) developed tetranucleotide markers at 10–15 cM spacing. These spacings are average and are not evenly distributed. The standard sets of markers are called *mapping sets*. Using lod scores, a value of 3 is considered to be significant. When 2 genes are tested, the lod score probability may be corrected for more accuracy; it should be $3 + \log 2 \approx 3.3$, and in case of say 20 genes, it should be $3 + \log 20 \approx 4.3$. Sib pair data sets may be evaluated with the aid of the χ^2 procedure and the appropriate degrees of freedom, but general probability may be determined by using lod $= (\chi^2)/4.605$. ($2 \ln 10 \approx 4.605$, and converts the lod scores to χ^2 with 1 degree of freedom. Lod score of 3 corresponds to a P value of 0.001, and $\chi^2 \approx 4.605 \times 3 \approx 13.83$.) Jianfeng Xu, et al. (1998) present the justification for the very high significance level of 0.001 as follows. If the human genome is 3000 cM and it is divided into sixty 50 cM segments, and the studied locus is in one of them, then the chance for the location of this gene is $1/60 \approx 0.02$. That is, under this assumption, the a priori chance of linkage for any single locus is ~2%. According to Bayes' theorem with a lod score of 3, the posterior probability for linkage is about 95%, a conventional limit for level of significance. Usually these linkage data are loaded with false positive results and the best statistical procedures have

not been agreed upon or generally accepted. Directed genomic screening may identify *locational candidate regions* where the investigated genetic difference is likely to be situated. As a general rule, large relative risk, determined by λ_S, is very helpful in locating genes. *See* Bayes' theorem; candidate gene; chi square; λ_S; lod score; mapping; microarrays; microsatellite; minisatellite; physical mapping.

genomic stress Such as dissimilar genetic backgrounds in hybrids, in vitro cell culture, etc., may activate dormant transposable elements and cause genetic instability. *See* somaclonal variation; transposable elements.

genomic subtraction Method that removes all of the sequences present in a deletion mutant from wild-type DNA but retains the wild-type DNA sequences corresponding to the deletions by denaturing a mixture of wild-type and biotinylated mutant DNA. Allowing the mix to reassociate, the biotinylated sequences are subtracted by several repeated cycles of binding to avidin-coated polystyrene beads (which have great affinity for biotin). The remaining (nonbiotinylated) DNA is wild type and contains only the sequences that were deleted in the mutant but present in the wild type; it can then be amplified by PCR and studied by standard techniques of sequencing. This method also permits the isolation of genes affected by the deletion (caused by, e.g., ionizing radiation). *See* avidin; biotinylation; DNA sequencing; gene isolation; PCR; physical mutagens; RDA; RFLP subtraction; subtractive hybridization. (Kingsley, P. D., et al. 2001. *Dev. Growth Differ.* 43[2]:133.)

genomotyping Hybridizes the DNA of a particular strain/isolate to the genome of a sequenced standard line to assess the difference between the two.

genophore Gene string not associated with large amounts of protein (bacterial chromosome). (Ris, H. & Chandler, B. L. 1963. *Cold Spring Harbor Symp. Quant. Biol.* 18:1.)

genotoxic chemical Causes gene mutation, chromosomal aberration, and cancer. Genotoxic stress activates cell cycle checkpoints to allow time for repair if possible. The most recommended tests involve bacterial mutation assays, in vitro tests for chromosomal damage using mammalian cells (rodent hematopoietic cells), and in vitro assays of mouse lymphoma $tk^{+/-}$ cells (MLA). Molecular effects of genotoxic chemicals can be assessed by single nucleotide polymorphism analysis. *See* cell cycle; checkpoint; databases; environmental mutagens; gene-tox; mutagen assays; pharmaceuticals; SNIP. (Müller, L., et al. 1999. *Mutation Res.* 436:195.)

genotype Genetic constitution; the full set of genes.

genotype elimination Statistical algorithms are used for the identification of genotypes that are inconsistent with the pedigree information. (O'Connell, J. R. & Weeks, D. E. 1999. *Am. J. Hum. Genet.* 65:1733.)

genotypic frequency *See* Hardy-Weinberg theorem.

genotypic mixing After infecting a cell with viruses of different genotypes, in a single viral capsid more than one type of viral DNA may be included. *See* rounds of matings.

genotypic risk ratio (GRR) Total number of offspring affected/twice the number of affected homozygotes. *See* displacement; empirical risk; genetic hazards; genetic risk; risk.

genotypic segregation *See* trinomial distribution.

genotypic value Quantitative genetics term indicating the genetically determined component (G) of the phenotypic variation; phenotypic value $(P) = G + E$, where (E) stands for environmental variation. *See* additive effects; breeding value; midpoint.

genotyping Identifying the genotypic constitution at one or more loci by genetic, molecular, immunological, or any other means using cells, tissues, or whole organisms. Quite commonly, RFLP, mini- and microsatellites, trinucleotide repeats, single-nucleotide polymorphism, and PCR are used. High-thoroughput methods have been developed by immobilizing DNA on silicon chips and using Maldi. Statistical methods based on inheritance are available for the detection of genotyping errors (Douglas, J. A., et al. 2002. *Am. J. Hum. Genet.* 70:487; Sobel, E., et al. 2002. *Am. J. Hum. Genet.* 70:496). *See* DNA chips; genotype; MALDI; microarray hybridization; microsatellite; minisatellite; PCR; RFLP; SNIP; trinucleotide repeats. (Tang, K., et al. 1999. *Proc. Natl. Acad. Sci. USA* 96:10016; Ranade, K., et al. 2001. *Genome Res.* 11:1262; Beaulieu, M., et al. 2001. *Nucleic Acids Res.* 29:1114; Wolfe, J. L., et al. 2002. *Proc. Natl. Acad. Sci. USA* 99:11073.)

gens (plural gentes) Organisms with shared relations (a subrace).

GENSCAN Gene predictor program. (Burle, C. & Karlin, S. 1997. *J. Mol. Biol.* 268:78.)

gent algorithm Generates contigs from optical mapping data. *See* contig; optical mapping. (Mathe, C., et al., 1999. *J. Mol. Biol.* 285:1977.)

gentamicin Aminoglycoside antibiotic (see formula).

Gentamicin sulfate.

genus Taxonomic category usually including several species of common descent, e.g., *Drosophila* (genus) *melanogaster* (species), the fruitfly most commonly used in genetics studies. Some genera are monotypic, however, inasmuch as they consist of a single species, e.g., *Arabidopsis*. *See* species.

geographical race   Topologically separate population with distinctive gene frequencies.

geographic isolation   Populations cannot exchange genes because physical distance or other physical factors (mountain ranges, lakes, etc.) keep them apart. *See* speciation.

geological age   *See* evolutionary clock.

geological-evolutionary time periods ($\sim$ millions of years ago) Formation of the Earth—4600—origin of Life—3000—Cambrian—600—Ordovician—450—Silurian—410—Devonian—345—Carboniferous—280—Permian—225—Triassic—190—Jurassic—135—Cretacious—65—Eocene—36—Oligocene—Miocene—13—Pliocene—3—Pleistocene—0.01 → Recent. *See* Archeozoic; Cenozoic; evolution, prebiotic; evolution of the genetic code; Mesozoic; origin of life; Paleozoic; Pterozoic.

geometric mean   *See* mean.

geometric progression   Series of elements increasing by the same factor, e.g., 2, 6, 18, 54 (i.e., by a factor of 3 in this example). *See* arithmetic progression.

George III   Mad king of England (1738–1820) who might have been a victim of porphyria. *See* porphyria.

geotropism   Growth influenced by gravity; positive geotropism directed toward (+) and away (−) from gravity. Plant roots grow downward (+) and the shoots upward (−).

GEP   *See* guanine nucleotide exchange protein.

gephyrin (93 kDa)   Peripheral nervous system membrane protein binding the inhibitory β-subunit of the motor neural glycine receptor to tubulin in the cytoskeleton. It is also used for a cofactor that regulates molybdenum-dependent enzymes. *See* cytoskeleton; neuron; tubulin. (Sola, M., et al. 2001. *J. Biol. Chem.* 276:25294.)

geranyl pyrophosphate   Precursor of farnesyl pyrophosphate. Two molecules of farnesyl pyrophosphate then join by the pyrophosphate end; through the elimination of both pyrosphosphates, squalene is formed. Squalene is then cyclicized to form lanosterol before being converted into cholesterol. *See* cholesterols; prenylation.

Gerbich (Ge blood group)   Distinguished by its encoding β- and γ-sialoglycoproteins (glycophorins). These red blood cell membrane proteins are suspected of being the receptors of the *Plasmodium falciparum* merozoite (malaria-causing protozoan). *See* blood group; malaria. (Mayer, D. C., et al. 2001. *Proc. Natl. Acad. Sci. USA* 98:5222.)

gerbil   *Gerbillus cheesmani* $2n = 38$; *Gerbillus gerbillus* $2n = 43$ male, 42 female.

germ   Pathogenic microorganism or an initial cellular structure capable of differentiation and development into a special organ or organism.

german   Closely related, such as having the same parents. *See* cousin german.

german measles   *See* rubella virus.

germarium   Location of the pro-oocytes, which through mitotic divisions give rise to the oocysts, and one of them becomes the oocyte. *See* karyosome; oocyte, primary.

germ cells   Reproductive (sex) cells of eukaryotes such as spores, eggs, and sperm. The spores frequently come about by nonsexual processes such as the conidia of fungi and may not function similarly to sex cells. The egg and sperm are direct or indirect products of meiosis that have undergone a process of differentiation without division, e.g., the spermatozoa of animals arise from the spermatids, or the sperm of plants are formed by postmeiotic division of the microspore nuclei. The eggs of animals arise by an additional division of the haploid secondary oocytes. The egg of plants is formed through three divisions from one of the haploid megaspores. *See* conidia; egg; gametogenesis; megaspore; microspore; sperm; spore.

germinal center   Group of naive (uncommitted) B cells. When activated by a specific antigen, they may develop into memory B cells after antigen selection or become plasma cells. In the presence of interleukin-1,-10 and CD40 ligands, they become memory B cells. By removal of CD40 ligands, the cells differentiate into plasma cells. A rapidly growing center also includes antigen-specific helper T cells. CD21 may be required for the B cells to survive in the germinal center. *See* antigen; B lymphocyte; CD21; CD40; clonal selection; memory, immunological; OBF; plasma cell, somatic hypermutation; T lymphocyte. (Schebesta, M., et al. 2002. *Curr. Opin. Immunol.* 14:216.)

germinal choice   Idea that parents should not necessarily rely on their own gametes for producing offspring but adopt eggs or sperm or even fertilized eggs from superior gene pools as a practical measure of positive eugenics. *See* ART; eugenics; in vitro fertilization; sperm bank.

germinal mutation   Occurs in the germline, gonads, or gametes. *See* gamete; germline; gonad.

germinal vesicle   Large nucleus of the amphibian oocyte. This nucleus contains the three eukaryotic RNA polymerases and can transcribe exogenous (microinjected) DNA. The oocyte then translates the mRNAs into a variety of proteins. *See* in vitro translation systems.

germ layers   Gastrulation forms the inner-most layer, *endoderm*; the surface layer, *ectoderm* (epithelium); and the in-between mesenchyme cell layer, *mesoderm*. Some embryologists attribute the differentiation to the neural crest. *See* gastrula; neural crest.

germline Cell lineage that contributes to the formation of the gametes. In the majority of animals, the germline is determined very early in the zygote, although the embryonic stem cells have pluripotent capability. In mice the germ cells originate from extra-embryonic ectoderm under the influence of an inducible transmembrane protein encoded by the *fragilis* gene. Then gene *stella* is expressed in the cells that are restricted to the germline. The latter gene represses homeobox genes in the cells and thus they retain pluripotency (Saitou, M., et al. 2002. *Nature* 418:293). According to some views, plants do not have germline — certainly not in the sense of animals — because the generative cell lineage is not set aside definitely in early development and plant cells may retain totipotency for almost the entire life of the individuals. Nevertheless, by fate maps the cell lineages giving rise to megaspores and microspores of plants can be traced to origin. In *Drosophila*, for the development of the germline, the product of the *nanos* (*na*) gene locus is essential. If mutation occurs in the germline, the genetically mosaic tissue may produce different gametes. Some mutations that appear recessive in the somatic tissues may display reversal and function as dominant. In such cases, selection is possible before formation of the gametes. *See* cell lineages; *Drosophila* life cycle; gametogenesis; genetically effective cell number; germ plasm; gonads; morphogenesis in *Drosophila*; somatic embryogenesis; stem cells; Keimbahn. (Lin H. 1997. *Annu. Rev. Genet.* 31:455; Saffman, E. E. & Lasko, P. 1999. *Cell. Mol. Life Sci.* 55:1141; Extavour, C. & García-Bellido, A. 2001. *Proc. Natl. Acad. Sci. USA* 98:11341; Crittenden, S. L., et al. 2002. *Nature* 417:660.)

germline transcript (sterile RNA) Not translated into protein. These specific guanine-rich RNAs are transcribed from the immunoglobulin heavy-chain S (switch) sequences in the B lymphocytes. These RNAs of 1 to 10 kb in length, containing repeats of 20 to 100 bp, anneal with the cytosine-rich DNA template. Although they have a similar overall structure, the sterile transcripts are specific for each switch sequence preceding a heavy-chain gene, and each mediates in cis-position class switching of a specific heavy-chain gene. It has been hypothesized that these RNA-DNA hybrids are the recognition sites for the endonuclease that cuts the DNA double strands in the process of class switching. *See* antibody gene switching; cis arrangement; immunoglobulins. (Tracy, R. B., et al. 2000. *Science* 288:1058.)

germ plasm Development (in *Drosophila*) begins with the formation of the primordial germ cells, also called pole cells. The syncytial nuclei congregate at the posterior segment of the pole. Cellularization begins after about 2 hours. During gastrulation, the germ cells move to the embryonic gonad and form the germline stem cells. In both males and females, after four rounds of cell divisions, 16 cells are formed. In the male, all 16 contribute to sperm formation. In the female, these 16 cells remain interconnected, but only one becomes an oocyte and the other 15 become polyploid nurse cells and nourish the oocyte. The oocyte proceeds with meiosis. About 80 maternal (somatic) follicle cells surround the oocyte and nurse cells. The development of the germ plasm is controlled by the interacting products of a series of genes. *See Drosophila*; germline; morphogenesis in *Drosophila*.

germplasm (Keimplasma) Sum of the genetic determinants transmitted through the gametes to the progeny. In a broader sense it is used for the designation of a collection of genotypes of organisms for plant and animal breeding resources. *See* genotype; germ plasm.

gerontology Clinical, biological, and sociological study of aging. *See* aging; apoptosis; Hayflick's limit.

Gerstmann-Sträussler disease (GSD) Chromosome 20p12-pter-dominant brain disease with substantial similarities to Creutzfeldt-Jakob disease. There are some apparent differences inasmuch as in GSD there are numerous multicentric tuft-like plaques in the cerebral and cerebellar cortex, basal ganglia and the white matter of the brain. GSD appears to have a greater recurrence risk than Creutzfeldt-Jakob disease. *See* Creutzfeldt-Jakob disease; encephalopathies; encephalopathy bovine spongiform; prion; scrapie.

gestation Time from fertilization of the ovum (ova) to delivery of the newborns in viviparous animals. The average days of gestation: opossum 13, hamster 17, mouse 19, rat 21, rabbit 31, giant kangaroo 39, dog 61, cat 63, guinea pig 68, sow 114, sheep and goat 151, Virginia deer 215, Rhesus monkey 164, chimpanzee 238, woman 267, cow 284, mare 340, elephant 624. There may be substantial deviations from these averages. Some of the differences in the literature data are due to either biological or developmental variations, or the information indicates the time between ovulation and birth. *See* hatching time in poultry.

gestational drive (green beard effect) Maternal genes recognizing and favoring special genes of the offspring already during gestation; favoring or disfavoring a genetic constitution may lead to consequences somewhat similar to meiotic drive. Population geneticists do not generally accept the concept. *See* meiotic drive.

g factor *See* human intelligence.

GFAP Glial fibrillary acidic protein. It affects myelination of the peripheral nerve cells and brain, and its defect causes long-term depression. *See* depression; leukemia inhibitory factor; myelin. (Headley, S. A., et al. 2001. *J. Comp. Pathol.* 125[2–3]:90.)

g force *See* centrifuge.

GFP *See* green fluorescing protein.

GGA Protein that sorts mannose phosphate receptors (MPR) into vesicles budding from the trans-Golgi network (TGN). The proteins are eventually delivered to endosomal and lysosomal compartments. GGA is composed of a VHS (VPS27, Hrs, STAM) domain at the NH_2 end and a GAT domain that flexibly hinges to a GAE domain at the carboxyl end. GGA is moved to the trans-Golgi membrane after the GAT (transporter) domain interacts with the ARF-GTP (ADP-ribosylation factor–guanosine triphosphate) complex on the TGN membrane. The VHS domain binds the acidic cluster dileucine motif (ACLL) of MPR. GGA recruits a clathrin triskelion at the GAE hinge and γ-synergin (controlling clathrin-coated vesicle traffic) and the endosome fusion regulator protein rabaptin 5. *See* endocytosis; lysosome; mannose phosphate receptor; trans-Golgi network; triskelion. (Tooze, S. A. 2001. *Science* 292:1663.)

G$_h$ G protein with GTP-binding signaling function and transglutaminase activity. *See* G protein.

GH   Growth hormones such as hGH (human, encoded in 17q22-q24) or rGH (rat). *See* GHRH; GHRHR; growth hormone releasing hormone; hormone-response elements; hormones; pituitary dwarfness.

ghost   (1) Empty phage capsid without its genetic material. (2) electronic noise.

ghost QTL   Erroneus localization result obtained by QTL analysis. *See* QTL.

ghrelin   Acetylated 28-amino-acid secretagogue produced in the hypothalmus that releases growth hormone from its receptor. It promotes feeding and is an antagonist of leptin. Ghrelin regulates neuropeptide Y and agouti-related protein neurons. *See* agouti; growth hormone, pituitary; leptin; secretagogue. (Inui, A. 2001. *Nature Rev. Neurosci.* 2:551.)

GHRH   Growth hormone-releasing hormone, 20q11.2. It stimulates the release of growth hormones from the pituitary. Antagonists of GHRH receptors suppress cancerous proliferation. Somatostatin inhibits growth hormone secretion. *See* animal hormones; brain, human; GH; pituitary; somatostatin.

GHRHR   Growth hormone-releasing hormone receptor, 7p15-p14. It results in dwarfism. Several variants are known. *See* dwarfism; GH. (Szepesházi, K., et al. 2001. *Endocrinology* 142:4341.)

gi   Identification number in the GenBank database that is used in addition to the accession number. This permits closer identification of subsequently discovered variations in a particular sequence to which a string of gi's may be added as new information becomes available for that particular DNA. *See* accession number; asn.1; GenBank; identifier syntax.

GI$_{50}$ Chemical dose that provides 50% growth inhibition, e.g., for a certain cancer cell line.

giant axonal neuropathy (GAN, 16q24)   Recessive sensory and motor disease of the central and peripheral nervous system. Its onset is in early childhood and usually causes death by late adolesence of curly-haired individuals. It causes swelling of the axons due to a defect in the protein gigaxonin affecting the axonal cytoskeleton. A similar disease afflicts some German Shepherd dogs. *See* neuropathy.

giant chromosomes   Polytenic chromosomes and lampbrush chromosomes. *See* lampbrush chromosomes; polytenic chromosomes; salivary gland chromosomes.

giant platelet syndrome (Bernard-Soulier syndrome, 22q11.2, 17pter-p12)   Caused by deficiency of a major platelet glycoprotein (glycoprotein Ib-β, GP1BB), resulting in a bleeding disorder. *See* May-Hegglin anomaly; thrombocytopenia; thrombophilia.

GIB   *See* Genome Information Broker.

Gibberella fujikuroi   Plant-pathogenic fungus that produces the plant hormones, gibberellins, by its normal metabolism. *See* plant hormones.

Gibberellic acid.

gibberellin   *See Gibberella fujikuroi*; plant hormones. (Rojas, M. C., et al. 2001. *Proc. Natl. Acad. Sci. USA* 98:5838; Richards, D. E., et al. 2001. *Annu. Rev. Plant Physiol. Mol. Biol.* 52:67; Olszewski, N., et al. 2002. *Plant Cell* 14:S61.)

gibbon   *See* Pongidae; primates.

Giemsa stain   Contains azure II, azure-eosin, glycerol, and methanol. The dark bands appear to be low in GC and the light bands are rich in GC content. *See* chromosome banding; G banding; rye. (Niimura, Y. & Gojobori, T. 2002. *Proc. Natl. Acad. Sci. USA* 99:797.)

Gierke's disease   *See* glycogen storage disease type I.

GIFT   Gamete intrafallopian transfer. It is a method of artificial insemination. *See* ART; artificial insemination.

giga   Prefix for 10^9 size or quantity.

GigAssembler   Algorithm suitable for preparing the human genome working draft including about 88% of the 400,000 initial contigs. *See* contig; genome projects; human genome. (Kent, W. J. & Haussler, D. 2001. *Genome Res.* 11:1541.)

Gilbert syndrome   Very common in human chromosome 2–dominant hyperbilirubinemia, similar to Crigler-Najjar syndrome, and probably controlled by genes allelic to it. *See* Crigler-Najjar syndrome; Dubin-Johnson syndrome; hyperbilirubinemia.

Gilles de la Tourette syndrome   *See* Tourette disease.

Gin   Invertase. *See* invertases.

ginger (*Zingiber officinale*, $2n = 2x = 22$)   Perennial rhizome spice. It dilates blood vessels, relieves pain, reduces flatulence, increases perspiration, and is a stimulant. *See* phenolics.

Ginkgo biloba   Ornamental tree in the United States. Its leaves are considered an herbal medicine for neurological disorders associated with aging such as Alzheimer disease, hearing and memory loss, attention deficit, etc. Its flavonoids appear to be effective scavengers of free radicals. Microarray hybridization reveals a higher level of tyrosine/threonine phosphatase and other mRNAs involved in upregulation of activity in the brain cortex of mice upon consuming leaf extracts. (Watanabe, C. M. H., et al. 2001. *Proc. Natl. Acad. Sci. USA* 98:6577.)

GIP   (1) G-protein subunit and a potential oncoprotein. *See* G protein; oncoprotein. (2) Glucose-dependent insulinotropic polypeptides mediate insulin secretion. (Hinke, S. A. 2001. *Biochim. Biophys. Acta* 1547:143.)

G$_i$ protein   Member of the trimeric G-protein family; it activates adenylate cyclase and thus opens K^+ channels. The $\beta\gamma$ subunits activate the ERK/MAPK signal transduction path through tyrosine kinase. This pathway responds positively to RAS and is antagonized by RAP1. *See* adenylate cyclase; G proteins; ion channels; RAP1; RAS; signal transduction.

giraffe (*Giraffa camelopardalis*) $2n = 30$; *Okapia johnstoni* $2n = 45$.

girdle bands   Concentric rings of thylakoids. *See* chloroplasts; thylakoids.

GIRK   G-protein-gated inwardly rectifying K^+ channel. It is a heterotrimeric guanine nucleotide-binding protein. *See* G proteins; ion channels. (Seeger, T. & Alzheimer, C. 2001. *J. Physiol.* 535[pt 2]:383.)

GISH   Genomic in situ hybridization. It may identify chromosomes in species hybrids and reveal crossovers among homoeologues. *See* FISH; genome; homoeologous chromosome; in situ hybridization; see color plate in Color figures section in center of book.

Gitelman syndrome (16q13)   Hypocalciuria, hypomagnesemia, and hypertension. *See* Bartter syndrome; hypertension; hypoaldosteronism; hypokalemia; Liddle syndrome.

GITR   Glucocorticoid-induced tumor necrosis factor receptor-related, 1p36.3. It regulates cell proliferation, differentiation, and survival. Its ligand is AITR (activation-inducible TNFR). *See* TNF. (Nocentini, G., et al. 2000. *DNA Cell Biol.* 19[4]:205.)

Glanzmann's disease   Variety of blood platelet anomalies determined by autosomal-recessive genes. The overall symptoms include bleeding under the skin (ecchymosis), tiny, round, and flat purplish (later yellow or blue) spots under the skin caused by blood release (petechia), bleeding of the teeth and gums (gingiva), nosebleeds (epistaxes), gastrointestinal bleeding, excessive uterine bleeding (menorrhagia), and bleeding from the uterus at irregular intervals (metrorrhagia). The platelets may appear normal, yet their number is reduced (thrombocytopenia). Sometimes the size of the platelets increases, their shape becomes abnormal, and they appear isolated rather than aggregated. *See* hemophilias; platelet abnormalities; von Willebrand disease.

GLAST Na^+-dependent transporters of glutamate and aspartate; GLASTs may have 68% homology with another glutamate transporter, GLT. *See* transporters. (Gegelashvili, G., et al. 2001. *Progr. Brain Res.* 132:267.)

glaucoma   May be controlled by autosomal-dominant or -recessive genes and may be manifested at birth, during juvenile years, or in adults. The incidence of the different forms may vary from 10^{-4} to a couple of percent in the general population, usually at higher risk in adult life. The most general features are opacity of the eye lens caused by a gray gleam on the iris and increased intraocular pressure, which eventually distorts vision. In the early stages or in mild forms, the anterior chamber of the eye is open (open-angle glaucoma). This stage may pass into an intermittent form that may be transient but can last for several months; eventually the angle becomes closed, resulting in great pressure and swelling of the cornea accompanied by substantial pain. If untreated, total blindness may follow.

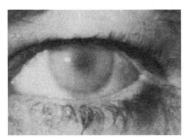

Glaucoma. (From Bergsma, D., ed. 1973. *Birth Defects. Atlas and Compendium*. National Foundation–March of Dime.)

Testing eye pressure before the visible onset of the condition may monitor it. The penetrance and expressivity of this disease is highly variable. The gene (GLC1A) coding for juvenile open-angle glaucoma (JOAG) was assigned to human chromosome 1q23-q25. GLC1B is at 2cen-q13 and GLC1C at 3q. The GC3B (buphtalmos) locus is at 1p36. The gene encodes the trabecular (supportive connective tissue) meshwork — inducible glucocorticoid response (TIGR) or myocilin. The dominant glaucoma at 6p25 encodes a forkhead-type transcription factor. For early detection of glaucoma, the endothelial leukocyte adhesion molecule (ELAM - 1, 1q23 - q25) test has been suggested. *See* Axenfeld-Rieger anomaly; eye diseases; FKH. (Jacobson, N., et al. 2001. *Hum. Mol. Genet.* 10:117.)

GLC   *See* gas liquid chromatography.

Gle1   *See* export adaptors; RNA export.

Gleason score Classification of prostate cancer on the basis of histology with predictive value for progression. (Gleason, D. F. 1992. *Hum. Pathol.* 23:273.)

GLEEVEC (Glivec) Tyrosine kinase inhibitor anticancer drug (Capdeville, R., et al. 2002. *Nature Rev. Drug. Discovery* 1:493.)

GLGF repeat DHR domain or PDZ domain. (Tochio, H., et al. 2000. *J. Mol. Biol.* 295:225.)

gliadin *See* glutenin; zein.

glial cell (neuroglia) Can be astrocytes, oligodendrocytes, or microcytes; the first two have supportive roles, and the latter phagocytizes the waste products of the nerves. Glial cells modulate synaptic transmission through an acetylcholine-binding protein. *See* acetyl choline; FGF.

glioma Tumor of the tissues supporting the nerve cells (astrocytes), but it may spread. The glioma may be benign, yet the malignant forms lead to rapid death. The most active glioma (glioblastoma multiformis) develops from an interaction of Ras and Akt in mice. Gliomas seem to secrete glutamate, which activates the NMDA receptors and further facilitates the expansion of the tumor. Surgery, radiation treatment, and chemotherapy, or gene therapy with adenoviral vectors carried herpes simplex virus thymidine kinase (HSVTK) and ganciclovir are used, but neither gives entirely satisfactory results. *See* adenovirus; Akt; cancer gene therapy; ganciclovir; GLI oncogene; NMDA; RAS. (Lam, P. Y. P. & Breakfield, X. O. 2001. *Hum. Mol. Genet.* 10:777; Holland, E. C. 2001. *Nature Rev. Genet.* 2:120; Takano, T., et al. 2001. *Nature Med.* 7:1010.)

GLI1 oncogene (glioma) Has been located to human chromosome 12q13. It is highly amplified in gliomas. GLI2 in chromosome 2q14 appears homologous to the *Krüppel* gene of *Drosophila*, encoding a DNA-binding protein and regulating embryo morphogenesis. Similarly, GLI2 is expressed in embryonal carcinomas but not in late developing ones. GLI3 (7q13) is apparently not an oncogene, but it is involved in Greig syndrome and in Pallister-Hall syndrome. Other homologous genes were also found in the human genome—altogether six loci in five different chromosomes. Some of the homologues (GLI4, 8q24.3) are denoted as HKR (human Krüppel). Gli transcription factors are suspected in the transduction of sonic hedgehog signals. Carboxy-terminal deletions in Gli3 facilitate its association with SMADs. *See* DNA-binding protein domains; glioma; Greig's cephalopolysyndactyly syndrome; *hedgehog*; Krüppel; nevoid basal cell carcinoma; oncogenes; Pallister-Hall syndrome; polydactyly; Rubinstein-Taybi syndrome; SMAD; *sonic hedgehog*; syndactyly. (Kim, Y.-S., et al. 2002. *J. Biol. Chem.* 277:30901.)

glnA Bacterial glutamine synthase.

glnAp2, glnAp1 Major and minor glutamine synthase promoters, respectively, in bacteria.

global genetic effect Involves most or all of the genome.

global single-cell reverse transcription–polymerase chain reaction (GSC RT-PCR) The aim of the method is to determine differences in gene expression among individual cells within a population of cells. This may be of interest for determining the process of metastasis, changes in gene expression during development, etc. A description of the procedure can be found in Brailo, L. H., et al. 1999. *Mutation Res. Genomics* 406:45. Although much difference is detectable among different cells, the components of the procedure also introduce substantial variations. *See* microarray hybridization, SAGE.

globin Ancestral protein molecule that diverged over a billion years ago into the oxygen-carrying muscle protein, myoglobin, and into the respiratory hemoglobins of the red blood cells. The neuroglobin, encoded at human chromosome 14q24, is expressed predominantly in the brain. *See* haptoglobin; hemoglobin; LCR; leghemoglobin; myoglobin. (<globin.cse.psu.edu>.)

GLOBO H Glycoceramide present in breast, ovarian, gastric, pancreatic, endometrial, prostate, and small cell lung carcinomas. It may be employed in cancer vaccines. *See* cancer gene therapy. (Keusch, J. J., et al. 2000. *J. Biol. Chem.* 275:25315.)

globoid cell leukodystrophy *See* Krabbe's leukodystrophy.

globoside Glycosphingolipid with the most common structure: acetylgalactoseamine-galactose-galactose-glucose-ceramide. *See* ceramides; sphingolipids. (Puri, V., et al. 2001. *J. Biol. Chem.* 154:535.)

globozoospermia Round-headed spermatozoa. Developmental anomaly caused by the loss of α'-subunit of casein kinase II. This enzyme has many substrates and it is involved in numerous metabolic controls. (Larson, K. L., et al. 2001. *J. Androl.* 22[3]:424.)

globulin Salt-soluble proteins with many diverse cellular functions.

glomerulocystic kidney disease, hypoplastic, familial (GCKD, 17 cen-q21.3) Dominant mutations in the hepatocyte nuclear factor-1-β gene, causing chronic renal failure, renal cysts, and diabetes-like symptoms. *See* HNF; kidney diseases.

glomerulonephritis Autosomal-dominant kidney disease associated with very sparse hairs and red lesions due to dilation of the blood vessels (telangiectasis). This disease (membranoproliferative glomerulonephritis) is frequently associated with reduced levels of C3 complement component. More recent information indicates that the FcγR (fragment crystalline gamma receptor) of the antibody molecule is the most critical factor in the disease. Dominant IgA nephropathy (6q22-q23) occurs at a frequency of 1×10^{-3} and may cause death in

~20% of the afflicted despite dialysis. *See* antibody; complement; hair; immunoglobulins; kidney diseases; skin diseases; telangiectasis.

glomerulosclerosis, focal and segmental, familial

Involves increased urinary protein excretion and decreasing kidney function or even morbid kidney defects. The α-actinin gene at human chromosome 19q13.1 may be one of the causes for the stronger-than-normal binding of this protein to filamentous actin. Another dominant locus is in chromosome 11q22-q24 and a recessive steroid-resistant NPHS2 gene has been assigned to 1q25-q31. The latter locus encodes the transmembrane protein podocin. *See* actin; actinin; kidney diseases.

gloves

Frequently recommended for laboratory work when handling hazardous material or when contamination by hands must be avoided. Remember that surgical latex gloves easily develop invisible holes and permit unseen contamination of the hands. (Mercury penetrates latex-disposable gloves in 15 seconds.) Latex gloves may cause (serious) allergic reactions to about 10% of regular users and food allergies may aggravate the condition. Longer than 15-minute use of a latex glove may result in leakage. Organic solvents damage some plastic gloves and they may develop holes easily. For most operations, neoprene gloves provide the greatest safety. For very hazardous material, the use of double gloves may be advisable. Handwashing after removal of the gloves is recommended. *See* laboratory safety.

GLT   *See* GAST.

glucagon

Polypeptide hormone secreted by the α-cells of the pancreas when blood glucose sinks below a certain level. The hormone then increases the concentration of blood sugar by breaking down glycogen with the cooperation of epinephrine. *See* animal hormones; cAMP; diabetes mellitus; epinephrine.

glucan   Polymer (repeating units) of glucose.

glucanase   Glucan-digesting enzyme. *See* glucan; host-pathogen relation.

glucocorticoid

Kidney cortex hormone that regulates carbohydrate, lipid, and protein metabolism, muscle tone, blood pressure, the nervous system, etc. It inhibits the release of adrenocorticotropin, slows down cartilage synthesis, mitigates inflammation, allergy, and various immunological responses. Cortisol (hydroxycortisone) is an important natural glucocorticoid, whereas dexamethasone is a synthetic product that is two orders of magnitude more potent than cortisol. Glucocorticoid-mediated immunosuppression involves activation of the IκBα gene and an increase of its cytoplasmic protein product. When the nuclear regulator factor NF-κB is active (because of the expression of TNF), its inhibitor, the IκBα protein, is degraded, and NF-κB moves into the nucleus and activates the immune system. Dexamethasone—in contrast to the natural glucocorticoids—causes an increased transcription of IκBα, and NF-κB translocation to the nucleus is inhibited, leading to less nuclear NF-κB and reduction of inflammation because the immune system is suppressed. Familial and sporadic glucocorticoid deficiencies are caused by defective adrenocorticotropic hormone receptors. Deficiency of the glucocorticoid receptor (94 kDa, encoded at 5q31) causes cortisol and dexamethasone resistance. Melanocortin unresponsiveness is due to receptor deficiency at 18p11.2. The glucocorticoid receptor is an indispensable transcription factor, and it can attach to naked DNA as well as to nucleosomal structures. *See* adrenocorticotropin; allergy; apoptosis; calreticulin; cortisol; Cushing syndrome; dexamethasone; GRE; IκB; immunophilins; immunosuppression; NF-κB; opiocortin; stress.

glucocorticoid response element (GRE)

Located generally about 100 to 2,000 nucleotide pairs upstream from the transcription initiation site (the human growth hormone-response element is within the transcribed region). These elements, such as mammary tumor virus (MTV), metallothionein (MTIIA), tyrosine oxidase (TO), and tyrosine amino transferase receptor, respond to different activating proteins. Despite their differences in structure, they share a consensus: CGTACANNNTGTTCT. *See* DNA looping; hormone-response elements; mammary tumor virus; metallothionein; regulation of gene activity; tyrosine aminotransferase. (Herrlich, P. 2001. *Oncogene* 20:2465.)

glucogenic amino acid

Can be converted into glucose or glycogen through pyruvate (alanine, cysteine, glycine, serine, tryptophan), α-ketoglutarate (arginine, glutamine, histidine, proline), succinyl CoA (isoleucine, methionine, threonine, valine), fumarate, (phenylalanine, tyrosine), and oxaloacetate (asparagine, aspartate). *See* amino acids.

glucokinase (GK)

Phosphorylates glucose to form glucose-6-phosphate. Heterozygosity for GK mutation in the fetus may cause mild hyperglycemia and may reduce insulin secretion, resulting in reduced intrauterine growth. In case of maternal glucokinase mutation, hyperglycemia stimulates fetal insulin secretion and growth. *See* insulin.

gluconeogenesis

Synthesis of sugars from noncarbohydrate precursors such as oxaloacetate, pyruvate, citrate, and malate.

glucose (glycose)

6-carbon sugar (dextrose), an aldohexose. Besides being a source of energy, it induces and represses many genes. *See* galactose (for formula).

glucose effect

Form of catabolite repression; as long as glucose is available in the nutrient medium, the synthesis of enzymes involved in the utilization of other carbohydrates is prevented. The preferential growth on, e.g., glucose is followed by a temporary pause before the utilization of another carbon source, which is commonly called *diauxic growth*. Glucose may act at three levels: (1) it inhibits the uptake of inducer molecules by relying on the dephosphorylated component of the phosphoenolpyruvate-dependent glucose phosphotransferase; (2) it lowers the level of cAMP and its receptor and indirectly activates adenylate cyclase; (3) it increases the level of catabolites that repress the synthesis of inducible enzymes. In fungi, the mechanism of glucose effect may be mediated through the function of hexokinase. In yeast, *SNF1* (sucrose nonfermenting), encoding a transactivator protein (protein threonine/serine kinase) gene, can relieve *SUC* and *GAL* glucose repression. The Mig1/CREA zinc-finger DNA-binding protein, Glc7, protein phosphatase, and the Tup1 general suppressor have also been implicated in regulation. Two glucose-signaling loci (*gsf1* and *2*) affect the glucose repression of *SUC2* and *Gal10*. Glucose suppression has been analyzed in prokaryotes and lower and higher eukaryotes — animals and plants. The *PRL1* locus of *Arabidopsis* encodes an α-importin WD protein that regulates glucose/sucrose sensitivity as well as hormone responses in plants. *See* catabolite repression; feedback control; *GAL*; repression; *SUC2*; SW1; transactivator; Tup1; WD-40; zinc finger. (Ronne, H. 1995. *Trends Genet.* 11:12; Németh, K., et al. 1998. *Genes & Development* 12:3059; Stülke, J. & Hillen, W. 1999. *Curr. Opin. Microbiol.* 2:195; Rolland, F., et al. 2002. *Plant Cell* 14:S185.)

glucose-galactose malabsorption (GGM)   *See* SGLT.

glucose induction   In the cell membrane of yeast, *SNF* and *RGT1* genes monitor glucose. Most likely, glucose causes a conformation change in these proteins by attaching to their N-terminal domains outside the cell membrane. Both are transmembrane proteins with their C-terminus tail within the cytoplasm. That tail probably recruits the Hxt glucose transporters. The transcriptional suppressor Zn-finger protein, Rgt1, represses *HXT* glucose transporter genes and the SCF[Grr1] complex inhibits Rgt1 (regulator of transport) when a low concentration of glucose appears in the culture medium. (SCF is an acronym for Skp1, Cdc53, and Cdc34; it includes an F-box protein. Grr is a Cdc34-dependent protein factor of ubiquitination of cyclins.) Then HXT genes are activated. When sucrose is increased beyond a certain level, the Mig1 suppressor system becomes active. When the concentration of glucose becomes high, Rgt1 turns into an activator of HXT1. *See* Cdc34; Cdc53; F-box; glucose effect; glucose transporters; SCF; Skp; SNF. (Vaulont, S., et al. 2000. *J. Biol. Chem.* 275:31555.)

glucose-6-phosphate dehydrogenase   First enzyme in the pentosephosphate pathway that converts G-6-P into 6-phosphoglucone-δ-lactone. The final product of the pathway is D-ribose-5-phosphate, and NADPH is also generated. Although about 90% of the cellular glucose in mammals is converted to lactate by glycolysis, 10% is driven through the pentose phosphate path, and this is the principal reaction for providing the erythrocytes with NADPH for the reduction of glutathion. The deficiency of the enzyme caused by Xq28-chromosomal genes was first identified as a hemolytic anemia caused by the antimalarial drug 8-aminoquinoline. Most of the afflicted individuals are essentially asymptomatic until exposed to drugs such as certain analgesics, sulfonamides, antimalarial drugs (atabrine), quinine, etc., or until they are afflicted by certain other diseases. G-6-P dehydrogenase deficiency is widespread in human populations, probably because the heterozygotes and hemizygous males are protected against falciparum malaria by a 46–58% reduction of the infectious disease. Heterozygotes (XX) may display lyonization. In the Jewish populations of Kurdistan, Caucasus, and Iraq, the frequency of the defect reached 58.2, 28.0 and 24.8%, respectively, whereas in geographical areas free of malaria it was generally less than 2%. Cavalli-Sforza and Bodmer

estimated that G-6-P dehydrogenase deficiency conveyed an extremely high 0.15% selective advantage against malaria. A similar sequence is situated in human chromosome 17 and it may be a pseudogene. *See* analgesic; atabrine; glutathion; glycogen storage diseases; glycolysis; malaria; pentose phosphate pathway; selection coefficient; selection conditions. (Tishkoff, S. A., et al. 2001. *Science* 293:455. <http://www.rubic.rdg.ac.uk/g6pd/>.)

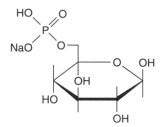

Glucose-6-phosphate.

G-6-PD cell-deficient red blood cell peripheral "BITE".

glucose-phosphate isomerase   *See* phosphohexose isomerase.

glucose repression   *See* glucose effect.

glucose tolerance test   *See* diabetes.

glucose toxicity   Normally insulin regulates the physiological range of glucose in the cells. When the level of glucose increases for a longer period of time, glucose toxicity results. Glucose may generate reactive oxygen species (ROS). Antioxidants, as well as binding transcription factors PDX-1/STF and RIPE-3b1 to the insulin promoter, may increase insulin production and reduce toxicity. (Shimoi, K., et al. 2001. *Mutation Res.* 480–481:371.)

glucose transporter   GLUT (12p13.3) 49 kDa protein involved in moving glucose. GLUT2 (3q26.1-q26.3) is another solute/sugar carrier (Fanconi-Bickle syndrome). GLUT1 (1p35-p31.3) mediates sugar transport to the brain across the blood-brain barrier membrane. GLUT4 (17p13) defect seems to be involved in the resistance to insulin in diabetes. GLUT5 (1p36.2) is a fructose transporter. *See* BBB; diabetes. (Brown, G. K. 2000. *J. Inherit. Metab. Dis.* 23:237.)

glucosidase (GCS1)   Enzyme digests 1,2-N-linked glycoproteins; it is encoded in human chromosome 2p13-p12. *See* acid maltase; Gaucher disease; Pompe diseases.

glucoside   When D-(+) glucose is treated with an alcohol (methanol) and HCl, methyl D-(+)glucoside is formed, which still has one methyl group yet its properties resemble those of an acetal. Acetals may be formed from aldehydes, and they are common in different plants. Cardiac glucosides present in plants such as *Digitalis, Scilla,* etc., have cardiotonic effects (strengthen heart function) and are used as medicine. Many of the plant glucosides are highly toxic and cause anorexia (loss of appetite), nausea, vomiting, salivation, diarrhea, headache, drowsiness, delirium, hallucinations, and possibly death. Glycosides linked to cyanides also occur in common food plants such as beans, apricot, almond seed, etc. Forage plants such as Sudan grass, white clover, etc., may contain enough cyanide to kill a 50 kg animal if it eats 1 to 2 kg fresh plant material. Through plant breeding efforts, the synthesis of the glucoside (lotoaustralin) may be blocked or the production of the enzyme linamarase may reduce the toxicity. *See* cyanide; lotoaustralin. (Tattersall, D. B., et al. 2001. *Science* 293:1826.)

glucosylation   Attaching glucose to another molecule. Defective N-glycosylation is the cause of mucolipidosis II and impacts the immune system. Glycosylation has many important consequences for plant metabolism. *See* congenital disorders of glycosylation; mucolipidodosis. (Lowe, J. B. 2001. *Cell* 104:809.)

glucuronic acid   Derivative of uronic acid (a derivative of glucose). It is present in glucosaminoglycans. *See* GUS; mucopolysaccharidosis.

glume   Lowermost bract of the grass florets. The glume is generally free from the fruit; in some cases, however, it may be firmly associated with the kernels.

GluR   *See* glutamate receptor.

GLUT   Insulin-dependent glucose transporters encoded by the SLC genes. GLUT may be homologous to the cJun amino-terminal kinase-interacting protein JIP. MAPK81P1 (11p11.2-p12), a potential SLC transactivator, may be a major gene for type II diabetes. There are also several other glucose transporters. *See* diabetes; insulin; MAPK; MODY. (Doege, H., et al. 2001. *Biochem. J.* 359[pt2]:443.)

glutamate ($HOOCCH[NH_2]CH_2CH_2CONH_2$)   Uncharged derivative of glutamic acid. It has a key role as a nitrogen donor in the cell. Glutamate neurotransmitter activates the glutamate receptors (iGluR) regulating ion uptake and (mGluR) nerve synaptic strength and frequency. Mitochondrial glutamate facilitates insulin secretion. *See* amino acids; glutamate synthase; glutamate synthetase; glutamine; neurotransmitter.

glutamate decarboxylase deficiency disease (GAD) Pyridoxine-dependent epilepsy. The two enzymes require the cofactor pyridoxal phosphate. These enzymes convert glutamic acid into γ-aminobutyric acid (GABA), which controls neurotransmission in vertebrates and invertebrates. The phenotype is autosomal recessive (GAD1 at 2q31, GAD2 at 10p11.23). *See* amino acid metabolism; epilepsy; GABA.

glutamate dehydrogenase (M_r 330,000)   Catalyzes oxidative deamination of glutamate in the mitochondria, resulting in the formation of α-ketoglutarate. The reaction requires NAD^+ or $NADP^+$ as cofactors, and it is regulated allosterically by GTP and ADP. Then in turn α-ketoglutarate and ammonia may again form glutamate. If the concentration of NH_3 is low, glutamate dehydrogenase cannot function to an appreciable extent. In such a case, NH_3 plus glutamate are converted to glutamine by nonadenylylated glutamine synthetase. In the presence of a high amount of NH_3, *glutamine synthetase* is adenylylated and becomes inactive; in this form it represses its own synthesis (autoregulation). In its nonadenylylated state (when the level of ammonia is low), it represses *glutamate dehydrogenase* instead. From glutamine and α-ketoglutarate, glutamate can be synthesized by *glutamate synthase* in the presence of $NADPH + H^+$. Glutamate synthase also serves as an inducer for tryptophan permease, which together with tryptophan transaminase may contribute to glutamate synthesis. In its nonadenylylated state, glutamine synthetase activates the histidine utilization operon (*hut*). This operon yields glutamate and ammonia. In humans, a small multienzyme family codes this enzyme (GLUD); its level is relatively high in the brain. The principal and functional *GLUD1* is in human chromosome 10q23. This gene is homologous to mouse locus *Glud-2* in chromosome 14. *See* autoregulation; glutamate synthase; olivopontocerebellar atrophy; UTase.

glutamate formiminotransferase   Autosomal-recessive deficiency of this enzyme leads to the accumulation of formiminoglutamate and folic acid in the urine and in the serum, causing physical and mental retardation. *See* amino acid metabolism; mental retardation.

$$\text{formimino} - OOC - CH - (CH_2)_2COO^-$$
$$| $$
$$NH$$
$$|$$
$$HC = NH$$

glutamate oxaloacetate transaminase (GOT2)   Encoded in human chromosome 16q21, but the protein is mitochondrially located. In many plants and lower animals, the enzyme is mitochondrially coded. Pseudogenes have been found at two locations in human chromosome 1 and in chromosome 12. *See* aspartate aminotransferase, mitochondrial; mtDNA; tyrosine aminotransferase.

glutamate-pyruvate transaminase (GPT1)   Catalyzes the reversible reaction:

$$HOOCCH_2CH_2COCOOH \ + \ CH_3CH(NH_2)COOH$$

α-ketoglutaric acid   L-alanine

$$\leftrightarrows HOOCCH_2CH_2CH(NH_2)COOH \ + \ COOHCOCH_3$$

L-glutamic acid   pyruvic acid

The soluble enzyme is encoded in human chromosome 8q24.2-qter. Cytosolic and mitochondrial forms exist. It is also called alanine aminotransferase (AAT1). *See* alanine aminotransferase; amino acid metabolism; glutamine.

glutamate receptor (GluR)   Cation channels mediating the postsynaptic current in the central neurons. Certain mutations in GluR-B subunits lead to increased calcium uptake and concomitant seizures if, e.g., the position of 586 arginine prevents editing of pre-mRNA. The glutamate receptors are tetrameric. GluR genes with 16 to 63% homology to animal GluRs have been identified in both monocot and dicot plants with a role in light signal transmission. *See* ion channels; neurotransmitters. (Borges, K. & Dingledine, R. 2001. *J. Biol. Chem.* 276:25929.)

glutamate synthase   Catalyzes the reaction that leads to α-ketoglutarate + glutamine + NADPH + H$^+$ → 2 glutamate + NADP$^+$. The result of the combined action of glutamate synthetase and glutamate synthase in bacteria is α-ketoglutarate + NH$_4^+$ + NADPH + ATP → L-glutamate + NADP$^+$ + ADP + Pi. *See* autoregulation; glutamate dehydrogenase; glutamine.

glutamate synthetase   In *E. coli*, it is a ca. 800,000 M$_r$ protein containing flavin, iron, and S^{2-}. *See* autoregulation; glutamate dehydrogenase; glutamate synthase; glutamic acid; glutamine.

glutamate transporter   *See* GLAST.

glutamic acid   HOOC · CH(NH$_2$) · CH$_2$ · CH$_2$ · COOH (L(+), amino-glutaric acid).

glutaminase (GLS)   Enzyme converting glutamine into glutamic acid; it has been mapped to human chromosome 2q32-q34. It is activated by phosphate and it may affect the neurotransmitter role of glutamate. *See* amino acid metabolism; glutamic acid; glutamine.

glutamine   HOOC-CH(NH$_2$)-(CH$_2$)$_2$-C(O)NH$_2$. *See* glutamic acid.

glutamine amidotransferases   Group of enzymes with two domains: one binds glutamine and the other binds another molecule. After cleaving ammonia from glutamine, they transfer it to the other substrate, generally in the presence of ATP.

glutamine repeat diseases   *See* dentatorubral pallidoluysian atrophy; fragile sites; Huntington's chorea; Kennedy disease; Macho-Joseph disease; olivopontocerebral atrophy; trinucleotide repeat.

glutamyl ribose-5-phosphate glycoproteinosis   ADP ribose protein hydrolase deficiency resulting in proteinuria and neurological disorders. It is also regarded as a lysosomal storage disease. It may be X-linked. *See* lysosomal storage diseases.

glutamyl-tRNA synthetase (QARS)   Enzyme charging cognate tRNA with glutamic acid; it is encoded in human chromosome 1q32-q34. *See* aminoacyl tRNA synthetase.

glutaraldehyde   *See* fixatives.

glutaredoxin   Catalyzes NADPH-dependent reduction of disulfides usually in a complex with glutathione and glutathione reductase. *See* DsbA; thioredoxin.

glutaricacidemia (GA)   GAI autosomal-recessive (19p13.2) glutaryl-CoA dehydrogenase deficiency results in an increase in glutaric acid in the blood and in the urine, causing neurodegenerative disorders. GAIC encoded in human chromosome 4q32-qter involves deficiency in the electron transfer flavoprotein oxidoreductase. GAIIA (15q23-q25) also causes the excretion of lactic, ethylmalonic, isovaleric, and different forms of butyric acids. Similarly, an X-linked (Xq26-q28) acyl-CoA dehydrogenase deficiency results in the abnormal excretion of glutaric and other organic acids. *See* aminoacidurias; glutaricaciduria.

glutaricaciduria   Autosomal-recessive glutaryl-CoA dehydrogenase deficiency leading to the accumulation of glutaric acid in the urine, degeneration of the nervous system, and impairment of muscle functions. Limiting amino acid intake may alleviate the symptoms. An autosomal-dominant form (15q23-q25) was identified as a defect in an electron transfer flavoprotein. Some glutaricacidemias are also called glutaricaciduria, e.g., glutaryl-CoA dehydrogenase deficiency (GAI, 19p13.2). Glutaricaciduria IIC (GAIIC) was assigned to 4q32-qter. *See* aminoacidurias; glutaricacidemia; neuromuscular diseases.

glutathione γ-L-glutamyl-L-cysteinylglycine is a reducing agent and protects SH groups in proteins. About 10% of the blood glucose is oxidized to 6-phosphogluconate by glucose-6-phosphate dehydgrogenase (G6PD) using NADP$^+$, and the reducer NADPH keeps glutathione reduced. Deficiency of G6PD results in the destruction of red blood cells and thus anemia. Glutathione is indispensable for development. Protozoa with anaerobic metabolism lack glutathione and mitochondria. *See* glucose-6-phospate dehydrogenase. (Meister, A. 1988. *J. Biol. Chem.* 263:17205; Spector, D., et al. 2001. *J. Biol. Chem.* 276:7011.)

glutathione peroxidase (GPX1)   Assigned to human chromosome 3p21.3 (earlier it was assigned to 3q11). Its deficiency causes hemolysis and jaundice. The frequency of the GPX1 gene is >0.5 in Mediterranean Jewish populations, but it is <0.2 in Northern Europeans. Locus GPX2 is in 14q24.1, GPX2 is in 5q32-q33.1, and GPX4 is in 19p13.3. The ailment may also be caused by a selenium-deficient diet. *See* glutathione reductase; glutathione synthetase deficiency; hemolytic anemia.

glutathione reductase (GSR)   The gene was located to human chromosome 8p21. Its deficiency results in hemolytic anemia. Thioredoxin substitutes for GSR in insects. *See* hemolytic anemia.

glutathione-S-transferases (GST) Family of enzymes metabolizing and detoxifying mutagens and carcinogens (some alkylating agents, cisplatin, carbonyl, peroxide, and epoxide groups). GST 3 was assigned to 11q13; GST2 to 6p12; GST1, GST4, and GST5 all at 1p13.3. GSTPL (glutathione transferase-like enzyme) is encoded at 12q13-q14. These enzymes, despite different locations of the coding units, show homology. GST is also used for protein labeling. It is extremely stable and facilitates the solubilization of fused proteins. *See* cisplatin; multidrug resistance.

glutathione synthetase deficiency Form of human chromosome 20q11.2 recessive hemolytic anemia and/or 5-oxyprolinuria. It may also result in excess metabolic pyroglutamic acid in the urine and a variety of ailments. GST2 (γ-glutamylcystein synthetase) gene, assigned to human chromosome 6p12, also causes hemolytic anemia. *See* anemia; glutathione; hemolytic anemia.

glutathionuria (GGT) Recessive defect (human chromosome 22q11.1-q11.2) in γ-glutamyl transpeptidase enzyme causing accumulation of glutathione in the urine.

gluten Mixture of several seed proteins in cereals. The main fractions are alcohol-soluble gliadin and alkali-soluble glutenin. The proportion of the components is genetically determined and defines nutritional value and baking quality. *See* glutenin; zein.

glutenin About half of the seed storage protein in wheat; soluble in 70% ethanol and alkali but insoluble in water. It is a polymer of extremely large molecular weight, up to the tens of millions. Its composition bears similarity to the muscle protein titin, comprising about 27,000 amino acid residues. The similarities based on (PEVK) proline, glutamate, valine, and lysine sequences may be attributed to the fact that both proteins require great elasticity in bread dough. It was indirectly selected by humans to retain gas bubbles in the dough to return to the original position after extension. In wheat, gliadin occurs with glutenin. The former conveys resistance to extension while the latter provides the softness and viscosity of the dough. *See* celiac disease; gliadin; gluten; *Triticum*. (Kobrehel, K., et al. 1992. *Plant Physiol.* 99:919.)

glycan Polysaccharide.

glycerol (CH_2OH — $CHOH$ — CH_2OH) Intermediate in carbohydrate and lipid biosynthesis.

glycerol kinase deficiency (GKD, Xp21-p21.2) It causes physical and mental retardation, osteoporosis, myopathy, eye defects, hyperglycerolemia. Chromosomal deletions may overlap with several other genes in the region of the X chromosome. *See* contiguous gene syndrome. (Gaudet, D., et al. 2000. *Am. J. Hum. Genet.* 66:1558.)

glycerophospholipid Fatty acids are esterified to glycerol and a polar alcohol is linked to it by a phosphodiester bond. Glycerophospholipids are parts of cell membranes (synonymous with phosphoglycerides).

glycine biosynthesis Glycine (NH_2CH_2COOH) is synthesized by hydroxymethyltransferase from serine ($HOCH_2CH(NH_2)COOH$) while tetrahydrofolate is converted to N^5, N^{10} methylene tetrahydrofolate. The transferase gene has been located to human chromosome 12q12-q14, whereas the tetrahydrofolate cyclases are in human chromosomes 8q21 and 18-qter. Glycine is synthesized alternatively from CO_2 and NH_4 by glycine synthase in the liver of vertebrates. *See* glycinemia, ketotic; hyperglycinemia.

Glycine max (soybean) Leguminous plant (basic chromosome number 20). The seed contains 20 to 23% oil and its protein content (meal) may exceed 40%. It is one of the most important sources for vegetable oil products and textured proteins for human food. It is also used as a supplement to animal feed mixtures.

glycinemia, ketotic (PCC) Caused by two genes at two human chromosomal locations (PCCB at 3q21-q22 and PCCA at 13q32). The biochemical defect is propionyl-CoA carboxylase deficiency. This enzyme's primary known role is the generation of D-methyl-malonyl-CoA, which is epimerized into the L form and subsequently by a mutase with vitamin B_{12} cofactor to succinyl-CoA. These processes concomitantly somehow produce ketosis, hypoglycemia, and hyperglycinemia. The symptoms are growth retardation, vomiting, lethargy, protein intolerance, low level of neutrophilic leukocytes, reduction in platelet number, etc. *See* amino acid metabolism; glycine biosynthesis; hyperglycinemia; ketoacidosis; methylmalonicaciduria.

glycocalyx Carbohydrate-rich membrane glycoprotein-lipid layer of prokaryotic and eukaryotic cell surfaces.

glycoform Proteins with differences in glycosylation. *See* glycosylation.

glycogen Main storage polysaccharide in animal cells. About 7% of the wet weight of the liver is glycogen and it is present in the muscle cells. It is branched at every 8 to 12 residues. As needed, it is hydrolyzed into glucose to supply energy with the aid of enzymes associated with its granular form. Glycogen is synthesized from glucose-6-phosphate by first changing it into glucose-1-phosphate by phosphoglucomutase. Then UDP-glucose pyrophosphorylase converts G-1-P and UTP into UDP-glucose and pyrophosphate (PP_i). Glycogen synthase then converts UDP — glucose into glycogen. *Glycogen synthase a* is the dephosphorylated active form of the enzyme, whereas phosphorylated *glycogen synthase b* is inactive. The reaction requires a primer of $\alpha 1-4$ polyglucose and the protein glycogenin. The branching is generated by branching enzymes amylo-$(1{\rightarrow}4)$ to $(1{\rightarrow}6)$ transglycolase or glycosyl-$(4{\rightarrow}6)$ transferase. Glycogen metabolism is regulated by glucagon and insulin in the liver and mainly by epinephrine and insulin in the muscles. The level of glucagon is regulated by cAMP. *Glycogen synthase kinase-3* regulates glycogen and protein synthesis by insulin, modulates transcription factor AP-1 and CREB, and regulates dorsoventral patterning of embryogenesis, and apoptosis. *See* Akt; AP; CREB; diseases; epinephrine; insulin; PTG. (Weston, C. R. & Davis, R. J. 2001. *Science* 292:2439.)

glycogenosis Used to designate glycogen storage diseases. *See* glycogen storage.

glycogen storage diseases The following hereditary defects have been identified in the synthesis and catabolism of glycogen: (1) *von Gierke's disease* (type I glycogen storage disease, 17q21) involves a deficiency of glucose-6-phosphatase determined by an autosomal-recessive gene. The patients develop liver enlargement (hepatomegaly as indicated by the extended abdomen; see photo at below) and a subnormal level of blood sugar (hypoglycemia), increased levels of ketone bodies (acetone) in tissues and fluids (ketosis), as well as high amounts of lactic and uric acids in the blood. (2) *Type II glycogen disease* (Pompe disease, GAA, chromosome 17q25.2-q25.3) is determined by an autosomal-recessive condition causing a deficiency of lysosomal α-1,4-glucosidase (acid maltase). Infants develop excessive enlargement of the heart (cardiomegaly) because of the deposition of glycogen in the lysosomes. By age 2, they succumb to cardiorespiratory failure. The defect can be diagnosed prenatally with amniocentesis. A milder form of the disease is also known with prolonged survival. Intravenous injection of the normal GAA gene in an adenovirus vector construct significantly alleviated the disease in a mouse model. (3) *Type III glycogen disease* (*see* Forbes disease) is also caused by autosomal-recessive (1p21) mutations. The basic physiological defects involve the glycogen debranching process. The symptoms are not as severe as in type II disease and the patients may survive longer. Some of the symptoms may be somewhat alleviated with age. 4. *Type IV disease* involves a 3p21 recessive defect of the glycogen branching enzymes. Progressive destruction of the liver cells is accompanied by an increase of the connective tissues and liver substance (cirrhosis). Increase in size of the liver and spleen and accumulation of fluids in the abdominal cavity (ascites) result in death before age 2. 5. In *type V McArdle's disease* (chromosome 11q13), the homozygosity of the autosomal-recessive glucose-6-phosphate translocase gene causes variable symptoms accompanied by glycogen accumulation. Phosphorylating activity in the muscle tissues is deficient. Painful cramps with physical exercise are the first symptom with an onset around age 20. There is no hypoglycemia or increase of lactate in the blood, but some patients excrete myoglobins in the urine. 6. *Type VI* (chromosome 14q21-q22) patients accumulate glycogen and some have reduced phosphorylating activity. 7. *Type VII disease* (Tarui disease), determined by chromosome 12q13.3 recessive genes, resembles type V disease, but the patients have reduced phosphofructokinase activity. 8. *Type VIII* glycogen storage disease is caused by a Xp22.2-p22.1 chromosomal recessive gene and thus affects primarily males. It is based on a leukocyte phosphorylase b

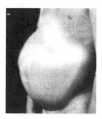

Photo by courtesy of Dr. C. Sterm.

activation deficiency. Some glycogen diseases involve multiple enzyme defects. These diseases are frequently associated with muscle weakness and various other adverse effects. *See* acid maltase deficiency; epilepsy; glucose-6-phosphate dehydrogenase; glycogen; neuromuscular disease.

glycogen synthase *See* glycogen.

glycolipid Lipid with a carbohydrate group.

glycolysis Catabolic pathway from carbohydrates to pyruvate. *See* Embden-Meyerhof pathway; pentose monophosphate shunt.

glycome Glycosylated proteins of the cell. *See* glycosylation.

glycophorin 131-amino-acid transmembrane glycoprotein. Serological glycophorin assays have been developed to detect somatic mutations. The glycophorin-spectrin/actin bridge determines membrane shape and stability. *See* actin; spectrin. (Gerber, D. & Shai, Y. 2001. *J. Biol. Chem.* 276:31229.)

glycoprotein Protein with covalently linked carbohydrate(s).

glycosaminoglycan (mucopolysaccharide) Heteropolysaccharide alternating N-acetylglucosamine + uronic acid and N-acetylgalactosamine + uronic acid (glucuronic acid). This family of compounds includes chitins, chondroitin sulfate, heparan, heparin, hyaluronic acid, keratans, and keratin. *See* exostosis; mucopolysaccharidosis; proteoglycan. (Constantopolous, G. & Dekaban, A. S. 1975. *Clin. Chim. Acta* 59[3]:321.)

glycose Generic name of monosaccharides, e.g., glucose, fructose, mannose, etc.

glycosidase Digests glycosidic bonds and transfers glycosyl moieties from a donor sugar to an acceptor of another sugar or molecule(s).

glycosidic bond Sugar linked to alcohol or purine, or pyrimidine or sugar, through an oxygen or nitrogen atom.

glycosome Peroxisome (microbodies) filled with glycolytic enzymes. *See* glycolysis; microbody.

glycosphingolipid Present in the plasma membrane rafts and caveolae and has an important role in differentiation and development. *See* caveolae; RAFT; sphingolipid.

glycosuria Incompletely recessive defect in glycose reabsorption by the kidney resulting in a high sugar level in the urine. *See* disaccharide intolerance; phlorizin.

glycosylase Enzyme involved in excision of damaged purines and pyrimidines from the sugar-phosphate backbone of DNA. Different enzymes work on different bases. The uracil-DNA glycosylases (human gene UNG, chromosome 12q23-q24)

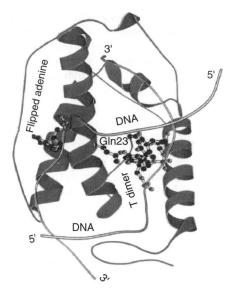

T4-pdg (pyrimidine dimer glycosylase/AP lyase, a TDG) X-ray crystal structure as a ribbon diagram. The DNA is distorted and the flipped-out adenine and the thymine dimer are shown as ball and sticks. The 3′ and 5′ indicate the directions of the DNA strands. Gln23 is a mutant amino acid residue. (Courtesy of R. S. Lloyd. From McCullough A. K., et al. 1999 *Annu. Rev. Biochem.* 68:255.)

remove uracils that are formed by spontaneous or induced deamination of cytosine to avoid U-G mispairing, potentially leading to GC→AT transitions. It works in the nucleus and the mitochondria as well. Thymine-DNA glycosylase (TDG/UDG, 12q24.1) is one of the most efficient of these repair enzymes. The enzyme pushes and pulls out the improper uracil nucleotide from the major groove of the DNA. Subsequently, TDG excises U, AP endonuclease cleaves the DNA backbone, deoxyribophosphodiesterase removes the 5′-phosphate group, DNA polymerase β replaces the correct nucleotide, and ligase finishes the job. It has been estimated that in a human cell 100 to 500 cytosine residues are deaminated daily. MUG (mismatch-specific uracil DNA glycosylase, human chromosome 12) removes uracil/thymine when mispaired with guanine. Human hSMUG1 operates primarily at single strands of DNA during replication and transcription. MBD4 glycosylase (3q21) may remove mismatched U or T nucleotides. The 3-methyladenine-DNA glycosylase (gene AAG/MPG, chromosome 16p-telomere) works on N-3 and N-7 methylation adducts of purines (including hypoxanthine) and cyclic adducts. Pyrimidine hydrate-DNA glycosylase removes damaged or altered pyrimidines. OGG1 (oxoguanine glycosylase, 3p25.3/3p26.2) removes 8-oxoguanine across cytosine. MYH (1p32.1-p34.3) glycosylase excises adenine when misincorporated across oxoguanine. Formamidopyrimidine-DNA glycosylase (NTHL1, 16p13.3-p13.2) excises oxidatively damaged purines such as 8-oxyguanine, 8-hydroxyguanine, thymine glycol, and cytosine glycol but requires the cofactor XPG (xeroderma pigmentosum G, 13q33). This DNA glycosylase also removes deaminated 5-methylcytosines that are common in eukaryotic DNA. All of these excision repair enzymes maintain the working conditions of the human cells; each suffers more than 10,000 damages each day. The yeast or *E. coli* glycosylases have similar functions, but the proteins involved are different in size. *See* adduct; AP endonucleases; base flipping; cyclobutane ring; DNA repair; endonuclease III; endonuclease VIII; excinucleases; pyrimidine dimer; RAD27; transition; X-ray repair. (McCullough, A. K., et al. 1999. *Annu. Rev. Biochem.* 68:255; Hollis, T., et al. 2000. *Mutation Res.* 460:201; Ischenko, A. A. & Saparbaev, M. K. 2002. *Nature* 415:183.)

glycosylation Attachment of sugars to proteins either through a hydroxyl group of serine or threonine (O-glycosylation, Ser[Thr]-O-GlcNAcylation) or to the amide group of an asparagine (N-glycosylation). Glycosylated proteins have many different types of cellular functions. O-glycosylation occurs in proteins of the nuclear pore, RNA polymerase II, transcription factors, oncoproteins (tumor suppressors), chromatin proteins, microtubule-associated proteins, cytoskeletal-binding proteins, tyrosine phosphatase, SV40 T antigen, estrogen receptors, etc. Some antibiotics (tunicamycin) interfere with the process. Glycosylation increases the stability of proteins and may facilitate antigen recognition, appropriate folding, signal transduction, nerve function, etc. It plays an important role in the function of the immune system and several other health-related metabolic functions. *See* glycoform; sialic acid. (Rudd, P. M., et al. 2001. *Science* 291:2370; Lübke, T., et al. 2001. *Nature Genet.* 28:73.)

glycosyltransferase Enzyme adding glucose to proteins and lipids involved in the formation of lipopolysaccharides used for the bacterial cell wall. The ABO blood group alleles also encode glycosyltransferases (also the B gene product adds galactose). These enzymes shape the cell surface, determine cell-to-cell contacts, play some role in cancer, and have an important role in various sphingolipidoses. *See* ABO blood group; sphingolipidoses. (Cosgrove, D. J. 1999. *Annu. Rev. Plant Physiol. Plant Mol. Biol.* 50:391.)

glyoxylate cycle Converts acetate into succinate and finally into carbohydrate. *See* Krebs-Szentgyörgyi cycle.

glyoxysome Vesicle in plant seeds. A special type of microbodies (peroxisomes) in plants mediates the conversion of fatty acids to succinic acid to produce peroxiacetyl CoA and glucose through the glyoxylate cycle. *See* glyoxylate cycle; microbody.

glypican Transmembrane protein with phosphatidyl inositol distal to the membrane and heparan sulfate outside but proximal to the membrane. Glypican-1, -3, and -5 are encoded at human chromosomes 2q35-q37, Xq26, and 13q31-q32, respectively. *See* heparan sulfate; phosphatidyl-inositol; Simpson-Golabi-Behmel syndrome; syndecan. (Filmus, J. & Selleck, S. B. 2001. *J. Clin. Invest* 108[4]:497.)

GLYT Glycine-specific transporters to the nervous system; they may be inhibitory neurotransmitters through ligand-gated Cl⁻ channels activated by glycine or may modulate glutamate-mediated neurotransmission. *See* ion channels; transporters. (Hanley, J. G., et al. 2000. *J. Biol. Chem.* 275:840.)

GM *See* GMO.

GM-CSF Granulocyte macrophage colony-stimulating growth factor, a lymphokine. *See* G-CSF; granulocyte; lymphokines; macrophage; MCSF; M-CSF. (Collins, S. J., et al. 2001. *Blood* 98:2382.)

GMENDEL Computer program for analysis of segregation and linkage. (*J. Hered.* 81:407 [1990].)

GM1-gangliosidosis *See* gangliosidosis type I.

GM2-gangliosidosis *See* Sandhoff disease; Tay-Sach disease.

GM3-gangliosidosis *See* gangliosidosis type III; Sandhoff disease.

GMHT Genetically modified herbicide-tolerant plants. They have an agronomic advantage by facilitating the elimination of undesirable weeds from crops, but they may adversely affect the population of wild birds that feed on weed seeds. *See* GMO.

GMO Genetically modified organisms; a name actually used for transgenic plants and animals. On the basis of fear that the products may cause allergic reactions (e.g., Brazil nut albumins in transgenic soybean), consumers may oppose genetically modified (GM) food. The modified organisms develop antibiotic resistance (in case antibiotic resistance genes were used in the transformation vectors) or the transgene may be transcribed and translated into harmful substances. Lectins, alkaloids in the GMO, may create human or environmental hazards (e.g., transmitting glyphosat herbicide resistance by cross-pollination to weeds [*Cruciferae*], harming useful insects [e.g., neuropteran lacewings] or other species such as Monarch butterfly). Ecologists expressed concerns about the potential selective advantage of genetically modified plants in the natural environment. A 10-year study involving rape, maize, beet, and potato carrying various transgenes found, however, that in general these plants are no more invasive than conventional crops (Crawley, M. J., et al. 2001. *Nature* 409:682). Transgenic rice producing more β-carotene and accumulating more iron may not entail any danger but can reduce malnutrition, anemia, and some blindness in underdeveloped areas of the world. It is worth considering that the transgenic organisms may be easier, cheaper, and safer to produce because of resistance to pests and diseases and may curtail hunger in poor areas. The extensive use of chemical pesticides may be reduced. Regulations and/or labeling of the products and further research to clarify the cost/benefit dilemmas will be needed.

Monarch butterfly [*Danaus*] larva.

Rainbow trout (*Onorhynchus*) or salmon carrying engineered growth hormone genes may grow dramatically larger,

especially the nondomesticated forms, which have not been subjected to selection for increased productivity. One must keep in mind that a type of genetic modification, selection, raised the sugar content of beets from ~2% to ~20% from the middle of the 18th century to the present time. Similarly purposeful plant breeding increased maize production from about 1.25 metric tons/hectare to ~15 tons since the 1930s. The "green revolution" doubled cereal production since 1960. Nevertheless, about 40,000 children still die daily from malnutrition-related diseases. Obviously, technological progress will have trade-offs. The only question is whether it is worth it. There was a general fear in the 1970s about the use of recombinant DNAs even for laboratory purposes. Most of these fears turned out to be unfounded, but certain types of genetic engineering (e.g., using toxin genes) can be carried out only in a highly controlled environment and only when there is a, justified need. The arguments against and for genetically modified organisms must be based on scientifically validated facts rather than on political views or preconceived notions. Existing information indicates relatively fast (100–150 days) decomposition of most (60–70%) of the Bt toxin in soil and even in plant tissues. Paracelsus of Hohenheim (1493–1541), the "Luther of Medicine," remarked: "Guilty is he who does not know it properly and who does not apply it properly." The genetic and ethical problems relevant to inheritable genetic modification of humans can be accessed on the Web: <www.aaas.org/spp/dspp/sfrl/germline/main.htm>. *See* Asilomar conference; *Bacillus thüringiensis*; biohazards; biotechnology; Bt; gene therapy; GMO; input trait; nuclear transplantation; pest eradication; recombinant DNA and biohazards; stem cells; targeting genes; xenotransplantation. (Wolferbarger, L. L. & Phifer, P. R. 2000. *Science* 290:2088; Quist, D. & Chapela, I. H. 2001. *Nature* 414:541; Dale, P. J., et al. 2002. *Nature Biotechn.* 20:567; Hare, P. D. & Chua, N.-H. 2002. *Nature Biotechn.* 20:575; <www.nbiap.vt.edu>; <www.usia.gov/topical/global/biotech>; <www.colostate.edu/programs/lifesciences/TransgenicCrops>.)

GMP Guanosine monophosphate.

GMS Genomic mismatch scanning is a method designed to scan large genomic DNA samples for differences in order to identify alterations, e.g., those responsible for hereditary disease. The principles are as follows: Two DNA samples (diseased and healthy) are digested with restriction endonuclease. Fragments of one of the samples are methylated. Then both samples are denatured and allowed to hybridize. From reannealed DNA, only hybrid strands are subjected to further study (i.e., one of the two strands is methylated but the other is not). These hybrids are exposed to bacterial mismatch repair enzymes that recognize mismatches and nick the unmethylated strand at that site. The nicked strands are removed and the intact duplexes are retained. These would be expected to include the desired marker(s). The method is very elegant in principles but cannot yet be applied to the very complex human genome. Single genetic regions can, however, be studied with the aid of array hybridization. *See* array hybridization; genetic screening; mismatch repair; RDA. (Mirzayans, F. & Walter, M. A. 2001. *Methods Mol. Biol.* 175:37.)

gnotobiota Microbes (animals and plants) associated with laboratory animals.

GnRHA Gonadotropin-releasing hormone agonist. When administered at a constant rate, it shuts down mammalian reproductive functions and induces a condition resembling menopause. It can be employed as a fertility controlling agent but must be supplemented with periodic treatments with other hormones to prevent menopause-like side effects. It can be used to save an implanted ovum or zygote by preventing ovulation. GnRHA has many other medical applications. *See* ART; gonadotropin-releasing factor; in vitro fertilization; menopause; menstruation. (Smitz, J., et al. 1992. *Hum. Reprod. Suppl.* 1:49.)

GNRP Guanine nucleotide releasing protein. When activated by receptor tyrosine kinase in the signal transduction pathway, an RAS protein switch is turned on. *See* RAS; signal transduction. (Marshall, M. 1995. *Mol. Reprod. Dev.* 42[4]:493.)

GO Dormant stage of cell divisions in fission yeast. *See* cell cycle; *Schizosaccharomyces pombe*.

goat (*Capra hircus*) $2n = 60$. It was probably the first large herbivorous domesticated animal. (MacHugh, D. E. & Bradley, D. G. 2001. *Proc. Natl. Acad. Sci. USA* 98:5382.)

goat-sheep hybrids Domesticated sheep (*Ovies aries*, $2n = 54$) can be impregnated by domesticated goat (*Capra hircus*, $2n = 60$), but the hybrid embryos rarely develop normally; occasionally some hybrids grow up. *See* animal species hybrids. (Hancock, J. L., et al. 1968. *J. Reprod. Fertil. Suppl.* 3:29; Ilbery, P. L., et al. 1967. *Aust. J. Biol. Sci.* 20:1245.)

GOGAT Glutamine-2-oxoglutarate transferase. *See* nitrogen fixation.

goiter, familial Actually a collection of various metabolic anomalies involving enlargement of the thyroid gland, which may become obvious by viewing the neck. The defect may involve various dominant or recessive mutations in the thyroglobulin gene. The thyroglobulin gene (TG) is in human chromosome 8q24, extending to about 300 kb genomic DNA and containing 37 exons and large introns. The dimeric thyroglobulin protein has a molecular weight of ca. 660,000. This protein is iodinated at tyrosine residues to form mono- and diiodotyrosines.

Thyroxine is a tetraiodothyronine, but also *triiodothyronine* is formed upon activation by peroxidase. The iodinated proteins are transported by the blood, increasing the metabolism and regulating the function of the nervous system, kidneys, liver, and heart. *Hyperthyroidism* occurs with overproduction of iodinated thyroglobulin hormones, resulting in goiter, fast heart rate, fatigue, muscular weakness, heat intolerance and sweating, tremor, and emotional instability. Excessive secretion of thyroid hormones is called *Graves'* or *Basedow* disease, with susceptibility controlled by several sites (14q31, 6p21, 2q33, 20q13, Xq21). The new Graves' disease maps to 18q21. The latter condition may or may not be genetic, although its frequency may be quite high (0.008). The basic defect may involve autoimmunity of the receptor of the hormone. *Hypothyroidism* is the consequence of underproduction of the thyroid hormone, resulting in fatigue, lethargy, low metabolism, cold sensitivity, and menstrual problems in females. This condition may lead to *cretinism*, which is most commonly caused by failure of releasing *thyrotropin*, the glycoprotein thyroid-stimulating hormone of the anterior pituitary. Cretinism also means an arrest of physical and mental development caused by this hormonal deficiency.

Hypothyroidism may lead to deafness. Defects in deiodination of iodotyrosines may cause hypothyroidism. *Permanent congenital hypothyroidism* has a prevalence of $\sim 3-4 \times 10^{-4}$ newborns, and unless it is caused by hypothalmic or pituitary defects, it is accompanied by overexpression of thyroid-stimulating hormone and lower-than-normal thyroid function or thyroid dysgenesis. Thyroid therapy is required within the first couple of months to prevent neurological damage (cretinism). In some cases, mutation of the Pax8 gene at human chromosome 2q12-q14 has been detected. PAX8 seems to be required for the differentiation of endoderm primordia into thyroxin-producing follicular cells. Goiter-type diseases are known in the majority of mammals. Thyroxine-binding globulin is encoded in human chromosome Xq28 and a thyroxin-binding serum globulin is autosomal. The multinodular goiter with 5:1 female:male ratio has been assigned to Xp22. *See* animal hormones; hyperthyroidism; NIS; PAX; Pendred syndrome; thyroid-stimulating hormone; tyrosine. (Tomer, Y., et al. 1997. *J. Clin. Endocrinol. Metab.* 82:1645; Vaidya, B., et al. 2000. *Am. J. Hum. Genet.* 66:1710.)

Goldberg-Hogness box *See* Hogness box.

Goldenhar syndrome Autosomal-dominant and -recessive forms with different expressions of facial and other developmental deformities.

G$_{olf}$-protein Trimeric G protein that stimulates cAMP in the control of olfactory neurons. *See* G proteins; olfactogenetics; olfactory.

Golgi apparatus Flat vesicles (cisternae) containing cellular storage and transport material involved in glycosylation, sulfation, proteolysis, etc., in animals. Although the model shown on next page indicates transport in one direction, evidence is accumulating for transport by the cisternae from the cell membrane toward the endoplasmic reticulum. *Coat protein I* (COPI) and SNARE seem to play important roles in the transport. The homologous structures in plants are frequently called dictyosomes. Some of the Golgi structures are located next to the endoplasmic reticulum and are called *cis-Golgi*; others are at a distance (*trans-Golgi*). In these vesicles, some proteins are modified after the completion of their synthesis in the endoplasmic reticulum.

The Golgi complex is inherited by fragmentation of the elements into small vesicles, which were thought to be randomly distributed during mitosis. However, according to more recent observations, the distribution may not be entirely random. The fragments aggregate around the mitotic spindle pole and the motor proteins pull them into the daughter cells. After cytokinesis, Cdc2 supposedly phosphorylates the p115 receptor and then the Golgi structure is reconstituted of the fragments probably via the endoplasmic reticulum. The exact mechanisms are not generally agreed upon, however. *See* cell structure; cis-Golgi; dictyosome; endoplasmic reticulum; RAB

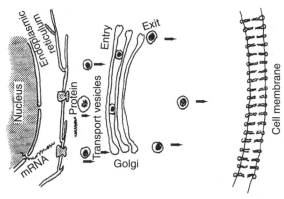

Transport function of the Golgi apparatus. From the nucleus through the nuclear pores, mRNA is reaching the ribosomes sitting on the endoplasmic reticulum (ER). The synthesized protein may enter the lumen of the er with the assistance of a signal peptide transfer particle–mediated system. The proteins may emerge in transport vesicles to enter the Golgi at the cis end and exit at the bulbous ends of the stacked membrane vesicles. In the golgi, the proteins are glycosylated and modified posttranslationally.

oncogene; SNARE; trans-Golgi network. (Allan, B. B. & Balch, W. E. 1999. *Science* 285:63; Roth, M. G. 1999. *Cell* 99:559; Müsch, A., et al. 2001. *EMBO J.* 20:2171; Allan, V. J., et al. 2002. *Nature Cell Biol.* 4:E236.)

Gombo syndrome Autosomal-recessive growth retardation with eye, brain, skeletal, and mental defects. *See* growth retardation.

Gomori's (Gömöri's) stain Used primarily for histological localization of phosphatases and lipases in sectioned specimens by the light microscope. *See* histochemistry; stains.

gonadal dysgenesis Failure of normal differentiation of the gonads (ovary, testis). It is a common cause of sterility in aneuploids. Gonadal dysgenesis of XY chromosomal constitution occurs in mammalian females. They have "streak gonads" and fail to develop the secondary sexual characteristics. Gonadal neoplasias are frequent in these individuals. It has been shown that the testis-determining factor resides in a Y-chromosomal segment and either deletion or base substitution may lead to an inactive human SRY (Yp11.3) product, a DNA-binding protein involved in testis determination. Transfection of the TDY (the mouse homologue of SRY) DNA into XX mouse has induced male development. Gonadal dysgenesis may also occur in XX females, which have a higher-than-normal level of gonadotropins and underdeveloped male gonads. In XY females (GDXY, Xp22.11–21.2), gonadal dysgenesis causes multiple developmental anomalies and hypermuscular appearance. Premature ovarian failure (2p21-p26) is a mutation in the follicle-stimulating hormone receptor. The cause of the dysgenesis may reside either in autosomal-recessive genes or in sex chromosomes. Mutation in the human desert hedgehog (12q12-q31.1) may lead to partial gonadal dysgenesis and neuropathy. *See* campomelic dysplasia; FSH; gonad; hedgehog; hermaphroditism; H-Y antigen; pseudohermaphroditism; sex reversal; Smith-Lemli-Opitz syndrome; SOX; SRY; Swyer syndrome; testicular feminization; Turner syndrome.

gonadoblastoma Rare type of neoplasm containing germ cells, immature Sertoli-like cells, and cells resembling the granulosa cell of ovarian follicles. Mutation in a 1–2 Mb fragment encoding the testis-specific protein (TSPY) near the centromere in the short arm of the human Y chromosome may be responsible for gonadoblastoma.

gonadotrophin Gonadotropin.

gonadotropin Group of hormones that regulate gonadal and placental functions. *See* GnRHA; MSAFP.

gonadotropin-releasing factor *See* luteinizing hormone-releasing factor.

gonadotropin-releasing hormone agonist *See* GnRHA.

gonads Organs of gametogenesis, such as the ovary and testis. In *Drosophila*, the male gonad includes about 15 cells and in the female about 12 cells that further proliferate during embryonic differentiation; there are 60–110 primordial germ cells by the late instar stage. In mice, the primordial germ cells appear 7 days after mating (dpc). In the female, the primordial germ cells stop mitosis 13–15 dpc and meiosis is initiated immediately. Spermatogenesis begins 5–6 days after birth, and on the 9th day, proleptotene spermatocytes appear. By day 18, haploid spermatids are formed followed by spermiogenesis. The general pattern of gonadal development is similar in mammals as sketched at the below. The testes and the ovary differentiate from sex-neutral structures (shown in the middle). From the Müllerian ducts, the fallopian tube develops, and the uterus develops at its base while the primitive gonad is converted to the ovary. From the Wolffian duct, the vas deferens and the seminal vesicles at its base are formed, and the primitive gonad is converted into the testis. On this sketch (for saving space), only half of the female and male sexual apparatus are shown at right and left, respectively. At the undifferentiated stage, the steroidogenic factor, SF-1, Wilms' tumor 1 zinc-finger protein, and homeobox gene Lim-1 are important. During the sexual differentiation stage, the Y-chromosomal testis-determining gene, SRY, and the autosomal male sex differentiation factors, SOX9, glucoprotein WNT4 (female), WT1 (male), SF-1, a general sexual differentiation factor, and DMRT1 play key roles. The Müllerian inhibitory substance inhibits testosterone synthesis in the female, whereas it is promoted in the male. Dax-1 protein (originally

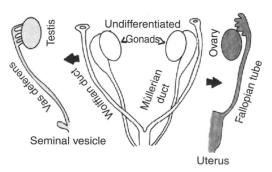

(The diagram was redrawn after H. Eldon Sutton & R. P. Wagner. 1985. *Genetics. A Human Concern.* Macmillan, New York.)

considered an ovary-promoting substance) at a higher dosage suppresses male development. InsL3 protein has a role for the gubernaculum (a ligament), facilitating the descent of the testes into the testicular bag, the scrotum. Numerous other proteins are also involved. *See* Müllerian duct, Wolffian duct, gametogenesis, cell cycle, germline, dpc, mismatch repair, sex, sex reversal, GDNF. (Roberts, L. M., et al. 1999. *Am. J. Hum. Genet.* 65:933; Koopman, P. 2001. *Curr. Biol.* 11:R481; Mackay, S. 2000. *Int. Rev. Cytol.* 200:47.)

gonidia Specialized asexual reproductive cells.

gonochorism Normally the species has separate male and female individuals. *See* dioecious; hermaphrodite.

gonocytes Progenitors of the spermatogonia. *See* gametogenesis; spermatogonia.

gonosomes Sex chromosomes in distinction from the autosomes. *See* autosome; sex chromosome.

Gonzalo of Spain Great-grandson of Queen Victoria of England, who inherited from her grandmother, Beatrice, the classic X-chromosomal hemophilia gene and died by hemorrhage in an automobile accident at age 20. *See* anticoagulation factors; hemophilia; Queen Victoria.

good genes (1) Genes that are advantageous for the individual or for evolution (eugenics). (2) Genes with good penetrance and expressivity, thus facilitating analysis of their inheritance and function. *See* expressivity; penetrance.

goodness-of-fit test *See* chi-square.

Goodpasture syndrome Autosomal-recessive autoimmune reaction of the basement membrane of the renal glomeruli and the lung. The basic defect is in the α-chain of collagen type 4. Although some cases have indicated familial occurrence, most likely the patients were exposed to similar (viral, bacterial, hydrocarbon) environmental factors. *See* autoimmune disease; basement membrane; collagen.

goose *Anser anser*, $2n = 80$.

gooseberry (*Ribes* spp) Tart berry fruits; $2n = 2x = 16$.

gopher Genetic information databases accessible through Internet electronic networks. Software for gopher is free and can be obtained with FTP (file transfer protocol) at <gopher@boombox.micro.umn.edu>. (The name comes from a ground squirrel.)

G$_0$ protein Subunit of trimeric G protein; it activates K$^+$ channels and shuts down Ca^{2+} ion channels. Mutations in the gene encoding it cause behavioral anomalies in *Caenorhabitis* similar to those caused by a defect in the serotonin receptor. The main symptoms are hyperactivity, premature egg laying, and male impotence due to defects in neuronal and muscle functions. *See* G$_\alpha$; ion channels; serotonin; signal transduction.

Gordon syndrome (PHA2A, 1q31-q42) Involves hypertension and high salt concentration in the blood with a normal filtration rate in the kidneys. It is also called pseudohypoaldosteronism II. Similar disorders occur at 17q21-q22 and 12p13. *See* hypertension; Liddle syndrome; pseudohypoaldosteronism.

gorilla *See* Pongidae; primates.

Gorlin-Chaudhry-Moss syndrome Very rare autosomal-recessive craniofacial dysostosis (head malformation), excessive hairiness, heart and lung defects (patent ductus), and hypoplasia (reduced growth) of the female external genitalia. *See* craniosynostosis; hypertrichosis.

Gorlin-Goltz syndrome *See* nevoid basal cell carcinoma.

Gossypium (cotton) Member of the Malvaceae family of plants. Economically the most important are the long staple upland species, *G. hirsutum* ($2n = 4x = 52$), which produces 95% of the cotton fibers; *G. barbadense*, an extra-long staple (Sea Island, Egyptian) cotton (also a tetraploid), contributes about 5% of the world fiber. There are 30 diploid species. *G. herbaceum* and *G. arboreum* carry the *A* genome and are the only diploids with spinnable lint. The *B* genome is represented by North African and Cape Verde Islands species. The *C* genome occurs in Australian diploids. *D* genome plants occur in Mexico, Peru, the Galápagos Islands, and the United States. *E* genome species occur in North Africa, Arabia, and Pakistan. The *F* genome is represented by a single African species. The new world tetraploids contain the *A* and *D* genomes. Most of the cottons are naturally cross-pollinated but tolerate inbreeding. The various genomes are distinguished primarily on the basis of chiasma frequencies and the number of univalents in the species hybrids, although some chromosome morphological differences also exist. (<http://www.tigr.org/tdb/tgi.shtml>.)

GO units *See* gene ontology.

gout Complex hereditary disorder of the joints, leading to arthritis caused by overproduction and/or underexcretion of uric acid. In autosomal-recessive gout, glucose-6-phosphatase is deficient. In X-linked gout, hypoxanthine-guanine phosphoribosyl transferase deficiency exists. Some gouts are associated with increased turnover of nucleic acids. Autosomal-dominant and polygenic forms are known. Gout may be asymptomatic initially, but at later stages the joints and kidneys may become permanently injured. (See swollen gouty fingers at left.) The first sign is pain in the great toe, but it may spread to other

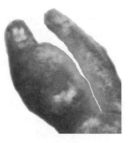

Gouty fingers.

parts of the foot and also to the wrists and other body parts. The prevalence may vary in different populations from 0.2 to 10%. The serum urate level may vary from 6 to over 9 mg/100 mL serum. Generally fewer women than men suffer from it, but in women the gout may be more severe and destructive. If the diet is very low in proteins, gout may not appear.

Uric acid crystals (urate) activate the Hageman factor in the viscous fluid (synovia) of the joints, which in turn sets into motion a series of events leading to inflammation. In chickens that lack the Hageman factor or in dogs with a suppressed number of leukocytes (leukopenia), the inflammatory reaction fails, indicating the role of these factors in gout attack. After the first attack, gout symptoms apparently disappear for weeks or many months but return with greater strength. Chronic arthritic gout may produce ulcerating tophi (a chalky urate deposit) in the joints and may cause severe deformation of the affected area. Urates may also be deposited in kidneys, cartilage, and bone tissues. Tophi may be present at the fingertips, palms, soles, eyelids, nasal cartilages, and in the eye. Rarely it is observed in and on the penis, the aorta, the heart wall (myocardium), valves, the tongue, the entrance of the larynx (epiglottis), and vocal cords. Urate deposits occur in between the vertebral disks and cartilages. There is very little or any urate in the spinal cord or in the nervous system. In the kidney, medulla urate crystals may accumulate and kidney stones may form (lithiasis) in 20 to 40% of affected persons.

Gout is frequently associated with obesity, and hyperuricemia is common in the case of diabetes mellitus. Hyperlipoproteinemia and high triglyceride levels are common in gout. Alcoholism may aggravate hyperuricemia. Serum urate levels are about the same in people of European origin, North American Indians, Hawaiians, Japanese, and Chinese. In some Polynesiansm, Australian aborigines, and South American Indians, the urate level may be higher. Overproduction of uric acid is correlated with the availability of L-glutamine and phosphoribosyl-1-pyrophosphate, which are rate-limiting precursors in purine biosynthesis. Uric acid is dramatically overproduced in the case of (partial) deficiency of hypoxanthine-guanine-phospho-ribosyltransferase (HPRT). Glucose-6-phosphatase and glutathione reductase also increase uric acid synthesis. Exposure to lead may increase the occurrence of gouty arthritis due to inflammation of the kidney (saturnine gout). Starvation, Down syndrome, and psoriasis (a skin disease causing silvery scaling and plaques) may increase uricemia. Acute attacks of gout may be successfully treated with colchicine and allopurinol, an inhibitor of xanthine oxidase. Both of these compounds may be highly toxic. Gout due to genetically determined factors is called primary gout; secondary gout is the result of ingestion of certain chemicals and drugs. Many famous historical persons were afflicted by gout: Medici, Newton, Darwin, Luther, Calvin, Benjamin Franklin, Cotton Mather (who reported plant hybrids in America in 1721), and others. *See* antihemophilic factors; colchicine; Hageman factor; Lesch-Nyhan syndrome. (Chen, S. Y., et al. 2001. *Metabolism* 50:1203.)

gp In general, the abbreviation for glycoprotein; the gp is usually followed by a number.

gp Gene of phage, e.g., the first gene of λ-phage entering the capsid is *gpNu*. *See* lambda phage.

gp39 Same as the CD40 ligand.

gp120 HIV glycoprotein activates B lymphocytes with receptors carrying variable heavy-chain (V_H3) immunoglobulins. *See* B lymphocytes; HIV; immunoglobulins.

gp130 ~101 kDa (without glycosylation) subunit of the interleukin-6 family receptors, encoded at 5q21. It is a signal transducer chain for IL-6, IL-11, LIF, OSM, and CNTF. *See* APRF; interleukins. (Chow, D.-C., et al. 2001. *Science* 291:2150.)

gp190 ~121 kDa (without glycosylation) subunit of various cytokine receptors, encoded at 5p12-p13.

GPA Genes of yeast homologous to Gα cDNAs involved in mammalian G-protein coding. GPA1 protein is 110 and GPA2 is 83 amino acids longer at the N termini than the mammalian proteins. GPA1 may be involved in mating signal transduction; GPA2 controls cAMP level. GPA1 (α-subunit of G protein) plays a negative role (growth arrest) in mating signal transduction, whereas *STE4/STE18* (β, γ subunits) are responsible for a positive transducing signal (enhancement) for mating. *See* cAMP; G proteins; mating type determination in yeast; STE.

GPCR (G-protein-linked receptors) *See* signal transduction.

G6PD Glucose-6-phosphate dehydrogenase deficiency is responsible for one type of hemolytic anemia in humans; it is controlled by a sex-linked recessive gene (map location X28). It catalyzes the reaction $G6P + TPN^+ + H_2O \rightleftarrows$ 6-phosphogluconic acid $+ TPNH + H^+$. *See* glucose-6-phosphate dehydrogenase; glutathione; malaria; Zwischenferment.

G4 phage Single-stranded DNA phage (5,507 bases), ~67% related to φX174. *See* bacteriophages; φX174. (Godson, G. N., et al. 1978. *Nature* 276:236.)

G0 phase State of a pause for the cell before it enters the G_1 phase and until divisional activities start again after mitosis. *See* cell cycle.

G1 phase First phase of the cell cycle following mitosis (*C* value = 2). *See* cell cycle.

G2 phase Phase following DNA replication during the cell cycle (*C* value = 4). *See* cell cycle. (O'Connell, M. J., et al. 2000. *Trends Cell Biol.* 10[7]:296.)

GPI anchor Glycosyl-phosphatidylinositol cell surface membrane protein.

G1ps G1 (gap 1) Presynthetic phase (preceding S phase) of mitosis. *See* cell cycle; mitosis.

GPR Coreceptors of HIV and SIV. *See* acquired immunodeficiency syndrome.

G protein Guanine nucleotide-binding protein that serves as an intermediary in biological signaling pathways. The signal is received by *receptors* and the *G proteins* forward it by mediation of a different number of intermediaries to the *effectors* that regulate genes in response to the signals. G proteins are activated by aluminum fluoride and the α-subunit can be ADP-ribosylation mediated by bacterial toxins (cholera, pertussis).

The large G proteins are heterotrimeric (α, β, γ) and control the opening and closing of the signal transduction pathways by changing the attached GDP$\rightleftharpoons$GTP. In the GTP-associated form, they have a key role in signal transduction from receptors to effectors. There are also low-molecular-weight small G proteins with a single (α) subunit. G protein (G_s) is involved in the regulation of the level of the enzyme adenylyl cyclase and thus cAMP- and cAMP-dependent protein kinase. The G_i form is involved in the inhibition of adenylate cyclase; G_{iia2} is required for insulin function. The light-activated GTPase activity is mediated by the G_t protein, also called transducin. G proteins stimulate the hydrolysis of phosphoinositides with the aid of phospholipase C. cAMP degradation by cyclic nucleotide phosphodiesterase is also mediated by G proteins and indirectly by Ca^{2+}. G proteins regulate ion channels. When a proper ligand binds to a transmembrane receptor, the trimeric G protein dissociates into a $\beta\gamma$ and an α-subunit. The α-subunit stimulates adenylyl cyclase, and first the transition of GDP to GTP and later the transition of GTP to GDP through mild GTPase activity. In the G-GDP state, reassociation of the three subunits follows.

G-proteins regulate Ca^{2+} metabolism and indirectly control allosteric effector proteins. Several human diseases are associated with defects in G proteins (pituitary tumors, McCune-Albright syndrome, Albright hereditary osteodystrophy, puberty precocious) or with defects in the G-protein receptors (hypercalcemia, hypercalciuria, hyperparathyroidism, diabetes insipidus, retinitis pigmentosa, color blindness, glucocorticoid deficiency, opiate addiction, retinitis pigmentosa, hypertension, myocardial ischemia, chronic heart disease). Dozens of genes are involved with coding and regulation of G-protein subunits scattered among several human chromosomes. Many of the G-protein-coupled receptors are without introns, suggesting that retrogenes code for them. Mutations in exons may alter splicing sites remote from the mutation and thus create new variants. G proteins have important roles in both animal and plant cells; however, plants have much fewer genes encoding the trimeric G-protein subunits. Several human diseases implicate G-protein malfunctions. *See* adenylate cyclase; calmoduline; cAMP; cell cycle; cholera toxin; G_α-; G_i-; G_s-; G_o-; G_q-; G region; G' region; G'' region; GTPase; G_t- protein; pertussis toxin; phosphodiesterase; receptor; retrogene; rhodopsin; signal transduction. (Sprang, S. R. 1997. *Annu. Rev. Biochem.* 66:639; Bockaert, J. & Pin, J. P. 1999. *EMBO J.* 18:1723; Farfel, et al. 1999. *New Engl. J. Med.* 340:1012; Dohlman, H. G. & Thorner, J. 2001. *Annu. Rev. Biochem.* 70:703; Knoblich, J. A. 2001. *Cell* 107:183; Peterson, Y. K., et al. 2002. *J. Biol. Chem.* 277:6767; Assmann, S. M. 2002. *Plant Cell* 2002. 14:S355; Chalmers, D. T. & Behan, D. P. 2002. *Nature Rev. Drug Discovery* 1:599; <http://tGRAP.uit.no/>.)

GP32 protein (of phage T4) Required for (1) configuration of the single-strand DNA (ssDNA) to accommodate the replisome, including DNA polymerase; (2) melting adventitious secondary structures; (3) protecting ssDNA from nucleases; and (4) facilitating homologous recombination. *See* replication fork; replisome.

G_q-protein Member of the trimeric G-protein family; activates phospholipase C-β and responds to acetylcholine. *See* acetylcholine; acetylcholine receptors; G proteins; phospholipase; signal transduction.

G-quadruplexes Are four-stranded guanine-rich structural elements of the telomeres. *See* telomeres. (Parkinson, G. N., et al. 2002. *Nature* 417:876.)

G quartet Guanine-rich nucleotide sequences may form four-stranded complexes stabilized in Hogsteen structures. The G quartets may be formed in phosphorothioate octamers such as S-$T_2G_4T_2$ or from other sequences like $GTG_2TG_3TG_3TG_3T$. These quartets or double quartets may be potent viral inhibitors. The latter may block HIV1 integrase.

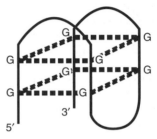

The guanine quartets form hydrogen bonds and may be stacked at the ends of the chromosomes. (After Moine, H. & Mandel, J.-L. 2001. *Science* 294:2487.)

Guanine-rich sequences are common in the telomeres, transcription regulatory regions, immunoglobulin switching areas, etc. Antisense RNAs containing G quartets seem to suppress MYC and MYB cellular oncogenes, HIV integrase, etc., to various degrees. G quartets may regulate the translation of mRNA transcript of the fragile X gene and thus control synaptic activity of neurons in the brain. *See* antisense technologies; fragile X chromosome; HIV; Hogsteen pairs; immunoglobulins; oncogenes. (Horvath, M. P. & Schultz, S. C. 2001. *J. Mol. Biol.* 310:367.)

Graafian follicle Small sac-like structures on the ovary of mammals containing a mature egg (secondary oocyte). The release of the egg is called ovulation and afterward the follicle is transformed into a corpus luteum. *See* corpus luteum; luteinization.

gradient centrifugation Technique of separation of cells, subcellular organelles, and macromolecules on the basis of their density and shape by centrifugation. High-speed centrifuges may separate the larger particles; ultracentrifuges are used for macromolecules. The medium of separation may be sucrose, percol, cesium salts, etc. The material is placed either on top of the medium, which is made in various concentrations in steps—i.e., 60% sucrose, layered on top with 40%, then 20% solutions are placed in the centrifuge tube; or cesium salts may be used at an average density of the macromolecule. In the latter case, during high-speed centrifugation the medium forms a continuous density gradient. In either case, the material will accumulate either at the top of the step (layer), which has a higher density than the substance to be separated, or it will accumulate as a band in the medium that corresponds to the density of the macromolecule (DNA, ribosomes, viral particles). *See* buoyant density; DNA density; ultracentrifuge.

grading up   An animal breed is repeatedly backcrossed with males of another, more desirable livestock to improve its productivity and/or quality. *See* gain.

gradualism   Evolution is supposed to proceed by slow acquisition of adaptive mutations in the Darwinian sense. *See* Darwinian evolution; punctuated evolution.

Graeco-Latin square   Experimental design similar to the Latin square. Three or more variates, e.g., A, B, C, are tested separately under three or more different treatments, e.g., 1, 2, 3, and the results are usually evaluated by analysis of variance. One such arrangement is shown in the box below. *See* analysis of variance; factorial experiments; Latin square.

$$\begin{array}{ccc} A_1 & B_2 & C_3 \\ B_3 & C_1 & A_2 \\ C_2 & A_3 & B_1 \end{array}$$

graft   Transplantation of plant or animal tissues by surgical means.

graft hybrid   Chimera produced by fusion of two genetically different cells in tissues. The followers of Mitchurin, Lysenko, and Glushchenko's group of Soviet ideologues postulated nonchimeric-type graft hybrids. They also called them vegetative hybrids and claimed that grafting alters the hereditary material of both graft and scion. These claims were not reproducible by appropriate methods of experimentation, and several of the results were due either to ignorance or deliberate deception. *See* acquired characters; grafting; lysenkoism; Mitchurin; Soviet genetics.

grafting   (1) Transfer of one piece of tissue or an organ from one place to another within the body or to another body. Horticulturists have practiced grafting of plants as a means of propagation. Grafted roses and other ornamentals, as well as fruit trees, assure the maintenance of genetic uniformity in the grafts where multiplication by seed would produce a heterogeneous offspring because of heterozygosity at multiple loci. Some grafts are horticulturally advantageous because the root stock may be resistant to soil-borne pests and can secure crops in more valuable varieties of *Vitis vinifera* grapes. *Vitis rotundifolia*, wild grape stocks, may be 20-fold more resistant to the *Dactylosphaera vitifolii* root parasite than the standard varieties. Grafting may be used to propagate inviable plants on appropriate stocks and to study the physiological interactions between scion and stock in such complex processes as flowering response, etc. Macromolecules such as viruses may move from stock to scion through plasmodesmata. mRNA may move through the phloem and mutant phenotype may be expressed in the scion. *See* graft hybrids. (Kim, M., et al. 2001. *Science* 293:287.) (2) Grafting is practiced in modern medicine by transplanting skin, kidney, liver, heart, and other organs. Allografts are generally incompatible with the host immune system. The immune response is controlled by a large number of genes that are part of the major histocompatibility gene families. Experimental studies on tissue transplantation are carried out with inbred strains of mice (*See* congenic resistant). The histocompatibility genes are codominant and F_1 hybrids between different inbred lines may accept grafts from both parents, whereas the two parental lines may be incompatible with each other. F_1 hybrid generally also accepts grafts from the later-generation offspring. The incompatibility is inherited in a Mendelian manner and three-fourths of the F_2 individuals are compatible with one or the other parent; one-fourth is not. If the number of independent histocompatibility loci is (n), $(3/4)^n$ = the number of histocompatible individuals in F_2, and in a backcross it is $(1/2)^n$. There are some confounding factors, however. Within some highly inbred lines, skin grafts from male to female may be rejected but not from female to male. This may be due to male-specific antigens encoded by the Y chromosome. Also, tumor tissues may be rejected when skin grafts are accepted because of tumor-specific antigens. Heterotopically (placed to a nonregular position) transferred hearts and kidneys may be accepted when skin grafts are rejected. Even apparently accepted grafts may produce a very low level of antibodies. Allogeneic inhibition may occur, i.e., parental animals fail to accept transplants from their offspring, but the offspring may accept the transfer from that parent. Most of these principles of grafting were derived from studies on inbred mice strains. Grafts have another interest for medicine with the discovery of the regenerative capacity of stem cells. *See* allogeneic; cell therapy; HLA; microcytotoxicity assay; mixed lymphocyte reaction; organ culture; stem cells; therapeutic cloning; zoonosis. (Quisenberry, P. J., et al. 2001. *Ann. NY Acad. Sci.* 938:54.)

graft inheritance   *See* cortical inheritance.

graft rejection   Manifestation and result of histoincompatibility between transplanted tissues: host-versus-graft disease. *See* cytotoxic T cells; graft-versus-host disease; HLA; MHC; microcytotoxicity assay; mixed lymphocyte reaction; therapeutic cloning.

graft-versus-host disease (GvH)   May arise when the grafted tissue damages the host because of the immune reaction. GvH may also be beneficial in eradicating the residual leukemia cells through the mediation of T cells and the HLA molecules of the major histocompatibility system. *See* graft rejection; HLA. (Kärre, K. 2002. *Science* 295:2029.)

gramene (rice and other grasses database)   <http//www.gramene.org>.

gram molecular weight   Grams of a compound equal to its molecular weight in moles.

gram negative/gram positive   Classification of bacteria depending on retention of the gram stain (gentiana violet after an iodine stain, then extracted by acetone or alcohol).

The outer membrane of the gram-positive bacteria (stain blue-purple) does not have lipopolysaccharides, but they are present in the membrane of gram-negative bacteria (stain pink-red). The cell wall of gram-positive bacteria contains peptidoglycans and teichoic acid. *Gram-positive bacteria*: Streptococci, Staphylococci, Pneumococci, Corynebacterium, Mycobacterium, Bacillus anthracis, *B. cereus*, Listeria, Actinomyces, Streptomyces, etc. *Gram-negative bacteria*: Neisseria, Enterobacteriaceae (*E. coli*, Salmonella, Shigella), Haemophilus, Bordatella, Yersinia, *Vibrio cholerae*, Pseudomonas, Brucella, Proteus, Campylobacter, Legionella, etc. *See* peptidoglycan; teichoic acid.

gram stain   *See* bacteria; gram negative/gram positive.

grana (sing. granum)   Dark green pile of flattened membrane vesicles thylakoids) in the chloroplasts. *See* chloroplast; chloroplast genetics.

Granum.

grandchildless-knirps syndrome   In *Drosophila*, maternal effect genes cause embryonic lethality by eliminating pole cells and one or more abdominal segments. *See* maternal effect genes; morphogenesis; pole cells.

granddaughter design   Analysis of genetic linkage of quantitative loci to (usually) DNA markers among the granddaughters. The markers are identified in grandsires and sons, but quantitative analysis is carried out on the daughters of sons. *See* least squares; maximum likelihood method applied to recombination; QTL; RFLP. (Weller, J. I., et al. 1990. *J. Dairy Sci.* 73:2525; Cappieters, W., et al. 1998. *Genetics* 149:1547.)

grande   Wild-type cells of yeast in comparison with the petite mitochondrial mutants deficient in respiration. *See* mtDNA; petite.

granin   Calcium-binding acidic proteins (21 to 76 kDa) in the Golgi network. Their function is processing secreted proteins, and they are subject to processing by converting into biologically active peptides. The granin consensus bears similarity to breast cancer gene proteins BRCA1 and BRCA2. *See* breast cancer; Golgi. (Rosa, P. & Gerdes, H. H. 1994. *J. Endocrinol. Invest.* 17[3]:207.)

Grantham's rule   From the synonymous codons, highly expressed genes preferentially use those that have a pyrimidine at the third position of the triplets. *See* codon usage; genetic code. (Grantham, R., et al. 1981. *Nucleic Acids Res.* 9:43.)

granule exocytosis   *See* cytotoxic T cells.

granulocyte-macrophage colony-stimulating growth factor   GM-CSF (18–30 kDa monomeric glycoprotein, encoded at human chromosome 5q21-q32) activates the cells of the granulocyte pathway. Binding of GM-CSF to its receptor dimerizes the receptor and leads to activation of the Jak-STAT pathway of signal transduction. A small nucleotide, SB 247464, may serve as a nonpeptidyl inducer of the oligomerization of the receptor. It may be involved in the maturation or function of special antigen-presenting cells. *See* antigen-presenting cell; lymphokines; neutropenia; signal transduction. (Tian, S. S., et al. 1996. *Blood* 88:4435; Roth, M. D., et al. 2000. *Cancer Res.* 60:1934; Trapnell, B. C. & Whitsett, J. A. 2002. *Annu. Rev. Physiol.* 64:775.)

granulocytes (polymorphonuclear leukocyte) Specialized white blood cells such as neutrophils, eosinophils, and basophils. They contain numerous lysosomes and secretory vesicles (granules) and play an important role in the defense system of the animal body. *See* C/EBP; eosinophils; lysosome; neutrophils.

granulolysin   *See* cytotoxic T cell.

granulomatous disease, chronic (CGD) Group of X-chromosomal (Xp21.1) recessive conditions involving chronic infections, based on defects in NADPH-oxidase subunits of the neutrophils and other phagocytotic leukocytes. If the normal enzyme is activated, it generates superoxide, which is converted to antimicrobial hydrogen peroxide. The 91 kDa membrane glycoprotein, a phagocyte oxidase (gp91^{phox}, p47^{phox}, p40^{phox}, p67^{phox}), is a part of the cytochrome b system. The autosomal (16q24)-recessive type is deficient in cytochrome b α-subunit (CYBA) and the neutrophil cytosol factor deficiency (NCF1) form is located in chromosome 7q11.23. A third CGD (NCF2) was assigned to 1q25. The Duchenne muscular dystrophy gene may involve CGD and several other genes have a bearing on the disease. *See* contiguous gene syndrome; hydrogen peroxide; leukocyte; McLeod syndrome; muscular dystrophy; neutrophil; superoxide dismutase. (Grizot, S., et al. 2001. *J. Biol. Chem.* 276:21627.)

granum in chloroplasts (plural grana)   Multilayered thylakoids appear as dark "grains" in the chloroplasts when viewed by the light microscope. *See* chloroplast genetics; chloroplasts.

granzymes   Cytotoxic T-cell serine proteases and perforin responsible for apoptosis by activating the precursor (CPP32) of the protease cleaving poly(ADP-ribose) polymerase. Granzyme B is activated in T cells; in the active form, cytolytic CD8$^+$ T cells (CTL) destroy infecting particles. Besides granzymes A and B, perforin is important for the action of CTLs. Granzymes cleave lamins and may be responsible for cytolysis. *See* apoptosis; caspase; cathepsin; CTL; ICE; lamins; perforin. (Kam, C. M., et al. 2000. *Biochim. Biophys. Acta* 1477:307; Zhang, D., et al. 2001. *Proc. Natl. Acad. Sci. USA* 98:5746.)

grape   *Vitis vinifera*, $2n = 38$; *Muscadina* species, $2n = 2x = 40$. *See* resveratrol.

grape, seedless Normal diploid, but a gene prevents the division of the embryo, presumably because of a shortage of hormones (stenospermocarpy). *See* seedless fruits.

grapefruit *Citrus paradisi*, $2n = 18, 27, 36$.

grasses (cultivated for herbage) Blue grass (*Poa pratensis*) $2n = 36-123$; Italian ryegrass (*Lolium multiflorum*) $2x = 14$; meadow fescue (*Festuca pratensis*) $2x = 14$; orchardgrass (*Dactylis glomerata*) $4x = 28$; perennial ryegrass (*Lolium perenne*) $2x = 14$; smooth brome (*Bromus inermis*) $8x = 56$; tall fescue (*Festuca arundinacea*) $x = 7, 2n = 42$; timothy (*Phleum pratense*) $6x = 42$.

grasshopper (*Orthoptera*) Suitable objects of cytological and evolutionary investigations because of the large size of chromosomes and variable numbers ($n = 13$ to 57 in males that are of XO constitution); *Melanopus differentialis* $2n = 24$. The variation in chromosome number is supposed to be due to chromatin reorganization rather than to polyploidy.

gratuitous inducer Substrate analog of an inducible enzyme that may trigger transcription of the gene concerned, such as IPTG (isopropyl thiogalactoside) for the *Lac* operon, although it is not metabolized by the *z* gene of the operon. *See* inducer; *Lac* operon. (Horton, N., et al. 1997. *J. Mol. Biol.* 265:1.)

grauzone Female meiosis regulatory WD-type zinc-finger protein distantly related to Cdc20 that binds to the cortex promoter during meiosis and early embryo development. Meiosis fails when the protein mutates. *See* CDC20; DNA-binding proteins; WD-40. (Harms, E., et al. 2000. *Genetics* 155:1831; Chu, T., et al. 2001. *Genesis* 29:141.)

Graves disease *See* goiter.

gravitropism Tendency of plant organs such as roots to grow in the direction of the terrestrial gravitation. The mechanism of this response is unclear, although amyloplasts and other cytoplasmic characteristics have been suggested as possible receptor sites. *See* phototropism; statolith. (Kato, T., et al. 2002. *Plant Cell* 14:33.)

gravity Either hypo- or hypergravity may cause chromosomal damage in human cells.

gray crescent Pale area in some amphibian eggs, opposite to the sperm entry; at this point, the dorsal parts will be initiated. *See* dorsal.

gray matter Butterfly-shaped neuronal tissue of the hippocampus; it is surrounded by the axonal *white matter*. Its anomalies may lead to psychomotor retardation and seizures that are resistant to anticonvulsant therapy. Some of the hereditary infantile seizures respond dramatically to large doses of vitamin B_6 (pyridoxine). Frontal gray matter differences are under genetic control and increased size appears positively correlated with cognitive abilities. *See*

brain, human; neuron. (Thompson, P. M., et al. 2001. *Nature Neurosci.* 4:1253.)

Gray units Of ionizing radiation, 1 Gy = 100 rad absorbed dose. *See* R; rad; rem; Sievert.

GRB Growth factor receptor–bound protein. Vertebrate adaptor protein with SH2- and SH3-binding domains; it is a downstream receptor kinase. It mediates the activation of guanine nucleotide exchange (GTP$\rightleftharpoons$GDP) on RAS, a homologue of the *Drosophila* protein, DRK. *See* DRK; SH2; SH3; signal transduction. (Jahn, T., et al. 2001. *J. Biol. Chem.* 276:43419; Kessels, H. W. H. G., et al. 2002. *Proc. Nat. Acad. Sci. USA* 99:8524.)

GRE Glucocorticoid receptor element. Situated upstream from the TATA box gene regulatory tract. *See* glucocorticoid; glucocorticoid response element. (Herrlich, P. 2001. *Oncogene* 20:2465.)

Greek key Protein configuration where β-sheets are connected across the end of a barrel. *See* barrel.

green beard effect Idea that some unique traits are favored by the parents' altruistic behavior during evolution. In general, other individuals of the population, regardless of whether they carry it themselves, recognize the "green beard" gene (gene product). *See* gestational drive; kin selection. (Dawkins, R. 1976. *The Selfish Gene*. Oxford Univ. Press, New York; Nee, S. 1989. *J. Theor. Biol.* 141:81.)

green fluorescent protein *See* aequorin; drFP583; Renilla GFP.

green revolution Development of new plant (cereal crop) varieties that permit more intensive agricultural practices (use of higher doses of fertilizers, irrigation, etc.) because of shorter and stronger stems and improved disease resistance, thus resulting in 2–3–fold increases in grain yield. (Khush, G. S. 2000. *Nature Rev. Genet.* 2:815.)

G region Consensus N-K-X-D *see* amino acid symbols in GTP-binding proteins that interacts with guanine in GTP. *See* G proteins.

G′ region, GTP-binding protein With highly conserved consensus D-X-X-G-Q (*see* amino acid symbols), involves GTPase function and may affect oncogenicity. *See* G proteins; signal transduction.

G″ region In RAS, interacts with GTP through the E-T-S-A-K (*see* amino acid symbols) consensus. In some G proteins, H-(F/M)-T-C-A(T/V)-D-T (*see* amino acid symbols) may be the corresponding functional area. *See* G proteins.

Greig's cephalopolysyndactyly syndrome (GCPS) Dominant in the short arm of human chromosome 7p13. It involves polysyndactyly and malformation of the head without mental defects. Molecular analysis indicates that the anomaly is concerned with GLI3 oncogen, a CREB-binding protein. The protein product of the *cubitus inrruptus locus* of *Drosophila,* involved in the regulation of limb development, is also a homologue. *See* cubitus interruptus; GLI oncogene; hedgehog; morphogenesis in *Drosophila*; Pallister-Hall syndrome; polydactyly; polysyndactyly; Rubinstein-Taybi syndrome.

grey matter   *See* gray matter.

GRIP   Glutamate receptor interacting protein. Contains seven PDZ domains; interacts with the C end and links AMPA to other proteins. GRIP lacks catalytic functions. *See* AMPA; CARM. (van Beeren, H. C., et al. 2000. *FEBS Lett.* 481:213.)

Griscelli syndrome   Recessive 15q21 mutation causing anomalous pigmentation and T lymphocyte and macrophage function aberrations (hemophagocytic syndrome). The basic defect is in the RAB27A guanosine triphosphate-binding protein. Defects at the same site may involve also myosin 5a, a motor protein. The former lesion involves immune defects, the latter is concerned with neurological impairment. *See* RAB. (Anikster, Y., et al. 2002. *Am. J. Hum. Genet.* 71:407.)

GRM   General regulator of mating type in yeast in cooperation with PRTF. *See* mating type determination in yeast; PRTF.

gRNA   Guide RNA. In the kinetoplast, mitochondrial DNA of some protozoa the pan-edited primary RNA transcripts substantially modified by U additions or deletions, but some short sequences (50–100 bases) may remain homologous to the primary transcript. These sequences are apparently anchored to the 3′ end and thus pairing may get started. Additional homology may occur in the middle (20–30 bases) and at the 5′ end (ca. 10 bases). These gRNAs may serve as templates for editing. The process requires a series of enzymatic steps. Free uridine triphosphates are the source of the inserted U and they are added to the 3′ ends generated by the enzymatic cleavage. *See* kinetoplast; *Leischmania*; mtDNA; pan editing; RNA editing; *Trypanosoma*. (Müller, U. F., et al. 2001. *EMBO J.* 20:1394; Bloom, D., et al. 2001. *Nucleic Acid Res.* 29:2950.)

G8 RNA   *See* thermal tolerance.

GROα   *See* melanoma growth-stimulating factor.

GroEL   Homotetradecameric chaperonin composed of 57 kDa subunits of three functional domains each, arranged as a hollow cylinder of two stacked rings with a seven-fold symmetry in *E. coli.* It binds to the smaller GroES molecule. GroEL and GroES are encoded in the same operon of *E. coli.* Although the information for folding resides in the primary structure of proteins, the GroEL-GroES complex facilitates the realization of this potential. *See* chaperone; chaperonin; protein folding. (Feltham, J. L. & Gierash, L. M. 2000. *Cell* 100:193; Farr, G. W., et al. 2000. *Cell* 100:561.)

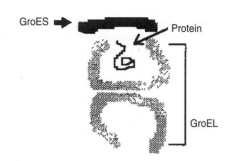

GroES   10 kDa monomer.

groucho   In *Drosophila*, it is a corepressor protein somewhat limited in scope and is responsible for extra bristles and ocelli above the eyes of the flies. It can act with hairy, engrailed, and dorsal, although the interacting domains in hairy and engrailed are different. Its mammalian homologue is TLE1 and is structurally and functionally related to Tup1 of yeast. *See* corepressor; dorsal; engrailed; hairy; ocellus; Tup1.

groundnut (peanut, *Arachis hypogea*)   About 40–70 species, $2n = 2x = 20$; some have higher ploidy. Some people are allergic to peanut protein that binds immunoglobulin (IgE) in the intestinal mucosa, resulting in histamine release, which may cause contraction of the smooth muscles of the airways and an anaphylactic reaction. It should be promptly counteracted by epinephrine to prevent serious consequences that may include death. (Burow, M. D., et al. 2001. *Genetics* 159:823.)

ground state   Stable, normal, unexcited form of an atom, molecule, or gene.

group selection   May occur when the behavior of individuals influences its own fitness and the fitness of related individuals and selected by Nature accordingly. *See* kin selection; nepotism.

group transfer potential   Ability of a compound to donate an activated group (e.g., phosphate or acyl).

grow   Hamster gene activated by mitogens. *See* KC; MGSA; N51.

growth   *See* cell growth; exponential growth; growth curve.

growth-associated kinase   M-phase histone-1 kinase; functions at its peak in mitotic M phase, but its activity ebbs at other phases; it is also active during meiosis. *See* histones; mitosis.

growth cone   Tip of growing axons. *See* axon.

growth curve   Cell multiplication may start at an exponential rate under ideal conditions for proliferation, then it reaches a stationary phase (growth flattens) and only maintenance of the cell population takes place (S curve). The exponential growth is so named because an exponent of the base 2 mathematically can define the growth. Thus, after 10 divisions of a cell, the expected number is 2^{10}. In case the initial cell number was 100, after exponential growth the number of cells 10 generations later would be $2^{10} \times 100 = 1,024 \times 100$. Alternatively, the growth may decline and the level of the population will decrease. In higher organisms, such growth curves can be observed only in isolated cell cultures. In differentiated tissues, the growth has structural limitations.

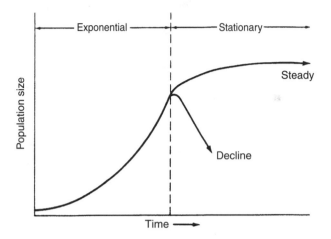

growth factor *See* cell cycle; EGF; erythropoietin; FGF; growth hormone, pituitary; IGF-I; IL-2; IL-3; NGF; PDGF; TGF.

growth hormones *See* animal hormones; plant hormones.

growth hormone, pituitary Gene complex is in human chromosome 17q23-q24 and encodes a 190-amino-acid protein, human growth hormone (hGH), somatotropin, a chorionic growth hormone (191-amino-acid residue with ≈85% homology to hGH), and a growth hormone−like protein (GHL, 22 kDa). The expression of the hormone gene complex is regulated by the 33 kDA growth hormone transcription factor (GHF1) and a growth factor-response protein (GFRP1). hGH is released by a 44-amino-acid growth hormone-release factor (GHRF) encoded at chromosome 20p12. *See* brain, human; Gapo syndrome; ghrelin; pituitary; pituitary dwarfness; Rowley-Rosenberg syndrome; secretagogue.

growth retardation Reduction in the rate of increase in cell size, multiplication, differentiation, and development. *See* GAPO syndrome; Gombo syndrome; retardation; Rowley-Rosenberg syndrome.

GRP **(1)** Heat-shock glycoproteins of the HSP family. GrpE is not itself a chaperone but as an ADP-ATP exchange factor it is part of the DnaJ-DnaK chaperone complex of prokaryotes and the replication machinery of phage lambda. Its homologues are present in prokaryotes as well as in the mitochondria of yeast, insects, and mammals. The mammalian Grp94 (member of the Hsp90 family of proteins) chaperones are a small number of proteins suspected to be involved in antigen presentation and tumor rejection. *See* antigen processing and presentation; DnaK; Droe1; endoplasmic reticulum; heat-shock proteins;

HSP; Hsp90; Mge1. **(2)** General receptors of phosphoinositides. *See* phosphoinositides.

Grunstein-Hogness screening Involves in situ lysis of bacterial colonies on nitrocellulose filters (or other membranes) and noncovalent attachment of the probe DNA to that support medium. *See* Benton-Davis plaque hybridization; probe. (Elvin, P., et al. 1988. *Br. J. Cancer* 57:36.)

GSC RT-PCR *See* global single-cell reverse transcription polymerase chain reaction.

GSD Genetically significant dose. Regarding mutagenic exposure. *See* doubling dose; genetic hazards; genetic load; mutation, spontaneous; mutation rate.

GSH Reduced glutathione. *See* glutathione.

GSK3β Glycogen synthase kinase 3β. Protein encoded by *Drosophila* gene *zeste white* (z^{W3}, chromosome 1-1); homologous proteins are present in other animals. It is assumed that GSK mediates a step in the intestinal polyposis carcinogenic pathway. It also regulates global protein synthesis. *See* conductin; GBP; polyposis, adenomatous intestinal; translation initiation.

GSM Genetic sexing mechanism is a method of insect control by producing translocation between the Y and X chromosomes. The Y translocation serves as a dominant selectable marker and reduces the fertility of the female. *See* autosexing; genetic sterilization.

gsp c-oncogene; its product is the α-subunit of G proteins. *See* G-, c-oncogene.

GSP Gene-specific primer. *See* c-oncogene; directed mutation.

G$_s$ protein Stimulatory G protein; when bound to GTP, it stimulates the activity of adenylate cyclase, the membrane-bound enzyme that generates cAMP. G$_s$ has α-, β-, and γ-subunits, the GTP/GDP-binding site being on the α-subunit. When GDP is at the nucleotide-binding site, adenylate cyclase activity ceases. Displacement of GDP and replacement by GTP (mediated by the hormone epinephrine) restores the active form. At this stage, the α-subunit with bound GTP dissociates from β and γ. *See* adenylate cyclase; G proteins; signal transduction.

G$_{ST}$ Index of genetic diversity similar to F$_{ST}$. *See* F$_{ST}$. (Nei, M. 1973. *Proc. Natl. Acad. Sci. USA* 70:3321.)

GST *See* glutathione-S-transferase.

GSTB Genome Sequence Data Bank. It maintains nucleotide sequence information on genes and clones. *See* GenBank; NCBI.

GT-AG rule (Chambon's rule) The first two and the last two nucleotides of introns are GT and AG, respectively; some exceptions are known. *See* intron; exon.

G test Basically a goodness-of-fit test, but instead of calculating the χ^2 in the common way, the probability of p is divided by $\hat{p} = L$ (likelihood), and $G = 2\ln(L) = 2(\ln 10)\log L$, and the distribution is approximated by the χ^2 distribution in large samples. Instead of G, sometimes the symbol I (information) is used. *See* chi square; information. (Sokal, R. R. & Rohlf, F. J. 1969. *Biometry*. Freeman, San Francisco.)

GTBP G/T mismatch-binding protein is encoded in human chromosome 2 and its mutations lead to genetic instability at single nucleotide sites. *See* DNA repair; hereditary nonpolyposis colorectal cancer; mismatch.

GTF General transcription factors. Proteins that are required for the initiation of transcription by RNA polymerase I (TFIs), RNA polymerase II (TFIIs), and RNA GTP polymerase III (TFIIIs). *See* transcription factors.

GTP Guanosine triphosphate.

GTP

HO—P—O—P—O—P—O

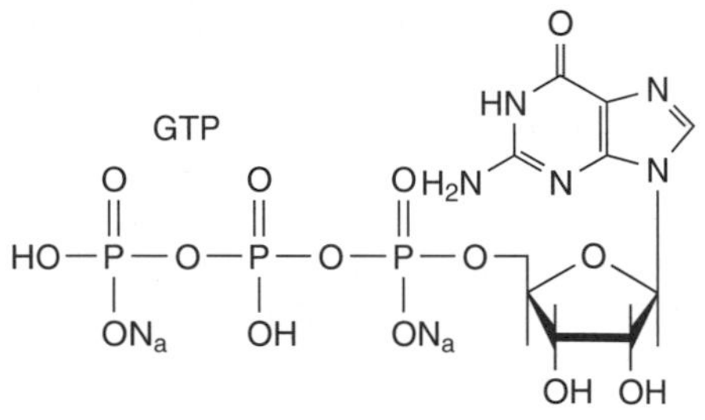

GTPase $\sim$60 proteins mediating the conversion of GTP into GDP. These enzymes regulate translation, signal transduction, cytoskeletal organization, vesicle transport, nuclear import, protein translocation across membranes, etc. Two different GTPases may modulate each other's activity. *See* Arf; dynamins; GAP; GEF; G proteins; RAB; RAC; RAN; RAS; RASA; RHO; SAR. (Yang, Z. 2002. *Plant Cell* 14:S375.)

GTPase-activating protein (GAP) Increases GTP hydrolysis to GDP by several orders of magnitude in signal transduction. *See* signal transduction.

GTP-binding protein superfamily Includes transitional factors, transmembrane signaling proteins, Ras proteins, and tubulins. *See* signal transduction.

GTP cyclohydrolase deficiency (14q22.1-q22.2) Recessive deficiency of guanosine triphosphate cyclohydrolase I results in hyperphenylaninemia because tetrahydrobiopterin is not converted into dihydroneopterin triphosphate by a process requiring GTP. The disorder involves low urinary pterine, serotonin, and dopamine levels. The afflicted individuals show convulsions, muscular hypotonia of the trunk, and hypertonia of the limbs. Oral L-erythrotetrahydrobiopterin may alleviate some of the symptoms. *See* biopterin; dopamine; hyperphenylalaninemia; hypertonia; hypotonia; phenylalanine; pteridines; serotonin.

G$_t$-protein (transducin) Member of the trimeric G-protein family; it activates cGMP phosphodiesterase in photoreceptors. *See* G proteins; rhodopsin.

Guam disease Autosomal-dominant complex syndrome discovered in Guam that displays the characteristics of amyotrophic lateral sclerosis, parkinsonism, and dementia. Environmental conditions such as low calcium and magnesium in the diet and consumption of Cycas plants seems to favor toxic metal accumulation in the central nervous system and appears to promote the onset. *See* neurodegenerative diseases.

guanidinium chloride Used in molecular genetics similarly to guanidinium isothiocyanate for isolation of undegraded RNA. *See* RNA extraction.

guanidinium isothiocyanate Used for the isolation of RNA. It breaks up cells, dissociates nucleoproteins, and inactivates tough RNase enzymes (at 4 M solutions) in the presence of the reducing agent β-mercaptoethanol. *See* RNA extraction.

guanine Purine base in DNA and RNA. *See* glycosylases.

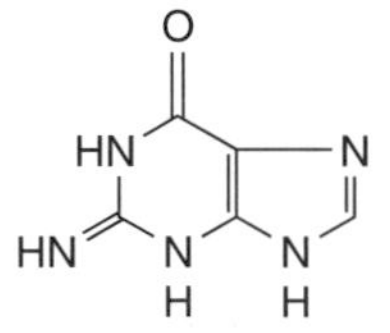

Guanine.

guanine methyltransferase Methylates the mRNA cap. *See* cap.

guanine nucleotide-binding protein (16q13, 6q21.3, 20q13.2, 7q21, 22q11.2, 1p13, 3p21, 19p13) Mediate signal transduction on G proteins.

guanine nucleotide exchange protein (11p23.3, 19q13.3) Catalyzes the reaction GTP⇌GDP. The large Rho family of G proteins generally contains a Dbl (MCH2) domain and usually a pleckstrin homology (PH) domain. *See* brefeldin; G protein; MCH2; Rho.

guanine nucleotide-releasing protein (GNRP, 11q13) Hydrolyzes GTP bound to G proteins into GDP. *See* signal transduction.

guanosine Nucleoside of guanine.

guanosine tetraphosphate, guanosine pentaphosphate (ppGpp, pppGpp) Effectors of the stringent response. *See* stringent response. (Chatterji, D. & Ojha, A. K. 2001. *Curr. Opin. Microbiol.* 4:160.)

guanylate cyclase Mediates the formation of cyclic guanosine monophosphate (cGMP).

guanylic acid Guanine nucleotide.

guanylyl transferase Attaches GTP to the mRNA cap. *See* cap; GTP.

guard cell *See* stoma.

guard hypothesis Postulates the requirement for a specific protein to activate the plant resistance gene when it encounters an avirulence gene of a pathogen, and it guards against the suppression of the plant defense mechanism by any bacterial effector.

guava (*Psidium guajava*) Subtropical, tropical, small allogamous fruit tree; $2n = 2x = 22$.

guessmer Usually thirty 7-base-long synthetic oligonucleotides representing limited degeneracy and using neutral bases (inosine) at sites of ambiguity. The nucleotide sequence is generated on the basis of information of amino acid sequences in the protein. This label can be used for screening for specific coding sequences (genes). If the codons were picked at random, the synthetic sequence would represent at least 76% homology by chance, but by considering codon usage of the organism, the homology may be over 90%. The probes are labeled with the aid of polynucleotide kinase or primer extension with the Klenow fragment. *See* primer extension; probe. (O'Farrell, P. A., et al. 1997. *Biochem. Biophys. Res. Commun.* 239:810.)

guest peptide *See* CD tagging.

guest RNA *See* CD tagging.

guest tag *See* CD tagging.

guide RNA Chaperones the alignment of splicing by attaching either to the intron or to the exon sequences of the transcript. *See* gRNA; intron; RNA editing; splicing. (Kabb, A. L., et al. 2001. *Nucleic Acids Res.* 29:2575.)

Guillain-Barré syndrome Sporadic or familial (autosomal-dominant?) demyelinating neuropathy arising after infection by *Campylobacter jejuni*. *See* Campylobacter.

guinea pig *Cavia porcellus*, $2n = 64$.

Guinea pig.

Gunther disease *See* porphyria (erythropoietic porphyria, 10q25.2-q26.3).

GUS The tetrameric glycoprotein acid hydrolase, β-glucuronidase enzyme (gene), is frequently used as a reporter for the in vivo testing of promoters, identifying site-specific expression or monitoring the excision of transposable elements. Several substrates of the enzyme are useful for releasing a blue color upon activity of GUS. Deficiency of the enzyme in mammals leads to lysosomal storage diseases. *See* gene fusion; lysosomal storage diseases; reporter gene. (Jefferson, R. A., et al. 1987. *EMBO J.* 6:3901; Schenk, P. M., et al. 2001. *Plant Mol. Biol.* 43:399.)

gustatory Involving taste sensation. *See* taste.

gustducin *See* taste.

gut Gastrointestinal tract or the developmentally primitive (early) digestive tract composed of fore-, middle- and hindgut sections. The mammalian gut endoderm forms different "buds," giving rise to the liver, lung, pancreas, thyroid, and gastrointestinal tissues. The developmental fates are determined by the additional growth factors recruited.

Guthrie test Detects phenylketonuria, because if the blood contains phenylalanine, the analog β-2-thienylalanine does not interfere with the growth of *Bacillus subtilis*. *See* genetic screening; phenylketonuria.

gutless vector Usually a viral vector almost without viral genes.

guttation Water ascending through the xylem vessels of plants may drop from the leaves when the relative humidity increases. *See* cohesion tension; transpiration.

GVG Transcription factor constructed of the yeast GAL4 DNA-binding domain, the transactivation domain of herpes virus VP16, and the hormone-binding domain of the glucocorticoid receptor. *See* dexamethasone; galactose utilization; glucocorticoid response elements; VP16.

Gy (1) Gray units of radiation; 1 Gy = 100 rad. *See* rad. (2) Billion years of geological time.

gymnosperm (Coniferophyta) Plants with seeds that are not enclosed in an ovary. Typical representatives are the pine trees ($2n = 24$).

gymnothecium Fruiting body of some ascomycetes fungi; it may cause skin infections. *See* ascogonium; cleistothecium; perithecium.

gynander Gynandromorph.

gynandromorph Sex mosaic (part male/part female); same as gynander. Gynandromorphs are the result of the loss of one of the X chromosomes during the development of *Drosophila* and other organisms where the XO chromosomal constitution leads to the development of male phenotypic characteristics. The loss of the X chromosome reveals the recessive alleles present in the remaining homologue. These sex mosaic individuals can be exploited for fate mapping. The right side of the fly shows the male body characteristics and has a ruby eye and *broad wing* because the sector is X0. The left

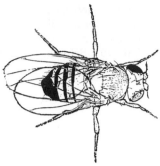

Gynadromorph of *Drosophila*. From Morgan, T. H., et al. 1925. *Bibliographia Genet.* 2:1.

side appears like females (XX). *See* fate mapping; lyonization; variegation. (Szabad, J. & Fajszi, C. 1982. *Genetics* 100:61.)

gynecomastia   Increased development of the mammary gland of males caused by estrogen accumulation and/or reduction of testosterone. Deficiency gene encoding hydroxysteroid dehydrogenase III (9q22) and increased expression of cytochrome P450 (CYP, 15q21.2) aromatase subunit may be responsible for pseudohermaphroditism and gynecomastia. X-linked inheritance with male transmission has also been suggested. A transient mild form may not be abnormal during puberty. *See* animal hormones; aromatase; Kennedy syndrome; Klinefelter syndrome; pseudohermaphroditism male; steroid dehydrogenase/hydroxysteroid dehydrogenase.

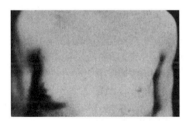

Gynecomastia [photo by courtesy of Dr. C. Stern].

gynodioecious   The population consists of both hermaphroditic and female individuals generally determined by nuclear and mitochondrial genes. *See* hermaphrodite. (Taylor, D. R., et al. 2001. *Genetics* 158:833.)

gynoecium   Carpels and enclosed structures by it in the flowers. *See* fruits. (Ferrándiz, C., et al. 1999. *Annu. Rev. Biochem.* 68:321.)

gynogenesis   Reproduction by parthenogenesis, i.e., the sperm does not fertilize the egg but stimulates the cleavage of the unreduced egg (pseudogamy). Also, embryos developed by transfer of male pronuclei into the egg, and thus diploid, are called *gynogenones*, in contrast to *parthenogenones* (gynogenotes) that arise by parthenogenesis. *See* androgenesis; apomixis; EP; parthenogenesis.

Gypsy   Somewhat diverse ethnic group migrating from the Indian subcontinent northward and southward, presumably before the 9th century, to Asia and to Egypt and from there to most of the Northern Hemisphere, although they are now found all over the world. Their Indian origin is asserted by orally transmitted legends. Linguistic evidence indicates Sanskrit roots. Y-chromosomal and mtDNA information support Indian origin. Their ethnic identity has been preserved by cultural and genetic isolation. J. B. S. Haldane (1935) used ABO blood-type frequencies to show that the Hungarian Gypsies are more closely related to some Eastern Indian populations than to those of Hungary, although some of them lived in that country since the early 15th century. They prefer to be called Roma. *See* ethnicity. (Kalaydjieva, L., et al. 2001. *Eur. J. Hum. Genet.* 9[2]:97; Gresham, D., et al. 2001. *Am. J. Hum. Genet.* 69:1314.)

gypsy retroposon   *See* copia.

gyrase   DNA topoisomerase II that reverses the direction of coiling in DNA, resulting in negative supercoiling. *See* DNA replication; supercoiling; topoisomerase; transcription. (Gellert, M., et al. 1979. *Proc. Natl. Acad. Sci. USA* 76:6289; Kirchhausen, T., et al. 1985. *Cell* 3:933; Williams, N. L. & Maxwell, A. 1999. *Biochemistry* 38:13502.)

> "William Curtis, British botanist, described a weed called *Arabidopsis thaliana* as having 'no particular virtues or uses'. More than 200 years later, he could not have been proved more wrong." *The Guardian*, UK, quoted after Jane Alfred, Editor of Nature Rev. Genet. 2001, 2:86.

A historical vignette.

H

h (1) Planck constant. Energy quantum of radiation that relates it to the frequency of the oscillator that emitted it. $E = h\nu$, where E is the energy quantum, ν is its frequency, numerically 6.624×10^{-27} erg/sec. (2) Human homologue of a gene or protein standing in front of the symbol.

H Heavy chain of a double-stranded DNA

H1, H2A, H2B, macroH2A, H3, H4 H1 and H5 are variants of H1 histones. *See* histones.

H-2 The major histocompatibility gene cluster in the mouse is located to chromosome 17, proximal to the centromere within a segment of about 1.3 cM, consisting of about 2,000 kb DNA. They encode cell surface glycoproteins that have a major role in recognition and immune response to foreign antigens. The gene order in this cluster encoding class I, class II, and class III– and class I–like polypeptides is

$$K - A - E - C2 - Bf - SLP - OH - C4 - TNF - D - D2$$

$$- D3 - D4 - L - Q - T - T1a - \text{centromere.}$$

The transplantation antigens are the class I proteins coded for by genes K, D, and L. The class II proteins, encoded by genes A and E, occupy the surface of B and T lymphocytes and the macrophages and participate in cell immune responses. The class III proteins coded for by genes C2, Bf, SLP, OH, and C4 are the complement proteins of the serum involved in the lysis of foreign material after the recognition by the antibody. The Q and T loci determine the so-called differentiation antigens present in the blood cells. TNF is the tumor necrosis factor gene. *See* antibody; HLA; lymphocytes; TNF. (Fischer, K., et al. 1997. *Annu. Rev. Immunol.* 15:851.)

h^2 Symbol of heritability; it is derived historically from Sewall Wright's definition of heritability as the ratio of the standard deviations of the additive and phenotypic variances, $h^2 = V_A/V_P$. Heritability is not a squared entity, and h^2 stands for heritability, not for its square. *See* correlation; heritability; heritability estimation in humans; intraclass correlation; offspring-parent regression.

HA *See* hemagglutinin.

HAART Highly active antiretroviral therapy.

habitat Area in nature where an organism(s) occurs naturally. *See* sink habitat; source habitat.

habituation (accoutumance à l'auxine, anergie à l'auxine) Plant tissues after prolonged culture may dispense of the continued reliance on exogenous auxins for proliferation. This alteration does not involve somatic mutation, yet it bears a similarity to oncogenic transformation. In animal cells, the SV40 T-antigen loss after a period of time may still not cease proliferation in the absence of the oncoprotein. *See* oncoprotein; somatic mutation; SV40; tissue culture; transformation, oncogenic; tumor. (Gautheret, R. J. 1955. *Rev. Gen. Bot.* 62:1.)

Habrobracon *See* wasp.

Habsburg lip Protruding lower lip and the undershot lower jaw was transmitted among male and female members of this European dynasty up to the 20th century. The coin shown here represents Maximilian I (1459–1519). (Thompson, E. M. & Winter, R. M. 1988. *J. Med. Genet.* 25:838.)

Habsburg lip.

HAC (1) hyperpolarization activated channels. *See* I_h. (2) Human artificial chromosome. Critical elements (telomeres, centromeres, and replicator) must be present. HACs can eventually be used to ferry desirable genes to human cells for medical purposes. A useful human artificial chromosome should be much smaller than the smallest chromosome in the natural genome in order to transfer genes into human cells, yet it should be large enough to carry large human or mammalian genes and some regulatory sequences. Minichromosomes of ∼4 Mbp appear stable, but those below 2.5 Mbp seem unstable. The useful HAC is expected to be stable through mitosis and possibly even through meiosis. This requires a minimal functional centromere and telomeres. The centromeric DNA (α-satellite) is variable in length in the different human chromosomes, but in order to function it may have to comprise ∼150 kbps. It should not incite adverse immunological reaction in the recipient cell. For experimental manipulation, it is desirable that the HAC would be easily transferable to any type of cell, including, e.g., yeast cells. *See* α-satellite; BAC; DT40; HAEC; human artificial chromosome; MAC; minichromosome; PAC; vectors; YAC. (Henning, K. A., et al. 1999. *Proc. Natl. Acad. Sci. USA* 96:7125; Shen, M. H., et al. 2000. *Curr. Biol.* 10:31; Csonka, E., et al. 2000. *J. Cell. Sci.* 113[pt18]:3207; Mejia, J. E., et al. 2001. *Am. J. Hum. Genet.* 69:315.)

Hac1 Basic leucine-zipper protein is a transcription factor that binds to the UPR element in the promoter of the UPR genes. *See* unfolded protein response.

Hae II Restriction enzyme with recognition site $^A_G GCGC\downarrow^T_C$.

***Hae* III**   Restriction enzyme with recognition site GG↓CC.

HAEC   Human artificial episomal chromosome was constructed by using the replicational origin of the Epstein-Barr virus. Such a construct may carry over 300 kb inserts and may be maintained in human cells without integration (as an episome) in the genome. *See* episome; Epstein-Barr virus. (Wade-Martins, R., et al. 2000. *Nature Biotechnol.* 18:1311.)

HAEM   Iron porphyrin; occurs in different forms. *See* heme.

Haem.

haematopoietic growth factor   *See* IL-3.

haematoxilin   *See* hematoxilin, hematopoietic.

haemochromatosis   *See* hemochromatosis.

Hageman factor   Protein (M_r ca. 80,000) present in the blood plasma and serum of the majority of mammals but absent in dolphins, killer whales, and birds. It is involved in the pathway of blood coagulation; it also affects vascular permeability, dilates blood vessels, contracts smooth muscles, provokes pain, promotes the migration of leukocytes, induces fibrinolysis (dissolution of fibrin), etc. *See* antihemophilic factors; Hageman trait. (Schousboe, I., et al. 1999. *Thromb. Haemost.* 82:1041; Gaffney, P. J., et al. 1999. *Haemostasis* 29:58.)

Hageman trait   Controlled by an autosomal-recessive gene causing deficiency of the Hageman factor in the blood. Normally the individuals lacking this factor do not show any disease symptom, although blood coagulation is slow in the laboratory. No therapy is required, yet in case of surgery it is advisable to keep appropriate blood or plasma at hand. *See* Hageman factor.

Hailey-Hailey disease (HHD, benign pemphigus)   Autosomal-dominant skin disease involving vesicle formation generally on the neck, groin, and armpits. The benign disease is precipitated by infection by the fungus *Candida albicans*, but antifungal, antibacterial drugs may also initiate it. Mutation in an ATP-powered ion pump sequestering calcium into the Golgi apparatus causes HHD. *See* pemphigus; skin diseases.

hair   Human hair has a high- and a low-sulfur protein; the former is about 40% of the hair proteins. In some animal hairs, a high-tyrosine protein also occurs. The hair develops from multipotent clonogenic keratinocytes similar to multipotent stem cells. *See* alopecia; baldness; catenins; De Lange syndrome; glomerulonephritis; hair-brain syndrome; hair color; hair whorl; hairy ears; hairy elbows; hairy nose; hairy palms and soles; hypertrichosis; hypotrichosis; keratin. (Fuchs, E., et al. 2001. *Developmental Cell* 1:13; Jave-Suarez, L. F., et al. 2002. *J. Biol. Chem.* 277:3718.)

hair-brain syndrome (trichothiodystrophy, BIDS)   Autosomal-recessive brittle hair, low intelligence, short stature, reduced fertility, and reduced cystine-rich protein in the hair and nails. *See* hair; stature; trichothiodystrophy.

hair color, human   Strikingly blonde and red hair colors (prevalence about 2% or less) appear to be autosomal recessive, but the latter may have some expression in the heterozygotes. Red hair is hypostatic to brown and black. Brown hair appears to be autosomal dominant and it is closely linked to green eye color. Dark hair appears to be dominant. First hair of a baby may not be concordant with that of later years. Relative proportions of the reddish pheomelanin and the black eumelanin pigments determine hair color. Their level is controlled by melanocyte-stimulating hormone (MSH) and its receptor (MC1R). Environmental factors (temperature, sunshine, diseases) may also transiently affect the color. Graying of the hair proceeds usually by aging; however, precocious graying may be determined by dominant genes, and it may be a symptom shared by several syndromes such as Book, Waardenburg, and Werner syndromes; pernicious anemia (a vitamin B_{12} deficiency) may also cause it. Actually, the inheritance of human hair is determined by many loci. In human chromosome 19 alone, there are 6 loci homologous to fur color genes of the mouse. *See* aging; albinism; forelock, white; hypostasis; pigmentation in animals. (Sturm, R. A., et al. 1998. *Bioessays* 20[9]:712; Flanagan, N., et al. 2000. *Hum. Mol. Genet.* 9:2531; Healy, E., et al. 2001. *Hum. Mol. Genet.* 10:2397.)

hairpin   Double-stranded structure in nucleic acids brought about by folding back of palindromes like a hairpin. At the end where the arrow is pointed → ⊂ the unpaired structure may be digested by S_1 single-strand-specific nuclease. *See* palindrome; pseudohairpin; S_1 nuclease.

hair whorl   May be clockwise, which is autosomal dominant, over counterclockwise rotation. *See* hair.

hairy ears   Occur primarily in male humans; thought to be due to a Y-chromosomal (holandric) gene(s) or to two genes in the homologous segments of the X and Y chromosomes. Most likely it is determined by autosomal-dominant factors and sex-influenced inheritance. *See* holandric genes; sex influenced.

[Photo is the courtesy of Dr. Curt Stern.]

hairy elbow   Autosomal-dominant hairiness on the elbows associated with short stature. *See* hair.

hairy nose Autosomal-dominant (?) hairs on the nose tip; onset after puberty in the male only. *See* hair.

hairy palms and soles Apparently autosomal-dominant, male transmitted, site-specific hairiness. *See* hair.

Haldane's mapping function $(1 - e^{-m})0.5 = y$ (recombination fraction), where m is the number of exchanges and e is the base of the natural logarithm. Hence $(1 - e^{-m}) = 2y$, and $e^{-m} = 1 - 2y$, and $m = -\ln(1 - 2y)$; the corrected map distance estimate is $x = m/2$ because each exchange produces maximally 50% recombination. This formula does not take into account interference and thus frequently overestimates map distances. The graph below permits reading directly the recombination frequencies corrected by Haldane's mapping function. Different organisms may require different mapping functions (Stahl, F. W., *Genetic Recombination*. Freeman, San Francisco, CA). *See* coefficient of coincidence; Kosambi's mapping function; mapping, genetic; recombination frequencies.

Haldane-Muller principle The seriousness of the consequences of a mutation is not its deleterious effect because a dominant lethal mutation is eliminated immediately but its probability of death. Thus moderately disadvantageous mutations may be maintained for many generations and in each generation they may adversely affect the fitness and eliminate a number of individuals. *See* genetic load.

Haldane's rule When in the F_1 offspring of two different animal races one sex is absent or sterile, that sex is the heterogametic one. The cause is not an imbalance between autosomes and sex chromosomes; rather it is due to the general fact that lethality is usually completely recessive, whereas deleterious mutations may show a series of less debilitating effects that may also be expressed in the heterozygotes. Another interpretation in some species is the "faster male" theory, i.e., the sex-specific male fertility genes evolve faster than those responsible for female fertility. It has been suggested that the reduction in the heterogametic class is due to sex-transforming genes. This mechanism may play some

role, but it is apparently insignificant. Another interpretation points to dosage compensation, which may break down in crosses between distantly related species. This interpretation is contradicted by the fact that Haldane's rule is observed in ZW sex determination systems, albeit dosage compensation is not evident. Furthermore, in mammals, the X-chromosome inactivation in the females (*See* Lyon hypothesis) and the hypoactivation of the X chromosome in the hermaphroditic *Caenorhabditis* are kinds of female dosage compensation. X-Y and Y-autosome incompatibilities may account, to some extent, for the sterility of species hybrids but not for the viability. Meiotic drive has also been invoked as a possible cause without general acceptance. There are some examples of cytoplasmic effects, which are independent of heterogamety, although they may affect heterogametic individuals (*See* infectious heredity; segregation distorter). With rare exceptions, Haldane's rule is generally valid. *See* sexchromosomes; sex determination; hybrid inviability; hybrid sterility. (Laurie, C. C. 1997. *Genetics* 147:937; Haldane, J. B. S. 1922. *J. Genet.* 12:101; Nesbit, R. E., et al. 2002. *Genetics.* 161:1517.)

half chromatid Involves only one of the strands of the DNA double helix in the chromatid. The expression of mutation is delayed by one division if it occurs in a half chromatid and it may result in somatic sectors after replications. *See* chromatid.

half conversion In yeast, the tetrad contains 2 A, 1 a and 1 A/a heteroduplex sectorial spores. *See* gene conversion.

half-life In general, the time required for the decay of one-half of a compound; the decay, e.g., of a promutagen or procarcinogen may lead to the formation of even more active mutagens and/or carcinogens. The half-life of H^3 and C^{14} is 12.4 and 53,700 years, respectively, and that of P^{32}, P^{33}, I^{131}, and I^{125} is 14.3, 25.4, 8, and 60 days, respectively. The half-life $T_{1/2} = 0.693/\lambda$, and λ = the characteristic disintegration constant for the specific isotope. The radioactivity is gradually decreasing by time and the correlation between the number of half-lives passed and the chart next page shows the amount of radioactivity still remaining. *See* evolutionary clock; isotopes.

Haldane's mapping function. Read the point of intersection between the recombination fraction observed (ordinate) line and the solid line curve corresponding to the mapping function and project the point to the bottom line representing map units. For example, 0.3 recombination corresponds to about 45−46 map units. (After Haldane, J. B. S. 1919. *J. Genet.* 8:299.)

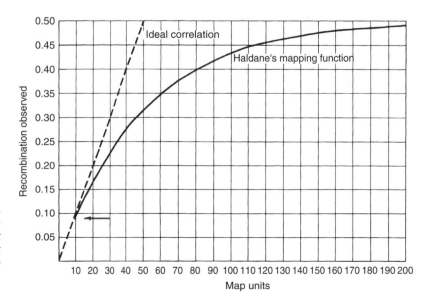

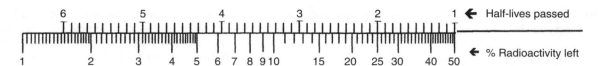

half-mutant Hugo de Vries' term for the new types that arose relatively frequently in *Oenotheras* with translocation rings. The half-mutants produced normal (translocation ring) and sterile progenies due to recombination between the differential segment and its homologue in another chromosome within the ring, resulting in smaller rings and other chromosomal changes. When the noncrossover original translocation ring was recovered in the egg and sperm, the offspring was the same as the original complex heterozygotes. *See* complex heterozygote.

half-sibs Share only one biological parent; they are half-sisters or half-brothers. *See* sibling.

half-tetrad analysis Tetrad analysis is feasible in a limited number of organisms where the four products of meiosis can separately be recovered either as a tetrad or after a postmeiotic division as an octad, a phenomenon common is the ascomycetous fungi. In *Drosophila*, half-tetrad analysis can be carried out in the presence of attached X chromosomes. In this case, the products of meiosis have either two attached X chromosomes or are nullisomic for the X chromosome. X-chromosomal genes are inherited as a block unless recombination takes place. Flies heterozygous for attached X-chromosomal genes can become homozygous only after crossing over and produce two different nonparental gametes from a single meiocyte. This type of recombination provided the first direct evidence that genetic recombination takes place at the 4-strand stage of meiosis. Half-tetrad analysis has been adopted to a plant (*Medicago*) case using four RFLP markers in situations when restitution nuclei were observed at a high frequency. Three of the markers were linked

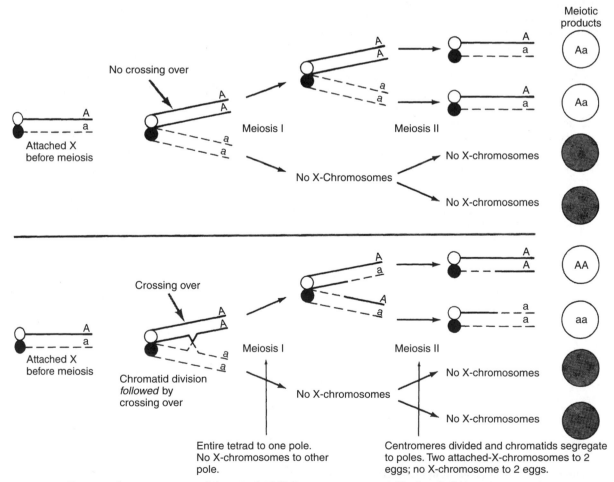

Because the centromeres of the attached X chromosomes are stably fused, they cannot pass to opposite poles during anaphase I. Segregation of the chromatids is equational at anaphase II. Instead of four viable meiotic products, only two X-disomic and two lethal, nullisomic gametes are formed. If the two attached chromosomes carry different alleles, all the noncrossover gametes will be the same and only the crossovers will be of two nonparental types. (After Shult, E. E. and Lindegren, C. C. 1959. *Can. J. Genet. Cytol.* 1:189.)

(chromosome 1) and the fourth was independent (chromosome 6). The analysis permitted the localization of the centromere. Trisomics may also be used for half-tetrad analysis. Molecular analysis of individual secondary oocytes by PCR may reveal gene positions. *See* attached X chromosomes; disomic; gene-centromere distance; nullisomic; restitution nucleus; tetrad analysis. (Zhao, H. & Speed, T. P. 1998. *Genetics* 150:473.)

half-translocation Half of an original reciprocal translocational event. Such a situation occurs, e.g., when a presumably reciprocal interchange takes place between the proximal region of an X chromosome and the small 4th chromosome of *Drosophila*, but one of the translocated strands is not recovered because it gets lost during segregation of chromosomes into the functional gamete. *See* translocation, chromosomal.

half-value layer Reducing the transmission of radiation to half.

Hallermann-Streiff syndrome Most likely autosomal-recessive bird-like face, long, narrow nose, cataracts, sparse hairiness, occasionally teeth by birth and proportionate dwarfism. *See* dwarfs; tooth.

Hallervorden-Spatz disease (neuroaxonal dystrophy) Human chromosome 20p12.3-p13-located recessive brain atrophy accompanied by involuntary movement and early death. The brain accumulates iron in the basal ganglia probably due defects in the panthotenate kinase gene. *See* ferritin; Lewy body; synuclein. (Zhou, B., et al. 2001. *Nature Genet.* 28:345.)

halo assay A substance is placed on a of a microbial culture plate and inhibition of growth or reverse mutations appear as a circle around the spot.

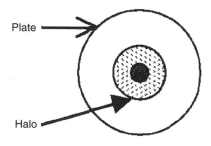

halophyte Salt-tolerant or -resistant plant species.

halothane gene Controls malignant hyperthermia syndrome in pigs (in chromosome 6). Halothene (2-bromo-2-chloro-1,1,1-trifluoroethane) is an anesthetic applied by inhalation. *See* PSE.

halter Balancing organ. *See* morphogenesis [26] in *Drosophila*.

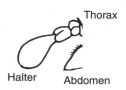

HALVORSON 5 Nutrient concentrate for yeast g/L $(NH_4)_2SO_4$ 20, K_2HPO_4 43.5, succinic acid 29, $CaCl_2.2H_2O$ 1.99, $MgSO_4.7H_2O$ 5.11, *trace elements* 5 mL [containing mg/0.5 L $Fe_2(SO_4)_3$ 307, $MnSO_4$ 280, $ZnSO_4.H_2O$], pH 4.7. This medium is usually diluted five times.

HAMA Human antimouse antibody is a human antibody response against murine monoclonal antibodies. HAMA complexes are usually rapidly cleared from the human body and may incite allergic reactions or even anaphylactic shock. Engineered rodent antibodies may overcome the problems. Human antibodies are better because they are compatible with the complement system and ADCC. *See* ADCC; complement; humanized antibody; monoclonal antibody.

HAMARTIN *See* epiloia.

hamartoma Tissue overgrowth (of darker color).

Hamiltonian path Directed graph if and only if there is a sequence of a one-way path beginning with v_{in} (vertex in) and ending at v_{out}, entering every other vertex exactly once. This mathematical method may serve as the basis for designing a molecular (DNA) computer that may assist in developing very complex combinatorial approaches to manipulating macromolecules, e.g., designer enzymes. *See* DNA computer; RNA computer. (Liu, Q., et al. 2000. *Nature* 403:175.)

Hamilton's rule *See* inclusive fitness.

hammerhead *See* ribozymes.

hamster *Cricetulus griseus* (Chinese [gray] hamster) $2n = 22$; *Mesocricetus auratus* (Syrian [golden] hamster), $2n = 44$.

hand clasping Some authors claim that it is genetically determined whether the left- or right-hand fingers (the latter more common among females) are on top when the hands are clasped. The control may be autosomal dominant or polygenic. *See* handedness.

handcuffing Products of plasmid replication associated—incA and —incC (incompatibility factors of replication) that cause antiparallel pairing of the two DNAs, preventing replication until the 'handcuffs' are disrupted. (Chattoraj, D. K. 2000. *Mol. Microbiol.* 37:464.)

handedness About 93% of the human population is right-handed; left-handedness is higher in younger than in older age groups, indicating greater longevity of right-handed individuals. The inheritance of handedness is unclear; it was attributed to polygenic control and it has been suggested that it is due to a homozygous-recessive state, the heterozygotes being ambidextrous (able to use both hands with equal facility). In *Arabidopsis* left — right asymmetry of leaf twisting is controled by the intradimer interface of α-tubulins 4 and 6 (Thitamadee, S., et al. 2002. *Nature* 417:193).

hand-foot-genital syndrome (hand-foot-uterus syndrome, HFU, 7p15-p14.2) Rare dominant malformations of hands, feet, fingers, toes, vaginal septum and uteral anomalies in girls, hypospadias in boys, common infections of the urinary tract, and other variable symptoms. The mutation affects a HOXA13 gene. *See* homeotic genes; hypospadias.

hanging-drop slide The material to be microscopically examined hangs from the cover slip into the concavity of the slide in a drop of a solution.

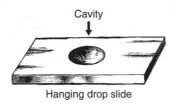

Hanging drop slide

Hap Positive control protein of transcription (46 kDa), attached to one or two sites in the DNA. It generally associates with heat-shock protein70 (Hsp70). Its murine homologue is BAG-1.

haplodiploidy In some social insects (Hymenoptera), the males (drones) are haploid (because they develop from unfertilized eggs), whereas the females (queen and workers) are diploid (because they hatch from fertilized eggs). Consequently, usually full sisters (progeny of a single mating) are more closely related to each other and less closely related to their brothers than to their daughters and sons. In the wasp *Bracon hebetor*, the females are normally heterozygous for genes of sex determination. Inbreeding in such populations may result in homozygosity of the sex-determining genes and male diploidy. In *Nasonia vitripennis*, there is a "selfish" supernumerary chromosome, PSR (paternal sex ratio chromosome). When PSR males produce offspring with normal females, only haploid males occur because the paternal chromosome set, except the PSR chromosome, is eliminated. When PSR males mate with triploid females, the offspring become diploid males. *See* sex determination. (Smith, N. G. 2000. *Heredity* 84[pt 2]:186.)

haplogroup Group of haplotypes that share some sequence variation. *See* haplotype.

haploid Contains only a single complete set(s) of chromosomes (*see* monoploid). The gametes are haploid; the meiotic products of polyploids are often called *polyhaploid*. Many of the fungi are haploid during most of their life cycle, except after fusion of the nuclei in dikaryotes preceding meiosis (*see* fungal life cycle; meiosis). Haploid individuals among diploids show up spontaneously at low frequency. Haploid lines are very rare in animals, although haploid frog cell lines have been used

in cultures. In the spider mite *Brevipalpus phoenicis*, most of the populations are haploid females caused by infection by the bacteria *Cytophaga-Flavobacterium-Bacteroides*. Curing the females of the bacterium by antibiotics, they are converted to male individuals (Weeks, A. R., et al. 2001. *Science* 292:2479).

The spontaneous frequency of haploids varies among different lines of the same species. A high frequency of haploids occurs in crosses involving the wild barley, *Hordeum bulbosum*, and certain wheat varieties crossed with the grass *Aegilops caudata (Triticum dichasians)*. In these crosses, one of the genomes (e.g., *H. bulbosum*) is eliminated at high frequency. Haploid plants have been obtained in many species in artificial culture of immature pollen or microspores and anthers. The microspores can be cultured without separation from the anthers, although separation of the pollen sacs may improve their development because the anther walls may contain growth inhibitors. From these haploid cells, haploid plants can be regenerated. Successful embryogenesis is generally easier if the cultures are initiated at the uninuclear stage of the microsopore. Haploid plants may be obtained actually through two different routes from microspores: either by so-called *direct androgenesis* when the haploid male cells are guided through embryogenesis directly, or the haploid initial cells can be converted to callus, and plants are generated from the callus (indirect androgenesis). The latter procedure is less desirable because during callus growth spontaneous chromosome doubling and other types of chromosomal anomalies frequently occur. If the microspores are exposed to colchicine or other agents that block the mitotic spindle, chromosome doubling may be induced.

Haploid cell cultures have great advantages for mutation studies because all of the recessive mutations are detectable without the masking effect of the dominant alleles at the loci. Similar advantages are available in the use of hypoploid animal cell cultures, such as derived from XY males where X-chromosomal genes are present in a single dose or, e.g., in the culture of Chinese hamster ovary cells (CHO) where individual chromosomes are spontaneously eliminated. Although effective screening methods are available in animal cells (BUdR, antibiotic, temperature), regeneration of animals from isolated cells cannot be accomplished in culture, except from stem cells or via nuclear transplantation. Haploids may have great advantages in plant breeding because by doubling the chromosome numbers in a single step, 100% homozygosity results. Ordinarily, by self-fertilization or inbreeding, six to eight cycles result in only 98 to 99% homozygosity ($0.5^6 \cong 0.0156$ heterozygosity). S. S. Chase developed a successful method for selective isolation of haploid maize. Female flowers, recessive for several markers easily recognizable in the kernels, are pollinated by the corresponding dominant stocks (*A, B, Pl, R*). Kernels that fail to show the paternal markers in the endosperm are discarded because they originated by unintended selfing. The seedlings displaying the dominant paternal markers are also discarded because they are most likely diploid. The true haploids thus arise from fertilization of the endosperm nucleus by the dominant sperm, but the egg that develops into an embryo without fertilization yields seedlings with maternal traits only and (thus haploid) can be recognized. After doubling their chromosome number, homozygotes are obtained. Some caution may be required in critical studies because through spontaneous mutation some variations may arise in these otherwise completely homozygous lines. Haploidy may also be induced by pollination with heavily irradiated pollen or by other means with damaged pollen. Some genotypes display a proclivity for spontaneous androgenesis. *See* androgenesis; anther culture; antibiotic resistance; apomixis; bromouracil; conversion; doubled haploid; embryo culture; embryogenesis, somatic; selective medium; sex determination. (Hall, D. W. 2000. *Genetics* 156:893.)

haploidization   Reduction of the chromosome number to the haploid level.

haploid-specific gene   Turned on in response to mating factors (yeast). The responding consensus is 50 bp upstream of the translation initiation site. *See* consensus; mating type determination in yeast; translation; upstream.

haploinsufficient   The gene in a single dose (such as in hemizygotes) cannot assure its normal function and may even be lethal. *See* Chotzen syndrome; DiGeorge syndrome; familial hypercholesterolemia; hemizygous ineffective; PCR-mediated gene replacement; polycystic kidney disease; Turner syndrome; Waardenburg syndrome.

haplontic   During most of its life the organism is haploid. *See* haploid.

Haplopappus gracilis   Composite plant with only two pairs ($n = 2$) of good-size chromosomes. Due to self-incompatibility genes, its culture is somewhat inefficient.

haplotype   Set of genes in each chromosome of the genome inherited ordinarily as a bloc. (The term was most commonly used in immunogenetics.) Originally it represented the haploid set of genes of the MHC (multiple histocompatibility) antigens. It is generally assumed that individual genes within a haplotype would display collinearity with homologous genes of other haplotypes of the same species. This expectation may not be valid, however, since it has been shown that e.g., at and around the *bz* locus of maize substantial nucleotide heterogeneity occurs (Fu, H. & Dooner, H. K. 2002. *Proc. Natl. Acad. Sci. USA* 99:9573). *See* HLA. (Gabriel, S. B., et al. 2002. *Science* 296:2225.)

haplotype analysis   Infers the relative position of genes and DNA markers by assuming a minimum number of crossing overs along the chromosome. Haplotype analysis may be very useful for the identification of human disease genes. *See* crossing over; linkage disequilibrium. (Daly, M. J., et al. 2001. *Nature Genet.* 29:229; Johnson, G. C., et al. 2001. *Nature Genet.* 29:233.)

HAPMAP   Based on haplotype patterns of SNIPs. *See* haplotype analysis; SNIPs.

happiness   Most likely genetically determined because the correlation between monozygotic and dizygotic twins was found to be 0.44 and 0.08, respectively. It has been suggested that happiness may be determined by the D4 dopamine receptor and unhappiness is related to the control of serotonin metabolism. *See* dopamine; serotonin.

HAPPY   Haploid genome equivalent and polymerase chain reaction is an in vitro method of mapping DNA fragments. The fragments are generated from genomic DNA by irradiation and classified by size with the aid of pulsed-field gel electrophoresis. The fragments are distributed into a 96-well panel. The panel members are amplified by PCR, then screened for specific new STS markers. LOD scores test cosegregation. The procedure is very efficient. The disadvantage of the methods—which may later be overcome—is that the preamplification of the genomic DNA results in new STS markers flanked by interspersed repeat elements and are thus not identical to known STS markers; they reflect the distribution of interspersed repeats rather than the standard maps. *See* lod score; pulsed-field gel; PCR; radiation hybrids; STS. (Dear, P. H. & Cook, P. R. 1993. *Nucleic Acids Res.* 21:13; Walter, G., et al. 1993. *Nucleic Acids Res.* 21:4524.)

happy puppet syndrome   Abandoned, derogatory name of Angelman syndrome. *See* Angelman syndrome.

hapten   Small molecule that can act as an antigen in association with a protein (carrier). Alone, haptens are only antigenic but not immunogenic. The hapten carrier is the basis of the immune response to the complex. *See* affinity labeling; affinity maturation; antigen; immune system.

Hapsburg lip   *See* Habsburg lip.

haptoglobin   Mammalian serum protein composed of two α- and two β-chains. The $\alpha 1$ chain contains 84 amino acids and differs from the $\alpha 2$ chain, which is nearly double the size in the presence of the Hp^2 allele due to a duplication in an intercalary segment, presumably brought about by an unequal crossing over event during its evolution. The $\alpha 2$ chain occurs only in humans and it is thus most likely of relatively recent origin. This protein is attached to hemoglobin and has a role in recycling heme. The HP gene was located to human chromosome 16q22. A number of electrophoretic variants have been identified. The frequency of the gene responsible for the $\alpha 1$ chain varies a great deal in ethnic populations. *See* globin; hemoglobin; plasma proteins; unequal crossing over.

hard disk   Permanently sealed computer disk that operates faster than floppy disks and has a large capacity. *See* disk.

harderoporphyria   Variant form of coproporphyria where the relative amount of excreted coproporphyrin is less in favor of the other intermediate in heme biosynthesis, harderoporphyrin. *See* coproporphyria; porphyria. (Lamoril, J., et al. 2001. *Am. J. Hum. Genet.* 68:1130.)

hard heredity   Inheritance determined by a permanent genetic material such as DNA and RNA, rather than the diffuse hypothetical gemmules, pangenes, etc., hypothesized before the acceptance of Mendelian genetics. *See* gemmules; Mendelian laws; pangenesis.

hardware Physical equipment such as the computer. *See* software.

Hardy-Weinberg equilibrium Genotype and gene frequencies remain constant from generation to generation because there is random mating between the individuals and neither selection nor mutation or migration affect the composition of the population. *See* Hardy-Weinberg theorem.

Hardy-Weinberg theorem For one allelic pair, $p^2 + 2pq + q^2$, where p is the frequency of the dominant allele and q is the frequency of the recessive allele $(1 - p)$. If the genotypic frequencies are available, the allelic frequencies can be derived because the two types of homozygotes have two copies of the alleles, whereas the heterozygotes have one of each. In the case of three alleles at a locus such as in the ABO blood group, the frequency of the i^O recessive allele is

$$r = \sqrt{\frac{O}{N}}.$$

Since the combined frequencies of i^O and I^A is $r^2 + 2pr + p^2$, the frequency of the i^A allele is

$$p = \sqrt{\frac{A + O}{N}} - r,$$

and therefore the frequency of the allele, by subtraction, becomes $q = 1 - p - r$. The O, A, and B are the actually observed numbers of the representatives of the blood groups and N is the population size. In the case of trisomy, the various possible genotypes will be given by $(p_1A_1 + p_2A_2)^3$ after the expansion of this binomial. *See* ABO blood group; allelic frequencies.

hare *Lepus americanus* $2n = 48$; *Lepus towsendii* $2n = 48$.

harelip Hereditary cleft of the upper lip. The maxilla (upper jaws) and palate (the partition of the oral and nasal cavity) may also be affected. The incidence in the general population varies between 0.04 and 0.08%, and the recurrence risk among brothers and sisters is about 0.2%. It is somewhat more common among males than females. Harelip is a sex-influenced trait. *See* cleft palate; sex influenced; Van der Woulde syndrome.

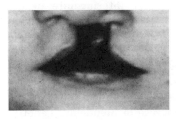

[photo after R. J. Gorlin, Univ. Minnesota.]

Harlequin chromosome staining Results in different coloration of sister chromatids as the result of one or two cycles of replication in the presence of the nucleoside analog 5-bromodeoxyuridine (BrDU). The chromatids replicated in the presence of BrDU absorb less fluorescent stain Hoechst 33258 than the chromatid that has replicated in the presence of thymidine. On a black-and-white photo negative, the chromatid free of BrDU appears lighter than the one that incorporated BrDU; in the print it appears darker. Thus, chromatids containing one, two, or no BrDU can be distinguished. Therefore, such a staining permits the detection of sister-chromatid exchange. *See* bromouracil; sister chromatid exchange. (Rachel, A. J., et al. 1991. *Mutation Res.* 264:71; Jordan, R., et al. 1999. *Biotechniques* 26:532.)

Hartnup disease (11q13) Autosomal-recessive disorder involving photosensitivity, rash, cerebellar ataxia (impaired muscle coordination by the brain), and aminoaciduria. The uptake of methionine and tryptophan and to some extent of other neutral amino acids by the intestines is reduced. It is diagnosed by urine analysis for increase in neutral amino acids. Its prevalence is about 4×10^{-5}. *See* ataxia telangiectasia; light sensitivity diseases; tryptophan.

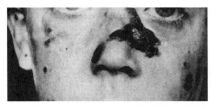

Photograph from Bergsma, D., ed. 1973. *Birth Defects*. March of Dimes Foundation.

harvest index Proportion of the economically directly usable productivity of crops, e.g., grain versus straw. It has also been defined as the ratio of dry weight of the harvestable organs to the total dry weight.

Harvey murine sarcoma virus (transformation gene) Originally derived from rats and it is found to encode a 21 kDa oncoprotein (p21) or RAS. The human homologue was mapped to chromosome 11p14.1. *See* p21; RAS.

HARWEY *Drosophila* transposable element (7.2 kb). *See* transposable elements.

HAT *See* coactivator; histone acetyltransferase; histones; nucleosomes.

HAT medium Of animal cell culture, contains hypoxanthine, aminopterin, and thymidine and has been extensively used for the isolation of bromodeoxyuridine- and azaguanine-resistant mutants and complementary fused cell lines by the rationale outlined in the figure. *See* aminopterin; azaguanine; bromodeoxyuridine; selective isolation; diagram next page.

hatching time, poultry The eggs of chicken hatch in 3 weeks; turkey, goose, and duck eggs require 4 weeks of incubation.

Alternative pathways of nucleic acid biosynthesis. The upper, de novo, pathway can be selectively blocked by aminopterin, an inhibitor of dehydrofolate reductase (an enzyme essential for the biosynthesis of thymidylate). In case of such a block, pyrimidines and purines may still be synthesized through the salvage pathway from nucleosides. TK (Thymidine kinase) can make thymidylic acid and hgprt (hypoxanthine guanine phosphoriboyl transferase) can supply guanylic acid. In case *TK* is inactivated by mutation (*TK⁻*), the cells become resistant to 5-bromodeoxyuridine because the cells cannot convert it into a nucleotide analog. Similarly, if hgprt is inactivated (*hgprt⁻*) by mutation, the guanine analog, 8-azaguanine, cannot be incorporated into dna and thus the cell will be resistant to it. Therefore, on HAT medium both bromodeoxyuridine- and aza-guanine-resistant mutant cells can selectively be isolated.

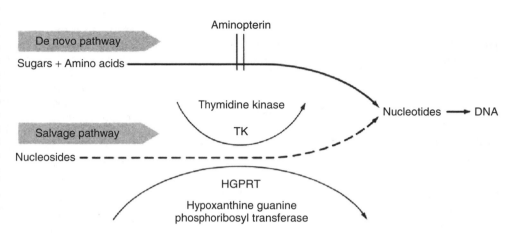

hausp Herpesvirus-associated ubiquitin-specific protease.

haustorium Organs of parasites (e.g., fungi) that penetrate the periplasmic space of the host cells for the purpose of absorbing nutrients. In parasitic plants (*Striga asiatica*), quinone may serve as a signal to the expression of the expansin gene and the cellulose fibers are modified to form haustoria. (Voegele, R. T., et al. 2001. *Proc. Natl. Acad. Sci. USA* 98:8133.)

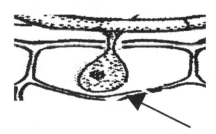

Haustorium grows through epidermis.

hawkinsinuria (4-α-hydroxyphenylpyruvate hydroxylase deficiency) Hawkinsin is a 2-L-cystein-S-yl-1,4-dihydroxycyclohex-5-en-1-yl acetic acid. On a restricted phenylalanine and tyrosine diet, the dominant condition much improves. It is a childhood disease.

Haw River syndrome Dentatorubral-pallidoluysian atrophy.

hay fever *See* allergy.

Hayflick's limit Human cells usually die in culture after 50 to 60 or fewer cell cycles. Cancer cells are ordinarily not subjected to such a limit by senescence. *See* apoptosis; hybridoma; immortalization; senescence. (Shay, J. W. & Wright, W. E. 2000. *Nature Rev. Mol. Cell Biol.* 1:72.)

Haynaldia villosa Diploid wild grass (2n = 14) carrying the V genome. It can directly be crossed with tetraploid wheats; the AABBVV hybrids can be backcrossed with hexaploid wheat (AABBDD) and AABBDD + 7V additions can thus be generated. *See* addition lines; *Triticum*.

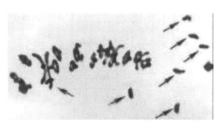

Haynaldia addition chromosomes marked by arrows. (Courtesy of Dr. E. R. Sears.)

Hay-Wells syndrome (3q27) Dominant ankyloblepharon-ectodermal defects such as cleft lip/palate. Mutations in p63 may be involved. *See* ankyloblepharon; p63. (McGrath, J. A., et al. 2001. *Hum. Mol. Genet.* 10:221.)

hazard function $\hat{h}(t) =$ (number of individuals exposed within time t)/(number of survivors within the same period of time).

hazardous chemicals *See* biohazards; chemicals, hazardous; environmental mutagens.

hazelnut (*Corylus* spp) Monoecious shrub with edible nuts; $2n = 2x = 28$.

HBV Hepatitis B virus. *See* hepatitis.

HCK *See* SRC kinase family of oncogenes.

HCP Nonreceptor tyrosine phosphatase. *See* B lymphocyte.

HC-Pro   *See* posttranscriptional silencing.

HCT   *See* CDH1.

HDA1, HDA2, HDA3   Histone 3, histone 4 deacylating enzymes of yeast. *See* DHAC; histone deacetylation.

Hdj2   Mammalian homologue of DnaJ, and it may chaperone translocation into the Golgi apparatus and the nucleus. *See* chaperones; DnaK; Golgi apparatus.

HDL   *See* high-density lipoprotein; Tangier disease.

hDNA   Used in two ways: human DNA or heteroduplex DNA. *See* heteroduplex.

H-DNA   (1) Protonated molecule; apparently without natural biological role. *See* DNA types. (2) May be formed as a triplex tract when either pyrimidines (PY) or purines (PU) abound in the DNA single strands. The PY-rich strand may fold back and pair with the PU-rich tract or vice versa. The 3′ halves (PY3′) contribute preferentially to the triplex. *See* nodule DNA; trinucleotide repeats; triplex.

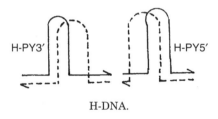

H-DNA.

head/face/brain defects   *See* Aarskog syndrome; Angelman syndrome; Borjeson syndrome; Coffin-Lowry syndrome; craniometaphyseal dysplasia; cranioorodigital syndrome; de Lange syndrome; fragile X chromosome; holoprosencephaly; Langer-Giedion syndrome; Miller-Dieker lissencephaly; Opitz-Kaveggia syndrome; Opitz syndrome; otopalatodigital syndrome; Prader-Willi syndrome; Smith-Lemli-Opitz syndrome; Walker-Wagner syndrome.

headful rule   Bacteriophages replicate their DNA in concatamers, but the phage head (capsid) has a limited storage capacity; therefore, the molecules have to be cut to "headful" length. *See* concatamer; lambda phage. (Droge, A. & Tavares, P. 2000. *J. Mol. Biol.* 296:103; Coren, J. S. & Sternberg, N. 2001. *Gene* 264:11.)

hearing deficit   *See* deafness.

Heartbreaker (*hbr*)   Nontranslated, short (314 bp) insertion element in several grasses (maize and other cereals). It is present in ~4,000 copies in the maize genome. It is suspected to be responsible for a significant portion of the polymorphism observed. *See* MITE; transposable elements, plants. (Zhang, Q., et al. 2000. *Proc. Natl. Acad. Sci. USA* 97:1160.)

heart disease   Affects about 1% of the population and only a little less of newborns; about 15% it is lethal. Congenital heart disease frequently occurs as a component of various syndromes. About 2% of the cases are caused or precipitated by environmental effects and various diseases (alcoholism, lithium, thalidomide, retinoic acid [a derivative of vitamin A], trimethadione [anticonvulsant drug], viral infections (rubella), maternal diabetes and phenylketonuria, trisomies (21, 8, 13), Turner syndrome, deletions by DiGeorge syndrome, asplenia, patent ductus, Holt-Oram syndrome, Ellis–van Creveld syndrome, mucopolysaccharidosis, Pompe's disease, endocardial fibroelastosis, coarctation of the aorta, Noonan syndrome, LEOPARD syndrome, Fallot's tetralogy, mitral prolapse, myotonic dystrophy, Jervell and Lange-Nielsen syndrome, Opitz-Kaveggia syndrome, etc. *See* aneurysm; atherosclerosis; atrial septal defect; cardiovascular disease; coronary heart disease; hypertension; hypertrophic cardiomyopathy; mitochondrial diseases in humans; nemaline myopathy; supravalval aortic stenosis. (Srivastava, D. & Olson, E. N. 2000. *Nature* 407:22; Chien, K. R. 2000. *Nature* 407:227; Nicol, R. L. & Olson, E. N. 2000. *Annu. Rev. Genomics Hum. Genet.* 1:179; Molkentin, J. D. & Dorn, G. W. 2001. *Annu. Rev. Physiol.* 63:391; *Nature* [2002], 415:198–240.)

heat repeat   Repeated element first discovered in the proteins: huntingtin, translation elongation factor EF3, the A subunit in protein phosphatase 2A (PP2A) and TOR1, a target for rapamycin. Similar repeats have since been found in different proteins of several species. *See* EF3; huntingtin; rapamycin; TOR. (Andrade, M. A., et al. 2001. *J. Struct. Biol.* 134:[2–3]:117.)

heat shock element   *See* HSE.

heat shock proteins (hsp)   Most commonly molecular chaperones that under heat stress inhibit or prevent denaturation of other proteins by binding to interactive surfaces and thus securing that these chaperoned proteins maintain their structure required under normal, nonstress conditions. The production of heat-shock proteins under this type of stress is very rapid because the *hsp70*, *hsp26*, and *hsp27* genes of *Drosophila* initiate transcription in anticipation of heat by a halted 20- to 40-nucleotide transcript that can then be readily elongated. Some hsp proteins may have other functions as well. Hsp110 is present in prokaryotes and eukaryotic organelles in response to stress. Hsp90 family members bind ATP and are involved in autophosphorylation, but in animals they are immunophilins (binding to immune-suppressive antibiotics such as cyclosporin). Hsp70 members are made in all types of organisms and their main role is mediating the folding of nascent proteins. The heptameric hsp60 molecules participate in the molding of proteins using ATP ⇆ ADP for energy. The hsp40 proteins may be involved in sorting polypeptides. The smaller hsp proteins may form aggregates before acting on other molecules. Prokaryotes do not contain the very smallest hsp proteins. The *hsp* genes are classified into families on the basis of the molecular weight of the proteins (in kDa) encoded by them. A generalized structure of an Hsp70 protein:

N-end	ATPase Fragment		Peptide-Binding Fragment		COOH Terminal
1	~385		~550–600		~650

For the small heat-shock proteins, *see* sHsp. Under high-temperature stress, heat-shock proteins may accumulate to 15–20% of the total cellular proteins. Their formation is tightly regulated by both positive and negative controls. In *E. coli*, the heat signals activate the $E\sigma^{32}$ RNA polymerase and the rpoH operon, whereas misfolding activates the $E\sigma^E$ (σ^{24}) operon. Besides elevated temperature (5–10°C above normal), heat-shock responses may be elicited by toxic chemicals, metabolic inhibitors and analogs, microbial or viral infections, various injuries, cancer, aging, and developmental events. During heat shock, normal protein synthesis may be selectively repressed. *See* chaperone; cold shock; DnaJ; elongation factors; Hsc; Hsc66; HSE; HSP; Hsp70; Hsp90; immunophilin; sHsp; σ; spastic paraplegia; thermal tolerance; transcription factors, inducible. (Nagao, R. T. & Gurley, W. B. 1999. *Inducible Gene Expression in Plants*, Reynolds, P. H. S., ed., p. 97. CABI; Wallingford, UK, Macario, A. J. & Conway-Macario, E. 2000. *Int. J. Clin. Lab. Res.* 30[2]:49; Queitsch, C., et al. 2002. *Nature* 417:618; Srivastava, P. 2002. *Annu. Rev. Immunol.* 20:395.)

heavy chain *See* antibody; DNA heavy chain; Watson-Crick model.

heavy chain, DNA *See* DNA heavy chain.

Hebbian mechanism Old hypothesis proposed by Donald Hebb (1949) for the memory function of the brain as a series of excitations of axons resulting in firing of other nerve cells and long-term potentiation. *See* memory. (Rao, R. P. & Sejnowski, T. J. 2001. *Neural. Comput.* 13:2221.)

Hec 80 kDa coiled-coil nuclear protein. It interacts with Smc1 and Smc2 and apparently controls chromosome segregation. *See* spindle.

hedgehog *Erinaceus europaeus*, $2n = 48$.

hedgehog (*hh*, 3-81) Segment polarity-type embryonic lethal mutation in *Drosophila*. The product of the wild-type allele of *hh* encodes a signaling molecule that is processed by autoproteolysis into two active species. By the action of the C-terminal domain of the protein, it is cleaved into two and the N-terminal signaling domain covalently binds a cholesterol moiety. This binding is required for neutralizing the effects of Patched and Smoothened proteins that would block the signaling pathway. The Cholesteryl-Hedgehog–Patched–Smoothened complex displays differences in tissue distribution and instructs adjacent cells to express the organizing signal encoded by the *decapentaplegic* (*dpp*) locus. The Hh protein is secreted under the control of the En (engrailed) protein. The *en* locus is continuously expressed in the posterior part of the embryo. The anterior part of the

embryo expresses the *ci* (*cubitus interruptus*) locus, which encodes a zinc-finger-binding protein of the Gli family of transcription factors. If *ci* is not expressed, the *hh* gene product shows up and posterior compartment properties appear without the expression of the *en* signal. Increased levels of *ci* products induce the expression of *dpp* independently from *hh*. Expression of the normal *ci* product in the anterior cells results in limb development by limiting the expression of *hh* to posterior cells and mediating the ability to respond to the protein signal of the *hh* locus. Ci transduces the Hh signal by activating the *dpp* and *ptc* (*patch*) genes. The Patch gene product and cyclic AMP-dependent protein kinase A interfere with inappropriate expression of *dpp* if Hh product is not available. Ci product at a low level represses *dpp* and at a higher concentration it appears to be an activator of the same gene. Hh induces both decapentaplegic (dorsal compartment) and wingless (ventral compartment), and these two modulate each other's function to assure normal axial development.

The product of *patched* (*ptc*) gene is the receptor of Hh and *smoothened* (*smo*) is a signaling component of Ptc. Hip (hedgehog-interacting protein, ~78 kDa) has a somewhat similar role as Ptc. The human gene EXT-1 (and the apparently homologous *Drosophila* gene *Ttv*) seems to regulate the embryonic movement of the hh protein. EXT encodes a protein apparently involved in cell-surface glycosaminoglycan (GAG) synthesis. The *hh* gene has at least three homologues in humans and other vertebrates: Sonic hedgehog (SHH, 7q36, patterns the neural tube, early gut endoderm, posterior limb buds); Indian hedgehog (*Ihh*, 2q33-q35, expressed primarily in cartilage); and Desert hedgehog (*Dhh*, 12q13.1, expressed primarily in the testes). The *hh* signaling system is present in many invertebrates and vertebrates, but it is absent from *Caenorhabditis*. Upon autocatalytic processing, Hh releases a 19 kDa ligand with cholesterol linked to its C-end. It is assumed that Hh works in *Drosophila* as a raft for intracellular transport. Basal cell carcinoma and medulloblastoma cancers are frequently associated with mutation of Hh. *See* cholesterol; cubitus interruptus; Greig's cephalopolysyndactyly syndrome; glycosaminoglycan; holoprosencephaly; morphogenesis in *Drosophila*; nevoid basal cell carcinoma; Rubinstein syndrome; *sonic hedgehog*; tumor suppressor pathway. (McMahon, A. P. 2000. *Cell* 100:185; Kalderon, D. 2000. *Cell* 103:371; Ingham, P. W. 2001. *Science* 294:1879. Bale, A. F., 2002. *Annu. Rev. Genomics Hum. Genet.* 3:47.)

HEI (Hybrid element insertion). The left end of one hybrid dysgenesis-causing P element of *Drosophila* in *trans* position can move with the left end of another P element as a unit and cause viable exchanges or inviable rearrangements. The exchange products may also be due to hybrid excision and single-strand repair synthesis (HER). The HEI model may account for male recombination in *Drosophila*. *See* DNA repair; hybrid dysgenesis; male recombination. (Tanaka, M. M., et al. 1997. *Genetics* 147:1769.)

HeLa cell line Immortalized human cancer cell line originated in 1951 from a highly malignant cervical carcinoma of patient Henrietta Lacks, who died the same year, but the line has been maintained all over the world indefinitely. Many of the HeLa cell lines have been contaminated — by error or by deception — with cells of a different origin. With improved forensic laboratory techniques, these can be identified by high probability. *See* immortalization. (O'Brien, S. J. 2001. *Proc. Natl. Acad. Sci. USA* 98:7656; Masters, J. R. 2002. *Nature Rev. Cancer* 2:325.)

helical twist   Angle between neighboring DNA base pairs; it is within the range of 24° and 51° with a mean of 36.1 ± 5.9.

helicases   Enzymes unwinding the double helix of a nucleic acid for replication and transcription, repair, recombination, and in reactions associated with binding and hydrolysis of DNA. About 60 helicases have been identified; some unwind DNA and RNA hybrids. *E. coli* bacterium encodes about 12 different helicases, and all organisms have several types. The activity of helicase may be coordinated with DNA polymerase, and the complex may travel at about 1,000 nucleotides/second. The RecQ family of helicases is implicated in mutator functions and chromosomal rearrangements such as those occurring in Bloom syndrome, Cockayne syndrome, xeroderma pigmentosum B and D, and others. *See* ABC excinuclease; Bloom syndrome; branch migration; chromosomal rearrangements; Cockayne syndrome; DEAD box; DNA replication in prokaryotes; recA; recombination, molecular mechanism in prokaryotes; Rep; Rothmund-Thompson syndrome; unwinding protein; Werner syndrome; xeroderma pigmentosum. (Patel, S. S. & Picha, K. M. 2000. *Annu. Rev. Biochem.* 69:651; von Hippel, P. & Delagoutte, E. 2001. *Cell* 104:177; van Brabant, A. J., et al. 2000. *Annu. Rev. Genomics Hum. Genet.* 1:409; Mohaghegh, P. & Hickson, I. D. 2001. *Hum. Mol. Genet.* 10:741; Enomoto, T. 2001. *J. Biochem.* 129:501.)

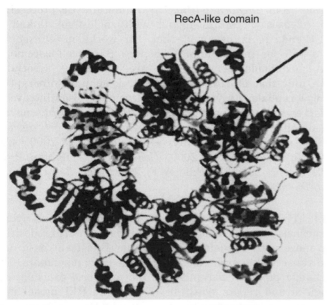

RecA-like domain

Phage T7 hexameric helicase motor protein. (From Waksman, G., et al. 2000. *Nature Struct. Biol.* 7:20.)

Helicobacter pylori   Bacterium (strain 26695) with a completely sequenced genome of 1,667,867 bp and 1,590 ORF (*Nature* 388:539 [1997]). Strain J99 is also sequenced (1,643,831 bp; *Nature* 397:176 [1999]). The latter has 24,036 bp smaller genome. The genic difference between the two strains is about 6–7% and half of these are clustered in a hypervariable

region. These common pathogens are tolerant for high acidity and responsible for peptic ulcers and possibly for gastric carcinoma but may also be harmless. The carcinogenic effect may be brought about by increased gastric acidity and interleukin-1. *H. pylori* can produce an antibacterial peptide (a cecropin) bearing similarity to ribosomal protein RpL1. A protein-protein interaction map using yeast two-hybrid assay has been developed for over 46% of the proteome. *See* antimicrobial peptides; *E. coli*; ORF. (Rain, J.-C., et al. 2001. *Nature* 409:211; Ernst, P. B. & Gold, B. D. 2000. *Annu. Rev. Microbiol.* 54:615; Del Giudice, G., et al. 2001. *Annu. Rev. Immunol.* 19:523; <http://genolist.pasteur.fr/>.)

helitron   *See* rolling circle.

helix (of macromolecular structure)   Coil of a three-dimensional ribbon; the DNA helix is similar to a staircase, with the steps corresponding to the bases and the rails of the staircase representing the sugar-phosphate backbone of the two polynucleotide chains; thus, it is not a simple spiral. *See* DNA; Watson and Crick model.

helix destabilizing protein (RF-A)   Single-strand-binding protein instrumental in DNA replication. *See* binding proteins; DNA replication; RF-A.)

helix-loop-helix   Polypeptides each with three-partite structures; two of the helices are connected through a loop in each component of the dimer; the third helix of the components is rich in basic amino acids and this end binds to DNA. The monomers thus appear as HOOC-helix-loop-helix positively charged helix-NH$_2$. They recognize the CANNTG (E-box) sequence in the DNA. The H-L-H proteins may form dimers and recognize two different neighboring DNA-binding sites. The proteins may saddle into the major groove of the DNA through their positively charged amino acids in the α-helix of the NH$_2$ terminus. Basic helix-loop-helix proteins (>250) control transcription of genes involved in a great variety of cellular functions. *See* DNA-binding protein domains; helix-turn-helix motif; regulation of gene activity. (Lodent, V. & Vervoort, M. 2001. *Genome Res.* 11:754.)

helix-turn-helix motif   Parts of regulatory proteins of prokaryotes and homeodomains of eukaryotes. One α-helix (*recognition helix*) fits into the major groove of the DNA and the other is positioned at a right angle above it, allowing interaction with other proteins. The other monomer of the dimeric structure binds to the next major groove along the DNA. There are large varieties of these proteins in bacteria and eukaryotes, yet they contain a conspicuous symmetrical structure formed of antiparallel β-sheets or α-helices separated by a turn of several amino acids. H-T-H motifs occur in the homeodomain proteins in various repressors (λ *cI* and *cro* proteins, in the *E. coli* catabolite activator [CAP] proteins, etc.). *See* DBP; DNA-binding protein domains; DNA grooves; lac operon; lambda phage; monomeric; protein structure; regulation of gene activity.

Helminthosporium maydis   Fungus is the causative agent of the maize disease, southern corn leaf blight. Plants with T (Texas) cytoplasmic male sterility are extremely susceptible to the disease and suffer serious damage. *See* cms; cytoplasmic male sterility.

helper plasmid   *See* binary vector.

helper T cell   *See* T cell.

helper virus   Provides the functions that a defective virus particle lacks.

hemagglutinin   Protein that causes agglutination of the red blood cells, e.g., antibodies, lectins, certain viruses (influenza, mumps, etc.). Variations in hemagglutinins due to amino acid substitutions are important factors in viral infectivity and the ability of the cells to make antibodies against them. *See* antibody; antigen; fusigenic liposome. (Skehel, J. J. & Wiley, D. C. 2000. *Annu. Rev. Biochem.* 69:531.)

hemangioblast   Mesodermal stem cells giving rise to blood vessels and hematocytoblasts (hemocytoblasts). *See* angiogenesis; blood; hematocytoblast.

hemangioma   Several forms of neoplasias of blood vessels under autosomal-recessive gene control in humans. *See* angioma.

hematocytoblast (hemocytoblast)   Totipotent stem cells of blood. *See* angiogenesis; blood; hemangioblast; stem cell.

hematopoiesis   Blood formation. It is a homeostatic process as new cells are formed and old ones are removed by apoptosis. After a certain age, the blood volume does not expand. The major regulatory factors are cytokines. Growth factors include the Kit ligand and various macrophage factors. *See* G-CSF; GM-CSF; interleukins; M-CSF. (Metcalf, D. 1998. *Stem Cells* 16:314.)

hematopoietic receptor   Binds hormone (GH, PRL, EPO, G-CSF) or cytokine (interleukin) ligands and is involved in cellular signaling. These receptors may be homodimers (α-chain) and-or heterooligodimers (α- and β-chains).

hematoxylin   Histological and cytological stain in different formulations. (Delafield's hematoxylin, Heidenehain hematoxylin, iron hematoxylin, etc.). *See* stains.

heme   Iron or magnesium-porphyrin prosthetic group in hemoglobin, cytochromes, and chlorophyll, respectively. *See* haem.

hemeralopia   *See* day blindness.

hemerythrin   Circulatory transport, nonheme iron protein in some nonvertebrates. *See* heme; hemocyanin; hemoglobin. (Kurtz, D. M. 1992. *Adv. Comp. Env. Physiol.* 13:151.)

hemibiotrophy   The parasite requires live cells for infection, but later in can grow on dead ones.

hemicellulose   Polymer of neutral polysaccharides present in the plant cell wall matrix.

hemiclonal (hybridogenetic species)   *Rana esculenta* (European edible green frog) arises from the hybrid of *Rana ridibunda* × *R. lessonae*. In *R. esculenta*, one of the parental genomes is eliminated before DNA synthesis and after duplication of the remaining genome meiosis takes place; thus, the gametes are produced from only one of the genomes. Backcrossing to the parental species whose genome was eliminated then reforms the hybrid. In case of exposure to ionizing radiation, *R. esculenta* individuals may lose up to ~50% of their DNA, thus making it possible to measure the extent of radiation fallout with the aid of flow cytometry. The *Poeciliopsis* fishes and the *Bacillus rossius-grandii* stick-shape insects are also hybridogenetic. The latter may reproduce by parthenogenesis, gynogenesis, or androgenesis. *See* androgenesis; gynogenesis; parthenogenesis. (Giorgi, P. P. 1992. *Nature* 357:444; Vinogradov, A. E. & Chubinishvili, A. T. 1999. *Genetics* 151:1123.)

hemiglobin   *See* methemoglobin.

hemimetabolous   The insect hatches as a miniature adult and grows to the imago size. *See* holometabolous; imago.

hemimethylated   Only one of the two DNA strands is methylated. *See* demethylation; methylation of DNA.

hemin (ferriporphyrin chloride, $C_{34}H_{32}ClFeN_4O_4$)   Used for the treatment of porphyria and the preparation of rabbit reticulocyte lysate. *See* haem; porphyria; rabbit reticulocyte in vitro translation.

hemin-regulated inhibitor   *See* HRI.

hemizygous   Gene(s) present in a single dose in an otherwise diploid cell or organism (e.g., X-chromosome-linked genes in an XY male or XO cells). *See* dosage compensation.

hemizygous ineffective   Special class of recessive mutations in allopolyploids. These recessive alleles are not expressed in monosomics (hemizygotes), and in case of nullisomics, the dominant wild type is expressed for these loci. Their expression requires two doses of these recessive alleles and never displays pseudodominance. Thus, the hemizygous ineffective alleles resemble recessive suppressor mutations in diploids. Recessive mutations in allopolyploids are generally of this type because the homoeologous loci cover up the mutations at the other corresponding genes. *See* allopolyploid; homoeologous alleles; monosomic; nullisomic; pseudodominance; suppressor gene. (Sears, E. R. 1972. *Symp. Biol. Hung.* 12:72.)

hemochromatosis (HFE/HLAH)   Disease of iron accumulation, accompanied by cirrhosis (fibrous condition) of the liver, diabetes, dark pigmentation of the skin, heart abnormality, and cancer. If diagnosed early by determining plasma iron and ferritin (a red 80,000 M_r serum protein with two iron-binding sites to transport iron, also called siderophilin), it is curable by lowering the iron level through venesection (phlebotomy), i.e., letting out blood by cutting the vein. The recessive gene HFE is located within the HLA-H complex in human chromosome 6p21.3. The frequency of the gene is about 6 to 10% with an incidence of homozygosity of 2–3 per 1,000 births. The HFE2 locus is at 1q and HFE3 is at 7q22 encoding (TFR2) transferrin receptor-2. Neonatal hemochromatosis (8q21.3) has a very complex expression due to modifying effects. The solute carrier family member, ferroportin (FPN1, 2q32), may be involved in dominant hemochromatosis. *See* anemia; ferritin; HLA; liver cancer; Menke's disease; metal metabolism and disease; transferrin; Wilson disease. (Njajou, O. T., et al. 2001. *Nature Genet.* 28:213.)

hemocyanin Copper-containing protein in several invertebrates that is used to bind O_2 in a way similar to hemoglobin. *See* hemerythrin; hemoglobin; keyhole limpet hemocyanin. (van Holde, K. E., et al. 2001. *J. Biol. Chem.* 276:15563.)

hemocytometer Special microscope slide with compartments of known exact volume when a coverglass is placed over it. It is used for the microscopic counting of the number of blood cells or any other suspended cells or protoplasts in a certain volume.

The ruling on the hemocytometer slide showing the compartments.

hemoglobin Oxygen-transporting heme proteins (M_r 64,500) in the red blood cells. The four polypeptide chains are attached to four heme prosthetic groups (Fe_2^+ state). The human adult hemoglobin consists of two α- (141 amino acids) and two β- (146 amino acids) polypeptide chains, whereas in the early embryo the polypeptide composition is $\zeta\zeta\varepsilon\varepsilon$. The ζ (zeta) is α like; the ε (epsilon) is β-like. By the 8th week of gestation, the embryonal hemoglobin is replaced by fetal hemoglobin with a structure of $\alpha\alpha G_\gamma A_\gamma$ (the latter ones are β-like). Just before, birth, these two γ-chains are replaced by β- and δ-globin chains. By 6 months after birth, 97–98% of hemoglobin A (HbA) is $\alpha\alpha\beta\beta$ and about 2% is $\alpha\alpha\delta\delta$ (HbA2); a very small amount is still fetal hemoglobin (HbF). Thus, there are two gradual developmental switches in the β-like genes but only one in the α-like genes. These changes are programmed as the sites of the synthesis shift from the yolk sac of the embryo to the liver, spleen, and bone marrow of the fetus and finally to the bone marrow in adults. The two gene families, of hemoglobin, α and β, include α, γ, ζ and ∂ (θ, theta, unknown function), located to human chromosome 16p13, and $\beta, \delta, \varepsilon$, and γ are at chromosome 11p15.5.

Both the α-family of genes ($\rightarrow \zeta, \psi\zeta, \psi\alpha, \psi\alpha, \alpha2, \alpha1, \partial$) and the β-family ($\rightarrow \varepsilon, G\gamma, A\gamma, \psi\beta, \delta, \beta$) include pseudogenes ($\psi$) and are transcribed in the order shown by the arrow. Hundreds of different amino acid substitutions and deletions are known in the globin chains and some of them lead to hemoglobinopathies, i.e., blood diseases. One of the most famous is a substitution of valine for glutamic acid at residue 6 of the β-chain (hemoglobin S). This is responsible for a hydrophobic change on its surface, resulting in an abnormal quaternary association of the subunits. When the oxygen level

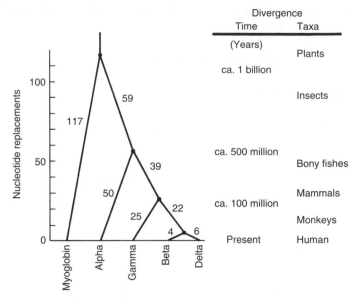

Evolution of hemoglobins. The numbers of the presumed nucleotide replacement in the genetic code during evolution are indicated by the numbers at each node. The time of divergence is estimated on average amino acid replacement data and evolutionary clocks. (After Fitch, W. M. & Margoliash, E. 1970. *Evol. Biol.* 4:67.)

is reduced, the subunits polymerize into a linear array of fibers that alter the normally doughnut-shaped red blood cells to assume a sickle shape; therefore, it is called *sickle cell anemia* in the homozygous condition. In the heterozygous state, it is called the *sickle cell trait*. The sickling erythrocytes are inefficient in oxygen transport. In the arteries, normal hemoglobin is 95% saturated with oxygen; in the venous blood—on the return to the lung—the saturation is about a third less. Over 3 million Americans carry this abnormal HbB gene and there are many millions more all over the tropical and subtropical regions of the world where malaria is common. The heterozygotes are apparently at a selective advantage when infested with the *Anopheles* mosquito that spreads the causative protozoan, *Plasmodium falciparum*. A milder form of sickling, expressed only in the homozygotes, is caused by another mutation, called hemoglobin C (HbC, β6Glu$\rightarrow$Lys). HbC also protects against falciparum malaria. Other variants are HbD and HbE.

Many other amino acid substitutions that have been identified on the basis of altered electrophoretic mobility or amino acid substitutions at precisely identified residue sites in either of the chains may not involve a disease. In the M hemoglobins, the oxygenated molecule has hereditary deficiencies of NADH-methemoglobin reductase activity and remains largely charged with oxygen. The stability of the hemoglobin molecules depends mainly on the close fit of the hemes to the globin chains that would be thermolabile at 50°C. Mutations affecting the tight fit of the heme pocket may cause the formation of methemoglobin or the loss of the heme. Such unstable hemoglobins are, e.g., the Hb Zürich in which at β-chain residue 63 His changes to Arg. Hb Köln β 98 has a Val$\rightarrow$Met substitution. The increase in positive charge (e.g., Hb Zürich) favors the loss of heme. In case of Hb Hammersmith β 42, Phe$\rightarrow$Ser substitution, there is no change in charge, yet instability occurs. Instabilities have also been associated with α-chain substitutions, e.g., in Hb Boston 58 His$\rightarrow$Tyr.

Other anomalies include *hereditary persistence of high fetal hemoglobin*. In the latter case, numerous types of variations exist. It appears that deletions involving a regulatory region of 3.5 kb at 5′ to the δ gene may be a very important factor for the high expression of fetal hemoglobin. In one form, all the cells are equally affected, whereas in the *heterocellular* hereditary persistence of fetal hemoglobin (HHPFH), only some of the cells display this anomaly. In Lepore hemoglobins, $\delta\beta$ or $\beta\delta$ fusion products were observed, indicating the possibility of unequal crossing over between these DNA sites. Another type of hemoglobin anomaly involves the slow rate or lack of synthesis of one hemoglobin chain, resulting in thalassemias. The HBA gene is located to human chromosome 16p13.33-p13.11, whereas the HBB, HBD, and HBG genes are in 11p15.5. Senescent or damaged erythrocytes are disposed of by macrophages in the bone marrow to protect from the oxidative and toxic effects of heme. Heme is converted into bilirubin and iron. Hemoglobins are ubiquitous in animals and are highly variable. A different type (truncated) of hemoglobin occurs in prokaryotes, protozoa, algae, and higher plants but is absent in archaea and metazoa. *See* anemia; globin; methemoglobin; sickle cell anemia; thalassemia. (Hardison, R. C. 1996. *Proc. Natl. Acad. Sci. USA* 93:5675; Watts, R. A., et al. 2001. *Proc. Natl. Acad. Sci. USA* 98:10119; Modiano, D., et al. 2001. *Nature* 414:305; Wittenberg, J. B., et al. 2002. *J. Biol. Chem.* 277:871 <http://globin.cse.psu.edu>.)

hemoglobin, fetal, persistence of   The condition may be restricted to some of the cells (heterocellular) or it involves all of the cells (pancellular). The fetal hemoglobin is a normal protein; only the persistence is a disorder. It may have point mutations or various types of deletions. The DNA may have either the G-γ gene without the presence of A-γ, -δ, -β, or it may retain the G-γ gene and the A-γ, -δ, -β, or it may have the G-γ and the A-γ but the δ and β are missing. The phenotype depends primarily on the genes present rather than on the missing ones. *See* GATA; hemoglobin. (Ikunomi, P., et al. 2000. *Gene* 261:277.)

hemoglobin evolution   Molecular genetics evolved from early studies (late 1940s) showing that the globin genes can be obtained in pure forms relatively easily. Over 90% of the soluble proteins in the blood plasma are globins. The still nucleated erythrocytes and the reticulocytes synthesize mainly globin mRNA. Amino acid sequence variants were detected in these proteins at the beginning of protein sequencing. Since globin genes occur in animals and in plants, evolutionary studies became attractive. The number of amino acids in globins of plants and animals is fairly close: soybean leghemoglobin 143, seal myoglobin 153, human-α 141, human-β 146. Also, there are sequence homologies in the primary structure of the chains that could be the evolutionarily quantitated with aid of estimated average replacement data. In vertebrates, the globin genes have two introns at around the 30th amino-acid-coding region and another around the 100th. In plants there are three introns: two at about the same location and a third near the middle in between. The size of the introns varies substantially from 99 triplets in the middle intron of the leghemoglobin to 4,800 in the first intron of the seal myoglobin. In vertebrates, the heme binding is in the second exon, closer to the 3′-end. The residues required for tetramer association are in exon 3. *See* evolutionary clock; evolutionary tree; figure on p. 530; globin;

hemoglobin; leghemoglobin; myoglobin. (Hardison, R. 1998. *J. Exp. Biol.* 201 [pt 8]:1099.)

hemoglobinopathies   Diseases that involve hemoglobins. *See* anemia; hemoglobin, fetal persistence of; hemoglobins; hemophilia; methemoglobin; paroxysmal nocturnal hemoglobinuria; sickle cell anemia; thalassemia.

hemoglobin switching   During development, various different hemoglobin genes are activated, resulting in the formation of the different fetal and eventually adult types. *See* hemoglobins. (Ristaldi, M. S., et al. 2001. *EMBO J.* 20:5242.)

hemolin   Immunoglobulin-like molecule in insects. *See* immunoglobulins.

hemolymph   Blood-like fluids in insects and other invertebrates. Hemolymph has also a role in the defense against pathogens.

hemolysin   Proteins secreted by pathogenic bacteria, causing the formation of pores in mammalian cell membranes and the dissolving of red blood cell membranes. *See* erythrocyte; lysin.

hemolysis   Disruption of the membranes of erythrocytes by antibodies, hemolysin enzyme, and chemicals. *See* hemolytic disease.

hemolytic anemia   Disease may occur due to several causes such as emphysema, sensitivity to high temperature, Rh blood type, and other autoimmune diseases, etc. *See* anemia; autoimmune diseases; glutathione peroxidase; glutathione reductase; glutathione synthetase deficiency; hemolysis; hexokinase deficiency; phosphohexose isomerase; pyruvate kinase deficiency.

Hemolytic disease   *See* erythroblastosis fetalis; Rh blood type; Su blood type.

hemolytic uremic syndrome (HUS)   Hemolytic anemia, hypertension, and acute kidney failure are the main clinical symptoms. In the dominant form (1q32), complement factor H is defective. *See* complement.

hemophilia   Hereditary bleeding disease caused by a deficiency of antihemophilic blood-clotting protein factors. The classic *hemophilia A* (deficiency of antihemophilic factor VIII, encoded at Xq28) has a prevalence of about 1/10,000. The estimated mutation rate is about $2–3 \times 10^{-5}$. Somatic mosaicism may occur in 25% of patients (Leuer, M., et al. 2001. *Am. J. Hum. Genet.* 69:75). The most famous case of hemophilia involved the descendants of Queen Victoria of England. She was heterozygous for the gene (probably through a new mutation) and transmitted it through marriage to the Russian, German, and Spanish royal families (See pedigree chart).

The inept handling of the social problems by Tzar Nicholas II of Russia has been attributed partly to his preoccupation with and worries about the affliction of his son, Tzarevitch

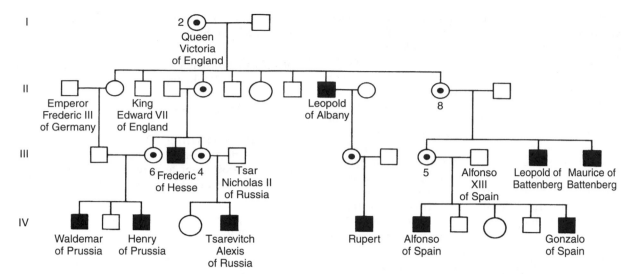

The family of queen Victoria of England. Hemophiliac male descendants are represented by black squares. The circles with a dot inside stand for the carrier females, 8: Beatrice, 7: Alice, 6: Irene, 5: Victoria, 4: Alexandra, 3: granddaughter Alice.

Alexis, by hemophilia. Thus, a single recessive gene might have contributed indirectly to the largest upheaval in the social order of the world. The so-called *Christmas disease* (named after the family Christmas [deficiency of antihemophilic factor IX]), hemophilia B, is also an X-chromosome-linked disease (Xq27.1-q27.2). Hemophilia B is about one-fifth to one-tenth as frequent as the classic hemophilia. These diseases are generally expressed in the males (because of X-chromosomal hemizygosity). Heterozygotes also have generally lower amounts of the proteins responsible for the condition, and thus the carriers can be identified in most cases. The estimated average mutation rate for hemophilia B is $\mu = 7.73$, for males $\nu = 18.8$ and for females $u = 2.18$ per gamete per generation, $\times10^{-6}$ (Green, P. P., et al. 1999. *Am. J. Hum. Genet.* 65:1572). Transitions are 7.3×10^{-9}, transversions 6.9×10^{-9}, deletions/insertions 3.2×10^{-10}. (Gianelli, F., et al. 1999. *Am. J. Hum. Genet.* 65:1580.)

Temporarily effective therapy involves treatment with antihemophilic factor. For gene therapy — using Moloney retroviral or adenovirus vectors — the B domain of factor VIII is deleted from the cDNA and only about a 170 kDa section is used from the total of 293 kDa. The transcription of factor VIII sequences may be repressed presumably by MAR action. Gene therapy has also been explored for factor IX (Christmas disease), and the adeno-associated vector carrying the factor IX gene into skeletal muscles had beneficial effects and no toxicity in phase I clinical trials. The latter has the advantage of being a much smaller protein (454 amino acids). In certain cases, the therapy is difficult because patients may produce significant amounts of the so-called circulating anticoagulant (IgG-type antigen), which counteracts the beneficial effects of the clinically transfused blood or the supplied blood factors. Prenatal diagnosis is potentially feasible after 10 weeks by DNA analysis and physical mapping or blood samples can be studied for factor VIII. *See* adeno-associated virus; adenovirus; afibrinogenemia; antihemophilic factors; blood clotting pathways; clinical trials; coumarin-like drug resistance; dysfibrinogenemia; fibrin-stablizing factors; gene therapy; Glanzmann's disease;

Hageman trait; hemostasis; MAR; parahemophilia; platelet abnormalities; prothrombin deficiency; pseudohemophila; PTA deficiency; retroviral vectors; Stuart factor; thrombopathia; thrombopathic purpura; transition; transversion; vitamin K–dependent clotting factors; von Willebrand's disease; X-chromosomal linkage. (*New England J. Med.* [2001], 345[14].)

hemopoiesis   Same as hematopoiesis.

hemorrhage   Bleeding.

hemostasis   Checking or arrest of blood flow. Hereditary diseases involved in hemostasis are hemophilias, platelet abnormalities, Nishimina factor, Tatsumi factor, Fletcher factor, Dynia factor, Flood factor, thrombocytopenia, May-Hegglin anomaly, and a number of bleeding anomalies characterized by abnormal bleeding such as Fanconi's disease, Meekrin-Ehlers-Danlos syndrome, von Willebrand disease, osteogenesis imperfecta, hemorrhagic telangiectasia, Osler-Rendu-Weber syndrome, pseudohemophilia, and stroke. (McEver, R. P. 2001. *Thromb. Haemost.* 86:746.)

hemp   *See Cannabis sativa.*

henbane   *Hyoscyamus niger* ($2n = 33$, 34) member of the *Solanaceae* family of plants requiring both vernalization and long photoperiods for flowering. It is a source of hyoscyamine and scopolamine alkaloids (ca. 0.04% in the leaves). The Egyptian henbane *Hyoscyamus muticus* has a higher alkaloid content (0.5%) and also contains hyoscipicrin and choline.

Henbane leaf and flower.

These alkaloids are used as smooth muscle relaxants and sedatives. Hyoscyamine (synonym atropine) has a low lethal oral dose of 5 mg/kg in humans, LD_{50} in mice intravenously 95 mg/kg. *See* alkaloids; vernalization.

Henderson-Hasselbalch equation $pH = pK + \log[(A^-)/(HA)]$, where pH is the hydrogen ion concentration, pK is an equilibrium constant (1/K), (A^-) is a proton acceptor, and HA is a proton donor. This equation expresses the relationship between pH and the ratio of acid to base in a solution.

Hennigian cladistics Basically a version of the parsimony method of analyzing evolutionary pathways. *See* cladistics; maximum parsimony; parsimony. (Mishler, B. D. 1994. *Am. J. Phys. Anthropol.* 94:143.)

Henrietta Lacks *See* HeLa cell line.

Henry of Prussia Great-grandson of Queen Victoria of England afflicted with hemophilia. *See* hemophilia.

Hensen's node (primitive knot) Group of cells in the primitive streak contributing to the formation of the head, notochord, and endoderm of the embryo. *See* endoderm; notochord; organizer; primitive streak. (Shilo, B. Z. 2001. *Cell* 106:17.)

HEPADNA virus (hepatotropic, DNA virus) Liver-targeting viruses (~3 kb) that replicate via an RNA intermediate. *See* hepatitis.

HEPA filter High-efficiency particulate air filter traps particles, notably microbes, of the air.

heparan sulfate Repeated disaccharide units composed of glucosamine linked to uronic acid or to (sulfated) glucuronic acid or (sulfated) L-iduronic acid. Heparan sulfates may be present in mucopolysaccharides and thus in the extracellular matrix. Heparan sulfate proteoglycans are coreceptors in cell adhesion, motility, proliferation, differentiation, and morphogenesis and regulate tumor growth and metastasis. Bacteria, viruses, and parasitic protozoa bind to the cell-surface heparan sulfate proteoglycans. The enzyme heparanase can break down the heparan sulfate meshwork, facilitate the metastasis of cancer cells, and aid in angiogenesis. Heparanase inhibitors are therefore expected to interfere with the development and invasiveness of cancer. *See* angiogenesis; exostosis; glucuronic acid; glypican; iduronic acid; metastasis, mucopolysaccharidosis; Simpson-Golabi-Behmel syndrome; syndecan; Wingless. (Bernfield, M., et al. 1999. *Annu. Rev. Biochem.* 68:729; Liu, D., et al. 2002. *Proc. Natl. Acad. Sci. USA* 99:568.)

heparin Mucopolysaccharide consisting of repeated units of uronic acid (glucosamine) and glucuronic acid disulfate. It is secreted into the bloodstream primarily by the liver and it has anticoagulant effects. Heparin is employed as an anticoagulant medicine, and in the molecular biology laboratory it is used (with or without dextran) for in situ hybridization. In Southern hybridization, heparin is used as a blocking agent on nitrocellulose (but usually not on nylon) membranes. Heparin may inhibit metastasis of cancer cells. *See* heparan sulfate; in situ hybridization; metastasis; Southern hybridization.

heparin cofactor II (HCF2, 22q11) Several cofactors facilitate the anticoagulant function of heparin. *See* thrombophilia.

heparinemia Accumulation of heparin in the blood that often causes problems in blood clotting and thus bleeding. *See* blood clotting pathways; hemostasis; platelet anomalies.

hepatitis Neonatal giant cell hepatitis (8q21.3) is a hemochromatosis due to a deficiency in steroid biosynthesis. *See* hemochromatosis; steroid dehdrogenase.

hepatitis B virus (HBV) Member of the hepadnavirus family. The double-stranded DNA genome is ~3.2 kb. HBV is replicated through a process of reverse transcription: DNA $\Rightarrow$ RNA $\rightarrow$ single-stranded DNA $\Rightarrow$ partially double-stranded DNA. The viral DNA may be episomal or it may integrate into the cellular DNA of the host cell by random nonhomologous recombination. The virus does not kill the cell upon integration; rather it causes a stable transformation and replicates along with the chromosomes. Infection may increase liver cancer development 5- to 30-fold or more. It is endemic in Southeast Asia, tropical Africa, and along the Amazon River in South America. Transmission is by body fluids (oral ingestion, blood transfusion, sexual contacts, nursing) of the infected persons. Incubation period is 1–6 months. The disease involves fever, vomiting, jaundice, arthritis, etc. Some individuals completely recover but remain carriers and may have a high chance of developing cirrhosis or cancer of the liver. There are about 250 million chronic HBV carriers worldwide and about 1 million die annually from complications caused by the infection. The development of effective DNA immunization seems to be a desirable goal. Currently, vaccines based on viral subunits are being used. The HAV and HCV viruses are single-stranded RNA viruses. *See* immunization, genetic; liver carcinoma; retroid virus. (Ganem, D. & Schneider, R. J. 2001. *Fundamental Virology*, Knipe, D. M. & Howlery, P. M. eds., p. 1285. Lippincott Williams & Wilkins, Philadelphia.)

hepatitis C virus (HCV) Member of the Flaviviridae family, has a positive-strand RNA genome of ~9.5 kb nucleotides, encoding a polyprotein of 3,010–3,033 residues. A serine protease processes its polyprotein and this is required for its replication. An estimated 1–3% of the world population is infected by it. Chronic infection may lead to liver diseases and cancer. Immunization is not available because of the great variety of mutant forms present in the infected individuals. Currently, interferons (IF) are used as a medication, but many isolates are resistant to it because the E2 viral coat protein can inhibit the RNA-inducible protein kinase (PKR). IF normally activates PKR that would block the translation initiation factor eIF2α. Thus, E2 conveys resistance to interferon by facilitating protein synthesis in the cell in the presence of IF. Other pathways of resistance may also exist. HCV can replicate in cell cultures when adaptive mutation(s) occurs. *See* eIF2α; PKR; positive-strand virus. (WHO. 1999. *J. Viral Hepat.* 6:35; Barbato, G., et al. 2000. *EMBO J.* 19:1195.)

hepatis delta virus (HDV) Has a closed circular RNA genome of 1.7 kb packaged into folded and base-paired rods. It codes for two proteins that are edited from a single RNA transcript. The editing is carried out by a *double-stranded RNA adenosine deaminase*. This virus cannot be packaged without the hepatitis B virus. *See* hepatitis B virus.

hepatocyte Type of liver cell; they are arranged in folded sheets facing blood-filled spaces (sinusoids). The hepatocytes are responsible for synthesis, degradation, and storage of many substances. The detoxification in the hepatocytes, with the assistance of cytochrome P450, may also produce ultimate mutagens and carcinogens as part of the process. Hepatocytes also secrete the bile that mediates the absorption of fats. *See* hepatoma; scatter factor.

hepatocyte growth factor (HGF) Mitogen and morphogen that controls processes in liver and placental development. Its receptor is a heterodimeric transmembrane protein tyrosine kinase. It has numerous functions such as organ regeneration, angiogenesis, and metastasis of tumors. It is a "scatter factor" because it dissociates epithelial cells and stimulates cell motility. *See* metastasis; mitogen; morphogen; scatter factor. (Cao, B., et al. 2001. *Proc. Natl. Acad. Sci. USA* 98:7443.)

hepatoma Liver tumor; originally the transition stage between the generally benign adenoma and the malignant carcinoma of the liver. Ovarian hormones suppress hepatoma development; therefore, male mice and men show a 5-fold higher incidence of hepatocarcinomas. *See* chi elements; liver cancer.

hepatomegaly Enlargement of the liver. *See* glycogen storage diseases; lipodystrophy.

hepatotoxicity Toxicity to the liver.

hepes (N-2-hydroxyethypiperazine-N'-2-ethanesulfonic acid) Used for the preparation of buffers in the pH range 7.2–8.2.

heptamer *See* immunoglobulins.

her Mutation converts X0 *Caenorhabditis* males into females.

HER Hybrid excision and repair. *See* HEI.

HER2 Synonymous with ERBB2 (avian erythroblastosis leukemia viral oncogen homologue 17q21.2); *See* ERBB1. Also Her-2/ErB-2 protein tyrosine kinase associated in breast cancer with the kinase CHK/MATK (19p13.3). The HER-2

gene also encodes the autoinhibitor herstatin. *See* ETS; tumor-associated antigen. (Tzahar, E., et al. 1997. *EMBO J.* 16:4938.)

HER3 (c-Erb-b3) Human epidermal growth factor receptor tyrosine kinase. (Singer, E., et al. 2001. *J. Biol. Chem.* 276:44242.)

herbaceous Plant tissue without woody components.

herbal Old books with description of plants considered mainly for spice or medicinal use.

herbarium Museum collection of dried plant specimens that are classified, identified, and described.

herbicide Plant growth-regulating chemical. The first herbicides were synthetic auxins (dichlorodiphenyl trichloroacetic acid, 2,4-D). Some of the general-type weed killers, such as 3-amino triazole, were very potent human carcinogens, although they did not respond as positive in the majority of short-term mutagen-carcinogen assay systems. Some herbicides kill plants as germination inhibitors (preemergent weed killers such as atrazine). Atrazine interferes with electron transport in photosystem II (photosynthesis) and may disrupt sex hormones. About 20 species of weeds have developed resistance to atrazine since its introduction to agricultural practice in the 1950s. Atrazine has been replaced by glyphosat, an inhibitor of the biosynthesis of aromatic amino acids by blocking enolpyruvylshikimate-3-phosphate synthase (EPSP). No plant species has developed resistance to glyphosat so far. (Glyphosat is also effective against some protozoa, e.g., *Plasmodium falciparum*.) Sulfonylureas are selective herbicides of dicotyledonous plants and inhibit the acetolactate synthase enzyme (ALS) in the branched pathway (leucine, valine, isoleucine). By highly selective isolation, system-resistant mutants were obtained. The nonselective herbicide gluphosinate-ammonium (phosphinothricin, BASTA) specifically inhibits the glutamine synthetase enzyme (GS). The plants accumulate highly toxic levels of ammonia. The dinitroanilin herbicides (trifluarin, oryzalin) depolarize cytoskeletal tubulins. Three base mutations in the host target molecule result in resistance against these herbicides. Herbicide research took advantage of transformation techniques by developing bacterial EPSP transgenic crop plants (cotton, sorghum, maize, alfalfa, canola, tomato, sugarbeet) to make them resistant to glyphosat while the weeds can be eliminated. The resistance of cultivated plants against sulfonylureas could also be enhanced 4-fold by insertion of the ALS gene. The high level of activity of these enzymes permits the transgenic plants to escape death. BASTA resistance was engineered into plants by transformation with a gene from *Streptomyces* bacteria that inactivates the herbicide by acetylation, and toxic levels of ammonia do not accumulate. Hydantocidin, a spironucleoside, binds to the regulation site of adenylosuccinate synthetase and thus blocks purine biosynthesis in the cells.

There is always the possibility that the herbicide-resistance transgene can escape from the crop plants to related weed species by cross-fertilization through the pollen. This risk can be greatly reduced if the resistance transgene is inserted into chloroplast DNA. The large majority of plants do not transmit plastids through the pollen. Another potential advantage of chloroplast transformation is the increased number transgene copies per cell. Transformation of chloroplasts and mitochondria (by biolistic methods) is somewhat more difficult, however, than introducing foreign genes by agrobacterial vectors into the plant cell nucleus. *See* acetyl-CoA carboxylase; *Agrobacterium*; bialaphos; biolistic transformation; spironucleoside; transformation; transformation of organelles; tubulins; vectors. (Palumbi, S. R. 2001. *Science* 293:1786; <http://www.dnr.state.wi.us/org/land/er/invasive/info/herbicides.htm>.)

herceptin Anti-HER-2 antibody to control the development of breast and other cancers. *See* breast cancer; erbB; HER-2.

hereditary Biologically inherited because of the genetic constitution of the family. Being hereditary does not imply perfect transmission to and expression in the progeny. The observed phenotype depends on the gene concerned and the environmental factors modifying their expression. The gene(s) responsible for a phenotype may be transmitted without expression under some conditions(s) and the condition may still be hereditary. *See* congenital; epigenesis; expressivity; familial; heritability; penetrance; reaction norm.

hereditary nonpolyposis colorectal cancer (HNPCC, MSH2 chromosome 2p16 and in MLH1 chromosome 3p21) Accompanied by a high rate of mutation in microsatellite sequences. The mutability is caused by deficiency in mismatch repair controlled by four loci, MSH2, MSH6, GTBP, MLH1, and to a lesser extent by PMS1 and PMS2, homologous to microbial enzymes MutS and MutL. Hypermethylation of the human MLH1 promoter is a frequent cause of insufficient repair. Its prevalence is about 2×10^{-3}. The hereditary nature of HNPCC is determined by the Amsterdam criteria. *See* Amsterdam criteria; colorectal cancer; Gardner syndrome; gastric cancer; microsatellite; mismatch repair; Muir-Torre syndrome; polyposis; trinucleotide; Turcot syndrome. (Dela Clapelle, A. & Peltoraki, P. 1995. *Annu. Rev. Genet.* 29:329; Scott, R. J. 2001. *Am. J. Hum. Genet.* 68:118.)

hereditary tyrosinemia *See* tyrosinemia.

heredity Study of storage, transmission, and expression of genetic information. *See* genetics; inheritance; reverse genetics.

heregulin (glial growth factor/neuroglin differentiation factor/NDF) Transmembrane protein tyrosine kinase receptor ligand (44 kDa, encoded at human chromosome 8p22-p11) present in breast carcinoma and fibrosarcoma cell lines. It binds to the NEU/ERBB2 oncogene. *See* ERBB1 [ERBB2]; oncolytic virus. (Lee, H., et al. 2001. *Cancer Res.* 61:4467.)

HERG Human inward rectifying K^+ ion channel responsible for LQT2 syndrome, encoded in chromosome 7q35-36. *See* ion channels; Jervell and Lange-Nielson syndrome; LQT1; LQT3. (Wang, H., et al. 2002. *Cancer Res.* 62:4843.)

heritability (h^2) Generally defined as the ratio of the additive genetic variance and the phenotypic variance (narrow-sense heritability). In some cases, only the ratio of the total genetic variance and the phenotypic variance can be determined (broad-sense heritability). Heritability can be estimated by offspring-parent regression, intraclass correlation, and by special methods in humans. Heritability below 0.25 is considered low and above 0.75 is high. Generally heritability of genes barely affecting fitness (e.g., spotting of the fur coat) is higher than for those traits that are important for reproductive success (fertility). The broad-sense heritability of intelligence displays variations during development. In children it may be 40–50%, but by adolescence it may climb to 60–70% or higher. The estimates of heritability are valid only for the population that provided the information. In different populations, different sets of polygenes may exist and they may provide different heritability estimates. In experimental animals and plants, generally the narrow-sense heritability is used because it is much more predictive. *See* behavior genetics; correlation; h^2, heritability, broad sense; heritability, narrow sense; heritability estimation in humans; intraclass correlation; QTL; realized heritability.

Selected heritability estimates in various animals and plants

HUMANS		CATTLE		MAIZE	
schizophrenia	0.75	white spots	0.95	plant height	0.51
epilepsy	0.50	milk production	0.43	kernel number	0.40
MOUSE		conception rate	0.03	yield	0.29
tail length	0.60	SWINE		ear number	0.20
litter size	0.15	litter number	0.20	SOYBEAN	
		weaning weight	0.10	maturity	0.75
DROSOPHILA				SOYBEAN	
abdom. bristles	0.50	CHICKEN		plant height	0.62
egg number	0.20	egg weight	0.75	oil percent	0.55
SHEEP		egg number	0.25	seed weight	0.54
wool length	0.55	viability	0.10		

heritability, broad sense Total genetically determined fraction of the phenotypic variance. *See* genetic variances; variance.

heritability, narrow sense Genetically determined fraction of the phenotypic variance, excluding nonfixable interactions such as those due to overdominance. *See* heritability; heritability, broad sense.

heritability estimation, human May be important in the study of various anthropometric traits and other polygenically determined conditions such as weight, height, behavior, various types of mental illness, epilepsy, heart disease, diabetes, etc. Since monozygotic twins are expected to be identical genetically, any variation between them is supposed to be environmental. Dizygotic twins are of the same age exactly but genetically as different as any other sibs. Thus, heritability (h^2) is calculated frequently as

$$\frac{\text{variance of dizygotics - variance of monozygotics}}{\text{variance of dizygotics}} = h^2$$

or

$$\frac{\text{percent monozygotic concordance} - \text{percent dizygotic concordance}}{100 - \text{dizygotoc concordance}} = h^2$$

or using the correlation coefficients (r) of monozygotic twins reared together (r_{mzt}) and reared apart (r_{mza})

$$\frac{(r_{mzt}) - (r_{mza})}{1 - (r_{mza})} = h^2$$

The interpretation in all of these cases has some limitations. The estimate is valid only if the variance is of the additive type because direct estimation, in the absence of controlled matings, cannot be carried out. In the presence of dominance variance, the genetic determination will be underestimated in the first two formulas. Because the common assortative matings in human populations leads to positive correlation between the parents, it overestimates heritability. The heritability estimates may be biased if major genes are also involved. Furthermore, environmental variation may not be the same for monozygotic (identical) twins as for dizygotic ones. Generally, it has been difficult to find enough twins for precise comparisons. Also, it must be kept in mind that in humans, just as in other organisms, heritability measured in one population, even for the same trait, may not be valid for another population. *See* confidence interval; correlation; h^2; heritability, broad sense; heritability, narrow sense; intraclass correlation; monozygotic twins; standard error; variance.

heritable translocation tests as bioassays in genetic toxicology

Animals exposed to mutagens are tested for sterility or semisterility and examined cytologically for multivalent association during diakinesis to metaphase I in the spermatocytes. Since many of these translocations are transmitted to the progeny, this procedure permits an assessment of the degree to which particular mutagenic agents increase the genetic load of mammals, potentially also of humans. *See* chromosome breakage in bioassays for genetic toxicology.

Hermansky-Pudlak syndrome

Rare, recessive human chromosome 10q23 disease, although in some endemic populations (Puerto Rico, Switzerland) its prevalence is $5-6 \times 10^{-4}$. Another locus is at 3q24. It involves pigmentation defects (ocular albinism, freckles but inability to get tanned), predisposition to bruising and bleeding, ceroid storage defects, large and abnormal melanocytes, lower platelet count, and defective lysosomes. Survival is usually limited to $20-25$ years. The defect seems to involve a transmembrane protein. This syndrome bears similarities to Chédiak-Higashi syndrome. *See* albinism; ceroid; Chédiak-Higashi syndrome; lysosomal storage diseases; lysosome; melanocyte; platelet. (Anikster, Y., et al. 2001. *Nature Genet.* 28:376; Huizing, M., et al. 2001. *Am. J. Hum. Genet.* 69:1022.)

hermaphrodite

Both male and female sex organs are present in the same individual. If the same plant bears both male and female flowers but individual flowers are either male (pollen producer) or female (egg producer), it is called monoecious. Hermaphroditism is the most common form of sexual differentiation in plants, but normally it is very limited in the animal kingdom (to flatworms, nematodes, some annelids and crustaceans, etc.). *True hermaphroditism*: the same individual develops both ovarian and testicular structures. Its frequency is low in mammals, including humans. The majority of human true hermaphrodites have

46XX constitution and about one-fourth are is 46XY; the remaining groups have sex chromosomal mosaicism and are considered to be males by appearance until puberty. In the majority of cases, *pseudohermaphroditism* is observed, i.e., the sex chromosomal constitution does not match the gonadal phenotype. The chromosomally XY individual appears feminine in many ways or the XX individual appears virile. Pseudohermaphroditism may arise by mutation in or translocation of the SRY gene, mutation in the Müllerian duct inhibitor substance gene, defects in the androgen receptor, or deficiency in the testosterone 5α-reductase. *See* freemartins; gonadal dysgenesis; intersex; Müllerian ducts; pseudohermaphrodite male; sex determination; sex ratio; sex reversal; testicular feminizaton; Wolffian ducts. (Mittwoch, U. 2001. *J. Exp. Zool.* 290:484.)

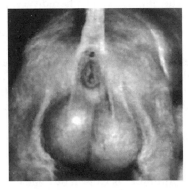

Photo of a hermaphrodite pig by courtesy of Dr. L. Lojda.

herpes

Family of double-stranded (linear) DNA viruses with $120-250$-kbp genetic material (HSV-1; ca. 75 genes); the capsid is made of about 30 polypeptides. The viruses replicate in the cell nucleus that protrudes through the nuclear membrane into the endoplasmic reticulum. There are more than 100 types of herpes viruses and many are serious pathogens; some benign forms are also common in humans. The herpes virus associated with Kaposi sarcoma encodes a chemokine (vMIP-II) that binds to a broad spectrum of chemokine receptors and thereby can block the entry of other viruses (e.g., HIV) into the cells. Type 1 is responsible for cold sores and type 2 for obnoxious, painful eruptions in the genital area. A member of the family, the Epstein-Barr virus, is responsible for nasopharyngeal (nose-throat) carcinoma. The cytomegalovirus that usually causes mild symptoms is also a herpes virus as well as equine and gallid herpes (the latter is responsible for Marek's disease of poultry [discovered by József Marek, Hungarian veterinary professor]). Herpes simplex virus 1 (HSV-1) is used as a therapeutic genetic vector to deliver to and express genesin in the central nervous system, for the selective destruction of cancer cells, and for prophylaxis against HSV and other infectious agents. The carrying capacity of the HSV vector is about $15-30$ kb DNA. The *E. coli* origin of replication permits its propagation in bacteria. For successful propagation in eukaryotic cells, helper HSV-1 is used, but it contains either mutations or deletions to avoid its lethal cytotoxic effects. Nevertheless, reversions may occur, but in the best helper stocks backmutation is in the 10^{-6} to 10^{-7} range. pHSV amplicon may also be propagated with the assistance of cosmid vectors, which may contain all of the essential genes of the virus in fragments but may lack the packaging signal. Therefore, lethal infectious virions cannot be reconstituted, yet assure they the propagation of the pHSV vector in various types of cells. The newer types of pHSVs may accommodate several genes, although generally passengers larger than 15 kb are undesirable. Also, several cell-specific promoters have been employed to facilitate more precise targeting. Another version of the pHSV carries a functional IE3 gene required for the production of the ICP protein (infected cell protein) regulators of transcription and translation. This gene is deleted in the helper virus and it can propagate only by complementation with pHSV. As a consequence, a higher proportion of pHSV is obtained after transfection. Another construct of pHSV contains the inverted terminal repeats (ITR) of the adeno-associated virus (AAV) as well as the AAV *rep* gene. Such a construct integrates specifically into human chromosomes at site 19q13 and behaves as a stable provirus without disrupting any essential function of the human genome. *See* adeno-associated virus; amplicon; chemokines; cleft palate; cytomegalovirus; Epstein-Barr virus; gene therapy; HVEM; Kaposi sarcoma; LAT; Marek's disease; shingles; viral vectors; VP16. (Boehner, R. E. & Lehman, E. R. 1997. *Annu. Rev. Biochem.* 66:347; Brune, W., et al. 2000. *Trends Genet.* 16:254; Albà, M. M., et al. 2001. *Genome Res.* 11:43.)

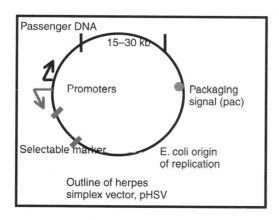

Outline of herpes simplex vector, pHSV

HERS disease   *See* glycogen storage diseases type VI.

Hershey-Chase experiment   Demonstrated that only the T2 phage DNA uptake was necessary to reproduce phage in *E. coli* and the phage coat protein was not required. Thus proved that DNA is the genetic material of this bacteriophage (Hershey, A. D. & Chase, M. 1952. *J. Gen. Physiol.* 36:39.)

Hershey circle   Nicked phage λ DNA circle in the bacterial host. *See* nick.

Hertz (Hz)   Unit of the number of cycles in alternating electric current. In the United States, the standard is 60 Hz. 1 megahertz (MHz) is 1 million pulses/second. Modern personal computers employ 300–600 or even much higher MHz microprocessors.

HERV   Human endogenous retrovirus. *See* endogenous virus. (<http://herv.img.cas.cz>.)

hesitation, transcriptional   *See* pausing, transcriptional.

HeT-A   Polyadenylated 6 kb retroposons (no LTR) specific for the telomeres of *Drosophila*, belonging to the family of LINE elements. The 2.8 kb open-reading frame encodes a zinc-finger protein but no reverse transcriptase. Thus, for transposition it is expected to depend on this function coded elsewhere in the genome. *See* LINE; retroposon; TART; telomere; zinc finger.

heteradelphian   Conjoined twins or other rare teratological anomalies such as extra limbs without a hereditary basis.

heteroalleles   Nonidentical alleles that may recombine (involve different nucleotide sites in a cistron). *See* allele; cistron.

heterobrachial   The two arms of the chromosomes are not identical in length. *See* chromosome arm; chromosome morphology; isobrachial.

Heterobrachial.

heterocaryon   *See* heterokaryon.

heterocatalysis   Catalytic process resulting in a product different from the starting material, e.g., the function of enzymes yields a compound different from the substrate or the enzyme. *See* autocatalytic function.

heterocellular   Body or tissue mixed of different cells.

heterochromatin   Parts of the chromosome that stain dark even during interphase (heteropycnotic). *Constitutive heterochromatin* remains in a highly condensed state in all cells of an organism. Heterochromatin usually reduces recombination and transposon insertion and may affect the expression of adjacent genes (position effect). It contains repetitive and methylated DNA that is usually not transcribed and translated; it is assumed to contain monotonous satellite DNA that is generally rich in AT sequences. It is frequently localized to both sides of the centromeres and to the telomeres.

The centromeric heterochromatin of *Drosophila* contains blocks of satellite DNA interrupted at 20 to 200 kb middle repetitive sequences and by single complete transposable elements. Certain chromosomal regions, such as most of the Y chromosomes and the B chromosomes, are heterochromatic. Constitutive heterochromatin can be seen as distinctly and characteristically located cross-bands of the chromosomes after stained with Giemsa or other stains.

Facultative heterochromatin may be in a relaxed state when it is expressing some information in cells at certain developmental stages but not under other conditions. The mammalian X chromosome, displaying dosage compensation, becomes heteropycnotic in the additional copies, e.g., in the normal females one of the two X chromosomes is heterochromatic, although both of the X chromosomes can be expressed in the males when they are present in a single dose (XY), or in the XO females the single X chromosome is not heteropycnotic. In XXX trisomic individuals, two are heteropycnotic and in XXXX superfemales three are heteropycnotic and show two or three Barr bodies, respectively, whereas in the normal females (XX) there is only one Barr body. Facultative heterochromatinization in the two X chromosomes permits the expression of one or the alleles of a gene associated with these chromosomes, resulting in a variegation called *lyonization*.

The function of heterochromatin has been investigated since the 1930s, but no general function could be assigned to it, probably because there are different functional properties of this material collectively identified as heterochromatin. Most commonly, regulatory and structural roles were attributed to heterochromatin. Lyonization proves some regulatory role. Position effect shows that genes transferred to heterochromatic regions can be silenced. This silencing may be permanent in stable position effect; presumably the transposed gene lacks the appropriate environment (promoter) for transcription. In case of variegation-type position effect (PEV), it was assumed that some binding proteins functioning between the homologous chromosomes have a trans-sensing role. The role of heterochromatin in crossing over remains confusing because both enhancing and suppressing effects have been observed. It has been suggested but not demonstrated conclusively that heterochromatin is a polygenic site, but the question remains of whether some of the redundant genes are heterochromatinized.

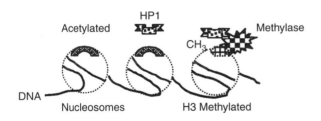

It came to light recently that centromeric heterochromatin (but not telomeric) has a role in the disjunction of achiasmate chromosomes. The GAGA transcription factor has been implicated in stimulating the variegation-type position effect of heterochromatin. The *kl*, *k*, and *cry* fertility genes of *Drosophila* seem to be in the Y-chromosomal heterochromatin, and the ribosomal gene repeats, *bb* (bobbed), can be found in both the X and Y chromosomes. The enhancer of the second chromosomal (2-54) segregation distorter (*Sd*) locus (Esd) and the activator of *Sd*, responder (*Rp*, 2-56.61), were localized in heterochromatin. The euchromatic abnormal ovule (*abo*, 2-44) mutations can be normalized by the *ABO* heterochromatic repeats in the X and Y chromosomes. About 13 autosomal genes have been associated with heterochromatin. The best-known component is heterochromatin-associated protein (HP1), which assists the expression of other heterochromatin genes. Methylation of histone H3 lysine creates a binding site for HP1 and contributes to gene silencing by the recruitment of histone methylase. At the amino end there is about a 50-amino-acid sequence (*chromodomain*) reminding of the Polycomb protein of *Drosophila*. The *chromoshadow*

domain is located at the C-terminus that is apparently involved in the organization of heterochromatic complexes. *See* achiasmate; banding; Barr body; boundary elements; bromodomain; centromere; chromodomain; cohesin; GAGA; histone deacetylase; histone methyltransferases; insulator; knob; locus control region; lyonization; nucleosome; PCNA; PEV; position effect; prochromosome; RAP polygenes; RIGS; satellite DNA; Sbf1; SET; silencer, telomeric silencing. (McCombie W. R., et al. 2000. *Cell* 100:377; Henning, W. 2000. *Chromosoma* 108:1; Henikoff, S. 2000. *Biochim. Biophys. Acta* 1470:1; Nielsen, A. L., et al. 2001. *Mol. Cell* 7:729; Grewal, S. I. & Elgin, S. C. 2002. *Current Opin. Genet. Dev.* 12:178.)

heterochromosomal recombination Involves a segmental exchange between repeats within nonhomologous chromosomes. *See* homologous recombination; illegitimate recombination.

heterochronic expression Of genes, controls timing of developmental events.

heterochronic mutant Expresses its genetic information with a timing pattern different from that of the wild type or it regulates development by a (+) or (−) control. Heterochrony has both developmental and evolutionary significance. Heterochrony also bears resemblance to homeotic mutations affecting spatial morphology rather than temporal changes. Heterochronic mutations may orchestrate cell lineage patterns, timing of hormone production, and/or the phasing of the developmental program(s). By studying heterochrony of development, evolutionary patterns in different species can be revealed. *See* cell lineages; homeotic genes.

heterochronic RNA Regulates the transition from one developmental phase to another. These RNAs occur in the majority of species with some exceptions, e.g., *E. coli*, *Saccharomyces cerevisiae*, and *Arabidopsis*. In *Caenorhabditis*, the 22-nucleotide *lin-4* RNA controls the transition from the first to the second larval stage. For the progress from the late larval stage to adult cell fates, the 21-nucleotide *let-7* RNA is required. These two RNAs are not homologous to each other, but they are complementary to the 3′ untranslated sequences of protein-coding genes and negatively affect their expression. These RNAs have single or multiple homologous sequences in humans, *Drosophila*, zebrafish, etc. *See* RNAi. (Pasquinelli, A. E., et al. 2000. *Nature* 408:86.)

heteroclitic antibody Binds to an antigen that is different but similar to the one which induced its formation.

heterocylic amines Include mutagens and carcinogens. Some are formed during high temperature cooking and frying of meats and fish.

heterocyst Terminally differentiated cell, e.g., the nitrogenase-protecting cells of filamentous cyanobacteria. *See* nitrogen fixation. (Wolk, C. P. 1996. *Annu. Rev. Genet.* 30:59.)

heterodimer Protein with two different polypeptide subunits.

heterodisomy *See* uniparental disomy.

heterodox chromosomes Special structures such as polytenic, lampbrush, sex, etc., chromosomes.

heteroduplex Base-paired polynucleotide chains with the two strands of different origins; may not be entirely complementary. In case the heteroduplex area includes different alleles, postmeiotic segregation may occur, resulting in aberrant ascospore octads or sectorial colonies from single (haploid) spores. *See* Meselson-Radding model of recombination.

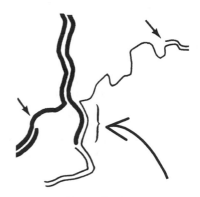

Heteroduplex.

heteroduplex analysis (heteroduplex tracking, HTA, HMA, QHTA) Method of mutation detection. DNA sequences containing even a single mismatch may move more slowly in the electrophoretic gel when compared with a corresponding homoduplex. *See* electrophoresis; mismatch; mutation detection; SNIP.

heteroencapsidation (transencapsidation) The viral coat and the enclosing genetic material do not match by origin.

heterofertilization Polar nucleus and egg of plants fertilized by genetically different sperms, and therefore the egg and the endosperm become nonconcordant. *See* embryo sac; gametophyte.

heterogametic sex Produces both X (Z) and Y (W) chromosome-containing gametes, or gametes with X and without X if the diploid has the chromosomal constitution XO. *See* MR; sex determination.

heterogamy Disassortative mating; alternation between sexual and parthenogenetic modes of reproduction. *See* disassortative mating; homogamy; parthenogenesis.

heterogeneity A population or family is segregating for genes of one or more loci. Appearance of phenocopies may confound the genetic conclusions. *See* phenocopy.

heterogeneous nuclear RNA *See* hnRNA.

heterogenote *See* endogenote.

heterograft The donor and the recipient of the graft are different species. *See* allograft; graft; grafting in medicine; isograft.

heterohybrid DNA One of the annealed strands is methylated; the other is not.

heterohybridoma (trioma) Obtained by fusing human lymphoid tumor cell with mouse myeloma cell. The purpose of generating such cells is to obtain cell lines, which produce more and more stable antibodies than the lines obtained by fusion with rodent cells alone. The production of human monoclonal antibodies may have several problems. There are ethical obstacles to immunize humans with disease markers. It is difficult to harvest appropriate immune cells from spleen or lymph nodes, and when peripheral blood is used, it contains relatively few B lymphocytes. Furthermore, the B blood cells carry IgM on their surface and therefore display low-affinity IgM antibodies. In addition, the human myeloma cells proliferate slowly and may express their own immunoglobulins, thus making purification difficult. Using murine myeloma cells may be only a partial solution, although they may not secrete their own antibody; but they may cause the loss of human chromosomes, resulting in the inability of human antibody production. *See* Epstein-Barr virus; hybrid; hybridoma; immunoglobulins; lymphocytes; monoclonal antibody; myeloma; somatic cell hybrids. (Jessup, C. F., et al. 2000. *J. Immunol. Methods* 246:187.)

heteroimmune When a lysogenic bacterium has a normal prophage and the λdgal transductant, and the two have different repressors, the transduction is heteroimmune. One species of animals was immunized by another species or the antigen (of any type) used evokes a pathological change. *See* HFT; lambda; λdgal; specialized transduction. (Yoshida, Y. & Mise, K. 1984. *Microbiol. Immunol.* 28[4]:415.)

heterokaryon More than one nucleus per cell of more than one type.

Cell.

This is of common occurrence in fungi when two different haploid cells undergo plasmogamy or when somatic cells of other organisms fuse and the cell fusion is not followed by fusion of the nuclei. *See* dikaryon; fungal life cycle; homokaryon; somatic cell hybrids. (Beadle, G. W. & Coonradt, V. L. 1944. *Genetics* 29:291.)

heterokaryon incompatibility Fusion of opposite mating-type (A and a) vegetative hyphae results in growth inhibition or cell death. *See* fungal incompatibility; heterokaryon; hypha; incompatibility.

heterokaryon test Two marked cells are fused and subsequently uninucleate cells are selected. If the nuclei are not fused and recombined but the uninucleate progeny still carries both parental markers, the markers must be carried by the cytoplasm.

heterolabel   Sister chromatid exchange can be identified if the two sister chromatids are different by harlequin staining. *See* harlequin staining of chromosomes; sister chromatid exchange.

heterologous probe   Used for the identification, localization, isolation, and cloning of specific genes in an organism by employing a labeled (radioactive, fluorescent) nucleotide sequence of presumably similar structure and function from another species. *See* probe.

heteromorphic bivalent   Morphologically distinguishable members of a homologous chromosome pair. *See* bivalent.

Heteromorphic bivalents.

heteromultimeric   Proteins with nonidentical subunits encoded by different genes.

heteroplasmy   The extranuclear genetic material within a eukaryotic cell is not homogeneous but contains genetically different components in analogy to heterozygosis. Pathogenic mutations in the mtDNA occurred at a frequency of $\sim 1.3 \times 10^{-4}$ in England. Selection would be expected to eliminate inferior mitochondria, yet some variation (neutral?) seems to be maintained, although in the majority of cells or bodies homoplasmy seems prevalent. Heteroplasmy for mitochondrial encephalopathy was detectable in 87% of the affected tissue (skeletal muscles) but only 0.7% in the unaffected blood. Deletions of mtDNA are rarely transmitted to the offspring, but point mutations (MELAS) reappear by random drift. The exhumed remains of Tsar Nicholas of Russia and his brother Georgij Romanov's mitochondria displayed two populations of mtDNA with both C and T at position 16,169, respectively. Human diseases involving defective mitochondria in a heteroplasmic state may potentially be remedied temporarily by selective destruction of defective mtDNA using a specific restriction enzyme. *See* chloroplast; DNA fingerprinting; homoplasmy; mitochondria; mitochondrial diseases in humans; paternal leakage; Romanovs. (Chinnery, P. F., et al. 2000. *Trends Genet.* 16:500; Brown, D. T., et al. 2001. *Am. J. Hum. Genet.* 68:533; Srivastava, S. & Moraes, C. T. 2001. *Hum. Mol. Genet.* 10:3093.)

heteroplastidic   Within the same cell more than one type of chloroplast exists as a consequence of mutation in ctDNA or cell fusion. *See* chloroplast genetics; ctDNA.

heteroploid   The chromosome number deviates from the normal.

Heteroplastidic cells.

heteropolymer   Synthetic polynucleotide containing more than a single type of base.

heteropolysaccharide   Contains more than one kind of monosaccharide.

heteropycnosis   One chromosome or part of it is darkly stained when the rest of the chromosome absorbs little or no stain at a particular stage. Dark staining at interphase is an indication of tight coiling and being in an inactive state. *See* Barr body; heterochromatin; pycnosis. (Zaccharias, H. 1990. *Chromosoma* 99:24.)

heteroscedasticity   Heterogeneity (inequality) of the variances in a group of samples.

heteroselection   In a population, selection for heterozygotes may prevail and improve its ability to respond to new environmental challenges. *See* homoselection; selection.

heterosis   Heterosis is frequently defined as the superiority of the F_1 hybrids over the midparent value. This definition expresses heterosis as Σdy^2, where d is the dominance effect at all loci and y is the difference of gene frequency between the two parental populations. This mathematical definition assumes that only dominant genes are responsible for heterosis. In agricultural practice, the breeders and growers require that the hybrids surpass in performance any known parental types, and overdominance and epistasis may also be involved (*See* hybrid vigor). Heterosis was known for ages and was systematically exploited in agricultural practice by cross-pollination between open-pollinated, distinct varieties of maize. Since the 1920s, inbred lines of maize have been used for the production of single-cross hybrids. A single-cross is the F1 generation of two inbreds. Inbreeding reduces vigor of the inbreds, but the single-cross may surpass not just the parental inbred lines but also the open-pollinated varieties from which the inbreds were isolated by several years of selfing in plants or brother-sister mating in animals. During the process of inbreeding, a progressive move toward homozygosity takes place. The different inbreds become homozygous for different alleles of the population. Therefore, not all inbreds are capable of contributing valuable genes to show heterosis. The breeders must select for inbred lines that produce superior hybrids, i.e., they have good combining ability. *Specific combining ability* means that a certain inbred contributes to valuable hybrids only with a particular other inbred line. *General combining ability* implies that an inbred yields high in combination with many other inbreds.

Heterosis was most successfully exploited in maize breeding where it was impractical and expensive to use F1 hybrids for commercial production. In the late 1910s, it was discovered that *double-crosses* were also very productive and it was much more economical to generate seeds in large quantities for agronomic production. A double-cross is the intercross of two F1 lines such as $(A \times B) \times (C \times D)$. In order to obtain 100% hybrid seeds, the plants had to be detasseled; the male inflorescence, the tassel, had to be cut off before shedding pollen, and when the female inflorescence (silk) was receptive, it had to be pollinated. This manual operation was very inefficient and costly. When cytoplasmic male sterility and fertility restorer genes were discovered, cross-pollination could be carried out economically on a large scale. Before hybrid corn was introduced into commercial production, the yield per hectare in the United States was about 1,400 kg. By the time the use of hybrid corn became practically universal, the yield increased 5- to 10-fold. Heterosis is now exploited in many other plant species, including some autogamous species such as tomatoes. Hybrid vigor is commercially utilized in the poultry and pork industry. Both positive and negative heterosis of molecules has also been revealed. Negative heterosis is generally hybrid inviability in interspecies or laboratory-bred offspring. The latter has been exploited for genetic sterilization of pests. *See* cytoplasmic male sterility; genetic sterilization; hybrid vigor; inbreeding; maize; overdominance and fitness; tassel. (Gowen, J. W., ed. 1952. *Heterosis.* Iowa State College Press, Ames, IA; Li, S. L. & Rédei, G. P. 1969. *Theor. Appl. Genet.* 39:68; Rhodes, D., et al. 1992. *Plant Breed. Revs.* 10:53; Comings, D. E. & MacMurray, J. P. 2000. *Mol. Genet. Metab.* 71:19.)

heterospecific   Belonging to another species.

heterosporic   Produces both micro- and macrospores. *See* macrospore; microspore.

heterostyly   The anthers in a flower are at a height different (generally lower) from that of the stigma in order to avoid self-pollination (inbreeding). *See* incompatibility alleles.

heterotaxy   Partial asymmetries in the placement (situs) of single visceral organs, e.g., different organs are oriented independently. *See* connexin; isomerism; situs inversus viscerum.

heterothallism   In lower eukaryotes (fungi) it is comparable to dioecy in higher plants, or bisexuality in animals, i.e., the plus and minus mating-type spores are carried by different individuals (thalli, colonies). This definition does not require that the species must have sexual organs (which many fungi lack), nor are these spores immediate meiotic products. In the case of *relative heterothallism*, the union of genetically different nuclei is favored. *See* homothallism.

heterotopia   Misplaced location of a tissue or organ. Hereditary nodular, periventricular heterotopia (Xq28) in the cerebral cortex is a serious health problem that results in epilepsy. The basic defect is in filamin-A synthesis. *See* filamins.

heterotopic graft   A piece of tissue is transplanted to a location different from its original site. *See* graft.

heterotroph   Cannot meet all of its requirements from inorganic nutrients.

heterotropic   An allosteric enzyme needs more than its substrate for modulation. *See* allostery; modulation.

heterozygosity, average   Varies in an outbreeding population in *Drosophila* as well as in humans. Many genes are homozygous, whereas heterozygosity at some other loci may vary from 0.10 to over 0.70. In a random mating population, heterozygosity (H) for a locus can be estimated:

$$H = 1 - \sum_{i=1}^{k} x_i^2$$

where x_i = frequency of the ith allele, k = the number of alleles, e.g., for the frequencies of the alleles A1 (0.4), A2 (0.2), A3 (0.3), and A4 (0.1), then $H = 1 - (0.4^2 + 0.2^2 + 0.3^2 + 0.1^2) = 1 - 0.3 = 0.70$. In case the size of the population is very small, the following formula may give a somewhat better estimate:

$$H = 2n/(2n - 1)(1 - \Sigma \hat{x}_i^2)$$

where $\hat{x}_i$ = the frequency of the ith allele and n = the number of alleles studied. *See* counseling genetic; utility index of polymorphic loci in genetic counseling.

heterozygosity-fitness correlation   Model assumes that heterozygosity is a major contributor to fitness in a population. An opposing assumption is that fitness is due to the association between certain marker genes and selectively advantageous loci. *See* fitness. (David, P. 1998. *Heredity* 80:531.)

heterozygote   Individual with different alleles at one or more gene loci. Segregation identifies it genetically. DNA sequencing may identify it molecularly. *See* homozygote.

heterozygote advantage   The fitness (the reproductive value) of the heterozygotes exceeds that of both types of homozygotes in a population, and this may lead to balanced polymorphism. *See* fitness; heterosis; hybrid vigor; inbreeding coefficient; selection coefficient.

heterozygote proportions   In the F_2 of a monofactorial cross, the genotypic proportions are 1AA:2Aa:1aa. If propagation is continued by selfing, the frequency of homozygotes will increase (inbreeding) and that of the heterozygotes will decrease, as discovered by Mendel. The proportion of each homozygote (AA or aa) relative to the heterozygote (Aa) will be in compliance with the formula: $[(2^{n-1}) - 1]:[2]$, where n stands for the number of filial generations. Applied to $F_2:[(2^{2-1}) - 1]:[2] = 1:2$; and to $F_6:[(2^{6-1}) - 1]:[2] = 31:2$. Note that the F_1 was not produced by selfing. The proportion of both types of homozygotes combined relative to the heterozygotes will be 2:64 at one locus. The same results can be obtained another way: 0.5^{n-1}; thus, the frequency of heterozygotes at a single locus in F_6 will be $0.5^{6-1} = 0.03125$, which is the same as 2/64 obtained above. In case the average effective population size (N_e) is known, the heterozygosity remaining after n generations can be determined as $[1 - 1/(2N_e)]^n$. *See* effective population size; inbreeding; inbreeding, progress of.

heterozygous   In a diploid or polyploid, the alleles at a locus are not identical; similar conditions may exist in haploids (merozygotes) carrying a duplication or a plasmid with the same locus. *See* homozygous; Mendelian segregation.

HETS   Heterozygotes.

He-T sequence   Repetitive DNA in the heterochromatin and near the telomeres of *Drosophila*. *See* heterochromatin; telomere.

heuristc search   Sequential estimation of the shortest branches of an evolutionary tree starting with a single individual, then choosing others. It is usually carried out with the PAUP computer program. *See* evolutionary tree; exhaustive search; PAUP.

Hevea brasiliensis (rubber plant)   Major source of latex in plants grown mainly in Asia and West Africa for the production of natural rubber (a polymer of cis-1,4 polyisoprene $[C_5H_8]_n$). The origin of the plant is South America. The basic chromosome number $x = 18$, but 9 has also been suggested. Some of the species are either diploid, $2n = 36$, or triploid, $2n = 54$.

hexaploid   Has six basic sets of chromosomes (genomes) in its cells (chromosome number $= 6x$). Hexaploids are generally allohexaploids, i.e., each pair of the sets has a different evolutionary origin, e.g., in the somatic cells of the common bread wheat (*Triticum aestivum*), $2n = 42$, there are 6×7 chromosomes of AABBDD genomic composition. Autohexaploids would have problems with multivalent association, resulting in unequal disjunction and sterility. *See* allopolyploid; polyploidy.

hexasomic   One particular chromosome is present in six copies within a cell. *See* aneuploid.

hexokinase   Phosphorylates glucose to glucose-6-phosphate using an ATP donor. HK1, HK2, and HK3 genes are in human chromosomes 10q22, 2p12, and 5q35.2, respectively. HK deficiency may lead to hemolytic anemia.

hexon   Capsomer with six neighbors in the viral capsid. *See* capsomer; penton.

hexosaminidase A and B   Lysosomal enzymes involved in the breakdown of ganglioside sphingolipids, forming about 6% of the membrane lipids in the gray matter of the brain and present in smaller amounts in other tissues. *See* gangliosides; lysosome; Sandhoff's disease; Tay-Sachs disease.

hexose   Sugar with 6-carbon backbone, e.g., glucose or fructose.

hexose monophosphate shunt   *See* pentose phosphate pathway.

Hfr   High-frequency recombination strain of a bacterium. The sex element (F plasmid) is integrated into the bacterial chromosome, resulting in about a 1,000-fold more efficient transfer of the bacterial chromosome with the integrated F^+ element into the F^- recipient cell. The transfer of the chromosome makes recombination possible, but unlike in eukaryotic crossing over, only one of the recombinant strands is recovered. The sex factor and the bacterial chromosome are transferred in a unidirectional manner. The F element can be integrated into the bacterial chromosome at different map positions and it may be transferred either clockwise or counterclockwise (*See* diagram). The transfer involves a process of a rolling circle-type replication. The genes closest to the initiation point of transfer are transferred first in proportion to their distance from that point. The transfer point is generally in the middle of the sex element, so for the recipient cell to receive the intact F requires the transfer of the entire Hfr plasmid. The rate of transfer depends on the nature of a particular Hfr element. The complete transfer of the entire bacterial genome plus the F element is about 4.7 million nucleotides and requires about 90–100 minutes. *See* bacterial recombination frequency; conjugation; conjugation mapping; F plasmid; *mob*; rolling circle replication. (Hayes, W. 1953. *Cold Spring Harbor Symp. Quant Biol.* 18:75.)

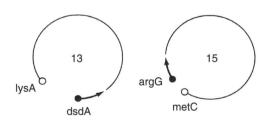

Two different Hfr groups, 13 and 15. The arrows on the dots indicate the direction of the transfer and the gene first transferred. The open circles show the position of the markers last transferred. In *E. coli*, more than two dozen Hfr strains are known.

HFS cell   Human fibroblast cell transformed by replication/transcription origin–deficient SV40 DNA and produces T antigen (Tag). Thus, HFS cells can provide helper functions for replication to high-copy-number minireplicons containing only viral replicational origins. *See* replicon; SV40.

HFT   High-frequency transducing lysate. Normal lambda phage and the defective λdgal phage can transduce bacteria. Upon induction, both λdgal and normal phages can be released at about equal frequencies because the wild-type phage can provide the information missing in the defective λ DNA. The two types of particles can be separated by CsCl density centrifugation because the density of the wild-type particles is about 1.7 g/cm³, whereas that of the defective particles may be as low as 1.3. *See* high-frequency lysate; lambda phage; LFT; specialized transduction.

HGF Hybridoma growth factor; hepatocyte growth factor. Its receptor is the product of the Met oncogene. Hepatocyte growth factor may interfere with the neoplastic activity of the c-myc oncogene in mice. *See* hepatocyte; hepatocytc growth factor; HNF; hybridoma; macrophage-stimulating factor; Met; Ron; signal transduction.

HGI Human gene index: <http://www.tigr.org/tdb/hgi/index. html>.

HGMD Human gene mutation database: <http://www. uwcm.ac.uk/uwcm/mg/hgmd0.html>.

HGP Human genome project. *See* genome projects. (<http://www.ncbi.nlm.nih.gov/genome/ guide>.)

HGPRT Same as HPRT. *See* hypoxanthine-guanine phos-phoribosyltransferase.

HhaI Restriction enzyme has the same recognition site as CfoI: GCG↓C.

HHV8 (KSHV) Human herpes virus 8 (~165 kb); it is consistently associated with Kaposi sarcoma. It has no in vitro transmission. It may inactivate the complement, induce IL-6, MIP-1α, β, RANTES, BCL2, and cyclin D2, and inhibit interferon signaling. *See* BCL2; Cd21/CR2; cyclin D; interferon; IL6; interferon receptor; Jak-Stat; Kaposi sarcoma; MIP-1; RANTES; signal transduction.

HIC (Hypermethylated in cancer, 17p13.3); zinc-finger protein regulating embryogenesis and differentiation. It is deleted in Miller-Dieker syndrome. *See* Miller-Dieker syndrome.

hickory (*Carya* spp.) Hardwood forest and shade trees, 2*n* = 32, 36.

hidden Markov model (HMM) Probabilistic model for a protein or RNA family. In an HMM graph, the nodes are "states" and the edges are "transitions." The total probability is the sum over all paths. *See* Markov chain statistics.

Hieracium Genus of the Compositae family with many weedy species in America and Europe. Several of them have euploid chromosome numbers (e.g., *H. japonicum* 2*n* = 18). The apomicts have chromosome numbers 2*n* = 27, 36, and 45. These apomicts, studied by Gregor Mendel upon the recommendation of Carl Nägeli, professor of Botany at the University of Munich, caused great worries to Mendel with regard to the validity of his discoveries about inheritance because the apomicts failed to segregate as expected. Thus, Nägeli, despite his great fame, almost thwarted the development of genetics with plants as he also argued against the existence of hard heredity in bacteria

Hieracium auranticum. Herbal illustration.

by advocating his theory of *pleomorphism*, meaning that these organisms lacked fixed genetic material and varied freely. If Nägeli's theory would have been accepted, it would have also foiled the development of bacteriology and medicine. *See* Mendel's laws; pleomorphism.

hierarchical shotgun sequencing Procedure based on mapped clones generated by BACs. The main advantage of this conservative approach versus the whole genome shotgun procedure is that several laboratories can work simultaneously on the project. The International Human Genome Sequencing Consortium relied on it. *See* shotgun sequencing.

HIF Hypoxia-inducible factor. Transcription factor binding to the 5'-ACGTG-3' element and involved in oxygen homeostasis under anoxic (low-oxygen) conditions by inducing glycolysis, erythropoiesis, and angiogenesis. The HIF1A-α subunit is encoded at 14q21-q24; the HIF2 (EPAS1)-α subunit is at 2p21-p16. In the presence of the wild-type allele of HIF, hypoxia and hypoglycemia reduce cell proliferation controlled by p53, p21, and Bcl-2 proteins, but inactivated HIF genes do not affect p27 and GADD153. Some tumors containing HIF display reduced apoptosis and increased proliferation under anoxia because of the different response of genes to the homeostatic effect of HIF. *See* apoptosis; Bcl-2; erythropoiesis; GADD153; glycolysis; homeostasis; hypoglycemia; p21; p27; p53. (Wenger, R. H. 2000. *J. Exp. Biol.* 203:1253: Ramírez-Bergeron, D. L. & Simon, M. C. 2001. *Stem Cells* 19:279; Semenza, G. L. 2001. *Cell* 107:1.)

high-density lipoprotein (HDL) Located in particles rich in protein and relatively low in cholesterol and cholesteryl esters. HDL is the benign lipoprotein; atherosclerosis is inversely related to the concentration of HDL. Lipoprotein lipase and endothelial lipase mutations may result in lower levels of HDL. The SR-BI scavenger receptor controls the amount of HDL and the concentration of cholesterol in the bile cells. The familial high-density lipoprotein deficiency syndrome (FHA, 9q22-q31) is characterized by a susceptibility to coronary heart disease because of the low level of HDL cholesterol in the blood. The ABC transporters are apparently responsible for the normal balance between cholesterol export and LDL import. Mutation of the ABC1 transporter accounts also for Tangier disease. *See* ABC transporters; atherosclerosis; cholesterol; hypertension; LDL; lipase deficiency; lipids; Tangier disease. (Sacco, R. L., et al. 2001. *JAMA* 285:2729.)

high-dose/refuge strategy Designed to prevent the development of resistance in insect populations against toxins synthesized by transgenic plants for their protection against insects. The resistance gene is engineered to a high degree of resistance and dominant inheritance. Any mutation for countering the resistance is expected to be recessive and thus not selected for effectively because in the heterozygotes the recessive allele is not expressed. Or even if the counterresistance is intermediate, the high toxin level is expected to be lethal or deleterious to them. The refuge strategy is expected to reduce the genetic survival (fitness) of the counterresistance. The refuge is the presence of nontoxic plants that allow the reproduction of susceptible insects. When there are substantial numbers of susceptible insects, the rare recessive counterresistant individuals will mate with them and there is a reduced probability for mating between two counterresistant individuals. Therefore, the counterresistant phenotypes will not be propagated and will eventually be eliminated. This strategy is very attractive theoretically, yet under the conditions of the natural environment it may have limitations. *See Cochliomya hominivorax*; genetic sterilization; insect resistance in plants. (Rauscher, M. D. 2001. *Nature* 411:857.)

high-energy bond Upon hydrolysis, the covalent linkage liberates large amounts of free energy, e.g., phosphodiester bonds of ATP and thioester linkage in acetyl coenzyme A. *See* phosphodiester; thioester.

high-frequency lysate When a temperate bacteriophage excises from the bacterial chromosome, it may pick up an adjacent site, e.g., bacteriophage λ may gain a *galactose* gene from *E. coli* map position 17, but because of the reciprocal recombination, a piece of its own genetic material required for lysogeny may be left behind. Thus, the new phage becomes λ*dgal* (lambda-deficient galactose). Such a phage particle may multiply vegetatively, but its infecting ability is reduced. When a helper phage is inserted into the bacterial chromosome next to λ*dgal*, the wild-type phage gene may compensate for the defect and a *double lysogenic bacterium* is produced. Upon induction (to liberate phage), a *high-frequency lysate* containing the two types of the phages is produced with high transducing ability. *See* lambda phage; specialized transduction. (Hartman, P. E. 1963. *Methodology in Basic Genetics*, Burdette, B. J., ed., p. 103. Holden-Day, San Francisco.)

high-frequency transducing lysate *See* HFT.

highly repetitive DNA Contains a high degree of redundancy and reassociates very rapidly after denaturation. *See* annealing; c_0t value; LINE; SINE.

high-lysine corn The seed proteins of cereals is composed of four major fractions in variable but frequently closely equal amounts: prolamine (zein in maize, gliadin in wheat), glutenin, albumin, and globulin. In maize, the genetically determined lysine content is highly variable and subject to increase and decrease by selection for high or low zein content, respectively. Two families of genes, *opaq* (*o*) and *floury* (*fl*), scattered over the maize map are particularly influential in reducing the low-lysine protein fractions with the concomitant increase (doubling or more) of the percentage of this essential amino acid and that of tryptophan as well. In some *fl2* lines, the level of methionine is also substantially higher. Particularly beneficial effects are due to genes *o2*, *o7*, *fl2*, and *fl3*. In wheat in mol/10^5 g protein, the lysine contents are ca. gliadin 5.0, glutenin 17.6, albumin 78.4, and globulin 98.0. Cereals with improved nutritional values are desirable for feeding of animals and more importantly for the production of cereal food for human populations suffering from malnutrition as a result of protein deficiency in the diet (kwashiorkor). *See* essential amino acids; kwashiorkor. (Coleman, C. E., et al. 1995. *Proc. Natl. Acad. Sci. USA* 92:6828; O'Quinn, P. R., et al. 2000. *J. Anim. Sci.* 78:2144; Zarkadas, C. G., et al. 2000. *J. Agric. Food Chem.* 48:5351.)

high-mobility group of proteins (HMG) Associated with functionally active chromatin. These proteins render the genes more sensitive to DNase and probably to RNase II. HMGs also promote the elongation of RNA transcripts. They are regulated by cell-cycle-dependent phosphorylation, affecting their ability to bind to DNA. HMG proteins are important for growth and development. A large number of transcription factors contain HMG-like domains. The specificity of these proteins varies, but a common feature is that they distort DNA structure and have an affinity for distorted DNA structures. The change in DNA electrophoretic mobility is correlated with this altered structure. Thus, HMGs have both regulatory and structural functions. Recurrent rearrangements of the HMGI-C group were detected in some benign tumors (lipomas). There are three groups of these proteins: HMGIC (12q15), HMGI, and HMGI(Y) encoded at 6p21. *See* cell cycle; chromatin; coactivator; DNA bending; elongation factors; FACT; lipomatosis; nonhistone proteins; SOX; transcription factors. (Bustin, M. 1999. *Mol. Cell. Biol.* 19:5237; Reeves, R., et al. 2001. *Mol. Cell. Biol.* 21:575.)

high-performance liquid chromatography (HPLC) A mixture of compounds is applied to chromatographic columns with strong ion exchange resins. The solvent is forced through the resin under pressure for rapid and sharp separation of the components of the mixture. The eluates are electronically scanned and identified. Denaturing high-performance liquid chromatography can be used to detect structural differences between mutant and wild-type molecules. *See* chromatography.

high-quality sequence Final stage of the contiguously sequenced genome that has an error rate of only $\sim 10^{-4}$. *See* first-draft sequence; genome projects.

high-throughput analysis Permits the study of large amounts of facts by the use of (generally) automated equipment.

Hill reaction Illuminated chloroplasts evolve oxygen and reduce an artificial electron acceptor (ferricyanide →

ferrocyanide). It was an important tool to study the mechanism of photosynthesis, namely that the evolved oxygen comes from water rather than from CO_2. This reaction demonstrated that isolated chloroplasts can perform part of the reactions and revealed the light-activated transfer of an electron from one substance to another against a chemical-potential gradient. *See* photosynthesis.

Hill-Robertson effect Selection at one locus may hinder the choice of a favorable allele at another linked gene in a population in the absence of recombination. *See* genetic drag; hitchhiking. (Hill, W. G. & Robertson, A. 1966. *Genet. Res.* 8:269.)

hilum Depression or pit where vessels and nerves enter an organ; the place ($\downarrow$) where the plant seed is connected to its stalk in the fruit.

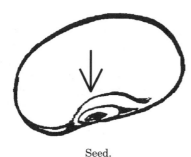

Seed.

him High incidence of male mutation. In *Caenorhabditis*, there is a high level of nondisjunction in XX hermaphrodites, thus producing X0 males. *See* Caenorhabditis; chromosomal sex determination; nondisjunction.

Himalayan rabbit Carries temperature-sensitive tyrosinase genes controlling pigmentation. The extremities, paws, ears, and tail, having lower blood circulation and concomitant lower body temperature, develop darker pigmentation. A similar pattern of pigmentation occurs in other rodents and in Siamese cats. Tyrosinase is a copper enzyme (also called polyphenol oxidase) involved in the formation of 3,4-dihydroxyphenylalanine (DOPA), which is responsible for the production of melanin in the hair and skin and darkening of wounded fruits and other plant tissues. *See* albinism; piebaldism; pigmentation of animals; Siamese cat; temperature-sensitive mutation.

*Hind*III   Restriction enzyme with recognition site A$\downarrow$AGCTT.

hindbrain (rhombencephalon) Caudal part of the brain including the cerebellum, pons (metencephalon), and medulla oblongata (myelencephalon). *See* brain, human.

HiNF Histone nuclear factor. A 48 K M_r protein identical to interferon regulatory factor IRF-2. *See* histones; IRF-2.

HINGE *See* antibody.

hinny She-ass ($2n = 62$) × stallion ($2n = 64$) hybrid. The reciprocal (mare × jackass) is called mule. Mules are easier to produce because the jackass willingly mates with the mare, but the stallion mates with the she-ass only under special circumstances (blindfolded). The hybrid's body more closely resembles the female parent as an apparent cytoplasmic influence. These sterile hybrids — known since the beginning of human civilization — may retain some sexual drive and occasionally fertility has been reported in backcrosses with either the jackass or the stallion. The jackass backcrosses are entirely sterile, but the backcrosses with stallions appear more normal. The advantages of the mule and hinny are that they are stronger yet as resistant to stressful conditions as the donkey, and they can thrive on much less feed than the horse. The adults are generally healthier than the horse and well adapted to work under primitive conditions. These hybrids do not require iron hoof plates (horseshoe). They are more obstinate than the horse, although they are shrewd and better self-disciplined under conditions scary to horses. *See* animal species, hybrids; mule. (Rong, R., et al. 1988. *Cytogenet. Cell Genet.* 47[3]:134.)

Mule.

HINT Autoproteolysis and protein-splicing domain. (Kroiher, M., et al. 2000. *Gene* 241:317.)

Hip 41.3 kDa cytosolic protein interacting with Hsc70 and homologous with DnaJ. *See* DnaK; Hsc; molecular chaperone interacting complex.

hip dislocation, congenital Autosomal-dominant complex with a prevalence of about 0.001/live birth. Generally involves laxity of joints.

Hippel-Lindau syndrome *See* von Hippel–Lindau syndrome.

hippocampus Ventral-temporal gray matter part of the brain including seven layers of tissues and the presumed site of memory-amnesia-learning deficit, etc. During mouse hippocampal development, 1,926 genes displayed dynamic changes by microarray analysis and seem to be involved in neuronal proliferation, differentiation, and synapse formation.

See brain, human; memory; microarray hybridization; synapse. (Mody, M., et al. 2001. *Proc. Natl. Acad. Sci. USA* 98:8862.)

hippopotamus (*Hippopotamus amphibius*) $2n = 36$.

HIR   Genes involved in histone and nucleosome assembly at 22q11.2. *See* histones. (Ray-Gallet, D., et al. 2002. *Mol. Cell* 9:1091.)

Hirschsprung's disease (megacolon)   Occurs in several forms involving megacolon (large and dilated terminal section of the intestine), microcephaly (reduced head size), short stature, and coloboma of the iris (the iris has two distinct colors); unequal development of the two sides of the face, obstructed anus (the terminal end of the intestine), and obstructed bladder were observed. The symptoms may not all occur in each individual and are attributed most commonly to autosomal-recessive mutation (chromosomal deletion 13q1-q32.1) and less frequently to autosomal-dominant mutation (human chromosome 10q11.2) in the RET oncogene (receptor tyrosine kinase), or even to multifactorial inheritance. The RET loss mutation may have dominant negative effects. A mutation at 9q31 appears to regulate the penetrance of RET. Mutations in the glial cell-derived neurotrophic factor (GDNF) encoded in human chromosome 5p may also have a minor role in the disease. Prevalence is about 1/5,000, but among sibs it is about 1/25. The chance for males to be affected is three to five times higher than females. Its incidence is high in Down syndrome (6%) and in piebaldism. This disease also occurs in horses and mice and probably in pigs. Endothelin-3 mutations may be responsible factors. In the complex Waardenburg-Hirschsprung disease called Waardenburg-Shah syndrome, mutations in the SOX10 gene have been detected. One form is controlled by the Smad-interacting protein (Sip1) encoded at 2q22. *See* aganglionosis; coloboma; diabetes mellitus; dominant negative; endothelin; eye diseases; hypogammaglobulinemia; piebaldism; psoriasis; RET oncogene; Shah-Waardenburg syndrome; Smad; SOX; stature in humans. (Wakamatsu, N., et al. 2001. *Nature Genet.* 27:369; Gabriel, S. B., et al. 2002. *Nature Genet.* 31:89.)

hirsute   Hairy.

HIS (HS2) oncogenes   Identified in mouse chromosomes 2 and 19 from leukemia cells. The human homologue was assigned to 2q14-q21. *See* leukemias; oncogenes.

***His* operon**   Coordinately regulated gene cluster. In *Salmonella typhimurium*:

gene order on the map: $E(2) \; I(3) \; F(6) \; A(4) \; H(5) \; B(7,9) \; C(8)$

$D(10) \; G(1)$

The numbers in parentheses after each gene indicate the order of the biosynthetic steps mediated by the genes. The substrates and products in the pathways are ordered below:
    (1) PHOSPHORIBOSYL PYROPHOSPHATE + ATP → **(2)** PHOSPHORIBOSYL-ATP → **(3)** PHOSPHORIBOSYL-AMP → **(4)** POSPHORIBOSYL-FORMIMINOAMINOIMIDAZOLE-4-CARBOXAMIDE RIBONUCLEOTIDE + GLUTAMINE →

(5) > ? →**(6)** IMIDAZOLE GLYCEROL PHOSPHATE + AMINOIMIDAZOLE-4-CARBOXIAMIDE RIBONUCLEO-TIDE → **(7)** IMIDAZOLEACETOL PHOSPHATE → **(8)** L-HISTIDINOL PHOSPHATE → **(9)** L-HISTIDINOL + PHOSPHATE → **(10) L-HISTIDINE**. *See* Alifano, P., et al. 1996. *Microbiol. Rev.* 60:44.

histamine   Decarboxylated histidine. A vasodilator (expands blood vessels) mediates the contraction of smooth muscles. It is released in large amounts as part of the allergic response. In allergic persons, an immunoglobulin E (IgE)–dependent histamine-releasing factor (HRF) is produced by lymphocytes. *See* allergy; basophils; hypersensitive reaction; mast cell. (Marek, J., et al. 2001. *Nature* 413:420.)

histidase (histidine ammonia lyase)   *See* histidinemia.

histidine kinase   Signal-transducing molecule of two components. One of the components, in response to environmental cues, autophosphorylates a histidine in the *catalytic* domain. Then the phosphate is transferred to an aspartate within the second component, which is a *response regulator*, binding to another cellular protein to induce a cellular response. The histidine kinase may be either a transmembrane signal receptor or a cytoplasmic protein receiving the signal indirectly from another transmembrane receptor. In yeast, the histidine kinase, SLN1, seems to be a transmembrane protein with the kinase domain in the NH_2 end region and it is joined to a response regulator domain at the COOH end. It appears that the sensor part is outside the cell and the kinase and the response regulator are within the cytoplasm. In yeast downstream of the histidine kinase (Sln1), the Ssk1 protein, normally phosphorylated by Sln1, is then inactive, but when it is not phosphorylated, it controls other proteins in the signal transduction pathway. Besides *SSK1*, the mitogen-activated *HOG1* gene also encodes another *SLN1* suppressor. Hog1 appears to be a homologue of MAP (mitogen-activated protein kinase) and Pbs2, which seems similar to the MAP kinase kinase (MAPKK). The *ETR1* (ethylene response) histidine kinase gene of the plant *Arabidopsis* is organized after the yeast pattern and this histidine kinase gene is a signal transducer for ethylene. The *CTR1* (constitutive triple response) gene downstream in the pathway encodes a serine-threonine kinase negative regulator of *ETR1*. *See* ethylene; histidine; histidine operon; MAP; phytohormones; signal transduction. (West, A. H. & Stock, A. M. 2001. *Trends Biochem. Sci.* 26[6]:369.)

histidinemia   Caused by a deficiency of histidase enzyme (histidine ammonia lyase) involved in the removal of a NH_3 from histidine. It frequently causes mental retardation and neurological abnormalities. Identification by enzyme assays is feasible prenatally. Its pattern of inheritance is not entirely clear; however, an autosomal-recessive gene was assigned to human chromosome 12q22-q23. Its prevalence is 1 in 20,000 to 40,000; however, in some isolated populations the heterozygote frequency was estimated to be as high as 3%. Histidinemia symptoms may also be caused by nongenetic factors. Histidase alleles were described in chromosome 10 of the mouse. *See* amino acid metabolism.

histidine operon   *See His* operon.

histidyl tRNA synthetase (HARS) Charges tRNA^{His} by histidine. Its gene is in human chromosome 5. *See* aminoacyl-tRNA synthetase.

histiocytoid lipidosis (familial cardiac lipidosis) Mitochondrial cardiomyopathy caused by accumulation of lipids in the heart muscle fibers. *See* cardiomyopathies; mitochondrial diseases in humans. (Reid, J. D., et al. 1968. *J. Pediatr.* 73:335.)

histiocytosis Heterogeneous autosomal-recessive lethal defect of the immune system involving uncontrolled activation of the T cells and macrophages, fever, enlargement of the spleen and the liver (hepatosplenomegaly), reduction in the number of blood cells (cytopenia), and neurological disorders. The HPLH1 gene was assigned to 9q21.3-q22 and to 10q21-q22. The erythrophagocytotic lymphohistiocytosis gene was mapped to 11p13 and to 11q25, respectively. The complex disease was also described as familial hemophagocytotic lymphohistocytosis (FHL, 10q21-q22) and familial histocytotic reticulosis. The basic problem in FHL appears to be due to a deficiency in perforin. *See* perforin. (Göransdotter Ericson, K., et al. 2001. *Am. J. Hum. Genet.* 668:590.)

histoblasts Cell groups that give rise to dorsal epidermis (tergites) and ventral epidermis (sternites), respectively. *See* epidermis.

histochemistry Determines tissue components on the basis of in situ analysis of their chemical reactions by color and/or by specific antibodies. Histochemical techniques are useful tools to study developmental mechanisms. *See* aequorin; Gomori's stain.

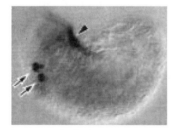

Histochemical localization of alkaline phosphatase enzyme in whole mount preparation of the annelide (*Tubifex*) embryo. (Courtesy of Professor T. Shimizu.)

histocompatibility *See* HLA (human leukocyte antigen); leukocytes.

histone acetylase *See* histone acetyltransaferases.

histone acetyltransferases (HAT) Acylate amino acid (lysine, serine) residues in the nucleosomes, contribute to remodeling of nucleosomal structure, and facilitate transcription. HAT has a preference for Lys5 and Lys12 or Lys16 (at dosage compensation) in histone 4. In the active chromatin, histone 2a is less acylated than in the inactive one. Acetylated histones permit better DNA access for transcription factors and other proteins modulating gene expression. Some histones are acetylated at the chromatidal location, but histone acetylation also takes place in the cytoplasm before entering the nuclei. The RNA polymerase II holoenzyme complex contains HAT activity. HATs may belong to the GNAT family of proteins (Gcn5-related acetyltransferases) of four conserved motifs. The MYST family (MOZ, Ybf2/Sas3, Sas2, Tip60) shares structural similarity with HATs. HATs may acetylate several nonhistone proteins. They play a role in development and in carcinogenesis. *See* ADA; bromodomain; CAC; CBP; chromatin remodeling; dosage compensation; E1A; elongator; GCN5; histones; *Msl*; NuA; nucleosomes; p300; PCAF; promoter; SAGA; TF$_{II}$ 230/250. (Imhof, A., et al. 1997. *Curr. Biol.* 7:689; Berger, S. L. 1999. *Curr. Opin. Cell Biol.* 11:336; Brown, C. E., et al. 2000. *Trends Biochem. Sci.* 25:15; Roth, S. Y., et al. 2001. *Annu. Rev. Biochem.* 70:81; Davie, J. R., et al. 1999. *Biochem. Cell Biol.* 77:265.)

histone code *See* histones.

histone deacetylase (HDAC) Component of the transcriptional suppression system in the inactive chromatin. Both HDAC1 and HDAC2 form complexes with DNA topoisomerase II. There are at least nine HDACs in the mammalian cells. The methyl-CpG-binding protein MeCP2 seems to be associated with histone deacetylase activity. The retinoblastoma protein may bind to the E2F transcription factors and the complex then apparently represses the promoters of cell-cycle S-phase genes. It may activate silenced genes such as those showing position effect by centromeric or telomeric heterochromatin, inactive X chromosomes in the imprinting mammalian females, and inactivated tumor suppressor genes. The steroid receptor coactivator (SRC-1) is required for the expression of these enzymes. Protein HDRP (HDAC-related protein) is associated with HDAC4, shares ~50% identity with the noncatalytic NH$_2$ domain of HDAC4 and HDAC5, and seems to be a transcriptional repressor. Histone deacetylation and acetylation are not limited to the nucleosomal site including the promoter gene, but both processes may have somewhat global effects. Blocking histone deacetylation by antisense construct in *Arabidopsis* resulted in pleiotropic expression of several genes. *See* chromatin remodeling; CpG; DHAC1; DNA methylation; E1A; E2F; HDA1; heterochromatin; histone; HOS; imprinting; lyonization; Mad; N-Cor; nucleosome; NuRD; RAR; REST; RPD; SAHA; signal transduction; Sin3/Rpd3; SMRT; SRC-1; topoisomerase; trichostatin; tumor suppressor gene; Xic; Xist. (Ahringer, J. 2000. *Trends Genet.* 16:351; Cress, W. D. & Seto, E. 2000. *J. Cell Physiol.* 184:1; Tanny, J. C. & Moazed, D. 2001. *Proc. Natl. Acad. Sci. USA* 98:415; Wade, P. A. 2001. *Hum. Mol. Genet.* 10:6693; Huang, X. & Kadanoga, J. T. 2001. *J. Biol. Chem.* 276:12497; Yang, W.-M., et al. 2002. *J. Biol. Chem.* 277:9447.)

histone fold The core histone octamer is composed of two tetramers (H2A, H2B, H3, H4) that share common motifs of two short α-helices and a long central helix connected by β-bridges. Each monomer dimerizes in a head-to-tail arrangement and interacts with the DNA. *See* histones; nucleosome. (Selleck, W., et al. 2001. *Nature Struct. Biol.* 8[8]:695.)

histone methyltransferases (HMT) Family of gene regulatory proteins. They repress gene expression and the cell-cycle transition G1→S by chromatin remodeling. SUV39H1 (Xp11.23, 412 amino acids) is homologous to the *Drosophila*

Su(var)3–9, which is a suppressor of variegation. Members of this family are ubiquitous in eukaryotes. SUV39H1 methylates Lys 9 in histone H3. It is associated with HP1, which is homologous to the two CBX-like proteins of 185 and 191 amino residues, respectively. HP1 targets the methyltransferase complex to the pericentromeric heterochromatin and to the promoters of genes to be silenced. SUV39H1 generally occurs in a complex with histone deacetylase required for methylation. SUV39H deficiency leads to chromosomal instability, increased risk of tumors, and male meiotic anomalies in mice. The retinoblastoma protein is also required for the activity of the complex, since it controls the expression of translation elongation factor E2F. G9a is a novel mammalian lysine-preferring HMTase. Like Suv39H1, the first identified lysine-preferring mammalian HMTase, G9a transfers methyl groups to the lysine residues 4, 9, and 27 of histone H3 at a 10–20–fold higher efficiency than Suv39H1, which methylates only lysine 9. G9a has an enzymatic nature distinct from Suv39H1 and its homologue H2. G9a was also localized in the nucleus but not in the centromeric domain where the Suv39H1 member of the family is found. Histone-4 methyltransferase (PRMT1) methylates Arg 3 followed by acetylation of the protruding H4 tails by p300. *See* chromatin remodeling; E2F; p300; position effect; retinoblastoma. (Nielsen, S. J., et al. 2001. *Nature* 412:561; Tachibana, M., et al. 2001. *J. Biol. Chem.* 276:25309; Wang, H., et al. 2001. *Science* 293:853; Peters, A. H. F. M., et al. 2001. *Cell* 107:323; Kouzarides, T. 2002. *Current Opin. Genet. Dev.* 12:198; Lachner, M. & Jenuwein, T. 2002. *Current Opin. Cell Biol.* 14:286.)

histone phosphorylation   Has important roles in chromatin remodeling and initiation of transcription. C-terminal phosphorylation of H2A histone follows exposure to ionizing radiation. H2A and H2B phosphorylation may be signals for caspases and apoptosis. H3 phosphorylation (by IpI1, AIR, NIMA) at Ser/Thr begins at the pericentromeric regions and spreads along the chromosome arms before mitosis. In the latter cases, it actually promotes chromosome condensation, although relaxation of coiling is required for transcription, which takes place during chromatin remodeling. *See* AIR; chromatin remodeling; histones; IpI1; NIMA; survivin; TAF$_{II}$230-250. (Jenuwein, T. & Allis, C. D. 2001. *Science* 293:1074.)

histones   Five basic (rich in arginine and lysine) DNA-binding proteins. H2A, H2B, H3, and H4 are parts of the nucleosome core and H1 (H5 in birds) is generally a linker between nucleosomes. The histone genes were highly conserved during evolution. The N-terminal domains of the core histones and the C-terminal domain of H2A protrude from the nucleosomes and provide the means for interacting with other proteins involved in genetic regulation and repair. MacroH2A variant is involved in the inactivation of the inactive X chromosome (lyonization). Phosphorylation of serine and acetylation of lysine residues in the N-terminal domain of core histones (H3, H4) contribute to modulation of transcription by the nucleosomal structure. Acetylation/deacetylation, phosphorylation, and methylation mediate the modulation. Methylation of tail lysines and arginines may result in activation or repression depending on the residue modified. Lys 4 methylation prevents the recruitment of histone deacetylase to the active coding region and protein Set1 protects against the deacetylase

(Bernstein, B. E., et al. 2002. *Proc. Natl. Acad. Sci. USA* 99:8695). Subsequently, a combination of proteins is attracted to the N end. It is assumed that these accessory proteins represent a "histone code" for the regulation of chromatin remodeling and transcription. The H2A tail contains a leucine zipper, which may be involved in interaction with the TATA box–associated proteins. The "deviant" histones may associate with the nucleosomal structure and exercise regulatory roles. The centromeric nucleosomes bind specific proteins (CENP-A, CENP-B, CENP-C) that are essential for proper chromosome segregation. The H1 and H5 linker histones in the nucleosomal structure contain a "winged helix"—a bundle of three α-helices attached to a three-stranded antiparallel β-sheet. The absence of H1 may slightly decrease the expression of some yeast genes (Hellauer, K., et al. 2001. *J. Biol. Chem.* 276:13587). This structure is found in sequence-specific regulator proteins such as the catabolite activator protein and the hepatocyte-activating factor HNF3. The HNF3 protein replaces the H1 linker within chromatin containing the serum albumin enhancer. Similarly, H3 may be replaced by CENP-A at the centromere. Histones also regulate telomeric sites and the silent mating-type locus. The yeast nucleosomes are compact and may not have H1, although an H1 gene was discovered in chromosome XVI. In *Tetrahymena*, H1 may act as a specific positive or negative regulator for some genes but does not have a general regulatory role in transcription, although it is believed that H1 stabilizes the higher order of chromatin structures. This information indicates the special role of histones and nucleosomes in gene regulation through chromatin structure. The histone genes lack introns; their mRNA is not polyadenylated. In the mammalian spermatozoa, protamines and transition proteins replace the somatic histones during maturation. *See* CCE; chromatin; chromatin remodeling; HiNF; histone deacetylase; histone acetyltransferase; H1TF2; HU; LCR; lyonization; nucleosome; promoter; RPD. (Santisteban, M. S., et al. 2000. *Cell* 103:411; Nakayama, J.-I., et al. 2001. *Science* 292:110; Faast, R., et al. 2001. *Curr. Biol.* 11:1183; Marmorstein, R., 2001. *Nature Rev. Mol. Cell Biol.* 2:422; Jenuwein, T. & Allis, C. D. 2001. *Science* 293:1074; Celeste, A., et al. 2002. *Science* 296:922; Ahmad, K. & Henikoff, S. 2002. *Mol. Cell* 9:1191; Nishioka, K., et al. 2002. *Mol. Cell* 9:1201. <http://genome.nhgri.nih.gov/histones/>.)

histone tail   *See* histone methyltransferase.

historical control   The concurrent controls are run along the experimental series. In some instances, concurrent control is impossible to use (e.g., some epidemiological studies, effects of environmental pollution, etc.) because there are no means to exempt the control from the overall consequences of the factors studied. For example, if we want to assess the effect of a particular compulsive vaccination system, there is no contemporaneous nonvaccinated cohort group for the comparison. In such cases, similar data preceding the experiment or control data collected elsewhere under similar conditions are used as standards for the comparisons. *See* cohort; concurrent control; control.

histotope   Site of the MHC molecule recognized by the T-cell receptor. *See* agretope; desetope; epitope; MHC; TCR.

hit-and-run technique Expected to target for mutation or replacement of only a single nucleotide or a very minute segment of a gene. The causative agent is not permanently required for the altered state. *See* targeting genes. (Skinner, G. G. 1976. *Br. J. Exp. Pathol.* 57[4]:361.)

hitchhiking Because of close linkage to genes of selective advantage, nonadvantageous genes may be maintained in a population. The effect of hitchhiking is inversely proportional to the frequency of recombination. *Directional hitchhiking* models assume that linked polymorphism is periodically eliminated from the population, yet it is replenished by mutation and genetic drift. *See* background selection; evolvability; genetic load; Hill-Robertson effect; linkage drag; linkage disequilibrium; Muller's ratchet; pseudo-hitchhiking. (Maynard-Smith, J & Haigh, J. 1974. *Genet. Res.* 23:23; Barton, N. H. 2000. *Philos. Trans. R. Soc. London B Biol. Sci.* 355:1553; Fu, Y. X. 1997. *Genetics* 147:915.)

HIT theory *See* target theory.

HIV-1, HIV-2 Lentiviruses and causative agents of acquired immunodeficiency disease that suddenly erupted in the 1950s. These two different yet related viruses evolved from the related Simian viruses (SIV) long present in about 26 African primate species. In wild primates, generally SIV infection does not lead to a devastating disease as in humans. HIV-1 is supposedly originated from chimpanzees, whereas HIV-2 originates from the sooty mangabeys, although the transmission from nonhuman primates to humans has taken place repeatedly from the 1930s on. *See* acquired immunodeficiency (AIDS); lentivirus; SIV; viral vectors. (Hahn, B. H., et al. 2000. *Science* 287:607.)

Hix Invertase. *See* invertases.

HKA Hudson, Kreitman & Aguadé test for evolutionary neutrality. *See* mutation, neutral.

HKR oncogene Renamed GLI. *See* GLI.

HKT High-affinity kalium transport. A membrane protein conferring the ability of potassium uptake. *See* ion channels.

HL60 Human granulocyte line. *See* granulocyte.

HLA Human leukocyte antigen. The H2 gene cluster determines the corresponding functions in mice. The development of congenic resistant lines in mice eventually permitted the determination of histocompatibility by serological tests rather than by tissue transplantation. The genes concerned with the determination of specificity are generally designated as the major histocompatibility complex (MHC). The *HLA class I* molecules have two polypeptide subunits. The 44 kDa highly variable heavy chain is coded within the *MHC* cluster in the short arm of human chromosome 6p21.31, whereas the invariable ~12 kDa β_2-microglobulin is encoded at another location (15q21-q22) in the genome. Paralogous HLA clusters appear in human chromosomes 1q21, 9q33-q34, and 19p131-p13.4. The heavy chain has three domains of about 90 amino

acids each, and one of them interacts with β_2-microglobulin. These proteins are trans-membranic with about 30 amino acids extending within the cell, and a larger portion is outside the cell. The heavy chain consists of three domains, $\alpha 1, \alpha 2$, and $\alpha 3$, encoded by genes HLA-A, HLA-B, and HLA-C. The class I polypeptides were originally designated as transplantation antigens. *HLA class II* antigens are heterodimeric and are composed of a 33 to 34 kDa α-chain and a 28 to 29 kDa β-chain. The human MHC class II genes are HLA-DR, HLA-DQ, and HLA-DP. All of the HLA genes are interrupted by several introns. The $\alpha 3$ and $\beta 2$-microglobulin of class I HLA genes and the $\beta 2$ and $\alpha 2$ polypeptides of class II polypeptides are homologous with the heavy-chain constant region of immunoglobulin M. β_2-microglobulin bears similarity to the V gene of immunoglobulins. Heterozygosity for both the class I and class II genes can improve protection against some viral infections. The HLA cluster occupies about 3,600 kb in the DNA, including interspersed genes not shown here:

The entire human HLA region has been sequenced and it includes 128 genes and 96 pseudogenes; some of the genes have paralogues in other chromosomes (*Nature* 401:921 [1999]). Within the class II region there are about 20 multiple repetitive-sequence families (retroposons, retrotransposons, human endogenous retroviruses [HERV]) occupying ~20% of the region. Retroelements seem to affect recombination within HLA. The DP cluster contains in opposite orientation two α and two β genes alternating; DQ also has the same arrangement of the two α and two β genes; DR has $\beta_1, \beta_2, \beta_3$ in the same left-to-right sequence and class I α genes in opposite orientation. Both class I and class II genes appear to have pseudogenes. The HLA genes have a large number of alleles, and alleles of some genes, e.g., A1 and B8, are in linkage disequilibrium, i.e., they are syntenic much more often than expected by random recombination.

The HLA genes display high polymorphism that may be correlated with the specificity of the immune reaction. In the N-termini of the A proteins, there may be 7% differences. In exons 2 and 3 of the A genes, the base substitutions generally result in amino acid replacements, whereas in exon 4 half of the base substitutions are silent at the protein level. The β-chains of DP, DQ, and DR are highly polymorphic, but the α-chains of DR and DP are conserved. The variations were suspected to be the result of gene conversion, but some of the variation implicates interallelic recombination.

The amount of HLA gene products varies substantially in different cells. B lymphocytes display many A and B antigens and less C antigens, whereas in T lymphocytes all three proteins are much less active. Interferons and tumor necrosis factors stimulate the expression of class I genes. Class II genes are expressed primarily on B lymphocytes, myelocyte-resembling cells, and macrophages. When activated, T lymphocytes as well as skin fibroblasts and some endothelial cells may express them. Class II gene transcription is induced only by γ-interferon. If mouse lymphocytes are transformed with HLA heavy-chain genes of class I, the heavy chain can function with murine β_2-microglobulin, and it is detectable by monoclonal antibodies against HLA determinants while in transgenic cells cytotoxic lymphocytes do not consistently attack HLA determinants. Genetically engineered HLA class I indicated that the first two external regions of the heavy chain (α_1 and α_2) are critical for encoding serological determination. Class II genes are expressed on the surface of mouse lymphocytes and the mouse cell can provide the invariant antigen chain.

Within the long DNA tract of the HLA cluster, several genes involved in disease susceptibility have been identified. 21-hydroxylase deficiency (in between DR and A) causes congenital adrenal hyperplasia (CAH1), hemochromatosis (HFE, either between B and A or distal to A), cirrhosis of the liver, diabetes, dark pigmentation, and heart failure. The gene for juvenile myoclonic epilepsy (JME; proximally to A) causes convulsions limited to certain areas and is possibly involved in ragweed sensitivity, etc. Tumor necrosis factors TNF-A and TNB-B, cachetin (hormone-like protein product of macrophages releasing fat and lowering the concentration of fat synthetic and storage enzymes), and lymphotoxin (lyses cultured fibroblasts) are also coded within the HLA cluster.

The most important functions of the complex are in immune recognition, defense against bacteria, and viruses, and they also have a number of other not entirely clarified roles. The glycoproteins encoded by the HLA genes are deposited on the surface of the majority of cells. These surface antigens are the ID cards of the individuals. The identity is determined by the inheritance of the specificity of the MHC alleles and not processed further as is the case with antibody transcripts. Class I gene products are located on cytotoxic T cells (CTC) that destroy the cells of the body when infected.

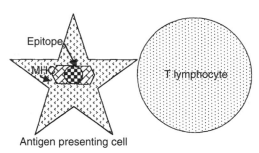

Epitope fragments of about a dozen amino acids are positioned in a cleft of the MHC molecule located on an antigen presenting cell are delivered to the T cell receptors and there destroyed by the enzymes secreted by the T cell.

Class II proteins are on the surface of B lymphocytes and macrophages involved in binding humoral (circulating) antigens. Antigens are separated into polypetides by cellular proteases, and the polypetides associate with specific compartments of the specific HLA protein molecule. CD8 (cluster of differentiation 8 protein, 34 kDa, encoded by a gene in human chromosome 2, product of differential processing of the transcript) becomes part of the T-lymphocyte cytotoxic or suppressor molecule in association with class I HLA proteins recognized by the T-lymphocyte receptor (TCR). The TCR protein is a complex of different polypeptide chains encoded at 14q11.2 (TCRα), 7q35 (TCR β), 14q11.2 (TCR δ), 11q23 (TCRε), and 7p15-p14 (TCRγ). CD4, a 55 kDa protein (encoded in human chromosome 12), is associated with TCR and with class II HLA protein on specific T lymphocytes. The CD8 and

CD4 proteins also act as signal transducers in association with CD3 antigens (multiple cistrons encoded at 11q23) and the helper T lymphocytes secrete lymphokines that activate B cells to secrete antibodies. They also turn on a protein tyrosine kinase to initiate signal transfers to the cell nucleus. The human MR1 gene locus (1q25.3), also called HLALS, encodes a polypeptide that has similarities to the class I major histocompatibility antigens. *See* adrenal hyperplasia; ankylosing spondylitis; antibody; CD3; CD4; CD8; congenic resistant lines; endogenous virus; *H*-2; hemochromatosis; lupus erythematosus; lymphochip; major histocompatibility complex; microcytotoxicity test; mixed lymphocyte reaction; myoclonic epilepsy, juvenile; psoriasis; ragweed sensitivity; Reiter syndrome; RFX; TAP; T cells; TCR. (Hill, A. V. 2001. *Lancet* 357:2037; Shiina, T., et al. 2001. *Genome Res.* 11:789; Hughes, A. L. & Yeager, M. 1998. *Annu. Rev. Genet.* 32:415; Ting, J. P.-Y., et al. 2002. *Cell* 109:[S02]:21.)

HLH   Amphipathic (both hydrophobic and hydrophilic sides) α-helices of proteins connected by a loop.

HLM oncogene   Mosaic of various types of retroviral elements found in avian and mammalian viruses. Oncogene HLM2 was assigned to human chromosome 1. *See* retrovirus.

HLP   Histone-like protein. *See* histones.

HMBA   Hexamethylene bisacetamide causes genome-wide transient demethylation in mouse erythroleukemia cells while the DNA is not replicating. *See* methylation of DNA.

HMCR   Human-mouse conserved region in the DNA.

HMG   *See* high-mobility group of proteins.

HMG box   High-mobility group domain of the SRY gene product and other proteins with high-affinity DNA binding but differing in sequence specificity. *See* SRY.

HMG-CoA reductase   *See* cholesterol, Niemann-Pick disease

HML and HMR   Left (α) and right (a) mating-type gene cassettes, respectively, in chromosome 3 of yeast. These elements are expressed only when transposed to the mating-type MAT site. Both of these loci are flanked by HML-E and HML-I and HME-E and HMR-I silencing elements, where binding sites exist for a combination of three proteins, ORC, a protein complex–originating replication, Rap1 and Abf. HMR-I has the ORC- and Abf-binding sites (ACS), but it has no Rap1-binding site. Rap1 binds the silencer proteins Sir3 and Sir4. ORC recruits Sir1. HMR-I has both an origin of replication and a silencing function. *See* Abf; mating type; mating type determination in yeast; ORC; Rap; Silencer.

hMLH1   Human DNA repair locus in chromosome 3p21 where mutation may lead to colorectal cancer. *See* DNA repair; hereditary nonpolyposis colorectal cancer.

HMM   Hidden Markov model. *See* Markov chain.

HMS   *See* copia.

HMSN Hereditary motor and sensory neuropathy. A group of hypomyelination diseases such as Charcot-Marie-Tooth disease, Dejerine-Sottas syndrome, and HNPP. *See* hypomyelinopathies.

HMS-PCI High-throughput mass spectrometric protein complex identification. (Ho, Y., et al. 2002. *Nature* 415:180.)

HNGFR High molecular nerve growth factor receptor. *See* LNGFR.

HNF Hepatocyte nuclear factor. Proteins are transcription factors of liver-specific genes involved in carcinogenesis, atherosclerosis, hyperlipidemia, insulin resistance, hypertension, blood clotting, etc. HNF-1α (human chromosome 12q22-t23, mouse chromosome 5) and HNF-4α ([TCF14] human chromosome 20q12-q13.1) regulate insulin secretion, and their defect may lead to non-insulin-dependent diabetes. HNF-3α (human chromosome 14q12-q13, mouse chromosome 12) is a negative regulator of HNF-1α and HNF-4α, whereas HNF-3β has a positive regulatory effect on HNF-1α, HNF-4α, and HNF-3α. HNF-3α appears to compete for the binding site of HNF-3β (human chromosome 20p11, mouse 2). HNF-3γ is in human chromosome 19q13-q13.4 and in mouse chromosome 7. HNF3 may bind to nonacetylated nucleosomes. *See* atherosclerosis; blood clotting; cancer; diabetes; glomerulocystic kidney disease, hypoplastic familial; hepatocyte growth factor; hyperlipidemia; hypertension; MODY; plasmin. (Shih, D. Q., et al. 2001. *Nature Genet.* 27:375.)

HNPCC Human nonpolyposis colon cancer. Hereditary nonpolyposis colorectal cancer. *See* colorectal cancer.

HNPP Hereditary neuropathy with liability to pressure palsy (human chromosome 17p11.2). It bears similarity to hereditary motor and sensory neuropathies (HMSN) represented by Charcot-Marie-Tooth syndrome and other hypomyelination disease, although it is distinguished from them in several ways. Its onset is during adolescence or shortly afterward. The hypomyelinated tomacula (sausage-like swellings) are diagnostic criteria. *See* Charcot-Marie-Tooth disease; HMSN; hypomyelination; MITE; MLE; neuropathy; palsy.

hnRNA Heterogeneous nuclear RNA; the thousands of diverse species of RNA found in the eukaryotic nucleus, including primary transcripts and pre-mRNA in various stages of processing. hnRNA includes introns and other transcribed but not translated RNAs. These RNAs may be associated with proteins of 34 kDa to 120 kDa size. There are about 20 different proteins within the particles. The six most common core proteins—A12, A2, B1, B2, C1, C2—occur in multiple copies within each globular aggregate. The complex takes a beads-on-string–like structure, with each globular structure (about 20 nm in diameter) containing 100 to 800 nucleotides and having a sedimentation coefficient of about 40S. A gene in human chromosome 19q13.3 codes the U1AP protein. U1-70K snRNP is the major antigen recognized by anti-(U1)RNP sera in autoimmune diseases. It has been estimated that only about 5% of pre-mRNAs are exported from the nucleus to the cytoplasm. Most of the rest are degraded. *See* autoimmune disease; hnRNP; KH domain; RNP; snRNA; U1-RNA. (Roy-choi, T. S. 1999. *Crit. Rev. Eukaryot. Gene Expr.* 9:107; Krecic, A. M. & Swanson, M. S. 1999. *Curr. Opin. Cell. Biol.* 11:363.)

hnRNP Heterogeneous ribonucleoprotein. These proteins regulate gene expression at posttranscriptional levels. Some of the recognition motifs are RNP1 octamers and RNP2 hexamers embedded in conserved regions of ~80 amino acids and present in 1 to 4 copies. The K-homology (KH) domains of about 60 amino acids may be present in up to 15 copies per protein. In the genome of *Arabidopsis*, 196 RNA recognition (RRM) proteins and 26 KH proteins were found. These proteins are more complex in this plant than those found in metazoa. *See* export adaptors; hnRNA; RNA-binding proteins. (Lorkovic, Z. J. & Barta, A. 2002. *Nucleic Acids Res.* 30:623.)

hobo *See* hybrid dysgenesis.

Hodgkin disease (HD) Malignant lymphoma with an unknown genetic determination because the familial nature of the condition is not sufficiently clear. W. F. Bodmer argued in favor of a gene linked with an HLA complex. In a two-marker case (A and A′), the distribution of 1AA:2AA′:1A′A′ would be expected in the progeny of heterozygotes without linkage. In a survey of 32 afflicted sib pairs, the actual proportions were 16:11:2, which is significantly different from 1:2:1 at the level of 0.005 probability. More recently, linkage to the pseudoautosomal region has been suggested. If it is assumed that the frequency of gene *a* (Hodgkin) is 0.01 and all *aa* individuals become afflicted but only 0.05 of the *aA* individuals develop the disease and none of the *AA* do, then only about 0.01 of the Hodgkin patients will be homozygotes for *aa*. In a case of 32 two-offspring families with two afflicted children, the expected frequency of *aa* was about 2.5×10^{-6}. This also assumes that the two-sib families may have $aA \times AA$ parents (about 4×10^{-3}) and the frequency of two-child-afflicted families is about 2.5×10^{-3}. These latter figures thus point to the possibility that in some populations the frequency of the recessive homozygotes may indicate high familial expression of the disease. *See* anaplastic lymphoma; ascertainment test; Epstein-Barr virus; Hardy-Weinberg theorem; HLA; IL-9; leukemia; lymphoma; MALT; pseudoautosomal. (Staudt, L. M. 2000. *J. Exp. Med.* 191:207.)

Hoechst stain 33258 Used for banding of AT-rich minor-groove DNA sequences in the chromosomes; it is also an antibiotic. *See* Harlequin staining of chromosomes; sister chromatid exchange.

HOG-1 (high osmolarity) Protein kinase of the MAPK family. *See* osmosis; p38; signal transduction.

Hogness (-Goldberg) box (TATA box) 7–8 base-pair region of conserved homology rich in TA, preceding the transcription initiation of mRNA by about 19–31 residues in the promoter region of eukaryotic genes:

Exceptionally, some promoter regions lack the TATA box; these are called *TATA-less promoters*, such as U1, U2, U4, and U5 RNA promoters. The TATA box is generally surrounded

TATA box

GC box — CCAT box in animals	−T A T A A A A−	PY A PY (transcription start)
or AGGA	82 97 93 85 63 83 50	
in some plant genes	approx. % of conservation	

by GC-rich sequences (proximal and distal sequence elements, PSE and DSE). In these U promoters, PSE and DSE are still present. Transcription by RNA polymerase II requires the association of the TATA box with a TATA-binding protein (TBP) or additional TATA-associated factors (TAF). *See* open promoter complex; Pribnow box; TAF; TBP; transcription factors. (Hernandez, N. 2001. *J. Biol. Chem.* 276:26733.)

hok-sok-mok *See* plasmid maintenance.

holandric gene Y-chromosome linked. The mammalian Y chromosome appears largely heterochromatic under the light microscope and it carries few genes. The H-Y antigen gene has been assigned to the proximal region of the long arm of Y, and the testis-determining factor, formerly called TDF, now SRF, is proximal to the centromere in the same arm in humans. The long arm also contains the pseudoautosomal region (PAS); this DNA sequence Yp (SMCY) is homologous to an X-chromosomal tract, Xp (SCX), the region where X and Y crossing over can occur. The gene for surface antigen MIC2Y was assigned to the euchromatic region Ypter-q1 of the Y chromosome. The homologue was assigned to an X-chromosomal band between Xp22.3 and Xpter. The azoospermia factor (AZF) Sp3 or HGM9 maps at the site of H-Y and may be identical with it. Genes controlling body height and tooth length were suspected to be in the Y chromosome. An arginosuccinate and an actin pseudogene were located to human Y chromosome. A gene for hairy ears was suspected to be in the Y chromosome, but its status is not resolved with certainty. No hereditary disease gene is linked to the Y chromosome, although in aneuploids (XYY, XO) it may cause defects. *See* actin; azoospermia; differential segment; hairy ear; heterochromatin; H-Y antigen; imprinting; pseudoautosomal; pseudogene; sex determination; SRF; surface antigen; Y chromosome. (Jobling, M. A. & Tyler-Smith, C. 2000. *Trends Genet.* 16:356.)

holandric inheritance Genetic transmission (only) through the male. *See* imprinting.

holin Protein product of the *t* gene of phages that conducts the lysozyme into the periplasmic space of the bacteria conducive to lysis. *See* endolysin; lysis; lysozyme; periplasma. (Wang, I.-N., et al. 2000. *Annu. Rev. Microbiol.* 54:799.)

holism View that organisms represent an integrated system of elements and mechanisms and that the integrated system is more than a collection of the parts; therefore, it cannot be properly understood on the basis of the separated components. *See* organismal genetics.

Holliday juncture Points were the polynucleotides forming the Holliday structure are exchanged during recombination. This juncture may be bound by Rad1 protein in the presence of Mg^{2+} and cut by this endonuclease. Rad1 appears to be the catalytic subunit of the Rad1/Rad10 endonuclease. *See* Holliday model; Holliday structure; RAD; recombination, molecular mechanism of. (Bond, C. S., et al. 2001. *Proc. Natl. Acad. Sci. USA* 98:5509; Lilley, D. M. J. & White, M. F. 2001. *Nature Rev. Mol. Cell Biol.* 2:433.)

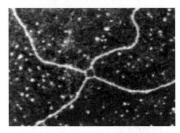

Electron micropgaph of a Holliday juncture. (Courtesy of Drs. H. Potter & D. Drechsler.)

Holliday model Of general recombination, is best explained by the figure on the next page. (After Potter, H. and Dressler, D. 1976. *Proc. Natl. Acad. Sci. USA:* 73:3000. The model was originally proposed by Holliday, R. 1974. *Genetics* 78:273.) This is the most widely accepted model of both prokaryotic and eukaryotic recombination. *See* Holliday juncture; Holliday structure; single-end invasion; figure on p. 553.

Holliday structure Recombinational intermediate of DNA displaying a four-strand cruciform arrangement (*see* Holliday model step I). Its resolution requires a specific endonuclease that takes place either by crossing over (flanking marker exchange) or gene conversion (noncrossover) without outside marker exchange results (step L). *See* cruciform DNA; model. (Allers, T. & Lichten, M. 2001. *Cell* 106:47.)

holocentric In several species of insects (*Lepidoptera, Hemiptera, Homoptera*), in the nematode *Caenorhabditis*, and in certain plants (*Luzula*), the spindle fiber attachment is not limited to the centromere (kinetochore), but the microtubules can be attached to many points along the chromosome, appearing as if the centromere were diffuse. In some genotypes of maize and rye, additional spindle fiber attachment sites (neocentromeres) accompany the major centromere. This characteristic may influence the distribution of the chromosomes to the poles (preferential segregation, nondisjunction). Species with holocentric chromosomes may be subjected to high doses of chromosome-breaking agents and still a more-or-less orderly anaphase distribution may take place. This feature has been exploited for biological control of *Lepidopteran* agricultural pests. The heavily irradiated individuals with broken chromosomes may survive and mate, but in their progeny the chromosomal fragments may fuse, leading to multiple translocations and lethal offspring. *See* centromere; centromere activation; genetic sterilization; neocentromere; screwworm control; translocation. (Dernburg, A. F. 2001. *J. Cell Biol.* 153[6]:F33; figure on page 554.)

Holyday model of recombination

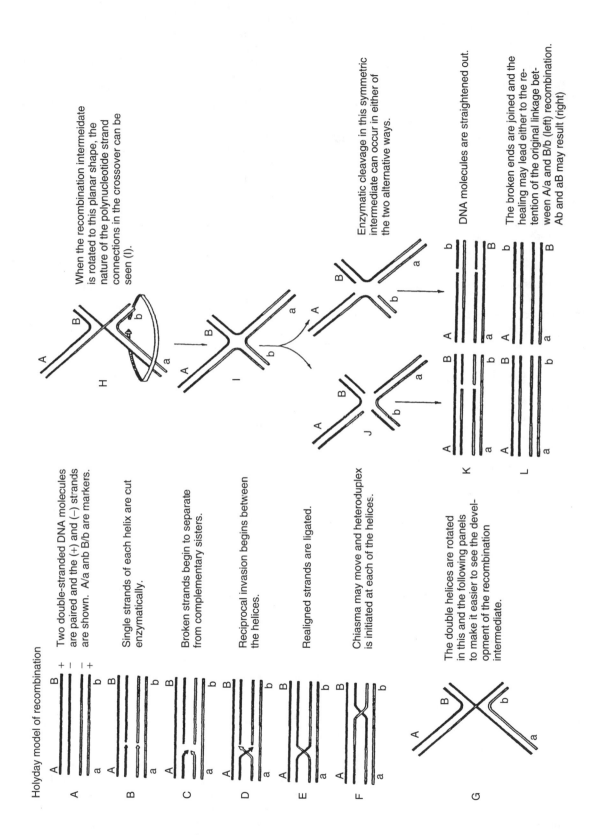

Two double-stranded DNA molecules are paired and the (+) and (−) strands are shown. A/a anb B/b are markers.

Single strands of each helix are cut enzymatically.

Broken strands begin to separate from complementary sisters.

Reciprocal invasion begins between the helices.

Realigned strands are ligated.

Chiasma may move and heteroduplex is initiated at each of the helices.

The double helices are rotated in this and the following panels to make it easier to see the development of the recombination intermediate.

When the recombination intermeidate is rotated to this planar shape, the nature of the polynucleotide strand connections in the crossover can be seen (I).

Enzymatic cleavage in this symmetric intermediate can occur in either of the two alternative ways.

DNA molecules are straightened out.

The broken ends are joined and the healing may lead either to the retention of the original linkage between A/a and B/b (left) recombination. Ab and aB may result (right)

553

Holocentric chromosome fragments. (From Hughes-Schrader, S. & Ris., H. 1941. *J. Exp. Zool.* 87:429.)

holoenzyme   Functionally complete enzyme, including cofactors. *See* apoenzyme.

hologenesis   View claiming that humans originated at many locations of the globe.

holography   Three-dimensional photography with the aid of split laser beams.

holokinetic   Holocentric.

holometabolous insect   Has a larval stage of development, and the larvae include imaginal disks, which are the initials of the adult body parts and from which all the appendages emerge according to a determined plan. *See* imaginal disk. (Yang, A. S. 2001. *Evol. Dev.* 3[2]:59.)

holoprosencephaly   Dominant with human chromosomal locations 7q36 (sonic hedgehog, SHH), 2p21, and 13q32 (ZIC2, a homologue of the *Drosophila* odd-paired, Opa, Zn-finger protein), but possibly other locations affect the expression of this syndrome (21q22.3, 18p11.3 [Niemann-Pick syndrome], 2q35 [Indian hedgehog], 12q13). The anomalies involve cleft lip, hypotelorism (abnormal distance between organs), defective head and face, and mental defects due to neural tube anomalies. SHH human gene transmembrane receptors are smoothened (7q32) and patched (9q22.3). TG-interacting factor (TGIF, 18p11.3), a homeodomain protein, interacts with Smad2, Nodal, and sonic hedgehog protein and represses the transcription of genes that control signaling to the neural axis pathway. Its incidence is 1/250 in conceptuses and 1/16,000 live borns. *See* cleft lip; conceptus; head/face/brain defects; *hedgehog*; megalin; Nodal; SMAD; Smith-Lemli-Opitz syndrome; *sonic hedgehog*; tumor suppressor pathway.

Holt-Oram syndrome   Dominant or sporadic arm, thumb, and heart malformation associated with human chromosome 12q24.1. The defective transcription factor involved is homologous with the *Drosophila* gene *Serrate* and the *Brachyury* mouse gene family of T box. The prevalence of the syndrome is 1×10^{-5}. *See* adactyly; *Brachyury*; heart disease; polydactyly; Tabatznik syndrome; T box; thrombocytopenia, sporadic. (Ghosh, T. K., et al. 2001. *Hum. Mol. Genet.* 10:1983.)

homeoallele   Alleles in the homoeologous chromosomes. *See* homoeologous chromosome.

homeobox   Conserved (183 bp) sequence within homoeotic genes. *See* homeotic genes.

homeodomain   Part of the protein that is coded for by the homeobox; it contains a DNA-binding helix-turn-helix motif protein domain. The homeodomain contains three α-helices and a flexible N-terminal arm. The third or recognition helix takes position in the major groove of the DNA. The N-terminus keeps contact with several bases in the minor groove of the DNA. The best-conserved part of the homeobox is the TAAT motif. The homeodomain genes determine the anterior posterior pattern of development and they are usually clustered in the genome. The homeodomain proteins are transcription factors. *See* anterior; DNA-binding protein domain; helix-turn-helix motif; homeobox; homeotic genes; morphogenesis in *Drosophila*; pseudogenes; posterior. (Gehring, W. J., et al. 1994. *Annu. Rev. Biochem.* 63:487; Banerjee-Basu, S. & Baxevanis, A. D. 2001. *Nucleic Acids Res.* 29:3258; <http://genome. nhgri.nih.gov/ homeodomain/>.)

homeogene   *See* homeotic gene.

homeogenetic induction   Cells or tissues start on a certain path of development by induced cells and continue producing the same cell types. (Tiedemann, H., et al. 2001. *Dev. Growth Differ.* 43[5]:469.)

homeologous   *See* homoeologous alleles; homoeologous chromosomes.

homeologous recombination   May take place between DNA (chromosome) strands that are similar but not entirely homologous. Recombination between *E. coli* and *Salmonella* with about 16% nonhomology is about 10^{-5} of that of recombination within the species. Mismatches within the species affect recombination in a less dramatic extent. In mouse 2/232, mismatches reduced recombination to about 5%. Mitotic recombination in budding yeast at a difference of about 17–27% resulted in reduced exchange by a factor of 13–180. Meiotic recombination is also much reduced in higher eukaryotes in case of sequence differences. *See* homologous recombination; illegitimate recombination.

homeopathy   Administering small doses of a medicine to a sick person that in larger doses would produce the same disease symptoms in a healthy individual. Homeopathy's origin goes back to Hippocrates (~4th–5th century B.C.) and Paracelsus (A.D. 16th century) and it was elaborated by Hahnemann (1767). Today, it is a generally discredited practice in the United States, although there are some adherents worldwide. (Barberis, L., et al. 2001. *J. Altern. Complement. Med.* 7[4]:337.)

homeosis   Changing a body part into the likeness of another body part. *See* homeotic genes.

homeostasis   Property of a system to maintain its composition by a flexible adjustment of the function of its genes (genetic homeostasis), or a physiological or developmental buffering capacity of cells or developing organisms under a range of conditions. *See* logarithmic stability factor; stress.

homeostasis, genetic   Property of a population to equilibrate its genetic composition and resist mutational changes.

homeotic gene Specifies an alternative competence for differentiation of a part of the body, e.g., in *Drosophila*, legs in place of antennae; in plants, petals in place of stamens, etc.; they contain a homeobox. Homoeosis (term coined by Bateson in 1894) was recorded by the ancient world (King Midas, 7th century B.C., grew 60-petaled roses). The discovery of the first homeotic gene, *bithorax*, (*bx*, 3.58.8), by Calvin Bridges in 1915, stimulated more interest in these genetically determined developmental anomalies. Subsequently, other such developmental genes were discovered both in *Drosophila* and all types of higher organisms. In *Caenorhabditis*, about 10% of the developmental genes contain homeoboxes. In the plant *Arabidopsis*, 35–70 homeogenes were estimated to exist. The homeotic genes are generally large complexes; the *BXC* complex occupies more than 300 kb and less than one-tenth of it codes for mRNA. They are interspersed with introns and intergenic DNAs required for the developmental regulation of the complex. In *Drosophila melanogaster*, the homeotic genes are basically continuous (only introns are wedged within); in *Drosophila viridis*, the *Ultrabithorax* complex is mapped to two different salivary chromosome bands in chromosome 2. The first molecular analysis conducted (in D. S. Hogness' laboratory, 1983) on the *antennapedia* complex, *ANTC* (3–47.5), revealed that this complex spans 335 kb and includes several transcription units. All homeotic genes have a 180 bp consensus sequence, the so-called *homeobox* that specifies the *homeodomains* of regulatory proteins. The structural features of the *Ant* gene of *Drosophila* represent the organization of the homeoproteins below.

Left: *Normal Drosophila* female, middle: *Antennapedia* mutant. Right: Homeotic shoot apex of *Arabidopsis* (Courtesy of Dr. K. Németh).

The 7–8 amino acids at the amino termini are rather well conserved across taxonomic groups, except the one represented by (?). The homeodomain of this protein may form three helical units between amino acids 1 (Ser) and 22 (Glu), between 28 (Arg) and 38 (Leu), and between 42 (Glu) and 58 (Lys). The best-conserved amino residues are outlined. This homeodomain region has homology to the helix-turn-helix motif of prokaryotic repressor proteins and to the MAT

α2 protein with repressor function at the mating-type site in yeast. These three helical regions may fold into a helix-turn-helix DNA-binding motif.

The homeoprotein may make (unspecific) surface contact with the phosphate backbone of the DNA in the major groove. The homeodomain's conserved residues, preceding the helixes, attach to the minor groove of the DNA, and for the more specific contacts, probably helix III is responsible. These homeoproteins regulate processes of differentiation and either point or frameshift mutations may abolish their DNA-binding abilities and alter the pattern of differentiation. Binding can also take place at more than one DNA sequence as long as some basic similarities are shared in the base sequences. Therefore, one homeoprotein may regulate more than a single gene, although to a different degree. Homeoboxes display great sequence similarities among different organisms. Mammals usually have four homebox gene clusters: A, B, C, and D. Homeobox *HOX2* of the mouse is in chromosome 11, but its human homologue is in human chromosome 17 (containing 9 homeobox genes separated by a few units of recombination in a 180 kb region). *HOX1* of the mouse is in chromosome 6, but its homologue is in human chromosome 7p along with seven other homeoboxes. *HOX1* is also homologous to the *ANTC* homeobox of *Drosophila*.

Hox genes in mice occur in chromosomes 11, 15, and 2. Each Hox/Hom contains a cluster of 9–11 genes with an average length of about 10 kb, and thus the clusters are about 100 kb each. The genes within the cluster show paralogy and they most likely arose by serial duplications during evolution. The genes within a paralogous group are also referred to as cognate. Some *trans-paralogous* genes (situated in chromosomes 6 and 2, respectively) of mice are capable of substituting for each other if one of them is deleted. Some similarities are apparent among the homeotic loci in different organisms, and these are called *orthologous* because they appear to have common evolutionary origin. Statistical surveys of *Drosophila*, mice, *Caenorhabditis*, humans, etc., indicate that the flanking areas of the homeotic gene clusters were conserved during evolution. Homeoboxes are directly or indirectly regulated by cis- and trans-acting elements. *See* collinearity; coordinate regulation; developmental pattern formation; helix-turn-helix; heterochronic mutation; mating-type yeast; morphogenesis; operon; orthologous; parahox genes; paralogous; paralogous loci; *Polycomb*; POU; syndactyly; Wolf-Hirschhorn syndrome. (Krumlauf, R. 1994. *Cell* 78:191; Ferrier, D. E. K. & Holland, P. W. H. 2001. *Nature Rev. Genet.* 2:33; Zákány, J., et al. 2001. *Cell* 106:207; Galant, R. & Carrol, S. B. 2002. *Nature* 415:910; Ronshaugen, M., et al. 2002. *Ibid.*, p. 914; <http://www.iephb.nw.ru/ hoxpro>.)

[MSSLYY?]-variable region-[IYPWM]-intron-[RKRGRQTYTRYQTLELEKQ⌡ acidic tail-COOH

NH₂ tract   Homeopeptide   1   10 →   helix I

FHFNRYLTRRRRIEIAHAHALC⌡
← 22   28 →   helix II   ← 38

LTERQIKIWFQNRRMHWHK⌡
42 →   helix III   ← 58

EN] 60 basic amino-acid-rich
60   Homeodomain

(For the identity of amino acids see amino acid symbols in protein sequences).

HOMER 186-amino-acid glutamate receptor-binding protein with a single PDZ domain. *See* AMPA; GABA; PDZ. (Sal, C., et al. 2001. *Neuron* 31:115.)

homing Moving of insertion or other mobile elements within or into a genome.

homing endonuclease Encoded within mitochondrial and nuclear introns and catalyzing the movement of introns and inteins. The best-studied representatives of these enzymes is I-Sce. It is mitochondrially encoded within the ω^+ yeast factor (21S rRNA). After being spliced out, it is translated into Sce protein. The VDE endonuclease is within the intron of a vacuolar membrane ATPase. They are transcribed and translated together, and when the protein is processed, the VMA1 ATPase and the VDE endonuclease are produced by evicting VDE and splicing the amino- and carboxy-terminal amino acid sequences of VMA1. The *Sce* system can be used for site-specific chromosome breakage and repair studies after introduction with appropriate vectors into the cells. The event can be monitored in a *S1neo* construct, where the *neo* antibiotic resistance gene is interrupted by an 18 bp *I-Sce I* sequence and therefore *neo* is not expressed. The *I-Sce I* sequence is flanked by a CATG duplication:

S1neo≈≈≈≈CATG≈≈CATG. After *I-Sce I* cleavage, two types of *neo* genes are found: *neo*≈≈≈≈CATG expressed and *neo*≈≈≈≈Δ≈≈ not expressed because of the Δ deletion. Other types of constructs can also be produced that can restore *neo* activity by cleavage and repair. *I-Sce I* can be introduced into the cells by electroporation and can bring about nonhomologous repair. The double-strand breaks generated by the homing endonuclease increase gene targeting up to 1,000-fold. They induce site-specific breaks within the 14–44 base-pair recognition sequence. Four families of homing endonucleases are distinguished. Although basically different from the *Cre/loxP* or *FLP/FRT*, these systems can be used for similar purposes. *See* chromosomal rearrangements; *Cre/loxP*; *FLP/FRT*; inteins; intron homing; marker exclusion; mitochondrial genetics; site-specific recombinases; super-Mendelian inheritance. (Dujon, B., et al. 1989. *Gene* 82:115; Jurica, M. S. & Soddard, B. S. 1999. *Cell. Mol. Life Sci.* 55:1304; Edgell, D. R. & Shub, D. A. 2001. *Proc. Natl. Acad. Sci. USA* 98:7898.)

Hominidae Family of humans, *Homo sapiens*, *Homo erectus*, *Homo habilis*, and other most closely related, now extinct, genera that are more highly evolved than the closest family of the *Pongidae*, which includes the orangutan, chimpanzee, and gorilla. The brain size in modern humans varies. The Australian aborigines have a brain volume of over 1,200 cm^3, whereas that in *Homo erectus* was over

1,000, *Homo habilis* above 700, *Australopithecus* 400 to 500, gorilla 500, and orangutan and chimpanzee just below 400. Apparently, brain size has changed very little in the last 300,000 years. Sequencing 53 intergenic, nonrepetitive DNA segments (24,234 bp) from humans, chimpanzee, gorilla, and orangutan revealed a very low average sequence divergence. The divergence is for human–chimpanzee 1.24%, human–gorilla 1.62%, and chimpanzee–gorilla 1.63%. On the basis of a molecular evolutionary clock, humans separated from chimpanzees 4.6 to 6.2 million years ago. (Chen, F.-C. & Li, W.-H. 2001. *Am. J. Hum. Genet.* 68:444.)

The taxonomic classification and evolutionary descent of *Primates* is not entirely clear. The majority of anthroplogists favor the idea that *Homo sapiens* evolved in Africa (about 1 million years [Myr] ago), then spread to Asia and Europe in relatively recent time. One tree of hominid descent is shown below. Human evolution from the primates will be revealed by the completion of the genome projects. It is already known that the DNA base composition of humans and chimpanzees is almost 99% identical. Therefore, the morphological and brain differences may be accounted for by different regulation of the basically identical genetic material. The lower human chromosome number appears to be due to chromosome fusion (e.g., human chromosome 2). A pericentric inversion of human chromosome 18 is the major difference from the corresponding ape chromosome. Rearrangements frequently alter regulation of genes. The human genus (Hominini) has been reclassified on the basis of new criteria; because of the inherent problems of scanty anthropological remains, modifications are likely to be proposed. *See* Eve, foremother of molecular mtDNA; genome projects; hologenesis; inversion; out-of-Africa; Primates; Y chromosome. (Wood, B. & Collard, M. 1999. *Science* 284:65; Stone, A. C., et al. 2002. *Proc. Natl. Acad. Sci. USA* 99:43; Wood, B. 2002. *Nature* 418:133; <www.mnh.si.edu/anthro/humanorigins>.)

hominoid Includes the gibbons, great apes, and hominids (and extinct species). *See* great apes; hominids; pongidae; primates.

homoalleles Differ at the same nucleotide site of a codon, and therefore only four different alleles can be produced, containing at a particular site either A or T or C or G; within the nucleotides, recombination is impossible. Such an allele can be changed by intracodon (between nucleotides) recombination or base substitution. *See* heteroalleles; homoeoalleles.

homocaryon *See* homokaryon.

homocitrullinuria *See* urea cycle.

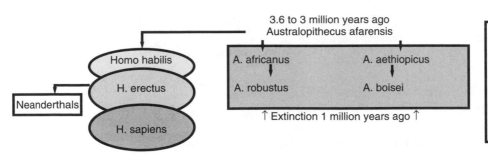

homocystinuria   May be due to different causes: (i) recessive deficiency of cystathionine synthetase (human chromosome 21q22.3); (ii) defects in vitamin B_{12} metabolism; (iii) poor intestinal absorption of B12; (iv) deficiency of methylenetetrahydrofolate reductase (1p36.3). Within group (i), different forms were also found; some responded to vitamin B_6 (pyridoxine). The phenotypes may resemble those of Marfan syndrome. In some instances, deletion of the 145-amino-acid carboxy terminal of cystathionine synthase alleviates the metabolic problems. It seems that this segment negatively controls the catalytic activity and is under the regulation by AdoMet. The general symptoms involve dislocation of the eye lens, thromboembolism (obstruction of the blood vessels), bone abnormalities, mental retardation in two-thirds of the afflicted, psychological disorders, etc. Prevalence is $\sim 2 \times 10^{-6}$, although in Ireland it is much more common. The diagnostic test for elevation of homocysteine and methionine in the urine is carried out by the cyanide-nitroprusside reaction. Prenatal and carrier identification is practical. Types (ii) and (iii) respond favorably to vitamin B_{12} (hydroxycobalamin) and cultured cells require methionine. Type (iv) individuals respond favorably to pyridoxine and folic acid. Defects in methionine synthase reductase (MTRR, 5p15.3-p15.2) may also increase the blood homocystine level. In Down syndrome, the level of cystathionine and cysteine is increased, but the level of homocysteine is reduced and folate-dependent methionine synthesis is diminished. Folate deficiency may occur. *See* AdoMet; amino acid metabolism; aminoacidurias, coronary heart disease; cystathionuria; cystinosis; cystinuria; Down syndrome; genetic screening; hyperhomocysteinemia; hypertension; Marfan syndrome; methionine biosynthesis; vitamin B_{12} deficiency. (Janošik, M., et al. 2001. *Am. J. Hum. Genet.* 68:1506.)

homoeoalleles   Evolutionarily and functionally closely related genes in polyploid species. *See* homoalleles.

homoeobox   *See* homeobox; homeotic genes.

homoeologous alleles   Occur in the homoeologous chromosomes of allopolyploid species; also called homoeoalleles.

homoeologous chromosomes   Nonidentical yet related chromosomes derived from a common ancestor. Despite some evolutionary divergence, they show partial homology like the A, B, and D genomes of wheat. *See* homologous chromosomes; nullisomic compenstion.

homoeotic   Related by evolutionary descent but modified during the course of evolution; similar but not entirely identical (recent usage is generally homeotic, although homoeotic would be the etymologically correct spelling).

homogametic sex   XX produces only X-chromosome-containing gametes, in contrast to the heterogametic XY individual, which can have both X- and Y-chromosome-bearing gametes. The homogametic sex can be either female (XX) or male (ZZ). *See* chromosomal sex determination, heterogametic.

homogamy   Mating between similar types or self-fertilization in plants. *See* assortative mating; autogamy; heterogamy.

homogeneity test   Can be used to determine whether different sets of data are statistically homogeneous enough to consider them to be part of the same population and whether the information is homogeneous enough to permit pooling. This test is basically a chi-square procedure, but the use of the Yates correction is not allowed. Without testing the homogeneity of separate sets of experiments, the information should not be pooled even when the combined data fit well to a null hypothesis. The use of this test is best explained by a tabulated example shown below. *See* chi square; null hypothesis; Yates correction.

homogeneously stained region (HSR)   Due to amplification in cancer cells, extended chromosomal bands are detectable by light microscopy. *See* double minutes.

Homogeneity test (using pea data of Mendel)

Family 1 ($n = 36$)	D	R	df $\otimes$	chi^2	Family 3 ($n = 97$)	D	R	df $\otimes$	chi^2
(1) observed numbers	25	11			(1) observed numbers	70	27		
(2) expected (3:1)	27	9			(2) expected (3:1)	72.25	24.25		
(3) difference (1)-(2)	2	2			(3) difference (1)-(2)	2.75	2.75		
(4) square of difference	4	4			(4) square of difference	7.563	7.563		
(5) divide (4) by (2)	0.148	0.444	1	0.592	(5) divide (4) by (2)	0.105	0.312	1	0.417
					Families				
Family 2 ($n = 39$)	D	R	df $\otimes$	chi^2	Combined ($n = 172$)	D	R	df $\otimes$	chi^2
(1) observed numbers	32	7			(1) observed numbers	127	45		
(2) expected (3:1)	29.25	9.75			(2) expected (3:1)	129	43		
(3) difference (1)-(2)	2.75	2.75			(3) difference (1)-(2)	2	2		
(4) square of difference	7.563	7.563			(4) square of difference	4	4		
(5) divide (4) by (2)	0.259	0.776	1	1.035	(5) divide (4) by (2)	0.031	0.093	1	0.124

Homogeneity test	Chi Squares	Degrees of Freedom	Probability*
Total of the 3 families	2.044	3	
Combined data	0.124	1	0.72
HOMOGENEITY (difference of the above lines)	1.92	2	0.38

*Use chi-square table (here chi-square charts were used). $\otimes$df = degrees of freedom.

homogenote *See* endogenote.

homogenotization Production of homozygosity for knockout either by targeting both chromosomes in the somatic cells or culturing the heterozygous knockout cells under highly selective conditions favoring the knockout chromosome. *See* knockout; targeting genes.

homogentisic acid *See* alkaptonuria.

homograft There is no known genetic difference between the transplanted and host tissues.

homohistont Nonchimeric tissue derived from a chimera. *See* chimera.

homohybrid DNA Annealed product of two methylated or unmethylated DNA sequences.

homoiogenetic induction Passed on by cells that have been previously induced. (Nieuwkoop, P. D. 1999. *Int. J. Dev. Biol.* 43[7]:615.)

homokaryon *See* dikaryon.

homologous chromosomes Contain the same cytological gene loci and form bivalents in meiosis. The homology may not be perfect and the difference may be up to several hundred kilobases. The nonhomologous chromosomes may display substantial telomeric homology among several chromosomes. In molecular evolutionary terms homology may exist among certain chromosomes or chromosomal regions or sites of different taxonomic groups if they carry similar nucleotide sequences. (homoeologous chromosomes, Mefford, H. C., et al. 2001. *Hum. Mol. Genet.* 10:2363.)

homologous genes Carry out basically the same function and descended from common ancestors, yet their primary structure may not be entirely the same. *See* analogous genes. (<http://www.ncbi.nlm.nih.gov/HomoloGene/>.)

homologous incompatibility The pathogen cannot cause disease because a particular race-specific gene of the host conveys resistance to it.

homologous pathogen The pathogen is genetically qualified to cause disease on the particular host.

homologous proteins Occur in different species and display similar structure and function, such as the various globins, the majority of metabolic enzymes, etc.

homologous recombination Genetic exchange between essentially identical chromosomes (polynucleotide chains). Homologous recombination can be used for gene disruption and studying the consequence of lack of function of a particular locus. A plasmid is constructed that carries other nucleotide sequences such as an antibiotic resistance gene in between the flanking sequences of the target gene. The flanks permit homologous pairing and recombination and may take place within the boundary of an interrupted DNA stretch (e.g., the target):

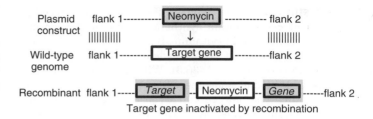

Such a manipulation is useful in animal cell cultures, and the resistance to the antibiotics can select the recombinant cells. Extremely rare double crossovers can thus be recovered. Cutting within the target by an appropriate restriction endonuclease can enhance homologous recombination by about 10-fold. The frequency of somatic homologous recombination in mammalian cells may vary within the range of 10^{-6} to 10^{-8} per cell per generation. The term *homologous recombination* gained a new meaning because classical genetics always recognized recombination as homologous. In yeast, homologous recombination usually requires a homology of at least 30 bp. Single-stranded DNA of about 30 nucleotides long may promote recombination by promoting annealing to complementary sequences. *See* Cre/loxP; crossing over; ectopic recombination; excision vector; FLP/FRT; heterochromosomal recombination; illegitimate recombination; I-Sce1; recombination; recombination molecular mechanisms eukaryotes; recombination molecular mechanisms prokaryotes; targeting genes. (Hiom, K. 2001. *Curr. Biol.* 11:R278; Sonoda, E., et al. 2001. *Proc. Natl. Acad. Sci. USA* 98:8388; Bollag, R. J., et al. 1989. *Annu. Rev. Genet.* 23:199.)

homolog Has similar primary and three-dimensional structure (3D) and function. In homologous proteins, the 3D sequence is commonly better preserved during evolution than the amino acid sequence. Thus, they have common motifs. Similarity does not necessarily mean homology. Homology indicates relationship by common evolutionary descent. *See* orthologous loci; paralogous loci.

homolog-scanning mutagenesis Applicable to gene families coding for structurally related proteins. Homologous domains of the proteins (7 to 30 amino acids) are substituted for each other. In the substituted domain, amino acids are replaced by another amino acid, e.g., alanine. Then the binding of the substituted domains to the protein receptor (e.g., growth hormone, protein kinase) is analyzed. Such an analysis may reveal any change in receptor binding and function. A simpler procedure is charged-to-alanine scanning mutagenesis in which blocks of amino acids (4 to 8) are replaced by alanine. *See* cysteine-scanning mutagenesis; Kunkel mutagenesis; oligonucleotide-directed mutagenesis; site-specific mutation. (Vik, S. B., et al. 1988. *J. Biol. Chem.* 263:6599; Kunkel. T. A., et al. 1991. *Methods Enzymol.* 204:125; Griffith, K. L. & Wolf, R. E. 2002. *J. Mol. Biol.* 322:237.)

homology Similarity based on nucleotide sequences in the DNA and RNA or amino acid sequences in the protein. It also indicates evolutionary relationship. Information on homology of human and yeast genes can be obtained at <http://www.ncbi.nlm.nih.gov/XREFdb>. The discovery of homology sometimes appears puzzling. *Saccharomyces* yeast contains a gene homologous to the *NifS* gene of nitrogen-fixing *Azotobacter*, although yeast does not fix nitrogen. Studies have revealed that this gene actually inserts sulfur into metal-sulfur centers of metalloenzymes using pyridoxal phosphate as a cofactor. The reliance on homology for understanding genetic functions is of great importance. Yeast, *Drosophila*, and *Arabidopsis* are relatively simple organisms and can genetically be manipulated by mutation, recombination, transformation, etc. They can help to shed light on how a series of genes function in a genomic context in other organisms, e.g., humans, where the application of these laboratory techniques (e.g., controlled mating) are limited or impossible. Evolutionists distinguish between *repetitive homology*, such as the multiplicity of legs in millipedes, and *nonserial homology*, e.g., the leaves of a plant. Usually a 25% or higher homology is considered evidence for a common evolutionary relationship. On the next page, the amino acid sequence homology of cytochrome-c protein is shown in 35 different organisms. The shading indicates complete identity at the alignments. (Courtesy of Margaret Dayhoff, ed. *Atlas of Protein Sequence and Structure*, 5. Natl. Biomed. Res. Found., Georgetown Univ., Washington, DC.) Alignment of nucleotide sequences provides more critical information because the amino acid sequences do not distinguish among synonymous codons. Among bacteria, the homologous genes, except members of operons, are not in identical orders. Apparently replication may rearrange the gene positions in the various genomes. A conservative estimate of the sequenced human and mouse genomes revealed 3,920 orthologous gene pairs. *See* analogy; congruence analysis; DNA sequence alignment; indel; orthology; paralogy; phylogenetic weighting; URF; xenology; table on page 560.

homology-dependent gene silencing *See* antisense technology; cosuppression; posttranscriptional gene silencing; RIP. (Cogoni, C. 2001. *Annu. Rev. Microbiol.* 55.381.)

homomeric protein   Built from identical subunits.

homomultimeric protein   Consists of more than two identical subunits.

homonomous   Similar in structure and function; the metameric body parts of insects have developed in a similar homonomous manner in different species. *See* homeobox; homeogenes; metamerism.

HOMOPHILA   Database that correlates human disease genes with their *Drosophila* homologues. (<http://homophila.scdsc.edu>.)

homoplasmy   Within a cell, tissue, or organism, the organellar genomes (mitochondrial, plastid) do not show genetic differences. Despite the usually large number of organelles per cell and the high rate of mutation, homoplasmy is the most common situation. A low level of heteroplasmy is difficult to detect, however, if the number of organelles is large. In cancer cells, a high degree of mutant mtDNA homoplasmy is common. Homoplasmy for mtDNA defects may cause lethality in the offspring of a clinically normal mother who is a carrier. *See* heteroplasmy; heteroplastidy; mitochondrial diseases in humans; mtDNA; sorting out. (Coller, H. A., et al. 2001. *Nature Genet.* 28:147; McFarland, R., et al. 2002. *Nature Genet.* 30:145.)

homoplastidic   The chloroplasts within a cell are genetically identical. *See* chloroplast genetics; heteroplastidy.

homoplasy   Parallel evolution (similarity is not based on common ancestry)—that is, two alleles or genes are identical or near-identical in state and/or in function, but they do not share common ancestry. *See* convergent evolution. (Collard, M. & Wood, B. 2001. *J. Hum. Evol.* 41[3]:167.)

homopolymer   Synthetic polynucleotide chain built from only one type of nucleotide, e.g., AAAA or CCCC, etc.

homopolymeric amino acids   May occur in proteins due to increased numbers of trinucleotide repeats. The best know among them are the polyglutamine stretches that lead to a variety of diseases. Although rare, polyleucine and polyalanine sequences are even more toxic. *See* trinucleotide repeats. (Dorsman, J. C., et al. 2002. *Hum. Mol. Genet.* 11:1487.)

homopolysaccharide   Built of one type of sugar subunit.

homoproline   *See* pipecolic acid.

Homo sapiens (man) $2n = 46$. *See* hominidae; human races; primates.

homoscedasticity   Homogeneity (equality) of the variances in a group of samples.

homoselection   In small populations, selection may favor homozygotes, which increases the specialization to the unique environmental niche. *See* heteroselection; selection.

homosexual   An individual is attracted to the same sex and identifies him-/herself as homosexual, develops sexual fantasies about the same sex, and practices homosexuality. Various studies have estimated the prevalence of homosexuality to be from 2 to 10% in human populations. Homosexual orientation occurs in other mammals, especially under conditions when the opposite sex is not available.

In some primitive human societies, magical powers are attributed to homosexuals; however, in the majority of societies and religions, homosexuality is disapproved. The ancient Greeks accepted it as an abnormal condition. Base reliefs in the tombs from the Fifth Dynasty of Old Kingdom Egypt discovered in the 1960s depict men in intimate poses expected only in conjugal relations (Reeder, G. 2000. *World of Archeol.* 32:193).

The majority of the individuals may begin same-sex orientation between ages 5 and 30, most commonly by puberty. The bases of homosexuality are not entirely clear. Homosexuality had been attributed to hormonal, psychological, anthropological, genetic, and moral causes or to a combination of part or all of these factors. The exact scientific study of homosexuality is difficult because generally a multitude of factors influence the development of all behavioral traits, and homosexuality may have different categories

Homology of the amino acid sequences of cytochrome C in 35 organisms.

from obligate homosexuality to bisexuality and primarily heterosexuality. Anatomical studies have attributed brain differences to homosexual orientation. Twin studies suggested higher concordance of this type of sexual orientation between monozygotic twins than between dizygotic ones. The observation that male homosexuals have more homosexual males than lesbian females in their kindred may indicate that the genetic bases of female homosexuality may not be identical to those of males but may also point to some environmental causes. One study involving 114 families of homosexual index cases found an increase of same-sex orientation among maternal uncles and cousins in comparison to paternal relatives, suggesting the possibility of sex-linked transmission of the gene(s) concerned. In 40 families where at least two homosexual males occurred, in approximately 64% of the sib pairs tested an apparent linkage was observed to the DNA marker Xq28 with a multipoint lod score of 4, indicating a higher than 99% probability for synteny. Another study (Rice, G., et al. 1999. *Science* 284:665) failed to confirm linkage to Xq28 by studying 52 gay male siblings. The heritability of male homosexuality was reported to be 0.50, but the inheritance of female homosexuality seems more complex.

Research workers debate some of this human information. There is some indication that fetal androgen levels may favor the development of both male and female adult homosexuality. In *Drosophila*, the *satori* (*sat*) mutants of males do not court or copulate with females but have a sexual interest in males. The locus co-maps and is allelic with *fru* at the 91B chromosomal band. The *fru*^{sat} flies lack the male-specific Lawrence muscle (MOL). The fru^{sat} protein expressed in some brain cells is probably a transcription factor with two zinc-finger domains. The female flies produce two double bonds (dienes) with 27 and 29 carbons, and these excite males. The male flies make monoenes (one double bond) of 23 and 25 carbons of the pheromones. The transformer gene (*tra*) regulates these two pheromones. When the *tra* gene is ubiquitously expressed, a mixture of the two pheromones is produced in the males and elicits homosexual courtship by normal males. The social status of human homosexuals is an ethical rather than a biological problem, yet the ethical solution may be facilitated by better biological information. *See* behavioral genetics; ethics; kindred; lod score; sex determination; sibling; twinning. (Hammer, D. H., et al. 1993. *Science* 261:321; Hu, S., et al. 1995. *Nature Genet.* 11:248; Yamamoto, D. & Nakano, Y. 1999. *Cell. Mol. Life Sci.* 56[7–8]:634.)

homosexual cross   The mitochondrial genome of yeast appears to have sex-factor-like elements ω^+ and ω^-. The crosses between $\omega^+ \times \omega^-$ are heterosexual, where as the crosses $\omega^+ \times \omega^+$ and $\omega^- \times \omega^-$ are homosexual. *See* mitochondrial genetics; mtDNA.

homothallism   The same individual (thallus) of lower eukaryotes produces both *plus* and *minus* or *A* and *a* mating-type spores that can fuse into a zygospore. Homothallism bears a similarity to both monoecy and autogamy in higher plants. These spores are not necessarily immediate meiotic products. *See* dioecious; heterothallism; monoecious; zygospore.

homotopic transplantation   Transfer of cells or tissue(s) to an identical site but in a different individual.

homotropic enzyme   Allosteric enzyme acting on the same substrate and usually regulated by the substrate. *See* allostery.

homozygosity in a randomly selected individual   at a locus for an allele, under equilibrium conditions between mutation and genetic drift, is approximately

$$F = \frac{1}{1 + 4N\mu}$$

where N = population size and, μ = mutation rate. The homozygosity estimator in a population is obtained by

$$\frac{2(H_w - H_b)}{1 + H_w - 2H_b}$$

where H_w = proportion of homozygosity within an individual and H_b = proportion of homozygosity between individuals. In outbreeding populations of dioecious species, homozygosity for the majority of loci is expected to be low. A study of short tandem repeats in human families in CEPH revealed that on the average more than 10 cM homozygous tracts occur. The longest, 77 cM segment, included 118 homozygous markers. Apparently there is a substantial degree of autozygosity or relatedness in these populations. The linkage disequilibrium can be preserved only in the absence of recombination and low mutation rate. Actually, the long homozygous segments also indicate that these families are of relatively recent origin and mutation, and selection did not yet have a chance to break up the linkage. *See* autozygosity; CEPH; genetic drift; genetic equilibrium; linkage disequilibrium. (Broman K. W. & Weber, J. L. 1999. *Am. J. Hum. Genet.* 65:1493.)

homozygosity mapping   Locates autosomal-recessive genes in consanguineous families by identity of descent in the pooled records. *See* consanguinity; IBD; mapping.

homozygous   In a diploid or polyploid, the alleles at a locus are identical; being a homozygote, e.g., *aa* or *aaaa* or *AA* or *AAAA*.

honey bee   *Apis mellifera* $2n$ = female 32, male n = 16. Diploid workers and $2n$ female (queen) and haploid males (drones, $1n$). The difference between the workers and the queen is developmental, due to a difference in nutrition. *See* social insects. (Michener, C. D. 2000. *The Bees of the World.* Johns Hopkins Univ. Press, Baltimore; <www.cyberbee.net>, <http://www.tigr. org/tdb/tgi.shtml>.)

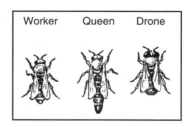

Worker	Queen	Drone

Hoogsteen pairing   Refers to nucleotide pairs, by an association different from that proposed by Watson and Crick. The A–T pairs have an 80° angle between the glycosylic bonds and a 8.6 Å distance between anomeric carbons (differing in configuration about C). In the *reversed Hogsteen* pairs, one base is rotated 180° relative to the other. Hoogsteen pairing takes place between nucleotides of a third strand of DNA in the major groove of the duplex. This happens when in the duplex one strand is polypurine and the other is polypyrimidine; the third strand is most commonly polypyrimidine in the triplex DNA. *See* triple helix formation; Watson and Crick model. (Soliva, R., et al. 1999. *Nucleic Acids Res.* 27:2248.)

DNA triple helix and Hoogsteen base pairing.

hop (*Humulus lupulus*)   Climbing dioecious plant with $2n = 20$ chromosomes, although forms (*H. japonicus*) with $2n = 16$ females and $2n = 17$ males have also been reported. Its main use is for brewing beer, but it may be mixed in bread in some areas of the world. The soft resins and essential oils of the flowers add a typical bitter flavor to the brew and have some preservative effect. It is related to hemp (*Moraceae*). Some extracts have pharmaceutical value.

hopeful monster   Neo-Darwinism assumes that evolutionary changes take place by accumulation of minor changes that favor fitness of an organism. An alternative suggestion raised the possibility that some major mutations conveyed distinctive alterations to an individual, transcending conventional taxonomic boundaries, and such an individual was called a hopeful monster. *See* gradualism; saltatory replication; saltation. (Richards, E. 1994. *Isis* 85[3]:377.)

hopping   Movement from one location to another by a transposable or insertion element. Also in translation when the peptidyl-tRNA passes further downstream to a similar or the same codon. *See* insertion elements; recoding; translational hopping; transposable elements.

Hordeum bulbosum   Wild barley; it exists in diploid and tetraploid forms ($x = 7$). It gained particular attention because when crossed with the common barley (*Hordeum vulgare*, $2n = 14$) or with wheat ($2n = 42$), its chromosomes are eliminated from the zygote, and thus haploids are produced at a high frequency. *See* chromosome elimination; haploids.

Hordeum vulgare   Cultivated barley ($2n = 14$). It is used for animal feed, human food, and for the industry to produce malt and beer. World production is about half of that of maize and about 40% that of wheat. It probably evolved from the wild *H. spontaneum* ($2n = 14$), a species with which it is readily crossable and forms fertile hybrids. *H. spontaneum* has two dominant genes (*Bt* and *Bt*$_1$) that makes the ear brittle. Two-row barley develops fertile flowers in the central part of the spikelets, whereas in six-row barley all three flowers are female-fertile under the control of gene *v*. In naked barley, the glumes (husks) are not attached to the kernel because of a gene. It is an autogamous species. *See* databases.

horizontal transmission   *See* transfer, lateral; transmission.

horizontal resistance   The host is resistant to all races of a pathogen.

hormesis   Increased growth by irradiation at low doses or by other stress factors. It is a controversial idea that very low doses of radiation make human cells more resistant to subsequent exposure to higher doses. *See* radiation. (Macklis, R. M. & Beresford, B. 1991. *J. Nucl. Med.* 32:350; Holzman, D. 1995. *J. Nucl. Med.* 36:13n Calabrese, E. J. 2002. *Rev. Mut. Res.* 511:181.)

hormonal effects on sex expression   Although sex determination is under the control of genes within the sex chromosomes, the expression of sex characteristics may be influenced by natural hormones or administered through medical treatment. Human females treated with steroid hormones to prevent miscarriage may give birth to females that may become somewhat masculinized after puberty. The bovine freemartins also display some virile features presumably caused by intrauterine exposure to male sex hormones. Genetically determined subnormal production of pituitary growth hormone (human gene assigned to chromosome 17q23-q24) may involve recessive sexual anomalies. Castration and ovariectomy lead to intersex phenotype that can be further enhanced by grafting ovaries into male or testes into female chickens. The plant hormone gibberellin may affect the sex ratio in some species and can restore fertility in some genetic dwarf plants. *See* animal hormones; freemartins; gonads; growth hormones; hormones; plant hormones; sex determination; testicular feminization.

Testosterone-treated hen develops rooster type comb.

hormone receptors   Located on the surface or within target cells and transmit their signals to the genes that are set into action when the signals reach them. In *Caenorhabditis elegans*, hormone receptors represent about 1.5% of the coding sequences of the entire genome.

The general structure of the Zn fingers of the steroid and thyroid (peptide) hormone receptors is shown below. Steroid hormones are not readily soluble in aqueous media like the blood and require special serum proteins to be carried to the plasma membrane. The glucocorticoid receptor is usually situated in the cytoplasm and moves to the nucleus after binding with a ligand. The estrogen and progesterone receptors are mainly in the nucleus, whereas the thyroid receptors are present only in the nucleus. The hormones combine with hormone receptor proteins. These complexes may then bind to other complexes of hormones and navigate to the hormone-response elements (HRE) in the upstream regulatory regions of the genes. The HREs are found in either the promoter or in the enhancer regions, where they regulate the transcription of particular genes. These HRE elements vary from hormone to hormone, but all have a consensus sequence of dyad symmetry,

sometimes palindromic and frequently separated by three or more nonconserved bases. The most important functional parts of the hormone receptors are the 66 to 68 conserved and basic amino-acid-rich tracts that bind to the DNA. The binding may require that a ligand be associated with the receptor or removal of the ligand-binding domain.

The DNA-binding region of the hormone receptor forms two zinc fingers by cross-linked to Zn. The two fingers are coded for by two separate exons, and the entire DNA-binding region is separated by introns from other coding regions of the receptor gene. The structure shown below represents the estrogen receptor, but it is characteristic for many transcription factors. Thus, the hormone receptors are hormone-inducible transcription factors. For the DNA binding, the 3 amino acids following the last Cys residue of the right finger appear most important. The palindrome-like structure of the HRE in the DNA indicates that the hormone receptor protein works as a dimer. The function of the receptor as a transcription factor requires hormone binding. The residues near the C terminus primarily determine hormone binding. This hormone-binding region is critical for specificity. If a chimeric protein is constructed that replaces this binding region with the amino acid sequences of another hormone receptor, then the chimeric protein activates the transcription of genes normally receptive to the first segments of the chimera in response to the latter hormone but not to the hormone that normally activates these genes. The receptor protein does not bind to HRE unless it is linked to the hormone. Yet if the hormone-binding domain is deleted, the receptor can bind to the cognate HRE.

When the human estrogen response element (HRE) is transfected into yeast along with a functional estrogen receptor gene, estrogen can promote the expression of yeast genes even if their normal UAP (upstream activating sequence) has been deleted. In some instances, all the components of transcriptional activation by hormones are present but no

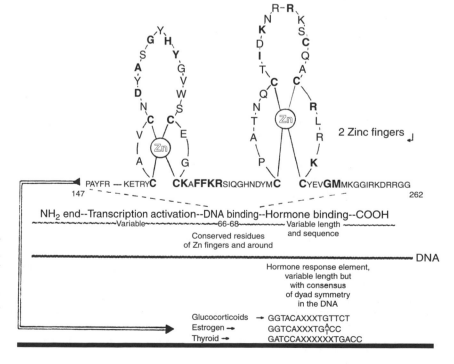

Zinc fingers of the estrogen receptor DNA-binding protein. The most conserved residues are shown in bold. Nuclear localization is specified by the carboxy terminus. The amino acid symbols in protein sequences are given by the single-letter code. Three different hormone-response consensus sequences are shown at the lower part of the diagram. The hormone-response element is generally about 200 bases upstream from the transcription initiation site.

activation is observed because an inhibitor may tie up the system and may prevent, e.g., dimerization of the receptor. These basic molecular genetic studies have found medical applicability. The proliferation of breast cancer depends on a continuous supply of estrogen. The drug tamoxifen can compete for the estrogen-binding site, but this complex is not capable of activating transcription. Thus, it may be used as an antineoplastic drug in combination with surgery, irradiation, and chemotherapy if clinical evidence indicates that the tumor has estrogen and progesterone receptors. The steroid analog, RU486, developed in France, is a progesterone analog (antiprogesterone) and can block the implantation of the fertilized ova because of the depletion of normally functional receptors. It is also beneficial for the treatment of endometriosis and leiomyoma. Thus, it is used as a morning-after pill for the prevention of pregnancy in some countries. The most commonly used birth-control pills contain a combination of estrogen and progesterone. Their elevated level shuts down the production of the pituitary hormones, thus preventing ovulation. See breast cancer; DNA-binding protein domains; endometriosis; hormone-response elements; hormones; leiomyoma; nuclear receptors; RU486; signal transduction; tamoxifen; testicular feminization; transcription factors. (Zhu, T., et al. 2001. *Cell Signal.* 13:599; Chang, C. & Stadler, R. 2001. *Bioessays* 23:619; Robinson-Rechavi, M., et al. 2001. *Trends Genet.* 17:554.)

hormone-response elements (HRE) Short DNA sequences flanking genes that respond rapidly to activation by steroid or peptide hormones. Steroid hormones, retinoic acid, thyroid hormone, and vitamin D_3 interact with ligand-activated transcription factors. The receptors for this steroid/nuclear receptor superfamily are bound to the interspaces of the "half-sites" (n) between the tandem repeats of six base pairs:

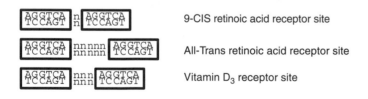

9-CIS retinoic acid receptor site

All-Trans retinoic acid receptor site

Vitamin D_3 receptor site

These six nucleotide pairs are unchanged, but the number of bases (n) between the boxes varies. These are recognized by heterodimers of receptor proteins. The binding domains may include Zn fingers. See estrogen response elements; ethylene; glucocorticoid response elements; nuclear receptors; RAR; RAX; regulation of gene activity, retinoic acid; thyroid hormone response elements; Zn fingers. (Klinge, C. M. 2001. *Nucleic Acids Res.* 29:2905.)

hormones Peptides (polypeptides), amino acid derivatives, and steroids synthesized in a gland of animals and carried by the blood to the site of their action(s) where they control the local function (first messengers). Animal hormones are, e.g., somatotropin (growth hormone), corticotropin (in adipose and kidney tissues), thyrotropin (stimulates thyroids), follitropin and lutropin (in gonads), prolactin and lipotropin (in mammary glands), insulin (regulates sugar metabolism), serotonin (in nervous system), testosterone, estrogen (in most

cells), progesterone (in uterus), prostaglandins (in smooth muscles), etc. Plants synthesize five groups of hormones that are quite different (except the brassinosteroids) in chemical nature from animal hormones. They primarily control cell elongation (auxins), cell divisions (cytokinins), germination, elongation (gibberellins), abscission of leaves and fruits, dormancy, germination (abscisic acid), ripening, and morphogenesis (ethylene). These plant hormones display numerous types of interactions and are involved in a complex manner in signal transduction. See animal hormones; brassinosteroids; hormone receptors; hormone-response elements; plant hormones; steroid hormones.

horotelic evolution See bradytelic evolution.

horse *Equus caballus*, $2n = 64$. Mongolian wild horse (*Equus przewalskii*) $2n = 66$. Mitochondrial DNA evidence indicates that domestication of horses must have occurred in numerous independent events. See hinny; map; mule; Prezewalsky horse. (*Genomics* 66:123 [2000].)

Przewalsky horse.

horsetail movement During the meiotic prophase of yeast, the nucleus — under the direction of the spindle pole body and with the aid of the microtubules — slowly oscillates, reminding of how the horse moves its tail. The telomeres are then clustered and the chromosomes are paired. (Hiraoka, Y. 1998. *Genes Cells* 3:405.)

horseradish (*Armoracia rusticana*) Perennial pungent condiment; $2n = 4x = 32$.

horsetail stage In fission yeast (but not in budding yeast) meiosis, before intimate pairing the 3 chromosomes aggregate to the spindle pole body (SPB) by their telomeres and are moved back and forth until the homologous regions find each other along the string. This stage is similar to the bouquet stage in the majority of eukaryotes. See bouquet; spindle pole body.

horse threadworm See Ascaris.

HOS1, HOS2, HOS3 Histone 3 and 4 deacetylase enzymes of yeast. See histone deacetylation.

host Organism that can enter into a particular relationship with another.

host cell reactivation DNA repair mechanisms operating in bacterial cells harboring defective bacteriophages. See DNA repair.

host-controled DNA modification   *See* restriction modification.

host-lethal genes   *See* colicins, killer strans, killer plasmids, pollen killer.

host-mediated assay   Tests whether the host metabolism can convert a compound (a promutagen) into a mutagen, detectable by the cells passed through the host, e.g., yeast, bacterial or animals cells injected into and withdrawn from the abdominal cavity of a mouse previously exposed to a potential mutagen. Forward or backmutation frequency in the cells is assessed after withdrawal from the abdominal cavity of the test animal, then plating the cells onto selective solid media or growing them in liquid media. Such a rapid assay was designed to be substituted for the slow and expensive direct mammalian assays of promutagens. *See* Big Blue®; bioassays in genetic toxicology; Muta™ Mouse. Diagram below.

host–pathogen relationship   Based on susceptibility or tolerance/resistance genes in the host and virulence and avirulence genes in the pathogen. The development of the disease depends on the appropriate genetic determination and expression of genes in the two organisms. In order to display resistance, the plants usually carry an *R* (dominant gene[s] for resistance) and the microorganism has an *avr* (avirulence) gene(s). The several disease-resistant plant genes (R) interact with RAR1 and the SGT1 complex of SKP1, the Cul1 subunits of SCF (Skp1-Cullin-F box), and have ubiquitin ligase function (Azevedo, C., et al. 2002. *Science* 295:2073). Other allelic combinations lead to disease susceptibility. In plant breeding, successful efforts were made to transfer disease-resistant genes from wild relatives to cultivated varieties — e.g., the *Ry* dominant genes of the wild potato, *Solanum stoloniferum*, convey resistance against the potato Y virus (PVY) or leaf rust (*Puccinia triticina*) from *Aegilops umbellulata* (now name *Triticum umbellulatum*) to the cultivated bread wheat, etc.

These classical methods of breeding for resistance had to overcome problems of linkage of the resistance to agronomically undesirable genes present in the primitive species. Some viral plant pathogens have lines with extra genomic *satellite RNA*, and the presence of the latter may suppress the expression of full-scale disease (*attenuation*). Disease symptoms may be prevented in some cases by the simultaneous presence of another related plant virus (*cross-protection*). It has been shown that 25-nucleotide viral RNA may target a 25 kDa protein involved in viral movement and may generate a posttranscriptional silencing signal and protection against other viruses. This phenomenon, also called *preimmunity*, can be exploited by transferring a coat protein gene of some of the RNA viruses through transformation into the plant species.

Another molecular plant breeding approach is to introduce antisense mRNA of the viral proteins into the plants to prevent their synthesis. This latter method may be coupled with a ribozyme expression system that would degrade the viral RNA. Plant resistance or tolerance to viral infection may be based on transcriptional of posttranscriptional silencing bearing similarities to cosuppression by transgenes. In some plants, the silencing may take place without transgenes.

Various genetic mechanisms are known to be involved in resistance against bacterial and fungal pathogens. The *hypersensitivity reaction* of plant tissues restricts the infection by death of the surrounding host cells, resulting in the liberation of antimicrobial cellular substances. A collection of chemically diverse substances called phytoalexins regulate the synthesis of plant cellular compounds that are involved in the defense mechanisms or are the consequences of the infection process. Various genes involved in the synthesis of phenolderivatives may be the regulatory targets. The activation or enhancement of the transcription of genes determining cell wall components (polysaccharides, lignin, suberin, saponin, etc.) may provide a barrier to infection. The activation of plant enzymes (*pathogenesis-related proteins*) such as various glucanases, chitinases, and proteases may lead to the continued breakdown of the cell walls of the pathogens,

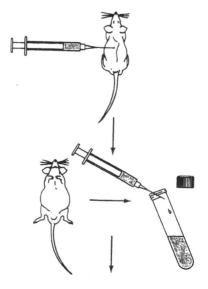

Microbial or cultured animal cells are injected intraperitoneally into a mammalian host for proliferation.

After 1 to 3 days, a test chemical dissolved in saline or in saline plus a solubilizer is injected subcutaneously once or more into the mouse or rat. The control animals are treated with the solvent solution only. This mutagenic exposure lasts for a few hours or a few days.

A sample of the injected cell population is withdrawn from the abdominal cavity into centrifuge tubes in preparation for assaying the mutagenic effectiveness of the test chemical in comparison with the control.

The recovered cells are plated on Petri dishes or incorporated into soft agar culture tubes to screen the mutants produced within the host as a consequence of the treatment.

General outline of host-mediated mutagen test. (Modified after Fischer, G. A., et al. 1974. *Mutation Res.* 26:501.)

thus facilitating the escape of potential hosts from microbial infection. Upon infection, the plants may produce reactive oxidative species (ROS) that may damage or destroy the infectious agents. Plants ectopically expressing ferritin may become tolerant to oxidative damage and pathogens (Deák. M., et al. 1999. *Nature Biotech.* 17:192).

Plants do not have an immune reaction to invading agents, yet some type of systemic resistance may be induced by exposure to necrosis-causing microorganisms with certain chemicals such as nicotinic acid derivatives, etc. Some microorganisms secrete or contain organic molecules (proteins, glucans, glucosamines, fatty acid derivatives, etc.) in their cell walls that provoke the defense mechanisms of plants, or the plants themselves may produce such molecules upon contact with the pathogens. These compounds are called *elicitors*. Elicitors come in much different chemical compositions and display substantial specificity for different hosts and pathogens. Many of these substances seem to reach the plant cell nucleus through the various ion channels or the signals are transduced through the membrane systems to activate the appropriate plant cells and functions. Several plant genes have been cloned that are involved in the disease-resistance-tolerance mechanisms (*RPS1, RPS2, Pto,* [*Pseudomonas syringae*], *Cf-9* [*Cladosporium fulvum*], L^6 [*Melampsora lini*], *N* [tobacco mosaic virus], *Hm1* [*Cochliobolus carbonum*], etc). It is believed that these genes activate the plant defense system through responding to the elicitors of the avirulent pathogens or by producing enzymes that degrade the fungal toxin. In some instances, inorganic antimicrobial compounds accumulate.

In *Theobroma cacao*, cyclooctasulfur accumulation in the xylem walls was observed in *Verticillium dahliae*–resistant plants. Salicylic acid is generally considered an agent for improving plant resistance probably through its stimulating action of hydrogen peroxide release from the cells. Plant mutants with a reduced ability to synthesize salicylic acid or phytoalexins become more susceptible to pathogens. These latter compounds are credited with systemic acquired resistance (SAR). The *Xa21* (*Xanthomonas oryzea*) gene of rice conveys resistance to this plant pathogen. DNA sequence analyses indicate a leucine-rich repeat characteristic for serine/threonine kinases. It has been suggested that this protein is a signal transducer to alert the plant cell defense system. In some pathogenic bacteria (*Pseudomonas*), the same virulence factors mediate pathogenicity in plants (*Arabidopsis*) and in animals (mouse). It is assumed that a signal transduction path in general mediates plant disease resistance. A pathogen generated ligand produced by an avirulence gene is recognized by an extra- or intracellular receptor, encoded by a plant receptor gene, which sets into motion the process of defense. Different plant species and different resistance reactions indicate the involvement of a leucine-rich repeat (LRR) and leucine zipper-binding sites shared by many resistance genes. The majority of isolated disease-resistant genes display a nucleotide-binding domain, a common feature of many ATP- and GTP-activated protein families involved in signal transduction. The nucleotide-binding leucine-rich repeat (NB-LRR) resistance genes (R) and pseudogenes are present in multiple (about 150) copies in *Arabidopsis*. They are present in all five chromosomes as singletons or multiple clusters. The R genes may recognize more than a single type of avirulence factors and protect against different types of

pathogens, e.g., fungi and viruses or viruses and nematodes. Great diversity may exist among the different R genes; some have lost the LRR sequences, yet they may be involved in some types of signal transduction. Plant disease-resistant genes may be classified into five structural groups and some of their domains reveal homologies to innate immunity genes of animals (e.g., the Toll-like receptors [TLR]). The criteria for classification of resistance genes involve their location or the location of some of their domains, e.g., cytoplasmic or transmembrane or membrane association or the combinations of these features.

The *Pto* serine/threonine protein kinase of tomato and the closely linked *Prf* gene seem to be binding directly to the *Pseudomonas syringae* avirulence gene (*AvrPto*) conveying susceptibility to the pathogen. A single amino acid substitution in the plant-resistant gene *Pto* may be sufficient to develop resistance to *Pseudomonas*. Several bacterial species rely on acyl-homoserine lactones (AHL) for quorum sensing and to select a host plant for invasion, eventually causing disease. If AHL can be inactivated by breaking special bonds of the molecule (quorum quenching), the disease symptoms can be prevented (Dong, Y.-H., et al. 2001. *Nature* 411:813).

Many of the putative resistance genes cloned may lack the stringent criteria to be the principal determinants of resistance to a particular pathogen. Some plant pathogenic fungi (*Helminthosporium victoriae*) can be infected with double-stranded RNA viruses (totivirus) and cause lytic disease to the fungus. In a wild relative of cultivated beets, a membrane protein was identified that conveys resistance against a nematode (a lower eukaryotic animal). The defense of animals relies on the innate and adaptive immunity systems. Posttranscriptional gene silencing may control resistance to viral diseases. Although insect resistance and herbicide resistance were successfully incorporated into plants by molecular engineering methods, disease-resistant transgenic plants appear to be of less economic success. *See* 2,5-A; alien substitution; antimicrobial peptides; antisense RNA; apoptosis; avirulence; biological control; caspase; chitin; coevolution; cosuppression; ethylene; Feys; Floor's model; guard hypothesis; hypersensitive reaction; immune system; infection; innate immunity; insect resistance in plants; jasmonic acid; methylation of DNA; octadecanoic acid; oleuropein; oxidative burst; parasitoid; pathogen-derived resistance; pathogenicity island; phytoplasma; plant defense; plantibody; plant pathogenesis; posttranscriptional gene silencing; PTGS; guorum sensing; RdRp; ribozyme; RNAi; Ros; salicylic acid; saponins; SAR; secondary metabolism; secretion machine; selection types; signal transduction; silencer; transformation; VIGS; virulence; wound response. (Feys, B. J. & Parker, J. E. 2000. *Trends Genet.* 16:449; Dangle, J. L. & Jones, J. D. G. 2001. *Nature* 411:826; Cao, H., et al. 2001. *Annu. Rev. Phytopath.* 39:259; Hulbert, S. H., et al. 2001. *Annu. Rev. Phytopath.* 39:285; Lengeling, A., et al. 2001. *Mamm. Genome* 12[4]:261; Asai, T., et al. 2002. *Nature* 415:977; Maleck, K., et al. 2002. *Genetics* 160:1661; Mackey, D., et al. 2002. *Cell* 108:743; Schneider, D. S. 2002. *Cell* 109:537; Lellis, A. D., et al. 2002. *Current Biol.* 12:1046; <http://www.ifgb.uni-hannover.de/extern/ ppigb/ppigb.htm>.)

host range Infectious agent (virus, bacterium, fungus) can grow in some but not in other individuals depending on the genotype of the agent and its target. (*See* da Silva, A. C. R., et al. 2002. *Nature* 417:459.)

host-range restriction The same oncoprotein may not transform different species because the host cellular machinery does not favor its function. Some viruses infect or fail to infect other species. (Oto, T. & Kawaoka, Y. 2000. *Vet. Microbiol.* 74[1–2]:71.)

host-resistance genes In mammals, these genes are generally quantitated on the basis of the length of survival after infection by viruses, bacteria, or protozoa or on the basis of the density of the pathogen in the infected foci. The inheritance is assessed on the basis of crosses with recombinant inbred strains in mice. Most of the resistance genes are linked to the H-2 complex (MHC), but other locations have also been identified as resistance sources. Several resistance genes have been cloned. Comparative analyses assist in the use of the information for human and livestock research. *See* host-pathogen relations; pathogenicity island. (Staskawicz, B. J., et al. 2001. *Science* 292:2285.)

hot spot Highly mutable site within a gene or high-frequency recombination site in chromosomes. Recombinational hot spots are generally expressed as cM/kb (map unit/kilobase). Within the a^1 gene of maize, the recombination frequencies varied between 5×10^{-3} and 8×10^{-2} cM/kb, whereas in the 140 kb distance between a^1 and *sh2*, 6×10^{-4} cM/kb was observed. Within the pseudoautosomal site (*PAR*) of *Drosophila*, 1 cM/53 kb was found. Within the mammalian histocompatibility locus, recombination frequencies of 1 cM/0.6 kb to 1 cM/400 kb have been reported. The frequency of recombination shows great variation along the length of the chromosomes. Usually, a number of hot spots are recognized and there is no equivalence between genetic and physical length. In the vicinity of minisatellites, recombinational hot spots may be found by molecular analyses. In human chromosome 22, recombination is increased between long tandem GT repeats. *See* chi elements; coefficient of crossing over; cold spot; HLA; mariner; MITE; recombination frequency; recombination hot spots. (Lichten, M. & Goldman, A. S. H. 1995. *Annu. Rev. Genet.* 29:423; Majewski, J. & Ott, J. 2000. *Genome Res.* 10:1108; Gerton, J. L. 2000. *Proc. Natl. Acad. Sci. USA* 97:11383.)

hot-start PCR Used to detect allele-specific changes in mtDNA. The reaction begins at relatively high temperatures, but temperatures are reduced during the cycling (touch-down PCR) in order to improve the specificity of the amplification. *See* PCR allele-specific; polymerase chain reaction; TULIPS-PCR. (Ailenberg, M. & Silverman, M. 2000. *Biotechniques* 29:1018.)

hot-stop PCR Method for quantitation of allele ratios. (Uejima, H., et al. 2000. *Nature Genet.* 25:375.)

housefly *Musca domestica*, $2n = 12$. Genome size bp/$n = 9 \times 10^8$.

housekeeping genes Functional throughout the life of a cell and in the majority of cells and tissues. In the promoters, the TATA and CAAT boxes are not present. Their 5′ CpG islands are not methylated. *See* asparagine synthetase; CAAT box; capping enzymes; CpG islands; methylation of DNA; TATA box, transcription, illegitimate.

HOVERGEN Homologous vertebrate genes. A database that facilitates identifying homology between nucleic acid and protein sequences. *See* databases; evolution. (<http://acnuc.univlyon.fr/start.html>.)

HOX Proteins control differentiation along different paths in the anterior-posterior axis of the body. Mammalian cells are endowed with 39 HOX genes in four clusters (HoxA to HoxD) that belong to 13 paralogous groups. The expression of the members is generally activated in $3' \to 5'$ direction within the group. *See* homeotic genes; PAX. (Gehring, W. J., et al. 1994. *Annu. Rev. Biochem.* 63:487; Moconochie, M., et al. 1996. *Annu. Rev. Genet.* 30:529; Shen, W., et al. 2001. *Mol. Cell Biol.* 21:7509.)

HOWDY Integrated human genome database. It permits cytogenetic localization. (<http://www-alis.tokyo.jst.go.jp/HOWDY/>.)

HP1 *See* heterochromatin.

***Hpa* II** Restriction enzyme recognition site is C↓CGG.

HPBP Human platelet basic protein is a member of a large cytokine family. *See* cytokines. (Zhang, C., et al. 2001. *Blood* 98[3]:610.)

HPFH *See* thalassemia.

HPFN Human plasma fibronectin is a cell adhesion molecule. *See* CAM. (Pouloouin, L., et al. 1999. *Protein Expr. Purif.* 17:146.)

HPLC High-performance liquid chromatography is suitable for the separation (among other molecules) of oligonucleotides of up to 20 residues by partitioning them between a stationary phase (such as a chromatography column) and a mobile phase (such as solvents), forming a gradient pumped through a system that is carefully monitored. *See* electrophoresis; high-performance liquid chromatography; sequenator.

HPRT Hypoxanthine phosphoribosyl transferase is controlled in humans by a recessive gene (map location Xq26-q27.2, 57 kb). Its deficiency causes Lesch-Nyhan syndrome involving choreoathetoses (involuntary, uncoordinated movements), spasticity (high muscular tension), mental retardation, tendency to self-mutilation, overproduction of uric acid, renal damage, etc. It can be detected prenatally. The molecular basis of the defect is an interruption of the salvage pathway of nucleotides because guanosine and hypoxanthantosine cannot be phosphorylated. Defect of this enzyme has been extensively exploited for in vitro selective isolation of fused mammalian cells and for the isolation of mutations. Prenatal diagnosis and carrier identification of defects is feasible. *See* epilepsy; HAT medium; HGPRT; Lesch-Nyhan syndrome; salvage pathway.

HPV Human papilloma virus. *See* bovine papilloma virus; papilloma virus.

hr Human recombinant in abbreviation, e.g., hrEGF, human recombinant epidermal growth factor.

HR Hypersensitive reaction. *See* host-pathogen relationships; hypersensitive reaction.

hRAS (HRAS) RAS protooncogene of humans. *See* RAS.

HRE *See* hormone-response elements.

HRF Homologous restriction factor (C8bp). *See* membrane attack complex.

HRI Hemin-regulated inhibitor, EIF2AK3, human chromosome 2p12. Hemin apparently interferes with the function of this specific protein serine/threonine kinase involved in translational activation; in the absence of hemin, the protein kinase is active and protein synthesis may be shut off, leading to cell death. Hemin inactivates it by forming disulfide-linked homodimers. Functionally it is similar to yeast GCN2 and PKR endoplasmic reticulum elongation factor kinase, PEK/PERK. *See* GCN2; GCN4; hemin; PKR; protein kinase. (Crosby, J. S., et al. 1994. *Mol. Cell Biol.* 14:3906.)

hRP-A Human homologue of the single-stranded-binding protein involved in genetic recombination. It is required for in vitro replication of SV40 DNA and growth. *See* SV40. (Baumann, P. & West, S. C. 1999. *J. Mol. Biol.* 291[2]:363.)

hs As a prefix, stands for *Homo sapiens*, e.g., hsDNA, human DNA, or hDNA.

Hsc Heat-shock cognate. A form of a heat-shock protein in the normal cytosol. *See* heat-shock protein; Hsp70.

Hsc66 615-amino-acid *E. coli* protein, encoded by *hscA*, member of the Hsp70 family. It is induced by cold shock It lacks the ATP-binding domain of Hsp70. It is cotranscribed with *hscB* (encoding a DnaJ-like protein) and *fdx* (encoding ferredoxin). *See* cold shock; DnaK; ferredoxin; heat shock. (Silberg, J. J., et al. 2001. *J. Biol. Chem.* 276:1696.)

hsdR Some mutations in *E. coli* that eliminate restriction function but not the modification in the restriction-modification system. *See* restriction enzyme; restriction modification. (Doronina, V. A. & Murray, N. E. 2001. *Mol. Microbiol.* 39:416.)

HSE Heat-shock elements located in the upstream region of eukaryotic heat-shock responding genes and recognized by heat-shock protein transcription factors. A typical heat-shock element is about a 14 bp palindromic sequence with a GAA core:

5′-CTN**GAA**NNTTCNAG-3′

3′-GANCTT**NN**AAGNTC-5′

See heat-shock proteins; Hsc; HSTF; transcription factors, inducible.

HSF Heat-shock transcription factor that activates the transcription of heat-shock proteins. *See* heat-shock proteins; HSE.

Hsj1a, Hsj1b Neuron-specific chaperones of the DnaJ family. Alternative splicing of the same RNA transcript produces them. They do not have intrinsic ATPase activity, yet they increase this function of Hsp70. *See* chaperone; DnaJ.

HSP Heat-shock protein. HSP110 of mammals is similar to the SSE proteins of yeast and is related to Hsp70. The 40 kDa mammalian chaperone Hsp40 facilitates the folding of polypeptides coming off the ribosomes. These are also stress proteins highly inducible by heat, although they may be formed constitutively. The sea urchin egg carries an HSP that serves as a receptor for sperm. Another related protein is Osp94/APG-1, which is formed in response to heat, high osmotic, and other stresses. GRP glycoproteins of diverse molecular sizes are also heat-inducible and function in the spleen, lymph nodes, and Peyer's patches in connection with immune responses. The HSP40 family members are cochaperones of the Hsp70 family of proteins resembling DnaJ. The HSP40 members interact with nucleic acid and proteins in concert with Hsp70, and occur in the nucleus, mitochondria, endoplasmic reticulum, and on the membrane surfaces, mediating chaperone functions (protein folding, translocation, renaturation of proteins of wrong conformation, proteolysis), signal transduction, and formation of macromolecular complexes. The Hsp100 (Clp) family of proteins is generally involved in the aggregation and disaggregation of proteins of different sizes. Hsp104 is another stress protein controlling the Psi+ prion-like protein of yeast. Hsp47 is a serpin family stress protein of vertebrates found in the endoplasmic reticulum, where it binds to procollagen. Hsp101 is a plant chaperone closely related to Hsp104. Hsp78 is a soluble mitochondrial chaperone of yeast. Hsp26 is a sHsp of yeast with a major role in stress and cytoskeletal control. *See* chaperones; chaperonins; Clp; conformation; constitutive gene; DnaJ; endoplasmic reticulum; heat-shock proteins; Hsc; HSE; Hsp70; Hsp83; Hsp90; HSTF; immune system; lymph nodes; Peyer's patches; pro-collagen; PSI+; serpin; sHSP; signal transduction; spleen.

HSP60 (CPN60, chaperonin 60) *See* chaperonins.

Hsp70 Heat-shock protein gene (activated by elevated temperature); it is a chaperone. These proteins are widely distributed among diverse species, prokaryotes, and eukaryotes, and have been found in various subcellular compartments. The proteins are endowed with multiple functional domains required for peptide and cofactor binding and have weak ATPase activity. Hsp70 functions involve the control of protein folding, lytic phage infection of bacteria, uncoating of clathrin-coated vesicles and thereby selective endocytosis, assistance of proteases in degradation of abnormal proteins, etc. ATP hydrolysis and polypeptide-binding activity of Hsp70 depend on two different domains of the protein. Hsp70 needs the cooperation of Hsp40 (DnaJ). The human homologues (HSPA) are encoded in chromosomes 6p21.3, 14q22, 14q24.1, 5q31.1, 9q34, 1q, 14q24, and at other locations. Mouse homologues were identified in chromosomes 17, 12, 2, and 18. Male mice lacking Hsp70 homologues are unable to produce spermatids and mature sperm and showed a dramatic increase of apoptosis of the spermatocytes. Female mice of the same constitution did not show meiotic disturbances or infertility. The 4 *SSA* genes of yeast are highly homologous (80%) members of this family of proteins (~70 kDa). *SSA2* is essentially constitutive; the others are heat-inducible. The SSA gene products negatively regulate their own and each other's synthesis. The *SSA* genes are essential for normal cell viability. The SSB gene products of yeast are associated with the ribosomes. The Ssc yeast proteins are in the mitochondrial matrix and envelope and aid in the translocation between the cytosol and the mitochondria; they are involved in general protein folding and lysis in that organelle. Ssc proteins are weakly inducible by heat. Ssh1 protein is encoded in chromosome 12 of yeast and located in the mitochondria, where it may be involved with DNA replication and protein processing. Ssi (100 kDa) protein controls endoplasmic reticulum traffic, particularly when the *KAR2* gene of yeast is defective. Kar2 protein (682 amino acids) in the endoplasmic reticulum of yeast mediates translocation and folding of proteins and nuclear fusion after mating. BiP is an essential endoplasmic reticulum chaperone present in fission yeast, several plant species, *Trypanosomas*, and mammals. The Hsc4 proteins of *Drosophila* are also Hsp70-related. Hsc3 in *Drosophila* is homologous to Kar2 and BiP. The mammalian Prp73 is a 73 kDa chaperone recognizing the S peptide (residues 1–20 of ribonuclease A). Pbp74 (74 kDa) peptide-binding protein in the mammalian and yeast mitochondria aids in peptide import and processing within the mitochondrial matrix. Stch chaperone (471 amino acids) is encoded in human chromosome 21q11.1, and the protein is located in the microsomal fraction of the cell and has ATPase activity. *See* apoptosis; ATPase; chaperone; clathrin; endocytosis; heat-shock proteins; HSE; HSTF; Prp73. (Hartl, F. U. 1996. *Nature* 381:571; Bukau, B. & Horwich, A. L. 1998. *Cell* 92:351.)

Hsp83   *See* Hsp90.

Hsp90   Chaperone protein family including heat-inducible and constitutive chaperones involved with the signal transduction kinases and steroid receptors. Homologues occur in prokaryotes (HtpG), *Drosophila* (Hsp83), mammals (Grp94), yeast, and plants. *See* chaperones; Grp94; heat-shock proteins. (Kumar, P., et al. 2001. *Cell Stress Chaperones* 6:78.)

HSR   *See* homogeneously stained region.

HSSB   Human single-strand-binding protein. *See* replication protein A.

HS site   *See* DNase hypersensitive site.

HSTF   Heat-shock transcription factor regulates about 20 heat-shock genes by binding cooperatively to more than one heat-shock element. It is activated by phosphorylation. *See* HSE; HSP. (Tanguy, R. M. 1988. *Biochem. Cell Biol.* 66[6]:584.)

HST oncogene   Assigned to human chromosome 11q13; originally identified in stomach cancer, but it was also found in Kaposi sarcoma. Its product is a heparin-binding protein with homology to fibroblast growth factor (FGF). It is also called K-FGF. *See* FGF; growth factors; Kaposi sarcoma; oncogenes.

Hst1p (homologue of Sir2)   Protein involved in mating-type determination in yeast. (Rusche, L. N. & Rine, J. 2001. *Genes Dev.* 15[8]:955.)

HSV   Herpes simplex virus. *See* herpes.

HTC   Database for high throughput of not yet definite genomic sequences.

HTDV   Human teratocarcinoma-derived virus is an endogenous human retrovirus family occurring in about 25–50 copies per genome, and about 10,000 solitary LTRs are present. They are well expressed in different organs. *See* retroviruses; solitary LTR.

hTERT   Catalytic subunit of human telomerase. *See* telomerase.

H1TF2   Histone transcription factor binding to the CCAAT box. *See* histones. (Martinelli, R. & Heintz, N. 1994. *Mol. Cell Biol.* 14:8322.)

HTF islands   1–2 kb sequences around the 5′ region of genes where the Hpa II restriction endonuclease cleaves tiny fragments because the cytidines are not methylated. At most other regions, because of methylation only larger fragments are cut. The mammalian genome may display 30,000 HTFs. (Bird, A. P. 1986. *Nature* 321:209; Zardo, G. & Caiafa, P. 1998. *J. Biol. Chem.* 273:16517.)

HTG   High-thoroughput genome. Division of gene banks permitting the search for unfinished nucleotide sequences. *See* databases; GenBank; genome projects.

HTLV   Human T-cell leukemia virus is the retroviral causative agent of some adult leukemias and partial paralysis (HTLV-1-associated myelopathy). The 40 kDa viral TAX protein (trans-acting viral factor) is apparently responsible for its carcinogenicity. *See* leukemia; myelopathy. (Overbaugh, J. & Bangham, C. R. M. 2001. *Science* 292:1106; Yoshida, M. 2001. *Annu. Rev. Immunol.* 19:475.)

HTML   Hypertext markup language. The computer program that can generate world wide web pages and links. HTTP (hypertext transfer protocol) is the program to use with the www. *See* www.

HtpG   *See* Hsp90.

HU   "Histone-like" prokaryotic protein involved in maintaining DNA structure and transposon integration. *See* histones.

hUBF   Human upstream binding factor is a transcription factor. *See* transcription complex; transcription factors.

Hudson, Kreitma, Aguadé method   *See* mutation, neutral.

HUGO   Human Genome Organization. Source of mammalian genomic data available in Macintosh Hypercard disks from HUGO Europe, One Park Square West, London NW1 4IJ, UK; phone: 44 71 935 8085; fax: 44 71 706 3272; <s.brown@sm.ic.ac.uk>.

Reference points of the human chromosomes (see caption on p. 571)

human artificial chromosome (HAC) Not yet equivalent to similar constructs in *Saccharomyces* or *Schizosaccharomyces*. The closest are the minichromosomes, which may be less than one-tenth of the size of the chromosome before truncation yet mitotically quite stable. Human minichromosomes can be generated in vitro by mixing alphoid, telomeric, and carrier DNAs. These structures that are transfected into human cells are maintained, segregate, and bind centromeric proteins. Current evidence indicates, however, that neither the alphoid DNA nor the CENP-B protein may be required for normal centromeres. Neocentromeres have been analyzed and apparently lack alphoid DNA. They are not suitable for the construction of human artificial chromosomes. Human artificial chromosomes may not function normally if they are introduced into somatic cells because under normal conditions the cells are disomic rather than trisomic, although the extra chromosome (HAC) would not carry genes already present in one or two doses in the genome. This problem might even further be aggravated if the cell undergoes meiosis. In cell culture, human artificial chromosomes have been used as genetic vectors to introduce specific genes and have demonstrated the potentials of complementing deficiencies of hypoxanthine guanine phosphoribosyl transferase. HACs may be exploited for the transfer of genes from humans to other mammals. Human artificial chromosome vectors equipped with chromosome 17 alphoid DNA generated 32–79% artificial minichromosomes, but the alphoid DNA of the Y chromosome generated only in 4%. The stability of the minichromosomes decreased when the guanine phosphoribosyl locus (HPRT1) was included. *See* α-satellite; alphoid DNA; BAC; disomic; HAC; minichromosome; neocentromere; PAC; trisomy; YAC. (Harrington, J. J., et al. 1997. *Nature Genet.* 15:345; Ikeno, M., et al. 1998. *Nature Biotech.* 16:431; Tyler-Smith, C. & Floridia, G. 2000. *Cell* 102:5; Mejía, J. E., et al. 2001. *Am. J. Hum. Genet.* 69:315; Mejía, J. E., et al. 2002. *Genomics* 79:297; Csonka, E., et al. 2000. *J. Cell Sci.* 113:3207.)

human chromosome maps Linkage and mapping information before the 1960s was obtained mainly by studying family pedigrees. This procedure was very inefficient because in human populations controlled mating is not feasible, and because of the small populations available, crossing-over frequencies could not be used in the same manner as, e.g., in *Drosophila*. Some of the gene locations were determined by deletion mapping, but until 1956 even the number of human chromosomes was uncertain. In the late 1960s and early 1970s, new chromosomal staining techniques made the recognition of chromosome bands possible, permitting the more precise location of genes relative to these beacons. During the 1960s, the culture of isolated mammalian cells was improved. Fusion of somatic cells was accomplished in 1960, followed in 1965 by the production of mammalian somatic cell hybrids. It was discovered that human chromosomes in human-mouse somatic cell hybrids are preferentially lost, facilitating the assignment of particular genes to specific human chromosomes. Only those human genes were expressed in the cultures that were present, and their expression ceased when the critical chromosome was eliminated. By induced breakage, the location of genes could be further specified when it was seen that the retention of particular cross-bands was essential for the expression of a particular gene.

In 1960, nucleic acid hybridization was discovered, and by the mid-1960s, nucleic acid hybridization was used for localizing genes in cultured human HeLa cancer cells. Somatic cell hybridization permitted the chromosomal assignment of three times as many genes within a few years as family analysis accomplished over half a century. By 1970, methods were developed to assign mouse DNA fragments to chromosomal location by in situ hybridization. In situ hybridization also facilitated the location of about twice as many genes than the pedigree analysis. The discovery of restriction endonuclease enzymes in the late 1960s and 1970s opened a new era in genetics of physical mapping, localizing cloned genes or just "anonymous DNA" sequences to chromosomal positions. Isolated DNA sequences could be ordered on the basis of the overlapping fragments by the use of chromosome walking and jumping first employed in *Drosophila* in 1983. The use of isolated genes from other species (*Drosophila*, mouse, yeast, etc.) as probes permits the identification of the position of homologous genes and DNA sequences. By the extension of these techniques, individual human genes can be isolated, cloned, and sequenced; thus, the technology is in use for the ultimate mapping of the human genome at

---

Reference points for mapping genes or DNA sequences to human chromosomes (p. 570) Each chromosome is numbered; 1 is the longest by cytological evidence. The metaphase length is somewhat ambiguous because the condensation may vary due to genetic and other causes, and length is not an absolutely consistent feature of the karyotype. The approximate minimal DNA content (Mb) of each chromosome is shown in parentheses under each diagram. The centromere is represented by constrictions and the pericentromeric region is diagonally hatched. The short arms (shown above the centromere) are designated by p (for petit) and the long arms are designated by q. Each arm has cross-bands corresponding to observations by Giemsa, Q and C banding techniques. The fine hatching indicates heterochromatic regions. Each chromosome, starting at the centromere, is divided into domains and subdomains designated by numbers. If we see that the location of TNFR-2 (encoding tumor necrosis factor 2, a 75 kDa protein) is at 1p36.3-p36.2, we know that it is in the terminal region of chromosome 1, either in the first black band or above. The telomeric region is usually designated by ter. Acrocentric chromosomes (13, 14, 15, 21, 22) and their heterochromatic areas and satellites are finely cross-hatched. The large number of genes are impossible to represent on a single page. The location of the physical markers occupied, by 1996 (*Nature* 380, suppl.), about 138 pages in fine print. (Illustration is based partly on the New Haven Human Gene Mapping Library Chromosome Plots. Number 4. HGM9.5, constructed by Spence, M. A. & Spurr, N. K.) By 2001, after the completion of the rough draft of the human genome, the length of the individual chromosomes required revision compared to the data seen in this illustration. Thus, the latest length in Mbp of the chromosomes is: *1* 212.2, *2* 221.6, *3* 186.2, *4* 168.1, *5* 169.7, *6* 158.1, *7* 146.2, *8* 124.3, *9* 106.9, *10* 127.1, *11* 1286, *12* 124.5, *13* 92.9, *14* 86.9, *15* 73.4, *16* 73.1, *17* 72.8, *18* 72.9, *19* 55.4, *20* 60.5, *21* 33.8, *22* 33.8, *X* 127.7, *Y* 21.8. These figures ignore the unsequenced gaps. (For publicly available genome data, see *Nature* 409:860 [2001].) Physical maps of the human chromosomes can be found in some detail in *Nature* 409:942–958 (2001). The cytogenetic map shown here is still widely used, but one must remember that its resolution is not sufficient for the precise location of the large number of genes, estimated to be about 35,000 to 40,000. Therefore, the same cytogenetic map sites often accommodate several functionally different genes.

the nucleotide level, to recognize the structural bases of functions, regulation, and organization of the genome. The detailed mapping is assisted by sophisticated computer programs to determine the order of several gene loci on the basis of the maximum likelihood principle. This information, sought by the human genome projects, will contribute to evolutionary information as well as to medical applications of gene therapy. The 2002 (Kong, A., et al. *Nature Genet.* 31:241) genetic map includes 5,136 microsatellites (AC/TG) markers based on 1,257 meiotic events. The "complete" nucleotide sequence of the human genome available by 2000 facilitated the determination of the position of all human genes. Although the physical map of the human genome is practically complete (*Science* 291, Feb. 15, 2001; *Nature* 409, Feb. 15, 2001), the full functional annotations will have to wait. The genetic and the nucleotide sequence-based physical maps are not entirely consistent because of technical ambiguities (DeWan, A. T. 2002. *Am. J. Hum. Genet.* 70:101). *See* chromosome banding; chromosome walking; deletion mapping; draft genome sequence; DNA sequencing; gene therapy; human genome; in situ hybridization; mapping; maximum likelihood; mouse; nucleic acid hybridization; pedigree analysis; probes; radiation hybrids; restriction endonucleases; RFLP; somatic cell hybridization. (*Nature* 402:489; *Nature* 402:311; human physical map: <http://www.ncbi.nlm.nih.gov/genemap>; cytogenetic map of morbid genes: <http://www.ncbi.nlm.nih.gov/htbin-post/Omim/getmap>; HOWDY.)

human evolution   Important tools gained by the application of molecular markers of mtDNA (of skeletal remains) and mtDNA and Y chromosomes to the surveying of living populations. These methods permit the separation of gene flow by the sexes. Commonly, in migrating population mixtures, immigrant men and native women's gene pools seem to prevail because the invading men eliminated or reduced the number of the native men but begot offspring with the women, and among the immigrants, females were not always represented equally to males. The information may be biased by the size of the admixing populations. *See* Eve foremother of mitochondrial DNA; evolution; hominidae; mtDNA; Y chromosome. (Jones, S., Martin, R. & Pilbeam, D., eds. 1992. *Human Evolution.* Cambridge Univ. Press, New York; <www.becominghuman.org>.)

human genetics   Basic genetics using human (cells or individuals) as the subject of study. The human genome contains approximately 30,000 to 75,000 (?) genes, and their physical mapping and sequencing is near completion. In general, the function of several physically identified genes is discovered weekly. Human genetics is relying more and more on the methods of reversed genetics and bioinformatics. *See* BodyMap; clinical genetics; HGMD; medical genetics; OMIM. (Hawly, R. S. & Mori, C. A. 1999. *The Human Genome: A User's Guide.* Academic Press, San Diego, CA; Jenssen, T. K., et al. 2001. *Nature Genet.* 28:21; Bodano J. L. & Katsanis, K. 2002. *Nature Rev. Genet.* 3:779; <http://www.ncbi.nlm.nih.gov/>; <http://hummolgen.de>; <www.ncbi.nlm.nih.gov/genome/guide/human>; <www.ensembl.org/genome/ central>.)

human gene transfer   The vectors can be replication-deficient retroviral or adenoviral ones. The retroviral vectors may integrate 9 kb passenger DNAs into the chromosomes and may destroy tumor suppressor genes or activate oncogenes if inserted nearby, and thus may permanently alter the genome either favorably by inserting the desired genes or undesirably as mentioned above. Adenoviral vectors have a carrying capacity of about 7.5 kb DNA. They enter the cells by special receptors. Adenoviral vectors do not integrate into the chromosomes and are not replicating indefinitely, but they have to be reapplied after a few weeks or months. They are well suited for in vivo use because they may be efficient in replicating and nonreplicating cells and may have high titer (10^{13} virion/mL). Adenoviral vectors may involve inflammation of the tissues and may encounter antivector cellular immunity. New vectors are continuously developed for greater efficiency.

Plasmid-liposome carriers also have been explored under in vivo conditions. Human gene transfer may be used to insert selectable markers into T cells, to stem cells, to tumor-infiltrating lymphocytes, to neoplastic cells in hematopoietic lines, and to sarcoma cells, etc. Gene transfer may have therapeutic goals. Most of the human gene transfer experiments were plagued with inconsistent results and the expectations based on animal model experiments were not entirely fulfilled. The vectors need further improvement regarding homing specificity (targeting), side effects (inflammation), elimination of the possibility of insertional mutations, and elimination of possible immunological reactions against the vectors. *See* adenovirus; cancer gene therapy; gene therapy; liposomes; transformation, genetic; vectors. (Factor, P. H. ed. 2001. *Gene Therapy for Acute and Acquired Diseases*. Kluwer, Boston.)

human genome   About 90% of the genome, 2.91 billion bp of the euchromatic portion, was sequenced by the whole-genome shotgun approach using the DNA of five individuals of different ethnicity, published by the Celera group on Feb. 16, 2001, and on Feb. 15 by the Human Genome Sequencing Consortium: 25% of the genome is in scaffolds of over 10 million base pairs and 90% is in assemblies of 1 million bp or more. Celera found 25,588 protein-coding transcripts, and computational analysis and mouse information indicated the presence of ~12,000 additional genes. The Consortium estimated ~31,000 protein-coding genes. About 75% of the DNA is intergenic. About 740 genes are transcribed into RNA but not translated into protein. Only 1.1% is in exons and 24% is in introns. On the average, there was 1 bp variation per 1,250 bp among genomes, and less than 1% of the SNPs altered the amino acid sequences in the proteins. Mate pairs using BAC clones arraigned the megabase size sequences. The high-throughput automation and computational technology made rapid progress possible beyond expectation. The gene density of the chromosomes varies and generally the genes are not distributed uniformly along the length of the DNA. There are more genes in the GC-rich tracts than in the AT-rich sequences. Chromosomes 17, 19, and 22 are relatively gene-rich, whereas X, 4, 18, 13, and Y are poor. Recombination frequencies per Mbp are higher in the telomeric region and also higher in females than in males. The first coding exons appear to have high CpG proportions. There is a high proportion of processed pseudogenes, especially for ribosomal proteins (67%), lamin receptors (10%), and others to a smaller extent. Repeated genes are contained in 107 blocks; 781 of them include five or more. Duplications are extensive over the entire genome, especially in chromosomes 2 and 14. The latter is duplicated in over 70%. Some of the genes responsible for diseases (bleeding disorders, developmental regulation, cardiovascular conduction) are in paralogs of duplicated segments. Interestingly, these duplications are similar to those in mice, indicating that their origin predated the evolutionary divergence. The nucleotide diversity based on SNIPs for the autosomes was found to be 8.94×10^{-4} and for the X chromosome 6.54×10^{-4}. The transition:transversion rate varies from 1.59:1 to 2.07:1, depending on the analytical methods. The distribution of SNIPs deviates from the Poisson expectation. In the coding sequences, the missense nucleotide substitutions are relatively rare (0.12–0.17%), indicating conservatism for the integrity of the proteins is favored by selective evolution. The intergenic and intron regions harbor most of the SNIPs. About 40% of genes had no identified function by 2001. The annotation remains of central interest (Hubboard, T. 2002. *Bioinformatics*, Suppl. 2:S140). Some inconsistencies and errors turned up on reexamination of the draft completed by 2000. The largest categories are nuclei acid metabolism (7.5%), transcription factors (6%), hydrolases (proteases), signal transduction (G proteins), receptors, kinases, phosphatases, transporters, cytoskeletal proteins, ion channels, motor proteins, and cell adhesion molecules. Human-*Drosophila* orthologs were 2,758 and human-*Caenorhabditis* orthologs appeared to be 1,523. The nucleic acid metabolic enzymes, polymerases, helicases, ligases, nucleases, and ribosomal proteins were best preserved during evolution. The lower animal genomes (*Drosophila*) are devoid of the human genes mediating acquired immunity. The human nervous system is controlled by a much larger number of genes than the similar functions of *Caenorhabditis* or *Drosophila*. Intracellular signaling, hemostasis, apoptosis, translation, and ribonucleoprotein genes are substantially expanded in humans compared to lower organisms. The number of genes is about only twice that of *Caenorhabditis*, because with the more elaborate regulatory system in humans, the higher complexity of functional needs can be taken care of. *See* acquired immunity; BAC; clone-based mapping; diversity; draft genome sequence; exon; fingerprinting; gene; intron; mate pairs; missense mutation; paralogous loci; Poisson distribution; rough draft; scaffolds in genome sequencing; shotgun sequencing; SNIPs; transition; transversion. (Venter, J. C., et al. 2001. *Science* 291:1304; Lander, E. S., et al. 2001. *Nature* 409:860; errata in *Nature* 412:565–566; Katsanis, N., et al. 2001. *Nature Genet.* 29:88; <http://www-alis.tokyo.jst.go.jp/HOWDY/>; human genome variations [HGVbase]: <http://hgvbase.cgb.ki.se>.)

Human Genome News   Monthly publication on the human genome project by the National Institute of Health and the U.S. Department of Energy. Information: Human Genome Management Information System, Betty K. Mansfield, Oak Ridge National Laboratory, 1060 Commerce Park, MS-6480, Oak Ridge, TN 37831; phone: 615-576-6669; fax: 615-574-9888; <bkq@ornl.gov.>; <http:www.ornl.gov/hgmis/>. Human genome information is also available at web sites: <http://www.ncbi.nlm.nih.gov/genome/central> and <http://www.ensembl.org/genome/central/>.

humanics   Combination of genetics and social sciences.

human immunodeficiency virus   *See* acquired immunodeficiency syndrome.

humanin   Short polypeptide that prevents neuronal death in Alzheimer disease. *See* Alzheimer diseases. (Hashimoto, Y., et al. 2001. *Proc. Natl. Acad. Sci. USA* 98:6336.)

human intelligence (IQ)   Not a simple qualitative trait, but it is generally defined as the composite index of a variety of scores in which each is expected to have a different genetic determination. Intelligent quotients (IQ) also vary, depending on the variety of test batteries used for the analysis, such as cognitive, verbal, mathematical, logical, etc. Since intelligence is such a complex trait, it must be under the control of many genes. All polygenic traits have genetic and environmental components; so does intelligence. As a consequence, it is impossible to clearly separate ability from achievement, and the tests generally involve the latter. Nevertheless, it is obvious that all traits are expressed on the basis of a genetic blueprint in the DNA, and intelligence cannot be an exception.

Shortly before the end of the 19th century, Francis Galton, the father of statistical genetics, came to the conclusion that the apparent mental abilities of parents and children are correlated. Galton found that 36% of the sons of the 100 most distinguished men were still eminent, but only 9.5% of their grandsons and 1.5% of their great-grandsons were such. Also, 23% of the brothers of eminent men were eminent. Since then, several studies have confirmed the existence of such general correlations. These correlations may be biased to a great degree because the environmental conditions of parental and offspring populations are highly correlated. Before World War I, intelligence quotients were introduced for standardized quantitative measurements of intelligence. The Binet test had been widely used for determining scholastic performance and predictions. According to these quotients, children who were scored according to the average of their age group were classified with a score of 100 and those whose performance corresponded to 2 years behind or 2 years ahead of their peers were assigned IQ values of 80 or 120, respectively. Within similar socioeconomic and educational groups, these figures were meaningful. Mono- and dizygotic twins reared together and apart, as well as comparisons of adopted children with foster and birth parents, were studied for the inheritance of IQ.

These studies have proven the existence of a significant hereditary component of the IQ indexes by objective measures. Some studies attribute great influence to maternal effects through the womb environment. The IQ values should be very cautiously considered if different ethnic or socioeconomic groups are compared. The developmental rate of individuals may vary a great deal. Some children are early or late "bloomers," and this condition limits the predictive value of the indexes. There can be no question that certain genes concerned with the nervous system are responsible for mental retardation. The IQ indexes should not be used, however, for ethnic or social group discrimination because the statistical ranges of individual IQ values are highly overlapping in human populations, e.g., of blacks and whites in the United States. Although the averages of the two groups displayed a 15% difference, such information does not serve any useful purpose (see figure next page, drawn on the basis of data by A. R. Jensen. 1998. *The 3 g Factor*. Praeger, Westport, CT). Humans can be judged only individually. The diverse individual values of athletes, businesspeople, laborers, physicians, geneticists, theoretical physicists, etc., are all important for the good function of human societies.

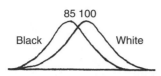

IQ data are statistical and do not have a rational neurobiological basis. Sex differences in mental test scores have been examined repeatedly in large representative populations. Although the overall differences between the sexes appear small, the males' abilities in mathematical and mechanical performance appeared higher, whereas in associative memory, reading comprehension, and perceptual speed, the females were at an advantage. The genetic meaning of these data is not clear. Males and females share the same chromosomal complement, except the Y chromosome, which by current knowledge contains minimal information beyond what is located in the pseudoautosomal segment and the genes related to sex determination and male fertility. Fragile sites in several human chromosomes are associated with trinucleotide repeats. The C residues of these variable-length repeats may be methylated to an extent and the increase of methylation is correlated with a decrease of intellectual abilities. The range of IQ of dulls, 20–84; feeblemindeds, 50–70; imbeciles, 20–50; and idiots, <20. These four lower types combined represent 2–3% of the general population but are substantially higher in consanguineous sibships, especially when one or both of the parents are subnormal.

IQ represents a good example of regression. Children of higher IQ people (IQ 140) display values of about 121, whereas of low IQ (85) people's offspring may exceed (92) the parent's. Although various IQ tests have been widely used for psychological and clinical studies, the general cognitive abilities can be better assessed by Spearman's *factor analysis* based on the *g* test (general factor test) developed at about the same time as the Binet test. The *g* test uses correlation matrices of various abilities, e.g., study of classics, modern languages, mathematics, pitch, music, etc. The basic idea is that one single ability is expected to be correlated with the diverse other abilities and the *g* score can express it all in a single numerical manner (*factor loading*). The *g* test generates a composite index that weights the separate tests by the overall correlation with all other component tests. Since the 1910s, Cyril Burt proposed the more complex *group factor* approach as a refinement of the *g* test. The IQ tests are numerical, but the final score is derived from the summation of the average individual scores (e.g., reading, memory, judgment, etc.).

The availability of methods for the study of quantitative gene loci and multivariate genetic analyses permits the study of genetic correlation among numerous genes. The heritability of cognitive abilities can be assessed with much better resolution. The progress in neurobiology (neuroimaging) sheds some light on the spatial organization and cooperation of the sites of these separate cognitive centers in the brain. The availability of chromosome and genome sequence information, along with microarray hybridization techniques, is expected to further reveal how the cognitive index of *g* is controlled by molecular factors. Since in a human population the individual genes may be represented by different alleles, the fine differences may be revealed by single nucleotide polymorphism (SNIP). With the progress of the proteome analysis, eventually better treatment may be feasible for the mentally retarded or those who are afflicted by affective disorders. The molecular genomics of behavior will certainly raise many ethical problems. The progress in biology and genetics must proceed even in some of the delicate areas. Application of the new technology requires, however, different types of moral and ethical decisions. One must keep in mind the lack of a generally agreed upon definition of what constitutes intelligence. *See* affective disorders; autuism; behavior genetics; behavior in humans; brain, human; cognitive abilities; covariance; DNA chips; dyslexia; eutelegenesis; FMR1 mutation; fragile sites; genius; heritability; imaging; mental retardation; microarray hybridization; multivariate analysis; myopia; polygenes; proteome; QTL; SNIP; Spearman rank correlation test; trinucleotide repeats. (Macphail, E. M. & Bolhuis, J. J. 2001. *Biol. Rev. Camb. Philos. Soc.* 76[3]:341; Dickens, W. T. & Flynn, J. R. 2001. *Psychol. Rev.* 108:346; Plomin, R. & Craig, I. 2001. *Br. J. Psychiatry*, Suppl. 40:s41. Bartels, M., et al. 2002. *Behavior Genet.* 31:132.)

humanized antibody   Chimeric molecule with the variable region of the mouse fused to the constant region of the human antibody. This molecular chimera retains its binding specificity and some characteristics of the human antibody. Some humanized antibodies have been tried as immunosuppressors of lymphatic, breast, and other tumors. Hyperchimerization and civilization are synonyms for humanized antibody. *See* antibody; HAMA; immune system; veneering of antibody. (Vaswani, S. K. & Hamilton, S. G. 1998. *Ann. Allergy Asthma Immunol.* 81[2]:105.)

human mutagenic assay   Humans cannot be subjected to mutagenic treatments, and mutagenic risks to human populations are determined largely by indirect means using microbial, animal, and plant bioassays of genetic toxicology. Other methods may involve epidemiological efforts encompassing dominant mutations, surveys of chromosomal aberrations in blood samples, testing of cell cultures for mutability, biochemical and molecular methods to assay genetic repair in cell culture, monitoring of changes in DNA by restriction fragment polymorphism using appropriate probes, SNIPs, etc. Increases in recessive mutations are difficult to detect because the human mating system does not favor homozygosis. Although the populations of Hiroshima and Nagasaki were exposed to very high doses of ionizing radiation as a consequence of the atomic explosion during World War II, no statistically significant increase in gene mutation could be detected, but developmental defects and the incidence of neoplasia increased. Similarly, the meltdown of the nuclear reactor in Chernobyl in 1986 caused a significant increase in birth defects and cancer; it is too early to tell whether these were only teratological effects or whether genetic causes are also involved. More recent data seem to indicate that genetic alterations were also caused. *See* atomic radiations; bioassays in genetic toxicology; host-mediated assays; mutation detection; mutation rate in humans; SNIP; substitution mutation. (<ariel.ucs.unimelb.edu.au:80/~cotton/glsdb.htm>.)

human origin   Anthroplogists and paleontologists are not in agreement on how *Homo sapiens* populated the world.

Some scientists claim that Europe was invaded from Africa 40,000–50,000 years ago during the Paleolithic period. Others favor the demic-diffusion hypothesis according to who during the Neolithic period (10,000 years ago) migrated to Europe from the Near East, bringing with them early agricultural practices. These waves of immigrants failed to interbreed with the earlier settlers. Others postulate that agriculture moved to Europe without actual migration of people. The different concepts were based on mtDNA, Y chromosomes, and classic gene frequencies. Ancient migration from Africa via Southeast Asia might have taken place southward to Australia and northward to China and Japan. Indians and Eskimos populated the Americas about 10,000 years ago through a land bridge across the Behring Strait from North Asia. *See* Eve foremother; hominoids; out-of-Africa; Y chromosome. (Lell, J. T. & Wallace, D. C. 2000. *Am. J. Hum. Genet.* 67:1367.)

human population growth   (Modified after Biraben, J. N. 1980. *Population* 4:1):

Years	400	1	500	1000	1500	2000
Millions of people	160	250	200	250	460	≈6000

Some calculations indicate a two-thirds probability that the world's population will not double during the 21st century. A probabilistic forecast indicates an 85% chance that the world population will cease growing by the end of the 21st century, and the probability is 60% that it will not exceed 10 billion by the end of this century. Then there will be a 15% decline in population. *See* age-specific birth and death rates. (Lutz, W., et al. 2001. *Nature* 412:543.)

human races   Distinguished by anthropologists on the basis of anthropometric traits. Geneticists delineate the races on the basis of gene frequencies shared within the group that are different from other "racial" populations. The main human races are Caucasoids, Mongoloids (including Chinese, Japanese, Koreans, and Native Americans, etc.), and Negroids. Khoisanoids or Capoids (Bushmen and Hottentots) and Pacific races (Australian aborigines, Polynesians, Melanesians, Indonesians) may also be distinguished. Many other subgroups within the larger ethnic groups may be classified. There is no genetic incompatibility among the various human races and there is no well-founded scientific evidence that interracial marriage would lead to the disruption of coadapted gene blocks, resulting in biological or mental deterioration in the offspring. The three major human races are genetically closely related as indicated by determination of evolutionary genetic distance. There is a controversy in the biomedical community as to what extent race would have any meaning for therapy. Different alleles are represented by different frequencies among ethnic groups and there may be differences in response to medication and a different need for social services. The human race is extremely closely related on the basis of DNA sequences to primates, primarily to chimpanzees, yet the genetic isolation is complete. *See* allelic frequencies; evolutionary distance; genetic isolation; *Homo sapiens*; interracial human hybrids; miscegenation; primates; racism. (Barbujani, G., et al. 1997. *Proc. Natl. Acad. Sci USA* 94:4516; Roychoudhury, A. K. & Nei, M. 1988. Human Polymorphic Genes: World Distribution. Oxford Univ. Press, New York.)

human subjects of experimentation   *See* bioethics. (<http://www.bioethics.gov>.)

humoral antibody   Made by the B lymphocytes and circulating in the bloodstream rather than being attached to the surface of T lymphocytes. *See* antibody; immune system.

humoral antigen   Secreted into the bloodstream. *See* B cell; immune system; T cell.

humunculus   17th–18th–century preformationist figment in which either in the human egg or sperm a miniature version of a human adult resides that eventually develops into a human embryo. *See* preformation.

humus   Decaying organic matter in soil.

Hunter syndrome   *See* mucopolysaccharidosis (MPS II).

Hunter-Thompson chondrodysplasia   *See* chondrodysplasia.

huntingtin   *See* Huntington's chorea.

Huntington   *See* Huntington's chorea.

Huntington's chorea   Dominant gain-of-function genetic disorder (chromosome 4p16.3) with a prevalence of 4 to 9×10^{-5} live births of Western European descent. The deletion of this chromosomal area leads to Wolf-Hirschhorn disease. It is a progressive degeneration of the basal ganglia (spinal cord) and the brain cortex, causing uncoordinated (choreic) movements and loss of mental abilities (dementia). It exists in the juvenile (akinetic rigid) form with an onset in the teen years or with late onset at ages 30 to 50. Expectancy of survival after the first symptoms appear is about 20 years. Motor disorders (jerkiness) in the carriers may be observed years before the onset of the disease. The late-onset form is generally inherited through the mother and the early-onset form through the father. Although the difference of this gene with a near-perfect penetrance was attributed initially to a mitochondrial cofactor, it appears to be determined by differential methylation (imprinting). Prenatal testing is feasible because of a very closely linked DNA marker, D4S10 (G8). Because the onset is frequently delayed beyond the reproductive period, early analysis of risk on the basis of linkage to this tight molecular marker is desirable.

Various biochemical alterations (GABA, glutamic acid decarboxylase, choline acetylase deficiency) may be associated with the disease. It has been shown that the huntingtin protein (≈350 kDa) in the basal ganglia and cerebral cortex displays 37 to 121 glutamine (CAG) repeats vs. the normal protein, which has only 6 to 34 repeats. Amino-terminal mutant fragments accumulating in the striatal neurons are the principal cause of the neuropathy. PKR binds preferentially to CAG repeats in mutant huntingtin RNA, and this may be important in pathogenesis. When the mismatch repair enzyme MSH2 is deficient, trinucleotide instability increases. Glucocorticoid receptor (GR) localizes the repeat into the nucleus. Deletion or mutation in the C-terminal of GR suppressed aggregation and nuclear localization. Surprisingly, mutation in the DNA-binding N-terminal domain increased aggregation and nuclear localization by GR. The genetic transmission of the normal range of the repeats is stable, whereas in the abnormal range it is unstable. The longer repeats hasten onset and increase severity of the condition. Homozygosity may result in embryonic lethality. The polyglutamine aggregates compromise cellular viability.

When in a mouse model the HD gene was turned off during development by an antibiotic in the drinking water of 18-week-old puppies, the extranuclear polyglutamine aggregates were gradually reduced or disappeared and concomitantly the neurological motor disorder was alleviated. The huntingtin protein resembles neuronal nitric oxide synthase. It has been assumed that it is a transcription factor. The mutant huntingtin and atrophin-1 seem to interfere with transcription through CREB-binding protein CBP. Increased histone deacetylase activity may play a role in the accumulation of the polyglutamine tracts and the neurodegeneration. The ~10 kb normal Huntington gene is transcribed into two mRNAs of 13.5 and 10.5 kb, and it is essential for embryo survival. The presence of the wild-type huntingtin is required for the normal production of the brain-derived neutrotrophic factor (BDNF), a requisite for the survival of striatal neurons in the brain. By neural transplantation, both the cognitive and the motor system deficits may be alleviated. Mice mutant for the huntingtin gene can be normalized by introduction of the wild-type HD transgene. *See* atrophin-1; CREB; DNA marker; epilepsy; fragile sites; genetic screening; junctophilin; mental retardation; mismatch repair; nitric oxide; PKR; trinucleotide repeats; Wolf-Hirschhorn syndrome. (Leavitt, B. R., et al. 2001. *Am. J. Hum. Genet.* 68:313; Peel, A. L., et al. 2001. *Hum. Mol. Genet.* 10:1531; Zuccato, C., et al. 2001. *Science* 293:493; Steffan, J. S., et al. 2001. *Nature* 413:739; Dunah, A. W., et al. 2002. *Science* 296:2238.)

HUPO   Human Proteome Organisation. *See* HUGO; proteome. (<http://www.hupo.org/>.)

HuR   Group of AU-rich element-binding proteins of the ELAV family. *See* AU-rich elements; ELAV.

Hurler syndrome (mucopolysaccharidosis I, MPS I, 4p16.3) Also called Scheie syndrome or Hurler-Scheie phenotype. It is due to a deficiency of α-L-iduronidase. The symptoms range from severe mental retardation to enlargement of the liver and spleen (hepatosplenomegaly, bone diseases (dysostosis multiplex), opacity of the cornea, hearing loss, and heart problems but usually normal intelligence and life span. Aminoglycoside antibiotics (geneticin, gentamycin) by suppression of nonsense mutation may alleviate the symptoms. *See* aminoglycosides; Hunter syndrome; hypertrichosis; iduronic acid; mucolipidosis; mucopolysaccharidosis. (Keeling, K. M., et al. 2001. *Hum. Mol. Genet.* 10:291.)

Hutchinson-Gilford syndrome (progeria)   Rare and very precocious aging of either recessive or dominant inheritance (human chromosome 1?). The disease is usually accompanied by multiple heat-labile protein defects. *See* aging.

***hut* operon**   Histidine utilization genes. Histidine is synthesized from three precursors—ATP, phosphoribosylpyrophosphate (PRPP), and glutamine—and it is involved in the regulation of other amino acids by yielding both glutamate and ammonia. *See* histidine operon. (Zalieckas, J. M., et al. 1999. *J. Bacteriol.* 181:2883.)

HUVEC   Human umbilical vein endothelial cells.

H-value paradox   Expresses the tendency of homology across a genome, and it may indicate a late specific expansion after losing some of the evolutionarily ancestral genes. *See* C-value paradox; N-value paradox. (Petrov, D. A. 2001. *Trends Genet.* 17:23.)

HVEM   Herpes virus entry mediator may block the entry of herpes virus into T cells but not into other cells. The 283-276 residue protein participates together with its receptor, LIGHT, in TNF function. *See* LIGHT; TNF.

HVR   Hypervariable regions in the DNA. *See* DNA fingerprinting; somatic hypermutation.

hyacinth (*Hyacinthus orientalis*)   Bulbous, monocot, fragrant spring flower, $2n = 16$.

Hyacinth.

hyaloplasm   Very finely granulated part of the cytoplasm. *See* cytoplasm.

hyalurodinase deficiency   Autosomal-recessive hyaluronuric acid storage disease. Short stature and excess hyaluronate in the fluid of the joints.

H-Y antigen (6p23-q12, structural gene)   Histocompatibility antigen controlled by another gene on the X chromosome (Xp22.3, HYR), the H-Y regulator. The HYA histocompatibility Y antigen is located in the long arm of the Y chromosome (Yq). It was recognized by rejection of skin grafts of male donors by female recipients, but the male recipients did not reject the grafts donated by females. The mice involved in these studies were inbred for many generations and were supposed to be isogenic, except for factors in the Y chromosome. Therefore, the rejection was attributed to a male-specific antigen. Thereupon it was hypothesized that the H-Y male-specific cell-surface antigen is also a male (testes) determining factor. Another gene, SMC, encoding an H-Y epitope on both the X and Y chromosomes, was located near the centromere in the long arm of the Y chromosome encoding an 11-peptide residue of the SMCY protein of 1,539 amino acids. The X chromosome has a homologue SMCX with about 200 amino-acid-site differences scattered along the entire length, except in the 11-residue H-Y antigen, where there is only a single amino acid difference between SMCY and SMCX. In the mouse, the H-Y controlling gene (*Hya*) in the short arm of the Y chromosome and the epitope is defined by the octamer Thr-Glu-Asn-Ser-Gly-Lys-Asp-Ile. Exceptional individuals with XY chromosomal constitution displayed female phenotype, and cytological analyses revealed a deletion at the tip of the short arm of the Y chromosome. Chromosomally XX exceptional males carried a translocated terminal segment of the short arm of the Y chromosome in one of the X chromosomes. This terminal segment was identified as the testis-determining factor (TDF) and thus the H-Y antigen has a true histocompatibility role, but it is apparently not responsible for testis differentiation. The H-Y antigen is the product of several genes. The current name of *TDF* is *SRY* (sex-determining region Y). *See* azoospermia; chromosomal sex determination; freemartins; gonadal dysgenesis; pseudohermaphroditism; sex determination; sex reversal; SRY; testicular feminization. (Wolf, U. 1998. *Cytogenet. Cell Genet.* 80:232.)

hybrid Progeny of two genetically unidentical parents. *See* F$_1$.

hybrid, asymmetric After somatic cell fusion, some chromosomes of one of the "parental" cells are eliminated. *See* alien addition; cell fusion; somatic cell genetics.

hybrid antibody Has more than one epitope-binding site because it is produced by genetic or chemical modification. *See* antibody; humanized antibody; hybrid hybridoma. (Chintalacharuvu, K. R., et al. 2001. *Clin. Immunol.* 101:21.)

hybrid arrested translation (HART) When an mRNA is hybridized with a cDNA, only those sequences can be translated in vitro that are not base-paired. The pairing prevents translation of the sequences. On this basis, the coding sequence for a particular polypetide can be identified. Hybridization of mRNA by antisense RNA also prevents translation. *See* antisense RNA; cascade hybridization; hybrid-released translation. (Paterson, B. M., et al. 1977. *Proc. Nat. Acad. Sci. USA* 74:4370; Nagy, E., et al. 1987. *Virology* 158:211.)

hybrid breakdown Reduced viability of the F$_2$ generation compared to F$_1$. The most likely cause is homozygosity of deleterious recessive genes or negative recessive epistasis. *See* epistasis.

hybrid capture Commercial (Digene Corp. Silverspring MD) DNA solution hybridization method for testing viral cancer risks. (Clavel, C., et al. 1999. *Br. J. Cancer* 80:1306; Lörincz, A. T. 1996. *J. Obstet. Gynecol. Res.* 22:629.)

hybrid depletion Identifies any cDNA that encodes a subunit of a multimeric protein. It is managed by preparing a mRNA pool coding for the protein of interest. The mRNA is hybridized to a cDNA cloned in a single-stranded vector and the hybrids are fractionated by CsCl equilibrium centrifugation providing the unhybridized, i.e., the antisense RNA at the bottom of the centrifuge tube. When it is injected into *Xenopus* oocytes along with the coding mRNA of the original pool, it may block the translation of the complementary mRNA. *See* antisense technology; density gradient centrifugation; mRNA.

hybrid DNA Heteroduplex. *See* heteroduplex.

hybrid dysgenesis Historical term for various genetic phenomena caused by transposable elements in *Drosophila*. It entails mutation and chromosomal rearrangements in hybrids of two genetic stocks. *P-M* system: The *P-strains* carry a transposable element and a suppressor of transposition. In the other strains (*M*), there is a genetic factor that derepresses the transposase and thus the hybrid becomes genetically unstable while both of the parental forms are stable. (The *P* and *M* originally indicated paternal and maternal conditions, respectively.) In *Drosophila*, over 30 different hybrid dysgenesis systems have been identified. The best known among them is the *P-M* system. The physical structure of a complete *P* element is shown in the diagram below.

Some *P* elements have internal deletion, duplications, and substitutions (such as *π*2, *P*c[ry]). The *P* element can be *autonomous* (transpose by its own power) and nonautonomous (requires a more complete [helper] element to move it). The

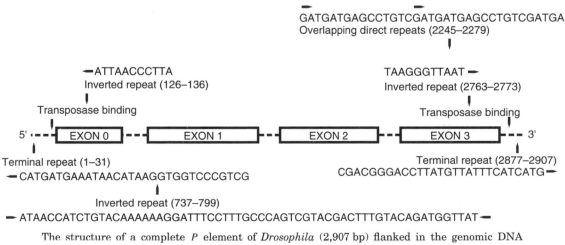

The structure of a complete *P* element of *Drosophila* (2,907 bp) flanked in the genomic DNA target site by 8 bp direct repeats. The numbers in parentheses indicate the nucleotide positions (beginning with the 5′ end). The horizontal arrows indicate the direction of the repeats; the vertical arrows point toward the positions in the linear sequence of the *P* element. (After Engels, W. R. 1989. *Mobile DNA*, Berg, D. E & Howe, M. M., eds., pp. 437–484. American Society of Microbiology, Washington, DC.)

autonomy of an element has been successfully tested by introduction through transformation of an in vitro engineered *P* element (*Pc[ry]*) carried on a plasmid along with the *rosy* gene (*rosy* is the structural gene for xanthine dehydrogenase [map position 3-52] and mutants have rosy eyes). The wild-type allele introduced into *ry⁻* homozygotes have normal eyes and are capable of moving into the *sn* (*singed*, map location 1-21, responsible for bristle [microchaetae] deformations), which can be easily monitored. The movement of *P* is controlled by the transposase function encoded by the four exons that extend to almost the entire length of the element (see diagram p. 577). The transposase begins transcription at base 85 and terminates at 2,696; thus, the transcript includes about 2.5 kb. The transposase enzyme is an 86.8 kDa protein. The *P* element can excise almost completely and leave behind the original genomic sequence or it may delete internal sequences, including sometimes even flanking nucleotides, involving genes with a total length rarely exceeding 7 kb. The imprecise excisions generate the defective elements. The frequencies of these excisions vary, ranging from 0.4 to nearly 2% per generation of the dysgenic flies. The mutability at the *sn* locus varies from 20 to 60% but may reach up to 90% when two reverse-oriented *P* elements (*double P*) are at the target site. The targets for insertion are not distributed at random and *P* is inserted by several orders of magnitude more frequently in the *sn* locus than into the alcohol dehydrogenase (*Adh*) gene. There is a tendency for *P* elements to become clustered. For insertion, the nontranslated upstream regions of genes are favored compared to the coding regions. Insertion into euchromatic sites is favored over heterochromatin. Interbands appear more likely targets than chromosomal bands.

P elements integrate with preference for the 5′-end of the genes. The integration does not seem to have base specificity; rather some structural properties of the DNA are the bases of choice for insertion. The 8 bp target site duplication created by the *P* insertion is situated within a 14 bp palindromic sequence. Transposition in the germline is much more frequent than in the somatic cells. The suppression of transposition of the *P* chromosomes may only partially be relieved in the strains designated as *M'*. These transposable elements induce a variety of genetic events, including recombination in the male *Drosophila*, mutation, and chromosomal rearrangements. The *P-M* system slightly boosts the effect of other mutagens. The frequency of X-chromosomal rearrangements was estimated to be 10% per generation, and the second breakpoint tends to stay within the same chromosome. The active transposable elements cause segregation distortion because the transmission of the *P* chromosomes is reduced compared to the *M* chromosomes. The transposons are associated with gonadal abnormality and (GD) sterility, particularly at temperatures above 27°C. The cytological sites and cis-effects may affect the activity of the *P* elements. Various mutations induced by *P* are subject to suppressors. This transposon, similarly to others, can be used for gene tagging and isolation, particularly its special constructs carrying selectable markers such as neomycin resistance so that they can be screened efficiently (smart ammunition). The element *pogo* (about 2.2 kb) is somewhat similar to *P*, and it has either 23 bp inverted terminal repeats and no target site duplication or a 21 bp inverted terminal repeat flanked by duplication of TA. The transposable element *hobo* (variable up to 3 kb) with up to 50 copies has 8 bp

target site duplication. Some (H) *hobo* elements are located in euchromatic regions and others are empty (E) sites. Reciprocal crosses do not activate *Hobo*, but its presence is associated with a high degree of instabilities. *HB* is a small (1.6 kb) element with 20 copies and 8 bp target site duplication. *HB* contains one reading frame of 444 bp that shares 25% homology with the amino acids of the *Tc* element of *Caenorhabditis elegans*.

The other best-studied transposable system of *Drosophila* that causes hybrid dysgenesis is the *I-R system*. The complete *I* element is 5.4 kb (5–15 copies/genome scattered among the chromosomes) and has many features of a retrotransposon (its transposition is via an RNA intermediate), but it does not have long terminal repeats. Thus, it is structurally quite different from *P* yet; some of its functions warrant its description. The counterpart of the *M* cytotype of the *P-M* system is the *R* (responsive) cytotype. Hybrid dysgenesis is observed in the crosses of *R* females with *I* males. These sterile/semisterile females are called SF (*stérilité femelle*), whereas the reciprocal nondysgenic ones are RSF. The female sterility of *I-R* does not involve gonadal anomalies (in contrast to GD in *P-M*), but hatching of the eggs is reduced. Eventually, the *R* strains may be converted to *I* by chromosome "contamination," i.e., accumulation of chromosomes derived from an *I* strain by crossing and segregation. *I*-factor activity involves mutations (recessive and dominant) that are frequently clustered, indicating their occurrence shortly before or at meiosis. The frequency of mutation varies at different loci and does not follow the same pattern as with *P*. The molecular structure of the complete *I* element of 5,371 bp is known. It does not have terminal repeats; however, four TAA reiterations are near the 3′-end of one strand and in the genomic DNA there are 12 bp duplications at the target site. One of the strands of *I* has open-reading frames (ORF) I (1,278 bp) and II (3,258 bp), separated by 471 bases. There is probably another ORF of 228 bp. The base sequences in ORF II are similar to viral and virus-like transposase reverse transcriptases. There are apparent coding sequences at the COOH-end for RNase H (ribonuclease digesting RNA in RNA-DNA hybrid molecules as required in reverse transcription).

Elements similar to *I* have been detected in the mammalian *L1*, *Drosophila* nonviral retrotransposable elements (*F* family), *R2* ribosomal insertions in silk worm, *Cin4* element in maize, and in the *ingi* elements of *Trypanosoma brucei*. Near the 3′ end of ORF II, the longest ORFs of *L1* and *Cin4* code the amino acid sequence Cys–Pro–Phe–Cys–Gln–Gly–Asp–Ile–Ser–Leu–Asn– His–Ile–Phe–Asn–Ser–Cys, which resembles the metal-binding domain of general transcription factor TFIIIa. ORF I has a sequence with some homology to the DNA-binding viral *gag* polypeptides (group-specific antigen). The *I* elements are most common near the centromeric regions. Mutations induced by *I* are stable and do not revert (unlike to *P*). Many of the *I* elements are truncated and show internal deletions and rearrangement. The *R* factor is quite complex, and it is determined by both nuclear and cytoplasmic regulatory components. Its role is to release *I* element expression. The *FB* family of transposable elements is complex; ca. 6.5 kb or smaller-size transposons cause a variety of genetic effects. They seem widespread in *Drosophila* but present in small copy number. They cause chromosomal rearrangement in 1/1,000 chromosomes. *See* copia; HEI; retroposons; transposable elements. (Engels, W. R. 1996. *Curr. Top. Microbiol. Immunol.*

204:103; Simmons, M. J., et al. 2002. *Proc. Natl. Acad. Sci. USA* 99:9306; Rio, D. C. 2002. Craig, N. L., et al., eds. p. 484 in Mobile DNA II. Amer Soc. Microbiol. Press, Washington, DC, USA.)

hybrid histocompatibility phenomenon *See* allogeneic inhibition.

hybrid hybridoma (quadroma) Fusion of two different hybridomas. *See* heterohybridoma; hybridoma. (Withoff, S., et al. 2001. *Br. J. Cancer* 84:1115.)

hybrid element insertion In meiotic recombination, it is induced by a P element of *Drosophila* at the site of recombination; most commonly, with either a deletion or duplication, a P element is retained. Two P elements in sister chromatids combine to form a hybrid element insertion at a nearby position. This process generates deletions flanking the existing P elements. Subsequent recombinational events contribute to additional variability.

hybrid inviability Postmating or zygotic mechanism of sexual isolation. The hybrids die either before sexual maturity or the offspring are sterile. *See* embryo culture; Haldane's rule; hybrid lethality; hybrid sterility; zygotic lethal. (Burke, J. M. & Arnold, M. L. 2001. *Annu. Rev. Genet.* 35:31.)

hybridization Crossing (mating) of genetically different individuals. Also, annealing DNA single strands with RNA or single-stranded DNAs of different origin (probe). The two nucleic acid strands must bear some homology to anneal. *See* c$_0$t curve; dot blot; in situ hybridization; Mendelian laws; Northern hybridization; Southern hybridization; South-Western method; Western hybridization.

hybridization arrest Antisense oligonucleotides selectively interfere with the function of a particular part of the genetic material by virtue of Watson-Crick or Hoogsteen base pairing. *See* antisense technologies; Hoogsteen pairing; Watson and Crick model.

hybridization probe Radioactively or fluorochrome (e.g., biotin) labeled nucleotide sequence that will hybridize with the complementary nucleotide sequences and identify the homologous tract(s) either on an extracted DNA or in situ in a chromosome. *See* biotinylation; probe; radioactive label.

hybrid lethality The parental forms are normal, yet hybrid embryos are aborted; this phenomenon is not uncommon when different species (with different chromosome numbers) are crossed. *See* Haldane's rule; hybrid inviability; hybrid sterility; zygotic lethal.

hybrid nucleic acid Double-stranded structure from two strands of different origin. *See* DNA; DNA hybridization; heteroduplex.

hybridogenetic Species hybrid containing, say, the A and B genomes and normally mating with one of the parental species (say B), but its gametes transmit only one of its genomes (say, A); thus the A genome is reproduced clonally while the B genome is added newly by each mating. *See* hemiclonal. (Marescalchi, O. & Scali, V. 2001. *Mol. Reprod. Dev.* 60[2]:270.)

hybridoma Myeloma (cancer) cell fused with spleen B lymphocyte, producing monoclonal (identical by origin) antibodies. The cancer cell assures the rapid proliferation and indefinite growth while the other component determines the specificity. Hybridomas have many applications in basic research and are very useful for the production of monoclonal antibodies and for the generation of lymphokines. *See* heterohybridoma; hybrid hybridoma; immortalization; lymphocytes; lymphokines; magic bullet; monoclonal antibody; senescence. (Springer, T. A., ed. 1985. *Hybridoma Technology in the Biosciences and Medicine*. Plenum, New York.)

hybridoma growth factor (HGF) *See* interferon β-2.

hybrid PCR product Can occur when the amplified sample is heterozygous or when related sequences are amplified with the same primer. *See* PCR.

hybrid released translation (HRT) Cloned DNA is attached to a membrane filter and annealed with mRNAs. After melting, the hybridized mRNA is eluted, then translated in an in vitro system (wheat germ or rabbit reticulocytes). The labeled polypeptides are then analyzed by electrophoresis. *See* hybrid-arrested translation. (Conlan, R. S., et al. 1995. *Plant Mol. Biol.* 28[3]:369.)

hybrid resistance *See* allogeneic inhibition.

hybrid-specific amplification (HAS) Procedure for the isolation of common fractions of two DNA samples avoiding the repeated seqence background. The principle of the procedure is basically the same as that of subtractive suppression hybridization. *See* subtractive suppression hybridization. (Lecerf, F., et al. 2001. *Nucleic Acids Res.* 29:e87.)

hybrid sterility The hybrid gonads or gametes are abnormal and incapable of normal sexual union. Usually, this phenomenon is based on misfunction of numerous genes. More commonly, the sterility affects the males or the male gametes and the females may be successfully backcrossed with normal males. *See* Haldane's rule; hybrid dysgenesis; hybrid inviabilty; hybrid lethality; zygotic lethal. (Orr, H. A. & Turelli, M. 2001. *Evol. Int. J. Org. Evol.* 55:1085.)

hybrid swarms Arise when the habitats of two related species are adjacent and mass outcrossing takes place. In such cases, the parental and the hybrid forms may not be easily recognized in the zone.

hybrid vigor The superior performance (growth, fitness) of hybrids has been observed for centuries before the birth of modern genetics. In its simplest case, hybrid vigor may be attributed to the presence of *complementary dominant genes*. If one parent is *AAbb* and the other is *aaBB*, their offspring are *AaBb*. The favorable dominant alleles are at both loci (*dominance theory of hybrid vigor*). Geneticists measure vigor by reproductive advantage and fitness. The fitness of a homozygous-recessive (R) class may be $w_{aa} = 1 - s$, where s is the coefficient of selection. The frequency of the homozygous recessives in a population may be determined by the rate of mutation (μ) from allele A to allele a. The average proportion of the recessive class is expected to be:

$$\hat{p}^2 = \mu/s$$

and the average frequency of recessive alleles is expected to be

$$\hat{q} = \sqrt{\mu/s}$$

The average fitness of the population then becomes

$$\hat{w} = 1 - s\hat{q} = 1 - s\sqrt{\mu/s} = 1 - \sqrt{s^2\mu/s} = 1 - \sqrt{s\mu}$$

If each of the n alleles contributes equally to the performance of the individual (additive effect), the average reduction of fitness caused by homozygosity of the recessive alleles in the populations is $n\sqrt{s\mu}$. For example, assuming that an organism has 10,000 gene loci and an average mutation rate of 10^{-5}, and the selection coefficient against the recessive alleles is 0.01, after substitution we obtain

$$n\sqrt{s\mu} = 10,000\sqrt{0.01 \times 0.00001} \cong 3.16$$

This indicates that inbreeding may reduce fitness of the population by a factor of about 3 compared to the situation when each locus has at least one dominant allele at all loci of a diploid organism.

On the average, this hypothetical example is in agreement with experimental observations on inbred and hybrid populations and lends support to the dominance theory of hybrid vigor (heterosis). If the midparent value, $m = 0.5(P_1 + P_2)$ and $0.5(P_1 - P_2) = d$ and $h = F_1 - m$, it is possible to predict the best performance to be expected by accumulating all of the favorable alleles in the inbreds in the case of perfect additivity of all genes and if the F_1 displays hybrid vigor.

$$P_{mx} = m + \frac{h}{\sqrt{H/D}}$$

where H = heterozygotes and D = homozygous dominants. Increased vigor of hybrids may also be caused by *overdominance* (superdominance), i.e., the heterozygote Aa is surpassing both AA and aa. Let us assume that the selection coefficient of each of the two classes of homozygotes (AA and aa) is -0.05. In such a case the three genotypic classes (according to the Hardy — Weinberg theorem) would occur in one generation of reproduction: AA (95):Aa (200):aa (95); if the population is mating at random, the allelic frequencies are equal (0.5) and the fitness of both types of homozygotes is also equal. Since the size of the population (the three classes combined) is 390 rather than 400 as expected without selection, the proportion of the surviving zygotes is 390/400 = 0.975, indicating 2.5% (10/400) reduction by a single cycle of reproduction. This reduction may not be significant for the majority of species that produce more offspring than the population of the habitat can maintain. The reduction in size may become, however, quite serious if overdominance occurs at several or many loci (see table). The table indicates that for most populations even a 0.5% disadvantage of the homozygotes at a larger number of loci may have very serious adverse consequences and would be hardly acceptable in herds of domesticated animals or in crop plants. The contribution of overdominance at single loci may be large enough to be of selective advantage in feral conditions or to improve the performance of agricultural species. *See* inbreeding; overdominance; selection coefficient; superdominance. (Barton, N. H. 2001. *Mol. Ecol.* 10:551.)

Reduction in population size at three different percentages of disadvantage of the homozygotes

Number of over-dominant loci	5%	1%	0.5%
10	0.77633	0.95111	0.97528
100	0.07952	0.60577	0.77856
1000	1.01×10^{-11}	0.00665	0.08183

hybrid zone   At the geographical boundary of two races, natural hybrids occur; as a result, speciation may ensue. *See* speciation. (Perry, W. L., et al. 2001. *Evolution* 55:1153.)

hydatidiform mole   Human hyperplasia resulting from an abnormal fertilization when the epithelial layer of the ovum is induced to proliferate into a tuft of cysts resembling a bunch of grapes. The karyotype of such structures is 46 (XX) and all of the chromosomes are derived from a diploidized sperm of 23 (X) constitution. *See* androgenesis.

hydra   About a 25–30 mm freshwater animal of about 100,000 cells. The freshwater hydra (*Hydra vulgaris attenuata*) is $2n = 32$. Propagates mainly by budding; a new clonal offspring is produced every 1.5 to 2 days. Sexual reproduction has a minor role; the production of a few eggs — predominantly by hermaphroditic means — takes a few weeks. It produces a variety of cell types (including a nerve system) and a differentiated body under the control of morphogens similar to those in other more complex animals. Some hydras are green because of symbiosis with algae and thus in the past were mistaken as plants. They have excellent abilities for regeneration. (Lohmann, J. U. & Bosch, T. C. 2000. *Genes Dev.* 14:2771; <www.ucihs.ici. edu/biochem/steele/ default.html>.)

Hydra.

hydrocarbons   Organic compounds containing only hydrogen and carbon; can be aliphatic (alkanes [paraffin], alkenes, alkynes, cyclic aliphatic) or aromatic. Although some of them are chemically rather inert (paraffin), the complex polynuclear hydrocarbons (benzo-a-pyrene, benzanthracenes, methylcholantrenes) are highly toxic, carcinogenic, and mutagenic.

They are present in combustion products. *See* cigarette smoke; environmental mutagens.

hydrocephalus (Xq28)   Disease of cerebrospinal fluid accumulation in the brain as a result of a defect in its secretion and absorption. It can be the symptom of physiological or mechanical lesions and it may be due to several autosomal-recessive (Dandy-Walker syndrome) occlusions of the openings of the fourth ventricle of the brain (Albers-Schönberg osteopetrosis [extreme bone density and bone proliferation]). Autosomal-dominant mutation (achondroplasia, osteogenesis imperfecta congenital type II, defects in bone formation), X-linked recessive (narrowing of a brain fluid channel following inflammation or bleeding), and X-linked dominant (orofacial-digital syndrome I, involving oral, digital, and mental abnormalities) are the genetic causes. Its prevalence is 0.01 to 1×10^{-3} births. It may be entirely sporadic or the recurrence risk when it is hereditary may vary from 15 to 50% within families, depending on the type of inheritance involved. Prenatal diagnosis may be feasible using ultrasound, brain tomography, or magnetic resonance imaging. *See* anencephaly; Arnold-Chiari malformation; Dandy-Walker syndrome; mental retardation; orofacial-digital syndrome; osteopetrosis; prenatal diagnosis; sonography; Walker-Wagner syndrome.

hydrocortisone   Glucucorticoid hormone, an antiinflammatory drug.

hydrogen bond   Weak bond between one electronegative atom and a hydrogen atom that is covalently linked to another electronegative atom, C or N. Hydrogen bonds tie together the polynucleotide chains of a DNA double helix (A = T and G ≡ C) and affect conformation of proteins. *See* hydrogen pairing; protein structure; Watson and Crick model. (Pauling, L. 1960. *The Nature of the Chemical Bond.* Cornell Univ. Press, Ithaca, NY; Luscombe, N. M., et al. 2001. *Nucleic Acids Res.* 29:2860.)

hydrogen hypothesis   *See* endosymbiont theory.

hydrogenosomes   Mitochondria-like organelles in *Trichomonads*. They are surrounded by double membranes and produce ATP from pyruvate or other molecules but have no DNA. In the ciliate *Nyctotherus ovalis,* living in the intestinal tract of cockroaches, hydrogenosomes were found that contain a minimal DNA genome, but the hydrogenase gene resides in the nucleus and protein encoded by it is imported into this organelle. *See* endosymbiont theory; mitochondria. (Dyall, S. D. & Johnson, P. J. 2000. *Curr. Opin. Microbiol.* 3[4]:404.)

hydrogen pairing   Secures the double-stranded form of DNA or RNA by establishing hydrogen bonds between the O atom attached to the C atom at position 4 (or position 6, depending on whether the American or the Beilstein numbering system is used for the pyrimidine ring) of thymine (uracil) and the NH2 group at the C atom of position 6 in adenine. The second hydrogen bond is formed between the hydrogen attached to the N at position 3 (or according to the Beilstein system 1) of the pyrimidine ring of thymine. Cytosine pairs with three hydrogen bonds with guanine between positions 4-6, 3-1, and 2-2 as shown by the figure next page. Cytosine and adenine cannot form hydrogen bonds unless

a tautomeric shift occurs. Similarly, the pairing of thymine (or analogs) with guanine requires another tautomeric shift. The tautomeric shift is an isomerization of the bases, changing the position of a hydrogen from 3N to O^4 on the thymidine molecule or moving one hydrogen from the N^6 position to the 1N position in adenine. As the formula at the left shows, the bases may undergo keto-enol transformations. *See* base analogs; base substitution; DNA; imino transformation; point mutation; steric-exclusion model; tautomeric shift. (Kool, E. T. 2001. *Annu. Rev. Biophys. Biomol. Struct.* 30:1.)

$$-C = C - \underset{\text{enol}}{OH} \quad \rightleftarrows \quad -\underset{\underset{H}{|}}{C} - \underset{\text{keto}}{C = O}$$

hydrogen peroxide (H_2O_2)   May act as an intracellular messenger in both plant and animal tissues. In plants, salicylic acid inactivates catalase, which may be a factor in activation of plant defense mechanisms to pathogens. In animal tissues, it may be involved in the activation of NF-κB transcription factor in the regulation of the immune reaction and immunosuppression. H_2O_2 may initiate apoptosis. When cells are stimulated by cytokines, phorbol esters and growth factors H_2O_2 may be released into the extracellular space. *See* granulomatous disease; glucocorticoids; host-pathogen relationship; immunosuppression; NF-κB; OH•; ROS; superoxide dismutase.

hydrogen pump   *See* proton pump.

hydrolase   Enzyme carrying out hydrolysis reactions. *See* hydrolysis.

hydrolethalus syndrome   Recessive 11q23-q25 hydrocephalus, polydactyly, and other developmental defects that have a prevalence $\sim 2 \times 10^{-4}$ in Finland.

hydrolysis   Splitting a molecule by inserting a molecule of water. One of the parts will obtain OH and the other H.

hydropathy index   Indicates the relative hydrophilic and hydrophobic properties of chemicals.

hydropathy plot   Used to determine the hydrophobic amino acid tracts in (membrane) proteins on the basis of the energy requirement for transfer into water from a nonpolar solvent. *See* membrane proteins. (Jayasinghe, S., et al. 2001. *J. Mol. Biol.* 312:927.)

hydrophilic   Readily miscible with water; it has polar groups.

hydrophobic   Insoluble in water, or poorly (if at all) soluble in water; it lacks polar groups. Also, (unjustified) fear of (drinking) water such as occurs in the viral disease called *rabies*, which is also called hydrophobia.

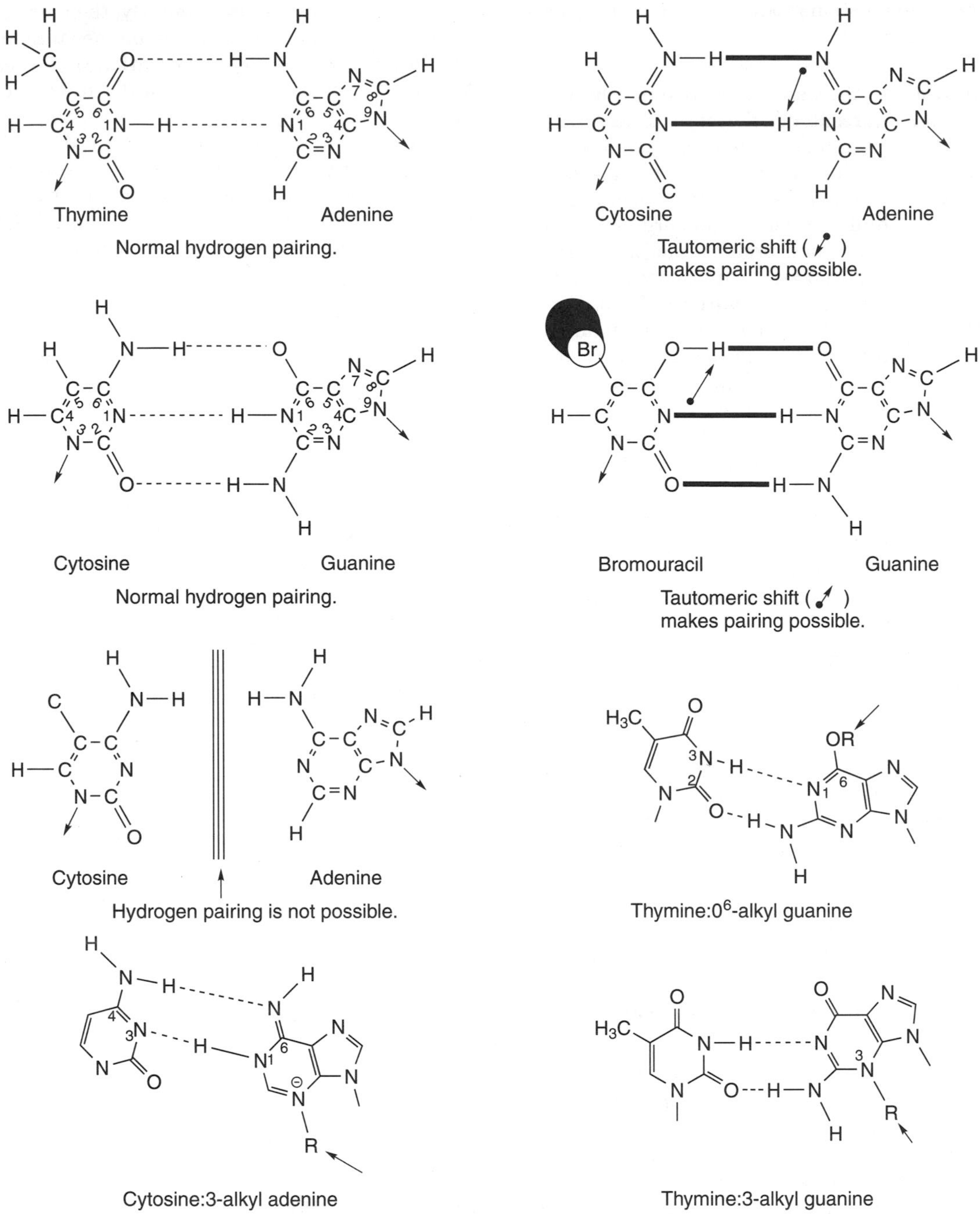

Thymine   Adenine

Normal hydrogen pairing.

Cytosine   Adenine

Tautomeric shift ( ) makes pairing possible.

Cytosine   Guanine

Normal hydrogen pairing.

Bromouracil   Guanine

Tautomeric shift ( ) makes pairing possible.

Cytosine   Adenine

Hydrogen pairing is not possible.

Thymine:0⁶-alkyl guanine

Cytosine:3-alkyl adenine

Thymine:3-alkyl guanine

Hydrogen bonding between DNA bases. Normal bonds are dashed lines. After tautomeric shift, or the hydrogen bonds are formed as represented by the heavy lines.

hydrophobic vacuum cleaner Mechanism of action of multidrug resistance transporters. *See* multiple drug resistance. (Sharom, F. J., et al. 2001. *Semin. Cell Dev. Biol.* 12[3]:257.)

hydrophobin Hydrophobic membrane material on the surface of aerial hyphae and on dikaryons of fungi. *See* dikaryon; hypha. (Wosten, H. A. 2001. *Annu. Rev. Microbiol.* 55:625.)

hydroponic culture Growing plants in salt solutions rather than in soil under semiaxenic conditions. (Hill, W. A., et al. 1992. *Adv. Space Res.* 12[5]:125.)

hydrops An edema (accumulation of fluids) may occur in kidney diseases or in erythroblastosis fetalis and in many other diseases. *See* erythroblastosis fetalis.

hydrops fetalis Prenatal anemia accompanied by fluid accumulation in the fetal body caused by failure to synthesize the α-chain of hemoglobin (an extreme form of thalassemia major) or by destruction of hemoglobin in other hemolytic anemias (rh). The idiopathic form is of spontaneous origin and others are caused by some immunological conditions. *See* adrenomedullin; erythroblastosis fetalis; hemoglobin; Rh; thalassemia.

hydroxyapatite Calcium phosphate hydroxide (may also contain silica gel). The phosphate residues of nucleic acids bind to calcium and thus double-stranded DNA binds to it stronger than the single-stranded molecules. On this basis, by adsorption chromatography, the two types of DNA can be separated from each other and DNA-RNA hybrids from RNA. The single-stranded molecules come off the columns by a low-molarity buffer and at higher molarity the double-stranded molecules can be eluted. Hydroxyapatite also participates in calcification of the extracellular matrix and in bone formation. Calcification of the arteries may lead to aortic failure and death.

hydroxyguanine One of the products of oxidative damage to the DNA that is generally repaired by base excision. *See* DNA repair. (Tuo, J., et al. 2001. *J. Biol. Chem.* 276:45772.)

hydroxylamine (NH$_2$OH) Antagonist of pyridoxalphosphate (PLP)-requiring enzymes; cleaves Asp-Gly linkages at high pH; blocks oxidation of H$_2$O but permits electron transfer through photosystems I and II from artificial donors. Its poisonous effect (α-effect) may be based on its high nucleophilic reactivity. For genetics, it is important that hydroxylamine reacts with carbonyl groups (C=O) of pyrimidines, specifically targets cytosine residues in nucleic acids, and generates hydroxylaminocytosine (a thymine analog), causing the transition of a G≡C base pair into a T=A pair and resulting in base-specific mutations. It is, however, a weak mutagen effective in prokaryotes but without much effect in higher eukaryotes. *See* base analog; base substitution; hydrogen pairing; transition mutation. (Freese, E. 1971. *Chemical Mutagens*, Hollaender, A., ed., p. 1. Plenum, New York.)

21-hydroxylase deficiency *See* adrenal hyperplasia.

hydroxyl radical (—OH•) Widespread in biological molecules; it can be formed from hydrogen peroxide in the presence of transition metals and may cause damage to macromolecules. *See* hydrogen peroxide; ROS; SOD; transition state.

5-hydroxymethyl cytosine Most common form of cytosine in T-even phage DNA. *See* 5-azacytidine; DNA base composition; DNA methylation.

3-hydroxy-3-methylglutaryl CoA lyase deficiency Leucine is degraded in six enzymatic steps into acetoacetic acid and in the process acetyl-CoA is generated. The last enzyme in this path is 3-hydroxy-3-methylglutaryl CoA lyase. Deficiency of this enzyme causes acidosis (accumulation of acids) and reduction of sugar in the blood (hypoglycemia). This potentially fatal disease is controlled by autosomal-recessive mutation. *See* amino acid metabolism; isoleucine-valine metabolic pathway; isovalericacidemia; methylcrotonylglycinemia; methylglutaconicaciduria.

hydroxytryptamine *See* serotonin.

hydroxyurea (NH$_2$CONHOH) Inhibitor of ribonucleoside reductases, thereby preventing the formation of deoxyribonucleotides from ribonucleotides. Consequently, it blocks DNA synthesis in the S-phase of the cell cycle and is also an antineoplastic agent and a strong poison. *See* DM chromosome; neoplasia; S phase.

hyena *Crocuta crocuta,* 2n = 40; *Hyena brunnea,* 2n = 40.

hygromycin B (Hyg, {5-deoxy-5-[[3-[4-[(6-deoxy-β-D-arabino-hexofuranos-5-ulos-1-yl)oxy]-3-hydroxyphenyl]-2-methyl-1-oxo-2-propenyl]amino]-1-2-O-methylene-D-neoinositol}) Antibiotic commonly used for screening transformed plant and animal cells by the expression of hygromycin phosphotransferase gene (*hph*) as a selectable marker (confering resistance). Hyg inhibits ribosomal ATPase. It is also an antihelminthic drug. *See* antibiotics. (Ganoza, M. C. & Kiel, M. C. 2001. *Antimicrob. Agents Chemother.* 45:2813.)

Hylandra suecica (formerly *Arabidopsis suecica*) Putative amphidiploid (2n = 26) of *Arabidopsis thaliana* (2n = 10) and *Cardaminopsis arenosa* (2n = 32). (Löve, Å. 1961. *Svensk. Bot. Tidskr.* 55:211.)

hymenium Fruiting body–forming tissue in fungi.

hyoscyamine Anticholinergic alkaloid that blocks neurotransmission through the parasympathetic system (originating in the brain and controlling the heart, head, neck, chest, abdomen, and pelvic organs). It is an alkaloid of the solanaceous species of plants *Hyoscyamus, Datura,* and *Atropa.* *See* alkaloids.

Hyoscyamus *See* henbane.

hyperactivity *See* attention deficit hyperactivity disorder.

hyperacute reaction (HAR) In xenotranslantation of organs to immunologically competent human tissues, the circulating natural antibodies immediately recognize the Galα1-3Gal antigens on the endothelial lining of the graft vascular tissues, which activates the recipient's complement. As a result, the transplant is destroyed within minutes to hours. *See* antibody; complement; grafting in medicine; xenograft; xenotransplantation. (Dawson, J. R., et al. 2001. *Immunol. Res.* 22[2-3]:165.)

hyperammonemia *See* acetylglutamate synthetase deficiency; carbamoylphosphate synthetase deficiency; glutamate dehydrogenase; ornithine aminotransferase deficiency; ornithine transcarbamylase deficiency; urea cycle.

hyperargininemia *See* argininemia.

hyperbetalipoproteinemia *See* hyperlipoproteinemia.

hyperbilirubinemia Excessive amounts of bilirubin in the blood causing jaundice, a common symptom of several diseases involving the destruction of red blood cells. It may be involved in indirect epistasis. *See* bilirubin; Crigler-Najjar syndrome; Dubin-Johnson syndrome; epistasis; Gilbert syndrome.

hypercalciuric hypercalcemia There are two forms: homozygous (dominant) primary severe neonatal hyperparathyrodism (NSHPT) at chromosome 3q13.3-q21 and heterozygous (recessive) hypocalciuric hypercalcemia (HHC1) at the same 3q13.3-q21 chromosomal location. The basic defect is attributed to G-protein receptors and the connected Ca^{2+} sensors that are involved in parathyroid hormone release from the cell. *See* G proteins; hypercalcemia-hypercalciuria; parathormone.

hyperchimerization Indicates the production of humanized antibodies by CDR transfer. *See* CDR; civilization; humanized antibody.

hypercholesterolemia *See* familial hypercholesterolemia.

hyperchromicity Single-stranded nucleic acids have increased UV absorption relative to the double-stranded molecules. *See* DNA; DNA denaturation; DNA thermal stability.

hyperferritinemia with cataracts Iron storage disease encoded at 19q13.3-q13.4. *See* ferritin.

hypergeometric distribution When a collection or population is sampled (selected) without replacement, the probabilities will change; unlike in the binomial or multinomial distributions where the basic probabilities are assumed to remain constant, the distribution can be characterized by the formula shown below. $T =$ the finite number of elements, $k =$ the exclusive and exhaustive classes of T, $N =$ the number of observations $= n_1 + n_2 + \cdots n_k$; $T = T_1 + T_2 + \cdots T_k$. *See* distributions; Fisher's exact test. (McDonald, J. W., et al. 1999. *Biometrics* 55:620.)

$$\frac{\begin{bmatrix} T_1 \\ n_1 \end{bmatrix} \begin{bmatrix} T_2 \\ n_2 \end{bmatrix} \cdots \begin{bmatrix} T_k \\ n_k \end{bmatrix}}{\begin{bmatrix} T \\ N \end{bmatrix}}$$

hyperglycemia Increase in blood glucose level.

hyperglycerolemia *See* glycerol kinase deficiency.

hyperglycinemia (hyperglycinuria, 13q32) May be the result of a deficiency of propionyl-CoA carboxylase I. It may cause protein intolerance, retarded development, swollen face, lethargy, frequent vomiting, low platelet counts, and other blood diseases. *See* glycine biosynthesis; glycinemia, ketotic; iminoglycinuria; thrombocytopenia.

hyperhaploid Gametes with more than the full basic chromosome number. *See* disomic.

hyperhomocysteinemia (MTHFR, 1p36.3) Caused by mutation in the 5,10-methylenetetrahydrofolate reductase (MTHFR) enzyme, and homozygotes for the condition have increased levels of homocysteine and an increased risk of cerebrovascular (brain), peripheral (vein), and coronary heart disease. Mutations in methionine synthase (1p43) may be responsible for both hyperhomocysteinemia and hypermethioninemia (Watkins, D., et al. 2002. *Am. J. Hum. Genet.* 71:143). Homocysteinemia may also be caused by other metabolic defects. *See* homocystinuria; tetrahydrofolate. (Chen, Z., et al. 2001. *Hum. Mol. Genet.* 10:433.)

hyper-IgE syndrome (Job syndrome) Immunodeficiency with high IgE, recurrent skin abscesses, and pneumonia is located in the short arm of human chromosome 4.

hyper-IgM syndrome (HIGM, HIM, X-linked immunodeficiency with hyper IgM) Chromosomal position Xq26. Although initially a defect in the B cells was suspected, recent evidence indicates a defect involving a ligand of CD40 in the helper T cells. Other symptoms (neutropenia, inflammations, etc.) have also been described. Usually the production of IgG, IgA, and IgE is reduced, but IgM and IgE are not affected. The clinical symptoms are respiratory infections, lymphoid hyperplasia, oral ulcers, autoimmunity, etc. *See* CD40; class switching; ectodermal dysplasia; immunoglobulins; ligand. (Weller, S., et al. 2001. *Proc. Natl. Acad. Sci. USA* 98:1166.)

hyperimmunoglobulinemia D (HIDS) Human chromosome 12q24 disease with strong resemblance to mevalonic aciduria, periodic fever, rash, diarrhea, swollen lymph nodes (adenopathy), and arthralgia (pain in joints). HIDS seems to be involved with a deficiency of mevalonate kinase. *See* mevalonic aciduria.

hyperinsulinemia *See* hypoglycemia.

hyperinsulinism *See* hypoglycemia.

hyperkalemic paralysis Group of dominant periodic paralysis diseases induced by a high level of potassium caused by human chromosome 17q23-q25 defects. *See* periodic paralysis.

hyperlipidemia The autosomal-dominant condition has similarities to familial hypercholesterolemia, but in this case hypercholesterolemia is not found in the offspring. Familial combined hyperlipidemia (FCHL) has a prevalence of 0.01 to 0.02 and is responsible for 10–20% of early-onset coronary heart disease. One gene reponsible for hyperlipidemia has been located to human chromosome 1q21-q23 in the area of the apolipoprotein A-II locus (APOA2). Generally, low-density lipoproteins accumulate. Early onset is relatively rare and results in hypertriglyceridemia. Its incidence is five times that of hypercholesterolemia. *See* apolipoprotein; CD36; cholesterol; coronary heart disease; familial hypercholesterolemia; familial triglyceridemia; HLP; hypertension; lipid; sterol. (Allayee, H., et al. 1998. *Am. J. Hum. Genet.* 63:577.)

hyperlipoproteinemia Caused by a dominant gene in human chromosome 19q13.2, coding normally for apolipoprotein E-d, a 299-amino acid polypeptide. A defect in this protein involves the accumulation of chylomicron (a small lipoprotein that normally transports dietary cholesterols and triglycerides from the intestines to the bloodstream). As a consequence of the defect, the conditions for coronary heart disease may develop. Several forms of this disease are usually distinguished. Some types (IV) are induced by high carbohydrate diet, alcohol, uremia, glycogen storage diseases, and steroid contraceptives. The recessive form of type I disease is associated with human chromosome 8p22. It involves large amounts of chylomicron accumulation even on a normal diet but disappears on a fat-free diet. It may not lead to early atherosclerosis. Lipoprotein lipase activity is deficient in this recessive form. The lipase gene encodes a 475-amino-acid protein, including a 27-residue leader peptide. *See* abetalipoproteinemia; apolipoprotein; atherosclerosis; betalipoprotein; cholesterol; chylomicron; coronary heart disease; lipase; lipoprotein; sex hormones; triaglyceride.

hyperlysinemia (7q31.3) Autosomal-recessive phenotype caused by a defect in the enzyme lysine:α-ketoglutarate reductase. This enzyme is a bifunctional complex of saccharopine dehydrogenase, controlling a step subsequent to α-ketoglutarate reduction in lysine degradation. The complex is also called α-aminoadipic semialdehyde synthase. The defect in this enzyme causes the accumulation of lysine in the blood, resulting in physical and mental retardation. Hyperlysinemia may occur if the transport of lysine into the mitochondria fails, and dibasic aminoaciduria also involves excessive urinary excretion of lysine. *See* amino acid metabolism; dibasic aminoaciduria; lysine biosynthesis. (Sacksteder, K. A., et al. 2000. *Am. J. Hum. Genet.* 66:1736.)

hypermorphic mutant Overexpresses a particular trait. *See* gain-of-function mutation; hypomorphic.

hypermutation (somatic hypermutation) Common phenomenon in the variable (V) region-coding sequences of immunoglobulin (Ig) genes. If a κ-chain promoter is inserted upstream of the constant region (C), it may cause mutational events in the C and V regions, indicating that transcription may be required for somatic hypermutation. Some other genes (e.g., BCL-6) may also undergo hypermutation if an Ig enhancer is introduced in B cells. The rate of mutation has been estimated to be six orders of magnitude higher than the average somatic rate of mutation in other genes. If the variable κ-immunoglobulin gene segment is replaced by heterologous sequences (prokaryotic *neo*, *gpt*, or β-globin), the rate of hypermutations does not decrease. Although many of the hypermutational events appear to be point mutations, a substantial fraction of the mutations involve double-strand breaks in the B cells that have completed the cell cycle. *See* affinity maturation; antibody; antibody gene switching BCL; enhancer; immune response; immunoglobulins; junctional diversification; neomycin; somatic hypermutation; transposable elements; transposon. (Muramatsu, M., et al. 2000. *Cell* 102:553; Kinoshita, K. & Honjo, T. 2001. *Nature Rev. Mol. Cell Biol.* 2:493.)

hypermutation, germinal May occur in the germinal centers under stress. This mechanism may generate random mutations, including adaptationally (evolutionarily) beneficial ones at high frequencies. *See* directed mutation. (Toellner, K. M., et al., 2002. *J. Exp. Med.* 195:383.)

hypernephroma (adenocarcinoma of kidney) Most commonly it involves rearrangement (translocations) of the short arm of human chromosome 3 or loss of 3p14.2. Trisomy or tetrasomy 7 may also be associated with kidney carcinomas. *See* renal cell carcinoma; von Hippel-Lindau syndrome.

hyperornithinemia *See* urea cycle.

hyperoxauria *See* oxalosis.

hyperparathyroidism (HRPT1, 1q25-q31, MEN1, 11q13) Dominant adenoma of the parathyroid gland. The condition is caused by a defect in thyrotropin receptor A and G-protein receptor. Another dominant form is characterized by multiple bone tumors on the jaws. An autosomal-recessive form affecting newborns is also known. Mutations in the calcium-sensing receptor (Casr) may upset parathyroid hormone production and may cause familial hypocalcemia or hypocalciuric hypercalcemia. The parathyroid hormone receptor is at 3p22-p21.1. *See* adenoma; goiter; G-protein; hypocalcemia; multiple endocrine neoplasia; parathormone.

hyperphenylalaninemia Autosomal-recessive condition attributed to reduced phenylalanine hydroxylase activity or dihydropteridine reductase deficiency. Hyperphenylalaninemia is also caused by a deficiency of pterin-4-α-carbinolamine dehydratase (10q22). *See* GTP cyclohydrolase I deficiency; phenylalanine; phenylketonuria; pteridines.

hyperplasia Abnormal increase of normal cells in a normal tissue. *See* neoplasia.

hyperploid Contains extra chromosomes beyond the normal number.

hyperpolarization Negative shift in the electric potential in a cell membrane. *See* I_h.

hyperprolinemia Type II (1p36) is caused by autosomal deficiency of the enzyme Δ'-pyrroline carboxylate dehydrogenase and consequently accumulation of proline and glycine in the blood (the mechanism is unclear). The patients generally show some degree of mental retardation and convulsions. In type I (22q11.2) disease, proline oxidase is deficient and renal problems occur with or without mental defects. *See* amino acid metabolism; iminoglycinuria; proline biosynthesis.

hypersensitive reaction (HR) At the place of infection by fungi, bacteria, and viruses, plant cells may suddenly die in a limited area and thus stop the spread of the infection and convey resistance to the host. Hypersensitivity reactions in case of fungi are elicited by the hyphal cell membrane, particularly by the fatty eicosapentaenoic (EPA) and arachidonic (AA) acids, and are inhibited by β-1,3- and β-1,6-glucans. The availability of EPA and AA is determined by lipoxigenase activity. Peroxides or other reactive oxygen intermediates (ROI) may have a role in the development of HR and may be signal transducers. In mammalian macrophages, ROI is usually aided by nitric oxide (NO) in attacking invading bacteria. In plants, HR seems to be activated by an oxidative burst and increase in NO. In addition, NO may initiate a signaling pathway through phenylalanine ammonia lyase and salicylic acid, leading to the activation of different disease-resistance genes. HR development may be the cause or consequence of the alteration of ion channel functions. In bacteria, *hrp* (hypersensitivity and pathogenesis) genes and mutants were identified. They interact with various plant products, e.g., the plant flavonoids (acetosyringones) initiate the activation of the virulence cascade of *Agrobacterium*, although these bacteria do not produce a typical HR and an early report could not be generally confirmed. The harpin 44 kDa regulatory protein has been implicated in HR expression. Bacterial avirulence (avr) genes may suppress HR, depending on particular host species. In animals, the hypersensitivity is usually called allergy. In the *immediate response*, the allergens bind to the surface of IgE and the following degranulation causes the release of histamines and interleukins, resulting in inflammation. In late-phase (4–24 h) inflammation response, leukotrienes and platelet-activating factors are secreted. The hypersensitive reaction is functionally comparable to apoptosis in animals, although it may program cell death by somewhat different mechanisms. Similarly to animal apoptosis, the mitochondria of plants also control cell death. The pro-apoptotic Bax protein promotes plant cell death. BAX-inhibitor-like proteins have been found in plants. Plants also have caspase-like enzymes, but their role in HR is not clear compared to the critical role of caspases in animals. The inner mitochondrial membrane enzyme called alternative oxidase (AOX) is absent from animal mitochondria but present in plants. Its suppression by antimycin results in hypersensitivity and cell death. In contrast, activation of AOX may reduce the size of necrotic spots. Some plastid proteins may also mediate HR. In plants, the MAPK pathway is activated during the hypersensitive reaction with the concomitant generation of hydrogen peroxide. *See* allergy; apoptosis; histamine; host-pathogen relation; hydrogen peroxide; interleukins; leukotrienes; MAPK; nitric oxide; phenylalanine ammonia lyase; platelet activating factor; porin; ROS; salicylic acid. (Lam, E., et al. 2001. *Nature* 411:848; Ren, D., et al. 2002. *J. Biol. Chem.* 277:559.)

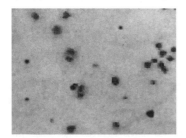

Hypersensitive necrotic spots.

hypersensitive site Includes nucleotide sequences that are readily cut by endonuclease because these tracts are (relatively) free of chromosomal proteins. These sites are generally found in front of transcribed genes and supposedly facilitate the attachment of the transcriptase enzyme. *See* regulation of gene activity.

hypersensitivity Term used in mutagen (carcinogen) testing to indicate the percentage of compounds classified as carcinogens (mutagens) among all compounds tested by a system. Also, it designates atopic (hereditary) allergy manifested within minutes after exposure to an allergen. *See* allergy; DTH. (Kleinjan, D. A., et al. 2001. *Hum. Mol. Genet.* 10:2049.)

hypertelorism Abnormally long distance between two organs or organ parts in the body.

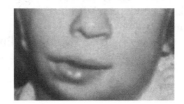

Hypertelorism.

hypertension Probably under the control of about 50 genes or less. In rats, genetic analysis identified at least two genes with major effects: BP/SP-1 and BP/SP-2 (blood pressure/sodium pump) in rat chromosome 10 (human chromosome 17q23). BP/SP-1 is probably linked closely to gene ACE1 (angiotensin-converting enzyme). Hypertension has an important physiological component associated with lithium-sodium countertransport, thus depending on dietary factors. RHO-associated Ca^{2+}-sensitive protein kinase seems to have a regulatory role. Hypertension occurs in about a third of humans and is frequently evoked by kidney diseases. Hypertension is the most common cause of heart disease, especially among black Americans (33% vs. whites 25%). (*See* cardiovascular disease; coronary heart disease.) Various familial disorders are frequently associated with hypertension: coarctation of the aorta, polycystic kidney disease, Alport syndrome, pheochromocytoma, neurofibromatosis, aldosteronism, hypoaldosteronism, hyperthyroidism, homocystinuria, Wilms' tumor, familial hypertriglyceridemia, hyperlipidemia, hydroxysteroid dehydrogenase deficiency (11βHSD). Blood pressure may vary from 80 mm (Hg) or less at diastole (expansion of the heart ventricles) to a very high 200 mm at systole (contraction of the ventricles). The average normal blood pressure in adults is 80 at diastole and 120 at systole. In children it is lower; in older people it is usually higher. An elevated salt diet is normally conducive to hypertension as the heart releases an atrial natriuretic peptide. The formation of this peptide is regulated by a guanyl cyclase A receptor (GC-A). Disruption of the GC-A gene in mice results in high blood pressure irrespective of the amount of salt in the diet. Mutations in 11-β-hydroxysteroid dehydrogenase (16q22) may lead to hypertension. *See* aldosteronism; angiotensin; brachydactyly; DC36; debrisoquine; Gitelman syndrome; Gordon syndrome; hyperlipidemia; hypotension; LDL; Liddle syndrome; mineral corticoid syndrome; nitric oxide; pulmonary hypertension; RHO; stroke. (Garbers, D. L. & Dubois, S. K. 1999. *Annu. Rev. Biochem.* 68:127; Lifton, R. P., et al. 2001. *Cell* 104:545.)

hyperthermia (malignant hyperthermia, MHS) May involve only an increase of the skin temperature or a general elevation of the body temperature (hyperpyrexia) occurring after anesthesia. The susceptibility is under dominant control in humans, but it is recessive in light-skinned pigs. Gene MHS1 is in human chromosome 19q13.1, also encoding a ryanodine receptor (RYR1); MHS2 is at 17q11.2-q24; MHS3 encodes a subunit of a voltage-dependent Ca^{+2} channel at 7q21-q22; MHS4 is at 3q13.1; MHS5 is at 1q32; MHS6 is in 5p. Several other genes also play various roles in hyperthermia. *See* cold hypersensitivity; halothane gene; ion channels; ryanodine; temperature-sensitive mutation. (Robinson, R. L., et al. 2000. *Ann. Hum. Genet.* 64:307.)

hyperthyroidism An autosomal-dominant defect may be caused by an inadequate response of the thyrotropin-secreting cells to the pituitary thyroid-stimulating hormone (TSH). As a consequence, high levels of thyroid hormones appear, and goiter, increased pulse rate, fatigue, nervousness, sweating, heat intolerance, and other symptoms develop. In a recessive form (human chromosome 14), a deficiency of the TSH receptor is caused by an insert of an 8-amino-acid sequence near the NH_2 end of the protein. A human chromosomal site (22q11-q13) is coding for a thyroid autoantigen, but it is not the receptor as once assumed. *See* cardiovascular disease; goiter; hormones; thyroid hormone response element; TRE.

hypertonia Excessive tension of the muscles; they are hard to stretch.

hypertonic The salt concentration of this type of solute is high enough to draw out water from a cell. *See* hypotonic; isotonic.

hypertranscription Mechanism of dosage compensation in the male *Drosophila* by upregulation of gene transcription in the single X chromosome, thus making the phenotypes alike in males (XY) and females (XX). It is mediated by the MSL proteins, which seem to be required for the accumulation of histone H4, acetylated at lysine 16 in the X of the male. Apparently, MLE and MSE proteins direct an acetyltransferase to the X-chromosomal histone. *See* dosage compensation. (Gorman, M. & Baker, B. S. 1994. *Trends Genet.* 10:376; Ruiz, M. F., et al. 2000. *Genetics* 156:1853.)

hypertrichosis Excessive hairiness. In the autosomal-dominant hypertrichosis universalis, hair abundantly covers the entire body until the end of infancy; in another form, gum disease is also present. In an autosomal-recessive form, the excessive hairiness is accompanied by nerve disease (neuropathy). Xq24-q27.1-linked hypertrichosis affects males more than females; lyonization causes patchy appearance of the hairs. The condition is very rare; only about 50 cases have been described. Excessive hairiness does occur, however, as part of several syndromes such as Hurler, de Lange, Coffin-Sirius, Lawrence-Seip, Schinzel-Geidion, and Gorlin-Chaudhary-Moss. *See* atavism; hair; hirsute; lyonization.

hypertriglyceridemia Increase in hepatic very-low-density lipoprotein secretion. Carnitin appears to regulate hypertriglyceridemia. *See* carnitin; hyperlipidemia.

hypertrophic cardiomyopathy Autosomal-dominant heart defect in the β-myosin heavy chain, the myosin light chain, or troponin. *See* cardiomyopathy, dilated; cardiomyopathy, hypertrophic familial; heart diseases; myosin; troponin.

hypertrophy Overgrowth generally with larger-than-normal cells.

hypervalinemia (valinemia) Accumulation of valine in the urine and blood plasma because of a defect in the enzyme valine transaminase caused by an autosomal-recessive factor. Symptoms include drowsiness and vomiting. Prenatal diagnosis is feasible. *See* isoleucine-valine biosynthetic pathway.

hypervariable sites In the light and heavy chains, antibody molecules are responsible for their high specificity in recognizing different antigens. They are short polypeptide loops and are also called complementarity-determining regions (CDRs). *See* antibody.

hyperzincemia Causes dwarfism, anemia, hypogonadism, etc., due to an abnormally high level of Zn in the blood. *See* acrodermatitis enterpathica.

hypha Fungal filament; cylindrical structural unit of the mycelia. Hyphae are surrounded by a wall and filled with cytoplasm unless they are vacuolated. They elongate by growing at the tips (apex). When branching hyphae aggregate, they may form a standing-up mycelium, called *coremium*, or horizontal strands may form *rhizomorphs*. The spherical or irregular aggregates functioning as enduring (dormant) bodies are *sclerotia*. Hyphae aggregating in pseudoparenchyma tissue are *stroma*. Hyphae may be involved in plasmogamy and function in some sort of sexual function (*somatogamy*), although they may not be sexually specialized. *See* fungal life cycles, heterokaryon incompatibility. (Xiang, Q., et al. 2002. *Genetics* 160:169.)

hypoaldosteronism (adrenal hyperplasia) The two genes responsible for the recessive 11-β-hydroxylase deficiency have been located to human chromosome 8q21, and a defect is responsible for the accumulation of 11-deoxycorticosterone and consequently for hypertension and other hormonal abnormalities. The same enzymes are involved in the hydroxylation of 18-hydroxysteroids and 17-hydroxysteroids. These genes are members of the P450 (cytochrome) enzyme coding family. Affected females show masculinization and males exhibit precocious puberty because of the accumulation of steroids. Some of the aldosterone deficiency mutations are allelic to the aldosterone-overproducing defect (aldosteronism), indicating the presence of a multifunctional protein. *Pseudohypoaldosteronism*, which is caused not by aldosterone deficiency but by the recessive deficiency of the mineral corticoid receptor, is encoded at human chromosome 4q31.1 or q31.2. This condition is characterized by salt wasting in the urine and responds favorably to salt administration. Hypoaldosteronism defects appear to be quite common among Oriental (Persian) Jews. Pseudohypoaldosteronism type 1 may be due to a defect in an epithelial sodium channel (ENaC), resulting in hyperkalemic (high in potassium) acidosis and salt wasting. The α-subunit of the channel is encoded in human chromosome 12p13.1-ter, whereas the β- and γ-subunits are coded by chromosome 16p12.2-p13.11. The most common basis of pseudohypoaldosteronism is, however, mutation in the mineral corticoid receptor gene (MRL). *See* aldosteronism; Bartter syndrome; cardiovascular diseases; Gitelman syndrome; ion channels; Jews and disease; Liddle syndrome; mineral corticoid syndrome; P-450; pseudohypoaldosteronism.

hypo-α-lipoproteinemia (11q23.3, 9q22-q31) One of the several lipid metabolism disorders contributing to coronary heart disease. *See* cardiovascular diseases; lipoproteins; Tangier disease. (Kort, E. N., et al. 2000. *Am. J. Hum. Genet.* 66:1845.)

hypo-β-liporoteinemia (2p23, 3p21-p22) Involves dominant, reduced levels of apolipoprotein B, which may result in increased levels of blood cholesterol and atherosclerosis, yet individuals with this anomaly may have a prolonged life span. *See* abetaliproteinemia; apolipoproteins; atherosclerosis; betalipoprotein; cardiovascular diseases; hyperlipoproteinemia.

hypoblast Precursor of the mesoderm and endoderm. *See* endoderm; mesoderm.

hypocalcemia Reduced amount of calcium in the blood in several diseases involving defects in ion channels. Autosomal-dominant hypocalcemia (CASR, 3q13.3-q21) is a mutation in the calcium-sensing receptor of the parathyroid gene (PTH, 11p15.3-p15.1) or in PTH itself. *See* hypocalciuric hypercalcemia; ion channels; parathormone.

hypocalciuric hypercalciuria (HHC2, 19p13.3) The calcium level in the blood and urine is elevated. In some cases, it involves bone and blood vessel disorders. The basic defect seems to be in a G-protein receptor. The phenotype is very similar to that of hypercalciuric hypercalcemia at 3q13.3. *See* G-proteins; hypercalciuric hypercalcemia.

hypochlorate (Ca[ClO]$_2$) Calcium and sodium hypochlorites (hypo) are among the oldest and most effective oxidative sterilizing agents. Calcium does not leave alkalic residues after washing and it is more useful as an antiseptic for live tissues. *See* sterilization.

hypochondriasis Type of affective disorder when illness is imagined on the basis of irrelevant signs. *See* affective disorders.

hypochondrogenesis Connective tissue disorder caused by a mutation in collagen.

hypochondroplasia This autosomal phenotype resembles achondroplasia, but the tibia (shin bone) and the head are rather normal. The fingers are short, but the hand is not three-pronged. This gene appears to be allelic to that responsible for achondroplasia. It may be caused by a defect in the function of fibroblast growth factor receptor 3 (FGFR3) at human chromosome 4p16.3. *See* achondrogenesis; achondroplasia; dwarfism; fibroblast growth factor; receptor tyrosine kinase; stature in humans.

hypochromicity, nucleic acid In double-stranded molecules, the free rotation of the bases is hindered, resulting in reduced optical density in UV light compared to single-stranded molecules (hyperchromicity). *See* DNA; hyperchromicity.

hypocotyl Section of a plant embryo or plant situated between the cotyledon attachment point and the radicle or root, respectively.

hypocretin *See* orexin.

hypodontia (adontia) Autosomal-dominant condition of lack or underdevelopment of up to 6 teeth. (See also teeth, dental non-eruption, dentinogenesis imperfecta, dental ankylosis, tooth agenesis, olygodentia)

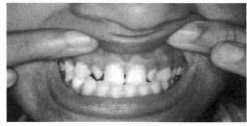

Hypodontia. (Fom Bergsma, D. ed. 1973 Birth Defects. Atlas and Compendium. By permission of the March of Dimes Foundation.)

hypogammaglobulinemia/common variable immunodeficiency Complex immunodeficiency with uncertain familial determination. *See* Epstein-Barr virus; gammaglobulinemia; Hirschsprung disease; immunodeficiency.

hypoglycemia Involves lower than normal blood sugar content and may result in shaking, cold sweat, low body temperature, headache, irritability, and eventually coma. An autosomal-recessive gene may cause hypoglycemia by a deficiency of glycogen syntethase in the liver. Another recessive gene in human chromosome 1p31 may cause acyl−coenzyme A−dehydrogenase deficiency and hypoglycemia. Mutations in the sulfonylurea receptor gene may cause hyperinsulinemia (human chromosome 11p14-p15.1) and consequently hypoglycemia. Hypoglycemia may result indirectly from a large number of different ailments. *See* epilepsy; ion channels; sulfonylurea.

hypogonadism Less than normal function of the gonads (ovary and testes); it is frequently associated with retardation in growth and mental abilities. Mutation in the DAX-1 human gene may result in adrenal defects such as adrenal hypoplasia and hypogonadotropic hypogonadism (deficiency in the hormone gonadotropin and gonadal hypofunction). Mutation in the gonadotropin-releasing hormone receptor (4q13.1-q21.1) may also be responsible. *See* adrenal hypoplasia; Kallmann syndrome.

hypohidrosis Reduced ability to sweat caused by autosomal-recessive defect of the sweat glands.

hypokalemia Autosomal-recessive (16q13) low potassium and magnesium levels in the body, causing neuromuscular abnormalities. *See* Bartter syndrome; Gitelman syndrome; hypoaldosteronism; periodic paralysis; pseudoaldosteronism.

hypolactasia *See* disaccharide intolerance; lactose intolerance.

hypomagnesemia (11q23)   Dominant renal defect involving wasting of Mg^{2+} and frequently involving Ca^{2+} excretion due to a mutation in the Na^+, K^+-ATPase γ-subunit.

hypomelanosis Ito   Pale skin spots in whorls or patches of tan associated with diverse anomalies, indicating that this syndrome is a mixed bag, often associated with breakage of different chromosomes. *See* incontinentia pigmenti; pigmentation defects; skin diseases.

hypomorphic   The product of the gene is below the level of that of the wild type. *See* leaky mutant.

hypomyelination *See* hypomyelinopathies.

hypomyelinopathy   Involves defects in the myelin coat of the nerves such as congenital hypomyelinating neuropathy. *See* Charcot-Marie-Tooth type 1 disease; Dejerine-Sottas syndrome; Egr; HMSN; HNPP; neuropathy; sensory neuropathy.

hyponasty *See* epinasty.

hypophosphatasia   Either a *dominant* mutation expressed in adults as a deficiency of the liver (general) alkaline phosphatase gene in human chromosome 1p36.1-p34 or the *recessive* mutation in the same locus appears as infantile hypophosphatasia. The adult phenotype involves early loss of teeth and bowed legs like in rickets. The level of intestinal alkaline phosphatase is normal. In some forms, the stature is somewhat shorter. The infantile type is manifested before birth, involving severe skeletal anomalies, increased levels of phosphoethanolamine and inorganic pyrophosphate in the urine, and higher levels of pyridoxal phosphate in the serum. Death may occur within a year. In some instances, infusion of normal blood plasma resulted in prolonged normalization. It was suggested that a cofactor of the enzyme is missing. *See* dwarfism; hypophosphatemia; rickets; stature in humans.

hypophosphatemia   Dominant bone disease coded in human chromosome Xp22 region, resulting in a low level of phosphate in the blood and no or minimal response to vitamin D. The defect may also involve abnormal phosphate absorption. Its prevalence is 2×10^{-4}. *Autosomal-dominant hypophoshatemic rickets* (ADHR, 12p13.3) is due to a mutation in gene FGF23; it is resistant to vitamin D. The X-linked PHEX gene encodes a neutral endopeptidase and is responsible for *hypophosphatemic rickets* (XLH, Xp22.2-p22.1). Heterogeneous *autosomal hypophosphatemic bone disease* (HBD) and autosomal-recessive *hereditary hypophosphatemic rickets with*

hypercalciuria (HHRH) both respond to vitamin D. *See* exostosis; FGF; hypophosphatasia; rickets; spermine. (Sabbagh, Y., et al. 2001. *Hum. Mol. Genet.* 10:1539.)

hypoplasia   Underdevelopment of an organ. *See* aplasia.

hypoploid   Contains less than the full set of chromosomes. *See* autopolyploid; nullisomic.

hypoproconvertinemia   Recessive bleeding disease caused by a deficiency of blood-clotting factor VII encoded in human chromosome 13q34-qter. Afflicted individuals may also be deficient in antihemophilic factor X. *See* antihemophilic factors; proconvertin.

hypospadias   The urethra opens at the lower part of the penis; it has autosomal-dominant and 1-recessive transmission. Prevalence is close to 0.3% with a heritability of about 0.57. Hypospadias may accompany several syndromes, e.g., hypertelorism (22q11.2), steroid α-reductase 2 pseudohermaphroditism (2p23), androgen receptor defects (Xq11-q12), Wilms' tumor (11p13), mutation in the HOXA13 gene (7p15-p14.2), McKusick-Kaufman syndrome (20p12), etc. *See* penis; steroid dehydrogenase.

hypostasis   Condition in which expression of a gene is masked by an epistatic gene. *See* epistasis.

hypotension   In contrast to hypertension, it involves low blood pressure. Some of the cases may be determined by identical loci but different alleles. The autosomal-dominant orthostatic form (*Shy-Drager syndrome*) is characterized by incontinence, anhidrosis (absence of sweating), ataxia, tremor, and low norepinephrine levels in the plasma. *Pseudohypoaldosteronism* (*PHA-1*) is autosomal dominant and involves serious dehydration, salt wasting, high levels of potassium, and a form of hyperaldosteronism. The mutation seems to affect the genes controlling the same ion channel as in Liddle syndrome. *Gitelman syndrome* is a human chromosome 16 recessive condition that involves a defect in the NaCl cotransporter and consequently salt wasting. *See* adrenergic receptors; aldosteronism; hypertension; Liddle syndrome; mineral cortical syndrome. (Zuscik, M. J., et al. 2001. *J. Biol. Chem.* 276:13738.)

hypothalmus (forebrain)   Part of the brain where vision, visceral activities, water balance, temperature, sleep, etc., control centers are located. *See* brain, human.

hypothesis   Supposition or multiple alternative suppositions (hypotheses) are used to explain certain experimental data and generally statistical approaches are used to decide which of the alternatives have the greatest probability to be true. One must keep in mind that the statistical methods do not prove cause-effect relationships or mechanisms; only the degree of chance or likelihood is indicated. The *working hypothesis* is used as guidance to design experimental procedures to test the most likely mechanism involved. In experimental science, such as genetics, only the testable hypotheses have any value. *See* chi square; genetic risk; likelihood; maximum likelihood; null hypothesis; probability.

hypothyroidism *See* goiter.

hypotonia Weak muscle tension. It is a typical feature of Down syndrome, Prader-Willi syndrome, cri-du-chat syndrome, and various other chromosomal anomalies and trisomies.

hypotonic The concentration of salts in hypotonic media is lower than the osmolality of the cells; therefore, water from the cells may be drawn out into the media. *See* hypertonic; hypotonic; osmolality; osmotic pressure.

hypotrichosis Rare autosomal-recessive reduction of hairs on the face and absence of pubic hairs even after puberty. Baby hair is shed shortly before or after birth. Congenital atrichia (baldness) or hypotrichosis of Marie Unna was mapped to 8p22-p21, but the latter does not seem identical with congenital atrichia. Hypotrichosis simplex is a rare dominant baldness affecting both sexes and was mapped to 6p21.3. Recessive gene for hypotrichosis associated with juvenile macular dystrophy encodes P-cadherin at 16q22.1. *See* alopecia; cadherin; hair; hypertrichosis; macular corneal dystrophy. (Sprecher, E., et al. 2001. *Nature Genet.* 29:134.)

hypotrophy Less than normal growth of a tissue or organ caused by inadequate nourishment.

hypouricemia Type of nephrolithiasis. In the Dalmation coachhund type, the kidneys do not reabsorb urate. In other cases, the decrease of urate in the urine is accompanied by excessive excretion of calcium. *See* kidney diseases; nephrolithiasis.

hypoviruses Persistent and hard-to-transmit infectious agents in fungi causing no symptoms. However, when introduced into the chestnut blight fungus (*Cryphonectria parasitica*) by electroporation, the synthetic transcripts of such viruses can reduce the virulence of the fungus, thus indicating a means of biological control. *See* symbionts, hereditary.

hypoxanthine Purine base very similar to guanine, except that from the C2 position NH_2 is removed. It can also be formed by deamination at the 6 position of adenine. Its nucleoside is (confusingly) inosine and its nucleotide is called inosinic acid. It may occur in some anticodons of tRNA. When guanine is deaminated into hypoxanthine through chemical mutagens (nitrous acid), lethal mutation occurs because hypoxanthine cannot support nucleic acid replication. The enzyme hypoxanthine-guanine phosphoribosyl transferase (deficient in Lesch-Nyhan syndrome) can phosphorylate both of these purines in the salvage pathway of nucleic acid. *See* ADA; HAT medium; HPRT; inosine; nitrous acid mutagenesis; salvage pathway; wobble.

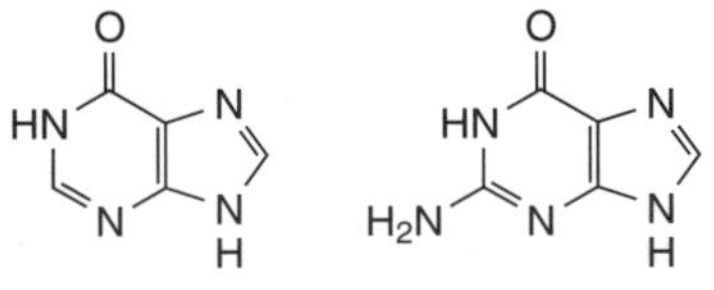

Hypoxanthine   Guanine

hypoxanthine-guanine phosphoribosyl transferase (HGPRT) Enzyme that donates the ribose phosphate moiety to the purine bases hypoxanthine and guanine from 5′-phosphoribosyl-1-pyrophosphate, thus forming the corresponding nucleotides. Another enzyme, adenosine phosphoribosyl transferase, synthesizes adenylic acid from adenine. HGPRT is the key enzyme in the salvage pathway of nucleic acid synthesis. *See* HAT medium; Lesch-Nyhan syndrome; salvage pathway.

hypoxia Low oxygen concentration in the environment of cells or tissues. The effectiveness of ionizing radiation in induction of mutation or chromosome breakage is reduced under low oxygen concentration. Hypoxia may trigger the formation of double minutes and chromosomal instability. Hypoxia promotes the binding of hypoxia-inducible factors to the hypoxia-response element in the vascular endothelial growth factor promoter, thus stimulating angiogenesis. Deletion of the hypoxia-response element may cause motor neuron degeneration that resembles the anomaly of amyotrophic lateral sclerosis. Hypoxia is a factor in tumorigenesis. *See* angiogenesis; double minutes; fragile sites; oxygen effect; VEGF. (Jiang, H., et al. 2001. *Proc. Natl. Acad. Sci. USA* 98:7916; Harris, A. L. 2002. *Nature Rev. Cancer* 2:38.)

hypoxic gene Expressed primarily under anoxia. *See* amyotrophic lateral sclerosis; anoxia.

hysterectomy Surgical removal of the uterus. *See* uterus.

hysteresis (1) Time lag between two processes. (2) Lowering of the point of freezing without lowering the temperature required for melting; antifreeze proteins may regulate it. *See* antifreeze proteins; temperature-sensitive mutation.

On May 8, 1900, William Bateson wrote in the *J. Roy. Hort. Soc.* 25:54:

"An exact determination of the laws of heredity will probably work more changes in man's outlook on the world, and in his power over nature, than any other advance in natural knowledge that can be foreseen.

There is no doubt whatever that these laws can be determined. In comparison with the labour that has been needed for other great discoveries it is even likely that the necessary effort will be small."

A historical vignette.

IAA Indole acetic acid. *See* plant hormones.

Indole-3-acetic acid.

IAM *See* infinite allele mutation model; microsatellite; minisatellite; SMM model.

IAP (1) *See* chloroplast import. (2) Intracisternal (within closed compartments) A particles are retroviral elements in the mouse genome capable of transposition. Normally they are not active, but demethylation of their long terminal repeats increases their transcription 50–100–fold. Male germ cells are demethylated only shortly before the formation of the spermatogonia. Female germs are demethylated a little longer but only during the period from primordial germ cells to growing oocytes. The spermatogonia and the oocytes are heavily methylated. IAP elements that are not found in humans. *See* Prader-Willi syndrome. (Vogler, C., et al. 2001. *Pediatr. Res.* 49[3]:342.) (3) Inhibitor of apoptosis controls apoptosis by binding or interfering in other ways with caspase or procaspase activation. At its N-terminus it contains one or several BIR motifs and its C-terminus has a RING-finger domain for recruiting caspase. IAPs have ubiquitin ligase activity and may autoregulate their degradation by proteasomes. The Smac/DIABLO protein binds to IAP and promotes caspase activity. Binding proteins REAPER, HID, and/or GRIM promote caspase activity presumably by neutralizing IAP. *See* apoptosis; BIR; caspase; DIABLO; proteasome; RING finger; Smac; ubiquitin. (Verhagen, A. M., et al. 2001. *Genome Biol.* [2: reviews 3009]; Sharief, M. K. & Semra, Y. K. 2001. *J. Neuroimmunol.* 119:350.)

IARC International Agency for Research on Cancer, Lyon, France; publishes reviews on carcinogens as *Scientific Publications of IARC*. *See* cancer; databases; environmental mutagens. (<http://www.iarc.fr/>.)

iatrogenic Adverse effect(s) resulting from or concomitant to medical treatment.

IBD Identical by descent, i.e., an allele inherited from an ancestor is exactly the same as that of the particular ancestor and not the result of a new mutation identical by state (IBS). IBD-APM is identical by descent for affected pedigree member. *See* inbreeding coefficient.

ibid. Abbreviated form of the Latin adverb *ibidem*, meaning the same place. It is sometimes used in citations when another paper appears at the same place, i.e., in the same journal.

IBIDS *See* ichthyosis.

I blood group (Ii system) I and i are universal erythrocyte antigens that exhibit alteration during development but minimal polymorphism. The synthesis of I/i antigens results from the cooperation of glycosyltransferases on common substrates and there is not a single diagnostic immunodeterminant sugar specific for the blood type. It may be associated with autoimmune hemolytic anemias. *See* ABO blood type; blood types; hemolytic anemia; hemolytic disease.

IbpA, IbpB Inclusion body proteins are small heat-shock protein chaperones of prokaryotes. *See* sHSP. (Kitagawa, M., et al. 2000. *FEMS Microbiol. Lett.* 184[2]:165.)

IBS *See* IBD, identity by descent; identity of state.

IC$_{50}$ Median inhibitory concentration (of a drug or other chemical compound).

ICAD Inhibitor of CAD. *See* AIF; CAD; caspase. (Sakahira, H., et al. 1999. *J. Biol. Chem.* 274:15740.)

ICAM (CD54, 19p13.3-p13.2) Intercellular adhesion molecules are proteins that bind (epithelial) cells together through integrins. The concentrations of ICAM-2 in uropod projections, rather than evenly distributed, are special targets for cytotoxic lymphocytes (CTL). The targeting is assisted by the cytoskeletal membrane linker protein ezrin. ICAM-1 protects against allograft rejection. *See* cadherin; CD proteins; DC-SIGN; integrin; LFA; N-CAM; T cell; uropod. (Springer, T. A. 1990. *Nature* 346:425; Sultan, P., et al. 1997. *Nature Biotechnol.* 15:759.)

ICAT Isotope-coded affinity tag is a protein analytical procedure that may be substituted for two-dimensional gel electrophoresis. The cells are grown under two different conditions and a biotin-linked isotope labels the cysteine residues of the proteins. The extracted protein is digested and the labeled peptides are subjected to affinity chromatography. The peptides are further analyzed by high-performance liquid chromatography/mass spectrometry. The differences in expression of the genes are shown by the differences at the level of the isotope. *See* gel electrophoresis; MALDI/TOF/MS; proteomics; two-dimensional gel electrophoresis. (Smolka, M. B., et al. 2001. *Anal. Biochem.* 297:25; Han, D. K., et al. 2001. *Nature Biotechnol.* 19:946; Turecek, F. 2002. *J. Mass Spectrom.* 37:1.)

iccosomes Dendritic cells package antigen:antibody complexes into budding structures that shed from their surface. *See* dendritic cell. (Terashima, K., et al. 1992. *Semin. Immunol.* 4[4]:267.)

ICE Interleukin-1β-converting enzyme, caspase 1 in human chromosome 11q22. A cysteine protease activated during Fas-mediated apoptosis. Its family includes CPP32 and

Ich-1 proteases. ICE is a mammalian homologue of *ced-3*. *See* apoptosis; caspase; CPP32; Fas; granzyme B; IL-1; interleukins; RNKP-1. (Druilhe, A., et al. 2001. *Cell Death Differ.* 8[6]:649.)

I-cell disease *See* mucopolysaccharidosis.

Ice man (Ötzi) Mummy discovered frozen in the Tyrolean Alps in 1991; its age was estimated to be 5,100 to 5,300 years. Some claims have been made that it may be a scientific hoax, but detailed analysis negates this possibility. Apparently, a 10-genome-equivalent quantity of DNA was recovered per gram of its tissue, constituting only redundant sequences. One DNA sequence of the hypervariable region of the mitochondrial DNA indicated close similarities to central and northern European populations. X-ray analysis revealed the cause of death was murder by an arrow flint. *See* ancient DNA; mtDNA; mummies. (Williams, A. C., et al. 1995. *Biochim. Biophys. Acta* 1246:98.)

ICER Inhibitor of CRE. Basic leucine-zipper domain protein of the CREB family lacking the C-terminal transactivation domain. It is inducible by FSH. *See* CRE; CREB; FSH; leucine zipper. (Trocme, C., et al. 2001. *J. Neurosci. Res.* 65[2]:91.)

ICF See immunodeficiency; methylation of DNA.

ICH1 Protease implicated in apoptosis. Its N-terminus is homologous to caspase 2. *See* apoptosis; caspase. (Zeng, Q., et al. 2000. *J. Comp. Neurol.* 424:640.)

ichthyosis Noninflammatory keratosis of the skin appearing in different forms and under different genetic controls, often associated with syndromes. In the *autosomal-dominant* ichthyosis vulgaris (1q21), the scaling of palms and soles appears during the first 3 months after birth. It may be caused by a deficit in keratohyalin, a precursor of filaggrin and an element of keratin fibers. A number of different variations of the dominant types exist. The *autosomal-recessive* lamellar form (14q11.2, keratinocyte transglutaminase, TGM1) has great variations involving redness of the skin, brittle hair, blisters, liver disease, physical and mental retardation, etc. There are other recessive lamellar ichthyosis genes in human chromosome 2 (LI2, 2q33-35) and LI3 is located at 19p12-q12. The recessive forms are generally more serious than the dominant ones. The harlequin type (so named because of the diamond-shape 4–5-cm-diameter scaly, horny spots, like the pattern on the robes of harlequins) may cause death within the first week after birth. Some recessive forms involve hair loss, progressive neural defects, enlarged liver, and kidney defects. In Sjögren-Larsson syndrome (17p11.2), the frequency of the recessive gene responsible for the disease in a county of Sweden appeared to be 0.01 and that of the carriers 0.02, while the prevalence was 8.3×10^{-5}. Fatty aldehyde dehydrogenase deficiency, neurological disorders, and keratosis accompanied this latter form of ichthyosis. Another form of recessive ichthyosis is connected to triglyceride storage anomalies. In the autosomal-recessive ichthyosis (14q11), the activity of keratinocyte transglutamase activity is much reduced. X-linked ichthyosis (Xp22.32)–based steroid sulfatase deficiency may involve asymmetric malformations of the lung, thyroid, several nerves, etc., and hypoplasia at

the same side as the ichthyosis. Another X-linked (distal part of Xp) group showed deficiency of placental steroid sulfatase. Ichthyosis is also called IBIDS. *See* Chanarin-Dorfman disease; collagen; erythroderma; keratosis; lyonization; skin disease; trichothiodystrophy.

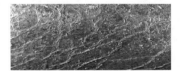

Keratosis of the skin.

ICOS Inducible co-stimulator, a human homologue of CD28, is a homodimeric protein (M_r 55–60K) that enhances T-cell proliferation and lymphokine secretion. It upregulates cell-to-cell interaction, antibody production by B cells, and the production of IL-10 (but not IL-2) in response to foreign antigens. The B7-related protein (B7RP) expressed on B cells and ICOS without interacting with the B7-CD28 system stimulates adaptive immune response. ICOS$^{-/-}$ mice are deficient in immunoglobulin class switching and germinal center formation. CD40 and its ligand may compensate for ICOS defects. ICOS seems to influence T_H-1 and T_H-2 lymphocytes and may be partly responsible for graft rejection. *See* B7 protein; CD28; CD40; CTLA; germinal center; IL-2; IL-10; immunoglobulin; T-cell receptor. (McAdam, A. J., et al. 2001. *Nature* 409:102; Özkaynak, E., et al. 2001. *Nature Immunol.* 2:591 and following articles.)

icosahedral Body with 20 facets like the capsids of some viruses. *See* isometric phage.

ICR (1) Compounds synthesized by the International Cancer Research Institute, Philadelphia, PA, are acridines with alkylating side chains causing frameshift mutations. *See* mutagens. (2) Internal control regions through the zona pellucida. *See* A box. (3) Insulator control region. *See* enhancer competition; insulator.

ICSI Intra cytoplasmic sperm injection. Some people's and animal's spermatozoa are unable to penetrate the egg. In such cases, the sperm may need to be injected mechanically into the ooplasm. This technology can be used for genetic transformation. There are some risks of mechanical or chemical (in the delivery fluid) injury to the egg. The genetic risk is that the injected sperm is not competing and a less viable or defective one may fertilize. ICSI makes it possible for men with a very low sperm count to father children. Unfortunately, the genetically determined low sperm count is transmitted to the offspring. *See* ART. (Ma, S., et al. 2001. *Fertility & Sterility* 75:1095; Faddy, M. J., et al. 2001. *Nature Genet.* 29:131.)

ID (1) *See* idant. (2) Integrating database is formed in the GenBank containing information on accession number, ASN.1, gi, bioseq. *See* accession number; ASN.1; bioseq; GenBank; gi.

idant Higher hierarchical genetic structure as conceived during the 19th century. Biophores (~alleles) are aggregated into ids (~loci) and the ids form idants (~chromosomes).

idaxozan *See* clonidine.

IDC Idiopathic dilated cardiomyopathy. *See* cardiomyopathy, dilated.

IDDM Insulin-dependent diabetes mellitus. *See* diabetes mellitus.

identical by descent *See* IBD.

identical by state *See* IBD.

identical twins *See* twinning.

identifier Sequences within introns or noncoding 3′ regions are involved in genetic regulation. The identifier sequences are transcribed by RNA polymerase III, may define chromatin regions, and facilitate the transcription by RNA polymerase II as promoters are made more accessible to soluble trans-acting molecules. *See* introns; regulation of gene activity; RNA polymerase; trans-acting element. (Mellon, S. H., et al. 1988. *Nucleic Acids Res.* 16:3963.)

identifier syntax In the BLAST search program, indicates the source of the information, e.g., GenBank, EMBL, Swiss-Prot, etc. *See* gi.

identity by descent (IBD) Shared genes in a pedigree were derived from a common ancestor and not from an identical mutation (identity by state).

identity gene Controls developmental specifications for an organ. (See Science (2002) 256:297–316.)

identity index *See* evolutionary distance.

identity of state When the coefficient of inbreeding is determined, the identity of an allele must be specified as *identity by descent* (i.e., passed on by a common ancestor), not by coincidentally occurring mutation. Mutation may generate only from *identity by state*. *See* coefficient of inbreeding; consanguinity.

idiogram Diagram of the chromosome set, including all essential morphological features of the chromosomes, e.g., arm ratio, satellites, banding pattern, etc.

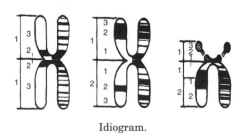

Idiogram.

idiomorphs Opposite mating type determining gene loci in fungi. (Arie, T., et al. 2000. *Mol. Plant Microbe Interact.* 13:1330.)

idiopathic Spontaneous, i.e., the cause or origin is unknown.

idiopathic torsion dystonia (ITD) Dominant (human chromosome 9q) neurological disorder ($\approx$30% penetrance) with onset before the late 20s, first affecting the limb muscles, and later the effects may spread. Its frequency ($1.6-5 \times 10^{-4}$) is higher among Ashkenazi Jews than in other populations. *See* dystonia.

idiopathic ventricular fibrillation (IVF) Common cause of fatal heart disease involving arrhythmia. Missense or splice-donor mutations in cardiac sodium channel gene SCN5A may be responsible for the risk of developing this condition.

idioplasm 19th-century hypothetical concept about the physical basis of heredity.

idiotope Determinant of antigenic specificity. *See* antibody; antigen.

idiotype Originally (in 1884) it was coined for identifying the entire complex of genetic determinants in a cell, but in that sense it is no longer in use. In modern immunogenetics it means the idiotopes distinguishing one group of antibody-producing cells from other groups of immunoglobulin-producing cells. Thus, the idiotype represents the specificity of all the idiotopes because the antigens bind within or at the idiotopes to the variable region of the antibody. The recurrent (also called public, major, or cross-reactive) idiotypes regularly appear during immune response, but the private (or minor) idiotypes may or may not appear even in genetically identical individuals. *See* allotype; antibody; anti-idiotype; epitope; HLA; immunoglobulins; internal image immunoglobulin; isotype; paratope. (Jerne, N. K. 1985. *Science* 229:1057.)

idiotype exclusion In a family of genes, only one gene product is expressed. *See* allelic exclusion; locus exclusion.

idling reaction In amino acid–starved cells, synthetic activity is reduced (*see* stringent control) and the ribosomes are uncharged with aminoacylated tRNAs. Under such conditions, uncharged tRNAs may be attached to the ribosomes and block any residual protein synthetic activity as the idling reaction. *See* protein synthesis. (Bilgin, N., et al. 1992. *J. Mol. Biol.* 224:1911.)

iDNA Initiator DNA. At the beginning of DNA replication, the polymerase α/primase synthesizes iDNA before polymerase δ takes over the chain elongation, *See* DNA replication; replication fork. (Law, A., et al. 1998. *Nucleic Acids Res.* 26:919.)

Id protein Interferes with differentiation by blocking the access of basic helix-loop-helix transcription factors to the E box in the DNA. In mice there are four *Id* genes and knocking out all of them is lethal, but when at least one allele is functional, premature withdrawal of the neuroblasts from the cell cycle, failure of normal angiogenesis, and vascularization occur. Overexpression of Id2 results from transcription activation of the Myc family of oncoproteins. The progression of the cell cycle induced by Myc requires the inactivation of Rb (retinoblastoma) protein by the dominant-negative Id2. Id1 opposes the effect of p16^{INK4q} in senescence. *See* angiogenesis; E box; helix-loop-helix; Myc; p16^{INK4a}; retinoblastoma; senescence. (Alani, R. M., et al. 2001. *Proc. Natl. Acad. Sci. USA* 98:7812.)

iduronic acid Epimer (a diastereomer [stereoisomers without being mirror images]) of glucuronic acid and dermatan sulfate, heparan sulfate, and heparin. *See* glucuronic acid; heparin; Hurler syndrome; mucopolysaccharidosis.

I element *See* nonviral retrotransposable elements retroposon; retrotransposon.

IES *See* internally eliminated sequences.

IEX Immediate early expressed protein may block apoptosis induced by FAS and TNF. Its function is mediated by NF-κB. *See* apoptosis; FAS; NF-κB; TNF. (Segev, D. L., et al. 2001. *J. Biol. Chem.* 276:26799.)

IF-1, IF-2, IF-3 Protein synthesis initiation factors in prokaryotes and eukaryotic organelles. IF-1 is a ribosome dissociation factor, and it cooperates with IF-2 and IF-3. IF2-1 and IF2-2 facilitate the binding of tRNA^{fMet} to the AUG translation initiation codon on the 30S ribosomal subunit. In *E. coli*, IF2-1 is a 97.3 kDa single polypeptide with a GTP-binding domain in its center. IF2-2 is 79.7 kDa. IF3 is 20.7 kDa, is a ribosome subunit antiassociation factor, and is a facilitator of tRNA^{fMet} and initiator codon interaction. In the eukaryote *Saccharomyces*, there is a (*FUN1* gene product yIF2) homologue to IF2. Eukaryotic eIF2 is a three-subunit functionally similar but structurally different protein. yIF2, however, is present in an evolutionarily similar form from *Archaebacteria* to eukaryotes. *See* eIF; protein synthesis; Shine-Dalgarno sequence. (Carter, A. P., et al. 2001. *Science* 291:498.)

IFGT Irradiation and fusion gene transfer. Human chromosomes are fragmented by irradiation (~3 Krad) and then allowed to hybridize somatically with rodent chromosomes. From such a cell culture, the majority of the human chromosomes and fragments are eliminated, but fragments that recombine with the chromosome are retained. The coretention of human markers sheds light on their physical proximity within the human chromosomes. *See* radiation hybrid; somatic cell hybrids (chromosome assignment). (Walter, M. A. & Goodfellow, P. N. 1995. *Mol. Biotechnol.* 3[2]:117.)

IFN *See* IFR; interferons.

IFN-gamma Same as MAF, a lymphokine.

IFR Interferon regulatory factor. Negative regulator of interferon. When it binds to CCE, it activates histone 4 transcription involved in the progression from the G1 to S phase of the cell cycle. *See* CCE; cell cycle; HiNF; histones; interferons. (Nakaya, T., et al. 2001. *Biochem. Biophys. Res. Commun.* 283:1150.)

Ig Immunoglobulin (with additional letters to identify types). *See* antibody; immune system; immunoglobulins.

Igα and Igβ Immunoglobulin-associated proteins that form disulfide-linked heterodimers (Igα-Igβ). They belong to a large family of antigen receptor-associated signal transducers and they have a tyrosine-containing cytoplasmic motif. B-cell activation requires this dimer for triggering the Src and Sky family kinases for receptor phosphorylation. These proteins have an additional regulatory role. The heterodimer is essential for the differentiation of B cells. *See* CD40; immunoglobulins; lymphocytes; Sky; Src. (Rudolph, A. K., et al. 1981. *Eur. J. Immunol.* 11:[6]:527.)

IGC *See* speckles intranuclear.

IGF *See* insulin-like growth factor.

IGI Integrated gene index is a list of all genes revealed in a genome. *See* IPI.

IGM Inheritable genetic modification of humans. *See* GMO.

ignorant DNA Molecular by-product of coincidental amplification; may be slightly harmful but may also be advantageous under certain circumstances, e.g., rRNA, satellite sequences, etc. *See* junk DNA; satellite; selfish DNA. (Epplen, J. T. 1988. *J. Hered.* 79[6]:409.)

IGS *See* intergenic spacer.

IgSF Immunoglobulin superfamily. The cell adhesion molecules (CAM) commonly contain immunoglobulin domains. These domains have a core with two face-to-face arranged β-sheets and other highly variable regions. *See* CAM; immunoglobulins; protein structure.

I$_h$ (I$_f$) Hyperpolarization-activated cation channels (HAC) in the cell membrane activated by cAMP- and cGMP-controlled Na$^+$/K$^+$ ions. Such a process determines the pacemaker activity of active neurons and heart cells. The three cloned HACs contain 863, 910, and 779 amino acids, respectively. *See* hyperpolarization; ion channels.

IHF Integration host factor. Heterodimeric host protein required for site-specific recombination. It may have some function in transcription and replication. It is a member of the high-mobility group proteins and its function may bear some similarity to that of histones in eukaryotes. *See* high-mobility group proteins; intasome; recombination; site-specific recombination. (Holbrook, J. A., et al. 2001. *J. Mol. Biol.* 310:379.)

Ii Invariant chain of the major histocompatibility complex participating in assembly of the class II molecules in the endoplasmic reticulum. This chain is subsequently degraded by lysosomal proteinases, and amino acids 81–104 are retained as CLIP (class II–associated Ii peptide) before the foreign antigen is loaded onto the T lymphocytes, and it is removed by the chaperone H-2M. *See* HLA; major histocompatibility complex. (Frauwirth, K. & Shastri, N. 2001. *Cell Imunol.* 209[2]:97.)

iIF-3P Translation elongation factor stimulating the preinitiation complex on the 40S ribosomal subunit. *See* eIF; PIC; protein synthesis.

Ii gene *See* ABO blood group.

IIH Shorthand for TFIIH. *See* transcription factors.

I$_{KACh}$ Inwardly rectifying acetylcholine regulated K$^+$ channel activated by G$_i$ protein; inhibits the opening of the voltage-gated Ca^{2+} channels and thus regulates nerve and muscle functions. *See* G$_i$ protein; ion channels.

IκB Regulatory protein binding to transcription factor NF-κB, keeping it in the cytoplasm until it is degraded by proteolysis after being phosphorylated, then releasing the transcriptional activator NF-κB to migrate into the nucleus. Inactivation of IκB usually requires stress signals or pathogenic attack. Ankyrin repeats characterize the various types of IκB and these are also present in several morphogenetic factors of *Drosophila* (e.g., cactus). Epstein-Barr virus nuclear antigen contains glycine-alanine (GA) repeats, which are inhibitory to cis-MHC class I−restricted antigen presentation. Thus, it prevents the activation of cytotoxic T cells (CTL) and the virus may evade immune recognition. Insertion of GA repeats into the IκBα chain protects it from ubiquitination and thereby the activation of the NF-κB transcription factor. Phosphorylation and proteolytic cleavage are key instruments in the separation of IκB from the NF-κB complex, providing a chance for the two NF-κB subunits to migrate into the nucleus and serve there as transcription factors. Phosphorylation of IκB is carried out by morphogenesis factor- and cytokine-activated kinases IKK-α and -β isozymes (M$_r$ 70−90 K), respectively. IKK is part of a large complex including an associated scaffold protein (IKAP, 150 K M$_r$). IκB is modified by the ubiquitin-carrier protein SUMO-1 and is therefore not ubiquitinated and is protected by degradation by proteasomes. Whereas ubiquitination is favored by phosphorylation, SUMO-1 is inhibited by phosphorylation. IκB-Ras1 and 2 proteins degrade the IκBβ subunit at a lower rate than the α-subunit. *See* ankyrin; antigen-presenting cell; FADD; glucocorticoid; IKK; immunosuppression; MAPKKK; morphogenesis in *Drosophila*; NIK; NF-κB; PEST; PIC; RAS; regulation of gene activity; RIP; signal transduction; TNF; TRADD; ubiquitin. (Imbert, V.,

et al. 1996. *Cell* 86:787; Pahl, H. L. 1999. *Oncogene* 18:6853; Bottero, V., et al. 2001. *J. Biol. Chem.* 276:21317.)

Ikaros Transcriptional regulator of lymphocytes. It is associated with pericentromeric heterochromatin. *See* hetrochromatin; lymphocytes. (Trinh, L. A. 2001. *Genes Dev.* 15:1817; Lopez, R. A., et al. 2002. *Proc. Natl. Acad. Sci. USA* 99:602.)

IKK IκB protein serine kinases. IKKα (IKK1) responds to morphogenetic signals in the activation of NF-κB, whereas IKKβ responds to proinflammatory cytokines and controls NF-κB, which in turn regulates inflammation and apoptosis. IKKγ is a modulatory subunit of NF-κB. Aspirin inhibits IKKβ. *See* Akt; aspirin; cyclooxygenase; ectodermal dysplasia; IκB; NAK; NF-κB. (Senftleben, U., et al. 2001. *Science* 293:1495.)

IL Interleukin. (O'Neill, L. A. J. & Bowie, A., eds. 2001. *Interleukin Protocols*. Humana Press, Totowa, NJ.)

IL-1 Lymphocyte-activating factor. IL-1α (17 kDa) and -β (also about 17 kDa) chains are encoded in human chromosome 2q13 at very close linkage. IL-1ra is a cytokine receptor antagonist of IL-1. IL-1 receptors (IL-1R) are encoded close by in human chromosome 2. IL-1 mediates Ser/Thr phosphorylation of various proteins (cytoskeleton) and Hsp27. IL-1-activated Ser/Thr kinase is IRAK and the specific transduced molecule is TRAF6. These molecules are elements of the NF-κB cascade. The IL-1 system is involved with the hematopoietic cells, the neuroendocrine system, and the central nervous system. It generally cooperates with TNF. IL-1 has clinical implications in inflammations, blood coagulation, osteoporosis, metastasis, neurodegenerative and autoimmune diseases. *See* dendritic cell; ICE; interleukins; IRAK; mental retardation; NF-κB; prostaglandin; TNF; Toll. (Bomford, R. & Henderson, B., eds. 1990. *Interleukin-1*. Elsevier, New York.)

IL-2 T-cell growth factor, TCGF, a ~15.5 kDa glycoprotein, encoded at 4q26-q27. It promotes the proliferation and differentiation of lymphocytes. The discovery of IL-2 made possible the longtime culture of T cells. IL-2Rβ and IL-2Rγ chain subunits are members of the cytokine receptor superfamily. IL-2Rβ (human chromosome 22q11.2-12) and -γ (Xq13) chains are shared by the other IL receptors. Human IL2Rα (chromosome 10p14-p15) has only 13 amino acids in its cytoplasmic domain and it may not have a significant role in signal transduction. The IL-R receptor complex has a prominent role in various signal transduction pathways. Mice with an IL-2 receptor chain β (IL-Rβ) show excessive differentiation of B cells into plasma cells,

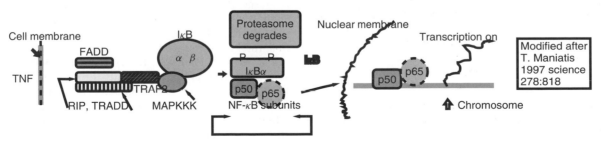

The action path of the IκB.

excessive amounts of immunoglobulins G1 and E, and the production of autoantibodies, resulting in hemolytic anemia. Defects in the widely shared IL-2Rγ cause severe combined immunodeficiency (SCID) in humans. It is synthesized primarily by CD4$^+$-T$_H$ lymphocytes. The IL-2 gene is activated by a Jun kinase at the 5′-untranslated region and modulated at 3′-untranslated sequences. IL-2 may serve as an adjuvant in genetic immunization. See anemia; AP-1; autoantibody; B cells; cytokines; ICOS; immunization, genetic; immunoglobulins; interleukins; JUN; memory, immunological; NF-AT; NF-κB; Oct-1; RAP1; SCID; T cells. (Lotzova, E. & Herberman, R. B., eds. 1990. *Interleukin-2 and Killer Cells in Cancer*. CRC Press, Boca Raton, FL.)

IL-3 Hematopoietic growth factor is a 28 kDa glycoprotein secreted by CD4$^+$ lymphocytes, encoded in human chromosome 5q31. IL-3 is within a gene cluster IL-3–CSF2–IL-13–IL-4–IL-5→centromere. IL-3 plays an important role in the development of immunity to parasites by stimulating the formation of hematopoietic effector cells. Jointly with IL-13, it inhibits inflammation and boosts humoral immunity and IgE responses. See allergy; CIS; hematopoiesis; IL-13; interleukins. (Craddock, B. L., et al. 2001. *J. Biol. Chem.* 276:24274.)

IL-4 B-cell growth factor, ~18 kDa glycoprotein, encoded at human chromosome 5q31.1. It stimulates the formation of IgG4 and IgE. Its receptor is encoded at 16p12.1-p11.2. Activation of gene expression by IL-4 requires phosphorylation of tyrosine, homodimerization, and nuclear translocation of STAT6. IFN-β and -γ inhibit IL-4-induced activation of STAT6 and STAT6-induced genes partly by inducing SOCS-1. See asthma; IFN; IL-3; IL-13; immunoglobulins; interleukins; SOCS; STAT. (Spitz, H. 1992. *IL-4: Structure and Function*. CRC Press, Boca Raton, FL.)

IL-5 Eosinophil differentiation factor encoded at human chromosome 5q31. IL-5Rα is expressed on B lymphocytes, eosinophilic and basophilic granulocytes. It is involved in the differentiation of B cells by its association with the syntenin protein that binds transcription factor Sox4. See allergy; IL-3; lymphocytes; Sox; syntenin. (Geijsen, N., et al. 2001. *Science* 293:1136.)

IL-6 (IFN-β2/BSF-2) Interleukin group of proteins (21–28 kDa), encoded in human chromosome 7p21. It binds to upstream DNA sequences in some genes. Its receptor, IL-6R, is a transmembrane protein. IL-6 formation is induced by infection and primarily mediates humoral immune reactions. The IL-6 promoter contains GRE, AP-1, SRE, and other transcription factor-binding sites. The IL-6 receptor (IL-6R) is trimeric and includes gp130 (918 amino acids, encoded at 5q21), IL-6Rα chain (80 kDa, encoded at 1q21), and IL-6. See AP-1; CRE; gp130; GRE; interleukins; SRE. (Marsch, J., et al., eds. 1992. *Polyfunctional Cytokines: IL-6 and Lif*. Wiley, New York; MacKiewicz, A., et al., eds. 1995. *Interleukin-6-Type Cytokines*. Ann. New York Acad. Sci., vol. 762.)

IL-7 (lymphopoietin, ~25 kDa) Stimulates lymphocyte precursors in the bone marrow. It is indispensable for normal proliferation of T cells. If the α-chain of the IL-7 receptor is defective in mice, D-J joining is normal in immature B cells, but recombination of the distantly situated heavy-chain V gene segments is progressively impaired in proportion to their distance from D.-J. IL-7 receptor ligands seem to signal for recombination by regulating the access of recombinase to DNA. IL-7 promotes embryonic neurons. See B cell; immunoglobulins; T cell. (Tan, J. T., et al. 2001. *Proc. Natl. Acad. Sci. USA* 98:8732.)

IL-8 Activates neutrophils, cell migration, adhesion, and inflammation. See melanoma growth-stimulatory factor; SDF. (Xie, K., 2001. *Cytokine Growth Factor Rev.* 12[4]:375.)

IL-9 (32-39-kDa glycoprotein) Promotes (mast) cell proliferation and differentiation and is the erythroid colony-stimulating factor. The IL-9 receptor (522 amino acids) is encoded in the pseudoautosomal region of Xq28. But pseudogenes are found at 9p34, 10p15, 16p13.3, and 18p11.3, as well as in the X and Y chromosomes. IL-9R appears to be constitutively expressed in Hodgkin disease and seems to have a role in asthma. IL-9 appears to stimulate myeloid leukemia. See asthma; Hodgkin disease; leukemia; mast cells; pseudoautosomal; pseudogene. (Demoulin, J. B., et al. 2001. *Cell Growth Diff.* 12[3]:169.)

IL-10 (CSIF) 18.5 kDa protein encoded at 1q31-q32 and produced primarily by T$_H$ cells; acts as a cytokine synthesis inhibitor, yet it may stimulate the growth of mast cells and CD8$^+$ T cells. IL-10 has a role in autoimmune diseases; protozoan, fungal, and bacterial infections; and various carcinomas. Its receptor (hIL-10R, 90–110 kDa) is similar to interferon receptors and is encoded at 11q23.3. There is also a second receptor gene in human chromosome 21. IL-10 signals through the Jak-Stat and Tyk2 pathways. Gene therapy for the reduction of inflammation seems promising. See ICOS; mast cell; osteopontin; signal transduction; T cell; Tyk2. (Moore, K. W., et al. 2001. *Annu. Rev. Immunol.* 19:683; Xing, Z. & Wang, J. 2000. *Curr. Pharm. Des.* 6:599.)

IL-11 (23 kDa, encoded at 19q13.3) Controls hematopoiesis in the bone marrow. Many of its functions overlap with those of IL-6. The ~43 kDa IL-11 receptor, IL-11Rα1 chain, is encoded at 9p13 close to IL-11Rα2. They recruit the gp130 and gp190 glycoproteins and activate the Jak/Tyk cytoplasmic tyrosine kinases. See gp130; gp190; hematopoiesis. (Kaye, J. A. 1996. *Stem Cells* 14 Suppl. 1:274.)

IL-12 (interleukin-12) Heterodimeric cytokine, one of the most potent stimulators of T helper cells (T$_H$), natural killer cells (NK), and B lymphocytes; increases the production of interferon-γ (IFN-γ). Macrophages, neutrophils, dendritic cells, and B lymphocytes produce IL-12. The light chain (p35) is encoded at human chromosome 3p12-q132 and the heavy chain (p40) is at 5q31-q33.1. The deficiency of the IL-12 receptor (IL12R) causes susceptibility to infections. Microbial lipoproteins stimulate IL-12 production by macrophages and the process is mediated by Toll-like receptors. Nitric oxide synthase (NOS2) is important in mediating signaling to IL-12. IL-12 is a very promising tool in controlling HIV, Leischmaniasis, malaria, tuberculosis, schisostomiasis, and other infectious diseases, although some side effects (toxic shock syndrome, atherosclerosis) may limit its application. See osteopontine. (Lotze, M. T., et al., eds., 1996. *Interleukin 12*,

Ann. New York *Acad. Sci.* volume 795; Picard, C., et al. 2002. *Am. J. Hum. Genet.* 70:336.)

IL-13 The 131- or 132-amino-acid protein is encoded at human chromosome 5q31. It is a monocyte and B-lymphocyte regulator synergistic with IL-2. It is functionally related to IL-4, although the homology between the two is ∼20%. IL-13, unlike IL-4, is not a T-cell proliferation factor. IL-13 receptor α1 subunit is encoded at Xq13. *See* asthma; IL-3. (Chiaramonte, M. G., et al. 1999. *J. Immunol.* 162:920.)

IL-14 Produced by activated T cells and B-cell lymphoma. *See* lymphoma; T cell. (Ford, R., et al. 1995. *Blood* 86:283.)

IL-15 ∼114 amino acid T-cell growth factor, encoded at human chromosome 4q13, produced by monocytes and epithelial cells. IL-15 induces the proliferation of activated T cells and B cells. In cooperation with IL-2, it boosts the production of IFNγ and TNFα. *See* IFN; memory immunological; monocyte; T cell; TNF. (Perera, L. P. 2000. *Arch. Immunol. Ther. Exp.* 48[6]:457.)

IL-16 (LCF, lymphocyte chemoattractant) 130-amino-acid lymphokine (M_r 13,500), secreted by activated CD8$^+$ cells, binds to T cells by the CD4 receptor, and suppresses HIV and SIV. *See* HIV; SIV; T cell. (Cruikshank, W. W., et al. 2000. *J. Leukoc. Biol.* 67[6]:757.)

IL-17 (CTLA-8) Interleukin with homology to ORF13 of herpes virus Samiri (HVS). The human IL-7 is a 20 to 30 kDa homodimeric protein. It regulates cytokine and (leukemia) oncogene mRNA stability. *See* CTLA. (Shi, Y., et al. 2000. *J. Biol. Chem.* 275:19167.)

IL-18 (IGIF, interferon γ-inducing factor) Produced by peripheral blood mononuclear cells, dendritic cells, intestinal epithelial cells, osteoblastic stroma cells, and upon stimulation by monocytes and macrophages (Kupffer cells). It has two different immunoglobulin receptors. IL-18 enhances inflammatory responses, but in cooperation with IL-12 it is antiallergic. IL-18 activates natural killer lymphocytes, T_H cells, and the production of interferon-γ. IL-18 is a defense molecule against bacteria, viruses, some fungal pathogens, and protozoa. Along with IL-12 and INFγ, it may cause tumor regression. Its therapeutic use is hampered by the toxicity of cytokines and by causing inflammatory and autoimmune reactions. *See* autoimmune disease; cytokines; interferons; killer cell; Kupffer cell; monocytes; T_H. (Nakanishi, K., et al. 2001. *Annu. Rev. Immunol.* 19:423.)

IL-20 176-amino-acid homologue of IL-10. Its overexpression in transgenic mice is lethal. IL-20 binding activates STAT3 in keratinocytes. IL-20 is upregulated in psoriasis. *See* keratin; psoriasis; STAT. (Rich, B. E. & Kupper, T. S. 2001. *Curr. Biol.* 11:R531; Blumberg, H., et al. 2001. *Cell* 104:9.)

IL-21 IL-21 and its receptor (IL21R) regulate clonal expansion of natural killer lymphocytes. (Parrish-Novak, J., et al. 2000. *Nature* 408:57.)

ILK Integrin-linked protein kinase phosphorylates protein kinase B, Akt, and cyclins *See* Akt; cyclins; integrin; protein kinases. (Yamaji, S., et al. 2001. *J. Cell Biol.* 153:1251.)

illegitimate child Begotten by a male other than the legal father. This status can now be identified with great certainty by DNA fingerprinting. The mother seems always to be certain. Illegitimacy, contrary to some myths, does not endow the "love baby" with better abilities, despite some famous examples (Leonardo da Vinci, Francis Bacon, etc.). Actually, out-of-wedlock children generally suffer physical and emotional stress. Illegitimate offspring in old pedigrees may confound recombination frequencies, especially at tight linkage. Illegitimacy may cause problems in prenatal diagnosis of some disorders and may frustrate genetic counseling. *See* DNA fingerprinting; genetic counseling; prenatal diagnosis.

illegitimate insertion Insertion elements may be incorporated into the genome by a process that does not require complete or even substantial homology unlike the general recombination events that usually have the requisite of homology between the sites of recombination. Restriction enzymes may facilitate integration when it is nonhomologous but do not promote homologous insertion. *See* illegitimate recombination. (Manivasakam, P., et al. 2001. *Nucleic Acids Res.* 29:4826.)

illegitimate pairing Synapsis between not entirely identical strands, which may lead to aberrant crossover products. *See* oblique crossing over, unequal crossing over.

illegitimate recombination Recombination when the synaptic strands are not entirely homologous. Insertion and transposable elements are integrated by nonhomologous recombination at their target site that shows minimal homology. *See* DNA ligases; ectopic recombination; homeologous recombination; insertion element; Ku; nonhomologous recombination; RecA-independent recombination; recombination; synapsis; transposable elements. (Ehrlich, S. D., et al. 1993. *Gene* 135:161; Hanada, K., et al. 2001. *J. Bacteriol.* 183:4964.)

illumination Various types of units of measurement are in use. Most common is Lux = 1 standard new candela from 1 meter distance per 1 meter^{-2}, or it is measured in watt or joule units. *See* candela; joule; watt.

ILTs (LIR, MIR, CD85, human chromosome 19q) Surface receptors on monocytes, macrophages, dendritic cells, and B lymphocytes containing immunoglobulin-like domains. Most of them carry ITIM sequences in the cytoplasmic side. ILT1 and ILT7 lack ITIM and are associated with ITAM containing FcεRIγ membrane adaptor proteins. *See* FcεRIγ; ITAM; ITIM. (Volz, A., et al. 2001. *Immunol. Rev.* 181:39.)

image analyzer (image processor) Uses video cameras with high light sensitivity attached to the microscope. The camera is connected to a computer that can further enhance the image going to the camera. It thus can electronically enhance (by digitalization) and process the image and detect details that cannot be perceived by the human eye by direct viewing through the microscope. The electronic system may show the picture in "false colors," i.e., selected by the operator rather than the natural color that the object has. The use of such devices can free the image from background "noise" and greatly enhance the clarity. In addition, the operator's eyes are not stressed. The tissue-specific distribution and expression of the bacterial luciferase transgene can be monitored by microchannel plate–enhanced photon counting analyzers. This setup consists of a microscope, image processor, TV monitor, and computer. The heart of the equipment can be outlined:

MICROSCOPE → photocathode → microchannel plates

→ phosphor screen → DISCRIMINATING VIDICON

This system can resolve a single photon. By comparison, a single native bacterial cell releases about 10,000 photons. An image analyzer (e.g., VAX Station II/GP4) is also used for scanning DNA fingerprint autoradiograms. *See* luciferase; microscopy.

image clones Gridded sequences verified (Research Genetics <http://www.resgen.com>) or not-sequence-verified (Genome Systems <http://genomesytems.com/>) human and other gene clones and gene clusters suitable for microarray hybridization. *See* microarray hybridization; UniGene.

image processing *See* image analyzer.

imaginal disk After fertilization of the *Drosophila* egg, the zygote nucleus begins dividing within a common cytoplasm and forms a syncytium (multinucleate protoplasm) without actual separation by cell membranes. After about 2 hours and about 13 divisions, the cellular blastoderm stage is reached (about 6,000 cells). Then the cells move to the periphery of the embryo and the central space is occupied by the yolk. Within less than a day, the larva is hatched and this developmental stage (three instar steps) lasts for about 4 days. Inside the larva, some groups of cells, distinguishable by location shape and size—the *imaginal disks* (50,000 cells)—are set aside for serving as initials for the various organs and structures of the adult organism. The larva does not use the imaginal disks and removal of them does not kill the organism. As the larva is metamorphosed into the pupa, most of the larval tissues disintegrate to support the development of the differentiation from the disks. During pupation, from these "prefabricated elements," the imago is assembled by cell and tissue fusions. From the anterior (front) part, the head; from the middle region, the thorax; and from the posterior (hind) part, the abdomen are formed. If the imaginal disks are lifted from their original position after the third instar stage (about 5 days) and grafted into a new location of another larva, these disks develop into extra eye, wing, leg, or other structures, depending on which imaginal disk has been chosen for the surgery not on the host tissue. Thus, in the imaginal disks the developmental determination is completed much before the onset of morphological and functional differentiation. For the correct differentiation, the transplant must be placed in a larva and not into an adult. In the abdomen of an adult, the disks only proliferate. After a series of transfers through adults, the original determination of the disks may change, and when the proliferated disk tissue is transplanted into a larva, an antennal disk may give rise to a leg instead. This altered course of differentiation is called *transdetermination*. The path of the transdetermination is not accidental or random, e.g., a genital disk may develop into a proboscis (mouth part) or may pass through some indirect steps in a certain order such as genitalia → proboscis → antenna → leg. A wing from the genital disk cannot be formed directly but may arise through an antenna or leg. The process of transdetermination is only a change in competence for differentiation without a mutational change, whereas in homeotic mutants one structure is replaced or altered after a change in the DNA. *See* antenna; blastoderm; determination; development; *Drosophila*; fate maps; homeotic genes; homeotic mutants; morphogenesis; proboscis. (Ramirez-Weber, F.-A. & Kornberg, T. B. 2000. *Cell* 103:189.)

imaging New technologies (fMRI, PET) permit scanning whole bodies to monitor the turning on of genes and follow metabolic processes (including activities in the brain) without sacrificing the animals. *See* image analyzer; nuclear magnetic resonance spectroscopy; tomography.

imago Adult form of an insect, male or female. *See* *Drosophila*; imaginal disk.

Imago.

imbibition Absorption of water or other liquids.

IMC Intramolecular chaperone. *See* chaperone.

Imerslund-Grasbeck syndrome Megaloblasatic anemia (MGA1).

imidazole Present in many organic molecules and may mediate tautomerism.

Imidazole.

iminoaciduria Proline and hydroxyproline use the semirecessive renal glycine reabsorption mechanism and in its failure these amino acids accumulate in the urine. *See* aminoacidurias.

imino form Of an amino acid, arises from an amino form according to the reaction below:

$$NH_2-C-COOH + flavin \longrightarrow NH=C-COOH + flavin\ H_2$$

Amino   α-Imino

Similar transformation can take place on nucleotides and can facilitate unusual hydrogen pairing between bromouracil and guanine or cytosine and adenine. *See* hydrogen pairing.

iminoglycinuria Excessive proline and hydroxyproline in the urine. This is probably due to heterozygosity of hyperglycinuria. *See* hyperglycinemia.

imitatory epigenotype Theory that assumes functional units of the phenotype are genetically modular and integrated within the group and relatively independent of the rest of the phenotype. It can be studied by the statistical methods used for QTL. *See* QTL. (Mezey, J. G., et al. 2000. *Genetics* 156:305.)

immediate early genes Of a virus, are first turned on after infection without any requirement for synthesizing virally encoded proteins. *See* delayed early genes; late genes.

immortalization A cell in culture would not senesce but would proliferate indefinitely. Normally, cultured cells will die, but cancerous transformation or fusing normal cells with cancer cells makes them "immortal." Immortalization and tumorous growth are, however, under separate genetic control. Cells of the germline are also practically immortal because of their transmission from generation to generation until the extinction of the strain. *See* cancer; hybridoma; p16^{INK}; telomeres. (Tevethia, M. J. & Ozer, H. L. 2001. *Methods Mol. Biol.* 165:185.)

immune clearance Destruction of infecting foreign cells by the phagocytotic activity of various special lymphocytes. Its major sites are in the liver and to a lesser degree in the spleen, but other tissues with phagocytotic activity may carry it out. The C3 complement and its receptors, C5b-9 and IgG, are important factors. *See* complement; immunoglobulins.

immune deviation In non-atopics, postnatal exposure to inhalant allergens may redirect the fetal immune response toward the T$_H$1-like cytokine pattern. *See* allergen; allergy; asthma; atopy.

immune evasion Some viruses may inactivate the MHC class I antigen presentation system and thus escape the attention of cytotoxic T lymphocytes (CTL). Killer cells (NK) may still destroy the invader. Herpes simplex viruses may inactivate or interfere with TAP transporters and various subunits of proteasomes. In HIV-infected human cells, the peptide epitopes of the virus may mutate, and rather than being a ligand, they may serve as T-cell receptor antagonists, resulting in inactivation of HIV-specific T cells. Some adenoviral glycoproteins immobilize MHC class I molecules to the endoplasmic reticulum. Some cytomegaloviruses (CMV) make proteins that bind to MHC class I molecules and "dislocate" the heavy chains into the cytosol (from the endoplasmic reticulum); thus, proteasomes destroy them. Another trick of some viruses is to mediate the internalization and thus inactivation of MHC I proteins. CMVs may produce their own MHC class I homologue proteins (UL18, m144), which help evading natural killer cell attacks on the virus. These protein homologues may regulate the immune system in various other ways. CMV may also downregulate

the synthesis of MHC type II molecules involved in the B-lymphocyte defense system. An IL-10 homologue encoded by the Epstein-Barr virus (EBV) may interfere with T-cell function. Viral IL-10 homologues may subvert IL-12-regulated interferon, cytokine, and TAP production. Some viruses may interfere with the process of apoptosis used by the organ to pathogen-infected cells before the maturing infective viral particles can be released. *See* epitope; immune system; MHC. (Kavanagh, D. G., et al. 2001. *J. Exp. Med.* 194:967.)

immune homeostasis After an immune reaction, the immune system is reset to the preimmunization state and can respond to a new antigen. *See* immune system.

immune privilege Grafts to certain sites (eyes, testis, brain) may be protected from rejection caused by the constitutive expression of the Fas ligand (FasL). Other factors may be blood tissue barriers, direct drainage of the tissue fluid into the blood, absence of efferent lymphatics, potent immunosuppressive environment (TGFβ), reduced expression or induction of MHC, neuropeptides, α-melanocyte-stimulating hormone, vasoactive intestinal peptide, calcitonin gene-related peptide (CGRP), and membrane-bound inhibitors of complement activation and fixation. Immune privilege may be necessary for pregnancy, avoiding certain diseases, graft rejection, etc. *See* calcitonin; complement; Fas; MHC; TGF. (Rall. G. F., et al. 1995. *J. Exp. Med.* 182:1201; O'Connell, J., et al. 2001. *Nature Med.* 7:321.)

immune response *See* immune system; MHC.

immune suppression Removal or inactivation of the antigen. This can be achieved by destroying the target cells by CTL, B cells, macrophages, or killer cells or by cytokines and other immunosuppressive agents. Contrasuppression prevents suppression and allows for the development of immunity. *See* CTL; immune tolerance; killer cell; veto cell. (Takemoto, S. K. 2000. *Clin. Transpl.* 481.)

immune surveillance *See* immunological surveillance.

immune synapse On the surface of natural killer T cells (NK), inhibitory immunoglobulin-like receptors (KIR1, KIR2) occur that cause clustering of HLA-C molecules at the surface of the target cells. At the target and the NK cells, HLA-KIR forms rings around intercellular cell adhesion molecules and lymphocyte-associated antigen-1 for about 20 minutes. Multiple synapses may occur as the T cells invade the selected target. Synapse formation requires reorganization of the cytoskeleton of the T cells and the antigen-presenting cells. *See* agrin; antigen; antigen-presenting cell; cytoskeleton; HLA; killer cell; KIR. (Khan, A. A., et al. 2001. *Science* 292:1681; Dustin, M. L. & Coper, J. A. 2000. *Nature Immunol.* 1:23; van der Merwe, P. A. & Davis, S. J. 2002. *Science* 295:1479.)

immune system Complex defense organization of vertebrate animals. The main cellular components of the immune system are the lymphocytes and macrophages. The B lymphocytes synthesize the antibodies that react and destroy the foreign antigens. The T lymphocytes generate the antigen receptor–MHC protein complex, which specifically recognizes individual antibodies and is involved in the destruction of foreign antigens. Lower animals may use phagocytosis as a defense. Plants do not have an immune system. The immune system protects the body from microbial infections and from noninfectious foreign macromolecules such as proteins, polysaccharides, and tissue grafts.

Smaller foreign molecules (*haptens*) may also trigger the immune system primarily when associated with proteins. The molecules that elicit the immune response are called antigens. The antigens stimulate the formation of antibodies and the antibodies are the actual defense molecules. The immune reaction is extremely specific, and minute differences between antigens, such as single amino acid substitutions or isomers, may be specifically recognized. The immune system distinguishes between the body's own antigens and those of extraneous origin. The absence of immune response to the body's self antigens is an *acquired immunological tolerance*. The immature immune system may learn how to ignore even a foreign antigen if exposed to it during a very early stage. Immunological tolerance may be produced in later stages of development with immunosuppressive drugs by very high or repeated exposure to very low amounts of the foreign antigen. Also, altering the so-called *antigen-presenting cells* may affect tolerance. These antigen-presenting cells bind, process, and combine foreign antigens with class I and II proteins of the HLA complex (MHC). T lymphocytes can be made tolerant to foreign antigens more readily than B cells. In some rare *autoimmune diseases*, even self-discrimination may break down with very serious consequences for the individual because the defense system of the body may destroy its own tissue. The key players in mounting an immune response are the approximately two trillion lymphocytes (white blood cells) of the human body.

The *B lymphocytes* are involved in the *humoral* (circulating in the blood) immune response. The B cells make antibodies. These are called B cells because they are made in the bursa Fabricius (intestinal pouches) in birds. In humans, the B cells originate from the stem cells of the fetal yolk sac, then in the liver and finally in the bone marrow (humans do not have bursa). The pre-B cells appear in the fetal liver by the 8th week of gestation and contain μ-immunoglobulin chains in their cytoplasm but not on their surface About 2 weeks later, B cells appear. By the 13th week, B cells have μ-immunoglobulin on their surface. By the 12th week, B-cell production shifts to the bone marrow. IgD (immunoglobulin-δ) appears by the 14th week in the spleen, lymph nodes, and blood. After the 14th week, the HLA antigens begin to appear. As the antibody gene rearrangements proceed, all other immunoglobulin genes may be expressed. The proliferation and differentiation of B cells requires the presence of antigens on their surface and some soluble factors obtained from T cells. T_H (helper) and T_S (suppressor) cells regulate the development of B cells. Activation of helper T cells requires the presence of APC cells (*antigen-presenting cells*), HLA class I and II proteins, and interleukin-1. T_H cells secrete some mediators (e.g., lymphokines).

The activated lymphocytes and mediators then induce the formation of receptors followed by proliferation and differentiation of both T and B lymphocytes. The suppressor lymphocytes may prevent the induction of the immune response by exposure to a wide variety of molecules. The *T lymphocytes* (shaped in the thymus) are important representatives of the immune system; their primordia develop by the 5th to 6th week of human gestation, and the first mature T cells appear by the 9th to 10th week. The T cells are involved in *cell-mediated* immune response, i.e., they react with antigens bound to their surface. As the maturation of T cells proceeds, CD antigens (CD1, CD2, CD4, etc.) appear on their surface and identify T-cell subsets. By the 16th to 20th week, the spectrum of subsets reaches that of the adults. T cells usually respond to antigens in association with the HLA complex (MHC, major histocompatibility complex). With CD8 antigens, T cells respond to class I proteins of the HLA complex, whereas CD4 T cells are limited to HLA class II proteins (DP, DQ, DR). This phenomenon is called *MHC restriction*. Maternal IgG may cross the placenta after the 16th week and provides early immune protection for the fetus. Fetuses and newborns may not respond much to foreign cells because the lymphocyte development may be arrested when contacted by antigens and antiidiotype antibodies. IgM and IgA are normally (in the absence of infection) not present in substantial amounts in newborns. Although after birth the maternal IgG tends to be eliminated, it is gradually replaced by the infant's own antibodies. By age 3 years, IgG levels become sufficient. Before the end of the first year, IgM levels reach those of adults, but IgA levels rise very slowly, reaching adult levels only by age 9 to 12 years.

The appearance of the immune reaction is the result of a developmental process. The *virgin T cells* (that have never been exposed earlier to a specific antigen) become *effector cells*, i.e., they begin to proliferate and produce either the cell-mediated (T-cell) or the humoral (B-cell) response. Some other lymphocytes are also induced to proliferation and differentiation. But this time they do not participate in the immune response; instead, they become *memory cells* that will "remember" the same antigen by virtue of their differentiation and assume the effector cell role at a later similar exposure (*clonal mechanism of secondary immune response*). The immune response is mounted relatively slowly (in days or weeks) when first-time exposed to a particular antigen occurs (*primary immune response*), but on subsequent exposures the response is much faster because of the availability of memory cells. This phenomenon is the basis of immunization (vaccination) as it is used in medicine. In order to generate an immune response, the lymphocyte receptors must be exposed to an antigen. This is usually followed by the signals of cytokines and other costimulator molecules (CD40, CD80, CD86 on the antigen-presenting cells), resulting in the clonal proliferation of specific lymphocytes and memory cells. Upon contact of CD28 and CD80, CD86 initiates the immune response and the CTLA-4 receptor with CD80; CD86 terminates the T-cell immune reactions. CD28 also stimulates the synthesis of Bcl antiapoptosis proteins or B cells. The C3d component of the complement in association with the CD21 complement receptor is the costimulator. In case either antigen cannot reach the lymphocytes or the costimulatory response is defective, there will be no immune response. The immune system is largely associated with lymph nodes and the spleen, and thus represents systemic responses. Some of the immune reactions are, however, localized to specific tissues such as the gut epithelia or intraepithelial lymphocytes.

In insects, different types of peptides, such as cecropin (35–39 amino acids), defensin, attacin, diptericin (82 amino acids), drosopcin, metnikowin, drosomycin, andropin, and

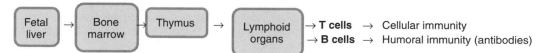

Development of the immune system and its major components

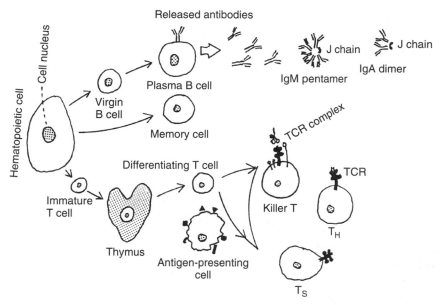

The B and T lymphocytes in mature individuals are generated by the hematopoietic cells in the bone marrow. The virgin B cells become either plasma cells or memory cells upon the encounter of a foreign antigen and lymphokines. As a response, the plasma cells manufacture antibodies that are first embedded in the plasma membrane of the B cells, and when the membrane is "saturated," they are released as circulating antibodies. Antibody IgM polymerizes into A pentamer and joins A J Peptide chain. The IgA antibody forms a dimer and A J Polypeptide chain braces it. The memory cells do not become active until follow-up exposure to the particular antigen. Each of the antibody-producing cells makes a single type of antibody and its clonal progeny makes the same antibody. Different antibodies are made by different B-cell clones. The immature T cells differentiate into T cells within the thymus. The thymus is a bilobate organ that develops in the embryo and it reaches its maximal function during puberty, then begins a gradual, slow decline, accompanied by a decrease of T-lymphocyte production and a weakened immune response. The mature T cells are of three main types: *killer cells* (cytotoxic T lymphocytes, CTL), *helper T cells* (T_H), and *suppressor T cells* (T_S). The specificity of the T cell function depends on the surface antigens carried by the antigen-presenting cells. The TCR (T-cell receptor) has substantial structural and functional homology with the immunoglobulin proteins. The TCR complex is formed by the participation of the MHC protein, encoded by the HLA complex and a series of cell membrane proteins. The killer cells destroy the mebranes of the invading cells, followed by lysis of their contents.

lysozyme, mediate the humoral defense system. In addition, cellular response (phagocytosis) may be available. A genome-wide survey of *Drosophila* (13,197 genes tested) using microarray analysis revealed that microbial infection induced 230 genes and repressed 170. *See* allergy; antibody; antigen; antigen-presenting cell; apoptosis; asthma; autoimmune disease; B cells; blood cells; bystander activation; CD21; CD28; CD40; CD80; complement; hapten; HLA; host-pathogen relations; immune evasion; immune system disease; immune tolerance; immunoglobulins; innate immunity; interleukin; leukalexins; lymphocyte; lymphocyte homing; lymphokines; memory, immunological; MHC; molecular mimics; phagocytosis; surrogate chains; T cells; TCR; xenotransplantation. (Abbas, A. K. & Janeway, C. A. 2000. *Cell* 100:129; Janeway, C. A., Jr. 2001. *Proc. Natl. Acad. Sci. USA* 98:7461; Germain, R. N. 2001. *Science* 293:240;

DiGregorio, E., et al. 2001. *Proc. Natl. Acad. Sci. USA* 98:12590; <http://sdmc.krdl.org.sg:8080/fimm/>; see diagram next page.)

immune system diseases See agammaglobulinemia; arthritis; autoimmune diseases; celiac disease; DiGeorge syndrome; immunodeficiency; lupus erythematosus; Reiter syndrome; Sjögren syndrome; Wiskott-Aldrich syndrome.

immune tolerance Failure of the immune system to respond to antigen(s) by supposedly suppressing alloreactive T cells. Neonates were supposed to display immune tolerance; however, newer studies indicate that their immune response may be different. Immune tolerance is regulated either by anergy or by suppression. The suppression may be mediated by orally supplied low doses of antigens or by clonal anergy

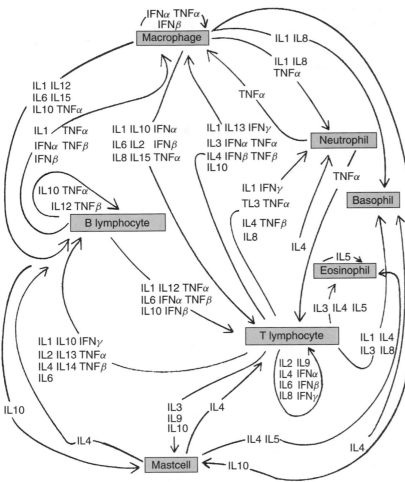

IL = INTERLEUKIN, INF = INTERFERON, TNF = TUMOR NECROSIS FACTOR

Communications in the immune system.

when high doses of the antigen are provided. Oral tolerance may extend to tolerance of the gut immune system and thus can present systemic immune reaction to ingested proteins. The brain-blood barrier prevents lymphocytes from entering the central nervous system. Peripheral immune tolerance may include T-cell-activated repression by the secretion of TGF-β. It has been somewhat of a puzzle why allogeneic fetuses are not rejected by the maternal immune system. The mother's body may reject organ transplants from the child because of the paternal genes of the offspring. Notwithstanding, the same mother can carry the child to term. It has been hypothesized that either physical separation or antigenic immaturity of the fetus or some sort of immunological inertness of the mother toward the conceptus may be responsible. It appears that the maternal T cells specific for paternal MHC class I alloantigens are reduced during pregnancy. The observation that the concentration of tryptophan is lower in the maternal-fetal interface tissues led to the experimentally supported assumption that tryptophan may be required for T-cell proliferation. The level of tryptophan is reduced by indoleamine 2,3-dioxygenase (IDO). Inhibition of IDO by higher doses of the tryptophan analog 1-methyltryptophan reduced the maintenance of allogeneic concepti in mice apparently by curtailing the chances for T-cell-mediated rejection. Loss of autoimmunity may occur after infection by pathogens due to molecular mimicry, epitope spreading, or bystander activation.

B-cell tolerance may be lost after removal of the self-antigen or when T-cell help is provided at the initial exposure but not if the T cell is available after the exposure to the self-antigen. *See* alloreactive; anergy; antigen; bystander activation; conceptus; epitope; immune system; immunity; immunosuppression; incompatibility; molecular mimics; mucosal vaccines; Rh blood group; self-tolerance; T cell; T-cell receptor; TGF; thymus. (Krensky, A. M. 2001. *Pediatr. Nephrol.* 16:675; Godnow, C. C. 2001. *Lancet* 357:2115; Robertson, S. A. & Sharkey, D. J. 2001. *Semin. Immunol.* 13[4]:243; Moffett-King, A. 2002. *Nature Rev. Immunol.* 2:656.)

immunity Protected state against biological (microbial) or other agents. *Natural immunity* is the readily emerging incompatibility between donor and recipient tissues. The xenoreactive natural (human) antibodies (XNA) react with the Galα1-3Gal terminal antigens present on the endothelial lining of blood vessels of the xenografts of the majority of mammals (except humans and some primates). This binding then activates the human complement system. The graft *complement decay accelerating factor* (DAF), *membrane cofactor proteins* (MCP), and CD59 cannot neutralize the host natural immunity reaction and thus rejection may be initiated. Further injury to the graft is inflicted by the natural killer cells of the host. Although NK cells normally carry *inhibitory receptors* that normally prevent attack on MHC class I molecules, this system may fail with the MHC system of the graft. *Acquired immunity* develops as a reaction of the immune system of the organism. The transfer of antibodies or activated lymphocytes from another body convey s *passive immunity*. *Natural immunity* means that the cells are not susceptible to a particular organism, e.g., humans are not infected by wheat rust. *Genetic or familial immunity* indicates that a group of individuals are resistant to or free of a particular type of infection. *Maternal or intrauterine immunity* is when humoral antibodies are passed through the placenta into the fetus. *Cell-mediated immunity* is the result of T-lymphocyte activation by adoptive transfer (adoptive immunization) of an immune donor to syngeneic recipients. *Humoral immunity* is provided by the antibodies secreted by the B lymphocytes. *Neonate immunity* is due to passing IgG immunoglobulins from the mother's milk with the assistance of the FcRn receptor. At the newborn stage, the monocytes and the macrophages have limited activity, resulting in the reduction of MHC class II proteins, costimulatory molecules, antigen processing, and cytokines. Natural killer cells are present but not very active. *Phage immunity* designates the condition when a lysogenic bacterium cannot be superinfected by another bacteriophage. *See* antibody; antigen-presenting cell; complement; cytokines; FcRn; humoral antibody; immune response; immune tolerance; killer cells; lymphocyte; MHC; mucosal immunity; superinfection; syngeneic; xenotransplantation.

immunization   Induction of immunity. It can be active immunization by introducing specific antigens into the body to promote the formation of a specific antibody or passive immunization by introducing antibodies into the body that may convey temporary immunity. During vaccination, live (attenuated) or killed microbes are introduced into the body either through the bloodstream or by oral means. The bonding subunit of *E. coli* heat-labile enterotoxin (LT-B) is a very effective immunogen. Bacterial polysaccharide vaccines are not effective in babies before age 18–24 months, although protein conjugates may be effective at the age of a few weeks. Attenuated live viral or bacterial vaccines at this stage are usually not advisable. The results of DNA immunization at the early stages of life are still debated.

Transgenic plants encoding LT-B express the foreign peptides and are able to bind the natural ligand after oligomerization. Mice fed by antigens produced by transgenic plants could orally be immunized. After intercrossing, tobacco plants transgenic for murine antibody κ-chain, hybrid immunoglobulin A–G heavy chain, immunoglobulin J chain, and a rabbit secretory component, respectively, produced segregating progeny that simultaneously expressed all four polypeptides. The polypeptide chains were successfully assembled within the transgenic hybrid into a functional high-molecular-weight secretory immunoglobulin that could recognize streptococcal surface adhesion molecules. Interestingly, in plants this process could take place in a single cell, whereas in animals it requires two cell types. The results indicate the possibility of using immunoglobulins manufactured by plants for oral vaccination. In reactive immunization the antigen is so highly reactive that a chemical reaction takes place at the site of the combining antibody. This mechanism may enhance catalytic antibody chemistry. The common type of immunization does not result in covalent interaction between antigen and antibody. Although sharks can synthesize a variety of immunoglobulins (somewhat different from higher mammals), they cannot be immunized. *See* antibody; immunization, genetic; immunization, in vitro; immunoglobulins; intercellular immunization; intracellular immunization; memory, immunological; transformation, genetic (transformation of plants); vaccines.

immunization, genetic (DNA vaccination)   Involves introducing into an animal (or into a specific animal tissue) a gene or expression gene library encoding a particular antigenic protein(s) by biolistic transformation or plasmid vectors, resulting in antibody production. The transgene is introduced into muscle tissue or into the skin. The latter, especially when associated with lymphoid tissues, is more suitable because it harbors more antigen-presenting cells. This method does not involve the risk of live or attenuated pathogen vaccines. Targeting antigen-ligand (L-selectin or CTLA-4) constructs to the lymph nodes or to antigen-presenting cells (dendritic cells, macrophages, and lymphocytes) enhances the effectiveness of DNA immunization by 100-to 1,000-fold. They may elicit Th1 or CD4+ and CD8+ T-cell responses. The presence of special DNA sequences (IL-2) may serve as *adjuvants* and substantially magnify the reaction. Poly G:C, poly I:C, or palindromic (5′-GACGTC, 5′-ADCGCT, 5′AACGTT) sequences may effectively induce the production of interferons and interleukins, potent stimulants of the killer effects of the lymphocytes. Coinjection of vectors containing mutant caspases stimulate T- and B-lymphocyte responses. DNA vaccination appears promising against infectious diseases, malaria, autoimmune diseases, and cancer. The goal of this procedure for cancer is not prevention but rather the cure from tumors already present. A potential advantage of DNA vaccination versus protein vaccines is the continuous expression of the antigen. DNA vaccines are expected to be processed for immune recognition by the cell's special protein degradation system, have high temperature stability, and the cost may be low. A possible slight risk is that the transgenic products may have adverse side effects such as inducing autoimmune reaction or immune tolerance. DNA immunization appears to be substantially less effective as aging progresses. In a rat model, orally administered adeno-associated virus vaccine containing the NR1 subunit of the NMDA receptor appeared to generate strong antiepileptic and neuroprotective effects. *See* adeno-associated virus; antibody; autoimmune disease; biolistic transformation; cancer; CTLA-4; epilepsy; gene therapy; IFN; IL; IL-2; immune tolerance; immunization; library; malaria; monoclonal antibody; NMDA receptor; polynucleotide vaccination; selectins; T cell; viral vectors. (Xu, M., et al. 2000. *Trends Biotech.* 18:167; Maloy, K. J., et al. 2001. *Proc. Natl. Acad. Sci. USA* 98:3299; Schadendorf, D. 2002. *Semin. Oncol.* 29:503; <http://www.genweb.com/Dnavax/dnavax.html>.)

immunization, intracellular   *See* immunization, genetic.

immunization in vitro   B lymphocytes are cultured in vitro in the presence of cytokines and growth factor complexes in order to produce antibody. It is of somewhat limited significance; however, it may be particularly useful when antibody production is elicited against hazardous antigens. *See* immunization.

immunoadsorbtion   Binding cognate antibody to antigen in order to facilitate the separation of a specific type of antigen or antibody from a mixture. In *extracorporeal immunoadsorption*, the blood is pumped out of the body and allowed to react with an affinity column, then may be returned to the body after this clearance, e.g., removal of radioactively labeled antibody. *See* immunofiltration.

immunoblotting   Separating proteins by gel electrophoresis and identifying an appropriate component by the specific monoclonal antibody labeled by fluorochrome or radioactivity. *See* electrophoresis; Western blot.

immunocontraceptive   Vaccination of animals with zona pellucida proteins and adjuvants prevents fertilization to a substantial extent. *See* contraceptives.

immunocytochemistry   Locates antigens in the cell with the aid of labeled antibodies through microscopic examination. *See* immunolabeling.

immunodeficiency   May have ~80 causes; some immunodeficiencies involve milder effects, whereas others are lethal. Immunodeficiency is frequently caused by defects in the lymphocyte differentiation system (thymus). Some of the immunodeficiencies are parts of other syndromes such as Down syndrome, sickle cell anemia, ataxia telangiectasia, glycoprotein deficiency, glucose-6-phosphate dehydrogenase deficiency, immunoglobulin imbalance (deficiency of IgA and IgG but increased IgM), defects of the HLA histocompatibility system, and some are caused by nongenetic events such as infection (HIV) or drugs and various allergies that may have a genetic component. Xq13.1-linked immunodeficiency with hyperimmunoglobulin M (IgM) but absence of IgG, IGA, and IgE is caused by a defect in the interleukin receptor IL-2R γc-chain and interaction between CD40 and CD 40L; it accounts for about half of all the cases. As a consequence, the T and B cells do not differentiate normally. ICF syndrome (human chromosome 20q11.2, immunodeficiency, centromere instability [centromere areas 1, 9, and 16] and facial anomalies) involves a defect in DNA methyl transferase 3B. Methylation at specific sites in the heterochromatin is essential for the maintenance of chromosome integrity (Hansen, R. S., et al. 2000. *Hum. Mol. Genet.* 9:2575.) Some immunodeficiencies are the consequences of medical treatment, e.g., cancer therapy or the use of immunosuppressive drugs in tissue transplantation. Jak-3 deficiency also causes lymphocyte problems. ADA (adenosine deaminase)-recessive defects are responsible for about 15–25% of the cases with severe immunodeficiency. RAG1/RAG2 (recombination activating genes) interfere with the generation of immunoglobulins and results in impaired lymphocyte receptors. Recessive mutation in the ZAP 70 gene (2q12) is detrimental for the development of T-cell receptors. Recessive mutations at 11q23 (CD3ε/CD3γ) are also a cause of immunodeficiency due to problems with immunoglobulins. Mutations in MHC class I (6p21.3) and class II peptides adversely affect T and B lymphocytes. *See* acquired immunodeficiency; ADA; agammaglobulinemia; bare lymphocyte syndrome; CD3; CD40 ligand; chronic granulomatous disease; DiGeorge syndrome; Duncan syndrome; Epstein-Barr virus; HLA; hyper-IgM syndrome; hypoglobulinemia; lymphoproliferative disease; methylation of DNA; Nezlof syndrome; RAG1/RAG2; RFX; severe combined immunodeficiency (SCID); Wiskott-Aldrich syndrome; ZAP-70. (<http:www.uta.fi/imt/bioinfo/idr/>.)

immunodeficiency, viral   *See* acquired immunodeficiency syndrome. (AIDS).

immunodominance   of the many possible epitopes of the antigen the cytotoxic lymphocytes (CTL) recognize only one or a few. *See* CTL; epitope. (Brehm, M. A., et al. 2002. *Nature Immunol.* 3:627.)

immuno-electron microscopy   Fluorescent antibodies label cell constituents and thus the location of proteins can be identified with electron microscopic resolution. *See* histochemistry; microscopy.

immunoelectrophoresis   Can be carried out by different procedures. *Diffusion*: Over the electrophoretically separated proteins, antisera (antibody or antibodies) are placed in a trough. The protein bands are permitted to diffuse in a radial manner from their original position in the gel while the antibody is diffusing vertically. At their position of reaction with each other an elliptical arc of precipitation is formed. *Rocket electrophoresis*: The antigen is placed in the wells of the gel that contain the antibody. After turning on the electric current, the antigen moves and forms a rocket-shape trail of precipitation as it reacts with the antibody. The length of the rocket indicates the amount of antigen applied to a well. *See* ELISA; rocket electrophoresis.

immunofluorescence   Identification of a protein (antigen) by *direct* adsorption to a specific antibody that has an attached fluorochrome. Alternatively, to the first (unlabeled) antibody a second cognate, fluorochrome-labeled antibody is added for the sake of immunostaining (*indirect* method). *See* antibody; antigen; fluorochromes.

immunogen   Substance that elicits immune response. *See* immune system.

immunogenetics   Concerned with the hereditary and molecular aspects of antigen and antibody systems and the immune response. *See* <imgt.cines.fr>.

immunoglobulins   Structural units of the antibody molecules. Each antibody is a heterotetramer consisting of two identical light immunoglobulin chains (either of two λ or two κ) and two identical heavy chains. The five heavy chains of humans and mice display considerable variations, but the light chains are rather constant. There are five classes (*isotypes*) of immunoglobulins—IgM, IgG, IgA, IgD and IgE—identified according to the heavy-chain subunits: μ, γ, α, δ, and ε The Ig heavy chains are glycoproteins containing about 15, 4, 10, 18, and 18% sugars in a molecular mass of 70, 50, 55, 62, and 70 kDa, for μ, γ, α, δ, and ε respectively. IgM is a pentamer containing five μ heavy chains, five light chains, and one J chain; its molecular mass is about 900 kDa.

The J polypeptide (ca. 20 kDa) synthesized within IgM-secreting cells is covalently attached between two Fc domains (see antibody) and it is supposed to initiate oligomerization. (The J polypeptide is not the product of the J genes located between the variable and constant gene clusters.) IgA antibody is either a monomer, dimer, or trimer with molecular weights 153, 325, and 580, respectively, and may have a J chain. Besides the J polypeptide, IgA may have a secretory component (SC) of 558 amino acids. The SC is picked up by the secretory

$$5'-L-V_n-V_1-D_1 \bullet D_{20}-JJJJ_{\sim\sim} \blacksquare \mu \blacksquare \delta \blacksquare \gamma_3 \blacksquare \gamma_1 \blacksquare \varepsilon_\psi \blacksquare \alpha_1 \blacksquare \gamma_\psi \blacksquare \gamma_2 \blacksquare \gamma_4 \blacksquare \varepsilon \blacksquare \alpha_2 - 3'$$

IgA dimers from the surface of epithelial cells and braces the IgA dimers, protecting them from proteolysis. The SC is part of the transport receptor and mediates the translocation of Ig and IgM into glandular secretions. IgG, IgD, and IgE antibody monomers are of 150, 180, and 190 kDa, respectively. IgD and IgE are of minor significance, although the letter mediates allergic reactions and its level is elevated hundreds of fold in chronic infections. Each of the five major types of heavy chains can be associated with either one of the light chains.

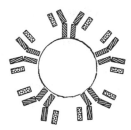

The pentameric structure of immunoglobulin M (IgM) joined by the J chain in the center. Note the absence of hinges.

The five classes of antibodies have somewhat specialized roles. IgG is the humoral, circulating antibody and it is the major class. It can cross the placenta and enter extravascular areas. IgE deals with the allergic reactions and IgA has the major role in defense against microbial infections. IgM and IgA have two subclasses; IgG and the λ-chains have four, designated as, e.g., IgM1, IgM2, and so on. Some of the IgG subclasses can cross the placenta. IgM is secreted in low amounts by B lymphocytes, but this is the first immunoglobulin made by newborns and it readily activates the complement cascade, thus representing one of the first lines of defense. IgA is produced mainly in the gastrointestinal system and plays a major role in mucosal immunity. Within these subclasses, *allotypes* are distinguished: 25 for human IgG, 2 for IgA, and 3 for the kappa (κ) chain. These allotypes represent antigenic markers on the immunoglobulin chains and are designated as, e.g., Gm1, Gm2...Am1...Km1... and so on for IgG, IgA, and κ-chain markers, respectively. The allotype variants represent amino acid substitutions at one or more sites. The characteristic series of allotype markers are inherited as gene blocs and are called *haplotypes*.

The genes encoding immunoglobulins are clustered in three supergene families. The heavy-chain genes are clustered at human chromosomal location 14q32.33. The κ-genes are at human chromosome 2p12, whereas the λ-gene family is in human chromosome 22q11.12. Within the approximately 7,000 kb heavy-chain region in the long arm of chromosome 14, the genes L (encodes the signal that leads the polypeptide to the endoplasmic reticulum), V_H (variable heavy), D (diversity), J (juncture), and the various constant heavy-chain genes C_μ to C_α are arranged in groups. The groups are separated by

long base sequences ($\sim\sim$) and spacers ($\blacksquare$) with switch signals (S) in front of the constant heavy-chain genes. The general organization of the immunoglobulin genes is very similar in all mammals, although they are located in different chromosomes:

(The ψ indicates pseudogenes.) Up front of this sequence, the basal promoter of the Ig heavy chains contains non-translated regulatory transcriptional elements, such as the dispensable *heptamer* consensus (5'-CTCATGA-3'), and 10 to 40 bp downstream, the indispensable *octamer* consensus (5'-ATGCAAAT-3'). The latter is 30 to 60 bp upstream from the TATA box, which is followed within about 20 to 30 bp by the transcriptional initiation site for the LVDJ and constant heavy-chain sequences, μ to α_2. (The orientation of the octamer is opposite the direction of transcription.) Actually, between the LVDJ region and constant heavy-chain genes, there are enhancer elements of the 5'-CAGGTGGC-3' motif and three core repeats of multiple GC sequences.

The κ-cluster is similarly organized in human chromosome 2:

$$5'-L-V_S--V_n_____JJJJJ-$$
$$-1 \ \kappa \ \text{constant gene group}-3'$$

The λ-genes are in human chromosome 22. The variable genes occur in six groups. Here the J genes are not clustered separately but situated in front of the six constant λ-gene groups. Some of the λ-genes are outside the clusters and may not be functional. The individual λ segments are quite variable in size. The V_S is one of the switching sequences explained below.

The base promoter of the κ-light chain contains a pentanucleotide consensus about 100 bp upstream from the transcription initiation site, then within −90 to −60 bp the octamer consensus (oriented in the direction of the transcription) follows. This does not have the heptamer shown at the heavy genes. The TATA box is about the same distance from the transcription initiation site as the heavy chains. There are enhancer elements, designated as κB (5'-GGAAAGTCCCC-3') and Eκ1 to Eκ3 (variants of the enhancer motif shown at the heavy chains), between the LVJ genes and the constant κ-gene group. The strongest enhancer is κB. Both the heavy and light gene enhancers act preferentially in B lymphocytes. The heavy-chain enhancers seem to be constitutive, whereas the light gene enhancers become active after the rearrangement of the genes (discussed below). Besides the enhancers, the immunoglobulin genes seem to have silencers of expression for nonlymphocyte chromosomes. Turning on the promoters requires transcription factors. One of them, the 60 kDa OTF-2, has specificity for the immunoglobulin enhancer consensus 5'-CAGGTGGC-3'. The 90 kDA OTF-1a general mammalian transcription factor is also present in the lymphocytes. The DNA-binding domains of these two factors are very similar, but their other domains are different. These enhancers also bind other types of proteins, and the only lymphocyte-specific enhancer appears to be the octamer. The light-chain-specific transcription factor is protein NF-κB, which binds to the 5'-GGGPu(C/T)TPyPy(C/T)C-3' motif. After the immunoglobulin light chain has undergone rearrangement preparatory to transcription, the pattern of the *nuclease-sensitive sites* in the promoter region is altered.

The variable-diversity regions of the light and heavy chains determine antibody specificity. The antigen has to fit to be complementary to the NH_2 end of the antibody. The three complementarity-determining regions (CDR1, CDR2, and CDR3) represent the hypervariable region of the antibody (that has the highest specificity for the antigen).

Although in the germline the complete array of all immunoglobulin genes is present, during development various rearrangements and gene elimination take place. Thus, the mRNA does not represent all of the genes all of the time in the somatic cells, but different ones may be represented, depending on their transcription, which is stimulated as the immune response unfolds. The heavy-chain genes can generate an enormous variety of polypeptides. The variable region is put together from a menu of hundreds of V_H, about 20 diversity (D) segments, and 5 or more joining segments (J_H). The V, J, and constant heavy gene clusters are separated by sequences containing 12 or 23 spacers, flanked by conserved heptamer (7mer) and nonamer (9mer) nucleotide tracts that apparently serve the purpose of rearrangements. The general organization of the switching sequences is diagrammed below. The heavy-chain genes are in mouse chromosome 12 and the corresponding human gene cluster is in chromosome 14. The light κ-genes of the mouse are in chromosome 6, whereas the homologous human genes are in chromosome 2. The λ-chain genes are in mouse chromosome 16 and in human chromosome 22. The 12 and 23 bp spacers are shown in bold numbers:

```
     9MER          7MER          7MER          9MER        7MER
V_H-<GGTTTTTGT - 23 - CACATGT>--<CACAGTG - 12 - ACAAAAACC>- D -<CACAGTG-12--⌐
 ∟--ACAAAAACC>--<GGTTTTTGT - 23 - CACATGT>--J--C_H     (mouse chromosome 12)
     9MER          9MER          7MER
```

Recombination is limited to the 12 and 23 base spacers (12/23 rule). The RAG (recombination-activating) gene products provide signals for the recombinational signal sequences (RSS) for double-strand breaks. The 12/23 and RAG signals are concerted, and mutation in one signal prevents cleavage at both. The blunt signal ends and coding ends then form a hairpin-like structure by transesterification with the assistance of a score of repair enzymes. The hairpin intermediates lead to the assembly of the antigen receptor sequences of the genes. This process has similarities to retroviral integration.

The switching recognition sites in the two mouse light-chain gene series are as follows:

```
V_κ-<CACAGTG - 12 - ACAAAAACC>------<GGTTTTTGT - 23 - CACATGT>--J--C_κ
V_λ ▬<GGTTTTTGT - 23 - CACATGT>--▬<CACAGTG - 12 - ACAAAAACC>--J--C_λ
```

The consensus sequences shown above are present in between each of the variable and joining genes in the clusters. Thus, rearrangements alone can generate enormous variability, particularly within the heavy and κ-clusters, which each have hundreds of variable region genes. The λ-cluster has only a very small number of variability genes; thus, it contributes minimally to the antibody arrays, yet in combination with the highly variable heavy chains it may produce about 100,000 different antibodies. The κ-heavy-chain assemblies are capable of generating more than a million types of antibodies. This assembly process is limited to the lymphoid cells. The T-cell receptor assembly occurs only in the T cells and the complete assembly of the Ig genes takes place in the B cells. The accessibility of the V(J)D recombinase to the complementarity region is regulated by the transcription system. The enhancer motif called E-box $\mu E3$ and other upstream regulatory elements control accessibility of the recombinase.

The V and J consensus sequences display opposite orientation. Opposite orientation means that the same nonamer (9MER) can be either $\frac{ACAAAAACC}{TGTTTTTGG}$ or $\frac{GGTTTTTGT}{CCAAAAACA}$ in the DNA double strands as the sequences are inverted horizontally and vertically ($\rightarrow\downarrow$). The types of sequences shown ensure that V and J genes do not recombine within the V or J group, but the V genes recombine with the J genes because one type of spacing can recombine only with another type of spacing. In the heavy gene cluster, a V gene is obligated to transpose to a D gene, which in turn can be relocated to a J gene. V and J genes have the same type of spacers (different from the D gene); thus, the relocation of a V gene to a J gene must involve a D gene. These types of relocation mechanisms are frequently called *recombination* and *recombinase systems*. One must keep in mind that these events are not crossing overs. The 12 and 23 base spacers apparently represent one or two turns, respectively, of the DNA double helix, and thus seem to indicate the rejoining mechanics. The V and J genes can also recombine by a breakage-reunion mechanism that may take place at the heptamers (7mers) at the ends of the two coding units, resulting in the elimination of the interjacent segment with the *signal ends*. The ends of the coding sequences are the *coding ends*. The signal ends may be joined to form a circular DNA structure. This circularization, followed by elimination, is apparently a major source of the rearrangements. In some instances, there is an inversion of a V gene relative to a J gene. In such a case, the intervening material may not be deleted, although the V gene may be inactivated. Some of the rearrangements are *nonproductive* because they occur at random and are in the wrong register. Since codons are triplets, two-thirds of the rearrangements may result in garbled sequences. Some nucleotides may be lost at the coding ends and some may be added (*N nucleotides*) by the enzyme deoxynucleotidyl transferase. At the coding ends, the 5′ terminus of one of the DNA strands may covalently fuse with the 3′ terminus of its complementary strand. When the resulting hairpin structure breaks, a protruding end of nucleotides may be formed that can serve as a template to generate an inverted terminal repeat of a few nucleotides (*P nucleotides*). The 96th codon at the end of the V_H gene is actually generated by a fusion between V and J genes. This is a critical point because the 96th amino acid is part of the antigen-binding region as well as the connection of the light and heavy chains. Both deletions and additions increase the variability of the antibody genes. The V_H regions can then combine with any of the five constant heavy chains ($\mu, \delta, \gamma, \varepsilon, \alpha$, and their subclasses, *isotypes*), and this is the source of another variation.

The expression of the immunoglobulin genes requires that the promoter be transposed to the vicinity of an enhancer (see base promoter structure above). The enhancer becomes normally active only in the B lymphocytes. Before any antigen is encountered, IgM- and IgD-class antibody production starts. In such virgin B cells, both immunoglobulins have identical variable regions, but they may differ in constant regions of the μ- and the δ-chains. Since B cells are diploid, genes only in one of the two homologous chromosomes can be expressed at a time, and only one type of rearrangement can function in a cell, resulting in *allelic exclusion*. The virgin B cells can then differentiate either into *plasma cells* or *memory cells* upon exposure to an antigen. The former becomes an immediate producer of antibody; the latter will be activated into plasma cells only upon subsequent exposure to the same antigen. The differentiation is aided by *lymphokines* (a variety of growth-regulating proteins), various T lymphocytes (T_H, T_S), and macrophages. Association of the virgin lymphocytes with an antigen triggers the mechanism of *isotype switching*. Isotype switching brings about the selection of the proper heavy-chain constant region by a process of transcription, although DNA deletions may also be involved. In the undifferentiated B lymphocyte, transcription begins at the heavy-chain gene leader sequence upstream and continues through the variable and diversity regions to the end of the δ-gene, passing through exons and introns. A polyadenylation signal follows the last exon of the transcribed constant region of the μ- and δ-genes:

5′ Promoter- Enhancers--V_n...V_1---D_{20}...D_1-J_n...J_1--S-μ --S- δ -S-γ --Sϵ -S- α 3′ DNA

Virgin B cell ...V_2---D_5-J_5----μ------δ.. primary RNA transcripts

Processing ...V_2 D_5 J_5 μ polyA and V_2 D_5 δ polyA two mRNAS

Translation V_2 D_5 J_5 μ and V_2 $D5$ δ IgM and IgD polypeptides

Actually, two types of μ chains exist at the early stage of the B lymphocyte: the μ_m chain (with a hydrophobic C terminus and alternative processing event) and μ_Σ The former is included in the lymphocyte membrane as a monomeric IgM antibody (2 light -2 μ-chains), whereas the secreted IgM becomes pentameric and adopts a 20 kDa J peptide with the composition of $(\mu_2\lambda_2)_5$J (the J is synthesized in the B lymphocyte but not coded in the constant heavy-chain cluster). After recombination and deletion occurs at one of the **S** switch points in the noncoding ca. 2 kb upstream tracts at the (GAGCT)nGGGGT motifs of the constant heavy-chain genes, the rearranged genes are transcribed. The introns are eliminated and the transcript is processed into polyadenylated mRNA. The mRNA is translated into the individual heavy-chain monomers:

Switching at γ V_2---D_5-J_5----γ Primary transcript
                                                 $\downarrow$
                                    IgG 50 kDA polypeptide
Switching at α V_2---D_5-J_5----α Primary transcript
                                                 $\downarrow$
                                    IgA 55 kDa polypeptide

In these examples, the same VD and J genes are shown; actually, any of the VDJ genes can be selected before switching. Additional variation is generated by somatic mutation during the proliferation of lymphocyte clones. The frequency of these mutational events (about 10^{-3} per base) appears to be higher than the usual mutation rate of other types of genes. After the heavy chain is finished, it may combine with any of

the two light-chain polypeptides. IgG and IgA are further polymerized to form the final antibody. The activity of B cells is terminated partly by binding the antigen to the secreted antibody. This event then prevents the binding of the antigens to the B lymphocyte receptors and thus the stimulation of immunoglobulin synthesis ceases. Some birds (ducks, geese, swans) produce immunoglobulin Y (IgY), which does not occur in mammals or in some other avian species, but it bears similarities to IgG and IgE. IgY is produced in larger and smaller forms; the latter is deficient in the crystalline fragment.

The completed polypetide chains are modified gly-cosylation chains (using covalently D-galactose, N-acetyl-D-galactosamine, N-acetyl-D-glucosamine, L-fucose, D-mannose, and acetylneuraminic acid) with their structure and function altered. IgG has only 3% carbohydrates, whereas other chains may have $3-4$ times as much. The glycans are mainly at the constant regions of the heavy chains, although the number and location of the glycosylation sites vary. Glycosylation affects the activation of the complements, their stability regarding proteolysis, how many antigen-binding sites are available (avidity), etc. In certain diseases (rheumatoid arthritis, tuberculosis, Crohn disease, Sjögren syndrome, scleroderma, some autoimmune conditions), IgG molecules lose their galactose. Agalactosyl IgG molecules decrease during pregnancy and their level is lower during the first 25 years of life, then increases again.

In humans and pigs, the proportion of the κ:λ chains is about 6:4, whereas in rodents the κ-chains are about 19 times more common than the λ-chains. Chickens have only λ-light chains, and in many other animals (bovines, horse) the λ-chain is preponderant. All vertebrates synthesize immunoglobulins, yet the antibodies of the lower animals (fishes, amphibians, reptiles, birds) are somewhat different from higher forms. In mammals, usually five types of immunoglobulins are made. There are, however, exceptions. Rabbits lack IgD; mice and rats have the five types. In cattle, sheep, pigs, and horses, IgM, IgG, IgA, and IgE occur. In cats, there are IgG, IgM, IgA, and IgE and several subclasses; dogs display all five types. In camels and llamas, the IgG molecules are built of heavy and light chains, whereas other immunoglobulins have only heavy chains and the variable domain of the light chain is missing. In invertebrates, there are some immune defense molecules. In *Drosophila*, the *Amalgam* (*ama*, $1-47.5$)-encoded cell adhesion proteins display some immunoglobulin-like domains. The fasciclin II glycoproteins involved in neuronal recognition of grasshoppers have five immunoglobulin-like domains. It is conceivable that the PapD prokaryotic protein involved in pilus assembly has some evolutionary relation to immunoglobulins. The similarity is not in the amino acid sequence but in the overall structure of the domains.

Extremely large phage antibody libraries can be generated in bacteria with the aid of two nonhomologous *Lox* sites in a phagemid vector. Exchange among variable heavy- and light-chain genes creates functional recombinants with a diversity of $\sim 3 \times 10^{11}$. Gene conversion generates additional diversity. *See* accessibility; affinity maturation; AID; antibody; antibody gene switching; B cell; CDR; class switching; complement; *Cre/LoxP*; DNA-PK; ELISA; hemolin; HLA; hybridoma; hypermutation; immune system; immunization; lymphocytes; macroglobulinemia; membrane segment; monoclonal antibody; multiple myeloma; RAG; repertoire shift; RSS; SCID; somatic hypermutation; surrogate chains; tailpiece secretory; T cell; T-cell receptors; TCR; terminal nucleotidyl transferase; translin; transposons (Tn3, Tn 5, Tn 7, Tn 10); V(J)D recombinase. (Bross, L., et al. 2000. *Immunity* 13:589; Arakawa, H., et al. 2002. *Science* 295:1301; <http://imgt.cines.fr:8104/textes>; <http://sdmc. krdl.org.sg:8080/fimm>.)

immunoglobulins in the human serum (mg/mL)   IgG1, 146 kDa (γ1): ~9; IgG2, 146 kDa (γ2): ~3; IgG3, 170 kDa (γ3): ~1; IgG4, 146 kDa (γ4): ~0.5; IgM, 970 kDa (μ): ~1.5; IgA1, 160 kDa (α1): ~3; IgA2, 160 (α2): ~0.5; IgD, 200 (δ): ~0.03; IgE, 200 kDa (ε): ~0.0001. Immunoglobulins also occur in milk, tears, genitourinary and lung secretions.

immunohistochemistry   *See* immunocytochemistry.

immunolabeling   *See* ELISA; immunocytochemistry; immunofluorescence; immunoscintography; immunosensor; immunostaining; monoclonal antibody; RIA.

immunoliposome   Liposome that is coated with target-specific antibodies to deliver, e.g., therapeutic agents to cancer cells. *See* bispecific antibody; cancer gene therapy; liposome.

immunological learning   The quality of the antibody improves as the clonal selection progresses. *See* clonal selection.

immunological memory   Survivors of a cell (individual) that mounted an immune response are more effectively protected in case of a subsequent infection. The mechanism of this protection may be based on maintenance of specific T or B cells even in the absence of the antigen, or some types of lymphocytes have the ability to remember the antigen. Recurrent low levels of infection may maintain some lymophocytes, or some regulatory networks of cytokines respond in case of a second infection. *See* B cells; immune system; lymphocytes; T cell. (Utzny, C. & Burroughs, N. J. 2001. *J. Theor Biol.* 211:393; Hu, H., et al. 2001. *Nature Immunol.* 2:705.)

immunological privilege   At some tissue sites (such as the central nervous system, maternal-fetal interface, adrenal cortex, testis, ovary, hair follicles, liver, etc.), immunological reaction is not elicited either by acquiring tolerance or by failure of the antigens to communicate effectively with other sites.

immunological surveillance   Considered one of defense mechanisms against cancerous body cells. The surface antigens of the transformed (cancer) cells are different from their normal counterparts. The immune system that continuously monitors the body for invading microorganisms and other foreign antigenic material (macromolecules, grafts, etc.) recognizes the cancer cells at an incipient stage and disposes of them with the aid of the immune system (Rousseau Merck, M. F., et al. 1996. *J. Exp. Med.* 183:725). Experimental data in support of this idea come from the fact that antibodies produced against mammary cancer preferentially recognize the metastatic cells without reacting with the normal cells. The CCR7 chemokine receptor is an organizer of the immune response. Cancerous transformation may be initiated by a single mutation, although additional events are required for the development of neoplasias. Mutation rates per cell are in the range of 10^{-9}, yet because the human body may have 4 to 5 orders of higher numbers of cells, cancer mutations may affect each person numerous times. Nevertheless, the incidence of cancer death is about 0.2 of all deaths. If no biological protection were available, all individuals would have cancerous transformation(s).

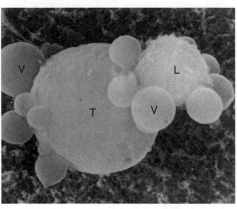

The small lymphocyte (L) is attacking a large tumor cell (T). As the cancer cell begins to lyse, first the microvilli dilate, then surface vesicles (V) appear, and eventually the cancer cell shows blebbing before destruction. (Courtesy of Dr. A. Liepins. See also Liepins, A., et al. 1978. *Cell. Immunol.* 36:331.)

The general validity of immunological surveillance for protection against cancer has been questioned because several types of cancers occurred at the same frequency in immune-compromised or genetically weak immune system animals as in normal individuals. In nude mice with a defective immune system, cancer incidence did not increase. When naive CD4$^+$ T cells specific for an antigen expressed by tumor cells were transferred into tumor-bearing mice, transient clonal expansion occurred early after transfer, accompanied by phenotypic changes associated with antigen recognition (Stavely-O'Carroll, K., et al. 1998. *Proc. Natl. Acad. Sci. USA* 95:1178). In transgenic mice where T lymphocytes and tumor-associated antigens could be monitored in some tumors, a degree of immune tolerance developed. More recent information (Ochsenbein, A. F., et al. 2001. *Nature* 411:10558) shows that (1) tumor-specific induction of protective cytotoxic T cells (CTLs) is contingent on how many tumor cells reach the secondary lymphatic organs early and stay there long enough; (2) diffusely invading systemic tumors can eliminate CTLs; (3) tumor cells located outside the lymphatic organs or not within the reach of T cells remain; (4) major histocompatibility class I molecules may not be protective. Killer cells, CD8$^+$ T cells, and activated macrophages express the stimulatory NKG2D lectin-like receptors, which may also contribute to the rejection of tumor cells. *See* cancer; cancer prevention; cancer therapy; CCR; chromosome breakage; genetic tumors; immune system; MHC; T cell. (Diefenbach, A., et al. 2001. *Nature* 413:165; Shastri, N., et al. 2002. *Annu. Rev. Immunol.* 20:463.)

immunological synapse   TCR engagement with antigen-presenting cells aided by integrin family adhesion molecules. *See* antigen-presenting cell; integrin; TCR.

immunological tests Extremely sensitive for the detection of a particular protein or other molecules that can form cross-reacting material (crm) with specific antibodies. Frequently, in case the quantity of the material is very low and standard biochemical assays are not sensitive enough for identification, immunoprobes are the choice for testing. *See* antibody preparation; immunofluorescence; immunoprobe; immunoscreening.

immunological tolerance *See* immune system.

immunomodulators Viral-encoded proteins regulating antigen presentation, regulators of cytokines, cytokine antagonists, and inhibitors of apoptosis, interfering with the functions of the complement. *See* antigen-presenting cell; apoptosis; complement; cytokines.

immunopanning Cell separation technique used in neurobiology. (Gard, A. L., et al. 1993. *Neuroprotocols* 2:209.)

immuno-PCR Polymerase chain reaction version resembling capture-PCR. It detects specific antibody-DNA conjugates with high sensitivity. *See* capture PCR; polymerase chain reaction. (Sano, T., et al. 1992. *Science* 258:120.)

immunophilins Two classes of proteins that bind immunosuppressants such as rapamycin and FK506 (FKBP) or cyclosporin. All known immunophilins display rotamase (peptidyl-prolyl cis-trans isomerase) activity in vitro. The 59 kDa member of the FKBP family is a component of the inactive glucocorticoid receptor. The immunophilins apparently interact with protein kinases of the signal-transducing paths and with the heat-shock protein 90, a chaperone. FKBP12 immunophilins bind to the GSGS domain of TGF β receptors and stabilize them in an inactive form. Activation occurs upon phosphorylation. *See* cyclophilin; cyclosporin; FK506; immunosuppressants; rapamycin; T cell. (Ivery, M. T. 2000. *Med. Res. Rev.* 20:452.)

immunoprecipitation The reaction of an antigen with a cognate antibody may lead to blood coagulation or to the selective precipitation of a protein. From a mixture of proteins, the interactive molecules (enzyme-substrate, proteins of signaling pathways, etc.) may be coprecipitated. *See* gel retardation assay; microcalorimetry; protein complexes; pulldown assay; Western blotting.

immunoprobe Or immunoblot; bacterial colonies are immobilized on a filter and a specific antibody is added. This antibody can bind the epitope of a second antibody or an antibody plus protein A, which may be labeled by a radioactive isotope (I^{135}) or a biotinylated molecule. The complex can then be detected by autoradiography on the dot blot or separated in SDS-polyacrylamide gel; the labeling identifies the substance of interest. *See* colony hybridization; DNA probe; probe; protein A; Western blot.

immunoproliferative disease, X-linked *See* Epstein-Barr virus.

immunoproteasome Used for degradation of foreign antigens by cytotoxic T cells. The proteins are ubiquitinylated and partially degraded by immunoproteasomes. Some of the subunits of the regular proteasome are replaced by other polypeptides induced via interferon-γ. This cytokine also induces the proteasome activator PA29, which facilitates antigen presentation through a more open proteasome structure. *See* antigen processing and presentation; interferon; proteasome; T cells. (Kloetzel, P.-M. 2001. *Nature Rev. Mol. Cell Biol.* 2:179.)

immunoscintigraphy By using a scintillation camera (capable of detecting the flashes emitted by radioactive isotopes), radioactively labeled monoclonal antibodies can be localized in the body or tissues by a three-dimensional image.

immunoscreening The product of a gene is identified on the basis of a cognate antibody.

immunosensor Solid-state apparatus capable of detection of antigen-antibody binding, based on either changes in mass or electrochemical or optical properties. It may be employed in clinical, environmental, or food analysis.

immunostaining Purified antibodies can be labeled by fluorochromes and their specific recognition sites can be visualized in situ with the aid of fluorescence light microscopy. Antibodies labeled by colloidal gold permit their analysis by electron microscopy.

immunostimulatory DNA (ISS) Contains CpG dinucleotides within short stretches of plasmid vehicles: 5'GACGTC-3', 5'-AGCGCT-3' or 5'AACG%%-3'. Such sequences promote the production of interferon-α and -β and interleukin-12. The significance of this finding for gene replacement therapy is that ISS may cause the production of proinflammatory cytokines and thereby downregulate gene expression. *See* cytokines; T cells; therapy. (Uchijima, M., et al. 2001. *Biochem. Biophys. Res. Commun.* 286:688.)

immunosuppressant Blocks or reduces the immune response by irradiation, specific antimetabolites, or specific antibodies. *See* calcineurin; cannabinoids; cyclosporin; immunophilin.

immunosuppression Activation of the immune system generally involves activation of cytokines and cell adhesion. Generally, in tumor cells the immune system is suppressed and this suppression may involve inhibition of lymphocytes (CTL, NK, B cells) by IL-2, IL-10, TGF-β, etc. Repression of this process requires the inhibition of transcription factors required to develop the key elements of the immune system. Glucocorticoids, prednisone (also a glucocorticoid), fungal cyclic oligopeptides, e.g., cyclosporin, mycophenolic acid ($C_{17}H_{20}O_6$), acidcyclophosphamide (carcinogen), azathioprine (an arthritis drug), cytarabin (cytosine analog), mercaptopurine (a purine analog), methotrexate (a folic acid antagonist), muromonab-CD3 (a murine monoclonal antibody [$IgG_{2\alpha}$] targeted to the lymphocyte membranes), etc., are used. The new suppressants may prevent stimulation of the T-cell receptors (TCR) or the costimulatory responses or may downregulate the amplification of the specific antigen-responding T cells. *See* CTL;

cyclophosphamide; cyclosporin; Galα1-3Gal; glucocorticoids; hyperacute reaction; IL-2; IL-10; immune system; immune tolerance; killer cell; methotrexate; TGF. (Kahan, B. D. & Koch, S. M. 2001. *Curr. Opin. Crit. Care* 7[4]:242.)

immunotherapy Includes immunization, use of immunopotentiators, immunosuppressants, hyposensitization to allergens, transplantation of bone marrow or thymus. In rats, heat-shock proteins prepared from the same cancer (but not from others) retarded the progression of the primary cancer, reduced metastasis, and prolonged life. Another possibility is to fuse antigen-presenting dendritic cells with carcinoma cells and the cell hybrids used for immunization syngeneic animals. When surgically removed cancer cells (surface antigens) were delivered to the same animal, the mouse body immunorejected the cancer. In some instances, both CD4+ and CD8+ T cells responded favorably and both primary tumors and metastatic cells were rejected. Immunotherapy of cancer may use IL-1, IL-2, IL-4, IL-5, and IL-6, receptor antagonists, interferon, tumor necrosis factor (TNFα), granulocyte-macrophage colony-stimulating factor (GM-CSF), or interferon (IFNγ) treatment to boost the host effectors and MHC class I and II molecules or apply tumor-specific antigens in order to activate cytotoxic T cells against the cancer cells. GM-CSF appeared particularly effective. *See* adoptive cellular therapy; bacillus Calmette-Guerin; cancer gene therapy; SEREX; vaccinia virus. (Chen, Z. N., et al. 2001. *Cell Biol. Int.* 25:1013; McLaughlin, P. M., et al. 2001. *Crit. Rev. Oncol. Hematol.* 40:53.)

immunotherapy, active specific Boosts immunogenic response by immunization with endogenous antigenic determinants of the cells. This is expected to result in immunological memory and thus have a long-lasting effect. *See* immune response; memory, immunological. (Pol. S., et al. 2001. *J. Hepatol.* 34:917.)

immunotherapy, adoptive (passive) *Ex vivo*−selected allogeneic transgenic donor lymphocytes are employed against viral infection or leukemia. After reintroduction, it may become necessary to select against the introduced lymphocytes in case host-graft incompatibility occurs. The negative selection requires activation of a special suicide gene. *See* allogeneic; *ex vivo*; leukemia; suicide vector. (Bathe, O. F., et al. 2001. *J. Immunol.* 167:4511.)

immunotoxin May be an antitoxin or specific antibody equipped with a bacterial, fungal, or plant toxin. The monoclonal antibody (Fab domains, Fv) provides the means of homing on the special target cell(s) of cancer (lymphoma, melanoma, breast and colorectal carcinomas) or graft rejection (bone marrow transplant) or T cells responsible for autoimmune disease (arthritis, lupus or HIV infected T cells.) In addition, it may carry cytokine and soluble receptors to assist targeting. The toxin (*Pseudomonas* exotoxin, diphtheria toxin, ricin, abrin, α-sarcin) then specifically destroys the target by inhibiting local protein synthesis without affecting other cells. Lysosome targeting (lysosomotropic) amines (NH4Cl), chloroquine, and carboxylic ionophores (monesin) protect the cells from some immunotoxins (e.g., diphtheria toxin) but make them more sensitive to others (e.g., to ricin). The clinical applicability of this therapy is still very limited. The toxins

may damage other cell types to some extent. *See* antibody engineering; antitoxin; magic bullet; monoclonal antibody. (Knechtle, S. J. 2001. *Philos. Trans. R. Soc. Lond. B Biol. Sci.* 356:681; Manzke, O., et al. 2001. *Med. Pediatr. Oncol.* 36:185.)

impact factor Scientometric index monitoring the citation frequency of "average articles" in particular journals within certain years (Garfield, E. 1972. *Science* 178:471). It is calculated by dividing the number of cited articles in a journal by the number of articles published in that journal during a period of 2 years. The information is available in the *Journal Citation Reports* in alphabetical order and by grouped fields. It is commonly used for the evaluation of the performance of individuals or departments because it indicates the impact of the publications. It is a useful tool of evaluation within a discipline, although frequently publishing methods are cited more frequently than data and theory. It is not entirely suitable for comparison across different disciplines because "glamorous" journals are commonly cited more frequently than traditional ones. The Institute of Scientific Information that does the tallying may also be affected by some human errors (*Nature* [2002], 415:101; ibid., 726; *ibid.*, 731). Nevertheless, it is probably the most objective tool for rating the prestige of a journal. The *Citation Index* is used for the evaluation of author impact of scientific papers. (Citation Index: <http://www.isinet.com/isi/products/citation/wos/>; <http://www.isinet.com/isi/hot/essays/journalcitationreports/7html>.)

impala (*Aepyceros melampus melampus*) $2n = 60$.

impaternate Originated by parthenogenetic reproduction. *See* parthenogenesis.

IMPDH Inosine-5′- monophosphate dehydrogenase is involved in lymphocyte replication. Mycophenolic acid (MPA), an approved immunosuppressive drug, is its potent inhibitor. (Desmoucelles, C., et al. 2002. *J. Biol. Chem.* 277:27036.)

imperfect flower Has either the male or the female sexual apparatus; it is monoecious or dioecious but not hermaphroditic. *See* dioecious; flower differentiation; hermaphrodite; monoecious.

Asparagus flowers (After J. A. Huyskes & J. Sneep).

imperfect fungi Have no known sexual mechanism of reproduction. *See* fungal life cycles.

impetigo Pus-forming skin infection caused by the plasmid-carrying *Staphylococcus aureus* bacteria.

implant Grafted addition to the body or an inserted artificial object or an implanted zygote.

implantation Attachment of the blastocyst to the lining of the uterus after about a week of fertilization and embedding into the endometrium (in humans). *See* blastocyst; uterus. (Paria, B. C., et al. 2002. *Science* 296:2185.)

importin α and β Protein factors mediating passage through the nuclear pore by binding the α-subunit to a nuclear localization sequence of a protein, Ran-GTP. *See* GTP; karyopherin; nuclear localization sequence; nuclear pore; NuMA; Ran; RNA export; TPX2; transportin. (Mingot, J.-M., et al. 2001. *EMBO J.* 20:3685; Gruss, O. J., et al. 2001. *Cell* 104:83.)

impotence Inability to initiate or maintain erection of the penis due to organic or psychological causes. *See* nitric oxide. (Renand, R. C. & Huaseb, H. 2002. *Nature Rev. Drug. Discovery* 1:663.)

imprinting The expression of behavioral or other traits may be influenced by the parental source of the chromosome, i.e., the paternal and maternal genomes may have different effects (imprinting) on the developing offspring because of the modification of an allele by a cis-element or different methylation of the sequence. Methylation also affects the organization of the chromatin. The generation of antisense transcript may be a means of imprinting (Runte, M., et al. 2001. *Hum. Mol. Genet.* 10:2687). The insulin-like growth factor gene (*IGF-2*) of mice transmitted through females is not transcribed (imprinted) in most of the tissues and only the one transmitted through the male is active. If the offspring receives a mutant copy of the gene through the male and a normal copy through the female, the heterozygote is crippled. The choroid plexus (the brain tissue secreting the cerebrospinal fluid) and the leptomeninges (the innermost of the three membranes covering the brain and the spinal chord) were not subject, however, to *IGF-2* gene imprinting in mice. *IGF-2* is a single-chain polypeptide and an autocrine regulator of hormone response and growth. Deletion of a silencer element from the mouse *Igf2* involves loss of imprinting. It appears that the methylation takes place in CG-rich islands of 200 to 1,500 base pairs and conspicuously several of the imprinted genes are either in chromosome 11p or 15q in humans, or in the mouse in chromosome 7. The maternally expressed gene, MEG3, is in human chromosome 14q.

In mouse chromosome 17 in the IGF-2 gene, a 113 bp methylation imprinting box was identified. In mice, imprinted genes are situated at nine regions in six autosomes. In human chromosome 15q11-q13, an *imprinting center* (IC) was found that is involved in epigenetic resetting of this 2 Mb domain. The IC is part of the promoter and the first exon of the small nuclear ribonucleoprotein particle peptide N (SNRPN) gene. When the untranslated *H19* mouse gene was disrupted, the *Ins-2* and *Igf-2* genes — 100 kb upstream of *H19* — were transmitted by the female. It has been suggested (but not verified) that the chromosomal choice of imprinting is determined by a competition for a nearby enhancer. *Igf* is preferentially expressed in the male because a germline-inherited methylation silences the promoter of the *H19* gene. The 5′ upstream region of *H19* contains an imprinting methylation signal (mark) in the male rodent. This ~42 bp element is conserved by evolution. In the offspring, 27% higher weight was observed compared to animals that received the same chromosome from their father. In this mouse chromosome 11, *mash-2* encodes a helix-loop-helix protein, and it is maternally expressed only in the placenta. If this gene is deleted, maternal lethality results, but there is no consequence if transmitted through the male. If the mouse conceptus receives two paternal or two maternal chromosomes 12, intrauteral death results. Uniparental conceptuses are also inviable. Normally, *Ins-2* (insulin) is expressed paternally in the embryo yolk but biparentally in the pancreas. The tissue specificity of imprinting of the insulin-like growth factor is determined by which of the four promoters of the gene was used. Several paternally inherited X-chromosomal genes are inactive in early embryonic tissues. An exception is the *Xist* gene, which is expressed only from the paternally derived X chromosome. In some cases, the demonstration of true imprinting is difficult because in human diseases the penetrance and expressivity of the genes may widely vary. Imprinted genes frequently carry special repeats, display unusual sex-specific rates of recombination, and the size of their introns is relatively short. Parthenogenesis may cause embryonic lethality in mice if the imprinted paternal genes are not expressed. During tumorigenesis, both the paternal and maternal copies of the IGF-2 gene are expressed. Imprinted genes usually replicate asynchronously from the rest of the gene pool. Although it appears that expressed genes replicate early, this rule does not seem to hold for imprinting because early replicating paternal genes may still be silent. Imprinting appears mainly in mammals; however, imprinting-like phenomena were observed in the plant *Arabidopsis* (Choi, Y., et al. 2002. *Cell* 110:33). In mammals, 0.1–1% of the genes may show some degree of imprinting, although in mice and humans only about three dozen imprinted genes were identified by 2000. (See Bartolomei, M. S. 2003. *Annu. Rev. Cell Dev. Biol.* 19, in press.)

There is no generally valid interpretation for the evolutionary utility of imprinting. It has been hypothesized the imprinting has a *dosage compensation* purpose. Loss of imprinting may predispose to sporadic colorectal cancer. Also, imprinting may permit the expression of an oncogenic allele. Alternatively, the *conflict theory* has been proposed. According to this theory, if the female produces offspring by more than one male during her period of fertility, the more vigorous pregnancy places more demand on the female and thus weakens the mother, thereby potentially harming future offspring sired by other males. To resolve this conflict, the mother turns on growth-promoting genes. The male silences genes that suppress growth. Some of the facts seem to support this conflict theory; others do not. According to the conflict hypothesis, there should be no imprinting in monogamous species. This expectation is not realized, however, in the monogamous *Peromyscus polionatus* rodents. Imprinted genes, because of monoallelic expression, may have a higher risk of contributing to tumorigenesis because a single recessive mutation may trigger the process. *See* Albright hereditary osteodystrophy; Angelman syndrome; ataxia; Beckwith-Weidemann syndrome; cosuppression; CTCF; diabetes mellitus; dosage compensation; enhancer competition; IGF; imprinting box; insulin-like growth factor; KIP2; lyonization; methylation of DNA; MYF-3; myotonic dystrophy; obesity; parent-of-origin effect; parthenogenesis;

polar overdominance; Prader-Willi syndrome; regulation of gene activity; Russel-Silver syndrome; snRNP; *Venture*; Wilms' tumor; Xist (Tsix). (Bartolomei, M. S. & Tilghman, S. M. 1997. *Annu. Rev. Genet.* 31:493; Pfeifer, K. 2000. *Am. J. Hum. Genet.* 67:777; Nakagawa, H., et al. 2001. *Proc. Natl. Acad. Sci. USA* 98:591; for linkage analysis: Strauch, K., et al. 2000. *Am. J. Hum. Genet.* 66:1945; Spencer, H. G. 2000. *Annu. Rev. Genet.* 34:457; Reik, W. & Walter, J. 2001. *Nature Rev. Genet.* 2:21; Mann, J. R. 2001. *Stem Cells* 19:287; Ferguson-Smith, A. C. & Surani, M. A. 2001. *Science* 293:1086; Bourchis, D., et al. 2001. *Science* 294:2536; Sleutels, F., et al. 2002. *Nature* 415:810.)

imprinting, behavioral Response learned during an early phase of an individual has a lifetime effect on animal behavior.

imprinting, molecular Biological molecules are coated by polymers preserving their three-dimensional structure (imprint) in order to facilitate their manipulation (breaking up at selective locations, fractionation from complex mixtures).

imprinting box Responsible for imprinting of genes situated in the region 15q11-q13 of the human chromosome (and in the homologous chromosome 7 segment of the mouse). The human imprinting box extends to ~200 bp of the promoter/exon 1 site of the small ribonucleoprotein polypeptide N (SNRPN, 15q12) gene and a 1 kb sequence about 35 kb upstream including the short regions of overlap (SRO) of Angelman syndrome (AS-SRO) and Prader-Willi syndrome SRO (PWS-SRO). The insulin-like growth factor receptor gene (IGF2R) can act alone as a methylation initiating imprinting box. *See* Angelman syndrome; imprinting; insulin-like growth factor; Prader-Willi syndrome. (Shemer, R., et al. 2000. *Nature Genet.* 26:440.)

inactive-X hypothesis *See* Lyon's hypothesis.

inborn error of metabolism Historical term for biochemical genetic defects. Generally, mutation in single genes blocks or changes the metabolic pathways at a single specific step. The consequences of the mutation may be alleviated either by providing the missing compound or by avoiding the supply of the accumulated precursors that cannot be further processed because of the defect in the enzymatic step. *See* auxotrophy; biochemical genetics; one gene — one enzyme theorem. (Garrod, A. E. 1902. *Lancet* 2:1616.)

inbred Line developed by continued inbreeding until the majority of the genes become homozygous. In mice, generally 20 generations of brother x sister (or parent x offspring) mating is employed to produce such lines. In species where self-fertilization is feasible (e.g., plants), 10 generations of inbreeding result in more than 0.999% homozygosity $(1-0.5^{10})$, unless the genes are very tightly linked in repulsion. *See* coefficient of inbreeding; coisogenic; congenic; linkage disequilibrium; subline, substrain.

inbreeding Mating among biological relatives, including self-fertilization, brother-sister mating and mating, with ancestors or offspring. *See* coefficient of inbreeding; inbreeding and death rates; inbreeding and population size; inbreeding in autopolyploids; inbreeding progress. (Fisher, R. A. 1949. *The Theory of Inbreeding.* Oliver & Boyd, Edinburgh.)

inbreeding and death rates Inbreeding results in homozygosity of deleterious and lethal genes; as a consequence, spontaneous abortions, infant mortality, and frequency of hereditary diseases will increase with this type of mating system. The frequency of the adverse consequences depends on the frequency of these undesirable genes in the population concerned and the degree of inbreeding. When infant mortality of first-cousin marriages and that of general-population marriages are compared, the frequency in the former is about double. Some of the variation within columns may also be due to random statistical error:

Ethnicity	First-cousin marriages %	General population %
Canadian	8.8	4.1
French	9.4	4.4
Japanese	6.2	3.9
Swedish	8.5	4.0

(After Fraser, F. C. & Biddle, C. J. 1976. *Am. J. Hum. Genet.* 28:522.)

In a more recent analysis, the excess death rate — up to age 10 — of the progeny of first-cousin marriages in Japan, Pakistan, India, and Brazil combined appeared to be 4.4%. *See* coefficient of coancestry; genetic load; incest; lethal equivalent.

inbreeding and linkage The probability of homozygosity after backcrossing to an inbred line varies according to the intensity of linkage and the number of backcrosses performed.

Recombination	Number of Backcrosses →	2	3	6	9	12
0.5		0.500	0.750	0.969	0.996	0.999
0.3		0.300	0.510	0.832	0.942	0.980
0.1		0.100	0.190	0.409	0.569	0.686

(After Klein, J. 1975. *Biology of the Mouse Histocompatibility-2 Complex.* Springer, Berlin.)

inbreeding and population size Inbreeding increases more in smaller populations than in large ones. The increase can be reduced by controlled mating, i.e., when the mating pairs are selected from different families or if an equal number of mates are selected from each family of the herd. The rate of inbreeding $(\Delta_F) = 1/2N_e$. If the actual size of the population is say 10, then $\Delta_F = 1/(2 \times 10) = 0.05$. In case the effective population size is only say 0.75 of the total population, then $\Delta_F = 1/2N_e = 1/(2 \times 10 \times 0.75) = 1/15 = 0.066$. The coefficient of inbreeding becomes $F_g = 1-(1 - \Delta_F)^g$, where F_g is the inbreeding coefficient of the gth generation and Δ_F is the rate of inbreeding. *See* inbreeding coefficient; inbreeding depression; inbreeding rate; population size, effective.

inbreeding autopolyploids Since autopolyploids have more alleles present per locus, homozygosity at a locus is achieved only after a larger number of generations. Many of the autopolyploid species reproduce by self-fertilization, and the deleterious consequences of inbreeding have been eliminated by natural selection. Since many of the crop plants are polyploid, plant breeders rely on crossing for improvement. In allopolyploids, the consequences of inbreeding may vary according to the species. The table below provides a comparison of the proportion of homozygotes in diploids and tetraploids of the initial genetic constitutions Aa, $Aaaa$, and $AAaa$, respectively, after five generations of self-fertilization.

(After Burnham, C. R. 1962. *Discussions in Cytogenetics*. Burgess, MN.)

Generation	Aa (diploid)	$Aaaa$ Chrom. segr.[1]	$Aaaa$ Max. equat.[2]	$AAaa$ Chrom. segr.[1]	$AAaa$ Max. equat.[2]
F_2	0.500	0.250	0.295	0.050	0.099
F_3	0.750	0.380	0.460	0.194	0.285
F_4	0.875	0.493	0.581	0.326	0.442
F_5	0.938	0.558	0.674	0.438	0.566
F_6	0.969	0.648	0.747	0.531	0.662

[1]Chromosome segregation indicates that the gene is absolutely linked to the centromere.
[2]Maximal equational segregation occurs when the gene segregates independently from the centromere.

inbreeding coefficient Probability that *two alleles at a locus in an individual* are identical by *descent* from a common ancestor, i.e., the chance that an individual is homozygous for an ancestral allele by inheritance (not by mutation). *Consanguinity* (coancestry) is a similar concept, but the coefficient of coancestry indicates the chances that *one allele in two individuals* would be identical by descent. F symbolizes the coefficient of inbreeding. The calculation of F is based on the fact that in a diploid at each locus there are two alleles and only one is contained in any gamete (either one in a particular egg or sperm). Thus, each individual has 0.5 chance for passing on a particular allele to a particular offspring. Examples make it simpler to illustrate the method of calculation required. Brother (X) and sister (Y) have two common parents (W) and (V). An offspring of the mating (X) × (Y) → (I) has a chance to inherit a gene from grandparent (W) through two routes (W) → (X) → (I) or (W) → (Y) → (I); therefore, its chances for homozygosity for one allele derived from (W) is $0.5^2 = 0.25$. In

other words, in the F_2, the chance is 1/4 for homozygosity for any allele according to Mendelian law. In a half-brother and half-sister, progeny grandparent (A) can transmit a particular allele to (I) either through (B) or (C) parents, and the inbreeding coefficient of (I) is $0.5^3 = 1/8$ because three individuals are involved in the transmission route (A), (B), and (C). Similarly, the inbreeding coefficient of other types of matings can be calculated as indicated on the chart. In half first-cousin mating individuals (C), (B), (A), (E), and (D) are involved in the transmission path, each with a 0.5 chance; thus, the coefficient of inbreeding becomes $0.5^5 = 0.03125 = 1/32$. In two generations of brother-sister matings (see scheme 6), the transmission of alleles may follow the routes [E-C-F, F-D-E], {E-C-A-D-F, F-D-B-C-E, E-D-A-C-F and F-C-B-D-E}, i.e., [2] and {4} paths of $[0.5]^3$ and $\{0.5\}^5$, respectively. Thus, the coefficient of inbreeding is $2[0.5]^3 + 4\{0.5\}^5 = 0.375 = 3/8$. If there are multiple paths through the same ancestor, all the paths through the shared ancestors must be included in the calculation with the precaution that within the same paths the same ancestor must be counted only once.

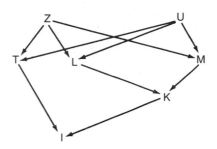

Let us illustrate the method with another example where (Z) and (U) are the common ancestors and again the inbreeding coefficient of individual (I) is sought. There are 2 routes through (Z): T-Z-L-K and T-Z-M-K and 2 paths through (U): K-M-U-T and K-L-U-T. Each of the 4 paths involves four ancestors contributing genes to (I). Therefore, the coefficient of inbreeding of (I) is $4(0.5)^4 = 0.25 = 1/4$. Under practical conditions of breeding, much more complicated schemes may be encountered, yet their solution can be sought on the basis of these much simpler examples. It is easier to determine

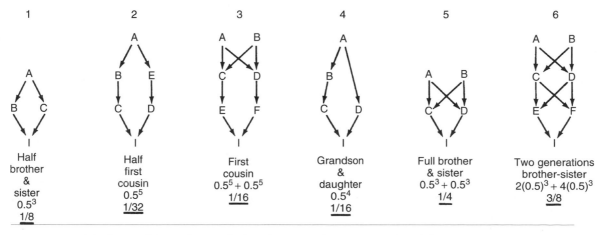

Calculation of the coefficient of inbreeding on the basis of the paths of allele transmission.

the loops of gamete contribution by working backward from the critical individual, (I) in this case. It is conceivable that the common ancestors are not completely unrelated, contrary to the assumption used in the calculations above, but they may have some degree of relatedness and their inbreeding coefficient, F_A (ancestral coefficient of inbreeding), must also be taken into account. Therefore, the general formula for the coefficient of inbreeding $F = \Sigma[(0.5)^n(1 + F_A)]$, where Σ is the sum of the paths through which an individual can derive identical alleles from the ancestors and $n =$ the number of individuals in the paths. The $1 + F_A$ is a correction factor for the inbreeding coefficient of the common ancestor in the path. The knowledge of the coefficient of inbreeding may be very important in a breeding project, but it may also be quite relevant to human families. Let us assume that the frequency of a recessive genetic disorder is $q^2 = 1 \times 10^{-6}$, and if the population is in genetic equilibrium, the frequency of that allele is $q = \sqrt{q^2} = \sqrt{0.000001} = 0.001$. Then the risk related parents have of an afflicted child is $q^2(1 - F) + q(F)$. Since the inbreeding coefficient of the offspring of first cousins is $F = 1/16 = 0.0625$ (see the first chart on the preceding page), after substitutions we get $(0.000001 \times 0.9375) + (0.001 \times 0.0625) = 0.000063$. Since 0.000063 is $\sim$63-fold higher than 0.000001 (the frequency of individuals with this affliction in the general population), the first-cousin parents take a >63-fold risk than unrelated parents of having an offspring afflicted with a hereditary disease that has a gene frequency of 0.001. In some cases, inbreeding may be detected by DNA fingerprinting. *See* coefficient of coancestry; consanguinity; DNA fingerprinting; F; fixation index; genetic load; inbreeding progress; inbreeding rate; relatedness degree. (Fisher, R. A. 1965. *The Theory of Inbreeding*. Academic Press, New York.)

inbreeding depression   When heterozygotes of normally outcrossing or dioecious or monoecious species are propagated by inbreeding and the deleterious recessive genes become homozygous, the viability, vigor, and fitness of the individuals and the population decline. The degree of depression varies according to species and the traits affected. *See* controlled mating; inbreeding and death rates; inbreeding coefficient; inbreeding progress; heterosis.

inbreeding progress   The proportion of homozygosity of any selfed or inbred generation for any number of allelic pairs can be computed by the formula $[1 + (2^g - 1)]^n$, where g is the number of generations selfed (note that, e.g., by F_5 there are 4 selfings because the F_1 is the result of crossing) and n is the number of allelic pairs involved. Example of expansion of the binomial in case of 3 pairs of alleles in F_5:

$$[1 + (2^4 - 1)]^3 = [1 + (16 - 1)]^3 = 1^3 + 3(1)^2(15)$$
$$+ 3(1)(15)^2 + 1(15)^3$$

or rewritten

$$1^3(15)^0 + 3(1)^2(15)^1 + 3(1)^1(15)^2 + (1)^0(15)^3$$

where the first exponent in each term indicates the number of heterozygous genes and the second exponent shows the number of homozygous genes in each class of individuals. In this example, 1 will be heterozygous for all 3 loci and homozygous for none (3×15). And 45 will be heterozygous for 2 loci and homozygous for 1, (3×15^2), i.e., 675 will be heterozygous for 1 locus and homozygous for 2. And 1×15^3, i.e., 3,375 will be heterozygous for none of the 3 loci but will be homozygous for all 3 in a total population of 4,096 ($1 + 45 + 675 + 3,375 = 4,096$). *See* binomial; coefficient of inbreeding; fixation index; heterozygote proportions.

inbreeding rate   Determined as $\Delta_F = 1/(2N_e)$, where $N_e =$ effective population size. If the actual size of a population size is 10, then $\Delta_F = 1/(2 \times 10) = 0.05$. In this computation the effective population number was considered to be equal to the total number. Under practical circumstances, this is rarely the case. If we suppose that the effective size is only three-fourths of the actual number, then $\Delta_F = 1/(20 \times 0.75) \cong 0.066$, a higher fraction. The coefficient of inbreeding can be also calculated as $F_g = 1 - (1 - \Delta_F)^g$ where $g =$ the number of generations of inbreeding and $F =$ the inbreeding coefficient. Accordingly, after 25 generations, $F_{25} = 1 - (1 - 0.066)^{25} = 1 - (0.934)^{25} = 0.819$, whereas if 10 were the effective size, $F_{25} = 0.723$ in this hypothetical case. *See* effective population size; fixation index; inbreeding coefficient; inbreeding progress.

incenp   *See* sister chromatid cohesion.

incest   Legally prohibited sexual intercourse between close (biological) relatives. Thus, incest is primarily a legal, ethical, and moral concept. From the viewpoint of genetics, the consanguinity of the parents is considered and the legal restrictions may not be sufficient to avoid the deleterious consequences for the progeny of such matings. In ancient Egypt the pharaohs frequently (and legally) married their sisters, which may explain the large number of child mummies at the burial places. Relatively scarce data are available on children of first-degree relatives (parent × child, brother × sister), yet it is clear that about 40% suffer more or less severe physical and/or mental defects. Interestingly, in populations where uncle-niece and cousin-cousin marriages have been practiced for centuries, the number of birth defects is not as high as would be expected. Apparently, the inbreeding continuing for many generations purges the gene pool from the most deleterious alleles that are maintained at random mate selection. *See* coefficient of coancestry; coefficient of inbreeding; genetic load.

inchworm model   During transcription the RNA polymerase is flexibly connected to the template. The front-end domain is tightly associated with the DNA ($\sim$10 bases), which is followed by a loose association ($\sim$15–20 bases), including the catalytic domain, and that is followed by another tight association ($\sim$10 bases) to the transcript. Thus, the movement of the polymerase displays a variable discontinuous pattern. A somewhat similar mechanism was revealed as *end-to-end template switching*. When the nucleotide supply is low, the polymerase can jump from a single-strand template to 7–9 bases of another linear double-stranded DNA. The upstream electrostatic interaction involves the C-end of the β-subunit, whereas the downstream interaction is by the N-end of the β-subunit of the polymerase. The polymerase ternary complex may also move backward and the catalytic site may be involved in the cleavage of the RNA strand. *See* promoter clearance; protein synthesis; RNA polymerase; TCF; transcript elongation; transcription. (Uptain, S. M., et al. 1997. *Annu. Rev. Biochem.* 66:117.)

incidence   Frequency of occurrence of a genetic alteration or disease in a population. *See* prevalence.

incipient species   Group of organisms in the process of speciation. *See* evolution; speciation.

inclusion body   Protein aggregate in *E. coli* cells expressing at a high rate some foreign gene(s) in the presence of amino acid analogs or antibiotics. Inclusion bodies appear to be the products of protein processing and defective polypeptide degradation. These proteins are visible by phase-contrast microscopy and can be concentrated by centrifugation of lysed or sonicated cells. Some RNA viruses replicate within cytoplasmic inclusion bodies. *See* phase-contrast microscope; sonicator. (Rattenholl, A., et al. 2001. *Eur. J. Biochem.* 268:3296.)

inclusion body myopathy (cytoplasmic body myopathy)   Late-onset nonprogressive muscle weakness. The muscle fibers contain microscopically visible inclusions in the cytoplasm. Both autosomal-dominant and -recessive inheritance have been reported. Dominant inclusion body myopathy 3 (IBM3, 17p13.1) displays rimmed vacuoles and ophthalmoplegia. The recessive IBM2 (9p12-p13) shows amyloid-like inclusions. The latter is relatively frequent among Jews of Kurdish and Iranian origin. This disease is caused by a mutation in UDP-N-acetylglucosamine-2-epimerase/N-acetylmannosamine kinase. *See* amyloid; Lewy body; ophthalmoplegia. (Eisenberg, I., et al. 2001. *Nature Genet.* 29:83.)

inclusive fitness   Type of altruistic behavior. Parents defend their children – at their own risk — for the sake of maintenance of their genes, which the offspring shares with them. This altruistic behavior is positively correlated with the degree of relationship. According to Hamilton's rule, the closer the genetic relationship, the greater is the altruism. R. A. Fisher and J. B. S. Haldane discussed this phenomenon earlier. Thus, siblings — on the average — are expected to share half of their genes and fitness, whereas cousins share only one-fourth. *See* altruistic behavior; fitness; inbreeding coefficient; kin selection. (Sundstrom, L. & Boomsma, J. J. 2001. *Heredity* 86[pt 5]:515.)

incompatibility   *See* ABO blood group; cytoplasmic incompatibility; erythroblastoma; fungal incompatibility; heterokaryon incompatibility; histocompatibility; immune tolerance; incompatibility alleles; maternal tolerance; plasmid incompatibility; Rh blood group; S alleles.

incompatibility, vegetative (somatic incompatibility)   Blocks hyphal fusion and formation of heterokaryons. *See* fungal incompatibility; heterokaryon; hypha.

incompatibility alleles   Self-incompatibility in a large number of plant species (tobacco, clovers, crucifers, fescue, beets, cherry, etc.) may prevent self-fertilization or formation. The number of different incompatibility alleles may run into the hundreds in some species. A plant pollen carrying a particular incompatibility allele may not successfully develop a pollen tube in the stylar tissue of the same genetic constitution but may successfully fertilize another plant of a different allelic constitution. A sperm with an S1 allele is thus incompatible with a stylus of S1S1 type but may fertilize an S2 egg. In an S1S2 heterozygote, neither S1 nor S2 sperm may be successful, but they have no barriers in S3S3 plants. In some species, compatibility may be determined by the sporophytic tissue, and in an S1S2 stylus, if the S2 allele is dominant, the S1 pollen may be successful and produce S1S1 seed. In some cases, the S1S2 pollen of a tetraploid plant may be compatible with an S1S1 egg if the S2 allele is dominant. In *Brassicaceae*, the S alleles extend to several hundred kb and include several closely linked transcriptional units, often referred to as the S haplotype.

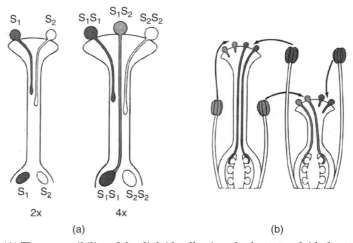

(A) The compatibility of the diploid pollen in a duplex tetraploid plant. (B) incompatibility associated with dimorphism of style and stamen length. (After Linskens, H. F. & Kroh, M. 1967. *Encyclopedia of Plant Physiology.* vol. 18. Springer. Berlin, Germany.)

Incompatibility may also be based on different timing of pollen release and stigma receptivity. Compatible combinations may arise through induced mutations. Heterostyly (different height of the stylus and stamen) may prevent self-fertilization. S-specific glycoproteins and ribonuclease enzymes cause self-incompatibility. In *Brassicaceae*, two tightly linked genes that mediate incompatibility encode the S-locus receptor serine/threonine protein kinase (SRK) and the secreted glycoprotein (SLG, 431 amino residues). It is assumed that a pollen-born ligand ties SLR and SLG into a signaling complex that prevents the germination or the growth of the pollen tube on the stigma or in the style. It is an interesting observation that an SRK incompatibility protein provides protection against *Pseudomonas syringae* infection. In *Solanaceae*, the S locus encodes another type of glycoprotein with ribonuclease activity (S-RNase), and this ribonuclease seems to inhibit the growth of incompatible pollen tubes. *See* ABO blood group; apomixis; fungal incompatibility; gametophyte; genetic load; male sterility; mentor pollen effect; Rh blood group; self-incompatibility. (Wang, X., et al. 2001. *Plant Physiol.* 125:1012.)

incompatibility load   *See* genetic load.

incompatibility mother-fetus May be genetically determined, such as in erythroblastosis fetalis in case of the wrong Rh allele combinations, or to some extent in other blood groups. The incompatibility may be mediated by activation of the complement. Decay-accelerating factor (DAF) inactivates the C3 complement component convertase protein that activates C3. Membrane cofactor protein (MCP) is required for degradation of activated C3 and C4. In murines, a regulatory protein suppresses C3 deposition, but if it becomes homozygous for an inactive mutation, fetal loss may ensue. *See* complement; decay-accelerating factor; erythroblastosis fetalis; genetic load; killer cells; MCP.

incompatibility plasmids Utilize the same system of replication and cannot coexist. When they are introduced into the bacterial cell, they have to compete with each other and one is eliminated. Plasmids (phages) that carry the same replicon belong to the same incompatibility group. RNA I, RNA II, and Rop protein determine incompatibility. There are more than 30 plasmid compatibility groups. *See* RNA I. (Miller, C. A. & Cohen, S. N. 1993. *Mol. Microbiol.* 9:695.)

incomplete digestion Reaction with restriction enzymes is terminated before all potential cleavage sites are cut. A larger variety and some large-size fragments are cut because some neighboring sequences are not cleaved apart. *See* restriction enzymes.

incomplete dominance Or semidominance, observed when the expression of the gene does not entirely mask or prevent the expression of the recessive allele at the same locus in a hybrid. *See* dominance; epistasis.

incomplete linkage Recombination takes place between or among the syntenic genes in question. *See* crossing over; linkage; recombination; synteny.

incongruence May be construed as evidence of horizontal gene transfer because the variation at certain loci is higher than that of the flanking regions. *See* transfer, lateral. (Farris, J. S., et al. 1995. *Cladistics* 10:315.)

incontinentia pigmenti (Bloch-Sulzberger syndrome, IP1) Human X-linked (Xp11) dominant "marble-cake-like" dark pigmentation on skin of the trunk, generally preceded by an inflammation. It may begin soon after birth and may fade by age 20. The anomaly may be associated with eye, tooth, bone, and heart anomalies. The majority of IP2 cases are apparently due to mutation/chromosomal breakage at the gene NEMO (NF-κB essential modulator)/IKKγ (IκB kinase-γ) closely linked to antihemophilia factor VIII at Xq28. Mutation/chromosome breakage in both IP1 and IP2 is male lethal before or around birth, whereas the heterozygous females survive to adulthood and the severity of the symptoms varies. Some of the rare males with IP1 may be mosaics or of XXY constitution. *See* antihemophilic factors; ectodermal dysplasia; erythrokeratodermia variabilis; hypomelanosis of Ito; NF-κB; NEMO; pigmentation defects. (Aradhya, S., et al. 2001. *Hum. Mol. Genet.* 10:2171.)

incorporation error Mechanism of mutation in which a nucleic acid base analog or a wrong base is inserted into the nucleic acid during replication. As a consequence, one base pair replaces another. Thus, the meaning of the codon may change and appears as a visible mutation if it leads to an amino acid substitution at a critical site in the protein. Example: During replication, 5-bromouracil is inserted into DNA at a C site, resulting in a BrU—G base pair. During the next replication, a BrU—A pair is formed, and at a subsequent replication a T=A pair is substituted at a site where originally a C≡G pair existed. *See* base substitution mutations; hydrogen pairing; replication error. (Freese, E. 1963. *Molecular Genetics*, Taylor, J. H., ed. p. 207. Academic Press, New York.)

incremental truncation for the creation of hybrid enzymes (ITCHY) *See* iterative truncation.

incross Hybridization between two strains that have the same genetic background.

indel Insertion or deletion in the DNA nucleotide sequences. By the simple example below, the alignment score can be illustrated. $Pr = p^3 q^2 r^1$, where p is the probability of identity (match), q is the probability of substitution (mismatch), and r is the probability of an indel. The alignment score can be derived as follows:

$S' + \log Pr = 3(\log p) + 2(\log q) + (1(\log r))$ and $S = S'$ -nlog $s = S' - 6(\log s)S = $ a constant satisfying $\log(p/s) = 1$. And $S = 3 - 2\mu - 1\delta$, where $\mu = \log(q/s)$ and $\delta = \log(r/s)$ and $S = $ number of identities $-\mu$ number of substitutions $-\delta$ number of indels. Computer programs based on high-level mathematics that cannot be presented here can resolve the task. *See* databases; DNA sequence information. (Waterman, M. S., Joyce, J. & Eggert, M. 1991. *Phylogenetic Analysis of DNA Sequences*, Miyamoto, M. M. & Cracraft, J., eds., pp. 59–89. Oxford Univ. Press, New York.)

independence Two events are independent when the occurrence of one does not affect the chance of occurrence of the other. Genes at a distance of 50 map units or more segregate independently; the sex of the first child is (normally) independent from the sex of the next one; if two pennies are tossed they can land on heads or tails independently from each other unless they are defective or biased.

independence test *See* association test (contingency chi-square).

independent assortment Alleles of different loci (nonallelic genes) may reassort freely in the gametes and therefore segregate independently in zygotes in the absence of linkage.

The independent assortment of the alleles is one of the most essential discoveries of Mendel and is frequently called Mendel's third law. *See* Mendelian laws.

independent events   Do not affect or influence each other.

indeterminate inflorescence (raceme)   The main axis can elongate indefinitely, but the branches terminate in a flower bud. *See* raceme.

index   Alphabetical or other ordered set of files, symbols, or numbers distinguishing particular things in an array. Examples: allele a^1, the "1" distinguishes this allele among all other a alleles; or *NK3* homeobox 3 of *Drosophila*; or *adp^{fs}* of an *adipose* allele conveying female sterility; or the second asymmetric leaf locus as_2 (*AS-2*) of *Arabidopsis*.

index case   *See* proband (propositus, proposita).

INDEXERs   Actually an amplification system for specific DNA fragments from whole-genome digests without a need for cloning. It uses restriction enzymes, class IIs. The nonidentical cohesive ends can then be selectively modified by ligation to synthetic oligodeoxyribonucleotides with the corresponding complementary ends. This permits the introduction of PCR and sequencing primer sites and labels the genomic digest into small fragments. An advantage over cloning is that in fragment losses, rearrangements can be avoided without cloning, probes, or libraries. The procedure is applicable to small prokaryotic and large eukaryotic genomes for analyzing sequence-tagged sites, restriction mapping, RFLP, sequencing, and DNA diagnostics. *See* PCR; restriction enzymes class IIs; RFLP; sequence-tagged site. (Unrau, P. & Deugau, K. V. 1994. *Gene* 145:163; Guilfoyle, R. A., et al. 1997. *Nucleic Acids Res.* 25:1854; Sibson, D. R. & Gibbs, F. E. M. 2001. *Nucleic Acids Res.* 29[19]:e95.)

index locus   *See* polymorphism information content (PIC).

index value   Concept used for selection in animal breeding. A score weights each trait, and these scores are summed in an index value. The use of this index alleviates the danger of selecting for one particular trait only, which would jeopardize the overall success of the program because disease susceptibility or low fertility, etc., accompany frequently high performance. *See* gain, selection. (Falconer, D. S. 1960. *Introduction to Quantitative Genetics.* Ronald, New York.)

indicator mice   Transgenic for a recombinase (*Cre* or *Flp*). When it is crossed to another transgenic line carrying a construct with an FRT sequence in front of the structural gene *LacZ*, on Xgal medium it develops a blue color only when the recombinase flips out the FRT stop codon. The test thus indicates the functionality of the recombinase and its utility for targeting. *See* Cre/LoxP; Flp/FRT; targeting genes; Xgal.

indirect diagnosis   Identification of a gene by linkage rather than by direct evidence.

indirect end labeling   Determines the distance of a DNase-hypersensitive site from a restriction enzyme cleavage site. The chromatin is first digested with a Klenow fragment of DNAase I, then isolated and treated with a restriction endonuclease. The double digest is separated by electrophoresis and probed with a sequence adjacent to the restriction site. The size of the fragment generated by the double cuts indicates the distance of the DNAase hypersensitive site from the site of restriction. This procedure can thus localize the sites in the DNA where transcription may be initiated because the hypersensitive sites are correlated with the position of transcriptionally active genes. *See* DNase hypersensitive site; Klenow fragment; restriction endonuclease. (Li, S., et al. 2000. *Methods* 22[2]:170.)

indirect suppression   Some suppressor mutations do not correct the primary change in the gene; rather they modify the translation process and thereby suppress the expression of the mutation. *See* suppressor mutation.

indole   Heterocyclic compound present in many organic associations in biological materials. When excited by ultraviolet light, indole displays a characteristic fluorescence spectrum suitable for its rapid detection. Among other roles, it is a precursor of tryptophan synthesis: anthranylate → indole glycerol phosphate → indole + serine → tryptophan.

Indole.

indole acetic acid (IAA)   Plant hormone synthesized from tryptophan via the pathway Trp→indole-3-acetaldoxime→ indole acetonitrile→IAA or Trp→indolepyruvic acid→indole-3-acetaldehyde→IAA. *See* IAA. (Leyser, O. 2002. *Annu. Rev. Plant Biol.* 53:377.)

indophenoloxidase   *See* superoxide oxidase.

induced fit   Of enzymes (ribozymes), happens when the conformation is so modified that the activity improves; this may be caused by binding to a ligand or substrate. *See* DNA-binding proteins; key-lock; *Lac* operon; ligand.

induced helical fork   After the binding protein contacts a few bases of the DNA, it keeps the double helix apart. *See* binding proteins; Watson and Crick model.

induced mutation   Obtained by exposure to a mutagen, and presumably generated by the mutagen rather than by an incidental spontaneous event. *See* mutation; spontaneous mutation.

induced replisome reactivation   *See* replication-restart.

inducer   Substrate or an analog of a substrate of an enzyme prevents a repressor protein from attaching to the promoter (operator) of a gene and thus facilitates its expression.

See activator proteins; gratuitous inducer; inducible gene expression; induction; repression.

inducible enzyme Presence of substrate or substrate analog is required for their synthesis. *See* Ara operon; *Lac* operon. (Monod, J. & Audureau, A. 1946. *Ann. Inst. Pasteur* 72:868.)

inducible gene expression Required in many instances to stimulate the expression of a particular gene (transgene) in a specific tissue or particular cells. Inducible promoters are useful tools in biotechnology as they can be employed for turning on/off genes in response to special physiological or developmental factors. *See* metallothionein; split-hybrid system; tetracycline; three-hybrid system; transactivator; transcriptional activators; two-hybrid method; VP16.

induction Phage induction indicates the facilitation of the transition from the prophage stage to the lytic phase. Induction of enzymes sets into motion the catalytic activity. The assembly of the preinitiation complex of transcription induces gene expression. Embryonic development is induced by the transmission of the various exogenous and endogenous signals. *See* enzyme induction; morphogenesis in *Drosophila*; organizer; photomorphogenesis; prophage; regulation of gene activity; signal transduction; transcription.

induction, developmental The fate of a cell or tissue is affected by the interaction of the embryonic cell or tissue with its neighbors. *See* embryonic induction.

Induction of a lysogenic bacterium To liberate phage particles by first inducing a change from prophage to a vegetative state of the phage. *See* prophage; zygotic induction.

induction of an enzyme Initiates the synthesis of new enzyme molecules by the presence of an inducer that may be the substrate or an analog of the substrate of that enzyme. *See* derepression; gratuitous inducer.

indusium Membrane-type layer over the sorus (sporangial cluster) of ferns.

industrial melanism As industrialization (coal-burning pollution) increased (in Britain), the dark variants (dominant) of the black-peppered moth (*Biston betularia*) increased as a selective trend to camouflage the insect on the soot-covered tree barks. *See* adaptation; *Biston betularia*; natural selection. (Kettlewell, H. B. D. 1961. *Annu. Rev. Entomol.* 6:245.)

infantile amaurotic idiocy *See* Tay-Sachs disease.

infection Invasion of a host by a virus or another organism. The invader may establish a mutually beneficial or neutral relation with the host (symbiosis). Most frequently, pathogenic consequences result. The microbial agent generally subverts the host defense system on the surface and adheres to the special receptors. Usually, they engage the cytoskeleton to facilitate penetration and disable the phagocytotic mechanisms of the host cell. In order to maintain itself, the infective agent must overcome the immune system of the host. After the invasion, the pathogen may redirect the host metabolism to its own interest and evoke disease symptoms. If the host quickly succumbs to the disease, a frequency-dependent selection may eventually work against the invader or by the time the host may develop resistance. In *latent infection*, the virus remains in a few copies and a few viral proteins are expressed, so the host defense is not evoked temporarily. At an opportune time, the virus may, however, enter a lytic stage. *See* frequency-dependent selection; host-pathogen relationship. (Knodler, L. A., et al. 2001. *Nature Rev. Mol. Cell Biol.* 2:578; Hill, A. V. S. 2001. *Annu. Rev. Genomics Hum. Genet.* 2:373; Kazmierczak, B. I., et al. 2001. *Annu. Rev. Microbiol.* 55:407.)

infectious center Spot from where infectious phage or bacteria can be produced.

infectious diseases They are not hereditary as it was assumed before the era of genetics, yet susceptibility may be genetically determined. (Cooke, G. S. & Hill, A. V. 2001. *Nature Rev. Genet.* 2:967.)

infectious drug resistance Drug-resistance genes are carried on conjugative plasmids of bacteria. *See* conjugation, bacterial; plasmid.

infectious heredity *See* plasmid; prions; segregation distorter; symbionts, hereditary; Wolbachia.

infectious nucleic acid May be a purified viral DNA or RNA that may propagate in the host cell and code subsequently for viral particles.

infectious protein *See* encephalopathies; kuru; prion.

infertility May have various causes such as anatomical abnormalities of the sexual organs (hermaphroditism, dysgenesis, testicular feminization, polycystic ovary disease). Infectious diseases, hormonal abnormalities, malfunction of CREM, psychological factors, organic diseases, medications, alcoholism or other substance abuse, malnutrition, and chromosomal defects (trisomy, translocations, inversions, deletion, duplications, aneuploidy) may cause infertility. Hereditary abnormalities (cystic fibrosis, mental retardation, Kallman syndrome, Kartagener syndrome, myotonic dystrophy) may involve infertility. In the U.S., about 15% of couples are involuntarily infertile. About 20–50% of human infertility cases involve the males; 70–90% of them have defects in spermatogenesis or spermiogenesis. After age 50, semen volume and sperm concentration gradually decreases (in some cases, increases), sperm motility is reduced, and fertility rate is generally lower. Male infertility may be due to an absence of a 10-repeat CAG microsatellite within the mitochondrial DNA polymerase γ-gene (Rovio, A. T., et al. 2001. *Nature Genet.* 29:261). About 38% of human females may be infertile. The consequence of the long-term use of fertility drugs on the incidence of ovarian cancer may not be entirely clear (Parazzini, F., et al. 2001. *Hum. Reprod.* 16:1372). Human females who smoke have earlier menopause, which may be caused by polycyclic aromatic hydrocarbon (PAH) exposure. PAH binds to its receptor (AHR) in the promoter of the BAX gene and promotes apoptosis of the egg, leading to infertility (Matikainen, T., et al. 2001. *Nature Genet.* 28:355). By tissue transplantation between two sterile male mouse lines, fertility may be restored due to complementation. *See* ART; asthenozoospermia; azoospermia; CBAVD; cell cycle; claudin-11; CREM; fecundity; fertility; fertilization; gametogenesis; microsatellite; miscarriage; PN-1; sex hormones; smoking; spermiogenesis; sterility. (Ogawa, T., et al. 2000. *Nature Med.* 6:29; Moore, F. L. & Reijo-Pera, R. A. 2000. *Am. J. Hum. Genet.* 67:543; Kidd, S., et al. 2001. *Fertility & Sterility* 25:237; Cooke, H. J. & Saunders, P. T. K. 2002. *Nature Rev. Genet.* 3:790.)

infiltration Introduction of various substances into biological tissues by diffusion, frequently facilitated by evacuation (under negative pressure).

infinite allele mutation model (IAM) Among the practically infinite number of genetic variations of nucleotide sequences, each mutation creates a new allele that did not exist earlier in the genome. It predicts that evolutionary alterations in (microsatellite) DNA occur either by addition or deletion of one copy (of tandem repeats) in a novel fashion. *See* microsatellite; stepwise mutation model. (Kimura, M. & Crow, J. F. 1964. *Genetics* 49:725.)

infinitesimal model When linkage is sought between quantitative trait loci (QTL) and other genetic markers, the analysis is conducted on the basis of a null hypothesis that a particular chromosome or chromosome segment segregates independently from a QTL. *See* interval mapping; null hypothesis; QTL.

inflorescence Cluster of flowers characteristic for taxonomic classification of plants.

influenza virus Group of most commonly spherical (120 nm) single-stranded RNA (12,900–14,600 nucleotides) viruses. The genome of the A strain is segmented and consists of eight molecules. When its density becomes high, defective-interfering particles containing deletions may slow down viral multiplication. Their major surface glycoprotein is hemagglutinin (560 amino acids), which is usually modified. Hemagglutinin after cleavage by host proteases mediates the attachment of the virus to the cell and the transfer of ribonucleoprotein into the cell. The other common surface glycoprotein, neuraminidase (460 amino acids), is anchored to the lipid membrane by its amino end. The C-type flu virus has another single-surface glycoprotein, HEF, which destroys the cellular receptor by neuraminate-O-acetylesterase. The viral M1 protein ($M_r \sim 28$ K) controls the nuclear traffic of the virus. The virus has several types designated by origin such as the Spanish, Hong Kong, and Russian strains or as type A (most common and reoccurring in 2–3-year cycles), type B (causes epidemics in 4–5-year cycles), and type C (sporadically occurring). In the type A virus, hemagglutinin HA1 plays an important role for infectivity. The nucleotide substitution in this domain is high (5.7×10^{-3} per site). At least 18 amino acids are critical for evading the host immune response. The expected mutation rate at these sites has a predictive value for the pharmaceutical industry for the preparation of inoculation for the year ahead. Spanish flu was particularly devastating in 1918 and killed 675,000 Americans, reducing the average life expectancy by 10 years. By an alternative reading frame, influenza virus A encodes and translates an 81-residue protein, PB1-F2, which promotes apoptosis. It appears that upon infection this mitochondrially localized protein kills the host immune cells (Chen, W., et al. 2001. *Nature Med.* 7:1306). Birds, horses, swines, and cats also have influenza-type infections by different viruses. The bird influenza virus became a threat to human populations only in 1997. The highly virulent flu strains usually develop from reassorted viruses of the human and avian types sometimes via the pig flu virus. Viral infection of the respiratory tract occurs with possible secondary infection by *Streptococcus*, *Staphylococcus*, and *Haemophilus* bacteria. *See* hemagglutinin; neuraminidase deficiency; reassortant. (Gibbs, M. J., et al. 2001. *Science* 293:1842; Bae, S.-H., et al. 2001. *Proc. Natl. Acad. Sci. USA* 98:10602.)

informatics System of databases and electronic retrieval. *See* bioinformatics; databases.

information (1) Obviously the greater amount of information is available about a population the easier and more reliable is the decision of the geneticist about a parameter of that population or populations. Statistically, the information

$$I_p = \frac{1}{Vp}$$

indicates that the total amount of information is inversely proportional to the variance (V) of the statistic employed. The calculation of the information for a particular set of data can be carried out by

$$I = \Sigma \left(\frac{1}{m} \left[\frac{dm}{d\theta} \right]^2 \right)$$

where m is the expectation in terms of parameter θ, and Σ is the sum of all classes. R. A. Fisher pointed out that maximizing the likelihood function provides an estimate of T, which has the limiting value of $1/nV_T = I$. The reciprocal of the variance of the maximum likelihood estimate permits the value assessment of other estimates. If the variances obtained by other methods are not $1/nI$, they do not give us the full possible information and thus are inferior to the maximum likelihood statistics. *See* maximum likelihood; variance. (Mather, K. 1957. *The Measurement of Linkage in Heredity*. Methuen, London.) (2) In statistics it is sometimes called support or lod score. *See* lod score.

informational macromolecules DNA, RNA, and proteins that can convey genetic, developmental, biochemical, and evolutionary instructions to a cell or organism.

information retrieval Procedures to obtain information from a set of stored data such as a specific nucleotide sequence in the databases. *See* databases. (Yandell, M. D. & Majoros, W. H. 2002. *Nature Rev. Genet.* 3:601.)

informative mating Reveals the inheritance or linkage relationship of a gene or allele.

informed consent   May be a dilemma of a genetic counselor regarding the information he/she may wish to withhold from the counselee because of psychological impact. Legally, all of the dangers that the professional evaluation indicates should be exposed to the individual within the legal limits of confidentiality. Action to be pursued requires informed consent. *See* bioethics; cancer gene therapy; confidentiality; counseling, genetic; gene therapy; genetic privacy; public opinion. (U.S. Office for Protection from Research Risks [OPRR]. 1993. Dept. of Health and Human Services, Washington DC; Greely, H. T. 2001. *Annu. Rev. Genet.* 35:785.)

informosome (masked RNP)   mRNA complexed with protein, thus acquiring a very low turnover rate and stability. (Spirin, A. S. 1994. *Mol. Reprod. Dev.* 38:107.)

ingi   *See* hybrid dysgenesis; I-R.

INGI   *See* p53.

ingression   Movement of cells from the surface into the inner region. *See* gastrulation.

INH   Protein complex containing at least six species that is isolated from oocytes; it inhibits the activation of pre-MPF. *See* maturation protein factor.

INHAT   Inhibitor of histone acetyltransferases CBP and PCAF. *See* CBP; PCAF.

inherency   Genes that are important for organogenesis in higher evolutionary forms usually have some comparable representatives in more primitive forms.

inheritance   Receiving genes from ancestors and passing them on to offspring. DNA codes these genes in eukaryotes and prokaryotes, whereas in some viruses the transmitted genetic material is RNA. The genetic material may be located in the nucleus (nuclear inheritance) or carried by the nucleoid in prokaryotes. The genetic material in mitochondria and chloroplasts mediates extranuclear inheritance. Prokaryotes and cytoplasmic organelles may also have plasmid vehicles of heredity. Contrary to some common loose wording, traits are not inherited; only the genes that determine their expression are transmitted. *See* acquired characters, inheritance; DNA; genealogy; genetics; genotype; heredity; pedigree; phenotype; prions; reverse genetics; RNA.

inheritance, cortical   *See* cortical inheritance.

inheritance, cultural   Information transfer by nonbiological means such as customs, traditions, behavior, etc. Cultural inheritance plays a significant role in the phenotype, but the genes of the organism determine its genetic significance. The genes involved may have different selective values. *See* fitness.

inheritance, delayed   *See* delayed inheritance.

inhibin   Antagonist of activin. Inhibins are glycoproteins (A and B) in the seminal and follicular fluids, and they inhibit the production of follicle-stimulating hormone and regulate gametogenesis, embryonic and fetal development, and blood formation (hematopoiesis). *See* activin; FSH.

inhibition   *See* inhibitor.

inhibition of transcription   Any inhibitor of the RNA polymerase protein can block transcription. Bis(1,10)-phenanthroline cuprous chelate ([OP]$_2$Cu$^+$) is one inhibitor. On its own, it is not gene-specific; however, it can cut oxidatively single-stranded DNA templates and it is suitable for mapping the transcription initiation sites. Gene-specific inhibition of transcription can be accomplished by antisense RNA, triple-helix formation, and DNA-binding polyamides. Gene-specific inhibition is feasible by targeting ([OP]$_2$Cu$^+$) to the promoter in an open transcription complex with the aid of template-specific oligonucleotides with ([OP]$_2$Cu$^+$) attached to the oligonucleotides at various positions at either ends or interstitially. The template strand is then interrupted by the ([OP]$_2$Cu$^+$) position, e.g., OP-5′-GUGGA-3′, 5′-GUGGA-3′-OP, or 5′-GU[OP]GGA-3′. The inhibition is most efficient with 5 nucleotides representing one-half turn of A or B DNA-type double helix. The preferred cleavage site is 2–3 nucleotides from the OP-linkage toward the 3′-end. 2′-aminouridine appears to increase the specificity of intercalation. The presence of RNA polymerase is essential for binding. *See* antisense RNA; DNA types; polyamides; RNAi; TFO; transcriptional repression; transcription corepressor; triple helix formation. (Milne, L., et al. 2000. *Proc. Natl. Acad. Sci. USA* 97:3136.)

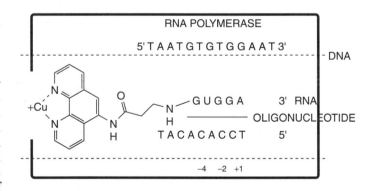

inhibitor   A substance that interferes with the *activity* of an enzyme versus a repressor that prevents the *synthesis* of the enzyme. *See* regulation of enzyme activity.

INI1   Integrase interacting protein tethers the retroviral (HIV) integrase enzyme and facilitates the integration at or near the DNase hypersensitive site of the eukaryotic chromosome. *See* Cre/loxP; DNase hypersensitive site; FLP; HIV; integrase; resolvase; retroviruses.

initial sequence contig   Assembly of overlapping sequences from a single clone.

initiation codon   First translated codon. In prokaryotes, it is most commonly AUG (90%) translated into formylmethionine, but GUG (8%) and UUG (1%) can also be used. AUU is rarely employed because IF3 discriminates against this *noncanonical* codon versus the above three *canonical* codons. In prokaryotes, the nonformyl AUG is prevented from initiation by a secondary structure in the mRNA and interaction between mRNA and ribosomal RNA. In eukaryotes, the AUG does not code for formylmethionine but for methionine. In some insect viruses, the initiator codon is CAA (glutamine) and an initiator tRNA is not required. The initiation codon is charged to a specific initiator tRNA. *See* aminoacyl-tRNA synthetase; elongation factors; modified bases; ribosome scanning; Shine-Dalgarno sequence; transcript elongation; translation initiation.

initiation complex Contains the small subunit of the ribosome with associated mRNA, aminoacylated tRNA, various initiation protein factors, and energy donor nucleotide triphosphates. In prokaryotes, the initiation complex, comprised of three single-polypeptide-chain proteins, has a mass of ~150 kDa. In eukaryotes, about 10 initiation factors comprising >25 polypeptide chains has an aggregate mass of ~1,200 kDa. Although both types of systems require the ternary complex of Ifs (initiation factors)•GTP•tRNA, several differences exist in the details of executing the functions. In eukaryotes, most commonly the ternary complex binds the 40S ribosomal subunit before mRNA binding, although the reverse sequence of events is possible. In prokaryotes, the chance is about equal for following either of theses possible routes of binding the 30S subunit. *See* preinitiation complex; protein synthesis. (Kimball, S. R. 2001. *Progr. Mol. Subcell Biol.* 26:155.)

initiation factor for transcription *See* eIF; IF; initiation complex.

initiation factors of protein synthesis Involved in initiation of translation. *See* eiF; IF; iIF. (Sonenberg, N., et al., eds. 2000. *Translational Control of Gene Expression.* Cold Spring Harbor Lab. Press, Cold Spring Harbor, NY.)

initiator (Inr) *See* promoter.

initiator codon Starting site of translation in mRNA; generally 5′-AUG-3′, but can be less frequently 5′-GAG-3′, 5′-GUG-3′, or 5′-GUA-3′. *See* protein synthesis; translation.

initiator tRNA Carries formylmethionine (prokaryotes) and initiation methionine (eukaryotes) to the P site of the ribosome to begin translation. This tRNA has a structure distinguished from the rest of the transfer RNAs. Besides AUG, this rRNA in bacteria may recognize GUG and UUG as a formylmethionine codon. Not all proteins begin with a methionine because of processing. Mutation in the anticodon of tRNA^{fMet} may start translation with amino acids other than formylmethionine or methionine. The eIF2 initiation protein distinguishes between the intiator and the elongation tRNA^{Met} on the basis of several criteria: the A1:72 base pair at the bottom of the amino acid acceptor stem, three G:C pairs in the anticodon stem, initiators that do not have the TψC in the T arm, A54 rather than T54 in the T arm, and A60 within the T loop replacing pyrimidine-60. In plants and fungi, a phosphoribosyl group at position 64 is attached to the 2′-OH of the ribose. Additional variations may exist in some species. *See* eIF2; initiation codon; IRES; protein synthesis; pseudoknot; ribosome; transfer RNA. (O'Connor, M., et al. 2001. *RNA* 7:969.)

INK (p^{INK}) Polypeptide inhibitors of cyclin-dependent kinases cause cell cycle G1-phase arrest. The INK family includes proteins p15, p16, p18, and p19; they bind cyclin D/CDK4. *See* cancer; CDK; cell cycle; p15; p16; p18; p19.

innate Inherited, congenital. *See* congenital.

innate immunity (natural immunity) Based on cell-surface receptors and other protein molecules encoded by the germline. These systems generally recognize carbohydrate structures, then stimulate the synthesis of various molecules such as cytokines, interleukins, and tumor necrosis factor. Some natural killer cells (neutrophils, macrophages) recognize inimical cells by lectin-like membrane receptors. This innate immunity is thus a rather fixed, rigid system in comparison to the acquired immunity mediated by immunoglobulins, which are greatly adaptable and variable. Innate immunity can shape the development of acquired immunity by interacting with it. The protein fragments cut up by the macrophages may be presented to the adaptive immune system represented by the B and T cells. It may guide the selection of antigens by lymphocytes and the secretion of cytokines by helper T lymphocytes. Innate immunity is the first line of defense by initiation of inflammation through recruiting phagocytic and bactericidal neutrophils and macrophages. The innate immune system recognizes foreign bodies with the aid of pattern-recognition receptors. The complement is part of innate immunity, but it also cooperates with the acquired immunity system. Innate immunity is also present in insects (*Drosophila*), although they lack the adaptive immunity system of vertebrates. The innate defense system of bacteria recognizes nonmethylated DNA and cleaves foreign DNA by restriction endonucleases. Mammals possess relatively few CG pairs and the C is commonly methylated within the PuPuCGPyPy sequence. The unmethylated bacterial, fungal, and insect CG pairs then activate macrophages, dendritic and B cells without normally attacking self-DNA. The genes of the acquired immune system are turned on through the mediation of the MEK signal transduction pathway and NF-κB. The specificity of the discrimination is attributed to the different Toll-like receptors, which recognize bacterial lipopolysaccharides, glycolipids, flagellin, etc. Another signaling route leads through DNA-PK. These two pathways may anastomose. The defense system in *Drosophila* relies on phagocytosis, proteolytic cascades, melanin formation, opsonization, and the synthesis of antimicrobial peptides. *See* acquired immunity; antibody; antimicrobial peptides; CD14; complement; DNA-PK; flagellin; host-pathogen relationship; immune system; interferon; lymphocytes; natural antibody; NF-κB; opsonins; signal transduction; *Toll*. (Aderem, A. & Ulevitch, R. J. 2000. *Nature* 407:782; Aderem, A. & Hume, D. A. 2000. *Cell* 103:993; Kimbrell, D. A. & Beutler, B. 2001. *Nature Rev. Genet.* 2:256; Roger, T., et al. 2001. *Nature* 414:920; Janeway, C. A., Jr. & Medzhitov, R. 2002. *Annu. Rev. Immunol.* 20:197.)

innervation Development of the nervous system in an organ or tissue.

innexin   Invertebrate gap-junction protein; functionally similar to connexins in vertebrates. *See* connexins; gap junction.

Inoculation loop.

inoculum   Usually a small microbial cell sample used for starting a culture.

inorganic pyrophosphatase   Cleaves off two molecules of phosphates from molecules.

inosine   *See* hypoxanthine.

Inosine.

inosinic acid   *See* hypoxanthine.

inositide   *See* phosphoinositides.

inositol   In cells it occurs generally as myoinositol as part of vitamin B complex. It is formed through cyclization from glucose-6-phosphate. Inositol is an indispensable constituent of some lipids (phosphoinositides). In some forms of diabetes it may accumulate in the urine. *See* diabetes; signal transduction.

Inositol.

inositol trisphosphate   *See* phosphoinositides.

in-planta transformation   *See* transformation, genetic.

input trait   In genetic engineering of plants the goal of the research is to facilitate the culture of the plants (e.g., increase resistance to herbicides, pathogens, parasites, cold, etc.). The *output traits* are higher nutrient content, modified plant products (e.g., lipids, fatty acids) manufacturing special

proteins (e.g., antibodies), and synthesis of special industrial raw material (e.g., silk fibroin, plastics, etc.). *See* genetic engineering.

Inr   *See* promoter.

insect control, genetic   *See* genetic sterilization; holocentric chromosomes.

insect control, physiological   Uses chemicals as well as biological agents such as viruses, bacteria, fungi, protozoa, and natural parasites. Chemicals include alkylating agents (e.g., methyl bromide) that alter the DNA, organophosphates and carbamates that inhibit choline esterases and thus nerve function, nicotine and derivatives that act in a similar manner to acetyl choline mimics. Arsenics inhibit glycolysis; cyanides poison the respiratory system. DDT and pyrethrins activate sodium channels; growth regulators may inhibit chitin synthesis, hormone synthesis, etc. The effectiveness of many compounds may be diminished by time because of mutation to resistance.

insect resistance in plants   Some plant species contain genes for insect tolerance and these are being incorporated into plant breeding material by conventional techniques. The most successful insect resistance gene, the *Bacillus thüringiensis* toxin gene, has been transformed into several species of dicots and provides almost complete defense when it is expressed under the control of efficient promoters. Pea plants transgenic for α-amylase inhibitor 1 and 2 are protected from the pea weevil (*Bruchus pisorum*). αAl-1 causes larval mortality at the first and second instar stages, whereas αAl-2 is responsible for blocking maturation of the larvae. Maize plants defend themselves against the armyworm (*Spodoptera exigua*) caterpillars by releasing sesquiterpene and indole. These volatile compounds encoded by the *stc1* and *Igl* genes attract parasitic wasps, which deposit their eggs and thus eventually destroy the caterpillars. Some of these volatile defense compounds (cis-3-hexen-1-ol, linalool, cis-α-bergamotene) perform multiple tasks, e.g., repelling herbivorous invaders, decreasing their rate of oviposition, and recruiting their natural predators. Spider mite–infected plants may release terpenoids (β-ocimene) that may stimulate defense-related genes in uninfected lima bean plants. Plants transgenic for the cyanogenic glucoside, dhurrin, become resistant to the flea beetle *Phylotreta nemorum*. *See* Bacillus thüringiensis; chitin; high-dose/refuge strategy; host-pathogen relations; lotaustralin; selection types; wound response. (De Moraes, C. M., et al. 2001. *Nature* 410:577; Farmer, E. E. 2001. *Nature* 411:854; Palumbi, S. R. 2001. *Science* 293:1786; Tattersall, D. B., et al. 2001. *Science* 293:1826; Schriber, J. M. 2001. *Proc. Natl. Acad. Sci. USA* 98:12328; Kessler, A. & Baldwin, I. T. 2002. *Annu. Rev. Plant Biol.* 53:299.)

Dhurrin.

insect viruses Diverse types. The enveloped Baculoviridae and Poxviridae have double-stranded DNA. The mosquito iridescent virus has nonenveloped dsDNA. The genetic material of the Parvoviridae is nonenveloped single-stranded DNA. The Reoviridae carry nonenveloped double-stranded RNA. The Flaviviridae are single-stranded, enveloped RNA viruses. The *Drosophila* C virus is a nonenveloped ssRNA virus. The Baculoviruses are important for engineering genetic vectors. *See* baculoviruses. (Friesen, P. D. & Miller, L. K. 2001. *Fundamental Virology*, Knipe, D. M. & Howley, P. M., eds., p. 443. Lippincott Williams & Wilkins, Philadelphia, PA.)

insemination by donor (DI) Therapeutic process to secure offspring in case of male infertility. In the U.S., ~30,000 births/year occur by these means. Generally, cryopreserved semen from sperm banks is utilized. To avoid potential consanguinity and the transmission of hereditary diseases, it is advisable to set a limit of no more than 10 inseminations per donor. If the donors are genetically screened, genetic abnormalities may be reduced from 2–5% to about half of that in the general population where DI is not used. *See* ART; in vitro fertilization; sperm bank. (Kuller, J. A., et al. 2001. *Hum. Reprod.* 16:1553.)

insertional inactivation Insertion of any type of DNA into an antibiotic resistance or other gene; it may inactivate the function (e.g., becomes antibiotic sensitive) because of the disruption of gene continuity. *See* insertional mutation; pBR322. (McClintock, B. 1951. *Cold Spring Harbor Symp. Quant. Biol.* 16:13; Koncz, C., et al. 1992. *Plant Mol. Biol.* 20:963.)

insertional mutation Insertion of a transposable genetic element within a gene disrupts its function and as a consequence lethal effects or altered functions are displayed. The insertion may modify the expression of the target genes by overexpression or suppression of the mutant phenotype. It has been shown that many of the insertions do not lead to observable changes in gene expression or their effect is minimal and only sequencing of the target loci reveals their presence. Labeled probes for the insert can selectively isolate the gene carrying the insertion. Although initially it was believed that the insertions occured at random sites, it is now documented that the preferred target sites are upstream of the promoters or other locations where no essential functions are disrupted. The target selection is mediated by integrase or transposase functions. In some cases, only DNA elements are involved in the insertions, but retrotransposons sever the target DNA, the free 3′-OH end primes the reverse transcription of the template RNA element, and the DNA is then inserted. Generally there is no homology-dependent pairing requisite for recombination except for the type 2 introns. The insertion requires protein cofactors such as IHF, HU, MuA, and MuB (an ATP-dependent activator of Mu-phage protein MuB). The IS10/Tn10 bacterial and the TC1/TC3 *Caenorhabditis* elements interact specifically with a target site consensus. TC transposase recognizes and is inserted at TA dinucleotides. In IS10 the interaction with the target catalyzes the excision of the element from the original location. Retroviral and some other elements preferentially choose nucleosomal sites where the DNA is bent. Tn7 preferentially inserts to a conserved attachment site in *E. coli* located outside the boundary of the *glmS* gene (glucosaminephosphate isomerase, map position 83). The yeast element Ty3 has high specificity for the promoters of RNA pol III–transcribed genes, whereas Ty1 and Ty5 display less stringent specificities but (although not exclusively) within the same region. Ty1 also prefers insertion when transcription is (potentially) active, whereas Ty5 selects transcriptionally inactive sequences. Therefore, Ty1 is well suited for the induction of insertional mutations. Transposition is generally targeted to sites outside the transposon, yet it may occur within the element and often leads to its destruction, but it may lead to the evolution of a new element. The maize transposable elements and the *Drosophila* P element move preferentially to nearby sites, although they may jump to other chromosomes. Agrobacterial vectors can produce insertional mutations in plants. The position of insertions can be defined by various polymerase chain reaction techniques. By the use of microarray hybridization, insertion libraries can be screened with the aid of capture PCR (Mahalingam, R. & Fedoroff, N. 2001. *Proc. Natl. Acad. Sci. USA* 58:7420) and on the basis of homologies to expressed sequence tags. *See* Ac-Ds; *Agrobacterium*; Alu family; *Caenorhabditis*; cancer gene therapy; capture PCR; DNA bending; EST; gene isolation; gene therapy; HU; hybrid dysgenesis; IHF; insertion elements; integrase; introns; LINE; microarray hybridization; REMI; T-DNA; Tn7; Tn10; transposase; transposition immunity; transposons; TRIPLES; Ty. (Vidan, S. & Snider, M. 2001. *Curr. Opin. Biotechn.* 12:96.)

Insertional mutation of *Arabidopsis* knocked out the stem, but fertility of the fruits is retained.

insertion elements DNA sequences generally shorter than 2,000 bases that can insert into any part of a genome (*see* transposons) are common in all organisms from prokaryotes to eukaryotes. Some of the bacterial insertion elements have the characteristics shown next page. They are not carrying any genetic information beyond that needed for insertion. Viruses also can act as insertion elements in cells. Insertion elements are the major factors of spontaneous mutability. According to some estimates, 5–15% of the spontaneous mutations in bacteria are caused by *IS* elements. The presence of *IS*1 may increase deletion frequency of the *gal* operon by 30- to 2,000-fold. *IS* elements cause chromosomal rearrangements. Many *IS* elements have a large number of potential target sites; others display clear preferences. The mechanism of transposition is either dependent on replication and the new copy is transposed or it simply involves a relocation of the existing element. The movement of the *IS* elements is affected by host genetic factors (DNA polymerase, gyrase, histone-like proteins, dam methylase, DnaA protein, and proteins mediating recombination).

Name	Size (bp)	Inverted Terminal Repeats (bp)		Target Duplication
*IS*1	768	18/23	(*E. coli*)	8–11
*IS*2	1,324	32/41	(*E. coli*)	5
*IS*3	1,258	29/40	(*E. coli*)	3
*IS*5	1,250	16	(*E. coli*)	4
*IS*10	1,329	17/22	(*Tn*10)	9
*IS*66	2,548	18/20	(*Agrobacterium tumefaciens*)	8

IS elements may not only alter mutation but also gene expression by their presence in the control regions of genes. In *Drosophila melanogaster*, the average insertion density (average number of large insertions/kb/chromosome) is about 0.004 and the average frequency of movement is 0.023. Some of the historical insertions were modified and became permanent regulatory elements of the genes. *See* copia elements; hybrid dysgenesis; illegitimate recombination; LIN; pathogenicity islands; retroposons; Ti plasmids; Tn; transposable elements; Ty. (Chandler, M. & Mahillon, J. 2000. p. 306 in Mobile DNA II, Craig, N., et al., eds., p. 109. *Am. Soc. Microbiol.*, Washington, DC; Mahillon, J., et al. 1999. *Res. Microbiol.* 150:675.)

insertion vectors   Have selectable markers and restriction enzyme sites on vectors and chromosomes where foreign DNA can be inserted. The insertion involves a single reciprocal recombination and may generate a duplication because the insert has homology to the target. The homology does not have to be complete.

insert restriction site into plasmid   Amplify the restriction endonuclease gene by PCR, then open up the plasmid by the restriction enzyme. After the primers are removed, insert the target DNA into the plasmid and ligate to circularize the plasmid.

in silico biology   Identification of genes in databases with the aid of bioinformatics. (Harris, S. B. 2002. *EMBO Rep.* 3:511.)

in situ hybridization (ISH)   The DNA double strands within the cells (chromosomes) are separated (denaturation) in cytological preparations on microscope slides and labeled (radioactively or by fluorochromes or immunoprobes), and complementary DNA or RNA strands (probes) are annealed. The cells are then visualized by microscopy as usual for cytological microtechniques and can either be autoradiographed or viewed through fluorescence microscopy to ascertain the position of the hybridized ↑ probe. Thus, chromosomal location of molecular markers can be determined. *See* chromosome painting; DNA hybridization; FISH; fluorochromes; GISH; immunoprobe; in situ PCR; nick translation; PRINS; probes; somatic cell hybrids. (Pardue, M. L. 1973. *Cold Spring Harbor Symp. Quant. Biol.* 38:475; Carpenter, N. J. 2001. *Semin. Pediatr. Neurol.* 8[3]:135.)

in situ PCR   Employs PCR technology within single cells. The DNA is amplified, then in situ hybridization is used. Treating the cells with reverse transcriptase before PCR enhances its utility. Such a procedure permits the detection of cellular genic activity, viral infection, and expression of newly introduced transgenes for gene therapy. *See* FISH; PCR; PRINS; RT-PCR. (Kher, R. & Baccalao, R. 2001. *Am. J. Physiol. Cell Physiol.* 281:C726.)

insomnia   *See* fatal familial insomnia.

InsP$_2$ (inositol-1,4-diphosphate)   Phosphoinositides.

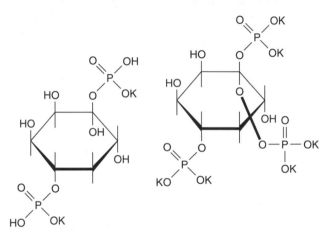

InsP$_2$ left and InsP$_3$ right.

InsP$_3$ (inositol-1,4,5-trisphosphate)   Global signaling molecule that liberates Ca^{2+} in the cytoplasm. *See* IP$_3$; phosphoinositides.

instability, genetic   Caused by a high rate of mutation, defects in genetic repair, recombination among repeated sequences (trinucleotide repeats, inverted repeats), palindromes, presence of transposable elements, illegitimate recombination, and double-strand breaks. *See* at-risk-motif; cancer; chromosomal aberration; chromosome breakage; DNA repair; mutagens; mutator genes; repeats, inverted; transposable element; trinucleotide repeats. (Myung, K., et al. 2001. *Nature* 411:1073.)

instar   Insect larvae between the processes of molting (shedding the outer cover layer). In *Drosophila*, three moltings take place after hatching and before pupation. The first and second instars last for ~1 day each, whereas the third instar lasts for ~2 days. *See Drosophila*.

instincts   Variety of innate behavioral patterns that are developing without learning, although learning may reinforce them, e.g., motherly love, nursing, fearing for life, conforming to some standards, etc. *See* aggression; behavior genetics; ethics; ethology.

Institutional Biosafety Committee Must approve genetic engineering projects regarding (1) the source of DNA used, (2) the nature of the inserted DNA, (3) the hosts and vectors contemplated, (4) the protein products expected, (5) the nature of the containment facilities and the compliance with the regulations of the National Institutes of Health. *See* recombinant DNA; recombinant DNA and biohazards.

instructive signal Receptor-mediated; involves a signal transduction cascade. *See* signal transduction.

insulator DNA element of 0.5 to 3 kbp that prevents interactions between enhancers and the target promoter, but it may/may not permit the expression of stably integrated transgenes. Insulators must be located between the enhancer and the promoter to act. When there is a spacer DNA between the enhancer and the promoter, the structural gene expressed as illustrated by a lightbulb is turned on (☼). When the spacer is replaced by an insulator element, the expression of the structural gene is much reduced or eliminated. The insulator may affect the organization of the chromatin or it may act as a regulator of transcription. Such elements are Fab-7 and Fab-8 or the HMR and HML elements controlling sex determination in yeast.

Insulator site in the chromosome.

Insulators may prevent position effect or they may display some sort of position effect and two cis-insulators may facilitate transcription rather than blocking it. The boundary between heterochromatin and euchromatin displays a distinct pattern of methylation of histone, H3. *See* boundary element; CTCF; enhancer; enhancer competition; histones; promoter; position effect; sex determination in yeast; silencer. (Sun, F.-L. & Elgin, S. C. R. 1999. *Cell* 99:459; Bell, A., et al. 1998. *Cold Spring Harbor Symp. Quant. Biol.* 63:509; Bell, A. C. & Felsenfeld, G. 1999. *Curr. Opin. Genet. Dev.* 9:191; Bell, A. C.,

et al. 2001. *Science* 291:447; Noma, K.-I., et al. 2001. *Science* 293:1150; Gerasimova, T. I. & Corces, V. G. 2001. *Annu. Rev. Genet.* 35:193; Oki, M. & Kamakaka, R. T. 2002. *Current Opin. Cell Biol.* 14:299.)

insulin One of the most important peptide hormones (MW ≈5,700) in the body and it is synthesized in the pancreas. The pancreas is a large gland behind the stomach. Inside, the Langerhans islets produce and secrete insulin into the blood. Insulin regulates glucose uptake into the muscles by glucose transporters and into the liver by glucokinase. In the muscles and the liver, with the assistance of glycogen synthase, glycogen is made. The breakdown of glycogen is mediated by the insulin-regulated glycogen phosphorylase. Glycolysis and acetyl coenzymeA synthesis is boosted by insulin through the enzyme phosphofructokinase and the pyruvate dehydrogenase complex. Fatty acid synthesis in the liver is promoted by acetyl-CoA carboxylase and neutral fat (triaglycerol) is stimulated by insulin with the aid of lipoprotein lipase.

The deficiency of insulin leads to hereditary diabetes mellitus. Diabetes is a very complicated disease controlled by several genes involved either in differentiation of the Langerhans islets or at various steps in the synthesis and regulation of the hormones. About 5 to 10% of the population in Western countries has a form of this disease. Of these, however, only 1/10 are insulin-dependent. Insulin-dependent diabetes (IDDM) is largely familial and occurs early in life. The non-insulin-dependent form of the disease (NIDDM) generally affects those who become diabetic after age 40. The latter has a relatively small hereditary component, and it is most common among obese individuals between ages 50 and 60. One of the most important clinical manifestations of the disease is hyperglycemia (excessive amount of sugar in the blood), but there are many other complex characteristics. Actually, about 60 known human diseases are accompanied with diabetes symptoms. A well-controlled diet and continuous supply of insulin manage early-onset diabetes. The treatment of the other forms (NIDDM) varies according to the causative metabolic defect.

The primary structure of insulin was first determined by the double Nobel laureate, F. Sanger, in 1953, the same year as the Watson-Crick model of DNA was published. These two events signaled the beginning of molecular biology. Insulin is synthesized as a pre-proinsulin equipped with a signal sequence that directs the molecules into the secretory vesicles. After removal of this signal peptide, proinsulin is formed and stored in β-cells. When the blood glucose level increases, insulin-specific peptidases process the protein into the functional final form. The A-chain of bovine insulin contains 21 amino acids and the B-chain has 30; the two chains are joined by disulfide bridges between A7 and B7 as well as A20 and B19 residues, and a third disulfide bridge is formed within the A-chain between residues 6 and 11 (*see* protein structure). The A-chain is the same in humans, dogs, pigs, and rabbits; the bovine B-chain is identical with that of pigs, goats, dogs, and horses. The function of insulin requires the presence of a receptor.

The insulin receptor protein is made of two α- and two β-chains. The identical α-chains bind insulin above the surface of the plasma membrane, whereas the two β-chains reach inside the cell through the membrane with their carboxy termini. Upon binding of insulin, the β-chains become a specific protein tyrosine kinase that first autophosphorylates and then phosphorylates other proteins in a cascading series of events. As a consequence, a number of enzyme activities are altered by phosphorylation of tyrosine and serine residues. These multiple reactions explain why diabetes occurs in many forms and is part of dozens of other syndromes under the control of a variety of genes and manifested in aneuploids (Turner syndrome, Klinefelter syndrome, Down syndrome). Insulin receptor substrate-1 tends to the integrin family of surface receptors and in this way integrin and growth factor signaling are connected.

The various forms may be under the control of autosomal-dominant genes (diabetes insipidous nephrogenic type II, diabetes mellitus juvenile with early onset). The latter one was attributed, in one case, to a substitution of serine for phenylalanine at the 24th position of the β-Chain. Only a single locus seems to encode the human insulin structural gene in chromosome 11p15.5. The human INS and the mouse *Ins* (chromosome 7) are expressed only from paternal alleles at some stages of development (imprinting). In rats and mice, two distinct loci are involved. Some mutations interfere with the proteolytic processing of proinsulin, resulting in its accumulation (hyperproinsulinemia) without diabetes symptoms.

Non-insulin-dependent diabetes was observed to be associated with a mutation of the glucose transporter-4 protein at residue 383 (Val [GTC] → Ile [ATC]) in some cases but not others. Vasopressin and oxytocin deficiency may also be involved in diabetes. These two octapeptides regulate muscle contraction and renal secretion. Diabetes is generally associated with abnormalities of kidney functions.

Autosomal-recessive inheritance appeared to account for juvenile-onset diabetes mellitus type I, and this locus was assigned to human chromosome 6 in close proximity to the HLA class II genes. This locus is probably a regulatory one, determining insulin susceptibility with a penetrance of 71% for the homozygotes and 6.5% for the heterozygotes. It was reported that if at position 57 the HLA-DQ-β-chain has Asp, most likely diabetes was absent, whereas among diabetics there was a high chance that this 57th residue was non-Asp. Other studies indicated that HLA genes DR3 and DR4 and HLA-DQ have predisposing effects on the expression of diabetes. Actually, it appeared that HLA-DQw1.2 allele was protective, but HLA-DQw8 increased the chances of developing insulin-dependent diabetes. Population genetic studies also confirmed that having a nonpolar amino acid at position 57 increased the chance to develop diabetes by a factor of about 30. X-chromosomal linkage was observed for nephrogenic diabetes insipidus type I, and it was suggested that this form of the disease is associated with a defect in the vasopressin 2 (V2) receptor.

Insulin for therapeutic purposes, up to recent times, was obtained exclusively from animal pancreas collected at the slaughterhouses. By the use of recombinant DNA, human insulin can industrially be produced with the aid of bacterial cultures and further processing. The industrial production of human insulin by genetic engineering is more successful than that of other proteins because it does not require glycosylation. In yeast, proinsulin mRNA is normally produced, but the translation of the message is very inefficient. The human insulin has a definite advantage for some patients who may be allergic to the slightly different animal protein.

Lower-than-normal insulin resistance is frequently associated with infections, diabetes, obesity, and some forms of cancer. Obesity and diet-induced insulin resistance may be reversed by salicylate or disruption IKKb (Yuan, M., et al. 2001. *Science* 293:1673). Insulin resistance is mediated by tumor necrosis factor α and it decreases tyrosine kinase activity in the insulin receptor by activating serine phosphorylation (Kahn, B. B. & Flier, J. S. 2000. *J. Clin. Invest.* 106:473). White adipocytes transcribe a 750-residue mRNA encoding the protein resistin (resistance to insulin). Neutralization of resistin enhances insulin-stimulated glucose uptake and active resistin reduces the uptake. *See* diabetes; HLA; IKK; insulin-like growth factor; insulin receptor substrates; obesity; resistin; S6 kinase; thiazolidinedione; TNF. (Bell, G. I., et al. 1980. *Nature* 284:26; Steiner, D. F., et al. 1985. *Annu. Rev. Genet.* 19:463; Lizcano, J. M. & Alessi, D. R. 2002. *Current Biol.* 12:R236.)

insulin-like growth factors (IGF-1, IGF-2)   Required for passage of the cell cycle from G1 to S phase; promote cell maintenance, metabolism, and cell division. The human genes for IGF-1 and IGF-2 are located in chromosome 11p15-p11, separated by about 12–13 kbp. The human IGF-2 gene is expressed only from the paternal chromosome. In mice, the IGF genes (chromosome 7) display tissue-specific imprinting and are also expressed from the paternal chromosome. Somatomedin, a mammalian second messenger, has insulin-like functions in regulating bone and muscle growth in conjunction with the pituitary hormone receptor. A minor 4.8 kb mRNA generates the prepro-IGF2, whereas the major 6 kb mRNA encodes a posttranscriptionally regulated IGF. The insulin-like growth factor receptor (IGF-1R) and the insulin receptor belong to the tyrosine kinase receptor family and regulate both normal and malignant cell proliferation. The paternally expressed IGF2 seems to control—like a QTL—muscle mass and fat deposition in pigs. *See* achondroplasia; growth factors; imprinting; insulin; insulin receptor; mannose-6-phosphate receptor; pituitary dwarfness; prostate cancer; QTL; signal transduction; Simpson-Golabi-Behmel syndrome; somatomedin. (LeRoith, D., et al. 1992. *Ann. Intern. Med.* 116:854; Jiang, F., et al. 2001. *Dev. Biol.* 232:414.)

insulinoma   A relatively benign pancreatic cancer leading to excessive secretion of insulin.

insulin receptor protein (IR)   Heterotetrameric membrane protein, a tyrosine kinase (related to the epidermal growth factor [EGFR] family); its deficiency is Donohue syndrome. The enzyme protein tyrosine phosphatase (PTP-1B) dephosphorylates IR, and, as a consequence, in diabetes type II disease, although the cells can make insulin, they are unable to respond to this hormone because of the expression of PTP-1B. Rabson-Mendenhall disease (INSR, 19p13.2) is due to IR deficiency. *See* diabetes; Donohue syndrome; GLUTs; insulin; insulin-like growth factor; obesity; Rabson-Mendenhall disease.

insulin-receptor substrates (IRS1, IRS2, IRS3, IRS4)   After multiple-site tyrosine phosphorylation, they bind to and activate phosphatidylinositol-3'-OH kinase and other proteins with SH2 domains. Defects in IRS2 seem to be responsible for insulin-independent diabetes. IRS1 and IRS2 are essential for normal embryonic and postnatal growth. The IRS$^{-/-}$ individuals by 30 days reach less than one-fourth of the wild-type weight and reproductive capability—especially of the females—is impaired. *See* diabetes; insulin; insulin-like growth factor; SH2.

intasome Bacterial nucleoprotein complex, including a negative supercoiled phage DNA wrapped around by several copies of the phage-encoded integrase and the bacterial integration host factor protein (IHF). *See* IHF; integrase. (Chen, H., et al. 1999. *J. Biol. Chem.* 274:17358; Esposito, D., et al. 2001. *Nucleic Acids Res.* 29:3955.)

integral membrane protein Protein in the membrane without covalent linkage to it; it is bound there by about two dozen uncharged and/or hydrophobic amino acids.

integrase Protein that specifically recognizes transposon, phage, or yeast integration sites and opens them up for insertion to take place. The integrase makes a staggered cut in the DNA and the enzyme is covalently bound to DNA via a catalytic tyrosine (Tyr^{342}). A Holliday intermediate is formed; after a second exchange (and removal of the 40 kDa protein), the recombination product is generated. There is great structural variation among the more than 60 integrases, yet a 4-residue catalytic unit is highly conserved. The position of the integrase (IN) in the Ty/gypsy group of transposons is 5′-LTR-gag-protease-reverse transcriptase-RNaseH-IN-LTR-3′, and among the Ty1/copia-like elements, 5′-LTR-gag-protease-**IN**-reverse transcriptase-RNaseH-LTR-3′. The same enzyme may catalyze both integration and excision as determined by recombination directionality factors (RDF). *See* att sites; copia; disintegration; Holliday model; Holliday structure; IHF; INI1; intasome; integration; integron; lambda phage (site 27815); retroviruses; transposable elements; Ty. (Hindmarsh, P. & Leis, J. 1999. *Microbiol. Mol. Biol. Rev.* 63:836; Craigie, R. 2001. *J. Biol. Chem.* 276:23213; Lewis, J. A. & Hatfull, G. F. 2001. *Nucleic Acids Res.* 29:2205.)

integrated circuit (chip) Electronic circuit within a single piece of semiconducting material. *See* semiconductors.

integrated map Based on the combined information on genetic linkage and physical mapping. *See* mapping, genetic; physical map; skeletal map; unified genetic map.

integrating vector *See* yeast integrating vector.

integration (1) DNA sequence inserted at both termini by covalent linkage into the host DNA. DNA structure (bends) affects the site of integration. HIV integrates nonrandomly into purified naked DNA or preferentially into CpG stretches

modified by cytosine methylation within distorted nucleosomal DNA. Murine leukemia virus (MLV) integrates nonrandomly, with preference into the major groove of nucleosomal DNA. Retroviruses most commonly select DNase hypersensitive sites and stretches involved in active transcription. Agrobacterial vectors are most commonly integrated into the nontranslated regions of potentially expressed genes; similar predilections were observed in bacterial transposons. The different yeast Ty elements also display preferences. Transfer of foreign DNA into mammalian genomes either by infection or transformation may result in genome-wide methylation at sites remote from those of the integration and may alter the pattern of gene expression in the recipient genomes. *See* DNase hypersensitive sites; HIV; methylation of DNA; MIP; MLV; RIP; T-DNA; transposons; Ty. (Müller, K., et al. 2001. *J. Biol. Chem.* 276:14271.) **(2)** In mathematics, integration is the finding of a function of which the integrand is a derivative or an equation among finite variables that is the equivalent of the integrated differential equation.

$$F(x) = \int f(x)\, dx$$

and

$$\frac{dF(x)}{dx} = f(x)$$

See derivative.

integration host factor *See* IHF.

integration plasmid Has homologous sequences to the chromosome. After transformation, it may be inserted in multiple copies into the yeast chromosome by recombination. Linearization of the plasmid may increase its efficiency of integration by 10–1000–fold.

integrative circuits in signal transduction Multiple kinases initiate a common response. *See* signal transduction.

integrative suppression Elimination of the manifestation a genetic defect by insertion of a normal copy of the gene. (Brasch, M. A. & Meyer, R. J. 1988. *Mol. Gen. Genet.* 215:139.)

integrator gene Hypothetical regulator and coordinator of eukaryotic genes. (Wadgaonkar, R., et al. 1999. *J. Biol. Chem.* 274:1879.)

integrins (ITG) Heterodimeric family of integral membrane proteins that stretch out to the intercellular space and form the extracellular matrix. In connection with fibronectin and other ligands and ICAM, integrins control cell adhesion and cell shape, development of *Drosophila* halteres, legs, and wings, signaling, intracellular Ca^{2+}, inositol and lipid metabolism.

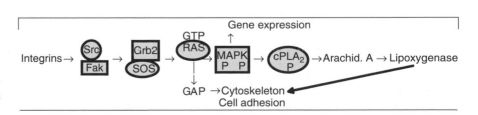

(After Clark, E. A. & Brugge, J. S. 1995. *Science* 268:233.) (For symbols, see individual entries; Arachid. A = arachidonic acid.)

Leukocytes express a variety of integrins that participate in inflammatory and immune responses. Integrins have α- and β-subunits that are translated separately, and their association is not by covalent linkage. Some of the integrin α- and β-subunit genes were assigned to human chromosome 2, whereas β-7 integrin appears to be in chromosome 12. α-6 integrin forms a 991-residue, extracellular, 23-amino transmembrane, and 36-amino-acid intracellular domain. Integrin is associated in the cells with a 59 K serine/threonine protein kinase containing four ankyrin-like repeats. The integrin-mediated signal transduction pathway is represented on page 627. This integrin-linked kinase (ILK) probably regulates integrin-mediated signal transduction. By binding to ligands, integrins, may control substratum adhesiveness and thus cell migration. Integrins integrate the extracellular signal systems with the cytoskeleton. Focal adhesion complexes containing β1 integrin, talin, actin, and vinculin are formed upon activation by the extracellular matrix. This complex is attached to the cytoskeleton and localized signal-transducing molecules. Preexisting mRNAs are targeted to cytoskeletal microcompartments where integrin controls their translation in the vicinity of the signal receptor sites. Integrins appear to be involved in fertilization by binding fertilin molecules. Integrins are involved in the control of the cell cycle and apoptosis. They signal to the RAS and ERK systems.

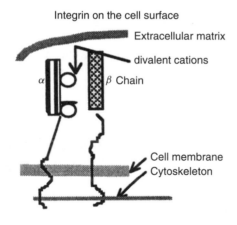

Integrin on the cell surface

See actin; angiogenesis; ankyrin; atresia; cadherin; calreticulin; CAM; caveolin; CD98; cell migration; cPLA₂; epidermolysis; ERK1; FAK; fertilin; fibronectin; GAP; Grb2; ICAM; invasin; lipoxygenase; MAPK; metastasis; myopathy; RAS; signal transduction; Src; SOS; talin; T-cell receptor; vinculin. (Clegg, D. O. 2000. *Mol. Cell Biol. Res. Comm.* 3:1; Calderwood, D. A., et al. 2000. *J. Biol. Chem.* 275:22607; <www.geocities.com/CapeCanaveral/9629>; <www.life.uiuc.edu/csb/integrins/index1.html>.)

integron   Mobile DNA element, a transposon with a cassette, flanked by a 5′ and a 3′ conserved sequence. The internal cassette can accommodate antibiotic resistance genes by site-specific recombinase (integrase, *intI*) located in the 5′ element and a closely linked *attI* (attachment) site. It contains several rightward (P2, P2, P4, and P5) and one leftward (P3) promoter sites and a ribosome-binding site. The 3′ element includes a conserved sulfonamide resistance gene. The origin of the sulfonamide resistance is based on the lack of sensitivity of dihydropteroate synthase to sulfonamide inhibition. Its origin is unclear since sulfonamides are synthetic antibiotics. Dihydropteroate is a precursor of folic acid. The integron is the vehicle of antibiotic resistance genes among various types of bacteria. The integrases bear 43–58% homology to the similar sequences of bacteriophages. *See* folic acid; gene cassette; integrase; R plasmid; shoufflons; transposon. (Rowe-Magnus, D. A., et al. 2001. *Proc. Natl. Acad. Sci. USA* 98:652.)

integument   Maternal somatic tissue layers that surround the ovule of plants and give rise to the seed coat; thus, the integument may show delayed inheritance. *See* delayed inheritance; megagametophyte development.

inteins   Elements inserted into proteins before sequence completion and removed afterward; they are protein-splicing elements. They are spliced out posttranslationally by autocatalytic proteolysis and ligation. Some inteins are site-specific endonucleases. They may also function as transposons and insert their coding sequences into inteinless genes. Inteins show different structures and they occur in prokaryotes, algal chloroplasts, and other lower eukaryotes. Inteins may be evidence of horizontal transmission of nucleotide tracts during evolution. *See* extein; intron; transmission. (Paulus, H. 2000. *Annu. Rev. Biochem.* 69:447; Liu, X.-Q. 2000. *Annu. Rev. Genet.* 34:61; Gogarten, J. P., et al. 2002. *Annu. Rev. Microbiol.* 56:263; <http://www.neb.com.neb/inteins.html>; <http://www.biomedgate.com/web/Databases_Resources/Proteins/>.)

intellectual property   *See* copyright; patent.

intelligence quotient (IQ)   *See* human intelligence.

intelligent design   Theory of organic evolution that does not accept either darwinism or creationism in its entirely and postulates that the development of the complexity of biochemical structures required an intelligent design. *See* creationism; darwinism; evolution.

intensifier   Animal or plant gene that intensifies (darkens) color.

interaction trap   Basically similar procedure as the two-hybrid method where interaction of two or more molecules permits a biological or physical observation. *See* two-hybrid method. (Toby, G. G. & Golemis, E. A. 2001. *Methods* 24:201.)

interaction variance   Due to epistasis between quantitative traits and the effect of the environment on gene expression. *See* analysis of variance; epistasis.

interactome   System in the proteome where proteins interact and cooperatively determine function(s). *See* gene product interaction; protein complexes; proteome. (Ito, T., et al. 2001. *Proc. Natl. Acad. Sci. USA* 98:4569; Ge, H., et al. 2001. *Nature Genet.* 29:482.)

interactor   In evolutionary context, an individual that interacts with the biotic and abiotic environment, and its traits impart reproductive success by the interaction. Its hereditary material is transmitted to the progeny.

interallelic complementation *See* allelic complementation.

interband region In the polytenic chromosomes, the relatively lighter-stained space between the characteristic darkly stained bands.

Bands and interband regions.

interbreeding Within a population, individuals of different genotypes may mate. *See* mating systems; random mating.

intercalating mutagens (such as acridines, some nitrogen mustards, etc.) Can insert within nucleotide sequences and cause frameshift mutations, short insertions, and deletions. Intercalation also causes separation of the base pairs, lengthening and untwisting of the double helix. *See* acridine dye; frameshift mutation; nitrogen mustards.

intercellular Situated in between cells. *See* intracellular.

intercellular immunization The antibody production is ectopic and its secretion depends on cells other than lymphocytes. The expression of immunoglobulin genes can be directed to specific cells by employing cell-type-specific promoters and enhancers in transformation. Such manipulations may block the expression of genes by the antibodies and may reveal the consequences for their function, for development or disorder, and possibly for therapy targeted with high specificity. Besides the transgenic approach, grafting of specific cells to an appropriate tissue may be employed. *See* ablation; ectopic; genetic engineering; immunization; intracellular immunization; lymphocytes; neuroantibody.

intercept *See* correlation (in a linear regression $a = Y - bx$).

interchange Reciprocal translocation of chromosomes. *See* translocations.

interchange, trisomic *See* trisomic, tertiary; trisomic analysis.

intercistronic region Number of nucleotides between the end of one gene and the beginning of the next one. *See* cistron.

intercross Mating between individuals (siblings) of the same parentage.

interfacial enzymology The enzymes act on substrates located on a surface and thus their activity is also regulated by the concentration of the substrate on the surface. Such enzymes may reside within or on cellular membranes.

interference, bacterial One strain excludes others from infection or colonization sites. Generally the interference is mediated by the inhibition of the synthesis of virulence factors or surface receptors. *See* RNA interference.

interference, chromatid *See* chromatid interference.

interference, chromosome One crossing over may either reduce (positive interference) or increase (negative interference) the occurrence of additional ones. *See* chromatid interference; coefficient of coincidence; coincidence; mapping function. (Zhao, H., et al. 1995. *Genetics* 139:1045; Browman, K. W. & Weber, J. L. 2000. *Am. J. Hum. Genet.* 66:1911; Tapper, W. J., et al. 2002. *Ann. Hum. Genet.* 66:75.)

interference, dominant Changes a transcriptional activator into a repressor. *See* activator protein; repressor.

interference, negative *See* coincidence.

interference RNA *See* RNAi; RNA interference.

interferon Specific cellular glycoprotein developed after viral infection or as a reaction to RNA or other compounds. Interferons have antiviral activity and possibly antitumor activity. They have three major forms: IFN-α, -β, and -γ. Interferons can be produced after transformation in a variety of cells (yeast, silkworm, mouse, hamster, etc.). Interferons produced by leukocytes predominantly contain the α-type. Lymphoblastoid cells (lymphocytes stimulated by an antigen) have 90% α- and 10% β-interferons. Induced fibroblasts mainly contain the β-chain. γ-interferon is produced by antigen- or mitogen-stimulated T lymphocytes. Interferon-induced protein genes have been located to human chromosomes 21, 10, 4, and 1. Interferons may interfere with protein synthesis by causing phosphorylation of eukaryotic peptide chain initiation factor eIF-2.

Interferon α (leukocytic interferon, IFNA1) genes (up to 30) were located to human chromosome 9p21-p13. One of these genes activated by the double-stranded viral replication intermediate is 2′-5′-oligoadenylate synthetase, which in turn activates latent ribonuclease L. RNase L degrades single-stranded RNA, the viral genome. Another induced enzyme is dsRNA-activated protein kinase R (PKR). PKR may mediate apoptosis and may assist in establishing persistent infections by several viruses. An *interferon* α-*receptor* (antiviral protein, AVP) was assigned to human chromosome 21q22. Translocations of INFA have been identified in leukemia patients. It has been claimed that intranasal use of interferon α may prevent the common cold. Natural IFN-producing cells (dendritic cell precursors) that express CD4 and major class II histocompatibility synthesize IFNA.

Interferon β-1 (fibroblast interferon/IFB1, human chromosome 9p21-pter) is structurally homologous to α-interferon. Patients with acute monocytic leukemia displayed translocations of IFB1 to chromosome 21, and the breakpoint was within this gene at about 17 cM from the site of the ETS-2 oncogene (chromosome 21q22.1-q22.3). Actually, in chromosome 9, several interferon genes occur. *Interferon* β-2 (IFNB2, human chromosome 7p21-p15) is induced by tumor necrosis factor (TNF) and interleukin (IL-1) when interferon β-1

(IFNB1) is not induced. It is identical with the B-cell differentiation factor (BSF2) and hybridoma growth factor. Interferon β-3 (IFNB3) is in human chromosome 8.

Interferon γ (IFNG, human chromosome 12q14) induces the expression of HLA class II genes. The induction is modulated by a factor in human chromosome 16 (probably a receptor) and by another in chromosome 21 that may control the transduction of the γ-interferon signal. In RAG (recombination-activating) cells, a human chromosome 6 factor was also required in human-rodent cell hybrids. IFNG induces the production of a 98-residue polypeptide (IP-10, chromosome 4q21) and other activating polypetides. Hereditary interferon-γ receptor deficiency increases the susceptibility to mycobacteria and *Salmonella* infections. The interferon regulatory factor (IRF) family includes the interferon consensus sequence-binding protein (ICSBP, expressed constitutively in B lymphocytes) and other transcription factors (IRFs). Their N-terminal region binds to DNA (IFN-stimulated responsive element, ISRE) and the C-terminal contains the regulatory sequences. *See* allergy; B cell; CD4; dendrocyte; granulocyte; hybridoma; interferon response element; interleukin; IRF; killer cell; leukemia; lymphokines; MHC; modulation; mycobacterium; PKR; receptor protein; signal transduction. (Stark, G. R., et al. 1998. *Annu. Rev. Biochem.* 67:227; Taniguchi, T. & Takaoka, A. 2001. *Nature Rev. Mol. Cell Biol.* 2:378; Sen, G. C. 2001. *Annu. Rev. Microbiol.* 55:255.)

interferon-induced proteins Located in different human chromosomes 1, 2, 4, 10, 12, and 21. Interferon-inducible cytokine IP-10 (human chromosome 4q21) is located close to the breakpoint associated with monocytic leukemia. (Monocytes are phagocytic mononuclear leukocytes that develop into macrophages in the lung and liver.) IP-10 has substantial homology to several activating peptides and it may control inflammatory responses. *See* cytokines; leukemia; macrophage.

interferon receptors IFNα and β share the same 63.5 kDa receptor. The IFNγ, IFNβ, and IFNω receptors are located in the same cluster (21q-q22.1). IFNγ receptor binds the ligand at the 245-amino-acid extracellular domain, whereas the 222-amino-acid intracellular domain is involved in signal transduction. *See* interferon; interferon regulatory factors; signal transduction (Jak-Stat).

interferon regulatory factors (IFR) Bind to the upstream regions of both α- and β-interferon genes and serve as transcription activators. IRF-1 and IRF-2 have an antagonistic effect. Thre are several additional IRFs. IRF1 binds to two regulatory elements (PRDI and II) in the IFNβ gene promoter. These elements also respond to Jun and NK-κB, which bind to the nearby PRDII and IV. They compete for the same cis-element. Both factors were assigned to human chromosome 5q23-q31. The IFR-2 protein is identical to HiNF. The IRF-E DNA-binding site has the consensus G(A)AAA(G/C)(T/C)GAAA(G/C)(T/C). Some IFRs (IURF-4, vIRF) are elevated in certain cancers. *See* CCE; cis-acting element; cytokines; HiNF; histones; interferon; Jun; leukemia; macrophage; NK-κB; transcription factors; upstream. (Taniguchi, T., et al. 2001. *Annu. Rev. Immunol.* 19:623.)

interferon sequence-response element (ISRE) AGTTT-CNNTTTCN[C/T]. *See* GAS; interferon; interferon receptors; Jak-Stat; signal transduction.

intergenic complementation Evidence that the two genes are not allelic and belong to separate loci. *See* allelic complementation; allelism test.

intergenic regions Represent the bulk of DNA, which contains in higher eukaryotes only a few percent of coding sequences. The noncoding sequences may have regulatory roles. A large fraction includes mobile elements with evolutionary significance. *See* introns; selfish DNA; transposable elements. (Kondrashov, A. S. & Shabalina, S. A. 2002. *Hum. Mol. Genet.* 11:669.)

intergenic spacers (IGS) Sequences between rRNA genes, which have very short transcripts (appear like feathers) but may be contributing to transcription initiation. *See* ITS; pol III; tRNA.

(Courtesy of Spring et al. 1976. *J. Microsc. Biol.* 25:107.)

intergenic suppressor One mutation suppresses the expression of another situated in a different locus. Most commonly the suppressor encodes a mutant tRNA. *See* suppression.

interkinesis *See* interphase.

interleukins Proteins secreted by white blood cells, phagocytes, and B lymphocytes. They are involved in stimulating growth and differentiation of lymphocytes concerned with the natural defense system of the body. *See* IL-1; IL-2; IL-3; IL-4; IL-16; lymphokines.

interlocking bivalents When another chromosome passes through the terminalizing chiasmata (ring bivalents), the nonhomologous chromosome may be trapped (interlocked) within the ring. *See* ring, bivalent. (Rasmussen, S. W. 1976. *Chromosoma* 54:245.)

Centromere of the chromosome

intermediary metabolism Enzymes within the cells produce energy from nutrients and use it to synthesize other compounds or organize cellular components.

intermediate filaments Ubiquitous 10 nm protein filaments abundant in the eukaryotic cells and encoded by at least 50 genes. They are composed of keratin, desmin, vimentin, neurofilament proteins, glial fibrillary acidic protein (GFAP), lamin, etc. Their anomalies may result in epidermolysis, keratosis, and possibly other skin diseases. *See* desmosome; epidermolysis; filament; keratin; keratosis; lamins; skin diseases; vimentin.

intermedin   Melanocyte-stimulating protein factor. *See* melanocyte.

internal control region (ICR)   *See* A box.

internal image antibody   Antiidiotypic antibody that binds to the antigen-binding site of the complementarity-determining region of an antibody rather than to the antigen. *See* antibody; antiidiotypic antibody; antigen; complementarity-determining region; immunoglobulin.

internal membranes   Membranes within the cell excluding the plasma membrane.

internal promoter   *See* promoter.

internal transcribed spacers   *See* ETS; ITS; tRNA.

internalins   Proteins mediating bacterial uptake by eukaryotic cells. The internalization requires cofactors such as cadherin, catenin, actin, and PI-3 kinase and the reorganization of the cytoskeleton. *See* actin; cadherin; catenin; cytoskeleton; phosphoinositides. (Schubert, W. D., et al. 2001. *J. Mol. Biol.* 312:7387.)

internally eliminated sequences (IES)   During the formation of the macronuclei of *Ciliates*, chromosome diminution and fragmentation occur. The deleted DNA involves repetitive sequences, but IES is not part of the repetitive sequences and could even include functional genes. *See Ascaris*; chromosome diminution; macronucleus; *Paramecia*.

International Prognostic Index (IPI)   Attempt by the European Society for Medical Oncology to predict the risk/survival of some cancers under standard conditions of treatment.

Internet   Complex system of interconnected electronic communication networks.

internode   Segment of plant stem between nodes (leaves).

interologs   Conserved protein-protein interactions facilitating the identification of protein networks. *See* genetic networks. (Matthews, R. L., et al. 2001. *Genomes Res.* 11:2120.)

interoperability   Different computer programs and computers can communicate with each other.

interorganismal genetics   *See* Flor's model.

interphase   Part of the cell cycle (boxed) $\boxed{\text{G1, S, G2,}}$ **M** $\boxed{\text{G1, S, G2,}}$ **M** when mitosis (M) or meiosis is not in progress. During interphase between mitoses, the cells are actively synthesizing DNA (S phase), and during the G phases other molecules are made and the cellular organelles divide. In meiotic divisions, DNA synthesis is limited only to repair and all of the DNA is made during the interphase preceding meiosis. *See* cell cycle; meiosis; mitosis.

interpolation, linear   *See* F_2 linkage estimation.

interracial human hybrids   The genetic distance based on gene frequencies is relatively small (see evolutionary distance) among the various human races, and despite the theoretical expectation of deleterious effects of breaking up coadapted genetic sequences, interracial hybrids apparently suffer no physical harm. Problems may arise, however, in cultural adaptation because the hybrids generally are classified socially with the minority race (whatever the majority is) and discrimination against minorities is not uncommon in all racial, ethnic, and cultural groups. Despite the great similarities of the genetic structure of all primates, no hybrids between humans and other primates have been verified. Somatic cells of all types of eukaryotes—including those of humans—can, however, be hybridized but cannot be regenerated into hybrid organisms. *See* human races; somatic cell hybrids.

interrogation, genetic   In-depth study of the function of a gene or group of genes.

interrupted mating   Stopping bacterial conjugation at definite intervals (by stirring the culture) in order to determine the order of gene transfer and establish map position on the basis of minutes required for transfer of particular genes from donor to recipient. The interruption stops transfer. *See* conjugation mapping.

intersectin   Endocytosis protein. *See* endocytosis.

intersex   The true intersex types have both male and female gonads, a very rare condition. More common are those that have either female or male gonads and chromosomal constitution, but they express, to various degrees, the secondary sexual characteristics normally unexpected for their chromosomal constitution. The intersex phenotype is determined by autosomal genes, and in species where the sex chromosome:autosome ratio determines sexuality, the aneuploids appear as intersex types. In *Drosophila*, the *tra* (*transformer*, chromosome 3-45) homozygotes of XX chromosomal constitution are sterile males. The *tra2* (chromosome 2-70) mutation in XX background has similar effects as *tra*; in XY background, the males look normal and behave normal, yet their sperm is not motile. The *dsx* (*doublesex*, 3-48.1) locus has numerous alleles. The homozygotes for their null alleles (in either XX or XY background) and the heterozygotes for the dominant alleles (in XX background only) are intersexes. The *ix* gene (*intersex*,

2-60.5) makes females intersex with reduced male and irregular female external genitalia. The *ix/ix* males look normal, but they are largely homosexual and the *ix/+* heterozygotes court females and young males but not adult ones. The *tra* and *dsx* genes regulate sex expression by alternative splicing of the RNA transcript involved in normal sexuality. In some insects, like the Gypsy moth (*Lymantria dispar*), the sex-determining genes have "strong" and "weak" alleles in some populations and the crosses regularly yield intersex individuals. The crosses between head lice (*Pediculus capitis*) and body lice (*P. vestimenti*) produce intersexes in F_2 and F_3. In angiosperm plants, 96 to 98% are hermaphroditic; among the 2 to 4% dioecious species, intersexes occur, depending on the number of each of the sex chromosomes they carry—e.g., in *Melandrium*, the XXY and XXXY males occasionally produce intersex flowers, but the XXXXY individuals are hermaphroditic. *See* anti-Müllerian hormone; gonads; gynandromorphs; hermaphroditism; homosexuality; introns; *Lymantria*; psedohermaphroditism; relative sexuality; sex determination. (Vaiman, D. & Pailhoux, E. 2000. *Trends Genet.* 16:488.)

interspersed repetitious DNA About 35% of the human genome, including various types of active or inactive transposable elements. *See* LINE; redundancy; SINE.

inter-SS PCR *See* cancer.

interstitial segment Chromosomal region between the centromere and the translocation break point. *See* translocation.

interval mapping Considers pairs of adjacent markers and maximizing the likelihood for quantitatively expressed gene loci (QTL) being in between. ELOD is the expected lod score, θ = recombination fraction, p = proportion of variance contributed by QTL, n = sample size. The calculation may be difficult under practical conditions because there may be more than one QTL per linkage group. An interval mapping based on the least squares method may be better. *See* ASP analysis; least squares; lod score; QTL. (Ott, J. 2001. *Advances Genet.* 42:125.)

$$\text{ELOD} = \left(\frac{1-2\theta}{1-\theta}\right)\frac{n}{2}\log\left(\frac{1}{1-p}\right)$$

intervarietal substitution Basically similar to alien substitution except the chromosomes belong to different varieties of the same species. *See* alien substitution.

intervening sequence (IVS) *See* intron.

intimate pairing (synapsis) Very close apposition of the chromosomes at the meiotic prophase that makes crossing over (gene conversion) and recombination possible. *See* recombination, molecular mechanisms, prokaryotes; recombination mechanisms, eukaryotes; recombination models; synapsis.

intimim Enterobacterial protein that mediates the attachment of the bacterium to the epithelial cells of the intestine and causes the erosion that facilitates the transport of proteins and intestinal inflammation as a consequence of the infection.

intine Inner layer of the wall of the pollen grain.

INT oncogene INT1 was assigned to human chromosome 12q13 and to mouse chromosome 15. The product of *Drosophila* gene *wingless* (*wg*) is a homologue and therefore the gene is sometimes mentioned as WNT. INT4 (human chromosome 17q21-q22, mouse chromosome 11) is also homologous to the *Drosophila wg* gene. In mammary carcinomas, INT1 is frequently the target site for insertion and inactivation. INT2 is a mammary tumor gene, and it was assigned to human chromosome 11q13; its only relation to INT1 is its presence in mammary tumors. INT2 product shows some relationship to fibroblast growth factor 6. INT3, another mammary tumor gene, encodes a transmembrane protein. *See* FGF; morphogenesis in *Drosophila*; oncogenes; transmembrane proteins; WNT1.

intrabody Antibody expressed by intracellular immunization in the target cells. *See* gene therapy; intracellular immunization; tumor vaccination. (Duff, R. J., et al. 2000. *Methods Enzymol.* 313:297.)

intracellular Within the cell(s).

intracellular clock Differentiation of particular cells into a particular type of tissue during embryonal development is controlled by the timing of the signal for differentiation received—e.g., an animal epithelium excised at the gastrula stage grafted into the eye disc of an embryo may differentiate into a neural tube, but if the same tissue if grafted to the same position a few hours later, it may differentiate into an eye lens.

intracellular immunization Involves the targeting of the antibody to a specific cellular compartment. This can be accomplished by signal peptide sequences specific for, e.g., the endoplasmic reticulum, mitochondria, cell nucleus, cytoplasm in general, membranes, etc. The function of proteins at the target sites may be modulated, leading to alteration of susceptibility to viral infection and replication, modifying cell surface receptors, altering light receptors, affecting mitosis, etc. *See* biotechnology; differentiation; intrabody. (Chames, P. & Baty, D. 2000. *FEMS Microbiol. Lett.* 189:1.)

intrachromosomal recombination May be responsible for deletions and duplications. *See* ectopic recombination; sister chromatid exchange; transposition.

intraclass correlation Form of analysis of variance used for the estimation of heritability on the basis of variances between and within classes, e.g., in the progeny of a larger number of different males mated to a smaller number of females, each of which has several offspring. Thus, a comparison can be made of the variance within the litter of a female mated (sired) to the same male (within "sires") and the variance among the total offspring of individual males mated to different females (between "sires"). The intraclass correlation of the sires is one-fourth of the heritability for the trait considered because each male contributes half of the genetic material to the offspring quantitatively analyzed, and these again will contribute only half of their chromosomes by their haploid sperm. The procedure of calculation is illustrated by a hypothetical example below. *See* correlation; heritability; variance; variance analysis. (Hill, W. G. & Nicholas, F. W. 1974. *Biometrics* 30[3]447.)

Procedure for calculating heritability based on intraclass correlation. The mean scores of the offspring are represented in the body of the table. (Y stands for individual or group measurements)

Males →	**A**		**B**		**C**		**D**	
Females ↓	Y_i	$(Y_i)^2$	Y_i	$(Y_i)^2$	Y_i	$(Y_i)^2$	Y_i	$(Y_i)^2$
E	2	(4)	3	(9)	3	(9)	4	(16)
F	3	(9)	2	(4)	4	(16)	3	(9)
G	3	(9)	3	(9)	4	(16)	2	(4)
H	4	(16)	2	(16)	3	(9)	4	(16)
I	3	(9)	4	(16)	4	(16)	6	(36)
J	5	(25)	2	(4)	5	(25)	5	(25)
Sum Y_i	20		16		23		24	
Sum Y_i^2		72		46		91		106

Sum (Σ) all   Sum $(\Sigma\Sigma)$ all n (all measurements) = 24,
  $Y_i = Y\ldots = 83$, $Y_i^2 = 315$, n_i (no. of families of females) = 6
Correction factor (C) = $Y^2\ldots/n = 83^2/24 = 6889/24 = 287$
Uncorrected Sum of Squares $(\Sigma Y_i^2/n_i)$
  $= (20^2 + 16^2 + 23^2 + 24^2)/6 = 1761/6 = 293.5$
Sum of Squares "Between Males" = $SS_S = (\Sigma Y_i^2) - C$
  $= 293.5 - 287 = 6.5$
Sum of Squares "Within Males" = $SS_P = \Sigma\Sigma Y_i - (\Sigma Y_i^2/n_i)$
  $= 21.5$
Mean Square (MS) is SS/df
(The lower index S stands for "sires"; the lower index P is
  for progenies)

	Analysis of Variance		
	df	SS	MS
Between Males (SS$_S$)	3	6.5	2.17
Within Males (SS$_P$)	20	21.5	1.075

$\hat{\sigma}^2{}_S = \dfrac{\text{MSS} - \text{MSP}}{n_i} = \dfrac{2.17 - 1.075}{6}$
  $= 0.1825$ (the male's variance component)
$\hat{\sigma}^2{}_P = \text{MSP} = 1.075$ (the progeny variance component)

The Males Intraclass Correlation (r_1), $(\hat{\sigma}^2{}_S/(\hat{\sigma}^2{}_S + \hat{\sigma}^2{}_P)\bullet$ is equal to 1/4 of the heritability; hence

$\qquad h^2 = 4\hat{\sigma}^2{}_S/\hat{\sigma}^2{}_S + \hat{\sigma}^2{}_P) = 4[0.1825/(0.1825 + 1.075)] = 0.58$

(heritability is denoted by h^2 for historical reasons, but it is not the
  second power of an entity)

intracytoplasmic sperm injection   *See* ICSI.

intra-Fallopian transfer   Of gametes (GIFT) or zygotes (ZIFT), infertility treatment procedures whereby the spermatozoa and the mature oocytes or in vitro–generated zygotes, respectively, are placed surgically into the fallopian tube of the female where fertilization and/or segmentation may proceed. These procedures have higher rates of success for conception than in vitro fertilization, but they also lead frequently to twinning if more than a single egg or zygote is used. *See* ART; in vitro fertilization; PROST; TET. (Klonoff-Cohen, H., et al. 2001. *Fertil. Steril.* 76:675.)

intragenic recombination   Rare event because alleles of a locus are very close to each other. Intragenic reciprocal

recombination is expected to yield wild-type and double mutant recombinants that can be verified only if flanking nonallelic markers are available. Ideally these outside markers should not be more than 5–10 map units at both sides of the locus. The mutant alleles are by m^1, m^2, and m^3, respectively, and p, t, and a are outside markers. The + sign indicates nonmutant. The following crosses are required to determine linear order of the m alleles in a simple case:

Test crosses	Reciprocal recombinant phenotypes
$\dfrac{p\ m^1\ t^+\ a}{p^+\ m^2\ t\ a^+}$ X $\dfrac{p\ m\ t\ a}{p\ m\ t\ a}$	$p\ m\ t\ a^+$ and $p^+\ m^+\ t^+\ a$
$\dfrac{p^+\ m^2\ t\ a^+}{p\ m^3\ t^+\ a}$ X $\dfrac{p\ m\ t\ a}{p\ m\ t\ a}$	$p^+\ m^+\ t^+\ a$ and $p\ m\ t\ a^+$
$\dfrac{p\ m^1\ t^+\ a}{p^+\ m^3\ t\ a^+}$ X $\dfrac{p\ m\ t\ a}{p\ m\ t\ a}$	$p\ m^+\ t\ a^+$ and $p^+\ m\ t^+\ a$

Among the recombinants, the m phenotype indicates double recessive alleles in the same strand, whereas the m^+ phenotype is an indication of recombination between the two recessive alleles present in the heterozygous parent that is test-crossed. According to these results, the order of the mutant alleles and the markers must be $\underline{p - m^3 - m^1 - m^2 - t - a}$ and none of the recombinant classes are supposed to be contaminants because the markers are consistent with the recombination events suggested. The double mutant recombinants may be further tested by recombination to yield the two mutant classes. Intragenic recombination is a rare event (~0.02 to 0.000001 or less) close to the range of mutation frequency. (Whittinghill, M. 1950. *Science* 111:377.)

intragenic suppressor   *See* suppressor gene.

intrasteric regulation   Internal sequences of protein that resemble substrate tracts (pseudosubstrate) act directly at the active site in contrast to the allosteric effects in which the allosteric molecule bears no similarity to the substrate and attaches to the protein at a site different from the active site. Intrasteric control is a means of autoregulation. *See* allosteric control; autoregulation. (Kobe, B., et al. 1997. *Adv. Second Messenger Phosphoprotein Res.* 31:29.)

intrauterine fertilization (IUI)   Sperm is deposited directly into the uterus, bypassing the cervix (the anterior part of the uterus, which may form a barrier to the passage of the sperm). This type of medical intervention may overcome human infertility. *See* ART.

intravasation   Entrance of an extraneous substance into the blood vessels. Metalloproteinases, receptor urokinase plasminogen activators (uPA), and their inhibitors may affect the process. Metastasis of cancer cells may depend on intravasation. *See* extravasation; metalloproteinases; metastasis; urokinase.

intrinsic rate of natural increase *See* age-specific birth and death rates.

intrinsic terminators Of transcription in prokaryotes, require no cofactors (are rho-independent) for the termination of transcription. *See* transcription termination in prokaryotes.

introgression Transfer of genes from one group of the species into another. The two populations may inhabit the same area (sympatric) or they have only occasional contacts because they live in a different area (allopatric). After the initial crossing, the mating continues generally only within the group; therefore, only a few of the "borrowed" genes are maintained. *Marker-assisted introgression* uses molecular tools to monitor the transfer/presence of the desired genetic regions, especially for quantitative trait loci. *See* marker-assisted selection; QTL. (Anderson, E. 1949. *Introgressive Hybridization*. Wiley, New York; Saetre, G. P., et al. 2001. *Mol. Ecol.* 10:737.)

introgressive hybridization Accomplishes introgression. *See* introgression.

intron homing (retrohoming) Process of insertion of an intron at a particular site within an intronless cognate sequence. Accordingly, some introns of yeast mitochondria also serve as mobile genetic elements besides having ribozyme function. The homing introns, mitochondrial or nuclear, are endonucleases with similarities to restriction enzymes, but their recognition sequence is much longer. Intron homing may utilize either the double-strand break-repair pathway common in genetic recombination and yielding reciprocal crossover and noncrossover products or a synthesis-dependent strand-annealing process that does not produce flanking crossing over. *See* introns; mtDNA; ribozyme. (Mohr, G., et al. 2000. *Genes & Development* 14:559; Belfort, M., et al. 2002. p. 761 in Mobile DNA II, ASM Press, Washington, DC.)

intronless paralogs Genes inserted into the genome by retrotransposition of processed mRNA. *See* processed gene; retrotransposon. (Schimenti, J. C. 1999. *Mamm. Genome* 10[10]:969.)

intron slippage (intron sliding) The position of introns (intron-exon boundary) may vary among homologous genes of evolutionarily related species by one or a dozen or more nucleotides. *See* introns. (Wistow, G. J. & Piatigorsky, J. 1990. *Gene* 96:263.)

introns Nucleotide sequences within a gene that are not represented in the mature mRNA transcripts of that gene (intervening sequences, IV). Introns are transcribed but not translated into protein that will not be part of the products of exon-coded genes or will not be included in the final RNA encoded by the gene. Introns separate the coding sequences of the exons. They occur in the majority of eukaryotic genes, including genes of mitochondria and chloroplasts, and also in viruses of eukaryotes, but they are exceptional in prokaryotic genes. In land-plant plastid DNA, ca. 20 introns are present and their size varies from 400 to over 1,000 bp. The average intron size in mice is about $1,800 \pm 300$; in humans about $3,000 \pm 550$. In the X chromosome, the intron size is substantially longer. In imprinted genes, the intron number, and particularly size, is much smaller.

The RNA maturases in mammals are actually intron sequences within protein genes and assist the splicing of pre-mRNA transcripts. In algae cpDNA, introns are quite variable in number and size; in *Euglena*, 149 introns were detected. In the red alga *Porphyra*, there were no introns in the cpDNA. In *Chlamydomonas*, the insertion site of an intron corresponds to an *E. coli* rRNA domain. Introns may appear as latecomers to eukaryotic organelle genes, since in multiple evolutionary lines the introns are nonhomologous either by sequence or location. The thymidylate synthetase genes of T4 bacteriophage has an intron however, and the archebacterial tRNA^{Leu} and the large ribosomal subunit genes have introns. One of the largest introns (64 kb) was found in the human thyroglobulin gene involved in the regulation of energy metabolism and in various forms of goiter disease. The overall profile of a gene can be represented as

5′-enhancer-promoter — transcription initiation site — leader

— exon◆**intron**◆exon-termination signal-3′

The origin of introns is unclear. Some arguments favor their ancient evolutionary presence preceding the divergence of eukaryotes from prokaryotes. Some introns may be located at critical regions of the gene, dividing it into functional domains or separating α-helices from β-sheets. It has been suggested that increased replication rate is inversely proportional with the number of introns. Indeed, introns are rare in prokaryotes and reduced in number in yeast.

Introns may regulate genes controlling complex developmental pathways. True, large homeotic genes of *Drosophila* seem to have more introns than other simple genes. The greater density of introns within genes may be promoted by sexual reproduction, which enables them to propagate in a selfish manner through gametes of both parents. Introns could protect against the deleterious consequences of recombination, because if such events occur within introns rather than within exons, no harm may ensue to the coding capacity. Recombination frequency is negatively correlated with intron length. Introns could also protect against illegitimate recombination by interrupting homeologous tracts at random sites. Although these problems are not yet resolved, the mechanics of transactions concerning introns have made substantial progress.

The number of introns varies a great deal (Llopart, A., et al. 2002. *Proc. Natl. Acad. Sci. USA* 99:8121). Some genes — e.g., histones, human α- and β-interferon genes — have no introns, whereas the γ-interferon gene has several. The large (~2 Mbp) human dystrophin gene has more than 60 introns. (Dystrophin is a muscle protein deficient in muscular dystrophy.) The average size of an intron generally varies between 75 and 2,000 bp, but some introns are several times as long or no more than about a dozen nucleotides. No introns occur in the 5S, 5.8S, U RNA, 7SL, and 7SK RNA genes. The number of introns is generally small in organisms with compact genomes like *Drosophila* and budding yeast, but fission yeast has a larger number. Introns make about 24% of the human genome. Interestingly, the mitochondrial genes of budding yeast have relatively more introns than the nuclear genes.

Introns are removed from the primary transcripts of the genes during processing. Removal of introns is essential for the expression of genes. Group II introns have six domains,

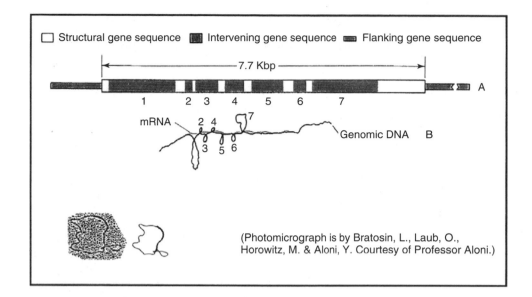

□ Structural gene sequence   ▦ Intervening gene sequence   ▭ Flanking gene sequence

(Photomicrograph is by Bratosin, L., Laub, O., Horowitz, M. & Aloni, Y. Courtesy of Professor Aloni.)

and domain 1 (D1) and D5 are essential for splicing. Thus, the mRNA becomes much shorter than the primary transcript. When the mRNA is hybridized to the genomic coding strand of DNA, the latter reveals loops at the positions where the introns were removed from the mRNA. The chicken ovalbumin gene contains seven introns and loops corresponding to their location, and length can be detected by electron microscopy (map by Dugaiczyk et al. 1979. *Stadler Symp.* 11:57, and interpretative drawing by Rédei, 1982). Alternative mRNAs can be obtained from the same DNA sequence by controlling the length, number, or pattern of exons used, depending on the site of (1) initiation of transcription, (2) the alternative sites of polyadenylation signals, or (3) the selective retention of particular exons in the mRNA. By these mechanisms the products of single genes can be diversified during development or differentiation. Alternative splicing may be accomplished by using more than a single promoter, resulting in alternative long or short transcripts, depending on the site of transcription initiation. Alternative polyadenylation signals at more than a single location downstream may truncate the transcript at the 3'-end. An example for mechanism (1): alternative promoters of the exons are bracketed; introns are in bold numbers and parentheses. Alternative splicing occurs in various eukaryotic genes and various viruses of eukaryotes. In the small genomes of the latter systems, a single transcript may permit the production of several proteins. In the determination of correct splicing, both cis- and trans-acting proteins cooperate. When the introns are removed, the exons are *spliced* together and the continuity of the mRNA is restored. There are two splice sites: the *upstream donor site* and the *downstream acceptor site*. When the invariant bases (shown in bold numbers below) are altered, splicing generally fails.

TATA1–[1]–(**1**)–TATA2–[2]–(**2**)–[3]–(3)–[4]–(**4**)–[5]–(**5**)–[6] **DNA**

Example of splicing mechanism (1):

    mRNA1 : –[1]–[4]–[5]–[6]– alternative transcripts

    mRNA2 : –[2]–[3]–[4]–[5]–[6]–

Example for alternative truncation (mechanism 2):

TATA–[1]–(**1**)–[2]–(**2**)–[3]–(**3**)–[4]–(**4**)–*poly A signal I*

   –[5]–[6]–(**6**)-*poly A signal II*–**DNA**

mRNA1 : –[1]–[2]–[3]–[4]–AAAAA or

mRNA2 : –[1]–[2]–[3]–[5]–[6]–AAAAA

Example for mechanism (3), alternative splicing between identical promoter and polyA signal:

TATA 1–[1]–(**1**)–[2]–(**2**)–[3]–(**3**)–[4]–(**4**)–[5]–(**5**)–[6]–(**6**)— **DNA**

mRNAs :– [1]–[2]–[3]–[4]–[6]–AAA, –[1]–[3]–[4]–[5]–[6]–AAA
        ↑   ↑   ↑   ↑

  –[1]–[2]–[4]–[5]–[6]AAA
    ↑   ↑

Other neighboring bases may also effect the efficiency of splicing. In animal nuclear genes, an A residue in the vicinity of the splice site is required, but its position does not have to be absolutely fixed relative to the splice site. In yeast, there is an absolute requirement for a conserved UACUAAC tract within 6 to 59 nucleotides upstream from the 3' splice signal. Some genes have more than a single splice site pair that facilitates alternative splicing and the assembly of different mRNAs. Although introns were originally regarded as junk DNA, it is known now that some introns are translated but have independent functions from the exons. Although some have roles in the processing of the transcript shared by the neighboring exons, others seem to have regulatory functions.

On the basis of the structural information about the splice sites, some special introns have been classified into group I and group II (see table next page). The principal characteristics of group I introns are their splicing does not always require a protein enzyme; rather a short internal sequence facilitates their folding and splicing is initiated by an extraneous guanosine or a phosphorylated form of it. Group II introns do not have the conserved internal sequences, yet they are capable of fold-back pairing; the splicing requires an intrinsic signal rather than an extraneous guanine. The spliced group II

2′ OH
|
5′ P-X-O-p-X-O-p-X-O-p-X-O-p-X-O-p-X-O-p-X-OH   Intron in the RNA

lariat →   **5′ P**
            2′-5′ phosphodiester bond in the lariat
         2′ O
            | Branch point
         X-O-p-X-Op-X-O-p-X-O-p

The intron-exon border sequences are conserved within groups of different classes of RNA transcripts (after Cech, T. 1986. *Cell* 44:207). The bold letters indicate constant bases, the ↓ indicates splice sites, Py stands for any pyrimidine, Pu for any purine, and X means any base

Intron	5′ Splice Junction	3′ Splice Junction
Common nuclear pre-mRNA	CPuG↓**GU**$_G^U$ AGU	(PY)$_n$**AG**↓X
Yeast nuclear pre-mRNA	↓**GUAUGU**	(PY)$_n$**AG**↓X
tRNA	X↓X	X↓X
[1]Introns Group I	U↓	G↓
[2]Introns Group II	↓**GUGCG**	(Py)$_n$**AU**↓
Euglena plastid mRNA	↓**GUG**$_U^C$ **G**	(Py)$_n$**A**$_C^U$↓

[1]Nuclear rRNA genes in some lower eukaryotes, mitochondrial and plastid rRNA genes.
[2]Yeast mitochondrial genes for cytochrome oxidase and cytochrome b.

introns form a *lariat* (similar to that of a tethering rope); the 5′ P-end forms a phosphodiester bond with the 2′-OH group of a nucleotide within the chain at some distance. The loop itself may have three nucleotides (GpApA). The splicing of group I introns requires *transesterification* (phosphodiester linkage exchanges). The transesterification takes place without first severing the bonds. The self-splicing is mediated by ribozymes, which are RNAs that have enzyme-like catalytic functions.

Ribozymes facilitate the formation of the proper configuration of the RNA transcript as well as possibly functioning similarly to endonucleases. Actually, the aI2 ribonucleoprotein of the *COX1* yeast mitochondrial gene catalyzes the cleavage of the DNA target recognized by complementarity of the sequences within the intron RNA. After cleaving the DNA, the aI2 protein reverse-transcribes the intron RNA using the 3′ end of the DNA as a primer. Actually, the aI2 RNA cleaves the sense strand of the DNA, whereas the aI2 protein cuts the antisense strand and the latter also boosts the activity of the aI2 ribozyme. The intron assumes a complex secondary structure by base pairing of some complementary sequences separated by nonpairing tracts. The extraneous guanosine makes a nucleophilic attack (an electron-rich molecule reacts with an electron-poor molecule that is willing to accept electrons) at the exon-intron boundaries, resulting in nucleotide chain breakage at the site and intron exclusion. Not all of the group I introns are able to self-splice without the help of *maturase* proteins encoded by some introns within some yeast mitochondrial genes. The self-splicing of group II introns is somewhat similar only to that of group I. In group II, there is no guanine participation. The 2′-OH group of an adenine within the intron initiates the nucleophilic attack. After appropriate

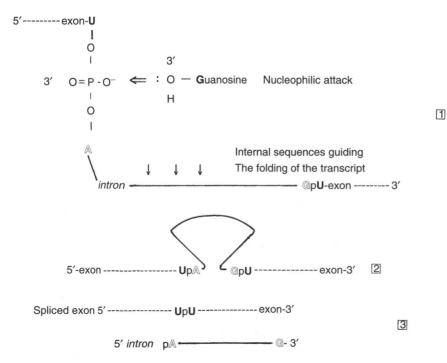

Self-splicing of a group I intron. (1) A guanosine (or guanylic acid) conducts a nucleophilic attack against a phosphate group near the point of juncture, and transesterification follows later at both the 5′ and 3′ junctions. This is accompanied by the recognition of internal sequences ↓↓↓ that guide the folding. (2) Exon termini are in bold and the ends of the intron are in outline letters. (3) Spliced exon and intron are displayed.

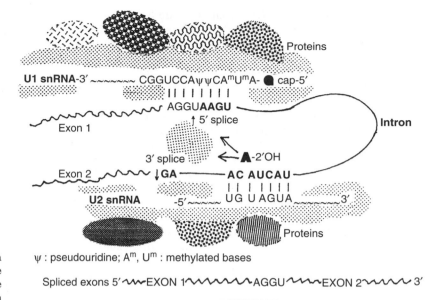

ψ : pseudouridine; A^m, U^m : methylated bases

U1 and U2 snRNAs facilitate the proper configuration of the folding of the transcript and thus identify the correct splicing sites. The A-2'OH nucleotide then leads the nucleophilic attack, which is followed by transesterification and ligation of the exon.

folding to bring the 5' and 3' junctures to each other's vicinity, the exon is spliced by transesterification and the intron is released as a lariat (*see* lariat). The primary RNA transcript of nuclear pre-mRNAs is capped. Cleavage takes place at the 5' end of the intron upstream of a ↓pGU pair as indicated by the arrow. The free 3' end of the intron then forms a lariat with an internal A residue; in another step, the intron with the lariat is released and the 3' end of the first exon is ligated to the 5' end of the next exon. This outcome is the same as shown above. The splicing of the nuclear pre-mRNAs requires, however, a more complex machinery involving spliceosomes.

The spliceosomes are assembled from short U-rich RNAs, ranging in size from 65 (U7) to 217 (U3) nucleotides, and a set of different proteins. The most common U RNAs in the mammalian nucleus are U1, U2, U4, and U6. U3 RNA is involved in RNA processing in the nucleolus. U7 RNA assists in the formation of the 3' end of the histone mRNAs that have no introns. The corresponding U RNAs are highly homologous even among taxonomically very different organisms such as humans and dinoflagellates. The homologous U RNAs among vertebrates are almost identical (ca. 95% similarity). The spliceosomes contain a common set of seven proteins (generally designated by capital letters [B, A', etc.]) or by molecular weight (e.g., 59K, 25K) and they may have a few specific proteins. About 1% or less of the introns use AU and AC as terminal nucleotides (rather than the most common GU and AG). In plants (crucifers), AU-AA introns may occur exceptionally.

Some of these introns also use U12 snRNA for processing. The intron-encoded U22 small nucleolar RNA facilitates the processing of 18S ribosomal RNA in *Xenopus* oocytes. U1 snRNA specifically binds to the 5'-splice site of the intron and protects this region from RNase attack, but it does not cleave at the splice site. The U2 snRNA complex does the same near the 3'-end of the intron around the adenosine residue whose 2'-OH is involved in the formation of the lariat. The spliceosomes are ellipsoid complexes of about 25 × 50 μm dimensions that actually bring together the splicing sites to make the *cut-and-paste* process possible. These bindings generally require some degree of complementarity, but it is not complete. Besides these two spliceosomes, others (U2, U4, U6) and a number of binding proteins may be involved in facilitating and stabilizing the binding.

The position of *introns in tRNA* genes is next to the third (3') base of the anticodon triplet, with a single base in between. Yeast has about 40 nuclear-coded tRNAs, and 10% of them have a single intron containing from 14 to 46 bases. This regular location is somewhat enigmatic because tRNA^Tyr genes of yeast with deleted intron sequences are still transcribed, processed normally, and function. In the primary transcript, the intron has, however, a triplet that is complementary to the anticodon. For the correct splicing it appears that the various loops and arms of the tRNA cloverleaf are important. The first step in splicing of tRNA transcripts is a cleavage by an endonuclease at the ends of the intron. In yeast, the 5' terminus of the exon

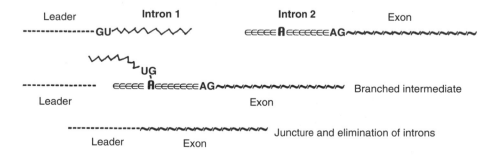

is then represented by an OH group and the 3′ terminus by a 2′,3′ cyclic phosphate. A cyclic phosphodiesterase breaks up the ring and generates a 3′-OH and a 2′-O-(PO₃) terminus. The 5′-OH end is phosphorylated by a kinase, requiring ATP. Thus, the termini become ready for joining by an *RNA ligase* enzyme. Essentially similar reactions take place in splicing nuclear tRNAs in mammals and plants.

The splicing mechanisms shown and discussed above involve splicing exons within the same primary RNA transcripts. In some instances, exons located originally in different transcripts, even from different chromosomes, may be spliced by *transsplicing*. Transsplicing can generate mRNAs from leader sequences remote to the exons. Transsplicing is common in the protozoa *Trypanosomes* where different exons can be joined to a single 35 bp leader sequence. (The exons are scattered among the approximately 100 chromosomes.) The 3′-end of the leader and the 5′-end of the coding exons can be joined. Since the 5′-end of the intron carries the generally conserved 5′-GU sequence, the remote other intron, preceding the remotely transcribed coding exon, terminates at 3′ with the conserved AG bases and also has an adenosine nucleophile somewhere in the vicinity of its 5′ terminus. Then the two originally remote introns can be brought together by a 5′-to-2′ bond as a branched molecule; subsequently, both introns are eliminated and the leader is joined to the coding exon as outlined in the diagram p. 641. Transsplicing also occurs in other eukaryotes, e.g., in the nematode *Caenorhabditis elegans* and in the chloroplast genes of plants. The open-reading frames of class II introns have sequences reminding to reverse transcriptases of the type of that retroposons. There is no evidence, however, that these would function as reverse transcriptases. Introns and spliceosome sites display evolutionary variations and are suitable for tracing phylogenetic relationships among species. Group II introns can be targeted for insertion into a ~14-nucleotide region of target DNA of prokaryotes. Introns inserted in the antisense orientation cannot splice and disrupt the gene. Introns inserted in the sense orientation may serve for selective regulation of gene expression when linked to inducible promoters. *See* alternative splicing; branch-point sequence; DEAD-box proteins; DEAH-box proteins; endonuclease; gene; intein; intron group III; ligase; mtDNA; *Neurospora* mitochondrial plasmids; ribozymes; RNA maturase; snoRNA; speckles, intranuclear; spliceosomal intron; splicing; spliceosome; SR motif; transcription. (Bassi, G. S., et al. 2002. *Proc. Natl. Acad. Sci. USA* 99:128; Clark, F. & Thanaraj, T. A. 2002. *Hum. Mol. Genet.* 11:451; <http://mcb.harvard.edu/gilbert.EID>; <http://isis.bit.uq.edu.au>; exon-intron database: <http://intron.bic.nus. edu.sg/exint/newexint/ exint.html>; organellar introns: <http://wnt.cc.utexas.edu/ ~ifmr530/ introndata/main.htm>.)

introns group I   Self-splicing introns of some ribosomal genes using mechanisms different from the pre-mRNA or the majority of tRNA genes. *See* introns for characterization; mitochondrial genetics for introns as mobile elements and plasmids; spliceosomal introns.

introns group II   Introns are large ribozymes in a few mitochondrial, chloroplast, fungal, and bacterial genes differing in structure and splicing mechanisms from the common introns (type I) of eukaryotes. They are also self-splicing and are capable of intron homing. *See* intron homing; introns; mitochondrial genetics; spliceosomal introns.

introns group III (twintron)   Relatively short, group II introns inserted within the boundary of group II introns in the cpDNA of *Euglena*. *See* ctDNA.

intron sliding   Alternative type of splicing used by cryptic splice sites generally caused by the presence of SINE elements. *See* introns; SINE; splicing. (Rogozin, I. B., et al. 2000. *Trends Genet.* 16[10]:430.)

intussusception   Insertion of interstitial tissue into the lumen of existing vessels.

invader   Molecular procedure for DNA and RNA quantification. It is suitable and highly sensitive discrimination between mutant and wild-type forms and SNPs. (Kwiatkowski, R. W., et al. 1999. *Mol. Diagn.* 4:353; Olivier, M., et al. *Nucleic Acids Res.* 30[12] e53.)

invagination   Folding inward.

Invagination.

invariance   Reciprocal value of the variance. *See* variance.

invasin   Integrin-binding protein of *Yersinia* bacteria assisting infection. *See* integrins; *Yersinia*.

invasive   Penetrating cells (and commonly causing their destruction). Also, migration of cancer cells. *See* metastasis; RAGE.

inverse PCR   *See* inverse polymerase chain reaction.

inverse polymerase chain reaction   Can be used for molecular analysis of flanking regions of a target:

Restriction fragment including flanks and target.

The fragment shown above is circularized by its cohesive ends and cut with a restriction enzyme within the target. As a consequence, a head-to-tail association of the flanks is produced:

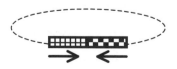

This sequence is amplified by PCR for nucleotide sequencing. The two flanks may be separated if an appropriate restriction endonuclease site is known at or near their boundary. *See* capture PCR (CPCR); PCR; tail PCR; target.

inversion Chromosomal segment turned around by 180°; (β) centromere.

Thus if the **NORMAL ARRANGEMENT** of the genes is: A B C D E∘F G H
after **PARACENTRIC INVERSION** the order becomes: A Ɑ Ɔ B E∘F G H
after **PERICENTRIC INVERSION** the sequence is: A B Ɔ E∘ꟻ G B A

Such chromosomal rearrangements require two breaks, both of them in one chromosome arm in case of *paracentric inversions*, and one in each arm across the centromere in case of *pericentric inversions*. Pericentric inversion may alter the arm ratio of the affected chromosome as seen above. Within a single chromosome more than one inversion may be present. These multiple inversions may be independent or may be overlapping, or a shorter inversion may be included within a longer one.

If the chromosomes are well suited for cytological analysis, the various types can be identified with the aid of a light microscope. Inversions as such may not have much phenotypic consequence, unless they cause "position effect," influencing the expression of the gene because they interrupt either the

Two-strand paracentric inversion heterozygote displays double chromatid bridge and two chromatid fragments. Because the chromatid tie keeps the crossover chromatids in the middle of the cell, they fail to be incorporated into the egg and thus usually female sterility is prevented. (Courtesy of A. H. Sparrow.)

coding or the nontranslated regulatory sequences. These effects may be serious in the relatively rare inversion homozygotes. During the early years of genetics, inversions were thought

Paracentric inversion heterozygotes produce 50% defective gametes when crossing over takes place within the inverted chromosomal segment. Since the frequency of crossing over depends on the length of the segment inverted, the frequency of the defective gametes varies from 0 to 50%. At the top, an inverted and a normal chromosome are shown. Such chromosomes can pair only if the inverted strands form a loop and thus gene-by-gene alignment becomes possible. The configuration of one inverted and one normal strand is shown on the second line at left. In the center of that line, the pairing and crossing over ($\Leftarrow$) are diagrammed. As a result of recombination, one of the crossover strands becomes dicentric and deficient for marker (D) and duplicated for marker (A/a). The other crossover strand becomes acentric, and deficient for A/A and duplicated for D/d. At meiotic anaphase I, the crossing over results in a chromatid bridge because the chromatids are tied together even when anaphase separation proceeds. The tie may hold together the crossover chromatids and they will not be able to get to either pole; rather they will remain in the middle of the metaphase plane, or if the tie breaks, depending on the site of the breakage, additional duplication and simultaneous deficiency may be generated. In animals and plants, after recombination of paracentric inversion heterozygotes, generally only viable (normal or inverted) chromosomes enter the egg and thus female sterility is usually not observed. In the males (where polarity does not occur during gametogenesis), if recombination takes place within the inversion, 25% of the gametes contain only normal chromosomes and 25% carry inverted chromosomes containing complete chromatin. But the other 50% result in duplication deficiency or deficiency of the genes, and such gametes are generally sterile or result in embryo lethality. Recombination in pericentric inversion heterozygotes does not result in a dicentric bridge, but half of both male and female gametes may become defective.

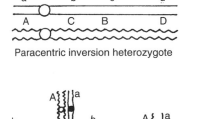

Paracentric inversion heterozygote

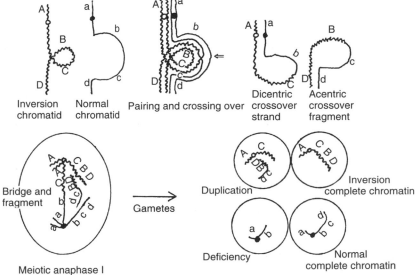

Inversion chromatid   Normal chromatid   Pairing and crossing over   Dicentric crossover strand   Acentric crossover fragment

Bridge and fragment

Meiotic anaphase I   Gametes

Duplication   Inversion complete chromatin

Deficiency   Normal complete chromatin

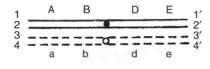

Inverted chromosome segment is marked with lowercase bold letters (chromatid ends are numbered).

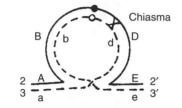

Chromosome pairing at the 4-strand stage displays a loop. Only two of the chromatids (involved in a a chiasma) are shown for the sake of simplicity.

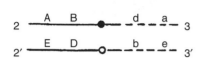

Crossing over in paricentric inversion heterozygotes does not result in a chromatid bridge and fragment, yet the recombinants have duplications and deficiencies.

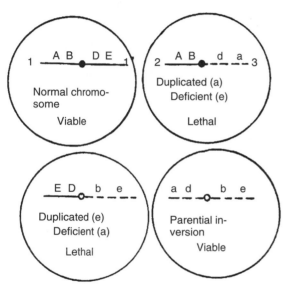

The four products of meiosis of pericentric inversion heterozygotes are shown at left. Two of the gametes carrying the intact parental chromosomes are viable; the other two are lethal because of duplications and deficiencies. There is no difference between the male and female gametes; unlike in paracentric inversion heterozygotes, they are affected equally. In a population the frequency of sterility depends on the frequency of recombination within the inverted segment.

Recombination in a heterozygte for a pericentric inversion.

to be "C factors," crossing-over inhibitors. Actually, inversions inhibit crossing over in inversion heterozygotes only in the vicinity of the breakage points where the rearrangement prevents intimate pairing of the chromatids. Crossovers are not observed primarily because the strands involved in recombination are usually not transmitted through the paracentric inversion females; thus, they do not generate defective gametes. The consequences for the sperm are not the same for paracentric inversion heterozygotes because the microspore tetrad is not linear and the crossover chromatids are not tied by the inversion bridge in such a way that they are not included in any microspore. Therefore, in the males all four products of meiosis could potentially be transmitted; however, 50% of the gametes are still defective.

In pericentric inversion, heterozygote duplication-deficient gametes are formed (without a bridge) in both females and males if crossing over takes place within the inverted segments during meiosis. Thus, crossing over may occur, but the crossover gametes or zygotes may not be viable. In animals, the consequence of duplication deficiencies may be different from that in plants. In plants, the defective gametes are usually prevented from fertilization because of inviability,

whereas in animals the defective sperm may be capable of fertilization, but the offspring resulting from such a mating may not survive. The cytological consequences of crossing over within inverted chromosome segments are diagrammed and shown by a photomicrograph p. 639. Crossing over within pericentric inversions has the same genetic consequence as that of paracentric inversions, namely, 50% of the gametes formed by meiocytes that have suffered recombination are duplication deficient. That is, some of the genes are present in the same strand twice and others are entirely absent, and therefore generally inviable. Inviability may be gametic or zygotic. In the gametophytes of plants, generally the former is the case, whereas in animals the latter is prevalent. Crossing over in pericentric inversion heterozygotes cytologically differs from that in paracentric inversions.

In the former, dicentric chromosomes and acentric fragments are not generated by recombination. Double crossing over within the same inverted paracentric segment involving two homologous chromatids does not produce defective gametes (see diagram on next page). Three-strand double crossing overs yield, however, both acentric fragments and dicentric chromosomes besides the two parental ones. Four-strand double

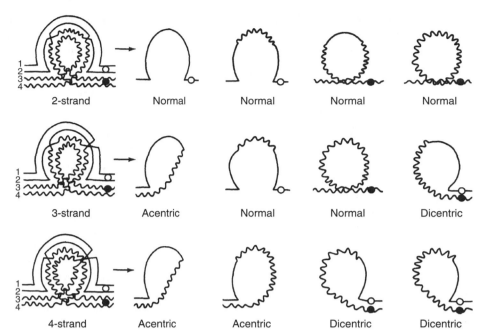

2-strand → Normal   Normal   Normal   Normal

3-strand → Acentric   Normal   Normal   Dicentric

4-strand → Acentric   Acentric   Dicentric   Dicentric

Double crossing over within paracentric inversions. Only two-strand double crossing over results in two complete chromosomes. Pericentric inversion heterozygotes do not yield dicentric strands, yet the overall genetic consequences of double recombination are the same as in pericentric inversion heterozygotes.

crossing over results in the formation of two acentric fragments and two dicentric chromosomes and generally all gametes thus become defective (see figure above).

Paracentric inversions may play an important role in speciation because they may cause hybrid male sterility and thus partial sexual isolation. W. S. Stone has estimated that during the evolution of the approximately 2,000 species of *Drosophila*, about 350 million paracentric inversions might have occurred, and from these tens of thousands have been permanently fixed, and the fate of another similar in number is still undecided. Pericentric inversion may also play a role in speciation because of the complete sexual isolation and the preservation of the inverted segment as supergenes refractory to recombination. The descendence of inverted chromosomes in natural populations can be determined on the basis of banded chromosomes or restriction fragment patterns (RFLP).

The frequency of inversions in human populations is less than 1% per birth. The actual rate of occurrence may be much higher because many the afflicted fetuses are spontaneously aborted. Inversions have been added to various mutation tester strains of *Drosophila* so that the new mutations can be identified within a particular chromosome without confounding the results by recombination because inversions eliminate the crossovers. Short inverted nucleotide repeats are found at the termini of insertion/transposable elements. Inverted duplication of nucleotide sequences form palindromes, facilitating the formation of double-stranded sequences in single-stranded nucleic acid chains; they have determining roles in conformation of the molecules, such as in tRNAs. In *Drosophila buzzatii*, evidence was obtained that indicated inversions may arise by recombination either within the target site duplications or between two copies of the *hobo* transposable element. Chromosomal (DNA) inversions would be expected to produce "retroproteins" with altered three-dimensional structure. The stability of these retroproteins may or may not differ from the natural sequence and may or may not affect the function of the proteins. *See* conformation; hybrid dysgenesis; mutation, chromosomal; palindrome; recessive lethal tests in

Drosophila; speciation; unbalanced chromosomal constitution. (Ranz, J. M., et al. 2001. *Genome Res.* 11:230.)

inversion, oligonucleotide The linkages are 3′3′ or 5′5′ and thus have inverted polarity within the normal nucleotide tract. They are very resistant to exonucleases and have a longer half-life than normal oligonucleotides, yet they may not disturb the Watson-Crick structure. These may be used for oligonucleotide therapy.

inversion, paracentric Involves only one arm of a chromosome. *See* inversion.

inversion, pericentric Inversion is spanning across the centromere. *See* inversion.

inversion, phage and bacteria One group of enzymes are invertases (transposases) mediating inversions between inverted terminal repeats of transposable elements. The other classes are the integrases. The inversion may affect an invertible promoter or the entire gene. The invertase of phage MU (*gin*) is positioned outside the inversible 3000 bp G region flanked by 34 bp inverted repeats. In one orientation, the host specificity genes Sv and U (→) permit the attachment of Mu to *E. coli*; when inversion (←) takes place, the phage can be adsorbed to other host bacteria. A similar system is found in the phase variation of *Salmonella*. *See* phase variation; resolvase; transposase. (Roth, J. R. & Schmid, M. B. 1981. *Stadler Symp.* 13:53; Johnson, R. C. 2002. p. 230 in Craig, N. L., et al., eds. Mobile DNA II. *Amer Soc. Microbiol. Press*, Washington, DC.)

inversion loop *See* inversion.

invertases Proteins (Cin, Hin, Gin) involved in viral transposition and bearing extensive homology with the N-terminal domains of the resolvases of transposons. In biochemistry the enzyme sucrase, which hydrolyzes sucrose to fructose and glucose, is also called invertase. *See* Cin; resolvase; Tn.

inverted repeat *See* direct repeat; repeat, inverted.

invertrons Linear mobile elements in plasmids and mitochondrial DNA. It seem that 5′-linked proteins encode DNA and RNA replication and integration functions. *See* *Agrobacterium*; transformation, genetic. (Hermanns, J. & Osiewacz, H. D. 1992. *Curr. Genet* 22[6]491.)

in vitro The reaction or culture is carried out within a "glass" vessel rather than in an intact cell or in natural culture conditions, respectively, such as in vitro culture, in vitro fertilization, or in vitro enzyme assay. *See* ART; cell genetics.

in vitro fertilization (IVF) Extracted mammalian eggs can be fertilized by competent sperm (AID or AIH) outside the body, then surgically implanted in the uterus. This procedure may have a higher than 40% chance for success and helps to overcome sterility barriers in many women with infertility problems caused by factors other than the eggs. In vitro fertilization produces normal babies, but the chance of having nonidentical twins is greatly increased if the gynecologist implants more than a single fertilized egg to assure reasonable success. IVF has applied significance in animal breeding. Eggs can be retrieved from the ovaries of cows in slaughterhouses, and after in vitro fertilization, can be reimplanted into any (even some sterile) cows. By this procedure more beef can be produced. There is an opportunity to obtain more offspring from genetically superior individuals of domestic animals by the use of surrogate mothers. It also has applications in wildlife maintenance of endangered species (in zoos) or in species in which the natural rate of reproduction is unsatisfactory. *See* ART; artificial insemination; GnRFA; ICSI; insemination by donor; preimplantation, genetic; test tube baby; twinning. (Elder, K. & Dale, B. 2000. *In Vitro Fertilization*. Cambridge Univ. Press, New York; Ozturk, O., et al. 2001. *Hum. Repr.* 16:1319.)

in vitro mutagenesis Mutation is produced in isolated DNA sequences, which is then reintroduced into the cells by transformation. *See* gene replacement; localized mutagenesis; site-specific mutagenesis; TAB mutagenesis; transformation, genetic. cassette mutagenesis, Kunkel mutagenesis, PCR-based mutagenesis, doping nucleotides. (Lehoux, D. E., et al. 2001. *Curr. Opin. Microbiol.* 4:515.)

in vitro packaging Recombinant DNA equipped with the phage-λ-*cos* sites and genes required for packaging (*origin* of replication and other sequences of about 4–6 kb at both *cos* neighborhoods) can accept inserts so that the total length will remain in the range of 37–52 kb, can be packaged into phage capsids that may infect *E. coli* and yeast cells, and can bring about transformation. Similar procedures are applicable to other systems. *See* cosmid vector; lambda phage. (Okimoto, T., et al. 2001. *Mol. Ther.* 4[3]:232.)

in vitro protein synthesis *See* in vitro translation systems.

in vitro translation *See* oocyte translation; rabbit reticulocyte translation assay; translation in vitro; wheat germ translation.

in vivo The process takes place in intact cells or in tissues of a live organism or cell. *See* ex vivo; in vitro.

involucre Whorl of bracts around an inflorescence. *See* bract.

involution Degenerative type of development; return to a more primitive or inactive state.

iodine stain For coloring starch (amylose) blue-black while amylopectin (dextrin) is colored red-brown (iodine 120 mg and potassium iodide 400 mg in 100 mL water).

iojap (*ij*) Nuclear mutations in maize located in chromosome 3L-90 (*ij1*) and in chromosome 1L (*ij2*), respectively, causing leaf striping because of defects in the development of plastids. The *ij* gene appears to be a specific mutator of extranuclear DNA and displays normal Mendelian inheritance, whereas the striping itself is maternally transmitted. Defects in plastid RNA polymerase were implicated in this variegation. (Silhavy, D. & Maliga, P. 1998. *Curr. Genet.* 33[5]340.)

ion Positively (cation) or negatively (anion) charged atom or radical. *See* electrolyte.

ion channels Pores with special passage specificity, e.g., a membrane protein when bound to acetylcholine permits the influx of sodium (sodium channel); a variety of ion channels exist (mechanically gated, voltage gated, ligand gated, cAMP gated, etc.). The anion-gated ion channels appear structurally different from most of the cationic channels inasmuch as they may have different subunit associations creating double or triple pores. The thousands of different odors activate the trimeric G protein, G_{olf}, which in turn activates adenylate cyclase, and the cAMP-gated cation channels open and transmit the signal to the brain. Some olfactory receptors utilize IP_3-gated ion channels. Cyclic guanosine monophosphate (cGMP)–gated channels mediate visual perception. Light rapidly induces the formation of guanylate cyclase, generating cGMP, and it is degraded by cGMP phosphodiesterase. The photoreceptors (rhodopsin pigment) are in the retina of the eye. Voltage-gated Ca^{2+} ion channels regulate the influx of calcium through the plasma membrane. Mechanical stimuli may affect Ca^{2+} and other ion channels. The L-type ion channels of the neurons may be shut off when the intracellular level of calcium increases beyond a certain point. The Ca^{2+} ions then serve as widespread intracellular messengers and regulate many diverse cellular functions, particularly the secretion of neurotransmitters. Their modulation is due to the $\beta\gamma$ subunits of the trimeric G protein. The glutamate receptors are permeable to Na^+, K^+, and Ca^{2+} and are gated by glutamate in eukaryotes and prokaryotes. The autosomal-dominant human disease *periodic paralysis* appears to be due to an amino acid substitution in the α-subunit of a sodium channel transmembrane protein. Mutation in sodium channel SCN5A may slow down myocardial conduction and may cause life-threatening cardial arrhythmias. Cystic fibrosis is due to a defect in transmembrane conductance regulator protein kinase A and the ATP-regulated chloride ion channel. In pancreatic β-cells, ATP-dependent K$^+$ channels are important for glucose-induced insulin secretion and targets of sulfonylureas used for oral treatment of non-insulin-dependent diabetes (NIDDM). Truncation of the sulfonylurea receptor (SUR) causes persistent hyperinsulinemic hypoglycemia of infancy and unregulation of insulin secretion in severe hypoglycemia. Channels can be inward or outward rectifying, depending on the predominant direction of the flow of the ions. The rectification is not always an intrinsic property of the channel protein, but accessory substances, spermidine, spermine, and other polyamines may control it. Glucocorticoid stress hormones may also regulate K channels. Ion channels regulate the expression of many genes. See ataxia; Bartter syndrome; calmodulin; color blindness; cystic fibrosis; diabetes; dihydropyridine receptor; glucocorticoid; G proteins; HERG; hyperthermia; hypoglycemia; IP$_3$; Jervell and Lange-Nielson syndrome; LQT; myokymia; myotonia; neurotransmitters; patch clamp; periodic paralysis; pyrethrin; rhodopsin; ryanodine; salt-tolerance; signal transduction; sulfonylurea. (Doyle, J. L. & Stubbs, L. 1998. *Trends Genet.* 14:92; Apse, M. P., et al. 1999. *Science* 285:1256; Hübner, C. A. & Jentsch, T. J. 2002. *Hum. Mol. Genet.* 11:2435; <phy025.lubb.ttuhsc.edu/Neely/ionchann.htm>.)

ion-exchange resins   Can be cation or anion exchangers. Their cross-linkage determines their use for separation of molecules of different sizes. There are a variety of ion exchange resins; they are made by the copolymerization of styrene and divinylbenzene and various other substances and are combined to produce the phosphocelluloses, diethylaminoethylcellulose (DEAE), carboxymethylcellulose (CMC), etc. They can be used for the separation and purification of monovalent ions or polyelectrolytes of high molecular weight.

ion exchangers, cell membrane   See ion channels.

ionic bond   Noncovalent attachment between a positively and a negatively charged atom.

ionization   Separation of molecules into ions. See ion.

ionization chamber   Measures the radioactivity of gases by the ionizations generated through molecular collisions. Electrodes collect the ions and the current (amplified and registered) is proportional to the radioactivity. See Geiger counter; radiation measurement; scintillation counter.

ionizing radiation   High-energy electromagnetic radiation causing intramolecular alterations (ion pairs) in organic material, thereby capable of inducing mutation and cancerous transformation in living cells. The maximum legal permissible occupational limit for human exposure should not exceed 0.5 mSv per year; the legal limit actually should be 0.2 mSv. 1 Sv (Sievert) = 100 rem (röntgen equivalent man); 1 rem = 1 rad of 250 kVp X-rays. See cosmic radiation; electromagnetic radiation; Gray units; physical mutagens; radiation effects; radiation hazard assessment; radiation measurement; radiation protection.

ionophores   Hydrophobic molecules involved in ion transport through cell membranes.

ionotropic receptors   Mediate the control of ion channels after binding the appropriate ligand. They — and the metabotropic receptors — have an important role in nerve synapses. See metabotropic receptor; synapse.

ion pump   Mediates ion transport through membranes by the use of energy (ATP). Ion pumps assure osmotic balance and convert ATP energy into electrochemical gradients that are utilized by metabolic pathways. See proton pump. (Dunbar, L. A. & Caplan, M. J. 2001. *J. Biol. Chem.* 276:29617.)

IP$_3$ (inositol 1,4,5-trisphosphate)   Derived from inositol phospholipid PIP$_2$; in response to external signals it releases Ca^{2+} from the endoplasmic reticulum. See InsP; Ipk1; Ipk2; olfactogenetics; phosphoinositides.

IP$_5$, IP$_6$ IP$_3$-derived signaling molecules; IP$_6$ along with other phosphoinositides and nuclear pore–associated proteins are required for mRNA export from the nucleus. See InsP; Ipk1; Ipk2; nuclear pore; RNA export.

IPCR   See inverse polymerase chain reaction.

IPEX   Immune dysregulation, polyendocrinopathy, enteropathy, X-linked is caused by mutation in the region of human chromosome Xp21-Xq13.3 and encodes the FOX3 protein involved in the regulation of transcription or RNA splicing.

IPI   Integrated protein index is an inventory of all the revealed proteins encoded by a genome. See IGI.

Ipl1   Yeast mitotic histone kinase.

Ipk1   Inositol polyphosphate kinase converts inositol-1,3,4,5,6-pentakisphosphate (IP$_5$) into hexakisphosphate (IP$_6$). IP$_6$ is probably a regulator mRNA export from the nucleus and it modulates the function of synaptic vesicles by interacting with synaptogamin. See Ipk2; phosphoinositides; RNA export; synaptogamins.

Ipk2   Inositol polyphosphate kinase converts inositol 1,4,5-trisphosphate (IP$_3$) to inositol-1,4,5,6-tetrakisphosphate (IP$_4$) and insositol-1,3,4,5,6-pentakisphosphate (IP$_5$). Ipk2 seems to be the same yeast enzyme that was named previously as Arg82, a pleiotropic kinase regulating sporulation, mating, stress, arginine metabolism, and transcription. IP$_4$ and IP$_5$ may be effectors for the processes. Ipk2p may also stabilize MCM1p, which seems to be required for Ipk2 function. See Ipk1; MCM; phosphoinositides.

IPMDH   Isopropylmalate dehydrogenase. α-ispropylmalate is a precursor of leucine and it is derived from α-ketoisobutyrate.

ipsilateral   Affecting only one side. See contralateral.

IPTG   Isopropyl-β-D-thiogalactoside, a gratuitous inducer analog of the *Lac* operon. See β-galactosidase; gratuitous inducer; *Lac* operon.

IQ *See* human intelligence.

Ir Immune-response genes is the old name of the HLA genes that encode the MHC complex. *See* HLA.

I-R *See* hybrid dysgenesis.

IRA1, IRA2 Negative regulators of RAS in yeast, antagonistic to CDC25; they are structurally related to GAP. *See* GAP; RAS. (Mitts, M. R., et al. 1991. *Mol. Cell Biol.* 11:4591.)

IRAK IL-1 activated serine/threonine kinase. Pelle-like interleukin receptor-associated serine/threonine kinase and adaptor of the MyD88 signaling pathway. *See* IL-1; interleukins; MyD88; NK-κB; Pelle; Toll. (Jensen, L. E. & Whitehead, A. S. 2001. *J. Biol. Chem.* 276:29037; Kobayashi, K., et al. 2002. *Cell* 110:191.)

Ire Yeast protein kinase and endoribonuclease, functionally homologous to mammalian JNK and SAPK. Ire1 is a transmembrane protein bound in the cytoplasm to TRAF2, an adaptor protein in signal transduction. Ire signals the stress in the endoplasmic reticulum (ER) when the proteins inside are not folded properly and activation of the chaperones is needed. ER stress is monitored by PERK. *See* chaperone; JNK; PERK; ribonuclease L; SAPK.

IRE Iron-responsive element is a 28-nucleotide 5′-UTR sequence in the ferritin mRNA; it is necessary for iron regulation. IRE sequences in the 3′-UTR of the transferrin receptor mRNA protect the mRNA from degradation if the iron level is low. Iron is an essential element for many biological functions, but beyond a certain level it may be very toxic because its reaction with oxygen forms hydroxyl radicals and damages macromolecules. *See* anemia; ferritin; hemochromatosis; ROS; transferrin; UTR. (Andrews, N. C. 2000. *Annu. Rev. Genomics Hum. Genet.* 1:175.)

IRES Internal ribosome entry site, also called RLP, may be present in circular viral RNAs (picornaviruses) and they can be translated on eukaryotic ribosomes. These viral RNAs (300–450 nucleotides) are not capped at the 5′-untranslated region and carry several AUG codons, showing specific sequences serving as ribosome landing pads. The 5S ribosomal protein interacts with the IRES site. The interaction with ribosomes requires the cellular eIF-4F eukaryotic translation initiation factor. IRES-type elements have been found in other viruses and in eukaryotes. The IRES-containing noncoding regions (5′NCR) are generally longer than the regular 5′-UTRs. The presence of IRES elements (9 or double of 9 nucleotides) permits the dicistronic transcription of genes and facilitates gene targeting, homologous recombination, and modification of gene expression. The use of promoterless vector constructs will position the IRES-carrying sequences within transcribed regions rather than into nontranslated regions of the genome. IRES elements are frequently borrowed from the encephalomyocarditis virus (EMCV) family. The advantage of this system is that it does not interfere with the regular Cap-mediated ribosome scanning. IRES-mediated translation is frequently characteristic for mRNAs involved in physiological stress responses and apoptosis. *See* apoptosis; capping enzymes; dicistronic transcription; eIF4F; eIF4G; gene fusion; picornaviruses; ribosome scanning; RLP; targeting genes; translation initiation. (Holci, M., et al. 2000. *Trends Genet.* 16:469; Pinkstaff, J. K., et al. 2001. *Proc. Natl. Acad. Sci. USA* 98:2770; Fukushi, S., et al. 2001. *J. Biol. Chem.* 276:20824; Hellen, C. U. & Sarnow, P. 2001. *Genes Dev.* 15:1593.)

IRF Interferon regulatory transcription factor; homologue of c25 rat factor. *See* c25; interferon; p53.

IRF4 *See* LSIRF.

iris (1) Circular pigmented membrane in front of the lens of the eye and in its center where the pupil is located. (2) Monocot genus of perennial flowers ($2n = 44$, *Iris germanica*).

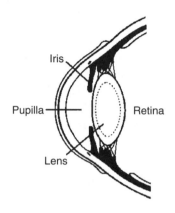

iris pattern The iris of the eye with potentially 266 distinguishable physical features provides a pattern suitable for individual identification when the image obtained by a video camera is analyzed by a computer. According to some views, this procedure is superior to any other personal identification methods, including fingerprinting or DNA fingerprinting. *See* fingerprinting of macromolecules; fingerprints. (Daugman, J. & Downing, C. 2001. *Proc. R. Soc. Lond. B. Biol. Sci* 268:1737.)

IRM Interference reflection microscopy is used to study the cell membrane and protein associations (e.g., in immunological synapse). *See* immunological synapse.

IRMA Immunoradiometric assay uses radiolabeled antibody to quantitate a particular antigen. In the *two-site IRMA*, two antibodies are used that bind different epitopes. *See* ELISA; epitope. (Frystyk, J., et al. 2001. *Growth Horm. IGF Res.* 11[2]:117.)

Iron Age About 3,000 to 4,000 years ago; the beginning of recorded history.

iron metabolism Several human diseases involve anomalies in iron metabolism. *See* aceruloplasminemia; anemia; ferritin; Friedreich ataxia; Hallervorden-Spatz disease; hemochromatosis; IRE; porphyria; sideroblastic anemia; transferrin. (Roy, C. N. & Andrews, N. C. 2001. *Hum. Mol. Genet.* 10:2181.)

irradiation *See* radiation.

IRS (1) Interferon-response factor. *See* interferon. (2) Insulin receptor substrate. *See* insulin receptor substrate.

IRS-PCR Interspersed repetitive sequence–polymerase chain reaction is used with radiation hybrids as a rapid test for identifying the hybrid nature of the putative hybrid cells; also, the PCR product is hybridized as a probe to a normal chromosomal complement and compared to the previously established map. IRS-PCR can be used to screen cosmid and artificial chromosome libraries. *See* PCR; radiation hybrid. (Himmelbauer, H., et al. 2000. *Nucleic Acids Res.* 28[2]:e7.)

IS *See* insertion elements.

Isadora 8.3 kb transposable element of *Drosophila*, generally present in eight copies.

I-Sce1 Unusual restriction endonuclease, encoded by yeast mitochondrial introns. It recognizes 18 bp sequences; thus, in the mouse genome it cuts only once per 7×10^{10} bp, i.e., less frequently than once in 10 times the size of the total mouse genome. In a modified form, it is useful for studying the consequences of double-strand breaks and the mechanism of homologous recombination. *See* double-strand break; homologous recombination; restriction endonuclease. (Rouet, P., et al. 1994. *Proc. Natl. Acad. Sci. USA* 91:6064.)

ischemia Involves restriction of the blood vessels. It may be a major cause of stroke. For gene therapy the introduction of antiapoptotic genes, interleukin-1 receptor antagonists, angiogenesis-activating vascular endothelial growth factor, superoxide dismutase, etc., have been considered. Among the genetic vectors tried, adenovirus and herpes virus appeared most promising by injection into the brain or into the spinal artery. *See* adenovirus; alcohol; apoptosis; herpes; interleukin; stroke; superoxide dismutase; VEGF.

ISGF-3α *See* APRF.

ISIS Isotype-specific inhibitory sequence. Various types of antisense constructs are effective in reducing or eliminating gene translation. *See* antisense technologies; isotype.

island model of populations A group of subdivided populations exchanges m alleles with each other regularly. They also have a chance to receive alleles at a constant $m - 1$ rate from an earlier population. Thus, the situation resembles that of a drift in a single population under migration pressure. The model assumes that each subpopulation has the same size and they are equidistant from each other. The continuous-model version considers equal density at any point and the discontinuous version assumes that the subpopulations are clustered at nodal points of a lattice. Each of these distributions may be one- or two-dimensional. The discontinuous model has also been called the stepping-stone model; the latter may also be three-dimensional. All variations of the basic model have been criticized because most of the basic stipulations may not be met under natural conditions. (Cavalli-Sforza, L. L. & Bodmer, W. F. 1971. *The Genetics of Human Populations*, p. 423, Freeman, San Francisco.)

isoacceptor tRNAs Group of different tRNAs accepting the same amino acid. The identity of a particular isoaccepting tRNA is determined by particular nucleotide sites within the group. The anticodon may serve as an identity site; however, the six serine tRNAs do not share a common anticodon base. Generally, nucleotide 73 is a discriminator base and the first base pairs within the acceptor stem. The long variable arm of tRNA^{Ser}, the extra G1:C73 bp in tRNA^{His}, and the G3:U70 bp in tRNA^{Ala} may serve for identity in *E. coli*. In other species, identity may be differently determined. *See* aminoacyl-tRNA synthetase; protein synthesis; tRNA; wobble. (Heyman, T., et al. 1994. *FEBS Lett.* 347[2–3]:143; Chaley, M. B., et al. 1999. *J. Mol. Evol.* 48[2]:168.)

isoallele Wild-type allele distinguishable only by special techniques. *See* allele. (Harris, M. J. & Juriloff, D. M. 1989. *J. Hered.* 80[2]:127; King, J. L. 1974. *Genetics* 76:607.)

isoallotype Variable antigens of one immunoglobulin (IgG) molecule subclass but invariant on the molecules of another subclass or subclasses. *See* allotype. (Delacroix, D. L., et al. 1986. *Mol. Immunol.* 23[4]:367.)

isoantigen Allelic variant of an antigen within the species. *See* alloantigen.

isobrachial The two arms of a chromosome are identical. *See* chromosome arm; chromosome morphology; isochromosome.

isochores Segments (generally larger than 200 kb) of the DNA of higher eukaryotes of rather homogeneous GC or AT content. Usually, one rather light and two or three different heavier components can be separated by buoyant density centrifugation in the presence of certain ligands, e.g., silver ions (Ag$^+$). In the mammalian genome, the GC-poor isochores represent about 62% and the GC-rich and GC-very-rich isochores are 22% and 3–4% of the genome, respectively. Genes within isochores have a base content characteristic for the isochores. In the high-GC isochores, the gene concentration is much higher than in the AT-rich areas. The isochores may affect codon usage and replication pattern. The cytologically identifiable Giemsa-stained bands are low in GC, the T bands are high, and the R bands occupy a position in between. Because of the high GC content of the T bands, their codon usage is biased. R bands and G/R borders are characterized by higher frequency of chromosomal exchanges (breakage and chiasmata). Viral and transposon integration sites are correlated with the base composition of the elements; sequences similar to their own are the prefered targets. The origin of isochores was attributed to natural selction or variation in mutational bias. Recent evidence supports the latter alternative. *See* chromosome banding; codon usage; CpG islands; transposons. (Bernardi, G. 2000. *Gene* 241:3; Galtier, N., et al. 2001. *Genetics* 159:907.)

isochromatid break   Damage simultaneously inflicted to both chromatids of a single chromosome. *See* radiation effects.

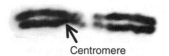

Centromere

Isochromatid breaks. (Courtesy of B. R. Brinkley).

isochromosome   Has two identical arms and generally comes about by misdivision of a telocentric chromosome. *See* isobrachial; misdivision; trisomic, secondary.

isodiametric   Its diameters (lines passing through its center) are the same in all directions.

isodisomy   *See* uniparental disomy.

isoelectric focusing   Electrophoretic separation technique based on the isoelectric point of the molecules to be separated. The isoelectric point is at a pH where a solute has no electric charge and thus does not move in the electric field. When, e.g., denatured protein mixture is placed in an electrophoretic gel that contains a pH gradient established by different buffers, the polypeptides migrate to their isoelectric zones. *See* electrophoresis; two-dimensional electrophoresis.)

isoelectric point   When a charged molecule in a solution shows no net electric potential because of the pH and consequently does not move in an electric field. *See* isoelectric focusing.

isoenzymes (isozymes)   Multiple, distinguishable forms (by primary structure [electrophoretic mobility], substrate affinity, reaction velocity and/or regulation) of enzymes that catalyze the same reaction. *Sorting isozymes* are targeted to specific subcellular compartments, e.g., mitochondria, chloroplasts, or to different organs. *See* isozyme.

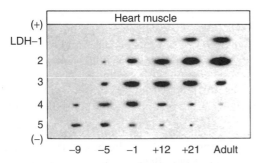

Lactate dehydrogenase isoenzyme profile in the heart muscles in the mouse shows dramatic changes during development. Nine days before birth, isozyme LDH-5 is predominant, and LDH-1, LDH-2, and LDH-3 are not detectable. In the adult animal, almost the opposite is true. LDH is a key enzyme in the pathway of converting sugars into amino acids, lipids, etc. The numbers at the bottom indicate days before (−) or after (+) birth. (From Markerts, C. L. and Ursprung, U. 1971. *Developmental Genetics*. Prentice-Hall, Englewood Cliffs, NJ.)

isofemale   Line descended from the same female ancestor that was originally infected by a cytoplasmically infected bacterium, e.g., in *Drosophila* by *Wolbachia*. *See* Wolbachia.

isoflavones   *See* phytoestrogens.

isoform   Due to some difference in amino acid composition, caused by mutation or alternative splicing of the RNA transcript, RNA editing, use of alternative promoters, and posttranscriptional or posttranslational processing, essentially the same protein is somewhat altered. Polypeptides in the cellular organelles may have basically similar catalytic functions as those residing in the cytosol or organ- or tissue-specific forms of enzymes. The human WT1 (Wilms' tumor) has 24 isoforms, which all share four C-terminal C_2H_2 zinc fingers and an N-terminal Pro/Gln-rich regulatory domain. *See* alternative splicing; splicing; Wilms' tumor. (Hastie, N. D. 2001. *Cell* 106:391.)

isogamy   The gametes forming a sexual union appear to be the same, e.g., in protozoa. *See* anisogamy; heterogametic; homogametic.

isogenic stocks   Their genes are represented by the same alleles at all loci. In rodents, for practical purposes, isogenicity is tested by skin grafts. Grafts within inbred lines are supposed to be successful but not in between different inbred lines. Grafts from parents to F_1 should be successful but not in the opposite direction. *See* congenic.

isograft   The genotypes of the donor and recipient tissues match. *See* allograft; grafting in medicine; heterograft.

isolabeling ^{3}H label (or other) in both daughter chromatids after one replication in ^{3}H-thymidine medium (or other labeling medium) at a certain tract(s) as the consequence of sister chromatid exchange. *See* sister chromatid exchange.

isolation, genetic   May be based on the presence of inversions in the population that in case of recombination yield defective gametes. Also, since recombination within inverted segments produces defective gametes, advantageous gene blocks may be preserved as "supergenes." Genetic isolation may be the first step in speciation. *See* incompatibility.

isoleucine-valine biosynthetic pathway   Steps 1 and 2 are controlled by identical enzymes in both pathways in several organisms and therefore genetic defects in either may generate nutritional requirements for both valine [$CH_3CH(CH_3)CH(NH_2)COOH$] and isoleucine [$CH_3CH(CH_2CH_3)CH(NH_2)COOH$] or the accumulation of intermediates. *Maple syrup urine disease* (MSUD) exists in different forms in humans and other mammals. It is caused by a block in the degradation, decarboxylation (step 2), and accumulation of leucine, isoleucine, and valine. Keto-methyl-valerate accumulation then causes the characteristic maple syrup odor ketohydroxy-methylvalerate ↔ dihydroxy-methylvalerate ↔ keto-methylvalerate ↔ **isoleucine**

| **1** | | **2** | | **3** |

ketohydroxy-isovalerate   ↔   dihydroxy-isovalerate   ↔   keto-isovalerate   ↔   **valine**

of the urine. One of the genes (MSUD II) was assigned to human chromosome 1p31 and a pseudogene to 3q24. The more serious aspects of the disease are physical and mental retardation, potential coma, and death. The disease is also controlled by nonallelic recessive genes (BCKDHA [branched-chain keto acid decarboxylase and dehydrogenase], locus is at 19q13.1-q13.2, the BCKDHB is at 6p22-p21). Gene A is responsible for the biosynthesis of the α-chain of the enzyme (BCKDHA) and locus B for the β-chain (BCKDHB). In one form of the type A disease, thiamin (10 mg/day) reduces the hyperaminoacidemia without dietary limitations. Apparently the larger subunit (M_r 46,500) of the enzyme is part of a mitochondrial protein complex. The enzyme complex also contains component E2 (M_r 52,000), which transfers the acyl group of the keto acid from the E1 component (protein A) to coenzyme A. The disease may be due to a single base-pair change, resulting in a tyrosine substitution at asparagine site 394 of protein A, or to deletions of several nucleotides. The defect is detectable prenatally, but carriers cannot be identified because of recessivity. The prevalence varies greatly in different ethnic groups from 3×10^{-5} to 6×10^{-3}. Defects in the transaminase reaction (step 3) may be controlled by a common glutamic-branched-chain amino acid transaminase. However, in humans, hypervalinemia (valinemia) is caused by a defect in a specific transaminase, resulting in lack of ability to catabolize valine into keto-isovalerate without affecting the level of leucine and isoleucine in the blood. MSUD III (7q31-q32) is caused by dihydrolipoamide dehydrogenase deficiency. Methyl-crotonyl-CoA carboxylase deficiency (3q25-q27) is a recessive defect in leucine catabolism without acidosis. Apparently spinal defects result in muscular hypotony and atrophy. Isovaleric acid CoA dehydrogenase deficiency (15q14-q15) is a ketoacidosis with isovaleric acidemia and is similar to MSUD. The clinical symptoms include retarded psychomotor activity, vomiting, and protein aversion. See genetic screening; hydroxymethylglutaricaciduria; 3-hydroxy-3-methyl-glutaryl CoA lyase deficiency; hypervalinemia; leucine metabolism; methacrylaciduria; methylacetoaceticaciduria; methylcrotonylglycinemia; methylglutaconicaciduria. (Nellis, M. M. & Danner, D. J. 2001. *Am. J. Hum. Genet.* 68:232.)

isolocus   Paralogous locus originated from duplication of the genome. *See* paralogous loci.

isomerase   Enzyme that interconverts enantiomorphs. *See* chirality; enantiomorph.

isomerism   Developmental anomaly displaying single organs, normally present asymmetrically in the body, in a position symmetrical on the two sides of the body axis. *See* heterotaxy; left-right asymmetry; situs inversus viscerum.

isomerization, strand   One crossed-over DNA strand changes into a two-strand crossover through rotation of the molecules. *See* Meselson-Radding model of recombination.

isomers   Different compounds that have the same molecular formula, but their structure is different. Their differences my be relatively subtle (D- and L-glucose) or may be as large as those between ethyl alcohol and methyl ether.

isometric phage   Enclosed in an icosahedral capsid.

Icosahedron.

isonymy   Individuals having the same family name have a probability of kinship. According to some suggestions, the frequency of isonymous couples multiplied by a factor of 1/4 may provide information on the coefficient of inbreeding in the population. One-fourth of the married pairs have identical family names before marriage because of the inheritance of the grandfather's name through two sibs and three-fourths have different surnames. The probability of isonymy for the first cousin is 1/4, for second cousins 1/16, and for third cousins 1/64. When multiplied by 1/4, these fractions may provide the coefficient of inbreeding, i.e., $(1/4 \times 1/4) = 1/16$, $(1/16 \times 1/4) = 1/64$, and $(1/64 \times 1/4) = 1/256$ for the three types of matings, respectively. (Lasker, G. W. & Mascie-Taylor, C. G. 2001. *Ann. Hum. Biol.* 28:546.)

isopentyladenine (6-[γ,γ-dimethylallylamino]purine) Posttranscriptionally modified base in tRNA; it is also a cytokinin plant hormone. *See* plant hormones; tRNA. (Takei, K., et al. 2001. *J. Biol. Chem.* 276:26405.)

isopeptidase   Deubiquitinating enzyme. *See* ubiquitin; UBP.

isoprene   2-methyl-1,3 butadiene unit of terpenoids, fragrances, rubber, etc. Isoprenylated proteins are anchored to the cell membranes. The role of the oncogene product, RAS, functions in carcinogenesis and signal transduction in isoprenylated form attached to a farnesyl pyrophosphate. *See* membrane proteins; plastoquinone; prenylation; RAS; signal transduction. (Lichtenthaler, H. K. 1999. *Annu. Rev. Plant Physiol. Plant. Mol. Biol.* 50:47.)

$$CH_3$$
$$|$$
$$CH_2 = C - CH = CH_2$$

isoprenoids   Include >23,000 diverse compounds in viruses to mammals such as cholesterol, steroid hormones, bile acids, retinoids, isopentenyl-tRNA, sphingomyelins, ecdysone, gibberellic acid, abscisic acid, brassinosteroids, carotenoids, rubber, etc. They may function as regulators of transcription, developmental processes involving the hedgehog family of proteins, meiosis, apoptosis, etc. *See* apoptosis; hedgehog; Niemann-Pick disease; SF-1. (Edwards, P. A. & Ericsson, J. 1999. *Annu. Rev. Biochem.* 68:157; Sharkey, T. D. & Yeh, S. 2001. *Annu. Rev. Plant Physiol. Plant Mol. Biol.* 52:407.)

isopropyl thiogalactoside   *See* IPTG.

isopycnic   Molecules with equal density. *See* buoyant density.

isopycnotic chromosome   Does not display heterochromatic regions. *See* heterochromatin.

isoschizomers   Restriction endonucleases with very similar recognition sites and their cleaved ends are identical (cohesive) and thus capable of joining each other, e.g.,

Bgl II	5′	A↓pGpApTpCpTA
	3′	TpCpTpApGp↑A
Bam HI	5′	G↓pGpApTpCpC
	3′	CpCpTpApGp↑G

Their termini after ligation are

GGATCT

CCTAGA

This juncture is no longer recognized, however, by either of the two enzymes because the bases at the left and right ends are incompatible with both Bgl II and Bam HI.

Other isoschizomers are

Hpa II	5′ C↓ÇGG	3′	does not cut when Ç is methylated
Msp I	5′ C↓CGG	3′	is indifferent to methylation

Isoschizomers SmaI (CCC↓GGG) and Xma I(C↓CCGGG) recognize the same sequence but cut it (↓) at a different position. *See* restriction enzymes; restriction modification.

isosteres   Have similar electron arrangements but different chemical structures; used for substrate analog designs.

isothermal   Being at identical temperature.

isotope-coded affinity tag (ICAT)   Two chemicals with biotin affinity and a thiol group are labeled with a heavy or light isotope, respectively. The experimental and the control proteins are reduced and derivatized with either the heavy or the light forms. The mixed samples are digested to obtain peptide fingerprints. The tagged fragments are analyzed by mass spectrometry after purification. Only the double peaks with 8 Da difference (i.e. the difference between the light and heavy label) are further characterized (Gygi, S. P., et al. 1999. *Nature Biotechnol.* 17:994). Non-isotopic labeling

techniques have also been developed. *See* mass spectrometer [MALDI/TOF/MS]; non-isotopic labeling, proteomics.

isotonic   The active salt concentration of this medium is the same as in the cell. A NaCl solution of 0.9% is isotonic with the human blood and can be used to temporarily maintain good osmotic conditions after substantial bleeding by injecting sterile solutio natrii chlorati isotonica into the (blood vessels) venae. *See* hypotonic; isotonic.

isotope discrimination   The heavier atoms may be used less effectively than the lighter ones. *See* isotopes.

isotopes   Two or more nuclides have the same atomic number; thus, they are the same elements but differ either in mass (stable isotopes such as hydrogen and deuterium) or radioactive isotopes (atom) that disintegrate by emission of corpuscular or electromagnetic radiation (α-, β-, or γ-rays). The latter ones are particularly useful in biology as radioactive tracers of minute amounts of labeled (H^3, C^{14}, P^{32}) nucleotides or (C^{14}, S^{35}, I^{125}) proteins. Stable isotopes (N^{15}, C^{13}, H^2) can be used as density labels for distinguishing old and newly synthesized molecules. The types of radiations and energies in million electron volts are H^3: β, 0.017–0.019, C^{14}: β 0.155, P^{32}: β 1.71, S^{35}: β 0.167, I^{131}: β 0.605, 0.250, and γ 0.164, 0.177, 0.284, 0.364, 0.625, I^{125}: γ 0.0355, Y^{90}: β 2.24, K^{42}: β 3.6, 2.4, and γ 1.5, Cs^{137}: β 0.518, 1.17, and γ 0.663, Co^{60}: β 1.56, γ 2.33, U^{238}: α 4.180, γ 0.045. Ra^{226} is generally understood to be radium; it has a long half-life (1,620 years) and thus is quite stable. It primarily emits α-radiation (helium nuclei) 4.750 MeV and γ-electromagnetic radiation 0.188 MeV. Radium is converted to a number of other isotopes. Among those, radon, the commonly present gas diffusing from the rocky soils, has a short half-life (3.825 days), yet it is usually replenished from the source. It emits α-radiation and may pose an environmental health hazard if its concentration reaches higher levels. Radium is no longer used for medical purposes and to a very limited extent for industrial purposes because of the hazards in handling. Luminous watch dials are now made from fluorochromes. *See* biotin; curie; fluorochromes; μC; nonisotopic labeling.

isotopic graft   A group of cells or a piece of tissue is transplanted to an equivalent position of another animal.

isotype   Closely similar immunoglobulins with differences in the constant region. Sometimes the word is used in a broader sense as similar in constitution or sequence. *See* allotype; idiotype; immunoglobulins.

isotype switching   Process of immunoglobulin gene rearrangement. *See* immunoglobulins.

isotypic exclusion   Either λ- or κ-light chains are used with any of the heavy chains for the formation of a particular antibody (both light chains cannot be used simultaneously). *See* antibody; immunoglobulins.

isovalericacidemia (IVD)   Controlled by a recessive gene in human chromosome 15q14-q15. The disorder is characterized by sweaty feet—like odor (butyric and hexanoic acid) of the urine, dislike of protein, vomiting, anemia, ketoacidosis, high isovaleric acid content of the blood, leading to injury of the nervous system. The biochemical lesion is in isovaleric acid CoA dehydrogenase deficiency. Several types of mRNAs of this gene have been identified due to differences in transcription and different mutations or deletions. *See* acetyl-CoA dehydrogenase deficiency; amino acid metabolism; isoleucine-valine biosynthetic pathway; Jamaican vomiting sickness.

isozygotic   Homozygous for all genes.

isozyme   *See* isoenzymes.

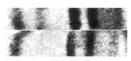

Arabidoposis thaliana. Fructose-1,6-diphosphate aldolase isozymes. *Hylandra suecica.*

ISP45   *See* mitochondrial import.

ISPCR   In situ polymerase chain reaction. *See* in situ PCR.

ISRE   Interferon sequence-response element. *See* signal transduction (Jak-STAT).

ISSR   Intersimple sequence repeats are repeats between nonrepeated sequences of chromosomes.

ISTR   Inverse sequence-tagged repeat technique is used for physical mapping of genomes. (Rohde, W. J. 1996. *J. Genet. Breed.* 50:249.)

ISWI   ATPase subunit of NURF and other chromatin remodeling proteins. *See* chromatin remodeling; nuclear receptors; nucleosome; Nurf. (Langst, G. & Becker, P. B. 2001. *J. Cell Sci.* 114:2561; Xiao, H., et al. 2001. *Mol. Cell* 8:531; Clapier, C. R., et al. 2002. *Nucleic Acids Res.* 30:649.)

ITAM   Immune receptor tyrosine-based activation motif. *See* BLK; ITIM; killer cell; lymphocytes; T-cell receptor.

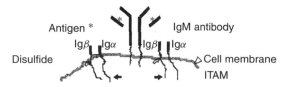

ITAMs are phosphorylated by Lck, Blk, Flyn, Lyn and recognized by Syk The phosphorylation and activation process may be initiated by crosslinking antigens of IgM molecules

ITD   *See* idiopathic torsion dystonia.

iteration   Involves repetitions; a commonly used form is the iterated integral in which we differentiate first with respect to one of the variables while holding the other constant and then in the result with respect to the other variable. For biological experiments the distribution may be fitted to a negative binomial, requiring a similar procedure in the calculation of the maximum likelihood of the values. For complex numerical iterations, computer programs may be required. *See* maximum likelihood; negative binomial.

iterative crosses   Used in breeding programs such as the three-way and double crosses employed for the testing or production of hybrid maize. *See* heterosis.

iterative truncation (ITCHY)   Incremental truncation for the creation of hybrid enzymes is a method for the generation of hybrid enzymes by fusing truncated N- or C-terminal fragment libraries. The DNA coding sequences are progressively reduced with the aid of exonuclease III before ligating the "single-crossover" hybrid library. The new coding sequence may be the same length as the "parental" ones except the contribution to the hybrid by the parents may vary. The components of the newly created protein may be as different as prokaryotic and eukaryotic. *See* DNA shuffling; protein engineering. (Ostermeier, M., et al. 1999. *Proc. Natl. Acad. Sci. USA* 96:3562; Lutz, S., et al. 2001. *Proc. Natl. Acad. Sci. USA* 98:11248.)

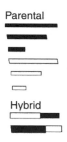

Parental

Hybrid

iterons   Multiple, short DNA repeats, which may bind to the plasmid replication protein and either initiate or inhibit replication. (Chattoraj, D. K. 2000. *Mol. Microbiol.* 37[3]:467.)

iteroparity   Reproduction takes place on more than a single occasion.

ITG   *See* integrin.

ITIM   Immunoreceptor tryrosine-based inhibitory motif. Typically the cytoplasmic domain of these molecules contains the six amino acids Ile/Val/Leu/Ser)-X-Tyr-X-X-(Leu, Val), where X can be any amino acid. A balance between activation (ITAM) and inhibition motifs determines the development of the immune system (NK). Clustering of these receptor sites is induced by ligands and an Src-type kinase phosphorylates the tyrosine residue. Such an event facilitates the recruitment of SH2 domain phosphatases such as SHP-1 and SHIP. PIR-B attenuates B-cell receptor activation in cooperation with SHP-1. Other targets are Sky, BLNK, BASH, phosholipase C, FcγR, etc. *See* BASH; BLNK; B lymphocyte receptor; DAP; FcγR; ITAM; killer cells; KIR; phospholipases; PIR; SH2; SHIP; SHP-1; Sky. (Ravetch, J. V. & Lanier, L. L. 2000. *Science* 290:84.)

ITK Nonreceptor tyrosine kinase of the Tec family; it signals to TCR. *See* TCR; Tec.

ITP Inosine triphosphate. *See* hypoxanthine; inosine.

ITR Inverted terminal repetition such as occurring in human adenovirus DNA. *See* adenovirus.

ITS Internal transcribed spacers are short sequences within eukaryotic pre-tRNA transcription units (5′ 18S - ITS - 5.8S - ITS - 28S 3′), and these clusters are separated by external transcribed spacers. Their analysis facilitates identification of different strains. *See* ETS; tRNA. (Fujita, S. I., et al. 2001. *J. Clin. Microbiol.* 39:3617.)

IU International unit is a quantity of various vitamins, hormones, enzymes, etc., that bring about a standard response as determined by the International Conference for Unification of Formulas.

IUCD (IUD) Intrauterine contraceptive device prevents implantation of the egg. Although it is an effective method of birth control, it may promote infection and may cause some discomfort.

IUGT In utero gene transfer would correct genetic defects of somatic cells by transfer of gene(s) into the unborn fetus. Its advantage would be that it could be applied early during embryonic development before irreversible damage can take place, e.g., in various neurological diseases (Tay-Sachs, Niemann-Pick, Lesh-Nyhan, Sandhoff, Leigh, leukodystrophies, gangliosidoses, immunological disorders, thalassemias, osteopetrosis). IUGT may less likely involve vector or cell rejection. Hematopoietic stem cells might be treated with more success. Animal experiments have provided encouraging results. Its drawbacks are the potential risk to the mother and fetus. Also, the vector DNA may cause undesirable insertions or mutations in the germline. *See* ART; gene therapy. (Heikkila, A., et al. 2001. *Gene Ther.* 8:784.)

IUI *See* ART; intrauterine fertilization.

Ivermectin Antibiotic against parasitic infections.

IVET In vivo expression technology selects specifically induced bacterial genes that are expressed when bacteria are committed to infection or passage through a host.

IVF *See* ART; in vitro fertilization.

IVIG Intravenous immunoglobulin gamma administration has polyclonal antiinflammatory effectiveness against autoimmune cytopenias, Guillain-Barré syndrome, myasthenia gravis, anti–factor VIII autoimmune disease, dermatomyositis, vasculitis (inflammation of the blood vessels), and uveitis (eye inflammation). *See* antihemophilic factors; cytopenia; Guillain-Barré syndrome; myasthenia; polyomyositis.

IVS Intervening sequences; same as introns. *See* introns.

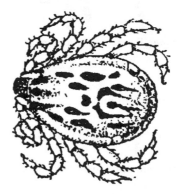

Dermatocentor marginatus (tick).

Ixodoidea Group of insects (ticks); common carriers of *Borrelia* infection, causing Lyme disease and viral infections, resulting in encephalitis. *See Borrelia*.

Professor T. H. Morgan in The American Naturalist 60, p. 490 (1926) "... except for the rare cases of plastid inheritance, the inheritance of all known characters can be sufficiently accounted for by the presence of genes in the chromosomes. In a word the cytoplasm may be ignored genetically".

A historical vignette

J

JAB Jun activation domain–binding protein. Coactivator of AP1 transcription factor by transactivating c-Jun and JunD. It interacts with the $\beta2$ subunit of LFA-1. It may turn off cytokine signaling. *See* AP; CIS; Jun; LFA; SOCS-box; transactivator. (Harding, T. C., et al. 2001. *J. Biol. Chem.* 276:4531.)

jackknifing Statistical device for the estimation of bias and variance of genetic parameters without providing essential estimates on the distribution of the estimates. The jackknife estimator of a parameter is shown at the below, where $\hat{\theta}$ is the usual estimator using the complete set of n observations. In the jackknife procedure, each sample member in turn is omitted from the data, thus generating n separate samples each of $n-1$ size. This method may be used for the estimation of the size of misclassification in conjunction with discriminant analysis. *See* bootstrap; discriminant function. (LaPointe, F. J., et al. 1994. *Mol. Phylogenet. Evol.* 3[3]:256.)

$$\tilde{\theta} = n\hat{\theta} - (n-1)\frac{\sum_{i=1}^{n} \hat{\theta}i}{n}$$

Jacobsen syndrome Dominant fragile site involving human chromosome 11q23.3 is within a distance of 100 kb to the CBL2 oncogene and CCG repeats. This trinucleotide repeat is called FRA11B. The CpG repeats are liable to methylation. The syndrome involves growth and psychomotor retardation, anomalies of the face, finger and toe development. *See* ataxia; dentatorubral-pallidoluysian atrophy; FMR1; fragile sites; Huntington's chorea; Kennedy disease; Machado-Joseph disease; trinucleotide repeats.

jackpot mutation Occurs early during the growth of a population and is represented by more copies than late-occurring mutations. The jackpot mutations may bias the calculations of the mutation frequency if not identified. *See* mutant frequency.

jackpot vessel In a series of dilutions or in a fluctuation test, one vessel has more than the expected number of cells caused either by a clump of cells or a preexisting mutation. *See* fluctuation test.

Jackson Laboratory Backcross DNA Panel Map Service Makes available DNA from the reciprocal mouse crosses (C57BL/6J x *Mus spretus*), characterized by SSLP markers, proviral loci, and several other sequences. Information: Lucy Rowe or Mary Barter, Jackson Laboratory, 600 Main St., Bar Harbor, ME 04608; phone: 207-288-3371, ext. 1687; fax: 207-288-5079; <lbr@aretha.jax.org> (L.R.) or <meb@aretha.jax.org> (M.B.).

Jackson-Lawler syndrome Keratosis of the skin; involves teeth at birth. *See* ichthyosis; keratosis.

Jackson-Weiss syndrome *see* Apert syndrome; Crouzon syndrome, Pfeiffer syndrome.

Jak kinases (Janus tyrosine kinases) Jak3 is required for the progression of the development of B lymphocytes. Jak kinases are involved in the transmission of interleukin signals. *See* interleukins; signal transduction by interferon signaling. (O'Brien, K. B., et al. 2002. *J. Biol. Chem.* 277:8673.)

Jak-Stat pathway Several Jak kinases and signal transducers and activators of transcription (STATs) regulate the signal transduction of interleukins and interleukin-mediated transcription. The pathway may be activated by interferons, phospholipase C (PLC), growth hormones, epidermal growth factor (EGF), and platelet-derived growth factor (PDGF). SOCS/JAB/SSI and CIS proteins exert negative control. This pathway is missing from *Caenorhabditis*. *See* CSF; EGF; JAB; PDGF; signal transduction; SOCS; SSI. (Schindler, C. & Darnell, J. E. 1995. *Annu. Rev. Biochem.* 64:621; Hilton, D. J. 1999. *Cell. Mol. Life Sci.* 55:1568; O'Shea, J. J., et al. 2002. *Cell* 109:S121; Schindler, C. W. 2002. *J. Clin. Invest.* 109:1133.)

Jamaican vomiting sickness Caused by the consumption of unripe ackee fruit, a common food for people of the island. The obnoxious component of the fruit hypoglycin A may reduce blood glucose content to the level of 10 mg/100 mL and may even cause death. The compound is a specific inhibitor of isovaleryl-CoA dehydrogenase. Isovaleric acid accumulates in the blood, causing depression of the central nervous system. The poisoning has an effect similar to human isovalericacidemia. *See* isovalericacidemia.

Jansky-Bielschowsky disease *See* ceroid lipofuscinosis.

Janus kinases Include the Jak kinases and Tyk2; they are nonreceptor tyrosine phosphorylating enzymes. *See* Jak kinases.

jarovization (yarowization) *See* vernalization.

jasmonic acid ([±]-1α,2β-[Z]-3-oxo-2-[2-pentenyl]cyclopentanacetic acid) Fatty acid derivative protease inhibitor in plants and activator of stress response genes in case of infection or wounding and a number of developmental processes. *See* plant defense; wound response. (Seo, H. S., et al. 2001. *Proc. Natl. Acad. Sci. USA* 98:4788; Turner, J. G., et al. 2002. *Plant Cell* 14:S153.)

jaundice May be caused by hyperbilirubinemia and is characteristic for several hereditary syndromes. *See* hyperbilirubinemia.

JAVA Commercially available computer language for various applications.

Java man Representative of *Homo erectus*, with small cranium (brain $\approx$815–1067 cm^2) and robust jaws, who lived about 100,000 years ago. *See* hominids.

J base β-D-glycosylhydroxymethyluracil occurs in the repetitive DNAs of protozoa. Its presence is correlated with a J-binding protein and with the epigenetic silencing of telomeric surface glycoprotein genes. These surface glycoproteins mediate antigenic variations of *Trypanosomas* and other related parasites. *See* antigenic variation; pyrimidine; *Trypanosomas*. (Sabatini, R., et al. 2002. *J. Biol. Chem.* 277:958.)

J chain   15 kDa polypeptide participating in the formation of antibody molecules. *See* immunoglobulins.

J chromosome   Chromosome moving toward the pole appearing like the letter J below:

JE   PDGF (platelet-derived growth factor) and serum-inducible cDNA. *See* PDGF.

Jefferson, Thomas   President's paternity. *See* Y chromosome.

jellyfish   (<www.ucis.uci.edu/biochem/steele/default.html>.)

Jervell and Lange-Nielsen syndrome   21q22.1-q22.2 and 11p15.5 recessive heart and auditory (deafness) syndrome. On electrocardiogram, the interval Q-T is prolonged. In this method the excitation of the heart atrium is denoted by the P wave, followed by the QRS complex of deflections and excitations (depolarization) of the ventricles, and the T waves indicate the repolarization of the ventricles. Fibrillations (uncoordinated arrhythmia) of the heart atrial muscles are also observed as a consequence of inadequacy of potassium and/or sodium ion channels. Sudden death may occur. *See* Beckwith-Wiedemann syndrome; deafness; electrocardiography; heart disease; HERG; ion channels; LQT; Ward-Romano syndrome. (Neyroud, N., et al. 1997. *Nature Genet.* 15:186.)

Jesuit model   There are more potential replicational origins than actually selected in eukaryotes. *See* replication bubble.

Jews and genetic diseases   *Ashkenazi Jews*: Common: Riley-Day syndrome, Tay-Sachs disease, Gaucher's disease, Niemann-Pick syndrome, diabetes mellitus, pentosuria, dystonia, about 1% of women carry deletions at various positions in the BRCA1 and BRCA2 breast cancer genes, Cohen syndrome, Canavan disease, PTA deficiency disease. Rare Occurrence: juvenile form of Gaucher's disease, glucose-6-phosphate dehydrogenase deficiency, Bloom syndrome. *Sephardic Jews*: Common: Mediterranean fever. Rare Occurrence: Tay-Sachs disease. *Oriental Jews of Persian origin*: hypoaldosteronisms, Dubin-Johnson syndromes appear relatively common. *Lybian Jewish populations*: Creutzfeldt-Jakob disease is disproportionally frequent. For these disease differences (gene frequencies) there is no generally valid explanation. It has been suggested that genetic drift in small isolated populations may be the cause. The fact that most of these diseases are based on mutations at different sites within the respective loci is at variance with this argument. The high frequency of Tay-Sachs, Gaucher, and Niemann-Pick diseases involves lysosomes, but how this could be the cause is unclear. Selective advantage of the heterozygotes specific for these particular populations has also been considered. *See* Amish; Ashkenazim; aspartoacylase deficiency; ethnicity, evolutionary distance; founder principle; human intelligence; Sephardic. (Adam, A. 1973. *Isr. J. Med. Sci.* 9:1383; Ostrer, H. 2001. *Nature Rev. Genet.* 2:891.)

J gene   *See* immunoglobulins; J chain.

JIL-1   Chromosomal kinase that may upregulate gene expression in the single Y chromosome of *Drosophila* males. *See* dosage compensation.

jimpy mice   Special strain of these animals with reduced rate of cerebroside synthesis resulting in neurological defects. *See* cerebroside.

jimson weed   *See Datura*.

Seed capsule of *Datura stramonium*.

JIP   JNK-interacting protein. *See* JNK.

JNK   Jun amino-terminal kinase acts on the amino terminal of Jun oncogenes and other transcription factors. It is the same as SAPK. They belong to the MAK family of protein kinases that are activated by stress (environmental stress, heat shock, tumor necrosis factor). SAPK appears to be inhibited by p21, a transforming protein. Activated JNK stimulates the transcriptional activity of AP1. JNK interacting protein-1 (JIP-1) causes the retention of JNK in the cytoplasm and thus inhibits JNK-regulated gene expression. JNK signaling activates CD4 helper T cells (T$_H$), which during clonal proliferation release interleukins and become T$_H$1 and T$_H$2 effector cells that mediate inflammatory responses. JNK is involved in the mitochondrial release of cytochrome c and the apoptotic path. *See* AP1; apoptosis; ASK1; aspirin; ATF2; ELK; interleukins; Ire; JUN; MAPK; MLK; NFAT; p21; Pyk; SAPK; T cell; TRAF. (Davis, R. J. 2000. *Cell* 103:239; Bagowski, C. P. & Ferrell, J. E., Jr. 2001. *Curr. Biol.* 11:1176; Weston, C. R. & Davis, R. J. 2002. *Curr. Opin. Genet. Dev.* 12:14.)

Jockey *see* nonviral retrotransposable elements

Joining of DNA *See* blunt-end ligation; cohesive ends; ligase.

joint probability When two events are independent from each other, the probability of their joint occurrence can be obtained by multiplication of the independent probabilities. The same rule applies to more than two independent frequencies. Independence means that the occurrence of one has no bearing on the occurrence of the other(s). *See* probability.

Jost factor *See* Müllerian inhibitory substance.

Joubert syndrome Heterogeneous autosomal-recessive developmental defect of the human brain (cerebelloparenchymal disorder, cerebellar vermis agenesis) was assigned to human chromosome 9q34.3.

Joule 1 joule $= 10^7$ ergs, the energy expended per 1 second by an electric current of 1 ampere in a resistance of 1 ohm; approximately 0.24 calorie.

Juberg-Marsidi syndrome Xq12-q21 mental retardation, growth and developmental anomaly is based on mutation in a helicase. The X-linked α-thalassemia and mental retardation seem to involve the same protein. *See* ATRX; mental retardation; thalassemia.

Judassohn-Lewandowsky syndrome (pachyonychia congenita, PC1, 17q12-q21, 12q13) Hereditary recessive keratosis of the nails (onchyogryposis), palm, sole, and mouth. It is due to mutation in keratin 16. *See* ichthyosis; keratosis.

Jukes-Cantor estimate of evolutionary divergence Is based on the number of nucleotide substitutions since the separation of two DNA sequences during evolution. D = $2\alpha t$ where D = distance, α = the probability (p) that one nucleotide is replaced in time t. The separation in time t = D/2α. *See* evolutionary distance. (Chen, F. C., et al. 2001. *J. Hered.* 92:481.)

Jumping Frenchman of Maine Rare and obscure apparently autosomal-recessive anomaly characterized by very rapid emotional reactions.

jumping genes Move in the genome because they are within transposons. *See* transposable elements.

jumping library Generated by circularizing large eukaryotic DNA fragments and cloning the junctions of the circle. The large fragments are obtained by using restriction enzymes that very rarely cut the DNA. *See* chromosome jumping; DNA library; linking library; slalom library. (Zabarovsky, E. R., et al. 1991. *Genomics* 11:1030.)

jumping translocations Involve one (donor) chromosome and multiple recipient chromosomes. Such unstable phenomena are common in cancers, mainly involving human chromosome 1. The breakpoints are generally in regions of repetitions such as centromeric, telomeric, and rRNA sequences. *See* Levy, B., et al. 2000. *Cytoget. Cell Genet.* 88:25; Padilla-Nash, H. M., et al. 2001. *Genes Chromosomes Cancer* 30:349.)

jump stations Collection of links for genetic and biological information regarding databases, journals news groups, etc. *See* databases (general directories).

JUN (*jun*) Avian fibrosarcoma oncogene homologue JUN-A is in human chromosome 1p32-p31 and in mouse chromosome 4. Its homologues are also present in other vertebrate species and it appears to be identical to a subunit of transcription factor AP-1. Along with the product of oncogene FOS, these homologues activate several genes. The products of JUN and FOS are bound together with a leucine zipper and at their carboxyl end they have a DNA-binding domain (5'-TGAGTCA-3'). They apparently form the C/EBP protein. JUN-B and JUN-D oncogenes are closely linked in mouse chromosome 8. JUN-B human homologue is in human chromosome 19p13.2. UV-irradiated mammalian cells may exit from the p-53 imposed block of the cell cycle by induction of Jun. *See* AP1; bZIP; C/EBP; FOS oncogene; JNK; oncogenes; signal transduction; UV. (Barr, R. K. & Bogoyewitch, M. A. 2001. *Int. J. Biochem. Cell Biol.* 33:1047.)

junction complex Assembly of the various types of junctions (tight junctions, adhesion belt, desmosome) within cells. *See* desmosome; gap junctions.)

junction of cellular networks Integrators of molecular signals coming from a different origin and regulated by the interconnections. cAMP may represent such a *junction* because it is affected in a positive or negative manner by a variety of signals. Phosphokinase A as a *node* may split the signals and direct them to multiple targets such as the cytoskeleton and cellular traffic, gene expression and cell growth, metabolism, ion channels, G-protein-coupled receptors of signal transduction, neuronal synapsis, etc. Another example of a node is Cdc42, which receives signals through receptor tyrosine kinases (RTK) and G-protein-linked receptors (GPCR), then sorts them by serum response factor (SRF) and p21 activated kinase (PAK), S6 kinase affecting transcription, translation, and cellular traffic. *See* Cdc42; coordinate regulation; G proteins; PAK; RTK; signal transduction; S6 kinase. (Jordan, J. D., et al. 2000. *Cell* 103:193; McCarty, D. R. & Chory, J. 2000. *Cell* 103:201; Vohradsky, J. 2001. *FASEB J.* 15:864.)

junctional diversification When immunoglobulin genes are recombined to generate specific antibodies, a few nucleotides may be lost or added to the recombining ends. *See* antibody; combinatorial diversification; immunoglobulins; RAG. (Wang, C., et al. 1997. *J. Immunol.* 159:757.)

junction sequence   *See* introns.

junctophilin   Junctional membrane complex protein. Junctophilin deficiency may lead to muscle and motor defects and Huntington disease–type anomalies. *See* Huntington's chorea. (Takeshima, H., et al. 2000. *Mol. Cell* 6:11; Holmes, S. E., et al. 2001. *Nature Genet.* 29:377.)

jungles   Chromosomal regions with high frequency of recombination (genes). *See* deserts.

juniper (*Juniperus communis*)   Evergreen woody species, $2n = 22$.

junk DNA   DNA that appeared without any obvious function, such as some introns and spacers, when the term was coined in 1980. Today, several introns are known with maturase and other functions. Some of the noncoding DNA is interspecifically conserved, indicating some kind of biological function. In animal chromosomes, nearly 97% of the DNA is noncoding and this "junk" DNA is predominantly intron material. *See* C-value paradox; LINE; selfish DNA; SINE; trinucleotide repeats. (Wong, G. K.-S., et al. 2000. *Genome Res.* 10:1672.)

Jurassic period   About 137 million to 190 million years ago; an era dominated by the dinosaurs and reptiles, although the ancestral forms of most vertebrates were also present and even primitive mammals began to appear.

juvebione   *See* juvenile hormone.

juvenile hormone   Secreted in the larval state; prevents precocious metamorphosis into the pupal stage of the insect. Juvenile hormones include ethylpolyprenyl components. Similar terpenes and terpene-related substances, e.g., juvebione (in balsam fir) and gossypol (in cotton), occur in plants and affect feeding insects. Synthetic hormones have been produced with similar physiological effects. *See* abscisic acid; allostatin; ecdysone; metamorphosis; molting; pupa. (Davey, K. G. 2000. *Insect Biochem. Mol. Biol.* 30[8–9]:663; Gilbert, L. I., et al. 2000. *Insect. Biochem. Mol. Biol.* 30[8–9]:614.)

juvenile mortality   Frequently the function of the consanguinity of the parents, e.g., stillbirth and neonatal death, if the parents were unrelated, was found in one study to be 0.044. If the parents were first cousins (consanguinity 1/16), it was 0.111. Similarly infant and juvenile death rates were 0.089 and 0.156, respectively. *See* coancestry; inbreeding; mortality.)

juvenile onset   Hereditary condition appearing in childhood. *See* diabetes mellitus.

juxtacrine signaling   The membrane-anchored growth factors and cell adhesion molecules are signaled through the juxtacrine mediators. *See* signal transduction.

JX$_2$ DNA   *See* PX DNA.

Carl Wilhelm von Nägeli made an effort to convince Mendel about the insignificance of his experiments with peas because he felt it inconceivable that segregation in plants should obey statistical rules. Similarly, he was the founder and unbending adherent of the theory of pleomorphism of bacteria. According to this idea, bacteria did not have a stable heredity, but would change from one form to another by a change in the environment. Although Mendel's theory was not understood by his contemporaries, pleomorphism was subjected to serious criticism; yet von Nägeli's obvious influence definitely hampered bacteriology. Dr. W. Migula, Professor at the College of Technology in Karlsruhe, gives a vivid account of the situation in his *System der Bakterien* (Fischer V1g., Jena, 1897, p. 215):

"When Nägeli says, p. 20, that 'Cohn [the founder of modern bacterial systematics in 1872] had established a system of genera and species, in which each function of the Schizomycetes [bacteria] is represented by a particular species; by this he expressed the rather widespread view exclusive to physicians. So far I have not come across any factual ground that could be supported by morphological variations or by pertinent definitive experiments.' When Nägeli still says this in 1877, one must either assume that he was unaware of the work of the preceding 5 years, or that he chose to ignore it on purpose because it did not fit to his theory."

A historical vignette

K

K Kinase.

K1 Non-nucleoidal methylation sites of *E. coli* transducer proteins spaced seven amino acid residues apart.

K$_a$ *See* dissociation constant.

Ka Thousand of years before present time.

K$_A$/K$_S$ Ratio of nonsynonymous and synonymous mutations. The former leads to amino acid replacement in the protein. The ratio thus indicates an adaptive change and has been used to measure molecular evolution. K_I substitution rate within introns; K_4 rate of synonymous substitution in fourfold degenerate sites. *See* degenerate code; Grantham rule; molecular evolution; mutation, beneficial; synonymous codon.

KABAT database *See* immunoglobulins.

kainate Cyclic analog of glutamic acid; also forms a synaptic receptor. Kainate may be a neurotoxin and it is antihelminthic. (Bailey, A., et al. 2001. *Eur. J. Pharmacol.* 431:305; Lauri, S. E., et al. 2001. *Neuron* 32:697.)

Kalanchoe Bryophyllum of several species with chromosome numbers varying from 34 to nearly 300 among them. They have been used for studies of development and differentiation and had been a favorite subject for investigations on the effects of agrobacteria on plants. *See Agrobacterium*.

Kalilo *See* killer plasmids. (Bok, J. W. & Griffith, A. J. 2000. *Plasmid* 43[2]:176.)

K-allele model Interprets mutation mechanisms. Accordingly there are K possible allelic states (at a gene or microsatellite) and each has a constant probability ($\mu/[K-1]$) to undergo mutation to any of the other $K-1$ allelic forms. In IAM, K is infinite. *See* IAM; microsatellite; minisatellite; SMM; TPM. (Vitalis, R. & Couvet, D. 2001. *Genetics* 157:9111.)

kallikrein Serine proteinase in the pancreas, saliva, urine, and blood plasma that cleaves kallidin (a kind of kinin) from globulin. It has a vasodilator and possibly some type of skin-irritating effect. (Diamandis, E. P., et al. 2000. *Trends Endocrinol. Metab.* 11:54.)

Kallmann syndrome Rare autosomal-recessive malfunction of the gonads resulting in infertility, lack of ability to smell (anosmia), cleft palate, and cleft lip. It is apparently caused by defects of the olfactory receptor neurons in the steroid hormone receptor(s) and in the gonadotropin-releasing hormone neurons. There is also an X-linked (Xp22.3) form of the disease. It has been suggested that the X-linked 14 exons code for a cell adhesion (fibronectin) molecule. *See* adrenal hypoplasia; gonadotropin-releasing hormone; hypogonadism; infertility; N-CAM; olfactogenetics. (Rugarli, E. I. 1999. *Am. J. Hum. Genet.* 65:943.)

KAM *See* K-allele model.

Kammerer, Paul (20th-century lamarckist) From 1905 on he reported hereditary changes in the coloration of salamanders and midwife toads by exposing the animals to yellow or dark background, respectively. The much-heralded experiments could not be confirmed under appropriate scrutiny and it turned out that the dark spots on the preserved specimens were marked by India ink rather than by genetic mechanisms. When the forgery came to light, Kammerer acknowledged the truth but denied personal fraud before committing suicide in 1926. The rumors were that an assistant or a janitor in Vienna, Austria, played a practical joke on him. In the Soviet Union where he was invited as a professor, the withdrawal of his reports was never acknowledged because his claims were in line with the then current political dogmas. *See* lamarckism. (Meinecke, G. 1973. *Med. Welt* 24[38]:1462, in German.)

kanamycin Aminoglycoside antibiotic frequently used as a selectable marker in genetic transformation. *See* aminoglycoside phosphotransferase; antibiotic resistance; antibiotics; geneticin; transformation, genetic; vector.

Kanamycin (R = NH$_2$; R$'$ = OH).

kangaroo *Macropus rufus*, $2n = 20$; rat kangaroo (*Potorous tridactylus apicalis*), $2n = 13$ in the male and 12 in the female.

Kanzaki disease (Schindler type II) Autosomal-dominant lysosomal glycoaminoacid storage disease with angiokeratoma but no neurological defects. *See* angiokeratoma; Schindler disease.

KAP Human phosphatase with specificity for Thr[160] in Cdk2. *See* Cdk2.

Kaplan-Meier estimator of survival (product limit estimator) Calculated in steps for the periods of time of the survivor numbers of a treated experimental population. The formula at the time of the first death (t_1): $(Y_1 - d_1)/Y_1$, where $Y(Y_i)$ stands for the residual number of animals at risk and d (or in general terms d_i) for the number of deaths. After the second death time (t_2), the formula is ($[Y_2 - d_2]/Y_2$), and so on, until

t_i, when it is $([Y_i - d_I]/Y_I)$. At each step a fraction is obtained. By multiplying these fractions, we obtain the survival estimate at t_n. The cumulative hazard $H_t = -\ln\Sigma(t)$. *See* genetic hazards; LD50; MELD; MTD. (Dubin, J. A., et al. 2001. *Stat. Med.* 20:2951.)

kapok (*Ceiba*) Southeast Asian bamboo fiber tree, $2n = 72$–88, with unknown basic chromosome number.

Kaposi's sarcoma (hemangiosarcoma) Shows red-purple nodules and plaques that become tumorous. It is due to autosomal-dominant genes, but it is also an opportunistic tumor because it is expressed mainly when some types of infections, such as *Pneumocystis* microorganisms, are present as in the case in AIDS and other immunodeficiencies. Herpes virus KSHV/HHV8 is generally present in the tissues in a latent form as episomes. In primary effusion lymphoma (PEL) cells, KSHV encodes LANA (latency-associated nuclear antigen). In PEL, LANA and KSHV are colocalized in dots of interphase nuclei and they are in a diffuse form on mitotic chromosomes. The higher frequency and aggressiveness of Kaposi sarcoma in AIDS is explained by the synergism between cellular basic fibroblast growth factor (FGF) and Tat enhancer protein of the virus. A novel herpes virus (HHV8/KSHV [Kaposi sarcoma–associated herpes virus]) has also been blamed for causing the disease. HHV8 infection seems to be mediated by an interaction between the RGD peptides of the viral envelope protein and $\alpha 3\beta 1$ integrins. The virus transactivates the promoter of the reverse transcriptase of the telomerase. *See* acquired immunodeficiency; FGF; herpes virus; HHV8; telomerase; transactivator. (Lagunoff, M., et al. 2001. *J. Virol.* 75:5891; Knight, J. S., et al. 2001. *J. Biol. Chem.* 276:22971; Akula, S. M., et al. 2002. *Cell* 108:407.)

κ *See* symbionts, hereditary.

κ **chain**   *See* immunoglobulins.

kappa-deleting element   Assists in the rearrangement of the immunoglobulin κ-chain. *See* immunoglobulins. (Seriu, T., et al. 2000. *Leukemia* 14[4]:671.)

kappa particles   Lysogenic bacterial symbionts that kill the sensitive hosts of *Paramecium aurelia*. *See Paramecium*; symbionts, hereditary. (Preer, L. B., et al. 1972. *J. Cell Sci.* 11[2]:581.)

Kar1   Yeast protein regulating nuclear migration (congression) during karyogamy. *See* cell cycle.

Kar2   Cytoplasmic assembly protein regulating nuclear fusion. *See* Hsp70; karyogamy; Sec63.

Kar3   84 kDa motor protein of the kinesin family in yeast operating in nuclear fusion and microtubule sliding prior to anaphase. *See* cell cycle; kinesin; mitosis.

karmellae   Nuclear-associated endoplasmic reticulum membrane components. (Koning, A. J., et al. 2002. *Genetics* 160:1335.)

Kartagener syndrome (dextrocardia, 19q13.3) Complex recessive syndrome of left-right inverted location of major visceral organs involving lack of cilial movement and sperm motility, as well as nasal polyps. The basic defect seems to involve mutation in the axonemal intermediate chain of dynein (DNAI1) or in the heavy chain (DNAH5). *See* asthenozoospermia; asymmetry of cell division; axis of asymmetry; dynein; left-right asymmetry; situs inversus viscerum. (Guichard, C., et al. 2001. *Am. J. Hum. Genet.* 68:1030; Olbrich, H., et al. 2002. *Nature Genet.* 30:143.)

karyogamy   Nuclear fusion in fungi following fusion of the cytoplasms of two cells, plasmogamy. The word $\kappa\alpha\rho\nu o\nu$ means kernel; $\gamma\alpha\mu\varepsilon\iota\nu$ means to marry. *See* fungal life cycles.

karyogram   Depiction of the karyotype. *See* karyotype.

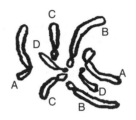

Karyogram of *crepis parviflora*.

karyokinesis   Division of the cell nucleus. *See* cell cycle; mitosis.

karyolymph   Fluid fraction of the cell nucleus in contrast to the particulate ones, e.g., chromosomes.

karyopherin/importin (α and β) Cytosolic protein that mediates nuclear traffic in cooperation with nucleoporin, with a GTPase and RAN. Importin-α (M_r 60 K) recognizes the nuclear localization sequences and importin-β (M_r 97 K) docks the nuclear pore complex. There are more than a dozen karyopherins in yeast and even more in higher eukaryotic cells. *See* CAS; chromosome maintenance region 1; GTPase; importin; nuclear localization sequences; nuclear pore; RAN; RNA export; transportin. (Yoshida, K. & Blobel, G. 2001. *J. Cell Biol.* 152:729.)

karyoplast   Nucleus surrounded by only a thin layer of cytoplasm and membrane. *See* cytoplast; transplantation of organelles.

karyosome   Mass of chromatin of the oocytes before metaphase of animal cells.

karyotheca   Synonymous with nuclear envelope. *See* nuclear membrane.

karyotype   Characteristics of a (mitotic) metaphase chromosome set by number, morphology, arm ratio, secondary constrictions, banding pattern, etc. *See* chromosome morphology; electrophoretic karyotyping; FISH; GISH.

The haploid chromosome set in the subgenus sophophora of *Drosophila*. The lowest chromosomes are the X. Note the gradual fusion of the arms and chromosomes from left to right. The X chromosome of *D. anassae* probably evolved from that of *D. melanogaster* by a pericentric inversion. In *D. willstonii*, the small dot-like chromosome seems to be incorporated into the biarmed X chromosome. (After Sturtevant, A. H. 1940. *Genetics* 25:337.)

Subobscura   Pseudoobscura   Melanogaster   Ananassae   Willistoni

karyotype evolution Chromosome number is a rather good characteristic of a species, although an identical number of chromosomes may not indicate any relationship.

The morphology of the chromosomes and arm ratio are frequently used to assess similarities, but substantial differences may exist within related groups and pericentric inversions may alter the relative position of the centromere. Centromere fusion and Robertsonian translocations may convert telocentric chromosomes into biarmed ones. Misdivision may generate telocentrics from biarmed chromosomes and at the same time change the chromosome numbers. Paracentric inversions may serve the purpose of speciation and the sequence of change in the pattern of chromosome bands (in polytenic chromosomes or specially banded chromosomes) may be traced by the techniques of light microscopy. Translocations and other types of chromosomal aberrations may also be followed in related species. The similarities of the karyotypes can be assessed on the basis of chiasma frequencies if the species are closely related enough to permit meiotic analyses. FISH and GISH are important tools of karyotyping. Chromosomal mapping of classical genetic markers or restriction fragments, sequence-tagged sites, and RAPDs may reveal the order of genes and nucleotide sequences and their evolutionary path. Karyotic changes (chromosome number and chromosome arm number) per evolutionary lineage per MY vary a great deal from 1.395 in horses to 0.025 in whales to 0.029 in other vertebrates (lizards, teleosts). *See* banding techniques; chromosomal aberrations; evolution of the karyotype; FISH; fruit fly; GISH; inversion; misdivision; polyploidy; polytenic chromosomes; RAPD; RFLP; sequence-tagged sites. (Graphodatsky, A. S., et al. 2001. *Cytogenet. Cell Genet.* 92[3–4]:243, Yu, K. & Ji, L. 2002. *Cytometry* 48:202.)

karyrhexis Fragmentation of the nucleus at cell death. *See* apoptosis.

kasugamycin Antibiotic that alters 16S rRNA and inhibits protein synthesis on the ribosome.

katanin ATPase of the AAA protein family that cleaves and disassembles microtubules. *See* AAA proteins; ATPase; microtubule. (Hartman, J. J. & Vale, R. D. 1999. *Science* 286:782.)

kazal (5q32) Serine protease inhibitor. Its defect may be responsible for chronic pancreatitis.

kb Kilobase, i.e., 1,000 bases.

kbp Kilobase pairs, thousand pairs of nucleotides in double-stranded nucleic acids.

KC Cytokine protein, homologous to N51, MGSA, and gro. *See* cytokines; gro; MGSA; N51.

KCNA Potassium voltage-gated ion channel diseases encoded at several loci in the short arm of human chromosomes 12 and 19 involving neurological disorders. *See* ion channels; LQT. (Charlier, C., et al. 1998. *Nature Genet.* 18:53.)

K_d *See* dissociation constant.

kDa Kilodalton (1,000 Da). *See* dalton.

KDEL (Lys-Asp-Glu-Leu) Amino acid sequence serving as a conserved carboxy-terminal peptide in many endoplasmic reticulum (ER) lumenal proteins involved in the traffic between ER and the Golgi apparatus. Proteins with KDEL are retained within the lumen. Several other motifs are also conserved in the transport systems. *See* antibody intracellular; ER; plantibody. (Gatti, G., et al. 2001. *J. Cell Biol.* 154:525.)

kDNA Kinetosome DNA. *See* kinetosome.

KDR Receptor for neuropilin. *See* neuropilin.

Kearns-Sayre syndrome *See* mitochondrial disease in humans; optic atrophy.

KEGG *See* Kyoto Encyclopedia of Genes and Genomes.

keel Two petals associated along the edge.

Keimbahn Germline; A. Weissmann, 1885.

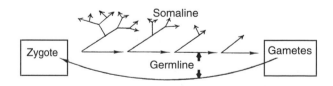

Keimplasma *See* germplasm.

Kell-Cellano blood group (KEL) The KEL antigen is a 93 kDa membrane glycoprotein associated with the cytoskeleton and is encoded in the human chromosome 7q32 area. Its precursor substance (Kx) is coded in the X chromosome (McLeod syndrome). Its mutations may cause "horny" appearance of the erythrocytes (acanthocytosis) and granulous inflammations in response to infectious and other factors. The frequencies of the KEL and Kx alleles in England were found to be between 0.0457 and 0.9543, respectively. *See* blood groups; McLeod syndrome. (Lee, S., et al. 2001. *J. Biol. Chem.* 276:27281.)

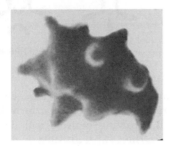

Horny erythrocyte.

kelp   Large brown algae.

Kelvin   Temperature scale is used primarily in thermodynamics; $0°C = 273°K$; the conversion between $C°$ and $K°$ is $C° = K° - 273$, e.g., $100°C = 373°K$ or $0°K = -273°C$

KENAF (*Hibiscus cannabinus*)   Warm-climate fiber crop; $2n = 2x = 36$.

Kennedy disease (SBMA)   Nonlethal spinal bulbar muscular atrophy, sensory deficiency, frequently with gynecomastia and impotence expressed primarily in adults. The recessive gene was mapped to Xq12. The basic defect is a CAG repeat increase (11–33 CAG in normal → 38–66 in SBMA) within the first exon of the androgen receptor (AR) in the spinal cord, brain stem, and sensory neurons. Glucocorticoid receptor (GR) localizes the repeat into the nucleus. Deletion or mutation in the C-terminal of GR suppressed aggregation and nuclear localization. Surprisingly, mutation in the DNA-binding N-terminal domain increased aggregation and nuclear localization by GR. Female heterozygotes for the repeats in AR are not affected and males with AR deletions do not show SBMA, although they are usually sterile and somewhat feminized. Increase in size is more common when the transmission is by an affected male than by a carrier female. Longer repeats increase the severity of the symptoms. (This syndrome is not named after President Kennedy's back ailment.) *See* androgen receptor; dihydrotestosterone; fragile sites; glucocorticoid; gynecomastia; neuromuscular disease; spinal muscular atrophy; testicular feminization; trinucleotide repeats.

keratin   Protein of the surface layer of skin, hair, nails, hoofs, wool, feather and porcupine quills, etc. Keratins may be high-sulfur, acidic matrix proteins or low-sulfur, basic fibrous proteins. These two types usually appear in pairs and are controlled in humans by autosomal-dominant genes. Point mutation in keratin genes in human chromosomes 12 and 17 may lead to various epithelial anomalies. *See* cathepsin; hair; keratosis.

A keratin fibril.

keratitis   Autosomal-dominant inflammation of the cornea.

keratoma (hyperkeratosis)   Formation of keratoses on the palms and other parts of the body. Palmoplantar keratoderma (17q12-q21) is a dominant epidermolytic condition due to mutation in the keratin 9 gene with potential carcinogenic consequences. Another palmoplantar keratoma (12q11-q13) is frequently elicited by fungal infection and does not cause epidermolysis. Striate palmoplantar keratoderma was located to 18q12.1-q12.2. Desmoplaquin (6p24) mutation is responsible for striate II keratoderma. *See* keratosis; Mal de Maleda.

keratosis   Either a wart-like flat or emerging (scaly) spot(s) that may become cancerous or a soft friable (sometimes colored), noninvasive benign skin lesion. Both may have a number of different forms and are under the control of autosomal-dominant genes. These skin lesions generally appear during adulthood, but some start in very early childhood and develop progressively; they may signal more serious conditions. Sunburn may lead to keratosis and squamous cell carcinoma if mutation occurs in p53. *See* Darier-White disease; dyskeratosis; FGF; ichthyosis; intermediate filaments; Jackson-Lawler syndrome; Judassohn-Lewandowsky syndrome; p53; pachyonychia; psoriasis; skin diseases.

KERMA   Kinetic energy released to the medium upon irradiation.

kermit   *Drosophila* transposable element (4.8 kb). *See* copia.

Wheat kernel with the embryo at base.

kernel   "Seed" (grain) covered by the pericarp (fruit) and not just with the seed coat, such as a wheat or maize kernel; in barley it may have also glumes (husks) attached.

k$_e$ test   Detects electrophilicity of chemicals and thus indicates their potential to react with DNA and cause mutation or cancer. *See* electrophile.

ketoacidosis   Occurs in several human diseases when ketones accumulate due to a defect in succinyl-CoA, 3-ketoacid CoA-transferase in patients with diabetes mellitus, Gierke's disease (type I glycogen storage disease), glycinemia, methylmalonic aciduria, and lactic aciduria.

ketogenic amino acids   Tryptophan, phenylalanine, tyrosine, isoleucine, leucine, and lysine can serve as percursors of ketone bodies (acetoacetate, D-3-hydroxybutyrate, acetone) formed primarily from acetyl coenzyme A if fats are degraded. *See* amino acid metabolism.

ketone Closely resembles aldehyde. Both have the same unsaturated carbonyl group.

$$R-C=O$$

with R above the carbon.

ketose Monosaccharide. The carbonyl group is a ketone such as fructose.

ketosis Condition with accumulation of ketone bodies, e.g., in diabetes mellitus. The C=O group is generally joined to two other C atoms in the ketone bodies. *See* ketone.

ketotic Showing ketosis. *See* ketosis.

ketothiolase *See* methylacetoaceticaciduria.

Keutel syndrome Human chromosome 12p12.3 area recessive defects involving lung, hearing, cartilage, and face. The gene encodes an 84-amino-acid transmembrane protein MGP with a 19-amino-acid signal peptide. It belongs to the family of extracellular matrix proteins. In all of these proteins, γ-carboxyglutamic acid is found. This residue potentiates high affinity for Ca, PO_4; and hydroxyapatite crystals and calcification of the extracellular matrix. *See* extracellular matrix; Singleton-Merten syndrome.

KeV (Kilo electron volt) 1,000 electron volts. *See* electron volt.

keyhole limpet hemocyanin (KLH) Carrier protein (from *Megathura crenulata*) that can be linked to synthetic amino acid sequences with 12 to 15 hydrophobic residues through an NN_2 or COOH-terminal cysteine. The synthetic peptides can be used to raise antisera against them and these may be advantageous because their recognition is independent of the conformation of the whole protein. *See* antiserum; hemocyanin; monoclonal antibody.

key-lock theory Enzymes are rigid molecules that can accommodate the substrate in the manner of jigsaw puzzles in contrast to the induced fit theory. *See* induced fit.

KGK Keratocyte growth factor is involved in epithelial cell development in wound repair.

KH module The K homology motif is present in the heterogeneous nuclear ribonuclear protein (RNP) K, protein, ribonuclease P, Mer splicing modulator of yeast, and fragile X product. A common feature of these domains is the binding of single-stranded RNA. *See* FMR1; fragile X; hnRNA; ribonuclease P; splicing. (Grishin, N. V. 2001. *Nucleic Acids Res.* 29:638.)

KID Kinase-inducible domain in transcription factors. *See* CREB; kinase; transcription factors.

Kidd blood group (Jk) Apparently associated with human chromosome 18q11-q12. The worldwide frequency of the Jk^a allele appears to be 0.5142 and that of the Jk^b is 0.4858. *See* blood groups.

kidney diseases May be caused by various environmental and genetic factors. *See* Addison disease; Alport syndrome; Bardet-Biedl syndrome; Dent disease; diabetes insipidus; glomerulocystic kidney disease, hypoplastic familial; glomerulonecrosis; glomerulonephritis; glomerulosclerosis, focal and segmental familial; hypernephroma; hypouricema; kidney stones; nephritis; nephritis, familial; nephrolithiasis; nephropathy juvenile hyperuricemic; nephrosialidosis; nephrosis; nephrotic syndrome, steroid resistant; oncocytoma; polycystic kidney disease; renal cell carcinoma; renal dysplasia; renal dysplasia and limb defects; renal hepatic-pancreatic dysplasia; renal tubular acidosis; retinal aplasia; urogenital dysplasia; xanthinuria.

kidney stones (nephrolithiasis) Accumulation of primarily oxalate crystals in the kidneys caused by intestinal malabsorption of dietary salts. Polygenic and autosomal-dominant control have been claimed to be responsible, although one form appears to be X-linked recessive (Xp11.22). In the X-linked form, the defect is caused by various mutations in an outwardly rectifying chloride ion channel. *See* Dent disease; ion channel; kidney diseases; nephrolithiasis.

KIF Motor protein superfamily. In cooperation with AP1 and other proteins, KIF facilitates vesicular transport between organelles. *See* AP1; kinesin. (Nagai, M., et al. 2000. *Int. J. Oncol.* 16:907.)

killer cells (NK, natural killer cell) Part of the immune system; they are large granular lymphocytes and they cooperate with other elements of the defense system to eliminate foreign organisms. Natural killer cells mount a cytotoxic reaction against invaders without prior sensitization. Similarly to other lymphocytes, they kill some virus-infected cells and tumor cells. NK cells are controlled (inhibited) by MHC receptors specific for class I molecules of the major histocompatibility complex. NK cells do not express conventional receptors for antigens, but their triggering is carried out by the surface molecules NKp46 (activating natural killer receptor p46, human chromosome 19), NKp30 (lymphocyte antigen 117, 6p21.3-p21.1), and NKp44 (lymphocyte antigen 95 [LY95], human chromosome 6) that are the natural cytotoxicity receptors (NCR). This system is coupled with signal transducers such as CD3ζ, FcεRIγ, and KARAP/DAP12 (killer cell−activating receptor-associated protein).

MIC proteins select the targets for destruction. In association with NKG2D protein, MICA/B proteins are a part of NK cell receptors; in association with other proteins (DAP10 [19q13.1], p85 [5q13], p110 [5q13]), they activate killing by NK cells. 2B4 (1q22) and NKG2D (12p13.2-p12.3) appear to be coreceptors. NK cell may kill allogeneic cells because of the absence of autologous MHC molecules and cells that display different antigens due to viral infection. Human NK class I receptors consist of an immunoglobulin domain containing lysine in the transmembrane section, whereas in mice the transmembrane portion of receptors resembles C-type lectins. The lectin-like heterodimer CD94/NKG2 recognizes HLA-E molecules. The killer cell inhibitory receptors (KIR, human chromosome 19q13.4) recognize HLA-B (KIR3D) or HLA-C (KIR2D).

The inhibitory receptors are characterized by ITIM (immunoreceptor tyrosine-based inhibitory motif). The ITIMs are phosphorylated but can attract and activate tyrosine phosphatases, resulting in suppression of the cytotoxic function of NK cells. Those cells, which have short receptors (without ITIM), remain phosphorylated and are killers. In case ITIMs are lacking or unavailable, proteins with ITAM (immunoreceptor tyrosine-based activation motif) can be phosphorylated by ZAP-70 and Syk kinases and thus become activated. One such ITAM protein is DAP12 ($M_r = 12$ K, 19q13.1). DAP12 also binds a killer cell–activating receptor (KIR2DS2). The activity of NK cells is increased by the secreted interferons, especially by γ-interferon, but IFNγ provides protection for normal cells against NKs. NK cells secrete chemokines MIP-1α and RANTES. They usually cooperate with cytotoxic T cells (CTL). They are often referred to as non-MHC-restricted cells because they can even attack cells that do not express MHC. The outermost layer of the human placenta lacks class I and class II MHC proteins, which is sufficient to protect the hemiallogeneic fetal cells against T cells, but this would not protect from NK killer cells. The trophoblast cells (extraembryonic ectodermal tissue) in contact with the placenta (the tissue connecting the maternal and fetal system) express the HLA-G class I molecules, which is sufficient to protect the pregnancy from some adverse immunological reaction. There are also mononuclear killer cells that have antibody-dependent cellular cytotoxic ability. Herpes viruses may produce MHC class I homologues that may interfere with NK cell defense systems. *See* allogeneic; anomalous killer cell; antibody; at least one hypothesis; autologous; blood cells; caspase; cytotoxic T cells; DAP; HLA; immune system; ITIM; lectin; lymphocytes; MHC; MICA/B proteins; Mic-1α; missing self hypothesis; monocyte; NKT cells; p85; p110; RANTES; Syk; T cells; ZAP-70. (Brown, M. G., et al. 1997. *Immunol. Rev.* 155:53; Biron, C. A., et al. 1999. *Annu. Rev. Immunol.* 17:189; Guidotti, L. G. & Chisari, F. V. 2001. *Annu. Rev. Immunol.* 19:65; Moretta, A., et al. 2001. *Annu. Rev. Immunol.* 19:197; Raulet, D. H., et al. 2001. *Annu. Rev. Immunol.* 19:291; Colucci, F., et al. 2002. *Nature Immunol.* 3:807.)

Killer genes Well-known examples of the antimorphic *Killer of prune* mutations (awd^K, *abnormal wing disc*; at the end of the long arm of chromosome 3) in *Drosophila* that are viable as homo- or hemizygotes but exert dominant killing effects on most of the hemizygous (male) third instar recessive *prune* mutations (*pn*; 1–0.8). The *awd* gene encodes a nucleoside diphosphate kinase and displays very high homology with the human gene NM23 (17q22) that encodes metastasis inhibitor proteins and is detectable at reduced levels in several malignant cancers. The *prune* alleles encode a 45 kDa protein and control a lower level of pteridine (drosopterin) pigments in the eye, resulting in brownish rather than reddish color of the wild type. Some mutant *prune* alleles do not respond lethally to the awd^K mutations; others respond only in a temperature-sensitive manner and some are lethal. The biochemical mechanism of the lethal interaction between *pn*

and awd^K is not clear. *See* (Timmons, L. & Shearn, A. 1997. *Adv. Genet* 35:207.) drosopterin; killer plasmids; Medea factor; pollen killer; temperature-sensitive mutation.

killer plasmids The mitochondrial *kalilo* plasmid (8.6-kb) has 1,338 bp inverted terminal repeats in *Neurospora intermedia* from Hawaii. At the 5′-end it is covalently linked to a 120 kDa protein. The plasmid has two nonoverlapping, opposite-orientation open-reading frames. ORF1 codes for a RNA polymerase (homologous to that of phage T7) and ORF2 is a DNA polymerase gene. Integration of this kalDNA into mtDNA causes senescence and death because it builds up at the expense of normal mtDNA. It is transmitted in heterokaryons. The *maranhar* plasmid is prevalent in *Neurospora crassa* from South Asia. In function, *maranhar* is similar to *kalilo*, although the proteins encoded are substantially different. In prokaryotes, several killer systems are known. In plasmid of R1 group, at least 13 hok/sok genes have been identified that kill cells not carrying the plasmid. *See* hok/sok; killer strains; mitochondrial disease in humans; mitochondrial genetics; mitochondrial plasmids; plasmid maintenance; pollen killer; senescence. (Schaffrath, R. & Meacock, P. A. 2001. *Yeast* 18:1239.)

killer strains Of *Paramecium*, harbor symbionts that release toxins lethal to sensitive strains. Bacteria harboring colicinogenic plasmids may destroy colicin-sensitive strains. In yeast, the double-stranded RNA viruses, L (4.6 kb) and M (two 1.8 kb dsRNA), result in the production of a killer toxin, affecting sensitive strains. These two viruses occur together because only the L strain encodes the capsid protein, whereas the actual toxin is encoded by the M genome. L and M do not exist outside the cells and are transmitted during mating. In some insects, *Wolbachia pipientis* bacterial infection of the males kills all of the offspring sired by these males. *See* colicins; *Paramecia*; symbionts, hereditary.

killer toxin Secreted by many yeast cells. The producer cells themselves are immune to the toxin, but on the sensitive strains a pore is bored that kills the cells. The K1 killer strains contain two double-stranded RNA viral genomes: M_1 dsRNA is 1.8 kb, encoding the toxin (42 kDa) and the immunity substance precursors; the larger L-A dsRNA (4.6 kb) replicates and maintains M_1. These virus-like particles are transmitted during cell divisions and mating. Several genes are required for the maintenance of MAK (maintenance of killer), SKI (superkiller), and KEX (killer expression, endopeptidase and subtilisin-like proteins). KRE (killer resistance) affects cell wall receptors and the function of the killer state. The viral killer toxin, TOK1, activates plasma membrane potassium channels. *See* colicins; killer virus of yeast; *Paramecium*; subtilisin.

killer virus of yeast Has double-stranded satellite RNA as genetic material. Diploid cells of **a**/α-mating type with the M dsRNA segment produce a protein toxin that is lethal to noninfected cells but not to infected cells; therefore, they are denoted as K^+ (killer) and R^+ (resistant), whereas the cells without this viral segment are K^- and R^-. The killer and resistance substances are produced by different processing of the same gene (*KIL-d*) product. The killer can be M1 or M2 type; the two are mutually incompatible (exclusion) and usually M1 prevails. The yeast cells generally contain a L-A dsRNA or a somewhat different helper viral genome. The **a** or the α-cell display defective killer variegation. After mating *KIL-d × KIL-d* haploids, the killer phenotype seems to *heal* in the diploids, but in the haploid progeny of these diploids the variegation reappears, although the defect remains mitotically stable. The *healing* seems to be the epigenetic result of nuclear fusion and it is apparently evoked by viral RNA. The *KIL-d* elements have prion-like properties. *See* killer toxin; mating type determination in yeast; prion. (Sesti, F., et al. 2001. *Cell* 105:637.)

kilobase (kb) 1,000 bases in nucleic acids.

kin selection Generally, natural selection favors the survival of the fittest individuals. In some instances, this principle may not be so obvious because altruism supports individuals for the benefit of the population sharing their genes. Selfless females may sacrifice themselves to predators in attempts to rescue their multiple offspring. In social insects (bees, ants, termites), only the queens and selected males reproduce, yet the workers and soldiers of the colonies protect the reproductive individuals even at the cost of their life. The fitness of these nonreproducing castes is measured by the success of their mating sibs. Actually, the survival of their genes is assured indirectly through these reproducing individuals. It is a nepotism motivated by natural selection. Olfactory stimuli frequently play a role in kin recognition. *See* altruistic behavior; fitness; green beard; inclusive fitness; male stuffing; olfactogenetics; selection. (Agrawal, A. F. 2001. *Proc. R. Soc. Lond. B Biol. Sci.* 268:1099.)

KIN17 Nuclear protein encoded at human chromosome 10p15-p14 and it is associated apparently with unrepaired DNA damage caused by ionizing radiation. *See* DNA repair; ionizing radiation. (Biard, D. S. F., et al. 2002. *J. Biol. Chem.* 277:19156.)

KIN28 Cyclin-dependent kinase of *Saccharomyces cerevisiae*; it phosphorylates the C-terminal domain of RNAP II to facilitate transcription. *See* CDK; RNAP.

KinA, KinB *Bacillus subtilis* kinases affecting sporulation regulatory proteins SpoA and SpoF, respectively. *See* sporulation.

kinase Enzyme that joins phosphate to a molecule. Genes encoding kinases and phosphatases may represent up to 15% of the genome controlling certain developmental pathways. Phosphorylation/dephosphorylation may alter the structure/activity of proteins. *See* protein kinases. (Bauman, A. L. & Scott, J. D. 2002. *Nature Cell Biol.* 4[8]:E203; Cheek, S., et al. 2002. *J. Mol. Biol.* 320:855.)

kindred Group of biological relatives with a determined pedigree. *See* pedigree analysis.

kinectin Membrane protein, binding intracellular vesicles to kinesin. *See* kinesin.

kinesin Cytoplasmic protein involved in moving vesicles and particles along the microtubule plus end; it assists segregation of chromosomes and transport of organelles (endosomes,

Golgi complex, lysosomes, mitochondria, nerve axons) by using energy derived from ATP hydrolysis. The ca. 380 kDa NH_2 domain has the motor function, whereas the COOH terminus probably binds to organelles or microtubules. The kinesins consist of two 120 kDa heavy chains and two 64 kDa light chains arranged in two (10 nm) globular ends connected by a linear molecule either N- or C-terminal or in the middle. Some kinesins may move in a plus or minus direction on the microtubules. The motor function may be at different sites of the sequences with a total length of 80 nm. Kinesins belong to the KIF family of motor-associated proteins. Kinesin motors transport cargo vesicles toward the plus end of the microtubules. The Ncd minus-end-directed motor proteins facilitate the segregation of chromosomes. *See* axon; bimC; centrosome; dynein; endocytosis (endosome); Golgi; kinectin; lysosome; microtubules; monastrol; myosin. (Kikkawa, M., et al. 2000. *Cell* 100:241; Kikkawa, M., et al. 2001. *Nature* 411:439.)

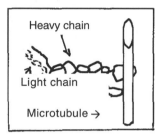

kinetic complexity Measured by the reassociation kinetics of denatured DNA; the increase in complexity (large number of unique diverse sequences) requires a longer time for reassociation. *See* c_0t curve; kinetics.

kinetics Analysis of reaction rates. The reactions may run to completion or may remain incomplete. *Zero order* of the reaction kinetics indicates that the velocity of the process is constant and independent from the initial concentration of the substrate. *First-order* kinetics indicates that only one substrate is involved (monomolecular reaction). Mutation rates and terminal chromosomal deletions below a certain dose of mutagens or clastogens also follow first-order kinetics. *Second-order* kinetics indicates bimolecular reactions, and ionizing radiation-caused chromosomal rearrangements (inversion, transposition, translocation) may also be in this category. *Multiple-order* kinetics may be involved with more than two reacting factors. *See* clastogen; DNA repair; LET; radiation effects.

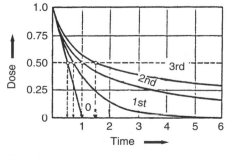

Reaction kinetics: dose (concentration) plotted against time. The 0-, 1st-, 2nd-, 3rd-order kinetics are marked on the curves.

kinetin (6-furfurylaminopurine) Cytokinin plant hormone. *See* plant hormones.

Kinetin.

Kinetochore Structural protein part of the centromere where spindle fibers are attached. The kinetochores associate with the microtubules that connect the chromosomes with the spindle pole. These microtubules pull mitotic/meiotic chromatids/chromosomes to the opposite poles. During interphase, microtubule-kinetochore association does not exist. In *Schizosaccharomyces pombe*, the formation of the spindle takes about 1.5 min at 36°C, whereas its elongation takes about 6.2 minutes. *See* cell cycle; CENP; centromere; microtubules; MTOC; Roberts syndrome; SKP1; SPB. (Skibbgens, R. V. & Hieter, P. 1998. *Annu. Rev. Genet.* 32:307; Kitagawa, K., et al. 1999. *Mol. Cell* 4:21; He, X., et al. 2001. *Cell* 106:195; Tanaka, T. U., et al. 2002. *Cell* 108:31; Shimoda, S. L. & Soplomon, F. 2002. *Cell* 109:9.)

kinetoplast The mitochondrial DNA in some protozoa (*Trypanosoma*) is organized into kDNA (kinetoplast DNA) concatenated circular molecules. The ca. 20–50 copies of the *maxicircular* genome (22 kbp) resembles the mtDNA of other species. The heterogeneous 5,000 to 10,000 copies of the *minicircles* are 0.5–2.8 kbp and represent about 95% of the kDNA. The sum of kDNA constitutes 7–30% of the total cellular DNA in the different *Trypanosoma* and *Leishmania* species. The maxicircle transcripts are subject to editing by adding or (rarely) by subtracting uridine from the primary transcript. Such editing, also called pan-editing, may cause some of the edited transcripts to have more than 50% uridine. The gene that encodes the transcript and is subject to pan-editing is called a *cryptogene* because the sequences of the original transcripts are almost concealed by this process. In the minicircles, short (50–100 base) sequences (guide RNA [gRNA]) remain complementary to the edited RNA. kDNA may have a major role in the life cycle of the protozoan while it is in the insect gut but not in the mammalian blood. The edited RNA of cytochrome bc₁ is translated. *See* gRNA; *Leishmania*; mtDNA; RNA editing; *Trypanosoma*. (Klingbeil, M. M., et al. 2002. *Molecular Cell* 10:175.)

kinetosome *See* kinetoplast.

kingdoms of taxonomy The definition varies among different authors: Prokaryotae, Archaea, Protista, Fungi, Plantae and Animalia or only Bacteria, Archaea and Eukarya.

King George (**III, 1762–1830**) **madness** Neurotic behavior likely caused by porphyria. *See* porphyria. (Hindmarsh, J. T. 1997. *Lancet* 349:364; Warren, M. J., et al. 1996. *Trends Biochem Sci.* 21[6]:229).

kinin Group of endogenous peptides acting on blood vessels, smooth muscles, and (injury-sensing) nerve endings. *See* kininogen.

kininogen Kinin precursor α2-globulins of either 100,000–250,000 M_r or 50,000–75,000 M_r are split by kallikrein to bradykinin and lasyl-bradykinin (kallidin), respectively. The brady-kinins regulate blood vessel contraction, inflammation, pain, and blood clotting. *See* eclampsia; kinin; Williams factor.

kink Distortion (twist) of the DNA double helix or in the secondary structure of RNA. The M_r 7K archaeon protein (Sac7d, Sso7d) binds to the minor groove of the DNA—(GCGATCGC)₂ and (GTAATTTAC)₂—and increases its melting temperature by kinking. (Oussatcheva, E. A., et al. 1999. *J. Mol. Biol.* 292:75, Klein, D. J., et al. 2001. *EMBO J.* 20:4214.)

kinship *See* coefficient of coancestry.

KIP2 (p57KIP2) Cyclin-dependent kinase inhibitor protein, encoded in human chromosome 11p and in mouse chromosome 7. At some developmental stages, it is expressed from the maternal chromosome. KIP1 (p27^KIP1) also has a closely related function in the cell cycle. *See* cancer; CDK; imprinting.

KIR Killer cell inhibitory receptor. *See* killer cell.

Kirsten-RAS (K-RAS) Oncogene of a rat sarcoma RNA virus at human chromosome 12p12.1. It encodes a 21 kDa membrane-associated protein involved in signal transduction. It is commonly responsible for pancreatic and other adenocarcinomas. *See* oncogenes; p21; RAS.

KIS Kirsten murine sarcoma virus oncogene.

KISS (1q32) 145/154 amino acid polypeptide (called metastin) suppresses metastasis of melanoma and breast cancer but does not affect tumorigenesis. It seems to be a ligand of a G-protein-coupled receptor. *See* metastasis; signal transduction. (Ohtaki, T., et al. 2001. *Nature* 411:613.)

kissing loop Dimerization site of HIV at nucleotides 248–271, and it facilitates the incorporation of two genomic RNAs into the virion. *See* acquired immunodeficiency.

kissing interactions Formed when unpaired nucleotides within two RNA hairpins (pseudoknots) pair with each other. These interactions stabilize the structures and may facilitate recognition by various ligands or metal ions. *See* pseudoknot. (Andersen, A. A. & Collins, R. A. 2001. *Proc. Natl. Acad. Sci. USA* 98:7730.)

KIT oncogene Homologue to the viral *v-kit* gene of a feline sarcoma. It is in human chromosome 4q12 and in chromosome 5 of the mouse. This protooncogene codes for transmembrane tyrosine kinase with homology to CSF1R and PDGF. The KIT gene was assumed to be responsible for piebaldism in humans and mice. KIT also signals to spermatogenesis and oogenesis. Gastrointestinal stromal tumors may be caused by gain-of-function mutations in the c-KIT gene. The receptor for KIT appeared to be encoded by the *W* (*white fur*) and PDGF genes of mice, located to about the same or identical chromosomal site as the KIT homologue. The KIT/stem cell factor receptor is involved in hematopoiesis, melanogenesis, and gametogenesis. Stem cell factor proliferation depends on phosphatidylinositol-3′-kinase (PIK). Mutation in the PIK-binding site causes male sterility but no female sterility. Another mouse locus, *Sl* (*Steel*), encodes MGF (mast cell growth factor), a ligand for the growth factor receptor. The KIT oncogene product is also required for the phasic contraction of the mammalian gut and the control of hematopoietic cells. *See* CD117; CSF1R; FLT; gain-of-function mutation; hematopoiesis; mast cells; microphthalmos; oncogenes; PDGF; piebaldism; PIK; receptor tyrosine kinase; stem cell factor; transmembrane proteins; tyrosine kinase. (Smith, M. A., et al. 2001. *Acta Haematol.* 105:143; Kitamura, Y., et al. 2001. *Mutation Res.* 477:165.)

KIX CREB-binding domain of CBP. *See* CBP; CREB.

Kjeldahl method Determines total nitrogen content in organic or inorganic material after hot sulfuric acid digestion in the presence of a catalyzator (Se, Hg). After titration of the distilled ammonia in the presence of phenolphthalein, 1 mL 0.1 N H_2SO_4 bound by the ammonia corresponds to 0.0014 g nitrogen, and the amount of nitrogen multiplied by 6.25 estimates protein content. *See* Bradford method (for protein); Lowry method.

KL Male fertility complex in the long arm of the Y chromosome of *Drosophila melanogaster*. *See Drosophila; KS*; 3 sex determination.

Klebsiella Gram-negative, facultative anaerobic enterobacterial genus (closely related to *E. coli*) with several species, widely present in nature (including hospitals) and capable of causing urinary, pulmonary (lung), and wound infections, but it has the desirable feature of fixing atmospheric nitrogen. *See* nitrogen fixation.

Klenow fragment Large fragment of bacterial DNA polymerase I lacking 5′-to-3′ exonuclease activity (located in the small fragment of the enzyme) but retaining the 5′ → 3′ polymerase and the 3′ → 5′ exonuclease functions. It is generated by cleavage with subtilisin and other proteolytic enzymes. It has been used for nick translation and Sanger's dideoxy method of DNA sequencing. *See* DNA polymerase I; DNA replication in prokaryotes; DNA sequencing; nick translation. (Kuchta, R. D., et al. 1988. *Biochemistry* 27:6716.)

KLH *See* keyhole limpet hemocyanin.

Klinefelter syndrome Caused most commonly by XXY chromosomal constitution, although the consequences are similar of XYY, XXXY, XXXXY, and some of the mosaicisms involving more than one X and Y chromosome. XXY condition affects about 1 to 2 boys among 1,000 births. XYY is somewhat less frequent, and the more complex types are much more rare. XYY males may be fertile, although the other symptoms are common with those of XXY. Underdeveloped testes (hypogonadism) and seminiferous ducts characterize Klinefelter syndrome. The afflicted individuals are generally sterile; although effeminate, they are heterosexual in behavior. Their height is usually above average and the limbs appear longer than normal. About half of them show increased breast size and they are about as likely to develop breast cancer as women. They are more likely to develop insulin-dependent diabetes and heart failure (mitral valve prolapse) than normal males. Some develop speech problems. Their intelligence is generally low, particularly in those having a higher number of sex chromosomes. Klinefelter individuals are frequently slow in development and have learning disabilities. Although they tend to be shy and immature in behavior, they were considered to be prone to violence, but this latter classification turned out to be based on false statistics. There is a higher incidence of Klinefelter syndrome in prison populations, but this is due to their mental deficiency. Klinefelter symptoms also occur in various other mammals. Among the sperm of human males in their 20s, the average frequency of XY sperm is 0.00075 and it gradually increases; in the 50s age group, it may be 0.0176, i.e., 160% higher. *See* gynecomastia; polysomic cells; sex chromosomal anomalies in humans; sex chromosomes; sex determination; sex mosaics; trisomy; XX males. (Klinefelter, H. F. Jr., et al. 1942. *J. Clin. Endocrin.* 2:615; Smyth, C. M. & Bremner, W. J. 1998. *Arch. Intern. Med.* 158:1309; Lowe, X., et al. 2001. *Am. J. Hum. Genet.* 69:1046.)

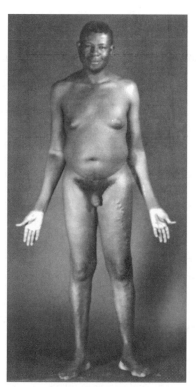

Klinefelter male. (Courtesy of Dr. K. L. Becker. By permission of the American Fertility Society.)

Klippel-Feil syndrome In the autosomal-dominant form it is associated with fusion of cervical (neck) vertebrae and malformation of head bones conducive to conductive and/or sensorineural hearing loss. The autosomal-recessive forms do not involve hearing deficit. *See* deafness.

Klotz test Estimates the statistical variance of two populations displaying identical median values. *See* median; variance. (Drinkwater, N. R. & Lotz, J. H. 1981. *Cancer Res.* 41:113.)

Kluyveromyces lacti Yeast species in which the first-time linear plasmids were found.

Kniest dysplasia (metatropic dwarfism) Autosomal-dominant disease concerned with the locus of collagen II α-1 polypetide (human chromosome 12q13.11-q13.2). The defect involves a deficiency of C propeptide that is required for normal fibril formation. The urine contains increased amounts of keratan sulfate. There are general problems with the cartilage. Specifically, the afflicted individuals cannot tightly close their fist, their palm is purplish, severe myopia (nearsightedness) appears, the retina is detached, and various types of bone defects including dwarfism may be evident. *See* collagen; connective tissue disorders; dwarfism; eye diseases; keratin; Stickler syndrome.

knirps (*kni*, map location 3-46) Zygotic gap mutation in *Drosophila*. The first seven abdominal segments are abnormal (fused/deleted), but the head, thorax, eighth abdominal segment, and tail are normal. The locus encodes a steroid/thyroid receptor protein that responds to various ligands such as the product of gene *Krüppel*. *See* gap genes; *Krüppel*.

knob Heterochromatic cytological landmark of a chromosome. Knobs have been used to identify the fate of particular chromosomes during meiosis and evolution. Knobs were implicated in preferential segregation in maize, affecting the frequency of recombination of syntenic markers. Knobs (altogether about 22) were observed on all chromosomes of maize. These are composed of tandem arrays of about 180 bp elements and may function as neocentromeres. Molecular evidence suggests that through inversion centromeric heterochromatin moved to the arm of the chromosome where it is composed largely of tandem repeats and retrotransposons. *See* chromosome knob; heterochromatin; neocentromere; preferential segregation. (Ananiev, E. V., et al. 1998. *Genetics* 149:2025.)

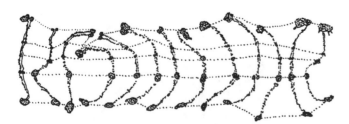

Chromosome knobs of the pachytene chromosomes of 13 different house crickets. The dotted lines connect homologies. (From Wenrich, D. H. 1916. *Bull. Mus. Comp. Zool. Harvard* 60:58.)

Knobloch syndrome (21q22.3) Retinal degeneration, meningocoele (protrusion of the meninges [spinal cord and brain membrane]), myopia (defect of focusing of the eye), etc.

knock-down Slowing down the expression of a particular gene either by transformation with mRNA-digesting ribozymes or more practically by the introduction of synthetic 18–25 nucleotide antisense constructs. This procedure may be practical even when only an EST is known. Generally low levels of phosphophorothioate are used to avoid toxicity and still protect against nucleases. *See* antisense technologies; EST. (Griffoni, C., et al. 2000. *Biochem. Biophys. Res. Commun.* 276:756.)

knock-in (Ki) Insertion of a functional copy or a domain into an inactive/active gene. Generally a vector cassette flanked by the *loxP* gene is employed. *See Cre/loxP*; knockout. (Golub, R., et al. 2001. *Eur. J. Immunol.* 31:2919.)

knock-out (KO) Inactivation of a gene by any means (e.g., deletion, insertion, targeted gene transfer) to determine the phenotypic, metabolic, behavioral, or other consequences, and to draw conclusions concerning its normal function. Removal of genes by site-specific recombination is expected to specifically knock out a discrete genic sequence. The consequence of the knockout may also depend on the neighboring genes. The general procedure involves growing embryonic stem cells (ES) of mice. The cells are then electroporated with a gene construct carrying a selectable marker (e.g., neomycin resistance) within the coding sequence and thus it is inactivated. The disrupted gene is flanked by sequences homologous to the target to facilitate homologous double crossing over. The cells are grown out on selective media and recombinants are isolated. These recombinants will not be able to carry out the normal function of the target gene because of the insert in the coding sequence. Eventually, the ES cells are injected into host blastocytes with a micromanipulator and transferred into the blastocoele. In case of success, the developing embryo becomes chimeric and some of the germline cells or the entire germline will carry the knockout and transmit it to some of the progeny. In 17 days after the transfer, knockout offspring may be obtained. The rate of success varies depending on the technical skills of the investigator and the mice concerned, but it may be quite high — over 50%. *See* excision vector; phenotypic knockin; RNA, double-stranded; site-specific mutagenesis; targeting genes. (Mak, T. W., ed. 1998. *The Gene Knockout Facts Book.* Academic Press, Orlando, FL; Thorneycroft, D., et al. 2001. *J. Exp. Bot.* 52:1593; Mansouri, A. 2001. *Methods Mol. Biol.* 175:397; <http://www.gdb.org/Dan/tbase/tbase.html>.)

knotted circle of DNA *See* concatenane.

A knotted circle.

knowledge discovery Methods of extracting meaningful information from sets of sequence data of macromolecules. *See* data mining. (Kurgan, L. A., et al. 2001. *Artif. Intell. Med.* 23[2]:149.)

knucklewalking African apes do not use their palm but their knuckles (flexed fingers) to lean on in quadrupedal movement.

Knudson's two-mutation theory In retinoblastoma, usually a germline mutation is followed by a somatic mutation(s) in order to express the cancer. The number of tumors in an individual indicates the rate of somatic mutation. *See* Armitage-Doll model; cancer; genetic tumors; Moolgavkar-Venzon model; retinoblastoma. (Hethcote, H. W. & Knudson, A. G., Jr. 1978. *Proc. Natl. Acad. Sci. USA* 75:2493; Knudson, A. G. 2000. *Annu. Rev. Genet.* 34:1.)

KO *See* knockout.

koala *Phascolarctos cinereus*, $2n = 16$.

Kobberling-Dunnigan syndrome *See* lipodystrophy.

Kohara map Of *E. coli*, is a restriction fragment map based on partial digests of 3,400 phage-lambda clones with eight restriction endonucleases. *See* cosmids; *E. coli*; RFLP; vectors. (Rudd, K. E. 1998. *Microbiol. Mol. Biol. Rev.* 62:985.)

Kolmogorov-Smirnov test Nonparametric statistical procedure for the analysis of frequency distributions. It may be used in genetics for different problems, e.g., analysis of gene or physical marker frequency distributions in different populations. The method is usually applied to expected and observed cumulative distributions. The differences are expressed as differences between relative cumulative frequencies. Usually, statistical tables are used to ascertain the maximum difference between observed and expected cumulative frequency at a certain level of significance. The largest absolute vertical deviation $(D) = \text{maximum} |F_s(x) - T_T(x)|$, where $F_s(x)$ is the sample cumulative distribution frequency (CDF) and $T_T(x)$ is the theoretical CDF. The procedure is simple to carry out, and computer programs are available. Details can be found in Hays, W. L. & Winkler, R. L. 1971. *Statistics: Probability, Inference and Decision*. Holt, Rinehart and Winston, New York; Sokal, R. R. & Rohlf, F. J. 1969. *Biometry*. W. H. Freeman & Co, San Francisco.

Kosambi's function $x = 0.25 \ln[(1 + 2y)/(1 - 2y)]$, where x is the recombination frequency corrected by the mapping function and y is the observed recombination fraction. Kosambi's formula considers "average" interference and has been used in different species, although interference may vary in different situations. *See* coefficient of coincidence; Haldane's mapping function; mapping, genetic; mapping function; recombination frequency. (Kosambi, D. D. 1944. *Ann. Eugen.* 12:172.)

kosmotrope *See* chaotrope.

Kozak rule In eukaryotes, the most efficiently translated mRNAs start with **AUGxG** sequence. Commonly optimal translation occurs with ACC**AUGG**. *See* ribosome binding; ribosome scanning. (Kozak, M. 1999. *Gene* 234[2]:187.)

Krabbe's leukodystrophy (globoid cell leukodystrophy) Rare chromosome 14q31 recessive disease, also called galactosyl ceramide lipidosis. The enzymatic basis of the condition is the deficiency of galactocerebroside β-galactosidase/galactosylceramidase. The onset of the disease is expected within the first half year of life and it is generally fatal by around age 2. Exceptionally its onset may be delayed to late childhood. The early symptoms are irritability, hyperactivity followed by lethargy, degeneration of the nervous system, resulting in blindness and deafness. Cerebrospinal fluid (CSF) proteins accumulate. In the white matter of the brain, myelin is reduced and infiltrated with globoid cells that are rich in galactosyl ceramides. The *twitcher* mouse is an enzymatically appropriate animal model of this disease. *See* Addison disease; galactosidase; metachromatic leukodystrophy; saposin; sphingolipidoses; sphingolipids. (Matsuda, J., et al. 2001. *Hum. Mol. Genet.* 10:1191.)

KRAS *See* RAS oncogene.

Krebs-Szentgyörgyi cycle (1) **Oxaloacetate** → (2) Citrate ⇔ cis-Aconitate (3) cis-Aconitate ⇔ Isocitrate (4) Isocitrate ⇔ α-Ketoglutarate (5) α-Ketoglutarate ⇔ Succinyl Co-enzyme A (6) Succinyl CoA ⇔ Succinate (7) Succinate ⇔ Fumarate (8) Fumarate ⇔ Malate (9) Malate ⇔ **Oxaloacetate**. This simplified outline does not show the energy donors and cofactors. This cycle is the most efficient path to generate energy. Also called tricarboxylic acid cycle and citric acid cycle. A modified form is the glyoxylate cycle where isocitrate lyase converts isocitrate to glyoxylate and succinate. Glyoxylate then reacts with Co-A to form malate.

Krev-1 Probably the same as Rap1. *See* Rap.

kringle Disulfide-linked, triple-looped protein domain present in some plasma proteins, apolipoproteins, plasminogens, serine proteases, phosphoglycerate kinase, thrombin, and hepatocyte growth factors. Some of the kringle proteins regulate angiogenesis, cancer, and metastasis. (Ozhogina, O. A., et al. 2001. *Protein Sci.* 10:2114.)

k-RNA (kinetoplast RNA) Mitochondrial RNA in *Trypanosomas*. *See* RNA editing; *Trypanosoma*.

KROX20 Serum-inducible primary-response gene with Zn-finger, originally from *Drosophila*. It controls the myelination of the peripheral neurons. *See* Egr; myelin; neuron; serum response element; zinc finger. (Turman, J. E., Jr., et al. 2001. *Dev. Neurosci.* 23[2]:113.)

Krox-24 *See* Egr; NGKI-A.

Krüppel (Kr, 2-107.6) Maternal effect gap gene of *Drosophila*, encoding a protein with regulatory Zn-finger domain interacting with Kni (knirps) and Hb (hunchback) proteins. Its monomers activate transcription in its vicinity by TFIIB and its dimer may repress transcription by interacting with the TFIIEβ subunit. Kr-like factor also controls G_1-S progression. In the human genome there are an estimated 600–700 Krüppel-like (KLF) regulatory proteins. All of them bind a very similar GT-box (CACCC element). *See* cell cycle; EKL; gap genes; Gli oncogenes; kni; morphogenesis in *Drosophila*; transcription factors. (Shields, J. M., et al. 1996. *J. Biol. Chem.*

271:20009; Chen, X., et al. 2001. *J. Biol. Chem.* 276:30423; Bieker, J. J. 2001. *J. Biol. Chem.* 276:34355.)

Kruskal-Wallis test Nonparametric method of analysis of variance by ranks. Each observation of the groups of treatments, genotypes, or phenotypes to be compared is ranked as shown in parentheses:

	Group 1		Group 2		Group 3	
	4	(1.5)	11	(7)	18	(11)
	7	(4)	12	(8.5)	30	(15)
	10	(5.5)	12	(8.5)	24	(12.5)
	4	(1.5)	10	(5.5)	24	(12.5)
	6	(3)	13	(10)	25	(14)
T_j		(15.5)		(39.5)		(65)

Total of $N = 15$ observations
T_j = sum of ranks of group j
T = total ranks of the groups
$T = 120$, i.e.,

$$T = \frac{N(N+1)}{2} = \frac{(15)(16)}{2} = 120$$

if the ranking is correct.

$$H = \frac{12}{(n(N+1))}\left(\sum_j \frac{T_j^2}{nj}\right) - 3(N+1)$$

$$= \frac{12}{15(16)}\left(\frac{15.5^2 + 39.5^2 + 65^2}{5}\right) - 3(16) = 12.25$$

(in the example above, note 12 is a constant). If there are ties in the ranking, the H value should be corrected by dividing H by the formula shown in the box. $1G$ = number of sets of tied observation, t_i = numbers of tied in any set of i. This correction may not make much difference if N is very small or the number of ties is very large. From the chi-square table (*see* chi-square table) we can determine for $J - 1 = 2$ degrees of freedom in the example shown above that the probability is less than 0.005 and that these three sets would be identical. Generally much larger samples are required to get meaningful results. It is considered to be a very powerful nonparametric test. See Mann-Whitney test; nonparametric tests.

$$1 - \left(\frac{\sum_{i}^{G}(t_i^3 - t_i)}{N^3 - N}\right)$$

KS Male fertility complex in the short arm of the Y chromosome of *Drosophila melanogaster*. See *Drosophila*; *KL*; sex determination.

KSHV Kaposi sarcoma–associated herpes virus. *See* HHV8; Kaposi sarcoma.

KSR Kinase suppressor of RAS is structurally related to RAF and has a similar role in cell proliferation and development of *Drosophila*. See CNK; RAF; RAS.

KSS1 Protein kinase of the MAPK family. Along with Fus3, it causes arrest of yeast cells in G1 prior to mating. *See* Fus3; MAPK; signal transduction; Ste.

Ku Heterodimeric (Ku70 [22q13] and Ku86 [2q35], kDa) serine/threonine protein kinases that bind to DNA in cooperation with transcription factors. Ku also interacts with the termini of DNA double-strand breaks and is the binding domain of DNA protein kinase (DNA-PK), whereas the catalytic subunit of DNA-PK is encoded at 8q1. During telomere replication, Ku is bound to a guanine-rich overhang in yeast. Ku has a role in DNA repair and recombination, including immunoglobulin V(D)J rearrangements. The deficiency of Ku70 leads to a defect in B-cell maturation and development of T-cell lymphoma. The XRCC5 human gene is also encoded at 2q35. Ku70 apparently mediates nonhomologous chromosomal end-joining in somatic cells and is essential, but it seems to be absent during the meiotic prophase. *See* Bloom syndrome; DNA-PK; DNA-PK silencer; DNA repair; immunoglobulins; ligase DNA; lymphoma; Mre11; NHEJ; nonhomologous end joining; p350; RAG; terminal deoxynucleotidyl transferase; transcription factors; XRCC. (Featherstone, C. & Jackson, S. P. 1999. *Mutation Res.* 434:3; Woodard, R. L., et al. 2001. *J. Biol. Chem.* 276:15423; Walker, J. R., et al. 2001. *Nature* 412:607; Li, G., et al. 2002. *Proc. Natl. Acad. Sci. USA* 99:832.)

Kugelberg-Welander syndrome Muscular atrophy expressed at infancy. It is determined either by a dominant or recessive factor in human chromosome 5q11.2-q13.3. See atrophy; neuromuscular disease.

Kunkel mutagenesis Template DNA is generated in a bacterial strain of *dut ung* constitution. It is defective in dUTPase (*dut*) and uracil-N-glycosylase (*ung*). Consequently, several uracil residues are incorporated into the single-stranded M13mp19 phage DNA and cannot be removed by DNA repair because of the defective glycosylase. A mutagenic nucleotide primer is added to this DNA. In the presence of all four deoxyribonucleotides, a new strand is synthesized using the U-containing single-strand DNA template of M13. After the new M13 DNA synthesis is completed, the molecule is transfected into wild-type *E. coli*, which gets rid of the U-containing strand and synthesizes DNA containing the mutant nucleotide; the resulting phage plaques contain this localized, directed mutation. *See* bacteriophages; glycosylases; homolog-scanning mutagenesis; localized mutagenesis; site-specific mutagenesis. (Kunkel, T. A. 1985. *Proc. Natl. Acad. Sci. USA* 82:488.)

Kupffer cell Liver macrophage-type cell. *See* IL-18; macrophage.

kurtosis Departure from the symmetrical (normal curve) frequency distribution by displaying excess or deficiency at the shoulders compared to the tails and the highest point of the curve (peakedness). At normal distribution $(x - \mu)^4/\sigma^4 \cong 3$.

A higher ratio indicates kurtosis. A *leptokurtic* distribution is characterized by a higher number of observations or measurements at the mean and at the tails. *Platykurtic* distribution displays the opposite, i.e., less at the mean and tails than expected by normal distribution. *See* moments; normal distribution; skewness. (Diagrams redrawn after Hyperstat.)

Leptokurtosis   Platykurtosis

kuru Infectious human chronic degenerative disease characterized by tremors, ataxia (lack of muscular coordination), strabismus (contorted visual axis in the two eyes), dysarthria (stemmering and stuttering), dysphagia (difficulties in swallowing), fasciculation (bundling of nerve and muscle tissues), etc. Generally, death results within a year after onset. It affects about 1% of the Fore women and their daughters and sons. It shows vertical transmission because in the Fore tribe New Guinian populations as a cannibalistic, religious ritual the females and their youngsters consume some flesh of dead relatives. It occurs sporadically in neighboring tribes due to stenosis, LEOPARD syndrome, or it may be viral or due to prions. *See* Creutzfeldt-Jakob disease; encephalopathies; LEOPARD syndrome; prion; stenosis. (Collins, S., et al. 2001. *J. Clin. Neurosci.* 8[5]:384.)

KUZ (Kuzbanian) *See* ADAM.

kV Kilovolt, 1,000 V. *See* volt.

KVLQT1 *See* Beckwith-Wiedemann syndrome.

kW kilowatt, 1,000 watts. *See* watt.

kwashiorkor Condition caused by malnutrition of humans on diets low in essential amino acids, particularly lysine, tryptophan, and methionine. The symptoms are emaciation with altered pigmentation in patches on the hair and skin. On dark spots of the limbs and back, the epithelial layer may be shed, showing the raw flesh as pink blotches. If it is coupled with a deficiency in caloric intake, the condition is further aggravated by losing both flesh and fats and is generally coupled with dehydration as well (marasmic kwashiorkor). It is a widespread anathema, particularly in the tropical and subtropical areas of the world with underdeveloped agriculture and political turmoil. The term's origin is an African Gold Coast (Ghanaian) language meaning pink boys, descriptive of the syndrome. This severe malnutrition primarily affects children. The condition is aggravated by infectious diseases (e.g., AIDS) and poisons in the food (e.g., aflatoxin). Therapy is a gradual return to balanced, nutritious diet. *See* aflatoxin; AIDS; essential amino acids; high-lysine corn.

Kwok Polymorphism (SNIPs, duplications, deletions, rearrangements) in ESTs, named after P-Y. Kwok, who first called to their usefulness for studying genomic variations. *See* EST; SNIP. (Kwok, P.-Y. 2002. *Hum. Mutation* 19:315.)

KYA Kiloyears ago. 1 ky = 1,000 years. *See* MY.

kynurenine Intermediate in tryptophan metabolism. In humans an autosomal-recessive deficiency of the enzyme kynurinase results in excessive amounts of xanthurenic acid in the urine. Xanthurenic acid is a tryptophan metabolite that also accumulates in case of a pyridoxal phosphate (vitamin B_6) shortage in the diet. *See* tryptophan.

Kynurenic acid.

Kyoto Encyclopedia of Genes and Genomes (<http://www.genome.ad.jp/kegg/>.)

W. Haacke assumed in 1893 (*Biol. Zbl.* 13:525) that the waltzing-walking traits of mice are located in cytoplasmic elements (centrosome) whereas coat color (whitegray) segregation is assured by the reductional division of the chromosomes. ("I do not know whether the number of chromosomes present in mice had been recorded, but this number would enable us to establish the possible combinations.") The fact that he was able to obtain experimentally all 16 combinations of these 4 traits seemed to indicate to him the validity of this interpretation. Although C. Correns and independently E. Baur (*Zeitschr. Ind. Abst. Vererb.-Lehre* 1:291 and 330, respectively) reported genuine cytoplasmic inheritance in 1909, and their findings were abundantly confirmed, John R. Preer, Jr., an eminent contributor to the field remarked: "Cytoplasmic inheritance is a little bit like politics and religion from several aspects. First of all, you have to have faith in it. Second, one is called upon occasionally to give his opinion of cytoplasmic inheritance and to tell how he feels about the subject." (P. 374 in *Methodology in Basic Genetics*, W. J. Burdette, ed., Holden-Day, San Francisco, 1963.)

A historical vignette

L

L1 (1) LINE1 (long interspersed repeat). *See* LINE. (2) *See* hybrid dysgenesis I-R. (3) Cell adhesion molecule with important role in neurogenesis. It contains fibronectin domains. *See* fibronectin.

La Autoimmune antigen transiently associated with pre-tRNAs and 5S rRNAs. It mediates both transcription initiation and termination by polIII. *See* pol III; ribosomal RNA.

labeling Attaching or incorporating a radioactive or fluorescent compound into a molecule that permits the recognition of the molecule itself in the cell or in the extract of the cells or any molecule with which it is hybridized, attached, or into which the label is inserted. *See* aequorin; biotinylation; FISH; fluorochromes; gene tagging; immunofluorescence; immunoprobe; nick translation; nonradioactive labels; probe; radioactive labeling; radioactive tracer.

labeling index Shows the fraction of cells that incorporate labeled nucleotides, i.e., the percent of S-phase cells in a tissue. The fraction of labeled cells in relation to DNA content permits a convenient estimate of the cells in G1 phase (1 DNA unit), G2 + M (2 DNA units), and in between 1 and 2, indicating S phase. The fractions of cells can be analyzed with the aid of an automatic cell sorter. *See* cell cycle; cell sorter; labeling.

labor Process of child delivery.

laboratory safety The most important requisite is to know the potential hazards of equipment and the biological and chemical materials to be employed. Develop plans how to cope with possible accidents and how to dispose of spillage, fumes, fire, etc. Most of the commercial suppliers provide safety information for chemicals. Use nonporous (neoprene) gloves. Sometimes using bare hands may be justified because accidental contact can be immediately sensed and proper washing can decontaminate the body. Fume hoods (with proper air exchange) must be used with chemicals that evaporate or sublimate. Appropriate sterilization and the use of certified laminar flowhoods can minimize biological hazards. Laboratory waste must be segregated for solids and liquids. Do not dump any chemical (mutagens and carcinogens) into drains, which may hurt plumbers and cause problems at wastewater treatment. Monitoring with radiation counters and appropriate shielding may prevent radiation hazards. Keep work benches clean. All laboratory personnel must be properly instructed about safety and checked regularly for compliance. Remember the admonitions of Paracelsus, the 15th-century physician and scientist: "Poison is everything and no thing is without poison. The dosage makes it either a poison or a remedy." *See* biohazards; chemical mutagens; environmental mutagens; gloves; ionizing radiation; radiation hazard assessment. (Fleming, D. O. & Hunt, D. L., eds. 2000. *Biological Safety*. ASM Press, Washington, DC; <http://www.cdc.gov/od/ohs>; <http://www.absa.org/>.)

labrum Anterior-most structure (mouth) of the head of arthropods; in general, morphology edges; lips are designated as such.

labyrinth Communication canal or cavity such as in the labyrinthine trophoblast connecting maternal and fetal tissues of the placenta, internal part of the ear, tubules in the kidney, etc.

LacI^q Mutant *lacI* that synthesizes about 10 times more repressor than the wild-type allele.

***Lac* operon** *E. coli* can utilize the milk sugar lactose by splitting it into glucose and galactose with the aid of the β-galactosidase enzyme encoded by the *Lac z* gene. This enzyme is not made unless lactose or one of its analogs is present in the culture medium (inducible enzyme) and even then not until there is glucose available. This particular metabolic response is under a very precise and complex genetic regulation in prokaryotes. When transcription begins, actually 3 genes are transcribed into a 3-cistronic RNA. The *Lac y* gene encodes a galactoside permease, a membrane protein that facilitates the uptake of the substrate for the galactosidase. The *Lac a* gene is transacetylase, which acylates the galactoside with the assistance of acetyl-coenzyme-A. The nucleotide sequence of this regulatory upstream region is shown on the next page.

For the sake of brevity, the map (at 8 min) is shown on top of the next page. When galactosidase is not synthesized, a repressor protein blocks the repressor-binding sequences, a definite tract within a section of the promoter region, a part of the operator gene. The product of the *Lac I* (1040 bp) gene is 152 kDa, a 4-subunit repressor protein. The transcription of the *Lac I* gene is separately regulated from the genes that it controls by suppression (negative control). Since the repressor binding site is within the operator gene where the transcriptase enzyme (RNA polymerase) is attached to carry out its function, it prevents the expression (transcription) of the downstream genes (z, 3510 bp; y, 780 bp; a, 825 bp) of the operon.

```
Lac i -⌈--|CAP-cAMP|-⌈operator-
              {repressor-binding sequences-
                 transcription initiation}⌋ ⌉- leader -z - y - a
         <---------------------Promoter---------------------->
```

The Lac Z protein is 125 kDA; *Lac y* and *Lac a* are both about 30 kDa. Because of the coordinated operation of iuxtapositioned genes, the system was named *operon*. The *Lac i* gene is transacting because its product can flow to the operator irrespective of where the gene is located within the cell; it can be in the vicinity of the operon or it can be carried by a plasmid. Furthermore, if there is an inactive *i⁻* gene next to the operon within the bacterial chromosome and the z, y, and a genes are transcribed. Introduction of a *i⁺*

(wild-type) suppressor gene in a plasmid, the transcription is blocked from this transposition. Also, it shows that the active form of i is dominant. The tetrameric repressor protein is normally a homotetramer, i.e., the four subunits are identical. If the cell has two different i genes that code for differently altered repressor monomers, the aggregate becomes a heterotetramer and may show *allelic complementation*. The i^{-d} gene product in the presence of the wild-type polypeptide imposes a conformational change on the repressor tetramer and renders it inactive by dominant negative complementation. This phenomenon indicates that the monomers alone are not functional, but their aggregate (quaternary structure) is the functional repressor. Base substitution mutations within the *Lac i* may abolish the ability of the protein to bind to the operator, then protein synthesis can go on constitutively (without a need for induction). Mutations may just reduce its binding, then without induction a reduced level of transcription can still go on. If the mutations affect the inducer-binding sites of the repressor, it may no longer bind the inducer and it will become a superrepressor mutation, i^S. The number of repressor molecules per cell is about 5 to 10.

Similarly, mutations within the operator region may alter the binding of the wild-type suppressor. If the repressor is bound to it rather than the RNA polymerase, the transcription of the three structural genes (z, y, a) cannot proceed. The repressor does not prevent the binding of the RNA polymerase to the promoter; it may even enhance the binding of the polymerase, but it blocks the initiation of transcription. Since the RNA polymerase is at its site before induction, as soon as the inducer makes contact with the repressor, transcription is initiated. The operator gene is unique because it does not have any product; it merely serves as the starter site for transcription if the RNA polymerase can attach to it. Therefore, the operator must always be in cis position, in front of the structural genes. The binding sequences within the operator shown below have inverted repeats (**bold**) and the inactivating mutations are shown below in ⓄⓊⓉⓁⒾⓃⒺ (Modified after Watson, J. D., et al. 1987. *Molecular Biology of the Gene.* Benjamin/Cummings, Menlo Park, CA):

TGG **AATTGT**GAGCGGATA **ACAATT**
ACC **TTAACT**CT CGCCTAT **TGTTAA**
| | | | |
T - TGTTA - - - C - - - T
T - ACAAT - - -G - - - A

The left side of the operator is more likely to render it unreceptive to repressor binding. If the repressor cannot bind to the operator, an operator-constitutive system emerges that is functional without induction (o^C). When the cells are grown without galactoside, the number of galactosidase molecules may be less than 5 per cell. After supplying galactose, within a couple of minutes, the number of galactosidase molecules increases by about a thousand-fold. If the inducer is used up, the system reverts to the uninduced state very rapidly because the half-life of the mRNA is only about 3 minutes, although the already synthesized proteins may linger on for a little longer. Induction can take place without β-galactoside if thiogalactosides, particularly the often-used isopropylthiogalactoside (IPTG), are provided. See formula at top of next page. These thiogalactosides induce the synthesis of the galactosidase enzyme, although the enzyme cannot use these analogs as substrates. Therefore, these are *gratuitous inducers*. The presence of the slight amounts of the permease protein in the noninduced cells is required to initiate the uptake that eventually jumpstarts the system. When the inducer is added to the system, it combines with the repressor and a change in the conformation prevents it from attaching to the repressor-binding sites in the operator region. The repressor protein has two essential sites: one that binds the inducer and the other that binds the operator. Although the binding of the repressor is not exclusively limited to the operator, it binds to this region by about 10 million–fold more effectively than to other sequences of the genome. When the cells take up the inducer, the unspecific binding to all over the genome does not change, but the binding to the operator is almost completely relieved.

The mechanisms discussed above (*negative control*) are only parts of the total regulatory system. We have to look at the role of the CAP-cAMP site upstream in the promoter. CAP stands for *catabolite activator protein*. It has also been called

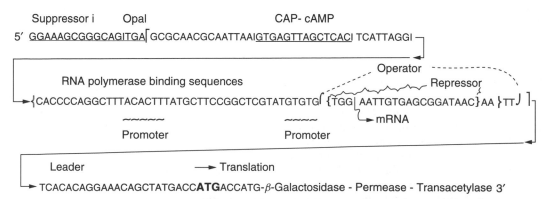

The structural and functional elements of the *Lac* operon of *E. coli*. The promoter region is enclosed within the signs ⌐ ¬ the | cat and cAMP binding site | is within the promoter and is shown underscored. The sequences occupied by the DNA-dependent RNA polymerase { are delimited } within this region. The two specific promoter elements are underlined ~~~~. The operator region ⌐ ⌐ includes the repressor-binding site {} and transcription begins within it ∟→. The leader sequence is followed by the ATG formylmethionine codon and continues into the transcripts of the β-galactosidase (z)–permease (p)–transacetylase (a) genes.

α-Lactose   Allolactose   Isopropyl
                                         β-D-thiogalactoside

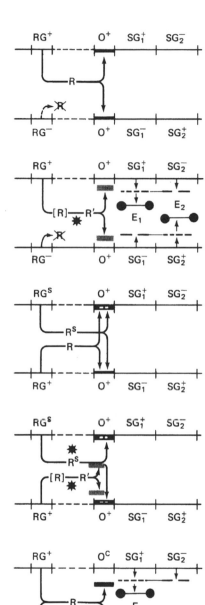

A simplified model of the operation of negative control in an operon like lac. RG⁺: wild-type allele of regulator gene; RG⁻: defective regulator gene; O⁺: wild-type operator gene; SG⁺: wild-type; SG⁻: mutant structural genes (responsible for enzyme proteins); ▬ repressor.

The genetic conditions are identical to those at A, but an inducer (✱) is provided. Now the repressor is neutralized (▦), and either functional (---) or defective (— —) transcripts are made depending on the structural genes. The nonmutant mRNA of the SG⁺ genes is translated into protein (●—●).

A mutation from RG⁺ to RGˢ leads to the production of a superrepressor protein (▭) which locks permanently into the operator. No mRNA and no protein are produced by the structural genes.

In the RG⁺/RGˢ heterozygotes, the product of the RG⁺ allele is receptive to the inducer (✱) and is prevented from blocking the operator, but the superrepressor product (▭) cannot be removed from the system. No mRNA and no protein are produced.

In the presence of two RG⁺ genes (wild-type repressor), one of the operator alleles mutates to Oᶜ (constitutive operator). In this case, the wild-type repressor protein is unable to bind to the Oᶜ region even in the absence of an inducer. Consequently, transcription may proceed on one sequence of the operon.

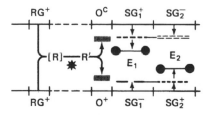

Lac operon has been a paradigm of genetic regulation since its inception. The diagram at the left illustrates some of the functional circuits involved by more general symbols. Of course not all genetic regulatory systems use exactly the same principles but several elements of it are valid for other regulations. (Modified after Jacob, F. & Monod, J. 1961. *Cold Spring Harbor Symp. Quant. Biol.* 26:193.)

The genetic constitution is identical to that diagrammed in E, but an inducer is provided for the system. Now transcription becomes possible on both sequences, and functional protein is made from the two wild-type structural alleles just as it is in the case illustrated in B.

cyclic AMP receptor protein (CRP, encoded by gene *crp* at map position 73 min). CAP is a 22.5 kDa dimeric protein and the subunits contain a DNA-binding and other transcription-activating sites. Thus, CAP interacts both with DNA at the evolutionarily conserved upstream site and with α-subunits of the RNA polymerase. The CAP protein becomes active upon forming a complex with cyclic adenosine monophosphate (cAMP). cAMP is formed from ATP by the enzyme adenylate cyclase (the encoding *cya* gene is at map position 84 min) and its formation is reduced by glucose. We discussed that as long as glucose is available for the cells, β-galactosidase and the companion enzymes are not formed in appreciable amounts. This phenomenon is *glucose effect* or by another name *catabolite repression*. The basis of this repression is that there is not enough cAMP to activate transcription by the CAP-cAMP complex. If the CAP-cAMP system is defective, the *lac* operon cannot function even if the repressor is not formed and the operator is constitutive. Therefore, the CAP-cAMP complex constitutes a *positive regulatory* element of the *lac* operon in contrast to the *negative regulatory i-o* system seen p. 671. We can conclude that this system uses great wisdom in managing cellular energies: it calls to duty the enzymes only when they are needed and turns off the synthesis as soon as they are no longer needed. One of the products of the galactosidase, glucose, reduces the synthesis of the enzyme that splits it off from lactose. The lac gene has been used extensively as a reporter in eukaryotic systems (Mills, A. A. 2001. *Genes Dev.* 15:1461). *See* allelic complementation; β-galactosidase (for assay); cAMP receptor protein; galactose; galactose utilization; galactosidase; helix-turn-helix motif; *Lac* repressor; suppression; transcription; translation. (Müller-Hill, B. 1996. *The lac Operon.* De Gruyter, New York.)

***Lac* repressor**   Protein that negatively regulates the *Lac* operon in the absence of lactose in the medium. When the repressor is combined with lactose (inducer) or isopropyl-β-D-thiogalactoside (gratuitous inducer), cAMP, and CAP, the transcription of the structural genes may be started. The bound repressor either inhibits binding of the RNA polymerase or prevents elongation of the transcript. The Lac repressor protein (LacR/LacI) has a small headpiece (binding domain) and a large core (regulatory domain). The intact 38 kDa repressor forms a homotetramer of 152 kDa. Each of the two LacR core dimers arrange the headpieces in such a manner that they bind maximally to the operator. Headpiece monomers bind only weakly. In the presence of an inducer (IPTG), the binding is reduced by three orders of magnitude. LacR also mediates DNA looping to make contacts at multiple sites. There are about 20 members of the family of Lac repressor proteins. The lactose, fructose, and raffinose repressors are tetramers, whereas the others are dimeric. The majority of the Lac repressor (*LacI*) family members are most effective if no other proteins bind to them. PurR (purine repressor), however, requires the presence of a corepressor (hypoxanthine and guanine) ligand. *See* cAMP receptor protein; hypoxanthine; *Lac* operon; looping of DNA; purine repressor. (Fried, M. G. & Daugherty, M. A. 2001. *J. Biol. Chem.* 276:11226.)

lactacystin   *Streptomyces*-produced inhibitor of proteasomes by affecting the amino-terminal threonine. It inhibits several proteases, the cell cycle, and causes neurite outgrowth

in neuroblastoma cells. It may inhibit the degradation of p53. *See* neurite; neuroblastoma; p53; proteasome; ubiquitin. (Yamada, Y., et al. 2000. *Eur. J. Haematol.* 64[5]:315.)

lactam   Is a tautomeric form of uracil. *See* penicillin; pyrimidin; tautomeric shift.

Lactam

lactamase   *See* β-lactamase.

lactase β-galactosidase.

lactase deficiency   *See* disaccharide intolerance.

lactate dehydrogenase (LDH)   Catalyzes the reaction: pyruvate + NADH + H$^+$ $\rightleftarrows$ lactate + NAD$^+$. Lactate is generally not utilized as such, but it is converted back to pyruvate. The rationale of the reaction is to regenerate NADH for glycolysis in the skeletal muscles and shift the metabolic burden from the muscle to the liver. In yeast, L-lactate dehydrogenase is also called cytochrome b$_2$. The mammalian enzyme is a tetramer consisting of four subunits, each with an approximate M_r of 33,500. These subunits are encoded by two separate genes and can be combined into five different isozymic forms: A_4B_0, A_3B_1, A_2B_2, A_1B_3, A_0B_4. In the skeletal muscles the A chains (synonymous with M) predominate, whereas in the heart the B chains (synonymous with H) predominate. The isozymic forms vary during development. In cancer cells, LDH isozymes are present in the fetal rather than in the adult form. In humans, the LDHA gene was assigned to chromosome 11p15.4 and LDHB to 12p12.1-12p12.2. The testes-specific LDHC protein is also encoded in the close vicinity of LDHA. The nucleotide and amino acid homology between human LDHA and LDHB varies between 68 and 75%. In the mouse, LDHC displayed 72–73% homology with LDHA, and in humans the similarity was just slightly higher between the two amino acid chains. In the Japanese population, LDHA and LDHB deficiencies occurred at frequencies of 0.19 and 0.16, respectively. These figures are supposed to be too high for other populations. LDHA deficiency in humans was associated with skin lesions, myoglobinuria, and fatigue. LDHA deficiency was found to be caused by 20 bp deletions in exon 6, but nonsense mutation GAG → TAG at codon 328 was implicated at another human deficiency. LDHB deficiency was found to be due to a replacement of the highly conserved arginine 173 by a histidine. *See* code genetic; dehydrogenase; isozymes.

lactic acid (CH$_3$CHOHCOOH, MW 90.08)   Formed from pyruvate through glycolysis or by fermentation of lactose (yogurt). Lactic acid is the main end product of sugar metabolism of lactic acid bacteria. It is formed in the skeletal muscles and oxidized in the heart for providing energy or it is again converted into glucose by gluconeogenesis. It is the most important preservative in silage fermented at moderate temperature. (Gladden, L. B. 2001. *Proc. Natl. Acad. Sci. USA* 98:395.)

lactic acidosis May be caused by the deficiency of dihydrolipoamide dehydrogenase (7q31-32). It may also be due to pyruvate decarboxylase deficiency. *See* Leigh encephalopathy; mitochondrial disease in humans.

lacticaciduria Due to an autosomal-recessive condition. Blood lactate and pyruvate levels are increased. In different forms of the disease, pyruvate carboxylase, pyruvate dehydrogenase, and phosphoenolpyruvate carboxykinase enzymes may be defective. Mental retardation, loss of hair, lack of muscle coordination, and infant death may occur. *See* ketoacidosis.

Lactococcus lactis Gram-positive bacterium of cheese making. The sequenced laboratory strain IL1403 contains 2,365,589 bp encoding 2,310 proteins; among them, 293 appears to be of prophage origin and it contains 43 insertion elements. *See* insertion element; prophage. (Bolotin, A., et al. 2001. *Genome Res.* 11:731.)

lactoferrin Iron-binding glycoprotein in milk, in other secretions, and in neutrophils. It defends the cells against infections and is a growth regulator that modulates killer cells of the immune system. It is also a peptide messenger, binding to conserved DNA sequences and activating transcription. *See* killer cells; messenger polypeptide. (Kanyshkova, T. G., et al. 2001. *Biochemistry* 66:1.)

lactose (milk sugar) Disaccharide of galactose + glucose. *See Lac* operon.

lactose intolerance *See* disaccharide intolerance.

lactose permease *See Lac* operon.

lactosyl ceramidosis Deficiency of β-galactosyl hydrolase involving a hereditary sphingolipidosis type of disease. *See* galactosidase; sphingolipids.

lacuna Hole in the bacterial lawn caused by the production of bacteriocin.

LacZ Gene for β-galactosidase enzyme (map position 8 min). Another locus of *E. coli* (*ebgA⁰*, map position 67 min) may evolve into a β-galactosidase gene if the locus at position 8 is deleted. At least two different mutations are required to acquire this enzyme activity, but it may evolve to this state through different paths. The new activity is based on an immunologically different protein from that of the *lacZ* product. *See* galactosidase; *Lac* operon; Xgal.

***lacZ* ΔM15** The amino terminal of the bacterial β-galactosidase gene is deleted. *See Lac* operon.

ladder Collection of oligonucleotides with known lengths that can be used as standards for identifying the length of

Part of a molecular ladder.

DNA fragments separated by electrophoresis. The synthetic ladders generally have a certain number of progressive base increments, e.g., 123, 246, 369.

LaFora disease *See* myoclonic epilepsy (6q24).

lagging strand Of DNA, facing the replication fork by its 5′ end, and the elongation can be accomplished only by adding 5′-P ends of the nucleotides to the 3′-OH ends by phosphodiester linkage. Therefore, it must be synthesized in pieces, in Okazaki fragments. *See* DNA replication; leading strand; nucleic acid chain growth; replication fork.

Lagotrix (woolly monkey) *See* Cebidae.

lag phase Of a culture, when growth is minimal and under favorable conditions it may be followed by exponential growth. *See* exponential growth.

lair Leukocyte receptors (encoded at human chromosome 19q13.4) containing two ITIMs. *See* ITIM.

LAK Lymphokine-activated killer cell is a special type of T lymphocyte of the immune system. *See* immune system; lymphocytes; lymphokines.

laloo Member of the Src family of protein tyrosine kinases that activates mesoderm formation in *Xenopus* using TGF-β and FGF signaling. *See* FGF; Src; TGF; *Xenopus laevis*.

Lamarckism Largely discredited evolutionary theory embodying the ideas of the French biologist J. B. de Lamarck (1774–1829). He proposed a comprehensive theory claiming that evolution proceeds by the inheritance of gradually acquired characters. He supposed that the use or lack of use of a body structure eventually leads to reinforcement or lapse of that trait by strive and direct environmental influence. The giraffes stretched their neck to reach the tree tops; thus, the *inner drive* and the circumstances contributed to their familiar shape. Contrarily, the current concepts such as the neo-Darwinian theory believe that the longer-neck animals could feed better, and thus through continuous selection for increased neck length their progeny had a selective advantage, facilitating the propagation of cumulative mutations that assured the survival of the best-adapted genotypes. Neo-Lamarckism is basically an identical dogma to Lamarckism, except it emphasizes the use or disuse idea rather than the inner drive or autogenesis aspects. *See* directed mutation; Kammerer; lysenkoism; Soviet genetics; transformation; neo-Darwinian evolution, (Aboitiz, F. 1992. *Med. Hypotheses* 38[3]:194.)

λ Lambda bacteriophage. *See* lambda phage.

lambda chain (λ chain) Immunoglobulin light chain. *See* immunoglobulins.

λDASH Replacement vector for the stuffer DNA fragment that carries multiple cloning sites on both sides of *red* and *gam* genes and appropriate promoter(s) specific for T3 or Ty RNA polymerase. Thus, the insertions can be transcribed without a need for recloning. *See* cloning; EMBL3; λFIX; lambda phage; stuffer DNA.

λdgal Lambda-phage-deficient but carries the galactose gene. *See* specialized transduction.

λFIX Replacement vector very similar to λDASH. *See* λDASH.

lambda phage (λ) Temperate bacteriophage, an obligate parasite. It belongs to the family of *lambdoid* phages including phages φ21, φ80, φ81, etc. Lambdoid phages are characterized by cohesive ends, ability to recombine, and inducibility by UV. Each λ-particle contains one double-stranded DNA molecule of ca. 49,502 bp in its icosahedral head (0.05 μm in diameter) that was completely sequenced by 1982. In its *E. coli* host, it can replicate either autonomously and produce ~100 progeny particles in 50 min at 37°C. Alternatively, it may insert into the host chromosome as a *prophage*, generally near the map position of the galactose (*gal*) operon. Phage λ usually does not kill all bacterial cells and therefore its plaques are turbid.

Gene *cI* is the principal phage gene enforcing the lysogenic state (λ being a provirus in the host chromosome). In addition, *cI* assures immunity to infection by other λ-phages. When *cI* is inactivated, the phage may enter *productive growth*, i.e., it is liberated from the host chromosome (*induction*) and begins autonomous replication, resulting in lysis of the host. Spontaneous induction occurs at a frequency of 10^{-3} per generation. Exposure to UV light may cause induction in most cells. Mutations *recA⁻* and *cI ind⁻* prevent induction by radiation as well as by spontaneous means. The *N* gene of the phage regulates both *cI* (the repressor of autonomous replication) and the morphogenetic genes involved in operations during the lytic lifestyle. The λ-genome may assume either a linear or circular form because at the ends (12 bp) of the DNA strands the complementary cohesive sites (*cos*) may open up to a linear form or circularize:

$$5'\ \text{pGGGCGGCGACCT}$$

$$\text{CCCGCCGCTGGAp}\ 5'.$$

In between these ends the genes (encoding about 50 proteins) are in functional units corresponding to their temporal sequence of expression. This relatively small genome transcribes genes in both strands of the DNA double helix. Thus, it has both left- and rightward *transcriptons*. The multiple promoters serve in the most efficient regulation of gene expression. The immunity region contains the major regulatory genes of the phage. After infection of the host cell or induction, transcription begins at either the leftward or rightward promoters. The leftward promoter, p_L, mediates the transcription of genes involved in recombination (integration and excision) under the control of the (12.2 kDa) *N* gene product, which may prevent transcription termination of genes using the p_L or p_R promoters. Some mutations (*ninL* and *ninR*) on both sides of *N* may make the system insensitive to the N protein. A balance of the action of its own and some host genes determine the fate of the behavior of the phage. For lysogeny, genes *cII* and *cIII* (in opposite promoters, p_R and p_L, respectively) activate promoters p_E (includes the *cI*) and p_L (includes gene *int*).

The *cI*, the λ-repressor, produces a 236-amino-acid protein that prevents the expression of genes involved in phage DNA

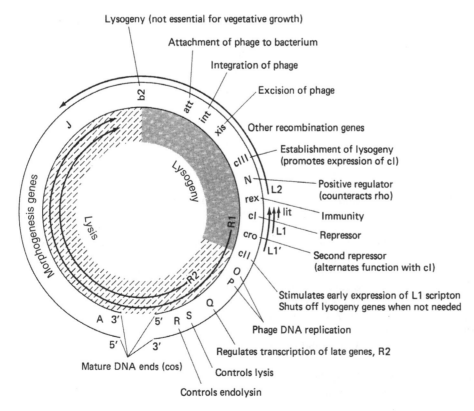

The major functional regions of phage lambda. The head and tail genes are essential for morphogenesis and for the packaging of foreign DNA into λ-phage vectors. The region between J and N (the stuffer) can be removed and replaced by foreign genes in the replacement vectors. Just left of the DNA replication genes is the phage immunity region. Before gene *N* is the left *nin* and after gene *P* is the right *nin* region controlling transcription termination. The *ori* gene is the replicational origin and transcription may proceed left- or rightward from *pL* and *pR*, respectively. There are additional promoters for the delayed early genes (*pI*, internal), for the transcription of genes located in the replaceable region, and *pE* and *pM* for the establishment and maintenance of *cI* repressor controlling the transition from lysogeny to morphogenesis in cooperation with other genes. The total length of the DNA in the wild-type phage is about 49.5 kbp. In order to assure packaging, the total DNA in the vectors cannot be less than 78% and not more than 105% of the wild-type genome. The region of the *A* to *J* genes (ca. 20 kb) and that from *pR* to *the cos* site (8 to 10 kb) must be retained for the viability of the particles.

synthesis and morphogenesis. It activates the synthesis of the Int and Xis proteins, which in turn mediate recombination; thus, the phage can integrate into the host chromosome as a prophage. In that state practically all λ-functions cease, except *rex A* and *cI*, which exclude superinfection by other phages (*immunity*), and replication of phage DNA becomes synchronized with that of the host. Actually, the prophage appears as an integral part of the host chromosome.

At this state the p_M maintenance promoter regulates concentration of the *cI* repressor. The recognition site of the lambda repressor is:

```
T A T C A C C G C C A G A G G T A
A T A G T G G C G G T C T C C A T
```

The structure of the lambda repressor. (From Alm, E. & Baker, D. 1999. *Proc. Natl. Acad. Sci., USA* 96:11305.)

The ability of λ to have a lysogenic state makes it a temperate phage. If for any reason *cI* is not functional, the bacterial plate shows clear plaques, indicating that the phages are in the reproductive growth phase.

The first gene turned on at the p_R promoter is *cro*, followed by *cII*, which has the product of λ-gene *P*, a DNA prepriming protein, with host proteins. The expression of gene *Q* prevents transcription termination of the genes involved in lysis and morphogenesis. The products of the *S* and *R* genes mediate lysis of the host cell, permitting the liberation of the phage progeny. Under the influence of protein Q, the pR' promoter assists in the transcription of all late genes and its role ends somewhere in the recombination-control region around the area of *b*. In the integrated state, the prophage is replicated under the control of the host as if it were an intrinsic part of the bacterial genome. Autonomous replication of phage DNA requires the function of the replicational origin (*ori*) where the O protein binds and replication originates. The replication is preceded by circularization of the phage DNA is with the mechanical assistance of the cohesive (*cos*) ends, which are then sealed by a DNA ligase. When the DNA is injected into the host cell first it replicates by the bidirectional theta (θ) replicational form and generates monomeric DNA circles. When the function of the *gam* gene is turned on, replication is switched to the bacterial recBC protein-dependent rolling circle type (sigma replication, σ). This replication results in the formation of linear concatamers attached by the *cos* ends. These concatamers are then stuffed into the preformed phage heads (encapsidation) as shown by the diagram on the next page.

The pR' promoter mediates the transcription of the head and tail genes. The head and tail components are made and assembled separately before joining into the mature phage particle (see illustration on next page). In this process, the cooperation of host proteins (groE) is needed. The concatenated λ-DNA molecules are filled into the prehead. The DNA was synthesized in head-to-tail molecules involving more than a single genome (concatamer). The protein products of the *Nu1* and *A* genes bind to the DNA not far from the left *cos* sites and are brought to the prehead. The *FI* gene product reels it into the head until the terminal *cos* site is found where the Nu1-A protein complex (*terminase*) cuts it off. The product of the *D* gene stabilizes the head shell and the FII protein completes the head that is now attached to the prefabricated tail assembly. The mature phage particle is equipped with injector mechanisms that facilitate the adsorption of the particles to the host cell and introduction of its genome through the host cell membrane.

The phage genes under the control of the *pL* promoter are often called *early genes*; the ones under *pR* are *delayed early* and those regulated through *pR'* are referred to as *late genes*. The two major promoters, *pL* and *pR*, may not necessarily stop at the termination signals represented by the arrowheads but may continue transcription and thus represent read-through. The regulation of the *N* gene product is accomplished by acting at the *nut* sites rather than at the terminator sequences (*tL* and *tR*). The *nutL* is located before the terminator less than 60 bp from *pL* and the *nutR* is about 250 bp from *pR*. The *nut* regions have a dyad symmetry, including *boxA* and *boxB*. The region of binding of the *Q* gene product is designated *qut*. The activities of the N and Q proteins are frequently called the *antitermination* function. The function of the λ *N* gene is also regulated by bacterial (N utilization) genes *nusA*, *nusB*, *nusE*, and *nusG*. NusA is a transcription factor; *nusG* encodes ribosomal protein 10. NusB protein and S10 dimer bind to the *boxA* sequences in bacteria. NusG assembles the other Nus proteins for binding to RNA polymerase. NusA can facilitate transcription termination by the N protein at intrinsic terminators, but all Nus proteins are required for stopping transcription in rho-dependent termination. When the Nus complex and other proteins associate with RNA polymerase, it becomes modified at the *nut* site and can pass through the terminator sites without stopping transcription. The polymerase core can associate either with the σ-subunit of the transcriptase or with the NusN complex. When the N protein replaces the σ-subunit of the RNA polymerase, the termination signals are ignored. The *qut* sites permit a change in transcriptase to work faster and avoid stoppage at the terminator sites, proceeding through the lytic phase into the vegetative growth stage.

Recombination of λ may be mediated by the closely linked genes *exo* (also called *redX*, *redα*, or *redA*) and the *bet* complex (also called *β*, *redβ*, *redB*). These two *red* complexes are involved in general recombination and are not dependent on the function of the host *recA* function. The *exo* gene codes for a 5′-exonuclease (M_r 24,000) that can convert a branched DNA structure to an unbranched nicked duplex by the process called *strand assimilation*.

The *bet* gene product (subunit M_r 28,000) binds to the exonuclease and promotes reannealing with the complementary DNA strand. The *gam* gene product (M_r 16,500) in a dimeric form binds to the bacterial host enzyme recBC and inhibits its activities. General recombination is affected by *chi* (χ) sites near *A*, between *I* and *J*, *xis* and *exo*, within *cII*, between *Q* and *S* genes. *Chi*s do not have any gene product; rather they offer a suitable

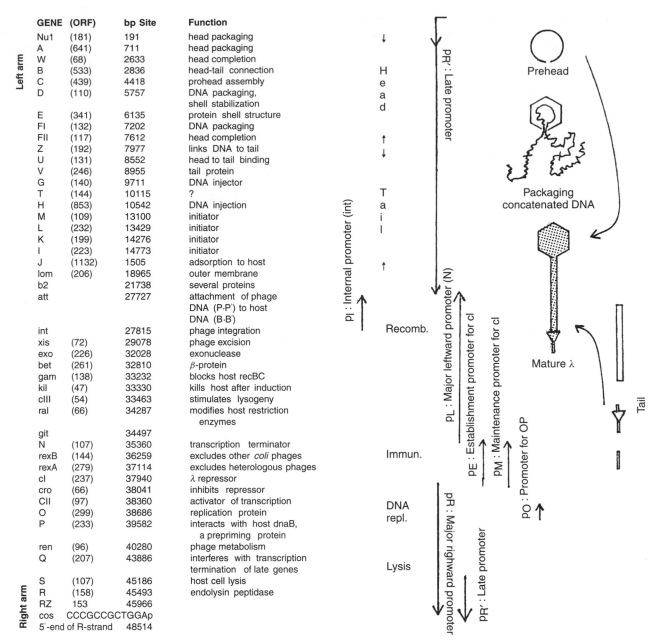

GENE	(ORF)	bp Site	Function
Nu1	(181)	191	head packaging
A	(641)	711	head packaging
W	(68)	2633	head completion
B	(533)	2836	head-tail connection
C	(439)	4418	prohead assembly
D	(110)	5757	DNA packaging, shell stabilization
E	(341)	6135	protein shell structure
FI	(132)	7202	DNA packaging
FII	(117)	7612	head completion
Z	(192)	7977	links DNA to tail
U	(131)	8552	head to tail binding
V	(246)	8955	tail protein
G	(140)	9711	DNA injector
T	(144)	10115	?
H	(853)	10542	DNA injection
M	(109)	13100	initiator
L	(232)	13429	initiator
K	(199)	14276	initiator
I	(223)	14773	initiator
J	(1132)	1505	adsorption to host
lom	(206)	18965	outer membrane
b2		21738	several proteins
att		27727	attachment of phage DNA (P·P') to host DNA (B·B')
int		27815	phage integration
xis	(72)	29078	phage excision
exo	(226)	32028	exonuclease
bet	(261)	32810	β-protein
gam	(138)	33232	blocks host recBC
kil	(47)	33330	kills host after induction
cIII	(54)	33463	stimulates lysogeny
ral	(66)	34287	modifies host restriction enzymes
git		34497	
N	(107)	35360	transcription terminator
rexB	(144)	36259	excludes other coli phages
rexA	(279)	37114	excludes heterologous phages
cI	(237)	37940	λ repressor
cro	(66)	38041	inhibits repressor
CII	(97)	38360	activator of transcription
O	(299)	38686	replication protein
P	(233)	39582	interacts with host dnaB, a prepriming protein
ren	(96)	40280	phage metabolism
Q	(207)	43886	interferes with transcription termination of late genes
S	(107)	45186	host cell lysis
R	(158)	45493	endolysin peptidase
RZ	153	45966	
cos	CCCGCCGCTGGAp		
5'-end of R-strand	48514		

The map of bacteriophage lambda showing the order of the major genes. ORF indicates the size of the open-reading frames and bp site stands for the number of nucleotides starting at the left arm. The base sites may vary in different λ-phages. The morphogenetic pathway is also outlined. (Data from Hendrix, R. W., et al., eds. 1983. *Lambda II*. Cold Spring Harbor Laboratory Press, Cold Spring Harbor, NY.)

site for an enhancement of recombination (about 5-fold, measured by burst size) through the RecBC pathway. Near the *chi* sites an octamer consensus exists: 5'-GCTGGTGG-3'. The stimulation of recombination by *chi* is not limited to its location, but it is effective over a considerable distance. *Chi* s act "dominant" because it is enough to be present in one of the recombining partners. When *chi* is present in only one of the partners, recombinants without active *chi* are much more frequent than the reciprocals (*nonreciprocality*). Enhancement of recombination to its left is greater than to its right side (*directionality*).

Recombination may be *site-specific*, requiring a special nucleotide sequence, the *primary att* sites of high efficiency, and the much lower-efficiency secondary *att* sites. The *att*

sites have common 15 bp core sequences that are 70 to 80% AT. The primary attP sites in the phage are represented as —POP'— and there is a corresponding site of homology in the bacterium attB (~~~BOB' ~~~). After recombination (insertion) at the crossing-over region (O), by the means of pairing as an inversion loop, in the bacterial cell there will be:

$$\sim\sim\sim BOP'-----POB' \sim\sim\sim$$

In excision the process is reversed. The insertion is mediated by λ-protein Int and the excision uses the products of λ genes *Int* and *xis*. Both processes require the bacterial integration protein factor (IHF, M_r 20,000) and other host gene products.

The coding region of int extends 84 to 1151 bp right from the crossing-over site. The M_r of the Int protein is about 40,000; at the amino terminus it is rich in basic residues and it works as a topoisomerase. The Xis protein shares amino acid sequences with Int. The *xis* product contains 72 amino acids with 25% Lys and Arg. Because of the site-specific recombination, λ-phage can mediate specialized transduction.

Phage λ has been used for the development of a number of genetic transformation vectors. Genes not exceeding 5% of the genome can be inserted at a single target site into the λ-DNA and can be propagated (insertional vectors). The nucleotide sequences between genes J and N can be deleted (representing about 25% of the genome) and replaced. The DNA can still produce nearly normal-size plaques as long as the total length of the DNA remains no less than 37 kb and not more than 52 kb (replacement vectors). In cosmid vectors it is essential to retain the *cos* sites plus 4–6 kb at both termini, including the origin of replication. The vectors so constructed can propagate foreign DNA exceeding 30 kb. *See* burst size; Charon vectors; chi elements; cosmid vectors; DNA-binding protein domains; helix-turn-helix; O-some; recombination; rho terminator; rolling circle; specialized transduction; theta replication. (Hendrix, R. W., et al. 1983. *Lambda*. Cold Spring Harbor Lab. Press; Cue, D. & Feiss, M. 2001. *J. Mol. Biol.* 311:233; Friedman, D. I. & Court, D. L. 2001. *Curr. Opin. Microbiol.* 4[2]:201.)

λ_S Risk of relatives/population risk may be used to determine the genetic part of a disease. A recessive disease is expected among the relatives of a proband at a frequency of 0.25. If the prevalence of this disease in the general population is say 0.0005, then $λ_S = 0.25/0.0005 = 500$. In case the disease is controlled by a dominant gene, then $λ_S$ will be larger. The higher $λ_S$, the greater the risk of recurrence. $λ_S$ larger than 2 indicates a significant genetic component in the disease. In case the disease is under the control of multiple loci, the risk may diminish in successive generations. *See* prevalence; risk; sibling.

λZap Carries the *lacIq* and tetracycline resistance in an F′ plasmid. This vector is very effective for cDNA cloning. Its cloning capacity is about 10 kb at multiple cloning sites. In the presence of helper M13 or f1 helper phages, the cloned DNA is excised and placed into a small plasmid to facilitate restriction mapping or sequencing of the insert. λZap is also suitable for making RNA transcripts of either strand using T3 or T7 transcriptases. *See* cloning site; filamentous phages; F′ plasmid; helper phage; lambda phage; restriction enzyme; sequencing of DNA.

lambdoid phage Closely related to λ-phage such as P22 of *Salmonella. See* P22. (Clark, A. J., et al. 2001. *J. Mol. Biol.* 311:657.)

lambert Unit of luminous intensity; 1 lumen per cm^2. The lumen is the unit of luminous flux emitted in a unit solid angle (steradian) by a uniform point source of 1 candela. *See* candela (candle).

lamella Thin plate or membrane sheet.

lamellipodium Sheet-like cellular extension involving actin and assisting cell movement.

laminin Large glycoprotein (~1 MDa) localized in the synaptic cleft between the neuronal basal lamina of the muscle cell sheath and the acetylcholine receptors. *See* acetylcholine; agrin; basement membrane; epidermolysis; synapse. (Koch, M., et al. 1999. *J. Cell Biol.* 145:605.)

lamin Intermediate filament proteins (encoded at 1q21). Upon lamin polymerization, the nuclear lamins are formed; during interphase, lamins support the nuclear membrane. A-kinase anchoring protein 149 (AKAP149) seems to recruit lamins to the nuclear envelope, and protein phosphatase 1 (PP1) modulates the membrane architecture during mitosis. Lamins also help to define embryonal polarity. *See* cardiomyopathies (CMD1A); intermediate filaments; lipodystrophy; muscular dystrophy (Emery-Dreifuss type). (Wilson, K. L., et al. 2001. *Cell* 104:647.)

L-amino acids (levorotatory amino acids)　Common natural amino acids. *See* D-amino acids.

LAMP-1, LAMP-2 Lysosome-associated membrane proteins. LAMP-2 deficiency is associated with Danon disease characterized by impaired degradation (autophagy) of some liver proteins and various muscle anomalies (cardiomopathy). *See* Danon disease. (Winchester, B. G. 2001. *Eur. J. Paediatr. Neurol.* 5, Suppl. A:11.)

lampbrush chromosome Giant chromosome in oocytes (mainly in amphibia), with conspicuous loops extending about 40 μm on a core resembling brushes used to clean the glass chimney of kerosene lamps. These loops are highly active in RNA synthesis, and although not polytenic, they are well visible under the light microscope. The photograph is a section of a long chromosome. (Courtesy of Dr. Joseph G. Gall.) A set of amphibian lampbrush chromosomes may display 10,000 loops alternating with condensed chromomeres. On the surface of the DNA loops, the nascent RNA transcripts may be visible. This high activity may precede meiosis and continues during it in order to secure a good supply to meet the later needs of the rapidly developing zygote. *See* giant chromosomes; newt. (Gall, J. G. & Murphy, C. 1998. *Mol. Biol. Cell* 9:733.)

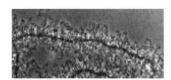

LANA Latency-associated nuclear antigen. *See* Kaposi sarcoma.

landrace Locally adapted variety of plants or breed of animals that may include a number of different genotypes but of similar phenotypes. Landraces are the products of an intuitive selection of the growers or husbandryman.

Lange-Nielsen syndrome (cardioauditory syndrome)　Involves heart arrhythmia and deafness, and sudden death. The gene responsible, KVLQT1, spans the region 11p15. This area includes Beckwith-Wiedemann syndrome. *See* Beckwith-Wiedemann syndrome; deafness; LQT.

Langer-Giedion syndrome (LGS, 8q24.11-q24.13)   Generally involves mental retardation, small brain (microcephaly), bulbous nose, sparse hair, emergences on the bones (exostosis), etc. The autosomal-dominant phenotype is based on deletions in the 8p22-8q24.13 chromosomal region. *See* deletion; exostosis; head/face/brain defects; mental retardation; trichorhinophalangeal syndrome.

Langerhans cell   Dendritic antigen-presenting cell with indented nucleus. Langerhans cells are present in the epidermis and some internal organs. *See* antigen-presenting cell; dendritic cell.

Langerhans islets   Cells with dentate nucleus in the pancreas, arranged in groups. They appear to have antigen-presenting abilities. *See* diabetes mellitus.

Langer mesomelic dwarfness   *See* dyschondrosteosis.

language   *See* speech and grammar disorder.

language, computer   Set of representations used by a computer program.

language impairment, specific (SLI)   About 4% of the English-speaking children of normal intelligence and average environment suffer from low language skills for their age. It is a polygenic disorder; however, chromosome 13q21, sites 16 and 19 are significantly involved. *See* speech and grammar disorder. (SLI Consortium. 2002. *Am. J. Hum. Genet.* 70:384.)

langurs   *See* Colobidae.

lanosterol   Precursor of cholesterol. *See* cholesterol; geranyl pyrophosphate.

laparoscopy   Examination or treatment of the interior of the abdomen by a special device, the laparoscope, which may be equipped with a laser source.

LARD   Apo-3 family TNF/NGF receptor. *See* Apo-3; NGF; TNF.

large embryo/offspring syndrome   Bovine and ovine embryos under unusual experimental conditions preceding the blastocyte stage increase in size and involve problems of fetal loss, difficulties in parturition (birth), and developmental defects. In vitro embryo culture, nuclear transplantation, high urea diet of the mother, etc., may be responsible for the anomalies. Mice produced by nuclear transplantation tend to be large and obese, but the condition is not transmitted to their offspring. *See* nuclear transplantation; tetraploid complementation. (Young, L. E. 1998. *Rev. Reprod.* 3:155; Tamashiro, K. L. K., et al. 2002. *Nature Med.* 8:262.)

large T antigen   *See* SV40.

lariat RNA   Formed during the splicing reaction of the primary transcript of eukaryotic genes. In the first step, the pre-mRNA is cut at the junction of exon 1 and the intron, resulting in a piece of RNA containing intron-exon 2; this "2/3 molecule" immediately forms a loop-like lariat. In the second step the intron-exon 2 junction is cut and exons 1 and 2 are ligated (spliced). Afterward, the intron is released and the procedure follows for the other junctions. *See* introns; splicing. (Ooi, S. l., et al. 2001. *Methods Enzymol.* 342:233.)

Laron-type dwarfism   *See* dwarfness; pituitary dwarfism.

Larsen syndrome (LRS1, 3p21.1-p14.1)   Dominant knee dislocation, prominent forehead, cylindrical fingers, short digits, flat nasal bridge. *See* bone diseases.

larva   Insect at the developmental stage after hatching from the egg. It resembles a worm more than an adult insect, and after pupation the imago emerges from it. During the larval stage, the germline and the imaginal discs are laid down. *See* Drosophila.

Drosophila larva.

Lascaux Cave   Located in the caves of Dordogne, southwestern France; reveals the remarkable art of the Cro-Magnons, originated about 15,000 years ago. These colored paintings disclose a great deal about the types of animals present in the area in the Paleolithic age and attest to the intelligence and artistic abilities of the early European ancestors. More recently, even older (35,000–40,000 years) ancient artwork and artifacts have been discovered in Europe (Grotte Chauvet) and Africa. The animal shown below displays great similarity to a present-day longhorn bull. By comparison with the modern breeds of cattle, the success of animal breeding can be assessed. (Valladas, H., et al. 2001. *Nature* 413:479.)

(Courtesy of the Caisse Nat. Mon. Hist. Sites, France.)

laser   Light amplification by stimulated emission of radiation equipment produces electromagnetic radiation in the infrared and visible spectrum by stimulation of atoms. The laser beam does not diffuse like that of an electric light. The radiation travels in the same direction at the same wavelength of a very narrow frequency band. Thus, it focuses all of the energy to a fine point. Lasers can produce radiation at many wavelengths and frequencies. An intense light source and electron currents can activate gases, semiconductors, and ions in solid material to produce coherent laser light. The lasers can be pulsed or continuous. The former produces extremely high peaks of power; the latter is highly stable and gives pure emission. Laser radiation is applicable to different scientific instruments (e.g., laser scanners), photochemistry (e.g., laser photolysis), identification of fluorochrome-labeled molecules, surgery, etc.

laser-capture microdissection May be used to trace the event in cellular differentiation. Cell cultures or tissue sections are placed on a polymer film activated by laser pulses. The shape of the cells and their DNA, RNA, and protein content are preserved. cDNA libraries can then be hybridized to thousands of genes to study their expression in health and disease or patterns of DNA methylation, and they can be used for biological or diagnostic purposes. They may be used for isolation of pure cell types — without injury to macromolecules — from microscopic tissue sections and transferred to laser-activated polymer films. The procedure permits molecular analysis (e.g., microarray hybridization) of tissue-, cell-specific alterations due to disease and/or therapy. *See* microarray, laser. (Emmert-Buck, M. R., et al. 1996. *Science* 274:998; Simone, N. L., et al. 1998. *Trends Genet.* 14:272; Kerjean, A., et al. 2001. *Nucleic Acids Res.* 29[21]:e106; Craven, R. A. & Banks, R. E. 2001. *Proteomics* 1:1200.)

laser scanning cytometry Analyzes DNA content of normal and cancer cells; it can be used to measure in situ hybridization by automation. (Darzynkiewicz, Z., et al. 1999. *Exp. Cell Res.* 249:1; LaSalle, J. M., et al. 2001. *Hum. Mol. Genet.* 10:1729.)

last-male sperm precedence In many insects (e.g., Drosophila) the female stores the sperm in her spermatheca before fertilization. When multiple insemination takes place by different males, the sperm of the last male preferentially sires the offspring due to sperm displacement and inactivation. The sperm elimination is controlled by the seminal fluid produced by the accessory gland. *See* certation; multiparental hybrid; multipaternate litter; sperm precedence. (Hooper, R. E. & Siva-Jothy, M. T. 1996. *Mol. Ecol.* 5:449.)

LAT Class of herpes virus genes of mainly unclear function. They are transcribed when the virus is maintained in a quiescent state bound to the nucleosomes in the neurons. LAT protein has an adapter function in TCR-mediated signal transduction. *See* herpes; signal transduction; TCR. (Aguado, E., et al. 2002. *Science* 296:2036.)

late genes Transcribed only later during the life cycle of an organism; viral genes involved in replication and in the synthesis of structural (coat) proteins (e.g., during the lytic phase of a bacteriophage). These genes use late promoters that are different from the early ones. *See* delayed-response genes; early genes.

latent period The infection has taken place, but the symptoms are not manifested yet; in phage biology the time

between injection of the phage DNA into the bacterium and the beginning of lysis. It is also the time from the initiation of a disease to the onset of symptoms. *See* burst.

late period Of phage development, starts with the beginning of DNA replication. *See* development; lambda phage.

lateral transmission Same as horizontal transmission. Lateral transmission is generally suggested when similar nucleotide sequences occur among species that may not be related by orthologous descent. Lateral transmission is common (1.5 to 14.5%) in bacteria by means of transformation by exogenous DNA, by phage-mediated transduction and conjugational transfer of plasmids. Antibiotic resistance and virulence genes are commonly transmitted laterally and may become important factors of population adaptation and evolution. On the basis of the sequenced genomes of eukaryotes, the lateral transfer of genes might have occurred during evolution. The lateral acquisition of the number of bacterial genes by the human genome has been estimated as 40 to over 200. A common objection to such claims is that convergent evolution of these genes or sequences may not be ruled out with great certainty. Furthermore, many of the supposedly horizontally acquired sequences in vertebrates were already present in nonvertebrate lineages (Stanhope, M. J., et al. 2001. *Nature* 411:940). In addition, the endosymbiotic acquisition of chloroplasts and mitochondria and DNA traffic among these organelles pose problems in identifying the lateral transfers. *See* convergent evolution; infectious, heredity; orthologous genes; transformation; transmission; transmission, lateral; transposable elements. (Garcia-Vallvé, S., et al. 2000. *Genome Res.* 10:1719; Andersson, J. O., et al. 2001. *Science* 292:1848.)

late-replicating chromosome Or chromosomal region, is heterochromatic and genetically not active. *See* Barr body; heterochromatin; lyonization.

lathyrism Caused by the presence of β-cyano-alanine and its decarboxylation product, β-aminopropionitrile in the seeds of the food and forage legume, *Lathyrus sativus*. Some other contaminating weed legumes may be major culprits in lathyrism. In some countries the ground seed is used for a filler in bread making. As a feed its amount should be kept below 10% of the ration. Cooking substantially reduces its adverse effect. These compounds affect the cross-linking of collagen and result in spasms, pain, paralysis of the lower extremities (paraplegia), abnormal sensitivity (hyperesthesia), burning sensation of the skin (paresthesia), curvature of the spine, rupture of the aorta, etc. *See* collagen. (Spencer, P. S., et al. 1993. *Environ. Res.* 62:106; Riepe, M., et al. 1995. *Nat. Toxins* 3:58.)

Lathyrus odoratus (sweet pea) Ornamental legume ($2n = 14$), a favorite object of early studies on the gene-controlled anthocyanin synthetic pathways. The *L* dominant allele (long) and the recessive ℓ allele (disc) determine the shape of the pollen. Although this is a gametophytic trait, the phenotype is determined by the genotype of the anther (sporophytic) tissue and thus is an example of delayed inheritance. *See* delayed inheritance; diagram on p. 679.

Latin square Square array of items in parallel columns and rows in such a manner that each must occur once but only once in each row and column. Such a design lends itself for the evaluation of the experiments by analysis of variance and it can be employed in agricultural field experiments, but it is useful for pharmacological, microbial, and other research where the partition of the data into relevant variances is very important. *See* analysis of variance; Graeco-Latin square.

```
B A C
A C B
C B A
```

Latin square layout design.

latrotoxin Large peptide neurotoxin binding to neurexins. *See* neurexin.

LATS Large tumor suppressor. It is a negative regulator of CDC2 and cyclin A. LATS1 deficiency in mice may produce soft-tissue sarcomas, ovarian tumors, and pituitary defects. *See* CDC2; cyclin A; pituitary; sarcoma. (Hori, T., et al. 2000. *Oncogene* 19:3101.)

lauryl sulfate *See* dodecyl sulfate sodium salt.

law Both civil and criminal law considers genetic principles and methods. Examples are regulating marriage, social policy, demographic factors, using dermatoglyphycs (fingerprinting), serological methods, DNA fingerprinting, patents, etc. The branch of genetics involved in legal matters is called forensic genetics. *See* forensic genetics; genetic privacy; patents.

lawn, bacterial Petri plate containing nutrient agar and growing bacterial cell colonies.

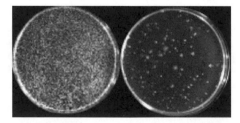

Bacterial lawn, heavy at left and sparse at right.

law of large numbers The increase in the size of a population assures the closer fit of the observed data to a valid null hypothesis or that the experimentally observed mean represents the true mean of the population dispersed according to the normal distribution. *See* mean; normal distribution; null hypothesis.

Lawrence-Moon syndrome Autosomal-recessive mental retardation involving retinal defects, underdeveloped genitalia, and partial paralysis.

Lawrence-Seip syndrome Lipoatrophy (loss of fat substance from under the skin); it is accompanied by enlargement of the liver, excessive bone growth, and insulin-resistant diabetes.

LB agar Add 1.5 to 2% agar to LB medium. *See* LB bacterial medium.

LB (Luria-Bertani) bacterial medium H_2O, deionized, 950 mL, bactotryptone 10 g, bacto yeast extract 5 g, NaCl 10 g, adjust pH to 7 with 5 N NaOH, fill up to 1 L.

LC50 Lethal concentration 50 of a chemical is expected to cause death in 50% of the treated cells or individuals of a population. *See* LCLo; LD$_{50}$; LDLo.

LCA oncogene Isolated from human liver carcinomas; assigned to human chromosome 2q14-q21. *See* oncogenes.

LCK Lymphocyte nonreceptor protein tyrosine kinase of the SRC family, located to human chromosome 1p35-p32. *See* ITAM; oncogenes; signal transduction; T cell; tyrosine kinase. (Holdorf, A. D., et al. 2002. *Nature Immunol.* 3:259.)

LCLo Lethal concentration low. The lowest concentration of a substance in air that causes death to mammals in acute (<24 h) or subacute or chronic (>24 h) exposure. *See* LDLo.

LC-MS Liquid chromatography–mass spectrometry is a bioanalytical pharmaceutical tool used in conjunction with quadrupole mass spectrometers. *See* proteomics; quadrupole. (O'Connor, D. 2002. *Curr. Opin. Drug Discov. Devel.* 5:52.)

LCR Locus control region is a nuclease-hypersensitive region far upstream or downstream or at other sites apart from the structural gene. Its presence is required for the expression of a particular gene locus and its role is probably opening of the chromatin for transcription. All the genes are in competition for the assistance of the LCR and those that are closer to it have an advantage. LCR may be at many kb distance from the locus it controls. The multiple genes appear to be transcribed alternately as the chromatin loops back and forth. In the LCR of the β-globin gene, there are 5 DNase I hypersensitive sites that synergistically activate the gene cluster (globin ε-Gγ-Aγ-δ-β) in a developmentally specific sequence. Hypersensitive site 4 is the most important for activation of transcription. Each of the hypersensitive sites binds transcriptional activators (E-box factors, GATA-1, NF-E2, CACCC-binding proteins [e.g., Sp1]), yet they also have some specificities. For the globin cluster, the most important role of LCR is the steady maintenance of transcription. The globin gene LCR is very well conserved among the various species. The CACCC box of the β-globin genes has an important role in the transcriptional regulation of the γ- and β-gene promoters. Sp1-related protein binds ubiquitously and Krüppel-like proteins (EKLF) bind to this box only in the erythroid lineage. Although CACCC boxes are present in both the γ- and β-promoters, in the γ

four repeating copies of a CCTTG sequence suppress EKLF binding. Thus, the functions appear to also be controlled by the structural/sequence context. The LCRs include enhancer elements. *See* chromatin; chromatin remodeling; E-box; EKLF; GATA; heterochromatin; looping of DNA; NF-E2; nuclease sensitive site; position effect; promoter; regulation of gene activity; Sp1. (Fraser, P. & Grosveld, F. 1998. *Curr. Opin. Cell Biol.* 10:361; Grosveld, F. 1999. *Curr. Opin. Genet. Dev.* 9:152; Johnson, K. D., et al. 2001. *Mol. Cell* 8:465; Ho, Y., et al. 2002. *Mol. Cell* 9:291.)

LD$_{50}$ Calculated dose of a substance expected to kill 50% of the exposed experimental population. For the determination of LD$_{50}$, rats have been used. This type of test will be phased out to avoid cruelty to animals. *See* LDLo.

LDB LIMB domain-binding factor. *See* Lim domain; NLI.

Ldj1, Ldj2 DnaJ-like farnesylated chaperones of higher plants. *See* chaperones; DnaK; prenylation.

LDL *See* familial hypercholesterolemia; low-density lipoproteins; sterol.

LDLo Lethal dose low is the lowest dose of a substance introduced by any route, except inhalation, over a period of time in one or more portions; has caused death to mammals. *See* LClo; LD50.

LDL receptor Mediates LDL endocytosis; it is encoded in human chromosome 19-p13.1-13.3. It is homologous with the EGF receptor. LDL-related protein (LRP5) defects cause a type of osteoporosis. *See* EGFR; LDL; low-density lipoproteins; lysosomes; osteoporosis. (Gong, Y., et al. 2001. *Cell* 107:513.)

L-DNase II Posttranslational product (from a serpin-like protease inhibitor or leukocyte elastase inhibitor) upon cytosolic acidification. It degrades DNA during apoptosis. *See* apoptosis; elastase; serpines.

LDP Long-day plant. *See* photoperiodism.

leader peptide Directs the translocation of proteins. In bacteria it consists of 16–26 amino acids involving a basic amino terminal, a polar central domain, and a nonhelical carboxy domain. The latter is essential for recognition by leader peptidase to cut it off. For mitochondrial import, the leader has 10–70 residues and it is rich in positively charged and hydroxylated residues. Similar leader peptides mediate the import of proteins from the cytosol to the chloroplasts or to the nucleus. The leader sequences do not enter the target organelles. *See* signal peptide. (Wang, A. H. & Yang, X. J. 2001. *Mol. Cell Biol.* 21:5992.)

leader sequence Nontranslated stretch of nucleotides at the 5′-end of mRNA. Some mRNAs lack leaders, e.g., lambda-phage cI gene. The tobacco mosaic virus carrying the translational enhancer Ω works more efficiently in the absence of a Shine-Dalgarno sequence. *See* mRNA; Shine-Dalgarno sequence. (O'Donnell, S. M. & Janssen, G. R. 2001. *J. Bacteriol.* 183:1277.)

leading strand Of DNA, faces the replication fork by the 3′-OH end and is extended by directly adding 5′-deoxynucleotidephosphates to that end after removal of the γ- and β-position phosphates from the nucleotide triphosphate precursors. *See* DNA replication; lagging strand; replication; replication fork.

leadzyme Ribozyme that can cleave itself at a specific phosphodiester site in the presence of lead ions (Pb^{2+}). The leadzyme has an asymmetric internal loop flanked by Watson-Crick hydrogen-paired nucleotides. The shaded section indicates the catalytic core. *See* DNA-zyme; ribozyme. (Doherty, E. A. & Doudna, J. A. 2000. *Annu. Rev. Biochem.* 69:597.)

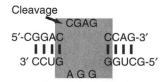

Diagram of the LZ4 leadzyme after Wedekind, J. E. & McKey, D. B. 1999. *Nature Struct. Biol.* 6:261.

leaf skeleton hybridization Procedure carried out on plant leaves infected by cauliflower mosaic virus vector with the same purpose as colony hybridization of bacteria, i.e., detecting foreign gene sequences in the CaMV vectors. *See* CaMV; colony hybridization. (Melcher, U., et al. 1992. *Arch. Virol.* 123[3–4]:379.)

leaky mutant Has an incomplete genetic block in a synthetic step. *See* genetic block; hypomorphic.

learning behavior Partly genetically controlled. It has been found that slow-learning mice have a reduced amount of protein kinase C in the hippocampus of the brain; thus, it appears that the PKC gene is a part of the polygenic system affecting this complex trait. Learning is potentiated by regeneration ability of the central nervous system. *See* Boolean algebra; brain, human. (Dubnau, J. & Tully, T. 1998. *Annu. Rev. Neurosci.* 21:407.)

learning disability *See* dyslexia; Tourett's syndrome.

least significant difference (LSD) Measure of the significance among a set of means compared. The standard error of the difference between two means can be determined on the basis of the student's *t* distribution using $\sqrt{2s^2/n}$ where s^2 is the standard error and *n* is the size of the population. The standard error of the difference is then multiplied by the minimal *t* value at a particular level of significance from the student's *t* table; any value that exceeds this product constitutes the LSD. *See* F distribution; student's *t* distribution; *t* value.

least squares Formula is the basis for the theory of regression (*b*),

$$b = \frac{S(y[X-x])}{S(X-x)}$$

where S = sum, y = one of the variates, x = other variate, $\bar{x}$ = mean of x. Least square methods are also used for the estimation of evolutionary distance. The smallest minimum sum of squared differences computed from paired data indicates the best topology for an evolutionary tree. *See* correlation; four-cluster analysis; minimum evolution method; neighbor joining method.

leaving group Displaced molecular group in a chemical reaction.

Leber optic atrophy *See* amaurosis congenita; eye diseases; mitochondrial disease in humans; optic atrophy.

lecithin Glycerol phospholipid, also called phosphatidyl choline.

lecithin:cholesterol acyltransferase deficiency (Norum disease, fish-eye disease) Either a relatively rare recessive Norum disease coded in human chromosome 16q22.1 and in mouse chromosome 8 or a dominant mutation at the same locus, which is fish-eye disease. This defect in lipid metabolism causes proteinuria, anemia, renal and heart defects. In Norum disease there is a general failure in esterification of cholesterol in high-density lipoprotein, whereas in fish-eye diseases the deficiency is more specific. The name comes from the opacity of the eye resembling that of boiled fish. *See* apolipoprotein; cardiovascular disease; lipoprotein.

lectins First identified as plant proteins that agglutinate erythrocytes by virtue of binding to surface sugars. Lectins also occur in invertebrate and vertebrate animals and may serve as ligands in the natural killer cells of mice. They play a role in general cell adhesion, management of peptides in the endoplasmic reticulum, and surface recognition and protection against bacteria and viruses. Lectins introduced into the potato by genetic engineering may have adverse effects on experimental animals if consumed. *See* cell adhesion; complement; concavalin; endoplasmic reticulum; killer cells; selectin. (Weis, W. I. & Drickamer, K. 1996. *Annu. Rev. Biochem.* 65:441.)

leech Hirundinaceous (blood-sucking) lower animal with only ≈350 nerve cells per ganglion. *See* ganglion. (Stent, G. S., et al. 1992. *Int. Rev. Nerurobiol.* 33:109.)

Leech.

LEF Lymphoid-enhancer-binding factor is a transcription factor regulating lymphocyte differentiation; it is a member of the high-mobility group of proteins. LEF interacts with β-catenin and regulates signal transmission to the nucleus. It binds to the TTCAAACC sequences in the TCRα enhancer region. LEF is similar to TCF, and this family of proteins also regulates axis formation with the aid of β-catenin-associated proteins. Catenin activation plays a role in human cancers. *See* catenins; Gardner syndrome; high-mobility group proteins;

lymphocytes; melanoma; pilomatricoma; TCF. (Liu, T., et al. 2001. *Science* 292:1718.)

left-handedness *See* handedness.

left-right asymmetry Widely found in all higher organism syndromes (prevalence $>1 \times 10^{-4}$). In some humans, the normal pattern of the location of organs is disturbed (Kartagener syndrome, situs inversus viscerum). The pattern of differentiation is under the control of a cascade of events that bear some similarities and differences in various organisms. The processes are under the control of genes and proteins with somewhat variable homology. N-cadherin appears to be an early regulator for establishing the asymmetry of the body. The initial signals emanate from the organizer where activins on the right side stimulate fibroblast growth factor (FGF-8), which in turn blocks the Cerberus (Cer)/Dan/gremlin/Caronte (CAR) family of genes/proteins. The homologous macromolecules are so named in different organisms. On the left of the organizer/primitive streak, Sonic hedgehog (SHH) acts in a direction opposite to FGF-8; it activates CAR. CAR then interferes with the bone morphogenetic protein-blocking action and NODAL (Nod)/Lefty-2/Activin is turned on ♙. Nod has a central morphogenetic function. On the left side, ACTIVIN RIIB/SMAD-2 blocks ♥ SnR (Snail related). ACTIVIN RIIB also turns on ♙ PitX2/NKX3.2/bagpipe. On the right side of the lateral-plate mesoderm, the mirror image of the processes goes on. What is activated on the left side is more or less attenuated on the right side.

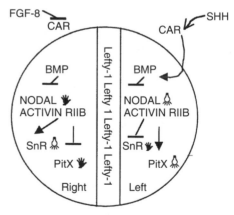

Some of the developmental circuitries involved in the determination of left/right differentiation.

In the midline, Lefty-1, stimulated by CAR, seems to stabilize the difference between the two sides of the mesoderm. Lefty-1 is highly homologous to Lefty-2. Although this simple model provides a generalized picture, actually the processes involve additional genes and proteins in an interacting cascade of events. The EGF (epidermal growth factor)–Cryptic protein 1 (224 amino residues in humans) complex interacts with Nodal and regulates laterality. The normal asymmetry of the organ plan is called *situs solitus*, whereas the reverse arrangement is *situs inversus*. Some other laterality differences are called isomerism. The overall normal pattern of left/right asymmetry is evolutionarily conserved in vertebrates. *See* Kartagener syndrome; organizer; Rieger syndrome; situs inversus viscerum. (Diagram modified after Esteban, C. R., et al. 1999. *Nature* 401:243; Capdevila, J., et al. 2000. *Cell* 101:9; Burdine, R. D. & Schier, A. F. 2000. *Genes & Development* 14:763; Garcia-Castro, M. I., et al. 2000. *Science* 288:1047; Mercola, M. & Levin, M. 2001. *Annu. Rev. Cell Dev. Biol.* 17:779; Hamada, H., et al. 2002. *Nature. Rev. Genet.* 3:103.)

leghemoglobin   Hemoglobin protein coded for by genes of leguminous plants. This protein accumulates in the root nodules formed by the presence of nitrogen-fixing bacteria. Leghemoglobin transfers oxygen to the electron transport system of bacteria and prevents the accumulation of toxic amounts of oxygen that would interfere with nitrogen fixation. *See* globin; hemoglobin; myoglobin. (Kawashima, K., et al. 2001. *Plant Physiol.* 125:641.)

Leghorn, White   The poultry breeds White Plymouth Rock and White Wyandotte have plumage determined by recessive genes *ii cc*. The White Leghorn, in contrast, has the *IICC* genotype, where *I* is a dominant suppressor gene of color (*C*). Thus, when it is crossed to another white breed, e.g., White Wyandotte, in the F_2 of their hybrids the segregation is 13 white or speckled and 3 black. The blacks have the *iiCC* or *iiCc* genotype.

Legionnaires' disease   The airborne, potentially lethal, opportunistic infection is caused by the bacterium *Legionella pneumophila*, although over 40 different species exist. Some amoebae may rarely be responsible for similar disease. (Swanson, M. S. & Hammer, B. K. 2000. *Annu. Rev. Microbiol.* 54:567.)

legit   Statistical concept worked out by R. A. Fisher (*Biometrics* 6:353 (1950)) showing the change in allelic frequencies in a cline. If the allelic frequency is known, the legit can be read from the table published in this *Biometrics* article. *See* cline; diffusion, genetic.

legume   Taxonomic group of plants (e.g., pea, beans) with an ability to accumulate atmospheric nitrogen in symbiosis with some nitrogen-fixing bacteria. *See* nitrogen fixation; Rhizobia.

Leigh's encephalopathy (LS, 19q13, 11q13, 9q34)   The autosomal-recessive disorders occur in multiple forms and are characterized by high pyruvate and lactate concentrations in the serum and urine. The biochemical findings may be caused by a defect of pyruvate carboxylase or a necessary cofactor of the enzyme, thiamin triphosphate (TTP). The latter deficiency may be brought about by the absence of thiamin pyrophosphate-adenosine triphosphate phosphoribosyl transferase. The pyruvate carboxylase gene appears to be in human chromosome 11. The nuclear-encoded (3q29 and 5p15) flavoprotein subunit of succinate dehydrogenase controls mitochondrial enzyme complex II, including succinate dehydrogenase. Mutation in the SURF1 gene (9q34) encoding cytochrome c oxidase is involved in some cases of Leigh syndrome. Mutation in the mitochondrial NDUF complex may cause leukodystrophy and myoclonic epilepsy. Gluconeogenesis may become defective, and necrotic brain lesions, heart and respiratory disorders may be present. Prenatal diagnosis is successful in some cases. A defect in the NADH-ubiquinone oxidoreductase Fe-S protein (NDUFS8, 11q13) or a mutation in the 18 kDa NADH-coenzyme Q reductase (NDUFS4, 5q11.1), the NADH-coenzyme-ubiquinone oxidoreductase Fe-S protein 7 (NDUFS7, 19p13), or the NADH-ubiquinone oxidoreductase flavoprotein 1 (DUFV1, 11q13) may also be the cause of Leigh syndrome. These proteins are components of the mitochondrial respiratory chain that includes more than 40 subunits of which only 7 are encoded in mtDNA. *See* encephalopathies; gluconeogenesis; lactic acidosis; mitochondrial diseases in humans; NDUF; neuromuscular diseases; Saguenay–Lac-Saint-Jean syndrome.

Leiner's disease   Deficiency of the C5 complement component resulting in low opsonization and various types of infections, primarily during childhood. *See* complement; opsonin.

leiomyoma   Generally benign tumors of the smooth muscles occurring in the uterus, genitalia, gullet, etc., controlled by autosomal-dominant genes and their translocations. Predisposition is coded in the long arm of human chromosome 1q42.3-q43. *See* cancer.

Leishmania   Protozoan closely related to *Trypanosoma*. In mammals, including humans, it causes skin (T_H-1 response) and visceral ailments (T_H-2 response) ranging from relatively mild to life-threatening types depending on the infectious species. Activation of macrophages through IFN-γ may control the disease through the formation of antimicrobial substances, e.g., NO. The amastigote form occurs in mammalian blood and the promastigote form in the gut of the sandfly, which spreads the infection. The virulence and transmission is controlled by lipophosphoglycan on the surface of the cells. The genome of this protozoan can now be effectively manipulated by molecular techniques. *See* IFN-γ; kinetoplast; macrophage; *mariner*; metacyclogenesis; nitric oxide; T_H; Trypanosoma. (Tamar, S. & Papadoupoulou, B. 2001. *J. Biol. Chem.* 276:11662; Sacks, D. & Kamhawi, S. 2001. *Annu. Rev. Microbiol.* 55:453, <http://www.tigr.org/tdb/tgi.shtml>.)

lek   Social mating group of animals.

lemma   Upper cover bract of the grass flower, frequently bearing an awn at the tip. *See* palea.

Wheat lemma with awn.

lemon (*Citrus limon*) $2n = 18$ or 36. *See* grapefruit; orange.

lemur   *Cheirogaleus major* and *C. minor*, $2n = 66$; *Hapalemur griseus griseus*, $2n = 54$; *Hapalemur griseus olivaceus*,

$2n = 58$; *Lemur catta*, $2n = 56$; *Lemur coronatus*, $2n = 46$; *Lemur fulvus albifrons*, $2n = 60$; *Lemur fulvus fulvus*, $2n = 48$; *Lemur macaco*, $2n = 44$; *Lemur variegatus subcinctus*, $2n = 46$. *See* primates; prosimii.

length mutation Either deletion, duplication, insertion, or any other alteration of the chromosome or nucleic acid sequence affecting the size of a tract. *See* chromosomal aberrations.

lenticel Structure on tree barks permitting passage of gaseous substances; lens-shaped glands at the base of the animal tongue.

lentigines (singular lentigo) Freckles. *See* LEOPARD syndrome; xeroderma pigmentosum.

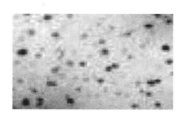

Lentigines.

lentil (*Lens culinaris*) Pulse; popular in the old world because of its high protein ($\approx 25\%$), $2n = 2x = 14$.

lentiviruses (slow viruses) Cytopathic single-stranded RNA retroviruses with similarity to HIV. They infect primates and domestic animals. The equine infectious anemia virus (EIAV) is contagious for horses. The Maedi visna virus (MVV) infects sheep. the Caprine arthritis encephalitis virus (CAEV) is pathogenic for goats. The bovine immunodeficiency virus (BIV) attacks cattle. The feline immunodeficiency virus (FIV) is a threat to cats. The simian immunodeficiency virus (SIV) may be found in apes and monkeys and may or may not affect them seriously, but probably led to the evolution of human immunodeficiency viruses HIV1 and HIV2. The HIV genome contains two identical repeats coupled to a tRNA primer for reverse transcription. The two terminal repeats are essential for replication, transcription, polyadenylation, and integration. Besides three minimally needed genes (*gag*, *pol*, *env*) for the family, they may contain additional genes (*see* acquired immunodeficiency syndrome). They are suitable for vector construction by pseudotyping and are promising for gene therapy. An advantage of these vectors is that they are suitable for transformation of most nondividing cells because of the ability to pass through cell membranes and nuclear pores, also using a system for nuclear localization factors. However, some HIV-based vectors may only infect activated cells. Some retroviral vectors can only infect cells during mitosis when the nuclear membrane transiently disintegrates.

Amphotropic, replication-deficient HIV vectors have been developed that retain the ψ-signal as well as parts of the gag and the RRE domain of Rev and Tat, facilitating packaging as well transcription and cytoplasmic transfer of the vector transcripts. Their use may involve the potential danger of self-replication and activation of protooncogenes by insertion in their vicinity. For gene therapy, the cells are transfected with three separate plasmids. One (the vector) contains the sequences required for infection and transfer of the gene selected. The packaging plasmid carries the elements (structural proteins but not the envelope protein) needed for vector production. The *env* gene must be avoided so that the vector does not infect $CD4^+$ T cells. The third plasmid carries a different envelope gene, encoding, e.g., the G glycoprotein of vesicular stomatitis virus. This protein assures stability and can be employed at high titer. Because the three vectors lack overlapping sequences, reconstitution of a virion by recombination is prevented. For the efficient transfection of different cell types, accessory proteins (Vpr, Vif, Tat) may be required. The 5′ LTR is needed as a promoter. A deletion in the 3′ LTR is incorporated into the 5′ LTR during reverse transcription and prevents the production of vector RNA. Such constructs prevent the formation of replication-competent lentivirions (RCL). *See* adeno-associated virus; cytomegalovirus; gene therapy; HIV; MoMuLV; pseudotyping; ψ; retroviral vectors; retroviruses; vesicular stomatitis virus. (Lever, A. M. L. 1999. *Gene Therapy Technologies, Applications and Regulations*, Meager, A. ed., p. 61. Wiley, New York; Tang, H., et al. 1999. *Annu. Rev. Genet.* 33:133; Schnell, T., et al. 2000. *Hum. Gene Ther.* 11:439.)

Lentz-Hogben test Ascertainment test.

leopard cat (*Felis bengalensis*) $2n = 38$.

LEOPARD syndrome Lentigines, electrocardiographic conduction abnormalities, ocular hypertelorism, pulmonary stenosis, abnormalities of the genitalia, retardation of growth, sensorineural deafness, and autosomal-dominant inheritance. The lentigines may turn neoplastic. The lesions in the PTPN11 gene are shared with the Noonan syndrome. *See* cancer; hypertelorism; lentigines; sensorineural; stenosis. (Digilo, M. C., et al. 2002. *Am. J. Hum. Genet* 71:389.)

Leopold of Albany Hemophilic son of Queen Victoria, who died at age 31 of hemorrhage after a fall. He transmitted this gene through her daughter, Alice, to her grandson, Lord Trematon, who died from hemorrhage after an auto accident. *See* hemophilia.

Leopold of Battenberg Hemophilic grandson of Queen Victoria who died of hemorrhage after surgery at age 33. *See* hemophilia.

lepidotrichia Bony structures of the fins of fishes.

```
LTR- ψ- GA- RRE- p-transgene - LTR
==============================
LTR: long terminal repeat
ψ - GA: packaging signals
RRE:  Rev-response element
p: promoter of transgene

ELEMENTS OF AN HIV VECTOR
```

Lepore   *See* thalassemia.

leprechaunism (Donahue syndrome)   Dwarfism caused by a defect in the insulin receptor.

leprosy (Hansen disease)   Neurological disease caused by *Mycobacterium leprae* (3,268,203 bp) invading the Schwann cells and macrophages in the peripheral nervous system, damaging the host immune system, and disfiguring the extremities, head, and other parts of the body. Only 49.5% of the genome codes for proteins, 27% is pseudogenes, and 23.5% is noncoding but may have some regulatory functions. It appears that during evolution this bacterium (compared to other mycobacteria) lost significant functional parts. A major human susceptibility locus is at 10p13. *See* Bacillus Calmette-Guerin; mycobacteria. (Cole, S. T., et al. 2001. *Nature* 409:1007; Young, D. & Robertson, B. 2001. *Curr. Biol.* 11:R381; Siddiqui, M. R., et al. 2001. *Nature Genet.* 27:439; Cole, S. T., et al. 2001. *Nature* 409:1007.)

leptin   Assumed to be a hormonal feedback signal (167 amino acids) produced by adipocytes. It acts on the hypothalmus and negatively regulates food intake and metabolic rate.

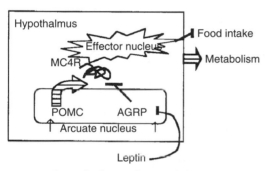

A simplified outline of the melanocortinergic pathway of leptin action. POMC = pre-proopiomelanocortin, AGRP = agouti-related protein, MC4R = melanocortin. (After Barsh, G. S., et al. 2000. *Nature* 404:644.)

Obesity may be the result of resistance to leptin. A deficiency or resistance to leptin causes increased food intake in mice and humans, whereas injection of leptin, the hormone encoded by the *ob* gene of mice, reduces body weight by reduction of fat tissues and controls glucose and lipid metabolism. Leptin has repressive effects on hepatic stearoyl-CoA desaturase-1, an enzyme catalyzing the biosynthesis of monounsaturated fatty acids (Cohen, P., et al. 2002. *Science* 297:240). Leptin also has angiogenic activity. Glucosamine, hyperglycemia, and hyperlipidemia may stimulate leptin synthesis. Recombinant human leptin corrects sterility caused by the *ob*/*ob* condition in female mice. The leptin receptor of mouse (*Ob-R*) gene encodes five alternatively spliced transcripts. Leptin-deficient humans are deficient in sexual development. Leptin signals through the STAT3 protein and modulates the activity of insulin. Leptin activates ATP-sensitive K channels and inhibits hypothalamic neurons (nucleus arcuatus hypothalami and nucleus paraventricularis hypothalami), but the arcuate nucleus may also stimulate the lateral hypothalamic neurons, and the parasympathetic and preganglionic neurons may receive mixed signals. Neuropeptide Y interacts with leptin and may promote appetite. Leptin affects the synthesis of α-melanocyte-stimulating hormone (α-MSH), which downregulates appetite. Melanin-concentrating hormone (MCH) is high in the brain of *ob* homozygous rodents. Dopamine-deficient mice do not feed even when leptin is missing. The OBR-A receptor may affect leptin transport. The suppressor of cytokine signaling (SOCS3) appears to be a factor in mediating resistance to leptin. The synthesis of leptin may be suppressed by CART (cocaine- and amphetamine-regulated transcript). Leptin may inhibit bone formation by acting through the central nervous system. The leptin receptor (OB-R) occurs in several isoforms because of alternative splicing of its RNA transcript. The leptin and insulin pathways may interact at several levels. Peroxisome proliferator-activated receptor γ (PPAR) coactivator (PGC-1) is an important regulator of energy output (thermogenesis). Administration of leptin to starving mice reversed the immunosuppressing effect of starvation. Leptin-deficient mice, however, are protected from T-cell-mediated hepatotoxicity because of the reduction of TNF-α and IL-18. Malnutrition and starvation appears to be a serious factor in susceptibility to infectious diseases. It is not known whether treatment with leptin under conditions of hunger would be desirable because it increases energy storage and expenditure, and may further aggravate malnutrition. *See* AGRP; angiogenesis; brain, human; CART; ciliary neurotrophic factor; ghrelin; IL-18; insulin; ion channels; melanocortin; melanocyte-stimulating hormone; muscarinic acetylcholine receptors; neuropeptide Y; obesity; opiocortin; orexin; SOCS-box; STAT; TNF. (Sierra-Honogmann, M. R., et al. 1998. *Science* 281:1683; Harris, R. B. S. 2000. *Annu. Rev. Nutr.* 20:45; Forbes, S., et al. 2001. *Proc. Natl. Acad. Sci. USA* 98:4233.)

leptokurtic   *See* kurtosis.

leptomycin   Antibiotic and nuclear export inhibitor of meiosis.

leptonema   *See* leptotene stage.

leptotene stage   Distinguished in meiosis when the chromosomes appear under the light microscope as single strands, although they have been doubled during the preceding S stage. The chromosomes are much relaxed and stretched out. The threads (leptonema) are usually quite tangled and their termini cannot be distinguished by light microscopy. Bead-like structures of localized condensations, the chromomeres, are usually visible. Photo shows late leptotene; courtesy of Dr. A. Sparrow. *See* chromomere; meiosis.

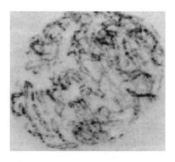

Leptotene stage.

Leri-Weill syndrome *See* dyschondrosteosis; pseudoautosomal.

lesbian Homosexual female. *See* homosexuality.

Lesch-Nyhan syndrome *See* HPRT.

Lespedeza bicolor Forage legume, $2n = 18, 20, 22$.

LESTR Leukocyte-expressed seven-transmembrane-domain receptor is the same as fusin.

LET Linear energy transfer is a measure of energy delivered by ionizing radiation in keV (kilo electron volts) per nanometer path within the target. Approximate LET values in biological material: gamma rays from ^{60}Co source (1.2 to 1.3 MeV): 0.3, hard x-rays (250 keV): 3.0, β-rays from ^{3}H (0.6 keV): 5.5, recoil proton from fast neutrons: 45, heavy nuclei from fission (α-particles): 5,000. *See* DNA repair; ionizing radiation; kinetics.

let-23 Gene of *Caenorhabditis* coding for a transmembrane receptor protein. *See* signal transduction.

lethal equivalent Combined genetic factors that are responsible for death, e.g., if three genes each are expected to reduce life expectancy by a chance of 0.33, their combined lethal equivalent value is ≈ 1, i.e., approximately the same as a gene that causes early death with 100% probability. However, the actual calculation must be carried out by a different procedure because no information can be obtained on the number of the individual's recessive sublethal genes. Thus, if it is observed that the infant mortality among the offspring of totally unrelated parents ($F = 0$) is 8%, and among the offspring of first cousins ($F = 1/16$) of the same population it is 13%. (F is the coefficient of inbreeding.) The difference $0.08 - 0.13 = 0.05$, i.e., 5%. Then $5\% \times 16 = 0.8$, i.e., 80%. Since among diploids two recessive alleles at a locus are required for expression, the frequency of lethal equivalent recessive genes is $\sqrt{0.8} < 0.89$; thus, the number of lethal equivalent recessive genes per gamete in this particular population is ≈ 0.89, close to 1 per gamete. The number of lethal equivalents in a population is Σsq, where q is the frequency of the recessives and s is the selection coefficient. *See* coefficient of coancestry; coefficient of inbreeding; genetic load; incest; selection coefficient. (Makov, E. & Bittles, A. H. 1986. *Heredity* 57[pt 3]:377; Anderson, N. O., et al. 1992. *Plant Breed. Rev.* 10:93.)

Lethal mouse embryo.

lethal factors Have been extensively used for the genetic analyses of developmental pathways because they permit studies on the consequences of arrest of differentiation at particular stages. Conditional lethal mutations, which survive under certain temperature regimes or on supplemented nutritive media, offer particular advantages because they can be analyzed more easily by biochemical methods. The presence of lethal factors may modify segregation ratios. Dominant and recessive lethal factors may not permit the expression of syntenic genes or only at low frequency depending on the extent of recombination. Therefore, genes on the homologous strand may appear in excess. Recessive lethal mutations may cause 2:1 segregation instead of 3:1 in F_2 (ll dead). *See* certation; meiotic drive; segregation distorter.

lethal mutation Normally fails to survive. In *Caenorhabditis*, only 20–35% of the genes are mutable to an obvious visible, lethal, or sterile phenotype and the majority of mutations appear like the wild type. This information resulted from the sequencing of the whole genome and has been interpreted by (partial) redundancy of the genetic material. In *Drosophila*, 24–30% of the genes are estimated to be vital. *See* conditional lethal mutations.

lettuce (*Lactuca sativa*) Composite salad crop with chromosome numbers $n = 8, 9$, and 17.

leucine metabolism Leucine [$(CH_3)_2CHCH_2CH(NH_2)CO$-OH] may be formed from isoleucine [$CH_2CH(CH_3)CH(NH_2)CO$-OH]. *See* hydroxymethylglutaricaciduria; 3-hydroxy-3-methylglutaryl CoA lyase deficiency; isoleucine-valine biosynthetic pathway; methylcrotonylglycinemia; methylglutaconicaciduria.

leucine-rich repeats 15–29 leucines assist ligand recognition in processes such as signal transduction, cell development, DNA repair, and RNA processing. *See* ligand; introns; leucine zipper; DNA repair.

leucine zipper Dimeric regulatory proteins of two subunits; at the COOH terminus, amino leucine occurs at every seventh position; at the amino-terminal domain, positively charged amino acids are found and this region is bound to DNA in a zipper-like manner. The leucine zippers may be homo- or heterodimeric and thus increase their specificity by a combinatorial control mechanism. The basic leucine zipper (bZIP) contains 25 conserved amino acids; of these, 9 substitutions disrupt function. The binding domain is usually 100 amino acids long. *See* binding proteins; diagram next page, DNA-binding protein domains; induced helical fork; Max; regulation of gene activity. (Rieker, J. D. & Hu, J. C. 2000. *Methods Enzymol.* 328:282; Palena, C. M., et al. 2001. *J. Mol. Biol.* 308:39.)

leucocyte (leukocyte) White blood cell. *See* blood.

leucyl tRNA synthetase (LARS) Charges tRNA^{Leu} by leucine. Its gene is in human chromosome 5. *See* aminoacyl-tRNA synthetase.

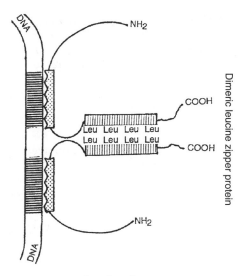

Leucine zipper.

leukalexins Released by effector cells, and they slowly degrade DNA of the target cells. *See* immune system. (Joag, S., et al. 1989. *J. Cell Biochem.* 39[3]:239.)

leukemia Cancer of the blood-forming organs characterized by an increase in the number of leukocytes and accompanied by enlargement or proliferation of the lymphoid tissues. It involves anemia and increasing general weakness and tiredness. Two main types are generally distinguished: *acute myelogenous* leukemia (AML/FPD, 21q22.1-q22.2) that primarily affects adults; *chronic lymphocytic* leukemia (CLL, 11q13.3), in which the circulating malignant cells are differentiated B lymphocytes. A low-grade B-cell malignancy involves chromosomal aberrations at 13q14, a site concerned with about 25–50% of CLL, the position of a tumor suppressor. Acute myelogenous leukemia (AML) may be preceded by haploinsufficiency of the core-binding factor α-subunit (CBFA2, 6p21). Acute myeloid leukemia involving inversion in human chromosome 16 is a fusion of the core-binding factor β (BGF/CBFB/PEBP2B) and the smooth-muscle myosin heavy chain (MYH11) that results in transcriptional repression. About half of AML cases involve cytological anomalies.

A defect in JunB may lead to AML. *Acute lymphoblastic leukemia* affects mainly children. The acute myeloid form frequently involves a translocation between human chromosomes 8 and 21 [t(8;21)(q22;q22)]. In chromosome 21, the breakpoints are generally within the same intron of the receptor for the granulocyte-macrophage colony-stimulating factor (CSF2R). At the same chromosome 21 location, a basic helix-loop-helix protein, class B1 (BHLHB1), is encoded as shown by the t(14;21)(q11.2;q22) translocations. Patients in relapse usually have activated Jak-2 protein tyrosine kinase. Tyrphostins (AG-490) effectively block leukemic cell growth in these cases by inducing apoptosis in the cancer cells without affecting normal blood-forming cells. *Chronic myeloid leukemia* (CML) is generally associated with a translocation between chromosomes 22q11.21 and 9q34.1. The translocation is within the BCR (break-point cluster region) of chromosome 22 and in chromosome 9; the ABL gene (Abelson murine leukemia virus oncogene) is affected. The translocation causes a fusion between the 5′-proximal region of one of the BCR-spanned loci and the ABL gene. The fused BCR-ABL region encodes a 210 kDa tyrosine kinase protein, that causes malignant transformation of leukocytes. Transplantation experiments in mice demonstrated hematological malignancies in the recipients of this protein. In the early stages of the disease, the use of the drug STI-571 specifically and effectively blocks the kinase and counteracts the malignancy. Treatment at advanced stages involves resistance in about a year.

Chronic lymphocytic leukemia type2 (BCL2) is a follicular lymphoma involving translocations between chromosome 18q21 (immunoglobulin heavy-chain gene J) and gene BCL in chromosome 18 and placing BCL under the influence of the IgJ_H enhancer, which leads to a low-grade malignancy. In another step, a translocation with chromosome 8 brought in the MYC c-oncogene (myelocytomatosis v-oncogene homologue in chromosome 8q24) and the double translocation caused a high-grade malignancy. Transfection with a BCL-2 construct resulted in oncogenic potential. BCL-2 product is an integral mitochondrial membrane protein (M_r 25,000). A human chromosome 19q13 oncogene apparently causes *B-cell leukemia/lymphoma (BCL3 or BCL4)*. Usually it involves translocations. *Chronic B-cell leukemia* (BCL5) also involves a translocation between BCL-2 and a truncated MYC, resulting in high-grade expression.

Chronic lymphatic leukemia (CLL or BCL1) is a clearly familial disease involving translocations (11;14)(q13;q32). Again, the translocation involves the joining of segments 3 and 4 of the heavy-chain Ig in chromosome 14. As a consequence, the immune system is weakened and the leukemia is often associated with other disturbances of the immune system (hyperthyroidism, pernicious anemia, rheumatoid arthritis, autoimmune disease). *Acute T-cell leukemia* (ATL/TAL/SCL) is due to the activation of the α-chain of the T-cell receptor (TCRA, chromosome 11p13) by translocations t(11;14)(p13;q11). The 14q11 band contains the variable region of TCRA and the 11p15.5 is the location of the HRAS (Harvey rat sarcoma) oncogene. *Lymphoid leukemia*, LYL1, involves translocations between chromosome 7 at the T-cell receptor β-chain gene (TCRB, 7q35) and the LYL1 gene, chromosome 19p13. *Myeloid/lymphoid mixed-lineage leukemia* (MLL/ALL/HRX) involves translocations of 11q23 and displays lymphoid myeloid phenotypes; it is present in infancy. The sequences of 11q23 involve a postulated transcription factor and it is frequently included in about 30 different translocations. Cleavage of the MLL gene (translocations) may be caused by inhibition of topoisomerase II by bioflavonoids (flavone, rutine, kaempferol, etc.). The MLL gene is an apparent regulator of a homeobox shared by other animals.

The very common B-cell precursor (BCP) leukemia responds favorably to the CD19-associated tyrosine kinase inhibitor, genistein. CD19 is a B-cell-specific TK receptor and its destruction leads to apoptosis of 99.9% of the leukemia cells. The *stem cell leukemia* (SCL) gene is normally involved in the differentiation of the hematopoietic cells, but it may form a complex with the oncogene product LMO-2 and GATA proteins, and the tetramer may act in the development of acute T-cell leukemia. In *acute promyelocytic leukemia* (APL), the translocation involves 15q21 (PML [promyelotic leukemia inducer])-17q11.2 (RARA [retinoic acid receptor α]). A translocation between 11q23 (PLZF [promyelotic leukemia Zn fingers]) and 17q12 (RARA) is another cancerous gene fusion transactivating p21 transforming protein. PLZF normally is a transcriptional repressor and regulates limb and axial skeletal patterning. PML and PLZF heterodimerize and thus form nuclear bodies. The oncogenic domain of PML is involved in the degradation of ubiquitinated proteins. PML-RARA is retinoic acid responsive, whereas PLZF-RARA is a retinoic acid–resistant leukemia. In animal models, N-methylnitrosourea (MNU), 3-methylcolanthrene (MCA), ionizing radiation, Moloney mouse leukemia virus, etc., may induce leukemia. *See* Abelson murine leukemia virus; ABL; apoptosis; BCR; C/EBP; colony stimulating factor; CREB; ELL; FMS; GATA; genistein; Hodgkin disease; homeobox; HTLV; immunoglobulins; interferon; Jun; leukemia inhibitory factor; LYT oncogene; MLV; Myb oncogene; MYC; myelodysplasia; NCoR; neurofibromatosis; non-Hodgkin lymphoma; OL1p53; p21; Philadelphia chromosome; PLZF; PML; RAR; RAS; SMRT; T cell; TCR; topoisomerase. (Ross, J. A. 2000. *Proc. Natl. Acad. Sci. USA* 97:4411; Burmeister, T. & Thiel, E. 2001. *J. Cancer Res. Clin. Oncol.* 127:80; Armstrong, S. A., et al. 2002. *Nature Genet.* 30:41; Staudt, L. M. 2002. *Annu. Rev. Med.* 53:303; Kelly, L.M. & Gilliland, D.G. 2002. *Annu. Rev. Genomics Hum. Genet.* 3:179.)

leukemia inhibitory factor (LIF) The human cytokine (241 amino acids, encoded in chromosome 22q12.1-q12.2) is involved in negative and positive regulation of myeloid leukemia cell lines, in the development of motor neurons, and in a wide variety of other functions. LIF causes the dimerization of LIFR (LIF receptor) and glycoprotein gp130. LIF activates transcription factors STAT3; BMP2 induces Smad1. The STAT and Smad families of proteins bind p300 and are involved in the differentiation of astrocytes. BMP causes the tetramerization of the bone morphogenetic protein receptors (BMPR1 and BMPR2) and the heterodimer phosphorylates Smad1. p300 most probably is involved with the acetylation of histones and thereby the facilitation of transcription. Its receptor, encoded at 5p12-p13, is shared with other cytokines. Inactive LIF does not affect embryonic development in mice, but it may be important for implantation of the egg. LIF$^{-/-}$ males are normal, but the females are sterile. *See* Abelson murine leukemia virus; APRF; astrocyte; bone morphogenetic protein; cytokines; GFAP; gp130; histone acteyl transferase; neurons; p300; Smad; STAT; TGF. (Stewart, C. L., et al. 1992. *Nature* 359:17; Cheng, J.-G., et al. 2001. *Proc. Natl. Acad. Sci. USA* 98:8680.)

leukocyte White blood cell. *See* blood.

leukocyte adhesion deficiency (LAD) LAD2 is a deficiency of CD15-fucose cell-surface glycoprotein, a ligand of selectins E and P. *See* fucosyltransferase; selectins.

leukocyte antigen *See* HLA.

leukocyte common antigen *See* CD45.

leukocytosis Activation of the leukocyte-producing system. It is mediated by the leukocyte mobilization factor (LMF) formed from the α-chain of the C3 complement component by enzymatic fragmentation. *See* complement; leukocytes.

leukodystrophy *See* Addison disease; Charcot-Marie-Tooth disease; cytotoxic T cell; Gaucher disease; Krabbe's leukodystrophy; metachromatic leukodystrophy; Palizaeus-Merzbacher disease.

leukoencephalopathy Group of neurological disorders. A dominant dementia, CADASIL, was located at 19p13.2-p13.1. Leukoencephalopathy with vanishing white matter and neurological deterioration is at region 3q27 encoding the ε- and β-subunits of the translation initiation factor eIF2B. Swelling and cysts in the brain characterize a third type. *See* CADASIL; eIF-2B. (Leegwater, P. A. J., et al. 2001. *Nature Genet.* 29:383.)

leukopenia Condition with a reduced number of leukocytes and neutrophils in the blood.

leukoplasts Plastids without carotenoids and chlorophylls.

leukosis Organic basis of leukemia. *See* leukemia.

leukotomy *See* lobotomy.

leukotrienes 20-carbon carboxylic acids with one or more conjugated double bonds and some oxygen substitutions. They are formed from arachidonic acid by lipoxygenase and control inflammatory and allergic reactions such as asthma, and indirectly obesity. Leukotrienes may activate the transcription factor *peroxisome proliferator-activated receptor* (PPAR) subunits that are involved in the regulation of enzymes of fatty acid oxidation, control of development, adipocyte differentiation, glucose metabolism, and indirectly in inflammatory reactions through macrophages. Leukotriene C4 synthase (LTC4S) is encoded in human chromosome 5q35. The cysteinyl leukotriene-1 receptor (CysLT$_1$) is encoded at Xq13-q21. *See* anemia; fatty acids; hemolytic anemia; integrin; lipoxygenase; peroxisome; PPAR; prostaglandins. (Funk, C. D. 2001. *Science* 294:1871.)

levorotatory The plane of polarized light is rotated counterclockwise.

Lewis blood group Characterized by the expression of the Le gene coding for α4-L-fucosyltransferase. Four different types can be characterized with the frequencies indicated: (1) Le^{a-b+}, 0.7 (genotype Le/le Se/se); (2) Le^{a+b-}, 0.20 (genotype Le, se); (3) 0.09 (genotype le,Se/se); (4) Le^{a-b-}, 0.01 (genotype le, se). The ABH blood group may cause less severe reactions in incompatible application in blood transfusion than ABO, yet it may cause serious hemolytic problems. The Le^y antigen is overexpressed in the majority of human carcinomas. *See* ABH antigen; ABO blood group; Bombay blood group; secretor.

Lewy body Round structure in neural cytoplasms composed mainly of α-synuclein. Lewy bodies occur in Parkinson disease, a neuron degeneration. *See* Parkinson's disease; synuclein; ubiquilin. (Gasser, T. 2001. *Adv. Neurol.* 86:23.)

Lewy neurites Neurons with α-synuclein deposits. *See* Parkinson's disease; synuclein.

lexA Gene of *E. coli* (map position 91) controls resistance/sensitivity to X-rays and UV; it is a repressor of all SOS repair operons. Its protein product is autoregulated. The RecA protein cleaves the LexA repressor protein (22 kDa), then the genes that it repressed become activated. *See* DNA repair. (De

Henestrosa, F., et al. 2000. *Mol. Microbiol.* 35:1560; Luo, Y., et al. 2001. *Cell* 106:585.)

lexitropsins Organic molecules with the ability to recognize G-C in the minor groove of DNA. There are three major types of lexitropsins: pyyrole, imidazole, and hydroxypyrrole. They are of potential use for targeting genes and controlling their function. *See* netropsin. (Goodsell, D. S. 2001. *Curr. Med. Chem.* 8:509.)

Leydig cells *See* Wolffian ducts.

LFA Lymphocyte-associated proteins belong to the integrin family present on T lymphocytes; promote cell adhesion and stimulate T-cell activation in association with ICAM, CD28, and myosin motor proteins. *See* antigen-presenting cell; CD28; ICAM; JAB; myosin; T-cell receptor.

LFT Low-frequency transducing lysate is generated when aberrant excision of a phage particle takes place from a single lysogen. *See* HFT; lysogen.

LGT Lateral gene transfer may occur during evolution. It is postulated on the basis that some genomes contain nucleotide sequences absent from close relatives but present in more distant ones. *See* lateral transmission; transmission.

LH Luteinizing hormone. *See* animal hormones; luteinizing hormone.

LHCP Light-harvesting chlorophyll protein complex. *See* chlorophyll-binding protein complex.

Lhermitte-Duclos disease *See* multiple hamartoma syndromes; PTEN.

LHX LIM homeobox genes of mammals and similar genes in invertebrates. *See* homeobox; LIM.

liability The appearance of many human traits cannot be explained by simple genetic mechanisms because they show up more frequently in certain human families than in others or in the general population, yet the proportions of afflicted individuals vary greatly. The manifestation of these traits is commonly explained by polygenic systems. Various environmental effects that make predictability of the manifestation quite difficult generally influence polygenic systems. The frequencies of diabetes, schizophrenia, hypertension, dental caries, peptic ulcer, various forms of cancers, etc., belong to these categories. Until a certain threshold is reached, the person is considered healthy, but beyond that point medical attention is necessary. The passing of the threshold requires special unidentified environmental conditions. Naturally, it is important to have some predictive ability regarding the liability of individuals to succumb to such diseases.

A relatively simple statistical procedure exists to assess the heritability of the liability based on the normal distribution of these traits. In order to proceed with the empirical calculation, the table below may help:

1	2	3	4	5	6	7	8	← Columns
$Q\rightarrow$	0.001	0.005	0.010	0.050	0.100	0.150	0.200	
$t\rightarrow$	3.090	2.576	2.326	1.645	1.282	1.036	0.842	
$z\rightarrow$	0.0037	0.1446	0.02665	0.10314	0.17550	0.23316	0.27996	

Q = the incidence of the trait, t = the student's t distribution with one tail of the normal distribution, and z = the truncation point at the t value. Let us assume that the incidence of the trait in the general population is $Q_p = 0.001$ and the incidence among the offspring of an afflicted individual is $Q_a = 0.1$. The truncation point in the general population is thus $t_p = 3.090$ (first column) and that in the afflicted family $t_a = 1.282$ (6th column). The mean liability of the affected parent $\mu^1 \cong z_a/Q_a = 0.0037/0.001 = 3.7$. The mean liability of the offspring of the affected parent is $\mu^2 \cong t_p - t_a = 3.090 - 1.282 = 1.808$. The heritability of the liability of the trait determined by offspring-parent regression is $h^2 = 2b_{\overline{OP}}$ (where regression is on one parent rather than on the midparent value) and $b_{\overline{OP}} = \mu^2/\mu^1 = 1.808/3.7 \cong 0.49$; hence, heritability, $h^2 = 2 \times 0.49 = 0.98$. This hypothetical value is obviously very high for heritability. We must keep in mind, however, that the trait has low penetrance for being controlled by a single gene and most likely it is under the control of multiple genes and is not expected to occur frequently because the threshold event is needed for its manifestation. Would have been the incidence of the trait in the general population $Q_p = 0.005$ and that of the afflicted family $Q_a = 0.500$, the heritability would have come out by this calculation as 0.65. The liability increases with the consanguinity of the parents. *See* consanguinity; correlation; GTL; heritability; normal distribution, penetrance; polygenic inheritance; recurrence risk; risk; student's t distribution, threshold trait; z.

libido Sexual drive. Libido is the motive for sexual contact, although it may exist even in the absence of sexual potency such as in sterile mules.

library, genetic Collection of cloned fragments, representing the entire genome (at least once). The construction of the library requires cutting up the genome by one or more restriction enzymes or mechanical shearing. Subsequently, the DNA fragments are ligated into appropriate cloning vectors and transformation is carried out in a suitable host (most commonly into *E. coli*). The screening of the cloned colonies requires nucleic acid hybridization, southwestern analysis, immuno-chemical procedures, recombinational assays, or genetic analysis and identification through one or more of these final steps. *See* gene bank; restriction endonucleases; southwestern method; transformation; vectors.

licensing factor (replicational licensing factor) *See* geminin; MCM.

Liddle syndrome (pseudoaldosteronism) Autosomal-dominant moderate hypertension encoded in human chromosome 16p13-p12. It is involved in the control of a Na^+ ion channel and renal sodium reabsorption. *See* aldosteronism; Bartter syndrome; degenerin; Gitelman syndrome; hypertension; mineral cortical syndrome.

LID domain Oligopeptide that covers the top layer of the 26S proteasome and is used for the recognition of ubiquitin, ubiquitin isopeptidase, and some proteins with homology to signal transduction complexes. *See* proteasome; ubiquitin.

LIF (1) *See* APRF; leukemia inhibitory factor. (2) Laser-induced fluorescence detector.

life beginning on Earth Latest evidence indicates ~3,800 Myr before present. *See* Myr.

life cycle Successive changes in generations of organisms, including the modes of reproduction. In higher organisms this includes the generation of gametes, fertilization and other modes of propagation, and the development of the adult forms. *See* alternation of generations; gametophyte; germline; sporophyte.

life expectancy *See* longevity.

life form domains Bacteria, archaea, and eukarya. *See* archaea; eukaryotes; prokaryotes.

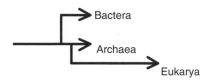

lifespan *See* longevity.

Li-Fraumeni syndrome (LFS) In some families several types of cancers (breast, sarcomas, brain, lung, laryngeal, adrenal cancers, melanomas, prostate cancer, gonadal germ cell tumor, and leukemias) are found with about 50-fold increased incidence compared to the general population. The assumption is that in these families alleles of the dominant p53 tumor-suppressor gene are segregating. LFS is encoded in human chromosome 17p13.1. In some LFS cases, the p53 gene appears normal, but various single-nucleotide substitutions occur in the CHK2 gene. The human gene is homologous to the budding yeast G2 checkpoint kinase RAD53 that normally phosphorylates protein CDC25C. *See* CDC25; cancer; checkpoint; Lynch cancer families; malignancy; *p53*; RAD53; substitution mutation; tumor suppressor genes.

lifting *See* phasmid.

ligand Molecule that can bind (by noncovalent bonds) to a receptor by virtue of special affinity. Ligands are transcription factors, growth factors, activators, DNA-binding proteins, hormones, neurotransmitters, antigens, morphogens, membrane receptors, etc. Target-guided combinatorial assembly

can screen ligands. In the first step, a set of potential binding elements, including common chemical linkage groups, is prepared. Second, their interaction with the selected target (enzymes, receptors, etc.) is determined. Third, the combinatorial library of linked binding elements is connected by flexible linkers. Fourth, this library is tested and classified for tightness of binding to the biological target. (Maly, D. J. 2000. *Proc. Natl. Acad. Sci. USA* 97:2419; <http://www.genome.ad.jp/dbget/ligand.html>.)

ligand-activated site-specific recombination The Cre recombinase is fused to a ligand-binding domain of the estrogen receptor (ER); thus, the recombinase may be activated by tamoxifen (but not by estradiol). A DNA sequence flanked by loxP sites can then be excised. *See* Cre/loxP; estradiol; estrogen; site-specific recombination; tamoxifen. (Feil, R., et al. 1996. *Proc. Natl. Acad. Sci. USA* 93:10887.)

ligase, DNA Enzymes that tie together ends of single polynucleotide chains. The *E. coli* enzyme requires 5′ phosphate and 3′ OH termini and NAD; the T4 enzyme requires ATP for the reaction. The T4 enzyme can ligate both cohesive and blunt ends, although the latter reaction is much slower, but monovalent cations (NaCl) and polyethylene glycol (PEG) can increase its efficiency. Human cells contain at least four ligase isozymes and all are ATP dependent. The efficiency of these ligases depends on the total concentration of the substrates and on the closeness of the ends to be ligated. Mammals have three known DNA ligases and they have an essential natural role in replication and recombination. Ligase I joins Okazaki fragments and seals exicision repair nicks. Ligase IV (3q22-q34) deficiency is lethal in mice and V(D)J joining is absent. Ligase IV is required for the repair of double-strand breaks and nonhomologous end joining. In the laboratory, DNA ligases are used for the sealing of DNA in vectors or joining of DNA molecules. Proliferating cell nuclear antigen (PCNA) after being loaded onto the DNA by replication factor C enhances ligase I activity up to 5-fold. *See* DNA-PK; Ku; NHEJ; recombination; replication; RF-C; vectors. (Tom, S., et al. 2001. *J. Biol. Chem.* 276:24817.)

ligase, RNA Mediates the joining of RNA 5′ and 3′ ends in the presence of ATP. RNA ligases have an important role in the final processing of RNA transcripts. There is a very large number of RNA ligase ribozymes that structurally belong to three classes and within the classes to several families. *See* ligase DNA; ribozyme. (Stage-Zimmermann, T. K. & Ehlenbeck, O. C. 2001. *Nature Struct. Biol.* 8:863.)

ligase chain reaction Two pairs of complementary oligonucleotides are added to both strands of denatured DNA at the immediate vicinity of each other and amplified. The ligation products are only useful for the signal amplification but not for amplifying other DNA copies. The method has wide applicability for diagnostics. (Rodriguez, H., et al. 2000. *Methods* 22:148.)

ligation Covalent joining of nucleotide ends by DNA or RNA ligase, respectively.

ligation, embryo Imposing a constriction on an embryo, e.g., *Drosophila*. The development of the thoracic (central) structures requires the cooperation of both poles. Ligation at the blastoderm stage interferes with the development of only one segment. Ligation has various medical applications. *See* morphogenesis; morphogenesis in *Drosophila*.

light   *See* electromagnetic radiations; HVEM.

light chain   *See* antibody; DNA heavy chain.

light-controlled reactions in plants   (Simoni, R. D. & Grossman, A. R. 2001. *J. Biol. Chem.* 276:11447.)

light-directed parallel synthesis   Light directs the simultaneous synthesis on solid support. The pattern of light masking activates different regions of the solid-state support and facilitates the chemical coupling reactions. Activation is the result of the removal of different photolabile protecting groups. After deprotection of the first set of building blocks (amino acids or nucleotides bearing the photolabile compound), the entire surface is exposed to light and the reaction takes place only in the light-exposed areas in the preceding step. Subsequently the substrate is illuminated by using another mask, then a second block is activated. The masking and illumination pattern determines the location and the ultimate product. This way, complex molecules can be fabricated in a combinatorial manner and the product may be useful for industrial and molecular biology studies. *See* array hybridization; DNA chips; microarray hybridization. (Lipshutz, R. J., et al. 1999. *Nature Genet.* 21, Suppl. 1:20; Barone, A. D., et al. 2001. *Nucleosides, Nucleotides Nucleic Acids* 20[4−7]:525.)

light harvesting chlorophyll protein complex   *See* CAB; LHCP.

light intensities   Of various emitters (lambert units [1 lambert − 1 new candela/cm^2/π]): sunshine at noon, 519,000; sun at the horizon, 1,885; moonlight, 0.8; Tungsten bulb of 750 watts, 7,500; mercury vapor light of 1,000 watts, 94,000. These are approximate average data. *See* electromagnetic radiation; illumination.

light microscopy   Used for the study of small biological specimens, biological tissues, cells, and subcellular organs down to a resolution of 0.2 μm. The most essential elements of the light microscope are the objective lenses for viewing details of different sizes. The nosepieces generally contain objectives with numerical aperture 0.1 (4×), 0.25 (10×), 0.65 (high dry, 40×), and 1.25 (oil immersion, 100×) lenses. Eyepieces are usually 10× and permit the viewing of the objects at about 1,000-fold larger, at maximum. The condensor focuses the light, coming from a special low-voltage illuminator. The microscope stand includes a stage for the slides on which the specimens are moved and various adjustment knobs for focusing and adjusting light intensities. Light microscopes may be equipped with other devices, e.g., camera stand or built-in camera, filters, etc. Common light microscopes require special handling of the specimens before viewing, e.g., fixation, staining, sectioning, etc. On the slides, the specimen is generally covered by a very thin (about 0.13−0.25 mm) cover glass (slips). For viewing natural specimens in three dimensions, stereomicroscopes are used that may permit magnifications within the range of 2× to 160×, generally with zooming capabilities. *See* atomic

force microscope: confocal microscopy; electron microscopy; fixatives; fluorescence microscopy; Nomarski; phase-contrast microscopy; resolution, optical; sectioning; stains.

light reactions   Can be carried out in light only. *See* DNA repair.

light repair (of DNA)   Splitting of pyrimidine (thymine) dimers by visible light-inducible enzymes. *See* DNA repair.

light response elements (LRE)   Nucleotide sequences binding transcriptional regulators of light receptor protein (e.g., phytochrome) genes. *See* hormone-response elements; photomorphogenesis; phytochrome.

light sensitivity diseases   *See* albinism; ataxia telangiectasia; Bloom syndrome; Cockayne syndrome; DNA repair; excision repair; Fanconi anemia; Hartnup disease; protoporphyria; RAD; Rothmund-Thompson syndrome; trichothiodystrophy; xeroderma pigmentosum.

light sensitivity in natural populations   Relatively large variations exist in *Arabidopsis*. (Maloof, J. N., et al. 2001. *Nature Genet.* 29:441.)

lignin   Rigid woody polymer of coniferyl alcohol (derived from phenylalanine and tyrosine) and related compounds occurring in plants along with cellulose. It is the industrial source of vanillin and dimethyl sulfoxide. Lignin is used in manufacturing certain plastics, rubber, precipitation of protein, etc.

ligule   "Tongue-like" and frequently hairy outgrowth on the upper and inner side of the leaf blade (at the leaf sheath) of grasses.

likelihood   Statistical concept dealing with a hypothesis based on experimental data; $\lambda = -2\ln(L_{H0}/L_{H1})$ likelihood ratio λ, H_0 and H_1 are the 2 hypotheses can be expressed as $L(H/R)$, where L is the likelihood, H is the hypothesis, and R is the experimentally obtained result. In contrast, probability is fixed by the fit of the data to a preconceived null hypothesis. The likelihood ratio LR = likelihood of data model/likelihood data of null hypothesis or the log likelihood ratio, i.e., $\log(LR)$. *See* lod; maximum likelihood.

lilac   (1) The plant *Syringa vulgaris* is an ornamental shrub ($2n = 46, 47, 48$). (2) Rodent fur color determined by epistatic action of certain gene combinations.

Liliaceae   Family of monocotyledonous plants; many of them have been exploited for cytological studies because of the large chromosomes, *Lilium* subspecies ($2n = 24$, ~1.8×10^{11} bp), onion (*Allium cepa*, $2n = 16, 32$), hyacinth ($2n = 16$), *Trillium* subspecies ($2n = 10$), *Bellevalia* $2n = 8, 16$). In 1928, John Belling counted 2,193 chromomeres in the pachytene chromosomes of *Lilium pardalinum* and believed this number represented the number of genes in the species. In *Fritillaria davisii* ($2n = 24$), the DNA content in the somatic nuclei was estimated to be 295 pg (approximately 50 times the amount in human somatic cells or ~1,000 times that of *Arabidopsis* or *Drosophila*). *See* chromomeres; Dalton; DNA; genome; measurement units.

LIM domain Cysteine-rich zinc-binding unit facilitating protein-protein interactions in signaling molecules, transcription factors, cytoskeletal proteins, motor neuron pathways, axon guidance, etc. LIM has the properties of an organizer. LIM proteins do not have a functional relationship beyond the common feature of protein-protein binding and may convey both inhibitory and activating functions. The RLIM corepressor inhibits the LIM homeodomain transcription factors by recruitment of the histone deacetylase complex. *See* adaptor proteins; cleft palate; CRIP; CRP; DNA-binding protein domains; histone deacetylase; LDB; LHX; LMO; NLI; organizer; PINCH; Williams syndrome. (Schmeichel, K. L. & Beckerle, M. C. 1994. *Cell* 79:211; Hobert, O. & Westphal, H. 2000. *Trends Genet.* 16:75.)

limb bud Embryonal cell group initiating limb development. NF-κB transcription factors and regulatory REL proteins are required. NF-κB seems to transmit fibroblast growth factor (FGF) signals between the ectoderm and the underlying mesenchyme cells during limb differentiation. *See* AER; FGF; NF-κB; organizer; ZPA. (Capdevila, J., et al. 2001. *Annu. Rev. Cell Dev. Biol.* 17:87.)

limb defects, human Can be isolated, but most commonly associated with particular syndromes. In some cases, entire limbs are absent (amelia) or there is a partial reduction (meromelia), frequently missing some parts between the extremities (hand or foot) and the trunk (phocomelia). Not uncommonly, extra fingers and toes are involved or fusion or deformations are observed. The defects may be caused by teratogenic factors, chromosomal defects, or mutations and may be accompanied by a variety of other symptoms of complex syndromes. *See* adactyly; ADAM complex; arthrogryposis; Chotzen syndrome; clubfoot; De Lange syndrome; EEC; exostosis; Holt-Oran syndrome; Majewski syndrome; Moebius syndrome; orofacial-digital syndromes; Pätau's syndrome; Poland syndrome; polydactyly; renal dysplasia and limb defects; Roberts syndrome; syndactyly; thrombo-cytopenia.

limb girdle muscular dystrophy *See* muscular dystrophy.

LIM kinase *See* Williams syndrome.

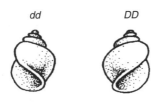

Limnaea.

Limnaea peregra Freshwater snail displaying delayed inheritance of its shell shape. *See* delayed inheritance; maternal effect genes.

limpet *See* keyhole limpet hemocyanin.

Lincoln, Abraham, President (1809–1865) Suspected of having had the relatively mild form of spinocerebellar ataxia, although earlier it was also suggested that he might have had Marfan or Marfanoid syndrome. *See* ataxia; Marfan syndrome. (Ranum, L. P., et al. 1994. *Nature Genet.* 8:280.)

lincomycin 6-(1-methyl-4-propyl-2-pyrrolidinecarboxamidol)-1-thio-o-erythro-D-galacto-octopyramide antibiotic of gram-positive bacteria and chloroplasts; inhibits protein synthesis in the 23S and 16S 70S ribosomal RNA genes. (Cséplö, A., et al. 1993. *Mol. Gen. Genet* 236:163.)

LINE (1) Genetic stock with defined characteristics. (2) L1: Long (≈5–7 kbp) interspersed repetitive DNA sequences, including retroposons generated by reverse transcription; they may occur in 10^4 to 10^5 copies in the eukaryotic genome. Sequencing of the human genome revealed 535 Ta and 415 pre-TA LI elements. LINE elements do not have long terminal repeats like the retrotransposons yet retain reverse transcriptase sequences that are frequently (>95%) nonfunctional. These are remnants of ancient retroviral insertions, but once they are inserted—because of the loss of the LTR (transposase) function—they generally remain at the position of insertion and can be used to trace phylogenetic descendance of various species. The elements have internal promoters for RNA polymerase. cDNA copies can be inserted at new locations. L1 elements are regulated by SOX-type proteins. In humans, L1s (mainly mariner type) are the most common retrotransposons and directly or indirectly represent about 15% of the genome. The majority of these elements are truncated and only 3,000 are of full length; ~40–50 are active retrotransposons. Several human pathological conditions involve LINE insertions. Blood coagulation protein VIII gene insertions were found to cause hemophilia. L1 endonuclease has limited specificity and thus the element can be inserted at many sites, including genes. L1 elements were identified in Duchenne muscular dystrophy (DMD), adenomatous polyposis cancer (APC), and β-globin genes. In mice, about 10% of the spontaneous mutations seem to be due to transposition. In humans, 1/600 mutations may be due to retrotransposition and 1/8 to 1/50 per 100 individuals may carry an endogenous insertion. The human LINE retroposons can also mobilize the transcripts of other genes and thus create processed pseudogenes. L1 elements appear to regulate the inactivation of the mammalian X chromosome (Bailey, J. A., et al. 2000. *Proc. Natl. Acad. Sci. USA* 97:6634) as postulated by the Lyon hypothesis. *See* Alu; antihemophilic factors; cis preference; globins; HERV; Lyon hypothesis; mariner; muscular dystrophy; p40; processed pseudogene; retroposon; retrotransposon; reverse transcriptase; SINE; SOX; Ta; Tα. (Sheen, F.-M., et al. 2000. *Genome Res.* 10:1496; Ostertag, E. M. & Kazazian, H. H., Jr. 2001. *Annu. Rev. Genet.* 35:501.)

LINE element

LINE 1 (L1) Major mouse LINE that can be used for DNA fingerprinting of YAC sequences from the mouse genome. *See* DNA fingerprinting; YAC.

Color Figures

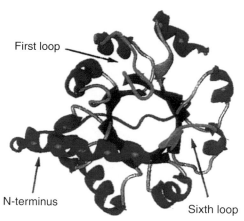

See Barrel. α/β Barrel folds of phosphoribosylanthranylate isomerase. The β-sheets in the center are colored purple. (From Gerlt, J. A. 2000. *Nature Struct. Biol.* 7:171, by permission.)

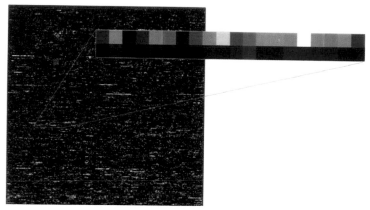

See Microarray hybridization, DNA chips. Different 25-mer oligonucleotide probes in an area of 1.28×1.28 cm. The array contains probe sets for more than 6,800 human genes, and the image was obtained after overnight hybridization of an amplified and labeled human mRNA sample. Image by courtesy of Affymetrix Inc. and from Fields, S., et al. 1999. *Proc. Natl. Acad. Sci. USA* 96:8825. (Copyright by the National Academy of Sciences, U.S.A 1999.)

See FISH. Human male metaphase chromosomes. In situ multiplex-fluorescence hybridization permitted each chromosome pair in different color. (Courtesy of Dr. Michael R. Speicher, Institut für Anthropologie und Human Genetik, LMU, München, Germany.)

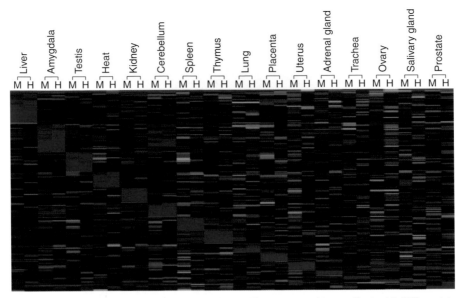

See Transcriptome. Transcriptome of 427 human and mouse gene pairs. Gene expression profile in 16 different types of tissues detected by Affymetrix GENECHIPs. Individual tissues express about 30 to 40% of the genes. Statistical analysis indicated that 78 to 82% of the genes were expressed differently in the mouse (M) and humans (H). The expression of the genes may be strictly tissue-specific. In this particular study 85 human genes were expressed only in the testis. (From Su, A. I., Cooke, M. P., Ching, K. A., Hakak, Y., Walker, J. R., Wiltshire, T., Orth, A. P., Vega, R. G., Sapinoso, L. M., Moqrich, A., Patapoutian, A., Hampton, G. M., Schultz, P. G. and John B. Hogenesch 2002. *Proc. Nat. Acad. Sci. USA* 99:4465–4470. Copyright 2002 National Academy of Sciences, U.S.A. Courtesy of Dr. J. B. Hogenesch, The Genomics Institute of the Novartis Research Foundation, San Diego, CA.)

See Transcriptional gene fusion vector. Transcriptional *in vivo* gene fusion of bacterial luciferase genes to a flower-bud-specific promoter of *Arabidopsis*. The highest level of expression is indicated in red (false color). The background is photographically cleared. (From Rédei, G. P., Koncz, C., Langridge, W. H. R., unpublished.)

lineage Line of descent from an ancestral cell or ancestral individual(s). *See* cell lineage.

linear energy transfer *See* LET.

linear function Set of data (variables, parameters) without powers or multiplication of the quantities, e.g., $3x_1 \pm 2x_2 \pm x_3$.

linearization Conversion of covalently closed circular nucleic acid to an open linear form.

linear programming Analysis or solution of problems when linear functions are maximized or minimized.

linear quartet In the embryo sac of plants, the four megaspores, and in the asci of some fungi, four spores arranged in the order in which they were formed during meiosis. *See* embryo sac; megaspore.

Linear spore quartet.

linear regression *See* correlation.

Lineweaver-Burk equation Reciprocal of Michaelis-Menten equation:

$$\frac{1}{v} = \frac{K_m + (S)}{V(S)} = \frac{K_m}{V}\left[\frac{1}{S}\right] + \frac{1}{V}$$

v = velocity constant of the reaction

$$K_m = \frac{[(E) - (ES)(S)]}{(ES)}$$

(S) = substrate concentration

V = maximal velocity of reaction

It provides a simpler solution for representing enzyme kinetics; $1/v$ is plotted against $1/(S)$ or $(S)/V$ against (S). *See* Michaelis-Menten equation and the Eadie-Hofstee plot.

linkage Association of genes within the same chromosome but can be separated by recombination. In test crosses, four phenotypic classes are expected in equal numbers if the segregation is independent (absence of linkage). These four classes can be designated AB (a), Ab (b), aB (c), and ab (d) for the n individuals. For the verification of linkage, chi-square tests (χ^2) can be used. First, the segregation for A can be examined,

$$\chi_1^2 = \frac{[a + b - c - d]^2}{n}$$

then in a second step the

$$\chi_2^2 \text{ (for B)} = \frac{[a + b + c - d]^2}{n}$$

is determined. The linkage

$$\chi_L^2 = \frac{(a - b - c + d)^2}{n}$$

Example: *AB*: 191, *Ab*: 37, *aB*: 36, *ab*: 203, Total 467. Without linkage in each class, 116.7 is expected in a test cross.

	χ^2	Degree of Freedom	Probability of Greater χ^2
Segregation for A—a	0.259	1	>0.5
Segregation for B—b	0.362	1	>0.5
Linkage	220.645	1	$<10^{-10}$
Total	221.266	3	

The segregation of each marker is very normal, but the linkage chi-square is extremely high, indicating that the segregation is not independent. Another simpler formula that does not take into account the transmission of the markers (p = parental, r = recombinant) is shown below:

$$\chi^2 = \frac{(\text{observed}_p - \text{expected}_r)^2}{(\text{observed}_p - \text{expected}_r)}$$

Similarly, for F_2, the chi-squares for the A and the B loci and for linkage can be determined for the

$$AB\left(\frac{9}{16}n\right), \quad Ab\left(\frac{3}{16}n\right), \quad aB\left(\frac{3}{16}n\right), \quad ab\left(\frac{1}{16}n\right)$$

classes designated as a, b, c, and d, respectively:

$$\chi_A^2 = \frac{[a + b - 3c - 3d]^2}{3n}, \quad \chi_B^2 = \frac{[a - 3b + c - 3d]^2}{3n} \quad \text{and}$$

$$\chi_L^2 = \frac{[a - 3b - 3c + 9d]^2}{9n}$$

Mather suggested the following χ^2 formula for the detection of linkage in F_2 in case the segregation is not normal:

$$\chi^2 = \frac{(a_1 a_4 - a_2 a_3)^2 n}{(a_1 + a_2)(a_3 + a_4)(a_2 + a_4)(a_1 + a_3)}$$

Fisher proposed to detect linkage by using repulsion and coupling information combined in case of abnormal transmission as shown below:

	ab	Ab	aB	AB	Sum
Repulsion	30	70	2	24	126
Coupling	519	119	12	349	999

Then add separately for the repulsion and coupling data the ab + AB and designate as α_1 and the Ab + aB as α_2, so α_1 represents the recombinants in repulsion but the parentals in coupling, and α_2 represents the parentals in repulsion and the recombinant in coupling. In the absence of linkage, α_1 and α_2 are expected to be near equal. A 2×2 contingency χ^2 (*See* association test) can then be set up and the χ^2 is computed with the null hypothesis so that there is no linkage:

$$\chi^2 = \frac{(868 \times 72 - 1341 \times 54)^2 1125}{126 \times 999 \times 203 \times 922} > 146$$

and for 1 degree of freedom it absolutely rules out independent segregation.

	α_1	α_2	$\sum$
Repulsion	72	54→	126
Coupling	131	868→	999
$\sum$		203 922	1125

When we want to estimate the probability of finding linkage all over the genome, the chance is relatively very small of finding it. The P value obtained by the χ^2 procedure can be converted to an actual estimate of linkage based on a Bayesian estimate:

$$\text{Probability of linkage} = 1 - \left(\frac{P}{P + f_{\text{swept}}}\right)$$

where P is the value obtained by χ^2 and f_{swept} is the fraction of the genome that appears to be linked around the marker within the so-called *swept radius*. The swept radius is the maximal distance on both sides of the marker between two loci with a certain probability. Usually a probability limit of 0.95 is chosen to be significant. Thus, the swept radius is (shown in parentheses) 10 cM (21%), 20 cM (35%), and 30 cM (44%), respectively, for this critical probability.

Estimating linkage in human populations is difficult because of the small family sizes and frequently there are other complications, such as paternity, penetrance, expressivity, etc. Classical human genetics devised a different formula (u statistics) for *double backcrosses* (AaBb × aabb) as $u_{11} = (a - b - c + d)^2 - (a + b + c + d)$. For *single backcrosses* (AaBb × Aabb): $u_{31} = (a - 3b - c + 3d)^2 - (a + 9b + c + 9d)$; and for the *progenies resembling* $F_2 : u_{33} = (a - 3b - 3c + 9d)^2 - (a + 9b + 9c + 81d)$; (a, b, c, and d indicate the four genotypes as indicated above.) The lod scores are used for family data. In case there is no difference between female and male recombination,

$$Z \text{ score} = \log \frac{P(F|\theta)}{P(F|0.5)}$$

and where there is a sex difference, the formula

$$log \frac{P(F|\theta', \theta'')}{P(F/(0.5)(0.5))}$$

is applicable. θ is the recombination fraction (θ' for males and θ'' for females). Linkage in humans is generally studied by somatic cell hybridization, radiation hybrids, or molecular techniques. For multipoint linkage analysis, computer programs are generally used such as Vitesse (O'Connel, R. & Weeks, D. E. 1995. *Nature Genet.* 11:402), MAPMAKER (Lander E. S., et al. 1987. *Genomics* 1:174), MENDEL (Lange, K. D., et al. 1988. *Genet. Epidemiol.* 5:471), JOINMAP (Stam, P. 1993. *Plant. J.* 3:739), or several others listed by Terwilliger, J. D., et al. 1994. *Handbook of Human Genetic Linkage*. Johns Hopkins Univ. Press, Batimore, MD. This book includes sources of availability. *See* affected-sib-pair method; Bayesian

theorem; chi square; coincidence; F_2 linkage estimation; interference; linkage group; lod score; mapping, genetic; mapping function; maximum likelihood method applied to recombination; meta-analysis of linkage; radiation hybrid; recombination frequency; recombination mechanisms; sex linkage; sex-linked lethal mutation; somatic cell hybrids; tetrad analysis. (<http://linkage.rockefeller.edu/>.)

linkage, complete Linkage is practically complete in the heterogametic sex in insects, e.g., in the majority of the species of *Drosophila*, male (XY), and in the silkworm female (WZ). Recombination in male flies has been observed, however, in the presence of transposable elements causing hybrid dysgenesis. Linkage is practically complete within inversions because the recombinant gametes or zygotes are lethal (balanced lethals). In case of multiple translocations involving all or most of the chromosomes, even genes situated in different chromosomes do not segregate independently, and this is characteristic for the complex heterozygotes. *See* complex heterozygote; hybrid dysgenesis; inversions; translocations. (Morgan, T. H. 1912. *Science* 36:719.)

linkage detection, nonparametric *See* affected sib-pair method. (Abreu, P. C, et al. 1999. *Am. J. Hum. Genet.* 65:847.)

linkage detection, tetrads The linear spore tetrads can be arranged as PD, TT, and NPD. At linkage, the frequency of the double recombinant class (NPD) must be smaller than the single recombinant (TT) class. If the deviation between TT and NPD is very small, a chi-square test may be necessary by using the formula:

$$\chi^2 = \frac{(PD - NPD)^2}{PD + NPD}$$

See chi-square; linkage; tetrad analysis.

linkage disequilibrium Certain groups of genes are syntenic in higher frequency than would be expected on the basis of unhindered genetic recombination. In random mating populations, linkage disequilibrium should disappear in time. The tighter the linkage between genes, the less the chance to reach linkage equilibrium. If two loci are independent, the frequency of their joint occurrence is the product of the two independent frequencies. A cause of persistent association may be based on their selective advantage as a group or recent introgression. With these possibilities in mind, the linkage disequilibrium determined for several loci may provide information on the chromosomal location of a gene. Disequilibrium is expected to be the greatest for a group of markers that are closest to the gene that needs to be assigned to a position. The quantitative measure of linkage disequilibrium

$$\delta = (p_D - p_N)/(1 - p_N)$$

where p_D and p_N represent the two different linkage-phase chromosomes. The decline (Δ_n) of linkage disequilibrium depends on recombination frequency (unless a particular gene block has a high selective advantage and linkage drag exists)

$$\Delta_n = \Delta_0(1 - r)^n$$

where r = recombination frequency, n = number of generations, and Δ_0 = the beginning of disequilibrium. Linkage disequilibrium may shed light on the origin or changes within and among populations. Hitchhiking may affect linkage disequilibrium. In human populations, the extent of linkage disequilibrium varies in different chromosomes and chromosomal regions, also depending on the closeness of the physical location of the alleles studied. Single-nucleotide polymorphisms (SNP) have been used successfully to map genes on the basis of linkage disequilibrium. Gene flow between distinct populations generates *admixture linkage disequilibrium*. Errors in genotyping may seriously affect the outcome of the estimate. Studied at the molecular level, the exchange hot spots are not distributed at random and the intensity of exchange may vary considerably. These facts may affect the outcome of the calculations of linkage disequilibrium. *See* association mapping; coupling; genetic equilibrium; Hardy-Weinberg equilibrium; hitchhiking; introgression; MALD; mitochondrial genetics; mutation dating; recombination, genetic; repulsion; sib TDT; SNP; SWT; syntenic; transmission disequilibrium test. (Huang, J. & Jiang, Y. 1999. *Am. J. Hum. Genet.* 65:1741; Reich, D. E., et al. 2001. *Nature* 411:199; Jorde, L. B. 2000. *Genome Res.* 10:1435; Pritchard, J. K. & Przeworski, M. 2001. *Am. J. Hum. Genet.* 69:1; Rannala, B. & Reeve, J. P. 2001. *Am. J. Hum. Genet.* 69:159; Jeffreys, A. J., et al. 2001. *Nature Genet.* 29:217; Morris, A. P., et al. 2002. *Am. J. Hum. Genet.* 70:686. linkage disequilibrium map: Dawson, E., et al. 2002. *Nature* 418:544.)

linkage drag Undesirable genes are preserved in a population on the basis of hitchhiking if no recombination occurs between the selected desirable and undesirable genes. *See* genetic load; Hill-Robertson effect; hitchhiking. (Hospital, F. 2001. *Genetics* 158:1363.)

linkage equilibrium *See* genetic equilibrium.

linkage genome screening Used to approximately locate genes with low penetrance and/or expressivity to chromosomes on the basis of the λ_s value and maximum likelihood. *See* maximum likelihood; λ_s.

linkage group Syntenic genes (situated in one chromosome) belong to a linkage group. Therefore, the number of linkage groups equals the number of chromosomes in the basic set. When the linkage information about a chromosome is incomplete, it is possible to see two or more linkage groups for a particular chromosome because gene clusters may recombine freely when the distance between them is 50 map units or more. If trisomic analysis is used, all syntenic genes must appear as a linkage group, irrespective of their dispersal along the chromosome. In multiple translocations (complex heterozygotes), several interchanged chromosomes are transmitted together and form superlinkage groups such as in the plants *Oenotheras*. Similar translocation complexes may exist in animals. In the termite *Kalotermes approximatus*, meiotic chains of 11 to 19 chromosomes have been observed, forming a single male-determining linkage group. *See* complex heterozygote; mapping; synteny; trisomic analysis.

linkage in autotetraploids *See* recombination in autotetraploids.

linkage in breeding In applied genetics, linkage may be very useful if advantageous traits are controlled by closely linked genes. The opposite is true if disadvantageous characters are syntenic. The breeder may benefit by knowing linkage with neutral, visible, or easily detectable chromosome markers that permit monitoring of the inheritance of sometimes hard-to-recognize quantitative traits. The male silkworm produces 25–30% more silk than the female. By having a dominant color (dark) gene in the Y chromosome, the less productive female-producing eggs (XY) can be separated by an electronic sorter before hatching and eliminated. In baby chickens, it is difficult to identify the males by genitalia, but the presence of the *B* (*Barring*) X-chromosome-linked gene reduces the color of head spots more effectively in two (XX, males) than in single dose (XY, females). Thus, by autosexing, the males can early be assigned to meat, whereas the females to egg production regimes.

Recombination Frequency	Genetic construction of the F_1		
	(+++/−−−) (−−−/+++)	(++−/−−+) (−−+/++−)	(+−+/−+−) (−+−/+−+)
0.000	62,500	0.000	0.000000
0.075	33,498	1.448	0.000063
0.225	8,134	57.787	0.410526
0.375	1,455	188.787	24.441630
0.450	523	234.520	105.094534
0.500	244	244.141	244.140625

The frequency of 6-fold double homozygous dominants for desirable genes located in two chromosomes. The numbers in the body of the table indicate the number of individuals carrying 12 advantageous alleles (+) in an F_2 population of 10^6 size. (After Power, L. 1952. In *Heterosis*, Gowen, J. W., ed. Iowa State College Press, Ames, p. 315.)

The use of RFLP techniques facilitates the recognition of linkage of quantitative trait loci with DNA markers. The knowledge of recombination frequency may facilitate the size planning of the populations for selecting a combination of desirable genes. The yellow seed-coat color gene in the flax is very tightly linked to a high quantity and high quality of oil. Unfortunately, low yields and susceptibility to disease are also linked in coupling. Therefore, it is practically difficult to produce a commercially acceptable flax variety with yellow seed coat. The availability of appropriate linkage information may facilitate the positional cloning of agronomically desirable genes. If the frequency of recombination is known, the expected frequency of double homozygous dominant offspring can be predicted by the following formulas: for coupling $0.25(1 - p)^2$ and for repulsion $0.25(p)^2$, where p = recombination frequency. Genetic transformation by cloned genes has an enormous advantage over classical breeding methods because it does not have to go through the tedious and long-lasting period of selection after crossing if the linkage is tight. *See* autosexing; chromosome substitution; gain; positional cloning; QTL; transformation.

linkage map *See* genetic map; mapping, genetic; physical map.

linked genes They are in the same chromosome and their frequency of recombination is less than 0.5. *See* linkage; recombination.

linker (1) Very short DNA sequences (commercially available) that can be added by blunt-end ligation to DNA termini to generate particular cohesive ends for insertion of passenger DNA fragments in molecular cloning or transformation. *See* blunt-end ligation. (2) In nucleosomes, 40–60 nucleotides (generally associated with H1, sometimes with H5 histone) that connect the cores of nucleosomes. *See* histones; nucleosome.

linker insertion Insertion of a DNA sequence in place of deleted sequences to test for specific functional units in the deleted region. *See* linker scanning.

linker scanning Molecular method for the identification of upstream regulatory elements. The procedure is as follows: In defined lengths of upstream DNA fragments, deletions are induced at different positions. The gaps are filled in such a way that the total original length of all fragments is neither reduced nor extended even by a single base pair. If the linker falls into the gap that was normally the site of a regulatory element in front of the gene, expression is reduced or abolished, serving as evidence that the site of the deletion (now the linker) involves an essential upstream element. *See* linker; promoteral. (Laumonnier, Y., et al. 2000. *J. Biol. Chem.* 275:40732.)

isomers. After a break, one strand can be rotated about the other and such a reaction may change one isomer into the other by the action of DNA topoisomerases. These topological processes are the requisites for several functional activities of DNA such as replication, recombination, transcription, and others. *See* isomer; supercoiled DNA; topoisomerase; writhing number. (Quian, H. & White, J. H. 1998. *J. Biomol. Struct. Dyn.* 16:663.)

linking number paradox In a nucleosome, the DNA is coiled by about 1 3/4 turns over the histone octamer and forms close to -2 superhelical turns. Yet when the histones are removed from the DNA-histone fiber, only -1 negative supercoiled turn is found. The explanation of this paradox may be that the DNA released from the histones becomes more tightly coiled than in the nucleosome-restricted form, i.e., the bound DNA has different structural periodicity than the free form. *See* histones; nucleosome; supercoiled DNA. (Prunell, A. 1998. *Biophys. J.* 74:2531.)

linseed *See* flax.

LINUX Bioinformatics program similar to that of the UNIX operating system.

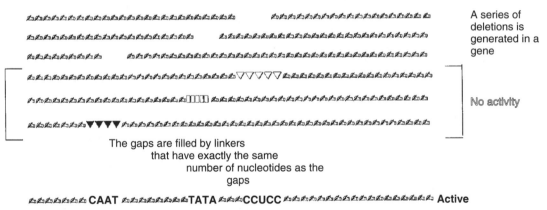

A series of deletions is generated in a gene

No activity

The gaps are filled by linkers
that have exactly the same
number of nucleotides as the
gaps

CAAT   TATA   CCUCC   Active

If the gaps/linkers involve essential regulatory elements of the gene,
expression is either reduced or lost and thus the position of these signals is revealed.

Linker scanning for the identification of the position of essential regulatory elements.

linking library The 5'-end of many genes is rich in NotI restriction enzyme recognition sites (GC↓GGCCGC). A NotI-linking clone containing, e.g., the transferrin receptor gene, can be used as a point of jumping over NotI sequences and thus mapping genes. *See* jumping library; slalom library. (Koshuba, V. I., et al. 1997. *FEBS Lett.* 419:181.)

linking number Indicates how many times one strand of DNA double helix crosses over the other strand in space. The linking number may reflect the degree of supercoiling in a closed DNA molecule. The linking number has two components: the writhing number (W) and the twisting number (T); a change in linking number $\Delta L = \Delta W + \Delta T$. A change in linking number requires the breakage of at least one strand. Molecules that are the same as their linking number are called topological

lion (*Panthera leo*) $2n = 38$; lion × panther (leopard) (*Panthera pardus*) hybrids are sterile.

lipase Enzyme that digests triaglycerols into fatty acids and mediates energy supply in the cell. *See* fatty acids; lipids; triaglycerol.

lipase deficiency *Hepatic lipase* (HL) dominant 15q2; *liposome lipase* (8p22) result in triglyceride-rich high- and low-density lipoproteins. It also occurs in a recessive form at the same locus. The deficiencies are called *chylomicronemia* and *hyperlipoproteinemia*, respectively. There is also an autosomal-dominant *hormone-sensitive lipase* (HSL). Pancreatic lipase deficiency and lysosomal acid lipase are autosomal recessive. The lipoprotein lipase gene (LPL, 10 exons) was mapped to human chromosome 8p22. Endothelial lipase normally reduces the level of HDL in the liver, lung, kidney, and placenta but not in the skeletal muscles. Lysosomal acid lipase (LAL, 10q24-q25) is Wolman disease. Lipase deficiency loci are also found at 15q21-q23 (hepatic triglyceride lipase, LIPC), 9q34.3 (carboxylester bile salt–dependent lipase, CEL), and 10q26.1 (pancreatic lipase, PNLIP). *See* HDL; lipids; Wolman disease. (Henderson, H. E., et al. 1996. *Biochem. Biophys. Res. Commun.* 227:189; Du, H., et al. 2001. *Hum. Mol. Genet.* 10:1639.)

lipid, neutral   Soluble only in very low-polarity solvents. *See* lipids; lipids, cationic.

lipid bilayer   Main structural element of membranes with two layers of hydrophobic lipid tails facing each other and the hydrophilic heads exposed at the outer parts. Proteins occur within the lipids and some transmembrane proteins pass through the lipid bilayer. *See* cell membrane; seven-membrane proteins.

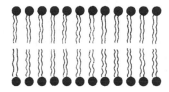

Model of lipid bilayer.

lipidoses   *See* lipid storage disease; lysosomal storage diseases; sphingolipidoses.

lipid peroxidation   May be a substantial source of mutagenic/carcinogenic alterations in DNA. *See* malondialdehyde.

lipids   Usually large molecules polymerized from fatty acids and associated frequently with sugar, protein, and inorganic components. Lipids are the principal material of membranes; they are insoluble in water but readily soluble in nonpolar organic solvents and detergents. The genetic analysis of lipids in higher organisms is quite difficult because of their indispensable function in maintaining cell viability. In *E. coli*, several mutations have been identified in the phospholipid biosynthetic pathways. *See* cell membranes; fatty acids; high-density lipoproteins; liposome; low-density lipoproteins; phosphoinositide; sphingolipids; sphingosine; steroids. (*Annu. Rev. Biochem.* Dowhan, W. 1997. 66:199; Kersten, S. 2001. *EMBO Reports* 2:282; <http://www.lipidat.chemistry.ohio-state.edu>.)

lipids, cationic   Synthetic molecules that form a bilayer and are suitable for the construction of nucleic acid delivery vehicles. They enhance the internalization of nucleic acid fragments 5–20–fold and that of antisense oligonucleotides up to three orders of magnitude. The cargo may pass through the nuclear pores into the nucleus, but its degradation is more likely than in liposomes. They are made of 12–18 carbon atom chains and single or multiple cations of amines. The vehicles frequently use DOTMA (N-[-{2,3-dioleyl-oxy}propyl]-N,N,N-trimethyl ammonium chloride), DOGS, and DOPE (dioleyl phosphatidylethanolamine). Cationic lipids are not allergenic,

are nontoxic, and can be administered through the vascular system. In case the vector is equipped with cell- or tissue-specific promoters, its expression may be successfully targeted. Monoclonal antibodies or other ligands to (tumor) cell-surface antigens may facilitate targeting. In addition, the cationic lipids may be immunostimulatory—a feature meaningful for cancer gene therapy. In some instances, DOPE and other lipids have reduced the expression of the gene that they carried. Various versions of liposome-mediated gene transfer are under clinical trials (<http://clinicaltrials.gov/ct/gui/c/r>). *See* cancer gene therapy; DOGS; gene therapy; liposomes. (Audouy, S. & Hoekstra, D. 2001. *Mol. Membr. Biol.* 18[2]:129.)

lipid storage disease   *See* lysosomal storage disease.

lipoate   Carrier of H atoms and acyl groups in α-keto acids.

lipo-chitooligosaccharide   *See* nodule.

lipodystrophy, familial   Complex of autosomal-dominant, -recessive, and X-linked anomalies of partial or predominant loss of lipid tissues, especially in some areas or a body and an increase at others. It may be associated with cystic angiomatosis (neoplasia of the blood vessels or lymphatic ducts), microphthalmia or anophthalmia (reduced eyes), mental retardation, and insulin-resistant diabetes. It is surprising that both obesity and lipodystrophy can cause diabetic symptoms. Seip-Berardinelli congenital lipodystrophy is at 9q34.1. Another locus at 11q13 is also responsible for the recessive Berardinelli-Seip syndrome involving much reduced adipose tissue and severe insulin resistance at birth. In addition, acanthosis nigricans, hyperandrogenism, hepatomegaly, and hypertriglyceridemia are typical. Kobberling-Dunnigan syndrome, a partial lipodystrophy (human chromosome 1q21-q22), involves defects in lamins and is encoded at the same location as the LMNA gene. *See* acanthosis nigricans; androgen; angioma; diabetes mellitus; eye diseases; hepatomegaly; lamin, obesity; SREBP; triaglycerols. (Magré, J., et al. 2001. *Nature Genet.* 28:365.)

lipofection   Genetic transformation by liposome transfer. *See* lipids, cationic; liposomes.

lipofuscinosis   *See* ceroid lipofuscinosis.

lipoic acid   *See* lipoate.

lipomatosis   Development of usually benign tumors in fat tissues. They may be under autosomal-dominant control and are frequently associated with breakpoints in human chromosome 12q13-q14. An autosomal-recessive lipomatosis of the pancreas (Schwachman-Bodian syndrome) involves frequent chromosomal breakage. *See* cancer; high-mobility group proteins.

lipopolysaccharide   *See* LPS.

lipoprotein   Lipid-protein (apolipoprotein) complex; transports hydrophobic lipids. *See* cholesterol; coronary heart disease; HDL; LDL.

lipoprotein lipase deficiency   *See* hyperlipoproteinemia.

liposarcoma, mixoid Usually involves the balanced translocations (t12;16)(q13;p11), but additional chromosomal aberrations also occur. The C/EBP transcription factor (also called CHOP, 12q13.1-q13.2) forms a fusion protein with the product of chromosome 16.

liposomes Delivery vehicles of macromolecules to cells. Within a protective coat of unilamellar vesicles consisting of phosphatidyl serine and cholesterol (1:1), they can carry DNA to the cell membrane, then after fusion, to the nuclear membrane, and eventually into the nucleus. The loading of the liposomes requires DNA in a buffer thoroughly mixed with the ether-lipid mixture. The ether is then removed. The delivery may be facilitated by polyethylene glycol and dimethyl sulfoxide. With appropriate coating (e.g., monoclonal antibody, DNA, viral vectors, drugs, etc.), they can be targeted to specific sites within the cell. Guanidinium-cholesterol cationic lipids (BGSC, BGTC) can be used for construction of eukaryotic vectors. Good results are obtained by the use of the cationic lipid DOTAP (dioleoyl trimethylammonium propane) and the so-called neutral helper lipid DOPE (dioleoyl phosphatidyl ethanolamine) mixtures and also with the mix of DOPC (dioleoyl phosphatidylcholine) and DOTAP. DMRIE (1,2-dimyristoyloxypropyl-3-dimethylhydroxyethyl ammonium bromide) promotes the transfer of reporter and therapeutic genes. DNA delivery is quite efficient when the linear cationic lipid monolayer is transformed into hexagonal lattices. Apparently, endocytosis mediates the uptake of liposomes. The entry of the liposomal content into the nucleus is not efficient, but it can be increased by four orders of magnitude if human papilloma virus enhancer elements are incorporated into the upstream regulatory region in the plasmid replicon. Liposome vehicles can deliver very large amounts of DNA (1 Mbp) and in contrast to viral vectors they do not evoke an immune response. Liposomes are also used for the delivery of vaccines and drugs. *See* cytofectin; deoxyribonucleotide-gated channels; DOGS; fusigenic liposome; immunoliposome; lipids; lipids, cationic; lipofection; vaccines. (Chesnoy, S. & Huang, L. 2000. *Annu. Rev. Biophys. Biomol. Struct.* 29:27.)

lipotropin Adenocortical peptides controlling lipolysis. *See* opiocortin.

lipoxygenase (LOX) Oxygenates arachidonic acid into leukotrienes. 15-Lipoxygenase integrates into membranes of cellular organelles and mediates the access of proteases to the integral and lumenal membrane proteins. As a consequence, in some special cells (e.g., erythrocytes, eye lens) mitochondria are degraded. 12-Lipoxygenase controls the angiotensin pathway and 5-lipoxygenase mediates leukotriene metabolism. Lipoxygenase-3 and lipoxygenase-12 mutations may be involved in non-bullous congerutal ichthyoriform erythrodema (17p13.1). *See* angiotensin; arachidonic acid; cyclooxygenase; erythroderma; integrin; leukotriene. (van Leyen, K., et al. 1998. *Nature* 395:392; Feussner, I. & Wasternack, C. 2002. *Annu. Rev. Plant Biol.* 53:275.)

liprin *See* synapse.

liquid crystal *See* mesogen.

liquid-holding recovery Type of genetic repair that takes place in irradiated cells if kept in liquid medium (and allowed to utilize energy) for 2 days before plating.

liquid scintillation counter *See* scintillation counters.

LIR (1) Long inverted repeat. DNA sequence of about 100 or more base pairs that is palindromic, or there may be an insertion between the inverted repeats. Such motifs are conducive to recombination or to deletion. *See* palindrome; repeat, inverted. (2) *See* ILT.

LISA Localized in situ amplification. *See* amplification; PCR; PRINS.

Lisch nodule Bloody spots (hamartomas) on the iris, characteristic sign of neurofibromatosis. *See* hamartoma; neurofibromatosis.

lissencephaly Norman-Roberts type (Xq22.3-q23) involves "smooth" brain, cerebellar hypoplasia, abnormal migration of cortical neurons, and abnormal connectivity of axons. The "double cortex" anomaly is at the same location. The 388 kDa RELN (reelin) protein is encoded at the same chromosomal site at 7q22. There are other diseases with lissencephaly. *See* Miller-Dieker syndrome.

Listeria monocytogenes 2.95 Mb bacterium causing abortion, heart/lung infections, diarrhea, and inflammation of the brain. It causes similar ailments in sheep and cattle. Its surface protein internalin binds to the intestinal transmembrane protein E-cadherin for infection. The host specificity of the different virulent forms depends on the amino acid sequence of the cadherin molecule. Mouse chromosomes 5 and 13 modify susceptibility to this bacterium. *See* cadherins. (Boyartchuk, V. L., et al. 2001. *Nature Genet.* 27:259.)

lithium Drug used for the treatment of manic depression. It causes various developmental anomalies in diverse organisms. It appears that it antagonizes the enzyme glycogen synthase kinase-3β. *See* manic depression.

lithography, photo *See* DNA chips.

litter Youngsters given birth to at the same time by a multiparous animal.

liver cancer Liver cell carcinoma, (LCC) is generally initiated upon the integration of hepatitis B virus in one or more chromosomal locations. Antitrypsin deficiency, hemochromatosis, and tyrosinemia may also be conditioning factors. When the viral infection takes place in an individual genetically susceptible to liver carcinoma, he/she may have an eventual incidence of 0.84 (males) and 0.46 (females) of the cancer. Without viral integration the chances for the cancer are 0.09 and 0.01, respectively, even when the genetic constitution is permissible for its development. One of the viral integration sites (17p12-p11.2) is near the p53 tumor suppressor gene. The flanking sequences are homologous to the autonomously replicating sequence (ARS1) required for the replication of the *Saccharomyces cerevisiae* integrative plasmids. Other integration sites identified are 11p13 and

18q11.1-q11.2. *See* antitrypsin; ARS; cancer; chi elements; hemochromatosis; hepatitis B virus; tumor suppressor gene; tyrosinemia; yeast vectors.

liverwort   *See* bryophytes.

L19 IVS (intervening sequence lacking 19 nucleotides)   395-nucleotide linear RNA left after 19 nucleotides were consumed by the *Tetrahymena* ribozyme intron originally 414 nucleotides long. This RNA molecule is capable of nucleotidyl transferase activity and can elongate various nucleotide sequences in a protein enzyme–type manner. Preferred substrates are the nucleotide sequences that can pair with guanylate-rich sequences in L19IVS, i.e., have C residues. *See* intron; ribozyme; *Tetrahymena*.

LKB   *See* polyposis hamartomatous.

LMO   Class of proteins with only LIM domain. *See* CRIP; CRP; LIM; PINCH.

LMP-2, LMP-7   Proteasomal subunits. *See* proteasome.

LM-PCR   Ligation-mediated PCR is used for the study of DNA double-strand breaks. *See* PCR.

LMYC   *See* MYC.

LNGFR   Low-molecular-weight nerve growth factor receptor. *See* HNGFR; nerve growth factor.

load   *See* genetic load.

loading test   Heterozygotes for many metabolic defects usually appear very close to normal. In case large amounts of the metabolite are given that the homozygotes are unable to process, the heterozygotes are also slower than normal to clear them from the system. Example: normal persons and phenylketonuria heterozygotes at fasting displayed phenylalanine (μmoles/mL) in the plasma, 0.067 ± 0.032 and 0.103 ± 0.029 (1.54), respectively, and 1 hour after an oral dose of 0.1 g phenylalanine/kg body weight, the observations were 0.55 ± 0.186 and 1.14 ± 0.18 (2.07), respectively. Histidine loading in case of heterozygosity for B_{12} or folate deficiency excessively increases the amounts of formiminoglutamic acid

in the urine. *See* folic acid; heterozygote proportions; His operon; phenylketonuria; vitamin B_{12} defects.

lobotomy   Surgery of the skull for (nowadays questionable) psychotherapeutic means during the first half of the 20th century before psychotherapeutic drugs were discovered.

localized determinants   In some species, different parts of the egg determine the formation of different blastomeres depending on what cellular portions they received during cleavage. *See* blastomere; cleavage.

localized mutagenesis   Involves in vitro and in vivo manipulations. An amber mutation in a phage can be corrected by synthesizing a short DNA sequence complementary to a section of ϕX174 DNA single strand having a single mismatch at one base (A) corresponding to the amber codon. The synthetic sequence will have C there correctly. The short synthetic DNA is then extended to a double-stranded form. After a cycle of replication, the corrected strand makes a correct complementary copy of itself, and its progeny is thus permanently corrected by localized mutagenesis. The other old single strand of the DNA will produce another amber mutant phage.

Similar methods can be applied to any DNA that can be transferred to cells by transformation. Site-directed transposon mutagenesis is feasible by identifying genes, which contain a P element within or near a particular gene. P elements are preferentially inserted at the 5′-ends of genes and thus gene-specific promoters can be placed there. *See* cysteine-scanning mutagenesis; gene replacement; homologue-scanning mutagenesis; hybrid dysgenesis; insertional mutagenesis; Kunkel mutagenesis; PCR-based mutagenesis; site-specific mutagenesis; TAB mutagenesis; targeting genes; figure below and next page.

locational candidate region   *See* genomic screening.

location score   Natural logarithm of the likelihood of the odds ratio for linkage. It is determined by multiplying the lod score by 4.6. *See* lod score.

An amber mutation is corrected by a short DNA sequence containing a cytosine rather than an adenine in the codon. Adding the appropriate nucleotides and through the use of a DNA polymerase, the synthetic nucleotide strand is completed and the strands are tied together by DNA ligase. The new single-stranded DNA has a single mismatch. After transfection of the phage to *E. coli* bacteria, each strand replicates, and both amber and wild-type ϕX174 genomes are made. By selective screening, the wild-type phages can be isolated as the result of this directed, local mutation. (modified after Itakura, K. & Riggs, A. D. 1980. *Science* 209:1401.)

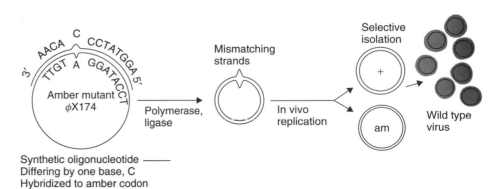

Synthetic oligonucleotide ——
Differing by one base, C
Hybridized to amber codon

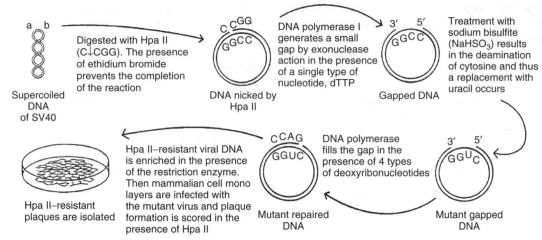

Local mutagenesis using restriction endonuclease gapping and in vivo mutagenesis followed by DNA polymerase correction of the gap. (After Shortle, D. & Nathans, D. 1978. *Proc. Natl. Acad. Sci. USA* 75:2170.)

locule   Fruiting body of fungi or the cavity of the ovary of plants occupied by the ovule or the place of the uterus of some mammals where the embryo is attached.

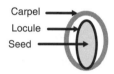

locus   Site of a gene in the DNA (or RNA) or the site of a gene in the chromosome.

locus content map   Displays the observations on the presence or absence of genes in chromosome fragments produced by ionizing radiation. This information may not be suitable for genetic mapping if contrasting alleles are not available. The data obtained can be confirmed by physical mapping. *See* radiation hybrids. (Teague, J. W., et al. 1996. *Proc. Natl. Acad. Sci. USA* 93:11814.)

locus control region   *See* LCR.

locus exclusion   Both alleles at a locus are prevented from expression in deference to an allele(s) at another locus. *See* allelic exclusion; idiotype exclusion.

locus heterogeneity   The same phenotype is produced by more than one combination of the genetic constitution.

LocusLink   Provides information about genes as a central hub. (<http://www.ncbi.nlm. nih.gov/LocusLink/.)

locust   *Locusta migratoria*, 2n = 23; B chromosomes occur.

lodicule   Scale-like structure at the base of the ovary of grasses; it mediates the opening of the flower when the pollen is shed in allogamous species.

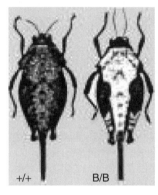

Locust (*Paratettix texanus*).

lod score   Expresses the relative probability of linkage compared to the odds against linkage (log odds) because the score is obtained by dividing the sum of logarithms indicating linkage by the sum of logarithms suggesting independent segregation. The principle briefly is

$$\text{Probability of linkage } (P) = \frac{\text{Recombination Frequency } \theta}{\text{Recombination Frequency } 1/2}$$

θ stands for any value of recombination, and 1/2 recombination indicates independent segregation. In case several linkage probabilities exist, the formula can be rewritten as

$$\text{Relative linkage } (P) = \frac{p_1(\theta_1)\,p_2(\theta_2)\ldots p_n(\theta_n)}{p_1(1/2)\,p_2(1/2)\ldots p_n(1/2)}$$

Hence the relative odds of linkage is expressed:

$$\text{Log}_{10}(\text{odds of linkage}) = \frac{\log p_1(\theta_1) + \log p_2(\theta_2) + \log p_n(\theta_n)}{\log p_1(1/2) + \log p_2(1/2) + \log_n(1/2)}$$

$$= \text{lod score}$$

The lod scores for all θ (recombination frequency) values summed up is expressed as $Z(\theta)$. This value may be empirically defined at various values of θ:

θ	θ^2	$(1-\theta)^2$	$2[\theta^2+(1-\theta)^2]$	$Z(\theta)$
0	0	1	2.00	0.30303
0.05	0.0025	0.9025	1.81	0.25768
0.10	0.0100	0.8100	1.64	0.21484
↓	↓	↓	↓	↓
0.50	0.2500	0.2500	1.00	0

Thus, a lod score if $\theta = 1/2$ is 0 because $\log 1 = 0$. A lod score >3 indicates approximately 95% probability of linkage, but lod score of >4 gives a probability of about 99.5%. A lod score of −2 indicates independent segregation. Lod scores between −2 and 3 are considered inconclusive. It is most useful for sequential data as available from (human) pedigrees. It is widely used for mapping restriction fragments (RFLP). The homogeneity of the lod scores among different families can be tested by chi-square analysis. The probability of two genes being independent may also be obtained by the formula based on Bayes' theorem, e.g., for the 22 human autosomes:

$$P(\theta) = (1/2) = \frac{21}{\lambda + 21}$$

where λ is the average value of $\lambda(\theta)$ and it is equal to antilog $Z(\theta)$. The chance that any particular unknown gene would be in a human autosome is 21/22 and the chance for another gene being in the same chromosome is 1/22. The computation based on maximum likelihood is somewhat complicated when multipoint crosses are evaluated, but computer programs (e.g., Mapmaker, Joinmap, LINKAGE, MENDEL, VITESSE, FAST-LINK, LODLINK) are available for linkage estimation. The lod score can be used for other biological comparisons, e.g., evolutionary relationships of nucleotide or protein sequences. *See* antilog; genomic screening; linkage; mapping, genetic; mod score; wrod score. (O'Connel, J. R. & Weeks, D. E. 1995. *Nature Genet.* 11:402; Lander, E. & Kruglyak, L. 1995. *Nature Genet.* 11:241; Nyholt, D. R. 2000. *Am. J. Hum. Genet.* 67:282; Ott, J. 2001. *Advances Genet.* 42:125.)

lognormal distribution   Logarithms of the data may approximate the normal distribution. Mutation rates may show this type of distribution and the variance $(\sigma^2) \sim \mu$ (mean).

loganberry   (*Rubus loganobaccus*)   Fruit shrub, $2n = 42$.

logarithm   Can be defined as $x = b^y$ and from this $y = \log_b x$, or y is the logarithm of x to the b base. The common (Briggs) logarithm uses base 10; the base of the natural logarithm is called e and it is 2.71828 (to five digits). Logarithms can be determined to any base and can be converted to each other, e.g., the natural logarithm, $\log_e$ (commonly designated also as ln), can be converted to common logarithms as $\log_{10} x = \log_e x (1/\log_e 10)$. *See* antilogarithm.

logarithmic growth   *See* exponential growth.

logarithmic stability factor   Measures the developmental homeostatic values of individuals of certain genotypes under different environmental conditions. LSF is calculated as the absolute difference between two logarithmic means. If LSF = 0, the developmental homeostasis is maximal. *See* homeostasis. (Li, S. L. & Rédei, G. P. 1969. *Theor. Appl. Genet.* 39:68.)

logic   Study of the methods of reasoning; it deals with propositions, their implications, possible contradictions, conversions, etc. *Symbolic logic* employs mathematical symbols for the propositions, quantifiers, mutual relations among propositions, etc. Logic is one of the most important tools of science for interpreting the experimental data. *See* syllogisms.

log phase   *See* exponential growth.

LOH   Loss of heterozygosity mediated by mutation (deletion) or somatic recombination or gene conversion. It is a common cause of cancer when a cancer suppressor gene is lost. *See* gene conversion; LOS; mitotic recombination; pseudodominance; tumor suppressor gene.

lollipop structure   When single strands of DNA carrying palindromic terminal repeats are allowed to fold back, the complementary ends pair while the noncomplementary sequences in between remain single-stranded and thus form a stem and loop configuration as shown at the below. Denatured insertion elements may also display lollipop structures if the complementary sequences are allowed to pair within the single strand. *See* insertion element; palindrome.

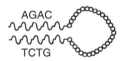

Stem loop.

Lon   Monomeric eukaryotic polypeptide with one of the domains associated with ATPase and functions as a chaperone and the other domain functions as a protease. *See* Clp.

long branch attraction   Placement in evolutionary tree sequences that are not truly relevant to their phylogenetic relationship. *See* evolutionary tree. (Qiu, Y. L., et al. 2001. *Mol. Biol. Evol.* 18:1745.)

long-day plant   Requires usually more than 12 hr daily illumination to be able to develop flower primordia. *See* photoperiodism; primordium; short-day plants.

longevity   Represents the length of life. *life span* is the biologically determined potential of an organism to live. *Life expectancy* is the actual, expected duration of life under certain conditions. The average human life expectancy in years in the U.S. in 1850 was 38.3 for males and 40.5 for females; by 1950, it became 66.5 and 73.0, respectively, and it has increased since by approximately 7–8 years. Raymond Pearl, an American pioneer of human population genetics, found (during the early years of the 20th century) the life expectancy at birth of sons of fathers whose death was before age 50, 50–79, and over 80 to be 47, 50.5, and 57.2 years, respectively. Despite regression, approximately similar life expectancy differences continued after age 40. Currently the average life expectancy is in the mid-70s in the U.S., and if the present trends continue, in about 50 years it may increase by about 10 years. At the present period, the estimated maximal human life span is about 120 years. Compared to the U.S. 1900 cohort, male siblings of centenarians were about 17 times as likely to attain age 100 themselves; while female siblings were at least 8 times as likely (Perls, T. T., et al. 2002. *Proc. Natl. Acad. Sci. USA* 99:8442).

In *Drosophila* increased egg-production, receipt of male accessory fluid and courtship shorten lifespan. Mating and courtship also shorten the life span of the flies. Caloric restriction in the diet or mutation in sodium dicarboxylate cotransporter or overexpression of a protein repair methyltransferase, extend the life span of the flies. The gene *Indy* (for I'm not dead yet) encodes a membrane protein that transports Krebs cycle intermediates. The hermaphrodite *Caenorhabditis* has a shorter life if mated, but this is independent from egg production or receipt of sperm. The males of this species are not affected by mating. Some single genes of *Caenorhabditis* (*daf, age-1*) may extend the life span 2–3 times. In this nematode, the life span is also regulated by sensory perception, and defects in the sensory neurons may extend the life span. Some suggestions were made regarding correlation between longevity and MHC alleles, apolipoprotein E and B, angiotensin-converting enzyme, upregulation of superoxide dismutase activity, etc. When the insulin-like growth factor (IGF-1) receptor DAF-2 reduces the level of DAF-16 (a forkhead/winged helix-type transcriptional regulator [FKH]) protein, the worm's life span is doubled. Besides DAF-16, DAF-12 (a nuclear hormone) has a beneficial role. This pathway (DAF-2) is under germline-controlled negative regulation. The somatic gonad also affects aging. In *Drosophila*, the loss of an insulin receptor substrate protein significantly extends the life span. *Drosophila* longevity is increased by 4-phenylbutyrate without apparent side effects (Kang, H.-L., et al. 2002. *Proc. Natl. Acad. Sci. USA* 99:838). The p66[Shc] protein (encoded by SHC gene) protects mammalian cells from oxidative stress (ROS) and may increase the life span of mice by 30%, and homozygosity for the Prop[df] and PitI[dw] mutation (affecting growth factor production) may expand the life span by 40 to 50%. In the fungus *Podospora anserina*, inactivation of subunit V of the mitochondrial COX5 gene channels respiration to an alternative pathway, decreases the production of reactive oxygen species, and strikingly increases longevity. In *Saccharomyces*, activation of the gene silencing protein Sir2 by NAD contributes to longevity. Longevity among taxonomic groups may be positively correlated with the size of the genome. *See* age and mutation; age-specific birth and death rates; aging; angiotensin; apolipoproteins; copulation; COX; dauer larva; FKH; insulin; menopause; MHC; mortality; NAD; ROS; SHC; silencer; superoxide dismutate. (Tuljapurkar, S., et al. 2000. *Nature* 405:789; Clancy, D. J., et al. 2001. *Science* 292:104; Chavous, D. A., et al. 2001. *Proc. Natl. Acad. Sci. USA* 98:14814; Larsen, P. L. & Clarke, C. F. 2002. *Science* 295:120; Perls, T., et al. 2002. *Current Op. Genet. Dev.* 12:362; <www.demog.berkely.edu/wilmoth/mortality/index.html>.)

Longhorn cattle Primitive form, used extensively as draft farm animal before the era of mechanization; it resembles the cattle depicted (15,000 years ago) by neolithic humans in European caves. *See* Lascaux cave.

Long interspersed nucleotide element *See* LINE.

long-patch repair Repair of longer mismatches. In humans, DNA polymerase β may have the major role, although pol δ and pol ε are also repair polymerases. Base excision and nucleotide excision repairs are short-patch repairs. *See* DNA polymerases; DNA repair; excision repair; mismatch repair; PARP. (Podlutsky, A. J., et al. 2001. *EMBO J.* 20:1477.)

long-term selection *See* artificial selection.

long-period interspersion Long sequences of repetitive DNA alternate with long unique sequences.

long QT syndrome (KCNQ1/LQT1, 11p15.5; LQT2, 7q35-q36) Involves voltage-gated potassium channel defects. *See* LQT; Ward-Romano syndrome. (Marx, S. O., et al. 2002. *Science* 295:496.)

long-term depression *See* memory.

long terminal repeat (LTR) Retroviruses at the two flanks may have 2–8 kb repeated nucleotide sequences; similarly a few hundred repeats are found at the ends of retrotransposons. *See* retroposons; retroviruses; solitary LTR; transposable elements.

long-term potentiation (LTP) Activation of synaptic junctions in the brain relating to memory. Nonassociative LTP is induced by high-frequency stimulation of the presynaptic termini. Associative-type LTP increases the probability that neighboring synapses on the same neuron will be strengthened if they are activated within milliseconds. This provides an opportunity to encode associations between different events if they occur concurrently. Ca^{2+}/calmodulin-dependent protein kinase (CaMKII) and cAMP regulate LTP. Inhibition of protein phosphatase-1 (PP1) can substitute for the cAMP function, which uses PP1 to gate CaMKII signaling. *See* calmodulin; cAMP; Hebbian mechanism; memory. (Malenka, R. C. & Nicoll, R. A. 1999. *Science* 285:1870.)

loop Single-stranded (unpaired) nucleic acid sequences (loops) alternating with double-stranded, paired regions in isolated molecules. Paired regions of the meiotic chromosomes of inversion heterozygotes, unpaired regions of normal chromosomes paired with deletions. The nondeleted strand pops out opposite the deletion. *See* deletion; inversion; lampbrush chromosome; lollipop; looping of DNA; tRNA.

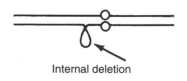

Internal deletion

loop domains model The DNA fibers form 5 to 100 kb loops that are attached to the nuclear matrix (protein) at the *matrix attachment region* (MAR). These domains of the chromatin are supposed to be transcriptional and replicational units. The matrix is also called the nuclear scaffold. MAR regions are expected to regulate the expression/silencing of genes *See* A box; chromatin; T box.

looping, DNA Brought about by protein binding to DNA at two different positions from 10 to thousands of nucleotides apart. The association of proteins with specific DNA sequences permits the DNA to fold back from a longer distance to the location of the gene promoter and thus regulates expression, recombination, and replication of the genetic material. *See* hormone-response elements; LCR; regulation of gene activity; transcription factors, inducible. (Irani, M. H., et al. 1983. *Cell* 32:783; Schleif, R. 1992. *Annu. Rev. Biochem.* 61:199; Geanocopoulos, M., et al. 2001. *Nature Struct. Biol.* 8[5]:432.)

loop-out loop-in mutagenesis Generation of deletions and replacement of the deleted segments with other sequences.

LOR Human low-copy repetitive sequences in about five copies per genome. *See* redundancy.

lordosis Curvature of the spine. It is generally applied to the condition when the curvature is expressed above the normal level, e.g., in some types of muscular dystrophy. *See* muscular dystrophy.

Lorisidae Family of prosimii. *Arctocebus calabrensis* $2n = 52$; *Galago crassicaudatus* $2n = 62$; *Galago demidovii* $2n = 58$; *Galago senegaliensis braccatus* $2n = 36, 37, 38$; *Perodicticus potto* $2n = 62$. *See* prosimii.

LOS Loss of heterozygosity mutation. In a heterozygote, one of the alleles (infrequently both) is lost. If the remaining allele is nonfunctional or suffers a secondary (somatic) mutation, the normal function of the gene is discontinued. *See* LOH.

loss-of-function mutation Generally results in a recessive phenotype, but only one-third or one-fourth of the *Drosophila* genes can be classified by phenotype as loss-of-function mutation because of redundancy. *See* gain-of-function mutation; LOH.

Lotka-Volterra formula Quantifies the competitive interactions between populations:

$$\frac{dN_1}{dt} = r_1 N_1 \left[1 - \frac{(N_1 + \alpha_{12} N_2)}{K_1} \right]$$

and

$$\frac{dN_2}{dt} = r_2 N_2 \left[1 - \frac{(N_2 + \alpha_{12} N_1)}{K_2} \right]$$

Subscripts 1 and 2 stand for the two populations (N); dN_1/dt and dN_2/dt are the calculus symbols for the rate of change; $1/K$ is the effect of an individual of species 1 on the growth of species 1; α_{21}/K_1 is the effect of an individual of species 2 on species 1; similarly, $1/K_2$ and α_{12}/K_2 stand for the influence of one individual of species 1 on species 2; and r is the constant of intrinsic rate of growth. (Drossel, B. & Mckane, A. 2000. *J. Theor. Biol.* 204[3]:467.)

lotoaustralin A glucoside linked with cyanide, making some white clover varieties toxic or lethal (if 1 to 2 kg is consumed) to animals (50 kg). This toxic effect has been circumvented by breeding white clovers deficient in the synthesis of glucoside and in linamarase. Several other plant species (Sudan grass, flax, some beans, almond, apricot) produce potentially toxic cyanides. *See* cyanogenic glucosides.

Lou Gehrig's disease *See* amyotrophic lateral sclerosis.

Louis-Bar syndrome Ataxia telangiectasia. *See* ataxia.

lovastatin (mevinolin/monacolin K) Inhibitor of the (3S)-hydroxy-methylglutaryl-coenzyme A reductase, which normally reduces its substrate (HMG-CoA) to mevalonate, a precursor of cholesterol. *See* cholesterol.

Lovastatin (acid).

low-complexity sequences Contain repetitions of the building blocks or tracts of repetitions.

low-copy repats May constitute 5% of the genome of 10 to 40 kb size and ~95–97% similarity. Because of recombination among these similar sequences, they may be responsible for genomic instabilities and disease (Stankiewicz, P., & Lukski, J. R. 2002. *Current Op. Genet. Dev.* 12:312.)

low-density lipoproteins (LDL) Form 22-nm-diameter particles containing bilayered vesicles about 1,500 cholesterol molecules esterified to long-chain fatty acids; form a vehicle of delivery of lysosomes from which the cholesterol is made available for membrane synthesis. If there is a defect in the synthesis of the receptor proteins (19p13.2, 12q13.1-q13.3, 1p34, 2q24-q31, 4p16.3, 11p12-p11.2), the cell may not draw cholesterol from the blood, resulting in atherosclerosis and eventually artery disease. LDL receptors are involved in high-affinity and broad specificity endocytosis such as the macrophage scavenger receptors that are supposed to mediate cell adhesion, host defense, and atherosclerosis. *See* Alzheimer disease; apolipoproteins; atherosclerosis; cholesterol; HDL; hypercholesterolemia; hypertension; lysosome; VLDL.

low-energy phosphate Compound that can release relatively low energy upon hydrolysis, e.g., glucose-6-phosphate.

low-frequency transducing lysate *See* LFT.

Lowe's oculocerebrorenal syndrome (OCRL) X-chromosome-linked (Xq26.1) eye defect, mental retardation, aminoaciduria, kidney anomaly, ricketts, etc. The defect is in an inositol-5-phosphatase, synaptojanin. *See* mental retardation; synaptojanin.

Lowry test (*J. Biol. Chem.* 193:265 [1951]). Quantitatively estimates total protein in tissue extracts using 1% CuSO4 (0.5 mL), 2.7% Na-K-tartrate (0.5 mL), 2% Na_2CO in 0.1 M NaOH (50 mL); add 5 mL to 1 mL sample protein (25–500 μ/mL), mix, allow to stand 10 min, then add 0.5 mL Folin & Ciocalteu reagent and mix; after 30 min, determine O.D. at 750 nm and compare to bovine serum albumin standard solution series (detergents interfere with the test). *See* Bradford method; Kjeldhal method.

Lox (*loxP*) *See* Cre / Lox.

L-phase Due to various shocks (temperature, osmotic, antibiotics, etc.), the bacteria may lose their walls possibly in a reversible manner, yet they can multiply.

LPS (endotoxin) Lipopolysaccharide, lipid-carbohydrate complexes such as exist on the walls of some bacteria. The infection of mammalian cells by gram-negative bacteria leads to the release of toxic LPS, which in turn activates TNF and other toxic cytokines. *See* cytokines; endotoxins; TNF.

LQT Cardiac (heart) disease involving left ventricular arrhythmia. Several genetic defects results in the same symptoms (SCN5A, LQT3 [chromosome 3p21], HERG [LQT2, chromosome 7], KVLQT1/KCNQ1 [LQT1, chromosome 11p15.5], and another in chromosome 4). All of these loci are involved in the control or regulation of muscle cell K^+ or Na^+ ion channels. *See* Beckwith-Wiedemann syndrome; cardiovascular diseases; electrocardiogram for LQT; ion channels; Jervell and Lange-Nielson syndrome; KCNA; Ward-Romano syndrome. (Bennett, P. B., et al. 1995. *Nature* 376:640; Schwartz, P. J., et al. 2001. *Circulation* 103:89; Keating, M. T. & Sanguinetti, M. C. 2001. *Cell* 104:569.)

LRE *See* light-response elements.

LRF Luteinizing hormone releasing factor. *See* corticotropin; luteinization.

LRR Leucine-rich repeat. *See* host-pathogen relations.

LSIRF (IRF4) Transcription factor specific for mature B and T cells.

LTα, LTβ Proteins of the TNF/NGF family encoded at the HLA gene cluster. *See* HLA; NGF; TNF.

LTD *See* memory.

LTP *See* memory.

LTR Long terminal repeats in movable genetic elements and oncogenic viruses. *See* retroviruses; transposable elements.

LUC Firefly luciferase gene. *See* luciferase.

luciferase Can be effectively used as a tissue- or developmental stage-specific reporter of gene expression. The luciferase of the firefly (*Photinus pyralis*) is a single polypeptide of 550 amino acids. In the presence of oxygen (NO), and ATP (or coenzyme A), it oxidizes luciferin, resulting in the emission of yellow-green light flashes, which can be monitored by luminometers, scintillation counters, or extended exposure to highly sensitive photographic film. Other insects also produce luciferases, which produce somewhat different light emission. The generation of the light flashes is part of the courtship process and reproduction in insects. The bacterial luciferases such as the one produced by *Vibrio harveyi* or other similar strains consist of two subunits A and B. The substrate of the bacterial luciferase enzymes is n-decyl aldehyde (decanal), and for the emission of light, reduced flavin mononucleotide ($FMNH_2$) is required. While the aldehyde (RCHO) is converted to acid (RCOOH) in the presence of oxygen (O_2), water (H_2O), FMN, and *light* is generated. The luminescent bacterium *V. fischeri* uses a quorum-sensing signal; combined with the enzyme LuxI, it activates LuxR, which in turn binds to specific activating domains in bacterial DNAs to turn on the genes required for light emission. About 20 additional genes cooperate with and modulate the expression of bacterial luciferase. The light can be monitored in the tissues without destruction by the use of a microchannel plate-enhanced photon-counting image analyzer that uses a video camera, an image processor, a TV processor, and computerized controls. Such equipment may even detect a single photon. A much less expensive luminometer can be used with ground tissues. *See* GUS; image analyzer; quorum-sensing. (Baldwin, T. O., et al. 1995. *Curr. Opin. Struct. Biol.* 5:798; Pazzagli, M., et al. 1992. *Anal. Biochem.* 204:315; Olsson, O., et al. 1988. *Mol. Gen. Genet.* 215:1.)

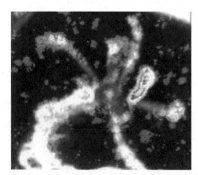

Expression of bacterial luciferase αβ genes in *Arabidopsis*. (From Rédei, G. P, Koncz, C, Langridge, W. H. R., unpublished.) See color plate in Color figures section in center of book.

luciferin Upon activation, it cleaves a pyrophosphate off ATP and luciferyl adenylate is formed. Upon the action of firefly luciferase—in the presence of O_2—light is emitted. Oxyluciferin is then generated in the presence of CO_2 and AMP, and subsequently luciferin is reformed. These reactions are suitable for monitoring gene expression and for the quantitation of ATP. *See* ATP; luciferase. (Wilson, T. & Hastings, J. W. 1998. *Annu. Rev. Cell Dev. Biol.* 14:197.)

Lucy 3-million-year-old *Australopithecus afarensis* skeleton of about 1 m tall (discovered in 1974 in Ethiopia) believed to be a representative of the origin of the human family tree. Although the skeleton was assumed to be that of a female (because of the small size), some paleoanthropologists argue that it is a male because the pelvis would not have been sufficient for child bearing. *See* hominids.

Ludwig effect Biological habitats are generally divided into numerous microniches; the species respond to these by selection.

Luft disease Rare human hypermetabolism caused by a defect in the coupling of oxidative and phosphorylation processes in the mitochondria. The genetic determination is not entirely clear; autosomal recessivity is suspected. *See* mitochondrial disease in humans.

lumen (1) Interior compartment of a membrane-enveloped structure in the cell. (2) Unit of luminous intensity. *See* lambert.

luminescence Light emission from cool sources such as excited gases (neon light), fluorescent tubes, television and computer screens, and bioluminescence of fireflies, glow worms, and certain fishes and bacterial luciferase. *See* luciferase.

lung *See* small cell lung carcinoma.

lupines (leguminous plants) Yellow lupine (*Lupinus luteus*, $2n = 46, 48, 52$); blue lupine (*L. angustifolius*, $2n = 40$); white lupine (*L. albus*, $2n = 30, 40, 50$) are crop plants used for "green manure" or in alkaloid-free forms as forage crops in acid soils. The perennial lupine (*L. polyphyllus* ($2n = 48$) is an attractive ornamental. The alkaloid-free lupine is one of the best examples of scientific plant breeding. *See* alkaloids).

lupus erythematosus Variety of skin and subcutaneous inflammations, each with a specific medical name. The genetic basis is not clear because some of the common symptoms are apparently caused by certain drugs (e.g., procaine anesthetics); in other cases, viral infection has been suspected because of similar symptoms observed on family dogs. Some indications are for the involvement of steroid hormone problems because of its incidence with Klinefelter syndrome. In some cases, along with the normal DNA, a low-molecular-weight DNA was associated with the disease. In other individuals, anti-RNA or antinuclear antibodies were identified. Lymphocyte defects were also shown in some afflicted individuals. Serum complement deficiencies involve a major risk of onset. C1q-deficient individuals carry a risk of ~90%. C1q is assembled from 18 polypeptide chains including the C1r and C1s serine protease homodimers. After C1q binds to an antibody, the proteases are activated; from C2 and C4 complement parts, C3 convertase is formed. Females are about 8-fold more likely to be afflicted by it than males in the age group 15 to 50, but later or earlier the relative risk differences are much less. Protein p21 appears to be responsible for preventing lupus-like diseases in females. The lifetime risk for a U.S. Caucasian female is about 0.15% and much less for blacks or Hispanics. A mouse model indicates that the secretion of BAFF protein by T cells or dendritic cells initiates the abnormal stimulation of B-cell proliferation. Furin protease cleaves off BAFF from the surface of the cells that secreted it. When BAFF goes into the bloodstream, it recognizes the B-cell receptors BCMA and TACI, and in cooperation with CD40, it activates B cells. Decoy receptors for BAFF may alleviate the inflammation. Antinuclear antibodies presumably due to deficiency of DNase1, which normally removes them, characterize systemic lupus erythematosus (SLE, 4p16-p15.2). As a consequence, immune reactions afflict blood vessel walls and joints, and arthritis symptoms develop. In lupus erythematosus, a molecular mimic of Asp/Glu-Trp-Asp/Glu-Tryr-Ser/Gly pentapeptide is antigenic to double-stranded DNA; it is also present in murine and human NMDA receptor NR2 subunits. As a consequence, neurons are subject to apoptosis and to this autoimmune disease. *See* autoimmune disease; B cell; CD40; complement; convertase; decoy receptors; dendritic cell; HLA; p21; procaine anesthetics; RoRNP; TACI; T cell. (Gray-McGuire, C., et al. 2000. *Am. J. Hum. Genet.* 67:1460; Yasutomo, K., et al. 2001. *Nature Genet.* 28:313; DeGiorgio, L. A., et al. 2001. *Nature Med.* 7:1189.)

Luria-Bertani medium *See* LB.

luteinization After ovulation, the ovarian follicle is converted into a corpus luteum (yellow body). *See* corpus luteum; Graafian follicle; luteinizing hormone-releasing factor.

luteinizing hormone pGlu-His-Trp-Ser-Tyr-Gly-Leu-Arg-Pro-Gly:NH₂. *See* animal hormones.

luteinizing hormone-releasing factor (LH-RF or LRF, gonadotropin-releasing factor) Hypothalamic neurohormone stimulating the secretion of pituitary hormones. It is involved in the control of mammalian fertility. *See* ART; corticotropin; GnRHA; in vitro fertilization; puberty, precocious. (Latronico, A. C. & Segaloff, D. L. 1999. *Am. J. Hum. Genet.* 65:949.)

Lutheran blood group Named after the first patient identified in 1945. The Lu(a⁻ b⁻) phenotype is determined by either dominant or recessive alleles of the gene in human chromosome 19cen-q13, distinguishable by hereditary pattern and by serotype. The frequency of Lu(a) allele is about 0.04 and that of Lu(b) is about 0.96. The Lu phenotype may be caused by dominant inhibitors identified as mice antigens A3D8 and A1G3 under the control of a locus in human chromosome 11p. This 80 kDa antigen apparently has different epitopes on the same protein molecule. The Lu system appears to be identical to the Auberger blood group. *See* blood groups; epitope.

Lux LUC; *See* luciferase genes of bacteria.

lux Unit of illumination, 1 Lux = 1 lumen/m². *See* lambert.

luxuriance Hybrid vigor displayed by the somatic tissues of reduced fertility or sterile hybrids of different species or genera. *See* hybrid vigor.

luxury genes Not involved in functions indispensable for the viability of a cell and common basic function but only in differentiated, special cells.

Luzula purpurea (Juncaceae, $n = 3$) And other related plant species, have holocentric chromosomes and polyploid as well as an aneuploid series. *See* aneuploid; holocentric; polyploid.

L virus *See* killer strains.

LW blood group Produces the rodent antirhesus antibody. *See* Rh blood group.

LXR Group of liver hormone receptors regulating cholesterol metabolism. *See* cholesterol.

lyase Enzymes catalyze additions or removal of double bonds.

Lychnis Synonyms *Melandrium* and *Silene*, a dioecious plant species.

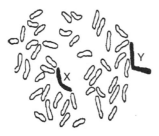

Lychnis male karyotype (Warmke, H. E., 1946. *Am. J. Bot.* 33:648).

lycopene ($C_{40}H_{56}$) Carotenoid pigment. *See* carotenoids.

Lyell syndrome toxic epidermal necrolysis, TEN Adverse reaction to drugs due to lytically active FAS ligand causing apoptosis. *See* FAS.

LYL Lymphoid leukemia oncogene, encoded in human chromosome 19p13.

Lymantria *Lymantria dispar*, $2n = 62$. It has been used for studies on sex determination. The heterogametic female (at left) and the homogametic male (at right) have different colors. The intersex forms may resemble the male more than the female and frequently display light-color sectors. Its English name is gypsy moth. *See* intersex.

Lyme disease (borreliosis) *See Borrelia*; *Ixodoidea*.

lymph Transparent (yellowish or sometimes pinkish) fluid filtered through the capillary blood vessels from blood.

lymphedema, hereditary I (Milroy disease, 5q35.3) Dominant with incomplete penetrance; displays edema generally below the waist. The basic defect is in the epidermal growth factor receptor/receptor tyrosine kinase or VEGFR-3. *See* EGFR; receptor tyrosine kinase; VEGF.

lymphedema distichiasis (FOXC2, Meige disease, 16q24.3) Failure of the lymph nodes to drain properly; eyelashes are in double and turn against the eyeball. Translocations involving site 16q24 also cause the disease. Another lymphedema (Milroy disease) involving the VEGFR-3 gene was mapped to chromosome 5q35.3. *See* distichiasis VEGF; eyelashes, long., (Finegold, D. N., et al. 2001. *Hum. Mol. Genet.* 10:1185.)

lymph nodes Small (1 to 25 mm) nodules in the body where lymphatic vessels, lymphocytes, and antigen-presenting cells accumulate. They have defensive roles by removing toxic and infectious agents. *See* antigen-presenting cell; Hodgkin disease; lymph; lymphocytes; lymphoma; T cell; T-cell receptor; thymus.

lymphoblast Lymphocyte progenitor, which after enlargement divides about four times daily and eventually forms a clone of about 1,000 lymphocytes. *See* lymphocytes.

lymphoblastic lymphoma Leukemia, a neoplasia of the lymph-producing cells (such as Hodgkin's disease). Its basis is a DNA-ligase deficiency. *See* Hodgkin's disease.

lymphochips Microarrays facilitating the recognition of genes that are turned on/off in the process of the development of the immune reaction. *See* microarray hybridization.

lymphocyte Cell produced in the fetal hepatocytes and in adults by bone marrow; may differentiate into T lymphocytes and B lymphocytes. In the thymus, the T-cell receptor immunoglobulins undergo rearrangement and the terminal nucleotidyl transferase generates additional diversity. The T lymphocytes are involved in cell-mediated immunity; the B cells are responsible for the humoral antibodies. During the development of B cells, the secreted and membrane-bound immunoglobulin (Ig) undergoes a series of events. The B-cell antigen receptor (BCR) is formed by complexing a membrane Ig with the heterodimer of Ig-α and Ig-β (*pro-B stage*). The complex now contains an extracellular Ig-like domain, a transmembrane domain, and a short cytoplasmic domain. The signal transducing part of the receptor contains a two-tyrosine motif in the α- and β- cytoplasmic domains. When the hematopoietic stem cell of the bone marrow becomes committed to B-cell development, the Ig gene undergoes a

Lymantria larva and imagos.

series of rearrangements. In the first step, a D_H (diversity heavy chain; *see* immunoglobulins) joins a J_H (juncture heavy chain) gene. One V_H (variable heavy) is attached to the D_H J_H segment in order to form a functional heavy chain. Under normal conditions, a functional μ-heavy chain is expressed (pre-BI stage). This transition from *pro-B* to *pre-BI* constitutes an important checkpoint, contingent on the formation of Ig-β.

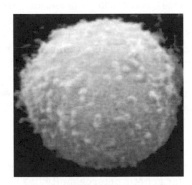

T lymphocyte.

The next step (*pre-BII*) is reached when temporary light-chain immunoglobulin attaches to the VpreBI globulin. Ig-κ chain undergoes a rearrangement and it joins a V-J$_\kappa$ (variable kappa light-chain gene), and the development thus arrives to the *immature B stage*. The λ5-VpreB chain is replaced by a κ-μ-BCR (B-cell receptor) complex, and the so-formed *mature B cell* emerges from the bone marrow in the presence of a functional Ig-α cytoplasmic domain. This B cell is then activated when it comes in contact with an antigen and secretes Ig. This requires the Ig-α cytoplasmic domain and the expression of a cytoplasmic (Btk) tyrosine kinase. For T-cell-dependent antigen response, Ig-α and Btk are not needed.

The tumor-infiltrating lymphocytes have potential use in cancer therapy. The differentiation and maintenance of the lymphocytes requires interleukin-7 (IL-7) and its receptor (IL-7R). The transcription factor Ikaros (a Zn-finger protein) is specific for lymphoid cell differentiation. This protein has binding sites in several other genes involved in lymphocyte development such as CD3-T-cell receptor, CD4, CD2, terminal nucleotidyl transferase, interleukin-2Rα, and NF-κB. In young adult mice, $2-4 \times 10^7$ new T cells are produced daily by 100 to 200 million thymocytes, but only a couple of percent of them enter the T-cell pool. The rest are disposed of by apoptosis. The survivors differentiate into CD4$^+$ CD8$^+$ T cells and subsequently to single-positive CD4$^+$ MHC class II—restricted and CD8$^+$ MHC class I—restricted T cells in the peripheral T-cell pool. Selective elimination of the inadequately activated T cell continues. The retained T cells undergo IL-2—stimulated clonal expansion after antigenic challenge. After the cease of the challenge, some T cells die and some others become memory T cells. *See* antibody; basophili; B cells; CD3; CD4; CD8; dendritic cell; eosinophil; germinal centers; immune system; immunoglobulins; lymphocyte homing; macrophage; memory, immunological; monocyte; neutrophil; NF-κB; T cells; TCR; T$_H$; TIL. (Israels, L. G. & Israels, E. D. 1999. *Stem Cells* 17:180.)

lymphocyte, natural selection of *See* lymphocyte, positive selection of.

lymphocyte degeneracy Ability of T cells to recognize and respond to different peptides (epitopes) carried by different MHC complexes despite the renowned specificity for molecules. Some degeneracy may also exist with B cells, which have Ig's of low affinity. *See* epitope; Ig; major histocompatibility complex; T cell.

lymphocyte homing The immune processes are systematically distributed through movement of lymphocytes from their origin (bone marrow for B cells and thymus and bone marrow for T cells) to lymph nodes, glands, tonsils, and spleen. In these organs microbial antigens are deposited and the naive B and T cells are exposed to them. That propels their differentiation into memory cells and effector cells, respectively. These differentiated cells then home on the bone marrow (B cells) or to extralymphoid effector sites (T cells). These processes are complicated and may go through multiple and reversible substeps. The passage from the blood vessels requires an interaction of the lymphocyte receptors with vascular ligands, activation by G-protein-linked receptors, and finally passage through the vessel wall (diapedesis). After stepping out and entering special tissues, they home on, e.g., B-cell follicles or T zones, during memory cell formation in germinal centers. They also seek sites of inflammation. In each of these niches, they are subject to various regulatory forces by attaching, e.g., to TGFβ-regulated integrin that directs them to intraepithelial sites by attachment to cadherin. They exchange signals with antigen-presenting cells, surface receptors, chemoattractant receptors, immunoglobulins, antigen receptors, chemokines, etc. Probably the RHO homologue of the RAS oncogene plays a central role in the signaling paths and the antigens also affect the trafficking. There is some sort of competition and orderly assignment of the various lymphocytes to specific sites. *See* antigen; antigen-presenting cell; cadherin; chemokines; germinal center; immunological; integrin; ligand; memory, immunoglobulins; receptors; RHO; signal transduction; TGFβ. (Wiedle, G., et al. 2001. *Crit. Rev. Clin. Lab. Sci.* 38:1.)

lymphoedema Dominant human chromosome 5q mutation in VEGFR-3 disables normal lymphatic functions. *See* vascular endothelial growth factor.

lymphoid organs The primary ones are the thymus and bone marrow; the secondary ones are the lymph nodes, spleen, tonsils, mucosal lymphoid tissues, and Peyer's patches. Antigens entering the body are ferried to these secondary lymphoid organs. After prospecting, B and T lymphocytes detect the presence of the antigens at these sites and antigen-specific B and T cells move in on the cues represented by chemokines. *See* chemokines.

lymphokines Proteins secreted by lymphocytes in response to specific antigens. They are involved in multiple ways in serological defense mechanisms of the body and in differentiation. Interferons (IF), interleukins (IL), granulocyte colony-stimulating factors (G-CSF), granulocyte-macrophage colony-stimulating factors (GM-CSF), macrophage activity factor (MAF or IFN gamma), T-cell replicating factor (TRF), migration inhibition factor (MIF), and tumor necrosis factor (TNF) belong to this large group. *See* chemokines; cytokines; cytotoxic T cells; immune system; interferons; interleukins;

rel; tumor necrosis factor. (Webb, D. R., et al. 1988. *Molecular Basis of Lymphokine Action.* Humana, Totowa, NJ.)

lymphoma Cancer of the lymphocytes, neoplasia of lymphoid tissues. In follicular B-cell lymphoma because of a chromosome 18 translocation, BCL-2 gene expression is aberrant and as a consequence apoptosis is reduced. *See* anaplastic lymphoma; Burkitt's lymphoma; Chédiak-Higashi syndrome; diffuse large B-cell lymphoma; Duncan syndrome; Hodgkin disease; leukemia; lymphoblastic lymphoma; MALT; Mantle cell lymphoma; non-Hodgkin lymphoma; translin. (Staudt, L. M. 2002. *Annu. Rev. Med.* 53:303.)

lymphopenia Recessive deficiency of lymphocytes (cytokines), an autoimmune anemia. *See* autoimmune disease; cytokines; immunodeficiency; lymphocyte.

lymphopoiesis (development of lymphocytes) Is a continuous process beginning at the hematopoietic stem cells. (See Gounari, F., et al. 2002. *Nature Immunol.* 3:489.)

lymphoproliferative diseases, X-linked (immunoproliferative disease, Xp25, XLP) A protein domain SH2D1A defect may be involved. This protein is also called SAP (for association with SLAM [signaling lymphocyte activation molecule, 1q23]) and inhibits SLAM signaling. *See* Canale-Smith syndrome; Duncan syndrome; Epstein-Barr virus; FAS; SAP; SH2; SLAM. (Morra, M., et al. 2001. *Annu. Rev. Immunol.* 19:657; Czar, M. J., et al. 2001. *Proc. Natl. Acad. Sci. USA* 98:7449.)

lymphotactin C-type cytokine, i.e., it lacks two of the four cystin residues characteristic for chemokines. It is expressed in $CD4^+$ T cells and $CD4^-$ and $CD8^-$ T-cell receptor $\alpha\beta^+$ thymocytes. It has no chemotactic activity for monocytes and neutrophils. *See* blood; chemokines; cytokines; lymphocytes; monocytes; T cells. (Ju, D. W., et al. 2000. *Gene Ther.* 7[4]:329; Cerdan, C., et al. 2001. *Blood* 97:2205.)

lymphotoxins Forms of lymphokines of the tumor necrosis factor family that cause the lysis of some cells, e.g., cultured fibroblasts, and are required for growth, differentiation, and regulation of the immune system. They are encoded within the human HLA gene cluster. *See* HLA; immune system; lymphokines; TNF. (Yin, L., et al. *Science* 291:2162.)

LYN Tyrosine protein kinase of the SRC family involved in the differentiation of B lymphocytes. In the nervous system, Lyn is associated with the AMPA (α-amino-3-hydroxy-5-methyl-4-isoxazole propionate) receptor. AMPA is a ligand-gated cation channel that mediates the fast component of the excitatory postsynaptic currents and it is also a cell-surface signal transducer. Lyn is activated by Ca^{2+} and Na^+ influx, and Lyn activates the MAPK signal pathway. *See* B lymphocyte receptor; BTK; Csk; ion channels; signal transduction; SRC. (Luciano, F., et al. 2001. *Oncogene* 20:4935.)

Lynch cancer families Groups of dominant genes responsible for the development of certain cancer syndromes such as endometrial cancer, adenocarcinoma of the colon, and others. The instability may be caused by defects of the genetic repair system. *See* colorectal cancer, Li-Fraumeni syndrome.

Lyon hypothesis In mammalian cells with more than one X chromosome, usually all but one are heteropycnotic (highly condensed and thus dark-stained at all stages) and form $n - 1$ Barr bodies. Heteropycnosis may not equally affect the entire length of the chromosome. Mary Lyon, British geneticist, suggested that the frequently observed mosaicism in these animals is the phenotypic consequence of the chromosomal behavior. The heteropycnotic chromosomes carry their genes in an inactivated state (possibly key nucleotides are methylated). Depending on which allele of a heterozygote is expressed, sectorial mosaicism is displayed. In order to carry out the inactivating switch, the X chromosome must have an intact *inactivation center*. During development, the X chromosomes may switch between their inactive and active states. These decisions are not made entirely by the X chromosomes because in human triploids with XXY constitution both X chromosomes remain active. A typical example is the tortoise-shell cat color that is observed almost exclusively in females heterozygous for yellow-black fur color. The exceptional (less than 1/500) tortoise-shell male cats are of XXY constitution (Klinefelter syndrome). Similar genetic heterogeneity has been observed in heterozygous females with two classes of lymphocytes: one with normal and the other without testosterone-binding capacity. The LINE (L1) elements appear to regulate the inactivation of the mammalian X chromosome (Bailey, J. A., et al. 2000. *Proc. Natl. Acad. Sci. USA* 97:6634) and may spread the lyonization along the X chromosome or even to translocated autosomal segments fused to the X chromosome. *See* Barr body; dosage effect; heteropycnosis; LINE; lyonization; methylation of DNA; regulation of gene activity. (McBurney, M. W. 1988. *Bioassays* 9:85.)

lyonization Variegation in mammalian females as predicted by the Lyon hypothesis. In an XX mammalian female, one of the X chromosomes remains in a condensed state, replicates its DNA asynchronously, and its genes are not transcribed after the blastocyst stage: 3.5–4.5 dpc in the trophectoderm and 5.5–6.5 dpc in the embryo cell initials of the female mouse. In the germline, the inactive X is reactivated at the beginning of meiosis (12.5–13.5 dpc; the average time of gestation in the mouse is 19 days). In the male, X-chromosomal inactivation is limited to the duration of meiosis presumably to restrict deleterious recombination with the Y chromosome. After fertilization and before implantation, inactivation recurs in the female. The other X chromosome displays a more open structure and its gene content is expressed. The majority of the genes in the inactive X replicate late, but those that escape inactivation (>20) replicate synchronously with the active X chromosome.

One of the X chromosomes is selected for inactivation by a chromosomal locus, *Xic* (X-chromosome inactivating center, Xq13). This *Xic* locus (450 kb) is responsible for *counting* (encoded within the 6th exon) the number of X chromosomes to be inactivated. In case of more than 2 X chromosomes, only one remains active normally. In addition, the *Xic* must be in cis-position to the genes to come under its influence (*spreading effect*). Thus, *Xic* also selects the chromosome to be inactivated (*choice function*). If the *XIC* locus is lost after inactivation has been initiated, the inactive state is maintained.

There is another functional site within the *XIC* locus, *Xce* (*X control element*). *Xce/Xce* homozygotes undergo normal, random inactivation. In *Xce* heterozygotes there is a higher chance that the chromosome remains active if it carries a strong *Xce* allele. The *Xce* locus seems to control nonrandom (skewed) inactivation. In monozygotic human twins, skewed distribution (nonconcordance) of inactivation was observed. Skewing may also be due to selection against mutant alleles or chromosomal abnormalities. The product of *XIC* is XIST (X inactive specific transcript) transcribed from the 15 kb site within XIC in only the inactive X chromosome after inactivation has taken place. In *XIST/Xist* heterozygotes, there is a large RNA transcript (ca. 15 kb) in the nucleus, apparently associated with the inactive X, and it cannot be translated. Thus, it appears that RNA mediates the regulation. During early embryogenesis, *Xist* expression precedes the inactivation of the X chromosome, and in XY males it is expressed only in the male germ tissues and only before meiosis.

A protein (not translated from the XIST transcript) is preferentially associated with the inactive X chromosome. This protein, mH2A (macroH2A), bears resemblance in its N-terminal third to H2A histone, but the rest of the molecule is different from the histones. The mH2A protein in females accumulates in dense *macrochromatin bodies* in proportion to the number of inactive X chromosomes. In marsupials, the inactive X chromosome is the paternally transmitted one. Similar observations were also made in the extraembryonic endoderm cells of some rodents and to some extent in humans. About 20 human X-chromosomal short-arm genes (especially in the vicinity of the pseudoautosomal region) escape inactivation in a discontinuous manner, e.g., Xg blood group, the closely linked steroid sulfatase locus (ichthyosis), and the tissue inhibitor of metalloproteinases (TIMP1). This region of the human X chromosome appears to be a more recent evolutionary addition (XAR) to the ancestral X-conserved region (XCR) and the coordination between the two has not been completed yet. (The marsupials do not have the XAR.) At the tip of the long arm of the human X, there is a very short pseudoautosomal stretch that carries one escapee and one inactivated gene. In the inactivation, differential methylation between the active X and the inactive X chromosome is implicated, although the overall methylation appears to be the same. In the inactive X chromosome, CpG islands in the 5' region are mostly methylated, whereas in the active ones they are not. The inactive X-chromosomal regions are hypoacetylated at

lysine residues of the core histones, whereas genes escaping inactivation are hyperacetylated.

Aging may reactivate some genes. Microarray hybridization applied to cloned X-chromosomal fragments representing 1,317 genes indicated that rather than being silenced, 53 genes displayed elevated activity, in the presence of multiple X chromosomes and many just escaped inactivation (Sudbrak, R., et al. 2001. *Hum. Mol. Genet.* 10:77). X-chromosomal inactivation is somehow related to imprinting. A deleted *Xist* allele in the maternal X chromosome does not affect the viability even of the adults, but the *Xist* allele deleted from the paternal X chromosome prevents survival of the embryo after implantation. In the mouse, the presence of two maternal X chromosomes (maternal disomy) is very detrimental to development of the embryo. In contrast, in humans maternal X disomy is not detrimental or even trisomy (XXX) or Klinefelter syndrome (XXY), irrespective of the origin of the X's, is of the same viability. Although X-chromosomal inactivation occurs in all mammals, the pattern of inactivation varies. The X chromosome of the mouse is acrocentric and it is not covered with Xist RNA, whereas the human X is submetacentric and has Xist. Nevertheless, the mouse and human genes display great homology. Microarray technologies may provide further insight into the factors subjected to or involved in inactivation/reactivation. *See* Barr body; dosage compensation; dosage effect; dpc; ectodermal dysplasia, anhidrotic; histones; ichthyosis; imprinting; Lyon hypothesis; metalloproteinases; methylation of DNA; pseudoautosomal; Tsix; Wiskott-Aldrich syndrome; Xg; *Xic*; *Xist*. (Heard, E., et al. 1997. *Annu. Rev. Genet.* 31:571; Avner, P. & Heard, E. 2001. *Nature Rev. Genet.* 2:59; Maxfield Boumil, R. & Lee, J. T. 2001. *Hum. Mol. Genet.* 10:2225.)

lyophilization *See* freeze drying.

lysate When a cell is lysed (dissolved), its contents are released; the lysis of bacterial cells may produce bacteriophage particles as lysate. *See* lysogeny.

Lysenkoism T. D. Lysenko (1898–1976) presided over the greatest scandal in cultural history. Beginning in the 1930s and with the culmination of his almost complete victory in 1948, he destroyed genetics and most of biology in the Soviet Union. His group of charlatans, misled ignorants, psychopaths, and common criminals brought havoc to biology in the Soviet Union and other countries in her political interest sphere. The Lysenkoists rejected Mendelian inheritance, cytogenetics, biochemical genetics, and statistics as "capitalist fraud." The movement traced its origin to Mitchurin, an uneducated railwayman who successfully practiced empirical plant breeding and attempted to interpret his observations without being familiar with scientific principles and facts. Lysenkoists denied the existence of genes, the role of "hard heredity," and instead claimed the inheritance of the phenotype and continuous change of heredity under the influence of nutrition, temperature, day length, and other environmental conditions. This was a revival of the ancient myths of pangenesis, lamarckism, inheritance of acquired characters, directed genetic change by agrotechnical methods (fertilizers, irrigation, etc.), grafting, vernalization (yarowisation), blood transfusion, etc. Although these ideas alone were disgraceful

Lyonization phenotype in rodents. (From Lyon, M. F. 1963. *Genet. Res.* 4:93.)

and appalling, the major problem was that Lysenkoism became a state political doctrine, imposed on and substituted for all scientific activities in biology, totally subjugating agriculture and making inroads even into medicine. Marx and Engels were supporters of lamarckism in the hope that the future of the human race might be improved by these doctrines. Stalin, the Soviet dictator, embraced the Lysenkoist ideas because they promised instant increased agricultural productivity by the application of vernalization, summer culture of potatoes, species transformation, graft hybridization, supplementary pollination, etc., at a period of famines caused by the senseless social experimentation of the Soviet bureaucracy. The promised results failed to come true and agricultural productivity further declined partly because of the application of Lysenkoism. The failures were then attributed to sabotage by the scientists and technicians. Therefore, after 1948, Stalin granted virtually unlimited power to Lysenko to purge the universities and research institutes from his and the "people's enemies." Thousands of scientists actually lost their lives either by execution or imprisonment. Geneticists who had enjoyed reasonably good support before 1948 and worldwide renowned scientists of the Soviet Union were persecuted, forced underground, or physically eliminated. Resurrection of genetics in Russia could come only after Stalin's death. *See* acquired characters; directed mutation; hard heredity; lamarckism; pangenesis; Soviet genetics; vernalization. (Medvedev, Z. A. 1969. *Rise and Fall of T. D. Lysenko*. Columbia Univ. Press, New York.)

lysergic acid   Psychomimetic alkaloid derivative. *See* ergot; psychomimetic.

lysidine   Cytidine with a lysine residue qualifying it for recognition of the AUA codon by tRNA. *See* aminoacylation; cytidine; genetic code; tRNA.

lysin   Complement-dependent lytic protein such as hemolysin, bacteriolysin, and the sperm lysin, which creates a hole in the egg envelope for penetration. Mutation in lysin may lead to reproductive incompatibility and speciation. *See* antibody; complement; speciation.

lysine biosynthesis ($NH_2[CH_2]4CH[NH_2]COOH$)   Some fungi and algae synthesize lysine from α-aminoadipic acid ($HO_2CCH[NH_2]CH_2CH_2CH_2COOH$). Bacteria, some fungi, and higher plants make lysine from ($HOOCCH(NH_2)(CH_2)_3$-$CH(NH_2)COOH$), diaminopimelic acid. Asparate semialdehyde is also a precursor of lysine. For mammals it is an essential amino acid; they rely on their diet to meet their need. Some food or feed may have amino acid levels that are too low to meet the requirements, then human malnutrition or low weight gain in animals results. Some mutant plants, e.g., high-lysine maize, may be used to avoid the nutritional problems. The degradation of lysine may take place in a number of different ways, and a common intermediate is α-ketoglutarate. *See* high-lysine corn; hyperlysinemia; kwashiorkor; lysine intolerance; lysine malabsorbtion. (Galili, G. 2002. *Annu. Rev. Plant Biol.* 53:27.)

lysine intolerance (lysinurin protein intolerance, LPI)   Recessive human disorder (14q11.2) resulting in periodic vomiting, growth retardation, muscle hypotonia, hepatosplenomegaly (liver and spleen enlargement), and possibly coma based on a defect in the membrane transporter (y^+LAT-1, also named SLC [solute carrier]) of basic (cationic) amino acids. *See* dibasic aminoaciduria; lysine biosynthesis.

lysine malabsorption   Causes excessive amounts of lysine in the urine and low levels in the serum due to a recessive autosomal mutation. *See* lysine biosynthesis.

lysinuria (lysinuric protein)   *See* lysine intolerance.

lysis   Disruption of the (bacterial) cell (before the release of a virus or plasmid from a cell) by spontaneous or any other means such as exposure to lysozyme, by boiling or alkaline treatment, or the dissolution of eukaryotic cells or cellular organelles by lysozyme or detergents. *See* lysis inhibition; rapid lysis; photo next page.

lysis from without   When too many phages (>20) attack a single bacterial cell, the bacterium may disintegrate by the perforations suffered.

lysis inhibition (LIN)   In case of shortage of bacterial cells (that the phage senses by some means), it postpones lysis for several hours and produces the maximal possible number of virions from that cell (e.g., 400 copies of T4). Lysis is controlled by the *r* genes of the phage. *See* rapid lysis.)

lysogen   Bacterium with phage integrated into its genome. In such a state the replication of the phage DNA is under the control of the host and the phage's own replication system is repressed (stable lysogen). Rarely, the suppression of the replication of the phage fails after insertion (abortive lysogeny) and the prophage cannot be replicated; it is lost by dilution during subsequent divisions of the host. *See* lambda phage. (McAdams, H. H. & Shapiro, L. 1995. *Science* 269:649; Kihara, A., et al. 2001. *J. Biol. Chem.* 276:13695.)

lysogenic bacterium   Can harbor temperate phage in an integrated state in the chrroromosome; after induction, phage particles can be formed and released. *See* lysis; prophage.

lysogenic immunity   *See* immunity in phage.

lysogenic repressor   Prevents the lysis of a lysogen. *See* lamda phage.

lysogeny   The phage coexists with the bacterium as a prophage. *See* symbionts, hereditary. (Bertani, G. 1958. *Advances Virus Res.* 5:151.)

lysophosphatidic acid (oleoyl-sn-glycero-3-phosphate)   Derived from L-α-phosphatidic acid dioleyl by the action of phospholipase A. It stimulates smooth muscle contraction and may affect blood pressure, cell adhesion, mitogenesis, etc. *See* endophilin; Pyk-2. (Hla, T., et al. 2001. *Science* 294:1875.)

lysopin   Octopine-type opine. *See* opines.

lysosomal storage diseases   Involve defects in lysosomal hydrolase enzymes (~50) and accumulation of the products of

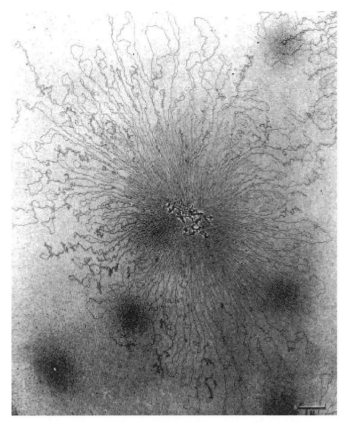

Electron micrograph of the folded chromosome of *E. coli* bacterium lysed and spread out on 0.4 m salt. (From Kavenoff, R. & Bowen, B. C. 1977. *Stadler Symp.* 8:159. Courtesy of Dr. Kavenoff.)

these hydrolases, which affect various organs in many ways. The cumulative prevalence may be 5×10^{-4}. Glycosphingolipid storage diseases are Tay-Sachs disease and Gaucher disease. According to a mouse model, N-butyldeoxynojirimycin may hinder the accumulation of the substrate for β-hexoseaminidase A and may alleviate the symptoms of Tay-Sachs disease. Enzyme replacement or modified autologous bone marrow transplantation have been attempted in some types of these diseases. Gene transfer into hematopoietic cells or into the central nervous system using herpes or adenovirus vectors is being studied. Some mucopolysaccharidoses may be treated by implantation of neoorgans, tissues of skin fibroblasts, secreting β-glucuronidase or α-L-iduronidase. *See* Coffin-Lowry syndrome; Fabry's disease; Farber's disease; gangliosidoses; Gaucher's disease; glucuronic acid; glutamyl ribose-5-phosphate glyco-proteinosis; glycoprotein storage disease; Hermansky-Pudlak syndrome; Hunter syndrome; Hurler's syndrome; iduronic acid; Kanzaki disease; Krabbe's leukodystrophy; lipase deficiency; lipidoses; lysosomes; mannosidosis; Maroteaux-Lamy syndrome; metachromatic leukodystrophy; miniorgan; min-iorgan therapy; mucolipidoses; mucopolysaccharidoses; neuroaminidase deficiency; neuromuscular disease; Sanfilippo syndrome; Schindler disease; sialidosis; Tay-Sachs disease; viral vectors.

lysosome   Cytoplasmic organelle (in eukaryotes) containing hydrolytic enzymes. The digestion is either by a phagocytosis-type mechanism or by selective ingestion and lysis of molecules. The latter process requires Mg-ATP, which are

molecular chaperones (heat-shock proteins 73 and 70). The receptor for the uptake may be a 96 kDa (Lamp2) lysosomal membrane protein. Clathrin plays a role in targeting proteins to the lysosomes by its KFERQ amino acid sequence motif. Within the lysosomes, in an acidic environment, inter- and intrachain disulfide bonds of the proteins are destroyed by the enzyme thioredoxin. *See* clathrin; endocytosis; LDL receptor; lysosomal storage diseases; lysozymes; phagocytosis; thioredoxin; Wolman disease. (de Duve, C. & Wattiaux, R. 1996. *Annu. Rev. Physiol.* 28:435; Dell'Angelica, E. C. & Payne, G. S. 2001. *Cell* 106:395.)

lysozyme   Enzyme capable of digesting bacterial cell walls. Lysozymes have an important role in phage infection and liberation. *See* lysosome.

LYST   *See* Chediak-Higashi syndrome.

lytic   Capable of initiating and performing lysis.

lytic cycle   Life cycle of the phage involving infection, growth (one-step growth), and lysis of the host to liberate the infectious phage particles. *See* lambda phage; one-step growth.

lytic infection   *See* lysis of bacteria.

LYT oncogene   Associated with lymphoblastic leukemias and non-Hodgkin-type lymphomas when the gene (in human chromosome 10q24) is translocated to chromosome 14q11 or 14q32. Genes neighboring the 14q32 area (TNG1, TNG2) activate T-cell leukemia genes TCL1 and TCL2 located at 14q23.1. The normal product of LYT is similar to that of NFKB (NF-κB) products, which are transcription factors. *See* Hodgkin's disease; leukemias; NFKB; REL oncogene.

Erwin Chargaff, the discoverer of Chargaff Rule in 1949, six years before the breaking of the DNA code, made the following statement (*Istituto Lombardo (Rend. Sc.)* 89:101). It must be remembered that his Rule was one of the cornerstones for the construction of the Watson & Crick model of the double helix.

"I believe, however, that while the nucleic acids, owing to the enormous number of possible sequential isomers, could contain enough codescripts to provide a universe with information, attempts to break the communications code of the cell are doomed to failure at the present very incomplete stage of our knowledge. Unless we are able to separate and to discriminate, we may find ourselves in the position of a man who taps all the wires of a telephone system simultaneously. It is moreover, my impression that the present search for templates, in its extreme mechanomorphism, may well look childish in the future and that it may be wrong to consider the mechanisms through which inheritable characteristics are transmitted or those through which the cell repeats itself as proceeding in one direction only."

A historical vignette.

M

M Abbreviation for mitosis, or in statistics for the mean. *See* mean; mitosis.

Ma (*mille mille annus*) Million years. *See* My.

MAb Monoclonal antibody.

Mab Mouse antibody.

MAC Mammalian artificial chromosome. *See* BAC; HAC; HAEC; PAC; SATAC; YAC. (Grimes, B. & Cooke, H. 1998. *Hum. Mol. Genet.* 7:1635.)

MAC-1 Oligonucleotide-binding heparin-like integrin on the surface of neutrophils, macrophages, and killer cells (NK). *See* blood; heparin; integrin; killer cell.

Macaca Rhesus Old World monkeys. *See* Cercopithecidae; cynomolgus; Rhesus.

macadamia nut (*Macadamia* spp) Delicious nut, $2n = 2x = 28$.

macaloid Clay colloid capable of adsorbing ribonuclease during disruption of cells; can be centrifugally removed from the RNA preparation after extraction by phenol. *See* RNA extraction.

macaroni wheat (hard wheat) *Triticum durum* ($2n = 4x = 28$) varieties containing the AB genome and taxonomically classified as a subgroup of *T. turgidum*. The endosperm of the commercially used varieties has more β-carotenes and therefore the milled products show yellowish color that is desirable for pastas. It is milled to a grainier product (semolina) that favors cooking quality. The durum wheats have higher protein and mineral contents than the soft wheats. The turgidum wheats, except the durums, are not used for food. *See Triticum.*

Macaroni wheat.

macerozyme Pectinase enzyme of fungal origin; it is used for the preparation of plant protoplasts in connection with a cellulase. *See* cellulase; protoplast.

MACH (caspase 8) *See* caspases.

Machado-Joseph syndrome (Azorean neurologic disease, spinocerebellar ataxia type 3, MJD) 14q32.1 chromosome-dominant defect of the central nervous system involving ataxia and other anomalies of motor control. An abnormally increased number of CAG repeats (from 13–36 in normal to 61–84 in disease) occurs in the gene locus. If the long CAG repeats are translated into polyglutamine, cell death may result. The neurodegeneration may be suppressed by the chaperone HSP70. *See* ataxia; fragile sites; muscular atrophy; spinocerebellar ataxia; trinucleotide repeats.

machine learning Study of computer algorithms in the interest of improving the study of scientific data. (Mjolsness, E. & DeCoste, D. 2001. *Science* 293:2051.)

macroarray analysis Mutations of unknown function are grown in the presence of compounds that make them sensitive if they are defective, e.g., cell wall–binding calcofluor dyes (fluorescent brightener 28/Tinopal) detect a group of yeast mutations defective in cell wall biogenesis. Mutations in several different gene loci involved in microtubule functions are sensitive to benomyl/benlate ($C_{14}H_{18}N_4O_3$, a fungicide and ascaricide). Genetic defects in sugar utilization may not grow on the nonfermentable glycerol if the mutation affected a certain domain of the protein product. Thus, this type of analysis can find NORFs involved in certain pathways. *See* genomics; microarray hybridization; NORF; synexpression.

macroconidium Large multinucleate conidium. *See* conidia; fungal life cycle; microconidia.

macrochromatin *See* lyonization.

macroevolution Major genomic alteration that produced the taxonomic categories above the species level. *See* evolution; microevolution.

macrogametophyte Megaspore. *See* gametogenesis in plants.

macroglobulinemia (Waldenström syndrome) HLA-linked immunodeficiency resulting in an increase in IgM in the blood, thrombosis, skin, nose, and gastrointestinal bleeding. *See* immunodeficiency; immunoglobulins; multiple myeloma.

macrolesion In genetic toxicology, visible alterations in the chromosomes. *See* Gene-Tox.

macrolide Type of antibiotic (including more than one keto and hydroxyl group such as erythromycin, troleandomycin) associated with glycoses. *Streptomyces* bacteria produce these compounds that inhibit the 70S ribosomes by binding to the large (50S) subunit and interfering with peptidyl transferase. *See* antibiotics; glycoses; maytansinoids; protein synthesis; ribosome; *Streptomyces*. (Retsema, J. & Fu, W. 2001. *Int. J. Antimicrob. Agents* 18 Suppl. 1:3.)

macromelic dwarfism (Desbuquois syndrome) Skeletal, digital anomaly.

macromere   Large blastomere. *See* blastomeres.

macromolecule   Molecule with molecular weight of several thousands to several millions such as DNA, RNA, protein, and other polymers.

macromutation   Genetic alteration involving large, discrete phenotypic change.

macronucleus   Larger type of polyploid nucleus in protozoa. While inheritance is mediated through the small micronucleus, the macronucleus is directing metabolic functions by being transcriptionally active, and the latter is responsible for cell phenotype. After the internally eliminated sequences (IES) are removed from the micronuclear DNA, the leftover tracts are joined into macronuclear-destined sequences (MDS) to form the macronucleus. The macronuclear genome is derived at sexual reorganization during conjugation. This involves various chromosomal rearrangements. Macronuclear differentiation includes site-specific fragmentation of the micronuclear chromosomes. A 15 bp site that is required for fragmentation marks each fragmentation site. The two ends of the fragments form telomeres and thus become *autonomously replicating pieces* (ARP). The macronucleus contains about 45 copies of the ~300 subchromosomal fragments except the rDNA, which is amplified by ca. 10,000 times. The macronucleus divides by fission rather than by mitosis, yet recombination between and within genes occurs. In the absence of recombination, groups of genes may stay together. This phenomenon is called *coassortment*. The genes of these polyploid macronuclei may sort out to the pure form of the initial heterozygous state by a process of *phenotypic assortment*. On the basis of the assortments, the genes may be mapped genetically. The macronuclei are formed by the fusion of two diploid micronuclei of the zygotes. *See* chromosome diminution; IES; *Paramecium*; *Tetrahymena*. (Katz, L. A. 2001. *Int. J. Syst. Evol. Microbiol.* 51[pt4]:1587.)

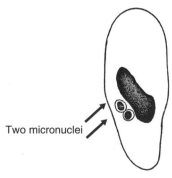

Paramecium macronucleus.

Two micronuclei

macronutrient   Required in relatively large quantities by cells.

macroorchidism   Condition of larger than normal testes. *See* fragile X.

macrophage (Mφ)   Phagocytic cell of mammals with accessory role in immunity. Macrophage cells are produced by the bone marrow stem cells as monocytes that enter the bloodstream. Within 2 days of circulation, they enter various tissues of the body and develop into macrophages. The macrophages may differ in shape, but generally they have single, large, round or indented nuclei, extended Golgi apparatus, lysosomes, and vacuoles for the ample storage of digestive enzymes. The cell membrane forms microvilli of various sizes suited for phagocytosis (engulfing particles) and pinocytosis (uptake of fluid droplets); thus, they are important as part of the cellular defense mechanism against foreign antigens, including tumor antigens. The macrophages have a wide array of receptors, including Fc, complement, carbohydrate, chemotactic peptide, and extracellular matrix receptors. Some macrophages phagocytize apoptotic cells. *See* antibody; apoptosis; granulocytes; immune system; phagocytosis.

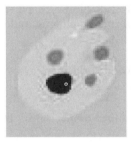

Macrophage ingesting bacteria.

macrophage activity factor (MAF)   Lymphokine that is identical to IFN-gamma. *See* interferon; lymphokine.

macrophage colony-stimulating factor (M-CSF, 5q33.2-q333)   Its receptor is the KIT oncogene. The receptor MCSFR is a protein tyrosine kinase. It promotes B-cell proliferation with the assistance of IL-7. *See* B-cell; colony-stimulating factor; FMS; G-CSF; GM-CSF; IL-7; KIT oncogene; RAS; signal transduction. (Csar, X. F., et al. 2001. *J. Biol. Chem.* 276:26211.)

macrophage-stimulating protein (MSP)   80 kDa serum protein that stimulates responsiveness to chemoattractants in mice, induces ingestion of complement-coated erythrocytes, and inhibits nitric oxide synthase in endotoxin- or cytokine-stimulated macrophages. MSP is structurally homologous to hepatocyte growth factor–scatter factor (HGF-SF), but its targets are different. MSP has been located to human chromosome 3p21, a site frequently deleted in lung and renal carcinomas. *See* HGF-SF; macrophage; Met; Ron; scatter factor. (Stella, M. C., et al. 2001. *Mol. Biol. Cell* 12:1341.)

macrospore   *See* megaspore.

macrothrombocytopathy   *See* giant platelet syndrome; thrombocytopenia.

macula   Spot or thicker area, particularly on the retina; it is often colored. *See* choroidoretinal degeneration; foveal dystrophy; macular dystrophy; macular degeneration; retina.

macular corneal dystrophy (MCD, 16q22) Involves progressive recessive opacity of the cornea. Two types may be distinguished on the bases of absence (MCDI) and presence (MCDII) of keratan sulfate in the serum. A third type (CHST6) has a defect involving corneal N-acetylglucoseamine-6-sulfotransferase, which has overlapping mutations within the two other types. *See* hypotrichosis; Stargardt disease.

macular degeneration Dominant-acting 6q25 deletion causing degeneration of the vitelline layer of the eye. The dominant 2p16-located diseases Malattia Leventinese and Doyne honeycomb retinal dystrophy are similar to some extent to macular degeneration and show drusen (bright speckles on the retina) due to mutation in epidermal growth factor (EGF)–containing fibrillin-like extracellular matrix protein 1 (EFEMP1). The Stargardt-like STGD3 locus is in human chromosome 6q14. This gene is responsible for ELOVAL4 (elongation of very long-chain fatty acid) and autosomal-dominant macular dystrophy. These diseases manifest with some variations in expression and onset. Age-related macular degeneration (ARMD) is a very common cause of blindness in the West. *See* eye diseases; macula; Stargardt disease.

macular dystrophy Autosomal-recessive (8q24) and X-linked forms with symptoms similar to macular degeneration. The 1.4 Mb vitelline macular dystrophy (VMD2) gene (human chromosome 11q11-q13.1) encodes a 585-amino-acid protein (bestrophin) named after the other name of the condition, Best's disease (autosomal dominant at 11q13). Its prevalence in the U.S. is about 3×10^{-5}. *See* eye diseases; macula; retinal dystrophy; Stargardt disease.

mad (*many abnormal discs*; 3-78.6) Homozygous disc and cell-autonomous lethal, affecting several morphogenetic functions in *Drosophila*.

Mad (*Mothers against decapentaplegic*) *Drosophila* gene controlling several developmental events in the fly. Its human homologue is SMAD1 (Sma in *Caenorhabditis*). SMAD is a TGF (transforming growth factor)/BPM (bone morphogenetic protein) cytokine family-regulated transcription factor involving serine/threonine kinase receptors. The Mad family of proteins can be found in a wide range of organisms. Upon dimerization with MAX, they function as transcriptional repressors. The repressor function is correlated with the recruitment of Sin3 protein binding to the Sin3-Mad interaction domain (Mad-Sid). SMAD is also called hMAD; MADR is its receptor. Mad is a component of the N-CoR/Sin3/RPD complex that mediates the repression of certain classes of genes. *See* cytokines; *decapentaplegic*; *dpp*; Dpp; histone deacetylase; MYC; RPD; serine/threonine kinase; SMAD; tumor growth factor.

MAD Multiwavelength anomalous dispersion analysis is a physical method for the determination of crystal structure of molecules.

MAD2, MAD1 (mitotic arrest deficient, MAD2L1, 4q27; MAD2L2, 1p36) Anaphase-regulating proteins that interact with estrogen receptor-β. They are essential for normal mitosis checkpoints and their defect may lead to chromosome instability, cancer, and embryonic lethality. *See* anaphase; BUB; cell cycle; spindle. (Skoufias, D. A., et al. 2001. *Proc.*

Natl. Acad. Sci. USA 98:4492; Martin-lluesma, S., et al. 2002. *Science.* 297:2267.)

Mad/Max Sequence-specific heterodimeric transcriptional repressors.

mad cow disease *See* encephalopathy.

MADS box Conserved motif of 56 residues in a DNA-binding protein of transcription factors involved in the regulation of *MCM1* (yeast mating type), *AG* (agamous homoeotic gene and a root morphogenesis gene of *Arabidopsis*), *ARG80* (arginine regulator in yeast), *DEF A* (deficient-flower) mutation of *Antirrhinum*, and SRF (serum response factor in mammals), regulating the expression of the c-fos protooncogene. The MADS box has amino-terminal sequence specificity and carboxyl dimerization domains. In addition, SRF recruits accessory proteins such as ELK1, SAP-1, and MCM1 and relies on MATα1 and MATα2, STE12, and SFF. *See* mating-type determination in yeast; MEF. (Jack, T. 2001. *Plant Mol. Biol.* 46:515.)

MAF Macrophage activity factor is a lymphokine, oncogene, and regulator of NF-E2 transcription factor. Heterodimers of Maf protooncogene family members promote the association with and the expression of NF-E2, whereas the homodimers are inhibitors of it. *See* homodimer; lymphokine 4; macrophage; NF-E2; oncogenes; transcription factors. (Swamy, N., et al. 2001. *J. Cell Biochem.* 81:535.)

MAFA (12p12-13) Inhibitory immune receptors on myeloid, mast, and natural killer cells.

magainin *See* antimicrobial peptides.

MAGE Melanoma antigens encoded by several genes and expressed in different tumors. In normal tissue, MAGE is limited to the testes and wound healing. In normal tissues, MART-1/Melan-A differentiation antigen represents the melanoma lineage. The immunogenic epitopes for MAGE-1 are EADPTGHSY and SAYGEPRKL; for MAGE-3: EVDPIGHLY, FLYGPRALV, and MEVDPIGHLY. Each has different HLA specificity. The MART-1 peptide epitopes are AAGIGILTV, ILTVILGVL, and GIGILTVL. *See* amino acid symbols in protein sequences; antigen; melanoma; tumor antigen. (Otte, M., et al. 2001. *Cancer Res.* 61:6682.)

MAGI-2 Membrane-associated guanylate kinase inverted-2 is a scaffold protein enhancer of PTEN. *See* PTEN. (Wu, X., et al. 2000. *Proc. Natl. Acad. Sci. USA* 97:4233; Vazquez, F., et al. 2001. *J. Biol. Chem.* 276:48627.)

magic bullet A specific monoclonal antibody is supposed to recognize only one type of cell-surface antigen (e.g., one on a cancer cell), growth factor receptor, or differentiation antigen. The antibody carries the *Pseudomonas* exotoxin, diphtheria toxin, ricin, maytansinoids (an extract of tropical shrubs or trees), or a radioactive element (such as Y^{90}) capable of selectively unloading these harmful agents at the target cell and thus killing the cancer cell without much harm to other cells. *See* ADEPT; antigen; diphtheria toxin; hybridoma; immunotoxin; isotopes; maytansinoid; monoclonal antibody;

receptor; ricin; vascular targeting. (Frankel, A. E., et al. 2000. *Clinical Cancer Rev.* 6:326.)

magic number The 64 triplet (4^3) combinations of the 4 nucleotides specify the 20 natural amino acids in all organisms. (Gamov, G. & Yčas, M. 1955. *Proc. Natl. Acad. Sci. USA* 41:1011.)

magic spot pppGpp and ppGpp nucleotides that serve as effectors of the stringent control. *See* stringent control. (Schattenkerk, C., et al. 1985. *Nucleic Acids Res.* 13:3635.)

magnetic relaxation switches Nanometer-size colloidal metal particles may be coupled to affinity ligands and used as chemical sensors. Highly uniform magnetic nanoparticles can be covalently and stoichiometrically attached to oligonucleotides, nucleic acids, peptides, proteins, receptor ligands and antibodies. These nanomagnetic probes can be assembled also into larger units. These superparamagnetic scale particles can efficiently dephase the spins of the surrounding water protons and enhance spin-spin relaxation times. This technology can thus detect at miniaturized scale DNA–DNA, Protein–Protein, Protein–Small Ligand and enzyme reactions. Thus, e.g. non-sense DNA sequences, green fluorescent mRNA can be identified even in cell lysates. (See Perez, J. M., et al. 2002. *Nature Biotechnol.* 20:816.)

magnetic resonance *See* nuclear magnetic resonance spectroscopy.

magnetoreception Response of organisms to the magnetic field of the Earth for behavior and orientation.

magnification Increase in the units of ribosomal genes, hypothesized to occur by extra rounds of limited replication or unequal sister chromatid exchange. At the *bobbed* (*bb*, 1-66.0) locus present in both sex chromosomes of *Drosophila* in about 225 copies organized as large tandem arrays, separated by non-transcribed spacers. The copy numbers of *bb* in wild population Y chromosomes may vary 6-fold. *See* microscopy; ribosomes.

MAGUK Membrane-associated guanylate kinases mediate nuclear translocation and transcription. *See* ELL; zona occludens. (McMahon, L., et al. 2001. *J. Cell Sci.* 114[pt 12]:2265.)

maize *Zea mays* belongs to the family of Gramineae (grasses) and the tribe Maydeae along with teosinte (*Euchlena mexicana*) and the genus *Tripsacum* (with a large number of species). The male inflorescence of maize is more similar to that of teosinte than the female inflorescence. The ear or maize carries generally 8 to 24 rows of kernels, whereas teosinte has only 2 rows. The teosinte ear is fragile; the maize ear is not. The seed is really a karyopse (caryopse), a single-seed fruit (kernel). The basic chromosome number $x = 10$. *Tripsacum* resembles some members of the *Andropogonaceae* family more than these two closer relatives and its basic chromosome number $x = 18$. Teosinte can be crossed readily with maize and the offspring is fertile, whereas *Tripsacum* is more or less strongly isolated sexually and their hybrids are not fully fertile. Teosinte genes display the same chromosomal and gene arrangements as maize in contrast to *Tripsacum*. All three genera have evolved in the Western Hemisphere.

Maize is one of the genetically most thoroughly studied plants. It is a monoecious species; under natural conditions it is allogamous. Approximately 500 kernels may be fixed on the cob for easy Mendelian analysis for a good number of endosperm, pericarp, and embryo and seedling characters. Mendelian genes have been identified at about 1,000 loci and more than 500 have been mapped. RFLP maps are also available. The characteristic pachytene chromosomes facilitate cytogenetic analyses. Several genes controlling meiosis have been identified. Cytogenetic studies with maize have contributed significantly to the understanding of chromosomal rearrangements. Transposable genetic elements were first recognized in maize.

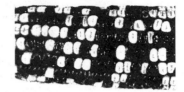

Segregation in maize.

Transformation is possible but requires either electroporation or biolistic techniques. The discovery and the commercial production of hybrid corn (heterosis) made an unprecedented increase in the food and feed supply. (Davis, G. L., et al. 1999. *Genetics* 152:1137; Doebley, J. 2001. *Genetics* 158:487; for genetic databases, contact by e-mail: <dhancock@teosinte.agron.missouri.edu> or <http://teosinte.agron.missouri.edu>; <http://zmdb.iastate.edu>; <http://www.tigr.org/tdb/tgi.shtml>.)

Majewski syndrome Autosomal-recessive phenotype that has many similarities to oral-facial-digital syndromes, particularly Mohr syndrome. It is distinguished from Mohr syndrome on the basis of laryngeal (throat) anomalies and polysyndactyly of the feet. *See* oro-facial-digital syndromes; polydactyly.

major gene Determines clear, qualitative phenotypic traits. *See* minor gene.

major groove As it turns the DNA helix displays two grooves in a 3.46 nm pitch; the wider one (almost two-thirds of the pitch) is the major groove. *See* DNA; hydrogen pairing; Watson and Crick model.

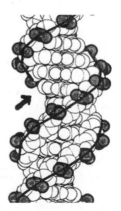

Major groove.

major histocompatibility complex (MHC in humans, H-2 in mice) Triggers the defense reactions of the cells against foreign proteins (invaders). Their existence was first recognized by incompatibilities of tissue grafts. MHC molecules are transmembrane glycoproteins encoded by the HLA complex in humans and by the H-2 in mice. Class I molecules have three extracellular domains at the NH_2 end and the COOH terminus reaches into the cytosol. The extracellular domains are associated with β_2 microglobulin. Class II MHC molecules are formed from α- and β-chains without microglobulin. Class I heavy chain and β_2 microglobulin are translocated to the endoplasmic reticulum before their translation is completed; their assembly takes place there with the assistance of the chaperones, BiP (heat-shock protein) and calnexin (glycoprotein). MHC I molecules then associate with TAP and are ready to pick up foreign antigenic peptides. After a peptide is bound, the system is released. After passing through the Golgi where their attached carbohydrates may be modified, exocytosis moves them to the surface of the cell membrane, providing a chance for $CD8^+$ T cells (CTL) to react to them. At this stage, the cells that recognize self-antigens are eliminated by apoptosis. MHC molecules resemble immunoglobulins, and the class II molecules, especially the β-chains, are highly polymorphic (the HLA-DRB 1 locus has more than 100 alleles.). MHC molecules bind foreign antigens and present them to the lymphocytes. Murine CD1 proteins bear similarities to MHC molecules, but they present lipids and glycolipid antigens to the T cells. MHC class I proteins accumulate the cut pieces of peptides from inside the cells, whereas class II MHC proteins are attached to the pieces of antigens from outside the cell. Before a class II protein is loaded with an invader peptide, it carries a neutral (dummy) peptide called CLIP, which is then replaced by foreign pieces regulated by acidic conditions in the presence of DA. Class I molecules are expressed on practically all cells that T cells recognize. Class II molecules are found on $CD4^+$ B cells and other antigen-presenting cells. MHC class I peptides (CTL epitopes) stimulate $CD8^+$ cytotoxic T cells, and MHC class II peptides are recognized by $CD4^+$ T cells. T_H cells destroy any cell that presents the antigen with MHC II molecules. Helper T cells do not directly attack the invaders but stimulate the action of macrophages. MHC genes rely on transactivation by CIITA (non-DNA-binding) and RFX5, NF-X, NF-Y, and other DNA-binding transcription factor proteins. There is a great diversity among MHC molecules. Gene conversion may create variations in the 10^{-4} range in sperm. The histocompatibility system of rabbits is called RLA, in rats RT1, in guinea pigs GPLA. MHC I homologues of herpesviruses may disarm natural killer T cells. *See* antigen-presenting cell; antigen processing and presentation; apoptosis; bare lymphocyte syndrome; blood cells; CTL; DA; HLA; Ii; immune system; microglobulin; proteasome; RFX; self-antigen; TAP; T cells. (Beck, S. & Trowsdale, J. 2000. *Annu. Rev. Genomics Hum. Genet.* 1:117.)

majority class spores *See* polarized recombination.

malaria Infectious disease affecting 300–500 million people annually, causing 1.5–2.7 million deaths. It is caused by a species of the *Plasmodium* protozoa. Transmission is mediated by *Anopheles* mosquito bites, but transplacental infection or transfusion with contaminated blood may transmit malaria. This disease is prevalent in the subtropical and tropical areas of the world. The symptoms are chills, fever, sweating, anemia, etc. The attacks are recurring according to the major reproductive cycles of the parasite. Heterozygotes of sickle cell anemia are somewhat resistant to the disease, which explains the higher than expected frequency of this otherwise deleterious gene.

Plasmodium merozoites.

Plasmodium degrades hemoglobin by two plasmalepsin proteases. Two minor and one major quantitative trait loci have been identified in *Anopheles gambiae* that inhibit the development of *Plasmodium cynomolgi* B in the midgut of the mosquito. The human body defends itself by cytotoxic T cells. The human immune response cannot prevent the infection because the malaria-infected erythrocytes attach and slow down the maturation of the antigen-presenting dendritic cells and thus fail to properly activate the T lymphocytes. In some populations, HLA and other ligands may cooperatively downregulate this immune response. During the development of the parasite, the surface antigens are altered, providing immune evasion and chances of reinfection after a period of apparent curing.

Some success was obtained by vaccination before the parasite reaches the blood infection stage. Viral vector–delivered multigenic DNA vaccines encoding different epitopes and applied before the parasite reaches the blood stage may overcome the problems caused by antigenic variations. Transgenic mosquitos may fail to transmit the parasites (Ito, J., et al. 2002. *Nature* 417:452). First-time gravid humans are more susceptible to malaria because the parasite attaches to chondroitin sulfate A on the placenta. By subsequent pregnancies, a partially protective antiadhesion response develops. The size of the *Plasmodium falciparum* genome is $\sim$33 Mb ($\sim$82% A = T) located in 14 chromosomes. It has two organelle genomes: the mitochondrial is 5.9 kb (tandemly repeated) and $\sim$35 kb in the apicoplast, a plastid-like organelle. The technology of molecular biology greatly facilitates the identification of the biology of the protozoa and the development of more effective control measures. *P. falciparum*, the most fatal species, is rapidly developing resistance to the drug chloroquine and to several newer drugs. New drugs (fosmidomycin) that block the nonmevalonate pathway of isoprenoid biosynthesis in the apicoplast appear promising. The cerebral form of malaria attacks the brain and causes mental disorders, hyperthermia, and about half of the infections are lethal. Malaria may mitigate hepatitis B viral infections. *See* apicoplast; CD36; chloroquine; cytotoxic T cell; glucose-6-phosphate dehydrogenase; hemoglobin; HLA; immunization, genetic; Oct-1; *Plasmodium*; QTL; refractory genes; sickle cell anemia. (de Koning-Ward, T. F., et al. 2000. *Annu. Rev. Microbiol.* 54:157; Kappe, S. H. I., et al. 2001. *Proc. Natl. Acad. Sci. USA* 98:9895; *Nature* 2002; *Insight* 415:670–710; <www.malaria.org>.) *Anopheles* genome: *Science* 298, 4 oct. 2002.

malate dehydrogenase The product of MDH1 (human chromosome 2p23) is cytosolic, whereas the MDH2 enzyme (encoded in 7p13) is located in the mitochondria. *See* Krebs-Szentgyörgyi cycle; mitochondria.

malattia leventinese *See* macular degeneration.

MALD Mapping by admixture linkage disequilibrium. Mapping on the basis of linkage disequilibrium in a population. *See* linkage disequilibrium. (Collins-Schramm, H. E., et al. 2002. *Am. J. Hum. Genet.* 70:737.)

mal de Maleda (MDM, 8q24.3) Rare, recessive keratoderma, redness on the face, brachydactyly, and nail abnormalities. The basic defect is in a secreted Ly-6/uPAR domain−related protein (defined by disulfide bonding pattern between 8 and 10 cysteine residues). *See* brachydactyly; keratoma. (Fischer, J., et al. 2001. *Hum. Mol. Genet.* 10:875.)

MALDI/TOF/MS Matrix-assisted laser desorption ionization/time of flight/mass spectrometry can detect minute differences in masses of large DNA molecules (~2,000 nucleotides) in ~femtomole (10^{-15})-size samples. This technology is used for the characterization and identification of proteins and protein fragments. It detects posttranslational modifications and any alteration in protein mass during development and modulation of function. For proteomics, generally a tryptic peptide mixture is used. Trypsin cuts proteins at arginine and lysine residues and unknown proteins may be identified on the basis of matches with known proteins in the databases. The analytical procedure requires that the target molecules be coprecipitated with an excess matrix consisting of either α-cyano-4-hydroxycinnamic acid or dihydrobenzoic acid (DBH). The material is then dried on a metal surface and irradiated by a flanking nanosecond nitrogen laser at a wavelength of 337 nm to generate ionization. The sample is analyzed by a mass spectrometer. *See* CID; DNA chips; electrophoresis; genomics; protein chips; proteomics; SNIPS, mole. (Li, L., et al. 2000. *Trends. Biotechn.* 28:151; Mann, M., et al. 2001. *Annu. Rev. Biochem.* 70:437; Bennett, K. L., et al. 2002. *J. Mass Spectrom.* 37:179.)

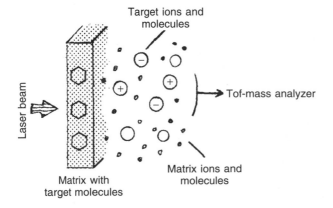

male gametocide Chemical that destroys male gametes and thus may facilitate cross-pollination in plants or may serve as a human birth-control agent. *See* birth control; crossing.

male gametophyte *See* gametophyte.

male recombination Normally absent in *Drosophila* (except when transposable elements are present; up to 1%) and in the heterogametic sex (females) of the silk worm. In case of the presence of a P element, most of the recombination is site-specific within a 4 kb tract near the element. Recombination then involves adjacent duplication and deletion of a few to >100 bp. Genetic markers can be mapped relative to known P-element sites. Mitotic recombination may occur, however, even in the absence of the meiotic one. In the plants maize and *Arabidopsis*, certain cases indicate reduced frequency of recombination in the megasporocytes compared to the microsporocytes. In rainbow trout, the average female:male recombination ratio is 3.25:1 but there is significant variation according to chromosomal regions and families. Generally, in most species the recombination frequency in the heterogametic sex is slightly lower. *See* HEI; hybrid dysgenesis; *MR*; recombination frequency.

male-specific phage Infects only those bacterial cells that carry a conjugative plasmid.

male sterility May be caused by various chromosomal aberrations, such as inversions, translocations, deficiencies, duplications, aneuploidy, polyploidy, etc., and by cytoplasmic factors. Generally, in plants male sterility is more common than female sterility because the pollen, the male gametophyte, has a more independent life phase than the megaspore and cannot rely on support by sporophytic tissues. In plants, male sterility may be caused by a mutation in the male gametophyte or it may be the result of abnormal development of the anthers or other parts of the flower, e.g., failure to release the normally developed pollen. A common cause of male sterility is an alteration of the mitochondrial DNA known as cytoplasmic male sterility (*cms*). Incompatibility between nuclear genes and certain cytoplasms as well as viral infection may also cause male sterility. Certain chemicals (mutagens) and chromosome-breaking agents (maleic hydrazide) may have gametocidic effects. Male sterility may result in transgenic plants when a tapetum-specific promoter drives a ribonuclease gene within the anthers, thus destroying the pollen. Fertility can be restored if the sterile plants are employed as female in crossing with a male carrying the transgene *barstar* inhibiting the ribonuclease. Other self-destroying — restoring combinations have been considered. A healthy human male ejaculates 25−40 million sperm each time, and if for any

50% sterile pollen.

reason this number is reduced to 20,000 or below, male infertility/sterility may result, although only a single sperm functions in fertilization. *See* asthenospermia; azoospermia; chromosomal defects; cytoplasmic male sterility; gametocides; KIT oncogene; oligospermia. (Hackstein, J. H. P., et al. 2000. *Trends Genet.* 16:565.)

male stuffing In some social insects the workers may force (and kill) males into empty cells in order to keep them from using the scarce food, thus making it available to the larvae. The relatedness among workers in single-mated queen colonies is 75%, whereas in males it is only 25%. Thus, this behavior is a form of kin selection. *See* kin selection.

malignant growth Defect in the regulation of cell division that may lead to cancer and may cause the spreading of abnormal cells. *See* cancer; cell cycle; metastasis; oncogenes.

malondialdehyde (MDA) Mutagenic carbonyl compound generated by lipid peroxidation. *See* adduct; lipid peroxidation; pyrimidopurinone.

malonic aciduria (16q24) Recessive deficiency in malonyl-CoA decarboxylase that causes abnormally large amounts of malonic acid in the urine. It involves developmental retardation, constipation, abdominal pain, and infantile death.

Malpighian tubules Excretory channels of arthropods that empty into the hindgut, performing kidney-like functions.

malsegregation The two homologous chromosomes are not recovered in 2:2 proportion at the end of meiosis. *See* meiotic drive; nondisjunction; segregation distorter.

MALT Mucosa-associated lymphoid tissue lymphomas are most common in the gastrointestinal tract and in non-Hodgkin lymphoma. MALT is frequently associated with translocations t(1;14)p22;p32 and with the apoptotic signaling gene BCL10. *See* apoptosis; BCL; Hodgkin disease; lymphomas.

MalT Specific activator of the bacterial maltose operon. *See* maltose.

Malthusian parameter Assumes that the age distribution in a population remains constant from one generation to the next. In case the age-specific birth and death rates would continue without any hindrance, the population growth would ultimately reach that level, frequently denoted by r. *See* age-specific birth and death rate; population growth. (Fisher, R. A. 1958. *The Genetical Theory of Natural Selection*. Dover, New York; Demetrius, L. 1975. *Genetics* 79:535.)

Malthusian theorem The population growth exceeds the rate of food production and thus jeopardizes the survival of humans. This 19th-century idea was contradicted by the fact that between 1961 and 1983 the available food calories per capita increased from 2,320 to 2,660 (over 7% in two decades). Unfortunately, this growth is slowing down again. Between 1984 and 1990, the increase was only about 1%. For about 2,000 years, the global human population grew by an annual rate of about 0.04%. Between 1965 and 1970, the rate increased to 2.1%, and in 1995 it was about 1.6% per year. *See* human population growth. (Black, J. A. 1997. *BMJ* 315:1686; Lee, R. D. 1987. *Demography* 24:443.)

maltose Glucopyranosyl glucose is a disaccharide. The Snf1 protein kinase controls the Mig1 transcriptional repressor of *SUC2* (sucrose), *GAL* (galactose), and *MAL* (maltose) genes. Induction of *MAL* requires a transcriptional activator. Maltose utilization is controlled by several operons. A common activator is 103-kDa MalT. Glucose and the global repressor *Mlc* repress *MalT*. The maltose transporter system is an integral part of maltose utilization. *See* catabolite repression; glucose effect; Snf. (Boos, W. & Böhm, A. 2000. *Trends Genet.* 16:404.)

MAMA Monoallelic mutation analysis. Mutations are screened in somatic cell hybrids of lymphocytes and hamster cells with the aid of DNA techniques. *See* mutation detection. (Jinneman, K. C. & Hill, W. E. 2001. *Curr. Microbiol.* 43[2]:129.)

Mammalian Comparative Mapping Database Part of the Mouse Genome Database URL: <http://www.informatics.jax.org>; questions, problems, etc., can be addressed to Mouse Genome Informatics Project: <mgi-help@informatics.jax.org>; phone: 207-288-3371, ext. 1900; fax 207-288-2516.

Mammalian Genome Monthly publication of the Mammalian Genome Society. Contact V. M. Chapman, Dept. Molecular and Cellular Biology, Roswell Park Cancer Institute, Elm & Carlton, Buffalo, NY 14263. *See* databases.

mammary tumor virus MTV of the mouse causes an estrogen-stimulated adenocarcinoma. The virus is transmitted through the milk. *See* glucocorticoid-response element. (Kang, C. J. & Peterson, D. O. 2001. *Biochem. Biophys. Res. Commun.* 287:402.)

manatee (*Trichechus manatus latirostris*) Large herbivorous aquatic mammal, $2n = 48$.

mango (*Mangifera*) Tropical fruit tree, $2n = 2x = 40$.

manic depression Psychological condition characterized by recurrent periods of excessive anguish (unipolar) or by manic depression (bipolar). The latter form is accompanied by hyperactivity and obsessive preoccupation with certain things or events. Depression and other affective disorders may involve 2 to 6% of human populations, although the incidence of unipolar depression may be as high as 21% among females and 13% among males during the entire lifetime. The concordance among monozygotic twins may be as high as 80%, whereas among dizygotic twins it is about 8%. It appears that the recurrence in families is higher with early-onset types. Also, bipolar types appear to have higher hereditary components. The genetic control of depression is unclear. X-linked-recessive and autosomal-dominant (chromosome 11p) genes have been implicated, but the majority of assignments were not well reproducible. These may be major genes, but other genes are also involved. The physiological bases may vary from defects in neurotransmitters to electrolyte abnormalities, etc. Commonly recommended therapy involves monoamine oxidase inhibitors, tranquilizers (Prozac), lithium, etc. *See* affective disorders; bipolar mood disorder; lithium; tyrosine hydroxylase.

mannans Yeast cell wall polymers of mannose. *See* mannose.

mannopine N2-(1′-deoxy-D-mannitol-1′-yl)-L-glutamine. *See* opines; Ti plasmid.

mannose Aldohexose. Left, D mannose; right, L mannose.

mannose phosphate isomerase (MPI) Zn^{2+} monomeric enzyme that converts mannose-6-phosphate into fructose-6-phosphate. It is coded in human chromosome 15q22.

mannose-6-phosphate receptor (MPR) Same as insulin-like growth factor; it plays a role in signal transduction, growth, and lysosomal targeting. LOH mutation of this receptor (human chromosome 6q26-q27) causes liver carcinoma. MPR is also a death receptor for granzyme B during CTL-induced apoptosis. *See* apoptosis; CTL; death receptor; GGA; granzyme; insulin-like growth factor; LOH; RAB; TIP47.

mannosidosis The recessive (α-mannosidase B) deficiency has been located to human chromosome 19cen-q12 (MAN2B1) and involves a large increase of mannose in the liver, causing susceptibility to infection, vomiting, facial malformations, etc. Another mannosidase defect at another autosome (4q22-q25, MANBA) caused excessive mannosyl-1-4-N-acetylglucosamine and heparan in the urine, apparently involving glycoprotein abnormalities and a variety of physical and mental defects.

Mann-Whitney test Powerful nonparametric method for determining the significance of difference between two normal-distributed populations. This method is useful for evaluating scores of samples even if the size of them is not identical. The procedure is illustrated by small samples; however, samples of $n > 20$ are preferred. The scores are ranked (T) as follows (in case of ties, the average ranks are assigned):

Popu-lations	I	II	II	I	I	II	II	Sum of I = T_I = 10
Scores	1	4	5	7	8	9	10	Sum of II = T_{II} = 18
Rank	1	2	3	4	5	6	7	$n_I = 3, n_{II} = 4$

The null hypothesis to be tested is that the distribution of the two populations (I and II) is identical. Then U is determined for sample I:

$$U_I = n_I\, n_{II} + \{[n_I(n_I + 1)]/2\} - T_I$$
$$= 3 \times 4 + (12/2) - 10 = (12 + 6) - 10 = 8.$$

In case the resulting U value is larger than $(n_I\, n_{II})/2$, calculate $U' = n_I\, n_{II} - U_I$. For large populations the sampling distribution for U is approximately normal and $E(U) = (n_I\, n_{II})/2$. The variance is determined $\sigma^2 = [n_I\, n_{II}(n_I + n_{II} + 1)]/12$; hence, the z value (the standard normal variate) = $[U - E(U)]/\sigma_U$ and the probabilities corresponding to z can be read from statistical tables of the cumulative normal probabilities; a few commonly used corresponding values for z are 1.65, 1.96, 2.58, and 3.29; the P values are 0.90, 0.95, 0.99, and 0.999, respectively. *See* null hypothesis; standard deviation, probability; Wilcoxon's signed rank test.

Mantle cell lymphoma (MCL), human translocation [11;14][q13;q32] or mutation at 11q22-q23 is Non-Hodgkin-type lymphoma. Derived from naive CD5$^+$B cells of the primary follicles or of the mantle zones of the secondary follicles. The segment deleted from chromosome 11q22-q23 includes the ataxia telangiectasia locus. The translocation juxtaposes cyclin D1 (CCND1) and the transcriptional control element of the immunoglobulin G gene, although other factors (c-Myc) are also required for the development of lymphoma. *See* anaplastic lymphoma; ataxia telangiectasia.

The caudal end of a manx cat.

Manx cat Tailless because of fusion, asymmetry, and reduction in size of one or more caudal vertebrae. A dominant gene that is lethal in homozygotes causes this phenotype. The Manx protein is also essential for the development of the notochord of lower animals. *See* brachyury; notochord.

MAO Monoamine oxidase enzyme is involved in the biosynthetic path of neurotransmitters from amino acids. Mutation in the 8th exon (amino acid position 936) converted the glutamine codon CAG to TAG (chain termination codon) and resulted in mild mental retardation, continued and impulsive aggression, arson, attempted rape, and exhibitionism in human males with this X-chromosomal recessive defect. The block of MAOA resulted in accumulation of normetanephrine (a derivative of the adrenal hormone epinephrine) and tyramine (an adrenergic decarboxylation product of tyrosine) and a decrease in 5-hydroxyindole-3-acetone. The heterozygous women were not affected behaviorally or metabolically; monoamine oxidase B level remained normal. Both enzymes are in human chromosome Xp11.2-p11.4 and the enzymes are located in the mitochondrial membrane. The enzymes may affect various psychiatric disorders such as Gilles de la Tourette syndrome, panic disorder, alcoholism, etc. *See* mitochondria; Norrie disease; Tourette syndrome.

MAP *See* *ASE1*; map, genetic; microtubule-associated proteins.

Map, genetic *See* mapping, genetic.

map, genetic, versus physical map *See* coefficient of crossing over; gene number. (Ashburner, M., et al. 1999. *Genetics* 153:179.)

map, metric On-scale ordered FISH map where cosmid clones can be positioned. *See* cosmid; FISH.

map, physical *See* physical map.

map-based cloning (positional cloning) Isolation of genes on the basis of chromosome walking and propagation usually by YAC and/or cosmid clones. After a genome sequence is completed, positional cloning will no longer be needed as long as the function of the gene is known. Positional cloning may be complicated by the fact that some phenotypes are affected by more than a single locus. *See* chromosome landing; chromosome walking; position effect. (Lukowitz, W., et al. 2000. *Plant Physiol.* 123:795; Tanksley, S. D. 1995. *Trends Genet.* 11:63.)

map distance Indicates how far syntenic genes are located from each other in the chromosome as estimated by their frequency of recombination; 1 map unit = 1% recombination = 1 centimorgan. The greater the distance between two genes, the higher is the chance that they are separated by recombination. A single recombination between two genes in a meiocyte produces maximally 50% recombination that is 50 map units. The distance between syntenic genes may exceed 50 map units several times; these longer distances are determined in a staggered manner, proceeding stepwise from left to right and right to left. In prokaryotes, using conjugational transfer and recombination, map distances are measured in minutes of transfer (*see* conjugational mapping). *See* mapping function; radiation hybrids; recombination frequencies.

map expansion The distance between two distant markers exceeds the sum of the distances of markers in between; it is commonly observed in gene conversion. *See* gene conversion. (Holliday, R. 1968. *Replication and Recombination of Genetic Material*, Peacock, W. J. & Brock, R. D., eds., p. 157. *Australian Acad. Sci.*, Canberra.)

MAPK Mitogen-activated protein kinases are distinct kinases responding to different environmental cues and set into motion in different development/physiological pathways. The family includes KSS1 (filamentous growth), HOG-1 (hypertonic stress), FUS3 (mating), Mpk1 (cell wall remodeling), SLT-2, sapk-1 (stress-activated protein kinase), FRS (FOS-regulating kinase), erk-1 (extracellular signal-regulated kinase), Smk1 (sporulation), etc. In yeast, association with the Ste12 transcriptional activator regulates specificity. Ste12 in combination with other proteins may bind to the pheromone-response element (PRE). Ste12 protein associates with the filamentation- and invasion-response element (FRES), including A Ste12

protein-binding site (TGAAACA), and a neighboring CATTCY sequences specific for the Tec1 (nonreceptor tyrosine kinase) transcription factor. The mating and filamentous growth pathways are initiated in a similar way, but for mating the FUS kinase is activated rather than the KSS. In the absence of phosphorylation, KSS inhibits the Ste12-Tec1 complex and also filamentous growth. The Dig proteins appear to be cofactors of the inhibitory path. In the inhibitory MAPKs (KSS) there is the MKI (MAP kinase insertion) site, which is remodeled upon phosphorylation and thus conversion into an activator. The specificity of the MAPKs is also secured by the complexes of recruited proteins, and each of these selects the appropriate MAPK. *See* anthrax; arrestin; Ask; FOS; JNJ; MAPKK; MEK; signal transduction; Ste. (Ito, M., et al. 1999. *Mol. Cell Biol.* 190:7539; Roberts, C. J., et al. 2000. *Science* 287:873; Pouysségur, J. 2000. *Science* 290:1515; Barsyte-Lovejoy, D., et al. 2002. *J. Biol. Chem.* 277:9896.)

MAP kinase (MPK) Family of serine/threonine protein kinases associated with mitogen activation (growth) and stress responses. These two paths of responses interact at various levels. They have a key role in signal transduction pathways. The p42 and p44 MPKs are also called ERK2 and ERK1, respectively. The MAP kinase family is activated by STE20, RAS, and Raf protein serine/threonine kinases. MAPK81P1 (11p11.2-p12) may be a transactivator of SLC2A2 gene encoding GLUT2 glucose transporter and one of the factors responsible for MODY diabetes. *See* cell cycle; diabetes; ERK; GLUT; JNK/SAPK; MAPK; MODY; p38; signal transduction; STE. (Cobb, M. H. 1999. *Progr. Biophys. Mol. Biol.* 71:479; English, J., et al. 1999. *Exp. Cell. Res.* 253:255; Chang, L. & Karin, M. 2001. *Nature* 410:37; Dong, C., et al. 2002. *Annu. Rev. Imunol.* 20:55.)

MAP kinase kinase (MAPKK) *See* Ste 7.

MAP kinase kinase kinase (MAPKKK) *See* Ste 11.

MAP kinase kinase kinase kinase (MAPKKKK) *See* Ste 20.

MAP kinase phosphatase (MPK) MPK-3 dephosphorylates phosphotyrosine and phosphothreonine and inactivates the MAP kinase family proteins. Binding activates it by its noncatalytic C-end to ERK2 without a need for phosphorylation. The homologous MPK-4 is activated by ERK2 but also by JNK/SAPK and p38. *See* signal transduction. (Zhang, T., et al. 2001. *Gene* 273:71.)

MAPKK Mitogen-activated (MAP) protein kinase kinase. This protein mediates signal transduction pathways by phosphorylating RAS, Src, Raf, and MOS oncogenes. When such an active kinase was introduced into mammalian cells, the AP-1 transcription factor was activated, the cells formed cancerous foci, and they became highly tumorigenic in nude mice, indicating that MAPKK is sufficient for tumorigenesis. *See* anthrax; NPK; signal transduction; tumor.

MAPKKK Mitogen-activated protein kinase kinase kinase. *See* TAK1.

maple (*Acer* spp.)   Hardwood trees, and the sugar maple is used for collecting syrup, $2n = 26$.

maple syrup urine disease   *See* isoleucine-valine biosynthetic pathway.

MAPMAKER 3.0   Software for constructing linkage maps using multipoint analysis in test cross and F_2. MAPMAKER/QTL is for quantitative trait loci. Available for Sun (Unix), PC (DOS), and Macintosh. Contact Eric Lander, Whitehead Institute, 9 Cambridge Center, Cambridge, MA 02142; fax: 617-258-6505; <mapmaker@genome.wi.mit.edu>.

Map Manager 2.5   Software for storing and organizing genetic recombination data and database for RI strains of mice. Information: K. F. Manly, Roswell Park Cancer Institute, Elm & Carlton, Buffalo, NY 14263. Phone: 716-845-3372; fax: 716-845-8169; <kmanly@mcbio.med.buffalo.edu>.

map unit   1% recombination = 1 map unit (m.u. or 1 centimorgan, c.m. or cM). In approximate kilobase pairs equivalent to 1 centimorgan in a few species: *Arabidopsis* ≈ 140; tomato ≈ 510; human $\approx 1,108$; maize $\approx 2,140$. The *Salamanders* have the largest known genetic map: 7,291 cM.

MAPCs (multipotent adult progenitor cells)   "Universal stem cells" present in small number in some adult tissues and have the capacity to produce other types of cells in a manner similar to embryonic stem cells. *See* stem cells. (Jiang, Y., et al. 2002. *Nature* 418:41; Schwartz, R. E., et al. 2002. *J. Clin. Invest* 109:1291.)

mapping   Establishing the sequential location of genes, restriction fragments, or PCR products. *See* databases; genome projects; PCR; physical maps; recombination; restriction endonuclease. (<http://linkage.rockefeller.edu/>.)

mapping, genetic   Mapping of chromosomes can be carried out on the basis of recombination frequencies of chromosomal markers, either genes or DNA markers such as RFLPs, RAPDs, etc., or molecular methods can be used in physical mapping (chromosome walking). As a hypothetical example, assume that genes *a*, *b*, *c*, *d*, and *e* are syntenic and the recombination frequencies between them are

a   0.06   *b*   0.04   *c*   0.06   *d*   0.16   *e*   0.18   *g*

The sum of the recombination frequencies is $0.06 + 0.04 + 0.06 + 0.16 + 0.18 = 0.50$, indicating that the segregation between *a* and *g* is independent. The results shown could be obtained between two genes at a time, but such a two-point cross would not have permitted the determination of the order of the genes relative to each other. For the determination of genes at left and genes at right, a three-point cross is required as a minimum, and multipoint crosses are even more helpful (*see* recombination frequencies). The results of a hypothetical three-point test cross are tabulated below.

According to the data, the number of recombinants in interval I is $10 + 2$ and in interval II it is $20 + 2$ (the double recombinants had recombination in both intervals I and II and their number must be added to the numbers observed). Thus, the frequency of recombination between *A* and *B* is $12/100 = 0.12$ and between *B* and *D* $22/100 = 0.22$. The number of recombinations between *A* and *D* is $10 + 20 + 4 = 34$ (the 4 is the double of the number of recombinants in intervals I and II because these represent double recombination events). Thus, the relative map positions are $A - B - D$. Had we found that the combined parental numbers were 68 but the recombinants *Abd* plus *aBD* 10 and *ABd* plus *abD* 2, and *AbD* plus *aBd* 20, we would have had to conclude that the gene order was $A - D - B$, because the lowest-frequency class ($0.10 \times 0.20 = 0.02$) must have been the double recombinants, and thus the gene order would have been $A - D - B$. The observed recombination frequencies may have had to be corrected by mapping functions because not all double crossovers might have been detected (*see* mapping functions). The recombination frequencies may be biased by interference when the frequency of double crossovers is either higher or lower than expected on the basis of the product of the two single crossovers (see coincidence, interference). The (corrected) recombination frequencies can be converted to map units by multiplication with 100 and 1 map unit (m.u. or centimorgan [cM]) is 0.01 frequency of recombination. Recombination frequencies can be estimated in F_2 by using the product ratio method (see separate entry). In the latter case, recombination frequencies can be calculated only between pairs of loci, yet from the data of two pairs involving three loci, the gene order can be determined. *See* chromosome walking; comparative map; consensus; crossing over; deletion mapping; F_2 linkage estimation; genomic screening; integrated map; Joinmap; linkage; Mapmaker; mapping functions; maximum likelihood method, recombination; physical maps; QTL; radiation hybrids; RAPD; recombination frequencies; RFLP; skeletal map; SNIPs; unified genetic map. (Ott, J. & Hoh, J. 2000. *Am. J. Hum. Genet.* 67:289; Grupe, A., et al. 2001. *Science* 292:1915.)

mapping, genetic, additivity of   Ideally it means that the distance between genes A–C is equal to the sum of the distance between A–B and B–C if the order of genes is A B C. Exceptions exist to this generally valid rule because of genetic interference. *See* coincidence; interference; mapping; genetic; recombination frequency.

mapping by dosage effect   If the activity of enzymes is proportional to their dosage and disomics can be distinguished from critical trisomics, genes (for the enzymes) located in a specific trisome can be identified and assigned to that specific

Phenotypic classes →	ABD	abd	Abd	aBD	ABd	abD	AbD	aBd
Number of individuals	34	34	5	5	10	10	1	1
		Parental		Recombinants interval I		Recombinants interval II		Recombinants intervals I + II
Total of 100		68			10		20	2

chromosome, or in the case of telotrisomics to a particular chromosome arm. Theoretically the enzyme activity is expected as follows if the locus is situated in the long arm of a particular chromosome:

```
----O--------  ----O--------  -----O----------  ------O---------  -----O---------
----O--------  ----O--------  -----O----------  ------O---------  -----O---------
               ----O--------  -----O            O---------  --------O---------
   2              3              2                 3                 4
```

In human trisomy 21, several genes show increased expression ranging from 1.21 to 1.61 relative to the normal disomic condition. Thus, in practice the dosage effect may not be perfectly additive, yet it may be clear enough for classification. *See* trisomics. (Carlson, P. S. 1972. *Mol. Gen. Genet.* 114:273.)

mapping function   Corrects map distance estimates from recombination frequencies when the recombination frequency in an interval exceeds 15–20% and double crossing overs are undetectable because of the lack of more densely positioned markers. *See* Carter-Falconer mapping function; coefficient of coincidence; count location models; Haldane's mapping function; Kosambi's mapping function; mapping; recombination frequency; stationary renewal process. (Zhao, H. & Speed, T. P. 1996. *Genetics* 142:1369.)

mapping panels   DNA sequences with known chromosomal location and can be used to locate unknown sequences to chromosomes. *See* radiation hybrid panels.

mapping in silico   Feeding phenotypic information to a computer program facilitates fast information gathering on the regulation and chromosomal locations of the multiple factors involved in a polygenic disease. *See* polygenic. (Grupe, A., et al. 2001. *Science* 292:1814.)

mapping set   *See* genomic screening.

MapSearch   Locates regions of a genomic restriction map that best resemble a local restriction, the so-called probe.

MapShow   Computer program displaying MapSearch alignments and drawing Probe-to-Map alignments in Sun Workstations. *See* mapping.

MAR   Matrix attachment region attaches chromatin loops to the nuclear matrix. The attachment region has a consensus of a so-called A box (AATAAATCAA) or T box (TTA/TAA/TTTA/TTT). The MARs are about 100 to 1,000 bp long and frequently include replicational origins and transcription factor binding sites. *See* chromatin; loop domains; mode; scaffold. (Stratling, W. H. & Yu, F. 1999. *Crit. Rev. Eukaryot. Gene Expr.* 9[3–4]:311; Pemov, A., et al. 1998. *Proc. Natl. Acad. Sci. USA* 95:14757; <http://www.futuresoft.org/MAR-Wiz>.)

MARANHAR   *See* killer plasmids; *Neurospora*.

MARCKS   Myristoylated alanine-rich C-kinase substrate. Protein substrates of the protein kinase C (PKC) involved in differentiation; they bind actin filaments. (Spizz, G. & Blackshear, P. J. 2001. *J. Biol. Chem.* 276:32264.)

mardel10   Supernumerary human chromosome 10 with a large deletion at the regular centromere. The deletion activates, however, a functional neocentromere at 10q25, which lacks α-satellite and CENP-B centromeric protein, although it shows some other centromere proteins. *See* centromere; human artificial chromosome; neocentromere. (Voullaire, L. E., et al. 1993. *Am. J. Hum. Genet.* 52:1153; Choo, K. H. A. 1997. *Am. J. Hum. Genet.* 61:1225.)

Marek's disease   Lymphoproliferative viral chicken disease. The growth hormone GH1 conveys resistance. *See* Epstein-Barr virus; herpes. (Liu, H.-C., et al. 2001. *Proc. Natl. Acad. Sci. USA* 98:9203.)

marfanoid syndromes   May resemble Marfan syndrome, but one of the forms has no ectopia lentis (displacement of the crystalline lens of the eye). Another form does not involve cardiovascular defects. The marfanoid-craniosynostosis is called Shprintzen-Goldberg syndrome and the anomaly is caused by mutation in the fibrillin-1 gene. *See* craniosynostosis syndromes; eye diseases; fibrillin; Marfan syndrome.

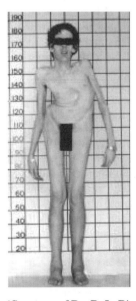

Marfan syndrome. (Courtesy of Dr. D. L. Rimoin, Los Angeles.)

Marfan syndrome (MFS, FBN1) Includes a tall, thin stature, long limbs and fingers; chest deformations are also common. The three most consistent defects are skeletal, heart-vein (cardiovascular), and eye (ectopia lentis) abnormalities. The disease may affect the development of the fetus and may be recognized in early development. Life expectancy in serious cases may not exceed 30 years. The cause of death is generally heart failure, but the defect may surgically be corrected in some cases. The penetrance appears very good, but the expressivity is highly variable. The symptoms frequently overlap with other anomalies, particularly with those of Ehlers-Dunlop syndrome, which involves a defect in collagen. The primary defect in MFS involves the elastic fiber system glycoprotein, fibrillin. This protein of the connective tissue contains repeats resembling sequences in epidermal growth factor (EGF) where the observed lesion leads to the identification of the basic molecular cause. Formerly a collagen defect was suspected. Several investigators confirmed a transversion mutation at codon 293 leading to CGC (Arg) → CCC (Pro) replacement. Similar molecular defects have been identified in *Drosophila, Caenorhabditis*, and cattle. This dominant gene has been assigned to human chromosome 15q21.1. Interestingly, mosaicism for trisomy 8 causes similar symptoms. The prevalence of MFS/FBN1 is about 1×10^{-4}, but this figure may not be entirely reliable because of the wide range of symptom manifestations. The recurrence risk is about 50%; 15–30% of the cases may be due to new mutation, which is the cause of the most severe cases, whereas the familial incidence generally entails milder symptoms. The estimated mutation rate is 4–5×10^{-6}. FBN1 may be dominant negative. *See* arachnodactyly; cardiovascular disease; coronary heart disease; connective tissue disorders; dominant negative; Ehlers-Dunlop syndrome; expressivity; fibrillin; Marfanoid syndromes; penetrance; transversion mutation. (Dietz, H. C. & Pyeritz, R. E. 1995. *Hum. Mol. Genet.* 4:1799.)

marijuana *See* cannabinoids; *Cannabis*.

mariner Probably the smallest transposable element in eukaryotes (1,286 bp). It has not been observed in *Drosophila melanogaster*, but it has been detected in African species of the *D. melanogaster subgroup*, *D. sechellia* (1–2 copies), *D. simulans* (usually 2 copies), *D. yakuba* (about 4 copies), *D. teissieri* (10 copies), and *D. mauritiana* (20 to 30 copies). A mariner-like element has been detected in soybeans.

Mariner contains 28 bp inverted terminal repeats and a single open-reading frame (1,038 bp) beginning with an ATG codon at position 172 and termination with an ochre (TAA). Overlapping AATAA bases may serve as polyadenylation signals.

The target in the untranslated leader of the w^{pch} is 5′-TGGCGTA↓TAAACCG-3′. The arrow marks the insertion and the TATA probably indicate the target site duplication. *Mariner* is different from other transposable elements inasmuch as it induces a high frequency of somatic sectors (4×10^{-3}) at the w^{pch} (white *peach*) locus. Germline mutation is about 2 to 4×10^{-3} — twice as high in males than females (no sex difference in somatic mutation). Before the transposable element was recognized, the somatic instability was attributed to the factor named *Mos* in chromosome 3. This element also causes dysgenesis, but it does not display the reciprocal difference observed in the *P-M* and *I-R* systems; however, *mariner* transmitted through the egg shows higher rates of somatic excisions. *Mariner* homologues occur in other species too, including humans, but the (*human*) sequences are pseudogenic, although they increase unequal crossing over in human chromosome 17p11.2-p12. The mariner-type DNA-based transposons are most common in the human genome, representing ~1.6% of it. The *mariner* sequences are also expressed in *Leishmania* and *Caenorhabditis*. *See* Charcot-Marie-Tooth syndrome; HNPP; hybrid dysgenesis; *Leishmania*; MITE; MLE; neuropathy; sleeping beauty; transposable elements; unequal crossing over. (Hartl, D. L., et al. 1997. *Annu. Rev. Genet.* 31:337; Zhang, L., et al. 2001. *Nucleic Acids Res.* 29:3566; Feschotte, C. & Wessler, S. R. 2002. *Proc. Natl. Acad. Sci. USA* 99:280.)

marker A genetic marker is any gene or detectable physical alteration in a chromosome (e.g., knobs, microsatellites) or cytoplasmic organelle used as a special label for that chromosome or chromosomal area. A molecular marker is a macromolecule (nucleic acid or protein) of known size and electrophoretic mobility used as a reference point in estimating the size of unknown fragments and molecules. In genomics, type I markers are the protein-coding genes, type II markers are the highly polymorphic microsatellites, and type IIIs are SNPs. *See* ladder; microsatellite; RAPD; RFLP; SNIPs.

marker-assisted selection Used for animal and plant breeding once linkage has been established between physical markers (RFLP, microsatellite loci) and others; economically desirable traits (e.g., disease resistance, productivity) that have low expressivity and/or poor penetrance because of the substantial environmental influences can be better identified. The physical markers should not have ambiguity of expression and thus may facilitate much faster progress in breeding. *See* expressivity; microsatellite; penetrance; RFLP; QTL. (Hospital, E. F. 2001. *Genetics* 158:1363; Davierwala, A. P., et al. 2001. *Biochem. Genet.* 39[7–8]:261; Hayes, B. & Goddard, M. E. 2001. *Genet. Sel. Evol.* 33[3]:209.)

marker effect By generalized transduction, theoretically any bacterial gene should be transferred by the transducing phage. As a matter of fact, some genes are transduced 1,000-fold better than others. The differences have been attributed to the distribution of *pac* sites in the bacterial chromosome. Recombination by transduction may vary along the length of the bacterial chromosome. Marker effects have been observed also in eukaryotic recombination. *See* generalized transduction; *pac* site.

marker exchange mutagenesis *See* targeting genes.

marker exclusion Occurs upon joint infection of bacteria by phages T4 and T2. In the progeny T2 genes are recovered in about 30% rather than 50% as expected. The apparent basis of the bias is that T4 harbors 13 sequence-specific endonucleases that selectively eliminate particular tracts from the other

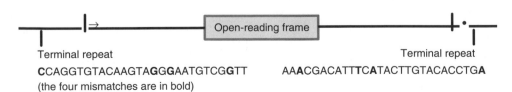

Terminal repeat

CCAGGTGTACAAGTA**GG**AATGTCG**G**TT

(the four mismatches are in bold)

Terminal repeat

AAACGACATTTCATACTTGTACACCTGA

Mariner element.

phage. *See* homing endonucleases. (Edgell, D. R. 2002. *Current Biol.* 12:R276.)

marker panels DNA probes or genes that cover portions of the genome regarding linkage. *See* linkage; probe.

marker rescue Integration of markers into normal DNA phage from mutagen-treated (irradiated) phage during mixed infection of the host; it is similar to cross-reactivation where normal phages can be obtained from defective phages by recombination. Alternatively, a wild-type DNA fragment can be inserted by recombination into a mutant one and the mutation can thus be mapped. Marker rescue is the most commonly used method of mapping in phages. *See* multiplicity reactivation; reactivation; Weigle reactivation. (Barricelli, N. A. & Doermann, A. H. 1961. *Virology* 13:460; Thompson, C. L. & Condit, R. C. 1986. *Virology* 150:10.)

marker transfer Actually a gene replacement by recombination.

Markov chain Monte Carlo algorithm Applies Bayesian inference for evaluation of the posterior distribution. It is also used in classical likelihood calculation. *See* Bayes' theorem; Monte Carlo method. (Gilks, W. R., et al., eds. 1996. *Markov Chain Monte Carlo in Practice.* Chapman and Hall, London; Larget, B. & Simon, D. L. 1999. *Mol. Biol. Evol.* 16:750.)

Markov chain statistics Sequence x_1, x_2 ... of mutually dependent random variables constitutes a Markov chain if there is any prediction about $x_{n|1}$. Knowing $x_1 \ldots x_n$ without loss x may be based on x_n alone. It is used in physical mapping of genomes and for ascertaining frequency distributions in populations. The *hidden Markov model* (HMM)–type analysis permits the identification of protein domains in peptide chains or in nucleotide sequences. *See* TB-parse. (Stephens, D. A. & Fisch, R. D. 1998. *Biometrics* 54:1334.)

marmoset New World monkey. *See* Callithrichicidae.

Marmoset.

marmota *Marmota marmota* $2n = 38$; *Marmota monax* $2n = 38$.

Maroteaux-Lamy syndrome *See* mucopolysaccharidosis.

Marquio syndrome *See* mucopolysaccharidosis.

marriage In every state of the United States and many other countries, marriages between parent and child, grandparent–grandchild, aunt–nephew, uncle–niece, and brother–sister are illegal. In about half of the states, first cousin and half-sibling unions are also prohibited. In some societies, consanguineous marriages are legal. *See* consanguinity; miscegenation.

Marshall syndrome *See* Stickler syndrome.

MARS model Developed to interpret the fate of mitochondria during the development of an organism. The acronym is derived from (1) accumulation of defective mitochondria, (2) accumulation of aberrant proteins, (3) effect of oxygen-free radicals and antioxidant enzymes, and (4) turnover of proteolytic scavengers. *See* aging; mitochondria.

MARS shield Symbol of male; in pedigrees, usually squares are used. *See* Venus mirror.

Mars shield.

marsupial Mammalian group of animals that carry their undeveloped offspring in a pouch, e.g., the kangaroos and other Australian species, also the North American opossum. The organization of the genetic material of marsupials differs in several ways from other placental mammals. Ohno's law is not entirely complied with inasmuch as genes in the short arm of the eutherian X chromosomes are dispersed in three autosomes. Marsupial Y chromosomes are extremely small, yet they contain testis-determining and other sequences homologous to other mammals, although the Y chromosome is not critical for sex determination. Their sex chromosomes also carry a pseudoautosomal region at their tip but do not form a synaptonemal complex and do not recombine. *See* eutheria; monotrene; Ohno's law; pseudoautosomal; SRY. (Marshall Graves, J. A. 1996. *Annu. Rev. Genet.* 30:233; Zenger, K. R., et al. 2002. *Gnetics* 162:321.)

MART-1 (Melan-A) *See* MAG.

marten (*Martes americana*) $2n = 38$.

Martin-Bell syndrome *See* fragile Xq27.3.

Martsolf syndrome Apparently autosomal-recessive cataract-mental retardation-hypogonadism. *See* cataracts; cerebro-oculo-facio-skeletal syndrome; hypogonadism; mental retardation. (Hennekam, R. C., et al. 1988. *Eur. J. Pediatr.* 147:539.)

MaRX Method of isolation of mammalian cells of a certain type. A DNA library is introduced into a packaging cell line (linX cells) with the aid of retroviral vectors. The virus-infected cells are then subjected to selection for the desired phenotype. Proviruses are recovered and used for the production of viruses and further screening. *See* DNA library; packaging cell line; provirus; retroviral vectors. (Hannon, G. J., et al. 1999. *Science* 283:1129.)

MAS *See* marker-assisted selection.

masculinization *See* sex reversal.

MASDA Multiplex allele-specific diagnostic assay is a mutation detection test capable of simultaneous detection of up to 100 nucleotide changes in multiple genes concerned with disease. *See* peptide nucleic acids; single-strand conformation polymorphism. (Shuber, A. P., et al. 1997. *Hum. Mol. Genet.* 6:337.)

MASH2 Mammalian helix-loop-helix transcription factor controlling extraembryonic trophoblast development but not that of the mouse embryo. *See* DNA-binding protein domains; MYF-3.

Mask Computer program that produces ambiguous files in order to protect confidential DNA sequences in EcoSeq programs.

masked mRNA Present in eukaryotic cells and cannot be translated until a special condition is met. *See* mRNA. (Spirin, A. S. 1996. In Hershey, J. W. B., et al., eds., *Translational Control*, p. 319. Cold Spring Harbor Lab. Press, Cold Spring Harbor, NY.)

masked sequences Of nucleic acids, are associated either with proteins or other molecules to protect them from degradation.

MAS oncogene MAS1 was assigned to human chromosome 6q27-q27; it encodes a transmembrane protein. *See* oncogenes; transmembrane protein.

Mas6p *See* mitochondrial import.

maspin Protease inhibitor of the serpin family; maspin sequences are frequently lost from advanced cancer cells. Maspin is an inhibitor of angiogenesis. *See* angiogenesis; serpin. (Maass, N., et al. 2000. *Acta Oncol.* 39:931; Futscher, B. W., et al. 2002. *Nature Genet.* 31:175.)

mass-coded abundance tagging (MCAT) Is a method of protein characterization for proteomic studies. It is based on differential guanidination of the C-terminal lysine residues in tryptic peptides and followed by capillary liquid chromatography–electrospray tandem mass spectrometry. *See* capillary electrophoresis, electrospray ms, proteomics, trypsin. (Cagney, G. & Emili, A. 2002. *Nature Biotechnol.* 20:163.)

mass spectrum When molecules are exposed to energetic electrons in a mass spectrometer, they are ionized and fragmented. Each ion has a characteristic mass-to-charge ratio, m/e or m/z. The m/e values are characteristic for particular compounds and provide the mass spectrum for chemical analysis. Quantitative mass spectrometry uses the incorporation of a stable isotope (N^{15}) derivative that shifts the mass of the peptides in a known extent. The ratio between the derivatized and underivatized target can be measured. *See* affinity-directed mass spectrometry; electrospray mass spectrum; genomics; laser desorption mass spectrum; MALDI; MS/MS; proteomics; quadrupole. (Oda, Y., et al. 1999. *Proc. Natl. Acad. Sci. USA* 96:6591; Mann, M., et al. 2001. *Annu. Rev. Biochem.* 70:437; Cohen, S. L. 2001. *Annu. Rev. Biophys. Biomol. Struct.* 30:67.)

mast cell Resides in the connective or hemopoietic tissues and plays an important role in natural and acquired immunity. Mast cells release TNF-α (tumor necrosis factor), histamines, and attract eosinophils (special white blood cells) that destroy invading microbes, especially if they have IgE or IgG (immunoglobulins) on their surface. They are responsible for the inflammation reactions in allergies. *See* IL-9; IL-10 immune system; immunoglobulins; TNF.

master chromosome Large circular genome within the mitochondria. *See* mtDNA.

master gene Has a major role in a range of functions. (Prior, H. M. & Walert, M. A. 1996. *Mol. Med.* 2:405; Silver, L. M. 1994. *Mamm. Genome* 5:S291.)

master molecule Regulates series of reactions in differentiation, involving several genes. *See* morphogenesis in *Drosophila*; neuron-restrictive silencer factor.

master-slave hypothesis Interpreted redundancy in the genomes by multiple copying of the "slaves" from the original "master" sequences based on the loop structure of the lampbrush chromosomes of the newt *Triturus*. *See* lampbrush chromosomes; redundancy. (Callan, H. G. 1967. *J. Cell Sci.* 2:1.)

MAT1 RING-finger protein subunit stabilizing cyclin H-CDK7 complex or CAK. *See* CAK; CDK; cyclin; four-hybrid system; RING finger. (Devault, A., et al. 1995. *EMBO J.* 14:5027.)

MAT cassette *See* mating type determination in yeast.

matched pairs *t*-test Checks the equality of the means of paired observations. The difference between the matched pairs is tested by the formula

$$t = \frac{\bar{d}}{s_d/\sqrt{n}}$$

$\bar{d}$ = the mean of the differences, s_d = standard deviation, and t is determined at $n - 1$ degrees of freedom from a t table. The null hypothesis is that the means of the paired observations are true. *See* mean; null hypothesis; standard deviation; student's t table.

mate killer Mu particle. *See* symbionts, hereditary.

mate pairs Randomly sequenced DNA fragments are fit together by their matching mate pair ends into a continuous sequence. *See* human genome; WGS.

maternal behavior Very complex trait regulated by hormones such as estradiol, progesterone, prolactin, oxytocin, and β-endorphin. The Mest/Peg1 (mouse chromosome 6, human chromosome 7q32) is expressed only from the paternally transmitted allele and its defect leads to altered maternal behavior. *See* hormones; imprinting.

maternal coordinate genes Expressed during oogenesis and determine positional information in the egg.

maternal effect genes Display delayed inheritance because only the offspring of the homozygous or heterozygous dominant females are affected; these females themselves may appear normal. Also, genes with products (RNA, protein) in the follicle and nurse cells that may diffuse into the oocytes and the embryo, and thus not just the zygotic genes, affect early development. *See* cadherin; delayed inheritance; imprinting; indirect epistasis; morphogenesis; transgenerational effect. (Evans, M. M. S. & Kermicle, J. L. 2001. *Genetics* 159:303.)

The Drosophila egg chamber cysts are surrounded by follicle cells and inside the diploid primary oocyte is shown with the often polyploid nurse cells. These are connected by cytoplasmic bridges. The maternal genes affect oogenesis and the early development of the embryo

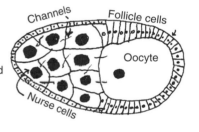

maternal embryo Develops from an unfertilized egg. *See* apomixis; parthenogenesis.

maternal genes *See* maternal effect genes. *Limnaea*;

maternal inheritance Genetic elements (generally extranuclear) are transmitted only through the female. *See* chloroplast genetics; doubly uniparental inheritance; mitochondrial genetics; mtDNA.

maternal tolerance The embryo expresses antigens that are supposed to be foreign to the maternal tissues, yet usually the embryo does not suffer immunological rejection. The cause of the tolerance appears to be due to the secretion of corticotropin-releasing hormone (CRH) and the activation of the proapoptotic Fas ligand (FasL) resulting in killing of activated T lymphocytes. Some female infertility may be caused by lack of an adequate level of CRH in the endometrium. *See* apoptosis; corticotropin-releasing factor; FAS; incompatibility; T cell. (Makrigiannakis, A., et al. 2001. *Nature Immunol.* 2:1018.)

maternity verification Rarely needed as the Romans held *mater certa*; in case of legal disputes, the methods of forensic genetics are available. *See* forensic genetics; paternity testing.

MATH Meprin and TRAF homology is a protein domain shared by metalloendopeptidases and TRAF proteins regulating the folding to activatable forms of the molecules. *See* meprin; TRAF.

mating, assortative *See* assortative mating.

mating, bacterial *See* conjugation.

Mating Bacteria.

mating, controlled *See* controlled mating.

mating, interrupted *See* conjugation mapping; interrupted mating.

mating nonrandom *See* assortative mating; autogamy (self-fertilization); controlled mating; inbreeding.

mating, physiological consequences In some species, the male seminal fluid may contain substances toxic to the female. In contrast, the seminal fluid of the male cricket (*Gryllus lineaticeps*) increases the life expectancy of the female by about a third and multiple matings (common in this species) almost double the fertility of the female. (Wagner, W. E., Jr., et al. 2001. *Evolution* 55:994.)

mating, random *See* mating systems; random mating.

mating plug Gelatinous material deposited at the vulva after mating of hermaphroditic *Caenorhabditis*. In mating between some hermaphroditic males, plugs may be visible at the head. *See* vaginal plug.

mating success (K)

$$\frac{\text{No. of females mated by mutants/No. of mutant males}}{\text{No. of females mated by wild-types/No. of wild-type males}}$$

Increasing success in mating in insect species may reduce the level of phenoloxidases in both sexes. Thus a major humoral immune component is weakened and the individuals become more susceptible to infection. Therefore copulation may adversely influence fitness. (*See* Rolff, J. & Siva-Jothy, M. T. 2002. *Proc. Natl. Acad. Sci. USA* 99:9916.)

enhancer (E) elements. The transposition activity of a *MAT* cassette requires the presence of nuclease hypersensitive sites in the flanking regions. The transposition at the *MAT* sites is initiated when the HO endonuclease makes a staggered cut at the *YZ*1 junction, generating strands with four base overhangs see below.

5'-GCTTT↓CGGCAACAGTATA-3' *MATa* 3'-CGAAAGGCG↑TTGTCATAT-5'	5'-ACTTCGCGC AACA↓GTATA-3' *MATα* 3'-TGAAGCGCG↑TTGTCATAT-5'

mating systems   Can be random mating, self-fertilization, inbreeding, assortative mating, or a combination of these. *See* Hardy-Weinberg theorem.

mating type   Designation of individuals with plus or *a* ("male") and minus or *α* ("female") labels when sex type in higher eukaryotes cannot be recognized yet genetically two types exist that do not "mate" within the group but in between groups; the diploid zygote subsequently undergoes meiosis and reproduces the mating types in 2:2 proportion. (Ferris, P. J., et al. 2002. *Genetics* 160:181.)

mating type, yeast   *Saccharomyces cerevisiae* yeast can exist in homothallic and heterothallic forms. The heterothallic yeast cells are haploids and are either of *a* or *α* mating type. The homothallic yeast cells are diploid and heterozygous for the mating-type genes (*a/α*). The diploid cells arise by fusion of haploid cells. Recognition of the opposite mating-type cells is mediated by the pheromones, the *α* factor (13 amino acids) and the *a* factor (12 amino acids) peptide hormones, respectively. The two types of cells are equipped with surface receptors for the opposite mating-type pheromones. The diploid cells lack these factors and receptors and do not fuse but may undergo meiosis and sporulate by releasing both *α* and *a* haploid cells.

The two mating types are coded by genes in chromosome 3 on the left and right sides of the centromere, respectively. These genes are clustered within the so-called mating-type cassettes and are silent at their positions named *HMLα* and *HMRa* locations. They are expressed when transposed to the mating-type site, *MAT*. At *MAT*, either the left *HMLα* or the right *HMRa* cassette can be expressed within a particular homologous chromosome. The *MAT* site is approximately 2 kb; it is about 187 kb from *HMLα* and 93 kb from *HMRa*. The overall structure of the regions is shown below. In the outline, the *MAT* site is shown empty, but in reality it is occupied by either the *HMLα* or *HMRa* cassette.

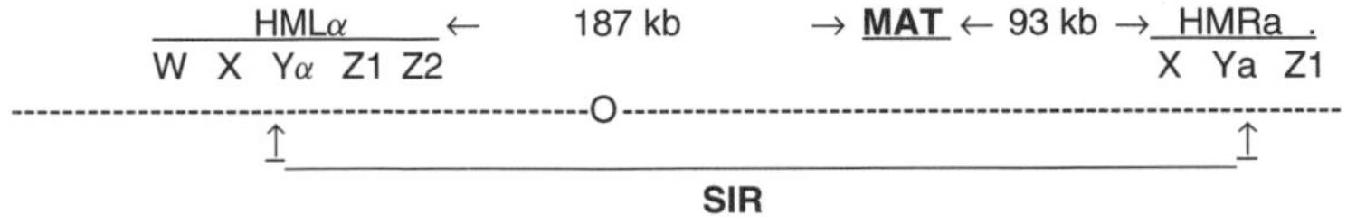

The *HO* gene is expressed only in the haploid cells but not in the diploids and only at the end of the G1 phase of the cell cycle. Both new cells have the same mating type and can transpose their mating-type gene only after the completion of the next cell cycle. The product of gene SIN1-5 represses the function of HO and the expression of gene *SW1-5* is required for the expression of *HO*. Mother cells selectively transcribe *HO*. In daughter cells, Ash1p suppressor of *HO* may accumulate as a result of asymmetric mRNA distribution. For the actual mating, the functions of other (sterility, *STE*) genes are also needed. *STE2* and *STE3* are cell-type-specific receptors of G proteins. *GPA1* encodes the Gα subunit, *STE4* the Gβ subunit, and *STE18* the Gγ subunit. After activation, the GβGγ subunits, regulate units of the pheromone-signaling pathway further downstream, including the kinases STE20, a series of MAP kinases (encoded by *STE11*, *STE7*, and *FUS/KSS* genes). The Ste5p protein (encoded by *STE5*) is presumed to be a scaffold for organizing all of the kinases.

Transposition involves pairing with the homologous Z sequences, followed by an invasion of a double-stranded receiving Y site and degradation of these Y sequences. Subsequently, the X site is invaded. New DNA synthesis takes place using the sequences of the invader molecule that is replacing the old DNA tract as a template. Integration is mediated in the pattern of a gene conversion mechanism as suggested by the Holliday model of recombination.

The *α* mating-type gene has two elements: *α*1 (transcribed from right to left), which induces the *α*-mating functions, and *α*2 (transcribed from left to right similarly to a), which keeps the expression of the a mating-type expression in check. The simultaneous expression of a1 and *α*2 represses *SIR* and other haploid-specific genes. These processes recruit proteins PRTF (pheromone receptor transcription factor) and GRM (general regulator of mating type), which recognize specific nucleotide sequences within the haploids and diploids. The *α*2 protein also interacts with MCM1 DNA binding and MADS box proteins. The binding of MATα2 with MCM1 represses a-specific genes

	HMLα	←	187 kb		→ **MAT** ← 93 kb →	HMRa	.
W	X Yα Z1 Z2					X Ya Z1	

SIR

Mating type determination in yeast.

When one of the silent complexes is unidirectionally transposed to the *MAT* site, the expressed *MATα* and *MATa*, respectively, are generated. At the original (left and right) locations, the Y*α* (747 nucleotides) and Y*a* (642 nucleotides) are kept in place and silent by the product of the *SIR1-4* gene (silent information regulator). Repression is also mediated by the autonomous replication sequences (ARS) and the pertinent

in haploids, but in diploid cells MATα2 heterodimerizes with MATa1 and thus haploid-specific genes are repressed (*see Nature* 391:660 for the structural interactions of these proteins).

Whether *HML* or *HMR* switching occurs depends on the surrounding sequences and it is not intrinsic to the elements. *Mat a* cells recombine with *HML* almost an order of magnitude more frequently than with *HMR*. *MATα* cells recombine in 80–90% of the cases with *HMR*. The switching is controlled by a 700 kb element 17 kb proximal to HML by a recombinational enhancer.

The expression of the mating-type genes involves a complex cascade of events. The mating-type protein factors interact with G-protein-like receptors situated in the cell membrane. These G proteins then transduce the mating signals through a series of phosphorylation reactions to transcription factors that control the turning on/off genes mediating the cell cycle, cell fusion, and conjugation. The industrial strains of yeast are frequently diploid or polyploid and may be heterozygous for mating type. In such a case they fail to mate. If they are plated on solid medium, they may sporulate and the chromosome number will be reduced. *See* Abf; cell cycle; heterothallic; HML and HMR sex determination; Holliday model; homothallic; MADS box; MCM1; ORC; pheromones; RAP1; rare mating; regulation of gene activity; *Schizosaccharomyces pombe*; signal transduction; silencer; Ty. (Haber, J. E. 1998. *Annu. Rev. Genet.* 32:561; Dohlman, H. G. & Thorner, J. W. 2001. *Annu. Rev. Biochem.* 70:703.)

MAT locus   Involved in the determination of mating type in yeast. *See* mating type determination in yeast.

matrilineal   Descended from the same maternal ancestor, e.g., mtDNA. *See* mtDNA.

matrix   Solutes in cells, organelles, or chromosomes, etc.; the *extracellular matrix* fills the space among the animal cells and it is composed of a meshwork of proteins and polysaccharides secreted by the cells. The viral matrix connects the genomic core with the envelope.

$$\begin{bmatrix} 1 & 4 & 7 \\ 2 & 5 & 8 \\ 3 & 6 & 9 \end{bmatrix} \begin{bmatrix} 2 & 1 \\ 4 & 6 \end{bmatrix}$$

matrix algebra   Deals with elements that are arranged as shown above. A matrix with r rows and c columns is of *order r x c* or an *r x c matrix*. If $r = c$, it is a *square matrix*. A matrix with one row is a *row vector* and a matrix with only one column is a *column vector*.

matrix-assisted laser desorption ionization/time of flight/mass spectrometry (MALDI-TOF)   Procedure for separation of DNA fragments mixed with a carrier that is subsequently painted on the surface of a solid face target. Laser desorbs and ionizes the fragments and acceleration in a mass spectrometer is used to determine fragment length. The MALDI-TOF procedure may even detect a nucleotide change in length of a DNA sequence (ca. 100 or perhaps 1,000 bp) or changes in microsatellite numbers. This technique can be utilized in DNA sequencing, discrimination among mutations of a gene, and identification of STS. It may eventually be used for clinical and diagnostic procedures. *See* laser; laser desorption mass spectrum; MALDI-TOF; mass-spectrum; microsatellite; proteomics; SNIPs; STS; TOFMS. (Shahgholi, M., et al. 2001. *Nucleic Acids Res.* 29:E91; Chu, J., et al. 2001. *Clin. Chim. Acta* 311:95; Nordhoff, E., et al. 2001. *Electrophoresis* 22:2844.)

matrix attachment region   *See* MAR.

matroclinous   The offspring resembles the mother because it developed either from an unfertilized egg or from an egg that underwent nondisjunction and carries two X chromosomes; or from a female with attached X chromosomes; or by imprinting or dauer modification; or due to sets of dominant genes or the failure of transmission of a particular chromosome through the sperm; or by nonnuclear (mitochondrial, plastid) genes. *See* chloroplast genes; mitochondrial genetics; mtDNA; *Rosa canina*.)

Matthiola (garden stock, wallflower)   Cruciferous ornamental ($2n = 14$) and an early object of genetic research. In floricultural use the presence of a lethal factor tightly linked to the simple flower character (S) has been of special interest. This causes the appearance of the ever-segregating full-flower trait (s) in 1:1 proportion rather than 3:1. By sophisticated breeding techniques and seed selection in trisomic offspring, the commercially available seed germinates and develops into nearly 100% full-flower/double-flower plants that are, however, completely sterile due to the recessive lethal factor. Thus, the seed supply is entirely dependent on commercial sources. (Kappert, H. 1937. *Ztschr. Ind. Abst.- Vererb.-lehre* 73:233; Roeder, A. H. K. & Yanofsky, M. F. 2001. *Developmental Cell* 1:4.)

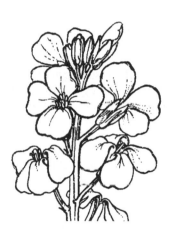

maturases   Proteins that mediate a conformational change in the pre-mRNA transcript and cooperate, in the splicing reactions. *See* introns; mitochondrial genetics.

maturation, DNA   Phage proteins cut the linear, continuous DNA into pieces that can be accommodated by the phage capsids.

maturation division   Meiosis.

maturation promoting factor   *See* MPF.

Maurice of Battenberg   Hemophiliac grandson of Queen Victoria of England; son of carrier daughter Beatrice. *See* hemophilias.

Mauriceville plasmid   *See Neurospora* mitochondrial plasmids.

MAX   A b/HLH/LZ (basic helix-loop-helix/leucine-zipper) protein hetero-oligomerizes with the MYC oncoproteins, and this state is required for malignant transformation by c-MYC. MAX alone lacks the transactivator domain of MYC. Its DNA recognition site is CACGTG. MAX may be orchestrating the biological activities of b/HLH/LZ transcription factors. The basic α-helices follow the major groove of the DNA in a *scissors grip. See* helix-loop-helix; leucine zipper; major groove; Mxi/Max; MYC; RFX.

Maxam-Gilbert method   *See* DNA sequencing.

maxicell Bacterial cell that lost most or all of its chromosomal DNA because of heavy irradiation by UV light. Therefore, maxicells do not replicate their DNA. The plasmid they contain may have escaped irradiation. It represents an appropriate replicon and can carry on replication of that plasmid and direct the synthesis of plasmid-coded proteins. This makes such cells ideal for the expression of the plasmid-born protein without a background of cellular proteins. Especially useful are those maxicells containing lambda vectors, which have sufficient expression of the phage repressor and thus do not permit λ-protein expression. *See* lambda phage; mindless; plasmids; replicon. (Jemiolo, D. K., et al. 1988. *Methods Enzymol.* 164:691.)

maxicircle Large mitochondrial genome. *See* mtDNA. (Carpenter, L. R. & Englund, P. T. 1995. *Mol. Cell Biol.* 15[12]:6794.)

maximal equational segregation Takes place when a gene is segregating independently from the syntenic centromere. It has particular significance in polyploids because it facilitates an increase of double (or multiple) recessive gametes and thus affects segregation ratios as a function of the map distance between genes and centromeres. *See* autopolyploids; polyploidy; synteny; trisomic analysis. (Mather, K. 1935. *J. Genet.* 30:53; Rédei, G. P. 1982. *Genetics.* Macmillan, New York.)

maximal parsimony *See* evolutionary tree.

maximal permissive dose *See* radiation hazard assessment.

maximization of gene expression Can be achieved by the selection or modification of optimal promoters or in prokaryotes by varying the bases immediately after the Shine-Dalgarno sequence or manipulation of the triplet preceding the first methionine codon. *See* regulation of gene expression; regulation of protein synthesis; Shine-Dalgarno.

maximum likelihood method applied to recombination
The justification for the use of the maximum likelihood principle in estimating recombination frequencies is that the value obtained has the smallest variance among all procedures. The estimation is based on the maximization of

$$\frac{n!}{a_1!a_2!\ldots a_t!}(m_1)^{a_1}(m_2)^{a_2}\ldots(m_t)^{a_t}$$

where n is the population size, $a_1\ldots a_t$ stands for the number of individuals in the different phenotypic or genotypic classes,

$m_1\ldots m_t$ represents the expected proportions of individuals in classes $1\ldots t$. After maximizing the logarithm of the likelihood (L) expression with respect to the recombination fraction (p), we have

$$L = C + a_1\log m_1 + a_2\log m_2 + \cdots a_t\log m_t$$

where C is a constant of the maximum likelihood that is eliminated upon differentiation:

$$\frac{dL}{dp} = a_1\frac{d\log m_1}{dp} + a_2\frac{d\log m_2}{dp} + a_t\frac{d\log m_t}{dp} = 0$$

For the coupling experiment table next page:

$$L = 4032\log\left(\frac{1}{2} - \frac{1}{2}p\right) + 149\log\left(\frac{1}{2}p\right)$$
$$+ 152\log\left(\frac{1}{2}p\right) + 4035\log\left(\frac{1}{2} - \frac{1}{2}p\right)$$

After maximization and differentiation:

$$\frac{dL}{dp} = \frac{4032}{1-p} + \frac{149}{p} + \frac{152}{p} - \frac{4035}{1-p} = 0 \text{ and}$$
$$p = \frac{149 + 152}{8368} \cong 0.03597$$

The standard error s_p is calculated:

$$-\frac{1}{V_p} = S\left(mn\frac{d^2\log m}{dp^2}\right)$$

Since

$$a\frac{d\log m}{dp}$$

was defined earlier, after a second differentiation and substitution (mn) for (a), we obtain

$$-\frac{1}{V_p} = -\frac{n}{2}\left(\frac{1}{1-p} + \frac{1}{p} + \frac{1}{p} + \frac{1}{1-p}\right) = \frac{n}{p(1-p)}$$
$$\cong \frac{8368}{0.034676} \cong 241,319$$

and hence

$$V_p = 0.000004143$$

and

$$s_p = \sqrt{V_p} \cong 0.00204$$

or by the general formula

$$s_p = \sqrt{\frac{p[1-p]}{n}}$$

Recombination in F_2 can also be estimated with the aid of the maximum likelihood principle and exemplified by a coupling phase progeny:

Test-cross examples:

	PARENTAL	RECOMBINANT	RECOMBINANT	PARENTAL	
Gametic genotypes →	AB	Ab	aB	ab	Σ
Observed in **coupling**	4032	149	152	4035	8368
Expected coupling	$\frac{1}{2}n(1-p)$	$\frac{1}{2}n(p)$	$\frac{1}{2}n(p)$	$\frac{1}{2}n(1-p)$	n
	RECOMBINANT	PARENTAL	PARENTAL	RECOMBINANT	
Gametic genotypes →	AB	Ab	aB	ab	Σ
Observed **repulsion**	638	21,379	21,096	672	43,785
Expected repulsion	$\frac{1}{2}n(p)$	$\frac{1}{2}n(1-p)$	$\frac{1}{2}n(1-p)$	$\frac{1}{2}n(p)$	n

F_2	PAREN-TAL	RECOMBI-NANT	RECOMBI-NANT	PAREN-TAL		
Phenotypic classes	AB	aB	Ab	ab	Σ	
Expectation	$\frac{n}{4}(2+P)$	$\frac{n}{4}(1-P)$	$\frac{n}{4}(1-P)$	$\frac{n}{4}P$	n	**(1)**
Observed	663	36	40	196	935	

$$L = 663 \log\left(\frac{1}{2} + \frac{1}{4}P\right) + 36 \log\left(\frac{1}{4} - \frac{1}{4}P\right)$$

$$+ 40 \log\left(\frac{1}{4} - \frac{1}{4}P\right) + 196 \log\left(\frac{1}{4}P\right) \quad \textbf{(2)}$$

Upon maximization:

$$\frac{dL}{dP} = \frac{663}{2+P} - \frac{36}{1-P} - \frac{40}{1-P} + \frac{196}{P} = 0 \quad \textbf{(3)}$$

This can be reduced

$$\frac{663(1-P)(P)}{2P - P^2 - P^3} - \frac{76(2+P)(P)}{2P - P^2 - P^3} + \frac{196(2+P-2P-P^2)}{2P - P^2 - P^3} \quad \textbf{(4)}$$

Common denominator omitted and multiply

$$663(P - P^2) - 76(2P + P^2) + 196(2 + P - 2P - P^2) \quad \textbf{(5)}$$

Multiplication completed:

$$663P - 663P^2 - 152P - 76P^2 + 392 + 196P - 392P - 196P^2 \quad \textbf{(6)}$$

Terms summed up:
$$392 + 315P - 935P^2 = 0.0001585$$
$$\text{(close to zero)} \quad \textbf{(7)}$$

Designate terms: c b a

The right side of eq. (7) can be determined only after solving the quadratic equation below:

$$P = -b \pm \frac{\sqrt{b^2 - 4ac}}{2a} = 315 \pm \frac{\sqrt{99225 + 1466080}}{1870}$$

$$= -315 \pm \frac{\sqrt{1565305}}{1870} = -315 \pm \frac{1251.1215}{1870} = \frac{-1566.1215}{1870}$$

$$= -0.837498; \text{ after changing sign, } P = 0.837498$$

Thus, $P = 0.837498$ and $\sqrt{P} = 0.9151492 = 1 - p$, and hence the recombination fraction $p = 1 - 0.9151492 = 0.0848508$. The variance of P,

$$V_P = \frac{2P[1-P][2+P]}{n[1+2P]} = 0.0004495,$$

where $n(= \Sigma) = 935$ and the variance of p,

$$V_p = \frac{V_P}{4P} = 0.0001342$$

and the standard error

$$s_p = \sqrt{V_p} = \sqrt{0.0001342} = 0.01158$$

Thus, the frequency of recombination between the two genes is ~0.085 ± 0.012. Data may be entered at step (6) to expedite routine calculations. *See* F_2 *linkage estimation; information; maximum likelihood principle; recombination frequency.* (Mather, K. 1957. *The Measurement of Linkage in Heredity.* Methuen, London, UK; Wu, R. & Ma, C.-X. 2002. *Theor. Population Biol.* 61:349.)

maximum likelihood principle Provides a statistical method for estimating the optimal parameters from experimental data. The best statistics for the computations provide the smallest variance, e.g., the variance of the median of a sample is $\frac{\pi\sigma^2}{2n}$, which is $\frac{\pi}{2} = 1.57$ times the size of the variance of the mean $(\bar{x})$. Therefore, the mean is a much better characteristic of the population than the median. The binomial probability is expressed as $\binom{n}{r} p^r (1-p)^{n-r}$ given the probability (p) that (r) events occur in a sample of (n).

The relative probability of r/n events for different values of (p) is called the *likelihood*. The procedure that facilitates finding a population parameter (θ) that maximizes the likelihood of a particular observation is a *maximum likelihood procedure*. If the dispersion of a population follows the normal distribution, the variance $V = \sigma^2$ is a maximum likelihood estimator of the distribution of that population. All other methods need to be compared with and tested against this method before their results can be accepted and used.

Naturally, all statistics provide only predictions and not direct proof regarding the biological mechanism concerned. Therefore, careful collection of data, replications, sufficient sample sizes, etc., are indispensable for accuracy and predictability. The maximum likelihood mandates that the choice of the parameter (θ) makes the likelihood, $L(X_1, X_2 \dots X_n|\theta)$ the largest value. Example: A random sample of 20 is obtained, and among them, say 12 belongs to a particular class. We can hypothesize that the true frequency of this class is either (I): $p = 0.6$ or (II): 0.5 or (III): 0.7. According to the normal distribution:

$$\text{(I) } \binom{20}{12} (0.6)^{12}(0.4)^8 = \frac{20!}{12!8!}(0.6)^{12}(0.4)^8$$
$$= 125{,}970 \times 0.002176782$$
$$\times 0.00065536$$
$$\cong 0.17971$$

$$\text{(II) } \binom{20}{12} (0.5)^{12}(0.5)^8 = \frac{20!}{12!8!}(0.5)^{12}(0.5)^8$$
$$= 125{,}970 \times 0.000244140$$
$$\times 0.00390625$$
$$\cong 0.12013$$

$$\text{(III) } \binom{20}{12} (0.7)^{12}(0.3)^8 = \frac{20!}{12!8!}(0.7)^{12}(0.3)^8$$
$$= 125{,}970 \times 0.013841287$$
$$\times 0.00006561$$
$$\cong 0.11440$$

Obviously, hypothesis (I) has the maximum likelihood to be applicable to this case. After this simple demonstration, we can generalize the likelihood function as

$$L(X_1, X_2 \dots X_N | p) = \binom{N}{r} p^r (1-p)^{N-r}$$

where X are the samples, $N =$ population size, $p =$ probability, and $r = 0, 1, \dots N$. The maximized likelihood is conveniently expressed by the logarithm of the likelihood function:

$$\log L = \log \binom{N}{r} + (r)\log(p) + (N-r)\log(1-p)$$

After differentiation to (p) and equating it to zero:

$$\frac{d}{dp} \log L = \frac{r}{p} - \frac{N-r}{1-p} = 0$$

After bringing it to the common denominator:

$$\frac{r(1-p) - (N-r)p}{p(1-p)} = 0$$

The denominator omitted:

$$r(1-p) - (N-r)p = 0 = r - rp - Np + rp$$

and hence $p = r/N$, and this is the *maximum likelihood estimator* of p. Similarly it can be shown that for a population in normal distribution the arithmetic mean of the sample

$$\left(\frac{\sum x_i}{N} \right)$$

is the maximum likelihood estimator of μ. The probability P for a multinomial distribution is

$$\frac{N!}{X!Y!Z!\dots} p^X q^X r^X$$

where $p, q, r \dots$ are the probabilities of $X, Y, Z \dots$ classes. Although we may not know these probabilities, we may have experimentally observed the classes (genotypes, alleles, etc.), and we can derive the likelihood function that permits the estimation of the parameters of p, q, etc. If in a random mating population the proportion of A is p^2, that of B is $2pq$ and that of C is q^2, we can write the likelihood function as

$$L = \frac{N!}{A!B!C!} (p^2)^A (2pq)^B (q^2)^C$$

from which after logarithmic conversion and differentiation we can obtain the value of

$$p = \frac{2A + B}{2N}$$

and

$$q = \frac{2C + B}{2N}$$

and the variance

$$V_p = \frac{pq}{2N}$$

For an in-depth treatment of maximum likelihood, mathematical statistics monographs should be consulted. The maximum likelihood method is widely used in decision-making theory. In genetics, it is most commonly used for the estimation of recombination and allelic frequencies. *See* information; maximum likelihood method applied to recombination frequencies; probability.

maximum parsimony   *See* evolutionary tree.

maxizyme   Dimeric ribozyme.

May-Hegglin anomaly (Dohle leukocyte inclusions with giant platelets) Asymptomatic dominant granulocyte and platelet disorder often resulting in thrombocytopenia located to chromosome 22q12.3-q13.1. The locus is about 0.7 megabase DNA. In the granulocytes, spindle-shaped cytoplasmic inclusions have been observed that appear to be the depolarization relics of ribosomes. Fechtner and Sebastian syndromes share the major characteristics and the chromosomal location (22q13.3-q13.2). Nonmuscle myosin heavy-chain IIA (MYH9) mutations appear to account for the three diseases. Alport syndrome shares some of the symptoms, although it is a different disease. *See* giant platelet; hemostasis; platelet anomalies. (Martignertti, J. A., et al. 2000. *Am. J.*

Hum. Genet. 66:1449; Kelley, M. J., et al. 2000. *Nature Genet.* 26:106.)

maytansinoids Extract of tropical trees or shrubs with an LDLo of 190 µg/kg as an intravenous dose for humans. *See* LDLo; magic bullet, formula below.

Maytansin (related compounds are rifamycin, streptovaricin, macrolides, etc).

M9 bacterial minimal medium 5 concentrated salt solution g/L H_2O, Na_2HPO_4. 7 H_2O 64, KH_2PO_4 15, NaCl 2.5, NH_4Cl 5 → 200 µL, 1M $MgSO_4$ 2 mL, glucose 4 g, fill up to 1 L after any supplement (e.g., amino acids) added.

MBD Methyl-binding domain proteins are members of the histone deacetylase complex, bind to DNA, and remodel the chromatin, resulting in gene silencing. *See* 5-azacytidine; histone deacetylase.

MBF *See* Mbp1; Swi.

Mbp Megabase pair, 1 million base pair.

Mbp1 Mitotic-binding protein. Components of MBF (microtubule-binding factor) with Swi6 mediate S-phase expression of the cell cycle. *See* cell cycle; SBF; Swi. (Iyer, V. R., et al. 2001. *Nature* 409:533.)

MBP Maltose-binding protein, encoded by gene *malE* (91 min) of *E. coli*.

MCA Metabolic control analysis is the study of complex enzyme systems in response to any changes in substrate(s). It may facilitate the detection of thresholds, potentially leading to disease. It may permit the classification of genes and proteins regarding their role in metabolic networks and thus may help drug discovery. *See* proteomics. (Cascasnte, M., et al. 2002. *Nature Biotechnol.* 20:243.)

McArdle's disease *See* glycogen storage disease type V.

McCune-Albright syndrome (GNAS1, 20q13.2) Pituitary neoplasia resulting from excessive secretion of growth hormone, caused by mutation and constitutive expression of the GTP-binding subunit (G_s) of a G protein. *See* G proteins; pituitary gland; pituitary tumor; securin.

MCF oncogene Synonymous with DBL and ROS. The human mammary carcinoma protooncogene was assigned to human chromosome Xq27. It encodes a serine-phosphoprotein (p66). *See* oncogenes; ROS.

Mch ICE-related proteases. *See* apoptosis; ICE.

MCH Melanin-concentrating hormone reduces appetite and increases metabolic rate. MCH encodes a neuropeptide precursor at 12q23; PMCHL1 at 5p14 and PMCHL2 at 5q13 are truncated versions of MCH. *See* leptin; melanin; obesity.

MCK Muscle-specific kinase. *See* MyoD.

McKusick-Kaufman syndrome (MKKS, 20p12) Recessive developmental anomaly including accumulation of fluids in the uterus and vaginal area (hydrometrocolpos), extra finger at the area of the little finger (postaxial polydactyly), heart disease, etc. Several other genes apparently map to the same chromosomal location, e.g., that of Bardet-Biedl (BBS) syndrome. The distinction between MKKS and BBS is by the three criteria named above. Some of the symptoms are shared by Ellis–van Creveld syndrome that is at another chromosomal location. The critical protein appears to be a chaperonin. *See* Bardet-Biedl syndrome; Ellis–van Creveld syndrome. (David, A., et al. 1999. *J. Med. Genet.* 36:599; Slavotinek, A. M., et al. 2000. *Nature Genet.* 26:15.)

McLeod syndrome (XK) Recessive human Xp21 region deficiency of the Kx blood antigen precursor. The symptoms vary because of an overlapping defect with closely linked genes, particularly CGD (chronic granulomatous disease). It may be associated with acanthocytosis, which is characteristic for abetaliproteinemia. *See* abetalipoproteinemia; contiguous gene syndrome; granulomatous disease, chronic; Kell-Cellano blood group.

MCM1 (licensing complex) Yeast DNA-binding protein, a product of the minichromosome maintenance gene also involved in the regulation of mating type; it controls the entry into mitosis. All eukaryotes apparently have at least six different MCMs assisting in DNA replication. MCM 2–7 appear to become activated in telophase before the next cell cycle (Dimitrova, D. S., et al. 2002. *J. Cell Sci.* 115:51). Some have helicase and DNA-dependent ATPase function. The licensing actually means that the chromatin must be subject to quality control to qualify for replication after mitosis. MCM prevents reentry into S phase. *See*

ARS; Cdc6; Cdc18; CDC19; CDC21; Cdc45/Cdc46/Mcm5; cell cycle; Cdt1; geminin; mating type determination in yeast; MCM3; reinitiation of replication. (Tye, B. K. 1999. *Annu. Rev. Biochem.* 68:649; Lee, J.-K. & Hurwitz, J. 2000. *Proc. Natl. Acad. Sci. USA* 98:54; Nishitani, H., et al. 2001. *J. Biol. Chem.* 276:44905.)

MCM3   Apparently the same as the replicational licensing factor (RLF) that appears in tight binding to DNA during interphase but is released during S phase. This factor assures that within a cell just one cycle of DNA replication occurs. This protein belongs to the family of MCM1 to MCM5 factors detected in yeast. *See* cell cycle; MCM1; replication licensing factor.

M component   Paraprotein.

MCP   Membrane cofactor protein CD46 regulates (along with other proteins such as DAF, factor H, and C-4-binding protein) complement functions and protects the cells from attacks by their own defense system. MCP and DAF control reproductive functions (spermatozoa, extrafetal tissues) besides infectious diseases and xenografts. These functions are encoded in human chromosome 1q3.2 region. MCP (14 exons) has four isoforms generated by alternative splicing of a single transcript. MCP shares a 34-amino-acid signal peptide with DAF (11 exons). The signal sequence is followed by *complement control protein repeats* (CCPR) where C3b and C4b complement components bind. Glycosylated residues and serine, threonine, and proline (STP) residues follow CCPR sequences. The other regions of MCP and DAF are different. *See* atherosclerosis; complement; decay accelerating factor; signal sequence; xenotransplantation. (Kemper, C., et al. 2001. *Clin. Exp. Immunol.* 124[2]:180.)

MCP-1   Monocyte chemoattractant protein controls (along with IL-8) adhesion of monocytes to the vascular epithelium. It is inducible by platelet-derived growth factor (PDGF). Mice lacking these receptors are more prone to atherosclerosis. *See* atherosclerosis; chemokines; monocytes. (Yamamoto, T., et al. 2001. *Eur. J. Immunol.* 31:2936.)

mcr   *See* methylation of DNA.

MCR   Mutation cluster region is a segment of a gene where mutations occur at a high frequency.

MCS   Multiple cloning sites. *See* polylinker.

M-CSF   *See* macrophage colony-stimulating factor.

M cytotype   *See* hybrid dysgenesis.

mdg   *See* copia.

Mdl1   Mitochondrial export protein of the AAA transporter family. *See* AAA proteins.

MDM2   Murine double-minute homologue 12q14.3-q15 is a cellular oncoprotein that can bind and downregulate p53 tumor suppressor, attach to the retinoblastoma suppressor, stimulate transcription factors E2F1 and DP1, and thus may promote tumorigenesis. MDM2 can suppress TGF-β effect without the inactivation of p53. MDM2 is regulated by RASA through the Raf/Mek/Map kinase pathway. Raf also activates p19^{ARF}, an inhibitor of MDM2. MDM2 involves TGF resistance in various tumor cell lines. MDM2 prevents transcriptional activation by Sp1, but the retinoblastoma protein displaces Sp1 from MDM2 and restores Sp1 transcriptional activity. MDM2 may stimulate ubiquitins. *See* apoptosis; ARF; cyclin G; DP1; E2F; oncogenes; p53; retinoblastoma; TGF; transcription factors; ubiquitins. (Johnson-Pais, T., et al. 2001. *Proc. Natl. Acad. Sci. USA* 98:2211.)

MDR   *See* multidrug resistance.

MDS   Macronuclear destined sequences. From the germline DNA during vegetative development of ciliates internal sequences are eliminated by the process of chromosome diminution and only the MDS is retained in the macronucleus. *See* chromosome diminution; macronucleus; *Paramecium*. (Prescott, D. M. 1997. *Curr. Opin. Genet. Dev.* 7:807.)

meal worm (*Tenebrio molitor*)   Sex is determined by a larger X and a smaller Y chromosome in this insect. *See* antifreeze protein.

mealybug   Member of the coccidea taxonomic group of animals, with the name reflecting the "mealy" appearance of the wax coat of the body of the insects. They received attention by the peculiarity of their chromosome behavior. During the cleavage divisions immediately after fertilization, all the chromosomes are euchromatic. After blastula, one-half of the chromosomes ($2n = 10$) become heterochromatic in the embryos that develop into males. At interphase, these heterochromatic chromosomes clump into a chromocenter. By metaphase, the heterochromatic and euchromatic sets are no longer distinguishable. In the males the first meiotic division is equational and during the second division the two types of chromosomes go to opposite poles. Two of the four nuclei are heterochromatic and two are euchromatic. The heterochromatic nuclei then disintegrate and the euchromatic cells proceed to spermiogenesis. The euchromatic set of the fathers later becomes the heterochromatic chromosomes of the sons. This was verified by X-raying the females and males. Only 3% of daughters of males irradiated by 16,000 R survived, but the sons were unaffected even after 30,000 R. Some sons survived even after 90,000 R exposure of the fathers. Thus, sex appears to be determined developmentally in these insects. *See* chromosomal sex determination. (Nur, U. 1967. *Genetics* 56:375; Palotta, D. 1972. *Can. J. Genet. Cytol.* 15:809.)

mean   The *arithmetic mean* $\bar{x}$ is equal to the sum (Σ) of all measurements (x) divided by the number (n) of all measurements, or

$$\bar{x} = \frac{\sum x}{n}$$

The *geometric mean* (G) is the nth root of the product of all measurements:

$$G = \sqrt[n]{x_1.x_2 \ldots x_n}$$

The *harmonic mean* (H) is the inverse average of the reciprocals of the measurements

$$H = \frac{n}{\Sigma[1/x]}$$

Examples:

$$\bar{x} = \frac{2+8}{2} = 5, \quad G = \sqrt[3]{2 \times 8 \times 4} = \sqrt[3]{64} = 4,$$

$$H = \frac{2}{[1/2]+(1/8)} = 3.2$$

The *weighted mean* is the calculated mean multiplied by the pertinent frequency of the groups in a population. *See* variance.

meander   Two consecutive β-sheets of a protein are adjacent and antiparallel. *See* protein structure.

mean lethal dose   Mutagens or toxic agents are denoted by LD_{50}. *See* LC50; LD_{50}; LDLo.

mean squares   Average of the squared deviations from the mean; it is obtained by dividing the sum of the squared deviations by the pertinent degrees of freedom. Basically, this is the estimated variance. *See* intraclass correlation; variance; variance analysis.

measurement units   *Length:* 10 ångström (Å) = 1 nanometer (nm), 1,000 nm = 1 micrometer (μm), 1,000 μm = 1 millimeter (mm), 10 mm = 1 centimeter (cm), 100 cm = 1 m. *Volume:* 1,000 microliter (μL) or λ = 1 milliliter (mL), 1,000 mL = 1 liter (L). *Weight:* 1,000 picogram (pg) = 1 nanogram (ng), 1,000 ng = 1 microgram (μg), 1,000 μg = 1 milligram (mg), 1,000 mg = 1 gram (g), 10 g = 1 dekagram (dg), 100 dg = 1 kilogram (kg). *Generally:* milli = 10^{-3}, micro = 10^{-6}, nano = 10^{-9}, pico (p) = 10^{-12}, fempto (f) = 10^{-15}, atto (a) = 10^{-18}, kilo (k) = 10^3, mega (M) = 10^6, giga (G) = 10^9, tera (T) = 10^{12}. *See* dalton; M_r.

MEC   *See* degenerin; ion channels.

MEC1   Kinase locus of yeast (member of the PIK family); it phosphorylates RAD53 and RAD9, signal transducers of DNA damage. Mec3 protein seems to regulate telomere length. Mec1 and Rad53 are also involved in the G1, S, and G2 checkpoint control. The Mec1 function can also be carried out by Teℓ1. The homologues are *SAD3, ESR1,* and the human gene is homologous to ΛT, responsible for ataxia telangiectasia. *See* ataxia; cell cycle; checkpoint; DNA replication; PIK; RAD; signal transduction; telomeric silencing. (Tercero, J. A. & Diffley, J. F. X. 2001. *Nature* 412:553; Lopes, M., et al. 2001. *Nature* 412:557.)

mechanism-based inhibition   *See* regulation of enzyme activity.

mechanosensory gene   Involved in the neurobiological control of proprioceptory sensations. The proprioceptory nerve terminals are located in the muscles, joints, tendons, and ears and perceive the information about movements, touch, balance, and hearing. The process converts mechanical forces into electrical signals through special ion channels.

mecillinam β-lactam-type antibiotic that targets the penicillin-binding protein 2 (PBP2) required for the elongation of the bacterial cell wall. *See* β-lactamase.

Meckel syndrome   Rare complex recessive syndrome (17q22-q23) with brain defects, cystic kidneys, and polydactyly. In Finnish populations it may occur in the 10^{-4} range. *See* neural tube defects; polydactyly.

MeCP1   Methyl-CpG-binding protein MBD1 binds methylated CpG sequences in the DNA and is part of a transcriptional repression complex along with histone deacetylase. It is encoded at human chromosome 18q21. *See* CpG islands; histone deacetylase; methylation of DNA.

MeCP2   Methyl-CpG-binding protein encoded at Xq28. It is involved with Rett syndrome autuism. *See* autism.

MED-1   Null promoter.

medaka (*Oryzias latipes*)   Small, fertile fish well suited for developmental studies because a genetically pigment-free (because of four recessive genes) and transparent stock is available that permits the direct visualization through a stereoscopic microscope of the major internal organs (heart, spleen, blood vessels, liver, gut, gonads, kidney, brain, spinal cord, eye lens, air bladder, etc.) in live animals. *See* zebrafish. (Wakamatsu, Y., et al. 2001. *Proc. Natl. Acad. Sci. USA* 98:10046.) Photo by courtesy of Dr. Yuko Wakamatsu.

Medaka female.

MEDEA factor   Maternal effect dominant embryonic arrest is a lethality factor transmitted through the egg cytoplasm, which kills the offspring, in, e.g., *Tribolium*, unless it inherits a rescuing M factor from either parent. *See* killer genes; *Tribolium*. (Beeman, R. W. & Friese, K. S. 1999. *Heredity* 82:529; Grossniklaus, U., et al. 1998. *Science* 280:446.)

median   Statistical concept indicating that equal numbers of (variates) observations are on its sides at both minus and plus directions. *See* mean; mode.

mediator   Assembly factors of RecA and like recombinases and single-strand DNA-binding proteins. *See* RecA. (Gasior, S. L., et al. 2001. *Proc. Natl. Acad. Sci. USA* 98:8411.)

mediator complex (Meds)   Group of ~20 or more proteins involved in the facilitation of transcription by RNA polymerase II in yeast and other eukaryotes (see Bjorklund, S., et al. 1999. *Cell* 96:759). These proteins share subunits and participate in a large variety of different complexes that have different functions in gene expression and developmental control. They may mediate chromatin remodeling and interact with various proteins (activators), general transcription factors, and directly or indirectly with RNA polymerase II. *See* activator proteins; chromatin remodeling; coactivator; DRIP; elongator; NAT; reinitiation; Srb; TAF; TBP; transcription factors. (Svejstrup, J. Q., et al. 1997. *Proc. Natl. Acad. Sci. USA* 94:6075; Gustafsson, C. M., et al. 1998. *J. Biol. Chem.* 273:30851; Myers, L. C. & Kornberg, R. D. 2000. *Annu. Rev. Genet.* 69:729; Gustafsson, C. M., et al. 2001. *Mol. Microbiol.* 41:1; Boube, M., et al. 2002. *Cell* 110:143.)

medical error   Kills 44,000 to 98,000 people in U.S. hospitals annually (Hayward, R. A. & Hofer, T. P. 2001. *J. Am. Med. Assoc.* 286:415.)

medical genetics   Genetics applied to medical problems. *See* clinical genetics; human genetics.

medicinal chemistry   Involved with drug design and development.

Mediterranean fever, familial (FMF)   Human chromosome 16p13 recessive disease with recurrent spells of fever, pain in the abdomen, chest, and joints and red skin spots (erythema). It is a type of amyloidosis. The basic defect involves a 781-amino-acid protein, pyrin. In some populations, the prevalence, gene frequency, and carrier frequency may be 0.00034, 0.019, and 0.038, respectively. *See* amyloidosis. (Schaner, P., et al. 2001. *Nature Genet.* 27:318.)

MEDLINE   Bibliographic system of the National Library of Medicine. It can be reached online as part of the MEDLARS database: <http://www4.ncbi.nlm.nih.gov/PubMed/>.

medRNA   Mini-exon-dependent RNA. *See* Trypanosoma brucei.

medulla   Inner part of organs, the basal part of the brain connecting with the spinal chord. *See* brain, human.

medulloblastoma (17p13.1-p12, 10q25.3-q26.1, 1p32)   Brain cancer of childhood. Its frequency in adenomatous polyposis increases by about two orders of magnitude. *See* Gardner syndrome; nevoid basal cell carcinoma.

Meekrin-Ehlers-Danlos syndrome   Connective tissue disorder.

MEF   Series of myocyte enhancer-binding factors that specifically potentiate the transcription of muscle genes and thus differentiation of various types of muscles and myoblasts. The MEF group belongs to the family of MADS domain protein. MEF2 is a Ca^{2+}-regulated transcription factor and it is actively transcribed during development of the central nervous system. MEF2 is activated when it dissociates from histone deacetylase in response to MAPK signals. Nur77 and Nor1 steroid receptors mediate apoptosis of T-cell receptors controlled by MEF2. MEF2C is involved in inflammation responses and it is stimulated by lipopolysaccharides of gram-negative bacteria. MEF2C transactivation is regulated through phosphorylation by p38. *See* apoptosis; gram-negative; MADS box; MAPK; MRF4; MYF5; myocyte; MyoD; Nor1; Nur77; p38; T cell. (Mora, S., et al. 2001. *Endocrinology* 142:1999.)

megabase (Mb)   1 million nucleic acid bases.

megadalton (Mda) 10^6 dalton; 1 da (or Da) = 1.661×10^{-24} g.

megalencecephalic leukoencephalopathy (MLC, 22qtel)   Recessive enlargement of the brain, defective motor functions (ataxia, spasms), and mental deterioration caused by defects in a transmembrane protein. Onset is within a year of birth and a slow progressive realization of the symptoms follows.

megaevolution   Process and facts of descent of higher taxonomic categories. *See* evolution; macroevolution.

megagametophyte   One of the four functional haploid products of female meiosis (megaspores) in plants, and it develops into the embryo sac. Its origin and most prevalent developmental paths are outlined in a figure at the gametophyte entry. *See* gametophyte (female).

megakaryocyte   Large cell in the bone marrow with large lobed nucleus. The cytoplasm of megakaryocytes produces the platelets. Megakaryocyte formation from stem cells is regulated by the cytokine receptor cMpl and its ligand, megakaryocyte lineage-specific growth factor (meg-CSF), which is homologous to erythropoietin and has both meg-CSF and thrombopoietin-like activities. *See* erythropoietin; platelet; thrombopoietin.

megalin   Cell surface apolipoprotein-B receptor. Its deficiency may lead to holoprosencephaly. *See* apolipoprotein; cholesterol; holoprosencephaly. (Barth, J. L. & Argraves, W. S. 2001. *Trends Cardiovasc. Med.* 1:26.)

megaloblast   Large, nucleated, immature cells giving rise to abnormal red blood cells.

megaloblastic anemia   Autosomal-dominant (human chromosome 5q11.2-q13.2) deficiency of dehydrofolate reductase (involved in the biosynthetic path of purines and pyrimidines) resulting in hematological and neurological anomalies. The symptoms may be alleviated by 5-formyltetrahydrofolic acid. A rare recessive type (MGA1, 10p12.1) is caused by intestinal malabsorption of vitamin B_{12}, due to defects in cubilin, the intrinsic factor (IF)-B_{12} receptor. Another form (TRMA, Rogers syndrome, 1q23.2-q23.3) responds favorably to thiamin. In chromosome 5p15.3-p15.2, methylcobalamine and methionine synthase reductase have been located. Megaloblastic anemia may be found in several other syndromes. *See* anemia; folic acid; megaloblast; phosphatase (ACP1); thiamin; transcobalamine deficiency.

megaspore   *See* gametogenesis; megagametophyte.

megaspore competition   Determines which of the four products of meiosis (megaspore) in the female (plant) becomes functional. It occurs only in a few species such as *Oenotheras*. *See* certation; gametogenesis; pollen.

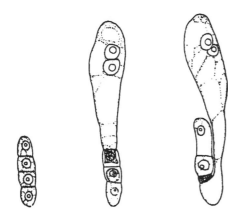

| Megaspore tetrad | TOP megaspore develops | BASAL megaspore develops |

Megaspore competition in the *Oenotheras* where normally the top spore of the tetrad is functional, but in case there is a deleterious gene in the top spore, the basal spore may compete with it successfully. (After Renner, O. from Goldschmidt, R. 1928. *Einführung in die Verebungswissenschaft*. Springer-Verlag, Berlin, Germany.)

megaspore mother cell   Diploid cell that produces the haploid megaspores through meiosis in the female plant. *See* gametophyte (female).

Basal megaspore.

megasporocyte   Same as megaspore mother cell. *See* gametogenesis; gametophyte.

MEI41   270 kDa *Drosophila* phosphatidylinositol kinase. When inactivated, meiotic recombination is reduced. *See* PIK.

meiocyte   Cell that undergoes meiosis. *See* meiosis.

meiosis   Two-step nuclear division that reduces somatic chromosome number ($2n$) to half (n) and is usually followed by gamete formation. (See diagram of meiosis next page.) Meiosis is genetically the most important step in the life cycle of eukaryotes. Meiosis proceeds from the 4C sporocyte (in diploids) and includes one numerically reductional and one numerically equational chromosome division.

Synapsis takes place at prophase I to metaphase I. Chiasmata may be visible by the light microscope during prophase I. Centromeres do not split at anaphase I and the sister chromatids are held together at the centromeres during the separation of the bivalents at anaphase I. At the completion of meiosis I, the chromosome number is reduced to half (2C), and by the end of meiosis II, the C-value of each of the 4 haploid daughter cells is 1. The major stages of meiosis are shown in the diagram below. These stages are more transitional than absolutely distinct. The nucleolus is not shown. The nuclear membrane is generally not discernible by the light microscope from metaphase to anaphase but reappears at telophase. Dashed and solid thin lines represent the spindle fibers. The genetic consequences of the meiotic behavior of the chromosomes are best detected in ascomycete fungi with linear tetrads.

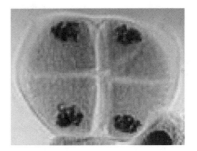

The four male meiotic products in plants.

The duration of meiosis generally much exceeds that of mitosis and the longest is the prophase stage. In the majority of plants, the completion of mitosis requires 1–3 hours, whereas meiosis may need 1–8 days. In yeast, meiosis takes place in about 7 hours. The stages most revealing for the cytogeneticist—pachytene (2–8), diplotene (0.5–1), metaphase I (1.5–2), and anaphase I (0.5–1)—generally require the number of hours indicated in parentheses. In human females, meiosis may stall at the late prophase stage, at dictyotene, and the subsequent divisions take place only before the onset of ovulations (and following fertilization), a period repeated approximately 13 times annually during about 40 years. The activation of meiosis requires C_{29} sterols in both males and females. In *Schizosaccharomyces pombe*, starvation initiates the switch from mitosis to meiosis. Under such conditions protein kinase Pat1/Ran1 is inactivated by the expression of the $mei3^+$ gene. The substrate of this kinase is the RNA-binding protein Mei2. Dephosphorylation of Mei2 protein also causes a change from mitosis into meiosis. Mei2 is required for premeiotic DNA synthesis and the polyadenylated meiRNA promotes the first nuclear division. Mei2 is localized in the cytoplasm in mitotic cells, but during meiotic prophase it has been visualized in the microtubule-organizing center. Meiosis in budding is controlled by about 150 genes, but meiosis affects the expression of over 10 times more loci. *See* anaphase; C amount of DNA; cell cycle; diakinesis; dictyotene; diplotene; gametophyte; interphase; leptotene; metaphase; mitosis; nucleolus; pachytene; Pat1; Ran1; sister chromatid cohesion; tetrad analysis; zygotene. (Zickler, D. & Kleckner, N. 1998. *Annu. Rev. Genet.* 32:619; Zickler, D. & Kleckner, N.

1999. *Annu. Rev. Genet.* 33:603; Rabitsch, K. P., et al. 2001. *Curr. Biol.* 11:1001; Davis, L. & Smith, G. R. 2001. *Proc. Natl. Acad. Sci. USA* 98:8395; Forsburg, S. 2002. *Mol. Cell* 9:703; Nakagawa, T. & Kolodner, R. D. 2002. *J. Biol. Chem.* 277:28019; Lemke, J., et al. 2002. *Am. J. Hum. Genet.* 71:1051.)

meiosis I First stage of meiosis when through reduction of the chromosome number the 4C amount of DNA in the meiocyte takes place and each of the two daughter nuclei has 2C amounts of DNA. *See* C amount of DNA; meiosis.

meiosis II Follows meiosis I and is basically an equational division of the chromosomes resulting in four daughter nuclei of the meiocyte. Each has only 1C amount of DNA. *See* C amount of DNA; meiosis.

meiotic drive Results in unequal proportions of two alleles of a heterozygote among the gametes in a population because certain meiotic products are not functional or are less functional and consequently the proportion of other gametes increases. In spite of the preferential transmission of the segregation distorters, the various populations display fewer carriers of the distorter than expected. Meiotic drive usually requires the *drive locus* (with driving and nondriving alleles) and the *target locus* (with sensitive and resistant alleles). These two loci are usually tightly linked in repulsion. Coupling the drive and sensitive alleles is expected to eliminate the system. These loci are usually within inversions and they involve the sex chromosomes more commonly than the autosomes, probably because the two sex chromosomes (X and Y) in *Drosophila* do not recombine, except in pseudoautosomal region. Meiotic drive may become a microevolutionary factor because it may alter gene frequencies. Meiotic drive may favor selectively disadvantageous gene combinations and thus may contribute to the genetic load of a population. Meiotic drive may be subject to genetically determined modification. Meiotic drive operates in the males or in the females but not in both. In mice, cytoplasmic factors may affect meiotic drive, and it has also been attributed to the effect of the paternal allele of the *Om (ovum)* locus by inducing the maternal allele to go preferentially to the

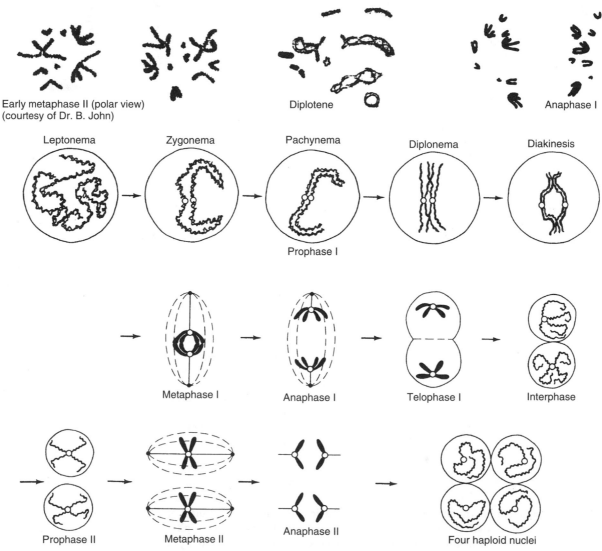

Early metaphase II (polar view)
(courtesy of Dr. B. John)

Diplotene

Anaphase I

Leptonema   Zygonema   Pachynema   Diplonema   Diakinesis

Prophase I

Metaphase I   Anaphase I   Telophase I   Interphase

Prophase II   Metaphase II   Anaphase II   Four haploid nuclei

A generalized course of meiosis represented by only one pair of chromosomes. Above a few characteristic photomicrographs of the male grasshopper *Chortippus paralellus*.

polar body after fertilization. Meiotic drive may be caused by deletions from the pericentromeric heterochromatin of the X chromosome—and Y chromosome–autosome translocations. X-linked insertions containing the 240 bp rRNA intergenic spacer or the rRNA genes may restore normal conditions. The failure of pairing in the pseudoautosomal region causes meiotic drive and XY nondisjunction. Meiotic drive may be detected by sperm typing. *See* brachyury; certation; genetic load; killer strains; megaspore competition; mitotic drive; polarized segregation; polycystic ovarian disease; preferential segregation; pseudoautosomal; rRNA; segregation distorter; sex ratio; sexual dimorphism; sexual selection; spacer DNA; sperm typing; symbionts, hereditary; transmission. (Hurst, G. D. & Werren, J. H. 2001. *Nature Rev. Genet.* 2:597.)

meiRNA Polyadenylated meiotic RNA cooperates with meiotic nonphosphorylated protein Mei2 to promote premeiotic DNA synthesis and nuclear division. *See* meiosis; Pat1; Ran. (Ohno, M. & Mattaj, I. W. 1999. *Curr. Biol.* 9:R66.)

MEIV Multilocus exchange with interference and viability is a statistical model of recombination and achiasmate segregation in tetrads. (Zwick, M. E., et al. 1999. *Genetics* 152:1615.)

MEK Member of the extracellular signal-regulated kinase (ERK) family. *See* MAPK; MP1; signal transduction. (Widmann, C., et al. *Mol. Cell Biol.* 18:2416.)

MEKK 196 kDa protein serine/threonine kinase. MEK kinase (i.e., a kinase kinase). *See* MEK; signal transduction. (Yujiri, T., et al. 1998. *Science* 282:1911.)

Melandrium (synonymous with *Lychnis, Silene*) Dioecious plant ($2n = 22 + $ XX or $22 + $ XY, Caryophyllaceae. *See* intersex; *Lychnis*. (Lardon, A., et al. 1999. *Genetics* 151:1173; Lebel-Hardenack, S., et al. 2002. *Genetics* 160:717.)

MELANIE II Computer software package that can match two-dimensional protein gel data to information in database. *See* databases.

melanin *See* agouti; albinism; piebaldism; pigmentation of animals.

melanism Increased production of the dark melanin pigment. *See Biston betularia*; industrial melanism; melanin.

melanocortin Synthesized as a complex pre-pro-opiomelanocortin that by processing contributes to the formation of adrenocorticotropin hormone (ACTH), melanocyte-stimulating hormone (MSH), and β-endorphin. MSH regulates the brain melanocortin-4-receptor (MC4R) and thereby leptin. Haploinsufficient, morbid autosomal (18q29) mutation in MC4R leads to obesity in the carriers. The agouti and related gene products are natural antagonists of the MC4R ligand MSH. *See* ACTII; agouti; endorphin; haploinsufficient; leptin; melanocyte stimulating hormone; opiocortin; pigmentation of animals. (Huszar, D., et al. 1997. *Cell* 88:131; Kistler-Heer, V., et al. 1998. *J. Neuroendocrinol.* 10:133; Chen, A. S., et al. 2000. *Nature Genet.* 26:97.)

melanocyte Produces melanin. *See* melanin.

melanocyte-stimulating hormone (MSH) Exists in forms α-MSH, β-MSH, and γ-MSH. These adenocortical hormones regulate melanization and energy utilization. *See* agouti; leptin; opiocortin; pigmentation of animals; syndecan. (Haskell-Luevano, C., et al. 2001. *J. Med. Chem.* 44:2247.)

melanoma (CMM1, 1p36; CMM2, 9p21) Forms of cancer arising in the melanocytes or other tissues. The most prevalent form appears as a mole of radial growth of reddish, brown, and pink color with irregular edges that penetrate into deeper layers as they progress. It may originate in dark freckles on the head or other parts of the body. Excessive exposure to sunlight may condition its development, although autosomal-dominant genes determine susceptibility to melanoma. Three alleles of the melanocortin receptor (MC1R, 16q24.3) double the cutaneous malignant melanoma (CMM1) risk for red-haired individuals. Melanoma may be one of the most aggressively metastatic cancers. The development of melanoma is regulated by the melanoma mitogenic polypeptide, encoded by GRO human gene in chromosome 4q21. Melanoma-associated antigen ME49 appears in the early stages of this cancer, and it is coded by an autosomal-dominant locus (MLA1) in human chromosome 12q12-q14. Melanoma-associated antigen MZ2-E is coded for by another autosomal-dominant locus. Melanoma-associated antigen p97 is a member of the iron-binding transferrin protein family and is coded by an autosomal-dominant gene, MAP97 (MF12), at human chromosome 3q28-q29. An autosomal-dominant gene in human chromosome 15 encodes the melanoma-specific chondroitin sulfate proteoglycan expressed in melanoma cells. In cultured melanoma and metastatic tissues, a mutant CDK4/CDKN2A (9p21) protein was found that was unable to bind the p16^{INK4a} protein and thereby interferes with normal regulation of the cell cycle inhibitor. In human chromosome 11, several presumptive melanoma suppressors have been found. Many of the hereditary melanomas are attributed to G → 34T mutation at this locus. Because of this mutation, an AUG initiation codon is created at site −35 and the resulting 4 kDa translation product shows no homology to CDKN2A/p16. Deletions of p16 may frequently be responsible for malignant melanoma. Catenin seems to be a transcriptional coactivator of Tcf and Lef and seems to affect melanoma progression. Approximately 5 to 10% of the melanoma patients have at least one afflicted family member. Nonmelanoma skin cancer in fair-skinned redheads may be associated with the melanocortin-1 receptor (MCR1). Genetic hybrids between the species of the platyfishes (*Xiphophorus*) are prone to develop melanoma. *See* Apaf; cancer; catenins; CDKN2A; freckles; Lef; MAGE; melanocyte; MEL oncogene; metastasis; p16; survivin; Tcf. (Mellado, M., et al. 2001. *Curr. Biol.* 11:691; van der Velden, P., et al. 2001. *Am. J. Hum. Genet.* 69:774.)

melanoma growth-stimulatory factor (GROα) Interleukin-related protein like SDF. *See* chemokines; CXCR; IL-8; SDF. (Wang et al. 2000. *Oncogene* 19:4647.)

melanosome Specialized pigment-containing animal organelle produced by melanocytes. Melanosomes include tyrosinase and related proteins, GP100, etc., and bear resemblance in function to lysozymes. *See* albinism; lysozymes; melanocyte; pigmentation of animals. (Kishimoto, T., et al. 2001. *Proc. Natl. Acad. Sci. USA* 98:10698; Marks, M. S., et al. 2001. *Nature Rev. Mol. Cell Biol.* 2:738.)

MELAS syndrome *See* mitochondrial disease in humans.

melatonin Hormone synthesized in the pineal gland that controls reactions to light, diurnal changes, seasonal adjustment in fur color, aging, sleep, reproduction, etc. It is a scavenger of oxidative radicals and it protects from ionizing radiation without substantial side effects. It is synthesized from serotonin by serotonin N-acetyltransferase and is regulated by cAMP. *See* circadian rhythm; radioprotectors; serotonin; sulfhydryl. (Reiter, R. J., et al. 2001. *Nutr. Rev.* 59[9]:286.)

MELD Overlapping, merged DNA fragments.

MELD$_{10}$ Mouse equivalent lethal dose. *See* LD50.

melibiose Galactopyranosyl-glucose, a fermentable disaccharide.

Melibiose.

MEL oncogene Isolated from human melanoma, although its role in melanoma is unclear; it was assigned to the broad area of human chromosome 19p13.2-q13.2. *See* melanoma; oncogenes; p16^{INK4}.

melt-and-slide model The dual functions of DNA polymerase I are carried out by pol I occupying the duplex primer template site for the polymerase action, and for the editing reaction the DNA melts (strands separate), unwinds, and single-strand DNA is transferred to the exonuclease site of the polymerase enzyme. *See* DNA repair; DNA replication.

meltdown, mutational Can occur when the mutation rate is high, the genetically effective population size is small, and genetic drift is high. It may lead to extinction. *See* effective population size; extinction; genetic drift. (Zeyl, C., et al. 2001. *Evolution* 55:909.)

melting Breakdown of the hydrogen bonds between paired nucleic acid strands *See* breathing of DNA; denaturation; Watson and Crick model.

melting curve of DNA Higher temperatures cause progressively higher disruption of the hydrogen bonds between DNA strands; it is also affected by the base composition of the DNA because there are three hydrogen bonds between $G \equiv C$

and two between A=T. *See* c$_0$t curve; hydrogen pairing; melting temperature; renaturation.

melting temperature The temperature where 50% of the molecules is denatured (T_m); the DNA strands may be separated (depending on the origin of the DNA, $G \equiv C$ content, solvent, and homology of the strands) generally above 80°C and melting may be completed below 100°. *See* C$_0$t curve; hyperchromicity; melting curve of DNA.

meltrin Metalloproteinase protein mediating the fusion of myoblasts into myotubes. *See* ADAM; metalloproteinase; myotubes. (Inoue, D., et al. 1998. *J. Biol. Chem.* 273:4180.)

memapsin Cloned and sequenced aspartic protease with β-secretase activity. *See* secretase.

membrane Lipid protein complexes surrounding cells and cellular organelles and forming intracellular vesicles. *See* cell membranes.

membrane attack complex (MAC) C5b can be converted to C5b6 and then to C5b67 by binding to C6 and C7 complement components, then with C8 and C9, resulting in the formation of MAC. This complex can protect the cells against certain foreign intruder cells, but it must be regulated to protect the membrane system of the cells. The so-called homologous restriction factor (HRF) is a 65 kDa glycoprotein bound to cell membranes through glycosylphosphatidylinositol (GPI) can perform the task. A smaller (20 kDa) immunoglobulin G (IgG1), also called HRF20 (CD59), and MIRL (membrane inhibitor of reactive lysis) function in a similar manner. Paroxysmal nocturnal hemoglobinuria is a mutational defect in GPI anchoring of HRF20 to the hematopoietic membrane. *See* complement; immunoglobulins; paroxysmal nocturnal hemoglobinuria; phosphoinositides. (Linton, S. 2001. *Mol. Biotechnol.* 18:135.)

membrane channel Permits passive passing of ions and small molecules through membranes. *See* cell membrane; ion channel.

membrane filter May be used for clarification of biological or other liquids, trapping macromolecules, for exclusion of contaminating microbes, for Southern and Northern blotting, etc. The filters may be cellulose, fiberglass, or nylon and may have been specially treated to best fit the purpose.

membrane fusion Maintains subcellular compartments. The process requires ATPases (NSF), accessory proteins (SNAP), integral membrane receptors of SNAP (SNARE), GTPases (RAB). and additional proteins. The SNARE function may be transient. In the fusion of vacuoles, Ca^{2+} /calmodulin regulates membrane bilayer mixing in the final steps. Protein phosphatase 1 (PP1) has an essential role in membrane mixing. Infection by viral pathogens, penetration of the egg by the sperm, vesicular transport, etc., all involve fusion of membranes. *See* ATPase; protein phosphatases; RAB; SNAP; SNARE. (Eckert, D. M. & Kim, P. S. 2001. *Annu. Rev. Biochem.* 70:777; Jahn, R. & Grubmüller, H. 2002. *Current Opin. Cell Biol.* 14:488.)

membrane potential Electromotive force difference across cell membranes. In an average animal cell inside, it is 60 mV relative to the outside milieu. It is caused by the positive and negative ion differences between the two compartments.

membrane proteins May be *integral* parts of the membrane structure and cannot be released. The *transmembrane* proteins are single amino acid chains folded into (seven) helices spanning across the membrane-containing lipid layers. The latter have a hydrophobic tract that passes through the lipid double layer of the membrane and their two tails, one pointing outward from the membrane and the other reaching into the cytosol are hydrophilic. Some of the membrane proteins are attached only to the outer or to the inner layer of the membrane lipids, and they are called *peripheral membrane proteins*. The membrane proteins regulate the cell membranes and cell morphology by anchoring to the cytoskeleton, pH, ion channels, and general physiology of the cell. The structure of membranes can be analyzed with the aid of membrane mutants available in several organisms. *See* cell membranes; cytoskeleton; farnesyl pyrophosphate; myristic acid; prenylation diagram below. (Dalbey, R. E. & Kuhn, A. 2000. *Annu. Rev. Cell Dev. Biol.* 16:51.)

membrane segment Antigen receptor immunoglobulins possess a membrane-bound segment at the C-end of their heavy chains. The transmembrane part is composed of 25–26 highly conserved amino residues and the intracellular portion is of 25 amino acids in IgG and 14 in IgE and IgA in mice. Membrane-bound IgG, IgE, and IgA are involved in the stimulated B-cell receptors, whereas IgM and IgD are parts of the naive (immature) B-cell receptors. *See* immunoglobulins.

membrane-spanning helices *See* membrane proteins; seven-membrane proteins.

membrane transport Movement of polar solutes with the aid of a transporter protein through cell membranes. *See* cell membrane. (Kaback, H. R., et al. 2001. *Nature Rev. Mol. Cell Biol.* 2:610.)

meme (1) Unit of cultural transmission (imitation) that bears some similarities to the gene from which it is transmitted from generation to generation as cultural inheritance. (Boyd, R. & Richerson, P. J. 2000. *Sci. Am.* 283[4]:70.) (2) Replicator function of organisms that controls coevolution of genes and organisms. (Bull, L., et al. 2000. *Artif. Life* 6[3]:227.)

memory Information storage in the brain or in a computer. The mammalian brain deals with synaptic strength as memory. If synapses are used repeatedly, the strength is improved and long-term potentiation (LTP) takes place. The opposite of LTP is LTD (long-term depression). The latter may erase the effect of LTP. Acetylcholine, dopamine, and norepinephrine mediate different memory signals acting in the prefrontal cortex of the brain. Both of these mechanisms are triggered by the inflow of Ca^{2+}, regulated by an ion channel (NMDAR), glutamate-activated N-methyl-D-aspartate receptor channel. Multiple protein kinases (adenylate cyclase, CREB) appear to be involved in extremely complex manners. It is not entirely clear how discrimination between LTP and LTD is accomplished. Nerve growth factor (NGF) gene transfer to basal forebrain of rats resulted in recovery from age-related memory loss. There are some indications that memory is controlled by positive and negative signals. LTP seems to require protein synthesis, but long-term memory (LTM) may not.

Modern neurobiology uses a variety of techniques for exploring the mechanisms of memory and learning, including mutations, knockouts, transgenes, tomography, magnetic resonance imaging, etc. *Explicit* (declarative) *memory* is the remembrance of facts and resides in the hippocampal region of the brain. *Implicit* (nondeclarative) *memory* involves perceptual and motor skills (based on basal ganglia) that may be more widely distributed; memory of conditional fear appears to be in the corpus amygdaloideum (an almond-shaped part in the temporal lobe of the brain with connections to the hippocampus, thalamus, and hypothalamus). Social memory (parental care, etc.) in mice is based on olfactory functions controlled by oxytocin in the brain. *Drosophila* mutants *dunce* (encoding cAMP phosphodiesterase) and *rutabaga* (encoding adenylate cyclase) appear to have an ability to learn yet display short memory. The mutant *amnesiac* is defective in a

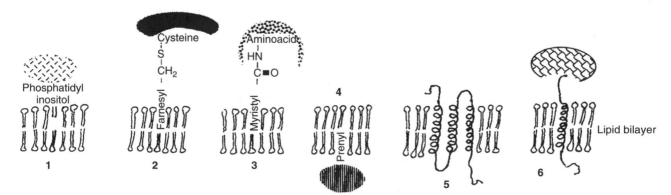

1. Globular protein attached to the outer surface by a glycosyl-phosphatidyl inositol anchor. **2.** The protein is attached to the outer surface by a thioether linkage between the sulfur of cysteine and a farnesyl molecule. **3.** Amino acid and myristil anchor join a protein to the outer surface of the membrane. **4.** Prenyl residue anchors the protein to the inner part of the membrane. **5.** Transmembrane protein chain passes through the membrane three or seven times. **6.** Transmembrane protein anchors another protein without covalent linkage.

pituitary peptide required for activation of adenylate cyclase, and mutation *linotte* is affected in a helicase-like function. In *Aplysia* and mice, CREB affects long-term memory but not short-term memory. All these pieces of information point to the role of cAMP in memory. Ca^{2+}/calmodulin-dependent protein kinase (CaMKII) is an important signaling molecule for memory. Dynamin, a microtubule-associated GTPase, is important for synaptic vesicle recycling. Recalling olfactory memory in *Drosophila* requires the function of the mushroom bodies, although this organelle is not necessary for learning or storage of information. *See* acetylcholine; adenylate cyclase; brain, human; CREB; dopamine; dynamin; Flyn; Hebbian mechanism; ion channels; knockout; long-term potentiation; mushroom body; NMDAR; norepinephrine; nuclear magnetic resonance spectroscopy; oxytocin; synapse; tomography; transgene. (Silva, A. Z., et al. 1997. *Annu. Rev. Genet.* 31:527; Zars, T., et al. 2000. *Science* 288:672; Sweatt, J. D. 2001. *Curr. Biol.* 11:R391; Waddell, S. & Quinn, W. G. 2001. *Science* 293:1271; Wadell, S. & Quinn, W. G. 2001. *Annu. Rev. Neurosci.* 24:1283.)

memory, immunological Immunological memory rests with the lymphocytes. There are three phases in its development: (1) activation and expansion of the CD4 and CD8 T cells, (2) apoptosis, and (3) stability (memory). During phase (1), lasting for about a week, the antigen selects the appropriate cells and an up to 5,000-fold expansion of the specific cells takes place. They also differentiate into effector cells. Between a week and a month, as the antigen level subsides, the T cells die and effector function declines. This is called activation-induced cell death (AICD). AICD is a defense against possible autoreactive function. The memory stage may then last for many years and responds very rapidly to low exposure of the reintroduced antigen. A certain level of antigen maintenance seems to be required to keep memory cells at a high level, but it appears that memory cells can be saved even in the absence of the antigen (Maruyama, M., et al. 2000. *Nature* 407:636). The conditions for CD4 and CD8 cell maintenance appear different. Memory B and T cells do not provide immediate protection against infection, but after infection they start rapid formation of effectors. The duration of the memory depends on the strength of immunization. The protection against peripheral reinfection is antigen-dependent. The long-term success of immunization depends on the presence of memory cells. Therefore, there is much significance in selecting T cells that may become memory cells. The division of CD8$^+$ memory T cells is slow and it can be increased by IL-15, but IL-2 has the opposite effect. Thus, these memory T cells are under the control of interleukin balance. The central memory cells have CCR7$^+$ receptors, whereas the effector memory T cells are CCR7$^-$. *See* affinity maturation; apoptosis; B cell; CCR; CD4; CD8; effector; germinal center; IL-12; IL-15; immune system; immunization; immunotherapy, active specific; lymphocytes; T cell; vaccines. (Mackay, C. R. & von Andrian, U. H. 2001. *Science* 291:2323; Sprent, J. & Tough, D. F. 2001. *Science* 293:245; Fearon, D. T., et al. 2001. *Science* 293:248; Sprent, J. & Surh, C. D. 2002. *Annu. Rev. Immunol.* 20:551.)

memory, molecular Heritable specific pattern of gene expression.

MEN (1) Multiple endocrine neoplasia. MEN1 (dominant) was located to human chromosome 11q13, responsible for the production of a 610-amino-acid protein, Menin, encoded by 10 exons predisposing to pancreatic islet cell tumors. MEN1 homozygosity is lethal. MEN2 is encoded at 10q11.2 and puts individuals at risk of thyroid cancer. RET tyrosine kinase is defective. *See* endocrine neoplasia; ganglioneuromatosis; RET oncogene. (Crabtree, J. S., et al. 2001. *Proc. Natl. Acad. Sci. USA* 98:1118.) (2) Mitotic exit network includes cyclin-dependent kinase (CDK) inactivators, CDC15, CDC14, CDC5, DBF, TEM1, etc. *See* FEAR. (Asakawa, K., et al. 2001. *Genetics* 157:1437.)

menaquinone (vitamin K_2)   Synthesized by bacteria. Menaquinone is an electron carrier. *See* vitamin K.

menarche   First menstruation event followed by a period of about 3 years of no ovulations before the regular menstruation begins and continues until menopause. *See* menopause; menstruation.

Mendelian laws The term was first used by Carl Correns (1900), one of the rediscoverers of these principles, which he named (1) Uniformitäts- und Reziprozitätsgesetz, (2) Spaltungsgesetz, and (3) Unabhängige Kombination. *First Law*: Uniformity of the F_1 (if the parents are homozygous) and the reciprocal hybrids are identical (in the absence of cytoplasmic differences). *Second Law*: independent segregation of the genes in F_2 (in the absence of linkage). *Third Law*: independent assortment of alleles in the gametes of diploids. Thomas Hunt Morgan (1919) also recognized three laws of heredity: (1) free assortment of the alleles in the formation of gametes, (2) independent segregation of the determinants for different characters, and (3) linkage — recombination. In some modern textbooks, only two Mendelian laws are recognized, but this is against the tradition of genetics that the first used nomenclature is upheld. Mendel himself never claimed any rules as such to his credit. He did not observe any linkage among the seven factors he studied in peas, although he had less than 1% chance for all factors segregating independently. This was called "Mendel's luck." If he would have found linkage, it would probably have been recorded in his notes. Unfortunately, after his death his successor at the abbey, Anselm Rambousek, disposed of most of the records. After he experimented with *Hieracium*, an apomict (unknown at that time), he doubted the general validity of his discoveries. Yet before ending his career he stated: "My scientific work has brought me a great deal of satisfaction, and I am convinced that I will be appreciated before long by the whole world." That appreciation began in 1900 — 16 years after his death — and continues today. *See* epistasis; Mendelian segregation; modified Mendelian ratios.

Mendelian population Collection of individuals that can share alleles through interbreeding. *See* population genetics.

Mendelian segregation It can be predicted for independent loci on the basis of the table below:

Number of different allelic pairs	1	2	3	4	n
Kinds of gametes and number of phenotypes in case of dominance	2	4	8	16	2^n
Number of phenotypes (in case of no dominance) and number of genotypes	3	9	27	81	3^n
Number of gametic combinations	4	16	64	256	4^n

Mendelian segregation ratios may show only apparent deviations in case of epistasis. Reduced penetrance or expressivity may also confuse the segregation patterns and in such cases it may be necessary to determine the difference between male and female transmission. *See* certation; epistasis; modified Mendelian ratios; penetrance, expressivity; segregation distorter.

Mendelizing   Segregation corresponds to the expectations by Mendelian laws. *See* Mendelian laws; Mendelian segregation.

Ménière disease (COCH, 14q12-q13)   Late-onset, dominant nonsyndromic deafness, although in some cases vertigo (dizziness-like sensation of whirling of the body or the surroundings) and tinnitus (ringing inside the ears) may occur periodically. In the majority of cases, mutation at nucleotide position C-T^{208} results in proline^{51} → serine substitution in the COCH protein. *See* deafness.

MENIN1   Tumor suppressor protein (binding JunD) encoded at 11q13 by the gene responsible for multiple endocrine neoplasia. *See* endocrine neoplasia; Jun. (Guru, S. C., et al. 2001. *Gene* 263:31.)

meninges   Three membranes (pia, arachnoid, dura maters) surrounding the brain and the spinal cord.

meningioma   Usually slow proliferating brain neoplasias classified into different groups on the basis of anatomical features. Generally meningiomas involve the loss of human chromosome 22 (hemizygosity) or part of its long arm or some lesions at 22q12.3-q13 where the SIS oncogene, responsible for a deficit of platelet-derived growth factor (PDGF), is located. Chromosomes 1p, 14q, and 17 have also been implicated. *See* cancer; meninges; neurofibromatosis; SIS.

meningocele   *See* spina bifida.

Menkes syndrome (MNK, kinky hair disease)   The gene is situated in the centromeric area of the human X chromosome (Xq12-q13). The phenotype involves hair abnormalities, mental retardation, low pigmentation, hypothermia, and short life span. Apparently, the defect is in the malabsorption of copper through the intestines, resulting in copper deficiency of the serum. The prevalence is in the 10^{-5} range or less. It is detectable prenatally, and the heterozygotes

can be identified, although its inheritance is apparently recessive. Lysyl oxidase and other copper-dependent enzyme levels (tyrosinase, monoamine oxidase, cytochrome c oxidase, ascorbate oxidase) are reduced in afflicted individuals. In the mouse homologue, Mottled-Bridled, the nonexported copper is tied up by metallothionein and the afflicted individual dies within a few weeks after birth. Occipital Horn syndrome may be an allelic variation of MNK and of cutis laxa. *See* acrodermatitis; collagen; cutis laxa; Ehlers-Danlos syndrome; hemochromatosis; mental retardation; metallothionein; Wilson disease.

menopause   End of periodic ovulation (menstruation) and fertility around age 50 in human females. Animals in the wild usually stay fertile in old age. The evolutionary cause of menopause is not known, but it has been hypothesized that it is a protection against the increase of chromosomal aberrations in old egg cells. An alternative hypothesis assumes that life beyond menopause may aid the rearing of the last offspring or grandchildren, and this conveys fitness by kin selection. *See* age-specific birth and death rates; andropause; dictyotene stage; kin selection; menarche; menstruation; porin.

menses   *See* menstruation.

menstruation   Monthly discharge of blood from the human (primate) uterus in the absence of pregnancy. If the egg is not fertilized, it dies, and the endometrial tissue of the uterus is removed amidst the bleeding. Fertilization takes place within the oviduct. About 3 days are required for the egg to reach the uterus through the oviduct, where it is implanted within a day or two, and about a week after being fertilized. Fertilization may occur if coitus takes place about 2 weeks after the beginning of the last monthly menstruation. The calendar rhythm method of birth control relies on knowledge of this receptive period. Unfortunately, its effectiveness is not very high. *See* hormone receptors; menarche; ovulation; sex hormones.

mental retardation   Collection of human disabilities caused by direct or indirect genetic defects and acquired factors such as diverse types of infections (syphilis, toxoplasma coccidian protozoa), viruses (rubella, human immunodeficiency virus, cytomegalovirus, herpes simplex, coxsackie viruses), bacteria (*Haemophilus influenzae*, meningococci, pneumococci), etc., mechanical injuries to the brain pre-, peri-, and postnatally, exposure to lead, mercury, addictive drugs, alcoholism or deprivation of oxygen during birth, severe malnutrition, deficiency of thyroid activity, social and psychological stress, etc. An estimated 2 to 3% of the population suffers from mild (IQ 50–70%) or more or less severe (IQ below 50%) forms of it. Special education programs can help an estimated 90% of cases. Approximately 10% of the human hereditary disorders have some mental-psychological debilitating effects.

Autosomal-dominant hemoglobin H disease–associated mental retardation due to a lesion in the α-globin gene cluster with chromosomal deletion and without it have been observed. Other cases of mental retardation were also observed involving autosomal-dominant inheritance caused by breakage in several chromosomes. Autosomal-recessive inheritance was involved in mental retardation associated with head, face, eye, and lip abnormalities, hypogonadism, diabetes, epilepsy, heart and kidney malformations, and phenylketonuria. X-chromosome-linked mental retardation was observed as part of the syndromes involving the development of large heads, intestinal defects, including anal obstructions, seizures, short statures, weakness of muscles, obesity, marfanoid appearance, etc. In some cases, "kinky hair" syndrome (Menkes syndrome), apparently caused by abnormal metabolism of copper and zinc, also involved mental retardation.

A fragile site in the X chromosome (Xq27-q28) apparently based on a deficiency of thymidine monophosphate caused by an insufficient folate supply is associated with testicular enlargement (macroorchidism), big head, large ears, etc. A defect in the IL-1 receptor accessory protein, encoded at Xp22.1–21.3, affects learning ability and memory. The transmission of the fragile X sites (FRAX) is generally through normal males. The carrier daughters are not mentally retarded and generally do not show fragile sites. In the following generation, about a third of the heterozygous females display fragile sites and become mentally retarded. This unusual genetic pattern was called the Sherman paradox and it is interpreted by some type of a premutational lesion. The premutation ends up in a genuine mutation only after transmitted by a female who already had a microscopically undetectable rearrangement.

The risk of the sons was estimated to be 50% from mentally retarded heterozygous females, 38% from normal heterozygous mothers, and 0% from normal transmitting fathers. The probability of these sons being a mentally sound carrier was estimated as 12, 0, and 0%, respectively. The risk of the daughters of the same mothers to become a mentally affected carrier was calculated to be 28, 16, and 0%, and to become a mentally normal carrier was estimated to be 22, 34, and 1%, respectively. The chance of mental retardation for the brother of a proband whose mother has no detectable fragile X site may vary from 9 to 27%; among first cousins this is reduced to 1–5%. It was proposed (Laird, 1987. *Genetics* 117:587) that the expression of the fragile X syndrome is mediated by chromosomal imprinting. The imprinting can, however, be erased by transmission through the parent of the other sex. The fragile X syndrome is apparently caused by localized breakage and methylation of CpG islands at the site (Bardoni, B. & Mandel, J.-L. 2002. Current Op. Genet. Dev. 12:284). Currently the most reliable diagnosis of this condition is based on DNA probing.

Besides the fragile X syndrome, autosomal and sex-chromosomal trisomy and chromosome breakage associated with translocations may be contributing factors of mental retardation. Also, mutations causing metabolic disorders (phenylketonuria, homocystinuria, defects in the branched-chain amino acid pathway [maple syrup urine disease], anomalies in amino acid uptake [Hartnup disease]), defects involving mucopolysaccharides (Hunter, Hurler, and Sanfilippo syndromes), gangliosidoses and sphingolipidoses (most notably Tay-Sachs disease, Farber's disease, Gaucher's disease, Niemann-Pick disease, etc.), galactosemias, failure of removal of fucose residues from carbohydrates (fucosidosis), defects in acetyl-glucosamine phosphotransferase (I-cell disease), defects in HGPRT (Lesch-Nyhan syndrome), hypothyroidism, a variety of defects of the central nervous system, and other genetically determined conditions may be responsible for mental retardation. The incidence, establishment of genetic risks, and possible therapies are as variable as the underlying causes.

Mental retardation is defined as borderline: IQ $\approx$ 70–85; mild: IQ $\approx$ 50–70; moderate: IQ $\approx$ 35–50; severe: IQ $\approx$ 25–35; profound: IQ $\leq$ 20.

In utero radiation exposure may cause mental retardation; 140 rad during the first 8–15 weeks may result in such damage in 75% of the fetuses and in 46% of the irradiated at any stage. *See* Apert syndrome; Apert-Crouzon disease; aspartoacylase deficiency; autism; Bardet-Biedl syndrome; biotinidase deficiency; CADASIL; cerebral gigantism; ceroid lipofuscinosis; Coffin-Lowry syndrome; Cohen syndrome; craniofacial dystosis; craniosynostosis; De Lange syndrome; double cortex; dyslexia; FMR1 mutation; focal dermal hypoplasia; fragile sites; glutamate formimino transferase deficiency; head/face/brain defects; heritability; human behavior; human intelligence; Huntington disease; hydrocephalus; IL-1; Juberg-Marsidi syndrome; Langer-Giedion syndrome; Laurence-Moon syndrome; Lowe syndrome; Menke syndrome; mental retardation, X linked; Miller-Dieker syndrome; muscular dystrophy; myotonic dystrophy; neurodegenerative diseases; neurofibromatosis; Noonan syndrome; oligophrenin; Opitz-Kaveggia syndrome; PAK; periventricular heterotopia; Prader-Willi syndrome; psychoses; QTL; Roberts syndrome; Rubinstein syndrome; Russel-Silver syndrome; Seckel dwarfism; serpines; Smith-Lemli-Opitz syndrome; Smith-Magenis syndrome; spina bifida; tetraspanin; trinucleotide repeats; tuberous sclerosis; Walker-Wagner syndrome; West syndrome; Wilms' tumor.

mental retardation, X-linked (MRX)

These hereditary defects occur in a variety of forms: (1) MRXS with diplegia (bilateral paralysis), (2) associated with psoriasis (skin lesions), (3) with lip deformities, obesity, and hypogonadism, (4) Renpenning type with short stature and microcephaly, (5) with seizures (EFMR), (6) with Marfan syndrome-like habitus, (7) with fragile X-chromosome sites, etc. Several of the MRX disorders map in the human Xq28 region and in this general area apparently 12 genes are located. At Xp22.1–21.3 there is the IL1RAPL (IL-1 receptor accessory protein) gene, expressed in the hippocampus. Translocations involving Xq26 and 21p11 (ARHGEF6) involve defects in a guanine exchange factor for Rho GTPases. Some of the sequences observed at Xp22 are also found at Yq11.2, where infertility and azoospermia factors are situated. *See* Coffin-Lowry syndrome; fragile X chromosome; Juberg-Marsidi syndrome; Lowe's syndrome; Marfan syndrome; mental retardation; oligophrenin; Rett syndrome; tetraspanin. (Fukami, M., et al. 2000. *Am. J. Hum. Genet.* 67:563; Chelly, J. & Mandel, J.-L. 2001. *Nature Rev. Genet.* 2:669.)

Menthas

Group of dicotyledonous species of plants of various (frequently aneuploid) chromosome numbers. *M. arvensis*: $2n = 12, 54, 60, 64, 72, 92$; *M. sylvestris*: $2n = 24, 48$; *M. piperita*: $2n = 34, 64$. They were the source of menthol (peppermint camphor) and other oils used in cough drops, nasal medication, anti-itching ointments, candy, liquors, etc. Menthol appeared noncarcinogenic, although doses (above 1 g/kg) may cause 50% death in laboratory rodents when administered subcutaneously or orally.

mentor pollen effect

Simultaneous application of dead or radiation-damaged compatible (mentor) pollen with incompatible pollen may in some instances help to overcome the incompatibility of the latter and fertilization may result. *See* incompatibility alleles. (Stettler, R. F. 1968. *Nature* 219:746.)

menu Of a computer, lists the various functions you can choose. The menu bar on top of the monitor screen displays the titles of the menus available.

meprin Metalloendoproteinase with α- and β-subunits. The β-subunit is a kinase-splitting membrane protease. The α-subunit stays in the endoplasmic reticulum until its transmembrane and cytoplasmic domains are cleaved off, then it moves out. (Ishmael, F. T., et al. 2001. *J. Biol. Chem.* 276:23207.)

MEPS Minimal efficient processing segment is the identical nucleotide tract length required for efficient initiation of recombination. *See* recombination mechanisms in eukaryotes.

Mer Human T-cell protooncogene-encoded receptor tyrosine kinase. It mediates phagocytosis and apoptosis of thymocytes. *See* apoptosis; phagocytosis; receptor tyrosine kinases; TCR; thymocytes.

MER (1) From the Greek $\mu\varepsilon\rho\sigma\sigma$, part. Used as octamer, e.g., indicating it is built of 8 units (octamer would be a Latin-Greek hybrid to be avoided even by geneticists). (2) Medium reiteration frequency sequence; ~35 copies/human genome. *See* redundancy.

mercaptoethanol Keeps SH groups in a reduced state and disrupts disulfide bonds while proteins are manipulated in vitro. *See* DTT; thiol.

mercaptopurine Purine analog inhibiting DNA synthesis and therefore cytotoxic.

merged sequence contig Overlapping initial sequence contigs are merged. *See* contig; initial sequence contig.

mericlinal chimera The surface cell layers are different from the ones underneath just like in the periclinal chimeras, but the difference is that the different surface layer does not cover the entire structure but only a segment of it. *See* chimera; periclinal chimera. (Jørgensen, C. A. & Crane, M. B. 1927. *J. Genet.* 18:247.)

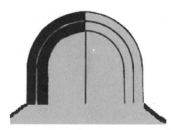

meristem Undifferentiated plant cells capable of production of various differentiated cells and tissues, functionally similar to the *stem cells* of animals. *See Arabidopsis* mutagen assay;

flower differentiation; stem cells. (Weigel, D. & Jürgens, G. 2002. *Nature* 415:751; Nakajima, K. & Benfey, P. N. 2002. *Plant Cell* 14:S265.)

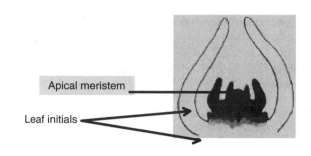

meristic traits Quantitative traits that can be represented only by integers, e.g., the number of kernels in a wheat ear or the number of bristles on a *Drosophila* body. *See* quantitative traits.

merit, additive genetic Same as the breeding value of an individual.

MERLIN Moesin, ezrin, radixcin-like protein. The same as schwannomin. *See* ERM; neurofibromatosis. (Bretscher, A., et al. 2000. *Annu. Rev. Cell Dev. Biol.* 16:113.)

7mer-9mer Seven- or nine-base-long conserved sequences in the vicinity of the V-D-J (variable, diversity, junction) segments of immunoglobulin genes in the germline DNA. *See* immunoglobulins.

merodiploid *See* merozygous.

merogenote *See* merozygous.

merogone Fragment of an egg.

meromelia *See* limb defects in humans.

merosin *See* muscular dystrophy.

merotelic attachment Capture of single kinetochores by microtubules from both centrosomes. It may cause aneuploidy. *See* aneuploidy; centrosome; kinetochore. (Stear, J. H. & Roth, M. B. 2002. *Genes & Development* 16:1498.)

merozoite *See Plasmodium.*

merozygous Prokaryote, diploid for part of its genome (merogenote). Prokaryotes are functionally haploid, but transduction or a plasmid may add another gene copy into the cell. *See* conjugation; transduction. (Wollman, E. L., et al. 1956. *Cold Spring Harbor Symp. Quant. Biol.* 21:141.)

Mer⁻ phenotype   Mammalian cell defective in methyl-guanine-O^6-methyltransferase.

MERRY   *See* mitochondrial diseases in humans.

MES   Maternal effect sterility is due to recessive genes causing developmental defects of the germ cells in the offspring.

Meselson-Radding model of recombination   Explains gene conversion (occurring by asymmetric heteroduplex, symmetric heteroduplex DNA) and crossing over occurring from one initiation event as indicated by the data of *Ascobolus* spore octads. In yeast, the aberrant conversion tetrads arise mainly from asymmetric heteroduplexes as suggested by the Holliday model. Symmetric heteroduplex covers the same region of two chromatids, whereas asymmetric heteroduplex means that the heteroduplex DNA is present in only one chromatid. The heteroduplexes can be genetically detected very easily in asci containing spore octads. In the absence of heteroduplexes, the adjacent (haploid) spores are identical genetically. If the heteroduplex area carries different alleles, the two neighboring spores may become different after postmeiotic mitosis. Actually, heteroduplexes may be detectable in yeast (that forms only four ascospores) by sectorial colonies arising from single spores. Branch migration indicates that the exchange points between two DNA molecules can move and eventually they can reassociate in an exchanged manner in both DNA double helices involved in the recombination event. Rotary diffusion indicates that the joining between single strands can take place by movement of the juncture in either direction, thus making the heteroduplex shorter or longer. *See* recombination, molecular mechanisms; recombination models.

Meselson-Stahl model   Proved that in bacteria DNA replication is semi-conservative. See semi-conservative replication. (Meselson, M. & Stahl, F. W. 1958. *Proc. Natl. Acad. Sci. USA* 44:671.)

mesenchyma   Unspecialized early connective tissue of animals that may give rise to blood and lymphatic vessels. *See* mesoderm.

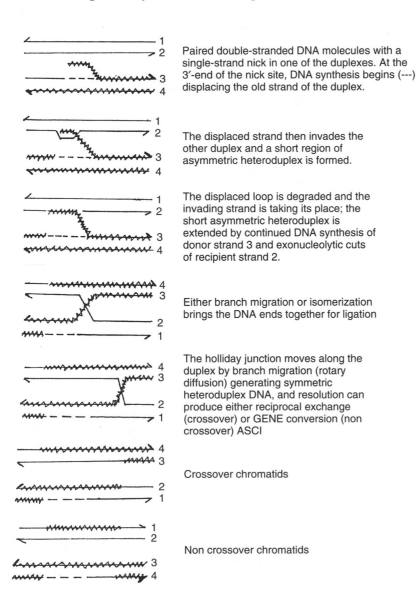

Paired double-stranded DNA molecules with a single-strand nick in one of the duplexes. At the 3′-end of the nick site, DNA synthesis begins (---) displacing the old strand of the duplex.

The displaced strand then invades the other duplex and a short region of asymmetric heteroduplex is formed.

The displaced loop is degraded and the invading strand is taking its place; the short asymmetric heteroduplex is extended by continued DNA synthesis of donor strand 3 and exonucleolytic cuts of recipient strand 2.

Either branch migration or isomerization brings the DNA ends together for ligation

The holliday junction moves along the duplex by branch migration (rotary diffusion) generating symmetric heteroduplex DNA, and resolution can produce either reciprocal exchange (crossover) or GENE conversion (non crossover) ASCI

Crossover chromatids

Non crossover chromatids

Meselson-Radding model of recombination (1975. *Proc. Natl. Acad. Sci. USA* 72:358).

mesocarp Middle part of the fruit wall. *See* endocarp; exocarp.

mesoderm Middle cell layer of the embryo developing into connective tissue, muscles, cartilage, bone, lymphoid tissues, blood vessels, blood, notochord, lung, heart, abdominal tissues, kidney, and gonads. Members of the transforming growth factor (TGFβ) family of proteins regulate mesoderm development. *See* eomesodermin; morphogenesis; TGF. (Furlong, E. E. M., et al. 2001. *Science* 293:1629.)

mesogen (liquid crystal) Some compounds may exhibit transitions between crystalline and liquid forms and can be manipulated by rotation or external electric fields causing rotation. They have various applications (in cellular phones, pocket computers, etc), and because of these properties they may be used for detection of binding of ligands to specific molecules. Heat-responding mesogens may generate optical anisotropy and thus may transduce optical signals, facilitating molecular diagnostics without invasive procedures.

mesokaryote Organism that occupies some kind of a middle position between prokaryotes and eukaryotes. Mesokaryotes are endowed with cytoplasmic organelles like plant cells, but their nuclear structure is similar to prokaryotes. The amount of chromosomal basic proteins in the nucleus is low; the chromosomes are attached to the membrane, yet they develop a nuclear spindle apparatus. Several microtubules pass through the nuclear membranes; in the majority of species, they pull the chromosomes to the poles with the membrane without being attached to the chromosomes. In the dinoflagellate *Cryptecodinium cohnii*, 37% of the thymidylate is replaced by 5-hydroxymethyluracil. More than half of the DNA is repetitious. The vegetative cells appear to be haploid and thus mutations can be readily detected. Both homo- and heterothallic species are known. *See* eukaryote; prokaryote. (Hamkalo, B. A. & Rattner, J. B. 1977. *Chromosoma* 60:39.)

mesolithic age About 12,000 years ago when domestication of animals and agriculture started. *See* neolithic; paleolithic.

mesomere Blastomere of about medium size. *See* blastomeres.

mesophyll Parenchyma layers of the leaf blade (below cuticle and epidermis).

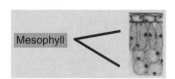

mesophyte Plants avoiding extreme environments such as wet, dry, cold, or hot.

mesosome Invaginated membrane within a bacterial cell.

mesothorax Middle thoracic segment of insects, bearing legs and possibly wings. *See Drosophila.*

Thoracic area and legs, wings unshaded.

Mesozoic Geological period in the range of 225 to 65 million years ago; age of the life and extinction of dinosaurs and several other reptiles. *See* geological time periods.

messenger polypeptide Extracellular signaling molecule that can pass the cell membrane, enter the cell nucleus, and recognize in the DNA a special sequence motif, then activate transcription. These messengers are different from hormones or other signaling molecules because they do not need membrane receptors, special ligands, series of adaptors, or phosphorylation to be activated and becoming nuclear coactivators of gene expression. *See* lactoferrin.

messenger RNA *See* mRNA.

mestizo Offspring of Hispanic and American Indian parentage. *See* miscegenation; mulatto.

meta-analysis, linkage Used when the information in an experiment is inadequate for drawing definite conclusions. In such cases, data from other comparable experiments are pooled in the best possible way. Usually the formula of R. A. Fisher (Statistical Analysis for Research Workers) is used, where P = the probability of obtaining as different or more extreme information under the assumption that there is no linkage, m = the data sets. The χ^2 is calculated by 2 degrees of freedom. Meta-analysis may be used for other synthetic purposes when the information available is inconclusive. Its conclusions, therefore, may have to be regarded with some reservations. Meta-analysis may reveal substantial variation/heterogeneity in the genetic association of disease symptoms. *See* chi square; linkage. (Allison, D. B. & Heo, M. 1998. *Genetics* 148:859; Ioannidis, J. P. A., et al. 2001. *Nature Genet.* 29:306.)

$$\chi^2_{2m} = -2 \sum_{i=1}^{m} \ln(P_i)$$

metabolic block Nonfunctional enzyme (due to mutation in a gene or to an inhibitor) prevents the normal flow of metabolites through a biochemical pathway.

metabolic pathway Series of sequential biochemical reactions mediated by enzymes under the control of genes. Genetic studies greatly contributed to their understanding along with the use of radioactive tracers. *See* auxotrophy; MCA; radioactive tracer. (<http://wit.mcs.anl.gov/WIT2>;

<http:emp.mcs.an.gov>; <http://expasy.ch>; <http://ecocyc.PangeaSystems.com/ecocyc>.)

metabolism Enzyme-mediated anabolic and catabolic reactions in cells. *See* BRENDA; EXProt; MCA; MetaCyc. (compound/enzymes/reactions: <http://www.genome.ad.jp/ligand/>.)

metabolite Product of metabolism. *See* metabolism.

metabolite engineering Transformation of certain genes into a new host may lead to production of substances that the organisms never produced before or can synthesize larger quantities or different qualities of certain proteins. Examples are production of indigo or human insulin in *E. coli*, novel antibiotics, expressing antigens of mammalian pathogens in plants, or changing the pathway of diacetyl formation in yeast to acetoin production and thus shortening the lagering process in brewing beers, etc. *See* genetic engineering; protein engineering. (Kholodenko, B. N., et al. 2000. *Metabol. Eng.* 2[1]:1.)

metabolome Complete set of (low-molecular-weight) intermediates in a cell or tissue metabolism. (Tweeddale, H., et al. 1998. *J. Bacteriol.* 180:5109, Fiehn, O. 2002. *Plant Mol. Biol.* 48:155.)

metabolon Supramolecular-associated complex of sequential metabolic enzymes. (Reithmeier, R. A. 2001. *Blood Cells Mol. Dis.* 27:85.)

metabotropic receptor Binding of a ligand initiates intracellular metabolic events. Many of them have transmembrane regions and may respond to second messengers such as cAMP, cGMP, or inositol triphosphate. *See* ionotropic receptor. (Ramaekers, A., et al. 2001. *J. Comp. Neurol.* 438[2]:213.)

metacentric chromosome The two arms are nearly equal in length. *See* chromosome morphology.

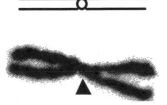

Centromere in middle.

metachondromatosis Autosomal-dominant multiple exostoses particularly on hands, feet, knees, and limb bones without deforming the bones or joints. It may disappear in time. *See* exostosis; fibrodysplasia.

metachromasia The same stain colors different tissues in different hues.

metachromatic leukodystrophy (MLD) Sulfatide lipidosis. Two distinct forms have been identified that are due to two recessive alleles of a gene: mutations in *A* and *I*, in human chromosome 22q13.31-qter. Due to the deficiency arylsulfatase A, cerebroside sulfate accumulates in the lysosomes. The accumulation of galactoside-sulfate-cerebrosides in the plasma membrane and particularly in the neural tissues (myelin) causes a progressive and fatal degeneration in the peripheral nerves, liver, and kidneys. As a consequence, failure of muscular coordination (ataxia), involuntary partial paralysis, hearing and visual defects, as well as lack of normal brain function, arise after 18 to 24 months of age and usually cause death in early childhood. In the juvenile form of the disease, the symptoms appear between ages 4 and 10 years. There is also an adult type of the disease with an onset after age 16 that involves schizophrenic symptoms. The mutations involve either a substitution of tryptophan at amino acid residue 193 or at threonine 391 by serine, or a defect at the splice donor site at the border of exon 2. MLD has been observed in animals. The reduced activity of the enzyme can be identified in cultured skin fibroblasts of heterozygotes and prenatally in cultured amniotic fluid cells of fetuses. *See* arylsulfates; Krabbe's leukodystrophy; lysosomal storage diseases; prenatal diagnosis; saposin; sphingolipidoses; sphingolipids.

metacline hybrids Were called in *Oenotheras* the exceptional progeny that occurred only in the reciprocal crosses due to the difference in transmission through egg and sperm of the different complex translocations. *See* certation; complex heterozygotes; megaspore competition.

MetaCyc Database of 445 metabolic pathways involving 1,115 enzymes in 158 organisms. (<http://ecocyc.org/ecocyc/metacyc.html>; <http://bioinformatics.ai.sri.com/ptools/>.)

metacyclic *Trypanosoma* Lives in the salivary gland of the insects, which spread sleeping sickness. *See Trypanosoma*.

metacyclogenesis Differentiation of the promastigote of *Leishmania* into a highly infective form in the sandfly. *See Leishmania*.

metafemale Has more than the usual dose of female determiners; it may be XXX. *See Drosophila*; triplo-X.

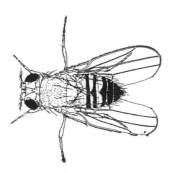

Euploid Metafemale *Drosophila*.

metagenesis Sexual and asexual generations alternate.

metalloprotein The prosthetic group of the protein is a metal, e.g., hemoglobin. *See* hemoglobin; zinc finger. (<http://metallo.scripps.edu>.)

metalloproteinases May be cell-surface endopeptidase enzymes mediating the degradation of the extracellular matrix, cartilage formation, the release of tumor necrosis factor α, a cytokine involved in inflammatory reactions, in embryogenesis, cell migration, etc. Membrane-bound metalloproteinase (matrix metalloproteinase, MMP3, stromelysin 1, 11q13) deficiencies cause craniofacial anomalies, arthritis, dwarfism, and other defects due to collagen and connective tissue problems. MMP11 (stromelysin III, 22q11.2) is overexpressed in metastatic breast cancers. Cancer tissues are associated with increased protease activities and it is supposed that these activities facilitate the bursting out of the tumor from the normal cell milieu and increase angiogenesis. MMP-9 (gelatinase, 20q11.2-q13.1) is a contributor to skin carcinogenesis. MMP8 I is collagenase I (11q21-q22); MMP2 is collagenase type IV (16q13), mediating endometrial breakdown during menstruation and possibly facilitating carcinogenesis by promoting angiogenesis. MMP1 is also a collagenase (11q22-q23). MMP26 (matrilysin 2, 11p15) is involved in tissue healing and remodeling. MMP7 (11q21-q22) is a relatively short uterine matrilysin protein. MMP15 (16q13-q21) and MMP16 (8q21) are membrane enzymes with roles in normal physiological as well as pathogenic processes at special organs. Metalloproteinase inhibitors are explored for cancer therapy. *See* ADAM; arthritis; bone morphogenetic protein; collagenase; disintegrin; meltrin; night blindness (Sorsby syndrome); stromelysin; TACE. (Nagase, H., et al. 1992. *Matrix*, Suppl. 1:421; Bode, W., et al. 1999. *Cell. Mol. Biol. Life Sci.* 55:639; Sternlicht, M. D. & Werb, Z. 2001. *Annu. Rev. Cell Dev. Biol.* 17:463; Coussens, L. M., et al. 2002. *Science* 295:2387; Egeblad, M. & Werb, Z. 2002. *Nature Rev. Cancer* 2:163.)

metallothionein SH-rich metal-binding protein in mammals. Its main function is detoxification of heavy metals. The promoter is activated by the same heavy metal that the protein product binds. This promoter is very useful for experimental purposes because structural genes attached to it can be turned on and off by regulating the amount of heavy metal in the drinking water of transgenic animals. It is encoded in human chromosome 16q13. *See* Menke's disease; metals; transgenic. (Vasák, M. & Hasler, D. W. 2000. *Curr. Opin. Chem. Biol.* 4:177.)

metal metabolism and disease *See* aceruloplasminemia; acrodermatitis enteropathica; coproporphyria; glutathione peroxidase; hemachromatosis; hyperzincemia; Menkes syndrome; metalloproteinases; metallothionein; occipital horn syndrome; porphyria; protoporphyria, erythropoietic; selenium-binding protein; Wilson syndrome.

metals May contribute oxidative damage to the DNA and may hinder repair and enhance radiation damage. Beryllium, chromium, and lead salts may enhance radiation damage. *See* metallothionein.

metamale *See Drosophila*; supermale.

metamerism Anterior-posterior segmentation of the body of annelids and arthropods. In chemistry it is rarely used for a type of structural isomerism when different radicals of the same type are attached to the same polyvalent element and give rise to compounds possessing identical formulas.

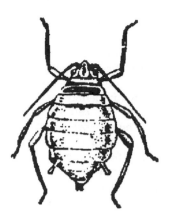

Metamerism of the body.

metamorphosis Distinct change from one developmental stage to another, such as from larva to adult or from tadpole to toad. *See Drosophila*.

metanomics Using gas chromatography and mass spectrometry to trace metabolic changes upon mutation or environmental changes. *See* gas-liquid chromatography; mass spectrum.

metaphase Stage in mitosis and meiosis when the eukaryotic chromosomes have reached maximal condensation and spread out on the equator of the cell (metaphase plane), and their arm ratios and some other morphological features can be well recognized. In meiosis I, the bivalent chromosomes may be associated at their ends if chiasmata took place during prophase. The ring bivalents indicate crossing over between both arms, whereas rod bivalents are visible when crossing over is limited to only one of the two arms. *See* chromosome rosette; meiosis; mitosis; ring bivalent.

metaphase arrest May take place if the nuclear division is blocked by toxic agents or when two separate kinetochores are joined by translocation. Homologous recombination does not lead to metaphase arrest, indicating kinetochore tension may be responsible for the event. *See* meiosis.

metaphase plane Central region of the cell where the chromosomes are located during metaphase. Often incorrectly called metaphase plate (but no plate is involved). *See* meiosis; mitosis.

metaplasia One cell type gives rise to another, and in the tissue the two may be adjacent. Normally these two cell types are not expected to occur together. The phenomenon is attributed to different activation of a particular cell(s). *See* stem cells. (Tosh, D. & Slack, J. M. W. 2002. *Nature Rev. Mol. Cell Biol.* 3:187.)

metapopulation Large population composed of smaller, local populations. *See* coalescent; F. (Hanski, I. & Gilpin, M. E., eds. 1997. *Metapopulation Biology: Ecology, Genetics, and Evolution*. Academic Press, San Diego.)

metastable Potentially transitory state; it can change to a more or less stable form.

metastasis Spread of cancer cells through the bloodstream, thus establishing new foci of malignancy in any part of the body. The invasiveness of cancer cells requires an active state of the integrin system, cell-surface gelatinases (collagenases, proteases), so that they can penetrate the extracellular matrix of the target cells. On the cell surface, precursors of the gelatinases are found that are proteolytically activated by met-alloproteinases. The plasminogen activator proteases include urokinase and tissue-type activators. Plasminogen activator inhibitor (PAI1) is also required. IL-18 and TNF-α may promote cell adhesion and metastases by upregulating vascular cell adhesion molecule (VCAM-1) synthesis. Also, ICE may facilitate metastasis after processing the precursors of IL-1β and IL-18 proinflammatory cytokines. The CCR gene product appears to suppress metastasis by apoptosis of small cell lung carcinoma and melanoma cells. CXCR4 and CCR7 are highly expressed in breast cancer cells and metastases. It seems that these chemokine receptors and chemokines play an important role in metastasis. The direction of invasiveness may depend on the organs/tissues where these receptors are expressed. TIP30 kinase, which appears to be the same as CC3, upregulates some apoptotic genes by phosphorylating the C-terminal domain of the largest subunit of DNA-dependent RNA polymerase II. Metastasis of melanoma cells may be initiated by an increase in the expression of fibronectin, RhoC, and thymosin β4 as visualized by microarray hybridization. *See* anoikis; apoptosis; cancer; CCR; CD44; collagen; contact inhibition; CXCR; DAP kinase; extracellular matrix; fibronectin; ICE; IL-1; IL-18; intravasation; Kiss; malignant growth; metalloprotein; oncogenes; plasminogen activator; PRL-3; RAGE; Rho; saturation density; thymosin; TNF; urokinase. (Al-Mehdi, A. B., et al. 2000. *Nature Genet.* 6:100; Müller, A., et al. 2001. *Nature* 410:50; Liotta, L. A. & Kohn, E. C. 2001. *Nature* 411:375; Trusolino, L. & Comoglio, P. M. 2002. *Nature Rev. Cancer* 2:289; Chambers, A. F., et al. 2002. *Nature Rev. Cancer* 2:563.)

metatarsus Middle bones beyond the ankle but preceding the toes, or in insects the basal part of the foreleg distal to the tibia but proximal to the tarsal segments and the claw. It carries the sexcombs in the male *Drosophila*. *See Drosophila*.

metaxenia Physiological modification of maternal tissues of the fruit in plants by the genetically different embryo, e.g., in green hybrid apples the exocarp (skin) may become reddish if the pollen carries a dominant gene for red color. *See* fruit; xenia.

metazoa All animals with differentiated tissues; thus, protozoa are excluded.

methacrylateaciduria β-hydroxyl-isobutyryl CoA deacylase deficiency involving the catabolism of valine leads to urinary excretion of cysteine and cysteinamine conjugates of methacrylic acid and to teratogenic effects. Methacrylic acid [$CH_2=C(CH_3)COOH$] is a degradation product of isobutyric acid [$(CH_3)_2CHCOOH$] and the amino acid valine [$(CH_3)_2CH(NH_2)CHCOOH$]. *See* isoleucine-valine biosynthetic pathway.

Methanococcus jannaschii Bacterium with a genome of 1.67×10^6 bp DNA and 1,738 ORF. *Methanococcus thermoautotrophicum's* genome is 1.75×10^6 and has 1,855 ORF. *See* missing genes; ORF. (Bult, C. J., et al. 1996. *Science* 273:1058.)

methemoglobin (ferrihemoglobin, hemiglobin) Hemoglobin with the ferroheme oxidized to ferriheme; has impaired reversible oxygen-binding ability. Mutation and certain chemicals (ferricyanide, methylene blue, nitrites) may increase the normally slow oxidation of hemoglobin. Reduction of methemoglobin may be accomplished either through the Embden-Meyerhof pathway or through the oxidative glycolytic pathway (pentose phosphate pathway). The principal methemoglobin-reducing enzyme (22q13.31-qter) is NADH-methemoglobin reductase, which may be adversely affected by autosomal-recessive genetic defects. Methemoglobinemia may also be caused by a deficiency of cytochrome b5 (18q23). As a consequence, lifelong cyanosis (bluish discoloration of the skin and mucous membranes) results following the accumulation of reduced hemoglobin in the blood. Males are apparently more affected. This anomaly may be corrected with the reducing agents methylene blue or ascorbic acid. NADH-methemoglobin reductase is often called diaphorase. *See* Embden-Meyerhof pathway; hemoglobin; hexose monophosphate shunt; NADH; pentose phosphate pathway.

methionine adenosyltransferase deficiency (MAT, 10q22, 2p11.2) Homocystinuria and tyrosinemia may cause hypermethioninemia. The most direct cause of hypermethioninemia is an autosomal-recessive defect in methionine adenosyltransferase. Besides these genetically determined causes, methionine accumulation may be due to prematurity at birth or the overactivity of cystathionase when cow milk (rich in methionine) is fed, etc. Hypermethioninemia itself may not have very serious consequences. Methionine adenosyl transferase catalyzes the biosynthesis of S-adenosylmethionine, a major methyl donor in the cell. *See* AdoMet; amino acid metabolism; homocystinuria; methionine biosynthesis; tyrosinemia.

methionine biosynthesis Methionine is biosynthesized from homocysteine by methionine synthase using N^5-methyltetrahydrofolate as a methyl donor:

$$\underset{\text{homocysteine}}{HSCH_2CH_2CH(NH_2)COOH} \rightarrow \underset{\text{methionine}}{CH_3SCH_2CH_2CH(NH_2)COOH}$$

Methionine can then be used by methionine adenosyltransferase to generate S-adenosylmethionine and subsequently S-adenosylhomocysteine, which after hydrolyzing off adenosine again yields homocysteine. *See* AdoMet; amino acid metabolism; homocystinuria; methionine adenosyltransferase deficiency.

methionine malabsorbtion Under the control of an autosomal-recessive gene, and it results in the excretion of α-hydroxybutyric acid (with its characteristic odor) in the urine. Mental retardation, convulsions, diarrhea, respiratory problems, and white eye accompany the condition. *See* methionine biosynthesis.

methionine synthase reductase (MTRR, 5p15.3-p15.2) Its deficiency may lead to megaloblastic anemia, hyperhomocyteinemia and hypomethioninemia, and meiotic nondisjunction of human chromosomes. (Leclerc, D., et al. 1998. *Proc. Natt. Acad. Sci. USA* 95:3059.)

methionyl tRNA synthetase (MARS) Charges tRNA^{Met} by the amino acid methionine. The MARS gene is in human chromosome 12. *See* aminoacyl-tRNA synthetase.

methotrexate (amethopterin) Folic acid antagonist and inhibitor of dihydrofolate reductase. It is an inhibitor of thymidylic acid synthesis. It is used as a selectable agent in genetic transformation. A mutant dehydrofolate reductase may improve the resistance to methotrexate and may be advantageous for cancer chemotherapy. *See* amethopterin; DB (double-minute) chromosome; folic acid; transformation, selectable. (Goodsell, D. S. 1999. *Stem Cells* 17:314.)

methylacetoacetic aciduria (11q22.3–23.1) 3-oxothiolase deficiency in the degradation of isoleucine, resulting in 2-methyl-hydroxybutyric acid, 2-methylacetoacetic acid, tiglylglycine, and butanone in the urine. Tiglic acid [CH$_3$CH:C(CH$_3$)CHO] is trans-2-methyl-2-butenoic acid. *See* isoleucine-valine biosynthetic pathway.

methylase Enzymes in bacteria protect the cell's own DNA from type II restriction endonucleases by transferring methyl groups from S-adenosyl methionine to specific cytosine or adenine sites within the endonuclease recognition sequence (cognate methylases). When eukaryotic DNAs are transferred to *E. coli* cells by transformation for cloning, their methylation pattern may be lost because the methylation system of the prokaryotic cell is different from that of the eukaryote. *E. coli* does not methylate the C in a 5'-CG-3', but the dam methyltransferase methylates A in the 5'-GATC-3' sequence and the dcm methyl transferase methylates the boxed C in the 5'-C $\boxed{\text{C}}$ $^{\text{A}}_{\text{T}}$GG-3' group. Such methylations may change the restriction pattern of cloned DNAs depending on whether the particular restriction enzyme can or cannot digest methylated DNA. *See* methylation, DNA; methylation-specific PCR. (Cheng, X. & Roberts, R. J. 2001. *Nucleic Acids Res.* 29:3784.)

methylated DNA-binding proteins Bind methylated DNA and recruit additional proteins; the methylome silences transcription of DNA. *See* methylation, DNA. (Nan, X., et al. 1998. *Nature* 393:311; Ng, H. H., et al. 1999. *Nature Genet.* 23:58.)

methylation, DNA (DNMT) In many eukaryotes 1–6% or more of the bases in DNA are methylcytosine. In T2, T4, and T6 bacteriophage DNA, 5-hydroxy-methylcytosine occurs in place of cytosine. Methylation of other bases such as thymine (= 5-methyluracil), adenine, and guanine may also occur. Some of the alkylations of DNA bases lead to mutations by base substitutions. Methylation protects DNA from most of the restriction endonucleases (*see* restriction endonuclease types). In the majority of *E. coli* strains, two enzymes are responsible for DNA methylation, *dam* and *dcm* methylase; *dam* methylates adenine at the N^6 position within the sequence 5'-GATC-3'. This sequence occurs at the recognition sites of a number of frequently employed restriction enzymes (Pvu I, Bam HI, Bcl I, Bgl II, Xho II, Mbo I, Sau 3AI, etc). Mbo I ($\downarrow$GATC) and HpaII (C$\downarrow$CGG) are sensitive to methylation, but Sau 3AI and MspI, respectively, are not, and their recognition sites are identical (isoschizomers); therefore, when the DNA is methylated, the latter ones still can be used. For several restriction enzymes to work, the DNA must be cloned in bacterial strains that do not have the *dam* methylase.

Mammalian DNA is not methylated at the N^6 position of adenine; therefore, Mbo I is always supposed to work, as well as Sau 3AI. The DNAs of eukaryotes are most commonly methylated on C nucleotides in $^{\text{CG}}_{\text{GC}}$ sequences. The dcm methylase methylates the internal C positions in the sequences 5'-CCAGG-3' and 5'-CCTGG-3'; this methylation interferes with cutting by EcoRII [$\downarrow$CC(A/T)GG] but not by BstN I, although at another position [CC$\downarrow$(A/T)GG] of the same sequence. *E. coli* strain K also has methylation-dependent restriction systems that recognize only methylated DNA: *mrr* (6-methyladenine), *mcrA* [5-methyl-C(G)], *mcrB* [(A/G)5-methyl C]. Mammalian DNA with extensive methylation at 5-methyl C(G) is, e.g., restricted by *mcrA*. Once the DNA is methylated, this feature may be transmitted to the following cell generation(s) by an enzyme — *maintenance methylase* — although methylation is usually lost through the meiotic cycle.

In bacteria, the expression of methylated genes may be reduced by a factor of 1,000, but in mammals the reduction may be of six orders of magnitude. Methylation in the promoter region usually prevents transcription initiation but not RNA chain elongation of that gene in mammals. In mice, *Dnmts3a* and *Dnmts3b* are necessary for embryonic survival (*b*) or development after birth (*a*). The double mutants (*a, b*) cannot develop beyond gastrulation. *Dnmt1* is incapable for de novo methylation; its role is maintenance methylation of the unmethylated strand generated by replication across the methylated strand in the double helix. Inactivation of DNMT1 does not affect the status of methylation of CpG doublets in human cancer cells involved in the silencing of tumor susceptibility genes in several types of cancers. In melanoma and several solid tumors, about 40% of the promoter of the p16$^{\text{INK4a}}$ gene is hypermethylated. In endometrial, colon, and gastric cancer cells, the microsatellite instability is accompanied by up to 70–90% methylation. The DAP kinase gene in Burkitt's lymphoma may be completely methylated. Mutation in the human DNMT3B gene is responsible for the rare immunodeficiency, ICF. ICF (20q11.2) involves immunodeficiency, pericentromeric hypomethylation, and instability in lymphocytes and facial malformations. In some fungi (*Ascobolus*, *Neurospora*), peptide chain elongation may also be inhibited. The inhibition of initiation is attributed to the reduced binding of transcription factors to methylated DNA. Also, methylation-dependent DNA-binding proteins (MDBP) may suppress transcription. Methylation by Dnmt1 may affect histone deacetylation and chromatin remodeling.

Proteins MeCP1 and MeCP2 mediate silencing of gene expression due to methylation of CpG. MeCP1 effect on silencing depends on the density of methylation near the promoter. MeCP2 binds only single methylated CpG pairs and can act at a distance from the transcription factor-binding sites. The effect of MeCP2 is apparently not gene specific; rather it is global. It has been suggested that genomic imprinting is caused by differential methylation. In mice, demethylation or lack of MeCP2 may prevent normal completion of embryonic development. In *Arabidopsis* plants, demethylation to about one-third of the normal level caused either by a DNA (*ddm1*) demethylation mutation or by introducing an antisense RNA of the cytosine methyltransferase (MET1) by transformation caused alterations in the morphogenesis and developmental time of the plants.

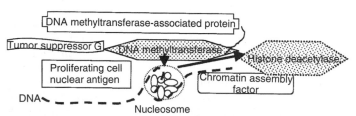

Some of the major factors controlling gene silencing. The proliferating cell nuclear antigen (PCNA) and the chromatin assembly factor (CAF) cooperate in building the nucleosomal structure. DNA Methyltransferase (DNMT) is assisted by the DNA Methyltransferase associated protein (DMAP) and the tumor suppressor gene 101 (TSG) methylates primarily cytidine residues and silences genes. Histone deacetylase (HDAC) joins the silencing system.

Demethylation of *XIST* leads to its expression, but that inactivates both X chromosomes in the female and the single X in the male, an obviously lethal condition. In some organisms with small genomes (*Caenorhabditis*), methylation of the DNA is very low; thus methylation, may not have a general developmental regulatory role. In *Drosophila* DNA, 5-methylcytosine is detectable only at very low frequency and mainly in the embryo, but interestingly the genome encodes two proteins that resemble cytosine DNA methyltransferase and methyl-CpG-binding-domain proteins.

If 5-methylcytosine is deaminated, thymine results, and a $C \equiv G$ pair may suffer a transition mutation to T=A. It seems that the genome of higher eukaryotes, including humans, includes 35% or more active or silent transposable elements. It had been suggested that methylation of infective (inserted) DNA is part of the eukaryotic defense system. Actually, most of the methylated cytosine in mammals is in the parasitic transposable elements. This methylation suppresses their transcription and the $C \rightarrow T$ mutation leads to the formation of pseudogenes. In the small invertebrate chordate *Ciona intestinalis*, nonmethylated transposons and normally methylated genomic sequences were detected (Simmen, et al. 1999. *Science* 283:1164). Not uncommonly, the DNA of cancer cells is undermethylated at C residues, indicating the demethylation of their parasitic sequences (SINE, LINE, Alu, etc.) and leading to the destabilization of the genome. The DNA methylase (methyltransferase) enzymes, therefore, have been supposed to be means to defend the genome against the deleterious effects of the infective transposable/viral elements. Although this hypothesis is in agreement with many observations, it does not seem to be of general validity, particularly for the methylation of plant transposable elements.

The epigenetic state of methylation can be transferred in the ascomycete *Ascobolus* by a mechanism resembling or related to recombination. After fertilization, the methyl moieties are generally removed from the CpGs and an unmethylated state is maintained through the blastula stage. Some of the genes involved in tumorigenesis display an increased methylation on aging. Housekeeping genes stay unmethylated, whereas the methylation of tissue-specific genes varies by tissues. Reduced methylation causes developmental anomalies in plants and animals. The maternal genomes in haploid and diploid gynogenetic one-cell mammalian embryos are always methylated. The polar bodies are always methylated. The paternal genomes in haploid or diploid androgenetic embryos are demethylated. Triploid digynic embryos show two methylated maternal and one demethylated paternal chromosome sets. In the diandric triploid embryos, the methylation pattern is the opposite (Barton, S. C., et al. 2001. *Hum. Mol. Genet.* 10:2983). Changes in methylation are apparently not required for the regulation of development of zebrafish. Methylation of the normally barely methylated *Drosophila* DNA reduces viability. In the embryonic tissues of mice, CpA and CpT are also methylated to some extent, not just CpG. The methylation pattern of cancer cell DNA is usually altered (the promoter of tumor suppressor genes is heavily methylated), but the overall extent of methylation is lower. Methylation of the promoter may interfere with the attachment of the transcription factors. The silencing effect of methylation may be associated with the simultaneous deacetylation of the nucleosomes. The gene silencing by methylation in *Arabidopsis* is controlled by the 2001-amino-acid product of the *MOM* gene. This protein contains a sequence with similarity to the ATPase region of SWI2/SNF2 protein family, which is involved in chromatin remodeling. Inactivating *MOM* by antisense technology releases heavily methylated genes from the silenced state. *See* alkylation; Alu; ascomycete; 5-azacytidine; base flipping; cancer; carcinogenesis; chemical mutagens; chromatin remodeling; CpG islands; cross-linking; DAM; diandry; epigenesis; histone deacetylase; HMBA; housekeeping genes; immunodeficiency; imprinting; integration; LINE; MeCP; methylation resistance; methylation-specific PCR; methylome; methyltransferase; MIP; paramutation; PCNA; regulation of gene activity; RIP; RLSG; silencing; SINE; Sp1; transposable elements; transposition. (Bird, A. P. & Wolffe, A. P. 1999. *Cell* 99:451; Robertson, K. D. & Wolffe, A. P., 2000. *Nature Rev. Genet.* 1:11; Baylin, S. B., et al. 2001. *Hum. Mol. Genet.* 10:687; Reik, W., et al. 2001. *Science* 293:1089; Aoki, A., et al. 2001. *Nucleic Acids Res.* 29:3506; Bender, J. 2001. *Cell* 106:129; Cervoni, N., et al. 2002. *J. Biol. Chem.* 277:25026.)

methylation, RNA 2′-O-methyladenosine, 2′-O-methylcytidine, 2′-O-methylguanosine, 2′-O-methyluridine, and 2′-O-methylpseudouridine are minor nucleosides in RNA. The cap of mRNA is a 7-methyl guanine, and an N^6-methyladenosine occurs near the polyadenylated tract in mRNA. *See* capping enzyme. (Santoro, R. & Grummt, I. 2001. *Mol. Cell* 8:719.)

methylation interference assay Detects whether a binding protein can attach to the specific DNA sites and thus provides information on the binding site and on the protein.

The analysis is carried out by combining DNA and binding proteins followed by treatment with methylating enzyme. If the protein binds to a specific guanine site(s), that base will not be methylated. Piperidine breaks DNA at bases modified by methylation, and sites protected from methylation by bound protein are not cleaved by piperidine. *See* methylation, DNA. (Shaw, P. E. & Stewart, A. F. 2001. *Methods Mol. Biol.* 148:221.)

methylation resistanc

methylation-specific
is usually detected by
restriction endonucleases
detects methylated CpG
The DNA is treated wit
nonmethylated cytosines
designed to represent origi
methylated and unmethyl
then be sequenced to comp
samples. *See* methylation,
(Velinov, M., et al. 2001. *G*

methylator Simultaneou
different cancer suppresso
DNA; tumor suppressor.

3-methylcrotonyl glycin
of a mitochondric enzym
of leucine, β-methylcroton
exons, 3q25-q27). The β-s
(MCCB, 17 exons). As a
recessive condition, muscl
urinary overexcretion of 3
hydroxyisovaleric acid, occur.
to biotin because this vitamin
amino acid metabolism; 3-hy
isoleucine-valine metabolic p
methylglutaconaciduria. (Gal
Hum. Genet. 68:334.)

5-methylcytosine Commor
otes. Deamination of 5-methylc
dine is one of the most commor
for 20% of all human point
binding domain protein MBD4
from the mismatched sites and
(Millar, C. B., et al. 2002. *Science* 297:403.) *See* 5-azacytidine; 5-hydroxymethyl cytosine.

5-Methylcytosine.

methyl-directed repair *See* mismatch repair.

methylene blue Aniline dye, for microscopic specimens, an indicator of oxidation-reduction, an antiseptic, an antidote for cyanide and nitrate poisoning.

5,10-methylenetetrahydrofolate dehydrogenase (MTHFD1) An enzyme in folate and purine biosynthesis is encoded in human chromosome 14q24. *See* folic acid.

methylenetetrahydrofolate reductase (MTHFR, 1p36.3) Its deficiency is responsible for homocystinuria and may affect human chromosomal nondisjunction. *See* homocystinuria; nondisjunction.

methylglutaconicaciduria Autosomal-recessive condition caused by a deficiency of 3-methylglutaconyl-CoA hydratase, an enzyme mediating one of the steps in the degradation of leucine. The patients may develop nerve disorders such as partial paralysis, involuntary movements, eye defects, etc. In some cases, there is a marked increase of methylglutaric and methylglutaconic acid (an unsaturated dicarbonic acid) in the body fluids. Leucine administration may exaggerate the symptoms. 3-methylglutaconicaciduria (MGA) type III is encoded at 19q13.2-q13.3. *See* amino acid metabolism; endocardial fibroelastosis; 3-hydroxy-3-methylglutaric aciduria; isoleucine-valine metabolic pathway; isovalericacidemia. (Anikster, Y., et al. 2001. *Am. J. Hum. Genet.* 69:1218.)

methylgreen pyronin Histological stain; coloring DNA blue-green and RNA red. *See* stains. (Brachet, J. 1953. *Quart. J. Microscop. Sci.* 94:1.)

methylguanine-O^6-methyltransferase Enzyme that reverses the alkylation of this base, and it is thus antimutagenic and anticarcinogenic. Cells defective in the enzyme (Mer$^-$, Mex$^-$) are extremely sensitive to DNA-alkylating agents. Mutant enzymes with increased methylguanine-O^6-methyltransferase activity may decrease cell sensitivity to alkylating mutagens. *See* antimutator; methylation, DNA. (Lips, J. & Kaina, B. 2001. *Mutation Res.* 487:59; Zhou, Z.-Q., et al. 2001. *Proc. Natl. Acad. Sci. USA* 98:12566.)

methylimidazole *See* pyrrole.

methyljasmonate Fragrance of jasmine and rosemary plants; it is a proteinase inhibitor.

methylmalonicaciduria There are several forms of the metabolic disorder: methylmalonic-CoA mutase deficiency (MUT), in human chromosome 6p21, and another is caused by a defect in the synthesis of adenosyl-cobalamin (cblA, vitamin B_{12}), a necessary cofactor in the biosynthesis of succinyl-CoA from L-methylmalonyl-CoA by MUT. A third type of methylmalonic aciduria is due to a defect in the enzyme epimerase (racemase) that converts D-methylmalonyl-CoA to the L form. This pathway can be represented as

Propionyl-CoA		D-methylmalonyl-CoA		L-methylmalonyl-CoA		Succinyl-CoA
	Carboxylase		Epimerase		Mut B_{12}	

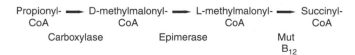

In these disorders, methylmalonic acid and glycine may accumulate in the body fluids, and the affected individuals may show serious (growth and mental retardation, acidosis [keto acids in the blood]) or almost no adverse effects. A high-protein diet (valine, isoleucine) may aggravate the condition. Administration of vitamin B_{12} may alleviate the problem in some cases. *See* amino acid metabolism; ketoaciduria; methylcrotonylglycemia; vitamin B_{12} defects.

methylmercuric hydroxide (MMH) May be added to the electrophoretic agarose running gel of RNA, and when it is stained with ethidium bromide in 0.1 M ammonium acetate, the color of the RNA is enhanced. It is also used to treat mRNA for preventing the formation of secondary structure during the synthesis of the first strand of cDNA. Note that MMH is an extremely toxic volatile compound. *See* cDNA; electrophoresis.

methylmethanesulfonate Powerful alkylating agent and mutagen/carcinogen. *See* ethylmethanesulfonate.

methylome Factors involved in methylation of DNA. *See* methylated DNA-binding proteins; methylation, DNA.

methylphosphonates Oligonucleotide analogs used for antisense operations. They are readily soluble in water and resistant to nucleases. The oligonucleotide methylphosphonates form stable complexes with both RNA and single- and double-stranded DNA. They have antiviral and anticarcinogenic effects. *See* antisense technologies; mixed backbone oligonucleotide. (Schweitzer, M. & Engels, J. W. 1999. *J. Biomol. Struct. Dyn.* 16[6]:1177.)

CH₃P(O)(OH)₂
Methylphosphonic acid

methyltetrahydrofolate cyclohydrolase deficiency
See phosphoribosylglycinamide formyltransferase, affects folate and purine metabolism.

methylthioadenosine phosphorylase (MTAP) Encoded in human chromosome 9p21 area, and its defect or deletion is characteristic for many malignant tumors. It may be associated (linked) to a tumor suppressor activity of CDK-4. Methylthioadenosine is abundant in some human tissues and it is an important donor for methylation. *See* CDK; methylation, DNA. (Hori, Y., et al. 1998. *Int. J. Cancer* 75:51.)

methyltransferase, DNA (dnmt1, dnmt1-b, 19p13.3-p13.2) Responsible for the methylation of CpG sites. These are encoded in humans by the same gene locus, but the transcript is spliced alternatively. The activity of these enzymes is increased during the initiation and progression of carcinogenesis, but the methylation of DNA in tumors is altered and usually reduced. The 1,620-amino-acid mammalian dnmt is essential for embryonic development of the mouse. Methyltransferases have an important role in regulation of gene activity, restriction-modification in bacteria, mutagenesis, DNA repair, cancer, imprinting, chromatin organization, and lyonization. DNA methylating enzymes are scarce in some insects (*Drosophila*), which do not have methylation in the genetic material. *See* cancer; methylation, DNA; methylguanine-O^6-methyltransferase, immunodeficiency, restriction-modification, (Adams, R. L. P. 1995. *Bioassays* 17:139; Bestor, T. H. 2000. *Hum. Molec. Gen.* 9:2395; Kiss, A., et al. 2001. *Nucleic Acids Res.* 29:3188.)

methylviolet Aniline dye for bacterial microscopic examination.

MET oncogene Hepatocytic growth factor receptor gene in human chromosome 7q31. Its α-subunit is extracellular; its β-subunit is an extra- and transmembrane protein. It is a tyrosine kinase as well as a subject for tyrosine phosphorylation. The receptor of HG-SF is the product of Met. *See* hepatocyte growth factor; oncogenes; papillary renal cancer; tyrosine kinase.

METREE Computer program package for inferring and testing minimum evolutionary trees; designed by the Institute of Molecular Evolution, Pennsylvania State University, Philadelphia. *See* evolutionary tree.

METRO Message transport organizer is a mechanism and center of sorting out molecules within the developing oocyte. *See* morphogenesis; RNA localization.

metronidazole (2-methyl-5-nitro-1-imidazole ethanol) Radiosensitizing agent and mutagen for chloroplast DNA; a suspected carcinogen. *See* chloroplast genetics, formula.

Metronidazole.

MeV Mega electron volt or million electron volt. *See* electron volt.

mevalonicaciduria (MVK) Recessive (human chromosome 12q24) defect with huge increase of mevalonic acid in the urine caused by a defect in mevalonic acid kinase. *See* hyperimmunoglobulinemia.

Mex⁻ Methylation excision minus. Deficient in methyltransferase DNA repair.

Mex67 Protein carrying leucine-rich nuclear localization signal; serves an export adaptor. *See* export adaptor; nuclear localization sequence.

MF1 5′ to 3′ exonuclease. *See* DNA replication, eukaryotes.

MFG Moloney murine leukemia retrovirus-based vector.

MGD *See* mouse genome database.

Mge1   26 kDa GrpE homologue in *Saccharomyces* mitochondria. *See* Grp.

MGMT   *See* methylguanine-O⁶-methyltransferase.

MGSA   Melanoma growth-stimulating activity. *See* gro; KC; N51.

MGT   MGMT group of enzymes; they are encoded in human chromosome 10q and protect the cells against the genotoxic, recombinogenic, and apoptotic effects of O^6-methylguanine with the aid of mismatch repair. When MGMT is expressed at a high level, the formation of O^6-methylguanine is substantially reduced. *See* Mer⁻; MGMT. (Kaina, B., et al. 2001. *Progr. Nucleic Acid Res. Mol. Biol.* 68:41.)

MHC   (1) Major histocompatibility complex is involved in immunological reactions; it is controlled by linked multigene families (HLA) determining cell surface antigens and thus cellular recognition. *See* HLA; immune system. (2) Myosin heavy chain.

MHC restriction   *See* immune system.

MIC   Minimal inhibitory concentration. *See* micRNA.

MICA/B protein   Intracellular protein that marks (cancer) cells for destruction by natural killer cells. *See* killer cells.

micelle   Round body of (protein) substances surrounded by lipids.

Michaelis-Menten equation   Measures enzyme kinetics in a process:

$$(E) + (S) \overset{k1}{\underset{k2}{\Leftrightarrow}} (E) \overset{k3}{\rightarrow} \text{Product(s)} + (E)$$

where k_1, k_2, k_3 are constants of the reactions, (E) = enzyme, (S) = substrate concentration, and

$$v = \frac{V(S)}{K_m + (S)} \quad \text{or} \quad k_m = (S)\left[\frac{V}{v} - 1\right]$$

where v is the velocity of the reaction when half of the substrate molecules are combined with the enzyme, V = the maximum velocity, and

$$K_m = \frac{[(E) - (ES)(S)]}{(S)}$$

is the Michaelis-Menten constant. *See* Linweaver-Burk plot.

Michel syndrome (oculopalatoskeletal syndrome)   Autosomal-recessive multiple defect involving the eyelid, opacity of the cornea, cleft lip and palate, defects of the inner ear and spine column, etc. It causes complete deafness. *See* deafness; eye diseases.

micRNA   Messenger RNA-interfering RNA. This RNA is transcribed on short sections of the complementary strand DNA and it prevents gene expression. *See* antisense RNA; RNAi. (Mizuno, T., et al. 1984. *Proc. Natl. Acad. Sci. USA* 81:1966.)

Microarray hybridization

Cloned DNA placed in the wells of microtiter tray. After amplification by PCR and purification, the samples (~5 nL) are spotted by a robot pipetter onto specifically coated microscope slides.

Cellular RNA transcripts of sample and an appropriate reference are labeled by two different fluorochromes and hybridized under stringent conditions to the clones.
The slides are irradiated by laser and the emission is measured by confocal laser microscope. The monochrome images are transfered to computer software, which generates pseudocolor images. The color intensities as well as intensity ratios of sample and reference (the normalized ratio) are determined, and the difference in expression ratio reveals how the genes are expressed under the two different conditions.

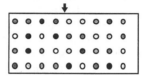

This slide is a black/white simulation of a colored picture where different light intensities may be represented by different colors.

microarray   *See* DNA chips; microarray hybridization.

microarray hybridization   Microtiter tray with wells containing DNA to which fluorochrome-labeled-RNA can be hybridized. After incubation and scanning, the amounts of the mRNAs derived from the same tissue can be quantitated (by fluorescence intensity). A microarray tray looks similar to the pattern shown above. Its size may be as small as 18 × 18 mm. The thousands of genes expressed on a single tray may display different fluorescent colors according to the fluorochrome labeling of the probe in a spot test. Such an analysis may identify the simultaneous expression of large sets of genes at a particular developmental or disease/health stage and thus permit a functional study at the entire genome level. Most commonly the colors shown in the publications represent the intensity of hybridization (red the highest).

If the tissues are, e.g., from healthy and diseased sources, on the basis of the level of expression, the genetic cause of the disease can be inferred. Depending on the type of fluorescent used, the RNA sample may vary from 10 μg to 50 ng and the number of cells needed to extract this much material may vary from >1 to 10⁹. The analysis may begin with cDNA clones, and PCR-amplified samples are transferred to glass plates or nylon filters where appropriate probes are hybridized to the arrays. Usually fluorochrome labeling is used for the glass slides, whereas phosphor imager (P³³, half-life ~25 days) is employed with the nylon. The data are evaluated by image processing after the information is fixed in JPEG or GIF forms,

Probe array

Image

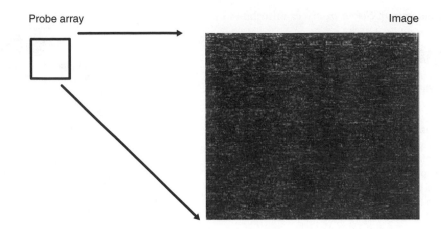

then statistical and biological methods are applied. (JPEG, TIFF, and GIF are Adobe PhotoShop® computer formats.) The information reveals the function of many genes in a context in different tissues, developmental or pathological states, and may be used for simultaneous characterization of many tumor biopsies. Sophisticated computer-based analysis of the biological meaning of the information, *data mining*, permits the identification of genes, which are expressed simultaneously or in certain tissues, in a specific state of a disease or diseases, or appear epistatic or respond to a particular condition. Microarray interrogated with short synthetic RNA oligomers permits the analysis of polymorphism at ~2% accuracy in up to 40,000 genes per microarray—more than enough for a genome the size of yeast. By the use of hierarchial clustering algorithms, parameterization, or profiling methods, gene expression can be mapped—not unlike the methods of genetic or physical mapping, but the clustered functions independent from the physical location of the genes may indicate the cotranscribed compartments and interaction of gene products. The pattern of expression reflects the dynamic network of timing, the physiological and developmental processes. The microarrays may be used to study the coregulation of genes and for the exploration of the global or particular effects of mutation or repression of single genes. Furthermore, the coexpression patterns of pathogens and host cells can be revealed.

The *spotting method* outlined above uses a single clone for the analysis of each mRNA. The GeneChip® Expression (Affymetrix, 3380 Central Expressway, Santa Clara, CA 95051) employs about 16 pairs of specific oligonucleotide probes to interrogate each transcript. (The description below is based on information and images provided by Affymetrix and Gene Microarray Shared Resource, Ohio State University.) The latter procedure is better suited for reducing and identifying nonspecific hybridization and background effects and therefore it is more sensitive and more accurate. A *target sequence* of ~600 nucleotides of gene is selected from a public database. About 16 to 20 *probes* are generated to match the sequences of the database. A pair of 25-base probe is of two types. The *perfect match* (PM) is entirely identical to a tract of the DNA. The *mismatch probe* (MM) contains a single mismatch in the middle, but otherwise it is the same as the PM. The rationale for using an MM probe is to have a control for nonspecific hybridization. These probes are synthesized on a GeneChip, a small glass plate. The synthesis is a light-directed process (photolithographic fabrication) yielding a large quantity of accurate probes in an economical manner.

From the biological sample, biotinylated mRNA is prepared, fragmented by heat, and applied to a *probe array*, which is a 1.28 × 1.28 cm glass surface held on a small tray. Hybridization is allowed for about 16 hr. The extent of hybridization is ascertained by the fluorescence intensity as detected by a laser scanner.

From a small segment of image, the intensity of the hybridization (i.e., the identity of the target and the probe pair) may have the alternative matches shown below. White corresponds to high hybridization intensity, black no measurable hybridization signal. Intermediate shades or colors correspond to intermediate signals. When the Perfect Match and the Mismatch signals are inconsistent, the hybridization is not detectable (see diagram below, Appendix II).

Microarray Suite (MAS) software manages the Affymetrix GeneChip experiments. The relative expression of a transcript is determined by the difference between each probe pair (PM minus MM) and averaging the difference over the whole probe set (*Avg Diff*). The term *Abs Call* for a probe set is a qualitative measure based on three different determinations collected by MAS 4.0. The Absolute Call for a probe set can be A for nondetectable, M for marginal, and P for present. *Diff Call*, difference call, is the qualitative call for a probe set representing the outcome of one array set compared to another. There are five possibilities: I increased, MI moderately increased, NC no change, MD moderately decreased, and D decreased call. Microarray hybridization is an essential tool in proteomics, for the study of genetic networks that play an important role in development and in studies of the reaction to drugs.

Perfect match
Mismatch

Signal detectable

Undetectable

Undetectable

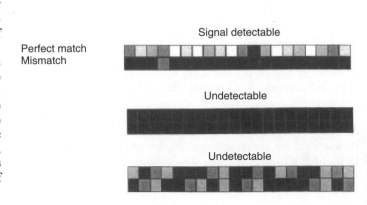

The impact of stress or cancer on gene expression may become amenable to interventions. The simultaneous effect of various drugs on a family of genes can be assessed. The availability of these methods is completely revolutionizing information gathering on the expression of genes and may be exploited for the study of the simultaneous expression of many genes involved in a pathway or related pathways. A number of factors may affect the efficiency of hybridization, such as temperature, base composition ($A=T$ versus $G\equiv C$), and even at the same composition, the actual sequences of the bases, secondary structures, sequence length, distribution of mismatches, etc. Unexplained variations have been seen with the present-day steadily improving technologies. An alternative approach for monitoring gene expression pattern optically measures light emission of transcriptional gene fusions of luciferase structural genes to promoters of operons or regulons. Gene expression (mRNA) and protein levels can be analyzed under a variety of conditions and an integrated picture can be derived of metabolic networks. The same basic principle has been applied to an array of animal cells, which are transfected with a variety of cDNAs, and the expression of the transgenes that affect cellular physiology is monitored. *See* activity-based protein profiling; base-calling; cluster analysis; DNA chips; electrospray MS; epistasis; genetic network; genomics; global single-cell reverse transcription–polymerase chain reaction; IMAGE clones; interrogation, genetic; laser desorption MS; light directed parallel synthesis; luciferase; lymphochips; macroarray analysis; operon; protein arrays; regulon; SAGE; SOM; support vector machine; synexpression; tissue microarray; TOGA; transcriptional gene fusion; two-hybrid method. (Ermolaeva, O., et al. 1998. *Nature Genet.* 20:19; *Nature Genet.* 21(1) Supplement; *Nature Biotechn.* 17:974; Scherf, U., et al. 2000. *Nature Genet.* 24:236; Van Dyk, T. K., et al. 2001. *Proc. Natl. Acad. Sci. USA* 98:2555; Ideker, T., et al. 2001. *Science* 292:929; Ziauddin, J. & Sabatini, D. M. 2001. *Nature* 411:107; computations: Quackenbush, J. 2001. *Nature Rev. Genet.* 2:418; Zhao, L. P., et al. 2001. *Proc. Natl. Acad. Sci. USA* 98:56321; Schulze, A. & Downward, J. 2001. *Nature Cell Biol.* 3:E190; Yang, Y. H. & Speed, T. 2002. *Nature Rev. Genet.* 3:579, *Nature Genet. Suppl.* [2002], 32:461–552; <www.ncbi.nlm.nih.gov/geo>; <http://www.ebi.ac.uk/microarray/>; temporal gene expression software:<genomethods.org/caged>; <http://industry.ebi.ac.uk/~alan/MicroArray>; <http://www.nhgri.nih.gov/>; <http://www.nhgri.nih.gov/DIR/LCG/15K/HTMLO/>; collection of yeast data: <http://transcriptome.ens.fr/ymgv>; <www.ohsu.edu/gmsr/amc/tech.shtml>.) Atlas human cDNA. Please see color plate in Color figures section in center of book.

microbe　Small organisms like the eukaryotic fungi, algae, and protozoa and the prokaryotic blue-green algae, bacteria, and viruses. The 2–3 billion microbial species exhibit enormous genetic diversity and are largely unknown. They are frequently associated with disease. The identification of particular microbes may be quite difficult by morphology or culture methods. Broad-range polymerase chain reaction, representational difference analysis, and expression library screening for specific sequences are feasible and greatly improve identification by molecular means. *See* microarray hybridization; PCR broadbase; RDA. (<http://pbil.univ-lyon1.fr/emglib/emglib.html>, <http://locus.jouy.inra.fr/micado>.)

microbial safety index　Logarithm of the reciprocal number of surviving microbes after a procedure of sterilization. Sterilization in principle is the destruction of all infective agents, but in practice a small fraction of $\sim 10^{-6}$ may survive the treatment.

microbiome　Collective genomes of the very large number of microbes that inhabit the human body.

microbodies　*See* peroxisomes.

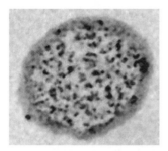

Microbody.

microcalorimetry　Detects the minute change in heat energy resulting from molecular interactions, associations, and dissociations. Two procedures are used: *differential scanning calorimetry* (DSC) of the temperature changes while proteins are unfolding or *isothermal titration calorimetry* (ITC), recording the temperature changes while the solutions are mixed. *See* immunoprecipitation; surface plasmon resonance.

microcell　Micronucleus, a piece of chromatin, a chromosome, or a few chromosomes surrounded by a membrane. *See* micronucleus.

microcell hybrid　Contains a single (e.g., human) chromosome in a complete other (e.g., mouse) genome. The hybrid, using deletions, may permit the identification of specific functions associated with segments of the critical chromosome. *See* somatic cell genetics. (Cao, Q., et al. 2001. *Cancer Genet. Cytogenet.* 129[2]:131.)

microcephaly　Abnormal smallness of the head; generally involves mental retardation. It is a condition due to various genetic and environmental causes (e.g., X-ray, heat [febrile] exposure of the fetus). The incidence of the autosomal-recessive form is about 2.5×10^{-5}. A deletion of chromosomal segment 1q25-q32 or mutation in 1q31-32 may be the cause of severe cases. Recurrence risk among sibs was estimated to be 0.19, but the risks may vary depending on the cause of the defect. It is generally accompanied by other abnormalities. Microcephaly with normal intelligence characterizes the autosomal-recessive Nijmegen breakage syndrome, which is associated with chromosomal instability, immunodeficiency, and radiation sensitivity. *See* cerebral gigantism; craniofacial dysostosis [Crouzon syndrome], hydrocephalus; mental retardation. (Pattison, L., et al. 2000. *Am. J. Hum. Genet.* 67:1578.)

microchaetae (pl.)　Hairs of insects. *See* chaetae.

microchannel plate detector   Analytical equipment for the detection of radioactive labels in proteins separated by 2-dimensional gel electrophoresis. *See* two-dimensional gel electrophoresis. (Richards, P. & Lees, J. 2002.*Proteomics* 2:256.)

microchimerism   Survival of donor cells in the recipient after transplantation of foreign tissues or organs. Preliminary evidence indicates that the rate of survival of pig cells in humans appears to be as low as 10^{-5}. *See* PERV; xenograft; xenotransplantation. (Johnson, K. L., et al. 2001. *Arthritis Rheum.* 44:2107.)

microchromosomes   Uniformly very small chromosomes of the avian genome rich in GC content. They replicate ahead of the larger (macro)chromosomes. Their density of genes appears very high, probably because their introns are short. *See* human artificial chromosome; intron. (McQueen, H. A., et al. 1996. *Nature Genet.* 12:321.)

microcin   Enterobacterial heptapeptide (Acetyl−Met−Arg−Thr−Gly−Asn−Ala−Asp−X) inhibiting protein synthesis where the first amino acid is acetylated and X is an acid-labile group. It is encoded by 21 bps and thus appears to be one of the smallest translated genes. *See* gene size.

microcinematography   Time-lapse motion photorecording of living material. Successive frames delayed in real time are projected at normal speed, giving the sensation as if the events, movements of cells or chromosomes, etc., would have taken place at an accelerated time sequence and thus, e.g., the progress of mitosis requiring 1−2 hr can be seen in motion in minutes. *See* Nomarski; phase-contrast microscopy. (Matter, A. 1979. *Immunology* 36[2]:179.)

micrococcal nuclease   From *Staphylococcus aureus*, degrades DNA (with preference to heat-denatured molecules) and RNA and generates mono- and oligonucleotides with 3′-phosphate termini.

microconidia   Small uninucleate conidia. *See* conidia; fungal life cycles; macroconidium.

μC (microcurie) 3.7×10^4 dps (disintegration/second). *See* Curie; isotopes.

microcytotoxicity assay   Detects allelic variants of MHC proteins. Cells are usually labeled by the green fluorescein diacetate and exposed to two different antibodies. The immunologically reacting cells are lysed and their nuclei are stained red by propidium iodide. Thus, the two alleles are distinguished in vitro. (Wahlberg, B. J., et al. 2001. *J. Immunol. Methods* 253:69.)

microdeletion   Loss of a segment too short to be visualized by light microscopy.

microdot DNA   Molecular version of steganography (concealing text) used to communicate secret spy messages. The messages are represented in nucleotide sequences generated by PCR using 20-base forward and reverse primers. The encrypt code may be represented by base triplets, e.g., CGA for the letter A, CCA for B, etc. The sequences within the PCR products can be hidden within the total human DNA or a mixture of DNAs of different organisms from which they can be fished out despite the enormous complexity of the mix. This total mixture, including the secret PCR message in ~10 ng, would not occupy a much larger microdot than about a full stop (.) on a filter paper. Primers may permit amplification of the message, and it can be subcloned and sequenced. If the cipher is known, the message is readable in English (or any other language) by those who are familiar with the primers, but it is virtually impossible to decipher it by others. *See* PCR. (Taylor Clelland, C., et al. 1999. *Nature* 399:533.)

microencephaly   Abnormal smallness of the brain caused by developmental genetic blocks or degenerative diseases. *See* microcephaly.

microevolution   Minor variation within species that may lead to speciation. *See* evolution; macroevolution. (Hendry, A. P. & Kinnison, M. T. 2001. *Genetica* 112:1.)

microfilaments   Actin- and myosin-containing fibers in the cells serving as part of the cytoskeleton. They mediate cell contraction, amoeboid movements, etc. *See* cytoskeleton.

microfusion   Fusion of protoplast fragments with intact protoplasts to generate cybrids (generally by electroporation). *See* cell fusion; cybrid; somatic hybridization.

microglia   Mesodermal cells supporting the central nervous system. Microglia are a class of monocytes capable of phagocytosis. They can bind β-amyloid fibrils (present in Alzheimer plaques) through scavanger receptors, resulting in the production of reactive oxygen species and leading to cell immobilization and cytotoxicity toward neurons. *See* Alzheimer disease; monocyte. (Giulian, D. 1999. *Am. J. Hum. Genet.* 65:13; Nakajima, K. & Kohsaka, S. 2001. *J. Biochem.* 130:169.)

β₂-microglobulin   Class I MHC α-chain is noncovalently associated with this polypeptide that is not encoded by the HLA complex. The α_3 domain and the microglobulin are similar to immunoglobulins and they are rather well conserved, in contrast to the α_1 and α_2 domains, which are highly variable. Its defect results in renal amyloidosis. *See* amyloidosis; HLA; TAP. (Hamilton-Williams, E. E., et al. 2001. *Proc. Natl. Acad. Sci. USA* 98:11533.)

micrognathia   Abnormally small jaw.

microgonotropen (MGT)   Inhibitor of transcription factor−DNA interactions. Transcription factors usually bind to the major groove of the DNA, whereas MGTs associate primarily with the minor groove but to some extent with the major groove. Efficient MGTs may regulate gene expression and may be of therapeutic value. *See* DNA grooves; transcription factors. (Wemmer, D. E. & Dervan, P. B. 1997. *Curr. Opin. Struct. Biol.* 7:355; White, C. M., et al. 2001. *Proc. Natl. Acad. Sci. USA* 98:10590.)

micrograph (photomicrograph, electron micrograph) Photograph taken through a microscope (light or electron microscope). *See* microscopy.

microinjection Delivery method of transforming DNA or other molecules into animal or other cells by a microsyringe. This procedure is not considered a highly efficient method of transformation in plants. An advantage is, however, that the delivery can be targeted to cells but not into chromosomal location unless gene targeting is used. *See* caged compounds; gene transfer by microinjection; targeting genes; transformation animals.

microlesion In genetic toxicology, denotes microscopically undetectable change in the genetic material.

micromanipulator Mechanical device usually employing glass needles or microsyringes to carry out dissections or injections of cells while viewed under the microscope.

micromere (small micromere) Small cells in the vegetal pole arising from the eight-cell blastomere and giving rise to the coelom. *See* blastomere; coelom; vegetal pole.

micronucleus Reproductive nucleus of *Infusoria*, as distinguished from the vegetative macronucleus; a small additional nucleus containing only one or a few chromosomes in other taxonomic groups. In some organisms, broken chromosomes may be visible as micronuclei. *See* microcell; micronucleus formation as a bioassay; *Paramecium*; *Tetrahymena*.

micronucleus formation as a bioassay Micronuclei are formed when broken chromosomes known as chromosomal fragments fail to be incorporated into the daughter nuclei during cell division. Also, the damage to the spindle apparatus may result in the appearance of micronuclei. These phenomena have been exploited for testing mutagenic agents that specifically cause these types of genetic damage to animal and plant cells. Such assays can be done in cultured cells, but in vivo assay of meiotic plant cells or mammalian bone marrow polychromatic erythrocytes has also been used. *See* bioassays in genetic toxicology. *Tradescantia*. (Riccio, E. S., et al. 2001. *Environ. Mol. Mutagen.* 38:69.)

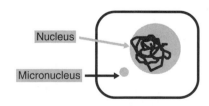

micronutrient Required for nutrition in small or trace amounts. This class of approximately 40 different compounds includes vitamins, minerals, etc. The deficiency of folate, vitamins B_{12}, B_6, niacin, C, and E, etc., has been suggested to be responsible for chromosomal damage and certain types of tumors. (Ames, B. N. 2001. *Mutation Res.* 475:7, and ff. articles in the same issue.)

microorganisms *Prokaryotic* → bacteria, *eukaryotic* → protozoa, fungi, algae.

Dominant eyeless mutant (*see* normal eye at morphogenesis in *Drosophila*).

microphthalmia *See* microphthalmos.

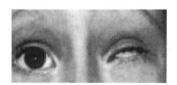

microphthalmos (nanophthalmos) Genetically determined (dominant and recessive) forms involve (extreme) reduction of the eye(s). In some instances, it does not involve additional defects. Its frequency in the general population is low, about 0.004%, and the incidence among Caucasian sibs is about 12–14% in the recessive form. Transformation of mice with diphtheria toxin genes attached to eye-specific (γ-crystalline, a globulin of the lens) or pancreas-specific elastase I, a collagen-digesting enzyme, promoted this developmental condition (ablation). Microphthalmia-associated transcription factor (MITF) mutations transform fibroblasts into melanocytes. Stimulation of melanoma cells by Steel factor (Sl) activates a MAP kinase that phosphorylates MITF, resulting in the transactivation of a tyrosinase pigmentation promoter. In humans, ~5 loci control the condition. Single amino acid replacement mutations at 14q24.3 of the retinal homeobox gene CHX10 have been identified. Dominant colobomatous microphthalmia was assigned to 15q12-q15. *See* ablation; anophthalmos; coloboma; eye diseases; Kit oncogene; Steel factor; Waardenburg syndrome. (Planque, N., et al. 2001. *J. Biol. Chem.* 276:29330.)

micropia *See* copia.

microplasmid (miniplasmid) *See* πVX; recombinational probe.

microprojectile *See* biolistic transformation.

micropropagation Plant regeneration from somatic, usually apical, meristem cells by in vitro techniques. It may be useful for rapid propagation of rare plants or nonsegregating hybrids or secure virus-free stocks. *See* embryo culture; synthetic seed. (Evans, D. A., et al., eds. 1983. *Handbook of Plant Cell Culture.* Macmillan, New York.)

micropyle Pore of the plant ovule between the ends of the integuments through which the pollen tube (sperms) reaches the embryo sac. The pore on the ovules of arthropods (and some other invertebrates) serving for the penetration of the sperm.

MicroRNA (micRNA) About 22-nucleotide-long regulatory molecules of diverse sequence. *See* RNAi; RNA noncoding. (Ambros, V. 2001. *Cell* 107:823; Hutvágner, G. & Zamore, P. D. 2002. *Science* 297:2056.)

microsatellite Mono-, tetra-, or hexanucleotide repeats distributed at random (10–50 copies) in the eukaryotic nuclear and mitochondrial chromosome can be used for constructing high-density physical maps and for the rapid screening for genetic instability, evolutionary relationships, detection of cancer in bladder cells found in the urine, etc. Although it was initially assumed that the alterations at the microsatellites follow a stepwise pattern and therefore are very useful for following changes within populations or between populations, these expectations were not entirely realized. Because the variability depends on the length of the arrays, there may be size constraints to the expansion, mutations may increase the flanking regions, etc. The microsatellite region may expand (most commonly) or contract and thus result in genetic instabilities. Microsatellite loci may mutate at a frequency of ~0.8% per gamete, more frequently than in somatic cells. Some estimates for eukaryotes are in the 10^{-4} to 10^{-5} range, making them useful for linkage analysis using PIC.

The rate of mutation varies according to organisms, microsatellite length and site, etc. In bacteria, the mutation rate in repeats appears two orders of magnitude higher. In general, longer repeats display higher mutation rates. The length of the microsatellite repeats appears longer in vertebrates than in *Drosophila*. In the majority of the cases, the expansion of the loci does not change the linkage phase of the flanking markers and therefore gene conversion is implicated. In humans, one ($\geq$4 bp) microsatellite per 6 kbp genomic DNA was estimated. Three-quarters of the repeats are A, AC, AAAN, or AG. The CA and TG repeats are distributed at random in the genome and used most commonly for linkage studies. In the human X chromosome, 3 and 4 base repeats occur in every 300–500 bp. The most common poly(A)/poly(T) repeats are not well suited for genetic studies because of PCR instability. A large-scale survey indicates that most of the microsatellites were generated as a 3′ extension of retrotranscripts and may serve as pilots to direct integration of retrotransposons.

Defective mismatch repair may be caused by mutations in the human gene MBD4/MED1 at 3q21-q22, leading to carcinomas with microsatellite instability. The *perfect microsatellite* represents one repeated motif without any insertion of a different base. The *imperfect microsatellite* has repeats with one base different from the main type. The *interrupted microsatellite* contains several repeated bases that are different from the pattern of the rest of the repeat structure. The *compound microsatellite* includes two or more different types of repeats in adjacent microsatellites. In *Saccharomyces*, the poly(AT) sequences are common. Prokaryotes display relatively few repeats and they are mainly poly(A)/poly(T). About 30% of the human microsatellites are conserved in murines. The microsatellites appear to have regulatory roles, but some may be transcribed and translated as coding trinucleotides (e.g., CAG polyglutamine tracts). Some of the untranslated upstream repeat elements are conserved among related species and may serve as enhancers by binding transcription factors. The enhancer activity may be only moderate but may increase or decrease by high copy number. Discrete repeat lengths may carry specific adaptive value in some populations. Diversity in microsatellite repeats may be generated by replicational slippage, unequal crossing over (although the reciprocal products are not always present), gene conversion, additions, and deletions. The microsatellites show a tendency of association with the nonrepetitive portion of the genome. *See* cryptically simple sequences; DNA fingerprinting; enhancer; hereditary nonpolyposis colorectal cancer; IAM; KAM; minisatellite; mismatch repair; PCR; PIC; slippage; slip-strand mispairing; SMM; stutter bands; TRM; trinucleotide repeats; unequal crossing over; VNTR. (Graham, J., et al. 2000. *Genetics* 155:1973; Tóth, G., et al. 2000. *Genome Res.* 10:967; Morgante, M., et al. 2002. *Nature Genet.* 30:194.)

microsatellite mutator (MMP) Increases frameshift and other mutations in G-rich microsatellites. *See* microsatellite; minisatellite.

microsatellite typing In case of linkage disequilibrium, the inheritance of microsatellite markers can be followed on PCR-amplified DNA. For example, the CAAT repeat in the human tyrosine hydroxylase genes (chromosome 11p15.5) using the Z alleles displayed the pattern of segregation below in the ethidium bromide–stained nondenaturing gel. (After Hearne, C. M., Ghosh, S. & Todd, J. A. 1992. *Trends Genet.* 8:288). *See* DNA fingerprinting; microsatellite; paternity testing; PIC. (Calbrese, P. P., et al. 2001. *Genetics* 159:839.)

microscopy Using a special optical device for viewing objects that are not clearly discernible by the naked eye. *See* atomic force microscopy; atom microscopy; confocal microscopy; Cryo-EM; dark-field microscopy; electron microscopy; fluorescence microscopy; light microscopy; Nomarski; phase-contrast microscopy; resolution; scanning electron microscopy; scanning tunneling; stereomicroscopy. (Sharpe, J., et al. 2000. *Science* 296:541; Jain, R. K., et al. 2002. *Nature Rev. Cancer* 2:266.)

microsomes Membrane fragments with ribosomes and enzymes obtained after grinding up eukaryotic cells and

Father	Mother	Sib-1	Sib-2	Sib-3	Paternal grandfather	Paternal grandmother	Alleles
▮▮▮	▮▮▮		▫▫	▮▮▮	▮▮▮		Z
					▮▮▮		4
		▮▮▮	▮▮▮			▮▮▮	12
▮▮▮			▮▮▮	▮▮▮		▮▮▮	16

The ▮ symbolize gel bands and the ▪ stand for homozygosity of the Z allele).

Microsatellite typing.

separating the cellular fractions by centrifugation. After about $9,000 \times g$ force (10 min), the microsomes (S9) remain floating while other cellular particulates become sediment. For the Ames genotoxicity bioassay, generally Sprague-Dawley rat livers are used. The rats are previously fed in the drinking water with polychlorinated biphenyl (PCB), Araoclor 1254, a highly carcinogenic substance (requires special caution of disposal), in order to induce the formation of the P-450 monooxygenase-activating enzyme system associated with the endoplasmic reticulum. *See* activation of mutagens; Ames test; carcinogen; centrifuge.

microspectrophotometer Spectrophotometer that can measure monochromatic light absorption of microscopical objects. *See* spectro-photometer.

microsporangium Sac that contains the microspore tetrad of plants and the microspores of fungi and some protozoa. *See* microspore.

microspore (small spore) Immature male spore of plants that develops into a pollen grain. *See* gametogenesis; megaspore; meiosis; microspore mother cell; microsporocyte; microspores.

microspore culture In vitro method for the production of haploid plants by direct or indirect androgenesis. *See* diagram on next page.

microspore dyad Microspore mother cell after the end of the first meiotic division is divided into two haploid cells (the microspore dyad). *See* meiosis.

microspore mother cell gives rise to the microspores of plants that mature into pollen. *See* microsporocyte.

microsporidia (Archezoa) Represent an evolutionary branch of anaerobic protists that diverged from the main line of Eukarya before the acquisition of mitochondria. *See* evolution.

microsporocyte Cell within which meiosis takes place and the microspores develop. *See* gametogenesis; microspore.

microsporogenesis *See* meiosis in plants.

microsporophyll Leaf on which microsporangia develop in lower plants.

microsurgery Dissection or other surgical operations carried out under the microscope, generally with the aid of a micromanipulator. *See* micromanipulator.

microtechnique Procedure used for the preparation of biological specimens for microscopic examination (involving fixation, staining, sectioning, squashing, etc.).

microtome Instrument that by means of various (sliding or rotating or rocking) motions cuts serial thin (usually within the range of $1-20\ \mu\mathrm{m}$) sections of the embedded or frozen specimens to be examined by light or electron microscopy. The cutting edge may be steel or glass. Electron microscopy also requires sectioning of the specimens, usually employing a diamond knife. *See* embedding; sectioning; smear; stains.

microtubule Various types of long cylindrical filaments of about 25 nm in diameter within cells built by polymerization of α- and β-tubulin proteins. Microtubules are hollow tubulin filaments of the spindle apparatus of the dividing nuclei, elements of the cytoskeleton, cilia, flagella, etc. The energy for polymerization is provided by hydrolysis of GTP to GDP. The beginning of the polymerization is called nucleation, and in animal cells it begins at the centrosomes. Microtubule elongation is a polarized process. Microtubules move around chromosomes and various protein complexes according to the blueprints of differentiation. The nerve axon microtubules are oriented with the plus end (their growing point) away from the cell body, whereas in epithelial cells they point toward the basement membrane. In fibrobasts and macrophage cells, the microtubules originate in the center of the cell and the plus end faces the outer regions. The transport function requires the cooperation of motor proteins (kinesins, dynein) that push the protein complex organelles on the microtubule trails.

Microtubules are somewhat unstable molecules, and antimitotic drugs such as colchicine or colcemid may block their growth. Taxol (an anticancer extract of yew plants) stabilizes the microtubules and arrests the cell cycle in mitosis. After polymerization, the microtubules undergo modification, e.g., a particular lysine of α-tubulin may be acylated and tyrosine residues removed from the carboxyl end. Microtubule-associated proteins (MAP) mediate the maturation of microtubules. MAPs aid the differentiation of nerve axons and dendrites that are loaded with microtubules. Microtubules move various organelles such as the chromosomes during nuclear divisions in the cells with the assistance of the proteins kinesin and dynein. The positioning of the microtubules is controlled by the cortical protein Kar9 and the Bim1/EB1 microtubule-stabilizing protein. EB1 is also a suppressor of the *adenomatous polyposis coli* tumors. The cilia and flagella involved in movements are built of bundles of microtubules. The microtubule protein complexes bind ATP at their "head" and associate with organelles by their "tail." The "plus" end of microtubules is where tubulin subunits are added rapidly and the "minus" end is where the addition is more slow. There are several protein factors that move the tubulins and associated structures. In prophase the duplicated centrosomes are separated by bimC, KAR3, and cytoplasmic dyneins (450–550 kDa, motility $75\ \mu\mathrm{m}$/minute). The bimC family members (120–135 kDa with $1-2\ \mu\mathrm{m}$/min motility) have different names in the different species (KLP61F and KRP_{130} in *Drosophila*, Cin8 in *Saccharomyces*, cut7 in *Schizosaccharomyces*). The KAR3 family (65–85 kDa, motility $1-15\ \mu\mathrm{m}$/min) is involved in spindle stabilization during metaphase. During prometaphase, the microtubules are captured by the kinetochore with the assistance of MCAK (mitotic centromere associated kinesin). The chromosomes congregating at the metaphase plane are moved by KIF4 (140 kDa) and related proteins that are involved in chromosome alignment during metaphase. During anaphase, CENP-E and the KAR3 family of proteins propel the movement of chromosomes toward the poles. MKLP, bimC, and cytoplasmic dynein proteins mediate the elongation of

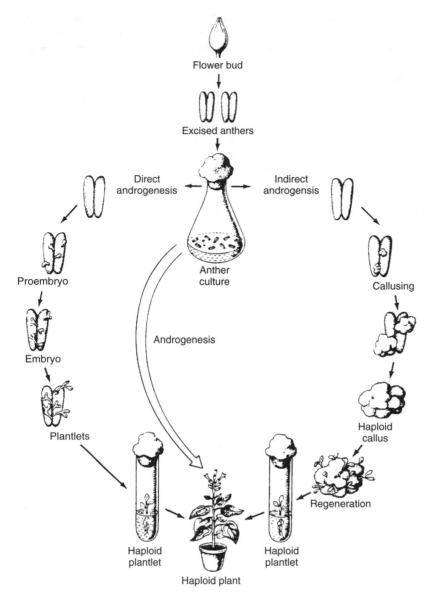

Flower bud

Excised anthers

Direct androgenesis

Indirect androgensis

Anther culture

Proembryo

Callusing

Androgenesis

Embryo

Haploid callus

Plantlets

Regeneration

Haploid plantlet

Haploid plantlet

Haploid plant

Outline of microspore culture. After Reinert, J. and Bajaj, Y. P. S. *Plant Cell and Tissue Culture*, p. 251. Springer, New York.

the spindle fibers. *See* centromere; centrosome; colchicine; dynamic instability; dynein; filaments; katanin; kinesins; meiosis; mitosis; polyposis adenomatous, intestinal; spindle; tau; taxol; treadmilling; tubulin; vinblastin. (Nogales, E. 2000. *Annu. Rev. Biochem.* 69:277; Downing, K. H. 2000. *Annu. Rev. Cell Dev. Biol.* 16:89; Popov, A. V., et al. 2001. *EMBO J.* 20:397; Schuyler, S. C. & Pellman, D. 2001. *Cell* 105:421; Lloyd, C. & Hussey, P. 2001. *Nature Rev. Mol. Cell. Biol.* 2:40.)

microtubule-associated protein (MAP) Controls stability and organization of microtubules. *See* dynein; kinesin; microtubule.

microtubule organizing centers Areas in the eukaryotic cells from which the microtubules emanate and grow, such as the mitotic centers (poles) that give rise to the mitotic spindle. *See* mitosis; MTOC; spindle.

microvilli Small emergences on the surface of various cells in order to increase their surface. The microvilli of the

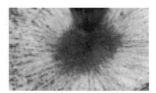

Microtubule organizing center.

chorion may be sampled for amniocentesis in prenatal genetic examinations. *See* amniocentesis.

microwave radiation Electromagnetic radiation in the $\sim 10^9$ to $\sim 10^{11}$ nm range. Genetic effects are difficult to separate from heat effects. In the 2–3 Ghz range, ambiguous mutagenic effects were reported, and the mutagenicity of the radiations could not be confirmed. Microwave radiation can damage cell membranes.

MIDA1 Helix-loop-helix–associated protein of mammals. It is inactive in erythroleukemic cells. *See* DNA-binding protein domains; erythroleukemia; helix-loop-helix; Zuotin.

midbody   After division, microtubule fragments (midbodies) may be detected in animal cells connecting the two daughter cells by a refringent structure. *See* microtubules. (Piel, M., et al. 2001. *Science* 291:1499.)

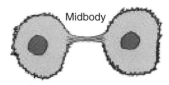

Midbody

midbrain   Middle part of the brain. Degeneration of the motor neurons in this region may be responsible for Parkinson disease. *See* brain, human; Parkinson disease.

middle lamella   Material (mainly pectin) that fills the intercellular space in plants.

middle repetitive DNA   Made up of relatively short repeats dispersed throughout the genome of (higher) eukaryotes. *See* LINE; redundancy; SINE.

midgut   Middle portion of the alimentary tract of insects and other invertebrates.

midkine   *See* pleiotrophin.

midparent value   *See* breeding value (midpoint).

midpoint   *See* breeding value.

MIF   Macrophage inhibitory factor is a proinflammatory pituitary factor; it may override glucocorticoid-mediated inhibition of cytokine secretion. MIF binds to cytoplasmic protein Jab1, which induces the phosphorylation of c-Jun and through it the activity of AP-1. Jab1 also helps the degradation of p27^{Kip1}. *See* AP1; cytokines; glucocorticoid; Jun; p27; septic shock. (Froidevaux, C., et al. 2001. *Crit. Care Med.* 29, Suppl. 7:S13.)

mifepristone   (11-[44-(dimethylamino)phenyl-17-hydroxy-17-(1-propynyl)-(11β,17β)-estra-4,9-dien-3-one)   Antiprogestin targeted to the progesterone receptors. This receptor family includes the glucocorticoid, mineralocorticoid, androgen, estrogen, and vitamin D receptors. receptors; RU486. (Ho, P. C. 2001. *Expert Opin. Pharmacother.* 2:1383; Zalányi, S. 2001. *Eur. J. Obstet. Gynecol. Reprod. Biol.* 98[2]:152.)

migraine   Neurological anomaly causing recurrent attacks of headaches, nausea, light and sound avoidance. It may or may not be preceded by an aura (subjective sensation). Migraine may be triggered by dietary factors (monosodium glutamate, tyramine and phenylethylamine in chocolates, citrus fruits, or certain cheeses) in persons with low levels of phenolsulfotransferase activity. (This enzyme catalyzes the conjugation of sulfate of catecholamines and phenolic drugs.) Its duration may be minutes to days. The sexual migraine pops up during sexual activity, commonly at or near orgasm. It may more frequently affect females (18–24%) than males (6–12%) to a very variable degree. Migraine is relatively rare in children (~4%) and its onset usually begins after age 30. In males it usually ceases after age 45, but in women it may continue well beyond menopause. It is generally attributed to multiple genes. Familial hemiplegic migraine (FHM) is, however, a rare autosomal-dominant condition coded in human chromosome 19p13 as is episodic ataxia type 2. CADASIL has been localized to the same chromosomal area and it involves migraine. It appears that the basic defect is in a brain-specific Ca^{2+} ion channel α-1 subunit translated from 47 exons of the CACN1A4 gene. Spinocerebellar ataxia type 6 and episodic ataxia are allelic to the latter gene. Some migraines are associated with mtDNA-encoded MELAS syndrome. Another migraine susceptibility locus appears to be at Xq24. *See* CACN1A4; CADASIL; ion channels; mitochondrial diseases in humans; spinocerebellar ataxia. (Guida, S., et al. 2001. *Am. J. Hum. Genet.* 68:759; Wessman, M., et al. 2002. *Am. J. Hum. Genet.* 70:652.)

migration   *See* gene flow.

migration, nuclear   Mediates polarization of cell division and provides direction for cell growth. (Bloom, K. 2001. *Curr. Biol.* 11:R326.)

migration, population   *See* gene flow.

migration inhibition factor   MIF is a lymphokine. *See* immune system; lymphocytes; lymphokines.

Mik1   Mitotic kinase and an inhibitor of Cdc2. Mik1 accumulates in S phase and may mediate the transition from S phase to mitosis. *See* Cdc2; cell cycle; Wee1.

MILC   (maximum identity length constrast):   It is a statistical method for genetic analysis of multifactorial diseases. It looks for an excess of identity of parental haplotypes transmitted to the affected offspring as compared to non-transmitted haplotypes. *See* haplotype. (Bourgain, C., et al. 2002. *Ann. Hum. Genet.* 66:99.)

Miller-Dieker syndrome   Characterized by smooth brain (lissencephaly I, AFAH1B1), more like in the early fetus, defects in other internal and external organs, mental retardation, and death before age 20. Apparently deletions in area 17p13.3 are responsible for the recessive phenotype. It is a cell-autonomous disease inhibiting neuronal migration. Other genes involving hydrocephaly and severe brain lesions may cause lissencephaly II (DCX, X-linked double cortin). The gene PAF (lipid platelet-activating factor) encodes a subunit of brain platelet-activating acetylhydrolase. *See* deletion; head/face/brain defects; HIC; lissencephaly; malformations; Walker-Wagner syndrome; Walker-Warburg syndrome.

miller units   *See* β-galactosidase (activity measurement).

millets (*Eleusine, Pennisetum*)   Arid climate grain crops. The cultivated *E. coracana* is $2n = 4x = 36$, tetraploid. *P.*

americanum is $2n = 2x = 14$ diploid. The common millet (*Panicum miliaceum*) is an old grain crop; $2n = 4x = 36$. The foxtail millet (*Setaria italica*) is $2n = 2x = 18$ is mainly a hay crop.

MIL oncogene Avian representative of the RAF murine oncogene, a protein serine kinase; it is also related to the murine leukemia virus (MOS). *See* MOS; oncogenes; RAF.

MIM *See* mitochondrial import.

mimicry Process and result of protective change in the appearance of an organism that makes it resemble the immediate environment for better hiding, or imitating the features of other organisms that are distasteful or threatening to the common predators. *See* Batesian mimicry; *Biston betularia*; industrial melanism; molecular mimics; Müllerian mimicry. (Mallet, J. 2001. *Proc. Natl. Acad. Sci. USA* 98:8928.)

mimicry, macromolecular Some proteins may mimic nucleic acids in shape, structure, and to some extent function. Such mimicry may protect a phage from the host restriction endonucleases. (Nissen, P., et al. 2000. *EMBO J.* 19:489; Walkinshaw, M. D., et al. 2002. *Mol. Cell* 9:187.)

mimicry, structural Pathogenic microorganisms produce virulence factors that are molecular mimics of host proteins so that they may evade the host defense system. *See* molecular mimics. (Stebbins, C. E. & Galán, J. E. 2001. *Nature* 412:701.)

mimics (1) Individuals that develop mimicry as a form of adaptation and evasion of predators. (2) Genes that control practically the same phenotype yet are not allelic.

mimotope Conformational mimic of an epitope without great similarity in amino acid residues or as a response to microbial anti-DNA antibodies. *See* epitope. (Wun, H. L., et al. 2001. *Int. Immunol.* 13[9]:1099; Mullaney, B. P., et al. 2002. *Comp. Funct. Genomics* 3:254.)

mineral corticoid syndrome (AME) Causes hypertension without overproduction of aldosterone. This syndrome is activated by cortisol. 11β-hydroxysteroid dehydrogenase converts cortisol to cortisone and thus activates the mineral corticoid receptor. Patients are deficient in this enzyme, which is inhibited by glycyrrhetinic acid (enoxolone [$C_{30}H_{46}O_4$], present in licorice). The mineral corticoid receptor and other steroid receptors regulate the activity of many genes. *See* aldosteronism; hypertension; Liddle syndrome; nuclear receptor. (Pearce, D., et al. 2002. *J. Biol. Chem.* 277:1451.)

mineral requirements of plants Nine macro elements (H, C, O, N, K, Ca, Mg, P, S) and seven micro elements (Cl, B, Fe, Mn, Zn, Cu, Mo). Under some circumstances, other elements may also be beneficial. *See* embryo culture.

minicell DNA-deficient bodies surrounded by a cell wall (in bacteria). Because they have no DNA, they cannot incorporate labeled precursors either into RNA or protein. In case the minicells descended from parents with plasmids, they may contain DNA and thus can make RNA and direct protein synthesis depending on the nature of their DNA. *See* maxicells.

minichromosome In eukaryotic viruses (SV40, polyoma virus), the histone-containing, small nucleosome-like structure of genetic material; also in eukaryotes, an extra chromosome with extensive deletions. Such minichromosomes can be generated by the insertion of human telomeric sequences (TTAGGGG)$_n$ between the centromere and the natural telomere, eliminating the sequences distal to the insertion point. Human mini-Y chromosomes have about 32.5 to 4 Mb compared with the normal Y chromosome of 50–75 Mb. Minichromosomes may be generated by removal of nonessential distal genes from each arm. Neocentromeres may serve as regular centromeres for human artificial chromosomes. These minichromosomes permit the analysis of the role of different sections of the chromosomes. It may become feasible to extend the analysis of mammalian chromosomes in yeast cells. Artificial chromosomes may be exploited as large-capacity cloning vectors. *See* human artificial chromosome; MCM1; neocentromere. (Saffery, R., et al. 2001. *Proc. Natl. Acad. Sci. USA* 98:5705.)

minichromosome maintenance factor *See* MCM.

mini-F Basic replicon of the bacterial F plasmid. *See* F plasmid; replicon.

minigels Used for the separation of small quantities (10–100 ng DNA), in small fragments (<3-kb) on about 5×7.5 cm slides (10–12 mL agarose), in a small gel box (ca. 6×12 cm) for 30 to 60 min at 5 to 20 V/cm. *See* electrophoresis.

minigene Some of its internal sequences are deleted in vitro before transfection. *See* transfection.

minihelix of tRNA Consists only of the TΨC arm and the amino acid acceptor arm and may be aminoacylated by some aminoacyl-tRNA synthetases according to the rule of the operational RNA code. The minihelix is considered to be the most ancestral part of the tRNA. *See* aminoacyl-tRNA synthetase; operational RNA code; transfer RNA; tRNA. (Nordin, B. E. & Schimmel, P. 1999. *J. Biol. Chem.* 274:6835.)

minimal genome size Smallest number of genes required for survival (replication) in a specific milieu for a free-living organism. Based on transposon inactivation experiments of the genome of *Mycoplasma genitalium*, the estimate was ~265–350. Some viruses have only 4 genes, but they are genetic parasites. *See* gene number; genome; *Mycoplasma*.

minimal medium Provides only the minimal (basal) menu of nutrients required for maintenance and growth of the wild type of the species. *See* complete medium.

minimal promoter Includes the most essential sequences to facilitate transcription of genes (basal promoter) without some other regulatory elements such as enhancer or transcriptional activator. *See* basal promoter; enhancer; promoter; transcriptional activators.

MINIMAL RESIDUAL DISEASE (MRD)   Even after apparently successful chemotherapy of cancer, some residual cancerous cells may persist. Their detection is important to develop treatment for the prevention of relapse. Molecular techniques (PCR) are applicable for this goal. *See* PCR. (Kim, Y. J., et al. 2002. *Eur. J. Hematol.* 68:272.)

minimal tiling path   Tightly overlapping set of bacterial vector clones suitable for sequencing eukaryotic genomes. *See* tiling.

minimization   *See* crossing over.

mini mu   Deletion variants of phage Mu cloning vehicles that still carry the phage ends, a selectable marker, and replicational origin. *See* Mu bacteriophage.

minimum description length (MDL)   Principle frequently used in characterizing macromolecular sequence information. The best principle uses the least number of bits for the theory and the data. L(T) indicates the complexity of the theory by the number of bits that encode the theory. L(D/T) is the number of bits required to define the data in connection with the theory and reveals the consistency of the data with the theory. *See* bit.

minimum evolution methods   Use an estimate of a branch length of an evolutionary tree construct on the basis of pairwise distance data calculated by various mathematical algorithms. The most plausible tree should provide the smallest sum of total branch length. *See* algorithm; evolutionary distance; evolutionary tree; four-cluster analysis; least square methods; neighbor joining method.

miniorgan (neo-organ) therapy   Ex vivo genetically modified group of cells capable of synthesis of immunotoxins, angiogenesis inhibitors, hormones, ligands, or other proteins and enzymes delivered to target cells/tissues in vivo with the purpose of correcting acquired or genetic disorders. In case the promoters can be regulated, the supply of the gene product(s) can be adjusted according to the need. *See* angiogenesis; cancer gene therapy; ex vivo; gene therapy; immunotoxin; promoter; retroviral vectors. (Bohl, D. & Heard, J. M. 1997. *Hum. Gene Ther.* 8:195; Rosenthal, F. M. & Kohler, G. 1997. *Anticancer Res.* 17[2A]:1179.)

miniplasmid   *See* π VX microplasmid; recombinational probe.

miniprep   Small-scale quick preparation of DNA from plasmids or from other sources. (Ferrus, M. A., et al. 1999. *Int. Microbiol.* 2[2]:115.)

minireplicon   Vector consisting of a pBR322 replicon, a eukaryotic viral replicational origin (SV40, Polyoma), and a transcriptional unit. These vectors can be shuttled between *E. coli* and permissive mammalian cells. Also, deficient replicons containing only the replicational origin. *See* replicon; shuttle vector. (Roberts, R. C. & Helinski, D. R. 1992. *J. Bacteriol.* 174:8119.)

minisatellite   In eukaryotic genomes, short (14–100 bp), tandem, highly polymorphic repeats occur at many locations with repeat arrays of 0.5–30 kb. In forensic work, the minisatellites are used for DNA fingerprinting. They are supposed to be products of replicational errors (slippage) and localized amplifications. Their high variability is probably due to frequent unequal crossing over and duplication deficiency events. Gene conversion may also expand or contract minisatellites. These sequences are highly variable. The variations are associated with diabetes mellitus and various types of cancers. The minisatellite mutation rate in the human germline was estimated as ~5.2%, although it may vary at different loci and may be different in the two sexes. Minisatellites are used as RFLP or PCR probes in physical mapping or for characterization of populations. In contrast to humans, minisatellite intra-allelic mutation rate is quite low 5×10^{-6} per sperm (Bois, P. R. J., et al. 2002. *Genomics* 80:2). *See* DNA fingerprinting; gene conversion; IAM; microsatellite; MVR; PCR; RFLP; SINE; small-pool PCR; SMM; TPM; trinucleotide repeats; unequal crossing over; VNTR. (Vergnaud, G. & Denoeud, F. 2000. *Genome Res.* 10:899.)

minisegregant   Budlike extrusions of animal cells with pinched-off DNA.

MinK   15K protein that forms potassium ion channels in association with other proteins.

MINOCYCLINE   Tetracycline type antibiotic capable of passing the blood-brain barrier and it is a neuroprotective by inhibition of caspases. It may delay the progression of neurodegenerative diseases (Zhu, S., et al. 2002. *Nature* 417:74.)

minor grove, DNA double helix   Marked by arrow below.

Minor grove of DNA.

minor histocompatibility antigen   Has some role in immune reactions, but it is not coded in the HLA region. *See* HLA; MHC. (Dazzi, F., et al. 2001. *Nature Med.* 7:769.)

MINOS   1,775-base-pair transposable element of *Drosophila hydei* with 255 bp inverted terminal repeats and two nonoverlapping open-reading frames. *See* transposable elements, animals. (Zagoraiou, L., et al. 2001. *Proc. Natl. Acad. Sci. USA* 98:11474.)

minus end   Of microtubules or actin filaments, is less liable for elongation. *See* plus end.

minus position, nucleotide Indicates the upstream distance from the first translated triplet of the transcript of a gene. *See* RNA polymerase; triplet code.

minutes Approximately 60 dominant mutations in *Drosophila* that slow down the development of heterozygotes; lethal when homozygous. They are defective in ribosomal proteins.

MIP Methylation induced premeiotically is a mechanism of gene silencing that apparently causes a stall of the RNA polymerase before completing the transcription of fungal genes. *See* RIP.

MIP-1α Macrophage inflammatory protein is a chemokine-mediating virus and other microbial-induced inflammation related to RANTES. It belongs to the family of FK506-binding proteins. *See* acquired immunodeficiency syndrome; blood cells; FK506; peptidyl-prolyl isomerases; RANTES. (Matzer, S. P., et al. 2001. *J. Immunol.* 167:4635.)

MIPS Database of functional genomes and proteomes. (Mewes, H. W., et al. 2002. *Nucleic Acids Res.* 30:31; <http://mips.gsf.de>.)

MIR Mammalian-wide interspersed repeat, $\sim 12-30 \times 10^4$ copies/primate genome. They may regulate the expression of genes, alternative splicing, polyadenylation sites, and evolution. *See* redundancy. (Matassi, G., et al. 1998. *FEBS Lett.* 439:63.)

Mirabilis jalapa (four-o'clock) Ornamental plant and early object of inheritance, $2n = 58$. *See* figure below.

Mirabilis jalapa.

MIS Müllerian inhibitory substance. *See* gonads; Müllerian duct.

miscall *See* base calling.

miscarriage Loss of pregnancy. It occurs spontaneously in about 15–20% of pregnancies at least once during the life of the human female. Repeated miscarriage before 20 weeks of pregnancy occurs in 0.5–2% of women and it may be caused by environmental factors, extrinsic factors (infections), uterine anomalies, hormonal problems, chromosomal aberrations, autoimmune reactions or other immune problems (Rh), or metabolic dysfunction (folic acid deficiency, hyperhomocystinemia, defects in nitric oxide synthase, etc.). (Tempfer, C., et al. 2001. *Hum. Reprod.* 16:1644.)

miscegenation Sexual relations between partners of different human races. In the majority of human tribes, marriage was generally limited to within the tribes; however, marriage by "capture" existed in ancient societies where the conquerors in war abducted females. Miscegenation was applied particularly to marriage between whites and blacks in the United States and in some South American societies. There is no genetic justification against interracial marriage. Although marriage between blacks and Orientals was not prohibited by any law, marriage between Caucasians and blacks was unlawful in about 15 states of the U.S. until 1967 when the Federal Court ruled that the choice to marry resides with the individual. Racial and social (cast) discrimination in marriage still exists in many underdeveloped countries and in backward communities. *See* marriage; mestizo; mulatto; racism. (Hulse, F. S. 1969. *J. Biosoc. Sci. Suppl.* 1:31.)

mischarged tRNA Linked to wrong amino acids. *See* aminoacyl-tRNA synthetase; protein synthesis.

miscoding *See* ambiguity in translation.

misconduct, scientific Fabrication, falsification, or embellishing data, use of inadequate statistics or techniques, making unjustified conclusions, plagiarism, omitting relevant facts concerning the experimental data or the literature, misrepresenting previous or competing publications, claiming undue credit, or any other unethical behavior. *See* ethics; publication ethics; scientific misconduct.

misdivision of the centromere Is a vertical rather than longitudinal division. It generates two telochromosomes from one biarmed chromosome. The telochromosomes may open up to isochromosomes and may undergo misdivision, again forming telochromosomes. *See* telochromosome.

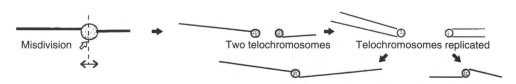

The cycles of bi-armed → telochromosome → isochromosomes → telochromosomes may continue.

misexpression of a gene *See* ectopic expression.

misincorporation During DNA replication, a normal nucleotide or an analog is placed into the growing strand at a site where correct pairing (e.g., A=T, G≡C) is not available. Such an event may lead to base substitution in the following step of replication, e.g., A + T → A : C→G≡C, resulting in mutation unless mismatch repair corrects it. *See* base substitution; DNA repair; mismatch repair. (Freese, E. 1963. *Molecular Genetics*, Taylor, J. H., ed., p. 207. Academic Press, New York.)

misinsertion DNA polymerase binds to a correctly matched 3′ end of a primer where right and wrong dNTP substrates compete for insertion and occasionally the wrong may succeed, resulting in misinsertion. *See* DNA editing; dNTP.

mislocalization Gene products are normally localized according to a specific pattern within the tissues, cells, or at subcellular sites. Deviations from the normal pattern may lead to disease. (Sutherland, H. G. E., et al. 2001. *Hum. Mol. Genet.* 10:1995.)

mismatch There is one or more wrong (noncomplementary) base(s) in the paired nucleic acid strands. Mismatches are generally identified (localized) by the use of flurochrome-labeled probes or they can be labeled by nanoparticle gold and analyzed on the basis of reduction silver (in the presence of hydroquinine at pH 3.8) by ordinary flatbed scanners with high sensitivity. Chemical cleavage of mismatch (CCM) is a hydroxylamine/osmium tetroxide-based (HOT) analysis of sequence variability in DNA. This procedure has been improved by adapting fluorescence techniques. *See* transition mismatch; transversion mismatch. (Ellis, T. P., et al. 1998. *Hum. Mutat.* 11:345.)

```
ATCG A GTCA
TAGC C CAGT
```

mismatch extension DNA polymerase binds to either a matched or mismatched primer end, and the mismatch is extended in replication. *See* mismatch.

mismatch repair (MMR) Excision repair that removes unpaired or mispaired bases and replaces them through unscheduled DNA synthesis with correct base pairs. The long-patch mismatch repair may have to correct tracts of hundreds of nucleotides. Mutations in *E. coli* gene *mutL* (homodimer, subunits ~68 kDa) and *mutS* (~95 kDa subunits) increase genetic instabilities. In yeast, defects in *PMS1, MLH1*, and *MLH2* may increase genetic instability 100- to 700-fold because of deficient mismatch repair. Mismatch repair deficiency may enhance mutation rate by 1–2 orders of magnitude, depending on the background (Ji, H. P. & King, M.-C. 2001. *Hum. Mol. Genet.* 10:2737). Similar are the consequences of deletions of RTH1, encoding a 3′ → 5′ exonuclease. The G/T-binding protein (GTBP, 100K) and hMSH2 (160K, homologue of bacterial mismatch-binding protein) are essential for mismatch recognition in human cells.

In fission yeast, the mismatch repair enzyme, exonuclease I, reduces the mutation rate. Defects in the bacterial methyl-directed mismatch repair system may also enhance mutability. The repair system is directed to DNA strands by methylation of adenine in the d(GATC) sequences. Since newly synthesized strands are not methylated, this is the criterion for recognition by the repair system (MutH, MutL, MutS, ATP). The mismatch may be located kilobases away from the d(GATC) tract. Interaction of this complex with heteroduplexes activates the MutH-associated endonuclease that responds to an initial sequence discontinuity or nick. Excision from the 5′ side of the mismatch requires the RecJ exonuclease or exonuclease VII, and digestion from the 3′ end needs exonuclease I and a helicase to unwind the strands because these proteins hydrolyze only single strands. RecJ and exo VII require that the unmethylated GATC should be downstream from the defect. Exo I usually does not have this precondition. The replacement synthesis requires DNA polymerase III and other polymerases not to work in this system. In the eukaryotic systems, the MutL homologues are PMS1 in yeast and PMS1; PMS2 in mammals. The yeast PMS1 system corrects G-T, A-C, G-G, A-A, T-T, and T-C mismatches, but C-C or long insertion/deletion sequences were barely repaired if at all.

The human repair system fixes 8–12 pair mismatches and C-C. A MutS homologue in mammals is called GTBP (G-T-binding protein). Defects in PMS1, MLH1, MSH2, and MSH3 enhance the mutation rate in yeast mitochondria, increase somatic mutability up to three orders of magnitude at certain loci (e.g., canavanine resistance), and destabilize (GT)$_n$ sequences. Mutations in the yeast Rth1 5′ → 3′ exonuclease also increase mutability, particularly of plasmid-encoded genes. MutS function is required for the normal progression of meiosis and the maintenance of female gonads. The yeast homologues of bacterial MutS and MutL not only correct errors in replication but block recombination between not entirely homologous (homeologous) sequences. Heterocomplexes of the mismatch repair genes have an important role in meiotic recombination. Defects in mismatch repair result in increased frequency of recombination and mutation, indirectly affecting evolution. Double-strand DNA breaks may be repaired by gene conversion. Mismatch repair deficiency may increase the susceptibility to some methylating drugs or ionizing radiation. *See* cisplatin; DNA repair; double-strand break; excision repair; gene conversion; gonads; hereditary nonpolyposis colorectal cancer; homeologous recombination; Huntington's chorea; incongruence; Lynch cancer families; microsatellite; MRD; Muir-Torre syndrome; mutator genes; short patch repair; slippage; unscheduled DNA synthesis; Walker boxes. (Nakagawa, T., et al. 1999. *Proc. Natl. Acad. Sci. USA* 96:14186; Oblomova, G., et al. 2000. *Nature* 407:703; Lamers, M. H., et al. 2000. *ibid.* 701; Buermeyer, A., et al. 1999. *Annu. Rev. Genet.* 33:533; Harfe, B. D. & Jinks-Robertson, S. 2000. *Annu. Rev. Genet.* 34:359; Evans, E. & Alani, E. 2000. *Mol. Cell. Biol.* 20:7839; Junop, M. S., et al. 2001. *Molecular Cell* 7:1; Wang, H. & Hays, J. B. 2002. *J. Biol. Chem.* 277:26136, 26143.)

Mis12/Mtw1 Protein that controls kinetochore orientation in yeast. *See* kinetochore.

mispairing Occurs when the homology between paired nucleotide sequences is imperfect and illicit binding takes place between nucleotides.

misreading   A triplet is translated into an amino acid different from its standard coding role. *See* ambiguity in translation.

misrepair   *See* DNA repair; SOS repair.

misreplication   *See* error in replication.

missense codon   Inserts in the polypeptide chain an amino acid different from that encoded by the wild-type codon at the site. *See* nonsense codon.

missense mutation   Change in DNA sequence that results in an amino acid substitution in contrast to mutations in synonymous codons where the change involves only the DNA but not the protein. *See* nonsense mutation.

missense suppressor   Enables the original amino acid to be inserted at a site of the peptide in the presence of a missense mutation. *See* suppressor tRNA. (Benzer, S. & Weisblum, B. 1961. *Proc. Natl. Acad. Sci. USA* 47:1149.)

missing genes   In the archaebacterium *Methanococcus janaschii*, 4 of the 20 aminoacyl-tRNA synthetases were not detected in the completely sequenced genome. It has been hypothesized that glutamine and asparagine are incorporated into polypeptides as transamidated derivatives of glutamate and aspartate. Furthermore, cysteine was assumed to be inserted as a trans-sulfurated serine and tRNA^{Lys} synthetase function was replaced by a protein quite dissimilar to known aminoacyl-tRNA synthetases. *See* aminoacyl-tRNA synthetase. (Bult, C. J., et al. 1996. *Science* 273:1058.)

missing link   Lack of transitional forms in evolution in between two species of which one was/is supposed to have descended from the other.

missing self hypothesis   One of the functions of natural killer lymphocytes (NK) is to recognize and eliminate cells that do not express class I MHC (major histocompatibility complex) molecules. *See* killer cells; MHC. (Ljunggren, H. G. & Karre, K. 1990. *Immunology Today* 11:237.)

missplicing   Incorrect splicing. *See* splicing.

mistranslation   *See* ambiguity in translation; misreading.

MIT⁻   General designation of mitochondrial point mutations.

Mitchurin   19th- to 20th-century Russian plant breeder who contributed a large number of improved varieties, mainly fruits, to Russian and then to Soviet agriculture. His career is often compared to that of the American Burbank. There was a very important difference, however. Burbank produced over 100 varieties in U.S. agriculture without governmental pretensions that he made novel basic scientific discoveries. Mitchurin, on the other hand, published undigested and misunderstood theories and became the official forefather

of lysenkoism, a state-supported charlatanism. *See* Burbank; lysenkoism.

MITE   Mariner insect transposon-like element, miniature inverted-repeat transposable element. A 1,457 bp DNA sequence in the vicinity of MLE containing a 24 bp tract with homology to Mos1 and supposedly responsible for the high frequency of unequal crossing over, resulting in a duplication appearing as Marie-Charcot-Tooth disease, and in the complementary deletion appearing as HNPP. Similar MITE elements occur in rice, maize, sugarcane, and other plants and control transcription. *See* heartbreaker; HNPP; hot spot; Marie-Charcot-Tooth disease; mariner; MLE; Mos1; unequal crossing over. (Casa, A. M., et al. 2000. *Proc. Natl. Acad. Sci. USA* 97:10083.)

mitochondria   Cellular organelles 1–10 μm long and 0.5–1 μm wide surrounded by double membranes. The outer membrane has a diameter of 50–75 and the inner 75–100 Å. The latter forms the structures, called cristae. Mitochondria are frequently associated with the cytoskeleton, which may control the distribution of this organelle within the cell and its transmission during cytokinesis.

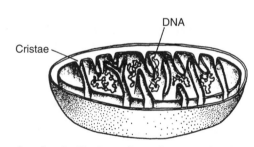

Diagram of a longitudinal section of a mitochondrion showing the cristae and the DNA strings in between them. According to electron microscopic view, the cristae are tubular rather than lamellar structures.

The outer membrane is associated with monoamine oxidase and an NADH–cytochrome-c reductase. Cardiolipin, phosphatidyl inositol, and cholesterol are major compounds associated with the inner membrane. Mitochondria have an important role in the generation of ATP and in electron transport. ATP synthesis requires oxidative phosphorylation. The mitochondria generally encode cytochrome b, cytochrome oxidase (COX) subunits, ATPase, and NADH dehydrogenase. All human cells except the erythrocytes contain mitochondria. In human mitochondria, 13 out of the 78 polypeptides involved in electron transport are coded for by mitochondrial DNA.

Mitochondria are the major source of reactive oxygens (ROS) in the body. In humans, about 300 nuclear genes control mitochondrial functions. The larger mitochondrial genomes of fungi and plants have a larger coding capacity. The mitochondrial DNA and the prokaryotic-type ribosomes in this organelle are capable of independent protein synthesis, although they may not have all their necessary tRNA genes within the mt genome. The majority of the proteins within the mitochondria are synthesized in the cytosol and imported. This import also depends on the function of cytosolic chaperones such as Hsp70 and Hsp60. In yeast and some other organisms, under some circumstances, a single giant mitochondrion is formed that spreads all over the cell and can break up into smaller organelles. Mitochondria are usually transmitted only through the eggs. Exceptions exist, however. In a human cell, about 100–1,000 mitochondria may be found, but a single mouse oocyte has $\sim 1 \times 10^5$. Each may contain 10 or more DNA molecules and about 1,000 peptides of which only 13 are coded by mtDNA. In addition, the mtDNA genome (16.5×10^3 bp in higher animals) encodes 2 rRNAs and 22 tRNAs. (In other organisms, the number of genes in the mitochondria may vary between 5 and 60.) Mitochondria are involved in the production of $\sim 80\%$ of cellular ATP, carry out respiration, synthesize amino acids, nucleotides, lipids, heme, and regulate inorganic ion channels and apoptosis.

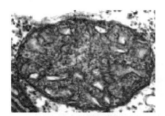

A plant mitochondrion.

During mammalian oogenesis, the number of mitochondria may be 4,000–200,000, but their number may be much reduced during early development of the embryo. The number of mitochondria seems to be regulated by mtTFA, and mtTFA is regulated by a nuclear regulatory factor (NRF-1). The large number of mitochondria within mammalian and other higher eukaryotic cells was probably necessitated by the limited capacity of DNA repair within the mitochondria. The large number can compensate for defective organelles. There is apparently no nucleotide excision repair (NER) in the mammalian mitochondria, but base excision repair (BER) by the use of glycosylases has been detected. Mismatch repair (MMR) also exists in mitochondria. It is not clear whether recombinational repair has a role in mitochondria. The evolutionary origin of the mitochondria has been interpreted by syntrophic hypotheses assuming the fusion of a primitive amitochondrial eukaryote with a *Clostridium*-like eubacterium or a *Sulfolobus*-like archaebacterium. Genome signatures, the selective advantage of expanded energy metabolism, and the possibility that *Clostridium* provided an opportunity for the development of the nucleus and the cytoplasm of the eukaryotic cell support the latter hypothesis. Mitochondrial DNA analysis (Finnilä, S., et al. 2001. *Am. J. Hum. Genet.* 68:1475) is a very important tool for evolutionary and demographic studies. *See* aging; amitochondriate; apoptosis; aspartate aminotransferase; CSGE; cytoplasmic male sterility; double uniparental inheritance; endosymbiont theory; Eve, foremother of mitochondrial DNA; Hsp60; Hsp70; hydrogenosome; membrane proteins; mitochondrial abnormalities, plant; mitochondrial diseases, human; mitochondrial genetics; mitochondrial import; mitochondrial plasmids; mtDNA; mtTFA; *Neurospora*; organelle compatibility; organelle evolution; organelle sequence transfer; paternal leakage; petite colony mutations; photorespiration; polar granules; promitochondria; respiration; rho factor; Rickettsia; ROS; signature, genomic; sorting out; Tid50;

Ydj. (<http://www3.ebi.ac.uk/Research/Mitbase/mitbase.pl>; <http://www.gen.emory.edu/mitomap.html>.)

mitochondria, cancer and   Dysfunction of apoptosis may lead to abnormal cell proliferation and thus to cancer. The apoptosis-blocking Bcl-2 and other proteins are localized in the mitochondrial membrane. The oxidative processes in the mitochondria may cause mutagenic lesions, leading to tumor initiation and progression. *See* apoptosis; cancer.

mitochodria, cryptic   Are double-strand membrane-enclosed structures in microsporidia that were supposed to lack mitochondria. Microsporidia are intracellular parasites of some animals and protists. The cryptic mitochondria are about 1/10 in size of mitochondria of animals and contain only a few proteins. The majority of the mitochondrial functions were apparently lost in these parasites as a consequence of retrograde evolution. *See* retrograde evolution, hydrogenosomes. (Roger, A. J. & Silberman, J. D. 2002. *Nature* 418:827)

mitochondrial abnormalities, plant   The cytoplasmic male sterility, widespread in many species of plants and the *nonchromosomal striped* mutations of maize have proven mitochondrial defects. The latter ones have deletions either in the cytochrome oxidase (*cox2*), NADH-dehydrogenase (*ndh2*), or ribosomal protein (*rps3*) mitochondrial genes. The mtDNA defects apparently originate by recombination between repeats (6–36 bp) at different locations within the genome. Recombinations involving larger repeats may give rise to normal mtDNA as well as duplication-deficiency molecules. The phenotype (green and bleached stripes on the leaves, stunted growth, etc.) might suggest plastid involvement. A similar mt-coded enzyme causes mitochondrial defects in other plants as well as in animals. *See* cytoplasmic male sterility; killer plasmids; mitochondria; mitochondrial disease in humans; mitochondrial genetics; mitochondrial plasmids; senescence. (Newton, K. J., et al. 1996. *Dev. Genet.* 19:277.)

mitochondrial complementation   When mitochondria with long deletions (4,696 bp), including the site of cytochrome oxidase c (COX), are lost, mice do not show any disease when COX$^+$ mitochondria are introduced. Within single cells, apparently only normal mitochondrial function can then be detected, indicating genetic complementation. The results promise the possibility of gene therapy by mtDNA. *See* mitochondrial heterosis. (Nakada, K., et al. 2001. *Nature Med.* 7:934.)

mitochondrial control regions   Encompass the areas where replication of the heavy strand (control region I, bp 16,040–16,400 in humans) and light strand (control region II, bp 39–380 in humans) takes place. The control regions display greater variation than the rest of the molecule and therefore these regions are frequently examined for base variations in evolutionary and population studies. *See* D loop; DNA replication, mitochondria; DNA 7S; mtDNA.

mitochondrial diseases, human   Before the 1970s, no human disease was attributed to mtDNA. Approximately

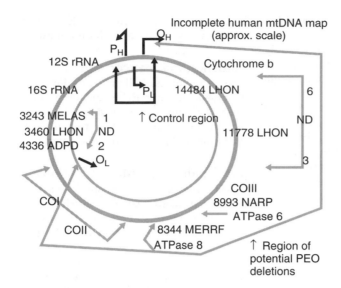

Incomplete human mtDNA map (approx. scale)

O_H: Origin of heavy-chain replication, P_H: promoter of heavy-chain replication, P_L: light strand promoter, ND: NADH-coenzyme Q oxidoreductase subunits, ADPD: Alzheimer disease late onset, CO: cytochrome oxidase subunits, NARP: neuropathy, ataxia, retinitis pigmentosa syndrome, MELAS: mitochondrial myopathy, encephalopathy, lactice acidosis, stroke-like syndrome (associated mainly with mutation in the tRNA^{Leu}), LHON: Leder's hereditary optic neuropathy, MERRF: myoclonic epilepsy and ragged-red fibers, PEO: progressive external ophthalmoplegia (most commonly deletions)

0.001 fraction of newborns have a mitochondrial disease. The role of mitochondrial DNA was definitely proven only after molecular analysis became practical. The inheritance pattern is not always clear because of heteroplasmy. *Chloramphenicol* resistance in human cells is caused by several mutations in the mitochondrial DNA-encoded 16S ribosomal RNA genes. Leber hereditary optic atrophy (LHON), a progressive disease (most commonly single-base-pair substitution [G(A(Arg→His]) resulting in wasting away of the eyes [optic atrophy]), abnormal heartbeat accompanied by neurological anomalies characterizes this disease (involving 5 NADH dehydrogenases at different sites in the mtDNA) that it appears a human X-chromosomal factor has a role in determining susceptibility to the mitochondrial defect. In LHON models of mice and dogs, adeno-associated virus vector carrying the gene for the retinal pigment epithelium photoreceptor successfully restored visual function (Acland, G. M., et al. 2001. *Nature Genet.* 28:92).

Kearns-Sayre syndrome (KSS/PEO), a progressive external ophthalmoplegia (paralysis of the eye muscles), is characterized by pigmentary degeneration of the retina, cardiomyopathy (inflammation of the heart muscles), and other less specific symptoms caused either by deletion or base substitution mutation in nucleotide 8,993 in the mtDNA. Deletions in 30–50% of the afflicted are flanked by the 5′-ACCTCCCTCACCA direct repeats and show the loss of nucleotides 8,468–13,446 and a less frequent deletion between the repeats 5′-CATCAACAACCG at positions 8,468–16,084. The larger deletions may simultaneously affect other mitochondrial genes. KSS most commonly involves heteroplasmy, and its occurrence is frequently sporadic. The basic defect is either in a protein phage T7 gene-4-like pimase/helicase of DNA (Spellbrink, J. N., et al. 2001. *Nature Genet.* 28:223) or in DNA polymerase γ, located at 15q22-q26 (Van Goethem, G., et al. *Nature Genet.* 28:211). In both cases, multiple mutations/deletions occur in the mtDNA.

One form of *myoclonic epilepsy*, MERRF (shock-like convulsions), is caused by a point mutation in the mtDNA-coded tRNA^{Lys}. *Mitochondrial myopathy* (muscle/eye degeneration), *encephalopathy* (brain degeneration), *encephalomyopathy* (MTTL2) mutation in tRNA^{Leu}, *lactic acidosis* (accumulation of lactic acid in the blood), and *stroke-like episodes* (MELAS syndrome, *mitochondrial encephalopathy with lactic acidosis and stroke-like symptoms*) are due to a mtDNA-encoded tRNA^{Leu} base substitution (A→G) at nucleotide 3,243 and at other sites and by mutations in tRNA^{Cys} and tRNA^{Val}. The same tRNA^{Leu} mutation occurs in about 20% of patients with the recessive autosomal *progressive external ophthalmoplegia* (PEO) and in some cases of diabetes mellitus (tRNA^{Leu}).

Pearson marrow-pancreas syndrome is caused by mtDNA deletions affecting subunit 4 of NADH dehydrogenase, subunit 1 of cytochrome oxidase, and subunit 1 of ATPase. This rare disease is usually heteroplasmic and frequently fatal at infancy. *Oncocytoma*, responsible primarily for benign solid kidney tumors, loaded densely with mitochondria, has deletions in subunit 1 of cytochrome oxidase. In general, more human males are affected by mitochondrial mutations; 85% of *Leber's optic dystrophy* cases are found in males. Alzheimer disease (tRNA^{Gln}) and Parkinson disease are associated with defective mitochondrial energy metabolism, reduced sperm motility, and reduced fertility. *Leigh syndrome* (an ATP synthase defect at nucleotide 8,527–9,207) is a progressive encephalopathy of children accompanied by over 90% mutant mitochondria in the blood, muscle, and nerve cells, although the inheritance appears autosomal.

Several myopathies are associated with mutations in various tRNAs of the mitochondria (tRNA^{Phe}, tRNA^{Ile} [cardiomyopathy], tRNA^{Glu} [cardiomyopathy], tRNA^{Met}, tRNA^{Ala}, tRN^{Asn}, tRNA^{Cys} [ophthalmoplegia], tRNA^{Tyr}, tRNA^{Ser}, tRNA^{Asp}, tRNA^{Lys} [MERRF syndrome], tRNA^{Gly}, tRNA^{Arg} [LHON], tRNA^{Ser}2, tRNA$^{Leu(CUN)}$ [skeletal myopathy], tRNA^{Glu}, tRNA^{Thr} [cytochrome b subunits]). A decrease in the number of mitochondria caused by genetic factors or drugs may lead to disease. *Aminoglycoside sensitivity* (modest doses of streptomycin) may lead to hearing defects due to mutation in the 12S rRNA gene. About 5×10^{-5} fraction of the human population is hypersensitive to chloramphenicol and may become anemic from the drug. Deletion of the COX (cytochrome oxidase) gene may result in *myoglobinuria*. *Autosomal-dominant progressive external ophthalmoplegia* (adPEO, CPEO) is a muscle weakness affecting primarily the eyes; it is caused by a mutant gene in human chromosome 10q23.3-q24.3, which is responsible for multiple deletions.

Hereditary spastic paraplegia is encoded by chromosome 16q, but the 795-amino-acid paraplegin protein is located in the mitochondria. Some of the late-onset neurodegenerative diseases apparently have mitochondrial components.

Human *colorectal tumors* frequently display purine transition mutations and appear homoplasmic. *Hereditary paraganglioma* is caused by a mutation at 11q23 in the SHDS gene encoding a small subunit of cytochrome b involved in succinate-ubiquinone oxidoreductase (sybS). This disease shows vascularized benign tumors in the head and neck due to defects in the carotid body (the main artery to the head), which senses oxygen levels in the blood. The mtDNA contains 37 genes. Most of them are transcribed from the heavy chain, but 9 are read in opposite direction from the light chain of the DNA and they are thus somewhat overlapping. The human mitochondria encode only 13 respiratory chain proteins, although they may harbor about 1,000 proteins. Some human diseases or syndromes (hypotonia [reduced muscle tension], ptosis [dropping down eyelids], ophthalmoplegia [eye muscle paralysis], high level of lactate in the blood serum, liver defects) may be associated with a reduced level of mtDNA.

Genetic counseling with mitochondrial disease is difficult because the transmission of the heteroplasmic conditions is irregular. Treatment of mitochondrial diseases has appeared rather elusive. The mitochondrial respiratory chain involves the multisubunit complexes (I) NADH-UQ oxidoreductase, (II) succinate dehydrogenase, (III) UQ-cytochrome c oxidoreductase, (IV) cytochrome c oxidase, and (V) ATP synthase. Blocking complex I by 1-methyl-4-phenylpyridinium (MPP$^+$) mimics Parkinson disease and LHON. Inhibition of complex II by 3-NPA (3-nitropropionic acid) causes the symptoms of Huntington's chorea. Cyanide and azide inhibition of complex IV resembles the expression of Alzheimer disease. The effect of oligomycin on complex V involves neuropathy, ataxia, and retinitis pigmentosa–like symptoms. There are a number of diseases that are not encoded by mitochondrial DNA and neither are the proteins localized in the mitochondria, yet they affect mitochondrial functions. Mitochondrially determined disease due to ATP deficiency can be alleviated (Manfredi, G., et al. 2002. *Nature Genet.* 30:394) by introduction into the nucleus the functional mitochondrial gene. *See* aging; Alzheimer disease; amyotrophic lateral sclerosis; antibiotics; apoptosis; atresia; chondrome; colorectal cancer; diabetes mellitus; dimethylglycine dehydrogenase; epilepsy; Friedreich ataxia; heteroplasmy; homoplasmy; Leigh's encephalopathy; mitochondria; mitochondrial abnormalities, plant; mitochondrial gene therapy; mitochondrial genetics; MNGIE; mtDNA;

mtPTP; myoglobin; myoneuralgastrointestinal encephalopathy; NARP syndrome; ophthalmoplegia; optic atrophy; oxidative DNA damage; Parkinson disease; pleiotropy; protoporphyria; senescence; transmitochondria; Wilson disease; Wolfram syndrome. (Wallace, D. C. 1992. *Annu. Rev. Biochem.* 61:1175; Smeitink, J. & van den Heuvel, B. 1999. *Am. J. Hum. Genet.* 64:1505; Acland, G. M., et al. 2001. *Nature Genet.* 28:92; Thornburn, D. R. & Dahl, H. H. 2001. *Am. J. Med. Genet.* 106:102; Steinmetz, L. M., et al. 2002. *Nature Genet.* 31:400; <http://www.mips.biochem.mpg.de/proj/medgen/mitop/>.)

mitochondrial DNA *See* mtDNA.

mitochondrial export *See* polar granules.

mitochondrial gene therapy Several mitochondrial diseases became ascertained and studied since the 1970s. They may include (1) nuclearly encoded gene products imported to the mitochondria, (2) strictly mitochondrially encoded defects, and (3) gene products under dual control. Correcting the defect of proteins encoded by the nucleus and translated in the cytoplasm has about the same requisites as gene therapy in general. There are more hurdles to overcome when the therapeutic DNA sequences are intended for transport into the mitochondria with the aid of cytoplasm-located vectors. The first requirement is the construction of appropriate signal peptides to mediate the import. There is experimental evidence for the feasibility of targeting a hairpin-shape DNA into the mitochondrial matrix with the aid of an appropriate transit/signal peptide (see diagrom below). The transforming vector must use the coding system particular for the mitochondria, which may be different in different organisms. The transcript and the translation product must be suitable for processing within the mitochondria by its own protein-synthesizing machinery. The mitochondrial ribosomes are also different from the cytosolic ones. The promoter must respond to the regulatory system of the mitochondria and the expression of the gene should be optimally modulated in that environment. The maintenance replication of the introduced transgene must be secured for both somatic divisions and for the maternal germline. Problems may arise if the defective, mutant polypeptide interferes with the assembly of the functional protein complex even in the presence of a correct product of the transgene. Since the majority of mtDNA mutations involve deletions, duplications, and rearrangement of the mitochondrial genome, some diseases may not be amenable to exogenously delivered corrections. The mitochondrial genome in higher organisms is present in polyploid forms and the rules of mitochondrial segregation within the cells are not entirely known. There is evidence that in the relatively very simple yeast system the mitochondrial ATPase 8 defect could be corrected by a genetic vector product in the cytoplasm. It has been reported that the β-subunit of the mitochondrial ATP synthase fused to the bacterial *CAT* (chloramphenicol acetyltransferase), driven by the cauliflower mosaic virus CaMV 35S promoter, and introduced into tobacco (*Nicotiana plumbaginifolia*) by an agrobacterial vector was expressed mainly in the mitochondria but not in the chloroplasts. The introduction of transgenes by biolistic methods to both mitochondria and chloroplasts has been achieved repeatedly. Defects of mtDNA frequently result in faster replication of the mutation. This type of anomaly may be corrected by antisense technologies. Antisense technology may also silence a mitochondrial gene with harmful effects. Eventually, gene replacement techniques may become practical. *See* antisense technologies; ATPase; biolistic transformation; gene replacement; gene therapy; genetic code; mitochondrial anomalies, plant; mitochondrial complementation acid; mitochondrial diseases, human; mitochondrial genetics; peptide nucleic acid; signal peptide; transformation, plant; transit peptide. (Murphy, M. P. & Smith, R. A. 2000. *Adv. Drug Deliv. Rev.* 41:235.)

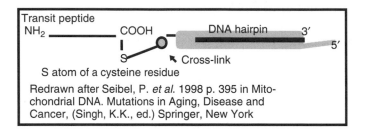

Transit peptide
NH$_2$ ———————— COOH   DNA hairpin   3'
                                                        5'
S atom of a cysteine residue   ↖ Cross-link

Redrawn after Seibel, P. *et al.* 1998 p. 395 in Mitochondrial DNA. Mutations in Aging, Disease and Cancer, (Singh, K.K., ed.) Springer, New York

mitochondrial genetics

Mitochondrial DNA is generally transmitted only by the egg cytoplasm. Wild-type yeast mtDNA is inherited biparentally. The hypersuppressive petite mutants of yeast are transmitted, however, only uniparentally. Segregation of mitochondrial genes is followed by the cell phenotype because individual mitochondria are not amenable to direct genetic analysis. (Molecular analysis can identify, however, differences among mitochondria.) Some of the genes are specific for the mitochondria and others, e.g., the cytochrome oxidase complex, share subunit coding with the nucleus. This sharing may vary in different evolutionary groups, which seems to indicate interchanges between the nuclear, mitochondrial, and chloroplast genomes. Despite the large number of mitochondria (>100/cell) and mtDNA molecules (4–6 per nucleoid), the homoplasmic condition may be obtained at a much faster rate than expected on the basis of random sorting out. In mice, it has been estimated that the number of segregating mitochondrial units is about 200 and the rapid sorting out was attributed to random genetic drift. (Some estimate the number of sorting mammalian mtDNAs to be >50,000.) The sorting out of the mitochondria—according to studies of yeast mutants of the MDM (mitochondrial distribution and morphology) group—is controlled by proteins of the cytoskeleton and integral proteins of the outer mitochondrial membrane, such as a dynamin-like protein. The ubiquitin-protein ligase suppressor, Rsp5p, and Ptc1p (serine/threonine phosphatase) seem to be involved in the regulation of mitochondrial transmission. In mouse oocytes, ~10^5 mtDNA molecules may occur, yet it appears that a relatively small number of mitochondria (20? to 200?) are transmitted by the egg cytoplasm to the zygote. This bottleneck facilitates a relatively rapid sorting out. Generally, 20 cell generations of the zygotes are used for appropriate mitochondrial typing followed by the use of selective media. In mammals, ~100 mitochondria may be transmitted by the spermatozoa; these, however, are destroyed within less than a day in the egg cytoplasm.

In natural populations of higher eukaryotes, mitochondrial recombination is considered to be rare because mitochondria are usually inherited through the egg, although paternal leakage may occur. The highest frequency of recombination in heteroplasmic cells is usually 20–25% because recombination between identical molecules also takes place and multiple rounds of matings occur within longer intervals. In humans and chimpanzees, the decline of linkage disequilibrium of mitochondrial markers was interpreted as evidence for recombination. In *Schizosaccharomyces*, the distribution of the mitochondria is mediated by microtubules. It has been suggested that mitochondria move along the cytoskeleton with a combination of motor proteins. Apf2 protein seems to be involved in the control of segregation and recombination of mtDNA in budding yeast.

Recombinational mapping is generally useful within short distances (≈1 kb). Genetic maps of mtDNA are generally constructed by physical mapping procedures. Also, the relative position of the genes may be ascertained by codeletion-coretention analyses. These methods resemble the deletion mapping of nuclear genes. Genes simultaneously lost by deletion, or retained in case of deletion of other sequences, must be linked. Mitochondrial genes have been mapped by polarity. It was assumed that the yeast mtDNA carried some sex-factor-like elements, ω^+ and ω^-, respectively. The omega plus (ω^+) cells appeared to be preferentially recovered in polar crosses ([ω^+] × [ω^-]). In these so-called heterosexual crosses, the order of certain genes, single and double recombination, and negative interference were observed. The exact mechanism of the recombination was not understood. The ω^+ factor turned out to be an intron within the 21S rRNA gene with a transposase-like function. Nonpolar crosses have also been used to determine allelism. In the latter case, all progeny were parental type and recombination was an indication of nonallelism. Linkage and recombination could be detected in nonpolar crosses, but mapping was impractical. Complementation tests can be carried out in respiration-deficient mutations on the basis that in "nonallelic" crosses respiration is almost completely restored within 5–8 hours, whereas recombination may produce wild types in 15–29 hours. Different loci are complementary and appear unlinked. Some loci may not complement each other. Mitochondrial fusion and complementation of mitochondrial genes in human cells has also been reported (Ono, T., et al. 2001. *Nature Genet.* 28:272). Homologous recombination has been extensively studied in yeast by the Flp/FRT system. Recombination of mtDNA in vertebrate cybrids does not readily occur. In case of fusion of haploid cells, gene conversion may take place.

Transformation of mitochondria requires the biolistic method or implantation of cell hybrids with mutant mitochondria. Unfortunately, only a fraction of the mutant mitochondria are transmitted and only through the female. Usually cotransformation by nuclear and mitochondrial DNA sequences is employed. The nuclear gene (*URA3*) is used for selection of transformant cells. The ρ^- cells cannot be selected because they are defective in respiration. They are crossed with mt$^-$ cells that carry a lesion in the target gene and among the recombinants the transformants may be recovered. It is possible that the wild-type *URA3* gene is transferred from the mitochondrion into the *ura3* mutant nucleus, whereas the cytochrome oxidase gene (*cox2*) is maintained in the nucleoid.

The large subunit of mitochondrial ribosomal RNA may house a mobile element, ω^+. This element (a group I intron) may show polar transmission and can be used to a limited extent for recombination analysis as already mentioned. It encodes 235 amino acids and represents *Sce*I endonuclease. A *Chlamydomonas smithii* intron (from the *cytb* mitochondrial gene) can move into the same but intronless gene in *C. reinhardtii*. Several other group I introns have similar mobility and the introns usually share the peptide LAGLI-DADG, a consensus also conserved in some maturases. Other group I introns display the GIY-10/11aa-YIG pattern. The VAR-1 yeast gene (ca. 90% A + T) encoding the small mt ribosomal subunit is also an insertion element. Other mobile elements with characteristic G + C clusters are common features of these insertion elements. Group II introns present in various fungi are mobile. Splicing/respiration–deficient mutants become revertible if the defective intron is removed. Similar mobile elements can be found in the cpDNA of *Chlamydomonas* algae. Mitochondrial plasmids may be circular or linear. Some of the circular mt plasmids display structural similarities to group I introns, and their transcripts are reminiscent of reverse transcriptases (resembling protein encoded by group II introns).

The probability (P) of segregation of mtDNA molecules under the assumption that each of the cells carries 10 rings of mtDNA, and among them the number of mutants may be 0, i (1) to j. The bracketed expressions represent the binomial probability functions.

$$P_{1-5} = \frac{\begin{bmatrix} 10 - 2i \\ 5 - j \end{bmatrix}\begin{bmatrix} 2i \\ j \end{bmatrix}}{\begin{bmatrix} 10 \\ 5 \end{bmatrix}}$$

(From Kirkwood, T. B. L. & Kowald, A. 1998, p. 131 in *Mitochondrial DNA Mutations in Aging, Disease and Cancer* (Singh, K. K., ed.), Springer-Verlag, New York.)

In yeast, nuclear genes *MDM1* (involved in the formation of nontubulin cytoplasmic filaments), *MDM2* (encodes Δ9 fatty acid desaturase), and *MDM10* (determines mitochondrial budding) control the transmission of mitochondria to progeny cells. The subunit(s) of the mitochondrial transcriptase and the mtTFA transcription factor are coded in the yeast nucleus.

Nuclear and mitochondrial promoters share cis elements. It seems that mitochondrial signals regulate nuclear genes controlling mitochondrial functions. Partially reduced intermediates of NADH dehydrogenase and coenzyme Q (ubiquinone) are held responsible for the production of ROS. In human diseases such as cancer, ischemic heart diseases, Parkinson disease, Alzheimer disease, and diabetes, ROS products have been implicated. Mitochondrial ROS activates mammalian transcription factors NF-κB, AP, and GLUT glucose transporters. The majority of the prokaryotic mutagens are apparently effective in causing mutations in mtDNA. The ROS molecules generated within the mitochondria may be responsible for a variety of alterations in mtDNA.

Deletions and duplications besides point mutations may occur spontaneously during the processes of aging. Mutation rates per mitochondrial D loop have been estimated to be 1/50 between mothers' offspring, but other estimates used for evolutionary calculations are 1/300 per generation or 3.5×10^{-8} per site per year, although the DNA polymerase γ appears to have low error rates. Some of the (recessive?) mitochondrial mutations are apparently nondetectable without molecular analysis because functionally the mitochondria are polyploid and their numbers in the cells are commonly very high. Because of the large number of cells, the large number of mitochondria per cell, and the multiple copies of the mtDNA per mitochondria, chances for mutation are high within a multicellular organism. Within a single individual, different mitochondrial mutations may occur and their spectrum may vary in different tissues. By 1998, nearly 1,000 mitochondrial point mutations were known, and 65 were involved in human disease. Mutations in mtDNA (5×10^{-7}/base/human generation) are much higher than in nuclear DNA (8×10^{-8}). The mutation rate/bp in both mtDNA and nuclear DNA varies substantially from region to region. About 40% of the mtDNA deletions involve 7,000–9,000 bp and 20% encompass 4,000–5,000 bp. Their distribution is biased inasmuch as 95% of the deletions affect the region between bp 3,300 and 12,000.

In plants, recombination repeats occur that may generate additional variation by interchanges. In somatic cell hybrids of plants, the mitochondria may recombine and nonparental sequences and duplications of genes may be generated. The pseudogenic sequences are also attributed to recombination. Early studies indicated an apparent lack of genetic repair systems in the mitochondria. Recent studies detected, however, several proteins (glycosylases, excinucleases, mismatch repair enzymes, Rec A–like proteins, etc.) that may mediate genetic repair in the mitochondria. *See* Alzheimer disease; AP; atresia; binomial probability; biolistic transformation; bottleneck effects; chloroplast genetics; conplastic; cybrid deletion mapping; DNA repair; double uniparental inheritance; dynamin; endosymbiont theory; Eve, foremother of mitochondrial DNA; GLUT; heteroplasmy; homoplasmy; interference; introns; ischemic; killer plasmids; mapping function; mating rounds; maturase; mismatch repair; mitochondria; mitochondrial diseases, human; mitochondrial gene therapy; mitochondrial import; mitochondrial mutation; mitochondrial plasmids; mitochondrial suppressor; MSS; mt; mtDNA; mutations, cellular organelle; mutation; spontaneous; NF-κB; organelle transformation; Parkinson disease; paternal leakage; petite colony mutants; physical mapping; recombination repeat; replicative segregation; RNA editing; Romanovs; ROS; Ru maize; senescence; sorting out; transmitochondria. (*Mutation Res.* 1999. 434[3]; Elson, J. L., et al. 2001. *Am. J. Hum. Genet.* 68:145; Tully, L. A. & Levin, B. C. 2000. *Biotechnol. Genet. Eng. Rev.* 17:147; <http//www.gen.emory.edu/mitomap.html>; <http://www.ncbi.nlm.nih.gov/PMGifs/Genomes/organelles.html>; <http://megasun.bch.umontreal.ca/ogmp/projects/projects.html>; http://www.mips.biochem. mpg.de/proj/medgen/mitop/>.)

mitochondrial heterosis It was claimed that the presence of different mitochondria may lead to complementation and increased vigor. *See* hybrid vigor; mitochondrial complementation. (Sarkissian, I. V. & Srivastava, H. K. 1973. *Basic Life Sci.* 2:53.)

mitochondrial import Transport into the mitochondria must pass two cooperating membrane layers. The outer membrane carries four receptors (Tom37, −70, −20, −22). The transmembrane import channel is built of at least six proteins; the transmembrane translocation system uses at least three proteins (Tom40, −6, 7). The Tom5 forms the link (a relay) between the outer and inner channel proteins. The inner membrane import channel proteins (Tim) are Mas6p (MIM23), Sms1p (MIM17), and Mpi1p (MIM44/ISP45). The mitochondrial heat-shock 70 protein (Hsp70), the MIM44 complex, and ATP play a central role in import. It appears that MIM44 first binds the incoming unfolded polypeptide chain as it is passing the entry site; it is then transferred to Hsp70 as ATP dissociates the complex. The polypeptide moves further and binds again to the complex, eventually traversing the inner membrane. Tim10p and Tim12p mediate the import of the multispanning carriers into the inner membrane. Tim12 is bound to Tim22. Tim23 passes proteins through the inner membrane. Although the mitochondria contain their own protein synthetic machinery (ribosomes, tRNAs), the majority of mitochondrial proteins are encoded by the nucleus and translated in, or imported from, the cytosol. Imported cytosolic tRNAs may correct mutations of mitochondrial tRNAs. The import system is also nuclearly encoded. The general import factors include the Hsp proteins, cyclophilin 20, ADP/ATP carrier (AAC), and proteases. The role of the imported assembly facilitator proteins is more specialized. Yeast genes *SCO1, PET117, PET191, PET100, OXA1, COX14, COX11*, and *COX10* are such facilitators. The latter two are actually involved in heme biosynthesis. The COX10 product farnesylates protoheme b. *See* Brownian ratchet; chloroplast import; cyclophilin; farnesyl; heat-shock proteins; heme; Hsp; Hsp70; mitochondria; mitochondrial gene therapy; mitochondrial genetics; MSF; mtDNA; Ydj. (*Annu. Rev. Biochem.* 66:863; Bauer, M. F., et al. 2000. *Trends Cell Biol.* 10:25; Scheneider, A. & Maréchal-Drouard, L. 2000. *Trends Cell Biol.* 10:509; Wiedemann, N., et al. 2001. *EMBO J.* 20:951; Rehling, P., et al. 2001. *Crit. Rev. Biochem. Mol. Biol.* 36[3]:291; Kovermann, P., et al. 2002. *Mol. Cell* 9:363; Pfanner, N. & Wiedemann, N. 2002. *Current Opin. Cell Biol.* 14:400; Neupert, W. & Brunner, M. 2002. *Nature Rev. Mol. Cell Biol.* 3:555.)

mitochondrial mapping   *See* mitochondrial genetics.

mitochondrial mutations   Occur about 10 times more frequently than in the nuclear genes. This may be due to inadequate DNA repair. The mtDNA is "naked" (free of histones) and there is a relative abundance of free oxygen radicals in this organelle. Large deletions are common in human mitochondrial DNA encompassing the 5 kb deletion (mtDNA^{4977}) between nucleotide positions 8470−8482 and 13447−13459, respectively. Some deletions may encompass even longer segments of the mtDNA and cover the position of more than a single mitochondrial gene. These deletions are most common in muscle tissues and the brain and frequently occur by aging. The average mutation rate in the elderly appeared to be 2×10^{-4}/bp in mtDNA (Lin, M. T., et al. 2002. *Hum. Mol. Genet.* 11:133). These other deletions usually occur between direct repeats of 13 to 5 nucleotides. Some of the deletions may involve only single bases. The deletion mutants and their relative extent are characterized by various PCR procedures. The data on increased frequency of point mutations during aging are somewhat ambiguous. Apparently, deletions and duplications, however, accumulate by aging. Sometimes it may be difficult to ascertain that a particular mutation occurs in the nuclear or mitochondrial DNA. In cases of ambiguity, the problem may be resolved by transferring a new nucleus (karyoplast) into an enucleated cytoplasm (cytoplast). If the expression of the mutation is limited to the donated cytoplast, the mitochondrial origin of the mutation can be proved in animal cells. In plant cells the plastids may complicate the identification. In cancer cells the mtDNA mutations occurred 19 to 229 times as abundantly as in the nuclear p53 gene. Interestingly, the cancer cells were largely homoplasmic for the mtDNA mutations, indicating that the mutation had selective advantage. With the techniques available, the complete sequence of the mtDNA can be determined in single cells. Deficiency for uracil-DNA glycosylase in yeast results in mitochondrial mutator phenotype. *See* aging; homoplasmic; mitochondrial diseases in humans; mutation rate; oxidative DNA damage; oxygen effect; p53; PCR; petite colony mutants; somatic cell hybrids. (Inoue, K., et al. 2000. *Nature Genet.* 26:176; Taylor, R. W., et al. 2001. *Nucleic Acids Res.* 29(15):e74; Jacobs, H. T. 2001. *Trends Genet.* 17:653; Chatterjee, A. & Singh, K. K. 2001. *Nucleic Acids Res.* 29:4935.)

mitochondrial myopathy   *See* mitochondrial diseases in humans.

mitochondrial plasmids   Circular or linear; occur in the mitochondria of some cytoplasmically male sterile lines of maize plants and relatives. Their sizes vary between 1.4 and 7.4 kb. The main types are S, R, and D. S2, R2, and D2 are the same, and R1 and D1 are apparently also identical. S1 appears to have emerged as a recombinant between R1 and R2. The *cms*-S plants carry the S1 and S2 plasmids. During the formation of S1, a terminal part of R1 (R*) was lost and inserted at two sites in the mtDNA. The S elements can be either integrated or free mitochondrial episomes. The S2 element encodes (*URF1*) a protein somewhat homologous to a viral RNA polymerase, whereas another (*URF3*) appears to be homologous to a DNA polymerase. The *cms*-C and *cms*-T nucleoids are free from these plasmids, whereas the R1 and R2 sequences (from RU) are integrated in the N nucleoids. Another 2.3 kb plasmid (or a 2.15 kb derivative) is homologous to tRNA^{Trp} and tRNA^{Pro} in cpDNA, and also represents the only functional tRNA^{Trp} in the mitochondrion. *See* cpDNA; cytoplasmic male sterility; endosymbiont theory; episome; killer plasmids; mitochondria; mitochondrial genetics; *Neurospora* mitochondrial plasmids; RU maize; senescence; tRNA. (Bok, J. W. & Griffith, A. 2000. *Plasmid* 43:176.)

mitochondrial proteins   Most of the mitochondrial proteins are imported from the cytosol, but the ~17 kbp mammalian mitochondrial DNA transcribes, then the organelle translates 13 polypeptides, and the ~337 kbp *Arabidopsis* mtDNA encodes 32 proteins. *See* mitochondrial diseases; mitochondrial import. (Kenmochi, N., et al. 2001. *Genomics* 77:65.)

mitochondrial recombination   Generally inferred from observations of linkage disequilibrium. The problem is controversial, however. *See* mitochondrial genetics. (Wiuf, C. 2001. *Genetics* 159:749; Innan, H. & Nordborg, M. 2002. *Mol. Biol. Evol.* 19:1122.)

mitochondrial suppressor Enzyme complexes involved in oxidative phosphorylation are encoded by cooperation among nuclear (*PET*) and mitochondrial genes (*oli/ATP9*). Mutation in, e.g., *AEP2* nuclear gene regulating *oli1* mRNA stability may prevent the formation of a functional subunit 9 of ATP synthase. Mutation in the 5'-untranslated region of *oli1* may, however, suppress the mutation in *aep2*. *See* mitochondrial genetics. (Chiang, C. S. & Liaw, G. J. 2000. *Nucleic Acids Res.* 28:1542; Alfonzo, J. D., et al. 1999. *EMBO J.* 18:7056; Bennoun, P. & Delosme, M. 1999. *Mol. Gen. Genet.* 262:85; Chen, W., et al. 1999. *Genetics* 151:1315.)

mitochondrion *See* mitochondria.

mitochondriopathies *See* mitochondrial diseases in humans.

mitogen Collective name of substances stimulating mitosis and thus cell proliferation (such as growth factors).

mitogen-activated protein kinases *See* MAPK.

mitogenesis Processes leading to cell proliferation.

mitokinesis Hypothetical mechanism of orderly distribution of mitochondria between two cells during cell division possibly using the cytoskeleton as a vehicle.

mitomycin C ($C_{15}H_{18}N_4O_5$) Antibiotic, antineoplastic agent after activation, causing DNA crosslinks, enhancing mitotic recombination, inhibitor of DNA synthesis, etc.

MitoNuc Database of nuclear-encoded mitochondrial proteins. *See* mitochondrial genetics. (<http://bighost.area.ba.cnr.it/mitochondriome>.)

mitosis Nuclear division leading to identical sets of chromosomes in the daughter cells. Mitosis assures the genetic continuity of the ancestral cells in the daughter cells of the body. It involves one fully equational division preceded by a DNA synthetic phase. There is normally no (synapsis) pairing of the chromosomes. The centromeres that split at metaphase separate at anaphase and the chromosomes relax after moving to the poles in telophase. The centromere-spindle association is regulated by a number of kinases (Mad1, Bub1, Msp1). Before anaphase takes place, a number of proteins (Pds1, Scc1, Ase1) regulating sister chromatid cohesion are degraded by proteasomes under the control of the cyclosome (APC). The critical features of mitosis are diagrammed on the next page and comparable photomicrographs are also shown. Thus, mitosis is different from meiosis where there is one reductional and one numerically equational division. During meiotic prophase, the homologous chromosomes (bivalents) are synapsed and chiasmata may be observed. At the first meiotic division, the centromeres do not split, and the undivided centromeres separate at anaphase I. Meiosis reduces the chromosome number to half, in contrast to mitosis, which preserves the number

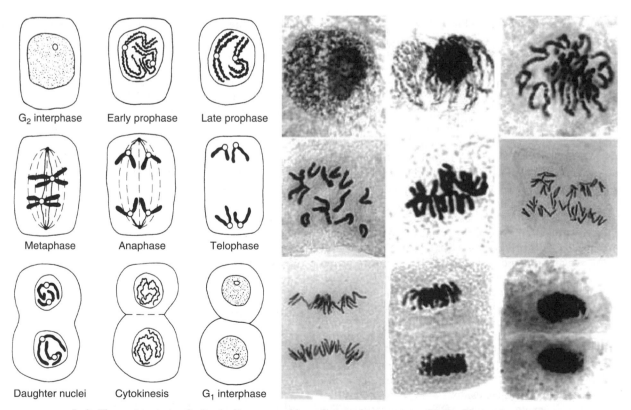

Left: The major steps of mitosis diagrammed by only two chromosomes. Right: Photomicrographs of mitosis in barley $2n = 14$. (Courtesy of T. Tsuchiya.) Top right to left: Interphase, early prophase, late prophase, middle; early metaphase, metaphase, early anaphase, Bottom: Early telophase, late telophase, two daughter nuclei in the two progeny cells.

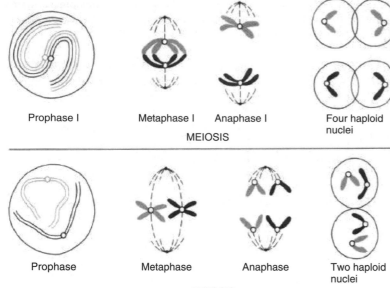

| Prophase I | Metaphase I | Anaphase I | Four haploid nuclei |

MEIOSIS

| Prophase | Metaphase | Anaphase | Two haploid nuclei |

MITOSIS

Meiosis
One numerical reduction & one numerically equational division
Synapsis & chiasma at prophase
Centromeres do not split at metaphase and separate undivided at anaphase I
Chromosome number reduced

Mitosis
One fully equational division
No pairing & no chiasma
Centromeres split in metaphase and separate at anaphase
Chromosome number maintained in daughter cells

Meiosis and mitosis compared by cytological behavior.

of chromosomes in the daughter cells. A comparison between mitosis and meiosis is diagrammed next page.

Mitosis and meiosis are the genetically most important processes of eukaryotic cells (organisms). Mitosis maintains genetic continuity through the development of an organism from conception to the end of its life. Somatic mutation may bring about changes in the exact continuity, but this usually affects only a small fraction of the genes. The subsequent mitotic divisions provide the precise mechanism for the maintenance of such mutations. Mitotic and meiotic cell cycles show some differences in molecular mechanisms. In mitosis, before the S phase, the two key cyclins, Clb5 and Clb6, are kept inactive by the *cyclin-dependent kinase inhibitor protein* Sic1. Two cyclins (Cln1 and 2), along with Cdc28/Cdc2, degrade Sic. In case Clb5 and Clb6 are nonfunctional, the initiation of the S phase is delayed until cyclins CLB1−4 are activated at a subsequent stage. In meiosis, Cdc28/Cdc2 is not required and its role is taken over by a similar protein, SPF (S-phase-promoting factor), encoded by the yeast gene *IME2*. Note that only a single S phase is required for meiosis I, and meiosis II does not require the synthesis of new DNA because it involves only the segregation of the sister chromatids after reduction. *See* Cdc2; CDC6; cdc10; Cdc18; cdc22; cell cycle; CENP; condensin; cyclosome; lamins; MCM; meiosis; mitotic exit; nucleolar organizer; nucleolar reorganization; nucleolus; proteosome; sister chromatid cohesion.

The mitotic cell cycle requires DNA replication. The prereplication complex in yeasts includes Cdc18/CDC6 and the minichromosome maintenance proteins (MCM). For meiosis in fission yeast, these proteins are apparently not mandated.

Also, the vegetative replication checkpoint genes are not active when there are problems during S-phase events. Cdc2 kinase, Cdc10 transcription factor, and Cdc22 functions were needed for meiosis. Mutation in DNA polymerases α and ε and ligase (Cdc17) functions delayed meiotic S phase. Mutation in DNA polymerase δ (Cdc6) and in GEF (Cdc24), similarly to mitosis, reduced the number of divisions completed. *See* separin; sister chromatid cohesion. (Forsburg, S. L. & Hodson, J. A. 2000. *Nature Genet.* 25:263; Tóth, A., et al. 2000. *Cell* 103:1155; Nasmyth, K. 2001. *Annu. Rev. Genet.* 35:673; Georgi, A. B., et al. 2002. *Curr. Biol.* 12:105; Lemke, J., et al. 2002. *Am. Hum Genet.* 71:1051.)

mitospore Meiotic product of fungi ready for mitotic divisions.

mitostatic Stopping or blocking the mitotic process. Many anticancer agents are mitostatic.

mitotic apparatus Subcellular organelles involved in nuclear divisions such as the spindle (microtubules), centromere (kinetochore), centriole, poles, and CENP.

mitotic center *See* centrosome.

mitotic chromosomes Chromosomes undergoing mitosis.

mitotic crossing over Recombination in somatic cells. Recombination is generally a meiotic event (*see* crossing over),

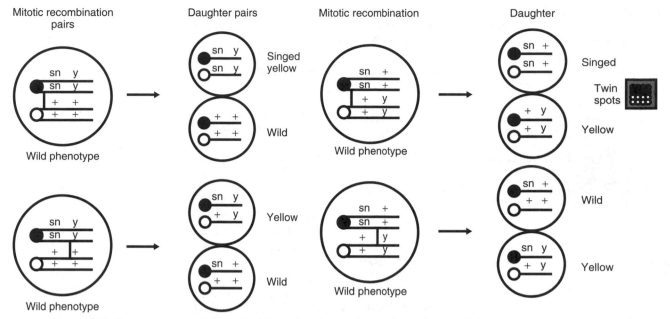

Mitotic recombination pairs	Daughter pairs	Mitotic recombination	Daughter

Mitotic crossing over (at sign |) with markers in coupling and repulsion and the consequences of recombination. (On the basis of experiments of Curt Stern, 1936. *Genetics* 21:625.)

but in some organisms the chromosomes may associate during mitosis, which may be followed by genetic exchange of the linked markers. Although this may take place spontaneously, several agents capable of chromosome breakage (radiation, chemicals) may enhance its frequency. For the detection of somatic recombination in higher eukaryotes, the markers must be cell and tissue autonomous, i.e., they should form sectors (spots) at the locations where such an event has taken place. In the first experiments with *Drosophila*, the chromosomal construct diagrammed below was used. If exchange took place between *sn* and *y*, then only a yellow sector (homozygous for *y*) was formed. *Twin spots* were observed only when in repulsion; the exchange took place between the proximal locus and the centromere.

Somatic recombination takes place in the *Drosophila* male, although this usually does not happen in meiosis. The characteristics of somatic recombination are that exchange is between two chromatids at the four-strand stage, but instead of reductional division, as in meiosis, the centromeres are distributed equationally. Mitotic recombination is a relatively rare event in higher eukaryotes. Somatic recombination can be studied in plants when the chromosomes are appropriately marked, and generative progeny can be isolated from the sectors or by tissue culture techniques. In fungi (lower eukaryotes) where longer diploid phase exists (*Aspergillus*), mitotic recombination can be used even for chromosomal mapping because some of the parental and crossover products can selectively be isolated. The meiotic and mitotic maps are collinear, but the relative recombination frequencies in the different intervals vary (see diagram on below).

Some research workers assumed that the mechanisms of meiotic and mitotic recombination are the same. Homology of the chromosome is required in both meiotic and mitotic recombination (Shao, C., et al. 2001. *Nature Genet.* 28:169). Several facts, however, cast some doubt on this view. In general, caution must be taken in interpreting genetic phenomena

in somatic cells unless classical progeny tests or molecular information can confirm the assumptions. *See* mitotic mapping; mitotic recombination as a bioassay in genetic toxicology; parasexual mechanisms; twin spot. (Pontecorvo, G. 1958. *Trends in Genetic Analysis*. Columbia Univ. Press, New York.)

mitotic drive   Stepwise expansion of trinucleotide repeats. *See* anticipation; meiotic drive; trinucleotide repeats. (Khajavi, M., et al. 2001. *Hum. Mol. Genet.* 10:855.)

mitotic exit   After metaphase, the mitotic spindle moves the chromosomes during anaphase to the poles of the cell and eventually the two daughter cells are formed. PDS and Clb5 protein components of the anaphase inhibitory complex inhibit the formation of the spindle apparatus. Both PDS and Clb5 need to be degraded by the activation of Cdc20 to prepare the path for exit from mitosis. PDS degradation activates Cdc14 phosphatase. Clb5 kinase degradation facilitates dephosphorylation of Sic1, Cdh1, and various cyclin kinases by Cdc14, thus permitting the transition from anaphase into G1 interphase. *See* Cdc14; Cdc20; Cdh1; Clb5; FEAR; mitosis; PDS; Sic1. (Bardin, A. J. & Amon, A. 2001. *Nature Rev. Mol. Cell Biol.* 2:815.)

mitotic index   Fraction of cells in the process of mitosis at a particular time.

mitotic mapping   Genetic recombination during mitosis is generally a very rare event in the majority of organisms. In some diploid fungi, its frequency may reach 1 to 10% that of meiosis. As a consequence of ionizing irradiation and certain chemicals, the frequency of mitotic exchange may increase. On the basis of mitotic recombination, genetic maps can be constructed. Although the order of genes is the same in mitotic and meiotic maps, the recombination frequencies may be quite

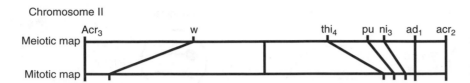

Comparison of a segment of the meiotic and mitotic maps of the fungus *aspergillus nidulans* at identical scales. (Redrawn after Pritchard, R. H. 1963. *Methodology in Basic Genetics*, Burdette, W. J., ed., p. 228. Holden-Day, San Francisco.)

different. *See* mitotic crossing over; mitotic recombination as a bioassay in genetic toxicology; parasexual mechanisms, diagram above.

mitotic nonconformity Genetic change in fungi due to chromosome translocations.

mitotic recombination *See* mitotic crossing over; mitotic mapping.

mitotic recombination as a bioassay in genetic toxicology Has been used in *Saccharomyces cerevisiae* yeast. In a diploid strain (D5) heterozygous for *ade2-40* and *ade2-119* alleles, twin spots are visually detectable in case of mitotic crossing over. Homozygosity for *ade2-40* has an absolute requirement for adenine and involves formation of red color. Homozygosity for *ade2-119*, a leaky mutation, results in pink coloration. Since the two genes are complementary, the heterozygous cells do not require adenine and the colonies are white. Any environmental or chemical factor that promotes mitotic recombination is thus detectable. This procedure has not been extensively used in genetic toxicology programs. *See* bioassays in genetic toxicology; mitotic crossing over.

mitotic spindle *See* spindle; spindle fibers.

mitoxantrone ($C_{12}H_{28}N_4O_6$)   Antineoplastic drug with ability to break DNA double strands.

mitral cells Can make junctions in a beveled fashion.

mitral prolapse Buckling of the heart atrial leaflets as the heart contracts, resulting in backward flow (regurgitation) of the blood. It is caused by an autosomal-dominant allele and it is a common congenital heart anomaly affecting 4–8% of young adults, particularly females. It may accompany various other syndromes such as Marfan syndrome, Klinefelter syndrome, osteogenesis imperfecta, Ehlers-Danlos syndrome, fragile X syndrome, and muscular dystrophies. In an autosomal-recessive form, ophthalmoplegia (paralysis of the eye muscles) is also involved. *See* heart disease. (Towbin, J. A. 1999. *Am. J. Hum. Genet.* 65:1238.)

mixed backbone oligonucleotide (MBO)   Composed of phosphorothionate backbone and 2′-O-methyloligoribonucleotides or methylphosphonate oligodeoxyribonucleotides. Such an antisense construct is resistant to nucleases and forms stable duplexes with RNA. *See* antisense technologies.

mixed-function oxidases Generally, flavoenzymes that oxidize NADH and NADPH; in the process, they may activate promutagens and procarcinogens. *See* microsomes; monooxygenases; P-450; procarcinogen; promutagen.

mixed lymphocyte reaction (MLR) For the detection of histoincompatibility in vitro. The principle of the test: Lymphocytes of two individuals are mixed. One (donor) serves as *responder* and the other as *stimulator*. Proliferation of the antigen-presenting cells is blocked by irradiation or by mitomycin C. CD4 T cells proliferate as they recognize foreign MHC II molecules, and the measure of their proliferation is assessed by the incorporation of H^3-thymidine. Cytotoxic CD8 T cells recognize differences primarily in MHC I molecules, and the extent of the killing reaction is detected by using Cr^{51} (sodium dichromate)-labeled cells. In 7 days, the reaction may detect histoincompatibility between two persons. *See* HLA; lymphocytes; MHC; microcytotoxicity test.

mixoploid Chromosome numbers in the different cells of the same organism vary.

MKI MAP kinase insertion factor. *See* MAPK.

MKK Homologue of MEK. *See* MEK.

MKP-1 Dephosphorylates Thr^{183} and Tyr^{185} residues and thus regulates mitogen-activated MAP protein kinase involved in signal transduction. *See* MAP; PAC.

Mle (*maleless*)   *mle* is located in *Drosophila* chromosome 2-55.2, *mle3* in *Drosophila* chromosome 3.25.8, and similar genes are present in X chromosomes. The common characteristics that homozygous females are viable but the homozygous males die. The underdeveloped imaginal discs of *mle3* may develop normally if transplanted into wild-type larvae. MLE protein appears to be an RNA helicase, and its amino and carboxyl termini may bind double-stranded RNA. Ribonuclease releases MLE from the chromosomes without affecting MSL-1, MSL-2, and RNA. MLE and MSL proteins appear to control dosage compensation. *See* dosage compensation; *Msl*.

MLE Mariner-like element is a transposable element in *Drosophila* that is also present in the human genome where it is responsible for a recombinational hot spot in chromosome 17. *See* Charcot-Marie-Tooth disease; HNPP; mariner; MITE.

MLH Muscle enhancer factor (MLH2A) is a MADS box protein inducing muscle cell development in cooperation with basic helix-loop-helix proteins. *See* bHLH; MADS box; MyoD.

MLH2 *See* hereditary nonpolyposis colorectal cancer.

MLH3 Mismatch repair gene with frequent variation in hereditary nonpolyposis cancer. *See* hereditary nonpolyposis colorectal cancer; mismatch repair. (Wu, Y., et al. 2001. *Nature Genet.* 29:137.)

MLINK Computer program for linkage analysis.

MLK Mixed-lineage group kinases are members of the MAP kinase family of phosphorylases that mediate signal transduction. *See* MAP kinase; signal transduction. (Sallo, K. A. & Johnson, G. L. 2002. *Nature Rev. Mol. Cell Biol.* 3:663.)

MLL Mixed-lineage leukemia (All-1, Htrx). The protein encoded in human chromosome 11q23 has four homology domains: the A-T hook region, DNA methyltransferase homology, zinc fingers, and *Drosophila trithorax (trx,* 3-54.2) homology. *See* leukemia; methyltransferase; SET.

MLS Maximum likelihood lod score or multipoint lod score is a linkage analysis method (in human families). MLS increases when the proportion of relatives carrying an allele identical by descent is higher than expected on the basis of the degree of relatedness and independent segregation. *See* lod score; maximum likelihood; maximum likelihood method applied to recombination; relatedness degree.

MLV *See* Moloney mouse leukemia virus.

MLVA Multilocus variable-length repeat analysis is basically the same as VNTR, and it is used for the identification of related but different bacterial strains. *See* VNTR.

mm Prefix for mouse (*Mus musculus*) protein or DNA, e.g., mmDNA.

MMP (1) Microsatellite mutator phenotype. *See* microsatellite mutator. (2) Matrix metalloprotease. *See* metalloproteinases.

MMR *See* mismatch repair.

MMTV (1) Murine mammary tumor virus. (2) *See* Mouse mammary tumor virus (MTV).

M-MuLV Moloney murine leukemia virus.

Sialic acid.

MN blood group The human chromosomal location is 4q28-q31. For the M blood type, α-sialoglycoprotein (glycophorin A), and for the N type, δ-peptide (glycophorin B) is responsible. Glycophorin-deficient erythrocytes are resistant to *Plasmodium falciparum* (malaria). The En(a⁻) blood group variants are also lacking glycophorin A. The M and N alleles are closely linked to the S/s alleles, and the complex is often mentioned as the MNS blood group. Allelic combination frequencies in England were MS (0.247172), Ms (0.283131), NS (0.08028), and Ns (0.389489). *See* blood groups.

mnemons L. Cuénot's historical (early 1900s) term for genes.

MNGIE Mitochondrial neurogastrointestinal encephalopathy is an autosomal-recessive disease involving myopathy with ragged red fibers, reduced activity of respiratory chain enzymes, etc. The basic defect is attributed to a deficiency of TK2 (a thymidine kinase), which normally phosphorylates, also deoxynucleosides of other pyrimidines. The reduction of enzyme activity apparently affects the maintenance of mtDNA. The TK2 gene has been located to human chromosome 16q22 (earlier to chr. 17), and more recently MNGIE was assigned to 22q13.32-qter. *See* mitochondrial diseases in humans; mtDNA.

Methylnitrosoguanidine.

MNNG N-methyl-N'-nitro-N-nitrosoguanidine is a monofunctional alkylating agent and a very potent mutagen and carcinogen (rapidly decomposing in light). It methylates O^6-position of guanine. Cell lines exist that are highly resistant to the cytotoxic effects of MNNG, but they are even more sensitive to mutagenic effects. *See* alkylating agents; mutagens.

MNU N-methyl-N-nitrosourea is a monofunctional mutagen and carcinogen forming O^6-methylguanine. *See* alkylating agents; monofunctional; mutagens, ethylnitrosourea below.

Ethylnitrosourea.

MØ *See* macrophage.

MO15 CDK-activating kinase, related to CAK. Association of MO15 with cyclin H greatly increases its kinase activity toward Cdk2. *See* CAK; CDK; Cdk2; cell cycle; cyclin.

MoAb *See* monoclonal antibody.

mob Bacterial gene that facilitates the transfer of bacterial chromosomes or plasmids into the recipient cell. In order to transfer the plasmids, there is a need for a cis-acting *nick* site and a *bom* site. At the former, the plasmid is opened up (nicked) and the *bacterial origin of mobilization (bom)* makes the conjugative transfer possible. *See* conjugation; Hfr.

mobile genetic elements Occur in practically all organisms, and they represent different types of mechanisms and serve diverse purposes. They share some common features. They are capable of integration and excision from the genome (like the temperate bacteriophages), or move within the genome (like the insertion and transposable elements). These elements may fulfill general regulatory functions in normal cells (such as the switching of mating-type elements in yeast). Phase variation in *Salmonella*, antigenic variation as a defense system in bacteria (*Borrelia*) and protozoa (*Trypanosomas*), parasitizing plant genomes by agrobacteria, etc., may depend on such elements. They have a role in the generation of antibody diversity in vertebrates by transposition of immunoglobulin genes. *See* LINE; organelle sequence transfers; pathogenicity islands; retroposons; retrotransposons; SDR; SINE; transposable elements; transposons.

mobility shift assay When two molecules (DNA-protein, protein-protein) bind, upon electrophoretic separation the mobility is retarded. *See* gel retardation assay. (Filee, P., et al. 2001. *Biotechniques* 30:1944.)

mobilization Ribosome binding to mRNA initiates polysome formation; the process of conjugative transfer; also the release of a compound in the body for circulation.

mobilization, plasmid Transfer of conjugative plasmids to another cell. *See* conjugation.

Möbius syndrome Congenital partial or full paralysis and dysfunction of a cranial nerve, face and limb malformations, possible mental retardation. A dominant locus was found at 3q21-q22 and another dominant on the long arm of chromosome 10. Recessive and X-linked inheritance was also suspected.

modal *See* mode.

mode Value of the variates (class of measurements) of a population that occurs at the highest frequency. *See* mean; median.

model Represents the essential features of a concept with minimum detail. Model organisms are frequently of no economic interest beyond making research efficient, fast, and less demanding in labor and funds, yet they make possible the understanding of basic genetic and molecular mechanisms common in other biological systems. *Escherichia coli*, yeast, *Neurospora*, *Drosophila*, *Sciaras*, *Caenorhabditis*, zebrafish, mouse, *Arabidopsis*, *Vicia faba*, etc., each have some special advantage for the study of particular problems, although they may be inferior tools for others. Since DNA sequence and proteome information is becoming available, generalizations on functions, development, and evolution is much facilitated.

model, genetic In human genetic analysis, specified by the type of inheritance, e.g., dominant/recessive, autosomal/sex-linked, penetrance/expressivity, frequency of phenocopies, mutations, allelic frequencies, pattern of onset, etc.

model-free analysis Human genetics term indicating no need for a genetic model because the persons and ancestors have already been genotyped. *See* genotyping; model, genetic.

modeling Physical, mathematical, and/or hypothetical construction for the exploration of the reality or mechanism of theoretical concepts. *See* Monte Carlo method; simulation.

model organisms Relatively simple biological systems. They facilitate experimental studies because of ease of manipulation, short life cycle, small genomes, etc. Such models are *E. coli* for prokaryotes, *Saccharomyces cerevisiae* for fungi and other eukaryotes, *Caenorhabditis elegans* and *Drosophila melanogaster* for lower animals and humans, mice for higher animals and humans, and *Arabidopsis thaliana* for higher plants. Using certain types of research methods is immoral or otherwise objectionable in humans (e.g., controlled mating, testing of new drugs and mutagens, genetic engineering), and mouse models are indispensable and invaluable. (Reinke, V. & White, R. K. 2002. *Annu. Rev. Genomics Hum. Genet.* 3:153)

modem (modulator/demodulator) Links the computer to another computer through a telephone line, e.g., fax modem (sends printed and graphic information) or e-mail modem sends and retrieves information from a mainframe computer to other computer operators through the information networks such as BITNET, INTERNET, and various online services.

modification Methylation (in molecular biology). *See* methylase; methylation of DNA.

modified bases Occur primarily in tRNA and are formed mainly by posttranscriptional alterations. The most common modified nucleosides are ribothymidine, thiouridine, pseudouridine, isopentenyl adenosine, threonylcarbamoyl adenosine, dihydrouridine, 7-methylguanidine, 3-methylcytidine, 5-methylcytidine, 6-methyladenosine, inosine, etc. These modified bases have an important role in the function of tRNA, e.g., 1-methyladenosine in the $T\psi C$ loop at position 58 facilitates the translation from the $tRNA_1{}^{Met}$ initiation codon of yeast. *See* initiation codon; transfer RNA. (Anderson, J., et al. 2000. *Proc. Natl. Acad. Sci. USA* 97:5173.)

modified mendelian ratios Observed when the products of genes interact. These situations in case of two loci can be best represented by modified checkerboards using the zygotic rather than the gametic constitutions at the top and at the left side of the checkerboards (below). At the top, a standard constitution is shown where for each locus the genotypes are 1*AA*, 2*Aa*, 1*aa* and 1*BB*, 2*Bb*, 1*bb*, respectively, etc. Although in the boxes only the relative numbers of the phenotypes (genotypes) are shown (as a modification of the 9:3:3:1 Mendelian digenic ratio), their genetic constitution can be readily determined from the top left checkerboard. These schemes assume complete dominance. Additional variation in the phenotypic classes occurs in case of semidominance or codominance and the involvement of more than two allelic pairs. In common usage these modifications are mentioned as gene interactions, but actually the products of the genes do interact.

Sometimes it is not very easy to distinguish between two segregation ratios within small populations and statistical analysis may be required. Example: Let us assume that we

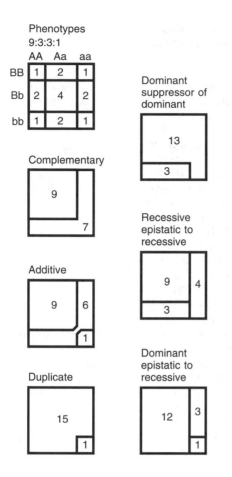

Phenotypes
9:3:3:1

	AA	Aa	aa
BB	1	2	1
Bb	2	4	2
bb	1	2	1

Complementary

9
7

Additive

9 6
1

Duplicate

15
1

Dominant
suppressor of
dominant

13
3

Recessive
epistatic to
recessive

9 4
3

Dominant
epistatic to
recessive

12 3
1

wish to ascertain whether the segregation is 3:1 or 9:7. With 3:1, the standard error of the recessives is

$$\sqrt{\frac{3n}{16}},$$

and with 9:7, it is

$$\sqrt{\frac{63n}{256}}.$$

Since deviations from the expectation can be either + or −, the statistically acceptable misclassification of 0.025 can rely on the deviate of 0.05. From a table of the normal deviates (*see* normal deviate), the expected number of recessives must be 1.9599 times the standard error or higher. Thus, for 9:7

$$(7/16n - \text{recessives}) = 1.9599\sqrt{\frac{63n}{256}}$$

and for the 3:1 segregation the expected number is

$$r - 1/4(n) = 1.9599\sqrt{\frac{3n}{16}}.$$

By adding, we get

$$n([7/16] - [1/4] = 1.9599\sqrt{n}\left(\sqrt{\frac{63}{256}} + \sqrt{\frac{3}{16}}\right)$$

and

$$\sqrt{n} = \frac{16}{3}\left\{1.9599\left[\left(\frac{7.9373}{16}\right) + \left(\frac{1.7321}{4}\right)\right]\right\} = 9.7121,$$

and $n = 9.7121^2 = 94.32$; therefore, about 95 individuals permit a distinction between the two hypothetical segregations at a level of 0.025 probability. In a similar way, other segregation ratios can be tested. *See* codominance; genotype; Mendelian segregation; normal deviate; phenotype; Punnett square; semidominance.

modifier gene Affects the expression of another gene. *See* epistasis. (Nadeau, J. 2001. *Nature Rev. Genet.* 2:165; Burghes, A. H. M., et al. 2001. *Science* 293:2213.)

mod score Maximized lod score over recombination fractions (θ) and genetic models (ϕ). *See* lod score; model, genetic.

modulation Reversible alteration of a cellular function in response to intra- or extracellular factors.

module Single or multiple motif containing structural and functional units of macromolecules (proteins, DNA, RNA) and possibly small molecules. Their function is not a simple sum of that of the isolated components. Modules may be spatially isolated in the cell, e.g., ribosomes at the complex site of polypeptide synthesis, or may be only functionally separated as, e.g., a particular signal transduction pathway. Modules have the ability, however, for hierarchical interaction. One evolutionary advantage of the modular organization is that a mutation in a modular component may alter the function of that module without an upset in the whole system of the organism. Module components involved in a central role of cell biology, e.g., histones, must be well conserved across phylogenetic ranges. *See* motif; Williams syndrome.

MODY Maturity-onset diabetes of the young. Apparently dominant forms of familial diabetes with an onset at or after puberty but generally before age 25. Human chromosomal location is 20q11.2 (MODY1), 7p13 (MODY2), or 12q24 (MODY3), and it is a heterogeneous disease. MODYs are responsible for about 2–5% of non-insulin-dependent diabetes. MODY1 is responsible for the coding of HNF-4α, a hepatocyte nuclear transcription factor, a member of the steroid/thyroid hormone receptor family, and an upstream regulator of HNF-1α. MODY2 encodes glycolytic glycokinase, which generates the signal for insulin secretion. MODY3 displays defects in hepatocyte nuclear factor HNF-1a, a transcription factor that normally transactivates an insulin gene. SHP/NRPB2 (1p36.1) modulates the expression of HNF-4α and contributes to obesity. *See* diabetes insipidus; diabetes mellitus; GLUT.

Moebius syndrome The major characteristic is facial paralysis of the sixth and seventh cranial nerves and often limb deformities and mental retardation. The dominant disorder was located to human chromosome 13q12.2-q13. The recurrence rate is below 1/50. *See* limb defects; mental retardation; neuromuscular diseases; periodic paralysis.

Mohr syndrome *See* orofacial digital syndrome II.

Mohr-Tranebjaerg syndrome (DDP) Recessive deafness encoded in Xq21.3-Xq22, involving poor muscle coordination, mental deterioration, but without blindness. In the same chromosomal region there is another transcribed region with symptoms similar to DDP but with blindness, DFN-1. *See* deafness.

MOI *See* multiplicity of infection.

molar solution 1 gram molecular weight compound dissolved in a final volume of 1 L.

mole (1) *Talpa europaea*, $2n = 34$, an insect-eating small underground mammal. (2) *See* fleshy (placental) neoplasia; gram molecular weight. (3) Abnormal tissue mass in the uterus.

Mole.

molecular beacon Molecular method for allele discrimination or monitor enzyme activity. (Tyagi, S., et al. 1998. *Nature Biotechnol.* 16:19; Rizzo, J., et al. 2002. *Mol. Cell. Probes* 16(4):277.)

molecular biology Study of biological problems with physical and chemical techniques and interpretation of the functional phenomena on macromolecular bases.

molecular breeding (1) Its goal is to develop highly efficient new vectors for genetic engineering. The procedure is reshuffling the coding regions of viral glycoprotein genes, which determine tropism of the proteins. This is followed by selection of vectors with exquisite specificity, improved gene transfer, etc. (2) Application of molecular biology methods for the development of new crops or animal stocks.

molecular chaperone heterocomplex (MCH) Includes heat-shock proteins Hsp90 and Hsp70, chaperone interacting proteins Hop, Hip, and p23, and peptidyl-prolyl isomerases FKBp51, FKBp52, and Cyp-40. *See* chaperones; chaperonins; cyclophilins; FK506; FKB; heat-shock proteins. (Bharadwaj, S., et al. 1999. *Mol. Cell Biol.* 19:8033.)

molecular clock *See* evolutionary clock.

molecular cloning Reproduction of multiple copies of DNA with the aid of a vector(s). (Sambrook & Russel, R. 2001. *Molecular Cloning*. Cold Spring Harbor Lab. Press.)

molecular combing Spreading and aligning purified, extended DNA molecules on a glass plate. Appropriate probes and optical tools quantitatively reveal the amplification or losses of the genome at high resolution. (Allemand, J. F., et al.

1997. *Biophys. J.* 73:2064; Gueroui, Z., et al. 2002. *Proc. Natl. Acad. Sci. USA* 99:6005.)

molecular computation Could be called reversed mathematics because DNA or protein sequence information is used in an attempt to solve complex mathematical problems. *See* DNA computer.

molecular disease The molecules involved have been identified, e.g., in sickle cell anemia a valine replaces a glutamic acid residue in the hemoglobin β-chain. Numerous diseases have been explained in molecular terms.

molecular drive Copies of redundant DNA sequences are rather well conserved in the genomes, although divergence would be expected by repeated mutations. The force behind this tendency for uniformity within species has been named molecular drive or concerted evolution. The current literature uses mainly the latter term. *See* concerted evolution.

molecular evolution Studies the relationship of the structure and function of macromolecules (DNA, RNA, protein) among taxonomic groups. *See* evolutionary clock; evolutionary distance; evolutionary tree; evolution of the genetic code; K_A/K_S; origin of life; polymerase chain reaction; RNA world. (Nei, M. 1996. *Annu. Rev. Genet.* 30:371.)

molecular farming Use of transgenic animals or plants for the production of substances needed for the pharmaceutical industry or for other economic activities. *See* pharming; transgenic.

molecular genetics Application of molecular biology to genetics; it is a somewhat unwarranted distinction because genetics as a basic science must always use all of the best integrated approaches.

molecular hybridization Annealing two different but complementary macromolecules (DNA with DNA, DNA with RNA, etc.). *See* probe; c_0t curve.

molecular markers *See* ladder; molecular weight; RAPD; RFLP; VNTR.

molecular mimics May trigger an autoimmune reaction because they have sequence homology to bacterial or viral pathogens. It is assumed that viral infection may lead to the synthesis of antigens structurally of functionally resembling self-proteins. Naive T cells may then be activated by this antigen and may mount an immune response to self-proteins. Some proteins appear to mimic the structure of nucleotides, e.g., domain 4 of EF2 peptide elongation factor mimics the anticodon stem and loop of tRNA. Similarly, eukaryotic release factors mimic tRNA. *See* autoimmune diseases; bystander activation; immune response; immune tolerance; structural mimics; tRNA. (Putnam, W. C., et al. 2001. *Nucleic Acids Res.* 29:2199.)

molecular modeling Aimed at determining the three-dimensional atomic structure of macromolecules. Database and software information is available at <http://www.ncbi.nlm.nih.gov/Structure/>; <http://www.pdb.bnl.gov>.

molecular plant breeding Uses DNA markers to map agronomically desirable quantitative traits by employing the techniques of RFLP, RAPD, DAF, SCAR, SSCP, etc. The purpose is to incorporate advantageous traits into crops. *See* QTL.

molecular plumbing Generation of blood vessels by tissue culture techniques for the purpose of replacing defective organs.

Molecular weight The relative masses of the atoms of the elements are the atomic weights. The sum of the atomic weights of all atoms in a molecule determines the molecular weight of that molecule (MW). In the determination of the relative mass, the mass of C^{12} is used, which is approximately 12.01. Relative molecular weight is usually abbreviated as M_r. The molecular weight of macromolecules is generally expressed in daltons, 1 Da = 1.661×10^{-24} grams. Molecular weights are determined by a variety of physicochemical techniques. In gel electrophoresis of DNA, fragments are compared with sequenced (known base number) restriction fragments of λ-phage generated by *HinD* III (125 to 23,130 bp); *HinD* III–*Eco* RI double digests (12 to 21,226 bp); *Eco* RI (3,530 to 21,226 bp) or pUC18 plasmid cleaved by *Sau* 3AI (36 to 955 bp); $\phi X 174$ digested by *Hae* III (72 to 1,353 bp); commercially available synthetic *ladders* containing 100 bp incremental increases from 100 to 1,600 or several other sizes. For the large chromosome-size DNAs studied by pulsed-field gel electrophoresis, T7 (40 kb), T2 (166 kb), phage G (758 kb), or even larger constructs obtained by ligation are used. For protein electrophoresis, bovine serum albumin (67,000 M_r), gamma globulin (53,000 and 25,000 M_r), ovalbumin (45,000 M_r), cytochrome C (12,400 M_r), and others can be employed. *See* ladder.

molecule Atoms covalently bound into a unit.

Moloney mouse leukemia oncogene (MLV) Integrates at several locations into the mouse genome; an integration site has been mapped to human chromosome 5p14. Moloney leukemia virus-34 (Mov34) integration causes recessive lethal mutations in mice. The homologue to this locus is found in human chromosome 16q23-q24. The Mos oncogene encodes a protein serine/threonine kinase. It is a component of the cytostatic factor CSF and regulates MAPK. *See* CSF; MAPK; oncogenes.

Moloney mouse sarcoma virus (MSV, MOS) The c-oncogene maps to the vicinity of the centromere of mouse chromosome 4. The human homologue MOS was assigned to chromosome 8q11-q12 in the vicinity of oncogene MYC (8–24). Breakpoints at human 8;21 translocations have been found to be associated with myeloblastic leukemia. It has been suspected that band 21q22, critical for the development of Down syndrome, is responsible for the leukemia that frequently affects trisomic individuals for chromosome 21. The cellular mos protein (serine/threonine kinase) is also required for the meiotic maturation of frog oocytes. If the c-mos transcript is not polyadenylated, maturation is prevented. Overexpression of c-mos causes precocious maturation, and after fertilization, cleavage is prevented but disruption of the gene may lead to parthenogenetic development of mouse eggs. This protein is active primarily in the germline and may cause ovarian cysts and teratomas; it may also cause oncogenic transformation in somatic cells. *See* cell cycle; c-oncogene; Down syndrome; leukemia; polyadenylation; teratoma.

MOLSCRIPT Computer program for plotting protein structure. *See* protein structure. (*J. Appl. Crystallog.* 24:946.)

molting Shedding the exoskeleton (shell) of insects during metamorphosis (developmental transition stages). *See Drosophila.*

molybdenum Cofactor of several enzymes in prokaryotes and eukaryotes. In mammals, xanthine dehydrogenase, aldehyde, and sulfite oxidase require it. An autosomal-recessive molybdenum cofactor (molybdopterin) deficiency/sulfite oxidase deficiency involves neonatal seizures, neuronal damage, and is a rare lethal condition between ages 2 and 6. Two open-reading frames in human chromosome 6p encode the protein. *See* nitrate reductase.

moments Expectations of different powers of a variable or its deviations from the mean, e.g., the first moment is $E(X)$ = the mean, the second moment is $E(X^2)$, and the third is $E(X^3)$. The first *moment about the mean*: $E(X - E[X])^2$; the second $E(X - E[X])^3$. The third moment about the mean (if it is not 0) indicates skewness; the fourth moment reveals kurtosis. The moment of the joint distribution of X and Y is the covariance. *See* correlation; covariance; kurtosis; skewness.

MoMuLV Moloney murine leukemia virus is a retrovirus. It is frequently used for vector construction. It is potentially useful for transformation of both rodent and human cells. Integration is very efficient because the viral genome is retained in the reverse-transcribed DNA form of the virus. It can be targeted to rapidly dividing and not to postmitotic cells. The *gag, pol,* and *env* protein genes are dispensable and ensure that the virus is incapable of autonomous replication, but the ψ-packaging signal is required. If the target cells do not have the ψ, the virus does not spread to other tissues. This is a particular advantage when tumor cells are targeted but normal cells are spared. For example, herpes simplex virus thymidine kinase (*HSV-tk*)–containing viral vectors convey ganciclovir sensitivity only to tumor cells (which divide), and 10–70% may then be selectively killed by this cytotoxic drug. Similar vector designs can be applied to some degenerative diseases. (Parkinson disease, Huntington's chorea, Alzheimer disease, and other neurodegenerative diseases). The protective gene can be transfected into cultured cells and grafted back onto the target tissue of the same patient. *See* ganciclovir; gene therapy; retroviral vectors.

monastrol Protein causing monopolar (rather than the normal bipolar) spindle in dividing cells. It blocks kinesin-related proteins (such as those in the bimC family, e.g., Eg5). *See* bimC; kinesin. (Kapoor, T. M., et al. 2000. *J. Cell Biol.* 150:975.)

Mondrian   Type of representation of microarray of RNA transcripts resembling the "Broadway boogie-woogie" by the Dutch painter Piet Mondrian (1872–1944). It identifies open-reading frames within the sequenced genomic DNA. (Penn, S. G., et al. 2000. *Nature Genet.* 26:315.)

Mongolism (mongoloid idiocy)   Now rejected name of Down syndrome, human trisomy 21. Besides the epicanthal eyefold, other—nonmongoloid—features are more characteristic for the condition. *See* Down syndrome; trisomy.

monilethrix   Autosomal-dominant (human chromosome 14q) and possibly autosomal-recessive baldness (alopecia) due to defects of the keratin filaments. *See* baldness; filaments.

monitor   A video monitor receives and displays information directly received by a computer; or a television monitor that accepts broadcast signals; monitoring: keeping track of something.

monoallelic expression   In a diploid (or polysomic) cell or individual, only one of the alleles is expressed, such as the genes situated in one of the mammalian X chromosomes or in case of maternal or paternal imprinting. Actually, some lower organisms (*Plasmodium, Trypanosoma*) also express only one allele of some genes of the diploid. *See* allelic exclusion; imprinting; lyonization; *Plasmodium*; *Trypanosoma*.

monoamine oxidase   *See* MAO; Norrie disease.

monobrachial chromosome   Has only one arm. *See* chromosome morphology; telocentric.

Monobrachial chromosome.

monocentric chromosome   Has a single centromere as it is most common.

monochromatic light   Light of a single color, practically light emission with a single peak within a very narrow wavelength band.

monocistronic mRNA   Codes for a single type of polypeptide. It is transcribed from a single separate cistron (not from an operon). Eukaryotic genes are usually monocistronic. The class I genes, 18, 5.8, and 28S rRNAs, including spacers, are transcribed as a single unit containing one of each gene's pre-rRNAs and are subsequently cleaved into mature rRNA. The 5S RNA genes are transcribed from separate promoters by polIII. *See* cistron; ribosome.

monoclonal antibody (MAb)   Single type produced by the descendants of one cell and specific for a single type of antigen. MAbs are generated by injecting purified antigens (immunization) into mice, then isolating spleen cells (splenocytes, B lymphocytes) from the immunized animals and fusing these cells, in the presence of polyethylene glycol (a fusing facilitator), with bone marrow cancer cells (myelomas) that are deficient in thymidine kinase (TK) or HGPRT. This process assures that the nonfused splenocytes rapidly senesce in culture and die. The unfused myeloma cells are also eliminated on HAT medium because they cannot synthesize nucleic acids either by the de novo or salvage pathway. The selected myelomas do not secrete their own immunoglobulins and thus hybrid immunoglobulins are not produced. Single hybridoma cells are cultured in multiwell culture vessels and screened for the production of specific MAbs by the use of radioimmunoassays (RIA) or by enzyme-linked immunosorbent assay (ELISA). Most of the hybridomas will produce many types of cells and are of limited use; a few, however, may be more specific, and these are rescreened for the specific type needed—e.g., animals immunized with melanoma cells produce HLA-DR (human leukocyte antigen D-related) antibodies, melanotransferrin (a 95 kDa glycoprotein), or melanoma-associated chondroitin proteoglycan (heteropolysaccharides, glucosaminoglycans attached to extracellular proteins of the cartilage). Monoclonal antibodies have been used effectively for identification of tumor types in sera and histological assays.

Some MAbs have been successfully applied for direct tumoricidal effect. Radioactively labeled monoclonal antibodies of melanoma-associated antigens (MAA) have been used for imaging (immunoscintigraphy) and detecting melanoma cells in the body when other clinical and laboratory methods have failed. The "magic bullet" approach of combining specific monoclonal antibodies (prepared from individual cancer patients) against a specific cancer tissue with Yttrium[90] isotope (emitting β-rays of 2.24 MeV energy and 65 hr half-life), ricin (LD50 for mice by intravenous administration 3 ng/kg), the deadly diphtheria toxin, or tumor necrosis protein (TNF) is expected to home on cancer cells and destroy them. This scheme has not been proven successful in clinical trials so far (*see* abzymes). Another newer approach is to separately clone cDNAs of a variety of light and heavy chains of the antibody and allow them to combine in all possible ways, thus producing a combinatorial library of antibodies against all present and possible, emerging future epitopes. These antibodies can then be transformed into bacteria by using λ-phage or filamentous phage such as M13. The phage plaques can be screened with radioactive epitopes. The advantage of using filamentous phage is that they display the antibodies on their surface and permit screening in a liquid medium that enhances the efficiency by orders of magnitude. Although monoclonal antibodies do not completely fulfill all the (naive) therapeutic expectations, they still remain a power tool of biology. Monoclonal antibodies cannot be produced against self-antigens or against less than about 1 kDa antigens. In the latter case, high-molecular-weight carrier proteins such as bovine serum albumin or limpet hemocyanin are used. A typical yield of MAb from hybridoma cells is 10–100µg/10⁶ cells/day. Currently, mammalian cells are used mainly to produce IgG and obtain posttranslational modifications (e.g., glycosylation) essential for the maintenance of the conformation of the immunoglobulin effector function. For transient expression of antibody genes, COS cells are most commonly employed. For stable expression, myeloma cell lines SP2/0, NS0, and Chinese hamster ovary (CHO) cells are used. Antibodies can be produced in transgenic mammals, insects, and plants. *See* antibody; antibody, polyclonal; B cells; bispecific monoclonal antibody; CHO; combinatorial library; COS; diphtheria; ELISA; epitope; genetic engineering; HAMA; HAT medium; heterohybridoma; HGPRT; hybridoma; immune system; immunization, genetic; immunoglobulins; keyhole limpet hemocyanin; LD50; lymphocytes; MAb; melanoma; phage display; plantibody; quasimonoclonal; RIA; ricin; senescence; somatic cell hybrids; T cells; TK; toxin; transformation, transient. (Ritter, M. A. & Ladyman, H. M., eds. 1995. *Monoclonal Antibodies. Production, Engineering and Clinical Application.* Cambridge Univ. Press, New York.)

monoclonal antibody therapies   Open a variety of applications in medicine, although at the moment they do not offer perfect solutions. Reocclusion of blood vessels after angioplasty caused by unwanted aggregation of the platelets may be alleviated by the 7E3 antibody fragments of the chimeric mouse-human Fab. Single- or two-chain fragments of the antibody variable region and the urokinase fusion protein (scFv-uPA) or tissue plasminogen activator (tPA) conjugated to antifibrin antibody have been designed for thrombolysis. Monoclonal antibodies may be generated against interleukins (IL-1), TNF, and cell adhesion molecules (selectins, integrins) to block inflammatory responses after wounding or autoimmune diseases (rheumatoid arthritis, septic shock), T cells, T-cell receptors, etc. *See* abzymes; ADEPT; antibodies, intracellular; antiviral antibodies; autoimmune diseases; cancer gene therapy; fibrin; gene therapy; immunotoxins; integrin; magic bullet; plantibody; platelet; rheumatic fever; selectins; septic shock; T cell; T-cell receptor; thrombin; tissue plasminogen activator; vascular targeting. (Baselga, J. 2001. *Eur. J. Cancer* 37 Suppl. 4:16.)

monocotyledones   Plants that form only one cotyledon such as the grasses (cereals).

monocytes   Mononuclear leukocytes that become macrophages when transported by the bloodstream to the lung and liver. *See* atherosclerosis; leukocyte; macrophage; MCP-1; microglia.

monoecious   Separate male and female flowers on the same individual plant. *See* autogamous; outcrossing; protandry; protogyny.

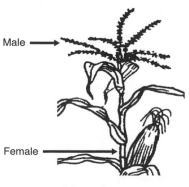

Male →

Female →

Monoecious.

monofactorial inheritance   A single (dominant or recessive) gene (factor) determines the inheritance of a particular trait. Although this term has some value for classification of inheritance, in most cases the phenotype is the function of a number of genes with different alleles that respond to a variety of internal and external factors at the same time. Thus, strictly monogenic inheritance may not exist. The so-called monogenic human diseases generally are expressed as syndromes. *See* epistasis; Mendelian laws; pleiotropy; reaction norm; syndrome. (Ming, J. E. & Muonke, M. 2002. *Am. J. Hum. Genet.* 71:1017.)

monofunctional alkylating agent   Has only a single reactive group.

monogenic heterosis   Overdominance, superdominance.

monogenic inheritance   Monofactorial inheritance.

monogerm seed   Contains only a single embryo. The fruit of some plants (e.g., sugarbeet) is frequently used for propagation and usually it contains multiple seeds. This fruit may, however, be genetically modified to contain a single seed or mechanically fragmented to become monogerm. The agronomic advantage of the monogerm seed is that the emerging seedlings are not crowded and the labor-consuming thinning may be avoided or is at least facilitated.

monogyne   Social insects with a single functional female (queen) in the colony. *See* polygyne.

monohybrid   Heterozygous for only one pair of alleles. *See* monofactorial inheritance.

monoisodisomic   In wheat, $20'' + i1''$, $2n = 42$ ($'' =$ disomic, i $=$ isosomic).

monoisosomic   In wheat, $20'' + i'$, $2n = 41$ ($'' =$ disomic, $' =$ monosomic, i $=$ isochromosome)

monoisotrisomic   In wheat, $20'' + i2'''$, $2n = 43$, ($'' =$ disomic, i $=$ isosomic, $''' =$ trisomic).

monokine   Lymphokine produced by monocytes and macrophages. *See* lymphokine.

monolayer   (1) Noncancerous animal cell cultures grow in a single layer in contact with a solid surface. (2) Single layer of lipid molecules. *See* cell culture; tissue culture.

monomer   One unit of a molecule (which frequently has several in a complex); a subunit of a polymer (protein).

monomorphic locus   Represented by one type of allele in a population.

monomorphic trait   Represented by one phenotype in a population.

mononucleosis Caused by cytomegalovirus; infectious mononucleosis is the result of activation of the Epstein-Barr virus. *See* cytomegalovirus; Epstein-Barr virus.

monooxygenases Introduce one atom of oxygen into a hydrogen donor, e.g., P450 cytochromes, and have varied functions in cells in normal development and as detoxificants, converting promutagens and procarcinogens into active compounds. *See* mixed-function oxidases; P450.

monophyletic Organisms have evolved from a single line of ancestry. *See* polyphyletic.

locus or chromosomal region. Monosomics can produce both monosomic and nullisomic gametes. *See* chromosome substitution; monosomic analysis; nullisomics.

monosomic analysis Very efficient in Mendelian analysis of allopolyploids such as hexaploid wheats. Monosomic individuals can produce monosomic and nullisomic gametes; their proportion is different in males and females. The proportion depends on the individuality of the particular chromosomes. On the average, the gametic output and zygotic proportion of selfed monosomics of wheat are as shown below:

	Sperms	
	Monosomic (0.96)	Nullisomic (0.04)
Eggs ↓	**Zygotes**	
Monosomic 0.25	Disomic ≈ 24% (0.25 x 0.96)	Monosomic ≈ 1% (0.25 X 0.04)
Nullisomic 0.75	Monosomic ≈ 72% (0.75 x 0.96)	Nullisomic ≈ 3% (0.75 x 0.04)

monoploid Has only a single basic set of chromosomes. *See* haploid.

monopolar spindle Pulls all chromosomes to one pole.

monoprotic acid Has only a single dissociable proton.

monosaccharide Carbohydrate that is only one sugar of basic units $C_nH_{2n}O_n$ and thus can be diose, triose, pentose, hexose, etc., depending on the number of C atoms; monosaccharides can also be aldose (e.g., glucose, ribose) or ketose sugars (e.g., fructose, deoxyribose).

monosaccharide malabsorption (22q13.1) Controlled by a sodium/glucose transporter at the intestinal brush border. Glucose/galactose malabsorption may be remedied by fructose.

monosodium glutamate ($HOOCCH(NH_2)CH_2CH_2COONa$. H_2O) Used as a flavor enhancer (0.2–0.9%) in salted food or feed or for repressing the bitter taste of certain drugs or as a medication for hepatic coma. *See* Chinese restaurant syndrome.

monosomic The homologous chromosome(s) is represented only once in a cell or all cells of an individual. The gametes of diploids are monosomic for all chromosomes. A monosomic individual of an allopolyploid has $2n - 1$ chromosomes. About 80% of the human $45 + X$ monosomics involve the loss of a paternal sex chromosome. Medical cytogeneticists frequently call deletions partial monosomy, but this is a misnomer because such a condition is hemizygosity for a particular

Very few nullisomic sperm are functional, whereas the majority of the eggs are nullisomic because most of the monosomes (in the absence of a partner) remain and get lost in the metaphase plane; the eggs will receive only one representative from each chromosome that has paired during prophase I. The monosomics can be used to assign genes to chromosomes. If the proper genetic constitution is used, cytological testing may not be required. On the basis of the table and the box below, it is obvious that 75% of the F_1 will be of recessive phenotype if the genes are in the chromosome for which the female is monosomic. In case the recessive is in the monosomic female, only 3% will be recessive in the F_2 offspring. *See* nullisomy. (Morris, R. & Sears, E. R. 1967. *Wheat and Wheat Improvement*, Quiasenberry, K. S. & Reitz, L. P., eds. Am. Soc. Agron., Madison, WI; Sears, E. R. 1966. *Hereditas Suppl.* 2:370; Tsujimoto, H. 2001. *J. Hered.* 92:254; Rédei, G. P. 1982. *Genetics. Macmillan*, New York.)

A		a
	x	a
Monosomic female with dominant allele		Disomic male with recessive allele

monospermy Fertilization by a single sperm.

monotelodisomic In wheat, $20'' + t2'''$, $2n = 43$ ($''$ = disomic, t = telosomic, $'''$ = trisomic).

monotelomonoisosomic In wheat, $20'' + t' + i'$, $2n = 42$, ($''$ = disomic, $'$ = monosomic, t = telosomic, i = isosomic).

monotelo-monoisotrisomic In wheat, $20'' + (t + i)1'''$, $2n = 43$, ($'' =$ disomic, $''' =$ trisomic, $t =$ telosomic, $i =$ isosomic).

monotelosomic In wheat, $20'' + t'$, $2n = 41$ ($'' =$ disomic, $' =$ monosomic, $t =$ telosomic)

monotocous species Produces a single offspring by each gestation.

monotrene Belonging to the taxonomic order Monotremata, a primitive small mammalian group including the spiny anteater and the duck-billed platypus. They lay eggs and nurse their offspring within a small pouch through a nippleless mammary gland. They also show other organizational similarities to marsupials. *See* marsupials.

monotypic A taxonomic category is represented by one subgroup, e.g., a monotypic genus includes a single species.

monoubiquitin The surface of the ubiquitin may be ordained for separate multiple functions, e.g., for degradation or endocytosis. The various monoubiquitins are specialized to one function only, although this function may vary, e.g. it may involve regulation of histones H2A and H2B, endocytosis, or budding of the retroviral Gag polyprotein. *See* endocytosis; histones; retrovirus; sumo; ubiquitin. (Hicke, L. 2001. *Nature Rev. Mol. Cell Biol.* 2:195.)

monozygotic twins Develop from a single egg fertilized by a single sperm; therefore, they should be genetically identical, except mutations that occur subsequent to the separation of the zygote into two blastocytes after implantation into the wall of the uterus. Although MZ twins are generally identical genetically, their birth weight may be different because of differences in intrauteral nutrition due to the path of blood circulation. Developmental malformations may affect only one of the MZ twins. The concordance in susceptibility to infectious diseases has been investigated, but the information is not entirely unequivocal. Multifactorial diseases are more concordant in MZ twins than among dizygotic twins. *See* concordance; cotwin; discordance; dizygotic twins; heritability estimation in humans; twinning; zygosis.

Monte Carlo method Computer-assisted randomization of large sets of tabulated numbers; can be used for testing against experimentally obtained data for determining whether their distribution is random. The Monte Carlo method is used for simulation when the analysis is intractable or too complex. *See* Markov chain; modeling; simulation. (Roederer, M., et al. 2001. *Cytometry* 45:47; Roederer, M., et al. Ibid., 37.)

Moolgavkar-Venzon model Revised version of Knudson's two-mutation model of carcinogenesis. It considers that after the first mutation not all the cells will survive and allows for differential growth of the cells before malignant transformation occurs. *See* Armitage-Doll model; Knudson's two-mutation hypothesis. (Holt, P. D. 1997. *Int. J. Radiat. Res.* 71[2]:203.)

Moore-Federman syndrome Dwarfism with stiff joints and eye anomaly. *See* dwarfism.

mooring sequence 11-nucleotide element anchored downstream to base C to be RNA edited (CAAUUUU-GAUCAGUAUA). *See* RNA editing.

moose North American: *Alces alces*, $2n = 70$.

morality Socially accepted principles and guidelines for the distinction of right from wrong in human behavior, usually based on customs and generally required by society to abide by. Some moral rules are present in all human societies; these rules may have, however, distinct variations. *See* behavior genetics; ethology.

morbidity Diseased condition or the fraction of diseased individuals in a population. *See* mortality. (Petronis, A. 2001. *Trends Genet.* 17:142.)

morgan 100 units of recombination, 100 map units, usually the centimorgan, 0.01 map unit, is used. *See* map unit.

morphactins Various plant growth regulators.

morphallaxis *See* epimorphosis; regeneration in animals.

morphine Opium-type analgesic and addictive alkaloid; methylation converts it to the lower-potency codein. *See* alkaloids.

morphoallele Gene involved in morphogenesis. *See* morphogenesis.

morphogen Compound that can affect differentiation and/or development, and can correct the morphogenetic pattern of a mutant that cannot produce it if it is supplied by an extract from the wild type. The exact chemical nature of the morphogens is generally unknown, however, some hormones and proteins have these properties. The morphogen is either a transcription factor or a type of transcriptional regulator. Their control involves concentration gradients over a threshold level. Generally, the initial morphogen signal recruits other similar signals and signal receptors (encoded by genes) that act in positive or negative manners and determine cell fates, bringing about a differentiational event(s). The activin signal (a member of the transforming growth factor-β family) spreads in a passive gradient about 300 μm or 10 cells diameter within a few hours in the vegetal cells of animals. *See* activin; cross-talk; cue; morphogenesis; selector genes; signal transduction; vegetal pole. (Pagès, F. & Kerridge, S. 2000. *Trends Genet.* 16:40; Tabata, T. 2001. *Nature Rev. Genet.* 2:620; Gurdon, J. B. & Bourillot, P.-Y. 2001. *Nature* 413:797.)

morphogenesis Process of development of form and structure of cells or tissues and eventually the entire body of an organism beginning from the zygote to embryonic and adult shape. Morphogenesis is mediated through morphogens in response to inner and outer factors. Morphogenesis takes place in three phases: determination, differentiation, and development. Determination is a molecular change preparing the cell (or virus) for competence for differentiation. Differentiation is the realization of molecular and morphological structures that determine the differences among cells endowed with identical genetic potentials. The events of differentiation are coordinated in sequences of development. These steps generally occur in this order yet may be running in overlapping courses for different aspects of morphogenesis. Morphogenetic events vary in the different organisms because this is the basis of their identity, yet some basic principles are common to all. The start point of morphogenesis is difficult to pinpoint because these events run in cycles of the generations (what comes first: the egg or the hen?). The life of an individual begins with fertilization of the egg by the sperm and the formation of the zygote. The zygote has all the genes that it will ever have (*totipotency*), yet many of these genes are not expressed at this stage. (New genes may be acquired during the life cycle by mutational conversion of existing ones or by transformation and transduction.) Also, there are *maternal effect* genes that control oogenesis and affect the zygote from outside without being expressed at this stage within the embryo's own gene repertory. Morphogenesis cannot be explained by one general set of theories, e.g., *gradients of morphogens* or the *signal relay* system. Apparently evidence for either one can be found in different morphogenetic pathways. Many structures in higher eukaryotes are under the control of partially redundant gene families. In different tissues, different sets of transcription factors or transcriptional activators and enhancers may orchestrate the expression of different cell types in a combinatorial manner adapted to a particular organ. The cytoskeleton and cell adhesion may determine cell shape. Signals originating outside or inside the cell may act as morphogens and may trigger morphogenetic changes by their concentration gradients along their distribution path. Alternatively, it has been suggested that signals induce a change in one type of cell, influencing other cells through a relay of series of cells. Signals transmitted to metalloproteinases, collagenase-1, and stromelysin-1 mediated by the RAS family (Rac) of GTP-binding proteins are also involved in cell morphogenesis, *See* anchor cell, CAM, clonal analysis; compartmentalization; cue; cytoskeleton; development; developmental genetics; *Drosophila*, founder cells; Hensen's node; homeotic genes; imaginal disks; left-right asymmetry; metalloproteinases; morphogenesis in *Drosophila*; morphogenetic furrow; morphogens; mRNA targeting; Notch; oncogenes; organizer; pattern formation; primitive streak; Rac; RAS; RNA localization; segregation; asymmetric; selector genes; signal transduction. (Madden, K. & Snider, M. 1998. *Annu. Rev. Microbiol.* 52:687; Chase, A. 2001. *Bioessays* 23:972.)

morphogenesis, *Drosophila* Has been studied the most extensively with genetic techniques. Oogenesis requires four cell divisions within the oocyst (see illustration at maternal effect genes), resulting in the formation of 16 cells, 1 oocyte, and 15 nurse cells; the latter ones become polyploid and are surrounded by a single layer of somatic (diploid) follicle cells. The nurse cells are in communication with the egg through cytoplasmic channels. Both the nurse cell genes (somatic maternal genes) and the egg (germline maternal genes) influence the fate of the zygote through morphogens. The oocyte itself is transcriptionally not active. These maternal effect genes determine the polarity of the zygote. The anterior-posterior gradients of the morphogens account for the future position of the head and tail, respectively. Genes *gurken* (*grk*, 2–30) and *torpedo* (2–10) play an important role in anterior-posterior polarity and later during development in dorsoventral determination. Dorsoventral determination is responsible for the sites of the back and belly, respectively. Mediolateral polarities are involved in the determination of the left and right sides of the body. The larvae and the adults develop from several compartments: 1 for the head, 3 for the thorax, and 9 for the abdomen, formed during the blastoderm stage of the embryo.

Mutations of the maternal genes mentioned in the box next page usually cause recessive embryo lethality, although the homozygous mothers are generally normal. The males are usually normal and fertile. The molecular basis of their actions is known in a few cases, e.g., the amino terminal of the DL protein is homologous to the C-REL (avian reticuloendothelial viral oncogene homologue) protooncogene present in human chromosome 2p13-2cen and in mouse chromosome 11. The N terminus of *dorsal* is homologous to the product of gene *en* (*engrailed*, 2–62) that is also expressed during gastrulation in stripe formation of the embryo. The carboxyl terminus of the *sna* gene product appears to contain five zinc-finger motifs indicating DNA-binding mechanisms of transcription factors.

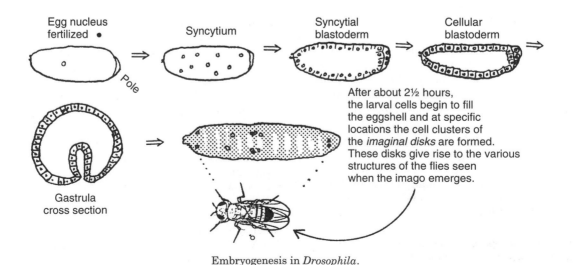

Embryogenesis in *Drosophila*.

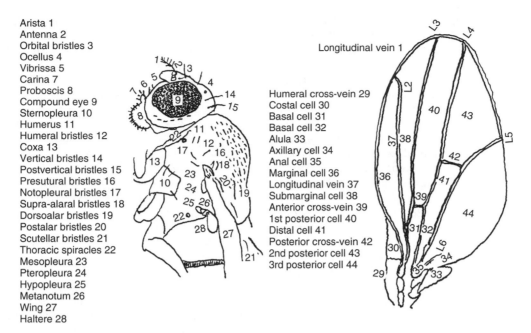

Arista 1
Antenna 2
Orbital bristles 3
Ocellus 4
Vibrissa 5
Carina 7
Proboscis 8
Compound eye 9
Sternopleura 10
Humerus 11
Humeral bristles 12
Coxa 13
Vertical bristles 14
Postvertical bristles 15
Presutural bristles 16
Notopleural bristles 17
Supra-alaral bristles 18
Dorsoalar bristles 19
Postalar bristles 20
Scutellar bristles 21
Thoracic spiracles 22
Mesopleura 23
Pteropleura 24
Hypopleura 25
Metanotum 26
Wing 27
Haltere 28

Longitudinal vein 1
Humeral cross-vein 29
Costal cell 30
Basal cell 31
Basal cell 32
Alula 33
Axillary cell 34
Anal cell 35
Marginal cell 36
Longitudinal vein 37
Submarginal cell 38
Anterior cross-vein 39
1st posterior cell 40
Distal cell 41
Posterior cross-vein 42
2nd posterior cell 43
3rd posterior cell 44

Head and wing landmarks of the Drosophila body.

Somatic maternal effect lethal genes: in *Drosophila* 1 *tsl* (*torsolike*, chromosome 3−71) controls the anterior-posteriormost body structures (labrum, telson). Genes 2 *pip* (*pipe*, 3−47) and 3 *dl* (*dorsal*, 2−52.9) eliminate ventral and lateral body elements, and their product is homologous to the *c-rel* proto-oncogene, a transcription factor homologue, NF-κB. 4 *ndl* (*nudel*, 3−17) exaggerates dorsal elements of the embryos, 5 *wbl* (*windbeutel*, 2−86) controls dorsal epidermis, and the 6 *sna* (*snail*, 2−51) strong alleles eliminate most mesodermal tissues. 7 The *gs* (*grandchildless*, 1−21) and *gs(2)M* (chromosome 2) cause blockage of the embryos of normal-looking homozygous females before the blastoderm stage at temperatures above 28.5°C.

Maternal and germline genes (A very incomplete list. Numbers in parentheses indicate map locations)

{8} *ANTC* (*Antennapedia complex*, 3−47.5) contains elements *lab* (*labial*), *pb* (*proboscipedia*), *Dfd* (*Deformed*), *Scr* (*Sex combs reduced*), and *Ant*, affecting head structures in anterior-posterior relations. The *lab* and *pb* elements share cuticle protein genes. Elements *fushi tarazu (ftz)*, *zerknüllt* (*zen*), and *z* (*zen-2* [*zpr*]) affect segment numbers and *bicoid (bcd)* functions as a maternal effect gene eliminating anterior structures (head) and duplicating posterior elements (telsons). The product of *bcd* also binds RNA and acts as a translational suppressor of the *caudal* (*cad*) protein product. The *bcd* and *zen* genes are included in a 50 kb transcription unit called *Ama* (*Amalgam*). These latter four do not have homeotic functions. Scr regulates the segmental identity of the anterior thorax, and the posterior part of the head is under the control of *Scr*, *Dfd*, *pb*, and *lab*. The entire complex encompasses 355 kb genomic DNA with multiple exons. Eight of the gene products are transcription factors and *Ant*, *Scr*, *Dfd*, *pb*, and *lab* have homeotic functions. The *Antp*

(*Antennapedia*) gene has both lethal and viable loss-of-function and gain-of-function recessive and dominant alleles controlling structures anterior to the thorax and the thoracic segments. The first observed mutations converted the antennae of the adult into mesothoracic legs. The gene has two promoters and four transcripts that control spatial expression differently, relying also on alternate splicing. The *bcd* gene is situated within *zen*, and *Ama* is a maternal lethal affecting head and thorax development. The strong alleles in the females replace the head and thorax of the embryos with duplicated telsons.

Injection of *bcd*[+] cytoplasm into the embryo (partially) remedies the topical alterations brought about by mutant alleles. The RNA transcript sticks to the anterior pole of the embryo and forms a steeply decreasing gradient in the posterior direction. This *bcd* gradient is regulated by genes exu, swa, and stau (see them below), and the gradient may be eliminated by mutations in these genes. In *bcd*⁻ embryos, the anterior activity of *hb* is eliminated and replaced by mirror image posterior *hb* stripes. The four exons are transcribed in either a long, complete RNA (2.6 kb) or a short one (1.6 kb) with exons 2 and 3 spliced out. The protein contains homologous tracts to the nonmaternal effect genes *prd* (*paired*, 2−45, involved in the control of segmentation) and *opa* (*odd paired*, 3−48, deleting alternate metasegments). Exon 3 contains the homeodomain with only 40% homology to other homeoboxes. The C termini of the *bic* protein appears to be involved in transcription activation by binding to five high-affinity upstream sites of *hb* (TCTAATCCC). *Dfd* (*Deformed*) is a weak homeotic gene with recessive and dominant lethal alleles affecting the anterior ventral structures of the head and occasionally also thoracic bristles may appear on the dorsal part of the head. It is composed of five exons coding for a 586-residue protein. Gene *ftz* (*fushi tarazu* [segment deficient in Japanese]) has recessive late embryo lethal and dominant and viable regulatory alleles affecting genes in the *BXC* (*Bithorax complex*), *Ubx* (*Ultrabithorax*, 3–58.8), involved in the control of the posterior thorax and abdominal segments. The general characteristic of *ftz* is the pair-rule feature, i.e., in the mutants the even-numbered abdominal and nerve cord segments are deleted (or fused). The striped pattern of the abdomen is controlled within a 1 kb tract upstream of the beginning of transcription, whereas a more distal upstream element regulates the central nervous system and an even more distal tract is required for the maintenance of the striped pattern. The homeobox is within the second of the two exons of this gene. The *ftz Rpl* mutations may transform the posterior halteres into the posterior wing while the *ftz Ual* mutations convert patches of the first adult abdominal segment into a third abdominal segment-like structure. (The latter two types are not embryonic lethal as the others.) The *lab* (*labial*) mutations are embryonic lethal because of the failure of head structures. The protein product of the gene contains *opa* (*odd paired*, 3–48; deletes alternate metasegments) as well as a homeodomain, although it does not display homeotic transformations. Gene *pb* (*proboscipedia*) may convert labial ("lip") portions into prothoracic leg structures or antennae. From the nine exons, exons 4 and 5 contain the homeobox, and in exon 8 there are again *opa* sequences. The gene products (RNA and protein) are localized to the general area affected by the mutations. Null mutations in gene *Scr* (*Sex combs reduced*) are embryonic lethals. Homeotic transformations involve the labial and thoracic areas. Dominant mutation reduces the number of sex comb teeth. Gene *z2* (*zen-2*) has no detectable effects on development. Gene *zen* (*zerknüllt*) mutations may involve embryo lethality, and the products may be required for postembryonic development.

{9} *arm* (*armadillo*, 1–1.2): cell lethal at the imaginal disc stage because the posterior part of each segment is replaced by an anterior denticle belt. Transcripts have been found in all parts of the larva.

{10} *bcd* (*bicoid:*) see *AntC*.

{11} *bic* and *Bic* (*bicaudal*, 2–67, 2–52, 2–52.91) genes affect the anterior poles of the embryo by replacing these segments with posterior ones in opposite orientation. *Bic* apparently encodes a protein homologous to actin, which is part of the cytoskeletal system.

{12} *btd* (*buttonhead*, 1–31) mutations fail to differentiate the head.

{13} *BXC* (*Bithorax complex*, 3–58) is a cluster of genes that determine the morphogenetic fate of many of the thoracic and abdominal segments of the body. The second thoracic segment, which develops the second pair of legs and a pair of wings, is the most basic part of the complex. The genetic map appears as follows:

abx bx Cbx **Ubx** bxd pbx iab2 **abd-A** Hab iab3 iab4 Mcp

iab5 iab6 iab7 **Abd-B** iab8 iab9

The entire complex is organized into three main integrated regions. *Ubx* (*Ultrabithorax*) is responsible for parasegments PS 5–6, *abd-A* (*abdominal-A*) defines the identity of PS7-13, and *Abd-B* (*Abdominal-B*) is expressed in PS10-14. In the *Ubx* domain the *anterobithorax-bithorax* (*abx-bx*) region specifies PS5 and *bithorax-postbithorax* (*bxd-pbx*) defines PS6. In the *abd-A* region are *iab2*, *iab3*, and *iab4* elements, and in *Abd-B* are *iab5* to *iab9* elements. Mutations in *iab* (*infraabdominal*) tracts cause the homeotic transformation of an anterior segment to a more anterior abdominal (A) segment (e.g., A2 → A1 or A3 [or more posterior ones] → A2, etc.). Mutations *abx* (*anterobithorax*) cause changes in thoracic (T) and abdominal (A) segments: T3 → T2, *bx* (*bithorax*): T3 → T2, *Cbx* (*Contrabithorax*): T2 → T3, *Ubx* (*Ultrabithorax*): A1 + T3 → T2 and T2 + T3 → T1, *bxd* (*bithoraxoid*): A1 → T3, *pbx* (*postbithorax*): T3 → T2, *abd-A* (*abdominal-A*): A2 to A8 → A1, *Hab* (*Hyperabdominal*): A1 + T3 → A2, *Mcp* (*Miscad-astral pigmentation*): A4 & A5 to an intermediate between A4 & A5, *Abd-B* (*Abdominal-B*): A5, A6, A7 may be weakly transformed into anterior forms. Most of these changes involve only some structures in the segments, but additional alterations may also occur.

{14} *cact* (*cactus*, 2–52) mutations reduce the dorsal elements and enhance ventral structures. This gene encodes a homologue of the Iκ-B protein, which forms a complex with a product of *dl*, a NF-κB homologue transcription factor, which is released from the complex upon phosphorylation of the *cact* product. After entering the cell nucleus, it may participate in the activation of its cognate genes.

{15} capu (cappucino, 2–8) mutations may be lethal. It causes somewhat similar alterations as *stau* in addition to making pointed appendages on the head.

{16} *ci^D* (*cubitus interruptus-*, Dominant, 4.0): the wing vein 4 is twice interrupted proximal and distal to the anterior cross-vein. In homozygotes, the anterior portions of the denticle belts are duplicated in a mirror image manner in place of the posterior parts, and they are lethal.

{17} *dpp* (*decapentaplegic*, 2–4.0, [old name was *ho*]) is a complex locus with multiple developmental functions (it is homologous to BMP, TGF-β). The haploid-insufficient *Hin/+* condition is dominant embryonic lethal because of defects in gastrulation. The *Hin-Df* (deficiency) and *hin-r* (recessive) are also lethal. The *hin-emb* is an embryonic lethal mutation. It is complementary to the *shv* (*shortvein*) and *disk* (*imaginal disk*) region genes in the same complex. The *shv-lc* recessive larva-lethal mutants complement all *disk* alleles except for mutant *Hin* and *hin-r*. Mutants *shv-lnc* do not complement *disk*, *Hin*, or *hin-r*, and *shv-p* alleles are viable and complement the *disk* mutants but not *Hin* or *hin-r*. Another mutation in the *shv* region, *Tg* (*Tegula*), causes the roof-like appearance of the wings. This mutation is complementary to all *dpp* genes. The *disk* group of mutations is either viable or lethal and may affect the eyes, wings, haltere, genitalia, head, imaginal disks, etc. The *dpp* gene apparently acts as a regulator of mesodermal genes and it encodes a secreted protein of the TGF-β family transforming growth factor involved in mammalian cancerous growth.

{18} *ea* (*easter*, 3–57) maternal effect lethal with loss- or gain-of-function effects on the dorsal, mesodermal, or lateral structures, depending on the alleles. The mutants may be rescued by injection of normal cytoplasm.

{19} *ems* (*empty spiracles*, 3–53), the interior of the breathing orifices, is partially missing, and it is embryo lethal.

{20} *en* (*engrailed*, 2–62): Some point mutations and chromosome breaks are viable. Others display pair rule defects; adjacent thoracic and abdominal segments fuse.

{21} *eo* (*extra organs*, 1-[66]) in the homozygous lethal embryos causes head defects and a ventral hole.

{22} *eve* (*even skipped*, 2–58), lethal segmentation and head defects, pair rule effects. Its expression is reduced in *h* mutants and *en* segments do not appear. The expression of *Ubx* protein is high in the odd-numbered parasegments 7 to 13 rather than in every segment from 6 to 12; the *ftz* segments are disrupted.

{23} *exu* (*exuperentia*, 2–93) replaces the anterior part of the head with an inverted posterior midgut and anal pit (proctodeum).

{24} *fs(1)K10* (*female sterile*, 1–0.5) and a whole series of other *fs* genes in chromosome 1 may have both specific and overlapping effects. Expression may depend on cues from the oocyte. Eggs of homozygous females are rarely fertilized, but if they are, the gastrulation is abnormal and the anterior ends are dorsalized.

{25} *fu* (*fused*, 1–59.5): Veins L3 and L4 are fused beyond the anterior cross-vein with the elimination of the latter. Heterozygous daughters of homozygous mothers have a temperature-sensitive segmentation problem that is not observed in reciprocal crosses.

{26} *ftz* (*fushi tarazu*); See *AntC* above (also in a separate entry).

{27} *gd* (*gastrulation defective*, 1–36.78) causes dorsal and ventral furrowing of the gastrula-stage embryos.

{28} *gsb* (*gooseberry*, 2–107.6) is homozygous lethal because the posterior parts of segments are deleted and the anterior parts are duplicated in mirror-image fashion.

{29} *gt* (*giant*, 1–0.9) increased-size larvae, pupae, and imagos based on increased cell size. DNA metabolism is abnormal, and both viable and lethal alleles are known that affect the entire embryo in many ways. The protein product appears similar to that of *opa*.

{30} *h* (*hairy*, 3–26.5) displays extra microchaetae along the wing veins, membranes, scutellum, and head. It is also a pair rule gene and affects the expression of *ftz*. It regulates the expression of genes in the ASC (*achaete-scute* complex, 1.-0.0) involved in the control of hairs and bristles. The major gene products are located in the posterior and adjacent anterior parts of segment primordia.

{31} *hb* (*hunchback*, 3–48) alleles have different effects: class I alleles lack thoracic and labial segments: class II mutants retain the prothoracic segment; class III alleles retain the labial parts; class IV mutations prevent the formation of mesothoracic segments; class V alleles cause various gaps as well as segment transformations. These alleles are transcribed from two different promoters and produce up to five different transcripts. The products of this locus interact with those of *kr* and *ftz*. The product of *nos* activates some of the *hb* alleles. In cooperation with *hb*, dMi-2 proteins repress *Polycomb* (see *Polycomb*).

{32} *hh* (*hedgehog*, 3–81): In homozygous embryos a posterior-ventral portion of each segment is removed and the anterior denticle belt is substituted for in a mirror image. The embryos may not have demarcated segments. The gene has two activity peaks: during the first 3–6 hr and at 4–7 days of development (see *hedgehog* in seperate entry).

{33} *kni* (*knirps*, 3–46): These are lethal zygotic gap mutants. The shorter transcript is expressed only until the blastoderm stage, but the longer one is expressed even after gastrulation. The NH_2 end of the protein is homologous to one of the vertebrate nuclear hormone receptors. The *kni* box, a Zn-finger domain, is homologous to parts of the products of genes *knrl* (*knirps related*, 3-[46]) and *egon* (*embryonic gonad*, 3-[47]).

{34} *Kr* (*Krüppel*, 2–107.6) mutants show gaps in thoracic and abdominal segments and other anomalies, and are lethal when homozygous. The protein product is similar to the Zn-finger domain of transcription factor TFIIIA. It interacts with transcription factors TFIIB and TFIIEβ. In monomeric form it is an activator; in dimeric form at high concentration it represses transcription. This protein binds to the AAGGGGTTAA motif upstream of *hb*. It also affects other maternal effect genes. The GLI oncogene is a homologue.

{35} *nkd* (*naked cuticle*, 3–47.3): Denticle bands are partially missing; germ bands are shortened and thus lethal.

{36} *nos* (*nanos*, 3–66.2) is active in the pole cells. Transport of its product in an anterior direction is required for the normal abdominal pattern, and it is essential for the normal development of the germline. It is a maternal lethal gene. Its product represses that of *hb* in the posterior part of the embryo. Deficiency of both *hb* and *nos* is conducive, however, to normal development.

{37} *oc* (*ocelliless*, 1–23.1): Some alleles are viable, although the ocelli are eliminated; others (*otd*) involve lethality because of neuronal defects.

{38} *odd* (*odd skipped*, 2–8): embryonic lethal because the posterior parts of the denticle bands are replaced; the anterior parts are in mirror image fashion in T2, A1, A1, A3, A5, and A7.

{39} *opa* (*odd paired*, 3–48): Alternate metasegments are genetically ablated. Denticle bands of T2, A1, A3, A5, A7, and naked cuticles of T3, A2, A4, and A6 are absent. The product of *en* is lost, but that of *Ubx* increases in even-numbered parasegments.

{40} *osk* (*oskar*, 3–48): Homozygous females and males are fertile, but the embryos produced by homozygous females are defective in the pole cells and consequently also in abdominal segments; affects *BicD* and *hb* expression.

{41} *phl* (*pole hole*, 1-0.5) blocks the formation of anterior-posterior end structures as well as the entire 8th abdominal segment. Phenotype is similar to that caused by *tso*. It is the *raf* oncogene in *Drosophila*.

{42} *pll* (*pelle*, 3–92) gene causes maternal embryo lethality by preventing the formation of ventral and lateral structural elements.

{43} *prd* (*paired*, 2–45): In strong mutants, the anterior parts of T1, T3, A2, A4, A6, and A8 and posterior parts of T2, AS1, A3, A5, and A7 are absent.

{44} *run* (*runt*, 1-[65.8]; syn. *legless* [*leg*]): pair rule embryo lethal; eliminates the central mesothoracic and uneven-numbered abdominal denticle belts. The deletions are accompanied by duplication of the more anterior structures. The wild-type allele appears to regulate the expression of *eve* and *ftz*.

{45} *slp* (*sloppy paired*, 2.8): Parts of the naked cuticle are missing from T2, A1, A3, A5, and A7 in an irregular way, causing lethality.

{46} *sna* (*snail*, 2–51): Embryonic lethals causes dorsalization and reduction or elimination of most of the mesodermal tissues. The C terminus of the encoded polypetide has five Zn-finger motifs.

{47} *snk* (*snake*, 3–52.1) is a maternal lethal gene with dorsalizing effects. The encoded polypeptide contains elements homologous to serine proteases.

{48} *spi* (*spitz*, [*spire*], 2–54) is embryonic lethal, blocking the development of anterior mesodermal tissues.

{49} *spz* (*spätzle*, 3–92): Maternal lethal alleles accentuate dorsal structures.

{50} *stau* (*staufen*, 2–83.5) ablates pole cells and other anterior structures and some abdominal segments; causes *grandchildless-knirps* syndrome. Staufen protein binds RNA; along with Prospero RNA, it is asymmetrically partitioned into the ganglion mother cells (GMC) and they determine neuroblast cell fate.

{51} *swa* (*swallow*, 1–15.9) is expressed in the nurse cells. The product is transported to the oocyte and into the blastoderm until gastrulation, leading to nuclear division problems, head and abdominal defects. In *swa* homozygotes, the *bcd*$^+$ products are disrupted.

{52} *tl* (*Toll*, 3–91) females heterozygous for the dominant or homozygous for the recessive *tl* mutant alleles produce lethal embryos that are defective in gastrulation and dorsal or ventral structures. The gene product is an integral membrane protein. Toll-like proteins are receptors of interleukin signals in the innate immunity system of mammals.

{53} *tll* (*tailless*, 3–102) deletes several posterior structures (Malpighian tubules, hindgut, telson), but brain and other anterior structures are also missing. Its expression is required for the manifestation of pair rule genes *h* and *ftz* in the 7th abdominal segment and for site specificities of *cad* (*caudal*, 2-[55] involved with head, thorax, and abdominal structures and regulation of *ftz*), *hb* (see above), and *fkh* (*fork head*, 3–95, involved with homeosis in both anterior and posterior structures of this nonmaternal embryonic lethal). Gene *tll* has negative effects on genes *kni*, *Kr*, *ftz*, and *tor*.

{54} *tor* (*torso*, 2-[57]) locus has both loss-of-function and gain-of-function alleles. The former type of alleles eliminate anterior-most head structures and segments posterior to the 7th abdominal segment. The latter-type alleles are responsible for defects in the middle segments and enlargement of the most posterior parts of the body. The gene is expressed in the nurse cells, oocytes, and early embryos. The expression of *ftz* is reduced in the gain-of-function *tor* mutants, whereas *phl* mutations are epistatic to *tor*. With the exception of the NH$_2$ terminus, the protein is homologous to the growth factor receptor kinases of other organisms. It is concentrated in both pole cells and at the surface cells. Apparently, the product of this gene receives and transmits maternal information into the interior of the embryo.

{55} *trk* (*trunk*, 2–36) mutants lack anterior head structures as well as segments posterior to the 7th abdominal band.

{56} *Tub* (*tubulin* multigene families, scattered in chromosomes 2 and 3): The *αTub* genes are responsible for the production of α-tubulin and are apparently active in the nurse cells. The transcripts accumulate in the early embryo and in the ovaries, controlling the mitotic and meiotic spindle and cytoskeleton. The β-tubulin genes are expressed in the nurse cell, early embryos, and various structures and organs.

{57} *tud* (*tudor*, 2-[97]): The germline autonomous mutants display the "*grandchildless-knirps*" phenotype and lack pole cells, yet about 30% of the embryos survive, becoming sterile adults.

{58} *tuf* (*tufted*, 2–59): Segment boundaries are duplicated in a mirror image, and other parts are deleted and the neuronal pattern is altered.

{59} *twi* (*twist*, 2–100): Mutants are embryo lethal with defects in mesodermal differentiation. The embryos are twisted in the egg case. Mutations at the *dl*, *ea*, *pll*, and *Tl* loci prevent the expression of *twi*. The *twi* polypeptides are homologous to DNA-binding myc proteins.

{60} *Ubx* (*Ultrabithorax*, 3–58.8): See BXC.

{61} *vas* (*vasa*, 1-[64]): affects the pole region and segmentation (*grandchildless-knirps* syndrome). The protein product is homologous to murine peptide chain translation factor eIF-4A.

{62} *vls* (*valois*, 2–53): The phenotype is very similar to that caused by *vas*.

{63} *wg* (*wingless*, 2–30): Visible viable and lethal alleles control the segmentation pattern and imaginal disk pattern (wings and halteres). Its protein product is homologous to the mouse mammary oncogene *int-1* (INT1/Wnt). It appears that *frizzled* (*fz*) is the receptor of *wg*. The signal then may be transmitted to *Dsh* and *Arm* (β-catenin). Transcription of *En* is turned on with the assistance of other factors. *See* wingless; wnt.

{64} *zen* (*zerknüllt*), a segment of *ANTC* {8}, affecting segmentation and dorsal structures of the early embryo. The maternal germline mutations are genetically identical in the unfertilized egg and their gene products generally cause lethality in the egg or in the homozygous embryos. The heterozygous mothers are semisterile, and the males are usually fairly normal in appearance and function. The molecular bases of a few such lethal genes are known, e.g., the *phl* gene appears to be homologous to the *v-raf* protooncogene; the *phl* gene is the *Drosophila raf* gene that encodes a serine-threonine kinase protein in humans and mice. The *snk* locus encodes a protein that appears to have a calcium-binding site at the NH$_2$ end with homology to several serine proteases at its C terminus. The product of the *ea* gene has some homology to an extracellular trypsin-like serine protease. Specific cytoplasm extracted from some (e.g., *osk*, *tor*, *ea*, *bcd*) unfertilized normal eggs or from normal embryos when injected into the mutant cytoplasm end may rescue the embryos that develop into sterile adults. The 923-amino-acid protein encoded by *tor* has no homology to other known proteins in the NH$_2$ end, but the rest of it is similar to a growth factor receptor tyrosine kinase, and a hydrophobic segment appears to be associated with the cell surface membrane. The product of the *Tl* locus is an integral membrane protein with both cytoplasmic and extracytoplasmic domains containing 15 repeats of leucine-rich residues that resemble yeast and human membrane proteins. The cytoplasmic domain is homologous to the interleukin-1 receptor (IL1R), the heterodimeric platelet glycoprotein 1b, coded in human chromosome 2q12 and mouse chromosome 1 near the centromere.

Although all of the mutants assigned to different chromosomal locations have different molecular functions, the phenotypic manifestation may not necessarily distinguish this. The majority of the morphogenetic-developmental mutations are pleiotropic (e.g., affect the pole cells and cause the *grandchildless-fushi tarazu* syndrome). Many display epistasis and indicate the complex interactions of the regulatory processes (e.g., the expression of *twi* may be prevented by mutations at *dl, ea, pll*, and *Tl*). The genes may be expressed at a particular position, but their products form a diminishing gradient (e.g., *bcd, nos*). The DNA-binding protein encoded by *bcd*, depending on its quantitative level, qualitatively regulates the transcription of, e.g., *hb*. Some of the genes display the so-called gap effect. They eliminate particular body segments, e.g., *kni, Kr*, and *hb*. The so-called pair rule genes may eliminate certain body segments and replace them with others (e.g., *ftz, eve*) or eliminate half of the segments and fuse them together in pairs (e.g., *en*). Several of the genes, particularly those in the huge *BTC* and *ATC* clusters, may display homeotic effects. Although a great deal of information has been gathered on these genes, more specific knowledge is required for understanding the precise functions (especially the interacting circuits) of the morphogenetic processes. With the aid of microarray hybridization, information on the interacting systems will greatly expand. It appears that these genes are frequently expressed preferentially at a particular position. Their mRNA or protein product is then spread in a gradient to the sites of the required action. In some instances, only the RNA is spread and the protein is made locally. The position of morphogenetic function is controlled by a hierarchy of signals. These genes are expressed differently in different time frames and in coordinated sequence. The coordination is provided by the interaction of gene products. The same protein may turn on a set of genes and turn off others, depending on the local concentration of the products.

According to their main effects, some of these genes may be classified into groups; others are more difficult to place because they affect several stages and different structures. (The horizontal line stands for the embryo axis and the arrows or gaps illustrate the typical sites of action; the best representative genes are bracketed):

MATERNAL ANTERIOR GENES: →___ 1, 10 see [8] 23, 51,

MATERNAL POSTERIOR GENES: ___ ← 6, 11, [36], 40, 50, 57, 61, 62

MATERNAL END SEGMENT: ___ ← 1, 24, [54], 55,

MATERNAL DORSOVENTRAL: ___↓↑___ 2, [3] 4, 5, [14], 17, 18, 42, 46, 49, 52, 59, 63

ZYGOTIC GAP: _ __ __ __ __ 29, [31], [33], [34], 53

ZYGOTIC PAIR RULE: __ _ __ __ [22], 26, [30], 38, 39, [14], 45

ZYGOTIC SEGMENT POLARITY: __ →← __ →← __ 9, 16, [20], 25, 28, 32, 35, 43, [63]

HOMEOTIC GENES: genes within the *Antennapedia* complex {8}, and the *Bithorax complex* {13}

The development of wing veins is controlled by several genes. Locus *vn* (*vein*, 3–16.2, *Vein*, 3–19.6) disrupts longitudinal vein L4, the posterior cross-vein, and sometimes L3; *ri* (*radius-incompletus* 3–46.8) interrupts L2; mutations in *px* (*plexus*, 2–100.5) produce extra veins. *N* (*Notch*, 1–3.0) complex (*Ax, Co, fa, l(1)N, N, nd, spl*), with a very large number of (dominant and recessive) alleles, is homo- and hemizygous lethal and removes small portions of the ends of the wings; it also affects hairs and embryo morphogenesis, thickening of veins L3 and L5, and the hypertrophied nervous system. *Notch* gene homologues are present in vertebrates. The N complex codes for a protein with EGF-like repeats. *Ax* (*Abruptex*) homozygotes reduce the length of longitudinal vein L5 and commonly L4, L2, and sometimes L3. The various *Ax* alleles are either positive or negative regulators of *Notch*. The Notch signals are modulated by *fringe*, and this gene determines the dorsal-ventral boundaries. *Co* (*Confluens*) causes thickening of the veins. The *fa* alleles affect the eye facets and are noncomplementing to the *spl* gene, which also causes rough eyes and bristle anomalies. Gene *nd* (*notchoid facet*) is homozygous viable and displays some of the characteristics of the other genes within the complex. *E(spl)* (*Enhancer of split*, 3–89) mutation became known as exaggerating the expression of *spl* (enhancer in this context does not correspond to the term *enhancer* as used in molecular biology). This gene produces 11 similar transcripts. They share homologies with *c-myc* oncogene (a helix-loop-helix) protein and with the β-transducin G-protein subunit known to involve signal transduction. *Dl* (*Delta*, 3–66.2) causes thickening of the veins (and a number of other developmental defects) and is responsible for a protein that has an extracellular element with nine repeats resembling EGF, an apparent transmembrane, and an intracellular domain with apparently five glycosylation residues. *Egfr* (*epidermal growth factor receptor* [synonyms *top* {*Torpedo*}, *Elp* {*Ellipse*}, *fbl* {*faint little ball*}, 2–100) genes cause embryonic lethality and a number of other developmental effects including extra wing veins, eyes, etc.

The genes *wg* (see {63}) and *dpp* (see {17}) are involved in anterior-posterior specifications of the embryo and thus also in wing formation. Several other genes affect wing and vein differentiation. Molecular evidence permits the assumptions that Dsh (dishevelled), a cytoplasmic protein, is one of the receptors of the Wg-protein signal. Dsh binds to N (Notch, a transmembrane protein) and inhibits its activity. When N binds to Delta, it activates Su(H), suppressor of hairless (H). Su(H) then moves to the cell nucleus and operates as a transcription factor. The Wg signal can also activate the Shaggy-Zeste white (Sgg-Za3) serine-threonine kinase, and the phosphorylation of Armadillo may lead to Wg-dependent gene expression. The level of Armadillo may be elevated by binding Wg to a member of the frizzled family (Dfz2). This *frizzled* (*fz*, 3–41.7) appears to be another receptor of Wg. A diffusible protein product of *optomotor-blind* (*omb*, 1–{7.5}) that was initially identified on the basis of locomotor activity is also required for the development of the distal parts of the wing within the *wg-dpp* system.

amx (*almondex*, 1–27.7) is a locus with multiple functions including eyes (some alleles complement the *lozenge* mutants). Most relevantly, in the mutants there is hyperplasia of the central-peripheral nervous system and a concomitant reduction in epidermogenesis.

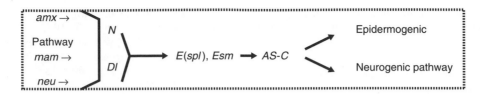

A general picture of neurogenesis is emerging. (Modified after Campos-Ortega & Knust. 1992. *Development*, Russo, V. E. A., et al., eds. p. 347. Springer Verlag, New York):

mam (*master mind*, 2–70.3) affects the eyes, but mutations lead to neural hyperplasia and epidermal hypoplasia.

neu (*neural*, 3–50) mutations cause neural hyperplasia and epidermal hypoplasia.

N and *Dl* have been briefly described above. Both have protein products with epidermal growth factor (EGF-like) repeats. EGF has growth-promoting signals and has the ability to bind appropriate ligands.

E(spl): The wild-type alleles encode a protein with helix-loop-helix motifs characteristic for transcription factors and display similarities to one subunit of the trimeric G proteins, which have key roles in several signal transduction pathways.

ASC (*achaete-scute complex*, 1-0.0) controls sensilla and micro- and macrochaetae that are sensory organs of the flies and correspond to the peripheral nervous system. The ca. 100 kb region contains four major distinguishable areas: *ac* (*achaeta*), *sc* (*scute*), *l(1)s* (*lethal scute*), and *ase* (*asense*). All four reduce and alter the pattern of the sensory organs. The dominant components, the *Hw* (*Hairy wing*) mutations, increase hairyness. Another regulatory mutation has been named *sis-b* (*sisterless b*). The complex includes nine transcription units. Four of them appear to be transcription factors because of the helix-loop-helix motifs.

svr (*silver*), *elav* (*embryonic lethal abnormal vision*), and *vnd* (*ventral nervous system defective*) are all affected in the nervous system and located at 0 position of the X chromosome just like *ASC*. A large number of other genes at various locations are also involved in the nervous system. Even in a relatively simple organism such as *Drosophila*, ~14,000 genes exist and their functions are too complex to be represented by simple models. *See* clonal analysis; developmental genetics; *Drosophila*; gap genes; homeotic genes; imaginal disks; morphogens; oncogens; pattern formation; RNA localization; selector genes; signal transduction. (Morisato, D. & Anderson, K. V. 1995. *Annu. Rev. Genet.* 29:371; Mann, R. S. & Morata, G. 2000. *Annu. Rev. Cell Dev. Biol.* 16:243.)

morphogenetic field   Embryonal compartment capable of self-regulation.

morphogenetic furrow   Embryonic tissue indentation marking the front line of the differentiation wave. Signal molecules mediate its progression. *See* daughter of sevenless; morphogenesis; morphogenesis in Drosophila.

morpholinos   Antisense oligomers that block cell proliferation, interfere with normal splicing of pre-mRNAs, and generate aberrant splicing. They are highly specific and immune to nuclease (RNase H). They are suitable for the inactivation (knockdown) of targeted genes. (Summerton, J. 1999. *Biochim. Biophys. Acta* 1489:141.)

Morpholine.

morphology   Study of structure and forms.

morphometry   Quantitative study of shape and size of a body or organ. *See* QTL.

morphosis   Morphological alteration; a phenocopy rather than a mutation. *See* epigenetic; phenocopy.

Morquio disease   *See* gangliosidosis type I.

mortality   Condition of being mortal, i.e., subject to death. The rate of mortality is computed as the average number of deaths in a particular (midyear) population. An important factor of (particularly) infant death rate is the coefficient of inbreeding. First-cousin marriages (inbreeding coefficient 1/16) approximately double infant mortality. The rate of human mortality calculated as the number of deaths per 1,000 population may vary substantially in different parts of the world. In 1955–59, it was 9.4 in the U.S., 11.6 in England, in and 7.8 in Japan. In some other parts of the world, it was double or higher. *Extrinsic mortality* is an age and condition-independent concept. *See* age-specific birth and death rates; aging; inbreeding coefficient; juvenile mortality; longevity; morbidity.

morula   Mass of blastomeres before implantation of the zygote onto the uterus.

Mos   *See* mariner.

MOS (*mos*)   *See* MITE; Moloney mouse sarcoma virus oncogene.

mosaic   Mixture of genetically different cells or tissues. Mosaicism is generally the result of somatic mutation, change in chromosome number, deletion or duplication of chromosomal segments, mitotic recombination, nondisjunction, sister chromatid exchange, sorting out of mitochondrial or plastid genetic elements, infectious heredity, intragenomic reorganization by the movement of transposons or insertion elements, lyonization, gene conversion, lyonization, etc. A mosaic hybrid may be the result of codominance of the parental alleles. Some breeds, however, display homozygosity for mosaicism. Introduction of foreign DNA into the cells by transformation or gene therapy may also be the cause of mosaic tissues. Somatic mutation may be of medical importance and may lead to oncogenic transformation and the expression of cancer. Somatic reversion of recessive genes causing disease, e.g., reversion of adenosine deaminase (ADA) or tyrosinemia (FAH) genes, followed by selective proliferation of the normalized sector, may alleviate the disease. Loss of the extra chromosome in trisomics or nondisjunction in monosomics may restore the normal chromosome number. When a mutation is induced only in a single strand of DNA, its expression may be delayed, but in mice fur patches may occur in the heterozygotes; these animals are called *masked mosaics*. *See* allophenic; lyonization; sex chromosomal anomalies in humans; variegation. (Extavour, C. & Garcia-Bellido, A. 2001. *Proc. Natl. Acad. Sci. USA* 98:11341; Yousonffian, H. & Pyeritz, R. E. 2002. *Nature Genet.* 3:748.)

Mosaic hybrid.

mosaic genes Contain exons and introns. *See* exons; introns.

mosaic theory *See* suicide vectors.

Mosolov model In the eukaryotic chromosome an extremely long DNA fiber is tightly packed into a very compact chromosome. One of the many existing models is shown here. Each line represents an elementary chromosome fiber including the nucleosomal structure at various stages of folding. E: conceptualizes the chromomeres. The elementary fiber is about 25 Å in diameter and forms a tubular coil (solenoid). D: tightly coiled coils of the metaphase chromosomes. *See* chromosome coiling; chromosome structure; nucleosome; packing ratio; SMC; diagram above right.

mosquito *Culex pipiens*, $2n = 6$. *See Aedes aegypti*; malaria. (<klab.agsci.colostate.edu>; <http://www.tigr.org/tdb/tgi.shtml>.) (Atkinson, P. W. & Michel, K. 2002. *Genesis* 32:42.)

mossy fibers Microtubules on nerve axons.

motheaten Mutation in the SH2 domain of a tyrosine phosphatase of mice leading to autoimmune anomalies. *See* autoimmune disease; SH2; tyrosine phosphatase. (White,

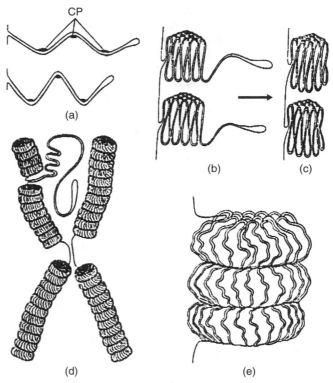

(d)   (e)

(Diagram from Kushev, V. V. 1974. *Mechanisms of Genetic Recombination*. By permission of the Consultants Bureau, New York.)

E. D., et al. 2001. *J. Leukoc. Biol.* 69:825; Hsu, H. C., et al. 2001. *J. Immunol.* 166:772.)

mother-fetus compatibility *See* immune tolerance; incompatibility; Rh blood groups.

motif Small structural domain or sequence of amino acids or nucleotides present in different macromolecules. Motifs are generally preserved during evolution and usually convey some functionality. The probability of a motif is a matrix of probabilities. The column number in the matrix is the length (N) of the motif. The number of rows is 4 in DNA (the 4 kinds of nucleotides) and 20 in protein (the 20 common amino acids). Row i and column j specify the probability for finding a special appearance (instance) of a motif. *See* matrix algebra; module; MP score; protein domains. (<http://www.bioinf.man.ac.uk/dbbrowser/PRINTS/>.)

motif-trap technology Random fragments of DNA fused to fluorescent protein are screened and their location in the cells is monitored. The procedure may facilitate the identification of the function of the unknown sequences. *See* aequorin. (Cutler, S. R., et al. 2000. *Proc. Natl. Acad. Sci. USA* 97:3718.)

motilin 22-amino acid well-conserved peptide hormone regulating motility of the intestinal tract. (Coulie, B., et al. 2001. *J. Biol. Chem.* 276:35518.)

motor proteins Can move along microtubules and actin filaments or macromolecules by deriving energy from the hydrolysis of energy-rich phosphates such as ATP. They transport various molecules and vesicles within the cell. Motor proteins mediate some of the processes involved in establishing body plans, such as left-right, dorsoventral, and anterior-posterior differentiation. Representatives include myosin, kinesin, dynein, helicases, bimC, etc. The MyoVa motor is involved in melanosome transport and organization of the endoplasmic reticulum in Purkinje cells. *See* actin; albinism; anaphase; caveolae; cytoskeleton; duty ratio/duty cycle; dynein; kinesin; microtubule; myosins; Purkinje cells. (Karcher, R. L., et al. 2001. *Science* 293:1317.)

Mountjack chromosomes after Wurster & Benirschke.

mountjack (*Muntiacus muntjack*)   The male is $2n = 7$. The female is $2n = 6$, but the *Muntiacus reevesi* is $2n = 46$.

mouse   (1) Of a computer is a pointer device. (2) *Mus musculus*, $2n = 40$. Rodent belonging to the subfamily Murinae, including about 300 species of mice and rats. They are extensively used for genetics and physiological studies because of the small size (25–40 g), short life cycle (10 weeks), life span of about 2 years, gestation 19 days, 5–10 pups/litter, and practically continuous breeding. The genome is $\sim 1.8 \times 10^6$ kDa in $2n = 40$ chromosomes. In a m^2 laboratory space, up to 3,000 individuals can be studied annually. Very detailed linkage information is available. According to the 1996 map, the genetic length-based 7,377 genetic marker including RFLP and other markers is 1,360.9 units. The average spacing between markers was 400 kb. A nucleotide sequence draft became available in 2002 and the final annotated sequence is expected by 2006. Transformation, gene targeting, and other modern techniques of molecular genetic manipulations are well worked out. It is very important that mice are being used as human genetic models for immunological, cancer, and other human diseases. About 85% of the autosomal gene repertory of the mouse displays conserved synteny with that in the human genome. The X-chromosomal genetic structure is practically identical in humans and mice. *Peromyscus* wild mice are $2n = 48$. In mice, Robertsonian translocations are common and therefore the chromosome number may vary in different populations and in different cultures. *See* animal models; databases; *Encyclopedia of the Mouse Genome*; Mouse Genome Database; Mouse Genome Informatics Group; *Portable Dictionary of the Mouse Genome*. (Mapping information in *Nature* 380:149; radiation hybrid map: *Nature*

Genet. [1999] 22:383; YAC-based physical map: *Nature Genet.* [1999] 22:388; Festing, M. F. W. 1979. *Inbred Strains in Biomedical Research.* Oxford Univ. Press, New York; electronic information sources: <http://www.informatics.jax.org>; <http://www.genome.wi.mit.edu> or by "help" to <genome_database@genome.wi.mit.edu>; <http://genome.rtc.riken.go.jp>; BodyMap, nude mouse: <http://www.rodentia.com/wmc/index.html>; <www.informatics.jax.org/silver>; knockout database, TBASE: <http://www.jax.org/tbase>; Mouse Atlas: <http://genex.hgu.mrc.ac.uk>; sequencing information, Celera Genomics: Mouse Sequence Consortium: <www.ncbi.nlm.nih.gov/genome/seq/MmProgress.shtml>; <http://mouse.ensembl.org/>; cDNA clones: <http://fantom.gsc.riken.go.jp/db/>; GXD, *Adv. Genet.* 35:155; Joyner, A. L., ed. 2000. *Gene Targeting. A Practical Approach.* Oxford Univ. Press, Oxford, UK, mouse chromosome 16 versus human genome: Mural, R. J., et al. 2002. *Science* 296:1661; Gregory, S. G., et al. 2002. *Nature* 418:743.)

Mouse Genome   Newsletter by Journal Subscriptions Dept., Oxford Univ. Press, Walton St., Oxford, OX2 6DP, UK.

Mouse Genome Database (MGD)   Integrates various types of information, mapping, molecular phenotypes, etc. Contact: Mouse Genome Informatics, Jackson Laboratory, 600 Main St., Bar Harbor, ME 04609; tel.: 207-288-3371, ext. 1900; fax: 207-288-2516; <mgi-help@informatics.jax.org> for knockout. *See* databases, TBASE.

Mouse Genome Informatics Group   Electronic bulletin board. Information: Mouse Genome Informatics User Support, Jackson Laboratory, 600 Main St., Bar Harbor, ME 04609; tel.: 207-288-3371, ext. 1900; fax: 207-288-2516; <mgi-help@informatics.jax.org>.

mouse lymphoma test for genotoxicity (MLA)   Employs thymidine kinase heterozygous ($TK^{+/-}$) and homozygous ($TK^{-/-}$) mouse lymphoma cells (L5178Y) and classifies the treated cultures for chromosomal aberrations and colony morphology. *See* bioassays in genetic toxicology. (Hozier, J., et al. 1981. *Mutation Res.* 81:169; Clements, J. 2000. *Mutation Res.* 455:97.)

mouse mammary tumor virus (MMTV)   Causes mammary adenocarcinomas. The virus is transmitted to the offspring by breast-feeding. If the virus is transposed within 10 kb distance to the *Wnt-1* oncogene (homologue of the *Drosophila* locus *wingless* [*wg*]), the insertion may activate the oncogene because of the very strong enhancer in the viral terminal repeat. *See* DNA-PK; pattern formation, retroviruses.

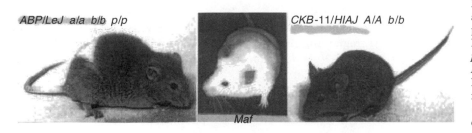

ABP/LeJ a/a b/b p/p   CKB-11/HIAJ A/A b/b

Maf

Mice have a very large number of spontaneous and induced variations involving morphological, physiological, biochemical, and behavioral traits. Left: *pink-eyed dilution, A/a Tyrp1^b/Tyrp1^{b}bt/bt p/p*, chromosome 7–28.0. Middle: Kit ligand, *Mgf^{Sl-pan}/Mgf^{Sl-pan}*, chromosome 7 10–57.0. Right: tyrosine related, A/A Tyrp1^b/ Tyrp1^b, chromosome 4–38. (Courtesy of Dr. Paul Szauter, <http://www.informatics.jax.org/mgihome/other/citation.shtml>.)

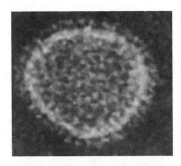

Mouse mammary tumor virus.

Mov34 *See* Moloney mouse leukemia virus oncogene.

movable genetic elements *See* mobile genetic elements; transposable elements; transposons.

movement proteins Their synthesis is directed by plant viruses in order to spread the infectious particles through the plasmodesmata with the aid of microtubules. *See* microtubules; motor proteins; plasmodesma.

MOWSE Peptide mass database (http://www.hgmp.mrc. ac.uk/Bioinformatics/Webapp/ mowse/mowsedoc.html>).

MOZ Monocytic leukemia zinc-finger domain. *See* CREB; leukemia.

M6P Mannose-6-phosphate. The M6P proteins are transmembrane proteins in the trans-Golgi network. *See* endocytosis; Golgi apparatus.

Mpl Regulator of megakaryocyte formation. *See* megakaryocyte.

MP1 MEK partner 1 is involved in the activation of MEK and ERK in the signal transduction pathway. ERK; MEK; *See* signal transduction. (Schaeffer, H. J., et al. 1998. *Science* 281:1668.)

Mpd 36 kDa yeast protein in the endoplasmic reticulum with disulfide isomerase activity. *See* Eug; PDI.

MPD Maximal permitted dose. *See* radiation effects; radiation hazards; radiation threshold.

MPF Maturation protein factor/M-phase-promoting factor contains two subunits: a protein kinase (coded for by the $p34^{cdc2}$ gene) and B cyclin. Activation takes place (probably by phosphorylation at threonine 161 of the $p34^{cdc2}$ protein) during M phase and deactivation is mediated by degradation of the cyclin subunit during the rest of the cell cycle. MPF is deactivated by phosphorylation at Thr-14 and Tyr-15 amino acid residues. These sites are dephosphorylated probably by the product of gene $p80^{cdc25}$ or a homologue before the onset of mitosis. *See* APC; cell cycle; cyclin; p34^{cdc},2; protein kinases; signal transduction. (Frank-Vaillant, M., et al. 2001. *Dev. Biol.* 231:279; Taieb, F. F., et al. 2001. *Curr. Biol.* 11:508.)

M13 phage *See* bacteriophages; DNA sequencing.

M phase Part of the cell cycle in which the structure, movement, and separation of chromosomes or chromatids are visible and the nuclear divisions are completed. *See* meiosis; mitosis.

M phase histone-1 kinase *See* growth-associated kinase.

M phase promoting factor (MPF) Maturation-promoting factor.

MPI minimal protein identifier serves as an index of protein identity on the basis of proteomics characteristics. *See* proteomics.

MPK2 *See* p38.

Mp1p *See* mitochondrial import.

Mps Kinetochore-associated kinase required for maintaining the anaphase checkpoint until the microtubules are attached to the kinetochore. It acts before the anaphase-promoting complex. Mps1 also participates in the duplication of the centrosome in cooperation with Cdk2. *See* anaphase; APC; Cdk2; centrosome; checkpoint; kinetochore. (Abrieu, A., et al. 2001. *Cell* 106:83; Fisk, H. A. & Winey, M. 2001. *Cell* 106:95.)

MP score Mean sum of pairs score for a column in an alignment of motifs—i.e., the value of the SP score divided by the total number of, e.g., amino acids. The SP score is the sum of pair scores for a column in a multiple alignment. *See* motif.

MPT Mitochondrial permeability transition indicates protein release from mitochondria and apoptosis; it may be detected by ICAT. *See* apoptosis; ICAT. (Bruno, S., et al. 2002. *Carcinogenesis* 23:447.)

MQM Marker-QTL-marker. *See* QTL.

M$_r$ Relative molecular weight. relative molecular mass of a molecule compared to that of the mass of a C^{12} carbon atom. This is different from the gram molecular weight traditionally used in chemistry. *See* dalton.

MR Male recombination factor of *Drosophila* (map location 2–54) is apparently a defective P element (*see* hybrid dysgenesis) that cannot move, but it can facilitate the movement of other P elements. Besides causing recombination in the male, it induces many of the symptoms of hybrid dysgenesis, including chromosomal aberrations, high mutation rate, and mitotic exchange. *See* hybrid dysgenesis; male recombination.

MRCA Most recent common ancestor is an evolutionary concept for divergence or coalescence. *See* coalescence; evolutionary tree.

MRD Mismatch repair detection. DNA fragments are cloned and inserted into bacterial plasmids and bacteria are transformed by the heteroduplex constructs. If there is a mismatch, the growing bacterial colonies become white, whereas in case of no mismatch the colonies are blue. The reporter gene is *LacZα*. The template for repair is a hemimethylated double-stranded DNA. Mismatches activate repair and the unmethylated strand is degraded, whereas the methylated strand becomes the template for repair. The method can scan up to 10 kb fragments for mismatches that are below five nucleotides in length. A revised procedure more suitable for high-throughput analysis uses the Cre/lox recombinase, a tetracycline-resistance (Tet^R) marker, and a streptomycin-sensitive (Str^S) marker. Two vectors are constructed. One of them carries a 5 bp deletion for the *Cre* gene. In other respects, the two vectors are identical. DNA fragments cloned in the active Cre$^+$ vector are propagated in a bacterial strain, which is dam methylase–free. The two clones are made only for use as standard panels for testing human DNA samples. Human DNA fragments from each individual to be tested are amplified, then pooled. Linearized methylated DNAs from the Cre deletion and the Cre$^+$ unmethylated DNA are combined in one vessel and single-stranded PCR-amplified DNA is added. Subsequently, Taq ligase is added to generate hemimethylated, closed heteroduplex circles. The remaining linear strands are removed by exonuclease III digestion. The heteroduplexes are cloned into an *E. coli* strain (carrying Str^R in its chromosome) carrying an F' plasmid with Tet^R and Str^S cassettes, each flanked by two lox sites. The DNA strands without mismatch replicate normally in the bacterium and both types of plasmids survive. The Cre$^+$ protein mediates recombination between the two lox sites, and Tet^R as well as the Str^S genes are removed. The bacterial cell thus becomes tetracycline sensitive, but because of the presence of the chromosomal Str^R gene, it will thrive on streptomycin but not on tetracycline media. The mismatch in the heteroduplex will be repaired and the unmethylated Cre$^+$ strand will be degraded. The inactive Cre strand will, however, be used as a template for replication and thus the intact Tet^R and Str^S cassettes will stay functional; cells that carry them remain tetracycline resistant and streptomycin sensitive. The tetracycline- and streptomycin-resistant cells are propagated on petri plates and their restricted DNA content can be assayed by gel electrophoresis. *See* Cre/lox; dam methylase; electrophoresis; lactose; mismatch repair; PCR; restriction enzymes; streptomycin; tetracycline. (Faham, M. & Cox, D. R. 1995. *Genome Res.* 5:474; Faham, M., et al. 2001. *Hum. Mol. Genet.* 10:1657.)

Mre11   Mediates double-strand break repair in somatic cells in cooperation with Ku. In meiosis it is expressed without Ku and apparently mediates repair of the recombined DNA strands. Its defects may cause symptoms of Nijmegen breakage syndrome and ataxia telangiectasia. *See* ataxia telangiectasia; DNA repair; Ku; Nijmegen breakage syndrome. (Petrini, J. H. 2000. *Curr. Opin. Cell Biol.* 12:293, Costanzo, V., et al. 2001. *Mol. Cell* 8:137.)

MRF   Member of the family of muscle proteins. *See* MEF; MYF5; MYOD; myogenin.

MRI   Magnetic resonance imaging. *See* magnetic resonance.

mRNA   Messenger RNA carries genetic information from DNA for the sequence of amino acids in protein. Its half-life in prokaryotes is about 2 min; in eukaryotes, 6–24 hr, or it may survive for decades in trees. In individual cells, the life of mRNA may vary by an order of magnitude (Grunberg-Manago, M. 1999. *Annu. Rev. Genet.* 33:193). Unstable mRNAs even in eukaryotes may last only for a few minutes. mRNA is produced by a DNA-dependent RNA polymerase enzyme using the sense strand of the DNA as a template, and it is complementary to the template. mRNA is derived from the primary transcript by processing, including the removal of introns (in eukaryotes). The size of mRNA molecules varies a great deal because the size of the genes encoding the polypeptides is quite variable. Upstream of the coding sequences of mRNA there are several regulatory sequences (G box, CAAT or AGGA box, etc.) that are important for transcription, but they are not included in mRNA. Transcription begins by the recognition and attachment of the transcriptase to a TATA box and the assembly of the transcription complex. Eukaryotic mRNA is capped with a methylated guanylic acid after transcription. Preceding the first amino acid codon (Met in eukaryotes and fMet in prokaryotes), there is an untranslated leader sequence that helps the recognition of the ribosome. In prokaryotes, the leader includes the Shine-Dalgarno sequence, which assures a complementary sequence on the small ribosomal unit. In eukaryotes, such a sequence is not known; however, usually there is a $\boxed{\text{AG—CC\textbf{AUG}G}}$ preferred box around the first codon and the ribosomal attachment is relegated to "scanning" for it in the leader. The structural genes then follow, which in eukaryotes have been earlier spliced together from a highly variable number of exons. In eukaryotic mRNA, there are untranslated sequences also at the 3' end, including a polyadenylation signal (most commonly AAUAAA) to improve stability of mRNA by the posttranscriptional addition of over 100 adenylic residues. This signal is used for discontinuing transcription. In prokaryotes, a rho protein-dependent palindrome, a rho-dependent GC-rich palindrome, or a polyU sequence serves the same purpose. The number of mRNA molecules per cell varies according to the gene and the environment. In yeast cells, under good growing condition, 7 mRNA molecules were detected with a half-life of about 11 minutes, indicating that the formation of one mRNA molecule required about 140 seconds. The maximal transcription initiation rate for mRNA in yeast was found to be 6–8 seconds. The level of mRNA is frequently used for protein assessment. The obtained estimate may be biased, however, because proteins may be in a dynamic state subject to maturation and degradation, and many proteins are regulated by alternative splicing of mRNA and by posttranslational processing. *See* aminoacyl-+RNA synthetase; cap; decapping; degradosome; enhancer; G-box; GC-box; Hogness box; introns; leader sequence; monocistronic mRNA; mRNA degradation; mRNA tail; mRNA targeting; open promoter complex; operon; polyadenylation signal; polyA polymerase; polynucleotide phosphorylase; Pribnow box; regulation of gene activity; regulon; ribonuclease III; ribonuclease E; RNA noncoding; RNA polymerase II; RNA surveillance; Shine-Dalgarno box; transcription complex; transcription factors; transcription termination, eukaryote; transcription termination, prokaryote; UAS; up-promoter. (Brenner, S., et al. 1961. *Nature* 190:576; Maquat, L. E. & Carmichael, G. G. 2001. *Cell* 108:173; <http://biochem.otago. ac.nz/Transterm>.)

M RNA   *See* cowpea mosaic virus (CPMV).

mRNA CAP   *See* cap.

mRNA, transgenic   Result of trans-splicing of the transcripts. *See* trans-splicing. (Finta, C. & Zaphiropoulos, P. G. 2002. *J. Biol. Chem.* 277:5882.)

mRNA circularization   The 5′ and 3′ ends of messenger RNA my be joined through the transcription initiation polypeptides eIF-4 G and eIF-4E and PABp, and this structure favorably modulates translation initiation, possibly preventing the truncation of the message and the polypeptide. Rotavirus mRNA, which has a GUGACC rather than a poly(A) tail, uses a special binding protein, NSP3, and probably employs this structure to suppress host protein synthesis. *See* PABp; polyadenylation signal. (Mazumder, B., et al. 2001. *Mol. Cell Biol.* 21:6440.)

mRNA degradation   Not an incidental random process, but it is under precise genetic control. The decay in eukaryotes may begin by shortening or removal of the poly(A) tail, followed by removal of the Cap and digestion by 5′ → 3′ exoribonucleases (encoded by *XRN1, HKE1*). The decay by 3′ → 5′ exonuclease is of minor importance. The poly(A) tail is removed by the PAN 3′ → 5′ exonuclease but requires PABP (poly(A)-binding protein) and other proteins for activity. Decapping is mediated by pyrophosphatases. Degradation is predominantly cytoplasmic, although it may take place in the nucleus before the transfer to the cytoplasm. Most frequently, the attacked mRNAs have a termination codon 50–55 nucleotide upstream of the last exon-exon junction. AU-rich elements (AURE) and other factors in the downstream regions regulate the decay process. Decay regulatory purine-rich elements (180–320 base) reside within the coding region. Histone mRNAs lack poly(A) tails and their stability depends on a 6-base double-stranded stem and a 4-base loop; it also depends on the 50 kDa SLBP (stem-loop-binding protein) and other similar proteins. Destabilization of these mRNAs depends on the process of translation and probably on the association of the stem-loop proteins with the ribosomes; it also appears to be autoregulated by histone(s). mRNA degradation is initiated by defects in the translation process of the or by encountering nonsense codons, wrong splicing, upstream open-reading frames (uORFs), and transacting protein factors in the cytoplasm and the nucleus. Steroid hormones, growth factors, cytokines, calcium, and iron affect mRNA stability. Cytokine and protooncogene mRNAs are degraded after binding the AUF1 protein to the 3′ untranslated AU-rich sequences. This AU factor complexes with heat-shock proteins, translation initiation factor eIF4G, and a polyA-binding protein. When AUF1 is dissociated from eIF4G, AU-rich mRNA is degraded. Exo- and endoribonucleases, regulated by various factors may also be involved in processing and in protecting mRNA. Viral infections may rapidly destabilize host mRNAs without affecting rRNAs and tRNAs. Premature termination of translation may account for Fanconi anemia, Duchenne muscular dystrophy, Gardner syndrome, ataxia telangiectasia, breast cancer, polycystic kidney disease, desmoid disease, etc. *See* Cap; eIF4G; endonuclease; exonuclease; exosome; heat-shock proteins; histones; mRNA; mRNA surveillance; mRNA tail; NMDi nonsense codon; polyA mRNA; RNAI; transcription factors; transcription termination; ubiquitin; URS. (Ross, J. 1995. *Microbiol. Rev.* 59:423; Wilson, G. M., et al. 2001. *J.* Biol. Chem. 276:8695; Wilusz, C. J., et al. 2001. *Nature Rev. Mol. Cell Biol.* 2:237; Chen, C.-Y., et al. 2001. *Cell* 107:451; van Hoof, A. & Parker, R. 2002. *Current Biol.* 12:R285.)

mRNA display   Entirely in vitro technique for the selection of peptide aptamers to protein targets. The binding of the aptamers does not require disulfide bridges or special scaffolds (e.g., antibody), yet the affinities are comparable to those of monoclonal antibody–antigen complexes. The polypeptides are linked to their mRNA in vitro while stalling the translation on the ribosomes with the aid of puromycin, which attaches to the 3′ end of mRNA. The mRNA-peptide fusions are then purified and selected in vitro. The procedure is highly efficient and permits selection in libraries of about 10^{13} peptides. It obviates the need for transformation, a disadvantage of phage display. *See* aptamer; phage display; puromycin. (Wilson, D. S., et al. 2001. *Proc. Natl. Acad. Sci. USA* 98:3750.)

mRNA entrapment   The molecule is kept inactive by a ternary complex. (Schalx, P. J., et al. 2001. *J. Biol. Chem.* 276:38494.)

mRNA Leader   *See* leader sequence.

mRNA surveillance   (nonsense-mediated decay, NMD) RNA transcripts containing premature stop codons or other defects are more liable to destruction by specific proteins than normal ones. Prematurely terminated mRNA molecules may cause about one-third of human hereditary diseases. In human cells, three proteins — hUpf1 (cytoplasmic ATP-dependent RNA helicase), hUpf2 (perinuclear), and hUpf3b (nuclear export-import signals shuttling between the nucleus and cytoplasm) — are involved. The NMD complex is apparently already formed in the nucleus at the exon-exon junctions and triggers destruction of mRNA in the cytoplasm beginning downstream of the translation termination site. In *Caenorhabditis*, seven genes (*Smg*) are involved in NMD. Drugs in cultured cells of patients can inhibit the NMD pathway and the nonsense transcripts can be stabilized. The drug-induced changes are then revealed in normal and test systems with the aid of microarray hybridization. By identifying NMD inhibition, human disease genes can be recognized. Map location and biological function of the disease genes can be determined even when no a priori information is available. *See* microarray hybridization; mRNA; mRNA degradation. (Hilleren, P. & Parker, R. 1999. *Annu. Rev. Genet.* 33:229; Lykke-Andersen, J., et al. 2000. *Cell* 103:1121; Noensie, E. N. & Dietz, H. C. 2001. *Nature Biotechnol.* 19:413.)

mRNA tail   Mature mRNAs of eukaryotes are generally tailed by ca. 200 adenylate residues. This is not the end of the primary transcript; transcription may continue by 1,000 or more nucleotides beyond the end of the gene. Polyadenylation requires that the transcript be cut by an endonuclease, then a poly-A RNA polymerase attaches to the poly-A tail, which is probably required for stabilization of the mRNA. Several of the histone protein genes do not have, however, a poly-A tail. Other histone mRNAs not involved with the mammalian cell cycle and histone mRNAs of yeast and *Tetrahymena* are polyadenylated. The common posttranscriptional polyadenylation is signaled generally by the presence near the 3′-end of an AATAAA sequence followed by two dozen bases downstream by a short GT-rich element. Polyadenylation takes place within the tract bound by these two elements. Within most gene tracts, AATAAA occur at more upstream locations, but they are not used for poly-A tailing. Several genes may have alternative polyadenylation sites, however, and thus can be used for the translation of different molecules, e.g., for the membrane-bound or secreted immunoglobulin, respectively. The polyadenylation signal may also have a role in signaling the termination of transcription, no matter how much farther downstream it takes place. In the nonpolyadenylated histone genes there is a 6-base pair palindrome that forms a stem for a 4-base loop near the 3'-end of the mRNA, and it is followed farther downstream by a short polypurine sequence. The latter may pair with a U7 snRNP that facilitates termination. The U RNA transcripts of eukaryotic RNA polymerase II are not polyadenylated either. The formation of an appropriate 3'-tail requires that it would be transcribed from a proper U RNA promoter and the transcript would have the 5' trimethyl guanine cap. The 3'-end of U1 and U2 RNAs is formed by the signal sequence $GTTN_{0-3}AAAPU\genfrac{}{}{0pt}{}{PUPU}{PYPY}AGA$ (PU any purine, PY any pyrimidine) near the end. *See* decapping; polyadenylation signal; polyA polymerase; transcription termination. (Hilleren, P., et al. 2001. *Nature* 413:538.)

mRNA targeting In order for proteins required for differentiation and morphogenesis to be available at the location needed, the mRNA generally carries a relatively long 3'-UTR (untranslated region). These long sequences appear to provide means for binding multiple proteins while being transported to the target sites. Additional "zip codes" may be located in the 5'-untranslated sequences. During the process, the mRNA is bound and stabilized at the "ordained" sites from which the translated product may diffuse (in a gradient). *See* compartmentalization; differentiation; morphogenesis. (Jansen, R.-P. 2001. *Nature Rev. Mol. Cell Biol.* 2:247.)

mRNA turnover *See* mRNA degradation.

mRNP Messenger ribonucleoprotein is a repressed mRNA. It is found in the eukaryotic cytoplasm. *See* mRNA.

MRP **(1)** Mitochondrial RNA processing is an RNase that cleaves the RNA transcribed on the H strand of mtDNA at the CSB elements. MRP frequently designates mitochondrial ribosome proteins. *See* chondrodysplasia, McKusick type; CSB; DNA replication mitochondria; mtDNA; Rex; ribonuclease P; ribosomal protein; ribosomal RNA. (Ridanpaa, M., et al. 2001. *Cell* 104:195.) **(2)** Matrix representation with parsimony is a method for constructing composite phylogenetic trees based on published data. The source phylogenetic information is encoded as a series of binary characters that represent the branching pattern of the original trees. The data matrix of the phylogenies is evaluated by parsimony and integrated into a composite tree. *See* evolutionary tree; maximum parsimony. (Bininda-Emonds, O. R. 2000. *Mol. Phylogenet. Evol.* 16:113.)

mrr *See* methylation, DNA.

MRX *See* mental retardation, X-linked.

MS *See* multiple sclerosis.

MSAFP Maternal serum α-fetoprotein analysis may detect prenatally chromosomal aneuploidy (trisomy) and open neural tube defects between 15 and 20 weeks of pregnancy. Oxidation products of estradiol and estrone (estriol) and chorionic gonadotropin generally accompany the high level of this protein. *See* Down syndrome; fetoprotein; gonadotropin; trisomy; Turner syndrome. (Ochshorn, Y., et al. 2001. *Prenat. Diagn.* 21:658; Miller, R., et al. 2001. *Fetal. Diagn.* 16[2]:120.)

msDNA Multicopy single-stranded DNA is formed in bacteria, which contain a reverse transcriptase, and it is associated with this enzyme. This DNA has a length of 86 to 162 bases and it may be repeated a few hundred times. The 5'-end of the msDNA is covalently linked by a 2'-5'-phosphodiester bond to an internal G residue, thus forming a branched DNA-RNA copolymer of stem-and-loop structure. In some cases, because of processing, msdRNA does not form a branched structure and it is only a single-stranded DNA. The 5'-region of msDNA is part of internal repeats within the copolymer. The system is transcribed from a single promoter and thus constitutes an operon. *See* retron; reverse transcription. (Lampson, B., et al. 2001. *Progr. Nucleic Acid Res. Mol. Biol.* 67:69.)

msdRNA *See* msDNA. (Lima, T. M. & Lim, D. 1997. *Plasmid* 38:25.)

MSF Mitochondrial import-stimulating factor selectively binds mitochondrial precursor proteins and causes ATP hydrolysis. The MSF-bound precursor is made up of at least four proteins: Mas-20p, -22p, -70p, and -37p. *See* mitochondria; mitochondrial import. (Hachiya, N., et al. 1993. *EMBO J.* 12:1579.)

MSH *See* hereditary nonpolyposis colorectal cancer; melanocyte-stimulating hormone.

MSH2 (2p22-p21) Homologue of the bacterial MutS mismatch repair gene. *See* mismatch repair.

MSI Microsatellite instability. *See* microsatellite.

mSin Proteins are transcriptional corepressors. *See* corepressor; nuclear receptors. (Ayer, D. E., et al. 1996. *Mol. Biol. Cell* 16:5772.)

MSK Enzyme of the MAPK family; closely related to RSK and involved in phosphorylation of histone 3 and thereby in chromatin remodeling. *See* chromatin remodeling; RSK.

Msl Male-specific lethal. At least five of these genes exist in *Drosophila* that along with *Mle* (maleless) assure that a single X chromosome in the male carries out all of the functions at approximately the same level (dosage compensation) as two X chromosomes in the female. *Msl-1* (2–53.3) and *Msl-2* (2–9.0) are located in the 2nd chromosome, but similar genes are found along the X chromosome. Normally the male chromatin is highly enriched in a histone-4 monoacetylated at lysine-16 (H4Ac16). Mutation at this lysine alters the transcription of several genes. Mutation in *Msl* genes prevents the accumulation of H4Ac16 in the male X chromosome. By containing a zinc-binding RING finger motif, Msl-2 protein may specifically recognize X-chromosomal sequences and in this way distinguish between X and autosomes. The *Msl* complexes contain at least two RNAs on the X: roX1 (3.7 kb) and roX2 (0.6 kb). These RNAs, similarly to *Xist* RNAs in mammals, spread along over 1 Mbp of the chromosome, although not as broadly as *Xist*, which may span over >100 Mbp. *See* histone acetyltransferase; dosage compensation; *Mle*; ring finger; *Xist*. (Larsson, J., et al. 2001. *Proc. Natl. Acad. Sci. USA* 98:6273; Smith, E. R., et al. 2001. *J. Biol. Chem.* 276:31483.)

MS/MS   Tandem mass spectrometer. It can be used as an automatable tool for the diagnosis of amino acid sequence in proteins, etc. *See* mass spectrometer; proteomics.

Msp I   Restriction endonuclease with recognition site C↓CGG.

MS2 phage   Mainly single-stranded icosahedral RNA bacteriophage of about 3.6 kb with four genes completely sequenced. *See* bacteriophages. (Fiers, W., et al. 1976. *Nature* 260:500; Bollback, J. P. & Huelsenbeck, J. P. 2001. *J. Mol. Evol.* 52:117.)

Mss   Mammalian suppressor of Sec4 is a guanosine-nucleotide exchange factor regulating RAS GTPases. *See* GTPase; Rab; RAS; Ypt1.

MSUD   Meiotic silencing by unpaired DNA. DNA unpaired in meiosis silences all homologous sequences including genes that are paired. *See* cosuppression; quelling. (Shiu, P. K. T., et al. 2001. *Cell* 107:905.)

MSV *(msv)*   *See* Moloney mouse sarcoma virus.

MSX   Muscle segment homeobox (*msh*). Pleiotropic loci in humans (MSX1, 4p16.1) and MSX2 (5q34–q35) and homologous genes in mice control cranoifacial (skull, teeth, etc.) development. *See* craniosynostosis; pleiotropy; tooth-and-nail dysplasia. (Milan, M., et al. 2001. *Development* 128:3263; Cornell, R. A. & Ohlen, T. V. 2000. *Curr. Opin. Neurobiol.* 10:63.)

MTA *a*   Yeast mating type *a* gene. *See* mating type determination in yeast.

MTA *α*   Yeast mating type *α* gene. *See* mating type determination in yeast.

MTD   Maximum tolerated dose of a treatment that does not affect the longevity of the animal (except by cancer if the agent is a carcinogen) or does not decrease its weight under long exposure by more than 10%. MTD has been used in classification of carcinogens, but in some instances at high doses some chemicals appear carcinogenic yet under the conditions of normal use they may not pose a risk. Another potential problem is that genetic differences among test animals and humans may affect the sensitivity or susceptibility to the agents. *See* bioassays in genetic toxicology. (Leung, D. H. & Wang, Y. G. 2002. *Stat. Med.* 21:51.)

mtDNA   Mitochondrial DNA. In mammals it generally consists of 5–6 small 16.5×10^3 bp mtDNA rings. The yeast mtDNA genome is circular and 17–101 kb. In *Paramecium*, the mtDNA is linear and 40 kb. In about one-third of the yeasts and many other species (protozoa, fungi, algae), mtDNA may be linear. In plants it varies in the range of 200 to 2,500 kbp and occurs in variable size of mainly circular molecules, although smaller linear mtDNAs also exist and show inverted terminal repeats. The mtDNA genome of *Plasmodium falciparum* is only 6 kbp; that of the plant *Arabidopsis* contains 366,924 bp. More than 80% of plant mtDNA is noncoding, whereas in protists only ~10% is noncoding. The human mtDNA genome contains 16,596 base pairs and it is similar to that of other mammals, encoding 13 proteins and transcribing 2 ribosomal genes and a minimal (22) set of tRNAs. The 13 proteins are ATP synthetase subunits 6 and 8, cytochrome oxidase subunits I, II, III, apocytochrome b, NADH dehydrogenase subunits 1–6, and 4L. Among the metazoan mtDNAs some variations exist in gene number, and the base composition of the coding strands also varies. *Reclinomonas* protozoans encode 97 genes by mtDNA. *Arabidopsis* borrows 2 tRNAs from the chloroplasts. The number of mitochondria in the eukaryotic cells may run into hundreds to thousands. The buoyant density of the mtDNA is surprisingly uniform among many species; it varies between 1.705 and 1.707 g/mL.

In the marine mussels (*Mytilus*) there is an M (transmitted by the male to the sons) and an F (transmitted by the female to sons and daughters) mitochondrial DNA, resulting in a high degree of heteroplasmy among the males.

The replication of mouse mtDNA generally proceeds through two origins of replication, thus forming D loops, a tripartite structure of two DNAs, and a nascent RNA strand. Replication of the heavy strand begins at the single origin (O_H) and continues until the origin of the light chain (O_L) is reached. Then the "lagging" strand replication is initiated in the opposite direction (for an overview of the human mtDNA map, see p. 770). When studied by two-dimensional gels, restriction fragments, including the replication forks, reveal structures that are called "Y arc". In vertebrates, only a single DNA polymerase (polymerase γ) is involved in replication. (Iyengar, B., et al. 2002. *Proc. Natl. Acad. Sci. USA* 99:4483.) This enzyme has two subunits: the larger (~125 to ~140 kDa) is involved in polymerization, and the smaller (35–40-kDa) subunit may have a role in the recognition of the DNA replication primer and may function as a processivity clamp. The smaller, the accessory subunit, shares structural homology with aminoacyl-tRNA synthetases. The large subunit has a $3' \rightarrow 5'$-exonuclease activity, assuring the high fidelity of DNA synthesis. Replication in mitochondria requires a number of accessory proteins. The origin of replication of the light chain is at some distance from that of the heavy strand. The replication is regulated by the mtTFA protein, which assures proper maintenance of DNA and regulates transcription. The mutation rate of mtDNA, despite the exonuclease function of polymerase γ, is an order of magnitude higher than that of nuclear DNA. This higher rate has been attributed to oxygen generated by oxidative phosphorylation. DNA repair may take place by nucleotide excision and mismatch repair and alkyltransferase, but some aflatoxin B1, bleomycin, and cisplatin damage as well as pyrimidine dimers cannot be repaired efficiently.

In animals, the heavy strand codes for 2 rRNAs, 14 tRNAs, and 12 polypeptides of the respiratory chain (ATP synthase, cytochrome b, cytochrome oxidase, 7 subunits of NADH dehydrogenase). Plant mtDNA may encode 3 rRNAs and 15–20 tRNAs. The heavy chain of mammalian mtDNA may be transcribed in a single polycistronic unit and subsequently cleaved into smaller functional units. There are also abundant shorter transcripts of the heavy chain of mtDNA. The light DNA strands of vertebrates are transcribed into 8 tRNAs and 1 NADH dehydrogenase subunit. tRNA genes are scattered over the genome. The small mammalian mtDNA is almost entirely functional, and only a few (3–25) bases are between the genes without introns, and some genes overlap. The promoter sequences are very short and are embedded in noncoding regions. The human mtDNA employs two forms of transcription factors that bear structural similarities to rRNA modifying enzymes. These two factors (TFBM1 and TFBM2, the latter being much less efficient) bind to the core RNA polymerase (PolRMT), which can recognize the two promoter sites (LSP and HSP). TFAM is the protein that binds upstream to both promoters, unwinds the DNA template and facilitates bidirectional transcription (Shoubridge, E. A. 2002. *Nature Genet.* 31:227). The upstream leader sequence is minimal and the typical eukaryotic cap is absent. In addition to the AUG initiator codon, AUA, AUU, and AUC may start translation as methionine codons. The genetic code dictionary of mammalian and fungal (yeast) mitochondria further differs from the universal code. The UGA stop codon means tryptophan; the AUA isoleucine codon represents methionine in both groups; the CUA leucine codon in yeast mtDNA spells threonine (but in *Neurospora* mtDNA it is still leucine); and the AGA and AGG arginine codons are stop codons in mammalian mtDNA. Other coding differences may still occur in other species. Some mtDNA genes have no stop codons at the end of the reading frame, but a U or UA terminates the transcripts after processing, possibly becoming a UAA stop signal by polyadenylation. Since mitochondria have only 22 tRNAs, the anticodons must use an unusual wobbling mechanism. The 2 codon-recognizing tRNAs can form a G*U pairing and the 4 amino-acid codon sets are base-paired by 2 nucleotides only, or the 5′ terminal U of the anticodon is compatible with one of the other 3′ bases of the codon. Nuclear mRNA may not be translated in mitochondria because of the differences in coding. Mitochondrial translation does not use the Shine-Dalgarno sequence for ribosome recognition in contrast to prokaryotes. Rather, it depends on translational activator proteins that connect the untranslated leader to the small ribosomal subunit. Plant mitochondria utilize the universal code.

The fission yeast and *Drosophila* mtDNAs are just slightly larger (19 kb) than those of mammals. The mtDNA of the budding yeast is about 80 kb, yet its coding capacity is about the same as that of mammalian mtDNA. Large mitochondrial DNA is rich in AT sequences and contains introns. Some yeast strains have the same gene in long form (with introns) and in other strains in short form (without intron). The introns may have maturase functions in processing the transcripts of the yeast genes that harbor them (cytochrome b) or they may be active in processing the transcripts of other genes (e.g., cytochrome oxidase). The 1.1 kb intron of the 21S rDNA gene contains coding sequences for a site-specific endonuclease that

facilitates its insertion into some genes lacking this intron as long as they contain the specific target site (5′-GATAACAG-3′). Thus, this intron is also an insertion element.

For transcription of vertebrate mtDNA, the transcription factor mtTFA (a high-mobility-group protein) must bind upstream (-12 to -39 bp) to the promoter to unwind the DNA for the mtRNA polymerase, which binds downstream. The distance between these two binding sites is important.

The *Saccharomyces* mtDNA genes are replicated from 7 or more scattered replicational origins, each containing 3 GC-rich segments separated by much longer AT tracts. Transcription is mediated by a mitochondrial RNA polymerase (similar to T-odd phage polymerases) that is coded, however, within the nuclear DNA. The conserved 5′$^{T}_{A}$-TTATAAGTAPuTA-3′ promoter is positioned within 9 bases upstream from the transcription initiation site. Unlike the much smaller mammalian mtDNA genes, yeast genes have more or less usual upstream and 3′ sequences. The much more compact mammalian or fission yeast genes do not use upstream regulatory sequences (UTRs). In mtDNA, a 13-residue sequence embedded in tRNA$^{Leu(UUR)}$, a 34 kDa protein (mTERM), is bound, and the complex is required for the termination of transcription. Transcription initiation also requires—besides sc-mtTFA—the protein sc-mtTFB. It seems that the primers for the heavy-chain transcription are synthesized on the light chain of mtDNA. Mammalian light chains contain three conserved sequence blocks (CSBs) serving apparently as primers for the transcription of the heavy chain. These seem to be analogous to the GC-rich segments of yeast. In yeast, for the transcription mtDNA heavy strand, an R loop is also formed. Near the origin of the heavy chain there is an RNase specific (RNase MRP) for processing the 3′-OH ends of the origin-containing RNAs. The rRNA genes, unlike in the nuclear genomes, are separated by other coding sequences. Yeast mtDNA encodes at least one ribosomal protein. In mammals, all mitochondrial ribosomal proteins are coded in the nucleus and imported into this organelle. The majority of the other proteins are also imported. Nuclear proteins mediate the translational control locus by specific or global manners.

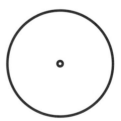

The relative size of a plant (*Arabidopsis*) mtDNA molecule (outer circle) and that of a mammal (human, inner circle).

The transport of proteins through the mitochondrial membrane takes place in several steps. The NH$_2$ terminus passes into the inner membrane through the protein import channel. Mitochondrial heat-shock protein 70 (mtHsp70) stabilizes the translocation intermediate in an ATP-dependent process. The traffic may be in two directions, and it may be a passive transport. The mtDNA of plants is much larger than that of other organisms, varying between 208 and 2,500 kbp. The mtDNA of *Arabidopsis* is 366,924 bp and contains 57 genes. The mtDNA of plants frequently exists in multiple-size groups. This DNA is interspersed with large (several thousand kb) or smaller (200–300 kb) repeated sequences. Recombination between these direct repeats of a "master circle" may generate in smaller, subgenomic DNA rings stoichiometric proportion. In species without these repeats (e.g., *Brassica hirta*), subgenomic recombination does not occur. The very small mtDNA molecules are called plasmids. Plasmids *S*-1 and *S*-2 share 1.4 kb termini. These plasmids may integrate into the main mtDNA.

The variety and the number of repeats may vary in the different species, and some species (*Brassica hirta, Marchantia*) may be free of recombination repeats. Recombination repeats may contain one or more genes. The repeats display recombination within and between the mtDNA molecules. These events then generate chimeric sequences, deficiencies, duplications, and a variety of rearrangements. Recombination between direct repeats tosses out DNA sequences between them and generates smaller circular and possibly linear molecules. The 570 kb maize mtDNA "master circles" may have 6 sets of repeats, whereas in the T cytoplasm the repeats are quite complex. The size of the repeats may be 1 to 10 kb. Not all of them promote recombination, and even the recombination repeats may vary in different species. Plant mitochondria may contain single- and double-stranded RNA plasmids. The latter may be up to 18 kb in size. The single-stranded RNA plasmids may be replication intermediates of the double-stranded ones. The base composition of some is different from that of the main mtDNA and appears to be of foreign (viral) origin.

It has been reported that passage of cells through in vitro culturing (tissue culture) may incite rearrangements in the mtDNA genome. This phenomenon may be the result, however, of different amplification of preexisting alterations. Formation of cybrids may also result in new combinations of the mtDNAs (*see* somaclonal variation). The mitochondrial protein complexes may be organized by chaperone-like mitochondrial proteases.

Some of the DNA sequences in the plant mitochondria are classified as promiscuous because they occur in the nuclei and in the chloroplasts. These promiscuous DNA sequences may contain tRNA (tRNA^{Pro}, tRNA^{Trp}) and 16S rRNA genes. mtDNA has been inserted into human nuclear DNA several times during evolution and is detectable mostly as pseudogenes. It has been estimated that mtDNA sequences are translocated into the nuclear chromosomes at the high frequency of 1×10^{-5} per cell generation. With the aid of biolistic transformation, any type of DNA sequence can be incorporated into organellar genetic material, including mtDNA. Plant mitochondrial tRNAs may be encoded by mtDNA, nuclear DNA, or plastid DNA directly, or by plastid DNA inserted into mtDNA. Plant mitochondria may code for some mitochondrial ribosome proteins. These ribosomal proteins may be different in different plant species. The liverwort mtDNA genome is very large and contains at least 94 genes.

Human mtDNA also used for evolutionary studies to trace the origin of the present-day human population to a common female ancestral line, the so-called phylogenetic Eve. Restriction fragment-length polymorphism carried out on the mtDNA of 147 people, representing African, Asian, Australian, Caucasian, and New Guinean populations, indicated an evolutionary tree by the maximal parsimony method. Accordingly, it appeared that Eve lived about 200,000 years ago in Africa because that population was the most homogeneous in modern times and other populations shared the most common mtDNA sequences with these samples. The advantage of mtDNA for these studies was that mitochondria are transmitted through the egg and thus recombination with males would not alter its sequences. Later studies have shown, however, that in mice (and probably in humans) a small number of mitochondria are transmitted through the sperm. Therefore, some of the conclusions of the original study regarding the time and population scales may require revision; the basic ideas about the human origin may be correct. The unique descendance of males can be traced through the Y chromosome.

In vitro in heteroplasmic cells the mutant mtDNA replication could be inhibited by peptide nucleic acid complementary to the mutant sequences. Mitochondria depend on either the salvage pathway for deoxyribonucleotides or on special transporters to provide these DNA precursors. Human mitochondria have only two deoxyribonucleoside kinases. Thymidine kinase phosphorylates both deoxythymidine and deoxycytidine. Deoxyguanosine kinase phosphorylates both deoxypurine nucleosides. *See* aflatoxin; ancient DNA; bleomycin; bottleneck effect; chaperone; chondrome; cisplatin; cryptogene; cytoplasmic male sterility; D loop; DNA fingerprinting; DNA repair; DNA replication, mitochondrial; DNA 7S; Eve; evolution by base substitution; heat-shock proteins; heteroplasmy; intron homing; introns; kitetoplast; maximal parsimony; methyltransferase; mismatch repair; mitochondria; mitochondrial abnormalities; plants; mitochondrial control regions; mitochondrial disease in humans; mitochondrial genetics; mitochondrial import; MSF; mtTFA; organelle sequence transfers; pan editing; passive transport; paternal leakage; *Plasmodium*; petite colony mutations; peptide nucleic acid; plastid male transmission; polymerase chain reaction; RFLP; R loop; RNA editing; RNA 7S; RU maize; spacers, sublimon; Y chromosome. (The original human mtDNA sequences [Anderson, S., et al. 1981. *Nature* 290:457] require correction as given by Andrews, R. M., et al. 1999. *Nature Genet.* 23:147; Lang, B. F., et al. 1999. *Annu. Rev. Genet.* 33:351; Richards, M., et al. 2000. *Am. J. Hum. Genet.* 67:1251; Holt, I. J., et al. 2000. *Cell* 100:515; Moraes, C. T. 2001. *Trends Genet.* 17:199; for the evolutionary sequences of mtDNA, see <http://megasun.bch.umontreal.ca/gobase/>.)

MTF mouse tissue factor is a membrane protein initiating blood clotting.

mtFAM Mitochondrial transcription factor (Tfa, TCF6) is a high-mobility-group nuclear gene product controlling transcription and replication in the mammalian mitochondria and possibly affecting transcription in the nucleus. It is also indispensable for embryogenesis. Heart-specific local mutation (deletion) in mtFAM in mice leads to the expression of heart anomalies appearing in Kearns-Sayre syndrome in humans. *See* high-mobility-group proteins; mitochondrial diseases in humans; mitochondrial genetics; mtDNA.

MTHF *See* photolyase.

Mtj1 Transmembrane murine chaperone with homology to DnaJ and Sec63. *See* DnaJ; Sec63.

MTOC microtubule-organizing centers are formed early in the 16-cell stage within the cyst that gives rise to the primary oocyte and nurse cells. *See* centromere; centrosome; maternal effect genes; microtubule-organizing center; oocyte; spindle; spindle pole body. (Rieder, C. L., et al. 2001. *Trends Cell Biol.* 11:413.)

MtODE Mitochondrial AP lyase that cleaves double-stranded DNA at 8-oxoguanine sites and repairs oxidative damage. *See* AP lyase; oxidative DNA damage. (Croteau, D. L., et al. 1997. *N. j. Biol.Chem.* 272:27338.)

mtPTP Mitochondrial transition pore is an opening on the mitochondrial inner membrane through which mitochondrial proteins can be released to the cytosol and may initiate, e.g., apoptosis. *See* AIF; apoptosis.

M-TROPIC (e.g., virus) Homes on macrophages. *See* tropic.

MTS1 Multiple tumor suppressor is a gene that encodes an inhibitor (p16) of the cyclin-dependent kinase-4 protein. When it is missing or inactivated by mutation, cell division may be out of control. MTS1 is implicated in melanoma and pancreatic adenocarcinoma. *See* melanoma; p16; pancreatic adenocarcinoma; tumor suppressor.

MTT dye reduction assay Measures cell viability by spectrophotometry.

mtTF Mitochondrial transcription factor. *See* mtDNA; transcription complex; transcription factors.

MtTGENDO Oxidative damage-specific mitochondrial AP lyases. It cleaves TG mismatches in double-stranded DNA. It does not recognize 8-oxodeoxyguanine or uracil sites. *See* AP lyase; MtODE; oxidative DNA damage. (Stierum, R. H., et al. 1999. *J. Biol. Chem.* 274:7128.)

Mu (*Mutator*) Transposable element system of maize increases the frequency of mutation of various loci by more than an order of magnitude (10^{-3}–10^{-5}). About 90% of the induced mutations induced carry a *Mu* element. Their copy number in the genome may be 10–100. The element comes in various sizes, but the longest ones are less than 2 kb. The shorter elements appear to be originated from the longer ones by internal deletions. There are relatively long (0.2 kb) inverted repeats at the termini, and 9 bp direct repeats are adjacent to them. The inverted terminal repeats are conserved among the at least five different classes of *Mu* elements, although the sequences in between them may be quite different. The *Mu* elements appear to transpose (in contrast to other maize transposable elements) by a replicative type of mechanism. The two best studied forms, *Mu1* (1.4 kb) and *Mu1.7* (1.745 kb), were identified in circular extrachromosomal states. When *Mu* is completely methylated, the mutations caused by it become stable; less-than-complete methylation of the element and some other sequences (e.g., histone DNA) is associated with mutability . *Mu* is regulated by MuDR regulatory transposon (4.9 kb) carrying the *mudrA* and *mudrB* genes with transcripts of 2.8 kb and 1.0 kb, respectively. MuDR element frequently suffers deletions limited to the MURB region. Mutator-like elements (MULEs) with substantial variations in structure and size have been discovered in *Arabidopsis*. These 22 mutator elements are normally dormant in Arabidopsis, but a mutation that decreases DNA methylation (DDM1) activates them. *See* controlling elements; hybrid dysgenesis; insertional mutation; methylation of DNA; transposable elements. (Singer, T., et al. 2001. *Genes Dev.* 15:591; Miura, A., et al. 2001. *Nature* 411:212.)

Mu bacteriophage 37 kbp temperate bacteriophage. Its linear double-stranded DNA is flanked by 5 bp direct repeats. Transcription of the phage genome during the lytic phase requires the *E. coli* RNA polymerase holoenzyme. During replication it may integrate into different target genes of the bacterial chromosome and thus may cause mutations (as the name indicates). Mu may integrate in more than one copy. If the orientation of the prophage is the same, it may cause deletion; or if the orientation of the two Mu DNAs is in reverse, inversion may take place by recombination between the transposons. The phage carries three terminal elements at both left and right ends where recombination with the host DNA takes place. There is a transpositional enhancer (internal activation sequence) of about 100 bp at 950 bp from the left end. This left enhancer–right complex is called LER. The 75 kDa transposase binds to the two ends and to the enhancer. A complex of nucleoproteins, the transposome, mediates the transposition. The transposome includes 4 subunits of the MuA transposase, and each contains a 22 bp recognition site. These recognition sites — in case there is a shortage of Mu DNA — may recruit non-Mu DNA and then can transpose it. In the presence of bacterial-binding proteins (HU and IHF), a stable synaptic complex (SSC) is formed. Then Mu is cleaved at the 3′-ends by the transposase and they form the cleaved donor complex (CDC). Transesterification at the 3′-OH places Mu into the host DNA. In the strand transfer complex (STC), the 5′-ends are still attached to the old flanking DNA, but the 3′-ends are joined to the new target sequences and a cointegrate is generated by replication, or nucleolytic cleavage separates Mu from the old flanks, then the gaps are repaired and the transposition is completed. *See* cointegrate; mini-Mu; mutator phage; transposition site; transposons. (Bukhari, A. I. 1976. *Annu. Rev. Genet.* 10:389; Abbes, C., et al. 2001. *Can. J. Microbiol.* 47[8]:722; Goldhaber-Gordon, I., et al. 2002. *J. Biol. Chem.* 277:7694, Pathania, S., et al. 2002. *Cell* 109:425.)

mucins Glycosylated protein components of the mucosa that lubricate the intestinal tract. Mutation in the genes involved may lead to cancer. (Velcich, A., et al. 2002. *Science* 295:1726.)

Muckle-Wells syndrome Dominant (1q14) inflammatory disease accompanied by fever, abdominal pain, and urticaria (red or pale skin eruptions). Progressive nerve deafness and amyloidosis may follow. *See* amyloidosis; cold hypersensitivity; deafness; urticaria familial cold.

mucolipidoses (ML) Include a variety of recessive diseases connected to defects in lysomal enzymes. ML I is a neuroaminidase deficiency; ML II is an N-acetylglucosamine-1-phosphotransferase deficiency (human chromosome 4q21-q23). This enzyme affects the targeting of several enzymes to the lysosomes. Congenital hip defects, chest abnormalities, hernia, and overgrown gums but no excessive excretion of mucopolysaccharides is observed. ML III is apparently allelic to ML II, although the mutant sites seem to be different. MP III (pseudo-Hurler polydystrophy) has similarities to Hurler syndrome (*see* mucopolysaccharidosis), although the basic defects are not identical. MP III A is a heparan sulfate sulfatase deficiency (*see* mucopolysaccharidosis; Sanfilippo syndrome A), whereas MP IIIB is basically Sanfilippo syndrome B (*see* mucopolysaccharidosis). MP IV (19p13.2-p13.3) is a form of sialolipidosis with typical lamellar body inclusions in the endothelial cells that permit prenatal identification of this disease, which may cause early death. *See* Hurler syndrome; mucopolysaccharidosis.

mucopolysaccharidosis (MPS)   Hurler syndrome (*MPS I*) recessive 22pter-q11 deficiency of α-L-iduronidase (IDUA) resulting in stiff joints, regurgitation in the aorta, clouding of the cornea, etc. The latest chromosomal assignment of MPS1 is 4p16.3. The IDUA protein (about 74 kDa) includes a 26-amino-acid signal peptide. The *MPS II* (*Hunter syndrome*, I-cell disease) is very similar to *MSP I*, but it is located in human chromosome Xq27-q28. The symptoms in this iduronate sulfatase deficiency are somewhat milder and the clouding of the cornea is lacking. Dwarfism, distorted face, enlargement of the liver and spleen, deafness, and excretion of chondroitin sulfate and heparitin sulfate in the urine are additional characteristics. The following MPSs are controlled by autosomal-recessive genes and the deficiencies involve: *MPS IIIA* (Sanfilippo syndrome A; human chromosome 17q 25.3), heparan sulfate sulfatase; *MPS IIIB* (Sanfilippo syndrome B, human chromosome 17q21), N-acetyl-α-D-glucosaminidase; *MPS IIIC*, acetylCoA:α-glucosaminide-N-acetyl-transferase; *MPS IIID* (Sanfilippo syndrome D, 12q14), N-acetylglucosamine-6-sulfate sulfatase; *MPS IVA* (Morquio syndrome A, 16q24.3), galactosamine-6-sulfatase; *MPS IVB* (Morquio syndrome B), β-galactosidase; *MPS VI* (Maroteaux-Lamy syndrome, 5q11-q13), N-acetylgalactosamine-4-sulfatase (arysulfatase B); *MPS VII* (Sly syndrome, 7q21.11), β-glucuronidase; *MPS VIII* (DiFerrante syndrome), glucosamine-3-6-sulfate sulfatase. Hurler syndrome and Sly disease may be treated by implantation of neo-organs, tissues of skin fibroblasts secreting β-glucuronidase, or α-L-iduronidase, respectively. *See* arylsulfates; coronary heart disease; deafness; eye diseases; geleophysic dysplasia; glucuronic acid; heparan sulfate; iduronic acid; miniorgan; mucolipidoses.

mucosal immunity   The mucosal membranes (in the gastrointestinal tract, nasal and respiratory passages, and other surfaces that have lymphocytes) are supposed to trap 70% of the infectious agents. The mucosal cells rely on immunoglobulin A (IgA) rather than IgG used by the serum. In the gastrointestinal system Peyer's patches are located that trap infectious agents and pass them to the antigen-presenting T and B cells. The B cells generate the IgA antibodies. Concomitantly, the released antigen may stimulate the formation of IgG. Providing attenuated forms of the pathogen (e.g., *Vibrio cholerae*) orally can trigger the mucosal immunity system. Another approach is to deliver the antigens by bacteria that more effectively stimulate both IgA and IgG production simultaneously. Other delivery systems may use a biodegradable poly(DL-lactide-co-glycolide), PLG, or liposomes. The oral polio vaccine has been quite successful because it provides lasting protection, but most oral vaccines

are short in this respect. *See* antigen; antigen-presenting cell; B cell; immune system; immunoglobulins; Peyer's patch; plantibody; T cell; vaccines. (Simmons, C. P., et al. 2001. *Semin. Immunol.* 13[3]:201.)

mucosulfatidosis   Multiple sulfatase deficiency with excessive amounts of mucopolysaccharides and sulfatides in the urine. It results in abnormal development and early childhood death. Some of the symptoms are shared by leukodystropy, Maroteaux-Lamy, Hunter, Sanfilippo A, and Morquio syndromes as well as by ichthyosis.

Mud   Transposon modified (derived) from bacteriophage Mu. *See* Mu bacteriophage.

MudPIT   Multidimensional protein identification technology is a method for the preparation of complex protein mixtures for mass spectrometric analysis by digesting the cell lysate by endoproteinase lysC and then further purified. *See* mass spectrum; MALDI-TOF; proteomics. (Smith, R. D., et al. 2002. *Proteomics* 2:513.)

Muenke syndrome (nonsyndromic coronal craniosynostosis)   Primarily a head bone fusion disorder that sometimes affects finger and toe development. It is generally caused by a mutation affecting the proline[250] site in the fibroblast growth factor receptor 3 (FGFR3) gene at 4p16.3. Its incidence is higher in women than in men. *See* craniosynostosis syndromes.

Muir-Torre syndrome   Familial autosomal-dominant disease involving skin neoplasias apparently due to hereditary defects in the genetic (mismatch) repair system. *See* colorectal cancer; Gardner syndrome; hereditary nonpolyposis colorectal cancer; mismatch repair; polyposis hamartomatous.

mulatto   Offspring of white and black parentage. *See* miscegenation.

mulberry (Morus spp)   Fruit tree; $x = 14$. *M. alba* and *M. rubra* are $2n = 28$, whereas the Asian *M. nigra* is $2n = 38$.

mule   *See* hinny.

Mulibrey dwarfism   *See* dwarfism (Mulibrey nanism).

Mulibrey nanism (MUL, 17q22-q23)   Autosomal anomaly involving all or some of these characteristics: low birth weight, small liver, brain, eye, growth, triangular face, yellow eye dots, etc. The basic defect is in a zinc-finger protein (RBCC) containing a ring finger, a B box, and a coiled-coil region. It appears to be a peroxisomal disorder. *See* B box; coiled coil; peroxisome; ring finger; stature in humans; zinc finger. (Kallijärvi, J., et al. 2002. *Am. J. Hum. Genet.* 70:1215.)

Müllerian ducts   Gonadal cells begin to develop before the mouse embryo becomes 2-weeks old. From the unspecialized primordial cells in the male the Wolffian ducts develop and in the female the Müllerian ducts. Initially, however, both sexes form both of these structures and appear bisexual, but later, according to sex, one or the other degenerates. The Müllerian-inhibiting substance (MIS, Jost factor) causes the degeneration of the Müllerian ducts. This is a member of TGF-β (transforming growth factor family of proteins) or AMH (anti-Müllerian hormone), and it is encoded in human chromosome 19p13-p13.2. A defect in the AMH receptor causes pseudohermaphroditism with uterine and oviductal tissues in males observed in PMDS (persistence of Müllerian duct syndrome). MIS regulates NFκB signaling and breast cancer cell growth in vitro. In mammals, Wnt-4 signaling is required for ovarian morphogenesis. *See* gonad; pseudohermaphroditism; sexual dimorphism; SF-1; SRY; TGF; Wolffian ducts. (Segev, D. L., et al. 2001. *J. Biol. Chem.* 276:26799.)

Müllerian inhibitory substance (Jost factor) *See* Müllerian duct.

Müllerian mimicry Two monomorphic species share morphological similarities signaling a defense (e.g., distastefulness, poisonousness) to predators and both benefit from the trait. Similar mimicry occurs in various species. *See* Batesian mimicry. (Kapan, D. D. 2001. *Nature* 409:338.)

Muller 5 technique *See Basc.*

Muller's ratchet Genetic drift can lead to the accumulation of deleterious mutations, particularly in asexual populations. By each mutation the ratchet (a toothed wheel) may click by one notch. The expected time of losses of individuals with a successive minimal number of mutations depends on the absolute number of individuals with a minimal number of mutations: $N_m = q_m$, where q is their expected frequency and N is the effective population size. Muller's ratchet may operate during the evolution of transposons and retroviruses; fix shorter sequences than the initial elements; and establish elements that would depend on trans-acting elements for transposition. In case the number of nondeleterious mutations is small (n_0), the equilibrium value $n_0 = Ne^{-\mu/s}$, where $N =$ effective population size, $\mu =$ expected number of deleterious mutations per genome, $s =$ selection coefficient, and $e =$ base of natural logarithm. Conversely, the rapid mutations of retroposons may eliminate elements that would lose their transposase and thus can escape Muller's ratchet. Deleterious mutations may be eliminated by reversions or compensating new mutations or by genetic recombination. *See* genetic drift; retrovirus; transposable elements; Y chromosome. (Gabriel, W. & Bürger, R. 2000. *Evolution Int. J. Org. Evolution.* 54:1116, Gordo, I. & Charlesworth, B. 2000. *Genetics* 156:2137.)

multibreed Population including pure-bred and cross-bred groups. *See* cross-breeding; pure breeding.

multicase (multiple computer-automated structure evaluation) Relies on information of molecular fragments as descriptors of potential biological (carcinogenic) activity of chemicals. It compares the structure-activity relationship among many compounds in order to find commonality. *See* biophore; CASE; SAR. (Cunningham, A. R., et al. 1998. *Mutation Res.* 405:9.)

multicellular organisms The cells of an individual are coordinated for different functions and situated closely enough to ensure interaction. A bacterial colony is formed from many cells, but these cells are not coordinated even when they display patterned growth. The multicellular condition apparently evolved repeatedly during evolution. Evolution of multicellularity involved signal transmission and reception and localized expression of specialized genes. Their main advantages are in feeding and dispersion. *See* quorum sensing. (Kaiser, D. 2001. *Annu. Rev. Genet.* 35:103.)

multicompartment virus Each individual viral particle may carry only part of the total genome and complementation between the particles can provide the full function. Such viral strategy permits a combinatorial advantage to the virus.

multicomponent virus Its genome is segmented, i.e., its genetic material is in several pieces similarly to the chromosomes in eukaryotic nuclei. *See* bacteriophages.

multicopy plasmids Have several copies per cell. *See* plasmid types.

multidrug resistance (MDR/ABCB1) Mediated by the multidrug transporter (MDT, 1,280 amino acids) phosphoglycoprotein that regulates the elimination (or uptake) of chemically different drugs from mammalian (cancer) cells in an ATP-dependent manner. In cancer cells, the MDR gene is generally amplified as episomes and double minutes. The MDR gene may be used for gene therapy by protecting the bone marrow from the effects of cytotoxic cancer drugs. The MDT gene controls a very broad base of drug resistance. MRP (multidrug resistance–associated protein) is a glutathione conjugate protein belonging to the same family as MDR, and it may be expressed in cell lines where MDR function is usually limited. The bacterial multidrug resistance gene, *LmrA*, is very similar to the human gene, and it is expressed in human cells when introduced by transformation. Besides gene mutation, MDR can be acquired by rearrangement of the genome (aneuploidy) in cancer cells. *See* ABC transporters; aldehyde dehydrogenase; amplification; chemosensitivity; DHFR; double minute; episomes; glutathione-S-transferase; MGMT; multiple drug resistance; P-glycoprotein; SOD. (Hipfner, D. R., et al. 1999. *Biochim. Biophys. Acta* 1461:359; Rosenberg, M. F., et al. 2001. *J. Biol. Chem.* 276:16076; Duesberg, P., et al. 2001. *Proc. Natl. Acad. Sci. USA* 98:11283.)

multifactorial cross The mating is between parents that differ at multiple gene loci.

multiforked chromosomes In bacteria, DNA replication may start again before the preceding cycles of DNA synthesis have been completed, thus displaying multiple replication forks. *See* DNA replication; replication, bidirectional; replication fork. (Diagram from N. Sueoka 1975. *Stadler Symp.* 7:71.)

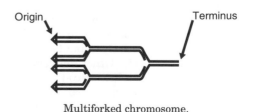

Multiforked chromosome.

multifunctional proteins   *See* one gene–one enzyme.

multigene family   Clusters of similar genes evolved through duplications and mutations and display structural and functional homologies. Some members of the families may be at different locations in the genome and some may be pseudogenes. *See* pseudogenes.

multigenic   *See* polygenic inheritance.

multilocus probe (MLP)   In DNA fingerprinting, alleles of more than one locus are simultaneously examined. *See* DNA fingerprinting.

multilocus sequence typing   Detects allelic variations (mutations and recombination) within about 450 bp internal sequences of seven bacterial housekeeping genes. Since the integrity of housekeeping gene function is generally vital, most of the detected alterations are neutral. *See* DNA typing; housekeeping genes. (Feil, E. J., et al. 2000. *Genetics* 154:1439.)

MULTIMAP   Computer program for linkage analysis using lod scores. *See* lod score.

multimeric proteins   Have more than two polypeptide subunits.

multinomial distribution   May be needed to predict the probability of proportions:

$$P = \frac{n!}{r_1!r_2!\ldots r_z!}(a)^{r_1}(b)^{r_2}\ldots(z)^{r_z}$$

where P is the probability; $r_1, r_2 \ldots r_z$ stand for the expected numbers in case of $a, b \ldots z$ theoretical proportions, and n = total numbers. Example: In case of codominant inheritance and expected 0.25 AA, 0.50 AB, and 0.25 BB, the probability in an AB × AB mating that we would find F2 among 5 progeny 2AA, 2AB, and 1 BB is:

$$P = \frac{5!}{2!2!1!}(0.25)^2(0.50)^2(0.25)^1 = \frac{120}{4}(0.0625)(0.25)(0.25)$$
$$= 30(0.00391) \approx 0.117.$$

See binomial distribution; distribution.

multiparental hybrid   Can be generated by fusion of two or more different embryos of different parentage at the (generally) 4–8 cell stage, then reimplanted at (generally) the blastocyst stage into the uterus of pseudopregnant foster mothers. Such hybrids may appear chimeric if the parents were genetically different and can be used advantageously for the study of development. *See* allophenic; pseudopregnant.

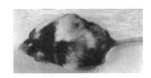

(Photo modified after R. M. Petters & C. L. Markert.)

multiparous   Animal species that usually gives birth to multiple offspring by each delivery; a human female that gives birth to twins at least twice.

multipaternate litter   Produced if the receptive multiparous female mates during estrus with several males. *See* estrus; last-male sperm preference; superfetation.

multiple alleles   More than two different alleles at a gene locus. The number of different combinations at a particular genetic locus can be determined by the formula $[n(n + 1)]/2$, where the number of different alleles at the locus is n. *See* allelic combinations.

multiple birth   *See* twinning.

multiple cloning sites (MCS)   *See* polylinker.

multiple crossovers   Genetically detectable only when the chromosomes are densely marked. In the absence of interference, the frequency of multiple crossovers is expected to be equal to the products of single crossovers. *See* coincidence; mapping; mapping function.

multiple drug resistance (MDR)   Based on several mechanisms such as active detoxification system, improved DNA repair, altered target for the drugs, decreased uptake, increased efflux, and inhibition of apoptosis. The MDR genes control multiple drug resistance in mammals. The 27-exon MDR1 locus (human chromosome 7q21.1) encodes a phosphoglycoprotein and controls drug transport and drug removal (hydrophobic vacuum cleaner effect). A repressor binding to a −100 to −120 GC-rich sequence regulates the MDR1 gene. At −70 to −80 is a Y box (binding site of the NF-Y transcription factor). This area overlaps with the binding site of Sp1. There is also a 13 bp sequence (IN) surrounding the transcription initiation site and promoting the accuracy of transcription. If tumors have high expression of this gene, the prognosis for chemotherapy is not good. Upon drug treatment, the activity of the MDR protein may increase several-fold. MDR activity can be suppressed by some calcium channel blockers, some antibiotics, steroids, detergents, antisense oligonucleotides, anti-MDR ribozymes, etc. Non-P-glycoprotein-mediated MDR has also been restricted by MRP (a multiple drug resistance–related glycoprotein), encoded in human chromosome 16p13.1. The MDR gene targeted to special tissues of transgenic animals by tissue-specific promoters may protect these tissues (e.g., bone marrow) during cancer chemotherapy. Also, MDR transgenic cells may display selective advantage and thus be advantageous during the curing process. *See* cancer therapy; chemotherapy; multidrug resistance; promoter; Sp1; transcription factors. (Gottesman, M. M., et al. 1995. *Annu. Rev. Genet.* 29:607; Litman, T., et al. 2001. *Cell. Mol. Life Sci.* 58[7]:931.)

multiple endocrine neoplasia (MEN)   *See* endocrine neoplasia.

multiple epiphyseal dysplasia (MED)   Osteochondrodysplasia causing short stature and early-onset osteoarthritis.

Mutations in a cartilage oligomeric matrix protein (a pentameric 524 kDa glycoprotein, COMP, 19p13.1) and in collagen IX (COL9A2, 1p33-p32.2, COL9A3, 20q13.3) cause MED. *See* collagen; COMP; Ehlers-Danlos syndrome; osteoarthritis; osteochondromatosis; pseudoachondroplasia; thrombospondin. (Briggs, M. D., et al. 1995. *Nature Genet.* 10:330.)

multiple hamartoma syndrome (MHAM, PTEN, MMAC1) Hamartomas are groups of proliferating, somewhat disorganized, mature cells occurring under autosomal-dominant control on the skin, in breast cancer, thyroid cancer, mucous membranes, and gums but my be found as polyps in the colon or other intestines. The lesions may become malignant. The symptoms may be associated with a number of other defects. Cowden disease was assigned to human chromosome 10q22-q23. Bannayan-Riley-Zonona syndrome resembles Cowden disease and maps to the same location. The molecular basis is a dual-function phosphatase with similarity to tensin. Another form involving megalocephaly and epilepsy symptoms is Lhermitte-Duclos disease. *See* Bannayan-Riley-Zonona syndrome; breast cancer; cancer; PTEN; tensin. (Backman, S. A., et al. 2001. *Nature Genet.* 29:396; Kwon, C.-H., et al. Ibid., p. 404.)

multiple hit The mutagen causes mutations at more than one genomic site. *See* kinetics.

multiple myeloma Bone marrow tumor–causing anemia, decrease in immunoglobulin production, frequently accompanied by the secretion of Bence-Jones proteins. Multiple myeloma may be genetically determined as it may appear in a familial manner. In some forms that have independent origin, the protein may show some variations within the same family or even within the same person. It can also be induced in isogenic mice by injection of paraffin oil into the peritoneal cavity. Such animals may produce large quantities of the light-chain immunoglobulin that can be subjected to molecular analysis. The globulins are not necessarily monoclonal, although from single transformation essentially similar molecules are expected. Multiple myeloma is also called amyloidosis encoded by a gene in human chromosome 11q13. Translocation of the IgH (14q32) immunoglobulin H gene to the amyloidosis locus or to fibroblast growth factor receptor 3 (4p16.3) may evoke multiple myeloma. Lymphocyte-specific interferon regulatory factor 4 (6p25-p23) may also be involved. The cyclin D1 protein, mediating cell cycle events and the expression of oncogenes, is also at the 11q13 location. *See* Bence-Jones proteins; cyclin D; immunoglobulin; interferon; macroglobulinemia; monoclonal antibody. (Kuehl, W. M. & Bergsagel, P. L. 2002. *Nature Rev. Cancer* 2:175.)

multiple sclerosis (MS) Disease caused by a loss of myelin from the nerve sheath or even defects of the gray matter. It is called multiple because it frequently is a relapsing condition involving incoordination, weakness, and abnormal touch sensation (paresthesia) expressed as a feeling of burning or prickling without an adequate cause. The exact genetic basis is not clear. It may be determined polygenically or it may be recessive — in both cases with much reduced penetrance. It was suggested that it may be associated with defects of the HLA system, but it has not been proven. There is molecular evidence that in some forms α-B-crystallin, a small heat-shock protein, is formed as an autoantigen. Viral initiation has also been considered. MS is basically an autoimmune disease. The controls may reside in the thymus where the T lymphocytes develop. Damage in the white matter of the central nervous system may be responsible. It appears that MS is induced by damage to the blood-brain barrier (BBB) caused by inflammatory cytokines (TFN, IFN-γ). Among close relatives the incidence may be 20 times higher than in the general population. The concordance between monozygotic twins is about 30%. HLA DR2 carriers have a four-fold increased relative risk. Females have almost twice the risk of males in the relapsing type, but male affliction is slightly more common in progressive MS. The prevalence in the U.S. is about 1/1,000, but it varies by geographical areas. The onset is between ages 20 and 40. Some of the symptoms may be shared with other diseases, and conclusive identification requires laboratory analysis of myelin. Synonym: disseminated sclerosis. MS is under the control of several genes, yet linkage to 5p14-p12 with a lod score of 3.4 was found. Pelizaeus-Merzbacher disease, a late-onset multiple sclerosis–like disorder with an autosomal-dominant expression, has frequently been confused with MS. Two principal regions in human chromosomes 17q22 and 6p21 (in the HLA area) epistatically control the susceptibility to MS. *See* Addison-Schilder disease; autoimmune disease; BBB; CD45; concordance; epilepsy; heat-shock protein; HLA; IFN; lod score; MHC; myelin; neurological diseases; neuromuscular diseases; Pelizaeus-Merzbacher syndrome; TFN; twinning. (Klein, L., et al. 2000. *Nature Med.* 6:56; Smith, T., et al. 2000. *Nature Med.* 6:62; Barcellos, L. F., et al. 2001. *Nature Genet.* 29:23; Schmidt, S., et al. 2002. *Am. J. Hum. Genet.* 70:708; Keegan, B. M. & Noseworthy, J. H. 2002. *Annu. Rev. Med.* 53:285.)

multiple translocations *See* translocation complex.

multiple tumor suppressor *See* MTS1.

multiplex amplifiable probe hybridization Uses PCR and electrophoresis for the detection of small deletions and duplications within genes. *See* (White, S., et al. 2002. *Am. J. Hum. Genet.* 71:365.)

multiplexing Several pooled DNA (or other) samples are sequenced/processed/analyzed simultaneously to expedite the process. *See* genome project.

multiplication colony Expansion by random mating of the foundation stock of inbred rodents for experimental use. *See* foundation stock; inbred.

multiplication rule *See* joint probability.

multiplicative effect Alleles at more than one gene locus together have more than a simply additive contribution to the phenotype. *See* additive effects.

multiplicity of infection A single bacterial cell is infected by more than one phage particle. *See* double infection.

multiplicity reactivation   When bacterial cells are infected with more than one phage particle inactivated by heavy doses of DNA-damaging mutagens, the progeny may contain viable viruses because replication and/or recombination restores functional DNA sequences. *See* Weigle reactivation.

multipoint cross   More than two genes are involved.

multipolar spindle   During mitosis more than two poles exist under exceptional conditions, e.g., aneuploidy caused by fertilization by more than one sperm. Such a condition may be the result of centrosome defects in animals. Multiple centrosomes may be formed in case the tumor suppressor gene p53 is not functioning. *See* centrosome; mitosis; p53; spindle.

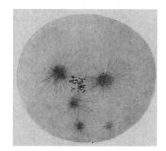

Multipolar spindle in a starfish egg.

multipotent   Capable of regeneration of different kinds of cells or tissues. *See* pluripotency; generation.

multiregional origin   *See* out-of-Africa hypothesis.

multisite mutations   Occur generally when an excessively large dose of a mutagen(s) is applied. Frequently deletions or other chromosomal aberrations are included. *See* chromosomal aberration; deletion; mutation.

multistranded chromosomes   Contain more than two chromatids such as in the polytenic chromosomes produced by repeated replication without separation of the newly formed strands. At the early period of electron microscopic studies of ordinary chromosomes, apparent multiple strands were observed, and it was assumed that each chromosome has

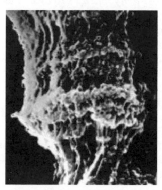

Scanning electron micrograph of a segment of polytenic salivary gland chromosome. (Courtesy of Dr. Tom Brady.)

many parallel strands of DNA. This assumption could not be validated by subsequent investigations. Each chromatid has only a single DNA double helix that is folded to assure proper packaging. *See* Mosolov model; packing ratio; polytenic chromosomes.

multivalent   Association of more than two chromosomes in meiosis I. *See* meiosis.

multivariate analysis   Has many uses in genetics when a decision is needed in classifying syndromes with overlapping symptoms and the diagnosis requires quantitation. It may also be used to classify populations with similar traits. The assumption is that the variates $X_1 \ldots X_k$ are distributed according to a multivariate normal distribution. The variance of X_i, σ_{ii} and the covariance of X_i and X_j are presumed to be identical in the two populations, but σ_{ii} and σ_{ij} are not the same from one variate to another or one pair of variates to another pair, respectively. The difference between the two means for X_i is $\delta_I = \mu_{2i} - \mu_{1i}$. Hence, the *linear discrimination function* $\Sigma L_i X_i$ provides the lowest probability of incorrect classification. The L_i values will be determined according to the procedure shown below. The δ/σ is maximized for minimization of the chance of misclassification. The *generalized squared distance* is $\Delta^2 = (\Sigma L_i \delta_i)^2 / \Sigma\Sigma L_i L_j \sigma_{ij}$ and L_i is obtained by a set of equations: $\sigma_{11}L_1 + \sigma_{12}L_2 \cdots + \sigma_{1k}L_k = \delta_1, \sigma_{k1}L_1 + \sigma_{k2}L_2 + \cdots + \sigma_{kk}L_k = \delta_k$. *See* covariance (at correlation entry); discriminant function; Euclidean distance.

multivariate normal distribution $F(x_1, x_2 \ldots x_q) = (2\pi)^{-q/2} |\Sigma|^{1/2} \exp{-1/2(x - \mu)' \Sigma^{-1}(x - \mu)}$.

MUM   Maximum unique match of the nucleotides in a DNA sequence.

mummy   Dried animal or human body preserved by chemicals and/or desiccation. It may contain protein (blood antigens) and DNA sequences to carry out limited molecular analysis using PCR technology. *See* ancient DNA; ancient organisms; ice man.

MUNC18   Mammalian homologue of the Unc18 protein of *Caenorhabditis* that binds syntaxin. *See* syntaxin.

mung bean nuclease   Single-strand-specific endonuclease.

Mu printing   Analysis of the pattern of integration sites of Mu phage in large populations of bacteria. *See* insertional mutation; Mu bacteriophage.

Murashige & Skoog medium (MS1)   Suitable for plant tissue culture for growing callus and different plant organs. Composition mg/L: NH_4NO_3, 1650; KNO_3, 1900; $CaCl_2.2H_2O$, 440; $MgSO_4.7H_2O$, 370; KH_2PO_4, 170; KI, 0.83; H_3BO_3, 6.2; $MnSO_4.4H_2O$, 22.3; $ZnSO_4.7H_2O$, 8.6; $Na_2MoO_4.2H_2O$, 0.25; $CuSO_4.5H_2O$, 0.02; $CoCl_2.6H_2O$, 0.025; ferric EDTA, 43; sucrose 3%, pH 5.7; inositol, 100; nicotinic acid, 0.5; pyridoxine.HCl, 0.5; thiamin.HCl, 0.4; indoleacetic acid (IAA), 1–30; kinetin, 0.04–10. Microelements, vitamins, and hormones may be prepared in a stock solution and added before use. For kinetin, other cytokinins may be substituted such as 6-benzylamino purine (BAP) or isopentenyl adenine (or its nucleoside); for IAA, naphthalene acetic acid (NAA) or 2,4-D (dichlorophenoxy acetic acid) may be substituted or a combination of the hormones may be used in concentrations that are best suited for the plant and the purpose of the culture. For solid media, use agar or gellan gum. Heat-labile components are sterilized by filtering through 0.45 μm syringe filters. Variations of this medium are commercially available as a dry powder ready to dissolve, but the pH needs to be adjusted. *See* agar; Gamborg medium; gellan gum; syringe filter. (Murashige, T. & Skoog, F. 1962. *Physiol. Plant.* 15:473.)

murine   Pertaining to mice (or rats).

muscarinic acetylcholine receptors   Activated by the fungal alkaloid, muscarine, a highly toxic substance that causes excessive salivation (ptyalism, sialorrhea), lacrimation (shedding tears), nausea, vomiting, diarrhea, lower than 60 pulse rate, convulsions, etc. Deficiency of the M3 receptor decreases appetite in mice and the animals stay lean. Antidote: atropine sulfate. *See* signal transduction.

muscular atrophy   Peroneal muscular atrophy may be classified as (1) demyelinating form and sensory neuropathy, (2) axonal motor and sensory neuropathy, and (3) distal hereditary neuronopathy or distal spinal muscular dystrophy. *See* Charcot-Marie-Tooth disease; dentatorubral-pallidoluysian atrophy; Kennedy disease; Machado-Joseph syndrome; olivopontocerebellar atrophy; spinal muscular atrophy; trinucleotide repeats; Werdnig-Hoffmann disease.

muscular dystrophy   Collection of anomalies primarily involving the muscle, controlled by X and autosomal-recessive or -dominant genes. The most severe form is *Duchenne muscular dystrophy* (DMD) in human chromosome Xp21.2 (12q21). Its transmission is through the females because the affected males do not reach the reproductive stage. Similar defects have been observed in animals. The prevalence is about 3×10^{-4}. The frequency of female carriers is about twice as high as the affliction of males. The onset is around age 3. Within a few years, the affected persons fail to walk and usually die by age 20. Mental retardation is common in this disease. The earliest diagnosis may be made by an abnormally high serum creatine phosphokinase (CPK) level at birth. Prenatal diagnosis is feasible and successful in about 90% of the cases. The dystrophin gene is one of the largest human genes; more than 2 megabases (about half of the size of the entire genome of *E. coli*) may be involved. This gene includes many introns with an average size of about 16 kb, whereas about 79 exons average in size of about only 50 kb. In a tissue-specific manner, it may choose among 8 promoters located at different positions. The full-length dystrophin is ~427 kDa, but the R dystrophin is 260, the B 140, the S 114, and the G 71 kDa. The pertinent promoter used 5′ downstream determines the length of the protein. All have the same C-terminal domain, except the 40 kDa apodystrophin-3, which has a truncated C-terminal

synthrophin-binding domain. The gene in the muscles appears to use a promoter other than in the brain. The majority of the cases are deletions and other chromosomal defects, yet gene mutations (predominantly frameshifts) are common because of the large size of the genes. Intragenic recombination frequency was estimated to be high at 0.12. Dystrophin connects F-actin through its N-terminus, and its C-terminus binds to a complex of glycoproteins; through them to the extracellular matrix and the sarcolemma.

In DMD the dystrophin protein may be entirely missing, whereas in the milder form of the disease, the *Becker type* (BMD), the dystrophin protein is just shorter. With antisense technology (in a mouse model using 2′-O methyl oligoribonucleotides), skipping of exon 23 induction altered the reading frame and the mRNA was translated into a protein resembling that of the Becker-type dystrophin. Gene therapy significantly mitigated the symptoms of the disease (Mann, C. J., et al. 2001. *Proc. Natl. Acad. Sci.* 98:42). The frequency of BMD is about 0.1 of that of DMD, and patients may live to age 35 and thus may have children. Apparently allelic genes code for BMD and DMD. Deletions in the DMD gene may affect genes proximal to the centromere: CGD (chronic granulomatous disease), XK (McLeod syndrome), or XP6 (retinitis pigmentosa), and distally GKD (glycerol kinase) or AHC (adrenal hyperplasia).

The two forms of *Emery-Dreifuss dystrophy* affecting the shoulder muscles (scapulohumeral dystrophy) are autosomal (1q21-q23, encoding lamins A and C) and X-linked (Xq28, encoding emerin). In *limb-girdle muscular dystrophy* (LGMD, 3 dominant, 8 recessive), the defects are in the sarcoglycan subunits; α-sarcoglycan is encoded in chromosome 17q12-q21, the β-subunit is coded in chromosome 4q12, whereas the γ-subunit was localized to 13q12. The human chromosome 4q45 region harbors the genes, which are misregulated in FSHD (*facioscapulohumoral muscular dystrophy*). LGMD1A (human chromosome 5q) and LGMD2 (human chromosome 1q11-q21) are both the dominant forms. LGMDC (3q25) involves a defect in the plasma membrane protein caveolin-3. Chromosomes 2p13, 5q22.3−31.3, and 15q15 also encode LGMD subunits. Recessive (chromosome 9q31) *Fukuyama congenital muscular dystrophy* (FCMD) has a prevalence of ~10^{-4} in Japan. It involves polymicrogyria (numerous small convolutions of the brain) due to defects of migration of the neurons and involves mental retardation. The mutation is caused by a ~3 kbp insertion of a retrotransposal tandem repeat at the 3′-untranslated sequences of a gene encoding a secreted protein, fukutin.

Recessive *Miyoshi myopathy* (2q13.2) has an early adulthood onset due to a defect in the 6.9 kb cDNA encoding the dysferlin protein (named after dystrophy and fertility in *Caenorhabditis*). *Oculopharyngeal muscular dystrophy* (OPMD), a late adulthood−onset disease of swallowing, drooping eyelids, and limb weakness, is caused by 8−13 expansion of a (GCG)6 tract specifying polyalanine encoded by the poly(a)-binding protein 2 gene (PABP2) at human chromosome 14q11-q13. Symptoms of muscular dystrophies display considerable variation in onset and severity of expression. The face and shoulder (fasciscapulohumeral) dystrophy is limited to the named body parts. The gene was assigned to human chromosome 4q35. The congenital dystrophy gene causes deficiency in laminin; α2-chain (merosin) around the muscle fibers was located to human chromosome 6q22-q23. *Ullrich* (scleroatonic)*muscular dystrophy* is a recessive defect in collagen 6 (COL6A1, COL6A2, 21q22.3, COL6A3, 2q37). Other types of muscular dystrophies may accompany symptoms of other human diseases. Dystrophin is a rod-like cytoskeletal protein normally localized at the inner surface of the sarcolemma (the muscle fiber envelope). Dystrophin is attached within the cytoplasm to actin filaments and to the membrane-passing β-subunit of the dystroglycan (DG) protein, and the α-subunit of DG joins the basal lamina through the laminin protein (encoded in chromosome 6q22-q23). In the membrane, DG is associated with the β- and γ-subunits of sarcoglycan (SG); α-sarcoglycan is extracellularly attached to the other two subunits.

In some LGMD patients, there is a defect in the muscle-specific protease calpain-3 encoded in chromosome 15q15. (LGMD disease is also called severe childhood autosomal-recessive muscular dystrophy or SCARMD.) In some of the SCARMDs, the defect was associated with adhalin, a 50 kDa sarcolemma dystrophin-associated glycoprotein encoded in human chromosome 17q. Bone marrow cells that are normally precursors of cartilage, bone, and some parenchyma cells may differentiate into myotubes. They have been considered as genetic therapeutic agents for the treatment of muscular dystrophy. Carriers of dystrophies can be *biochemically normalized* when from overexpressed positive cells the protein is diffused to the nearby negative cells. In case of *genetic normalization*, degenerated myonuclei are replaced by nuclei from dystrophin-positive *muscle satellite cells* (muscle stem cells).

Gene therapy is possible by the injection of integrating vectors carrying the normal genes into muscle satellite cells that may fuse with the muscle fibers. Adenovirus vectors, however, do not integrate into the host nuclei and replicate only in extrachromosomal form. Therefore, after a variable period of time, they may be lost. Retroviral vectors integrate and are stably expressed in the chromosomes. They can be introduced in vivo or ex vivo. Direct injection of supercoiled plasmid DNA may also be successful for transformation, although such DNAs are not integrated into the host genome and frequently display variable expression. Another therapeutic approach is the injection of fusogenic in vitro cultured myoblast cells that may cure the host's defect in a mosaic pattern. Upregulation of the utrophin gene (6q24, its encoded protein is 65–80% homologous to dystrophin) has also been considered as a remedy. The utrophin gene is normally active only during embryonal development, then falls dormant. *See* adenovirus; antisense technologies; apoptosis; caveolin; Charcot-Marie-Tooth disease; collagen; contiguous gene syndrome; dysferlin; dsytrophin; emerin; gene therapy; lamins; mitochondrial diseases, human; muscular atrophy; myoblast; myotonic dystrophy; myotubes; neuromuscular diseases; neuropathy; retroviral vectors; RFLP; sarcolemma; viral vectors; Walker-Wagner syndrome. (Koenig, M., et al. 1989. *Am. J. Hum. Genet.* 45:498; Arahata, K. 2000. *Neuropathology, Suppl. Dystrophin S34-S41*; Burton, E. A. & Davies, K. E. 2002. *Cell* 108:5; Durbeej, M. & Campbell, K. P. 2002. *Current Op. Genet. Dev.* 12:349; Chamberlain, J. S. 2002. *Hum. Mol. Genet.* 11:2345.)

mushroom Fruit body of basidiomycete fungi. The mushroom is generally a genetic mosaic arising from mycelial fusion of two to several genetically compatible colonies. *See* basidiomycetes; fungal life cycles.

mushroom body Central nerve (neuropil) complex in the brain; the pairs of mushroom bodies are implicated in olfactory memory recall and elementary cognitive functions. *See* memory; olfactogenetics. (McGuire, S. E., et al. 2001. *Science* 293:1330.)

musical talent Pitch perception appears to be associated with an auditory area of the brain (asymmetry of the planum temporale). The immediate and correct recognition of musical pitch, an auditory tone without an external reference, is a rare autosomal-dominant factor in the human population and is called perfect or absolute pitch (AP). The heritability of musical ability appears high (70–80%). The development of this ability is favored by early musical training. Whales and several species of birds use music somewhat similarly to humans (Gray, P. M., et al. 2001. *Science* 291:52). The neurobiological bases of human music perception and performance is discussed by Tramo, M. J. 2001. *Science* 291:54. *See* amusia; Bach; dysmelodia; pitch; prosody; Strauss. (Baharloo, S., et al. 1998. *Am. J. Hum. Genet.* 62:224; Ibid., 67:755 [2000]; Peretz, I., et al. 2002. *Neuron* 33:185; <www.provide.net/~bfield/abs_pitch.html>.)

Music of Macromolecules <Ndbserver.rutgers.edu/NDB/archives/MusicAtlas/index.html>; <geneticmusic.com/dnamusic>.

mustard gas (dichloroethyl sulfide, $[ClCH_2CH_2]_2S$) Radiomimetic, vesicant poisonous gas used for warfare in World War I. *See* nitrogen mustards; radiomimetic.

mustards (family of cruciferous plants) The taxonomy is not entirely clear. The white mustard (*Sinapis alba*) is $2n = 24$. The black mustard (*Brassica nigra*) is $2n = 2x = 16$ and supposedly the donor of its genome to the Ethiopian mustard (*B. carinata*) is $2n = 34$, $(2x\,[8+9])$, an amphidiploid that received 9 chromosomes from *B. oleracea* ($2n = 18$). The brown mustard (*B. juncea*) $2n = 36(2x\,[8+10])$ has one genome ($x = 8$) of the black mustard and another genome ($x = 10$) from *B. campestris* ($2n = 20$). The $38(2x\,(9+10)$ chromosomes of the rapes and swedes descended from *B. oleracea* ($2n = 18$) and *B. campestris* ($n = 20$). Sometimes *Arabidopsis* ($2n = 10$) is called mustard, since mustard also means crucifer.

mutability Indicates how prone a gene is to mutate; it indicates a genetic instability.

mutable gene Has a higher than usual rate of mutation. *See* DNA repair; mismatch repair; mutator; mutator genes; transposable elements.

mutagen Physical, chemical, or biological agent capable of inducing mutation. *See* biological mutagens; chemical mutagens; distal mutagen; effective mutagen; efficient mutagen;

environmental mutagens; mutagen, direct; mutagen, indirect; mutagen activation; physical mutagens; promutagen; proximal mutagen; supermutagen; triple helix formation; ultimate mutagen.

mutagen, direct Acts by modifying DNA (or RNA bases) by, e.g., alkylation or deamination, cross-linking DNA strands, and breaking the nucleic acid backbone.

mutagen, indirect Damages DNA by its metabolic products (activation of promutagens), increases reactive free radicals (ROS), affects (reduces) apoptosis, increases recombination, mutates oncogenic suppressors, etc.

mutagen assay *See* Ames test; autosomal-dominant assays; autosomal-recessive assays; Basc; bioassays in genetic toxicology; ClB.

mutagen sensitivity Is determined by DNA repair and the metabolic enzymes degrading or activating mutagens/promutagens. *See* DNA repair, promutagen. (Tuimula, J., et al. 2002. *Carcinogenesis* 23:1003.)

mutagenesis Induction and procuring of mutations either by mutagenic agents in vivo or by the use of reversed genetics. *See* cassette mutagenesis; directed mutation; insertional mutation; insertion elements; localized mutagenesis; mutation induction; oligonucleotide-directed mutagenesis; REMI; reversed genetics; RID; transposable elements. (Jackson, I. J. 2001. *Nature Genet.* 28:198.)

mutagenesis, site-selected *See* directed mutation; localized mutagenesis.

mutagenesis, site-specific *See* directed mutation; localized mutagenesis; RID; site-specific mutation; targeting genes.

mutagenicity and active genes Experimental data indicate that replicating or actively transcribed genes are preferred targets of mutagens, probably because of the decondensation of the genetic material.

mutagenicity of electric and magnetic fields A large body of the published positive results lack rigorous experimental verification and reproducibility. Some newer data do not apparently suffer from these shortcomings. (McCann, J., et al. 1998. *Mutation Res.* 411:45.)

mutagenic potency Difficult to determine for many agents that have low mutagenicity. Therefore, the genetic and carcinogenic hazards of many compounds are unknown, and possibly harmless agents may have been found mutagenic in some studies when actually their harmful effect may

not be verified by others. There are other agents that may be mutagenic and/or carcinogenic but may have escaped attention. Also, there is no perfect way to quantitate mutagenic effectiveness, particularly for human hazards, because of the different quantities to which humans may be exposed. As an example, the mutagenicity in the Ames *Salmonella* assay of a few compounds is listed in the table below. Interesting to note that ethylmethane sulfonate, probably the most widely used mutagen, is only ninth on this list, and nitrofuran, a food preservative, exceeds its effectiveness as a mutagen on a molar basis more than 3,000-fold. *See* Ames test; bioassays in genetic toxicology; environmental mutagens; mutagen assays.

Compound	Mutations /nmole	Compound	Mutations/ nmole
Caffeine	0.002	Ethidium bromide	80
EDTA	0.002	Sodium azide	150
Sodium nitrite	0.010	Acridine ICR-170	260
Ethylmethane sulfonate	0.160	Nitrosoguanidine	1375
Captan (fungicide)	25.000	Aflatoxin B-1	7057
Proflavine	38.000	Nitrofuran (AF-2)	20800

(After McCann, J., et al. 1975. *Proc. Natl. Acad. Sci. USA* 72:5135.)

mutagenic specificity Base analogs (5-bromouracil, 2-aminopurine) affect primarily the corresponding natural bases. The target of hydroxylamine is cytosine. Alkylating agents preferentially affect guanine. There are no simple chemicals, however, that would selectively recognize a particular gene or genes. With the techniques of molecular biology, specifically altered genes can be produced by synthesis and introduced into the genetic material by transformation. Gene replacement by double crossing over can substitute one allele for another. *See* Cre/Lox; gene replacement; knock-in; knockout; localized mutagenesis; synthetic genes; TAB mutagenesis; targeting genes; transformation, chimeraplasty, homolog-scanning mutageneses, TFO, zinc-finger nuclease.

Mutagen Information Center *See* databases

mutagens-carcinogens A very large fraction of mutagens (genotoxic agents) are also carcinogens, but not all carcinogens are mutagenic. The difference is based on the biology of the two events. Mutation is a single-step alteration in DNA/RNA and carcinogenesis is a multistep process. *See* bioassays in genetic toxicology. (Zeiger, E. 2001. *Mutation Res.* 492:29.)

MUTA™ Mouse Commercially available transgenic strain with the *LacZ* insertion useful for testing mutagens/carcinogens. The transgene is extracted from mice and introduced into λ-phage and subsequently into *E. coli* bacteria. In the presence of Xgal substrate, the mutant bacterial colonies appear colorless on a blue background because of the inactive galactosidase. Actually, mutations inside the mouse are ascertained outside the animals for convenience of detectability. *See* β-galactosidase; Big Blue®; Lac operon; Xgal. (Nohmi, T., et al. 2000. *Mutation Res.* 455:191.)

mutant   An individual with mutation.

mutant enrichment   *See* filtration enrichment; mutation detection; screening.

mutant frequency   Frequency of mutant individuals in a population, disregarding the time or event that produced the mutation; it is thus a concept different from mutation frequency. *See* jackpot mutation; mutation rate.

mutant hunt   Inducing /collecting/isolating particular mutants for a purpose.

mutant isolation   Basically similar procedures can be used in various microorganisms. The isolation of mutations is greatly facilitated if selective techniques are available. Revertants of auxotrophs can grow on minimal media, whereas the resistants (antibiotics, heavy metals, metabolite analogs, etc.) can be selectively isolated in the presence of the compound to which resistance is sought. Herbicide-resistant plants can be selected in the presence of the herbicide (*see* mutation detection). Auxotrophic animal cells could be isolated in the presence of 5-bromodeoxyuridine because the wild-type cells that incorporated the nucleoside analog after exposure to visible light—because of breakage of the analog-containing DNA—are inactivated. The nongrowing mutant cells fail to incorporate the analogs and thus stay alive; after they are transferred to supplemented media, they may resume growth. Feeding the cells allyl alcohol can isolate alcohol dehydrogenase mutations. The wild-type cells convert this substance to the very toxic acrylaldehyde and are killed, whereas the alcohol dehydrogenase inactive

mutants (microorganisms, plants) cannot metabolize allyl alcohol and selectively survive. *See* Ames test; filtration enrichment; fluctuation test; penicillin screen; replica plating; reverse mutation; selective medium, diagram and photo next page.

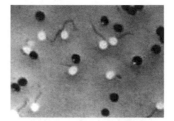

Allyl alcohol selection of alcohol dehydrogenase mutant maize pollen. The black is dead. The germinating white pollen is resistant. (Courtesy of M. Freeling, photo by D. S. K. Cheng; see also *Nature* 267:154.)

mutarotation   Change in optical rotation of anomers (isomeric forms) until equilibrium is reached between the α- and β-forms.

mutase   Enzyme mediating the transposition of functional groups.

mutasome   Enzyme complex (Rec A, UmuC-UmuD', pol III) involved in the error-prone DNA replication in translesion. *See* DNA repair; translesion pathway.

mutasynthesis   Biochemically altered mutant of an microorganism may produce modified metabolites either directly or from a precursor analog; the new product (e.g., antibiotic) may be more useful as a drug.

mutation   Heritable change in the genetic material; the process of genetic alteration. *See* auxotrophy; equilibrium mutations; forward mutation; genetic load; mutagen; mutagen assays; mutagenic potency; mutagenic specificity; mutant, isolation; mutation, beneficial; mutation, chromosomal; mutation, neutral; mutation, useful; mutation detection; mutation frequency; mutation in cellular organelles; mutation in human

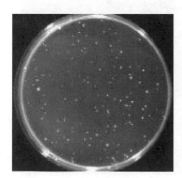 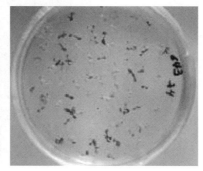 

Isolation of *Salmonella* revertants, kanamycin-resistant mutants, and thiamine-revertant *Arabidopsis* (from left to right).

mutant isolation in microorganisms

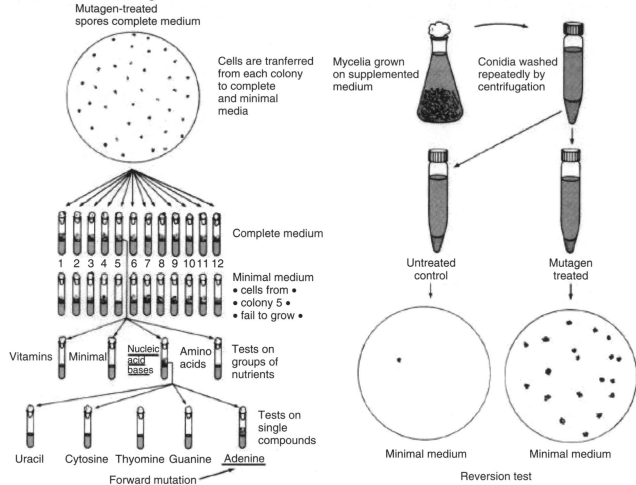

General scheme for the nonselective isolation of mutations in haploid organisms. mutagen-treated haploid cells are propagated on a complete medium (containing vitamins, amino acids, nucleic acid bases, etc.). Each colony on the petri plate (top left) is the progeny of a single cell (because of the great dilution used for plating). Inocula are transferred to both complete and minimal media (without organic supplement). Auxotrophic mutants fail to grow on minimal media but may grow on nutrients that satisfy their requirement. Thus, specific forward mutations can be isolated and identified. Reversion tests are carried out as outlined at the right part of the diagram. Only the revertants of the auxotrophic mutants will grow on minimal media.

The thiamine biosynthetic pathway is controlled by single genes as shown:

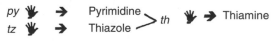

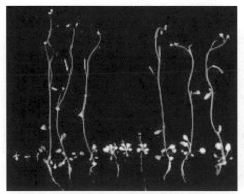

The thiazole-requiring mutants of *Arabidopsis* fail to grow on basal or pyrimidine media but display normal growth on thiazole or thiamine. (From left to right: Basal, Thiazole, Pyrimidine, Thiamine media.)

populations; mutation in multicellular germline; mutation pressure; mutation rate; mutator genes; polymorphism; reverse mutation; somatic hypermutation; somatic mutation; spontaneous mutation; transposable elements, diagram and photo next page.

mutation, adaptive *See* mutation, beneficial.

mutation, age of The age of a mutation may be inferred on the basis of breaking up the linkage disequilibrium of the ancestral haplotype by additional mutations and recombination. The task is linking the present haplotype to the haplotype of the coalescence. The identification of the ancestral haplotype is, however, generally problematic. There is no way to determine with certainty the constitution of the original haplotype. One approximation is the study of closely linked markers. The probability that the haplotype stays ancestral during the period elapsed since the mutation occurred can be statistically inferred. If we consider that p is the probability of the ancestral haplotype in a very large population in which all lineages are basically independent, then $G = -\ln(p)/r$, where G is the generations and r is the frequency of mutations and recombination. *See* coalescent; genealogy; haplotype; linkage disequilibrium. (Reich, D. E. & Goldstein, D. B. 1999. *Microsatellites*, Goldstein, D. B. & Schlötterer, C., eds., p. 129. Oxford Univ. Press, Oxford, UK; Slatkin, M. & Rannala, B. 2000. *Annu. Rev. Genomics Hum. Genet.* 1:225.)

mutation, beneficial When a new mutant appears in the population with a reproductive success of 1.01 (selective advantage 0.01), the odds against its survival in the first generation are $e^{-1.01} \cong 0.364$. Its chances for elimination by the 127th generation will be reduced to 0.973 compared to the probability of a neutral mutation that has a 0.985 chance of extinction. Ultimately, at a selective advantage of 0.01, its chance of survival will be 0.0197. According to the mathematical argument, even mutations with twice as great fitness than the prevailing wild type have a high (<13%) probability of being lost, $e^{-2} \cong 0.1353$, during the first generation. Under normal conditions the mutants' selective advantage is generally very small and the probability of its ultimate survival is $(y) = 2s$. The chance of its extinction is $(l) = 1 - 2s$. In order for a mutant to have a better than random chance (0.5) for survival, the following conditions must prevail:

$$(1 - 2s)^n < 0.5 \text{ or } (1 - 2s)^n > 2; \text{ hence, } -n \log_e(1 - 2s) > \log_e 2$$
or approximately $-n(-2s) > \log_e 2$, and therefore $n > \log_e 2 /(2s)$, i.e., $0.6931/(2s)$

If $s = 0.01$ and $n =$ number of mutations, n must be larger than $0.6931/(2 \times 0.01) = 34.655$ — that is, a mutation with 0.01 selection coefficient must occur approximately 35 times to be ultimately accepted and fixed. If the spontaneous rate of mutation at a locus is 1×10^{-6}, a population of 35 million is required for providing an adequate chance for the survival of a mutant of that type. Since the proportion of advantageous mutations is probably no more than 1 per 10,000, very few new mutations have a real chance of survival. This may be attributed to the fact that during the long history of evolution of organisms, the majority of possible mutations

have been tried, and the good ones have been adopted by the species. The chances of new mutations are favored, however, when environmental conditions, such as agricultural practice, change in pests, pathogens, predators, etc., take place for which historical adaptation does not exist. *See* adaptive evolution; fitness; mutation, neutral; mutation rate; selective advantage; selective sweep.

mutation, chromosomal Partial chromosomal losses (deletion, deficiency), duplications, and rearrangements of the genetic material (inversion, translocation, transposition) are considered chromosomal mutations. The various types of aneuploids (nullisomy, monosomy, trisomy, polysomy, etc.) or increase in the number of genomic sets (polyploidy) can also be classified as chromosomal mutations.

mutation, cost of production Varies a great deal from organism to organism and depends on the developmental stage when the mutagen is applied. Mutation induction at the gametic stage may be less expensive than at the multicellular (diploid) germline stage. In higher plants such as *Arabidopsis*, mutagens are generally applied to the mature seed and thus two generations are required for the isolation of mutants for further experimental studies. The cost of raising these two generations may not be equal and both contribute to the final cost. Although in large M_2 families there is an increased probability for recovering the mutations induced, from the viewpoint of cost-effectiveness a large number of families with minimal size is desirable. In case of recessive mutations, depending on the (n) number of individuals in M_2 families, the probability of recovery (P) of a mutant is at $n = 1(P = 0.25)$, $n = 2(P = 0.437)$, $n = 4(P = 0.683)$, $n = 8(P = 0.899)$, $n = 16(P = 0.989)$, and $n = 24(P = 0.998)$ in case of heterozygosity of the M_1 generation derived from a single cell. It is thus obvious that increasing the M_2 size from 1 to 24 involves the increase of probability of recovery only from 0.25 to 0.998.

CALCULATION OF THE COST OF MUTATION

M_2 family size →	1	2	4	8	24
$M_1 + M_2$ size →	24 + 24	12 + 24	6 + 24	3 + 24	1 + 24
Mutant Expected →	6	5.244	4.098	2.697	1

Cost M_1:M_2 ↓	Cost of 1 Mutation in Arbitrary Units Under the Conditions Shown Above and in the Left-Side Column of This Tabulation				
1:1	8	**6.845**	7.321	10.011	25
1:2	12	**11.442**	13.177	18.910	49
1:3	**16**	16.018	19.034	27.809	73
2:1	12	9.153	**8.785**	11.123	26
3:1	16	11.442	**10.249**	12.236	27
4:1	20	13.783	**11.713**	13.384	28

The cost of both generations must be included along with the effectiveness of recovery of the mutations induced. The table indicates that M_2 family sizes larger than 4 individuals generally increase the labor and the cost. The recovery of mutations is most effective if selective techniques are applicable. Unfortunately, in some cases (morphological mutants) this may not be possible. (Rédei, G. P., et al. 1984. *Mutation, Cancer and Malformation*, Chu, E. H. Y. & Generoso, W. M., eds., p. 295. Plenum, New York.)

mutation, dating of origin G (number of generations since the emergence of the mutation) can be determined in a population using the information of linkage disequilibrium (δ) concerning closely linked genetic or molecular markers and their known recombination frequencies (θ):

$$G = \frac{\log[1 - Q/(1 - pN)]}{\log(1 - \theta)} \times \frac{\log \delta}{\log(1 - \theta)}$$

where $\delta = (pD - pN)/(1 - pN)$ and pD and pN represent the two different linkage phases of the homologous chromosomes. Q (probability) $= (1 - [1 - \theta]^G)(1 - pN)$. (Guo, S. W. & Xiong, M. 1997. *Hum. Hered.* 47[6]:315; Colombo, R. 2000. *Genomics* 69:131.)

mutation, neutral A neutral mutation is neither advantageous nor deleterious to the individual homo- or heterozygous for it. Its chance of immediate loss is determined by the first term of the Poisson series, $e^{-1} :\cong 0.368$ (*see* evolution, non-Darwinian). Consequently, its chance for survival is $1 - 0.362$. The chance for its extinction during the second generation is $e^{-0.632} \cong 0.532$ and by the third generation it is $e^{-(1-0.532)} \cong 0.626$ ($e = 2.71828$). In general terms, the extinction of any neutral mutation is e^{x-1}, where x is the probability of its loss in the preceding generation. According to R. A. Fisher, by the 127th generation the odds against the survival of a neutral mutation are 0.985 and may eventually reach 100%. Several evolutionists, notably M. Kimura, have argued statistically against the conclusion of Fisher (and the Darwinian theory of the necessity of selective advantage). Accordingly, if the rate of mutation for a locus is μ, the population size is N, and the organism is diploid, the number of new mutations occurring per generation per gene is $2N\mu$. The chance for random fixation is supposed to be $(1/2)N$ because the new allele is represented only once among the total of $2N$. Accordingly, the probability of fixation is expected to be $(2N\mu) \times [(1/2)N] = \mu$, indicating that the rate of incorporation of a new neutral mutation into the population is equal to the mutation rate. The average number of generations required for fixation of a neutral mutation (according to Kimura and Ohta) is expected to be $4N_e$, where N_e = the effective population size. Thus, if the effective population size $= 1,000$ individuals, it takes 4,000 generations for a neutral mutation to be fixed in the population. In *Caenorhabditis elegans*, 65–80% of the mutations do not have a visible phenotype and thus do not seem to have much effect on fitness. In some human families, a 28 kb apparently neutral deletion was detected in chromosome 15q11-q13 area involved in imprinting (Buiting, K., et al. 1999. *Am. J. Hum. Genet.* 65:1588). On the basis of nucleotide sequences, the neutrality can be statistically estimated by the HKA formula (Hudson, R. R., et al. 1987. *Genetics* 116:153) at the constant-rate neutral model under specified conditions:

$$\chi^2 = \sum_{i=1}^{L}(S_i^A - \hat{E}[S_i^A])^2/\hat{V}ar[S_i^A] + \sum_{i=1}^{L}(S_i^B - \hat{E}[S_i^B])^2/\hat{V}ar[S_i^B]$$
$$+ \sum_{i=1}^{L}(D_i - \hat{E}[D_i])^2/\hat{V}ar[D_i]$$

where L indicates the locus (loci, l) in species A and B sequenced. S_i^A and S_i^B stand for the number of nucleotide sites that are polymorphic at locus i in the number of gametes of species A and B, respectively. D_i indicates the number of differences at locus i between a random sample of gametes of species A and B, respectively. $\hat{E}$ and $\hat{V}ar$ stand for the estimates of expectation and variance, respectively. In the past, sequence variation in mitochondrial DNA has been considered as neutral, but this assumption may not be generally valid. *See* adaptive evolution; codon usage; effective population size; fitness; imprinting; mutation, beneficial; mutation rate; non-Darwinian evolution; radical amino acid substitution; Tajima's method. (Hudson, R. R., et al. 1987. *Genetics* 116:153; Kimura, M. & Ohta, T. 1977. *J. Mol. Evol.* 9:[4]:367; Gerber, A. S., et al. 2001. *Annu. Rev. Genet.* 35:539.)

mutation, spontaneous Occurs at a low frequency, and its specific cause is unknown. The rate of spontaneous mutation per genome may vary. In bacteriophage DNA, 7×10^{-5} to 1×10^{-11}; in bacteria, 2×10^{-6} to 4×10^{-10}; in fungi, 2×10^{-4} to 3×10^{-9}; in plants, 1×10^{-5} to 1×10^{-6}; in *Drosophila*, 1×10^{-4} to 2×10^{-5}; in mice for seven standard loci, 6.6×10^{-6} per locus; in humans, 1×10^{-5} to 2×10^{-6} rates have been reported. The total number of detectable spontaneous mutations per live birth had been estimated as 10–3%, but these estimates are not very accurate because of the uncertainties of identifying very low levels of anomalies. In RNA viruses, 1×10^{-3} to 1×10^{-6} per base per replication has been estimated. Amino acid substitutions in proteins may occur at the rate of 10^{-4}/residue. The error rate of the reverse transcriptase is extremely high because the enzyme does not have an editing function. The induced rate of mutation may be three orders of magnitude higher, but it varies a great deal depending on the locus, the nature of the mutagenic agent, the dose, etc. Mutation rate per mitochondrial D loop has been estimated to be 1/50 between mothers and offspring but other estimates used for evolutionary calculations are (1/300) per generation. Mutation rate in mitochondria on the basis of phylogenetic studies was estimated as $\sim 0.17 \pm 0.15 - 2.2$/site/Myr to $1.35 \pm 0.72 - 1.98$/site/Myr at the 95% confidence level. Mutation rate per base in the human mitochondria has been estimated as 5×10^{-7}. *See* confidence interval; mutation rate, induced; Myr. (Sankaranarayanan, K. 1998. *Mutation Res.* 411:129.)

mutation undetected *See* mutation spectrum.

mutation, useful Although the majority of the new mutations have reduced fitness and are rarely useful for agricultural or industrial applications, some have obvious economic value. Spontaneous and induced mutations have been incorporated into commercially grown crop varieties by improving disease and stress resistance, correcting amino acid composition of proteins, and eliminating deleterious chemical components (erucic acid, alkaloids, etc.). Most of the natural variation in the species arose by mutation during their evolutionary history, and many have been added to the gene pool of animal herds and cultivated plants. Many of the floricultural novelties are induced mutants. Some of the animal stocks have accumulated single mutations; others such as the platinum fox are based on a single dominant mutation of rather recent occurrence. Genetic alterations in industrial microorganisms contributed very significantly to the production of antibiotics. (Demain, A. L.

1971. *Adv. Biochem. Eng.* 1:113; Sakaguchi, K. & Okanishi, M., eds. 1980. *Molecular Breeding and Genetics of Applied Microorganisms.* Academic Press, New York; Quesada, V., et al. 2000. *Genetics* 154:421.)

Dark platinum fox mutant. (From Mohr, O. & Tuff, P. 1939. *J. Heredity* 30:227.)

mutational bias   Microsatellites tend to change to longer repeat sequences; thus, the bias is in favor of longer repeat tracts, although long repeats may not recombine if mismatches are included that impede chromosome pairing. *See* microsatellite.

mutational delay   There is a time lag between the actual mutational event and the phenotypic expression because recessive mutations may show up only when they become homozygous. In organelles, the mutations must sort out before becoming visible, etc. *See* premutation; sorting out.

mutational dissection   Analysis of a biochemical, physiological, or developmental process with the aid of mutations in the system.

mutational distance   Number of amino acid or nucleotide substitutions between (among) macromolecules that may indicate their time of divergence on the basis of a molecular clock. *See* evolutionary clock; evolutionary distance.

mutational load   *See* genetic load.

mutational spectrum   Array and frequency of the different mutations that have been observed under certain conditions or in specific populations. The spectrum of auxotrophic mutations is much lower in most eukaryotes (except yeast and other fungi) than in lower and higher photoautotrophic organisms. *See* mutation spectrum.

mutation and allelic frequencies   *See* equilibrium mutations.

mutation and DNA replication   Mutation is most commonly caused by base substitution as replicational error. In multicellular eukaryotes, the detection of the time of mutations is technically difficult or impossible, although dormant seeds exposed to ionizing radiation regularly display mutations in the progeny. These mutations are due, however, to chromosomal breakage or to the production of reactive chemicals that may act during subsequent cell divisions.

In prokaryotes, tests can be devised for the detection of mutations occurring during the stationary phase. Using a conditional lethal system for selection, the cells may survive under both restrictive and permissive conditions. Mutations arising under restrictive growth conditions are not expected to survive but may live under permissive conditions. Thus, bacterial cultures in the stationary phase are expected to lose all mutants that occurred during the preceding exponential growth phase. When the culture is shifted to permissive conditions, all of the immediately detectable mutations must have their origin in mutations without replication. *See* base substitution; conditional lethal; growth curve; permissive condition. (Freese, E. 1963. *Molecular Genetics*, Pt. I, Taylor, J. H., ed., p. 207. Academic Press, New York.)

mutation bias   Microsatellite sequences undergo frequent unequal recombination or replicational error, but the changes increase when the size of the tracts is short and decrease when longer than a particular size (~200), because the longer tracts are likely to include imperfections. This bias may be maintained in the population by selective forces. *See* unequal crossing over; microsatellite. (Udupa, S. M. & Baum, M. 2001. *Mol. Genet. Genomics* 265:1097.)

mutation clearance   Removal of disadvantageous mutations from a population. The emergence of sex and recombination facilitate the process.

mutation detection   Depends on the general nature of the organisms. In haploid organisms (bacteria, algae, some fungi), haploid cells (microspores, pollen, sperm), or hemizygous cells (e.g., Chinese hamster ovary cell cultures), for easily visible somatic markers in heterozygotes, the mutations are readily detectable. In diploid or polyploid cells, only the dominant mutations can immediately be observed, and for the detection of recessive mutations, a more elaborate procedure is required (*see* mutation in the multicellular germline). The most effective methods involve selective screening. Molecular methods are also available, but these are generally not suitable for large-scale screening. When single-stranded DNA is subjected to electrophoresis in nondenaturing gel, a changed pattern is detectable by SSCP. Heteroduplexes can be resolved by instability in nondenaturing gradient gels (DGGE). Also, heteroduplexes move differently in nondenaturing gels. Cleavage of heteroduplexes by chemicals or enzymes may detect mutant sites. Polymorphism at a single locus can be detected by automated analysis using PCR and fluorescent dyes coupled with a quencher (*see* polymorphism). Mutation may be detected by chemical modification of the mismatches. Carbodiimide generates electrophoretically slow-moving DNA containing mismatched deoxyguanylate or deoxythymidylates. Hydroxylamine modifies deoxycytidylate and osmium tetroxide modifies deoxycytidylate and deoxythymidylate in such a way that the DNA at these residues becomes liable to cleavage by strong bases. Ribonuclease A cleaves at mismatches (depending on the context) in RNA-DNA hybrids. The Mut Y glycosylase of *E. coli* excises adenine from A-G and with somewhat less efficiency from A-C mispairs. In *E. coli*, a methyl-directed mismatch system works primarily up to 3-nucleotide insertions or deletions, but longer sequences are barely if at all corrected, and C-C pairs are ignored.

The MutH/L/S multienzyme complex misses only 1% of the G-T mispair-induced cuts at nearby GATC sequences. The repair is identified by PCR, which also may be a source of replicational error. The latter errors can be estimated as $f = 2lna$, where f = the estimated fraction of mutations within the sequence, l = the length of the amplified sequence, n = the PCR cycles, and a = the polymerase-specific error rate/nucleotides (a for Taq polymerase is within the range of 10^{-6} to 10^{-7}). Direct sequencing may provide the most precise information at the highest investment of labor. *See* alanine scanning mutagenesis; allele-specific probe for mutation; Ames test; ARMS; bioassays in genetic toxicology; cancer (INTER-SS PCR); comet assay; DGGE; DNA sequencing; EMC; footprinting, genetic; gel electrophoresis; hemiclonal; heteroduplex analysis; host-mediated assays; mutant isolation; mutation in human populations; padlock probe; PCR; polymorphism; replica plating; selective screening; sex-linked lethal tests; SNIPS; specific locus tests; SSCP; substitution mutation.

mutation equilibrium   *See* equilibrium mutations.

mutation frequency   Mutation rate. *See* mutant frequency (which is different from mutation frequency).

mutation in cancer cells   *See* cancer.

mutation in cellular organelles   Cellular organelle mitochondria and plastids are generally present in multiple copies per cell, and within individual organelles generally several copies of DNA molecules exist. Therefore, it generally is not expected that the mutations would be immediately revealed. For their visible manifestation, they have to "sort out," i.e., the mutations may not be visible until single organelles become "homogeneous" regarding the mutation, and single cells would be "homoplasmonic" regarding that particular organelle. Although this process is frequently claimed to be stochastic, the direct observation does not seem to support the assumption. Organelles divide by fission, and the daughter organelles are expected to stay in the vicinity of the parental organelle within the viscous cytosol. Thus, their location is not merely by chance, and the progeny organelles tend to remain clustered unless the plane of the cell division separates them. Therefore sorting out may be a relatively fast process. Mutation in tRNA^Leu (MTTL1) in lung carcinoma cybrid cells may be higher than 95% and suppressor mutations in mitochondrial DNA have been identified in only about 10% of the total mtDNA. *See* chloroplast genetics; chloroplasts; mitochondrial diseases in humans; mitochondrial genetics; mtDNA; sorting out; suppressor gene; suppressor tRNA.

mutation induction   *See* cassette mutagenesis; chemical mutagens; directed mutation; environmental mutagens; localized mutagenesis; mutagen; physical mutagens.

mutation in human populations   Difficult to study directly because the random or assorted mating system does not favor the identification of recessive mutations. Therefore, geneticists generally rely on the relative increase of sentinel phenotypes such as new dominant mutations, hemophilia, muscular dystrophy, and cancer, which may be used as epidemiological indicators for an increase in mutation. Also, sister chromatid exchange, chromosomal aberrations, and sperm motility assays may reveal mutations. Molecular methods have become more available, such as RFLP, RAPD, and PCR, and have changed the previously held view based on the fact that no transmitted genetic effects of radiation were clearly detected by the traditional methods in human populations. Mutational hazard for humans is generally inferred from mouse data, which include induced mutations. The immature mouse oocytes are insensitive to radiation-induced mutation but quite likely to be killed by 60 rads of neutrons and 400 R of X-rays or γ-rays. Maturing or mature mouse oocytes, on the other hand, are very susceptible to mutation by acute radiation, although the sensitivity to low-dose irradiation is 1/20 or less. In contrast, the immature human oocytes are not susceptible to killing, but their susceptibility to mutation is not amenable to testing. The estimated mutational hazard of mouse oocytes at various stages ranged from 0.17 to 0.44 times that in spermatogonia, indicating the lower hazard to females than to males. In industrialized societies, chemical mutagens constitute the major hazard, and because of their variety and potency, their effects are difficult to assess, especially at low levels of exposure. Mutation rates can also be estimated on the basis of base substitutions in synonymous and nonsynonymous codons. It may be assumed that the majority of nonsynonymous substitutions are more or less deleterious, whereas the synonymous substitutions are neutral. It has been estimated that a total of 100 new mutations occur in the genome of each human individual and that the rate of deleterious mutations per generation per diploid human genome is ~1.6. *See* atomic radiations; base substitution mutation; bioassays in genetic toxicology; diversity; doubling dose; genetic load; human mutation assays; mutation, neutral; mutation rate (undetected mutation); RBE; specific locus mutations test; synonymous codon. (<http://mutview. dmb.med.keio.ac.jp>, <ariel.ucs.unimelb.edu. au:80/~cotton/ glsdb.htm>.)

mutation in multicellular germline   Reveals the numbers of cells in that germline at the time the mutation occurred, e.g, if a plant apical meristem is treated with a mutagen at the genetically effective 2-cell stage, in the second generation of this chimeric individual the segregation for a recessive allele will not be 3:1 but 7:1. This is because one of the cells will produce 4 homozygous wild-type individuals; the other cell will yield 1 homozygous wild type + 2 heterozygotes (= 3 dominant phenotypes) and 1 homozygous recessive mutant ([4 + 3]:1). In case the germline contains 4 cells, the segregation ratio is expected to be 15:1 (1/16). In case it consists of 8 cells, the proportion is 31:1. *See* GECN; mutation rate. (Rédei, G. P.,

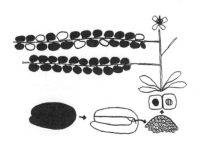

et al. 1984. *Mutation, Cancer and Malformation*, Chu, E. H. Y. & Generoso, W. M., eds., p. 295. Plenum, New York.)

mutation load see genetic load

mutation pressure Repeated occurrence of mutations in a population.

mutation pressure opposed by selection Mutations are frequently prevented from fixation by chance alone (*see* mutation, beneficial; mutation, neutral; non-Darwinian evolution). An equilibrium of selection pressure and mutation pressure may be required for the new mutations to have a chance for survival. Mathematically, the frequency of homozygous mutants is $q^2 = \mu/s$, where μ is the mutation rate and s is the coefficient of selection. *See* allelic fixation.

mutation rate Frequency of mutation per locus (genome) per generation. This calculation is relatively simple if the cell population is haploid (prokaryotes, most of the fungi, or when the gametes are mutagenized). Mutation rate can be expressed also as alteration per nucleotide or as mutation per replicational cycle. If mutation takes place in the multicellular germline of higher eukaryotes, only indirect procedures can be used. The genetically effective cell number (GECN) in the germline must be determined (*see* genetically effective cell number). The level of ploidy must also be known. Thus, mutation rate in these germline cells (R) is

$$R = \frac{\text{number of independent mutational events}}{\text{survivors} \times \text{GECN} \times \text{ploidy}}$$

The standard error of mutation rates (s_m) can be computed as $\sqrt{\mu[1-\mu]/n}$, where μ = the mutation rate observed and n = the size of the population. Calculating the mutation rate on the basis of survivors may pose problems if two different agents are compared. For example: If we use 10,000 haploid cells as a concurrent control and find 10 mutations, the spontaneous mutation rate would be 10/10,000 = 0.001. If we expose another 10,000 cells to a mutagen that is lethal to 5,000 cells but again we obtain 10 mutations, the calculated apparent induced mutation rate is 10/5,000 = 0.002, when actually the treatment may not have induced any mutation but only reduced survival. In case the mutagen is very potent and in each genome or family multiple mutations may occur, it may become very difficult to distinguish the multiple mutations from the single ones without further genetic analysis.

Since the distribution of mutations follows Poisson distribution, in such a case the average mutation rate may be better determined on the basis of the size of the fraction of the population that shows no mutation at all. This is the zero class of the Poisson series, $e^{-\mu}$. If, e.g., the fraction of mutations of the population is 0.3, then zero class is $1 - 0.3 = 0.7 = e^{-\mu}$. Hence, $-\mu = \ln 0.7 = -0.3566$ and $\mu \cong 0.36$. In yeast, for the determination of spontaneous backmutation rate, the formula of von Borstel may be used. Accordingly, $M = e^{(N_0/N)} - m_b/2C$, where N = the total number of compartments, N_0 = the number of compartments without reversions, m_b = the average number of mutants in the inoculum, and C = the average number of cells in the compartment after the stoppage of growth. If no mutations are found in an experimental population, it does not necessarily mean that the mutation rate is zero. An approximation to the possible mutation frequency may be made. If we assume perfect penetrance and normal distribution of undetected mutations, we may further assume that in a population of n genomes the frequency of mutations is $(1 - q)$ at a probability of P. In order to obtain an estimate of q, we have to solve the equation

$$(1 - q)^n = (1 - P)/2$$

Hence

$$\hat{q} = 1 - \sqrt[n]{[1 - p]/2}$$

As an example, after arbitrary substitution of 10,000 for n and 0.99 for P

$$\hat{q} \approx 1 - \sqrt[10000]{0.01/2} \approx 1 - 0.9995 \approx 5.3 \times 10^4$$

That is, if we observed no mutations at all, there is a high probability that actually more than 5 may have occurred under these conditions but we have missed them by chance. It is also possible that the rate of mutation in this case is much below the 10^{-4} range or it may even be zero.

Mutation rate in human populations in case of dominance may be estimated:

$$\frac{\text{number of sporadic cases}}{2 \times \text{number of individuals}} = \mu.$$

The number of individuals stands for the number of total populations studied and 2 is used to account for diploidy. Frequently, it may be necessary to use a correction factor for penetrance or viability. Recessive X-linked mutations can be estimated on the basis of the afflicted males and the formula becomes $\mu = (1/3)s(n/N)$, where n = the number of new mutations, N = the size of the population examined, and s = the relative selective value and/or penetrance. (The 1/3 multiplier is used because the females have 2 and the males have 1 X chromosome.)

Estimation of the rate of recessive mutations is very difficult in human populations because the detection of homozygotes would require controlled mating (inbreeding). The majority of spontaneous mutations among humans occur in the males (achondroplasia, acrodysostosis, Marfan syndrome, oculodentaldigital syndrome, Pfeiffer syndrome) because more cell divisions take place in the male germline than that of the female. Some analyses indicate four times higher mutation rates in males than in females and lower base substitutions in the human X chromosomes than in the autosomes. Some other studies claim a much smaller excess of mutations in human males (Crow, J. F. 2000. *Trends Genet.* 16:525). On the basis of sequencing the human genome, the male:female substitution rate $\alpha_m : \alpha_f = 2.1$. Mutability increases with paternal age. In some human diseases (e.g., Duchenne muscular dystrophy, neurofibromatosis) there is no large sex bias, however. Actually, deletions may be more frequent in females than in males (presumably because of the long dictyotene stage). In point mutation there is a reverse tendency. Human geneticists may also have great difficulty in identifying mutations that have symptoms overlapping several syndromes. For single cases, multivariate statistics may be used, but for many this procedure may be arduous.

Precise calculation of mutation rate is almost intractable by classical techniques but may be solved by molecular methods. DNA sequencing indicates genetic variations in the range of about 1 per kb. Many of these are synonymous at the protein level or do not alter the visible phenotype. Calculating mutation rate may not be accurate if the incidence of the mutation is considered at a particular age but potential loss of the mutations by abortus or neonate death is not considered. The mutation rate on the basis of single-nucleotide polymorphism in a population for the Y chromosomes and autosomes has been estimated within the range of 1.9×10^{-9} to 5.4×10^{-9}, and for mitochondria 3.5×10^{-8} per site per year. Other estimates for average nucleotide changes/human genome are higher (2.5×10^{-8}, ~175/diploid genomes/generation). Single-nucleotide replacements appear an order of magnitude higher than length mutation. Both transitions and transversions are most common at CpG sites. The average deleterious mutation rate (U) in humans appears to be 3 or less. In the HIV1 retrovirus, mutation rate per nucleotide per generation was estimated to be as high as 1 per 10^{-5} to 10^{-4}.

Mutation rate in bacteria has been estimated by various means.

$$\text{Rate} = \ln 2(M2 - M1)/(N2 - N1)$$

Alternatively,

$$R = 2 \ln 2 \left(\frac{M2}{N2} - \frac{M1}{N1} \right) \Big/ g$$

where $M1$ and $M2$ are the number of mutant colonies at time 1 and 2, respectively; $N1$ and $N2$ are the corresponding bacterial counts; $\ln$ = natural logarithm; g = number of generations. In order to obtain reliable estimates on the rates, the culture must be started by large inocula to avoid bias due to mutation at the early generations when the population is still small, and the experiments must be maintained over several generations under conditions of exponential growth. The majority of *Caenorhabditis* genes do not mutate to visible, lethal, or sterile phenotypes. Some double mutants may display, however, mutant phenotype. Therefore, the true mutation rate can be determined only by DNA sequencing. *See* band-morph variants; base substitution; chromosome replication; discriminant function; diversity; DNA repair; error in replication; fluctuation test; F_{ST}; hemophilia; mitochondrial mutation; multivariate analysis; mutation, spontaneous; polymorphism; SNIP; substitution mutation; synthetic lethal. (Crow, J. F. 2000. *Nature Revs. Genet.* 1:40; Nachman, M. W., et al. 2000. *Genetics* 156:297; Kumar, S. & Subramanian, S. 2002. *Proc. Natl. Acad. Sci. USA* 99:803.)

mutation rate, evolution of   Determined by the presence of mutator and antimutator factors. *See* antimutator; mutator.

mutation rate, induced   Using N-ethyl-N-nitrosourea in mice induced at specific loci, 1.5×10^{-3} mutation per locus was observed. This rate may, however, be an overestimate for other eukaryotes, and the range appears to be about 10^{-4} or less. The induced mutation rate per cGy per locus in mice was estimated to be 2.2×10^{-7} and in *Drosophila* $1.5\text{-}8 \times 10^{-8}$. It appeared that in mice the females had lower rates than the males, although the frequencies are greatly affected by the developmental stages. *See* error in replication; Gy; MNU; mutation rate, spontaneous; supermutagens.

mutation rate in fitness   In *Drosophila*, about 0.3/genome/generation was estimated as the frequency of deleterious mutations. In *E. coli*, the rate of deleterious mutations per cell was estimated to be 0.0002. This is larger than three orders of magnitude difference. The *Drosophila* genome is about 20-fold larger than that of *E. coli*, and during a generation approximately 25 divisional cycles take place compared to 1 in the bacterium. If we take the liberty of making adjustments for these differences, $0.0002 \times 20 \times 25 = 0.1$, the deleterious mutation rate in the eukaryote and the prokaryote falls within close range. Mutation rate in fitness is very difficult to estimate and in *Caenorhabditis* it appears three orders of magnitude lower than in *Drosophila*. (Zeyl, C., et al. 2001. *Evolution Int. J. Org. Evolution* 55:909.)

mutation scan   System of mutation detection.

mutation screening   Selective isolation of mutation(s).

mutation spectrum   Indicates the range of mutations observed in a population under natural conditions or after exposure to different types of mutagens. The theoretical expectations would be that the common genetic material would mutate at the same rate at identical nucleotides in different organisms or under different conditions. This expectation is not met because some mutagenic agents have specificities for certain bases. Others act by breaking the chromosomes, depending on their organization and the physical and biological factors present. Also, genes present in multiple copies per genome may suffer mutational alterations, but this may not be observed by classical genetic analysis, although molecular methods may detect them. In general, obligate auxotrophic mutations are very limited in higher plants when the screening uses selective culture media. (Li, S. L., et al. 1967. *Mol. Gen. Genet.* 100:77; Reich, D. E. & Lander, E. S. 2001. *Trends Genet.* 17:502.)

Sample of the spectrum of morphological mutations in *Arabidopsis* expressed at the rosette stage grown under 9 hours of daily illumination. Columbia wild type is at the top right corner.

mutation tester stock, Oak Ridge   Includes the following mouse genes: (1) agouti, (2) brown, (3) albino, (4) pink-eyed dilution, (5) dilute, (6) short ear, (7) piebald. (Peters, J. & Lewis, S. E. 1991. *Mutation Res.* 249:323.)

mutator genes May be functioning on the basis of an abnormal level of errors in DNA synthesis (error-prone DNA polymerase III) or an abnormally low level of genetic repair due to a defect of proofreading exonuclease in bacteria. The actual repair DNA polymerase in bacteria is pol I, which also has $5' \rightarrow 3'$ polymerase and double- and single-strand $3' \rightarrow 5'$ exonuclease capability. MutS protein recognizes DNA mismatches and MutL protein scans for nicks in the DNA, then through exonuclease action slices back the defective strand beyond the mismatch site and facilitates the replacement of the erroneous sequences with new and correct ones. In bacteria, MutH protein recognizes mismatches not by the proofreading system of the exonucleases but by the distortion of newly made DNA molecules. The recognition of the new strands depends on the not-yet-methylated A or C sites within GATC sequences in the new strands. Once the distortion is found, the mismatched base can be selectively excised. If either of these repairs fails, mutator action is observed.

Any defects in the bacterial gene *dam* (DNA methylase) may also result in high mutation rates because the correct base may be excised and replaced erroneously. Some of the so-called mutator genes of the past were actually insertion elements that moved around in the genome and caused mutation by inactivation of genes through disrupting the coding or promoter sequences. The mutator genes *mutA* and *mutC* of *E. coli* cause the A• T→T• A and G• C→T• A transversions in the anticodon of two different copies of tRNA genes that normally recognize the GGU and GGC codons. As a consequence, Asp replaces Gly at a rate of 1 to 2%. Defective methyl-directed mismatch repair (gene MutS) occurs in 1–3% of *E. coli* and *Salmonella*, raising the mutation rate to antibiotic resistance (Rif [rifampicin], Spc [spectinomycin], and Nal [nalidixic acid]) to hundreds of fold, depending on the strain of bacteria. Besides polymerase and repair defects, increase in mutation rate may be caused by anomalies of the replication accessory proteins such as RPA/RAF, PCNA, and RAD27/Rthp/Fen-1. High mutator activity may increase evolutionary adaptation in bacteria.

Once the particular strain is well established, the high mutator activity is expected to decrease in that habitat. *See* amino acids; anticodon; antimutators; DNA polymerases; DNA repair; DNA replication, prokaryotes; insertional mutation; mismatch repair; Mu; mutation detection; mutations, cellular organelle; PCNA; proofreading; RAD; RAD27; RPA; transposable elements; transversion; tRNA. (Giraud, A., et al. 2001. *Science* 291:2606; Shah, A. M., et al. 2001. *J. Biol. Chem.* 276:10824; Shinkai, A. & Loeb, L. A. 2001. *J. Biol. Chem.* 276:46759; Shaver, A. C., et al. 2002. *Genetics* 162:557.)

mutator phage The best characterized is the Mu bacteriophage. The Mu particle includes a 60 nm isosahedral head and a 100 nm tail (in the extended form) also containing base plates, spikes, and fibers. The phage may infect enterobacteria (*E. coli, Citrobacterium freundii, Erwinia, Salmonella typhyimurium*). Its DNA genetic material consists of 33 kb (α) and 1.7 kb (β) double-stranded sequences separated by a 3 kb essentially single-stranded G-loop (specifies host range). It also has variable-length (1.7 kb) single-strand split ends (SE). Besides the coat protein genes, the *c* gene is its repressor (prevents lysis), *ner* (negative regulator of transcription), *A* (transposase), *B* (replicator), *cim* (controls superinfection [immunity]), *kil* (killer of host in the absence of replication), *gam* (protein protects its DNA from exonuclease V), *sot* (stimulates transfection), *arm* (amplifies replication), *lig* (ligase), *C* (positive regulator of the morphogenetic genes) and *lys* (lysis). Upon lysis, 50 to 100 page particles are liberated. The Mu chromosome may exist in linear and circular forms. Mu can integrate at about 60 locations in the host chromosome with some preference. At the position of integration, 5 bp target-site duplications take place. The integration events cause insertional mutation in the host. Mu causes host chromosome deletions, duplications, inversions, and transpositions. These functions require gene *A*, the intact termini of the phage, and replication of the phage DNA. Another related phage, D108, has several DNA regions that are nonhomologous. Its host range is the same as that of Mu. *See* bacteriophages; insertion elements; Mu bacteriophage; temperate phage. (Nakai, H., et al. 2001. *Proc. Natl. Acad. Sci. USA* 98:8247.)

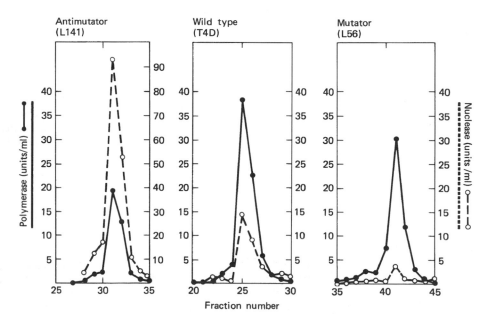

Mutator and antimutator activity may depend on the balance between DNA polymerase and nuclease functions. (From Muzyczka, N., et al. 1972. *J. Biol. Chem.* 247:7116.)

mutein   Mutant protein.

muton   Outdated, no longer used term meaning the smallest unit of mutation. Since 1961, it has been known that single nucleotides or nucleotide pairs are basic units of mutation. (Benzer, S. 1957. *The Chemical Basis of Heredity*, McElroy, W. D. & Glas, B., eds. p. 70 Johns Hopkins Univ. Press, Baltimore, MD.)

mutual exclusiveness   For example, alternative alleles at particular genetic sites in a haploid cannot exist simultaneously.

mutualism   Mutually beneficial or alternatively a selective situation for increased exploitative association of organisms. *See* commensalism; symbionts. (Curie, C. R. 2001. *Annu. Rev. Microbiol.* 55:357.)

M virus   *See* killer strains.

MVR   Minisatellite variant repeat. Within different-length minisatellite allele pairs of the same size, different alterations may occur at different sites and generate isoalleles:

Locus pair 1 ▲▲▲▲▲▲▲▲▲▲▲
            ▲▲▲▲▲▲▲▲▲▲▲

Locus pair 2 ▼▼▼▼▼▼▼▼▼▼▼▼▼▼
            ▼▼▼▼▼▼▼▼▼▼▼▼▼▼

These hypervariable alterations (represented by the triangles) may be identified with the aid of restriction enzyme digestion and electrophoresis and can yield the profile of an individual. The majority of mutations show polarity, i.e., the alterations are most common at the end of the minisatellite site. Many of the mutations generate alleles of different lengths as a result of recombination of the parental alleles. Mutation rate is variable depending on alleles, but apparently it is independent from the length of the repeat as long as homologies are not significantly violated. *See* DNA fingerprinting; minisatellite. (Junge, A. 2001. *Forensic. Sci. Int.* 119:11.)

Mxi1   Tumor suppressor gene at human chromosome 10q24−26. *See* MYC; prostate cancer. (Wang, D. Y., et al. 2000. *Pathol. Int.* 50[5]:373.)

Mxi/Max   Heterodimeric specific transcriptional repressors. *See* E box; Mad/Max; Max. (Billin, A. N., et al. 2000. *Mol. Cell Biol.* 20:9945.)

Mx Proteins   70−80 kDa interferon-inducible GTPases of the dynamin family interfere with the replication and transcription of negative-strand RNA viruses. Mx1: myxovirus (influenza) resistance; it is located at human chromosome 21q22.3. *See* dynamin; replicase. (Regad, T., et al. 2001. *EMBO J.* 20:3495.)

my   Million years during the course of evolution; mya means million years ago. *See* kya.)

myasis (myiasis)   Infestation of a live body with fly larvae (maggots) such as occurs as a consequence of screw worm, a common pest of southern livestock. *See* genetic sterilization.

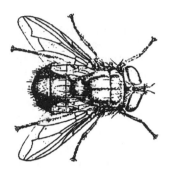

An adult screw worm fly. It has an average life span of 3 weeks and may travel 100 to 200 kilometers, but it can survive only during the warm winters such as exist in Mexico or in the southern U.S. (From Stefferud, A., ed. 1952. *Insects. Yearbook of Agriculture*. USDA, Washington, DC.)

myasthenia (gravis, MG, 17p13)   Generally involves muscular (eye, face, tongue, throat, neck) weakness, shortness of breath, fatigue, and the development of antibodies against acetylcholine receptors. The genetic determination is ambiguous; generally it appears autosomal recessive. The infantile form lacks the autoimmune feature, although it also includes a defect in the acetylcholine receptor. Anticholinesterase therapy and immunosuppressive drugs may be helpful. Newborns of afflicted mothers may be temporarily affected through placental transfer. *See* epistasis, indirect; IVIG; neuromuscular diseases; rapsyn.)

MYB oncogene   Avian myeloid leukemia (myeloblastosis) oncogene; its human homologue was assigned to chromosome 6q21-q23. Its product is a transcription factor. The same gene is also called AMV (*v-amv*). The MYB gene product is translated into ~75 kDa protein in immature myeloid cells and its activity is substantially reduced as differentiation proceeds. The Myb protein, with a nuclear protein-binding leucine zipper in the regulatory domain, recognizes the 5′-PyAAC(G/Py)G- 3′ core sequence. In case the C- or N-terminus or both are deficient, it becomes a potent activator of leukemia, although overexpression of the entire Myb may also be oncogenic. The Myb oncogene regulates hematopoiesis. The Myb A protein is required for normal spermatogenesis and mammary gland development. Using antisense oligodeoxynucleotides for codons 2−7 may reduce its transcription. The same treatment may be effective against chronic myelogenous leukemia. Myb homology domains occur in multiple copies in both monocotyledonous and dicotyledonous plants. *See* AMV; antisense technologies; hematopoiesis; leukemias; oncogenes.

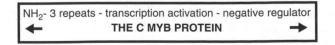

NH₂- 3 repeats - transcription activation - negative regulator
◄————   THE C MYB PROTEIN   ————►

mycelium   Mass of fungal hyphae. *See* hypha.

mycobacteria   Gram-positive bacteria that cause tuberculosis (killing 3 million people annually) and leprosy (*Mycobacterium leprae*, 3.27 Mb, 1604 genes), respectively. Their cell wall has low permeability and therefore they are rather resistant to therapeutic agents. *Mycobacterium tuberculosis* H37Rv genome is 4,411,529 bp and includes about 4,000 genes (see *Nature* 393:537 [1998]; see also corrections in *Nature* 396:190 [1998]). Tuberculosis is most commonly treated by isoniazid (INH). As a result, saturated hexacosanoic acid (C26:0) accumulates on a 12 kDa acyl protein carrier. This complex associates with β-ketoacyl carrier protein synthase (KasA). When mutations occur in the latter, INH resistance develops. Apparently, INH acts by inhibiting KasA. Nitroimidazopyran drugs are promising against tuberculosis. The cure of tuberculosis generally requires strict multidrug therapy to avoid the emergence of drug resistance, which is a very serious threat nowadays. *Mycobacterium vaccae* and *Bacillus Calmette-Guerin* (BCG, also a mycobacterium) produce cross-reactive material against *M.l.*, and *M.v.* is suited for vaccine production against leprosis. *M.l.* cannot efficiently be propagated in vitro for vaccine production, and only the nine-banded armadillo would be a suitable animal host. The virulence of mycobacteria seems to be controlled by the *erp* (exported repetitive protein) gene. Some major (2q32-q35, 15q11-q13) and apparently several minor genetic factors determine the susceptibility to tuberculosis and other diseases. Gene UBE3A, a ubiquitin ligase is located in the imprinted region of chromosome 15. *See Bacillus Calmette-Guerin*; leprosy; vaccine. (Abel, L. & Casanova, J.-L. 2000. *Am. J. Hum. Genet.* 67:274; Glickman, M. S. & Jacobs, W. R., Jr. 2001. *Cell* 104:477; Russel, D. G. 2001. *Nature Rev. Mol. Cell Biol.* 2:569; Cervino, A. C. L., et al. 2002. *Hum. Mol. Genet.* 11:1599, <http://genolist.pasteur.fr/>.)

mycology   Study of the entire range of mushrooms (fungi).

MYC oncogene (8q24.12-q24.13)   Named after the myelocytomatosis retrovirus of birds from which its protein product was first isolated. It is a widely present DNA-binding nuclear phosphoprotein in eukaryotes. c-MYC is expressed only in dividing cells and it is an inhibitor of terminal differentiation. C-MYC induces the expression of the cell cycle–activating kinase, CDK4. Besides the *v-myc* gene, the MH-2 virus carries the *v-mil* gene, which encodes serine/threonine kinase activity and invariably causes monocytic leukemia. Most retroviruses in this group cause only the formation of nonimmortalized macrophages requiring growth factors for proliferation. The cellular forms of the oncogenes are specially named after the type of cells in which they are found. NMYC occurs in neuroblastomas (and retinoblastomas). Homologues of this gene were assigned to human chromosome 2p24 and to mouse chromosomes 12 and 5. The LMYC genes are involved in human lung cancer where one was located to chromosome 1p32 and two to the mouse chromosomes 4 and 12. Alternative splicing and polyadenylation produce several distinct mRNAs from a single gene. Metastasis is favored by the presence of a 6 kb restriction fragment of the DNA. In human chromosome 7, another form, the MYC-like 1 gene, was detected with a 28 bp near-perfect homology to the avian virus. In both humans and mice,

translocation sequences were identified between the Myc gene (human chromosome 8) and several immunoglobulin genes encoding the heavy κ- and λ-chains. PVT1 has an activation role in Burkitt's lymphomas (caused by the Epstein-Barr virus) and plasmacytomas (neoplasia). Myc/MAX heterodimer binds to the CDC25 gene and activates transcription after binding to the DNA sequence CACGTG. This binding site is recognized by a number of MYC-regulated genes. Frequent targets of Myc activation are the cycline genes as well as ornithine decarboxylase, lactate dehydrogenase, and thymosine β4. When cellular growth factors are depleted, Myc can induce apoptosis with the cooperation of CDC25. MYC, in cooperation with RAS, mediates the progression of the cell cycle from G1 to the S phase through induction of the accumulation of active cyclin-dependent kinase and transcription factor E2F. MAD (Mxi1; 10q24-26) protein holds MYC in check. MAD forms a heterodimer with MAX (MAD-MAX), and this successfully competes with MYC-MAX heterodimers for transcription, thus controlling malignancy. Downregulation of the human ferritin gene by MYC leads to cellular proliferation. Myc expression may mediate Fas-FasL (Fas-ligand) interaction and an apoptosis pathway through FADD. The BIN1 protein interacts with MYC and serves as a tumor suppressor. *See* apoptosis; BIN1; Burkitt's; cancer; CDC25; CDF; CDK; E box; E2F; FADD; Fas; ferritin; Gardner; Id protein; immunoglobulins; MAD; MAX; oncogenes PVT; RAS. (Henriksson, M. & Lüscher, B. 1996. *Adv. Cancer Res.* 68:109.)

mycophenolic acid (MPA)   *See* IMPDH.

mycoplasma   Parasitic and/or pathogenic bacteria. They may also contaminate animal cell cultures. *Mycoplasma genitalium* was the first free-living organism to be completely sequenced (by 1995), containing 580,070 bp DNA, 480 open-reading frames, and 37 RNA transcribing genes; many of its known genes (127) are involved in translation. *Mycoplasma pneumoniae* has $\sim 0.82 \times 10^6$ bp DNA genome and 679 genes. *See* gene number; phytoplasmas. (Glass, J. I., et al. 2000. *Nature* 407:757; Flynn, J. L. & Chan, J. 2001. *Annu. Rev. Immunol.* 19:93; <http://genolist.pasteur.fr/>.)

mycorrhiza   Symbiotic association between plant roots and certain fungi. *See* nitrogen fixation. (Smith, K. P. & Goodman, R. M. 1999. *Annu. Rev. Phytopath.* 37:473.)

mycosis   Infection or disease by fungi. The infection may be superficial, subcutaneous, and systemic by inhalation of the spores. Most commonly, fungi cause skin diseases. Immune-compromised individuals are particularly susceptible to mycoses.

mycotoxins   Fungal metabolites that damage DNA. The best known are aflatoxins, sterigmacystin, ochratoxin, zearalenone, and some penicillin toxins. *See* aflatoxin.

MyD88   Myeloid (bone marrow–like) differentiation primary response factor, encoded in human chromosome 3p22-p21.3. Adaptor protein in the Toll–NF-κB signaling pathway. *See* NF-κB; Toll/TRAF6.

myedema (hyperthyroidism)   Attributed to hypothyroidism caused by autosomal-recessive/dominant conditions. It involves dry skin with wax-like deposits. Some HLA genes and autoimmune causes have been implicated. *See* Graves disease.

myelencephalon   Part of the brain of the embryo that develops into the medulla oblongata (the connective part between the pons [brain stem]) and the spinal cord. *See* brain.

myelin   Lipoprotein forming an insulating sheath around nerve tissues. *See* Marie-Charcot-Tooth disease; Pelizaeus-Merzbacher disease. (Greer, J. M. & Lees, M. B. 2002. *Int. J. Biochem & Cell Biol.* 34:211.)

myeloblast   Precursor (formed primarily in the bone marrow) of a promyelocyte and eventually a granular leukocyte. It usually contains multiple nucleoli. *See* leukocyte; Wegener granulomatosis.

myeloblastin   Serine protease, specific for myeloblasts controlling growth and differentiation and myelogenous leukemia. *See* leukemia.

myelocyte   Precursor of the granulocytes, neutrophils, basophils, and eosinophils.

myelocytoma   *See* myeloma.

myelodysplasia   Recessive form of leukemia generally associated with monosomy of human chromosome 7q. The critical deletion apparently involves chromosomal segment 7q22-q34. *See* leukemia.

myeloma   Bone marrow cancer. Myeloma cells are used to produce hybridomas. *See* Bence Jones proteins; hybridoma; monoclonal antibody.

myelomeningocele   *See* spina bifida.

myelopathies   Diseases involving the spinal chord and myelination of the nervous system. About 15 different human anomalies are involved. *See* Charcot-Marie-Tooth disease; Dejerine-Sottas neuropathy.

myelopoiesis   Formation of granulocytic and monocytic cell lineages. *See* granulocyte; monocyte.

myelosuppression   Interferes with bone marrow function and blood cell formation. It is a most common target of cancer chemotherapy drugs. These drugs are aimed at highly proliferative cells and the bone marrow produces 4×10^{11} cells daily for long periods of time.

MYF-3   Myogenic factor was located to human chromosome 11-p14, and its mouse homologue, *MyoD*, is in chromosome 7. It controls muscle development, may be subject to imprinting, and may possibly be involved in the formation of embryonic tumors. The protein is a helix-loop-helix transcription factor (MASH2), and it is a mammalian member of the *achaete-scute* complex of *Drosophila*. *See* imprinting; Mash-2; morphogenesis in *Drosophila*).

Myhre syndrome   *See* dwarfism.

myleran (1,4-di[methanesulfonoxy]   butane)   Clastogenic (causing chromosome breakage) alkylating mutagen. Also called busulfan.

myoadenylate deaminase   *See* adenosine monophosphate deaminase.

myoblast   Muscle cell precursor. Myoblasts can be well cultured in vitro and reintroduced into animals where they fuse into myofibers. They can be used as gene or drug delivery vehicles. *See* vectors.

myocardial infarction   Obstruction of blood circulation to the heart; may be accompanied by tissue damage. Infarcted (dead/decaying) heart muscle cells may be replaced by transplantation of regenerating bone marrow cells. *See* coronary heart disease.

myoclonic epilepsy   Occurs in different forms. The recessive juvenile EJM gene is situated within the boundary of the HLA gene complex in the short arm of chromosome 6 (Janz syndrome). It is characterized by generalized epilepsy with onset in early adolescence. Mutations in a subunit of the sodium channel SCN1A gene (2q24) cause slow growth from the 2nd year of life, the individuals often become ataxic, and later cause speech problems. Myoclonic epilepsy associated with ragged-red fibers is characterized by epileptic convulsions, ataxia, and myopathy (enlarged mitochondria with defects in respiration). The defect was attributed to an adenine → guanine transition mutation at position 8344 in human mitochondrial DNA involving the tRNA^{Lys} gene. Human chromosome 6q24 recessive (EPM2A) myoclonous epilepsy (LaFora disease) shows up at about age 15 and results in death within 10 years. LF disease involves defects in a protein tyrosine phosphatase. Myoclonus epilepsy Unverricht and Lundborg (EPM1) is 21q22.3 chromosome recessive with onset between ages 6 and 13, beginning with convulsions and turning within a few years into shock-like seizures (myoclonus). The latter disease is caused by mutation in the cystatin B (cysteine protease inhibitor) gene. EPM1 disease involves a large insertion (600–900 bp) consisting of 12 bp repeats (CCCCGCCCCGCG). The cases appeared to be initiated by premutational changes of 12–13 repeats and usually by maternal transmission the alleles appeared stable. In the majority of cases of paternal transmission, the repeats increased. In some cases, an 18mer repeat was observed at the 5′ region of the promoter and a 15mer repeat at the 3′ region. The 18 and 15mer repeats also included six and four T and A bases, respectively. Benign familial adult epilepsy maps to 8q23.3-q24.1. Some human myoclonus diseases are apparently under dominant control. A *myo*clonus disease in cattle has an apparent defect in glycine-strychnine receptors. In mice the spastic mutation in chromosome 3 affects the α-1 glycine receptors, whereas the α-2 receptor defect is X-linked. *See* epilepsy; ion channels; mitochondrial disease in humans; premutation; trinucleotide repeats.

myoclonus Sudden, involuntary contractions of the muscle(s) like in seizures of epilepsy or sometimes as a normal event during sleep. Myoclonus dystonia, a dominant disease, is located to 7q21. See dystonia; myoclonic epilepsy; sarcoglycan.

MyoD Muscle-specific basic helix-loop-helix protein that regulates muscle differentiation and the cessation of the cell cycle. Its DNA-binding consensus is CANNTG in the promoter or enhancer of the controlled genes. MyoD may activate p21 and p16 and promotes muscle-specific gene expression. When MyoD is phosphorylated by cyclin D1 kinase (Cdk), it may fail to transactivate muscle-specific genes. See Cdk; cell cycle; cyclin; enhancer; helix-loop-helix; MEF; Myf3; myogenin; p16; p21. (Tedesco, D. & Vesco, C. 2001. Exp. Cell Res. 269:301.)

myofibril Muscle fiber made of actin, myosin, and other proteins (bundle of myofilaments).

myogenesis Muscle development.

myogenin Protein involved in muscle development. MyoD activates myogenin, and overcoming the inhibitor of DNA-binding (helix-loop-helix) gene *Id* and other muscle gene transcription and differentiation, is set on course. See MEF; MRF; MYF5; MyoD. (Sumariwalla, V. M. & Klein, V. H. 2001. Genesis 30[4]:239.)

myoglobin Single polypeptide chain of 153 amino acids attached to a heme group; functions in the muscle cells to transport oxygen for oxidation in the mitochondria. Recurrent myoglobinuria may result from a microdeletion in mitochondrially encoded cytochrome C oxidase (COX) subunit III. Some Antarctic fishes do not have myoglobin, and homozygous myoglobin knockout mice function in a practically normal manner. See hemoglobin, mitochondrial disease in humans.

myo-inositol Active form of inositols. They are widely used as phospholipid head groups. They are essential for signaling, membrane trafficking, etc. See embryogenesis, somatic; inositol; phosphoinositides; phytic acid.

myokymia Hereditary spasmic disorder of the muscles caused by potassium channel defects. See ion channels.

myoneurogastrointestinal encephalopathy (mitochondrial myoneurogastrointestinal encephalopathy, MNGIE, 22q13.32-qter) Recessive defect in a nuclear gene; however, mitochondrial DNA aberrations my also accompany it. The onset is usually after the second decade of life and the clinical symptoms involve atrophy of the muscles (ophthalmoplegia), multiple neural disorders, lactic acidosis, etc. See mitochondrial diseases in humans; ophthalmoplegia.

myopathy Collection of diseases affecting muscle function, controlled by autosomal-dominant, autosomal-recessive, X-linked, and mitochondrial DNA. Besides the weak skeletal muscles, ophthalmoplegia (paralysis of the eye muscles) usually accompanies it. The recessive homozygotes for myotonic myopathy are also dwarf and have cartilage defects and myopia. Two types are known with carnitine palmitoyle transferase deficiency (CPT1; mutation at 1pter-q12 [CPT2] involves myoglobinuria, especially after exercise or fasting). Another recessive myopathy is based on succinate dehydrogenase and aconitase. Phosphoglycerate mutase deficiencies (human chromosome 10q25 and 7p13-p12) also involve myoglobinuria. X-linked forms display autophagy (cytoplasmic material is sequestered into lysosome-associated vacuoles) or slow maturing of the muscle fibers, or swelling and hypertrophy in the quadriceps muscles, respectively. Insertion or deletion or frameshift mutations in the ITGA7 integrin gene cause congenital myopathy (human chromosome 12q13). The Xq28-linked tubular myopathy is caused by defects in the myotubularin tyrosine phosphatase. Dominant and recessive base substitution mutations in the skeletal muscle actin, ACTA1 gene (1q42), involve thin muscle fibers and severe hypotonia, muscle weakness, feeding and breathing difficulties, and most commonly death during the first few months after birth. Occasionally, some afflicted individuals may survive to adulthood. See actin; adenosine monophosphate deaminase; Batten-Turner syndrome; cardiomyopathies; carnitine; cartilage; desmin; inclusion body myopathy; integrin; mitochondrial disease in humans; muscular dystrophy; myopia; myositis; Sbf1; troponin.

myopia Caused by the increased length of the eye lens in the front-to-back dimension, and focusing the refracted light in front of the retina (nearsightedness). Most forms are under polygenic control with rather high heritability (around 0.6). Autosomal-dominant, autosomal-recessive (infantile), and X-linked forms have also been suggested. Some studies indicate significant positive correlation between myopia and intelligence. Myopia may be concomitant with various syndromes. See eye diseases; human intelligence.

myosins Contractile proteins that form thick filaments in the cells, hydrolyze ATP, bind to actin, and account for the mechanics of muscle function, organelle movement, phagocytosis, pinocytosis, cell movement, RNA transport, phototransduction, signal transduction, etc. Myosins are activated by kinases (MLCK) and deactivated by phosphatases. MLCK is inhibited by PAK. Members of the 11 families of myosin are represented in amoebas, insects, mammals, and plants. The majority of myosin motors move in one direction toward the plus (+) end of actins. A conformational change at the actin-binding site (converter) of myosin class VI may facilitate movement of the lever in the minus (-) direction. This class of myosin has a 50-amino acid insertion at the converter region. See deafness; filament; kinesin; microfilament; motor proteins; myofibril; PAK; phagocytosis; pinocytosis; RAC; spindle; Usher syndrome. (Homma, K., et al. 2001. *Nature* 412:831; <www.mrc-lmb.cam.ac.uk/myosin/ myosin.html>.)

myositis   Name of some myopathies. *See* myopathy.

myostatin gene (human chromosome 2q32.1 recessive, GDF8) Its inactivation (partial deletion) substantially increases muscle development in mammals. This mutation is found in the Belgian Blue and Piedmontese cattle breeds. Transforming growth factor β is involved. The condition is also named double muscling and growth/differentiation factor 8. (Lee, S. J. & McPherron, A. C. 2001. *Proc. Natl. Acad. Sci. USA* 98:93067.)

myotonia   Recessive myotonia congenita is apparently in human chromosome 7 and there is also a dominant form (Thomsen disease). Both involve difficulties in relaxing the muscles. The various forms (M. fluctuans, M. permanens, paramyotonia congenita, K^+-activated myotonia) may involve Na^+ ion channel malfunctions. *See* ion channels; periodic paralysis.

myotonic dystrophy (DM) Dominant human disorder expressed as wasting of head and neck muscles, eye lens defects, testicular dystrophy, speech defects, frontal balding, and frequently heart problems. It is controlled by a dominant gene (DMPK) at the centromere of human chromosome 19q13.2-q13.3 (Steinert disease). DM2 is at 3q. The DMPK-related kinase, DK serine/threonine kinase, is at 1q41-q42. The prevalence is highly variable. It seems that DM may be a contiguous gene syndrome. In some isolated populations, it may occur in 1/500 to 1/600 proportion, whereas in other populations the occurrence is about 1/25,000. The manifestation of the symptoms is enhanced in subsequent generations, and more are found in children of affected mothers than affected fathers. This observation may be due, however, to anticipation. The condition is dominant and it is quite polymorphic. Molecular evidence indicates that the DNA sequences concerned are unstable in the 3'-untranslated tracts downstream of the last exon of the protein kinase gene. The difficulties involving nucleosome assembly in DNA containing CTG triplet repeats in the 3'-untranslated region of a serine/threonine kinase gene seem to be concerned. It appears that the CUG repeats of the RNA transcript bind proteins (CUG-BP) that interfere with the proper splicing of the transcript. A troponin protein has been implicated with the binding. On the basis of linkage with chromosome 19 centromeric genes, prenatal tests are feasible. Mutation frequency was estimated within the $1.1-0.8 \times 10^{-5}$ range. Ribozyme-mediated trans-splicing of the trinucleotide repeats has been tried to reduce the expansion. A second myotonic dystrophy gene (DM2) was mapped to a 10 cM region of human chromosome 3q21 (ZNF9). The latter does not display CTG expansion (~15 kb) seen in DM1 but large expansion (~44 kb) of the tetranucleotide repeat (CCTG). The expanded mRNAs are not distributed normally to the cytoplasm and are not translated into the needed protein in sufficient amounts or interfere with the function of nuclear RNA-binding proteins. In the brain, hyperphosphorylated short tau protein may occur. *See* anticipation; contiguous gene syndrome; Duchenne periodic paralysis; fragile sites; imprinting; mental retardation; muscular dystrophy; Pompe's disease; prenatal diagnosis; Schwartz-Jampel syndrome; tau; transsplicing; trinucleotide repeats; troponin; Werdnig-Hoffmann disease. (Liquori, C. L., et al. 2001. *Science* 293:864; Sergeant, N., et al. 2001. *Hum. Mol. Genet.* 10:2143.)

myotubes   Aggregate of myoblasts into a multinucleated muscle cell. *See* acetylcholine; cadherin; fertilin; integrin; meltrin.

myristic acid (tetradecanoic acid, $CH_3(CH_2)_{12}COOH$) Natural 14-carbon fatty acid without double bonds. Myristoylation is common in oncoproteins, protein serine/threonine and tyrosine kinases, protein phosphatases, in the α-subunit of heterotrimeric G proteins, transport proteins, etc. Cell membranes can be targeted by myristoylation of nascent proteins with the aid of myristoyl-CoA:protein N-myristoyltransferase enzyme. *See* fatty acids; G protein; kinase. (Farazi, T. A., et al. 2001. *J. Biol. Chem.* 276:39501.)

Myt1   *See* Cdc2.

myxobacteria   Slime-secreting bacteria, which may form cysts that contain lots of cells.

myxoviruses   Group of RNA viruses. *See* RNA viruses.

MZ   Monozygotic twin. *See* twinning.

H. J. Muller was the second geneticist recipient of the Nobel Prize, on 31 October 1946, for his research on the influence of X-rays on genes and chromosomes. At the Cold Spring Harbor Symposia on Quantitative Biology (9:163, in 1941) he stated:

"We are not presenting ... negative results as an argument that mutations cannot be induced by chemical treatment."

"... it is not expected that chemicals drastically affecting the mutation process while leaving the cell viable will readily be found by our rather hit-and-miss methods. But the search for such agents, as well as the study of the milder, 'physiological' influences that may affect the mutation process, must continue, in the expectation that it still has great possibilities before it for the furtherance both of our understanding and our control over the events within the gene."

A historical vignette

N

n Gametic chromosome number, $2n$ = the zygotic chromosome number. *See* genome; polyploid; x.

N *See* Newton.

N₂, N₃ ... Nₙ Backcross generations. *See* B.

N51 Serum-inducible gene, identical to KC, MGSA, and *gro*.

NAA α-naphthalene acetic acid is a synthetic auxin. *See* plant hormones; somatic embryogensis.

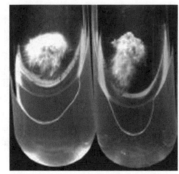

Plant callus transferred to naphthaleneacetic acid medium develops abundant roots.

NAADP Nicotinic acid adenine dinucleotide phosphate may coordinate agonist-induced Ca^{2+} signaling and other metabolic processes. *See* agonist.

β-NAADP.

NAC Nascent polypeptide-associated complex *See* protein synthesis.

NACHT ATPases involved in apoptosis and MHC gene activation. *See* apoptosis; MHC; NTPase. (Koonin, E. V. & Aravind, L. 2000. *Trends Biochem. Sci.* 25[5]:223.)

NAD β-nicotinamide-adenine dinucleotide (diaphorase): a cofactor of dehydrogenation reactions and an important electron carrier in oxidative phosphorylation. DIA1 (cytochrome b5 reductase) is in human chromosome 22q13-qter, DIA2 in chromosome 7, and DIA4 in 16q22.1.

NADH Reduced form of NAD. *See* NAD.

NADP⁺ Nicotine adenine dinucleotide phosphate is an important coenzyme in many biosynthetic reactions. The enzyme-reduced nicotinamide-adenine dinucleotide oxidase is an important producer of reactive oxygen species in the cell, and its deficiency may lead to chronic granulomatous disease. *See* granulomatous disease; NAD; NADH; Rossmann fold.

nail-patella syndrome (Turner-Kieser syndrome, Fong disease) Malformation or absence of nails, poorly developed patella (a bone of the knee), defective elbows and other bone defects, kidney anomalies, collagen defects, etc. The dominant locus was assigned to human chromosome 9q34.1 encoding an LIM-domain protein. *See* collagen; LIM domain; patella aplasia-hypoplasia.

NAIP Neuronal apoptosis inhibitory protein (5q12.2-q13.3). Its partial deletion leads to spinal muscular atrophy. *See* apoptosis; spinal muscular atrophy. (Crocker, S. J., et al. 2001. *Eur. J. Neurosci.* 14:391.)

NAIS Nucleotide analog interference suppression. A short nucleotide sequence is ligated to a special RNA site to reveal the functional consequences of this modification, e.g., for selfsplicing of introns. (Ryder, S. P. & Strobel, S. A. 1999. *J. Mol. Biol.* 291:295; Szewczak, A. A., et al. 1998. *Nature Struct. Biol.* 5:1037.)

naïve Unaffected, e.g., a B lymphocyte that has not yet been exposed to a foreign antigen.

NAK NF-κB-activating kinase is an IKK kinase that mediates NF-κB activation in response to growth factors and phorbol ester tumor promoters that stimulate protein kinase C-ε. *See* IKK. (Tojima, Y., et al. 2000. *Nature* 404:778.)

naked DNA Not embedded in protein; pure DNA.

nalidixic acid (1-ethyl-1,4-dihydro-7-methyl-4-oxo-1,8-naphthydrine-3-carboxylic acid is an antibacterial agent (inhibits DNA synthesis); used also in veterinary medicine against kidney infections.

nanism *See* dwarfism; Mulibrey nanism.

nanobacteria (nanobacteria) $\sim 0.1\,\mu$m mineralized particles isolated from fetal bovine serum (FBS), human serum, and kidney and dental pulp. The living nature of these particles and their nucleic acid content could not be unambiguously confirmed. (Cisar, J. O. 2000. *Proc. Natl. Acad. Sci. USA* 97:11511.)

nanocrystals, semiconductor   May be used similarly to fluorochromes. They can be excited at any wavelength shorter than their emission peak. Many sizes of nanocrystals can be excited by one wavelength, resulting in the simultaneous production of different colors. *See* FISH; fluorochromes; quantum dot. (Holmes, J. D., et al. 2001. *J. Am. Chem. Soc.* 123:3743; Hamad-Schifferli, K., et al. 2002. *Nature* 415:152.)

nanoelectrospray mass spectrometry   *See* electrospray.

nanoparticles   Usually less than 1 μm colloidal capsules of biodegradable or nonbiodegradable vehicles for the protected delivery of molecules into the cells of an organism. They are usually internalized by endocytosis and are transported to the lysosomes, where the contents may or may not be saved from degradation. This technology may facilitate the detection of DNA with extremely high selectivity. (Park, S.-J., et al. 2002. *Science* 295:1503.)

nanopore technology   Upcoming development for extremely rapid detection of single-stranded DNA, RNA, and nucleotides. It may be applicable to sequencing these macromolecules. (Deamer, D. W. & Akeson, M. 2000. *Trend Biotechn.* 18:147; *Science* 290:1524 ff.)

nanotechnology   Generally uses atomic force microscopy or scanning tunneling microscopy to manipulate objects at the level of individual molecules with the aid of special tweezers, pipettes, and nanolithography dip pens. This technology may be exploited for the manipulation of gene chips. Construction of branched DNA structures and other novel DNA motifs, including those that appear during crossing over, etc. *See* atomic force microscopy; DNA chips; scanning tunneling microscopy. (Seeman, N. C. 1998. *Annu. Rev. Biophys. Biomol. Struct.* 27:225; <www.pa.msu.edu/cmp/csc/nanotube.html>.)

naphthaleneacetic acid (naphthylacetic acid)   *See* NAA.

naphthylacetic acid   *See* NAA.

CH₂COOH

1-Naphthylacetic acid.

narcolepsy (17q21)   Pathological frequent sleep, commonly associated with hallucinations, loss of muscle tone, and paralysis. It may be associated with other syndromes such as catalepsy. Its sporadic incidence is about 2×10^{-3}. Deficiency of hypocretin (orexin, hypothalamus-specific neuroexcitatory peptides), the cerebrospinal fluid, is the suspected cause. *See* apnea; autoimmune diseases; cataplexy; HLA; orexin; sleep. (Mignot, E., et al. 2001. *Am. J. Hum. Genet.* 68:686.)

naringenin   Trihydroxyflavonone; with naringin it causes the bitter flavor in grapefruit.

NARP syndrome   Complex disease involving neuronal defects, ataxia, seizures, feeblemindedness, and developmental retardation caused by a mutation in subunit 6 of mitochondrial ATPase. *See* mitochondrial diseases in humans.

narrow-sense heritability   *See* heritability.

NarX   *E coli* kinase affecting nitrate reduction by regulator protein NarL. *See* nitrate reductase; nitrogen fixation. (Wei, Z., et al. 2000. *Mol. Plant Microbe Interact.* 13:1251.)

NAS   nonsense-associated altered splicing may be executed by skipping an exon, choosing alternate splice sites, or insertion of an intron as a consequence of stop codons. *See* ESE; intron; NMD; nonsense codon; splicing. (Liu, H.-X., et al. 2001. *Nature Genet.* 27:55; Li, C. M., et al. 2001. *Eur. J. Hum. Genet.* 9[9]:685.)

NASBA   Nucleic acid sequence–based amplification. An RNA strand in the presence of a primer is amplified by reverse transcriptase. Then the RNA strand is removed by RNase H and the resulting cDNA, using RNA primers, is amplified by (T7) RNA polymerase into sufficient quantities of RNA. *See* LCR; PCR; RT-PCR. (Borst, A., et al. 2001. *Diagn. Microbiol. Infect. Dis.* 39[3]:155.)

nascent   Just being born or synthesized, e.g., the mRNA still associated with the ribosomes. The molecule is not combined yet with any other molecule and may be highly reactive.

nasopharyngeal carcinoma   *See* Epstein-Barr virus.

NAT   Negative regulator of activated transcription. About a 20-polypeptide complex in human cells including homologues of yeast proteins (Srb, Med, Rgr, CDK8) negatively regulating transcription of RNA polymerase II. A defect or deficiency of the NAT2 gene may increase cancer risk. *See* Mediator; *RGR*; Srb. (Gadbois, E. L., et al. 1997. *Proc. Natl. Acad. Sci. USA* 94:3145.)

native   Original to a particular region, a natural form of a substance.

native conformation   Natural active structural form of a molecule.

natural antibody   Produced spontaneously without deliberate immunization. Immunoglobulins M, G, and A may be involved. May assist in antigen uptake and antigen presentation by B lymphocytes through complement. *See* antibody; complement; immune reaction; immmunoglobulins; innate immunity.

natural immunity   *See* immunity.

natural killer cell (NK)   *See* killer cell; T cell.

natural selection   Action of forces in nature that maintain or choose the genetically fittest organisms in a habitat. The *fundamental theorem of natural selection* is "The rate of increase in fitness of any organism at any time is equal to its genetic variance in fitness at that time" or mathematically expressed $\Sigma\alpha\,dp = dt\,\Sigma\,\Sigma'(2pa\alpha) = Wdt$, where α = the average effect on fitness of introducing a gene, a = the excess over the average of any selected group, W = fitness, and $dt\,\Sigma'(2pa\alpha)$ is the sum of increase of average fitness due to the progress of all alleles considered (according to R. A. Fisher, 1929). *See* cost of evolution; fitness; selection. (Fischer, R. A. 1958. *The Genetical Theory of Natural Selection.* Dover, New York, Edwards, A. W. F. 2002. *Theor. Population Biol.* 61:335.)

nature and nurture   Hereditary (nature) and environmental factors (nurture) that cooperatively mold the actual appearance of an organism. Francis Galton first used this term in science in 1874 (although W. Shakespeare in *The Tempest* (1612) speaks of "a born Devil, on whose nature nurture can never stick"). According to R. Woltereck (1909), the genes assure a "reaction norm" upon which nutrition, climate, education, etc., act and form the phenotype. The relative impact of nature and nurture varies depending on the attributes. Some traits and faculties are under almost complete genic determination; others are more influenced by the environment in the broad sense. *See* heritability.

Navajo neuropathy   Apparently rare disease involving loss of pain and temperature sensation in the limbs due to poor myelination of the nerve fibers.

navel oranges   Commonly seedless but being "navel" may not rule out seed development, since some navel citruses produce a normal number of seeds. The Washington navel orange and the Satsuma mandarin are completely pollen sterile, which is the cause of the seedlessness when foreign pollen is excluded. The navel oranges (and grapefruits) have two or three whorls of carpels, resulting in a fruit-in-fruit appearance. This abnormal carpel formation may also interfere with pollination. *See* orange; seedless fruits.

NBM paper   *See* diazotized paper.

NBT   Nitroblue tetrazolium used as a chromogen (0.5 g in 10 mL 70% dimethyl formamide).

NC   Negative cofactors interfere (may also promote) with TFIIB transcription factor binding to the preinitiation complex. *See* DSTF; PIC; preinitiation complex; TBP; transcription factors. (Cang, Y. & Prelich, G. 2002. *Proc. Natl. Acad. Sci. USA* 99:12727.)

N-CAM   neural cell adhesion molecule is a Ca^{2+}-independent immunoglobulin-like protein that binds cells together by homophilic means (i.e., by being present on neighboring cells). N-CAM may activate NF-κB. NCAM1 is encoded at 11q23.1, NCAM2 at 21q21, and L1CAM at Xq28. The latter mutation involves symptoms similar to Kallmann syndrome and MASA syndrome (mental retardation), aphasia (speech and writing defect), walking anomaly, and abnormal thumb position. Other cell adhesion defects may be coded at other locations. *See* ICAM; Kallmann syndrome; neurogenesis; NF-κB.

NCAM (N-CAM)   Neural cell adhesion molecule. *See* N-CAM. (Thomaidou, D., et al. 2001. *J. Neurochem.* 78:767.)

NCNI   (National Center for Biotechnology Information) maintains nucleotide sequence information on genes and clones. *See* databases; GenBank; GSDB.

NCoA   *See* signal transduction.

N-CoR   Protein of M_r 270 K binding to the ligand-binding domain of thyroid hormone and retinoic acid receptors by means of its carboxy-terminal. The binding of this protein to these hormone receptors mediates a ligand-independent transcriptional repression. These hormone receptors are transcriptional repressors for their target genes without their cognate ligands. N-CoR seems to be associated with SIN3 and its binding proteins. *See* chromatin remodeling; Cor-box; histone deacetylase; SANT; signal transduction; Sin3; SMRT. (Guenther, M. G., et al. 2001. *Mol. Cell Biol.* 21:6091.)

NCp7   Viral nucleocapsid protein that serves as a chaperone for the folding of the RNA and enhances recombination of two single-stranded RNAs within the capsid. (Takahashi, K., et al. 2001. *J. Biol. Chem.* 276:31274.)

NCR   Natural cytotoxic receptors. *See* killer cells.

N-DEGRON   Degradation signal of a protein, the NH_2-terminal amino acids, and an internal lysine residue of the substrate protein. Several ubiquitin molecules form a multiubiquitin chain at the lysine. The degradation may be mediated by a G-protein. *See* degron; G proteins; N-end rule; PEST; ubiquitin. (Suzuki, T. & Varshavsky, A. 1999. *EMBO J.* 18:6017.)

NDF   *See* heregulin.

nDNA   Nuclear DNA.

NDUFV   NADH-ubiquinone-oxidoreductase flavoprotein is part of the mitochondrial oxidoreductase system and is responsible for some of the symptoms in Alexander's and Leigh's diseases. *See* Alexander's disease; Leigh's encephalopathy; mitochondrial disease in humans.

Neanderthal (Neandertal) people   Hominids who lived about 30,000 to 200,000 years ago in France, southern Germany, and the Middle East. They had large jaws and heavy bones and may have developed later (about 30,000 years ago) into the Cro-Magnon men, who showed more similarity to present-day humans. Their precise relationship to *Homo sapiens* is not known, but DNA evidence indicates a dead-end detour from human evolution. A mtDNA (extracted from fossil bones) analysis revealed 22–36 bp difference from modern humans, whereas the same mtDNA among pairs of humans varies within the range of 1–24. The difference from chimpanzees and gorillas is much greater than from modern humans. The common ancestors of the Neanderthals and modern humans date back an estimated 550,000 to 690,000 years. *See* hominidae. (Ovchinnikov, I. V., et al. 2000. *Nature* 404:490; Scholz, M., et al. 2000. *Am. J. Hum. Genet.* 66:1927; Krings, M., et al. 2000. *Nature Genet.* 26:144; <www.neanderthal-modern. com>.)

nearest-neighbor analysis Technique to determine base sequences in oligonucleotides, a procedure of historical interest for DNA sequencing, but it gained new usefulness in designing DNA-binding ligands for antisense therapy. The DNA is synthesized with 5'-P^{32}-labeled nucleotides. The sequence is then digested with micrococcal endonuclease and by spleen phosphodiesterase. In the digest specific 3'-P^{32} mononucleotides and oligonucleotides are found and thus indicating which base was nearest to the original radioactively labeled nucleotide (e.g. adenylic acid). Nearest-neighbor analysis of protein complexes may reveal the patterns of subunit association. *See* antisense RNA; endonuclease; phosphodiesterase. (Josse, J., et al. 1961. *J. Biol. Chem.* 236:284; Crevel, G., et al. 2001. *Nucleic Acids Res.* 29:4834.)

NEB *See* copia.

nebenkern (paranucleus) Cellular body; generally a mitochondrial aggregate resembling the cell nucleus by microscopical appearance situated in the flagellum of the spermatozoan.

nebularine 9-β-D-ribofuranosyl-9H-purine is a natural product of some fungi and streptomyces with an antineoplastic effect. Its subcutaneous LD50 for rodents varies from 220–15 mg/kg, depending on the species. *See* LD50.

Nebularine.

nebulin Actin-associated large protein in the skeletal muscles, built of repeating 35-residue units. *See* titin.

necrophagous Referring to organisms that thrive by eating dead tissues.

necropsy (autopsy) Examination of the body after death.

necrosis Death of an isolated group of cells or part of a tissue or tumor. It generally involves inflammation in animals because of the release of cell components toxic to other cells. Necrosis is different from apoptosis, which involves elimination of cells no longer needed, and it is normally a programmed cell death. *See* apoptosis; hypersensitive reaction.

necrotic Referring to necrosis. *See* necrosis.

necrotroph The pathogen kills the host and feeds on its dead tissue. *See* biotroph.

nectar Sweet plant exudate fed on by insects and small birds.

nectins In four isoforms represent an immunoglobulin-like cell-cell adhesion system, which organizes adherens junctions cooperatively with the cadherin-catenin system. *See* adherens junction. (Mizoguchi, A., et al. 2002. *J. Cell Biol.* 156:555.)

NEDD ICE-related proteases. *See* apoptosis; ICE. (Murillas, R., et al. 2002. *J. Biol. Chem.* 277:2897.)

neddylation Degrades protein(s) with the aid of NEDD. *See* COP.

Needleman-Wunsch algorithm Finds optimal global alignment of macromolecule building blocks. *See* algorithm. (Laiter, S., et al. 1995. *Protein Sci.* 4:1633.)

negative binomial $(q - p)^{-k}$ after expansion

$$p_x = [(k + (x - 1)] \frac{R^x}{q^k},$$

where $p = m/k$, m = mean number of events, $q = 1 + p$, px = probability for each class, $R = p/q = m/(k + m)$, x = number of events/class, and k must be determined by iterations using z_i scores and approximate k values until the z_i becomes practically zero. The detailed procedure cannot be shown here (*see* Rédei, G. P. & Koncz, C. 1992. *Methods in Arabidopsis Research*, Koncz, C., et al., eds., p. 16 World Scientific, Singapore). The negative binomial distribution resembles that of the Poisson series. It may provide superior fit to data that are subject to more than one factor, each affecting the outcome according to the Poisson distribution, e.g., mutations occur according to the Poisson series, but their recovery follows an independent Poisson distribution, etc. *See* distributions; Poisson distribution.

negative complementation Intraallelic complementation when the polypeptide chain translated on one cistron of a locus interferes with the function of the normal polypeptide subunit(s) encoded by the same or another cistron of the multicistronic gene. Thus, actually there is an interference with the expression of the locus concerned. *See* allelic complementation; cistron; complementation; hemizygous, ineffective. (Garen, A. & Garen, S. 1963. *J. Mol. Biol.* 7:13.)

negative control Prevention of gene activity by a repressor molecule. The gene can be turned on only if a ligand molecule (frequently the substrate of the enzyme, which the gene encodes) binds to the repressor, resulting in moving the repressor from the operator site or from an equivalent position. *See* DNA-binding domains; DNA-binding proteins; *lac* operon; lambda phage; *tryptophan* operon.

negative cooperativity Binding of a ligand or substrate to one subunit of a multimeric protein precludes the binding to another. (Horovitz, A., et al. 2001. *J. Struct. Biol.* 135[2]:104.)

negative dominant *See* dominant negative; negative complementation. (Müller-Hill, B., et al. 1968. *Proc. Natl. Acad. Sci. USA* 59:1259.)

negative feedback *See* feedback inhibition.

negative interference *See* coefficient of coincidence; interference; rounds of matings.

negative numbers in nucleotide sequences Indicate position of bases upstream of the position (+1) where translation begins. *See* upstream.

negative regulator Suppresses or reduces transcription or translation in a direct or indirect manner. *See* Lac operon; lambda phage; NAT. (Maira, S. M., et al. 2001. *Science* 294:374.)

negative selection of lymphocytes see positive selection of lymphocytes.

negative staining Used for the study of macromolecules. The electron microscopic grid with the specimen on it is exposed to uranyl acetate or phosphotungstic acid, which produces a thin film over it except where the macromolecule is situated. The electron beam illuminates the noncovered macromolecules while the metal-stained parts appear more dense. This gives a negative image of viruses, ribosomes, or other complex molecular structures. *See* electron microscopy; stains.

negative-strand viruses Their genomic and replicative intermediates (plus strand) exist as viral ribonucleoprotein (RNP). They require viral RNA polymerase and precise 5′ and 3′ termini for both replication and packaging. *See* positive-strand virus; replicase. (Pekosz, A., et al. 1999. *Proc. Natl. Acad. Sci. USA* 96:8804.)

negative supercoil Double-stranded DNA molecule twisted in the opposite direction as the turn of the normal right-handed double helix (e.g., in B DNA). *See* supercoiled DNA; Z DNA.

neighbor joining method Relatively simple procedure for inferring bifurcations of an evolutionary tree. Nucleotide sequence comparisons are made pair-wise and the nearest neighbors are expected to display the smallest sum of branch lengths. *See* DNA likelihood method; evolutionary distance; evolutionary tree; Fitch-Margoliash test; four-cluster analysis; gene tree; least square methods; population tree; protein likelihood method; transformed distance; unrooted evolutionary trees. (Saitou, N. & Nei, M. 1987. *Mol. Biol. Evol.* 4:406; Romano, M. N. & Weigendt, S. 2001. *Poult. Sci.* 80:1057.)

neighborliness Statistical method for the estimation of the position of two taxonomic entities in an evolutionary tree. *See* evolutionary tree; Fitch-Margoliash test; transformed distance. (Charleston, M. A., et al. 1994. *J. Comput. Biol.* 1[2]:133.)

Neisseria gonorrhoeae, N. meningitidis Gram-negative bacteria. Strain B (2,272,351 bp) and strain A (2,184406 bp) of *N. meningitidis* are responsible for inflammation of the brain membranes (meninges), and blood poisoning (septicemia) has been completely sequenced. Serogroup A causes large epidemics in sub-Saharan Africa, whereas B and C are responsible for sporadic outbreaks worldwide. *Neisserias* display great abilities for antigenic variation, making vaccine development difficult. The sequenced genomes may facilitate defense measures against these pathogens. *See* antigenic variation; vaccines. (Merz, A. J. & So, M. 2000. *Annu. Rev. Cell Dev. Biol.* 16:423; Tettelin, H., et al. 2000. *Science* 287:1809; Parkhill, J., et al. 2000. *Nature* 404:502.)

NELF Negative protein factor of mRNA elongation by polymerase II. *See* DRB; DSIF; TEFb; transcript elongation. (Ping, Y. H. & Rana, T. M. 2001. *J. Biol. Chem.* 276:12951.)

nemaline myopathy Collection of muscle fiber gene defects (nemaline = thread or rod-like). The majority of the muscles may be affected and the severity may vary from intrauterine death to relatively mild anomalies. The heart muscles are usually not affected. Dominant tropomyosin-3 is encoded at 1q22-q23. Recessive alpha-skeletal muscle myofilament disease gene (actin myopathy ACTA1) is at 1q42.1. Recessive nebulin gene (2q22) encodes the filaments in the sarcomeres. *See* heart diseases. (North, K. N., et al. 1997. *J. Med. Genet.* 34:705.)

NEMO NF-κB essential modulator (Xp28) is the γ-subunit of IKK. *See* IKK; incontinentia pigmenti; NF-κB. (Courtois, G., et al. 2001. *Trend Mol. Med.* 7:427; Aradhya, S., et al. 2001. *Hum. Mol. Genet.* 10:2557.)

N-end rule The half-life of a protein is determined by the amino acids at the NH_2 end. *See* destruction box; N-degron; protein degradation within cells. (Rao, H., et al. 2001. *Nature* 410:955.)

neobiogenesis Idea that living organisms arose from organic and inorganic material. It is also used to describe the formation of new organelles. *See* spontaneous generation.

neocarzinostatin (NCS) Naturally occurring enediyne antibiotic. The NCS chromophore specifically attacks a single residue in a two-base DNA bulge. It attacks HIV type I RNA and other viruses. *See* bulge; enediyne; HIV. (Maeda, H., et al. eds. 1997. *Neocarzinostatin: The Past, Present, and Future of an Anticancer Drug*. Springer, New York.)

Precocious neocentromeres at metaphase I in 5 chromosome pairs. (After M. M. Rhoades.)

neocentric *See* neocentromere.

neocentromere Extra spindle-fiber attachment site in the chromosomes in eukaryotes of certain genotypes, different from the regular centromere position yet containing tandem repeats of satellite DNA in plants. Human neocentromeres are not repetitive and do not contain the 171 bp α-satellites present in the regular human centromeres. Human neocentromeric DNA sequences do not share commonality; therefore, they may be the result of epigenetic alterations rather than basic sequence specificity. The proteins and functions associated with human neocentromeres are shared with the same of the regular centromeres. Neocentromeres occur at low frequencies. Some genetic constitutions (e.g., in maize and rye) promote neocentromere formation. Neocentromeres are not surrounded by heterochromatin. From the maize neocentromeres, centromeric protein CENP-C is absent. *See* centromere; centromere activation; holocentric; human artificial chromosomes; knob; preferential segregation. (Warburton, P., et al. 2000. *Am. J. Hum. Genet.* 66:1794; Hiatt, E. N., et al. 2002. *Plant Cell* 14:407; Amor, D. J. & Choo, K. H. A. 2002. *Am. J. Hum. Genet.* 71:695.)

neo-Darwinian evolution   Evolution is attributed to small random mutations accumulated by the force of natural selection. Characters acquired as a direct adaptive response to external factors have no role in evolution. *See* darwinism; directed mutation; fitness; mutation, beneficial; mutation, neutral; selection. (Matsuda, H. & Ishii, K. 2001. *Genes Genet. Syst.* 76[3]:149.)

neo-Lamarckism   *See* Lamarckism.

Neolithic   About 7,000 years ago when humans turned to agricultural activity from hunting and gathering. *See* mesolithic; paleolithic.

neomorphic   Mutation that displays a new phenotype of any structural or other change that evolved recently. (Muller, H. J. 1932. *Proc 6th Int. Congr. Genet.* 1:213.)

neomycin (neo, $C_{23}H_{46}N_6O_{13}$)   Group of aminoglycoside antibiotics. *See* antibiotics; geneticin; kanamycin; neomycin phosphotransferase.

Neomycin.

neomycin phosphotransferase (NPTII)   *See* aminoglycoside phosphotransferase; aph(3′)II, *neo^r*; kanamycin resistance.

neonatal tolerance   *See* immune tolerance.

neo-organ   *See* mini organ.

neoplasia   Newly formed abnormal tissue growth such as a tumor; it may be benign or cancerous. *See* cancer; carcinogens; oncogenes.

neo^r   Neomycin resistance gene encoding the APH(3′)II enzyme. *See* aminoglycoside phosphotransferases; APH(3′)II; neomycin.

neo-sexchromosome   Translocation involving an autosome and the X or Y chromosome; some have binding sites for the MSL (male-specific lethal) proteins. *See* MSL.

neoteny   Retention of some juvenile (or earlier-stage) function in a more advanced stage.

neo-X-chromosome   *See* neo-sexchromosome.

neo-Y-chromosome   *See* neo-sexchromosome.

nephritis, familial (nephropathy)   Autosomal dominant without deafness or eye defects. It closely resembles Alport syndrome. Elevated blood pressure, proteinuria, and only microscopically detectable blood cells in the urine precede kidney problems. In one form, immunoglobulin G accumulates in the serum and the expression of the disease is promoted by dietary conditions (e.g., high gluten). *See* Alport's disease; kidney disease.

nephrolithiasis (kidney stones)   May be caused by CLCN5 (Xp11.22) voltage-gated chloride channel and by Dent disease, which maps to the same cytological position and causes a similar disorder. Human chromosomes 12q12-q14, 1q23-q24, 10q21-q22, and 20q13.1-q13,3 may harbor additional genes affecting the disease. Kidney stones (calcium oxalate, calcium phosphate, uric acid, ammonium magnesium sulfate [struvite], and cystine) may afflict about 10% of Western populations, with substantial variations according to genetic factors, diet, and climatic conditions. *See* Dent disease; kidney diseases. (Ombra, M. N., et al. 2001. *Am. J. Hum. Genet.* 68:1119.)

nephronophthisis (3q22)   Autosomal recessive kidney disease characterized by anemia, passing of large amounts of urine (polyuria), excessive thirst (polydipsia), wasting of kidney tissues, etc. It is the most common cause of renal failure in children. *See* diabetes. (Otto, E., et al. 2002. *Am. J. Hum. Genet.* 71:1161.)

nephropathy, juvenile hyperuricemic (16p21.2)   Dominant hyperuricemia (excessive amounts of uric acid in the blood), elevated serum creatinine but reduced uric acid excretion, gout, renal malfunction. *See* kidney disease; nephritis, familial.

nephrosialidosis   Autosomal-recessive kidney inflammation caused by oligosaccharidosis. *See* Hurler syndrome; lysosomal storage diseases.

nephrosis, congenital Autosomal-recessive inflammation of the kidney. It can be detected prenatally by the accumulation of α-fetoprotein in the amniotic fluid. The basic defect is in the basement membrane structure of the glomerular membranes. Its incidence may be as high as 1.25×10^{-4} in populations of Finns or those of Finnish descent. *See* basement membrane; fetoprotein; glomerulonephritis; kidney diseases; prenatal diagnosis.

nephrotic syndrome, steroid-resistant (SRN1, 1q25-q31) Recessive fetal kidney disease due to a defect in the NPHS2 gene encoding podocin (42 kDa), an integral membrane protein. In mice, the expression is most evident in the kidneys, brain, and pancreas. Concomitant massive proteinuria causes neonatal death. *See* kidney diseases. (Boute, N., et al. 2000. *Nature Genet.* 25:125; Putaala, H., et al. 2001. *Hum. Mol. Genet.* 10:1.)

nephrotome Part of the mesoderm that contributes to the formation of the urogenital tissues and organs.

nepotism in selection Some disadvantageous individuals or groups may be favored by selection in case they promote the fitness of the reproducing groups that share genes with them or the ability to favorably recognize kin. *See* indirect fitness; selection. (Mateo, J. M. & Johnston, R. E. 2000. *Proc. R. Soc. Lond. B Biol. Sci.* 267:695.)

neprilysin Endopeptidase that normally degrades amyloid β peptides but not in Alzheimer disease, then the brain plaques develop. *See* Alzheimer disease. (Iwata, N., et al. 2001. *Science* 292:1550.)

NER (Nucleotide exchange repair) *See* DNA repair, excision repair.

Nernst equation

$$\left(\frac{RT}{zF} \ln \frac{C_o}{C_i} = V \right)$$

expresses the relation of electric potential across biomembranes to ionic concentration at both sides; V = equilibrium potential in volts, C_o and C_i-outside and inside ionic concentrations, $R = 2$ cal mol^{-1}°K^{-1}, T = temperature in K (Kelvin), $F = 2.3 \times 10^4$ cal V^{-1} mol^{-1}, z = valence, and ln = natural logarithm. *See* chemosmosis.

nerve cell *See* neuron.

nerve function The nervous system includes the central (brain and spinal cord) and the peripheral systems. The central nervous system in humans contains an immensely large number of cells ($\sim$30 billion in the cerebellum). The elements of this complex system are well coordinated. The neurons communicate with each other through the synapses of the dendrites. The signals are transmitted through the neurotransmitters and received by the receptors. The information is passed along the axons. The impulses may be forwarded in two opposite directions electrically by changes

in the polarity of the cell membranes. The membranes are equipped with ion channels. The nervous system begins to develop at the embryo stage and continuously expands up to the adult stage and beyond. Mutation in the genes, rearrangement of the chromosomes, infectious diseases, drugs, and malnutrition, and changes in metabolism due to various causes may affect this development. Sleeping, learning, emotional events, pain, and all types of normal changes in external or internal events may bring about fluctuations and alterations in the function of the system. Differences may be based on the genetic constitution of the individual, age, sex, etc. Generally, monozygotic twins respond very similarly and narrow-sense heritability may be quite high (22–88%). Modern neurobiology can monitor and map in vivo the activity in different parts of the brain or in the peripheral system by the use of electroencephalography, positron emission tomography, functional magnetic resonance imaging, and other methods. (Kennedy, M. B. 2000. *Science* 290:750; Kandel, E. R. & Squire, L. R. 2000. *Science* 290:1113.)

nerve growth factor (NGF) Stimulatory factor of growth and differentiation of nerve cells. NGF promotes neuronal survival partly with the aid of CREB transcription factor activation of antiapoptotic gene (Bcℓ-2) activity. Structurally it is different from ciliary neurotrophic factor. It is present in small amounts in various body fluids. NGF is a hexamer composed of α-, β-, and γ-subunits, but only the β-subunit is active for the ganglions. The active component of mouse submaxillary factor has a dimeric structure of two 118-amino-acid residues (MW 13 kDa). The mouse gene for α is in chromosome 7; β is in chromosome 3. The human gene for β-NGF is in chromosome 1p13. It acts primarily on a tyrosine kinase receptor. NGF receptors (NGFR) share similarities to TFNRs (tumor necrosis factor receptors). *See* apoptosis; Bcℓ-2; ciliary neurotrophic factor; CREB; growth factors; neurogenesis; neuropathy; protein 4.1N; signal transduction; TNFR. (Saltis, J. & Rush, R. A. 1995. *Int. J. Dev. Neurosci.* 13:577.)

NES Nuclear export signal. *See* chromosome maintenance region 1; RNA export.

nested genes Partially overlapping genes sharing a common promoter. The structural genes may be read in different registers. This is a very economical solution for extremely small genomes $\phi \times 174$ for multiple utilization of the same DNA sequences within different reading frames. *See* contiguous gene syndrome; knockout; overlapping genes; recoding. (Turner, S. A., et al. 2001. *J. Bacteriol.* 183:5535.)

nested primers The product of the first PCR amplification internally houses a second primer in order to minimize amplification of products by chance. *See* PCR; primers. (Menschikowski, M., et al. 2001. *Anal. Cell Pathol.* 22[3]:151.)

Net1 Inhibitor of Cdc14; it moves Cdc14 into the nucleolus and removes inhibitory phosphates from APC. Net 1 anchors Sir2 in the nucleolus. *See* APC; Cdc14; Sir. (Shou, W., et al. 2001. *Mol. Cell* 8:45.)

Netherton syndrome (NS, 5q32) Recessive ichthyosis, hair shaft defect, skin allergy (atopy), hayfever, and high serum IgE. The gene colocalizes with LEKTI (lymphoepithelial Kazal-type inhibitor), a serine protease inhibitor (SPINK5). *See* ichthyosis. (Walley, A. J., et al. 2001. *Nature Genet.* 29:175; Lauber, T., et al. 2001. *Protein Exp. Purif.* 22:108.)

netrins Secreted by neuronal target cells and assist homing in the proper nerve axons under the guidance of Ca^{2+}. Netrins serve as maintenance factors. *See* axon; collapsin; colorectal cancer (DCC); *frazzled*; semaphorin; tenascin; UNC-6. (Kennedy, T. E., et al. 1994. *Cell* 78:425.)

netropsin Organic molecules that recognize A-T base pairs in the minor groove of DNA. *See* lexitropsin. (Wemmer, D. E. 2000. *Annu. Rev. Biophys. Biomol. Struct.* 29:439; Wang, L., et al. 2001. *Biochemistry* 40:2511.)

network Associated proteins most likely interact. Thus if the function of one protein is known, the function of its associated partner can be inferred. On this basis a network of interacting proteins can be determined. *See* genetic network. (Schwikowski, B., et al. 2000. *Nature Biotechnol.* 18:1257.)

NEU ERBB2. *See* ERBB1.

neural code Represents the various stimuli as sensory experiences such as visual, olfactory, gustatory, mechanical, and auditory qualities as well as intensities, frequencies, and spatial relations. (Fairhall, A. L., et al. 2001. *Nature* 412:787.)

neural crest Ectodermal cells along the neural tube cleave off from the ectoderm and migrate through the mesoderm to form the peripheral nervous system, pigment cells, and possibly other tissues such as the thyroid and adrenal glands, connective tissues, heart, eye, etc. *See* germ-layers.

neural plate Thickened notochordal overlay of nerve cells that develop into neural tubes.

neural tube Central nervous system of the embryo derived from epithelial cells of the neural plate. *See* floor plate.

neural tube defects Occur at a frequency of 1–2/1,000 births. It results from the failure of closing (normally by the 4th week of human pregnancy) of the neural tube. The recurrence risk is usually 3–5%. In about 70% of the cases, folic acid administration may prevent it. At the molecular level, the defect may be caused by mutation in the Slug Zn-finger transcription factor. Protein Bcℓ 10 seems important for neural tube closure and lymphocyte activation. *See* anencephaly; hydrocephalus; Meckel syndrome; microcephaly; spina bifida.

neuraminic acid Pyruvate and mannosamin-derived 9-carbon amino sugar. Its derivatives, such as sialic acid, are biologically important. *See* neuraminidase deficiency; sialic acid; sialidase; sialidoses; sialiduria.

neuraminidase deficiency (sialidosis) Recessive autosomal lysosomal storage disease with multiple and variable characteristics. The basic common defect in the various forms is a deficiency of the sialidase enzyme. Sialidase (6p21.3) cleaves the linkage between sialic acid (N-acetyl-neuraminic acid) and a hexose or hexosamine of glycoproteins, glycolipids, or proteoglycans. The uncleaved molecules are then excreted in the urine. In sialurias, free sialic acid is excreted. For the normal expression of the 76 kDa enzyme, the integrity of the structural genes in human chromosome 10pter-q23 and a 32 kDa glycoprotein coded by chromosome 20q13.1 are required. The deficiency is characterized by cherry red muscle spots, progressive myoclonus (involuntary contraction of some muscles), loss of vision, but generally normal intelligence. *See* influenza virus; lysosomal storage diseases; neuraminic acid. (Lukong, K. E., et al. 2001. *J. Biol. Chem.* 276:17286.)

neuregulins (NDF [neuron differentiation factor], GGF [glial growth factor], ARIA [acetylcholine receptor-inducing activity]) Human chromosomes 8p22-p11 (NRG1) and 10q22 (NRG3) encode protein signals of the epidermal growth factor family (EGF) that activate acetylcholine receptor genes in synaptic nuclei. They also have multiple effects on various processes of differentiation. Some neuregulins belong to the tyrosine kinase transmembrane receptor family (NRG2, 5q23-q33). *See* acetylcholine; argin; EGF; heregulin; synapse. (Frenzel, K. E. & Falls, D. L. 2001. *J. Neurochem.* 77:1.)

neurexins Great variety of nerve surface proteins generated by alternative splicing of three genes. They may function in cell-to-cell recognition and interact with synaptotagmin, latroxin, and neuroligins. *See* latrotoxin; neurexophilin; neuroligin; splicing; synaptotagmin. (Sugita, S., et al. 1999. *Neuron* 22:489.)

neurexophilin Small neurexin-α-binding molecule with probable signal function. *See* neurexin. (Missler, M., et al. 1998. *J. Biol. Chem.* 273:34716.)

neurite Extension from any type of neuron (axon and dendrite).

neuritis Nerve inflammation, increased sensitivity or numbness, paralysis, or reduced reflexes caused by one or more (polyneuritis) nerves defects. *See* neuropathy.

neuroantibody Antibody secreted by the nerve cells. Although antibody secretion is the duty of the plasma cells, many other types of cells of animals (and plants) are capable of antibody production when the proper immunoglobulin genes are present and expressed. The degree of secretion varies a great deal. *See* antibody; immunoglobulins; lymphocytes; plasma cell. (Ruberti, F., et al. 1993. *Cell. Mol. Biol.* 13:559.)

neuroaxonal dystrophy, late, infantile (Hallervorden-Spatz disease, NBI1, 20p13-p12.3) Recessive nerve degenerative disease with onset generally between ages 10 and 20 and death before age 30. It involves involuntary movements, speech defects, difficulties with swallowing, and progressive mental deterioration. Brown discoloration is visible in the brain (substantia nigra, globus pallidus) after autopsy. Both J. Hallervorden and H. Spatz discredited themselves during the era of the Third Reich by being involved in murderous human experimentation. Therefore, it has been suggested that the eponymous designation of the disease be changed to NBI1.

Another similar but separate disease is also called infantile neuroaxonal dystrophy. *See* neurodegenerative diseases.

neuroblast Cell that develops into a neuron. *See* neurogenesis.

neuroblastoma Nerve cell tumor located most frequently in the adrenal medulla (in the kidney). An MYC oncogene seems to be responsible for its development. Deletions of the short arm of human chromosome 1 (NB, 1p36.3-p36.2) inactivate the relevant tumor suppressor gene. *See* cancer; ERBB1; MYC; tumor suppressor gene.

neurod Helix-loop-helix regulatory protein of pancreas development. Its defect may lead to diabetes mellitus II and defects of the sensory neurons. *See* diabetes mellitus; helix-loop-helix. (Kim, W. Y., et al. 2001. *Development* 128:417.)

neurodegenerative diseases *See* affective disorders; Alzheimer disease; amyotrophic lateral sclerosis; dementia; encephalopathies; fatal familial insomnia; Friedreich ataxia; FTDP-17; Guam disease; Huntington chorea; Lewy body; mental retardation; neuroaxonal dystrophy; Parkinson disease; Pick disease; prions; synuclein; trinucleotide repeats; Wilson disease.

neuroectoderm Contributes to the formation of the nervous system. *See* neurogenesis.

neuroendocrine immunology Study of the interactions among the central-peripheral nervous system, endocrine hormones, and immune reactions.

neuroepithelioma *See* Ewing sarcoma.

neurofibromatosis Autosomal-dominant (NF-1, incidence $\sim 3 \times 10^{-4}$), near the centromere of human chromosome 17q11.2, and another gene (NF-2) in the long arm of chromosome 22q12.2. It affects the developmental changes in the nervous system, bones, and skin; causes light brown spots and soft tumors (associated with pigmentation) over the body, and in NF-2 particularly in the Schwann cells of the myelin sheath of neurons; involves mental retardation, etc. The NF-2 protein is called schwannomin or merlin and it is practically absent from schwannomas, meningiomas, and ependymomas (nerve neoplasias). In some instances, schwannomin is broken down by calpain. Schwannomin is also a tumor suppressor and interacts with the actin-binding site of fodrin and thus with the cytoskeleton. The literature distinguishes several forms of this syndrome. The loss of the NF1 protein activates the RAS signaling pathway through the granulocyte macrophage colony-stimulating factor and makes the cells prone to develop juvenile chronic myelogenous leukemia. NF1 mutations also inactivate p53. NF-1 activates an adenylyl cyclase coupled to a G protein and is deeply involved in the regulation of the overall growth of *Drosophila*. A pseudogene (NF1P1) was located in human chromosome 15. Estimated mutation rate $1-0.5 \times 10^{-4}$. The neurofibromin protein (guanosine triphosphatase activating, GAP) is encoded by NF-1 in a 350 kb DNA tract with 59 exons. About 5–10%

of the affected individuals have ~ 1.5 kb microdeletions due to unequal crossing over occurring mainly during female meiosis. *see* actin; ataxia; café-au-lait spot; calpains; cancer; Cushing syndrome; cytoskeleton; ependymoma; epiloia; eye disease; fodrin; GAP; GCSF; hypertension; Lisch nodule; leukemia; meningioma; mental retardation; merlin; p53; RAS; schwannoma; tumor suppressor; von Recklinghousen disease. (Ars, E., et al. 2000. *Hum. Mol. Genet.* 9:237; Nguyen, R., et al. 2001. *J. Biol. Chem.* 276:7621; Gutman, D. H., et al. 2001. *Hum. Mol. Genet.* 10:1519.)

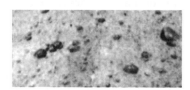

Photo modified from Dr. Curt Stern, orig. from Dr. V. McKusick.

neurogastrointestinal encephalomyopathy (MNGIE) Human chromosome 22q13-qter external ophthalmoplegia with drooping eyelids (ptosis), abnormality of the intestines, defects of the nervous system and the muscles, lactic acidosis (accumulation of lactic acid), and slender body. The mitochondria in the skeletal muscles are structurally defective and the mtDNA show deletions. The activity of the respiratory enzymes is reduced. The basic defect appears to involve nuclearly encoded thymidine phosphorylase controlling maintenance of mtDNA. *See* encephalopathies; mitochondrial diseases in humans; mtDNA; ophthalmoplegia.

neurogenesis The nervous system has a great deal of similarity among all animals. The *neuroblasts* (the cells that generate the nerve cells) develop from the ectoderm. The *neural tube* (originating by ectodermal invagination) gives rise to the *central nervous system* composed of *neurons* and the supportive *glial* cells. The differentiated neural cells do not divide again. The *neural crest* produces cells that eventually migrate all over the body and form the *peripheral nervous system*. A neuron contains a dense cell body from which the *dendrites* (reminding to the root system of plants) emanate. From the neurons extremely long *axons* may emanate, which at the highly branched termini make contact with the target cells by *synapses*. The dendrites and axons are also called by the broader term *neurites*.

Initially, the various components of the system develop at the points of migration and subsequently they are connected into a delicate network. The point of growth is named *growth cone*, which manages a fast expansion, generating some web-like structures (*microspikes or filopodia* organized into *lamellipodia*). The neurites are frequently found in *fascicles*, indicating that several growth cones travel the same track across tissues. This motion is mediated by cell adhesion molecules (N-CAM, cadherin, integrin). When the migrating growth cones reach their destination, they compete for the limited amount of *neutrophic factor* (NGF) released by the target cell and about a half of them starve to death.

The nerve cells function at the target by their inherent *neuronal specificity* rather than by their positional status. When the branches of several axons populate the same territory of control, they are trimmed back and a process

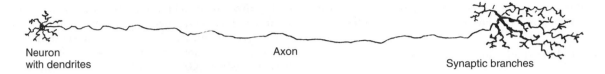

Neuron
with dendrites   Axon   Synaptic branches

of activity-dependent synapse elimination disposes of some of them. Due to this process, the synaptic function becomes more specific. The migration of the motor neurons from the central nervous system toward the musculature begins in the 10th hour of embryo development in *Drosophila*. It has been found that the membrane-spanning receptor tyrosine phosphatases DLAR, DPTP99A, DPT69D, and others determine the "choice point" when the neurons head toward the muscle fiber bundle. Apparently, after receiving the appropriate extracellular signal, they dephosphorylate the relevant messenger molecules. Deleting these surface molecules, the axons lose their guidance system. When the neurons break out of the nerve fascicles, they are homing to the muscles by the attraction of fasciclin III proteins in the muscle membranes. At this stage, the muscles secrete chemorepellent semaphorin II and thus the neurons settle down to form eventual synapses. The late bloomer protein (LBP) mediates the slowing down of the growth cone and the formation of the synapse. The neurotransmitters are then released by the terminal arbors. The agrin protein secreted by the nerve cells binds to heparin and α-dystroglycan and causes the clustering of the acetylcholine receptors. Other molecules still not identified are likely to be involved. Genetic study of neurogenesis is best defined in *Drosophila*, and *Caenorhabditis*. In the latter, the nervous system contains only about 300 cells and a large array of mutant genes (many cloned) are available. In *Drosophila*, the large *achaete-scute complex* (*AS-C*, in chromosome 1–0.0) and several other genes cooperate in neurogenesis. The differentiation and function of the systems appear to be operated by cell-to-cell contacts rather than by diffusing molecules. In most of the mutants, this communication is disrupted. For a long period of time, regeneration of neuronal tissues was unsuccessful. Recently, experiments with EAK, RGD, and RAD peptides (the letters stand for the amino acid symbols in protein sequences) promoted the formation of microscopic fibers on nerve cells, giving some hope that eventually problems with regeneration can be resolved for therapeutic purposes. *See* amino acid symbols in protein sequences; cadherin; ciliary neurotrophic factor; integrin; morphogenesis; N-CAM; nerve growth factor; neuron; neuron-restrictive silencer factor; signal transduction. (Reh, T. A. 2002. *Nature Neurosci.* 5:392.)

neurogenetics Study of the genetic bases of nerve development, behavior, and hereditary anomalies of the nervous system. (<http://www.hbp.scripps.edu/Home.html>.)

neurogenic ectoderm Set of cells that separates from the epithelium and forms the neurons by moving into the interior of the developing embryo. *See* gastrulation.

neurogenin Basic loop-helix transcription factor promoting the expression of neuronal genes. Its overexpression may inhibit glial differentiation. *See* glial cell; neuron. (Sun, Y., et al. 2001. *Cell* 104:365.)

neurohormone Secreted by neurons, such as vasopressin and gastrin. *See* gastrin, gonadotropin-releasing factor; vasopressin. (Maestroni, G. J. 2000. *Ann. NY Acad. Sci.* 917:29.)

neuroleptic Changes effected by the administration of antipsychotic drugs.

neuroligins Nerve cell-surface proteins in the brain, encoded by three genes. They bind to neurexins. *See* neurexins. (Bolliger, M. F., et al. 2001. *Biochem. J.* 356[pt 2]:581.)

neurological disorders *See* affective disorders; Alzheimer disease; Down syndrome; encephalopathy; epilepsy; gangliosidoses; hypomyelination; lysosomal storage disease; mental retardation; mucopolysaccharidoses; multiple sclerosis; neurofibromatosis; neuromuscular diseases; neuropathy; Parkinson disease; phenylketonuria; trinucleotide repeats.

neuromedin Peptide widely available in the central nervous system and the gut. It stimulates smooth muscles, blood pressure, controls blood flow, regulates adrenocortical functions, and appears to control feeding. (Kojima, M., et al. 2000. *Biochem. Biophys. Res. Commun.* 276:435.)

neuromodulin (GAP-43) Protein abundant at the nerve ends, may be involved in the release of neurotransmitters, and is a negative regulator of secretion at low levels of Ca^{2+}. *See* neurotransmitter; synaptotagmin. (Slemmon, J. R., et al. 2000. *Mol. Neurobiol.* 22:99.)

neuromuscular diseases Include a variety of hereditary ailments with the common symptoms of muscle weakness and lack of movement control. The specific and critical identification is often quite difficult because of the overlapping symptoms. *See* abetalipoproteinemia; aminoacidurias; amyotrophic lateral sclerosis; aspartoacylase deficiency; ataxia; atrophy; brain malformations; Brody disease; carnosinemia; cerebral palsy; Charcot-Marie-Tooth disease; chromosome defects; diastematomyelia; dystonia; dystrophy; gangliosidoses; glutaric aciduria; glycogen storage diseases; Kennedy disease; Kugelberg-Welander syndrome; Leigh encephalopathy; lipidoses; Moebius syndrome; multiple sclerosis; muscular dystrophy; myasthenia; myotonic dystrophy; palsy; Parkinson disease; Refsum disease; sphingolipidoses; spinal muscular atrophy; Werdnig-Hoffmann syndrome; Zellweger syndrome. (<www.neuro.wustl.edu>.)

neuron Nerve cell capable of receiving and transmitting impulses. In an adult human body, there are about 10^{12} neurons connected by a very complex network. For a long time, regeneration of neural cells was not expected. Recently, pluripotent neural stem cells have been identified that can give rise to neurons, myelinating cells, and other types of cells. *See* neurogenesis.

neuronopathy *See* muscular atrophy.

neuron-restrictive silencer factor (NRSF/REST/XBR) Protein that can bind to the neuron-restrictive silencer element (NRSE) by virtue of eight noncanonical zinc fingers. NRSE sequences were detected in at least 17 genes expressed only in the nervous system. It is a repressor of some neural-specific genes in neural and nonneural tissues, but in case of its derepression it may induce the expression of neural-specific genes in nonneural cells without converting these cells into neural cells. *See* master molecule; silencer; Zn-finger. (Kuwahara, K., et al. 2001. *Mol. Cell Biol.* 21:2085.)

neuropathy Noninflammatory disease of the peripheral nervous system. Sensory neuropathies may be associated with deficiency of α-methylacyl-CoA racemase (AMACR). This peroxisomal enzyme mediates the conversion of pristanoyl-CoA and C27-bile acyl-CoA to their (s)-stereoisomers. *See* Charcot-Marie-Tooth disease; Egr; giant axonal neuropathy; HNPP; hypomyelination; Krox; Navajo neuropathy; pain insensitivity; Riley-Day syndrome; sensory neuropathy 1.

$$CH_3CH(CH_2)_3CH(CH_2)_3CH(CH_2)_3CHCH_3$$

with CH_3 groups attached.

Pristane.

neuropeptide Signaling molecule (peptide) secreted by nerve cells. (Hansel, D. E., et al. 2001. *J. Neurosci. Res.* 66:1.)

neuropeptide Y (NPY, encoded in mouse chromosome 4) 36-amino-acid neurotransmitter that is supposed to modulate mood, cerebrocortical excitability, hypothalmic-pituitary signaling, cardiovascular physiology, sympathetic nerve function, and increased feeding behavior. NPY receptor–deficient mice become very sensitive to various pain signals, but this is not fully supported by the evidence obtained with NPY mutant mice; however, several NPY receptors exist. *See* CART; ghrelin. (Niimi, M., et al. 2001. *Endocrine* 14[2]:269.)

neurophysin Group of soluble carrier proteins (M_r about 10,000) of vasopressin and oxytocin and related hormones secreted by the hypothalamus. *See* brain, human; diabetes insipidus; hypothalamus; oxytocin; vasopressin. (Assinder, S. J., et al. 2000. *Biol. Reprod.* 63:448.)

neuropil (neuropile) Bunch of dendrites, axons in cells, and neuroglia in the gray matter of the brain.

neuropilin-1 (NRP) 130 kDa transmembrane receptor that directs axonal guidance in cooperation with semaphorin; it may be coupled with VEGF and aids angiogenesis. *See* angiogenesis; axon; KDR; semaphorin; VEGF. (Marin, O., et al. 2001. *Science* 293:872.)

Neurospora crassa (red bread mold) An ascomycete ($n = 7$, DNA 4×10^7 bp) was introduced into genetic research more than half a century ago for the exact study of recombination in ordered tetrads. Its original home is in tropical and subtropical vegetation. The asexual life cycle is about 1 week, whereas the sexual cycle requires about 3 weeks. The two mating types (*A* [5 kb] and *a* [3 kb]) are determined by a single locus. It was the first fungus to be used for centromere mapping on the basis of the frequency of second-division segregation in the linear asci of eight spores. The first auxotrophic mutants were induced in *Neurospora*, and these experiments led to the formulation of the one-gene–one-enzyme hypothesis that formed the cornerstone of biochemical genetics. Although it is a haploid organism, the availability of heterokaryons permits the study of dominance and allelic complementation. This was the first eukaryote in which genetic transformation with DNA became feasible. Currently, about 10^4 to 10^5 transformants can be obtained per μg DNA.

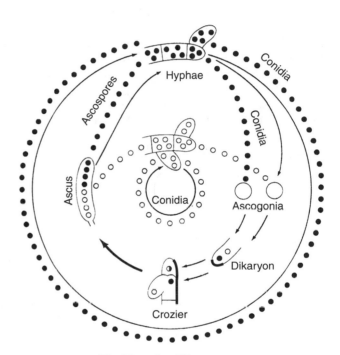

The life cycle of *Neurospora*.

Neurospora (and other ascomycetes) may display RIP (repeat-induced point mutation) observed at high frequency when duplicated elements are introduced into the nuclei (by transformation), resulting in GC→AT transitions and eviction of the duplications, as well as chromosomal rearrangements. Over 800 genes have been mapped and about one-third have been cloned. The genetic maps of *Neurospora* are generally shown in relative distances only because the different *rec* genes present in the various mapping strains may alter recombination frequencies by an order of magnitude. Mitochondrial mutants are known. The respiration-deficient *poky* mutants are somewhat similar to the petite colony mutants of budding yeast. The *stopper* mutants are also 350–5,000 bp deletions and are unable to make protoperithecia. Related species used for genetic studies are *N. sitophila* and *N. tetrasperma*. *See* channeling; diagrammed life cycle; fungal life cycles; *Neurospora* mitochondrial plasmids; RIP; tetrad analysis. (Perkins, D. D., et al. 2000. *The Neurospora Compendium*. Academic Press, San Diego,

CA; <http://www.fgsc.net/outlink.html>; <http://www.mips.biochem.mpg.de/proj/neurospora/>; <http://www.tigr.org/tdb/tgi.shtml>.)

Neurospora mitochondrial plasmids *Neurospora crassa* and *N. intermedia* strains may harbor 3.6–3.7 kb mitochondrial plasmids with a single open-reading frame encoding a 81 kDA reverse transcriptase protein. The proteins appear similar to the class II intron-encoded reverse transcriptase-like elements. The Mauriceville and Varkud plasmid DNAs apparently use a 3'-tRNA-like structure as a primer. The latter two plasmids integrate into the mitochondrial DNA at the 5'-end of the major plasmid transcript. *N. intermedia* contains a smaller Varkud satellite plasmid (VSP) that may be either linear or circular single-stranded RNA. The VSP plasmid bears resemblance to class I introns. *See* introns; mitochondrial plasmids. (Kevei, F., et al. 1999. *Acta Microbiol. Hung.* 46[2–3]:279.)

neurotactin Membrane-bound chemokine encoded in human chromosome 16q. It is overexpressed with inflammation of brain tissues. *See* chemokine. (Pan, Y., et al. 1997. *Nature* 387:611.)

neurotoxin Destroys nerve cells. *See* toxins.

neurotransmitters Several types of signaling molecules (acetylcholine, noradrenaline, glutamate, glycine, serotonin, γ-aminobutyric acid, catecholamines, neuropeptides, nitric oxide, carbon monoxide, adenosine, etc.) used for information transmission of neuronal signals. These signals can be excitatory (e.g., glutamate, acetylcholine, serotonin) or inhibitory, such as gamma-aminobutyric acid and glycine, and may open Cl⁻ channels. The secretion is activated by electric nerve impulses in response to extracellular signals. The signals are transmitted to other cells by chemical synapses across the synaptic cleft by attaching to the transmitter-gated ion channels. Although it was once believed that individual neurons can release only a single type of neurotransmitter, information now indicates that both glycine and GABA can be simultaneously released by spinal interneurons, such as the neurons between the sensory and motor neurons. *See* AMPA; calmodulin; complexin; DARPP; dopamine; GABA; ion channels; signal transduction; synapse; synaptic cleft; transmitter-gated ion channel; voltage-gated ion channel. (Matthews, G. 1996. *Annu. Rev. Neurosci.* 19:219; Lin, R. C. & Scheller, R. H. 2000. *Annu. Rev. Cell Dev. Biol.* 16:19.)

neurotrophins (NT) Growth and nutritive factors for the nerves, but may also potentiate nerve necrosis. Their receptor is either a glycoprotein (LNGFR) or trk. Neurturin, a structurally NT-related protein, also activates the MAP kinase signaling pathway and promotes neural survival. *See* aging; FOS oncogene; LNGFR; MAP; NGF; trk. (Huang, E. J. & Reichardt, L. F. 2001. *Annu. Rev. Neurosci.* 24:677; McAllister, A. K., et al. 1999. *Annu. Rev. Neurosci.* 22:295.)

neurovirulence Viral infection attacking the nerves. *See* virulence.

neurturin *See* GDNF; neurotrophins.

neurula Embryonic stage at which the development of the central nervous system begins.

neurulation Formation of the neural tube (the element that gives rise to the spinal chord and the brain) from the ectoderm during gastrulation. The process depends on multiple genes. A nucleosome assembly (NAP)-like protein, Nap1/2 (Nap1 like 2), plays a specific role. *See* ectoderm; gastrula. (Colas, J. F. & Schoenwolf, G. C. 2001. *Dev. Dyn.* 221[2]:117.)

neutered Ovariectomized (ovaries surgically removed) female. *See* castration; oophorectomy; spaying.

neutrality, conditional The difference among QTLs is distinguishable conditionally.

neutrality index, proteins

$$\frac{no.\ polymorphic\ replacement\ sites\ /\ no.\ fixed\ replacement\ sites}{no.\ polymorphic\ synonymous\ sites\ /\ no.\ fixed\ synonymous\ sites}$$

neutralizing antibody Covers the viral surface and prevents binding of the virus to cell-surface receptors and thus interferes with infection. IgG1 subtype protects against mucosal infection and infection by HIV1 and SIV. *See* acquired immunodeficiency; antibodies; SIV. (Grossberg, S. E., et al. 2001. *J. Interferon Cytokine Res.* 21[9]:743.)

neutral-loss scan Ions can decompose into both charged and uncharged (neutral) fragments. The products can be analyzed by mass spectrometry. The loss of the charged part affects the peaks in the mass spectrum of, e.g., proteins and peptides. *See* CID. (Vrabanac, J. J., et al. 1992. *Biol. Mass Spectrom.* 21[10]:517.)

neutral mutation *See* mutation, neutral.

neutral petite *See* petite colony mutants.

neutral substitution Amino acid changes in the proteins that have no effect on function.

neutron *See* physical mutagens.

neutron flux detection Personal monitoring may be based on microscopic examination of proton recoil tracks on a film or dielectric materials (i.e., which transmit by induction rather than by conduction) such as plastic or glass. Precise measurement of neutron flux is more complex. *See* radiation measurement.

neutropenia, chronic Autosomal-dominant or -recessive defect of the neutrophils. *See* neutropenia, cyclic; neutrophil.

neutropenia, cyclic Rare blood disease in humans and dogs. In humans it appears to be autosomal dominant (19p13.3), whereas in dogs it is recessive. In 21-day cycles in humans (12 in dogs), fever, anemia, and a decrease in neutrophils, granulocytes, eosinophils, lymphocytes, platelets, and monocytes recur and enhance the chances of infection. In some instances, this childhood disease is outgrown by adulthood. Granulocyte colony-stimulating factor may alleviate the condition. Neutrophil elastase (ELA2, 240 amino acids), a serine protease mutation, may account for the condition. *See* elastase; endocardial fibroblastosis; granulocyte colony-stimulating factor.

neutrophil   One of the specialized white blood cells (leukocyte). Generally, neutrophils are distinguished by irregular-shaped, large nuclei and granulous internal structure (polymorphonuclear granulocytes). They contain lysosomes and secretory vesicles, enabling them to engulf and destroy invading bodies (bacteria) and antigen-antibody complexes. They produce reactive oxygen species (superoxide, singlet oxygen, hydroxyl radical, N chloramines, etc.) and cationic antimicrobial proteins (defensins). *See* basophils; eosinophils; granulocytes; macrophage. (Reeves, E. P., et al. 2002. *Nature* 416:291.)

nevirapine   Nonnucleoside inhibitor of HIV-1 reverse transcriptase. It is much less expensive than AZT. *See* acquired immunodeficiency syndrome; AZT; TIBO. (De Clercq, E. 2001. *Curr. Med. Chem.* 8:1529.)

nevoid basal cell carcinoma (NBCCS, Gorlin-Goltz syndrome)   Quite frequently caused by new dominant mutation in human chromosome 9q31, 9q22.3, or 1p32. Patients generally have bifid ribs, bossing on the head, teeth and craniofacial defects, poly- or syndactyly, and reddish birthmarks that may become neoplastic. These cutaneous neoplasias frequently become horny but are sporadic and benign basal cell carcinomas (BCC) in the majority of cases, although they may become metastatic. The skin is very sensitive to ionizing radiation and in response develops numerous new spots (nevi). The more serious form is basal cell nevoid syndrome (BCNS). This is the most common type of cancer in the U.S. and annually about 750,000 cases occur. The major culprit in this disease is a homologue of the *Drosophila* gene *patch* (*ptc* or *tuf* [*tufted*] in chromosome 2–59). *Ptc* encodes the receptor for the product of the sonic hedgehog (*Shh*) gene, and a mutational event prevents Ptc from inhibiting SMO (smoothened), a seven-span transmembrane protein, which is a factor in carcinogenesis. Gli1 transcription factor signals the expression of BCC in the basal cells. This segment polarity locus encodes a transmembrane protein and the wild-type allele represses the transcription of members of TGF-β (transforming growth factor). *Drosophila* gene *hh* (*hedgehog*, 3–81) has the opposite effect, i.e., it promotes the transcription of TGF proteins. Homologues of these genes exist in other animals, including mice. The human homologue has now been cloned by the use of a mouse probe of *ptc*. Nucleotide sequencing revealed that a 9 bp (3-amino-acid)

duplication occurred in the afflicted member of one of the families, and in another kindred an 11 bp deletion was associated with expression of BCNS. Other developmental genes may also contribute to the expression of these genes. *See* cancer; Gli; *hedgehog*; morphogenesis in *Drosophila*; *Sonic hedgehog*; TGF-β. (Bale, A. E. & Yu, K. 2001. *Hum. Mol. Genet.* 10:757.)

nevus   Autosomal-dominant red birthmark on infants that usually disappears in a few months or years. Autosomal-dominant *basal cell nevus syndrome* is a carcinoma. It causes bone anomalies of diverse types, eye defects, sensitivity to X-radiation, etc. *See* cancer; carcinoma; skin diseases; vitilego.

Newcastle disease   Avian influenza that may be transmitted to humans. The 15,186-nucleotide viral RNA genome has been completely sequenced. *See* oncolytic viruses.

Newfoundland   Rare blood group.

newt (*Triturus viridescens*)   Salamander; $2n = 2x = 22$; its lampbrush chromosomes have been extensively studied. *See* lampbrush chromosome.

Triturus.

newton (N)   Physical unit of force; $1\,N = 1\,kg{\cdot}m{\cdot}sec^{-2}$. In molecular kinetics, usually pN (picoN, 10^{-12} N) is used.

Newton, Isaac (1642–1727)   Mathematician, physicist, and natural philosopher. Afflicted by gout. *See* gout.

nexin   Protein connecting the microtubules with a cilium or flagellum. *See* PN-1.

nexus   Junction.

Nezelof syndrome   Autosomal-recessive T-lymphocyte deficiency immune disease. The affected individuals lack cellular immunity, while display humoral immunity. The development of the thymus is abnormal. *See* immunodeficiency.

NF   Nuclear factors required for replication. NF-1 (similar to transcription factor CTF) binds to nucleotides 19–39 of adenovirus DNA and stimulates replication initiation in the presence of DBP. NF-2 is required for the elongation of replicating intermediates. NF-3 (similar to OTF-1 transcription factor) binds nucleotides 39–50 adjacent to the NF-1-binding site and stimulates replication initiation. The NFs bind to the same DNA sequence, yet they display somewhat different functions in the regulation of development. *See* CTF; DBP; OTF.

NF-1   Human neurofibromatosis involves proteins homologous to GAP and IRA. NF is a CCAAT-binding protein. *See* CAAT box; GAP; IRA1; IRA2; neurofibromatosis.

NF-AT   Nuclear factor of activated T cells is a family of transcription factors (NF-AT$_c$ and F-AT$_p$) that activate the interleukin (IL-2) promoter or the dominant negative forms block IL-2 activation in the lymphocytes. CD3 and CD2 antigens activate NF-AT jointly but not separately and they are suppressed by cyclosporin (CsA). For the nuclear import of cytoplasmic NF-AT, it is dephosphorylated by calcineurin at the nuclear import site, at the nuclear localization signal (NLS). When the level of calcineurin is lowered, NF-AT is rephosphorylated and exported from the nucleus, and the transcription regulated by it may cease. Crm1 export receptor mediates the rephosphorylation process at the nuclear export site (NES). Calcineurin binding to NF-AT may prevent Crm1 binding to NF-AT. The different members of the NF-AT family have sequence differences at their NLS and NES regions and may be acted upon by different kinases. These transcription factors, AP1 and Fos-Jun, cooperatively bind to DNA. NF-AT$_c$ transcription factor has an important role in the differentiation of heart valves and the septum. Its nuclear localization is blocked by FK506, an inhibitor of calcineurin. *See* AP; calcineurin; cyclophyllin; cyclosporin; FK506; FOS oncogene; immune system; interleukin; JUN; lymphocytes; NF-κB; T cell; transcription factors. (Rao, A., et al. 1997. *Annu. Rev. Immunol.* 15:707.)

NFAT   Family of phosphoprotein transcription factors that mediate the production of cell-surface receptors and cytokines and regulate the immune response. Its nuclear localization is facilitated by phospholipase C (PLC-γ) through signals by the cell membrane receptor. Then the phosphoinositide pathway is activated and NFAT moves into the nucleus after dephosphorylation. *See* NF-AT; phosphoinositides; rel; T cell. (Graef, I. A., et al. 2001. *Curr. Opin. Genet. Dev.* 11:505; Macian, F., et al. 2001. *Oncogene* 20:2476; Crabtree, G. R., et al. 2002. *Cell* 109:S67.)

NF-E2   Transcription factor with a basic leucine zipper domain; it regulates specific genes. Its binding site in erythroid-specific genes is GCTGAGTCA and in β-globin genes the NF-E2-AP1 motif (a subset of an antioxidant-response element) is G̲C̲T̲G̲A̲G̲T̲C̲A̲TGATGAGTCA. *See* DNA-binding protein domains; LCR; Maf; NRF. (Francastel, C., et al. 2001. *Proc. Natl. Acad. Sci. USA* 98:12120.)

NF-κB   Nuclear factor kappa binding. A *Drosophila* homologue is *Dorsal*. Transcription factor family dimeric with an REL oncoprotein, and they are specific for the IκB (immunoglobulin kappa B lymphocytes, *Drosophila* homologue Cactus) proteins. NF-κB is activated by interleukin 1 (IL-1R1) signaling cascade with the aid of TNF, hypoxia, and viral proteins. NF-κB normally stays inactivated in the cytoplasm by being associated with inhibitor IκB. The activation dissociates the two proteins and IκB is degraded before NF-κB moves to the nucleus, where it binds to the GGGACTTTCC

consensus and activates the transcription of *Twist*, which is involved in the establishment of embryonal germ layers and may inhibit *bone morphogenetic protein-4* (*Drosophila* homologue *decapentaplegic*). The target genes are involved either with cellular defense or differentiation such as IL-2, IL-2 receptor, phytohemagglutinin (PHA), and phorbol ester (PMA) synthesis. NF-κB suppresses caspase-8 and thus apoptosis is mediated by tumor necrosis factor (TNF-α) and its receptor (TNFR). Na salicylate, aspirin, glucocorticoids, immunosuppressants (cyclosporin, rapamycin), nitric oxide, etc., inhibit NF-κB involved with inflammation and infection processes. In arthritis, lupus erythromatosus, Alzheimer disease, HIV, influenza infection, and several types of cancer, the level of NF-κB is elevated. The function REL subunit is necessary for the activation of TNF and TNF-α-dependent genes and for NF-κB protection against apoptosis. Inhibition of NF-κB translocation to the nucleus enhances apoptotic killing by ionizing radiation, some cancer drugs, and TNF. Thus, managing NF-κB may assist in drug therapy of cancer. *See* Alzheimer disease; apoptosis; autoimmune diseases; bone morphogenetic protein; glucocorticoid; HIV; IκB; IKK; immunosuppressants; interleukins; morphogenesis in *Drosophila*; NF-AT; NFKB; NIK; nitric oxide; phorbol-12-myristate-13-acetate; REL; signal transduction; T cell; TFD; TNF. (Baldwin, A. S., Jr. 1996. *Annu. Rev. Immunol.* 14:649; Ghosh, S., et al. 1998. *Annu. Rev. Imunol.* 16:225; Martin, A. G., et al. 2001. *J. Biol. Chem.* 276:15840; De Smaele, E., et al. 2001. *Nature* 413:308; Tang, G., et al. 2001. *Ibid.*, 313; Ghosh, S., et al. 2002. *Cell* 109:S81; Larin, M., et al. 2002. *Nature Rev. Cancer* 2:301.)

NFKB   Nuclear factor kappa B (NF-kB, NF-κB) transcription factors. NFKB1 is in human chromosome 4q23 and NFKB2 in human chromosome 10q24. The former codes for a protein p105 and the latter for p49. Both are regulators of viral and cellular genes. IKKβ is essential for IκB phosphorylation and its subsequent degradation, leading to the activation of NFKB1. IKKα seems to be involved in the activation of NFKB2 through phosphorylation reactions. They have homology to the REL retroviral oncogene, the product of the *Drosophila* maternal gene *dl*, and some regulatory proteins of plants. NF-κB is selectively activated by nerve growth factor using the neurotrophin p75 receptor (p75^{NTR}), a member of the tyrosine kinase family, located in the Schwann cells. NF-κB usually exists within the cell in association with the inhibitory molecule IκB. Their dissociation (induced by TNF, IL-1, etc.) permits NF-κB to move into the nucleus. NF-κB regulates the expression of cytokine genes, acquired immunodeficiency, cancer metastasis, rheumatoid arthritis, inflammation, and other processes. Protein kinase θ mediates NF-κB activation via T-cell receptor and CD28. *See* Akt; IκB; IKK; morphogenesis in *Drosophila*; NF-κB; oncogenes; Schwann cell. (Senftleben, U., et al. 2001. *Science* 293:1495; Cludio, E., et al. 2002. *Nature Immunol.* 3:958.)

NF-X, NF-Y   CAAT-binding transcription factors. *See* major histocompatibility complex; RFX. (Stroumbakis, N. D., et al. 1996. *Mol. Cell Biol.* 16:192; Linhoff, M. W., et al. 1997. *Mol. Cell Biol.* 17:4589.)

NG   Nitrosoguanidine. *See* MNNG.

NGF   *See* nerve growth factor.

NGFI-A   Mitogen-induced transcription factor (probably the same as egr-1, zif/268, Krox-24, and TIS8). (Slade, J. P. & Carter, D. A. 2000. *J. Neuroendocrinol.* 12[7]:671.)

NGFI-B   *See* nur77; TIS1.

NHEJ   Nonhomologous end joining is a repair process for limited homology chromosomes broken in both chromatid strands by mutagenic agents. Proteins mediating the NHEJ repair pathway include XRCC4, DNA-PK, Ku, and DNA ligase IV. Several of the NHEJ complexes are telomere bound and participate in the maintenance of telomere length. *See* DNA-PK; DNA repair; Ku; ligase DNA; XRCC. (Haber, J. E. 2000. *Trends Genet.* 16:259; Barnes, D. E. 2001. *Curr. Biol.* 11:R455.)

NHR   Nuclear hormone receptor. *See* hormone receptors; nuclear receptors.

niacinamide (nicotinamide)   Vitamin. *See* nicotinic acid.

Niacinamide.

nibrin   *See* p95.

nicastrin   Transmembrane glycoprotein that associates with presenilins and participates in the generation of the amyloid-β-peptide fragment (Aβ) contributing to Alzheimer disease. *See* Alzheimer disease; presenilins; secretase. (Fagan, R., et al. 2001. *Trends Biochem. Sci.* 26[4]:213.)

niche   (1) Small depression on a surface or a special area of nature favorable for the species. (2) Involved in transdetermination and self-renewal of the stem cell. The Jak-Stat signaling pathway mediates in *Drosophila* spermatogenesis from stem cells. *See* signal transduction; stem cells; transdetermination. (Spradling, A., et al. 2001. *Nature* 414:98; Kiger, A. A., et al. 2001. *Science* 294:2542; Trulina, N. & Matunis, E. 2001. *Science* 294:2546.)

Nicholas II   Czar of Russia and the distressed father of a hemophiliac son, Alexis. Some historians suggested that this monogenic disease (transmitted by mother Alexandra, granddaughter of Queen Victoria of England) was one important factor in the disability of Nicholas II to deal with the social problems of his country. Major historical upheavals led to his murder and to communist rule for about three-quarters of a century. A single gene altered the history of about one-fourth of the world's population. *See* antihemophilic factors; hemophilia; Romanov.

nick   Disruption of the phosphodiester bond in one of the chains of a double-stranded nucleic acid by an endonuclease. *See* endonuclease; phosphodiester bond.

nicked circle   Circular DNA with nicks. *See* nick.

nicking enzyme   *See* nick.

nick translation   Restriction enzyme cuts DNA. DNA polymerase I enzyme of *E. coli* can attach to nicks and add labeled nucleotides to the 3′ end while slicing off nucleotides from the 5′ end, thus moving (translating) the nicks through this nucleotide labeling. This is a process of replacement replication. *See* DNA replication; Klenow fragment. (Rigby, P. W. J., et al. 1977. *J. Mol. Biol.* 113:237.)

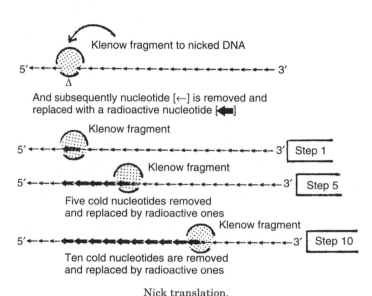

Nick translation.

Nicotiana (x = 12)   Genus in the *Solanaceae* family includes over 60 species. The majority are of new world origin, although some are native in Australia and the South Pacific islands. About 10 species have been used for smoking and alkaloid production. Some tobaccos are ornamentals. Economically the most important is *N. tabacum* (2n = 48). The latter is an amphidiploid presumably of *N. syvestris* (2n = 24) and either *N. tomentosiformis* (2n = 24) or *N. otophora* (2n = 24). Synthetic amphidiploids with the latter resemble less *N. tabacum* than *N. tomentosiformis*. *Sylvestris* x *otophora* amphiploids are more fertile than those of *sylvestris* x *tomentosiformis*. *N. tabacum* has been widely used for cytogenetic analysis; monosomic and trisomic lines are available. This species has been extensively studied by in vitro culture techniques, including regeneration of fertile plants from single cells, because it is very easy to establish cell and protoplast cultures and carry out transformation by agrobacterial vectors. This was also the first higher plant species in which Nina Fedoroff could express foreign transposable elements (*Ac* of maize) in the Jeff Schell laboratory. For manipulations involving isolated cells, most commonly the SR1 (streptomycin-resistant chloroplast mutation of P. Maliga) is used. This stock originated from the variety Petite Havana that is relatively easy to handle because of its small size. Mendelian experiments are difficult to conduct with *N. tabacum* because of its allotetraploid nature. Single-gene mutations or chromosomal markers are generally not available. Antibiotic-resistant chloroplast mutations (streptomycin, lincomycin) have been produced in *Nicotiana* and serve as tools to demonstrate recombination of the plastid genome. Transformation of chloroplast genes has been accomplished both by the biolistic methods and polyethylene glycol treatment of protoplasts. *N. plumbaginifolia,* one of the diploid species, offers some advantages for Mendelian analysis but it is less suitable for cell cultures and regeneration. *See* tobacco.

| Nicotiana tomentosa 2*x* = 24 | Nicotiana tabacum 4*x* = 48 | Nicotiana sylvestris 2*x* = 24 |

(Figures of flower shape after Goodspeed, T. H. 1954. The Genus Nicotiana. Chronica Bot. Waltham, MA.)

nicotine (*β*-pyridyl-*α*-N-methylpyrrolidine) Contact poison insecticide and an indispensable metabolite for nerve and other functions. Nicotine is considered an addictive substance. Nicotine is metabolized to cotinine (an antidepressant) by the wild-type enzyme CYP2A6. Individuals with the inactive alleles CYPA6*2 and CYPA6*3 are less likely to become smokers, or if they smoke they consume less tobacco products. The carriers or homozygotes for the null alleles are also less likely to be affected by diseases common among smokers (cancer, Alzheimer disease, Tourette syndrome, ulcerative colitis, etc.) Therapeutic treatment targeted to the wild-type enzyme may be a means of reducing smoking addiction. *See* alkaloids; smoking. (Pianezza, M. I., et al. 1998. *Nature* 393:750.)

nicotine adenine dinucleotide *See* NAD+.

nicotine adenine dinucleotide phosphate *See* NADP+.

nicotinic acetylcholine receptors Ion channel–linked receptors in the skeletal muscles and presynaptic neurons. Thus, nicotine and/or acetylcholine activate the nicotine-acetylcholine-regulated calcium channels, resulting in the release of glutamate to its receptor in the postsynaptic cells of the hippocampus. This modulation of synaptic transmission may play an important role in learning, memory, arousal, attention, and information processing. In Alzheimer diseases the nicotinic cholinergetic transmission degenerates. After smoking a cigarette, about 0.5 μM nicotine may be delivered within 10 seconds to the brain and lungs. *See* nicotine; signal transduction; smoking. (Itier, V. & Bertrand, D. 2001. *FEBS Lett.* 504[3]:118.)

nicotinic acid (niacin) Vitamin PP, coenzyme component. *See* niacinamide.

Nicotinic acid.

NIDDM Non-insulin-dependent diabetes mellitus. *See* diabetes mellitus.

nidus Origin of a process; well; pit; shallow depression.

Niemann-Pick disease (NP, sphingomyelin lipidosis) Types C (18p11-q12), A, B, and E (11p15.4-p15.1) are known as autosomal-recessive hereditary disorders of sphingomyelinase deficiency. The differences among these types involve the level of activity of the enzyme and the onset of the symptoms. The most common form, type A is identified as a severe enlargement of the liver, degeneration of the nervous system, and generally death by age 4. Phosphorylcholine ceramide (sphingomyelin) accumulates in the lysosomes. Type B is also deficient in sphingomyelinase, and since the nervous system is not affected, the patients may live to adulthood. Type C is a milder form of type A with low activity of the enzyme, and the afflicted persons may survive up to age 20. In C1, the accumulation of LDL-derived cholesterol is observed as a consequence of a defect in the NPC-1 protein, a permease with similarity to 3-hydroxy-3-methyl-glutaryl coenzyme A reductase. This protein displays homology to the Patch morphogen receptor of *Drosophila*. The NPC-2 type of the disease is due to a deficiency in HE1 (14q24.3), a lysosomal protein. In type D, the symptoms are similar to those in type C and sphingomyelin accumulates, yet the activity of the enzyme appears close to normal. The latter type apparently involves cholesterol transport from the lysosomes. In type E disease, the nervous system remains normal and sphingomyelin accumulation is limited to some organs. Type E may not be directly determined genetically. A human acid sphingomyelinase gene has been mapped to chromosome 11p15.1-p15.4. Prevalence of NP is ~7 × 10^{-6}. *See* cholesterol; epilepsy; hedgehog; LDL; SCAP; sonic hedgehog; sphingolipidoses; sphingolipids; SREBP. (Sun, X., et al. 2001. *Am. J. Hum. Genet.* 68:1361; Millat, G., et al. 2001. *Am. J. Hum. Genet.* 69:1013.)

Nieuwkoop center Position of the early embryo for dorsal and bilateral differentiation. *See* chordin; noggin; Wnt.

nif *See* nitrogen fixation.

nigericin (antibiotic of *Streptomyces*) Facilitates K+ and H+ transport through membranes and activates ATPase in the presence of a lipopolysaccharide; promotes the processing of interleukin-1*β* precursors. *See* ATPase; interleukins; ionophore. (Cascales, E., et al. 2000. *Mol. Microbiol.* 38:904.)

night blindness (nyctalopia) Defective vision in dim light is caused by several choroid and retinal defects. The X-linked phenotype (Xp11.3) is distinguished from the autosomal type by frequent association with myopia (nearsightedness). The autosomal-dominant Sorsby syndrome involves retinal degeneration caused by mutation in the tissue inhibitor metalloproteinase-3 gene (TIMP3). Mutation in the leucine-rich proteoglycan, nyctalopin (481 amino acids), encoded by gene NYX (Xp11.4), apparently disrupts retinal interconnections. *See* color blindness; day blindness; Oguchi disease; optic atrophy; retinitis pigmentosa; Stargardt disease. (Pusch, C. M., et al. 2000. *Nature Genet.* 26:324.)

nightshade (*Atropa belladonna*) Solanaceus alkaloid-producing plant, $2n = 72$. *See* burdo.

Nightshade plant.

NIH National Institutes of Health, Bethesda, MD, USA.

Nijmegen breakage syndrome (NBS) Rare 8q21 recessive phenotype closely resembling the characteristics of ataxia telangiectasia with the exception of the absence of ataxia, telangiectasia, and elevated levels of α-fetoprotein. It involves chromosomal instability, increased radiation damage, microcephaly, retarded growth, immunodeficiency, and increased chances for (breast) cancer. The 50 kb gene encodes a 754- amino-acid protein. Ataxia telangiectasia kinase phosphorylates the NBS protein. The NBS protein may be translated in an alternative manner, resulting in a somewhat modified phenotype. *See* ataxia telangiectasia; fetoprotein; Mre11; p95. (Zhao, S., et al. 2000. *Nature* 405:473; Wu, X., et al. 2000. *Nature* 405:477; Moser, R. S., et al. 2001. *Nature Genet.* 27:417; Williams, B. R., et al. 2002. *Current Biol.* 12:648.)

NIK NF-κB-inducing kinase. *See* IKK; NEMO; NF-κB; NκB.

NIL Near isogenic line. *See* inbred.

NIMA Mitosis-specific protein kinase. *See* cell cycle; histone phosphorylation; parvulins.

NINA *Drosophila* cyclophilin mediating rhodopsin maturation in the endoplasmic reticulum. *See* cyclophilin; rhodopsin.

ninhydrin Triketohydrindene hydrate, a reagent for ammonia. Upon heating, amino acids liberate ammonia, which reduces ninhydrin and produces a blue color. This reagent is used for very sensitive colorimetric estimation of amino acids or as a spot test on paper or thin-layer chromatograms. *See* paper chromatography; thin-layer chromatography.

NIS Sodium (Na)-iodide symporter is a transmembrane carrier protein responsible for transporting also radioactive iodine into the thyroid and breast epithelial cells, causing cancer. NIS activity is correlated with pregnancy and the onset of lactation. Iodine is essential for normal development. Radioactive iodine has been used to treat Basedow/Grave's disease. Regulating NIS activity may be medically useful. *See* atomic radiation; goiter; isotopes.

NISH Nonisotopic in situ hybridization. *See* FISH; in situ hybridization.

Nishimine factor Special hemostatic factor required for the generation of thromboplastic (blood-clot-forming) activity. *See* antihemophilic factors.

nisin Z Lanthionin-containing peptide antibiotic produced by some strains of the bacterium *Lactococcus lactis*. (Lanthionine is bis[2-amino-2-carboxyethyl] sulfide.) It makes pores in the membranes of gram-positive bacteria. It is highly effective and resistance against it has not been found. *See* antimicrobial peptides. (Breukink, E., et al. 2000. *Biochemistry* 39:10247.)

nitrate reductase Used in prokaryotes to reduce nitrate to nitrite under the conditions of anaerobic respiration. The enzyme is associated with molybdenum and it is essential for normal function. Nitrate reductase is generally also associated with cytochrome b. Plants and fungi reduce nitrate before it can enter into the amino acid synthetic path. Nitrate reductase mutations are generally selected on chlorate media. *See* chlorate; molybdenum; nitrogenase; nitrogen fixation. (Heath-Pagliuso, S., et al. 1984. *Plant Physiol.* 76:353; Lejay, L., et al. 1999. *Plant J.* 18:509.)

nitric oxide (NO) Endothelial cells may make and release NO in response to liberation of acetylcholine by the nerves (NOS1, 12q24.2) in the blood vessel walls. The hepatic expression gene, NOS2A, is at 17cen-q11.2. The endothelial enzyme, NOS3, is encoded at 7q36. NOS4 is a chondrocyte enzyme. Nitric oxide synthase isozymes (NOS) produce NO from L-arginine. NO activates potassium channels through a cGMP-dependent protein kinase. Then the smooth muscles of the veins become relaxed and the blood flow is boosted. This mechanism is the basis of penile erection and it the basis of the action of the antiimpotence drug Viagra (Bivalacqua, T. J., et al. 2000. *Trends Pharmacol. Sci.* 21:484). In vivo gene transfer of the endothelial isoform of nitric oxide synthase may also alleviate erectile dysfunction in animal models. NO contributes to the activation of macrophages and neutrophils in the body's defense reaction. NO may also cause cytostasis during differentiation. It is called a physiological messenger of the cardiovascular, immune, and nervous systems, memory, and learning. Deficiency of NO may cause pyloric stenosis, depression, hypotension, inflammation, aggressive behavior, recurrent miscarriage, etc. NO may contribute to the formation of reactive nitrogen species (RNS); reacting with $O_2\bullet^-$ it may generate $ONOO^-$ (peroxynitrite) radicals and may contribute to the formation of peroxides. NO may have disinfectant effects against fungi and viruses. $ONOO^-$ can oxidize sulfhydryl groups, peroxidize lipids and nitrate, or deaminate DNA bases that may cause cytotoxicity and mutation. NO regulates angiogenesis and thus may have both promoting and suppressive effects on tumor growth. Atherosclerosis, pulmonary hypertension, pyloric stenosis, stroke, multiple sclerosis, etc., may be related to NO metabolism. NOS expression/activity is controlled by steroid sex hormones. NO controls metabolism of metals. NO gene therapy has been considered to relieve hyperoxic lung injury and pulmonary hypertension. Unfortunately, the technology is not yet satisfactory because NO is inhibitory to the adenoviral vectors used (Champion, H. C., et al. 1999. *Circ. Res.* 84:1422). The 1998 Nobel Prize in medicine recognized the discovery of the physiological role of nitric oxide. *See* aggression; AKT oncogene; cGMP; chemical mutagens; circadian rhythm; depression; hypersensitivity reaction; hypotension; ion channels; luciferase; olfactogenetics; peroxynitrite; pyloric stenosis. (Stuehr, D., et al. 2001. *J. Biol. Chem.* 276:14533; Ganster, R. W., et al. 2001. *Proc. Natl. Acad. Sci. USA* 98:8638.)

nitrification Conversion of ammonium into nitrate.

nitroblue tetrazolium ($C_{40}H_{30}Cl_2N_{10}O_6$) *See* BCIP; NBT.

nitrocellulose filter Can be pure nitrocellulose or of cellulose acetate and cellulose nitrate mixtures. It can be used for immobilization of RNA in Northern blots, Southern blotting of DNA, replica plating and storage of λ-phage or cosmid libraries, bacterial colonies, Western blotting, antibody purification, etc. Nucleic acids generally have superior binding to nylon filters. (Thomas, P. S. 1980. *Proc. Natl. Acad. Sci. USA* 77:5201.)

nitrogen (N_2 = 28.02 MW) Odorless and colorless gas that becomes liquid at $-195.8°C$ and solidifies at $-210°C$; four-fifths of the volume of air is nitrogen.

nitrogenase *See* nitrogen fixation.

nitrogen cycle *See* nitrogen fixation.

nitrogen fixation Mechanisms of incorporating atmospheric nitrogen (N_2) into organic molecules with the aid of the nitrogenase enzyme complex present in a few microorganisms (*Clostridia, Klebsiella, Cyanobacteria, Azotobacters, Rhizobia*, etc.). The first step in the process is the reduction of N_2 to NH_3 or NH_4. It has been estimated that microbes annually reduce about 120 million ton of atmospheric nitrogen. Atmospheric nitrogen can be converted into ammonia industrially at high temperature and high atmospheric pressure (for the production of fertilizers and explosives). Many organisms are capable of using ammonia and can oxidize it into nitrite (NO_2^-) and nitrate (NO_3^-) by *nitrification*.

Bacterial nitrogen fixation primarily relies on two enzymes: the tetrameric *dinitrogenase* (M_r 240,000), containing 2 molybdenum, 32 ferrum, and 30 sulfur per tetramer; and the dimeric *dinitrogenase reductase* (M_r 60,000) with a single Fe_4-S_4 redox center and two ATP-binding sites. Nitrogen fixation begins by the reduced first enzyme. The reduction is the job of this second enzyme that hydrolyzes ATP. The role of ATP is the donation of chemical and binding energy. The nitrogenase complex uses different means to protect itself from air (oxygen) that inactivates it. In the symbiosis between *Rhizobia* and legumes, leghemoglobin, an oxygen-binding protein, is formed in the root nodules for protecting the nitrogen fixation system. In *Rhizobium leguminosarum* and related species, the genes required for nitrogen fixation and nodulation are in the large conjugative plasmids. The large plasmid of *Rhizobium* species NGR234 contains 536,165 bp, including 416 open-reading frames; 139 of these genes seem to be unique in any organism. In *R. meliloti*, the corresponding genes are within the bacterial chromosome. The fast-growing (colony formation 4–5 days) *Bradyrhizobium* species are common in the tropical areas, whereas in the moderate climates the various *Rhizobia* species

are found on the legumes. These two main groups cannot utilize N_2 in culture but only in the nodules. *Azorhizobia* can use N_2 without a symbiotic relation. The nodule formation requires a special interaction between the bacteria and the host. A series of nodulation factors have been identified; among others, *ENOD40* encoded a 10-amino-acid oligopeptide, a member of the TNF-R family. A homologue of this gene has been found in nonlegumes. The *Rhizobium leguminosarum* (*Vicia*) nodulation genetic system may be represented as

T N M	L EF	**D**	A B C	I J
→	←←←	←	←←	←	→→→→→

The *nod* (nodulation) genes of the bacteria are located in a plasmid within the bacterial cell The *nodD* expression is induced by root exudates (the flavonoids eryodyctyol, genistein). This is similar to the infection process of plants by *Agrobacteria* (related to *Rhizobia*) where the *virulence* gene cascade is induced by flavonoids (acetosyringone). As a consequence of turning on the *nod* system, signals are transmitted to the root hairs to curl up and develop into nodules inhabited by bacteria. The different legume species produce different inducers and the bacteria may have different genes for response, and this interaction specifies host and bacterial functions. Once within a root hair, the infection spreads to neighboring cells. The bacteria are enveloped by plant membranes and form *bacteroids*. Nodule formation assures the right supply of oxygen to the bacteroids and at the same time protects the nitrogenase system from oxygen that is detrimental to it.

Rhizobium bacteria within a soybean root nodule. (Courtesy of Dr. W. J. Brill.)

Nitrogen fixation is controlled by the *nif* and *fix* genes, encoding the FixLJ proteins. The FixL protein is a transmembrane kinase that is activated by low oxygen tension. This kinase then phosphorylates the FixJ protein, which in turn switches on *FixK* and *nifA* genes. The product of the latter interacts with an upstream element to promote the transcription of *fix* and *nif* operons. Nodule formation and nitrogen fixation depend on complex circuits between bacterial and (over 20) plant genes and environmental stimuli. There is a rather high degree of host specificity for nodulation. *R. leguminosarum bv. viciae* hosts are only pea, vetch, sweet pea, and lentil. *R. meliloti* nodulates sweet clover, alfalfa, soybean, and *Trigonella*. *Bradyrhizobium japonicum* uses soybean and cowpea (*Vigna sinensis*). Some other *Rhizobia* may have a larger range of hosts. The relationship between the host and bacteria is mutually advantageous and the symbiotic system has enormous economic value in maintaining soil fertility and crop productivity. It would be desirable to transfer the genes required for nitrogen fixation into crop plants. Some of the homologues of the nodulin genes are already present in plants (leghemoglobin). The *nif* genes of *Klebsiella* had been transferred to plants, but they failed to express. To overcome this problem, transformation of the (prokaryote-like) chloroplasts was considered, but it was not possible to protect the O_2-sensitive nitrogenase from the oxygen evolution concomitant with photosynthesis. Plants reduce nitrate to nitrite by nitrate reductase (NR). Then nitrite reductase (NiR) converts nitrite to ammonium. Glutamine synthetase (GS) and glutamine-2-oxoglutarate aminotransferase (GOGAT) incorporate the nitrogen into organic compounds (glutamate, asparaginate, etc.). The *Mesorhizobium loti* chromosome (~7 Mb) and its two plasmids (~0.35 Mb and ~0.208 Mb) and the *R. meliloti* (*Sinorhizobium meliloti*) chromosome (3.65 Mb) and two of its plasmids (1.35 Mb and 1.68 Mb) have been completely sequenced (Kaneko, T., et al. 2000. *DNA Res.* 7:331; and Galibert, F., et al. 2001. *Science* 293:668, respectively). *See* Agrobacteria; autoregulation; bacteria; glutamine synthetase; nodule; symbiosome. (Freiberg, C., et al. 1997. *Nature* 387:394; van de Sande, K. & Bisseling, T. *Essays in Biochemistry*, Bowles, D. J., ed., p. 127. Portland Press, London; Spaink, H. P., Kondorosi, Á. & Hooykas, P. J. J., eds. 1998. *The Rhizobiaceae.* Kluwer, Dordrecht; Christiansen, J., et al. 2001. *Annu. Rev. Plant Physiol. Mol. Biol.* 52:269; *Proc. Natl. Acad. Sci USA.* 2001. 98:9877–9894; Endre, G., et al. 2002. *Nature* 417:962; Geurts, R. & Bisseling, T. 2002. *Plant Cell* 14:S239.)

nitrogen mustards Radiomimetic alkylating agents. The general formula, where Al means alkyl groups, is trifunctional, with three chlorinated alkyl groups. When it is monofunctional, it has one chlorinated alkyl group; the bifunctional has two such groups. The nitrogen mustard family of antineoplastic and immunosuppressive compounds includes cyclophosphamide, uracil mustard, melphalan, chlorambucil, etc. *See* mustards; sulfur mustards. (Ross, W. C. F. J. 1962. *Biological Alkylating Agents.* Butterworth, London.)

$$Cl-Al-N \begin{array}{c} Al-Cl \\ \\ Al-Cl \end{array}$$

Nitrogen mustard.

$$\begin{array}{c} CH_3 \\ \\ CH_3 \end{array} N-N=O + NaCl + H_2O$$

N-nitrosodimethylamine.

nitrogenous base Purine or pyrimidine of nucleic acids.

nitrosamines Highly toxic carcinogens and mutagens. Lethal dose low (LDLo) of nitrosodimethylamine for mouse, orally 370 mg/kg, intraperitoneally 7 mg/kg; for dogs orally 20 mg/kg. *See* chemical mutagens; ENU.

nitrosomethyl guanidine (NNMG) Very potent alkylating agent, mutagen and carcinogen. It is light sensitive and potentially explosive. *See* chemical mutagens.

NITROSOGUANIDINE

nitrosourea *See* ENU.

nitrous acid mutagenesis (HNO_2, MW 47.02) *See* chemical mutagens.

NK *See* killer cell.

NKG2A (12p13) Inhibitory immune receptor of NK and T cells.

NKG2C, NKG2D NKG2E Activating receptors. HLA-E is the ligand.

NKT cells Similar in function to NK cells; share receptors. They have a regulatory function for lymphocytes by promptly releasing large amounts of IFN-γ and IL-4. *See* interferon; interleukin; killer cells. (Seino, K.-I., et al. 2001. *Proc. Natl. Acad. Sci. USA* 98:2577.)

N50 length Largest length such that 50% of all base pairs are contained in contigs of this length or larger. In the sequenced human genome, N50 appeared to be 82 kb. *See* contig.

NLI Nuclear LIMB domain-binding factor. *See* LDB; Lim domain.

NLS *See* nuclear localization sequence.

NMD Nonsense-mediated mRNA decay. *See* mRNA surveillance.

NMDA (N-methyl-D-aspartate) receptor Located in the postsynaptic membranes of the excitatory synapses of the brain. Activated NMDA receptors control the Ca^{2+} influx and excitatory nerve transmission. Tyrosine kinases and phosphatases regulate NMDA. *See* AMPA; CaMK; DARPP; Eph; lupus; nociceptor; PDZ domain; PTK; PTP; synapse. (Klein, R. C. & Castellino, F. J. 2001. *Curr. Drug Targets* 2:323.)

NMDAR *See* NMDA receptor.

N-methyl-N′-nitro-N-nitrosoguanidine　*See* MMNG.

NMR　*See* nuclear magnetic resonance spectroscopy.

NMYC　*See* MYC.

N nucleotides　*See* immunoglobulins.

Noah's Ark hypothesis　The low level of divergence among the different human populations indicates recent origin of the modern humans and the relatively fast replacement of the antecedent populations. *See* Eve foremother; Y chromosome. (Brookfield, J. F. Y. 1994. *Curr. Biol.* 4:651.)

nociceptin (orphanin)　*See* opiate;

nociceptor　Pain receptor involved in injuries. It is regulated by NMDA receptor and substance P neurotransmitters. The P2X$_3$ ATP receptor mediates ATP-gated cation channels and pain sensation. *See* capsaicin; NMDA; opiate. (Zubieta, J.-K., et al. 2001. *Science* 293:311; Julius, D. & Basbaum, A. I. 2001. *Nature* 413:203.)

nocturnal enuresis (bedwetting)　Dominant gene has been assigned to human chromosome 13q13-q14.3. The condition generally afflicts children under age 7, then gradually improves and usually disappears later. Its cause is the improper regulation of antidiuretic hormone and social maladjustment. *See* antidiuretic hormone.

NOD　Kinesin-like protein on the surface of oocyte chromosomes of *Drosophila* that stabilizes the chromosomes to stay on the prometaphase spindle. A similar protein, Xklp1, is found in *Xenopus* oocytes. *See* kinesin; nondisjunction; oocyte; spindle. (Clark, I. E., et al. 1997. *Development* 124:461.)

nodal　Vertebral embryonic signaling protein resembling TGF. It is involved in the control of mesoderm and endoderm differentiation, anterior-posterior determination, specification of the left-right axis, etc. *See* endoderm; left-right asymmetry; mesoderm; organizer; TGF.

node　(1) Knot- or swelling-like structure; the widened part of the plant stem from which leaves, buds, or branches emerge. (2) Crossover site in the DNA. *See* internode; junction of cellular network.

Nodes.

nodule　Small roundish structure. The root nodule of legumes harbors nitrogen-fixing bacteria. Rhizobial nodulation genes determine the formation of lipo-chitooligosaccharide signals that specify the Nod (nodulation) factors (NF). (See nitrogen fixation; recombination nodule. (Spaink, H. P. 2000. *Annu. Rev. Microbiol.* 54:257.)

nodule DNA　Formed when two H-DNAs (purine-rich *H-DNA-PU 3′ and pyrimidine-rich H-DNA-PY 3′) are combined into a structure shown below. *See* H-DNA; recombination nodule.

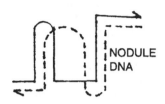

nodulin　Plant protein in the root nodules of leguminous plants. *See* nitrogen fixation.

Nod V　*Rhizobium* kinase affecting nodulation regulator protein NodW. *See* nitrogen fixation. (Loh, J., et al. 1997. *J. Bacteriol.* 179:3013; Stacey, G. 1995. *FEMS Microbiol. Lett.* 127:1.)

noggin　Bone/chondrocyte morphogenetic protein encoded at 14q22-q23; noggin mutations at the 17q22 gene cause brachydactyly, deficiencies of interphalangeal (finger and toe) joints, and bone fusion (multiple synostoses). *See* bone morphogenetic protein; organizer. (Marcelino, J., et al. 2001. *Proc. Natl. Acad. Sci. USA* 98:11353; Brown, D. J., et al. 2002. *Am. J. Hum. Genet.* 71:618.)

NOGO　Myelin-associated group of neurite growth inhibitors that prevent repair of damage and regeneration in the central nervous system and the spinal cord. (Grandpre, T. & Strittmatter, S. M. 2001. *Neuroscientist* 7[5]:377.)

noise　Random variations in a biological, mechanical, or electronic system or disturbance and interference in any of these systems. (Ozbudak, E. M., et al. 2002. *Nature Genet.* 31:69.)

nomad　Specially constructed vectors built of fragment modules in a combinatorially rearranged manner. These vectors allow sequential and directional insertion of any number of modules in an arbitrary or predetermined way. They are useful for studying promoters, replicational origins, RNA processing signals, construction of chimeric proteins, etc. (Rebatchuk, D., et al. 1996. *Proc. Natl. Acad. Sci. USA* 93:10891.)

nomadic genes　Dispersed repetitive chromosomal elements with a high degree of transposition. *See* copia; insertion elements; retroposon; retrotransposon; transposon.

Nomarski differential interference contrast microscopy　Phase shift introduced artificially in unstained objects to cause a field contrast that is due to interference with diffracted light. As a result, either dark-field (phase contrast) or bright-field contrast (Nomarski technique) can be obtained depending on the direction of the light by a quarter-wave phase shift in special microscopic equipment. In colored microscopic specimens the phase shift (and contrast of the image) is brought about by the differential staining by microtechnical dyes. *See* confocal scanning; electron microscopy; fluorescene microscopy; phase-contrast; resolution, optical.

nomenclature *See* databases; gene nomenclature assistance; gene symbols.

nonallelic genes Belong to different gene loci and are complementary. *See* allelism.

nonallelic noncomplementation Relatively rare phenomenon when recessive alleles of different gene loci fail to be complementary in the heterozygote despite the presence of a dominant allele at both loci. Such a situation occurs when the two gene loci encode physically interacting products. The mechanism may involve a particular *dosage* problem. Reduced dosage at one locus still supports the wild phenotype, but simultaneous reduction of dosage at another gene may not permit the expression of the wild phenotype. Such may be the case when a ligand and its receptor mutate at the same time. An alternative mechanism may be based on poisoning the expression of one mutation by another, e.g., one of the loci controls polymerization of another product. The first mutation is innocuous as long as the second locus is fully expressed, but reduction of the level of the product of the second gene may have deleterious consequences. Actually, physical interaction of the products of the two loci is not a requisite for the phenomenon, although most commonly physical interaction occurs between the two proteins. *See* allelic complementation; allelism test. (Yook, K. J., et al. 2001. *Genetics* 158:209.)

nonautonomous controlling element Transposable element that lost the transposase function and can move only when this function is provided by a helper element. A nonautonomous controlling element is *Ds* or *dSpm* of maize, whereas the helper element may be *Ac* or *Spm*, respectively. *See* *Ac-Ds*; *Spm*; transposable elements. (McClintock, B. 1955. *Brookhaven Symp. Biol.* 8:58; McClintock, B. 1965. *Brookhaven Symp. Biol.* 18:162.)

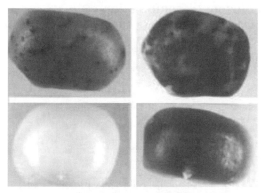

Active Spm transposase (top maize kernels) and inactive transposase (bottom endosperms) in either solid deeply colored (A) or inactivated (a-m^1) states. (Courtesy of Barbara McClintock.)

nonchromosomal genes Genes that are not located in the cell nucleus. It is more logical to call them extranuclear genes, such as the genes in mitochondria and plastids.

noncoding sequences Do not specify an amino acid sequence or a tRNA or rRNA. They are generally clustered (e.g., in human chromosome 5q31) in noncoding intergenic regions (~45%), in introns, in the nontranslated upstream (promoter) and downstream sequences, and sometimes overlap with coding sequences. Their numbers may run to hundreds of thousands in the mammalian genomes and may be well conserved (CNS) in some areas among species. They regulate various types of interleukins (IL-4, -5, -13) and are well conserved between mice and humans. The regulation has some cell specificity (T$_H$2) and may selectively affect genes at a distance over 120 kb. *See* coding sequence; interleukins; T cell. (Loots, G. G., et al. 2000. *Science* 288:136; Cliften, P. F., et al. 2001. *Genome Res.* 11:1175.)

noncompetitive inhibition Enzyme inhibition is not relieved by increasing the concentration of the substrate. *See* inhibitor; regulation of gene activity; repression.

nonconjugative plasmids Do not have transfer factors (*tra*) and are not transferred by conjugation. *See* conjugation; plasmid, conjugative; plasmid mobilization.

noncovalent bond Not based on shared electrons; these are weak bonds individually, but many of them may result in substantial interactions. *See* covalent bond; hydrogen bond.

noncrossover recombinant Recombination most commonly takes place by crossing over at the four-strand stage of meiosis or rarely at mitosis. About half of the gene conversion events do not involve crossing over of outside markers, yet the syntenic markers are recombined. Similarly, repair of double-strand breaks may produce recombination without crossing over. *See* crossing over; crossover; DNA repair; gene conversion; mitotic crossing over; recombination. (Allers, T. & Lichten, M. 2001. *Cell* 106:45.)

noncyclic electron flow Light-induced electron flow from water to NADP$^+$ as in photosystem I and II of photosynthesis. *See* drift, genetic; photosynthesis; Z scheme.

non-Darwinian evolution Evolution by a random process rather than based on selective value. *See* Darwinian evolution; evolution, non-Darwinian; mutation, neutral. (Matsuda, H. & Ishii, K. 2001. *Genes Genet. Syst.* 76[3]:149.)

nondirectiveness Principle of genetic counseling that only the facts are disclosed, but the decision is left to the persons seeking information. *See* counseling, genetic.

nondisjunction One pair of chromosomes goes to the same pole in meiosis and the other pole has neither of them; in mitosis, the movement of both sister chromatids

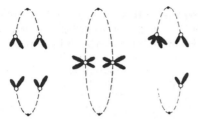

Left: Normal disjunction, Center: Normal metaphase, **Right**: Mitotic nondisjunction. (After Lewis, K. R. & John, B. 1963. *Chromosome Marker*. Little Brown, Boston.)

to the same pole. Nondisjunction can occur at meiosis I when the two nondisjoining chromosomes are homologues (bivalent) or at meiosis II when the nondisjoining elements are sister chromatids.

Nondisjunction in a normal euploid cell is called *primary nondisjunction*; when the event takes place in a trisomic cell, it is called *secondary nondisjunction*. Normal disjunction of meiotic chromosomes seems to have the prerequisite of chiasma (chiasmata). In the absence of chiasma, the likelihood of nondisjunction increases. In case nondisjunction occurs after chiasma, the chiasma seems to the be localized to the distal position of the bivalent. The human X chromosome fails to disjoin in the achiasmatic case, whereas the autosomes may be nondisjunctional even when distal exchanges take place. Meiosis II nondisjunction occurs when chiasmata take place in the near-centromeric region. In achiasmatic meiosis, orderly disjunction depends on the centromeric heterochromatin. It is interesting to note that in human females nondisjunction increases by advancing age, although the recombinational event determining chiasmata takes place prenatally. These facts require the assumption that the proteins (spindle and motor proteins) regulating normal chromosome segregation become less efficient during later phases of development. Nondisjunction occurs more frequently in the *Drosophila* and human females than in the males. In natural populations of flies, frequencies for X chromosomes varied from 0.006 to 0.241. Defects in maternal methylenetetrahydrofolate reductase (MTHFR) may create an increased risk for human chromosome 21 nondisjunction (James, S. J., et al. 1999. *Am. J. Clin. Nutr.* 70:495). Both MTHFR and methionine synthase reductase (MTRR) may affect nondisjunction of the human sex chromosomes and some autosomes. *See* aneuploidy; chromokinesin; Down syndrome; him; methionine synthase reductase; methylene tetrahydrofolate reducates; monosomy; nod; nullisomy; trisomy. (Bridges, C. B. 1916. *Genetics* 1:1, 107; Hassold, T. J., et al. 2001. *Am. J. Hum. Genet.* 69:434.)

nondivisible zones Flank the prokaryotic terminus of replication, and inversions within the zones are either deleterious or impossible.

nonessential amino acids Not required in the diet of vertebrates because they can synthesize them from precursors (alanine, asparagine, aspartate, cysteine, glutamate, glutamine, glycine, proline, serine, tyrosine). *See* amino acids; essential amino acids.

nonheme iron proteins Contain iron but no heme group.

nonhistone proteins /Proteins in the eukaryotic chromosomes. In contrast to histones, which are rich in basic amino acids, nonhistone proteins are acidic and are frequently phosphorylated. These proteins include enzymes, DNA- and histone-binding proteins, and transcription factors. Since they have regulatory roles, they are a heterogeneous group varying in tissue-specific and developmental-stage-specific manners. Along with DNA and histones, they are part of the chromatin. *See* chromatin; high-mobility group proteins; histones; illustration below (Wilhelm, F. X., et al. 1974. *Nucleic Acids Res.* 1:1043; Earnshaw, W. C. & Mackay, A. M. 1994. *FASEB J.* 8:947; Perez-Martin, J. 1999. *FEMS Microbiol. Rev.* 23:503.)

non-Hodgkin lymphoma *See* anaplastic lymphoma; lymphoma.

nonhomologous Chromosomes do not pair in meiosis, and their nucleotide sequences do not show substantial sequence similarities, except at the telomeric regions. (Mefford, H. C. 2001. *Hum. Mol. Genet.* 10:2363.)

nonhomologous end-joining (NHEJ) Mechanism of V(J)D recombination of immunoglobulin genes used to repair double-strand breaks in not completely homologous chromosomes, catalyzed by Ku. KU70/Ku80 recruit DNA-PK, RAD50, and other proteins to the broken ends. NHEJ repair frequently involves insertion of extraneous DNA into the double-strand break in yeast, plant, and mammalian cells. *See* DNA50; DNA-PK; DNA repair; double-strand breaks; end-joining; immunoglobulins; Ku; NHEJ. (Essers, J., et al. 2000. *EMBO J.* 19:1703; Pospiech, H., et al. 2001. *Nucleic Acids Res.* 29:3277; Lin, Y. & Waldman, A. S. 2001. *Nucleic Acids Res.* 29:3975.)

nonhomologous recombination Genetic exchange between chromosomes with less than the usual similarity in base sequences. The genetic repair system of the cell may work

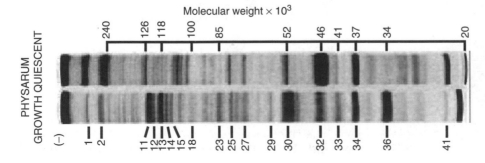

Phenol-soluble acidic protein profiles, separated by electrophoresis from *Physarum* at two developmental stages. (From Lestourgeon, A., et al. 1975. *Arch. Biochem. Biophys.* 159:861.)

against such recombination. *See* DNA ligases; heterochromosomal recombination; illegitimate recombination; Ku; mismatch repair. (Derbyshire, M. K., et al. 1994. *Mol. Cell Biol.* 14:156.)

Nonidet   Nonionic detergent used in electrophoresis.

nonisotopic labeling   *See* biotinylation; FISH; fluorochromes; nanocrystal semiconductor; nonradioactive labeling; quantum dot.

nonlinear tetrad   The ascus contains the spores in a random group of four. *See* tetrad analysis; unordered tetrad.

Nonlinear tetrad.

non-Mendelian inheritance   Extranuclear genetic elements are involved or gene conversion alters the allelic proportions. *See* chloroplast genetics; cortical inheritance; extranuclear genes; gene conversion; imprinting; meiotic drive; mitochondrial genetics; paramutation.

non-MHC-restricted cell   *See* killer cells.

nonorthologous gene displacement   In evolutionarily related species a particular function is encoded by genes with a dissimilar nucleotide sequence. *See* orthologous; paralogous.

nonparametric linkage test   *See* affected sib-pair method.

nonparametric tests   Do not deal with parameters of the populations (such as means); they estimate percentiles of a distribution without defining the shape of the distribution by the means of parameters. Nonparametric tests are simple and often they are the only ways of statistical estimations. *Wilcoxon's signed-rank test* and the *Mann-Whitney test* substitute for a *t*-test by using paired samples. The *Spearman rank-correlation test* determines correlation between two variables. The *chi-square* goodness-of-fit test, the *association test*, and many other frequently used tests in genetics are nonparametric. *See* Kruskal-Wallis test; parametric methods in statistics; robustness; statistics. (Kruglyak, L., et al. 1996. *Am. J. Hum. Genet.* 58:1347.)

nonparental ditype (NPD)   Tetrad or octad of spores in an ascus that display only two types of recombinant spores. *See* tetrad analysis.

nonpermissive condition   Regime where a conditional mutant may not thrive; the growth of the cell is not favored under these circumstances. *See* conditional lethal; temperature-sensitive mutation.

nonpermissive host   Cell that does not favor autonomous replication of a virus; it may not prevent, however, the integration of the virus into the chromosome, where the virus may initiate neoplasic transformation.

nonpermuted redundancy   Repeated sequences occur at the ends of T-even phage DNAs in the same manner in the entire population, e.g., $\boxed{1234\ldots1234}$ $\boxed{1234\ldots.1234}$. *See* permuted redundancy.

nonphotosynthetic quenching   Required in plants when the quantity of absorbed photosynthetic light exceeds the level that the plant can utilize under a particular culture or environmental condition. The protein concerned is a component of the chlorophyll-binding photosystem II complex. It protects the cells against reactive oxygen damage. *See* photosystem; Z scheme.

nonplasmid conjugation   Some transposons (Tn1545, Tn916) are capable of conjugal transfer at frequencies of about 10^{-6} to 10^{-8}. These larger transposons are very promiscuous and transfer DNA by a nonreplicative process to different bacterial species. *See* bacterial recombination; conjugation; plasmids; restriction modification; transposable elements. (Macrina, F. L., et al. 1981. *J. Antimicrob. Chemother.* 8 Suppl. D:77.)

nonpolar crosses (homosexual crosses)   In mitochondrial crosses some genes appeared to segregate in a polar manner and others apparently failed to display such polarity. The polarity was attributed to ω^+ and ω^- conditions in yeast mitochondria. Actually, ω is an intron within 21S rRNA genes. *See* mitochondrial genetics; mtDNA. (Linnane, A. W., et al. 1976. *Proc. Natl. Acad. Sci. USA* 73:2082.)

nonpolar molecule   Molecule lacks dipole features, repels water, and therefore has poor solubility (if any) in water-based solvents.

nonpolyposis colorectal cancer   *See* hereditary nonpolyposis colorectal cancer.

nonproductive infection   The virus DNA is inserted into the chromosome of the eukaryotic host and it is replicated as the host genes, but no new virus particles are released. In such a situation, the viral oncogene may initiate the development of cancer rather than the destruction of the cell as in productive infection. *See* oncogenes; temperate phage. (Butler, S. L., et al. 2001. *Nature Med.* 7:631.)

nonprogressor   Does not follow the expected course, e.g., a person infected by a virus does not develop all the symptoms of the disease.

nonradioactive labels   Biotinylated probes are used most frequently for DNA. Biotin 16-ddUTP (dideoxy-uridine triphosphate) contains a linker between biotin, and the 5′-position of deoxyuridine triphosphate is used for 3′-end labeling of oligonucleotides with the aid of nucleotidyl terminal transferase. Biotin-16-dUTP (deoxy-uridine-5-triphosphate) can replace thymidylic acid in nick translation. Then a signal-generating complex is added, containing streptavidin (protein with strong affinity for biotin) complex and peroxidase. After addition of peroxide and diamino-benzidine tetrahydrochloride substrates, a dark precipitate results if the probe is present. Some procedures use photobiotin® or digoxigenin-conjugated compounds (toxic) or other chemoluminiscents. See aequorin; chromosome painting; FISH; fluorochromes; immunolabeling.

Digoxigenin.

Photobiotin (N-biotinyl-6-aminocaproic hydrazide.)

nonrandom mating   Mate selection is by choice, either by the organism or by the experimenter. See assortative mating; autogamy; inbreeding; mating systems.

nonreciprocal recombination   See gene conversion.

nonrecurrent parent   Not used for backcrossing of the progeny. See backcross; recurrent parent.

nonrepetitive DNA   Consists of unique sequences such as the euchromatin of the chromosomes that codes for the majority of genes and displays slow reassociation kinetics in annealing experiments. It does not involve many redundancies such as SINE and LINE. See c_0t curve; c_0t value; euchromatin; LINE; SINE; unique DNA.

nonribosomal peptides   Built of amino acids by nonribosomal peptide synthetases; natural products such as penicillins or cyclosporin, etc. Iterated modules in these megasynthetases determine the identity and sequence of the amino acids. Each module activates a specific amino acid by means of coupled domains. An adenylation domain generates an aminoacyl-O-adenosine monophosphate, which is then joined to a phosphopantetheinyl group by thioester linkage of a peptidyl carrier. The chain elongates with the aid of acyl-S-enzymes. Polyketide synthases and nonribosomal peptide synthetases

synthesize some bacterial and fungal secondary metabolites. See aminoacyl-tRNA synthetases; polyketides; protein synthesis. (Silakowski, B., et al. 2001. Gene 275:233; Patel, H. M. & Walsh, C. T. 2001. Biochemistry 40:9023.)

nonsecretor   See ABH antigens; secretor.

nonselective medium   Any viable cell, irrespective of its genetic constitution, can use it in contrast to a selective medium containing, e.g., a specific antibiotic in which only the cells resistant to the antibiotic can survive. See selective medium.

nonsense-associated altered splicing   See NAS.

nonsense codon   Stop signal for translation. About one-third of the mutations in human disease involve new nonsense codons. See code, genetic; nonsense mutation.

nonsense-mediated decay   mRNA surveillance mechanism that may destroy mRNAs with a premature translation termination signal. See mRNA degradation; mRNA surveillance; translation termination. (Kim, V. N., et al. 2001. Science 293:1832.)

nonsense mutation   Converts an amino acid codon into a peptide chain terminator signal (UAA, ochre; UAG, amber; UGA, opal). See code, genetic; genetic code.

nonsense suppressor   Allows the insertion of an amino acid at a position of the peptide in spite of the presence of a nonsense codon at the collinear mRNA site. See suppressor tRNA.

non-small-cell lung carcinoma suppressor (NSCLCS) By deletion analysis it has been located to an $\sim$700 kb region in 11q23. See small cell lung carcinoma.

nonspecific binding   Involves adhesion to a surface. See cross-reaction.

nonspecific pairing   See chromosome pairing; illegitimate pairing.

nonsyndromic   Few or no other symptoms than the main characteristic of a condition appears in the diseases. See syndrome.

nonsynonymous mutation   Results in amino acid replacement in the protein.

nontranscribed spacer (NTS) Nucleotide sequences in between genes that are not transcribed into RNA, e.g., sequences between ribosomal RNA genes.

nonviral retroelements   There is a class of nonviral retrotransposable elements in Drosophila and other organisms (Ty in yeast, Cin in maize, Ta and Tag in Arabidopsis, Tnt in tobacco). These probably transpose by RNA intermediates but lack terminal repeats, although they encode reverse transcription-type functions. They contain AT-rich sequences at the 3′-termini and are frequently truncated at the 5′-end. Some of these elements are no longer capable of movement because of their extensive (pseudogenic) modification. *Drosophila*: *D* is a variable-length element with up to 100 copies and a variable number of target site duplications. *F* of variable length in 50 copies with 8–22 bp target site replication. The longest element of the *F* family is about 4.7 kb. The *Fw* element (3,542 bp) contains two long open-reading frames (ORF). The longer of these bears substantial similarity to the polymerase domains of retroviral reverse transcriptases. The other transcript (ORF1) codes for a protein that has similarities to the DNA-binding sections of retroviral gag polypeptides. The *FB* family of transposons of *Drosophila* is not related to the *F* family mentioned here (*see* hybrid dysgenesis). *G* transposable element is of variable length up to 4 kb in 10 to 20 copies with 9 bp target site duplication. *G* elements are similar to the *F* family inasmuch as they have polyadenylation signals and poly-A sequences at the 3′-end. *G* elements insert primarily into centromeric DNA. *Doc* is of variable length (up to 5 kb) with 6 to 13 bp target site duplication. The element has a variable 5′-terminus but a conserved 3′-terminus. *Jockey* is of variable length, up to 5 kb, in 50 copies. It has incomplete forms named *sancho 1, sancho 2,* and *wallaby. I* is of variable length, up to 5.4 kb, with 0 to 10 complete copies plus about 30 incomplete elements with a variable number of target site duplications (usually 12 bp). *See* copia; hybrid dysgenesis. (Higashiyama, T., et al. 1997. *EMBO J.* 16:3715; Schmidt, T., et al. 1995. *Chromosome Res.* 3[6]:335.)

Noonan syndrome (male Turner syndrome, female pseudo-Turner syndrome, 12q24.1) Apparently caused by an autosomal-dominant gene with an incidence of $1-2.5 \times 10^{-3}$. The symptoms are variable and complex, yet there is some resemblance to Turner syndrome females, but this also affects males. Mental retardation, heart and lung defects (stenosis), etc., frequently accompany it but no visible chromosomal abnormality. Missense mutation in the PTPN11 gene encoding nonreceptor protein tyrosine phosphatase is responsible for over 50% of the cases. *See* face/heart defects; heart disease; LEOPARD syndrome; stenosis; Turner syndrome. (Tartaglia, M., et al. 2001. *Nature Genet.* 29:465.)

nopaline Dicarboxyethyl derivative of arginine is produced in plants infected by *Agrobacterium tumefaciens* (strain C58). The nopaline synthase (*nos*) gene is located within the T-DNA of the Ti plasmid. The bacteria also have a gene for the catabolism of nopaline (*noc*). This and other opines serve bacteria with a carbon and nitrogen source. The *nos* promoter and tailing have been extensively used in the construction of plant transformation vectors. *See* Agrobactrium; octopine; opines; T-DNA.

Nopaline.

NOR *See* nucleolar organizer.

NOR1 Steroid receptor. *See* Nur.

noradrenaline *See* animal hormones.

norepinephrine (noradrenaline) *See* animal hormones.

NORF Not annotated open-reading frame. *See* annotation; ORF.

normal deviate The difference between two estimates divided by the standard deviation. The normal deviates have been tabulated by R. A. Fisher (see Fisher, R. A. & Yates, F. 1963.*Statistical Tables*. Hafner, New York) and the probabilities can be conveniently read. An example: The difference between two means = 4.5. The standard deviation of the difference of the means is $s_d^2 = s_1^2 + s_2^2 = 3^2 + 1^2 = 9 + 1 = 10$; hence, $s_d = \sqrt{10} = 3.16$. Thus, 4.5/3.16 = 1.42 = the normal deviate.

The probability of each entry is found by adding the column heading to the value of the far-left column at the appropriate line. This table is incomplete and is used only for illustrating the procedure of the calculation.

Probability →	0.00	0.001	0.002	0.01	0.02	0.05	0.08	0.09
0.0	∞	3.2905	3.0902	2.5758	2.3263	1.9599	1.7507	1.9954
0.1				1.6449	1.5982	1.4395	1.3408	1.3106

In order to find the probability sought, the value nearest to the normal deviate calculated is to be located in the body of the table above. The number on the second line under 0.05 (in the heading) is 1.4395, not too far from our estimated normal deviate of 1.42. Therefore, the probability will be at least 0.05 + 0.1 = 0.15. Thus, the difference between the two means of this example is statistically not significant. If we would have had a normal deviate of 1.96 or higher (on the first line), the probability would have exceeded 0.05 + 0.0 = 0.05, a minimal value of significance by statistical conventions. Interpolations for the tables may be required. *See* normal distribution; z distribution.

normal distribution Derived from binomial rather than experimental data. The normal distribution is continuous, yet it can be represented by the binomial distribution $(p + q)^k$, where k is an infinitely large number. The normal distribution takes the shape of a *bell curve*; however, a variety of bell curves may exist depending on the two critical parameters: μ (mean) and σ (standard deviation). The *mean* corresponds to the center of the bell curve, and σ measures the spread of the variates. The *normal distribution probability density function* is represented by the formula

$$f(x) = Z = \frac{1}{\sqrt{2\pi\sigma^2}} e^{-(Y-\mu)^2/2\sigma^2}$$

$f\mu = 0$ and $\sigma = 1$ then

$$f(z) = \frac{1}{\sqrt{2\pi}} e^{-z^2/2}$$

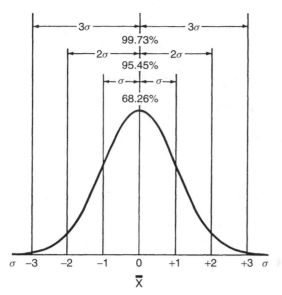

The normal distribution is represented by a normal curve, characterized by the standard deviation σ and within ± 1, ± 2 and $\pm 3\sigma$, 68.26, 95.45, and 99.73% of the population is expected around the mean, e.g., if the population is 100, the mean is 30, and the standard deviation = 5, approximately 68 individuals will be within the range of 30 ± 5.

where Z indicates the height of the ordinate of the curve, $\pi = 3.14159$, $e = 2.71828$, and Y = variable

Although in absolute sense perfect normal distribution is only a mathematical concept, experimental data may fit in a practically acceptable manner and can be used for the characterization of the available set of data. See mean; standard deviation.

normalization   Process in gene discovery by which the mutant gene is used to identify the wild-type (normal) sequences. Normalization selectively reduces the level of representation of multicopy gene products and thus facilitates the discovery of specific gene functions among the large pool of mRNAs. Before normalization, a typical cell expresses 1,000 to 2,000 different abundant messages at the level of >500 copies per cell. Middle abundant messages before normalization may appear in 15–500 copies per cell. The ca. 15,000 rare messages may appear in less than 5 copies per cell. Normalization is carried out by the generation of cDNA from a single tester RNA and a single cDNA with the aid of poly(dT) primer. The use of the poly(dT) primer results in long poly(dA/dT) sequences that become tangled and cause the loss of many templates during normalization and subtraction. Alleviation of this problem is expected by the use of anchored oligo(dT) primers. See gene discovery; subtractive cloning. (Wang, S. M., et al. 2000. *Proc. Natl. Acad. Sci. USA* 97:4162.)

normal solution   Each liter (L) contains a gram molecular weight equivalent quantity (quantities) of the solute; thus, it can be 1N or 2N…5N, etc. That is, in 1 L of the solvent there is a solute equivalent in gram(s) atom of hydrogen. Thus, 1 normal HCl (MW = 36.465, analytical-grade hydrochloric acid) requires 36.465 g HCl in 1 L water. (The commercially available HCl contains about 38% HCl [specific gravity about 1.19]; thus, the 1N solution should have approximately 80.62 mL of the reagent in a total volume of 1 liter.) The

molecular weight of sulfuric acid (H_2SO_4) is 98.076 and thus 98.076/2 = 49.038 gram/L is required for the 1N solution because H_2SO_4 has 2 hydrogen atom equivalents. The solutions are generally prepared in volumetric flasks. See mole.

Norrie disease   X-linked (Xp11.2-p11.4) retinal neoplastic disease (pseudoglioma) often complicated by diverse other symptoms, cataracts, microcephalus, etc. The same region includes two monoamine oxidase (MAOA and MAOB) genes. Deletion of this site may result in neurodegeneration, blindness, hearing loss, and mental retardation. MAO defects have been associated with several psychiatric disorders. See contiguous gene syndrome; eye diseases; MAO.

Northern blotting   Electrophoresed RNA is transferred from gel to specially impregnated paper to which it binds. It is then hybridized to labeled (radioactive or biotinylated) DNA probes, followed by autoradiography or streptavidin-bound fluorochrome reaction for identification. This procedure is suitable for the study of the expression profile of a large number of open-reading frames under different conditions. See amino-benzyloxymethyl paper; autoradiography; biotinylation; North-Western blotting; Southern blotting; Western blotting. (Brown, A. J. P., et al. 2001. *EMBO J.* 20:3177.)

Norum disease   See lecithin: cholesterol acyltransferase deficiency.

Norwalk virus   Single-stranded, positive-sense, ~7.7 kb RNA virus within a shell of 180 copies of a single ~5.6 kDa protein organized in an icosahedron. It is responsible for 96% of nonbacterial gastroenteritis cases in the U.S. Its cultivation is difficult. See icosahedron. (Belliot, G. M. 2001. *J. Virol. Methods* 19[2]:119.)

nosocomial   Hospital-originated condition, e.g., infection acquired in a hospital.

nosology   Subject area of disease classification.

NOT (CDC39)   Global regulator complex of yeast. See CDC39. (Maillet, L. & Collart, M. A. 2002. *J. Biol. Chem.* 277:2835.)

Notch (*N*)   *Drosophila* gene locus, map location 1–3.00. Homozygotes and hemizygotes are lethal; the wing tips of heterozygotes "notched" (gapped) and thoracic microchaetae are irregular. The expression is greatly variable, however, in the independently obtained mutants. Homozygotes may be kept alive in the presence of a duplication containing the normal DNA sequences. All of the mutants, whether they are homozygotes or hemizygotes, display aberrant differentiation in the ventral and anterior embryonic ectoderm. The nervous system is abnormal. The genetic bases of the mutations are deletions, rearrangements, or insertions. Homologues of the *Notch* gene are also found in vertebrates (including humans) and the transmembrane gene product is involved in cell fate determination. The 300-kDa Notch protein is a single-pass membrane receptor for several ligands in various developmental pathways. The extracellular domain includes

36 tandem epidermal growth factor (EGF)–like repeats and three cysteine-rich repeats. The intracellular part includes six tandem ankyrin repeats, a glutamine-rich domain, and a PEST sequence. *N* activation requires proteolytic cleavage of the intracellular domain of the encoded protein product and its association with the protein complex called CSL (Cbf1, Su[H], Lag-1). The transcription of several genes may ensue. In *Drosophila*, the two most important proteins that interact with the extracellular domain of N are Delta and Serrate, both single-pass transmembrane proteins (the equivalents of the latter in vertebrates are Delta and Jagged and LAG-2 and APX-1 in *Caenorhabditis*). Fringe proteins display fucose-specific β1,3 N-acetylglucosaminyltransferase activity, elongate the fucose residues on the epidermal growth factor–like sequence repeats of Notch, and modulate signaling. The primary target of N signaling is the gene *Enhancer of split* (*E[spl]*) in *Drosophila*, where it encodes a basic helix-loop-helix nuclear protein. The transcription factor Suppressor of hairless (Su[H]) is a major effector (in mammals CBF1/RJBk and in the nematode LAG-1 perform the same task). Several extracellular and intranuclear proteins are also involved in the regulation of Notch. Notch orchestrates up and down modulation of numerous proteins. Notch signaling initiates the metamerism of the somites. The Mesp2 basic helix-loop-helix protein contributes to the rostral (anterior) specificity, whereas presenilin is required for caudal differentiation. Both mediate the action of the Delta ligands of Notch. Notch affects practically all types of morphogenetic/developmental processes such as those of the central and peripheral nervous systems, eyes, spermatogenesis, oogenesis, muscle development, heart, imaginal disk, apoptosis, proliferation, etc. *See* ADAM; ankyrin; CADASIL; *Drosophila*; EGF; helix-loop-helix; knirps; morphogenesis; PEST; presenilins. (Baonza, A. & Garcia-Bellido, A. 2000. *Proc. Natl. Acad. Sci. USA* 100:2609; Allman, D., et al. 2002. *Cell* 109:S1.)

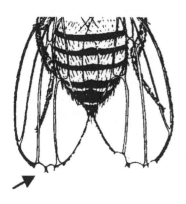

notochord Rod-shape cell aggregate that defines the axis of development of the animal embryo, giving rise to the somites that develop into the vertebral column in higher animals. *See* chordoma; organizer; somite.

notum Dorsal part of the body segments in arthropods. *See* dorsal.

NPC Nuclear pore complex is involved with the import of proteins into the nucleus. It is a 125 Mda complex embedded into the nuclear envelope. It is associated with cytoplasmic proteins, importin α and β, Ran (small guanosine triphosphatase), Nup214/CAN, and nuclear transport factor 2 (NTF2). *See* nuclear pore. (Walther, T. C., et al. 2001. *EMBO J.* 20:5703; Komeili, A. & O'Shea, E. K. 2001. *Annu. Rev. Genet.* 35:341.)

NPH Nucleoprotein helicase disrupts the interaction between U1A and the spliceosome; unwinds double-stranded RNA and protein associations. *See* DExH; snRNA; spliceosome. (Jankowsky, E., et al. 2001. *Science* 291:121.)

Npi Peptidyl-prolyl isomerase recognizing the nuclear localization sequences of the FKBP family. *See* FKBP; nuclear localization sequence; PPI.

NPK Plant MAPKKK required for the activation of the MAPK path involved in the regulation of auxin-responsive transcription. *See* MAPK; MAPKKK.

NPP-1 *See* PP-1.

NPTII *See* amino glycoside phosphotransferase; aph(3′)II; neomycin phosphotransferase.

NPXY Protein amino acid sequence: Asn-Pro-X-Tyr. *See* amino acid symbols in protein.

nr Prefix for rat (*Rattus norvegicus*) protein or DNA, e.g., nrDNA.

NRE1 *See* DNA-P.

NRF (NF-E related factor) *See* NF-E.

NRSF *See* neuron-restrictive silencer factor.

NRY Nonrecombinant part of the **Y** chromosome. *See* Y chromosome.

NSF National Science Foundation, U.S. agency for supporting basic research and science. (2) N-ethylmaleimide-sensitive factor is a vesicle transport ATPase component of the SNAP (soluble NSF attachment protein) in the fusion complex of vesicles. This then binds to the membrane receptor SNARE or tSNARE (target membrane SNARE). ATPase NSF and SNAP can also disrupt the SNARE, synaptobrevin, syntaxin ternary complex anchored to the lipid bilayer of the membrane. These play a role in the transport between the endoplasmic reticulum and the Golgi apparatus and neurotransmitter release. SNAPs are involved in the regulation of transcription of snRNA by RNA polymerases II and III. *See* caveolae; membrane fusion; RAB; SNARE; snRNA; synaptic vessel. (Yu, R. C., et al. 1999. *Mol. Cell* 4:97; May, A. P., et al. 2001. *J. Biol. Chem.* 276:21991.)

N-Ethylmaleimide.

nsL-TP Nonspecific lipid transfer proteins.

N-TEF *See* TFIIS.

N-terminus (amino end)   End of a polypetide chain where translation starts.

NTP Nucleotidetriphosphates.

NTPase (nucleotide triphosphatase, 14q14)   Divalent cation-dependent transmembrane enzymes.

NtrB Kinase in the bacterial nitrogen fixation system affecting NtrC. *See* nitrogen fixation.

NtrC Protein regulates the bacterial nitrogen assimilation signaling path by phosphorylation. *See* nitrogen fixation; signal transduction.

NTS *See* nontranscribed spacer.

NuA3, NuA4 Nucleosomal acetyltransferase of histone 3/histone 4, respectively. *See* histone acetyltransferase. (John, S., et al. 2000. *Genes & Development* 14:1196.)

nuage Large cytoplasmic inclusions in the cells destined to become the germline. (Ikenishi, K. 1998. *Dev. Growth. Differ.* 40:1.)

NU body Unit of a chromosome fiber containing $8+1$ molecules of histones and about 240 nucleotide pairs of DNA. *See* nucleosome.

nucellar embryo   Develops from diploid maternal nucellus. *See* apomixis; nucellus.

nucellus Maternal tissue of the ovule surrounding the embryo sac of plants. *See* embryo sac; megagametophyte development.

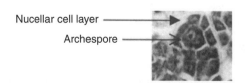

Nucellar cell layer surrounding the prominent archespore.

nuclear dimorphism In *Ciliates*, there is the genetically active micronucleus and the much larger macronucleus responsible for the metabolic functions of the cell but not for its inheritance. *See Paramecium*.

nuclear envelope   *See* nuclear membrane.

nuclear export sequences (NES)   Leucine-rich tracts of proteins and the Ran GTPase complex but no GTPase activity. *See* GTPase; nuclear localization sequences; nuclear pore; Ran. (Ossareh-Nazari, B., et al. 2001. *Traffic* 2[10]:684.)

nuclear family Parents and children living in the same household. The *extended family* includes other relatives of the household. These are actually social terms. *See* family.

nuclear fission Splitting of heavy atoms to elements, e.g., uranium may thus produce barium, krypton, etc., resulting in the liberation of a great amount of thermal energy and causing the split nuclei to fly apart at great velocity. *See* isotopes; nuclear fusion; nuclear reactor.

nuclear fusion Atomic energy can be liberated either by nuclear fission or by fusing lighter atomic nuclei (e.g., hydrogen) into heavier ones; during the thermonuclear reaction process, a huge amount of energy is generated. If this reaction could be made slower (rather than explosive), humanity could get access to vast amounts of inexpensive energy. *See* nuclear fission, fertilization.

nuclear import   *See* nuclear export; nuclear pore.

nuclear inclusions In diseases caused by trinucleotide repeats, polyglutamine proteins may accumulate in the neurons, causing degeneration. *See* trinucleotide repeats. (Ross, C. A. 1997. *Neuron* 19:1147; Chai, Y., et al. 2001. *J. Biol. Chem.* 276:44889.)

nuclear lamina Three polypeptides form a fibrous mesh within the cell nucleus attached to the inner nuclear membrane and participate in the formation of the nuclear pores. They anchor chromosomes to the membrane and control the dissolution of the nuclear membrane during mitosis. *See* mitosis; nuclear pores. (Guillemin, K., et al. 2001. *Nature Cell Biol.* 3[9]:848.)

nuclear localization sequences (NLS)   Direct the movement of proteins imported into the nucleus through ATP-gated pores. The targeting proteins (also called nuclear localization factor, NLF) usually contain one or more basic amino acid clusters. In the SV40 T antigen, there is a [126]PKKKRKV[132] at the location shown by the superscripts. In the nucleoplasmin, there are two bipartite clusters KRPAAIKKAGQAKKKK. Proteins so equipped are transported to the nucleus by karyopherin/importin proteins. *See* cofilin; export adaptors; export signals; footprinting; importin; ion channels; karyopherin; NF-κB; nuclear export sequences; nuclear pore; nuclear proteins; nucleoplasmin; Ran; second cycle mutation; T antigen; transportin. (Post, J. N., et al. 2001. *FEBS Lett.* 502:41.)

nuclear magnetic resonance spectroscopy (NMR) Physical method for studying three-dimensional molecular structures. Electromagnetic radiation is pulsed at small 15–20 kDa proteins in a strong magnetic field. This results in a change in the orientation of the magnetic dipole of the atomic nuclei. When the number of protons and neutrons is not equal in an atomic nucleus, they display a spin angular momentum. In the strong magnetic field, the spin is aligned, but it becomes misaligned in an excited state as a consequence of the radio frequencies of electromagnetic radiation. When the protons and neutrons return to the aligned state, electromagnetic radiation is emitted. This emission displays characteristics dependent on the neighbors of the atomic nuclei. Thus, from the emission spectrum it is feasible to estimate the relative position of hydrogen nuclei in different amino acids of the protein. If the primary structure of the polypetide is known from amino acid sequence data, the three-dimensional arrangement of the molecules can be determined. NMR technology—unlike X-ray crystallography—does not require that the material be in a crystalline state. NMR has found its use for the analysis of medical specimens and also in plant development and infection of tissue by pathogens. Relatively thick (500 μm) tissue slices can be studied by this nondestructive method. With the aid of color video processors, the distribution of water content can be followed and photographed. Magnetic resonance imaging (MRI) may permit viewing targets deep in the body and visualization of enzymes within a living organism (tadpole) without destruction. NMR is becoming a tool of proteomics. *See* circular dichroism; electromagnetic radiation; imaging; proteomics; Raman spectroscopy; X-ray crystallography. (Yee, A., et al. 2002. *Proc. Natl. Acad. Sci. USA* 99:1825; Walter, G., et al. 2000. *Proc. Natl. Acad. Sci. USA.* 97:5151; Ratcliffe, R. G. & Shachar-Hill, Y. 2001. *Annu. Rev. Plant Physiol. Plant Mol. Biol.* 52:499; Palmer, A. G., III. 2001. *Annu. Rev. Biophys. Biomol. Struct.* 30:129; Yee, A., et al. 2002. *Proc. Natl. Acad. Sci. USA* 99:1825; <http://www.bmrb.wisc.edu>.)

nuclear matrix Actin-containing scaffold extended over the internal space within the nucleus that participates in and supports various DNA functions, including replication, transcription, processing transcripts, receiving external signals, chromatin structure, etc. *See* actin; scaffold. (Lepock, J. R., et al. 2001. *Cell Stress Chaperones* 6[2]:136; <http://transfac.gbf.de/SMARtDB/>.)

nuclear matrix-associated bodies Number of different proteins involved in controlling cellular proliferation and oncogensesis. *See* PML. (Zuber, M., et al. 1995. *Biol. Cell.* 85:77; Carvalho, T., et al. 1995. *J. Cell Biol.* 131:45.)

nuclear membrane Surrounds the cell nucleus of eukaryotes. The pores of this membrane facilitate the export and import of molecules. The nuclear membrane seems to disappear during the period of prophase-metaphase and is reformed again during anaphase and telophase. The assembly of the envelope is induced by Ran and enhanced by RCC1. Dynein and dynactin proteins mediate the breakdown of the membrane. The Brr6 integral membrane protein surrounds the nuclear pore and regulates the transport through the pore. *See* dynactin; dynein; lamins; mitosis; nuclear pore; Ran; RCC1. (de Bruyn Kops, A. & Guthrie, C. 2001. *EMBO J.* 20:4183; Aitchison, J. D. & Rout, M. P. 2002. *Cell* 108:301; Burke, B. & Ellenberg, J. 2002. *Nature Rev. Mol. Cell Biol.* 3:487.)

nuclear pores Perforations in the nuclear membrane, defined on the inner side by the nuclear pore complex, which consists of eight large protein granules in an octagonal pattern and is associated with nuclear lamina. The pores selectively control the in-and-out traffic through their about 100 nm channel. An active mammalian cell nucleus may have 3,000–4,000 nuclear pore complexes. The nuclear pore may have a mass of 125 MDa and is built of about 50–100

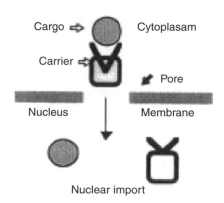

polypeptides (some of them in replicates). Small molecules may have passive passage, but larger ones (up to 25 nm in diameter) require active transport. In a minute, about 100 ribosomal proteins and 3 ribosomal subunits may pass through a pore. The transport requires appropriate recognition sites without clear consensus sequences. The major import factors are *importin α* and *β*; *α* picks up the molecules to be imported and *β* eases them through the pore by *Ran GTPase*; protein *pp15* provides the energy. The translocation may require effectors and other energy sources. After delivery, the import complex disengages and the two importin subunits quickly return to the cytoplasm. The export of RNA—generally following trimming and splicing—requires binding to special nuclear export sequences of export proteins. This process may be mediated by the Rev factor that has binding sequences to HIV-1 transcripts and may allow unspliced RNA through.

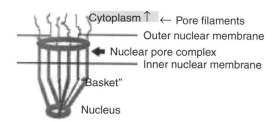

Schematic details of the nuclear pore. (Modified after *Science* 279:1129 [1998].)

Human cells have hRip (human Rev-interacting protein) or Rab. Viral RNAs, mRNA, tRNA, rRNA, and snRNA may each have somewhat different export system variants. The general transcription factor TFIII has some features that may qualify it for the viral Rev functions. The capped (by $m^1GppN-5'$) RNA ends are apparently joined in a cap-binding CBP80 and CBP20 protein complex (CBC) for export. The M9 region of the heterologous nuclear ribonucleoprotein (hnRNAP) also appears to be involved in RNA trafficking through the nuclear pores. The Ca^{2+} content of the nucleus may regulate the traffic of intermediate-size molecules (>70 kDa) across the membrane.

Some of the molecules exported from the nucleus to the cytoplasm are very large such as ribosomal nucleoprotein components. These molecules can pass through the pore only in extended forms. Mutational analysis, particularly in yeast, revealed a great deal about the protein components of the nuclear traffic. Some small molecules (proteins and RNA) may traffic by passive diffusion. The yeast proteins Mlp1 and Mlp2, localized to the basket filaments, play a role in active transport. These two proteins, as well as the associated Ku70 protein, nuclear pore protein 145 (Nup145), and Rap1, regulate the transcription of the genes situated in the telomeric region. *See* BR; cap; CIITA; CRM1; export signal; hnRNA; importin; karyopherin; NPC; nuclear localization sequence; nucleocytoplasmic interactions; nucleoporin; nucleus; RAN; Rap; RCC; RCC1; Rev; RNA export; RNA transport; RNP; transcription factors; transportin. (Fabre, E. & Hurst, E. 1997. *Annu. Rev. Genet.* 31:277; Mattaj, I. W. & Englmeier, L. 1998. *Annu. Rev. Biochem.* 67:265; Nakielny, S. & Dreyfuss, G. 1999. *Cell* 99:677; Route, M. P., et al. 2000. *J. Cell Biol.* 148:635; Wente, S. R. 2000. *Science* 288:1374; Ribbeck, K. & Görlich, D. 2001. *EMBO J.* 20:1320; Rout, M. P. & Aitchison, J. D. 2001. *J. Biol. Chem.* 276:16593; Cyert, M. S. 2001. *J. Biol. Chem.* 276:20805; Enninga, J., et al. 2002. *Science* 295:1523; Nemergut, M. E., et al. 2002. *J. Biol. Chem.* 277:17385; Ishii, K., et al. 2002. *Cell* 109:551; <http://www3.shinbiro.com/~virbio/index.htm>.)

nuclear proteins   Function within the nucleus. *See* nuclear localization signals.

nuclear reactor (atomic reactor)   Splits uranium or plutonium nuclei, and the neutrons in chain reactions split additional atomic nuclei. Cadmium and boron rods that can absorb neutrons regulate the process. For moderation (slowing down the neutron), heavy water and pure graphite are used. This makes the splitting more efficient. Only U^{235} is fissionable, but more than 99% of naturally occurring uranium is U^{238}, making enrichment mandatory. By capturing one neutron and releasing two electrons, in some reactions U^{238} and thorium^{232} yield fissionable plutonium^{239} and U^{233}. The liberated thermal energy can thus be used to drive water vapor turbines to generate electric energy for peaceful purposes. In some countries (France, Hungary), very substantial parts of electric energy are provided by atomic power plants. Atomic reactors also produce radioactive tracers used in basic research and medicine. By combining U^{235} and plutonium (Pu^{239}), *atomic bombs* may be produced. In *hydrogen bombs*, uranium or plutonium ignites a fusion process between deuterium (heavy hydrogen isotope [D] of atomic weight 2.0141) and tritium (H^3), generating helium in the process yielding millions of degrees of heat. All nuclear reactions generate some fallout of radioactive isotopes that may pose serious genetic danger to living organisms. The use of nuclear energy for peaceful purposes, under carefully shielded and monitored conditions, may be justified as a compromise between environmental protection and sustainable industrial society. The use of thermonuclear weapons does not have any justification. *See* atomic radiations; cosmic radiation; fallout; isotopes; plutonium; radiation effects; radiation hazard assessment; radiation protection.

nuclear receptors   Provide links between signaling molecules and the transcriptional system. They use several domains with different functions. The A/B domains are receptive to transactivation. The conserved C domains with Zn fingers bind DNA (usually after dimerization). The D domain is a flexible hinge. The E domain binds ligands (hormones, vitamin D, etc.),

dimerizes, and controls transcription. The role of the F domain is not quite clear. Fusing specific transactivator and receptor domains to nuclear receptors enhances their efficiency. The nuclear receptor superfamily includes more than 150 transcription factors. *See* coactivators; corepressor; estrogen receptor; glucocorticoid-response elements; hormone-response elements; ligand; NHR; orphan receptors; receptor; retinoic acid; transactivation diagram next page. (Weatherman, R. V., et al. 1999. *Annu. Rev. Biochem.* 68:559; Anke, C., et al. 2001. *Annu. Rev. Biophys. Biomol. Struct.* 30:329; Rosenfeld, M. G. & Glass, C. K. 2001. *J. Biol. Chem.* 276:36865; Chawla, A., et al. 2001. *Science* 294:1866; Xu, W., et al. 2001. *Science* 294:2507; Xu, H. E., et al. 2002. *Nature* 415:813.)

nuclear RNA   *See* hnRNA, transcript.

nuclear shape   Influencedly the organization of the cytoskeleton and nuclear matrix proteins, and these factors also impact gene expression. *See* cytoskeleton. (Thomas, C. H., et al. 2002. *Proc. Natl. Acad. Sci. USA* 99:1972.)

nuclear size and radiation sensitivity   Directly proportional except in polyploids, where multiple copies of the genes may greatly delay the manifestation of the damage. Although the larger nucleus is a more vulnerable target for damaging agents, the multiple copies of the same gene may compensate for each other's function. *See* physical mutagens; radiation effects; radiation vs. nuclear size. (Sparrow, A. H., et al. 1961. *Radiat. Bot.* 1:10.)

nuclear targeting   *See* nuclear localization sequences, receptor-mediated gene transfer.

nuclear traffic   *See* nuclear localization sequences; nuclear pore; RNA transport.

nuclear transfer   *See* nuclear transplantation.

nuclear transplantation   The nuclei of plant and animal cells can be inactivated by UV or X-radiation, then the nucleus can be replaced with another. Alternatively, the nucleus of one cell is evicted (enucleation, leading to the formation of a cytoplast) by cytochalasin (a fungal toxin) and the nucleus (karyoplast) is saved. Another cell is subjected to a similar procedure and the enucleated cytoplast is saved (its karyoplast is discarded). Then by reconstitution, the nucleus of cell #1 is introduced into the cytoplast of cell #2. In successful cases, the transplanted nucleus is functional and can direct the normal development of the cell into an organism (frog), indicating totipotency of nuclei harvested from different cell types. Similar transplantation procedures have been successfully employed for various mammals, including sheep and cattle, where viable progeny has also been obtained when the oocytes with transplanted nuclei were implanted into ewes or cows after several in vitro passages. The donor nuclei were obtained from still totipotent embryonic tissues.

In 1997, the transfer to enucleated metaphase II–stage eggs a mammary cell nucleus of a 6-year-old ewe resulted in a normal lamb (Dolly). Before transfer, the mammary cells were cultured at low-concentration (0.5%) serum to force the cells to exit from the cell cycle into G_0 stage. This step was required

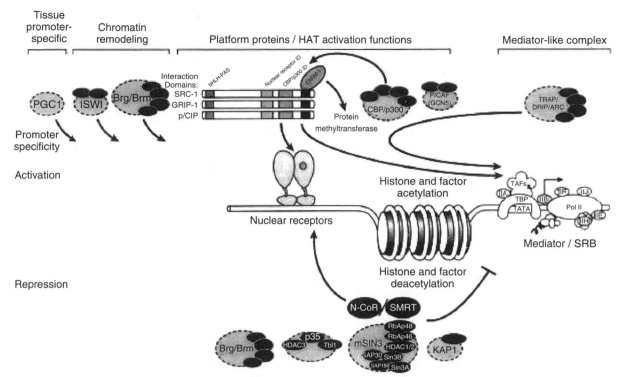

The activation and repression circuits tied to the nuclear receptor. PGC1: peroxisome proliferator-activated receptor-γ coactivator; ISWI: ATPase subunit of NURF, a chromatin remodeling protein, Brg/Brm: DNA-dependent human ATPase active in SWI/SNF-like manner in chromatin remodeling; SRC-1: hormone receptor coactivator that enhances the stability of the transcription complex controlled by the progesteron receptor; GRIP: glutamate receptor interacting protein that contains seven PDZ domains, interacts with the C end, and links AMPA (a glutamate receptor targeting protein) to other proteins; CARM-1: mediates transcription when recruited to the steroid receptor coactivator (SRC-1) and cofactors p300 and P/CAF; p/CIP: signal transducer and coactivator protein interacting with p300 (CREB), and it regulates transcription and somatic growth of mammals; GCN5: yeast transcriptional coactivator; TRAP/DRIP/ARC: multiprotein complex regulating transcription; TAFs: TATA box−associated transcription factors; TBP: TATA box−binding protein; IIA, IIB, IIE, IIF, IIJ, IIH: general transcription factors; *Pol*II: DNA-dependent RNA polymerase; MEDIATOR: complex of several proteins facilitating transcription; SRB: stabilizes the polymerase and its association with transcription factors; N-CoR: proteins are negative regulators of transcription; SMRT: silencing mediator of retinoic a thyroid hormone receptors; mSIN: proteins are transcriptional corepressors; RbAp: proteins of the WD-40 family of widely present regulators of chromatin, transcription, and cell division; HDAC: histone deacetylase proteins; Sin 3: repressor proteins; SAP: stress-activated regulatory proteins; p35: CDK5 protein; Tbi1: thyroxin-binding regulator. (Courtesy of Dr. Michael G. Rosenfeld. By permission of the American Society for Biochemistry and Molecular Biology.)

to assure compatibility of the donor nucleus (reprogramming) with the egg, avoiding DNA replication, which could result in polyploidy and other types of chromosomal anomalies. It was assumed that the success of reprogramming or remodeling of the donor nucleus was due to the presence of appropriate transcription factors and DNA-binding proteins. Nucleosomal ATPase ISWI may affect the course remodeling of the nucleus. The TATA box−binding protein was released from the nuclear matrix. Electrical pulses facilitated the uptake of the nucleus. The identity of the donor and recipient genotypes was verified by using microsatellite markers. The possibility of transplantation of nuclei from mature tissues and raising viable offspring after reintroducing the reconstituted egg into incubator females was the first example of cloning of an adult mammal. Mitochondrial DNA evidence was not included in the original report. In cattle, actively dividing embryonal fibroblasts were successful nuclear donors. Essentially similar

transplantation experiments produced viable offspring in rhesus monkeys (Meng, L., et al. 1997. *Biol. Reprod.* 57:454). Microinjected nuclei from cloned embryonic stem cells also resulted in normal mice, showing that from a single cell several identical offspring can be obtained. The sheep experiment indicated unexpected shortening of the telomeres of the cloned lamb similar to what happens during aging. This nuclear senescence did not occur in the cloned calves. Actually, older donor nuclei returned to a juvenile state and appeared normal. The cells of the clones displayed an extended life span (Lanza, R. P., et al. 2000. *Science* 288:665). Mice cloned from somatic cells have a shorter life because of pneumonia and hepatic failure (Oganuki, N., et al. 2002. *Nature Genet.* 30:253). The obesity of cloned mice is not inherited in the progeny (Tamashiro, K. L. K., et al. 2002. *Nature Med.* 8:262).

When nuclear transplantation takes advantage of genetically modified donors, special genes can be targeted for recombination and thus insertion or knockout. One must keep in mind that "perfect" cloning of males may not be feasible because the nuclear transfer takes place into an egg that may have different mitochondrial DNA. Also, if the eggs are collected from different females, even if the nuclei are obtained from a single individual, the offspring may not be entirely identical. In cattle, a low degree of mitochondrial heteroplasmy had been generated by nuclear transplantation. Propagation of embryos by splitting the cleavage-stage blastomeres presumably produces entirely identical embryos because in this case not only the cell nucleus but also the cytoplasm is cloned. These experiments foreshadowed the possibility of eventual cloning of all other mammals, including humans. Therefore, new political and ethical concerns surfaced regarding potential manipulation of the human race. Biologically the clones are not different from monozygotic twins. The manipulations, however, may cause abnormal development and—according to some observations—premature aging unless the procedure is perfected. In cloned bovine embryos, the various developmental anomalies were attributed to abnormal methylation, i.e., the methylation pattern of the donor nucleus persisted (Kang, Y.-K., et al. 2001. *Nature Genet.* 28:173). Frequently, respiratory and circulatory problems arise, and the fetus and the newborns may display abnormally large size. Using nuclei of embryonic stem cells involves fewer developmental problems than using nuclei from adult tissues. Some observations indicate subtle abnormalities even when the cloned animals are apparently normal.

Cloning of animals may contribute to the development of improved livestock and may facilitate medical research. Cloning embryonic stem cells obtained with the aid of nuclear transplantation from an individual may facilitate tissue and organ replacement for the same individual without the danger of immunological rejection. For xenotransplantation of cloned pig tissues or organs into humans, the removal or blocking of the porcine α-1,3-galactosyl transferase cell-surface antigen may prevent immunological rejection by humans. In the past, homozygosity of animal herds and flocks was pursued by inbreeding, which may also lead to reduced vigor (inbreeding depression) that is avoided by cloning in the short range, but in the long range appropriate mating schemes are required for sexual reproduction. Nuclear transplantation may rescue endangered species. Nuclei prepared from dead female mouflons (*Ovis orientalis musimon*) injected into enucleated sheep (*Ovis aries*) oocytes and then transfer of the blastocyst-stage embryos to domesticated sheep foster mothers produced apparently normal mouflon. *See Acetabularia*, allopheny; cell cycle; cloning; cytochalasin; cytoplast; differentiation; ethics; G_0; GMO; inbreeding coefficient; knockout; large embryo/offspring syndrome; mating systems; rejection; reprogramming; stem cells; totipotency; transplantation of organelles. (Poljaeva, I. A., et al. 2000. *Nature* 407:86; Humpherys, D., et al. 2001. *Science* 293:95; Loi, P., et al. 2001. *Nature Biotechnol.* 19:962; Lanza, R. P., et al. 2002. *Science* 294:1893; Wilmut, I. 2002. *Nature Med.* 8:215.)

nucleases   *See* DNase; endonuclease; exonuclease; restriction enzymes; ribonucleases.

nuclease hypersensitivity   *See* nuclease-sensitive sites.

nuclease-sensitive sites   In the transcriptionally active chromatin, some short DNA tracts are more easily digested by nuclease enzymes than the rest of the chromatin. These hypersensitive regions indicate that in order to be accessible to RNA polymerase and transcription factors, the DNA must be in a particular open conformation. A well-conserved 5'-CCGGNN-3' repeat sequence seems to exclude nucleosomes. The longer the repeat, the more effective is the nucleosome exclusion. Nuclease hypersensitive sites seem to be absent in regions where there is no potential transcription. *See* chromatin remodeling; nucleosomes; regulation of gene activity. (Li, G., et al. 1998. *Genes Cells* 3[7]:415.)

nucleation   In general usage it means the formation of an initial critical core in a process. During polymerization, a smaller number of monomers assemble in the proper manner, followed by a rapid extension of the polymer.

nucleic acid positive selection   Isolates and enriches desired types of nucleic acid sequences. The desired (tracer) sequences are digested by restriction endonucleases that generate cohesive ends. The rest of the nucleic acids (driver) are exposed to sonication (or the ends may be dephosphorylated), so sticky ends are not expected. Thus, mainly the tracer-tracer sequences are annealed when the mixture is treated with a ligase enzyme. *See* cohesive ends; genomic subtraction; ligase DNA; RFLP subtraction; sonicator; subtractive cloning.

nucleic acid bases   In DNA, adenine (A), guanine (G), thymine (T), and cytosine (C); in RNA, uracil (U) is comparable to (T). Besides these main bases, hydroxy-methyl cytosine (in bacteriophages) and methyl cytosine (in eukaryotes) are regular components of DNA.

To a lesser proportion, other methylated bases may occur in both DNA and RNA. In transfer RNAs, additional minor purine and pyrimidine bases may be found such as dimethyladenine, hypoxanthine, isopentenyl adenine, kinetin, zeatin, pseudouracil, 4-thiouracil, etc. The minor bases are modified after biosynthesis of the regular nitrogenous bases. *See* illustration on top of next page.

nucleic acid chain growth   Secured by adding 5'-triphosphates of nucleosides to the replicating system. The additions take place at the 3'-OH end of the ribose (deoxyribose) by joining only the α-phosphate, and the β- and γ-phosphate groups (pyrophosphate) are split off. Nucleic acid chains can grow only at the 3'-OH terminus. The first nucleotide in the chain may retain all three phosphates. The single-stranded DNA chain elongates in the same manner as the RNA strand. The only difference between the two is that in the DNA (solid arrow) there is 2' deoxyribose, whereas in the RNA at the same position there is OH. *See* DNA replication in prokaryotes;, DNA sequencing; replication fork; transcript elongation. *see* illustration on next page.

nucleic acid homology   Based on the complementarity of the bases in the nucleic acids. Complementary bases can anneal by the formation of hydrogen bonds, e.g., in two single-stranded nucleic acid chains such as A-T-**A-G-C**-T-G and G-C-**T-C-G**-T-A only the bases represented by bold letters are complementary. *See* hydrogen bond.

Nucleic acid bases.

Nucleic acid chain growth.

nucleic acid hybridization Annealing single strands of complementary DNAs or DNA with RNA. *See* c_0t curve; DNA hybridization; in situ hybridization; nucleic homology; Southern blotting. (Marmur, J. & Lane, D. 1960. *Proc. Natl. Acad. Sci. USA* 46:453; Doty, P., et al. 1960. *Proc. Natl. Acad. Sci. USA* 46:461; Hall, B. D. & Spiegelman, S. 1961. *Proc. Natl. Acad. Sci. USA* 47:137.)

nucleic acid probe *See* probe.

nucleic acids Deoxyribonucleic acid (DNA) and ribonucleic acid (RNA); polymers of nucleotides. *See* DNA; RNA.

nuclein Crude nucleic acid–containing preparation first obtained by Friedrich Miescher in 1868 and identified as a common constituent of the nuclei in pus cells and other tissue cells and yeast. *See* nucleic acid.

nucleobindin (calnuc) Ca^{2+} -binding protein in the cytosol and the Golgi apparatus; it interacts with the α-5 helix of the $G_{ai}3$-mediating signal transduction. *See* Golgi apparatus; G proteins; signal transduction. (Kubota, T., et al. 2001. *Immunol. Lett.* 75[2]:111.)

nucleocapsid Viral genome, including an inner protein layer that is not part of the viral envelope protein.

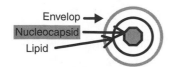

Envelope, Nucleocapsid, lipid.

nuclecytoplasmic factory   There are thousands of about 50 nm centers in the eukaryotic nucleus where several DNA-dependent RNA polymerases are turned on and active in different transcription units. The factory forms a cloud of RNA loops. (Cook, P. R. 2002. *Nature Genet.* 32:347.)

nucleo-cytoplasmic interactions   Products of chromosomal genes may affect the expression of organellar genes and organellar gene products may influence the expression of nuclear genes. These interactions are facilitated through the complex structure of the nuclear pore. *See* chloroplast genetics; cytoplasmic male sterility; mitochondrial genetics; nuclear pore; restorer genes. (Adam, S. A. 1999. *Curr. Opin. Cell Biol.* 11:402.)

nucleoid   In a region at the prokaryotic cell membrane where the cell's DNA is condensed. This DNA is not surrounded, however, by membrane as is the case with the eukaryotic nucleus. The DNA in the mitochondria and chloroplasts is organized in a similar manner without a special envelope; however, the organellar genetic material is present in multiple copies. *See* chloroplast; chondriolite; differentiation of plastid nucleoids; mtDNA; prokaryote.

nucleolar chromosome   Has a nucleolar-organizing region where the ribosomal genes are located. *See* nucleolar organizer; nucleolar reorganization; nucleolus; ribosomal RNA.

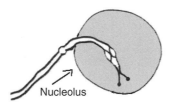

Nucleolar organizer region maize chromosome 6.

nucleolar dominance   In allopolyploid species, usually only one of the parental set of ribosomal RNA genes is expressed. *See* nucleolus. (Pikaard, C. S. 2000. *Trends Genet.* 16:495.)

nucleolar localization signals   Target proteins to the nucleolus. Usually they contain basic amino acid motifs. *See* nuclear localization sequences.

nucleolar organizer   Location of the highly repeated genes responsible for coding ribosomal RNA; it also assembles the products of their transcription until utilization. Morphologically the nucleolar-organizing regions (NOR) are identified by secondary constrictions. The number of nucleolar-organizing regions per genome varies from one to several. The human genome has nucleolar organizers in five chromosomes (13, 14,

15, 21, 22). *See* nucleolus; ribosome; rRNA; satellited chromosome. (Hourcade, D., et al. 1973. *Cold Spring Harbor Symp. Quant. Biol.* 38:537.)

nucleolar reorganization   The nucleolus becomes dispersed beginning with the prophase. It is invisible by the light microscope by metaphase and anaphase, it starts reorganization by telophase, and it finishes the reorganization by the end of the G1 phase. *See* mitosis; nucleolus.

nucleolin   110 kDa nucleolar protein coating the ribosomal transcripts within the nucleolus. Nucleolin accumulates in the cells at the chromosomes and is phosphorylated before mitosis. It has numerous functions: chromatin decondensation, transcription and processing mRNA, cell proliferation, differentiation, apoptosis, shuttling between the nucleus and the cytoplasm, binding lipoproteins, laminin, growth factors, the complement inhibitor J, etc. *See* nucleolus. (Westmark, C. J. & Malter, J. S. 2001. *J. Biol. Chem.* 276:1119.)

nucleolomics   Proteomic inventory of the nucleolus. (Dundr, M. & Misteli, T. 2002. *Mol. Cell* 9:5.)

nucleolus   Body (1–5 μm) associated with the nucleolus organizer region of one or more eukaryotic chromosomes at the coding region of ribosomal RNAs. Besides RNA, the nucleolus contains various proteins and is the site of the production of ribosomal subunits. In the human nucleolus, at least 271 proteins occur (Andersen, J. S., et al. 2002. *Curr. Biol.* 12:1). It has a larger variety of proteins than those incorporated into the ribosomes. By electron microscopy in the nucleolar organizer region, paler fibrillar regions are distinguished where the DNA is not transcribed. The darker fibrillar elements indicate the rRNA positions, and a granular background contains the ribosomal precursors. Although in some species there are several nucleolar organizer regions, the number of nucleoli may be smaller because of fusion. During nuclear division, the nucleoli gradually disappear from view (as rRNA transcription tapers off) to again be reformed beginning in telophase. Apparently, during the stage of no-show the ribosomal components are not destroyed but only dispersed to the surface of the chromatin bundle. *See* Cdc14; cell structure; coiled body; nucleolar chromosome; nucleolar organizer; nucleolar reorganization; nucleolin; paraspeckles; ribosome; rRNA; satellited chromosome. (Pederson, T. 1998. *J. Cell Biol.* 143:279; Nomura, M. 1999. *J. Bacteriol.* 181:6857; Filipowicz, W. & Pogacic, V. 2002. *Current Opin. Cell Biol.* 14:319.)

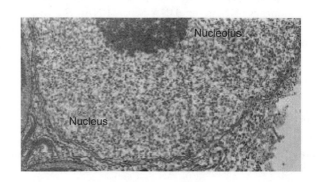

nucleolytic Function of nucleases cutting phosphodiester bonds.

nucleomorph Vestigial nucleus-like structure in some algae, enclosed within the *chloroplast endoplasmic reticulum*. The nucleomorph of the chlorachniophytes is only 380 kb, the smallest among eukaryotes. It contains three linear chromosomes, with subtelomeric rRNA genes at both ends. Two protein genes are cotranscribed. The 12 introns are very short (18–20 bp), and the spacers are too. Most of the genes are involved with the maintenance of the nucleomorph. In the cryptomonads, the nucleomorph is somewhat larger (550–600 kb), yet it has only three small chromosomes. In some organisms there are three or four layers of membranes, indicating that they absorbed the nucleomorph(s) of the symbiont(s) and retained it as a plastid(s) enclosed by the chloroplast. *See* chloroplast. (Douglas, S., et al. 2001. *Nature* 410:1091; Gilson, P. R., et al. 2002. *Genetica* 115:13.)

nucleophile Electron-rich group that donates electrons to electron-deficient (electrophile) carbon or phosphorus atoms. *See* electrophile.

nucleophilic attack Reaction between a nucleophile and an electrophile. *See* electrophile; nucleophile.

nucleophosmin (NPM) *See* anaplastic lymphoma.

nucleoplasm Solutes within the eukaryotic cell nucleus.

nucleoplasmin Acid-soluble proteins mediating the assembly of histones and DNA into chromatin; it is present in the nucleoplasm and in the nucleoli and carries out chaperonin functions. It removes sperm-specific basic proteins from the pronucleus after fertilization and replaces them by H2A and H2B histones. *See* chromatin; histones. (Andrade, R., et al. 2001. *Chromosoma* 109[8]:545; Dutta, S., et al. 2001. *Mol. Cell* 8:841.)

nucleoporin Structural elements of the nuclear pore. The Nup98 and Nup96 and the Nup98-Nup96 complex are translated from an alternately spliced mRNA. These two proteins are posttranslationally processed by autoproteolysis (without a need for proteases). The nuclear pore complex of budding yeast is built of 30 nucleoporins. *See* karyopherin; nuclear pore. (Allen, N. P., et al. 2001. *J. Biol. Chem.* 276: 29268.)

nucleoprotein Nucleic acid associated with protein.

nucleoside Purine or pyrimidine covalently linked to ribose, deoxyribose, or to another pentose.

nucleoside diphosphate kinase Mediates the transfer of the terminal phosphate of a nucleoside 5'-triphosphate to a nucleoside 5'-diphosphate.

nucleoside monophosphate kinase Transfers the terminal phosphate of ATP to a nucleoside 5'-monophosphate.

nucleoside phosphorylase deficiency Dominant human chromosome 14q22 anomaly involving T-cell immunodeficiency and neurological problems. *See* T cell.

nucleosome Histone octamer is wrapped around by about $1\frac{3}{4}$ times by DNA (core particle) and these nu bodies are connected with a 40–60-nucleotide-long DNA linker with either histone 1 or histone 5 (see histones). The total size of the nucleosomes is different in different species and may vary between 146 and 250 nucleotide pairs. Organization similar to the nucleosome core is found in dTAF$_{II}$ proteins associated with TFIID transcription factors. Before transcription, the nucleosomal structure is remodeled. The 2000 kDa SWI/SNF complex generates DNase hypersensitive sites by loosening the association of DNA with histones and permitting the entry of transcription factors. This function of SWI/SNF is only transient and requires ATP. There is another nucleosome remodeling factor (NURF, an ATPase) that remains associated with the nucleosomes.

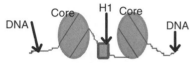

A dark-field electron micrograph (260,000×) of a nucleosome string of chicken. (Courtesy of Drs. Olins, A. L. and Olins, D. E.)

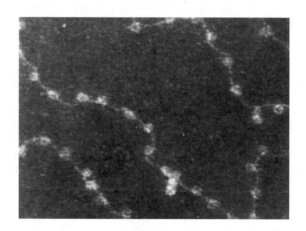

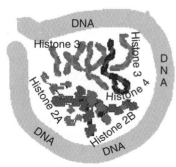

The nucleosome core particle containing four types of histone molecules wrapped around by 146 DNA nucleotides. The dimensions are not on exact scale, and peptide chains joining the polypeptide helices and attaching to the DNA are not shown. (Redrawn in a simplified form after Luger, K., et al. 1997. *Nature* 389:251.)

The compact yeast nucleosomes apparently lack histone 1. Studies indicate that H1 may be attached not only to the linker portion of the DNA but may be situated within the coiled section. Also, the role of the histones, considered earlier only to block transcription, in some cases may be actually slightly stimulative to transcription. The intact nucleosomal structure in place prevents the function of the DNA-dependent RNA polymerase; therefore, the histone octamer is temporarily displaced and reestablished after the polymerase passes through. In the active euchromatin, the available lysine sites of H4 histone are acetylated, whereas in heterochromatin the acetylation is minimal. The chromatin accessibility factor (CHRAC) facilitates the access to chromatin and the assembly of the nucleosomal structure. CHRAC is made of five subunits including the ATPases of NURF (topoisomerase II?), ISWtI, and Acf1, an accessory factor (Eberharter, A., et al. 2001. *EMBO J.* 20:3781.)

For the assembly of the nucleosomal H4-H3 structure by chromatin assembly factor1 (CAF1), lysine residue 5, 8, or 12 of H4 requires acetylation. CAF1 is a complex of p150, p60, and p48 proteins in humans. This complex may vary, however. In several organisms, nucleosome assembly protein 1 (NAP1) apparently joins histones 2A and 2B. The function of the histone acetyl transferase (HAT) enzyme may be a requisite for the initiation of transcription. Histone deacetylase (HDA) proteins, on the other hand, may prevent gene expression. In *Drosophila* and yeast heterochromatin, which is transcriptionally inert, only lysine 12 is acetylated. It appears thus that acetylation at Lys12 controls the silencing of genes. Heterochromatin packaging in yeast is mediated by RAP1 (repressor/activator protein). The various SIR (silencing information regulator) proteins mediate silencing. The effect of histones is not necessarily global in silencing of genes. Depletion of histone 4 in yeast may actually reduce the expression of some telomere-proximal genes and has little influence on many others. In yeast, deletion of the N-terminal 4−23 residues of histone 4 may reduce GAL1 transcription 20-fold, indicating that this region facilitates the unfolding of the nucleosome structure for the initiation of transcription (Durrin, L. K., et al. 1991. *Cell* 65:1023). For the binding of HNF3, acetylation of the nucleosome is not required (Cirillo, L. A. & Zaret, K. S. 1999. *Mol. Cell* 4:961). When only lysine 16 is acetylated, nucleosome assembly does not take place. The nucleosomal structure is assembled at the replication fork as an initial step in the maturation of chromatin. *See* ACF; ASF1; BRG1; CHRAC; chromatin remodeling; chromatosome; DNase hypersensitive site; heterochromatin; high-mobility group of proteins; histone acetyl transferase; histone deacetylase; histone fold; histones; NURF; ORC; RAP; RCAF; regulation of gene activity; rhabdomyosarcoma; RSF; signal transduction; silencer; SRC-1; SWI; transcription factors. (Workman, J. L. & Kingston. R. E. 1998. *Annu. Rev. Biochem.* 67:545; Wolffe, A. P. & Kurumizaka, H. 1998. *Progr. Nucleic Acids Res. Mol. Biol.* 61:379; Lomvardas, S. & Thanos, D. 2001. *Cell* 106:685; Lucchini, R., et al. 2001. *EMBO J.* 20:7294; Jacobs, S. A. & Khorasanizadeh, S. 2002. *Science* 295:2080; Ray-Gallet, D., et al. 2002. *Mol. Cell* 9:1091; Becker, P. B. & Hörz, W. 2002. *Annu. Rev. Biochem.* 71:247; Ahmad, K. & Henikoff, S. 2002. *Cell* 111:281.)

nucleosome phasing Nucleosome positions are not entirely random along the length of DNA, but small variations exist and serve as controls of transcription and packaging of eukaryotic DNA. *See* nucleosome. (Sykorova, E., et al. 2001.

Chromosome Res. 9[4]:309; Kiyama, R. & Trifonov, E. N. 2002. *FEBS Lett.* 523:7.)

nucleosome remodeling *See* chromatin remodeling.

nucleotide Purine or pyrimidine nucleosides with one to three phosphate groups attached.

nucleotide analog interference mapping (NAIM) Identifies atoms, chemical groups, ligand binding, and active sites in RNA. (Ryder, S. P. & Strobel, S. A. 2002. *Nucleic Acids Res.* 30:1287.)

nucleotide biosynthesis *purine biosynthetic path*: PHOSPHORIBOSYLAMINE → GLYCINAMIDE RIBONUCLEOTIDE → FORMYLGLYCINAMIDE RIBONUCLEOTIDE → FORMYLGLYCINAMIDINE RIBONUCLEOTIDE → 5-AMINOIMIDAZOLE RIBONUCLEOTIDE → 5-AMINOIMIDAZOLE-4-CARBOXYLATE RIBONUCLEOTIDE → 5-AMINOIMIDAZOLE-4-N-SUCCINOCARBOXAMIDE RIBONUCLEOTIDE → 5-AMINO-IMIDAZOLE-4-CARBOXAMIDE RIBONUCLEOTIDE → 5-FORMAMIDOIMIDAZOLE-4-CARBOXAMIDE RIBONUCLEOTIDE → INOSINATE. From the latter ADENYLATE is formed through adenylosuccinate, and GUANYLATE is synthesized through xanthylate. Free purine bases may be formed by the hydrolytic degradation of nucleic acids and nucleotides.

The *pyrimidine biosynthetic pathway*: N-CARBAMOYLASPARTATE → DIHYDROOROTATE → OROTIDYLATE → URIDYLATE → CYTIDYLATE. From uridylate THYMIDYLATE is made by methylation. *See* DNA replication; orotic acid; salvage pathway.

nucleotide chain growth At the initial position, the 5′ phosphates are retained; at subsequent sites, nucleotide monophosphates are attached to the 3′ position of the preceding ribose or deoxyribose after two phosphates of the triphosphonucleotides have been removed. The nucleotide chain always grows in the 5′ → 3′ direction. *See* nucleic acid chain growth.

nucleotide diversity τ = the average number of nucleotide differences between two sequences randomly chosen, and it permits the estimation of polymorphism within a species and divergence among species. In humans, the average τ was estimated to be 0.063% ± 0.036%; these values are about an order of magnitude lower than those determined for *Drosophila* from large populations. *See* diversity; evolutionary distance.

nucleotide excision repair *See* DNA repair; NER.

nucleotide flipping Removal of a base from a stack of nucleotides in a chain such as occurs in nucleotide exchange repair. *See* DNA repair; excision repair.

nucleotide sequencing *See* DNA sequencing.

nucleotide substitution *See* base substitution.

nucleotide triplet repeat *See* trinucleotide repeat.

nucleotidyl transferase Transfers nucleotides from one substance to another. *See* DNA polymerase; RNA polymerase; terminal nucleotidyl transferase. (Ranjith-Kumasr, C. T., et al. 2001. *J. Virol.* 75:8615.)

nucleus Genetically most important organelle (5–30 μm) in the eukaryotic cell surrounded by a double-layer membrane (ca. 25 nm) that encloses the chromosomes, proteins, and RNA. The nuclear membrane is equipped with well-organized pores for transport of macromolecules in both directions. *See* chromosomes; nuclear pore; nucleolus; nucleomorph.

nuclides Atoms with a characterized atomic number, mass, and quantum. There are almost 1,000 nuclear species, and about 40 are natural radioactive nuclides. By bombardment with radioactive energetic particles, many additional ones have been generated in the laboratory. *See* mass spectrometry; radionuclides.

nude mouse Genetically hairless; lacks thymus and thymic lymphocytes. It is commonly used in immunogenetic research. Because of the weakened immune reaction, these animals do not reject xenografts. Also, they are very useful for testing carcinogens because of the lack of immune surveillance that may eliminate the transformed cells. They are particularly advantageous for testing skin carcinogens because of the hairless skin. *See* alopecia; HLA; immune reaction; immunological surveillance; lymphocytes; mouse; xenograft.

Nude mouse.

Nu end Of the DNA; enters the phage capsid first. *See* packaging of DNA.

null allele Nonexpressed allele. It is commonly a deletion. *See* deletion.

null hypothesis Assumes that the difference is null between the actually observed and the theoretically expected data. Statistical methods are then used to test the probability of this hypothesis. Obviously, if the data observed do not fit the null hypothesis, it may be false to conclude that the null hypothesis is not valid. The right procedure is to determine what is the probability that the data might comply with the expectation. *See* maximum likelihood; probability; significance level; *t*-test.

nullichiasmate Does not show crossing over or recombination.

nulliplex Polyploid or polysomic individual that at a particular locus has only recessive alleles. *See* duplex; quadruplex; simplex; triplex.

nullisomic Cell or individual lacks both representatives of a pair of homologous chromosomes. Nullisomy is viable only in allopolyploids where the homoeologous chromosomes can compensate for the loss. In the photo below, A, B, and D denote the genomes, and the numbers indicate the particular chromosome within the three series of 7. The bottom right ear represents the normal hexaploid. Nullisomy may come about by selfing monosomics or by nondisjunction at meiosis I or II. However, in diploids it results in lethal gametes. Nullisomy is a normal condition for the Y chromosome in females (XX), whereas nullisomy for the X chromosome is lethal. In allohexaploid wheat, nullisomy has on average only 4% transmission through the male, whereas about 75% of the eggs of monosomics are nullisomic. The cause of the high frequency of nullisomic eggs is that during meiosis I the univalent chromosome (of monosomics) fails to go to the pole and is thus lost. *See* allopolyploid; genome; monosomic; monosomic analysis; nullisomic compensation; sex determination; *Triticum*. (Sears, E. R. 1959. *Handbuch der Pflanzenzüchtung*, Vol. 2, p. 164, Kappert, H. & Rudorf, W. eds. Parey, Berlin, Germany.)

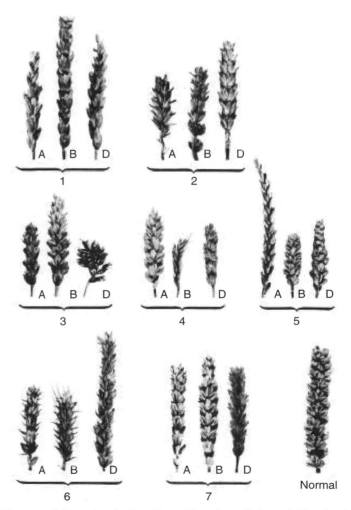

The complete set of the 21 nullisomics of hexaploid wheat, Chinese Spring.

nullisomic compensation Allopolyploids can survive as nullisomics, but it is a deleterious condition. If, however, they are made tetrasomic for another homologous chromosome, their condition is ameliorated because of some degree of restoration of the genic balance. If they are made tetrasomic for another (nonhomeologous) chromosome, their condition is further aggravated. The response to an added chromosome varies according to the specific chromosome. The compensation may occur spontaneously by occasional nondisjunction. If such a nondisjunction takes place in the germline, the tissue receiving the compensating homologous chromosome will be at an advantage in producing gametes and there is a higher chance for improved fertility. *See* dosage effect; homoeologous; monosomic; nullisomics; tetrasomic. (Sears, E. R. 1954. *Res. Bull. 572 Missouri Agric. Exp. Sta* Columbia, MO.)

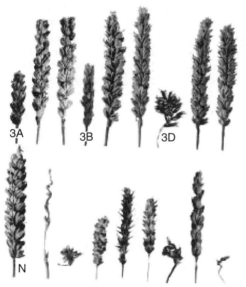

Nullisomic compensation *TOP ROW*: <u>Nullisomic 3A</u>, Nulli 3A-TETRA 3B, Nulli 3A-TETRA 3D, <u>Nulliisomic 3B</u>, 3A, Nulli 3B-TETRA 3A, Nulli 3B-TETRA 3D, <u>Nullisomic 3D</u>, Nulli 3D-TETRA 3A, Nulli 3D-TETRA 3B *BOTTOM ROW*: **normal** hexaploid (N), Nulli 2B-TETRA 4D, Nulli 4B-TETRA 5A, Nulli 5D-TETRA 4A, Nulli 6D-TETRA 1A, Nulli 7A-TETRA 1B, Nulli 7A-TETRA 4D, Nulli 7A-TETRA 6B, Nulli 3A-TRISOMIC 4A. Obviously the corresponding homoeologous chromosomes compensated for the entire loss of that chromosome, but the nonhomoeologous addition even aggravated the condition. (Courtesy of Professor E. R. Sears.)

nullizygous Loss of both alleles in a diploid. *See* nullisomic.

null mutation Entirely eliminates the function of a gene; it may be a deletion.

nullosomic Nullisomic (so used mainly by some human cytogeneticists).

null promoter Lacks TATA box and Initiator element. The transcription may begin at multiple start-site sequences (MED-1). *See* base promoter; core promoter; promoter.

NuMA Nuclear mitotic apparatus (centrophilin) is a nonhistone protein of about 250 kDa. It is present in the interphase nucleus and accumulates at the poles of the mitotic spindle until anaphase. Together with dynein and dynactin, NuMA tethers microtubules in the spindle pole, and they assure the assembly and stabilization of the spindle pole. *See* dynein; mitosis;, nonhistone proteins; spindle fibers; spindle. (Gobert, G. N., et al. 2001. *Histochem. Cell Biol.* 115[5]:381; Gordon, M. B., et al. 2001. *J. Cell Biol.* 152[3]:425.)

numerator Genetic element that "counts" the number of X chromosomes in sex determination and for dosage compensation. *See* dosage compensation; sex determination.

numerical aperture (NA) Of a microscope lens, determines the efficiency of the objective. The optical resolution of a dry lens with 0.75 NA at green light is about 0.5 μm and the depth of focus is about 1.3 μm. An oil immersion lens of 2 mm focal length and 1.3 NA has a resolution of 0.29 μm and a focal depth of 0.4 μm. *See* resolution, optical.

numerical taxonomy Classification of organisms into larger, distinct categories on the basis of quantitative measurements.

NUMTS (nuclearly located mitochondrial DNA sequences) These tracts may be only 100 nucleotides in length or up to 270 kb. In *Arabidopsis* plants 75% of the mtDNA is present in the nucleus. In humans the 300–400 numts may correspond to 0.5% to 88% of the mtDNA. Analysis of numts confirmed that in the oral polio vaccines (produced with the aid of chimpanzee kidney tissue cultures) macaque mtDNA sequences occur but not chimpanzee mtDNAs. Thus the hypothesis that the AIDS was initiated from simian virus, SIV by the use of polio vaccines does not seem likely *See* acquired immunodeficiency; mtDNA; organelle sequence transfer. (Vartanian, J.-P. & Wain-Hobson, S. 2002. *Proc. Natl. Acad. Sci. USA* 99:7566.)

Nup214 *See* nuclear pore.

Nup475 Transcription factor similar to TIS11 but differs in amino sequence at the NH$_2$ and COOH termini. *See* TIS; transcription factors.

nur77 (TR3) Ligand-binding transcription factor including steroid and thyroid hormone receptors (similar to NGFIB and TIS1). Its level is high in apoptotic lymphocytes but not in growing T cells. It permeabilizes the mitochondrial membrane when it migrates from the nucleus to the mitochondria. *See* apoptosis. (Langlois, M., et al. 2001. *Neuroscience* 106:117; Sohn, Y. C., et al. 2001. *J. Biol. Chem.* 276:43734.)

NuRD Histone deacetylase complex of ~2 MDa containing at least seven subunits. *See* histone deacetylase; Sin3. (Ahringer, J. 2000. *Trends Genet.* 16:351.)

NURF Four-protein complex mediating the sliding of the nucleosomes using ATP hydrolysis for energy. *See* ISWI; nuclear receptors; nucleosomes. (Xiao, H., et al. 2001. *Mol. Cell* 8:531.)

nurse cells In insect ovaries, 15 (generally polyploid) nurse cells surround the oocyte within the follicles and their gene products affect and have a morphogenetic role in the differentiation of the embryo at the early stages of development. *See* dumping; morphogenesis; oocyte.

nurture Nutritional (and other environmental) factors that affect the manifestation of the hereditary properties (nature). Twin studies are used in human genetics for the separation of the two components of the phenotype. The differences between identical twins permit the quantitation of the extent of the influence of nurture. *See* twinning.

nusA, nusB Lambda bacteriophage genes involved in the regulation of RNA chain elongation. *See* DSIF; lambda bacteriophage; transcript elongation; TRAP.

NUT Negative regulator element of transcription of the Mediator family. *See* Mediator.

nutmeg (*Myristica fragrans*) Evergreen, dioecious spice plant; $2n = 6x = 42$.

nutritional mutant *See* auxotroph; mutation.

nutritional therapy Humans cannot synthesize the essential amino acids and depend on the diet for a steady supply. Similarly, there may be a dependence on an exogenous (dietary or medicinal) supply of vitamin C or other vitamins or minerals, etc. Some epileptics may benefit from the administration of pyridoxin. Various hereditary defects are known in folic acid metabolism. Hereditary fructose intolerance, galactosemia, and lactose intolerance can be kept in check by limiting the supply of these carbohydrates in the diet or infant formulas. Phenylketonurics must avoid phenylalanine consumption. *See* disaccharide intolerance; epilepsy; fructose intolerance; galactosemia; phenylketonuria.

N value paradox Concept similar to the C-value paradox, indicating that the complexity of an organism is not directly proportional to the number of its genes because the same genome may code for a much larger proteome and the complexity may also depend on genetic networks. (See C-value paradox; proteome. (Claverie, J. M. 2001. *Science* 291:1255.)

nvCJD New variant of Creutzfeldt-Jakob disease. *See* Creutzfeldt-Jakob disease.

nyctalopia *See* night blindness.

NYMPH (nympha) Sexually immature stage between larvae and adults of some arthropods such as ticks (*Ixodes*). *See* Ixodiodia.

nympha of Krause *See* clitoris.

nymphomania Excessive sexual drive (abnormally long estrus) in the mammalian female based on hormonal disorders and usually accompanied by reduced fertility in mares and cows. The condition may have a clear hereditary component. *See* estrus.

nystagmus Involuntary eye movement (displayed by some albinos). This condition may be controlled by autosomal-recessive, -dominant, or X-linked inheritance and may be associated with parts of some syndromes. *See* achromatopsia; eye diseases. (Gottlob, I. 2001. *Curr. Opin. Ophthalmol.* 12[5]:378.)

nystagmus myoclonous Nystagmus accompanied by involuntary movement of other parts of the body. It is a rare congenital anomaly. *See* myoclonus; nystagmus.

nystatin Antibiotic produced by a *Streptomyces* bacterium. It is effective against fungal infections; it is also used as a selective agent in mutant isolation of yeast and other fungi. It primarily kills the growing cells. (Arikan, S. & Rex, J. H. 2001. *Curr. Opin. Investig. Drugs* 2[4]:488.)

NZCYM bacterial medium H_2O 959 mL, casein hydrolysate (enzymatic, NZ amine) 10 g, NaCl 5 g, Bacto yeast extract 5 g, casamino acids 1 g, $MgSO_4 \cdot 7\ H_2O$ 2 g, pH adjusted to 7 with 5 N NaOH and filled up to 1 L. *See* Bacto yeast extract; casamino acids.

NZM medium Same as NCZYM but without casamino acids.

Hermann J. Muller, recipient of the Nobel Prize in 1946, commented in 1931:

> "It is too late to protest that the choice of our own genes was determined by the sheer caprices of a generation now dead. But it is not too late for us to make sacrifices to the end that the children of tomorrow will start life with the best equipment of genes that can be gathered for them ... but it must also be remembered that a prime condition for an intelligent and moral choice of genes is an intelligent and moral organization of society."

(Quoted by G. Pontecorvo in 1968, *Annu. Rev. Genet.* 2:1.)

A historical vignette

O

O Replicational origin. *See* bidirectional replication; replication; replication fork.

oak (*Quercus* ssp) Forest and ornamental trees with great morphological variety, $2n = 24$. The delineation of some of the numerous species may be difficult because of the not uncommon spontaneous hybridization.

O antigen *See* ABO blood group.

oats (*Avena* ssp.) Major cereal crop with somewhat reduced acreage, since farm mechanization diminished the number of horses used in agriculture. The cultivated species (*A. sativa*) is an allohexaploid, but diploid and tetraploid forms are also well known. Basic chromosome number, $x = 7$.

OAZ Ornithine decarboxylase antizyme. Multi-Zn-finger protein affecting both the BMP-Smad and olfactory Olf signaling pathways. *See* BMP; olfactogenetics; Smad; zinc finger. (Hata, A., et al. 2000. *Cell* 100:229.)

OBA Office of Biotechnology Activities. Information about regulations can obtained by <http://www4.od.nih.gov/oba>.

obesity Accumulation of excessive body weight (primarily fat) beyond the physiologically normal range. Differences in predisposition to obesity are long recognized by animal breeders, and the different breeds of swine have large (over 100%) differences in fat content per body weight. Obesity is a health problem in humans because diabetes mellitus, hypertension, hyperlipidemia, heart disease, and certain types of cancer appear to be associated with it. In mice, obesity is regulated by a gene *ob* in chromosome 6, sequenced in 1994. It has been suggested that the 167-amino-acid protein product synthesized in the adipose (fat) tissues is secreted into the bloodstream and regulates food intake through signaling to the hypothalamus. Reduction of this gene product or specific lesions to the ventromedial (basal-central) region of the hypothalamus stimulates food consumption and reduces energy expenditures.

Some experimental data point to the *db* (*diabetes*) gene (mouse chromosome 4) as a receptor for the *ob*-encoded factor, leptin. The *tubby* gene of mice (chromosome 7) also causes maturity-onset obesity, insulin resistance, vision and hearing deficits. Tubby is activated by signal transduction from G-protein-linked receptors. Phospholipase C (PLC) releases Tubby from phosphatidylinositol-4,5-bisphosphate of the plasma membrane, then triggers its movement to the cell nucleus, where it functions as a transcription factor. The *fat* mutation in mice has a later onset than *ob*. In *ob/ob* mice, the serum insulin level decreases with an increase of blood glucose level. In *fat/fat* mice, exogenously supplied insulin decreases the serum glucose level. The fat mice store 70% of their insulin as proinsulin. Apparently, the *fat* gene causes a deficiency in carboxypeptidase, an enzyme that normally processes proinsulin. The protein tyrosine phosphatase-1B gene (PTP-1B) of mice is a negative regulator of insulin signaling. Mutational loss of PTP-1B activity results in decreased phosphorylation of the insulin receptor. Insulin receptor knockout protects against obesity (Blücher, M., et al. 2002. *Developmental Cell* 3:25). Consequently, the mutant animals gained much less weight than the animals that had the wild-type allele. Apparently, upon phosphorylation of the insulin receptor, glucose uptake and weight gain are promoted.

In humans, obesity has been attributed to both dominant and recessive genetic factors, with environmental (diet) factors contributing to about 40% of the variation in obesity. *Mahagony* (*mg*) locus has wide-ranging pleiotropic effects by suppressing obesity of the *agouti-lethal-yellow* locus. The *mg* gene encodes a transmembrane-signaling receptor of 1,428 amino acids with homology to the human attractin protein produced by T lymphocytes and cross-talking between the immune system and melanocortin. Major genes seem to determine 40% of the human variation in body and fat mass. There was some indication of greater effect of either human maternal or paternal body weight on the obesity of the progeny. There are some problems with how the geneticist should measure such a complex trait as obesity, e.g., by body mass, fat mass, visceral adipose tissue amounts, metabolic rate, respiratory quotient, insulin sensitivity, etc. The majority of human obesity factors were implicated mainly on the basis of mice models and putative linkage of quantitative trait loci (QTL). A major susceptibility locus was detected in the short arm of human chromosome 10 (MLS 5.28) and minor quantitative factors appeared in chromosomes 2 (lod score 2.68) and 5 (lod score 2.93).

Neuropeptide Y (NPY) appears to be a stimulant of food intake and an activator of a hypothalamic feeding receptor (Y5). cAMP-dependent protein kinase (PKA) also plays an important role in obesity. This holoenzyme is a tetramer containing two regulatory (R) and two catalytic (C) subunits. The catalytic function is phosphorylation of serine/threonine and the regulatory units slow down the enzyme when the level of cAMP is low. A knockout of the RIIβ regulatory subunit leads to stimulation of energy expenditures in mice, and they remained lean even on a diet that normally was conducive to obesity. In the inner membrane of the mammalian mitochondria, body heat is generated by uncoupling oxidative phosphorylation. Uncoupling proteins UCP1 (4q23), UCP2 (11q13), and UCP3 (11q13) also regulate obesity to some extent. Generally, UCP2 is associated with a somewhat reduced tendency for obesity (Esterbauer, H., et al. 2001. *Nature Genet.* 28:178). These proteins regulate energy balance and cold tolerance. Mice mutants in the uncoupling protein (ICP) have increased food intake, but because of the increase in the rate of metabolism, they do not become obese. Melanocyte regulatory factors (POMC, α-MSH, MC3-R, MC4-R) and bombesin receptor-3 (BRS-3) are also modulators of energy balance and thus obesity and associated diseases. MC4-R regulates food intake and likely energy use, and MC3-R affects the efficiency of the feed and the storage of fat. Mice mutants for both of these hormones eat excessively (because of the MC4-R deficiency), store fat excessively (because of MC3-R deficiency), and become quite obese.

Perilipin (an adipocyte protein) modulates hormone-sensitive lipase (HSL) activity. HSL hydrolyzes triacylglycerol, which stores energy in the cell. Deficiency of perilipin protects against obesity. Altogether, close to 30 genetic sites are known to involve human obesity. Obesity may be controlled by appropriate exercise regimens and control of food intake. Antiobesity drugs may target appetite and intestinal fat absorption, increase energy expenditures, stimulate fat mobilization, or decrease triglyceride synthesis. Some of the drugs of the dexfenfluramine family (inhibit serotonin reuptake and stimulate its release) may reduce obesity by 10%, but may have life-threatening side effects for some people. The newer drug sibutramine does not stimulate serotonin release and is considered safe. Current research explores the mechanism of leptin action (response of the brain to it) and the level of leptin biosynthesis or degradation. Cholecystokinin hormone receptor stimulants and glucagon-like peptides may reduce food intake. Fatty acid synthase (FAS) inhibitors may reduce food intake by inhibiting the removal of FAS from the cells, thus keeping the level of malonyl coenzyme A high. Malonyl-CoA (an appetite inhibitor) is generated from acetyl-CoA with the aid of acetyl-CoA carboxylase. *See* Alström syndrome; anorexia; attractin; Bardet-Biedl syndrome; body mass index; bombesin; bulimia; cachexia; cholecystokinin; ciliary neurotrophic factor; diabetes; GATA; glucagon; G proteins; hypertension; insulin; insulin receptor; leptin; lipodystrophy; lod score; MCH; melanocortin; melanocyte-stimulating hormone; MLS; muscarinic acetylcholine receptors; neuropeptide; orexin; paternal transmission; phosphoinositides; Prader-Willi syndrome; QTL; resistin; serotonin; triaglycerols; ZAG. (Barsh, G. S., et al. 2000. *Nature* 404:644; Robinson, S. W., et al. 2000. *Annu. Rev. Genet.* 34:255; Spiegelman, B. M. & Flier, J. S. 2001. *Cell* 104:531; Brockmann, G. A. & Bevova, M. R. 2002. *Trends Genet.* 18:367; Unger, R. H. 2002. *Annu. Rev. Med.* 53:319; Czech, M. P. 2002. *Mol. Cell* 9:695; <http://obesitygene.pbrc.ed/>.)

OBF   Oct-binding factor (synonyms BOB.1, OCA-B) regulates the lymphocyte-specific oct sequence in the transcription promoter of the immunoglobulin genes; it is required for the development of germinal centers. *See* germinal center; immunoglobulins; oct.

objective lens   Microscope lens next to the object to be studied. *See* light microscopy.

obligate   Restricted to a condition or necessarily of a type, e.g., obligate parasite, obligate anaerobe, etc. The latter can only thrive in the absence of air.

oblique crossing over   In case of adjacent (tandem) duplications in homologous chromosomes, pairing may take place in more than one register and crossing over may yield unequal products:

```
A A A A        A A A↓A  .              A A A A A
A A A A        A A↑A A    }            A A A
normal         oblique                 two, unequal
pairing        pairing and             recombinant
               crossing over           products
```

See unequal crossing over.

O blood group   *See* ABO blood group.

obsessive-compulsive disorder (OCD)   Type of schizophrenic behavior with some resemblance to Tourette syndrome. A major dominant gene with minor modifiers determines it without difference in the two sexes. *See* schizophrenia; Tourette syndrome. (Hanna, G. L., et al. 2002. *Am. J. Med. Genet.* 114:541.)

OCA-B   same as OBF or Bob. *See* OBF.

Occam's razor (Ockham's razor)   Philosophical precept of William Ockham (1280–1349), rebellious clergyman and venerabilis inceptor (= reverend innovator): *"pluralites non est ponenda sine necessitate,"* meaning that multiple alternatives should not be offered in logical argumentation; the simplest yet adequate explanation should be chosen. *See* maximal parsimony.

occipital horn syndrome (cutis laxa)   *See* cutis laxa; Menkes syndrome.

occluded virus particle   The virus is surrounded by proteinacious material that protects it from the adverse environment, e.g., when the insect host dies and decomposes. When the insect eats plant material, the alkaline gut fluid dissolves the occlusion and the infectious (baculovirus) particles are released. The lipoprotein viral envelope then fuses with the gut cell walls and the nucleocapsids are transmitted to the cytoplasm, then to the cell nucleus. *See* baculoviruses. (Hu, Z., et al. 1999. *J. Gen. Virol.* 80[pt 4]:1045.)

occlusion   Transcription from one promoter reduces transcription from a downstream promoter. *See* downstream; promoter; transcription.

occupational hazard   Presence of genotoxic (carcinogenic) agents at the workplace. Monitoring may use urine analysis for chemicals, sister chromatid exchange, abnormal sperm count or deformed sperm, or SNIPS, etc. *See* epidemiology; mutation detection.

Occupational Safety and Health administration   *See* OSHA.

occurrence risk   Chance that an offspring of a certain couple will express or become a carrier of a gene. *See* genetic risk; recurrence risk.

ocellus (plural ocelli)   Simple light sensor (eyelet, eyespot) on the top of the head of insects (see tiny arrows) behind the compound eyes. *See* compound eyes; morphogenesis in *Drosophila*; ommatidium; rhabdomere.

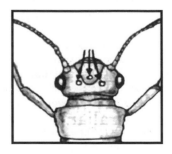

Ocellus.

ochre Chain-terminator codon (UAA).

ochre suppressor Mutation in the anticodon of tRNA that permits the insertion of an amino acid at the position of a normally chain-terminating UAA RNA codon; ochre suppressors frequently suppress amber (UAG) mutations too. *See* code, genetic; nonsense codon; suppressor tRNA.

ochronosis Blue pigmentation in alkaptonuria. *See* alkaptonuria.

OCT Mammalian gene regulatory proteins (helix-turn-helix transcription factors) with octa recognition sequence: ATTTGCAT. Oct-1 and Oct-2 regulate B-cell differentiation. Oct3/4 mediates differentiation and dedifferentiation of embryonic stem cells. The Oct-6 transcription factor regulates Schwann cell differentiation. The presence of the Oct-1 allele may increase the susceptibility to the cerebral form of malaria 4-fold. *See* B cell; immunoglobulins; lymphocytes; malaria; OBF; Oct-2; OCTA; octamer; Schwann cell. (Pesce, M. & Schöler, H. R. 2001. *Stem Cells* 19:271.)

Oct-2 Lymphoid transcription factor similar to Oct-1; both respond to BOB.1/OBF.1 activators. *See* OBF; OCT.

OCTA Eight-base sequence (ATTTGCAT) in the promoter of H2B histone gene and some other genes. Several slightly different octa sequences are found in the promoter regions of different genes. (octo in Latin, οκτασ in Greek: number 8, Oct-1). *See* OCT.

octad Being of eight elements, e.g., the spores in an ascus if meiosis is followed by an immediate mitotic step, e.g., in *Neurospora*, *Ascobolus*, etc. *See* tetrad analysis.

octadecanoic acid (stearic acid) Inducible plant defense molecule against insects. *See* sphingolipids.

octamers Conserved key elements in the promoter of immunoglobulin genes where several transcription factors bind. *See* immunoglobulins; OCT; octa. (Matthias, P. 1998. *Semin. Immunol.* 10:155.)

octaploid Cell nucleus carrying eight genomes, $8x$. *See* polyploid.

octopine Derivative of arginine synthesized by a Ti plasmid gene (*ocs*) in *Agrobacterium* strain Ach 5. *See* *Agrobacterium*; opines.

Octopine.

ocular Microscope lens next to the viewer's eye. *See* objective.

ocular albinism *See* albinism.

ocular cicatricial pemphigoid Autosomal-dominant autoimmune disease of the eye and possibly of other mucous membranes. It may be associated with defects in the HLA system. *See* autoimmune disease; eye diseases; HLA.

oculocutaneous albinism Several forms of the recessive disease are known. Type I (11q14-q21) is tyrosinase-negative recessive, whereas type II (15q11.2-q121) and type III (9p23) are tyrosinase positive (brown). There appears to be another brown type at 15q. The most prevalent is type II, with an incidence of ~1/1,100 among the Ibos in Nigeria, ~1/10,000 among U.S. blacks, and ~1/36,000 among the general U.S. population. *See* albinisms.

oculodentodigital dysplasia Autosomal-dominant disorder involving defects in the eyes (microphthalmos), small teeth, and polydactyly or syndactyly. Its mutation rate shows a large paternal effect. *See* eye diseases; microphthalmos; polydactyly; syndactyly.

O.D. Optical density; indicates the absorption of light at a particular wavelength by a compound in a spectrophotometer. O.D. can be used to characterize a molecule, e.g. pure nucleic acids have maximal absorption at 260 nm but contamination and the solvent may alter the absorption pattern. *See* DNA measurements; extinction.

odds ratio (OR) Comparison of the effect of a treatment or exposure on a particular group (of certain genotype) versus the same treatment or exposure on a different group. The OR may reveal the response of the genotype to the particular exposure. *See* lod score.

odontoblast Connective tissue cell, forming dentin and dental pulp of the teeth.

odor-sensing *See* olfactogenetics; pheromones. (Hurst, J. L., et al. 2001. *Nature* 414:631; Jacob, S., et al. 2002. *Nature Genet.* 30:175.)

ODP Origin decision point is a checkpoint for the initiation of replication. It precedes the restriction point. *See* R point. (Wu, J. R. & Gilbert, D. M. 2000. *FEBS Lett.* 484:108.)

Oedipus complex According to a psychological theory, during adolescence a child may become more attached to the parent of the opposite sex, and if the condition persists, it may lead to neurotic behavior.

Oenothera (*Onagraceae*, $x = 7$) Several diploid and polyploid species have been used for cytological study of multiple translocations and the nature and inheritance of plastid genes. *See* complex heterozygote; gaudens; megaspore competition; translocation; zygotic lethal. (Clealand, R. E. 1972. *Oenothera: Cytogenetics and Evolution*. Academic Press, New York; Hupfer, H., et al. 2000. *Mol. Gen. Genet.* 263:581.)

oestrogen (estrogen, estradiol)　Steroid hormone. *See* animal hormones.

oestrus (estrus)　Periodically recurrent sexual receptivity concomitant with sexual urge (heat) of mammals, except humans. In mice, it lasts for ~3–4 days, depending on crowding, exposure to male pheromones or hormone treatment, and the daily light/dark cycles.

OFAGE　Orthogonal-field alternation gel electrophoresis is used for the isolation of the small chromosomal-size DNA of lower eukaryotes. *See* pulsed-field electrophoresis.

Offermann hypothesis　Supposed to provide a mechanism for recombination within a short chromosomal region that would appear to be intragenic, although the recombination just separated genes and exerted a position effect on their flanking neighbor. These flanking loci were not supposed to have any detectable phenotype themselves, except the position effect on the neighbor. This idea emerged in 1935; pseudoallelism was discovered in 1940 by C. P. Oliver. *See* pseudoallelism. (Carlson, E. A. 1966. *The Gene: A Critical History*. Saunders, Philadelphia, PA.)

offspring-parent regression　*See* correlation.

O-GlcNAc transferase (OGT)　Indispensable cellular enzyme mediating posttranscriptional glycosylation of many different proteins involved in regulatory functions. (Hanover, J. A. 2001. *FASEB J.* 15:1865.)

OGOD　One gene–one disorder is a hypothesis based on the analogy of the one gene–one polypeptide (one enzyme) theorem. The majority of the human (and animal) diseases cannot be reconciled with a single gene mutation, however, and the majority of the disease symptoms (syndromes) are under multigenic control, although particular genes may have a major effect. *See* behavior in humans; one gene–one enzyme theorem. (Plomin, R., et al. 1994. *Science* 264:1733; Ming, J. E. & Muenke, M. 2002. *Am. J. Hum. Genet.* 71:1017.)

Oguchi disease　Recessive human chromosome-2q mutation or deletion in the arrestin protein modulating light signal transduction to the eye or by defects in the arrestin and the rhodopsin kinase genes. Night vision is impaired, but otherwise vision is normal. *See* arrestin; eye diseases; night blindness; retinal dystrophy.

OH (hydroxyl radical)　Responsible for the oxidative damage of superoxide and hydrogen peroxide. *See* Fenton reaction; hydrogen peroxide; ROS; superoxide.

ohm ($\Omega = 1V/A$)　Resistance of a circuit in which 1 volt electric potential difference produces a current of 1 ampere.

Ohno's LAW　The gene content of the X chromosome is basically the same in all mammals. Some exceptions are in humans, marsupials, and the monotreme, *Platypus*, where genes in the short arm of the X chromosome may be of autosomal origin. There are other exceptions such as the chloride channel gene (*Cln4*) is autosomal in the mouse *Mus musculus*, but it is X-linked in *Mus spretus*. Similarly, the human steroid sulfatase (STS) gene is near the pseudoautosomal region, whereas in lower primates it is autosomal. The rationale of the law is that translocations between autosomes and X chromosomes would upset sex-determinational gene balance. In the X chromosomes of various mammals, sequence homologies are conserved. FISH probes have, however, detected some rearrangements. *See* FISH; sex determination. (Ohno, S. 1993. *Curr. Opin. Genet. Dev.* 3:911; Palmer, S., et al. 1995. *Nature Genet.* 10:472.)

oidia　Asexual fungal spores produced by fragmentation of hyphae into single spores.

oil immersion lens　Highest power objective lens of the light microscope. It is used with a special nondrying immersion oil available at different viscosities with a refractive index of about 1.5150 for D line at 23°C; it increases light-gathering power and improves resolution. *See* light microscope; numerical aperture; objective; resolution, optical.

oil spills　About 22 bacterial genera are known to have a genetically determined ability to degrade petroleum hydrocarbons. Although a procedure invented by A. M. Chakrabarty did not involve molecular genetic engineering but only the introduction of two different plasmids into *Pseudomonas* (*P. aeruginosa* and *P. putida*) to degrade several harmful products, it was the first U.S. patent (#4,259,444) issued in 1981 for unique microorganisms. *See* biodegradation; patent. (Díaz, M. P., et al. 2002. *Biotechnol. Bioeng.* 79:145.)

okadaic acid ($C_{44}H_{68}O_{13}$)　Inhibitor of PP1 and PP2a protein phosphatases. *See* PP-1.

Okayama & Berg procedure　Permits cloning of full-length mRNA genes. The mRNA is extracted from the postpolysomal supernatant of reticulocyte lysate of rabbits made anemic by phenylhydrazine injection. The globin mRNA is recovered in the alcohol precipitate of phenol extract or with the aid of a guanidinium thiocyanate method. The poly-A-tailed mRNA is annealed to plasmid pBR322, which is equipped with a poly-T attached to a SV40 fragment inserted into the vector. In a following step, an oligo-G linker is constructed, separated, and purified by agarose gel electrophoresis. Now cloning of the mRNA can be started. The poly-A tail is annealed to the poly-T end of the vector. Using reverse transcriptase, a DNA strand, complementary to the mRNA strand already in the plasmid, is generated. Poly-C tails are added to one strand of the plasmid vector and to the DNA strand of the RNA-DNA double strand using terminal transferase enzyme. The oligo-G linker is added to the oligo-C ends and the plasmid is made circular by DNA ligase. Then the mRNA strand is removed by the use of RNase H and replaced by a complementary DNA strand generated by DNA polymerase I. The construction is finished by ligation into a circular cloning vector. This new vector, containing the full-length cDNA, is transformed into *E. coli* cells for propagation. *See* cloning vectors; guanidinium thiocyanate; linker; phenylhydrazine; reverse transcriptase; ribonuclease

H; RNA extraction; terminal transferase. (Okayama, H. & Berg, P. 1982. *Mol. Cell Biol.* 2:161.)

Okazaki fragments   Short (generally less than 1 kilobase in eukaryotes and about 2 kb in prokaryotes) DNA sequences formed during replication (of the lagging strand) and subsequently ligated into a continuous strand. Okazaki fragments are needed because nucleic acid chains can grow only by adding nucleotide to the 3′-end and the lagging strand template does not allow continuous-chain elongation like the leading strand of DNA. In the generation of the Okazaki fragments, the following steps are found: (1) Polymerase α and primase synthesize an RNA primer for 3′ $\leftarrow \leftarrow \leftarrow$ growth on the lagging strand template. (2) RFC assists in binding the primer to the DNA, and in the displacement of polymerase α, (3) PCNA promotes the assembly of the replicator DNA polymerase δ complex. (4) RNase H, Fen1, and Dna2 endonuclease digest off the RNA primer under the control of replication protein RPA as DNA synthesis proceeds. (5) The gaps between the Okazaki fragments are filled by DNA. (6) DNA ligase joins the fragment into a continuous strand of DNA. In eukaryotes, the process is more complex than in prokaryotes. *See* alpha accessory factor; DNA ligase; DNA replication; PCNA; polymerase switching; primosome; processivity; Rad27/Fen1; RCF; replication fork; ribonuclease H. (Bae, S.-H. & Seo, Y.-S. 2000. *J. Biol. Chem.* 275:38022; Jin, Y. H., et al. 2001. *Proc. Natl. Acad. Sci. USA* 98:5122; Bae, S.-H., et al. 2001. *Nature* 412:456.)

OK blood group   Encoded in human chromosome 19pter-p13. This antigen is present on the red cells of chimpanzees and gorillas but not in rhesus monkeys, baboons, or marmosets. *See* blood groups.

okra (*Abelmoschus esculenta*)   Annual vegetable of the Malvaceae with about 30–40 species with variable chromosome numbers generally higher than $n = 34–36$.

OKT3   Monoclonal antibody capable of blocking interleukin production. *See* Oct.

oleuropein   Phenolic secoiridoid glycoside in the leaves of privet (*Ligustrum*). When activated by herbivores, oleuropein becomes a protein cross-linking, lysine-decreasing glutaraldehyde-like structure, an α,β-unsaturated aldehyde as a means of self-protection. *See* host-pathogen relations; plant defense. (Konno, K., et al. 1999. *Proc. Natl. Acad. Sci. USA* 96:9159.)

olfactogenetics   Concerned with the genetically determined differences in (body) smell and the ability to recognize it. In the human brain, the olfactory bulbs ($\leftarrow$) situated under the frontal lobes interpret the olfactory signals. The mouse olfactory bulbs contain ~2 glomeruli for each olfactory receptor. The different olfactory receptors are segregated into groups of glomeruli. Scent is influenced by the chemical nature of secretions and to a great deal by the diet and the microflora of the body. Human polymorphism in olfactory responses is of concern for the cosmetics (perfume) industry. It has been claimed that the ability to distinguish between various odors is determined by the *H2* locus of mice (an analog of the human *HLA* complex). Some people are incapable of smelling, anosmic

for isobutyric acid or cyanide or urinary excretes of asparagus metabolites. The regulation of olfactory responses involves cAMP or phosphoinositide (IP$_3$)-regulated ion channels and G-protein (G$_\alpha$olf)-coupled receptor kinases. The olfactory memory in sheep is initiated by nitric oxide that potentiates the release of glutamate and GABA neurotransmitters, leading to an increase of cGMP in the mitral cells of the nose, the site of smell perception.

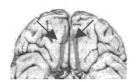

Olfactory bulbs in the brain ($\leftarrow$)

The olfactory memory formation has been localized in the mushroom body of the brain. In *Drosophila*, cAMP phosphodiesterase, encoded by gene *dunce* (*dcn*, chromosome 1–3.9), calcium-calmodulin-dependent adenylyl cyclase (encoded by *rutabaga* [*rut*, 1-{46}]), the catalytic subunit of cyclic AMP-dependent protein kinase A, and the α-integrin subunit encoded by *Volado* (*vol*, located in the X and the 2nd chromosome) are involved in the control of olfactory memory. There are also many other unidentified genes and proteins carrying out different olfactory functions. It appears that each neuron of the olfactory epithelium expresses only one olfactory receptor. In the human genome, ~1,000 olfactory receptor (OR) genes (encoding 7-transmembrane proteins) or pseudogenes have been identified. They are most common in chromosomes 7 and 17, but they are found in most of them, generally clustered as 6–138 genes. No olfactory receptor genes seem to be coded in human chromosomes 20 and Y.

Different olfactory sensory neurons in the nose express a different complement of the ORs and transmit the information through their axons to the olfactory bulbs in the brain. In fishes, approximately 100 genes are involved in olfactory functions. In rodents, there are about 1,500 but humans have about 1,000 only. A combinatorial use of the olfactory receptors permits the distinction of an almost indefinite variety of odors. The ligands of the receptors also contribute to sensory specificity. Different glomeruli respond qualitatively and quantitatively to the types and intensities of the odors. According to some estimates, olfactory receptors may represent 1% of all genes. It has been reported that the major histocompatibility complex is a major source of unique individual odors in animals, and women can detect differences among male odor donors with different MHC genotypes. This ability is dependent on the HLA allele inherited by the human female from her father (Jacob, S., et al. 2002. *Nature Genet.* 30:175).

In the vomeronasal organ — sensing pheromones — there are ~35 V1R and ~150 V2R receptor family members. The number of human OR genes exceeds 1,000, and many of them are pseudogenes. In animals, the major histocompatibility complex is a source of olfactory recognition of mating preferences and various other behavioral traits. The rodent ORs rarely, if at all, are pseudogenic. Natural odors are often mixtures of several components present in specific ratios. Therefore, an enormous number of odor (fragrance) signals may exist. Several OR genes have been cloned. Pheromones are perceived by >240 proteins of the vomeronasal system, including special 7-transmembrane proteins. Removal of the vomeronasal organ interferes with the pheromone response but not with other odor perceptions. The Ras-MAPK signal transduction pathway mediates odor perception and transmission of sensory signals to the *Caenorhabditis* olfactory neurons. In *Caenorhabditis*, deficiencies in the olfactory system seem to prolong life. *See Asparagus officinalis*; bisexual; brain human; Bruce effect; cAMP; cGMP; fragrances; IP$_3$; Kallmann syndrome; MHC; mushroom body; neurotransmitter; nitric oxide; odor sensing; pheromones; phosphoinositide; signal transduction; taste; vomeronasal organ. (Pilpel, Y., et al. 1999. *Molecular Biology of the Brain*, p. 93, Higgins, S. J., ed., Princeton Univ. Press; Buck, L. B. 2000. *Cell* 100:611; Glusman, G., et al. 2001. *Genome Res.* 11:685; Firestein, S. 2001. *Nature* 413:211; Mombaerts, P. 2001. *Annu. Rev. Genomics Hum. Genet.* 2:493; Young, J. M., et al. 2002. *Hum. Mol. Genet.* 11:535; <http://senselab.med.yale.edu/senselab/ordb>; <www.leffingwell.com>.)

olfactory   Concerned with smell sensation. *See* olfactogenetics.

2′-5′-oligoadenylate   Synthesized by 2–5A synthetases induced by interferons; they activate RNase L in defense of viral infection. *See* interferon; ribonuclease L.

oligodendrocyte   Nonneural cells that form the myelin sheath (neuroglia) of the central nervous system; they coil around the axons. They may become stem cells of the central nervous system. *See* neurogenesis. (Lu, Q. R., et al. 2002. *Cell* 109:75.)

oligodeoxyribonucleotide gated channel   Composed of a 45 kDa protein and involved in Ca^{2+}-dependent uptake of oligonucleotides through ~5 μm pore. Protein kinase C and a few organic molecules are inhibitors. *See* liposomes; protein kinases. (Salman, H., et al. 2001. *Proc. Natl. Acad. Sci. USA* 98:7247.)

oligodontia   Lack of the development of six or more permanent teeth is based on frameshift mutation in the PAX9 gene. *See* hypodontia; PAX; tooth agenesis.

oligodynamic action   Antimicrobial effect of trace amounts of heavy metals. *See* sterilization.

oligogenes   Small group of genes responsible for a particular trait; usually one has a major role. *See* breast cancer; polygenes; prostate cancer; QTL.

oligolabeling probes   Short (10–20 nucleotides), commonly radioactive or fluorochrome labeled synthetic probes for the identification of genes for isolation, labeling in gel retardation assays, screening DNA libraries, etc. *See* gel retardation assay; label; probe; variant detector assay.

oligomer   Polymer of relatively few units (amino acids, nucleotides, sugars, etc.) An oligomeric protein has a quaternary structure associated with noncovalent bonds.

oligonucleotide   Short nucleotide tract (about 15 to 30 units).

oligonucleotide-directed mutagenesis (Kunkel mutagenesis)   Synthetic oligonucleotide with the mutation sought is annealed to a single-stranded M13 phage DNA template with a number of uracil residues in place of thymine in a strain *dut*⁻ (dUTPase) and *ung*⁻ (uracil-DNA-glycosidase). The *E. coli* transformation medium should contain the four deoxynucleotide triphosphates, T4 DNA polymerase, and T4 DNA ligase to generate double-stranded circular DNA. M13 phages are selected with the mutation desired. The heteroduplex is introduced into a wild-type *dut* and *ung* strain to maintain the mutation. *See* site-specific mutagenesis. (Kunkel, T. A., et al. 1987. *Methods Enzymol.* 154:367.)

oligophrenia phenylpyruvica   Mental retardation due to phenylketonuria. *See* phenylketonuria.

oligophrenin   RAS-like GTPase protein (91 kDa) encoded in human chromosome Xq12. It is responsible for cognitive impairment. Similar mental retardation genes are scattered in the genome and afflict about 0.15–0.3% of males. Its level is higher in several cancerous tissues. *See* mental retardation; RAS. (Pinheiro, N. A. 2001. *Cancer Lett.* 172:67.)

oligoribonucleotide synthesis   *See* silyl-phosphite chemistry.

oligosaccharides   Consist of sugar residues such as glycans. They may be present in many metabolically, immunologically, structurally important molecules. Oligosaccharides may carry information for the folding of proteins in the endoplasmic reticulum. Their sequence can be determined by exoglycosidase-mediated digestion or by sequencing the amino acids of the protein. In both procedures electrophoretic (MALDI) analysis may be used. *See* electrophoresis; endoplasmic reticulum; folding; glycan; protein folding. (Billuart, P., et al. 1998. *Nature* 392:923; Lehrman, M. A. 2001. *J. Biol. Chem.* 276:8623.)

oligospermia   Low sperm content in the semen. Chromosomal rearrangement (translocations, inversions, ring chromosome, etc.) or aneuploidy may be responsible for the condition. *See* azoospermia; semen; sperm.

oligostickiness   Is a measure of the binding affinity of 12-base (dodeca-) oligonucleotides to the genome of a species. The affinity is characteristic for genomes and for chromosomes of a species and thus facilitates identification. It is called also as chromosome texture. (Nishigaki, K. & Saito, A. 2002. *Bioinformatics* 28:1153.)

oligozyme   Nuclease-resistant RNA oligomers (29–36 residues) that can cleave specific RNA sequences. *See* ribozyme. (Kitano, M., et al. 2001. *Nucleosides Nucleotides Nucleic Acids* 20:719.)

olive (*Olea europea*)   Oil-producing trees of about 30 species. The cultivated forms are $2n = 2x = 46$, although aneuploids have been identified.

olivopontocerebellar atrophy (OPCAI) Autosomal-dominant or -recessive variable types of expressions involve ataxia, paralysis, incoordination, speech defects, and brain and spine degeneration. In some forms, eye defects and other anomalies were also observed. Some cases displayed linkage to the HLA complex in human chromosome 6p21.3-p21.2; in other instances, such a linkage was not evident. Patients with this disease display 50% or less glutamate dehydrogenase activity. *See* ataxia; glutamate dehydrogenase; palsy.

OL(1)p53 Phosphorothioate oligonucleotide sequence (5′-d[CCCTGCTCCCCCCTGGCTCC]-3′) used as antisense DNA to suppress p53 function to pass the checkpoint into the S phase of the cell cycle as a potential treatment for acute myeloblastic leukemia. *See* antisense DNA; leukemia; Myb oncogene; p53; phosphorothioate. (Bishop, M. R., et al. 1997. *J. Hematother.* 6[5]:441.)

Ω (omega) Insertion element present in 0 to 1 copy per mitochondrion in yeast. *See* insertion elements; mitochondria; mtDNA.

ω-agatoxin Ion channel–blocking proteins from the *Agelenopsis* spider.

ω-conotoxins *Conus* snail inhibitors of calcium ion channels. *See* ion channels.

omega sequence Viral nucleotide sequence in the mRNA 5′-region of eukaryotic genes, enhancing translation.

Omenn syndrome *See* reticulosis, familial histiocytic.

OMIM Up-to-date catalog of autosomal-dominant, autosomal-recessive, X-linked, and mitochodrial genes of humans available through the Internet (<http://www.ncbi.nlm.nih.gov/>). The database provides information on the relevant literature of ("morbid") genes and connections to other databases. It is also available in book form: McKusick, Victor. 1997. *Mendelian Inheritance in Man.* The Johns Hopkins University Press, Baltimore, MD.

ommatidium (plural ommatidia) Self-sufficient element (facet) of the compound eye of arthropods, such as *Drosophila*. *See* compound eye; rhabdomere.

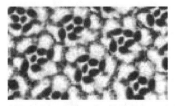

Drosophila ommatidia with eight photoreceptors inside each. There are about 800 ommatidia per eye.

ommochromes Insect eye pigments synthesized from tryptophan and the formation and condensation of hydroxykinurenine into xanthommatin, complexing with other components into brown pigment granules. *See* pigmentation in animals; pteridines; *w* locus.

OMNIBANK (OST) Collection of 80–700-nucleotide-long rapid amplification cDNA 3′-ends by PCR (RACE), generated from mouse embryonic stem cells for the purpose of identification of sequence-tagged mutations. *See* PCR; RACE; sequenced-tagged sites; stem cell. (Zambrowicz, B. P., et al. 1998. *Nature* 392:608.)

omnipotent *See* totipotent.

omp A Outer membrane protein A of the bacterial cell.

OmpC, OmpF Bacterial outer membrane proteins are regulated by kinase EnvZ in response to osmolarity (at low osmolarity by C and at high by F). OmpR is an outer membrane regulatory protein. OmpT is a serine protease, which cleaves cyclin A. *See* cell membrane; membrane proteins.

oncocytoma *See* kidney diseases; mitochondrial disease in humans.

oncogene antagonism therapy Transformation by vector constructs carrying tumor suppressor genes, dominant negative genes, suicide genes, antisense nucleotides, and toxin genes that may prevent tumor formation or disable tumor cells. *See* antisense technologies; cancer gene therapy; gene therapy; suicide vectors.

oncogene collaboration Some oncogenes do not transform cell cultures when singly applied, e.g., rat embryo fibroblast requires the simultaneous presence of RAS and MYC for complete transformation. *See* oncogenes; transformation, oncogenic.

oncogenes Of RNA viruses (v-oncogene) or similar genes in animal cells (c-oncogene), responsible for the first step in cancer initiation. The c-oncogenes (protooncogenes) perform a normal function in animal cells, but they may cause abnormal proliferation by activation or amplification or promoter/enhancer fusion (translocation), mutation, deletion, or inactivation. The transforming genes of DNA viruses do not have cellular counterparts and they induce tumors by interacting with tumor suppressor genes. An allelic form of an oncogene represents a gain of function that favors cancerous transformation. The primary target of the majority of oncogenes is the cell cycle. They deregulate the function of genes that normally control the initiation or progression of the cell cycle. The human genome contains about 30 recessive and more than 100 known oncogenes. For details, see ABL, AKT1, AMV, ARAF, ARG, BLYM, BMYC, CBL, DBL, ELK, EPH, ERBB, ERG, ETS, EVI, FES, FGR, FLT, FMS, FOS, GLI, HIS, HKR, HLM, HST, INT, JUN, KIT, LCA, LCK, LYT, MAS, MCF, MEL, MET, MIL, MYB, MYC, NGL, NMYC, OVC, PIM, PKS, PVT, RAF, RAS, REL, RET, RHO, RIG, ROS, SPI1, SEA, SIS, SK, SNO, SRC, TRK, VAV, YES, YUASA. Many of the oncogenic transformations are caused by cis-activation of protooncogenes by nononcogenic viruses. *See* amplification; cancer; carcinogens; CATR1; gene fusion; nonproductive infection; protein tyrosine kinases; protooncogene; retroviruses; tumor suppressor. (Kung, H. J., et al. 1991. *Curr. Top. Microbiol. Immunol.* 171:1; Liu, D. & Wang, L. H. 1994. *J. Biomed. Sci.* 1:65; Dua, K., et al. 2001. *Proteomics* 1:1191; <http://cgap.nci.nih.gov/>.)

oncogenic transformation Development of a cancerous state. It may begin by the loss of or suppression of tumor suppressor genes. *See* oncogenes; oncogenic viruses; oncoproteins. (Di Croce, L., et al. 2002. *Science* 295:1079.)

oncogenic viruses Can integrate into mammalian cells. Rather than destroying the host, they can induce cancerous proliferation of the target tissues by inhibiting the cellular tumor suppressor genes. The oncogenic viruses may have double-stranded DNA genetic material such as the adenoviruses (genome size ca. 37 kbp), Epstein-Barr virus (ca. 160 kbp), human papilloma virus (ca. 8 kbp), polyoma virus (ca. 5–6 kbp), and single-strand RNA viruses (retroviruses, 6–9-kb). *See* acquired immunodeficiency; adenoviruses; avian; Epstein-Barr virus; hepatitis B; Kaposi sarcoma; Kirsten; Moloney; papilloma virus; papova viruses; retroviruses; Rous sarcoma; SV40; tumor viruses.

oncolytic viruses Herpes simplex-1, Newcastle disease virus, reovirus, and adenovirus may selectively replicate in and destroy tumor cells without causing disease itself. Targeting selectively to tumor cells may be accomplished by fusing antibody fragments or erythropoietin or heregulin to the viral envelope protein. Erythropoietin recognizes its receptor on erythroid precursor cells. Heregulin is a nerve growth factor required specifically for breast cancer and fibrosarcoma cells. Engineering tumor-specific promoter to the viral genes may enhance tumor-specific viral gene expression. The herpes simplex or adenovirus then may directly lyse the tumor cells. New Castle disease virus may increase the sensitivity to tumor necrosis factor (TNF). Parvovirus may promote apoptosis of the cancer cells. The viral antigens bound to cellular MHC class I proteins may become targets for cytotoxic T lymphocytes (CTL). *See* adenovirus; apoptosis; cancer gene therapy; CTL; heregulin; herpes; Newcastle disease; ONYX-015; parvovirus; reovirus; TNF. (Wildner, O. 2001. *Annals Med.* 33[5]:291; Smith, E. R., et al. 2000. *J. Neuro-Onc.* 46[3]:268.)

Oncomouse® Trade name for a mouse strain prone to breast cancer and suitable for this type of research. *See* breast cancer. (Kerbel, R. S. 1999. *Cancer Metastasis Rev.* 17[2]:301.)

onconase Antitumor protein with ribonuclease activity to tRNA. *See* ribonuclease; tRNA. (Notomista, E., et al. 2001. *Biochemistry* 40:9097.)

oncoprotein Product of an oncogene responsible for initiation and/or maintenance of hyperplasia and malignant cell proliferation. *See* oncogenes.

oncostatin M *See* APRF. (Radtke, S., et al. 2002. *J. Biol. Chem.* 277:11297.)

one gene–one enzyme theorem Recognized that one gene is generally responsible for one particular biosynthetic step mediated through an enzyme. Somewhat more precisely stated is the one gene (cistron)–one polypeptide rule because some enzymatically active protein aggregates may be encoded by more than a single gene. There are also some other apparent exceptions, e.g., one mutation blocking the synthesis of homoserine may prevent the synthesis of threonine and methionine because homoserine is a common precursor of these amino acids. Also, in the branched-chain amino acid (isoleucine-valine) pathway, ketoacid decarboxylase and ketoacid transaminase enzymes control the pathways leading to both isoleucine and valine. As the information in genomics accrues, it becomes increasingly evident that one nucleotide sequence (depending on the multiple promoters and processing of the transcript) may carry out different (pleiotropic) functions. Sequencing and proteomic information of the human genome indicates that the same gene (DNA tract) spliced in three alternate ways and may be construed as one gene–three functions. *See* bifunctional enzymes; contiguous gene syndrome; homoserine; isoleucine-valine biosynthetic pathway; overlapping genes; pleiotropy. (Boguski, M. S. 1999. *Science* 286:543; Venter J. C., et al. 2001. *Science* 291:1304; Chen, J., et al. 2002. *J. Biol. Chem.* 277:22053.)

one gene–one polypeptide *See* one gene–one enzyme theorem.

one-hybrid-binding assay Basically a gene fusion assay where a transcriptional activation domain is attached to a particular gene. The function of such a "hybrid" may then be assessed by the expression of an easily monitored reporter gene (e.g., luciferase), depending on the signal received from the particular gene. *See* gene fusion; luciferase; split-hybrid system; three-hybrid system; two-hybrid assay. (Wilhelm, J. E. & Vale, R. D. 1996. *Genes Cells* 1[3]:317; Murakami, A., et al. 2001. *Nucleic Acids Res.* 29:3347.)

one-step growth Bacteriophages multiply within the bacterial cell until in one step, within less than 10 minutes during the "rise" period, all the particles are released. The temperate phages may have a longer period preceding the rise after infection because the phage DNA may become integrated into the bacterial chromosome, then replicates synchronously with the bacterial genes. Upon induction, the phage may switch

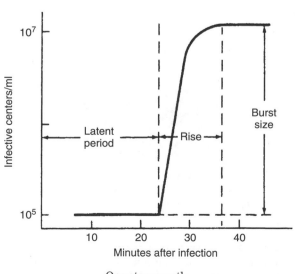

One-step growth.

to a lytic cycle, which begins with autonomous replication followed by liberation of the phage particles. The number of phage particles released in then called the burst size. *See* bacteriophages; phage life cycles. (Hayes, W. 1965. *The Genetics of Bacteria and their Viruses*. Wiley, New York.)

onion (*Allium* spp) The Alloideae subfamily of the lilies includes 600 species. *A. cepa*, $2n = 2x = 16$. Most of the North American species have $x = 8$. Polyploid species occur in different parts of the world with somatic chromosome numbers 24, 32, 40, 48. (Imai, S., et al. 2002. *Nature* 419:625.)

Allium, $x = 8$.

Onozuka R-10 *See* cellulase.

onset Stage (time) of expression of a genetically determined trait.

ontogeny Developmental course of an organism. *See* phylogeny.

ontology, genetic Seeks out the function and meaning of genes and genetic elements.

ONTOS *E. coli* computer data sets organized into an object-oriented database management system.

ONYX-015 Genetically modified adenoviral vector. It inactivates p53 tumor suppressor protein, facilitates the replication of oncolytic viral replication, and may selectively destroy cancer cells. *See* adenovirus; oncolytic viruses; p53. (Galanis, E., et al. 2001. *Crit. Rev. Oncol. Hematol.* 38[3]:177.)

oocyte, primary Has the same chromosome number as other common body cells (2n, 4C), but upon meiotic division each gives rise to two haploid *secondary oocytes* (n, 2C). One of the two, the smaller, is called the first polar body. By another division, the egg and three polar bodies (n, 1C) are formed. The egg may become fertilized, but the polar bodies do not contribute to the progeny and fade away. In human females, meiosis begins in the 4-month-old fetus and proceeds to the diakinesis (dictyotene) stage until sexual maturity. After puberty, in each 4-week cycle one oocyte reaches the stage of the secondary oocyte, then after completing the equational phase of meiosis (meiosis II), an egg is released during ovulation. Each of the two human ovaries contains about 200,000 primary oocytes. On the average, only about 400 eggs are produced during the entire fertile period of the human female, spanning on the average about 30 to 40 years from puberty to menopause. *See*

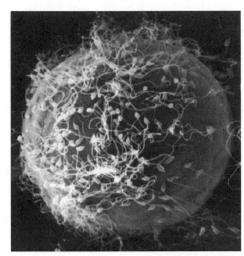

See photo: human oocyte with about 1,500 spermatozoa. (Courtesy of Dr. Mia Tegner.)

C amount of DNA; egg; fertilization; gametogenesis; meiosis; menopause; nurse cell; oocyte translation system; oogonium; oospore; spermatocyte; spermatozoan; *Xenopus* oocyte culture. (Johnstone, O. & Lasko, P. 2001. *Annu. Rev. Genet.* 35:365.)

oocyte, secondary *See* oocyte, primary.

oocyte donation Means to overcome infertility in women who for some reason (older age, genetic risks, etc.) do not want to or cannot conceive in the normal manner but can serve as a recipient of either her own ovum obtained earlier and preserved or an ovum from a donor. She can thus carry a normal baby to term. From a genetic viewpoint, it is important that the implanted ovum should be carefully analyzed to be as risk-free as possible. *See* ART; artificial insemination; surrogate mother. (Noyes, N., et al. 2001. *Fertil. Steril.* 76:92.)

oocyte micromanipulation The penetration of some types of disadvantaged sperm can be facilitated by mechanically opening an entry point through the zona pellucida (a noncellular envelope of the oocyte), thus facilitating sperm penetration (PZD). Alternatively, with the aid of a microneedle, the sperm can directly be deposited under the zona pellucida (SZI) into the space before the vitellus (egg yolk). *See* ART; in vitro fertilization; preimplantation genetics.

oocyte translation system mRNAs injected into the amphibian oocyte (*Xenopus*) nucleus (germinal vesicle) may be transcribed. In the cytoplasm, these exogenous messengers are translated and the proteins may be correctly processed, assembled, glycosylated, phosphorylated, delivered (targeted) to the proper location, etc. Similarly, injections of foreign DNA into the fertilized embryos may be replicated and inserted into the chromosomes. *See* translation in vitro. (Skerrett, I. M., et al. 2001. *Methods Mol. Biol.* 154:225.)

oogamy Fertilization of (generally) a larger egg with a (smaller) sperm. *See* oocyte.

oogenesis Formation of the egg. *See* gametogenesis. (Navarro, C., et al. 2001. *Curr. Biol.* 11:R162; Matzuk, M. M., et al. 2002. *Science* 296:2178; Schmitt, A. & Nebreda, A. R. 2002. *J. Cell Sci.* 115:2457.)

oogonium   Female sex organ of fungi fertilized by the male gametes. Also called zygote.

oogonium, animal   Primordial female germ cell enclosed in a follicle by the term of birth of the individual and becomes the oocyte. *See* gametangium; gametogenesis.

ookinete   Protozoan zygote at a motile stage within the malaria host mosquito.

oophorectomy   Surgical removal of the ovary. The same as ovariectomy, neutering, spaying. *See* castration.

ooplasm   Egg cytoplasm. Transfer of 5–15% of ooplasm from a donor may facilitate pregnancies after certain cases of in vitro fertilization. The transfer involves mitochondria and mRNA. *See* ART. (Brenner, C. A., et al. 2000. *Fertility and Sterility* 74:573.)

oöspore   Fertilized egg in fungi; it is either dikaryotic or diploid and frequently covered by a thick wall. *See* oogonium.

OPAL   Chain-terminator codon (UGA). *See* code, genetic.

opaque (*o*)   Genes in maize (several loci); *o-2* gained particular attention because in its presence the prolamine and zein-type proteins are reduced and the lysine content of the kernels increases. Thus, the nutritional value improves significantly and is important for some parts of the world where corn may be the main food staple. *See* floury; kwashiorkor.

open promoter complex   Partially unwound promoter (the DNA strands separated) to facilitate the operation of RNA polymerase. This separation is supposed to be the result of the attachment of the transcriptase to the promoter. The TATA box of the promoter is a logical place for the attachment of the pol enzyme because there are only two hydrogen bonds between A and T in contrast to the three bonds between G and C; thus, separation of the double helix is easier. This is followed by initiation of transcription. After the attachment of the RNA elongation proteins, the σ-subunit of the bacterial pol enzyme is evicted and transcription proceeds. Transcription factor TFIIB has a 7 bp recognition element immediately upstream of the TATA box; TFIIB and TBP are required for the formation of the preinitiation complex for RNA polymerase II. *See* closed promoter complex; Hogness box; nucleosome; PIC; pol; Pribnow box; PSE; RAD25; regulation of gene activity; reinitiation; TAF; TBP; transcription factors. (Uptain, S. M., et al. 1997.

Annu. Rev. Biochem. 66:117; Ranish, J. A., et al. 1999. *Genes Dev.* 13:49.)

open reading frame (ORF)   Nucleotide sequence between an initiation and a terminator codon. In higher organisms, most commonly one of the two DNA strands is transcribed into functional products, although there are open-reading frames in both strands. *See* initiation codon; nonsense codon.

open system   Exchanges material and energy within its environment.

operand   What is supposed to be operated (worked) on.

operational concepts   Frequently employed in genetics for providing an explanation when the underlying mechanism was not fully understood, but from the visible behavior a conceptualization was possible in agreement with what was known. Examples: T. H. Morgan defined the gene as the unit of function, mutation, and recombination before the nature of the genetic material was discovered. F. H. C. Crick concluded on the basis of frameshift mutagenesis — before the genetic code was experimentally determined — that the genetic code most likely uses nucleotide triplets.

operational RNA code   Sequence/structure-dependent aminoacylation of RNA oligonucleotides that are devoid of an anticodon. The specificity and efficiency is generally determined by a few nucleotides near the amino acid acceptor arm. *See* aminoacylation; aminoacyl-tRNA synthetase; minihelix of tRNA; transfer RNA. (Schimmel, P., et al. 1993. *Proc. Natl. Acad. Sci. USA* 90:8763; de Pouplana, L. R. & Schimmel, P. 2001. *J. Biol. Chem.* 276:6881.)

operational taxonomic unit (OTU)   *See* character matrix.

operator   Recognition site of the regulatory protein in an operon or possibly in other systems such as suggested for controlling elements (transposable elements) of maize. *See* *Ara* operon; *Lac* operon; operon; transposable elements.

operon   Functionally coordinated group of genes producing polycistronic transcripts. Operons were discovered in prokaryotes and they are exceptional in eukaryotes. Similar organization occurs in the homeotic gene complexes of eukaryotic organisms, such as *ANTP-C* and *BX-C* of *Drosophila*. Many genes of the nematode (ca. 15%), *Caenorhabditis* seem to be coordinately regulated and transcribed into polycistronic RNA that is processed into mRNA by transsplicing (Blumenthal, T., et al. 2002. *Nature* 417:851.) *See Arabinose* operon; coordinate regulation; dual-gene operons; *His* operon; homeotic genes; *Lac* operon; morphogenesis in *Drosophila*; SL1; SL2; transsplicing; *Tryptophan* operon; überoperon. (Hodgman, T. C. 2000. *Bioinformatics* 16:10; transcriptional regulation and operon organization in *E. coli*: <http://www.cifn.unam.mx/Computational_Biology/regulondb>.)

ophthalmoplegia   Autosomal-dominant phenotypes (incidence ~1 × 10⁻⁵, encoded at 10q23.3-q24.3, 3p14.1-p21.2,

4q34-q35) involve defects in moving the eyes and the head and some other variable symptoms. The 4q locus encodes a tissue-specific adenine nucleotide (ADP/ATP) translocator (ANT) and controls mtDNA integrity. Autosomal-recessive ophthalmoplegic sphingomyelin lipidosis appears to be allelic to the Niemann-Pick syndrome gene and is associated with mitochondrial DNA mutations. *See* eye disease; inclusion body myopathy; Kearns-Sayre syndrome; mitochondrial diseases in humans; myopathy; neurogastrointestinal encephalomyopathy; Niemann-Pick disease.

opiate Opium-like substance. Opiates regulate pain perception and pain signaling pathways and mood. *Endogenous opiates*, enkephalins and endorphins, were isolated from the brain and the pituitary gland, respectively. They contain a common 4-amino-acid sequence and bind to the same cell-surface receptors as morphine (and similar alkaloids). Nociceptins (orphanin) are 17-amino-acid antagonists of the opioid receptor-like receptor. Opioids are opiate-like, but they are not derived from opium. Nocistatin with an evolutionarily conserved C-terminal hexapeptide blocks pain transmission in mammals. Opioids activate the expression of FAS, which upon binding its ligand (FasL) promotes apoptosis of lymphocytes and thereby the immune system. Opioids may affect the immune system by suppressing cytokine synthesis. They modulate stress responses, learning and memory, metabolism, and may lead to addiction, etc. The major types of opiums are plant alkaloids. *See* apoptosis; cytokines; dynorphin; endorphin; enkephalin; FAS; immune system; nociceptor. (Formula after Massotte, D. & Kieffer, B. L. 1998. *Essays Biochem.* 33:65.)

Morphine: $R^1 = R^2 = H$
Codeine: $R^1 = CH_3$, $R^2 = H$
Heroin: $R^1 = R^2 = CH_3 - CO$

opines Synthesized in crown-gall tumors of dicotyledonous plants under the direction of agrobacterial plasmid genes. The bacteria use these opines as carbon and nitrogen sources. The octopine family of opines includes octopine, lysopine, histopine, methiopine, and octopinic acid. The nopaline group includes nopaline and nopalinic acid. Agropines are agropine, agropinic acid, mannopine, mannopinic acid, and agrocinopines. *See Agrobacteria*; nopaline; T-DNA octopine. (Petit, A., et al. 1970. *Physiol. Vég.* 8:205.)

opiocortin Prohormone (pro-opiocortin) translated as a precursor of several corticoid hormones, cut by proteases, and processed into corticotropin, β-lipotropin, γ-lipotropin, α-MSH (melanin-stimulating hormone), β-MSH, and β-endorphin as shown below. *See* animal hormones; pigmentation of animals; POMC. (Lowry, P. J. 1984. *Biosci. Rep.* 4[6]:467; De Wied, D. & Jolles, J. 1982. *Physiol. Rev.* 62:976; Challis, B. G., et al. 2002. *Hum. Mol. Genet.* 11:1997.)

opisthotonus Motor protein (myosin) dysfunction resulting in spasms and backward pulling of the head and heels while the body seems to move forward. *See* Usher syndrome.

Opitz-Kaveggia syndrome Xp22-linked phenotype involving large head, short stature, imperforate anus, heart defect, muscle weakness, defect in the white matter of the brain (corpus callosum), and mental retardation. *See* head/face/brain defects; heart disease; mental retardation; Opitz syndrome; stature in humans.

Opitz syndrome (G syndrome, BBB syndrome, 22q11.2) Apparently autosomal-dominant anomaly with complex features such as hypertelorism (the paired organs are unusually distant), defects in the esophagus (the passageway from the throat to the stomach), hypospadias (the urinary channel opens in the underside of the penis in the vicinity of the scrotum), etc. In an autosomal-recessive form, it shows polydactyly, heart anomaly, triangular head, failure of the testes to descend into the scrotum, and suspected deficiency of the mineralocorticoid receptor. An Xp22 gene encodes the 667-amino-acid Mid1 (midline) Ring-finger protein and its mutant forms may interfere with microtubule function. *See* corticosteroid; head/face/brain defects; Opitz-Kaveggia syndrome. (Liu, J., et al. 2001. *Proc. Natl. Acad. Sci. USA* 98:6650.)

opossum (marsupials) *Caluromys derbianus* $2n = 14$; *Chironectes panamensis*: $2n = 22$.

opsins Photoreceptor protein of the retina, but nonvisual photoinduction opsin exists in the pineal gland. The chromophore of opsins is either 11-cis-retinal or 3-dehydroretinal. The red-green color vision depends on these molecules. The red- and green-sensitive pigments differ mainly in amino acids at sites 180, 277, and 28, respectively. The red-sensitive alanine, phenylalanine, and the green-sensitive serine, tyrosine, and threonine, respectively, are most common. In hominids and Old World monkeys, two X-chromosomal genes encode these

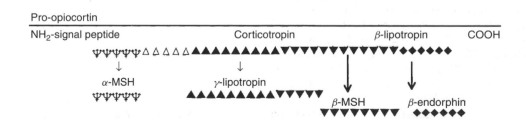

pigments. In all mammals (with the exception of the New World monkeys) there is only one locus encoding either red or green opsins. In humans, additional sites have minor effects. In pigeons, five retinal opsin genes have been distinguished. *See* color blindness; color vision; pineal gland; retinoic acid; rhodopsin. (Yokoyama, S. & Radlwimmer, F. B. 1999. *Genetics* 153:919.)

opsonins Trigger phagocytosis by the scavenger macrophages and neutrophils. These substances bind to antigens associated with immunoglobulins IgG and IgM and facilitate the recognition of the antigen-antibody complexes by the defensive scavenger cells. Also, they may bind to the activated complement of the antibody and assist in the recognition of the cell-surface antigens and thus mediate their destruction. *See* antibody; complement; immunoglobulins; macrophage; neutrophil. (Moghimi, S. M. & Patel, H. M. 1988. *FEBS Lett.* 233:143.)

opsonization Coating foreign invaders of the cells by opsonins to facilitate their destruction by phagocytosis. *See* opsonin; phagocytosis. (Mevorach, D. 2000. *Ann. NY Acad. Sci.* 926:226.)

optical density *See* O.D.

optical mapping May be used for ordering restriction fragments of single DNA molecules. The fragments are stained by fluorochrome(s) and the restriction enzyme-generated gaps can be visualized. The contigs are assembled automatically by using the Gent algorithm. *See* FISH; Gent algorithm; mapping, genetic; physical mapping. (Lin, J., et al. 1999. *Science* 285:1558; Aston, C., et al. 1999. *Trends Biotechnol.* 17:297.)

optical rotatory dispersion Variation of optical rotation by wavelength of polarized light; it depends on the difference in refractive index between left-handed and right-handed polarized light. The rotation is measured as an angle. It is very similar to circular dichroism. *See* base stacking; circular dichroism.

optical scanner Device that generates signals from texts, diagrams, pictures, electrophoretic patterns, autoradiograms, etc., which can be read or printed out with the assistance of a computer.

optical tweezer Consists of a special laser beam associated with a microscope system. It facilitates the manipulation of cell membranes, protein folding, chromosomes, etc. (Hayes, J. J. & Hansen, J. C. 2002. *Proc. Natl. Acad. Sci. USA* 99:1752.)

optic atrophy Determined by autosomal-dominant, -recessive (early-onset types), or X-linked or mitochondrial defects of the eye, ear, and other peripheral nerve anomalies. OPA1 gene in 28 exons encodes a dominant optic atrophy at human chromosome 3q28-q29. Its product is a 960-amino-acid dynamin-like protein localized in the mitochondria. The prevalence of OPA1 is 5×10^{-4}. *See* Behr syndrome; color blindness; dynamin; eye diseases; Kearn-Sayre syndrome; Leber optic atrophy; methylglutaconic aciduria; mitochondrial disease in humans; myoclonic dystrophy; night blindness; Wolfram syndrome. (Pesch, U. E. A., et al. 2001. *Hum. Mol. Genet.* 10:1359.)

opus *See* copia.

OR *See* lod score; odds ratio.

oral bacterial films The human oral cavity and the gut may be inhabited my more than 500 different taxa. These microorganisms may cause tooth decay, gingival bleeding, and other health problems. *See* Helicobacter. (Kolenbrander, P. E. 2000. *Annu. Rev. Microbiol.* 54:413; Hooper, L. V. & Gordon, J. I. 2001. *Science* 292:1115.)

oral-facial-digital syndrome *See* orofacial-digital syndrome.

orange (*Citrus aurantium*, $2n = 18$) Familiar fruit tree. Botanically the peel of the fruit is the pericarp, containing the fragrant oil glands at the lower face. The juice sacs are enclosed by the carpels, containing the seeds. *See* navel orange.

ORC Origin recognition complex is a six-subunit complex (including Rap1 and Abf1 silencers) required before DNA replication can start in eukaryotic cells. In yeast, ORC binds to the autonomously replicating sequence (ARS). During a cell cycle, replication can be initiated only once, but hundreds of sites of replicational origin exist. ORC may place the nucleosomes at the DNA replication initiation site to facilitate the process. The replication cannot start before the *origin of licensing* is created in the M phase with the participation of the MCM protein(s). This is followed by the *origin of activation* in the S phase. Both of these steps are controlled by ORC. Protein subunit ORC2 (Orp2 in fission yeast) apparently interacts with Cdc2, Cdc6, and Cdc18 proteins, which regulate replication. ORC is also required for silencing of the *HMRa* and *HMLα* loci of yeast involved in mating-type determination. The ORC homologue of *Drosophila* is DmORC2. Budding yeast genome appears to have ~429 replication origins. *See* Abf1; ARS; Cdc2; Cdc18; cell cycle; HML; HMR; mating type determination in yeast; MCM; nucleosomes; Rap1; replication; replication licensing factor. (Lipford, J. R. & Bell, S. P. 2001. *Molecular Cell* 7:21; Vashee, S., et al. 2001. *J. Biol. Chem.* 276:26666; Dhar, S. K., et al. 2001. *J. Biol. Chem.* 276:29067; Gilbert, D. M. 2001. *Science* 294:96; Wyrick, J. J., et al. 2001. *Science* 294:2357; Fujita, M., et al. 2002. *J. Biol. Chem.* 277:10345.)

orchard grass (*Dactylis glomerata*) Shade- and drought-tolerant forage crop, $2n = 28$.

orchids (*Orchideaceae*, $2n = 20$, 22, 34, 40) Monocotyledonous tropical ornamentals.

Ord 55 kDa chromosomal protein with some role in chromatid cohesion. *See* sister chromatid cohesion. (Bickel, S. E., et al. 1998. *Genetics* 150:1467.)

order Taxonomic category above *family* and below *class*, e.g., order of Primates within the class Mammals.

ordered tetrad   The spores in the ascus represent the first and second meiotic divisions in a linear sequence such as ⊙⊙⊙⊙. *See* tetrad analysis.

ORESTES   Contraction of the words open-reading frames + EST (expressed sequence tags). The procedure is aimed at sequencing midportions of the genes in contrast to ESTs, which deal with either the 5′ or 3′ ends. ORESTES may help in genome annotation. By 2001, 700,000 ORF tags were available for the definition of the human proteome. *See* EST; ORF; proteome. (de Souza, S. J., et al. 2000. *Proc. Natl. Acad. Sci. USA* 97:12690; Camargo, A. A., et al. 2001. *Proc. Natl. Acad. Sci. USA* 98:12103; Stupka, E. 2002. *Curr. Opin. Mol. Ther.* 4:265.)

orexin A and B (hypocretin)   Appetite-boosting polypetide synthesized in the lateral hypothalamus area of the brain. Their aberrant splicing (caused by insertional mutation) may lead to narcolepsy. Their G-protein-coupled receptors are Hcrtr-1 and -2. Hcrtr-2 was assigned to human chromosome 17q21. Orexins apparently also regulate sleep-wake cycles. *See* leptin; narcolepsy; obesity. (Scammell, T. E. 2001. *Curr. Biol.* 11:R769.)

ORF   Open-reading frame, the nucleotide sequences between the translation initiator and the translation terminator codons, e.g., AUG → UAG. *See* cORF.

ORFOME   Collection of all defined open-reading frames of an organism. (Harrison, P. M., et al. 2002. *Nucleic Acids Res.* 30:1083.)

organ   Body structure destined to a function.

organ culture   Growing organs in vitro to gain insight into function, differentiation, and development. Plant organ cultures, such as propagating roots, stem tips, embryos, etc., under axenic conditions on synthetic media have been known for decades. Research interest has turned to generating human tissues and organs as replacements in case of injuries or disease. Organs generated from a patient's own tissues avoid some of the problems of graft rejection. Since 1997, the U.S. Food and Drug Administration has approved the clinical use of cultured cartilage. Laboratory production of blood vessels, bladder, cardiovascular tissues, and eventually kidneys and livers is expected. Usually, the researchers employ biodegradable polymer scaffolds (polyglycolic acid, polylactide) to permit the development of thicker cell layers permeable to nutrients. *See* grafting in medicine; stem cells; tissue culture.

organelle   (1) Intracellular bodies, such as the nucleus, mitochondria, and plastids. (2) Specialized protein complexes.

organelle genetics   *See* chloroplast genetics; mitochondrial genetics; sorting out. (<http://www.ncbi.nlm.nih.gov/PMGifs/Genomes/organelles.html>.)

organelles   Membrane-enclosed cytoplasmic bodies such as the nucleus, mitochondrion, plastid, Golgi, and lysosome. *See* organelle sequence transfer.

organelle sequence transfers   During evolution, apparently sequences homologous among the major organelles were transferred in the direction shown below. In budding yeast, mtDNA sequences may regularly be transferred to the nucleus during double-strand break repair. In the 2nd chromosome of *Arabidopsis*, 135 genes appear to be of chloroplast origin. In the centromeric region of the same chromosome of one *Arabidopsis* ecotype, ~618 kb appears identical to part of the mitochondrial genome (Stupar, R. M., et al. 2001. *Proc. Natl. Acad. Sci. USA* 98:5099). The organelle genomes (mitochondrial, plastidic) are adopted by initial symbiosis. The complete genomes of the originally free-living organisms were not retained during evolution; some were lost and others were redistributed among the organelles. Some of the genes were apparently sequestered into the organelles in order to assure a homeostatic balance in the redox potential. Enhanced production of reactive oxygen by metabolic accident might kill the sensitive cells unless the damage is readily corrected at the origin. A survey of 277 genera of angiosperm plants indicated that the ribosomal protein gene, *rps10*, was transferred from the mitochondrion to the nucleus at a very high rate during evolution; this move probably still continues in plants but not in animals. There are interactions between organellar and nuclear functions without the actual transfer of genetic material (Traven, A., et al. 2001. *J. Biol. Chem.* 276:4020.) Microarray hybridization revealed that in petite strains of yeast (devoid of mitochondrial DNA), the expression of several nuclear genes (citrate synthase, lactate dehydrogenase, etc.) is altered. The petite strains displayed increased resistance to heat shock and pleiotropic drug resistance. *See* chloroplasts; double-strand break; mobile genetic elements; mtDNA; numts. (Race, H. L., et al. 1999. *Trends Genet.* 15:364; Adams, K. L., et al. 2000. *Nature* 408:354; Blanchard, J. L. & Lynch, M. 2000. *Trends Genet.* 16:315; Hedtke, B., et al. 1999. *Plant J.* 19:635; Adams, K. L., et al. 2002. *Plant Cell* 14:931; Martin, W., et al. 2002. *Proc. Natl. Acad. Sci. USA* 99:12246. <http://megasun.bch.umontreal.ca/gobase/gobase.html>.)

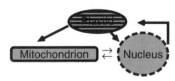

organic   Carbon-containing compound or something associated with a metabolic function.

organic evolution   Historical development of living beings of the past and present times. *See* geological time periods.

organismal genetics   Study of inheritance in complete animals and plants by biological means; does not employ molecular methods or the tools of reversed genetics. *See* interorganismal genetics; reversed genetics.

organizer (Spemann's organizer)   The dorsal lip of the blastopore (an invagination that eventually encircles the vegetal pole of the embryo) becomes a signaling center for differentiation. The formation of the organizer is preceded by induction of the mesoderm cell layer, resulting in the expression of organizer-specific homeobox genes and

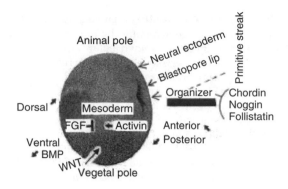

transcription of genes coding for signal molecules. This is followed by recruitment of the neighboring cells into axial mesoderm and neural tissues. Several of the nuclear genes responsible for the component of the organizer have been identified by genetic and molecular means. The organization is controlled by proteins such as Wnt that determine body axis and anterior-posterior development. Activin and fibroblast growth factors (FGF) involve signal reception to organize the formation of the mesoderm or noggin that controls dorsal/ventral differentiation. Chordin affects the development of the notochord, and follistatin is an antagonist of activin, etc. Chordin and noggin are required for the development of the forebrain. These proteins apparently bind to and antagonize the BMP protein of vertebrates (homologous to the decapentaplegic of *Drosophila*). The organizer may show some additional variations in different species. In a diagram it is not possible to correctly represent the complexity of embryonal differentiation, which involves thousands of genes directly and indirectly. The discovery of the organizer by Hans Spemann and co-workers in the 1920s is considered a major milestone in experimental embryology. The Nobel Prize was awarded to Spemann in 1935. This line of research initiated a new series of inquiries into chemical embryology and stayed in the focus of contemporary molecular biology. Many of the genes involved in embryonic induction, regulation of differentiation, and development have been cloned. The discovery of new techniques based on genetics, immunology, radioactive tracers, fluorochromes, microarray hybridization, scanning electron microscopy, etc., combined with genetic transformation, will answer the most basic problems of development and the nature of disease, cancer, etc. *See* activin; animal pole; BMP; bone morphogenetic protein; decapentaplegic; epiblast; FGF; follistatin; gastrula; homeobox; homeotic genes; induction; LIM; morphogenesis; noggin; pattern formation; primitive streak; signal transduction; Spemann's organizer; vegetal pole. (Harland, R. & Gerhart, J. 1997. *Annu. Rev. Cell Dev. Biol.* 13:611; De Robertis, E. M., et al. 2000. *Nature Revs. Genet.* 1:171; Shilo, B.-Z. 2001. *Cell* 106:17.)

organogenesis Development of organs from cells differentiated to special purposes.

ori Origin of replication. *See* DNA replication; ori_T; ori_V; plasmid; replication.

oriC Gene controlling the origin of replication in *E. coli* binds the replication proteins, DnaA and DnaB. *See* DnaA;

DnaB; DNA replication in prokaryotes. (Margulies, C. & Kaguni, J. M. 1996. *J. Biol. Chem.* 271:17035.)

orientation selectivity The majority of transposases work well only in one orientation. Accessory proteins associated with the transposases choose this orientation. It is synonymous with directionality of transposons. *See* transposable elements; transposons.

origin of life Prebiotic evolution laid down the foundations for the origin of life (*see* evolution, prebiotic; evolution of the genetic code). The following steps might have taken place: (1) generation of organic molecules; (2) polymerization of RNA, capable of self-replication and heterocatalysis (cf. ribozymes); (3) peptide synthesis with the assistance of RNA; (4) evolution of translation and polypetide-assisted replication and transcription; (5) reverse transcription of RNA into DNA; (6) development of DNA-RNA-protein auto- and heterocatalytic systems; (7) sequestration of this complex into organic micellae formed by fatty acid–protein membrane–like structures; (8) appearance of the first cellular organisms about 3–4 billion years ago; (9) development of photosynthesis and autonomous metabolism. According to some estimates, the starter functions of life might have been carried out by as few as 20–100 proteins. The extensive duplications in all genomes may be an indication that the early enzymes did not have strong substrate specificity. The minimal cellular genome size might have been comparable to that of *Mycoplasma genitalium*, 580 kb long and coding 482 genes. The time required to develop from the 100 kb genome to a primitive heterotroph cyanobacterium with 7,000 genes might have been about 7 million years. *See* exobiology; evolution of the genetic code; geological time periods; spontaneous generation, unique or repeated. (Nisbet, E. G. & Sleep, N. H. 2001. *Nature* 409:1083; Rotschild, L. J. & Mancinelli, R. L. 2001. Ibid.:1092; Carroll, S. B. 2001. Ibid.:1102.)

origin of replication Starting point of replication during cell proliferation.

Origin of Species Title of a book by Charles Darwin, first published in 1859. This book is one of the most influential works on human cultural history. Darwin recognized the role of natural selection in evolution, although he did not understand the principles of heredity (this book was published 7 years before the paper by Mendel). With the development of modern concepts of heredity, cytogenetics, population genetics, and molecular biology, the seminal role of this book generally became recognized and appreciated even from a sesquicentennial distance. *See* evolution; darwinism.

origin recognition element (ORE) DNA-binding site of proteins that initiate replication. *See* replication fork.

ori$_T$ Origin of transfer of the bacterial plasmid. Its A = T content is higher than the surrounding DNA. It has recognition sites for a number of conjugation proteins. Contains promoters for the *tra* genes localized in such a way that all of them be transferred only after the complete transfer of the plasmid. The transfer of the single strand proceeds in $5' \rightarrow 3'$ direction in all known cases, including the transfer of T-DNA to plant chromosomes. The 3' end can accept added nucleotides. After the transfer has terminated, the plasmid recircularizes. All these processes require specific proteins. *See* bom; conjugation; T-DNA.

ori$_V$ Origin of vegetative replication used during cell proliferation of bacteria.

ornithine (NH$_2$[CH$_2$]$_3$CH[NH$_2$]COOH) Nonessential amino acid for mammals. It is synthesized through the urea cycle. *See* urea cycle.

ornithine aminotransferase deficiency (hyperornithinemia, OAT) Ornithine is derived from arginine, N-acetyl glutamic semialdehyde, or glutamic semialdehyde, and carbamoylphosphate synthetase converts it to citrulline. The decarboxylation of ornithine (a pyridoxalphosphate [vitamin B$_6$]-requiring reaction) yields polyamines such as spermine and spermidine. OAT deficiency causes ornithinemia, a 10–20-time increase of ornithine in blood plasma, urine, and other body fluids. It also results in degeneration of eye tissue and tunnel vision and night blindness by late childhood. A restricted ornithine diet may prevent eye defects due to deficiency of the γ-chain. The OAT locus (21 kb, 11 exons) is in human chromosome 10q23qter and its mutation blocks the metabolic path between pyrroline-5-carboxylate and ornithine, Pseudogenes are at Xp11.3-p11.23 and Xp11.22-p11.21. *See* amino acid metabolism; hyperornithinemia; ornithine decarboxylase; ornithine transcarbamylase deficiency; pseudogene; urea cycle.

ornithine decarboxylase (ODC) The dominant allele (human chromosome 2p25, mouse chromosome 12, in *E. coli* 63 min) controls the ornithine → putrescine reaction, and its activity is very sensitive to hormone levels. There is a second ODC locus in human chromosome 7, but it has much reduced function. Elevated levels of ODC may indicate skin tumorigenesis. *See* amino acid metabolism; antizyme; ornithine aminotransferase deficiency.

ornithine transcarbamylase deficiency (OTC) This X-linked (Xp21.1) enzyme, normally expressed primarily in the liver mitochondria, catalyzes the transfer of a carbamoyl group from carbamoyl phosphate to citrulline; ornithine is made while inorganic phosphate is released. The defect may lead to an accumulation of ammonia (hyperammonemia) because carbamoyl phosphate is generated from NH$_4^+$ and HCO$_3^-$ (in the presence of 2 ATP). The high level of ammonia may cause emotional problems, irritability, lethargy, periodic vomiting, protein avoidance, and other anomalies. Na-benzoate, Na-phenylacetate, and arginine may medicate the plasma ammonium level. The incidence of OTC deficiency in Japan was found to be about 1.3×10^{-3}. In mice, an OTC deficiency mutation is responsible for the *spf* (sparse fur) phenotype. This single-gene metabolic defect may be corrected by somatic gene therapy. Unfortunately, one of the initial attempts using a very high dose of adenoviral vector (6×10^{14} particles/kg) ended with an unexpected fatal immune reaction to a patient in 1999. *See* amino acid metabolism; channeling; hyperammonemia; ornithine aminotransferase; ornithine decarboxylase.

Carbamoyl phosphate

ornithine transporter *See* urea cycle.

orofacial-digital syndrome (OFD) OFD I is apparently a dominant Xp22-linked or autosomal-recessive syndrome. The dominant form is lethal in males. OFD I shows hearing loss and polydactyly of the great toe, mental retardation, face and skull malformation, cleft palate, brachydactyly (short fingers and toes), etc. OFD II (Mohr syndrome) displays polysyndactyly, brachydactyly (short digits), and lobate tongue. OFD III and OFD IV include mental retardation, eye defects, teeth anomalies, incomplete cleft palate, hexadactyly, hunchback features, etc. OFD IV is distinguished on the basis of tibial dysplasia (shinbone defects). OFD V (Váradi-Papp syndrome), in addition, shows a nodule on the tongue and neural defects, etc. *See* limb defects; Majewski syndrome; mental retardation; polydactyly. (Ferrante, M. I., et al. 2001. *Am. J. Hum. Genet* 58:569.)

orosomucoid There are two orosomucoid (serum glycoprotein) genes in humans: ORM1 and ORM2 in 9q31-q32. (Yuasa, I., et al. 2001. *J. Hum. Genet.* 46[10]:572.)

Orotic acid monohydrate.

orotic acid Precursor in the de novo pyrimidine synthesis: N-CARBAMOYLASPARTATE $\overset{1}{\to}$ L-DIHYDROOROTATE $\overset{2}{\to}$ OROTATE $\overset{3}{\to}$ OROTIDYLATE $\overset{4}{\to}$ URIDYLATE $\overset{5}{\to}$ URIDYLATE (UTP) $\overset{6}{\to}$ CYTIDYLATE (CTP). The enzymes mediating the reactions numbered are: (1) ASPARTATE TRANSCARBAMOYLASE, (2) DIHYDROOROTASE, (3) DIHYDROOROTATE DEHYDROGENASE, (4) OROTATE PHOSPHORIBOSYLTRANSFERASE, (5) OROTIDYLATE DECARBOXYLASE, (6) KINASES, (7) CYTIDYLATE SYNTHETASE. *See* oroticaciduria.

oroticaciduria Recessive deficiency of the enzyme orotidylate decarboxylase, encoded at human chromosome 3q13, or a deficiency of orotate phosphoribosyl transferase. The rare human diseases have counterparts in other higher eukaryotes. The clinical symptoms are anemia with large immature erythrocytes and urinary excretion of orotic acid. Supplying uridylate and cytidylate alleviates the symptoms. Homologous mutations have been detected in cattle. *See* orotic acid.

orphan genes Open-reading frames in yeast (about one-third of the ORFs) that cannot be associated with known functions (FUN genes) and do not seem to have homologues in other organisms. Their number is expected to be smaller as information accrues. *See* fun genes; ORF; *Saccharomyces*. (Enmark, E. & Gustafsson, J. A. 1996. *Mol. Endocrinol.* 10:1293; Brenner, S. 1999. *Trends Genet.* 15:132.)

orphanin   *See* opiate.

orphan receptor   Has either no known ligand or has the ligand-binding domain but lacks the conserved DNA-binding site. *See* nuclear receptors; receptor. (Xie, W. & Evans, R. M. 2001. *J. Biol. Chem.* 276:37739.)

orphon   Former member of a multigene family but at a site separate from the cluster, and contains one coding region, or it is a pseudogene. *See* gene family; pseudogene.

orthogenesis   Evolution in a straight line toward a goal or in a predetermined path.

orthogonal-field alternation gel electrophoresis   *See* OFAGE.

orthogonal functions   Can be employed for comparisons between observed data that have not affected each other.

orthologous loci   Genes in the direct line of evolutionary descent from an ancestral locus, e.g., 1,000 genes in *Bacillus subtilis* are similar to those in *Escherichia coli*, indicating the common ancestry. Among the eukaryotes, *Saccharomyces cerevisiae* and *Caenorhabditis elegans*, ~57% of the protein pairs of the two organisms are represented by just one from each of these organisms. Intermediary metabolism (in 28%), DNA and RNA synthesis and function (in 18%), protein folding and degradation (in 13%), transport and secretion (in 11%), and signal transduction (in 11%) are carried out by genes preserved through evolution from common origin in these two eukaryotes. *See* duplications; evolution of proteins; genome projects; homologue; nonorthologous gene displacement; orthologous proteins; paralogous loci. (<http://www. ncbi.nlm.nih.gov/COG>.)

orthology   Common ancestry in evolution. *See* orthologous loci; paralogy.

orthostitchies   *See* phyllotaxis.

orthotopic   In normal position. *See* ectopic.

oscillator   Molecular mechanism that generates the circadian rhythm. *See* circadian rhythm.

oscillin   Protein formed in the egg after the penetration by sperm. It is involved with the oscillation of the level of Ca^{2+} and is apparently involved in triggering the early development of the embryo. (Nakamura, Y., et al. 2000. *Genomics* 68[2]:179.)

OSHA   Occupational Safety and Health Administration. It provides information about and protects against potential hazards involved in laboratory and industrial operations.

Oskar (*osk*, 3-48)   *Drosophila* homozygous females and males are normally viable and fertile. The embryos produced by the homozygous females lack pole plasm, the abdominal segments do not form, and the embryos die. The *osk* mRNA appears in the posterior of the embryo. Kinesin I facilitates the transport of oskar and Staufen mRNAs along the microtubules. The Mago Nashi (Kataoka, N., et al. 2001. *EMBO J.* 20:6424) and Y14 proteins (Hachet, O. & Ephrussi, A. 2001. *Curr. Biol.* 11:1666) are involved in the splicing of pre-mRNA (Palacios, I. M. 2002. *Curr. Biol.* 12:R50).

Osler-Rendu-Weber syndrome   *See* telangiectasia, hemorrhagic.

osmiophilic   Readily stains by osmium or osmium tetroxide.

osmium tetroxide (OsO_4)   Fixative for electron microscopic specimens. *See* fixatives; microscopy.

osmosis   Diffusion of water through a semipermeable membrane toward another compartment (cell) where the concentration of solutes is higher. The osmo-sensing signal seems to be mediated by Jnk protein kinase, a member of the mitogen-activated large family of proteins in mammals. In bacteria, a histidine kinase sensor (EnvZ) and a transcriptional regulator (OmpR) are involved. In yeast, a similar mechanism is implicated, also involving HOG-1. *See* histidine kinase; HOG-1; HSP.

osmotic pressure   In two compartments separated by a semipermeable membrane the solvent molecules pass toward the higher concentration of solutes. This flow may be prevented by applying high pressure to the compartment toward which the flow is directed. Osmotic stress may regulate the expression of genes as a homeostatic control. (Xiong, L., et al. 2002. *J. Biol. Chem.* 277:8588.)

osmotin   Protein in plants that may accumulate at high salt concentrations.

o-some   Nucleosome-like structure of lambda phage involved in DNA replication. It is formed from dimeric λO (lambda origin) protein-coding sequences and four inverted repeats. During the beginning of replication, the dimer of λP (λ promoter) and a hexamer of DnaB form oriλ ~ λO~ (λP~DnaB)$_2$, which gives rise to the pre-primosomal complex. *See* DNA replication, prokaryotes; lambda phage; nucleosome; primosome. (Zylicz, M., et al. 1998. *Proc. Natl. Acad. Sci. USA* 95:15259.)

OSP   *See* allele-specific probe for mutation.

OSP94   *See* HSP.

ossification of the longitudinal filaments of the spine   Common human disorder (over 3% above age 50) caused by ectopic bony development of the spinal muscles. In mice, defects in the nucleotide pyrophosphatase (*Npps*, chromosome 10), which regulates soft-tissue calcification and bone mineralization by producing inorganic phosphate, seems to be one major factor.

osteoarthritis   Common human affliction involving degeneration of the cartilage, caused by the replacement of the [α1(II)] collagen by αII chains with reduced glycosylation. Gene loci at 4q26, 7p15, 2q12, 11q, and others are involved. *See* arthritis; collagen.

osteoblast   Cell involved in bone production from mesenchymal progenitor. It is positively controlled in mice mainly by gene CBFA-1. Gene Hoxa-2 and leptin are negative regulators. *See* osteoclast. (Olsen, B. R., et al. 2000. *Annu. Rev. Cell Dev. Biol.* 16:191; *Science* 2000. 289:1501–1514.)

osteocalcin (1q25-q31, γ-carboxyglutamic acid)   Abundant, noncollagen-type protein in the bones dependent on Ca^{2+}. It is also involved in blood coagulation.

osteochondromatosis (dyschondroplasia, osteochondrodysplasia)   Inhomogeneous group of autosomal-dominant bone and cartilage defects encoded at 8q24 and 11p11-p12 and possibly at other sites. *See* spondyloepiphyseal dysplasia; spondyloepimetaphyseal dysplasia.

osteochondroplasias (osteochondrodysplasias)   Cartilage disorders caused by defects in different collagen genes. *See* chondrodysplasias; collagen.

osteoclastogenesis inhibitory factor (OCIF)   *See* osteoprotegrin.

osteoclasts   Multinucleate cells mediating bone resorption. The osteoclast differentiation factor (ODF) receptor is TRANCE/RANK. *See* bone morphogenetic protein; osteoblast; Paget disease; TRANCE. (Karsenty, G. 1999. *Genes & Development* 13:3037; Väänänen, K. H., et al. 2000. *J. Cell Sci.* 113:377; Motickova, G., et al. 2001. *Proc. Natl. Acad. Sci. USA* 98:5798.)

osteogenesis imperfecta (OI, 11q12-q13, [neonatal lethal] 17q21.3, [perinatal] 7q22.1)   Rare (prevalence 1/5,000–1/10,000) autosomal-dominant or -recessive disorders of collagen, resulting in abnormal bone formation due to mutation at numerous loci (ca. 17) that replace, e.g., a glycine residue by a cysteine. Such a substitution disrupts the Gly-X-Pro tripeptide repeats in collagen, that assures the helical structure of the molecules. Type I (chromosome 17) represents postnatal bone fragility, blue coloring of the eye white (sclera), and ear problems. Type II (Vrolik type) generally involves dominant negative lethality about birthtime due to a variety of bone defects. Most of the cases are considered new dominant mutations. Type III may result in prenatal bone deformities and the survivors are crippled. Hearing loss may accompany the bone symptoms. This may be either recessive or dominant. Type IV has much milder expression (17q21.3) of similar symptoms as the other classes and may not prevent survival.

See collagen; connective tissue disorders; dominant negative; hydrocephalus; mitral prolapse. (Millington-Ward, S., et al. 2002. *Hum. Mol. Genet.* 11:2201.)

osteolysis   Group of diseases characterized by bone resorption and osteoporosis controlled by several different loci. *See* bone diseases.

osteolysis, familial expansile   *See* Paget disease.

osteonectin (5q31.3-q32)   Glycin-rich protein involved in bone proliferation, formation of the extracellular cell matrix, repair, morphogenesis, etc.

osteopenia   Reduced bone mass. It may be due to defects in zinc transporter, ZNT5 resulting in defective maturation of osteoblasts to osteocytes (Inoue, K., et al. 2002. *Hum. Mol. Genet.* 11:1775.)

osteopetrosis   Occurs in different forms controlled by autosomal-dominant or -recessive genes in humans. The recessive form at 11q13 may be due to defects in the a3 subunit of osteoclast-specific proton pump encoded by ATP6i. Osteopetrosis (increased bone mass) with renal tubular acidosis is based on deficiencies in the carbonic anhydrase B or CA2 enzyme. The disease makes the calcified bones brittle. It generally involves mental retardation, visual problems, reduced growth, elevated serum acid phosphatase levels, and higher pH of the urine because of the excretion of bicarbonates and less acids in the urine. In the recessive Albers-Schönberg disease (16p13.3, 1p21), osteopetrosis causes reduction in the size of the head, deafness, blindness, increased liver size, and anemia. Defects in a colony-stimulating factor (CSF) are inferred from mouse experiments. A mild form of osteopetrosis is also known. A severe lethal form affects the fetus by the 24th week and may cause stillbirth. In mice and humans, the loss of chloride ion channel (CIC-7, 16p13.3) leads to dominant osteopetrosis and Albers-Schönberg disease. *See* carbonic anhydrase; cathepsins; colony-stimulating factor; ion channels; osteoporosis; TRANCE. (Lazner, F., et al. 1999. *Hum. Mol. Genet.* 8:1839; Kornak, U., et al. 2001. *Cell* 104:205; Bénichou, O., et al. 2001. *Am. J. Hum. Genet.* 69:647; Sobachi, C., et al. 2001. *Hum. Mol. Genet.* 10:1767.)

osteopontin   Glycoprotein produced by the osteoblasts. It is encoded in human chromosome 4q21-q25 at the SPP1 locus. Its expression is repressed by Hoxc-8, but Smad1 prevents Hoxc-8 from binding to the osteopontin promoter and facilitates its transcription. Osteopontin is also called Eta-1 (early T-lymphocyte activation factor 1) because it is a ligand of CD44 and plays a role in cell-mediated immunity. Its deficiency severely impairs the immunity of mice to some herpes simplex virus 1 and bacterial (*Listeria monocytogenes*) infections. The deficiency interacting with CD44 reduces the expression of IL-10. A phosphorylation-dependent interaction of Eta-1 and its integrin receptor stimulates IL-12. Osteopontin normally affects cell migration, calcification, immunity, and tumor cell phenotype. *See* CD44; herpes; homeotic genes; IL-12; integrin; L-10; osteoblast; Smad. (Agnihotri, R., et al. 2001. *J. Biol. Chem.* 276:28261.)

osteoporosis (17q21.31-q22.05, 7q22.1, 7q21.3) Abnormal thinning of the bone structure because of reduced activity of the osteoblasts to make bone matrix by using calcium and phosphates. The frequency of fracture of the vertebrae may increase up to ~30 times in postmenopausal women. Steroid and other hormones and vitamin D regulate this condition. The autosomal-recessive juvenile form (IJO) may be caused by defects in bone formation, and the problems may be alleviated spontaneously by adolescence or may be treated successfully with steroid hormones. Another apparently autosomal-recessive form in infancy involves eye defects (pseudoglioma) and possibly problems of the nervous system. The collagen genes COLIA1 and COLIA2 have been implicated. An extracellular noncollagen proteoglycan (biglycan) deficiency in mice may lead to osteoporosis-like phenotype. A high-vegetable diet may retard bone resorption. *See* ABL oncogene; aging; Cbl; collagen; estradiol;, hormone-receptor elements; LDL; Src.

osteoprotegrin (osteoclastogenesis inhibitory factor, OPG/OCIF) Receptor of the TNF family proteins. The ~29 kb human gene encodes an inhibitor of TRAIL and other TNF ligands. It regulates bone formation, and the mutant form may lead to early onset of osteoporosis and arterial calcification, i.e., calcium deposits in the veins. *See* osteoporosis; TNF; TRAIL. (Bucay, N., et al. 1998. *Genes Dev.* 12:1260.)

osteosarcoma Autosomal-recessive, usually malignant bone tumor with an onset in young adults. It is commonly part of a complex disease. People affected by bilateral (but not those afflicted by unilateral) retinoblastoma show a high tendency to develop bone cancer. *See* retinoblastoma; sarcoma.

osteosclerosis Abnormal hardening of the bones. *See* cathepsins; osteopetrosis.

ostiole (ostium) Small (mouth-like) opening on various structures such as fungal fruiting bodies or internal organs.

ostrich (*Struthio camelus*) The zoological name reflects the ancient belief that this huge bird descended from a misalliance of the sparrow and the camel. It differs from other birds by having only two toes. Also, while other birds copulate by bringing together their cloacas (the combined opening for urine, feces, and reproductive cells), the male ostrich everts a penis-like structure from the cloaca.

OTF-1, OTF-2 Binding proteins recognizing the consensus octamer 5′-ATGCAAAT-3′ and facilitating transcription in several eukaryotic genes. OTF-1 is identical to the NF-3 replication factor of adenovirus. *See* binding proteins; NF.

otopalatodigital syndrome (OPD) Human chromosome Xq28 semidominant, variable expression deafness, cleft palate, broad thumbs, and great toes. *See* cleft palate; cranioorodigital syndrome; deafness.

otosclerosis Autosomal-dominant hardening of the bony labyrinth of the ear, resulting in lack of mobility of the structures and thus conductive hearing defect. It may be a progressive disease starting in childhood and fully expressed in adults. Its incidence as hearing loss is about 0.003 among U.S. whites and about an order of magnitude less frequent among blacks. OTSC1 is at 15q25-q26 and OTSC2 is at 7q34-q36. *See* deafness.

otter *Amblonyx cinerea*, $2n = 38$; *Enhydra lutris*, $2n = 38$.

OTTO Gene predictor program based on integrated multiple evidence such as homology, EST, mRNA, RefSEq, etc. *See* EST; mRNA; RefSEq.

OTU Operational taxonomic unit is a particular population or a species used in evolutionary tree construction. *See* character matrix; evolutionary tree.

Ötzi *See* ice man (named after the Ötzthal where the mummy was found).

ouabain (3[{6-deoxy-α-L-mannopyranosyl}oxy]1,5,11α,11,19-pentahydroxy-card-20(22)-enolide) Cardiotonic steroid capable of blocking potassium and sodium transport through cell membranes; a selective agent for animal cell cultures. *See* steroids. (Aizman, O., et al. 2001. *Proc. Natl. Acad. Sci. USA* 98:13420.)

Ouchterlony assay *See* antibody detection.

outbreeding Opposite of inbreeding; the mating is between unrelated individuals (crossbreeding, allogamy). *See* allogamy; autogamy; inbreeding; protandry; protogyny.

outbreeding depression Coadapted and different subpopulations can no longer interbreed successfully because of divergence. *See* coadapted genes; speciation.

outcrossing Pollinating an autogamous plant by a different individual or strain or mating between animals of different genetic constitution. *See* incross; protandry; protogyny.

outgroup At least two species that can be used to distinguish an ancestral from a derived species and for the rooting of phylogenetic trees. Characters (physical or molecular) shared between the ingroup and the outgroup are considered to be ancestral. *See* evolutionary tree; PAUP.

outlier Gene that is more divergent from the typical members of a gene family than expected. These apparently represent highly specialized functions. In general, in statistical handling of data, the extreme values that appear inconsistent with the bulk of the data are called outliers. *See* gene family.

out-of-AFRICA Hypothesis suggests that the origin of the human race was in Africa. It is based on mtDNA composition showing that more differences exist in this respect in Africa than other parts of the world, supposedly because a few founders (highly conserved mtDNA) emigrated, then spread and evolved relatively recently into the majority of existing human racial groups. At the niche of origin, however, a longer evolutionary period permitted greater divergence. This hypothesis has been further supported on the basis of linkage disequilibrium between an Alu deletion at the CD4 locus in chromosome 12 and short tandem repeat polymorphisms (STRP) used as nuclear chromosomal markers. The two markers are separated by only 9.8 kb. The mapping data seem to indicate that 1 cM of the human genome corresponds to about 800 kb. The Alu deletion was mainly associated with a single STRP in Northeast Africa and non-African populations sampled from 1,600 individuals from European, Asian, Pacific, and Amerindian groups. In contrast, in sub-Saharan Africa, a wide range of STRP markers were with the Alu deletion. These data also indicate that migration from Africa took place relatively recently, an estimated 102,000 to 313,000 years ago. This estimate is in relatively close agreement with the age of the first human fossil records in the Middle East, dated to be 90,000–120,000 years old. *Homo sapiens* diverged from *H. erectus* an estimated 800,000 or more years ago. The European continent was settled from the Levant about 40,000 years ago in several waves during the Paleolithic, Mesolithic, and Neolithic eras. The subsequent waves of migration were expected to replace the preceding ones (*replacement theory*). Some anthropologists do not accept this theory of human evolution and suggest *multiregional origin* for modern humans. Analysis of the human Y chromosome lends support to the African origin of male descent and to the same general region as the EVE foremother theory suggested for females. *See* Alu; ancient DNA; CD4; EVE foremother of mitochondrial DNA; founder principle; Garden of Eden; hominids; mtDNA; Y chromosome. (Quintana-Murci, L., et al. 1999. *Nature Genet.* 23:437; Harpending, H. & Rogers, A. 2000. *Annu. Rev. Genomics Hum. Genet.* 1:3611; Barbujani, G. & Bertorelle, G. 2001. *Proc. Natl. Acad. Sci. USA* 98:22; Adcock, G. J., et al. 2001. *Proc. Natl. Acad. Sci. USA* 98:537; Templeton, A. R. 2002. *Nature* 416:45; African population genetics: Tishkoff, S. A. & Williams, S. M. 2002. *Nature Rev. Genet.* 3:611.)

output trait *See* input trait.

ovalbumin Nutritive protein of the eggwhite of chickens ($M_r \approx 4,500$); each molecule carries a carbohydrate chain. The gene is split by seven introns into eight exons and its transcription is induced by estrogen or progesterone. In the presence of the effector hormone, mRNA has a half-life of about 1 day, but without it, its persistence may be reduced to 20%. The gene is part of a family of three. *See* intron; oviduct.

ovalocytosis *See* acanthocytosis; elliptocytosis.

ovarian cancer May be due to mutation of the OVC oncogene (9p24). The expression of several other genes is also altered. The same genes affect a large variety of other cancers. Ovarian cancer may be associated with breast cancer and may increase when tamoxifen medication is employed. Mutation in the so-called ovarian cancer cluster region (OCCR, nucleotides 3059–4075 and 6503–6629) of the BRCA2 increases the relative risk of ovarian cancer by 0.46–0.84. Ovarian germ-cell cancer involves the germ cells in the ovaries and ~5% of the cases are due to genetic causes. Hormone (estradiol family) replacement therapy of menopausal women may increase the risk. It has been suggested that the use of oral contraceptives may lower the risk. A microarray hybridization profile seems to provide an early diagnosis. *See* breast cancer; sex hormones; tamoxifen. (Welsh, J. B., et al. 2001. *Proc. Natl. Acad. Sci. USA* 98:1176; Thompson, D. & Easton, D. (Breast Cancer Linkage Consortium). 2001. *Am. J. Hum. Genet.* 68:410.)

ovariectomy (oophorectomy) Removal of the ovary. *See* castration; neutering; spaying.

ovariole *See* ovary.

ovary (ovarium) Contains the ovules of plants, the egg-producing organ of animals, a female gonad. The ovaries of *Drosophila* contain *ovarioles*, each with two or three germline stem cells at the tip associated with *terminal filament cells* with the duty of somatic signaling. In *Caenorhabditis*, asymmetric differentiation of germ-cell lineage is not observed; rather, the distal mitotic stem cells undergo meiosis through adulthood. *See* cytoblast; gametogenesis; germarium; germline; gonad; pole cells.

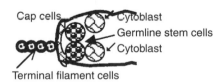

OVC oncogene Discovered in an ovarian cancer in a human chromosome with bromosomes 8–9 fusions. *See* oncogenes.

overdominance The *Aa* heterozygote surpasses both *AA* and *aa* homozygotes. Deleterious recessive genes may occur in heterozygotes at a greater frequency (at linkage disequilibrium) than expected because of association with neutral genes and is called *associative overdominance*. *Pseudooverdominance* occurs when two genes are closely linked and the aB and Ab combinations simulate overdominance when actually only dominance is contributing to increased performance. *See* heterosis; hybrid vigor; overdominance and fitness; polar overdominance; superdominance.

overdominance and fitness Assume that in case of overdominance the selection against the two homozygotes at a locus is 0.05. Thus, instead of Hardy-Weinberg proportions,

Allelic complementation (= overdominance) of temperature-sensitive pyrimidine mutants of *Arabidopsis*. Top and bottom rows are homozygous mutants, in the middle row: the F$_1$ hybrids. (From Li, S. L. & Rédei, G. P.)

the population will be *AA* 95, *Aa* 200, and *aa* 95. The proportion of the surviving zygotes will be 390/400 = 0.975. This means a 2.5% reduction in the size of the population in a single reproductive phase. It may not be very serious in case of a single locus and large populations, but in case of 10, 100, and 500 loci, it would be $0.975^{10} \cong 0.776$, $0.975^{100} \cong 0.0795$ and $0.975^{500} \cong 0.0000032$, respectively. Most likely, such a great reduction in population size could not be tolerated. Thus, overdominance may be advantageous only at one or a very small number of loci. *See* allelic complementation; fitness; heterosis; hybrid vigor; monogenic heterosis. (Rédei, G. P. 1982. *Cereal Res. Commun.* 10:5.)

overdrive   Consensus of 5'-TAAPuTPyNCTGTPuTNTGTT-TTGTTTG-3' in the vicinity of the T-DNA right border of some octopine plasmids, facilitating the transfer of T-DNA into the plant chromosomes. The VirC1 protein binds this region. In the *overdrive mode* of RNA polymerase, the system does not respond to protein signals of transcription termination or pause. *See* Agrobacterium; antitermination; RNA polymerase; T-DNA; terminator; transcription factors; virulence genes of *Agrobacterium*. (DeVos, G. & Zambryski, P. 1989. *Mol. Plant-Microbe Interact.* 2:43.)

overepistasis   Outdated term for overdominance.

overgrowth   Increase of cell size and/or number in neighboring tissues.

overhang   A double-stranded nucleic acid has a protruding single-strand end: _____

overlapping code   Historical idea about how neighboring codons may share nucleotides. *See* overlapping genes. (Ycas, M. 1969. *The Biological Code*. North-Holland, Amsterdam.)

overlapping genes   Occur in the small genomes of viruses, eukaryotic mitochondria, and rarely in eukaryotic nuclear genes. Sequences of the genetic material may be read in different registers; thus, the same sequences may be representing two or more genes in a complete or partial overlap because they do not need equal amounts of proteins from the overlapping genes. The means for achieving this

goal is either stop codon read-through or translational frame shifting on the ribosome. Murine leukemia virus (MLV) and feline leukemia virus (FeLV) use the first alternative. In HIV virus, the RNA transcript makes a small loop, and in the sequence 5' UUU UUUA GGGAAGAU-LOOP-GGAU, a one-base slippage (framed) takes place. This allows the normally UUA recognizing tRNALeu to pair with UUA (leucine). Next, in place of GGG (glycine), an AGG (arginine) was inserted; thus, the stop codon was thrown out of frame, permitting the synthesis of a fusion protein (*gag* + *pol*, i.e., envelop + reverse transcriptase). This frameshift translation is generally stimulated by a pseudoknot located 5–9 nucleotides (3') from the shift site. After the translation, the pseudoknot unfolds and the RNA returns to a linear shape. Such a ribosomal frameshifting occurred only in a fraction (11%) of the ribosomes, yet as a consequence, instead of a 10–20 envelope:1 reverse transcriptase protein, the proportion changed to 8 envelope:1 reverse transcriptase (*gag* + *pol*) production. Genome sequencing data in *Drosophila* indicate more than 30 examples of gene overlaps (by 1999). The majority seem to be transcribed from the opposite polarity DNA strand. Within a ~3 Mb region of the long arm of chromosome 2 of *Drosophila*, 12/17 were transcribed from the antiparallel strand. Deletions encompassing overlapping genes usually result in sterile or lethal phenotypes. *See* contiguous gene syndrome; deletion; frameshift; fuzzy logic; gag; knockout; nested genes; ϕX174; pol; pseudoknot; recoding; regulation of gene activity; retroviruses; reverse transcription; translational hopping; tRNA. (Pedersen, A. M. & Jensen, J. L. 2001. *Mol. Biol. Evol.* 18:763; Besemer, J., et al. 2001. *Nucleic Acids Res.* 29:2607; Barrette, I. H., et al. 2001. *Gene* 270:181.)

overlapping inversions   Part of one chromosomal inversion is included in another inversion. *See* inversion; salivary gland chromosomes.

Two overlapping inversions in a heterozygous salivary gland cell of *Drosophila*. (Remember, the salivary gland chromosomes display somatic pairing.) (From Dobzhansky, T. and Sturtevant, A. H. 1938. *Genetics* 23:28.)

overlay binding assay   Used to determine interactions between protein subunits (on a membrane) after labeling of the SDS-PAGE separated proteins, still retaining the critical conformation. *See* SDS-polyacrylamide gels. (Kumar, R., et al. 2001. *J. Biosci.* 26[3]:325.)

overproduction inhibition   Highly active promoter of a transposase may actually reduce the rate of transposition of a transposable element. *See* transposable elements.

overwinding Generates supercoiling in the DNA. *See* supercoil.

oviduct Passageway through which the externally laid eggs are released (e.g., birds); the channel through which the egg travels from the ovary to the uterus (e.g., mammals). *See* ovalbumin; ovary; uterus.

Ovis aries (sheep, $2n = 54$) Forms a fertile hybrid with the wild mouflons, but the sheep $\times$ goat (*Capra hircus*, $2n = 60$) embryos rarely develop in a normal way, although some hybrid animals have been obtained. *See* nuclear transplantation.

ovist *See* preformation.

ovotestis Gonad that abnormally contains both testicular and ovarian functions. *See* hermaphrodite.

ovulation Release of the secondary oocyte from the follicle of the ovary that will be followed eventually by the formation of the egg and the second polar body. Bone morphogenetic protein 15 (BMP15, Xp11.2-p11.4), a member of the TGFβ family of proteins, is expressed specifically in the oocytes and is essential for normal follicular development and ovulation. Heterozygosity for a mutation (in sheep) in the gene may lead to twinning, but homozygosity for the mutation impairs follicular growth beyond the primary stage and results in female sterility. *See* fertility; gametogenesis; twinning. (Richards, J. S., et al. 2002. *Annu. Rev. Physiol.* 64:69.)

ovule Megasporangium of plants in which a seed develops; the animal egg enclosed within the Graafian follicle. *See* Graafian follicle; nucellus.

ovum (plural ova) Egg(s) ready for fertilization under normal conditions.

ox (plural oxen) Castrated male cattle. *See* cattle.

OX40 Member of the tumor necrosis factor receptor family. *See* TNF. (Rogers, P. R., et al. 2001. *Immunity* 15:445.)

oxalosis (hyperoxaluria) Autosomal-recessive forms involve excretion of large amounts of oxalate and/or glycolate through the urine caused either by a deficiency of 2-oxoglutarate (α-ketoglutarate): glyoxylate carboligase or a failure of alanine:glyoxylate aminotransferase or serine:pyruvate aminotransferase. The accumulated crystals may result in kidney and liver disease. The glycerate dehydrogenase defects also increase urinary oxalates and hydroxypyruvate. Hydroxypyruvate accumulates because the dehydrogenase does not convert it to phosphoenolpyruvate. It is also called peroxisomal alanine:glyoxylate aminotransferase deficiency (AGXT) located to human chromosome 2q36-q37. Hyperoxaluria II (9cen) is a deficiency of glyoxylate reductase/hydroxypyruvate reductase. *See* glycolysis; peroxisome.

oxi3 Mitochondrial DNA gene responsible for cytochrome oxidase.

oxidation Loss of electrons from a molecule.

oxidation-reduction Reaction transferring electrons from a donor to a recipient.

oxidative burst In plants, resistance to microbial infection may be mediated by the production of reactive oxygen species (ROS) such as superoxide anion ($O^{\bullet-}$), OH^-, H_2O_2, etc. The oxidative burst may induce the transcription of defense genes in the plant tissue infected by avirulent or nonpathogenic agents. The reactive oxygen may kill the cells along with the invader and may modify cell wall proteins, which then reinforce the barrier to the infectious microbes. Several genes encoding proteins with functional similarity to RAC, an RAS family of proteins involved in signal transduction, mediate these reactions. *See* host-pathogen relations; RAC; ROS. (Martinez, C., et al. 2001. *Plant Physiol.* 127:334; Siddiqi, M., et al. 2001. *Cytometry* 46[4]:243.)

oxidative deamination For example, O replaces an NH_2 group of cytosine (C), resulting in a conversion of C $\rightarrow$ uracil. *See* chemical mutagens II; nitrous acid mutagenesis.

oxidative decarboxylation *See* pyruvate decarboxylation complex.

oxidative DNA damages Occurs when ionizing radiation or oxidative compounds hit cells. The major class of damages involves the formation of thymine glycol (isomers of 5,6-dihydroxy-5,6-dihydrothymine), 5-hydroxymethyluracil, 5,6-hydrated cytosine, 8-oxo-7,8-dihydroguanine, 2,2-diamino-4-[(2-deoxy-β-D-erythropentofuranosyl)amino]5(2H)-oxazolone and its precursor, cross-linking between DNA and protein, and elimination of the ribose. 8-oxoguanine represents about 10^{-5} of the mammalian guanine in DNA. 8-oxoguanine can be incorporated and pair in the DNA with A and C residues. If 8-oxoguanine is not repaired, A=T $\rightarrow$ G$\equiv$C transitions or G$\equiv$C $\rightarrow$ T=A transversions occur. The oxidative damages are usually repaired by exonucleases, endonucleases, glycosylases, and other repair enzymes in prokaryotes and eukaryotes. In mice, mutation of the MMH (MutM homologue) repair enzyme results in a 3- to 7-fold increase in 8-hydroxyguanine. *See* abasic site; DNA repair; MtGendo; MtODE; oxidative deamination; photosensitizer; ROS; transition mutation; transversion mutation. (Nunez, M. E., et al. 2001. *Biochemistry* 40:12465; Kaneko, T., et al. 2001. *Mutation Res.* 487:19; Vance, J. R. & Wilson, T. E. 2001. *Mol. Cell Biol.* 21:7191; Aitken, R. J. & Krausz, C. 2001. *Reproduction* 122[4]:497.)

oxidative phosphorylation (OXPHOS) ATP is formed from ADP as electrons are transferred from NADH or $FADH_2$ to O_2 by a series of electron carriers. Such processes take place on the inner membrane of the mitochondria with the assistance of cytochromes as electron carriers. This is the main mechanism of ATP formation. OXPHOS minus yeast mutants are petite and fail to grow on a nonfermentable carbon source and cannot convert an intermediate of the adenine (ade^-) biosynthetic path into a red pigment. About 20 different mutations in nuclear genes are known that control various functions of mtDNA. Actually, 82 structural subunits of mitochondrial genes are encoded in the nucleus versus the 13 coded by mtDNA. *See* ATP; FAD; mitochondria; mitochondrial diseases in humans; NADH; petite colony mutants. (Saraste, M. 1999. *Science* 283:1428; Shoubridge, E. A. 2001. *Hum. Mol. Genet.* 10:2277.)

oxidative stress   Exerted by free radicals (reactive oxygen species, ROS, singlet oxygen, nitric oxide) and peroxides and hydroxyl species through the action of enzymes involved in mixed-function oxidation and autooxidation (P450 cytochrome complex, xanthine oxidase, phospholipase A_2). These reactions play an important role in mutagenesis, aging, mitochondrial functions, signal transduction, etc., and are considered to affect neuronal degeneration in Alzheimer and Parkinson diseases and in amyotrophic lateral sclerosis. Transcription of protooncogenes may be regulated by redox systems. Glutathione and thioredoxin-dependent enzymes provide some protection. *See* catalase; glutathione; peroxidase; redox reaction; ROS; superoxide dismutase; thioredoxins. (Carmel-Harel, O. & Storz, G. 2000. *Annu. Rev. Microbiol.* 54:439; Rabilloud, T., et al. 2002. *J. Biol. Chem.* 277:19396.)

oxidizing agent   Acceptor of electrons.

oxindoles   Protein tyrosine kinase inhibitors. They inhibit fibroblast growth factor and other receptor tyrosine kinases. Chemically they are, e.g., 3-[4-(formylpiperazine-4-yl)benzyl-idenyl]-2-indolinone and 3-[(3-(carboxyethyl)-4-methylpyrrol-2-yl)methylene]-2-indolinone. (Cane, A., et al. 2000. *Biochem. Biophys. Res. Commun.* 276:379.)

8-oxodeoxyguanine   *See* abasic site; oxidative DNA damage.

O
‖
C
H—N₁ 6 5 N
      | 7
H₂N—C₂ 3 4 8 C=O
      N 9 N
         |
         H

8-Oxoguanine.

oxoprolinuria   *See* glutathione synthetase deficiency.

OXPHOS   Oxidative phosphorylation is a mitochondrial process by which molecular oxygen is combined with the electron carriers NADH and FADH₂ by the enzymes of the respiratory chain and mediate the $ADP + P_i \rightarrow ATP$ conversion. *See* oxidative phosphorylation.

oxygenases   *See* mixed-function oxidases.

oxygen effect   In the presence of air or oxygen, the frequency of chromosomal aberrations induced by ionizing radiation increases in comparison to conditions of anoxia (lack of air in the atmosphere). Bacterial irradiation experiments indicate that the oxygen must be present before or at the radiation pulse for reducing survival, but after the pulse its effect rapidly decreases. Glutathione level in

the cells protects against oxygen effect. Exogenous thiols {cysteine, cysteinamine, WR2721 (Amifostil, [(2-(3-amino propylamino)ethyl-phosphorothioic acid]) are also protective. Nitroaromatic compounds (e.g., antiprotozoan metronizadol) and 5′-halogen-substituted pyrimidines (bromouracil) are sensitizers under anoxia. The higher mutability of mtDNA relative to that of the nucleus is attributed to the presence of reactive oxygen in this organelle. *See* mtDNA; physical mutagens; radiation effects. (Thoday, J. M. & Read, J. M. 1947. *Nature* 160:608; Schulte-Frohlinde, D. 1986. *Adv. Space Res.* 6[11]:89; Hsieh, M. M., et al. 2001. *Nucleic Acids Res.* 29:3116.)

oxyntic cells   Secrete acid (HCl) in the lining of the stomach.

oxytocin   Octapeptide stored in the pituitary that regulates uterine contraction and lactation. It is synthesized in the hypothalamus with other associated proteins. They are assembled in the neurosecretory vesicles and transported to the nerve ends in the neurohypophysis (posterior lobe of the pituitary), then may be secreted into the bloodstream. The vasopressin, oxytocin (12 kb between them), and neurophysin genes are linked within the arm of human chromosome 20pter-p12.21. They are transcribed from opposite strands of the DNA. It appears that pre-proarginine-vasopressin-neurophysin are transcribed jointly and posttranslationally separated by proteolysis. *See* animal hormones; memory; neurophysin; vasopressin (antidiuretic hormone). (Breton, C., et al. 2001. *J. Biol. Chem.* 276:26931.)

ozone (O_3)   Bluish, highly reactive form of oxygen. It is a disinfectant, and its presence in the atmosphere protects the earth from the excessive ultraviolet radiation coming from the sun.

On July 9, 1909 (more than two decades earlier than the Neurospora work), F. A. Janssens, Professor at the University of Louvain, presented his theory of chiasmatypy in the journal *La Cellule* (25:389–411):

"In the spermatocytes II, we have in the nuclei chromosomes which show one segment of two clearly parallel filaments, whereas the two distal parts diverge ... The first division is therefore reductional for segment A and a and it is equational for segment B and b ... The 4 spermatids contain chromosomes 1st AB, 2nd Ab, 3rd ab, and 4th aB. The four gametes of a tetrad will thus be different ...

"The reason behind the two divisions of maturation is thus explained ... The field is opened up for a much wider application of cytology to the theory of Mendel."

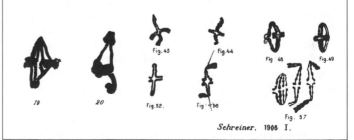

A historical vignette.

P

P (1) Parental generation. (2) *See* probability. (3) Polyoma, a regulatory DNA element in the viral basal promoter. *See* polyoma; Simian virus 40.

³²P Phosphorus isotope. *See* isotopes.

P1 Double-stranded DNA, temperate *E. coli* phage. It is also a vector used in DNA sequencing with a load capacity of ~80 kb.

P₁, P₂ Designations of the parents, homozygous for different alleles, at the critical locus (loci) in a Mendelian cross. *See* allelic combinations; gametic arrays; genotypic segregation; Mendelian laws.

p (petit) Short arm of chromosomes, also denoting frequencies. *See* q.

p13suc1 Yeast cell cycle activating enzyme binding to the cyclosome. *See* cell cycle; cyclosome. (Simeoni, F., et al. 2001. *Biochemistry* 40:8030.)

p14 *See* ARF.

p15^INK4B Inhibitor of CDK4 (encoded in human chromosome 9p21) appears to be an effector of TGF-β, a protein known to control the progression from the G1 phase of the cell cycle to S phase. *See* cancer; cell cycle; p16^INK4; p18; p19; TGF. (Seoane, J., et al. 2001. *Nature Cell Biol.* 3:400.)

p16 (*MTSI* [multiple tumor suppressor], CDKN2) Cell cycle gene (in human chromosome 9p21) that has a major role in tumorigenesis. In 50% of the melanoma cells, it is deleted; it is mutated in 25%. Over 70% of bladder cancer cases are associated with deletions of 9p21 in both homologous chromosomes; in head and neck tumors 33%; in renal and other cells, usually this tumor suppressor gene is lost. It normally restrains CDK4 and CDK6. *See* ARF; CDK4; CDK6; cell cycle; melanoma; pancreatic adenocarcinoma. (Serrano, M., et al. 1993. *Nature* 366:704.)

p16 Weak ATPase and a packaging protein of φ29 bacteriophage. It interacts with viral packaging RNA (pRNA) and phage portal protein and assists in pumping double-stranded DNA into the phage head. p16 and other packaging components are not found within the head. *See* packaging of DNA. (Ibarra, B., et al. 2001. *Nucleic Acids Res.* 29:4264.)

p16^INK4 (CDKN2A) Protein inhibitory to CDK4/CDK6; thus, it appears to be a tumor suppressor because it inhibits the progression of the cell cycle from G₁ to S. It may also direct the cell toward senescence rather than neoplasia. Both p16^INK4

and p19^INK4 are transcribed from the same 9p21 locus but from alternative alleles. *See* cancer; CDK4; cell cycle; Ets oncogene; Id protein; p18; p19; PHO81; senescence; tumor suppressor. (Quelle, D. E., et al. 1995. *Cell* 83:993; Wang, W., et al. 2001. *J. Biol. Chem.* 276:48655.)

p18^INKC Cell cycle inhibitors by blocking cell cycle kinases CDK4 and CDK6. *See* cancer; cell cycle; p15; p16; p19. (Blais, A., et al. 2002. *J. Biol. Chem.* 277:31679.)

p19^Arf *See* Arf.

p19^INK4d Cell cycle inhibitor of CDKs. *See* ARF; cancer; cell cycle; p15; p16; p18.

p21 Transforming protein of Harvey murine sarcoma virus. A Ras-gene-encoded 21 kDa protein binds GDP/GTP and hydrolyzes bound nucleotides and inorganic phosphate. This protein is involved in signal transduction, cell proliferation, and differentiation; p21 controls the cyclin-dependent kinases (Cdk4, Cdk6, Cdk2), binds to DNA polymerase δ-processivity factor, and inhibits in vitro PCNA-dependent DNA replication but not DNA repair. In the absence of p21, cells with damaged DNA are temporarily arrested at the G2 phase, followed by S phases without mitoses. Consequently, hyperploidy arises and apoptosis follows. Gene *p21* is under the control of p53 protein and retinoblastoma tumor suppressor gene RB. MyoD regulates expression of the p21 Cdk inhibitors (p21Cip/WAF1) during differentiation of muscle cells and nonmuscle cells. Withdrawal from the cell cycle does not require the participation of p53. RAF and RHO activate protein p21^Cip/Waf, leading to inhibition of cell cycle transition into the S phase. Actually, RAS activates the serine/threonine kinase Raf, which may facilitate the transition to the S phase, but upon excessive stimulation by RAF and RHO, p21^Cip/Waf has the opposite effect. p53 is also an independent activator of p21^Cip/Waf. After mitosis, p21 is expressed at the onset of differentiation, but it may again be downregulated at later stages of differentiation due to proteasome activity. p21 may not have an absolute requirement for induction by MyoD. The N-terminal domains of p27 and p57 provide other antimitogenic signals. *See* apoptosis; cancer; Cdk; cell cycle; hyperploidy; mitosis; MyoD; p27; p53; p57; PCNA; proteasome; RAS; RASA. (Prall, O. W. J., et al. 2001. *J. Biol. Chem.* 276:45433; Bower, K. E., et al. 2002. *J. Biol. Chem.* 277:34967; Wu, Q., et al. 2002. *J. Biol. Chem.* 277:36329.)

p23 Component of the steroid receptor complex with Hsp90. *See* Hsp; molecular chaperone interacting complex; steroid hormones. (Knoblauch, R. & Garabedian, M. J. 1999. *Mol. Cell Biol.* 19:3748; Munoz, M. J., et al. 1999. *Genetics* 153:1561.)

p24 Family of evolutionarily conserved small integral membrane proteins that form parts of the COP transport vesicles or regulate the entry of cargo into the vesicles. It also mediates viral infection. *See* COP transport vesicles; protein sorting. (Blum, R., et al. 1999. *J. Cell Sci.* 112[pt 4]:537; Hernandez, M., et al. 2001. *Biochem. Biphys. Res. Commun.* 282:1.)

p27 (p27^Kip1) Haplo-insufficient tumor suppressor protein that inactivates cyclin-dependent protein kinase 2. Its mutation (homo- or heterozygous) results in increased body and organ size and neoplasia in mice. The wild-type allele as a transgene may retard cancerous proliferation. Inactivation of p27^Kip1 is triggered by phosphorylation and mediated by the proteasome complex. *See* cancer; cancer gene therapy; Cdk2; cell cycle; KIP; Knudson's two-mutation theory of cancer; p21; p38; proteasome; SKP; transgene. (Mohapatra, S., et al. 2001. *J. Biol. Chem.* 276:21976; Malek, N. P., et al. 2001. *Nature* 413:323.)

p34^cdc2−2 Gene coding for the catalytic subunit of MPF in *Schizosaccharomyces pombe* (counterparts, *CDC28* in *Saccharomyces cerevisiae*, *CDCHs* in humans have 63% identity with *cdc2*; these genes are present in all eukaryotes). The gene product is a serine/tyrosine kinase and its function is required for the entry into the M phase of the cell cycle. If prematurely activated, it may cause apoptosis. *See* cell cycle; MPF. (Shimada, M., et al. 2001. *Biol. Reprod.* 65:442; Nigg, E. A. 2001. *Nature Rev. Mol. Cell Biol.* 2:21.)

p35 (Cdk5 regulatory subunit) Homologue of the baculoviral (survival) protein and cyclin-dependent regulator of neural migration and growth. It blocks apoptosis in a variety of eukaryotic cells. It colocalizes in the cells with RAC and Pac-1. Truncation of p35 results in a very stable p25 protein that is present in Alzheimer disease tangles. *See* Alzheimer disease; baculoviruses; CDK; Pac-1; RAC. (Tarricone, C., et al. 2001. *Mol. Cell* 8:657; Lin, G., et al. 2001. *In Vitro Cell Dev. Biol. Anim.* 37[5]:293.)

p38 (MPK2/CSBP/HOG1) Stress-activated protein kinase of the MEK family. It accelerates the degradation of p27^Kip1, and it phosphorylates H3 histone. p38 is one of the factors required for the initiation of G2/M checkpoint after UV radiation. *See* checkpoint; MAP kinase; MEF; MEK; UV. (Schrantz, N., et al. 2001. *Mol. Biol. Cell* 12:3139; Bulavin, D. V., et al. 2002. *Current Opin. Genet. Dev.* 12:92.)

p40 Tumor suppressor encoded in human chromosome 3q, produced by alternative splicing of p51. Also an L1 RNA transcript (of ORF II)-binding protein required for retrotransposon movement. *See* L1; ORF; p51; p53;

retrotransposon. (Hess, S. D., et al. 2001. *Cancer Gene Ther.* 8[5]:371; Henning, D. & Valdez, B. C. 2001. *Biochem. Biphys. Res. Commun.* 283:430.)

p42 *See* MAPK.

p50 N-terminus of the p105 light chain of NF-κB encoded at 4q23-q24. *See* NF-κB. (Yamada, H., et al. 2001. *Infect. Immun.* 69:7100.)

p51 Cell proliferation inhibitor protein related to p73 and encoded at 3q28. p51A is 50.9 kDa and p51B is 71.9 kDa. *See* p40; p53; p73. (Guttieri, M. C. & Buran, J. P. 2001. *Virus Genes* 23:17.)

p52^SHC RAS G-protein regulator protein; it is regulated through CTLA-4-SYP associated phosphatase. *See* CTLA-4; RAS; SYP. (Joyce, D., et al. 2001. *Cytokine Growth Factor Rev.* 12:73.)

p53 (TP3, 17p13.1) Tumor suppressor gene when the wild-type allele is present, but single base substitutions may eliminate suppressor activity and the tumorigenesis process may be initiated. p53 is a tetramer with separate domains for DNA binding, transactivation, and tetramerization. Its product binds to specific DNA sequences, activates transcription from promoters with p53 protein-binding sites, and represses transcription from promoters lacking p53-binding sites. It promotes annealing of DNAs, inhibits replication, controls G1- and G2-phase checkpoints, leads to apoptosis or just blocks cytokinesis if the DNA is damaged, interferes with tumorous growth, maintains genetic stability, and reduces radiation hazards by its regulatory role in the cell cycle. For the maintenance of G2 arrest after DNA damage, the presence of p21 is also required. Protein p53 binds to a somewhat conserved consensus and it is phosphorylated at serine 315 residues by CDK proteins during S, G2, and M phases of the cell cycle but not at G1, although p53 controls an important G1 checkpoint. It also controls proteins p21, p27, and p57. p53 protein binds to the four copies of its consensus in DNA (5′-PuPuPuGA/T-3′). The C-terminal domain controls tetramerization and the N-terminal domain is responsible for transcriptional activation and for the regulation of downstream genes. One study indicated that at least 34 different transcripts were induced by p53 more than 10-fold, although there was heterogeneity in the response. Coactivators TAFII40, TAFII 60, and other TATA box−binding factors mediate its transcriptional activation. When the first six exons of the gene are deleted, mRNA is still translated into a C-terminal protein fragment. Such a mutation enhances tumor suppression but leads to premature aging in mice (Tyner, S. D., et al. 2002. *Nature* 415:45). p53 is encoded in human chromosome 17p13.105-p12; in about

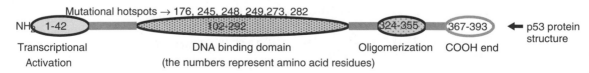

p53 tumor suppressor.

half of human tumors, the normal allele is altered. Protein 53 plays a central role in cellular metabolism and its expression is induced by many factors such as oncogene expression, chemotherapy, oxidative stress, hypoxia, etc. Topoisomerase I is a p53-dependent protein. p53-induced apoptosis may require transcriptional activation or it may occur in the absence of RNA or protein synthesis. The activation of p53 may be followed by FAS transport from Golgi intracellular stores without a need for synthesis of FAS.

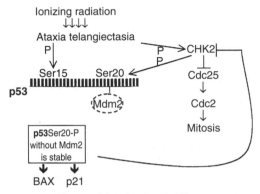

Some of the circuits of p53

Ionizing radiation damage to the ataxia telangiectasia protein results in the phosphorylation of the serine 15 residue of p53 and checkpoint 2 (CHK2), which in turn phosphorylates residue 20. The latter point is normally occupied by Mdm2, a negative regulator of p53 that is now dislodged by the phosphorylation of site 20. The p53 ser20-p protein is now stabilized, despite the radiation, and can induce BAX (a porin with antiapoptosis function) and p21, a suppressor of mitosis. The blocking of CHK2 then prevents its inhibition of cell division cycle proteins (Cdc25 and Cdc2). The process to mitosis is then facilitated.

Protein p73 has functions similar to those of p53. p33^INGI, encoded at human chromosome 13q34, cooperates with p53 by protein-protein interaction in repressing cellular proliferation and the promotion of apoptosis. DNA-dependent protein kinase (DNA-PK) activates p53 in case the DNA is damaged. When amino acid site 376 is dephosphorylated by ionizing radiation, protein 14–3–3 binds to p53 and increases its ability for DNA binding. Still other proteins such as IRF may be involved with p53 and other cooperating proteins. p53 activity may be affected by oncoproteins RAS and MYC. Chemotherapeutic agents, UV light, and protein kinase inhibitors may also activate p53. p53 is reversibly blocked by pifithrin-α (2[2-imino-4,5,6,7-tetrahydrobenzothiazol-3-yl]-1-polyethanone) and may protect from the undesirable side effects of anticancer therapy without causing new tumors in the absence of p53 function. When 33,615 unique human genes were tested by cDNA microarrays, 1,501 genes responded to p53 (Wang, L., et al. 2001. *J. Biol. Chem.* 276:43604). Pharmacological compounds (CP-31398, CP-257042, etc.) have been selected by large–scale screening that could stabilize the DNA-binding domain of mutant p53 and activate its transcription as well as slow tumor development in mice. In some cancers, p53 may increase sensitivity to chemotherapeutic agents, but in others it does not affect them. NADH quinone oxidoreductase may stabilize the p53 protein. Germline mutations occur in about 1×10^3 of the Caucasian populations, but in about half of the sporadic cancers they are somatically mutated. (*Science* magazine declared p53 the molecule of the year in 1993.) *See* annealing; apoptosis; ARF; ataxia telangiectasia; cancer; Cdc; cell cycle; CHK; DAP kinase; DNA-PK; E2F; GADD45; IRF; lactacystin; MDM2; p21; p27; p40; p51; p57; p63; p73; papilloma virus; porin; protein 14-3-3; ribonucleotide reductase; siRNA substitution mutation; TAF; TBP; transactivator; tumor suppressor gene. (Vogelstein, B., et al. 2000. *Nature* 408:307; Vousden, K. H. 2000. *Cell* 103:691; Asher, G., et al. 2001. *Proc. Natl. Acad. Sci. USA* 98:1183; Johnson, R. A., et al. 2001. *J. Biol. Chem.* 276:27716; Olivier, M., et al. 2002. *Hum. Mut.* 19:607; Martinez, C. A., et al. 2002. *Proc. Natl. Acad. Sci. USA* 99:14849.)

p55 (TNFR1) Tumor necrosis factor receptor of Fas. *See* Fas; TNF. (Dybedal, I., et al. 2001. *Blood* 98:1782; Longley, M. J., et al. 2001. *J. Biol. Chem.* 276:38555.)

p56^chk1 Protein kinase and a checkpoint for mitotic arrest after mutagenic damage inflicted by UV, ionizing radiation, or alkylating agents. DNA damage results in the phosphorylation of this protein in yeasts. Phosphorylation may prevent mitotic arrest, yet the cells may die later; p56 is not involved in DNA repair. Phosphorylation is required for other checkpoint genes to become/stay functional. *See* cell cycle; DNA repair. (Feigelson, S. W., et al. 2001. *J. Biol. Chem.* 276:13891.)

p57 (p57^Kip2) Antimitogenic protein; its carboxyl end assures nuclear localization; the amino end is involved in the inhibition of CDK proteins. *See* CDK; cell cycle; KIP; p21; p27. (Thomas, M., et al. 2001. *Exp. Cell Res.* 266:103.)

p60 Binds to Hsp70 and Hsp90 and chaperones the assembly of the progesterone complex. *See* animal hormones; chaperone; Hsp; Hsp70; progesterone. (Mukhopadhyay, A., et al. 2001. *J. Biol. Chem.* 276:31906.)

p63 Member of the p53 tumor suppressor gene family, encoded at 3q27-q29. The gene expresses at least six transcripts involved with transactivation of p53 and p73, DNA binding, and oligomerization. It controls ectodermal (limb, craniofacial, and epithelial) differentiation. *See* EEC syndrome; Hay-Wells syndrome; p53; p73. (van Bokhoven, H., et al. 2001. *Am. J. Hum. Genet* 69:481; van Bokhoven, & Brunner, H. G. 2002. *Am. J. Hum. Genet.* 71:1.)

p65 Component of the NF-κB complex. It can be exploited advantageously for gene activation in a chimeric construct with a mutant progesterone receptor ligand-binding domain of gene GAL4. *See* GAL4; Gene-Switch; NF-κB. (Burcin, M. M., et al. 1999. *Proc. Natl. Acad. Sci. USA* 97:355.)

p70/p86 Ku autoantigen. *See* DNA-PK; Ku.

p70^s6k Phosphorylates S6 ribosomal protein at serine/threonine residues before translation. Also called S6 kinase. *See* p85^s6k; signaling to translation; S6 kinase; translation initiation. (Harada, H., et al. 2001. *Proc. Natl. Acad. Sci. USA* 98:9666.)

p73 Has homology in amino acid sequence to p53 protein and is encoded in human chromosome 1p36.3. Similarly to p53, it regulates apoptosis and antitumor activity (upon E2F1 induction), hippocampal dysgenesis, hydrocephalus, immune reactions, and pheromone sensory pathways. It affects proliferation, although in a somewhat different manner, but its loss does not lead to tumorigenesis in mice. p73 may compete with p53. This protein may have antiapoptotic effects in neurons. *See* apoptosis; E2F; p53; p63. (Sasaki, Y., et al. 2001. *Gene Ther.* 8:1401; Stiewe, T. & Putzer, B. M. 2001. *Apoptosis* 6:1447; Melino, G., et al. 2002. *Nature Rev. Cancer* 2:605.)

p75 Nontyrosine kinase receptor protein, TNFR 2 (tumor necrosis factor receptor 2). It is a Fas receptor. *See* Fas; TNF. (Hutson, L. D. & Bothwell, M. 2001. *J. Neurobiol.* 49[2]:79; Wang, X., et al. 2001. *J. Biol. Chem.* 276:33812.)

p80^{sdc25} Protein phosphatase that activates the p34^{cdc2}-cyclin protein kinase complex by dephosphorylating Thr^{14} and Tyr^{15}. *See* cell cycle; Ku. (McNally, K. P., et al. 2000. *J. Cell Sci.* 113 [pt 9]:1623.)

p85^{s6k} Phosphorylates S6 ribosomal protein before translation at serine/threonine sites. Also called S6 kinase. p85 protein is involved in a p53-dependent apoptotic response to oxidative damage and natural killer activation. p85 phosphoinositide 3-kinase mediates developmental and metabolic functions. *See* p70^{s6k}; phosphoinositides; S6 kinase; translation initiation. (Fruman, D. A., et al. 2000. *Nature Genet.* 26:379.)

p95 Fas; involved in apoptosis, and its mutation leads to Nijmegen breakage syndrome and other double breakage of the chromosomes. *See* acrosomal process; APO; apoptosis; Fas; Nijmegen breakage syndrome.

p97 About M$_r$ 600 ATPase; mediates membrane fusion. (Hirabayashi, M., et al. 2001. *Cell Death Differ.* 8[10]:977.)

p105 *See* p50.

p107 Retinoblastoma protein-like regulator of the G1 restriction point of the cell cycle. *See* cell cycle; pocket; restriction point; retinoblastoma; tumor suppressor. (Charles, A., et al. 2001. *J. Cell Biochem.* 83:414.)

p110^{Rb} Protein encoded by the retinoblastoma (Rb) gene. When not fully phosphorylated, it interferes with the G$_0$ and G$_1$ phases of the cell cycle by inhibition of E2F transcription factor. *See* cell cycle; E2F; killer cells; retinoblastoma; tumor suppressor. (DeCaprio, J. A., et al. 1988. *Cell* 54:275.)

p115 Monomeric GTPase with a specific guanine exchange factor (GEF) for RHO (p115 Rho GEF). *See* GEF; GTPase; RHO. (Wells, C. D., et al. 2001. *J. Biol. Chem.* 276:28897.)

p125^{FAK} Focal adhesion kinase, a nonreceptor tyrosine kinase. *See* CAM. (Yurko, M. A., et al. 2001. *J. Cell. Physiol.* 188:24.)

p130 Retinoblastoma protein-like regulator of the G1 restriction point of the cell cycle. p130^{Cas} is involved in the organization of myofibrils, actin fibers, and anchorage dependence of cultured cells. *See* CAS; pocket; restriction point; retinoblastoma; tumor suppressor. (Tanaka, N., et al. 2001. *Cancer* 92:2117.)

p160 Family of transcriptional coactivators such as SRC-1, GRIP1/TIF2, and p/CIP. They modify chromatin structure by methylating some of the histones. *See* chromatin remodeling; nuclear receptors; p/CIP. (Mak, H. Y. 2001. *Mol. Cell Biol.* 21:4379.)

p300 (CBP) Cellular adaptor protein preventing the G$_0$/G$_1$ transition of the cell cycle; it may activate some enhancers and stimulate differentiation. It is also a target of the adenoviral E1A oncoprotein. Its amino acid sequences are related to CBP, a CREB-binding protein. Nuclear hormone receptors interact with CBP/p300 and participate in gene transactivation. PCAF is a p300/CBP-associated factor in mammals; is the equivalent of the yeast Gcn5p (general controlled nonrepressed protein), an acetyltransferase working on histones 3 and 4 (HAT A) and thus regulating gene expression. p300 functions as a coactivator of NF-κB. p300 binds PCNA. In several human cancers, p300 mutations were identified, indicating that the protein is a tumor suppressor. *See* adenovirus; bromodomain; chromatin remodeling; CREB; E1A; histone acetyltransferase; histone methyltransferases; NF-κB; PCNA. (Lin, C. H., et al. 2001. *Mol. Cell* 8:581.)

p350 DNA-dependent kinase; it is likely a basic factor in severe combined immunodeficiency and it may be responsible for DNA double-strand repair, radiosensitivity, and immunoglobulin V(D)J rearrangements. In association with KU protein, it forms a DNA-dependent protein kinase. *See* DNA-dependent protein kinase; immunoglobulins; kinase; KU; severe combined immunodeficiency. (Chan, D. W. 1996. *Biochem. Cell Biol.* 74:67.)

P450 (CYP) Family of genes coding for cytochrome enzymes involved in oxidative metabolism. They are widely present in eukaryotes and scattered around several chromosomes. All mammalian species have at least eight subfamilies. The homologies among the subfamilies are over 30%, whereas the homologies among members of a subfamily may approach 70%. These cytochromes possess monooxygenase, oxidative deaminase, hydroxylation, and sulfoxide-forming activities. The proteins are generally attached to the microsomal components of homogenized cells (endoplasmic reticulum [fragments]), often called S9 fraction. Some of these enzymes (subfamily IIB) are inducible by phenobarbital. Their expression may be tissue-specific, predominant in the liver, kidney, or intestinal cells. Mammalian P450 cytochrome fraction is generally added to the *Salmonella* assay media of the Ames test in order to activate promutagens. The pregnane X receptor (PXR) is activated by a variety of compounds and is thus responsible for the activation of different drugs involved in mutation, cancer, and interaction with other drugs. One member of the P450 series is involved in the regulation of the synthesis of the 6th class of plant hormones, brassinosteroids. P450 enzymes require the cofactor Nad or NADPH, and their activity is favored by the presence of peroxides as oxygen donors. By mutagenesis, more industrially useful P450 variants are being produced. P-450 (CYP1A1) dioxin and aromatic compound-inducible P450 maps to human chromosome 15q22-qter. CYP1A2 is phenacetin O-deethylase. Phenacetin is an analgesic and antipyretic carcinogen. CYP2D (22q13.1) is a debrisoquin 4-hydroxylase. Debrisoquine is a toxic antihypertensive drug. CYP51 (7q21.2-q21.3) is lanosterol 14-α-demethylase, a sterol biosynthetic protein. *See* Ames test; analgesic; antipyretic; brassinosteroids; cyclophilin; cytochromes; hypoaldosteronism; NAD; peroxide; steroid hormones. (Fujita, K. & Kamataki, T. 2001. *Mutation Res.* 483:35; Ingelman-Sundberg, M. 2001. *Mutation Res.* 482:11.)

PABp   Poly(A)-binding protein (~72-kDa) is the major protein that binds to the poly(A) tail of eukaryotic mRNA and converts it to mRNAP. Pab1p connects the mRNA end to the eIF-4H subunit of the eukaryotic peptide initiation factors eIF-4G and eIF-4F. It contains four RRM motifs. PABp interacts with PAIP, a translational coactivator protein in mammals. *See* binding proteins; eIF-4F; eIF-4G; mRNA circularization; mRNA decay; mRNAP; mRNA tail; polyadenylation signal; ribosome scanning; RRM; translation initiation; translational termination; Xrn1p. (Kozlov, G., et al. 2001. *Proc. Natl. Acad. Sci. USA* 98:4409.)

pac   Site in the phage genome where terminases bind and cut during maturation of the DNA before packing the DNA into the capsid. *See* packaging of the DNA; terminase.

PAC   Phage artificial chromosome. P1 phage PAC carries about 100–300 kb DNA segments. Most PAC vectors lack selectable markers suitable for mammalian cell selection but can be retrofitted by employing the Cre/loxP site-specific recombination system. *See* BAC; YAC. (Poorkaj, P., et al. 2000. *Genomics* 68:106.)

PAC-1   Dephosphorylates Thr[183] and Tyr[185] residues and thus regulates mitogen-activated MAP protein kinase involved in signal transduction. The Pac1 nuclease removes the 3′ external transcribed spacers from the nascent rRNAs in cooperation with RAC. *See* MAP; MKP-1. (Boschert, U., et al. 1997. *Neuroreport* 8:3077; Spasov, K., et al. 2002. *Mol. Cell.* 9:433.)

PACAP   Pituitary adenylyl cyclase- activating polypeptide-like neuropeptide is a neurotransmitter at the body wall neuromuscular junction of *Drosophila* larvae. It mediates the cAMP-RAS signal transduction path. *See* RAF; RAS; signal transduction. (Kopp, M. D., et al. 2001. *J. Neurochem.* 79:161.)

pachynema   Literally "thick thread" of chromosomes at early meiosis when the double-stranded structure of the chromosomes is not distinguishable by light microscopy because the chromatids are tightly appositioned. Also, the two homologous chromosomes are closely associated unless structural differences prevent perfect synapsis. If a pair of chromosomes is not completely synapsed by pachytene, they will not pair later either. In pachytene, the chromosomal knobs and chromomeric structure are visible and can be used for identification of individual chromosomes. After

pachytene, the synaptonemal complex is dismounted and the chromosomes progressively condense. In case the chromosomes are defective at this stage, the Red1 (required for chromosome segregation) and Mek1 proteins serve as checkpoint controls by preventing further progress of meiosis. Normally MEK kinase phosphorylates Red. Phosphatase Glc7 dephosphorylates Red. More than two dozen other proteins (named differently in different organisms) are also involved in pachytene controls. *See* chiasma; chromomere; meiosis; MEK; pachytene analysis; synapsis. (Bailis J.M & Roeder, G. S. 2000. *Cell* 101:211; Roeder, G. S. & Bailin, J. M. 2000. *Trends Genet.* 16:395.)

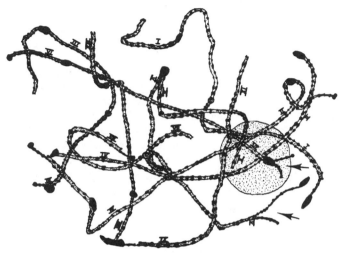

Naturalistic drawing of the 10 pachytene chromosome pair of a teosinte X maize hybrid. Note (←) unpaired ends of chromosomes V, VII, and some terminal and near-terminal knobs. (Courtesy of Dr. A. E. Longley, see also 1937. *J. Agric. Res.* 54:835.)

pachyonychia   Rare autosomal-dominant keratosis of the nails and skin. *See* keratosis.

pachytene analysis   Study of meiotic chromosomes at the pachynema stage when cytological landmarks, chromomeres, and knobs are distinguishable by the light microscope, and chromosomal aberrations (deletions, duplications, inversions, translocations, etc.) can cytologically be identified and correlated with genetic segregation information. The pachytene analysis of plants is analogous to the study of giant chromosomes in dipteran flies and other lower animals. The bands of the (somatic) salivary chromosomes are tightly appositioned chromomeres in these endomitotic chromosomes. *See* chromomere; endomitosis; meiosis; recombination nodule; salivary gland. (McClintock, B. 1931. *Missouri Agric Exp. Sta. Bull.* 163; Carlson, W. R. 1988. *Corn and Corn Improvement*, Agr. Monogr. 18, ASA-CS-SSA, p. 259. Madison, WI.)

pachytene stage   The chromosomes form pachynema. *See* pachynema.

packaging, cell line (for retroviral vectors)   For the replication of the vector the viral proteins gag, pol, and env are required, but they are deleted from the vectors to prevent the production of disease-causing virions. The packaging signal ψ is retained in the vector. Another solution is to insert these viral genes into host chromosomes or remove the packaging signal (Ψ [psi]) from the helper virus and delete the 3′-LTR. In neither case could the production (by two recombinations) of replication-competent virions be completely eliminated. Thus the nucleic acid (with the transgene) can be packaged, although the virions are defective. An improved construct removed LTRs from the structural genes and replaced them with heterologous promoters and polyadenylation signals. The *gag* and *pol* genes are placed on a plasmid different from the one that carries the *env* gene. In the packaging cell lines these two are inserted at different chromosomal sites. Also, if the number of cell divisions is limited, the chance of recombination between vector and helper is reduced. In an improved packaging system, a stop codon is engineered into the *gag* reading frame to prevent the assembly of a fully competent virus. In the packaging cell lines, the appropriate envelope protein for the intended target (ecotropic or amphotropic) should be present in the helper virus (pseudotyping) to ensure optimal transfection. The envelope protein may need modification in order to ensure the proper targeting to the intended types of cells. Antibodies specific for certain cell-surface antigens or against particular receptors may be employed. Although some of these procedures appear very attractive, they may not always be equally efficient. These technical problems are obviously attracting serious research efforts. *See* amphotropic retrovirus; ecotropic retrovirus; pseudotyping; retroviral vectors; viral vectors. (Thaler, S. & Schnierle, B. S. 2001. *Mol. Ther.* 4[3]:273.)

packaging, phage DNA

λ-phage gene A recognizes the cos sites; gene D assists in filling the head (capsid); genes W, F, V, ILK, and GMH assemble the phage from prefabricated elements and act in the processes shown diagrammatically below.

The DNA that first enters the phage capsid has the Nu end and the opposite end (the last) is the R end. *See* development; lambda phage; p16. (Smith, D. E., et al. 2001. *Nature* 413:748; Kindt, J., et al. 2001. *Proc. Natl. Acad. Sci. USA* 98:13671.)

packing ratio

The DNA molecule is much longer than the most extended chromosome fibers. The packing ratio was defined as the proportion of the DNA double helix and the length of the chromosome fibers. In the human chromosome complement, the packing ratio was estimated to be more than 100:1 at metaphase. The length of the *Drosophila* genome at meiotic metaphase was estimated to be 7.8 μm and the length of a chain of 3,000 nucleotides is approximately 1 μm. The *Drosophila* genome contains about 9×10^7 bp; hence, the total length of DNA within the *Drosophila* genome is about 30,000 μm, indicating a packing ratio of 3846:1. The packing ratio indicates some of the problems encountered by the eukaryotic chromosomes in condensing an enormous length of DNA to a small space and still replicating, transcribing, and recombining it in an orderly manner. To illustrate the

problems in a trivial way: Many eukaryotes have the same packing problem as folding a 2.5-km (1.6 mi)-long thread into a 2.5 cm (1 inch) skein. In prokaryotic-type DNA—such as without nucleosomal structure—the excessive amount of plasmid DNA forms liquid crystalline molecular supercoils. *See* Mosolov model; supercoiled DNA; lysis. (DuPraw, E. J. 1970. *DNA and Chromosomes*. Holt, Rinehart and Winston, New York; Cook, P. R. 2002. *Nature Genet.* 32:347.)

paclitaxel   *See* taxol.

pactamycin   Inhibitor of eukaryotic peptide chain initiation.

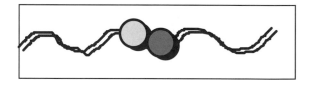

Pactamycin.

padlock probe

Contains two target-complementary segments connected by linker sequences. Hybridization to target sequences brings the two ends close to each other so that they can be covalently ligated. The circularized probes are thus catenated to the DNA (≈) like a padlock (OO). Such probes permit high-specificity detection, distinction among similar target sequences, and can be manipulated without alterations or loss. By using circularizable or circularized allele-specific probes and primers and rolling circle amplification, mutations in short genomic sequences can be detected. (After Lizardi, P. M., et al. 1998. *Nature Genet.* 19:225.)

Alternatively, for rolling circle amplification, two primers were used. After the first primer (P1) initiates the replication, the second (P2) primer is bound to the tandem repeats, and both primers generate repeats in opposite directions: O or N using either the (+) or (-) strands, respectively. In 90 minutes, at least 10^9 copies of the circles are generated, making it possible to

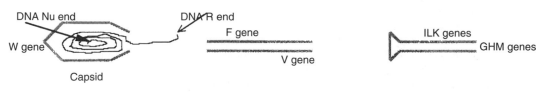

Packing of phage DNA.

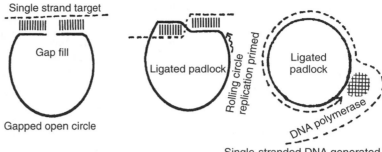

Single strand target

Gap fill

Gapped open circle

Ligated padlock

Rolling circle replication primed

Ligated padlock

DNA polymerase

Single-stranded DNA generated

The probe at left, interrupted by a 6- to 10-base gap, hybridizes to the target DNA. The gap is filled with an allele-specific or DNA polymerase-generated sequence. After ligation, it generates a closed duplex padlock. A complementary (18-base) primer was then employed with a DNA polymerase. The original target DNA is not shown.

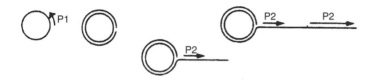

detect very rare somatic mutations. Rolling circle amplification can detect gene copy number single-base mutations and can quantify the transcribed mRNA (Christian, A. T., et al. 2001. *Proc. Natl. Acad. Sci. USA* 98:14238). The procedure generated replication products hybridized to either fluorescein- or Cy3-labeled deoxyribonucleoprotein-oligonucleotide (DNP) tags, respectively. (The tag was anti-DNP immunoglobulin M [IgM]). This process of condensation of amplification circles after hybridization of encoding tags is called CACHET. The procedure also permits the identification of single-copy genes by epifluorescence microscopy. *See* DNA polymerases; fluorochromes; immunoglobulins; microscopy; mutation detection; probe; rolling circle. (Baner, J., et al. 2001. *Curr. Opin. Biotechnol.* 12:11; Roulon, T., et al. 2002. *Nucleic Acids Res.* 30:[3]:e12.)

PAGE   Polyacrylamide gel electrophoresis.

Paget disease   Two autosomal-dominant forms have been described involving cancer of the bones or of the anogenital region (the region of the anus and genitalia). The disease is an anomaly of osteoclastogenesis. BDB1 gene was located to 6p21.3 and PDB2 (also called familial expansile osteolysis, FEO) to 18q21-q22. Mutations in the tumor necrosis factor receptor (TNFR) seem to be involved and affect the signaling by NF-κB (RANK, receptor activator of nuclear factor κB). Additional loci have been mapped to 5q35-qter (PDB3) and 5q31 (PDB4), 2q36 (PDB5) and 10p13 (PDB6). *See* NF-κB; osteoclast; TNFR. (Laurin, N., et al. 2001. *Am. J. Hum. Genet.* 69:528.)

PAH   (1) Poly aromatic hydrocarbons. The majority of them are carcinogenic. (2) Paired amphipathic helix motif may mediate protein-protein interactions in regulating enzyme functions. *See* amphipathic.

1,2,5,6-Dibenzanthracene

PAI   Plasminogen activation inhibitor. *See* plasminogen activator. (Eilers, A. L., et al. 1999. *J. Biol. Chem.* 274:32750.)

pain insensitivity   Controlled by defects causing hereditary sensory neuropathies. In the dominant form, the dorsal ganglia are degenerated. In recessive neuropathy, the loss of myelinated A fibers cause touch insensitivity. Congenital pain insensitivity with anhidrosis (CIPA, 1q21-q22) involves a defect of the nerve growth factor receptor (TRKA); in congenital insensitivity to pain without anhidrosis, the small myelinated A-delta fibers are defective. *See* anhidrosis; neuropathy; Riley-Day syndrome; sensory neuropathy 1; TRK. (Mardy, S., et al. 2001. *Hum. Mol. Genet.* 10:179; Cheng, H.-Y. M., et al. 2002. *Cell* 108:31.)

pain sensitivity   May be traditionally treated with analgesics. Gene therapy by introduction of genes producing analgesic substances (catecholamines, enkephalins) or antinociceptive peptides are potential molecular approaches. The capsaicin or vanilloid receptor (VR1) controls heat-gated ion channels with response to low temperature (~43°C) stimuli, whereas the VRL-1 receptor responds to about 52°C. The vanilloid channel receptor is induced by protein kinase C. The transcriptional repressor DREAM constitutively suppresses prodynorphin in the neurons of the spinal chord. When DREAM is knocked out, there is still sufficient expression of dynorphin, but there is a strong reduction in pain sensitivity. *See* analgesic; catecholamines; DREAM; dynorphin; endorphin; enkephalins; nociceptor; protein kinase. (Samad, T. A., et al. 2001. *Nature* 410:471; Costigan, M. & Woolf, C. J. 2002. *Cell* 108:297; Mantyh, P. W., et al. 2002. *Nature Rev.* Cancer 2:201.)

paired box genes   *See* PAX.

paired-end sequence   Product of the first sequencing of both ends of a cloned DNA tract. (Zhao, S., et al. 2000. *Genomics* 63:321.)

paired *t*-test   *See* matched pairs test; student's *t* distribution; *t* value.

pairing (synapsis)   Intimate association of the meiotic chromosomes mediated by several protein factors. In prokaryotes, the RecA and RecT proteins have an important role; in yeast and humans, the Rad52 protein carries out similar functions. *See* base pair; hydrogen pairing; meiosis; pachytene analysis; Rad; RecA; RecT; somatic pairing; synapsis; tautomeric shift; zygotene. (Kagawa, W., et al. 2001. *J. Biol. Chem.* 276:35201.)

pairing alkylated bases   *See* alkylation.

pair rule genes   Determine the formation of alternating segments in the developing embryo. A similar segment pattern, although with variations, occurs in other insects. *See* *engrailed*; *fushi tarazu*; *knirps*; metamerism; morphogenesis in *Drosophila*; *Runt*.

pair-wise likelihood score   Estimates the potential relationship between pairs of individuals on the basis of allele sharing. (Smith, B. R., et al. 2001. *Genetics* 158:1329.)

PAK   p21-activated kinase. Serine/threonine kinases activated by GTPases, Rac, and Cdc42. Pak3 regulates Raf-1 by phosphorylating serine 338 in rats. Paks regulate the actin cytoskeleton. A nonsyndromic mental retardation (human chromosome Xq22, yeast homologue is *STE20*) prematurely terminates PAK3 transcription. *See* actin; apoptosis; Cdc42; cytoskeleton; GTPase; mental retardation; nonsyndromic; p21; p35; PDK; Rac; raf. (Xia, C., et al. 2001. *Proc. Natl. Acad. Sci. USA* 98:6174.)

PAL   Phenylalanine ammonia lyase.

palea   Inner, frequently translucent bract around the grass flower.

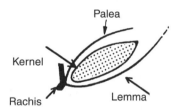

Paleolithic (old stone age)   More than 20,000 years ago, it marked the beginning of human tool formation and cave artistry by Cro-Magnon humans. *See* geological time periods; Lascaux; mesolithic; Neanderthal; neolithic.

paleontology   Deals with the relics of past geological periods. Its methods and materials are used for the study of the evolution of biological forms. *See* geological time periods; paleolithic age.

Paleozoic   Geological period between about 225 and 570 million years ago. During the later part of this period, land plants, amphibians, and reptiles appeared. *See* geological time periods.

palindrome   Region of a DNA strand where complementary bases are in opposite sequence, such as ATGCAC*GTGCAT. Palindromes may come about by inverted repeats of sections of the double-stranded DNA where these sequences of the opposite strands read the same forward and backward. Upon folding of these sequences in a single strand, they can assume structures with paired bases. Palindromic sequences in the DNA reassociate very rapidly because the complementary

bases are in close vicinity. A simple palindromic word is MADAM; it reads the same from left to right or from right to left. Palindromic sequences are often unstable; recombination within palindromes results in deletions and duplications. The restriction enzyme recognition sites are palindromic. *See* insertion elements; inverted repeats; RecA-independent recombination; restriction enzyme; stem and loop. (Leach, D. R. 1994. *Bioessays* 16:893; Nasar, F., et al. 2000. *Mol. Cell. Biol.* 20:3449; Zhu, Z.-H., et al. 2001. *Proc. Natl. Acad. Sci. USA* 98:8326.)

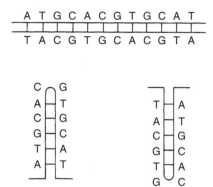

Palindromic DNA and possible pairing within single strands of it. (From Flavell, R. B. & Smith, D. B. 1975. *Stadler Symp.* 7:47.)

palingenesis   Regeneration of lost organs and parts or reappearance of evolutionarily ancestral traits during ontogeny. According to Ernst Haeckel (1834–1914), the ontogeny recapitulates the phylogeny. *See* ontogeny; phylogeny.

palisade cells   Oblong and arranged in a row; the large palisade parenchyma cells are below the upper epidermis of plant leaves and loaded with chloroplasts.

Palisade cells, lower layer.

Pallister-Hall   syndrome   Postaxial polysyndactyly (PAPA1), encoded in human chromosome 7q13, at the same location as Greig syndrome. The gene is also called GLI3 and its product has a homology to the *Drosophila* Krüppel Zn-finger protein. A similar syndactyly was recorded by French scientist Maupertuis in the Prussian royal court in Berlin in 1756. *See* DNA-binding protein domains; EEC syndrome; Greig cephalopolysyndactyly; polydactyly; Smith-Lemli-Opitz syndrome.

palm print   *See* Down syndrome; fingerprint; simian crease.

palomino   Horse with light tan fur color and flaxen mane and tail.

The genetic constitution of palomino is: *AAbbCCDd*.

PALS *See* alternative splicing.

palsy (paralysis) Cerebral palsy may be caused by physical injuries or may be part of the symptoms of diverse genetic syndromes. *See* syndrome; tau.

PAM *See* evolutionary clock.

PAMAM *See* polyamidoamine dendrimers.

PAN Genus of chimpanzees. *See* hominidae; primates. (Gagneux, P. 2002. *Trends Genet.* 18:327.)

pancreatic adenocarcinoma Cancer of the pancreas, frequently associated with loss or defect of DCC, or p53 or MTS1 oncogene suppressors. It was also attributed to mutations in codon 12 of the c-K-ras (Kirsten RAS) gene encoded at human chromosome 12p12. A dominant susceptibility locus was assigned to 4q32-q34. *See* DCC; Kirsten-Ras; p16; p21; p53. (Eberle, M. A., et al. 2002. *Am. J. Hum. Genet.* 70:1044.)

pancreatitis, hereditary Autosomal-dominant (7q35) gene (80% penetrance and variable expressivity) has an onset before the teen years, appearing as abdominal pain and other anomalies. The basic defect is in a cationic trypsinogen. *See* trypsin.

panda (*Ailurus fulgens*) $2n = 36$.

pandemic Infection by a microbe or virus is spread over large areas (countries, continents).

PAN editing Adding U residues to the primary transcripts of mtDNA, thus causing extensive posttranscriptional changes in RNA. *See* kinetoplast; RNA editing.

Paneth's cells Secretory epithelial cells expressing defensin. *See* defensin. (Ghosh, D., et al. 2002. *Nature Immunol.* 3:583.)

pangenesis Ancient misconception about heredity, originated in the Aristotelian epoch and periodically revived during the centuries. Charles Darwin has also interpreted inheritance as pangenesis. Accordingly, all the information expressed during the life of the individuals is transported to the gametes from all parts of the body. Thus, pangenesis is the means of the inheritance of all, including the acquired characters. *See* acquired characters; lysenkoism.

panicle Inflorescence of a compound raceme structure such as oats. *See* raceme.

panmictic index *See* fixation index; panmixis.

panmixis Random mating; in a population there is an equal chance for each individual to mate with another of the opposite sex. *See* Hardy-Weinberg theorem.

panning Use of antibody affinity chromatography or ELISA for the separation of specific molecules (in analogy to the gold-washing pans of the gold hunters). *See* affinity chromatography; ELISA; phage display. (Chen, G., et al. 2001. *Nature Biotechn.* 19:537.)

panspermia Theory claiming that life has originated at several places in the universe and spread to earth by meteorites or by other means. *See* origin of life.

panther (*Panthera pardus*, leopard) $2n = 38$: a feline species; in captivity, it may be crossed with a lion, but no mating is known to take place in the wild where conspecific sexual partners are available.

pantothenic acid Precursor of coenzyme A.

$$O=C-CH_2-CH_2-N\overset{H}{-}C\underset{O}{\overset{H}{-}}C\underset{OH}{\overset{CH_3}{-}}C\underset{CH_3}{\overset{H}{-}}CH_2OH$$

Pantothenate.

pantropic Can affiliate with many different types of tissues.

PAP (purple acid phosphatases) Ubiquitous metallophosphoesterase proteins with phosphatase, exonuclease, 5′-nucleotidase, etc. functions. (*See* Li, D., et al. 2002. *J. Biol. Chem.* 277:27772.)

PAPA syndrome (pyogenic sterile arthritis and acne and familial recurrent arthritis, 15p) It may involve also ulcerative skin lesions (pyoderma gangrenosum). It may be caused by defects in the 15-exon CD2-binding protein1 (CD2BP1). *See* CD2. (Wise, C. A., et al. 2002. *Hum. Mol. Genet.* 11:961.)

papain Member of a family of proteolytic enzymes with an imidazole group near the nucleophilic SH group, and the former plays a role as a proton donor to the cleaved-off part. Papain cleaves immunoglobulin G into three near-equal-size fragments, which helped in clarifying the structure of antibodies. *See* antibody; calpain; immunoglobulins; proteolytic.

Papanicolaou test *See* Pap smear.

papaya (*Carica papaya*) Melon-like edible fruit, latex-producing small tree with four genera and all $2n = 2x = 18$; it is the source of the proteolytic enzyme papain. *See* papain.

PapD Gram-negative bacterial chaperone (28.5 kDa) delivers the components of the pilus from the periplasm. *See* chaperones; gram negative/gram positive; periplasma; pilus.

paper chromatography Technique for the separation of (organic) molecules in filter paper by applying the mixture in a spot or band at the bottom of the paper and allowing an appropriate solvent to be sucked up, thus carrying the components at a different speed (to a different height) so that they can be separated. The components become visible by their natural color or by the application of specific reagents. A large variety of different modifications were worked out in one or two dimensions, in ascending and descending ways. Nowadays paper chromatography is not used very much. *See* affinity chromatography; column chromatography; high-performance liquid chromatography; ion exchange chromatography; Rf value; thin-layer chromatography.

papillary renal cancer, hereditary Based on the MET oncogene at human chromosome 7q31, encoding a hepatocyte growth factor receptor. *See* HGF; MET; receptor tyrosine kinase.

papillary thyroid carcinoma Caused by the RET oncogene. It accounts for about 80% of the thyroid cancers that have a prevalence in the 10^{-5} range. Its incidence is higher in females than males. *See* RET.

papilloma Premalignant neoplasia displaying epithelial and dermal finger-like projections.

papilloma virus (HPV, human papilloma virus) Double-stranded DNA ($\approx 5.3 \times 10^6$ Da or $\sim$8-kb) virus causing animal and human warts and squamous carcinomas in mice. HPV-16 and -18 are frequently present in cervical cancer. The E6 viral protein of HPV-16—through a ubiquitin path—is prone to degrade the p53 tumor suppressor if at amino acid position 72 there is an arginine rather than a proline. The E7 protein of strain HPV-18 degrades another tumor suppressor protein RB. HPV has been used as a genetic vector. *See* cervical cancer; p53; papova viruses; retinoblastoma; tumor suppressor. (Wolf, J. K. & Ramirez, P. T. 2001. *Cancer Invest.* 19:621.)

Papillon-Lefèvre syndrome *See* periodontitis.

papova viruses Large class of (oncogenic) animal viruses of double-stranded, circular DNA includes the polyoma viruses, the bovine papilloma virus, Simian virus 40 (SV40), etc., used as genetic vectors for transformation of animal cells. Also, they have been studied extensively by molecular techniques to gain information on structure and function. *See* papilloma virus; polyoma; Simian virus 40. (Soeda, E. & Maruyama, T. 1982. *Adv. Biophys.* 15:1.)

PAPS 3'-phosphoadenosine-5'-phosphosulfate is a sulfate donor in several biochemical reactions involving cerebrosides, glycosaminoglycans, and steroids. It is generated by the pathway ATP + sulfate → adenosine-3'-phosphosulfate (APS) + pyrophosphate, APS + ATP → PAPS + ADP. Mutations at 10q23-q24 (spondyloepimetaphyseal dysplasia, SEMD) locus, encoding the PAPSS2, cause short bowed limbs, large knee joints, brachydactyly, and curved spinal column (kyphoscoliosis). *See* bone diseases.

Pap (Papanicolaou) smear Cytological test for premalignant or malignant conditions (used primarily on smears obtained from the female urogenital tract). It also detects papilloma virus infections. *See* malignant growth; papilloma virus.

PAR (1) Pseudoautosomal region, which is where recombination may take place between the X and Y chromosomes. The ends of both the short arm (PAR1) and the long arm (PAR2) have pseudoautosomal regions. *See* lyonization; pseudoautosomal; X chromosome; Y chromosome. (Dupuis, J. & Van Eerdewegh, P. 2000. *Am. J. Hum. Genet.* 67:462.) (2) Protease-activated receptors are seven-transmembrane G-protein-coupled receptors that mediate thrombin-triggered phosphoinositide hydrolysis. PARs may also activate proteases and protect the airways. PAR is a cofactor of PAR4. Par2 is a trypsin receptor. *See* phosphoinositides; thrombin. (Kamath, L., et al. 2001. *Cancer Res.* 61:5933.)

parabiosis Two animals joined together naturally, such as Siamese twins, or by surgical methods; can be used to study the interaction of hormones, transduction signals, etc., between two different individuals. Intrauterine parabiosis develops immune tolerance.

paracellular space Intercellular space in the tissues.

paracentric inversion *See* inversion, paracentric.

Paracentrotus lividus Sea urchin, extensively studied by embryologists. *See* sea urchins.

paracrine effect Ligand (e.g., hormone) is released by a gland and affects neighboring cells.

paracrine stimulation One type of cell affects the function (such as proliferation) of another (nearby) cell. *See* autocrine. (Janowska-Wieczorek, A., et al. 2001. *Stem Cells* 19:99.)

paradigm Model or example to be followed.

paraganglioma *See* mitochondrial diseases in humans.

paraganglion Cells originating from the nerve ectoderm flanking the adrenal medulla; they are darkly stained by chromium salts. These cells may form a type of phaeochromocytoma tumor that secretes excessive amounts of epinephrine and norepinephrine. *See* SHC oncogene.

paragenetic Phenotypic alterations not involving hereditary mutation.

parahemophilia Determined by homozygosity of semidominant autosomal genes. The symptoms involve bleeding similar to the conditions observed in hemophiliacs; bleeding from the uterus (menorrhagia) several days following childbirth. The physiological basis is a deficiency of proaccelerin, a protein factor (V) involved in the stimulation of prothrombin synthesis. Therapy requires blood or plasma. *See* antihemophilia factors; hemophilia; hemostasis; prothrombin deficiency.

parahox genes Hox-like genes in clusters, separate from the hox genes, and originated by duplication from an ancestral protohox gene. *See* homeotic genes.

parainfluenza viruses Group of immunologically related but distinguishable pathogens responsible for some respiratory diseases. *See* Sendai virus.

PARALLEL CASCADE IDENTIFICATION It is nonlinear systems modelling approach. In biology, it can be used to predict long-term treatment response for cancer on the basis of small differences of gene expression levels. (*See* Korenberg, M. J. 2002. *J. Proteome Res.* 1:55.)

parallel substitution Various organismal lineages may display similar or different nucleotides at a number of sites. The chance of these substitutions at a site (p) in (n) lineages can be predicted on the basis of the binomial distribution, $(p + [1 - p])^n$, and upon expansion, e.g., for $n = 5$ it becomes $p^5 + 5p^4(1 - p) + 10p^3(1 - p)^2 + 10p^2(1 - p)^3 + 5p(1 - p)^4 + (1 - p)^5$, and the same change per any two lines is p^2. *See* evolution and base substitutions; evolutionary substitution rate; evolutionary tree; parallel variation.

parallel variation Within a taxonomically close group or even distantly related groups of organisms, similar mutations may occur during evolution. *See* parallel substitution. (Vavilov, N. I. 1922. *J. Genet.* 12:47; Pagel, M. 2000. *Brief Bioinform.* 1[2]:117.)

paraloci Have the same properties as pseudoalleles. *See* pseudoalleles.

paralogous loci Originated by duplication that was followed by divergence. *See* duplication; evolution of proteins; gene family; isolocus; nonorthologous gene displacement; orthologous loci; tetralogue. (Yamamoto, E. & Knap, H. T. 2001. *Mol. Biol. Evol.* 18:1522.)

paralogy Evolution by duplication of a locus. *See* homologue; orthologous loci; orthology.

Paramecium Unicellular protozoan. Normally reproduces by binary fission, i.e., a single individual splits into two. Each cell has two diploid micronuclei and a polyploid macronucleus. At fission the micronuclei divide by mitosis while the macronucleus is simply halved. These animals also have sexual processes (*conjugation*). Two of the slipper-shaped cells of opposite mating type attach to each other and proceed with meiosis of the micronuclei. Only one of the four products of meiosis survives in each of the conjugants. Each

of these haploid cells divides into four cells (gametes). One of these gametes (male) is passed on into the other conjugating partner through a *conjugation bridge* and fuses with a haploid gamete (female). This is a reciprocal fertilization, resulting in diploid nuclei in the conjugants. Subsequently, the pair separates into two *exconjugants*. The macronucleus disintegrates in both. The diploid zygotic nuclei undergo two mitoses and form four diploid nuclei each. Two of the four nuclei function as separate micronuclei of the cells, whereas the other two fuse into a macronucleus that becomes polyploid and is responsible for all metabolic functions and for the phenotype. Besides this sexual reproduction (conjugation), *Paramecia* may practice self-fertilization (*autogamy*). Meiosis takes place and the one surviving product divides twice by mitoses. Two of these identical cells then fuse and form two diploid, isogenic micronuclei. If the conjugation lasts longer, cytoplasmic particles may be transferred through the conjugation bridge. Chromosome numbers may be 63–123. *See* Ascaris; chromosome diminution; cortical inheritance; internally eliminated sequences; killer strains; macronucleus; symbionts hereditary. (Sonneborn, T. M. 1974. *Handbook of Genetics*, Vol. 2, p. 469, King, R. C., ed. Plenum, New York; Prescott, D. M. 2000. *Nature Rev. Genet.* 1:191.)

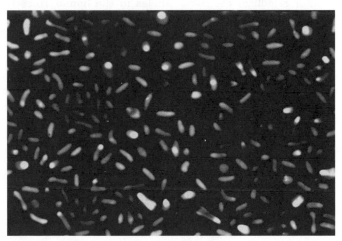

Paramecium aurelia (500 x) cells with bright and nonbright kappa particle symbionts. The symbionts are bacteria. The bright particles contain the so-called R (Refractive) bodies, which are bacteriophages. The nonbright kappa can give rise to bright, indicating lysogeny. The kappa-free (*kk*) paramecia are sensitive to the toxin produced by the bright particles and may be killed. The *K*, killer stocks are immune to the toxin. (From Preer, J. R., et al. 1974. *Bacteriol. Rev.* 38:113; photo by C. Kung; courtesy of Dr. J. R. Preer.)

parameter Quantity that specifies a hypothetical population in some respect or a variable to which a constant value is attributed for a specific purpose or process. Statistics usually denote parameters by Greek letters and Latin letters indicate the computed values.

parameter alpha *See* alpha parameter.

parametric methods, statistical Involve explicit assumptions about population distribution and parameters such as the mean, the standard deviation of the normal distribution, the p

parameter of the *Bernoulli process*, etc. *See* Bernoulli process; nonparametric statistics; normal distribution; robustness.

paramutation *Paramutable* allele becomes *paramutant* in response to a *paramutagenic* allele if the two are in heterozygous condition. The alteration is similar but not identical to the paramutagenic allele. Both *paramutability* and paramutagenic functions are allele-specific. In contrast to gene conversion, paramutation may take place at low frequency in the absence of a paramutagenic allele. At the *R* locus of maize, partial reversion of the paramutant may happen, but this has not been observed at the *B* locus of maize. The paramutant phenotype at the *R* locus may vary, but at the *B* locus the phenotype appears to be uniform. The exact mechanism of this heritable alteration is not fully understood. It appears that the level of transcription is reduced at the paramutant allele compared to that in the paramutable one. Apparently, hypermethylation is involved at the *R* locus of maize; at the *B* locus, involvement of methylation has not been detected. At the *pl* locus, the paramutation seems to result in a genetic alteration of the chromatin structure, which affects the regulation of the expression of the gene during development. Although paramutation has been considered to be an endogenous mechanism, it appears that in the promoter region of the two *r* alleles in the homozygotes the *doppia* (CACTA) transposable elements are present within the 387 bp σ-region that is intercalated between the two S elements in opposite orientation. (These elements are called S because they are responsible for anthocyanin coloration of the seed by this complex locus. The elements responsible for coloration of the plant were named P.) Paramutation is not a general property of all genes, although similar phenomena have been observed at a few other genes in maize and other plants. This phenomenon seems to violate the Mendelian principle that alleles segregate during meiosis independently

rrR ^{r}
dark mottled
cross *rr* × *R′ R′*

rrR St
stippled coming from
testcross *rr* × *R′ R^{st}*

rrR ^{r}
Light paramutant
testcross *rr* × *R′ R^{st}*

rrR St
from the cross
rr × *R^{st} R^{st}*

Paramutation results in reduced pigmentation in the triploid aleurone of maize. The R^R homozygotes are fully colored (when all other color-determining alleles are present). The *r* homozygotes are colorless. The *rrR^r* genotype is responsible for the dark mottled aleurone. R^{st} causes paramutation (stippling) of the R^r paramutable allele that may be manifested in different grades. In the crosses the pistillate parents are shown first, left. (Courtesy of Brink, R. A. See also 1956. *Genetics* 41:872.)

and during the process no "contamination" takes place. *See* blending inheritance; copy choice; cosuppression; directed mutation; epigenesis; gene conversion; graft hybrid; localized mutagenesis; pangenesis; position effect; presence-absence hypothesis; RIP; tissue specificity; transvection. (Brink, R. A. 1960. *Quart. Rev. Biol.* 35:120; Chandler, V. L. 2000. *Plant Mol. Biol.* 43[2-3]:121; Lisch, D., et al. 2002. *Proc. Natl. Acad. Sci. USA* 99:6130; Stam, M., et al. 2002. *Genetics* 162:917.)

paramyotonia Periodic paralysis (gene located in human chromosome 17). *See* myotonia.

paramyxovirus Single-stranded RNA viruses with a genome of 16–20 kb. Members of this group cause human mumps and respiratory diseases in humans and other animals, including birds and reptiles. *See* RNA viruses.

paranemic coils The two components of the coil can be separated from each other without any entanglement as one can easily pull apart two spirals that were pushed together after they were wound separately, i.e., they are not interlocked. *See* plectonemic coils.

paraneoplastic neurodegenerative syndrome *See* autoimmune diseases.

paranoia Psychological disorder in a more (paranoia) or less severe (paranoid) state. The major characteristics are delusions of persecution (delusional jealousy, erotic delusions) or less frequently feelings of grandiosity. It differs from schizophrenia that generally the rest of the personality and mental capacity may remain normal. Frequently, however, paranoid schizophrenia may occur. The precipitating factors are insecurity, frustration, physical illness, drug effects, etc. There is also an apparently undefined genetic component. *See* affective disorders; schizophrenia.

paranome Genes within gene families. *See* gene family.

paraoxonase (PON1, 7q21.3) May be associated with high-density lipoproteins in the blood plasma. It may protect against coronary heart disease by destroying oxidized lipids responsible for inflammation. It may detoxify organophosphate pesticides (parathion, chloropyrifos [Dursban]). *See* arylesterase; cholinesterase; HDL; pseudocholinesterase. (Brophy, V. H., et al. 2001. *Am. J. Hum. Genet.* 68:1428.)

parapatric speciation Groups of organisms inhabiting an overlapping region become sexually isolated. *See* allopatric; sympatric.

paraphyletic group Does not include all descendants of the latest common ancestor.

paraplegia Paralysis of the lower part of the body; it may be hereditary. *See* Pelizaeus-Merzbacher disease; Silver syndrome; spastic paraplegia.

paraplegin *See* mitochondrial disease in humans.

paraprotein Abnormally secreted normal or abnormal protein, e.g., the Bence-Jones protein in myelogenous myeloma. It is also called the M component. *See* Bence-Jones protein.

paraquat Artificial electron acceptor of photosystem I; it is a lung toxicant. It may produce oxidative stress by indirect production (through cellular diaphorases) of superoxide radicals. *See* diaphorase; diquat; photosystem I; ROS; superoxide.

pararetrovirus Its genetic material is double-stranded DNA, but it is replicated with the aid of an RNA molecule, e.g., in hepadnaviruses and caulimoviruses. They may occur in many copies in higher eukaryote genomes. *See* animal viruses; cauliflower mosaic virus; hepatitis B virus; plant viruses; retroviruses. (Richert-Poggeler, K. R. & Shepherd, R. J. 1997. *Virology* 236:137; Gozuacik, D., et al. 2001. *Oncogene* 20:6233.)

Parascaris Group of nematodes. *See* Ascaris.

parasegment Unit of a metameric complex consisting of the posterior part of one segment and the anterior part of another in insect larval and subsequent stages. *See* morphogenesis.

paraselectivity Apparent (but not real) selectivity in pollination among plants.

parasexual mechanism of reproduction Somatic cell fusion and mitotic genetic recombination. The processes bear similarities to those common at sexual reproduction, but they do not involve sexual mechanisms. *See* cell fusion; mitotic recombination; somatic cell genetics. (Pontecorvo, G. 1956. *Annu. Rev. Microbiol.* 10:393.)

parasitemia The blood contains parasites, e.g., *Plasmodium*. *See* Plasmodium; thalassemia.

parasitic Lives on and takes advantage of another live organism. *See* biotrophic; parasitoid. (<http://www.ebi.ac.uk/parasites/parasite-genome.html>.)

parasitic DNA Selfish DNA.

parasitoid Lives on another organism and eventually destroys it like some wasps and viruses. Upon attack and wounding by some insects some plants synthesize and emit host- and parasite-specific volatile compounds that attract parasitoid wasps, which in turn may destroy the insects. *See* parasitic.

paraspeckles Formed from RNA-binding proteins in the cell nucleus within the interchromatin nucleoplasmic space usually at the periphery of the nucleolus and in the vicinity of the nuclear speckles. *See* speckles. (Fox, A. H., et al. 2002. *Curr. Biol.* 12:13.)

parasterility Caused by incompatibility between genotypes that may be fertile in other combinations. *See* Rh blood group; self-incompatibility alleles.

parastichies Imaginary helical line in phyllotaxis. *See* phyllotaxis (phyllotaxy).

parathyroid hormone (11p15.3-p15.1) Produced by the parathyroid gland next to the thyroid gland. It is a regulator of calcium and phosphate metabolism (mediated by cAMP) primarily in the bones, kidneys, and digestive tract. A recessive hypoparathyroidism was mapped to Xp27. *See* enchondromatosis; hypercalcemia-hypocalciuria; hyperparathyroidism.

paratope Epitope-binding site of the antibody Fab domain. *See* antibody; epitope.

paratransgenic Insect that has transgenic symbionts inhabiting its gut. *Rhodnius prolixus* carries the actinomycete bacterium *Rhodococcus rhodnii* with which it has a symbiotic relationship. *R. prolixus* is a blood-sucking arthropod, vector of *Trypanosoma cruzi*, which is responsible for Chagas disease. When *R. rhodni* is transformed by cecropin A, a 38-amino-acid antimicrobial peptide derived from the moth *Hyalophora cecropia*, the peptide diffuses into the insect and lyses *Trypanoma cruzi* within the insect without serious damage to *R. rhodni*, thus effectively curtailing the propagation of the protozoan. This is a more attractive defense than using chemical pesticides. *See* Chagas disease; CRUZIGARD; transgenic; *Trypanosoma*. (Beard, C. B., et al. 2001. *Int. J. Parasitol.* 31:621.)

parcelation Relative lack of pleiotropic effects between two sets of nonoverlapping traits.

parenchyma (1) In plant biology, it means storage cells near isodiametric *spongy parenchyma*, closer to the lower surface of the leaves, or the *palisade parenchyma*, consisting of one or two layers of columnar cells with their long axis perpendicular to the upper epidermis. Both types of tissues contain conspicuous intercellular space. (2) In zoology, parenchyma cells are the functional units rather than the network of an organ or tissue. *See* palisade cells.

parens patriae State or community right to intervene against individual rights or beliefs and protect the interest of a person against potentially serious or actually life-threatening conditions, e.g., compulsory immunization, genetic screening, prohibition of incest, etc.

parental ditype *See* tetrad analysis.

parent-of-origin effect May be due to the differences in the cytoplasm, differential transmission of defective chromosomes through the two sexes, differences in trinucleotide-repeat expansions, endosperm:embryo chromosomal differences in the reciprocal crosses in case of polyploids and imprinting. *See* imprinting; trinucleotide repeats. (Morrison, I. M. & Reeve, A. E. 1998. *Hum. Mol. Genet.* 7:1599; Haghighi, F. & Hodge, S. E. 2002. *Am. J. Hum. Genet.* 70:142.)

parietal Situated on the wall or attached to the wall of a hollow organ.

Paris classification Of human chromosomes, standardized (in 1971) the banding patterns and classified them by size groups; it is basically very similar to what it is used today. Current maps show, however, only one of the two chromatids. *See* Chicago classification; Denver classification; human chromosomes.

parity (1) In gene conversion, the process can go equally frequently in the direction of one or the other allele. (2) In human biology, the condition in which a woman has borne offspring. *See* gene conversion. (Fogel, S., et al. 1971. *Stadler Symp.* 1–2:89.)

parity check In a digital system, reveals whether the number of ones and zeros is odd or even.

parkin *See* Parkinson's disease.

parking Set of rules for seeding in which no seed overlaps with any other seedlings. It is an iterative procedure that may be used at the early phase in genome sequencing. Each iteration sequences a new portion of a nonoverlapping piece of DNA. Aside from nonoverlaps, the sequences are chosen at random. *See* genome projects; seeding. (Roach, J. C., et al. 2000. *Genome Res.* 10:1020.)

Parkinson disease (PD, PARK) Shaking palsy generally with late onset; however, juvenile forms also exist. PD may include mental depression, dementia, reduced olfactory abilities, and deficiency of several different substances, notably dopamine, from the nervous system. The genetic determination of the heterogeneous symptoms is unclear; autosomal-dominant, -recessive, X-linked, polygenic, and apparently only environmentally caused phenotypes have been observed. An early-onset PD was located to human chromosome Xq28 and another (PARK6) to 1p35-p36. An autosomal-dominant form is in chromosome 22. Another locus encoding spheres of protofibrils of α-synuclein was assigned to human chromosome 4q21-q23. The α-synuclein gene is apparently responsible for only a minor fraction of parkinsonism. A gene has been located to human chromosome 17q21. A low-penetrance (40%), late-onset ($\sim$60 years) gene is at 2p13. An autosomal-dominant juvenile parkinsonism gene (1,395-bp ORF) encoding the 465-amino-acid *parkin* protein was located to 6q25.2-q27. Parkin is a ubiquitin protein ligase. Its prevalence is about 0.001 and it may be 0.01 over age 50. In some cases, a missense mutation in the carboxy-terminal hydrolase L1 (UUCH-L1), a component of the ubiquitin complex localized in the Lewy bodies, is responsible for PD. Not all forms of PD show Lewy bodies. Loss of dopaminergic neurons in the substantia nigra is usually associated with the disease. Glial cell line–derived neurotrophic factor (GDNF) has nutritive effects on the dopaminergic nigral neurons. In autosomal-recessive juvenile PD, a parkin-associated endothelin receptor-like (Pael-R) accumulates apparently because a misfolded Pael-R is not degraded if there is a defect in parkin. Parkin has several other substrates, which explains the complicated etiology of the different forms of PD (Imai, Y., et al. 2001. *Cell* 105:891; Shimura, H., et al. 2001. *Science* 293:263).

Gene therapy using lentivirus vector carrying the GDF gene injected into the brain (striatum and substantia nigra) of old or young monkeys pretreated with a nigrostriatal degeneration-inducing agent (1-methyl-4-phenyl-1,2,3,6-tetrahydropyridine [MPTP]) reversed the functional deficits and prevented degeneration, respectively. Anticholinergics (blocking choline), dopamine, and glial cell–derived neurotrophic factor (GDNF) treatments may be somewhat beneficial. An alternative approach may be using gene therapy by expressing transfected tyrosine hydroxylase or aromatic amino acid decarboxylase in the striated muscle cells. These enzymes can produce dopamine, yet the response is below expectations. Transplantation of fibroblasts equipped for secretion of BDNF and GDNF into the brain appears promising in animal models. Electric shocks localized to the globus pallidus (a medial part of the brain) had beneficial effects in some cases. Caspase-3 may be a conditioning factor in the apoptotic death of dopaminergic neurons in PD. Embryonic stem cells may develop into dopamine-producing neurons in the brain of the mouse and seem to be promising for cell-replacement therapy of Parkinson disease (Kim, J.-H., et al. 2002. *Nature* 418:50). *See* BDNF; caspase; dopamine; GDNF; Lewy body; mitochondrial disease in humans; neuromuscular diseases; parkinsonism; stem cells; subtantia nigra; synuclein; tau; tyrosine hydroxyls; ubiquitin. (Dawson, T. M. 2000. *Cell* 101:115; Kordower, J. H., et al. 2000. *Science* 290:767; Valente, E. M., et al. 2001. *Am. J. Hum. Genet.* 68:895; Vaughan, J. R., et al. 2001. *Ann. Hum. Genet.* 65:111; Lansbury, P. T., Jr. & Brice, A. 2002. *Current Op. Genet. Dev.* 12:299.)

Parkinsonism Secondary symptom caused either by drugs, inflammation of the brain (encephalitis), Alzheimer's disease, Wilson's disease, or Huntington's chorea, etc. *See* Parkinson's disease; tau.

paromomycin ($C_{23}H_{45}O_{14}N_5$) Aminoglycoside antibiotic. It may cause translational errors by increasing the initial binding affinity of tRNA. Oral LD_{50} in mice is 1,625 mg/kg. *See* aminoglycoside antibiotics; phenotypic reversion.

parotid gland Salivary gland; the proline-rich parotid glycoprotein is encoded in human chromosome 12p13.2.

proxysm Recurring events such as convulsions (but most commonly normal conditions in between); sudden outbreak of disease.

paroxysmal nocturnal hemoglobinuria Dominant human chromosome 11p14-p13 susceptibility of the erythrocytes to destruction by the complement because of a deficiency in protectin (HRF20/CD59) and DAF. *See* angioneuritic edema; CD59; complement; DAF; hemoglobin; membrane attack complex; protectin.

PARP (poly[ADP-ribose] polymerase) Enzyme involved in surveillance and base excision repair of DNA. It is cleaved by an ICE-like proteinase. Its deficiency increases the sensitivity to radiation damage, recombination, and sister chromatid exchange. *See* apoptosis; ICE; tankyrase; telomeres. (Bauer, P. I., et al. 2001. *FEBS Lett.* 506[3]:239; Lavrik, O. I., et al. 2001. *J. Biol. Chem.* 276:25541.)

PARS (poly[ADP-ribose] synthetase) Attaches ADP-ribose units to histones and to other nuclear proteins. It is activated when DNA is damaged by nitric oxide.

parsimony *See* evolutionary tree; maximal parsimony.

parsing Resolve it to parts or components. *See* exon parsing, pars means part in Latin.

parsley (*Petroselinum crispum*) Roots and leaves are used for flavoring; $2n = 2x = 22$.

parsnip (*Pastinaca sativa*) Root vegetable; $2n = 2x = 22$.

parthenocarpy Fruit development without fertilization. It may have horticultural application by producing seedless apple varieties such as Spencer Seedless or Wellington Bloomless. The gene responsible in these apples is homologous to *pistillata* of *Arabidopsis*. *See* apomixia; flower differentiation in *Arabidopsis*; parthenogenesis; seedless fruits. (Yao, J., et al. 2001. *Proc. Natl. Acad. Sci. USA* 98:1306.)

parthenogenesis Embryo production from an egg without fertilization. Parthenogenesis may be induced in sea urchins by hypotonic media or in some amphibia by mechanical or electric stimulation of the egg. In some fishes, lizards, and birds (turkey), it occurs spontaneously. Parthenogenesis in animals is most common among polyploid species. Parthenogenetic individuals produce only female offspring. On theoretical grounds, parthenogenesis may be disadvantageous because it deprives the species of elimination of disadvantageous mutations on account of the lack of recombination available in bisexual reproduction. Parthenogenesis may cause embryonic lethality in mice if the imprinted paternal genes are not expressed. Parthenogenesis is not known to occur in humans; however, it may exist as a chimera when after fertilization the male pronucleus is displaced to one of the blastomeres and then the maternal chromosome set in the other blastomere is diploidized. Many plant species successfully survive by asexual reproduction as an evolutionary mechanism. Parthenogenesis in plants is called apomixis or apomixia. The process of asexually reproducing plant populations that appear to be preponderant under conditions marginally suitable for the species is called *geographic parthenogensis*. *See* apomixia; gynogenesis; parthenocarpy; RSK. (Mittwoch, U. 1978. *J. Med. Genet.* 15:165; Cibelli, J. B., et al. 2002. *Science* 295:819.)

parthenote Individual developed from an egg that is stimulated to divide and develop in the absence of fertilization by sperm. *See* apomixia; parthenogenesis.

partial digest The reaction is stopped before completion of nuclease action and thus the DNA is cut into various size fragments, some of which may be relatively long because some of the recognition sites were not cleaved. *See* restriction enzyme.

partial diploid *See* merozygote.

partial dominance Incomplete dominance; semidominance.

partial linkage The genes are less than 50 map units apart in the chromosome and can recombine at a frequency proportional to their distance. *See* crossing over; recombination.

partial trisomy Only part of a chromosome is present in triplicate. *See* trisomy.

particulate inheritance Modern genetic theory that inheritance is based on discrete particulate material (written in nucleic acid sequences) transmitted conservatively rather than according to the pre-Mendelian theory of pangenesis. It claimed that the hereditary material is a miscible liquid subject to continuous changes under environmental effects. According to the particulate theory of genetics, genes are discrete physical entities that are transmitted from parents to offspring without blending or any environmental influence, except when mutation, gene conversion, or imprinting occurs. *See* blending inheritance; gene conversion; imprinting; mutation; pangenesis.

particulate radiation *See* physical mutagens.

partitioning (1) Segregation. Distribution of plasmids and/or the bacterial chromosome(s) into dividing bacterial cells. It may be a passive process mediated by the attachment of the DNAs to the cell membrane. It may be carried out by a bacterial tubulin-like protein FtsZ. The process is controlled by plasmid and bacterial genes. The bacterial chromosome or some of the plasmids seem to have a centromere-like protein that may bind to the cell poles and to 10 copies of a sequence situated along a 200 kb region near the replicational origin. The loss of the chromosome in the new cells is less frequent than 0.003. The segregation of plasmids present in multiple copies is more complex, Mechanisms ("addiction" system) exist to resolve plasmid dimers and to ensure that each cell will have at least one copy of a plasmid. The two-component addiction module includes a stable toxin and a labile antitoxin. The plasmid-free cells may be eliminated (Engelberg-Kulka, H. & Glaser, G. 1999. *Annu. Rev. Microbiol.* 53:43). *See* addiction module; cell division; plasmid maintenance; segregation. (Hiraga, S. 2000. *Annu. Rev. Genet.* 34:21; Gordon, G. S. & Wright, A. 2000. *Annu. Rev. Microbiol.* 54:681; Draper, G. L. & Gober, J. W. 2002. *Annu. Rev. Microbiol.* 56:567.) (2) In statistics, breaking down the variances among the identifiable experimental components or chi-square in a compound to reduce the quantity of the residual or error variance. *See* analysis of variance.

parturition Labor of child delivery.

parvoviruses Nonenveloped, icosahedral (18–25 nm), single-stranded DNA ($\sim$5.5 kb) viruses. The group includes the densoviruses of arthropods, the autonomous lytic parvoviruses, and the adeno-associated viruses. *See* adeno-associated virus; autonomous parvovirus; icosahedral; oncolytic viruses. (Lukashov, V. V. & Goudsmit, J. 2001. *J. Virol.* 75:2729.)

parvulin Very small monomeric 92-amino-acid prolyl isomerase of *E. coli* involved in protein maturation. Similar proteins occur in yeast (Ess1, 19.2 kDa), humans (Pin1, 18 kDa), and *Drosophila* (dodo, 18.3 kDa). Ess1 may not have, however, isomerase activity, but Pin1 does. In the absence of Ess1, the nuclei fragment and growth ceases; dodo apparently has a similar function to Ess1 and is interchangeable. Pin regulates mitotic progression by interacting with CDC25 and NIMA. *See* CDC25; NIMA; peptidyl-prolyl isomerases; PPI. (Rulten, S., et al. 1999. *Biochem. Biophys. Res. Commun.* 259:557.)

PAS domain   Shared motif of proteins involved in the regulation of the circadian rhythm of the majority of eukaryotes. PAS domain serine/threonine kinases regulate several different signaling pathways. *See* circadian rhythm. (Rutter, J., et al. 2001. *Proc. Natl. Acad. Sci. USA* 98:8991.)

PASA   Special PCR procedure by which chosen alleles can be amplified if the primers match the end of the allele. *See* PCR. (Smith, E. J. & Cheng, H. H. 1998. *Microb. Comp. Genomics* 3:13; Shitaye, H., et al. 1999. *Hum. Immunol.* 60:1289.)

Pascale triangle   Represents the coefficients of individual terms of expanded binomials: $(p+q)^n$:

$$1p^n + \frac{n}{1!(n-1)!}p^{n-1}q + \frac{n!}{2!(n-2)!}p^{n-2}q^2$$

$$+ \cdots + \frac{n!}{(n-1)!1!}p^{n-(n-1)}q^{n-1} + 1q^n$$

n→	1	2	3	4	5	6	7	8	9	10
	1	1	1	1	1	1	1	1	1	1
	1	2	3	4	5	6	7	8	9	10
		1	3	6	10	15	21	28	36	45
			1	4	10	20	35	56	84	126
				1	5	15	35	70	126	210
					1	6	21	56	126	252
						1	7	28	84	210
							1	8	36	120
								1	9	45
									1	10
										1
SUMS	2	4	8	16	32	64	128	256	512	1024

n→	11	12	13	14	15	16	17	18	19	20
	1	1	1	1	1	1	1	1	1	1
	11	12	13	14	15	16	17	18	19	20
	55	66	78	91	105	120	136	153	171	190
	165	220	286	364	455	560	680	816	969	1140
	330	495	715	1001	1365	1820	2380	3060	3876	4845
	462	792	1287	2002	3003	4368	6188	8568	11628	15504
	462	924	1716	3003	5005	8008	12376	18564	27132	38760
	330	792	1716	3432	6435	11440	19448	31824	50388	77520
	165	495	1287	3003	6435	12870	24310	43758	75582	125970
	55	220	715	2002	5005	11440	24310	48620	92378	167960
	11	66	286	1001	3003	8008	19448	43758	92378	184756
	1	12	78	364	1365	4368	12376	31824	75582	167960
		1	13	91	455	1820	6188	18564	50388	125970
			1	14	105	560	2380	8568	27132	77520
				1	15	120	680	3060	11628	38760
					1	16	136	816	3876	15504
						1	17	153	969	4845
							1	18	171	1140
								1	19	190
									1	20
										1
SUMS	2048	4096	8192	16384	32768	65536	131072	262144	524288	1048576

The pascal triangle represents the coefficients of individual terms of expanded binomials. The exponent of the binomial is n. The figures display a symmetrical hierarchy. The frequency of a particular class can be readily calculated because the sum of the coefficients is displayed at the bottom of the columns. Mendelian segregation follows the binomial distribution.

Since genetic segregation is expected to comply with binomial distribution, the coefficients indicate the frequencies of the individual phenotypic or (in case of trinomial distribution) genotypic frequencies. *See* binomial distribution; trinomial distribution.

passenger DNA   DNA inserted into a genetic vector.

passenger proteins   Include the inner centromeric protein (INCENP), which is a substrate for Aurora, the Aurora B kinase, the TD-60 autoimmune antigen, the inhibitor-of-apoptosis protein Survivin/BIR-1. These proteins are situated at the centromeres and move the spindle at late metaphase and anaphase. *See* Aurora; centromere; mitosis; Survivin. (Bishop, J. D. & Schumacher, J. M. 2002. *J. Biol. Chem.* 277:27577.)

passive immunity   Acquired by the transfer of antibodies or lymphocytes.

passive transport   Does not require special energy donor for the process. *See* active transport.

Pasteur effect   Fast reduction of respiration (glycolysis) if O_2 is added to fermenting cells.

Pasteurella multocida   Pathogenic bacterium causing cholera in birds, bovine hemorrhagic septicemia, atrophic rhinitis (inflammation of the nasal mucosa) in pigs. Humans may be infected by it through cat or dog bites. The sequenced genome of 2,257,487 bp contains ~2,014 coding sequences. *See Yersinia.* (May, B. J., et al. 2001. *Proc. Natl. Acad. Sci. USA* 98:3460.)

pasteurization   Reducing (killing) the microbe population in a material by heating at a defined temperature for a specified period of time. *See* aseptic; autoclaving; axenic.

PAT1 (Ran1)   Protein kinase required for the continuation of mitotic division in fission yeast. Its inactivation triggers the switch to meiosis. *See* meiosis; Ran1.

patatin   Glycoprotein-like storage protein in potato with lipid acid hydrolase and esterase activity. It inhibits pests larvae. It constitutes ~40% of the soluble proteins in the tubers. *See* potato. (Hirschberg, H. J., et al. 2001. *Eur. J. Biochem.* 268:5037.)

Pätau's syndrome   Caused by trisomy for human chromosome 13. This is one of the few (X, Y, 8, 18, 21, 22) trisomies that can be carried to term, but it generally leads to death within 6 months because of severe defects in growth, heart, kidney, and brain. It is accompanied by face deformities (severe hare lip, cleft palate), polydactyly, clubfoot, defects of the genital systems, etc. Definite identification is carried out by cytological analysis, including FISH, with the available chromosome-13-specific probes. An old designation of trisomy 13 was trisomy D because chromosome 13 belonged to the D group of human

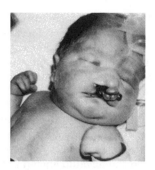

Pätau syndrome. (Courtesy of Dr. Judith Miles.)

chromosomes. *See* aneuploidy; clubfoot; hare lip; polydactyly; trisomy.

Patch *(Ptc)*   *See* hedgehog; sonic hedgehog.

PATCH (patched duplex) Resolution of a recombination intermediate (Holliday junction) without an exchange of the flanking markers (can be gene conversion). *See* Holliday model L.

patch clamp technique An electrode is tightly pressed against the plasma membrane so that the flow of current through a voltage-gated ion channel can be measured. It is also used for the sensitive in situ study of neurotransmitters. *See* ion channels.

patch mating Actin patches of the cytoskeleton can be used in yeast to quantify the number of viable diploid cells in the presence of silencer genes that regulate the expression of mating types. Actin patches are detectable for a duration of about 10 seconds at sites of polarized growth, then rapidly disappear. *See* mating type determination in yeast. (Smith, M. G., et al. 2001. *J. Cell Sci.* 114 [pt 8]:1505.)

patella aplasia-hypoplasia (PTLAH, 17q21-q22) Absence or reduction of size of the kneecap. The symptoms also occur in various syndromes such as Coffin-Siris syndrome and trisomy 8 syndrome. *See* Coffin-Siris syndrome; nail-patella syndrome.

patent The so-called gene patents do not protect the DNA sequence itself; rather the process of manipulation is the object. Gene "ownership" only prevents the use or selling of a particular sequence without permission. In general, according to U.S. patent laws, the patent is protected for 17 years from the date of issue. A requisite for patenting is that the subject of the patent application must be new, nonobvious, and practically useful, e.g., a probe for a gene. Natural DNA sequences are not patentable, but purified or isolated recombinant molecules or parts of a vector are patentable. A further requirement is "enablement," i.e., a trained person, after reading the patent description, can use the "invention" without further research. In October 1998, the first U.S. patent was awarded to an EST. Once a patent is issued, even further originally undisclosed applications are protected. Also, after isolating a full-length open-reading frame by the use of an STS, another person may obtain a patent but not without the

permission of the "inventor" of the patented STS. During the period of time of the patent, the patent-holder can prevent anybody from using it, including those who invented the same procedure independently or even those who improved of the procedure to such an extent that the second invention meets the requirements for patenting. The inventor is obligated by law to disclose the invention in sufficient technical detail that anybody with the proper expertise could use it. The fact that an invention was arrived at under federal financial support does not exclude patentability, but the inventor must report the patentable invention to the sponsoring agency.

The intention of the government is that the invention should be used at maximum benefit to the public, which can usually be achieved most effectively by commercial private enterprises. The patenting of the outcome of genetic research may be harmful to science if the investigators keep the ongoing work secret until it becomes patentable. If the discovery is published through the proper means of scientific communications prior to the patent application, it is disqualified from patenting. It is generally easier to patent a product than a process. Natural products (e.g., proteins) are usually not patentable unless they are modified in some way, are different from the natural product either in structure or function, and these properties were not generally known. A DNA sequence, identified as the coding unit for a genetic disease or a genetic marker in its vicinity, may be patentable, but a cloned gene that may be used for translating a protein may not be patentable. DNA markers are patentable only if their direct use can be determined.

The concept of patenting biological material raises several moral objections, but it is defended by the biotechnology industry because it takes hundreds of millions of dollars for the completion of such projects, and without the financial means, these investigations cannot be maintained. Between 1981 and 1995, a total of 1,175 human DNA sequences were patented. If the subject of the patent has been published or in use more than 1 year prior to the date of the patent application, it will not be approved. If another person can prove that he/she invented the object before the date of the publication by others, the person may still be entitled to a patent. The patent laws vary in different countries, and new legislation may take place at any time. The European Union is now approving patents for human genes and transgenic animals and plants. Laboratory assays, reagents and procedures, including computer programs may also be patentable.

An alternative to patenting is Trade Secret Protection. One means of preventing that any other person would patent an invention is public disclosure, e.g. publication in sufficient details (e.g. in a scientific journal) so another party would not be able to claim priority for the invention that is one of the requisites for patenting. Publications may not necessarily provide an effective lasting protection. Patent infringement usually does not entitle the patentee for more financial compensation than the reasonably calculated loss of royalty or profit caused by the infringement. It must be also verified that the original patent description did not include deceptive assertions (Arnold, B. E. & Ogielska-Zei, E. 2002. *Annu. Rev. Genomics Hum. Genet.* 3:415). *See* Cohen-Boyer patent on recombinant DNA; EST; SNP; STS. (Eisenberg, R. S. 1992. *Gene Mapping*, p. 226, Annas, G. J. & Elis, S., eds. Oxford Univ. Press, New York; DNA-based patents: Robertson, D. 2002. *Nature Biotechnol.* 20:639; <http://geneticmedicine.org>

or <http://208.201.146.119/>; patents in general: <http://www-sul.stanford.edu/depts/swain/patent/pattop.html>; <http://www.uspto.gov>; <www.bustPATENTS.COM/>.)

patent ductus arteriosus (6p12) The ductus arteriosus connects the lung artery and the aorta and shunts away blood from the lung of the fetus. Normally it fades away after birth. In ~1/2000 cases, this does not happen and the duct stays open, causing heart defects. The disease may be caused by fetal rubella infection and apparently by autosomal-dominant genes. The 6p12-p21 dominant *Char syndrome* involves patent ductus, facial anomalies, and an abnormal fifth digit of the hand. The basic problem has been traced to TFAP2B neural crest−related transcription factor that does not bind properly to its target. Risk of recurrence in an affected family is about 1–2%. General incidence is less than 10% of that. *See* risk. (Zhao, F., et al. 2001. *Am. J. Hum. Genet.* 69:695.)

paternal leakage Transmission of mitochondrial DNA through males. Generally, mitochondria are not transmitted through animal sperm. In mice, apparently male transmission of mitochondria is within the range of 10^{-5}. In interspecific mouse crosses, paternal mitochondria are transmitted, but they are eliminated during early embryogenesis or later during development. Heteroplasmy is rare. The role of transmission of mitochondria in humans is not clear. Some cytological observations may indicate the incorporation of the midpiece of the sperm (containing mitochondria) into the egg. Genetic evidence for human paternal transmission of mitochondria is rare. (Schwartz, M. & Vissing, J. 2002. *New England J. Pled.* 347:576.) In some mollusks (mussel), there is a strong biparental inheritance of mtDNA. In *Mytilus*, the paternal and maternal mtDNA displays 10 to 20% nucleotide divergence. The females transmit just one type of mitochondria to sons and daughters, whereas the males transmit a second type of mtDNA genome to sons. Biparental transmission of mtDNA may also occur in interspecific crosses of *Drosophila*. In *Paramecia,* mitochondria may be transmitted through a cytoplasmic bridge. In fungi, the transfer is maternal, although in some heterokaryonts cytoplasmic mixing may take place. In some slime molds, mtDNA transmission is also mating-type dependent. In *Physarum polycephalum*, different *matA* alleles regulate mtDNA transmission, but a plasmid gene may also be involved and recombination can take place between mtDNAs. In *Chlamydomonas* algae, several genes around the mating-type factors were implicated. *See* doubly uniparental inheritance; Eve, foremother; heteroplasmy; mitochondrial disease in humans; mitochondrial genetics; mtDNA; *Paramecium*; plastid male transmission; plastid genetics. (Eyre-Walker, A. 2000. *Philos. Trans. R. Soc. Lond. B Biol. Sci.* 355:1573; Shitara, H., et al. 1998. *Genetics* 148:851; Yang, X. & Griffith, A. J. 1993. *Genetics* 134:1055; Meusel, M. S. & Moritz, R. F. 1993. *Curr. Genet.* 24:539.)

paternal transmission Paternal transmission of certain genes may be caused by imprinting. Some of the human insulin and the insulin-like growth factor alleles may be preferentially inherited through the paternal chromosome and cause early-onset obesity. *See* imprinting; obesity; paternal leakage. (Le Stunff, C., et al. 2001. *Nature Genet.* 29:96.)

paternity exclosure Based on genetic paternity tests. *See* DNA fingerprinting; paternity testing; Y chromosome.

paternity testing Frequently required in civil litigation suits, but it may have significance for medical, population, immigration, archaeological, and other cases. The laboratory procedures are generally the same as used for DNA fingerprinting. Here, as in DNA fingerprinting in general, the exclusion of paternity is simple and straightforward, but the determination of identity may pose more difficulties because in the multilocus tests more than 10% of the offspring may show one band difference and 1% may show two, due to mutation. Therefore, Penas and Chakraborty (*Trends Genet.* 10:204 [1994]) recommended the formula shown in the box below:

$$PI = \frac{\binom{N}{U} \mu^{U}(1-\mu)^{N-U}}{\binom{n}{U} X^{n-U}(1-X)^{U}}$$

PI = paternity index, μ = mutation rate, X = band-sharing parameter, N = total number of bands per individual, n = number of test bands, U = number of bands not present in the alleged father. In rare instances (mistakes at maternity wards), similar tests may be necessary to prove maternity. The biological father of a child—even if the paternity can be accurately proven—cannot assert paternal rights against the will of the mother if she was/is married to another man (Hill J. L. 1991. *New York University Law Review* 66:353). When "the child is born to a mother who is single or part of a lesbian couple, law does permit the biological father to assert his paternal rights, even if he clearly stated his intention prior to conception to have no relation" (Charo, R. A. 1994. *The Genetic Frontier. Ethics, Law and Policy*, Frankel, M. S. & Teich, A., eds. Am. Assoc. Adv. Sci., Washington, DC). *See* DNA fingerprinting; forensic genetics; forensic index; microsatellite typing; surrogate mother; utility index; Y chromosome.

path coefficient Method of Sewall Wright has been worked out for studying the paths of genes in populations mathematically and by diagrams and genetic events determining multiple correlations. Here it is not possible to discuss meaningfully the mathematical foundations, but one type of graphic application for determining some relations between offspring and parents can be found under F and inbreeding coefficient. *See* correlation; inbreeding coefficient. (Wright, S. 1923. *Genetics* 8:239; Wright, S. 1934. *Ann. Math. Stat.* 5:161.)

pathogen Organism (microorganism) capable of causing disease to another.

pathogen-derived resistance Plants may be protected against certain pathogens by the transgenic expression of viral coat proteins, other proteins, antisense sequences, satellite and defective viral sequences. *See* host-pathogen relations; plantibody.

pathogenesis-related proteins (PR) Variety of acidic or basic proteins synthesized in plants upon infection with pathogens. The chitinases and glucanases apparently act by damaging the cell wall of fungi, insects, or even bacteria. *See* host-pathogen relations; SAR.

pathogenic Capable of causing disease. (*See* Hill, A. V. S. 2001. *Annu. Rev. Genome. Human. Genet.* 2:373.)

pathogenicity island (PAI) Group of genes in a pathogen involved in the determination and regulation of pathogenicity. In *Helicobacter pylori*, these islands are delineated by 31 bp direct repeats (DR) and acquired by horizontal transfer. Commonly the same genes are present between the ends of this large insert and the chromosomal genes in both pathogenic and nonpathogenic species of the same group of bacteria. Frequently, the insert is adjacent to a tRNA 3′ sequence or a codon for an unusual amino acid. The DNA inserts encode a rather specific secretory system (type III) and elements of transport and bacterial surface effectors located next to host cell receptors. The PAI may carry insertion elements, integrases, and transposases and their sequences may be unstable. Their location may change within the same bacterium. This organization is conducive to effective subversion of the host defense system. The size of the pathogenicity islands may vary from 10 to 200 kb or more. The base composition of the islands and the codon usage may differ from that of the core DNA. In some bacteria, only a single PAI occurs; in others, there are several. Similar mechanisms operate in both animals and plants. *See* cholera toxin; codon usage; *Helicobacter pylori*; host-pathogen relations; integrase; secretion system; transmission; transposase. (Hacker, J. & Kaper, J. B. 2000. *Annu. Rev. Microbiol.* 54:641.)

pathogenicity islet Similar in some functions to pathogenicity islands, but their size is much smaller, 1 to 10 kb.

pathogen identification The food industry may need rapid and highly sensitive methods for the detection of live pathogens in various products. In case of viable *E. coli* cells, this is feasible by infection with compatible bacteriophages carrying bacterial luciferase inserts. The genes in the phages are expressed only in live bacteria, and if they are present, even a single bacterial cell emitting light may be detected with a high-powered luminometer or by a microchannel plate-enhanced image analyzer. *See* luciferase, bacterial.

pathovar Plant varieties or species that share disease susceptibility/resistance genes.

patrilocality Anthropological term indicating that males more frequently bring in mates from outside their location than moving to the location of females.

patristic distance Sum of the length of all branches connecting two species in an evolutionary tree. *See* evolutionary tree.

patroclinous Inheritance through the male such as the Y chromosome, androgenesis, or fertilization of a nullisomic female with a normal male. Through nondisjunction, the chromosome to be contributed by the female is eliminated, some of the gynandromorphs, sons of attached-X female *Drosophila*, fail to express maternal genes etc. *See* attached X; gynandromorph; nondisjunction.

patronymic Designation indicating the descent from a particular male ancestor, e.g., Johnson, son of John or O'Malley descendant of Malley. Common family names may assist in isolated populations to establish relationships. This analysis may be improved by studies of Y-chromosomal molecular markers. *See* isonymy; Y chromosome.

pattern formation during development Developmental patterns may begin by intracellular differentiation (animal pole, vegetal pole, yolk), positional signals between cells, and intracellular distribution of the receptors to various signals. Fibroblast growth factor and transforming growth factor β apparently have major roles as epithelial and mesoderm induction signals. The *Drosophila* gene *fringe* (*fng*) is involved with mesoderm induction and in wing embryonal disc formation. Juxtaposition of *fng*-expressing and nonexpressing genes seems to be required for establishing the dorsal ventral boundary of wing discs. The gene *fng* is expressed in the dorsal half of the wing disc, whereas *wingless* (*wg*, 2−30) is limited to the dorsal-ventral boundary. Gene *hedgehog* (*hh*, 3−81) is expressed in the posterior half, and *decapentaplegic* (*dpp*, 2−40) is detected in the anterior-posterior boundary. Anterior-posterior patterning in *Drosophila* is affected by the *trithorax* (*trx*) and *Polycom* (*Pc-G*) family members such as *extra sex combs* (*esc*). The homologue of the latter, *eed* (embryonic ectoderm development), controls anterior-posterior differentiation in mice. Homologues to these genes have been identified in other animals and humans as well. In *Xenopus*, it appears that the FNG protein is translated with a signal peptide (indicating that it is secreted); the proFNG peptide is terminated with a tetrabasic site for proteolytic cleavage, and after this processing, it is ready for normal function. *Wingless* of *Drosophila* is homologous to *Wnt1* mouse mammary tumor gene. The branching pattern of trachea and lung, respectively, in *Drosophila* and mammals is controlled primarily by the fibroblast growth factor signaling pathway. This is used reiteratively in repeated sequences of a branching. At each stage, different feedback and other control signals provide the specifications (Metzger, R. J. & Krasnow, M. A. 1999. *Science* 284:1635). Developmental pattern formation is also under genetic control in plants (Lee, M. M. & Schiefelbein, J. 1999. *Cell* 99:473). The progress has been much slower, however, because the plant tissues and cells are more liable to dedifferentiation and redifferentiation. Mutants have been obtained with clear differences in morphogenesis, but with the exception of flower differentiation and photomorphogenesis, much less is known about the molecular mechanisms involved. *See* cell lineages; fibroblast growth factor; flower differentiation; homeotic genes; MADS box; morphogenesis; morphogenesis in *Drosophila*; photomorphogenesis; RNA localization; signal transduction. (Malakinski, G. & Bryant, P., eds. 1984. *Pattern Formation: A Primer in Developmental Biology*. Macmillan, New York.)

PAU genes *See* seripauperines.

pauling *See* evolutionary clock.

PAUP Phylogenetic analysis using parsimony is a computer program for the analysis of evolutionary descent on the basis of molecular data. *See* evolutionary distance; evolutionary tree.

PAUSE RNA polymerase, I, II, and III do not operate continuously at the same rate. Due to various causes, their transcription may hesitate, then resume synthesis. A minimal functional element of PAUSE-1 is $TCTN_xAGAN_3T_4$, where $x = 0$, 2, or 4. Various elongation proteins such as ELL, Elongin, and transcription factor TFIIF may mediate pausing. *See* attenuator region; Nus [see lambda phage]; σ. (Ogbourne S. M. & Antalis, T. M. 2001. *Nucleic Acids Res.* 29:3919.)

pausing, transcriptional (hesitation) Discontinuity of the transcriptional process by all RNA polymerases. As a consequence, there is a heterogeneity of the transcripts because of the differences in recognition of modulating factors such as attenuation, transcription factors TFIIF, ELL, silencers, nus (λ-phage), antitermination signal, etc. Paucity of a nucleotide(s) or too high concentration of it may slow down transcription. RNA polymerase often pauses before a GTP is incorporated. Pause signals have been detected in both the template and the nontemplate DNA strands. Generally, hairpin structures (RNA base-pairing) favor pausing, although secondary structures may not be the sole cause. DNA sequences 16–17 bp downstream of the pause may alter the conformation of the polymerase and the pause. Even the nontemplate strand may have an effect. *See* arrest, transcriptional. (Davenport, R. J., et al. 2000. *Science* 287:2497.)

PAX Paired-box homeodomains are so called because they include two helix-turn-helix DNA-binding units. Several PAX proteins are known to be encoded in at least five different chromosomes and they mediate the development of the components of the eyes in insects (compound eyes) and humans, teeth, the central nervous system, the vertebrae, the pancreas, and tumorigenesis. The 130-residue paired domain binds DNA and functions as a transcription factor for B cells, histones, and thyroglobulin genes. The *Pax5* gene encodes the BSAP transcription factor. Mutation in *Pax5* arrests B cells at the pro-B stage. BSAP also may promote the expression of CD19 and indirectly IgE synthesis. BSAP may block the immunoglobulin heavy-chain 3′-enhancer and isotype switching and the formation of the pentameric IgM antibody. The activator motifs of BSAP display about 20 times higher binding affinity to the DNA than the repressor motif, yet the activator or repressor function depends primarily on the context of the motif. The level of BASP is high in the pre-B and immature B-cell stages, and after the antigen signal has arrived, its level is greatly diminished by signals from IL-2 and IL-5. Overexpression of BSAP results in its repressor activity. PAX6 (11p13) mutations cause absence of the iris of the eye (aniridia) without elimination of vision but other general neurodevelopmental problems. *See* animal models; aniridia; B cell; DiGeorge syndrome; FKP; goiter, familial; helix-turn-helix motif; histones; homeotic genes; hox; immunoglobulins; integrin; isotype switching; rhabdosarcoma; thyroglobulin; Waardenburg syndrome; Wilms' tumor. (Balczarek, K. A., et al. 1997. *Mol. Biol. Evol.* 14:829; Chi, N. & Epstein, J. A. 2002. *Trends Genet.* 18:41; Pichaud, F. & Desplan, C. 2002. *Current Op. Genet. Dev.* 12:430; <http://www.hgu.mrc.ac.uk/Softdata/PAX2/>; <http://www.hgu. mrc.ac.uk/ Softdata/PAX6/>.)

P blood group Controlled by two nonallelic loci. The nonpolymorphic P blood group is located in human chromosome 6 and encodes globoside, whereas the polymorphic P1 locus in human chromosome 22 encodes paragloboside. The frequency of the P gene in Sweden was found to be 0.5401 and that of P1 was 0.4599. According to other studies, the frequency of P among Caucasoids is about 0.75. The P1 blood type facilitates bacterial attachment to the epithelial cells of the urinary tract and kidney. Therefore, infections are more common. Some P alleles raise the risk of abortions, and others may increase the chances of stomach carcinomas. Some of the literature refers to P as P1 and P1 as P2. *See* blood groups; globoside. (Stroud, M. R. 1998. *Biochemistry* 37:17420.)

Pbp74 *See* Hsp70.

pBR322 Nonconjugative plasmid (constructed by Bolivar & Rodriguez) of 4.3 kb can be mobilized by helper plasmids because (although it lost its mobility gene) it retains the origin of conjugal transfer. It is one of the most versatile cloning vectors with a completely known nucleotide sequence and over 30 cloning sites. It carries the selectable markers ampicillin resistance and tetracycline resistance. Insertion into these antibiotic resistance sites permits the detection of the success of insertion because the inactivation of these target genes results in either ampicillin or tetracycline sensitivity. Although direct use of this 20-year-old plasmid has diminished, pBR322 components are present in many currently used vectors. *See* Amp; plasmid; tetracycline; vectors. (Bolivar, F., Rodriguez, R. L., et al. 1977. *Gene* 2:95; see diagram on the next page).

PBS Phosphate buffered saline. *See* saline.

PBSF Pre-B-cell growth-stimulating factor is a ligand of CXCR controlling B-cell development and vascularization of the gastrointestinal system. *See* CXCR; lymphocytes. (Egawa, T., et al. 2001. *Immunity* 15:323.)

PBX1, PBX2 Transcription factors involved in B-cell leukemias, encoded in human chromosomes 1q23 and 3q222, respectively. *See* leukemia.

PCA (principal component analysis) Multivariate statistical method that separates the original variables into independent variables and associated variances. It may be useful for the interpretation of microarray data. (*See* Méndez, M. A., et al. 2002. *FEBS Lett.* 522:24.)

P/CAF Human acetyltransferase of histones 3 and 4. *See* bromodomain; chromatin remodeling; histone acetyltransferases; INHAT; nucleosome; p300; signal transduction; $TAF_{II}230/250$. (Blanco, J. C., et al. 1998. *Genes Dev.* 12:1638.)

PCB Polychlorinated biphenyl is an obnoxious industrially employed carcinogen. A *Pseudomonas* enzyme may break it down. *See* environmental carcinogens; sperm.

PCD *See* apoptosis.

pCIP Coactivator protein that interacts with p300 (CREB) and regulates transcription and somatic growth of mammals.

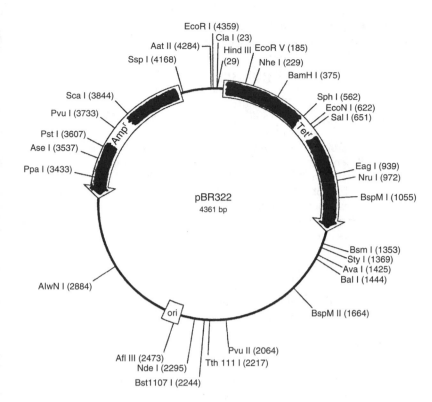

EcoR I (4359)
Cla I (23)
Aat II (4284)
Hind III
Ssp I (4168)
(29)
EcoR V (185)
Nhe I (229)
BamH I (375)
Sca I (3844)
Pvu I (3733)
Sph I (562)
EcoN I (622)
Sal I (651)
Pst I (3607)
Ase I (3537)
Ppa I (3433)
Amp'
Tet'
Eag I (939)
Nru I (972)
BspM I (1055)
pBR322
4361 bp
Bsm I (1353)
Sty I (1369)
Ava I (1425)
Bal I (1444)
AlwN I (2884)
ori
BspM II (1664)
Afl III (2473)
Pvu II (2064)
Nde I (2295)
Tth 111 I (2217)
Bst1107 I (2244)

Illustration from Pharmacia Biotech inc., by permission.

See nuclear receptors; signal transduction. (Wang, Z., et al. 2000. *Proc. Natl. Acad. Sci. USA* 97:13549.)

PCL   Putative cyclin. *See* cyclin.

P1 cloning vectors   Have carrying capacity of up to 100 kbp DNA; thus, they fall between lambda and YAC vectors. *See* vectors. (Park, K. & Chattoraj, D. K. 2001. *J. Mol. Biol.* 310:69; Grez, M. & Melchner, H. 1998. *Stem Cells* 16 Suppl. 1:235.)

PCNA   Proliferating cell nuclear antigen is an auxiliary protein (a processivity factor) in pol δ and pol ε functions in eukaryotes. It has a similar role in DNA replication in general, and in the cell cycle and repair. Its function is similar to that of the β-subunit of the prokaryotic pol III; it provides a "sliding clamp" on the DNA to be replicated. Binding to p21 may inhibit PCNA replicative function. PCNA also binds other proteins such as cyclin D, FEN1/Rad27/MF1, DNA ligase 1, GADD, DNA methyltransferase, DNA repair proteins XPG, MLH, and MSH. Mutations in PCNA alter the conditions for nucleosomal assembly by interacting with CAF. Under such conditions silencing by heterochromatin is reduced or lost. Some mutations in RFC may compensate for defects in PCNA. *See* ABC excinuclease; CAF; cell cycle; cyclins; DNA polymerases; DNA replication; eukaryotes; excision repair; heterochromatin; ligase DNA; methylation of DNA; mismatch repair; p21; Rad27/Fen1; RFC; sliding clamp. (Karmakar, P., et al. 2001. *Mutagenesis* 16:225; Ola, A., et al. 2001. *J. Biol. Chem.* 276:10168; López de Saro, F. J. & O'Donell, M. 2001. *Proc. Natl. Acad. Sci. USA* 98:8376.)

PCR   *See* polymerase chain reaction.

PCR, allele-specific   Used to screen a population for a particular allele-specific mutation, e.g., mutations responsible

for MELAS in aging human mitochondrial DNA. One of the most common mutations in this anomaly involves the transition A → G at site 3243. If a primer containing the complementary base C is used, the mutant sequence from appropriate tissues is successfully amplified and can be detected by gel electrophoresis. The same C primer does not generate a substantial quantity of the fragment using the wild-type template. Thus, by this procedure the approximate frequency of this allele-specific mutation can be determined. *See* hot-start PCR; mitochondrial disease in humans; polymerase chain reaction; transition mutation. (Ugozzoli, L. & Wallace, R. B. 1992. *Genomics* 12:670.)

PCR, broad-based   Uses primers that amplify a broad base of genes, e.g., the microbial rRNA genes or a group of viral genes common to the majority of related species. In order to facilitate molecular identification of pathogens, RDA and other procedures may supplement the analysis. *See* RDA.

PCR, competitive   Used for quantifying DNA or RNA. The competitor nucleic acid fragment of known concentrations in serial dilutions is coamplified with another (the experimental) nucleic acid of interest using a single set of primers. The beginning quantity of the experimental molecules is estimated from the ratio of the competitor and experimental amplicons obtained during the PCR procedure that are supposedly amplified equally. The quantity of the unknown DNA is determined by the equivalence point (EQP), where the experimental nucleic acid and the competitor show the same signal intensity, indicating that their amounts are the same. A simplified new version is described by Watzinger, F., et al. 2001. *Nucleic Acids Res.* 29(11):e52.

PCR, discriminatory   Detects small mismatches or point mutations. *See* diagram next page.

```
Wild type              Mismatch mutant
GGCGTGTGAACTG          GGCGTGTGGTCTG
|||||||||||||          |||||||||||||
CCGCACACTTGAC          CCGCACACCAGAC

PCR↓                   PCR↓

primerTGTGAA  →        primerTGTGAA  →
CCGCACACTTGAC          CCGCACACCAGAC

Product made           No product made
```

Discriminatory PCR.

PCR, doc Degenerate oligonucleotide-primed polymerase chain reaction and capillary electrophoresis of DNA is a random amplification technique combined with analysis on microchips. *See* capillary electrophoresis; PCR. (Cheng, J., et al. 1998. *Anal. Biochem.* 257:101.)

PCR, electronic *See* electronic PCR.

PCR, multiplex Employs a multiple set of primers for amplification in a single reaction batch. (Broude, N. E., et al. 2001. *Proc. Natl. Acad. Sci. USA* 98:206.)

PCR, overlapping Uses two sets of primers; each has complementary sequences at the 5′-end. Two separate PCRs are carried out, then the products are purified by gel to remove the unincorporated primers. A second PCR process uses only the outside primer pairs and the two primary products are joined. *See* PCR.

PCR, real-time reverse transcription (*See* Seeger, K., et al. 2001. *Cancer Res.* 61:2517.)

PCR, transcriptionally active (TAP) (1) Specific primers amplify the gene of interest. (2) Mixtures of DNA fragments are equipped with promoter and terminator elements, then can be used for transfection in a suitable plasmid. They can also be inserted into plasmids by homologous recombination. TAP products can be used as DNA vaccines and generate antibodies against the encoded genes. The procedure is suitable for the generation of hundreds or thousands of transcriptionally active genes for genomic/proteomic studies. (Liang, X., et al. 2002. *J. Biol. Chem.* 277:3593.)

PCR-based mutagenesis Any base difference between the amplification primer will be incorporated in the future template through polymerase chain reaction. Actually only half of the new DNA molecules would contain the alteration present in the original amplification primer unless a device is used, e.g., the undesired strand would be made unsuitable for amplification and therefore lost from the reaction mixture. The method may include multiple-point mutations, small insertions, or deletions. It is possible that the amplification may result in other nucleotide alterations as a result of the error-prone Taq polymerase. *See* DNA shuffling; local mutagenesis; polymerase chain reaction; primer extension; small-pool PCR; VENT. (Nelson, R. M. & Long, G. L. 1989. *Anal. Biochem.* 180:147.)

PCR-LSA Polymerase chain-reaction amplification is a method for the localization of SNIPS. *See* RRS; SNIPS.

PCR-mediated gene replacement By mitotic recombination the procedure replaces particular genes with an identifiable marker of a neutral phenotype. A 20-base-pair unique sequence tract tags each of these lines. On the basis of hybridization of the PCR products to a tag sequence, it is possible to quantitate the altered cell lines in a population. When one of the target genes is deleted in diploid yeast, the other is still expected to be functional. Some of them, however, will display a defective phenotype due to haploinsufficiency. These "heterozygotes" may also display increased sensitivity to drugs and may be used for pharmaceutical research. *See* haploinsufficiency. (Giaever, G., et al. 1999. *Nature Genet.* 21:279.)

PCR targeting *See* targeting genes.

PC-TP Phosphatidylcholine transfer protein mediating transfer of phospholipids between organelles within the cell.

P cytotype *See* hybrid dysgenesis.

PD-1 Inhibitory immune receptor (2q37.3) on B, T lymphocytes and natural killer cells.

PDECGF Platelet-derived endothelial cell growth factor.

pDelta *See* γδ.

PDF Electronic publishing software readable with the aid of Adobe Acrobat reader. Frequently used by journals available through the Internet.

PDGF *See* platelet-derived growth factor.

PDGFR Platelet-derived growth factor receptor.

PDI Protein disulfide isomerase is a cofactor of protein folding mediated by chaperones. *See* chaperone; Erp61; Eug; Mpd; PPI.

PDK Phosphoinositide-dependent kinase is part of the MAPK, RSK signaling pathway. *See* Akt; MAPK; phosphoinositides; PIK; RSK. (Toker, A. & Newton, A. C. 2000. *Cell* 103:185.)

PDS Anaphase inhibitory protein that must be degraded with the assistance of CDC20 component of APC before the cell cycle can exit from mitosis regulated also by the activity of Cdc14 phosphatase. *See* APC; Cdc14; Cdc20; cell cycle; checkpoint; mitotic exit; sister chromatid cohesion. (Salah, S. M. & Nasmyth, K. 2000. *Chromosoma* 109:27.)

pDUAL *See* γδ.

PDZ domains Post-synaptic density, disc-large, zo-1. Approximately 90 amino acid repeats are involved in ion channel and receptor clustering and linking effectors and receptors.

PDZ-domain proteins are involved in the regulation of the Jun N-terminal kinase pathway, postsynaptic density (PSD) proteins at glutamate-ergic synapses, Rho-activated citron protein function, visual signaling, etc. The *Drosophila* gene, *scribble* (*scrib*), encodes a multi-PDZ-domain protein and in cooperation with a leucine-rich protein controls apical polarization of the embryo. *See* AMPA; citron; effector; HOMER; ion channels; Jun; NMDAR; receptor; Rho; signal transduction. (Harris, B. Z. & Lim, W. A. 2001. *J. Cell Sci.* 114[pt 18]:3219; Hung, A. Y. & Sheng, M. 2002. *J. Biol. Chem.* 277:5699.)

pea (*Pisum* ssp) Several self-pollinating vegetable and feed crops: the Mendel's pea is *P. sativum*, and others are $2n = 2x = 14$. *See Pisum*; photograph shows segregation for smooth and wrinkled within a pod.

peach (*Prunus persica*) $x = 7$; the true peaches are diploid.

peacock's tail Evolutionary paradigm when a clear disadvantage (like the awkward tail) turns into a mating advantage because of the females' preference for the fancy trait, thus increasing the fitness of the males that display it. *See* fitness; selection; sexual selection.

pea comb Comb characteristic of poultry of *rrP(P/p)* genetic constitution. *See* walnut comb.

peanut *See* groundnut.

pear (*Pyrus* spp) About 15 species; $x = 17$ and mainly diploids, triploids, or tetraploids. It is very difficult to hybridize it with apples, but it can be crossed with some *Sorbus*. *See* apple.

Pearson marrow pancreas syndrome *See* mitochondrial disease in humans.

pebble *See* scaffolds in genome sequencing.

pectin Polygalacturonate sequences alternated by rhamnose and may contain galactose, arabinose, xylose, and fucose side chains. Molecular weight varies from 20,000 to 400,000. Its role is intercellular cementing of plant cells. Acids and alkali may cause its depolymerization.

pedicel Stalk of flowers in an inflorescence. *See* peduncle.

pedant Protein extraction, description, and analysis tool. (<http://www.dl.ac.uk/CCP/CCP11/conferences/gpc_v/frishman.html>.)

pedigree analysis Generally carried out by examination of pedigree charts, used in human and animal genetics where the family sizes are frequently too small to conduct meaningful direct segregation studies. The pedigree chart displays the lines of descent among close natural relatives. Females are represented by circles, males by squares, and if the sex is unknown, a diamond (◇) is used. The same but smaller symbols or a vertical or slanted line over the symbol indicates abortion or stillbirth. For spontaneous abortions, triangles may be used. Individuals expressing a particular trait are represented by a shaded or black symbol, and in case they are heterozygous for the trait, half of the symbol is shaded. When an unaffected

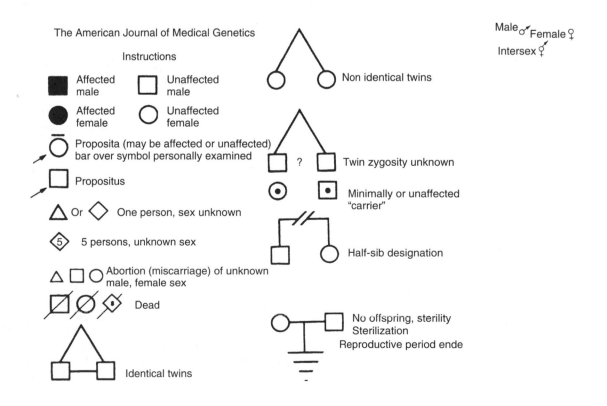

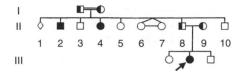

A pedigree chart.

female is the carrier of a particular gene, there is a dot within the circle. In case segregation for traits needs to be illustrated in the pedigree, the individual displaying both traits may be marked by a horizontal and vertical line within the symbol or only by a horizontal or vertical line, respectively. Horizontal lines connect the parents, and if the parents are close relatives, the line is doubled. The progeny is connected to the parental line with a vertical line and the subsequent generations are marked by Roman numerals at the left side of the chart, I (parents), II children, III grandchildren, and so on. Twins are connected to the same point of the generation line. If they are identical, a horizontal line connects them to each other. The order of birth of the offspring is from left to right and may be numbered accordingly below their symbol. An arrow to a particular symbol indicates the proband, the individual who first became known to the geneticist as expressing the trait. The appropriate symbols of adopted children may be bracketed. If a prospective offspring is considered at risk, broken lines draw the symbol. A horizontal line connected by a vertical line to the "parental" line may indicate lack of offspring by a couple. Infertility may be represented by doubling a horizontal line under and connected to the male or female symbol, respectively. Egg or sperm donors (in case of assisted reproductive technologies [ART]) are indicated by a D and surrogate mothers by an S within the symbols. Males may be represented by a *Mars shield* and females by a *Venus mirror*, and intersexes by the sign in the box shown on preceding page. *See* ART. (Bennett, R. L., et al. 1995. *Am. J. Hum. Genet.* 56:745.) (*Am. J. Med. Genet.* pedigree chart is reprinted by permission of John Wiley & Sons, Inc. ©)

pedogenesis   Egg production by immature individuals such as larvae.

peduncle   Stalk of single-standing flowers ($\rightarrow$); bundle of nerve cells *See* pedicel.

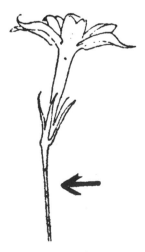

Peduncle.

PEG   (1) Polyethylene glycol may be liquid or solid and comes in a range of different viscosities (200, 400, 600, 1,500, etc.). It facilitates fusion of protoplasts, uptake of organelles, precipitation of bacteriophages, plasmids and DNA, promoting end labeling, ligation of linkers, reduction of immunogenicity when attached to humanized antibody, etc. (2) Paternally expressed gene. *See* imprinting.

PEG-3   Progression-elevated gene-3. In nude mice it upregulates carcinogenesis in progress via activation of VEGF. It can be blocked by antisense technology. *See* antisense technology; VEGF.

PEK   *See* HRI.

pelargonidine   *See* anthocyanin.

Pelargonium zonale (geranium)   Ornamental plant. Some variegated forms transmit the nonnuclear genes through the sperm, whereas in the majority of plants the plastids are transmitted only through the egg. *See* chloroplast genetics; chloroplasts; uniparental inheritance.

P element   *See* hybrid dysgenesis.

P-element vector   Constructed from the 2.9 -kb transposable element P of *Drosophila* equipped with 31 bp inverted terminal repeats. The gene to be transferred is inserted into the element, but in order to generate stable transformants, the transposase function located in the terminal repeats is disabled. Functional transposase is provided in a separate helper plasmid (pπ25.7wc). Such a binary system permits the separation of the two plasmids and the screening of the permanent transgenes if a selectable marker is included. Both plasmids are mixed in an injection buffer and delivered into preblastoderm embryos. The various P vectors have been widely used in gene tagging, induction of insertional mutation, and exploration of functional genetic elements in *Drosophila* and some other insects. *See* hybrid dysgenesis; transposon vector. (Sullivan, W., et al. 2000. *Drosophila Protocols.* Cold Spring Harbor Lab. Press, Cold Spring Harbor, NY.)

Pelger(-Huet) anomaly   Autosomal-dominant condition in humans as well as rabbits, cats, etc., characterized by fewer (1.1–1.6) than normal (2.8) nuclear lobes in the granulocytic leukocytes. It may be a mild anomaly, but it may be associated with other more serious ailments. The prevalence varies from 1×10^{-3} to 4×10^{-4}. Similar phenotypes were described as autosomal recessive or X-linked.

Pelizaeus-Merzbacher disease   Xq22 chromosomal-recessive leukodystrophy that accumulates a proteolipoprotein (PLP, a 276-amino-acid integral membrane protein) of the endoplasmic reticulum and the surface protein DM20 (26.5 kDa). The defect involves alternative splicing of the same mRNA. Duplications and deletion may be responsible for the disorder. The clinical symptoms are defective myelination (dysmyelination) of the nerves and defective interaction between oligodendrocytes and neurons, pathogenesis of the central nervous system, and impaired motor development with an onset before age 1. Mutation in the same gene encoding PLP is responsible for X-linked spastic paraplegia type 2 (SPG-2), and the difference is in the degree of hypomyelination and motor dysfunction. Hereditary spastic paraplegia alleles were assigned to 8p, 16p, 15q, and 3q27-q28. The corresponding defect in mice displays *jimpy* and the myelin-deficient (*msd*) phenotype. *See* Charcot-Marie-Tooth disease; leukodystrophy; myelin; spastic paraplegia.

pelle   Serine/threonine kinase, involved in dorsal signal transduction. *See* IRAK.

peloric   Circular symmetry of the flower in contrast to the bilateral symmetry of the wild type first described in *Linaria* by Linnaeus. Homologous mutations occur in *Antirrhinum* as shown in the figure. This variation of floral symmetry of *Linaria* is due to the different methylation of a gene, *Lcyc*. Top: the wild-type *Antirrhinum* flower of bilateral symmetry. Below: the *cycloidea* mutant with radial symmetry (pelory). *See* epigenesis; methylation of DNA; superman.

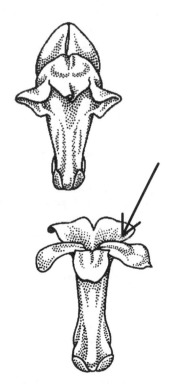

(Illustration is the courtesy of Professor Hans Stubbe.)

pelota   *Drosophila* gene involved in sperm function. *See* azoospermia.

pemphigus   Collection of skin diseases with the general features of developing smaller or larger vesicles of the skin that may or may not heal and in extreme cases may result in death. Autosomal-dominant familial pemphigus vulgaris is an autoimmune disease of the skin and mucous membranes. In the majority of cases, HLA-DR4 is involved. This anomaly is particularly common among Jews in Israel. *See* autoimmune disease; desmosome; Hailey-Hailey disease; HLA; skin diseases.

Pendred syndrome   Recessive (7q31, PDS) thyroid anomaly and neurosensory deafness. The locus encodes pendrin, an anion transporter, a presumed sulfate transporter localized in the cell membrane, and a bicarbonate secretion in the kidney. Recent evidence also indicates chloride and iodide transport. This locus is responsible for about 1–10% of genetically determined hearing loss. *See* deafness; goiter. (Royaux, I. E., et al. 2001. *Proc. Natl. Acad. Sci. USA* 98:4221.)

penetrance   Percentage of individuals in a family that express a trait determined by a gene(s). The genetic basis of this phenomenon is poorly understood and may cause serious problems in genetic counseling. *See* expressivity.

penicillin   Antibiotic originally obtained from *Penicillium* fungi. The arrow points to the reactive bond of the β-lactam ring. *See* antibiotics; β-lactamase; lactam; *Penicillium*.

Basic structure of penicillins.

penicillin screen   Was used for mass isolation of auxotrophic microbial mutations that failed to grow in basal media (in contrast to the wild type); therefore, the presence of the antibiotic did not lead to their death (in contrast to the wild type). After transfer to complete (or appropriately supplemented) media, the auxotrophs grew and thus were selectively isolated. *See* mutant isolation; replica plating; selective medium. (Davis, B. D. 1948. *J. Amer. Chem. Soc.* 70:4267; Lederberg, J. & Zinder, N. 1948. *J. Am. Chem. Soc.* 70:4267.)

penicillinase   *See* β-lactamase.

penicillin-binding proteins   *See* PBP.

penicillin enrichment   *See* penicillin screen.

Penicillium notatum (fungus) $x = 5$.

Penicillium conidiophore with conidia.

penis   Male organ of urinary excretion and insemination (homologous to the female clitoris). It contains the corpus spongiosum through which the urethra passes. Above that are the corpora cavernosa, which becomes extended when erection takes place due to enhanced blood supply to this elastic tissue as a consequence of NO (nitrogen monoxide) gas flow to the muscles of the blood vessel wall initiated by acetylcholine. The release of acetylcholine is controlled by steroid hormones. Cyclic GMP-dependent kinase is an essential enzyme for the maintenance of the extended state of the corpus cavernosum. Injection of prostaglandin E1 blocks cGMP-degrading phosphodiesterase and facilitates erection. Some animals (snakes, lizards, crustaceans, and insects) have two penises (virgae). In *Euborellia plebeja* (Dermatoptera), both are functional. *See* acetylcholine; acetylcholine receptors; animal hormones; baculum; cGMP; hypospadias; nitric oxide; prostaglandin. (Kamimura, Y. & Matsuo, Y. 2001. *Naturwiss.* 88:447.)

pentaglycines   *See* bacteria.

pentaploid   Its cell nucleus contains five genomes ($5x$). Pentaploids are obtained when hexaploids ($6x$) are crossed with tetraploids ($4x$). The pentaploids are generally sterile or semifertile because the gametes generally have an unbalanced number of chromosomes. *See* polyploids; *Rosa canina*.

penton   Capsomer with five neighbors in the viral capsid. *See* capsomer; hexon.

pentose   Sugar with 5-carbon-atom backbone, such as ribose, deoxyribose, arabinose, xylose.

pentose phosphate pathway   Glucose-6-phosphate + $2\,NADP$ + H_2O → ribose-5-phosphate + $2\,NADPH$ + $2\,H^+$ + CO^2, i.e., the conversion of hexoses to pentoses generates NADPH, a molecule that serves as a hydrogen and electron donor in reductive biosynthesis. *See* Embden-Meyerhof pathway; Krebs-Szentgyörgyi cycle.

pentose shunt   Pentose phosphate pathway.

pentosuria   Autosomal-recessive nondebilitating condition characterized by excretion of increased amounts of L-xylulose (1–4 g) in the urine because of a deficiency of the NADP-linked xylitol dehydrogenase enzyme. In Jewish and Lebanese populations, the frequency of the gene was about 0.013–0.03. *See* allelic frequencies; gene frequency.

PEPCK   Phosphoenolpyruvate carboxykinase, a regulator of energy metabolism.

pepper (*Capsicum* ssp)   Exists in a great variety of forms, but all have $2n = 2x = 24$ chromosomes. Some wild species are self-incompatible and the cultivated varieties also yield better if they have a chance for xenogamy. *See* self-incompatibility; xenogamy.

pepsin   Acid protease formed from pepsinogens. It has a preference for the COOH side of phenylalanine and leucine amino acids.

pepstatin ($C_{34}H_{63}N_5O_9$)   Protease (pepsin, cathepsin D) inhibitor.

peptamer   Exposed loop on the surface of a carrier protein; it is thus protected from degradation and its conformational stability is improved.

peptidase   Hydrolyzes peptide bonds. In humans, the peptidase gene PEPA is in chromosome 18q23, PEPB in 12q21, PEPC in 1q42, PEPD in 19cen-q13.11, PEPE in 17q23-qter, PEPS in 4p11-q12, and the tripeptidyl peptidase II (TPP2), a serine exopeptidase, is in 13q32-q33. (<http://www.merops.ac.uk>.)

peptide bond   Amino acids are joined into peptides by their amino and carboxyl ends ↑ (and they lose one molecule of water).

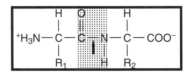

peptide elongation   *See* aminoacylation; cycloheximide; aminoacyl-tRNA synthetase; elongation factors (eIF); protein synthesis; ribosome; tmRNA.

peptide initiation   *See* pactamycin; protein synthesis.

peptide mapping   Separation of (in)complete hydrolysates of proteins by two-dimensional paper chromatography or by two-dimensional gel electrophoresis for the purpose of characterization. The distribution pattern is the map or fingerprint characteristic for each protein.

peptide mass fingerprints   The protein is first cleaved by a sequence-specific protease such as trypsin, analyzed by MALDI-TOF, compared with protein sequences with similar lysyl or arginyl residues of the same mass. On this basis, matching proteins even in a mixture can be identified. Modified proteins are detectable on the basis of peptide sequence with an incremental mass due to, e.g., a phosphogroup. *See* MALDI; proteomics; trypsin. (Mann, M., et al. 2001. *Annu. Rev. Biochem.* 70:437; Pratt, J. M., et al. 2002. *Proteomics* 2:157.)

peptide nucleic acid (PNA)   A nucleic acid base (generally thymine) is attached to the nitrogen of a glycine (or other amino acids) by a methylene carboxamide linkage in a backbone of aminoethylglycine units. Such a structure can displace one of the DNA strands and binds to the other strand. They are DNA mimics. This highly stable complex has similar uses as the antisense RNA technology. PNA may be used to inhibit excessive telomerase activity in cancer cells. Homopyrimidine PNA may invade homopurine tracts in double-stranded DNA and may form triplex DNA and interfere with transcription. PNA may also be used to screen for base mismatches and small deletions or base substitution mutations. Peptide nucleic acid complementary to mutant mtDNA selectively inhibits the replication of mutant mtDNA in vitro. PNA has been suggested to be the first prebiotic genetic molecule rather than RNA. PNA may be useful for delivering genes to the mitochondria. PNA-DNA hybrids may be identified by binding of the dye 3,3′-diethylthiadicarbocyanine and may be used for the rapid detection of mutations of clinical importance. *See* antisense DNA; antisense RNA; Hoogsteen pairing; mitochondrial gene therapy; mt DNA; RNA world; TFO. (Corey, D. R. 1997. *Trends Biotechnol.* 15:224; Chinnery, P. F., et al. 1999. *Gene Ther.* 6:1909; Wilhelmsson, L. M., et al. 2002. *Nucleic Acids Res.* 30:[2]:e3.)

PNA backbone resembles that of DNA. The letter **B** stands for nucleic acid bases.

peptide processing   *See* posttranslational processing.

peptide sequence tag   *See* electrospray ms.

peptide transporters   TAP; ABC transporters.

peptide vaccination   Synthetic polypeptides corresponding to CTL epitopes may result in cytotoxic T-cell-mediated immunity but in some instances may enhance the elimination of antitumor CTL response. *See* cancer prevention; CTL; epitope; immunological surveillance. (Vandenbark, A. A., et al. 2001. *Neurochem Res.* 26:713.)

peptidoglycan   Heteropolysaccharides cross-linked with peptides constituting the bulk of the bacterial cell wall, especially in gram-positive strains. *See* gram negative.

peptidyl site   *See* protein synthesis; P site; ribosome.

peptidyl-prolyl isomerases (PPI)   Mediate the interconversion of the cis and trans forms of peptide bonds preceding proline. PPI genes are in human chromosomes 4q31.3, 6p21.1, 7p13, and the mitochondrially located at 10q22-q23. This family includes cyclophilins, FKBs, and parvulins. *See* cyclophilins; FK506; parvulin. (Shaw, P. E. 2002. *EMBO Rep.* 3:54.)

peptidyl transferase   Generates the peptide bond between the preceding amino acid carboxyl end (at the P ribosomal site) and the amino end of the incoming amino acid (at the A site of the ribosome). It is a ribozyme and the catalytic function resides in the 23S ribosomal RNA. Essential function is attributed to adenine 2451. *See* macrolide; protein synthesis; ribosome. (Polacek, N., et al. 2002. *Biochemistry* 41:11602.)

percent identity plot (PIP)   Macromolecular sequence map displaying the percentage of identity between two sequences. (<bio.cse.psu.edu>.)

perdurance   Persistence and expression of the product of the wild-type gene even after the gene itself is no longer there.

perennial   Lasts through more than 1 year.

perfect flower   Has both male and female sexual organs, i.e., it is hermaphroditic. *See* flower differentiation; hermaphrodite.

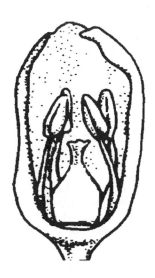

perforin   Encoded at 10q22 Pore-forming protein (homologous to component 9 of the complement) that establishes transmembrane channels; it is stored in vesicles within the CD8$^+$ cytotoxic T cells (CTL). These vesicles also contain serine proteases. Perforin mediates apoptosis by permitting killer substances to slowly enter the cell. Perforin may have a role in T-cell-mediated destruction of pancreatic β-cells in diabetes mellitus type I. *See* apoptosis; caspase; complement; diabetes mellitus; fragmentin-2; granzymes; histiocytosis; T cells.

perfusion   Addition of liquid to an organ through its internal vessels in vitro or in vivo.

perianth   Designates both sepals and petals of the flowers. *See* flower differentiation.

pericarp   Fruit wall (maternal tissue) developed from the ovary wall such as the pea pod; outer layer of the wheat or maize kernels. The *Arabidopsis* silique, the peel of citruses, the skin of apple, the shell of the nuts, etc., are similar but are exocarps. The outer layer of the common barley "seed" is not part of the fruit wall, but it is a bract of the flower.

pericentric inversion   *See* inversion, pericentric.

pericentromeric region   Highly redundant tract (<1 kb to ~85 kb); transition zone between the genic region of the chromosomes and the satellite heterochromatin. *See* centromere; heterochromatin; satellite; DNA. (Horvath, J. E., et al. 2001. *Hum. Mol. Genet.* 10:2215.)

perichromatin fibers   Active genes occupy the surface of specific compartments in the interphase nucleus (chromosome territories) and represent the perichromatin fibers. *See* chromatin; SR motif.

periclinal chimera   Contains genetically different tissues in different cell layers. *See* chimera; mericlinal chimera.

pericycle   Root tissue between the endodermis and phloem. *See* endodermis; phloem; root.

peridium   Covering of the hymenium or the hard cover of the sporangium of some fungi. *See* hymenium; sporangium.

perinatal   Period after 28 weeks of human gestation to 4 weeks after birth.

perinuclear space   ~20 to 40nm space between the two layers of the nuclear membrane. *See* nucleus.

periodic acid-Schiff reagent (PAP)   Tests for glycogen, polysaccharides, mucins, and glycoproteins. It breaks C-C bonds by oxidizing near hydroxyl groups, forms dialdehydes, and generates red or purple color.

periodicity   Number of base pairs per turn of the DNA or the number of amino acids per turn of an α-helix of a polypetide chain. *See* protein structure; Watson and Crick model.

periodic paralysis (PP, KCNE)   Group of autosomal-dominant human diseases manifested in periodically recurring weakness accompanied by low blood potassium level (hypokalemic periodic paralysis, defect in the α-subunits of Ca^{2+} channel) or in other forms with high blood potassium level (hyperkalemic periodic paralysis, paramyotonia). The latter types were attributed to base substitution mutations in a highly conserved region of the α-subunit of a transmembrane sodium channel protein. In another type of the disease, the blood potassium level appeared normal and the patients responded favorably to sodium chloride. The MiRP2 potassium channel defect is also associated with PP. *See* hyperkalemic periodic paralysis; ion channel; Moebius syndrome; myotonia. (Abbott, G. W., et al. 2001. *Cell* 104:217.)

periodontitis   Several diseases involving inflammation of the gingiva, especially at the base of the teeth and the alveolae, the bone support of the teeth. It is usually associated with keratosis of the palms and soles. About 30% of the human population is affected by it. In the juvenile form (encoded at 4q11-q13), both milk and permanent teeth may be lost in early childhood. The disease is the result of bacterial infections (~500 different species may inhabit the human mouth). Papillon-Lefèvre syndrome (11q14, prevalence $1-4 \times 10^{-6}$) is based on a deficiency of cathepsin C, a dipeptidyl peptidase I. In the similar autosomal-recessive periodontitis, deficiency of IL-1 is suspected. Similar symptoms may occur in Ehlers-Danlos syndrome type VIII. *See* cathepsin; Ehlers-Danlos syndrome; IL-1; keratosis. (Travis, J., et al. 2000. *Adv. Exp. Med. Biol.* 477:455.)

peripheral nervous system   Resides outside the brain and the spinal chord.

peripheral proteins   Bound to the membrane surface by hydrogen bonds or by electrostatic forces. *See* membrane proteins.

peripherin (retinal degeneration slow protein)   *See* retinal dystrophy.

periplasma   Bacterial cell compartment between the cell wall and cell membrane. In *E. coli*, the Sec family of proteins mediate translocation across the periplasmic and outer membranes. Extracellular stress response factor σ^E regulates the assembly of the outer membrane. The two-component Cpx seems to be involved in assembly of the pilus. *See* pilus; Sec; two-component regulatory system. (Danese, P. N. & Silhavy, T. J. 1998. *Annu. Rev. Genet.* 32:59; Raivio, T. L. & Silhavy, T. J. 2001. *Annu. Rev. Microbiol.* 55:591.)

peristalsis   Contraction of muscles of tubular structures (e.g., intestines) propelling the content.

peristome   Fringe of teeth at the opening of the sporangium of mosses or the buccal (mouth) area of ciliates.

perithecium   Fungal fruiting body of disk or flask shape with an opening (ostiole) for releasing the spores. A perithecium of *Neurospora* contains about 200 asci. Its primordium is called protoperithecium. *See* apothecium; ascogonium; ascus; cleistothecium; gymnothecium; *Neurospora*; tetrad analysis.

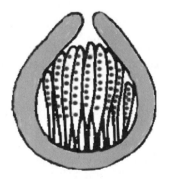

Perithecium.

periventricular heterotopia   Human X-chromosomal mental retardation and seizures caused by anomalies of the brain cortex. The neurons destined for the cerebral cortex fail to migrate. The mutation involves the filamin gene (FLN1, Xq28) encoding an actin-cross-linking phosphoprotein. *See* double cortex. (Sheen, V. L., et al. 2001. *Hum. Mol. Genet.* 10:1775.)

perk   PKR-like ER kinase is a phosphorylating enzyme in the endoplasmic reticulum similar to the mammalian RNA-dependent protein kinase (PKR). It is interferon-inducible and activated by double-stranded RNA. PERK and the phosphorylation of eIF2α inhibit the initiation of translation.

See eIF2 α; Ire; PKR; S6; unfolded protein response. (Kumar, R., et al. 2001. *J. Neurochem.* 77:1418.)

perlecan Heparan sulfate proteoglycan that interacts with the extracellular matrix and growth factor receptors, affecting signal transduction. *See* heparan sulfate; proteoglycan. (Knox, S., et al. 2002. *J. Biol. Chem.* 277:14657.)

***per* locus** The period locus in *Drosophila* (map location 1–1.4, 3B1–2) controls the circadian and ultradian rhythm, thus affecting eclosion (hatching), general locomotor activity, courtship, intercellular communication, etc. The mutations do not seem to affect the viability of the individuals involved; only the behavior is altered. When per^S (caused by a base substitution mutation in exon 5) in the brain is transplanted into per^{01} mutants (nonsense mutation in exon 4), causing short ultradian rhythm and multiple periods, some flies may be somewhat normalized. The locus has been cloned and sequenced and seems to code for a proteoglycan. The Per protein forms a heterodimeric complex with the Tim (*timeless* gene) protein and they jointly autoregulate transcription. Tim is degraded in the morning in response to light, resulting in the disintegration of the complex that is reformed again in the dark in a circadian oscillation. In mammals, three *mPer* loci have been identified that control the circadian clock. *See* circadian; proteoglycan; ultradian.

PERL SCRIPT Assembles and merges sequences from different DNA libraries. *See* DNA library; PHRAP.

permafrost Soil layer in cold regions that remains permanently frozen even when the top may thaw.

permanent hybrid *See* complex heterozygote.

permease Enzyme involved in the transport of substances through cell membranes. *See* ion channels; membrane channels; membrane potential; membrane transport.

permissible dose *See* radiation hazard assessment.

permissive condition State at which a conditional mutant can survive or reproduce. *See* conditional mutation.

permissive host A cell permits (viral) infection and/or development.

permutation Generating all possible orders of **n** numbers, and it can be obtained by the factorial n!, e.g., the factorial of 4, $4! = 4 \times 3 \times 2 \times 1 = 24$. *See* combination; variation.

permutation test Assesses the association of QTLs with multiple (molecular) markers in a randomized array. *See* QTL.

permuted redundancy A collection of redundant sequences occur in the phage population at the termini of phage DNA that start and end with permuted sequences of the same nucleotide sequence, e.g., 1234....1234, 2341....2341, 3412....3412, etc. This arrangement is characteristic for T-uneven phages, e.g., T1, T3, T5, etc. *See* nonpermuted redundancy.

Perodictus *See* Lorisidae.

peroxidase Heme protein enzymes that catalyze the oxidation of organic substances by peroxides. Glutathione peroxidase (and selenium) deficiency may cause hemolytic disease. Several peroxidase genes have been located in the human genome: GPX1 in 3q11-q12, GPX2 in 14q24.1, GPX3 in 5q32-q33, GPX4 (in testes) in chromosome 19; a rare eosinophil peroxidase (EPX) may compromise the immune system, a thyroid peroxidase deficiency (2p25) interferes with thyroid function. *See* eosinophil; hemolytic disease; immune system; oxidative stress.

peroxidase and phospholipid deficiency Autosomal-recessive anomaly (17q23.1) of the eosinophils involving the enzyme defects of peroxidase and reduced staining of phospholipids. *See* microbody; Refsum disease; Zellweger syndrome.

peroxides Display the $-O-O-$ linkage. Organic peroxides participate in activation and deactivation of promutagens, mutagens, procarcinogens, and carcinogens and in many other physiological reactions. Peroxides are formed by the breakdown of amino acids and fatty material in the cell and may inflict serious damage. According to some views, spontaneous mutation may be caused to a great extent by these regular components of the diet. Therefore, eating rancid food may pose substantial risk. *See* catalase peroxisomes; environmental mutagens; P450; peroxidase; promutagen; ROS.

peroxins Peroxisome proteins. PEX1 (7q21-q22), PEX10 (human chromosome 1), and PEX13 (2p15) are peroxisome biogenesis proteins. *See* microbodies; Refsum disease; Zellweger syndrome. (Walter, C., et al. 2001. *Am. Hum. Genet.* 69:35.)

peroxiredoxins Antioxidant enzymes present in prokaryotes and eukaryotes. They control signal transduction, apoptosis, HIV infection, etc.

peroxisomal 3-oxoacylcoenzyme A thiolase deficiency (pseudo-Zellweger syndrome) Autosomal-recessive disease assigned to human chromosome 3p23-p22. *See* adrenoleukodystrophy; microbody; Zellweger syndrome.

peroxisome 0.15–0.5- μm-diameter bodies in eukaryotic cells containing oxidase and catalase enzymes. The peroxisomes synthesize ether phospholipids by dihydroxyacetone-phosphate acetyltransferase (DHAPAT, human chromosome 1q42) and alkyl dihydroxyacetone-phosphate synthase (ADHAPS, 2q31). In a human cell the number of peroxisomes varies from less than 100 to more than 1,000. The peroxisomes have indispensable roles in fatty acid β oxidation, phospholipid and cholesterol metabolism. Fatty acid granules are also named microbodies. The peroxisome biogenesis disorders (PBD) are recessive lethal diseases in variable forms. The most extreme form is Zellweger syndrome; Refsum disease and adrenoleukodystrophy are milder, and rhizomelic chondrodysplasia punctata involves bone defects. Peroxisome mutations have been identified in yeast (PAS). Some rodent carcinogens increase the number of peroxisomes, but in humans these agents do not appear to be carcinogenic. Peroxisomal protein Pex2 controls photomorphogenesis in *Arabidopsis*. *See* adrenoleukodystrophy; chondrodysplasia; glyoxisome; microbodies; oxalosis; peroxidase and phospholipid deficiency; peroxin; peroxisomal 3-oxo-acylcoenzyme A thiolase deficiency; PPAR; Refsum disease; Zellweger syndrome. (Gould, S. J. & Valle, D. 2000. *Trends Genet*. 16:340; Sacksteder, K. A. & Gould, S. J. 2000. *Annu. Rev. Genet*. 34:623; Titorenko. V. I. & Rachubinski, R. A. 2001. *Trends Cell Biol*. 11:22; Thai, T.-P., et al. 2001. *Hum. Mol. Genet*. 10:127; Titorenko, V. I. & Rachubinski, R. A. 2001. *Nature Rev. Mol. Cell Biol*. 2:357; Purdue, P. E. & Lazarow, P. B. 2001. *Annu. Rev. Cell Dev. Biol*. 17:701; Hu, J., et al. 2002. *Science* 297:405, <www.peroxisome.org>.)

peroxynitrite (ONOO⁻/ONOOH)   Diffusion-limited product of nitric oxide with superoxide. It is strongly oxidizing and toxic. *See* nitric oxide; peroxides; superoxide.

personality   Can be characterized by five main groups of features: (1) *extraversion*, being outgoing or the lack of it, ability to lead and sell ideas versus reticent and avoiding company (heritability about 0.71); (2) *neuroticism*, emotional versus stable, worrisome or selfassured (heritability about 0.21); (3) *conscientiousness*, well organized versus impulsive, responsible or irresponsible, reliable or undependable (heritability 0.38–0.32); (4) *agreeableness*, empathic or unfriendly, warm versus cold, cooperative versus quarrelsome, forgiving versus vindictive (heritability about 0.49); (5) *openness*, insightful or lacking intelligence, imaginative versus imitative, inquisitive or superficial. These heritability estimates vary a great deal, however, and may be very different in some populations. On the basis of twin studies, several investigators concluded that overall close to 50% of the variance could be attributed to additive or nonadditive genetic determination. *See* affective disorders; behavior, humans; behavior genetics; heritability in humans; human intelligence.

perturbogen   Short peptides or protein fragments that can disrupt specific biochemical function in the cell.

pertussis toxin   Produced by *Bordetella* bacteria, responsible for whooping cough. The toxin stimulates ADP ribosylation of the Gα₁ subunit of a G protein in the presence of ARF; thus, GDP stays bound to the G protein, adenylate cyclase is not inhibited, and K⁺ ion channels do not open. As a consequence, histamine hypersensitivity and reduction of blood glucose levels follow. *See* adenylate cyclase; ADP; ARF; cholera toxin; GDP; G protein; signal transduction. (Alonso, S., et al. 2001. *Infect. Immun*. 69:6038.)

PERV   Porcine endogenous retrovirus exists in >50 copies/pig chromosome complement and has been feared to endanger humans with xenotransplantation of pig organs. So far the limited information indicates minimal risk relative to the potential benefits. PERVs can be transferred to mice by xenotransplantation. *See* nuclear transplantation; xenograft; xenotransplantation. (Specke, V., et al. 2001. *Virology* 285:177.)

PEST   Proline (P)-glutamate (E)-serine (S)-threonine (T)-rich motif is in the carboxyl domain of IκB and other proteins (Ubc) involved in the stimulation of proteolysis. *See* IκB; NF-κB; proteasome; Ubc; ubiquitin.

pest eradication by genetic means   *See Bacillus thüringiensis*; genetic sterilization; host-pathogen relations.

Bacillus thüringiensis toxin transgene is lethal to worms (right), but the wild-type plants (left) were destroyed. (Courtesy of Professor Marc Van Montagu, Rijksuniversiteit Gent.)

pesticide mutagens   *See* environmental mutagens.

pesticin   Toxin of *Pasteurella* bacteria

pestilence   Infectious epidemic of disease.

PET   *See* tomography.

Petaflop computer   Extremely powerful supercomputer. "Peta" comes from the Latin word *peto* ("I move forward") and in computer jargon "flops" designate floating operations. This new hardware may be capable of performing 1 quadrillion flops/second, which is more than 10⁶ times the efficiency of the best desktop computers.

petals   Generally the second whorl of modified leaves from the bottom of the flower. Frequently, they are quite showy because of their anthocyanin or flavonoid pigmentation. The petal number is a taxonomic characteristic, although petal number may be altered by homeotic mutations converting the anthers and/or pistils into petals and appearing as sterile double flowers of floricultural advantage. *See* flower differentiation; flower pigments; homeotic mutants. (Roeder. A. H. K. & Yanofsky, M. F. 2001. *Developmental Cell* 1:4.)

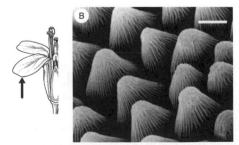

Petals of *Arabidopsis* are shown at left. At right: the adaxial ridge of the petal as viewed by scanning electron micrography, which reveals the beauty that the naked eye cannot see. (From Bowman, J. L. & Smyth D. R. 1994. *Arabidopsis: An Atlas of Morphology and Development*, Bowman, J. L., editor. By permission of Springer-Verlag, New York.)

petiole   Stalk of a leaf. *See* figure next page.

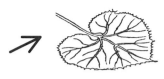

Petiole.

petite colony mutants Of yeast, form small colonies because they are deficient in respiration (OXPHOS minus) and lethal under aerobic conditions. *Vegetative petites* (ρ^-) are caused by (large) deletions in mitochondrial DNA; *segregational petites* are controlled by nuclear genes at over 200 loci. The mitochondrial mutations occur at high (0.1 to 10%) frequency. Using ethidium bromide as a mutagen, their frequency may become as high as 100%. The mitochondrial petites fail to transmit this character in crosses with the wild type except one special group, the *suppressive petites*, which may be transmitted at a low frequency in outcrosses with the wild type. In yeast, the A+T content of the normal mitochondrial DNA is about 83%; in some of the mitochondrial mutants, the A+T content may reach 96% because the coding sequences are lost and only the redundant A+T sequences are retained and amplified, so the mtDNA content is not reduced. *Hypersuppressive petite* mutants have short (400−900 bp) repeats that share 300 bp (*ori* and *rep*) sequences with the wild type that are necessary for replication. *Neutral petites* produce wild-type progeny when outcrossed to the wild type. Yeast cells that have normal mitochondrial function make large colonies and are called *grande*. Cells that can dispense with mitochondrial functions are sometimes called petite-positive, whereas those that absolutely need mitochondria are called petite-negative. Inactivation of an ATP and metal-dependent protease (Yme1p) associated with the inner membrane of the mitochondrion can convert the positives to negatives, and the presence of the Yme1p function may have the opposite effect. Yme is not universally present in all yeasts. *See* mitochondria; mitochondrial mutations; mtDNA; oxidative phosphorylation. (Sager, R. 1972. *Cytoplasmic Genes and Organelles*. Acad. Press, New York; MacAlpine, D. M., et al. 2001. *EMBO J.* 20:1807.)

Petri plate Flat glass or disposable plastic culture dish for microbes or eukaryotic cells.

Petunia hybrida ($2n = 28$) *Solanaceae*; predominantly self-pollinating, but allogamy also occurs. It has been used extensively for cell, protoplast, and embryo culture, intergeneric and interspecific cell fusion, genetic transformation, and genetic control of pigment biosynthesis. It has a good number of related species.

Peutz-Jeghers syndrome *See* polyposis hamartomatous.

PEV Position effect variegation *See* heterochromatin; position effect; RPD3.

pexophagy Sequestration to and engulfing of peroxisomes into vesicles. *See* peroxisome.

Peyer's patches Aggregated lymphatic nodes. Peyer's patches mediate the uptake of macromolecules, antigens, and microorganisms through the gut epithelium. These plaques are instruments of mucosal immunity. B cells are required for the normal functions of Peyer's patches. *Salmonella typhi* infection may cause perforation of Peyer's patches. *See* mucosal immunity.

PFAM is a database of over 3,000 protein families and domains, multiple sequence alignments and profile hidden Markov models (Bateman, A., et al. 2002. *Nucleic Acids Res.* 30:276, <http://pfam.wustl.edu>).

Pfeiffer syndrome Includes autosomal-dominant bone malformation affecting the head, thumbs, and toes (acrocephalosyndactyly), the autosomal-recessive headbone (craniostenosis), and heart disease. The origin is primarily paternal. The latter type seems to cosegregate with fibroblast growth factor receptor 1 (FGFR1) in human chromosome 8p11.2-p11.1. Another locus in chromosome 10q26 represents a fibroblast growth factor receptor, FGFR2. FGFR3 is located in chromosome 4p16.3 and its mutation is concerned with hypochondroplasia. Mutations in all three genes involve Pro→Arg replacements at identical sites, 253. This syndrome is allelic to Crouzon and Jackson-Weiss syndromes. Some of the mutations represent gain of functions. *See* achondroplasia; Alpert syndrome; craniosynostosis syndromes; Crouzon syndrome; fibroblast growth factor; gain-of- function; hypochondroplasia; Jackson-Weiss syndrome; receptor tyrosine kinase.

PfEMP1 Group of *Plasmodium falciparum* protein ligands expressed on the surface of infected red blood cells. They mediate cell adhesion (virulence factors) but may incite host immune reaction. PfEMP displays antigenic variation to evade this response. Other pathogenesis proteins of the parasite are rifins. *See* antigenic variation; Plasmodium; rifin. (Flick, K., et al. 2001. *Science* 293:2009.)

PFGE Pulsed-field gel electrophoresis separates very large nucleic acid fragments or even small chromosomes. The megabase-size fragments can be used for physical mapping of large chromosomal domains (PFG mapping). *See* pulsed-field gel electrophoresis.

pfu Plaque-forming unit. The number of phage particles/mL that can invade a bacterial lawn, then after reaching about 10^7 particle numbers, a clear spot appears on the petri plate where the bacterial cells have been lysed. *See* plaque.

PG *See* prostaglandins.

PGA Phosphoglyceric acid, a 3-carbon product of photosynthesis. *See* Calvin cycle; C3 plants; photosynthesis.

PGC Gynogenetic/parthenogenetic cell, the primordial cell of female gonads. *See* gonad; gynogenesis; parthenogenesis.

PGC1 (PPAR-γ-coactivator) Regulator of transcription, body heat production, mitochondrial biogenesis, and other processes. *See* nuclear receptor; PPAR. (Tsukiyama-Kohara, K., et al. 2001. *Nature Med.* 7:1102.)

PGD Preimplantation genetic diagnosis can be carried out for some human genetic disorders by, e.g., PCR or FISH (for fragile X, aneuploidy, etc.) at the stage of a few cells in the embryo. The disorders that have been identified by PCR include cystic fibrosis, Tay-Sachs disease, Lesh-Nyhan syndrome, Huntington chorea, Marfan syndrome, ornithine transcarbamylase deficiency, Fanconi anemia, etc. By using *in vitro* fertilization, the diagnosis may permit development of a human offspring of a certain (disease-free) genetic constitution. *See* ART; FISH; PCR. (Bickerstaff, H., et al. 2001. *Hum. Fert.* 4[1]:24; Simpson, J. L. 2001. *Mol. Cell. Endocrinol.* 183 Suppl. 1:S69; Findlay, I., et al. *Mol. Cell. Endocrinol.* 183 Suppl. 1:S5.)

PGK-neo Commonly used transformation cassette for gene knockout where the neomycinphosphotransferase gene (*neo*) is fused to the phosphoglycerate kinase (PGK) promoter. *See* knockout; vector cassette. (Scacheri, P. C., et al. 2001. *Genesis* 30[4]:259.)

P-glycoprotein 170 kDa product of the human multidrug resistance gene (MDR-1, 7q21.1) that exports different (mainly hydrophobic) toxic substances from the cells in an ATP-dependent manner. *See* multidrug resistance.

PGM *See* phosphoglucomutase.

P granule Serologically definable elements in the cytoplasm of animal cells at fertilization that segregate to the posterior part of the embryo where stem cell determination takes place. During embryogenesis, the P granules (RNA) may segregate asymmetrically into the blastomeres that produce the germline. (Harris, A. N. & MacDonald, P. M. 2001. *Development* 128:2823.)

PgtB Bacterial kinase that phosphorylates regulator protein PgtA. *See* kinase; protein kinases.

pH −log (H$_3$O$^+$), negative logarithm of hydronium ion concentration; pure water at 25°C contains 10^{-7} mole hydronium ions; solutions of acids could contain 1 mole and solution of bases 10^{-13} moles per liter. The pH meters measure the electrical property of solutions proportional to pH. pH 7 is neutral; below pH 7 is acidic; above pH 7 is alkalic (basic). The pH of body fluids and tissues is regulated by the function of the ion channels. *See* ion channels.

PH Pleckstrin homology domain. *See* pleckstrin.

PHA Phytohemagglutinin, a lectin of bean (*Phaseolus vulgaris*) plants; agglutinates erythrocytes and activates T lymphocytes. *See* agglutination; erythrocyte; lectins; lymphocytes.

Phaeochromocytoma *See* pheochromocytoma.

phaeomelanin Mammalian pigment. *See* pigmentation of animals.

phage (bacteriophage) Virus of bacteria. *See* bacteriophages; development; phage life cycle.

phage conversion Acquisition of new properties by the bacterial cell after infection by a temperate phage. *See* temperate phage.

phage cross *See* rounds of matings.

phage display Filamentous bacteriophages (M13, fd) have a few copies (3–5) of the protein III gene at the end of the particles. This protein controls phage assembly and adsorption to the bacterial pilus. When short DNA sequences are inserted into gene III (g3p), the encoded protein may be displayed on the surface of the particles. In case variable-region fragments of antibody genes are inserted into the protein III coding sequences, specific antigens may be screened. The peptides can be separated with antibody affinity chromatography (panning). Enormous arrays of recombinant libraries become available by repeated screening. The g3p product and the Fvs (fragments of variability) can be separated proteolytically or by inserting a stop codon between g3p and Fv. A huge combinatorial library of soluble epitopes may be generated by the insertion of a large array of nucleotide sequences. The specificity of the antibodies can be further manipulated by mutation (random or targeted), error-prone polymerase chain reactions, recombination, chain shuffling, i.e., trying out various light- and heavy-chain combinations, synthetic CDR sequences, etc. Similarly, a variety of different antigens may be displayed on the surface of protein III and can be used to screen for cognate antibody. Phage display may be of applied significance for the pharmaceutical industry because extremely large numbers of variants (up to 10^8 to 10^{10}) of monoclonal antibodies can be selectively isolated and tested. For in vitro testing, the two-hybrid method may be employed. The protein-protein interaction may then be studied in mammalian cells and screening techniques can be developed to isolate the cells that can neutralize the cytotoxic virus. The use of phagemid vectors may enhance the efficiency of the procedure. This procedure may facilitate the isolation of novel receptors, ligands, antibodies, anticancer reagents, transport proteins, signal transduction molecules, transcription factors, etc. Phage display technique may substitute for the construction of hybridomas. It can also be used for typing blood, for various diagnostic procedures, etc. A T7 phage display system permits the selection of RNA-binding regulatory proteins. *See* affinity chromatography; antibody engineering; CDR; combinatorial library; epitope screening; filamentous phages; hybridoma;

Protein III ↓   Cognate Antigen

Antibody fragment

monoclonal antibody; mRNA display; phagemid; pilus; two-hybrid methods. (Smith, G. P. & Petrenko, V. A. 1997. *Chem. Rev.* 97:391; Danner, S. & Belasco, J. G. 2001. *Proc. Natl. Acad. Sci. USA* 98:12954; Arap, W., et al. 2002. *Nature Med.* 8:121; <http://www.sgi.sscc.ru/mgs/gnw/aspd>.)

phage ghost   Empty protein shell of the virus.

phage immunity   Lysogenic bacterium that carries a prophage cannot be infected by another phage of the same type. *See* prophage; zygotic induction.

phage induction   Stimulates the prophage to leave a site in the bacterial chromosome and become vegetative. Physical and chemical agents may be inducive. *See* mutagens; UV light; zygotic induction.

phage life cycle   See diagram below. (Böhm, J., et al. 2001. *Curr. Biol.* 11:1168.)

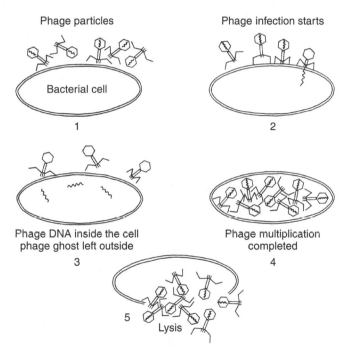

Phage particles

Phage infection starts

Bacterial cell

1

2

Phage DNA inside the cell
phage ghost left outside
3

Phage multiplication
completed
4

5

Lysis

Phage life cycle (Redrawn after the illustration provided by Drs. Simon, L. D. & Anderson, T. F. Institute of Cancer Research, Philadelphia, PA.)

phagemids   Genetic vectors that generally contain the ColE1 origin of replication and one or more selectable markers from a plasmid and a major intergenic copy of a filamentous phage (M13, fd1). When cells carrying such a combination are superinfected by a filamentous phage, they triggers a rolling circle-type replication of the vector DNA. This single-stranded product is used for sequencing by the Sanger-type DNA sequencing system, for oligonucleotide-directed mutagenesis, and as strand-specific probes. The phagemids can carry up to 10 kb passenger DNA. Their replication is fast (in the presence of a helper); they can produce up to 10^{11} plaque-forming units (pfu)/mL bacterial culture. Their stability is comparable to conventional plasmids. They obviate subcloning

the DNA fragments from plasmid to filamentous phage. The most widely used phagemids contain parts of phage M13 and pUC, πVX^c, and pBR322 vectors. *See* DNA sequencing; pfu; phasmid; plasmovirus; pUC; vectors. (Sambrook, J., et al. 1989. *Molecular Cloning*. Cold Spring Harbor Lab. Press, Cold Spring Harbor, NY; Connell, D., et al. 2002. *J. Mol Biol.* 321:49.)

phage morphogenesis   *See* development; one-step growth; phage life cycle.

phage mosaic   May be generated by phage display expressing different molecular structures on the surface of a filamentous phage. *See* phage display.

phage therapy   Bacteriophages are bacteria-eating systems. D'Hérelle, the discoverer of phages, attempted therapeutic use in poultry as well as for humans with considerable success. With the discovery of antibiotics, the interest in phage therapy ebbed. Another cause of the decline of interest was the discovery of phage resistance in bacteria. Also, bacteria encode restriction/modification systems of defense. The human body may react immunologically against the phages. Despite some technical problems, studies indicate feasibility of this type of therapy. (Summers, W. C. 2001. *Annu. Rev. Microbiol.* 55:437; Schuch, B., et al. 2002. *Nature* 418:884.)

phagocytosis   Special cell (phagocyte) engulfs a foreign particle microorganism (cell debris) and eventually exposes it to lysosomal enzymes for the purpose of destroying it. In lower animals, this mechanism substitutes for the immune system. The phagocytosis pathway is controlled by a battery of *Ced* genes (and homologues, e.g., Dock180 in humans) during apoptosis. The CD14 human glycoprotein on the surface of macrophages recognizes and clears apoptotic cells. The major phagocyte receptors are CR3 (binds opsonized C3bi complement fraction) and the Fc gamma receptor, FcγR (binds immunoglobulin G). Both processes require reorganization of the cytoskeleton under the control of RAC or RHO G proteins, respectively. *See* antibody; apoptosis; complement; lysosomes; macrophage; opsonins; pinocytosis; RAC; RHO. (Underhill, D. M. & Ozinsky, A. 2002. *Annu. Rev. Immunol.* 20:825.)

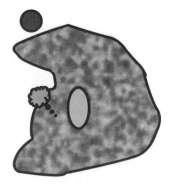

Phagocytosis.

phagosome   Body (vesicle) surrounded by a phagocyte membrane. *See* endocytosis; phagocytosis.

phakatomoses Diseases that display spots on the body, such as neurofibromatosis, tuberous sclerosis, von Hippel–Lindau syndrome, nevoid basal cell carcinoma, Cowden disease, Peutz-Jeghers syndrome, and polyposes. (Tucker, M., et al. 2000. *J. Natl. Cancer Inst.* 92:530.)

phalanges Three bones in fingers and toes (at left) with the metacarpal bone

Metacarpus.

Metacarpus.

phalloidin Amanotoxin, similar to but faster in action than amanitin. When labeled with fluorescent coumarin phenyl isothiocyanate, it is suitable to identify filamentous actin in the cells. It is extremely toxic. *See* α-amanitin; amatoxins. (Vetter, J. 1998. *Toxicon* 36:13.)

phallus Penis; the symbol of generative power; the fetal anlage of the penis and clitoris. *See* anlage; clitoris; penis.

phantom mutation Artifact of DNA sequencing; can be filtered out by statistical procedures. (Bandelt, H.-J., et al. 2002. *Am. J. Hum. Genet.* 71:1150.)

pharate Larva/adult emerging from the puparium.

pharmaceutical Chemical agent used for medical purposes. Data collected on 352 marketed drugs (excluding anticancer agents, nucleosides, steroids, and peptide-based formulations, which are known to affect DNA), 101 (28.7%) had at least one positive indication for genotoxicity. Four types of tests were used: bacterial mutagenesis, in vitro cytogenetics, in vivo cytogenetics, and mouse lymphoma assay. One must keep in mind that carcinogenicity may involve routes not testable by these methods. Also, the laboratory assays are not 100% reliable. *See* bioassays in genetic toxicology; combinatorial chemistry; genotoxic chemicals. (Snyder, R. D. & Green, J. W. 2001. *Mutation Res.* 488:151.)

pharmacogenetics Study of the reaction of individuals of different genetic constitutions to various drugs and medicines. Most of the differences are monogenic. Polymorphic genes frequently determine drug metabolism, drug transporters, and drug responses of the body. Pharmacogenetics is also the study of simultaneous drug responses by many genes. On the basis of these responses, drugs with selective effects can be developed. (Roses, A. D. 2001. *Hum. Mol. Genet.* 10:2261; Kuehl, P., et al. 2001. *Nature Genet.* 27:383; Roses, A. D. 2002. *Nature Rev. Drug Discovery* 1:541; <http://www.pharmgkb.org/>; key therapeutic targets in proteins and nucleic acids: <http://xin.cz3.nus.edu.sg/group/ttd/ttd.asp>.)

pharmacokinetics Study of the absorption, tissue distribution, metabolism, and elimination (ADAME) as a function of time of biologically relevant molecules.

pharming Production of pharmacologically useful compounds by transgenic organisms. *See* transgenic.

PHAS-1 Heat-stable protein ($M_r \approx 12,400$); when it is not phosphorylated, it binds to peptide initiation factor eIF-4E and inhibits protein synthesis. Its Ser^{64} site is readily phosphorylated by MAP, then no longer binds to eIF-4E, and protein synthesis may be stimulated. *See* eIF-4E; MAP.

phase-contrast microscope Alters the phase of light passing through and around the objects, permitting visualization without fixation and/or staining. *See* confocal microscopy; electron microscopy; fluorescene microscopy; microscopy, light; Nomarski.

phaseolin ($C_{20} H_{18} O_4$) Antifungal globulin in bean (*Phaseolus*).

phase variation Programmed rearrangement in several genetic systems. The flagellin genes of the bacterium *Salmonella* display it at frequencies of 10^{-3} to 10^{-5}. The flagellar protein has two forms: H1 and H2. The *H1* gene is a passive element. When *H2* is expressed, no H1 protein is made. When *H2* is switched off, the H1 antigen is made. The expression of *H2* is regulated by the expression of the *rh1* repressor (repressing the synthesis of the H1 protein) and the promoter of *H2*. This promoter is about 100 bp upstream from the gene and it is liable to inversion, then *H2* and *rh1* are turned off. Such an event switches on the synthesis of H1 protein. Reversing the inversion flip switches back to H2. Hin recombinase, which is very similar to the invertases or recombinases of phage Mu or Cin from phage P1, catalyzes the inversions. They can functionally substitute for each other. Hin binds to the *hixL* and *hixR* recombination sites. Additional genes are also involved in fine-tuning. Somewhat similar mechanisms control the host-specificity genes of phage Mu and the mating type of budding yeast. *See* antigenic variation; cassette model; flagellin; mating type determination in yeasts; regulation of gene activity; *Trypanosoma*. (Hughes, K. T., et al. 1988. *Genes Dev.* 2:937; Snyder, L. A., et al. 2001. *Microbiology* 147:2321.)

phasing codon Initiates translation (such as AUG) and determines the reading frame. *See* genetic code; reading frame.

phasmid (phage-plasmid) Plasmid vector equipped with the *att* site of the lambda phage, thus enabling the plasmid to participate in site-specific recombination with the λ-genome, resulting in incorporation of plasmid sequences into the phage (*lifting*). Because it contains both λ and plasmid origins of replication, it may be replicated either as a plasmid or as λ. *See* lambda phage; phagemid; vectors. (Briani, F., et al. 2001. *Plasmid* 45:1.)

Ph1 chromosome *See* Philadelphia chromosome.

phenacetin ($C_{10}H_{13}NO_2$) Analgesic and antipyretic drug, as well as a carcinogen.

phene Observable trait that may or may not have direct genetic determination. *See* gene.

phenetics Taxonomic classification on the basis of phenotypes.

phenocopy Phenotypic change that mimics the expression of a mutation. *See* epigenetic; epimutation; morphosis; phenotype.

phenodeviate Individual of unknown genetic constitution displaying a phenotype attributed to various genic combinations within the population.

phenogenetics Attempts to correlate the function of genes with phenotypes.

phenogram *See* character matrix.

phenolics Compounds containing a phenol ring such as acetosyringone, hydroxyacetosyringone, chalcone derivatives, phenylpropanoids, and some phytoalexins. These compounds may excite or suppress the *vir* gene cascade of *Agrobacterium* and may affect the response to plant pathogenic agents. Capsaicin, ginger, and resveratrol may be anticarcinogens due to their antioxidative properties and may promote apoptosis *See* acetosyringone; *Agrobacterium*; apoptosis; capsaicin; chalcones; ginger; phytoalexins; rasveratrol; virulence genes of *Agrobacterium*, wound response. (Nicholson, R. L. & Hammerschmidt, R. E. 1992. *Annu. Rev. Plant Path.* 30:369.)

phenology Study of the effects of the environment on live organisms.

phenome Group of organisms with shared phenotypes.

phenomenology Description of facts as observed without metaphysical interpretation; concept that behavior depends on how a person interprets reality rather than what is the objective reality.

phenotype Appearance of an organism that may or may not represent the genetic constitution. The 1014 human (disease) genes displayed 1,429 distinguishable phenotypes (~141%).

phenotype microarrays Automated analysis of phenotypic expression of hundreds of genes on microplates can be used to monitor the consequences of knockouts or other genetic alterations. *See* knockout; microarray hybridization. (Bochner, B. R., et al. 2001. *Genome Res.* 11:1246.)

phenotypic assortment *See* macronucleus.

phenotypic knockout Somatic gene therapy that neutralizes intracellular harmful mechanisms. *See* gene therapy; knockout.

phenotypic lag A period of time may be required for gene expression after transformation or mutagenic treatment. *See* premutation; transformation, genetic. (Ryan, F. J. 1955. *Am. Nat.* 89:159.)

phenotypic mixing Mixed assembly of viral nucleic acids and proteins upon simultaneous infections by different types of viruses. Therefore, the coat protein properties of the virions do not match the viral genotype by serological or other tests. (Hayes, W. 1965. *The Genetics of Bacteria and Their Viruses.* Wiley, New York.)

phenotypic plasticity Adaptive property of an organism enabling it to take advantage of local conditions without evolving a particular function and the expense of another function. *See* canalization; homeostasis, genetic.

phenotypic reversion Apparent restoration of the normal expression of a mutant gene; it is not inherited, however. Aminoglycoside antibiotics (paromomycin, geneticin, etc.) may successfully compete with the translation termination factors in eukaryotes in cases when the mRNA carries a stop-codon mutation. As a result, some of the polypeptide chains are not terminated/truncated but completed in the presence of the drug. Transposable elements may also alter gene function without causing mutation in the gene. Phenotypic reversion may be exploited for correcting the genetic defects in some diseases. *See* aminoglycosides; G418; paromomycin; phenocopy; suppressor tRNA; translation termination. (Gause, M., et al. 1996. *Mol. Gen. Genet.* 253:370; Franzoni, M. G. & De Castro-Prado, M. A. 2000. *Biol. Res.* 33:11; Biedler, J. L., et al. 1975. *J. Natl Cancer Inst.* 55:671.)

phenotypic sex May not reflect the expectation based on the sex-chromosomal constitution. *See* hermaphrodite; testicular feminization.

phenotypic stability factor Measure of developmental homeostasis calculated by the ratios of the quantitative expression of a parameter (gene) under two different environmental conditions. *See* homeostasis; logarithmic stability factor. (Lewis, D. 1954. *Heredity* 8:334.)

phenotypic suppression Apparently normal but nonhereditary phenotype brought about by translational error due to environmental effects and/or drugs. *See* error in translation.

phenotypic value In quantitative genetics, the mean value of a population regarding the trait under study. It is generally represented as the P value. *See* breeding value.

phenotypic variance *See* genetic variance.

phenylalanine ($C_9H_{11}NO_2$) Essential water-soluble, aromatic amino acid (MW 165.19). Its biosynthetic path (*with enzymes involved in parentheses*): Chorismate → (*chorismate mutase*) > Prephenate → (*prephenate dihydratase*) → Phenylpyruvate → (*aminotransferase*, glutamate NH_3 donor) → Phenylalanine. *See* chorismate; phenylketonuria; tyrosine.

phenylalanine ammonia LYASE (PAL) Deaminates phenylalanine into cinnamic acid; thus, it is involved in the synthesis of plant phenolics. *See* phenolics.

phenylalanine hydroxylase (PAH, 12q24.1) Its deficiency leads to phenylketonuria.

phenylhydrazine ($C_6H_9ClN_2$) Hemolytic compound, but it is also used as a reagent for sugars, aldehydes, ketones, and a number of industrial purposes (stabilizing explosives, dyes, etc.). *See* hemolysis.

phenylketonuria (PKU, PAH) Gene was located to human chromosome 12q24.1. It is a recessive disorder that has a prevalence of about 1×10^{-4} (carrier frequency is about 0.02) in white populations. Thus, an affected person has about 0.01 chance of having an affected child in case of a random mate, but the recurrence rate in a family where one of the partners is affected and the other is a carrier it is nearly 0.5.

Mentally retarded heterozygous children of a phenylketonuric mother (indirect epistasis). (Courtesy of Dr. C. Charlton Mabry, 1963; by permission of the *New England Journal of Medicine* 269:1404.)

Its incidence is substantially lower among Asian and black people (one-third of that in whites). PKU is more frequent in populations of Celtic origin than in other Europeans. This has been interpreted as the result of natural selection because PKU heterozygosity conveys some tolerance to the mycotoxin, ochratoxin A, produced by *Aspergillus* and *Penicillium* fungi common in humid northern regions. Before the nature of this disorder and the method of treatment were identified, about 0.5 to 1% of the patients in mental asylums were afflicted by PKU. A deficiency of the enzyme phenylalanine hydroxylase and consequently the accumulation of phenyl pyruvic acid and a deficiency of tyrosine cause the disease:

PHENYLPYRUVIC ACID $\rightleftarrows$ PHENYLALANINE $\Rightarrow$ TYROSINE

For the identification of the condition, the Guthrie test has been used: β-2-thienylalanine was added to cultures of *Bacillus subtilis* containing blood of the patients. This phenylalanine analog is a competitive inhibitor of tyrosine synthesis. In the presence of excess amounts of phenylalanine, the bacterial growth does not stop, however. Since the genetically determined defect in the enzyme varies in different families, so does the severity of the clinical symptoms. The accumulation of phenyl pyruvic acid is apparently responsible for mental retardation and the musty odor of the urine of patients.

The reduced amount of tyrosine prevents normal pigmentation (melanin) and thus results in pale color. Relative normalcy can be established if the condition is diagnosed early and

dietary restrictions for phenylalanine are implemented. The restriction of phenylalanine must start as early as possible (before birth if feasible) and continue until at least age 10 and before pregnancy, during pregnancy, and during breastfeeding or possibly for life to avoid harm to the nervous system. Phenylketonuria of the mother may damage the nervous system of a genetically normal fetus through placental transfer (indirect epistasis). Because of the multiple metabolic pathways involving phenylpyruvic acid, besides the deficiency of phenylalanine hydroxylase, other genes and conditions may cause similar clinical symptoms. Phenylalanine hydroxylase activity requires the availability of the reduced form of the cofactor 5,6,7,8-tetrahydro-biopterin that is made by the enzyme *dihydrobiopterin reductase* from 7,8-dihydrobiopterin. The dehydrobiopterin reductase enzyme is coded in human chromosome 4p15.1-p16.1. A defect in this enzyme also causes phenylketonuria symptoms, but lowering the level of phenylalanine in the diet does not alleviate the problems. Another form of phenylketonuria is based on a deficiency in dihydrobiopterin synthesis. Prenatal diagnosis can be carried out by several methods. Mutations at various related metabolic sites in mice may serve as a model for studying phenylketonuria. *See* alkaptonuria; amino acid metabolism; epistasis; genetic screening; Guthrie test; hyperphenylalaninemia; mental retardation; one gene−one enzyme theorem; phenylalanine; prenatal diagnosis; tyrosinemia. (Ledley, F. D., et al. 1986. *New Eng. J. Med.* 314:1276; Gjetting, T., et al. 2001. *Am. J. Hum. Genet.* 68:1353.)

phenylpropanoid *See* phenolics; phytoalexins.

phenylthiocarbamide tasting (PTC) The major incompletely dominant gene appears to be in human chromosome 7. About 30% of North American whites and about 8−10% of blacks cannot taste the bitterness of this compound. Persons affected by thyroid deficiency (athyreotic) cretenism (mental deficiency) are nontasters. Phenylthiocarbamide (phenylthiourea) has been used for classroom demonstration of human diversity, but it should be kept in mind that it is a toxic compound (LD_{50} oral dose for rats 3 mg/kg and for mice 10 mg/kg). *See* LD_{50}. (Guo, S. W. & Reed, D. R. 2001. *Ann. Hum. Biol.* 28[2]:111.)

phenylthiourea *See* phenylthiocarbamide tasting.

pheochromocytoma Bladder-kidney carcinoma overproducing adrenaline and noradrenaline. The disease may be caused by a mutation in the von Hippel−Lindau gene, neurofibromatosis 1, RET protooncogenes, or multiple endocrine neoplasia gene MEN2. Mutations in the subunits of mitochondrial succinate dehydrogenase in the long arm of human chromosome 11 also may be involved. *See* adenomatosis; animal hormones; endocrine neoplasia; MEN; neurofibromatosis; paraganglion; RET; SHC; succinate dehydrogenase; von Hippel−Lindau syndrome. (Astuti, D., et al. 2001. *Am. Hum. Genet.* 69:49; Maher, E. R. & Eng, C. 2002. *Hum. Mol. Genet.* 11:2347.)

pheresis Medical procedure of blood withdrawal. After fractionation, some fractions are reintroduced. Such a protocol may use stem cells, transfect them with a vector, or apply them to chemotherapy, and eventually place them back in the body of the same individual.

pheromones Various chemical substances secreted by animals and cells for the purpose of signaling, thus generating certain responses by members of the species, such as sex attractants, stimulants, territorial markers, or other behavioral signals and cues. The pheromones and their receptors in rodents are controlled by about 100 genes and the signals are transmitted through G-protein-associated pathways. The role of pheromones in humans may not be generally agreed on. The pheromone receptor genes in rodents include *V1r* and *V2r*; apparently there are no human homologues. *See* mating type determination in yeast; olfactogenetics; vomeronasal organ. (Kohl, J. V., et al. 2001. *Neuroendocrinol. Lett.* 22[5]:309).

Ph gene (pairing high) Approximately 700 Mb sequence that controls selective pairing in hexaploid wheat. In its presence homoeologous chromosomes do not pair. It is in chromosome 5B. Plants nullisomic for this chromosome display multivalent associations in meiosis. A similar gene, Ph2, is in chromosome 3D. Additional less powerful genes also regulate chromosome pairing. *See* homoeologous; nullisomic; Triticum. (Sears, E. R. 1969. *Annu. Rev. Genet.* 3:451; Martinez-Perez, E. 2001. *Nature* 411:204.)

φ (phi) Symbol of some phages.

phialide Fungal stem cells from which conidia are budded.

Philadelphia chromosome The long arm (q34) of human chromosome 9, carrying the *c-abl* oncogene, is translocated to the long arm (q11) of chromosome 22, carrying site *bcr* (breakpoint cluster region). The *bcr-abl* gene fusion is then responsible for myelogenous (Abelson) and acute leukemia as a consequence of the translocation and fusion. In the acute form, a 7.5 kb mRNA is translated into protein p190, and in the myelogenous form an 8.5 kb mRNA is translated into a chimeric protein p210. The fusion protein is a deregulated tyrosine kinase acting on hematopoietic cells. It causes leukemia-like oncogenic transformation in mice. Synthetic antisense phosphorothioate oligonucleotides ([S]ODN) complementary to the 2nd exon of BCR or to the 3rd exon of the ABL of the fused genes temporarily blocked the proliferation of chronic leukemic cells without harming the normal cells in a mouse model. The outcome of such a therapy could be improved by simultaneously targeting the c-Myc oncogene with an antisense construct. Further effectiveness was observed by exposing the cells to a low concentration of mafosfamide, an antineoplastic drug that promotes apoptosis, or to cyclophosphamide. *See* ABL; antisense technologies; apoptosis; BCR; cancer gene therapy; cyclophosphamide; hematopoiesis; leukemia; transresponder. (Saglio, G., et al. 2002. *Proc. Natl. acad. Sci. USA* 99:9882.)

phlebotomous Blood-sucking (insect) or bloodletting surgical procedure.

phloem Plant tissue involved in the transport of nutrients; it contains sieve tubes and companion cells, phloem parenchyma, and fibers. *See* parenchyma; root; sieve tube.

phlorizin Dichalcone in the bark of trees (Rosaceae); it blocks the reabsorption of glucose by the tubules of the kidney and causes glucosuria. *See* chalcones; disaccharide intolerance.

PHO81 Yeast CDK inhibitor homologous to p16^{INK4}. *See* CDK; p16^{INK4}.

PHO85 Cyclin-dependent kinase of *Saccharomyces cerevisiae*. *See* CDC28; CDK; KIN28; PHO81.

phoA Gene for alkaline phosphatase.

phobias Exist in different forms, all characterized by unreasonable avoidance of objects, events, or people. It may amount to seriously morbid mental illness. Apparently a duplication in human chromosome 15q24-q26 is associated with one form and with laxity of the joints. (Gratacós, M., et al. 2001. *Cell* 106:367.)

phocomelia Absence of some bones of the limbs proximal to the body. It may occur as a teratological effect of various recessive and dominant human genetic defects or as a consequence of teratogenic drugs, e.g., thalidomide use during human or primate pregnancy. *See* limb defects in humans; Roberts syndrome; teratogen; thalidomide.

PHOGE Pulsed homogeneous orthogonal-field electrophoresis is a type of pulsed-field gel electrophoresis, within the range of 50 kb to 1 Mb DNA, permitting straight tracks of a large number of samples. *See* pulsed-field gel electrophoresis.

PhoQ *Salmonella* kinase, affecting regulator of virulence PhoP. *See* *Salmonella*; virulence.

PhoR Phosphate assimilation regulated by PhoR kinase upon phosphorylation of regulator PhoB.

phorbol ester Facilitator of tumorous growth by activating protein kinase C. *See* carcinogen; PMA; procarcinogen; protein kinases; TPA.

phorbol 12-myristate-13-acetate (PMA) *See* phorbol ester; PMA.

phosphatases In animals, both acid and alkaline phosphatases are common; in plants, acid phosphatases are found. Some of the phosphatases have high specificities and have an indispensable role in energy release in the cells. A series of nonspecific phosphatases carry out only digestive tasks. In humans, the erythrocyte- and fibroblast-expressed acid phosphatase (ACP1) isozymes are coded in chromosomes 2 and 4. It has been suggested that these enzymes split flavin mononucleotide phosphates. In megaloblastic anemia, the ACP1 level is increased. Tartrate-resistant acid phosphatase type 5 (TR-AP) is an iron glycoprotein of 34 kDa (human chromosome

15q22-q26), and it is increased in the spleen in Gaucher disease. Lysosomal acid phosphatase (ACP2) is in human chromosome 11p12-p11. Alkaline phosphatase (ALPL) is present in the liver, bone, kidney, and fibroblast. It is often called the nontissue-specific phosphatase (human chromosome 1p36-p34) and is deficient in hypophosphatasias. The alkaline phosphatase ALPP is in the placenta (human chromosome 2q37), and several allelic forms have been identified. A similar alkaline phosphatase is present in the testes and the thymus; the gene is also at the same chromosomal location, but its expression is highly tissue-specific. *See* dual-specificity phosphatase; Gaucher's disease; hypophosphatasia; hypophosphatemia; megaloblastic anemia; serine/threonine and tyrosine protein phosphatases.

phosphatidylinositol (1,2-diacyl-sn-glycero-3-phospho[1-o-myoinositol]) Cell membrane phospholipid. Phosphatidylinositol transfer protein is required for vesicle budding from the Golgi complex. Phosphatidylinositol-3,4,5-trisphosphate activates protein kinase B. *See* Golgi apparatus; phosphoinositides; PIK; PKB; pleckstrin domain; PTEN. (Bourette, R. P., et al. 1997. *EMBO J.* 16:5880; Abel, K., et al. 2001. *J. Cell Sci.* 114:2207.)

3′-phosphoadenosine-5′-phosphosulfate *See* PAPS.

phosphodiesterases Exonucleases. The snake venom phosphodiesterase starts at the 3′-OH ends of a nucleotide chain and splits off nucleoside-5′-phosphate. The 3′-phosphate terminus does not lend the nucleotide chain for its action. The spleen phosphodiesterase, on the other hand, generates nucleoside-3′-phosphate molecules by splitting on the other side of the nucleotides. Phosphodiesterase converts cyclic AMP into AMP or cGMP into GMP. *See* phosphodiester bond.

phosphodiester bond Attaches the nucleotides into a chain by hooking up the incoming 5′-phosphate ends to the 3′-hydroxy tail of the preceding nucleotide. R^1 and R^2: nucleosides; O: oxygen; H: hydrogen; P: phosphorus. *See* Watson-Crick model.

$$5'-R^1-3'-O-P\overset{\overset{\textstyle O}{\|}}{\underset{\underset{\textstyle O^-}{|}}{-O}}-5'-R^2-3'$$

phosphofructo-2-kinase/fructose-2,6-bisphosphatase (PFKFB, PFRX, Xp11.21) Bifunctional enzyme encoded in the X chromosome of humans and rodents. The PFKFB3 locus is in 10p15-p14. The PFKFB4 enzyme is in 3p22-p21.

phosphofructokinase M (glycogen storage disease VII, 12q13.3) Phosphofructokinase in the muscles (PFKM). It deficiency may cause muscle cramps and myoglobinuria. Lactate production is reduced and fatigue develops after exertion. *See* glycogen storage diseases; myoglobin.

phosphofructokinase platelet type (PFKP, 10p15.3-p15.2) Expressed in the platelet but displays 71% identity of amino acid sequence of the muscle type and 63% identity with the liver enzyme.

phosphofructokinase X (PFKX) Chromosome 2–encoded enzyme expressed in the fibroblasts and the brain.

phosphofructose kinase 1 (PFK-1, PFKL, phosphofructokinase) Enzyme catalyzes the formation of fructose-1-6-bisphosphate from fructose-6-phosphate in the presence of ATP and Mg^{2+}. PFKL (liver enzyme) is encoded in human chromosome 21q22.3. The tetrameric enzyme may exist in five different forms due to the random association of the products of two different loci.

phosphofructose kinase 2 (PFK-2) Mediates the formation of fructose-2,6-bisphosphate formation from fructose-6-phosphate. It enhances the activity of fructosephosphate 1 enzyme by binding to it and inhibits fructose-2,6-bisphosphatase; therefore, it enhances glycolysis. *See* glycolysis.

phosphoglucomutase (PGM) Enzyme that catalyzes the reaction

GLUCOSE-1-PHOSPHATE ↔ GLUCOSE-6-PHOSPHATE

Phosphoglucomutase proteins are homologous in structure through the animal kingdom. In humans there are several PGM enzymes, some with multiple allelic forms with characteristic patterns, and they are reasonably stable. Therefore, PGM is used in forensic genetics for personal identification of samples up to 6 months old. Their human chromosomal locations are PGM1 (1p31), PGM2 (4p14-q12), PGM3 (6q12), PGM5 (9p12-q12). *See* forensic genetics.

phosphogluconate oxidative pathway Pentose phosphate pathway.

3-phosphoglycerate dehydrogenase deficiency (PHGDH, 1q12) Results in recessive serine biosynthetic defect, microcephaly, and neurological defects and seizures. *See* serine.

phosphoglyceratemutase deficiency *See* glycerophospholipid; myopathy.

phosphoglyceride *See* glycerophospholipid.

phosphohexose isomerase (PHI) Catalyzes the glucose-6-phosphate ↔ fructose-6-phosphate conversions. It is encoded in human chromosome 19cen-q12. Its defects result in dominant hemolytic anemia. *See* anemia; hemolytic anemia.

phosphoinositide-3-kinase *See* phosphoinositides; PIK.

phosphoinositides Inositol-containing phospholipids. They play an important role as second messengers, and in phosphorylated/dephosphorylated forms they participate in the regulation of traffic through membranes, growth, differentiation, oncogenesis, neurotransmission, hormone action, cytoskeletal organization, platelet function, and sensory perception. The signals converge on phospholipase C (PLC, 20q12-q13.1). It hydrolyzes phosphatidylinositol-4,5-bisphosphate (PtdInsP$_2$) into inositol trisphosphate (InsP$_3$) and diacylglycerol (DAG). InsP$_3$ regulates Ca^{2+} household and DAG activates PLC. InsP$_3$ levels also regulate pronuclear migration, nuclear envelope breakdown, metaphase-anaphase transitions, and cytokinesis. Cytidine diphosphate-diacylglycerol synthase (CDS) is required for the regeneration of PtdInsP$_2$ from phosphatidic acid. CDS is a key regulator in the G-protein-coupled phototransduction pathway. Pleckstrin homology domains selectively bind phosphoinositides. *See* DAG; inositol; IP$_2$ (InsP$_2$ for formula); IP$_3$ (InsP$_3$ for formula); *myo*inositol; phosphatidylinositol; phospholipase; PIK; pleckstrin; signal transduction. (Czech, M. P. 2000. *Cell* 100:603; Vanhaesebroeck, B., et al. 2001. *Annu. Rev. Biochem.* 70:535; Sato, T. K., et al. 2001. *Science* 294:1881; De Matteis, M. A., et al. 2002. *Current Opin. Cell Biol.* 14:434.)

phosphoinositide-specific phospholipase Cδ Signal transducer. It generates the second messengers inositol-1,4,5-triphosphate and diaglycerol. *See* second messenger; signal transduction.

phospholipase (PL) A, D, C Each splits specific bonds in phospholipids. PLC-β generates diacylglycerol and phosphatidylinositol 2,4,5-triphosphate from phosphatidylinositol 4,5-bis-phosphate. These second messenger molecules play roles in signal transduction. PLC-γ is activated by receptor tyrosine kinases and one of its homologues is the SRC oncoprotein. PLA is present in mammalian inflammatory exudates. Form A2 is coded by human chromosome 12; PLA2B is coded by chromosome 1. Phospholipase C is coded in human chromosomes at the following locations: PLCB3 (11q11), PLCB4 (20p12), PLCG2 (16q24.1). *See* Ipk1; Ipk2; phosphoinositides; serine/threonine phosphoprotein phosphatases; signal transduction; SRC. (Rhee, S. G. 2001. *Annu. Rev. Biochem.* 70:281; Wang, X. 2001. *Annu. Rev. Plant Physiol. Mol. Biol.* 52:211.)

phospholipid Lipid with phosphate group(s). *See* lipids; liposome.

phosphomannomutase deficiency Rare defect of glycosylation displaying large differences in expressivity. It involves inverted nipples, fat pads, strabismus, hyporeflexia (sluggish responses), mental retardation, hypogonadism, and early death. (Grünewald, S., et al. 2001. *Am. J. Hum. Genet.* 68:347.)

phosphomonoesterase Phosphatase digesting phosphomonoesters, such as nucleotide chains. *See* phosphodiester bond.

phosphonitricin (Basta) *See* herbicides.

phosphoramidates Used in antisense technologies by modification of the sugar-phosphate backbone of oligonucleotides. *See* antisense technologies; trinucleotide-directed mutagenesis; formula above right. (Jin, Y., et al. 2001. *Bioorg. Med. Chem. Lett.* 11:2057; Faria, M., et al. 2001. *Nature Biotechnol.* 19:40.)

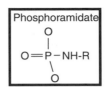

phosphoramidate/ink jotting Rapid method of DNA analysis. (Cooley, P., et al. 2001. *Methods Mol. Biol.* 170:117.)

phosphorelay *See* two-component regulatory system.

phosphorescence *See* fluorescence; luminescence.

phosphoribosylglycinamide formyltransferase Human chromosome 21q22.1 dominant. It controls purine/pyrimidine biosynthesis and folate metabolism.

phosphoribosylpyrophosphate synthetase (PRPS1, Xq22-q24) Enzyme of the purine/pyrimidine salvage pathway. Its deficiency may cause hyperuricemia (excessive amounts of uric acid in the urine), deafness, and neurological disorders.

phosphorimaging Detection of radioactive labels in tissues by phosphorescence.

phosphorolysis Glycosidic linkage holding two sugars together is attacked by inorganic phosphate and the terminal glucose is removed (from glycogen) as α-D-glucose-1-phosphate.

phosphorothioates Analogs of oligodeoxynucleotides. They are used in antisense technology. Their attachment to the 3'-end inhibits the activity of nucleases that attack RNA from that end. They can bind to proteins but do not stimulate the activity of RNase H (phosphorodithioate-modified heteroduplexes may stimulate RNase H), inhibit translation, and are taken up by cells relatively easily. Some of the effects of these molecules are not based on their antisense properties (e.g., binding to CD4, NF-κB, inhibition of cell adhesion, inhibition of receptors, etc.). Phosphorothioate-modified nucleotides (one of the oxygens attached to P is replaced by S) are used for in vitro mutagenesis to protect the template strand from nucleases while the strand to be modified is excised before resynthesis in a mutant form. *See* antisense DNA; antisense RNA; antisense technologies; CD4; NF-κB; OL(1)p53; ribonuclease H. (Sazani, P., et al. 2001. *Nucleic Acids Res.* 29:3965.)

phosphorylase b kinase Enzyme that phosphorylates two specific serine residues in *phosphorylase b*, thus converting it into *phosphorylase a* upon the action of cAMP-dependent protein kinase (protein kinase A). Phosphorylase b kinase mediates glycogen breakdown. This enzyme is a tetramer and for activation the two regulatory subunits (R) must be separated from the two catalytic subunits to be able to function. The dissociation is mediated by cAMP through A-kinase. The δ-subunit is calmodulin. *See* A-kinases; calmodulin; cAMP; cAMP-dependent protein kinase; epinephrine. (Brushia, R. J. & Walsh, D. A. 1999. *Front. Biosci.* 4:D618.)

phosphorylases (kinases) *See* A kinases; calmodulin; Jak kinase; phospholipase C; phosphorylase B; phosphorylase B kinase; serine/threonine kinases; serine/threonine phosphoprotein phosphatases; signal transduction; tyrosine protein kinase.

phosphorylation Adding phosphate to a molecule. It may play an important role in signal transduction, and depending on which of several potential sites phosphorylated the function of some transcription factors, it may be altered. *See* kinase; oxidative phosphorylation; phosphorylases. (Whitmarsh, A. J. & Davis, R. J. 2000. *Cell. Mol. Life Sci.* 57:1172.)

phosphorylation potential (Δg_p) Change in free energy within the cell after hydrolysis of ATP.

phosphoserinephosphatase Hydrolyzes O-phosphoserine into serine; it is encoded in human chromosome 7p15.1-p15.1.

photoaffinity tagging The labels may be radioactive or fluorescent and bind to certain compounds by noncovalent bonds upon illumination. *See* Knorre, D. G., et al. 1998. *FEBS Lett.* 433:9.

photoaging Skin collagens and elastin are damaged by the ultraviolet light–induced metalloproteinases, resulting in wrinkling of the skin similar to what occurs during aging. These enzymes are upregulated by AP-1 and NF-κB transcription factors. *See* aging; AP-1; collagen; elastin; NF-κB.

photoallergy Immunological response to a substance activated by light.

photoautotroph Organism that can synthesize in light from inorganic compounds all its required organic substances and energy. The majority of green plants are photoautotrophic. By introduction of a glucose transporter gene into obligate photoautotrophic alga, the organism could be converted to light-independent growth on glucose. (Zaslavskaia, L. A., et al. 2001. *Science* 292:2073.)

photochemical reaction center Site of photon absorption and initiation of electron transfer in the photosynthetic system. *See* photosynthesis.

photodynamic effect Photosensitivation, photodestruction. A dye or pigment absorbs light, converts the energy to a higher state, and exerts specific effects. Photodynamic effects may have various therapeutic applications. (Langmack, K., et al. 2001. *J. Photochem. Photobiol. B.* 60:37.)

photoelectric effect Has very wide applications of modern technology (television, computers, and other electronic instruments). Atoms may emit electrons when light hits a suitable target. When X-rays hit a target, very high-energy photoelectrons may be generated.

photogenes Chloroplast DNA-encoded proteins involved in photosynthesis. One of the most studied, *photogene 32*

(*psbA*), codes for a 32 kDa thylakoid protein involved in electron transport in photosystem II. Also, it binds the herbicide atrazine. By removing or altering this binding site, one can obtain plants resistant to the weedkiller through molecular genetic manipulations. *See* herbicides; photosynthesis. (Rodermel, S. R. & Bogorad, L. 1985. *J. Cell Biol.* 100:463.)

photography Has special requirements in the laboratory depending on the objects. Cell cultures in petri plates can be best photographed through macrolenses (for extreme closeups, use extension rings or teleconverters) and using highly sensitive color films comparable to Kodak Gold 400. To eliminate reflection, the blue photoflood lamps should be adjusted at about 45° angles. Agarose gels can be photographed with a Polaroid camera mounted on a copying stand and using high-speed (ASA 3000) films. Ultraviolet light sources of the longer wavelength are less likely to damage the DNA. The contrast can be enhanced by the use of orange filters on the camera (such as Kodak Wratten 22A). Note that ultraviolet light is dangerous to the skin and particularly to the eyes. Wear gloves, goggles, and a long-sleeve shirt. For photomicrography, built-in automatic exposure meters are very advantageous if frequently used. Otherwise, numerous exposures at the proper color temperature are necessary. For photocopying and editing, halftone image computers with (color) scanners can be used. The resolution of the digital cameras may not be satisfactory for all biological applications, although they are very convenient and the high pixel units are very powerful.

photolabeling Adding photoactivatable groups to proteins, membranes, or other cellular constituents in order to detect their reaction path. The labels are generally small molecules that are stable in the dark and highly susceptible to light without photolytic damage to the target. They are stable enough to permit analytical manipulations of the sample. Synthetic peptides containing 4'-(trifluoromethyl-diazirinyl)-phenylalanine or 4'-benzoyl-phenylalanine, etc., have been used to analyze biological structures (membranes, proteins, etc.). *See* green fluorescent protein; luciferase.

photolithography Modification of more than a century-old printing process. A solid plate is coated with a light-sensitive emulsion and overlaid by a photographic film, then it is illuminated. An image is formed after exposure of the plate to light. A similar principle has been adapted to visualizing DNA sequences for large-scale mapping, fingerprinting, and diagnostics. The process is also used for the synthesis of nucleotide probes. *See* DNA chips; microarray hybridization. (Barone, A. D., et al. 2001. *Nucleosides Nucleotides Nucleic Acids* 20[4−7]:525.)

photolyase Repair enzyme (M_r 54,000) that splits pyrimidine dimers into monomers upon absorption of blue light. In *E. coli*, two chromophores assist the process: 5,10-methenyltetrahydrofolate absorbs the photoreactivating light; 8-hydroxy-5-deazariboflavin and the energy are then transferred to $FADH_2$, although it also absorbs some energy. The excited $FADH_2^*$ transfers the energy to the dimer, and while $FADH_2$ is regenerated, the dimer splits up, the recipient member of the dimer breaks down, and monomeric pyrimidines are formed. A second cofactor, 5,10-methenyl-tetrahydrofolylpolyglutamate (MTHF), may be the light harvester. It is interesting that the blue light photoreceptor cryptochromes of plants bear substantial similarities to the bacterial photolyase; its cofactors are also the same, yet the exact role of photolyases in plant DNA repair is unclear. The cyclobutane photolyase does not split the pyrimidine-pyrimidinone (6-4) photoproducts. The 6-4 photolyases are under the control of two different genes. Topical application of photolyase and light to sunburned human skin may alleviate the symptoms by repair of the DNA damage. *See* base flipping; cryptochrome; cyclobutane ring; direct repair; DNA repair; photoreactivation; pyrimidine dimer; pyrimidinone. (Tanaka, M., et al. 2001. *Mutagenesis* 16:1; Komori, H., et al. 2001. *Proc. Natl. Acad. Sci. USA* 98:13560.)

photolysis Degradation of chemicals or cells by light.

photomixotrophic Organism that can synthesize some of its organic requirements with the aid of light energy, whereas for others it depends on supplied organic substances.

photomorphogenesis Light-dependent morphogenesis. Light affects the growth and differentiation of plant meristems (photoperiodism), plastid differentiation, and directly or indirectly many processes of plant metabolism. Certain stages in photomorphogenesis can be reached at low-intensity (fluence) illumination (or even in darkness) such as the formation of proplastids and etioplasts. Other steps, such as the full differentiation of the thylakoid system and photosynthesis-dependent processes, require a high fluence rate and critical spectral regimes (red and blue). Several genes involved in the control of plastid development have been identified in *Arabidopsis* and other plants. The *lu* mutation is normal green at low light intensity, but it is entirely bleached and dies at high light levels. Wild-type plants can make etioplasts in the dark, but the *de-etiolated (det1)*, *constitutive photomorphogenesis (cop1* and *cop9)* mutants develop chloroplasts in darkness. The Cop9 complex includes eight subunits, forming a signalosome in plants, and a homologue is also found in animals.

The *gun (genome uncoupled)* mutations grow normally in the dark but do not allow the development of etioplasts into chloroplasts. Various pale *hy (high-hypocotyl)* mutants deficient in phytochrome make light green plastids, indicating that phytochrome is not a requisite for plastid differentiation to an advanced stage. The *blu (blue light uninhibited)* class of mutants is inhibited in hypocotyl elongation by far red light. The *HY4* locus of *Arabidopsis* encodes a protein homologous to photolyases, and the recessive mutations are insensitive to blue light for hypocotyl elongation.

Mutants were identified with no response to blue light and some with a very high blue light requirement for curvature. Most of these light responses appear to be mediated by signal transduction pathways. The chlorophyll-b-free yellow-green mutants (*ch*) display chloroplast structure, appearing almost normal by electron microscopy. Several mutations defective in fatty acid biosynthesis and/or photosynthesis are rather normal in photomorphogenesis. Some mutants are resistant to high CO_2 atmosphere, and actually normal chloroplast differentiation requires high CO_2. Other mutants can be protected from bleaching only at 2% CO_2 atmosphere.

The *Arabidopsis* nuclear mutants of the *im (immutans)* type display variegation under average greenhouse illumination, but they are almost normal green under low light intensity and short daily light cycles, whereas at high-intensity continuous illumination they are almost entirely free of leaf pigments. Under the latter condition, by continuous feeding of an inhibitor or repressor of the de novo pyrimidine pathway, the leaf pigment content may increase 20-fold. In these variegated plants, the green cells have entirely normal chloroplasts, whereas the white cells lack thylakoid structure. The azauracil-treated plants display fully functional, although morphologically altered, thylakoids. An insertional mutation at the *ch-42 locus (cs)* identified a thylakoid protein essential for normal greening of the plants without abolishing cell viability. The *PRF (pleiotropic regulatory factor)* locus tagged by a T-DNA insertion controls several loci involved in photomorphogenesis. The product of the gene is a subunit of the G-protein family. The *det2, cyp90, cop, fus, dim axr2*, and *cbb* dwarf mutations develop their characteristic phenotypes because of defects in the brassinosteroid pathway. The nuclear gene *chm (chloroplast mutator)* induces a wide variety of plastid morphological changes due to extranuclear mutation (Rédei, G. P. 1973. *Mut. Res.* 18:149). *See* brassinosteroids; circadian; COP; photoperiodism; phototropism; proteasome; phytochrome; signal transduction. (Wada, M. & Kadota, A. 1989. *Annu. Rev. Plant Physiol. Plant Mol. Biol.* 40:169; von Arnim, A. & Deng, X.-W. 1996. *Annu. Rev. Plant Physiol. Plant Mol. Biol.* 47:215; Quail, P. H. 2002. *Nature Rev. Mol. Cell Biol.* 3:85.)

photon Quantum of electromagnetic radiation with zero rest mass and energy *h* times the radiation frequency. Photons are generated by collisions between atomic nuclei and electrons and other processes when electrically charged particles change momentum.

photoperiodism Response of some species of plants to the relative length of the daily light and dark periods. Besides the length of these cycles, the spectral properties and the intensity of the light are also important. Responses of the plants include the onset of flowering, vegetative growth, elongation of the internodes, seed germination, leaf abscission, etc. *Short-day, long-day*, and *day-neutral* plants are commonly distinguished on the basis of the critical day length, or in the latter category, by the lack of it.

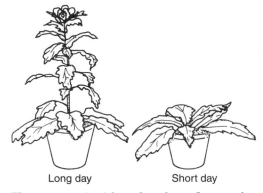

Long day   Short day

Henbane (*Hyoscyamus niger*) long-day plants flower only under long daily light periods (after appropriate cold treatment). (Courtesy of Professor G. Melchers.)

The geographic distribution of plants is correlated with their photoperiodic response. In the near-equatorial regions, short-day species predominate, whereas in the regions extending toward the poles, long-day plants are common. The onset of flowering of short-day plants is promoted by 15–16 hours of dark periods, whereas in long-day plants the flowering is accelerated by continuous illumination or by longer light than daily dark cycles. The critical day length is not an absolute term; it varies by species. Usually, there is a minimum number of cycles to evoke the photoperiodic response. The most important photoreceptor chromoprotein is *phytochrome*. The effect of phytochrome is affected by different plant hormones. Typical long-day plants are henbane (*Hyoscyamus*) spinach, *Arabidopsis* (without a critical day length), the majority of grasses and cereal crops (wheat, barley, oats), lettuce, radish, etc. Typical short-day plants are Biloxi soybean, cocklebur, aster, chrysanthemum, poinsettia, dahlia, etc. In the majority of species, the photoperiodic response is controlled by one or a few genes. *See* cryptochromes; flower evocation; photomorphogenesis; phototropism; phytochrome; vernalization. (Jackson, S. D. & Prat, S. 1996. *Plant Physiol.* 98:407; Amador, V., et al. 2001. *Cell* 106:343; Quail, P. H. 2002. *Current Opin. Cell Biol.* 2002. 14:180.)

photophosphorylation   ATP formation from ADP in photosynthetic cells.

photoreactivation   Elimination of the harmful effects of ultraviolet irradiation by subsequent exposure to visible light (which activates enzymes, splitting up the pyrimidine dimers in the DNA). With a few exceptions, e.g., *Haemophilus influenzae*, most organisms possess light-activated repair enzymes. The majority of mammals do not have efficient photoreactivation systems, except the marsupials. *See* DNA repair; error-prone repair; excision repair; glycosylases; light repair; photolyase dark repair. (Kelner, A. 1949. *J. Bacteriol.* 48:5111; Tuteja, N., et al. 2001. *Crit. Rev. Biochem. Mol. Biol.* 36(4):337; Sancar, G. B. 2000. *Mutation Res.* 451:25.)

photoreceptors   The human eye has very sensitive rod cells that mediate black and white vision and the less sensitive cone cells are for color vision. *See* CRX; metalloproteinases; phototropism; phytochrome; rhodopsin; S-cone disease; *sevenless*.

photoreduction   In photosynthetic cells, light-induced reduction of an electron acceptor.

photorespiration   Oxygen consumption in illuminated plants used primarily for oxidation of the photosynthetic product phosphoglycolate; it also protects C3 plants from photooxidation. *See* Calvin cycle; C3 plants; respiration. (Wingler, A., et al. 2000. *Philos. Trans. R. Soc. Lond. B Biol. Sci.* 355:1517.)

Photorhabdus luminescens   Gram-negative enterobacterium maintains a mutualistic association with insect-feeding nematodes. When the nematodes invade the insects, the bacteria are released, kill the host by the toxin, emit light, and make the cadaver luminescent. The toxins (tca and tcd) are potential insecticide, fungicide, and antibacterial agents somewhat similar to those of *Bacillus thüringiensis*. *See Bacillus*

thüringiensis. (Ehlers, R. U. 2001. *Appl. Microbiol. Biotechnol.* 56:623; Szállás, E., et al. 1997. *Int. J. Syst. Bacteriol.* 47:402.)

photosensitizer   May increase the oxidative damage to DNA. The action of photosensitizers may involve initial electron or hydrogen transfer to the DNA by the excited photosensitizer, followed by the generation of free radicals. Alternatively, they generate singlet oxygen that interacts with DNA, then produces peroxidic intermediates. Most commonly, guanine suffers lesions. *See* oxidative DNA damages.

photosynthesis   Uses light energy for the conversion of CO_2 into carbohydrates with the assistance of a reducing agent such as water. *See* Calvin cycle; chlorophyll-binding proteins; C3 plants; C4 plants; photosystems; thermotolerance; Z scheme. (Matsuoka, M., et al. 2001. *Annu. Rev. Plant Physiol. Mol. Biol.* 52:297; Xiong, J. & Bauer, C. E. 2002. *Annu. Rev. Plant Biol.* 53:503.)

photosystems   In photosynthesis, photosystem I is excited by far red light (~700 nm) while photosystem II requires higher energy red light (~650–680 nm). Photosynthesis in bacteria that do not evolve oxygen uses only photosystem I. Upon absorption of photons, photosystem I liberates electrons that are carried through a cascade of carriers to $NADP^+$, which is reduced to NADPH. The departure of electrons generates a "void" in the P700 photoreaction center of photosystem I, which is then filled by electrons produced through the splitting of water molecules in photosystem II. The overall reaction flow is

$$2H_2O + 2\,NADP^+ + 8\,photons \rightarrow O_2 + 2\,NADPH + 2\,H_+$$

Mutants of *Chlamydomonas* alga lacking photosystem I survive as long as the actinic light (beyond violet) reaches 200 microeinsteins per m^2/second. *See* antenna; CAB; chloroplast; LHCP; Z scheme. (Pakrasi, H. B. 1995. *Annu. Rev. Genet.* 29:755; Guergova-Kuras, M., et al. 2001. *Proc. Natl. Acad. Sci. USA* 98:4437; Jordan, P., et al. 2001. *Nature* 411:909; Chitnis, P. R. 2001. *Annu. Rev. Plant Physiol. Plant Mol. Biol.* 52:593; Szabó, I., et al. 2001. *J. Biol. Chem.* 276:13784; Rhe, K.-H. 2001. *Annu. Rev. Biophys. Biomol. Struct.* 30:307; Saenger, W., et al. 2002. *Current Opin. Struct. Biol.* 12:244.)

phototaxis   Movement of organisms (plants, animals, and microbes) in response to light.

phototransduction   Transmission of light signals mediating gene expression. A scaffold protein (InaD in *Drosophila*) assembles the components of the light transduction pathway. *See* retinal dystrophy; rhodopsin; signal transduction.

phototroph   Organism that uses light to generate energy and synthesizes its nutrients from inorganic compounds by using this energy.

phototropin   Flavoprotein photoreceptor for plant phototropism. *See* flavoprotein.

phototropism   Reaction of an organ or organism to light apparently involving more than a single photoreceptor.

In *Arabidopsis*, the phytochromes and two complementary cryptochrome mutations (*CRY1*, *CRY2*) have been identified. Inactivation of both is required to eliminate phototropic response. It was suggested that one of the receptors is a membrane protein with autophosphorylating ability. Additional genes (*NPH1*, *NPH2*, *NNPH3*, *RPT*, *NPL1*) are required for processing the responses after perception of the signals. *See* cryptochromes; gravitropism; photoreceptors; phototropin; phytochromes. (Briggs, W. R. & Liscum, E. 1997. *Plant Cell Environ.* 20:768; Quail, P. H. 2002. *Current Opin. Cell Biol.* 2002. 14:180.)

phragmoplast Hollow-looking ring- or barrel-like structure formed near the end of mitosis in the middle plane of plant cells before the *cell plate* appears, separating the two daughter cells. *See* mitosis. (Gu, X & Verma, D. P. 1996. *EMBO J.* 15:695; Zhang, Z., et al. 2000. *J. Biol. Chem.* 275:8779.)

PHRAP Frequently used DNA sequence alignment programs. A quality score of $10^{X/10} \approx 30$ corresponds to an accuracy of 99.9% regarding the base sequence. *See* base-calling; CONSED; PHRED. (Harmsen, D., et al. 2002. *Nucleic Acids Res.* 30:416; <http://www.phrap.org>.)

PHRED Automated base-calling computer program. *See* PHRAP; PolyPhred; base-calling.

phycobilins Highly fluorescent photoreceptor pigments in blue-green, red, and some other algae. They contain a linear tetrapyrrole prosthetic group for light harvesting, bile pigments, and an apoprotein. This family of pigments includes the blue phycocyanins, the red phycoerythrins, and the pale blue allocyanins. These pigments may form phycobilisome attached to the photosynthetic membrane. Phytochromes are also related pigments. *See* light-harvesting protein; phytochrome. (Wu, S. H., et al. 1997. *J. Biol. Chem.* 272:25700.)

phycocyanin Pigment of blue-green algae. *See* phycobilins.

phycoerythrin Red pigment of red algae. *See* phycobilins; phycocyanin.

phycomycetes Fungi with some algal characteristics. *Ph. blakesleeanus* is easy to grow with 4 days of asexual cycles and about 2 months of sexual cycles. It forms heterokaryons ($n = 14$) and can be subjected to formal genetic analyses, although the tetrads may be irregularly amplified. Transformation is feasible. Well suited for physiological and developmental studies.

phyletic evolution Gradual emergence of species in a line of descent. The gaps in the fossil records are supposed to be due to accidents in preservation of the intermediate forms.

phyllody Developmental anomaly of conversion of floral parts into leaves, generally after infection by pathogens.

phylloquinone Composed of a p-naphthokinone and a phytol radical. It catalyzes oxidation-reduction reactions in plants. *See* vitamin K.

phyllotaxy Consecutive leaves of plants do not occur above each other. Quite commonly single leaves are at opposite position (unless they occur in whorls). This arrangement makes sense for optimal utilization of light. In many plants, the leaves may not alternate in 180°, but they may be arranged in any other determined pattern. This pattern is called phyllotaxy. If the leaves are opposite to each other, the phyllotaxy is one-half. A common phyllotactic index is two-fifths (144°). That is, if the leaves are positioned by this index, leaf #1 will be followed by #2 at 144°, then #3 will take the place in a spiral at 288°, i.e., it will be above #1 (because 288:144 = 0.5 and 0.5 × 360 = 180), and so on. The arrangement of the fruits on the stem may also be caused by such an obliquity following either clockwise or counterclockwise directions. *See* embryogenesis in plants. (Hake, S. & Jackson, D. 1995. *ASGSB Bull.* 8[2]:29.)

Phyllotaxy.

phylogenetic analysis Uses pathogen strain DNA comparisons in forensic science for identifying the source of infection, e.g., the retroviral DNA in case of HIV. *See* acquired immunodeficiency; DNA fingerprinting; forensic genetics.

phylogenetic profile method Studies the correlations of inheritance of pairs of proteins among various species. These proteins are not necessarily homologous, but they appear to be linked functionally. *See* rosetta stone sequences. (<http://dip.doe-mbi.ucla.edu>.)

phylogenetic tree Graphically represents the phylogeny of organisms. Trees have been constructed in the past on the basis of morphology, sequences of single genes, or sequences of entire genomes. Similarity between two organisms can also be determined by dividing their total number of genes by the number of genes they have in common. Phylogenetic analysis based on molecular information greatly increases the precision of map construction. *See* evolutionary tree. (Madsen, O., et al. 2001. *Nature* 409:610; Murphy, W. J., et al. 2001. *Nature* 409:614.)

phylogenetic weighting DNA sequence information from various taxa is included in the phylogenetic tree in decreasing order of relationship. Thus, alignment from distant relatives should not precede alignment of closer relatives. This procedure prevents confounding similarity and descent. *See* DNA sequence alignment; evolutionary tree; homology; maximum parsimony. (Robinson, M., et al. 1998. *Mol. Biol. Evol.* 15:1091.)

phylogenomics Uses evolutionary information to infer function of genes.

phylogeny Evolutionary descent of a species or other taxonomic groups. *See* evolution; ontogeny; speciation. (Huelsenbeck, J. P., et al. 2001. *Science* 294:2310; <http://phylogeny. harvard.edu/treebase>; <http://www.herbaria.harvard.edu/

treebase>; <http://www.bioinformatik.de/; <http://beta.tolweb.org/tree/>.)

phylum First main category of the plant, animal, and other kingdoms.

Physarum polycephalum Single-cell slime mold that displays physiological dioecy. The cell forms a plasmodium, i.e., the nuclei divide without cell division and thus the cell becomes multinucleate. In the early embryos, only S and M phases of the cell cycle are detectable.

Physcomitrella patens Moss with a principal life phase as a haploid gametophyte. It can be used for the production of various mutants, parasexual research, transformation, and the study of plant hormones on developmental processes and various tropisms. (*See* Schaefer, D. G. 2002. *Annu. Rev. Plant Biol.* 53:477.)

physical containment *See* containment.

physical map Map where the genome is ordered in DNA fragments or nucleotide sequences rather than in units of recombination. The first physical maps were constructed in bacteriophages with small genomes. The DNA of phage P4 was cleaved completely by restriction endonuclease EcoRI into four fragments that could be separated by electrophoresis according to size:

After incomplete digestion for 5 minutes, larger fragments were also detected that contained fragments A + B + C, C + B, and C + D, but no fragment appeared with the size B + D. The cause of the absence of B + D must have been that B and D were not adjacent in the circular DNA. Therefore, the sequence of the fragments in the chromosome could have been only A – B – C – D.

The much larger polyoma genome was mapped by a different procedure. With a single EcoRI cut, the circular DNA was linearized, and that cut was designated as the zero coordinate of the map. HinDIII cut the circle into two fragments: A — 55% and B — 45%. HpaII produced eight fragments: a = 27%, b = 21%, c = 17%, d = 13%, e = 8%, f = 7%, g = 5%, and h = 2% of the total genome. When EcoRI and HpaII cleaved the DNA, fragment b (21%) was not detected by electrophoresis, but instead two new fragments of 1% and 20% were found. Obviously, the EcoRI cut was 1% from one end and 20% from the other end of fragment b. In a following step, the HindIII-generated A fragment was digested by HpaII and fragments c, d, e, g, and h were found again (17 + 13 + 8 + 5 + 2 = 45); two pieces of 3% and 7% were also obtained. When the HinDIII fragment of 45% length was exposed to HpaII, fragment f remained intact, but two other fragments of 18% and 20% were recovered. Therefore, the fragments could be pieced together as follows:

HindIII A:	7%	-	45%	-	3%
	part of a				part of b
HindIII B:	18%	-	7%	-	20%
	part of b		f		part of a

Incomplete digestion of A by HpaII produced fragments a + c, c + e, e + d, h + g, and g + b; therefore, the polyoma DNA appeared as b–f–a–c–e–d–h–g with the zero coordinate in b and g near the 100 coordinate.

Larger genomes such as *E coli*, yeast, or of higher eukaryotes are generally pieced together by a chromosome walking type of procedure using overlapping fragments generated by several restriction endonucleases, e.g.:

	1	**2**	**3**
fragments generated by enzyme A:	abcde	fghijklmn	oprstuvwz
	4	**5**	
fragments generated by enzyme B:	cdefghi	jklmnoprst	

will be tied into the order 1, 2, 3 on the basis of hybridization of 4 with 1 and 2, and hybridization of 5 with 2 and 3 but not 5 with 1 or 4 with 3. In the initial steps, YAC clones are generally used because they cover large segments of the genomes. Cosmid clones usually follow this and eventually large continuities (contigs) are established without gaps. By the employment of anchors, fragments with genetically or functionally known sites, the physical map can be correlated with the genetic map determined by recombination frequencies; thus, *integrated maps* are generated. The individual fragments can then be sequenced and maps of ultimate physical resolution can be obtained. *See* anchoring; chromosome walking; contigs; cosmids; dynamic molecular combing; electronic PCR; EcoRI; FISH; genomic screening; HindIII; HpaII; integrated map; PCR; restriction enzymes; RFLP; SAGE. (Bhandarkar, S. M., et al. 2001. *Genetics* 157:1021.)

physical mutagens The most widely used forms are *electromagnetic*, ionizing radiations such as X-rays, and γ-rays emitted by radioisotopes. The most commonly used radiation sources for the induction of mutation by γ-rays are cobalt^{60} (Co^{60}) and cesium^{137} (Cs^{137}). *Particulate radiations* such as those produced by atomic fission are also ionizing. Ionization is the dislodging of orbital electrons of the atoms. The particulate (corpuscular) radiation source is uranium^{235}, which releases neutrons, uncharged particles (slightly heavier than those of the hydrogen atom) with very high penetrating power and the ability to release about 15 times as much energy along their path as hard X-rays (of short wave length and high energy). The *fast neutrons* have energies between 0.5 and 2.0 MeV (million electron volt). The *thermal neutrons* have a much lower level of energy (about 0.025 eV) because they have been "moderated" by carbon and hydrogen atoms. Radioactive isotopes also emit *β-particles* (electrons). Their level of energy and penetrating power depend a great deal on the source; H^3 (tritium) has a very short path (about 0.5 μm) and P^{32} is much more energetic (2,600 μm). Beta-emitters are rarely used for mutation induction. They can, however, be incorporated directly into the genetic material by using radioactively labeled precursors or building blocks of nucleic acids and thus are capable of inducing localized damage. The degree of localization depends on the effective path length. Uranium^{238} emits *α-particles* (helium nuclei), releasing thousands of times more energy per unit track than X-rays. Because of the very low penetrating power, α particles can be stopped by a couple of sheets of cells, in contrast to X-rays and gamma-rays, which require heavy concrete or lead shielding. Alpha-radiation, because of its high energy per short path, can very effectively destroy chromosomes. The most common genetic effect of all ionizing radiations is chromosome breakage, particularly deletions.

Another physical mutagen is *ultraviolet (UV)* radiation. It causes excitation in the biological material rather than ionization. Excitation may raise the orbital electrons to a higher level of energy from which they return to the ground state very shortly. UV radiation sources are commonly mercury or cadmium lamps (black light, germicidal and sun lamps). UV radiation is included in natural sunlight, especially in clean air of the higher mountains. Near ultraviolet light, UV-B (290–400 nm), may be present in the emission of fluorescent light tubes, and in the presence of sensitizers it may be genetically effective on a few layers of cells. The most common genetic effect of UV light is the production of pyrimidine dimers.

The effect of radiation on cells and organisms may be *direct*, i.e., the radiation actually hits the target molecules, or it may be *indirect*, i.e., the radiation produces reactive molecules in the intra- or extracellular environment, and these in turn cause genetic and/or physiological damage. Exposure to high temperature may enhance mutability. If radiation is received during DNA replication, damage is more likely than in the dormant state. Generally hydrated cells and tissues are more sensitive to ionizing radiation than dry or nonmetabolizing cells. *See* atomic radiations; carcinogens; chemical mutagens; chromosomal mutation; cosmic radiation; DNA repair; electromagnetic radiation; genetic sterilization; genomic subtraction; LET; maximal permissive dose; nuclear reactors; radiation effects; radioisotopes; ultraviolet light; X-rays. (Hollaender, A., ed. 1954–56. *Radiation Biology*. McGraw-Hill, New York.)

physiology   Discipline dealing with the functions of living cells and organisms.

phytanic acid   20-carbon, branched-chain fatty acid is formed from the phytol alcohol ester of chlorophylls and it is degraded by β-oxidation into propionyl-, acetyl-, and isobutyryl-CoA. Deficiency of this oxidation leads to Refsum disease in humans. *See* peroxisome; Refsum diseases.

phytic acid (inositol hexaphosphoric acid)   Combined with Ca^{2+} and Mg^{2+} salts, phytins commonly present in plant tissues. Phytate also ties up iron in plant tissues and limits its availability for human nutrition unless it is degraded by phytase. *See myo*inositol.

phytoalexins   Generally relatively low-molecular weight-yet diverse compounds synthesized through the phenylpropanoid pathway. They were attributed to defense systems against various plant pathogens. They are now considered mainly as consequences of infection rather than active defense molecules. *See* host-pathogen relation; phenolics. (Hammerschmidt, R. 1999. *Annu. Rev. Phytopath.* 37:285.)

phytochromes   Five regulatory proteins with alternating absorbance peaks in red and far-red light. Through their absorbance peaks (red [R] 660 nm and far-red [FR] 730)

they control various photomorphogenic processes, such as short- and long-day onset of flowering, hypocotyl elongation, apical hooks, pigmentation, etc. These chromoproteins are homodimers of 124 kDa subunits and a tetrapyrrole complex joined covalently through a cystein residue at about one-third distance from the NH_2 end. The molecule exists in two conformations corresponding to the R and FR absorption states. The interconversion between these states is mediated very rapidly by light of R and FR emission peaks. In etiolated plant tissue, the inactive P_r conformation may constitute up to 0.5% of the protein. The transition from the P_r conformation into the active P_{fr} form also entails the degradation of this receptor. The apoproteins, coded by different genes (*PHYA* and *PHYB*) in *Arabidopsis*, may have only about 50% homology in amino acid sequences, although they bind the same chromophore. The specificity of PhyA (far-red) and PhyB (red) resides in the N-termini. Phytochrome can induce and silence the expression of genes in a specific selective manner. The transcription of the phytochrome genes is also light regulated; R light reduces the transcription more effectively than FR. Phy-A perceives continuous FR, whereas phy-B responds to continuous red light.

Phytochrome chromophore. (After Metzler, D. E. *Biochemistry* Academic Press, New York.)

Phytochrome B is also a photoreceptor in the circadian rhythm. Phytochrome phy-C is a light-stable molecule. Phytochrome A appears to be serine/threonine kinase. SPA1 (suppressor of phy-A), a WD-protein with sequence similarity to protein kinases, mediates photomorphogenic reactions among other factors. The phytochrome responses are under complex genetic regulatory systems involving light-response elements, transcription factors, and components of the signal transduction circuits. PIF3 (phytochrome-inducing factor) is a basic helix-loop-helix protein that attaches to the nonphotoactive C-terminus of phytochromes A and B and mediates their conversion into active forms. PIF3 also binds to a G-box in the promoter, thus regulating transcription. Nucleoside diphosphate kinase 2 (NDPK2) preferentially binds to the red-light-activated form of phytochrome and appears to have a role in eliciting light responses. In photomorphogenic responses, phytochromes interact with cryptochromes. Although phytochrome is known as a ubiquitous plant product, the yeast *Pichia* also synthesizes phytochromobilin (PΦB), a precursor of this plant chromophore. A phytochrome-like protein (Ppr) has been identified in nonphotosynthetic prokaryotes (*Deinococcus radiodurans, Pseudomonas aeruginosa*). In *Rhodospirillum centenum* purple photosynthetic bacterium, a photoreactive yellow (PYP) pigment has been identified with a central domain resembling phytochromes. In the cyanobacteria, the circadian input kinase (CikA), a bacteriophytochrome, mediates the circadian oscillations. *See* brassinosteroids; cryptochromes; G box; photomorphogenesis; photoperiodism; phycobilins; signal transduction; WD-40. (Neff, M. M., et al. 2000. *Genes & Development* 14:257; Martinez-Garcia, J. F., et al. 2000. *Science* 288:859; Smith, H. 2000. *Nature* 407:585; Bhoo, S.-H., et al. 2001. *Nature* 414:776; Nagy, F. & Schäfer, E. 2002. *Annu. Rev. Plant Biol.* 53:329.)

phytoestrogens Estrogen-like plant products, such as the isoflavones (genistein, daidzein). They can take advantage of animal estrogen receptors and regulate gene expression similarly to other estrogens. Isoflavones can thus be used in hormone replacement therapies to alleviate postmenopausal symptoms and for other selective modulation purposes of estrogen receptors. *See* estradiol; estrogen receptor. (An, J., et al. 2001. *J. Biol. Chem.* 276:17808; Yellayi, S., et al. 2002. *Proc. Natl. Acad. Sci. USA* 99:7616.)

phytoextraction *See* bioremediation.

phytohemagglutinin *See* PHA.

phytohormones *See* plant hormones.

Phytophtora Group of plant pathogenic fungi. (<http://www.ncgr.org/pgc/index.html>.)

phytoplankton Aquatic, free-flowing plant.

phytoplasmas (Mollicutes, 530–1,350 kbp circular DNA) Minute, round (200–800 μm or filamentous) bacteria without cell wall, infecting the phloem cells of plants and causing disease. The symptoms vary from yellowing to sterility, stunting, and heavy branching, They resemble mycoplasmas of animals but cannot be cultured in cell-free media. Sucking insects causing economic loss in vegetables and trees propagate them. Phytoplasma infection may be exploited for gain by floriculture to obtain bushier poinsettias (Lee, I.-M., et al. 1997. *Nature Biotechn.* 15:178). Phytoplasmas may be identified by DNA-DNA hybridization and serological means. *See* mycoplasma; phyllody. (Lee, I. M., et al. 2000. *Annu. Rev. Microbiol.* 54:221.)

phytoremediation *See* bioremediation.

phytosulfokines (PSK) PSK-α is a sulfated pentapeptide and PSK-β is a tetrapeptide. They are cell proliferation–promoting compounds of plants.

phytotron Plant growth chamber system with maximal physical regulation facilities.

π *See* diversity.

Pi Inorganic phosphate.

pI (pH$_I$) Isoelectric point. *See* isoelectric focusing.

PIBIDS *See* trichothiodystrophy.

PIC (1) Preinitiation complex. Proteins associated with RNA polymerase before transcription. *See* chromatin remodeling; open promoter complex; TBP; transcript elongation; transcription factors. (He, S. & Weintraub, S. J. 1998. *Mol. Cell. Biol.* 18:2876; Tsai, F. T. & Sigler, P. B. 2000. *EMBO J.* 19:25; Soutoglou, E. & Talianidis, I. 2002. *Science* 295:1901.) (2) *See* polymorphic information content. (3) Ubiquitin-like protein associated by RanGAP. *See* RanGAP; sentrin; SUMO; ubiquitin; UBL.

Pick disease (FTDP-17, frontotemporal dementia and parkinsonism) Chromosome 17q21.11 dominant behavioral, cognitive, and motor disease involving variable loss and atrophy of the frontal and temporal part of the brain caused by defects in splicing of the Tau microtubule-associated protein. The mutations responsible for the conditions occur in exon 10 of TAU or in its 5′-splicing site, resulting in duplications in Tau mRNA. *See* dementia; Parkinsonism; tau.

picornaviruses Their single-stranded RNA genomes of about 7.2 to 8.4 kb (ca. 2.5 to 2.9×10^6 Da) are transcribed into four major polypeptides. Their RNA transcript lacks the 5′ cap in the mRNA that is characteristic for other eukaryotic viruses. *Enteroviruses* are a group of mostly unsymptomatic intestinal viruses. The paralytic *poliovirus* may also belong to this group. *Cardioviruses* are responsible for myocarditis (causing inflammation of the heart muscles) and encephalomyelitis (inflammation of the brain and heart). *Rhinoviruses* (in over 100 variants) are responsible for the common cold and other respiratory problems in humans and animals. *Aphtoviruses* cause foot-and-mouth disease in cattle, sheep, and pigs and occasionally infect people. *Hepatitis virus* may also be classified among the picornaviruses. *See* animal viruses; coxsackie virus; IRES; papovaviruses; polio virus. (Knipe, D. M., et al., eds. 2001. *Fundamentals of Virology*. Lippincott Williams & Wilkins, Philadelphia, PA.)

PIDD p53-inducible death domain protein that promotes apoptosis. *See* apoptosis; death domain; p53.

PIE Polyadenylation inhibition element. *See* polyadenylation signal.

piebaldism In animals it is the result of hypomelanosis (low melanin), generally restricted to spots of the body; white spots are on a black background. It may be a mutation of the KIT oncogene (4q12) or other factors. *See* albinism; Himalayan rabbit; Hirschprung disease; KIT oncogene; melanin; mouse; nevus; pigmentation in animals; vitilego.

Piebald rat.

Pierre-Robin syndrome Autosomal-recessive defect involving the tongue (glossoptosis), small jaws (micrognathia), and sometimes cleft palate and syndactyly of the toes. In an autosomal-dominant form, reduced digit number (oligodactyly) is also found. There is an X-linked form involving clubfoot and heart defects. Another X-linked form increases the number of bones in the digits (hyperphalangy).

piezoelectric mechanism Under pressure crystalline material may generate electricity. In response to alternative electric current, mechanical stress, expansion, and contraction may take place in matter. This latter property has been exploited for insertion of cell nuclei into eggs after the destruction of the original egg nucleus. This type of nuclear transplantation may achieve cloning of higher animals. *See* nuclear transplantation.

PIF Proteolysis-inducing factor.

pig (*Sus crofa*) $2n = 38$. The domesticated breeds are the descendants of the crosses between the European wild boar and the Chinese pigs. They can still interbreed with the wild forms of similar chromosome numbers. The wild European pig is $2n = 36$. The Caribbean pig-like peccaries (Tayassuidae) are $2n = 30$. There are about 300 domesticated pig breeds. Sexual maturity begins by about 5–6 months and the gestation period is about 114 days. Pigs are a multiparous species with a litter size of 4–12. By adult, somatic cell nuclear transplantation live clones can be produced. *See* Polejaeva, I. A., et al. 2000. *Nature* 407:86; nuclear transplantation. <http://www.tigr.org/tdb/tgi.shtml>; <http://www.toulouse.inra.fr/lgc/pig/hybrid.htm>.

pigeon *Columbia livia*, $2n = 80$. Great morphological variations among the various breeds of pigeons caught Darwin's attention, who made a few crosses between "pure races" and observed some "Mendelian" patterns.

PiggyBac Cabbage moth (*Trichoplusa ni*) transposon-derived transformation vector of several different insect species. It is 2.5 kb with 13 bp inverted terminal repeats and contains a 2.1 kb open-reading frame. Its specific target is TTAA. Frequently, a green fluorescent protein marker is used for easy detection. *See* GFP; open-reading frame; transposon; transposon vector. (Handler, A. M., et al. 1998. *Proc. Natl. Acad. Sci. USA* 95:7520.)

pigmentation, animal In mammals, tyrosine is the primary precursor of the complex black pigment melanin. The enzyme tyrosinase (located in the melanosomes) hastens the oxidation of dihydroxyphenylalanine (DOPA) into dopaquinone, which is changed by a nonenzymatic process into leukodopachrome. Leukodopachrome is an indole derivative oxidized by tyrosinase into an intermediate of 5,6-dihydroxyindole. After another step of oxidation, indole-5,6-quinone is formed. Coupling the latter to 5,6-dihydroxyindole is the first step in additions of further dihydroxyindole units in the process of polymerization to melanin. When cysteine is combined with dopaquinone through a series of steps, the reddish pigments of the hair and feathers are formed. The different pigments may also have other adducts at one or more positions to yield the various colors. In the formation of the eye color of insects, tryptophan is a precursor to the formation of formylkynurenine → kynurenine → hydroxykynurenine → ommin, ommatin. The catabolic pathway of amino acids contributes to the formation of guanine and through the latter to pteridines, which contribute to the coloration of insects, amphibians, and fishes and serve as a light receptor. Xanthopterin and leucopterin account for the yellow and white pigmentation of butterflies, sepiapterin is found in the eyes of *Drosophila*, and biopterin is found in the urine and liver of mammals. The degradation of the heme group yields a linear tetrapyrrole from which the bile pigment biliverdin and ultimately bilirubin diglucuronide is synthesized, which is secreted into the intestines and may accumulate in the eyes and other organs, causing jaundice when the liver does not function normally. Oxidized derivatives of bilirubin, urobilin, and stercobilin color the urine. During the early years of genetics, mutations were detected that block the biosynthetic paths of these pigments, thus contributing to an understanding of how genes affect the phenotype. The color of the skin in humans is determined by its melanin content. Phaeomelanin is a reddish pigment and eumelanin is black. The former is responsible for light skin and red hair color and it potentially generates free radicals, thus possibly making the individual susceptible to UV damage. Eumelanin provides protection against UV. The melanocyte-stimulating hormone (MSH) and its receptor (MC1R) regulate the relative proportion of these two melanins. In mice, about 100 genes are known that control pigmentation. Differences in pigmentation of the human skin in various geographic areas of the world seem to be correlated with the degree of exposure to ultraviolet radiation. *See* agouti; albinism; chorismate; eye color in humans; hair color; Himalayan rabbit; melanin; melanocyte-stimulating hormone; opiocortin; phenylalanine; pigmentation in plants; Siamese cat; tanning; tryptophan; tyrosine. (Price, T. & Borntrager, A. 2001. *Curr. Biol.* 11:R405.)

pigmentation defects *See* Addison disease; albinism; erythermalgia; Fanconi anemia; focal dermal hypoplasia; hematochromatosis; hypomelanosis; incontinentia pigmenti; LEOPARD syndrome; neurofibromatosis; pigmentation in animals; polyposis hamartomatous; skin diseases; tuberous sclerosis; Waardenburg syndrome.

pigment epithelium-derived factor (PEDF) Potent inhibitor of angiogenesis of the retina. Its defect leads to opacity of vision and blindness. *See* angiogenesis; angiostatin; endostatin; thrombospondin.

pIgR Polymeric immunoglobulin receptor. *See* antibody polymers.

P$_{II}$ Proteins (involved in bacterial glutamine synthesis) accelerate hydrolysis of NtrC in the presence of NtrB and ATP in limiting N supply and 2-ketoglutarate level. P$_{II}$ uridylylation permits the increase of NtrC-phosphate level and an increase in transcription from the glnAp2 promoter. In an excess of N supply, PII is not altered, resulting in no NtrC buildup and glnA2 activation ceases. *See glnAp; NtrB, NtrC.*

PI3K *See* PIK.

PIK Phosphatidylinositol kinases (PI[3]K). Preferentially they phosphorylate the 3 and 4 positions on the inositol ring. PIK-catalyzed reaction products (PtdIins) are second messengers. They participate in meiotic recombination, immunoglobulin V(D)J switches, chromosome maintenance and repair, progression of the cell cycle, etc. The mouse Pik3r1 regulatory gene encodes proteins p85α, p55α, and p50α. p55/p50 are essential for viability. Their defect may lead to immunological disorders and cancer. In ovarian cancer, an increase of PIK3CA and increased PIK activity were detected. Inactivation of its γ-subunit may lead to invasive colorectal cancer in mice. PI3Kγ may signal to phosphokinase B or to MAPK. PI3K is negatively controlled by PTEN. PIK-related kinases are TOR, FRAP, TEL, MEI, and DNA-PK. Its inhibitor is wortmannin. The nuclear GTPase PIKE enhances PIK activity and it is regulated by protein 4.1N. *See* cell cycle; DNA repair; immunoglobulins; MEC1; phosphatidylinositol; phosphoinositides; protein 4.1N; PTEN; second messenger; wortmannin. (Kuruvilla, F. G. & Schreiber, S. L. 1999. *Chem. Biol.* 6:R129; Katso, R., et al. 2001. *Annu. Rev. Cell Dev. Biol.* 17:615.)

PI 3 kinase *See* phosphoinositide 3 kinase.

pileus (1) Umbrella-shaped fleshy mushroom fruiting body. (2) Membrane that may be present on the head of newborns.

pilin Protein material of the pilus. *See* pilus.

pilomatricoma Usually benign, calcifying skin tumors, densely packed by basophilic cells. They develop into hair follicle–like structures. Their origin is attributed to mutation in LEF/β-catenin. *See* basophil; catenins; follicle; LEF.

PILRα Inhibitory receptor of myeloid cell encoded at human chromosome 7q22. *See* ITIM.

pilus Bacterial appendage that may be converted into a conjugation tube through which the entire replicated chromosome or part of it is transferred from a donor to a recipient cell. It may serve as a protein conduit. In pathogenic enterobacteria (*Neisseria gonorrhoea, Vibrio cholerae*) and in some types of *E. coli*, the so-called pilus type IV may be formed. It facilitates bacterial aggregation (bundle-forming pilus, BFP, encoded by a 14-gene operon). The expression of the LEE (enterocyte effacement) element enhances the association of the bacteria with the mucous intestinal membranes and triggers diarrhea. The pilin protein may undergo antigenic variation to escape host defenses. *See* antigenic variation; conjugation; conjugation mapping; mating, bacterial; PapD; pilin; shoufflon. (Jin, Q. & He, S.-Y. 2001. *Science* 294:2556.)

pimento (*Pimento dioica*) Also called allspice. Tropical dioecious spice tree; $2n = 2x = 22$.

PIM oncogene In human chromosome 6p21-p12 and in mouse chromosome 17. The gene is highly expressed in blood-forming (hematopoietic) cells and myeloid cells and overexpressed in myeloid malignancies and some leukemias. The human protein is a serine/threonine kinase. *See* oncogenes; serine/threonine kinases.

pIN Promoter of the transposase gene of a transposon. There are two GATC sites involved in *dam* methylation within pIN. *See dam;* RNA-IN.

PIN1 Peptidyl-prolyl cis/trans-isomerase in human cells. It is important for protein folding assembly and/or transport. Its deficiency leads to mitotic arrest; its overproduction may block the cell cycle in G2 phase. It interacts with NIMA kinase. *See* cell cycle; NIMA; parvulin.

pinch Group of proteins with LIM and additional domain(s). *See* CRP; LIM domain; LMO.

pineal gland Site of melatonin synthesis and photoreception in the brain. *See* brain; melatonin; opsins; Rabson-Mendenhall syndrome.

pineapple (*Ananas comosus*) Monocotyledonous tropical or subtropical plant ($2n = 50, 75, 100$). The flowers and bracts sit on a central axis and form fleshy fruits. The lack of seeds is caused by self-incompatibility of commercial varieties, but they develop seeds if allowed to cross-pollinate by other varieties. *See* seedless fruits.

pines (Pinus ssp) Trees; all 94 species are $2n = 2x = 24$. *See* spruce.

pinna Ear lobe, the lobe of a compound leaf or frond. *See* hairy ear.

pinocytosis Formation of ingestion vesicles for fluids and solutes by the invagination of membranes of eukaryotic cells. *See* phagocytosis.

pinosome Small cytoplasmic vesicle originating by invagination of the cell membrane.

PIN*POINT Protein position identification with a nuclease tail. In vivo method to ascertain the position of the critical promoter-binding proteins involved in the LCR. Fusion proteins with an unspecific nuclease tail are studied for how the cleavage position affects the expression of the gene(s). *See* LCR. (Lee, J.-S., et al. 1998. *Proc. Natl. Acad. Sci. USA* 95:969.)

PinPoint assay Identifies single-nucleotide polymorphism (SNIP). The polymorphic DNA site is extended by a single nucleotide with the aid of a primer annealed immediately upstream to the site. The extension products are analyzed by MALDI-TOF mass spectrophotometry. *See* MALDI-TOF;

primer extension; SNIP. (Haff, L. A. & Smirnov, I. P. 1997. *Genome Res.* 4:378.)

PIP Phosphatidylinositol phosphate. *See* phosphoinositides; PIP2; PIP3.

PIP2 Phosphatidylinositol (4,5)-bisphosphate is involved (with PIP3) in mediating the inositol phospholipid signaling pathway and in the activation of phospholipase C (PLC). PIP_2 also controls the ATP-regulated potassium ion channel (K_{ATP}) by binding to the intracellular C-domain of the channel protein and interfering with the binding of ATP. Since K_{ATP} channels affect pancreatic β-cells, vascular and cardiac muscle tone, they may have relevance for human diseases, e.g., diabetes. Pleckstrin homology domains selectively bind phosphoinositides. *See* diabetes; InsP; ion channels; phosphoinositides; PITP; pleckstrin. (Martin, T. F. 2001. *Curr. Opin. Cell Biol.* 13:493.)

PIP3 Phosphoinositol-3,4,5-trisphosphate. Intracellular messenger and stimulator of insulin, epidermal growth factor, etc., by adding another phosphate to PIP2 and activating PKB. *See* InsP; PKB; PTEN. (Hinchliffe, K. A. 2001. *Curr. Biol.* 11:R371.)

pipecolic acid (homoproline) Intermediate in lysine catabolism. An increase of pipecolic acid (hyperpipecolathemia/hyperpipcolicacidemia) in the blood plasma and urine leads to an increase in the size of the liver (hepatomegaly), resulting in growth retardation, vision defects, and demyelination of the nervous system.

Pipecolic acid.

PIPES Piperazine-N,N′-bis(2-ethanesulfonic acid) is a buffer within the pH range of 6.2–7.3.

PIR Protein information resource. *See* MIPS. (<http://pir.georgetown.edu/>.)

PIR-A, PIR-B Immunoglobulin-like regulatory molecules (activator/inhibitor) on murine B cells, dendritic cells, and myeloid cells. A single gene encodes Pir-B, whereas a multigene family encodes the six Pir-A proteins. *See* ITIM. (Dennis, G., Jr., et al. 1999. *J. Immunol.* 163:6371.)

PISA Protein in situ assay. PCR-generated DNA fragments are transcribed and translated in a cell-free protein expression system on a coated microtiter plate where the protein was immobilized. Single-chain antibody fragments and luciferase have been successfully arrayed. *See* PCR. (He, M. & Taussig, M. J. 2001. *Nucleic Acids Res.* 29(15):E73.)

pistil Central structure of flowers (gynecium) consisting of the stigma, style, and ovary. *See* flower differentiation; gametophyte female; gametophyte male.

pistillate Flower or plant that carries the female sexual organs. A female parent in plants.

Pisum sativum (pea) Legume ($2n = 14$). It played an important role in establishing the Mendelian principles of heredity and it contributed further information on genetics. Curiously, the famous "wrinkled" gene of Mendel turned out to be an insertional mutation. *See* pea.

pit (1) Indentation. (2) Stony endocarp of some fruits, e.g., plums.

PITALRE (cdk9) *See* acquired immunodeficiency; TEFb. (Darbinian, N., et al. 2001. *J. Neuroimmunol.* 121:3.)

pitch Length of a complete turn of a spiral (helix) and the translation per residue is the pitch divided by the number of the residues per turn. In a keratin alpha helix, it is 0.54 nm$/3.6 = 0.15$ nm. (2) Dark black residue after distillation. (3) Physiological response of the ear to sound depending on the frequency of vibration of the air. Perfect/absolute pitch is the ability to recognize musical notes. *See* musical talent; prosody.

pith Parenchyma tissue in the core of plant stems, e.g., elderberry (*Sambucus*).

pithecia (saki monkey) *See* Cebidae.

PITP Phosphatidylinositol transfer proteins (35 and 36 kDa) are required for the hydrolysis of PIP_2 (phosphatidyl-inositol bis-phosphate) by PLC (phospholipase C). In a GTP-dependent signal pathway, PITP is required by epidermal growth factor (EGF) signaling. *See* EGF; GTP; phosphoinositides; PIP; PIP_2. (Cockcroft, S. 1999. *Chem. Phys. Lipids* 98:23.)

PI-TR Phosphatidylinositol transfer protein involved in transfer of lipids among organelles within cells.

PITSLRE Members of a cyclin-dependent protein kinase family involved in RNA transcription or processing. They are associated with ELL2, TFIIF, TFIIS, and FACT. *See* ELL; TFIIS; transcription factors. (Trembley, J. H., et al. 2002. *J. Biol. Chem.* 277:2589.)

pituitary (hypophysis) Located at the base of the brain and connected to the hypothalamus (a ventrical part of the brain). The anterior part secretes the pituitary hormones and the posterior part stores and releases them. *See* brain, human. (Fauquier, T., et al. 2001. *Proc. Natl. Acad. Sci. USA* 98:8891; Scully, K. M. & Rosenfeld, M. G. 2002. *Science* 295:2231.)

pituitary dwarfism Due to recessive mutation, deletion, or unequal crossing over in the gene cluster, containing somatotropin and homologues in human chromosome 17q22-q24. Administration of somatotropin may restore growth. The defect may also be in the hormone receptor (human chromosome 5p13.1-p12, mouse chromosome 15). In these cases, the growth hormone level may be high (Laron types of dwarfisms). The level of somatomedin (insulin-like growth factors) may also be low. Somatomedin is a peptide facilitating protein binding and it has insulin-like activity. In either case, dwarfism may result. Dominant-negative mutations in IGHD2 (isolated growth hormone deficiency) are also known. *See* binding protein; dwarfism; GH; growth hormone, pituitary; hormone receptor; insulin-like growth factor; pituitary gland; stature in humans. (Machinis, K., et al. 2001. *Am. J. Hum. Genet.* 69:961.)

pituitary hormone deficiency, combined familial Fails to normally produce one or more of the hormones: growth hormone (HGH), prolactin, thyroid-stimulating hormone (TSH), because of mutation in the POU1F1 gene, whereas mutation in the PROP1 gene cannot produce luteinizing hormone (LH) and follicular-stimulating hormone (FSH). *See* animal hormones.

pituitary tumor (GNAS1, 20q13.2) Caused by autosomal-dominant mutations in the α-chain of a G-protein (G_s). This protein is also called gsp (growth hormone-secreting protein) oncoprotein. The human securin, mediating sister chromatid cohesion, has substantial sequence homology with the pituitary tumor-transforming gene. Securin may block sister chromatid separation and thus can be responsible for chromosome loss or gain, which are common characteristics of tumors. *See* G protein; McCune-Albright syndrome; sister chromatin cohesion.

PI vector Contains packaging site (*pac*) and allows about 115 kb to be packaged. It infects *E. coli* at a pair of *lox P* recombination sites at which the *Cre* recombinase circularizes DNA inside the host cell. *See* vectors.

πVX Microplasmid (902 bp) containing a polylinker and an amber suppressor for tyrosine tRNA. It can be used for cloning eukaryotic genes. *See* recombinational probe.

pixel Picture element in the computer that represents a bit on the monitor screen or in video memory. *See* bit; byte.

pK$_a$ Negative logarithm of the dissociation constant K_a; stronger acids have a higher pK_a, whereas weaker acids have a lower pK_a. The dissociation of weaker acids is higher and that of stronger acids is lower.

PKA Protein kinase A (activated by cAMP). There are two types: PKA-I and PKA-II; they share a common catalytic subunit (C) but distinct regulatory subunits, RI and RII. RI/PKA-I positively controls cell proliferation and neoplastic growth. RII/PKA-II controls growth inhibition, differentiation, and cell maturation. RI is detectable in many types of cancers. Antisense methylphosphonate RNA of the RI$_\alpha$ subunit has caused arrest of proliferation of cancer cells without toxicity to normal cells. *See* antisense technologies; cocaine; export adaptors; protein kinases.

PKB Protein kinase B is a serine/threonine kinase, the same as Rac or Akt. It is activated by phosphatidylinositol-3,4,5-trisphosphate by binding to its pleckstrin homology domain. *See* CaM-KK; phosphoinositides; pleckstrin domain; protein kinases.

PKC Protein kinase C. *See* protein kinases.

PKD *See* polycystic kidney disease.

PKI Protein kinase I is a small protein that attaches to the catalytic subunits of heterotetrameric PKA. With the aid of its nuclear localization sequence (NES), it sends the complex to the nucleus. *See* export adaptors; nuclear localization sequence; PKA.

PKR Double-stranded RNA-dependent serine-threonine protein kinase involved in NF-κB signaling. One of the most important targets of PKR is the eIF-2A translation factor and thus protein synthesis. It may control cell division and apoptosis and may serve as a tumor suppressor. Translation is required for viral infection of mammalian cells. Viral infection may trigger the activation of PKR as a defense against infection through shutting off protein synthesis. The PKR active cell may succumb to apoptosis, but the animal may survive. Several viruses (adenovirus, vaccinia virus, HIV-1, hepatitis C, poliovirus, SV40, etc.) use various mechanisms to inhibit activation of PKR by interfering either with its dimerization or RNA binding, or regulation of eIF-2A, etc. PKR preferentially binds mutant huntingtin protein in Huntington disease. *See* apoptosis; eIF-2A; Huntington's chorea; interferon; NFκB; oncolytic virus; PERK; reovirus. (Kaufman, R. J. 1999. *Proc. Natl. Acad. Sci. USA* 96:11693.)

PKS oncogenes In human chromosomes Xp11.4 and 7p11-q11.2. These genes display very high homology to oncogene RAF1 and apparently encode protein serine/threonine kinases. *See* oncogenes; *raf*.

PKU *See* phenylketonuria.

PLAC Plant artificial chromosome. *See* artificial chromosome; YAC.

placebo Presumably inactive substance used in parallel but to different individuals in order to serve as a concurrent (unnamed) control for testing the effect of a drug. In some instances, the placebo has positive effects not because of physical or chemical properties but dur to expectation-caused dopamine release, e.g., in Parkinson disease. *See* concurrent control. (de la Fuente-Fernández, R., et al. 2001. *Science* 293:1164; Ramsay, D. S. & Woods, S. C. 2001. *Science* 294:785.)

place cells Locations in the brain for the firing of the specific nerve cells.

placenta (1) Maternal tissue within the uterus of animals that is in the most intimate contact with the fetus through the umbilical chord. Most commonly, the placenta is located on the side of the uterus; the placenta previa is situated at the lower part of the uterus in the zone of dilation. It may be involved in a painless hemorrhage in the eighth month of pregnancy. The latter situation may be correlated with the age of the mother. During pregnancy the placenta of eutherian mammals includes both maternal and zygotic tissues in close association (feto-maternal interface). The interaction between these two types of tissues is essential for normal embryo development and viability of the conceptus. In normal pregnancy the uterus is invaded by the cytotrophoblasts (the nutritive cells of the conceptus), but defects in the cell adhesion system may adversely affect the pregnancy and may lead to eclampsia. *See* eclampsia; imprinting; incompatibility. (Zhou, Y., et al. 1993. *J. Clin. Invest.* 91:950; Georgiades, P., et al. 2001. *Proc. Natl. Acad. Sci. USA* 98:4522.) (**2**) Also, the wall of the plant ovary to which the ovules are attached.

placode Heavy embryonal plate of the ectoderm from which organs may develop. *See* AER; ectoderm; germ layer; neural crest; organizer; ZPA.

PLAD Pro-ligand-binding assembly domains aggregate the (death) receptors before binding the ligands. *See* death receptors.

plagiary *See* ethics; publication ethics.

plague Has been used to loosely define widespread, devastating diseases. Strictly, the term applies to infection by the *Pasteurella pestis* (*Yersinia pestis*) bacterium. The disease may occur in three main forms: *bubonic* plague (most important diagnostic feature is swelling lymph nodes, particularly in the groin area), *pneumonic* plague (attacking the respiratory system), and *septicemic* plague (causing general blood poisoning). Many of its symptoms are overlapping with other infectious diseases. It used to be called "black death" on account of the dark spots appearing in largely symmetrical necrotic tissue with coagulated blood. The bacilli spread to human populations from rodents by fleas, but infections also occur through cough drops of persons afflicted by pneumonic plague. Various animal diseases are also called plague (pestis), but, except for those in rodents, they are caused by other bacteria or viruses. Pasteurellosis can be effectively treated with antibiotics, although some strains become resistant to a particular type of antibiotic (streptomycin, chloramphenicol). Eradication of rodent pests is the best measure of prevention. During the great epidemics in the 14th century, an estimated 25 million victims were claimed by the disease. Sporadic occurrence is known even today in the underdeveloped areas of the world. (Parkhill, J., et al. 2001. *Nature* 413:523.)

plakin >200 kDa dimeric, coiled coil, actin-binding proteins forming molecular bridges between the cytoskeleton and other subcellular structures. They also bind microtubules. *See* cytoskeleton; filaments; microtubule.

plakoglobin 83 kDa protein localized to the cytoplasmic side of the desmosomes. *See* adhesion; desmosome.

Planck constant (h) Constant of energy of a quantum of radiation and the frequency of the oscillator that emitted the radiation. $E = h\nu$, where E = energy, ν = its frequency; numerically 6.624×10^{-27} erg^{-sec}.

plankton Collective name of many minute free-floating water plants and animals.

plant breeding Applied science involved in the development of high-yielding food, feed, and fiber plants. It is also concerned with the production of lumber, renewable resources of fuel, and many types of industrial raw products (such as latex, drugs, cosmetics, etc.). A major goal of plant breeding is to improve the nutritional value, safety, disease resistance, and palatability of crops. Plant breeding and technological improvements in agriculture resulted in nearly a 10-fold increase in maize production, and wheat yields doubled in the 20th century. Plant breeding is based on population and quantitative genetics and biotechnology. (Mazur, B., et al. 1999. *Science* 285:372.)

plant defense Against herbivores, is mediated by the signaling peptide *systemin* activating a lipid cascade. Membrane linolenic acid is released by the damage and converted into phytodienoic and jasmonic acids, structural analogs to the prostaglandins of animals. As a consequence, tomato plants produce several systemic *wound-response proteins*, similar to those elicited by oligosaccharides upon pathogenic infections. Mutation in the octadecanoic (fatty acid) pathway blocks these defense responses. *See* fatty acids; host-pathogen relations; jasmonic acid; oleuropein; prostaglandins; systemin.

plant disease resistance *See* host-pathogen relation; plant defense.

plant hormones Auxins, gibberellins, cytokinins, abscisic acid, brassinosteroids, and ethylene. Polypeptide hormones play roles in the defense systems of plants (Ryan, C. A., et al. 2002. *Plant Cell* 14:251). The natural *auxin* in plants is indole-3-acetic acid (IAA), but a series of synthetic auxins are also known such as dichlorophenoxy acetic acid (2,4-D), naphthalene acetic acid (NAA), indole- butyric acid (IBA), etc. Auxins are involved in cell elongation, root development, apical dominance, gravi- and phototropism, respiration, maintenance of membrane potential, cell wall synthesis, regulation of transcription, etc. The bulk (≈95%) of IAA in plants is conjugated through its carboxyl end to amino acids, peptides, and carbohydrates. The conjugate regulates how much IAA is available for metabolic needs, although some conjugates may be directly active as a hormone. Enzymes have been identified that hydrolyze the conjugates. Over the developing tissues, auxins show concentration gradients, indicating their role in positional signaling similarly to animal morphogens. The conjugates may transport IAA within the plant. *Gibberellic acid* and gibberellins control stem elongation, germination, and a variety of metabolic processes. *Cytokinins* occur in a wide variety of forms such as kinetin, benzylaminopurine (BAP), isopentenyl adenine (IPA), zeatin, etc. Their role is primarily in cell division, but they regulate the activity of a series of enzymes. Regeneration of plants from dedifferentiated cells requires a balance of auxins and cytokinins. *Abscisic acid* and terpenoids control abscission of leaves and fruits, dormancy and germination of seeds, and a series of metabolic pathways. *Ethylene* has been recognized as a bona fide plant hormone. It is involved in the control of fruit ripening, senescence, elongation, sex determination, etc. The hormone-type action of *brassinosteroids* in controlling elongation and light responses was recognized by genetic evidence in 1996. *See* abscisic acid; brassinosteroids; ethylene; gibberellic acid; hormones; indole acetic acid; kinetin; signal transduction; zeatin. (Kende, H. 2001. *Plant Physiol.* 125:81; Mok, D. W. S. & Mok, M. C. 2001. *Annu. Rev. Plant Physiol. Mol. Biol.* 52:89, <www.plant-hormones.bbsrc.ac.uk>.)

plantibody Antibody synthesized by plants. Modified immunoglobulin produced in transgenic plants carrying the genetic sequences required for the site recognition of the viral coat or other proteins. The yield of the plantibody molecules is very high, up to 1% of the soluble plant proteins. The modification of the immunoglobulin usually involves the elimination of the constant region of the heavy chain while retaining the variable region. The plant antibodies are usually formed as single chains (ScFv). Other modifications for solubility and tissue-specific expression may be introduced. The plantibodies are also modified by intrinsic plant mechanisms (N-glycosylation) within the endoplasmic reticulum. Unfortunately, plant tissue lacks β1,4-galactosyltransferase, which is required for the synthesis of mammalian-like glycans. By transformation, the gene of this enzyme has been transferred into tobacco plants and it functions normally. Retention and excretion of ScFv immunoglobulin molecules is increased if the KDEL amino acid sequence is present in the polypeptide chain. Plant-produced antibodies may find biomedical application in humans and animals. Transgenic plants may produce large quantities of IgA and IgG-IgA.

Other components of the immunization system may thus be synthesized with single plants after combining the genes through classical crossing procedures. By eating IgA-secreting plant tissues, protection is expected through mucosal immunity or may protect against dental caries. If the antibody is expressed in seed tissues, it can be stored at room temperature (perhaps for years) without a loss of the variable region of the antibody and its antigen-binding ability. *See* antibody; host-pathogen relations; immunization; KDEL; monoclonal antibody; monoclonal antibody therapies; mucosal immunity; plant vaccine; ScFv. (Bakker, H., et al. 2001. *Proc. Natl. Acad. Sci. USA* 98:2899.)

plant pathogenesis Plant pathogens pose risks for agricultural, horticultural, and forest plants and may damage natural habitats of different organisms—plants as well as animals. Several plant pathogens and saprophytes may pose human health hazards, especially for immunologically compromised individuals. *See* host-pathogen relation. Vidaver, A. K. & Tolin, S. 2000. *Biological Safety*, pp. 27–33, Fleming, D. O. & Hunt, D. L., eds. ASM, Washington, DC.)

plant vaccines Transgenic plants may express immunogenic proteins. Consumption of the plant tissues may protect humans and animals against bacterial or viral diarrhea, or plant-synthesized immunoglobulins may protect against *Streptomyces mutans*, which are responsible for dental caries and gum disease. Hepatitis B surface antigen (HBsAg), Norwalk virus capsid protein (NVCP), *E. coli* heat-labile enterotoxin B subunit (LT-B), cholera toxin B subunit (CT-B), and mouse glutamate decarboxylase (GAD67) have been propagated in tobacco and potato tissues, respectively. So far these edible vaccines did not have clinical use. *See* immunoglobulin; plantibody; TMV; transformation, genetic; vaccines. (Daniell, H., et al. 2001. *J. Mol. Biol.* 311:1001; Ruf, S., et al. 2001. *Nature Biotechnol.* 19:870.)

plant viruses Vary a great deal in size, shape, genetic material, and host specificity. The majority of them have single-stranded positive-strand RNA as genetic material and are either enveloped or not. Reoviridae may have several double-stranded RNAs, and Cryptovirus carries two double-stranded RNAs. Cauliflower (Caulimo) virus has double-stranded DNA, whereas Geminiviruses have single-stranded DNA genetic material. The size of their genome usually varies between 4 and 20 kb, and their coding capacity is at least four proteins. The 5'-end may form a methylguanine cap or it may have a small protein attached to it. The 3'-end may have a polyA tail or may resemble the OH end of the tRNA. Approximately 600–700 plant viruses have been described. *See* CaMV; cap; geminivirus; polyA tail; TMV; tRNA; viroid; viruses. (Knipe, D. M., et al., eds. 2001. *Fundamental Virology*. Lippincott Williams & Wilkins, Philadelphia, PA; Harper, G., et al. 2002. *Annu. Rev. Phytopathol.* 40:119; Tepfor, M. 2002. ibid. 467.)

PLAP VECTOR *See* axon guidance.

plaque Clear area formed on a bacterial culture plate (heavily seeded with cells) as a consequence of lysis of the cells by virus; turbid plaques indicate incomplete lysis. *See* lysis.

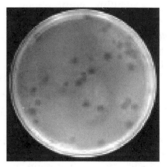

T3 bacteriophage plaques on petri plate heavily seeded by bacteria. (Courtesy of Dr. C. S. Gowans).

plaque-forming unit Number of plaques per mL bacterial culture.

plaque hybridization *See* Benton-Davis plaque hybridization.

plaque lift Plaques are marked on bacteriophage plates and overlaid by cellulose nitrate films. After denaturation and immobilization of the plaques on the filter, they are hybridized with probes to identify recombinants and are returned to the saved master plate for obtaining plugs of interest. The procedure generally requires repetition in order to isolate unique single recombinants. *See* colony hybridization. (Frolich, M. W. 2000. *Biotechniques* 29:30.)

plasma Fluid component of the blood in which the particulate material is suspended. The blood plasma is free of blood cells, but clotting is not allowed during its isolation; it contains platelets, which harbor animal cell growth factors. *See* cytoplasm; cytosol; PDGF; platelets; serum.

plasmablast Precursor of plasmacyte or precursor cell of the lymphocytes.

plasma cell (plasmacyte) B lymphocytes can differentiate into either memory cells or plasma cells; the latter secrete immunoglobulins. *See* immune system; immunoglobulins; lymphocytes.

plasmacytoma Cancer (myeloma) of antibody-producing cells.

plasmagene Nonnuclear genes (mitochondrial, plastidic, or plasmid). *See* chloroplast genetics; mitochondrial genetics.

plasmalemma Membrane around the cytoplasm or the envelope of the fertilized egg.

plasma membrane Envelops all cells. *See* cell membranes.

plasma proteins Proteins in the blood plasma. The major components are serum albumin, globulins, fibrinogen, immunoglobulins, antihemophilic proteins, lipoproteins, α_1 antitrypsin, macroglobulin, haptoglobin, and transfer proteins such as transferrin (iron), ceruloplasmin (copper), transcortin (steroid hormones), retinol-binding proteins (vitamin A), and cobalamin-binding proteins (vitamin B_{12}). The lipoproteins carry phospholipids, neutral lipids, and cholesterol esters. In addition, there is a great variety of additional proteins present in the serum.

plasmatocyte Macrophage-like elements in the insect hemolymph. *See* hemolymph; macrophage.

plasmid Dispensable genetic element that can propagate independently and can be maintained within the (bacterial) cell; may be present in yeast and mitochondria of a number of organisms. The plasmids may be circular or linear double-stranded DNA. Conjugative plasmids possess mechanisms for transfer by conjugation from one cell to another. Nonconjugative plasmids lack this mechanism and are therefore preferred for genetic engineering because they can be confined more easily to the laboratory. During evolution some of the advantageous plasmid genes are expected to incorporate into the chromosomes and the plasmids may be lost. The persistence of the plasmids may be warranted by their ability to disperse genetic information horizontally. Plasmids also occur in the organelles of higher and lower eukaryotes. *See* cryptic plasmids; curing of plasmids; pBR322; pUC; transposon, conjugative; Ty; vectors. (Summers, D. K. 1996. *The Biology of Plasmids*. Blackwell; Thomas, C. M., ed. 2000. *The Horizontal Gene Pool: Bacterial Plasmids and Gene Spread*. Harwood Press, Durham, UK.)

plasmid, amplifiable Continues replication in the absence of protein synthesis (in the presence of protein synthesis inhibitor). *See* amplification.

plasmid, chimeric Engineered plasmid carrying foreign DNA.

plasmid, conjugative Carries the *tra* gene, promoting bacterial conjugation; can be transferred to other cells by conjugation and can mobilize the main genetic material of the bacterial cell. *See* conjugation; F plasmid.

plasmid, cryptic Has no known phenotype.

plasmid, monomeric Present in a single copy per cell.

plasmid, multimeric Has multiple copies per cell.

plasmid, 2 μm 6.3 kbp circular DNA plasmid of yeasts present in 50–100 copies per haploid nucleus. It carries two 599 bp inverted repeats separating 2,774 and 2,346 bp tracts. Recombination between the repeats results in A- and B-type plasmids. Its recombination is controlled by gene *FLP* and its maintenance requires the presence of *REP* genes. *See* yeast. (Scott-Drew, S. & Murray, J. A. 1998. *J. Cell Sci.* 111:1779.)

plasmid, nonconjugative Lacks the *tra* gene required for conjugative transfer but has the origin of replication; therefore, when complemented by another plasmid for this function, it can be transferred. *See* conjugation.

plasmid, promiscuous Has conjugative transfer to more than one type of bacteria.

plasmid, recombinant Chimeric. It carries DNA sequences of more than one origin.

Recombinant plasmid.

plasmid, relaxed replication May replicate to 1,000 or more copies per cell.

plasmid, runaway replication Replication is conditional, e.g., under permissive temperature regimes the plasmid may replicate almost out of control, whereas under other conditions the plasma number per cell may be quite limited.

plasmid, single copy May have a single or very few copies per cell.

plasmid, stringent multicopy May grow to 10 to 20 copies in a cell.

plasmid addiction Loss of certain plasmids from the bacterial cells may lead to an apoptosis-like cell death, called postsegregational killing or plasmid addiction. *See* apoptosis.

plasmid incompatibility Plasmids are compatible if they can coexist and replicate within the same bacterial cell. If the plasmids contain repressors effective for inhibiting the replication of other plasmids, they are incompatible. Generally, closely related plasmids are incompatible and they thus belong to a different incompatibility group. The plasmids of enterobacteria belong to about two dozen incompatibility groups. Plasmids may also be classified according to the immunological relatedness of the pili they induce (such as F, F-like, I, etc.). The replication system of the plasmids defines both the pili and the incompatibility groups. Cells with F plasmids may form F sex pili, R1 plasmids belong to FII pili group, etc. *See* enterobacteria; F⁺; F plasmid; incompatibility plasmids; pilus; R plasmids.

plasmid instability   Indicates difficulties in maintenance caused by a defect(s) in transmission, internal rearrangements, and loss (deletion) of DNA. *See* cointegration.

plasmid maintenance   In prokaryotes, is secured either by the high number of copies or in low-copy-number plasmids by a mechanism reminding to some extent that of the centromere is mitosis of eukaryotes. The proteic plasmid maintenance system operates by the coordination of a toxin and an unstable antidote. When the labile antidote decays, the toxin kills the cells that do not have the plasmid. The antidote may be a labile antisense RNA that keeps the toxin gene in check. Plasmid-encoded restriction modification may also be involved. The modification system may go under an effective level in the plasmid-free cells, then the genetic material falls victim to the endonuclease. One example in *E. coli* is the *hok* (host killing)-*sok* (suppressor of killing)-*mok* (modulation of killing) system of linked genes. *See* antisense RNA; killer plasmids; partition; restriction modification. (Gerdes, K., et al.

1997. *Annu. Rev. Genet.* 31:1; Møller-Jensen, J., et al. 2001. *J. Biol. Chem.* 276:35707.)

plasmid mobilization   May take place by bacterial conjugation. Plasmid vectors use the gene *mob* (mobilization) if they do not have their own genes for conjugal transfer. Some plasmids may rely on *ColK* (colicin K, affecting cell membranes), which nicks plasmid pBR322 at the *nic* site, close to *bom* (basis of mobility). Mobilization proceeds from the nicked site (base 2254 in pBR322). Plasmids lacking the *nic/bom* system, e.g., pUC, cannot be mobilized. (Chan, P. T., et al. 1985. *J. Biol. Chem.* 260:8925.)

plasmid rescue   This procedure was designed originally for transformation with linearized plasmids of *Bacillus subtilis* that normally do not transform these bacteria. The linearized plasmid can be rescued for transformation in the presence of the *RecE* gene if recombination can take place.

The linearized monomeric plasmid can then carry any *in vitro* ligated passenger DNA into cells. If the host cells carry a larger number of plasmids (multimeric), special selection is necessary to find the needed one. Plasmid rescue has also been used for reisolation of inserts (plasmids) from the genome of transformed cells of plants. Reisolation requires appropriate probes for the termini of the inserts to permit recognition, then they are recircularized and cloned in *E. coli*. They have at least one selectable marker and an origin of replication compatible with the bacterium. The cloned DNA inserts or fragments are inserted into M13 phage for nucleotide sequencing. This permits the identification of any changes

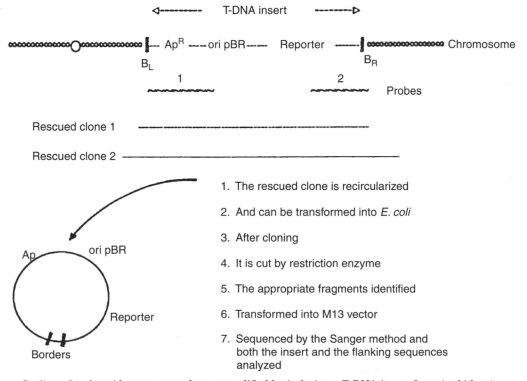

Outline of a plasmid rescue procedure exemplified by isolating a T-DNA insert from *Arabidopsis*. Ap = ampicillin-resistance gene (Apᴿ) of the pBR322 plasmid, oripBR = origin of replication of the pBR322 plasmid present in the plant-transforming vector, reporter is hygromycin resistance, | left (Bₗ) and | right (Bᴿ) border sequences of the T-DNA. (After Koncz, C., et al. 1989. *Proc. Natl. Acad. Sci. USA* 86:8467.)

that took place in the original transforming DNA and permits an analysis of the flanking sequences of the target sites as well. A number of different variations of the procedure have been adopted in prokaryotes, microbes, animals, and plants. *See* DNA sequencing; Rec; T-DNA. (Perucho, M., et al. 1980. *Nature* 285:207.)

plasmid shuffling The general procedure in yeast first disrupts the particular gene in a diploid strain. After meiosis, the cells can be maintained only if the wild-type allele is carried on a replicating plasmid (episome). Then mutant copies of that particular gene are introduced into the cell on a second episome and exchanged (shuffled) for the wild-type allele. The phenotype of any of the mutant alleles can be studied in these cells that carry the disrupted (null) allele. (Sikorski, R. S., et al. 1995. *Gene* 155:51; Zhao, H. & Arnold, F. H. 1997. *Nucleic Acids Res.* 25:1307.)

plasmid telomere Linear plasmids require exonuclease protection at the open ends. The problem may be resolved by capping with proteins or forming a lollipop-type structure by fusing the ends of the single strands as shown above at left. *See* telomere.

Loops   Protein

plasmid vehicle Recombinant plasmid that can mediate the transfer of genes from one cell (organism) to another. *See* vectors.

plasmin (fibrinolysin) Proteolytic protein (serine endopeptidase) with specificity of dissolving blood clots, fibrin, and other plasma proteins. For its activation, urokinases (tissue plasminogen activator) are required. Plasmin may be used for therapeutic purposes to remove obstructions in the blood vessels. *See* CAM; plasminogen; plasminogen activator; urokinase. (Lijnen, H. R. 2001. *Ann. NY Acad. Sci.* 936:226.)

plasmin inhibitor deficiency (PLI, AAP) Encoded in human chromosome 18p11-q11 as a recessive gene involved in the regulation of fibrinolysin. *See* plasmin.

plasminogen Precursor of plasmin. *See* angiostatin; plasmin; plasminogen activator.

plasminogen activator (PLAT) Cleaves plasminogen into plasmin; it is encoded in human chromosome 8q11-p11. The plasmin activator inhibitor (PLANH1) is encoded in human chromosome 7q21-q22 and PLANH2 at 18q21.1-q22. The plasminogen activator receptor was localized to 19q13.1-q13.2. *See* plasmin; plasminogen; PN-1; streptokinase; urokinase.

plasmodesma About 2 μm or larger channels connecting neighboring plant cells lined by extension of the endoplasmic reticulum. Functionally they correspond to the gap junctions of animal cells. Various molecules and signals, including viruses, may move through these intercellular communication channels. Plasmodesmata are subject to temporal and spatial regulation. *See* gap junctions. (Zambryski, P. & Crawford, K. 2000. *Annu. Rev. Cell. Dev. Biol.* 16:393; Hake, S. 2001. *Trends Genet.* 17:2; Haywood, V., et al. 2002. *Plant Cell* 14:S303.)

plasmodium Syncytium of the amoeboid stage of slime molds (such as in *Dictyostelium*).

Plasmodium One of several parasitic coccid protozoa causing malaria-type diseases in vertebrates, birds, and reptiles. A single *Plasmodium falciparum* ($n = 14$, ~23 Mb, ~5, 286 proteins) parasite simultaneously transcribes multiple *var* genes (at several chromosomal locations), encoding the erythrocyte membrane protein (PfEMP-1) that binds to the vascular endothelium and red blood cells. Functionally related genes tend to be clustered in subtelomeric regions of the chromosomes. Alignment of these *var* genes in heterologous chromosomes at the nuclear periphery may facilitate gene conversion and promotes diversity of antigenic determinants and adhesive phenotypes. Such a mechanism aids the evasion of the host immune system. The parasite invades the erythrocytes and destroys the host cells through the formation of merozoites (mitotic products), thus spreading to other cells. The merozoites may develop into gametocytes (gamete-forming cells) that infect blood-sucking mosquitos, then are transformed into sporozoites (the sexual generation) that are transmitted through insect bites to the higher animal host. The invaders first move to the liver where merozoites are formed and then go back to the erythrocytes; thus, the cycle continues. *Plasmodium falciparum* causes falciparum malaria. *P. malariae* is responsible for the *quartan*, or 4th-day recurring malaria. The protozoan contains two double-stranded extranuclear DNA molecules. The circular DNA resembles mitochondria, whereas the second molecule bears similarities to ctDNA and contains 68 genes. Mutation in a single gene (*pfmdr1*) encoding the P-glycoprotein homologue, Pgh1, may result in resistance to several antimalarial drugs of which some may be chemically related. Transformation of the gene encoding SM1 peptides into the *Anopheles* vector may render the insect resistant to *Plasmodium* infection. *See* chloroplast; gene conversion; malaria; mtDNA; PfEMP1; rifins; sex determination; thalassemia. (Fidock, D. A., et al. 2000. *Mol. Cell* 6:861; Ito, J., et al. 2002. *Nature* 417:452; <http://PlasmoDB.org>; <http://www. tigr.org/tdb/tgi.shtml>; sequenced P. falciparum genome: *Nature* 419: Oct. 30, 2002; Anopheles genome *Science* 298 Oct 4, 2002.)

plasmogamy Fusion of the cytoplasm of two cells without fusion of the two nuclei, thus resulting in dikaryosis. Plasmogamy is common in fungi but may occur in fused cultured cells of plants and animals. *See* fungal life cycle; cell genetics.

plasmolemma (plasmalemma) Plant cell membrane; the ectoplasm of the fertilized egg of animals.

plasmolysis Shrinkage of the plant cytoplasm caused by a high concentration of solutes (salt) outside the cell, resulting in loss of water. The cytoplasm separates from the cell wall.

plasmon Sum of nonnuclear hereditary units such as exists in mitochondrial and plastid DNA. *See* chloroplasts; mtDNA; plastome.

plasmon-sensitive gene *See* nucleocytoplasmic interaction.

plasmotomy Fragmentation of multinucleate cells into smaller cells without nuclear division.

plasmovirus Bears some similarity to phagemids, but in this case a retrovirus is combined with an independent vector cassette containing various elements. The envelope gene of the Moloney provirus is replaced by a transgene to prevent infective retroviral ability and does not regain it by chance recombination with another retrovirus. Such a construct can express transgenes and can multiply within the target cells and provide a tool for cancer therapy. *See* cancer therapy; phagemid; retrovirus; transgene; vectors; viral vectors. (Morozov, V. A., et al. 1997. *Cancer Gene Ther.* 4[5]:286.)

plasticity The ability of cells perform tasks not normally found in differentiated cells. *See* MAPCs; stem cells.

plastid Cellular organelle of plants containing DNA. It may differentiate into chloroplasts, etioplasts, amyloplasts, leucoplasts, or chromoplasts. *See* apicoplast; chloroplast; chloroplast genetics; ctDNA; plastid male transmission; plastid number per cell.

plastid male transmission Generally the genetic material in the plastids is transmitted only through the egg cytoplasm, but in a few species of higher plants (*Pelargonium, Oenothera, Solanum, Antirrhinum, Phaseolus, Secale,* etc.) a variable degree of male transmission takes place. Biparental transmission of plastid genes (about 1%) may occur in the alga *Chlamydomonas reinhardi*. The male transmission of these nucleoids is controlled by one or two nuclear genes. The nucleoids of the plastid and mitochondria of the male are usually degraded, or if they are included in the generative cells of the male, most commonly they fail to enter the sperm or are not transmitted to the egg cytoplasm. In contrast to the angiosperms, in conifers (pines, spruces, firs) the plastid DNA is usually transmitted through the males. In some interspecific hybrids, exclusively paternal, exclusively maternal, and biparental transmission were also observed. In other conifer crosses, mtDNA is transmitted maternally. In redwoods, the transfer is paternal. It has been claimed that the destruction of paternal ctDNA in females is carried out by a restriction enzyme while the maternal ctDNA is protected by methylation. Others implicate a special nuclease C. *See* chloroplast; ctDNA; genetics; mtDNA; paternal leakage. (Diers, L. 1967. *Mol. Gen. Genet.* 100:56; Avni, A. & Edelman, N. 1991. *Mol. Gen. Genet.* 225:273; Sears, B. 1980. *Plasmid* 4:233.)

plastid number per cell In the giant cells of *Acetabularia* algae, there may be 1 million chloroplasts, but in the alga *Chlamydomonas* there is only one per cell. In higher plants, the number of plastids vary according to the size of the cells, about 30–40 in the spongy parenchyma to about twice as many in the palisade parenchyma. *See* ctDNA; plastid.

plastochrone Pattern of organ differentiation in time and space.

plastocyanin Electron carrier in photosynthesis between cytochromes and photosystem I. *See* Z scheme. (Ruffle, S. V., et al. 2002. *J. Biol. Chem.* 277:25692.)

plastome Sum of hereditary information in the plastids. *See* chloroplast genetics; ctDNA.

plastome mutation Mutation in the plastid (chloroplast) DNA. *See* chloroplast genetics; mutation in cellular organelles.

plastoquinone Isoprenoid electron carrier during photosynthesis. *See* isoprene.

plate Petri dish containing a nutrient medium for culturing microbial or plant cells. The cell plate divides the two daughter cells after mitosis.

plate incorporation test Most commonly used procedure for the Ames test when the *Salmonella* suspension (or other bacterial cultures), the S9-activating enzymes, and the mutagen/carcinogen to be tested are poured over the bacterial nutrient plate in a 2 mL soft agar. After incubation for 2 days at 37°C, the number of revertant colonies is counted. *See* Ames test; spot test.

platelet abnormalities *See* giant platelet syndrome; Glanzmann's disease; Hermansky-Pudlak syndrome; May-Hegglin anomaly; platelets; thrombopathia; thrombopathic purpura.

platelet activating factor (PAF) Inflammatory phospholipid. PAF acetylhydrolase may be a factor in atopy. *See* atopy; platelets.

platelet-derived growth factor (PDGF) Mitogen secreted by the platelets, the 2–3-micrometer-size elements in mammalian blood; originated from the megakaryocytes of the bone marrow and concerned with blood coagulation. PDGF controls the growth of fibroblasts, smooth muscle cells, nerve cells, cell migration in the oocytes, etc. This protein bears substantial homologies to the oncogenic product of the simian sarcoma virus, the product of the KIT oncogene, and CSF1R (it also activates other oncogenes, such as c-fos). PDGF is required for the healing of vascular injuries, and in these cases the expression-induced Egr-1 (early growth response gene product) may bind to the PDGF β-chain promoter after displacing Sp1. PFGF- and insulin-dependent S6 kinase (pp70^{S6k}) are activated by phosphatidylinositol-3-OH kinase. Its receptor (PDGFR) is a tyrosine kinase. *See* CSF1R; growth factors; oncogenes; phosphatidyl inositol; platelets; signal transduction; S6 kinase; Sp1. (Betsholtz, C., et al. 2001. *Bioessays* 23[6]:494; Duchek, P., et al. 2001. *Cell* 107:17.)

platelets Originate as cell fragments or "minicells" (without DNA) from megakaryocytes of bone marrow. Their function is in blood clotting and repair of blood vessels; they secrete mitogens. Platelet abnormalities may cause stroke, myocardial infarction (damage of the heart muscles), and unstable angina (sporadic, spasmic chest pain). *See* blood; blood serum; megakaryocyte; platelet-derived growth factor. (Prescott, S. M., et al. 2000. *Annu. Rev. Biochem.* 69:419.)

plating efficiency   Percentage of cells or protoplasts that grow on a petri plate. The relative plating efficiency compares the fraction of growing cells in a treated series to that of an appropriate control.

platyfish (*Xiphophorus*/*Platypoecilus*)   Tropical fishes with complex sex determination. WX, WY, and XX are females; XY and YY are males. The pseudoautosomal region seems to be long. Their melanocytes frequently turn into melanoma. *See* melanocyte; melanoma; pseudoautosomal; sex determination.

platykurtic   *See* kurtosis.

platysome   Nucleosome core (when it was thought of as a flat structure). *See* nucleosome.

playback   The number of nonrepetitive sequences in a DNA can be determined by saturation of single-strand DNA with RNA of unique sequences. The kinetics of saturation, R_0t (by analogy to C_0t), is then determined. The annealed fraction is generally a small percent of the eukaryotic DNA that is highly redundant. To be sure that the RNA hybridizes to only the unique DNA sequences, in the DNA-RNA hybrid molecules the RNA is degraded enzymatically and the remaining DNA is subjected to a reassociation test to determine its C_0t curve. This "playback" then reveals whether all the isolated DNA has only genic DNA and is not redundant. Such studies may assist in estimating the number of housekeeping genes plus the genes that were transcribed when the RNA was collected. *See* c_0t; gene number; housekeeping genes.

PLC   Phospholipase C. *See* phospholipase.

pleated sheets   Relaxed β-configuration polypeptide chains hydrogen bonded in a flat layer. *See* protein structure.

pleckstrin domain   Approximately 100-amino-acid length; occurs in many different proteins such as serine/threonine kinases, tyrosine kinases, and the substrates of these kinases, phospholipase C, small GTPase regulators, and cytoskeletal proteins. Pleckstrin domains may participate in various signaling functions; they bind phosphatidylinosotol 4,5-bisphosphate. Pleckstrin is a substrate of protein kinase C in activated platelets. *See* adaptor proteins; desensitization, PH; phosphatidylinositol; phosphoinositides; platelets; PTB; SH2; SH3; SHC; WW. (Lemmon, M. A. & Ferguson, K. M. 1998. *Curr. Top. Microbiol. Immunol* 228:39; Rebecchi, M. J. & Scarlata, S. 1998. *Annu. Rev. Biophys. Biomol. Struct.* 27:503.)

plectin   500 kDa keratin of the cytoskeleton encoded in human chromosome 8q24. *See* epidermolysis [bullosa simplex]; keratin.

plectonemic coils   The two coils were wound together; therefore, they can be separated only by unwinding rather than by simple pulling apart like the paranemic coils. The DNA double helix represents plectonemic coils. *See* paranemic coils.

pleiomorphic   Displays variable expression (without a genetic basis for the special changes).

pleiomorphic adenoma   Salivary gland tumor caused by human chromosome breakage points primarily at 8q12, 3p21 and 12q13−15. The translocation t(3;8)(p21;q12) results in swapping the promoters of PLAG1, a Zn-finger protein encoded in chromosome 8, and β-catenin (CTNNB1), and activation of the oncogene. *See* β-catenin; zinc finger.

pleiotrophin (PTN)   18 kDa heparin-binding cytokine inducible by platelet-derived growth factor (PDGF). It is 50% identical with retinoic acid−inducible midkine, which like PTN is also a growth and differentiation factor. PTN reduces cell colony formation; it interacts with receptor protein tyrosine phosphatase, tumor growth, angiogenesis, and metastasis. *See* angiogenesis; metastasis; PDGF; retinoic acid. (Meng, K., et al. 2000. *Proc. Natl. Acad. Sci. USA* 97:2603.)

pleiotropy   One gene affects more than one trait; mutation in various elements of the signal transduction pathways, in general transcription factors, or in ion channels may have

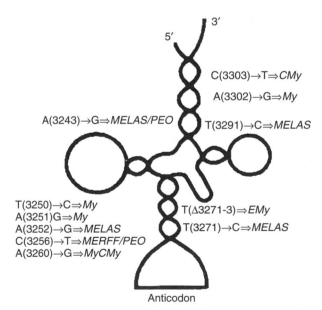

The diagram displays the mutations in the human mitochondrial Leu tRNA[UUR]. The first letter indicates the base that is changed, in parentheses is the nucleotide number at the physical map; after the → the substituted base is given; after ⇒ the diseases described under *mitochondrial diseases in humans* are identified with abbreviations. (Redrawn after C. T. Moraes. 1998. *Mitochondrial DNA Mutations in Aging, Disease and Cancer*, p. 167, Singh, K. K., ed. Springer, New York.)

pleiotropic effects. The existence of pleiotropy was questioned with the emergence of the one gene–one enzyme theory. It was inconceivable on that basis that one tract of DNA could code for more than a single function ("Pleiotropism non est... that is the dogma," p. 161 in *Genetics*. 1959. Sutton, E. H., ed. Josiah Macey Found, New York). It has been shown, however, that mutation at different sites within single mitochondrial tRNA genes may lead to several different human diseases. The complete sequence of the *Drosophila* genome shows that ~13,601 genes encode ~14,113 transcripts, indicating a minimum of nearly 4% of the genes display pleiotropy. *Antagonistic pleiotropy* claims that evolution does not work against variations, which adversely affect individuals after the completion of the reproductive stage of life, and the alternative genotypes display opposite phenotypes. *See* alternative splicing; epistasis; mitochondrial diseases in humans; signal transduction; transcription factors; two-hybrid method.

pleomorphism   Carl Wilhelm Nägeli's 19th-century suggestion claiming lack of hard heredity in bacteria and that they simply exist in a variety of pliable forms. This idea held back the development of bacterial genetics, although physicians like Robert Koch and the taxonomist W. Migula sharply criticized it and stated that it ignored the facts known by the 1880s. *See Hieracium.*

plesiomorphic   Trait in the more primitive state among several evolutionarily related species. *See* apomorphic; symplesiomorphic; synapomorphic.

pleura   Serous (moist) membrane lining the lung or insects' thoracic cavity.

plexins   Receptors for semaphorins. *See* semaphorins.

ploidy   Represents the number of basic chromosome sets in a nucleus. The haploids have one set (x), the diploids two (xx), autotetraploids four (xxxx), and so on. *See* polyploidy.

PLTP   Phospholipid transfer protein mediates the exchange of HDL cholesteryl esters with very-low-density triglycerides and vice versa. *See* CETP; cholesterol; HDL.

PLUM (*Prunus*)   Basic chromosome number $x = 7$, but a variety of polyploid forms exist. (Bliss, F. A., et al. 2002. *Genome* 45:520.)

plumule   Embryonic plant shoot-initial.

pluralism   Indicates sympatric speciation in evolutionary biology. *See* sympatric.

pluripotency   Has the ability to develop into various but not necessarily all types of tissues. Embryonic stem cells (from the inner mass of blastocysts), embryonic germ cells (primordial cells of the gonadal ridge), and the mesenchymal stem cells of the bone marrow possess pluripotency. The good cultures may grow for more than 70 doublings ($2^{70} \geq 10^{20}$) and may be free of chromosomal defects. *See* stem cells; totipotency. (Donovan, P. J. & Gearhart, J. 2001. *Nature* 414:92.)

plus and minus method   Early version of DNA sequencing using dideoxy analogs of nucleosides (+ batch) during replication; after the analog was incorporated to a site, T4 exonuclease failed to continue degradation. In the minus (−) batch, the synthesis stopped depending on which single nucleotide was omitted (the precursor mixture containing only three deoxyribonucleotides). Thus, nucleotide sequences of specific ends and length were generated, and the fragments of different lengths were analyzed by electrophoresis. The Sanger et al. (1977. *Proc. Natl. Acad. Sci. USA* 74:5463) method and its improvements replaced the plus and minus method. *See* DNA sequencing. (Sanger, F., et al. 1975. *J. Mol. Biol.* 94:441.)

plus end   The Preferential growing end of microtubules and actin filaments. *See* minus end.

plus strand   Of the single-stranded DNA or RNA of a virus, is represented in the mature virion, whereas the minus strand serves as a template for the transcription (replication) of the plus strand and the mRNA. In most cases, the plus strands are synthesized far in excess of the minus strands.

PLUS STRAND ——— MINUS STRAND ‑‑‑‑‑

The plus-strand viral genomic RNA serves directly as mRNA. *See* replicative form, RNA replication.

plutonium (Pu)   Metallic fissile element (atomic number 94, atomic weight 242) produced by neutron bombardment of uranium (U^{238}) during the production of nuclear fuel and used for making nuclear weapons. Radioactive Pu powers some of the heart pacemakers. Thus, the wearers, family members, and surgeons are exposed to some radiation, generally below 1.28 Sv per person per year, a little more than the average natural background (the doses are additive, however). If Pu particles are inhaled (the most common type of ingestion), the element may affect the lung and may eventually be preferentially deposited in the skeletal system, causing bone cancer by the emission of X- and γ-rays. Pu^{238} has a half-life of 86.4 years. It propels some space vehicles. Pu^{239} has a half-life of 24.3×10^3 years and primarily targets the bone marrow. Other Pu isotopes have an even much longer half-life. The level of Pu may be detected by radioactivity in the urine and by instruments placed on the body. Appropriate instruments can detect as low as 4 nCi (nanocurie) values. *See* atomic radiation; curie; isotopes; radiation hazard assessment.

Plx1   Kinase that phosphorylates the amino-terminal domain of Cdc25. *See* Cdc25.

Plymouth Rock   Recessive white-feathered breed of chickens with the genetic constitution of *iicc*. The dominant *I* gene is a color inhibitor and *C* symbolizes color. *See* Leghorn White; White Wyandotte.

PLZF   Zinc-finger protein encoded in human chromosome 11q23.It normally represses the promoter of cyclin A, but a transposition of RARα (retinoic acid receptor) results in transactivation of the cyclin A gene and may be involved in the initiation of cancer. *See* cyclin A; leukemia (acute promyelotic leukemia); RAR; transactivator; transcriptional activator.

PMA *See* phorbolesters; phorbol 12-myristate-13-acetate.

Phorbol.

pMB1 *See* ColE1.

PMDS Persistence of Müllerian duct syndrome. *See* Müllerian ducts.

PME Point mutation element in which the 3'-UTR regulatory proteins may bind and cause developmental switching.

PMF Peptide mass fingerprinting is a method of rapid identification of proteins without sequencing but using mass spectrometry information. *See* MALDI. (Jonsson, A. P. 2001. *Cell Mol. Life Sci.* 58:868.)

PML Promyelotic leukemia inducer is a putative zinc-finger protein, encoded in human chromosome 15q21. Formerly this gene was called MYL. There are about 10–20 PML bodies of ~0.3–1 μm per mammalian nuclear matrix. In acute promyelotic leukemia, these bodies become disorganized as the PML-RARα oncogenic complex is formed. PML bodies are associated with caspase- and FAD-induced apoptosis. Overexpressed PML also promotes apoptosis but without enhanced caspase-3 activity. In the absence of PML (PML$^{-/-}$) the cells become resistant to ionizing radiation. *See* apoptosis; leukemia; nuclear matrix; PLZF; POD; RAR.

PMS1 (2q31-q33), PMS2 (7q22) Increased postmeiotic segregation in yeast and increased colorectal cancer (or Turcot syndrome) in humans due to mismatch repair deficiency. *See* colorectal cancer; mismatch repair; Turcot syndrome.

PN-1 (protease nexin) 43 kDa inhibitor of serine proteases (thrombin, plasminogen activator). It is involved in the development of embryonic organs (cartilage, lung, skin, urogenital system, and nervous system). PN-1 is abundant in the seminal vesicle and its dysfunction leads to male infertility. *See* claudin-11; infertility; nexin; plasminogen activator; thrombin; urokinase. (Murer, V., et al. 2001. *Proc. Natl. Acad. Sci. USA* 98:3029.)

pN *See* Newton.

PNA *See* peptide nucleic acid.

Pneumococcus *See Diplococcus pneumoniae.*

Pneumocystis carinii Group of pathogenic ascomycetes with special susceptibility in immune-compromised individuals (e.g., AIDS patients) and rodents. It carries about 3,740 genes in the about 8 Mb genome. It reproduces asexually and sexually. *See* acquired immunodeficiency; ascomycete. (Kolls, J. K., et al. 1999. *J. Immunol.* 162:2890.)

PNPase *See* polynucleotide phosphorylase.

P nucleotides *See* immunoglobulins.

pocket Motif of the retinoblastoma (RB) tumor-suppressor protein family that binds to viral DNA-coded oncoproteins. Binding of RB to the E2F family of transcription factors blocks transcription that is needed for the progression of the cell cycle. The pocket proteins share this retinoblastoma (RB) motif. *See* cell cycle; E2F1; p107; p130; retinoblastoma; transcription factors; tumor suppressor. (Botazzi, M. E., et al. 2001. *Mol. Cell Biol.* 21:7607.)

POD PML oncogenic domain. *See* PML.

podophyllotoxin (epipodophyllotoxin) Antimitotic plant product.

podosome Actin-Containing adhesion structures on human primary macrophages controlling migration and immune reactions. *See* actin; immune reaction; macrophage.

Podospora anserina $n = 7$ is a genetically well-studied ascomycete.

Pof *(Painting of fourth)* *Drosophila* protein that binds only to the small 4th chromosome.

pogo *See* hybrid dysgenesis.

poikilocytosis Hemolytic anemia with variable-shape red blood cells. The defect is due to a reduction of ankyrin-binding sites or mutation in spectrin. *See* anemia; ankyrin; spectrin.

poikiloderma atrophicans (poikiloderma telangiectasia) *See* Rothmund-Thompson syndrome.

point mutation Does not involve detectable structural alteration (loss or rearrangement of the chromosome) and is expected to involve base substitutions. The point mutation rate per locus in eukaryotes is about 10^{-5} and may vary from locus to locus and among various organisms. The rate per nucleotides of a locus is in the range of 10^{-8}. *See* substitution mutation. (Krawczak, M., et al. 2000. *Hum. Mut.* 15:45.)

poise *See* stoke; viscosity.

poison sequence May be present in the genomes of some RNA viruses and thus even their cDNA cannot be cloned in full length in bacterial hosts. The problem may be overcome by

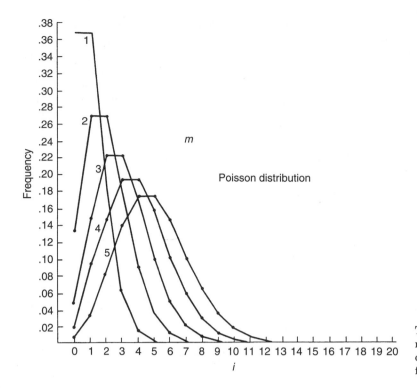

The Poisson distribution. Each curve corresponds to a numbered m value. The i classes represent the distribution of each mean value (m) with the ordinate indicating the frequencies.

propagating it in segments. (Brookes, S., et al. 1986. *Nucleic Acids Res.* 14:8231.)

Poisson distribution   Basic array of an extreme form of the normal distribution that is found when in large populations rare events occur at random, such as mutation. The general formula is e^{-m} $(m^i/i!)$, and expanded, $e^{-m}(m^0/0!,$ $m^1/m!,$ $m^2/2!\ldots m^i/i!)$, where e = base of natural logarithm ($\cong 2.718$), m = mean number of events, i = the number by which a particular m is represented at a given frequency, ! = factorial (e.g., 3! = 3 × 2 × 1, but 0! = 1). *See* negative binomial.

poky (synonym: *mi-1*)   Slow-growing and cyanide-sensitive respiration defective mitochondrial mutation in *Neurospora*. The basic defect appears to be a four-base deficiency of the 15-bp consensus at the 5′-end of the 19S rRNA of the mitochondria. Because of this defect, a further upstream promoter is used, making the transcript longer but during processing shorter RNAs are made. It is analogous to the petite colony mutations of budding yeast. *See* mtDNA; petite colony mutation; stoppers. (Akins, R. A. & Lambowitz, A. M. 1984. *Proc. Natl. Acad. Sci. USA* 81:3791.)

pol (bacterial RNA polymerase)   Synthesizes all the bacterial and viral RNAs in the bacterial cells. Its subunits are $\alpha\alpha\beta\beta'$ and σ. The σ-subunit identifies the promoter sequences and is required for the initiation of transcription within the cell. After about a half-dozen nucleotides were hooked up, it dissociated from the other subunits and further polymerization continued with the assistance of elongation protein factors. *See* eukaryotic RNA polymerases; pol I; pol II; pol III; transcription.

pol I   (1) Prokaryotic DNA polymerase where the polymerase (Klenow fragment) and exonuclease functions are located about 30 Å distance apart in a subunit, and editing (removal of wrong bases) follows the melt and slide model. It has a major role in prokaryotic repair and in the extension of Okazaki fragments for joining them into a contiguous strand by ligase. It adds 10–20 nucleotides/second to the chain and thus is much slower than pol III. *See* DNA ligase; DNA replication; Klenow fragment; melt and slide model; pol III; replication fork. (2) RNA polymerase involved in the synthesis of ribosomal RNA (except 5S rRNA) in eukaryotes. By endonucleolytic cleavage it generates the 3′ end of rRNA from longer transcripts. The upstream control regions for transcription initiation (binding proteins) vary from species to species. Human RNA pol I requires an activator UBF (upstream binding factor) and promoter selectivity factor SL1, including TBF (TATA box-binding protein) and associated subunits, TAF$_I$ 110, TAF$_I$ 63, and TAF$_I$ 48. The former two keep contact with the promoter, whereas TAF$_I$ 48 interacts with UBF and prevents RNA pol II from using this promoter site. *See* ribosome. (Reeder, R. H. 1999. *Progr. Nucleic Acid Res. Mol. Biol.* 62:293; Grummt, I. 1999. *Progr. Nucleic Acid Res. Mol. Biol.* 62:109.)

pol II   Prokaryotic DNA polymerase; functions are not completely defined so far. It has a role in DNA repair. *See* DNA repair.

pol II   RNA polymerase transcribes messenger RNA and most of the snRNAs of eukaryotes with the assistance of different transcription factors. Nine or ten of its subunits are very similar to other polymerases; pol II has 4–5 smaller unique subunits. The two largest subunits are very similar in the three eukaryotic RNA polymerases and are also similar to prokaryotic subunits. It is most sensitive to α-amanitin inhibition (0.01 μg/mL). The site of sensitivity is in the largest 220 kDa polypeptide. This large subunit is activated by phosphorylation. At the carboxy terminal, there are 26 (yeast), 40 (*Drosophila*), or 52 (mouse) heptapeptide (Tyr-Ser-Pro-Thr-Ser-Pro-Ser) repeats. These repeats are essential for function. Ser and Thr residues may be phosphorylated. Phosphorylation of the carboxy-terminal domain may affect the promoter specificity of the enzyme. The C-terminus of (CTD) of the large subunit is instrumental in the processing of the 3′-end of the transcript and the termination of transcription downstream of the polyA signal. CTD does not seem to affect initiation of transcription, but it also mediates the response to enhancers. This enzyme is different from RNA pol I and RNA pol III inasmuch as it requires a hydrolyzable source of ATP for the initiation of transcription. RNA pol II is different from the other polymerases in the requirement for a large array of special transcription factors that modulate the transcription of thousands of proteins. *See* α-amanitin; regulation of gene activity; transcription factors. (Cramer, P., et al. 2001. *Science* 292:1863.)

pol III   (1) Prokaryotic DNA polymerase where the α-subunit carries out the replication function and the ε-subunit is involved in editing (exonuclease) activity. It has a major role in replication of the leading and lagging strands. The replication has a speed of ≈1 kb/sec. There are only about 10–20 copies of the 10 subunit holoenzyme/cell. *See* core polymerase; DNA replication; replication fork; replisome. (2) RNA polymerase involved in the synthesis of transfer RNA, 5S rRNA, 7S rRNA, and U6 snRNA in eukaryotes. Transcription of pol III is higher during S and G2 phases of the cell cycle than in G1. Many neoplastic cells display high pol III activity, indicating that protein synthesis is demanded for tumorous growth. The RET protein appears to be a suppressor of increased pol III activity. *See* La; ribosomal RNA; ribosomes; tRNA. (Geiduschek, E. P. & Tocchini-Valentini, G. P. 1988. *Annu. Rev. Biochem.* 57:873; Huang, Y & Maraia, R. J. 2001. *Nucleic Acids Res.* 29:2675.)

pol α   DNA polymerase (encoded in fission yeast by gene *pol1/swi7*) replicating nuclear DNA (lagging strand) in cooperation with the primase of eukaryotes. *See* DNA polymerases; lagging strand; primase; replication fork.

Poland syndrome   Autosomal-dominant defect with low penetrance. The inheritance pattern is complicated by the teratogenic effects of diverse exogenous factors. It is characterized by fusion of fingers (syndactyly), short fingers, anomalies of the chest and sometimes other muscles. *See* limb defects; penetrance; syndactyly; teratogenesis.

polar   Hydrophilic, i.e., soluble in water; molecules with polarized bonds.

polar body   *See* gametogenesis in animals.

polar body diagnosis   The genetic constitution of the polar body is tested prenatallly by molecular techniques. *See* prenatal diagnosis.

polar bond   Covalent, yet the electrons are more firmly tied to one of the two molecules and therefore the electric charge is polarized.

polar coordinate model   Polar coordinate model of regeneration states that when cells are in nonadjacent positions, the process of growth restores all intermediate positions by the shortest numerical routes. The shortest intercalation mandates that small fragments may undergo duplication and large fragments may require regeneration. The position of each cell on a collapsed cone (the idealized primordium) is specified by the radial distance from a central point at the tip of the cone and the circumferential position on the circle defined by the radius of the base. *See* distalization. (Held, L. I. 1995. *Bioessays* 17:721.)

polar cytoplasm   Situated in the posterior (hind) portion of the fertilized egg cell. *See* pole cells.

polar granules   Present in the posterior pole region of insect eggs; have maternal effects and a germ cell specification role during embryogenesis. These granules are the mitochondrially coded 16S ribosomal RNA large subunits (mtRNA) exported from the organelle. *See* animal pole; morphogenesis in *Drosophila*; RNA localization.

polarimeter   Measures the rotation of the plane of polarized light.

polarisome   Defines polarity within a cell with the aid of several proteins. *See* Weiner, O. D. 2002. *Current Opin. Cell Biol.* 14:196.

polarity, embryonic   Required for differentiation and requires asymmetric cell divisions. In *Caenorhabditis*, the PAR proteins (serine/threonine kinase) control embryonic polarity and a nonmuscle-type myosin II heavy-chain protein (NMY-2) is a cofactor of this polarity. In *Drosophila*, the major body axes, primarily the anterior-posterior polarity, are controlled by the gurken-torpedo gene products, but other genes are also involved. Polarity may be achieved either by the asymmetric distribution of proteins or mRNA. *See* BUD; differentiation; morphogenesis in *Drosophila*; polar cytoplasm; RNA localization. (Drees, B. L., et al. 2001. *J. Cell. Biol.* 154:549; Wodarz, A. 2002. *Nature Cell Biol.* 4:E39.)

polarity of hyphal growth   *See* diagram below.

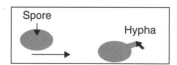

Polarity of hyphal growth.

polarity mapping   *See* mapping mitochondrial genes.

polarization   Distortion of the electron distribution in one molecule caused by another. *See* bouquet of chromosomes.

polarized differentiation   Basis of morphogenesis, chemotactic response, response to pheromones, etc. Polarized differentiation and growth is typical for neural and microtubule growth, for the pollen tubes, and for the roots of plants. (Hepler, P. K., et al. 2001. *Annu. Rec. Cell Dev. Biol.* 17:159.)

polarized light Exhibits different properties in different directions at right angles to the line of propagation. Specific rotation is the power of liquids to rotate the plane of polarization.

polarized recombination *See* polarized segregation.

polarized segregation May be brought about by meiotic anomalies, e.g., in maize plants heterozygous for some knobbed chromosomes (and syntenic markers) are preferentially included into the basal megaspore. Polarized segregation has been observed as a result of gene conversion, e.g., in *Ascobolus immersus* alleles of the *pale* locus in the cross

$$\frac{188w^+}{188^+w}$$

segregated in both cases 6:2, but in the first case the results were $(4[188] + 2[w^+]) : 2(w)$, whereas in the cross

$$\frac{w137^+}{w^+137}$$

the conversion asci were $(4[w] + 2[137]) : 2(w^+)$. The genetic order of these alleles was 188 w137. Thus, in the first cross *white* was in the minority class, whereas in the second cross it was part of the majority class. *See* gene conversion; map expansion; meiotic drive. (Whitehouse, H. L. K. & Hastings, P. J. 1965. *Genet. Res.* 6:27.)

polarizing microscope Uses a *polarizer* (a Polaroid screen) in front of the light beam and an *analyzer* over the eyepiece (permitting rotation). The anisotropic specimens (having a difference in transmission or reflection depending on the angle of light) will display optical contrast. *See* microscopy.

polar molecule Generally soluble in water; the distribution of the positive and negative charges is not even, thus resulting in a polarized effect.

polar mutation May be a base substitution (nonsense mutation), insertion, frameshift, or any chromosomal alteration that affects the expression of genes downstream in the transcription-translation system. *See* frameshift mutation. (Jacob, F. & Monod, J. 1961. *Cold. Spring Harbor Symp. Quant. Biol.* 26:193.)

polar nuclei In the embryo sac of plants, formed at the third division of the megaspore. After they have fused ($n + n$) and been fertilized by one sperm (n), they give rise to the triploid ($3n$) endosperm nucleus. *See* embryo sac; megagametophyte.

polarography Electrochemical measurement of reducible elements.

Polaroid camera Was developed in the 1940s. Many uses have been found in biological laboratories because it can provide almost immediate negative or positive images for recording observations such as of electrophoretic gels. The combined developing and fixing solution is contained in between the exposed negative film and the receiving film or paper. When the storage "pod" bursts under pressure of pulling, the processing is carried out within the camera. For some tasks, digital cameras are even better suited for fast imaging.

polaron Part of a locus within which gene conversion (or recombination) is polarized. *See* gene conversion; polarized segregation. (Whitehouse, H. L. K. & Hastings, P. J. 1965. *Genet. Res.* 6:27.)

polar overdominance Unusual type of inheritance, i.e., mutants heterozygous for the dominant *callypige* gene of sheep (chromosome 18) display the (CLPG) allele only when inherited from the males but not from the females. The phenotype is a muscular hypertrophy resulting from the cis-regulation of four imprinted genes. *See* imprinting; overdominance. (Charlier, C., et al. 2001. *Nature Genet.* 27:367.)

polar transport Certain metabolites move in only one direction in the plant body, e.g., the auxins under natural conditions are synthesized in the tissues over the ground, then move toward the roots.

pol β Eukaryotic DNA repair polymerase. *See* DNA polymerases.

pol δ Eukaryotic DNA polymerase (replicating the leading strand) of the nuclear chromosomes. *See* DNA polymerases; replication fork.

pol δ₂ pol ε. *See* DNA polymerases.

pole cells Localized in the posteriormost part of the cellularized embryo, pole cells eventually give rise to the germline. *See* germline.

pol ε Eukaryotic DNA polymerase (*cdc20*) with a repair role. *See* DNA polymerases.

pol γ DNA polymerase replicating eukaryotic organelle DNA. *See* DNA polymerases; θ-type replication.

polio viruses Icosahedral single-stranded RNA viruses with about 6.1 kb RNA in a total particle mass of about 6.8×10^6 Da. Type 1 was responsible for about 85% of the poliomyelitis (infantile paralysis) cases before successful vaccination (live oral, Sabin or inactivated, Salk) became widely used in developed countries. These small RNA viruses are highly mutable because their genetic material lacks repair systems. The three serotypes produce a cell-surface receptor (PVR) by alternative splicing of its transcript. Susceptibility to polio virus was located to human chromosome 19q12-q13. Mice are very resistant to this virus because they lack the membrane receptor for the infection. Polio virus can now be assembled in the laboratory. *See* IRES; picornaviruses. (Cello, J., et al. 2002. *Science* 297:1016.)

POLLED Dominant/recessive gene in goats and cattle responsible for lack of horns/intersexuality. The forkhead transcription factor (FOXL2)—also responsible for blepharophimosis—may be involved. In goats, a 11.7 kb deletion at

1q43 (homologous to human band 3q23) normally encodes two mRNAs. The FOXL2 transcript is homologous with the human blepharophimosis syndrome gene. *See* blepharophimosis. (Crisponi, L., et al. 2001. *Nature Genet.* 27:159; Pailhoux, E., et al. 2001. *Nature Genet.* 29:453.)

pollen Male gametophyte of plants developing from the microspores by two postmeiotic divisions. The first division results in the formation of a vegetative and a generative cell. The round vegetative cell directs the elongation of the pollen tube growing through the pistil toward the ovule. The crescent-shaped generative cells may divide before or after the shedding of the pollen grains. One of them fertilizes the egg and gives rise to the diploid embryo; the other fuses with the diploid polar cell in the embryo sac and contributes to the formation of the endosperm. The pollen tube elongates quite rapidly; it may grow 15 cm in just 5 to 15 hours. A protein that is glycosylated in that tissue regulates the pollen tube elongation. In allogamous species, a single individual may shed over 50 million pollen grains, whereas in autogamous species the number of pollen grains per anther may not exceed a couple of hundred. Since the pollen grain is haploid and may be autonomous (gametophytic control), it may express its genetic constitution independently from the genotype of the anther tissues (e.g., waxy pollen, various color or sterility alleles). In some instances, however, the morphology of the pollen grain is under sporophytic control. Since the pollen is a more independent product than the megaspore, it is more likely to suffer from genetic defects for which the surrounding tissues cannot compensate; therefore, pollen sterility is more common in plants than female sterility. Pollen sterility may not necessarily affect, however, the fertility of the individuals because of the abundance of functional pollen grains in case of heterozygosity for the defects. Under normal atmospheric conditions (high humidity) and high temperature, the viability of the pollen is maintained (depending on the species) for a few minutes or for several hours. In a refrigerator at low humidity, the viability of the pollen can be extended substantially. Freeze-dried and properly stored pollen of several species retains its ability to fertilize for years. Insects may carry viable pollen for long distances. According to a study, in rye populations cross-pollination (mediated by wind) may

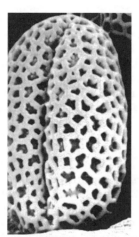

The sculptured surface of the mature pollen of *Arabidopsis*. (From Craig, S. & Chaudhury, A. 1994. in *Arabidopsis*: An Atlas of Morphology and Development. Bowman, J. L., ed. Courtesy of Bowman, J. L. By permission of Springer-Verlag, New York.)

occur to 50% at 100 m distance, to 20% at about 400 m distance and only to 3% at 600 to 700 m. Other studies reported only 7% cross-pollination in rye at a distance of 20 m. The prevailing environmental conditions (humidity, temperature, wind, etc.) and the quantity of the pollen influence the spread and viability. These problems have gained new interest with the use of genetically engineered crops, which are opposed by some environmentalists. The extracellular matrix of the pollen contains proteins that recognize species specificity and efficient pollination (Myfield, J. A., et al. 2001. *Science* 292:2482). These proteins are lipid-binding oleosins and lipases. *See* allogamy; autogamy; cross-pollination; gametogenesis; gametophyte; microsporogenesis; pollen tetrad; self-incompatibility.

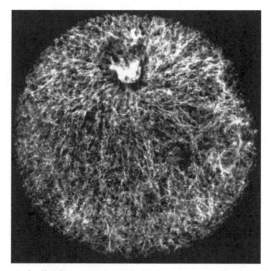

Fluorescent-phalloidin staining of the F-actin cytoskeleton in maize pollen. The bright spot on top is the pollen tube initial. (Courtesy of Dr. Chris Staiger. See Gibbon, B. C., et al. 1999. *Plant Cell* 11:2349.)

pollen competition *See* certation.

pollen-killer Or spore-killer genes in wheat, tomato, and tobacco, render the pollen incapable of functioning effectively in fertilization and may cause segregation distortion. *See* killer genes; killer plasmids; killer strains; pollen tube competition; segregation distorter.

pollen mother cell Microspore mother cell; microsporocyte. *See* gametogenesis.

pollen sterility Inability of the pollen to function in fertilization. It can frequently be detected by poor staining of the pollen grains with simple nuclear stains (acetocarmine, acetoorcein, etc.). Deletions, translocations, and inversion heterozygosity generally result in pollen sterility. Mitochondrial plasmids may also be responsible for some types of male sterility. *See* certation; cytoplasmic male sterility; fertility restorer genes; gametophyte; pollen.

pollen tetrad Four products of a single male meiosis. The components of the pollen tetrad may not stick together and may shed in a scrambled state. In some instances (*Salpiglossis*,

Elodea, some orchids), the tetrads remain together, however, in a way similar to the unordered tetrads of fungi. In *Arabidopsis*, induced mutations (*qrt1, qrt2, quartet*) cause the four pollen grains to stay together because of the alteration of the outer membrane of the pollen mother cell. Each tetrad may then fertilize four ovules. *See* tetrad analysis.

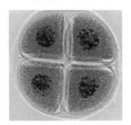

Pollen tetrad.

pollen tube *See* pollen; synergid. (Palavinelu, R. & Preuss, D. 2000. *Trends Cell Biol.* 10:517.)

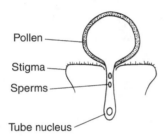

pollen tube competition *See* certation.

pollination Transfer of the male gametophyte to the stigmatic surface of the style (ovary). *See* allogamy; autogamy; gametophyte.

pollinium Mass of pollen sticking together; may be transported as such by the pollinator insects or birds.

Pollitt syndrome *See* trichothiodystrophy.

pollution Spoiling the environment by the release of unnatural, impure, toxic, mutagenic, carcinogenic, or any other undesirable and unaesthetic material, or disturbing nature by sound, odor, heat, or light. Pollution may cause mutation, cancer, and various other diseases.

POLO 577-amino-acid serine/threonine protein kinase of *Drosophila* required for mitosis. During interphase it is predominantly cytoplasmic, but at the end of prophase it associates with the chromosomes until telophase. *See* FEAR. (Llamazares, S., et al. 1991. *Genes Dev.* 5:2153.)

polyacrylamide *See* electrophoresis; gel electrophoresis.

Polyacrylamide gel electrophoresis.

polyadenylation signal Endonucleolytic processing of the primary transcript of the majority of eukaryotic genes is followed by posttranscriptional addition of adenylic residues downstream of the structural gene. The consensus signal for the process is 5'-AAUAAA-3' in animals and fungi. About half of the plants use the same signal. The rest rely on diverse signals. In eukaryotes, the number of added A residues might vary from 50 to 250. Polyadenylation is under the control of several genes. Besides the poly(A) signal, the RNA transcript of eukaryotes contains a CA element (PyA in yeast) and a GU-rich downstream element.

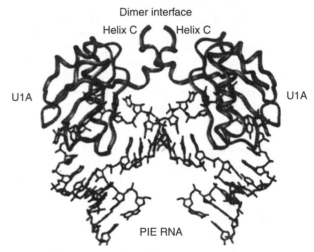

Poly(A) polymerase is regulated by inhibition (or stimulation) by the polyadenylation inhibition RNA element (PIE). PIE forms a trimolecular complex including the two u1a protein molecules. The U1A consists of a four-stranded β-sheet and two α-helices. The two C helices of the protein interface. The PIE RNA, which contains two asymmetric internal loops is separated by four Watson-Crick-paired nucleotides. (From Puglisi, J. D. 2000. *Nature Struct. Biol.* 7:263.)

The *positioning element* binds the *polyadenylation specificity factor* (CPSF), which is a tetrameric protein consisting of 160, 73, 100, and 33 kDa subunits, to AAUAAA. The *cleavage-stimulating factor* (CstF) is a trimeric protein of 64, 77, and 50 kDa subunits and binds to the GU-rich element of the RNA. The *polyadenylation polymerase* protein (PAP) binds downstream of the CPSF-binding sites. The *cleavage factors*, CFI and CFII, are positioned upstream of the GU-rich element and they terminate the mRNA. The polyadenylation complex of yeast is somewhat different. PABP (poly-A-binding protein, 70 kDa) regulates mRNA stability, translation, and degradation. CPEB (cytoplasmic element-binding protein), maskin, and cyclin B1 are regulators of polyadenylation and transcription of some mRNAs. Poly(A)-specific ribonuclease (PARN) is a cap-interacting 3' exonuclease. In cooperation with the cap, it mediates deadenylation from the cis position or at low concentration it may be inhibitory to deadenylation. From the trans position it inhibits deadenylation if its concentration is high. The poly(A) tail in synergy with the mRNA cap stimulates the initiation of translation. The poly(A)-binding protein (PABP) interacts with eIF4G of the eIF4F complex and eIF4E interacts with the cap and stabilizes mRNA. PABPs are located both in the nucleus and in the cytoplasm. In the human testes a specific poly(A)-binding protein occurs that is absent from other tissues (Féral, C., et al. 2001. *Nucleic Acids Res.* 29:1872). Deadenylation of the tail initiates mRNA decay, and when less than 10 A residue are left, an exonuclease attacks the RNA in the 5′ → 3′ direction (Martínez, J., et al. 2001. *J. Biol. Chem.* 276:27923).

In prokaryotes, rarely a few (14–60) adenine residues are also found at the mRNA 3′-terminus in 1 to 40% of the cases. In bacteria, *host factor q* (Hfq) plays a role similar to PABP. It stimulates the elongation of the polyA tail by poly(A) polymerase I (PAP) and protects against exoribonuclease attack. Some adenine sites are found in about 30% of both the early transcripts (transcribed by host polymerase) and late transcripts (transcribed by viral polymerase). Sometimes the poly-A sequence has other bases interspersed and may also be located within coding sequences. An interspersed long poly-(A) sequence was detected in chloroplast RNA transcripts. In mitochondria, the poly-A tract (35–55 A residues) is directly attached to the termination codon without an untranslated sequence after the endonucleolytic cleavage of the polycistronic transcript. In liver cancer, mitochondria tails of hundreds of A*s* have been observed. In prokaryotes, two similar (36 and 35 kDa) poly-A polymerases with overlapping functions have been identified. Polyadenylation of RNA in bacteria regulates plasmid replication and the degradation of RNAI. In *Archaea*, short poly-A tracts exist. Cordycepin (3′-deoxyadenosine) is an inhibitor of polyadenylation. *See* cleavage stimulation factor; eIF4; mRNA circularization; mRNA tail; PABp; polyA polymerase; RNA I; U1 RNA. (Hirose, Y & Manley, J. L. 1998. *Nature* 395:93; Sarkar, N. 1997. *Annu. Rev. Biochem.* 66:173; Beaudoing, E., et al. 2000. *Genome Res.* 10:1001; Mendez, R. & Richter, J. D. 2001. *Nature Rev. Mol. Cell Biol.* 2:521; Wang, L., et al. 2002. *Nature* 419:312.)

poly(ADP-ribose) polymerase   DNA-binding enzyme, but it appears to have no indispensable function.

PolyA⁺ elements   Transposons without long terminal repeats but poly-A sequences at the 3′-OH end. The RNA elements are usually mobilized via a DNA transcript with the aid of their encoded reverse transcriptase. Such elements are L1 (LINE) in mammals, the TART of *Drosophila*, the TRAS1 of silkworm, or the Ty5 telomere-specific elements of yeast. Other polyA⁺ elements (L1, I, and fungal and plant elements) can target a variety of other sites. *See* hybrid dysgenesis; TART; transposable elements.

polyamides   Contains N-methylimidazole and N-methyl-pyrrole amino acids with a high affinity for specific DNA sequences and may regulate the transcription similarly to

DNA-binding proteins. *See* binding proteins; inhibition of transcription; lexitropsin; netropsin. (Maeshima, K., et al. 2001. *EMBO J.* 20:3218.)

polyamidoamine dendrimers (PAMAM)   Highly branched, soluble, nontoxic molecules with amino groups on their surface. They are suitable for attaching antibodies, various pharmaceuticals, and DNA to this surface. They are effective vehicles for transfection. (Gebhart, C. L. & Kabanov, A. V. 2001. *J. Control Release* 73:401.)

polyamines   Various protein molecules derived in part from arginine and present in cells in millimolar concentrations, yet they have an important role in RNA and DNA transactions, replication, supercoiling, bridging between strands, binding phosphate groups, biosynthesis, degradation, etc. Typical polyamines are spermine, spermidine, putrescine, etc. *See* antizyme; lexitropsins. (Coffino, P. 2001. *Nature Rev. Mol. Cell Biol.* 2:188; van Dam, L., et al. 2002. *Nucleic Acids Res.* 30:419.)

polyA mRNA   Eukaryotic mRNAs posttranscriptionally polyadenylated at the 3′ tail before leaving the nucleus, and subsequently in the cytoplasm the tail may be reduced to 50–70 residues or further extended to hundreds. Polyadenylation improves stability and efficiency of translation in cooperation with mRNA cap. The polyA tail and the mRNA cap seem to cooperate in the initiation of translation. PolyA tail is frequently added to bacterial RNA. The addition of polyA tail accelerates the decay of RNA I of *E. coli*. The majority of eukaryotic viruses (except areana- and reoviruses) also produce a poly A tail. In *Drosophila*, the length of the poly(A) tail may be correlated with the function in differentiation of the mRNA. The regulatory mechanism of polyadenylation is interchangeable between mice and *Xenopus*. *See* capping enzymes; eIF; mRNA degradation; mRNA tail; PABP; polyadenylation signal; RNA I. (de Moor, C. H. & Richter, J. D. 2001. *Int. Rev. Cytol.* 203:567.)

polyandry   Form of polygamy involving multiple males for one female. It may have the advantage of reducing the relatedness within colonies of social insects, thereby increasing fitness. *See* fitness. (Tregenza, T. & Wedell, N. 2002. *Nature* 415:71.)

polyA polymerase (*PAP*)   Adds the polyA tail posttranscriptionally to the eukaryotic mRNA and antisense RNA transcripts. In yeast, at least two other genes, *RNA14* and *RNA15*, are involved in the processing of the 3′-end of pre-mRNA. *E. coli* also encodes at least two PAP enzymes. PolyA polymerase also facilitates the degradation of mRNA because it provides single-strand tails for polynucleotide phosphory-lase. *See* mRNA tail; polyadenylation signal; polynucleotide phosphorylase. (Dickson, K. S., et al. 2001. *J. Biol. Chem.* 276, 41810; Steinmetz, E. J., et al. 2001. *Nature* 413:327.)

polyaromatic compounds   Include various procarcinogens and promutagens such as benzo(a)pyrene, dibenzanthracene, methylcholanthrene, etc. *See* polycyclic hydrocarbons.

polyA tail   *See* polyadenylation signal; polyA mRNA.

Polybrene® (hexadimethrine bromide) Polycation used for introduction of plasmid DNA into animal cells; it is also an antiheparin agent and an immobilizing agent in Edman degradation. *See* animal cells; Edman degradation; heparin; transformation, genetic.

polycentric chromosome *See* neocentromeres.

polychlorinated biphenyl (PCB) Highly carcinogenic compound and an inducer of the P-450 cytochrome group of monooxygenases. It had been used in electrical capacitors, transformers, fire retardants, hydraulic fluids, plasticizers, adhesives, pesticides, inks, copying papers, etc. *Pseudomonas* sp. KKS102 is capable of degradation of PCB into tricarboxylic acid cycle intermediates and benzoic acid. *See* carcinogen; microsomes; P-450; S-9. (Ohtsubo, Y., et al. 2001. *J. Biol. Chem.* 276:36146.)

polychromatic Stainable by different dyes or displaying different shades when stained.

polycistronic mRNA Contiguous transcript of adjacent genes, such as those that exist in an operon, but it may be formed in short genes of eukaryotes, e.g., oxytocin. The *Trypanosomas* produce multicistronic transcripts. *See* *Caenorhabditis*; operon; oxytocin; *Trypanosoma*.

polyclonal antibodies Produced by a population of lymphocytes in response to antigens. These are not as homogeneous as the monoclonal antibodies. *See* monoclonal.

Polycomb (*Pc*, chromosome 3–47.1) *Drosophila* gene is a negative regulator of the *Bithorax* (*BXC*) and *Antennapedia* (*ANTC*) complexes. The homozygous mutants are lethal and the locus (and its homologues in vertebrates [*M33* in mice]) is involved in the repression of homeotic genes, which control body segmentation. *Pc* is a member of a group (*Pc-G*) of repressors of homeotic genes. Although *Pc* is located in the euchromatin, it is involved in the silencing of genes by heterochromatin. Insertion into the 5th exon of *M33* caused male → female sex reversal. *Pc* is required for the activation of other silencing elements and its mutation may lead to derepression of these elements. The suppressive effect of *Pc* may be associated with chromatin remodeling and histone deacetylation. The Polycomb group of proteins forms a large complex and the TATA box–binding proteins, Zeste, and others are associated with the general transcription machinery. *See* Antennapedia; Bithorax; chromatin remodeling; chromodomain; histone deacetylase; homeobox; homeotic genes; morphogenesis in *Drosophila*; sex reversal; *SWI*; TBP; transcription factors; trithorax; *w* locus; *zeste*. (Breilling, A., et al. 2001. *Nature* 412:651; Simon, J. A. & Tamkun, J. W. 2002. *Current Opin. Genet. Dev.* 12:210.)

polycross Intercross among several selected lines to produce a synthetic variety of a crop (Tysdal, H. M., et al. 1942. *Alfalfa Breeding*. Nebraska Agric. Exp. Sta. Res. Bull. 124, Lincoln, NE.)

polycyclic aromatic hydrocarbons (PAH) Generally carcinogenic and mutagenic compounds. They become more active during the process of the attempted detoxification by the microsomal enzyme complex. PAHs are the products of burning organic material (coal, charbroiling, smoking, etc.). Mice oocytes exposed to PAHs suffer apoptosis by activation of BAX. *See* apoptosis; BAX; benzo(a)pyrene; carcinogens; environmental mutagens; mutagens; PAH; procarcinogens; promutagens. (Matikainen, T., et al. 2001. *Nature Genet.* 28:355.)

polycystic kidney disease (PKD) Two main types, and several variations exist within each type. The adult-type dominant (ADPKD) is apparently controlled by the short arm of chromosome 16p13.31-p13.12. It involves fragility of the blood vessel walls. In ARPKD, the basic defect is in a Ca^{2+}-permeable nonselective cation channel. About 15% of the APKD cases are due mutation in the gene (PKD2) encoding polycystin. There is also another gene (PKD1) and these are involved in the proliferation of epithelial cells lining the cyst cavity, thickening of the basement membrane, fluid secretion and protein sorting (Bukanov, N. O., et al. 2002. *Hum. Mol. Genet.* 11:923). The autosomal recessive (ARPKD) generally has an early onset. Both forms occur at frequencies of 0.0025 to 0.001. Even the late-onset type may be detectable early by tomography. The symptoms vary and involve kidney disease, cerebral vein aneurism (sac-like dilatation), underdeveloped lungs, liver fibrosis, growth retardation, etc. The dominant type can be identified with high accuracy using chromosome 16p13 DNA probes, but less than 10% of the cases are due to genes not in chromosome 16. The autosomal-recessive form is at an unknown location and it can be identified after the third trimester by ultrasonic methods because the kidneys are enlarged. Genetic transmission of the dominant and recessive diseases is very good. One polycystic kidney (PKD1, 4300-amino-acid integral membrane glycoprotein) locus was assigned to 6q21-p12, and sequences were also found in 2p25-p23 and 7q22-q31 that are homologous to polycystic kidney disease of the mouse. There is a PKD2 locus in 4q21-q23, and this is similar in function to PKD1. PKD2 interacts with PKD1, and PKD2 interacts with the Hax-1 protein-binding F-actin, suggesting that the system affects the cytoskeleton. Thus, defects in PKD2 may be one of the causes of cyst formation in the kidney, liver, and pancreas. Infantile-type recessive PKD is also called Caroli disease. The ARPKD locus encodes a 968-amino-acid protein that forms six transmembrane spans with intracellular amino and carboxyl ends. It appears to be a voltage-activated Ca^{2+} (Na^+) channel protein. PKD1 may involve haploinsufficiency. The PKD1 homologue in *Caenorhabditis* (*LOV-1*) controls sensory neurons required for male mating behavioral steps. *See* cardiovascular disease; genetic screening; haploinsufficient; hypertension; ion channels. (Pei, Y., et al. 2001. *Am. J. Hum. Genet.* 68:355.)

polycystic lipomemembranous osteodysplasia with sclerosing leukoencephaly (PLOSL, Nasu-Hakola disease, 19q13.1) Recessive psychosis turning into presenile dementia and bone cysts limited to the wrists and ankles. Prevalence in Finland is 2×10^{-6}. The basic problem is a loss of function of the TYROBP/DAP12 tyrosine kinase-binding transmembrane protein, an activator of killer lymphocytes. *See* killer cells.

polycystic liver disease (PCLD, 19p13.2-p13.1) Dominant, often accompanies polycystic kidney disease. It involves fluid-filled cysts on the liver.

polycystic ovarian disease (Stein-Leventhal syndrome) Generally involves enlarged ovaries, hirsuteness, obesity, lack of or irregular menstruation and an increased level of testosterone, high luteinizing hormone:follicle-stimulating hormone ratios, and infertility. It appears to be due to an autosomal factor, yet 96% and 82% of the daughters of affected mothers and carrier fathers, respectively, developed the symptoms, indicating a meiotic drive-like phenomenon. Deficiency of 1-α-ketosteroid reductase/dehyrogenase (9q22) may cause polycystic ovarian disease as well as pseudohermaphroditism with gynecomastia in males. *See* Graafian follicle; gynecomastia; infertility; luteinization; meiotic drive; pseudohermaphroditism.

polycythemia (PFCP) Autosomal-dominant proliferative disorder of the erythroid progenitor cells, resulting in an increase in the number of red blood cells and in vitro hypersensitivity to erythropoietin. *See* erythropoietin.

polydactyly Presence of extra fingers or toes. In *postaxial* polydactyly (the most common type), the extra finger is in the area of the little finger, and in *preaxial* cases this malformation is on the opposite side of the axis (thumb) of the palm or foot. The various types of polydactyly may be determined by autosomal-recessive or -dominant genes, and their expression is usually part of other syndromes. Crossed polydactyly indicates coexistence of postaxial and preaxial types with discrepancy between hands and feet. Synpolydactyly is caused by an expansion of the normal 15 GCG trinucleotides to 22–29. *See* adactyly; diastrophic dysplasia; ectrodactyl; Ellis–van Creveld syndrome; focal dermal hypoplasia; Greig's cephalopolysyndactyly syndrome; *hedgehog*; Majewski syndrome; Meckel syndrome; Opitz syndrome; orofacial-digital syndromes; Pallister-Hall syndrome; Pätau's syndrome; polysyndactyly; Rubinstein-Taybi syndrome; syndactyly.

Polydactyly (From Bergsma, D., ed. 1973. *Birth Defects. Atlas and Compendium.* By permission of the March of Dimes Foundation.

polyembryony More than one cell of the embryo sac develops into an embryo in plants. *See* adventive embryos; embryo sac.

polyethylene glycol (PEG) Viscous liquid or solid compound of low toxicity, promoting fusion of all types of cells. PEG is widely used in the textile, cosmetics, paint, and ceramics industries. *See* PEG.

polygalacturons Complex carbohydrates in the plant cell wall.

polygamy Having more than one mating partner. In Western human societies, it is illegal, but in some others it is still acceptable that men have more than one wife simultaneously. Polyandry or polygyny is common practice in animal breeding, but it may be objectionable to humans on moral grounds. In the U.S., the polygamy laws are applied to all citizens, irrespective of religious affiliation or cultural tradition.

polygenes A number of genes are involved in the control of quantitative traits. *See* gene number in quantitative traits; QTL.

polygenic inheritance Determined by a number of nonallelic genes involved in the expression of a single particular trait (such as height, weight, intelligence, etc.). Polygenic inheritance is characterized by counting and measurements, and the segregating classes are not discrete but display continuous variation. *See* chaos; complex inheritance; QTL; quantitative genetics. (Tanksley, S. D. 1993. *Annu. Rev. Genet.* 27:205; Klose, J., et al. 2002. *Nature Genet.* 30:385.)

polygenic plasmids Obtained when two plasmids carrying identical genes cointegrate. Such plasmids may have merit in genetic engineering if the genes show a positive dosage effect for anthropocentrically useful traits. *See* cointegration.

polygeny One male has more than a single female mate. In *sororal polygeny*, the females are sisters. *See* effective population size; polygamy.

polyglutamine diseases *See* trinucleotide repeats.

polygyne Social insect colonies with more than a single queen. *See* monogyne.

polyhaploid Has half the number of chromosomes of a polyploid. The gametes of polyploids are polyhaploid. *See* polyploidy.

polyhedrosis virus, nuclear (BmNPV) About 130 kbp DNA baculovirus of the silkworm (and other insects). It has been used (after size reduction) as a 30 kb cloning vector and it may propagate in a single silkworm larva about 50 µg DNA . *See* baculoviruses; silkworm; viral vectors.

polyhybrid Heterozygous for many gene loci.

polyhydroxybutyrate (PHB) Bacterial polymer that can be manufactured by transgenic plants and is biodegradable.

poly I-G A DNA strand containing more cytosine is called the heavy chain of a DNA double helix because it binds more of the polyI-G (inosine-guanosine) sequences. Ultracentrifugation in CsCl separates these DNA heavy strands. *See* density gradient centrifugation; DNA heavy chain; inosine; ultracentrifuge.

polyisoprenyl phosphates Intermediates in cholesterol biosynthesis. They play a role in immune system signaling. *See* cholesterols; immune system.

polyketenes Polymers of $CH_2=C=O$ (ketene). Their biosynthesis is related to fatty acids. Several antibiotics (tetracycline, griseofulvin, etc.) contain ketenes. *See* antibiotics.

polyketides Various naturally occurring compounds built from residues that each usually contribute two carbon atoms to the assembly of a linear chain of which the β-carbon carries a keto group. These keto groups are frequently reduced to hydroxyls. The remaining keto groups, called polyketides, at many of the alternate carbon atoms form the chain. The polyketide synthesis pathway resembles that of the fatty acid path. Flavonoids, mycotoxins, antibiotics, etc., occurring in angiosperms to bacteria qualify for the polyketide collective name. Polyketide synthetases generate the precursors of erythromycin, rapamycin, and rifamycin antibiotics. *See* epothilone; lovastatin. (Khosla, C., et al. 1999. *Annu. Rev. Biochem.* 68:219.)

polykinetic chromosome Has centromeric activity at multiple sites. *See* neocentromeres.

polymerase Enzyme that builds up large molecules from small units, such as the DNA and RNA polymerases generate DNA and RNA, respectively, from nucleotides. *See* pol.

polymerase accessory protein (RF-C) Essential part of the DNA replication unit in SV40. *See* SV40.

polymerase chain reaction (PCR) Method of rapid amplification of DNA fragments when short flanking sequences of the fragments to be copied are known. The reaction begins by denaturation of the target DNA, then primers are annealed to the complementary single strands. After adding a heat-stable DNA polymerase, such as Taq or Vent/Tli (originally the less thermostable Klenow fragment of polymerase I was used), chain elongation proceeds starting at the primers. The cycles are repeated 20–30 times, resulting in over a million-fold ($2^{20} = 1,048,576$) replication of the target. The actual rate of replication may be less (80%) than the theoretically expected rate.

The amplified DNA can be subjected to molecular analysis such as preimplantation analysis, genetic screening, prenatal analysis, sperm typing, gene identification, etc. The error frequency for the Klenow fragment is about 8×10^{-5}, for Taq 10^{-5} to 10^{-4}, for Tli 2 to 3×10^{-5}. PCR amplification can be performed with a variety of

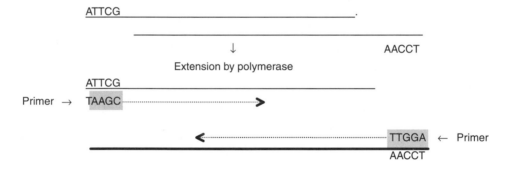

polylinker DNA sequence with several restriction enzyme recognition sites (multiple cloning sites, MCS) used in construction of different cloning or transformation vehicles (plasmids), e.g., TTCTAGAATTCT sequence has an overlapping XbaI (TCTAGA) and an EcoRI recognition site (GAATTC); thus, linking it to DNA may generate both types of cloning sites. *See* cloning sites; pUC; restriction enzymes; vectors.

poly(L-lysine) Polycation that can form complexes with negatively charged DNA and mediate gene transfer using retroviral vectors. In case the polycation has bound specific ligands, it can be targeted to special cell types. Without such a complex, the viral vector would have no target specificity. Some of the polycationic delivery systems are cytotoxic and/or may be subject to lysosomal degradation. *See* transformation, genetic. (Putnam, D., et al. 2001. *Proc. Natl. Acad. Sci. USA* 98:1200.)

polymarker test *See* DNA fingerprinting.

polymer Large molecule composed of a series of covalently linked subunits such as amino acids, nucleotides, fatty acids, carbohydrates, etc. *See* DNA; protein.

mechanical devices including chemical amplification on a microchip where the 20 cycles may be completed in only 90 seconds. All types of technical information and references are available at <http://apollo.co.uk/a/pcr>. *See* ancient DNA; capture PCR; DNA fingerprinting; double PCR and digestion; genetic screening; hot-start PCR; immuno-PCR; in situ PCR; INTER-SS PCR; inverse PCR; methylation-specific PCR; molecular evolution; PCR, allele-specific; PCR-based mutagenesis; PCR broad-base; PCR-LSA; PCR overlapping; preimplantation genetics; prenatal analysis; primer extension; PRINS; RAPDS; recursive PCR; reverse ligase-mediated polymerase chain reaction; RNA-PCR; RT-PCR; small-pool PCR; sperm typing; tail PCR; thermal cycler; tissue typing; touch-down PCR; vectorette. (Mullis, K. B. & Faloona, F. A. 1989. in *Recombinant DNA Methodology*, Wu, R., et al., eds., p. 189. Academic Press, San Diego, CA; Innis, M., et al., eds., 1990. *PCR Protocols: A Guide to Methods and Applications.* Academic Press, San Diego, CA.)

polymerase switching DNA replication is initiated by the polymerase α/primase complex, but subsequently chain elongation is continued by eukaryotic polymerase δ. Polymerase ε

may also have some role in initiation and elongation. *See* DNA polymerases; primase; processivity; replication fork.

polymery   Several genes cooperate in the expression of a trait. *See* polygenes.

polymorphic   A trait occurs in several forms within a population. The polymorphism may be balanced and genetically determined. *See* balanced polymorphism; polymorphism; RFLP; SNP.

polymorphic information content (PIC)   Used to identify and locate a hard-to-define marker locus. If the alleles of the marker locus are codominant, then PIC is the fraction of the progeny (the informative offspring) that cosegregates by phenotype with an index locus. The index locus (which is used for the detection of linkage with marker alleles) has two alternative alleles: a wild type and a dominant (mutant) allele. The marker locus is polymorphic for dominant (genetic or physical [nucleotide sequences]) alleles. Only those progenies are informative where the index locus is homozygous in one of the parents and the other parent is heterozygous for the marker. The converse constitutions are not informative. In case both parents are heterozygous at the marker locus, only half of the offspring are informative.

$$\text{PIC} = 1 - \sum_{i=1}^{n} p_i^2 - \left(\sum_{i=1}^{n} p_i^2 \right)^2 + \sum_{i=1}^{n} p_i^4$$

where p_i = frequency of the index allele and i and n are the number of different alleles. The PIC values may vary theoretically from 0 to 1. A hypothetical example: Four A alleles occur in a population with frequencies $A^1 = 0.2$, $A^2 = 0.1$, $A^3 = 0.15$, and $A^4 = 0.55$. After substitution,

$$\text{PIC} = 1 - (0.2^2 + 0.1^2 + 0.15^2 + 0.55^2)$$
$$- (0.2^2 + 0.1^2 + 0.15^2 + 0.55^2)^2$$
$$+ (0.2^4 + 0.1^4 + 0.15^4 + 0.55^4)$$

Thus, PIC = $1 - 0.375 - 0.140625 + 0.0937125 \approx 0.578$, and in this case almost 58% of the progeny are informative. Usually PIC values of 0.7 or larger are required for showing good linkage. The larger the number of the marker alleles, the more informative is the PIC. *See* microsatellite typing. (Da, Y., et al. 1999. *Anim. Biotechnol.* 10:25.)

polymorphism   Morphologically different chromosomes, or different alleles at a gene occur, or variable-length restriction fragments are found within a population. Polymorphism can now be detected by automated molecular techniques. During PCR amplification of a gene, one or more fluorescent reporter probes are attached to the 5′-end, and slightly downstream or at the 3′-end a quencher substance(s) is added. During amplification, the quencher may be cleaved by the Taq polymerase if it hybridizes to an amplified segment. The cleavage of the quencher enhances the fluorescence of the reporter fluorochrome. The samples placed in a 96-well plate can be scanned at three wavelengths in about 5 minutes. The procedure may be sensitive enough to detect a single base difference. In human DNA sequences, there is ca. 1 variation/500 bp. About 15% of the polymorphism involves insertions or deletions. At least 100 chromosomes are usually examined for base substitution before the alteration is considered a polymorphism. The average estimated nucleotide polymorphism in human populations is $\sim 8 \times 10^{-4}$. The diversity is variable at different loci and affected by several factors. *See* balanced polymorphism; blood groups; clone validation; fluorochromes; linkage disequilibrium; microsatellite; mutation; mutation detection; PCR; RLP; SNP. (Reich, D. E., et al. 2002. *Nature Genet.* 32:135.)

polymorphonuclear leukocyte (PMN)   *See* granulocytes; leukocyte.

polymyositis   Inflammation of muscle tissues that may lead to rheumatoid arthritis, lupus erythematosus, scleroderma, Sjögren syndrome, or neoplasia. Polymyosites are caused by two autoantigens: PMSCL1 and PMSCL2. Dermatomyositis is a form affecting the connective tissues. Polymyositis as such is not under direct genetic control. *See* IVIG. (Wang, H.-B. & Zhang, Y. 2001. *Nucleic Acids Res.* 29:2517.)

polyneme   Linear structure includes more than one strand, e.g., polytenic chromosomes (salivary gland chromosomes) may have 1,024 (2^{10}) parallel strands.

polynucleotide   Nucleotide polymer hooked up through phosphodiester bonds.

polynucleotide kinase (PK)   Phosphorylates 5′ positions of nucleotides in the presence of ATP such as $\text{ATP} + \text{XpYp} \xrightarrow{\text{PK}}$ p-5′XpYp + ADP (where X and Y are nucleotides), and can heal nucleic acid termini with ligase assistance. *See* ligase. (Wang, L. K. & Shuman, S. 2001. *J. Biol. Chem.* 276:26868.)

polynucleotide phosphorylase (PNPase)   Generates random RNA polymers [$(\text{NMP})_n$] — without a template — from ribonucleoside diphosphates (NDP) and releases inorganic phosphate (P_i):

$$(\text{NMP})_n + \text{NDP} \rightarrow (\text{NMP})_{n+1} + P_i.$$

It degrades mRNA from the 3′-end.

polynucleotide vaccination   Inoculation by subcutaneous, intravenous, or particle bombardment-mediated transfer of specific viral or other nucleotides/nucleoproteins to develop an immune response. The immune reaction is generally low. *See* immunization, genetic.

polyoma   Neoplasia induced by one of the polyomaviruses. The globoid (icosahedral) mouse polyoma viruses (a papovavirus of 23.6×10^6 Da) contain double-stranded circular DNA (4.5 kb). The BK and JC viruses infect humans. *See* papovaviruses. (Cole, C. N. & Conzen, S. D. 2001. *Fundamental Virology*, Knipe, D. M. & Howley, P. M., eds., p. 985. Lippincott Williams & Wilkins, Philadelphia, PA.)

polyp   Outgrowth on mucous membranes such as may occur in the intestines, stomach, or nose. A polyp may be benign, precancerous, or cancerous. *See* Gardner syndrome; PAP; polyposis, adenomatous.

polypeptide Chain of amino acids hooked together by peptide bonds. *See* amino acids; peptide bond; protein synthesis.

polyphenols Catechol-related plant products causing the formation of melanin-like brown color. The polyphenols in tea (theaflavin, catechins) apparently have antimutagenic and anticarcinogenic effects. *See* thea.

polypheny The same gene(s) can determine alternative phenotypes in response to internal or external cues, e.g. the queens and workers in social insects. Some older dictionaries and glossaries equate it with pleiotropy but this does not conform to current usage.

polyphosphates Linear polymers of orthophosphates (n up to 100 or more) present in all types of cells with roles similar to ATP, in metal chelation, bacterial competence for transformation, mRNA processing, growth regulation, etc. *See* chelation; competence of bacteria. (Kulaev, I. & Kulakovskaya, T. 2000. *Annu. Rev. Microbiol.* 54:709.)

$$^-O-\underset{\underset{O^-}{|}}{\overset{\overset{O}{\|}}{P}}-O\left[-\underset{\underset{O^-}{|}}{\overset{\overset{O}{\|}}{P}}-O\right]_n-\underset{\underset{O^-}{|}}{\overset{\overset{O}{\|}}{P}}-O^-$$

PolyPhred Computer program that automatically detects heterozygotes for single-nucleotide substitutions by fluorescence-based sequencing of PCR products at high efficiency. It is integrated by the Phred, Phrap, and Consed programs. *See* Consed; PCR; Phrap; Phred; SNIP. (Nickerson, D. A., et al. 1997. *Nucleic Acid. Res.* 25:2745.)

polyphyletic Organism (cell) originated during evolution from more than one line of descent. A polyphyletic group may contain species that are classified into this group because of convergent evolution. *See* convergence; divergence.

polyploid crop plants The most important polyploid crop plants include alfalfa (4x), apple (3x), banana (3x), birdsfoot trefoil (4x), white clover (4x), coffee (4x, 6x, 8x), upland cotton (4x), red fescue (6x, 8x, 10x), johnsongrass (8x), cultivated oats (6x), peanut (4x), European plum (6x), cultivated potatoes (4x), sugarcane (*x), common tobacco (4x), bread wheat (6x), and macaroni wheat (4x). Most of these are apparently allopolyploids. *See* allopolyploid.

polyploidy Having more than two genomes per cell. Definitive identification of polyploidy requires cytological analysis (chromosome counts), although many of the polyploid plants display broader leaves, larger stomata, larger flowers, etc. Polyploidy regulates the expression of individual genes in a positive or negative manner. A yeast study (using microarray hybridization) found the level of expression of some genes remained the same in haploid and tetraploid cells, whereas the expression of some cyclin genes decreased

with tetraploidy and a gene associated with cell adhesion was greatly overexpressed with tetraploidy. Polyploidy may permit separate evolutionary paths for additional gene copies. *See* allopolyploid; alpha parameter; autopolyploid; chromosome segregation; inbreeding autopolyploids; maximal equational segregation; microarray hybridization; tetrasomic; trisomy. (Otto, S. P. & Whitton, J. 2000. *Annu. Rev. Genet.* 34:401.)

Autotetraploid (top) and diploid (bottom) flowers of *Cardaminopsis petraea*.

polyploidy in animals Rare and limited mainly to parthenogenetically reproducing species (e.g., lizards). It also occurs in bees, silkworm, and other species. Some cells in special tissues of the diploid body may have an increased chromosome number as a normal characteristic. Among mammals, tetraploidy was found in the red visacha rat, *Tympanoctomys barrarae* ($2n = 112$). The rarity of polyploidy in animals is attributed to its incompatibility with sex determination and dosage compensation of the X chromosome. *See* honey bee; parthenogenesis; silkworm. (Zimmet, J. & Ravid, K. 2000. *Exp. Hematol.* 28:3; Wolfe, K. H. 2001. *Nature Rev. Genet.* 2:333.)

polyploidy in evolution Common in the plant kingdom, but the majority of polyploid species are allopolyploid. Some of the single-copy genes of invertebrates are, however, detectable up to four copies in vertebrates. A survey of plants indicated only 38% of polyploid species in the Sahara region, 51% in Europe, and 82% in the Peary Islands, thus showing an increasing trend toward the North. *See* allopolyploid. (Otto, S. P. & Whitton, J. 2000. *Annu. Rev. Genet.* 34:401; Wu, R., et al. 2001. *Genetics* 159:869.)

polyposis, juvenile Early-onset polyposis that frequently turns malignant. It is caused by a defect at the carboxyl terminal of the SMAD4/DPC4 (552 amino acids) protein, encoded in human chromosome 18q21.1. In a trimeric association, SMAD4 is involved in the TGF-β-signaling pathway. Although some of the symptoms are similar to other hamartomas, the Cowden disease gene (PTEN, phosphatase and tensin homologue) is encoded in chromosome 10 and Peutz-Jeghers syndrome is coded for in chromosome 19. *See* DCC; multiple hamartomas; polyposis hamartomatous; PTEN; SMAD; TGF. (Howe, J. R., et al. 2002. *Am. J. Hum. Genet.* 70:1357.)

polyposis adenomatous, intestinal (APC) Controlled by autosomal-dominant genes responsible for intestinal, stomach (Gardner syndrome), or other types (kidney, thyroid, liver, nerve tissue, etc.) of benign or vicious cancerous tumors. The various forms are apparently controlled by mutations or deletions in the 5q21-q22 region of the human chromosome and represent allelic variations. Retinal lesions (CHRPE) are associated with truncations between codons 463 and 1,387; truncations between codons 1,403 and 1,528 involve extracodonic effects, etc. In addition, it is conceivable that this is a *contiguous gene* region where adjacent mutations affect the expression of polyposis.

By the use of single-strand conformation polymorphism technique, DNA analysis may permit the identification of aberrant alleles prenatally or during the presymptomatic phase of the condition. The situation is further complicated, however, by the possibilities of somatic mutations. The *Min* gene of mice appears to be homologous to the human APC, thus lending an animal model for molecular, physiological, and clinical studies. The expression of *Min* is also regulated by the phospholipase-encoding gene *Mom1*, indicating the involvement of lipids in the diet. Polyposis may affect a very large portion of the aging human population, especially females. Certain forms of polyposis may affect the young (juvenile polyposis). Regular monitoring by colorectal examination is necessary for those at risk. Symptoms of bloody diarrhea and general weakness are usually too late for successful medical intervention.

Molecular genetic information suggests that vertebrates use the same pathway of signal transduction as identified by *Drosophila* genes: *porcupine* (*porc*, 1.59) → *wingless* (*wg*, 2-30.0) → *dishevelled* (*dsh*, 1-34.5) → *zeste white3* (z^{w3}, 1.1.0) → *armadillo* (*arm*, 1-1.2) → cell nucleus. The normal human APC gene appears to be either a negative regulator (tumor suppressor) or an effector, acting between z^w and the nucleus. When it mutates, either it can no longer carry out suppression or it may become an effector. The product of *dsh* also appears to be a negative regulator of z^w. When the *zeste* product, glycogen synthase kinase (GSK3β), is inactive, the *arm* product (catenin) is associated with the APC product and a signal for tumorigenesis is generated. Alternatively, when no signal is received, GSK phosphorylates and activates a second binding site on APC for catenin, but it causes the degradation of catenin and thus no tumor signal is generated. The APC protein may act as a tumor gene by docking at its COOH end with a human homologue of the *dlg1* (*disc large*, 1-34.82) of *Drosophila*. The *Dlg* product belongs to the *membrane-associated guanylate kinase* protein family analogous to the proteins in vertebrates sealing adjacent cell membranes (tight junction). *Dlg* is also considered a tumor gene. Although the molecular information reveals a number of mechanisms of action, it is not clear which one is being used or if multiple pathways are involved in polyposis. APC/FAP has a prevalence of about 1×10^{-4}. The EB1 protein binds the APC protein and it is situated on the microtubules of the mitotic spindle, serving as a checkpoint for cell division. *See* animal models; cancer; catenin; contiguous gene syndrome; cyclooxygenase; effector; Gardner syndrome; GSK3β; hereditary nonpolyposis colorectal cancer; microtubule; polymorphism; polyposis, juvenile; polyposis hamartomatous; single-strand conformation; spindle; tight junction; Turcot syndrome.

polyposis hamartomatous (Peutz-Jeghers syndrome, PJS) Chromosome-19p13.3 rare dominant overgrowth of mucous membranes (polyp), especially in the small intestine (jejunum) but also in the esophagus (the canal from the mouth to the stomach), bladder, kidney, nose, etc. Melanin spots may develop on lips, inside the mouth, and fingers. Ovarian

and testicular cancer were also observed. The susceptibility to this cancer is due to a deletion in a serine/threonine protein kinase gene (LKB). This gene signals to VEGF. *See* cancer; colorectal cancer; Gardner syndrome; Muir-Torre syndrome; multiple hamartomas; pigmentation of the skin; polyposis, adenomatous intestinal; VEGF. (Hemminki, A., et al. 1998. *Nature* 391:184; Sapkota, G. P., et al. 2001. *J. Biol. Chem.* 276:19469; Ilikorkala, A., et al. 2001. *Science* 293:1323; Bardeesy, N., et al. 2002. *Nature* 419:162.)

polyprotein Contiguously translated long-chain polypeptide that is processed into more than one protein.

polypurine Stretch of purine residues in nucleic acids.

polypyrimidine Sequence of multiple pyrimidines (mainly Us) in nucleic acids adjacent to the 3′-splicing site. Py-tract-binding proteins (PTB) recognize them, e.g., the essential splicing factor U2AF[65], the splicing regulator sex-lethal (*Sxl*), etc. *See* introns; sex determination; splicing. (Le Guinier, C., et al. 2001. *J. Biol. Chem.* 276:43677.)

polyribosome Polysome. *See* protein synthesis.

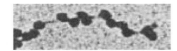

Polyribosome.

polysaccharide Monosaccharides joined by glycosidic bonds (e.g., starch, glycogen).

polysome Multiple ribosomes are held together by mRNA. The ovalbumin polysomes comprise an average of 12 ribosomes, and one peptide initiation takes place in every 6−7 seconds if all the required factors are functioning normally. The average polysome size for globin is ∼5 ribosomes (1 ribosome/∼90 nucleotides). Pactamycin may be an inhibitor of translation initiation and cycloheximide may interfere with peptide chain elongation. *See* cycloheximide; mRNA; pactamycin; ribosome, transcription; translation.

polysome display Polysomes are isolated and screened by affinity of the nascent peptides on an immobilized specific monoclonal antibody. The mRNA of the enriched pool of polysomes is reverse-transcribed into cDNA and amplified by PCR. The amplified template may be cloned and translated in vitro. The procedure is highly efficient for the screening of large, specific peptide pools. *See* cDNA; PCR; reverse transcription; translation in vitro. (Mattheakis, L. C., et al. 1994. *Proc. Natl. Acad. Sci. USA* 91:9022.)

polysomic cell Some chromosomes are present in more than the regular number of copies. The polyploids are polysomic for entire genomes. *See* aneuploidy; polyploidy.

polysomy Some of the chromosomes in a cell are present in more than the normal numbers. Examples of these cases in humans are 48,XXXX, 48,XXXY, 49,XXXXX, and 49,XXXXY. *See* nondisjunction; polyploid; trisomy.

polyspeirism   One cell makes several types of related molecules, e.g., different chemokines. (Montovani, A. 2000. *Immunology Tody* 21(4): 199.)

polyspermic fertilization   More than a single sperm enters the egg. Because each sperm may provide a centriole, multipolar mitoses may take place, resulting in aneuploidy and abnormal embryogenesis. *See* fertilization.

polysyndactyly   Encoded by the HOXD13 gene at human chromosome 2q31-q32. The amplification of the alanine codons (CCG, GCA, GCT, GGC) leads to expanded (25 to 35) alanine residues in the protein. Some polysyndactyly is due to mutation in the GLI3 gene at 7p13. *See* Pallister-Hall syndrome; syndactyly; trinucleotide repeats.

Polysyndactylic toes.

polytenic chromosomes   Composed of many chromatids (e.g., in salivary gland cell nuclei) because DNA replication was not followed by chromatid separation. The polytenic chromosomes in the salivary gland nuclei of diptera may have undergone 10 cycles of replication ($2^{10} = 1,024$) without division and may have over 1,000 strands. The polytenic chromosomes in the salivary glands are extremely long. A regular feature is the very close somatic pairing. Also, they are all attached at the chromocenter. The polytenic chromosomes have been extensively exploited for analysis of deletions, duplications, inversions, and translocations. The characteristic banding pattern was used as a cytological landmark for identification of the physical location of genes. Rarely, polyteny occurs in some specialized plant tissues (antipodals). *See* giant chromosomes; salivary gland chromosomes; somatic pairing.

Polytenic chromosomes of *Allium ursinum*. (Courtesy of G. Hasischka-Jenschke.)

polytocous species   Produce multiple offspring by each gestation. *See* monotocous.

polytomy   Multifurcating rather than bifurcating analysis of phylogenetic relations. *See* evolutionary tree. (Walsh, H. E., et al. 1999. *Evolution* 53:932.)

polytopic protein (multispanning)   Traverses the plasma membrane several times.

polytropic retrovirus   *See* amphotropic retrovirus.

polytypic   Species includes more than one variety or subtype.

pol ζ   Eukaryotic DNA polymerase without exonuclease activity. It is a repair enzyme inasmuch as it can bypass pyrimidine dimers more efficiently than polα. It is insensitive to 200 μM aphidicolin (and in this respect it is similar to pol β and pol γ) and to dideoxynucleotide triphosphates (which inhibit pol β and pol γ). It is moderately sensitive to 10 μM butylphenylguanosine triphosphates. It is relatively inactive with salmon sperm DNA or primed homopolymers. *See* DNA polymerases.

POMC   Pre-pro-opiomelanocortin. *See* ACTH; melanocortin; opiocortin.

pomegranate (*Punica granatum*)   Mediterranean fruit tree, $2n = 2x = 16$ or 18.

Pompe's disease   *See* glycogen storage diseases.

pongidae (anthropoid primates [hominoidea])   *Gorilla gorilla gorilla* $2n = 48$; *Hyalobates concolor s* (gibbon) $2n = 52$; *Hylobates lar* (gibbon) $2n = 44$; *Pan paniscus* (pygmy chimpanzee) $2n = 48$; *Pan troglodytes* (chimpanzee) $2n = 48$; *Pongo pygmaeus* (orangoutan) $2n = 48$; *Symphalangus brachytanites* $2n = 50$. *See* primates.

PO-PS copolymers   Phosphorothioate-phosphodiester copolymers are used for antisense technologies. *See* antisense RNA.

POP'   Symbolizes the ends of the temperate transducing phage genome integrating into the bacterial host chromosome. The corresponding bacterial integration site is BOB', and after integration (recombination) the sequence becomes BOP' and POB', respectively. *See* attachment sites.

poplar (*Populus* ssp)   Includes cottonwood trees, $2n = 2x = 38$. (Cervera, M.-T., et al. 2001. *Genetics* 158:787.)

pop-out, chromosomal   Originates by intrachromatid reciprocal exchange between direct repeats and excises one of the repeats (the pop-out) but may retain the other member of the duplication. *See* intrachromosomal recombination; sister chromatid exchange.

Pop1p   Protein component of ribonuclease P and MRP. *See* MRP; ribonuclease P.

poppy (*Papaver somniferum*)   Its latex is a source of opium. The plant is grown for its oil-rich seed as a food and for pharmaceutical purposes. Basic chromosome number $x = 11$; diploid and tetraploid forms are known.

population   Collection of individuals that may interbreed and freely trade genes (Mendelian population, deme), or it may be a closed population that is sexually isolated from other groups that share the same habitat. *See* Hardy-Weinberg theorem; population equilibrium.

population critical size   *See* critical population size.

population density   Number of cells or individuals per volume or area.

population effective size (N_e)   Number of individuals in a group or within a defined area that actually transmit genes to the following reproductive cycles (offspring). Each breeding individual has a 0.5 chance of contributing an allele to the next generation, and $0.5 \times 0.5 = 0.25$ is the probability of contributing two particular alleles. The probability that the same male contributes two alleles is $(1/N_m)0.25$, and for the same female it is $(1/N_f)0.25$, where N_m and N_f are the number of breeding males and females, respectively. The probability that any two alleles are derived from the same individual is $0.25N_m + 0.25N_f = 1/N_e$, and N_e is computed as $4N_mN_f/(N_m + N_f)$. *See* founder principle; genetic drift; inbreeding and population size. (Wright, S. 1931. *Genetics* 16:97.)

population equilibrium   *See* Hardy-Weinberg theorem.

population genetics   Study of the factors involved in the fate of alleles in potentially interbreeding groups. The individuals within these groups (demes) may actually reproduce by random mating or selfing or by the combination of the two within this range. Population genetics can be entirely theoretical and involves developing mathematical formulas for predicting the allelic frequencies, the effect of various factors that affect these frequencies, and the historical paths of the genes and factors as they emerge, become established, or disappear, form equilibria, or remain unstable during microevolutionary periods. Experimental population genetics conducts biological studies in the sense of the theoretical framework. Population genetics thus deals with the consequences of mutation, genetic drift, migration, selection, and breeding systems and is one of the most important approaches to experimental (micro) evolution. It provides the theory for many human genetics, animal and plant breeding research efforts. The availability of molecular information greatly advanced the resolving power of population genetics. Mitochondrial (maternally transmitted) and Y-chromosomal (paternally transmitted) markers are effective tools for studying the dynamics and history of human populations. *See* DNA chips; microsatellites; minisatellites; mtDNA; SNIPS; Y chromosome. (Population modeling software: <www.trinitysoftware.com>.)

population growth, human $P_t = P_0(1 + r)^t$, where $P_0 =$ the population at time 0, $r =$ rate of growth, and $t =$ time. It can also be calculated by $P_t = P_0e^{rt}$, where $e =$ the base of the natural logarithm. *See* age-specific birth and death rates; human population growth; Malthusian parameter.

population structure   Endemic by subpopulation groups. The dispersal of the subdivisions reflects adaptive genetic differences, gene flow, natural selection pressure, and sometimes genetic drift. *See* endemic; genetic drift; natural selection; population genetics.

population subdivisions   Smaller, relatively separated breeding groups with restricted gene flow among them. *See* gene flow; migration.

population tree   Constructed on the basis of genes frequencies among populations, indicating their evolutionary relationship. *See* evolutionary tree; gene tree.

population wave   Periodic changes in effective population size. *See* founder principle; gene flow; population size, effective; random drift.

porcupine man (ichthyosis histrix)   Dominant form of hyperkeratosis. *See* ichthyosis; keratosis.

porin   Voltage-dependent anion channel. It is opened by Bax and Bak proapoptotic proteins and closed by the antiapototic Bcl-x$_L$. Bax and Bak permit the exit of cytochrome c from the mitochondria and thus facilitate apoptosis by the activation of caspases. In case of IL-7 deficiency and an increase in pH over 7.8, the conformation of Bax is altered and the protein moves from the cytoplasm to the mitochondria, facilitating apoptosis. The Bcl-2 protein, localized to the mitochondrial membrane, normally suppresses the release of cytochrome c. Bax deficiency extends the ovarian life span into advanced age of mice. Normally the ovarian follicles fade by menopause in women and at similar developmental stages in mice. Degradation of Bax by the proteasomes may protect against the apoptosis overprotective effect of Bcl-2 and reduce cancer cell survival. For drug therapy of epithelial cancer, the state of BAX vs. Bcl-2 may be significant. *See* apoptosis; cytochrome c; hypersensitive reaction; ion channels. (Suzuki, M., et al. 2000. *Cell* 103:645; Gogvadze, V., et al. 2001. *J. Biol. Chem.* 276:19066.)

porphyria   Collective name for a variety of genetic defects involved in heme biosynthesis, resulting in under- and/or overproduction of metabolites in the porphyrin-heme biosynthetic pathway. These diseases may be controlled by recessive or dominant mutations. The affected individuals may suffer from abdominal pain, psychological problems, and photosensitivity. Autosomal-dominant acute *intermittent porphyria* (human chromosome 11q23-ter) is caused by a periodic 40–60% reduction in porphobilinogen deaminase enzyme, resulting in insufficient supplies of tetrapyrrole hydroxymethyl bilane, which is normally further processed nonenzymatically into uroporphyrinogen I. It was speculated that the famous Dutch painter van Gogh was a victim of this rare disease. Prevalence is in the range of 10^{-4} to 10^{-5}. Exogenous effects such as barbiturates, sulfonamides, alkylating agents, and many other drugs, alcohol consumption, poor diet, various infections, and hormonal changes generally elicit periodic attacks. An *adult* *type* of (hepatocutaneous) porphyria controlled by another human gene locus (1p34) involves light sensitivity and liver damage by the accumulation of porphyrins caused by uroporphyrinogen decarboxylase deficiency. The general effect may be less severe than in intermittent porphyria. The rare congenital *erythropoietic porphyria* (CEP) is the result of a defect in the enzyme uroporphyrinogen III cosynthetase controlled by a recessive mutation in human chromosome 10q25.2-q26.3. Laboratory identification is generally based on urine analysis for intermediates in the heme pathway. Porphyrias also affect various mammals. Defects in the porphyrin pathways are involved in several types of pigment deficiency mutations of plants. *Variegate porphyria* is caused by a defect of protoporphyrinogen oxidase (PPOX), with symptoms basically similar to those of intermittent porphyria. This dominant disease has low penetrance. Its prevalence is very high (about 3×10^{-3}) in South African populations of Dutch descent; it apparently represents the founder effect. The mental problems of King George III of England (reigned during the U.S. War of Independence) were also attributed to variegate porphyria. *See* coproporphyria; founder effect; heme; light sensitivity diseases; porphyrin; skin diseases.

HO$_2$C·CH$_2$ H$_2$C·CO$_2$H

Protoporphyrin.

porphyrin   Four special pyrroles joined into a ring; generally with a central metal, like iron in hemoglobin or in chlorophylls with magnesium. *See* coproporphyria; heme; porphyria.

porphyrinuria   *See* porphyria.

porpoise   *Lagenorhynchus obliquidens*, $2n = 44$. *See* dolphins.

Portable Dictionary of the Mouse Genome   Data on ~12,000 genes and anonymous DNA loci of the mouse, homologues in other mammals, recombinant inbred strains, phenotypes, alleles, PCR primers, references, etc. Can be used on Macintosh, PC in FileMaker, Pro, Excel, and text formats. Accessible through the Internet (WWW, Gopher, FTP), CD-ROM, or on floppy disk. Information: R. W. Williams, Center for Neuroscience, University of Tennessee, 875 Monroe Ave., Memphis, TN 36163. Phone: 901-448-7018; fax: 901-448-7266; e-mail: <rwilliam@nb.utmem.edu>.

portable promoter   Isolated DNA fragment, including a sufficient promoter that can be carried by transformation to other cells and may function in promoting transcription. *See* gene fusion; promoter; transformation.

portable region of homology   Insertion and transposon elements may represent homologous DNA sequences and can recombine. The recombination may then generate deletions,

cointegrates or insertions and inversions. These events can take place even in RecA$^-$ hosts. *See* cointegrate; deletion; inversion; targeting genes; Tn*10*.

positional cloning   *See* chromosome landing; chromosome walking; map-based cloning.

positional information   Provided to some cells by signal transducers in a multicellular organism and has an important influence on differentiation and development. *See* differentiation; morphogenesis.

positional sensing   Provides information for specific differentiation functions. *See* morphogenesis.

position effect   Change in gene expression by a change in the vicinity of the gene. The new expression may be *stable* or variable (*variegation-type position effect*). Stable position effect is observed when promoterless structural genes are introduced by transformation and the transgene is expressed with the assistance of a "trapped" promoter that is regulated differently than the gene's natural (original) promoter. Variegated position effect (PEV) is more difficult to interpret by molecular models. It has been assumed that heterochromatin affects the

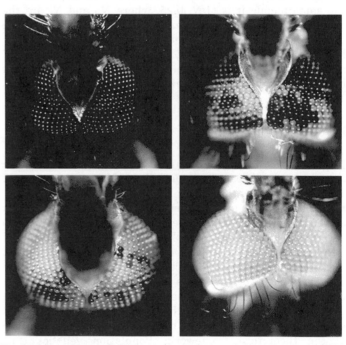

Duplication of the wild-type (p^+) allele into heterochromatic DNA results in the (p) eyes variegated expression of p^+ in the malaria mosquito *Anopheles gambiae*. (Courtesy of Dr. Mark Benedict, original photograph by James Gathany, CDC.)

intensity of somatic pairing and variations in somatic association and variations in cross-linking between the homologues by binding proteins bring about the silencing. The *trithorax-like* gene of *Drosophila* encodes a GAGA homology transcription factor that enhances variegation-type position effect (PEV) by decondensation of chromatin. The mosaicism may also be the result of the spontaneous and random derepression of the promoter in the presence of an activator. The telomeric isochores have been implicated in position effect (TPE). Position effect may be observed by altering the site or distance of the locus-control region. In *Drosophila*, over 100 genes were found that affect variegation-type position effect (PEV). It has been hypothesized that these genes control DNA packaging. Many cancers develop after translocations or transpositions, indicating the significance of position effect on the regulation of growth. Transposable elements may also cause position effect. Position effects may even be exerted from long distances (2 Mb) and may be difficult to distinguish the position effect from gene mutation within the target gene. Such cases may complicate positional cloning. Position effect also occurs in yeasts and other organisms. Some human genetic disorders are due to position effect. *See* cancer; chromosomal rearrangements; chromosome breakage; developmental regulator effect variegation; Dubinin effect; epigenesis; heterochromatin; histone methyltransferases; isochores; LCR; locus-control region; mating type determination in yeast; Offermann hypothesis; paramutation; positional cloning; regulation of gene activity; RIGS; RPD3; silencer; transposable elements. (Kleinjahn, D. J. & Heyningen, V. 1998. *Hum. Mol. Genet.* 7:1611; Baur, J. A., et al. 2001. *Science* 292:2075; Ahmad, K. & Henikoff, S. 2001. *Cell* 104:839; Csink, A. K., et al. 2002. *Genetics* 160:257; Monod, C., et al. 2002. *EMBO Rep.* 3:747.)

position-specific scoring matrix (PSSM) Represents amino acids in specific positions in a sequence alignment. It can be used for scanning proteins with matches to this tract. *See* PWM. (Gribskov, M., et al. 1987. *Proc. Natl. Acad. Sci. USA* 84:4355.)

position weight matrix *See* PWM.

positive control Gene expression is enhanced by the presence of a regulatory protein (in contrast to negative control, where its action is reduced). The arabinose operon of *E. coli* is a classic example. The regulator gene *araC* produces a repressor (P_1) in the absence of the substrate arabinose. If arabinose is available, P_1 is converted to P_2 (by a conformational change), which is an activator of transcription in the presence of cyclic adenosine monophosphate (cAMP). While the negative control (P_1) is correlated with low demand for expression, the activator (P_2) appears in response to the demand for a high level of expression. In general cases, the addition of an activator protein to the DNA makes normal transcription possible, but adding a special ligand to the system removes the activator and the gene is turned off. *See* arabinose operon; autoregulation; catabolite activator protein; *lac* operon; negative control; regulation of gene activity.

positive cooperativity Binding of a ligand to one of the subunits of a protein facilitates the binding of the same to other subunits.

positive interference *See* coincidence; interference.

positive/negative selection May be used to isolate cloned constructs containing the desired integrated sequence (positive selection). Negative selection is expected to eliminate integration sites containing the entire vector inserted at nontargeted sites and vector components that have no interest in cloning. Negative selection is usually less efficient—if possible at all—than positive selection. Positive selection in case of hypoxanthine/guanine phosphoribosyl transferase marker one may use hypoxanthine, aminopterin, thymidine (HAT) chemicals, whereas in the same experiment for negative selection 6-thioguanine or 5-bromodeoxyuridine may be used.

positive selection Generally indicates selection of a desirable type in a population rather than elimination of the undesirable phenotype/genotypes. *See* selection.

positive selection of lymphocytes Is a process of matruration of these cells into functional members of the immune system. In contrast, the negative selection eliminates, by apoptosis, early lymphocytes with autoreactive receptors. *See* immune system; lymphocytes.

positive selection of nucleic acids Isolates and enriches desired types of nucleic acid sequences. The desired (tracer) sequences are digested by restriction endonucleases that generate cohesive ends. The rest of the nucleic acids (driver) are exposed to sonication (or the ends may be dephosphorylated), so sticky ends are not expected. Thus, mainly the tracer-tracer sequences are annealed when the mixture is treated with a ligase enzyme. *See* cohesive ends; genomic subtraction; ligase DNA; RFLP subtraction; sonicator; subtractive cloning.

positive-strand viruses Their genome is also an mRNA. Upon transcription, they may directly produce infectious nucleic acid. *See* mRNA; negative-strand virus; plus strand; replicase.

positive supercoiling The overwinding follows the direction of the original coiling, i.e., it takes place rightward. *See* negative supercoil; supercoiling.

post-adaptive mutation Supposed to arise de novo in response to the conditions of selection. Actually, postadaptive mutation may not be found if the data are well scrutinized. *See* directed mutation; preadaptive mutation.

posterior Pertaining to the hind part of the body or behind a structure toward the tail end.

posterior probability *See* Bayes theorem.

postgenome analysis Study of the experimental, informatics of the sequential function (metabolic pathways), and interactions of genes and their products. (<http://www.genome.ad.jp>, <http://www.genome.ad.jp/kegg/comp/GFIT.html>.)

postmeiotic segregation Takes place when the DNA is a heteroduplex at the end of meiosis. Among the octad spores of ascomycetes this may result in 5:3 and 3:5 or other types of aberrant ratios instead of 1:1. Postmeiotic segregation may be an indication of failures in mismatch or excision repair. *See* DNA repair; gene conversion; tetrad analysis.

postprandial After consuming a meal, e.g., protein anabolism modifies protein synthesis due to the change in the amino acid pool or in insulin supply after eating (postprandially).

postreduction Segregation of the alleles takes place at the second meiotic division. *See* meiosis; prereduction; tetrad analysis.

postreplicational repair *See* DNA repair; unscheduled DNA synthesis.

PostScript Computer application to handle text and graphics at the same time. The PostScript code determines what the graphics look like when printed, although they may not be visible on the monitor screen.

postsegregational killing *See* plasmid addiction.

posttranscriptional gene silencing (PTGS) The transcript of a transgene is degraded before translation takes place and thus its expression is prevented. Also, it may be a defense mechanism against viruses in plants. The viral gene may be integrated into the chromosome and duly transcribed, yet it is not expressed. In addition, since the replication of the virus is mediated through a double-stranded RNA that has been found to be a potent inhibitor, it is conceivable that both the plant defense and the transgene silencing rely on similar mechanisms. In some plant species, the potyviruses, tobacco etch virus, and cucumber mosaic virus may produce a *helper component protease* (HC-Pro) and may inactivate this plant defense by degradation. HC-Pro may have another role. When a plant is infected simultaneously by two different viruses, one of them promotes the vigorous replication of the other, and the latter by its production of HC-Pro eventually facilitates the spread of the first type of the virus, thus enhancing the symptoms of the viral disease. In some of the silenced plant cells, a 25-nucleotide-long antisense RNA has been detected that seems to inactivate the normal transcript or infectious viral RNA. According to other studies, the ~25-nt RNA sequence apparently conveys specificity for a nuclease by homology to the substrate mRNA. Several types of hairpin structures of RNAs involving sense and antisense sequences and introns appeared to silence viral genes in plants very effectively. A calmodulin-related plant protein (rgs-CAM) may also suppress silencing. *See* cosuppression; homology-dependent gene silencing; host-pathogen relations; methylation of DNA; plant viruses; RNAi; silencing. (Bass, B. L. 2000. *Cell* 101:235; Jones. L., et al. 1999. *Plant Cell* 11:2291; Waterhouse P. M., et al. 2001. *Nature* 411:834; Mitsuhara, I., et al. 2002. *Genetics* 160:343.)

posttranscriptional processing The primary RNA transcript of a gene is cut and spliced before translation or before assembling into ribosomal subunits or functional tRNA; it includes removal of introns, modifying (methylating, etc.) bases, adding CCA to tRNA amino arm, polyadenylation of the 3′ tail, etc. *See* opiotropin. (McCarthy, J. E. G. 1998. *Microbiol. Mol. Biol. Revs.* 62:1492; Bentley, D. 1999. *Curr. Opin. Cell Biol.* 11:347.)

post-transcriptional operons Hypothesis that functionally related genes may be regulated post-transcriptionally as groups by mRNA-binding proteins that recognize common sequence elements in the untranslated 5′ and 3′ subsets of the transcripts. This conclusion is based on findings that mRNA-binding proteins recognize unique subpopulations of mRNAs, the composition of these subsets may vary depending on conditions of growth and the same mRNA occurs in multiple complexes. These conserved *cis* elements were named USER (untranslated sequence elements for regulation) codes. These systems may permit plasticity during developmental processes or responses to drug treatment. *See* operon, genetic networks. (Keene, J. D. & Tenenbaum, S. A. 2002. *Mol. Cell* 9:1161.)

posttranslational modification Enzymatic processing of the product of translation, the newly synthesized polypeptide chain. This may include proteolytic cleavage, glycosylation, phosphorylation, conformational changes, assembly into quaternary structure, etc. *See* conformation; protein structure; protein synthesis; proteomics. (Németh-Cawley, J. F., et al. 2001. *J. Mass. Spectrom.* 36:1301.)

posttransplantational lymphoproliferative disease (PTDL) Epstein Barr virus–infected B cells may continue to proliferate after engraftment because the immunosuppressive therapy required to maintain the graft inhibits cytotoxic T lymphocytes. Bone marrow transplantation may alleviate the problems. *See* CTL; Epstein-Barr virus; immunosuppression.

postzygotic *See* prezygotic.

postzygotic isolation Arises when the taxa diverge in allopatric evolution from the common ancestor by accumulation of different nondeleterious mutations. Although the divergent forms are well adapted, their hybrids may be inviable or sterile because of the negative effects of the alleles in a shared background. *See* allopatric speciation. (Orr, H. A. & Turelli, M. 2001. *Evolution* 55:1085.)

potassium-argon dating Based on the conversion of K^{40} into Ar^{40}, a stable gas. It is used for dating rocks over 100,000 years old.

potato (*Solanum tuberosum*) Has 170 to 300 related species with the basic chromosome number $x = 12$. Species with diploid, tetraploid, and hexaploid chromosome numbers are found in nature. Cultivated potatoes originated from *Solanum andigena* in Central America, where they produce tubers under short-day conditions. The majority of modern varieties are day-neutral and develop tubers under long-day conditions. Cultivated potatoes are usually cross-pollinating species, but many also set seeds by selfing. Generally, seed progeny are very heterogeneous genetically. Potatoes are rarely propagated by seed as a crop. Diploid relatives are usually self-incompatible, whereas polyploids may set seeds by themselves. Among the cultivated groups, tuber color may vary from white to yellow to deep purple. The chemical composition of the tubers varies widely, depending on the purpose of the market. Besides being a popular vegetable, the potato is an important source of industrial starch. The related species carry genes of agronomic importance (disease, insect resistance, etc.) that have not yet been fully exploited for breeding improved varieties. Application of molecular techniques of plant breeding seems promising. *See* patatin. (<http://www.tigr.org/tdb/tgi.shtml>.)

potato beetle (*Leptinotarsa decemlineata*, $n = 18$) One of the most devastating pests of agricultural production of potatoes. Plants transgenic for the δ-endotoxin of *Bacillus thüringiensis* are commercially available. *See* *Bacillus thüringiensis*; potato.

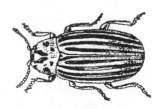

Potato beetle.

potato leaf roll virus Has double-stranded DNA genetic material.

Potocki-Shaffer syndrome *See* exostosis.

potocytosis Moving ions and other molecules into cells by caveola vehicles. *See* caveolae.

POU Region with several transcriptional activators of 150–160 amino acids (including a homeodomain) involved with a large number of proteins controlling development. The acronym stands for a prolactin transcription factor (PIT), a ubiquitous and lymphoid-specific octamer-binding protein (OTF), and the *Caenorhabditis* neuronal development factor (Unc-86). *See Caenorhabditis*; deafness; homeodomain; octa; transcription factor; unc. (Ryan, A. K. & Rosenfeld, M. G. 1997. *Genes Dev.* 11:1207; Bertolino, E. & Singh, H. 2002. *Mol. Cell.* 10:397.)

pOUT Strong promoter opposing pIN and directing transcription to the outside end of an insertion element. *See* pIN; RNA-OUT.

power of a test Algebraically it is $1 - \beta$, where β = type II error. This test reveals the probability of rejecting a false null hypothesis and accepting a correct alternative. The experimenter needs $1 - \beta$ to be as large as possible by reducing β to a minimum. To improve the power, increase the size of the experiment (population). In case the size cannot be increased, a more powerful test (statistics) should be chosen. *See* error types; significance level.

pox virus Group of oblong double-stranded DNA viruses of 130–280 kbp. Some of them parasitize insects; others in the family are the chicken pox, cowpox (vaccinia), and smallpox viruses. Their transmission is by insect vectors or by dust or other particles. Engineered pox virus vectors that are not able to multiply in mammalian cells may have the ability to express passenger genes without the risk of disease. Due to the success of vaccination, smallpox as a disease has been eradicated and vaccination against it is no longer necessary except in case of terrorist attack (Halloran, M. E., et al. 2002. *Science* 298:1428). Pox virus–based vectors are used orally to protect wildlife (red fox) from rabies and to protect chickens from the Newcastle virus. Recombinant canarypox virus is employed for the protection of dogs and cats against distemper, feline leukemia, equine influenza, etc. Highly attenuated derivatives expressing rabies virus glycoprotein, Japanese encephalitis virus polyprotein, or seven antigens of *Plasmodium falciparum* are used for safe and effective vaccination. *See* malaria; *Plasmodium falciparum*. (Moss, B. & Shisler, J. L. 2001. *Semin. Immunol.* 13:59; Takemura, M. 2001. *J. Mol. Evol.* 52[5]:419; Enserink, M. 2002. *Science* 296:1592.)

POZ Protein-protein interaction domain of zinc finger–containing transcriptional regulatory proteins. *See* zinc finger.

PP-1, PP-2 Protein serine/threonine phosphatases that are inhibited by okadaic acid. PP-1 may be associated with chromatin through the nuclear inhibitor of PP-1 (NIPP-1).

PP enzymes play key roles in many cellular processes. *See* DARPP; okadaic acid.

pp15 Protein factor required for nuclear import. *See* membrane transport; RNA export.

pp125^{FAK} *See* CAM.

PP2A Proline-directed protein serine-threonine phosphatase dephosphorylates proteins in the MAP pathway of signal transduction and thus balances the effect of kinases. PP2A subunit B56 regulates β-catenin signaling and several metabolic processes. PP2A is very sensitive to okadaic acid. *See* catenins; cyclin G; MAP; MAP kinase phosphatase; okadaic acid; signal transduction; Sit.

PPAR Peroxisome proliferator-activated receptor (17q12) is a transcription factor in the adipogenic (fat synthetic) pathways. The three types, α, γ, and δ, show different distribution in human tissues and associate with different ligands. PPARα is the target for the fibrates that reduce triglycerides. Type α also acts as a transcription factor for several genes affecting lipoprotein and fatty acid metabolism. PPARγ is a (3p25) regulator of glucose, lipid, and cholesterol metabolism and may be sensitized by thiazolidinediones (TZD). It offers some hope for the treatment of diabetes mellitus type II (IDDM). The PPARγ 12Ala allele is associated with a small yet significant reduction in risk for diabetes type II. PPARγ agonists have controversial promoting and suppressing effects on polyposis of the colon and other cancers. In human thyroid carcinoma, PAX8–PPARγ1 has been observed. *See* diabetes mellitus; dizygotic twins; farnesoid X receptor; hypertension; leptin; leukotrienes; obesity; PAX; peroxisome; polyposis; thiazolidinedione. (Lowell, B. B. 1999. *Cell* 99:239; Kersten, S., et al. 2000. *Nature* 405:421; Willson, T. M., et al. 2001. *Annu. Rev. Biochem.* 70:341.)

pPCV Plasmid plant cloning vector; designation (with additional identification numbers and/or letters) of agrobacterial transformation vectors constructed by Csaba Koncz.

ppGpp *See* discriminator region.

P1 phage *E. coli* transducing phage and vector with near 100 kb carrying capacity. (Lehnherr, H., et al. 2001. *J. Bacteriol.* 183:4105.)

P22 phage Temperate bacteriophage of *Salmonella typhimurium*; its genome is about 41,800 bp. (Vander Byl, C. & Kropinski, A. M. 2000. *J. Bacteriol.* 182:6472.)

PPI Peptidyl prolyl isomerase is an endoplasmic reticulum bound protein assisting chaperone function. There are three PPI families: cyclophilins, FK506, and parvulins. *See* chaperone; PDI. (Dolinski, K. & Heitman, J. 1997. *Guidebook to Molecular Chaperones and Protein Folding Catalysis*, Gething, M. J., ed., p. 359. Oxford Univ. Press, Oxford, UK.)

P1 plasmid Cloning vector with a carrying capacity of about 100 kb. *See* P1 phage; vectors. (Bogan, J. A., et al. 2001. *Plasmid* 45[3]:200.)

ppm   Parts per million.

PP2R1B   At human chromosome 11q22-q24 encodes the β isoform of the PP2A serine/threonine protein phosphatase. The gene displays alterations (LOH) in a variable fraction of lung, colon, breast, cervix, head and neck, and ovarian cancers and melanoma; it is thus a suspected tumor suppressor gene. *See* LOH; tumor suppressor gene. (Mumby, M. C. & Walter, G. 1993. *Physiol. Rev.* 73:673.)

PPT   Palmitoyl-protein thioesterases hydrolyze long-chain fatty acyl CoA, and PPT1 may cleave cysteine residues in the lysosomes. Its deficiency may lead to Batten disease. *See* Batten disease.

Prader-Willi syndrome (Prader-Labhart-Willi syndrome) Very rare (prevalence 1/25,000) dominant defect involving poor muscle tension, hypogonadism (hyperphagia [overeating]), obesity, short stature, small hands and feet, mental retardation, compulsive behavior by teenage, expressed from the paternal chromosome. The recurrence risk in affected families is about 1/1,000. This and cytological evidence indicates that the condition is caused in about 60% of the cases by a chromosomal breakage in the so-called imprinting center (IC) in the long arm of human chromosome 15q11.2-q12. The same deletion (4–5 Mbp or sometimes shorter) when transmitted through the mother results in Angelman syndrome. At the breakpoints, the HERC2 gene (encoding a very large protein) may be repeated. The repeats may then recombine and generate the deletions. See two chromosomes below with a different number of repeats, as detected by FISH. In some cases, there is no deletion but a mutation in a ubiquitin protein ligase gene (UBE3A). Mutations in the proximal part of IC lead to Angelman syndrome and in the distal part to Prader-Willi syndrome. Molecular studies indicated in many cases the missing (uniparental disomy) or silencing (imprinting) of paternal DNA sequences in patients.

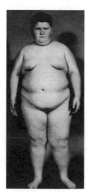

Prader-Willi syndrome at age 15. (From Bergsma, D., ed. 1973. *Birth Defects. Atlas and Compendium*. (By permission of the March of Dimes Foundation.)

The deletions of this syndrome usually involve the promoter of a snRPN gene resulting in the silencing (imprinting) of flanking genes (ZNF127 encoding a Zn-finger protein, NDN [necdin], IPW, and PAR) on either side. Lack of expression of snRNP is the most reliable clinical criterion

for the syndrome, although snRPN does not appear to be the only major pathogenic factor in the syndrome. Also, exon 1 (1,920 bp) includes more than 100 5′-CG-3′ and 5′-GC-3′ dinucleotides liable to methylation. Among the 19 methyl-sensitive restriction enzyme sites within the telomeric region, were completely methylated in this syndrome, but none of these sites were methylated in Angelman syndrome. A 2.2 kb spliced and polyadenylated RNA is transcribed 150 kb telomerically to snRPN in human chromosome 15q11.2-q12 and the homologous mouse chromosome 7 region. The transcript is not translated, however. This gene (IPW) is not expressed in individuals with Prader-Willi syndrome and therefore it is called imprinted in Prader-Willi syndrome. In the mouse gene *Ipw*, multiple copies of 147 bp repeats are found with retroviral transposon (IAP) insertions. *See* Angelman syndrome; disomic; head/face/brain defects; IAP; imprinting; imprinting box; obesity; snPRN. (Fulmer-Smentek, S. B. & Francke, U. 2001. *Hum. Mol. Genet.* 10:645.)

Duplications in the Prader-Willi syndrome. (Redrawn from Amos-Landgraf, J. M., et al. 1999. *Am. J. Hum. Genet.* 65:370.)

pRB   Retinoblastoma protein. *See* retinoblastoma.

preadaptive   Trait or mutation that occurs before selection would favor it, but it becomes important when the conditions become favorable for this genotype. *See* adaptation; fluctuation test; postadaptive mutation.

prebiotic   Before life originated. *See* evolution, prebiotic.

Precambrian   *See* Cambrian; geological time periods; Proterozoic.

precise excision   Genetic vector or transposon leaves the target site without structural alterations; the initially disrupted gene or sequence can return to the original (wildtype) form.

precursor ion scanning   Powerful technique in proteomics in connection with MS/MS and TOF. *See* MS/MS; TOFMS. (Steen, H., et al. 2001. *J. Mass Spectrom.* 36:782; Hager, J. W. 2002. *Rapid Commun. Mass Spectrom.* 16:512.)

predetermination   The phenotype of the embryo is influenced by the maternal genotypic constitution, but the embryo itself does not carry the gene(s) that would be expressed in it at that particular stage. *See* delayed inheritance; maternal effect genes.

predictive value True estimate of the number of individuals afflicted by a condition on the basis of the tests performed in the population.

predictivity Of an assay system, is, e.g., the percentage of carcinogens correctly identified among carcinogens and noncarcinogens by indirect carcinogenicity tests based mainly on mutagenicity. *See* accuracy; bioassays for environmental mutagens; sensitivity; specificity.

Predictome Database of protein links and networks. *See* genetic networks.

predictor gene Its expression signals differences among phenotypically similar but functionally different forms of malignancies. *See* cancer classification.

predisposition Susceptibility to disease.

preferential repair Transcriptionally active DNA is repaired preferentially. *See* DNA repair.

preferential segregation Nonrandom distribution of homologous chromosomes toward the pole during anaphase I of meiosis. It may constitute a genetic load if a harmful combination of genes (gene blocks) is preferentially included in the gametes. *See* meiotic drive; neocentromere; polarized segregation. (Rhoades, M. M. & Dempsey, E. 1966. *Genetics* 53:989; Buckler, E. S., et al. 1999. *Genetics* 153:415.)

prefoldins (PFDN) Molecular chaperones built as hexamers from the α- and β-subunits and four β-related subunits in eukaryotes. Prefoldin 1 was assigned to human chromosome 5, prefoldin 4 to chromosome 7. Prefoldins may be required for gene amplification in tumors. *See* chaperone. (Siegert, R., et al. 2000. *Cell* 103:621.)

preformation Absurd historical idea supposing that an embryo preexists in the sperm (spermists) or in the egg (ovists) of animals and plants rather than developing by epigenesis from the fertilized egg. *See* epigenesis. (Richmond, M. L. 2001. *Endeavour* 25[2]:55.)

pregenome RNA Replication intermediate in retroid viruses. *See* retroid virus.

P region of GTP-binding proteins Shares the G-X-X-X-X-G-K-(S/T) motif (*see* amino acid symbols) and suspected to involve the hydrolytic process of GTP-binding and several nucleotide triphosphate-utilizing proteins. *See* GTP-binding protein superfamily.

pregnancy, unwanted Estimated frequency in the human population of the whole world is between 35 and 53 million per year. *See* abortion, medical; pregnancy test.

pregnancy test Pregnancy is the formation of a fetus in the womb; there are about 40 pregnancy tests based on chemical study of blood and urine or other criteria. The currently used tests rely on estrogen level. *See* Aschheim-Zondek test.

Pregnenolone Precursor in the biosynthesis of several steroid hormones: CHOLESTEROL PREGNENOLONE → PROGESTERONE → ANDROSTENEDIONE → TESTOS – TERONE → ESTRADIOL. These steps are mainly under the control of several cytochrome P450 (CYP) enzymes, and their deficiency or misregulation leads to pseudohermaphroditism, hermaphroditism, and various other anomalies of the reproductive system. *See* steroid hormones.

preimmunity *See* host-pathogen relation.

preimplantation genetics Detects genetic anomalies either in the oocyte or in the zygote before implantation takes place. This can be done by molecular and biochemical analyses and cytogenetic techniques. The status of the egg—in some cases of heterozygosity for a recessive gene—may be determined prior to fertilization by examining the polar bodies. Since the first polar bodies are haploid products of meiosis, if they show the defect, then presumably the egg is free of it. The purpose of this test is to prevent transmission of identifiable familial disorders. The technology permits selection for sex of the embryo, but this is ethically controversial. *See* ART; gametogenesis; in vitro fertilization; micromanipulation of the oocyte; PGD; polymerase chain reaction; sperm typing; (Delhanty, J. D. 2001. *Am. J. Hum. Genet.* 65:331; Wells, D. & Delhanty, J. D. 2001. *Trends Mol. Med.* 7:23; Bickerstaff, H., et al. 2001. *Hum. Fertil.* 4:24.)

preinitiation complex *See* open promoter complex; PIC.

pre-mRNA (pre-messenger RNA) Primary transcript of the genomic DNA, containing exons and introns and other sequences. *See* hnRNA; introns; mRNA; posttranscriptional processing; RNA editing; RNA processing.

premutation Genetic lesion that potentially leads to mutation unless the DNA repair system remedies the defect before it is visually manifested. Premutational lesions lead to delayed mutations. UV irradiation or chemical mutagens with indirect effects (that is, the mutagen requires either activation or it induces the formation of mutagenic radicals, peroxides) frequently cause premutations. Incomplete expansion of trinucleotide repeats may also be considered premutational. *See* chromosomal mutation; chromosome breakage; mental retardation; point mutation; telomutation; trinucleotide repeats. (Auerbach, C. 1976. *Mutation Research*. Chapman and Hall, London, UK.)

prenatal diagnosis Determines the health status or distinguishes among the possible nature of causes of a problem with a fetus before birth. The results of cytological or biochemical analysis permit the parents to be psychologically and medically prepared if there is a problem. Although chromosomal abnormalities cannot be remedied, advance preparations can be made for metabolic disorders (e.g., galactosemia). Similarly, fetal erythroblastosis may be prevented. In case of very severe hereditary diseases, abortion may be an option if it is morally acceptable to the parents and does not conflict with existing laws. Prenatal diagnosis is now available for more than

100 anomalies. Until recently, prenatal diagnosis mainly required amniocentesis or sampling of chorionic villi. In some instances, the maternal blood can now be scanned for fetal blood cells and the DNA of the fetus can be examined by the use of polymerase chain reaction. *See* amniocentesis; ART; chorionic villi; DNA fingerprinting; echocardiography; fetoscopy; galactosemias; genetic screening; hydrocephalus; MSAFP; polymerase chain reaction; preimplantation genetics; PUBS; RFLP; sonography. (Weaver, D. D. & Brandt, I. K. 1999. *Catalog of Prenatally Diagnosed Conditions.* Johns Hopkins Univ. Press, Baltimore, MD.)

prenylation   Attachment of a farnesyl alcohol in thioester linkage with a cysteine residue located near the carboxyl terminus of the polypeptide chain. The donor is frequently farnesyl pyrophosphate. Cytosolic proteins are frequently associated with the lipid bilayer of the membrane by prenyl lipid chains or through other fatty acid chains. Prenyl biogenesis begins by enzymatic isomerization of isopentenyl pyrophosphate ($CH_2=C[CH_3]CH_2CH_2OPP$) into dimethylallyl pyrophosphate ($[CH_3]_2C=CHCH_2OPP$). These react to form geranyl pyrophosphate ($[CH_3]_2C=CHCH_2CH_2C[CH_3]=CHCH_2OPP$). Geranyl pyrophosphate is then converted into farnesyl pyrophosphate as shown below.

Members of the RAS family proteins involved in signal transduction, cellular regulation, and differentiation are prenylated at cysteine residues of the COOH-terminus. Prenylation determines the cellular localization of these molecules. Cellular fusions are mediated by prenylated pheromones. The cytoskeletal lamins attaching to the cellular membranes are farnesylated. Prenylation of the C-termini of proteins is generally mediated by farnesyltransferase, a heterodimer of 48 kDa α- and 46 kDa β-subunits. Squalene is a precursor of cholesterol and other steroids. *See* abscisic acid; cytoskeleton; lamin; lipids; pheromone; RAS.

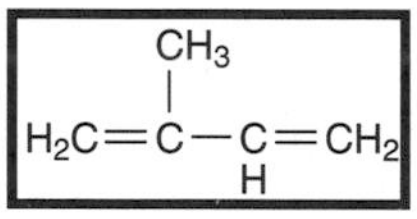

Isoprene units.

prepatent   Period before an effect (e.g., infection) becomes evident.

prepattern formation   Distribution of morphogens precedes the appearance of the visible pattern of particular structures. *See* morphogen. (Chiang, C., et al. 2001. *Dev. Biol.* 236:421.)

prepriming complex   Number of proteins at the replication fork of DNA involved in the initiation of DNA synthesis. *See* DNA replication; replication fork.

preprotein   The molecule has not yet completed its differentiation (trimming and processing).

prereduction   The alleles of a locus separate during the first meiotic anaphase because there is no crossing over between the gene and the centromere. *See* meiosis; postreduction; tetrad analysis.

pre-rRNA   Unprocessed transcripts of ribosomal RNA genes; they are associated at this stage with ribosomal proteins and are methylated at specific sites. The cleavage of the cluster begins at the 5′ terminus of the 5.8S unit and proceeds to the 18S and 28S units. *See* ribosomal RNA; ribosome; rrn; rRNA.

presence-absence hypothesis   Advocated by William Bateson during the first decades of the 20th century as an explanation for mutation. The recessive alleles were thought to be losses, whereas the dominant alleles were supposed to be the presence of genetic determinants. Similar views in a modified form have been maintained decades later and have been debated in connection with the nature of induced mutations. *See* genomic subtraction; null mutation. (Bateson, W., et al. 1908. *Rep. Evol. Comm. R. Soc. IV*, London, UK.)

presenilins   Proteins associated with precocious senility such as presenilin 1 (S182/AD3) encoded at human chromosome 14q24.3 (442 amino acids) and presenilin 2 (STM2/AD4, 467 amino acids encoded at 1q31-q34.2) proteins of Alzheimer's disease. Presenilins are integral membrane proteases. Presenilin 1 and presenilin 2 increase the production of β-amyloid either directly or most likely by their effect on secretases. They may also promote apoptosis. p53 and p21^{WAF-1} promote inhibition of presenilin 1, which may encourage apoptosis and tumor suppression as well. Presenilin 1 is associated with β-catenin and in the complex β-catenin is stabilized. Mutations in presenilin 1 may destabilize β-catenin and the latter is usually degraded in Alzheimer disease. Thus, mutation in presenilin 1 may predispose to early-onset Alzheimer disease. Presenilin 2 contains a domain that is similar to that of ALG3 (apoptosis-linked gene) and inhibits apoptosis. Presenilin 1 may also affect various (nonneurodegenerative) cancer-related

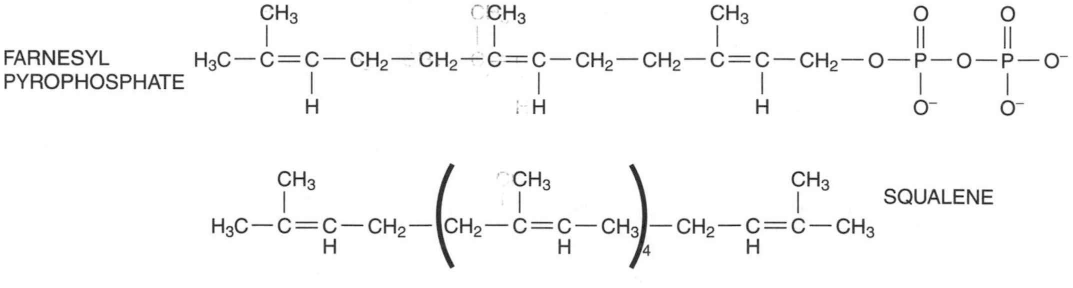

Two molecules of farnesyl pyrophosphate are converted into 30-C squalene in the presence of NADPH.

pathways. Presenilins are involved in the processing of the transmembrane domain of amyloid precursor proteins (APP), and they are essential for normal embryonal development. Presenilins also control the transduction of Notch signals. A presenilin locus exists in the third chromosome (77A-D) of *Drosophila melanogaster*. *See* Alzheimer disease; apoptosis; calsenilin; catenins; nicastrin; Notch; p21; p53; prion; secretase; ubiquilin. (Sisodia, S. S., et al. 1999. *Am. J. Hum. Genet.* 65:7; Baki, L., et al. 2001. *Proc. Natl. Acad. Sci. USA* 98:2381; Wolfe, M. S. & Haass, C. 2001. *J. Biol. Chem.* 276:5413.)

PRESENT Expressed open-reading frames during particular times or conditions when analyzed by microarrays. *See* microarray hybridization; open-reading frame.

presenting Behavioral signs shown by the female that indicate receptivity to mating.

presequence Generic name for signal peptides and transit peptides.

presetting Penchant of a transposable element to undergo reversible alteration in a new genetic milieu. It may be caused by methylation of the transposase gene. *See* Ac-Ds; Spm.

presymptotic diagnosis Identification of the genetic constitution before symptom onset. *See* genetic screening; prenatal diagnosis.

pre-tRNA *See* tRNA.

prevalence (K) Incidence of a genetic or nongenetic anomaly or disease in a particular human population. Percent of hereditary diseases caused by presumably single nuclear genes in human populations: autosomal dominant 0.75, autosomal recessive 0.20, X-linked 0.05. Multifactorial abnormalities account for about 6% of genetic anomalies. In case the general prevalence of the diseases in a population is x, the expected expression among sibs for autosomal dominant is $\frac{1}{2x}$, for autosomal recessives $\frac{1}{4x}$, and for multifactorial control $\frac{1}{\sqrt{x}}$. *See* mitochondrial diseases in humans.

prey *See* two-hybrid system.

prezygotic DNA molecule in the prokaryotic cell before recombination (transduction or transformation); after integration it becomes postzygotic.

PriA Replication priming protein. *See* DNA replication; primase; replication fork.

Pribnow box (TATA box) 5′-TATAATG-3′ (or similar) consensus preceding the prokaryotic transcription initiation sites by 5–7 nucleotides in the promoter region at about −10 position from the translation inititation site. There is another conserved element (extended promoter) separated by 17 bp in prokaryotes at −35 (TTGACA). The eukaryotic homologue of the Pribnow box is the Hogness box. *See* Hogness box;

open promoter complex; σ. (Gold, L., et al. 1981. *Annu. Rev. Microbiol.* 35:365.)

pride Living and mating community of animals under the domination of a particular male(s).

primary cells Taken directly from an organism rather than from a cell culture.

primary constriction Centromeric region of the eukaryotic chromosome.

Primary constriction ▼.

primary nondisjunction *See* nondisjunction.

primary response genes Their induction occurs without the synthesis of new protein but requires only preexisting transcriptional modifiers such as hormones. *See* secondary response genes; sign transduction.

primary sex ratio Ratio of males:females at conception. *See* sex ratio.

primary sexual characters Female and male gonads, respectively. *See* secondary sexual characters.

primary structure Sequence of amino acid or nucleotide residues in a polymer.

primary transcript RNA transcript of the DNA before processing has been completed. *See* pre-mRNA; pre-rRNA; processing.

primase Polymerase-α/primase synthesizes a 30-nucleotide RNA primer for the initiation of replication of the lagging strand of DNA. In prokaryotes, the primosome protein complex fulfills the function. In *E. coli*, gene *DnaG* (66 min) encodes it and it is associated with the replicative helicase. In eukaryotes, the ~60 and ~50 kDa subunits of DNA polymerase-α represent the primase. The latter complex is associated with proteins and forms a mass of ~300 kDa. The primases prime any single-stranded DNA, but they are much more effective at specific sequences. DnaG recognizes the 5′-CTG-3′ trinucleotide and synthesizes a 26–29 nucleotide RNA. The mouse primase works at ~17 sites that share either 5′-CCA-3′ or 5′-CCC-3′ at about 10 nucleotides downstream from the priming initiation site at the 3′-end. The active template is usually rich in pyrimidines. In eukaryotes, the primer is directly transferred to the DNA pol α without dissociating from the template. Primase inhibitors (cytosine or adenosine arabinoside, 2′-deoxy-2′-azidocytidine, etc.) have therapeutic potentials. Priming is assisted by several binding proteins. *See* DNA polymerases; DNA replication; Okazaki fragment; polymerase switching; PriA; primosome; replication fork. (Keck, J. L., et al. 2000. *Science* 287:2482; Arezi, B. & Kuchta, R. D. 2000. *Trends Biochem.* 25:572; Frick, D. N. & Richardson, C. C. 2001. *Annu. Rev. Biochem.* 70:39; Augustin, M. A., et al. 2001. *Nature Struct. Biol.* 8:57.)

primates   Taxonomic group that includes humans, apes, monkeys, and lemur. The following belong to the higher primates, also called anthropoidea or simians: old world monkeys (cercopitheoidea) such as the *Macaca*, *Cercopythecus*, etc.; hominoidea (chimpanzee [*Pan*], gorilla [*Gorilla*], orangutan [*Pogo*]) and humans, also the now extinct early evolutionary forms. The anthropoidea also include the new world monkeys (ceboidea). The lower primates or prosimians mean the genera of the lemur, galago, etc. According to data of D. E. Kohne, et al. (1972. *J. Hum Evol.* 1:627), on the basis of thermal denaturation of hybridized DNA, the numbers in millions of years of divergence (and the percent of nucleotide difference) of various primates from humans was estimated to be: chimpanzee 15 (2.4), gibbon 30 (5.3), green monkey 46 (9.5), capuchin 65 (15.8), galago 80 (42.0). Humans have substantially lower variations in the DNA than the great apes, chimpanzees, and orangutans (Kaessmann, H., et al. 2001. *Nature Genet.* 27:155). The taxonomic tree of primates can be outlined as PRIMATES: *I.Catarrhini.* IA1. Cercopithecidae (Old World Monkeys). IA1a. Cercopithecinae, IA1b. Colobinae. IA1c. Cercopithecidae. IB. Hominidae (Gorilla, Homo, Pan, Pongo). IC. Hylobatidae (Gibbons). *II.Platyrrhini* (New World Monkeys): IIA. Callitrichidae (Marmoset and Tamarins). IIA1. Callimico. IIA2. Callithrix. IIA3. Cebuella. IIA4. Callicebinae. IIA5. Cebinae. IIA6. Pitheciinae. *III Strepsirhini* (Prosimians) IIIA. Cheirogalidae. IIIA1. Cheirogaleus. IIIA2. Microcebus. IIIB. Daubentoniidae (Ayeayes). IIIB1. Daubentonia. IIIC. Galagonidae (Galagos). IIIC1. Galago. IIIC2. Otolemur. IIID. Indridae (Lorises). IIID1. Indri. IIID2. Propithecus (Sifakas). IIIE. Lemuridae (Lemurs). IIIE1. Eulemur. IIIE2. Hapalemur. IIIE3. Lemur. IIIE4. Varecia. IIIF. Loridae (Lorises). IIIF1. Loris. IIIF2. Nycticebus IIIF3. Perodicticus. IIIG. Megalapidae. IIIG1. Lepilemur. *IV.Tarsii* (Tarsiers). IVA. Tarsiidae (Tarpsiers). IVA1. Tarsius. *See* apes; Callithricidae; Cebidae; Cercopithecideae; Colobidae; evolutionary tree; Hominidae; *Homo sapiens*; human races; Pongidae; prosimii. (DeRousseau, C. J., ed. 1990. *Primate Life History and Evolution.* Wiley-Liss, New York; Enard, W., et al. 2002. *Science* 296:340; <www.primate.wisc.edu/pin>.)

primatized antibody   Chimeric antibody constructed using the variable region of monkey antibody linked to the human constant region. *See* antibody, chimeric.

primer   Short sequence of nucleotides (RNA or DNA) that assists in extending the complementary strand by providing 3'-OH ends for the DNA polymerase to start transcription. In some viruses (hepadna viruses, adenoviruses), the replication of viral DNA and some cases viral RNA is primed by proteins. The 3' OH group of a specific serine is linked to a dCMP and a viral enzyme drives the reaction. Replication may proceed from both ends of the linear molecules without being in the same replication fork. *See* nested primers; PCR; primase; Vpg. (<http://www.genome.wi.mit.edu>.)

primer extension   An RNA (or single-strand DNA) is hybridized with a single-strand DNA primer (30–40 bases) that is 5'-end-labeled. Generally, the primers are complementary to base sequences within 100 nucleotides from the 5'-end of mRNA to avoid heterogeneous products of the reverse transcriptase, which is prone to stop when it encounters tracts of secondary structure. After extension of the primer by reverse transcriptase, the length of resulting cDNA (measured in denaturing polyacrylamide gel electrophoresis) indicates the length of the RNA from the label to its 5'-end. When DNA (rather than RNA) is used as a template, DNA-DNA hybridization must be prevented. The purpose of the primer extension analysis is to estimate the length of 5' ends of RNA transcripts and identify precursors of mRNA and processing intermediates. The cDNA so obtained can be directly sequenced by the Maxam-Gilbert method or by the chain termination methods of Sanger if dideoxyribonucleoside triphosphates are included in the reaction vessels. *See* chimeric proteins; DNA sequencing; PCR-based mutagenesis; posttranscriptional processing; primary transcript. (Reddy, V. B., et al. 1979. *J. Virol.* 30:279; Sambrook, J., et al. 1989. *Molecular Cloning.* Cold Spring Harbor Lab. Press.)

primer shift   Used for confirmation that a PCR procedure indeed amplified the intended DNA sequence. For this purpose, a primer different from the one first employed is chosen and attached to the template a couple of hundred bases away from the position of the first one. After the completion of the PCR process, the amplified product is supposed to be as much longer as the difference between the position of the first and the second primers if the amplification involved the intended sequence. Such a procedure may be used when a DNA sequence corresponding to a deletion is amplified. *See* PCR; primer.

primer walking   Method in DNA sequencing whereby a single piece of DNA is inserted into a large-capacity vector. After a shorter stretch has been sequenced, a new primer is generated from the end of what was already sequenced and the process is continued until sequencing of the entire insert is completed. *See* DNA sequencing. (Zevin-Sonkin, D., et al. 2000. *DNA Seq.* 10[4–5]:245; Kaczorowski T. & Szybalski, W. 1998. *Gene* 223:83.)

primitive streak   Earliest visible sign of axial development of the vertebrate embryo when a pale line appears caudally at the embryonic disc as a result of migration of mesodermal cells. *See* differentiation; Hensen's node; morphogenesis; organizer. (Ciruna, B. & Rossant, J. 2001. *Developmental Cell* 1:37.)

primordium   Embryonic cell group that gives rise to a determined structure.

primosome   Complex of prepriming and priming proteins involved in replication of the Okazaki fragments. It moves along with the replication fork in the opposite direction to DNA synthesis. *See* DNA replication; Okazaki fragment; primase; replication fork. (Marsin, S., et al. 2001. *J. Biol. Chem.* 276:45818.)

Primula (Primrose)   Ornamental plant. *P. kewensis* ($2n = 36$) is an amphidiploid of *P. floribunda* ($2n = 18$) and *P. verticillata* ($2n = 18$).

PRINS Primed in situ synthesis is an in situ hybridization technique bearing some similarities to other methods of probing (e.g., FISH). The PRINS procedure uses small oligonucleotide (18–22 nucleotides) primers from the sequence of concern. After the primer is annealed to denatured DNA (chromosomal or other polynucleotides), a thermostable DNA polymerase is employed to incorporate biotin-dUTP or digoxygenin-dUTP. The procedure is very sensitive to mismatches (because the primer is short) and a mismatch at the 3′-end may prevent chain extension. The concentration of the primer (C) = Ab$_{260}/\varepsilon_{max}$ x L, where Ab$_{260}$ = absorbance at 260 nm, ε_{max} = molar extinction coefficient (M^{-1}), and L = the path length of the cuvette of the spectrophotometer. The molar extinction coefficients are determined as ε_{max} = (number of A × 15,200) + (number of T × 8,400) + (number of G × 12,010) + (number of C × 7,050)M^{-1} (A = adenine, T = thymine, G = guanine, C = cytosine). PRINS are useful for many purposes, including determination of aneuploidy, DNA synthesis, viral infection, etc. *See* biotinylation; extinction; FISH; in situ hybridization; LISA; nonradioactive label; PCR. (Hindkjaer, J., et al. 2001. *Methods Cell Biol.* 64:55.)

PrintAlign Computer program for graphical interpretation of fragment alignments in physical mapping of DNA. *See* physical map.

PrintMap Computer program that produces a restriction map in PostScript code. *See* PostScript.

PRINTS Database for the analysis of the hierarchy of protein families on the basis of fingerprints. *See* protein families. (<http://www.bioinf.man.ac.uk/dbbrowser/PRINTS/>.)

prion (PrP^C, PrP^{Sc}, PrP*) Infective glycoprotein particles responsible for degenerative brain diseases such as scrapie in sheep; BSE in cattle; kuru, Creutzfeldt-Jakob disease, Gerstmann-Sträussler syndrome, and fatal familial insomnia in humans; it may also be found in Alzheimer disease. Prions are transmitted among various animal species, although the expression may require a longer lag. Mutations in the gene may result in prion potentiation. Nonfamilial Creutzfeldt-Jakob disease may be traced to infections by gonadotropins, human growth hormones extracted from cadavers, grafts, or improperly sterilized medical equipment contaminated by prions.

On the basis of the degree and extent of glycosylation, about 400 prions have been distinguished. They appear like virus particles but are free of nucleic acid. It appears that a normal protein is structurally modified—the α-helical structure is largely converted into β-sheets—leading to the formation of these autonomous disease-causing proteins. In order to develop prion disease in mice, the organism must have PrP^C, and if it is absent, the animals become resistant to scrapie and show normal neuronal functions. Also, microglia (cells that surround the nerves and phagocytize the waste material of the nerve tissue) must be present. If microglia are destroyed by L-leucine-methylester, the neurotoxic PrP fragment containing amino acids 106–126 does not harm the neurons. The transition from the normal PrP^C → PrP^{Sc} (the insoluble scrapie prion) conformation involves changes in amino acid residues 121–231 involved in two antiparallel β-sheets and in three α-helices.

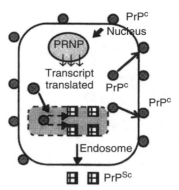

The normal PrP^C protein is encoded in the nucleus by the PRNP gene and after transcription the RNA transcript is translated in the cytoplasm. Some of the PrP^C molecules decorate the surface of the nerve cells and others may be sequestered into the endosomes or lysosomes. Within these compartments a conformational alteration may take place and the infectious PrP* or PrP^{Sc} protein molecules are released. These altered molecules may then infect other normal cells and initiate a process of degenerative protein accumulation. The conformational changes may be caused by mutations in PRNP and other genes located in several human chromosomes. (Modified after Weissmann, C. 1999. *J. Biol. Chem.* 274:3.)

Monoclonal antibody 15B3 discriminates between the two forms and may help in the diagnosis of prion diseases or perhaps cure (Peretz, D., et al. 2001. *Nature* 412:739). On the basis of experimental observations, it has been hypothesized that the toxicity of this protein is based on increased oxidative stress. The inactivation of the *PrP* gene in mice does not lead to an immediate deleterious condition, but by the age of 70 weeks an extensive loss of the Purkinje cells (large neurons in the cerebellar cortex) takes place and the animals have problems with movement coordination (ataxia). The disrupted *PrP* genes make them resistant to prions. On the basis of some genetic tests, it was concluded that the time of incubation of the mouse scrapie is controlled by allelic forms of a separate gene (*Sinc*/*Prni*). Molecular evidence indicates, however, that codons 108 and/or 109 of the *Prp* gene control incubation. In the mouse, there is a second *PrP* locus 16 kb downstream. This *Prnd* (d for Doppelgänger [alterego in German]) is truncated at the amino-end domain and encodes only 179 amino acids. In case the *PrP* (Prnp) exons are deleted, Doppelgänger exons can be spliced into the PrP mRNAs. In ataxic animals, this intergenic splicing is highly expressed. Apparently the manifestation of ataxia, loss of Purkinje cells, and degeneration of cerebellar granule cells is correlated with the alteration of a ligand-binding site.

A mutant form of PrP, ^{Ctm}PrP, a transmembrane protein, can also cause prion disease in the absence or presence of PrP^{Sc}. The latter may modulate the synthesis of the transmembrane form. PrP^C may be involved in signal transduction in nerve function. PrP^{Sc} apparently binds plasminogen that selectively imparts neurotoxicity to the prion protein.

In budding yeast, two nonnuclear elements (*URE3* and *PSI*) appear (among others) to be the infectious prion forms of the Ure2p protein, which is also a regulator of nitrogen catabolism. When Urep was overexpressed in wild-type strains, the frequency of occurrence of *URE3* increased 20–200-fold. If the overexpression of Urep was limited only to the amino ends of this protein, the frequency of occurrence of *URE3* increased

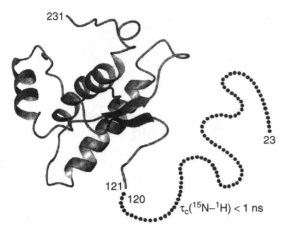

The nuclear magnetic resonance–revealed structure of the PrP^C protein. The amino end displays an ~100 residue flexible sequence that is modified when PRP^{Sc} is formed. (Modified after Rick, R., et al. 1997. *FEBS Lett.* 413:282. By Permission of Elsevier Science and Authors.)

6,000 times. The carboxyl domain of Urep seemed to carry out nitrogen catabolism, whereas the amino end induced the prion formation. Both *URE3* and *PSI* are the prion-causing forms of nuclear genes *URE2* and *SUP35*, respectively. The *URE2* gene is involved in the control of utilization of ureidosuccinate as a nitrogen source; the *SUP35* nuclear gene encodes a subunit (eRF3, eukaryotic release factor) of the yeast translation termination complex. Mutations in both of the nuclear genes involve derepression of nitrogen catabolism that is normally repressed by nitrogen. The propagation of *URE3* and *PSI* depends on *URE2* and *SUP35* nuclear genes, respectively. Guanidine-HCl blocks the propagation of PSI$^+$. In vitro the Sup35 protein may show prion-like properties. Normally, translation terminates at a stop codon by an interaction between Sup35 and other proteins such as Sup45. If the Sup35 proteins aggregate, they may assume prion conformation, the translation continues beyond the stop codon, and an additional protein sequence is formed. Five oligopeptide repeats at the N-terminus of Sup35 stabilize the aggregated form. The human PrP repeat (PHGGGWGQ) can substitute for the yeast peptides (Parham, S. N., et al. 2001. *EMBO J.* 20:2111). This feature of prions allows the development of diversity and may have evolutionary significance. Structurally, neither *URE3* nor *PSI* are similar to the mammalian PrP protein, indicating that there is more than one way for prions to arise. In the fungus *Podospora anserina*, the heterokaryosis incompatibility locus (*Het*) also makes prion-like proteins. The infectious forms of the normal Prp are also called PrP* and the PrP^{Sc} is designated as PrP^{res} (protease-resistant prion). PrP^C and PrP^{Sc} appear to be conformational isomers. PrP* is the misfolded pathological core form. The yeast prion *PSI*$^+$ can be reversibly removed, "cured" to *psi*$^-$ 100% in 7–8 generations when exposed to guanine hydrochloride or methanol. These denaturants induce the expression of chaperones, giving further support to the notion that the prion functions are based on conformational changes. Evidence indicates that the PrP^C → PrP^{Sc} transition may involve the chemical thiol/disulfide exchange between the terminal thiolate of PrP^{Sc} and the disulfide bond of a PrP^C monomer and not only a conformational change (Welker, E., et al. 2001. *Proc. Natl. Acad. Sci. USA* 98:4434). The protein

chaperones HSP104 and to a lesser extent HSP70 can affect the expression and transmission of *PSI*$^+$ and its conversion to *psi*$^-$.

When the *URE2* and *SUP35* genes or the N-terminal domain of their products are deleted, the *URE3* and *PSI*$^+$ elements permanently disappear. These yeast proteins are different from each other and from the prion proteins of higher eukaryotes, except the N-terminal region where homology exists. When fused to the rat glucorticoid receptor protein, the NH$_2$ domain of SUP35 can interact with the endogenous Sup35 protein, and it undergoes a prion-like change of state. The self-replication requires a conformational conversion of initially unstructured Sup35 protein. Thus, the prion-like behavior is transmissible to another protein (Derkatch, I. L., et al. 2001. *Cell* 106:171). More recently, additional yeast proteins (RNQ1, NEW1) with prion-like properties have been identified (Tuite, M. F. 2000. *Cell* 100:289; Derkatch, I. L. 2000. *EMBO J.* 19:1942). The vCJD (variant of Creutzfeldt-Jakob disease) prions appear to have either single amino acid differences or differences in glycosylation that may also be the cause or consequence of conformational differences. The differences in electrophoretic mobility of the protease-digested prions are expected to shed light on the problems of tracing the transmission of prions from cattle to humans or among different animal species. The PrP gene in humans is in chromosome 20p12 and encodes 253 amino acids by a single exon. The corresponding mouse gene is in chromosome 2. Other mammalian genes display very substantial homologies, although they may be transcribed by up to three exons. The NH$_2$ end of the protein displays an 8-amino-acid repeat consensus (PHGGGW) in 5–6 copies, depending on the species. Deletions in these repeats do not involve disease symptoms. Short conserved amino acids downstream of the last repeats are important for PrP^C → PrP^{Sc} conversion. Another unique feature of PrP is an alanine-rich tract (AGAAAAGA). The transmission of prions among different species prolongs the incubation period. Mice lacking the gene for PrP^c cannot develop the disease even when inoculated. Also, despite the fact that immunodeficient mice may accumulate some plaques upon scrapie infection, they fail do develop the disease. It appears that human individuals who might have been exposed to the same BSE source may not all respond with the development of the disease.

PrP → PrP^{Sc} conversion by infection with a prion from another species (heterotypic conversion), especially when the inoculum is small or the inoculation occurs rarely, is less likely. Wild-type mice that were brain infected with hamster prions did not develop scrapie; although a low level of maintenance of the hamster protein was detectable and reintroduced into hamsters, encephalitis followed. It is known that in Prp-deficient mice the immune system eliminates the PrP^c. It is also conceivable that in mice the hamster PrP^{Sc} is immunologically tolerated. The presence of the PrP gene is a requisite for the development of PrP^{Sc} protein. PrP^{Sc} exists in multimeric rather than monomeric form, but PrP may become part of the interacting PrP^{Sc} molecular network. There are indications that prion and DNA interaction may modulate the harmful aggregation of the protein (Cordeiro, Y., et al. 2001. *J. Biol. Chem.* 276:49400). Hamster-adapted prion protein heated up to 600°C for 5 to 15 min (actually ashed) still retained some infectivity and points to the role of an inorganic template in the replication of scrapie. Heating to 1000°C abolished all activity. In case of relatedness between these two proteins, PrP^{Sc} may easily facilitate the conversion to prion. The expression of

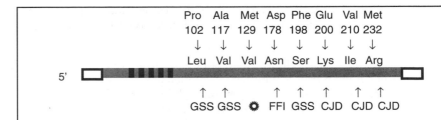

	Pro	Ala	Met	Asp	Phe	Glu	Val	Met
	102	117	129	178	198	200	210	232
	↓	↓	↓	↓	↓	↓	↓	↓
	Leu	Val	Val	Asn	Ser	Lys	Ile	Arg

5′

↑ ↑   ↑ ↑ ↑   ↑ ↑
GSS GSS ☀ FFI GSS CJD CJD CJD

- GSS is Gerstmann-Sträussler syndrome, FFI stands for fatal familial insomnia, CJD is for Creutzfeldt-Jakob disease.

The open boxes represent 5′ NH$_2$ and 3′ COOH regions of the proteins, respectively. The numbers indicate amino acid positions and are not shown on scale. ☀ indicates that the Val replacement at position 129 alone is not accompanied by prion disease but mutation at 178 associated with mutation to Val may cause FFI whereas the same mutation at 178 and methionine at 129 may lead to CJD. The black bars stand for 5 octa (nucleotide) repeats in the wild type and the same 5 octas may be repeated 9, 10, 11 or 12, times in CJD and 13 times in GSS.
(Redrawn after Weissmann, C. 1999. J. Biol. Chem. 274:3.)

Mutations in the prion gene.

the PrP^{Sc} may require chaperones. One such protein was named X, but its role is unclear. According to some views, the "protein-only" mechanism requires further proof, although all current evidence indicates a "protein-only" basis. Thus, the existence of prions seems to be an exception to the "nucleic doctrine." The prion diseases may be familial, with an onset at about 50 years of age in humans. The sporadic forms are attributed to dominant somatic mutations. From cultured scrapie-infected mouse (but not of hamster) neuroblastoma cells, the branched polyamines (polyamidoamide dendrimers, polypropyleneimine, polyethyleneimine) purged PrP^{Sc} prions at nontoxic concentrations. By 2002, no real cure or preventive measures emerged for prion diseases. There are some positive cues that proline-rich oligopeptides may restore the conformation of PrP^{Sc} to normal PrP. Preliminary results indicate that the lymphotoxin-β receptor may delay the onset of the symptoms temporarily in mice. It seems that the complement component C3 is important for the prions to attach to the follicular dendritic cells, which mediate infection. Antibodies generated against the μ-chain of PrP are a promising approach for the prevention of pathogenesis. *See* chaperones; Creutzfeldt-Jakob disease; curing plasmids; encephalopathies; fatal familial insomnia; Gerstmann-Sträussler disease; kuru; plasmin; presenilin; protein structure; PSI$^+$; tau. (*Annu. Rev. Genet.* 31:139; *Annu. Rev. Biochem.* 67:793; Prusiner, S. B., ed. 1999. *Prion Biology and Diseases*. Cold Spring Harbor Lab. Press, Cold Spring Harbor, NY; Umland, T. C. 2001. *Proc. Natl. Acad. Sci. USA* 98:1459; Heppner, F. L., et al. 2001. *Science* 294:178; Baskakov, I. V., et al. 2002. *J. Biol. Chem.* 277:21140; Kanu, N., et al. 2002. *Current Biol.* 12:523; Uptain, S. M. & Lindquist, S. 2002. *Annu. Rev. Microbiol.* 56:703.)

prisoner's dilemma Game theory applicable to interpretation of pairwise competition between two types of organisms using conflicting strategies. The two may cooperate, or one or the other may "defect" for selfish reasons and exploit the other. Consequently, the fitness may decrease to $1 - s_1$. In case both of them defect (are uncooperative), the population has to pay a cost (c), and the fitness becomes $1 - c$. In case the defector gains a fitness advantage $(1 + s_2)$, it may invade the cooperator's territory. If c is high $([1 - c] < [1 - s_1])$, a stable polymorphism may result. This theory is applicable to studies on economic activities and to other fields. (Page, K. M. & Nowak, M. A. 2001. *J. Theor. Biol.* 209:173; Neill, 2001. *J. Theor. Biol.* 211[2]:159.)

private blood groups Collective name of various blood groups with low frequencies compared to *public blood* group systems that occur frequently.

privilege *See* immune privilege.

PRL *See* prolactin.

PRL-3 (PTP4A3, 8q24.3) 22 kDa tyrosine protein phosphatase situated at the cytoplasmic membrane. Its elevated expression is associated with metastasis of colorectal cancer. *See* colorectal cancer; metastasis. (Saha, S., et al. 2001. *Science* 294:1343.)

PRM Pattern recognition proteins are involved in the regulation of transcription.

proaccelerin Labile blood factor (V); its deficiency may lead to parahemophilia and excessive bleeding during menstruation or after surgery or bruising. *See* antihemophilic factors.

probability Statistical measure of chance on a scale from 0 to 1, inclusive: 0 means the lack of chance for an event to occur and 1 indicates a certainty that it will occur. Values expressed as decimal fractions indicate the intermediate chances. The probability function indicates the value of a frequency predicted from the observations related to the parameter. *Simple probability* reveals the chance of a single event; *compound probability* is the chance of multiple events. When two events are independent, their *joint probability* is the product of their independent probabilities. *Alternate probability* exists in case of sex in dioecious species when an individual is either female or male and no intermediates are considered. Probability does not absolutely prove or disprove a point; it simply indicates the chance of its occurrence. *See* binomial probability; conditional probability; likelihood; maximum likelihood.

proband   Person(s) through which a family study of the inheritance of a human trait is initiated (also called propositus if male, or proposita if female). Determining the pattern of inheritance on the basis of families chosen by probands may display an excess of affected individuals relative to Mendelian expectations because of the bias in population sampling. *See* ascertainment test; pedigree analysis.

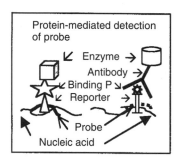

probe   Labeled nucleic acid fragment used for identifying or locating another segment by hybridization. Similarly, immunoprobes using primarily monoclonal antibodies, enzyme probes preceding page, or enzymes linked to antibodies (on the right) can be employed. The probe binds a reporter protein and another binding protein (binding p). For enzymatic detection of a probe, most commonly alkaline phosphatase or horseradish peroxidase is used. The tissue is incubated with the appropriate substrate of the enzyme and the colored precipitate formed through its action identifies its location. *See* heterologous probe; histochemistry; immunoprobe; labeling; nick translation; padlock probe; recombinational probe; synthetic DNA probes.

probe arrays   Oligonucleotides immobilized on silicon wafers in order to simultaneously study the functions of many genes. *See* DNA chips; microarray.

ProbeMaker   Computer program converts DNA sequence files in FASTA format to digital restriction maps used for MapSearch Probes.

probit   A cumulative normal frequency distribution is represented by an S curve. A cumulative curve can become a straight line by *probit transformation*. We may represent the probability scale in units of standard deviations. Thus, the 50% point is the 0 standard deviation, the 84.13 unit becomes +1, and the 2.27 point becomes the −2 standard deviation. The cumulative percentages are also called *normal equivalent deviates* (NED). If the ordinates are in NED units and we plot the cumulative normal curve, a straight line results. Probits are thus the NEDs with 5.0 added, so we do not get negative values for the majority of deviates. The probit value of 5.0 indicates a cumulative frequency of 50% and a probit value of 6.0 means a cumulative frequency of 84.13%, whereas probit 3.0 indicates a cumulative frequency of 2.27%. Probit value tables are available (Fisher, R. A. &

Yates, F. 1963. *Statistical Tables*. Hafner, New York). Probit transformations are frequently used for dosage mortality responses to chemicals, indicating the regression of cumulative mortalities on dosage. The graphs can be plotted on probit papers with the abscissa on a logarithmic scale.

proboscis   Tubular snout (nose-like emergence) on the head such as the feeding apparatus of *Drosophila*, elephant trunk, snout of tapirs, shrews, etc. *See* morphogenesis in *Drosophila*.

procaine anesthetics   Benzoic acid derivatives with local numbing of nerves or nerve receptors.

procambium   Primary meristem that gives rise to the cambium and the primary vascular tissue of plants. *See* cambium; meristem; root.

procapsid   Empty capsid precursor of phage into which the DNA can be packaged. *See* development; phage.

procarcinogen   Requires chemical modification to become carcinogenic. *See* activation of mutagens; carcinogen; phorbol esters.

procaryote   *See* prokaryote.

procentriole   Immature centriole. Upon maturing, it becomes the anchoring site of the spindle fibers, cilia, and flagella. *See* centriole; centromere; spindle fibers.

process, genetic   Gene products-mediated changes to reach a certain goal in the cell.

processed genes   Obtained by reverse transcriptase from mRNA. Therefore, they are free of all elements (e.g., introns) removed during processing of the primary transcript. They are widely expressed, highly conserved, short and low in GC. *See* cDNA; intron; primary transcript.

processed pseudogene (retropseudogene)   Similar to an mRNA, lacks introns, and may have a polyA tail, yet it is nonfunctional. Faulty reverse transcription of an mRNA may have produced processed pseudogenes. *See* cDNA: LINE; processed genes; pseudogene; reverse transcriptases. (Gonçalves, I., et al. 2000. *Genome Res.* 10:672.)

processing   Trimming and modifying the primary DNA transcripts into functional RNAs, or cutting and modifying polypeptide chains prior to becoming enzymes or structural proteins. *See* posttranslational processing; primary transcripts; protein synthesis.

processivity   Defines the number of nucleotides added to the nascent DNA chain before the polymerase is dissociated from the template. The processivity for *E. coli* DNA polymerase I, II, and III is 3–200, >10,000, and >500,000, respectively. *See* clamp-loader; DNA polymerases; error in aminoacylation; polymerase switching; replication fork.

processor   Data processing hardware or a computer program (software) that compiles, assembles, and translates information in a specific programing language.

prochiral molecule   Enzyme substrate after attaching to the active site undergoes a structural modification and becomes chiral. *See* active site; chirality.

prochloron   *See* evolution of organelles.

prochromatin   State of the chromatin that is conducive to transcription.

prochromosome   Heterochromatic blocks detected during interphase. In this interphase nucleus of *Arabidopsis*, on below the centromeric heterochromatin was stained by fluorescent isothiocyanate and it displayed a yellow-green color. (Courtesy of Drs. J. Maluszinszka & J. H. Heslop-Harrison.) *See* Barr body; heterochromatin; mitosis.

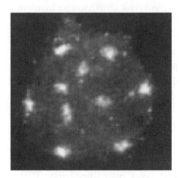

Prochromosomes of *Arabidopsis*.

procollagen   Precursor of collagen. *See* collagen.

proconsul   Fossil ape that lived about 17–23 million years ago.

proconvertin   Antihemophilic factor VII. Its deficiency may lead to excessive bleeding and hypoproconvertinemia. *See* antihemophilic factors; hypoproconvertinemia.

proctodeum   Invagination of the embryonal ectoderm where the anus is formed.

procyclic   *Trypanosoma* is in the gut of the intermediate host (tse-tse fly) and at this stage it is not infectious to higher animals. *See* metacyclic *Trypanosoma*.

prodroma   Ominous sign(s) of a looming disease.

prodrug   Processable to a biologically active compound. *See* activation of mutagens; ADEPT; suicide vector.

producer cell   Infected cell continuously produces recombinant retrovirus.

productive infection   The virus is not inserted into the eukaryotic chromosome, can propagate independently from the host DNA, and can destroy the cell while releasing progeny particles. *See* lysis.

product-limit estimator   Based on a number of conditional probabilities, e.g., the probability of survival after surviving for 1 day, then for the next day, and so on. Where $\hat{S}(t)$ is the survival function at subsequent times, r_j = the number of individuals at risk at time $t_{(j)}$ and d_j = the number of individuals involved in the event at risk time $t_{(j)}$.

$$\hat{S}(t) = \prod_{j|t_{(j)} \leq t} \left(1 - \frac{d_j}{r_j}\right)$$

product ratio method   *See* F_2 linkage estimation.

product rule   *See* joint probability.

proembryo   Minimally differentiated fertilized egg.

profile   Nucleotide or amino acid sequence probability motif. *See* motif.

profilin   Mediates actin polymerization. *See* actin; Bni1; cytoskeleton. (Carlsson, L., et al. 1977. *J. Mol. Biol.* 115:465.)

proflavin   Acridine dye capable of inducing frameshift mutations. *See* acridine dye; frameshift mutation. (Brenner, S., et al. 1958. *Nature* 182:933.)

progenitor   Ancestor or an ancestral cell of a lineage. Unlike stem cells, progenitor cells may lose their ability of self-renewal, yet they retain their mitotic ability and may generate one or more types of differentiated cells. *See* stem cells. (Reya, T., et al. 2001. *Nature* 414:105; Weissman, I. L., et al. 2001. *Annu. Rev. Cell Dev. Biol.* 17:387.)

progeny test   Procedure for determining the pattern of inheritance. *See* Mendelian laws; Mendelian segregation.

progeria   *See* aging.

progesterone   *See* animal hormones; estradiol; progestin; steroid hormones; testosterone.

Progesterone
(progestin)

progesterone receptors (PR) Assembled with the cooperation of at least eight chaperones including Hsp40, Hsp70, Hsp90, Hip, p60, p23, FKBPs, and cyclophilins. PRs are transcriptional regulators of progesterone-responsive genes. (Hernandez, P., et al. 2002. *J. Biol. Chem.* 277:11873.)

progestin Steroid hormone; medication for the prevention of repeated spontaneous abortion. When added to estrogen, it reduces the risk of endometrial cancer. *See* progesterone.

program (1) Set(s) of instructions in computer language (software) that permits the user to carry out specified tasks. (2) In biology, the development proceeds according to a genetically determined pattern realized by environmental effects.

programmed cell death *See* apoptosis.

progression Process involved in oncogenic transformation after the initial mutation of a protooncogene changes into an active oncogene. *See* cancer; phorbol esters.

prohibitin 30 kDa tumor-suppressor protein localized mainly in the mitochondria, although it is encoded at human chromosome 17q21. The well-conserved protein is present in other mammals, *Drosophila*, the plant *Arabidopsis*, and several microbes (*Pneumocystis carinii*, the cyanobacterium *Synechocystis*).

projectin Myosin-activated protein kinase. *See* myosin.

projection formula Modeling configurations of groups around chiral centers of molecules. *See* chirality.

prokaryon Prokaryote.

prokaryote Organism without membrane-enveloped (cell nucleus) genetic material (e.g., in bacteria). The majority of prokaryotic bacteria have circular double-stranded DNA chromosomes. However, *Borrelia, Streptomyces*, and *Agrobacterium tumefaciens* have linear chromosomes. *See Agrobacterium tumefaciens; Borrelia;* cell comparisons; Streptomyces.

prolactin (PRL) 23 kDa mitogen that stimulates lactation and the development of the mammary glands. Prolactin receptors are present on human lymphocytes, and prolactin may form complexes with IgG subclasses. A prolactin-releasing peptide was identified in the hypothalamus. *See* brain; immunoglobulins; lymphocytes; mitogen. (Mann, P. E. & Bridges, R. S. 2001. *Progr. Brain Res.* 133:251.)

prolamellar body Crystalline-like, lipid-rich structure in immature plastids that upon illumination develops into the internal lamellae of the proplastids and into the thylakoids of the chloroplasts. *See* chloroplast.

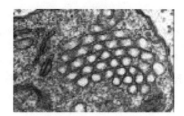

Prolamellar body.

prolamine *See* high lysine corn; zein.

proliferating cell nuclear antigen *See* PCNA.

proliferation Multiplication of cells or organisms. In cells, it may be caused by cytotoxic agents that may first induce cell death, then regenerative growth, or it may be the result of the action of mitogens. *See* cancer; mitogen.

proline biosynthesis Proceeds from glutamate through enzymatic steps involving glutamate kinase, glutamate dehydrogenase, and finally Δ'-pyrroline-5-carboxylate is converted to proline by pyrroline carboxylate reductase. In some proteins, e.g., collagen, prolyl-4-hydroxylase generates 4-hydroxyproline from proline. The latter enzyme is coded in human chromosomes 10q21.3-q23.1 (α-subunit) and 17q15 (β-subunit). *See* amino acid metabolism; hyperprolinemia.

prolog Database management and query system in physical mapping of DNA.

prolyl isomerase *See* immunophilin; PPI.

promastigote *See Trypanosoma*.

prometaphase Early metaphase. *See* mitosis.

promiscuous DNA Homologous nucleotide sequences occurring in the various cell organelles (nucleus, mitochondrion, plastid). They are assumed to owe their origin to ancestral insertions during evolution. *See* insertion elements. (Ayliffe, M. A., et al. 1998. *Mol. Biol. Evol.* 15:738; Lin, Y. & Waldman, A. S. 2001. *Nucleic Acids Res.* 29:3975.)

promiscuous plasmids *See* plasmids, promiscuous.

promitochondria Organelles in anaerobically grown (yeast) cells that can differentiate into mitochondria in the presence of oxygen. *See* mitochondria.

promoter Site of binding of the transcriptase enzyme (RNA polymerase), transcription factor complexes, and regulatory elements, including the ribosome-binding untranslated sequences. The promoter is usually (basal promoter) situated in front of the genes, although pol III may rely on both upstream and downstream promoters. The promoters of the 5S and tRNA genes are internal. The arrangement of the promoter used by pol II is outlined below. The promoters used by RNA polymerase II may encompass several hundred nucleotides in yeast, but in higher eukaryotes they may extend to several thousand bases. In yeast, UAS (upstream activating sequences) and URS (upstream repressing sequences) are regular binding sites. The transcription start site is usually within a stretch of 30 to 120 nucleotides downstream of the TATA box.

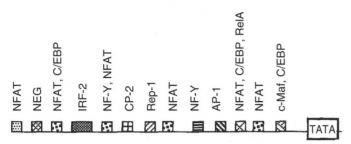

The promoter usually includes several regulatory boxes to which protein factors are recruited. (Modified after Guo, J. et al. 2001. *J. Biol. Chem.* 276:48871.)

At the ends of the genes, insulators (boundary elements) separate the genes or the used promoter from the others. The DNase hypersensitive site(s) (also called locus-control region) may permit the attachment of sequence-specific transcriptional activators, making the gene competent for transcription. The competence may involve histone acetylation.

Among 1,031 human protein-coding genes, the Pol II–like enzymes commonly (~32%) use a TATA box in prokaryotes and eukaryotes. The TATA box ca. 25 bp upstream from the initiation point of transcription is usually surrounded by GC-rich tracts (97%). Near the transcription initiation site (−3 to +5), there may be an initiator (Inr, 85%) with an average type of sequence: $(Pyrimidine)_2 CA(Pyrimidine)_5$. Many eukaryotic genes do not have Inr, but the TATA box directs the initiation. CAAT box is also a frequent (64%) element in the promoter. Some large eukaryotic genes utilize more than one promoter and the transcripts may vary. Some housekeeping genes and RAS genes do not use the TATA box. DNA-dependent RNA polymerase I synthesizes ribosomal RNAs; it has a core sequence adjacent to the transcription initiation site and upstream regulator-binding sites (UCE). Pol III promoters facilitating the transcription of tRNA usually have split promoters with an A-box and a B-box about 40 bases apart and situated inside the transcription unit 20 and 60 bases downstream from the transcription initiation site. The pol III promoter of some U RNAs has, however, a TATA box 30–60 bases upstream from the transcription initiation site and further upstream a proximal sequence element (PSE) near the TATA box. Synthetic promoters can be constructed with increased activity. The Promoter Scan II program identifies pol II promoters in genomic sequences and is available through the Internet: <http://www.cbs.umn.edu/software/software.html>.

Promoters (→) may be of different types and some genes may rely on multiple promoters:

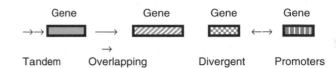

(arrows symbolize promoters; boxes stand for structural genes.)

The Signal Scan can be used to find transcription factor-binding sites by using TFD, TRANSFAC, and IMD databases. *See Arabinose* operon; basal promoter; chromatin remodeling; closed-promoter complex; complex promoter; core promoter; divergent dual promoter; divergent transcription; DPE; enhancer; histone acetyltransferase; insulator; *Lac* operon; LCR; minimal promoter; open-promoter complex; pol I; pol II; pol III; portable promoter; promoter, inducible; promoter clearance; promoter trapping; regulation of gene activity; TAF; TATA box; TBP; transcription complex; transcription factors; *Tryptophan* operon; UAS; URS. (Chalkley, G. E. & Verrijzer, C. P. 1999. *EMBO J.* 18:4835; Suzuki, Y., et al. 2001. *Genome Res.* 11:677; Pilpel, Y., et al. 2001. *Nature Genet.* 29:153; promoter tissue-specific: <http://www.epd.isb-sib.ch>; plant promoters: <http://sphinx.rug.ac.be:8080/PlantCare/cgi/index.html>.)

promoter, extended   *See* Pribnow box.

promoter, inducible   Turning on genes in response to biological, chemical, or physical signals. *See Lac*; metallothionein.

promoter, tissue-specific   Permits transcription of genes only or mainly in specific tissues. *See* promoter; tissue specificity.

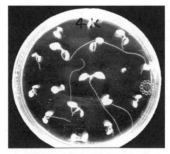

Tobacco seedlings segregating for kanamycin resistance on a root tissue-specific promoter. Transformation was made by a promoter-less vector construct. The non-transgenic plants cannot grow roots on kanamycin medium. (From Y. Yao & G. P. Rédei.)

promoter, tumorigenesis   Environmental substances or gene products that guide a group of precancerous cells toward

enhancer - PROMOTER - leader - exons - introns - termination signal - polyadenylation signal - downstream regulators

TRANSCRIPTION FACTOR-BINDNG SITES, DNase HYPERSENSITIVE SITE, TATA BOX, TRANSCRIPTION START malignant growth. The promoters themselves do not initiate cancer. *See* cancer; carcinogenesis; conversion; phorbol esters.

promoter bubble   *See* promoter clearance.

promoter clearance   The RNA polymerase complex (promoter bubble) starts moving forward from the promoter as the first ribonucleotides are transcribed. RNA polymerase can synthesize a few bases without leaving the promoter site, but tension subsequently develops that discontinues the contact between the DNA and the RNA polymerase. The movement may be represented by an inchworm model, a moving domain model (the translocation involves the entire transcription box with minimal stretching), or a model without a flexible polymerase that is tilted along the axis of the DNA. *See* inchworm model; replication bubble. (Pal, M., et al. 2001. *Mol. Cell. Biol.* 21:5815; Liu, C. & Martin, C. T. 2002. *J. Biol. Chem.* 277:2725.)

promoter conversion (Pro-Con)   Changing the promoter to a heterologous one.

promoter escape   *See* promoter clearance.

promoter interference   May occur when two genes are placed under separate controls within a single viral vector, e.g. the strong promoter within the long terminal repeat (LTR) may suppress the function of an internal promoter irrespective of its orientation. This problem may be overcome by utilizing an IRES for the second gene in the common transcript:

LTR——1st Gene——IRES——2nd Gene——. *See* IRES.

promoter occlusion   In retroviral elements with direct LTR repeats, the promoters at the 3′-end are inactivated and prevented from binding enhancers or transcription factors because they cannot facilitate transcription due to their wrong orientation. *See* enhancer; LTR; transcription factors.

promoter swapping   Exchange of promoter by, e.g., reciprocal chromosome translocation. *See* pleiomorphic adenoma; translocation.

promoter trapping   *See* gene fusion; promoter; transcriptional gene fusion vectors; translational gene fusion vectors; trapping promoters. (Medico, E., et al. 2001. *Nature Biotechnol.* 19:579.)

promutagen   Requires chemical modification (activation) to become a mutagen. *See* activation of mutagens; mutagen.

promyelocytic body (PML)   Product of the promyelocytic leukemia gene. It mediates the degradation of ubiquitinated proteins and is regulated by the nucleolus. PMLs also contain TRF1 and TRF2 telomeric proteins required for the maintenance of telomeres. PMLs may be targeted by viruses. They may act as growth and tumor suppressors and may mediate apoptosis. *See* apoptosis; leukemia; telomerase; ubiquitin.

pronase   Powerful general (nonspecific) proteolytic enzyme isolated from *Streptomyces*.

pronucleus   Male and female gametic nucleus to be involved in the sexual union.

proofreading   Bacterial DNA polymerase I (and analogous eukaryotic enzymes) can recognize replicational errors and remove the inappropriate bases by editing 3′–5′ exonuclease function. In case the editing function is diminished by mutation, mutator activity is gained. In case of gain in editing function, antimutator attributes are observed. In bacteria, proofreading is also performed by the *dnaQ* gene encoding the ε-subunit (an exonuclease) of the DNA polymerase III holoenzyme. The base selection is carried out by the product of gene *dnaE*. The enzymes MutH, MutL, and MutS and the corresponding homologues in higher organisms repair mismatches. The fidelity of replication due to the combined action of the sequentially acting bacterial genes was estimated to be in the range of 10^{-10} per base per replication. During the process of translation, the EF-Tu•GTP→EF-Tu•GDP change releases a molecule of inorganic phosphate (P_i) and allows a time window to dissociate the wrong tRNA from the ribosome. A similar correction is made by the aminoacyl synthetase enzyme by virtue of its active site specialized for this function. DNA polymerase η lacks exonuclease function required for proofreading, but correction is still accomplished by recruiting an extrinsic exonuclease to the error site. *See* ambiguity in translation; DNA polymerase I; DNA polymerase III; DNA polymerases; DNA repair; error in aminoacylation; error in replication; exonuclease; proofreading paradox; protein synthesis. (Friedberg, E. C., et al. 2000. *Proc. Natl. Acad. Sci. USA* 97:5681; Livneh, Z. 2001. *J. Biol. Chem.* 276:25639; Shevelev, I. V. & Hübscher, U. 2002. *Nature Rev. Mol. Cell Biol.* 3:364.)

propagule   Part of an organism that can be used for propagation of an individual by asexual means.

propeller twist in DNA   Surface angle formed between individual base planes viewed along the C^6–C^8 line of a base pair.

properdin (Factor P)   Serum protein of 3–4 subunits (each ca. 56 kDa, encoded in human chromosome 6p21.3). It is an activator of the complement of the natural immunity system by stabilizing the convertase. *See* complement; convertase; immune system. (Perdikoulis, M. V., et al. 2001. *Biochim. Biophys. Acta* 1548:265.)

prophage   Proviral phage is in an integrated state in the host cellular DNA and it is replicated in synchrony with the host chromosomal DNA until it is induced, thus becoming a vegetative virus. *See* lambda phage; lysogeny; prophage induction; temperate phage.

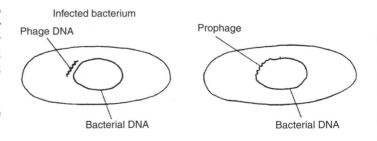

prophage induction Treating the bacterial cells by physical or chemical agents that cause the moving of the phage into a vegetative lifestyle, resulting in asynchronous, independent replication from the host and eventually lysis and liberation of phage. *See* lysogeny; prophage; zygotic induction.

prophage-mediated conversion The integrated prophage causes genetic changes in the host bacterium, and it is expressed as an altered antigenic property, etc.

prophase *See* meiosis; mitosis.

propionicacidemia *See* glycinemia, ketotic; isoleucine-valine biosynthetic pathway; methylmalonicaciduria; tiglic-acidemia. (Chloupkova, M., et al. 2002. *Hum. Mut.* 19:629.)

propionyl-CoA-carboxylase deficiency *See* glycinemia, ketotic.

proplastid Young, colorless plastid without fully differentiated internal membrane structures; it may differentiate into chloroplast. *See* chloroplasts; etioplast.

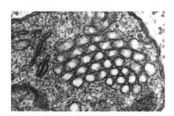

Proplastid.

proportional counters Used for measuring radiation-induced ionizations within a chamber. The voltage changes within are proportional to the energy released. It may be used for measuring neutron and α-radiations with an efficiency of 35–50%. The equipment must be calibrated to the radiation source. *See* radiation hazard assessment; radiation measurement.

propositus (proposita) *See* pedigree analysis; proband.

propyne ($CH_3-C\equiv CH$) Alkyne used as a modifier at the C-5 position of pyrimidines in antisense oligonucleotides frequently in combination with other modifications such as phosphorothioate. *See* antisense technologies.

Prosimii (prosimians) Suborder of lower primates, including Galago, Lemur, Tarsius, Tupaia, and Lorisidae. *See* Lemur; Lorisidae; primates.

PROSIT Protein sequence database searchable by PRO-SCAN (Bairoch, A., et al. 1977. *Nucleic Acid. Res.* 25:217; <http://pbil.univ-lyon1.fr/pbil.html>; PROSITE for uncharacterized proteins: <http://www.expasy.org/prosite/>.)

prosody Inability of sensing or expressing variations of the normal rhythm of speech. It seems to be independent of processing musical pitch. *See* amusia; musical talent; pitch.

prosome Small ribonucleoprotein body. It is identical with the ~20S multifunctional protease complex of the proteasome of eukaryotes and prokaryotes. *See* proteasome.

PROST Pronuclear state embryo transfer is basically very similar to intrafallopian transfer of zygotes, but the zygote is at a very early stage. *See* ART; intrafallopian transfer.

prostacyclins May be derived from arachidonic acid or prostaglandins, regulate blood platelets, cause vasodilation, and they are antithrombotic. *See* prostaglandins; thrombosis.

prostaglandins Long-chain fatty acids in different mammalian tissues with hormone-like, muscle-regulating, inflammation-regulating, and reproductive functions. They exist in several forms. They occur in the majority of cells and act as autocrine and paracrine mediators. Fever development is controlled by prostaglandin E_2 and EP_3 receptor. Prostaglandin synthesis is regulated by cyclooxygenases. Prostaglandin E (cyclopentenone prostaglandins) appears to be an inhibitor of IκB kinase. *See* animal hormones; autocrine; cyclooxygenases; eicosanoids; IκB; leukotrienes; paracrine. (Rudnick, D. A., et al. 2001. *Proc. Natl. Acad. Sci. USA* 98:8885.)

prostanoids Bioactive lipids such as prostaglandins, prostacyclin, and thromboxane. Aspirin-like drugs may inhibit prostanoid biosynthesis, reduce fever and inflammation, and interfere with female fertility. *See* prostacyclins; prostaglandins.

prostate (prostata) Gland in the animal (human) male surrounding the base of the bladder and the urethra. Upon ejaculation, it injects its content (acid phosphatase, citric acid, proteolytic enzymes, etc.) into the seminal fluid. *See* prostate cancer; PSA. (<http://www.pedb.org>; <http://www.pedb.org>.)

prostate cancer (HPC) About 9–10% of American males eventually develop this malignancy. The autosomal-dominant gene has a high penetrance: about 88% of the carriers become afflicted by age 85. High testosterone levels may increase the chances for this cancer. A reduced level of testosterone may not slow down advanced prostate cancer growth and metastasis. Metastatic prostate cancer cells may show high levels of caveolin-1 and a reduced amount of testosterone. Caveolin-1 antisense RNA promoted apoptosis and increased testosterone. A metastasis suppressor gene, KAI1, in human chromosome 11p11.2, has been identified. The KaI1 protein appears to contain 267 amino acids with four transmembrane hydrophobic domains and one large hydrophilic domain. This glycoprotein is expressed in several human tissues and in rats. A negative regulator of the MYC oncogene, MXI1 (encoded in human chromosome 10q24-q25), is frequently lost in prostate cancer. In the chromosome 10pter-q11 region, a prostate cancer suppressor gene, causing apoptosis of carcinoma, has been detected from loss of heterozygosity mutations (LOH). A major susceptibility locus was identified in human chromosome 1q24-q25 and at Xq27-q28. Candidate genes are expected in human chromosomes 3p, 4q, 5q, 7q32, 8p22-p23 and 8q, 9q, 10p15 (KLF6), 13q, 16q, 17p11, 18q, 19q12, and 20q13. Insulin-like growth factor (IGF-1) levels may be predictors of prostate cancer risks before cancerous growth is observed, but some other data are at variance with the claim. In prostate tumors, the prostate-specific cell-surface antigen (STEAP), human chromosome 7p22.3, is highly expressed in different organs and tissues except the bladder. A predisposing gene in 17p has been cloned. Prostate stem cell antigen (PSCA) may be the target for immunological therapy. The relative risk of first degree is 1.7–3.7 or more. Prostate cancer is the second most frequent cause of cancer mortality in the U.S., but it is very rare in other animals, except dogs. *See* antisense technology; apoptosis; cancer; caveolin; Gleason score; insulin-like growth factor; MYC; PSA; testosterone; tumor suppressor gene. (Ostrander, E. A. & Stanford, J. L. 2000. *Am. J. Hum. Genet.* 67:1367; Xu, J., et al. 2001. *Am. J. Hum. Genet.* 69:341; Stephan, D. A., et al. 2002. *Genomics* 79:41.)

prosthesis   Any type of mechanical replacement of a body part, such as artificial limbs, false teeth, etc.

prosthetic group   Nonpeptide group (iron or other inorganic or organic group) covalently bound (conjugated) to protein to assure activity.

PROT Na^+/Cl^--dependent proline transporter that also transports glycine, GABA, betaine, taurine, creatine, norepinephrine, dopamine, and serotonin in the brain. *See* transporters.

protamine   Basic (arginine-rich) protein occurring in the sperm, substituting for histones. It controls both condensation and decondensation of DNA by anchoring to it at about 11 bps. After fertilization, it is removed. In the somatic cells, protamines constitute less than 5% of the nucleus. The majority of mammals have only a single protamine, but mice and men have two. If protamines are deleted or missing, functional sperm is not produced because of haploinsufficiency. *See* haploinsufficiency; histones; transition protein. (Cho, C., et al. 2001. *Nature Genet.* 28:82.)

protandry   The pollen is shed in monoecious plants before the stigma is receptive. *See* monoecious; protogyny; self-sterility; stigma.

protanope   *See* color blindness.

protease (proteinase)   Enzyme that hydrolyzes proteins at specific peptide bonds. For *protease 3* see antimicrobial peptides.

protease inhibitors   For example, leupeptin, antipain, and soybean trypsin inhibitors are credited with anticarcinogenic effects and potential cures for asthma and schistosomiasis. Proteases process the primary proteins into their functional role in viral/microbial or other systems. If this processing is prevented, infectious or other agents may or not have reduced adverse effects to the cells. *See* AIDS; asthma; cancer; carcinogen; schistosomiasis.

proteasomes   Tools of degradation of proteins. ATP-dependent ubiquitinated proteins process intracellular antigens into short peptides, which are then transported to the endoplasmic reticulum with the aid of TAP and are responsible for MHC class I–restricted antigen presentation. Proteasomal polymorphism is partially determined by LMP2 and LMP7 genes encoded within the MHC class II region in the vicinity of TAPs that are upregulated by interferon-γ. The 26S ($\sim$2,500 kDa) proteasomes ($\sim$31 subunits) are hollow cylinders engulfing ubiquitinated proteins and degrading them with proteases. The lid and the base are each 19S (890 kDa). The $\sim$20S (720 kDa) middle-section barrel of the proteasomes contains multiple peptidases. Their active site is at the hydroxyl group of the N-terminal threonine in the β-subunit. The PA proteins are proteasome activators. The 26S proteasome is associated with at least 18 ancillary and essential proteins (PSM proteins, including ATPase), and many of these are genetically mapped to different human chromosomes. Chymostatin, calpain, leupeptin, etc., are inhibitors. The proteases of the 20S proteasome are activated by the heptameric 11S regulators, which also control the opening of the barrel-shaped structure. Proteasomes have ubiquitin-independent functions such as the degradation of excess amounts of ornithine decarboxylase, a key enzyme in polyamine biosynthesis. The proteasomes have important—although not fully understood—roles in differentiation and development by mediating protein turnover. In case the proteasome function is inhibited or lost, inhibitor-resistant cells that have a compensating mechanism for proteasome function may grow out of the culture. The Cop9 signalosome of *Arabidopsis* is functionally a homologue to the lid element of the proteasome. It appears that various elements of the proteasome complex are coregulated by the RPM4 putative transcription factor. *See* antigen-presenting cell; antigen-processing; Clp; DRiP; immune system; immunoproteasome; LID; MHC; N-end rule; photomorphogenesis; polyamine; signalosome; Skp1; TAP; tripeptidyl peptidase; ubiquitin. (Voges, D., et al. 1999. *Annu. Rev. Biochem.* 68:1015; Bochtler, M., et al. 1999. *Annu. Rev. Biophys. Biomol. Struct.* 28:295; Kloetzel, P.-M. 2001. *Nature Rev. Mol. Cell Biol.* 2001. 2:179; Ottosen, S., et al. 2002. *Science* 296:479.)

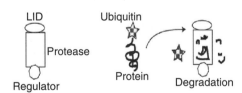

protectin (CD59)   Protein component of the complement encoded at 11p13. *See* complement; paroxysmal nocturnal hemoglobinuria. (Kawano, M. 2000. *Arch. Immunol. Ther. Exp.* 48[5]:367.)

protein   Large molecule (polymer) composed of one or more identical or different peptide chains. The distinction between protein and polypetide is somewhat uncertain; generally a protein has more amino acid residues (50–60) and therefore can fold. In animal cells, there are about 1×10^5 protein species. *See* amino acid sequencing; protein structure; protein synthesis. (Westbrook, J., et al. 2002. *Nucleic Acids Res.* 30:245; <http://www.rcsb.org/pdb/>; <http://www.ncbi.nlm.nih.gov/Entrez/structure.html>; <http:// pir.georgetown.edu/>.)

protein 14-3-3 Family of 28–33 kDa acidic chaperone proteins named after their electrophoretic mobility. These proteins occur in many forms in different organisms and have roles in signal transduction (RAS-MAP), apoptosis, exocytosis, and regulation of the cell cycle (checkpoint) and oncogenes. They generally bind to phosphoserine/threonine domains. *See* CaM-KK; Cdc25; cell cycle; chaperone; checkpoint; Chk1; p53. (Muslin, A. J. & Xing, H. 2000. *Cell Signal.* 12[11–12]:703; Masters, S. C. & Fu, H. 2001. *J. Biol. Chem.* 276:45193; Tzivion, G. & Avruch, J. 2002. *J. Biol. Chem.* 277:3061; Sehnke, P. C., et al. 2002. *Plant Cell* 14:S339.)

protein A Isolated from *Staphylococcus aureus*; it binds the Fc domain of immunoglobulins without interacting with the antigen-binding site. It is used both in soluble and insoluble forms for the purification of antibodies, antigens, and immune complexes. *See* antibody; immunoglobulins.

protein arrays Used in a manner analogous to microarrays of DNA. On specially treated microscope slide samples, proteins are lined up and exposed to other proteins or to drug molecules or molecular fragments in order to assess their interaction. This new procedure is expected to be useful for analytical purposes and particularly for the development of new drugs. *See* microarray hybridization; protein chips. (Avseenko, N. V., et al. 2001. *Anal. Chem.* 73:6047; Brody, E. N. & Gold, I. 2000. *J. Biotechnol.* 74:5.)

proteinase A Endopeptidase involved in protein folding. *See* endopeptidase; protein folding.

proteinase K Proteolytic enzyme frequently used to remove nucleases during the extraction of DNA and RNA. With appropriate heat treatment, any DNase associated with it can be safely removed. *See* protease.

protein assays *See* Bradford method; Kjeldhal method; Lowry test. (For analysis with single-cell resolution: Zhang, H. T., et al. 2001. *Proc. Natl. Acad. Sci. USA* 98:5497.)

protein C (2q13-q14) Vitamin K–dependent serine protease that selectively degrades antihemophilic factors Va and VIIIa; it is thus an anticoagulant. *See* anticoagulation; antihemophilic factors; protein C deficiency; thrombin; thrombophilia.

protein C deficiency (thrombotic disease) Human chromosome 2q13-q14 dominant. It may be a life-threatening cause of thrombosis. *See* protein C; thrombosis.

protein chips Protein mixture (e.g., serum) applied to an about 1 mm^2 surface containing a "bait" that is an antibody; a specific receptor or other kind of specific molecule that selectively binds a particular protein (tagged by fluorescent dye), thus facilitating its isolation even when it is present only in minute amounts. Or recombinant proteins are immobilized on chips, then putative interacting proteins (cell lysates) are applied to them. The unbound material is removed by washing and the bound ones are analyzed by mass spectrometry or phage display, or the two-hybrid method may be used. These procedures can efficiently handle a huge number of samples and bear a similarity to DNA chips. *See* DNA chips; electrospray; ELISA; gene product interaction; MALDI; mass spectrum; microarray hybridization; phage display; proteomics; two-hybrid method. (Zhu, H., et al. 2001. *Science* 293:2101.)

protein clock *See* evolutionary clock.

protein complexes Usually play an important role in protein and cellular function. Their study requires enrichment of the complex either by chromatography, coimmunoprecipitation, coprecipitation by affinity-tagged proteins, and SDS-PAGE separation of the components before additional analytical techniques are employed. One study involving 1,739 yeast genes, including 1,143 human homologous genes, revealed 589 protein assemblies. Among these, 51% included up to 5 proteins; 6% more than 40 proteins; 4%, 31–40; 6%, 21–30; 15%, 11–20; and 18% displayed interactions among 6–10 proteins. The technology did not reveal interaction of very short duration. Obviously, within the cells even more proteins interact. *See* genetic networks; immunolabeling; immunoprecipitation; mass spectrometry; SAGE; SDS-PAGE; TAP; two-hybrid method. (Gavin, A.-C., et al. 2002. *Nature* 415:141; Ho, Y., et al. 2002. *Nature* 415:180; <http://www.binddb.org/>.)

protein conducting channel Membrane passageway for proteins that interacts with the membrane protein and lipid components, *See* ABC transporters; protein targeting; Sec61 complex; SRP; TRAM; translocase; translocon. (Spahn, C. M., et al. 2001. *Cell* 107:373.)

protein conformation *See* conformation.

Protein Data Bank (PDB) Archive of macromolecular structures. *See* protein structure. (<http://www.pdb.org/>.)

protein degradation *See* endocytosis; lysosomes; major histocompatibility complex; proteasome; TAP.

protein degradation within cells Endogenous proteins are digested primarily by the proteasomes and exogenous proteins are cleaved mainly by the lysosomal system, although the compartmentalization is not rigid. *See* lysosome; N-end rule; proteasome.

protein design Computer program exists to design new proteins for physicochemical potential function and stereochemical arrangements using combinatorial libraries of amino acids. (Dahiat, B. I. & Mayo, S. L. 1997. *Science* 278:82.)

protein domains Generally formed by folding of 50–350 amino acid sequences for carrying out particular functions. Small proteins may have only a single domain, but larger complexes may have multiple modular units. The alternations of α-helices and β-sheets constitute a characteristic *motif*. Two β-sheet motifs are shown in black and white, respectively, below. The compact motifs are generally covered by polypeptide loops. Domain similarities among proteins from different organisms indicate a possible functional relationship (homology) of those proteins. *See* α-helices; binding proteins; helix-loop-helix; helix-turn-helix; motif; protein structure β sheets; zinc finger. (Ponting, C. P. & Russell, R. R. 2002. *Annu. Rev. Biophys. Biomol. Struct.* 31:45; <http://SMART.embl-heildelberg.de>; <www.ebi.ac.uk/interpro>.)

protein engineering   Constructing proteins with amino acid replacements at particular domains and positions (e.g., substrate-binding cleft, catalytic and ligand-binding sites, etc.) or adding a label or another molecule, etc., to explore their effect on function. *See* directed mutation; DNA shuffling; iterative truncation. (Tao, H. & Cornish, V. W. 2002. *Curr. Opin. Chem. Biol.* 6:858; Brennigan, J. A. & Wilkinson, A. J. 2002. *Nature Rev. Mol. Cell Biol.* 3:964.)

protein families   Share structural and functional similarities; generally share more than 30% sequence identity. *Superfamilies* catalyze the same chemical reaction or different overall reactions that share common mechanistic properties (partial reaction, intermediate or transition state) and 20 to 50% sequence identity. *Suprafamilies* are homologous enzymes that catalyze different reactions. *See* gene family; PRINTS. (Enright, A. J., et al. 2002. *Nucleic Acids Res.* 30:1575; Aravind, L., et al. 2002. *Current Opin. Struct. Biol.* 12:392; <http://pfam.wustl.edu/>; <http://pir.georgetown.edu/gfserver/ proclass.html>; <http://www.biochem.ucl.ac.uk/bsm/cath>; <http://www.protomap.cs.huji.ac.il>; <http:P//stash.mrc-imb.cam.ac.uk/SUPERFAMILY/>.)

protein folding   The majority of proteins fold to acquire functionality, although some (mainly) surface proteins do not require folding. Glycosylation in the endoplasmic reticulum may affect conformation of proteins. The folding is determined by the amino acid sequence; however, other factors (chaperones) may be needed to facilitate the process. Prokaryotic proteins (which are generally smaller, 200–300 amino acid residues) fold correctly only after the completion of the entire length of the amino acid chain. Eukaryotic proteins (usually on the average over 400–500 residues) may fold the separate domains in a sequential manner during their translation. Both prokaryotic and eukaryotic proteins may start folding before their translation is completed, i.e., cotranslationally. Therefore, fusion proteins can also fold, which might have been of evolutionary advantage. There is evidence that α-helices fold faster than β-sheets. Local interactions may facilitate speedier folding. Certainly many factors may affect the rate of folding, and the rate among different proteins may be 9 orders of magnitude. Disease may be due to misfolding of protein(s). *See* calnexin; calreticulin; chaperones; chaperonins; conformation; endoplasmic reticulum; folding; GroEL; protein structure; protein synthesis; Sec61 complex; trigger factor. (Bukau, B., et al. 2000. *Cell* 101:119; Baker, D. 2000. *Nature* 405:39; Parodi, A. J. 2000. *Annu. Rev. Biochem.* 69:69; Klein-Seetharaman, J.,

et al. 2002. *Science* 295:1719; Hartl, F. U. & Hayer-Hartl, M. 2002. *Science* 295:1852; Myers, J. K. & Oas, T. G. 2002. *Annu. Rev. Biochem.* 71:783.)

protein function   Generally determined by biochemical and genetic analyses such as enzyme assays, two-hybrid system, etc. Many proteins are involved in complex functions and interact with several other proteins. These complex functions can be inferred from a known role of proteins in evolutionarily different organisms, from amino acid sequence information, by the rosetta stone sequences, the correlation of mRNA expression, and gene fusion information from sequence data. *See* microarray hybridization; rosetta stone sequences; two-hybrid system.

protein G   Immunoglobulin-binding (IgG) streptococcal extracellular cell-surface protein.

protein H   Streptococcal IgG-binding protein. *See* immunoglobulins.

Protein Information Resource (PIR)   <http://pir.georgetown.edu>; <htpp://www.mips.biochem.mpg.de>; the largest and most comprehensive source is the Swiss-Prot: <http://www.expasy.ch/>. *See* databases.

protein interactions   *See* gene product interaction. (Bock, J. R. & Gough, D. A. 2001. *Bioinformatics* 17:455; <http://web.kuicr.kyoto-u.ac.jp/~vert/bibli/bock01.pdf>.)

protein intron   *See* intein.

protein isoforms   Closely related polypeptide chain family, encoded by a set of exons that share a structurally identical or almost identical subset of exons. *See* family of genes.

protein kinase   Phosphorylates one or more amino acids (frequently threonine, serine, tyrosine) at certain positions in a protein, and thus two negative charges are conveyed to these sites, altering the conformation of the protein. This alteration involves a change in the ligand-binding properties. The catalytic domain of this large family of enzymes is usually 250 amino acids. The amino acids outside the catalytic domains may vary substantially, specify the recognition abilities of the different kinases, and serve in responding to regulatory signals. Since the 1970s, hundreds of protein kinases have been discovered that can be classified into serine/threonine, TGF-β (transforming growth factor), tyrosine (EGF (epidermal growth factor) receptor), PDGF (platelet-derived growth factor) receptor, SRC (Rous sarcoma oncogene product), Raf (product of the Moloney and MYC oncogenes), MAP kinase, cell cyclin-dependent kinase (Cdk), cell division cycle (Cdc), cyclic-AMP- and cyclic-GMP-dependent kinases, myosin light chain kinase, Ca^{2+}/calmodulin dependent kinases, etc. Protein kinase R (PKR, dsRNA-dependent protein kinase) downregulates protein synthesis in virus-infected cells. In the N-terminal region, two double-stranded RNA-binding domains activate PKR by binding to dsRNA and recruit it to the ribosome where it phosphorylates the eukaryotic elongation factor eIF2α. The consensus sequences for a few protein kinases are shown below:

Protein kinase A	(?)-Arg-(Arg/Lys)-(?)-(Ser/Thr)-(?)
Protein kinase G	(?)-{[Arg/Lys] 2x or 3x}-(?)-(Ser/Thr)-(?)
Protein kinase C	(?)-([Arg/Lys] 1-3x)-([?] 0-2x)-(Ser/Thr)-([?]0-2x)-(Se/[Thr]1-3x)-(?)
Ca^{++}/calmodulin kinase II	(?)-arg-(?)-(?)-(?)-(Ser-Thr)-(?)
Insulin receptor kinase	Thr-Arg-Asp-Ile-Tyr-Glu-Thr-Asp—Tyr-Tyr-Arg-Thr
EGF receptor kinase	Thr-Ala-Glu-Asn-Ala-Glu-Tyr-Leu-Arg-Val-Arg-Pro

(?) indicates any amino acid; the numbers after the amino acid with an x indicate how many times it may occur.

The majority of protein kinases require phosphylation in their activation loop to perform their function. *See* cAMP-dependent protein kinase; EGF; epinephrine; kinase; MAP; MYC; obesity; PDGF; phosphorylase b kinase; PKB; RAF; signal transduction; TGF. (Plowman, G. D., et al. 1999. *Proc. Natl. Acad. Sci. USA* 96:13603; Ung, T. L., et al. 2001. *EMBO J.* 20:3728; Cohen, P. 2002. *Nature Cell Biol.* 4:E127; Huse, M. & Kuryan, J. 2002. *Cell* 109:275.)

protein L *Peptostreptococcus* bacterial protein binding to the framework of immunoglobulin κ-chains. *See* framework amino acids; immunoglobulins.

protein length Shows great differences among individual molecules by the number of amino acids. There is a statistically significant increase along the advancement in the evolutionary rank, e.g., in archaebacteria 270 ± 9, in bacteria 330 ± 5, and in eukaryotes (budding yeast and *Caenorhabditis*) 449 ± 25. Some of the mammalian proteins are huge, e.g., dystrophin.

protein likelihood method Used to determine evolutionary distance when the organisms are not closely related and the nonsynonymous base substitutions are higher than the synonymous ones. In such cases, the protein method may provide more reliable information. *See* DNA likelihood method; evolutionary distance; evolutionary tree; Fitch-Margoliash test; four-cluster analysis; least square methods; transformed distance; unrooted evolutionary tree. (Whelan, S. & Goldman, N. 2001. *Mol. Biol. Evol.* 18:691.)

protein machine Multimolecular interacting systems such as metabolic circuits, intracellular signal transduction, or cell-to-cell communication. These systems are operated under process control strategies involving integrated feedback control. The input and output of the circuits or modules are coordinated to assure the normal or adaptive function of the cell or organism. *See* feedback control; microarray hybridization. (Baines, A. J., et al. 2001. *Cell Mol. Biol. Lett.* 6:691; Tobaben, S., et al. 2001. *Neuron* 31:987.)

protein 4.1N Binds to the nuclear mitotic apparatus protein NuMA, a nonhistone protein that is associated with the mitotic spindle. 4.1N regulates the antimitotic function of the nerve growth factor NGF. *See* NGF; PIK. (Kontragianni-

Konstatonopoulos, A., et al. 2001. *J. Biol. Chem.* 276:20679; Scott, C., et al. 2001. *Eur. J. Biochem.* 268:1084.)

proteinoid Polymerized mixture of amino acids formed during the prebiotic stage of evolution (or simulated conditions in the laboratory). Proteinoids may resemble primitive cells and display fission-like phenomena. *See* prebiotic.

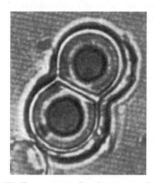

Proteinoid (From S. W. Fox., 1964. *BioScience* 14[12]:13; © Amer. Inst. Biol. Sci.) ℓ.

proteinosis Anomalous accumulation of protein at particular structures of the body.

protein phosphatases Remove phosphates from proteins. They include enzymes that reverse the action of protein kinases and have an important role, together with the kinases, in signal transduction. *See* membrane fusion; protein kinases. (Barford, D., et al. 1998. *Annu. Rev. Biophys. Biomol. Struct.* 27:133.)

protein-protein interaction Mediates structural and functional organization of the cells. The knowledge of these processes reveals the essential nature of the biology of organisms. The two-hybrid method may reveal the pairwise interactions, and by sequential and systematic analysis of the interacting systems, the metabolic modules can be identified. *See* gene product interaction; microarray hybridization; two-hybrid method.

protein purification Disrupt cells → separate subcellular organelles by differential centrifugation → wash by buffer the separated bodies → treat the fraction(s) needed by denaturing agents → dialyze to remove the denaturing agent → use reducing agents for protection → concentrate → remove the unneeded or improperly folded protein fractions by ion-exchange chromatography, gel filtration, immunoaffinity, isoelectric focusing, high-performance liquid chromatography, or other steps → the wanted pure protein. Quantitate the amount or yield of the protein obtained by UV absorption or by the Lowry or Bradford methods. Each of these steps may need detailed operations. *See* Bradford method; Lowry test; UV spectrophotometry of proteins.

protein repair Can be managed with the assistance of chaperones. If the refolding is not feasible, proteolytic enzymes destroy the chaperones either directly or by the mediation of ubiquitins. Nascent polypeptides that are transcribed from truncated mRNAs without a stop codon acquire a C-terminal oligopeptide (Ala, Ala, Asn, Asp, Glu, Asn, Tyr, Ala, Leu, Ala, Ala, or a variant), encoded by an *ssrA* transcript. The ssrA is a 362-nucleotide tRNA-like molecule that can be charged with alanine. The addition of the peptide tag takes place on the ribosome by cotranslational switching from the truncated mRNA to the ssrA RNA. The tagged polypeptide chain is degraded in the *E. coli* cytoplasm or periplasm by carboxyl-terminal-specific proteases. The Clp chaperone recognizes the peptides by the ssRA tag of AANDENYALAA and targets the proteins to the ClpX and ClpA ATPases. *See* amino acids; chaperone; Clp; DNA repair; periplasm; protease; ssRA; tmRNA; ubiquitin. (Wawrzynow, A., et al. 1996. *Mol. Microbiol.* 21:895.)

protein-RNA recognition   Almost all RNA functions involve RNA-protein interactions such as regulation of transcription, translation, processing, turnover, viral transactivation, and gene regulatory proteins in general, tRNA aminoacylation, ribosomal proteins, transcription complexes, etc.

protein S (PROS)   Human chromosome 3p11 vitamin K–dependent plasma protein that prevents blood coagulation and is a cofactor for protein C. Its deficiency and dysfibrinogenemia are genetically determined causes of thrombosis. *See* anticoagulation; antithrombin; APC; dysbrinogenemia; protein C; thrombophilia; thrombosis.

protein sequencing   *See* amino acid sequencing.

protein sorting (protein traffic)   Mechanism by which the polypeptides synthesized on the ribosomes in the endoplasmic reticulum reach their destination in the cell through secretory pathways by transport with the aid of endocytotic vesicles. *See* clathrin; COP transport vesicle; endocytosis; Golgi apparatus; RAFT. (Tormakangas, K., et al. 2001. *Plant Cell* 13:2021.)

protein splicing   *See* intein.

protein structure   The *primary structure* means the sequence of the amino acids. The *secondary structure* is formed by the three-dimensional arrangement of the polypeptide chain. The polypeptides may form an α-helix or β-conformation when hydrogen bonds are formed between pleated sequences of the same chain. The latter type of conformation is frequently found in internal regions of enzyme proteins and in structural protein elements such as silk fibers and collagen. The α-helixes

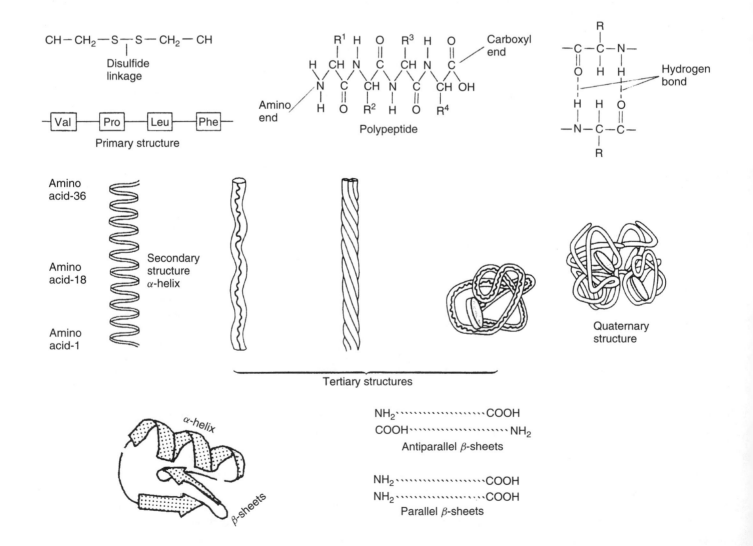

are commonly represented by cylinders, whereas the β-sheets are represented by ribbons with arrows frequently at their ends. The *tertiary structure* involves a folding either to globular or various types of rope-like structures of the polypeptide chain, which already has a secondary structure. Mature proteins may also be formed from multiple identical or different polypeptide subunits, and this type of association is the *quaternary structure*. The quaternary structure frequently includes some other molecule(s). For special biological functions, more than one protein may be joined in *supramolecular complexes* such as myosin and actin in the muscles, histones, and various nonhistone proteins in chromatin. The primary structure is genetically determined and all additional structural changes flow from this primary structure, although trimming, processing, and association with prosthetic groups may be involved. Proteins with greater than 30% sequence homology generally assume the same basic structures. Protein structure information can be obtained through the Protein Data Bank, ASTRAL, or SCOP: <http://biotech.embl-heidelberg.de8400/>; <http://biotech.embl-ebi.ac.uk:8400/>; <http://www.ncbi.nlm.nih.gov/entrez/query.fcgi?db = Structure)>; <http://www.ncbi.nlm.nih.gov/Entrez/structure.html>; <http://www.imb-jena.de/IMAGE.html>; <http://guitar.rockefeller.edu.tools/>; <http://astral.stanford.edu/>; <http://scop.mrc-lmb.cam.ac.uk/scop/>; <http:www.rtc.riken.go.jp/jouhou/Protherm/protherm.html>. *See* databases; electron density map of proteins; molecular modeling; MOLSCRIPT; protein domains; protein synthesis; x-ray diffraction analysis. (Goodsell, D. S. & Olsen, A. J. 2000. *Annu. Rev. Biophys. Biomol. Struct.* 29:105; Marti-Renom, M. A., et al. 2000. *Annu. Rev. Biophys. Biomol. Struct.* 29:291; Koonin, E. V., et al. 2002. *Nature* 420:218.)

protein synthesis Has many basic requisites and a large number of essential regulatory elements. It intertwines with all cellular functions. The blueprint for protein synthesis in the vast majority of organisms (DNA viruses, prokaryotes, and eukaryotes) is in the nucleotide sequences of the DNA code. In RNA viruses, the genetic code is in RNA. The viruses do not have, however, their own machinery for the actual synthesis of protein; rather, they exploit the host cell for this task. The genetic code specifies individual amino acids by nucleotide triplets using one or several synonyms for each of the 20 natural amino acids. The triplet codons are in a linear sequence of the nucleic acid genes. In the organisms with DNA as the genetic material, the process of transcription produces a complementary RNA sequence from one or both strands of the antiparallel ($\rightleftarrows$) strands of the DNA. The double strands unwind and the RNA polymerases synthesize a complementary RNA copy of the sequence in the DNA. In the single-stranded DNA and RNA viruses, the DNA or RNA may serve the purposes both of being the genetic material and the transcript for protein synthesis. In cellular organisms, three main classes of RNAs are made — messenger RNA (mRNA), transfer RNA (tRNA), and ribosomal RNA (rRNA) — and all three are indispensable for protein synthesis. In addition to these RNAs, a large number of proteins are required for the transcription process (transcription factors), the organization of the ribosomes (50–80 ribosomal proteins), the termination of transcription, the activation of tRNAs, etc. A broad overview (without details) is shown in the figure next page. Some of the details of the transcriptional process are different in prokaryotes from those in eukaryotes. In

the latter group, one DNA-dependent RNA polymerase is responsible for the synthesis of all RNAs. In eukaryotes, pol I synthesizes rRNAs with the exception of 5S and 7S rRNA; pol II transcribes mRNA and the small nuclear RNAs (snRNA); and pol III synthesizes tRNAs and 5S and 7S rRNA. The primary RNA transcripts must be processed to functional-size molecules in all categories that may require splicing and other posttranscriptional modifications (capping, formylation, etc.).

In prokaryotes, the processes of transcription and translation are *coupled*, i.e., as soon as the chain of mRNA unwinds from DNA, it is associated with the ribosomes and protein synthesis begins. In eukaryotes, when mRNA is released from its DNA template, it moves into the cytosol, where protein synthesis takes place. The fate of mRNA can be monitored by electron microscopy in both groups. The figures show the elongation of RNA and protein strands. The first products of both display long strands, and the short ones indicate the stage and place where they were started. The ribosomes are captured by mRNA and form an association of multiple units in the form called *polysomes*. Prokaryotic mRNA is directed to the proper position in the 30S ribosomal subunit by the Shine-Dalgarno nucleotide sequence within an 8- to 13-base area upstream from the initiation codon. In eukaryotes, such a sequence does not exist and mRNA is simply scanned by the ribosome until the first methionine codon is found. These ribosomal units then slide from the 5'-end of mRNA toward the 3'-end. Thus, the amino end of the polypeptide chain corresponds to the 5'-end of mRNA. The ribosomes in both prokaryotes and eukaryotes are composed of a small and a large subunit. The size of these units is somewhat different in the two major taxonomic categories. The small and large subunits of the ribosomes jointly form two compartments: the so-called P (peptidyl-tRNA-binding) and A (aminoacyl-tRNA-binding) sites. A newer *hybrid-states model* of the translational process is described under the ribosomes entry. The ribosomes actually do not look like those shown in the diagrams because they are three-dimensional and have a more elaborate structure. Before protein synthesis (translation) begins and the primary structure of mRNA is translated from the nucleotide triplet codon into the singular amino acid word language of the protein, the tRNA molecules must be charged with amino acids. This process is called activation of tRNA. (*See* aminoacyl-tRNA synthetase.)

The amino-acid-charged methionine-tRNA (tRNA^{Met}) in eukaryotes and the formylated tRNA^{fMet} in prokaryotes seek out the cognate codon in mRNA at the P site of the ribosome through the complementary anticodon. This event requires the presence of protein initiation factors and GTP as an energy source. The GTP is cleaved to GDP + inorganic phosphate (P_i), thus liberating some of the needed energy. The elongation factor proteins and GTP and GDP complexes also police the system to prevent the wrong charged tRNA from going to an A site (proofreading function). Actually, a similar correction mechanism is carried out earlier in the process by one of the active sites of the aminoacyl synthetase (activating) enzyme that usually dissociates the amino acid–tRNA link in case of a misalliance. With the double-checks available, misincorporation of amino acids is approximately in the 10^{-4} range. Protein synthesis in the mitochondria and chloroplasts is essentially patterned after the prokaryotic systems.

The 5'-base of the anticodon triplet may not be the exact and conventional base, yet it may function normally (see wobble). The two subunits of the ribosomes are combined and the

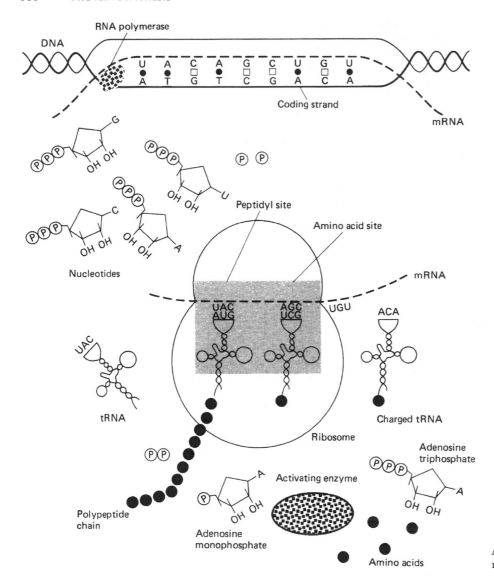

An overview of the protein synthesizing machinery.

second charged tRNA can land at the A ribosomal site. The carboxyl end of methionine forms a peptide bond with the amino terminus of the next incoming amino acid at the A site. This process is mediated by the enzyme peptidyl transferase. A 23S rRNA in the large subunit (a ribozyme) is responsible for this transferase function, not a protein. Again, energy donors and elongation protein factors cooperate in the process of peptide chain growth (*see* initiation and elongation factors IF, eIF, EF, EF-T, EF-Tu). When each peptide bond is completed, tRNA is released and recycled for another tour of duty. The *open-reading frame* of the gene is terminated by a nonsense or chain-termination codon. When the ribosome slides to this point, mRNA is released from the ribosomes with the assistance of release factors (*see* transcription termination in eukaryotes; transcription termination in prokaryotes). Protein synthesis proceeds at a rather rapid rate; it has been estimated that in *E. coli* 50–200 amino acids may be incorporated into peptides in 5–10 seconds. The process is slower in eukaryotes (3–8 per seconds).

The ribosomes have an important role in the regulation of protein synthesis. It appears that the availability of active ribosomes is controlled at the level of the transcription of rRNA genes. In most of the cases, the number of ribosomes is not a limiting factor of translation. Some of the bacterial ribosome proteins have dual roles, participating in transcription and translation (Squires, C. L. & Zaporojets, D. 2000. *Annu. Rev. Microbiol.* 54:775). When the supply of ATP and GTP is adequate, rRNA genes are activated for transcription. In case the level of these nucleotide triphosphates is low, rRNA transcription is reduced or halted. Abundance of free ribosomal proteins may feedback-inhibit ribosomal production. The ribosome-associated Rel-A protein may mediate the formation of ppGpp from GTP (and possibly from other nucleotides). Then ppGpp may shut off rRNA and tRNA synthesis by binding to the promoter of RNA polymerase or to its antitermination signal.

Some of the nascent peptides are segregated into the endoplasmic reticulum through the Sec61 conductance opening of the large subunit of the ribosomes. Within the endoplasmic reticulum the translation continues and the protein is folded by the appropriate chaperones. In prokaryotes, only the completed polypeptide chains are folded, whereas in eukaryotes the separate domains of the large polypeptides are folded as the chain grows. The dimeric NAC (nascent-polypetide-associated complex) interacts with the emerging polypeptide chains before 30 or fewer residue long chains

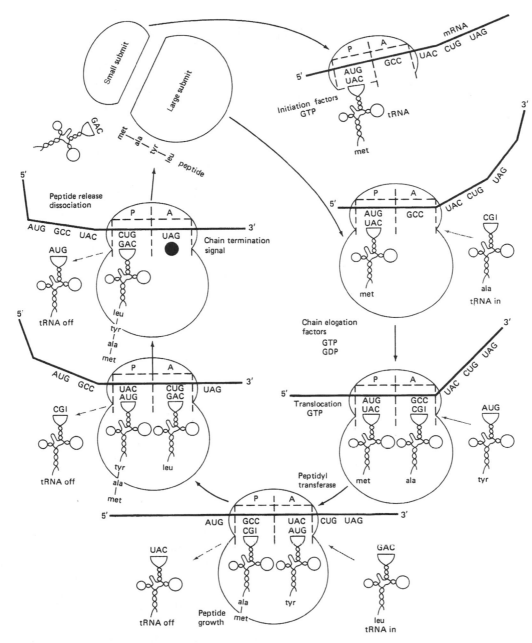

Classical model of genetic translation on the ribosomes.

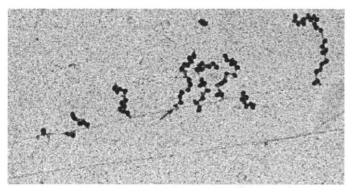

Transcription and translation coupled in *E. coli*. The thin thread is the DNA; the dark round structures are polysomes. The transcriptase attachment → is indicated. (From Hamkalo, B. A., et al. 1974. *Stadler Symp.* 6:91.)

are formed and protects the nascent chain from becoming associated with other cytosolic proteins until the signal peptide fully emerges; then the signal recognition particle (SRP) crosslinks to the polypeptide. The purpose of NAC is to assure that the polypeptide is oriented to the proper SRP and the endoplasmic reticulum. Alternatively, if the protein does not carry a signal peptide, the nascent chain may be folded by chaperones such as heat-shock proteins Hsp40, Hsp70, and TRiC. The completed amino acid sequences, the polypeptides, must then be converted to biologically active forms. This posttranslational process may involve trimming (removal of some amino acids), proteolytic cleavage, folding to a tertiary structure, aggregation of different polypeptide chains to form the quaternary structure, addition of prosthetic groups (such as heme, lipids, metals), and other non-amino-acid residues such as acyl, phosphate, methyl, isoprenyl, and

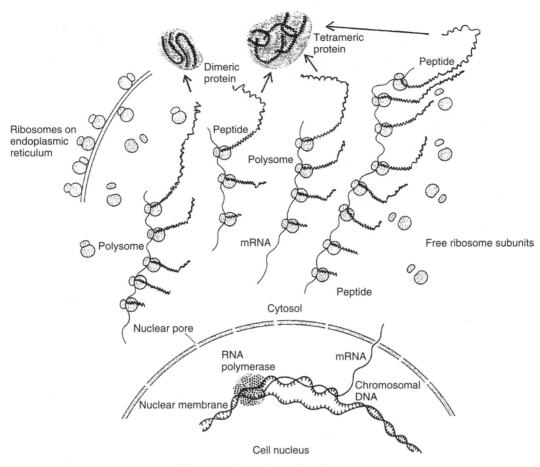

Overview of translation in eukaryotes.

sugar groups. *See* ambiguity in translation; aminoacyl-tRNA synthetase; antibiotics; cap; chaperone; code, genetic; discriminator region; eIF; EF-TuGTP; elongation initiation factors; E site; heat shock; introns; mRNA; nonribosomal peptides; polysome; prenylation; protein folding; regulation of gene activity; rho factor; ribosome recycling; ribosomes; RNA polymerases; rRNA; Sec61 complex; Shine-Dalgarno sequence; signaling to translation; signal peptides; signal sequences; SRP; tmRNA; toxins; transcription complex; transcription factor; transcription initiation; transcription termination; transit peptide; translation, nuclear; translation initiation; translation in vitro; TRiC; tRNA; wobble. (Sonenberg, N., et al. eds. 2000. *Translational Control of Gene Expression.* Cold Spring Harbor Lab. Press, Cold Spring Harbor, NY, Fredrick, K. & Noller, H. K. 2002. *Mol. Cell* 9:1125.)

protein synthesis, chemical   Building proteins by nonbiological means these from peptide domains. The peptides may be synthesized by these methods or organic chemistry, then ligated, spliced, and folded in order to assure some specific function. (Dawson, P. E. & Kent, S. B. H. 2000. *Annu. Rev. Biochem.* 69:923.)

protein synthesis inhibitors   *See* antibiotics; interferons; toxins.

protein targeting   Can be cotranslational, i.e., newly synthesized proteins are delivered to specific sites (endoplasmic reticulum) in the cell before the chain is completed, or posttranslational when the transport takes place after the polypeptide is completed. *See* protein conducting channel; signal hypothesis; signal sequence recognition particle; TRAM; translocon. (Bachert, C., et al. 2001. *Mol. Biol. Cell* 12:3152; Zaidi, S. K., et al. 2001. *J. Cell Sci.* 114:3093; Takayama, S. & Reed, J. C. 2001. *Nature Cell Biol.* 3:E237.)

protein trafficking   *See* protein sorting.

protein transduction   Introduction of protein into the bloodstream or organs for experimental or therapeutic purposes. This procedure is usually limited to small-size (<600 Da) molecules. When, however, the 120 kDa β-galactosidase was fused to an 11-amino-acid NH_2 domain of the Tat protein of HIV and introduced into the intraperitonial cavity of the mouse, the protein was detected in a biologically active form in several organs including the brain. *See* AIDS; BBB; galactosidase; protein targeting. (Embury, J., et al. 2001. *Diabetes* 50:1706.)

protein transport   *See* protein sorting.

protein truncation test   May be used to detect the effects of several mutations that do not permit the completion of a polypeptide chain. The gene is transcribed by using polymerase chain reaction, the RNA is translated in vitro, and the polypetide is analyzed in SDS minigels. *See* PCR; rabbit reticulocyte in vitro translation; SDS-polyacrylamide gel. (Lutz, S., et al. 2001. *Nucleic Acids Res.* 29:E16.)

protein tyrosine kinases (PTK) Phosphorylate tyrosine residues in some proteins. This function is frequently coded for by v-oncogenes of retroviruses, but cellular oncogenes and other proteins may be involved and control signal transduction and other cellular processes such as cell proliferation and differentiation. Cytosolic tyrosine kinases preferentially phosphorylate their own SH2 domains or related SH2 domains with hydrophobic amino acids at key positions, e.g., Ile or Val at -1 and Glu, Gly, or Ala at the $+1$ position. Receptor tyrosine kinases prefer Glu at the -1 position. These preferences specify their signaling role. The RET oncogene's receptor tyrosine kinase product can shift substrate specificity, thereby causing multiple endocrine neoplasia. Quercetin, genistein, lavendustin A, erbstatin, and herbimycin are natural plant products and inhibitors of these enzymes. *See* endocrine neoplasia, multiple; protein kinases; receptor tyrosine kinase; SH2; signal transduction; tyrosine kinase. (Hubbard, S. R. & Till, J. H. 2000. *Annu. Rev. Biochem.* 69:373; Blume-Jensen, P. & Hunter, T. 2001. *Nature* 411:355.)

protein tyrosine phosphatase *See* tyrosine phosphatases.

protein zero Major part of the nerve cell myelin sheath of vertebrates. Its defect may lead to neurological anomalies.

proteoglycan Heteropolysaccharides with a peptide chain attached through O-glycosidic linkage to a serine or threonine residue. Such molecules are enzymes, animal hormones, structural proteins, basement membranes, cellular lubricants (such as mucin), extracellular matrix proteins, and the "antifreeze proteins" of antarctic fishes. They control plant and animal growth, differentiation, development, and signal transduction. *See* amyloids; antifreeze proteins; glycosaminoglycan; glypican; syndecan. (Selleck, S. B. 2000. *Trends Genet.* 16:206.)

proteolipid protein Major part of myelin in the brain. *See* myelin.

proteolytic Hydrolyzing peptide bonds of proteins. *See* proteasome; ubiquitin.

proteome All the cellular proteins encoded by the cellular DNA. Proteome is the protein complement of the genome. The genome is very stable (except rare mutations) and it is the same in practically all cells of an organism. The proteome displays variations according to the developmental stage, organs, metabolic rate, health of the organism, etc. Since the proteins are organized and expressed in interacting systems, their study may be very complicated. While the genome does not reveal the detail of the function of a cell, proteomics has exactly this goal. The immediate products of the genome, the RNA, is frequently processed in more than one way (alternative splicing and combinatorial assembly) to be translated into more than a single type of polypeptide. The translated product can be further modified by trimming, docking, forming multimeric associations, ligand recruitment, phosphorylation and/or dephosphorylation, acetylation, glycosylation, and various other epigenetic mechanisms. Because of alternatives in transcription (using different promoters and processing of the transcripts), there are generally substantially more proteins than genes in the cells. The proteins also have various regulatory roles at the levels of replication, transcription, translation, etc. The amount and kind of RNAs are correlated with the amount of polypeptides, yet this correlation is variable. Proteins may undergo substantial posttranslational modifications. Although the genome is essentially constant, the encoded proteins may display great variations during differentiation and development. There are no well-established procedures fit for all proteins such as DNA sequencing after cloning, PCR, or microarray hybridization. Two-dimensional gel electrophoresis is powerful for the separation of thousands of proteins, and monoclonal antibody techniques can be used for the localization of proteins. Although definitive information on the proteome may not come easily, it should permit an insight into the function of cells, organisms, evolution, and disease that cannot be matched by other means. The size of the human proteome much exceeds that of the number of genes determined by sequencing the genome. The size of the human proteome has been estimated by the formula $N_{CDS} = f_1.f_2.N_{genes}$, where f_1 is the proportion of nonpseudogenic genes and f_2 is the ratio of the total number of protein-coding transcripts to the total number of genes, including those that are spliced alternatively. The estimates so obtained also vary within a wide range (see Harrison, P. M., et al. 2002. *Nucleic Acids Res.* 30:1083). *See* ACESIMS; core proteome; electrospray; genetic network; genome; genomics; ICAT; MALDI/TOF/MS; metabolic pathway; microarray hybridization; monoclonal antibody; MS/MS; networks; protein chips; TOGA; transcriptome; two-dimensional gel electrophoresis; two-hybrid method. (<http://www.expasy.ch>; on subscription basis: SOS protein recruitment; <http://www.proteome.com.databases/index.html>; yeast proteome: <http://www.infobiogen.fr/services/virgil>; <http://www.incyte.com>; Ito, T., et al. 2001. *Proc. Natl. Acad. Sci. USA* 98:4569; Walhaut, A. J. M. & Vidal, M. 2001. *Nature Rev. Mol. Cell Biol.* 2:55; Harrison, P. M., et al. 2002. *Nucleic Acids Res.* 30:1083; Auerbach, D., et al. 2002. *Proteomics* 2:611; Rost, B. 2002. *Current Opin. Struct. Biol.* 12:409; Burley, S. K. & Bonnano, J. B. 2002. *Annu. Rev. Genomics Hum. Genet.* 3:243; Appendix II-8, Appendix II-9, Appendix II-10.)

proteomic profiling Uses chemical labels for the identification of active groups of enzymes in complex mixtures and attempts the identification of the functional role of these groups of proteins. The procedure may reveal the role of protein arrays in the development of disease and may suggest targets for intervention. (Adam, G. C., et al. 2002. *Nature Biotechnol.* 20:805.)

proteomics Study of the system of proteomes, the modules of metabolism as they carry out cellular functions of the organisms. The new technologies detect the composition/structure of proteins, isoforms, conformational changes, modulatory alterations during development, posttranscriptional and posttranslational modifications (phosphorylation, glycosylation), interactions with other proteins or drugs, etc. With low mass tolerance, e.g., 10 ppm, single proteins can be identified in a mixture among thousands of molecules. Proteomics has modified the basic approach to investigating biological function. At one time, the experimental design was based on hypotheses. With the aid of proteomics technologies, more direct approaches are possible based on the simultaneous expression patterns of interacting genetic networks. *Expression proteomics* analyzes proteins of the cells by two-dimensional gel electrophoresis (Wagner, K., et al. 2002. *Anal. Chem.* 74:809). *Cell-map proteomics* is interested in the interaction between/among proteins at various phases of the cell function (Blackstock, W. P. & Weir, M. P. 1999. *Trends Biotechnol.* 17[3]:121). *Functional proteomics* targets specific functions rather than the entire proteome (Graves, P. R. & Haystead, T. A. 2002. *Microbiol. Mol. Biol. Rev.* 66:39). *Structural proteomics* seeks an understanding of protein function on the basis of three-dimensional analysis and modeling (Norin, M. & Sundstrom, M. 2002. *Trends Biotechnol.* 20:79). Liquid chromatography, two-dimensional polyacrylamide gel electrophoresis, and tandem mass spectrometry are important tools of proteomics on a large scale. Proteomics is concerned not only with the variability and interactions of proteins but may assist in modifying proteins for new types of interactions. The α-carboxyl group and preceding residues at the C-end of polypetides may offer a useful target for modifications. The PDZ and TPR domains are well qualified for interactions with the C-termini and may facilitate temporal and spatial interactions, degradation, neuronal signaling, and other functions (Chung, J. J., et al. 2002. *Trends Cell Biol.* 12:146). The proteome data are expected to be much more complex than those of the genome sequences. The number of proteins and their isoforms far exceeds that of the number of genes. There is a need to develop computer programs that can properly assist in interpreting the mountain of information. One of the most complete sources of information on the *E. coli* metabolic system is at <http://ecocyc.org/>. The increasing amount of information is becoming impossible to integrate and advanced computer models are indispensable. Proteomic information has an important impact on applied biology such as medicine, drug development, and agriculture. *See* annotation; FTMS; genetic networks; genomics; laser-capture microdissection; LC-MS; MALDI; MCA; MS/MS; NMR; nucleolomics; peptide mass fragments; posttranslational modification; proteome; quadrupole; two-dimensional gel electrophoresis; two-hybrid system. (Mann, M., et al. 2001. *Annu. Rev. Biochem.* 70:437; MOWSE, *Trends in Biotechnology* 19[10]:Suppl. 2001; Fraunfelder, H. 2002. *Proc. Natl. Acad. Sci. USA* 99 Suppl. 1:2479; Altman, R. B. & Klein, T. E. 2002. *Annu. Rev. Pharmacol. Toxicol.* 42:113; Regnier, F. E., et al. 2002. *J. Mass Spectrom.* 37:133; Laurell, T. & Mako-Varga, G. 2002. *Proteomics* 2:345; Auerbach, D., et al. 2002. *Proteomics* 2:611; Petricoin, E. F., et al. 2002. *Nature Rev. Drug Discovery* 1:683; <www.ebi.ac.uk/interpro>; <https://www.incyte.com/proteome/databases>; <http://dip.doe-mbi.ucla.edu/>.) Appendix II-8, and Appendix II-9, Appendix II-10.

proterozoic (Precambrian) Geological period 570 million to 5 billion years ago. Aquatic forms of living systems appeared during this era. *See* geological time periods.

Proteus syndrome Involves gigantism of parts of the body probably caused by lipomatosis (abnormally large local fat accumulation). The genetic control is unclear. *See* PTEN. Cohen, M. M., Jr. 1993. *Am. J. Med. Genet.* 47:645.

prothallium Haploid gametophyte generation of ferns.

prothrombin deficiency Caused by autosomal-recessive, semidominant defects in the formation of anticoagulation factor VII, Stuart factor, Christmas factor, and prothrombin. The human gene for prothrombin was assigned to chromosome site 11p11-q12. These proteins have similar proteolytic properties and the synthesis of all four depends on the presence of vitamin K. The patients have a bleeding tendency similar to hemophiliacs. Hereditary deficiency of factor VII itself is rare, but it may be fatal if bleeding affects the central nervous system. Stuart factor deficiency has symptoms similar to those in deficiency of factor VII. All of these conditions can be treated by transfusion with blood plasma. *See* antihemophilia factors; coumarin-like drug resistance; hemophilia; vitamin K dependence.

protist General term for single-cell eukaryotic organisms. The *Monera*, including bacteria, blue-green algae, and viruses, are sometimes called protists, although they are prokaryotes.

protocell Abiotic ancestor of living cells under prebiotic conditions. *See* origin of life.

protochlorophyll Precursor of chlorophyll ($C_{55}H_{70}O_5N_4Mg$); if the magnesium is removed, protophaeophytin results. The NADPH:protochlorophyllide oxidoreductases in the prolamellar body of the etioplast are required for the establishment of the photosynthetic apparatus (deetiolation) and for photoprotection in plants. *See* chloroplast; etioplast; NADP; photomorphogenesis; photosynthesis.

protogyny In monoecious plants, the stigma is receptive before the pollen is shed. *See* monoecious; protandry; self-incompatibility; stigma.

protomer Polypeptide subunit of an oligomeric protein encoded by a cistron of a gene. *See* cistron; oligomer.

proton Positive nucleus of the hydrogen atom. The proton carries a positive charge equal to the negative charge of an electron, but its mass is 1,837 times larger.

proton acceptor Anion capable of accepting protons. *See* anion; proton.

proton donor Acid.

protonema Filamentous stage in the formation of the gametophyte of mosses.

protonoma Red color–insensitive color blindness; an X-chromosomal anomaly. *See* color blindness.

proton pump Mediates transport or exchange of protons across cellular membranes. Energy is usually supplied by ATP or light. *See* ion pumps; proton. (Ferreira, T., et al. 2001. *J. Biol. Chem.* 276:29613.)

proto-oncogenes Cellular c-oncogenes, which after genetic alteration(s) may initiate or predispose to cancerous transformation. They generally have their counterparts in oncogenic viruses (v-oncogenes). Also, they may be involved in processes of signal transduction in a variety of organisms in fungi, plants, and animals. *See* carcinogenesis; cell cycle; oncogenes; signal transduction; tumor suppressors.

protoperithecium *See* ascogonia; perithecium.

protoplasia Formation of a new tissue.

protoplasm "Live" content of a cell.

protoplast Cell surrounded by the cell membrane but stripped of the cell wall, generally by a combination of pectin and cellulose-digesting enzymes. Under appropriate conditions, protoplasts may be regenerated into normal cells and intact plants. Bacterial protoplasts are generally called spheroplasts and may have some parts of the cell wall still attached. *See* cellulase; macerozyme; pectinase.

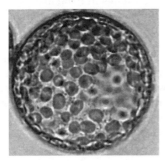

Plant protoplast From Durand, J., et al. 1973. *Z. Pflanzenphysiol.* 69:26.

protoplast fusion Protoplasts may fuse in the presence of polyethylene glycol (and some other agents). The fusion may take place within sister cells or with the cells (protoplasts)

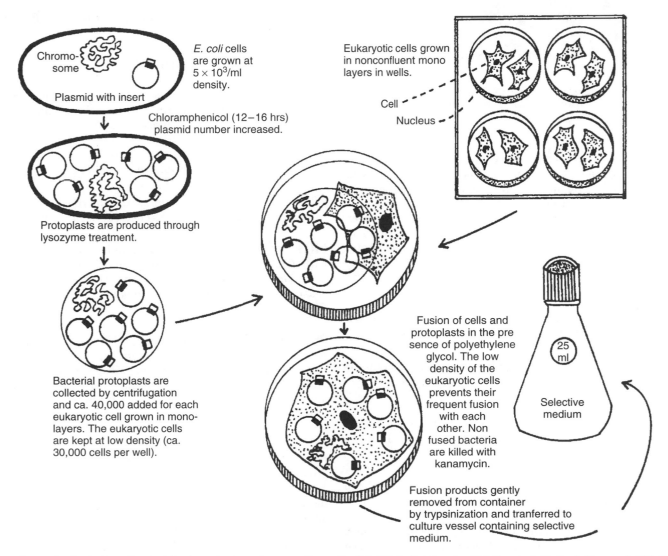

Transformation by fusion of bacterial spheroplasts and mammalian cells. (Modified after Sandra-Goldin, et al. *Methods Enzymol.* 101:402.)

of any taxonomically distant organisms such as mammalian and plant cells. These somatic hybrids, unlike the zygotes derived from the fusion of eggs and sperm, contain all the contents of the two cells, nuclei and cytoplasm, although some cytoplasmic organelles may be lost eventually. In certain rodent-human cell hybrids, even the human chromosomes may be eliminated; similar observations are available for carrot and parsley cell hybrids. When the genetic differences between the fused protoplasts is large, the fused cells may not divide or may not divide continuously. Somatic hybrids between related species may, however, behave like allopolyploids and form fertile or sterile hybrids after regeneration. Fusion of animal cells with bacterial spheroplasts is shown on the previous page. *See* cell fusion; polyethylene glycol.

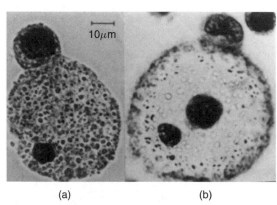

(a)   (b)

Human HeLa cells attached to tobacco protoplast (A); the HeLa nucleus (larger) inside the tobacco cell (B). (From Jones, C. W., et al. 1976. *Science* 193:401.)

protoporphyria, erythropoietic   Autosomal (human chromosome 18q21.3)-dominant (or -recessive) disease involving light-sensitive itching and inflammation of the skin. The porphyrin level of the blood may increase by over 16-fold, to 1 g/100 mL. The excess protoporphyrin is deposited in the liver, causing potentially serious damage. The basic defect probably involves a deficiency (10 to 25%) of the mitochondrially located ferrochelatase (FECH). *See* light sensitivity defects; mitochondrial disease in humans; porphyria. (Todd, D. J. 1994. *Brit. J. Derm.* 131:751.)

protoporphyrin   Organic part of heme consisting of four pyrroles joined by methylene bridges. *See* heme.

protosilencer   Alone it is incapable of silencing genes or its silencing effect is minimal, but it can reinforce and maintain the function of silencers. *See* silencers.

protostomes   Organisms that develop the mouth from the blastopore such as annelids, mollusks, and arthropods. *See* blastopore.

prototroph   Genotype that has a wild-type nutritional requirement. *See* autotroph; auxotroph.

protozoa   Unicellular animals, mainly free-living (such as the *Paramecia*). Some are, however, parasitic (such as the *Gaillardias*, which frequently contaminate drinking water sources). The *Trypanosomas* and *Leishmanias* cause potentially lethal infections in animals and humans. *See Leischmania; Trypanosoma.* (For the genetic nomenclature of *Tetrahymena* and *Paramecia*, see *Genetics* 149:459.)

provenance/provenience   Origin of a genetic stock.

provirus   DNA sequence in the eukaryotic chromosomal DNA that is a reverse transcriptase product of a retroviral RNA. *See* prophage; retroviruses; reverse transcription.

proximal   Situated in the vicinity of a reference point; e.g., a gene near the centromere is proximal, versus another that is in the direction of the telomere, and is thus called distal. In conjugational transfer of bacteria, the marker that is transferred before another is proximal. *See* centromere; conjugation mapping; telomere.

proximal mutagen   Chemical that has been activated into a mutagenic substance; it may not yet have reached its most reactive state. *See* activation of mutagens; chemical mutagens; promutagen; ultimate mutagen.

PRP   RNA-splicing factor component of the U-snRNP complex. *See* splicing.

PrP   Protease-resistant protein *See* prion.

Prp73   Mammalian chaperone binding to the first 20 residues (S peptide) of ribonuclease A; stimulates the uptake of polypeptides by lysosomes. *See* Hsp70; lysosome; ribonuclease A.

Prp20p   Yeast homologue of RCC1. *See* RCC.

PRR   (1) Postreplication repair. *See* DNA repair. (2) Positive regulatory region. *See Arabinose* operon; negative regulation.

PRTF   Pheromone receptor transcription factors cooperating with GRM (general regulator mating factor) in the determination of mating type. *See* mating type determination in yeast; pheromone; *Schizosaccharomyces pombe*. (Tan, S. & Richmond, T. J. 1990. *Cell* 62:367.)

Przewalsky horse   Mongolian wild horse, but apparently it can be found (~1,200) only in captivity, although its reintroduction into the wild in Mongolia and China is underway. All existing individuals have descended from the 13 animals captured about a century ago. Its chromosome number is 2n = 66, yet it makes viable hybrids with the domesticated species. *See* horse.

PSA   Prostate-specific antigen; M_r 33,000-kallikrein-type protease glycoprotein (APS) encoded at human chromosome 19q13. High levels of this protein in the serum may be an indication of prostatic carcinoma. The level of PSA varies a great deal. It is high after ejaculation and may provide a false positive indication of cancer. It may serve as a target for cancer gene therapy. The six-transmembrane epithelial antigen of the prostate (STEAP, 7p22.3) is also elevated in prostate cancer. Hepsin (transmembrane serine protease) and pim-1 (serine/threonine kinase) levels are strongly correlated with prostate cancer as detected by tissue microarray analysis. *See* cancer gene therapy; prostate cancer; tissue microarray. (Berry, M. J. 2001. *N. Eng. J. Med.* 344:1373; Dhanasekara, S., et al. 2001. *Nature* 412:822.)

PSD-95 Family of membrane-associated guanyl kinases; they also anchor K⁺ channels by their PDZ domains. *See* GTP; ion channels.

PSE (1) Proximal sequence element. *See* Hogness box. (2) Pale, soft exudative meat is controlled in pigs by the *Halothane* gene.

pseudoachondroplasia Dominant human chromosome 19p12-p13.1 gene mutation controlling the cartilage oligomeric matrix protein (COMP); it is responsible for short stature. *See* achondroplasia; COMP; multiple epiphyseal dysplasia. (Hecht, J. T., et al. 1995. *Nature Genet.* 10:325; Briggs, M. D. & Chapman, K. L. 2002. *Hum. Mut.* 19:465.)

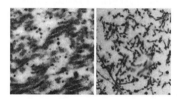

Left: Pseudoachondroplasiac, right: normal extracellular cartilage matrix.

pseudoaldostertonism (Liddle syndrome) Human chromosome 4 hypertension associated with hypoaldosteronism, hypokalemia, reduced renin, and angiotensin. *See* aldosteronism; angiotensin; hypokalemia; renin.

pseudoalleles Cluster of not fully complementing genes, separable by recombination. Pseudoalleles, e.g., a^1 and a^2, when heterozygous in trans position $a^1a^+//a^+a^2$, show mutant phenotype, whereas in cis position $a^1a^2//a^+a^+$ are complementary (wild type) except when dominant alleles are involved. Since these alleles are closely linked, in order to be able to prove that recombination takes place (rather then mutation), the pseudoalleles must be genetically marked by flanking genes preferably less than 10 m.u. apart from the locus. *See* complex locus; morphogenesis in *Drosophila*; step allelomorphism. (Carlson, E. A. 1959. *Quart. Rev. Biol.* 34:33.)

pseudoaneuploid The chromosome number appears aneuploid, but it is not truly the case. Only centromere fusion or misdivision of the centromeres causes the changes in numbers. *See* B chromosomes; misdivision; Robertsonian translocation.

pseudoautosomal (PAR) Genes located in both telomeric regions of the X and Y chromosomes (∼2.6 Mbp at the short arm [PAR1] and a similar PAR2 site in the long arm in the human genome) where recombination can take place; consequently, despite the sex-chromosomal location, sex linkage is not obvious. A gene for schizophrenia was suggested to be pseudoautosomal. SYBL1, encoding a synaptobrevin-like gene, is present in both X and Y chromosomal PAR regions, and it displays lyonization in the X chromosome and inactivation in the Y. The pseudoautosomal boundary is apparently spanned by a (depending on the species) 5′- or 3′-truncated gene. Short-stature (SHOX1/SHOXY), Leri-Weill dyschondrosteosis, and Hodgkin disease genes are all located in the PAR at Xpter-p22.32. The SHOX2 gene is at 3q25-q26. *See* autosome; differential segment; Hodgkin disease; holandric genes; IL-9; lyonization; short syndrome; syntagmin. (Ciccodicola, D., et al. 2000. *Hum. Mol. Genet.* 9:395; Cormier-Daire, V., et al. 1999. *Acta Paediatr.* 88 Suppl. 55.)

pseudobivalent Associated chromosomes are not homologous. *See* illegitimate pairing; synapsis.

pseudoborder DNA sequences in certain agrobacterial vectors or within the cloned foreign DNA may cause deletions and rearrangements within the T-DNA inserts in the transgenic plants. *See* T-DNA; transformation, genetic.

pseudocholinesterase deficiency (CH1, BCHE) Dominant (human chromosome 3q26.1-q26.2) breathing difficulty (apnea) after treated with the muscle relaxant suxamethonium (succinylcholine chloride), a drug used for intubation, endoscopy, cesarean section, etc., as an adjuvant to anesthesia. Several allelic forms respond differently to drugs. Individuals with a defective enzyme may be particularly sensitive to cholinesterase inhibitor insecticides (parathion). The frequency of the gene varies a great deal in different populations. In Eskimos, the frequency of the gene controlling the deficiency may be higher than 0.1; in other populations, it may be less than 0.0002. The BCHE2 form was assigned to 2q33-35 and the same enzyme was suggested to 16p11-q23.

pseudodiploidy Retroviral particles, because after infection only a single provirus is detected in the host. Normally, retroviruses carry two RNA genomes associated by base pairing at several sites, particularly at the 5′ end. It is assumed that the two copies are maintained for the purpose of assured survival and possible repair by recombination. They also contain tRNAs that prime replication. Other RNAs (5S, 7S, and cellular mRNA fragments) may be included. *See* retroviruses.

pseudodominance When a heterozygote loses the dominant allele, the recessive allele is uncovered (expressed) because of the lack of the dominant allele. Treating heterozygotes with mutagens (e.g., ionizing radiation) that cause deletions can readily induce pseudodominance. Before such experiments are conducted, it is advisable to place flanking genetic markers in the chromosome carrying the recessive markers to be able to rule out recombination and reversions. Segregation after somatic recombination may be a common cause of pseudodominance. Loss of heterozygosity is a frequent cause of oncogenic transformation. *See* deletion; LOH; mitotic crossing over; oncogenic transformation; segregation.

pseudoextinction Disappearance of a species by evolution into another form.

pseudogamy Apomictic or parthenogenetic reproduction. *See* apomixia; parthenogenesis.

pseudogene Has substantial homology with (clustered) functional genes of eukaryotes, but it is inactive because of numerous mutations that prevent its full expression and it may no longer available for transcription. The number of pseudogenes is variable in different species. Organisms with small genomes (e.g., *Drosophila*) have very few and it appears that some organisms eliminated the DNA sequences from their genome that are no longer functional. Pseudogenes may make the estimation difficult of the number of genes on the basis of incomplete sequences and lack of functional information. *See* C-value paradox; gene relic; processed pseudogene. (Harrison, P. M., et al. 2001. *Nucleic Acids Res.* 29:818; Avise, J. C. 2001. *Science* 294:86; Echols, N., et al. 2002. *Nucleic Acids Res.* 30:2515.)

pseudohairpin Overall structure is folded back, yet there is not full complementarity along the strand.

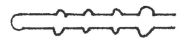

Pseudohairpin.

pseudohemophilia Bleeding disease, distinct from hemophilia; it is caused by some abnormalities of the platelets. *See* hemophilia; hemostasis; platelet anomalies.

pseudohermaphroditism *See* hermaphrodite.

pseudohermaphroditism, male Determined by a gene in human chromosome 17q12-q21. It is responsible for the deficiency of 17-ketosteroid reductase/17-β-hydroxysteroid dehydrogenase and consequently for feminization in prepubertal males and gynecomastia and virilization after puberty when the enzyme is usually expressed. The affected individuals may be surgically assisted to develop into a sterile female phenotype (by removal of the hidden testes) or into a male phenotype by reconfiguration of the external male genitalia. Infertility cannot be corrected, however. Recessive mutations in the luteinizing hormone receptor gene (LHB, 19q13.32) may also be responsible. The condition may be due to a deficiency of steroid 5-α-reductase (SRD5A2, 2p23). The SRDA1 isozyme encoded at 5p15 does not appear to be involved in this disorder. The afflicted XY individuals may have a blind vagina and a rudimentary hypospadias penis but no gynecomastia. They may produce viable sperm, although they may sire offspring only by intrauterine insemination because of underdeveloped prostate and seminal vesicles. Several defects in steroid biosynthesis may cause male pseudohermaphroditism. The 17,20 desmolase deficiency is most likely X-chromosome linked. Lipoid adrenal hyperplasia (8p11.2) responsible for complex defects in cortisol or aldosterone may cause even life-threatening conditions. Luteinizing hormone/choriogonadotropin receptor (LHCCGR, 2p21) may cause abnormalities of the Leydig cell differentiation in XY and possibly in XX individuals. Methemoglobinemia and deficiency of cytochrome b5 (18q23) may also cause pseudohermaphroditism. *See* adrenal hyperplasia; adrenal hypoplasia; androgen insensitivity; anti-Müllerian hormone; cytochromes; gynecomastia; hermaphroditism; hypospadias; infertility; luteinization; methemoglobin; Müllerian ducts; polycystic ovarian cancer; Reifenstein syndrome; testicular feminization; Wilms' tumor.

pseudohitchhiking Adaptive mutations near neutral loci may simulate genetic drift. *See* hitchhiking.

pseudo-Hurler syndrome *See* mucolipidosis.

pseudohypha (in *Saccharomyces cerevisiae*) Formation occurs by nutrient (N) deficiency and may cause polarized growth on the surface of the agar medium, favoring a delay in mitosis and precocious entry into meiosis. The pseudohyphal growth is symmetric and synchronous in comparison to regular budding, which is asymmetric and asynchronous. Cyclins 1 and 2 promote pseudohyphal growth, whereas cyclin 3 is inhibitory in yeast. Alternative controls exist. Protein Ste12 and the MAP kinase signal transduction pathways also regulate hyphal growth. Filamentous growth is a requisite for pathogenicity of *Ustilago maydis* and *Candida albicans*. *See* candidiasis; CDK; cyclin; MAP; Ste; *Ustilago maydis*.

pseudohypoaldosteronism (PHA; 1q31-q42, 17p11-q21, 12p13, 16p13-p12) Hyperkalemic, hyperchloremic acidosis and hypertension. The genes at chromosomes 17 and 1 encode a threonine/serine kinase, WNK4, localized in the tight junctions. The disease in this protein is due to missense mutations. In chromosome 12, the cytoplasmic WNK1 is encoded and the defect is due to large intronic deletions that boost the expression of the protein. Both of these proteins are in the distal nephron (a basic morphological and functional unit of the kidney) responsible for potassium and pH homeostasis. These two anomalies are dominant. The recessive PHA in chromosome 16 encodes subunits of an epithelial Na^+ ion channel. *See* aldosteronism; Gordon syndrome; hyperkalemic; hypertension; hypoaldosteronism; intron; ion channels. (Wilson, F. H., et al. 2001. *Science* 293:1107.)

pseudohypoparathyroidism *See* Albright hereditary osteodystrophy.

pseudoknot Formed when a stem-and-loop RNA structure is bound at the base of the loop by hydrogen bonds or by a ligand, resulting in a two-stem two-loop stacking. The actual configurations of the pseudoknots may vary. Pseudo-half-knots form only a single loop. Such structures may modulate RNA functions and can be exploited in designing highly selective drugs. Some insect RNA viruses that use CAA (glutamine) rather than AUG (methionine) for translation initiation and do not require an initiator tRNA but apparently rely on a pseudoknot formed between a 15–43 nucleotide upstream loop and the sequence immediately preceding the CAA codon. Pseudoknots initiate translational frameshifting in overlapping genes. *See* antisense RNA; overlapping genes; repeat, inverted; TFO. (Kim, Y.-G., et al. 1999. *Proc. Natl. Acad. Sci. USA* 96:14234.)

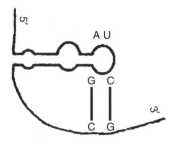

pseudolinkage Linkage due to translocation between nonhomologous chromosomes. *See* affinity.

pseudolysogen Lyses the bacterial cells slowly as if it would be lysogenic. *See* lysogeny.

Pseudomonas Bacteria include several species that degrade oil spills, polycyclic hydrocarbons, benzene, and other pollutants. *See* biodegradation; oil spills. (Coates, J. D., et al. 2001. *Nature* 411:1039.)

Pseudomonas aeruginosa 6.3 million bp bacterium and an opportunistic human parasite. It is the most common cause of death in cystic fibrosis, but it is involved in some pneumonias and other infections (urinary tract, burn victims, etc.). It also grows on soil and plant and animal tissues. This gram-negative bacterium is highly resistant to antibiotics and disinfectants. Close to 10% of its genes are regulatory, and the large number of its putative pump proteins explains its resistance to drugs. *See* cystic fibrosis. (Stover, C. K. 2000. *Nature* 407:959.)

Pseudomonas **exotoxin** Kills by irreversible ribosylation of ADP and subsequent inactivation of translation elongation factor, EF-2. Its applied significance is the potential for cancer therapy. *See* toxins.

Pseudomonas tabaci Bacteria causing "wild-fire" disease (necrotic spots) on tobacco leaves. The symptoms may be mimicked by methionine sulfoximine, a methionine analog.

Wildfire disease spots (Courtesy of Dr. Peter. Carlson.)

pseudomosaic May occur in a sample of amniocentesis caused by the conditions of culture rather than the genetic/chromosomal condition of the fetus.

pseudo-overdominance In a population certain phenotype(s) may appear in excess of expectation because of the close linkage of the responsible gene to advantageous alleles. Also, QTL loci may appear overdominant if they are relatively closely linked and display heterosis because the QTL mapping techniques cannot determine the map positions with great accuracy and the molecular function of the genes involved is not known. *See* fitness; hitchhiking; interval mapping; overdominance; QTL.

pseudoplasmodium Migrating slug of cellular slime molds. *See Dictyostelium*.

pseudopodium *See* amoeba.

pseudopregnant Female (mice) mated with vasectomized males, then implanted with blastocyst-stage embryos derived from other matings. *See* allopheny; vasectomy.

pseudoqueen In social insects (bees, ants, termites), one worker (XX) may become a fertile pseudoqueen after the loss of the queen of the colony. This type of development is promoted by special feeding (royal treatment) of the original worker caste insects. *See* honeybee.

pseudorecombinant Reassortment of two viral genome components from different viruses transmitted by the same insect vector.

pseudoreplication The samples are not independent replicates and the conclusion based on them may statistically be not reliable.

pseudoreversion Apparent backmutation caused by an extra-site suppressor mutation. *See* reversion.

pseudorheumatoid dysplasia Rare recessive cartilage defect due to mutation in the cysteine-rich secreted protein gene family. *See* arthritis; rheumatic fever.

pseudosubstrate Molecule with similarity to an enzyme substrate, but it is actually an inhibitor, and special regulators are required for its removal so that the enzyme is permitted to access its true substrate. *See* intrasteric regulation; substrate.

pseudotemperate phage It has a lysogenic cycle yet does not have a stable prophage state, e.g., the PBS1 transducing phage of *Bacillus subtilis*. *See* lysogeny; prophage.

pseudotrisomic Actually disomic, but one of the chromosomes is represented by two telocentric chromosomes. Each represents a chromosome arm; thus, two telocentrics + one normal chromosome occurs. *See* telocentric chromosome; trisomy.

pseudotype Virus carrying foreign protein on its envelope may expand the normal host range.

pseudotyping If two types of viruses invade the same cell, genetic material of one may slip into the capsid of the other. This type of packaging permits the introduction of the viral genome into a host, which otherwise would be incompatible with the virion. This phenomenon may be taken advantage of during the construction of viral vectors and helper viruses. The ability for a virus to infect a certain type of cell depends on the interaction between the viral glycoprotein and the nature of the cell-surface receptors. The vesicular stomatitis virus viral envelope glycoprotein (VSV-G) is highly fusigenic for a wide range of cell types and organisms. Thus, it can be employed for pseudotyped viral vectors to expand their effective host range. Similarly, the hemagglutinating paromyxovirus of Japan (HVJ) and other viruses can also be used. *See* amphotropic; ecotropic; packaging cell lines; pseudovirus; retroviral vectors. (Mazarakis, N. D., et al. 2001. *Hum. Mol. Genet.* 10:2109; Peng, K. V., et al. 2001. *Gene Ther.* 8:1456.)

pseudouridine (ψ) Pyrimidine nucleoside (5-β-ribofuranosyluracil) occurs in the T arm of tRNA by posttranscriptional modification of a uracil residue. Pseudouridine has been found in ribosomal RNAs and snRNAs. The modification is mediated by the nucleolar ψ-synthase with the assistance of other proteins. A requisite for the process is that a small nucleolar RNA (snoRNA) carrying a single-stranded H box (ANANNA) and a ACA-3' box must pair with the target RNA at an about 12 or less region of complementarity. After the enzyme gains access to the U site, the N1–C1' bond in a uracil is severed. After a 180° rotation, the C5 position becomes available for the formation of a new bond. Thus, the N1 and N3 sites may become readily available for hydrogen pairing and pseudouridine can bind easier in inter- or intramolecular reactions. *See* ψ for formula; snoRNA; tRNA. (Bortolin, M.-L., et al. 1999. *EMBO J.* 18:457; Hoang, C. & Ferré-D'Amaré, A. R. 2001. *Cell* 107:929.)

Pseudouridine.

pseudovirion (pseudovirus) Contains nonviral DNA within the viral capsid and can thus be used to unload foreign DNA into a cell if a helper virus is provided. *See* capsid; virion. (Liu, Y., et al. 2001. *Appl. Microbiol. Biotechnol.* 56:150; Ou, W. C., et al. 2001. *J. Med. Virol.* 64:366.)

pseudowild type Displays a wild phenotype because a mutation at a site different from the mutant locus that it masks—but most commonly a duplicated segment—compensates for the original and still present recessive mutation. In *Neurospora*, it occurs at a much higher frequency than expected by backmutation. It may also be due to a suppressor mutation. (Mitchell, M. B., et al. 1952. *Proc. Natl. Acad. Sci. USA* 38:569.)

pseudoxanthoma elasticum (PXE, 16p13.1) Autosomal-recessive or -dominant disorders of an ABCC6 (multiple drug resistance) transporter caused by degenerative changes in the skin (peau d'orange = orange rind), veins, eyes, intestines, etc., resulting in heart disease and hypertension. The defect involves dysplasia of elastin fibers and it affects the skin, retina, arteries, teeth, etc. *See* ABC transporters; coronary heart disease; hypertension; skin diseases. (Le Saux, O., et al. 2001. *Am. J. Hum. Genet.* 69:749.)

pseudo-Zellweger syndrome *See* peroxisomal 3-oxoacyl-coenzyme A thiolase deficiency; Zellweger syndrome.

psi (ψ) Pseudouridine, and also the packaging signal in retrovirions. The packaging signal located at the S' LTR repeat and reaches into the upstream end of the gag gene. It is not translated. *See* ψ; retroviral vectors; retrovirus; tRNA pseudouridine loop.

PSI$^+$ Yeast prion, an extrachromosomal protein suppressing nonsense codons. It functions in collaboration with the nuclear gene *SUP*35. Overexpression of this gene induces the formation of PSI$^+$, probably by a conformational change in the protein. Cells deleted in the amino-terminal region of Sup35 are resistant to PSI$^+$. Expansion of imperfect oligopeptide repeats in Sup35 (PQGGYQQYN) and in PrP (PHGGWGQ) seems to be responsible for the abnormality. Overexpression of Hsp104 heat-shock protein cures the cells from PSI$^+$. *See* Hsp; prion. (Masison, D. C., et al. 2000. *Curr. Issues Mol. Biol.* 2:51; Jensen, M. A., et al. 2001. *Genetics* 159:527.)

P site Peptidyl site on the ribosome where the first aminoacylated tRNA moves before the second charged tRNA lands at the A site as the translation moves on. The binding of tRNA to the 30S ribosomal subunit appears to be controlled by guanine residues at the 966, 1,401, and 926 positions in the 16S rRNA. *See* A site; protein synthesis; ribosome. (Feinberg, J. S. & Joseph, S. 2001. *Proc. Natl. acad. Sci. USA* 98:11120; Schäfer, M. A., et al. 2002. *J. Biol. Chem.* 277:19095.)

Psi vector *See* E vector.

PsnDNA 150–300 bp pachytene DNA sequences flanking 800–3,000 bp internal chromosomal segments in eukaryotes. The two short and the central DNA sequences are called PDNA (pachytene DNA). The PsnDNAs are supposed to be nicked by an endonuclease after homologous small nuclear RNA (snRNA) and a nonhistone protein (PsnProtein) have opened the sequences to the action of the enzyme. These molecules appear only during late leptotene to pachytene and are assumed to mediate recombination. *See* crossing over; meiosis; snRNA; ZygDNA. (Stern, H. & Hotta, Y. 1984. *Symp. Soc. Exp. Biol.* 38:161.)

psoralen dye Can combine with the DNA connecting nucleosomal core particles. After irradiation with near-ultraviolet light, cross-linking between the two DNA strands occurs. Psoralen-conjugated triple helix-forming oligonucleotides have been used to induce site-specific mutations in COS cells at very high frequency. Some celery stocks may contain higher-than-normal amounts of psoralen. *See* COS cell; site-specific mutation; triple helix formation. (Cimino, G. D., et al. 1985. *Annu. Rev. Biochem.* 54:1151; Luo, Z., et al. 1997. *Proc. Natl. Acad. Sci. USA* 97:9003; Oh, D. H., et al. 2001. *Proc. Natl. Acad. Sci. USA* 98:11271.)

Psoralen.

psoriasis (PSOR) Scaly type of skin defect determined either by dominant genes of reduced penetrance or polygenic inheritance involving relatively few genes. Its incidence is common in Caucasian populations (1–3%), but it is much less frequent in Orientals (Eskimos, Native Americans, and Japanese). Recurrence rate may vary (8–23% among first-degree relatives) depending on the type involved. If both parents are affected, the recurrence among children may reach up to 75%. The psoriasis haplotype appears to include HLA-BW 17 and HLA-A 13 genes. Some observations indicate that bacterial superantigens may trigger psoriasis. Psoriasis susceptibility genes have been assigned to 19p13.3, 3q21, 1q21, 17q25, 4qter, 14q31-q32, 6p21, and 20p. Linkage with other chromosomes is less certain. Psoriasis increases the risk of basal cell carcinoma. Microarray analysis revealed upregulation of transcription of at least 161 genes in psoriasis. Some of the transcripts are modulated in other skin diseases. *See* dermatitis, atopic; Hirschsprung disease; HLA; ichthyosis; IL-20; keratosis; nevoid basal cell carcinoma; skin diseases. (Bhalerao, J. & Bowcock, A. M. 1998. *Hum. Mol. Gen.* 7:1537; Bowcock, A. M., et al. 2001. *Hum. Mol. Genet.* 10:1793.)

P{*Switch*} *See* Gene-Switch; hybrid dysgenesis.

psychomimetic Drugs affect the state of mind in a manner similar to psychoses. *See* ergot; psychoses; psychotropic drugs.

psychoses Group of mental nervous disorders with variable genetic and environmental components. *See* affective disorders; attention deficit hyperactivity disorder; autism; dyslexia; IQ; manic depression; paranoia; schizophrenia; Tourette's syndrome.

psychotherapy Treatment/support provided for transient or lasting emotional and behavioral disorders. It may involve verbal support or chemical medication. Genetic counselors need to be familiar with the verbal support option. *See* counseling, genetic.

psychotropic drugs Affect the state of mind. They are used as medicine in various types of psychoses and may be very beneficial (e.g., lithium, valium, etc.) if applied under medical monitoring. Possible adverse side effects vary by the chemical nature of the drug and may include heart disease, birth defects, addiction, etc. *See* psychomimetic; psychoses.

psychrophiles Organisms that grow under low temperatures. *See* antifreeze proteins.

PTA deficiency disease Controlled by incompletely dominant (4q35) genes. Plasma thromboplastic antecedent protein deficiency is involved, which results in unexpected bleeding after tooth extraction or various surgeries. Nose bleeding (epistaxis) is common, but uterine bleeding (menorrhagia) or blood in the urine (hematuria) is rare. The carrier frequency in Ashkenazi Jewish populations is about 8.1%. *See* antihemophilia factors; pseudohemophilia.

PTB Phosphotyrosine-binding domain is present in proteins involved in signaling. *See* pleckstrin; SCK; SH2; SH3; signal transduction; WW.

PTC (1) *See* phenylthiocarbamide. (2) Papillary thyroid carcinoma; a variant of the RET oncogene-caused neoplasia. *See* RET.

PtdInsP$_2$ *See* phosphoinositides.

PTEN Phosphatase and tensin deleted in chromosome 10 (10ter-q11, 10q24-q26, 10q22-q23); MMAC1 (mutated in multiple advanced cancer). Tumor suppressor involved in brain, prostate, breast cancers, multiple hamartomas (Lhermitte-Duclos disease/Cowden syndrome), Bannayan-Zonona syndrome, and other cancers. It inhibits cell migration and cell adhesion and dephosphorylates FAK, serine, threonine, and tyrosine residues in proteins. The primary target of PTEN appears to be phosphatidylinositol-3,4,5 trisphosphate (PIP3) and acts as a tumor suppressor by promoting apoptosis. In vivo PTEN may act as a lipid phosphatase, and this function may be essential for tumor suppression. The protein (tyrosine, serine/threonine) phosphatase activity may not be important for tumor suppression. Some cancer cells (glioma, prostate, breast cancer) may be reverted to normalcy by the addition of PTEN. The catalytic domain identity motif is HCXXGXXRS/T. The two α-helix domains flanking the catalytic domain are encoded in its exon 5 and must be intact for proper function. The tensin homology domain enables the recognition of the cell adhesion system (actin, integrin, FAK, Src). In mice, the *Pten*$^{+/-}$ heterozygotes are subject to autoimmune disease and FAS-mediated apoptosis. The normal FAS function can be restored by the administration of phosphatidyl inositol 3 kinase. PTEN has an influence on cyclin D1 and signal transduction. Mutations in PTEN may be found in Proteus syndrome or Proteus-like syndrome. *See* apoptosis; Bannayan-Zonona syndrome; FAK; multiple hamartomas syndrome; phosphatidylinositol; PIK; PIP2; PIP3; polyposis, juvenile; Proteus syndrome; tensin; tumor suppressor. (Di Cristofano, A. & Pandolfi, P. P. 2000. *Cell* 100:387; Wen, S., et al. 2001. *Proc. Natl. Acad. Sci. USA* 98:4622; Maehama, T., et al. 2001. *Annu. Rev. Biochem.* 70:247; Waite, K. A. & Eng, C. 2002. *Am. J. Hum. Genet.* 70:829.)

pteridines Purine derivatives involved in coloring of insect eyes, wings, amphibian skin, etc. Pteridines may be light receptors. Reduction in tetrahydrobiopterin and related amines may be responsible for nervous disorders. *See* GTP cyclohydrolase I deficiency; ommochromes; photoreceptors; rhodopsin. (Blau, N., et al. 1998. *J. Inherit. Metab. Dis.* 21:433.)

Pterin.

PTG Protein-targeting glycogen forms complexes of phosphatases, kinases and glycogen synthase with glycogen. *See* glycogen; kinase.

PTGS Posttranscriptional gene silencing presumably by degradation of mRNA or inactivation of infectious (viral) RNA.

Recent evidence indicates the presence of a 25-nucleotide-long antisense RNA in the silenced cells. *See* epigenesis; methylation of DNA; posttranscriptional gene silencing; RIGS; RNA surveillance; RNAi.

ptilinum Inflatable head of the larva emerging from the puparium that cyclically is inflated/deflated to pry open the puparium by a wedging type of operation.

PTK Protein tyrosine kinase involved in regulation of signal transduction and in growth and differentiation of cells. *See* protein kinases.

ptosis Drooping eyelid(s). *See* blepharophimosis; epicanthus.

PTP *See* tyrosine phosphatase.

PTPRC Protein tyrosine phosphatase receptor type C.

PU.1 (PU1) Transcription factor in blood-forming cells regulating the differentiation of macrophages, B lymphocytes, and monocytes; it belongs to the ETS family of oncogenes. *See* ETS; lymphocytes; macrophages; monocytes. (DeKoter, R. P. & Singh, H. 2000. *Science* 288:1439; Lewis, R. T., et al. 2001. *J. Biol. Chem.* 276:9550.)

puberty Time of sexual maturation accompanied by the appearance of secondary sexual characteristics such as facial hair in males, breast enlargement in females, etc.

puberty, precocious Autosomal-dominant disorders occur in two forms: (1) isosexual, when sexual maturation in both males and females takes place before age 10 and 8.5, respectively, and may be even much earlier, especially in females; (2) male-limited form in which testosterone production seems to be independent from gonadotropin-releasing hormone production. The disorder is associated with a defect in the luteinizing hormone receptor. *See* animal hormones; G-proteins; hormonal effects on sex expression; luteinizing hormone-releasing factor.

PubGene Human gene-to-gene co-citation index involving 13,712 named human genes. (Jenssen, T.-K., et al. 2001. *Nature Genet.* 28:21.)

publication ethics Subject to the same common sense rules as any other principle of ethics. The detailed guidelines in *Human Reproduction* (2001), 16:1783–1788, contain specific valid points. *See* ethics.

public blood systems *See* private blood groups.

public opinion In the underdeveloped world with inadequate educational systems, superstitions greatly affect people's view on all aspects of life and society. In culturally and technically advanced nations, newspapers, television, and Internet resources may influence public opinion to a great degree. Application of scientific principles is commonly decided by legislative action. In a democratic society, the citizens' view must necessarily be considered. The dilemma of how well informed is the general public or the legislative/governmental system regarding the implications of scientific principles is an important problem. In a survey in England, the public indicated that automobiles are safer than trains. The actual statistics indicated, however, that the safety of trains is about 100 times better. People generally believe that atomic power plants expose the public to unnecessary health and genetic risks. The hazards of burning fossil fuels or using wood fireplaces are much less frequently considered, although they generate carcinogenic and mutagenic emissions. Very often, even the scientists are unable to predict the future consequences of the scientific achievements they brought about, as was apparent by the consensus reached on recombinant DNA by the historical Asilomar Conference. The problems of using genetically modified organisms, cloning, and stem cell applications cannot be resolved by political approaches. The problems created by technology and science can be resolved only by better scientific research. *See* atomic radiation; criticism on genetics; gene therapy; GMO; informed consent; recombinant DNA and biohazards; stem cells.

PUBS Percutaneous (through the skin) umbilical blood sampling, a method of prenatal biopsy for the identification of hereditary blood, cytological, and other anomalies. *See* amniocentesis; prenatal diagnosis.

Puccinia graminis *See* stem rust.

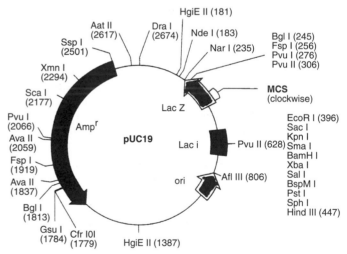

(The diagram is the courtesy of CLONTECH Laboratories Inc., Palo Alto, CA.)

pUC vectors Small (*pUC12/13* 1680 bp, *pUC18/19* 2686 bp) plasmids containing the replicational origin (*ori*) and the *Amp*ʳ gene of pBR322, and they carry the *LacZ'* fragment of bacterial β-galactosidase. Z' indicates that within this region there is a multiple cloning site (MCS) for recognition by 13 restriction enzymes. The orientation of the MCS is in reverse in pUC18 relative to pUC19. Genes inserted into *Lac* may be expressed under the control of the *Lac* promoter as a fusion protein. Most commonly, the insertion inactivates the *Lac* gene and white colonies are formed in Xgal medium rather than blue when the gene is active. The pUC vectors can be used with JM105 and NM522 *E. coli* strains. *See* filamentous phages; *Lac*, vectors; *Xgal*. (Messing, J. 1996. *Mol. Biotechnol.* 5:39).

PUF PROTEINS Control mRNA stability by binding to the 3′-untranslated end. (Wickens, M., et al. 2002. *Trends Genet.* 18:150.)

puff Swollen area of polytenic chromosomes active in transcription. Puffing is induced by expression of transcription factor genes regulated by steroid hormones (ecdysone).

Ecdysone formation comes in sequential pulses and thereby sequential activation of genes involved in metamorphosis of insects can be visualized at the level of the giant chromosomes. The puffs represent active transcription at particular genes. The pattern of puffing shifts along the salivary gland chromosomes during development and/or activation and the RNA extracted from the puffs reflect the differences in the base sequences of the genic DNA. Puffing has also been described in the rare polytenic chromosomes of some plant species, e.g., *Allium ursinum* or *Aconitum ranunculifolium*. These have been observed in specialized tissues of the chalaza or in the antipodal cells. *See* ecdysone; giant chromosomes, figure below. (Beermann, W. 1961. *Verh. Dtsch. Zool. Ges.* 1961:44; Mok, E. H., et al. 2001. *Chromosoma* 110:186.)

pufferfish, Japanese (*Fugu rubripes*) Small vertebrate with about 365 Mbp DNA, i.e., only somewhat more than one tenth of that of most mammals, and therefore it is suitable for structural and functional studies at the molecular level. More 95% of the genome have been sequenced by 2002. About 1/3 of the genome is genic and repetitive sequences occupy less than 1/6th. *Spheroides nephelus* is also used for studies of control of gene expression. *Tetraodon nigroviridis* DNA sequences have been used for the determination of human gene number. (Crollius, H. R., et al. 2000. *Genome Res.* 10:939; Aparicio, S., et al. 2002. *Science* 297:1301.)

pull-down assay Expected to reveal interacting proteins. One of the proteins is attached to agarose beads and thus immobilized. Then the test protein is added and the mixture is incubated to allow time for forming some links. Subsequently, the mix is centrifuged. If there is a binding between them, both proteins are found in the pellet and interaction is assumed. *See* immunoprecipitation. (Brymora, A., et al. 2001. *Anal. Biochem.* 295:119.)

pullulanase Secreted *Klebsiella* enzyme (~117 kDa) that cleaves starch into dextrin. It occurs in the endosperm of cereals and other plant tissues and is regulated by thioredoxin. *See* thioredoxin. (Schindler, I., et al. 2001. *Biochim. Biophys. Acta* 1548:175.)

pulmonary emphysema Increase in size of the air space of the lung by dilation of the alveoli (small sac-like structures) or by destruction of their walls. Smoking may cause it.

pulmonary hypertension (PPH, FPPH) Characterized by shortness of breath, hypoxemia, and arterial hypertension caused by the proliferation of endothelial smooth muscles and vascular remodeling. It is a 2q33 dominant disorder with reduced penetrance. Various drugs (such as the banned antiobesity drug fen-phen [fenfluramine + phentermine]) may trigger it. The basic defect is in gene BMPR2 (bone morphogenetic protein receptor II). Haploinsufficiency may cause it. The consequence is inappropriate regulation by the serine/threonine kinases of the phosphorylated Smad proteins, leading to inadequate maintenance of blood vessel integrity. *See* bone morphogenetic protein; haploinsufficient; hypertension; Smad. (Machado, R. D., et al. 2001. *Am. J. Hum. Genet.* 68:92.)

pulmonary stenosis *See* stenosis.

pulmonary surfectant proteins *See* respiratory distress.

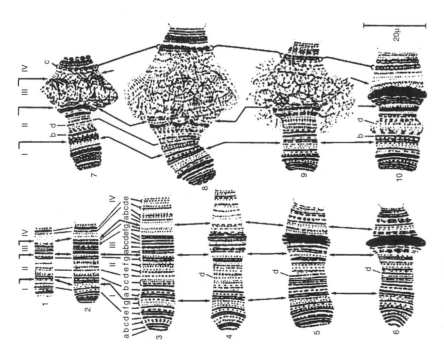

Selective activity of genes during development of the dipteran fly *Rhynchosciara angelae* is reflected in the puffing pattern of the salivary gland chromosomes. Lowercase letters designate bands. Roman numerals indicate regions of the chromosomes. (After Breuer, M. E., and Pavan, C. 1954. By permission from Kühn, A. 1971. *Lectures on Developmental Physiology*. Springer-Verlag, New York.)

Grown without label	Replication in ^{3}H	Labeled chromosomes replicated in ^{3}H-free medium		
		no exchange	sister chromatids exchanged	
				Cytological observation
				Interpretative drawing of the distribution of the radioactive label ■
			→	Interpretation of the replication of the DNA helices in the two chromatids ^{3}H . . . . .

Autoradiographic analysis of the replication of the DNA in chromosomes by the pulse–chase procedure. (Drawn after Taylor, J. H., et al. 1957. *Proc. Natl. Acad. USA* 43:122.)

pulse-chase analysis Exposes cells to a radioactive compound such as ^{3}H-thymidine (pulse) and examines the labeling of chromosomes in some cells. The culture is then transferred to nonradioactive thymidine, allowed to complete a division (chased to another stage), thus again enabling the study of the label distribution to determine its fate in the cells. The experiment permitted for the first time the valid conclusion that DNA replication is semiconservative. *See* radioactive label; radioactive tracer; figure above.

pulsed-field gel electrophoresis (PFGE) Procedure combining static electricity, alternating electric fields with gel electrophoresis for the separation of DNA of entire chromosomes of lower eukaryotes such as of yeast and *Tetrahymena* or large DNA fragments cloned in YAC vectors of any genome cut by rare-cutting restriction enzymes. *See* CHEF; FIGE; OFAGE; PHOGE; RGE; TAFE; YAC. (Mulvey, M. R., et al. 2001. *J. Clin. Microbiol.* 39:3481.)

PUMA (1) (*Felis concolor, Puma concolor*): $2n = 38$. (2) p53 upregulated modulator of apoptosis. *See* apoptosis; p53.

pump Various transmembrane proteins mediating active transport of ions and molecules through biological membranes. *See* sodium pump.

punctuated equilibrium *See* punctuated evolution.

punctuated evolution Theory that evolution would follow alternating periods of rapid changes and relatively stable intervals (punctuations). Natural selection of beneficial mutations appears after some intervals and spreads over the population. *See* speciation, beneficial mutation; gradualism; hopeful monster; neutral mutation; shifting balance theory. (Gould, S. J. & Eldredge, N. 1993. *Nature* 366:223; Elena, S. F., et al. 1996. *Science* 272:1797; Voigt, C., et al. 2000. *Adv. Protein Chem.* 55:79.)

punctuation codons (UAA, UGA, UAG) Terminate translation of mRNA.

Punnett square Permits simple prediction of the expected pheno- and genotypic proportions. It is a checkerboard where on top and at the left column the male and female gametic output is represented and in the body of the table the genotypes are found. If, e.g., the heterozygote has the genetic constitution of *Aa, Bb*, the gametes and genotypes will be AB, Ab, aB, and ab. In case of linkage and recombination, the actual frequency of each type of gamete must be used to obtain the correct genotypic proportions in the body of the checkerboard. *See* Mendelian segregation; modified Mendelian ratios.

MALE GAMETES →	AB	Ab	aB	ab
↓ FEMALE GAMATES				
AB	AB AB	AB Ab	AB aB	AB ab
Ab	Ab AB	Ab Ab	Ab aB	Ab ab
aB	aB AB	aB Ab	aB aB	aB ab
ab	ab AB	ab Ab	ab aB	ab ab

pupa Stage in insect development between the larval stage and the emergence of the adult (imago). *See* Drosophila; juvenile hormone.

Drosophila pupa.

puparium   Case in which the *Drosophila* (and other insect) pupa develops for about 4 days after hatching of the egg, and in another 4 days the imago emerges. *See Drosophila.*

pure breeding   Homozygous for the genes considered.

pure culture   Involves only a single organism. *See* axenic culture.

pure line   Genetically homogeneous (homozygous), and its progeny is expected to be identical with the parental line unless mutation occurs. (Johannsen, W. 1909. *Elemente der exakten Erblichkeitslehre.* Fischer, Jena, Germany.)

purine   Nitrogenous base composed of a fused pyrimidine and imidazole ring; the principal purines in the cells are adenine, guanine, xanthine, and hypoxanthine (but theobromine, caffeine, and uric acid are also purines).

5′,8-purine cyclodeoxynucleosides   Formed in two diastereoisomers by exposure of DNA to reactive oxygen species. The cyclopurines may cross-link the C-8 adenine or guanine and the 5′ position of 2-deoxyribose. These diastereoisomers may block DNA replication and are cytotoxic. Commonly excision repair may not correct the damage, although the xeroderma pigmentosum A protein may cut at both flanks and excise them. These types of damaged nucleosides may accumulate by time and result in progressive neurodegeneration in xcroderma pigmentosum patients. *See* cyclobutane; excision repair; xeroderma pigmentosum. (Kuraoka, I., et al. 2000. *Proc. Natl. Acad. USA* 97:3837.)

5′,8-cyclo-2′-deoxyadenosine phosphate.

purine repressor (PurR)   Member of the *Lac* repressor family of proteins regulating 10 operons involved in the biosynthesis of purine and affecting to some extent 4 genes controlling de novo pyrimidine synthesis and salvage. Its ca. 60 amino acids, the NH_2 domain, bind to DNA, and its ca. 280-residue COOH domain binds effectors and functions in oligomerization. *See Lac* repressor; salvage pathway. (Moraitis, M. I., et al. 2001. *Biochemistry* 40:8109.)

purity, gamete   One of the most important discoveries of Mendel. At anaphase I of meiosis of diploids, the bivalent chromosomes segregate, and at anaphase II, the chromatids separate. Therefore, in the gametes of diploids only a single allelic form of the parents is present with rare exceptions, e.g.,

nondisjunction and polyploids. *See* gene conversion; meiosis; Mendelian laws; nondisjunction.

Purkinje cells   Large pear-shape cells in the cerebellum are connected to multibranched nerve cells traversing the cerebellar cortex. In the heart, they are tightly appositioned cells that transmit impulses. *See* cerebellum; motor proteins.

puromycin   Antibiotic inhibiting protein synthesis by binding to the large subunit of ribosomes; its structure resembles the 3′-end of a charged tRNA. Therefore, it can attach to the A site of the ribosome to form a peptide bond, but it cannot move to the P site and thus causes premature peptide chain termination. *See* antibiotics; signaling to translation.

Puromycin.

pushme–pullyou selection   Positive-negative selection system to isolate engineered chromosomes in somatic cell hybrids, that have retained the segment positively selected for and lost the regions selected against. (Higgins, A. W., et al. 1999. *Chromosoma* 108:256; Trimarchi, J. M. & Lees, J. A. 2002. *Nature Rev. Mol. Cell. Biol.* 3:11.)

PV16/18E6   Human papilloma virus oncoprotein. *See* oncoprotein. Burkitt lymphoma and murine lymphocytomas; may have an activating role for MYC that is in the same chromosome. *See* Burkitt lymphoma; MYC; oncogenes.

PWM   Position weight matrix is used for identification of and search for functional nucleotide sequences, which are highly degenerate, e.g., TATA boxes in the promoters. PWM reflects the frequency of the four nucleotides (A, T, G, C) in an aligned set of different sequences sharing a common function. After it has been determined in well-characterized core promoter regions, the PWM can be used to scan for TATA boxes in anonymous nucleotide sequences. The similarities between PWM and specific sequences and the matching value (within an accepted range) is determined and called a signal. Bucher (*J. Mol. Biol.* 212:563) used a PWM for TATA box GTATAAAGGCGGGG, and when the best fit was designated as 0, the majority of "unknown" TATA boxes scored within 0 to -8.16. However, some of these might be false positives. *See* anonymous DNA segment; core promoter; position-specific scoring matrix; TATA box. (Audic, S. & Claverie, J.-M. 1998. *Trends Genet.* 14:10.)

PX (phox)   125-amino-acid module present in a variety of proteins involved in binding phosphoinositides.

P2X₁   Receptor for ATP in ligand-gated cation channels. P2X is a component of the contractile mechanism of the vas deferens muscles that propel the sperm into the ejaculate during copulation. Its defect entails ~90% sterility, although without apparent harm to the male or the female mice. *See* congenital aplasia of the vas deferens; ion channels; vas deferens.

PX DNA   Four-stranded molecule where the parallel helices are held together by reciprocal recombination at every site of juxtaposition. Its topoisomer is JX₂ and it contains adjacent helices, but there is no reciprocal exchange at the contact points. (Yan, H., et al. 2002. *Nature* 415:62.)

Pxr   *See* SXR.

pycnidium   Hollow spherical or pear-shaped fruiting structure of fungi producing the pycnidiospores, which are released through the top opening, the ostiole. *See* stem rust.

pycnodysostosis   Rare autosomal-recessive (1q21) human malady characterized by defects in ossification (bone development), resulting in short stature, deformed skull with large fontanelles (soft, incompletely ossified spots of the skull common in fetuses and infants), and general fragility of the bones. The primary defects appear to be in cathepsin K, a major bone protease, although interleukin-6 receptor has also been implicated. *See* cathepsins; cleidocranial dysostosis; Toulouse-Lautrec.

The famous French artist Henri Toulouse-Lautrec (1864–1901) might have suffered from this malady, and his self-portrait reveals some of the characteristics of the malformations. The exact nature of his condition cannot be diagnosed, but it is known that his parents were close relatives. (By permission of the St. Martin Press, New York.)

pycnosis (pyknosis)   Physiological effect of ionizing radiation on chromosomes expressed as clumping or stickiness. It is dose-dependent, and the late prophase stage of irradiation is most effective in causing it. Anaphase proceeds, but the chromosomes have difficulties in separation, display chromatin bridges, and may break up into fragments. *See* acinus; bridge; heteropycnosis; karyorrhexis.

Pygmy   Central African human tribe of about 100,000 has an average height of 142 cm. In comparison the average height of Swiss and Californian's is 167–169 and 170–172 cm, respectively. The Pygmies do not respond to exogenous stomatotropin, but the concentration of serum somatomedins in the adolescent Pygmies is about a third below that in non-Pygmies of comparable age. Although the shortness of Pygmies appears recessive, intermarriages indicate polygenic

determination of height. *See* dwarfism; nanism; somatomedin; somatotropin; stature in humans.

PYK2   Protein tyrosine kinase links Src with Gᵢ and G_q-coupled receptors with Grb2 and Sos proteins in the MAP kinase pathway of signal transduction. Lysophosphatidic acid (LPA) and bradykinin stimulate its phosphorylation by Src. Overexpressing mutants of Pyk or the protein tyrosine kinase Csk reduces the stimulation by LPA, bradykinin, or overexpressed Grb2 and Sos. *See* CAM; Csk; Gᵢ; G_q; Grb2; kininogen; lysophosphatidic acid; MAP; signal transduction; Sos; Src. (Felsch, J. S., et al. 1998. *Proc. Natl. Acad. Sci. USA* 95:5051; Sorokin, A., et al. 2001. *J. Biol. Chem.* 276:21521.)

pyknosis   *See* pycnosis.

pyloric stenosis   Smaller-than-normal opening of the pylorus, the lower gate of the stomach, which separates it from the small intestine (duodenum). It does not appear to have independent genetic control, but it is part of some syndromes. It affects males five times as frequently as females; the overall incidence for both sexes is about 3/1,000 births. About 20% of the sons of affected females display this anomaly, but only about 4% of the sons if the father has the malady. It may be caused by a deficiency of neuronal nitric acid synthase. *See* imprinting; nitric oxide; sex influenced.

PYO   Personal years of observations; a term used in medical and clinical genetics.

pyocin   Bacteriotoxic protein produced by some strains of *Pseudomonas aeruginosa* bacteria. *See* bacteriocins.

pyramidal cells   Excitatory neurons in the cerebral cortex. *See* brain; neuron.

pyrene   Fluorochrome, frequently used as a bimolecular excimer (excited dimer). *See* FRET.

pyrenoid   Dense, refringent protein structure in the chloroplast of algae and liverworts associated with starch deposition. *See* chloroplast.

pyrethrin (pyrethroids, permethrin)   Insecticides are natural products of *Pyrethrum* (*Chrysanthemum cineraiaefolium*) plants (Compositae). They affect the voltage-gated Na⁺ ion channels. Humans may have severe allergic reactions to pyrethrins. *See* ion channels.

pyrethrum (*Chrysanthemum* spp)   Source of the natural insecticide pyrethrin with basic chromosome number $x = 9$. Some species are diploid, tetraploid, or hexaploid. *See* pyrethrin.

pyridine nucleotide   Coenzyme containing a nicotinamide derivative, NAD, NADP.

pyridoxine (pyridoxal)   Vitamin B₆ is part of the pyridoxal phosphate coenzyme instrumental in transamination reactions. An apparently autosomal-recessive disorder involving seizures is caused by pyridoxin deficiency because of a deficit in glutamic acid decarboxylase (GAD) activity and consequently insufficiency of GABA, which is required for the normal function of neurotransmitters. Administration of pyridoxin caused cessation of seizures. The GAD gene has been located in the long arm of human chromosome 2. An autosomal-dominant regulatory pyridoxine kinase function has also been identified in humans. *See* epilepsy.

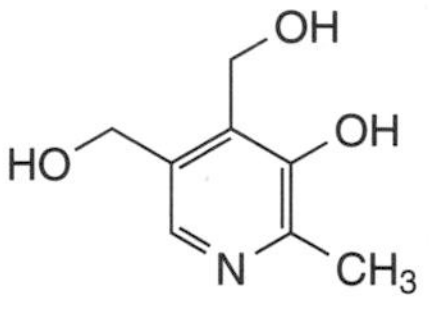

Pyridoxine.

pyridoxine dependency   May be manifested as autosomal-recessive seizures with perinatal onset (around birth).

pyrimidine   Heterocyclic nitrogenous base such as cytosine, thymine, or uracil in nucleic acids but also the sedative and hypnotic analogs of uracil, barbiturates, and derivatives. Pyrimidine biosynthesis may follow either a de novo or a salvage pathway. Some of the pyrimidine moieties, e.g., of thiamin, are biosynthesized through a route different from that of nucleic acid pyrimidines. *See* de novo synthesis; J base;

Uracil
(2,4-dioxypyrimidine)

Thymine
(5-methyl-2,4-dioxypyrimidine)

Cytosine
(2-oxy-4 aminopyrimidine)

5-Methylcytosine

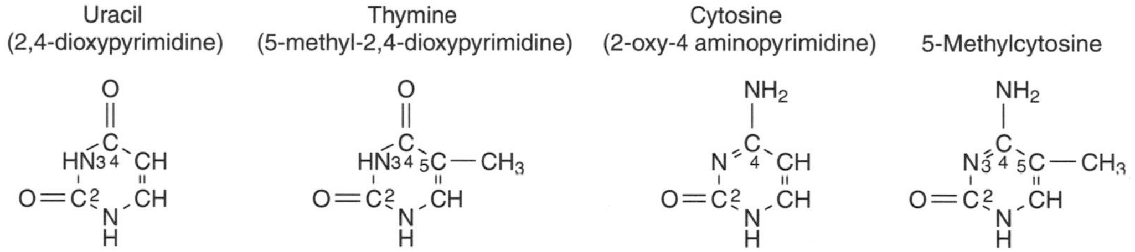

pseudouracil; salvage pathway; thiouracil. (Fox, B. A. & Bzik, D. J. 2002. *Nature* 415:926.)

pyrimidine dimer   Cross-linked adjacent pyrimidines (thymidine or cytidine) in DNA causing a distortion in the involved strand, thus interfering with proper functions. It is induced by short-wavelength UV irradiation. The thymidine dimers may be split by visible light-inducible enzymatic repair (light repair) or by excision repair (dark repair). *See* CPD; cyclobutane ring; DNA repair; genetic repair; glycosylases; photolyase; photoreactivation; physical mutagens; pyrimidine-pyrimidinone photoproduct. (Otoshi, E., et al. 2000. *Cancer Res.* 60:1729.)

pyrimidine dimer N-glycosylase   DNA repair enzyme that creates an apyrimidinic site. Then the phosphodiester bond is severed and a 3'-OH group is formed on the terminal deoxyribose. Exonuclease $3' \rightarrow 5'$ activity of the DNA polymerase splits off the new 3'-OH end of the apyrimidinic site. After this, the replacement-replication−ligation process repairs the former thymine dimer defect. *See* DNA repair;

glycosylase; pyrimidine dimer. (Piersen, C. E., et al. 1995. *J. Biol. Chem.* 270:23475.)

pyrimidine 5′-nucleotidase deficiency (P5N)   May cause hereditary hemolytic anemia as the pyrimidines inhibit the hexose monophosphate shunt in young erythrocytes. There are two isozymes of which P5NI is most commonly the cause of anemia. *See* anemia; pentose phosphate pathway. (Marinaki, A. M., et al. 2001. *Blood* 97:3327.)

pyrimidine-pyrimidinone photoproduct   Pyrimidine dimer involving a 6-4 linkage between thymine and cytosine. *See* cis-syn dimer; cyclobutane; Dewar product; photolyase; translesion pathway. (Vreeswijk, M. P., et al. 1994. *J. Biol. Chem.* 269:31858.)

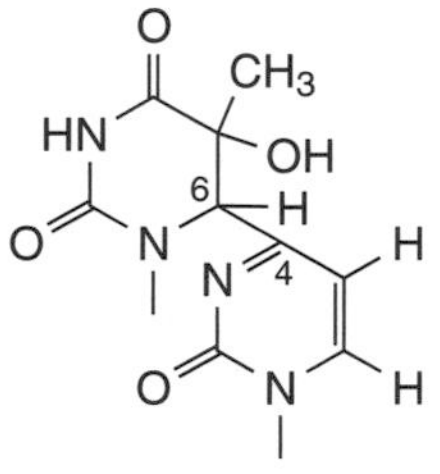

pyrimidone   Hydroxypyrimidine.

pyrimidopurinone   Malondialdehyde-DNA adduct derived from deoxyguanosine. *See* adduct:

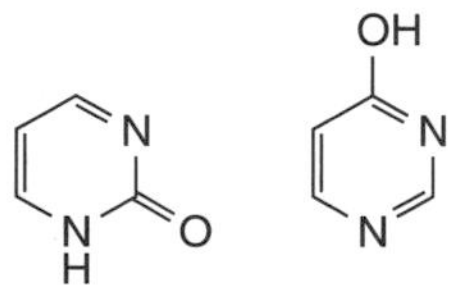

2-hydroxypyrimidine, right. 4(6)-hydroxypyrimidine left.

pyronin   Histochemical red stain used for the identification of RNA.

pyrosequencing   Used for the analysis of the nucleotide sequence of less than 200-base-long DNA strands for the detection of mutational alterations. It uses the enzymes DNA polymerase, sulfurylase, firefly luciferase, and apyrase. The incorporation of the nucleotides (which are not labeled) in the growing end is monitored by light flashes in a single tube.

Electrophoresis is not used. Nucleotide triphosphates add to the reaction in the sequence. Visible light is generated and detected when pyrophosphate is released during incorporation from the nucleotide triphosphates with the cooperative effects of sulfurylase and luciferase. This is a very fast procedure and may be automated. *See* apyrase; sulfurylase, luciferase. (Ronaghi, M. 2001. *Genome Res.* 11:3; Marziali, A. & Akeson, M. 2001. *Annu. Rev. Biomed. Engr.* 3:195; Fakhrai-Rad, H., et al. 2002. *Hum. Mut.* 19:479.)

pyrrole   Saturated five-membered heterocyclic ring such as found in protoporphyrin. Pyrrole-imidazole polyamides may bind to specific DNA of the transcription factor TFIIIA and regulate the transcription of 5S RNA. N-methylimidazole (Im)-N-methylpyrrole (Py) may target G≡C and Py-Im targets the C≡G base pairs, respectively. The Py-Py combination is specific for T=A and A=T. *See* heme; porphyria; porphyrin.

pyrrolizidine alkaloids (petasitenine, senkirkine)   Occur in several plant species (*Tussilago, Heliotropium*, etc.), some of which are used as food or medicinal plants, but they are mutagenic/carcinogenic. They also occur in some moths and convey protection against predators. (Ober, D. & Hartmann, T. 1999. *Proc. Natl. Acad. Sci. USA* 96:14777.)

pyrrolysine   The 22nd amino acid encoded in Archaea and Eubacteria by the stop codon UAG. *See* amino acids; genetic code. (Hao, B., et al. 2002. *Science* 296:1462; Srinivasan, G., et al. 2002. *Science* 296:1459.)

pyruvate dehydrogenase complex   Contains three enzymes — pyruvate dehydrogenase, dihydrolipoyl transacetylase, and dihydrolipoyl dehydrogenase — and the function of the complex requires the coenzymes thiamin pyrophosphate (TPP), flavine adenine dinucleotide (FAD), coenzyme A (CoA), nicotinamide adenine dinucleotide (NAD), and lipoate. The result of the reactions is oxidative decarboxylation, whereby CO_2 and acetyl CoA are formed. *See* oxidative decarboxylation. (Zhou, Z. H., et al. 2001. *J. Biol. Chem.* 276:21704.)

pyruvate kinase deficiency   Recessive (human chromosome 1q21-q22, PK1) hemolytic anemia actually caused by two enzymes that are the products of differential processing of the same transcript or chromosomal rearrangement. In the presence of some tumor promoters, hepatic pyruvate kinase activity decreases. *See* anemia; glycolysis; hemolytic anemia.

pyruvic acid   Ketoacid ($CH_3COCOOH$) formed from glycogen, starch, and glucose under aerobic conditions (under anaerobiosis it is reduced to lactate, and NAD^+ is formed). *See* Embden-Meyerhof pathway; pentose phosphate shunt.

PyV   Polyoma virus.

PYY$_{3-36}$ Is a neuropeptide Y (NPY)-like but it is a gastrointestinal hormone that inhibits food uptake. *See* obesity; leptin; neuropeptide Y. (Batterham, R. L., et al. 2002. *Nature* 418:650.)

PZD   *See* micromanipulation of the oocyte.

> "...alle essentiellen Merkmale...epigenetisch sind, und das die Determinierung ihrer Specificität durch den Kern erhalten." Theodor Boveri. 1903. *Roux' Arch. Entwickl.-Mech. Org.* 16:340–363.

A historical vignette.

Q

q Long arm of chromosomes. *See* p.

Q banding Chromosome staining with quinacrines that reveals cross-bands. Because of the availability of newer microtechniques, generally this procedure is no longer used. *See* chromosome banding; quinacrine mustard. (Caspersson, T. G., et al. 1971. *Hereditas* 67:89.)

Q-β RNA bacteriophage of a molecular weight of about 1.5×10^6 Da. Q-β replicase is an RNA-dependent RNA polymerase that synthesizes the single-stranded RNA genome of the phage without an endogenous primer. The replicase can use both the + and − strands as a template, and therefore it amplifies the genome rapidly. It is a heterotetramer consisting of one viral-encoded and three host polypeptides. *See* plus strand; replicase. (Munishkin, A. V., et al. 1991. *J. Mol. Biol.* 221:463.)

Q2 domain *See* CREB.

QSTAR Pulsar™ Quadrupole time-of-flight mass spectrometer. *See* MALDI; proteomics; quadrupole. (Steen, H., et al. 2001. *J. Mass Spectrom.* 36:782.)

QTL Quantitative trait loci control the expression of complex traits such as weight, height, cognitive ability, etc. Their expression is usually not strict and even in the absence of the critical genes a quantitative trait may appear under the influence of extrinsic factors, yet their expression is more likely when the appropriate alleles are present. Their physical presence may be traced by restriction fragments separated by electrophoresis because the DNA is independent of extrinsic factors. Their cosegregation is identified and can be used for improving quantitative traits for plant breeding purposes and for genetically defining behavioral traits and other polygenic characters. QTLs can be genetically mapped by several procedures, most commonly using the principles of maximum likelihood for statistical analysis. In a backcross generation, the phenotype (ϕ_i) and genotype (g_i) relations are expressed as $\phi_i = \mu + bg_i + \varepsilon_i$, where g_i corresponds to the homozygous and heterozygous dominants of the QTL, and its value may vary between 1 and 0. The mean of ε_i (a random variable) = 0 and its variance is σ^2. The values of μ, b, and σ^2 are unknown. The genotypic value of Qq and other contributors to the quantitative trait is μ, and b is the effect of a substitution of another allele at the quantitative trait loci. The statistical procedures shown below were adapted from Arús, P. & Moreno-González, J. 1993. *Plant Breeding*, pp. 314, Hayward, M. D., et al. eds. Chapman & Hall, London, New York:

The likelihood function

$$Lg_i(\mu, b, \sigma^2) = \frac{1}{\sqrt{2\pi\sigma^2}} e^{-\frac{(\phi_i - \mu - bg_i)^2}{2\sigma^2}}$$

and the likelihood that all individuals will be in the flanking parental marker classes (k) $M_1M_1M_2M_2$, $M_1M_1M_2m_2$, $M_1m_1M_2M_2$, and $M_1m_1M_2m_2$ will be L_k $(\mu, b, \sigma^2) = \Pi_i$

$[P_i(1)L_i(1) + P_i(0)L_i(0)]$, where $P_i(1)$, and $P_i(0)$ are the probabilities that QQ and Qq quantitative genes will be in the recombinant classes, respective of the flanking markers concerned. The maximum likelihood estimates for incomplete data can also be determined (Dempster, A. P., et al. 1977. *J. R. Stat. Soc.* 39:1). The likelihood for all observations is $L(\mu, b, \sigma^2) = \Pi_k L_k(\mu, b, \sigma^2)$. For the determination of the LOD score, to ascertain that the information obtained is real rather than a false, spurious conclusion, the following equation has to be resolved:

$$\text{LOD} = \frac{\log L(\mu, b, \sigma^2)}{L(\mu_0, b_0, \sigma_0^2)}$$

One must keep in mind that the estimates are as good as the data collected. Large populations and genes with greater quantitative effects improve the chances of finding linkage. For estimating linkage information from multiple marker data, computer assistance is required; various programs are available. For mapping QTL in humans, generally sib pairs are used. Statistical analysis indicates that choosing extreme discordant pairs makes the analysis more efficient. Just because of this selection, QTL estimates may be loaded with errors of under- and overestimation. Allison, D. B., et al. (2002. *Am. J. Hum. Genet.* 70:575) discuss methods useful for eliminating errors. Although most commonly single quantitative traits are analyzed at a time, the simultaneous study of multiple traits may be desirable. Quantitative traits generally are not expressed in isolation and we may have to face epistasis of multitraits. It is highly desirable that QTL would be amenable to isolation and cloning. *See* Frary, A., et al. 2000. *Science* 289:85 to better understand their function. Least squares, Bayesian, and nonparametric methods are also available. *See* ASP analysis; Bayes' theorem; BLUP; bootstrap; complex inheritance; cosuppression; gene block; infinitesimal model; interval mapping; least squares; liability; LOD score; mapping, genetic; nonparametric tests; RFLP. (Darvasi, A. 1998. *Nature Genet.* 18:19; Kao, C.-H. 2000. *Genetics* 156:855; Mackay, T. F. C. 2001. *Nature Rev. Genet.* 2:11; Flint, J. & Mott, R. 2001. *Nature Rev. Genet.* 2:437; Mackay, T. F. C. 2001. *Annu. Rev. Genet.* 35:303; Dekkers, J. C. M. & Hospital, F. 2002. *Nature Rev. Genet.* 3:22; Korstanje, R. & Paigen, B. 2002. *Nature Genet.* 31:235; Feingold, E. 2002. *Am. J. Hum. Genet.* 71:217.)

quadrant Consists of four parts, e.g., a tetrad.

quadratic check Used for testing two genes presumed to be required for phytopathogenic infection in the manner shown below:

	Low Pathogenicity	High Pathogenicity
Plant Reaction Low →	NO INFECTION	INFECTION
Plant Reaction High →	INFECTION	INFECTION

Resistance in the plants is usually a dominant trait. *See* Flor's model.

quadriradial chromosome   May be produced by cross-linking mutagens.

Quadriradial.

quadrivalent   Four partially or completely identical chromosomes in a polyploid that display pairing, although of the four, at any particular position, only two can be synapsed. During meiosis they show quadrivalent association of four chromosomes. *See* bivalent; meiosis; synapsis.

quadroma (hybrid hybridoma)   Fusion product of two hybridomas. *See* hybridoma. (Lindhofer, H., et al. 1995. *J. Immunol.* 155:219.)

quadruplets   Four-fold twins. In the absence of the use of fertility increasing treatment, their expected frequency is about 1 in $(89)^3$, whereas the expectation for triplets and quintuplets is about $(89)^2$ and $(89)^4$, respectively. *See* twinning.

quadruplex   Tetraploid or tetrasomic with four doses of the dominant alleles at a locus. *See* autopolyploid; G quartet.

quadruplex DNA   Has four parallel and antiparallel strands in vitro and blocks replication. (Schaffitzel, C., et al. 2001. *Proc. Natl. Acad. Sci. USA* 98:8572.)

quadruplicate genes   Four genes conveying identical or similar phenotype but segregating independently in F_2 and displaying a dominant recessive proportion of 255:1.

quadrupole   Used in specific mass spectrometers for the tracking of ion density in proteomic analysis. The quadrupole is made up of four rods that permit the filtering of the mass that traverses them with the mediation of an oscillating electric field to obtain the mass spectrum. The amplitude of the electric field is scanned and recorded. Peptides are usually analyzed by triple quadrupole devices. The unit mass resolution is excellent with an accuracy of 0.1 to 1 Da. *See* MALDI; mass spectrum; proteomics. (Hager, J. W. 2002. *Rapid Commun. Mass Spectrom.* 16:512.)

quantitative gene numbers   Difficult to determine because environmental effects obscure the impact of genes with minor effects. Several statistical procedures have been worked out for approximation. The simplest one is as follows:

$$N = \frac{R^2}{8(s_1^2 - s_2^2)}$$

where N = gene number, R = the difference between the parental means, s_1^2 = variance of the F1, and s_2^2 = variance of the F2 generations. The most common view is that quantitative traits are determined by a large number of genes and each of them contributes only a little to the observed phenotype. The association between bristle numbers (a quantitative trait) and the *scabrous* locus of *Drosophila* indicated, however, that approximately 32% of the genetic variation in abdominal and 21% of the sternopleural bristle number was associated with DNA sequence polymorphism at this single locus. *See* gene number; QTL; quantitative trait. (Mather, K. & Jinks, J. L. 1977. *Introduction to Biometrical Genetics*. Cornell Univ. Press, Ithaca, NY; Jones, C. D. 2001. *J. Hered.* 92[3]:274.)

quantitative genetics   Study of genetic mechanisms involved with the expression of quantitative traits. Its techniques involve those of population genetics and biometry. *See* biometry; heritability; population genetics; QTL; quantitative trait; selection; statistics.

quantitative trait   Shows continuous variation of expression and can be characterized by measurement or by counting, in contrast to qualitative traits, which can be identified satisfactorily by simple description such as black or white. *See* dichotomous trait; gene titration.

quantitative trait loci   *See* QTL.

quantum   Unit to quantify energy. *See* photon.

quantum dot   Built of semiconductor, luminescent nanometer-size crystals (e.g., of zinc sulfide–capped cadmium selenide). They may bind organic and macromolecules and thus permit their tracing as stable, very bright, water-soluble, and noninvasive labels. *See* luminescence; nanocrystal semiconductor; nonisotopic labeling; semiconductor. (Han, M., et al. 2001. *Nature Biotechnol.* 19:631.)

quantum speciation   Rapid formation of a new species by selection and genetic drift. *See* genetic drift; selection.

quarantine   State of isolation and observation without any external contact, especially from infection.

quarter-power scaling   Biological scaling can be expressed by the formula $Y = Y_0 M^b$, where Y is a variable (e.g., life span or metabolic rate), Y_0 is a normalization constant, b is a scaling exponent, and M is body mass. Y_0 varies with the trait and type of an organism, b is practically constant 1/4 or multiples of it, e.g., blood circulation time and life span are $M^{1/4}$, whole organism metabolic rate is $M^{3/4}$, diameter of tree trunks and aortas $M^{3/8}$, etc. (West, G. B., et al. 1999. *Science* 284:1677.)

quartet Structure consisting of four elements. *See* G quartet.

quasi In various combinations, indicates almost, e.g., quasispecies means that its difference from other forms may not qualify it for the status of a separate species with certainty.

quasidominant Recessive inheritance is misclassified as dominant because the mating took place between a heterozygote and a homozygous recessive individual.

quasilinkage *See* affinity.

quasimonoclonal antibody Produced by mice heterozygous for the V(D)J IM-imunoglobulin heavy chain (Ig), and the other allele is nonfunctional. The functional κ-chain is also missing. When the heavy chain specific for the hapten 4-hydroxy-3-nitrophenyl acetyl could join any λ-chain, the antibody was monospecific, but somatic mutation and secondary rearrangements changed the specificity of 20% of the B-cell antigen receptors. Such a system can thus be used to study antibody diversity. *See* antibody.

quasispecies Small degree of genetic (nucleic acid) variation does not qualify it clearly for separate species status.

quaternary structure Aggregate of multiple polypeptide subunits into a protein or by cross-linking DNA strands into a joint structure. *See* cross-linking; protein structure.

queen Reproductive female in cast insect colonies such as exist in bees and ants. *See* pseudoqueen.

Queen Victoria *See* hemophilias; Romanovs, photo below.

quelling Gene or chromatin repeat–associated silencing of genes without methylation. It occurs when foreign DNA is introduced into plants or fungi by transformation. *See* cosuppression; iRNA; MSUD; sense suppression; silencer; transvection. (Maine, E. M. 2000. *Genome Biol. Rev.* 1[3]:1018.)

quenching Suppression of fluorescence, transfer of electrons or suppression of an activator by blocking the binding site of the activator or binding it to another protein, which prevents its binding to the activator-binding site in DNA.

Quételet index Essentially the same as body mass index: (weight)/(height)^2. L. A. J. Quételet, an astronomer and pioneer of biometry, showed in 1835 that human stature follows the normal distribution. Old textbooks of genetics referred to the principle of normal distribution of quantitative traits as Quételet's Law. According to the Quételet-Galton Law, when a quantitative trait did not follow the normal distribution, the role of heredity in the expression of the trait was questioned and the variation was attributed to environmental causes. *See* body mass index.

queuine Rare modified purine.

quick-stop Temperature-sensitive DNA replication mutant *dna* of *E. coli* stops DNA replication immediately when the temperature rises to 42°C from the permissive 37°C. *See* temperature-sensitive mutation. (Rangarajan, S., et al. 1999. *Proc. Natl. Acad. Sci. USA* 96:9224.)

quinacrine mustard (ICR 100) Light-sensitive, fluorescent compound used for chromosome staining. Quinacrine (atabrine) staining permitted the visualization of banding in human chromosomes. It caused particularly bright fluorescence of the long arm of the human Y chromosome and facilitated the recognition of the XYY karyotype. ICR 100 is

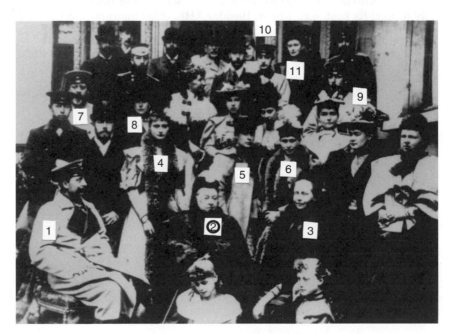

The most famous family of Queen Victoria of England, affected by hemophilia, at a reunion on April 23, 1894. (1) Kaiser Wilhelm II, Grandson, (2) **Queen Victoria**, (3) Daughter victoria, (4) Granddaughter Tsarina Alexandra, (5) Granddaughter Irene, (6) Granddaughter Alice, (7) Son and future king of England Edward VII, (8) Daughter Beatrice, (9) Son Arthur, (10) Granddaughter Marie, (11) Granddaughter Elizabeth. For the most likely genetic constitutions regarding hemophilia see under hemophilia. (Courtesy of the Humanities Research Center, Gernsheim Collection, University of Texas, Austin, TX.)

strongly mutagenic; it is also a highly toxic antihelminthic drug. *See* acridine dyes; Q banding.

1CR-stained Y chromosome.

Quinacrine mustard.

quintuplex *See* quadruplex.

quorum factors Signaling molecules in autoinduction. *See* autoinduction.

quorum sensing System of cell density–dependent expression of specific gene sets. In the luminescent bacteria *Vibrio fischeri* and *V. harveyi*, the quorum-sensing signal is acylated homoserine lactone (AHL). AHL triggers biofilm production in the infectious *Pseudomonas aeruginosa*, which protects the bacteria from antibiotics. AHL production is quite widespread among bacteria, including bacteria living on plant hosts. *P. aeruginosa*—in response to AHL—can produce, e.g., phenazine (mutagen involved in electron transport), an antibiotic that keeps away other bacteria and thus may protect its host, e.g., wheat. Besides AHL, other quorum-sensing signals have been detected in various bacteria and in fungi. *See* autoinduction; biofilm; multicellular. (Miller, M. B. & Bassler, B. L. 2001. *Annu. Rev. Microbiol.* 55:165; Fuqua, C., et al. 2001. *Annu. Rev. Genet.* 35:439.)

q.v (quod vide)   See it.

The majority of geneticists know that Carl Correns was one of the three rediscoverers of the Mendelian principles in 1900. Actually he named them as Mendel's Rules, and reported linkage in *Matthiola*. He was also one of the discoverers of cytoplasmic inheritance. In 1902 (Botanische Zeitung 60:64–82) he suggested a mechanism for crossing over nine years before Morgan's paper appeared in J. Exp. Zool. (11:365).

"We assume that in the same chromosome the two Anlagen of each pair of traits lie next to each other (A next to a and B next to b, etc.) and that the pairs of Anlagen themselves are behind each other. The picture is shown in Fig. 1. A, B, C, D, E, etc. are the Anlagen of parent I; a, b, c, d, e, etc. are those of parent II.

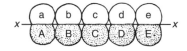

Through the usual cell- and nuclear divisions the same type of products are obtained as the chromosomes split longitudinally..."

When one pair contains antagonistic Anlagen, while the rest of the pairs are formed of two identical types of Anlagen, or the Anlagen are 'conjugated' as they are in *Matthiola* hybrids, which I have described, then further assumptions are *necessary...* Then AbCdE/aBcDe and aBcDe/AbCdE yield both AbCdE and aBcDe; ABcdE/abCDe and abCDe/ABcdE both ABcdE and abCDe, etc."

A historical vignette.

R (r, Röntgen, Roentgen) Unit of ionizing radiation (1 electrostatic unit of charge in $1 \, cm^3$ dry air at $0°C$ and 760 mm pressure; about 93 ergs/living cells). *See* cR; Gy; Rad; Rem; rep; Sv.

R1 Methylation sites in the cytoplasmic region of *E. coli* chemotaxis transducer proteins.

R1, R2 Ubiquitous retroposons in arthropod ribosomal RNA. *See* retroposon. (Perez-González, C. E. & Eickbush, T. H. 2002. *Genetics* 162:799.)

R2 *See* hybrid dysgenesis I–R.

*r*II *See* rapid lysis mutants of bacteriophages.

RA Rheumatoid arthritis. *See* autoimmune disease; rheumatoid fever.

rAAV Recombinant adeno-associated virus. *See* adeno-associated virus.

RAB RAS oncogene homologues (a guanosine triphosphatase) that regulate transport between intracellular vesicles (Golgi apparatus) and control endosome fusion. RAB was located to human chromosome 19p13.2. Rab3A functions in neural synaptic vessels. *See* EEA1; endosome; Golgi apparatus; Griscelli syndrome; GTPase; Mss; NSF; RNA export; RAS oncogene; Sec; SNARE; synaptic vessel; Ypt. (Zerial, M. & McBride, II. 2001. *Nature Rev. Mol. Cell Biol.* 2:107.)

rabbit *Oryctolagus cuniculus*, $2n = 44$; *Sylvilagous floridanus*, $2n = 42$.

rabbit reticulocyte in vitro translation Mammalian mRNA (extracted from cells or transcribed in vitro) can be translated into protein under cell-free conditions using lysates of immature red blood cells of anemic rabbits. Anemia is induced by subcutaneous injection of the animals for 5 days by neutralized 1.2% acetylphenylhydrazine solutions (HEPES buffer). After the larger white blood cells are removed by centrifugation, the red blood cells are lysed at $0°C$ by sterile double-distilled water. Then the endogenous mRNA is destroyed by micrococcal nuclease in the presence of Ca^{2+}. Without calcium, the nuclease does not work. The reaction is stopped by EGTA (ethylene glycol tetraacetic acid, which chelates calcium). Hemin ($C_{34}H_{32}ClFeN_4O_4$), dissolved in KOH, is needed for suppressing an inhibitor of eukaryotic translation initiation factor eIF-2. The translation mixture must contain spermidine or RNasin ribonuclease inhibitors, creatine phosphate (an energy donor), dithiothreitol (a reducing agent to prevent the formation of sulfoxides from the S-labeled amino acids), all normal amino acids (except the one that will carry the radioactive label), buffer, the radioactive amino acid (e.g., [^{35}S]methionine), the reticulocyte lysate, tRNAs, KCl and magnesium acetate (to enhance translation), and polyadenylate-tailed mRNA (to be translated into protein). All solutions must be made up with RNase-free material and the vessels should be made RNase-free. Incubation is at $30°C$ for 30 to 60 minutes. Before precipitating (by 10% trichloroacetic acid) the synthesized protein, the ^{35}S-methionine-tRNA is destroyed either by 0.3 N NaOH, or in case SDS-polyacrylamide gels are used for subsequent analysis, by pancreatic ribonuclease. Immunoprecipitation may also be employed for the analysis of the translation product. The amount of synthesized protein can be measured by scintillation counting. Rabbit reticulocyte lysates are also available commercially. Numerous variations of the procedure are available in laboratory manuals. Alternatively, wheat germ extract may be used for in vitro translation. *See* eIF-2; immunoprecipitation; polyA mRNA; scintillation counters; SDS polyacrylamide gels; translation repressor proteins; wheat germ translation system. (Olliver, L. & Boyd. C. D. 1998. *Methods Mol. Biol.* 86:221; Lorsch, J. R. & Herschlag. 1999. *EMBO J.* 18:6705.)

rabies Encephalomyelitis caused by an infection of a nonsegmented negative-strand RNA virus. The disease generally starts by inflammation and hyperactivity; death eventually follows. A wide range of wild (raccoons, foxes, mice) and domestic animals (dogs, cats) are susceptible to infection through saliva in bites. Humans are relatively resistant; there is only about 15% lethality without treatment. For prevention, attenuated virus or genetic immunization may be used. *See* encephalomyelitis; genetic immunization; positive strand; replicase; RNA viruses; segmented genomes; vaccination.

Rabl orientation Evidence by K. Rabl in the late 1800s for the continuity of the chromosomes inasmuch as the chromosomes emerge in very early prophase from premeiotic interphase in the same configuration as they entered anaphase and telophase, namely, the centromeres face the centrioles in close proximity. *See* coorientation.

Rabson-Mendenhall syndrome (19p13.2) Dominant insulin receptor defect (degradation) resulting in insulin-resistant diabetes. Hypertrophy of the pineal gland, dental and skin anomalies (acanthosis nigricans), and early lethality characterize the condition. *See* brain; diabetes; insulin receptor protein; pineal gland.

RAC (1) (same as Akt or PKB) Serine/threonine kinase member of the RAS protein family; transmits signals from the cell surface membrane to the cytoskeleton. When activated, RAC inhibits transferrin receptor-mediated endocytosis and regulates, with RHO, the formation of clathrin-coated vesicles and actin polymerization. It has an important role in RAS-mediated oncogenic transformation. RAC activates NADPH oxidase and thus free radical production as a defense against infections. Rac2 guanosine triphosphatase is selectively expressed in T_H1 lymphocytes (mediating cellular immunity) and in cooperation with NF-κB it induces IFN-γ promoter. Rac mediates the progression of the cell cycle. *See* cell membrane;

clathrin; cytoskeleton; endocytosis; IFN-γ; NF-κB; oxidative burst; p35; Pac; PAK; PKB; RAS; RHO; ROS; serine/threonine kinase; signal transduction; T cell; transferrin. (Mettouchi, A., et al. 2001. *Mol. Cell* 8:115; Hakeda-Suzuki, S., et al. 2002. *Nature* 416:438.) (2) Recombinant DNA Advisory Committee of the National Institute of Health (USA) overseeing the application of recombinant DNA technology.

raccoon (*Procyon lotor*) $2n = 38$.

race   Group within a species distinguished by several characteristics, such as allelic frequencies, morphology, etc. *See* ethnicity; evolutionary distance; human races; racism.

RACE   Rapid amplification of cDNA ends by PCR. *See* polymerase chain reaction. (Schafer, B. C. 1995. *Anal. Biochem.* 227:255.)

racemate   Mixture of D and L optical stereoisomers (enantiomorphs); the mixture then becomes optically inactive. All naturally synthesized amino acids are in the L form, but degradation generates the D enantiomorph. The degree of racemization of aspartic acid is faster than that of other amino acids. It has been used to determine the authenticity of ancient samples of DNA because the degradation of DNA and the racemization of amino acids, particularly Asp, indicates whether the spurious DNA is really ancient or just contaminant in the archaeological sample. In case the D/L Asp ratio exceeds 0.08, ancient DNA cannot be retrieved. The degradation depends on a number of factors, most notably the temperature to which the specimen had been historically exposed. The best preservation takes place in insects enclosed in amber (although there is some controversy about these samples). In specimens where the D/L Asp ratio was about 0.05, up to 340-bp-long DNA sequences could be detected using PCR technology. *See* carbon dating; enantiomorph; evolutionary clock.

raceme   Inflorescence with elongated main stem and with flowers on near equal-size pedicels.

Raceme.

rachis   Axis of a spike (grass ear) and fern leaf (frond).

racial distance   *See* evolutionary distance.

racism   Assumption of superiority of any particular ethnic group or groups and the consequently inferiority of some others. It advocates hatred and social discrimination on the basis of differences. The origin of racism can be traced back to prehistorical times for the purpose of exploitation of conquered or minority groups. Some forms of racism may be found in nearly all societies of Caucasians, Chinese, Japanese, blacks, etc.; even the Bible is not exempt from racist ideas, and

it has been frequently used as a justification by bigots. In the 19th century, the rise of the eugenics movement gave false scientific encouragement to racism, providing biological and ideological support for colonialism and social exploitation. Racist ideas were used to justify slavery. Racism culminated in the Third Reich of Hitler's Germany, resulting not just in discrimination and suppression but physical mass elimination of "non-Aryan" people and an aim to establish *Rassenhygiene* (race hygiene). Racism cannot be justified on the basis of any scientific evidence and it is morally unacceptable to enlightened societies. Human racial differences are based on a very limited number of genes, and the vast majority of genes are shared by all racial groups. Actually, the world's most successful societies exceled because of their multiracial and multicultural composition. Ample biological evidence supports the superiority of hybrids of mammalian and plant species. *See* eugenics; evolutionary distance; human intelligence; human races; hybrid vigor; miscegenation.

RACK   Receptor for activated C kinase. *See* C kinase.

RAD   Unit of ionizing radiation absorbed dose (100 ergs/wet tissue). *See* Gray; r; rem; Sievert.

RAD   Genes of yeast are involved in DNA repair and recombination. *See* ABC excinucleases.

RAD1   Yeast gene involved in cutting of damaged DNA in association with *RAD10* (ERCC1); its human homologue is XPF/ERCC4. *See* mismatch repair.

RAD2   Yeast gene involved in cutting DNA; its human homologue is XPG. *See* DNA repair.

RAD3   Yeast DNA helicase and a component of transcription factor TFIIH; its human equivalent is XPD. In yeast, *RAD3* regulates telomere integrity. A defect in a Rad3-like protein may be responsible for ataxia telangiectasia. *See* ataxia telangiectasia; DNA repair; telomeres.

RAD4   *See* RAD23.

Rad6   Protein with Ubc2 functions involved in both proteolysis and genetic repair in yeast. *See* DNA repair; Ubc2; ubiquitin.

RAD10   Yeast homologue of human gene ERCC1, participating in nucleotide excision repair. *See* mismatch repair; nucleotide excision repair.

RAD14   Yeast gene; its protein product binds to damaged DNA. The human homologue is XPA. *See* DNA repair.

Rad18   Yeast protein involved in genetic repair in association with Rad6. *See* Rad6.

RAD21   Controls double-strand break repair caused by ionizing radiation.

RAD23 Involved in nucleotide exchange repair. It interacts with the 26S proteasome by binding to the RAD4 repair protein. *See* DNA repair; proteasome; xeroderma pigmentosum.

Rad24 14-3-3 protein regulating nuclear export-import. *See* Chk1; protein 14-3-3.

RAD25 (ERCC) Helicase subunit of the general transcription factor TFIIH. It is credited with promoter clearance for the beginning of transcription after ATP hydrolysis and after the open promoter complex is formed. It is also a DNA repair enzyme. *See* DNA repair; helicase; open promoter complex; promoter clearance; regulation of gene activity; transcription factors.

RAD27/FEN1 (Rthp/Fen-1) 45 kDa $5' \to 3$ exonuclease/endonuclease removes the RNA primer from Okazaki fragments with the cooperation of other proteins such as RNA-DNA junction endonuclease, PCNA, and Dna helicase. The FEN-1/DNase IV protein of eukaryotes carries out the same functions as prokaryotic DNA polymerases beyond polymerization. For this function, eukaryotic cells rely on the PCNA-associated FEN-1. FEN-1 can also cut branched DNA molecules. *See* DNA replication in eukaryotes; flap nuclease; Okazaki fragment; PCNA. (Lieber, M. R. 1997. *Bioassays* 19:233; Debrauwère, H., et al. 2001. *Proc. Natl. Acad. Sci. USA* 98:8263.)

RAD28 Yeast homologue of the gene of Cockayne syndrome. *See* Cockayne syndrome.

RAD30 Encodes DNA polymerase η and it is homologous to *E. coli* DinB, UmuC, and *S. cerevisiae* Rev1.

RAD50, RAD51, RAD52, RAD53, RAD54, RAD55, Rad56, RAD57 Yeast genes involved in radiation sensitivity, DNA double-strand break, repair, and recombination. Rad51 and Rad52 proteins are the most essential for eukaryotic recombination. Overexpression of RAD51 and RAD52 reduces double-strand break-induced homologous recombination in mammalian cells (Kim, P. M., et al. 2001. *Nucleic Acids Res.* 29:4352). Replication protein A interacts with Rad proteins. The human breast cancer gene forms a complex with hRad50-p95-hMre11 proteins. Rad53, with chromatin assembly factor Asf1, mediates the deposition of acetylated histones H3 and H4 onto the newly replicated DNA. *See* chromatin assembly; DNA repair; nonhomologous end-joining; radiation sensitivity; recombination mechanisms, eukaryotes; replication; replication fork; replication protein A. (Masson, J.-Y., et al. 2001. *Proc. Natl. Acad. Sci. USA* 98:8440; Davis, A. P. & Symington, L. S. 2001. *Genetics* 159:515.)

RAD51 Gene of budding yeast regulates double-strand breaks and genetic recombination depending on ATP. The RAD51 protein bears similarity with the human protein (15q151) of similar functions. Disruption of *RAD51* in mice has embryonic lethal effects. *RAD51* is homologous with the bacterial gene *RecA* and bacteriophage T4 gene Uvsx, mediating strand exchange in genetic recombination. In the recombination function, RAD52 and its various yeast homologues (RAD55, RAD57, and other proteins) assist RAD1.

RPA (replication protein A) and its homologues, SSB (single-strand-binding protein) in bacteria and p32 protein in phage, prepare the broken ends of the DNA to find the proper sequences in the homologous chromosomes that may be suitable for joining. *See* Dmc1; DNA repair; RAD54; RecA1; recombination mechanisms in eukaryotes. (Fasullo, M., et al. 2001. *Genetics* 158:959; Yu, X., et al. 2001. *Proc. Natl. Acad. Sci. USA* 98:8419.)

RAD53 Yeast kinase gene encoding pRAD53 signal transducer and S-phase checkpoint controller; it is also called SAD1, MEC2, and SPK1. Rad53 is activated by a conserved protein, Mrc1 (mediator of replication checkpoint), in response to DNA damage. *See* DNA replication; MEC1. (Alcasabas, A. A., et al. 2001. *Nature Cell Biol.* 3:958.)

RAD54 Helicase function was suggested, but it appears that this protein is a DNA-dependent ATPase. It interacts with RAD1 scaffold and promotes homologous DNA pairing at the expense of ATP hydrolysis. *See* ATPase; helicase; RAD1. (Solinger, J. A. & Heyer, W.-D. 2001. *Proc. Natl. Acad. Sci. USA* 98:8447; Ristic, D., et al. 2001. *Proc. Natl. Acad. Sci. USA* 98:8454; Kim, P. M., et al. 2002. *Nucleic Acids Res.* 30:2727.)

radiation, acute The irradiation is delivered in a single dose at a high rate, in contrast to *chronic radiation*, when the same dose is administered during a prolonged time period. *See* physical mutagens; radiation effects.

radiation, adaptive *See* radiation, evolutionary.

radiation, background Includes all radioactive (ionizing) radiation in the environment arising from inadequately shielded X-ray machines, cosmic radiation, fallout, laboratory isotope pollution, radioactive rocks, or radon gas, increasing the dose delivered by medical treatment or other intended sources.

radiation, brain damage Actively dividing cells are most likely to suffer from ionizing radiation. The epidemiological data collected in the population of Hiroshima and Nagasaki indicated that the greatest susceptibility was during the first 8–15 weeks of the human fetus.

radiation, evolutionary Spread of taxonomic categories as a consequence of adaptation and speciation mediated by forces of selection, mutation, migration, and random drift.

radiation, indirect effects Radiation generates reactive radicals (e.g., peroxides) in the environment that in turn inflict biological damage. *See* radiation effects; target theory.

radiation, ionizing   *See* electromagnetic radiation; ionizing radiation.

radiation, natural   *See* cosmic radiation; isotopes.

radiation cancer   Many of the different cancers are associated with chromosomal rearrangement(s); ionizing radiation causes chromosomal breakage and rearrangements. Proximity of the X-ray-induced breakage sites favors rearrangements. Ultraviolet radiation may be responsible for the induction of skin cancer, especially if the body's genetic repair mechanism is weakened. According to estimates, 10 mSv may be responsible for 1 cancer death per 10,000 people. In the U.S., the permissible legal dose limit to the public is 0.25 mSv/year, but it should be reduced to 0.20 mSv/year. Ionizing radiation is used therapeutically against cancer. *See* DNA repair; excision repair; physical mutagens; radiation hazard assessment; radiation safety hazards; Sievert; ultraviolet radiation; xeroderma pigmentosum.

radiation chimera   (1) Antigenically different bone marrow transplant is harbored in a body after an extensive radiation treatment destroys or substantially reduces the immune reaction of the recipient. (2) Mutant sectors caused by radiation-induced mutations or deletions.

Chimeric dahlia flower, The consequence of radiation exposure. (Photograph of Dr. Arnold Sparrow. Courtesy of the Brookhaven National Laboratory, Upton, NY).

radiation, chronic   *See* radiation, acute.

radiation density   Generally measured by LET (linear energy transfer) values, i.e., the average amount of energy released per unit length of the tract. In case the density is low, the genetic damage is expected to be discrete. High LET radiation causes extensive damage along a very short path. *See* physical mutagens; radiation effects.

radiation doubling dose   *See* doubling dose.

radiation effects   Ionizing radiation may cause gross chromosome breakage (deletions, duplications, inversions, reciprocal translocations, isochromatid breaks, transpositions, change in chromosome numbers if applied to the spindle apparatus) or minute changes, including destruction of a single base in the nucleic acids or very short deletions involving only a few base pairs or oxidation of bases. The damage is often clustered. These effects depend on the quality of the ionizing radiation and the status of the biological material involved.

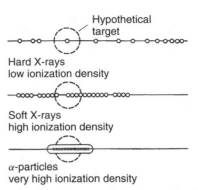

The pattern of ionization density along the track of hard and soft X-rays and α-particles. (After Gray, L. H. From Wagner, R. P. & Mitchell, H. K. 1964. *Genetics and Metabolism.* Wiley, New York.)

The physical effect of the radiation is frequently characterized by LET (linear energy transfer in keV/nm path), indicating the amount of energy released per unit tract as ionization and excitation in the biological target (*see* radiation density). Also, if the dose is delivered at a low rate, most of the damage may be repaired by the metabolic system of the cells. In prophase, the chromatid breaks may remain open for only a few minutes, but at interphase they may stay open substantially longer. The frequency of chromosome breakage is considerably increased in the presence of abundant oxygen, and anoxia has the opposite effect. Actively metabolizing cells are more susceptible to radiation damage than dormant ones (germinating seeds versus dry ones). Chromosomal aberrations requiring two breaks (e.g., inversions, translocations) occur by second-order kinetics, whereas the induction rate of point mutations displays first-order kinetics. Ultraviolet radiation causes excitation rather than ionization in the genetic molecules. The most prevalent damage is the formation of pyrimidine dimers, although chromosome or chromatid breaks may also occur. Radiation-induced malignant growth appears to be mediated by protein tyrosine phosphorylation followed by activation of the RAF oncogene. Suppressing cell proliferation due to interferon regulatory factor (IRF) that arrests the cell cycle may prevent accumulation of radiation-induced mutations. Independently from IRF, p53 may have a similar effect. Also, ionizing radiation may induce the transcription of p21, a cell cycle inhibitor, by a p53- and IRF-dependent mechanism. The adverse effect of radiation therapy may be circumvented by the application of pifithrin. The potential pathway of events after exposure to radiation is as follows: (1) initial hit; (2) excitation/ionization; (3) radical formation and other chemical reactions; (4) DNA/chromosome damage; (5) DNA repair or mutation; (6) cell death; (7) modifying physiological events; (8) teratogenesis/carcinogenesis/mutation fixation. Although for years radiation damage was attributed to damage within the exposed cell, evidence is accumulating (Azzam, E. I., et al. 2001. *Proc. Natl. Acad. Sci. USA* 98:473) for the damage to be communicated to neighboring cells through gap junctions with the mediation of connexin 43. As a consequence, in the neighboring cells the stress-inducible protein p21[Waf1] and genetic changes become detectable. The effect of radiation on a wide array of genes (Goss Tusher, V., et al. 2001. *Proc. Natl. Acad. Sci. USA* 98:5116) can be assessed by microarray hybridization. *See* Armitage-Doll model; DNA repair; doubling dose; interferon; ionizing radiation hazards; kinetics; Kudson's two-mutation theory; mental retardation; p21; p53; radiation hazard assessment; radiation measurement; RAF; signal transduction; target theory; ultraviolet light; UV; X-ray caused chromosome breakage. (Hollaender, A., ed. 1954–55. *Radiation Biology*. McGraw-Hill, New York, General References.)

radiation hazard assessment For human whole-body exposure of X- or γ-radiation, the minimal biologically detectable dose in mSv (milliSievert): no symptoms 0.01–0.05, chromosomal defects detectable by cytological analysis 50–250, physiological symptoms at acute exposure 500–700, vomiting in 10% of people 750–1,250, disability and hematological changes 1,500–2,000, median human lethal dose 3,000, but at doses above 4,500 the mortality is expected to be above 50%. Prolonged exposure below 1,000 Sv may cause leukemia and death. Single exposure of the spermatogonia to 0.5 Sv may block sperm formation. In mice, the LD_{50}/cGy was found for primary spermatocytes: 200; meiosis (leptotene–diplotene) 500; diplotene 800; diakinesis 900; secondary spermatocytes 1,000; spermatids 1,500; spermatozoa 50,000 (mouse data from Alpen, 1998). In mice, 1 Gy may kill the fertilized zygote, but the same dose may have no killing effect 5–8 days after conception, indicating the potentials of cell replacement. In humans, the most common response of the fetus is mental retardation by intrauterine exposure to 1–5 Gy. Carcinogenesis is well demonstrated by radiation exposure, but the course of the initial effect may be greatly affected by a variety of innate and environmental factors.

A single exposure of women to 3–4 Sv and 10–20 Sv over a longer time (2 weeks) may result in permanent sterility. A fetus in a pregnant woman should not be exposed to any radiation, but in case of an emergency, exposure should not exceed a total of 0.005 Sv; any person under age 18 should not receive an accumulated dose of more than 0.05 Sv/year. *Occupational exposure* maximal limits in mSv for whole body are 50, for lens of the eye 150, and for other specific organs or tissues 500. For cumulative exposure, the maximal limit should be below 10 mSv × age in years. For public exposure and educational and training exposure, the recommended maximal limit in mSv/year: for whole body 1; for eye, skin, and extremities 5.0. During spring 1996, the European Union (EU) lowered the permissible dose limits. Members of the public can be exposed to a maximum of 1 mSv each year (earlier limit was 5 mSv). Radiation industry personnel's limit of exposure is now 100 mSv in 5 consecutive years and an average limit of 20 mSv/year (previous limit was 50 mSv). These guidelines must be implemented within 4 years by EU member states.

The exposure by routine *medical* X-ray examination is supposed to be no higher than 0.04 to 10 mSv; by fluoroscopy or X-ray no higher than 25 mSv; by dental examination involving the entire jaws, not above 30 mSv. Nuclear medicine using radioactive tracers or positron emission tomography also involves some exposure. The replacement of ^{131}I by ^{123}I is desirable for thyroid analyses. Smoking tobacco may increase the exposure to ^{210}Pb and ^{210}Po. By comparison, in a normally operated nuclear power plant, the exposure for workers may be from 3 to 30 mSv/year. The exposure by living within a granite building may amount to 5 mSv/year and a transcontinental flight may involve an exposure of 0.03 mSv. Terrestrial radiation (^{40}K, ^{87}Rb, U, Th series, Rn) may also be a source to reckon with. A color television/video display set may deliver 0.001 mSv/year to the viewer if he/she stays

very close to the set; however, modern units are much safer (2–3 µSv/year). Airport luggage inspection may add 0.002 mrem, and smoke detectors 0.008 mrem to personal exposure. A plutonium-powered cardiac pacemaker may increase the radiation exposure of the wearer by 100 mrem. There might be no threshold below which ionizing radiation would have no effect. Current approximate incidence of mutation in live-born human offspring and the estimated increase per rem per generation in parentheses: autosomal dominant 0.0025–0.0075 (0.000005 – 0.00002), recessive 0.0025, X-linked 0.0004 (0.000001), translocations 0.0006 (0.0025), trisomy 0.0008 (0.000001). Approximately 5 Gy (500 rem) is considered the human lethal dose. The bacterium *Deinococcus radiodurans* can recover from doses as high as 30,000 Gy.

The very efficient recombinational repair in this prokaryote can explain this high radiation resistance. The chromosomes are just as well broken into pieces as other DNAs, but its genetic material exists in pairs, and in 12–24 hours, repair by recombination at the Holliday junctures restores their integrity. Even small doses, such as those delivered by therapeutic X-radiation, may increase by about one-third the number of broken chromosomes. Radiation by isotope treatment has a similar effect, depending on the dose and duration of the exposure. Eventually, the broken chromosomes, or at least some of them, may be eliminated from the body. Radiation sensitivity is generally positively correlated with the size of the genetic material, although in polyploids the damage may not be readily detectable because of gene redundancy. There is no universal consensus regarding the hazards involved in exposure to very low levels of ionizing radiation. *See* atomic radiation; BERT; DNA repair; doubling dose; Gy; hemiclonal; Holliday model of recombination; isotopes; mental retardation; radiation effects; radiation measurement; radiation protection; radiation sensitivity; radiation threshold; rem; Sv; X-ray chromosome breakage. (Dowd, S. B. & Tilson, E. R. 1999. *Practical Radiation Protection and Applied Radiobiology*. Saunders, Philadelphia [the book includes more than 100 Internet information addresses]; Sankaranarayanan, K. 1999. *Mutation Res.* 429:45; Mrázek, J. 2002. *Proc. Natl. Acad. Sci. USA* 99:10943; personal exposure: <www.umich.edu/~radinfo>.)

radiation hybrid (RH) Human chromosomes are broken into several fragments with 8,000 rad dose of X-rays. The irradiated cells are quickly fused to (with the aid of polyethylene glycol) somatic cell hybrids with Chinese hamster cells; thus, translocations and insertions into the hamster chromosomes are generated. The further apart are two human DNA markers, the higher the chance for a breakage. Thus, estimating the frequency of breakage, information is obtained about "recombination" in a manner analogous to classical genetic recombinational mapping. The recombination frequency in radiation hybrids varies between 0 and 1 (no recombination or the markers are always independent, respectively). In meiotic recombination, the maximal value is 0.5 for independent segregation. The formula and the recombination frequencies expressed in centiRays (cR) estimate the frequency of breakage. At 65 Gy, the estimated 1 cR ≈ 30 kb, at 90 Gy 1 cR ≈ 55 kb. $\theta = [(A^+B^-) + (A^-B^+)]/[T(R_A + R_B - 2R_AR_B)]$, where (A^+B^-) are the hybrid clones retaining A but not B and (A^-B^+) retain B but not A; T = the total number of hybrids and R_A, R_B stand for the recombinant fractions. The linkage analysis can be extended to more than two points. The retained fragments can also be analyzed by PCR procedures, and they are expected to carry neighboring sequences with them and thus provide information on physical linkage for sequences of about 10 megabases. The recombination process is dose-dependent. In one study, 50 Gy permitted the retention of an intact chromosome arm in 10% of the cases, whereas 40% had fragments of 3–30 Mb and 50% had fragments of 2–3 Mb. Using 250 Gy, less than 6% of the hybrids involved larger than 3 Mb pieces. If the generated fragments are intended for positional cloning, usually higher doses are used. The retention of fragments varies; centromeric pieces are more likely to be retained. The fragmented DNA can be further analyzed by probing with the aid of Southern blots, polymerase chain reaction, using sequence-tagged sites (STS), FISH, etc. The fragments are unstable unless they are fused with rodent chromosomes. The radiation hybrids may retain up to a dozen or slightly more fragments. An added special advantage is that genes without allelic variation can also be mapped. The radiation hybrid mapping method has been applied to plants. Single maize chromosome additions to oat lines permitted the resolution of 0.5 to 1.0 megabase sequences using 30 to 50 krad γ-rays. The hexaploid oat background assured the survival of (diploid) maize chromosome fragments. Radiation hybrid transcript maps are now available for mice, rats, humans, dogs, cats, and zebrafish and provide useful tools for evolutionary analyses. *See* addition lines; CentiRay; FISH; framework map; Gy; IRS-PCR; mapping, genetic; physical mapping; positional cloning; PRINS; probe; recombination; Rhalloc; Rhdb; RHKO; somatic cell hybrid; Southern blot; STS; θ (theta); WGRH. (Van Etten, W. J., et al. 1999. *Nature Genet.* 22:384; Riera-Lizarazu, O., et al. 2000. *Genetics* 156:327; Olivier, M., et al. 2001. *Science* 291:1298; Hudson, T. J., et al. 2001. *Nature Genet.* 29:201; Avner, P., et al. 2001. *Nature Genet.* 29:194; radiation hybrid map of the mouse genome: <http://wwww.ncbi.nlm.nih.gov/genemap>; <http://www. ebi.ac.uk/RHdb>.)

radiation hybrid panel Set of DNA samples containing radiation hybrid clones derived by fusion of human and rodent cells. *See* radiation hybrid; TNG.

radiation mapping *See* radiation hybrid. (<http://www-shgc.stanford.edu/ RH/>.)

radiation measurement *See* autoradiography; dosimeter film; dosimeter pocket; Geiger counter; ionization chambers; neutron flux detection; proportional counter; radiation hazard assessment; scintillation counter; thermoluminescent detectors.

radiation vs. nuclear size The harmful biological and genetic effects of ionizing radiations depend on the size of the cell nuclei, at the same dose. Larger nuclei present larger targets and suffer more damage than smaller ones. Haploid nuclei are more sensitive because the genes are generally present in a single dose. Polyploids are relatively less sensitive because of the multiple copies of the chromosomes. *See* ionizing radiation; physical mutagens; radiation effects; figure next page.

radiation particulate *See* physical mutagens.

radiation physiological factors The effect of radiation may be influenced by the species, age, developmental conditions, type of tissues and cells, metabolic state, genetic background, repair mechanisms, temperature, and chemical environment (presence of oxygen and other enhancing or protective compounds). Imbibed seeds, actively dividing cells, are much more sensitive to radiation damage than quiescent tissue or dry material. During pregnancy, radiation exposure should be avoided, especially during the early stages of gestation when cell division is most rapid. Developing children are at greater radiation risk than adults. Any condition with the possibility of diminished genetic repair increases the chances of chromosomal aberration and gene mutation. *See* physical mutagens; radiation brain damage; radiation hazard assessment; radiation sensitivity; radiation vs. nuclear size; target theory.

radiation protection Using isotopes, ventilation should be by exhaustion via seamless ducts through the buildings. Particulate material must be trapped in appropriate filters or washed (scrubbed) before environmental discharge. The source of radiation must be shielded. The transmission of the shielding material depends on the peak voltage of the X-ray or on the energy of the emitting isotope and the thickness of the shield. Attenuating material is characterized by half-value (HVL) or tenth-value layers (TVL).

Data on shielding effectiveness of commonly used radiation insulating material

kV X-ray	Lead in millimeters		Concrete in centimeters	
	HVL	TVL	HVL	TVL
50	0.05	0.16	0.43	0.15
100	0.24	0.8	1.5	5.0
200	0.48	1.6	2.5	8.25
500	3.6	11.9	3.5	11.5
1,000	7.9	26.0	4.38	14.5
4,000	16.5	54.8	9.00	30.00
10,000	16.5	55.0	11.5	38.25
^{60}Cobalt	6.5	21.6	4.75	15.50

Cracks, seams, conduits, filters, ducts, etc., should be kept under continued surveillance for possible leaks. Clothing (aprons, gloves, etc.) affords very limited protection. Radioactive waste must be disposed of according to government and local standards, whichever are stricter. Caution signs should identify radiation areas. "Radiation Area" is defined by the Occupational Safety and Health Standards of the U.S. as an area where a major portion of the body could receive in any hour a dose in excess of 5 millirem, or in any 5 consecutive days a dose in excess of 100 millirem. "High Radiation Area" means any accessible area to personnel in which there exists radiation at such levels that a major portion of the body could receive in any 1 hour a dose in excess of 100 millirem. For nonionizing electromagnetic radiation within the range of 10 MHz (megahertz, 10^7 cycles/sec) to 100 Ghz (gigahertz, 10^{11} cycles/sec), the energy density should not exceed 1 mW (milliwatt)/cm^2/0.1 hour. (Such radiations are within the realm of radio, microwave, and radar range.) Emergency plans must be prepared for spillage, cleanup, and fire. Never work

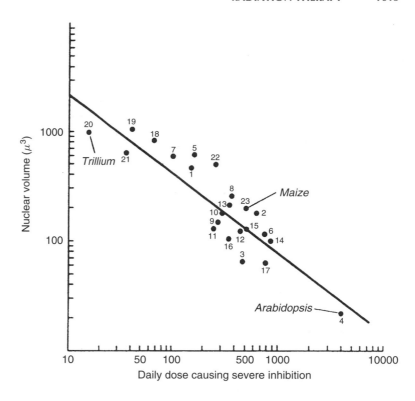

Radiation sensitivity in plants depending on nuclear volume. 1. *Allium cepa*, 2. *Anethum graveolens*, 3. *Antirrhinum majus*, 4. *Arabidopsis thaliana*, 5. *Brodiaea bridgesi*, 6. *Graptopetalum bartramii*, 7. *Haworthia attenuata*, 8. *Helianthus annuus*, 9. *Impatiens sultanii*, 10. *Luzula purpurea*, 11. *Nicotiana glauca*, 12. *Oxalis stricta*, 13. *Pisum sativum*, 14. *Raphanus sativus*, 15. *Ricinus communis*, 16. *Saintpaulia ionantha*, 17. *Sedum oryzifolium*, 18. *Tradescantia ohiensis*, 19. *Tradescantia paludosum*, 20. *Trillium grandiflorum*, 21. *Tulbaghia violacea*, 22. *Vicia faba*, 23. *Zea mays*. (From Sparrow, A. H., et al. *Radiation Bot.* 1:10.)

with any hazardous material before all personnel obtain sufficient training for safe storage, handling, and emergencies. *See* atomic radiation; electromagnetic radiation; ionizing radiation; isotopes; radiation effects; radiation hazard assessment; radiation measurement; sulfhydryl.

radiation-resistant DNA synthesis   Normally the ataxia telangiectasia gene/protein (ATM) is a target of the damaging effects of ionizing radiation. ATM activated by radiation results in the activation of the cell cycle (G1 → S) checkpoint kinase Chk2. Activation of Chk2 results in the phosphorylation of phosphatase Cdc25A at residue Ser[125]. Normally, Cdc25A prevents dephosphorylation of Cdk2, resulting in a transient stop of DNA synthesis at S phase. If Chk2 cannot bind or phosphorylate Cdc25A, radiation-resistant DNA synthesis proceeds. Thus, Chk2 acts as a tumor suppressor in ataxia telangiectasia. *See* ataxia telangiectasia; Cdc25; cell cycle; Chk2. (Falck, J., et al. 2001. *Nature* 410:842.)

radiation response   Deletions (single chromosomal breaks) and mutations occur with 1st-order kinetics, whereas the majority of chromosomal rearrangements (inversions, translocations) are proportional to the square of the dose and thus follow 2nd- or 3rd-order kinetics. *See* chromosomal aberrations; kinetics; oxygen effect; physical mutagens; radiation effects; radiation sensitivity; radiation vs. nuclear size.

radiation safety   *See* atomic radiation; cosmic radiation; radiation hazard assessment; radiation protection.

radiation safety standards   *See* radiation hazard assessment; radiation protection.

radiation sensitivity   Of DNA, depends on many diverse factors such as the degree of coiling, the status of the nuclear matrix, hydration, copper ions, OH scavengers, thiols, etc. In budding yeast, a genome-wide screen of 3,670 nonessential genes revealed 107 new loci that influence sensitivity to γ-rays; 50% of these yeast genes display homology to human genes. *See* aging; ataxia telengiectasia; chromosome breakage; *Deinococcus radiodurans*; DNA repair; microcephaly; nevoid basal cell carcinoma; radiation hazard assessment; radiation physiological factors; radiation protection; radiation response; radiation vs. nuclear size; xeroderma pigmentosum. (Bennett, C. B., et al., 2001. *Nature Genet.* 29:426.)

radiation sickness   Occurs when the human body or parts of it are exposed to ionizing radiation. The symptoms and hazards are dependent on dose. *See* radiation hazard assessment.

radiation therapy   Ionizing, electromagnetic radiation has antimitotic and destructive effects on live tissues and it is employed to suppress cancerous growth. The therapeutic effects of radiation in cancer therapy may not be a direct effect on the genetic material of the cancer cells. The effective target may be the surrounding (endothelial) cells that provide angiogenesis to satisfy the increased requirement of cancer cells for blood (Paris, F., et al. 2001. *Science* 293:293). Radiotherapy has been used to treat lymph nodes to suppress Hodgin's disease; radioactive isotopes can be applied by injection for localized radiation. Similar effects are expected by using magic bullets. Blood withdrawn from the body is irradiated by UV light and returned to the system. The level of radiocurability varies for tumors of different tissues from 2,000 to 3,000 rad for reproductive and nerve tissues, to 5,000 to 6,000 rad for lymphatic node and breast cancers, to 8,000 rad for melanomas and thyroid cancer. The proper function of the radiation source must be regularly monitored for the safety of patients and operators. *See* Hodgkin disease; lymph node; magic bullet; radiation effects; radiation hazard assessment. (Bharat, B., et al., eds. 1998. *Advances in Radiation Therapy*. Kluwer, Boston, MA.)

radiation threshold   The minimal harmful radiation dose is very difficult to determine because the visible physiological signs may not truly reflect the long-range mutagenic and carcinogenic effects. The maximal permissible doses for medical diagnostic or occupational radiation exposures reflect only conventional limits that have been revised many times as the sensitivity of physical and biological detection methods, as well as the instrumentation of delivery have improved. It is conceivable that no low threshold exists. *See* cosmic radiation; mutation frequency, undetected mutations; radiation hazard assessment; radiation response; radiation safety standards.

radical   Atom or group with an unpaired electron; a free radical.

radical scavenger   May combine with free radicals and reduce the potential harm caused by highly reactive molecules.

radical amino acid substitution   Occurs when nucleotide substitution alters the physicochemical property (e.g. charge, polarity, volume) of the mutant protein. By conservative substitution the physical/chemical characteristics are not altered. The proportion of radical/conservative ratios are positively correlated with the non-synonymous/synonymous replacements. Also, generally transversions are more likely to cause radical changes. *See* synonymous codons. (Zhang, J. 2000. *J. Mol. Evol.* 50:56; Dagan, T., et al. 2000. *Mol. Biol. Evol.* 19:1022.)

radicle   Seed or primary root of a plant embryo; the smallest branches of blood vessels and nerve cells.

Radin blood group (Rd)   Encoded in the short arm of human chromosome 1; its frequency is low.

radioactive decay   *See* isotopes.

radioactive isotope   *See* isotopes; radioactive label; radioactive tracer.

radioactive label   Compounds (nucleotides, amino acids) containing radioactive isotopes are incorporated into molecules to detect their synthesis, fate, or location (radioactive tracers) in the cells. Scintillation counters or autoradiography is used most frequently for detection. Geiger counters may also qualitatively detect the presence of these compounds.

For cytological analysis, isotopes are used that have a short path of radiation and display distinct, sharp marks on the film, e.g., tritium (^{3}H). For molecular biology, most commonly ^{32}P (in DNA, RNA), or ^{14}C and ^{35}S (in proteins), or ^{125}I (in immunoglobulins) is used. Radioactive labeling for genetic vectors or transgenes is not practical because at each subsequent division the label is diluted to about half. Labeling with integrated genetic markers is more useful because they are replicated along with the genetic material. *See* isotopes; nick translation; nonradioactive labels; Northern blotting; radioactive tracer; Southern blotting; Western blotting.

radioactive tracer   Radioactively labeled compound that permits tracing the biosynthetic transformations of the supplied chemical by determining when the radioactivity appears in certain metabolites after the supply. It also reveals what part of a later metabolite has acquired the label from the supplied substance. The availability of ^{14}CO$_2$ permitted tracing the path of photosynthesis, and through the use of various isotopes, the metabolism of pharmaceuticals and the role of hormones, etc., were determined. *See* radioactive label; radioimmunoassay.

radioactivity   Emission of radiation (electromagnetic or particulate) by the disintegration of atomic nuclei.

radioactivity dating   *See* evolutionary clock.

radioactivity measurement   *See* radiation measurements.

radioautography   Autoradiography.

radioimmunoassay (RIA)   The most commonly used isotope for radioimmunoassays, ^{125}I$^+$, is generated by oxidation of Na^{135}I by chloramine-T (N-chlorobenzene sulfonamide). It then labels tyrosine and histidine residues of the immunoglobulin without affecting the binding of the epitope and provides an extremely sensitive method for identifying minute quantities (1 pg) of the antigen in an experiment. ^{35}S-methionine or ^{14}C can radiolabel target proteins. In *competition RIA*, unlabeled target protein competes with a labeled antigen for binding sites on the antibody. The amounts of bound and unbound radioactivity are then quantified. In *immobilized antigen RIA*, unlabeled antigen is attached to a solid support and exposed to radiolabeled antibody. The amount of bound radioactivity measures the amount of specific antigen present in the sample. In *immobilized antibody RIA*, a single antibody is bound to a solid support and exposed to a labeled antigen. Again, the amount of bound radioactivity indicates the amount of the antigen present. In *double-antibody RIA*, one antibody is bound to a solid support and exposed to an unlabeled

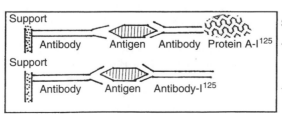

Support: diazobenzyoxymethyl cellulose or PVC

←Immobilized double antibody RIA

←Immobilized antigen RIA

antigen. After washing, the target antibody is quantitated by a second radiolabeled antibody (instead of the radiolabel, biotinylation can be used). This assay is very specific because it involves a step of purification. *See* antibody; immune reaction; immunoglobulins; isotopes; protein A. (Eleftherios, P., et al., eds. 1996. *Immunoassay*. Academic Press, San Diego, CA.)

radioisotope   *See* isotopes.

radioisotope dating   *See* fossil records.

radiomimetic   Agents, primarily alkylating mutagens, which may break single or both chromatids (isochromatids). Although it cannot be ruled out that the two chromatids break at the same place simultaneously, it is believed that isochromatid breaks are due to replicational events initially involving single-chromatid breaks. They mimic the effects of ionizing radiation. *See* alkylating agents; epoxides; ionizing radiation; nitrogen mustards; sulfur mustards. (Dustin, A. P. 1947. *Nature* 159:794.)

radiomorphoses   Morphological alterations in plants caused by ionizing irradiation during the life of the irradiated individuals. These effects may not be genetic.

radionuclide   Nuclide that may disintegrate upon irradiation by corpuscular or electromagnetic radiation. *See* nuclides.

radioprotectors   Protect against the harmful effect of ionizing radiation, e.g., sulfhydryl compounds, antioxidants, cysteine, cysteamine, amifostin, melatonin. (Maisin, J. R. 1989. *Adv. Space Res.* 9[10]:205.)

radiotherapy   *See* radiation therapy.

radish (*Raphanus sativus*)   Cruciferous vegetable crop; $2n = 2x = 18$. *See Brassica oleracea*; mustards; *Raphanobrassica*.

radix (root)   Multiplier of successive integral powers of a sequence of digits, e.g., if the radix is 4, then 213.5 means 2 times 4 to the second power plus 1 times 4 to the first power plus 3 times 4 to the zero power plus 5 times 4 to the minus 1 power.

radon (Rn) ^{219}Rn (An) is a member of the actinium series; ^{220}Rn (Tn) is a member of the thorium emanation group; ^{222}Rn is derivative of uranium, a heavy (generally accumulates in the basement), colorless radioactive noble gas formed from uranium (radium emanation)-contaminated rocks. It may pose health hazards in buildings with poor ventilation. In the U.S., an estimated 200 mrem is the average annual exposure/person due to radon. *See* cosmic radiation; radiation hazards; rem; WL. (Field, R. W. & Becker, K. 2001. *Radiat. Prot. Dosimetry* 95:75.)

raf   v-oncogene (cytoplasmic product is protein-serine/threonine kinase). The cellular homologue RAF1 is closely related to ARAF oncogenes. The v-raf is homologous to the Moloney murine leukemia virus oncogene. The avian MYC oncogene is the equivalent to the murine RAF. RAF1 has also been assigned to human chromosome 3, and a pseudogene, RAF2, is in chromosome 4. Human renal, stomach, and laryngeal carcinoma cells revealed RAF1 sequences. Raf may be recruited to the cytoplasmic membrane by a carboxy-terminal anchor (RafCAAX) and its activation is then independent from Raf, which is associated with the plasma membrane cytoskeletal elements, not with the lipid bilayer. RAF-1 phosphorylates MEK-1 kinase, which is involved in the signal transduction process of extracellular signal-regulated kinases. RAF induces NF-κB through MEKK1. Raf-1 is regulated by RKIP (Raf kinase-inhibiting protein). Phosphorylation at the appropriate sites activates/inactivates RAF. *See* ARA; MAP kinase; MEKK; Moloney; MYC; NF-κB; PAK; RAS; signal transduction; v-oncogene. (Chong, H., et al. 2001. *EMBO J.* 20:3716.)

RAF1   *See raf.*

raft   Association of sphingolipids and cholesterol ($\sim$50 nm in diameter) that mediates membrane traffic and cell signaling in mammals. Lipid rafts incorporate glycosylphosphatidyl inositol-anchored proteins, doubly acylated peripheral membrane proteins, cholesterol-linked proteins, and transmembrane proteins. Each raft carries no more than 10 to 30 proteins. *See* caveolae; cholesterols; protein transport; SFK; sphingolipids; TOR. (Langlet, C., et al. 2000. *Curr. Opin. Immunol.* 12:250; Brown, D. A. & London, E. 2000. *J. Biol. Chem.* 275:17221; Simons, K. & Toomre, D. 2001. *Nature Rev. Mol. Cell Biol.* 2:216.)

RAFTK   *See* CAM.

RAG1, RAG2   Recombination activating genes are closely linked (11p13) and encode the proteins of lymphocyte-specific recombination of the V(D)J sequences of the immunoglobulin genes. The functional part of RAG1 is the core sequence, whereas other tracts can be deleted without affecting recombination. Mutation in RAG results in inability to form functional antigen receptors on B and T cells and antibodies. The recombinational cleavage takes place between the *coding sequence* of the immunoglobulin genes and the so-called *recombinational signal sequence* (RSS) nucleotides. The recombination usually happens between the original coding sequences or it may take place by the rejoining of one coding end, the signal sequence, which originally belonged to the other coding end (hybrid joint), or the same coding and signal sequence end can be reunited (open-and-shut joint). Besides the RAGs, the joining reaction requires the double-strand repair protein XRCC4, a DNA-dependent protein kinase, and the Ku protein(s). An accessory protein, HMG1 or HMG2, is also required for recombination. Nucleotides may be added or deleted in each type of joining. RAG1 and RAG2 are actually transposons, but after the joining of the V(D)J ends, the RAGs are inactivated normally after the formation of the antibody genes. The RAG proteins seem homologous to the Tc1 transposon of *Caenorhabditis*. The RAG1 and RAG 2 genes are coordinately regulated by cell-type-specific elements upstream of RAG2. The 5′-promoter-upstream sequences in RAG2 regulate B and T cells differently. For the T cells, there are four T-cell receptors (TCR), and for the B cell to recombine there are three immunoglobulin loci. *See* antibody; combinatorial diversification; DNA-PK; hybrid dysgenesis;

immunoglobulins; junctional diversification; Ku; reticulosis, familial histicytic; RSS; TCR; V(J)D recombinase; XRCC4. (Schultz, H. Y., et al. 2001. *Molecular Cell* 7:65; Qiu, J.-X., et al. 2001. *Molecular Cell* 7:77; Raghavan, S. C., et al. 2001. *J. Biol. Chem.* 276:29126; Jones, J. M. & Gellert, M. 2001. *Proc. Natl. Acad. Sci. USA* 98:12926.)

RAGE Receptor for advanced glycation end product is a member of the immunoglobulin protein family on the cell surface. It interacts with multiple ligands and mediates homeostasis, development, inflammation, tumor proliferation, and the expression of some diseases (diabetes, Alzheimer disease). RAGE is also a receptor for amphoterin. In cooperation with p21^{Ras}, MAP, NF-κB, CDC42, SAP/JNK, and p44/p42 may alter cellular programming. Blocking the major factor complex of RAGE-amphoterin may thus prevent invasiveness of cancer and metastasis.

ragweed Mainly annual species of the genus *Ambrosia elatior* (Compositae) that is widespread in North America and Central Europe and causes pollen allergy (hay fever) of the nose and eye, skin irritations, and even asthma without hay fever. The susceptibility is genetically controlled by a locus, *Ir*, within the HLA complex. The antigen E contained in the ragweed pollen elicits the IgE antibody. *See* allergy; atopy; HLA; ragweed inflorescence.

RAIDD Adaptor protein that joins the ICE/CED-3 apoptosis effector molecules. *See* apoptosis; ICE.

RALA RAS-like protein, encoded in human chromosome 7. *See* RAS oncogene.

RALB RAS-like protein, encoded in human chromosome 13. *See* RAS oncogene.

raloxifene Lipid-like molecule that enters the cell nucleus, can bind to the special raloxifene-response element in DNA, and activates the tumor growth factor β3 gene. 17-epiestriol, an intermediate in the secretion of estrogen, activates raloxifene. Raloxifene is an antiestrogenic compound used in chemotherapy of breast cancer. Its advantage over the related compound tamoxifen is that it does not pose an appreciable risk for uteral cancer. *See* breast cancer; estradiol; estrogen receptor; hormone-response elements; tamoxifen. (Greenberger, L. M., et al. 2001. *Clin. Cancer Res.* 7:3166.)

ram Ribosomal ambiguity mutation causing a high rate of translational error. *See* ambiguity in translation.

RAM *See* rabbit antimouse immunoglobulin; random access memory in the computer.

Raman spectroscopy Similar to infrared spectroscopy, it uses the 10,000 to 1,000 nm spectral regimes to detect different rotational and vibrational states of molecules. If two atoms are far apart, their interaction is negligible. If they are very close, they may show repulsion. It is useful to obtain information on the physical state of nucleic acids. *See* FT-IR. (Thomas, G. J., Jr. 1999. *Annu. Rev. Biophys. Biomol. Struct.* 28:1.)

ramet Clonal descendant of a plant capable of independent reproduction. *See* genet.

ramie (*Boehmeria nivea*) Subtropical-tropical, monoecious fiber plant; $2n = 2x = 14$.

ramus Branch; it is used in various word combinations.

RAN RAS-like nuclear G protein (TC4). Guanosine triphosphatase (GTPase) required for import and export through nuclear pores and activation of the mitotic spindle. It participates as a switch of GTP → GDP in DNA synthesis and cell cycle progression. The proper balance between RanGTP and RanGDP determines the nucleo-cytoplasmic traffic. Cytoplasmic RanGTP may inhibit transport to the nucleus. The binding protein RanBP enhances the activity of RanGTP-activating protein RanGAP. RanBP1 is a nucleus-localized transporter binding well to RanGTP and weekly to RanGDP. RanBP2 is localized primarily in the cytoplasmic fibers of the nuclear pore complex. RAN also monitors the integrity of the tRNAs before their export to the cytoplasm. RAN processes the 7S RNA into the 5.8 ribosomal RNA of eukaryotes. *See* cell cycle; export adaptor; GAP; GTPase; importin; meiosis; nuclear localization signal; nuclear pore; Pat1; RAS; RCC; RNA transport; signal transduction; TPX. (Nachury, M. V., et al. 2001. *Cell* 104:95; Carazo-Salas, R. E., et al. 2001. *Nature Cell Biol.* 3:228; Seewald, M. J., et al. 2002. *Nature* 415:662.)

Rana Genus of frogs. Both the American leopard frog (*Rana pipiens*) and the European frog *R. temporaria* have $2n = 26$ chromosomes. The frogs are generally a more aquatic species than the toads. *See* Bufo; frog; toad; Xenopus.

random access memory (RAM) Storage of information that can be referred to at any order as long as the computer is turned on.

random amplified polymorphic DNA *See* RAPD.

random chromosome segregation The gene under study is absolutely or very closely linked to the centromere in polyploids. *See* maximal equational segregation.

random fixation *See* founder principle; random genetic drift.

random genetic drift   Change in gene frequency by chance. Such random changes are most likely to occur when the effective size of the population is reduced to relatively few individuals. Since random drift may occur repeatedly, eventually large changes may result in the genetic constitution of the population. Such changes are most likely when a few individuals migrate to a new (isolated) habitat. *See* effective population size; founder principle. (Whitlock, M. C. 2000. *Evolution Int. J. Org. Evolution* 54:1855.)

random mating   Each individual in the population has an equal chance to mate with any other of the opposite sex (panmixis). Random mating is assumed in the majority of principles of theoretical population genetics. The rules are based on the Hardy-Weinberg theorem, and on that basis the genetic structure of the population can be predicted as long as the allelic frequencies do not change or the change is negligible. Allelic frequencies may be altered primarily by selection, by migration, and if the size of the population is very small, by random genetic drift. In the short run, mutation does not affect allelic frequencies because of its rarity and because the chance of survival of the majority of new mutation is dim. In a population involving two different allelic pairs, the frequency of the mating genotypes and the genotypic proportion of their progenies derived from the binomial distribution by expanding $(p + q)^4$ are as follows:

MATES → (A1A1) × (A1A1)   (A1A1) × (A1A2)   (A1A2) × (A1A2)
(A1A2) × (A2A2)   (A2A2) × (A2A2)   (A1A1) × (A2A2)

Frequency → p^4	$4p^3q$	$6p^2q^2$	$4pq^3$	q^4
Progeny	**A1A1**	**A1A2**	**A2A2**	
	p^4	$2p^3q$	p^2q^2	
	$2p^3q$	$4p^2q^2$	$2pq$	
	p^2q^2	$2p^3$	q^4	
	4	8	4 → Sum = 16	

See Hardy-Weinberg theorem; mating systems; Pascal triangle.

random oligonucleotide primers   Used for synthesis of radioactive probes. Heterogeneous oligonucleotides can anneal to different and many positions along a nucleic acid chain. They can also serve as primers for the initiation of DNA synthesis. If the precursors are one type of radioactive (α-^{32}P)-deoxyribonucleotide (dNTP) and cold dNTPs, highly radioactive probes can be obtained. Single-stranded DNA templates can be copied by the aid of Klenow fragments of DNA polymerase I, or in case of an RNA template, reverse transcriptase can be used. The primers are usually short (6 to 12 bases) and can be generated either by DNA-ase digestion of commercially available DNA (from calf thymus or salmon sperm) or produced by an automatic DNA synthesizer. *See* nick translation; probe.

random sample   Withdrawn from a collection without any selection.

random walk   Physical theory of material distribution within media such as cell migration within connective tissues, Markov processes in DNA, diffusion in gas, liquids, and solids, etc. (Berg, H. C. 1993. *Random Walks in Biology*. Princeton Univ. Press, Princeton, NJ.)

range constraints   The number of repetitive units of a microsatellite has limitations because of viability or adaptability. *See* microsatellite.

RANK   Receptor activator of NF-κB. *See* TRANCE.

RANTES   Regulated on activation normally T-cell expressed and secreted. Chemoattractant of cytokines for monocytes and T cells. The chemokine receptors appear to be seven membrane proteins coupled to G proteins. RANTES is also involved with a transient increase of cytosolic Ca^{2+} and also Ca^{2+} release. The opening of the calcium channel increases the expression of interleukin-2 receptor, cytokine release, and T-cell proliferation. Thus, in addition to inducing chemotaxis, RANTES can act as an antigen-independent activator of T cells in vitro. RANTES and MIP-1 chemokines, along with the receptor CC CKR5 and fusin, are believed to suppress replication of HIV. *See* acquired immunodeficiency; CC CKR5; chemotaxis; cytokine; fusin; HIV; MIP-1a; T cell. (Alam, R., et al. 1993. *J. Immunol.* 150:3442; Casola, A., et al. 2001. *J. Biol. Chem.* 276:19715.)

RAP   *See* RNAIII/rnaiii.

RAP1A (1p13.3), RAP1B (12q14), RAP2 (13q34)   RAS-related eukaryotic proteins, but unlike RAS, they are localized on intracellular membranes. The guanine exchange factor of RAP1, Epac (exchange protein activated by cAMP), is activated by cAMP, and it bears homologies to the regulatory subunit of PKA. Rap1A is a suppressor protein of Ras-induced transformation. It has identical amino acid sequences with the effector region of Ras p21. RAP1 and SIR3 are transcriptional repressors of telomeric heterochromatin. RAP74 is a subunit of transcription factor TFIID and RAP74 is involved in binding to the serum-response element. RAP1 (repressor/activator protein) of yeast binds to upstream activator sequences (UAS) alone and in association with other proteins, activating many genes, in addition to silencing the mating type and the telomerase functions. Rap1 activates about 37% of the RNA polymerase II initiations in yeast. Rap1 is also a negative regulator of TCR-mediated transcription of the interleukin-2 (IL-2) gene. RAP1 may enhance meiotic recombination and opening up of the nucleosomal chromatin. RAP1 may stimulate the formation of boundary elements. Nerve growth factor (NGF) may signal to ERK (environmentally regulated kinases) either through RAS or RAP. *See* adherens junction; boundary element; cAMP; HML and HMR; IL-2; mating type determination; nucleosome; ORC; PKA; RAS oncogene; serum-response element; silencer; TCR; telomerase; transcription factors; TRF1. (Rousseau-Merck, M. F., et al. 1990. *Cytogenet. Cell Genet.* 53:2; Morse, R. H. 2000. *Trends Genet.* 16:51; Idrissi, F.-Z., et al. 2001. *J. Biol. Chem.* 276:26090.)

rapamycin (sirolimus)   Immunosuppressor that may block the cell cycle through the G1 phase by controling mitogen-activated signal transduction. It regulates the prokaryotic ribosomal protein S6 and the elongation initiation protein eIF-4E.

See cell cycle; eIF-4E; FK506; immunophilins; immunosuppressant; signal transduction; S6 kinase; TOR. (Cardenas, M. E., et al. 1998. *Trends Biotechn.* 16:427; Rohde, J., et al. 2001. *J. Biol. Chem.* 276:9583.)

RAPD (pronounce rapid)   Markers are generated by random amplified polymorphic DNA sequences using (on average) 10-base-pair primers and the PCR technique for the physical mapping of chromosomes on the basis of DNA polymorphism in the absence of "visible" genes. The map so generated may, however, be integrated into RFLP and classical genetic maps. *See* integrated map; physical mapping; polymerase chain reaction; sequence-tagged site. (Reiter, R. S., et al. 1992. *Proc. Natl. Acad. Sci. USA* 89:1477.)

rape (*Brassica napus*)   Oil seed crop, $2n = 38$. *See* canola; erucic acid.

Raphanobrassica   Man-made amphidiploid ($2n = 36$) of radish (*Raphanus sativus*, $n = 9$) and cabbage (*Brassica oleracea*, $n = 9$). *See* amphiploid; *Brassica oleracea*; radish.

raphe   Ridge on the seeds where the stalk of the ovule was attached; seam of animal tissues.

raphids   Needle-like crystals within plant cells (often of oxaloacetic acid).

raphilin   Peripheral membrane protein. It may bind RAB proteins in a GTP-dependent manner, may be phosphorylated by various kinases, and may bind Ca^{2+} and phospholipids. *See* RAB.

rapid lysis mutants   *r* mutants of bacteriophage rapidly lyse the infected bacteria and therefore the size of the plaques are much larger than the ones made by wild-type phage. *See* lysis inhibition; lysis; plaque lift.

Rapid lysis plaque.

Wild type plaque.

Rapp-Hodgkin syndrome   Anhidrotic ectodermal dysplasia with cleft lip and cleft palate. The latter symptoms do not always co-occur (mixed clefting). *See* anhidrosis; cleft palate; ectodermal dysplasia. (Neilson, D. E., et al. 2002. *Am. J. Med. Genet.* 108:281.)

rapsyn (43 kDa)   Peripheral membrane protein colocalized with the acetylcholine receptors at the neuromuscular synapsis. Mutations in rapsyn may lead to myasthenia. *See* acetylcholine; myasthenia. (Ohno, K., et al. 2002. *Am. J. Hum. Genet.* 70:875.)

RAR   Repair and recombination.

RAR, RARE   Retinoic acid receptor (element). RXR-α, -β, -γ retinoid-X receptors are transducers of ligand-activated morphogenetic and homeostasis signals. RAR and RXR can form homodimers but are usually found as heterodimers. They bind to the cognate hormone-response elements and increase the efficiency of transcription. Docosahexaenoic acid ($CH_3[CH_2CH=CH]_6[CH_2]_2CO_2H$) is an activator of RXR. RAR-α ligands can accomplish the binding of the RXR-RAR-α dimers to DNA, causing RXR activation and initiating the transcriptional activity of RAR-α. RAR is encoded in human chromosome 17q21. It may have a role in promyelocytic leukemia (PML) development when it associates with histone deacetylase and other cofactors. In PML-RARα patients, pharmacological doses of retinoic acid lead to cancer remission because of the almost normal differentiation of the hematopoietic (red blood–forming) cells, but in promyelotic leukemia zinc-finger (PLZF) patients this treatment is quite ineffective. Retinoic acid may downregulate telomerase activity. The retinoid system plays a role in differentiation, motor innervation of the limbs, skeletal development, and the development of the spinal cord. *RAR1* genes in plants play a defense role. *See* histone deacetylase; hormone-response elements; innervation; leukemia; PPAR; retinoic acid; Sin3; zinc finger. (Zhong, S., et al. 1999. *Nature Genet.* 23:287; Pendino, F., et al. 2001. *Proc. Natl. Acad. Sci. USA* 98:6662.)

RARE   RecA-assisted restriction endonuclease. At the site of restriction enzyme recognition in or near a locus, a triplex structure with an oligonucleotide is generated with the assistance of enzyme RecA. The genome is then enzymatically methylated and RecA is removed. Only the unprotected site will be cut by the restriction endonuclease. This procedure thus reduces the actual cleavage sites in the DNA. *See* DNA methylation; recA; restriction enzyme. (Ferrin, L. J. 2001. *Mol. Biotechnol.* 18[3]:233.)

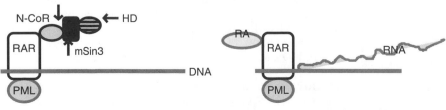

No transcription   Retinoic acid (RA) removed the inhibitory complex (N-Cor, mSin3, HA (histone deacetylase)

The action of RAR in promyelocytic leukemia. (Redrawn after Grignani, F., et al. 1998. *Nature* 391:815.)

rare-cutter Restriction enzyme with a longer DNA sequence (>8 nucleotides) recognition site. Therefore, it cleaves DNA much less frequently (because of the greater specificity for sites) than enzymes, which recognize only four bases in a sequence. *See* restriction enzyme.

rare-mating Cells of nonmating yeast strains are mixed with cells that are expected to mate under favorable conditions. Then low frequency of mating may take place and the progeny may be isolated by efficient selection. The rare-mating is presumed to be the result of mitotic recombination or nondisjunction of chromosome III in the nonmating strain. When normal mating fails, protoplast fusion may generate hybrids. The latter procedure results, however, in complex progeny. *See* mating type determination; protoplast fusion. (Spencer, J. F. & Spencer, D. M. 1996. *Methods Mol. Biol.* 53:39.)

RAS (p21^{ras}, 11p15.5) Protooncogene originally found in rat sarcoma virus; it codes for a monomeric GTP-binding protein in which point mutation mainly in codons 12, 13, 59, or 61 may lead to oncogenic transformation. The 12-site alterations in humans and the 61-site alterations in rodents are prevalent in tumors. The RAS proteins have an important role in transmembrane signaling. Protein Spred is an inhibitor of the RAS-MAP signaling pathway. The role of RAS may vary depending on cell type, from stimulation of adenylate cyclase to mating factor signal transduction and from proliferation to differentiation. The RAS protein becomes active only after prenylation by the 15-carbon farnesyl pyrophosphate. The prenylation is thioether formation with an amino acid, resulting in association of the protein with a membrane. RAS is one of the most important turnstile in signal transduction and one of the most common activated oncogene. The activation involves changing of the bound GDP into GTP. GAPs (GTPase-activating proteins) inactivate RAS by hydrolysis of GTP. In contrast, GNRPs (guanine nucleotide-releasing proteins) mediate the replacement of bound GDP by GTP and the activation of RAS. Receptor tyrosine kinases activate RAS either by inactivating GAP or activating GNRP. The guanine-nucleotide exchange reaction is mediated by the SOS (son of sevenless) protein, which has a GEF (guanidine exchange factor) domain. RAS is localized to the cell membrane.

After ligand binding, SOS attaches to the adaptor Grb2 protein containing an SH3 domain. Through the SH2 domain of Grb2, the complex binds to the phosphotyrosine residues of the signal receptor. The complex moving close to RAS, with the assistance of Cdc25, Sdc25, and GEF/GRF, respectively, makes it possible for the guanidine nucleotide to be released. The family RAS is represented in various human chromosomes: NRAS in 1p21, HRAS in 11p15, KRAS in 6p12-p11, RRAS in 19. Kras2 in mice also has a tumor-inhibitory feature (Zhang, Z., et al. 2001. *Nature Genet.* 29:25). On the basis of homology, three groups may be classified in mammals: (1) RAS, RAL, RRAS; (2) RHO; and (3) RAB. In *Drosophila*, there are *Ras1* in 3-49, *Ras2* in 3-15, and *Ras3* in 3-1.4 (the latter has a higher homology to *Rap1* and it is called *Rap1*). In yeast, the RAS homologues (*RAS1, RAS2*) are very closely related to the human protooncogenes and can be replaced by them. RAS protein is necessary for the completion of mitosis in association with other factors. Probably all eukaryotes carry RAS homologues. The p21 protein is a mobile RAS protein.

The RAS oncogene is generally active in tumorigenesis in the presence of the MYC or the E1A "immortalizing" oncogene products. In cooperation with RAS, MYC mediates the progression of the cell cycle from the G1 to the S phase through induction of the accumulation of active cyclin-dependent kinase and transcription factor E2F. The presence of RAS is necessary for tumor maintenance. The RAS promoter is high in GC and lacks a TATA box. The various RAS genes (human, mouse) may have over 50% difference at the nucleotide level, but the amino acid composition is highly conserved. RAS mutations have been detected in 90% of pancreatic adenocarcinomas, ~40–50% of colon adenocarcinomas, and other cancers. According to a genome-wide survey, RAS affects the expression of more than 250 genes. *See* adenylate cyclase; animal models; Cdc25; cell cycle; EF-Tu; farnesyl; GD; GEF; G-proteins; Grb2; GTPase; MYC; oncogenes; p21; prenylation; RAB; raf; RALA; RALB; RAP; RASA; retinoblastoma; RHO; Sdc25; SH2; SH3; signal transduction; SOS. (Zuber, J., et al. 2000. *Nature Genet.* 24:144; Johnson, L., et al. 2001. *Nature* 410:1111; Stacey, D. & Kazlauskas, A. 2002. *Current Opin. Genet. Dev.* 12:44; Quilliam, L. A., et al. 2002. *Progr. Nucleic Acid Res. Mol. Biol.* 71:391; Downward, J. 2003. *Nature Rev. Cancer* 3:11.)

RASA Guanosine triphosphate-activating RAS protein (21 kDa [p21]) encoded by human chromosome 5q13.3 and in mouse chromosome 13. *See* RAS oncogene.

Rasmussen's encephalitis *See* epilepsy.

raspberry (*Rubus* spp.) The majority of raspberries are diploid ($2n = 14$), but loganberry is $2n = 42$, and the blackberries exist with $2n = 28$, 42, and 56 chromosomes. Some wild blackberries are probably allopolyploids with $2n = 35$ and $2n = 84$; the latter is dioecious.

Rassenhygienie German term for negative eugenics. Its goal was to protect the "purity" of the Aryan (German) race enforced by the laws of the Nazi state. Between 1933 and 1945, resulted in 350,000 forced sterilizations, the mass murder of millions, and a ban on marriages between the genetically "fit" and "unfit," and persons whose ancestry included more than one-quarter Jews, Gypsies, or some other racial groups. *See* eugenics; racism. (Hubbard, R. 1986. *Int. J. Health Serv.* 16:227.)

rat *Rattus norvegicus*, $2n = 42$. A genetic linkage map has been published by Jacob, H. J., et al. *Nature Genet.* (9:63 [1995]). A radiation hybrid map of 5,255 markers: Watanabe, T. K., et al. 1999. *Nature Genet.* 22:27. The chromosome number of other rat species may be different. EST map: Scheetz, T. E., et al. 2001. *Genome Res.* 11:497. *See* EST map. (<http://ratEST.uiowa.edu>; <http://ratmap.gen.gu.se>; <http://www.tigr.org/tdb/tgi.shtml>; <http://rgd.mcw.edu/>.)

rate-limiting step Requires the highest amount of energy in a reaction chain or the slowest step in a metabolic path.

ratio labeling Using FISH cytological technology, different chromosomes may be labeled by different proportions of the same fluorochromes to distinguish individual chromosomes in the genome by color. *See* combinatorial labeling; FISH.

rationale   Logical basis of an act, process, or argument.

rationalize   Attempt to make something conform to reason. Sometimes apparent rationalization is applied in an effort to explain facts or ideas with inadequate justification.

Raynaud disease (hereditary cold fingers)   Familial periodic numb and white finger attacks. *See* vasculopathy.

Raynaud syndrome   Involves scleroderma, cyanosis, cold intolerance, chromosomal aberrations, and telangiectasia without a clear pattern of inheritance. *See* cyanosis; scleroderma; telangiectasia.

RB   Right border of T-DNA. *See* T-DNA.

Rb   *See* retinoblastoma.

R bands   Heat-denaturation-resistant chromosomal bands; half of them have telomeric sites. The bright-field R bands usually show the reverse Giemsa pattern. *See* C banding; chromosome banding; G banding; isochores. (Dutrillaux, B. & Lejeune, J. 1974. *Adv. Hum. Genet.* 5:119.)

R-banded chromosome.

RbAp   Proteins of the WD family of wide regulators of chromatin, transcription, and cell division. *See* WD-40. (Rossi, V., et al. 2001. *Mol. Genet. Genomics* 265:576.)

rBAT/4F2hc   Four-membrane-spanning proteins involved in membrane transport or regulation of transport of neutral and positively charged amino acids. *See* cystinuria; transporters. (Malandro, M. S. & Kilberg, M. S. 1996. *Annu. Rev. Biochem.* 65:305.)

rbc   Ribulose bisphosphate carboxylase/oxidase genes. *See* chloroplast genetics.

RBE   Relative biological effectiveness of radiation depends on a number of physical (type of radiation, wavelength, dose rate, temperature, presence of oxygen, hydration, etc.), physiological (developmental stage), and biological factors (species, nuclear size, DNA content, level of ploidy, repair system, etc.). The comparison usually relates to ^{60}Co gamma-radiation. *See* radiation effects; rem.

RBM   RNA-binding motif, also called RBMY, YRRM. A gene family only in the Y chromosome of mammals involved in male fertility. The HNRPG (hnRPG) gene located in human autosome 6p12 displays, however, ~60% homology to RBM. It is assumed that this autosomal locus was retrotransposed in an early ancestor to the Y chromosome. Similarly, in Xq26, sequences virtually identical to exon 12 of hnRPG were detected. Thus, it appears that the chromosome 6p12

sequence is a processed pseudogene of the Xq26 sequence and similar sequences were also found in chromosomes 1, 4, 9, and 11 — all retrotransposed supposedly from Xq26. BFLS actually displays hypogonadism and it was assigned to the same location as RBM. *See* azoospermia; Borjeson-Forssman-Lehmann syndrome (BFLS); boule; chromosome; DAZ; NRY; PABp.

R2Bm   Silkworm retroelement without long terminal repeats. *See* retrovirus; silkworm. (Luan, D. D., et al. 1993. *Cell* 72:595.)

RBMY   *See* RBM.

R bodies   Refractive bodies are temperate bacteriophages within the κ-particles of paramecia. *See Paramecium*; symbionts, hereditary.

RBTN (rhombotin)   Cystine-rich oncoprotein family (encoded in human chromosome 11) containing an LIM domain. *See* LIM domain. (Chan, S. W. & Hong, W. 2001. *J. Biol. Chem.* 276:28402.)

Rbx1   Ring-finger protein homologue of APC (anaphase-promoting complex) required for SCF- and VCB-mediated ubiquitination of Sic1 and probably other proteins. *See* APC; ring finger; SCF; Sic1; ubiquitin. (Carrano, A. C. & Pagano, M. 2001. *J. Cell Biol.* 153:1381.)

RCA   Regulators of complement activation. *See* complement; MCP.

RCAF   Replication-coupling assembly factor mediates chromatin organization into nucleosomes during replication. *See* ASF1; nucleosomes. (Tyler, J. K., et al. 1999. *Nature* 402:555.)

RCC   Renal cell carcinoma (human chromosome 3p14.2). RCC gene product frequently interacts with that of the von Hippel–Lindau (VHL) gene. *See* renal cell carcinoma; von Hippel–Lindau disease.

RCC1   Chromatin-bound guanine nucleotide release factor forming complexes with RAN (a G protein). Its deficiency interferes with cell cycle progression, chromosome decondensation, mating, RNA export, and protein import. It is a part of the nuclear pore complex. Its yeast homologue is Prp20p. *See* cell cycle; nuclear pores; RAN; RNA transport; TPX. (Hood, J. 2001. *Trends Cell Biol.* 11:321; Renault, L., et al. 2001. *Cell* 105:245.)

RcsC   *E. coli* kinase affecting capsule synthesis regulator RcsB. (Davalos-Garcia, M., et al. 2001. *J. Bacteriol.* 183:5870.)

RcsG   30.6 kDa protein with a C-terminal motif and an N-terminal sequence similar to that of DnaJ. In concert with RcsC/B, DnaK, and GrpE, it induces the *cps* capsule polysaccharide operon of *E. coli. See* DnaJ; RcsC.

RDA   Representational difference analysis is a genome-scanning procedure for the detection and identification of genetic markers representing disease, other genes, and chromosomal aberrations. Cellular DNA is cut by restriction endonucleases and the smaller fragments are amplified by PCR. Then DNA samples from affected and disease-free samples are denatured and the mixtures of the two samples are allowed to anneal. The sequences that do not match fail to hybridize and they are expected to be responsible for disease. The process is similar in principle to cascade hybridization. The relatively rapid mass screenings were expected to identify individuals liable to particular hereditary differences (diseases). The same procedure may be applicable to nondisease genes of eukaryotes. RDA and GDRDA procedures can be used to generate genetic maps in organisms with a paucity of chromosomal markers. *See* cascade hybridization; GDRDA; genetic screening; genomic subtraction; GMS; PCR; positional cloning; RNA fingerprinting. (Lisitsyn, N., et al. 1993. *Science* 259:946; Tyson, K. L., et al. 2002. *Physiol. Genomics* 9:121.)

rDNA   DNA complementary to ribosomal RNA. The ribosomal RNA genes are in the nucleolar organizer region of the eukaryotic chromosomes and there are multiple tandem repeats of transcriptional units consisting of 18S, 5.8S, 5S, and 26S RNAs. The mature rRNAs are cleaved from the large transcripts. *See* ribosome; rRNA. (Long, E. O. & Dawid, I. B. 1980. *Annu. Rev. Biochem.* 49:742.)

RdRP   RNA-directed RNA polymerase may be involved in gene silencing by synthesizing antisense transcripts from aberrant RNA, thereby causing PTGS. RdRP may convey virus resistance to plants. *See* antisense RNA; epigenesis; host-pathogen relation; PTGS; RNAi. (Cheng, J., et al. 2001. *Virus Res.* 80:41.)

reaction, chemical   Change in the atoms in or between molecules.

reaction intermediate   Generally a short life chemical in a reaction path.

reaction norm   Range of phenotypic potentials of expression of a gene or genotype. Usually, the genes do not absolutely determine the phenotype, but they permit a range of expressions, depending on the genetic background, developmental and tissue specificity conditions, and environment. *See* epigenesis; genotype; phenotype; regulation of gene activity. (Woltereck, R. 1909. *Verhandl. Dtsch. Zool. Ges.* p. 110.)

reactivation   *See* marker rescue; multiplicity reactivation; Weigle reactivation.

reactivity in hybrid dysgenesis   *See* VAMOS.

reading disability   *See* dyslexia.

reading frame   The triplet codons can be read in three different registers, starting with the first, second, or third; however, only one may spell the correct protein. *See* frameshift mutation; open-reading frame.

readout   DNA sequence recognition by proteins that may be *direct* (by hydrogen bonding) or *indirect* when the DNA conformation also plays a role.

read-through   Ribosome continues translation downstream of a stop codon. *See* autogenous suppression; translation termination.

read-through protein   Formed when a suppressor tRNA inserts an amino acid at a site where chain termination is normally expected because of a nonsense codon, thus producing a fusion protein from two different "in-frame" cistrons separated by a nonsense codon. Read-through may be brought about by mutation in the anticodon of a tRNA or modification of tRNA, e.g., selenocysteinyl-tRNA inserts selenocysteine into glutathione oxydase by recognizing the UGA (opal) stop codon. *See* gene fusion; transcriptional gene fusion; translational gene fusion; trapping promoters.

realized heritability   *See* gain.

real time   Actual time during which the physical process takes place.

reannealing (reassociation)   Double-stranded DNA can be heat denatured (strands separated) and can be restored to double-stranded form and reannealed when the temperature becomes lower than 60°C. *See* c_0t curve.

rearrangements   Structural changes of the chromosome(s), e.g., translocation, inversion.

reassociation kinetics   *See* c_0t curve.

reassortant   New virus strain emerging by combination of genes of two different strains, e.g., the pandemic influenza strains of 1957 and 1968 contain elements enabling the virus to replicate in humans and the avian segment. The hemagglutinin coding segment helps the avoidance of neutralizing antibodies of humans previously not exposed to the avian flu virus. *See* hemagglutinin; influenza virus; pandemic.

Rec8   Meiotic cohesin cleaved by separin before chiasma are resolved and meiotic anaphase I can proceed. In vertebrates, cohesin is removed during prophase, but Scc1 stays associated with the centromeres when separin cleaves this protein, thus facilitating the metaphase-anaphase transition. *See* cohesin; separin. (Buonomo, S. B., et al. 2000. *Cell* 103:387; Waizenegger, I. C., et al. 2000. *Cell* 103:399.)

rec⁻   Type of recombination-deficient mutation.

RecA protein   38.5 kDa polypeptide involved in homologous recombination by promoting pairing. It is a DNA-dependent ATPase that mediates strand exchange. RecA binds to single-stranded DNA and presynaptic nucleoprotein molecules mediate the pairing with the duplex DNA target. The paired DNA is then within a 25 Å hole. Within this cavity projecting toward the axis of the helix are mobile loops L1 and L2 representing the binding sites. The RecA protein expressed in transgenic plants substantially increases recombinational repair of DNA damage inflicted by mitomycin. The Archaea and eukaryotic homologues are RadA and Rad51, respectively. When equipped with the nuclear localization signal and transformed into tobacco cells, the RecA prokaryotic gene increased sister chromatid exchange by ~2–3-fold. *See* DNA repair; *RecA1*; RecA-independent recombination; recombination, molecular mechanism, prokaryotes. (Kowalczykowski, S. C. & Eggleston, A. K. 1994. *Annu. Rev. Biochem.* 63:991; Gourves, A.-S., et al. 2001. *J. Biol. Chem.* 276:9613; Bar-Ziv, R. & Libchaber, A. 2001. *Proc. Natl. Acad. Sci. USA* 98:9068; Robu, M. E., et al. 2001. *Proc. Natl. Acad. Sci. USA* 98:8211; Gasior, S. L., et al. 2001. *Proc. Natl. Acad. Sci. USA* 98:8411; Lusetti, S. L. & Cox, M. M. 2002. *Annu. Rev. Biochem.* 71:71.)

recA1 Recombination-deficient mutation of *E. coli* (map position 58 min) coding for a DNA-dependent ATP-ase, a 3522-amino-acid residue enzyme. Plasmids carrying it remain monomeric and do not form multimeric circles. When M13 vectors carry it, the foreign passenger DNA has fewer deletions. The recA protein mediates the association of double-stranded DNAs by synapsis mainly in the major groove but also in the minor groove of the DNA. The RecA-mediated pairing involves a triplex structure, i.e., along parts of the sequences double-stranded DNA associates transiently with a single strand of the other DNA molecule. The pairing of the DNA molecules may be plectonemic (intertwined) and thus may not require stabilization by proteins. (The paranemic coils are only juxtapositioned and require protein to keep them together.) Experimental information indicates that ATP hydrolysis is not required for the exchange between paired strands; rather, the removal of RecA requires ATP hydrolysis. In case the homology between the DNAs is not perfect, ATP is needed for the exchange. RecA is also involved in branch migration, but RuvAB and RecG proteins assist in the process. The extension of the DNA heteroduplex (at the rate of 2–10 bp/sec) in the 5′ → 3′ direction needs ATP hydrolysis. The length of the heteroduplex may become 7 kbp long. In both prokaryotes and eukaryotes, besides RecA (or homologues), a stimulatory exchange protein that binds single strands of the DNA (SSB) is required. The SSB monomers (1/15 base in ssDNA) facilitate synapsis between the heterologous strands. After exchange, RecA promotes DNA renaturation. The RecA homologues in yeast, mice, and humans are the RAD51 proteins and Mei3 in *Neurospora*. *See* branch migration; DNA repair; RAD; *recB*; recombination, molecular mechanisms, prokaryotes; RuvABC.

RecA-independent recombination (illegitimate recombination) May use three pathways: (1) simple replication

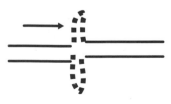

Palindrome.

slippage, (2) sister chromatid-associated replication misalignment, and (3) single-strand annealing. Single-strand anneal-

ing takes place after palindromic sequences within each strand fold into hairpin structures within the strands. When a nuclease (SbcCD) opens the cruciform palindromes resection, followed by annealing of the flanking repeats, deletion occurs. *See* illegitimate recombination; palindrome. (Bzymek, M. & Lovett, S. T. 2001. *Proc. Natl. Acad. Sci. USA* 98:8319.)

recB *E. coli* gene (map position 60 min), encoding a subunit of exonuclease and controlling recombination and genetic repair. *See* DNA repair; recombination models; recombination, molecular mechanism, prokaryotes.

RecBCD Enzyme (ribozyme) complex functioning in recombination of prokaryotes. RecBCD is a helicase and unwinds up to 42,300 bp per molecule. It is a strand-specific nuclease. The recognition sequence (also called χ) is 5′-GCTGGTGG-3′ and promotes recombination in its vicinity. The enzyme travels along one of the two strands of the DNA in the 3′ → 5′ direction. *See* DNA repair; recombination, molecular mechanism, prokaryotes; recombination models. (Jockovich, M. E. & Myers, R. S. 2001. *Mol. Microbiol.* 41:949.)

recC *E. coli* gene (map position 60 min), encoding a subunit of exonuclease V. It controls recombination and genetic repair. *See* DNA repair; recombination, molecular mechanisms. (Chen, H. W., et al. 1998. *J. Mol. Biol.* 278:89.)

recE Locus of Rac prophage (map position 30 min), encoding exonuclease VIII and promoting homologous binding between single-stranded and double-stranded DNAs. *See* RecA. (Muyrers, J. P., et al. 2000. *Genes Dev.* 14:1971.)

receptacle (1) Widened end of a flower stalk. (2) Container.

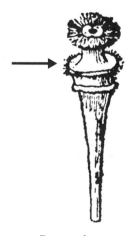

Receptacle.

receptor (1) Also called an operator, the site responding to the controlling element (transposase), as originally called by

Barbara McClintock. Components of the *Spm* transposable systems in maize. *See* transposable elements of maize receptors. **(2)** Protein that binds to ligands with cellular signaling functions. The receptors may be located within the plasma membrane (transmembrane proteins) or intracellularly. They bind ligands, which penetrate cells by diffusion. Some receptors are ligand-gated ion channels. The number of receptors in a cell may run into the hundreds. Several agonists and antagonists may affect their function. *See* adaptor proteins; cargo receptors; hormone receptors; ion channels; ligand; nuclear receptor; orphan receptor; receptor guanyl cyclase; receptors; receptor tyrosine kinase; receptor tyrosine phosphatase; serine/threonine kinase; signal transduction; T cell; TCR; transmembrane proteins; virus receptor. (Xu, L., et al. 1999. *Curr. Opin. Genet. Dev.* 9:140.)

receptor downregulation Epidermal growth factor (EGF)–binding receptors concentrate in coated pits after binding. They go into the lysosomes where degradation of the receptor and EGF takes place. The cell surface will have a reduced number of them because of receptor downregulation. *See* EGF; endocytosis.

receptor editing Modifies the antigen specificity of antigen receptors in the variable region of the antibody and may generate immune tolerance or antibody diversification by V(D)J recombination. This mechanism eliminates the autoreactive B cells when confronted with self-antigens. *See* antigen receptor; clonal selection; immune tolerance; immunoglobulins; V(D)J. (Kouskoff, V. & Nemazee, D. 2001. *Life Sci.* 69:1105.)

receptor guanylyl cyclase Transmembrane protein associated at the cytosolic end with an enzyme that generates cyclic guanosine monophosphate (cGMP). cGMP then activates cGMP-dependent protein kinase (G kinase), which phosphorylates serine/threonine residues in proteins. *See* cGMP; serine/threonine kinase. (Kusakabe, T. & Suzuki, N. 2001. *Dev. Genes Evol.* 211[3]:145.)

receptor-mediated endocytosis Very efficient delivery system of macromolecules (such as cholesterol) that adhere to coated pits and onto cellular organelles. *See* coated pits.

receptor-mediated gene transfer Cell-surface receptors may internalize their ligands by endocytosis. The ligand (peptides, lectins, sugars, antibody, glycoprotein, etc.) may form a conjugate with a polycation, e.g., polylysine. An expression vector plasmid may bind to a ligand-polycation conjugate. A fusogenic peptide or a disabled adenovirus may facilitate the entry of the complex into the cell by endocytosis transported with the aid of an endosomal vehicle. The DNA (gene) may be transferred from the endosome to the nucleus, where it may have a chance for expression. With the assistance of asialoglycoprotein receptor, the gene may be targeted to hepatocytes, or with a mannose receptor it may be targeted to macrophages. The transferrin receptor facilitates targeting erythrocytes; polymeric immunoglobulin receptors may aim the gene construct at the lung epithelia; other receptor and ligand combinations permit targeting to other

cells or tissues. The advantage of this type of transfection is that it is not infectious. The DNA carrying more than a single gene of most any size can be targeted. Cell division is not a requisite for expression. The transgene functions in the cytoplasm. Unfortunately, the level and duration of expression are variable. The system may elicit an undesirable immune reaction. *See* asialoglycoprotein receptor; endocytosis; macrophage; transferrin. (Varga, C. M., et al. 2000. *Biotechnol. Bioeng.* 70:593.)

receptor protein tyrosine phosphatase (RPTP) Signaling molecules required for cell development. They dephosphorylate negative regulatory C-terminal tyrosine residues of the Src family kinases. *See* signal transduction; Src. (Carothers, A. M., et al. 2001. *J. Biol. Chem.* 276:39094.)

receptor serine/threonine kinase Receptor of serine/threonine phosphorylating enzymes. It is the major type of plant receptor kinases, unlike in animals where receptor tyrosine kinases dominate. *See* serine/threonine kinase. (Choudhury, G. G. 2001. *J. Biol. Chem.* 276:35636.)

receptor tyrosine kinase (RTK) Binds protein tyrosine kinase enzymes such as the receptors for epidermal growth factor (EGF), insulin, insulin-like growth factor-1 (IGF-1), platelet-derived growth factor (PDGF), fibroblast growth factor (FGS), nerve growth factor (NGF), hepatocyte growth factor (HGF), vascular endothelial growth factor (VEGF), and macrophage colony-stimulating growth factor (M-CSF). RET proteins contain a cadherin-like, cysteine-rich extracellular domain. Defects of the RET protein family involve glial cell–derived neurotrophic factor problems, MEN2, Hirschsprung disease, familial medullary thyroid carcinoma, pheochromocytoma, hyperparathyroidism, and ganglioneuromatosis. Hepatocyte growth factor receptor defects are responsible for papillary renal cell carcinoma. Platelet-derived growth factor receptor changes activate the KIT oncogene. The insulin receptor anomalies lead to diabetes type II, leprechaunism, and Rabson-Mendenhall disease. Hereditary lymphedema receptor disease is due to vascular endothelial growth factor. Congenital pain with anhidrosis is caused by defects in the neurotrophin receptors. These receptors are transmembrane proteins. When the receptor becomes associated with the cognate phosphorylase enzyme, both the receptor and the target protein receive γ-phosphate groups from ATP at certain tyrosine residues. The phosphorylation results in dimerization or dimerization results in phosphorylation and activation. Activation increases the activity of RAS and subsequently the MAP kinases. Eventually, this leads to the expression of genes. The various regulatory proteins recognize different phosphorylated tyrosine residues in the receptor. Upon binding to their specific sites, they may also be phosphorylated on their own tyrosine residues and become activated. A cascade of events may follow that activate entire signaling pathways. The different receptors and the associated proteins may control the separate or interacting signaling pathways. Mutations or truncation of RTK that permits dimerization without the proper ligand may lead to carcinogenesis. The specificity of RTK (which is involved in diverse metabolic functions) is determined either by the strength or duration of the signal, or it may be qualitative. The different cell types may activate RTK signaling differently. In craniosynostosis (Crouzon, Pfeiffer, Apert, and Jackson-Weiss syndromes) and some dwarfness (hypochondroplasia, thanatophoric dysplasia), fibroblast growth factor receptor members of RTK are involved. *See* craniosynostosis syndromes; Eph; signal transduction; SIRP; tyrosine kinase. (Simon, M. A. 2000. *Cell* 103:13; Robertson, S. C., et al. 2000. *Trends Genet.* 16:265; Madhani, H. D. 2001. *Cell* 106:9; Haj, F. G., et al. 2002. *Science* 295:17080.)

receptor tyrosine phosphatase Binds protein tyrosine phosphates and splits off phosphate groups. (Bateman, J., et al. 2001. *Curr. Biol.* 11:1317.)

recessive Expression of a gene means that it is not visible in heterozygotes in the presence of the wild type or other dominant alleles of the locus. Recessivity is not necessarily an absolute lack of the expression of the gene (except in null alleles) because a very low level of transcription/translation may not be detectable by a particular type of study but may be observable by a finer analysis. *See* dominance; semidominance.

recessive allele Does not contribute to the phenotype in heterozygotes in the presence of the dominant allele. *See* pseudodominance.

recessive epistasis *See* epistasis; modified checkerboards.

recessive lethal Dies when homozygous, and can be maintained only as a heterozygote.

recessive lethal tests, *Drosophila* *See* autosomal-recessive lethal assay; *Basc*; *ClB* method; sex-linked recessive lethal.

recessive oncogenes Tumor-suppressor genes such as encoding p53. *See* p53; tumor suppressors.

recF *E. coli* gene (map position 82 min), also called *uvrF*, controls recombination and radiation repair. *See* DNA repair; recombination, molecular mechanisms. (Bidnenko, V., et al. 1999. *Mol. Microbiol.* 33:846.)

RecF Single- and double-strand-binding recombination protein. (Nakai, H., et al. 2001. *Proc. Natl. Acad. Sci. USA* 98:8247.)

recG *E. coli* gene (map position 82 min) controls recombination. *See* recombination, molecular mechanism. (Qourcelle, J. & Hanawalt, P. C. 1999. *Mutation Res.* 435:171.)

RecG Unwinds the leading and lagging strands at a damaged replication fork and thus may contribute to replication restart if it was stalled by the damage. *See* replication restart. (McGlynn, P. & Lloyd, R. G. 2001. *Proc. Natl. Acad. Sci. USA* 98:8227.)

recipient Bacterial cells of the F$^-$ state receive genetic material from the donor F$^+$ strains. Cell to which genetic material is transferred. *See* conjugation; transformation.

recipient site *See* donor site.

reciprocal crosses For example, A × B and B × A.

Reciprocal hybrids of *Epilobium hirsutum* Essen and *Epilobium parviflorum* Tübingen. Parents are $2n = 36$. In the cross at the left, an *E. parviflorum* female was crossed by an *E. hirsutum* male. The two plants at the right represent the reciprocal cross when the *E. hirsutum* female provided the cytoplasm. (From Michaelis, P. *Umschau* 1965 (4):106.)

In cases where cytoplasmically determined differences exist between the two parents, the F$_1$ offspring more closely resembles the female parent that usually transmits the cytoplasm. These reciprocal differences may persist indefinitely in the advanced generations. Although the reciprocal differences usually are most obvious in plants, animal hybrids, e.g., the mule and the hinny, are also easily distinguishable. *See* chloroplast genetics; mitochondrial genetics.

reciprocal interchange Reciprocal translocation of chromosomes.

reciprocal recombination Most common exchange between homologous chromatids at the 4-strand stage of meiosis in eukaryotes. In case of single crossing over in an interval, two parental types and two crossover strands are recovered. The exception is gene conversion where the exchange is nonreciprocal. In conjugational transfer in bacteria, the reciprocal products of the event are not recovered and their fate is unknown. In sexduction and specialized transduction, reciprocal recombination may also take place in bacteria. *See* conjugation; crossing over; recombination, molecular mechanisms, prokaryotes; sexduction; specialized transduction.

Parental	AB and ab
Reciprocal recombinants	Ab and aB

reciprocal selection *See* recurrent selection.

reciprocal translocation Segments of nonhomologous chromosomes are broken off and reattached to each other's place. As a consequence, generally 50% of the gametes of the translocation heterozygotes (formed by adjacent distributions) are defective because they do not have the correct amount of chromatin. *See* translocation.

RecJ Single-strand ($5' \rightarrow 3'$) exonuclease used in recombination of *E. coli*. (Hill, S. A. 2000. *Mol. Gen. Genet.* 264[3]:268.)

Rec-mutant Deficient in recombination and possibly altered in other functions of DNA. *See* Rec.

recoding Mechanism that may translate the same DNA sequence in more than one way. It is a common mechanism in viruses with overlapping genes. There are several other manners in which this can take place. Some genes utilize multiple promoters, and depending on the choice of their utilization, the same RNA may code for more than one protein. Frameshifting may take place, e.g., the mRNA may show slippage on the ribosome; a tRNA^{Leu} with an anticodon GAG may recognize CUUUGA in one frame and in a shift it inserts leucine (UUU) for 4 nucleotides: CUUUGA. Similar frameshifting cassettes may be determined by *E. coli* gene *SF2* and in other prokaryotes. In the TY3 transposable element of yeast, the GCG AGU U, instead of the Ala (GCG) and Ser (AGU), may read GCG **A** GUU Ala (GCG) and Val (GUU). The code words may be interpreted in different ways and stop codons may specify selenocysteine, tryptophan, and glutamine. The ribosome may also skip certain sequences, e.g., the T4 phage topoisomerase may bypass 50 contiguous nucleotides, and after the long frameshift it continues translation. Variants of phage-λ repressor and cytochrome b$_{562}$, when translated from mRNA without a stop codon, acquired an unusual COOH end. By cotranslational switches, the ribosome reading from the defective mRNA to the tRNA-like *ssrA* transcript translated into Tyr-Ala-Leu-Ala-Ala (the normal carboxyl end would have been very similar to Trp-Val-Ala-Ala-Ala). Recoding may be of importance in some human diseases, e.g., if in the cystic fibrosis transmembrane conductance regulator a glycine codon$_{542}$ or arginine codon$_{553}$ is replaced by UGA (opal) stop codon, the disease symptoms are alleviated compared with some missense mutations because this opal codon permits some read-through leakage. *See* cystic fibrosis; frameshift; fuzzy logic; overlapping genes; selenocysteine; set recoding; topoisomerase; Ty. (Shigemoto, K., et al. 2001. *Nucleic Acids Res.* 29:4079; Harrell, L., et al. 2002. *Nucleic Acids Res.* 30:2011.)

recoding signal Required for translational recoding. *See* overlapping genes; recoding.

recognition site, restriction enzyme *See* restriction enzyme.

recoil Bounce back; electromagnetic radiation recoils from glass and metal. *See* Compton effect.

recombinagenic May be involved in genetic recombination at an increased frequency.

recombinant Individual with some parental alleles that are reciprocally exchanged. *See* reciprocal recombination.

recombinant antibody (RAb) Genetically engineered. Recombinant antibodies usually include only the variable fragments, which are fused to some other proteins. The appropriate DNA fragments are amplified by PCR and cloned in *E. coli*, and a single peptide chain may contain the variable regions of both the light and heavy chains. Animal passage is not required. The antibody gene fragment can be fused to a bacterial signal sequence, enabling the direction of the molecule into the periplasmic space where chaperones can fold the engineered protein properly. The procedure may include selection by phage display. Since RAb is produced without an animal and in vitro, sources of contamination by pathogens can be eliminated. They are also monoclonal. Recombinant antibodies can be modified with the repertory of the tools of molecular biology and different properties can be added, e.g., the paratope can be specially targeted to tumor cells (bifunctional antibody). They can be obtained using human gene fragments, thus precluding an immune response against the RAb. *See* antibodies; immunoglobulins; monoclonal antibody; paratope; PCR; phage display; signal sequence. (Kortt, A. A., et al. 2001. *Biomol. Eng.* 18[3]:95; Karn, A. E., et al. 1995. *ILAR J.* 37[3]:132.)

recombinant congenic An outcross is followed by several generations of inbreeding in order to minimize the background genetic variations. *See* recombinant inbred strain panels.

recombinant DNA DNA that has been spliced in vitro from at least two sources with the techniques of molecular biology or that results from the replication of such molecules. From the viewpoint of safety regulations, synthetic DNA segments, which yield potentially harmful polynucleotides or polypeptides if expressed within cells, are subject to the same regulations as any harmful natural product. Transposable elements, unless they include recombinant DNA, are not subject to U.S. National Institute of Health recombinant DNA regulations. *See* cloning vectors; genetic engineering; restriction enzymes; splicing; transformation, genetic; vectors. (*Fed. Regist.* [1999] 64:25361; <http://www4.od.nih.gov/oba/rac/guidelines/guidelines.html>.)

recombinant DNA, evolutionary potentials Genes can be transferred by molecular biology techniques among organisms by means not routinely available in nature. However, during the process of natural evolution it cannot be ruled out that fragments of degraded DNA were taken up by direct transformation and exchanged between taxonomically unrelated species.

recombinant DNA and biohazards Feared by responsible scientists in the 1970s before the impact of techniques that could be fully assessed. Since then, evidence has accumulated showing that some of the worries were not entirely justified, except by the wisdom of caution with hitherto unknown and unused procedures. To avoid safety risks, various levels of containments were made mandatory to prevent accidental escape of the genetically engineered organisms. Certain types of gene transfers were entirely prohibited to avoid contagions and highly toxic products. Cloning vectors were constructed that would not survive outside the laboratory.

Bacterial strain X^{1776} (so designated in honor of the bicentennial anniversary of U.S. independence) had an absolute requirement for diaminopimelinic acid, an essential precursor of lysine absent from the human gut. Cloning bacterial hosts were made deficient for excision repair (*uvrB*), auxotrophic for thymidine (indispensable for DNA synthesis), and mutant for recombination (*rec⁻*) and conjugational transfer of plasmids to other organisms. If reversion frequency for any of, say five defects each, would be in the range of 1×10^{-6}, then the joint probability for simultaneous reversion for all five would be $(10^{-6})^5 = 10^{-30}$. Since the mass of a single *E. coli* cell is about 10^{-12} g, a 5-fold mutation could be expected in a mass of 10^{11} metric tons of bacteria. Obviously, such a mass of bacteria would not likely occur because the earth might not support them. For a comparison, the wheat production of the world in 1980 was only 4.5×10^8 tons, and the estimated mass of planet Earth is 10^{20} tons. To avoid any problem, nevertheless, government authorization is required in all countries where this technology is used for the release of any genetically engineered species (microbes, plants, animals) for the purpose of economic utilization. Objections to such carefully tested releases still occur based not so much on public concerns but on personal or political reasons and most commonly because of ignorance.

During more than 2 decades of practicing recombinant DNA technology, no major accident has happened, and with the guidelines available, none is expected. Before recombinant DNA experiments are initiated, the Institutional Biosafety Committees and Institutional Review Boards must approve the plans. Experiments involving cloning of toxin molecules with LD50 of less than 100 nanograms per kg body weight must be approved by the Office of Biotechnology Activities, National Institute of Health/MSC 7010, 6000 Executive Blvd., Suite 302, Bethesda, MD 20892-7010; tel. 301-496-9838. Such toxins are botulinum, tetanus, diphtheria, and *Shigella dysenteriae* neurotoxin. Specific approval is required for cloning in *Escherichia coli* K12 genes coding for the biosynthesis of toxic substances, which are lethal to vertebrates at 100 ng to 100 mg per kg body weight. Special review and approval by OBA and RAC are required for human experimentation or treatment. OBA sets specific guidelines for different risk categories. Working with plant and animal pathogens requires a permit by the U.S. Department of Agriculture. The U.S. National Institute of Health Guidelines classify the hazardous agents into four groups and name the pertinent agents according to group 1 is the least hazardous and 4 is the most dangerous. A principal investigator who is primarily responsible for the observance of the regulations must report all accidents or any potential hazardous events. *See* biohazards; cancer gene therapy; containment; gene therapy; GMO; Institutional Biosafety Committee; laboratory safety; OBA; RAC; recombinant DNA evolutionary potentials. (<http://www4.od.nih.gov/oba/rac/guidelines/guidelines.html>.)

recombinant inbreds (RI)   Generated for physical mapping of DNA by selfing F1 hybrid populations and selecting single-seed or animal progenies for about 8 generations, until only $(0.5)^8$ (≈ 0.0039) fraction remains heterozygous for a particular marker (linkage ignored). The parental lines are chosen on the basis of differences in their DNA sequences, and the map position of these physical markers can be determined genetically from the data by a combination of molecular and progeny tests. The calculation in animals is

as follows: R (the frequency of discordant individuals) is $R = (4r)/(1 + 6r)$, where r is the recombination in any single gamete. Because interference within very short distances is practically complete, the distance in cM is about $d = 100r$. The recombination fraction ($\hat{r}$) in function of the size of the sample (N) is $\hat{r} = i/(4N - 6i)$, where i is the number of discordant strains and $\hat{d} = 100 \times \hat{r}$ in cM. In plants, the frequency of recombinant monoploid gametes is calculated with basically the same formula: $r = R/(2 - 2R)$, where R is the frequency of homozygous recombinant diploid individuals. *See* congenic resistant lines of mice; congenic strains; RAPD.

recombinant inbred strain panels   Can be used for mapping in mice. Information: Jackson Laboratory Animal Resources, 600 Main St., Bar Harbor, ME 04609; phone: 1-800-422 MICE or 207-288-3371; fax: 207-288-3398.

recombinant joint   Site of connection of two molecules of DNA in a heteroduplex. *See* heteroduplex.

recombinant plasmid   Generated either from two different DNAs by the techniques of molecular biology or by spontaneous or induced genetic recombination. *See* plasmid.

recombinant vaccine   Produced by in vitro modifications of the genes/proteins; it does not carry the full complement of the infectious agent. *See* vaccine.

recombinase   Enzyme that mediates recombination. *See* *Cre/loxP*; *FLP/FRT*; *Rec*.

recombinase system   *See* immunoglobulins.

recombination   Process by which the linkage phase (coupling or repulsion) of syntenic genes is altered. Recombination is most common during meiosis, but mitotic recombination also takes place. The mechanism of the meiotic and mitotic events is not necessarily identical. Independent segregation and reassortment are outside the realm of this term, according to the original definition by H. A. Sturtevant, although some textbooks improperly include these too. Recombination can be precisely assessed with aid of sequenced genomes. The data show great (0 to 9 centimorgans per megabase) variation along each chromosome. The so-called desert sequences display low recombination and the jungle stretches display high recombination, which is represented graphically next page. *See* centimorgan; cold spot; coupling; *Cre/loxP*; flip-flop recombination; *FLP/FRT*; hot spot; linkage; linkage disequilibrium; rec; recombination, ectopic; recombination, homologous; recombination, molecular mechanisms of; recombinational probe; repulsion; sex circle model of recombination; site-specific recombination; STRP. (Cox, M. M. 2001. *Proc. Natl. Acad. Sci. USA* 98:8173; Sturtevant, A. H. 1913. *J. Exp. Zool.* 14:43; diagrams next page.)

recombination, homologous   *See* homologous recombination; recombination.

recombination, illegitimate   *See* illegitimate recombination.

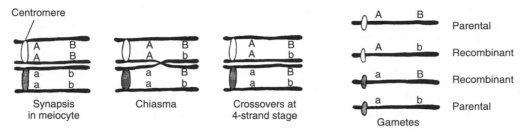

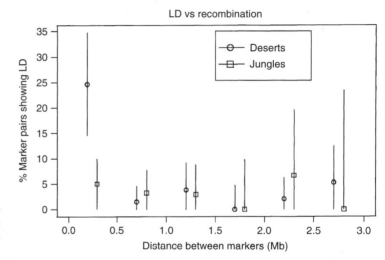

Cytological representation of recombination between homologous chromosomes in coupling phase. Each chiasma leads to crossing over and 50% recombination. The frequency of recombination depends, however, on the distance between the two loci considered. If crossing over takes place in all meiocytes between the bivalents, the frequency of recombination is 50%. If only half of the bivalents undergo crossing over in that particular interval, the recombination frequency will be 25% because, say in 4 meiocytes with 16 chromatids, $4/16 = 0.25$. If only two of the meiocytes display crossing over then $2/16 = 0.125$ is the frequency of recombination. The maximal recombination frequency by a single crossing over is 50%; the minimal may be extremely rare in case the linkage is tight.

Linkage disequilibrium (LD) among pairs of STRPs (microsatellite repeats) within human autosomal recombination deserts and jungles within 0.25 megabase intervals. (Courtesy of Dr. James L. Weber; see also Yu, A., et al. 2001. *Nature* 409:951.)

recombination, intragenic   *See* intragenic recombination.

recombination, modification of   The frequency of recombination may be altered by any means that affect chromosome pairing such as chromosomal aberrations, DNA inserts introduced through transformation, temperature (either low or high), physical mutagens, rarely by chemicals, *rec⁻* genes, etc. In the heterogametic sex of *Drosophila* and silkworm, meiotic recombination is usually absent, although mitotic recombination occurs. In animals, recombination may be more frequent in females than in males, and it is attributed to imprinting. In plants, when there is a sex difference in recombination, its frequency is commonly lower in the megaspore mother cell. *See* coincidence; imprinting; recombination, variation of. (Singer, A., et al. 2002. *Genetics* 160:649; Peciña, A., et al. 2002. *Cell* 111:173.)

recombination, molecular mechanisms of   The RecA protein (M_r 37,842) directs homologous pairing by forming a right-handed helix on the DNA, and it catalyzes the formation of DNA heteroduplexes. X-ray crystallography indicates that the DNA rests in the deep groove of this protein to facilitate the scanning for homologous sequences. The RecA protein is also involved in DNA repair function (SOS repair). It digests the LexA bacterial repressor and is indirectly instrumental in the derepression of over 20 genes involved in recombination and UV mutagenesis. The mechanism(s) of RecA activities can be studied by in vitro reactions. RecA can interact with three or four DNA strands by wrapping around the paired molecules. DNA-DNA pairing can also take place between linear and circular DNA. Strand exchange proceeds at a slow pace (2 to 10 base/sec) in a polar fashion (5′ to 3′). The transfer begins at the 3′-end of the duplex and is then transferred to a single-strand DNA. Homology is a requisite for the RecA-mediated reactions, yet it tolerates some mismatches or insertions (up to even 1,000 bases or more) that slow down the reactions. RecA can mediate pairing between two duplexes as long as there are short single-strand stretches or gaps. Low pH, intercalating chemicals, Z-configuration, and other structural changes of the DNA may alleviate the difficulties of binding two duplexes. The RecA protein is a low-efficiency ATPase. ATP hydrolysis is not an absolute requirement, however, for RecA activities in recombination but is more important for repair reactions. In the presence of ATP, the conformation of RecA is altered. In the nucleoprotein complex, the DNA is substantially underwound (the spacing between bases is extended from 3.4 Å to 5.1 Å). It is assumed that the paired DNA molecules are not just juxtapositioned but one molecule lays in the major groove of the other. The pairing may involve three or four strands.

DNA strand exchange requires that the RecA filament rotates along the longitudinal axis and the DNA molecules are "spooled" inside where they may form a Holliday junction (*see* Holliday model). ATP stabilizes the RecA–DNA association, and when ATP is split into ADT, the heteroduplex is released and RecA is recycled. Besides the RecA protein, recombination requires the presence of a single-strand binding protein (SSB), DNA polymerase I, DNA ligase, DNA gyrase, DNA topoisomerase I, and the products of genes *recB, recC, recD, recE, recF* (binding protein for single-strand DNA), *recG, recJ* (exonuclease acting on single-strand DNA), *recN, recO, recQ, RuvB* (helicases), *recR, ruvR, ruvB,* and *ruvC.* RuvC nicks the DNA at the point of strand exchange. RecBCD is a protein-RNA complex encoded by three genes (see p. 1024) that has the activities of (1) ATP-dependent double-strand exonuclease, (2) ATP-dependent single-strand exonuclease, (3) unidirectional DNA helicase, (4) site-specific endonuclease to nick four to six nucleotides dowstream of *chi,* a recombinational hot spot (5'-GCTGGTGG-3'). It was suggested that RecBCD generates 3'-tails that are utilized by protein RecA for DNA strand exchange. RecB and RecC mutations can be suppressed by *sbcA* and *sbcB* mutations. Mutations in *sbcA* lead to the activation of exonuclease VIII product of *recE.* Mutation in *scbB* inactivates exonuclease I, an enzyme that digests single-strand DNA, and its inactivation may assist the function of RecA in recombination (*see* models of recombination). The precise mechanism of how the Holliday junction (*see* Holliday model, steps I to L) is resolved is not clear, but endonuclease (RuvC) activity is postulated.

Bacteriophage T4 gene *49* encodes endonuclease VII, which under natural conditions splits branched DNA structures. Similarly, bacteriophage gene *3* encodes endonuclease I and cleaves branched DNAs. Some of the functions of the *ruv* operon of *E. coli* may be involved in the resolution of the Holliday junctions. *E. coli* also has in vivo systems in which the molecular mechanism of resolution of recombination intermediates can be studied. Covalently closed plasmid DNA, DNA polymerase I, and DNA ligase were transformed into *E. coli recA* mutants. Both monomeric and dimeric plasmid progenies were found, and the available markers permitted the conclusions that crossing over occurred in 50% of the progeny. Recombination is not limited to DNA, but viral RNA molecules can also recombine.

The molecular mechanisms of recombination in eukaryotes share many features with those of prokaryotes. It appears that double-strand breaks can stimulate homologous recombination within 1 kilobase of the site of the break or may affect recombination at a distance exceeding 30 kb. At the break, a *recombination machine* may gain entry, and as the machine moves, a heteroduplex of the DNA may form. At the broken ends, DNA replication may be primed and there is a potential for recombination. The recombination machine is a complex of many enzymes mediating the recombination process. *See* chi-elements; illegitimate recombination; recombination, RNA viruses; recombinational probe; recombination by replication; recombination mechanisms, eukaryotes; recombination models. (Camerini-Otero, R. D. & Hsieh, P. 1995. *Annu. Rev. Genet.* 29:509; Barre, F.-X., et al. 2001. *Proc. Natl. Acad. Sci. USA* 98:8189; West, S. C. 1992. *Annu. Rev. Biochem.* 61:603;

Cox, M. M. & Lehman, I. R. 1987. *Annu. Rev. Biochem.* 56:229; Smith, G. R. 2001. *Annu. Rev. Genet.* 35:243.)

recombination, RNA viruses   After coinfection of a cell by two different viruses, recombination may take place by template switching during replication; thus it is more like a copy choice than a breakage and reunion mechanism. The recombination can take place between homologous and non-homologous strands (illegitimate recombination). The latter mechanism may lead to deletions, duplications, and insertions. Among picornaviruses, the recombination frequency may be as high as 0.9 in case of high homology. Recombination in RNA viruses helps to eliminate disadvantageous sequences and can generate new variants. The estimated mutation rate per base is 6.3×10^{-4} and per genome about 5. The mutation rate is estimated as mutations per replication. *See* breakage and reunion; copy choice; illegitimate recombination; negative interference; reverse transcription. (Keck, J. G., et al. 1987. *Virology* 156:331; Kirkegaard, K. & Baltimore, D. 1986. *Cell* 47:433; Negroni, M. & Buc, H. 2001. *Annu. Rev. Genet.* 35:275.)

recombination, targeted   *See* Cre/loxP; FLP/FRT.

recombination, variations of   In the heterogametic sex of arthropods (male *Drosophila,* female silkworm), genetic recombination is usually absent or highly reduced. In the latter group of organisms, mitotic recombination occurs, however, and these premeiotic exchanges may account for the observation of recombinants. The most common cause of variation is the presence of *rec⁻* genes. In the Abbott stock 4A × Lindegren's wild-type crosses of *Neurospora,* postreduction frequency was found to be 4.6 ± 1.2, whereas in the Lindegren's stock it was 13 ± 1.2, and in the Emerson's × Lindegren's crosses 27.6 ± 3.7. L. J. Stadler, a pioneer of maize genetics, considered recombination as one of the most variable biological phenomena. *See* recombination, modification of; recombination frequency; recombination hot spots; tetrad analysis. (Browman, K. W., et al. 1998. *Am. J. Hum. Genet.* 63:861.)

recombination by replication   At the beginning of the 20th century, William Bateson suggested that recombination is basically associated with the process of replication. At that time, neither of these phenomena were sufficiently understood or could even be hypothesized meaningfully. On the basis of cytological evidence (1930s) for marker exchange accompanied by chromosome exchange and later evidence that DNA exchange and phage gene exchange were correlated, the generally accepted view was that recombination does not require replication. The discovery of gene conversion remained, however, a puzzling phenomenon, although it was observed that about 50% of the gene conversion events involved flanking marker exchange. The Holliday and other molecular models of recombination (during the 1960s and 1970s) permitted interpretation of classical crossing over and gene conversion without significant replication. Recently, it was found the mutation or loss of function of the PriA DNA replication protein blocked both replication and recombination in *E. coli.*

The SOS DNA repair activates a replication process that does not require the replicational *oriC* site or the normally functioning DnaA protein, but it needs RecA and RecBCD activities. It was assumed and subsequently demonstrated that double-strand breaks may be assimilated into the DNA, resulting in double D loops in the presence of nearby chi-elements. The *chi*-elements block nuclease activity and assist in the initiation of replication. It appears that PriA and other proteins of the primosome generate a replication fork at the D loop; relying on the DnaB helicase and the DnaG primase, replication and recombination can be turned on. Apparently, lagging strand synthesis begins by the replisome, and the lagging strand then primes the synthesis of the leading strand. Defective PriA may be compensated for by some elements of the primosome. The processes of double-strand break repair and recombination appear the same, with the exception that in repair only the defective region has to be corrected, whereas in recombination the entire strand must be replicated in order to recover the recombinants. There are some observations that indicate joint events of replicational repair and recombination also in eukaryotes. *See* breakage and reunion; chi-element; D loop; DNA repair; gene conversion; Holliday model; lagging strand; leading strand; recA; recBCD; recombination molecular models; reduplication theory; replication fork; replisome; SOS repair. (Michel, B., et al. 2001. *Proc. Natl. Acad. Sci. USA* 98:8181.)

recombination by transcription Some of the instabilities may be induced by RNA polymerase II, particularly between repeats of the eukaryotic chromosomes. Some yeast mutants may increase the process by over three orders of magnitude. (Gallardo, M. & Aguilera, A. 2001. *Genetics* 157:79.)

recombination cloning Based on the integration/excision mechanism of the lambda-phage and *E. coli* bacterium. Integration involves the λ-*attP* site within the bacterial *attB* site. The bacterial *attL* and *attR* sites flank the integrated phage genome. Excision reverses the process. Such a procedure can be used for the generation of vectors, with DNA-binding and activation domain sites facilitating the study of protein interaction by analogy of the two-hybrid method. *See* att sites; two-hybrid method. (Muyrers, J. P., et al. 2001. *Trends Biochem. Sci.* 26[5]:325.)

recombination frequency Linkage is generally noticed in F_2 when independent segregation of the genes does not occur. Two genes in the homologous chromosomes can be at two different arrangements: repulsion (*Ab/aB*) or coupling (*AB/ab*). Some people call the repulsion the trans and the coupling the cis arrangement. Recombination is most commonly calculated as the percentage of recombinants in a test cross population. This frequency is maximally 50% because at this value the frequencies of recombinant and parental chromosomes are equal, i.e., the segregation is independent. Linkage is usually first observed in F_2 by deviation of the phenotypic proportions from the expectations for independent segregation. Example:

	Phenotypic Classes Expected			
	AB	Ab	aB	ab
Independent Segregation →	9/16	3/16	3/16	1/16
Linkage, Repulsion →	less	more	more	less
Linkage, Coupling →	more	less	Less	more

Phenotypic classes in test crosses in two linkage phases and recombination:

	(A hypothetical case)			
	AB	Ab	aB	ab
Repulsion cross (*Ab/aB*) x *ab*	5	45	45	5
Coupling cross (*AB/ab*) x *ab*	45	5	5	45

Aleurone color (*C*) and shrunken endosperm (*sh*) genes of maize are closely linked in chromosome 9 of maize. On the two ears these markers are in different linkage phase. (From Hutchison, C. B. 1921. *J. Hered.* 12:76.)

The linkage phase does not affect the frequency of recombination, but it affects the frequency of the phenotypic classes. The frequency of recombination is the same in both cases $(5 + 5)/100 = 0.10 = 10\%$, as the table shows. In F_2, recombination frequencies cannot be calculated in such a simple way because in the heterozygotes the genetic constitution of the individual chromosome strands is concealed but may be revealed in F_3. Nevertheless, recombination frequencies can be calculated (*see* F_2 linkage estimation). Recombination takes place at the four-strand stage of meiosis (*see* exception of mitotic recombination). The bivalents pair and at the simplest case two chromatids exchange segments. The maximal frequency of recombination within a chromosomal interval is 50%. Recombination frequencies are generally converted to map units by multiplication with 100. The realistic conversion of recombination frequencies into map units generally requires the use of *mapping functions* because some of the recombinational events may not be detectable if the frequency of recombination between markers exceeds 15%. In physical measures one map unit has a different meaning in different organisms, depending on the size of the genome in nucleotides (nucleotide pairs) and the genetic length of the genome. Thus, 1 map unit in the plant *Arabidopsis* appears to mean about 150 kbp, in maize the same is about 2,140 kbp, and in humans about 1,100 kbp. In human male autosomes, the mean meiotic recombination frequency was estimated to be 8.9×10^{-3} per megabase by another study. Generally, smaller chromosomes display higher rates of recombination (cM/kb) not only among lower versus higher eukaryotes but also within one organism (yeast). The frequency of no recombination is a function (f) of the intensity of linkage and the population size: $f = (1 - r)^n$, where r is the recombination fraction and n is the number of test-cross progeny. Data obtained in maize (Fu, H., et al. 2002. *Proc. Natl. Acad. Sci. USA* 99:1082) indicate a much reduced frequency of recombination in regions containing methylated retrotransposons. *See* bacterial recombination; chiasma; F_2 linkage estimation; F_3 linkage estimation; hot spot; mapping; mapping function; maximum likelihood method applied to recombination frequencies; recombination, modification of;

recombination, variation of; sperm typing; test cross, product ratio method.

recombination frequency in bacteria *See* bacterial recombination.

recombination hot spots
Genetic recombinations do not occur uniformly along the physical length of the DNA (chromosomes). In *Arabidopsis*, 1 cM varied from 30 base pairs to >550,000 bp. Gene-rich regions display more exchanges than gene-poor sequences. Recombination is usually suppressed around or near the centromere or telomere. In wheat, 1 cM in a gene-rich region was estimated to be 118 kb, whereas it was 22,000 kb for gene-poor regions. In humans, 1 cM indicates 1 Mb, yet substantial variations are common. *See* coefficient of crossing over. (Faris, J. D., et al. 2000. *Genetics* 154:823.)

recombination in autotetraploids
Measuring linkage and recombination in autotetraploids is much more difficult than in diploids because of the multiplicity of the chromatids and alleles and because the segregation ratios are not simple to predict from genetic data without cytological information. The difficulties are almost insurmountable when the F_1 is a duplex or triplex and the genes are far apart. Even in close linkage, generally large populations are required. In case of

a coupling test cross, in simplex individuals the procedure is very similar to that of a test cross in diploids, as can be seen in an example. (After deWinton, D. & Haldane, J. B. S. 1931. *J. Genet.* 24:121):

Phenotypes observed in coupling

Parental	Recombinant	Recombinant	Parental	Total
SG	Sg	sG	sg	
336	215	210	353	1114

$$\text{Recombination frequency (p)} = \frac{Sg + sG}{\text{Total}}$$
$$= \frac{215 + 210}{1114} = \frac{425}{114} \cong 0.38$$

In case of repulsion, the calculation presupposes a knowledge of the possible gametic series that can be derived as shown in the figure below, and it is

$$(1)\frac{SB}{sb} : (p)\frac{SB}{sb} : (2-p)\frac{Sb}{sb} : (2-p)\frac{sB}{sb} : (1+p)\frac{sb}{sb}$$

The manipulation of autetraploids with the techniques available for classical genetics is impractical despite the

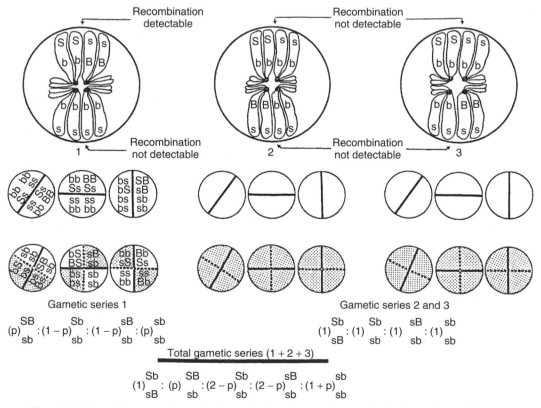

The derivation of the gametic series in a simplex tetrasomic case in repulsion. Recombination is considered only between loci *S* and *B*. In meiosis, three different quadrivalent associations are possible as shown at the top. Recombination is detectable only among the descendants of quadrivalent 1. The second row represents the different types of disjunctions at anaphase I, and the third row of circles shows the types of gametic tetrads formed. Gametes containing recombinant strands are open; gametes with parental strands are shaded.

theoretical framework, and molecular analyses are also still lagging.

Phenotypes Observed in Repulsion

SG (+Sg/sG)	Sg	sG	sg	Total
164	193	206	154	717

$$\text{Recombination frequency} = \frac{SG + sg}{\text{Total}} = \frac{2(1 + p)}{2(1 + p) + 2(2 - p)}$$

$$= \frac{318}{717} \cong 0.44$$

See alpha parameter; autopolyploids. (Luo, Z. W., et al. 2001. *Genetics* 157:1369; Wu, S. S., et al. 2001. *Genetics* 159:1339; Hackett, C. A., et al. 2001. *Genetics* 159:1819.)

recombination in vitro *See* staggered extension process.

recombination machine *See* recombination, molecular mechanisms.

recombination mechanisms, eukaryotes *YEAST*: The Sep 1 (strand exchange protein) 132 kDa fragment of a 175 kDa protein of yeast initiates the transfer of one DNA strand from a duplex to a single-stranded circle with 5′ to 3′ polarity without an ATP requirement. It also has 5′ to 3′ exonuclease activity and is probably required for the preparation of the 3′-end of single- and double-stranded DNA molecules for recombination. Mutation in Sep reduces mitosis, sporulation, meiotic recombination, and genetic repair. (The STPβ protein, encoded by gene *DST2/KEM1*, is probably identical to Sep 1.) One monomer of Sep 1 binds to about 12 nucleotides of single-stranded DNA. This requirement is reduced by the presence of the 34 kDa protein, which at a concentration of 1 molecule per 20 nucleotides reduces the requirement for Sep 1 to about 1/100. The DPA protein (120 kDa) of yeast controls DNA pairing and promotes heteroduplex formation in a nonpolar manner independently of ATP. It promotes single-strand transfer from double-stranded DNA to single-stranded circular DNA if the former has single-stranded tails. Protein STPα (38 kDa) increases 15-fold shortly before yeast cells are committed to recombination during meiosis. If the gene encoding it (*DST1*) mutates, meiotic recombination is greatly reduced without an effect on mitotic recombination. The *RAD50* gene product (130 kDa) has an ATP-binding domain and it binds stoichiometrically to duplex DNA. The *RAD51* gene product is homologous to the RecA protein of *E. coli* (*see* recombination mechanism, prokaryotes) and binds single- and double-stranded DNA. The *DMC1* (*disrupted meiotic cDNA*) gene product appears during meiosis; along with the product of *RAD51*, it performs functions similar to RecA in prokaryotes. Meiotic recombination in yeast is believed to involve double-stranded breaks of the DNA. The actual site of exchange may be 25 to 200 kb away from the prominent break (Young, J. A., et al. 2002. *Mol. Cell* 9:253). If DNA replication, which normally occurs

1.5 to 2 hours before double-stranded breaks, is blocked or delayed, recombination does not take place. *Drosophila*: Protein Rrp 1 seems to promote exchanges between single-stranded circular and linear duplex DNA. Its C-terminus has homology to *E. coli* exonuclease III and *Streptococcus pneumoniae* exonuclease A. *MAMMALIAN CELLS*: HPP-1 (human pairing protein with 5′ to 3′ exonuclease activity) binds to DNA and promotes strand exchange in the 5′ to 3′ direction, and it does not require ATP. Addition of the hRP-A (human single-strand binding) protein stimulates pairing about 70-fold and reduces the amount of HPP-1 requirement (cf. SF1 in yeast). The precise mechanism of how the Holliday junction (*see* Holliday model, steps I to L) is resolved is not clear, but endonuclease activity is postulated. Bacteriophage T4 gene *49* encodes endonuclease VII, which under natural conditions cuts branched DNA structures. Similarly, bacteriophage T7 gene *3* product encodes endonuclease I, which cleaves branched DNAs. In yeast, endonuclease XI (Endo XI, $\approx M_r$ 200,000, and other Endo proteins) was found in cells with mutations in the RAD genes and apparently cut cruciform DNA of the type expected by the Holliday juncture. *See* chiasma; databases; gene conversion; RAD1; RAG; recombination, molecular mechanisms in prokaryotes; recombination hot spots; recombination models; Sep 1; sex circle model; STPβ; synaptonemal complex. (Camerini-Otero, R. D. & Hsieh, P. 1995. *Annu. Rev. Genet.* 29:509; Baudat, F. & Keeney, S. 2001. *Curr. Biol.* 11:R45; Smith, G. R. 2001. *Annu. Rev. Genet.* 35:243.)

recombination minimization map Based on a skeletal map. The ordering relies on the smallest number of recombinations for single intervals. *See* skeletal map.

recombination models *See* Holliday model; Meselson-Radding model; Szostak et al. model.

recombination nodule Suspected site of recombination seen through the electron microscope as a densely stained structure 100 nm in diameter adjacent to the synaptonemal complex. There are early nodules seen at the association sites of the paired meiotic chromosomes and the late nodules are visible at pachytene when crossovers are juxtaposed. Noncrossovers do not show nodules after midpachytene. *See* association point; chiasma; crossing over; meiosis; pachytene; recombination; recombination, RNA viruses; synaptonemal complex. (Zickler, D., et al. 1992. *Genetics* 132:135; Anderson, L. K., et al. 2001. *Genetics* 159:1259.)

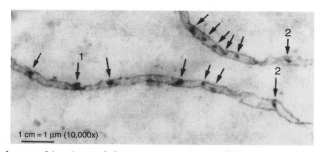

Early recombination nodules on two synaptonemal complexes of *Allium cepa* (onion). at positions marked 2 there is no synapsis yet. (Courtesy of Drs. L. K. Anderson and S. M. Stack.)

recombination probe One short probe is inserted into the 902 bp πVX miniplasmid containing a polylinker and the *supF* suppressor gene. Lambda-phage libraries also containing the miniplasmid construct are then propagated. If the phage carries a *supF*-suppressible amber mutation, recombination between sequences homologous to the probe can selectively be recovered by forming plaques on an *E. coli* lawn. The recombination may take place even in the absence of perfect homology; less than ca. 8% divergence may be tolerated. The very large populations may reveal recombination within 60-base or longer probes very effectively. *See* lawn; miniplasmid; πVX; *rec*; *Rec*; *supF*. (Perry, M. D. & Moran, L. A. 1987. *Gene* 51:227.) Figure below.

recombination repair *See* DNA repair.

recombination repeats Exist in the majority of plant mitochondrial DNAs in 1 to 6 pairs. Recombination can take place between/among these repeats, generating various subgenomic DNA molecules. These repeats (264 bp to >5 kbp in maize) may not be indispensable parts of the mtDNA, although they may contain genes for rRNA, cytochromes, etc. *See* mitochondrial genetics; mtDNA; rRNA. (Chanut, F. A., et al. 1993. *Curr. Genet.* 23:234.)

recombinator Cis-acting chromosomal sites promoting homologous recombination. *See* chi.

recombineering With the aid of a phage vector, large DNA molecules can be cloned into bacterial artificial chromosomes. PCR-amplified linear or double-stranded DNAs are introduced into targeting cassettes that have either short regions of homology at their end or single-stranded oligonucleotides. The inserted DNA may carry any type of mutation or modification, and there is no need for restriction enzyme cuts because the insertion is by homologous recombination. The method is simpler and safer than using the somewhat unstable YACs and can be applied to functional genomic studies of higher organisms, e.g., to mice. (Copeland, N. G., et al. 2001. *Nature Rev. Genet.* 2:769; Court, D. L., et al. 2002. *Annu. Rev. Genet.* 36:361.)

recombinogenic Agent (mutagen) that increases recombination.

recombinogenic engineering *See* recombineering.

recon Historical term for the smallest recombinational unit. Molecular genetics has shown that recombination can take place between two nucleotides within a codon. (Benzer, S. 1957. *The Chemical Basis of Heredity*, McElroy, W. D. & Glass, B. eds., p. 70. Johns Hopkins Univ. Press, Baltimore, MD.)

reconstituted cell Produced by fusing cytoplasts and karyoplasts. *See* cytoplast; karyoplast; transplantation of organelles.

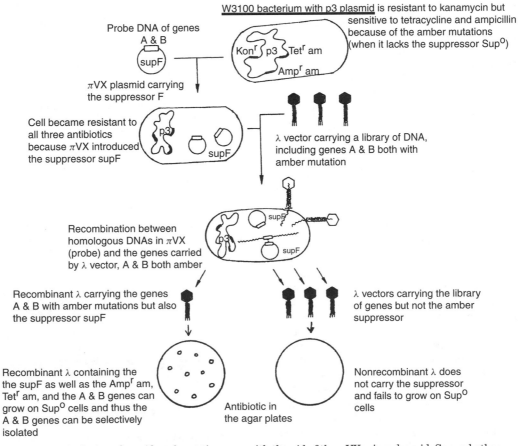

Selective isolation of specific eukaryotic genes with the aid of the πVX microplasmid. Several other plasmids have been constructed for similar purposes.

reconstituted virus A complete viral genetic material is introduced into an empty viral capsid, e.g., into the coat of tobacco mosaic virus (TMV) the genome of the related Holmes ribgrass virus (HRV) was introduced and the new particle expressed the characteristic functions of the donor RNA. This classic experiment proved that the genetic material can also be RNA. Influenza and other viruses can be reconstituted from cloned cDNAs. (Fraenkel-Conrat, H. & Singer, B. 1957. *Biochim. Biphys. Acta* 24:540; Neumamnn, G., et al. 1999. *Proc. Natl. Acad. Sci. USA* 96:9345.)

record True document of an observation or of a hypothesis, in each case explicitly stated.

recoverin *See* rhodopsin.

recovery, of DNA fragment from agarose gel The fragment is driven by electrophoresis onto DEAE cellulose membrane by cutting a slit in front of the band and placing the DEAE sliver in the slit. Alternatively, electroelution can be used or from low-melting-temperature agar the DNA can be extracted by phenol and precipitated by ammonium acetate in 2 volume ethanol. The DNA is collected by centrifugation. *See* DEAE cellulose; electrophoresis.

RecQ Family of helicase proteins. In case of mutation in the coding gene, chromosomal instability results in eukaryotes. In prokaryotes, *recQ* is involved in postreplicational repair. *See* chromosomal rearrangements; helicase; RecA; RecB. (Enomoto, T. 2001. *J. Biochem [Tokyo]* 129:501; Wu, X. & Maizels, N. 2001. *Nucleic Acids Res.* 29:1765; Cobb, J. A., et al. 2002. *FEBS Lett.* 529:43.)

RECQL RecQ-like protein in humans. *See* RecQ.

RecR Mediates DNA renaturation during recombination of *E. coli*. (Pelaez, A. I., et al. 2001. *Mol. Genet. Genomics* 265:663.)

recruitment For the initiation of transcription, some prokaryotic and eukaryotic genes require activators. The transcriptional complex will operate only if these activators are attracted to the transcriptional target. The GAL1 gene of yeast recruits the four units of the GAL4 activator at about 250 bases upstream to begin transcription. The GAL4 units are blocked, however, unless galactose is available in the culture medium. The activators make contact with some sites of the RNA polymerase subunits. In bacteria, typical activators are the CAP (catabolite activator protein) and the λ-repressor, which may bind to the σ^{70} subunit of the polymerase. In yeast, the transcription complex includes more than 30 different proteins. *See* activator proteins; transcription complex; transcription factors; two-hybrid method. (Francastel, C., et al. 2001. *Proc. Natl. Acad. Sci. USA* 98:12120.)

recruitment, exon Evolving genes may acquire coding sequences for functional domains by borrowing exons through recombination. The recruited DNA sequences may occur in several protein genes of different functions, e.g., the low-density lipoprotein (LDP) receptor (a cholesterol transport protein) has homology in eight exons with the epidermal growth factor (EGF) peptide hormone gene. *See* EGF; exon; LDP.

recruitment, gene Acquiring new genetic information through recombination or transfection (transformation). *See* recruitment of exons; transfection; transformation.

RecT Encoded by *recE*. It is involved in renaturation of homologous single-stranded DNA and pairing of DNA. *See* pairing; RecE.

rectification, inward Through a voltage-gated ion channel, the inward current exceeds the outward current. In case of outward rectification, the opposite is true. *See* ion channel.

recurrence risk Chance of having another child with the same defect by the same couple. *See* empirical risk; genetic risk; genotypic risk ratio; λ_S; risk.

recurrent parent Plant or animal is mated with selected lines in several cycles for one or several backcrosses. *See* recurrent selection.

recurrent selection (reciprocal recurrent selection) Variety of methods used for breeding superior hybrids of plants and animals of high productivity. The general procedure: Lines (inbred or not) A and B are crossed in a reciprocal manner, i.e., A males are crossed to B females and B males are mated with A females. The initial lines are expected to be genetically different to assure the sampling of different gene pools. In this manner, several lines are mated, not just A and B. The progenies are tested for performance and only the best parents are saved. The superior parents are then mated again to representatives of their own line. On the basis of their progeny, the parents are reevaluated. The mating cycle is repeated. Most commonly, each male is mated to several females of the other line in order to assure the availability of a sufficiently large population of offspring to be able to carry out statistically meaningful tests. The maintenance of the lines requires that the females be mated to selected males within their own lines. This procedure results in inbreeding but enhances the chances for further selection. Therefore, the performance of the selected parental lines is expected to decrease but that of the hybrids will increase. An alternative simplified method involves selection for combining ability in only one set of lines. Thus, line A is mated with a previously inbred tester that has a known combining ability and the selection is restricted to within line A. This latter modification provides faster initial progress but the final gain may be limited. *See* combining ability; diallele analysis; heritability; heterosis; hybrid vigor; QTL. (Hull. K. 1945. *Am. Soc. Agron.* 37:134.)

recursive partitioning Statistical approach for classifying information into alternative classes, e.g., normal or tumorous. It builds classification rules on the basis of feature information. The principle is as follows: Observations of *n* units represent a vector feature of measurement (vector means a single class matrix); or covariates (e.g., data from a type of condition) and a class label. Unlike linear discriminant analysis, recursive partitioning extracts homogeneous strata and constructs tree-based classification rules (strata means division of the data into parts). The information is partitioned into smaller and smaller samples (nodes) to facilitate critical discrimination between classes. *See* cancer classification; discriminant function; matrix, algebraic. (Zhang, H., et al. 2001. *Proc. Natl. Acad. Sci. USA* 98:6730.)

recursive PCR Method of DNA amplification. Synthetic oligonucleotide primers (50–90 bases) are used that have only terminal complementarity (17–20 bp). They are annealed at 52 to 56°C. The heating cycles are 95°C; the cooling is at 56°C. The Vent polymerase is used at 72°C. This thermostable polymerase has the capability not just for strand displacement but for exonuclease function, and it carries out proofreading; therefore, the fidelity of the amplification is very good. During the initial steps, each 3′-end is extended with the aid of the opposite strand as a template, and duplex sections are generated. In further cycles, one strand of the duplex is displaced by a primer oligonucleotide derived from one of the neighboring duplexes. During the last step, high concentrations of the terminal oligonucleotides assist in the amplification of the entire duplex. *See* polymerase chain reaction; Vent. (Prodromou, C. & Pearl, L. H. 1992. *Protein Eng.* 5:827.)

red *See* Charon vectors; lambda-phage.

RED613 Fluorochrome, a conjugate of R-phycoerythrin and Texas Red. Its excitation maximum is at 488 nm and emission at 613 nm. *See* fluorochromes.

RED670 Fluorochrome, a conjugate of R-phycoerythrin and a cyanine. Its excitation maximum is at 488 nm from an argon-ion laser, emission at 670 nm. *See* excitation.

red blood cell *See* blood; erythrocyte; sickle cell anemia.

red-green color blindness *See* color blindness.

redifferentiation Organ or organism formation from dedifferentiated cells, such as from callus. *See* callus.

Redifferentiation of leaves from *Arabidopsis* callus.

redox pair Electron donor and the oxidized derivative.

redox reaction *See* oxidation-reduction.

red queen hypothesis If a population does not continue to adapt at the same rate as its competitors, it will lose ecological niches where it can succeed, and if it stays put long enough, it may become extinct. The name RQ was adapted from Lewis Carrol (pen name of C. L. Dodgson) in 1872 in a fantasy story about a chess game, *Through the Looking Glass*. *See* adaptation; beneficial mutation; equilibrium in populations; extinction; genetic homeostasis; treadmill evolution. (Van Valen, L. 1973. *Evol. Theory* 1:1.)

reduced representation shotgun *See* RRS.

reducing sugar Its carbonyl carbon is not involved in a glycosidic bond and can thus be oxidized. Glucose and other sugars can reduce ferric or cupric ions, and this property is useful for their analytical quantitation (Fehling reaction).

reductant Electron donor.

reduction Gain of electrons.

reductional division At meiotic anaphase I, half of the chromosomes segregate to each pole and the two daughter cells have n number of chromosomes rather than $2n$ as the original meiocyte contained. In case of uneven numbers of crossing over between genes and centromeres, the numerical reduction of the chromosomes may not result in separation of the different pairs of alleles, i.e., the reduction does not extend to the alleles. The reductional division at meiosis assures the constant chromosome numbers in the species and serves as a basis for Mendelian segregation. *See* meiosis; postreduction; prereduction; tetrad analysis.

reductional separation In meiotic anaphase I, the parental chromosomes separate intact because there is no recombination between the bivalents. *See* equational separation.

reductionism Reducing ideas to simple forms or making efforts for explaining phenomena on the basis of the behavior of elementary units. This endeavor is frequently criticized because of the complexities of biological systems. One must also consider that without the analytical approach, science (molecular genetics) would not have progressed to the present stage.

reductive evolution During evolution, some obligate intracellular parasites (*Rickettsia, Chlamydia, Mycobacterium leprae*) lost a significant fraction (up to 76%) of the coding capacity of their genetic material. These organisms may still have many pseudogenes and apparently degraded nonfunctional DNA tracts. *See* leprosy; *Rickettsia*,

redundancy Repeated occurrence of the same or similar base sequences in the DNA or multiple copies of genes. The repeated gene sequences are considered to have duplicational origin. About 38–45% of the sampled (ca. one-third of all) proteins in *E. coli* are expected to be duplicated, and in the much smaller *Haemophilus influenzae* genome, which was completely sequenced by 1995, 30% appears to have evolved by processes involving duplications. In this small genome, some gene families were represented by 10 to over 40 members, whereas almost 60% of the genes appeared unique. In yeast, about 60% of the genes are redundant. In the large eukaryotic genomes, the repetitious sequences are represented by much larger fractions. The number of protein kinases in higher eukaryotic cells may reach 2,000 and that of phosphatases about 1,000. Theoretically, true redundancy should succumb to natural selection unless the rate of mutation is extremely low. In case the redundant genes have unique roles in addition to the shared functions, they are maintainable. Redundant genes are also saved when the *developmental error* rate is high. They may affect the fitness of an organism in a very subtle way and may not show clear independent phenotype. Also, they may be serving as insurance for loss or inactivation of other members of the gene family. Redundant genes may have the luxury of affording mutation to new functions. Mutation or deletion of redundant genes may not have phenotypic consequences. Some single-copy genes can also be removed without consequence for the phenotype. *See* inverted repeat; LINE; LOR; MER; MIR; polyploid; SINE; tandem repeat. (Goldberg, R. B. 1978. *Biochem. Genet.* 1–2:45.)

reduplication hypothesis At the dawn of Mendelism, William Bateson postulated that genetic recombination takes place by a differential degree of replication and different associations of genes after they separate at interphase rather than by breakage and reunion of synapsed chromosomes. *See* breakage and reunion; copy choice. (Bateson, W. & Punnett, R. C. 1911. *J. Genet.* 1:293.)

reelin Protein encoded by the *reeler* gene in mice and expressed in the embryonic and postnatal periods. It is similar to extracellular matrix serine proteases involved in cell adhesion. Reelin controls layering and positioning of neurons and mutations in gene impair coordination, resulting in tremors and ataxia. *See* ataxia; CAM. (D'Arcangelo, G., et al. 1995. *Nature* 374:719; Keshvara, L., et al. 2001. *J. Biol. Chem.* 276:16008; Quattroccchi, C. C., et al. 2002. *J. Biol. Chem.* 277:303.)

REF RNA-binding nuclear protein that facilitates the export of the spliced mRNA from the nucleus to the cytoplasm in cooperation with TAP. *See* TAP. (Le Hir, H., et al. 2001. *EMBO J.* 20:4987.)

Reference Library Database (RLDB) Cosmid, YAC, P1, and cDNA libraries for public use on high-density filters. Information: Reference Library Database, Imperial Cancer Research Fund, Room A13, 44 Lincoln's Inn Fields, London WC2A 3PX, UK; phone: 44-71269-3571; fax: 44-71-269-3479; INTERNET: <genome@icrf.icnet.uk>; <http://lecb.ncifcrf.gov/usefulURLs.html/>.

refractile bodies *Paramecia* may contain bacterial symbionts and the bacteriophages associated with them may appear as bright (refractile) spots under the phase-contrast microscope. *See* killer strains; *Paramecium*; symbionts, hereditary.

refractory genes Interfere with completion of the life cycle of parasites, e.g., *Plasmodium*, within the insect vector, e.g., mosquitos, and thereby can be exploited for the control of malaria and other diseases. *See* malaria; *Plasmodium*. (Yan, G., et al. 1997. *Evolution* 51:441.)

refractory mutation Genetic testing may not reveal its existence, although refractory mutation may lead to genetic disease, e.g., mutation in introns, promoters, or 3′-downstream regulatory sequences that control transcript levels. *See* genetic testing.

RefSeq Reference sequence in a few key genomes, transcripts, and proteins. (Pruitt, K. D. & Maglott, D. R. 2001. *Nucleic Acid Res.* 29:137; <http://www.ncbi.nlm.nih.gov/Locus Link/>.)

Refsum diseases (10pter-p11.2, 8q21.1, 7q21-q22, 6q23-q24) Autosomal-recessive disorders occurring in adult and early-onset forms. Both forms involve phytanic acid accumulation because of the deficiency of an oxidase enzyme residing in the peroxisomes. The symptoms include polyneuritis (inflammation of the peripheral nerves), cerebellar (hind part of the brain) anomalies, and retinitis pigmentosa. The early-onset form, in addition, displays facial anomalies, mental retardation, hearing problems, enlargement of the liver, lower levels of cholesterols in the blood, and accumulation of long-chain fatty acids and pipecolate (a lysine derivative). The symptoms of the infantile form overlap with those of Zellweger syndrome. *See* microbodies; phytanic acid; retinitis pigmentosa; Zellweger syndrome. (Mukherji, M., et al. 2001. *Hum. Mol. Genet.* 10:1971.)

regeneration, animal More limited than in plants where totipotency is preserved in most of the differentiated tissues. Regeneration can actually be classified into two main groups of functions: (1) the regular replacement of cells (e.g., epithelia, hairs, nails, feathers, antlers, production of eggs and sperms, etc.) in a wide range of animals; (2) the capacity to regenerate body parts lost by mechanical injuries. The latter type of regeneration may involve the formation of an entire animal from pieces of the body, such as what takes place by morphallaxis in, e.g., sponges, Hydra, flatworms, annelids (preferentially from the posterior segments), echinoderms, etc. A more limited type of regeneration is found in the higher forms. Arthropods may replace lost appendages of the body. The vertebrate fishes can replace lost fins, gills, or repair lower jaws. Some amphibians (salamanders, newts) readily regenerate lost limbs, tails, and some internal organs. The reptile lizards can reproduce lost tails, although the regenerated tail is not entirely perfect. Regeneration of feathers and repair of beaks may take place in birds. In mammals, lost blood cells may be replenished by activity of the bone marrow, or liver cells may regenerate new ones. More limited regeneration may occur in bone, muscle, skin, and nerve cells, but, unlike in plants, complete organisms cannot be regenerated from any part, except embryonic stem cells or possibly from other cells after special treatments. According to recent evidence, mesoderm, endoderm, and ectoderm cell lineages can be reprogrammed and, e.g., bone marrow cells can regenerate into nerve cells or muscle-derived cells and cells of the central nervous system can reconstitute other cell types. Some mouse strains (MRL) have an exceptional ability to regenerate heart tissues in vivo. *See* homeotic genes; nuclear transplantation; regeneration of plants; stem cells; transdetermination; transplantation of nuclei. (Leferovich, J. M., et al.

2001. *Proc. Natl. Acad. Sci. USA* 98:9830; Alsberg, E., et al. 2002. *Proc. Natl. Acad. Sci. USA* 99:12025.)

regeneration, plants Formation of new organs or entire organisms from dedifferentiated tissues or single cells. *See* callus; clone; dedifferentiation; embryo culture; embryogenesis, somatic; totipotency; vegetative reproduction.

reglomerate *See* aggregulon.

regression Measure of dependence of one variate on another in actual quantitative terms, in contrast to correlation, which uses relative terms from 0 to 1. Linear regression involves the independent variate to the first power. Quadratic regression involves the independent variate to the second power and cubic regression to the third power. *See* correlation for the calculation of regression coefficient; heritability; linear regression.

regulated gene The expression is conditional and affected by genetic and nongenetic factors. *See* constitutive genes; housekeeping genes; regulation of gene activity.

regulation, enzyme activity Enzyme activity is characterized by various measures of enzyme kinetics (*see* Eadie-Hofstee; Linweaver-Burk; Michaelis-Menten). The reaction is controlled by the quantity and/or activity of an enzyme. The quantity of the enzyme depends on protein synthesis/degradation controlled at the level of transcription, translation, processing of the protein, and its instability (*see* regulation of gene activity). The substrate of the enzyme may regulate the production of the enzyme protein (*see* attenuation; catabolite repression; *lac* operon; enzyme induction). *Feedback control* means that the accumulation of the product of an enzyme may shut down the operation of a pathway at any step preceding the final product. Feedback control may be simple or multiple, i.e., more than one enzyme may be affected either simultaneously or sequentially, or more than a single product of the pathway may act in a concerted manner (*see* feedback control). Feedback control may act either at the level of the synthesis (*feedback repression*) or by *inhibition* of the activity of a steady number of enzyme molecules. In general, the inhibitors are either *competitive* (bind to the enzyme and compete with the substrate for the active site) or *noncompetitive* (the inhibitors act by attaching to the enzyme at a site other than the active site yet lower enzyme activity [by allosteric effect]). *Uncompetitive inhibitors* operate by binding to the enzyme-substrate complex. *Suicide inhibitors* are converted by the enzyme into an irreversibly binding molecule that permanently damages the enzyme. The inhibitors may affect more than a single enzyme simultaneously. *Mechanism-based inhibitors* are highly specific to a single enzyme and as such have great significance for medicinal chemistry. Among these are the *antisense inhibitors* (*see* antisense RNA). *Allosteric enzymes* may also be stimulated (*modulated*) by allosteric compounds. The modulator may be *homotropic*, i.e., essentially identical structurally with the substrate or *heterotropic* in case of nonidentity with the substrate. The activity of an enzyme may require a proteolytic cleavage of the precursor protein, the *zymogen*. *See* allosteric control; allostery; feedback; protein synthesis; regulation of gene activity; signaling.

regulation, gene activity The various types of cells and differentiated tissues of an organism generally contain the same genetic material (*see* regeneration; totipotency), yet their differences attest that the genes must function quite differently in order to bring about the variety of morphological and functional differences. Genetic regulation accounts for this variety. Many genes are expressed in every cell because they determine the metabolic functions essential for life. Another group of genes is responsible for such generally required structural elements as membranes, microtubules, chromosomal proteins, etc. (*see constitutive* genes; *housekeeping*). Other genes are not constitutive, i.e., they are regulated in response to external and internal control signals; they are expressed only when they are called up for a duty. The latter group of genes are responsible for the differences within an organism.

PRETRANSCRIPTIONAL REGULATION: The expression of genes is regulated by several means, including the structural organization of the eukaryotic chromosome, although at one time it was thought that DNA associated with histones is or was not efficiently transcribed. Nucleosomal organization of the DNA may not prevent transcription, yet nucleosomal reorganization may be required for proper expression of the genes (*see* nucleosomes). For efficient gene transcription, chromatin remodeling (histone acetylation) is required. It has been known since the early years of cytogenetics that, e.g., heterochromatic regions of the chromosomes were not associated with genes that could be mapped by recombinational analysis. It appears that these tightly condensed regions of the chromosome are not suitable for transcription in general. The coiling of the chromosomes is also genetically regulated. Position effect indicates that gene expression is altered or obliterated by transposition into heterochromatin. Similarly, lyonization of the mammalian X-chromosome involves heterochromatinization and silencing of genes (*see* silencer). Insertion of normal genes (by transformation) into the condensed telomeric region (about 10^4 bp length) interferes with their expression (*see* heterochromatin; lyonization; position effect; telomeres). Gene expression also depends in some way on the presence of nuclease-sensitive sites in the chromatin. At these nuclease hypersensitive sites, apparently the DNA is not wrapped around so tightly and is more accessible for transcription initiation (*see* nuclease sensitive sites). The effects of the chromatin locale on the expression of genes is shown by the large differences in the production of a specific mRNA in different transgenic animals and plants carrying a particular gene inserted at different chromosomal locations (*see* LCR). Also, in order to make the gene accessible to transcription or replication, in bacteria negative supercoils are formed, then must be relaxed. In eukaryotes, DNA in Z conformation may be preferentially available for initiation of transcription (*see* supercoiling, Z DNA). Some genes are regulated by transposition; this mechanism is common in prokaryotes and eukaryotes to generate defense against the immune system of the host (*see* antigenic variation; phase variation), but it is also used for sex determination in yeast (*see* cassette model). At replication the four basic nucleotides are normally used; some nucleoside analogs (e.g., 5-bromodeoxyuridine) may be incorporated into the DNA with some consequences on gene expression. In the T-even (T2, T4, T6) phages, in place of cytosine, 5-hydroxymethyl cytosine is found as a protection against most of the restriction enzymes.

In eukaryotes, 5 to 25% of the cytosine residues are 5-methylcytosine. Genes with methylated cytosine are generally not transcribed *see* methylation, DNA; nuclear receptors; recruitment; SRB).

REGULATION OF TRANSCRIPTION AND TRANSCRIPTS: The cells have various options for the more direct regulation of transcription: (1) control of signal receptor and signal transmission circuits; (2) construct or take apart assembly lines geared to a particular function; (3) transcriptional control; (4) transcript processing and alternative splicing; (5) export of mRNA to the cytosol in eukaryotes. In prokaryotes and cellular organelles, a membrane does not enclose the genetic material, and transcription and translation are coupled. (6) Selective degradation of mRNA or a carboxypeptidase may cleave the transcription factors.

Nucleotide sequences in the DNA (structural gene) specify the primary structure of the transcripts. Upstream cis-elements (enhancers, promoters, and other protein-binding sequences) control the attachment and function of the DNA-dependent RNA polymerases (*see* pol I, pol II, and pol III RNA polymerases). Some eukaryotic genes may have more than one promoter; the tissue or cell type and the physiological conditions select the promoter to be used. Transcript length is dependent on the promoter element used. The upstream, nontranslated region contains binding sequences for further regulation of gene expression. The different upstream elements of the same gene may respond differently to cytokines, phorbol esters, and hormones (*see* hormone receptors; hormone-response elements). The enhancers may be positioned either upstream or downstream. Inducible genes receive cues through membrane

(*see* lambda phage). The consensus sequence of the budding yeast GAL4 upstream element (*see* galactose utilization) for the mating type α2 consensus (*see* mating type determination) and for the transcription factor GCN4 (*see* GCN4) regulate specific genes. In plants, the core sequence for a transcriptional activator protein is shown under the G-box element. The binding proteins have a short α-helix or a β-sheet that fits into the major groove of the DNA at the specific sequence motif (*see* helix-loop-helix; helix-turn-helix; leucine zipper; zinc finger). Specific activators may also regulate transcription. The activation may require a positive or negative control process (*see* arabinose operon; CAT; lac operon). Eukaryotic transcription requires the presence of a general transcription factor protein complex (*see* open transcription complex). Additional specific transcription factors may modulate transcription (*see* transcription factors, inducible). In the DNA, there are *response elements* or regulatory sequences that bind specific proteins, and the proteins may bind additional modules (*see* hormone-response elements; response elements). The inducible transcription factors in the eukaryotic nuclei help in the assembly of the transcription complex and activate or repress genes by assembling modules. The interacting elements are responsible for fine-tuning the metabolic pathways and regulating morphogenesis. These transcription factors may be syntenic with the genes they act on, and their number may vary depending on the gene concerned. The binding proteins may pile up in a specific way at the promoter after DNA looping brings them to that area. Also, the binding proteins may attract other molecules that act either in an activating or silencing manner. In an absolutely abstract form, this may be visualized with a few computer symbols:

receptors and transmitter cascades generally regulated by kinases and phosphorylases (*see* signal transduction). Downstream DNA nucleotide sequences control the termination of transcription. In eukaryotes, generally a polyA tail (exceptions are the histone genes) is added enzymatically without the use of a DNA template (*see* polyadenylation signals; transcription termination in eukaryotes and prokaryotes).

Gene expression begins by the initiation of transcription (*see* protein synthesis; transcription). The DNA displays some specific sequences in the major grooves of the double helix that are recognized by DNA-binding proteins (*see* lac operon, for the *E. coli lac* repressor-binding site and the CAP site for

The bacterial DNA-dependent RNA polymerase (*see* pol) attaches to the double-stranded DNA and generates an open-promoter complex, proceeding with transcription (*see* open-promoter complex). The bacterial RNA polymerase may rely on different σ-subunits for transcribing different bacterial or viral genes. In some instances, bacterial and eukaryotic genes also use activators of transcription to assist the RNA polymerase enzyme to generate the open promoter complex. These proteins may attach to the DNA at a region of some distance from the gene (enhancer) and looping may bring the protein to the promoter site (*see* looping of DNA). Actually, the likelihood for association of two DNA sites by looping reaches

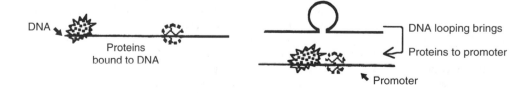

binding of the catabolite activator protein). In phage-λ, the *cI* repressor-binding element controls several genes by repression

an optimum at a distance of about 500 bp and it is much reduced when they are very close. Some of the enhancer DNA

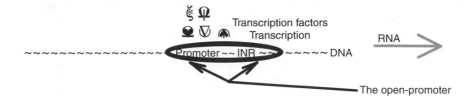

elements (binding sites for regulatory proteins) may be several thousands of nucleotides apart upstream or downstream of the structural gene (*see* enhancer). The various binding proteins (symbolized by ∪, ♥, ∩, Ψ, Ω, ♠, ζ, •, ∇) may associate with the general transcription factors and with each other in different combinations and numbers to activate or suppress, or modulate or silence, the gene.

The open-promoter complex includes the general transcription factors, RNA polymerase II, the TATA box, and the transcription initiator (INR). See diagrams above. These crude schematic figures cannot properly represent the interacting complexes that are required for turning on, turning off, and modulating expression as needed for orchestration of intricate processes such as the temporal and topological control of morphogenesis (*see* morphogenesis). The transcription factors regulate these processes, but the transcription proteins themselves are subject to regulation by metabolic and environmental cues. These processes include conformational changes, combinatorial assembly of subunits, ligand binding, phosphorylation and dephosphorylation, presence of inhibitors and activators (*see* signal transduction). In eukaryotes, there may be a need for chromatin remodeling in order for the activators and the TATA box-binding protein to access the DNA (*see* nucleosome). A histone acetylase or SWI/SNF complex may have to be recruited in preparation for transcription. In prokaryotes and eukaryotes, special control mechanisms have evolved for the termination of transcription (*see* transcription termination). Regulation of the transcriptional process and the turnover of the transcripts determines the quantity of the transcripts.

Many bacterial genes are organized into coordinated regulatory units employing negative, positive, or a combination of these two controls of transcription (*see* arabinose operon; *lac* operon). In these operons the genes are either exactly (*see tryptophan* operon) or with some modification (*see histidine* operon) arranged according to the order of the biosynthetic pathway. The amino acid operons also use *attenuation* for the control of the quantity of the transcripts for maximal economy (*see* attenuator region; tryptophan operon). The operons are characterized by coordinated regulation of the transcription of several genes belonging to the same transcriptional unit and transcribe them into a polycistronic mRNA. Eukaryotes usually do not produce polycistronic mRNAs, but the rRNA and tRNA transcripts are processed into functional units posttranscriptionally. Elements of a coordinated unit may not all be juxtapositioned (*see* arabinose operon; regulon). The small phage (*see* φX174) and retroviral genomes may have overlapping genes that specify more than one protein, depending on the transcription register (*see* overlapping genes; recoding; retroviruses). The need for the protein products of these overlapping genes transcribed with the aid of the same promoter may be not the same. Some proteins, e.g., viral coat proteins, may be needed in larger quantities than the replicase enzymes. Therefore, mechanisms have evolved to skip internal stop signals and produce some fusion proteins that assist in achieving this goal (*see* overlapping genes, recoding). Another means of regulation evolved in bacterial, plant, and animal viruses for the regulation of gene activity at different steps by the use of antisense RNA. This mechanism is being explored to develop specific drugs for the highly specific regulation of genes with minimal side effects or for the development of selective antimicrobial agents and more desirable crop plants without reshuffling the entire genome (*see* antisense technologies).

In prokaryotes, a short, transcribed stretch of nucleotides, the Shine-Dalgarno box, controls the attachment of mRNA to the small (30S) ribosomal subunit. For the same task, eukaryotes use ribosome scanning, i.e., the mRNA tethers a 40S ribosomal subunit and by reeling locates the first initiator codon. Eukaryotic 40S ribosomal subunits can enter circular mRNAs if they contain internal ribosomal entry sites (IRS).

The primary transcripts are generally not suitable for translation into a protein or for an RNA product (rRNA, tRNA). The transcripts are processed to mRNA and/or other RNA units. Introns are excised, and the sequences corresponding to exons are spliced and may even be transspliced with the cooperation of spliceosomes (*see* alternative splicing; exon; hnRNA; intron; snRNA; spliceosome). The splicing itself may be genetically and organ-specifically regulated. Transposition of the P element of *Drosophila* is relatively rare in the soma but is five times more common in the germline because one intron is not excised from the transposase transcripts in the somatic cells (*see* hybrid dysgenesis). Tissue specificity and function specificity of many proteins are controlled partly by alternative splicing (*see* immunoglobulins; sex determination). Mitochondrial RNA transcripts may also be modified by replacing C residues with U residues (*see* RNA editing).

The eukaryotic mRNAs are capped while still in the nucleus. The transcript is cut at the appropriate guanylic residue and is then modified (*see* cap; capping enzymes). Capping increases the stability of mRNA, facilitates its transport to the cytosol, and assists in the initiation of translation by being recognized by initiation protein factors eIF-4F, eIF-4B, etc. (*see* cap; eIF).

The tail of the eukaryotic mRNAs (with few exceptions, e.g., histones) is equipped with 50–250 adenylic units to increase their stability. Polyadenylation is controlled separately from transcription because a special enzyme adds these nucleotides after processing of the transcript. Generally, the genes carry a short A-rich consensus (*see* polyadenylation signal) in the DNA that instructs the RNA polymerase to terminate transcription after the enzyme passes through the signal and also indicates the need for polyadenylation. Eventually, the poly-A tail is reduced to about 30 A units. In eukaryotes, the 3′ tail may be specially regulated.

Some of the transmembrane proteins have a hydrophobic amino acid sequence in the section that is going to be located within the membrane, whereas the cytosolic end contains a longer hydrophilic carboxyl end. The positioning of the transmembrane proteins shows substantial variations, depending on their intrinsic properties. The transcript of the same coding sequences is differentially cut in such a manner as to assure terminus formation for the membrane-bound proteins, whereas a shorter hydrophilic end terminates the otherwise identical circulating immunoglobulin molecules.

After these intricate preparatory processes, the eukaryotic mRNA is transported to the cytosol through the nuclear pores. Prokaryotes do not have membrane-enclosed nuclei but only nucleoids anchored to the cell membrane, where the translation goes *pari passu* with transcription. (Carlson, M. 1997. *Annu. Rev. Cell Dev. Biol.* 13:1; Holstege, F. C., et al. 1998. *Cell* 95:717.)

POSTTRANSCRIPTIONAL REGULATION: The mRNA may be degraded before it can be translated into polypeptide chains. About half of the prokaryotic mRNAs may be degraded within 2–3 minutes after their synthesis. Eukaryotes have long-lived mRNAs, which usually last for at least three times longer, but in special dormant tissues of plants they may remain intact for many years. The degradation is mediated by special endonucleases that recognize mRNAs. Also, A-U sequences in the nontranslated downstream regions may remove the poly(A) tails, and thus in both cases stability is reduced.

Translation in eukaryotes begins with the transport of the capped mRNA outside the nucleus into the cytosol. The mRNA tethers several ribosomes and the polysomal structures are formed. Some mRNAs are equipped with a signal coding sequence, that codes for a special tract of 15 to 35 amino acids, which directs it toward the *signal sequence recognition particle* after only a few dozen amino acids are completed on the ribosome. The *signal peptide* then transports the nascent peptide chain into the lumen of the endoplasmic reticulum, Golgi vesicles, lysosomes, mitochondria, plastids, etc. This mechanism facilitates the subcellular localization of the emerging proteins at places where they are most needed and from where they may be diffused in a gradient as required for embryonic differentiation (*see* morphogenesis in *Drosophila*; signal peptide; signal sequence; signal sequence recognition particle). Various control mechanisms have been involved in generation of the protein products of genes: (1) translational control; (2) posttranslational modification of the polypeptides; (3) control of polypeptide assembly into proteins; (4) regulation of protein conformation; (5) compartmentalization of the proteins; (6) interaction of protein products and ribozymes; (7) feedback controls at the level of protein synthesis and function (*see* attenuation; induction; inhibition; repression; silencers, etc.) may be required before, during, and after the final protein products are made.

The state of phosphorylation of the eukaryotic initiation factor, eIF-2, is critical for the translation process. This protein may form a complex with guanosyl triphosphate (GTP), and it can then assist the attachment of the initiator tRNA^{Met} to the P site of the small subunit (40S) of the ribosome, scaning the mRNA until it finds a methionine codon (AUG). This occurs after the large ribosomal (60S) subunit joins the small subunit to form the 80S ribosome and at the same time one molecule of inorganic phosphate and the inactivated eIF-2 and GDP are released. Then eIF-2 can acquire another GTP and initiation goes on again (*see* protein synthesis).

Although all polypeptide chains start with a formylmethionine (prokaryotic) or methionine (eukaryotic), the final product is frequently truncated at both the amino and carboxyl termini.

Many proteolytic enzymes are translated as large units and become activated only after cleaving off certain parts of the original protein. Insulin is made as a pre-proinsulin that must be tailored in steps into pre-, then pro-insulin, and finally to insulin to become active. Several viral proteins, secreted hydrolytic proteins, peptide hormones, and neuropeptides are made as polyprotein complexes, and they have to be broken down into active units in the trans-Golgi network, in secretory vesicles, or even in the extracellular fluids to become fully functional. The formation of polyproteins appears to be justified as a protective measure against destruction in the cytosol until they can be sequestered and confined into some vesicles. The loaded vesicles then migrate to predetermined sites whereupon receiving the cognate signals they release the active protein. The signals can be chemical, physical (electric potentials), or topological. Actually, the release of the members of the polyprotein group may be selective regarding the site of release; different proteins can be released at different anatomical sites.

Some proteins are synthesized in separate polypeptide chains but must be folded and/or assume a quaternary structure, e.g, $\alpha\alpha\beta\beta$, and may even have to acquire a prosthetic group such as heme, a vitamin, or another organic or inorganic group(s). The folding in prokaryotes begins after completion of the chain. In eukaryotes, the folding may begin before finishing a polypeptide, and thus higher complexity is generated in the large proteins. The mRNA may be degraded before it can be translated into polypeptide chains. Proteins are very commonly acetylated after translation; carbohydrate side chains are added (glycoproteins), prenylated, and linked by covalent disulfide bonds; special amino acids (serine, threonine, tyrosine) are phosphorylated by kinase enzymes; lysine residues may be methylated; and extra carboxyl groups may be attached to aspartate and glutamate residues.

ENGINEERED REGULATION: Using a genetic vector, it is feasible to introduce into somatic cells a structural gene *A* for a protein of a special need. With the aid of another vector, one may also introduce gene *B* encoding its special transcription factor. The latter transcription factor gene is equipped with a promoter that responds to a specific drug (or to a special temperature or to any other conditional factor) regulating its transcription. Thus, supplying the drug at a variable dosage, the expression of gene *A* can be modulated by the controlled response of gene *B*. Such a system may permit fine-tuning of the controls and secure compensation for a genetic defect or improve productivity.

According to some estimates, about 2,000 different protein kinases and 1,000 phosphatases may exist in a higher eukaryotic cell. They must be regulated in time, space, and for other specificities. This regulation is an extremely complex task and is expected to be mediated by associations with modular, adaptor, scaffold, and anchoring proteins working in sequential cooperation through signal transduction pathways. The availability of complete nucleotide sequence information of prokaryotic and eukaryotic genomes, as well as microarray hybridization, permits the assessment of the simultaneous expression of thousands of genes. Eventually, with appropriate computer technologies the study of coordinated regulation of the function of entire genomes will become a reality. *See* attenuation; axotomy; cell cycle; chromatin; chromatin remodeling; DNA chips; DNA grooves; DNA looping; elongation factors; endoplasmic reticulum; genetic network; high-mobility proteins; insulator; LCR; microarray hybridization; open-promoter complex; polysome; protein synthesis; RAD25; rao; regulation of enzyme activity; regulation of transcription; RNA polymerase; serine/threonine phosphoprotein phosphatases; signaling to translation; signal transduction; SL1; TAF; TBP; transcription; transcriptional activator; transcriptional modulation; transcription complex; transcription factors; translation; translation initiation. (Tautz, D. 2000. *Curr. Opin. Genet. Dev.* 10:575; Lemon, B. & Tjian, R. 2000. *Genes Dev.* 14:2551; Rao, C. V. & Arkin, A. P. 2001. *Annu. Rev. Biochem. Eng.* 3:391; Emerson, B. M. 2002. *Cell* 109:267; reviews in Cell 108:439ff [2002]; Wang, W., et al. 2002. *Proc. Natl. Acad. Sci. USA* 99:16893; <http://transfac.gbf.de /TRANSFAC/>; Gene Resource Locator: <http://www.gene-regulation.com/pub/databases. html#transcompel>.)

regulator gene   Controls the function of other genes through transcription. *See* activator; coactivator; enhancer; operon; regulation of gene activity; silencer.

regulatory elements   Generally upstream (enhancer) sequences located within 100 to 400 bp from the translation initiation nucleotide (+1). They control cell and developmental specificities. Some enhancers may be located at much more distant positions and also downstream. The enhancer region provides binding sites for regulatory proteins. *See* basal promoter; regulation of gene activity; regulator gene; UAS.

regulatory enzyme   Allosteric or other modifications alter its catalytic activity rate, thus affecting other enzymes involved in the pathway.

regulatory sequence, DNA   Binds transcription factors and RNA polymerase, thus regulating transcription. *See* attenuator site; enhancer; open transcription complex; operon; transcription factors; UAS.

regulon   Noncontiguous set of genes under control of the same regulator gene. The different sections may communicate through looping of the DNA. *See* arabinose operon; looping of DNA; Manson; regulation of gene activity. (McGuire, A., et al. 2000. *Genome Res.* 10:744; Huerta, A. M., et al. 1998. *Nucleic Acids Res.* 26:55.)

Reifenstein syndrome (Xq11-q12)   XY chromosomal constitution, but there is insufficient androgen receptor production during fetal development. It displays male pseudohermaphroditism with hypospadias, hypogonadism, and gynecomastia, yet defective germ cells are present and fertility may be achieved by early treatment with testosterone. *See* androgen insensitivity; gynecomastia; hypogonadism; hypospadias; pseudohermaphroditism; testosterone.

reinitiation   Eukaryotic ribosomes can terminate an open-reading frame and initiate another downstream (at low efficiency). Reinitiation also takes place when the translation of one reading frame is completed and the process moves on to the next cistron. In an unfavorable nucleotide context, translation may be reinitiated not at the first AUG codon but at the next one downstream. Translation factor eIF2 may have an important role in the process. *See* cistron; eIF2; regulation of gene activity; translation; transcription. (Kozak, M. 1999. *Gene* 234:187; Park, H. S., et al. 2001. *Cell* 106:723; Kozak, M. 2001. *Nucleic Acids Res.* 29:5226.)

reinitiation of replication   The genome of eukaryotes replicates at many points along the chromosomes. To avoid chaos in the nucleus, restart of replication must be prevented. Reinitiation is prevented by cyclin-dependent kinases (CDKs) by phosphorylation of origin recognition complex (ORC), downregulation of Cdc6, and exclusion of the MCM2-7 complex from the nucleus. *See* Cdc6; CDK; MCM; ORC. (Nguyen, V. Q., et al. 2001. *Nature* 411:1068.)

reinitiation of transcription   Transcription factors and RNA polymerase must be reattracted to the promoter for a second cycle of transcription. Reinitiation appears to be a faster process than initiation. TFIID and TFIIA transcription factors do not leave the promoter when the rest of the transcription complex is released. The reinitiation intermediate includes TFIID, TFIIA, TFIIH, TFIIE, and the Mediator. Subsequently, the complete transcription complex, including activators, is reformed depending on ATP and TFIIH. *See* Mediator; preinitiation complex; transcription factors. (Hahn, S. 1998. *Cold. Spring Harbor Symp. Quant. Biol.* 63:181.)

reiterated genes   Present in more than one copy, possibly many times.

Reiter syndrome   Complex anomaly generally accompanied by overproduction of HLA-B27 histocompatibility antigen. The most common result is arthritis, inflammation of the eyes and the urethra (the canal that carries the urine from the bladder and serves also as a genital duct in males). The inflammations may be related to sexually transmitted and intestinal infections. *See* arthritis; connective tissue disorders; HLA; rheumatic fever.

rejection   Immune reaction against foreign antigens such as may be present in transfused blood or grafted tissue. Rejection of pig organs by humans and old-world monkeys is caused by the presence of α-1,3-galactosyl epitopes on the pig epithelia. During evolution, the rejecters lost the appropriate galactosyltransferase gene and as a consequence developed antibodies against the epitope of the foreign tissue transplant. This immune reaction cannot be satisfactorily mitigated through affinity absorption or complement regulators or other means of immunosuppression (drugs) even in transgenic animals. A better solution appears to be the inactivation of the gene and generation of clones by nuclear transfer into enucleated pig oocytes. When fully developed, this procedure may permit xenotransplantation of pig organs into humans who are suffering serious organ defects. *See* HLA; immune reaction; nuclear transplantation; transplantation of organelles; xenotransplantation. (Lai, L., et al. 2002. *Science* 295:1089; Prather, R., et al. 2003. *Theriogeneology* 59:115.)

rejoining   *See* breakage and reunion; breakage-fusion-bridge cycles.

relapsing fever   *See* Borrelia.

relatedness, degree of Indicates the probability in genetic counseling of sharing genes among family members. First-degree relatives such as a parent and child have half of their genes in common. Second-degree relatives such as a grandparent and grandchild have one-fourth of their genes identical. Population genetics prefers the use of mathematically simpler terms such as inbreeding coefficient, consanguinity, and coefficient of coancestry. *See* MLS; relationship coefficient.

relational coiling *See* chromosome coiling.

relationship, coefficient of $r = 2F_{IR}/\sqrt{(1 + F_I)(1 + F_R)}$, where F_I and F_R are the coefficients of inbreeding of I and R. If they are not inbred, F_I and F_R equal 0. *See* coefficient of inbreeding; relatedness degree.

relative biological effectiveness *See* RBE.

relative fitness *See* selection coefficient.

relative molecular mass (M_r) Expresses molecular weight relative to ^{12}C isotope (in 1/12 units). It is comparable to molecular weight in daltons, but it is not identical to molecular weight (MW) represented by the mass of the atoms involved. *See* Dalton.

relative mutation risk 1/doubling dose. *See* doubling dose; genetic risk.

relative sexuality Intensity of sexual determination may have degrees in some organisms. In extreme cases, a normally female gamete may behave as a male gamete toward a strong female gamete. *See* intersex; isogamy; pseudohermaphroditism.

relaxed circular DNA Not supercoiled because one or more nicks. *See* nick; supercoiled DNA.

relaxed control mutants (*relA*) Lost stringent control and continue RNA synthesis during amino acid starvation of bacteria. *See* fusidic acid; stringent control.

relaxed genomes Organelle DNAs are not replicated in lockstep with the nuclear genome. Their replication may be reinitiated during the cell cycle. The distribution of the organelles may not necessarily be equational during cytokinesis. *See* cytokinesis; stringent genomes.

relaxed replication control The plasmids continue to replicate when the bacterial division stops. *See* replication.

relaxin Water-soluble protein in the corpus luteum mediating the relaxation of the pubic joints and the dilation of the uteral cervix in some mammals. Its two receptors, LGR7 and LGR8, are heterotrimeric G-binding proteins and are widely distributed among organs, indicating their roles in diverse functions. *See* corpus luteum. (Hsu, S. Y., et al. 2002. *Science* 295:671.)

relaxosome DNA protein structure mediating the initiation of conjugative transfer of bacterial plasmids. It contains a *nic* site at the origin of transfer (*oriT*). Relaxase catalyzes the nicking and it becomes covalently linked to the 5′-end through a tyrosyl residue. Then a single strand is transferred to the recipient by a rolling circle mechanism. *See* conjugation; nick; rolling circle. (Xavier Gomis-Rüth, F., et al. 2001. *Nature* 409:637.)

relay race model of translation After passing a chain termination signal of an ORF, a ribosome does not completely disengage from the mRNA and may reinitiate protein synthesis if an AUG codon is within a short distance downstream. *See* ORF; regulation of gene activity; reinitiation; translation. (Ranu, R. S., et al. 1996. *Gene Expr.* 5[3]:143.)

release factor (RF) When translation reaches a termination codon, the release factors let the polypeptide go free from the ribosome. There are two direct release factors, in prokaryotes RF-1 (specific for UAG/UAA) and RF-2 (specific for UGA/UAA), and a third factor, RF-3, stimulates the activity of RF-1 and RF-2. RF-1 and RF-2 can discriminate between the termination and sense codons by 3 to 6 orders of magnitude effectiveness. In RF-1, a Pro-Ala-Thr tripeptide, and in RF-2, a Ser-Pro-Phe tripeptide, recognize the appropriate stop codon. The eukaryotic release factors, eRF and eRF-1, alone can recognize all three stop codons. RF-3 and eRF-3 are GTP-binding proteins. RF-3 is a GTPase on the ribosome in the absence of RF-1 and RF-2; eRF3 requires eRF-1 to act as a GTPase. ERF-1 may be sufficient alone for termination in yeast. It has been suggested that all release factors are homologous to elongation factor G, which mimics tRNA in its C-terminal domain. This would be the basis of the recognition of the RFs of the ribosomal A site. Class 1 release factor activity depends on the presence of Gly-Gly-Glu (GGQ) motif in the peptidyl transferase center for the release. The stop codons in the ciliates vary and so do the release factors. *See* EF-G; protein synthesis; regulation of gene activity; transcription termination. (Inagaki, Y. & Doolittle, W. F. 2001. *Nucleic Acids Res.* 29:921; Zavialov, A. V., et al. 2001. *Cell* 107:115; Ito, K., et al. 2002. *Proc. Natl. Acad. Sci USA* 99:8494.)

releasing factors Pituitary gland hormones are released under the influence of hypothalamic hormones. *See* animal hormones.

relics Genes with major lesions (insertions and deletions) in one or more components; they are similar to pseudogenes. *See* pseudogenes.

rel (REL) oncogene (2p13-p12, 11q12-q13) Turkey lymphatic leukemia oncogene, a transcription factor homologous with NF-κB. c-Rel as a homodimer or as a heterodimer with p50 or p52 is a strong transcriptional activator. In its absence or inactivation, the production of IL-3 and the granulocyte-macrophage colony-stimulating factor is impaired. Rel domains occur in several proteins such as NF-κB, NFAT, and ca. 12 others. *RelA* encodes guanosine tetraphosphate synthetase (ppGpp), *RelBE* in *E. coli* encodes the toxin-antitoxin proteins. The toxin severely inhibits bacterial growth as a stringent control, whereas the antitoxin is a repressor of the translation of the RelB toxin. *See* GMCSF; IL-3; morphogenesis in *Drosophila* {3}; NFAT; NF-κB; NFKB; oncogenes; p50; stringent response. (Gugasyan, R., et al. 2000. *Immunol. Rev.* 176:134; Christensen, S. K., et al. 2001. *Proc. Natl Acad, Sci. USA* 98:14328.)

REM (1) Roentgen equivalent man. It is the product of REB × rad. Generally 1 rem is considered to be 1 rad of 250 kV X-rays; 1 rem is 0.01 Sv (Sievert). *See* BERT; Gray; rad; REB; R unit; Sievert. (2) Ras exchanger motif required in signal transduction for interaction with RAS in GDP exchange for GTP using the GEF motif (Cdc25 homology catalytic unit) of SOS. *See* Cdc25; GEF; RAS; signal transduction; SOS.

REMI Restriction enzyme–mediated integration. An integrating vector is transformed into a cell in the presence of a restriction enzyme that facilitates insertion at the cleavage sites and may bring about insertional mutagenesis. *See* insertional mutation; restriction enzyme. (Thon, M. R., et al. 2000. *Mol. Plant Microbe Interact.* 13:1356.)

renal carcinoma, hereditary papillary (HPRC) Frequently caused by triplication (trisomy)/and or mutation of the MET oncogene, a cell-surface tyrosine kinase, encoded at human chromosome 7q31.

renal cell carcinoma (RCC) Most commonly involves translocation breakage points in human chromosome 3p, each representing a different type. The 3p14.2 region includes the gene for protein tyrosine phosphatase gamma (PTPγ). This region also contains a fragile site, FHIT (fragile histidine triad), and the von Hippel–Lindau syndrome gene. *See* fragile site; hypernephroma; papillary renal cell cancer; tyrosine phosphatase; von Hippel–Lindau syndrome. (Zanesi, N., et al. 2001. *Proc. Natl. Acad. Sci. USA* 98:10250.)

renal dysplasia and limb defects Autosomal-recessive underdevelopment of the kidney and the urogenital system accompanied by defects of the bones and genitalia. *See* kidney disease; limb defects.

renal dysplasia and retinal aplasia In the autosomal-recessive condition, kidney developmental anomaly is associated with eye defects. *See* eye disease; kidney disease.

renal glucosuria (16p11.2, 6p21.3) Dominant glycosuria; may not be related to diabetes.

renal-hepatic-pancreatic dysplasia (polycystic infantile kidney disease, ARPKD) Autosomal-recessive phenotypes include cystic (sac-like structures) kidneys, liver, and pancreas sometimes associated with other anomalies such as blindness. Polycystic kidney disease of adult type dominant (ADPKD, human chromosome 16) is associated frequently with internal bleeding or arterial blood sacs (aneurysm). *See* kidney disease.

renal tubular acidosis The 17q21-q22 dominant type I defect is primarily in the distal tubules with normal bicarbonate content in the serum. Type II is recessive, the defect is in the proximal tubules, and there is a low level of bicarbonate in the urine. In another recessive form — involving mutation in the B1 subunit of H$^+$-ATPase; in human chromosome 2cen-q13 — nerve deafness is present. A proximal type is X-linked recessive. Recessive distal tubular acidosis with normal hearing (rdRTA2) was assigned to gene ATP6N1B at 7q33-q34. The gene encodes an 840-amino-acid subunit of a kidney vacuolar proton pump. The excretion of ammonium is reduced and the urine pH is usually above 6.5, in contrast to types I and II, where it is around 5.5. Other variations have also been observed. *See* kidney diseases.

renaturation Complementary single DNA strands reform the double-strand structure by reannealing through hydrogen bonds. *See* c$_0$t curve; denaturation.

R end *See* packaging of λ-DNA.

Renilla GFP (from sea pansy) Green fluorescent protein with similarities to aequorin but with only one absorbance and emission peak, and its extinction coefficient is higher. *See* aequorin.

renin (chymosin, rennet) Protein hydrolase reacting with casein in cheese making. It is present in the kidneys and splits proangiotensin from α-globulin. *See* angiotensin; pseudoaldosteronism. (Kubo, T., et al. 2001. *Brain Res. Bull.* 56:23.)

Renner complex Chromosomal translocation complex that is transmitted intact. *See* translocation; translocation complex.

Renner effect *See* megaspore competition. (Renner, O. 1921. *Ztschr. Bot.* 13:609.)

reoviruses Double-stranded RNA viruses cause respiratory and digestive tract diseases and arthritis-like symptoms in poultry and mammals, but in humans the infection usually does not involve serious symptoms. The internal capsid particle transcribes (+)-strand copies from the 10 genomic segments. The transcript carries a cap and is exported to the cytoplasm of the infected cell. *See* cap; oncolytic virus; PKR; plus strand; rotaviruses. (Joklik, W. K. & Roner, M. R. 1996. *Progr. Nucleic Acid Res. Mol. Biol.* 53:249.)

rep Roentgen equivalent physical, a rarely used unit of X- and γ-radiation delivering the equivalent of 1 R hard ionizing radiation energy to water or soft tissues (≈93 ergs). *See* R.

Rep *E. coli* monomeric- or dimeric-binding protein and helicase. *See* binding protein; helicase; monomer. (Bredeche, M. F., et al. 2001. *J. Bacteriol.* 183:2165.)

REP Repetitive extragenic consensus of 35 nucleotides containing inverted sequences in the bacterial chromosome. There are over 500 copies of REP in *E. coli* in intergenic regions the at 3′-end of the genes. They are transcribed but not translated and appear to be the bacterial version of selfish DNA. *See* selfish DNA. (Herman, L. & Heyndrickx, M. 2000. *Res. Microbiol.* 151[4]:255.)

repair, genetic *See* DNA repair; unscheduled DNA synthesis.

repairosome Protein complex mediating DNA repair. *See* DNA repair.

repeat, direct Tandem duplication of the same DNA sequence. It may be present at the termini of transposable elements ABC–––ABC. The hexameric CeRep26 repeat of *Ceanorhabditis elegans* (TTAGGC) occurs at the telomeres and at many additional chromosomal regions. The 711 copies of CeRep11 are distributed over the autosomes, but only a single is in the X chromosome. *See* interspersed repeats; tandem repeat; transposable element; transposon.

repeat, inverted Double-stranded DNA carries inverted repeats such as the transposable elements. The single strands can fold back and form the *stem* (by complementarity) and *loop* (no complementarity) structure. In the sequenced *Caenorhabditis elegans*, inverted repeats represent 3.6% of the genome and occur on the average of once/4.9 kb; introns contain 45% of them and 55% are in intergenic regions. Inverted repeats may increase inter- and intrachromosomal recombinations by orders of magnitude and are responsible for a large part of genetic instabilities. Mutation in the *MRE11/RAD50/XRS2* and *SAE2* genes of yeast interferes with the repair of hairpins and contributes to instability of the genome. *See* LIR; tandem repeats; transposable elements. (Lobachev, K. S., et al. 1998. *Genetics* 148:1507; Waldman, A. S., et al. 1999. *Genetics* 153:1873; Lin, C.-T., et al. 2001. *Nucleic Acids Res.* 29:3529; Lobachev, K. S., et al. 2002. *Cell* 108:183.)

repeat, short tandem (STR) With the aid of polymerase chain reaction, STR can serve for individual (forensic) discrimination or for the identification of human cell lines, which may or may not be contaminated. *See* low-copy repeats; PCR. (Oldroyd, N. J., et al. 1995. *Electrophoresis* 16:334; Masters, J. R., et al. 2001. *Proc. Natl. Acad. Sci. USA* 98:8012.)

repeat, trinucleotide *See* fragile sites; trinucleotide repeats.

repeat-induced gene silencing *See* cosuppression.

repertoire, antigenic Complete set of antigenic determinants of the lymphocytes.

repertoire shift After a secondary immunization with a hapten following a primary immunization, the variable heavy/variable light (V_H/V_L) immunoglobulin genes display an altered spectrum of somatic mutations. *See* hapten; immunoglobulins. (Meffre, E., et al. 2001. *J. Exp. Med.* 194:375.)

repetitive DNA (repetitious DNA) Similar nucleotide sequences occurring many times in eukaryotic DNA. Some of these sequences represent transposable or retrotransposable elements; others, such as ribosomal genes, are called to duty when there is a special need for high gene activity, e.g., during embryonal development. More than 40% of the human genome appears highly or moderately repetitive and only about 3% may be genetically functional. *See* α-satellite DNA; cosuppression; LINE; microsatellite; minisatellite; pseudogenes; redundancy; SINE. (Britten, R. J. & Kohne, D. E. 1968. *Science* 211:667; Toder, R., et al. 2001. *Chromosome Res.* 9[6]:431; <http://www.girinst.org>.)

replacement theory *See* out-of Africa.

replacement vector By homology it recognizes, then replaces, a particular segment (gene) of the target. It has a pair of restriction enzyme recognition sites within the region of nonessential genes. Nonessential means that their removal and replacement do not impair packaging and propagation in *E. coli* by sequences of interest for the experimenter. *See* stuffer; vectors.

replica plating Has been designed for efficient selective isolation of haploid microbial mutants. Mutagen-treated cells are spread in greatly diluted suspension on the surface of complete medium and incubated to allow growth. Because of the dilution, each growing colony represents a single original cell (clone). Then impressions are made of this master plate on minimal medium where only the wild-type cells can grow. The absence of growth on the minimal media plates indicates that auxotrophs exist at the spots where no growth has been obtained. The impressions also represent a map of the colonies on the original complete-medium master plate. Thus, the experimenter can obtain cells from the original colonies and test them for nutritional requirements on differently supplemented media. This procedure permits the isolation of mutants and the identification of nutrient requirements. *See* mutant isolation; fig. next page. (Lederberg, J. & Lederberg, E. M. 1952. *J. Bacteriol.* 63:399.)

replicase RNA-dependent RNA polymerase enzyme of viruses encoded by viral RNA and packed to the progeny capsid, so upon entry to a cell, replication of the infective *negative-strand* RNA (influenza, stomatitisvirus [causing inflammation of mucous membranes]) forms the template for replication but does not code for viral proteins. Without the replicase, this negative strand would not be able to function. The *positive-strand* RNA viruses (e.g., poliovirus) are directly transcribed into the protein, including the replicase, and can be infectious in this form. DNA-dependent DNA polymerase enzymes may also be called replicases. *See* replication; RF; positive strand. (Tayon, R., Jr., et al. 2001. *Nucleic Acids Res.* 29:3576.)

replicating vector *See* transformation, genetic; yeast.

replication *See* chromosome replication; DNA replication; replication fork.

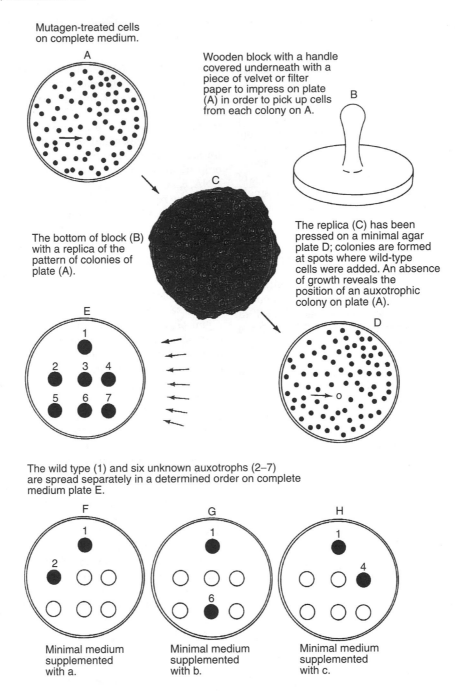

Mutagen-treated cells on complete medium.

Wooden block with a handle covered underneath with a piece of velvet or filter paper to impress on plate (A) in order to pick up cells from each colony on A.

The bottom of block (B) with a replica of the pattern of colonies of plate (A).

The replica (C) has been pressed on a minimal agar plate D; colonies are formed at spots where wild-type cells were added. An absence of growth reveals the position of an auxotrophic colony on plate (A).

The wild type (1) and six unknown auxotrophs (2–7) are spread separately in a determined order on complete medium plate E.

Minimal medium supplemented with a.

Minimal medium supplemented with b.

Minimal medium supplemented with c.

replication, bidirectional Mode of replication in bacteria and in the eukaryotic chromosome. Replication begins at an origin and proceeds in the opposite direction on both old strands of the DNA double helix. The helicase subunits encoded by the xeroderma pigmentosum genes XPB and XpD of the transcription factor TFIIH unwinds the DNA in both directions. An electron microscope reveals a θ (theta)-resembling structure of the circular DNA, whereas bubble-like structures are visible in the linear eukaryotic DNA. In prokaryotes, this replication is mediated by DNA polymerase III, and in eukaryotes, by a DNA polymerase α-type enzyme. Termination of replication in *E. coli* requires the 20-base-long Ter elements and the associated protein Tus (termination utilization complex, M_r 36 K). While replicating the template strand, T7 RNA polymerase can bypass up to 24 nucleotide gaps by making a copy of the deleted sequence using the corresponding nontemplate tract.

TerA, D, and *E* stop the replication in the anticlockwise direction; *TerC, B,* and *F* halt replication of the strand elongated clockwise. The Tus-Ter complex probably blocks the replication helicase. Similar mechanisms operate in most bacteria, but replication fork–arresting sites also exist in eukaryotes, including humans. *See* DNA replication, eukaryotes; DNA replication, prokaryotes; pol III; pol α; replication bubble; replication fork; θ replication; transcription factors; xeroderma pigmentosum. (Hiasa, H. & Marians, K. J. 1999. *J. Biol. Chem.* 274:27244; Abdurasidova, G., et al. 2000. *Science* 287:2023; Gerbi, S. A. & Bielinsky, A. K. 1997. *Methods* 13:271.) (Diagram after Kamada, K., et al. 1996. *Nature* 383:598.)

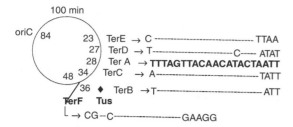

100 min

oriC / 84

23 / TerE → C ---------------------------------- TTAA
27 / TerD → T------------------------C---- ATAT
28 / Ter A → **TTTAGTTACAACATACTAATT**
48 34 / TerC → A---------------------------------- TATT
36 ♦ TerB → T---------------------------------- ATT
TerF Tus
└ → CG--C------------------GAAGG

replication, conservative Historical model of DNA replication, assuming that the two old (original) DNA strands produce two new copies, which then anneal to each other. That is, the double-stranded DNA is not composed of an old and a new strand as the current and experimentally demonstrated semiconservative replication mechanism shows. *See* semiconservative replication. (Delbruck, M. & Stent, G. S. 1957. *The Chemical Basis of Heredity*, McElroy, W. D. & Glass B., eds., p. 699. Johns Hopkins Press, Baltimore, MD.)

replication, defective virus Mutant for the replication function or lost genes required for producing infective particles. *See* replicase.

replication, dispersive Unproven old idea claiming that old and new double-strand DNA tracts alternate along the length of the molecule. *See* replication; replication fork. (Delbruck, M. & Stent, G. S. 1957. *The Chemical Basis of Heredity*, McElroy, W. D. & Glass B., eds., p. 699. Johns Hopkins Press, Baltimore, MD.)

replication bubble (replication eye) Indication of strand separation in a replicon. In a eukaryote nucleus, an estimated 10^3 to 10^5 replication initiations occur during each cell cycle without any reinitiation per site. This assures the maintenance of gene number. In yeast, the dynamics of chromosome replication can be studied with the aid of DNA microarrays. It appears that the two ends of each chromosome replicate rather synchronously, but the replication forks move quite differently in other regions. *See* DNA replication; FFA; geminin; ORC; promoter bubble; replication, bidirectional; replication fork; replication protein A; replicon. (Diffley, J. F. X. 2001. *Curr. Biol.* 11:R367; Raghuraman, M. K., et al. 2001. *Science* 294:115.)

replication during the cell cycle Eukaryotic DNA replication takes place predominantly during the S phase of the cell cycle, although some repair synthesis (unscheduled DNA synthesis) may occur at other stages. In prokaryotes, the replication is not limited to a particular stage, and DNA synthesis may proceed without cellular fission. Such a phenomenon (endoreduplication) is exceptional in eukaryotes and is most commonly limited to certain tissues, e.g., to the salivary gland chromosomes of insects (*Drosophila, Sciara*) or a rare non-repeating process (endomitosis) that doubles the number of chromosomes. The replication in eukaryotes is an oscillatory process tied to the S phase of the cell cycle. The process of replication shows some variations even among the different eukaryotes, and the process described below is modeled after that of *Saccharomyces cerevisiae* (the best known). During the G1 phase, the pre-origin-of-replication complex (pre-ORC), the ORC is assembled after the cyclosome (APC) proteases degrade the cyclin B–cyclin–dependent kinase (cyclin B-CDK). The cis-acting *replicator* element and the *initiator* proteins bind at each origin of replication (hundreds in eukaryotes). The replicator (0.5–1 kb) is a multimeric complex itself, and its indispensable component is the A unit, but B1, B2, and B3 are also used. A, B1, and B2 form the core of the replicator, and B3 is an enhancer that binds to the *autonomously replicating sequence* (ARS)–binding protein factor 1 (ABF1). The replicator (A + B1) hugs the ORC (origin recognition complex) composed of six subunits that form the hub of the replication process and attract other critical regulatory proteins. The initiation site in mammals may extend to 50 kb. At the origin, the nucleosomal structure is remodeled, and during S, G2 and early M-phase DNase hypersensitive sites are detectable that disappear before anaphase. It appears that protein Cdc7 is required for remodeling. CDC6 (or the homologous Cdc18 protein of fission yeast) is required in G1 or S phase for DNA synthesis (but the cells may proceed to an abortive mitosis and aneuploidy in its absence). CDC6 seems to be essential for the formation of the pre-ORC complex. Overexpression of this protein leads to polyploidy. Replication also requires a replication licensing protein (RLF) and members of the MCM (minichromosome maintenance) proteins. Cyclin-dependent kinases (CLB5 and CLB6) are required to establish the preinitiation complex, but after the assembly is completed, some of them may be degraded. Some cyclin-dependent kinases block the reinitiation of the complex until the cell passes through mitosis. Cyclin B5–cyclin-dependent kinase (Clb5-CDK) is inhibited by Sic1 (S-phase inhibitory complex), which is removed by ubiquitin-mediated proteolysis at the START point before the S phase is fired on. CDC34, CDC53, CDC4 SKP1, CLN1-Cdc28, CLN2-CDC28, and APC proteins promote ubiquitination. Initiation of DNA replication may also proceed through another pathway mediated by CDC7 and DNA-binding factor 4 (DBF4). *See* bidirectional replication; cell cycle; CLB; DNA replication, eukaryotes; DNA replication, prokaryotes; DNA replication in mitochondria; replication fork; replication protein A; reverse transcription; RNA replication; rolling circle replication; θ (theta) replication. Diagram at right above. (Waga, S. & Stillman, B. 1998. *Annu. Rev. Biochem.* 67:721; Kelly, T. J., et al. 2000. *Annu. Rev. Biochem.* 69:829.)

replication error Leads to base replacement and thus mutation. The rate of replicational errors of the different DNA polymerases varies, but it is partly compensated for by the editing function of the $3' \to 5'$ exonuclease activity of the polymerases. The error rate of the α-subunit of DNA polymerase III of *E. coli* is about 10^{-5}, but it is reduced by about two orders of magnitude by exonuclease subunit ε. The base substitution error of the *E. coli* DNA polymerase III holoenzyme is within the range 5×10^{-6} to 4×10^{-7}. The repair polymerase, pol I of *E. coli*, also has an error rate of about 1×10^{-5}, but the $3' \to 5'$ exonuclease activity again reduces the errors by two orders of magnitude. The T7 DNA polymerase has an error rate of 10^{-3} to 10^{-4}, but the repair system lowers it to 10^{-8} to 10^{-10}. The RNA polymerases of RNA viruses do not have proofreading and editing functions and their error rate may vary within the 10^{-3} to 10^{-4} range per nucleotide. The mitochondrial DNA polymerase γ has a base substitution error of $\sim 3.8 \times 10^{-6}$ to 2.0×10^{-6}. This appears to be one or two orders of magnitude higher than in the nucleus of mammals. The proofreading rate of the exonuclease depends a great deal on the availability of replacement nucleotides and therefore the difference in the final fidelity may not be as much. *See* DNA replication mitochondria; editing; mutation rate; proofreading; reverse transcriptases. (Kunkel, T. A. & Bebenek, K. 2000. *Annu. Rev. Genet.* 69:497; Johnson, A. A. & Johnson, K. A. 2001. *J. Biol. Chem.* 276:38090; Johnson, A. A. & Johnson, K. A. 2001. *J. Biol. Chem.* 276:38097; Pesole, G., et al. 1999. *J. Mol. Evol.* 48:427.)

replication eye   *See* DNA replication, eukaryotes; replication, bidirectional; replication bubble.

replication factor A   *See* helix-destabilizing protein; RF-A.

replication factor C   *See* RF-C.

replication factory   Replication machine. *See* DNA replication in prokaryotes.

replication fidelity   DNA replication error.

replication fork   Represents the growing region of DNA where the strands are temporarily separated.

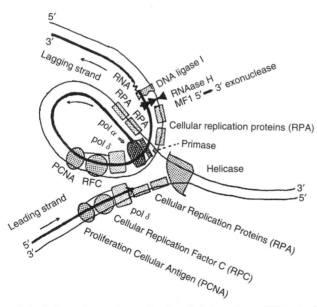

A model of the eukaryotic replication fork based on SV40 studies. thin lines: old DNA strands; heavy lines: new strands. The replication fork is opened at the replicational origin by helicases. The *cellular replication protein* (RPA) keeps the fork open and brings the pol α DNA polymerase complex to the replicational origin. After a short RNA primer is made (not shown on the diagram) at the beginning of the first okazaki fragment, *cellular replication factor C* (RFC) binds to the DNA and displaces pol α by pol δ. Then RFC, pol δ, and PCNA (proliferating cell nuclear antigen) form a complex on both leading and lagging strands and the two new DNA strands are replicated in concert. The synthesis of the leading strand is straightforward. The lagging strand is made of Okazaki fragments (by "backstitching") because the DNA can be elongated only by adding nucleotides at the 3'-OH ends. After an Okazaki fragment (100–200 bases) is completed, RNase H, MF1 exonuclease remove the RNA primer (jagged line) and DNA ligase I joins the fragment(s) into a continuous new strand. The long arrows indicate the direction of growth of the chain. (Redrawn after Waga, S. & Stillman, B. 1994. *Nature* 369:207.)

The simplest diagram of the replication is represented below with the new (thin line) facing the polymerase with the 3'-end. The leading strand is below and the lagging strand above. Actually, replication is a very complex process requiring two different polymerases (δ and α for the leading and lagging strands, respectively) and several proteins. *See* alpha accessory protein; DNA replication, eukaryotes; DNA replication, prokaryotes; GP32 protein; Okazaki fragment; PriA; primase; processivity; replication, bidirectional; replication bubble; replication licensing; replication machine. (Waga, S. & Stillman, B. 1998. *Annu. Rev. Biochem.* 67:721.)

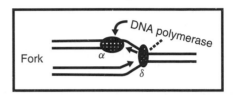

replication intermediate   *See* lagging strand; Okazaki fragment; replication fork.

replication licensing factor (RLF)   The initiation of replication requires two competency signals: the binding of RLF and an S-phase promoting factor. The sequential action of these two signals secures the accurate replication of the chromosomes. RLF has two elements: RLF-M and RLF-B. RLF-M is a complex of MCM/P1. RLF-M protein binds to the chromatin early during the cell cycle, but it is displaced after the S-phase. *See* cdt1; cell cycle; MCM1; MCM3; ORC; replication; replication bubble. (Chong, J. P. & Blow, J. J. 1996. *Progr. Cell Cycle Res.* 2:83; Nishitani, H., et al. 2001. *J. Biol. Chem.* 276:44905.)

replication machine   It has been shown that the *replication machine* of bacteria (*B. subtilis* and probably others) occupies a stationary central position in the cell and initially the twin PolC subunits are located in the replicational origin (O) of the bidirectionally replicating double-stranded DNA ring. The simultaneously replicating leading and lagging DNA strands are spooled through the twin machines. The machines contain the polymerase and several accessory proteins (also called replication factory). The eukaryotic replication bubbles are operated in a similar manner, but there are about 100 machines per nucleus, and each of these factories handles about 300 replication forks. *See* clamp-loader; DNA replication, eukaryotes; DNA replication, prokaryotes; replication fork. (Ellison, V. & Stillman, B. 2001. *Cell* 106:655; Bruck, I. & O'Donell, M. 2000. *J. Biol. Chem.* 275:28971; Turner, J., et al. 1999. *EMBO J.* 18:771.)

replication origin (o)   Point in the genetic material where replication begins. *See* CLB; replication bubble.

replication protein A (RPA, RFA, HSSB)   Complex of three different polypeptides (~ 70 kDa, human chromosome

17p13.3; ~32 kDa, 1p35; ~14 kDa, 7p22) binds most commonly to 20–25 nucleotides (with preference for pyrimidines) of single-strand DNA and may make the first step in DNA replication by participating in DNA unwinding by binding to A-T-rich sequences. The size of these subunits varies among different organisms. Despite the good homology, the subunits are antigenically not related and cannot be interchanged functionally. The binding to double-stranded DNA is 3–4 orders of magnitude less. RPA also binds to other proteins such as the primase subunit of DNA polymerase α and DNA repair proteins. The p53 cancer suppressor gene interferes with its binding to the replication origin. RPA plays a role in recombination and excision repair. It has affinity to xeroderma pigmentosum damage-recognition protein (XPA) and endonuclease XPG. RPA increases the fidelity of repair polymerases. During the cell cycle, RPA is phosphorylated by several kinases such as DNA-PK, CDK proteins, and others. *See* CDK; DNA-PK; DNA polymerases; DNA repair; DNA replication, eukaryotes; endonuclease; FFA; replication bubble; replication fork; RPA; xeroderma pigmentosum. (Wold, M. S. 1997. *Annu. Rev. Biochem.* 66:61; Patrick, S. M. & Turchi, J. J. 2001. *J. Biol. Chem.* 276:22630; Mass, G., et al. 2001. *Nucleic Acids Res.* 29:3892.)

replication restart Pathway of the SOS repair system of DNA by bypassing the mismatch and resulting in error-free replication. In bacteria, DNA polymerase II has an important role in the process. *See* DNA repair; translesion. (Rangarajan, S., et al. 1999. *Proc. Natl. Acad. Sci. USA* 96:9224.)

replication slippage May be one of the several mechanisms generating microsatellite diversity. *See* microsatellite. (Viguera, E., et al. 2001. *J. Mol. Biol.* 312:323.)

replication speed kbp/min: *E. coli* 45, yeast 3.6, *Drosophila* 2.6, toad 0.5, mouse 2.2.

replication timing Indicates whether chromosomal sequences replicate early or late during the cell cycle. Early tracts are usually more GC-rich and contain a larger number of genes than the AT-rich late replicating zones. Within the early/late transition regions, many cancer genes were found in human chromosomes 11q and 21q. *See* heterochromatin. (Watanabe, Y., et al. 2002. *Hum. Mol. Genet.* 11:13; Goren, A. & Cedar, H. 2003. *Nature Rev. Mol. Cell Biol.* 4:25.)

replicative form (RF) Double-stranded form of a single-stranded nucleic acid virus that generates the original complementary type of single-strand (+) nucleic acid. The necessity for the double-stranded replicative form is to generate a minus strand that is the template for the plus strand and is complementary to the sense molecule. This assures that all the progeny are identical and of one kind. *See* DNA replication; plus strand; RNA replication.

replicative intermediate *See* replication, intermediate; replicative form; RNA replication.

replicative segregation The newly formed cellular organelles display sorting out in the somatic cell lineages in case of mutation in organelle DNA. *See* cell lineage; sorting out.

replicative transposition *See* cointegrate; Mu bacteriophage; transposable elements; transposition.

replicator Origin of replication in a replicon. *See* ARS; replicon.

replichore Oppositely replicating half of the *E. coli* genome between the origin and the terminus of replication. *See* E. coli.

replicon Replicating unit of DNA. The size of the replicational unit varies a great deal. In *E. coli*: ≈4.7 Mbp, *Saccharomyces cerevisiae* (yeast): 40 kb, *Drosophila*: 40 kb, *Xenopus laevis* (toad): 200 kb, mouse: 150 kb, broad bean (*Vicia faba*): 300 kb. *See* DNA replication; minireplicon. (Jacob, F., et al. 1963. *Cold Spring Harbor Symp. Quant. Biol.* 18:329.)

replicon fusion *See* cointegration.

replisome Enzyme aggregate involved in the replication of DNA of *prokaryotes* (PriA, PrB, PriC, DnaC, DnaB, and other proteins). The DNA polymerase III holoenzyme consists of two functional enzyme units: one for the leading strand and the other for the lagging strand. The *polymerase core* contains one α-subunit (for polymerization), the ε-subunit ($3' \rightarrow 5'$ exonuclease for editing repair), and the θ-unit. It also includes a ring-like dimer of a β-*clamp* to hold on to leash the DNA strands and a five-subunit *clamp loader* γ-complex. There are two subunits that organize the two cores and the clamp loader into a pol III holoenzyme. The asymmetric replication of the leading and lagging strands is determined by the DnaC helicase, unwinding the double helix in front of the replisome. The helicase facilitates the hold of the complex onto the leading strand by the τ-unit to make a continuous extension of that strand. At the lagging strand the complex goes off and on, however, as the Okazaki fragments are made. *See* DNA polymerases; DNA replication in prokaryotes; GP32 protein; Okazaki fragment; replication fork in prokaryotes; replitase. (Benkovic, S. J., et al. 2001. *Annu. Rev. Biochem.* 70:181.)

replitase Replicational complex at the DNA fork in *eukaryotes*. *See* DNA polymerases; replication fork in eukaryotes; replisome. (Reddy, G. P. & Fager, R. S. 1993. *Crit. Rev. Eukaryot. Gene Expr.* 3[4]:255.)

replum Central membrane-like septum inside the silique (fruit) of cruciferous plants bearing the seeds. The carpels covering it were removed. Recessive pale mutant cotyledons show through the immature seed coats. At maturity the carpels dehisce at the base of the fruit.

reporter gene Structural gene with easily monitored expression (e.g., luciferase, β-glucuronidase, antibiotic resistance) that will report the function as differentiation progresses, or any heterologous or modified promoter, polyadenylation, or other signals attached to the gene by in vitro or in vivo gene fusion. *See* gene fusion; GUS; luciferase.

reporter ring According to the tracking concept of recombination between appropriate *res* (*resolvase*) points, during segregation, recombination of two DNA molecules would retain reporter rings catenated to one of the two DNA strands of the DNA recombination substrate during synapse and after its resolution in the product. The reporter rings were expected to be limited to one of the catenated product molecules, but the experimental data did not support this assumption. *See* resolvase; tracking.

representational difference analysis *See* RDA.

repressible Subject to potential repression. *See* repression.

repression Control mechanism interfering with the synthesis (at the level of transcription) of a protein. A general type of repression is attributed to histones tightly associated with DNA in the nucleosomal structure. When histones are deacylated, they reinforce repression by corepressors (N-CoR, mSin3). Histone acetyl transferases acylate histones with the assistance of pCAF (chromatin assembly factor), thus permitting recruitment of transcription factors to the gene. PCIP = p300/CEP cointegrator-associated protein (where CEP is a CREB-binding protein), CREB = cAMP-response element, CBP = CREB-binding protein, p300 = cellular adaptor and coactivator of some proteins. This scheme of repression is based on some eukaryotic systems. Other mechanisms are also known. Some repression mechanisms interfere with translation, e.g. threonyl-tRNA synthetase of *E. coli* represses its own synthesis by binding to the operator. tRNA^{Thr} serves an antirepressor and the balance between the two mechanisms

determines the level of translation (Torres-Larios, A., et al. 2002. *Nature Struct. Biol.* 9:343). *See Arabinose* operon; coactivator; feedback control; *Lac* repressor; MAD; regulation of enzyme activity; regulation of gene activity; repressor; signal transduction; transcription factors; *Tryptophan* operon. (Maldonado, E., et al. 1999. *Cell* 99:455.) (Diagram modified after Heinzel, T., et al. 1997. *Nature* 387:43.)

repressor Protein product of the regulator gene that interferes with transcription of an operon. The majority of the DNA-binding proteins bind the DNA by their α-helices, but some of the repressors (*met, arc, mnt*) bind by β-sheets or a combination of both (*trp*). Repression in eukaryotes has a variety of means for control. The short-range repressors are within 50–150 bp of the transcriptional activators. Since the promoters may be modulars (repeating units), the short-range silencers may only affect the nearest activators. This organization assures the expression of genes controlling segmentation along the axis of the developing embryo. The long-range repressors may silence the activators from several kb distance. The short-range repressor proteins are monomeric, whereas the long-range ones are multimeric. The repressors may be either gene-specific or more global. In the latter case, in bacteria the repressor binds to the σ-subunit of the RNA polymerase. *See Arabinose* operon; corepressor; gene switch; *Lac* operon; *Lac* repressor; morphogenesis in *Drosophila*; repression; tetracycline; transcription factors; *Tryptophan* operon. (Pardee, A., et al. 1959. *J. Mol. Biol.* 1:165; Hummelke, G. C. & Cooney, A. J. 2001. *Front. Biosci.* 6:D1186; Ryu, J.-R., et al. 2001. *Proc. Natl. Acad. Sci. USA* 98:12960.)

reproduction *See* allogamy; apomixia; asexual; autogamy; breeding system; clonal; conjugation; conjugation, *Paramecia*; cytoplasmic transfer; dioecious; fungal life cycle; genetic systems; hermaphroditism; incompatibility; life cycle; monoecious; parthenogenesis; social insects; vegetative reproduction; vivipary.

reproductive isolation Prevents gene exchange between two populations by a hereditary mechanism. Most commonly, reproductive isolation is caused by the inability of mating (premating isolation), but in some instances inviability or sterility of the offspring is the barrier (postmating isolation). In plants, chromosomal rearrangements most commonly lead to gametic disadvantage or sterility. In animals, translocated chromosomes are frequently transmitted at fertilization, but the duplication deficiency zygotes are inviable. Mating *Drosophila melanogaster* with *D. simulans*, a 3 kb DNA segment is inserted within the cyclin E locus, and causing male sterility and inviability but only a low degree of inviability or sterility in females. Actually, all of the *Drosophila* species show reproductive isolation from *D. melanogaster*. *See* drift, genetic; effective population size; founder principle; incompatibility; infertility; inversion; isolation, genetic; sexual isolation; speciation; transcription; translocation. (Harushima, Y., et al. 2001. *Genetics* 159:883.)

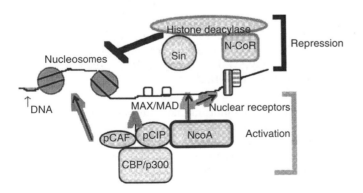

reproductive rate (R_0) Where ℓ_x = probability that a female survives to age x and m_x = expected number of female offspring produced by a female of age x. *See* age-specific birth and death rate; Malthusian parameter; population growth.

$$R_0 = \sum_{x=0}^{\infty} \ell_x m_x$$

reproductive success *See* fecundity; fertility; fitness.

reproductive technologies *See* ART.

reprogramming During development, the genome must be selectively turned on and off by methylation of nucleotides, acetylation and deacetylation of nucleosomes, recruiting general and specific transcription factors, etc. Beginning after fertilization of the egg, such reprogramming is a natural and indispensable process. Problems arise, however, when diploid nuclei of somatic cells are transplanted into eggs for the purpose of cloning. The transfer of nuclei from embryonic stem cells poses fewer problems, yet the in vitro culture conditions may not exactly duplicate the normal programming involved in natural fertilization of the haploid egg with the haploid sperm. Most commonly, the cloning involves developmental anomalies such as large offspring syndrome, various chromosomal anomalies, and inviability because of the lack of harmony between the donor nucleus and the recipient cytoplasm. *See* epigenesis; nuclear transplantation. (Rideout, W. M., III, et al. 2001. *Science* 293:1093; Håkelien, A.-M., et al. 2002. *Nature Biotechnol.* 5:460.)

reprolysin Metalloproteinase involved in regulation of morphogenesis. *See* bone morphogenetic protein.

reptation Theory about the movement of nucleic acid end-to-end in gels.

repulsion One recessive and one dominant allele are in the same member of a bivalent, such as *Ab* and *aB*. *See* coupling; linkage. (Bateson, W., et al. 1905. *Rep. Evol. Com. Roy. Soc.* II:1.)

RER Rough endoplasmic reticulum (endoplasmic reticulum with ribosomes sitting on it).

residue Elements of polymeric molecules, such as nucleotides in nucleic acids, amino acids in proteins, sugars in polysaccharides, and fatty acids in lipids.

resistance transfer factors (RTF) Plasmids that carry antibiotic or other drug-resistance genes in a bacterial host. They are capable of conjugational transfer. *See* plasmids.

resistin 12.5 kDa cysteine-rich protein hormone of the adipose tissues. The level of resistin is increased in genetically determined and diet-dependent obesity, and it may cause insulin resistance and type II diabetes. *See* adipocyte; diabetes mellitus; insulin. (Way, J. M., et al. 2001. *J. Biol. Chem.* 276:25651.)

RESites Related to empty sites. Genomic sites from which transposable elements moved out but left behind a footprint. Their flanking sequences are similar to insertion targets. (Le, Q. H., et al. 2000. *Proc. Natl. Acad. Sci. USA* 97:7376.)

resolution Depth of details revealed by the analysis.

resolution, optical Defines the ability of distinguishing between two objects irrespective of magnification. Magnification helps the human eye but does not improve the optical resolution. The power of resolution depends on the wavelength of the light and on the aperture of the lens used. (The achromatic lens is supposed to be free from color distortion; the apochromatic lens is free from color or optical distortions.) The unaided human eye may discern details larger than 100 μm; the light microscope with a good oil-immersion objective lens may resolve 0.2 μm; the lowest limit for an ideal electron microscope is 0.1 nm, i.e., 1 Å. Under practical conditions the resolution of the electron microscope is about 2 nm. The resolution of the light microscope is generally defined as

$$\frac{0.61\lambda}{n \sin \theta},$$

where λ is the wavelength of the light (for white light it may be about 530 nm), n stands for the refractive index of the immersion oil or air (when a dry lens is used), and θ is half of the angular width of the cone of the light beam focused on the specimen with the condensor of the microscope (sin θ is maximally about 1). Then sin θ is the *numerical aperture* of the lens; using oil immersion, its value may be 1.4. The immersion oil should be nonfluorescing, slow drying, and of right viscosity for vertical- or horizontal-inverted views. *See* confocal microscopy; electron microscopy; fluorescence microscopy; Nomarski; oil immersion lens; phase-contrast microscopy.

resolvases Endonucleases mediating site-specific recombination that are instrumental in resolving cointegrates or concatenated DNA molecules; transposon Tn3- and $\gamma\delta$-encoded proteins promoting site-specific recombination of supercoiled prokaryotic DNA containing replicon fusion, direct-end repeats, and internal *res* sites. *See* concatenate; EMC; $\gamma\delta$ element; phase variation; recombination, molecular mechanism; recombination, site-specific; reporter ring; site-specific recombination; TN3. (Kholodii, G. 2001. *Gene* 269:121; Croomie, G. A. & Leach, D. R. 2000. *Mol. Cell* 6:815.)

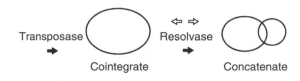

resolving power Ability of distinguishing between two alternatives or detecting an event expected to be very low, e.g., recombination between two alleles of a gene.

respiration Electrons are removed from the nutrients during catabolism and carried to oxygen through intermediaries in the respiratory chain. Oxygen is taken up and carbon dioxide is produced. *See* chlororespiration; fermentation; mitochondria; Pasteur effect.

respiratory distress syndrome Frequently accompanies premature birth due to deficiency of pulmonary surfactant protein A (SFTPA2, 10q22.2-q22.3) or PFTP3 (2p12-p11.2).

responder *(Rsp)* Component of the segregation distorter system in *Drosophila*, a repetitive DNA at the heterochromatic site in chromosome 2–62. *See* hybrid dysgenesis; segregation distorter.

response elements *See* hormone-response elements; inducible regulation of gene activity; transcription factors.

response regulator *See* two-component regulatory system.

restenosis (recurrent stenosis) Usually caused by therapeutic manipulation of the vascular system as an overcompensation of the injury repair, e.g., after angioplasty (surgical/balloon opening of stenosis) or coronary/peripheral bypass surgery. Cell cycle inhibitors, antisense technologies, and modifying the expression of particular genes by using genetic vectors to target these genes involved in the cell cycle have been considered for prevention of this very complex process. *See* antisense technologies; cell cycle; stenosis. (Gordon, E. M., et al. 2001. *Hum. Gene Ther.* 12:1277.)

restitution, chromosomal Rejoining and healing of broken-off chromosomes.

restitution nucleus Unreduced product of meiosis.

REST/NRSF RE1 silencing transcription factor/neural-restrictive silencing factor. Blocks the expression of genes in nonneural tissues. It is a component of the histone deacetylase complex. (Kojima, T., et al. 2001. *Brain Res. Mol. Brain Res.* 90:174.)

restorer genes *See* cytoplasmic male sterility, fertility restorer genes.

restriction endonuclease *See* restriction enzyme.

restriction enzyme Endonucleases cut DNA at specific sites (generally) when the bases are not protected (modified, usually, by methylation). Bacteria synthesize restriction enzymes as a defense against invading foreign DNAs such as phages and foreign plasmids. They may facilitate recombination and transposition. Three major types have been recognized. Type II enzymes are used most widely for genetic engineering. Type II enzymes have separate endonuclease and methylase proteins. Their structure is simple; they cleave at

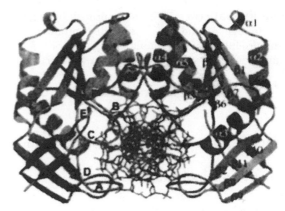

Crystal structure of *Bgl* II restriction endonuclease at 1.5 angström resolution. In the center is the DNA and α-helices and β-sheets embrace the DNA. (Courtesy of Aggarwal, A. K. From Lukacs, C. M., et al. 2000. *Nature Struct. Biol.* 7:134.)

the recognition site(s). The recognition sites are short (4–8 bp) and frequently palindromic; they require Mg^{2+} for cutting; the methylation donor is SAM. So far more than 3,000 type II restriction endonucleases have been identified. The type II proteins, unless they have a similar pattern of cleavage, have relatively low sequence homologies.

Type III enzymes carry out restriction and modification by two proteins with a shared polypeptide. They have two different subunits. Cleavage sites are generally 24–26 bp downstream from the recognition site. The recognition sites are asymmetrical 5–7 bp. For restriction, they require ATP and Mg^{2+}. For methylation, SAM, ATP, and Mg^{2+} are needed.

Type I enzymes are single multifunctional proteins of three subunits; cleavage sites are random at about 1 kb from specificity sites. Type I restriction enzymes are encoded in the *hsd* (host-specificity DNA) locus with three components: *hsdS* (host sequence specificity), *hsdM* (methylation), and *hsdR* (endonuclease). Their recognition sites are bipartite and asymmetrical: TGA-N8-TGCT or AAC-N6-GTGC. For restriction and methylation, SAM, ATP, and Mg^{2+} are required. Restriction enzymes may create protruding and receding ends, or the two ends may be of equal length, e.g.:

```
EcoRI – – – OH      PstI – – – OH       AluI – – – OH
– – – – – – p       – – – – p            – – – p
```

The Type II enzymes may be of high specificity, ambiguous, or isoschizomeric, and may be prevented from action on methylated substrates or may be indifferent to methylation. Some restriction enzymes such as McrA, McrBC, and Mrr actually cut only methylated DNA. Approximately 3,000 type II restriction enzymes with about 200 different specificities are known. Kilo- and megabase DNA substrates can rather precisely be cleaved by combining a DNA-cleaving moiety, e.g., copper: *o*-phenanthroline, with a specific DNA-binding protein (e.g., CAP). The complex thus cuts at the 5′-AAATGTGATCTAGATCACATTTT-3′ DNA site of CAP recognition. The specificity requires that the cutting moiety must be attached to an amino acid in such a way that it bends toward the selected target but not toward unspecific sequences. The IIS restriction endonucleases cleave DNA at a precise distance outside of their recognition site and produce complementary cohesive ends without disturbing their recognition site. *See* antirestriction; CAP; DNA methylation; hsdR; isoschizomers; nucleases; restriction modification; RNA restriction enzyme. (Roberts, R. J. & Macelis, D. 2001. *Nucleic Acids Res.* 29:268; Titheradge, A. J. B., et al. 2001. *Nucleic Acids Res.* 28:4195; Piungoud, A. & Jeltsch, A. 2001. *Nucleic Acids Res.* 29:3705; <http://rebase.neb.com>.) See table below.

restriction enzymes, class-IIS (Enases-IIS) Have 4–7 bp completely or partially asymmetric recognition sites and the cleavage site is 1–20 bp away from it. These enzymes are monomeric. They can be employed for precise trimming of DNAs, retrieval of cloned fragments, the assembly of genes, cleavage of single-stranded DNA, detection of point mutations, amplification, and localization of methylated bases. *See* indexer; restriction enzymes. (Szybalski, W., et al. 1991. *Gene* 100:13.)

restriction fragment Piece of DNA released after the digestion by a restriction endonuclease. The length of the fragments depends on how many nucleotides are situated between the two cleavage sites. From the same genomic DNA, the same enzyme generates fragments of different lengths because the nucleotide sequence varies along the DNA length. *See* restriction enzymes, restriction fragment number, RFLP.

restriction fragment length polymorphism *See* RFLP.

restriction fragment number Can be predicted on the basis of the number of bases at the recognition sites. Since the polynucleotide chain has four bases (A, T, G, C), four cutters can have $4^4 = 256$ bp average fragment length and six cutters $4^6 = 4,096$. The average frequency of these fragments is $0.25^4 = 0.0039$ and $0.25^6 = 0.000244$, respectively. These

---

Restriction endonucleases with recognition and cutting sites
(Isoschizomers: IS, Cutting site: ('), N: any base, ^{m}:Methylated base)

Aal IS Stu I	Bpu AI GAAGAC(N)$_{2/6}$	Hae III GG'CC	Pma CI IS: Bbr PI
Aat II GACGT'C	Bse AI T'CCGGA	Hgi AI IS: Asp HI	Pml IS Bbr PI
Acc I GT'(A,C)(T,G)AC	Bse PI IS: Bss H II	Hha I IS: Cfo I	Psp 1406 I AA'CGTT
Acc III IS: Mro I	Bsi WI C'GTACG	Hinc II IS: Hind II	Pst I CTGCA'G
Acs I (A,G)'AATT(T,C)	Bsi YI CC(N)$_5$'NNGG	Hind II GT(T,C)'(A,G)AC	Pvu I CGAT'CG
Acy I G(A,D)'CG(C,T)C	Bsm I GAATGCN'N	Hind III A'AGCTT	Pvu II CAG'CTG
Afl I IS: Ava II	CTTAC'GNN	Hinf I G'ANTC	Rca I T'CATGA
Afl II IS: Bfr I	Bsp 12861 IS: Bmy I	Hpa I GTT'AAC	Rsa I GT'AC
Afl III A'C(A,G)(T,C)GT	Bsp 14071 IS: Ssp BI	Hpa II C'CGG	Rsr II CG'G(A,T)CCG
Age IS: Pin AI	Bsp HI IS: Rca I	Ita I GC'NGC	Sac I GAGCT'C
Aha II IS: Acy I	Bsp LU11I A'CATGT	Kpn I GGTAC'C	Sac II IS: Ksp II
Aha III IS: Dra I	Bss HII G'CGCGC	Ksp I CCGC'GG	Sal I G'TCGAC
Alu I AG'CT	Bss GI IS: Bst XI	Ksp 632 I CTCTTC(N)$_{1/4}$	Sau I IS: Aoc I
Alw 44 I G'TGCAC	Bst 1107 I GTA'TAC	Mae II A'CGT	Sau 3A 'GATC
Aoc I CC'TNAGG	Bst BI IS: Sfu I	Mae III 'GTNAC	Sau 96 I G'GNCC
Aos I IS: Avi II	Bst EII G'GTNACC	Mam I GATNN'NNATC	Sca I AGT'ACT
Apa I GGGCC'C	Bst NI IS: Mva I, Eco RII	Mbo I IS: Nde II	Scr FI CC'NGG
Apo I IS: Acs I	Bst XI CCA(N)$_5$'NTGG	Mfe I IS: Mun I	Sex AI A'CC(A,T)GGT
Apy I IS: Eco RII, Mva I	Cel II GC'TNAGC	Mlu I A'CGCGT	Sfi I GGCC(N)$_4$'NGGCC
Ase I: IS: Asn I	Cfo I GCG'C	Mlu NI TGG'CCA	Sfu I TT'CGAA
Asn I AT'TAAT	Cfr I IS: Eae I	Mro I T'CCGGA	Sgr AI C(A,G)'CCGG(T,C)G
Asp I GACN'NNGTC	Cfr 10 I (A,G)'CCGG(T,C)	Msc I IS: Mlu NI	Sma I CCC'GGG
Asp 700 GAANN'NNTTC	Cla I AT'CGAT	Mse I IS: Tru 91	Sna BI TAC'GTA
Asp 718 G'GTACC	Dde I C'TNAG	Msp I C'C^{m}GG	Sno I IS: Alw 44 I
Asp EI GACNNN'NNGTC	Dpn I G^{m}A'TC	Mst II IS: Avi II	Spe I A'CTAGT
Asp HI G(A,T)GC(T,A)'C	Dra I TTT'AAA	Mst II IS: Aoc I	Sph I GCATG'C
Asu II IS: Sfu I	Dra II (A,G)G'GNCC(T,C)	Mun I C'AATTG	Ssp I AA'ATT
Ava I G'(T,C)CG(A,G)(A,G)G	Dra III CACNNN'GTG	Mva I CC'(A,T)GG	Ssp BI T'GTACA
Ava II G'G(A,T)CC	Dsa I C'C(A,G)(C,T)GG	Mvn I CG'CG	Sst I IS: Sac I
Avi II TGC'GCA	Eae I (T,C)'GGCC(A,G)	Nae I GCC'GGC	Sst II IS: Ksp I
Avr II IS: Bln I	Eag I IS: Ecl XI	Nar I GG'CGCC	Stu I AGG'CCT
Bal I IS: Mlu NI	Eam 11051 IS: Asp EI	Nci I CC'(G,C)GG	Sty I C'C(A,T)(A,T)GG
Bam HI G'GATCC	Ecl XI C'GGCCG	Nco I C'CATGG	Taq I T'CGA
Ban I G'G(T,C)(A,G)CC	Eco 47 III AGC'GCT	Nde I CA'TATG	Tha I IS: Mvn I
Ban II G(A,G)GC(T,C)'C	Eco RI G'AATTC	Nde II 'GATC	Tru 9 I T'TAA
Bbr PI CAC'GTG	Eco RII 'CC(A,T)GG	Nhe I G'CTAGC	Tth 111 I IS: Asp I
Bbs I IS: Bpu AI	Eco RV GAT'ATC	Not I GC'GGCCGC	Van 91 I CCA(N)$_4$'NTGG
Bcl I T'GATCA	Esp I IS: Cel II	Nru I TCG'CGA	Xba I T'CTAGA
Bfr I C'TTAAG	Fnu DII IS: Mvn I	Nsi I ATGCA'T	Xho I C'TCGAG
Bgl I GCC(N)$_4$'NGGC	Fnu 4 HI IS: Ita I	Nsp I (A,G)CATG'(T,C)	Xho II (A,G)'GATC(T,C)
Bgl II A'GATCT	Fok I GGATG(N)$_{9/13}$	Nsp II IS: Bmy I	Xma III IS: Ecl XI
Bln I C'CTAGG	Fsp I IS: Avi II	Nsp V IS: Sfu I	Xmn I IS: Asp 700
Bmy G(G,A,T)GC(C,T,A)'C	Hae II (A,G)GCGC'(T,C)	Pin AI AA'CCGGT	Xor II CGAT'CG

predictions would be valid, however, only if the distribution of the bases is random; it is not the case in the coding sequences, e.g., in the ca. 49.5 kb λ DNA, 12 EcoRI fragments would have been predicted, but only 5 are observed. Four-, six-cutter indicates that the enzyme cleaves the substrate at a 4- or 6-nucleotide-specified site. *See* restriction enzymes; restriction fragment.

restriction map   *See* RFLP.

restriction-mediated integration   *See* REMI.

restriction modification   The bacterial restriction enzymes are endonucleases and the modification enzymes are methyltransferases that recognize the same nucleotide sequence as the endonuclease and transfer a methyl group from S-adenosyl methionine to C-5 of cytidine or to cytidine-N^4 or to adenosine-N^6. Examples: HpaII cuts C↓CGG and methylates C^{m}CGG; TaqI cuts T↓CGA and methylates TGC^{m}A. The biological purpose of this complex is to destroy invading nucleic acids (phages) by cleaving the foreign DNA with the aid of the restriction endonuclease(s), i.e., to restrict the growth of the invader while protecting the bacterium's own genetic material by methylation. *See* antirestriction; methylation of DNA; methyltransferases; restriction enzymes. (Kobayashi, I. 2001. *Nucleic Acids Res.* 29:3742.)

restriction point   *See* cancer; cell cyle; checkpoint; commitment; R point.

restriction site   Site where the restriction enzyme cleaves. *See* cloning site.

restrictive conditions   Do not permit the growth or survival of some specific conditional mutants. *See* conditional mutation; permissive conditions.

restrictive transduction   *See* specialized transduction.

resveratrol (stilbene)   Antioxidant and antiinflammatory plant product. The beneficial effect of red wines is attributed to this phytoalexin. *See* lipooxygenase; phenolics; phytoalexins.

retardation   Slower than normal growth and development. *See* gel retardation assay; mental retardation.

reticulocyte   Immature enucleate red blood cell displaying a reticulum (network) when stained with basic dyes. *See* rabbit reticulocyte in vitro translation system.

reticulosis   Complex autosomal-recessive disease involving anemia, reduced platelet count, nervous disorders, immunodeficiency, etc. The symptoms may overlap with different types of leukemias. The prevalence is about 5×10^{-5}. Bone marrow transplantation and chemotherapy have been beneficial in some cases.

reticulosis, familial histiocytic   Involves spleen, liver, and lymph node enlargement caused by a mutation in the RAG1 or RAG2 gene at human chromosome 11p13. *See* RAG; Umenn disease.

retina   Inner layer of the eyeball connected to the optic nerve. See fig. at iris.

retinal   Vitamin A aldehyde. *See* vitamin A.

retinal dystrophy (retinopathies)   Caused by defects in the rod photoreceptor–retinal pigment epithelial complex. The incident light activates rhodopsin, which transmits the signal to the G-protein transducin, then to a phosphodiesterase (PDE), causing some reduction in the level of cGMP (required for the activity of the transducin) and the closure of Na$^+$ ion channels and cellular hyperpolarization. Arrestin and rhodopsin kinase regenerate the photoreceptor rhodopsin. Degeneration of the retinal pigment epithelium and the retinal rod receptors may result in blindness by failure at any step in this system. Low levels of cytosolic cGMP caused by mutation in the cGMP-activating protein may be the basis of amaurosis congenita. A defect of the photoreceptor-specific peripherin/RDS protein (located in the outer membranes of the rod and cone photoreceptors) results in retinal dystrophy, presumably because of the damage in anchoring the structures to the cytoskeleton. Another protein, ROM1 (rod outer membrane), homologous to peripherin, associates with a tetrameric form and represents the major outer part of the photoreceptor. Peripherin variants have been implicated in some forms of other eye diseases such as retinitis pigmentosa, macular dystrophy, and choroidal dystrophy. The peripherin-2 gene introduced by injection using adeno-associated viral vector can remedy the complex defect. Bietti corneoretinal dystrophy due to a 32 kDa fatty acid–binding protein defect that eventually causes night blindness has been assigned to 4q25-4qtel. Apoptosis may account for some retinal degenerations, which can be stopped by growth substances. *See* amaurosis congenita; apoptosis; arrestin; choroid; eye diseases; ion channels; night blindness; Oguchi disease; peripherin; phosphodiesterase; rhodopsin; signal transduction; Stargardt disease; transducin; Usher syndrome. (Rattner, A., et al. 1999. *Annu. Rev. Genet.* 33:89; Allikments, R. 2000. *Am. J. Hum. Genet.* 67:793.)

retinitis pigmentosa (RP)   Group of human autosomal recessives (84%), X-linked recessive (6%, Xp11.3 and Xp21.1), or autosomal dominant (10%, 13q14, 8q11-q13) or mitochondrial diseases entailing visual defects and blindness with an onset during the first two decades of life. Prevalence is in the 10^{-4} range. The Xp21 (RP6) gene may form a contiguous gene syndrome in that region. Only a small fraction of the pigmentary defect cases are associated with the rhodopsin receptor, a G-protein receptor (RGR). Mutation in the TULP1 gene, expressed only in the retina, may also involve RP. (Tulp-like proteins occur in vertebrates, invertebrates, and plants.) Several other diseases, e.g., some congenital deafness (Usher syndrome), hypogonadism–mental retardation, other neuropathies, and mitochondrial deficiencies may involve similar defects. Altogether about 14 genes may be involved in RP symptoms. Digenic retinitis pigmentosa is due to mutations in the unlinked loci of peripherin-2 (6p21-cen) and the ROM1 (retinal rod outer segment protein-1) in chromosome 11q13 (Loewen, C., et al. 2001. *J. Biol. Chem.* 276:22388). Mutations in the α-subunit of a cGMP phosphodiesterase (human chromosome 5q31.2-q34) gene, PDEA (phosphodiesterase A), may also cause retinitis pigmentosa. In autosomal-dominant disease, when there is single-amino-acid replacement in rhodopsin (His→Pro), ribozymes may discriminate against the mutant rat mRNA and destroy it when the photoreceptors are transduced by adenovirus-associated ribozyme constructs equipped with a rhodopsin promoter. The rhodopsin-like OPN2 gene was assigned to 3q21-q24. Recessive mutations at 1q31-q32.1 also cause RP12 due to photoreceptor degeneration. The latter gene is homologous to that of *crumbs* in *Drosophila*. MERTK receptor tyrosine kinase (2q14.1) may also be responsible for RP. A RP GTPase regulator-interacting gene (RPGIP1) was mapped to 14q11. Dominant mutations at 17p13.3, 1q21.1, and 19q13.4 encode pre-mRNA splicing factors. Inosine monophosphate dehydrogenase gene type 1 (chromosome 7q) deficiency may also be responsible for dominant RP. *See* choroidoretinal degeneration; contiguous gene syndrome; eye diseases; Lawrence-Moon syndrome; retinoblastoma; rhodopsin; ribozyme; Stargardt disease; Usher syndrome. (Bennett, J. 2000. *Curr. Opin. Mol. Ther.* 2:420; Phelan, J. K. & Bok, D. 2000. *Molecular Vision* 6:116; McKie, A. B., et al. 2001. *Hum. Mol. Genet.* 10:1555; Vithana, E. N., et al. 2001. *Mol. Cell* 8:375; Chakarova, C. F., et al. 2002. *Hum. Mol. Genet.* 11:87; Kennan, A., et al. 2002. *Hum. Mol. Genet.* 11:547.)

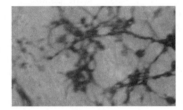

Part of the optic fundus of an eye with dark pigments indicating defects in the photoreceptor cell layer in retinitis pigmentosa. (Courtesy of the March of Dimes Foundation.)

retinoblastoma (RB)   Tumor arising from the retinal germ cells, a glioma of the retina. The overall incidence is about $1-6 \times 10^{-5}$ per birth, expressed within the first 2 years. About 40% of the cases are genetically determined. The estimated mutation rate is within the range of 10^{-5} to 10^{-6}. The bilateral form is generally familial, whereas unilateral cases may be due to new mutations. A dominant RB gene was assigned to human chromosome 13q14. The human RB1 allele may display meiotic drive, which is detectable by sperm typing. The mouse homologue Rb-1 is in mouse chromosome 14. Apparently, RB is more frequently traced to the paternal chromosome 13 than to the maternal one. The incidence of sporadic RB increases with parental age. It appeared that in RB cells tumor growth factor B was absent. The retinoblastoma protein (or homologues) may play a general role in tumorigenesis upon phosphorylation. RB is frequently associated with small cell lung carcinoma (SCLC), osteosarcoma, bladder cancer, breast cancer, leukemia, and other types of malignancies and esterase deficiency (in deletions). Retinoblastoma was the first type of cancer with recognized recessive inheritance in humans. About 5–10% of retinoblastomas are associated with deletions at chromosome 13q14 and rearrangements involving that site. RB-binding proteins (RBBP), with homology to the E7-transforming protein of a papillomavirus and to the large T antigen of SV40, have been identified. The normal allele of the retinoblastoma protein appears to restrain abnormal proliferation by limiting the activity of pol III and pol I. The retinoblastoma protein stimulates the transcription of several genes primarily by activating glucocorticoid receptors. RB has a decisive role at the G1 restriction point decisions (through RAS) in the cell cycle regarding differentiation or continuation of cell divisions. Transcription factor family E2F interacts with RB protein and regulates the cell cycle and cyclins. Retinoblastoma may be unilateral (the majority of nonhereditary cases) or bilateral (about two-thirds of hereditary cases). A defect in the RB gene may cause intrauteral death. The normal allele introduced into tumors by adenovirus vectors may slow down cancerous proliferation. *See* adenovirus; ARF; binding protein; CAF; cell cycle; cyclin; deletion; E2F; eye disease; glucocorticoid; histone deacetylase; MDM2; oncogenes; oncoprotein; p110^{Rb}; papilloma virus; papovavirus; pocket; pol III; RAS; restriction point; Simian virus 40; sperm typing; sporadic; tumor suppressor. (Nevins, J. R. 2001. *Hum. Mol. Genet.* 10:699; Chan, S. W. & Hong, W. 2001. *J. Biol. Chem.* 276:28402; <http://home.kamp.net/home/dr.lohman/>.)

retinoic acid   Carboxylic acid derivative of vitamin A; the aldehyde form is retinol. 11-cis retinal is the light-absorbing chromophore of the visual pigments (carotenoids). Retonic acids belong to a nuclear receptor family (RARE, RJR) and act as ligand-inducible transcription factors. Retinoids have a role in the anterior-posterior pattern of development of the body axis and limbs of vertebrates. They may also have anticancer effects. *See* opsin; RARE; transcription factors.

All-trans-retinol (Vitamin A).

retinoid   *See* RAR.

retinol   *See* retinoic acid.

retinopathy   Disease of the retina.

retinoschisis   Autosomal-dominant, -recessive, or X-linked (Xp22.3-p22.1) degeneration of the retina involving splitting of the organ. *See* eye diseases. (Wang, T., et al. 2002. *Hum. Mol. Genet.* 11:3097.)

RET oncogene   Rearranged during transfection, in human chromosome 10q11.2. In *Drosophila*, its homologue is *tor* (*see* morphogenesis in *Drosophila*). The protein product is a tyrosine kinase essential for the development of the nervous system; it is a signaling molecule for GDNF. Mutations at the RET locus may be responsible for familial medullary thyroid carcinoma (FMTC), multiple endocrine neoplasia (MEN2A and MEN2B), and Hirschsprung disease and may involve a dominant negative effect. The RET gene has five important domains cadherin-binding, cystein-rich calcium-binding, transmembrane, and two tyrosine kinase (TK) domains. The main course of the RET-activated signal pathway: RET receptor→Grb2→SOS→RAS→RAF→MAPKK →MAPK→NUCLEUS (transcription factors). *See* dominant negative; endocrine neoplasia; GDNF; Grb; Hirschsprung disease; MAPK; MAPKK; multiple endocrine neoplasia; oncogenes; papillary thyroid carcinoma; pheochromocytoma; RAF; RAS; SOS; TCR; tyrosine kinase; tyrosine receptor kinase. (Manie, S., et al. 2001. *Trends Genet.* 17:580.)

retroelements   *See* retroposon; retrotransposon.

retrogene   Pseudogene with transcriptionally active promoter. Retrogenes may be situated within the retroposon and have no introns. *See* pseudogenes; SINE.

retrograde   Backward. *See* anterograde.

retrograde evolution   Deletion of DNA sequences leads to adaption to new function. Many pathogens lost genes upon becoming pathogenic. Pathogens and symbionts generally show smaller genomes than nonpathogenic or nonsymbiotic relatives/ancestors. The reduction in genome size is probably the consequence of lack of need for maintenance of metabolic functions that are available in the host. Pseudogenes represent relics of genes that are no longer required.

retrograde regulation   The expression of nuclear genes is controlled by mitochondrial and chloroplast factors. (Surpin, M., et al. 2002. *Plant Cell* 14:S327.)

retrohoming   *See* intron homing.

retroid virus   Has double-stranded DNA genome (e.g., cauliflower mosaic virus, hepadnaviruses) and replicates the DNA with aid of an RNA intermediate. *See* cauliflower mosaic virus; hepatitis B virus; retroviruses.

retron   Responsible for the synthesis of msDNA. Retrons have several elements: the transcriptase *ret*, the coding regions of msDNA, and msdRNA. Retrons require RNase H. RNase HJ is required for the maintenance of the proper structure and termination of transcription. *See* msDNA; retronphage; reverse transcriptase; RNase H. (Lampson, B., et al. 2001. *Progr. Nucleic Acids Res. Mol. Biol.* 67:65.)

retronphage   Retrons that are parts of different proviruses, e.g., one in *E. coli* inserts within the selenocysteyl gene (*SelC*). *See* retron; selenocysteine.

retroposon   Transposable element mobilized through the synthesis of RNA, which is then converted to DNA by reverse transcription before integration into the chromosome. The retroposon may be a viral element or it may have originated from an ancient viral element. Retroposons are long (*see* LINE) and short (*see* SINE) interspersed elements, the copia elements in animals, and several other hybrid dysgenesis factors. They occur also in several species of plants. The majority of plant retroposons have lost their ability to move. Retroposons may be distinguished from retrotransposons by the former not having long terminal repeats. *See* copia; hybrid dysgenesis; retroviruses; reverse transcriptase; transposable elements. (Wilhelm, M. & Wilhelm, F. X. 2001. *Cell. Mol. Life Sci.* 58:1246.)

retroproteins   Have inverse folding. Although folding should theoretically depend only on the amino acid sequence, retroproteins may have altered stability. The majority of genetic inversions are detrimental, but the deleterious effect may not be the consequence of misfolding of proteins. *See* inversions.

retropseudogene   *See* processed pseudogene.

retroregulation   RNase III may degrade mRNA from the 3′-end; some mutations in temperate phages may prevent this degradation, thus permitting the translation of mRNA. This control operates from the end (downstream) forward (upstream) and is thus called retroregulation. *See* regulation of gene activity; retroviruses; ribonuclease III.

retrotransposon   Retrovirus-like transposable elements with long terminal repeats that change position within the genome as retroviruses; however, they lack an extracellular lifestyle. In the *Arabidopsis* transposon Tag1, a 98 bp 5′-terminal fragment containing a 22 bp inverted repeat and four copies of the AAACCX 5′ subterminal repeat is sufficient for transposition, but a 52 bp 5′-fragment with only one subterminal repeat is not. At the 3′-end, a 109 bp fragment containing four copies of the most 3′ TGACCC repeat, but not a 55 bp fragment, which has no subterminal repeats, is sufficient for transposition (Liu, D., et al. 2001. *Genetics* 157:817).

Retrotransposons are common in eukaryotic genomes and are apparently not distributed randomly in the chromosomes. The five *Ty* elements of *Saccharomyces* congregate in regions about 750 bp upstream of tRNA genes, and *Ty5* is found at the telomeres. Usually the sites of insertion are methylated and not transcribed, thus protecting the genome from insertional mutations. In a 280 kb region flanking the maize alcoholdehydrogenase gene (*Adh1*, chromosome 1L-128), 10 different retroelements were found crowded with repetition and inserted within each other. The repetitive elements of the maize genome largely represent retrotransposons and constitute at least 50% of the total nuclear DNA. The size of the repeats varies from 10 to 200 kb distributed throughout the genome. The maize retrotransposons with a very high copy number (10,000–30,000) usually do not cause insertional mutations. The elements with a small copy number (1–30) preferentially move into genic sequences. The *Arabidopsis* genome, which is about 1/20 that of maize, contains about 20 retrotransposons but only with 5–6 copy number. In *Vicia faba* plants with a genome size of 13.3 pg (about 1.3×10^{10} nucleotide pairs), 10% of the genome is retrotransposable elements, but in related species their number is much smaller. In *Allium* and other plants, the elements are mainly in the centromeric and telomeric heterochromatin regions, whereas in other organisms they may be dispersed. The plant retroposons (with the exception of the Tnt1 of

A DNA copy of a retrotransposon in the chromosome is replicated through an RNA transcript. The replication begins at the R segment of the 5′ long terminal repeat (LTR) and proceeds through segment U5 and the nonrepetitive internal region of the element (shown within the vertical lines) toward the 3′ boundary of the R segment of the long terminal repeat. The first strand of the DNA is synthesized by the reverse transcriptase, encoded within the retrotransposon, and it is primed by the CCA-3′-OH end of one or another kind of host tRNA that pairs by complementarity to the upstream LTR. The reverse transcription also employs a protein-tRNA complex that binds to the tRNA-protein site (tRNA-PPS). The second strand is primed at the SSP site (second-strand primer). At the target site, a duplication occurs.

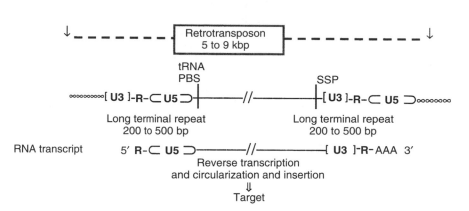

The integrase makes staggered cuts at the target and the single-strand overhangs are filled in by complementary bases, resulting in target-site duplications.

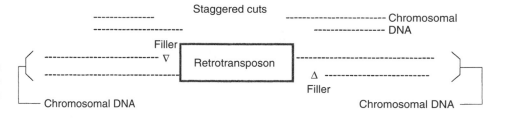

Element	5′Long terminal repeat	Internal domain	3′Long terminal repeat
Copia	**TGTTGGA...TACAACA** 276 bp	GGTTATGGGCCCAGTC...TTGAGGGGGCG 4190 bp	**TGTTGGA...TACAACA** 276 bp
TY912	**TGTTGGA...TTTCTCA** 334 bp	TGGTAGCGCCCTGTGCT...TATGGGTGGTA 5250 bp	**TGTTGGA...TTTCTCA** 334 bp
Tag-3	**TGTTGGA...GGTAACA** 514 bp	AGTGGTATCAGAGCCA....AAGGTGGAGAT 4190 bp	**TGTTGGA...GGTAACA** 514 bp

Retrotransposon and retrotransposon-like elements in *Drosophila* (*COPIA*), *Saccharomyces cerevisiae* (*TY 912*), and *Arabidopsis* (*TAG-3*) all make 5 bp target site repeats. Their long direct terminal repeats are of different length, yet they have highly conserved sequences (aligned in bold). Their internal domains of different length still have a few similarities, although *COPIA* is highly mobile, whereas *TAG-3* is no longer moving because its transposase gene underwent too many changes (pseudogenic). (After Voytas & Ausubel 1988. *Nature* 336:242.)

tobacco) fail to move because their transposase is pseudogenic. The major difference between retrotransposons and retroviruses is that the former do not produce envelope (Env) proteins. Retroposons and retrotransposons (retroelements) were expected to constitute about 5 to 10% of the human genome, but the sequencing data indicate a substantially higher proportion. *See* Line; retroposons; Sine; transposable elements. (Kumar, A. & Bennetzen, J. L. 1999. *Annu. Rev. Genet.* 33:479; <http://fly.ebi.ac.uk:7081/transposon/lk/melanogaster-transposon.html>.)

retroviral vectors Capable of insertion into the chromosomes of a wide range of eukaryotic hosts. Expected important features: (1) efficiency and selectivity for the target, (2) safety, (3) stable maintenance, (4) sufficient expression for the purpose employed. Generally the *gag, pol*, and *env* genes are removed, but all other elements required for integration and RNA synthesis and the ψ-packaging signal are retained. Usually, a selectable marker, neomycin (*neo*) or dehydrofolate reductase (*dhfr*), resulting in methotrexate resistance (MTX), bacterial hypoxanthine phosphoribosyltransferase (*hprt*), guanine-hypoxanthine-phosphoribosyltransferase (*gpt*), or mycophenolic acid (MPA) resistance, is inserted in such a way that its initiation codon (ATG) falls into the same place as the ATG of the group-specific antigen protein (*gag*) gene. Transcription will be initiated at the 5′-long terminal repeat (5′LTR) and translation will proceed from the ATG at the same place as that of *gag*:

----pBR322---- 5′ LTR PBS SD ψ ATG **SELECTABLE MARKER** 3′ LTR----pBR322----

pBR322: bacterial plasmid, 5′ LTR: long terminal repeat, PBS: binding site for primer to initiate first-strand DNA synthesis, SD: splicing donor site, ψ: packaging signal for the virion, ATG the first translated codon (Met) of the selectable marker (e.g., antibiotic resistance), 3′ LTR: long terminal repeat.

The generalized vector shown above is noninfectious and requires superinfection by a helper virus. There are several other types of vector designs. After the DNA has been packaged into a virus particle, the gene transfer is almost fully efficient as long as the cell has the appropriate receptor. The recombinant DNA integrates into the host as a provirus with the aid of the *integrase* protein encoded also by the *pol* gene. It is desirable to target the integrase to sites where the danger of insertional mutation is minimized in case the vector is used for therapeutic purposes. This goal may be difficult to achieve. Targeting to the sites of RNA polymerase III attachment with the aid of the yeast *Ty* transposons has been considered. This polymerase (pol III) transcribes ribosomal RNA genes that are present in multiple copies and thus pose a reduced risk for deleterious mutations. A better approach may be to use targeted homologous recombination.

The provirus is generally present in a single copy per cell, but this *producer cell* will proceed with the production of recombinant retrovirions. The production of the recombinant virus is much enhanced if initially the cell is coinfected by a wild-type proviral plasmid. Such a system has the disadvantage of the presence of a helper virus that competes with a pseudovirion, and it may also be hazardous by being pathogenic. The problems with the helper virus can be eliminated if it retains all the viral genetic sites except the packaging site (ψ) and thus cannot produce infectious particles.

Broad host range (amphotropic) viral vectors can be constructed by replacing a narrow-range (ecotropic) viral envelope protein with another of an amphotropic virus. The Env protein mediates the virus/vector entry into the cell generally through the Pit-2 receptor. Env determines the host range. The host range may be extended by pseudotyping, i.e., coinfection of the cells with two different viruses, which results in a mixed-envelope glycoprotein. Improved targeting of the retrovirus may also be achieved by attaching cell-specific ligands or antibodies to Env. Some of the Env proteins may elicit spongiform encephalomyelopathy. Caution must be taken because the vector or host-cell DNA and the helper virus genetic material may recombine and give rise to infectious particles. By genetic engineering, additional modifications have been introduced into the helper virus to prevent the formation of infectious single recombinants. This has been achieved by replacing, e.g., the 3′ LTR with a termination stretch of the SV40 eukaryotic virus DNA.

More useful are the *retroviral expression vectors*. They not only prove that the viral vector is present in the cell but can propagate desirable genes (e.g., growth hormone genes, globin genes, etc.). Retroviral vectors may carry strong promoters and enhancers within the LTR region and thus can overexpress some genes. They may also be employed for insertional mutagenesis because they can insert at different chromosomal locations and can be used as tools to study animal differentiation and morphogenesis. The most commonly used retroviral vectors were derived from the Moloney murine leukemia virus (MoMuLV). In some instances, within the host the retroviral vectors are silenced because the viral promoter in the 5′-LTR is methylated by host enzymes. This *promoter shutoff* may be avoided by using eukaryotic promoters. The inclusion of a locus-control region (LCR) of the host may boost the expression of the transgene. The use of insulators may protect the transgene from host repressors. Also, a heterologous promoter that may drive more efficient expression (SIN vector, double-copy vector) may replace the viral promoter. Retroviral vectors may permit the utilization of internal ribosomal entry sites (IRES) to drive the expression of more than one gene. By the use of tissue-specific promoters, the expression of the transgene may be limited to a certain cell type, e.g., in tumor cells. The promoter may be chosen for inducibility, so the expression may be regulated by the supply of glucocorticoids or a metal, or tetracycline, etc. Retroviral vectors are useful for

gene therapy because they integrate stably into human and animal cells and are normally transmitted through mitoses. *See* biohazards; cancer gene therapy; double-copy vector; epitope; E vector; foamy viral vectors; gene marking; gene therapy; inducible gene expression; insulator; IRES; laboratory safety; LCR; lentiviral vectors; MoMuLV; mycophenolic acid; packaging cell lines; pol III; pseudotyping; ψ; retroviruses; SIN vector; SV40; Ty; vectors; viral vectors. (Yee, J.-K. 1999. *Development of Human Gene Therapy*, Friedmann, T., ed., p. 21. Cold Spring Harbor Lab. Press.)

retroviruses (Retroviridae)   Include onco-, lenti-, and spuma viruses. In eukaryotes, they generally contain dimeric single-stranded RNA as the genetic material that is replicated through a double-stranded DNA intermediate with the aid of a reverse transcriptase enzyme. Retrovirus reverse transcription usually takes place after the viruses infect the host. In the human foamy virus (spumavirus), the infectious particles carry double-stranded DNA, indicating the reverse transcription precedes infection. Besides the polymerase gene (*pol*), they all carry group-specific antigen (*gag*) and envelope (*env*) protein genes. These three viral components are active in the trans position, whereas the other elements shown next page require the cis position. A provirus introduced into a cell by retroviral infection may become a retrotransposon (e.g., Ty element in yeast, copia in *Drosophila*, etc). It is characterized by long terminal repeats (LTR) of a few thousand nucleotides long. When the virus enters the host cell its RNA genetic material is converted into double-stranded DNA, and it may be covalently integrated into a host chromosome. The targets for integration spread all over the genome, yet transcriptionally active regions appear to be favored. Integration "hot spots" in chicken cells may be used by RSV a million-fold more often than expected by chance alone, and some sequences in mouse cells (e.g., the HGPRT gene) may be avoided. After infection of a cell by a single virion, thousands of viral particles may be produced in a day.

Retroviruses are considered to be diploid (dimeric) because they have a pair of genomes: two identical-size RNAs of 7 to 9 kb. All retroviruses minimally encode a protease, a polymerase, ribonuclease H activity of the reverse transcriptase, and an integrase function. A generalized structure of retroviruses is shown next page (individual types may display variations of this scheme).

The three-protein gag (group-specific antigen)—a polyprotein, pol (polymerase, reverse transcriptase), and env (envelope protein)—are transcribed in different, overlapping reading frames and then for RNA packaging into the virion. The genomic subunits are the same as the mRNA, i.e., they are (+) strands. The transcript RNA has a 7-methyl-guanylate group at the 5′-end, a 100 to 200 polyA tract at the 3′-end similar to eukaryotic mRNAs. The different retroviruses code for different proteins with known or still unidentified functions. The gag-pol polyprotein complex contains information for proteolytic activities that generate a *protease* from the carboxyl end of gag and some other proteins. *Reverse transcriptase* (RT) and *integrase* (IN) and a protease are generated by proteolysis from the translated *pol* gene product. The SU product of the *env* gene recognizes the retroviral receptors, which are transmembrane proteins of the host. Nucleocapsid proteins remain attached to proviral DNA and facilitate the integration of viral DNA into host DNA. Some murine leukemia viruses

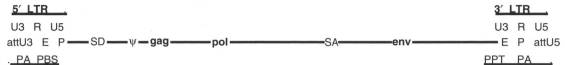

At the two ends of the viral genome are the _long terminal repeats_ (LTR) of 2 to 8 kb. At the left and right termini of the LTRs are *attU3* and _attU5_, respectively, for the attachment of the U3 (170 to 1,200 nucleotides) and U5 (80 to 120 nucleotides) direct repeats of the provirus in the host DNA. The *att* sequences at the 3′-end of U5 and at the 5′-end of U3′ contain usually imperfect, inverted repeats where viral DNA joins the host DNA at 2 nucleotides from these ends. The terminal repeats represented by *R* (10 to 230 nucleotides) are used for the transfer of the DNA during reverse transcription. *E*: transcriptional enhancer; *P*: promoter; *PA*: signal for RNA cleavage and polyadenylation; *PBS*: binding site for tRNA primer for first-strand DNA synthesis (different retroviruses use the 3′-OH end of different host tRNAs for initiation); *PPT*: polypurine sequences, which prime the synthesis of the second strand of DNA; *SD*: splice donor (the site where *gag*, *pol*, and *env* messages are spliced); *SA*: splice acceptor site (where the second splice site joins to the first [donor] site.)

can enter the cell nucleus only when the nuclear membrane breaks down during mitosis. Some of the lentiviruses can enter the nucleus with the assistance of integrase and other proteins. In the human foamy virus, the pol protein is translated by splicing mRNA that does not include the gag domain. Reverse transcriptase varies among the various retroviruses. In Rous sarcoma virus RT, there is an RNA- and DNA-directed polymerase, an RNase H, and a tRNA-binding protein, which works with either RNA or DNA primers and synthesizes up to 10 kDa molecules from single RNA priming sites. For transcription of the viral proteins, host RNA polymerase II is utilized. The host machinery translates the viral transcripts. The proteins may be processed in different ways by proteolysis to become functional.

Actinomycin D inhibits the replication on DNA but not on RNA template. Azidothymidine (AZT) inhibits polymerization and viral replication. RNase H activity removes RNA in both 5′ to 3′ and 3′ to 5′ and digests the cap, tRNA, and polyA tail. All retroviruses have an integrase protein derived from the C-end of the gag-pol polyprotein complex. Integrase (30 to 46 kDa protein with Zn-fingers) inserts the virus into the eukaryotic chromosomes:

with two single-stranded RNA copies through appropriate cellular receptors.

The retroviral genome has 10^5 times increased mutability compared with cellular genes. The retroviral genomes recombine with a 10 to 30% frequency during each cycle of replication. Integration may have a profound effect on cellular genes, may inactivate suppressor genes controlling cellular proliferation, or may activate the transcription of cellular genes with the same effect, thus initiating carcinogenesis.

Major types of retroviruses: (1) Bird's (avian) sarcoma and leukosis viruses such as Rous sarcoma virus (RSV), avian leukosis virus (ALV), Rous-associated viruses (RAV 1 and 2); (2) Reticuloendotheliosis viruses (hyperplasia of the net-like and endothelial tissues lining organ cavities), e.g., spleen necrosis virus (SNV); (3) Mammalian leukemia and sarcoma viruses, e.g., Moloney murine sarcoma virus (Mo-MSV), Moloney murine leukemia virus (MoMuLV), Harvey murine sarcoma virus (Ha-MSV), Friend spleen focus-forming virus (FSFFV), feline leukemia virus (FLV), simian sarcoma-associated virus (SSAV); (4) Mammary tumor viruses (MMTV); (5) Primate-type D viruses, e.g., Mason-Pfizer monkey virus (MPMV), simian retrovirus (SRV-1);

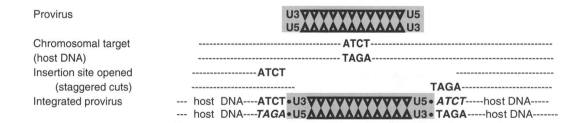

At the site of integration, there is a target duplication as the recessed ends of the target are filled in by complementary nucleotides (*in italics*). Inside the chromosomes of the host, the retroviral *provirus* still carries the LTRs at both ends. After transcription, viral RNAs are produced. Following proteolytic cleavage, the polyproteins can be converted into viral proteins with the assistance of cellular machinery. Viral particles (virions) may be assembled from viral RNA and viral proteins at the surface of the cell and the virions can exit from the cell membrane by "budding," thus infecting new cells

(6) Human T-cell leukemia-related viruses (HTLV-1, HTLV-2), simian T-cell leukemia virus (STLV), and bovine leukemia virus (BLV); (7) Immunodeficiency and lentiviruses, e.g., human immunodeficiency viruses (HIV1, HIV-2), Visna virus, simian immunodeficiency virus (SIV), caprine (goat) arthritis-encephalitis virus (CAEV), equine (horse) infectious anemia virus (EIAV) (classification of Varmus, H. & Brown, P. 1989). About 1% of the human genome includes retroviral sequences (human endogenous retroviruses, HERV). These elements presumably inserted themselves during evolution and have

been extensively modified. These HERV elements gave rise to the transposable elements or their pseudogenic forms in the modern genome. The endogenous viruses synthesize the Gag and Env proteins, enabling them to become infective but also possibly preventing their transposition. The Env protein may interfere with the viral receptors, thus limiting reinfection. HERV does not seem to have much significance for the genome at the present time; however, retrotransposition may lead to mutation and loss of gene function, including loss of cancer gene suppression. Typical retroviruses apparently do not occur in plants; however, the Athila elements of *Arabidopsis* and the SIRE-1 of soybean come close to retroviruses inasmuch as they harbor envelope-like genes. *See* animal viruses; cancer; copia elements; HTDV; hybrid dysgenesis; LINE; nucleocapsid; oncogenes; overlapping genes; pararetrovirus; plant viruses; retrogene; retroposon; retrotransposon; retroviral vectors; reverse transcription; tumor viruses. (Knipe, D. M. & Howley, P. M., eds. 2001. *Fundamental Virology*. Lippicott Williams & Wilkins, Philadelphia, PA.)

Rett syndrome (dominant, RTT, Xq28) Neurological disorder with onset after a period of normal early development. Loss of speech, loss of motor skills, seizures, and mental retardation are characteristic predominantly in girls. Prevalence is $\sim$1-2 $\times$ 10^{-4} among girls and is caused in 99.5% by new mutations occurring predominantly in the paternal X chromosome. The biochemical basis is missense or nonsense mutation in the MECP2 gene encoding a methyl-CpG-binding protein. *See* autism. (Wan, M., et al. 1999. *Am. J. Hum. Genet.* 65:1520; Meloni, I., et al. 2000. *Am. J. Hum. Genet.* 67:982; Trappe, R., et al. 2001. *Am. J. Hum. Genet.* 68:1093; Chen, R. Z., et al. 2001. *Nature Genet.* 27:327; Shahbazian, M. D., et al. 2002. *Hum. Mol. Genet.* 11:115.)

Rev Splicing element originally identified in viruses; it assists exports through the nuclear pore. *See* nuclear pore; RNA export.

REV1, REV3, REV7 Subunits of DNA polymerase ζ. REV1 is involved in DNA repair mutagenesis. *See* DNA polymerases; SOS repair; translesion. (Lawrence, C. W. & Hinkle, D. C. 1996. *Cancer Surv.* 28:21; Murakumo, Y., et al. 2001. *J. Biol. Chem.* 276:35644; Masuda, Y. & Kamiya, K. 2002. *FEBS Lett.* 520:88.)

reversal of dominance *See* dominance reversal.

reversed genetics (inverse genetics) Nucleic acids and proteins, etc., are first isolated and characterized in vitro by molecular techniques and subsequently their hereditary role is identified. Also, studying gene expression by introducing reporter genes, into cells by transformation as well as genes with truncated upstream or downstream signals, in vitro generated mutations, etc., thus determining the functional consequences of these alterations. Briefly, reversed genetics starts with molecular information, then deals with its biological role. Classical (forward) genetics recognizes genes when mutant forms become available, then studies their transmission, chromosomal location, mechanism of biochemical function, fate in populations, and evolution. Reversed genetics is sometimes called surrogate genetics.

See genetics; heredity; inheritance. (Masters, P. S. 1999. *Adv. Virus. Res.* 53:245.)

reverse dot blot *See* colony hybridization.

reverse endocrinology New hormones and ligands may be searched for and studied with the use of orphan receptors. *See* endocrinology; orphan receptors.

reverse ligase-mediated polymerase chain reaction (RL-PCR) when the beginning target mRNA is cleaved at a known location, the RL-PCR method generates a product of a predictable length in the presence of an appropriate linker and a nested primer. The linker is a probe for a synthesized strand. *See* nested primer; polymerase chain reaction; RACE. (Bertrand, E., et al. 1997. *Methods Mol. Biol.* 74:311.)

reverse linkage *See* affinity.

reverse mosaicism Secondary mutations restore wild-type function to a nucleotide sequence without returning to the wild-type nucleotide or amino acid sequence. Such

Wild type DNA ...	TTC.CTG.CTC.TGG.GCT...
Amino acids	F L L W A
Mutant DNA	TTC.CT**G.Cg**CTGG.GCT
Amino acids	F L **R** W A
Revertant DNA	TTC.CTG. tgC .TGG.GCT
Amino acids	F L C W A

somatic mosaicism may also occur by intragenic mitotic recombination, gene conversion frameshift mutation, or some type of compensatory sequence alterations in the gene as shown in the box above; the substituted nucleotides are in lowercase. The mutant is in bold. The revertant alteration is in shaded letters. (Data after Waisfisz, Q., et al. 1999. *Nature Genet.* 22:379.)

reverse mutation (backmutation) Change from mutant to wild-type allele, $a \rightarrow A$. In the experiment shown by the photo, thiamine prototrophs were selected on soil among thiamine auxotrophs. The thiamine mutants died in the absence of thiamine, but the revertant grew normally. The material was genetically marked at both flanks 5 and 9 map units, respectively, to verify that the apparent revertants were not contaminants. The progeny of the revertants were genetically analyzed. They segregated for auxotrophy and prototrophy in the proportion of 5:3 because at the time of the reversion the diploid germline consisted of two diploid cells. One of the cells remained homozygous for thiamine requirement, and the other became heterozygous and segregated for three wild-type (two heterozygotes) and for one homozygous for thiamine auxotrophy. *See* Ames test; mutant isolation; mutation; suppressor gene.

reverse transcriptases Enzymes that transcribe DNA on RNA template. An outline of the function of the enzymes within the protein coat using the diploid template is shown below. A very similar process is followed in vitro assays. Reverse transcriptases are commercially available from purified avian myeloblastosis (cancer of the bone marrow) cells or as cloned Moloney murine leukemia virus (Mo-MLV) gene product. The avian enzymes are dimeric and have strong reverse transcriptase and RNase H activities. The murine polymerase is monomeric and has only weak RNase H activity. Therefore, the murine enzyme is the choice when mRNA is transcribed into cDNA. Also, RNase H can degrade DNA and may reduce the efficiency of cDNA synthesis. The temperature optimum of the avian enzyme is 42°C (pH 8.3), and at this temperature the murine enzyme is already degraded. The pH optimum for the murine enzyme is 7.6. Both enzymes have much lower activity slightly below or above the pH optima. The avian enzyme more efficiently transcribes structurally complex RNAs. Reverse transcriptases are used for generating DNA from mRNA for vector construction or generating labeling probes for primer extension, and for DNA sequencing by the dideoxy chain termination method. Reverse transcriptase does not have an editing (exonuclease) function; therefore, it may make errors at the rate of 5×10^{-3} to 1×10^{-6} per nucleotide. This rate is orders of magnitudes higher than the error rate of most eukaryotic replicases. The HIV-1 reverse transcriptase can use either RNA or DNA as a template; in the latter case, it makes double-stranded DNA. *See* cDNA; central dogma; error

in replication; msDNA; retroviruses. Reverse transcriptase action is diagrammed below.

Reverse transcription of the retroviruses follows the generalized scheme. A single-stranded viral RNA (vvvv) serves as a template for the synthesis of the first-strand DNA ($\rightarrow\rightarrow\rightarrow$). The synthesis is primed by a tRNA attached to the PBS (primer-binding site of the retroviral [−] strand). The host tRNA is base-paired by 18 nucleotides to a sequence next to U5. The first-strand DNA (also called strong-stop minus DNA) is extended at the rate of about 2 kb per hour until the last part of the primer-binding site is copied. When the synthesis is extended, the copying of the second DNA strand ($\leftarrow\leftarrow\leftarrow$) begins. The process is practically the same in the cell and under in vitro conditions. DNA polymerases β and γ also have some reverse transcriptase function inasmuch as they can copy poly(rA) by using oligo-dT primer. *See* DNA polymerases; retroviruses; telomerase; transposon. (Whitcomb, J. M. & Hughes, S. H. 1992. *Annu. Rev. Cell Biol.* 8:275; Gao, G. & Goff, S. P. 1998. *J. Virol.* 72:5905; Vastmans, K., et al. 2001. *Nucleic Acids Res.* 29:3154.)

reverse translation If the protein is sequenced and the amino acids of a short segment are coded by nondegenerate or moderately degenerate codons, one may synthesize a few RNAs on the basis of the presumed codon sequences, and one of them may be complementary to the DNA. This short RNA sequence can be hybridized to the DNA that is coding for this particular protein. Thus, reversing the translation and generating an appropriate probe may isolate the gene. *See* probe; synthetic probe.

reverse two-hybrid system Monitors disruptions of protein-protein interactions. *See* two-hybrid system.

reversion Backmutation either at the site where the original forward mutation took place or at another (tRNA) gene that may act as a suppressor tRNA, or the reversion is caused by correcting frameshift. The possibility of reversion is frequently considered as evidence that the original forward mutation was not caused by a deletion. Suppressor mutations outside the mutant locus may restore, however, the nonmutant phenotype. Reversion may also take place when the mutation is caused by

---

1st retroviral RNA (-)   R   U5   PBS   gag   pol   env
                                        vvvvvvvvvvvvvvvvvvvvvvvvvvvvvvvvvvvvvvvvvvvvvvvvvvv
                          1st strand DNA $\rightarrow\rightarrow\rightarrow\rightarrow\rightarrow\rightarrow\rightarrow\rightarrow\rightarrow\rightarrow\rightarrow\rightarrow\rightarrow$ OH–tRNA primer
$\rightarrow$| 1st DNA synthesis continues after template switch to
vvvvvvvvvvvvvvvvvvvvvvvvvvvvvvvvvvvvv   2nd retroviral RNA template
                    U3   R
When the synthesis of the first strand reaches beyond the U3 site on the viral RNA template:
                    5'- U3 - R - U5---------------------------------------U3 - R - U5- PBS - 3'
RNAse H generates a purine-rich sequence that primes the synthesis of the second strand of the DNA starting exactly at the point corresponding to the left (5′) end of the LTR

                    vvvv$\leftarrow$ 2nd DNA strand synthesis
RNA primer ↑↑↑↑
generated by RNase H

Synthesis then continues through U3-R-U5 and PBS with the tRNA still attached until it arrives to the first modified base in the tRNA (plus strand strong stop DNA). Then RNase H removes the tRNA primer and after the two DNA strands become fully extended they anneal:

    5'$\rightarrow$ 3'
    3'$\leftarrow$ 5'
    U3 R U5   Double-stranded viral DNA   U3 R U5

a duplication and a deletion evicted the duplicated sequence. *See* Ames test; backmutation; base substitution; frameshift; reverse mutation; *sup*; suppressor; suppressor tRNA.

reversion assays, *Salmonella*, and *E. coli* in genetic toxicology   The *Salmonella* assay has been described by the Ames test. The most commonly used *E. coli* test employs strains WP2 and WP2$_{uvrA}$, which are deficient in genetic repair and are auxotrophic for tryptophan. They detect base substitution revertants, but the assay does not respond to most frameshift mutagens (unlike some of the *Salmonella* strains TA97, TA98, TA2637, and derivatives). The *E. coli* systems do not offer any advantage over that of the *Salmonella* assay of Ames. *See* Ames test; bioassays in genetic toxicology; mutation detection.

rex color   Of rodent (rabbit) hair, appears in the presence of the recessive fine fur gene, *r*, in certain combinations with black (*B/b*), agouti (*A/a*), and intensifier (*D/d*).

rexinoids   Agonists of RXR retinoid X receptor that regulate cholesterol absorption and bile acid metabolism/transport. *See* agonist; retinoic acid.

Rex1p, Rex2p, Rex3p   Exoribonuclease members of the ribonuclease D family. *See* ribonuclease D.

Reye syndrome   Nongenetic inflammation of the brain of infants that may cause fever, vomiting, coma, and eventually death. *See* acetyl-CoA dehydrogenase deficiency.

Reynaud's disease   Human complement deficiency that causes vascular problems and possibly paralyzes throat muscles. It is more common in females than males. *See* complement.

Rf   Fertility restorer genes in cytoplasmic male sterility. *See* cytoplasmic male sterility.

RF   (1) Release factor is a protein that mediates the release of the peptide chain from the ribosome after it recognizes the stop codons. *See* protein synthesis; release factor; transcription termination in eukaryotes; transcription termination in prokaryotes; translation termination. (Dontsova, M., et al. 2000. *FEBS Lett.* 472:[2–3]:213.) (2) Replicative form of single-stranded nucleic acid viruses (DNA or RNA) where the original single strand makes a complementary copy that serves as a template to synthesize replicas of the first (original) genomic nucleic acid chain. *See* plus strand; replicase. (Buck, K. W. 1999. *Philos. Trans. R. Soc. Lond. B Biol. Sci.* 354:613.)

host genome but can recombine with each other and generate new plasmids with multiple resistance factors. Because of the presence of multiple resistance factors, only simultaneous administration of multiple antibiotics may stop the multiplication of the bacteria. *See* antibiotics; plasmid; plasmid, conjugative; plasmid mobilization. (Watanabe, T. 1963. *Bacteriol. Rev.* 27:87; Falkow, S. 1975. *Infectious Multiple Drug Resistance*. Pion, London, UK; Patterson, J. E. 2000. *Semin. Respir. Infect.* 15[4]:299.)

RF-A   Replication factor A is a human single-strand DNA-binding protein, auxiliary to pol α and pol δ. *See* helix destabilizing protein; pol; replication; replication fork, eukaryotes.

RFA   Replication factor A is the same as replication protein A (RPA). *See* DNA replication, eukaryotes.

RF-C   DNA replication factor C, a primer/template-binding protein with ATP-ase activity; it has a primary role in replicating the leading-strand DNA in eukaryotes. RF-C loads PCNA on the DNA that tethers the DNA polymerase to the replication fork. RC-F is also called activator I. *See* PCNA; replication fork. (Mossi, R. & Hubscher, U. 1998. *Eur. J. Biochem.* 254:209.)

RFC   (also RF-C)   Cellular replication factor. *See* DNA replication, eukaryotes. (Schmidt, S. L., et al. 2001. *J. Biol. Chem.* 276:34792.)

RFLP   Restriction fragment-length polymorphism. Restriction endonuclease enzymes cut DNA at specific sites, thus generating fragments of various sizes in their digest, depending on the distances between available recognition sites in the genome. When base changes occurred at the recognition sites during mutation (during evolution), the length of fragments (within related strains) may have changed. After electrophoretic separation, a polymorphic pattern may be distinguished. These fragments may constitute codominant molecular markers for genetic mapping. Restriction-fragment maps can be generated by strictly physical means. If a small circular DNA is completely digested by a restriction enzyme yielding fragments, say A, B, C, D, and E, incomplete digestion with the same enzyme produces ABD, DB, AD, BC, and CE triple or double fragments, respectively, but never AB, BE, DC, or AC. Thus, the fragment sequence must be ADBCE because the double fragments must be neighbors. Another procedure is to digest by at least two enzymes and determine the overlaps by hybridization in a sequential manner. The overlapping fragments will indicate which fragments are next to each other, e.g.,

Fragments by enzyme 1
Fragments by enzyme 2

R factors   Resistance factors in bacterial plasmids that may make host bacteria insensitive to antibiotics and to normally bacteriotoxic drugs. They are common among gram-negative bacteria and are readily transmitted to other strains because the plasmids generally are endowed with transfer factors. These plasmids usually do not integrate into the bacterial

Restriction fragments can be used in genetic linkage analysis. They represent dominant physical markers because the DNA fragments can be recognized in heterozygotes. RFLP markers are useful for following the inheritance of linked genetic markers, which have variable expressivity and/or penetrance under unfavorable conditions. A *long-range restriction map* (macrorestriction map) represents the cutting sites of restriction enzymes along the chromosome. *See* physical map; restriction enzymes; restriction fragment length; restriction fragment number. (Sharma, R. P. & Mohapatra, T. 1996. *Genetica* 97[3]:313; Wicks, S. R., et al. 2001. *Nature Genet.* 28:160.)

RFLP marker   Restriction enzyme-generated DNA fragment that has been or can be mapped genetically to a chromosomal location and can be used for determining linkage to it. RFLP markers are codominant and always expressed. Their inheritance and recombination can be determined in relatively small populations. *See* integrated map; physical map; restriction enzyme; RFLP.

RFLP subtraction   Selective technique for the enrichment of particular polymorphic, eukaryotic genomic unique segments. Small restriction fragments are isolated and purified from one genome containing sequences that are in large fragments in a related genome of mice. By subtractive hybridization, the segments with shared sequences by both genomes are removed. Thus, small fragments unique to one of the other strain are obtained. These sequences then become mappable genetic markers. *See* genomic subtraction. (Rosenberg, M., et al. 1994. *Proc. Natl. Acad. Sci. USA* 91:6113.)

RFLV   RFLP variant. *See* RFLP.

Rf value   In paper or thin-layer chromatography, the distance from the baseline of the migrated compound divided by the distance of migration of the solvent (mixture). This Rf value, which is always less then 1, is characteristic for a particular compound within a defined system of chromatography. *See* paper chromatography; thin-layer chromatography.

$$Rf = \frac{\text{distance} - B}{\text{distance} - A}$$

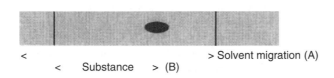

<   > Solvent migration (A)
     <   Substance   > (B)

RFX   Human DNA-binding protein that promotes dimerization of MYC and MAX, thus stimulating transcription. A group of transcription factors for the major histocompatibility complex is also designated as RFX. RFX binds cooperatively with NF-Y and X2BP. RFX has four complementation groups (CIITA [16p13], RFX5 [1q21.1-q21.3], RFXAP [13q14], RFXANK [19p12]). Their defects may be responsible for autosomal immunodeficiency syndrome. The RFX factor binds to X-boxes (5′-GTNRCC[0-3N]RGYAAC-3′), where N = any nucleotide, R = purine, and Y = pyrimidine. The human RFX1 is a helix-turn-helix protein that uses a β-hairpin (also called a wing) to recognize DNA. *See* bare lymphocyte syndrome; helix-turn-helix; immunodeficiency; MAX; MHC; MYC; NF-X; X2BP. (Katan-Khaykovich, Y. & Shaul, Y. 2001. *Eur. J. Biochem.* 268:3108.)

RGD   Amino acid sequence Arg-Gly-Asp in the extracellular matrix and in fibronectin is recognized by and bound to integrin. RGD peptides can activate caspase-3 and initiate the apoptotic pathway. *See* amino acid symbols in protein; apoptosis; fibronectin; integrin.

RGE   Rotating gel electrophoresis. the gel is rotated 90° at switching the cycle of the electric pulses. *See* pulsed-field gel electrophoresis.

rGH   Rat growth hormone is a thyroid hormone. *See* animal hormones; hormone receptors; hormone-response elements.

RGR   Yeast gene regulating transcription by RNA polymerase II. *See* NAT; RNA polymerase.

RGR   Retinal G-protein-coupled receptor is an opsin protein encoded at human chromosome 10q23. (Chen, X. N., et al. 1996. *Hum. Genet.* 97:720.)

R group (a radix)   Abbreviation for an alkyl group or any other chemical substitutions.

R locus of maize   Along with the *B* locus, it is involved in the activation of several genes of the anthocyanin biosynthetic pathway. These genes are separable by recombination and give rise to the phenotypes represented by the right-side kernels. P stands for plant color and S for seed color. In this diagram, the embryo represents the plant. The R locus includes a

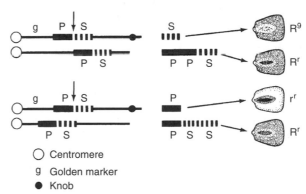

○ Centromere
g Golden marker
● Knob

large number of different alleles. A more detailed structure of the locus is described by May & Dellaporta. 1998. *Plant. J.* 13:247; Walker, E. L. 1998. *Genetics* 148:1973. Additional references are provided in these journals. *See* paramutation; tissue specificity.

RGS   Regulator of G-protein signaling is actually the same as GAP (GTP-ase activating protein). The different RGS proteins have various specificities for the $\alpha\beta$- and γ-subunits of the trimeric G proteins. In mammals, there are at least 19 members of this family of proteins with a common core, the RGS box. *See* conductin; GAP; G proteins; signal transduction. (Kehrl, J. H. & Sinnarajah, S. 2002. *Int. J. Biochem & Cell Biol.* 34:432.)

RH   *See* radiation hybrid.

rhabdomere   Rod-shaped element of the compound eye of insects. There are eight R1 to R8 neuronal photoreceptors in each of the ca. 20 cells of the *Drosophila* eyes containing about 800 ommatidial clusters. *See* compound eye; *Drosophila*; ommatidium; sevenless.

rhabdomyosarcoma   Type of cancer involving chromosome breakage in the Pax-3 gene 2q35 and 13q14 (Rhabdomyosarcoma-2) or other translocations involving chromosomes 3 and 11 (Rhabdomyosarcoma-1). These breakpoints may also be related to Beckwith-Wiedemann syndrome or WAGR syndrome. The malignant rhabdoid tumor is associated with deletions in human chromosome 22q11.2 encoding the homologue (hSNF5/INI1) of the yeast chromatin remodeling protein SWI/SNF. Embryonal rhabdomyosarcoma is due to a defect at 11p15.5. Isochromosome 3q has the same effect as mutation at the ATR (ataxia telangiectasia and rad3-related site), i.e., inhibiting MyoD (myogenesis), causing cell cycle abnormalities, and predisposing to cancer. *See* ataxia; Beckwith-Wiedemann syndrome; chromatin remodeling; histone; nuclease-sensitive sites; nucleosome; Pax; SWI; WAGR; Wilms' tumor.

Rhabdoviridae   Oblong or rod-shape (130–380 × 70–85 nm) single-stranded RNA (13–16 kb) viruses with multiple genera and wide host ranges. *See* CO_2 sensitivity in *Drosophila*.

Rhalloc   Sequences mapped by radiation hybrid methods by various mapping groups. *See* radiation hybrids.

Rh blood group   The name comes from a misinterpretation of the early study, namely that this human antigen would have the same specificity as that of rhesus monkey red cells. It is now known that this was incorrect. The animal antigen is different, but the name was not changed. Despite over half a century of research, the Rh antigen is not sufficiently characterized. The antigen may be controlled by three closely linked chromosomal sites, *C*, *D*, and *E*, and on this basis eight (2^3) different allelic combinations are conceivable, the triple recessive *cde* being a null combination. The eight combinations are also designated as *R* or *r* with superscripts: *CDe* (R^1 or $R^{1,2,-3,-4,5}$), *cde* (*r*, or $R^{-1,-2,-3,4,5}$), *cDE* (R^2 or $R^{1,-2,3,4,-5}$), *cDe* (R^O or $R^{1,-2,-3,4,5}$), *cdE* (*r″* or $R^{-1,-2,3,4,-5}$), *Cde* (*r′* or $R^{-1,2,-3,-4,5}$), *CDE* (R^Z or $R^{1,2,3,-4,-5}$), and *CdE* (r^y or $R^{-1,2,3,-4,-5}$). The first three of these occur at frequencies of about 0.42, 0.39, and 0.14, respectively, in England, and the others are quite rare. In some Oriental populations, R^1 (0.73%) and R^2 (0.19%) predominate, and the recessives have a combined frequency of about 2%. This is in contrast to Western populations, where they occur in over 40%. Clinically the most important is the D antigen, because 80% of the D⁻ individuals, in response to a large volume of D⁺ blood transfusion, make anti-D antibodies. The *d* alleles are amorph. The Rh genes are in human chromosome 1p. Additional regulatory loci have been identified in chromosome 3. For phenotypic distinction, antisera anti-D, anti-C, anti-E, anti-c, and anti-e are used. On the basis of the serological reactions, 18 phenotypes can be distinguished. Anti-D antibodies are usually immunoglobobulins of the G class (IgG) and develop only after immunization by Rh⁺-type blood. Anti-C antibodies are generally of IgM type, and they occur along with IgG after an Rh⁻ person is immunized with Rh⁺ blood. Anti-E

antibodies (IgG) are elicited in E⁻ individuals after exposure to E⁺ blood. Anti-c antibodies (IgG) occur in CDe/CDe individuals after transfusion with c⁺ erythrocytes. Anti-e antibodies are very rare (0.03). The major types of Rh antigens have several different variations. The Rh antigens are probably red blood cell membrane proteins. About 50 different Rh antigens have been identified. Rh deficiency may arise by the activity of a special suppressor gene in human chromosome 6p11-p21.1 or by CD47 protein, encoded at human chromosome 3q13.1-q13.2. The RG gene is situated at 1p34.3-p36.1.

It is clinically very significant that about 15% of Western populations are *cde/cde*. In Oriental populations, the frequency of this genotype is very low. These individuals — called Rh negatives — may respond with *erythroblastosis* when exposed to Rh-positive blood. If an Rh-negative female carries a fetus with Rh-positive blood type, antibodies against the fetal blood may be produced by the mother. This may then cause severe anemia with a high chance of intrauterine death and abortion. Generally, during the first pregnancy this hemolytic reaction is absent, but the chance in the following pregnancies is higher as sufficient immunization has taken place by the fetal blood entering into the maternal bloodstream. Thus, pregnancies of Rh⁻ females are monitored and appropriate serological treatment must be provided to prevent fetal erythroblastosis if antibody production is detected. Erythroblastosis may also occur if an Rh⁻ individual is transfused with Rh⁺ blood. The rodent antibodies responding to rhesus monkey red cells are now called the LW blood group. The physiological role of these antigens is ammonium transport. *See* antibodies; blood groups; erythroblastosis fetalis; immunoglobulins; schizophrenia. (Avent, N. D. 2001. *J. Pediatr. Hematol. Oncol.* 23:394; Stockman, J. A., III & de Alarcon, P. A. 2001. *J. Pediatr. Hematol. Oncol.* 23:385.)

Rhdb   Database containing the mapping information obtained by radiation hybrids. *See* radiation hybrids.

Rhesus blood group   *See* LW blood group; Rh blood group.

rhesus monkey (Macaque, *Macaca mulatta*, 2n = 42)   Representative of mainly Southeast Asian and North African species of long-tail monkeys. These small, intelligent animals have been used extensively for biological and behavioral studies. *See* Cercopithecidae; LW blood group; primates; Rh blood group.

rheumatic fever (rheumatoid arthritis, RA)   Ailment affecting mainly the connective tissues and joints, but it may cause heart and nervous system anomalies. The HLA region accounts for most of the susceptibility, but regions associated with other autoimmune diseases (lupus erythematosus, inflammatory bowel disease, multiple sclerosis, ankylosing spondylitis) are also implicated. The disease is complex because environmental and susceptibility factors heavily confound the direct genetic determination, e.g., certain streptococcal infections can precipitate rheumatic fever. The familial forms are attributed to dominant genetic factors. The susceptibility has been attributed to recessive genes. Several antigens have been identified that appeared to be more predominant within affected kindreds. One monoclonal antibody, D8/17, was present in 100% of the patients affected with the disease, whereas two other monoclonal antibodies showed up with 70 to 90% coincidence and with 17 to 21% presence even among the unaffected people. One susceptibility factor is linked to IL-3. Simultaneously blocking of both B- and T-cell receptors by the signaling molecule BlyS (B-lymphocyte stimulator) and TACI (transmembrane/T-cell activator and calcium-modulating and cyclophilin ligand interactor) prevented the development of arthritis in mice (Wang, H., et al. 2001. *Nature Immunol.* 2:632; Yan, M., et al. *ibid.* 638). *See* ankylosing spondylitis; arthritis; autoimmune disease; Coxsackie virus; HLA; IL-3; pseudorheumatoid dysplasia; rheumatoid. (Jawaheer, D., et al. 2001. *Am. J. Hum. Genet.* 68:927.)

rheumatoid   Resembling rheumatic condition. *See* rheumatic fever.

rhinoceros   *Ceratotherium simum*, $2n = 84$; *Rhinoceros unicornis*, $2n = 82$; *Diceros bicornis*, $2n = 134$.

Rhizobium   *See* nitrogen fixation.

rhizofiltration   *See* bioremediation.

rhizoid   Structure resembling plant roots.

rhizome   Underground plant stem modified for storage of nutrients and propagation.

Rhizome.

rhizomorph   *See* hypha.

RHKO   Random homozygous knockout. *See* knockout.

RhlB   Helicase of the DEAD-box family. *See* DEAD box; degradosome; helicase.

RHMAP   Radiation hybrid mapping is a multipoint radiation mapping procedure. It analyzes the minimal number of breaks (RHMINBRK) and may provide mapping information by the use of maximum likelihood procedure (RHMAXLIK) linkage information. *See* radiation hybrid. (*Am. J. Hum. Genet.* 49:1174.)

RHMAXLIK   *See* RHMAP.

RHMINBRK   *See* RHMAP.

RHO   GTPase homologue of the RAS oncogene. It relays signals from cell-surface receptors to the actin cytoskeleton. It regulates myosin phosphatase and Rho-associated kinase. In yeast cells, an RHO protein is involved in the stimulation of cell wall β $(1 \rightarrow 3)$-D-glucan synthase and the regulation of protein kinase C, and in mediation of polarized growth and morphogenesis. Actually, RHO is a subunit of the glucan synthase enzyme complex. Serine-threonine protein kinase and protein kinase N (PKN) are apparently activated by RHO. RHO also mediates endocytosis. In cooperation with RAF, RHO seems to induce p21$^{Waf1/Cip1}$ protein, which blocks the transition from the G_1 to the S phase of the cell cycle. In human chromosomes, RHOs are designated as ARH6: 3pter-p12; ARH12: 3p21; ARH9: 5q 31-qter. The RHO family includes Rac, Cdc42, RhoG, RhoE, RhoL, and TC10 proteins. An increase of RhoC activity accompanies metastasis of melanoma cells. Members of the RHO family of proteins are also involved in the regulation of photoreception and developmental events mediated by light. *See* Cdc42; citron; CNF; cytoskeleton; endosome; melanoma; metastasis; p21; photoreceptors; RAC; RAF; RAS oncogene; receptor; ROCK; ROK. (Kaibuchi, K., et al. 1999. *Annu. Rev. Biochem.* 68:459; Etienne-Manneville, S. & Hall, A. 2002. *Nature* 420:679.)

rho (ρ)   Designation of density. High $G + C$ content of DNA increases it; high $A + T$ content decreases it ($\rho = 1.660 + [0.098 \times \{G + C\}]$ fraction in DNA). The density is determined on the basis of ultracentrifugation in CsCl and refractometry of the bands. *See* buoyant density.

rhodamine b   Fluorochrome used for fluorescent microscopy; its reactive group forms a covalent bond with proteins (immunoglobulins) and other molecules. It is also a laser dye. Absorption maxima: 543 (355) nm. Caution: carcinogenic. *See* fluorochromes.

rho-dependent transcription termination   Actually none of the rho-dependent bacterial strains absolutely require this protein factor for termination. *See* rho factor; rho-independent; transcription termination. (Konan, K. V. & Yanofsky, C. 2000. *J. Bacteriol.* 182:3981.)

rhodopsin   Light-sensitive protein (opsin, $M_r \approx 28,600$, human chromosome 3q21-q24) coupled with a chromophore, 11-cis retinal, which isomerizes to all-trans-retinal immediately upon the receipt of the first photon. It functions as the light receptor molecule in the disks of the photoreceptive membrane of the photoreceptor cells of animal retina. Rhodopsin has seven short hydrophobic regions that pass through the endoplasmic reticulum (ER) membrane in seven turns. The amino end (with attached sugars) is within the ER lumen and the carboxyl end points into the cytosol. In the rod-shaped photoreceptor cells, rhodopsin is responsible for monochromatic light perception at low light intensities, and in the cone-shaped photoreceptor cells, color vision is mediated by it in bright light. The photoreceptor cells transmit a chemical signal to the retinal nerves, which then initiate the visual reaction series. When the receptor is activated, the level of cyclic guanylic monophosphate (cGMP) drops by the activity of *cGMP phosphodiesterase*, and it is quickly replenished in the dark by *guanylyl cyclase*. The activated opsin protein binds to transducin, an α_t G-protein subunit, which activates cGMP phosphodiesterase. When one single photon of light hits rhodopsin, through an amplification cascade, 500,000 molecules of cGMP may be hydrolyzed, 250 Na$^+$ channels may close, and more than a million Na$^+$ are turned back from entering the cell through the membrane within the time span of a second. In the dark, the sodium ion channels are kept open by cGMP; in the light, the channels are closed. When the sodium-calcium channels are shut, the intake of Ca^{2+} is reduced, which leads to the restoration of the cGMP level through the action of the *recoverin* protein, which cannot function well when it is bound to Ca^{2+}. Recoverin is a calcium sensor in the retinal rods. The rhodopsin gene has been assigned to human chromosome 3q21-qter and to mouse chromosome 6. *Drosophila* has three rhodopsin loci (*Rh2* [3-65], *Rh3* [3-70], *Rh4* [3-45]). In flies, the *nina* loci are also involved in the synthesis of opsins affecting the ommatidia and ocelli. *See* circadian rhythm; color blindness; color vision; G-proteins; ocellus; ommatidium; opsins; phytochrome; retinitis pigmentosa; retinoblastoma; signal transduction. (Yokoyama, S. 1997. *Annu. Rev. Genet.* 31:315; Palczewski, K., et al. 2000. *Science* 289:739; Bartl, F. J., et al. 2001. *J. Biol. Chem.* 276:30161; Sakmar, T. P. 2002. *Current Opin. Cell Biol.* 14:189.)

Rhoeo discolor   Ornamental plant with large chromosomes, $2n = 12$; genome $x = 14.5 \times 10^9$ bp.

Rhoeo disclor haploid set.

rho factor   Protein involved in termination of transcription in (rho-dependent) prokaryotes is about 46 kDa, and it is a hexamer (~275 kDa). For maximal efficiency, it is present in about 10% of the molecular concentration of the RNA polymerase enzyme. It is basically an ATP-dependent RNA-DNA helicase. Rho can stop elongation of the transcript only at specific termination sites in the RNA. In mitochondria, the mtTERM protein can stop transcription on both DNA strands. In yeast, the REB-1 protein terminates transcription and releases RNA from the ribosome. In mice, the TFF-1 protein terminates the action of RNA polymerase I, whereas the La protein controls RNA polymerase III. *See* antitermination; N-TEF; transcription termination in eukaryotes; transcription termination in prokaryotes. (Yu, X., et al. 2000. *J. Mol. Biol.* 299:1279; Kim, D.-E. & Patel, S. S. 2001. *J. Biol. Chem.* 276:13902.)

rho gene   Responsible for the suppressive petite (mtDNA) condition in yeast. *See* mtDNA.

rho-independent transcription termination   Also called intrinsic transcription termination. The original model visualized the involvement of an RNA hairpin followed by a 15-nucleotide T (thymidine-rich) region. The hairpin may be separated from the T sequences by a 2-nucleotide spacer. Since the *E. coli* genome has been fully sequenced, 135 terminators were identified and 940 putative terminators were found. Some of these are up to 60 nucleotides away from the 3′-end of the transcription units. *See* transcription termination. (d'Aubenton Carafa, Y., et al. 1990. *J. Mol. Biol.* 216:835; Lesnik. E. A., et al. 2001. *Nucleic Acids Res.* 29:3583.)

rhombocephalon   *See* hindbrain.

rhombomeres (neuromeres)   Metameric units of an eight-subdivision of the neuroepithelium of the hindbrain. *See* metamerism.

rho⁻ mutants   Of yeast, lost from their mitochondrial DNA most of the coding sequences and are very high in A + T content (the buoyant density of the DNA is low). *See* mtDNA.

rhubarb (Rheum spp)   About 50 species; $2n = 2x = 44$. It is an accessory food plant and some species are used as medicinal herbs (cathartic [laxative]).

Rhynchosciaras   Dipteran flies with very clearly banded polytenic chromosomes in the salivary gland nuclei. *See* polytenic chromosomes; *Sciara*.

RI   *See* recombinant inbreds.

RIA   *See* radioimmunoassay.

ribbon diagram, polypeptide structure   X-ray structure of an α-helix (at left) and a short β-sheet (at right, ending with an arrow). Usually, protein structure is more complex and contains several of these elements that form multiple domains. *See* protein structure.

ribocyte   Ancestral cell with RNA genetic material. *See* RNA world.

riboflavin (lactoflavin, vitamin B₂)   Vitamin precursor of flavin mononucleotide (FMN) and flavin adenine dinucleotide (FAD), oxidation coenzymes. Riboflavin is heat stable but rapidly decomposes in light. (Ritz, H., et al. 2001. *J. Biol. Chem.* 276:22273.)

Riboflavin.

riboflavin-retention deficiency   May prevent hatching of eggs in "leaky auxotrophic" chickens. The defect is not in

Tetraloop of 4 nucleotides.

ribonuclease (RNase) Occurs in a large number of specificities and digests various types of ribonucleic acids. The bovine pancreatic ribonuclease is a small (124 amino acids) and very heat-stable enzyme. The pancreatic ribonuclease was the first enzyme chemically synthesized in the laboratory. An autoradiogram shown below permits the distinction of the digestion patterns, separated in 20% polyacrylamide gel. NE: no enzyme control; T_1: G-specific enzyme; U_2: ribonuclease is A-specific; Phy M: *Physarum* enzyme M; specific for U + C, OH: random alkaline digest; B_c: *Bacillus cereus* enzyme with U + C specificity. On the left side, guanosine positions are indicated from the 5′-end; on the right side, the nucleotide sequences are shown as read from the gel. (Courtesy of P-L Biochemicals, Inc:) *See* angiogenins; RNases. (Condon, C. & Putzer, H. 2002. *Nucleic Acids Res.* 30:5339.)

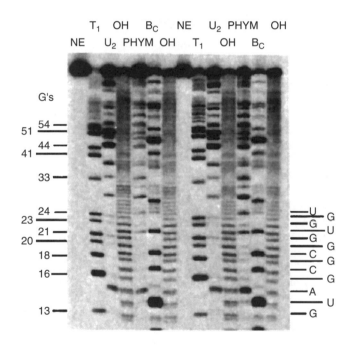

ribonuclease I Degrades RNA I. *See* RNA I. (Cunningham, K. S., et al. 2001. *Methods Enzymol.* 342:28.)

ribonuclease II Similar in action to ribonuclease D; its role is not just processing, but it can degrade an entire tRNA and mRNA molecule. *See* ribonuclease D. (Donovan, W. P. & Kushner, S. R. 1986. *Proc. Natl. Acad. Sci. USA* 83:120.)

ribonuclease III (RNase III) Homodimeric phosphodiesterase, an endonuclease cutting double-strand RNA from the 3′- or 5′-end. It cleaves prokaryotic and eukaryotic pre-rRNA at a U3 snoRNP-dependent site. RNase III controls maturation of cellular and phage RNAs and may determine the translation and half-life of mRNAs. In prokaryotes, its cleaving action may be restricted by antideterminants. In yeast, RNA tetraloops (AGNN) ⇒ are located 13–16 bp from the RNase III recognition sites. *See* antideterminant; snoRNP; trimming. (Grunberg-Manago, M. 1999. *Annu. Rev. Genet.* 33:193; Conrad, C. & Rauhut, R. 2002. *Int. J. Biochem. & Cell Biol.* 34:116.)

ribonuclease A Family of RNA-digesting enzymes, including pancreatic and brain ribonucleases, as well as the related eosinophil-derived neurotoxin (EDN), eosinophil cationic protein (ECP), and angiogenin involved in defense functions. *See* angiogenesis; eosinophil; Prp75. (Sheraga, H. A., et al. 2001. *Methods Enzymol.* 341:189.)

ribonuclease B Cuts at U + C sequences of RNA. (Zapun, A., et al. 1998. *J. Bi ol. Chem.* 273:6009.)

ribonuclease BN Exonuclease cutting tRNA. *See* exonuclease. (Callahan, C., et al. 2000. *J. Biol. Chem.* 275:1030.)

ribonuclease D Processes tRNA primary transcripts at the 3′-end into mature tRNA. *See* primary transcript; RNAi; tRNA.

ribonuclease E Cleaves RNAs with secondary structure within single-stranded regions rich in A and U nucleotides, such as RNA I. The N-terminal domain of 1,061 amino acids functions as an endonuclease involved in mRNA and rRNA processing and degradation in *E. coli* and other bacteria. This enzyme also shortens the polyA and polyU tails of RNA molecules. Its C-terminus may associate with a $3' \rightarrow 5'$ exoribonuclease and other proteins in the degradosome complex. *See* degradosome; endonuclease; *E. coli*; exonuclease; mRNA; protein repair; RNA I; tmRNA; tRNA. (Grunberg-Manago, M. 1999. *Annu. Rev. Genet.* 33:193; Walsh, A. P., et al. 2001. *Nucleic Acids Res.* 29:1864.)

ribonuclease G (CafA) Processes the 5′ end of 16S rRNA with RNase E. (Feng, Y., et al. 2002. *Proc. Natl. Acad. Sci. USA* 99:14746.)

ribonuclease H Digests RNA when paired with DNA, but it does not cut single-strand RNA, double-strand RNA, or double-strand DNA. RNase HI as an endonuclease can remove RNA primers (except one nucleotide) from the 5′-end of the Okazaki fragments. RNase H may also cleave "irrelevant sites," i.e., RNA that is imperfectly bound to DNA. RNase H can be used to prevent the translation of mRNA by recruiting it to phosphorothioated DNA. *See* antisense technologies; DNA replication, eukaryotes; Okazaki fragment; phosphorothioate; Rad27/Fen1; RNA I. (Wu, H., et al. 2001. *J. Biol. Chem.* 276:23547.)

absorption, but the vitamin is rapidly excreted by a genetic default, and the *rd/rd* eggs have only about 10 μg of the vitamin rather than the normal level of about 70 μg. If 200 μg is injected into the eggs before incubation, hatching occurs.

ribonuclease L   In dimeric form, it cleaves single-stranded RNA and its product may reduce viral replication in interferon-exposed cells. It may be induced by interferon and for activity it depends on 2′,5′-oligoadenylates. RNase L may be involved in apoptosis. *See* apoptosis; interferon; Ire; RNAi. (Stark, G. R., et al. 1998. *Annu. Rev. Biochem.* 67:227; Carpten, J., et al. 2002. *Nature Genet.* 30:181.)

ribonuclease P   Processes the 5′-end of transfer RNA transcripts (and cleaves some other RNAs). It may process some pre-tRNAs at the 3′-end. Its catalytic subunit is a ribozyme in bacteria, a 377-nucleotide RNA that can do the processing even without the ~120-amino-acid protein; however, the protein may enhance specificity and required for ribosomal translocation. For catalytic activity, the enzyme requires divalent cations (Mg^{2+}). The chloroplast enzyme is not a ribonucleoprotein but only a protein. The size of the protein subunits in bacteria is about 14 kDa, but in eukaryotes it may exceed 100 kDa. *See* external guide sequences; KH domain; MRP; ribozyme; RNA maturases; RNase. (Kurz, J. C. & Fierke, C. A. 2000. *Curr. Opin. Chem. Biol.* 4:553; Tous, C., et al. 2001. *J. Biol. Chem.* 276:29059; Gopalan, V., et al. 2002. *J. Biol. Chem.* 277:6759; Xiao, S., et al. 2002. *Annu. Rev. Biochem.* 71:165.)

RIBONUCLEASE R   is a RNase II homolog 3′ → 5′ exoribonuclease of *E. coli*. (*See* ribonuclease II, Cheng, A.-F. & Deutscher, M. P. 2002. *J. Biol. Chem.* 277:21624.)

RIBONUCLEASE S   enzymes are associated with self-incompatibility of plants (Ma, R.-C. & Oliviera, M. M. 2002. *Mol. Genet. Genomics* 267:71.)

ribonuclease T   Exonuclease of tRNA cutting at the amino-acid-accepting end (CCA). *See* tRNA. (Zuo, Y. & Deutscher, M. P. 1999. *Nucleic Acids Res.* 27:4077.)

ribonuclease T$_1$ Specific for G (guanine) linkages in RNA. (Kumar, K. & Walz, F. G., Jr. 2001. *Biochemistry* 40:3748.)

ribonuclease U1   Guanine-specific RNase. (Takahashi, K. & Hashimoto, J. 1988. *J. Biochem. [Tokyo]* 103:313.)

ribonuclease U2   Specific for A + U nucleotides in RNA. (Taya, Y., et al. 1972. *Biochim. Biophys. Acta* 287:465.)

ribonucleic acid   *See* RNA.

ribonucleoprotein (RNP)   Ribonucleic acid associated with protein. *See* RNP.

ribonucleotide   Contains one of the four nitrogenous bases (A, U, G, C), ribose, and phosphate. Building block of RNA. *See* deoxyribonucleotide.

ribonucleotide reductase (RNR)   Converts ribonucleotide di- and triphosphates into deoxyribonucleotide di- and triphosphates. It is required for DNA synthesis, the completion of the cell cycle, and malignancy. RNR proteins might have been instrumental in generating DNA in the RNA world. The allotetramer enzyme has a large subunit, R1, which regulates the maintenance of a deoxynucleotide triphosphate pool, and its level is constant throughout the cell cycle. The R2 subunit converts ribonucleotides to deoxyribonucleotides; it appears in G1 and vanishes at the early S phase. The 351-amino-acid R2 subunit is the product of the p53R2 gene, which is activated by p53. Gene p53R2 apparently has a DNA repair function. *See* cdc22; cell cycle; malignant; p53; RNA world. (Jordan, A. & Reichard, P. 1998. *Annu. Rev. Biochem.* 67:71; Tanaka, H., et al. 2000. *Nature* 404:42; Chimploy, K. & Methews, C. K. 2001. *J. Biol. Chem.* 276:7093.)

ribose   Aldopentose sugar present in ribonucleic acids with an OH group at both the 2′ and 3′ positions. Its deoxyribose form lacks the O at the 2′ position and it is present in DNA. *See* aldose.

D-Ribose.

ribose zipper   When two RNA strands of opposite polarity are situated in close vicinity to each other, hydrogen bonds may form between the 2′-OH groups of consecutive riboses in both strands. (Klostermeier, D. & Millar, D. P. 2001. *Biochemistry* 40:37.)

ribosomal DNA   Codes for ribosomal RNAs. *See* ribosome.

ribosomal filter   Is a hypothesis proposing that cis-regulatory sequences in the mRNA modulate its binding to the 40S ribosomal subunit by complementarity to the 18S or 28S ribosomal subunits or by affinity to specific ribosomal proteins. This binding may filter, i.e. influence the translation in a (+) or (−) manner and may be a factor in differential translation in a tissue- or development-specific manner. *See* ribosomes, translational control. (Mauro, V. P. & Edelman, G. M. 2002. *Proc. Natl. Acad. Sci. USA* 99:12031.)

ribosomal frame shifting (translational recoding)   *See* overlapping genes; recoding.

ribosomal genes   *See* ribosomes; *rrn*; rRNA.

ribosomal proteins   Generally designated with an S or L, indicating whether it is part of the small or large ribosomal subunit. The size of these 55 proteins in *E. coli* ranges from 6 to 75 kDa. They bind to the RNAs at specific binding sites either directly or through their association. In *E. coli*, the genes for these proteins are scattered among other genes in the chromosome. One of the bacterial ribosomal proteins is present in several copies, whereas the other ones occur only once per ribosome. In eukaryotes, the more than 80 ribosomal protein genes generally occur in a single or a few copies. The ribosomal proteins assure the proper structural conditions on the ribosomes for translation. About 35% of the bacterial ribosomes are proteins. The chloroplast ribosomal proteins are two-thirds imported from the cytoplasm, and even larger fractions of the mitochondrial proteins are coded by the nucleus. The number of ribosomal proteins in organelles is higher than in prokaryotes. The number of proteins in the mammalian mitochondrial ribosomes is about 85, and nearly all are imported. The number of ribosomal proteins in the large mitochondria of higher plants is about 65. Proteins bind only single-stranded sequences of RNA. See Diamond-Blackfan anemia; nucleolus; protein synthesis; ribosomal RNA; ribosomes. (Nomura, N., et al. 1984. *Annu. Rev. Biochem.* 53:75; Kenmochi, N., et al. 2001. *Genomics* 77:65; Uechi, T., et al. 2002. *Nucleic Acids Res.* 30:5369; Lecompte, O., et al. 2002. *Nucleic Acids Res.* 30:5382.)

ribosomal RNA   About 65% of the bacterial ribosomes are RNA. The 16S bacterial rRNA (1.54 kb) has short double-stranded domains and single-stranded loops. About 10 of the bases near the 3′-end are methylated. The 16S ribosomal RNA undergoes conformational changes (switches) before translation of the mRNA (Lodmell, J. S. & Dahlberg, A. E. 1997. *Science* 277:1262):

```
 885  887  890                      885  888 890
5'-G G G G A G-3'   ⟶         5'-G G G G A G-3'
3'-C U C-5'        ⟵         3'-C U C-5'
 912  910                            912  910
   Type I                              Type II
```

Actually, both types of base pairings have been found with physiological activity, although mutations that favored the type II conformation favored fidelity of the translation. Proteins S5 and S12 facilitate these switches, which seem to have a role in tRNA selection in algae and fungi.

The 23S rRNA (3.2 kb) carries about 20 methylated bases. The 18S mammalian rRNA (1.9 kb) has more than 40 and the 28S (4.7-kb) more than 70 methylations. The ribosomal RNAs provide not just a niche for translation, but they interact directly with translation initiation. The 16S rRNA cooperates with the anticodon at both the A and P sites, and the 23S rRNA interacts with the CCA end of the tRNA. The ribosomal exit channel (E) has a regulatory role in peptide elongation (Nakatogawa, H. & Ito, K. 2002. *Cell* 108:629). In the 23S rRNA, two guanine sites are universally conserved (G2252, G2253), and the cytosine 74 site of the acceptor end of the tRNA (CCA) is required for functional interaction at the P site of the ribosome for protein synthesis. Methylated sequences in the rRNA probably mediate the joining of the small ribosomal subunit to the large subunit after translation is initiated and hold on to the initiator tRNA^{fMet}. Some mutations in the

16S RNA may cause an override through the stop codon and failure of termination of the translation, and mutations in the 23S RNA may disturb the A and P ribosomal sites. The 23S ribosomal RNA has six domains. Of these domains, V, in an isolated form, can perform peptide elongation even better than the intact 23S molecule. It seems that ribosomal RNA alone (without protein) is required for peptide elongation. Ribosomal ribozymes mediate peptidyl transferase functions. The prokaryotic 23S ribosomal RNA contains the pentapeptide-coding minigene (GUGCGAAUGCUGACAUAAGUA) with a canonical ribosome-binding site and appears to mediate resistance to the antibiotic erythromycin. The 5S bacterial rRNA contains 120 nucleotides and binds 3 proteins (L25, L18, L5). 5S RNA is also present in eukaryotes, where it forms a complex with the L5 ribosomal protein. L25 binds to the E loop of 7 hydrogen-paired nucleotides stabilized by Mg^{2+}. The 3′-end of the small ribosomal subunit is highly conserved from prokaryotes to plants and mammals, e.g., in *E. coli*, <u>GAUCACCUCUUA</u>-OH; in yeast, GAUCA–UUA-OH; in maize, GAUCA–UUG-OH; and in rat, GAUCA–UUA-OH occur (the — signs are inserted for alignment). *See* aminoacyl-tRNA synthetase; class I genes; class III genes; gene size; introns; nucleolus; protein synthesis; ribosomal genes; ribosomal proteins; ribosomes; RNase; rrn; rRNA MRP; tmRNA; tRNA. (Noller, H. F. 1991. *Annu. Rev. Biochem.* 60:191; Liang, W.-Q. & Fournier, M. J. 1997. *Proc. Natl. Acad. Sci. USA* 94:2864; Gutell, R. R., et al. 2002. *Current Opin. Struct. Biol.* 12:301; Moore, P. B. & Steitz, T. A. 2002. *Nature* 418:229; <http://www.smi.stanford.edu/projects/helix/riboweb/kb-pub.html>; 5S ribosomal RNA: <http://rose.man.poznan.pl/5Sdata/>; small subunit: <http://silk.uia.ac.be/ssu>; diagnostics by 16S rDNA: <http://www.ridom.de>.)

ribosome   Provides the workshop and some of the tools for protein synthesis in all cellular organisms, including the subcellular organs, mitochondria, and chloroplasts. A yeast cell contains about 200,000–2 million ribosomes. The chloroplastic ribosome genes are situated in the characteristic inverted repeats, except in some *Fabaceae* and conifers. The prokaryotic and organellar ribosomes are similar and their approximate molecular weight is 2.5×10^6 Da, with a sedimentation coefficient of 70S. The eukaryotic ribosomes, excluding the organellar ones, have a molecular weight of about 4.2×10^6 Da, and they are ≈80S. The ribosomes have a minor and a major subunit. Both are built of RNA and protein:

The large subunit, 50S, is connected to the small subunit, 30S, by an RNA-protein bridge. The formation of the 80S ribosome requires the mediation of eIF-5. The mitochondrial ribosomes do not contain 5S subunits but chloroplasts do. The number of ribosomes may greatly increase when protein synthesis is very rapid. During early embryogenesis in

Transcription of ribosomal RNA in the nucleolar organizer region of the chromosomes of *Acetabularia*. The genes are separated by nontranscribed intergenic spacers. One transcription unit is about 1.7 μm. The ribosomal operons in *E. coli* use the proteins S4, L3, L4, and L13 for antitermination with functions similar to the *Nus* genes (Torres, M., et al. 2001. *EMBO J.* 20:3811). (The electron micrograph is the courtesy of Spring, H., et al. 1976. *J. Microsc. Biol. Cell* 25:107.)

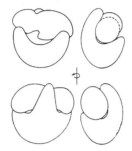

	Prokaryotic	Eukaryotic
<u>Small subunit</u>	30S	40S
rRNA types	16S (1.54 kb)	18S (1.9 kb)
protein, kinds of molecules	21	~33
<u>Large subunit</u>	50S	60S
rRNA types	23S (3.2 kb)	28S (4.7 kb)
	5S (0.12 kb)	5S (0.12 kb)
		5.8S (0.16 kb)
Protein, kinds of molecules	34	~49
Ribosomes, number/cell	15,000–70,000 (*E. coli*)	more and variable
Ribosomal gene number	in 7 operons	200 genes in *Drosophila*/genome

70S prokaryotic ribosome viewed from different angles. (Courtesy of Tischendorf, G. W., et al. 1975. *Proc. Natl. Acad. Sci. USA* 72:4870.) The ribosome structure is much more complex than shown above (see Brimacomb, R. *The Many Facets of RNA*, Eggleston, R. A., et al., eds. 1998, p. 41, Academic Press, San Diego, CA). By the use of crystallography, nuclear magnetic resonance, neutron diffraction, and cryo-electron microscopy, more detailed three-dimensional structures have been revealed. Also, by chemical footprinting, mutation, and other probes, the links between ribosomal components and ribosomal sites of tRNA and mRNA have been identified.

amphibia, the rRNA gene number may increase three orders of magnitude by a process of amplification, and the extra copies of the genes are sequestered into minichromosomes forming micronuclei. Their number in some higher plants regularly runs into thousands. The bacterial ribosomes are about 65% RNA and 35% protein. Several active centers on the ribosomes can be distinguished. The A and P sites receive the tRNAs. This area extends to both the small and large subunits. After unloading of the amino acids, the tRNA leaves the ribosome at the exit site (E) of the large subunit. The translocation factor EF-G seems to occupy a space in between the two subunits. The *classical* model of translation is discussed under the protein synthesis entry.

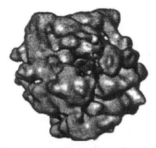

(Three-dimensional reconstruction of a ribosome. Courtesy of Dr. Wolfgang Wintermeyer.)

The newer *hybrid-states* model is given below. Elongation factor EF-TU may be located on the small subunit, but it communicates with EF-G. The EF-Tu.GTP.tRNA ternary complex is instrumental in the delivery of aminoacyl-tRNA to the A/T hybrid site of the peptidyl tRNA- ribosome complex. (A is the amino acid, P is the peptidyl, and T is the corresponding large subunit site.) In prokaryotes, the anticodon binds to the A site of the 30S ribosomal subunit. Hydrolysis of GTP is followed by the release of the elongation factor EF1A/EF-Tu. The CCA end of the amino-acid-charged tRNA moves to the A site of the large (50S) subunit. Peptidyl transferase then mediates peptide bond formation between the nascent peptide chain and the incoming amino acid. After the peptide bond is initiated, the peptidyl-tRNA is deacylated and transferred to the P site on the large subunit. The anticodon stays for a while at the A site of the small subunit while the CCA end is on the P site of the large subunit. Then the anticodon moves to the P site of the

large subunit and CCA goes to the E site of the large subunit. In the meantime, after the recognition of the cognate codon by the anticodon, the EF1A•GTP complex moves to the GTPase center of the ribosome, and EF1A•GDP and tRNA are released. After the translocation from the A site to the P site, the elongation factor EF1B mediates the conversion of the inactive EF1A•GDP into the active EF1A•GTP. This and other more recent models indicate that the movement on the ribosomes involves tRNA, but the peptidyl moiety moves very little.

Chloramphenicol, erythromycin, lincomycin, streptomycin, spectinomycin, kanamycin, hygromycin, etc., normally inhibit prokaryotic ribosomes. Eukaryotic ribosomes are sensitive to cycloheximide, anisomycin, puromycin, tetracyclines, etc. Ribosomes have an important role in the regulation of protein synthesis. It appears that the availability of active ribosomes is controlled at the level of the transcription of rRNA genes. When the supply of ATP and GTP is adequate, rRNA genes are activated for transcription. In case the level of these nucleotide triphosphates is low, rRNA transcription is reduced or halted. Abundance of free ribosomal proteins may feedback-inhibit ribosomal production. The ribosome-associated Rel-A protein may mediate the formation of ppGpp from GTP (and possibly from other nucleotides). Then ppGpp (or pppGpp) may shut off rRNA and tRNA synthesis by binding to the promoter of the RNA polymerase or to its antitermination signal. *See* antibiotics; A site; discriminator region; EF-G; EF-TU-GTP; E site; nucleolus; protein synthesis; P site; ribosomal proteins; ribosomal RNAs; ribosome recycling; Sec61 complex; transorientation hypothesis. (Green, R. & Noller, H. F. 1997. *Annu. Rev. Biochem.* 66:679; Venema, J. & Tollervey, D. 1999. *Annu. Rev. Genet.* 33:261; crystalline structure of small and large subunits of bacterial ribosomes: Clemons, W. M., Jr., et al. 1999. *Nature* 400:833; Ban, N., et al. Ibid. 841; Cate, J. H., et al. 1999. *Science* 285:2095; Wimberly, B. T., et al. 2000. *Nature* 407:327; Ban, N., et al. 2000. *Science* 289:905; Yusupova, G. Z., et al. 2001. *Cell* 106:233; Ogle, J. M., et al. 2001. *Science* 292:897; LaFontaine, D. L. J. & Tollervey, D. 2001. *Nature Rev. Mol. Cell Biol.* 2:514; Moss, T. & Stefanovsky, V. Y. 2002. *Cell* 109:545; Doudna, J. A. & Rath, V. L. 2002. *Cell* 109:153; Fatica, A. & Tollervay, D. 2002. *Current Opin. Cell Biol.* 14:313; <http://www.smi.stanford.edu/projects/helix/riboweb/kb-pub.html>; <http://rdp.cme.msu.edu>; small subunit <http://rrna.uia.ac.be/ssu/>; large subunit <http://rrna.uia.ac.be/lsu/>; <http://ribosome.fandm.edu/Department/Biology/Databases/RNA.html>.)

ribosome binding   *See* mRNA; ribosome; ribosome scanning; Shine-Dalgarno sequence.

ribosome binding assay   Used in the mid-1960s to identify several codons. RNA oligonucleotides bound to ribosomes attached to only those charged tRNA molecules that had the specific anticodons and carried the appropriate amino acids. This way, the relation between RNA codons and amino acids was revealed. *See* decoding; genetic code.

ribosome display   same as RNA-peptide fusions. (*See* display technologies, Hanes, J., et al. 2000. *Nature Biotechnol.* 18:1287.)

ribosome hopping   Bypassing the coding gaps in phage T4 genes with the assistance of a special protein factor. (Herr, A. J., et al. 2001. *J. Mol. Biol.* 311:445.)

ribosome recycling   After the termination of the translational process, the ribosome is disassembled into the small and large subunits. The posttermination complex in prokaryotes contains release factors RF1, RF2, and RF3; the 70S ribosome with the mRNA still attached to it; and a deacylated tRNA at the P site and an empty A site. This complex is split up by RRF (ribosome recycling factor) and the elongation factor G (EF-G) by GTP hydrolysis. RRF structurally mimics tRNAs, except the amino-acid-binding 3′-terminus, indicating that RRF interacts with the posttermination complex in a manner similar to how tRNA responds to the ribosome. *See* aminoacylation; A site; EF-G; protein synthesis; P site; ribosomes. (Kisselev, L. L. & Buckingham, R. H. 2000. *Trends Biochem. Sci.* 25:561; Inokuchi, Y., et al. 2000. *EMBO J.* 19:3788; Hirokawa, G., et al. 2002. *J. Biol. Chem.* 277:35847.)

ribosome scanning   Eukaryotic mRNAs do not have a Shine-Dalgarno consensus for ribosome binding. They are probably attached by the 5′-m^{7}G(5′)pp. The (5′)mRNA sequence reels on the ribosome until the initiator methionine codon is found. Circular viral or eukaryotic RNAs, if they contain internal ribosome entry sites (IRES), may be translated without a need for a free 5′-end. The RNA helicase eIF4E and the other translation initiation factor eIF4G, as well as the cap-binding protein eIF4E and the tail-binding protein Pab1, are instrumental in a cooperative manner in the initiation of translation. *See* Cap; dicistronic translation; eIF; elongation initiation; IRES; Kozak rule; protein synthesis; ribosome shunting; Shine-Dalgarno; translation. (Kozak, M. 1989. *J. Cell Biol.* 108:229; Sachs, A. B. 2000. *Cell* 101:243; Kozak, M. 2002. *Gene* 299:1.)

ribosome shunting   The long leader sequence of viral DNA (adenovirus, cauliflower mosaic virus, etc.) may contain several short open-reading frames that usually interfere with the translation of the downstream ORFs. These impediments may be bypassed (jumped) by the formation of some internal, e.g., stem-loop, structure. It seems that for the proper scanning of the ribosome, the ∼100 nucleotides at the 5′- and 3′-ends are most essential. *See* adenovirus; cauliflower mosaic virus; ORF; ribosome scanning; stem-loop. (Pooggin, M. M., et al. 2001. *Proc. Natl. Acad. Sci. USA* 98:886.)

riboswitch   In certain mRNAs there are receptor elements for target metabolites. Selective binding of the metabolite permits a conformational switch that may lead to modulation in the synthesis of protein. (Winkler, W. C., et al. 2002. *Proc. Natl. Acad. Sci. USA* 99:15098.)

ribothymidine   Thymine in the tRNA attached to ribose rather than to deoxyribose as it occurs in the DNA. *See* tRNA.

ribotype   RNA pool (similar to genotype for DNA); the information content of the RNA. It is different from the genotype because by differential processing, splicing, editing, etc., it may convey different meanings. The processed variations—during the course of evolution—may be integrated into the DNA genetic material with the aid of reverse transcriptases and become part of the hard heredity. (Herbert, A. & Rich, A. 1999. *Nature Genet.* 21:265.)

ribozyme   Catalytic RNA, possessing enzymatic activity such as splicing RNA transcripts, cleavage of DNA, amide and peptide bonds, polymerization and limited replication of RNA, etc. Thus, these ribonucleic acids carry out functions similar to those of protein enzymes. Ribozymes are generally metalloenzymes, commonly using Mg^{2+} for catalysis and stabilization. Some viral ribozymes do not require metal ions to cleave phosphodiester bonds. Most commonly, they cleave phosphodiester bonds, but they can also synthesize nucleotide chains. The ribozymes are generally large molecules, yet

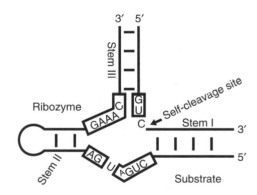

The critical sequences in the hammerhead ribozymes the boxed nucleotides are conserved

the shortest ribozyme is only UUU, and it acts on CAAA. Ribozymes commonly have an *internal guide* for substrate recognition near their 5′-terminus and a *splice site* (self-cleavage or catalytic site) where they cleave and splice the molecules. Frequently, ribozymes are classified into groups such as the hammerhead ribozymes that are used mainly by plant RNA viruses, the RNase P, the delta, group I, and hairpin ribozymes. The hammerhead ribozymes cut at UCX sequence if the neighboring sequences are complementary and pair. The hairpin ribozymes must have at least ∼50 nucleotides in the catalytic domain and ∼14 in the substrate domain. The two domains pair in a two-stem form separated with an unpaired loop of the ribozyme and substrate, most commonly containing a 5′-AGUC-3′ sequence. Cleavage is usually between A and G of the substrate.

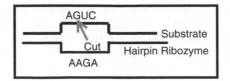

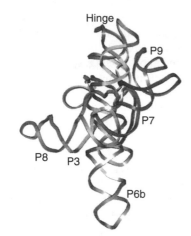

The crystal structure of 247-nucleotide *Tetrahymena* ribozyme of an rRNA intron. The conserved helical elements are identified by P1 to P9. (Courtesy of Dr. Tom Cech. From Golden, B. L., et al. 1998. *Science* 282:259.)

RNA-catalyzed RNA polymerization has also been identified. The substrate-binding by base-pairing and cleavage product release may be affected by various proteins. Although ribozymes may cleave molecules in trans, usually their efficiency is better for cis substrates. Ribozymes function as ligases and polynucleotide kinases, and in mRNA repair by transsplicing. Isomerases as well as self-alkylating catalysts have also been isolated from large pools (10^{14}) of diverse RNAs. From an evolutionary point of view, these diverse ribozyme functions lend support to the ideas of the prebiotic RNA world. Ribozymes can be engineered to recognize specific mRNAs, and by cleaving them the expression of a particular protein can be prevented. They have an advantage over protein enzymes because they are less likely to incite an immune reaction. Because of their small size, their introduction into the cell is facilitated with the aid of transformation vectors. Small RNA transcription units that may accumulate up to 106 copies per cell can propagate them. Transcription units for tRNA, U6 snRNA, have been used. Although these units produce high ribozyme titer in the cell, polymerase II transcribed units target the ribozymes more effectively to the desired location. In such pol II units, the ribozyme motif is inserted into the 5′-untranslated sequences. Evolution of ribozymes may be a much faster process than that of protein enzymes. A single polynucleotide sequence may fold into two different conformations and display two different catalytic activities such as the hepatitis delta virus self-cleaving ribozyme and a class III self-ligating ribozyme. These ribozymes may not share more than 25% (random) nucleotide identity, yet their folding pattern may satisfy the requirements of the two functions (Schultes, E. A & Bartel, D. P. 2000. *Science* 289:448). Both hammerhead and hairpin ribozymes were introduced into human cells infected with HIV and the ribozymes reduced

the level of the gag protein. Ribozyme gene therapy has been considered for malignancies caused by the human papilloma virus, Epstein-Barr virus, and hepatitis viruses. Ribozymes thus have various therapeutic potentials if an appropriate targeting system (e.g., retroviral, adenoviral vectors, cationic liposomes) is available. Ribozymes may inactivate tyrosine kinases, transcription factors, cell adhesion molecules, growth factors, telomerases, etc. Ribozymes may recognize complementary RNA targets. Problems may involve low efficiency of transfection, poor target recognition, transcriptional silencing, instability, etc. Ribozymes may be used similarly to antisense constructs but with the special advantage of catalytic activity. *See* alkylation; antisense technologies; cancer gene therapy; CBP2; cytofectin; deoxyribozyme; DNA-zyme; gene therapy; HIV; introns; kinase; leadzyme; ligase RNA; lipids, cationic; liposomes; peptidyl transferase; ribonuclease P; RNA restriction enzyme; RNA world; SELEX; transdominant molecules; transfection; viral vectors. (Wadekind, J. E. & Mckey, D. B. 1998. *Annu. Rev. Biophys. Biomol. Struct.* 27:475; Doherty, E. A. & Doudna, J. A. 2000. *Annu. Rev. Biochem.* 69:597; Ferbeyre, G., et al. 2000. *Genome Res.* 10:1011; Takagi, Y., et al. 2001. *Nucleic Acids Res.* 29:1815; Doudna, J. A. & Cech, T. R. 2002. *Nature* 418:222.)

ribulose 1,5-bisphosphate carboxylase oxidase (Rubisco) Chloroplast-located enzyme, generally the largest amount of protein in that organelle. Its large subunit is encoded and translated by the chloroplast system. The small subunit is coded for by a nuclear gene, translated in the cytosol, and imported into the plastids. The abundance of the small subunit affects the translation of mRNA of the large subunit. Rubisco is involved in early steps of photosynthesis and makes 3-phosphoglycerol through an intermediate. It is involved in oxidative and reductive carboxylation. In dinoflagellates, Rubisco is encoded by the nucleus. *See* chloroplast genetics; photosynthesis; Rubisco. (Suss, K., et al. 1993. *Proc. Natl. Acad. Sci. USA* 90:5514; Taylor, T. C., et al. 2001. *J. Biol. Chem.* 276:48159; Spreitzer, R. J. & Salvucci, M. E. 2002. *Annu. Rev. Plant Biol.* 53:449.)

rice (*Oryza*) Gramineae, $x = 12$, genome size ~430 Mbp; there are diploid and tetraploid species. Over 2,000 molecular markers are available in this small genome. Generation time 90–140 days. Two genomes have been sequenced in 2002. (Science 296:79 and 92) and seems to have more open reading frames (32,000–55,000) than other organisms. (Shimamoto, K. & Kyozuka, J. 2002. *Annu. Rev. Plant Biol* 53:399; Sasaki, T., et al. 2002. *Nature* 420:312; Feng, Q., et al. 2002. *Nature* 420:316; databases: <http://bioserver.myongji.ac.kr/ricemac.html>; <http://rgp.dna.affrc.go.jp/giot/INE.html>; <http://www.tigr.org/tdb/tgi.shtml>.)

Richner-Hanhart syndrome *See* tyrosine aminotransferase.

ricin Extremely toxic, ribosome-inactivating, dimeric toxin produced by the plant castor bean (*Ricinus*). It may have significance for cancer therapy research. *See* biological weapons; castor bean; magic bullet; RIP. (Lord, J. M., et al. 1991. *Semin. Cell Biol.* 2:15; Day, P. J., et al. 2001. *J. Biol. Chem.* 276:7202.)

ricinosome Protease precursor vesicle formed from the endoplasmic reticulum in senescing plant tissues. It contains large quantities of a 45 kDa cystine endoprotease and other proteins required for apoptosis. *See* apoptosis; protease. (Schmid, M., et al. 2001. *Proc. Natl. Acad. Sci. USA* 98:5353.)

rickets Anomalies in bone development caused by defects of calcium and phosphorus absorbtion and/or vitamin D deficiency. Human autosomal-recessive conditions may be caused by defects in the synthesis of calciferol (vitamin D) from sterols, and in such cases the dependency can be corrected by vitamin D₃. In some forms the receptor is defective and vitamin D cannot alleviate the hereditary condition (12q12-q14). Deficiency for pseudovitamin D (25-hydroxycholecalciferol-1-hydroxylase) is also at about the same chromosomal area. Hypophosphatemia (Xp22.2-p22.1) may cause vitamin D–unresponsive rickets. Rickets may then have multiple phenotypic consequences such as alopecia, epilepsy, etc. *See* Dent disease; hypophosphatasia; hypophosphatemia; spermine; vitamin D.

rickettsia Small rod-shape or roundish, obligate intracellular, gram-negative bacteria. They may carry typhus (typhoid fever, accompanied by eruptions, chills, headaches, and high mortality), spotted fever, and a tick-borne disease of cerebrospinal meningitis (brain inflammation) from animals to humans by infected arthropods (ticks, lice, fleas). The genome of *Rickettsia prowazekii* contains 1,111,523 bp and it shows the closest similarity to mtDNA of eukaryotes. *See* endosymbiont theory; mitochondria; mtDNA. (*Nature* 396:133 [1998] for complete physical map, reductive evolution; Ogata, H., et al. 2001. *Science* 293:2093.)

RID Random insertion/deletion is a technique by which certain bases of DNA at various positions can be deleted and/or replaced, thus generating mutations. *See* mutagenesis; targeting genes. (Murakami, H., et al. 2002. *Nature Biotechnol.* 20:76.)

RIDGE Region of increased gene expression. The highly expressed genes appear to be clustered in the chromosome as detected by SAGE. *See* SAGE.

Rieger syndrome Autosomal-dominant eye, tooth, and umbilical hernia syndrome. Chromosomal location (just as the Nazi discoverer of the disease) was controversial. Human chromosomes 21q22, 4q25, 13q14, and several others have been implicated. The basic cause is also unclear; epidermal growth factor, interleukin-2, alcohol dehydrogenase, and fibroblast growth factor deficiencies appeared to be involved in chromosome 4. The chromosome 4 gene (RIEG) has now been cloned and it encodes a transcription factor with similarities to the *bicoid* gene of Drosophila. Vertebrate homologues *Pitx*, *Potxlx*, and *Apr*-1 mediate the left-right development of visceral organs in concert with a number of other genes such as *Sonic hedgehog*, *Nodal*, etc. *See* eye diseases; left-right asymmetry; morphogenesis in *Drosophila* {8}; tooth agenesis. (Saadi, I., et al. 2001. *J. Biol. Chem.* 276:23034.)

rifampicin Antibiotic that inhibits prokaryotic DNA-dependent RNA polymerase (but not the mammalian RNA polymerase) and replication in *E. coli* and other prokaryotes. Rifamycin has similar effects. *See* antibiotics; maytanisoids. (Campbell, E. A., et al. 2001. *Cell* 104:901.)

RIFIN Repetitive interspersed family. *Plasmodium falciparum* proteins in high copy number involving antigenic variation are instrumental in parasite infection. *See* antigenic variation; *Plasmodium falciparum*.

rigens Translocation complex in *Oenothera muricata*. If during meiosis this complex goes to the top end of the megaspore tetrad, it may overcome its topological disadvantage and this megaspore may develop into an embryo sac because the other complex, *curvans*, is not functional in the megaspores. *See* megaspore competition; *Oenothera*; zygotic lethal.

RIG oncogene Probably required for all types of cellular growth. It is active in a very wide variety of cancers. *See* oncogenes.

RIGS Repeat-induced gene silencing is an apparently epigenetic phenomenon caused by methylation of cytosine residues. An alternative process is that locally paired regions of homologous sequences are flanked by unpaired heterologous sequences in transgenic plants. Silencing and variegation occur when transgenes are inserted either in trans or cis position to heterochromatin. The degree of silencing is proportional to the distance between the transgene and the heterochromatin. *See* cosuppression; epigenesis; heterochromatin; methylation of DNA; position effect; RIP; silencing. (Selker, E. U. 1999. *Cell* 97:157.)

Riley-Day syndrome (dysautonomia) Human chromosome 9q31-q33 recessive neuropathy involving emotional instability, lack of tearing, feeding difficulties, unusual sweating, cold extremities, etc. The prevalence in Ashkenazi Jews is about $2-3 \times 10^{-4}$, but it is quite rare in other ethnic groups. Defects in the nerve growth factor receptor are suspected. *See* IκB; nerve growth factor; neuropathy; pain insensitivity. (Slaugenhaupt, S. A., et al. 2001. *Am. J. Hum. Genet.* 68:598.)

ring bivalent Has terminalized chiasmata in both arms and thus in early anaphase I the homologous chromosomes appear to be temporarily connected at the telomeric regions of the four chromatids. This is not a ring chromosome, however. *See* translocation ring.

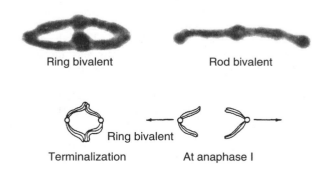

Ring bivalent   Rod bivalent

Ring bivalent

Terminalization   At anaphase I

ring canals Intercellular bridges during cytoblast differentiation. They transport mRNA and protein from nurse cells to oocytes. They are composed of actin, hts (hui-li tai shao), and kelch proteins. *See* cytoblast; maternal effect genes; RNA localization.

ring chromosome Circular chromosome without free ends (o) such as the bacterial chromosome, as the ring DNAs in mitochondria and plastids. Ring chromosomes may result by different types of chromosome breakage. Simultaneous breaks

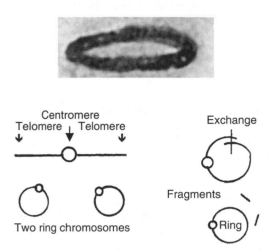

Left: A simultaneous breakage at the centromeres and the telomeres may result in a fusion between the broken ends and the generation of ring chromosomes from both of the arms of the normal chromosome generating two new ring chromosomes. **Right:** Exchange between the two ends yields a ring chromosome and two acentric fragments. For the sake of simplicity, the chromatids of the chromosomes are not represented in the diagrams.

across the centromere and the chromosome ends (telomeres) may result in fusion of the two broken termini generating one or two ring chromosomes. Crossing over between two ends of the same chromosome may give rise to a centric ring and acentric fragments. Sister chromatid exchange within a ring chromosome may result in a dicentric ring chromosome, which at anaphase separation may break at various points and generate unequal-size ring chromosomes and genetic instability. *See* dicentric ring chromosome; ring, bivalent; sister strand exchange; translocation ring. (Photo micrograph by Dr. D. Gerstel.)

ring finger Cysteine-rich amino acid motif such as Cys-X2-Cys-X(9−27)-Cys-X(1−3)-His-X2-Cys-X2-Cys-X(4−48)-Cys-X2-Cys. X stands for any amino acids in the numbers shown in parentheses. These are protein-protein, protein-membrane, protein-DNA-interacting elements involved in the regulation of transcription, replication, recombination, restriction, development, cancerous growth, etc. The ring fingers may also bind zinc and thus are related to zinc fingers. The name comes from the human gene ring — carrying such a motif — located in the vicinity of HLA. *See* DNA-binding protein domains; zinc finger. (Saurin, A. J., et al. 1996. *Trends Biochem. Sci.* 21:208.)

Ringer solution Prepared in somewhat different concentrations depending on the type of tissues used for as a sterilized

physiological salt solution in 100 mL water mg salts: NaCl 860, KCl 30, $CaCl_2$ 33; some formulations add $NaHCO_3$ 20, NaH_2PO_4, and glucose 200.

RIP **(1)** Recombination induced premeiotically or repeat-induced point mutation. When repeated (generally longer than 400 bp) DNA is introduced into fungi (*Neurospora* and others) by transformation or by other means, the replicated sequence may be lost premeiotically, or the duplications are methylated to reduce recombination, or mutations are induced. In *Ascobolus* and *Coprinus*, the repeats may be methylated without mutations following. The mutations are attributed to cytidylic acid methylation of 5′-CpG sequences as a defense against duplication introduced by transformation to silence the superfluous genetic material. The methylation results primarily in transition mutations, GC → AT. The distribution of the mutations in the DNA is not random, but most commonly they occur at 5′ of adenine sites; somewhat less frequently, they take place at 5′ to thymine or guanine but rarely at site 5′ to other cytosines. Generally within the same chromosome either C → T or G → A changes occur but not both. The majority of RIP mutations are missense or nonsense, but occasionally functional alleles arise. RIP mutations are frequently unstable and the longer duplications revert at a frequency of about 10^{-4} in the vegetative cells. *See* cosuppression; integration; methylation of DNA; MIP; position effect; RIGS; ripping; RNAi; suppressor genes. (Selker, E. U. 1997. *Trend Genet.* 13:296; Hsieh, J. & Fire, A. 2000. *Annu. Rev. Genet.* 34:187; Miao, V. P., et al. 2000. *J. Mol. Biol.* 300:249; Freitag, M., et al. 2002. *Proc. Natl. Acad. Sci. USA* 99:8802.) **(2)** Ribosome-inactivating protein is an antiviral protein in plants and animals with glycosylase activity. It may depurinate RNA of susceptible ribosomes, thus blocking protein synthesis. RIPs provide protection for plants against pathogens. *See* abrin; depurination; glycosylases; ricin; saporins. (Nielsen, K. & Boston, R. S. 2001. *Annu. Rev. Plant Physiol. Plant Mol. Biol.* 52:785.) **(3)** RNAIII inactivating peptide. *See* RNAIII/rnaiii.

Rip Regulated intramembrane proteolysis may control cell differentiation, lipid metabolism, and various proteins in prokaryotes and eukaryotes. The targets are proteins of the endoplasmic reticulum, sterol regulatory element-binding proteins (SREBP), amyloid precursor protein (APP), and Notch. *See* Alzheimer disease; Notch; SREBO. (Brown, M. S., et al. 2000. *Cell* 100:391.)

Rip1 *See* RNA export.

RI particles Formed in cold in vitro during the 30S ribosomal subunit reconstitution experiment of rRNA and about 15 proteins. Upon heating to assume the proper conformation, they become RI* particles. *See* ribosomal protein; ribosomal RNA; ribosome.

Ri plasmid Root-inciting plasmid of *Agrobacterium rhizogenes* can be used for genetic engineering similarly to the Ti plasmid of *Agrobacterium tumefaciens*. The bacterium is responsible for the hairy root disease of plants. Its T-DNA contains two segments. The right T-DNA (T_R) contains genes for the production of opines, mannopine, agropine, and auxin. The auxin genes are highly homologous to the comparable genes in the Ti plasmid of *Agrobacterium tumefaciens*. The left portion of the T-DNA (T_L) includes 11 open-reading frames with an organization similar to eukaryotic genes, but this segment is different from that of the Ti plasmid. *See* Ti plasmid. (Moriguchi, K., et al. 2001. *J. Mol. Biol.* 307:771.)

ripping   Process generated by RIP mutations. RIP may also generate point mutations primarily by G-C → A-T transitions. *See* RIP.

rise   *See* one-step growth.

risk   Combination of the degree of a hazard with its potential frequency of occurrence, e.g., if a recessive gene causes a particular malformation in 30% of the fetuses, the probability of its homozygosity among the progeny of heterozygous parents is 0.25. Thus, the risk of this malformation is $0.3 \times 0.25 \approx 0.075$, i.e., 7.5%. If an individual is known to be heterozygous for a recessive lethal gene and marries a first cousin, the risk that they will have a stillborn child may be as high as $0.5 \times 0.25 = 0.125 = 1/8$. If, however, the carrier marries an unrelated person from the general population where the frequency of this gene is only 0.005, the risk will be $0.5 \times 0.005 = 0.0025 = 1/400$. In some cases, the calculation of the genetic risk is not quite as simple. Let us assume that the penetrance of a dominant gene is 80%, but nontransmissible factors (somatic mutation not included into the germline, environmental effects) may also evoke the same symptoms; in this case, members of the family do not display the defect. Consider when new dominant germline mutations are responsible for 15% of the cases. In this instance, the ancestors are not affected. The offspring of such a mutant individual has, however, 0.8 (penetrance) × 0.5 (expected gametic transmission) = 0.4 chance for being affected. The risk of all their offspring for not being affected is $1 - 0.15 = 0.85$, and if not inherited (0 inheritance), the chance is $0.85 \times 0 = 0$. The probability of inherited (0.15) × penetrance (0.8) is $\cong 0.12$ (12%). Further considerations are necessary for proper genetic counseling if the proband already has normal, not-affected offspring. We designate the hereditary status of the proband, the probability of being a hereditary case, P(H) = 0.15, as specified above. The probability of the first child being normal (not affected), despite the defective parental gene, is P(N/H) = 1 − 0.4 = 0.6. The probability that the proband does not have this defective gene in the germline is $P(H^-) = 0.85$. The conditional probability that an offspring would be normal is $P(N/H^-) \cong 1$. From Bayes' theorem, the probability that the first child inherited but did not express the trait is

$$P(H/N) = \frac{P[H]P[N/H]}{P[H]P[N/H] + P[H^-]P[N/H^-]}$$
$$= \frac{[0.15][0.6]}{[0.15][0.6] + [0.85][1]} = \frac{0.09}{0.94} \cong 0.096$$

With 2 not-affected children, $P(N/H) = 0.6^2 = 0.36$, and n offspring, 0.6^n is the probability of the parent being normal in phenotype, although having the defective gene.

The probability of second child being normal, although carrying the defective gene, is

$$P(N/H) = 0.096 \times 0.4 \cong 0.0384,$$

and after substitution into the Bayes' formula,

$$P(H/N) = \frac{[0.15][0.0384]}{[0.15][0.0384] + [0.85][1]} = 0.0067$$

Attributable risk (AR) reveals the risk of genetically susceptible individuals relative to those who are not susceptible. It is estimated as $AR = P_{Aa}(1 - 2q[1 - P_{aa}])/(P_{Aa}[1 - 2q])$, where P is frequency and A and a are the dominant and recessive alleles, respectively, at a locus. The *relative risk* can also be estimated by contingency chi-square using an *association test*. *Absolute risk*: the excess risk of an agent causing a difference between exposed and unexposed populations. Usually, confounding factors also influence the risk, such as age, sex, addictions, etc., and then more elaborate statistical procedures are required. Life expectancy may be reduced by several factors in a complex manner (smoking a cigarette 10 min, accidents 95 days, obesity by 20% or 2.7 years, 1 mrem of radiation 1.5 minutes, medical X-rays 6 days). *See* association test; Bayes' theorem; confidence intervals; cosmic radiation; empirical risk; genetic hazards; genetic risk; genotypic risk ratio; λ_S; mutation in human populations; radiation hazard assessment; recurrence risk; utility index for genetic counseling.

RIZ   Retinoblastoma-interacting zinc-finger protein (220 kDa with 8 Zn-finger domains) has a common loss at human chromosomal site (1p36) and may be responsible for colorectal cancer, breast cancer, and endometrial neoplasias. RIZ also appears as a downstream effector of estrogen action. *See* colorectal cancer; effector; estradiol; retinoblastoma; zinc finger. (Steele-Perkins, G., et al. 2001. *Genes Dev.* 15:2250.)

RK   Rank of utility.

RK2 plasmids   Represent a family of broad host-range plasmids (56.4 kb) resistant to tetracycline, kanamycin, and ampicillin. Size and selectability of other members of the family vary. (Pogliano, J., et al. 2001. *Proc. Natl. Acad. Sci. USA* 98:4486.)

RKIP   Raf kinase inhibitor protein. *See* RAF.

RLDB   *See* reference library database.

RLGS   Restriction landmark genomic scanning. The goal of the procedure is to determine the methylation status of genes that may have undergone epigenetic changes such as oncogenic transformation or genes that are imprinted. It is based on DNA digestion by restriction endonuclease *Not*I, which cleaves methylated CpG islands. The fragments are then radioactively end-labeled at the *Not*I cut sites and further reduced in size by another restriction enzyme, e.g., *Eco*RV (GAT↓ATC). The fragments are separated by agarose gel electrophoresis and further cleaved with a third restriction enzyme, e.g., *Hinf*I (G↓ANTC), in the gel. The small fragments are subjected to electrophoresis in a second dimension in an acrylamide gel and the fragments containing the radioactive label are identified. The procedure holds promise for mass scanning of potential cancer genes. *See* electrophoresis; epigenesis; imprinting; restriction enzymes. (Costello, J. F., et al. 2000. *Nature Genet.* 25:132.)

RLK   Tec family kinase involved in T-cell receptor signaling. *See* T cell receptor; Tec. (Yang, W. C., et al. 2000. *Int. Immunol.* 12:1547.)

R loop   DNA strand displaced by RNA in a double-stranded DNA-RNA heteroduplex; also, the genomic DNA intron forms an R loop when the gene is hybridized with cDNA or mRNA. At the beginning of the replication of mtDNA, an R loop is formed that is synthesized on the light strand of mtDNA. This R loop is processed into primers for heavy-chain replication. *See* DNA replication, mitochondrial; primer. (White, R. L. & Hogness, D. S. 1977. *Cell* 10:177.)

R-LOOP   is a quasi three-stranded structure consisting of a double-stranded DNA and an single-stranded RNA, which displaces at a short section one of the DNA strands. Such a structure may occur primarily at the replication forks of DNA of prokaryotes and eukaryotes. *See* D loop. (Nossal, N. G., et al. 2001. *Mol. Cell* 7:31; Clayton, D. A. 2000. *Hum. Reprod.* 15 Suppl. 2:22; Tracey, R. B. & Lieber, M. R. 2000. *EMBO J.* 19:1055.)

RLP   Ribosome landing pad. *See* IRES.

RL-PCR   *See* reverse ligase-mediated polymerase chain reaction.

RME1   Inhibitor of yeast meiosis and sporulation. *See* mating type determination. (Shimizu, M., et al. 1998. *Nucleic Acids Res.* 26:2320.)

RMSA-1   Regulator of mitotic spindle assembly). Protein that is phosphorylated only during mitosis and is a substrate for Cdk2 kinase; it is required for the spindle assembly. *See* Cdk; spindle. (Yeo, J. P., et al. 1994. *J. Cell Sci.* 107:1845.)

RNA   Ribonucleic acid is a polymer of ribonucleotides. There are three main classes of RNAs in the cell: mRNA, which provides the instructions for protein synthesis; various ribosomal RNAs; and tRNAs. Other RNAs are involved in splicing, editing, posttranscriptional modification, ribonucleoproteins that insert proteins into membranes and mediate telomere synthesis, replication priming RNAs, inhibitory RNAs (RNAi), ribozymes, and noncoding RNAs involved in dosage compensation and imprinting. *See* dosage compensation; imprinting; mRNA; noncanonical bases; nucleic acid chain growth; replication; ribozymes; RNA I; RNA editing; RNAi; rrn; rRNA; telomerase; tRNA; Xist. (<http://prion.bchs.uh.edu/>.)

RNA, catalytic   *See* ribozyme.

RNA, double-stranded   Usually a minor fraction of the total cellular RNA (unlike in DNA). It appears that if double-stranded RNA with sequences homologous to an open-reading frame of a gene is introduced into the cell, antisense RNA more

effectively blocks the function of that gene (in *Caenorhabditis*). It is a surprising fact that even at very low abundance this RNA is a highly effective inhibitor. A double-stranded RNA-binding protein is apparently required for gene activation in mammalian germ cells. dsRNA is common in virus-infected cells and it is an inducer of interferon. It inhibits the transcription of about one-third of the genes and stimulates the rest to a variable degree. A substantial fraction of the pairing of RNA molecules is not in compliance with the Watson-Crick model. Hoogsteen and "sugar edge" pairing also occur. The edge indicates the relative orientation of the glycosidic and hydrogen bases. The four RNA bases (A, C, G, U) may associate in 4^2 matrices. These steric associations may be evolutionarily preserved better than the base positions themselves and facilitate long-range RNA—RNA interactions, create binding sites for proteins and small ligands. *See* antisense RNA, targeting genes, RNAi, Watson-Crick model, Hoogsteen pairing. (Geiss, G., et al. 2001. *J. Biol. Chem.* 276:30178; Leontis, N. B., et al. 2002. *Nucleic Acids Res.* 30:3497.)

RNA, heterogeneous   *See* hnRNA.

RNA, noncoding   There are many mRNA-like yet untranslated molecules in the cell. They are polyadenylated and spliced but lack a long open-reading frame. Presumably they have a regulatory or signal function. The tRNAs, ribosomal RNAs, snRNAs, and snoRNAs are noncoding. *See* microRNA; RNAi. (Erdmann, V. A., et al. 2001. *Cell. Mol. Life Sci.* 58:960; Eddy, S. R. 1999. *Curr. Opin. Genet. Dev.* 9:695; Storz, G. 2002. *Science* 296:1260; Kiss, T. 2002. *Cell* 109:145; Fahey, M. E., et al. 2002. *Comp. Funct. Genomics* 3:244; Tjaden, B., et al. 2002. *Nucleic Acids Res.* 30:3732. <http://biobases.ibch.poznan.pl/ncRNA>.)

RNA, ribosomal   *See* ribosomes.

RNA, small   *See* snRNA.

RNA, ubiquitous   *See* U RNA.

RNA I   Untranslated bacterial RNA controlling the maturation of RNA II that serves as a primer for plasmid DNA synthesis. RNA I and RNA II are synthesized on opposite DNA strands. RNA I binds to RNA II, thereby preventing its folding into a cloverleaf necessary for the formation of a stable DNA:RNA hybrid between RNA II and plasmid DNA. This binding is promoted by the Rop protein (63 amino acid residues) coded for by 400 bases downstream from the origin of replication. A single G → A transition mutation in Rop or upstream may contribute to plasmid amplification. RNA I, RNA II, and Rop also control plasmid incompatibility.

```
                          RNase H
        RNA II→  →  →  →  →  →  →    Replication
────────────────────────┃──────────────────────
RNA I ←              ori                    ← Rop
```

RNase H cuts off the preprimer section and prepares the primer for actual DNA synthesis. RNA I may be polyadenylated, then its decay is hastened similarly to mRNA. RNA I and RNA II may interact initially by base pairing between their seven-nucleotide complementary loops. Rom/Rop protein may also bind to the transient complex and assures a more stable duplex of the two RNAs, causing failure of replication initiation. *See* plasmid; polyadenylation; RNA polymerase; ROM. (Mruk, I., et al. 2001. *Plasmid* 46:128.)

RNA II *See* RNA I.

RNA binding proteins Modify RNA structure locally or globally; may affect RNA trafficking, mRNA biosynthesis, translation, splicing, polyadenylation, differentiation, and diseases. The length and composition of the binding domains may vary. A human RNA-binding protein is encoded in chromosome 8p11-p12 (RBP-MS). *See* DNA-binding protein domains; hnRNP. (Burd, C. G. & Dreyfuss, G. 1994. *Science* 265:615; Dreyfuss, G., et al. 2002. *Nature Rev. Mol. Cell Biol.* 3:195; Hall, K. B. 2002. *Current Opin. Struct. Biol.* 12:283.)

RNA cap *See* cap; capping enzymes.

RNA chaperone *See* chaperones.

RNA codons *See* genetic code.

RNA computer *See* DNA computer. (Faulhammer, D., et al. 2000. *Proc. Natl. Acad. Sci. USA* 97:1385.)

RNA-dependent RNA polymerase Replicates the RNA genome of viruses. RNA-directed RNA polymerase, primed by siRNA, amplifies the interference by RNAi. *See* RNAi. (Ahlquist, P. 2002. *Science* 296:1270.)

RNA-driven reaction In a DNA-RNA hybridization experiment, RNA is far in excess compared to single-stranded DNA. This assures that at all potential annealing sites hybridization will take place. *See* DNA hybridization; nucleic acid hybridization.

RNA editing Means of posttranscriptional or cotranscriptional altering of the RNA transcript (mRNA, tRNA, rRNA, 7 SLRNA). It is a very common process in the mitochondria of *Trypanosomes* (*see Trypanosoma brucei*). A separately transcribed 40–80-base "guide RNA" with homology to the 5′-end of the RNA to be modified pairs with the target. Then uracil residues from the 3′-end tract of the guide are transferred into the target sequences. This editing thus changes the content of the message and the amino acids of the translated protein. In the mitochondria of this protozoan, thousands of *U* nucleotides may be inserted into different pre-mRNAs. *U*s may also be removed at a 10-fold lower frequency. It has been hypothesized that the *U* replacements are provided by the 3′-end of the gRNA, but experimental evidence indicates that they come from free UTPs.

In the mitochondria of plants, in about 10% of the transcripts, U may replace C or C replaces U. Editing of 4 to 25 RNAs may take place in the chloroplasts. Editing appears to be rare and limited in higher animals. One C-residue deamination in apolipoprotein B results in a U

replacement, the creation of a stop codon, and consequently a truncated protein. Another similar deamination in the middle of the transcripts alters the permeability of a Ca^{2+} channel. Thus, editing produces two different mRNAs from one. Although $C \rightarrow U$ is the common change, $U \rightarrow C$ may also occur exceptionally. Simple deamination, addition, or cotranscriptional errors, such as stuttering of the polymerase, may also bring about the changes.

RNA editing takes place in different RNA viruses. In HIV-1 $G \rightarrow A$, editing also occurs besides $C \rightarrow U$. Mammalian nuclear RNA editing may involve the deamination of adenosine into inosine in the double-stranded pre-mRNA of the glutamate-receptor subunits. The enzyme responsible for the process is dsRAD (double-stranded RNA adenosine deaminase, also called DRADA or ADAR). ADAR1 and ADAR2 proteins are associated with spliceosomal components of a 200 S large ribonucleoprotein complex (lnRNP). Deficiency of ADAR in *Drosophila* results in behavioral anomalies during the advanced developmental stages. In the tRNA^{Asp} of marsupials, the GCC anticodon is found that can recognize only the glycine codon. In 50% of these tRNAs, the middle base is edited to U, and thus the regular Asp anticodon is generated. The marsupial mitochondria also have the regular tRNA^{Gly} with anticodon GGN that recognizes all 4 glycine codons, but the edited codon can match up with only 2 of the glycine codons. It is a puzzling observation that among only 22 tRNAs there are 2 for glycine (normal and edited).

RNA editing occurs through the plant kingdom (with few exceptions) in mitochondria as well as in chloroplasts, albeit at a lower frequency in the latter. In the plastids, there are about 25 editable sites, whereas in the mitochondria their number may exceed 1,000. RNA editing may generate new initiation and termination codons in plant organelles and thus new reading frames. No $U \rightarrow C$ editing was observed in the mitochondria or chloroplasts of gymnosperms or in the chloroplasts of angiosperms. The site of editing was apparently selected on the basis of the flanking sequences.

RNA editing occurs in nuclear genes and contributes to the regulation at an additional level. It appears that the neurofibromas are determined by editing in the neurofibromatosis gene (NF1). Editing may also regulate the mRNA of the serotonin-2C receptor. In the immune system, T-cell-independent B cells (B1) diversify their surface receptors by RNA editing. Defects in RNA editing may potentiate tumorigenesis. *See* ADAR; anticodon; apolipoproteins; B cell; DRADA; editosome; genetic code; gRNA; kinetosome; mooring sequence; mtDNA; RNA ligase; serotonin; stuttering; wobble; Z DNA. (Gott, J. M. & Emeson, R. B. 2000. *Annu. Rev. Genet.* 34:499; Raitskin, O., et al. 2001. *Proc. Natl. Acad. Sci. USA* 98:6571; Aphasizhev, R., et al. 2002. *Cell* 108:637; Bass, B. L. 2002. *Annu. Rev. Biochem.* 71:817.)

RNA enzyme *See* ribozyme.

RNA export From the nucleus, mRNA requires the presence of the nuclear export factor NES and a cellular cofactor Rip1/Rab. The NES function is part of the Gle1 yeast protein (M_r 62K). Gle1 interacts with Rip1 and nucleoporin (Nup 100) in the nuclear pore. The Rev splicing factor can substitute for the Gle1 function. Protein Aly of metazoans is involved in pre-mRNA slicing and mRNA export. The export of unspliced mRNA may be prevented by nuclear retention factors (RF); however, some intronless mRNAs can be exported and efficiently translated in the cytoplasm. Export of mRNA also requires PIP$_2$ and PIP$_3$. The abundant hnRNPs (hnRNP C and hnRNP K) have been implicated in mediating RNA export. Some TAP elements are apparently involved in the export. tRNA and U snRNA export may require members of the importin β-family of proteins. The export of tRNAs may be mediated by the aminoacyl-tRNA synthetases after transcript processing. This involves removal of the unnecessary 5′ and 3′ sequences, introns, addition to the 3′ end of the CCA sequence, and modification of some nucleotides. *See* aminoacyl-tRNA synthetase; cell-penetrating peptides; export adaptors; importin; Ipk1; nuclear export sequences; nuclear pore; REV; RNA transport; TAP; transport elements, constitutive. (Michael, W. M. 2000. *Trends Cell Biol.* 10:46; Zenklusen, D. & Stutz, F. 2001. *FEBS Lett.* 498:150; Sträßer, K., et al. 2002. *Nature* 417:304.)

RNA extraction   Essential requisite that RNase activity be eliminated or prevented during all operations. The glassware can be made RNase-free by baking for 8 hours or by chloroform washing. A 1% diethyl pyrocarbonate (DEPC [carcinogen!]) washing (2 hr, 37°C) may also be useful. RNase activity in the extraction media can be inhibited by vanadium or by the clay Macaloid. These are subsequently eliminated by water-saturated phenol extraction. RNases can be blocked by 4 M guanidium thiocyanate and β-mercaptoethanol. RNA is extracted from the tissues in a buffer containing a detergent (0.5% Nonidet) and a reducing agent (dithiothreitol). Proteins may be removed by proteinase K digestion. RNase-free DNase removes DNA. Finally, chilling in cold ethanol (containing Na acetate) precipitates RNA. The RNA is taken up in TE, pH 7.6 buffer. Its quantity can be measured spectrophotometrically at 260 nm. Several variations of these general procedures are being used to isolate RNA. *See* DNA extraction; RNase-free DNase; TE.

RNA factory   Complex associated with RNA polymerase II that carries out transcription, splicing, and cleavage-polyadenylation of the mRNA precursor. *See* mRNA. (McCracken, S., et al. 1997. *Nature* 385:357.)

RNA fingerprinting   The purpose is to identify the differential expression of the total array of genes that constitutes about 15% (or less) of all at a particular time in a mammalian genome. For this goal from a subset of mRNAs, partial cDNAs are amplified by reverse transcription using PCR. The short sequences are then displayed on a sequencing gel (differential display). Pairs of primers are selected in such a way that each will amplify 50 to 100 mRNAs. One of the primers (5′-TCA) is anchored to the TG upstream of the poly(A) tail of mRNA. This primer will recognize 1/12 (4!/2!) of the mRNAs with a different combination of the last two 3′ bases, omitting T as the penultimate base. The primer will then amplify only this subpopulation. As 5′ primers, 6 to 7 bp arbitrary sequences are used. Such a procedure can be used not just for molecular analysis of development, but eventually the genes producing the transcripts can be cloned. *See* differential display; DNA fingerprinting; fingerprinting of macromolecules; microarray hybridization; PCR; proteomics; RDA; reverse transcription; SAGE; TOGA. (Gill, K. S. & Sandhu, D. 2001. *Genome* 44:633.)

RNA G8   Contains about 300 nucleotides and is associated with the ribosomes in *Tetrahymena thermophila*. It is transcribed by RNA polymerase III and conveys thermal tolerance to the cells. *See* thermal tolerance.

RNA helicases   Unwind double-stranded RNA. They have two double-stranded RNA domains and have a cis-acting transactivation response (TAR) element binding and a dsRNA activated protein kinase (PKR) domain. TAR enhances the transcription of HIV-1. *See* acquired immunodeficiency; breast cancer; DEAD box; eIF4A; helicase. (Lüking, A., et al. 1998. *Crit. Rev. Biochem. Mol. Biol.* 33:259; Fujii, R., et al. 2001. *J. Biol. Chem.* 276:5445.)

RNAi   RNA-mediated genetic interference. Single- or double-stranded RNA formed or introduced into the cell may interfere with the translation of endogenous genes of *Caenorhabditis* or plants or *Drosophila*. The double-stranded RNA is at least an order of magnitude more potent. In the nematode, the most effective exogenous delivery to any part of the body is through the intestines. The effective pre-RNAi length is 1,000–2,000 nucleotides. RNAi is ATP-dependent, but it is not linked to mRNA translation. There is some evidence that the effect of RNAi is enzymatic (RNAi nuclease) and involves the degradation of mRNA rather than some kind of antisense mechanism. Nascent dsRNA, generated by RNA-dependent RNA polymerase, is degraded to eliminate the incorporated mRNA, and new cycles of dsRNAs are produced that yield new siRNAs (Lipardi, C., et al. 2001. *Cell* 107:297). The siRNA (small interfering RNA) combines with proteins that degrade the RNAs recognized by siRNA. In *Caenorhabditis* genes, *rde-1, -2, -3, -4* control this interference. RDE-1 and RDE-4 proteins seem to practice surveillance for the presence of double-stranded RNA, transposons, and RDE-2; RDE-3 (and MUT-7) degrades these RNA molecules as a defense mechanism. RNA interference and transposon silencing in *Caenorhabditis* is controlled by *mut-7* gene. The two strands of the targeted double-stranded RNA cleave into 21−23 (or 25)-nucleotide-long segments (siRNAs [short interfering RNAs] and stRNA [small temporal RNA]). siRNA degrades its target, whereas the regulatory stRNA represses translation of its target mRNA.

There appear to be two RNase III motif enzymes involved in processing from precursor RNAs: One, the initiator enzyme, Dicer, generates a ~22-nucleotide guide that marks the mRNA for further degradation by RISC (RNA-induced silencing complex), an effector ribonuclease complex (see figure). The interference may not be entirely specific for a gene, related genes may also be affected, and the effect may vary in degree in various tissues. The interference can also be transmitted to the progeny, although most probably posttranslational events are involved. The protein product of the gene is homologous with the 3′-5′-exonuclease domains of RNaseD and Werner syndrome protein. By establishing a library of DNA clones in a bacterium that produces double-stranded RNA (RNAi) and feeding it to postembryonic *Caenorhabditis*, the function of a large number of hitherto unknown ORFs could be identified by the silenced phenotype. Or in the case of genes controlling cell division, in vivo time-lapse differential interference contrast microscopy is used. In plants, the same mechanism can account for cosuppression and VIGS. In plants, two types of mechanisms account for viral RNA silencing: (1) The helper component-proteinase (HC-Pro) derived from the potyviruses

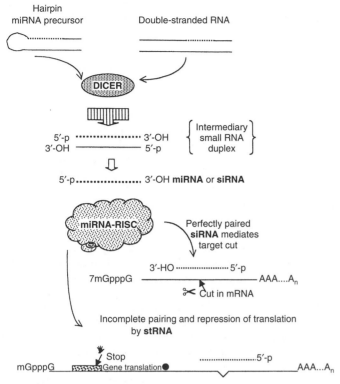

A model of the roles in gene silencing by miRNA, stRNA and RNAi
(Modified after Hutvágner, G. & Zamore, P.D. 2002 Science 297:2056)

is very efficient, can silence a broad range of plant viruses, and carries out transgene-induced as well as VIGS silencing. (2) The potato virus X (PVX) p25, is much less effective, and its action targets systemic silencing. Some endogenous proteins can also mediate silencing in plants and in *Caenorhabditis* or *Drosophila*. RNAi technology may be applicable also to human pathogenic viruses, such as HIV (Jacque, J.-M., et al. 2002. *Nature* 418:435). *See* antisense RNA; antisense technologies; cosuppression; epigenesis; heterochronic RNA; inhibition of transcription; Nomarski differential phase-contrast microscopy; posttranscriptional silencing; quelling; ribonuclease D; RNA interference; VIG; Werner syndrome. (Tabara, H., et al. 1999. *Cell* 99:123; Grishok, A., et al. 2000. *Science* 287:2494; Grishok, A., et al. 2001. *Cell* 106:23; Elbashir, S. M., et al. 2001. *Nature* 411:494; Vance, V. & Vaucheret, H. 2001. *Science* 292:2277; Hutvágner, G., et al. 2001. *Science* 293:834; Matzke, M., et al. 2001. *Science* 293:1080; Hammond, S. M., et al. 2001. *Science* 293:1146; Ruvkun, G. 2001. *Science* 294:797; Plasterk, R. H. A. 2002. *Science* 296:1263; Zamore, P. D. 2002. *Science* 296:1265; Mlotshwa, S., et al. 2002. *Plant Cell* 14:S289; Hutvágner, G. & Zamore, P. D. 2002. *Current Opin. Genet. Dev.* 12:225; Hannon, G. J. 2002. *Nature* 418:244; Tijsterman, M., et al. 2002. *Annu. Rev. Genet.* 36:489.)

RNAIII/rnaiii Bacterial virulence-controlling molecule induced by the RNA III-activating protein (RAP). The RIP protein that is produced in nonpathogenic strains competes for activation of rna III and the production of *Staphylococcus aureus* toxin. (Balaban, N., et al. 2001. *J. Biol. Chem.* 276:2658.)

RNA-IN Leftward transcript of the bacterial transposase gene in transposable elements transcribed from the pIN promoter. *See* pIN; RNA-OUT.

RNA interference In *Caenorhabditis* nematodes, fed on double-stranded RNA-containing bacteria, gene expression is transiently but specifically suppressed as long as dsRNA is available. The interfering dsRNA may be generated by inserting T7 phage RNA polymerase genes at the opposite ends of specific genes in a plasmid. Alternatively, a single copy of the polymerase may be used on inverted duplication of a specific gene. *See Caenorhabditis*; inhibition of transcription; RNAi. (Moss, E. G. 2001. *Curr. Biol.* 11:R772.)

RNA ligase Catalyzes the joining of RNA termini, such as generated during the processing of tRNA transcripts, in a phosphodiester bond, and may need ATP for the reaction. It is an essential enzyme for RNA editing. *See* ligase DNA; ligase RNA. (Stage-Zimmermann, T. K. & Uhlenbeck, O. C. 2001. *Nature Struct. Biol.* 8:863; Ho, K. C. & Shurman, S. 2002. *Proc. Natl. Acad. Sci. USA* 99:12709.)

RNA localization Embryonic development is an asymmetric process. The local distribution and enrichment of special RNAs play decisive roles. The RNA transcribed from the *bicoid* locus of *Drosophila* encoding a transcription factor specifies anterior cell fates in the oocytes and embryos. The *germ-cell-less* nuclear pore-associated factor participates in the definition of posterior development. The growth factor encoded by *gurken* affects anterior-dorsal differentiation, whereas the *prospero*-encoded transcription factor controls apical/basal development in neuroblasts. In mammals, the transcript of the β-actin genes is involved in the definition of the cytoskeleton, and it is localized primarily at the periphery of epithelial cells, fibroblasts, and myoblasts. Different RNAs appear at the predestined locations immediately following transcription. Some of the rather uniformly distributed RNAs are degraded, except at the locations where they exercise morphogenetic functions. Localization seems to be mediated by signals (usually a few hundred bases or less) generally in the untranslated 3'-end (3'-UTR) of mRNAs. mRNAs are not translated until properly localized. Translation without prior localization might cause developmental anomalies. Besides the cis-acting element (e.g., 3'-UTR), trans-acting factors encoded by other genes may affect localization (e.g., influencing the cytoskeletal motor proteins or RNA-binding proteins). Localization of RNA may also be assured by alternative splicing of the transcripts. The pattern of distribution may change as the microtubule-organizing center develops and motor proteins become available. At a later stage before the nurse cells decay, they dump large amounts of RNA into the oocyte through the ring canals. Later, ooplasmic streaming somewhat mixes up the previously laid-down distribution pattern. When the early syncytial blastoderm (~6,000 nuclei) is converted into a cellular blastoderm, anterior-posterior and dorsal-ventral polarization begins upon the expression of the genes, specifying the morphogenetic pattern. At this stage, control shifts toward the zygotic effect genes from the maternal effect ones. Another control involves the diffusion pattern of the morphogenetic gene products. *See* maternal effect genes; METRO; morphogenesis in *Drosophila*; motor proteins; MTOC; ring canal. (Bashirullah, A., et al. 1998.

Annu. Rev. Biochem. 67:335; Palacios, I. M. & St. Johnston, D. 2001. *Annu. Rev. Cell Dev. Biol.* 17:569; Bullock, S. L. & Ish-Horowicz, D. 2001. *Nature* 414:611; Saxton, W. M. 2001. *Cell* 107:707.)

RNA maturases   Ribosomal transcripts are processed to size by the U3, U8, U13 independently transcribed small RNAs (snoRNA, small nucleolar RNAs). U13-U14 snoRNAs and E3 are encoded by the introns of protein-coding genes participating in the process of translation. They are cotranscribed with pre-mRNA and removed during gene processing. By the intron of the U22 host gene (UHG), seven U RNAs are transcribed. These U RNAs display 12−15 base complementary sequences to rRNAs. *See* introns; ribonuclease P; splicing. (Delahodde, A., et al. 1989. *Cell* 56:431; Claros, M. G., et al. 1996. *Methods Enzymol.* 264:389.)

RNA-mediated gene inactivation   *See* RNAi.

RNA-mediated recombination   Thought to be involved in the exchange between the reverse transcript and the corresponding cellular allele. Ty element-mediated recombination is supposed to involve RNA. *See* Ty. (Derr, L. K., et al. 1991. *Cell* 67:355.)

RNA mimicry   *See* translation termination.

RNA mimics   2′-modified oligodeoxynucleotides such as the 2′-O-methyl-modified ones that enhance binding affinity to complementary RNA and are resistant to some nucleases. *See* antisense technology. (Putnam, W. C., et al. 2001. *Nucleic Acids Res.* 29:2199.)

RNA, MICRO (miRNA, mir)   Is a 22 nucleotide inhibitory RNA. *See* RNAi. (Zeng, Y., et al. 2002. *Mol. Cell* 9:1327.)

RNA nucleotides, modified   There are four basic nucleotides in the cell, but several others are produced by post-transcriptional modification as needed for tRNAs and various coenzymes. *See* modified bases. (<http://medlib.med.utah.edu/RNAmods>.)

RNA-OUT   Transcript originating from the strong pOUT promoter of bacterial transposable elements. It opposes pIN and directs transcription toward the outside end of the IS*10* element. *See* pIn; pOUT; RNA-IN; Tn*10*.

RNAP   RNA polymerases; in prokaryotes, there is only one DNA-dependent RNA polymerase, whereas in eukaryotes, RNA pol I, pol II, and pol III are found. *See* RNA polymerases.

RNA-PCR   Polymerase chain reaction may amplify rare RNAs after the RNA is reverse-transcribed into DNA. *See* polymerase chain reaction; reverse transcriptases.

RNA-PEPTIDE FUSIONS   Synthetic mRNAs can be fused to their encoded polypeptides when the mRNA carries puromycin, a peptidyl acceptor antibiotic at the 3′-end. After *in vitro* enrichment, proteins can be selected in a directed manner. *See* directed mutation; evolution. (Roberts, R. W. & Szostak, J. 1997. *Proc. Natl. Acad. Sci. USA* 94:12297.)

RNA plasmids   Exist in some mitochondria that are not homologous to mtDNA. *See* mtDNA.

RNA polymerase (RNAP)   DNA-dependent RNA polymerases synthesize RNA on DNA template. The T7 RNA polymerase is a relatively simple molecule (100 kDa). As a rule, this polymerase directs the incorporation of the first nucleotide in a template-directed manner with a single nucleotide primer. First it produces several pieces of short RNAs, then these 10−12 nucleotide units polymerize into an elongation complex and exit from the promoter. Between the DNA template and the 3′-proximal RNA transcript, at least 9-nucleotide-long hybrids are formed for efficient processivity during RNA elongation. The RNA products then separate from the template and the duplex of the DNA is restored. T7 phage promoter has a binding domain (−17 to −6) and an initiation domain (−6 to +6). After the binding domain of the polymerase recognizes the DNA, melting of the duplex takes place. During the elongation phase, about 200 nucleotides are added per second. The termination is rho-independent in T7 and the end forms a stem and loop structure and a stretch of U bases. The termination seems to be a reverse process of the initiation, i.e., a stable isomerization is followed by an unstable sequence.

In many respects, despite its simple structure, the phage enzyme functions similarly to more complex polymerases. In prokaryotes, a single polymerase synthesizes all cellular RNAs. The prokaryotic pol enzyme contains four subunits: $\alpha\alpha$, β, β', and σ. The large β-subunits are evolutionarily highly conserved and they are the main instruments of polymerization with the other subunits.

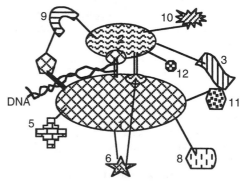

Schematic drawing of the 10/12 subunits of the yeast RNA polymerase II. Subunits 1, 5, and 9 grip the DNA below the active center of the enzyme. Pores below the active center form the entry of the nucleotides and the exit of the polymerized RNA. (Redrawn after Cramer, P., et al. 2000. *Science* 288:640.)

The prokaryotic RNA polymerases are about 500 kDA in size, whereas the T7 enzyme is only about 40 kDa. The *E. coli* α-subunits (36.5 kDa) recognize the promoter; the β' subunit (155.2 kDa) binds DNA, and β (150.6 kDa) is active in RNA polymerization. The σ is essential to start transcription in a specific way by opening the double helix for the action

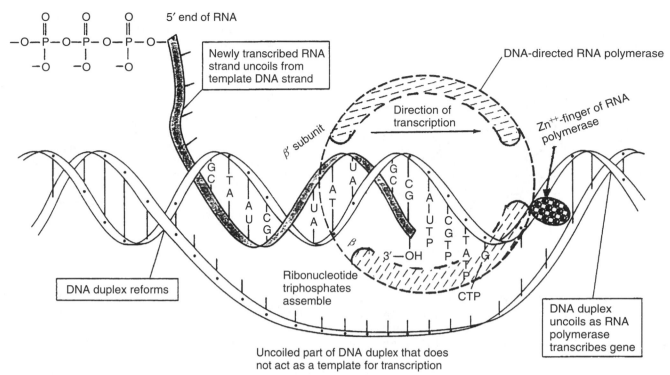

Conceptualization of the general process of transcription of RNA on DNA template from ribonucleoside 5′-triphosphates. The first 5′-nucleotide retains the three phosphates; the following ones split off a pyrophosphate and are then hooked up to the 3′-OH ends. In this process a ternary complex involving DNA, RNA, and protein is involved. The RNA-protein has been located to about 9 bp from the DNA fork into the transcription bubble where the DNA template forms an approximate 8 bp heteroduplex (HBS, heteroduplex-binding site). The RBS (RNA-binding site) together with HBS extends to approximately 14−16 RNA nucleotides from the 3′-OH end. About 7−9 nucleotides further downstream, RBS, HBS, and DBS (DNA-binding site) are situated and stabilize transcription. The DNA entry and the RNA exit sites on the polymerase are in close vicinity to each other. This diagram does not show the general transcription factors and the numerous other proteins regulating the process. (Diagram is modified after Page, D. S., *Principles of Biological Chemistry*. Willard Grant, Boston, MA; see also Nudler, E., et al. 1998. *Science* 281:424.) For the structure of initiation complex of the T7 enzyme at 2.4 Å resolution, see Cheetham, G. M. T. & Steitz, T. A. 1999. *Science* 286:2305.

of RNA polymerization. The eukaryotic organelles contain prokaryotic-type RNA polymerases. The eukaryotic RNA polymerase II — transcribing protein-encoding genes — has about 12 subunits and is somewhat larger but highly homologous across wide phylogenetic ranges. The two largest subunits are similar to the β-subunits of the prokaryotic enzyme. The carboxy-terminal domain (CTD) of the largest subunit of RNAP II carries amino acid repeats (YSPTSPS) liable to phosphorylation. Only the CTD unphosphorylated form participates in a transcription initiation complex, but it is phosphorylated when it begins to elongate the RNA transcript. The complex of transcription includes the Mediator and a total of about 60 proteins with a combined mass of 3.5 MDa. In an *E. coli* cell, there are about 2,000 core RNA polymerase molecules. The number of protein factors affecting transcription in *E. coli* is about 240–260.

In eukaryotes, there are three different DNA-dependent RNA polymerases (I, II, III) that display substantial homology. RNA polymerases replicate the genome of the RNA viruses. Mammalian RNA polymerase II may also carry out RNA-dependent RNA synthesis by switching to the RNA genome of hepatitis delta virus (Chang, J. & Taylor, J. 2002. *EMBO J.* 21:157). The viral enzymes — in contrast to the DNA polymerases — lack proofreading function and consequently their error rate may be within the 10^{-3} to 10^{-4} range per nucleotide, leading to the extreme diversity of the RNA viruses. The prokaryotic RNA polymerase moves along the DNA at a speed exceeding 10 nucleotides/second. Termination is basically the reverse process of transcription initiation. In a haploid yeast cell, there are 2,000 to 4,000 RNA polymerase II molecules and about 10 times more general transcription factor molecules are included with it. *See* antitermination; arrest, transcriptional; error in replication; error in transcription; inchworm model; mediator complex; nucleic acid chain growth; open-promoter complex; pausing, transcriptional; pol; processivity; promoter; protein synthesis; repression; SRB; terminator; transcript cleavage; transcription; transcription complex; transcription factors; transcription termination. (Archambault, J. & Friesen, J. D. 1993. *Microbiol. Rev.* 57:703; Barberis, A. & Gaudreau, L. 1998. *Biol. Chem.* 379:1397; Mooney, R. A. & Landick, R. 1999. *Cell* 98:687; Ishihama, A. 2000. *Annu. Rev. Microbiol.* 54:499; Cramer, P., et al. 2001. *Science* 292:1863; Gnatt, A. L., et al. 2001. *Science* 292:1876; Vassylyev, D. G., et al. 2002. *Nature* 417:712.)

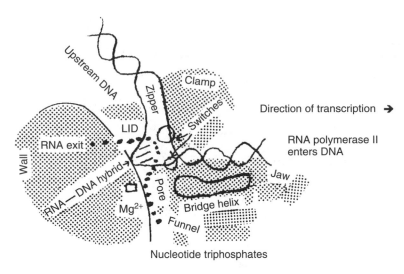

Nucleotide triphosphates

The structure of RNA polymerase II has been determined at 3.3 Å resolution. The schematic of the major functional elements is shown at left. The enzyme enters DNA at right and holds onto it by the element designated as *JAW*. The *CLAMP* domain keeps the protein on the DNA. The *Switches* turn on the function. A large protein subunit, the *WALL* causes an approximately 90° turn of the DNA template. This then facilitates the addition of nucleotides to the RNA. The RNA building blocks, the nucleotide triphosphates, enter the complex through the *FUNNEL*, which ends in A *PORE*. Magnesium (or other metals) is required for the polymerization. The RNA chain is symbolized by a dotted line. The DNA and RNA form A 9-base duplex. The *RUDDER* (not shown here) prevents the extension of the DNA — RNA hybrid to go beyond the 9 base pairs and facilitates the separation of the duplex. The synthesized RNA exits under the *LID*. The *BRIDGE HELIX* under the coding strand of the DNA apparently promotes the addition of nucleotides. (Modified after Gnatt, A. L., et al. 2001. *Science* 292:1876.)

RNA polymerase holoenzyme Complex of RNA pol II, transcription factors, other regulatory proteins including Srb, and a general transcriptional regulator of eukaryotes. *See* coactivators; RNA polymerase; TAF; TBP; transcription factors.

RNA polymerizations *See* pol.

RNA primer *See* DNA replication.

RNA processing Eukaryotic genes are in "pieces," although the DNA is continuous; in between the protein-coding nucleotide sequences (exons), nontranslated or not translated together with exons, additional nucleotide sequences, and introns occur. The long sequences are transcribed into a long RNA tract, but the introns are removed and the exons are spliced to make mRNA. Similar processing is carried out with rRNA and tRNA. The vast majority of prokaryotic genes do not require these processes. These cutoffs, mRNA, and primary transcripts constitute the pool of hnRNA (heterogeneous nuclear RNA). The primary RNA transcripts are coated with different proteins and thus form hnRNP, i.e., hnRNA-protein particles. These particles are instrumental in cutting the transcripts and splicing the exons into mRNA. After the splicing is completed, the methylated guanylic cap and the polyA tail are added, then the mRNA is exported into the cytosol. *See* exosome; hnRNA; introns; posttranscriptional processing; spliceosome. (Varani, G. & Nagai, K. 1998. *Annu. Rev. Biophys. Biomol. Struct.* 27:407.)

RNA-protein interactions Ubiquitous in all cells in the formation of ribosomes, spliceosomes, several ribozymes, posttranscriptional regulation, translation machinery (in tRNAs, elongation factors), and viral coat proteins. (Jones, S., et al. 2001. *Nucleic Acids Res.* 29:943.)

RNA replication Some bacteriophages, R17, MS2, fd2, f4, Qβ, M12, and some animal viruses such as the polio virus, the vesicular stomatitis virus, rhinoviruses (influenza viruses), and the vast majority of the plant viruses have RNA genetic material. This RNA is either single- or double-stranded. One of the best known is the Qβ replicase system.

The tetramer (210 kDa) is encoded by one viral (β-subunit about 65 kDa) and three bacterial genes (translation factors). The replicase enzyme does not have an editing exonuclease function. These replicases do not replicate or transcribe host RNAs. RNA replication is similar to the DNA replication inasmuch as all nucleic acid strands are elongated at the 3′-OH ends. (Retrovirus replication is discussed under reverse transcription.) Most of the bacterial RNA viruses and many animal viruses have the same genetic material as their mRNA (+ strand viruses). They use a replicative intermediate (RI) for the synthesis of the first new (−) strand. The (−) strand can then generate as many (+) strands as needed. Some viruses are, however, (−) strand viruses because their genetic material is not identical to mRNA but complementary to it. The polio virus (+) strand is the genetic material associated with a protein at the 5′-end through a phosphodiester linkage to a tyrosine residue. The (−) strand lacks this feature. Apparently, the OH group of tyrosine primes the synthesis of the (+) strand, but the (−) strand can get by without it. The reoviruses of vertebrates contain 8–10 short double-stranded RNA molecules. When the virus enters the host cell, the coat protein is shed and its RNA polymerase is activated, which works conservatively and asymmetrically. A virion enzyme copies the (−) strand and a new (+) is released. The original RNA duplex is conserved. On the new (+) strand, a (−) strand is made and the double-stranded RNA is reconstituted. *See* DNA replication; plus strand; replicase; replicative intermediate; retroviruses. (Tang, H., et al. 1999. *Annu. Rev. Genet.* 33:133; Knipe, D. M. & Howley, P. M., eds. 2001. *Fundamental Virology*. Lippincott Williams & Wilkins, Philadelphia, PA.)

RNA restriction enzyme Function exists in ribozymes. *See* restriction enzyme; ribozyme.

RNA 4.5S Bacterial RNA of 114 nucleotides. It targets signal peptide-equipped proteins to the secretory apparatus, where it forms the signal-recognition particle with protein Ffh. It also binds to translation elongation factor G. It is homologous to the eukaryotic 7S RNA. *See* elongation factors; Ffh; RNA 7S; secretion system. (Nakamura, K., et al. 2001. *J. Biol. Chem.* 276:22844.)

RNA 5S, RNA 5.8S *See* ribosomes. (<http://biobases.ibch.pooznan.pl/5SData>.)

RNA 6S Regulator of RNA polymerase of *E. coli*. (Wassarman, K. M. & Storz, G. 2000. *Cell* 101:613.)

RNA 7S Cytoplasmic class of RNA of ~300 nucleotides present in prokaryotes and eukaryotes. It participates in the function of RNA polymerase III and in the sequence signal recognition particle to direct proteins to the endoplasmic reticulum. 7S RNA, which is part of eukaryotic 35S pre-rRNA, is also a precursor of 5.8S ribosomal RNA, and the small GTPase RAN processes it. *See* DNA 7S; RAN; signal sequence recognition particle. (Luehrsen, K. R., et al. 1985. *Curr. Microbiol.* 12:69; Suzuki, N., et al. 2001. *Genetics* 158:613.)

RNA 10Sa (SsrA/tmRNA) When an mRNA is truncated at the 3′-end and has no stop codon, the ribosome may switch from one RNA to another to terminate the translation. The switching results in the generation of an 11-amino-acid residue at the carboxyl end that destines the protein for degradation. The last 10 amino acids are translated from a stable so-called 10Sa RNA (363 nucleotides). So far, these 10 amino acids have been found on murine interleukin-6 translated in *E. coli*, λ-phage *cI*, cytochrome b-562, and polyphenylalanine translated in polyU. 10Sa RNA is similar to a tRNA and is charged with alanine. This alanine is found as the first amino acid of the tag. This saves the ribosome from the "unfinished" mRNA and permits its quasinormal operation. The mechanism carries out the same task as ubiquitin does in eukaryotes. *See* lambda phage; tmRNA; translation, noncontiguous; ubiquitin. (Gillet, R. & Felden, B. 2001. *EMBO J.* 20:2966; Zwieb, C. & Wower, J. 2000. *Nucleic Acids Res.* 28:169; <http://www.indiana.edu/~tmrna>.)

RNA 20S Naked single-stranded RNA phage-like replicon of about 2,500 nucleotides in yeast encoding a 95 kd RNA polymerase-like protein. Its termini are 5′-GGGGC and GCCCC-3′ and therefore may circularize. (Rodriguez-Cusino, N., et al. 1998. *J. Biol. Chem.* 273:20363.)

RNase *See* ribonucleases.

RNase-free DNase Heat 10 mg RNase A per mL, 0.01 M Na-acetate (pH 5.2), at 100°C for 15 min, then cool, adjust pH to 7.4 (1 M Tris-HCl), and store at −20°C. *See* DNase free of RNase.

RNase MRP Ribonuclease that cleaves rRNA transcripts upstream of the 5.8 rRNA. In mitochondria, it cleaves the primers of DNA replication. *See* cartilage hair dysplasia; U RNPs. (Ridanpää, M., et al. 2001. *Cell* 98:195.)

RNA sequencing Can be done by digesting RNAs with different ribonuclease enzymes that cut the phosphodiester linkages at the specific nucleotides, then separate the fragments by gel electrophoresis (ca. 20% polyacrylamide) in one or two dimensions. The RNA is generally labeled by ^{32}P in order to autoradiograph the dried gels. T_1 ribonuclease (from *Aspergillus oryzae*) generates fragments with 3′-guanosine monophosphate ends. U_2 ribonuclease (from *Ustilago sphaerogena*) cleaves at purine residues (intermediates 2′,3′-cyclic phosphates). RNase CL 3 (from chicken liver) has about 16-fold higher activity for cytidylic than uridylic linkages. RNase B (from *Bacillus cereus*) cuts Up↓N and Ap↓N bonds. RNase Phy M (from *Physarum polycephalum*) has the specificity Ap↓N and Up↓N. When several of these enzymes are employed, from the separate electrophoretic patterns the sequences from the positions of the fragments with different termini can be deduced. In many ways, sequencing of RNAs with the aid of enzymatic breakage is very similar in principle to the Maxam and Gilbert method of DNA sequencing. RNA is sequenced also by first converting it by reverse transcriptase into cDNA, then the much easier DNA sequencing methods are used to determine the RNA nucleotide sequences. *See* DNA sequencing; ribonucleases.

RNA silencing *See* RNAi.

RNASIN® Ribonuclease inhibitor.

RNA 7SL About 300-base-long molecule forming the signal recognition particle (SRP) with six proteins of 10 to 75 kDa size. (Müller, J. & Benecke, B. J. 1999. *Biochem. Cell Biol.* 77:431.)

RNA splicing The primary transcript of RNA is much larger than required for particular functions; therefore, during processing it is cut up and the transcripts corresponding to introns are removed. The remaining pieces are reattached (spliced together) to form the mature RNA molecule. *See* introns; RNA processing. (Weg-Remers, S., et al. 2001. *EMBO J.* 20:4194.)

RNA structure (SCOR) Noncanonical base-base interactions. (<http://prion.bchs.uh.edu/1/>; <http://scor.lbl.gov>.)

RNA surveillance Monitors for the presence of RNAs that have stop-codon mutations, which might lead to the synthesis of truncated proteins. Nonsense mRNA is degraded by the NMD (nonsense-mediated decay) pathway. In yeast, the NMD process begins by recruiting the Upf1 protein to the ribosomes. Ubf1p can bind ATP and can also bind nucleic acids without ATP, can hydrolyze ATP in a nucleic acid–dependent manner, and it is an ATP-dependent 5′→3′ RNA/DNA helicase. Upf32p is probably a signal transducer from Ubf1 to Ubf3; the latter then enlists nonsense mRNA to the NMD complex. Ubf3p is located mainly in the cytoplasm, but it can shuttle between the nucleus and the cytoplasm. In yeast, proteins eRF1 and eRF3 termination factors are also required for the abortion of translation. The complex mediates the decay of the defective polypeptide if any is made. NMD systems operate in a wide range of eukaryotes and may have a role in some cancers and hereditary human diseases. *See* degradosome; exosome; Pab1p; Xrn1p. (Culbertson, M. R. 1999. *Trends Genet.* 15:74; Maquat, L. E. & Carmichael, G. G. 2001. *Cell* 104:173; Ishigaki, Y., et al. 2001. *Cell* 106:607; Frischmeyer, P. A., et al. 2002. *Science* 295:2258.)

RNA therapeutic May be used in medicine as a gene function inhibitory means such as in antisense technologies. Ribozymes may cleave pathogenic RNA molecules of infectious agents (e.g. HIV). Transsplicing of transcripts may correct defective functions.(e.g. sickle-cell disease, tumor suppressor transcripts). RNA may be used to prevent the binding of pathogenesis proteins to target (e.g. HIV-1 TAR, REV). Transfection of dendritic, antigen presenting cells with DNA that generates mRNA for tumor antigens may boost the effectiveness of cytotoxic T lymphocytes. *See* acquired immunodeficiency; antigen presenting cell; aptamer; CTL; decoy RNA; ribozyme; transsplicing; tumor antigen; tumor suppressor. (Sullenger, B. A. & Gilboa, E. 2002. *Nature* 418:252.)

RNA transcript   RNA copy of a DNA segment.

RNA transport   mRNA, snRNA, U3 RNA, and ribosomal RNA (but not tRNA) are exported from the nucleus by RCC1 (yeast homologue is PRP20/MTR1). This nuclear protein is a guanine nucleotide exchange factor for the RAS-like guanosine triphosphatase (GTPase). Some plant virus RNAs spread through plasmodesmata to a distance from the site of infection with the aid of protein transporters. Plants can move their own mRNAs through the phloem cells at great distances. *See* export adaptors; nuclear pore; phloem; plasmodesma; RAN; RCC1; RNA export; TAP; transport elements, constitutive. (Yang, J., et al. 2001. *Mol. Cell* 8:397.)

RNA trap   RNA tagging and recovery of associated proteins. The purpose of the procedure is to test the mechanism of action on gene expression exerted by enhancers, which are distant from a genic site. To this goal unprocessed RNA transcript of a specific locus with labeled oligonucleotides is hybridized to isolated cell nuclei. Horseradish peroxidase–conjugated antibodies are localized to the oligonucleotide probe. Peroxidase-activated biotinyl–tyramide covalently labels the electron–rich protein moieties by biotin in this region. From the sonicated cell fragments the biotin–conjugated chromatin is isolated by affinity to a streptavidin column. The specific sequences are then enriched by quantitative PCR. Enrichment of the locus control region and specifically the DNase hypersensitive sites indicated by immunofluorescence analysis that one of the hypersensitive sites becomes physically associated with the gene transcribed. This technology thus supports the role of DNA looping in gene function. *See* biotin; enhancer; LCR; PCR; regulation of gene activity; streptavidin; looping DNA. (Carter, D., et al. 2002. *Nature Genet.* 32:623.)

RNA trimming   *See* trimming.

RNA viruses   *See* animal viruses; ebola virus; MS2; paramyxovirus; plant viruses; reovirus; retroviruses; TMV; togavirus; viral vectors; viruses.

RNA world   Prebiotic era when RNA carried out auto- and heterocatalysis without DNA. There is evidence that ribozymes can catalyze even nucleotide synthesis, giving further support to the RNA world concept. There is evidence that RNA is involved in the catalysis of protein synthesis by (1) encoding amino acid sequence in protein, (2) activation of amino acids, (3) synthesis of aminoacyl-tRNA by a reaction analogous to the aminoacyl-tRNA synthetases, and (4) formation of peptide bonds. *See* autocatalytic function; heterocatalysis; origin of life; peptide nucleic acid; protein synthesis; ribozyme.

(Kumar, R. K. & Yarus, M. 2001. *Biochemistry* 40:6998; Joyce, G. F. 2002. *Nature* 418:214; Yarus, M. 2002. *Annu. Rev. Genet.* 36:125.)

RNKP-1   Homologue of ICE and fragmentin-2. *See* apoptosis; fragmentin-2; ICE.

RNP   Ribonucleoprotein; any type of RNA associated with a protein particle such as in the hnRNA or in the ribosomes. Proteins only bind single-stranded sequences of RNA. In recognition of the bases, the shape and charge distribution of RNA are also important. *See* hnRNA; ribosomes; RNA.

RNR   *See* ribonucleotide reductase.

RNS   Reactive nitrogen species. *See* nitric oxide; ROS.

ROAM mutation   Regulated overproducing alleles under mating signals. Activate yeast Ty elements by the influence of the MAT gene locus. Such a system may operate if the Ty insertion takes place at the promoter of a gene, and then the Ty enhancer may be required for the expression of that gene. These genes are expressed only in the *a* or *α* mating type cells but not in the diploid *a/α* cells. *See* mating type determination in yeast; Ty. (Rathjen, P. D., et al. 1987. *Nucleic Acids Res.* 15:7309.)

RO•, ROO•   Reactive oxygen, hydroperoxide radical. *See* ROS.

roan   Fur color (cattle, horse) with predominantly brown-red hairs interspersed with white ones. It is a common sign of heterozygosity for the *R* and *r* alleles of a gene locus. *See* codominance.

Roberts syndrome   Actually overlaps with the *SC phocomelia* (absence or extreme reduction of the bones of extremities located proximal to the trunk of the body) and with *TAR syndrome*, also involving *thrombocytopenia* (reduction in the number of blood platelets), mental retardation, cleft palate, etc. These three syndromes are autosomal recessive and apparently basically the same. They are caused by chromosomal instability. The prime suspect is a defect in the centromeric heterochromatin or the kinetochore itself. *See* chromosome breakage; Holt-Oram syndrome; mental retardation; thrombocytopenia; Wiskott-Aldrich syndrome.

Robertsonian translocations (Robertsonian change)   Two nonhomologous telocentric chromosomes fused at the centromere or more likely translocation between two nonhomologous acrocentric chromosomes. The outcome is a replacement of two telo or acrocentric chromosomes with one clearly biarmed chromosome. These translocated chromosomes may have preserved the centromeres of both acrocentrics and remain cytologically stable because one of the centromeres is inactivated. Robertsonian translocations are very common in mouse cell cultures, but they also occur in wild natural populations, resulting in an apparent change in chromosome morphology and numbers. If this translocation occurs, generally a minute piece of the acrocentric chromosome is lost, but being genetically inert, it has no consequence for fitness. Robertsonian

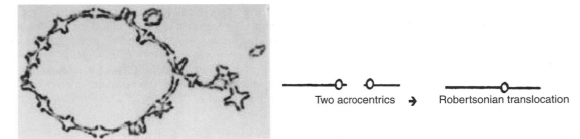

Late diakinesis/early metaphase I in the mouse, heterozygous for different Robertsonian translocations; 15 metacentrics in a superchain, 1 trivalent, 2 bivalents. The sex chromosomes are not involved in Robertsonian changes. (Courtesy of Capanna, E., et al. 1976. *Chromosoma* 58:341.)

translocation may affect 1/843 human neonates. *See* acrocentric; dicentric chromosome; fitness; telocentric; translocations. (Pardo-Manuel de Villena, F. & Sapienza, C. 2001. *Cytogenet. Cell Genet.* 92:342; Bandyopadhyay, R., et al. 2002. *Am. J. Hum. Genet.* 71:1456.)

Robinow syndrome (RRS) Rare autosomal-dominant phenotype usually involving short stature, normal virilization but micropenis, hypertelorism of the face, etc. In Robinow-Sorauf syndrome, the main characteristic is the flattened and almost doubled big toe. An autosomal-recessive Robinow syndrome includes the same facial and genital features, but multiple ribs and abnormal vertebrae are also present. The basic defect of the recessive RRS (9q22) is due to premature chain termination in an orphan receptor tyrosine kinase (ROR2). It is allelic to brachydactyly type B. *See* brachydactyly; Chotzen syndrome; limb defects; stature in humans.

Robo (*roundabout*) *Drosophila* gene determining the straight movement of axons along the embryonal axis and preventing crossing back over the midline once a crossover is made. Homologues exist in other eukaryotes. *See* axon. (Simpson, J. H., et al. 2000. *Cell* 103:1019.)

robustness Statistical concept indicating the justification of an assumption concerning the procedures applicable to the data at hand (e.g., normal distribution). A robust statistical method is less liable to violations of the assertion. Usually, parametric methods are more robust than nonparametric ones (because of, e.g., subjective scales). *See* nonparametric methods; parametric methods in statistics.

Rock RHO-associated protein serine/threonine kinase involved with microtubules of nuclear division. It induces the phosphorylation of cofilin by LIM-kinase. Caspase-3 removes an inhibitory domain of Rock 1, which then phosphorylates the light chain of myosin. This is the main cause of the blebs on cells undergoing apoptosis. *See* apoptosis; citron; cofilin; immunological surveillance for blebs; LIM; microtubules; myosin; RHO; ROK. (Sebbagh, M., et al. 2001. *Nature Cell Biol.* 3:346.)

rocket electrophoresis Type of immunoelectrophoresis in which antigens are partitioned against antisera. *See* antigen; antiserum; electrophoresis; immunoelectrophoresis. (Hansen, S. A. 1988. *Electrophoresis* 9:101.)

rocks *See* scaffolds in genome sequencing.

rodents (order Rodentia) A large number of species (mice, rats, hamsters, rabbits) have been extensively used for genetic research because of the small size of these multiparous mammals and short generation time. Mice and rats reach their sexual maturity in 1 or 2 months, and their gestation period is 19 and 21 days, respectively. They have been exploited as laboratory models for the study of development, cancer, antibodies, population genetics, behavior genetics, radiation, mutational responses, etc. Rodents are carriers of several human pathogens (bubonic plague [*Pasteurella pestis*], tularemia [*Pasteurella tularensis*], etc.). Several inbred strains of mice have contributed very significantly to the understanding of immunogenetics. *See* animal models; mouse.

roentgen (röntgen) Unit of ionizing radiation (X-rays). *See* Röntgen machine; R unit.

röntgen machine X-ray machine (invented by W. K. Röntgen), that produces ionizing radiation (used for induction of mutation, mainly deletions) for medical examination of the body. The dose delivered is measured by R, Rad, Rem, rep, Sv, Gy. *See* R; radiation effects; radiation hazard; radiation protection; radiation measurement; radiation threshold; roentgen.

Rogers syndrome (TRMA) *See* megaloblastic anemia (human chromosome 1q23).

rogue Off-type of unknown (genetic) determination.

ROI Reactive oxygen intermediates are by-products of oxidative metabolism of mitochondria and peroxisomes, and may be formed by ionizing radiation. *See* ROS. (Ono, E., et al. 2001. *Proc. Natl. Acad. Sci. USA* 98:759.)

ROK One of the multiple RHO-associated protein kinases regulating microtubules of the spindle. *See* citron; RHO; ROCK; spindle.

rolling circle Replication is common among circular DNAs (such as conjugative plasmids [e.g., F plasmid], double-stranded [λ-phage] and single-stranded phages [M13, φX174], amplified rDNA minichromosomes in amphibian oocytes). A protein nicks one of the DNA strands and remains attached to the 5'-end. The free 3'-OH terminus serves as the point of extension by DNA polymerase in such a way that the opened old strand is displaced from the circular DNA while the new strand is formed and immediately hydrogen–bonded to the old template strand. Thus, the rolling circle remains intact and may generate new single-stranded DNAs that may be doubled later. The displaced single strand may be formed just in a single unit length of the original duplex circle, or it may become a single- or double-stranded concatamer, or it may circularize in a single- or double-stranded form with the assistance of a DNA ligase to join the open ends.

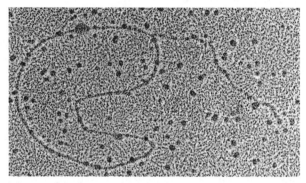

The (−) strand is the template for the (+) strand. The → indicates the direction of growth. The 5' end of the (+) strand attaches to the membrane (5'). (Diagram is the courtesy of Gilbert, W. & Dressler, D. 1968. *Cold Spring Harbor Symp. Quant. Biol.* 33:473. Electron micrograph is the courtesy of Dr. Nigel Godson.)

Rolling circle mechanisms have been identified in the transposition of bacterial insertion elements, but they are more common in eukaryotes. The transposons of eukaryotes that utilize rolling circle transposition are called helitrons. These elements may be quite abundant in eukaryotes but are hard to detect by the common computer programs because they do not generate target-site duplications like the majority of transposable elements. Many of them are nonautonomous in transposition. Helitrons do not have inverted terminal repeats. Their 5'-end begins with TC nucleotides and the 3'-end is CTRR (R stands for purine). Near the 3'-end, they have a characteristic 16–20-nucleotide palindrome, which does not have a conserved sequence, however. The *Arabidopsis*, rice, and *Caenorhabditis* helitrons may encode about 1,500 amino acids that embed a 5' → 3' helicase-like protein, a replication A protein, and some other gene products. For transposition, several host proteins are also required. Evolutionarily, helitrons may have originated from geminiviruses. *See* concatamer; conjugation; conjugation mapping; geminivirus; insertion element; padlock probe; palindrome; replication protein A. (del Pilar Garcillan-Barcia, M., et al. 2001. *Mol. Microbiol.* 39:494; Kapitonov, V. V. & Jurka, J. 2001. *Proc. Natl. Acad. Sci. USA* 98:8714; Feschotte, C. & Wessler, S. R. 2001. *Proc. Natl. Acad. Sci. USA* 98:8923.)

rolling circle amplification (RCA) Short DNA primer complementary to a segment of a circular DNA and an enzyme generate many single-stranded, concatameric copies of DNA in the presence of deoxyribonucleotide triphosphates. RCA can produce sufficient material for microarray analysis from specific locations. *See* concatamer; microarray hybridization. (Nallur, G., et al. 2001. *Nucleic Acids Res.* 29[23]:e118.)

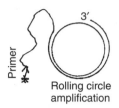

Rolling circle amplification

ROM RNA-one-Modulator. A protein of 63 amino acid residues affects the inhibitory activity of an antisense-RNA. A trans-acting inhibitor of plasmid replication is also named Rop/Rom. *See* antisense technology; Fanconi's anemia; RNA I. (Lin-Chao, S., et al. 1992. *Mol. Microbiol.* 6:3385.)

Roma (Romani)   *See* gypsy.

Romanovs   Last tsar of Russia. Nicholas II and his wife and three children were executed and buried near Ekaterinburg, Russia, on 16 July 1918, after the Bolshevik takeover. Seventy-five years later, the bodies were exhumed and the identities of the remains were determined from bone tissues on the basis of mitochondrial and nuclear DNAs. Tsarina Alexandra was granddaughter and Prince Philip (husband of the present queen of England, Elizabeth) is the great-grand son of Queen Victoria. Their mitochondrial DNA should be identical, as well as those of the three daughters of Alexandra. Forensic DNA analysis confirmed the expectation. Sex was determined from the bone samples by identifying the X-chromosomal amelogenin gene expressed in teeth enamel. It is rich in GC (51%) and codes for a proline-rich protein (24%). The nuclear DNA samples confirmed the identity. The tsar and his paternity and mtDNA tied him to his brother (by the exhumed remains of Grand Duke Georgij) on the basis of heteroplasmy. The four other skeletons did not belong to royal family members but were those of the physician and other people of the court. The two youngest children, Anastasia and Alexis, were not found in the grave. The purported identity of Anastasia with Anna Anderson was not confirmed by DNA analysis. *See* DNA fingerprinting; heteroplasmy Tsarevitch Alexis. (Gill, P., et al. 1994. *Nature Genet.* 6:130.)

ROMK Potassium ion channel family member; it is encoded in human chromosome 11 and alternative splicing generates its isoforms. *See* Bartter syndrome; ion channels; isoform; splicing.

Ron Cell membrane tyrosine kinase, the receptor of the macrophage-stimulating protein. Both are located in human chromosome 3p21. *See* HGF; macrophage-stimulating protein.

roo *See* copia.

roof plate   Organizing center for dorsal neural development in the embryo. *See* floor plate; organizer.

root   *See* a segment of the transverse cross-section below. Roots secrete a great variety of chemicals and transgenic plants may eventually be used to manufacture various needed chemicals in hydroponic cultures. The advantage of hydroponic cultures is that the purification of the secreted substances is easier. *See* hydroponic culture; transgenic. (Ryan, P. R. & Delhaize, E. 2001. *Annu. Rev. Plant Physiol. Plant Mol. Biol.* 52:527.)

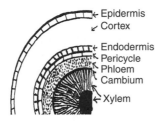

Longitudinal cross-section of a plant root.

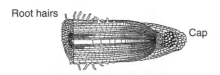

Segment of a root cross section.

root cap   Thin membrane-like protective shield at the tip of roots. Genetic ablation with the aid of transformation by diphtheria-toxin transgene, driven by a root cap–specific promoter, showed that the transgenic plants are viable without several root-tip cell layers and develop more lateral roots.

rooted evolutionary tree   Indicates the origin of the initial split of divergence. *See* evolutionary tree; unrooted evolutionary tree.

root nodule   *See* nitrogen fixation.

root pressure   Guttation of wounded stem caused by osmosis in the roots of plants.

Rop protein   *See* RNA I.

ROR   Retinoid-related orphan receptor. Animal hormone receptors regulating Bc*l* and thus the survival of lymphocytes, but they are also involved in the regulation of several other developmental processes. ROR-response elements (RORE) have the AGGTCA consensus, which is preceded by a 5 bp A/T-rich sequence. *See* Bc*l*; survival factors. (Jetten, A. M., et al. 2001. *Progr. Nucleic Acid Res. Mol. Biol.* 69:205.)

RoRNP   Ribonucleoprotein involving the so-called Y RNA (transcript of RNA polymerase III), the 60 kDa Ro60 protein, and in some cases the La protein (regulator of pol III). RoRNP has been identified in all eukaryotes and it appears to be the target of the autoimmune diseases of lupus erythematosus

and Sjögren syndrome. *See* autoimmune disease. (Labbe, J. C., et al. 1999. *Genetics* 151:143.)

ROS   Reactive oxygen species that may play a detrimental role in radiation damage, modification of DNA, degenerative human diseases, plant diseases, etc. Mitochondrial oxidative phosphorylation, lipid peroxidation, induced nitric acid synthase (NOS) may produce large amounts of reactive NO (nitric oxide), thus generating ROS. NO is a known deaminating mutagen. The defense against ROS damage is the production of antioxidants (vitamin C, vitamin E, glutathion, ferritin, β carotene). The cells may use SOD, catalases, and peroxidases to degrade ROS. Reactive oxygen may cause half of human cancer cases. Deficiency of ROS (by supplying 2-methoxyestradiol) may trigger the demise of leukemia cells because cancer cells have increased aerobic metabolism. *See* aging; Fenton reaction; host-pathogen relationship; hydrogen; hydroxyl radical; oxidative deamination; oxidative DNA damage; peroxide; RO•; ROI; ROO•; singlet oxygen; SOD. (Møller, I. M. 2001. *Annu. Rev. Plant Physiol. Plant Mol. Biol.* 52:561; Lee, D. H., et al. 2002. *Nucleic Acids Res.* 30:3566.)

Rosa spp   Ornamentals with $2n = 14, 21, 28$. *See Rosa canina.*

Rosa canina (dog rose)   Pentaploid species with 35 somatic chromosomes. Unlike other pentaploids, it is fertile. In meiosis, the plants produce 7 bivalents and 21 univalents. The univalents are lost at gametogenesis in the male, so the sperms contain only 7 chromosomes derived from the 7 bivalents. During formation of the megaspore, all 21 univalents and 7 chromosomes from the 7 bivalents are incorporated into the embryo sac. The addition of the 7 male and 28 female chromosomes to the zygote restores the somatic chromosome number of 35. The female contributes more chromosomes to the offspring; therefore, it is matroclinous. Recombination is limited to the 7 bivalents. Its breeding system is a unique mixture of generative and apomictic reproduction. *See* apomixis; matroclinous; pentaploids; univalent. (Gustafson, Å. 1944. *Hereditas* 30:405, illustration below.)

Rosa canina.

rosetta stone sequences   Aid in deciphering the function of simple polypeptide sequences. Prokaryotic proteins frequently carry out the same functions as the corresponding eukaryotic ones. In eukaryotes, the homologous proteins are frequently fused with other proteins that are required for their function, whereas in prokaryotes they may appear separately as single proteins. Once the function of the fused eukaryotic proteins is known, the function of both of the prokaryotic counterparts can also be inferred. An example: In *E. coli*, gyrase A and B are encoded separately, but in yeast they have homology to different domains of topoisomerase II. In a sequential manner, functional connections can be revealed to other proteins. *See* gene neighbor method; phylogenetic profile method.

rosette   Plant shoots with very much reduced internodes commonly found in dicots before the stem bolts after induction of flowering; any anatomical structure in animals arranged in a form resembling the petals of a rose.

Rosette.

ROS oncogene   In human chromosome 6q22. It is the c-homologue of the viral v-ros. It appears to be the same as MCF. *See* oncogenes.

Rossmann fold   NAD(P)-binding domain (GlyXXXGlyXGly) near the N-terminus, encoded by a large number of eukaryote and prokaryote genes. *See* NADP$^+$.

rostral   In the direction of the beak, mouth, or nose rather than toward the hind position.

r$_0$t   In an RNA-driven DNA-RNA hybridization reaction, the concentration of RNA × the time of the reaction (analogous to c$_0$t in reassociation kinetic studies with DNA). r$_0$t sheds information on the RNA complexity of different cells during development. *See* c$_0$t; RNA-driven reaction.

rotamase   Group of enzymes catalyzing cis-trans isomerization. *See* cyclophilin; immunophilins.

rotational diffusion   Membrane proteins travel within the membranes by rotation perpendicular to the plane of the lipid bilayer. *See* cell membrane.

rotaviruses (*Reoviriodae*)   Their genomes consist of 10–12 double-stranded RNAs, and each particle carries a single copy of this genome. The terminal sequences control replication and packaging. Through internal deletions, the RNAs may become aberrant, called DI RNA (defective interfering), resulting in lower infectious capacity. The rotaviruses may cause gastroenteritis (stomach and intestinal inflammation) and diarrhea in human babies and animals. *See* reovirus.

Rothmund-Thomson syndrome   Autosomal-recessive human disorder involving dermal (skin) lesions, dark pigmentation, light sensitivity, early cataracts, bone and hair problems,

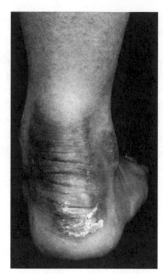

Rothmund syndrome skin lesions. (From Bergsma, D., ed. 1973. *Birth Defects. Atlas and Compendium*. (By permission of the March of Dimes Foundation.)

and premature aging. May lead to squamous (scaly) carcinomas. Autosomal-dominant genes determine some similar types of diseases. One form is assigned to human chromosome 8q24.3 and another to 17q25. Both genes encode helicases; the homologues control recombination in yeast. *See* cancer; helicase; light sensitivity diseases.

Rotor syndrome   *See* Dubin-Johnson syndrome.

rough draft, genome   Is a not complete sequence of a genome, yet it displays the sequences of about 90% of the euchromatic parts containing the coding units. *See* genome projects; human genome.

rough ER   *See* endoplasmic reticulum rough; RER.

rounds, mating   After bacteriophages have been replicated within the host cell, the newly formed molecules of vegetative DNAs may recombine with each other several times. Because of the multiple exchanges (in agreement with Poisson distribution expectations), it appears that the maximal recombination between markers cannot exceed 30–40%. *See* coefficient of coincidence; mapping function; negative interference. (Visconti, N. & Delbrück, M. 1953. *Genetics* 38:5.)

Rous sarcoma   Originally detected as a viral RNA cancer in chickens. The protooncogene homologue was detected in rats and other mammals. *See* oncogenes; RAS and homologues; TV/RCAS.

Rowley-Rosenberg   syndrome   Autosomal-recessive growth retardation, different from dwarfness, and characterized by aminoacidurias. *See* aminoacidurias; dwarfism.

RPA   Single-strand DNA-binding protein (of subunits of 70, 34, and 14 kDa) participating in replication, nucleotide excision repair, and repair of double-strand breaks by recombination. *See* BER; DNA replication, eukaryotes; NER; replication protein A; UNG2. (Davis, A. P. & Symington, L. S. 2001. *Genetics* 159:515.)

RPB Subunits (differently numbered) of RNA polymerase II. *See* RNA polymerase.

RPD3/Sin3 Histone deacetylase and a component of the Mad/N-CoR/Sin3/RPD transcriptional protein repressor complex. It may also block the position effect exerted by centromeric and telomeric heterochromatins. *See* histone; histone deacetylase; Mad; N-CoR; PEV; position effect; repression; signal transduction; Sin3. (Fazzio, T. G., et al. 2001. *Mol. Cell Biol.* 21:6450.)

R-phycoerythrin Phycobiliprotein fluorochrome isolated from algae. Maximal excitation at 545 and 565 nm, but it is also excited at 480 nm. Maximal emission at 580 nm; hence, the red color.

R plasmids Carry resistance factors (genes for antibiotic resistance and other agents).

rpo RNA polymerases such as A, B, C$_1$, C$_2$ in (organelles) plastids and mitochondria, resembling bacterial RNA polymerases. *See* RNA polymerase; σ.

R point (restriction point) Before the S phase, cells in the G1 phase pause and may or may not continue the cell cycle. Cancer cells bypass the restriction point and continue uncontrolled divisions. Cultured cells may require serum or amino acids to pass from G1 to S. *See* cell cycle. (Ekholm, S. V., et al. 2001. *Mol. Cell Biol.* 21:3256.)

r-proteins Ribosomal proteins. *See* ribosome.

RPTK ROS is a sperm receptor protein tyrosine kinase that may bind to the ZP3 (zona pellucida) protein of the egg matrix.

These proteins affect several cellular processes. *See* egg; fertilization; sperm. (Zeng, L., et al. 2000. *Mol. Cell Biol.* 20:9212.)

RRAS *See* RAS oncogene.

RRE Rev-response element is an RNA export adaptor promoting the export of HIV-1 transcripts from the nucleus. If an RRE is inserted into an intron (which is normally retained within the nucleus), even that can be exported to the cytoplasm. *See* acquired immunodeficiency; export adaptors. (Zhang, Q., et al. 2001. *Chem. Biol.* 8:511.)

rrn In *E. coli*, there are seven ribosomal transcription units, *rrn-A, -B, -C, -D, F, -G, -H*, including the 16S-23S-5S RNAs, spacers, and intercalated tRNA genes within the spacers. Maturation involves trimming of the cotranscript (cleavage by RNase III) and processing by other RNases, RNases P and D. The number of rRNA genes in eukaryotes is very variable and subject to amplification. At developmental stages of very active protein synthesis, the number of ribosomal genes may be amplified to several thousands, and in, e.g., the amphibian oocytes, may be sequestered into mininuclei. Eukaryotic rRNA genes are transcribed in a ca. 45S precursor RNA, containing the 18S-5.8S-28S (in this order) and spacer sequences. The pathway of trimming may vary among different species. 18S rRNA is methylated immediately at about 40 sites, and a few more methyl groups are added in the cytoplasm after maturation. 28S rRNA is methylated immediately after transcription at over 70 sites, and these methylated sites are saved during the process of maturation. The cleavage takes place at the 5'-side of the genes and between the spacers. 5.8S rRNA eventually associates with 28S rRNA by base pairing. *See* ribosomal proteins; ribosomal RNA; ribosomes; RNA maturases; stringent control; stringent response. (Hirvonen, C. A., et al. 2001. *J. Bacteriol.* 183:6305.)

rRNA Ribosomal RNA; structural component of the ribosomes; rRNAs may be preferentially amplified in the oocytes

Segment of a eukaryotic rRNA gene cluster in transcription. (Courtesy of Spring, H., et al. 1976. *J. Microsc. Biol. Cell.* 25:107.)

The DNA region of the rRNA and tRNA gene cluster (*rrn*) in *E. coli* is transcribed into longer than 30S primary transcripts, interrupted by spacers. The ribosomal and transfer RNA genes are clustered and cotranscribed in the order 5'-26S-23S-5S-3', and within the intergenic spacers the tRNA genes are situated. RNase III at the duplex stems trims the individual gene transcripts. The different loops contain ca. 1,600, 2,900, and 120 nucleotides, corresponding to the 16S, 23S, AND 5S RIBOSOMAL RNAs, respectively. (The 1,600 and 2,900 base sequences are also called p16 and p22, respectively.) RNase P further trims each of these rRNA precursors, and the tRNA precursors are processed by RNase P at the 5' and by RNase D at the 3' end.

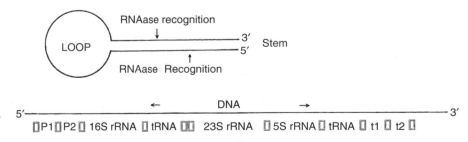

P1 and P2 : promoters, ▯ : spacers, t1 and t2 : termination signals (not on scale)

of amphibians and other organisms. *See* ribosomal RNA; ribosome.

RRS Reduced representation shotgun is a method for mapping single-nucleotide polymorphism in a genome. *See* SNIPS. (Altshuler, D., et al. 2000. *Nature* 407:513.)

RS domain Is rich in arginine (R) and serine (S); RS proteins are parts of the spliceosomes. *See* spliceosome.

RSC Remodel structure of chromatin. *See* chromatin remodeling.

RSF (1) *See* hybrid dysgenesis I-R system. (2) 400–500 kDa protein complex of a 325 kDa protein and the 135 kDa hSNF2h/ISWI. *See* chromatin remodeling; SNF2. (Labourier, E., et al. 1999. *Genes Dev.* 13:740.)

RSK Ribosomal protein S6 kinase (RSK-3 at Xp22.2-p22.1, RSK-2 at 6q27, RSK-1 in human chromosome 3; there is also an RSK-4). Epidermal growth factor–regulated protein kinase phosphorylating histone 3; RSK-3 defect is involved in Coffin-Lowry syndrome. MAPK-phosphorylated RSK prevents parthenogenetic development of unfertilized eggs. RSK integrates the MAPK and PDK1 signaling pathways. *See* chromatin remodeling; Coffin-Lowry syndrome; EGF; histones; MAPK; PDK; S6 kinase.

RSS Recombinational signal sequence. V(J)D recombinase mediates recombination only in gene segments flanked by tripartite recombination signal sequences consisting of a highly conserved heptamer (7mer), an AT-rich nonamer (9mer), and 12- or 23-base-long intervening nucleotides. *See* immunoglobulin; RAG; V(J)D recombinase.

R^{St} (*R-st*) Stippled, paramutable allele of the *R* locus of maize in the long arm of chromosome 10. *See* paramutation; R locus; photo.

R^{st} kernel.

R_{ST} Same as F_{ST}, but it is based on the variation in microsatellites. *See* F_{ST}; microsatellite.

r_0t value Measure of RNA-DNA or RNA-RNA hybridization; the product of the concentration of single-stranded RNA and time elapsed since the beginning of the reaction. *See* c_0t value.

RTF Resistance transfer factor. Bacterial plasmids carrying various antibiotic and other resistance genes. *See* conjugation in bacteria; plasmid mobilization; resistance transfer factors.

Rth *See* Rad 27.

RTK *See* receptor tyrosine kinase.

RTP Replication termination protein. Functionally but not structurally similar to Tus. *See* replication, bidirectional. (Gautam, A., et al. 2001. *J. Biol. Chem.* 276:23471.)

RT-PCR Reverse transcription-polymerase chain reaction. The purpose of the procedure is similar to PCR in general, in this instance to amplify the small amounts of RNA transcripts as cDNA. The reaction requires reverse transcriptase, mRNA, deoxyribonucleotides, and primers that can be random DNA sequences, oligodeoxythymidine or antisense sequences. The method is very sensitive and the RNA of a single cell can be amplified; thus, localized gene expression can be studied. Under well-controlled conditions, it can be semiquantitative. It is also used for clinical diagnostic purposes. *See* in situ PCR; PCR; RNA fingerprinting. (Barlič-Maganja, D. & Grom, J. 2001. *J. Virol. Methods* 95:101.)

rtTA Reverse transactivator tetracycline is basically a tTA system containing a nuclear localization signal at the 5′-end; efficiently binds the *tetO* operator only in the presence of tetracycline derivatives such as doxycycline or anhydrotetracycline. *See* tetracycline; tTA. (Pacheco, T. R., et al. 1999. *Gene* 229:125.)

rtTA-nls Same as rtTA.

RU486 (mifepristone) Pregnancy prevention drug. *See* hormone receptors. (Mahajan, D. K., et al. 1997. *Fertility & Sterility* 68:967; DeHart, R. M. & Morehead, M. S. 2001. *Ann. Pharmacother.* 35:707; Schulz, M., et al. 2002. *J. Biol. Chem.* 277:26238.)

RU maize Carry plasmid-like elements in their mitochondria, yet they are not male sterile. *See* cytoplasmic male sterility; mtDNA.

rubella virus (a toga virus) Causes German measles. Infection during early pregnancy may cause intrauterine death of the human embryo and/or developmental anomalies in the newborn. *See* teratogenesis.

Rubinstein syndrome (Rubinstein-Taybi syndrome) Dominant defects in the heart valve of the pulmonary aorta, collagen scars on skin wounds, enlarged passageway between the skull and the vertebral column. This condition has very low recurrence risk (1%) and about 0.2–0.3% of the inmates of mental asylums are afflicted by it. Haploinsufficiency for the CBP transcription factor seems to be involved in abnormal differentiation. Many afflicted individuals have breakpoints at chromosome 16p13.3, and this is the site of the human cyclic AMP response element-binding protein (CBP/CREB). *See* CBP; CREB; GLI3 oncogene; mental retardation. (Murata, T., et al. 2001. *Hum. Mol. Genet.* 10:1071.)

RUBISCO Ribulose bisphosphate carboxylase-oxygenase (M_r 550,000), a chloroplast enzyme. The eight large subunits (each M_r 56,000) are coded for by chloroplast DNA and the small subunits (each M_r 14,000) are under nuclear control. The carboxylase function catalyzes the covalent attachment of carbon dioxide to ribulose-1,5-bisphosphate, then splitting into two molecules of 3-phosphoglycerate. The oxygenase function mediates the incorporation of O_2 into ribulose-1,5-bisphosphate and the resulting phosphoglycolate reenters the Calvin cycle. *See* chloroplast; chloroplast genetics; photosynthesis; ribulose bisphosphate carboxylase. (Douce, R. & Neuberger, M. 1999. *Curr. Opin. Plant Biol.* 2:214; Spreitzer, R. & Salvucci, M. E. 2002. *Annu. Rev. Plant Biol.* 53:449.)

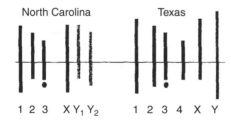

Genetic and electrophoretic evidence that the large subunits of RUBISCO are maternally inherited, whereas the small subunit is transmitted biparentally. Molecular studies mapped the large subunits in the chloroplast DNA. (After Chen, K., Gray, J. C. & Wildman, S. G. See also 1975. *Science* 190:1304.)

RULE 12/23 *See* immunoglobulins; RAG; V(D)J.

Rumex hastatulus North American herbaceus plant has variable chromosomal sex determination. The plants found in North Carolina have three pairs of autosomes: one X and two Y chromosomes. The form prevalent in Texas has four pairs of autosomes and one X and one Y chromosome. The middle line connects the positions of the centromeres. *See* chromosomal sex determination; sex determination.

(Redrawn after Smith, B. W. 1964. *Evolution* 18:93.)

runaway plasmids At lower temperature (30°C) they are present in relatively low copy numbers and do not interfere with the growth of the host cell. Above 35°C, their copy number raises substantially and so does the DNA they contain, but after about 2 hours (by the time they account for 50% of the DNA of the cell) they suppress cellular growth. *See* copy-up mutation; vectors. (Uhlin, B. E., et al. 1983. *Gene* 22:255.)

runaway replication Replication is not restricted in the normal manner. (Chao, Y. P., et al. 2001. *Biotechnol. Progr.* 17:203.)

R unit *See* r (Röntgen).

run-off transcription The inducer of gene activity (e.g., light signal) is withdrawn and the tapering off of transcription is monitored by incorporation of labeled nucleotides. (Delany, A. M. 2001. *Methods Mol. Biol.* 151:321.)

run-on transcription Once the transcription is turned on in isolated nuclei, it proceeds without further need for enhancers as measured by the incorporation of labeled nucleotides into mRNA. *See* regulation of gene activity; transcription. (Hu, Z. W. & Hoffman, B. B. 2001. *Methods Mol. Biol.* 126:169.)

Runt (*run*, 1–65) *Drosophila* pair-rule gene encoding a DNA-binding protein/transcription factor with homologies to, e.g., CBFA1, -2, -3 core-binding factor subunits of human and mouse genes. *See* leukemia; pair rule genes.

Rupert Hemophiliac grandson of Leopold of Albany (a hemophiliac himself), son of Queen Victoria. *See* hemophilias; Queen Victoria.

Russell-Silver syndrome (RSS) Most commonly autosomal recessive (7p11.2-p13), but X-linked or sporadic cases with low-birth-weight dwarfism, frequently with asymmetric body and limbs, deformed fingers, relatively large skull; mental retardation is common. The 7p site is closely associated with the location of the genes encoding growth factor receptor-binding protein 10 and the insulin-like growth factor-binding proteins 1 and 3. Maternal disomy, at least for a 35 Mb sequence, appears to account for the symptoms. *See* dwarfism; imprinting; stature in humans; uniparental disomy. (Hannula, K., et al. 2001. *Am. J. Hum. Genet.* 68:247; Nakabayashi K., et al. 2002. *Hum. Mol. Genet.* 11:1743.)

rust Disease of grasses (cereals) caused by *Puccinia* fungi. *See* host-pathogen relations. (Staples, R. C. 2000. *Annu. Rev. Phytopath.* 38:49; illustration below.)

Leaf rust (Puccinia triticina). (Courtesy of Dr. E. R. Sears.)

RUT *See* oestrus.

RuvABC Protein complex operating in the Holliday structure of recombination. RuvAB helicase/ATPase and the motor protein complex mediate branch migration (also through nucleosomes in eukaryotes) and replication. RuvC is an endonuclease. *See* AAA proteins; branch migration; endonuclease; Holliday juncture; recA; recombination, molecular mechanisms in prokaryotes. (West, S. C. 1997. *Annu. Rev. Genet.* 31:213;

Constantinou, A., et al. 2001. *Cell* 104:259; Yamada, K., et al. 2001. *Proc. Natl. Acad. Sci. USA* 98:1442.)

RVs *See* BIN.

RXR *See* RAR.

ryanodine (ryanodol 3-(1H-pyrrole-2-carboxylate)   Toxic extract (insecticide) from the new world tropical shrub *Ryania speciosa*. Regulates calcium ion channels in muscles. Mutation of deletion of the ryanodine receptor gene (RYR1, 19q13.1) may lead to malignant hyperthermia. *See* central core disease; hyperthermia; ion channels.

rye (*Secale cereale*) $2n = 2x = 14$ is an outbreeding crop plant used for the production of bread, biscuits, starch, and alcohol. Its taxonomy is somewhat controversial. Rye can be crossed with a number of other cereals; among them, the allopolyploid *Triticales* (wheat × rye hybrids, $2n = 42$ and $2n = 56$) are most notable. Addition lines and transfer lines carrying rye chromosomes or chromosomal segments have been made. It belongs to those exceptional grain crops where autotetraploid varieties have agronomic value. Trisomic lines are known. Some varieties harbor a variable number of B chromosomes. Rather unusually, some of the plastids are also transmitted through the pollen. *See* alien addition; alien substitution; chromosome banding; chromosome substitution; ergot; holocentric; transfer lines; *Triticale*. (<http://www.tigr.org/tdb/tgi.shtml>.)

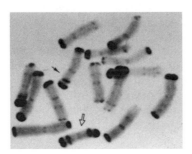

Giemsa-stained rye karyotype displaying an isochromosome ($\Rightarrow$) and the corresponding arm in a normal chromosome ($\rightarrow$). (Courtesy of Dr. Gordon Kimber.)

Rye ear.

ryegrass   *Lolium multiflorum* and *L. perenne*, both $2n = 14$.

Ryegrass spike.

At the January 6–8, 1909 meeting of the American Breeder's Association, the following paper was presented in Columbia, Missouri:

AMERICAN BREEDERS' ASSOCIATION.   365

WHAT ARE "FACTORS" IN MENDELIAN EXPLANATIONS?

By Prof. T. H. Morgan.

Columbia University, New York, N. Y.

In the modern interpretation of Mendelism, facts are being transformed into factors at a rapid rate. If one factor will not explain the facts, then two are invoked; if two prove insufficient, three will sometimes work out. The superior jugglery sometimes necessary to account for the results may blind us, if taken too naively, to the common-place that the results are often so excellently "explained" because the explanation was invented to explain them. We work backwards from the facts to the factors, and then, presto! explain the facts by the very factors that we invented to account for them. I am not unappreciative of the distinct advantages that this method has in handling the facts. I realize how valuable it has been to us to be able to marshal our results under a few simple assumptions, yet I cannot but fear that we are rapidly developing a sort of Mendelian ritual by which to explain the extraordinary facts of alternative inheritance. So long as we do not lose sight of the purely arbitrary and formal nature of our formulae, little harm will be done; and it is only fair to state that those who are doing the actual work of progress along Mendelian lines are aware of the hypothetical nature of the factor-assumption. But those who know the results at second hand and hear the explanations given almost invariably in terms of factors, are likely to exaggerate the importance of the interpretations and to minimize the importance of the facts.

A historical vignette.

S

S (such as in 5S RNA) *See* sedimentation coefficient.

s (1) *See* selection coefficient. (2) Standard deviation of a set of experimental observations. *See* σ parametric.

S8 Ribosomal protein with a binding site at the 597-599/640-643 at the hairpin of 16S rRNA. It is required for the assembly of the 30S small ribosomal subunit. *See* ribosomes.

Hairpin.

S-9 *See* Ames test; microsomes.

S³⁵, ³⁵S Sulfur isotope. *See* isotopes.

S49 Mouse lymphoma cell line.

SAA Serum amyloid A. *See* amyloidosis.

SAC Regulator of nuclear export and the cell cycle. *See* nuclear export factors; nuclear pore.

saccades Abrupt changes in the fixation of the eyes during scanning objects directed by the reflex center of the brain (superior colliculus).

saccharin Noncaloric sweetener; hundreds of times sweeter than sucrose. Oral LDLo for humans is 5 g/kg; it has been a suspected carcinogen and mutagen, but later studies have not confirmed its classification as a carcinogen. *See* aspartame; fructose; LDLo.

Saccharomyces cerevisiae Eukaryotic budding yeast with chromosome number $n = 16$; its genome size is about 1.2×10^7 bp, approximately three times that of the prokaryotic *E. coli*. Sequencing and knockout information contradict earlier estimates that less than 10% of its genome is repetitive. In chromosome III of 55 open-reading frames, only 3 appeared indispensable for growth on a rich nutrient medium. Of 42 other genes, only 21 displayed a phenotype. This information points to redundancy even in this small genome. This conclusion may be misleading, however, because some genes are called to duty only under specific circumstances. Its entire genome had been sequenced by 1996. The 5,885 open-reading frames are encoded by 12,068 kb. About 140 genes code for rRNA, 40 for snRNA, and 270 for tRNA. About 11% of the total protein produced by the yeast cells (proteome) has a metabolic function; 3% is involved in DNA replication and energy production, respectively; 7% is dedicated to transcription, 6% to translation; and 3% (ca. 200) are different transcription factors.

About 7% is concerned with transporting molecules. About 4% are structural proteins. Promoters, terminators, regulatory sequences, and intergenic sequences with unknown functions occupy about 22% of the genome. Many proteins are involved with membranes. In rich nutrient media, the doubling time is about $1\frac{1}{2}$ hours. The organism has regular meiosis and mitosis. The vegetative multiplication is by budding (budding yeast), i.e., the new (daughter) cell is formed as a small protrusion (bud) on the surface of the mother cell. Haploid cells may fuse to generate diploidy and the diploid cells may undergo meiosis (sporulation); the four haploid products are retained in an ascus as an unordered tetrad. The haploid cells may have *a* or α mating type. Although budding yeast is eukaryotic, it can be cultured much like prokaryotes, and thus it combines many of the advantages of both groups of organisms. Approximately 25–30% of the human genes have significant homology with a yeast gene. The present genome might have evolved through extensive duplications. *See* databases; fungal life cycles; gene replacement; mating type determination in yeast; mtDNA; *Schizosaccharomyces*; sequencing; tetrad analysis; YAC; yeast transformation; yeast transposable elements; yeast vectors. (Dwight, S. S., et al. 2001. *Nucleic Acids Res.* 30:69; <ftp://ftp.ebi.ac.uk/pub/databases/yeast>. or ftp://genome-ftp.stanford.edu/yeast/genome_seq>, protein coding sequences <ftp://ftp.ebi.ac.uk/pub/databases/lista>, <http://genome-www.stanford.edu/Saccharomyces>, Triples: <http://bioinfo.mbb.yale.edu/e-print/genome-transposon-nature/text.htm> Yeast introns: <http://www.cnrs-gif.fr/cgm/epissage_seraphin.html>.) Barnett, J. A. & Robinow, C. F. 2002. *Yeast* 19:151 and 745; Mackiewicz, P., et al. 2002. *Yeast* 19:619; diagram next page.

sacral agenesis (Currarino triad) Malformation of the caudal (tail) end of the notochord. Genes at human chromosomes 7q36 and 1q41-q42 may affect rare dominant expression. *See* Currarino triad; notochord.

SADDAN Severe achondroplasia with delayed development and acanthosis nigricans. *See* acanthosis nigricans; achondroplasia.

S-adenosylmethionine (SAM) Methyl donor for restriction-modification methylase enzymes, synonymous with Adomet. *See* methylation of DNA.

S-adenosylmethionine decarboxylase (AdoMetDC) Enzyme involved in the biosynthesis of spermidine and spermine. For its activity it is essential to contain a covalently bound

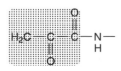

Pyruvoyle.

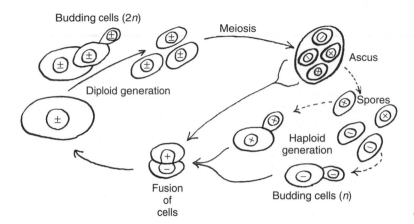

Budding cells (2n)

Meiosis

Ascus

Diploid generation

Spores

Haploid generation

Fusion of cells

Budding cells (n)

The life cycle of *Saccharomyces*.

pyruvoyle end group to the α-subunit of the dimeric enzyme. This enzyme is found in both prokaryotes and eukaryotes. *See* spermidine; spermine. (Li, Y.-F., et al. 2001. *Proc. Natl. Acad. Sci. USA* 98:10578.)

SAD mouse Animal model for human sickle cell anemia. It produces polymerized hemoglobin due to mutations in β-globin, Hb[SAD]. *See* sickle cell anemia. (Martinez-Ruiz R., et al. 2001. *Anesthesiology* 94:1113.)

Saethre-Chotzen syndrome *See* Chotzen syndrome.

SAF SKP-associated factors, F-box proteins. *See* F-box; SKP.

safety *See* chemicals, hazardous; cosmic radiation; gloves; environmental mutagens; laboratory safety; radiation hazard assessment; recombinant DNA and biohazards. (Fleming, D. O. & Hunt, D. L., eds. 2000. *Biological Safety. Principles and Practices*. ASM Press, Washington, DC.)

safflower (*Carthamus tinctorius*) Oil crop of warmer climates, $2n = 24$; other related species have $2n = 20$ or $2n = 44$ chromosomes.

SAGA Histone acetyltransferase complex of about 20 different proteins that interact with TBP (TFIID) and with gene-specific transcriptional activators. In yeast, SAGA and TFIID complexes appear to have some redundant functions. *See* bromodomain; chromatin remodeling; histone acetyltransferase; SNF/SWI; TBP; transcription; transcription factors. (Sterner, D. E., et al. 1999. *Mol. Cell. Biol.* 19:86.)

SAGE (1) Serial analysis of gene expression is a procedure that permits the analysis of the function of many genes by a sweeping procedure. The first step is to isolate all the mRNAs that are produced in a single organ at a particular developmental stage. By reverse transcription they are converted into cDNA. The 3′-ends are then tagged by biotin. The cDNAs are digested by a restriction enzyme and the end fragments are trapped on streptavidin beads. A second restriction enzyme is applied that cuts at least 9 bp from the fragments. Each short (9 bp or longer) tag is amplified by PCR and the tagged pieces are linked into a single DNA molecule. Then each tag is sequenced by an automatic sequencer and the tags are counted.

In this sweeping manner, 20,000 genes can be monitored in a month. The same effort would require years if conducted on separate genes. This procedure reveals not just the number of expressed genes in an organ but also the level of their activity. Some may be expressed in a single copy; others may be very active and are represented by multiple copies. *See* biotin; DNA chips; DNA sequencing, automated; electrospray MS; expressed sequence tag; genome project; genomics; laser desorption MS; microarray analysis; microarray hybridization, RIDGE; streptavidin. (Lash, A. E., et al. 2000. *Genome Res.* 10:1051; Polyak, K. & Riggins, G. J. 2001. *J. Clin. Oncol.* 19:2948; <www.sagenet.org>; <www.ncbi.nlm.nih.gov/SAGE>; <http://cgap.nci.nih.gov/SAGE>.) **(2)** Computer program for the analysis of (human) genetic data.

Sage Genie An innovative new bioinformatics tool based on SAGE but providing a single platform for acquiring, annotating and interpreting large sets of gene expression data. The new technology measures gene expression by the frequency of 3′ signature SAGE tags of 10 bases specific and unique to each transcript. The method permits the analysis of gene expression that are up- or down-regulated and allows automatic matching of SAGE tags to known transcripts. The incorrectly linked or occurring only once or those that were obtained by sequencing errors are filtered out from millions of tags. Then *confident SAGE tags* (CST) are obtained. The SAGE Genie permits *horizontal comparisons* (e.g. normal versus cancerous expression) and in addition *vertical comparisons* (e.g. expression profiles in different tissues or organs) under any desired conditions. This technology appears much simpler and has far greater specificity than microarrays. The SAGE Genie provides an automatic link between gene names and SAGE transcript levels accessible by the Internet: <http://cgap.nci.nih.gov/SAGE>. *See* SAGE; microarray hybridization. (Boon, K., et al. 2002. *Proc. Natl. Acad. Sci. USA* 99:11287.)

sagittal In the anterior-posterior body plan.

Saguenay–Lac-Saint-Jean syndrome (2p16) Rare, recessive, morbid cytochrome oxidase deficiency with symptoms similar to Leigh disease. *See* Leigh's encephalopathy.

SAHA Suberoylanilide hydroxamic acid is an inhibitor of histone deacetylase.

saimiri (squirrel monkey) *See* Cebidae.

sainfoin (*Onobrychis vicifolia*) Leguminous forage plant; $2n = 14$ or 28.

***Sal* I** Restriction endonuclease with recognition site G↓TCGAC.

salamander *Salamandra salamandra*, $2n = 24$; *Ambystoma mexicanum*, *A. tigrinum tigrinum*, $n = 14$. The estimated genome size of the ambystomas is 7,291 map units, the largest known. (Voss, S. R., et al. 2001. *Genetics* 158:735.)

salicylic acid (O-hydroxybenzoic acid) Painkiller, keratolytic and fungicidal agent; it also mediates the expression of disease defense-related responses in plants. Its methylsalicylate derivative, a volatile compound, may carry out airborne signaling after infection of plants by pathogens. The PAD genes encode lipase-like molecules involved in salicylic signaling. Salicylic acid activity may be modulated by desaturation. *See* aspirin; desaturase; host-pathogen relation; hypersensitivity reaction. (Dempsey, D. M., et al. 1999. *Crit. Rev. Plant Sci.* 18:547; Wildermuth, M. C., et al. 2001. *Nature* 414:562.)

Salicylic acid.

saline Water solution of NaCl; the "physiological saline" is 0.9% salt solution for humans.

salivary gland chromosomes Polytenic; because of their large size and clear landmarks, these chromosomes have been used extensively for cytogenetic analyses of *Drosophila* and other flies. The cultures for these studies should be less crowded and moist. Third instar larvae can be used as they crawl out of the medium before the cuticle hardens. The larvae are placed in aceto-orcein or into 7% aqueous NaCl on microscope slides. The larvae are held in place with a needle. Larvae are decapitated with a second needle placed behind the mouth parts, and the salivary gland is pulled out. The nuclei are stained in aceto-orcein on a clean slide in 5 to 10 min. The chromosomes may be spread by gentle pressure on the cover slide. After sealing at the edge with wax, they can be examined under a light microscope. *See* Drosophila; polytenic chromosomes. (Lifschytz, E. 1983. *J. Mol. Biol.* 164:17; *Cold Spring Harbor Symp. Quant. Biol.* 38 [1974].)

salivary gland chromosomes and mapping Because the salivary chromosomes display clear topological markers (bands), they can be used to locate deletions, duplications, inversions, translocations, and to associate mutant phenotypes with physical alterations. The genetic and cytological maps are collinear, yet they are not exactly proportional by distance. *See* coefficient of crossing over, *Drosophila*. (Painter, T. S. 1943. *J. Hered.* 25:465; Bridges, C. B. 1935. *J. Hered.* 26:60.)

Salla disease *See* sialic acid; sialuria.

S alleles Control self-sterility in plants. *See* incompatibility alleles; self-incompatibility.

salmon *Salmo gardneri*, $2n = 58-65$.

Salmonella Member of the enteric gram-negative bacteria; as such, it is related to *E. coli*, and its handling ease is similar to it. It is a human pathogen and even in the laboratory requires some caution when manipulated. Several of the related species contaminate food and feed supplies and create health hazards. F$^+$, F′, and Hfr strains are available. Its best known transducing phage is P22. The 4,809,037 bp genome of *Salmonella enterica* CT18 and two plasmids, pHCM1 (multiple-drug-resistance incH 218,159 bp) and the cryptic plasmid pHCM2 (106,516 bp), have been sequenced. It includes >200 pseudogenes and hundreds of insertions and deletions. *Salmonella enterica* serovar typhimurium LT2 contains 4,857 kb in the chromosomes and a 94 kb virulence plasmid. *See* Ames test; histidine operon; phase variation. (Parkhill, J., et al. 2001. *Nature* 413:848; McClelland, M., et al. 2001. *Nature* 413:852; <www.salmonella.org>.)

Salpiglossis variabilis (Solanaceae) Plant species with the rather unusual characteristics that the four pollen grains — products of a single meiosis — stick together and therefore can be used for tetrad analysis and gene conversion in higher plants. *See* gene conversion; tetrad analysis.

salpingectomy Sterilization of mammals by removal of the fallopian tube (salpinx) leading to the uterus. *See* birth control; sterilization in humans; tubal ligation; vasectomy.

saltation Unproven evolutionary proposition that species (and even higher taxonomic categories) arise by non-Darwinian, sudden, major alterations. *See* hopeful monster.

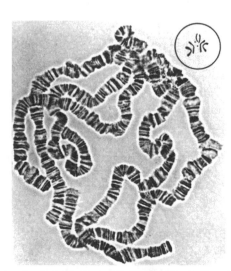

Salivary gland chromosomes of *Drosophila*. (Courtesy of Dr. H. K. Mitchell.) upper right (circled) a regular mitotic set of chromosomes at about the same scale. (Redrawn after Painter, T. S. 1934. *J. Hered.* 25:465.)

saltatory replication   Sudden amplification of DNA segments during evolution.

salt bridges   Noncovalent ionic bonds in multimeric proteins.

salt tolerance   Of plants, is regulated by HKT1 (high-affinity potassium [K+] transporter). This protein is actually a Na+ and K+ cotransporter and at a high K+ level results in a low level of Na+ uptake, thus conveying some salt tolerance. Ca²⁺ is beneficial for salt tolerance supposedly under the control of a protein sensor displaying about 50% similarity to calcineurin and neuronal calcium sensors. The Na+/H+ antiport protein (NH1X) mediates another salt control. Low proline content in the cells causes high salt sensitivity, whereas an increase in the level of sugar alcohols favors salt tolerance. Another potential mechanism for salt tolerance would be the improved elimination of salt from the cells. *Arabidopsis* plants carrying a mutant NHX1 gene displayed higher tolerance for sodium and, in addition, because of the accumulation of NaCl in the vacuoles, acquired better exploitation of the soil moisture through osmosis. Having salt tolerant crops may extend the usage of alkaline soils and may make it possible to use seawater for irrigation. *See* calcineurin; calcium signaling; ion channels. (Zhang, H.-X. & Blumwald, E. 2001. *Nature Biotechn.* 19:765; Zhu, J.-K. 2002. *Annu. Rev. Plant Biol.* 53:247; Shi, H., et al. 2003. *Nature Biotechnol.* 21:81.)

salvage pathway   Recycling pathway, in contrast to the de novo pathway, e.g., nucleotide synthesis from nucleosides after removal of the pentose followed by phosphoribosylation. *See* HAT medium; HGPRT; phosphoribosyl transferase.

SAM   *See* S-adenosyl-L-methionine; substrate adhesion molecules.

SAM68   SRC-associated in mitosis. A phosphoprotein and a target of SRC, FYN, and ITK kinases during mitosis. It also binds to the SH domains of the Grb adaptor protein of the cytoplasm and NCK in the nucleus. When SAM68 is tyrosine dephosphorylated, it binds RNA. A homologue of SAM68 is ÉTOILE. *See* signal transduction; Src; STAR.

sampling distribution   Probability distribution of a statistic estimated from a random sample and a certain size. The sampling distribution is somewhat different from the normal distribution characterized by the mean (μ) and the standard deviation (σ) inasmuch as the mean (M) has a standard deviation of $\sigma/\sqrt{n}$.

sampling error   Can occur when the population is too small or when few individuals are sampled.

$$n = \frac{2(z_\alpha - z_\beta)^2 \sigma^2}{(\mu_1 - \mu_2)^2},$$

where z_α is the normal deviate for a level of significance α; z_β is the normal deviate for β; $1 - \beta$ is the power required; σ^2 is the variance of each population with means μ_1 and μ_2 chosen or whenever the selection is not random. The proper sample size n can be statistically estimated as shown above. *See* effective population size; founder principle; genetic drift; normal deviate; normal deviation; Z distribution.

SANCHO   Nonviral retrotransposable element. *See* transposable elements.

Sandhoff's disease   Characterized either by the absence of both β-hexosaminidase α- and β-activity or by β-hexosaminidase β-subunit only. This autosomal-recessive hereditary defect has very similar symptoms to those of Tay-Sachs disease, which is caused by β-hexosaminidase α-subunit deficiency. β-hexosaminidase β-subunit defect blocks the degradation of β-galactosyl-N-acetyl-(1→3)-galactose-galactose-glucose-ceramide to β-(1→4)-galactosyl-galactose-glucose-ceramide and hexosaminidase-α normally converts GM₂ ganglioside (neuraminic-N-acetyl-galactose-4-galactosyl-N-acetyl-glucose-ceramide) into GM₃ ganglioside (neuraminic-N-acetyl-galactose-glucose) ceramide. The genes for β-hexosaminidase α and β are in human chromosomes 15q23-q24 and 5q13, respectively. Their structural similarity indicates evolution by duplication. *See* gangliosides; sphingolipidoses; sphingolipids; Tay-Sachs disease.

Sanfilippo syndrome   *See* mucopolysaccharidosis.

Sanger method of DNA sequencing   *See* DNA sequencing.

SANT   SWI3, ADA-2, N-COR, TFIIB. Chromatin remodeling complex sharing an ATPase and transcription and chromatin remodeling domains. *See* ADA; chromatin remodeling; N-COR; SWI; TFIIB.

Santa Gertrudis cattle   Originated by a cross between the exotic-looking Brahman × shorthorn breeds (*Bos taurus*). The Brahman cattle descended from *Bos indicus*. *See* cattle.

SAP-1   Stress-activated protein is involved in the activation of MAP kinases in binding to their serum-response factor (SRF). SAP is required for the activation of SLAM. *See* Elk; Epstein-Barr virus; lymphoproliferative diseases, X-linked; MADS box; MAP; SLAM; SRF. (Latour, S., et al. 2001. *Nature Immunol.* 2:681.)

SAPK   Stress-activated protein kinase of the ERK family. SAPK can be activated by SEK, a protein kinase related to MAP kinase kinases. The signaling cascade is targeted to JUN. *See* ERK; Ire; JNK; JUN; MAPKK; MEKK; signal transduction.

saponification   Alkaline hydrolysis of triaglycerols to yield fatty acids. *See* fatty acids; triaglyceride.

saponins   Glycosylated triterpenoid or steroidal alkaloids in plants that may be protective against fungal pathogens. *See* host-pathogen relation. (Bourab, K., et al. 2002. *Nature* 418:889.)

saporins   Plant glycosidases that remove adenine residues from RNA and DNA but not from ATP or dATP. *See* RIP.

saposins   (10q22.1)   Glycoproteins involved in the activation of galactosylceramidase. *See* leukodystrophy. (Qi, X. & Grabowski, G. A. 2001. *J. Biol. Chem.* 276:27010; Ahn, V. E., et al. 2003. *Proc. Natl. Acad. Sci. USA* 100:38.)

saprophytic   Lives on dead organic material. *See* biotrophic.

SAR   (1) Structure-activity relationship is a field of study in carcinogen, mutagen, and drug research. Structural modifications may or may not affect activity. It is important to know the decisive factors and how to design more effective drugs. *See* biophore; CASE; MULTICASE; TD$_{50}$. (2) Systemic-acquired resistance may be induced in plants by pathogens, salicylic acid, and etephon, a compound releasing ethylene. SAR-related genes may be members of regulons under common promoters binding specific transcription factors. Inactivation of MAP kinase 4 by the maize *Ds* transposon increased SAR and salicylic acid level and boosted the expression of pathogenesis-related proteins in *Arabidopsis*, although it reduced the size of the plants. *See* MAPK, desaturase; ethylene; host-pathogen relationship; salicylic acid. (Maleck, K., et al. 2000. *Nature Genet.* 26:403; Petersen, M., et al. 2000. *Cell* 103:1111; Maleck, K., et al. 2000. *Nature Genet.* 26:403.) (3) *See* scaffold.

SAR1   Low-activity GTPase related to Arf. It regulates the traffic between the endoplasmic reticulum and the Golgi apparatus. *See* Arf; endoplasmic reticulum; Golgi; GTPase. (Takai, Y., et al. 2001. *Physiol. Rev.* 81:153.)

SARA   Smad anchor for receptor activation. Recruits SMADs to the TGF-β receptor and thus regulates TGF signaling. *See* SMAD; TGF.

SarA   Pleiotropic staphylococcal accessory regulator of virulence. It binds to multiple AT-rich sequences. (Schumacher, M. A., et al. 2001. *Nature* 409:215.)

Saran Wrap   Thin sheet of plastic that clings well to most any surface and is suitable for covering laboratory dishes, gels, etc.

SAR by NMR   Structure-activity relationship by nuclear magnetic resonance-based methods. Essential and sophisticated method of synthetic drug design. Natural or synthetic molecules are screened and optimized analogs are synthesized to identify high-affinity ligands to develop effectively targetable drugs. *See* NMR; SAR.

sarcoglycans   (SGCA, 17q12-q21.33; SBCB, 4q12) Part of the dystrophin-glycoprotein complex in the sarcolemma. The complex protects the muscles and connects the cytoskeleton and the extracellular matrix. Sarcoglycan mutations may cause dystonia and myoclonous. *See* dystonia; dystroglycan; muscular dystrophy; myoclonous. (Zimprich, A., et al. 2001. *Nature Genet.* 29:66.)

sarcolemma   Membranes covering the striated muscle fibers. *See* caveolin; dystrophin.

sarcoma   Solid tumor tissue with tightly packed cells embedded in a fibrous or homogeneous substance; sarcomas are frequently malignant. *See* oncogenes; RAS; Rous sarcoma. (Bonnicelli, J. L. & Barr, F. G. 2002. *Curr. Opin. Oncol.* 14:412.)

sarcomere   Muscle units of thick myosin; thin actin filaments between two plate-like Z discs. These units are repetitive. *See* myosin; nebulin; titin.

sarcoplasmic reticulum   Membrane network in the cytoplasm of muscle cells containing a high concentration of calcium that is released when the muscle is excited.

sarcosinemia   Sarcosine (methylglycine, CH$_3$NHCH$_2$CO-OH) is normally converted to glycine (NH$_2$CH$_2$COOH) by the enzyme sarcosine dehydrogenase. A defect at this step increases the level of sarcosine in the blood and in the urine (hypersarcosinemia), and may result in neurological anomalies or may have almost no effect at all. Glutaric aciduria and defects in folic acid metabolism may also cause hypersarcosinemia. *See* folic acid; glycine.

sarkosyl   (N-lauroylsarcosine) Detergent for solubilizing membranes; also used for extraction of tissues.

SAS   Statistical analysis system. Computer software for the many types of data analyses.

SAS   Switch-activating site. *See Schizosaccharomyces pombe*.

SAT   *See* satellited chromosome.

SAT-DAC   Satellite-DNA-based artificial chromosome of mammals is expected to replicate independently from the genome and to express its gene content as a huge vector. It may also serve for transferring any gene into mammals (humans) without the risk of disrupting resident genes. SAT-DACs composed mainly of AT sequences may be readily isolated from the rest of the genome. Such a structure was stably inherited in mice. (Hadlaczky, G. 2001. *Curr. Opin. Mol. Ther.* 3:125.)

satellited chromosome   Carries an appendage to one arm by a constriction. The bridge between the main body of the chromosome and appendage is the site of the nucleolar organizer. It was originally named SAT as an abbreviation for *sine acido thymonucleinico*, i.e., a place where there was no detectable DNA (thymonucleic acid, as it was called in the 1930s) in that part of the chromosome. The appendages were also called trabants.

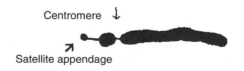

Centromere ↓

Satellite appendage

satellite DNA   DNA fraction with higher or lower density (during ultracentrifugal preparations) than bulk DNA. Generally it contains substantial repetitive DNA (in up to 5 Mb tracts). These satellite sequences generally pose problems in genome sequencing because of their instability in cloning vectors. Very little information is available regarding the function of SAT DNA. Dimeric oligopyrroles-imidazoles target adenine-thymine-rich satellite sequences and the scaffold-associated region (SAR) of *Drosophila*. Unexpectedly these polyamides induce gain- or loss-of-function phenotypes. It is assumed that these chemicals facilitate accessibility to chromatin. *See* α-satellite; heterochromatin; imidazole; polyamides; pyrrole; repetitious DNA; satellite; ultracentrifugation. (Janssen, S.,

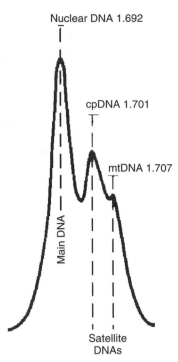

Satellite DNAs are identified by ultracentrifugation on the basis of their densities in CsCl.

et al. 2000. *Mol. Cell* 6:999; Janssen, S., et al. 2000. *Mol. Cell* 6:1013.)

satellite RNA Transcript of satellite DNA. *See* satellite DNA.

satellite virus Defective virus coexisting with another (helper) virus to correct for its insufficiency. The unrelated helper is required for infection.

Satsuma mandarin Pollen sterile citrus variety that produces seedless oranges if no foreign pollen can reach the flowers. This is also the characteristic for navel oranges, although some navel oranges have many seeds because abnormal carpel development may interfere with pollination, but they are not basically self-sterile. *See* seedless fruits.

saturated fatty acids All their chemical affinities are satisfied and they have a higher energy content than the unsaturated fatty acids that contain one or more double bonds. *See* cholesterols; fatty acids.

saturation density Of a mammalian cell culture, before contact inhibition takes place. *See* cancer; contact inhibition; malignant growth.

saturation hybridization One component in the nucleic acid annealing reaction mixture has excessive concentration to allow all possible sites of homology and hybridization to be found. *See* nucleic acid hybridization.

saturation mutagenesis Induces mutations at all available sites to reveal their relative importance. *See* linker scanning; localized mutagenesis; mutagenesis.

saturation of molecules Carbon-carbon attachments of single covalent bonds (no double bonds when entirely saturated). *See* saturated fatty acids.

***Sau* 3A** Restriction endonuclease with recognition sequence ↓GATC; *Sau* 96 I: G↓GNCC.

SAUR Small auxin-up RNA. RNAs encoded by several genes that are induced by auxins. *See* auxins. (Guilfoyle, T. J. 1995. *ASGSB Bull.* 8:39.)

SBD Single-strand DNA-binding domains of ~120 amino acids in the large subunits (and possibly in the small subunits) of the RPA protein of eukaryotes. *See* DNA replication in eukaryotes; RPA.

SBF *Saccharomyces*-binding factor is composed of Swi4 and Swi6. Along with MBF, (composed of Swi4 and Mbp1) it initiates the transcription of genes cyclin 1 (*CLN1*) and cyclin 2 (*CLN2*) required for activation of Cdc28, enabling the progress of the cell cycle from the G1 to S phase. *See* Cdc28; cell cycle; cyclin; MBF; Swi.

Sbf1 SET-binding factor 1 is a myotubularin-like protein (but without phosphatase activity) binding to SET domains. *See* myopathy; SET.

SBMA *See* Kennedy disease.

sc Indicating *Saccharomyces cerevisiae* (budding yeast) DNA, RNA, or protein as a prefix.

SC35 SR protein. *See* SR motif.

SCA *See* ataxia; ataxin; spinocerebellar ataxia.

scaffold Cytoskeleton of the cell or residual protein fibers left in the chromosome after the removal of histones. The bulk of the scaffold consists of two proteins, Sc1 (a topoisomerase II) and Sc2. The scaffold is attached to SAR (scaffold-attaching regions) of the chromatin. Cytoskeletal scaffolds facilitate the transport of various molecules along their network, securing them at the proper cellular positions. Artificial polymer scaffolds have been constructed and used as an alternative to viral vectors to deliver therapeutic proteins or gene constructs to cells with defective or deleted genes, releasing their cargo in their target environment. In animal models, implanted platelet-derived growth factor greatly enhanced vascularization. *See* cytoskeleton; genetic engineering; loop domains model; MAR; topoisomerase; WGS. (Dietzel, S. & Belmont, A. S. 2001. *Nature Cell Biol.* 3:767.)

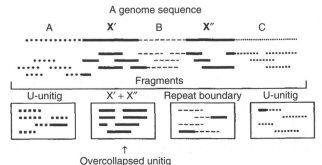

(Modified after Myers, E.W., et al. 2000. *Science* 287:2196.)

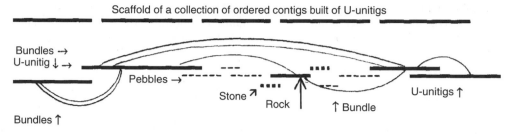

Scaffold of a collection of ordered contigs built of U-unitigs

(Modified after Myers, E.G., et al. 2000. *Science* 287:2196.)

scaffold in genome sequencing Set of ordered, oriented contigs assembled relative to each other by mate pairs in adjacent contigs see above. For assembling the over 99% accurate sequences of the *Drosophila* genome, Celera group used 8 Compaq Alpha ES40 computers with 32-gigabyte memory. In the assembly, the repetitive sequences pose real challenges because the overlaps can be *true overlaps* that belong together, or the overlaps can be parts of repeated sequences that may occur multiple times, and since they are scattered in the genomes, they do not belong together. Fragments whose arrangement is uncontested by overlaps from other fragments are *unitigs*. The unitigs that represent unique (nonrepetitive) sequences — although some extend into repeats — are called *U-unitigs*. In the diagram preceding page, X′ and X″ represent repeated sequences; in the box they are "overcollapsed" in a unitig because they consistently subassemble as the interior of the repeats. In the repeat boundary box, A and B or B and C do not overlap; they are computationally resolved and help in extending the assembly of the U-unitigs into scaffolds as shown above.

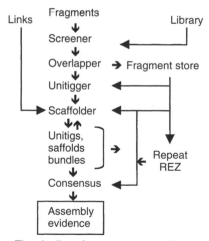

The pipeline of sequence assembly
(Modified after Myers, E.W., et al. 2000.
Science 287 2196.)

The overall strategy of the Celera operations.

If mate pairs and overlaps consistently appear in bundles, the ordering of the nonrepetitive euchromatic segments is facilitated. Usually, gaps appear between the U-unitigs and scaffolds. Smaller fragments called *rocks, stones*, and *pebbles* fill in these gaps. The shorter unitigs still have one or two mate links. The correct *tiling path* (a minimally overlapping DNA

fragment map spanning that length of the genome) is also verified by statistics. The *reads* (base sequences) are verified by multiple *base callings* (identifying the correct nucleotide in a sequence) and the sets are evaluated by Bayesian statistics.

The *Drosophila* genome was sequenced and assembled in 838 *firm scaffolds* (containing at least one U-unitig). As of 2000, the WGS still had 1,887 gaps with a total length of 2,322 Mbp varying in size up to 150 Mbp and zero. The number of U-unitigs was 7,164 (8.007 Mpb), rocks 1,787 (0.927 Mbp), stones 132 (0.118 Mbp), and pebbles 25,101. STS (sequence-tagged site) mapping and BAC/P1/cosmid clone tiling path (CTP) have validated the WGS map, yet some refinement will continue, especially in the heterochromatic regions (around the centromere, etc.). *See* BAC; Bayes' theorem; contig; cosmid; DNA sequencing; genome projects; mate pair; P1; STS; tiling; WGS. (Myers, E. W., et al. 2000. *Science* 287:2196.)

scaffold-mediated activation Involves the assembly of a molecular platform in response to death stimuli and the recruitment of procaspases. It is initiated through cell-surface receptors such as Fas receptors, TNFα receptors, and their bound ligands, leading to the formation of caspases and ending up in apoptosis. *See* apoptosis; caspase; TNF. (Earnshaw, W. C., et al. 1999. *Annu. Rev. Biochem.* 68:383.)

scaling *See* quarter-power scaling.

SCAM Substituted cysteine accessibility method is used for studying cysteine substitution and covalent modification on the structure-function relationship of proteins.

SCAMP Secretory carrier membrane proteins. Integral membrane proteins of secretory and transport vesicles, like the synaptic vessels, etc. (Wu, T. T. & Castle, J. D. 1998. *Mol. Biol. Cell* 9:1661.)

scanning, genetic Genetic scanning.

scanning electron microscopy (SEM) In contrast to transmission electron microscopy, the electron beam is reflected from the surface of the specimen coated with a heavy metal vaporized in a vacuum — a process called *shadowing*. As the electron beam scans the specimen, secondary electrons are reflected according to the varying angles of the surface of the object, generating a *three-dimensional image* corresponding to the grade of reflections. The maximal resolution is 50−100 times less than with the transmission electron microscopy, but the image can be highly magnified. It is an important technique for developmental studies. *See* electron microscopy; petals; scanning tunneling microscopy; SPM; stereomicroscopy.

scanning force spectroscopy (SFS)   Measures association and dissociation constants of biological molecules. *See* optical tweezer. (Bonin, M., et al. 2002. *Nucleic Acids Res.* 30[16]:e81.)

scanning mutagenesis   *See* homologue-scanning mutagenesis; linker scanning.

scanning of mRNA   Eukaryotic mRNA does not have a special consensus (such as the Shine-Dalgarno box in prokaryotes) for attachment to the ribosomes, so the mRNA leader adheres by its methylated guanylic cap to the ribosome, and the ribosome runs on it until it finds an initiation codon (generally AUG) to start translation. A preferred sequence, however, may be around the AUG codon AG−−−−−CC**AUG**G. The 3′ terminus of the 18S rRNA of mammals bears some similarity to the prokaryotic 16S terminus:

$$A^{Me2}A^{Me2}\ CCUGCGG\textbf{AA}GGAUGA-----UUA\text{-}3'\text{-}OH.$$

The presence of the m^7G-cap facilitates the initiation but is not absolutely indispensable. The average length of the 5′-untranslated (50−70 nucleotides) sequence and its structure may favor translation, but only large differences (very short tracts) substantially decrease initiation. A ~12-nucleotide hairpin secondary structure between the cap and AUG improves the efficiency of translation. Similarly, the polyA tail is advantageous for translation possibly by facilitating recycling of the ribosomes on the same mRNA. *Leaky scanning* indicates that either two initiator codons or only the second AUG is used for translation. In the first case, two different polypeptides may result. *See* eIF; initiation codon; Shine-Dalgarno sequence. (Kozak, M. 1989. *J. Cell. Biol.* 108:229; Samuel, C. E. 1989. *Progr. Nucleic Acid Res. Mol. Biol.* 37:127.)

scanning tunneling microscope (STM)   Can resolve biological molecules at the atomic level; its use for DNA sequencing has been proposed. With STM, vibrational spectroscopy is possible, permitting the analysis of molecules adsorbed on a surface. The vibrational energies may reveal adsorption sites, orientation, and adsorption changes. *See* electrospray MS; laser desorption; nanotechnology; SPM.

SCAP   SREBP cleavage-activating protein regulates cholesterol metabolism by promoting the cleavage of transcription factors SREBP-1 and -2 (sterol regulatory element-binding proteins). In low-sterol cells, discrete proteolysis cuts off the amino-terminal of SREBPs. As a consequence, these proteins enter the nucleus and activate the LDL receptor and cholesterol and fatty acid biosynthetic enzymes. The system can be studied by mutant CHO cells that either cannot synthesize cholesterols or LDL receptors in response to sterol depletion, or it is sterol resistant and cannot terminate the synthesis of sterols or their LDL receptor. *See* CHO; LDL; lipodystrophy; Niemann-Pick disease; sterol. (Shimano, H. 2001. *Progr. Lipid Res.* 40[6]:439.)

SCAR   (1) Sequence-characterized amplified region. Physical markers obtained by polymerase chain reaction−amplified RAPD bands. *See* amplification; PCR; RAPD. (Iturra, P., et al. *Heredity* 84:412.) (2) Suppressor of cAMP receptor. *See* cAMP.

SCARMD   *See* muscular dystrophy.

scatter factor (hepatocyte growth factor)   Its cellular responses are mediated by the Met tyrosine kinase receptor. It has multiple cell targets and is probably involved in mesenchymal-epithelial interactions, liver and kidney development, organ regeneration, metastasis, etc. *See* hepatocyte growth factor; macrophage-stimulating protein; metastasis. (Tacchini, L., et al. 2001. *Carcinogenesis* 22:1363.)

scattering   Deflection of electrons by collision(s). *See* Compton effect.

scavenger molecules   Clean up the cells from substances that are no longer needed.

SCC   Yeast integral membrane protein with partial structural homology to cyclophilins. *See* cyclophilins.

Scc1   *See* Rec8; separin.

SCD25   Suppressor of gene *cdc25* mutations in yeast; increases the dissociation of Ras•GDP but does not affect Ras•GTP. *See* cdc25; cell cycle.

SCE   *See* sister chromatid exchange.

SCF   (1) Skp1−CDC53−F-box protein is a complex of ubiquitination function. SCF protein complexes have a regulatory role in the cell cycle and development. *See* CDC53; CKS1; E1; E2; E3; F-box; glucose induction; Skp; ubiquitin; von Hippel−Lindau syndrome. (Patton, E. E., et al. 1998. *Trends Genet.* 14:237; Schwab, M. & Tyers, M. 2001. *Nature* 413:268; Zheng, N., et al. 2002. *Nature* 416:703.) (2) *See* stem cell factor.

ScFv   Single-chain fragment variable is a variable portion of the antibody molecule that may still be expressed, e.g., in plantibodies, bacteria, or other cells. *See* antibody; plantibody. (Norton, E. J., et al. 2001. *Hum. Reprod.* 16:1854.)

SCEUS   Smallest conserved evolutionary unit sequences reveal evolutionary similarities of the DNA in the genomes across taxonomic boundaries. *See* unified genetic map.

Scheie syndrome   *See* Hurler syndrome.

Schiff base α-amino groups of amino acids may react reversibly with aldehydes and form a Schiff base; these are labile intermediates in amino acid reactions. *See* Schiff reagent.

Schiff's reagent   Retains a blue color in the presence of aldehydes. Aldehydes are exposed to a fuchsin solution (0.25 g/L H_2O) and decolorized by SO_2. *See* aldehyde; fuchsin.

Schimke immuno-osseus dysplasia (SIOD) Apparently recessive disease displaying spondyloesipiphyseal dysplasia, lentigines, progressive immune and other defects. The basic anomaly is due to a mutation of a chromosomal matrix-associated protein (SMARCAL1, 2q34-q36) controlling chromatin remodeling. The gene contains 17 exons and encodes a 954-amino-acid protein. Variant forms exist. *See* chromatin remodeling; lentigines; spondyloepiphyseal dysplasia. (Boerkoel, C. F., et al. 2002. *Nature Genet.* 30:215.)

Schindler disease α-N-acetylgalactosaminidase deficiency (human chromosome 22), a lysosomal storage abnormality (?) leading to a neurological disease with onset before age 1 and progressive deterioration of motor and talking skills. *See* Kanzaki disease.

Schinzel-Geidion syndrome Autosomal-recessive malformation of the face, head, and heart, growth retardation, telangiectasia, and supernormal hair development. *See* hypertrichosis; telangiectasia.

Schinzel syndrome Autosomal-dominant ulnar-mammary syndrome shows complex symptoms including malformation of the hand, shoulder, and mammary glands, delayed puberty, obesity, etc. *See* ulna.

schistosomiasis State of disease of animals and humans seized by one or another species of the parasitic flatworms *Schistosoma* (fluke). The parasites infect the blood vessels through contact with contaminated waters in warm climates. The male carries the female in a ventral sac (gynecophoral canal). Intermediate hosts are snails and mollusks. *S. haematobium* is primarily a human parasite. *S. japonicum* infects several animals as well. The symptoms of the disease may vary according to the various species and may include systemic irritations, cough, fever, eruptions, tenderness of the liver, diarrhea, etc. The infected intestines, liver, kidneys, brain, etc., may be seriously damaged without medication. Therapeutic antimony derivatives are highly toxic to humans. The parasite is present in millions of people of the tropics including the ancient Egyptian mummies. A codominant locus in human chromosome 5q31-q33 provides some protection against *Schistosoma mansoni*. The interferon-γ receptor (IFN-γR1) locus (6q22-q23) also controls *S. mansoni* infection. *See* IFN. (<http://www.tigr.org/tdb/tgi.shtml>.)

Schistosoma haematobium male carrying the female.

schizocarp Fruit in which the carpels split apart in order to free the seeds.

schizophrenia (dementia praecox) Behavioral disorder. The afflicted individuals have difficulty distinguishing reality from dreams and imagination. Hallucinations and paranoid behavior, delusions, inappropriate emotional responses, lack of logical thought and concentration are common symptoms. The precise mechanism of inheritance is unknown, yet it occurs in about 13% of the children of afflicted parents, and 45% of the identical twins are concordant in this respect versus only 15% of dizygotic twins. The general incidence in the population varies from 1 to 4%. Both autosomal, pseudoautosomal single and multiple recessive and dominant loci have been implicated. There are some indications that schizophrenia genes are in chromosomes 1q32.2 (?), 5q33.2, 6q25.2, 6pter-p22, 7q11, 8p21-p22, 10p, 12q24, 13q34, 14q13, 9q21, 20, 22q12.1-q11.23 (proline dehydrogenase), 22q21, and 4q31 and in the long and short arms of chromosome 5. Some evidence indicates chromosomal sites 13q34 and 8p21-p22 as the most important for the determination of susceptibility. A schizophrenia locus has been assigned to human chromosome 1q21-q22 with a lod score of 6.5 (Brzustowicz, L. M., et al. 2000. *Science* 288:678). Various manifestations of the disease have strong environmental components. Maternal malnutrition during the first trimester, maternal influenza during the second trimester, perinatal complications, intrauterine fetal hypoxia, or maternal preeclampsia may substantially aggravate the risk (Tsuang, T. 2000. *Biol. Psychiatry* 47:210). The 22q11 site encodes also catechol-O-methyltransferase (COMT) and its deletion involves velocardiofacial syndrome and high incidence (20–30%) of psychiatric diseases, including schizopherenia. COMT controls the metabolism of catecholamine neurotransmitters (Shifman, S., et al. 2002. *Am. J. Hum. Genet.* 71:1296). Maternal-fetal incompatibility due to the presence of the Rh D protein in the pregnancy increases the chance for the psychiatric condition by a factor of ~2.6 (Palmer, C. G. S., et al. 2002. *Am. J. Hum. Genet.* 71:1312). Retroviral sequences have been detected in cerebrospinal fluid in ~28% of the patients with recent onset and in ~5% of patients with chronic affliction (Karlsson, H., et al. 2001. *Proc. Natl. Acad. Sci. USA* 98:4634).

MAO (monoamine oxidase) levels are reduced in afflicted individuals. MAO enzymes remove amino groups from neurotransmitters. The overproduction of dopamine (3,4-dihydroxyphenyl-ethylamine), a precursor of neurotransmitters, may also be suspected in schizophrenia. Chloropromazine (a peripheral vasodilator, antiemetic drug) and reserpine (alkaloid) may alleviate the psychological symptoms by inhibiting dopamine receptors. According to the glutamate-dysfunction hypothesis, the disease is caused by an imbalance between dopamine and glutamate or the glutamate receptor NMDA. In schizophrenia and other affective disorders, frequently an expansion of trinucleotide repeats is detectable. *See* affective disorders; catatonia, periodic; concordant; eclampsia; hypoxia; MAO; neurological disorders; NMDA; obsessive-compulsive disorder; paranoia; pseudoautosome; psychoses; Rh blood factors trinucleotide repeats; twinning. (Baron, M. 2001. *Am. J. Hum. Genet.* 68:299; Gurling, H. M. D., et al. 2001. *Am. J. Hum. Genet.* 68:661; Chumakov, I., et al. 2002. *Proc. Natl. Acad. Sci. USA* 99:13675.)

Schizosaccharomyces pombe (fission yeast) The three chromosomes are of 5.7, 4.7, and 3.5 Mb size. *S. pombe* has the lowest number of genes among eukaryotes: 4,944. (The cells are 7 × 3 μm. This eukaryote, an ascomycete, has both asexual (by fission) and sexual life cycles (each meiosis producing eight ascospores). Under good growing conditions, it reproduces by mitotic divisions. At starvation (for any factors of growth), the plus (P) and minus (M)—type cells fuse and meiosis follows. The mitotic cell cycle (2.5 h) has the typical G_1, S, G_2, and M phases. The G_2 phase takes 70% of the total time, whereas the other phases equally share the rest. Under severe nutritional limitations, instead of sexual development, the cells are blocked in either of the G phases; this dormant state is called GO.

The mating type is determined by which of the P or M alleles is switched (transposed) from their silent position to the *mat1* locus where they are expressed. Actually, both P and M genes have two alleles with a different number of amino acids in their polypeptide products: *Pc* (118), *Pi* (159) and *Mc* (181) and *Mi* (42). The c alleles (required for meiosis and conjugation) are transcribed rightward from the centromere when nitrogen is available, and the i alleles (required only for meiosis) are transcribed in the opposite direction in N starvation. The product of the *Pi* allele has a protein-binding domain, whereas that of *Mc* shows some homology to the *Drosophila Tdf* (testis-determining factor) and the mouse *Tdy*. Homothallic strains can switch between mating types, but the heterothallic ones are either P or M. The *MAT* site is comparable to the disk drive of a computer (or the slot of a tape player), where either the P or the M floppy disk (or tape cassette) is plugged in, which determines whether the mating type in the heterothallic strain will be P or M. The P (1,113 bp) and M (1,127 bp) sites are actually the storage sites for the P and M mating-type information, respectively.

The mating-type region in the right arm of chromosome II can be represented as

In about 25% of the cells, the *DSB* (double-strand break) near the *MAT* site is probably required for the chromosome to permit the insertion of a cassette. This breakage is probably transient and quickly restored so that the continuity of the chromosome is not compromised. According to data, the break at the *mat1* site is actually an artifact arising during purification of alkali-labile DNA due to a genetic imprint occurring while the lagging strand is synthesized. This imprint may reverse the *mat1* locus or introduce an origin of replication. The ≈15 kb L and K sequences are spacers where meiotic recombination is not observed. The H_1 (59 bp) and H_2 (135 bp) homology boxes flank the disk drive (*MAT*) and both floppy disks, whereas the H_3 (57 bp) occurs to the left of the P and M sequences only. It has been supposed that the reason why the P and M elements are silent at the storage sites is because of the H_3 presence there but not at the *MAT* site where they are expressed. The switching (transposition) is controlled by *SAS1* and *SAS2* (switch-activating sites) to the right of *DSB* (within 200 kb). In addition, at least 11 other trans-acting (*swi*) loci regulate switching (transposition). Mating-type determination in budding yeast (*Saccharomyces cerevisiae*) is also controlled by transposition, albeit in a different way, and the homology of the DNA sequences in the elements is low. Fission yeast has contributed to the study of many aspects of cell cycle control. Its genome sequence (Wood, V., et al. 2002. *Nature* 415:871)

has revealed 4,824 protein-coding genes, which is the smallest number in eukaryotes so far. Its promoter sequences are longer than those in budding yeast, indicating extended control of functions. About 43% of the genes contain introns. About 50 genes seem homologous to some extent to human genes that control disease, and half of these are cancer-related. *See* cell cycle; imprinting; mating type determination; *Saccharomyces cerevisiae*. (Vengrova, S., et al. 2002. *Int. J. Biochem. & Cell Biol.* 34:1031; <http:www.sanger.ac.uk/Projects/S_pombe/>; <http://pingu.salk.edu/~forsburg/pombeweb.hml>.)

Schneckenbecken chondrodysplasia Autosomal-recessive, lethal defect of the cartilage. It bears similarity to thanatophoric dysplasia.

Schwachman-Diamond syndrome Human chromosome 7q11. It involves recessive pancreatic lipomatosis and bone marrow dysfunction. (Goobie, S., et al. 2001. *Am. J. Hum. Genet.* 68:1048; Boocock, G. R. B., et al. 2003. *Nature Genet.* 33:97.)

Schwann cell Glial cell (forms myelin sheath for the peripheral nerves). Its differentiation requires the Oct-6 POU factor, Ca^{2+}/calmodulin kinase, MAPK, cAMP-response element-binding protein, and expression of *c-fos* and *Krox24* genes. ATP arrests the differentiation of the Schwann cells before myelination. *See* calmodulin; cAMP; fos; Krox-24; MAPK; myelin; Oct-1; POU.

Schwannoma *See* neurofibromatosis.

Schwartz-Jampel syndrome Chondrodystrophic myotonia (1p34-p36.1). It is a rare recessive failure to relax muscles. The syndrome is characterized by reduced height and skeletal dysplasia (abnormal development). The affection is due to an altered proteoglycan (perlecan) of the basement membranes. *See* basement membrane. (Arikawa-Hirasawa, E., et al. 2002. *Am. J. Hum. Genet.* 70:1368.)

Scianna blood group (Sc) Represented by antigenic groups Sc-1 and Sc-2 located in human chromosome 1. *See* blood groups.

Sciara Dipteran flies with polytenic, giant chromosomes in their salivary glands. The basic chromosome number in *Sciara coprophylla* is three autosomes and one X chromosome, but there are also the heteropycnotic so-called *limited chromosomes*, which are present only in the germline. They are eliminated from the nuclei during early cleavage divisions. The egg pronucleus contains three autosomes, an X chromosome, and one or more limited chromosomes. Sperm contributes three autosomes, two X chromosomes, and some limited chromosomes that are all of maternal origin. The first division of the spermatocyte is monocentric and separates the maternal chromosome set from the paternal one. Maternal chromosomes move to a single pole, whereas the paternal set is positioned away from the pole and never transmitted to the progeny. The single secondary spermatocyte displays an unusual, unequal-type division. The X chromosome divides longitudinally, both copies are included in the same cell, and only this cell survives. From the cleavage nuclei, the limited chromosomes are eliminated, then one of the X chromosomes is evicted from the cells that become males. Thus, the males become XO and the females are XX. *See* chromosomal sex determination; polytenic chromosomes; *Rhynchosciara*; salivary gland chromosomes; sex determination.

SCID   Severe combined immunodeficiency is a heterogeneous group of genetically determined diseases. It involves a defective V(J)D recombination of immunoglobulin genes. Several frameshifts, point mutations, or deletions in the IL-2Rγ (γc) chain (human chromosome Xq13) were found to be associated with SCID-X1. Actually, this part of the γ-chain is shared by the receptors of IL-4, IL-7, IL-9, and IL-15. Because of the deficiency, the T cells and the natural killer cells also become abnormal. The anomaly may involve the Jak/STAT and other signaling pathways. Jak3 defects cause SCID symptoms. Introduction into the lung of mice genetically engineered macrophages expressing IFN-γ provided marked protection against infection. IFN-γ upregulates the expression of major histocompatibility class I and class II molecules, which leads to the activation of killer lymphocytes. *See* adenosine deaminase deficiency; DNA-PK; gene therapy; IL-2; IL-4; IL-7; IL-9; IL-15; immunoglobulins; Jak/STAT; killer cell; MHC; RAG; severe combined immunodeficiency; signal transduction; T cell. (Wu, M., et al. 2001. *Proc. Natl. Acad. Sci. USA* 98:14589; French Gene Therapy Group 2003. *J. Gene. Med.* 5:82.)

science   Systematic study of natural phenomena with the explicit purpose of proving or negating a working hypothesis or hypotheses by experimental means. According to K. R. Popper (*Conjectures and Refutations*, p. 218. Basic Books, New York, 1962), for science the "criterion of potential satisfactoriness is thus testability, or improbability: only a highly testable or improbable theory is worth testing and is actually (and not merely potentially) satisfactory if it withstands severe tests — especially those tests to which we could point as crucial for the theory before they were ever undertaken. I refuse to accept the view that there are statements in science which we have, resignedly, accept as true merely because it does not seem possible … to test them." Science seeks to store, classify, and evaluate certainties. Research explores uncertainties to provide facts for science. Applied science seeks to find economic use of the principles discovered by basic or pure science. Gathering information is one of the tools of science. The information becomes science when it can be integrated into a proven theoretical framework. The framework may need modification as information accrues. *See* experiment, genetics.

scientific misconduct   According to the NSF (National Science Foundation), "fabrication, falsification, plagiarism, or other serious deviation from accepted practices." *See* ethics; misconduct, scientific; publication ethics. (*Federal Register* 18, March, 2002.)

scintigraphy   Photographic location of radionuclides within the body after introduction of radioactive tracers. *See* radioactive label; radioactive tracer.

scintillation counters   Can be liquid scintillation counters or crystal scintillation counters for solids. The counter is an electronic appliance where the sample is placed in a solution of organic compounds (cocktail). The radiation coming from the isotopes (even from the weak β-emitters) causes flashes in the fluorescing cocktail that are directed to a photoelectric cell. The cell then releases electrons that are amplified and registered (counted). Each flash corresponds to disintegration of an atom of the isotope, so the equipment displays (or prints out) the disintegrations per minute (dpm) or counts per minute (cpm), generally with background radiation subtracted. This information provides measures of the quantity of the label (or labeled compound). In the crystal scintillation counter, the radiation (usually energetic γ-rays, X-rays, or β-rays) emitted by the isotope hits a crystal of sodium iodide containing traces of thallium iodide, and again the disintegrations are registered similarly to the liquid scintillation counter. *See* dpm; isotopes; radiation hazard assessment; radioisotopic tracers; radiolabeling.

scission   Cuts in both strands of a DNA molecule at the same place. *See* nick.

scissors, molecular   *See* Cre/loxP; excision vector; targeting genes.

scissors grip   *See* Max.

Scj1   40 kDa budding yeast chaperone in the endoplasmic reticulum inducible by tunicamycin antibiotic but not by heat. *See* chaperones. (Nishikawa, S. & Endo, T. 1997. *J. Biol. Chem.* 272:12889.)

SCK   Protein with phosphotyrosine-binding domain but different from the SH2 domain of SRC. *See* PTB; SH2; SRC. (Kojima, T., et al. 2001. *Biochem. Biophys. Res. Commun.* 284:1039.)

SCLC   *See* small cell lung carcinoma.

sclerenchyma   Plant tissues with tough, hard cell walls.

scleroderma   Probably autosomal-dominant disease involving scaly hardening of the skin and increased frequency of chromosome breakage. *See* skin diseases.

scleroids   Cells with unusually hardened walls.

sclerosis   Hardening caused by inflammation or hyperplasia of the connective tissue.

sclerosteosis (SOST/BEER, 17q12-q21)   Rare recessive dysplasia of the bones displaying progressive and sometimes massive overgrowth distinct from osteopetrosis. It may distort the face and may be accompanied by syndactyly. Its highly conserved glycoprotein product, sclerostin, appears to be an antagonist of the bone morphogenetic protein and other members of the TGFβ family. It contains cystine-rich knots. *See* BMP; syndactyly; TGF. (Brunkow, M. E. 2001. *Am. J. Hum. Genet.* 68:577; Balemans, W., et al. 2001. *Hum. Mol. Genet.* 10:537.)

sclerotia   *See* ergot; hypha.

sclerotome Mesenchymal embryonic precursor of the vertebral column and ribs.

SCMRE Cis-acting element in the c-fos protooncogene. It is responsible for induction by some mitogens. *See* cis; FOS; mitogen; protooncogene.

scnDNA Single-copy nuclear DNA.

scoliosis The spine is not straight but laterally curved; it may be due to polygenic causes or it can be part of skeletal syndromes. Its incidence is about 1–6% of the adult human population. Kyphoscoliosis is a similar condition in mice. (Blanco, G., et al. 2001. *Hum. Mol. Genet.* 10:9.)

Scoliotic pig.

S-cone syndrome (ESCS, encoded by PNR/NR2E3 gene in human chromosome15q23) It is a night blindness and blue light hypersensitivity because of a defect in the retinal cones in the eye. *See* eye diseases.

SCOP Structural classification of proteins. (Murzin, A. G., et al. 1995. *J. Mol. Biol.* 247:536; Przytycka, T., et al. 1999. *Nature Struct. Biol.* 6:672; <http://scop.mrc-lmb.cam.ac.uk/scop>.)

scopolamine *See* alkaloids; *Datura*.

scotomorphogenesis Differentiation of plants in the absence of light, e.g., etiolated growth. *See* brassinosteroids.

SCP₂ Sterol carrier protein 2; probably the same as nsl-TP. *See* sterol.

SC phocomelia *See* Roberts syndrome.

scrapie *See* Creutzfeldt-Jakob disease; encephalopathies; kuru; prion.

SCRATCHY In vitro method of combinatorial protein engineering independent of sequence identity. It involves incremental truncation of protein-coding genes, then reshuffling and DNA fusion. The purpose is to improve protein function. (Lutz, S., et al. 2001. *Proc. Natl. Acad. Sci. USA* 98:11248.)

screening Selective classification of cell cultures for mutation or for special genes, conveying auxotrophy, antibiotic or other resistance, selecting antibodies by cognate antigens, plant populations for disease or chemical resistance, animal progenies for blood groups, etc. In the narrow sense, screening may not involve selective isolation, but special criteria are used to distinguish individuals in the growing population who have a special trait. *See* genetic screening; genetic testing. (Forsburg, S. L. 2001. *Nature Rev. Genet.* 2:659; Jorgensen, E. M. & Mango, S. E. 2002. *Nature Rev. Genet.* 3:356.)

screwworm *See Cochliomya hominivorax*; genetic sterilization; myasis.

Scribble *See* PDZ.

scripton (transcripton) Unit of lambda phage transcription. *See* lambda phage.

scRNA Small cytoplasmic RNA. *See* snRNA.

scRNAP Small cytoplasmic ribonucleoprotein.

scrotum Pouch containing the testes and accessory sex organs of mammals.

scutellum Single cotyledon of the grass embryo; S = scutellum, E = embryo.

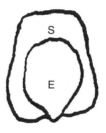

Scutellum.

S cytoplasm Present in some cytoplasmically male sterile lines. *See* cms.

SD *See Segregation distorter*; standard deviation.

Sdc25 Guanine-nucleotide release factor similar to Cdc25. *See* Cdc25; EF-Tu; RAS.

SDF-1 Stromal cell-derived factor is a chemokine and natural ligand of fusin. With its receptor CXCR-4 it mobilizes and promotes the proliferation of CD34⁺ cells. *See* CD34; chemokines; CXCR; fusin (LESTR). (Weber, K. S., et al. 2001. *Mol. Biol. Cell* 12:3074.)

S-DNA *See* slipped-structure DNA; trinucleotide repeats.

SDP *See* short-day plants.

SDR (1) Short dispersed repeats are organellar DNA sequences of 50–1,000 bp that may occur in direct or inverted forms and may represent more than 20% of the chloroplast genomes of *Chlamydomonas reinhardtii* alga and various land plants. Similar or shorter redundant sequences occur in the mitochondria of plants, animals, and fungi. In *Saccharomyces cerevisiae*, eight 200–300 bp *ori* and *rep* sequences and 200 G + C sequences of 20–50 bp may be clustered into several mtDNA gene families. Similar mobile G + C elements may occur in other fungi. SDRs have been exploited for forensic population studies. *See* chloroplasts; mobile genetic elements; mtDNA; organelle sequence transfers. (Seidl, C., et al. 1999. *Int. J. Legal Med.* 112[6]:355.) (2) *See* strain distribution pattern.

Sds Leucine-rich protein regulates that protein phosphatase 1C during mitosis of yeast. *See* protein phosphatases. (Peggie, M. W., et al. 2002. *J. Cell Sci.* 115:195.)

SDS Sodium dodecyl sulfate; detergent used for electrophoretic separation of proteins and lipids.

SDS-PAGE *See* SDS-polyacrylamide gel.

SDS-polyacrylamide gel Electrophoretic gel containing sodium dodecyl sulfate (SDS, also called sodium lauryl sulfate [SLS], detergents) and polyacrylamide. This medium dissociates proteins into subunits and reduces aggregation. Generally, the proteins are denatured with heat and a reducing agent before loading on the gel. The polypeptides become negatively charged by binding to SDS and are separated in the gel according to size (rather than by charge). On the basis of mobility, the molecular weight of the subunits can be estimated with the aid of appropriate molecular-size markers (ladder), but caution is required because glycosylated proteins may not reflect the molecular mass of the protein. The concentration of polyacrylamide determines the size of the polypeptides that can be separated. Polyacrylamides (bisacrylamide:acrylamide, 1:29) separate (kDa proteins) as follows: 15%, 12–43; 10%, 16–68; 7.5%, 36–94; 5%, 57–212. *See* electrophoresis; gel electrophoresis.

SDT test Nonparametric sign test used in pedigrees to compare the average number of candidate alleles between affected and nonaffected siblings. *See* sib TDT; TDT. (Rieger, R. H., et al. 2001. *Genet. Epidemiol.* 20[2]:175; <ftp://sph70-57.harvard.edu/XDT>.)

SE *See* standard error.

seal *Callorhinus ursinus*, $2n = 36$; *Zalophus californianus*, $2n = 36$; *Crystophora crystata*, $2n = 34$; *Erignatus barbatus* $2n = 34$; *Helichoereus grypus* $2n = 32$.

sea oncogene Avian erythroblastosis virus oncogene. The human homologue is at human chromosome 11q13 in very close vicinity to INT2 and BCL1. *See* oncogenes.

sea urchins *Strongylocentrotus purpuratus* and *Toxopneustus lividus*, both $2n = 36$, and other echinodermata have been favorable objects of cell cycle studies, fertilization, and embryogenesis. Their large-size eggs can be easily collected and handled in the laboratory. (<www.nhm.ac.uk/paleontology/echinoids/>.)

Sebastian syndrome *See* May-Hegglin anomaly.

Sec6/8 Multiprotein complex that mediates cell-to-cell contacts and transports vesicle delivery. *See* Golgi apparatus. (Matern, H. T., et al. 2001. *Proc. Natl. Acad. Sci. USA* 98:9648.)

Sec63 (NPL1/PTL1) 663-amino-acid yeast transmembrane chaperone protein with partial homology at the near N-end of DnaJ. It interacts with Kar2, Ces61, Sec71, and Sec72 proteins. *See* chaperones; DnaK; Kar2; Mtj1; Sec61 complex. (Young, B. P., et al. 2001. *EMBO J.* 20:262.)

SecA (α) protein Seven-component complex. Peripheral membrane domain of the translocase enzyme. It is the primary receptor for the SecB/preprotein complex by recognizing the leader domain of the preprotein. Hydrolyzes ATP and GTP; promotes cycles of translocations and preprotein release. *See* ARF; endoplasmic reticulum; exocytosis; membranes; Mss; protein synthesis; protein targeting; Rab; SRP; translocase; translocon; Ypt. (Hsu, S. C., et al. 1999. *Trends Cell. Biol.* 9:150.)

SecB (β) PROTEIN 17-subunit chaperone involved in translocation of preproteins by complexing, keeping them in the right conformation, and binding to the membrane surface of the endoplasmic reticulum. Recognizes both the leader and mature protein domains. *See* chaperones; endoplasmic reticulum; membranes; protein targeting; SRP; translocase; translocon. (Driessen, A. J. 2001. *Trends Microbiol.* 9:193.)

Sec61 complex Built of the three Sec subunits (α, β, γ) and other proteins; forms a protein-conducting channel across the endoplasmic reticulum (ER) membrane. It associates with the large subunit of the ribosome of prokaryotes and eukaryotes and transports some of the nascent proteins into the lumen of the ER. *See* chloroplast import; endoplasmic reticulum protein synthesis; ribosomes; Sec63; Sec proteins. (Mori, H. & Ito, K. 2001. *Proc. Natl. Acad. Sci. USA* 98:5128; Beckmann, R., et al. 2001. *Cell* 107:361.)

SECIS Selenocysteine tRNA regulatory element.

Seckel's dwarfism Bird-headed dwarfism (3q22.1-q24). Microcephalic autosomal-recessive condition with reduced intelligence. *See* dwarfism; microcephaly.

secondary constriction *See* nucleolar organizer; SAT; satellited chromosome.

secondary immune response Immune reaction conditioned by the memory cells when antigenic exposure occurs repeatedly. *See* immunological memory.

secondary lymphoid tissues Lymph nodes and spleen (in contrast to primary lymphoid tissues, bone marrow, and thymus). *See* immune system.

secondary metabolism Produces molecules that are not basic essentials for the cells, and their products occur only in specialized tissues, e.g., anthocyanin, hair pigments. Many of the secondary plant metabolites, e.g., phytoalexins, phenolics, flavanones, pterocarpan, chlorogenic acid, sesquiterpenes, diterpines, saponins, furanoacetylene, alkaloids, indole derivatives, etc., are defense molecules against microbial pathogens. (Dixon, R. A. 2001. *Nature* 411:843.)

secondary nondisjunction   *See* nondisjunction.

secondary response genes   Their transcription is preceded by protein synthesis; probably primary response genes are involved in their induction. They are stimulated by mitogens alone without cycloheximide. *See* mitogens; primary-response genes; signal transduction.

secondary sex ratio   Proportion of males to females at birth. *See* age of parents; primary sex ratio; sex ratio.

secondary sexual character   Usually accompanies the primary sexual characters but is not an integral part of the sexual mechanisms, e.g., facial hair in human males, red plumage of male cardinal birds, increased-size bosoms in females and higher pitch voice, etc. *See* accessory sexual characters; primary sexual characters.

secondary structure   Steric relations of residues that are next to each other in a linear sequence within a polymer such as α-helix, a pleated β-sheet of amino acids. *See* protein structure.

secondary trisomic (isotrisomic)   The third chromosome has two identical arms, originated by misdivision of the centromere or by the fusion of identical telochromosomes. *See* misdivision; telosome; trisomics.

second cycle mutation   Caused by the excision or movement of a transposable element, leaving behind a footprint that still causes some type of alteration in the expression of the gene. This type of alteration may be connected with the defective nuclear localization of a transcription factor. *See* nuclear localization sequences; transposable elements; transposon footprint.

second division segregation   *See* postreduction; tetrad analysis.

second-male sperm preference   *See* last-male sperm preference.

second messenger   Molecules with key roles in signal transduction pathways such as cyclic AMP, cyclic GMP, and others. Animal physiologists call hormones first messengers. In animal and plant cells, Ca^{2+} is also considered to play the role of second messenger. Inositol triphosphate is a second messenger. *See* cAMP; cGMP; G proteins; mRNA; PIK; PIP; signal transduction.

second site noncomplementation   *See* nonallelic noncomplementation.

second site reversion   It is actually a suppressor mutation at a site different from that of the original lesion, but it is capable of restoring the normal reading of mRNA. *See* compensatory mutation; reversion; suppressor gene; suppressor tRNA.

second strand synthesis   *See* reverse transcriptase.

secretagogue   Compound or factor that stimulates secretion. *See* ghrelin. (Pombo, M., et al. 2001. *Horm. Res.* 55 Suppl. 1:11.)

secretases α-secretase enzyme cleavage of APP (amyloid precursor protein) interferes with the production of α-amyloids, whereas cleavage by β- and γ-secretase contributes to the formation of amyloid plaques. γ-secretase activity requires the presence of the protein nicastrin, which also binds presenilin. There is an interaction among these and the fragment generated by β-secretase. The cytoplasmic tail of APP released intracellularly by secretase γ teams up with histone acetyltransferase and other proteins and may promote gene expression. Cleavage of APP by β-secretase at the N-end releases APPsβ ($\sim$100 kDa, N-terminal fragment) and a membrane-bound 12 kDa C-end fragment (C99). Cleavage by α-secretase generates the N-end APPsα and a membrane-bound 10 kDa piece (C83). C99 and C83 can further be split by secretases and yield 4 kDa Aβ (Alzheimer plaque material) and the harmless 3 kDa p3 peptides, respectively. γ-secretase also generates the 40-residue and in smaller proportion 42-residue Aβ fragments. The latter forms the tangled brain fibers and the 40 residues accumulate as brain plaques. β-secretase cuts at Asp1, Val3, and Glu11 and most commonly Aβ begins with Asp. Met$\rightarrow$Leu replacements favor the generation of amyloid plaques common in early-onset Alzheimer disease. BACE transmembrane aspartic protease is a β-secretase. *See* ADAM; Alzheimer disease; BACE; β-amyloid; memapsin; presenilins; TACE. (Kopan, R. & Goate, A. 2000. *Genes & Development* 14:2799; Cao, X. & Südhof, T. C. 2001. *Science* 293:115; Esler, W. P & Wolfe, M. S. 2001. *Science* 293:1449; Fortini, M. E. 2002. *Nature Rev. Mol. Cell Biol.* 3:673.)

secretion machine   Bacterial pathogens of animals and plants secrete about 20 different proteins that mediate infection of the host. Of these, 9 are conserved across phylogenetic boundaries and appear to have a universal mRNA targeting signal. *See* host-pathogen relation. (Lee, V. T. & Schneewind, O. 1999. *Immunol. Rev.* 168:241.)

secretion systems   Type III ($\sim$20 proteins) is a contact-dependent mechanism of transfer of bacterial pathogenicity island genes and toxins to other organisms. Bacterial plasmids may encode the pathogenicity island. Type IV/V: the autotransporters are the bacterial conjugation and the agrobacterial T-DNA transfer systems. Type II (12–14 proteins) is the general secretion system. The type I system requires three proteins and is encoded in pathogenicity islands or plasmids. *See* conjugation, bacterial; pathogenicity island; T-DNA. (Galán, J. E. & Collmer, A. 1999. *Science* 284:1322; Burns, D. L. 1999. *Curr. Opin. Microbiol.* 2:25; Cornelis, G. R. & Van Gijsegem, F. 2000. *Annu. Rev. Microbiol.* 54:735.)

secretion trap vector   Inserts a reporter gene, which is expressed in a transmembrane region. The constructs may carry a region of the CD4 gene fused in-frame to the 5'-end of the reporter, e.g., *LacZ*, which can then be identified by Xgal at the membrane. *See* CD4; Xgal. (Shirozu, M., et al. 1996. *Genomics* 37[3]:273.)

secretion vector   Besides an expressed structural gene, it carries a secretion signal to direct the gene product to the appropriate site. (Bolognani, F., et al. 2001. *Eur. J. Endocrinol.* 145:497.)

secretome   Type III secretion systems of bacteria serve as effectors for causing disease. *See* effector; secretion systems. (Guttman, D. S., et al. 2002. *Science* 295:1722.)

secretor   Secretes the antigens of the *ABH* blood group into the saliva. Fucosyltransferase, FUT, is the same gene locus as SEC/SE, but there is a difference in tissue-specific expression. *See* ABH antigen, fucosyltransferase. (D'Adamo, P. J. & Kelly, G. S. 2001. *Altern. Med. Rev.* 6[4]:390.)

secretory immunoglobulin A   IgA dimer with a secretory component. *See* immunoglobulins.

secretory proteins   Mainly glycoproteins that are released by the cell after synthesis, such as hormones, antibodies, and some enzymes.

secretory vesicle (secretory granule) Releases stored molecules, such as hormones, within the cell. Chromongranin A appears to control the biogenesis of the granules. *See* endocytosis; Golgi apparatus. (Kim, T., et al. 2001. *Cell* 106:499.)

sectioning   Generally required procedure for the preparation of biological specimens for histological examination. The material may need embedding before it is cut either by freehand or by microtomes. Some microtomes cut tissues frozen by CO_2. The 1–20 μm thin sections are subsequently placed on microscope slides and subjected to a series of manipulations (paraffin wax removal, dehydration, staining) before examination. *See* embedding; microtome.

sectored-spore colonies   Arise when the haploid spores carry heteroduplex DNA with different alleles in the heteroduplex region. *See* heteroduplex.

sectorial   Displays sectors, e.g., mitotic recombinant, sorting out of organelles, somatic mutation, etc. *See* chimeric; mosaic.

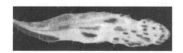

securins   Control the onset of the separation of sister chromatids during mitosis. The anaphase-promoting complex (APC) destroys securin before the sister chromatids can separate. Deletion of securin leads to chromosomal instability and is common in cancer cells. *See* APC; CDC2; separin; sister chromatid cohesion; spindle. (Sjögren, C. & Nasmyth, K. 2001. *Curr. Biol.* 11:991; Jallepalli, P. V., et al. 2001. *Cell* 105:445; Zhou, Y., et al. 2003. *J. Biol. Chem.* 278:462.)

SecY protein   Integral membrane component of bacteria involved in chaperoning the assembly of the membrane and some soluble proteins. *See* membrane. (Veenendaal, A. K., et al. 2001. *J. Biol. Chem.* 276:32559.)

SecY/E protein   Membrane-embedded domain of translocase enzyme consisting of SecY and SecE polypeptides. Stabilizes and activates SecA and facilitates membrane binding. *See* membrane; SecY.

SED   Spondyloepiphyseal dysplasia; bone diseases.

sedimentation analysis   *See* satellite DNA.

sedimentation coefficient   Rate by which a molecule sediments in a solvent. It is characterized by the Svedberg unit (S), which is a constant of 1×10^{-13}. S is derived from the equation $s = (dx/dt)/\omega^2 x$, where x is the distance from the axis of rotation in the centrifuge, ω is the angular velocity in seconds ($\omega = \theta/t$, where θ is the angle of rotation and t is time). At a constant temperature (20°C) in a solvent, s depends on the weight, shape, and hydration of a molecule. The S value is used for the characterization of macromolecules, e.g., RNA, such as 16S ($= 16 \times 10^{-13}$).

seed coat   *See* integument.

seed development   *See* embryogenesis in plants; endosperm.

seeding   Using a set of DNA fragments (in, e.g., BACs) as the beginning points for chromosome walking. *See* BAC; chromosome walking; genome projects; parking.

seed in genome sequencing   *See* genome projects.

seedless fruits   May be the result of different genetic mechanisms. Aneuploids, triploids, self-incompatibility or gametic sterility genes may be most commonly responsible for this condition. *See* bananas; naval orange; parthenocarpy; pineapple; seedless grapes; seedless watermelon; stenospermocarpy.

seedless grapes   Result of a gene that causes early embryo abortion, although fertilization occurs normally (stenospermocarpy). *See* seedless fruits.

seedless watermelons   Triploids (produced by crossing tetraploids with diploids). They are more convenient to eat. Their flavor and sweetness may make them superior to conventional varieties. *See* seedless fruits.

seed storage   Generally, viability (germination and survival) of seed can be maintained by storage below −20°C and low moisture content (<7–8%) well beyond the conditions of normal ambient temperature and humidity. *See* artificial seed; freeze drying. (Buitink, J., et al. 2000. *Proc. Natl. Acad. Sci. USA* 97:2385.)

segmental aneuploid   Contains an extra chromosomal fragment(s) in addition to the normal chromosome complement; it is a partial hyperploid. *See* aneuploid; hyperploid.

segmental interchange   *See* translocation chromosomal.

segmentation genes Control the polarity of body segments in animals. *See* metamerism; morphogenesis in *Drosophila*. (Zákány, J., et al. 2001. *Cell* 106:207; Dubrulle, J., et al. 2001. *Cell* 106:219.)

segmented genomes For example, the DNA of T5 phage is in four linkage groups; the RNA genetic material of alfalfa mosaic virus is in four segments.

segment polarity, body gene In *Drosophila*, mutations involved in the alteration of the characteristic body pattern often accompanied by inverted repetition of the remaining structures. By cell-cell communication they maintain the pattern imposed on them through subsequent developmental processes. *See* morphogenesis in *Drosophila*.

segregation Random separation of the homologous chromosomes and chromatids to the opposite pole during meiosis and carrying in them the alleles to the gametes. Independent segregation of nonsyntenic genes (or genes that are more than 50 map units apart within a chromosome) is one of the basic Mendelian rules. In the haploid products of meiosis of diploids, the 1:1 segregation of alleles may be identified. Sometimes 'x-segregation' is distinguished when each of the daughter cells carries one recombinant and one parental strand after mitotic crossing over at the four-strand stage, which is most common. At 'z-segregation' the distribution of the chromosomes into the daughter cells is biased inasmuch as one of them carries two parental strands and the other two recombinant strands. Mitotic nondisjunction has been called 'y-segregation'. *See* autopolyploid; chromosome segregation; epistasis; Mendelian laws; mitotic crossing over; nondisjunction; preferential segregation; segregation distorter; tetrad analysis.

segregation, asymmetric Differentiation and morphogenesis require that the progeny cells differ from the mother cell after cell division. This difference is brought about by the unequal distribution of cellular proteins. *See* morphogenesis in *Drosophila*.

segregation analysis Reveals the pattern of inheritance whether it is autosomal recessive, autosomal dominant, X-linked recessive or dominant, multifactorial, penetrance, or expressivity of a gene(s). This the first and sometimes laborious step, especially in human genetics, where controlled matings are not available, before gene frequencies, genetic risk, recombination, etc., can be meaningfully estimated.

segregation distorter Dominant mutation (*Sd*, map position 2–54 in *Drosophila*). When it is present, the homologous chromosomes (and the genes within) are not recovered in an equal proportion after meiosis. The second chromosomes that carry it are called SD and may be involved in chromosomal rearrangement and other (lethal) mutations. At the base of the left arm, they may have *E(SD)* (enhancer of *Sd*) or *Rsp* (Responder of *SD*, at the base of the right arm), and more distally *St(SD)* (stabilizer of *SD*) and other components of the system. *SD/+* males transmit the *SD* chromosome to about 99% of the sperm. When *Rsp* is in the homologous chromosome, *Sd* is preferentially recovered. *Sd* is actually a

tandem duplication of a 6.5 kb segment, and transformation by a 11.5 kb stretch of its DNA confers full *Sd* activity to the recipient flies. The sequence carries two nested genes, *dHS2ST*, a homologue of the mammalian heparan-sulfate 2-sulfotransferase, and *dRanGAP*, a guanine triphosphatase activator of the Ras-related Ran protein. It appears that the truncation of RanGAP is responsible for the poor transmission of the spermatids. Some of the other elements of the system act in a modifying manner and some may cause recombination in the male. In many species of insects, the infection of males by the bacterium *Wolbachia* kills the offspring of the infected males × uninfected females, but the viability of the infected eggs is normal. In mice, the transmission ratio distorter system (TRD) that impairs sperm flagellar motility and the *Tcr* (t complex responder) in cooperation with other genes located within chromosome 17 (t haplotype) affects chromosome segregation. The chromosome carrying the *Tcr* gene (even as a transgene at another location) enjoys high transmission in the presence of *Tcd*. The *Tcd* genes can act in both cis- and trans-position, whereas *Tcr* (80 kb protein kinase) acts in cis-position. *Tcr* represents a fusion between a part of the ribosome S6 protein kinase (*Rsk3*) and another gene of the microtubule affinity-regulating (MARK) Ser/Thr protein kinase family. The *Tcr* gene appears to have descended from a rearrangement between a member of the *Smok* (sperm motility kinase) family and a *Rsk* allele. Eventually, a practical method will be devised to increase one or the other sex by moving *Tcr* to the X or the Y chromosome of farm animals without relying on sperm sorting and artificial insemination. *See* certation; epistasis; gene conversion; heparan sulfate; infectious heredity; lethal factors; megaspore competition; meiotic drive; polarized segregation; pollen killer; preferential segregation; RAN; RAS; RSK; sex ratio; tetrad analysis; transmission disequilibrium. (Powers, P. A. & Ganetzky, B. 1991. *Genetics* 129:133; Schimenti, J. 2000. *Trends Genet.* 16:240; Pardo-Manuel De Villena, F. & Sapienza, C. 2001. *Mamm. Genome* 12:331.)

segregation index Gene number in quantitative traits, also called effective number of loci. *See* gene number in quantitative traits.

segregation lag The mutation or transformation is expressed only by third division of the bacteria until all chromosomes are sorted out. *See* phenotypic lag. (Angerer, W. P. 2001. *Mutation Res.* 479:207.)

segregation ratio Phenotypic (genotypic) proportions in the progeny of a heterozygous mating. *See* Mendelian segregation; modified Mendelian ratios; segregation.

segregation sterility A heterozygote produces unbalanced gametes. *See* gametophyte factor; hybrid dysgenesis; inversion; segregation distorter; translocation.

Seip disease *See* lipodystrophy, familial.

Seitelberger disease Degenerative encephalopathy (infantile neuroaxonal dystrophy). *See* encephalopathies.

seizure Sudden attack precipitated by a defect in the function of the nervous system. *Audiogenic seizures* are due to multifactorial mutations in the mouse upon exposure to loud high-frequency sound. *See* double cortex; epilepsy; periventricular heterotopia.

Sek1 (MKK) Tyrosine and threonine dual-specificity kinase involved in the activation of SAPK/JNK families of kinases. It also protects T cells from Fas and CD3-mediated apoptosis. *See* apoptosis; CD3; Fas; JNK; SAPK; T cell. (Yoshida, B. A., et al. 1999. *Cancer Res.* 59:5483.)

SelB Prokaryotic translation factor homologous to eIF-2A and eIF-2γ. *See* eIF-2A.

SELDI-TOF (surface enhanced laser desorption/ionization–time of flight spectrophotometry) Is a special form of MALDI–TOF procedure for the study of protein–protein interactions. (*See* MALDI–TOF, Bane, T. K., et al. 2000. *Nucleic Acids Res.* 30[14]:e69.)

selectable marker Permits the separation of individuals (cells) that carry it from all other individuals, e.g., in an ampicillin or hygromycin medium (of critical concentration) only those cells (individuals) can survive that carry the respective resistance genes (selectable markers). The aequorin gene equipped with an appropriate promoter may light up the tissue of expression in insects. Many different genes may serve as selectable markers. *See* aequorin.

selectins Cell-surface carbohydrate-binding, cytokine-inducible transmembrane proteins. They also bind to endothelial cells in the small blood vessels along with integrins and enable white blood cells and neutrophils to ooze out at the sites of small lacerations to combat infection. L-selectin facilitates the entry of lymphocytes into the lymph nodules by binding to CD34 cell adhesion molecules. ESL-1 selectin ligand is a receptor of fibroblast growth factor. The selectins contain an amino-terminal lectin domain, an element resembling epidermal growth factor, a variable number of complementary regulatory repeats, and a cytoplasmic carboxyl end. P (1q23-q25) and E (1q23-q25) selectins recruit T-helper-1 but not T-helper-2 cells to the site of inflammation. L-selectin may promote metastasis of cancer cells. *See* cell adhesion; cell migration; EGF; FGF; integrins; lectins; leukocyte adhesion deficiency; metastasis; TACE; T cell. (Dimitroff, C. J., et al. 2001. *J. Biol. Chem.* 276:47623.)

selection Unequal rate of reproduction of different genotypes in a population. *See* allelic fixation; fitness of hybrids; genetic drift; genetic load; mutation pressure and selection; natural selection; non-darwinian evolution; screening; selection and population size; selection coefficient; selection conditions; selection purifying; selection types. (Brookfield, J. F. Y. 2001. *Curr. Biol.* 11:388.)

selection, cyclic Different phenotypes are selected depending on seasonal variations in the environment.

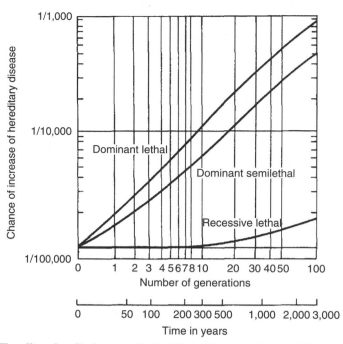

The effect of medical care on the incidence of human diseases. Thus, the improvement of medical services may lead to deterioration of the gene pool. If the initial frequency of a detrimental recessive allele is 0.001 or less, the consequences of the selection may not be evident for about 300 years. Even after thousands of years of selection, the increase of incidence is relatively modest. The frequency of dominant lethal or dominant semilethal alleles may increase much faster. (Redrawn after Bodmer, W. F. & Cavalli-Sforza, L. L. 1976. *Genetics, Evolution, and Man.* Freeman, San Francisco.)

selection, medical care The progress of effective medical care saves increasing numbers of human lives. Some of the saved individuals will have a chance to transmit deleterious genes to their offspring, so eventually some increase in detrimental alleles is expected. *See* selective abortion.

selection, natural *See* natural selection.

selection and population size In a very small population, chance (random drift) may be more important than the forces of selection in determining allelic frequencies. When the selection pressure becomes very large, even in small populations, there is a good chance for the favored allele to become fixed. *See* allelic fixation; genetic drift, in human populations; random drift. (Kreitman, M. 2000. *Annu. Rev. Genomics Hum. Genet.* 1:539.)

selection coefficient Measure of fitness of individuals of a particular genetic constitution relative to the wild type or heterozygotes in a defined environment. If the fitness is 0, the selection coefficient is 1 — that is, if an individual does not leave offspring (fitness is 0), the selection against it in a genetic sense is 1 (100%). The selection coefficient is generally denoted as (s) or (t), where the former indicates the selection coefficient of the recessive class and (t) indicates the selection coefficient of the homozygous dominants. The meaning and relation of fitness and selection coefficients can be illustrated best in table form as shown below in a population complying with the Hardy-Weinberg theorem:

Genotypes	AA	Aa	aa	Total
zygotic frequencies	p^2	$2pq$	q^2	1
fitness	w_1	w_2	w^3	
gametes produced	$(w_1) \times (p^2)$	$(w_2) \times (2pq)$	$(w_3) \times (q^2)$	1

In case we take the fitness of one genotype as unity (in this case, we choose the heterozygotes, e.g., $w_2 = 1$), the *standardized fitness* becomes $\frac{w_1}{w_2} = 1 - s$ and $s = 1 - \frac{w_1}{w_2}$. Similarly, $\frac{w_3}{w_2} = 1 - t$ and $t = 1 - \frac{w_3}{w_2}$.

See advantageous mutation; allelic frequencies; fitness; Hardy-Weinberg theorem.

selection conditions

Selection may operate at different levels beginning in meiosis (distorters, gametic factors) or at any stage during the life of the individual beginning with the zygote through the entire reproductive period. The intensity of the selection depends on the genes, the overall genetic constitution of the individual, and the environment, including the behavioral pattern of the population (e.g., protecting the young and infirm). The formulas representing the various means of selection were derived from the Hardy-Weinberg theorem $p^2 + 2pq + q^2 = 1$, and their use is exemplified.

The genotypic contributions to the *AA*, *Aa*, and *aa* phenotypes are p^2, $2pq$, and $q^2(1 - s)$, respectively, where s is the selection coefficient (*see* selection coefficient for derivation of s). The total contribution is reduced from 1 to $1 - sq^2$ because sq^2 individuals are eliminated by selection. Thus, the new frequency of the recessive alleles becomes q_1 and the

$$q_1 = \frac{q^2[1 - s] + pq}{1 - sq^2},$$

change in the frequency of q is

$$\Delta_q = \frac{q^2[1 - s] + pq}{1 - sq^2} - q,$$

which reduces to

$$\frac{sq^2(-q)}{1 - sq^2}.$$

The effectiveness of the selection depends on the frequency of the allele involved. Selection against rare recessive alleles may be very ineffective because only the homozygotes are affected. Selection most frequently works at the level of the phenotype, and the recessives may not influence the fitness of the heterozygotes. At low values of q, the majority of recessive alleles are in the heterozygotes and are thus sheltered from the forces of selection. Dominant, semidominant, and codominant alleles may be, however, very vulnerable if they lower the fitness of the individuals. As an example, let us assume that the frequency of a recessive allele is $q = 0.04$, and by using formula 4 in the table above the change in allelic frequency per generation is

$$\Delta_q = -\frac{[0.04]^2}{1 + 0.04} \cong -0.00154,$$

Formulas to calculate the change in allelic frequencies per generation at various conditions of selection. In case s or q is very small, omission of the sq product from the denominator may be of very little consequence for the outcome.

Type of Selection	Increase (+) or Decrease (−) in the Rate of Change of an Allele (Δq) per Generation
1. Selection against gametes	$-\dfrac{sq(1 - q)}{1 - sq}$
2. Differential selection in males and females (without sex linkage)	$q^2\left[1 - \frac{1}{2}(s_{\text{male}} + s_{\text{female}})\right]$
3. Selection at X-chromosome linkage	gametic in the heterogametic and zygotic in the homogametic sex
4. Selection against recessive lethals	$-\dfrac{q^2}{1 + q}$
5. Selection against the allele in absence of dominance	$-\dfrac{\frac{1}{2}sq(1 - q)}{1 - sq}$
6. Selection against dominant lethals	$+(1 - q)$
7. Partial selection against homozygous recessives in case of complete dominance	$-\dfrac{sq^2(1 - q)}{1 - sq^2}$
8. Partial selection against completely dominant alleles	$+\dfrac{sq^2(1 - q)}{1 - s(1 - q)^2}$
9. Selection against recessives in autotetraploids	$-spq^4$
10. Selection against intermediate heterozygotes	$-\dfrac{sq(1 - q)}{1 - 2sq}$
11. Selection against heterozygotes	$+2spq(q - \frac{1}{2})$
12. Selection against both homozygotes (heterozygote advantage)	$+\dfrac{pq(s_1p - s_2q)}{1 - s_1p^2 - s_2q^2}$

and after $n = 25$ generations, the initial frequency of the gene $q_0 = 0.04$ changes to

$$q_n = \frac{q_0}{1 + nq_0} = \frac{0.04}{1 + [25 \times 0.04]} = 0.02,$$

meaning that complete elimination of all homozygotes ($s = 1$) for 25 generations reduces only to half the frequency of that recessives allele. The initial zygotic frequency of $(0.04)^2 = 0.0016$ (1/625) will thus change to $(0.02)^2 = 0.0004$ (1/2,500).

If the same deleterious allele is semidominant and conveys a fitness of 0.5 relative to the homozygotes for the other allele, according to formula 5 in the table, the change of the frequency of this semidominant allele in a generation becomes

$$\Delta_q = -\frac{[1/2][0.5][0.04][0.96]}{1 - \{[0.5] \times [0.04]\}} \cong -0.0098$$

Thus, a semidominant allele with 0.5 selection coefficient will be selected against more than six times as effectively as a recessive lethal factor because $0.0098/0.00154 \cong 6.4$ in these examples.

The number of generations required to bring about a certain change in gene frequencies can be calculated in the simple case when the homozygous recessives are lethal, i.e., the selection coefficient is $s = 1$:

$$T_{\text{generations}} = \frac{q_0 - q_T}{q_0 q_T} = \frac{1}{q_T} - \frac{1}{q_0},$$

where q_0 is the initial frequency of the allele and q_T is its frequency after T generations. If it is assumed that the genotypic frequency is $(q_0)^2 = 0.0001$ and $q_0 = 0.01$, then the number of generations required to reduce the initial frequency to $q_T = 0.005$ is

$$T = \frac{1}{0.005} - \frac{1}{0.01} = 100.$$

After 100 generations, the frequency of the recessive lethal allele becomes $(q_T)^2 = (0.005)^2 = 0.000025 = 1/40,000$ compared to the initial frequency of 1/10,000. The effectiveness of selection is influenced greatly by the heritability of the allele concerned. In complex cases, more elaborate computations are required that cannot be illustrated here. *See* allelic fixation; balanced polymorphism; fitness of hybrids; gain; gametophyte; genetic load; heritability; mutation pressure opposed by selection; QTL; selection and population size; selection coefficient.

selection differential   *See* gain.

selection index in breeding   *See* gain.

selection intensity   *See* gain.

selection pressure   Intensity of selection affecting the frequency of genes in a population.

selection purifying   Acts against the heterozygotes of a new allele with lower fitness.

selection response   (heritability) × (selection differential) *See* gain; heritability.

selection types   (1) *Stabilizing* favors the intermediate forms that have the ability to survive under the most common but opposite conditions (such as cold and heat, draught and excessive precipitation). (2) *Directional selection* shifts the mean values of a population either toward higher or lower values than the current mean. (3) *Disruptive selection* breaks up the population into two or more subpopulations that each has an adaptive advantage in particular niches of a larger habitat. (4) *Frequency-dependent selection* favors an allele when it is relatively rare and may turn against it when it becomes abundant. Common examples are found in host-parasite, predator-prey relationships or in rsource utilization (Carius, H. J., et al. 2001. *Evolution* 55:1136).

When the number of predators increases beyond a point, there will not be enough prey to maintain the predators and their number will decrease. When animals overgraze in the natural habitat, ultimately the population decreases. Highly virulent viruses may outcompete the less aggressive types even when they may kill the host faster. *See* apostatic selection; artificial selection; competition; cyclic selection; fitness.

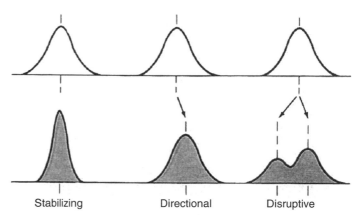

Top: Original frequency distribution of populations. Bottom: The shifts in distribution after selection. (After Mother, K. 1953. *Symp. Soc. Exp. Biol.* 7:66.)

selection value   *See* gain; selection index.

selective abortion   Termination of a pregnancy by precocious removal of the fetus from the womb if the condition of the mother or of the fetus medically justifies it and the legal system permits it. The genetic constitution or condition of the fetus may be tested with the aid of amniocentesis or sonography. From the viewpoint of genetics, selective abortion may pose biological problems. If all families would compensate for the abortions elected on the basis of genetic defects, the frequency of these defective genes might actually rise in the population because heterozygotes for genetic abnormalities would be assisted to leave offspring that could again transmit the undesirable genes to future generations, although they may not display the morbid trait. If all carriers would refrain from reproduction, the frequency of the deleterious genes might sink to the level of new mutations. Selective abortion may involve ethical, moral, and political problems, but these are beyond the scope of genetics. *See* abortion, medical; abortion, spontaneous; amniocentesis; bioethics; counseling, genetic; ethics; genetic screening; pregnancy, unwanted; selection and medical care; sterilization, humans.

selective advantage   In population genetics it is expressed by the relative fitness of bearers of (two) genotypes. Generally, the wild type has greater fitness (W_N) than the mutant type (W_M), and their relative fitness is $(W_N)/(W_M)$. W (fitness) is the reproductive success. Usually, $(W_N)/(W_M) = 1 - s$, where s is the selection coefficient indicating the disadvantage of the mutant type. In case the fitness of a genotype exceeds 1, it has an advantage in survival. *See* beneficial mutation; codon usage; fitness; selection coefficient.

selective fertilization   Because of their gene content, some sperm, may be at a disadvantage in competition with other sperm for penetrating the egg, or where multiple eggs or megaspore cells are formed their success depends on their genetic constitution. Therefore, the genetic segregation may deviate from the standard Mendelian expectation. *See* certation; gamete competition; megaspore competition; meiotic drive; sperm.

selective medium   Permits only the propagation of individuals or cells that carry a selectable marker, such as high or low temperature, antibiotic, drug resistance, etc. *See* selectable marker.

selective neutrality   Assumes that random drift is responsible for the allelic frequencies in a particular population. *See* allelic frequencies; random genetic drift.

selective peak   Determined by the genetic homeostasis of a population, i.e., the gene frequencies are maintained at this optimum as long as catastrophic changes in the environment do not occur. *See* adaptive landscape; homeostasis.

selective screening   *See* mutation detection; screening; selective medium.

selective sieve, extent of   Rate of substitution at a gene site/mutation rate for the gene.

selective sweep   Rapid establishment of advantageous alleles in a population. Such alleles may carry along other linked genes by a mechanism of hitchhiking. Less advantageous mutations may also be swept along with linked advantageous ones. The selective sweep may reduce the genetic variation in the population. *See* hitchhiking; mutation, beneficial; selection types; selective advantage. (Nurminsky, D. I., et al. 1998. *Nature* 396:572.)

selective value   *See* fitness; selection coefficient.

selectivity factor (SF)   Human general transcription factor homologous to TFIIB of other eukaryotes. *See* transcription factors.

selector genes   Supposed to specify segmental differences during morphogenesis, e.g., anterior/posterior or dorsal/ventral. After receiving morphogenetic signals, the selector genes may team up with the relevant transcription factor genes and recruit a number of specific activators to turn on the transcription of specific morphogenetic genes. The process of differentiation of an organ may take place by two major steps: The body position is specified, then the pattern of differentiation is determined. Selector genes are also used for genes that facilitate selection of transformed cells, e.g., neomycin, kanamycin, hygromycin resistance, and others. *See* differentiation; imaginal disc; morphogenesis; morphogenesis in *Drosophila*; signal transduction; figure below (Affolter, M. & Mann, R. 2001. *Science* 292:1080; Guss, K. A., et al. 2001. *Science* 292:1164.)

selenium-binding protein deficiency (SP56, 1q21-q22)   Causes neurological disorders and sterility due to the deficiency of a 56 kDa protein and other seleno proteins. Selenium may have an anticarcinogenic property. *See* selenocysteine. (Martin-Romero, F. J., et al. 2001. *J. Biol. Chem.* 276:29798.)

selenocysteine (SC, $C_6H_{12}N_2O_4Se_2$)   Reactive, very toxic, oxygen-labile amino acid. Selenophosphate is the donor of selenium for the synthesis of selenocysteyl-tRNA, which has the UCA anticodon, and SC incorporation is directed by the UGA (usually a stop) codon. In bacteria, a special mRNA fold immediately following UGA mediates the incorporation of SC carried by its own tRNA. The loop binds the SelB protein, which acts like a somewhat unusual elongation factor. In mammals, the loop (called SECIS) is locked at the end of the mRNA. Two proteins carry out the function of SelB. SBP2 binds the SECIS element, whereas eEFsec binds the selenocysteine tRNA. Thus, SC is the 21st natural amino acid. Selenocysteine-containing proteins (glutathione peroxidase, 5'-deiodinases, formate and other dehydrogenases, glycine reductase) may have substantially higher or lower catalytic activity and may have theoretical and applied interest. Selenoproteins may be scavengers of heavy metals and may favor survival of cultured neurons. *See* antideterminant; code, genetic; protein synthesis; SECIS; SelB. (Gladyshev, V. N. & Kryukov, G. V. 2001. *Biofactors* 14:87.)

selenoproteins   Appear to have roles in antioxidant defenses (protection against UV, peroxides) against tumor formation, viral resistance, thyroid and reproductive functions. The selenoprotein phospholipid, hydroperoxide glutathione peroxidase (PHHGPx), is an active, soluble enzyme during sperm maturation, but in mature spermatozoa it is enzymatically inactive and insoluble. There its role is structural, securing the stability of the midpiece where mitochondria are

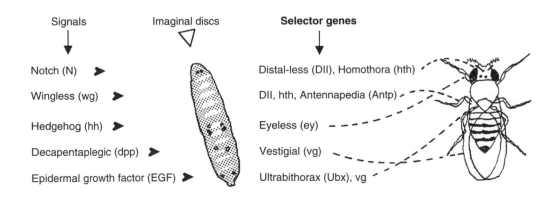

abundant. *See* glutathione peroxidase; selenocysteine; sperm; UV. (Copeland, P. R. & Driscoll, D. M. 2001. *Biofactors* 14:11.)

SELEX Systematic evolution of ligands by exponential enrichment is a method of aptamer selection from random oligonucleotide sequences (10^{15} to 10^{18}), resulting in the isolation of aptamers and ligands. The oligonucleotide library is synthesized by commercial DNA synthesizers and used in 20–30- or 200-base-long sequences ($4^{20} \sim 10^{12}$, whereas $4^{75} \sim 1.4 \times 10^{45}$ combinations). The phage R17 coat protein provides an excellent RNA recognition site. It binds to a loop (AUUA) and a four-base stem structure with a bulged A residue. The synthetic nucleotide sequence may use modified ribose residues such as 2′ F (fluoride) or 2′ NH_2 in the triphosphate nucleotides. This makes the oligonucleotides stable for several hours, whereas unmodified RNA may be degraded by nucleases immediately. By partition (gel electrophoresis, nitrocellulose filtration), the few good sequences must be separated from the large pool. The aptamers may be truncated to limit the minimal effective binding sites and may be further modified by extra nucleotide analogs for increased stability. The technology may be applied to the search for new drugs, diagnostic compounds, and ribozymes of diverse functions. *See* aptamer; gel electrophoresis; nitrocellulose filter; ribozyme. (Wilson, D. S. & Szostak, J. W. 1999. *Annu. Rev. Biochem.* 68:611; <http://wwwmgs.bionet.nsc.ru/mgs/systems/selex/>.)

self-antigen Antigens synthesized within the system that may stimulate autoimmune reaction by the T cells, although normally there is no immune response to the self-antigens. *See* autoantigen; immune tolerance; MHC; self-tolerance; T cells.

self-assembly Process of reconstituting a structure from components when there is enough information within the pieces to proceed without outside manipulation, e.g., the self-assembly of ribosomes from RNA and protein subunits.

self-compatibility Self-fertilization can take place and the offspring are normal. *See* incompatibility alleles; self-incompatibility.

self-cross Misnomer for self-fertilization. *See* cross.

self-deleting vector Lentiviral vector may integrate into the cell to deliver a gene (Cre) that is expressed in the target and subsequently deleted in a Creb-dependent manner. The undesirable effect of the vector is thus avoided. *See* Cre/LoxP; lentivirus; viral vectors. (Pfeifer, A., et al. 2001. *Proc. Natl. Acad. Sci. USA* 98:11450.)

self-destructive behavior *See* Lesh-Nyhan syndrome; Smith-Maganis syndrome.

self-fertilization Can take place between the gametes in hermaphroditic individuals; autogamy is the strictest means of inbreeding. The sperm of XO males preferentially fertilizes *Caenorhabditis elegans* hermaphrodites. *See* autogamy; *Caenorhabditis*; inbreeding. (Herlihy, C. R. & Eckert, C. G. 2002. *Nature* 416:320.)

self-immunity Mediated by protein factors that bind to, e.g., Moloney virus and thus prevent the integration into the sequences by another viral element. *See* cis-immunity; phage immunity.

self-inactivating/self-activating vector *See* E vector.

self-incompatibility Failure of fusion of male and female gametes produced by the same individual or lethality of the embryo formed by such fusion. Self-incompatibility may be controlled by either the sporophyte or the gametophyte. The incompatibility for effective pollination is determined by secreted glycoproteins (SLG, on the surface of the stigma) and membrane-bound receptor protein kinases (SRK, attached to the stigma cells). The stigma cells preferentially synthesize these proteins and only a low level of expression is found in the anthers of cruciferous plants. In several solanaceous plants (tomato, *Nicotiana alata*, *Petunia inflata*), an S (self-compatibility) RNase may attack in the style of the self-pollen. In poppy plants, a receptor on the pollen recognizes a stylar protein. The binding between this ligand and receptor leads to release of calcium ions, which in turn inhibit pollen-tube growth. In *Brassica* plants, self-incompatibility is overcome if the aquaporin gene is disrupted. This indicates that functional water channels are one of the requirements for the manifestation of self-incompatibility in *Brassica*. More recent information indicates that in the stigma cell membranes of crucifers the single *S* locus encodes an S-locus glycoprotein (SLG) and an S-receptor kinase (SRK). The latest evidence indicates that SRK and not SLG is responsible for the haplotype specificity; SLG reinforces the self-incompatibility. Transfer of the SRK and SCR genes from *Arabidopsis lyrata* to *A. thaliana* confers self-incompatibility to the latter species, which is normally autogamous (Nasrallah, M. E., et al. 2002. *Science* 297:247). The secretion of a cysteine-rich protein (SCR, encoded also at the *S* locus) by the self-pollen catalyzes the autophosphorylation of the SLG-SRK complex. The SRK protein then interacts with ARC1 (autorejection component) stigma protein and through a series of events the self-pollen is incapacitated on the stigma. Self-incompatibility is usually determined by a large array of alleles in the populations, and there are substantial differences among the species regarding the number of such alleles maintained depending on the effective population size. Heteromorphic self-incompatibility is based on mechanical barriers to self-fertilization. The stigma is located to a higher position in the flowers than the anthers, and the pollen usually does not reach the stigmatic surface. In such cases, artificial pollination is successful. Sporophytic incompatibility means that the incompatibility factors do not operate at the gametophyte (pollen-egg) level, but the pistil tissues prevent the growth of the pollen tube. *See* aquaporin; gametogenesis; gametophyte; HLA; incompatibility alleles; ligand; pistil; pollen; population effective size; RNA I; RNase; *S* alleles; unilateral incongruity. (Nasrallah, J. B. *Essays in Biochemistry*, Bowles, D. J., ed., p. 143. Portland Press, London; Hiscock, J. & Kues, U. 1999. *Int. Rev. Cytol.* 193:165; Dickinson, H. G. 2000. *Trends Genet.* 16:373; Kachoroo, A., et al. 2001. *Science* 293:1824; Takayama, S., et al. 2001. *Nature* 413:534; Tong, N. 2002. *Trends Genet.* 18:113; Kachroo, A., et al. 2002. *Plant Cell* 14:S227; Stone, J. L. 2002. *Quarterly Rev. Biol.* 77:17.)

selfing Self-fertilization; it is symbolized by $\otimes$.

selfish DNA Assumption for certain DNA sequences (introns, repetitive noncoding sequences, transposable elements) that they have no selective (adaptive, evolutionary) value for the carrier; therefore, the presence of such sequences is of no advantage to the cells concerned, and they are propagated only for selfish (parasitic) purposes. Some of the original selfish DNAs (1979–80) turned out to have some functions, e.g., as maturases, and others represent transposable elements that continuously reshape the genome and are thus significant for mutation and evolution. Alu sequences appear to be more common within the gene-rich GC regions, hinting at some regulatory functions. The minisatellite DNAs and the trinucleotide repeats are implicated in an increasing number of hereditary diseases. The majority of Y-chromosomal sequences of *Drosophila* do not seem to have any identifiable function, yet male fertility may be impaired if they are deleted. *See* Alu; copia; ignorant DNA; introns; junk DNA; plasmid addiction; REP; transposable elements; trinucleotide repeats. (van der Gaag, M., et al. 2000. *Genetics* 156:775; Hurst, G. D. D. & Werren, J. H. 2001. *Nature Rev. Genet.* 2:597.)

selfish replicons Small circular plasmids in eukaryotic nuclei (maize) without any apparent function beyond perpetuating themselves.

self-organizing map *See* cluster analysis.

self-protein *See* bystander activation; immune system; molecular mimics; self-antigen.

self-primed synthesis From single-strand DNA obtained through reverse transcription, one primer may be used at the 5′-end to produce the second strand by extension at the 3′-end in a hairpin-like structure. The synthesis by self-priming is slow.

Self-priming

self-renewal Ability of cells to produce stem cells in addition to dividing mitotically and generating progenitor cells. *See* progenitor; stem cells. (Smith, A. G. 2001. *Annu. Rev. Cell. Dev. Biol.* 17:435.)

self-replicating DNA Replication mechanism for short double-stranded molecules that do not require assistance by proteins. Similar mechanisms may have operated during prebiotic evolution. Self-replicating DNA can be reproduced in the laboratory. *See* replication; self-assembly.

self-replicating peptide Autocatalytic molecule capable of assembling amino acids into oligopeptides. The yeast transcription factor GCN4 leucine-zipper domain can promote its own synthesis of 15–17 amino acid residues. *See* GCN4; leucine zipper.

self-restriction Lymphocyte recognizes a foreign antigen bound to the self-MHC molecule. *See* lymphocyte; MHC.

self-tolerance Unresponsiveness of the immune system to self-antigens. *Central self-tolerance* may be caused by death of the lymphocytes when encountering autoantigens. Peripheral self-tolerance takes place among the mature lymphocytes in the peripheral lymphatic organs. *Clonal ignorance* fails to recognize the autoantigens because of their sequestration or the failure to stimulate the indispensable secondary signals, such as cytokines, etc. *See* at least one hypothesis; autoimmune diseases; immune system; immune tolerance; self-antigen. (Rubin, R. L. & Kretz-Rommel, A. 2001. *Crit. Rev. Immunol.* 21:29.)

selvin Rarely used unit of absorbed radiation dose. *See* Sievert.

SEM Scanning electron microscopy. *See* electron microscopy.

SEM-5 Homologue of *Grb2* in *Caenorhabditis* nematodes. *See* Grb2.

semaphorin Family of membrane-associated secreted protein factors required for axonal pathfinding in neural development. Transduction by semaphorin III is mediated by the neuropilin-1 receptor. Semaphorins also regulate the development of the right ventricle and the right atrium of the heart, as well as various cartilaginous and other tissues. Human semaphorins IV and V reside at the 3p21.3 chromosomal site. Semaphorin is deleted in small cell lung carcinoma. *See* axon; collapsin; fasciclin; netrin; neurogenesis; neuropilin; plexin; small cell lung carcinoma. (Tessier-Lavigne, M. & Goodman, C. S. 1996. *Science* 274:1123; Pasterkamp, R. J. & Verhaagen, J. 2001. *Brain Res. Brain Res. Rev.* 35:36.)

SEMD *See* PAPS.

semelparity The organism reproduces only once during its lifetime, e.g., *Palingea longicauda* (Ephemeroptera) or some marsupials, which die after one mating season.

semen Viscous fluid in the male ejaculate composed of the spermatozoa and secreted fluids of the prostate and other glands. The seminal fluid of *Drosophila* reduces the propensity of the females to mate with another male. Its higher quantity lowers the viability of the females. Thus, the semen per se may have a role in fitness. *See* testis; prostate.

semiconductors Materials of germanium, silicon, and others are characterized by increased electric conductivity as temperature increases to room temperature. These materials are called semiconductors because their conductivity is much lower than that of metals. Also, in metals the increase in temperature lowers conductivity. In the semiconductor material, the electronic motion is turned on through the crystal lattice structure. (The crystal lattice is a complex of atoms and molecules held together by electrons and atomic nuclei into an extremely large molecule-like structure.) The energy states are in so-called bands. When all the sites in an energy band are completely occupied by electrons, there is no flowing electric current because none of the electrons can accept increased energy even if exposed to an electric field of ordinary magnitude. This is a nonconductor state. When the energy gap between two bands is small, the electrons can be thermally excited into a conduction band and the electrons under the influence of an external electric source can initiate an electric current. This crystal state is an *intrinsic semiconductor*. The carriers of the current are called positive holes. Transistors (electronic amplifying devices utilizing single-crystal semiconductivity) operate by the principles of conduction electrodes and mobile positive holes. Industrially used semiconductors are *extrinsic semiconductors*, meaning that small amounts of other material is introduced into them, resulting in enhanced conductive properties. These devices are essential components of electronic laboratory equipment and communication systems, computers, and television sets.

semiconservative replication   Regular mode of DNA replication where one old strand serves as a template for the synthesis of a complementary new strand. These strands become the daughter double helix. *See* DNA replication; pulse-chase entry; replication; Watson and Crick model.

semidominant   The dominance is incomplete and therefore such genes may be useful because the heterozygotes can be phenotypically recognized. *See* codominance; incomplete dominance.

semigamy   Occurs when the egg and sperm do not fuse; rather they contribute separately to the formation of the embryo, which may become a paternal-maternal chimera. *See* androgenesis; apomixis; parthenogenesis.

semilethal   Genes reduce the viability of the individual and may cause premature death. *See* LD50; LDlo; lethal factors; lethal equivalent.

seminal fluid   *See* semen; sperm.

seminal root   Root of the embryo in plant seeds. *See* root.

semisterility   Indicates that in an individual some gametes or gametic combinations are not viable when others are normal. Semisterility is common after deletion and duplication in the offspring of inversion and translocation heterozygotes, but it may be caused by self-incompatibility, incompatible nonallelic combinations, cytoplasmic factors, fungal or viral infections, adverse environmental conditions, etc. *See* chromosomal aberrations; mtDNA.

semisynthetic compounds   Natural products that are chemically modified.

Semliki forest virus   Member of the alpha virus group. *See* alpha virus.

Sendai virus   Parainfluenza virus. In an ultraviolet light−inactivated form it has been used to promote fusion of cultured mammalian cells or uptake of liposomes by its modifying effect on the lipids of the plasma membrane. *See* alpha viruses; cell fusion; cell genetics; cell membranes; fusigenic liposome; polyethylene glycol.

senDNA   Mitochondrial DNA (ca 2.5 kb) excised from the first intron of the *cox1* gene (cytochrome oxidase) and amplified in *Podospora anserina*. This and similar structures appear to be responsible for aging in vegetative cultures. *See* aging; killer plasmids.

senescence   Process of aging of organisms. At the cellular level, it has a somewhat different meaning. Cell senescence indicates how many cell divisions are expected on the average from isolated mammalian cells. Generally, cell senescence is correlated with the age of the individual and organism from which it was explanted. Human fibroblast cells under normal conditions cease to proliferate after about ±50 divisions, although individual lineages may vary. Normal human mammary epithelial cells are different as they fail to senesce as fibroblasts do, but eventually they develop telomerase problems and chromosomal anomalies. It has been suggested that the activity of the telomerase enzyme slows down and causes this phenomenon. Tumor-suppresor protein p53, retinoblastoma protein (Rb1), and cyclin-dependent kinase (Cdk) inhibitors such as p21$^{CIP1/WAF1}$ and p16^{INK} are also involved. The disruption of p21$^{CIP1/WAF1}$ leads to an escape of senescence by human fibroblasts. Some rodent cell lines (glia oligodendrocyte precursor cells) may not senesce. Cancer cells and human fibroblast cells fused to cancer cells may divide indefinitely. When provided with an appropriate regime of phytohormones, plant cells may be maintained continuously and can even be regenerated into differentiated organisms. Senescence of plant cells is regulated by cytokinins, and the increase in cytokinin level inhibits the process of senescence. *See* aging; apoptosis; Cdk; cell cycle; embryogenesis, somatic; Ets oncogene; Hayflick's limit; hybridoma; Id proteins; killer plasmids; monoclonal antibody; p16^{INK4a}; p21; p53; plant hormones; retinoblastoma; senDNA; telomerase; telomeres; tissue culture. (Romanov, S. R., et al. 2001. *Nature* 409:633; Karlseder, J., et al. 2002. *Science* 295:2446; He, Y. & Gan, S. 2002. *Plant Cell*. 14:805.)

senescence, replicative   After a certain number of replications due to dysfunction of the telomeres, proliferation ceases. Out of control of this process may involve tumorigenesis.

sense codon   Specifies an amino acid. *See* genetic code.

sense strand   DNA strand that carries the same nucleotide sequences as mRNA, tRNA, and rRNA (of course, in the RNAs, U stands in place of T). It does not carry an absolute meaning because in some cases both strands are transcribed, although in context of a particular RNA it is correct. *See* coding strand; template strand.

sense suppression   *See* cosuppression; RNAi; quelling.

sensillum   Cuticular sensory element. *Sensilla campaniformia* are small circular structures along longitudinal veins of *Drosophila* wings. *See* Drosophila.

sensitivity   Percentage of correct identification of carcinogens on the basis of the mutagenicity or other rapid assay system. In DNA sequencing, the correctly predicted bases divided by total length of cDNA. *See* accuracy; bioassays for environmental mutagens; predictability; specificity of mutagen assays.

sensor gene Supposed to be responsible for perceiving a signal. *See* signal transduction.

sensory neural Affecting the nerve mechanism of sensing.

seta Bristles, stiff hairs of animals or plants.

sensory neuropathy 1 (HSN1) Dominant (human chromosome 9q22.1-q22.3) degenerative disorder of the sensory neurons and ulcerations and bone defects. A serine palmitoyle transferase subunit maps within the HSN1 gene and is expressed in the dorsal root ganglia. *See* ganglion; hypomyelination; neuropathy; pain in sensitivity. (Bejaoui, K., et al. 2001. *Nature Genet.* 27:261.)

sentinel phenotypes Used in human genetics to detect newly occurring mutations. These traits are supposed to be relatively easily detectable by direct appearance or clinical laboratory data can be obtained through routine examinations. Their frequencies are statistically evaluated for epidemiological information regarding possible increases in mutagenicity/carcinogenicity in an environment. *See* epidemiology; mutation in human populations. (Czeizel, A. 1989. *Mutation Res.* 212:3.)

sentrin Ubiquitin-carrier protein. *See* PIC; SUMO; ubiquitin; UBL. (Kahyo, T., et al. 2001. *Mol. Cell* 8:713.)

Sep 1 Pleiotropic eukaryotic (yeast) strand exchange protein. *See* recombination in eukaryotes; STPβ.

sepal Whorl of (usually green) leaves below the petals in the flower. *See* flower differentiation.

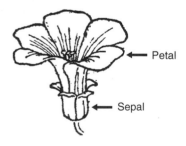

separins (Scc1, separase) Ubiquitous Esp1/Cut1-like proteins (~180–200 kDa) that are removed from their inhibitory association with securin by APC in order to separate the sister chromatids at anaphase. Separins may have endopeptidase (cysteine proteases) kinds of function. The Rec8 subunit of cohesin is cleaved by separin and consequently the meiotic chiasmata are resolved, and disjunction of the homologous chromosomes takes place during meiosis I. Separase also cleaves the kinetochore-associated protein Slk19 at the onset of anaphase. Slk stabilizes the anaphase spindle and assures the orderly exit from anaphase. *See* APC; chiasma; cohesion;

FEAR; kinetochore; mitosis; Rec8; Scc1; securin; sister chromatid cohesin; spindle. (Hauf, S., et al. 2001. *Science* 293:1320; Sullivan, M., et al. 2001. *Nature Cell Biol.* 3:771.)

Sephadex Ion exchanger or gel-filtration medium on cross-linked dextran matrix. *See* dextran; gel filtration.

Sephardic Jews who moved to Spain after the Roman occupation of Israel and then in the Middle Ages to Western European countries. *See* Ashkenazi; Jews and genetic diseases.

sepharose Anion exchanger of agarose matrix such as DEAE (diethylaminoethyl) sepharose.

septal Relating to the septum (dividing structure, a wall).

septate Separated by cross-walls (septa).

Septate (fungal mycelia).

septation *See* tubulins.

septic shock Bacterial lipopolysaccharide endotoxin-induced hypotension leading to inadequate blood supply to several organs. It is potentially fatal. It may be alleviated by neutralizing MIF. The inflammatory responses are amplified by the triggering receptors (TREM) on neutrophils and monocytes. *See* MIF. (Patel, B. M., et al. 2002. *Anesthesiology* 96:576.)

septins Rather ubiquitous (38–52 kDa GTP-binding) proteins in fungi to humans. They regulate cytokinesis and growth. A human septin gene is at 17q25.3. *See* cytokinesis. (McIlhatton, M. A., et al. 2001. *Oncogene* 20:5930.)

sequatron Automated high-performance DNA-sequencing apparatus (Hawkins, T. L., et al. 1997. *Science* 276:1887). *See* DNA chips; DNA sequencing; SAGE; sequenator.

Sequenase^{TM} Genetically engineered DNA polymerase. It combines the 85 kDa protein of phage T7 gene 5 and the 12 kDa *E. coli* thioredoxin protein (the latter keeps it associated with the template). The 3′ → 5′ exonuclease activity is suppressed. It synthesizes about 300 nucleotides per second, and it is used for DNA sequencing and oligolabeling. *See* DNA sequencing; oligolabeling probes.

sequenator Automated equipment that breaks up a protein sequentially, starting at the NH$_2$ terminus, into amino acids, identifying them by chromatography and thus determining their sequence. *See* amino acid sequencing.

sequence-based taxonomy <http://www.ncbi.nlm.nih.gov/Taxonomy/taxonomyhome.html>.

sequence skimming Long DNA fragment is probed with some known genes in order to test whether the probes have homology and thus a site within this long fragment is chosen at random. *See* probe. (Elgar, G., et al. 1999. *Genome Res.* 9:960.)

sequence space Number of possible sequences of a particular length.

sequence-tagged connector (STC) *See* genome project. (Mahairas, G. G., et al. 1999. *Proc. Natl. Acad. Sci. USA* 96:9739.)

sequence-tagged sites Single-copy DNA regions (100–500 bp) for which polymerase chain reaction (PCR) primer pairs are available and can be used for DNA mapping. *See* expressed-sequence tag; PCR; primer. (Venichanon, A., et al. 2000. *Genome* 43:47.)

sequencing *See* DNA sequencing; genome projects; protein sequencing; RNA sequencing.

SEQUEST Software package for the analysis of mass spectral data of proteins/peptides.

sequester Lay away or separate (into a compartment).

sequin Software tool for submitting nucleic acid sequence information to GenBank, EMBL, or DDBJ. It can be reached by <http://www.ncbi.nlm.nih.gov/Sequin/index.html>. *See* DBJ; EMBL; GenBank.

SER Smooth endoplasmic reticulum is an internal flat vesicle system in the cytoplasm involved in lipid synthesis. *See* RER.

SERCA Sarcoplasmic reticulum Ca^{2+} ATPase. *See* Brody disease; Darier-White disease.

SEREX Serological analysis of tumor antigens by recombinant expression cloning. Screens cancer patients' own sera for (autologous) tumor cells for antigens (cDNAs), which may be used for antibody-mediated immunotherapy. *See* cancer gene therapy; immunotherapy. (Okada, H., et al. 2001. *Cancer Res.* 61:2625.)

serine (Ser, S) Amino acid (β-oxy-α-amino-propionic acid, MW 105.09) that is soluble in water. RNA codons: UCU, UCC, UCA, UCG, AGU, AGC. Serine is derived from the glycolytic pathway:

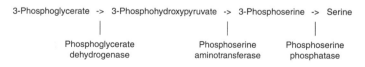

3-Phosphoglycerate -> 3-Phosphohydroxypyruvate -> 3-Phosphoserine -> Serine

Phosphoglycerate dehydrogenase   Phosphoserine aminotransferase   Phosphoserine phosphatase

Serine dehydratase enzyme, with the pyridoxalphosphate prosthetic group, degrades serine into pyruvate and NH_4^+. *See* amino acid metabolism; amino acids; oxalosis; 3-phosphoglycerate dehydrogenase.

serine kinase *See* MCF2 oncogene for serine phosphoprotein.

serine protease Degrades proteins in the extracellular matrix. *See* matrix.

serine/threonine kinase Phosphorylates serine and tyrosine residues in proteins. Their receptors are transmembrane proteins attached to the cytosolic carboxyl end of the receptor. *See* activine bone morphogenetic protein; membrane proteins; PIM oncogene; PKS oncogene; protein kinases; receptor guanylyl cyclase; signal transduction; SMAD; transforming growth factor β.

serine/threonine phosphoprotein phosphatases Remove phosphates from serine and threonine residues of proteins. *Protein phosphatase I* is inhibited by cAMP by promoting the phosphorylation of a *phosphatase inhibitor protein* through protein kinase A. *Protein phosphatase IIA* is the enzyme most widely involved in dephosphorylation of the products of serine/threonine kinases. *Protein phosphatase IIB* (calcineurin) is most common in the brain where Ca^{2+} activates it. *Protein phosphatase IIC* has only a minor role in the cells. The catalytic subunit of the first three is homologous, but they also contain special regulatory subunits. *Phospholipase C* (PLC) may be coupled to G-proteins, and upon its activation the level of Ca^{2+} increases. This cation mediates numerous cellular reactions. *See* phosphorylases; regulation of gene activity; serine/threonine kinases; signal transduction.

serine/threonine protein kinase *See* serine/threonine kinase.

seripauperines (PAU) Large group of proteins in eukaryotes. They are conspicuously low in serine and have amino-terminal signal sequences. Their function is still unknown. They are encoded at subtelomeric sites in all yeast chromosomes. *See* signal sequence. (Coissac, E., et al. 1996. *Yeast* 12:1555.)

seroconversion New antibody production against an antigen alters the serological state.

serodeme Particular type of antigen produced by a clone. *See* antigen.

serology Deals with antibody levels and with the reactions of antigens. *See* serum.

seronegative Fails to display antibodies against the antigen in question.

seropositive Reactive antibody to an antigen is present in the serum.

serotonin (5-hydroxytryptamine) Tryptophan-derived neurotransmitter modulates sensory, motor, and behavioral processes (including feeding behavior) controlled by the nervous system. Hydroxytryptamine 2B receptor regulates the cell cycle by interacting with the tyrosine kinase pathway through the phosphorylation of the retinoblastoma protein and activation of cyclin D1/Cdk4 and cyclin E/Cdk2. In concert with other proteins, cyclin D1 induces the MAPK pathway. Serotonin

Serotonin.

transporter (SERT) is a polytopic membrane transporter with 12 transmembrane domains. The 5-HT$_{3A}$ receptor is a high-conductance neuronal serotonin channel. *See* alcoholism; Cdk; cocaine; cyclins; MAPK; neurotransmitters; obesity; retinoblastoma; substance abuse.

serotype Distinguished from other cells by its special antigenic properties. *See* antigenic variation.

serovar Serotype.

serpentine receptors (hepathetical receptors) Seven-membrane spanning receptors. *See* seven-membrane proteins.

serpines Serine-protease inhibitors (14q32.1). When mutated they may be responsible for emphysema, thrombosis, and angioedema. Viral infection may counteract them. Accumulation of neuroserpin caused by mutation leads to familial encephalopathy with neuroserpin inclusion bodies (FENIB)—a dementia. There are about 500 serpins of 350–500 amino acid residues. They occur in all organisms from mammals to plants to viruses. *See* angioedema; antitrypsin; C1 inhibitor; emphysema; encephalopathy; Hsp (Hsp47); L-DNase II; maspin; thrombosis; (Atchley, W. R., et al. 2001. *Mol. Biol. Evol.* 18:1502; Silverman, G. A., et al. 2001. *J. Biol. Chem.* 276:33293; Crowther, D. C. 2002. *Hum. Mut.* 20:1; Lomas, D. A. & Correll, R. W. 2002. *Nature Rev. Genet* 3:759.)

serprocidin *See* antimicrobial peptides.

Sertoli cells *See* Wolffian ducts.

serum Clear part of the blood from which the cells and the fibrinogen are removed; the clear liquid that remains after blood clotting. The immune serum contains antibodies against specific infections. It differs from plasma, which is the nonparticulate portion of cells. *See* antibody production; plasma; serology.

serum dependence Animal cells may grow or differentiate only or preferentially in culture-containing serum.

serum response element (SRE) DNA tract that assures transcriptional activation in response to growth factors in the serum. *See* CArG box; SRE.

sesame (*Sesamum indicum*) Oil seed crop with about 37 related species; the cultivated form is $2n = 2x = 26$, but related species may have $x = 8$ and different levels of ploidy.

sesquidiploid Contains a diploid set of chromosomes derived from one parent and a higher-number set from the other. *See* allopolyploid; allopolyploid, segmental.

sessile Attached directly to a base without a stalk.

set motifs (Su[var3-9]-enhancer-of-zeste-trithorax) Originally named after the three *Drosophila* regulatory proteins where they occur and modulate chromatin structure and thus gene expression. So far at least 20 gene products display such a domain with an apparent role in epigenesis/development. The *polycomb* gene of *Drosophila*, yeast telomeric silencing, heterochromatin-mediated gene silencing, variegation position effects (PEV), as well as the Mll factors, involve SET domains. *See* chromodomain; Mll; Polycomb; Sbf1; *w* locus. (Baumbusch, L. O., et al. 2001. *Nucleic Acids Res.* 29:4319.)

set recoding *See* fuzzy inheritance.

Sevenless (*sev*) X-chromosomal gene (1–33.38) of *Drosophila* controlling the R7 rhabdomeres, thus altering photoreceptivity of the eye. The carboxy terminal of the protein product of the wild-type allele shows homology to the tyrosine kinase receptor of *c-ras*, *v-src*, and EGF. *See* BOSS; compound eye; daughter of sevenless; EGF; ommatidium; photoreceptor; RAS; rhabdomere; rhodopsin; signal transduction.

seven-membrane proteins (7tm) Integral parts of the plasma membrane that span the membrane by seven helices; they are important in signal receptor binding and in association with G-proteins. *See* G-protein; signal transduction; transmembrane receptors. (Pierce, M., et al. 2002. *Nature Rev. Mol. Cell Biol.* 3:639.)

seven-pass transmembrane proteins Seven-membrane protein.

severe combined immunodeficiency (SCID) Less frequent (0.00001–0.00005) autosomal disease than the X-linked agammaglobulinemia, but it is generally lethal before age 2. The thymus is abnormally small and therefore there is a severe deficiency of the T lymphocytes and sometimes the B lymphocytes. The afflicted infant cannot overcome infections. In some cases, viral infection may severely damage the thymus, and this nonhereditary disease may closely mimic the symptoms of SCID. The DNA-dependent kinase (p350) encoded in human chromosome 8q11 is most likely responsible for SCID-1. A gene in chromosome 10p of humans interferes with the V(D)J recombination system and thus prevents normal function of B and T lymphocytes (Moshous, D., et al. 2001. *Cell* 105:177). Several other gene loci may also be involved in the development of the disease. SCID mouse devoid of T and usually also B lymphocytes can accept human grafts and can be used to create a partial human immune system in the mouse. The most common form is X-chromosome-linked (Xq13). In about 40% of the cases, there is adenosine deaminase deficiency. T-cell deficiency may be treated by transplantation of hematopoietic cells. ADA may be corrected by gene therapy (Fischer, A., et al. 2001. *Immunity* 15:1). In some cases, transplantation of thymus tissues may result in improvement. *See* adenosine deaminase deficiency; agammaglobulinemia; DNA-PK; gene therapy; hypogammaglobulinemia; immunodeficiency; SCID.

Sewall Wright effect Genetic drift.

sex In eukaryotes, sex makes possible the production of two kinds of gametes and it is the requisite of syngamy. By recombination, sex facilitates selection of adaptive variation and provides a means for elimination of deleterious genes by promoting linkage equilibrium. In prokaryotes and viruses, sex is recombination. The majority of species reproduce sexually and the relatively few asexual species seem to represent dead ends in evolution. The Bdelloid rotifers are exceptional because they survived and evolved for 35–40 million years without sex. The *genetic sex* in the female is determined by the X chromosome(s), whereas the Y chromosome carries the male-determining genes. The *gonadal sex* is represented by the ovarian differentiation in the female and in the testicular differentiation in the male. *Somatic sex* is gonadally controlled. Female differentiation takes place — irrespective of the chromosomal constitution — if the gonads are removed during early fetal development. The anti-Müllerian hormone synthesized by Sertoli cells and fetal androgens (testosterone, androstenedione) synthesized by Leydig cells normally suppress female differentiation. The fetal female gonads have no effect on female somatic sex development. In the gonadless genital tract of both sexes, the Müllerian ducts are maintained, but the Wolffian ducts degenerate. Estrogen synthesized by the female may adversely affect the male-type differentiation. The anti-Müllerian hormone apparently blocks, however, an enzyme (aromatase) required for feminization and in this case the somatic sex shifts into the direction of masculinization. *See* copulation; gender; gonads; linkage; meiosis; Müllerian ducts; recombination; sex cell; sex determination; sex hormones; syngamy; Wolffian ducts.

sex, evolutionary significance It has been assumed that sex facilitates the purging of linked deleterious parental genes by recombination. Keightley and Eyre-Walker 2000 (*Science* 290:331) arrived at the conclusion that sex is not maintained by its ability to purge deleterious mutations. Sex may be disadvantageous for evolution. Acquisition of sex led to the evolution of diploidy, which is protective against the consequences of deleterious recessive mutations. *See* sex. (Rice, W. R. & Chippindale, A. K. 2001. *Science* 294:555; Kondrashov, F. A. & Kondrashov, A. S. 2001. *Proc. Natl. Acad. Sci. USA* 98: 12089.)

sex, phenotypic Phenotypic manifestation of the influence of the steroid sex hormones such as facial hair, increased phallic size in males, horns or special plumage in animals, and enlarged breast and mammary gland development in females. *See* animal hormones; sex determination.

sex allocation Variation in sex ratio in favor of males or females due to nonchromosomal sex-determining mechanisms such as those that exist in social insects, caused by colony size, mating behavior, and available resources. *See* sex determination; sex ratio.

sex bias in disease phenotype One sex is more likely to express the disease. *See* imprinting; Rett syndrome.

sex bias in mutation *See* mutation rate.

sex bivalent The X and Y chromosomes have homology only in the short common segment where they can pair and recombine. *See* pseudoautosomal.

sex cell Gamete that can fuse with another sex cell of an opposite mating type (sperm, egg) to form a zygote. *See* gamete; isogamy; mating type; zygote.

sex chromatin *See* Barr body.

sex chromosomal anomalies, humans Of various types; they may occur at a frequency of 0.002 to 0.003 of the births. *Females*: X0, XXX, XXXX, XXXXX, X0/XX, X0/XXX, X0/XXX/XX, XX/XXX, X0/XYY, XXX/XXXX, XXX/XXXX/XX. *Males*: XX, XYY, XXY, XXYY, XXXY, XXXYY, XXY/XY, XYY/XYYY, X0/XXY/XY, XXYYY/XY/XX, XXXY/XXXXY. Other more complicated types have been reported. The most common mechanism by which these anomalies occur is nondisjunction in meiosis and mitosis. The more complex-type mosaics (indicated by /) are the result of repeated nondisjunctional events. The X0 condition is called *Turner syndrome*, the XXX is *triplo-X*; XXY and other male conditions with multiple X and Y(s) are generally referred to as Klinefelter syndrome, along with the XX males, which have a Y-chromosome translocation to another chromosome. Similar sex-chromosomal anomalies have been identified in various other mammals. The X0 condition results in an abnormal female in humans and mice, but it is a normal male in grasshopper or *Caenorhabditis*, and it is an abnormal male in *Drosophila*. *See* chromosomal sex determination; gynandromorph; Klinefelter syndrome; testicular feminization; triplo-X; trisomy; Turner syndrome; XX males.

sex chromosome Unique in number and/or function to the sexes (such as X, Y or W, Z; *see* chromosomal sex determination). In the heterogametic sex, the X and Y chromosomes pair and may recombine in a relatively short terminal region, although in the heterogametic sex in insects recombination is practically absent, except when transposable elements function. *See* PAR.

sex circle model of recombination Basic tenets for fungi according to F. W. Stahl (1979, *Genetic Recombination*. Freeman, San Francisco, CA) are (1) any marker can recombine either by reciprocal exchange or by gene conversion; (2) close markers are more likely to recombine nonreciprocally; (3) gene conversion observes the principle of parity; (4) gene conversion is polar; (5) in half of the cases, gene conversion is accompanied by classical exchange of outside markers; (6) reciprocal recombination is always accompanied by exchange of outside markers; (7) conversion that does not involve outside exchange shows no interference of flanking genes; (8) gene conversion accompanied by outside marker exchange may involve interference; (9) conversion asci (5:3, 6:2) obey the principles listed under 1 to 8; (10) all markers (except deletions and a small fraction of conversion alleles) can segregate postmeiotically; (11) the very rare aberrant 4:4 conversion asci may be the result of two events. *See* gene conversion; recombination.

sex comb Special structures on the metatarsal region of the foreleg of *Drosophila* males. *See* Drosophila.

Sex comb.

sex controlled (sex influenced)   The degree of expression of a gene is determined by the sex (e.g., baldness is more common in human males than females). *See* Hirschsprung disease; Huntington disease; imprinting.

sex determination   Sex is usually determined in dioecious animals and plants by the presence of two X (female) and XY (male) chromosomal constitution, respectively. That is, the females are homogametic (i.e., the eggs all carry an X chromosome) and the males are heterogametic (i.e., they can produce sperm with either an X or Y chromosome). In some species—e.g., birds, moths—the females are heterogametic and the males are homogametic. In the nematode *Caenorhabditis,* some grasshoppers, and some fishes, the females are XX and the males are XO (single X). In *Drosophila,* the proportion of the X-chromosome(s) and autosomes (A sets) determines sex. Normally if the ratio is 1 X:2 sets of autosomes, the individual is male; if there are 2 Xs:2 sets of autosomes, the fly is female. All individuals with a sex ratio above 1 are also females and those with a ratio between 0.5 and 1 are intersexes. XO human and mouse individuals are females, however; irrespective of the number of X-chromosomes, as long as there is at least 1 Y chromosome, they appear male.

In hermaphroditic plants, the development of the gynoecia and androecea is determined by one or more gene loci. Actually, in *Drosophila* three major and some minor genes are known to control sex. *Sexlethal* (*Sxl,* 1–19.2) can mutate to recessive *loss-of-function* alleles that are deleterious to females but inconsequential to males. The dominant *gain-of-function* mutations do not appreciably affect the females but are deleterious to the males. The *Sxl* locus may produce 10 different transcripts. Three transcripts (4.0, 3.1 and 1.7 kb) are expressed at the blastoderm stage. Adult females have four transcripts (4.2, 3.3, 3.3, and 1.9 kb); the latter two are missing or reduced if the germline is defective. Adult males display three transcripts (4.4, 3.6, and 2.0 kb). The *Sxl* transcripts are alternatively spliced and functional in the female and are nonfunctional in the male. The *Sxl* gene product is apparently required for the maintenance of sexual determination and the processing of the downstream *tra* (*transformer,* 3–45) gene product. The Sxl protein (354 amino acids) controls alternative splicing of the *tra* premessenger RNA by binding to a polypyrimidine tract (UGUUUUUU) of a non-sex-specific 3′-splice site of one *tra* intron. This binding prevents the binding of the U2AF general splicing factor to the site and U2AF is forced to a female-specific 3′-splice site. The Sxl protein also binds to its own pre-mRNA and promotes its female-specific splicing.

The *Sxl* locus is regulated by other known genes: *da* (*daughterless,* 2–41.5) is a positive activator of *Sxl,* and it is suppressed by the gain-of-function mutations of the latter gene. The expression of *da*+ is necessary for the proper development of the gonads of the female in order to form viable eggs. In both sexes, the product of *da*+ is required for the development of the peripheral and central nervous system and the formation of the cells that determine the adult cuticle. Thus, the *da*+ gene has both maternal and embryonic influence. Females heterozygous for *da*[1] mutations produce sterile or intersex males and masculinize the exceptional daughters, which are homozygous for *male* (*maleless,* 2–55.2; [*male* is lethal to single X males but has no effect on XX females]). The DA gene product is a helix-loop-helix protein with extensive homology to the human kE2 enhancer (human chromosome 19p13.3-p13.2) of the κ-chain family of immunoglobulins. Chromosomally female (XX) flies homozygous for the third-chromosome recessive *tra* mutations become sterile males. XXY *tra*/*tra* individuals are also sterile males, but XY *tra*/*tra* males are normal males. A 0.9 kb transcript of the locus is female-specific and is required in the female, and another 1.1 kb RNA is present in both sexes, but no function is known and it is probably not essential. The splicing of the *tra* transcripts is controlled by *Sxl* gene products. When the 0.9 transcript is expressed in XY flies, the body resembles that of females.

Another *tra* locus (*transformer 2,* 2–70) regulates spermiogenesis and mating in normal males. Null mutations of *tra2,* when homozygous, transform XX females into sterile males. Actually, the *tra2* gene products seem to mediate the splicing of the *dsx* (*double sex,* 3–48.1) transcripts. Dominant mutations at the *dsx* locus, when heterozygous with the wild-type allele, change XX individuals into sterile males, but they have no effect on XY males. When homozygous, null alleles of *dsx* transform XX flies into intersexes. The recessive allele *dsx11* transforms XY flies into intersexes, and the null alleles change both XX and XY flies into intersexes. Germline sexual differentiation is not affected by the normal allele of this gene, but it is controlled by the X:autosome ratio. A 3.5 kb female-specific transcript is present in the larvae and adults. In the larvae, a 3.8 and a 2.8 kb male-specific transcript is detectable; by adult stage, a 0.7 kb RNA also appears. When homozygous, the *ix* (*intersex,* 2–60.5) mutations also change the XX flies into intersexes. Homozygous *ix* XY males appear normal morphologically, but their courtship and mating behavior is altered. Thus, sex determination in *Drosophila* appears to follow the cascade, and *fru* regulates mating behavior and sexual orientation through the *tra* and *tra2* genes:

$$fru$$
$$\rightarrow Sxl \rightarrow tra \rightarrow tra2 \rightarrow dsx \rightarrow ix$$

In summary, the X:autosome ratio is the trigger mechanism for the alternate sex developmental pathways. In the males, the *Sxl* and *tra* genes are expressed, but their transcript is not spliced to functional forms. The critical male sex-determining function is attributed to locus *dsx,* which in the wild type produces a protein that blocks the genes required for female development. In the females, with a chromosomal constitution of 2X:2A sets, a functional *Sxl* product is made that mediates the female-specific splicing of its own transcripts. The *Sxl* protein then mediates the splicing of the *tra* transcripts, leading to the synthesis of a Tra protein, which along with the Tra2 protein directs the female-specific splicing of the *dsx* transcript. The synthesized DSX protein blocks all the genes with functions that would be conducive to male development. Sex determination in *Caenorhabditis* is different from that in *Drosophila,* probably because the XX individuals are hermaphrodites and the nondisjunctional gametes lead to the development of the rare XO males. The level of expression of the known sex-determination genes is shown next page.

XX	high	low	high	low	high	low	high	**FEMALE**
X:A→	xol-1→	sdc-1→	her-1→	tra-2→	fem-1→	tra-1→		
ratio		|sdc-2		|tra-3	|fem-2			
fox-1		|sdc-3			|fem-3			
sex-1		|dpy 30						
XO	low	high	low	high	low	high	low	**MALE**

(Diagram modified after Kuwabara, P. E. & Kimble, J. 1992. *Trends Genet.* 8:164.)

Early in the pathway, *fox* (female X) acts as a numerator of the X-chromosomes, and five X-linked *dpy* (*dumpy*) alleles regulate dosage compensation. The hermaphroditic XX females of *Caenorhabditis* originally may produce sperm, then oocyte production is switched on. The *fem-3* gene turns on sperm production in the XX animals. The 3′-untranslated region of mRNA of the *fem-3* gene mediates the switch after the cytoplasmic-binding factor FBF protein binds to this region. Six *mog* genes are important regulators of female ⇌ male switching. In XX *Caenorhabditis*, the sex-determination complex protein (SDC-2) blocks the expression of the male-determining gene *her-1* and hermaphrodites are formed. SDC-2 recruits SDC-3, DUMPY (dpy), MIX-1 (mitosis and X), and other proteins to the X-chromosome, resulting in the reduced expression of X-chromosomal genes; thus, dosage compensation is realized. In XO males, *her-1* is transcribed, the SOC complex does not attach to the single X-chromosome, and all its genes are expressed normally.

Sex determination in mammals is much more complex and the pathway is not entirely clear. For a number of years, the H-Y antigen was thought to have a major role, but this did not turn out to be correct. A major critical difference was found in an 11-amino-acid segment of SMCX (structural maintenance chromosome X) and SMCY proteins encoded within the X and Y homology region. The genes DMRT1 and 2 in the short arm of human chromosome 9 have homologues in other mammals. Also in the Z chromosomes in chickens and alligators, *Drosophila*, and *Caenorhabditis,* higher expression is displayed in the male gonads than in the female ones appear to be basic regulators of sexual dimorphism. Sex chromosomal anomalies in humans generally lead to mental retardation. In several reptiles, sex is determined by the temperature to which the eggs are exposed during incubation in the sand. The actual manifestation of sex may be deeply affected by endocrine hormones directly or indirectly through environmental pollutants. In *Plasmodium* (causing malaria), induction of blood formation favors an increased production of the male parasite. *See* accessory sexual characters; amelogenin test; arrhenotoky; chromosomal sex determination; complementary sex determination; dosage compensation; F plasmid; freemartins; gynandromorphs; haploid, for sex determination in *Caenorhabditis*; hermaphrodite; Hfr; hormones in sex determination; H-Y antigen; intersex; mating type determination in yeast; mealy bug; mental retardation; *Mle*; *Msl*; numerator; *plasmodium*; *Rumex*; schisotomiasis; *Schizosaccharomyces pombe*; *Sciara*; sex, phenotypic; sex chromosomal anomalies in humans; sex determination in plants; sex hormones; sex plasmid; sex reversal; sex selection; social insects; SRY; testicular feminization; transexual; X-chromosome counting. (Meyer, B. J. 2000. *Trends Genet.* 16:247; Mittwoch, U. 2001. *J. Exp. Zool.* 290:484; Koopman, P. 2001. *Cell* 105:843; Vilain, E. 2000. *Annu. Rev. Sex Res.* 11:1; Goodwin, E. B. & Ellis, R. E. 2002. *Current Biol.* 12:R111; Hodgkin, J. 2002. *Genetics* 162:767.)

sex determination, plant In dioecious plants, sex determination is very similar to that in animals (*see* sex determination). In monoecious and hermaphroditic plants, sex is controlled without the presence of special chromosomes. A number of genes (nuclear, mitochondrial, and plastidic) involved in morphogenesis, phytohormone synthesis, and environmental responses determine the differentiation of the flowers, oogenesis (female), microsporogenesis (male), and thus sexuality. Genes are known that are similar to those of sex reversal in animals and feminize or masculinize, respectively, the monoecious or hermaphroditic flowers (e.g., *tassel seed, silkless* [in maize], *superman, gametophyte female* [in Arabidopsis], etc.). *Tasselseed 2* encodes a short-chain alcohol dehydrogenase involved in stage-specific floral organ abortion. Gibberellic acids, brassinosteroids, ethylene, chromosome-breaking agents, and mutagens may also influence the expression of sexual development as well as temperature regimes and other environmental factors. *See* flower differentiation; gametophyte; gametophyte factors; photoperiodism; phytohormones; self-incompatibility; vernalization. (Juarez, C. & Banks, J. A. 1998. *Curr. Opin. Plant Biol.* 1:68.)

sex differences Vary among phylogenetic groups in both morphology and function. In humans, the chromosomal constitution (XX, XY) is different. The X chromosomes except one, are generally inactivated in the soma. The ribosomal protein RPS4Y encoded by the Y chromosome is different from that of the X chromosome RPS4X. The relative level of hormones depends on sex; the ovaries produce more estrogen and progesterone (correlated with the incidence of breast cancer). Reduction of the natural supply of estrogen may involve reduced memory and may increase autoimmune disorders in females. Other phenotypic differences are obvious. *See* autoimmune diseases; estrogen; gender; lyonization; sex; sex determination.

sex differentiation *See* sex; sex determination.

sexduction (F-duction) Takes place when genes carried in the bacterial sex element (F′ plasmid) recombine with the bacterial chromosome. *See* conjugational mapping; F′ plasmid; Hfr; transduction. (Jacob, F., et al. 1960. *Symp. Soc. Gen. Microbiol.* 10:67; Lederberg, E. M. 1960. *Symp. Soc. Gen. Microbiol.* 10:115; diagram next page.)

sex factor Transmissible plasmid in bacteria that carries the fertility factor(s) F. *See* F′; F plasmid; Hfr.

sex hormones Have either an estrogenic (female) or androgenic (male) influence. These are steroids of the ovaries and placenta (estradiol, progesterone), or of the testes (testosterone), or of the adrenal cortex (cortisol and

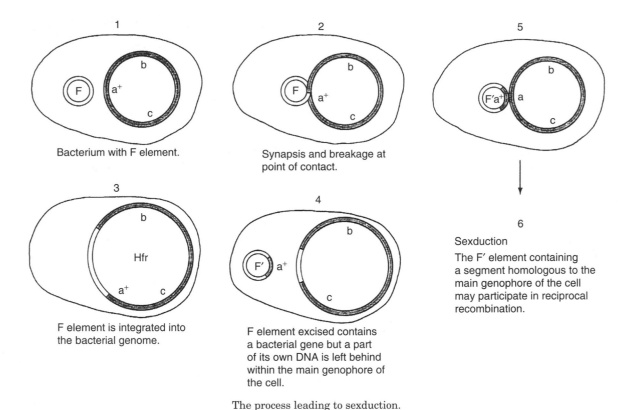

1

Bacterium with F element.

2

Synapsis and breakage at point of contact.

5

3

Hfr

F element is integrated into the bacterial genome.

4

F element excised contains a bacterial gene but a part of its own DNA is left behind within the main genophore of the cell.

6

Sexduction

The F′ element containing a segment homologous to the main genophore of the cell may participate in reciprocal recombination.

The process leading to sexduction.

aldosterol). Testosterone is required in females, although in smaller amounts. Androgens control the reproductive organs but also affect hair growth (beard) and the early death of hair follicles, causing preferential male baldness. Androgens promote bone and increased muscle growth. Some of the synthetic anabolic hormones without androgenic effects are used (illegally) by athletes to boost performance. Testosterones are precursors of estrogens. Estrogens are formed in female-specific organs and their targets include the mammary glands, bones, and fat tissues. Estrogen synthesis is regulated by the follicle-stimulating hormone (FSH) of the anterior pituitary. The pituitary luteinizing hormone mediates the release of the egg, and progesterone is required for the maintenance of pregnancy. The administration of exogenous estrogens and progestins inhibit ovulation and can be used as contraceptives. Other compounds act by prevention of the fusion of the sperm with the egg or implantation of the egg in the uterus, e.g., the drug RU486. Sperm production may be stopped by injection of progestin and androgen combinations or the inhibition of epididymal functions can prevent sperm maturation. Interfering with the release of enzymes requires breaking through the protective coat of the egg. Antiprogestins and antiestrogens and other inhibitors of steroid biosynthesis, nonpeptide anti-gonadotropin-releasing hormone antagonists may serve as female contraceptives. Steroid hormone-controlled sexual behavior is also mediated by neuronal activity. Prolonged use of steroid contraceptives or androgenic or anabolic steroids may increase the risk of liver, ovarian, and uterine carcinomas. There are several diseases or conditions (heart diseases, thromboses, embolism, diabetes, skin irritations, *Chlamydia* infection, etc.) in which contraceptive drugs are not permitted or are conditionally permitted. In some conditions (ovarian and endometrial cancer, uterine myoma, rheumatoid arthritis, etc.), oral contraceptives may be beneficial. *See* animal hormones; ART; epididymis; fertility; fertilization; hormone receptors; hyperlipoproteinemia; infertility; RU486; sex; transsexual.

sex influenced   The degree of expression of the trait is different in male and female individuals, e.g., facial hair in humans, color of plumage in birds, horns in deer, etc. *See* hare lip; Hirschsprung disease; imprinting; lupus erythematosus; pyloric stenosis. *See* imprinting; sex controlled.

sexing   Distinguishing female from male forms of animals. This procedure may be difficult in young birds because the genitalia may appear ambiguous to those who do not have special expertise (*see* autosexing). The avian males are homogametic (WW) and the females are heterogametic (ZW). Sexing also may be done by karyotyping of the tissues or Barr body detection. Molecular sexing may make the identification of possible sex on the basis of DNA markers from any tissue sample. On the Z chromosomes, there are both the CHD-W and the CHD-NW genes, whereas the W chromosome carries only the CHD-NW gene. The base sequences of these two genes are very similar, except in a short tract. When a restriction enzyme cuts within this segment of CHD-W, the females display three electrophoretic bands, but the males show only one.

A noninvasive sexing may be carried out on preimplantation embryos by inserting a green fluorescent transgene into the X chromosome. The male offspring of green fluorescent mammalian males will not display fluorescence. Such transgenic animals are apparently normal. *See* aequorin; autosexing; genetic sexing lines; sex determination.

sex linkage Various genes in the sex chromosomes are inherited with the transmission of those chromosomes. Sex linkage in females is generally partial because the two X-chromosomes may recombine. The recombination between the X- and Y-chromosomes is limited only to the homologous (pseudoautosomal) regions. In some insects (*Drosophila*, silkworm), recombination even between autosomes is usually absent in the heterogametic sex. *See* autosexing; criss-cross inheritance; crossing over; genetic equilibrium; hemophilia; pseudoautosomal; recombination frequency; recombination mechanisms. (Morgan, T. H. 1910. *Science* 32:120; Morgan, T. H. 1912. *Science* 36:719.)

sex-limited Expression of a trait is limited to one sex (e.g., lactation to females, Wildervanck syndrome.)

sex-linked lethal mutations: Served as the first laboratory test in *Drosophila* to quantitate mutation frequency and assess the mutagenic properties of physical and chemical agents. The old procedure was called the *ClB* test (*C* stands for crossover exclusion usually brought about by the presence of three inversions; *l* is a recessive lethal gene; *B* indicates the dominant *Bar eye* mutation). The principle of the techniques is diagrammed at *ClB*. If any new recessive mutation (?) takes place in the X chromosome of a male, then in the F₂ only females may occur because the original recessive *l* gene present in the inverted *Bl* chromosome will kill the hemizygous male progeny, and if a new lethal mutation occurs, it may kill (or much reduce the proportion) the other type of males in F₂. Instead of the *ClB* method, generally the improved Basc chromosome is used (*B*: *Bar*; *a*: *apricot* eye color [*w* locus]; *sc*: *scute* inversions). Any recessive lethal mutation in the X-chromosome of the grandfather's sperm results in the death of one of the grandsons. Rarely, some exceptional females are found, which are the result of unequal sister chromatid exchange in the inversion heterozygote mother. Somewhat similarly autosomal-recessive lethals can be detected in *Cy L/Pm* stocks. *See* autosomal-recessive lethal assay; *Basc*; bioassays in genetic toxicology; *Clb*.

sex mosaic The sex-chromosomal constitution in the body cells may vary in a sectorial manner. Typical examples are the gynandromorphs in insects, which have body sectors with both XX and XO constitution. Sex mosaicism occurs in humans with variable numbers of X- and Y-chromosomal sectors. The mosaicism is generally the result of nondisjunction or chromosome elimination. *See* gynandomorphs; nondisjunction; sex determination.

sex pilus *See* pilus.

sex plasmid Bacterial F plasmid. *See* F element; F plasmid.

sex proportion Proportion of male individuals in a population. *See* sex ratio.

sex ratio *Primary sex ratio* is the number of male conceptuses relative to that of females. *Secondary sex ratio* indicates the number of females:males at birth. *Tertiary sex ratio* states the ratio among adult males and females. Since XX females are mated with XY males, the proportion—just as in a test cross—should be 1:1. In the United States, at birth the proportion is about 105–106 males:100 females. In the West Indies, the proportion is about 1:1 or slightly more females. In China and Korea, the newborns are about 115 males to 100 females. Generally the female:male ratio shifts in favor of females with progressing age. By about 21–22, in the U.S., the female:male proportion becomes about 1:1, and because of the mortality differential of the sexes, by age 65 there are about 145 females for 100 males.

In dioecious plants the sex ratio may vary a great deal because of modifier genes and physiological factors (hormone supply). The problem of whether infanticide alters the sex ratio by selection has been repeatedly considered by population geneticists since the 1930s. Infanticide generally biases the childhood ratio against females. This might have the consequence that the genes of families producing males would be favored, would have greater fitness, and the secondary sex ratio would tend to be biased in favor of males. The problem is more complicated, however, in human societies because of the system of mating and the socioeconomical conditions have a substantial influence. In some *Drosophila* stocks, infection by filiform bacteria results in female offspring. The *sex-ratio* genes in the *Drosophila* X chromosome cause an excess of females in the progeny of the males carrying this gene. Usually, drive suppressors in the autosomes and in the Y chromosome balance the sex-ratio gene expression. Some of the gynandromorphs with very small XO sectors may also survive. Triploid intersexes live as well as females sex-transformed by *tra, ix, and dsx* genes (i.e., phenotypically males although XX).

The sex ratio in the Seychelles warbler's (*Acrocephalus sechellensis*) eggs may vary according to the availability of food supply; in low-food territories, they may have 77% male offspring, whereas at high food supply the proportion of sons may be only 13%. This difference is not due to different viability of the eggs. In some species of the *Cyrtodiopsis* flies in Malaysia, the sex ratio may be biased either toward the females or toward the males. In the cases of female bias, the males carry a *driver-X* chromosome that by some means eliminates from fertilization most of the Y-bearing sperms, resulting in predominantly female (XX) offspring. In the male-biased stocks, the Y chromosome carries a suppressor for the driver-X and actually somewhat increases the chances of function for the Y-bearing sperms, resulting in more than 50% male progeny. Irrespective of the actual mechanism of sex determination, natural selection tends to promote the 1:1 proportion of the two sexes, because as long as sexual reproduction is maintained, both males and females contribute to the offspring. *See* age of parents and secondary sex ratio; gynandromorph; hermaphroditism; infectious heredity; male-stuffing; meiotic drive; segregation distorter; sex determination; sex proportion; sex reversal; sex selection; spirochete; *Wolbachia*.

sex realizer Substance that determines whether male or female gonads will develop.

sex reversal Involves sex phenotypes that do not match the expectation based on chromosomal sex determination. It has been suggested that a certain number of trinucleotide repeats (glutamine) in the *Sry* gene might be responsible for sex reversal, but other studies do not confirm this mechanism in mice, and it appears that alteration in the function of autosomal regulator genes may be involved. There are apparently sex-determining autosomal (17q24.3-q25.1 and 9p24) factors that may cause sex reversal in 46XY individuals. Other autosomal and X-chromosomal genes (Xp21.3-p21.2) may also be responsible for sex reversal. In young mice with knockout for the estrogen receptors α and β ($\alpha\beta$ERKO), development of the sexual organs is near normal. By adult stage in the ovaries of the females, seminiferous tubule-like structures develop, Müllerian inhibitory substance is formed at an elevated level, and Sox9 protein has been found, indicating that estrogen receptors (ER) are essential for the maintenance of the normal ovarian phenotype. The $\alpha\beta$ERKO males displayed some spermatogenesis but also became sterile. In both males and females, αER is most essential for normal sexuality. Male-to-female sex change in mice seems to be controlled by fibroblast growth factor 9. A defect in the nuclear localization of SRY may lead to gonadal dysgenesis in humans. See adrenal hypoplasia, congenital; campomelic dysplasia; estradiol; FGF; gonadal dysgenesis; gonads; hermaphroditism; H-Y antigen; intersexes; knockout; Müllerian ducts; pseudohermaphroditism; sex determination; SF-1; SOX; SRY; Swyer syndrome; testicular feminization; *tra*; *Wingless*. (Ostrer, H. 2000. *Semin. Reprod. Med.* 18:41; Colvin, J. S., et al. 2001. *Cell* 104:875; Li, B., et al. 2001. *J. Biol. Chem.* 276:46480.)

sex selection   Possible with the use of cell sorters, followed by artificial insemination. The mammalian X-chromosome-bearing sperm has 2.8 to 7.5% more DNA than the Y-bearing ones. The sperm can be classified and selected with high efficiency and at high speed (18 million sperm/hr). It is used in animal husbandry for artificial insemination and its effectiveness is 85–95%. It would be technically feasible in human artificial insemination and ethically less objectionable than the preimplantation selection of fertilized eggs. See ART; cell sorter; segregation distorter; sex determination; sex ratio. (Garner, D. L. 2001. *J. Androl.* 22:519; Johnson, L. A. 2000. *Animal Repr. Sci.* 60–61:93; van Munster, E. B. 2002. *Cytometry* 47:192; Welch, G. R. & Johnson, L. A. 1999. *Theriogenology* 52:1343.)

sexual conflict   Enhanced reproductive success of one of the sexes reduces the fitness of the other, e.g., polyspermic fertilization. See polyspermic fertilization.

sexual differentiation   Realization of sex determination (gonads) and development of secondary sexual characters such as facial hair in men, differential plumage in birds, etc. (Wedell, A., et al. 2000. *Lakartidningen* 97:449; Burtis, K. C., 2002. *Science* 297:1135.)

sexual dimorphism   The two sexes are morphologically distinguishable. In some species, the differences appear only during later development or by the time of sexual maturity. The differences are not limited to morphology, but also various functions may differ. In the males, the language centers are localized in the left inferior frontal gyrus region of the brain; in females, both left and right regions are active. Conversion of testosterone to estradiol by neuronal tissue critically affects sexual differentiation of the brain. GABA and calcium-binding proteins are more abundant in the newborn

male rats relative to females. This difference appears to be a switch from excitatory action in the males to inhibitory signals in the females. GABA would increase phosphorylation of CREB at Ser[133] by protein kinase A, calcium-activated calmodulin kinase, ribosomal S6 kinase 2, and mitogen-activated kinase-activated protein kinase 2 in the brain of the male. This appears to be the initial signal to sexual dimorphism. Wnt-7a regulates sexual dimorphism of mice by controlling the Müllerian inhibitory substance. See autosexing; calmodulin; CREB; GABA; gonads; human intelligence; Müllerian duct; protein kinases; sex determination; *wingless*. (Auger, A. P., et al. 2001. *Proc. Natl. Acad. Sci. USA* 98:8059; Vincent, S., et al. 2001. *Cell* 106:399.)

sexual incompatibility   *See* incompatibility.

sexual isolation   Has significance in speciation by either preventing mating (isolation by life cycle, behavior or generative organs) between certain genotypes or because of gametic or zygotic death or inviability of their offspring. A very common form of sexual isolation is the inviability of the recombinants of chromosomal inversions. The cross-breeding at the incipient speciation may be prevented by pheromone genes responsible for the production of differences in the female cuticular hydrocarbons in *Drosophila*. See fertility; incompatibility; inversions; pheromone; speciation.(Majewski, J. 2001. *FEMS Lett.* 199:161; Fang, S., et al. 2002. *Genetics* 162:781.)

sexual maturity   Developmental stage at which reproductive ability is attained. It varies in different organisms (m: month, y: year): cat 6–12 m, cattle 6–12 m, chimpanzee 8–10 y, dog 9 m, elephant 8–16 y, horse 12 m, humans 12–14 y, mouse 1 m, rabbit 3–4 m, rat 2 m, sheep 6 m, swine 5–6 m. Sexual maturity is also affected by environmental factors. See gestation.

sexual orientation   *See* homosexual.

sexual reproduction   Production of offspring by mating of gametes of opposite sex or mating type.

sexual selection   Competition among mates of the same sex or gametes, or preferential choice of type of mate. The general purpose of the sexual selection is to find mates with a selective advantage for the offspring. In some instances, the actual value of the selected trait, e.g., the fancy tail of the peacock, may not be easily rationalized. The gynogen Sailfin Molly fish females reproduce clonally, yet they rely on sperm of heterospecific males to initiate embryogenesis; thus, it appears that the males do not contribute the progeny. Yet the sexual forms of the females prefer those males which mate with the gynogens. Consequently, the males exploited by the gynogens still benefit from the unusual sexual selection. Although sexual selection most frequently involves the selection of the males, in the sand lizard (*Lacerta agilis*) the females achieve selection. The females may copulate with several closely or distantly related males, but it appears that the share of offspring sired by the remotely or unrelated males is higher in the same clutch. Thus, the females apparently selected the sperm of the more distantly related males. Females of feral fowl may eject the sperm acquired through coerced mating by inferior males. Polymorphism exists in binding a protein that facilitates the attachment of the sperm to the egg, which may affect male selection by the egg. In some instances, the sexual selection is based on meiotic drive. In *Drosophila* species, mate recognition/preference is influenced by cuticular hydrocarbon composition. In guppy fish, sexual attractiveness of the males is positively correlated with ornamentation (encoded in the Y chromosome), but it is negatively correlated with survival. Asexual populations may have higher fitness than sexual ones because every individual has a chance to reproduce, especially when the deleterious mutation rate is higher in males than in females, as is most commonly the case. *See* assortative mating; certation; conflict, evolutionary; disassortative mating; gynogenesis; heterospecific; megaspore selection; meiotic drive; sexual dimorphism. (Swanson, W. J., et al. 2001. *Proc. Natl. Acad. Sci. USA* 98:2509; Snook, R. R. 2001. *Curr. Biol.* 11:R337; Knight, J. 2002. *Nature* 415:254; Pizzari, T. & Birkhead, T. R. 2002. *Biol. Rev.* 77:183.)

sexual swelling   Large, conspicuous reddish structure in between the vulva and the anus of the females of many Old World primates at the time of ovulation. It is a mating attractant for the males and it has a value in sexual selection. *See* anus; sexual selection; vulva.

sexuparous   Sexual production of offspring in species where parthenogenetic reproduction coexists with sexual reproduction. *See* parthenogenesis.

sex vesicle (XY body)   The meiotically paired mammalian sex chromosomes in the males may be heterochromatinized and form this special, visible structure. *See* heterochromatin; sex chromosomes.

SF   *See* hybrid dysgenesis I-R system; introns; splicing factor protein.

SF1   (1) Stimulatory factor. Yeast protein (33 kDa) reducing the binding requirement (at the concentration of 1 SF/20 nucleotides) of protein Sep 1 to DNA during recombination by two orders of magnitude. It probably has a role in DNA pairing. *See* synapsis. (2) Splicing transcription factor that represses transcription. (Goldstrohm, A. C., et al. 2001. *Mol. Cell Biol.* 21:7617.)

SF-1 (Ftz-F1, fushi tarazu factor homologue, nuclear receptor subfamily 5 group A member 1[NR5A1])   Steroidogenic factor (adrenal 4-binding protein [Ad4bp]) at human chromosome 9q33 (30 kb genomic DNA) regulates cytochrome 450 steroid 21-hydroxylase, aldosterone synthase, anti-Müllerian hormone (MIS), XY sex reversal, adrenal failure, and plays a key role in steroid biosynthesis. *See* fushi tarazu; Müllerian ducts; steroid hormones. (Whitworth, D. J., et al. 2001. *Gene* 277:209.)

S factor   Mitochondrial plasmid-like element in male sterile plants. *See* cms.

SF2/ASF   *See* SR motif of binding proteins; SR protein.

SFF   Cell cycle–regulating yeast protein. *See* cell cycle.

SFK   Group of tyrosine protein kinases. SFKs reside on the interior part of the cell membrane and respond to external signals. They are members of the Src family. When phosphorylated by Csk or other kinases, the proteins assume an inactive conformation and their activation requires phosphatases. Cbp attracts Csk to the membrane. The active SFK then phosphorylates other proteins. Rafts localized on the outer surface of the membrane modify SFKs. *See* CBP; Csk; RAFT; Src. (Vara, J. A., et al. 2001. *Mol. Biol. Cell* 12:2171.)

SFM   Serum-free medium.

Sfpi1   Protooncogene; probably the same transcription factor as PU.1 and Spi*1*. *See* PU.1; *Spi1*.

SGLT   Sodium/glucose cotransporter. Its defect results in glucose-galactose malabsorption (GGM), a potentially fatal neonatal (human chromosome 22) recessive disorder, unless the diet is sugar-free. (Xie, Z., et al. 2000. *J. Biol. Chem.* 275:25959.)

SH$_2$ *src* homology domain is an about 100-amino-acid-long binding site for tyrosine phosphoproteins. These phosphoproteins, such as SRC and ABL cellular oncogenes, phosphotyrosine phosphatases, GTPase-activating protein, phospholipase C, and Grb/Sem 5 adaptor protein, have an important role in signal transduction. *See* pleckstrin; PTB; SH3; signal transduction; SRC; WW.

SH$_3$ *src* homology domain is a binding site for the proline-rich motif (Arg-X-Leu-Pro-Pro-Z-Pro [the latter is Leu for the Src oncoprotein and Z is Arg for phosphoinositide kinase] or it can be X-Pro-Pro-Leu-Pro-X-Arg) in an adaptor or mediator protein in the signal transduction pathway through RAS. By binding, conformational and functional changes take place. The activity of the cellular SRC protein increases during normal and neoplastic mitoses. Protein p68 is closely related to the GAP-associated p62 and is bound to the SH3 domain of SRC. SH3 binding is specific in vivo yet of low affinity. *See* GAP; RAS; SH2; signal transduction; SRC.

The structure of the SH3 domain. (From Alm, E. & Baker, D. 1999. *Proc. Natl. Acad. Sci. USA* 96:11305.)

shadow bands   Stutter band.

shadowing   Electron microscopic preparatory procedure by which the surface of the specimen is coated with a vaporized metal such as platinum. The shadowed objects display a three-dimensional effect in scanning but even in some cases in transmission electron microscopy. *See* scanning electron microscopy.

Shah-Waardenburg syndrome (SOX10/WS4, 22q13; ED N3/ET3 20q13.2-q13.3) Involves the endothelin-3 signaling pathway; it has the combined symptoms of Hirschsprung disease and Waardenburg syndrome. The embryonic neural crest appears to have recessive (EDN3) or dominant (SOX10) defects. *See* endothelin; Hirschsprung disease; RET oncogene; SOX; Waardenburg syndrome. (Touraine, R. L., et al. 2000. *Am. J. Hum. Genet.* 66:1469.)

shaking (*shak*) Alleles at several chromosomal locations in *Drosophila* cause shaking of the legs under anesthesia to a variable degree, depending on the locus and allele involved. Some mutants may display hyperactive behavior in a temperature-dependent manner. Some may be viable; others are homozygous lethals. The *shakB* mutants may cause a defect in the synapse between the giant fiber neuron, postsynaptic interneuron, and dorsal longitudinal muscle and nerves operating the tergotrochanter (back-neck) muscle (*see* illustrations at *Drosophila*). The shaking may be caused by a defect in a protein of a potassium ion channel. *See* ion channel.

Shannon-Weaver index *See* diversity.

SHARP SMRT and histone deacetylase–associated repressor protein. Regulator of PPAR. *See* PPAR; SMRT. (Shi, Y., et al. 2002. *Proc. Natl. Acad. Sci. USA* 99:2613.)

shasta daisy (*Chrysanthemum maximum*) Ornamental plant; $2n \approx 90$.

SHC Adaptor protein involved in RAS-dependent MAP kinase activation after stimulation by insulin, epidermal growth factor (EGF), nerve growth factor (NGF), platelet-derived growth factor (PDGF), interleukins (IL-2,-3,-5), erythropoietin, granulocyte/macrophage colony-stimulating growth factor (CSF), and lymphocyte antigen receptors. It binds to tyrosine-phosphorylated receptors. When phosphorylated at tyrosine, it interacts with the SH2 domain of Grb2, which interacts with SOS in the RAS signal transduction pathway. The phosphotyrosine-binding (PTB) domain can also recognize tyrosine-phosphorylated protein, and the latter is similar to the pleckstrin homology domain, most likely binding acidic phospholipids of the cell membrane. SHC is also an oncogene involved in the development of pheochromocytoma neoplasias. *See* adenomatosis, endocrine multiple; CSF; EGF; insulin; interleukins; neoplasia; NGF; PDGF; pheochromocytoma; pleckstrin domain; signal transduction; SOS. (Ravichandran, K. S. 2001. *Oncogene* 20:6322.)

shearing Cutting DNA into fragments by mechanical means, e.g., by rapid stirring or brusque pipetting.

sheats (1) Any tube-like structure surrounding another. (2) Part of a leaf that wraps the stem.

sheep, domesticated (*Ovis aries*) $2n = 54$; some wild sheep have a higher number of chromosomes. A medium-density linkage map of 1,062 unique loci became available in 2001 that helps in mapping the cattle and goat genomes. (Maddox, J. F., et al. 2001. *Genome Res.* 11:1275; Cockett, N. E., et al. 2001. *Physiol. Genomics* 7:69.)

sheep hybrids Domesticated sheep (*Ovis aries*, $2n = 54$) form fertile hybrids with muflons, but the goat × sheep hybrid embryos rarely develop normally. *See* transplantation, nuclear. (Ruffing, N. A. 1993. *Biol. Reprod.* 49:1260.)

Sherman paradox Various recurrence risks among relatives caused by expansion of fragile X sites transmitted by nonsymptomatic males. *See* fragile X; mental retardation; recurrence risk; trinucleotide repeats. (Sherman, S. L., et al. 1985. *Hum. Genet.* 69:289; Fu, Y. H., et al. 1991. *Cell* 67:1047.)

shift Internal chromosomal segment generated by two break points translocated within the same or into another chromosome within a gap opened by a single break. It is a rare phenomenon. Shifting of the relative proportion of mitochondrial recombination products commonly occurs in plants. *See* reciprocal interchange; translocation; transposition.

shifting balance theory of evolution Polymorphism in a population is determined by a dynamic interplay of the forces of pleiotropy, epistasis, genotypic values, fitness, and population structure. Alternative ideas would be the discredited neo-Lamarckian internal drive or the well-documented neutral mutation concepts. *See* fitness; genotype; neo-lamarckism; neutral mutation; pleiotropy; polymorphism; population structure. (Wade, M. J. & Goodnight, C. J. 1991. *Science* 253:1015.)

Shigella Group of gram-negative enterobacteria causing dysentery (intestinal inflammation and diarrhea) in humans and higher monkeys. *See* primates.

shikimic acid Intermediate in aromatic amino acid biosynthesis.

Shikimic acid.

Shine-Dalgarno sequence Nucleotide consensus (AGGAGG) in the nontranslated 5′-region of the prokaryotic mRNA (close to the translation initiation codon) complementary to the binding sites of the ribosomes. The terminus of the 16S ribosomal RNA is generally

$A^{Me2}A^{Me2}$CCUGCGG**UUGGAUGA**<u>**CCUCC**</u>UUA-3′-0H.

The eukaryotic mRNAs do not have this sequence and the mRNA attaches to the ribosomes by means of ribosomal scanning. In the polycistronic messages, each cistron generally has a Shine-Dalgarno sequence. Mitochondrial and ribosome mRNAs generally but not always have this or a modified Shine-Dalgarno. *See* anti-Shine-Dalgarno; IF; initiation codon; polycistronic mRNA; ribosome scanning. (Shultzaberger, R. K., et al. 2001. *J. Mol. Biol.* 313:215.)

shingles Herpes virus responsible for chicken pox may emerge from a latent stage later in life when immunity has waned. It then causes shingles (sore eruptions) in parts of the body innervated by ganglions harboring the earlier latent virus. *See* ganglion; herpes.

SHIP Inositol phosphatase with SH$_2$ domain. SHIP proteins in a tyrosine phosphorylated form signal to hematopoiesis, cytokines, PTEN, etc. *See* cytokines; hematopoiesis; inositol; ITIM; PTEN; SH$_2$. (Rohrschneider, L. R., et al. 2000. *Genes & Development* 14:505.)

SHIV Simian immunodeficiency virus (SIV) engineered to carry an HIV coat protein and is thus capable of infection of, and symptoms of AIDS in macquaque monkeys. *See* acquired immunodeficiency syndrome; HIV; primates; SIV.

SHMOOs Mating projections in yeast.

SHOM Sequencing by hybridization to oligonucleotide microchips is one of the automated (robotic) procedures developed for nucleotide sequence diagnostics. *See* DNA chips; microarray hybridization. (Yershov, G., et al. 1996. *Proc. Natl. Acad. Sci. USA* 93:4913.)

shoot Plant part(s) above ground or a branch of a stem.

Shope papilloma Viral disease of rabbits causing nodules under the tongue. The double-stranded DNA virus of about 8 kbp has 49 mole percent G + C content. *See* bovine papilloma.

short-day plants Require generally less than 12–15 hr daily illumination for flowering; at longer light periods, they usually remain vegetative. *See* long-day plants; photoperiodism.

short dispersed repeats *See* SDR.

short patch repair Excision repair removing and replacing about 20 nucleotides. *See* DNA repair; excision repair; mismatch repair. (Mansour, C. A., et al. 2001. *Mutation Res.* 485:331.)

short syndrome Autosomal-recessive phenotype characterized by the initials of the SHORT acronym: short stature, hyperextensibility of joints and hernia, ocular depression, Rieger anomaly (partial absence of teeth, anal stenosis [narrow anus], hypertelorism [increased distance between organs or parts], mental and bone deficiencies, and teething delay). *See* dwarfism; pseudoautosomal; stature in humans.

shotgun cloning The DNA of an entire genome is cloned without aiming at particular sequences. From the cloned array of DNA fragments (library) the sequences of interest may be identified by appropriate genetic probes. *See* cloning; DNA library; DNA probe. (Matsumoto, S., et al. 1998. *Microbiol. Immunol.* 42:15.)

shotgun sequencing Random samples of cloned DNA, e.g., the segments of a cosmid are sequenced at random. *Whole-genome pairwise shotgun* procedure sequences paired ends of cloned DNAs of varying sizes fragmented into a larger number of contigs ordered with the aid of high-power computers. If there are still gaps between the contigs, they are filled in by 'finishing'. The *hierarchical shotgun sequencing* procedure is based on mapped clones generated by BACs. The *double-barrel shotgun* sequences the DNA from both ends. The short sequences are arranged into longer tracts by computers. The *full shotgun sequence* indicates that the cloned inserts have been covered about 8–10 times. *Half-shotgun coverage* is only 4–5-fold random sequence. *See* completion; contig; DNA sequencing; finishing; first-draft sequence; genome projects; human genome; scaffolds in genome sequencing; WGS. (Bankier, A. T. 2001. *Methods Mol. Biol.* 167:89.)

shoufflons Clustered (generally 6 to 7) recombination/inversion sites and a shoufflon-specific recombinase in bacteria. These elements determine the nature of the bacterial pili: thick, rigid, or thin. A typical shoufflon is plasmid R64, a 120.8 kb conjugative plasmid encoding streptomycin and tetracycline resistance and at least 49 genes in the 54 kb transfer region. Various types of shoufflons occur in different bacteria. *See* pilus; site-specific recombination. (Komano, T. 1999. *Annu. Rev. Genet.* 33:171.)

SHP-1 Synonymous with SH-PTP1, PTP1C, and HCP. Tyrosine phosphatase. It contains the SRC homology domain SH2. Upon activation of T cells, it binds to the kinase ZAP-70, resulting in increased phosphatase activity but a decrease in ZAP-70 kinase activity. It is a negative regulator of the T-cell antigen receptor activated by radiation stress. *See* ITIM; T cell; ZAP-70. (Kosugi, A., et al. 2001. *Immunity* 14:669.)

Shprintzen-Goldberg syndrome *See* Marfanoid syndromes.

shrew *Blarina brevicauda*, 2n = 50; *Cryptotis parva*, 2n = 52; *Neomys fodiens*, 2n = 52; *Notiosorex crawfordi*, 2n = 68; *Sorex caecutiens*, 2n = 42; *Suncus murinus*, 2n = 40.

shRNA Short heterochromatic RNA is instrumental in the formation of heterochromatin and epigenetic remodeling of chromatin. *See* heterochromatin, epigenesis. (Jenuwein, T. 2002. *Science* 297:2215.)

sHsp Small heat-shock proteins. Diverse ubiquitous proteins (15–80 kDa) formed in response to heat or other stress. Their transcriptional activation requires three inverted repeats of the NNGAAN motif (HSE) where the heat-shock transcription factor (HSF) binds. Their regulation may require other factors (e.g., estrogen, ecdysterone, etc.). The homology among the different types resides in the COOH-terminal half (α-crystalline domain). They may form large oligomers. Their role is heat and chemical protection of cells. Plant sHPS chaperones respond to various stresses to resist irreversible protein denaturation. *See* heat-shock proteins; ibp. (Haley, D. A., et al. 2000. *J. Mol. Biol.* 298:261.)

shuffling   *See* DNA shuffling.

shunting   During scanning, the 40S ribosomal subunits jump from region to region until they locate the translation initiation codon. *See* scanning; translational hopping.

shuttle vector   "Promiscuous" plasmid that can carry genes to more than one organism and can propagate the genes in the different cells, e.g., in *Agrobacterium*, *E. coli*, and plant cells. *See* cloning vectors; promiscuous DNA; transformation, genetic; vectors. (Perez-Arellano, I., et al. 2001. *Plasmid* 46:106.)

SI   Unit of absorbed dose (1 joule/kg) of electromagnetic radiation. It is generally expressed in gray (Gy) or sievert (Sv = 1 rem) units. Previously, rad (= 0.01 Gy) was used. *See* Becquerel; Curie; Gray; r; rem; Sievert.

sialic acid   Acidic sugar such as N-acetylneuraminate or N-glycolylneuroaminate. Sialic acids are present in gangliosides. Polysialic acid is involved in cell- and tissue-type differentiation, learning, memory, and tumor biology. The synthesis is mediated by polysialyl transferase under the regulation of neural cell adhesion molecules. The recessive sialic acid storage diseases (e.g., Salla disease, 6q14-q15) involve hypotonia, cerebellar ataxia, and mental retardation caused by a family of anion/cation symporters. The enzyme CMP-sialic acid hydroxylase changes N-acetylneuraminic acid into N-glycolylneuraminic acid. This enzyme is active in all mammals except humans, where the 6p22.3-p22.2 locus suffered a 92-base deletion in the 5′-region after the separation of humans from the primate lineage of evolution. *See* CAM; cytidylic acid; fusogenic liposome; gangliosides; gangliosidoses; lysosomal storage disease; neuraminidase deficiency; sialuria; sphingolipids; symport. (Aula, N., et al. 2000. *Am. J. Hum. Genet.* 67:832; Angata, T., et al. 2002. *J. Biol. Chem.* 277:24466.)

Sialic acid.

sialidase deficiency   *See* neuraminidase deficiency.

sialidosis   *See* neuraminidase deficiency.

sialolipidosis   *See* mucolipidosis IV.

sialuria (9p12-p11)   Caused by a semidominant/recessive gene defective in feedback sensitivity of uridine diphosphate N-acetylglucosamine 2-epimerase enzyme by cytidine monophosphate-neuroaminic acid. Those afflicted have defects in bone (dysostosis) and psychomotor (movement and psychic activity) development, and infantile death may incur. Salla disease (6q14-q15) is a sialic acid storage disease with mental and psychomotor retardation. *See* neuroaminidase deficiency.

siamese cat   Displays darker color at the extremities. Due to a temperature-sensitive gene, at slow blood circulation, more pigment develops at specific locations of the body, similarly to the Himalayan rabbits. *See* Himalayan rabbit; pigmentation of animals; temperature-sensitive mutation. *See* illustration below.

Siamese cat.

sib   Sibling. *See* full sib; half sib.

sibling   Natural children of the same parents. *See* genetic risk; genotypic risk ratio; λ_S; risk.

sibling species   Morphologically very similar and frequently share habitat, but they are reproductively isolated. *See* fertility; speciation; species.

sib pair method   *See* affected sib-pair method.

SIBPAL   Linkage analysis computer program for sib-pairs. *See* sibling. (Fann, C. S., et al. 1999. *Genet. Epidemiol.* 17 Suppl. 1:S151.)

sibship   Natural brothers and sisters. *See* kindred; sibpal.

sib TDT (s-TDT)   Transmission disequilibrium test to detect genetic linkage/association on the basis of analysis of close genetic markers and disease among sibs. *See* SDT. (Spielman, R. S. & Ewens, W. J. 1998. *Am. J. Hum. Genet.* 62:450.)

SIC1   Cell cycle S-phase cyclin-dependent kinase (CDK, Cdc28-Clb) inhibitor. *See* APC; CDC34; CDC6; cell cycle (START); mitotic exit. (Verma, R., et al. 2001. *Mol. Cell* 8:439.)

sickle-cell anemia   Human hereditary disease caused by homozygosity of a recessive mutation(s) or deletions in the hemoglobin β-chain gene. Heterozygosity causes the sickle cell trait. Under low oxygen supply, the red blood cells lose their plump appearance and partially collapse into sickle or odd shape because the abnormal hemoglobin molecules aggregate. The disease is not absolutely fatal, but crises may occur when the blood vessels are clogged. Complications may arise by poor blood circulation.

In the classical form of sickle cell disease, a valine residue in hemoglobin S replaces a glutamine residue of the normal beta chain (hemoglobin A). In hemoglobin C, a lysine replacement occurs at the same position, and this condition causes less severe clinical symptoms. Hemoglobin D and E are less common. This disease provided the first molecular evidence that mutation leads to amino acid replacement. Sickle cell anemia affects more than 2 million persons worldwide. About 10% of the population of African descent in the United States are carriers (heterozygous) for this mutation, and about 1/400 is afflicted with homozygosity at birth. In populations of European (except southern European) descent, the frequency of this mutant gene is about 1/20 of that among Mediterraneans and Africans. The high frequency of the genetic condition in areas of the world with a high infestation of malaria is correlated. Individuals without the sickle cell anemia gene have about 2–3 times higher chance of being infected by *Plasmodium falciparum*. The mutation is selectively advantageous by protecting heterozygotes against malaria.

Human red blood cells from a sickle cell anemia patient. **left**: in the presence of normal oxygen supply. **right**: at low oxygen supply. Normal red blood cells look like biconcave discs very similar to those at left here. The sickling cells are unable to hold oxygen and that condition is responsible for the disease. (Photographs are the courtesy of Cerami, A. & Manning, J. M. 1971. *Proc. Natl. Acad Sci. USA* 68:1180.) Carriers or sufferers can be unambiguously identified by electrophoretic separation of the blood proteins. Homozygotes for the normal blood protein (A-A), sickle cell anemia (S-S), hemoglobin C (C-C). In the heterozygotes (A-S, S-C AND A-C)—because of codominance—both types of parental proteins are detected. (From Edington, G. M. & Lehman, H. 1954. *Trans. R. Soc. Trop. Med. Hyg.* 48:332.)

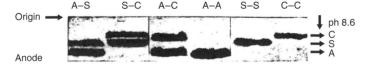

The globin gene cluster has been located to human chromosome 11p15. The order is $5' - \gamma G - \gamma A - \delta - \beta - 3'$. Correction of the defective allele is possible by transduction of the defective cells with retroviral or adenoviral vectors that can deliver the normal gene to the hematopoietic stem cells, but these viral vectors may have deleterious consequences for the body. There is a problem relating to the low expression of the transagene. Another method is to introduce chimeric DNA-RNA oligonucleotides into the lymphoblastoid cells with a correction for the β^S allele mutation brought about through gene conversion in the target cells. This procedure may eventually be clinically applicable. Another possible therapeutic approach involves correction of defective β-globin mRNA by the similar (fetal) normal (antisickling) protein transcript using a transsplicing ribozyme and generation of a normally functional transcript. When hemoglobin α- and β-genes were knocked out of mice and mated with animals transgenic for the human sickle cell gene, an animal model was generated for experimentation with this human disease. Attempts are being made to activate silent fetal hemoglobin genes—by urea compounds—at later developmental stages in order to compensate for the defective adult hemoglobin. *See* gene conversion; gene therapy; genetic screening; hemoglobin; introns; malaria; ribozymes; SAD mouse; sickle cell trait; thalassemia; transsplicing; viral vectors. (Pawliuk, R., et al. 2001. *Science* 294:2368; Vichinsky, E. 2002. *Lancet* 360:629.)

sickle cell trait Due to heterozygosity of the recessive mutation in the gene controlling the β-chain of hemoglobin. Normally these heterozygotes do not suffer from this condition, but under low oxygen supply, e.g., at high elevations, adverse consequences may arise. *See* hemoglobin; sickle cell anemia.

Side-arm bridge. (Courtesy of Dr. B. R. Brinkley & W. N. Hittelman.)

side-arm bridge Attachment of chromatids resembling chiasma, but actually it is only an anomaly in mitosis or meiosis, usually arising when the division has been disturbed by chemicals or radiation. *See* bridge.

siderocyte anemia (sideroblastic anemia) Anemia with erythrocytes containing nonhemoglobin iron; it may be controlled by autosomal-recessive or -dominant or X-linked genes based on defects in erythroid β-aminolevulinate synthase. *See* anemia.

siderophore Iron transporter. Transformation into rice plants highly efficient siderophore system genes (for nicotinamine aminotransferase) from barley facilitates iron uptake on alkalic soils and substantially improves growth. (Takahashi, M., et al. 2001. *Nature Biotechnol.* 19:466.)

siderosis Iron overload in the bloodstream.

SIE STAT-inducible element. *See* Jak-Stat pathway.

siemens (SI) Unit of conductance; 1 ampere/volt in a tissue of 1 ohm resistance. *See* ampere; ohm; volt.

sieve tube Plant food transporting tube-shaped, tapered, long cells. Sieve tubes may be connected by sieve plates.

sievert (Sv) Name for Sv (unit of absorbed dose equivalent [J/kg]) = 100 rem = 1 Sv. It is commonly used for measuring occupational radiation hazards. *See* Gray; R; rad; rem.

SIGLEs (19q13.3) Immune inhibitory receptors with sialic acid ligands. *See* sialic acid.

σ **(1)** Measure of superhelical density of DNA. (2) Subunit of prokaryotic RNA polymerase enzyme that is essential to start transcription in a specific way. This factor opens the double helix for the action of the RNA polymerase. Also, σ^{70} is involved in pausing of transcription. The σ^{70} recognition sites consist of two hexamers located at the -10 and -35 positions from the transcription start point. The σ^{38} subunits are used at the stationary phase of growth. σ^{28} is a minor subunit transcribing less than two dozen genes. σ^{54}, another minor subunit, binds to the promoter even in the absence of the core polymerase. In the synthesis of stress proteins, σ^{32} and σ^{24} are used. Usually, σ^{70} is released from the polymerase (RNAP) at the beginning of transcript elongation or shortly afterward. Some of the σ-elements stay on throughout elongation and regulate gene expression depending on the cellular circumstances. The σ-elements are complexed with anti-σ proteins when not in use. In some algae, the protein present in the chloroplast is encoded by the nucleus. In several plant species, the same polypeptide is coded for by the chloroplast DNA. *See* chloroplast; chloroplast genetics; DnaJ; open promoter complex; Pribnow box; RNA polymerase; rpo; transcription factors; UP elements. (Dartigalongue, C., et al. 2001. *J. Biol. Chem.* 276:20866; Marr, M. T., et al. 2001. *Proc. Natl. Acad. Sci. USA* 98:8972; Bar-Nahum, G. & Nudler, E. 2001. *Cell* 106:443; Kuznedelov. K., et al. 2002. *Science* 295:855; Mekler, V., et al. 2002. *Cell* 108:599.) **(2)** Parametric designation of standard deviation. *See* standard deviation; standard error. **(3)** Yeast transposable element. *See* Ty. **(4)** Viral infectious hereditary agent of *Drosophila*. *See* CO_2 sensitivity; infectious heredity. **(5)** 387 bp region intercalated between the two S elements in opposite orientation of the complex *R* locus of maize. *See* paramutation; *R* locus of maize; tissues specificity.

sigma factor Subunit of DNA-dependent bacterial RNA polymerase required for the initiation of transcription and promoter selection. *See* open-promoter complex; RNA polymerase; σ.

sigma replication *See* rolling circle.

σ^{S} (RpoS) Required for the expression of many growth-phase and osmotically regulated prokaryotic genes. *See* σ-subunit of RNA polymerase.

sigma virus Of *Drosophila*. *See* CO_2 sensitivity.

signal Molecular determinant emitted by an extracellular source or by an intracellular organizer. It is directed either to an adjacent tissue (vertical signal) or to adjacent cells of the same tissue (planar signal).

signal end *See* immunoglobulins.

signal joint *See* signal end.

signal hypothesis Postulated that the signal peptide of the nascent polypeptide chain guides it to the endoplasmic reticulum (and to other) membranes, where the signal peptidase cleaves it off. Subsequently, the peptide chain is completed within the lumen of the membranes. It is validated for transport through bacterial cell membranes, mitochondria, plastids, peroxisomes, etc. *See* protein targeting; signal peptides; transit peptide.

signaling Phosphorylation of enzymes mediated by second messengers leads to their activation through outfolding of the pseudosubstrate domains of the enzymes, thus opening the active sites for the true substrate. *See* signal transduction.

signaling molecule Alerts cells to the behavior of other cells and environmental factors.

signaling to translation Directed toward the 5'-untranslated region (UTR) and involves the translation of ribosomal proteins and elongation factors (eEF1A, eEF2). The target of the signals appears to be the 5'-terminal oligopyrimidine sequences (5'-TOP) and the UTR polypyrimidine tracts. Secondary structure formation with long UTRs may also be regulatory. Growth factors may phosphorylate the eukaryotic initiation factor eIF4E-binding protein, 4E-BP-1, and cause its dissociation from eIF4E. In order for translation to proceed, the initiation factors eIF4A, -4B, -4G, and -4E attach to the methylguanine cap and the secondary structure of RNA is untwisted by the helicase action of eIF4A. Phosphorylation by $p70^{S6k}$ kinase activates the S6 protein of the 40S ribosomal subunit, a process subject to enhancement by mitogens. This process is not a general requirement for translation, indicating that it only affects special genes with 5'-TOP and polypyrimidine tracts in their UTRs. Cycloheximide and puromycin are involved in the phosphorylation of S6. Both of these phosphorylations are inhibited by rapamycin. *See* cap; cycloheximide; F506; $P70^{S6k}$; protein synthesis; puromycin; regulation of enzyme activity; regulation of gene activity; secondary structure; signal transduction; 5'-Top. (Wilson, K. F. & Cerione, R. A. 2000. *Biol. Chem.* 381[5–6]:357.)

signalosome Molecular complex transmitting various cues. *See* signal transduction. (Lyapina, S., et al. 2001. *Science* 292:1382; Zhang, S. Q., et al. 2000. *Immunity* 12:301.)

signal peptidase *See* signal hypothesis.

signal peptides (signal sequence) 15- to 35-amino-acid-long sequences generally at the NH_2 terminus of the nascent polypeptide chains of proteins that have a destination for an intraorganellar or transmembrane location. They are made in eukaryotes and prokaryotes, but not all the secreted proteins possess one. At the beginning of the sequence, generally there are one or more positively charged amino acids, followed by a tract of hydrophobic amino acid residues that occupy about three-fourths of the length of the chain. This hydrophobic region may be required to pass into the lipoprotein membrane. The amino acid sequences among the various signal peptides are not conserved, indicating that the secondary structure is critical for recognition by the signal peptide recognition particles and for the function within the membrane. The eukaryotic signal sequences are recognized by the prokaryotic transport systems and the prokaryotic signal peptides can function in eukaryotes. After the passage of the nascent peptide has started, the signal peptides are split off by peptidases on the carboxyl end of (generally) glycine, alanine, and serine. Consequently, the majority of the proteins in the membrane or those transported through the membranes have the nearest downstream neighbor of one of these three amino acids at the amino end. One major characteristic of the signal peptide is cotranscriptional targeting, whereas transit peptides are targeted posttranslationally. The signal sequences may show polymorphism, which might result in incorrect targeting. *See* endoplasmic reticulum; leader peptide; signal sequence recognition particle; transit peptide. (von Heijne, G., et al. 1989. *Eur. J. Biochem.* 180:535; Watanabe, N., et al. 2001. *J. Biol. Chem.* 276:20474.)

signal recognition particle *See* signal sequence recognition particle; SRP.

signal sequence The amino terminal of some proteins signals the cellular destination of these proteins, such as the signal peptides. *See* signal peptide.

signal sequence recognition particle (SRP) In mammals, a complex of six proteins and an RNA (7SL RNA) that recognizes the *SRP receptor protein* on the surface of the endoplasmic reticulum and the *signal peptides* of the nascent proteins translated on the ribosomes. The signal peptides are associated with the endoplasmic reticulum (rough endoplasmic reticulum) and facilitate transport into the lumen of the Golgi apparatus and lysosomes (cotranslational transport). Signal recognition particle proteins have been located to human chromosomes 5q21, 15q22, 17q25, and 18. SRP binds to the signal

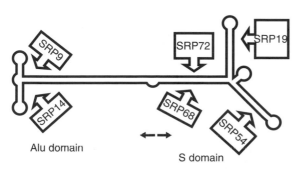

Schematic illustration of a signal recognition particle. The boxes represent the six proteins around the RNA. SRP9–SRP14 and SRP72–SRP68, respectively, form dimeric structures. (Modified after Weichenrieder, O., et al. 2000. *Nature* 408:167.)

peptide after about a 70-amino-acid chain is completed at the beginning of translation. The polypeptide chain elongation is somewhat relaxed until the SRP attaches to the SRP receptor. Then the SRP comes off the amino acid chain, elongation resumes its normal rate, and the entry of the chain through the membrane proceeds. In the meantime, a peptidase inside the endoplasmic reticulum cuts off the 15–30-amino-acid-long signal peptide sequences. *See* Golgi; lysosomes; protein conducting channel; protein synthesis; protein targeting; RNA 7S; signal peptides; signal sequence particle; TRAM; translocase; translocon. (Keenan, R. J., et al. 2001. *Annu. Rev. Biochem.* 70:755; Fulga, T. A., et al. 2001. *EMBO J.* 20:2338; Wild, K., et al. 2001. *Science* 294:598; Hainzl, T., et al. 2002. *Nature* 417:767; Pool, M. R., et al. 2002. *Science* 297:1345. <http://psyche.uthct.edu/dbs/SRPDB/SRPDB. html>.)

signal transduction System of proteins transforming various stimuli into cellular responses. The process requires four major categories of elements: signals, receptors, adaptors, and effectors.

SIGNALS: The extracellular signals interact with cell membrane receptors and make contacts with intracellular target molecules to stimulate a cascade of events, leading to the formation of effector molecules that turn genes on and off and control cellular differentiation in structure and time by regulating transcription. The *signals* are proteins, peptides, nucleotides, steroids, retinoids, fatty acids, hormones, gases (ethylene, nitric oxide, carbon monoxide), inorganic compounds, light, etc. The target cells accept the signals by special sensors called *receptors*. The receptors are generally specific proteins with high binding specificities positioned on the cell surface or within the plasma membranes; thus, they readily accept the signal *ligands*. The receptors may also be inside the cells, and the ligands may have to pass the cell membranes to reach them. Eventually, the instruction reaches the cell nucleus and the relevant genes. The *paracrine signals* are restricted in movement to the proper target that is generally nearby. The nerve cells communicate by *synaptic signals*. The *endocrine hormone* signals may affect distant targets in the entire body. The neurotransmitters are activated through long circuits of the nervous system by electric impulses emitted by neurons in response to the environment. The travel of the electric signals through the neurons is very fast, possibly meters per second. The neurotransmitters have only a few nanometers to pass and the process takes only a few milliseconds. The local concentration of the endocrine hormones is extremely low. In contrast, the neurotransmitters may be quite concentrated at a very small target area. The neurotransmitters also may very rapidly be removed either by reabsorption or by enzymatic hydrolysis. Generally, the hydrophobic signals persist longer in the cells than the hydrophilic ones. The membrane-anchored growth factors and cell adhesion molecules are signaled through the *juxtacrine* mediators.

RECEPTORS: The target cells respond by *receptor proteins*. These receptors are endowed with specificities regarding the signal they respond to. Also, the same signal may have different receptors in differently specialized cells. The interpretation and use of the signal within similar cells may vary. The signals may act in a combinatorial manner: Several signals together may be involved in the cellular decisions and influence the length and quality of the effect of a received signal. Despite substantial chemical differences of the signals (e.g., cortisol, estrogen, progesteron, thyroid hormones, retinoic acid, vitamin D) the various receptors may bind ligands that control closely related and interchangeable upstream DNA consensus elements through the signal transduction path involved in the regulation of the transcription of different genes. After bound to cognate hormones, steroid hormone receptors, may activate the transcription of the so-called *primary-response genes*. These proteins then repress the further transcription of the primary-response genes and turn on the transcription of *secondary-response genes* (*see* regulation of gene activity).

Receptors can be (i) *ion-channel* or (ii) *G-protein-* or (iii) *enzyme-linked* types. Group (i), also called transmitter-regulated ion channels, is involved in transmitting neuronal signals (*see* ion channels). Group (ii) receptors are transmembrane proteins of the so-called seven-membrane type associated with guanosine phosphate-binding G proteins (*see* G proteins). When GTP is bound to the G protein, a cascade of enzymes or other proteins may be activated or an ion channel may become more permeable. G-protein-linked receptors represent a large family of proteins, with more than 100 of them identified in a variety of eukaryotes. The receptors, generally monomeric and evolutionarily related proteins, respond

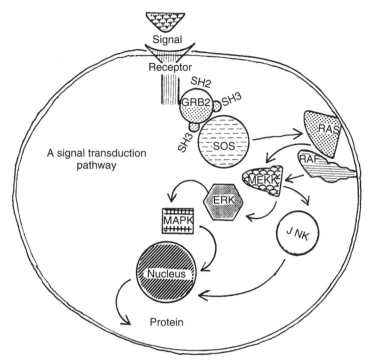

A signal transduction pathway leading through the G-protein RAS. This pathway controls cell division, differentiation, development of cancer, mating type, cell wall biosynthesis, and a variety of other processes. The signal can be a variety of molecules such as epidermal growth factor (EGF), nerve growth factor (NGF), or their homologues in various other animals such as *LET-23* in *Caenorhabditis or DER* in *Drosophila*. The RAS-mediated pathway also operates in fungi and plants. The process begins with the (mitogen) *signal* arriving to the double membrane of the cell (see at 12 O'clock). The signal is recognized by a *signal receptor protein*, which may vary from signal to signal. This is a transmembrane protein with a hydrophobic tract generally forming seven turns within the cell membrane. When the signal arrives, the receptor is activated. The protein tyrosine kinase receptors (e.g., SEV) recruit the *downstream receptor kinases* or adaptor proteins such as GRB (or homologues, e.g., SEM-5 in *Caenorhabditis*, DRK in *Drosophila*). The GRB proteins have an SH2 domain that is a binding site for tyrosine kinase proteins, and the SH3 domains are binding sites for proline-rich motif proteins. The SH domains were originally identified in the SRC protein (product of the rous sarcoma oncoprotein) and both are characteristic for mediators of the signal transduction path. The GRB protein then binds another mediator or adaptor protein, SOS (named after the product of a *Drosophila* gene, called *son of sevenless*). The *sevenless* gene encodes rhabdomere 7 light receptors in the eyes of the flies. In place of SOS, there may also be SHC (an oncoprotein of the pheochromocytoma tumor of the adrenal medulla or in paraganglia, thus causing increased secretion of the hormones epinephrine and norepinephrine). Upon the influence of the GRB-SOS complex, the membrane-bound RAS G-protein (see at 3 O'clock) becomes activated. The membrane association of RAS is mediated by a carboxy-terminal CAAX BOX, a signal for farnesylation, proteolysis, and carboxymethylation, and by the neighboring six lysine residues. RAS (named after rat sarcoma) is also a GTPase. RAS serves as a turnstile for a series of processes. It is shut down when the situation is RAS*GDP (RAS guanosine diphosphate) and it opens when it becomes RAS*GTP (RAS-guanosine triphosphate). RAS activation may be prevented by another gene, encoding a GTP-ase-activating protein (GAP). The RAF protein (its homologue in budding yeast is STE11, and BYR 2 in fission yeast) is also membrane bound by a CAAX BOX. RAF is a Ser-Thr kinase and phosphorylates MEK independently of RAS, a protein of the *extracellular signal regulated kinase* (ERK) family. The extracellular regulators can be growth factors (e.g., EGF, NGF), receptor kinases, and TPA (12-O-tetradecanoyl-phorbol-13-acetate). MEK kinase (MEKK) is phosphorylated on Thr and Ser residues by RAF. Protein MAPK (mitogen-activated protein kinase) may be capable of autophosphorylation and it may be phosphorylated by the ERK family of kinases at Tyr and Thr residues. MAPK homologues are KSS, HOG-1, FUS3, SLT-2, SPK-1, SAPK (a stress-activated kinase), FRS (FOS-regulating kinase), etc. MAPK may then activate the fos and jun oncogene complex also known as the AP-1 heterodimeric transcription factor of mitogen-inducible genes. The RAS to MAPK route may branch downstream through several effector proteins and may control the transcription of several different genes. The specificity of activation depends on combinatorial arrangements of the effectors. (The size, shape, or shading of the symbols was not intended to represent their structure.)

to a variety of signals, such as hormones, mitogens, light, pheromones, etc. Activation of group (iii) receptors may directly or indirectly lead to the activation of enzymes. These three different types of signal transduction may not be entirely distinct because the function of the ion channels may interact with the pathways mediated through G proteins and various kinases. Some of the receptors are *protein tyrosine phosphatases* (e.g., CD45 protein) and *serine/threonine phosphatases* residing within the membrane or in the cytosolic domain of transmembrane proteins or in the cytosol.

PATHWAYS: They may show a great variation depending on the signals and receptors involved. Ca^{2+} is a general regulator. It may enter nerve cell terminals through voltage-gated Ca^{2+} channels in the cell membranes and stimulate the secretion of neurotransmitters (*see* ion channels; voltage-gated ion channel). Alternatively, Ca^{2+} may have a more general role by binding to G-protein-linked receptors in the metabolism of inositol phospholipids, PIP (phosphoinositol phosphate) and PIP_2 (phosphoinositol bisphosphate). The specific trimeric G protein, Gq, is involved in the activation of *phospholipase C-β* that is specific for phosphoinositides and splits PIP_2 into inositol triphosphates and diaglycerol. Hydrolysis of PIP_2 yields IP_3 (inositol 1,4,5-triphosphate). The latter sets calcium free from the endoplasmic reticulum through IP_3-gated channels that are ryanodine receptors (*see* ryanodine). Upon further phosphorylation, IP_3 may give rise to IP_4 (inositol 1,3,4,5-tetrakisphosphate), which slowly yet

steadily replenishes cytosolic calcium. The calcium level in the cytosol rises and subsides in very short bursts according to how phosphatidyl-inositols regulate it (calcium oscillations). These oscillations still may assure increased secretion of second messengers and spare the cell from a constant level of the toxic Ca^{2+} in the cytosol. Besides pumping out Ca^{2+} shielded in the endoplasmic reticulum, diaglycerol and eicosanoids (arachidonic acid) may be produced.

Diaglycerol and the latter lipid derivatives may activate Ca^{2+}-dependent enzymes, serine/threonine kinases, which have a key role in activating proteins that mediate signal transduction (*see* protein kinases). *Protein kinase C* may activate cytosolic *MAPK* (mitogen-activated protein kinase) by phosphorylation and may phosphorylate a cytoplasmic inhibitor complex such as $I\kappa B + NF-\kappa B$. Thus, MAPK may phosphorylate DNA-binding proteins such as SRF (serum-response factor) and Elk (member of the ETS oncoprotein family) that are already sitting in the upstream regulatory regions of a gene(s), such as the serum-response element (SRE). Phosphorylation initiates transcription (*see* regulation of gene activity). The released protein factor NF-κB may migrate to the nucleus. By binding to its cognate DNA site, transcription is set into motion (if other factors are present). The response of the genes to the transducing signals depends on the number of regulatory proteins responding to the signal. Monomolecular reactions display a relatively slow response to the concentration of the signal molecules, whereas if the

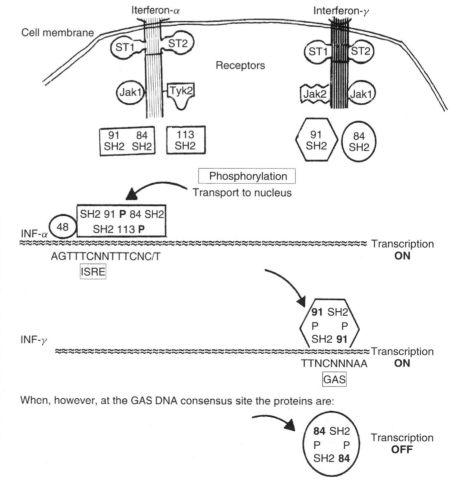

The JAK-STAT signal transduction pathway, based on interferon (IFN)-mediated receptors. *JAK* is a family of the Janus tyrosine protein kinase proteins, and *STAT* stands for signal transducer and activator of transcription. When the ligand binds to the signal-transducing receptors (ST), the receptor-attached *JAK kinases* modify the *STAT proteins*. After dimerization (using SH2 domains), they are transported directly to *ISRE* (interferon-α-response element) or to *GAS* (γ-interferon activation site) In the chromosomal DNA. These two elements vary, yet consensus sequences exist (as shown in the diagram). The 84, 91, 113, and 48 are proteins (in kDa), but additional ones may also be involved, depending on the nature of the signals received. The diagram does not display IFNGR1 and IIFNGR2 integral membrane proteins, which are also essential parts of the γ-receptor. (After Darnell, J. E., Jr., et al. 1994. *Science* 264:14125 and Heim, M. H., et al., 1995. *Science* 267:1347.)

number of effectors is multiple, the reaction to them may follow 3rd- or multiple-order kinetics. Similarly, a prompt response is expected if the signal activates one reaction (e.g., phosphorylation) while it deactivates an inhibitor or suppressor (e.g., by phosphatase action).

The enzyme-linked signal receptors do not need G-proteins. The transmembrane receptor binds the ligand at the cell surface and the cytosolic domain functions as an enzyme, or it associates with an enzyme and the transfer of the signal to the cell nucleus is more direct. An example is the cytokine-activated cell membrane receptor Jak *tyrosine kinases*, which when dimerized can combine with cytoplasmic STATs (signal transducers and activators of transcription) and chromosomal-responsive elements. In response to interferon or other cytokine signals, JAK phosphorylates tyrosine of the SH2 domains in a variety of STAT proteins. This may be followed by dimerization and transfer of these proteins to the nucleus, where they may turn on transcription of particular genes (see figure below). It has been estimated that about 1% of the human genes code for protein kinases.

In case of autophosphorylation or other reactions involving enzymes, which may bind their own products, the activity of the enzyme may be increased in the course of time with the increase in the number of product molecules through positive feedback. These kinases may be mainly serine/threonine or tyrosine specific. Some proteins may phosphorylate all three of these amino acids. The reliance on protein tyrosine kinases for signal transduction is rather general. Epidermal growth factor (EGF), nerve growth factor (NGF), fibroblast growth factor (FGF), hepatocyte growth factor (HGF), insulin, insulin-like growth factor (IGF), vascular endothelial growth factor (VEGF), platelet-derived growth factor (PDGF), macrophage colony-stimulating factor (M-CSF), etc., function with the assistance of transmembrane *receptor tyrosine kinases*. Upon the arrival of the ligand (signal), the receptor is dimerized either by cross-linking two receptors by the dimeric ligand or by inducing autophosphorylation and linkage of two cytosolic domains of the receptors. The different phosphorylated sites may bind different cytoplasmic proteins. The insulin itself is a tetramer ($\alpha\alpha\beta\beta$) and thus does not need

dimerization. After autophosphorylation, it phosphorylates an insulin receptor (IRS-1) at tyrosine sites, which may bind to other proteins that may become phosphorylated and may form different complexes, thus generating a variety of transcription factors. Alternatively, the *tyrosine kinase-associated receptors* themselves are not tyrosine kinases but associate with proteins of this capability. Some of the enzyme-linked receptors are *serine/threonine kinases* with specificities for these two amino acids. The phosphorylated tyrosine residues are binding sites for proteins with SH2 domains. The *receptor tyrosine phosphatases* may activate or inhibit the signal pathways by the removal of phosphate from tyrosine residues. The receptor guanylate (guanylyl) cyclases operate in the cytosolic domain of the receptors and function by serine/threonine phosphorylation in association with trimeric G proteins. The discoveries about signal transduction have changed the view about cellular functions and added a new dimension to biology by integrating reversed and classical genetics. Signal transduction mechanisms have a large variety of means to regulate diverse functions of metabolism, differentiation, and development. Different signaling molecules are organized into separate pathways. Different protein components appear to be regulated and coordinated into signaling complexes by SH, pleckstrin homology, phosphotyrosine, and PDZ (postsynaptic density, disc-large, zo-1) protein domains through protein-protein interactions.

PLANT SIGNALS: They are somewhat different from the signals in animals. The plant hormones, similarly to animal hormones, are signaling molecules, but most of them—except the brassinosteroids—are very different molecules. In plants, light and temperature signals (photoperiodism, vernalization, phytochrome) are very important for growth and differentiation. Salicylic acid is a signaling molecule for defense genes, etc. In plants, signaling depends a great deal on positional cues.

Several signal transduction pathways may be operational simultaneously within an organism and may interact at various levels. In addition, the pathways and components may operate differently in different cellular compartments (e.g., plasma membrane, cytosol, nucleus, organelles). The cytoskeleton network may serve in various capacities to

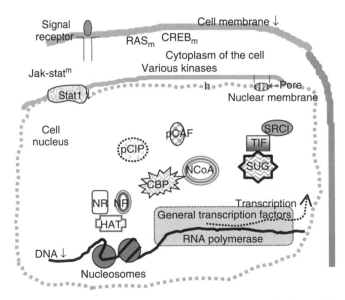

JAK-STAT: see more above

RAS: see more above

CREB: cAMP-response element-binding protein

pCIP: p300/CBP/cointegrator-associated protein

CBP: CREB-binding protein

pCAF: chromatin assembly factor

NCoA: nuclear receptor coactivator

SRCl, TIF, SUG: nuclear coactivator proteins

NR: nuclear receptors

HAT: histone acetyltransferase

Besides the molecules shown and named, several other proteins may be involved. Some of the proteins are activated only after binding with their ligands. The size or shape of the structures shown does not reflect the actual nature of these molecules.

Transactions inside the nucleus after the signals arrived.

direct the flow of reactions through the signaling pathways. The current view is that signaling proteins translocate in the cytoplasm and bind in a reversible manner and dynamic fashion (soft-wired signaling concept). The earlier idea was that receptors and other signaling proteins occupy fixed positions in the cell and second messengers mediate the connections with the aid of diffusion (hard-wired signaling). The introduction of GFP labeling now facilitates tracing of the signaling traffic. The interacting signals are usually expressed in a quantitatively variable manner. The understanding of how different cells work under the control of the genetic potentials and the environment will be the most challenging task of research on growth, differentiation, development, the nervous system, behavior, productivity, pathogenesis, evolution, etc. *See* adaptor proteins; AKAP79; arrestin; CBP; cell membranes; ciliary neutrotrophic factor; cross-talk; desensitization; diffusion; feedback; GEF; genetic network; GFP; G-proteins; histidine kinase; hormones; host-pathogen relationship; integrin; JAK-STAT pathway interferons; MLK; morphogen; morphogenesis; MPK; MPK phosphatase; NF-κB; nuclear pore; photomorphogenesis; photoperiodism; phytohormones; PP2A; regulation of gene activity; REM; salicyclic acid; selector genes; signaling to translation; SMAD; SUG; T cells; TIF; TRIP; vernalization. (Milligan, G., ed. 1999. *Signal Transduction. A Practical Approach.* Oxford Univ. Press, New York; Morris, A. J. & Malbon, C. C. 1999. *Physiol. Rev.* 79:1373; Hunter, T. 2000. *Cell* 100:113; Dohlman, H. G. & Thorner, J. W. 2001. *Annu. Rev. Biochem.* 70:703; Heldin, C.-H. 2001. *Stem Cells* 19:295; review articles in *Nature* 413:186–230 [2001]; Brivanlou, A. H. & Darnell, J. E., Jr. 2002. *Science* 295:813; for newer reviews: 2002. *Science* 296:1632 ff., Dorn, G. W. II & Mochly, D. 2002. *Annu. Rev. Physiol.* 64:407; Ernstrom, G. G. & Chalfie, M. 2002. *Annu. Rev. Genet.* 36:411; Pires-da Silva, A. & Sommer, R. J. 2003. *Nature Rev. Genet.* 4:39. signal transduction proteins: <http://www-wit.mcs.anl.gov/sentra>; <http://www.stke.org/>; <http://wit.mcs.anl.gov/WIT2/Sentra/>; <http://ecocyc.PangeaSystems.com/ecocyc>.

signal transfer particle PDGF associated with phospholipase C-γ, phosphatidyl inositol 3-kinase, and the RAF protooncogene product regulates signaling. *See* phosphatidyl inositol kinase; platelet-derived growth factor; RAF.

signature, evolutionary Within the various genomes (mammalian, mitochondrial, plants, prokaryotes, etc.) there appears to be a characteristic distribution of dinucleotide sequences, but it is different from that of the other species. (Campbell, A., et al. 1999. *Proc. Natl. Acad. Sci. USA* 96:9184.)

signature of a molecule Characteristic feature(s) convenient for identification. Discriminating sequences of DNA, RNA, or protein of organisms may serve as signatures.

signature-tagged mutagenesis (STM) Induction of mutation by insertion of plasmids, transposable elements, or passengers of specially constructed vectors into the genetic material. *See* insertional mutation; targeting genes; vectors. (Nelson, R. T., et al. 2001. *Genetics* 157:935; Shea, J. E., et al. 2000. *Curr. Opin. Microbiol.* 3:451.)

significance level Indicates the probability of error by rejecting a null hypothesis that is valid (type I error, α) or accepting one that is not correct (type II error, β). By convention, 5% (*, significant), 1% (**, highly significant), and 0.1% (***, very highly significant) levels are used most commonly. These are not sacrosanct limits. In field experiments with crops, the 5% level may be a satisfactory measure for comparative yields, but even 0.1% may not be acceptable for pharmaceutical tests because the chance of harming 1/1,000 persons is unacceptable. In general experimental practice, levels above 5% and below 0.1% are not considered meaningful, although they may have relevance for pharmacology. *See* goodness of fit; power of the test; probability; *t*-test.

sign mutations Frameshift mutations because an equal number of base additions (+) and deletions (−) at the gene locus may restore the reading frame but may not always restore normal function. *See* frameshift.

SIL Short insert library is generated by restriction cleavage of gap-bridging clones (used in the final stages of physical mapping) into 0.5 kb or smaller fragments to break up secondary structures of the DNA that complicate the determination of continuity in sequencing. *See* chromosome walking; physical mapping; restriction enzymes.

silencer Negative regulatory element reducing transcription of the region that involves the target genes. Its action bears similarity to the heterochromatic chromosomal regions, which reduce transcription of genes transposed to the elements vicinity. Sir proteins may interact with the amino-terminal of histones 3 and 4. Silencing is mediated by a combination of a protein(s) and the site where silencing takes place. There is evidence for Sir-generated heterochromatinization to interfere with the assembly of the components of the preinitiation complex (Sekinger, E. A. & Gross, D. S. 2001. *Cell* 105:403). For example, the *MATa* and *MATα* genes of yeast encode regulatory proteins that permit the expression of the *a*- and α-mating types, respectively, of the haploid cells and the nonmating phenotype of the sporulation-deficient *a*/α-diploid cells. When these genes are at the *HMLa* and *HMRα* sites, they are silenced until they are transposed to the *MAT* locus. The mating-type switch is catalyzed by a cut mediated through the HO endonuclease when the mating-type alleles are at the MAT locus but not at the *HMLa* and *HMRα* locations. This indicates that the silencing under the dual control of the repressed domains appears to extend to 0.8 kb proximal to the centromere from *HML-E* and the silencer protein and a specific site. Inactivation of *SIR2, SIR3,* and *SIR4* derepresses *HML* and *HMR*. These genes affect the telomeric position effect of other genes as well.

SIR2, SIR3, and SIR4 are involved in DNA repair and recombination in cooperation with the *HDF1* locus of yeasts (a *Ku* homologue). Mutations at the amino terminus of the *HISTONE 4* gene have a similar effect. Overexpression of *SIR2* causes hypoacetylation of this histone while *SIR3* mutations may alter the conformation of this histone bound to *HMR*. Loci *HML* and *HMR* are flanked by *HML-E* and *HML-I* silencer elements. These silencers are similar to the autonomously replicating elements of yeast involved in DNA synthesis and apparently also in silencing. *HML-E* is capable of repression only in the presence of *HML-I* and 0.4 kb distal from *HML-I* (based on Loo & Rine. 1994. *Science* 264:1768). *HMR-E* is a very potent silencer endowed with binding sites for ORC (origin recognition complex), Rap1 (a suppressor of RAS-induced replication), and Abf1 (another silencer) suppressors of the S phase of the cell cycle.

Evidence indicates that transcriptional silencing does not require DNA replication, although it seems to require some cell cycle events. Silencer elements are present in animal and plant systems. In plants, when multiple copies of a gene are introduced into the genome by transformation, all or most copies of the gene are inactivated (trans-inactivation). The mechanism of this phenomenon is unclear. It has been suggested that when the level of a particular RNA is increased, a degradative process is initiated by RNAi. This has been attributed to a defense mechanism, since the majority of plant viruses are RNA viruses. Some of the silencing appears, however, posttranscriptional. In fungi, silencing has been attributed to premeiotic methylation when multiple copies are present in cis-position. This view is supported by the longstanding knowledge that the repetitive sequences of heterochromatin are not expressed. Position effect has been known as a type of silencing. The reversible type of paramutation can be considered as a trans-inactivation mechanism. Silencing of genes may be accomplished by moving the region of the intact chromosome toward the centromere (position effect). Transposition to the vicinity of heterochromatin may also result in silencing. Sir1 protein is most common in the telomeric region of yeast chromosomes, but it may be present in the nucleolus. Sir3 and Sir4 are normally absent from the nucleolus, but they are present in the nucleolus if Sir2 is mutant. Although silencers have some similarities to insulators, the latter are different because they must be situated in between the enhancer and the target promoter. Hypermethylation of CpXpG nucleotides by chromomethylase 3 in *Arabidopsis* silences the expression of some genes. Its mutation restores wild phenotype to the epigenetically silenced genes and reactivates retrotransposons. *See* Abf; antisense technologies; cosuppression; CREB; dominant-negative mutation; enhancer; epigene conversion; epigenesis; heterochromatin; histone deacetylase; HML; HMR; Hs+1p; insulator; Ku; mating type determination in yeast; methylation of DNA; neuron-restrictive silencer factor; nucleosome; ORC; paramutation; PIC; position effect; posttranscriptional gene silencing, preinitiation complex; guelling; RAP1; RNAi; *Schizosaccharomyces pombe*; targeting genes; telomeric silencing; transcriptional gene silencing. (Guareente, L. 1999. *Nature Genet.* 23:281; Kirchmaier, A. L & Rine J. 2001. *Science* 291:646; Lindroth, A. M., et al. 2001. *Science* 292:2077; Sijen, T., et al. 2001. *Curr. Biol.* 11:436; Ogbourne, S. & Antalis, T. M. 1998. *Biochem. J.* 331:1; Moazed, D. 2001. *Mol. Cell* 8:489; Mlotshwa, S., et al. 2002. *Plant Cell* 14:S289; Béclin, C., et al. 2002. *Current Biol.* 12:684.)

Silene   *See Melandrium.*

silent mutation   Base-pair substitution in DNA that does not involve amino acid replacement in protein and entails no change in function. *See* mutation.

silent sites   Where mutations in the DNA base sequence have no consequence for function. *See* synonymous codons.

siliconization   Glassware is treated in a vacuum by dichlorodimethylsilane in order to prevent sticking of DNA molecules to the vessel, which would result in loss of recovery. *See* DNA extraction.

silique   Typical fruit of cruciferous plants; two carpels, dehiscing at the base at maturity, enclose the placentae, which sit in one row on each of the opposite sides of the replum.

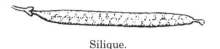

Silique.

silk fibroin   Protein rich in glycine and alanine residues arranged largely in β-sheets (β-keratin). It is synthesized within the silk gland to protect the pupa. The pupa is also called the chrysalis or cocoon. A fibroin gene is transcribed into about 10,000 long-life molecules of mRNA within a few days and they are translated several times into about a billion protein molecules. Each gland manufactures about 10^{15} fibroin molecules (300 μg) in 4 days. Actually, the gland is a single cell, but it contains polytenic chromosomes and thus the fibroin locus is amplified about a million-fold ($10^9 \times 10^6 = 10^{15}$ fibroins). In spiders, there is great diversity and conservation in the silk fribroin genes. Some of the spiders' silks are tougher than those of the silkworm and rival the best synthetic fibers. Spider silk can be synthesized in transgenic tobacco, potato plants, and transgenic mammalian cells. *See* polytenic chromosomes; silkworm. (Vollrath, F. & Knight, D. P. 2001. *Nature* 410:541; Scheller, J., et al. 2001. *Nature Biotechnol.* 19:573; Lazaris, A., et al. 2002. *Science* 295:472.)

silkworm (*Bombyx mori*, $2n = 56$)   One of the genetically best-studied insects. There are about 200 markers in the genetic map of ~900 cM. Its RAPD map (~2,000 cM) includes ~1,018 markers scattered over all chromosomes. Its genome contains a special type of transposable element, R2Bm, which is also present in some other insects. R2Bm has no long terminal repeats. It is inserted in the 28S rRNA genes only and encodes an integrase and a reverse transcriptase function within one protein molecule. The R2 protein nicks one of the DNA strands and uses it as a primer to transcribe its RNA genome, which is then integrated as a DNA-RNA heteroduplex. Subsequently a host polymerase synthesizes the second DNA strand. *See* autosexing; complete linkage; polyhedrosis virus; RAPD; silk fibroin; tetraploidy; transposon. <http://www.ab.a.u-tokyo.ac.jp/silkbase/>.)

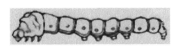

Silkworm larva.

Silver-Russel syndrome   *See* Russel-Silver syndrome.

Silver syndrome (Silver spastic paraplegia, SPG17, 11q12-q14)   Neurodegenerative disease involving amyotrophy (muscle weakness) in the hands. Hereditary dominant spastic paraplegia is a highly variable disease, and it is encoded in several other chromosomes and locations. *See* paraplegia; spastic paraplegia. (Patel, H., et al. 2001. *Am. J. Hum. Genet.* 69:209.)

silyl-phosphite chemistry   Oligoribonucleotide synthesis. (Agarwal, S., ed. 1995. *Methods in Molecular Biology*, p. 81. Humana, Totowa, NJ.)

simian   ape or monkey type. *See* hominidae; primates.

simian crease   It can be rarely observed (1–4%) on normal infants, but it is characteristic for human trisomy 21, De Lange, Aarskoog, and other syndromes. *See* Aarskoog syndrome; De Lange syndrome; Down syndrome for illustration.

simian sarcoma virus (SSAV)   Gibbon/ape leukemia retrovirus with a homologous element in human chromosome 18q21. The long terminal repeat (535 bp) appears to contain transcriptional control and signal sequences. Human chronic lymphatic-type leukemia seems to be associated with a break point of chromosome 18. *See* leukemia.

simian virus 40   Eukaryotic virus of a molecular weight of 3.5×10^6 with double-stranded, supercoiled DNA genetic material of 5243 bp. The DNA is organized into a nucleosomal structure that does not have H1 histone. The DNA around the nucleosome cores is 187 ± 11 bp, and the cores are separated by 42 ± 39 bp linkers. The viral particles are skewed icosahedral capsids and have 72 protein units. In primates, generally the virus follows a lytic lifestyle and the virions multiply in the cytoplasm, i.e., primates are *permissive hosts* for replication.

Occasionally, in humans the viral DNA integrates into the chromosomes. Such an event may lead to cancerous transformation. Rodent cells are *nonpermissive* hosts for viral replication, and the viral DNA integrates into the chromosomes, leading to cancerous tumor formation. Virus codes for early (t and T antigens) and late (VP1, 2, 3) viral proteins. Viral replication and transcription are bidirectional. In the nonpermissive host, only the early genes are expressed that are needed for replication of the genetic material before integration, but there is no need for the coat proteins. The integration can take place at different sites; therefore, it uses a mechanism of illegitimate recombination. The few integrated copies may be rearranged and may cause continued chromosomal rearrangement in the host. The infectious cycle

spans about 70 h. The joint replication and transcriptional origin (*ori*) area extends to about 300 bp and includes a rather sophisticated control system. The replication of SV40 DNA begins at the 27 bp palindrome of the *ori* site adjacent to a region consisting of 17 A-T base pairs. Next to it, on the side of the late genes, there are three other units of 22, 21, and 21 GC-rich repeats that also promote replication, although they are not absolutely essential to the process.

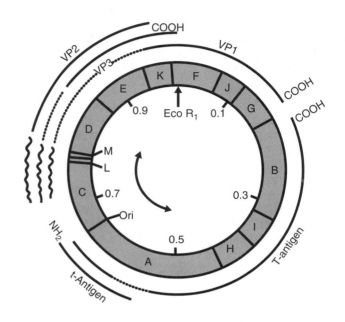

The SPH elements have an overlapping *octamer* that is present in other eukaryotic genes, and several other sequence motifs are similar in other promoters. The 72 elements include the 47 bp B and the shorter 29 bp A domains that are parts of the essential enhancer region. The A-T box is essentially a TATA box: 5′-TATTTAT-3′. For the start of replication, the large T has to bind to the Tag-binding sites. At the initiation when low amounts of Tag are available, binding begins at the T1 site located at the pre-mRNA region (right of *ori*). As more Tag becomes available, Tag binds to T2 and becomes an ATP-dependent helicase. With the cooperation of cellular proteins (DNA primase, polymerase, etc.), DNA replication proceeds. *Transcription*: For the function of *ori* in

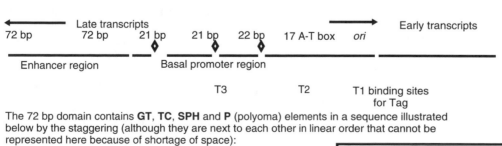

The 72 bp domain contains **GT**, **TC**, **SPH** and **P** (polyoma) elements in a sequence illustrated below by the staggering (although they are next to each other in linear order that cannot be represented here because of shortage of space):

GT II: 5′-GCTGTGGAATGT-3′
GT I: 5′-GGTGTGGAAATG-3′
TC I and **TC II**: 5′-TCCCCAG-3′
SPH - II: 5′-AAGTATGCA-3′
SPH - I: 5′-AAGCATGCA-3′
P: 5′-TTAGTCA-3′

Each tag binding motif has a minimum of two 5′-GAGGC-3′ of this pentanucleotide consensus transcription, the 17 bp A-T sequence is needed, although this TATA box does not affect the rate of transcription. The 21◇21◇22◇ sequences promote transcription and the 2 hexamers 5'-GGGCGG-3' within these elements are essential for transcription. Their orientation and inversion do not interfere with transcription. The 72 bp element of SV40 is a capable enhancer for mammalian, amphibian, plant, and fission yeast genes. The natural host of SV40 is the rhesus monkey (*Macaca mulatta*); in the laboratory, the kidney cell cultures of the African green monkey (*Cercopithecus aethiops*) are used primarily for its propagation. Single cells may produce 100,000 viral genomes after lysis. The general assumption is that SV40 does not cause human cancer, yet in many tumors its presence was detected by some laboratories. *See* COS cell; SV40 vectors. (Butel, J. S. & Ledniczky, J. A. 1999. *J. Natl. Cancer Inst.* 91[2]:119; see reviews in *Semin. Cancer Biol.* [2001], 11:5–85.)

SimIBD   Computer procedure to assess affected relative-pair calculations.

similarity index   *See* character index.

SIMLINK   Simulation-based computer program for estimating linkage information. *See* simulation; SLINK.

simple protein   Upon hydrolysis; only amino acids are produced.

simple sequence length polymorphism (SSLP)   Variations in microsatellite sequences can be used for DNA mapping. PCR primers are designed for the unique flanking sequences, and their length can be determined in the PCR products. *See* microsatellite; PCR. (Cai, W. W., et al. 2001. *Nature Genet.* 29:133.)

simplesiomorphy   Primitive features retained during evolution. It is not very useful to trace evolutionary development. *See* homology; synapomorphy.

simplex   Polyploid having only a single dominant allele at a particular gene locus; the other alleles are recessive at the locus. *See* duplex; quadruplex; triplex.

simplexvirus   Member of the herpes family of viruses infecting humans and other mammals. *See* herpes.

Simpson-Golabi-Behmel syndrome (Simpson dysmorphia, SGBS)   Human Xq26 (500-kilobase stretch) syndrome, encoding the GPC3 gene responsible for the synthesis of glypican (cell-surface molecules of heparan sulfate proteoglycans) and associated with insulin-like growth factor (IGF2). Individuals afflicted by this condition are generally very tall, usually have facial anomalies, heart and kidney defects, cryptorchidism, hypospadia, hernias, bone anomalies, and show susceptibility to cancer. Many of the symptoms involving overgrowth are shared with Beckwith-Wiedeman syndrome. *See* Beckwith-Wiedeman syndrome; IGF.

simulation   Representation of a biological system by a mathematical model generated frequently by a computer program. *See* modeling; Monte Carlo method. (<www.nrcam.uchc.edu>.)

Sindbis virus   Single-stranded RNA virus.

sine   Ratio of the side opposite (a) to an acute angle (A) of a right triangle and the hypotenuse (c): a/c, sine of angle A. *See* angular transformation; arcsine.

SINE   Short (0.5 kbp) interspersed repetitive DNA sequences that may occur over 100,000 times in the mammalian genomes. The B1 SINE of mice is 130 to 150 bp in length and constitutes nearly 1% of the genome; it is homologous to the human Alu sequences. The B2 SINE (~190 bp) constitutes about 0.7% of the mouse genome, but it apparently has no human homologue or its abundance is very low. RNA polymerase III transcribes the B2 SINE elements into short sequences that are not translated. The mouse gene carries an active RNA polymerase II promoter and can support transcription by pol II (Ferrigno, O., et al. 2001. *Nature Genet.* 28:77). The SINE elements are retroposons but lack reverse transcriptase function. SINE-type elements occur in all eukaryotes, including birds, fungi, insects, and higher plants. They may be pseudogenes of small RNA genes. The SINE sequences can also be used for fingerprinting and evolutionary studies. These are remnants of ancient retroviral insertions, but once inserted — because of the loss of the LTR (transposase) function — they remained at the position of insertion. *See* Alu family; DNA fingerprinting; LINE; retroposons; reverse transcriptase; transposable elements. (Cantrell, M. A., et al. 2001. *Genetics* 158:769; Weiner, A. M. 2002. *Current Opin. Cell Biol.* 14:343.)

singing ability   Has genetic determination. Some of the song genes of birds have been mapped for expression in different parts of the brain. (Marler, P. & Doupe, A. J. 2000. *Proc. Natl. Acad. Sci. USA* 97:2965.)

single burst experiment   Virus-infected bacterial population diluted and distributed into vessels in such a way that each vessel contains a single infected bacterial cell. (*See* Ellis, E. L. & Delbrück, M. 1939. *J. Gen. Physiol.* 22:365.)

single-cell analytical methods (chemical/physical)   *See* FISH; immunocytochemistry; immuno-electron microscopy; immunoelectrophoresis; immunofluorescence; MALDI/TOF/ MS. (Cannon, D. M., Jr., et al. 2000. *Annu. Rev. Biophys. Biomol. Struct.* 29:239. Slepchenko, B. M., et al. 2002. *Annu. Rev. Biophys. Biomol. Struct.* 31:423.)

single-copy plasmids   *See* plasmids.

single-copy sequence   DNA sequence containing nonredundant, genic portions.

single cross   *See* double cross.

single-end invasion (SEI)   Meiotic recombination intermediate during the transition from double-strand breaks to double-Holliday junction. SEIs are formed by strand exchange between one and then the other double strand. The appearance of SEI coincides with that of the synaptonemal complex. SEI is preceded by a nascent double-strand partner intermediate that differentiates into a crossover and a noncrossover type after the synaptonemal complex has formed. Strand exchange occurs relatively late after synapsis and recombination may be avoided between homeologous and structurally rearranged partners. *See* Holliday model; homeologous; synaptonemal complex. (Hunter, N. & Kleckner, N. 2001. *Cell* 106:59.)

single-gene trait Controlled by one gene locus and shows monogenic inheritance.

single-nucleotide polymorphism *See* SNIPS.

single-positive T cell Expresses either the CD4 or CD8 surface protein. *See* CD4; CD8; T cell.

single-strand annealing repair See SSA.

single-strand assimilation A single strand displaces another homologous strand, then takes its place during a recombinational event. *See* recombination, molecular models.

single-strand binding protein Binds to both separated single stands of DNA and thus stabilizes the open region to facilitate replication, repair, and recombination. *See* binding proteins; recombination, molecular mechanisms.

single-strand conformation polymorphism (SSCP) When small deletions or even single-base substitutions take place in one of the DNA strands of a gene locus, this alteration may be detectable by the electrophoretic mobility of the DNA in denaturing polyacrylamide gels. The two strands—the normal and the affected—may differ. If the individual is heterozygous for the amplified segment of the locus concerned, the electrophoretic analysis may indicate three or more band differences. In some cases, even the homozygotes may show multiple bands. With this method nearly all of the alterations are detected in fragments of 200–300 bp. *See* DGGE; dideoxy fingerprinting; gel electrophoresis; gene isolation; MASDA; mutation detection; polymerase chain reaction; (Orita, M., et al. 1989. *Proc. Natl. Acad. Sci. USA* 86:2766.)

SINGLET Gene that occurs only once in the genome.

singleton Singly occurring whole-body mutations; the spontaneous frequency in mice for seven standard loci is 6.6×10^{-6} per locus. *See* mutation rate.

Singleton-Merten syndrome Rare disease involving aortic calcification but defects in bone development.

singlet oxygen ($^1O_2$) Highly reactive O_2 molecule produced during inflammation by photosensitization in UV light, chemiexcitation in dark, decomposition of $NDPO_2$, etc. $^1O_2$ may be toxic to molecules in the cell, oxidize DNA and produce mutagenic 7-hydro-8-oxodeoxyguanosine. It may affect gene expression and carcinogenesis. *See* 8-oxodeoxyguanosine; ROS.

sink Storage of metabolites from where they can be mobilized on need.

sink habitat In which some individuals contribute less to the future generations than the average individual. *See* habitat; source habitat.

sinndakiss Receptor internalization signal.

Sinorhizobium *See* nitrogen fixation.

Sin3/RPD Repressor protein complex probably involved in chromatin remodeling by being recruited to histone deacetylase. It may be associated with SAP18, SAP30, and retinoblastoma-binding protein. *See* chromatin remodeling; histone deacetylase; Mad; NuRD. (Brubaker, K., et al. 2000. *Cell* 103:655.)

SIN vector Self-inactivating vector has a deletion in the U3 element of the 3′-LTR of the retroviral construct. After replication, it results in a deletion also in the 5′-LTR promoter and enhancer and prevents the transcription from the cell-specific internal promoter that may otherwise activate silent cellular oncogenes. This happens because the viral polymerase enzyme uses the 3′-U3 as a template for the replication of both 3′- and 5′-U3 sequences. A disadvantage of this construct is the generally slow replication. A high-efficiency heterologous promoter to enhance the expression of the transgene in the retroviral vector may also replace the deleted viral promoter. *See* double-copy vector; E vector; retroviral vector. (Gatlin, J., et al. 2001. *Hum. Gene Ther.* 12:1079.)

siphonogamy Immotile microgametes of higher plants are delivered to the archegonia through the elongating pollen tube. *See* embryo sac; pollen tube; zoidogamy.

Sipple syndrome *See* pheochromocytoma.

SIR *See* silencer.

sire Male mammal. The term is used primarily in animal breeding and applied animal genetics. *See* dam.

sirenomelia Developmental malformation showing fused legs and usually lack of feet.

SIRM Sterile insect release method. *See* genetic sterilization.

siRNA *See* RNAi.

sirolimus *See* rapamycin.

SIRPs Signal regulatory proteins (20p13). Members of this family inhibit signaling through tyrosine kinase receptors and represent immune inhibitory receptors expressed on macrophages or other blood cells. *See* macrophage; tyrosine kinase receptor. (Latour, S., et al. 2001. *J. Immunol.* 167:2547.)

sirtuin NAD-dependent histona deacetylase proteins of the Sir2 family. *See* silencer. (Grozinger, C. M., et al. 2001. *J.*

Biol. Chem. 276:38837; Pandey, R., et al. 2002. *Nucleic Acids Res.* 30:5036.)

SIS Simian sarcoma virus oncogene is in human chromosome 22q12.3-q13.1 and mouse chromosome 15. The SIS protein has high homology to the β-chain of the platelet-derived growth factor (PDGF), KIT oncogene, FOS oncogene, and colony-stimulating factor. *See* colony-stimulating factor; oncogenes; PDGF. (Liu, J., et al. 2001. *Nucleic Acids Res.* 29:783.)

Sis1 DnaJ structural homologue of budding yeast indispensable protein with multiple chaperone functions, including initiation of translation. *See* chaperones; DnaJ; DnaK.

sister chromatid cohesion Juxtaposition of the sister chromatids until the end of metaphase in mitosis and until the end of metaphase II in meiosis. The inner centromere proteins (INCENP) and the centromere-linking proteins (CLiP) provide the physical basis of the cohesion. The multiprotein cohesion complex, which binds most tightly to the centromere, is cohesin. A separation protein (separin, a cyteine protease) mediates sister chromatid cohesion, and the dissociation is achieved when the Scc1/Mcd1/Rad21 subunit of cohesin dissociates from the chromatids upon proteolytic cleavage. Rec8 is a component of the meiotic cohesin complex. Esp1 (separin) is tightly bound to the chromosomes by the anaphase inhibitor Pds1 (mammalian homologue securin). Pds1 is ubiquitinated by the triggering effect of the anaphase-promoting complex (APC) and Cdc20. The sister chromatids are closely juxtapositioned until anaphase, indicating the presence of intersister connector structures. Sister chromatid cohesion affects proper disjunction of the mitotic chromatids, but it appears important for meiotic recombination. In mitosis, the separation of the sister chromatids and the splitting of the centromere take place during the single anaphase. In meiosis, at anaphase I the sister chromatids separate, but the centromere does not separate until anaphase II. This timing is apparently under the control of a specific protein(s). For the orderly segregation of the sister chromatids during meiosis I, the protein monopolin is required in yeast. Dominant and recessive mutations have been identified in plants, animals, and yeast that are defective in chromatid cohesion. In yeast, centromeric element III (CDEIII) is essential for sister chromatid cohesion and for kinetochore function. *See* adherin; asynapsis; cell cycle; centromere; checkpoint; chiasma; cohesin; condensin; desynapsis; DNA polymerases; meiosis; mitosis; sister chromatid exchange; sister chromatids; synapsis. (Nasmyth, K., et al. 2000. *Science* 288:1379; Tóth, A., et al. 2000. *Cell* 103:1155; Carson, D. R. & Christman, M. F. 2001. *Proc. Natl. Acad. Sci. USA* 98:8270; Lee, J. Y., et al. 2001. *Annu. Rev. Cell Dev. Biol.* 17:753.)

sister chromatid exchange (SCE) Detectable in eukaryotic cells provided with 5-bromo-deoxyuridine for (generally) one cycle of DNA replication. Subsequently, at metaphase the chromosomes are stained with either the fluorescent compound Hoechst 33258 (harlequin staining) or according to a special Giemsa procedure. If sister chromatids are reciprocally exchanged, sharp bands appear in mirror image–like fashion. The frequency of sister chromatid exchange is boosted by about a third by potential carcinogens and mutagens. This method

has been successfully used in various animal and plant cells for identifying genotoxic agents. The data must be evaluated with care in comparison with the concurrent control because BrdU itself may break chromosomes under UV-B light. In *Saccharomyces cerevisiae*, molecular and genetic evidence are available for meiotic sister chromatid exchange. When one of the bivalents had a different number of ribosomal RNA repeats with an embedded LEU2 gene, duplication and deficiency of LEU2 and the repeats were detected. Similar observations were made with other chromosomes and markers. *See* bioassays in genetic toxicology; BrdU; chiasma; cohesin; crossing over; genotoxic; Giemsa staining; harlequin staining; ring chromosomes; sister chromatids; ultraviolet light. (Shaham, J., et al. 2001. *Mutation Res.* 491:71.)

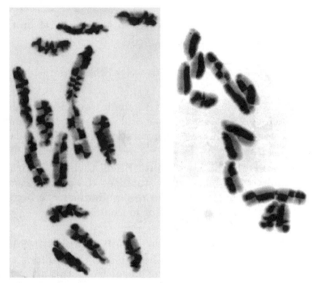

Sister chromatid exchange. right: untreated control. left: exposed to the alkylating compound thiotepa during DNA synthesis. (Courtesy of Professor B. A. Kihlman.)

sister chromatids Attached to the same side of the same centromere, but they seem to be coiled in opposite directions. Their separation in mitosis requires the activation of a proteolytic enzyme encoded by the *Cut2* gene in *Schizosaccharomyces pombe*. *See* chromatids. (Nasmyth, K. 2001. *Annu. Rev. Genet.* 35:673.)

Sister chromatids.

sister-strand exchange Sister chromatid exchange.

SIT (1) Family of protein phosphatases regulating diverse metabolic pathways. *See* PP2A. (2) Sterile insect technique. *See* genetic sterilization; GSM.

site-directed immunization *See* Immunization, genetic.

site-directed mutagenesis *See* Directed mutation; localized mutagenesis; targeting. (Storici, F., et al. 2001. *Nature Biotechnol.* 19:773.)

site-specific cleavage Of nucleic acids, is accomplished by restriction endonucleases. Some special RNases and oligonucleotide-phenanthroline conjugates may cut both strands of the DNA in the presence of Cu^{2+} and a reducing agent. $EDTA-Fe^{2+}$ may do the same if tethered to triplex molecules, albeit with low efficiency. In the presence of light, ellipticine attached to homopyrimidines may cleave a double helix within a triplex. *See* restriction endonucleases; tethering; triplex. (Gimble, F. S. 2001. *Nucleic Acids Res.* 29:4215.)

site-specific mutations Occur at particular nucleotides in the DNA and RNA, respectively. *See* alanine-scanning mutagenesis; base substitution; cassette mutagenesis; cystein-scanning mutagenesis; degenerate oligonucleotide-directed mutagenesis; gene replacement; homologue-scanning mutagenesis; Kunkel mutagenesis; localized mutagenesis; oligonucleotide-directed mutagenesis; PCR-based mutagenesis; site-specific recombination; TAB mutagenesis; targeting vector.

site-specific recombinases Resolvases that attach at the two-base staggered-cut sites. The enzyme is then covalently linked to the 5'-ends and the PO_4 of the DNA is covalently linked to the OH group of the recombinase. Subsequently, the broken DNA strand releases the deoxyribose hydroxyl group. PO_4 is joined to another deoxyribose OH group and the DNA backbone is reconstituted. The members of the integrase group of enzymes attach at sites 6–8 bases apart. The first breakage results in a Holliday juncture, which may lead to branch migration. After a second strand exchange and rotation isomerization (*see* Holliday model, steps H–J), the strands may be resolved either with an outside marker exchange (classical recombination) or in gene conversion (the constellation of the outside markers retained). It is conceivable that the broken ends are reconstituted without any change or deletions may take place, or the position of the broken ends is inverted by 180° resulting in what classical cytology called inversion. Resolvase and integrase reactions can be very specific for the sites, and the reaction is secured by the assistance of additional proteins that bring into contact only the appropriate DNA stretches. These two enzymes act only on supercoiled DNA. The integrase family of recombinase enzymes is more liberal in choice yet affected by various conditions. The Mu phage or the HIV integration does not require covalent association between the DNA and a protein. The phosphodiester bond of the donor DNA is hydrolyzed first to generate an OH group. Between this group and a phosphodiester group of the receiving DNA, the joining takes place and thus the strand is integrated. *See* Cre/loxP; FLP/FRT; Holliday juncture; homing endonucleases; integrase; phosphodiester linkage; resolvase; site-specific recombination; transesterification. (Woods, K. C., et al. 2001. *J. Mol. Biol.* 313:49.)

site-specific recombination Occurs when recombination is limited to a specific few nucleotide sequences. Homology may be present at the exchange region in both recombining molecules like at the integration-excision site of temperate phage. Or the specificity is limited to only one of the partners, like at the 25 bp termini of T-DNA or the direct and indirect repeats of transposable elements. In the latter cases, the recombinational target sites may have no or only minimal similarity. *See* chromosomal rearrangement; Cre/Lox; FLP/FRT; gene replacement; integrases; knockout; lambda phage; ligand-activated site-specific recombination; recombination; shoufflons; site-specific recombinase; switching; targeting genes; T-DNA. (Sauer, B. & Henderson, N. 1988. *Proc. Natl. Acad. Sci. USA* 85:5166; Pena, C. E., et al. 2000. *Proc. Natl. Acad. Sci. USA* 97:7760; Christ, N., et al. 2002. *J. Mol. Biol.* 319:305.)

sitosterolemia (phytosterolemia, STSL, 2p21) Rare, recessive hypercholesterolemia, resulting in more than a 30-fold increase of the level of this plant cholesterol in the plasma. Intestinal absorption of sterols is increased and the excretion of sterols into the bile is impaired. Initially it causes xanthomatosis and later premature coronary artery disease. Actually, two genes encoding sterolin 1 and sterolin 2 are involved in opposite orientation separated by a short interval. Sterolins apparently regulate sterol transport. *See* cholesterol; familial hypercholesterolemia; low-density lipoprotein; VLDL; xanthomatosis. (Lee, M.-H., et al. 2001. *Nature Genet.* 27:79; Lu, K., et al. 2001. *Am. J. Hum. Genet.* 69:278.)

situs inversus viscerum Malformation of mammals, including humans, where the internal organs such as the heart are shifted to the right side of the chest (thorax). It is frequently accompanied by chronic dilation of the lung passages (bronchi) and inflammation of the sinus. The latter disorder is also called Kartagener syndrome, which is characterized by immotility of sperm and cilia. The anomaly may by either autosomal or X-linked recessive. Its incidence in the general population may be about 1/10,000. In the mouse, the genes *iv* (chromosome 12) and *inv* (chromosome 4) disturb left-right axis formation and cause 50 and 100% manifestation of situs inversus, respectively. In the chicken, fibroblast growth factor (FGF8) mediates the determination of the right side and Sonic hedgehog (SHH) of the left side. In the mouse, FGF8 is instrumental in the left-side and SHH in right-side specification. *See* asymmetry of cell division; axis of asymmetry; dynein; FGF; heterotaxy; isomerism; Kartagener syndrome; left-right asymmetry; sonic hedgehog; ciliary dyskinesia. (Bartoloni, L., et al. 2002. *Proc. Natl. Acad. Sci. USA* 99:10282.)

SIV Simian immunodeficiency virus is a relative of HIV. *See* acquired immunodeficiency; HIV.

size Depends primarily on cell number and cell size and is developmentally and genetically determined. (Conton, I. & Raff, M. 1999. *Cell* 96:235.)

Sjögren-Larsson syndrome *See* ichthyosis.

Sjögren syndrome Autosomal-recessive autoimmune disease leading to the destruction of the salivary and lacrimal glands by the production of autoantibody against the SS-A (Ro RNA) and SS-B (La Sn RNA) particles. The autoantigens have been identified and purified. The Ro autoantigen appears to be encoded in human chromosome 19pter-p13.2. The La autoantigen may be involved with RNA polymerase III. The 120 kDa α-fodrin appears to be the critical autoantigen that elicits the disease. *See* autoimmune disease; fodrin; RoRNP.

SK Calcium-activated potassium ion channels. *See* ion channels.

7SK Small nuclear RNA (snRNA) of ubiquitous presence and involvement in the control of transcription. *See* snRNA. (Yang, Z., et al. 2001. *Nature* 414:317.)

skeletal map Uses only microsatellite marker data. *See* framework map; genetic map; integrated map; physical map; radiation mapping; recombination minimization map.

skewed distribution Data are not symmetrical around the mean; one of the extreme flanks is predominant. *See* kurtosis; normal distribution; graph below.

skewness Asymmetry in the distribution frequency of the data. *See* kurtosis; moments; normal distribution.

Skewness.

Ski Protein discovered at the Sloan Kettering Institute as a viral factor in tumorigenesis, Ski occurs in vertebrates and insects. Along with Sno (Si-related novel gene), it regulates the effect of Smad4 and Smad3 proteins, which may negatively control gene expression in response to the phosphorylation signals coming from TGF-β. It interacts with Skip, a transcriptional activator. *See* Smad; TGF. (Prathapam, T., et al. 2001. *Nucleic Acids Res.* 29:3469.)

S6 kinase (RSK, S6K) Collective name for cytosolic p70^s6k and the nuclear p85^s6k kinases that phosphorylate the S6 ribosomal protein before translation initiation. The supply of amino acids affects the process. Mutation in the genes results in reduced cell and body size. Mice deficient for S6K are glucose intolerant and hypoinsulinemic. The carboxyl end of S6 and the phosphorylation sites within are highly conserved from *Drosophila* to humans. *See* cell size; insulin; phosphatidylinositol; platelet-derived growth factor; S6 ribosomal protein; 5'-TOP; translation initiation. (Dufner, A. & Thomas, G. 1999. *Exp. Cell. Res.* 253:100.)

skin color *See* pigmentation of animals. (Sturm, R. A., et al. 1998. *Bioessays* 20[9]:712.)

skin diseases *See* acne; acrodermatitis; blisters; connexin; cutis, laxa; dermatitis; dyskeratosis; ectodermal dysplasia; eczema; epidermolysis; epithelioma; erythrokeratoderma variabilis; Fabry disease; familial hypercholesterolemia; focal dermal hypoplasia; Gardner syndrome; glomerulonephritis; ichthyosis; keratosis; light sensitivity; lupus erythematosus;

nevus; pemphigus; pigmentation defects; porphyria; pseudoxanthoma elasticum; psoriasis; Rothmund-Thompson syndrome; scleroderma; vitiligo; Werner syndrome.

SK oncogene Probably regulates tumor progression; it was assigned to human chromosome 1q22-q24.

skotomorphogenesis Morphogenesis without dependence on light. *See* deetiolation; photomorphogenesis.

SKP Cyclin A-CDK2-associated protein. Intrinsic kinetochore protein (22.3 kDa) widely conserved among species. It coordinates centromeres, centrosomes, and other cell cycle factors. Skp1p is a proteasome-targeting factor. Skp2^−/− mice are viable yet have a reduced growth rate, polyploid cells, and they accumulate cyclin E and p27^Kip1 proteins that they cannot efficiently eliminate during the S and G2 phases of the cell cycle. For degradation, SCF^Skp2 is required. Skp2 is upregulated in some epithelial carcinogenesis. *See* CDC4; CDK; cell cycle; cyclin A; F-box; kinetochore; proteasome; SCF; von Hippel–Lindau disease. (Nakayama, K., et al. 2000. *EMBO J.* 19:2069; Latres, E., et al. 2001. *Proc. Natl. Acad. Sci. USA* 68:2515.)

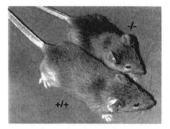

Skp2-disrupted mice are deficient in the F-box protein and SCF ubiquitin ligase has enlarged nuclei, multiple centrosomes, and reduced growth. They accumulate cyclin E and p27^Kip. (From Nakayama, K., et al. 2000. *EMBO J.* 19:2069.)

skunk *Mephitis mephitis*, $2n = 50$; *Spilogele putorius*, $2n = 64$.

SKY Spectral karyotyping. *See* spectral karyotyping.

Sky Cellular tyrosine kinase. It regulates B-cell development. *See* B lymphocyte. (Kishi, X. A., et al. 2002. *Gene* 288:29.)

SL1 Transcription factor complex of RNA polymerase I. It is a complex of the TATA-box-binding protein (TBP) and the three TATA-box-associated factors (TAF). The TBP protein binds exclusively either SL1 (RNA pol I) or TFIID (RNA pol II). In the case of RNA pol III, TFIIIB is required for the recruitment of the polymerase to the promoter complex. *See* pol I; pol II; TAF; TBP; transcription factors.

SL1, SL2 Spliced leader involved in transsplicing in *Caenorhabditis*. The 100-nucleotide leader donates its 5'-end 22 nucleotides to a splice acceptor site on the primary transcript. Transsplicing is very common (70%) among the nematode's genes. This mechanism is used for the coordinately regulated gene clusters transcribed in polycistronic RNA. The nematode operons use SL2, whereas other genes use SL1. *See* coordinate regulation; operon; transsplicing.

slalom library   Based on a combination of the principles of linking and jumping libraries. *See* jumping library; linking library. (Zabarovska, V. I., et al. 2002. *Nucleic Acids Res.* 30:[2]:e6.)

SLAM   Signaling lymphocyte activation molecule (CDw150) is a T-cell receptor protein (M_r 70K) of the immunoglobulin family, constitutively and rapidly expressed on activated peripheral blood memory T cells, immature thymocytes, and some B cells. It is a receptor for the measles virus. T cells carrying CD4$^+$ antigens produce increased amounts of interferon γ without an increase of interleukins 4 and 5. SLAM function is independent of CD28. *See* CD28; Epstein-Barr virus; interferon; interleukin; SAP; T cell. (Bleharski, J. R., et al. 2001. *J. Immunol.* 167:3174.)

SLAP (Fyb/Slap)   One of the adaptor proteins that regulates TCR-mediated signal transduction. Cbℓ has an inhibitory effect. SLAP interacts with Sky, ZAP-70, and LAT. *See* CBL; Fyb; signal transduction; TCR. (Peterson, E. J., et al. 2001. *Science* 293:2263.)

SLD   Yeast chromosomal replication protein acting after phosphorylation during the S phase. (Masumoto, H., et al. 2002. *Nature* 415:651.)

sleep   Circadian organization of rest after activity controlled by several neural genes. Apparently mutation of the human homologue of Per2 (2q) may be responsible for the familial advanced sleep-phase syndrome. The point mutation (Ser → Gly) is within the casein kinase Iε and alters the circadian clock. Sleep may have a weak role in the consolidation of memory. *See* apnea; circadian rhythm; memory; narcolepsy. (Siegel, J. M. 2001. *Science* 294:1058; Shaw, P. J., et al. 2002. *Nature* 417:287; Pace-Schott, E. F. & Hobson, J. A. 2002. *Nature Rev. Neurosci.* 3:591.)

Sleeping Beauty   Artificially constructed human mariner transposable element equipped with a salmon transposase function, enabling the otherwise nonmobile element to move in HeLa or other somatic cells by a cut-and-paste mechanism. Another type of transposon vector (pTnori) is outlined above. *See* cut-and-paste; HeLa; mariner; transposase. (Horie, K., et al. *Proc. Natl. Acad. Sci. USA* 98:9191; Izsvak, Zr., et al. 2002. *J. Biol. Chem.* 277:581.)

IR: inverted repeats, PRO: promoter, Tn5: bacterial transposon
NEO: selectable marker, ORI: origin of replication (pTnori)

sleeping sickness   Potentially fatal disease caused by *Trypanosomas*. The tse-tse fly spreads the disease. *See* *Trypanosomas*.

SLG   Self-compatibility locus-secreted glycoprotein. *See* Self-incompatibility.

sliding clamp   gp45 gene of phage T4 controls high speed of replication (processivity) by gp43 (DNA polymerase gene); gp44 and gp62 are the clamp loaders for gp45. The clamp hugs the DNA and slides along the duplex DNA. The replication complex is attached to the clamp. The cellular homologues for p45 are the β-units of the DNA polymerase III of prokaryotes and the PCNA of eukaryotes. The transcription may be coupled to replication and regulated by protein-protein and protein site-specific DNA interactions. In eukaryotes, the sliding clamp is PCNA. *See* clamp loader; DNA polymerases; PCNA; replication factor. (Fishel, R. 1998. *Genes Dev.* 12:2096; Trakselis, M. A., et al. 2001. *Proc. Natl. Acad. Sci. USA* 98:8368; Jeruzalemi, D., et al. 2001. *Cell* 106:417; Jeruzalemi, D., et al. Ibid., p. 429.)

SLINK   Computer program for estimating linkage information by a simulation approach. *See* SIMLINK.

slippage   Usually, when homopolymeric sequences are embedded in the template DNA strand, the RNA polymerase may synthesize RNA strands that are much longer (by a few to thousands of nucleotides) than the template. The slippage can be inhibited, however, when nucleotides, representing the next one to the homopolymeric stretch, are added. Frameshift mutations may be interpreted as the result of slippage. *See* attenuator region; microsatellite; mismatch repair; overlapping genes; replication slippage; slipping; unequal crossing over. (Viguera, E., et al. 2001. *EMBO J.* 20:2587.)

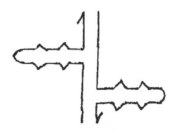

slipped-structure DNA (S-DNA)   Incomplete pairing within intrastrand folds (hairpins) in not exactly opposite position to each other. Such structure may form when there is a variable number of trinucleotide repeats in the DNA. *See* hairpin; trinucleotide repeats; diagram above.

slipping   Shifting of the translational reading frame. *See* hopping; recoding; slippage.

slip-strand mispairing   May cause replicational error if it is not corrected by repair, and it is frequently the cause of micro- and minisatellite instability. The slippage may cause additions to or deletions from the repeat sequences. This process is independent from the mechanism(s) of recombination (it is not unequal crossing over); flanking markers are not exchanged. In yeast, its frequency is about the same at mitosis and meiosis. Defects in the genes controlling replication and repair of the DNA may increase the instability and cause more mutation in microsatellites than in other types of DNA sequences. The packaging of DNA, temperature, methylation state, base composition, cell cycle stage, etc., may affect its frequency. Its rate may vary among different species because of differences in the replicase and mismatch repair enzymes. CTG repeats in the lagging strands are less stable than in the leading strand. *See* microsatellite; minisatellite; MVR; slippage; unequal crossing over. (Lewis, L. A., et al. 1999. *Mol. Microbiol.* 32:977.)

Slit Axon-repellent molecule with Robo as its receptor. *See* axon.

slithering Creeping-type motion toward each of the recombination sites. A recombinase enzyme within a supercoiled DNA molecule may mediate it in case of site-specific recombination. *See* site-specific recombination. (Huang, J., et al. 2001. *Proc. Natl. Acad. Sci. USA* 98:968.)

S locus *See* self-sterility alleles.

slot blot Binding cDNA or RNA onto slots → ● on membrane filters for the analysis of transcripts by hybridization to specific sequences.

slow component During a reassociation reaction of single-stranded DNA, the unique sequences anneal slowly. *See* annealing; c_0t value.

slow stop Bacterial mutant *dna* may slowly complete the replication underway but cannot start a new cycle at 42°C. *See* replication; strong-stop DNA.

SLP-76 76 kDa specific leukocyte protein adaptor. It binds to TCR and it is phosphorylated at tyrosines near the N-terminus. It provides SH2-binding sites for the VAV protein. SLP also binds the SH3 domain of Grb2. At the C-end, it associates with SLAP (SLP-76-associated phosphoprotein), resulting in activation of NF-AT. SLP-76 is essential for TCR activity and T-cell development in signaling pathways through the activity of phosphotyrosine kinases (PTK), but it is not required for macrophage and natural killer cells. *See* BASH; CD3; GADS; killer cell; macrophage; NF-AT; SH2; signal transduction; T-cell receptor; TCR genes; VAV Grb2. (Pivniouk, V. I., et al. 1999. *J. Clin. Invest.* 103:1737.)

7SL RNA RNA component in the signal recognition protein (SRP) complex. *See* Alu; signal sequence recognition particle.

SLS Sodium lauryl sulfate. *See* SDS.

SLT *See* specific locus mutations assay.

SLT-2 Protein kinase of the MAPK family. *See* MAPK; protein kinase; signal transduction.

slug (1) In general usage, a land mollusk. (2) *See* Dictyostelium.

Sly disease *See* mucopolysaccharidosis type VII.

***Sma* I** Restriction endonuclease; recognition site CCC↓GGG. *See* restriction enzymes.

SMAC Supra molecular activation cluster. CD4 T cells with antigen-presenting cells, receptors, and intracellular proteins form SMAC. (Potter, T. A., et al. 2001. *Proc. Natl. Acad. Sci. USA* 98:12624.)

Smac Second mitochondria-derived activator of caspases. Inhibits IAPs and facilitates apoptosis by caspase-3. It is homologous with Diablo. *See* apoptosis; DIABLO; IAP. (Zhang, X. D., et al. 2001. *Cancer Res.* 61:7339.)

SMAD Signal transducing proteins when stimulated by TGF-β can enhance gene transcription and tumor formation (TGF-β signaling). The Smad-binding element is GTCTAGAC. The facilitating effects of bone morphogenetic protein (BMP, acting via the serine/threonine kinase receptor) determine SMAD protein function and opposing epidermal growth factor (EGF, acting via the receptor tyrosine kinases). Smad2 is essential for the formation of the early embryonic mesoderm of mice. Smad3 is normally phosphorylated by TGF receptor TβRI and Evi-1 represses its transcriptional activator function. Smad4 controls mesoderm and visceral endoderm. Other SMADs associated with various ligands of the TGF family control the expression of genes involved in embryonal tissue differentiation. Smad3 and Smad4 cooperate with Jun/Fos (A1) and bind to the TPA-responsive gene-promoter elements. Smad2 and Smad3 are retained in the cytoplasm by SARA (Smad anchor for receptor activation). Pancreatic, colon, and other cancers are frequently associated mutations in SMAD2 and SMAD4/DPC4. SMADs 6 and 7 are modulators/inhibitors of signaling by some SMADs and their mutation may cause hyperplasia of the cardiac valves and other structural anomalies of the heart, as well as ossification of the aorta and high blood pressure in mice. The SMAD acronym was derived from the human SPA (spinal muscular atrophy genes, 5q12.2-q13.3) and the *Drosophila* gene *Mad* (*mothers against decapentaplegic*). Smad proteins are classified as R-Smads (receptor regulated), Co-Smads (common Smads), and I-Smads (inhibitory). *See* activin; AP1; bone morphogenetic protein; DPC4; EGF; Evi oncogenes; Gli; *Mad*; osteopontin; receptor tyrosine kinase; SARA; serine/threonine kinase; Ski; spinal muscular atrophy; TGF; TPA. (Wrana, J. L. 2000. *Cell* 100:189; Zauberman, A., et al. 2001. *J. Biol. Chem.* 276:24719; López-Rovira, T., et al. 2002. *J. Biol. Chem.* 277:3176.)

small-cell lung carcinoma (SCLC) Associated with a deletion of human chromosomal region 3p14.2; the susceptibility is dominant. It accounts for about one-third of all lung cancers. Lung cancer genes were also located to 3p21 and 3p25. Smoking may be the major cause of the development of this condition. Surgical remedy is usually not applicable because of rapid metastasis, but it generally responds to radiation and chemotherapy. Deregulation of the MYC oncogene is a suspected cause. A gene for fragile histidine triad (FHIT) was associated with SCLC and with some of the non-small-cell lung carcinomas (NSCLCs). The product of FHIT splits Ap$_4$A substrates asymmetrically into ATP and AMP. Its metastasis may be inhibited by CC3/TIP30. *See* AMP; ATP; cancer; metastasis; MYC; non-small-cell lung carcinoma suppressor; oncogenes; p53; semaphorin. (Zöchbauer-Müller, S., et al. 2002. *Annu. Rev. Physiol.* 64:681.)

small nuclear RNA   *See* snRNA.

small-pool PCR   Amplifies 20–100 molecules of minisatellite DNA from single individuals within a population and thus reveals a mutation rate that is $>10^{-3}$ in the sperm germline at a number of loci. *See* minisatellite; MVR; PCR. (Crawford, D. C., et al. 2000. *Hum. Mol. Genet.* 9:2909.)

small t antigen   *See* SV40.

S1 mapping   When genomic DNA is hybridized with the corresponding cDNA or mRNA, the nonhomologous sequences cannot find partners to anneal with, and the single-stranded loops can be digested with S1 nuclease. The remaining DNAs that formed a double-stranded structure can then be isolated by gel electrophoresis, or their position and length can be determined by autoradiography if appropriately labeled material was used. Thus, intron positions are revealed. *See* DNA hybridization; genomic DNA; introns; S1 nuclease diagram below. (Favaloro, J., et al. 1980. *Methods Enzymol.* 65:718; Dziembowski, A. & Stepien, P. P. 2001. *Anal. Biochem.* 294:87.)

SMART   Simple Modular Architecture Research Tool facilitates annotation of protein domains. *See* annotation; domain. (<http://smart.embl-heidelberg.de>.)

smart ammunition   *Drosophila* P-transposable element vectors with selectable markers to produce selectable (e.g., neomycin resistance) insertions. *See* hybrid dysgenesis; insertional mutation. (Engels, W. R. 1989. *Mobile DNA*, Berg, D. E. & Howe, M. M., eds., p. 437. Am. Soc. Microbiol., Washington, DC.)

smart cells   Generalized concept that cells (genes) have the ability to sense internal and external cues and respond to them in a purposeful manner such as shown in signal transduction.

smart linkers   Synthetic oligonucleotides with multiple recognition sites for restriction enzymes; they can be ligated to DNA ends to generate the desired types of cohesive ends. *See* blunt end; blunt end ligation; cloning vectors; cohesive ends.

Smart linkers can be cut at several sites with cohesive ends.

smart PCR   *See* polymerase chain reaction. (Villalva, C., et al. 2001. *Biotechniques* 31:81.)

SMC   Proteins involved in the structural maintenance of the chromosomes including condensation (increasing coiling) and segregation. The three Muk gene products of *E. coli* are functionally homologous. *See* chromosome coiling; cohesin; condensin; sex determination. (Ball, A. R. & Yokomori, K. 2001. *Chromosome Res.* 9[2]:85.)

smear   Preparing a soft specimen for microscopic examination by gentle spreading directly on the microscope slide. *See* microscopy; sectioning; squash.

Smg p21   Protein similar to Rap 1. *See* Rap.

SMGT (sperm-mediated gene transfer)   Sperm internalizes DNA and by artificial insemination transgenic pigs have been produced that carry the human decay accelerating (hDAF) gene. The efficiency of transformation is high (64%) and the expression is very good (83%). The presence of dDAF is expected to help in overcoming hyperacute rejection of xenotransplanted organs. *See* decay accelerating factor, xenotransplantation, transformation genetic. (Lavitrano, M., et al. 2002. *Proc. Natl. Acad. Sci. USA* 99:14230.)

Smith-Lemli-Opitz syndrome (SLOS/RSH)   High-prevalence (2×10^{-4}) autosomal-recessive anomaly involving microcephalus, mental retardation, abnormal male genitalia, polydactyly, etc., encoded in human chromosomes 11q12-q13 (type I) and 7q32.1 (type II). Prevalence in Northern European populations is 1×10^{-4} to 2×10^{-5}. It is caused by Δ^7-reductase deficiency in the cholesterol pathway and accumulation of 7-dehydrocholesterol. Deficiency of the Δ^{24}-reductase results in similar symptoms. The afflicted individuals show mevalonic aciduria. This condition may also involve holoprosencephaly. Some of the symptoms may overlap with those of Pallister-Hall syndrome. *See* cholesterol; chondrodysplasia; dwarfness; head/face/brain defect; holoprosencephaly; mental retardation; sonic hedgehog. (Wassif, C. A., et al. 2001. *Hum. Mol. Genet.* 10:555.)

Smith-Magenis syndrome   Involves head malformation, short brain, growth retardation, hearing loss, self-destructive behavior such as pulling off nails, putting foreign objects to ear, etc. Chromosome 17p11.2 region is generally deleted. *See* mental retardation; self-destructive behavior.

SMM   *See* IAM; stepwise mutation model; two-phase model.

smoking   Responsible for a wide variety of ailments such as heart disease, respiratory problems, cancer, etc., but it may decrease the risk of Parkinson's disease. The tobacco smoke of the mother may initiate cancer in the fetus. Although cancer may be induced by a variety of genotoxic agents in the environment, the smoking-induced alteration spectrum in the genetic material is different and thus can be distinguished from the effects of other agents. Tobacco smoke adducts induce a higher proportion of transversion mutations of the p53 gene

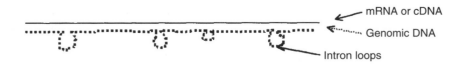

mRNA or cDNA
Genomic DNA
Intron loops

S1 mapping.

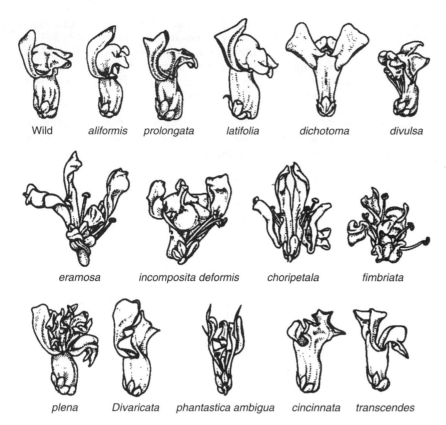

Wild   aliformis   prolongata   latifolia   dichotoma   divulsa

eramosa   incomposita deformis   choripetala   fimbriata

plena   Divaricata   phantastica ambigua   cincinnata   transcendes

Flower morphology mutants of snapdragon. (The original artwork is the courtesy of the late Professor Hans Stubbe.)

in the lung and increase loss of heterozygosity by deleting introns, particularly at the fragile site 3 (FRA3B) region, including FHIT (fragile histidine triad) in human chromosome 3p14.2. According to one report, 19/31 newborns of smoking mothers had the carcinogen 4-methylnitrosamino-1(3-pyridyl)-1-butanone in their urine.

The smoking habit is particularly prevalent in affective disorders. In the brain of smokers, the level of monoamine oxidase B (MAOB) is 40% lower relative to that in nonsmokers. MOAB degrades the neurotransmitter dopamine. The nicotinic acetylcholine receptors have, however, a very important role in cognitive processes of the brain. Tolerance to smoking seems to be influenced by diet and ethnic background. The carcinogenic effect of smoking tobacco is due primarily to specific N-nitrosamines. *See* affective disorders; chemical mutagens; dopamine; fragile site; infertility; intron; MAO; nicotine; nicotinic acetylcholine receptors; p53; Parkinson's disease; tobacco; transversion. (Hecht, S. S. 1999. *Mutation Res.* 424:127; Schuller, H. M. 2002. *Nature Rev. Cancer* 2:455; environmental exposure: Besaterinia, A., et al. 2002. *Carcinogenesis* 23:1171.)

smooth endoplasmic reticulum   Has no ribosomes on its surface. *See* endoplasmic reticulum; SER.

smooth muscle   Lacks sarcomeres; they are associated with arteries, intestines, and other internal organs, except the heart. *See* sarcomeres; striated muscles.

SMRT   Silencing mediator of retinoid and thyroid hormone receptors. It is also a corepressor of PPARδ. *See* animal hormones; nuclear receptors; PPAR; retinoic acid. (Becker, N., et al. 2001. *Endocrinology* 142:5321.)

smut   Infection of grasses by *Basidiomycete* fungi, causing a black carbon-like transformation of the inflorescence (by *Ustilago,* loose smut) or seed tissues (by *Tilletia,* covered smut).

snail   *Helix pomatia univalens,* $2n = 24$.

Snail   Family of zinc-finger transcription factors and a negative regulator of E-cadherin. *See* cadherin, (Betlle, E., et al. 2000. *Nature Cell Biol.* 2:84; Nieto, M. A. 2002. *Nature Rev. Mol. Cell Biol.* 3:155.)

snake venom phosphodiesterase   Releases 5′-nucleotides from the 3′-end of nucleic acids. *See* phosphodiesterases; phosphodiester bond.

SNAP   *See* membrane fusion; NSF; SNAREs.

snap-back   Inverted repeat sequence in nucleic acids. *See* lollipop structure; repeat, inverted.

snapdragon (*Antirrhinum majus*) $2n = 16$; a much employed dicotyledonous plant (*Scrophulariaceae*) for the study of mutation (transposable elements) and flower pigments. It is a popular ornamental. This plant has many beautiful flower morphology mutants as shown above. *See* flower morphology mutants; TAM. (<http://caliban.mpiz-koeln.mpg.de/~stueber/snapdragon/snapdragon.html>.)

SNARE   (1) Soluble N-ethylmaleimide-sensitive factor attachment protein receptor is a binding protein attaching vesicles (v-SNAREs) to target membranes (t-SNAREs). They mediate, among others, transport through the Golgi compartments. SNAREs mediate membrane fusions. SNARE seems to be activated by Ypt1p. The name has been attributed to the surgeons wire tools by which polyps and projections are removed or from bird traps (snares). *See* EEA1; exocytosis; Golgi apparatus; Ipk1; membrane fusion; NSF; RAB; Ypt, snare; synaptobrevin; synaptogamin; syntaxin; VAMP. (Bock, J. B. & Scheller, R. H. 1999. *Proc. Natl. Acad. Sci. USA* 96:12227; Peters, C., et al. 2001. *Nature* 409:581.)

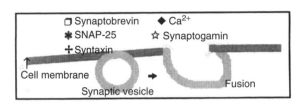

SNF Yeast genes are helicases involved in chromatin remodeling (*SNF2, SNF5, SNF6, SNF11*), and *SNF1* is an AMP-activated kinase. SNF1 senses depletion of ATP and an increase of AMP in the cell. This is a ubiquitous enzyme family involved in carbohydrate and lipid metabolism, phosphorylation of transcription factors, regulating stress responses in plants, etc. *See* bromodomain; chromatin remodeling; *SUC2*; *SWI*. (Eisen, J. A., et al. 1995. *Nucleic Acids Res.* 23:2723; Lo, W.-S., et al. 2001. *Science* 293:1142.)

SNIP Single-nucleotide polymorphism (SNP). Difference in a single nucleotide at a particular DNA site. SNIPs are used as genomic markers for human (or other) populations. The most common variation involves C↔T transition in CpG sequences. The analysis uses DNA chips or gel-based sequencing and biotin-labeled probes (VDA, variant detector array). A survey of 2,748 SNPs indicated a high degree of

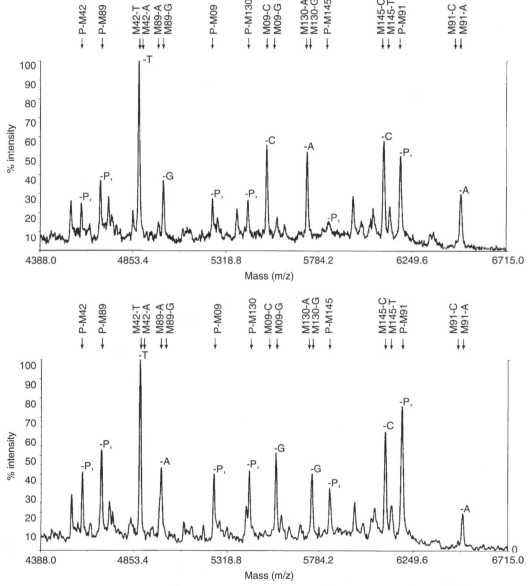

Multiplex primer extension products of the human Y chromosome of two individuals analyzed by MALDI-TOF mass spectrometry displaying allelic differences at the sites indicated at the top. -P indicate the primers and -A, -C, -G, -T stand for the nucleotides. (Courtesy of Silvia Paracchini, Barbara Arredi, Rod Chalk, and Chris Tyler-Smith, 2002.)

Normal sequence

| T G T G T G G G C T C C [TC] C C T G T T T C T G |

| Cys | Val | Gly | Ser | Ser | Leu | Phe | Leu |
| 1051 | 1052 | 1053 | 1054 | 1055 | 1056 | 1057 | 1058 |

Tangier disease

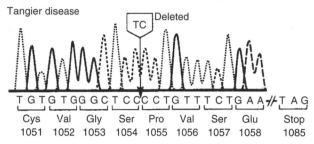

| T G T G T G G G C T C C C C T G T T T C T G A A // T A G |

| Cys | Val | Gly | Ser | Pro | Val | Ser | Glu | Stop |
| 1051 | 1052 | 1053 | 1054 | 1055 | 1056 | 1057 | 1058 | 1085 |

Two-nucleotide deletion in the ABC1 transporter gene results in Tangier disease. Note the frameshift. (Courtesy of H. Bryan Brewer, Jr. Modified after Remaley, A. T., et al. 1999. *Proc. Natl. Acad. Sci. USA* 96:12685.)

polymorphism (4.58×10^{-4}); mutation rate $\mu \sim 10^{-8}$ to range of about 10% of the confirmed SNPs. Other studies indicated one SNIP/600 base pairs in the human genome (Kruglyak, L. & Nickerson, D. A. 2001. *Nature Genet.* 27:235). The SNPs can be mapped to chromosomal location by radiation hybrid cell lines. If the SNP is not within the gene, recombination may lead to false positive identification. SNPs can be generated for the identification of the critical base substitutions responsible for human disease. The majority of SNPs occur in noncoding regions of the genome and are noninformative regarding human disease.

Mapping of SNIPS can be carried out by reduced representation shotgun sequencing (RRS) and locus-specific polymerase chain reaction amplification (LSA). When, however, many special cases of the same disease are analyzed, the significance of the base substitutions may be statistically or even causally determined. For population genetics and linkage studies, the SNPs are frequently classified into types I (involving nonsynonymous alterations regarding coding property and being a nonconservative change), II (within coding region and nonsynonymous yet conservative), III (in coding sequence but synonymous), IV (within the noncoding 5'-sequence), V (within the noncoding 3'-sequence), and VI (in other noncoding regions). Type I SNPs are most useful for genetic analyses because they have phenotypic (functional) characteristics. Preliminary information indicates that the majority of SNIP haplotypes (~80%) occur in all ethnic groups and only 8% are population-specific (Patil, N., et al. 2001. *Science* 294:1719). By March 2001, 2.84 million SNPs were deposited in the public databases, and they represented 1.64 million nonredundant mutations (Marth, G., et al. 2001. *Nature Genet.* 27:371). *See* ABC transporters; allele-specific probe; biotinylation; DASH; DNA chips; DNA repair; dynamic allele-specific hybridization; genotyping; linkage disequilibrium; LOS; MALDI/TOF/MS; MRD; padlock probe; PolyPhred; radiation hybrid; RRS; STS; Tangier disease. (Wang, D. G., et al. 1998. *Science* 280:1077; Sunyaev, S.,

et al. 2000. *Trends Genet.* 16:198; Mullikin, J. C., et al. 2000. *Nature* 407:516; Buetow, K. H., et al. 2001. *Proc. Natl. Acad. Sci. USA* 98:581; Grupe, A., et al. 2001. *Science* 292:1915; Roger, A., et al. 2001. *Genome Res.* 11:1100; Miller, R. D. & Kwok, P.-Y. 2001. *Hum. Mol. Genet.* 10:2195; Gut, I. G. 2001. *Hum. Mut.* 17:475; Werner, M., et al. 2002. *Hum. Mut.* 20:57; Kirk, B. W., et al. 2002. *Nucleic Acids Res.* 30:3295; Paracchini, S., et al. 2002. *Nucleic Acids Res.* 30[6]e27; Coronini, R., et al. 2003. *Nature Biotechnol.* 21:21; (human genetic variation: <http://www.ncbi.nlm.nih.gov/SNP>; <http://hgbase.interactiva.de>; <http://hgbase.cgr.ki.se>; cancer, SNP, frameshift; <http://lpgfs.nci.nih.gov:82/perl/snpbr>; <http://www.genome.wi.mit.edu/SNP/human/index.html>; <http://lpg.nci.nih.gov/html-cgap/validated.html>; disease genes: <bio.chip.org/biotools>.)

SNO oncogenes   Two SKI-related oncogenes. *See* oncogene; SKI.

snorbozyme   Ribozyme within the nucleolus that processes or degrades nucleolar RNA. (Samarsky, D. A., et al. 1999. *Proc. Natl. Acad. Sci. USA* 96:6609.)

snoRNA   Small nucleolar RNA assists maturation of ribosomal RNAs in the nucleolus, folding of RNA, RNA cleavage, base methylation, assembly of preribosomal subunits, export of RNP, etc. A family of snoRNAs of 10–21 nucleotides, complementary to the methylation sites of rRNA, guides the methylation within the nucleolus. Some of the snoRNA genes are situated in introns. The majority of snoRNAs are either box C/D or H/ACA snoRNA family members. All CD boxes contain fibrillarin and (spliceosomal) Snu13. *See* fibrillarin; introns; pseudouridine; RNA maturase; spliceosome. (Hirose, T. & Steitz, J. A. 2001. *Proc. Natl. Acad. Sci. USA* 98:12914; Song, X. & Nazar, R. N. 2002. *FEBS Lett.* 523:182.)

SNP   *See* SNIP.

snRNA   Small nuclear RNA is a low-molecular-weight RNA in the eukaryotic nucleus that is rich in uridylic residues. When associated with protein, it mediates the splicing of primary RNA transcripts and frees them from introns through the assistance of the lariat. Spliceosomal snRNP can be exported to the cytoplasm if it is appropriately capped and has the nuclear cap-binding complex, the export receptor CRM1/Xpo1, RanGTP, and the phosphorylated adaptor of RNA export (PHAX). *See* CRM1/SXPO1; export adaptr; hnRNA; introns; lariat; Ran; RNP; spliceosome; U1-RNA. (Ohno, M., et al. 2000. *Cell* 101:187; Kiss, T. 2001. *EMBO J.* 20:3617; *Gene Expr.* [2002] vol, 10 issue 1/2.)

snRNP   Small nuclear ribonucleoprotein (pronounced "snurp"). It is involved in the processing of RNA and the assembly of spliceosomes. *See* imprinting; KH domain; spliceosome. (Nagengast, A. A. & Salz, H. K. 2001. *Nucleic Acids Res.* 29:3841.)

S_1 nuclease   From *Aspergillus oryzae*, cleaves single-stranded DNA (preferentially) and single-stranded RNA. Double-stranded molecules and DNA-RNA hybrids are quite resistant to it unless used in very large excess. The enzyme produces 5′-phosphoryl mono- and oligonucleotides. S_1 has many applications in molecular biology: mapping of transcripts, removal of single-stranded overhangs from double-stranded molecules, analysis of the pairing of DNA-RNA hybrids, opening hairpin structures. Its pH optimum is 4.5 and this may cause unwanted depurination. *See* hairpin structure; nucleases; S1 mapping. (Kormanec, J. 2001. *Methods Mol. Biol.* 160:481.)

snurportin α-importin-like transport protein that handles snRPN import to the cell nucleus. *See* importin; nuclear pore; snRPN. (Paraskeva, E., et al. 1999. *J. Cell Biol.* 145:255.)

snurposomes Complex of the five snRNP particles that process RNA transcripts in the Cajal bodies. *See* Cajal body; coiled body; U RNAs. (Gall, J. G., et al. 1999. *Mol. Biol. Cell* 10:4385.)

SOB bacterial medium H_2O 950 mL, bactotryptone 20 g, bactoyeast extract 5 g, NaCl 0.5 g plus 10 mL of 250 mM KCl; pH is adjusted to 7 with 5 N NaOH and filled up to 1 L. Just before use, add 5 mL of 2 M $MgCl_2$.

SOC Store-operated channel. Group of plasma membrane ion channels controlling the release of ions stored in the lumen of the endoplasmic reticulum. *See* ion channels. (Ma. R., et al. 2001. *J. Biol. Chem.* 276:25759.)

SOC bacterial medium Same as SOB but contains glucose (20 mM). *See* SOB.

social Darwinism Application of the Darwinian views (survival of the fittest) to social order. Many anthropologists, sociologists, and ethicists rejected social Darwinism and portended that it was an attempt to justify inequalities, harsh competition without adherence to ethics, aggression, imperialism, racism, and unbridled capitalism as the necessity for the survival of the fittest. Social darwinism of the 19th century is no longer accepted in the developed world. *See* Darwinism; IQ; social engineering. (Rogers, J. A. 1972. *J. Hist. Ideas* 33:265.)

social engineering Utopistic idea that the genetic determination of individuals is unimportant in defining their abilities and realization, but education, welfare, medical service, etc., may determine how they will function in society. Therefore, the state and its institutions must actively control human life from cradle to death. Although compassion, education, caring, and many social safety nets are indispensable in a modern society, the significance of individuality cannot be ignored. *See* IQ; social Darwinism. (Graebner, W., 1980. *J. Am. Hist.* 67:612.)

social insects Bees, wasps, ants, and termites live in a colony and generally divide the various tasks among different casts such as workers (soldiers), queen, and drones. The queen (gyne), workers, and soldiers are diploid. The drones are different because they hatch from unfertilized eggs. The major differences between the queen and the workers is that the queen has predominantly 9-hydroxy-(E)2-decanoic acid and 9-keto-(E)2-decanoic acid in her mandibular glands. The workers have predominantly 10-hydroxy-(E)2-decanoic

acid. These pheromones then determine their functional roles in the colony. The workers and soldiers have ovaries and under certain circumstances (especially the soldiers) may produce haploid eggs. Cuticular hydrocarbons may regulate sex expression and convert workers into egg-laying females (gamergate), although they are usually smaller in size. Such a change in some groups may take place by physical contact of individuals transmitting hydrocarbon molecules. The status of soldiers has been questioned as being genetically equivalent with the workers. The role of the soldiers (not found in all species of social insects) is protection of the colony. *See* honey bee; male-stuffing; sex determination; wasp. (Bourke, A. F. 2001. *Biologist* 48[5]:205; Parker, J. D. & Hedrick, P. W. 2000. *Heredity* 85[pt 6]:530; Thorne, B. L. & Tranillo, J. F. A. 2003. *Annu. Rev. Entomol.* 48:283.)

sociobiology Study of the biology and behavior of social insects and other animal communities. *See* social insects. (Wilson, E. O. 2000. *Sociobiology: The New Synthesis*. Harvard Univ. Press, Cambridge, MA.)

SOCS-box Suppressor of cytokine signaling contains a protein-docking site(s) for protein-protein interaction and seems to be involved in transduction signal attenuation and ubiquitination. SOCS-$2^{-/-}$ mice display greatly increased body weight apparently caused by inappropriate regulation of the Jak/Stat signal transduction and the IGF-I pathways. *See* CIS; CSAID; cytokine; insulin-like growth factors; JAB; signal transduction; SSI-1; ubiquitin. (Zhang, J. G., et al. 2001. *Proc. Natl. Acad. Sci. USA*. 98:13261.)

SOD Superoxide dismutase. *See* amyotrophic lateral sclerosis; superoxide dismutase.

SODD *See* TNFR.

sodium azide (N_3Na) Inhibitor of respiration. It blocks electron flow between cytochromes and O_2. It is also a potent mutagen for organisms that can activate it.

sodium channel *See* ion channels.

sodium dodecil sulfate Sodium lauryl sulfate. *See* dodecyl sulfate sodium salt.

sodium pump Plasma membrane protein that moves Na^+ out and K^+ into the cells with the energy obtained by hydrolyzing ATP. It is also called Na^+-K^+ ATPase. *See* ion channels.

software Computer program that tells the hardware (the computer) what and how to carry out the instructions or applications.

Sog *See* bone morphogenetic protein.

solenoid structure Coiling electric conductor used for the generation of a magnetic field. By analogy, the coiled nucleosomal DNA fiber is frequently described as a solenoid, although it has no relation to electricity or magnetism. It only resembles those coils. The DNA solenoids are about 30 nm in diameter and contain about six nucleosomes per turn. They are packed with a large number of structural and catalytic proteins. *See* nuclear matrix; nucleosome.

solid-state control   Exercised by electric or magnetic means in solids, e.g., in a transistor. *See* semiconductors.

solitary LTR   Long terminal repeats that lost their internal (transposase) sequences by recombination and excision. *See* long terminal repeats; retroposon; retrotransposon; retrovirus. (Domansky, A. N., et al. 2000. *FEBS Lett.* 472[2–3]:191.)

solo elements   Terminal sequences (LTR) of retroposons that can exist in multiple copies without the coding sequences between the two direct repeats. *See* Ty.

soluble RNA   Somewhat outdated term for transfer RNA. *See* tRNA.

solute   Any substance dissolved in a solvent.

solution hybridization   Molecular hybridization in a liquid medium. *See* nucleic acid hybridization.

SOM   Self-organizing map is a type of mathematical cluster analysis that is particularly well suited for recognizing and classifying features in complex, multidimensional data of microarray hybridization. The method has been implemented in a publicly available computer package, GENECLUSTER, which performs the analytical calculations and provides easy data visualization. *See* cluster analysis; microarray hybridization. (Sinha, A. & Smith, A. D. 1999. *Microgravity Sci. Technol.* 12[2]:78; Haese, K. & Goodhill, G. J. 2001. *Neural Comput.* 13:595; Unneberg, P., et al. 2001. *Proteins* 42:460.)

soma   Body cells distinguished from those of sexual reproduction (germinal cells).

somaclonal variation   Genetic variation occurring at a frequency higher than spontaneous mutation in cultured plant cells. The causative mechanism is poorly understood. It is conceivable that the asynchrony between nuclear and cell divisions are accompanied by chromosomal damage. Also, there is evidence that movement of endogenous transposable elements is involved. The mobility is attributed to the stress imposed by the culture. (Kaeppler, S. M., et al. 2000. *Plant Mol. Biol.* 43[2–3]:179; diagram right above.)

somatic cell   Majority of the body cells (reproduced by mitosis), including those of the germline but not the products of meiosis, the sex cells (gametes). *See* cell genetics; cell lineages; germline; mitosis; parasexual mechanisms.

somatic cell hybrids   Formed through fusion of different somatic cells of the same or different species. Somatic cell hybrids contain the nucleus of both cells and all cytoplasmic organelles from both parents, in contrast to the generative

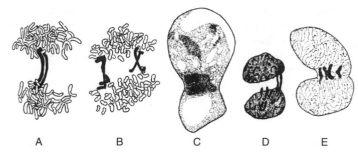

Mitotic anomalies in a clone of tobacco maintained in cell culture. (A), (B) anaphase bridges; (C) telophase bridge; (D) the same as (C) but enlarged nuclei; (E) early interphase with conjoined nuclei. Polyploidy is also of common occurrence. (From Cooper, L. S., et al. 1964. *Am. J. Bot.* 51:284.)

hybrids, in which mitochondria and plastids generally are not transmitted through the male.

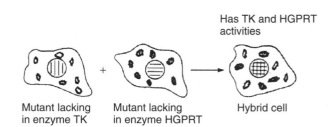

Selective isolation of animal somatic cell hybrids deficient in thymidine kinase and hypoxanthine-guanine phosphoribosyltransferase. On HAT medium only the complementary heterokaryons survive. (Modified after Ephrussi, B. & Weiss, M. C. 1969. *Sci. Am.* 220 [4]:26.)

For the fusion of cultured somatic cells, various special techniques are necessary. Most commonly, protoplasts are used and the fusion medium is polyethylene glycol (MW 1,300–1,600) 25 g, $CaCl_2.2H_2O$ 10 mM, KH_2PO_4 0.7 mM, glucose 0.2 M in 100 mL H_2O, pH 5.5 for plants. The best media may vary according to species. The isolation of somatic animal cell hybrids was greatly facilitated by using selective media (*see* HAT medium).

	Human chromosomes							
	1	2	3	4	5	6	7	8
Hybrid clones A	+	+	+	+	−	−	−	−
B	+	+	−	−	+	+	−	−
C	+	−	+	−	+	−	+	−

Chromosome assignment of genes on the basis of incomplete clones of mouse + human cell hybrids. A panel of 3 clones, each containing a set of 4 of 8 human chromosomes. If a gene is expressed uniquely in the clone (C) but not in (A) or (B), the locus must be in chromosome 7 Because only C carries chromosome 7. Additional panels are needed to test genes in the entire human genome. (After Ruddle, F. H. & Kucherlapati, R. S. 1974. *Sci. Am.* 231[1]:36.)

The fusion of animal cells is promoted by polyethylene glycol, attenuated Sendai virus, or calcium salts at higher

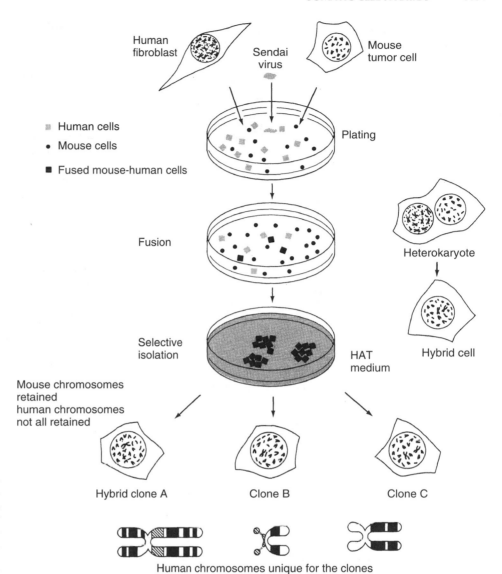

Human fibroblast

Sendai virus

Mouse tumor cell

Human cells

Mouse cells

Fused mouse-human cells

Plating

Fusion

Heterokaryote

Selective isolation

HAT medium

Hybrid cell

Mouse chromosomes retained human chromosomes not all retained

Hybrid clone A

Clone B

Clone C

Human chromosomes unique for the clones

Selective isolation of mouse + human cell hybrids. The fused cell usually retains all mouse chromosomes, but the human fibroblast chromosomes may be partially eliminated and in some clones only one human chromosome is maintained. (Modified after Ruddle, F. H. & Kucherlapati, R. S. 1974. *Sci. Am.* 231[1]:36.)

pH. Immediately after fusion, the somatic cells may become heterokaryotic, but eventually the nuclei may also fuse. The availability of fused cells along with the loss of one set of chromosomes or partial deletion of chromosomes makes possible the localization of many human genes (see figure above). In human genetics, most commonly mouse + human cells have been used. In such cultures, the human chromosomes are gradually eliminated. If, however, one of the mouse chromosomes carries a defective gene, the human chromosome carrying a functional wild-type allele must be retained in order for the culture to be viable in the absence of a particular supplement that is required for normal function.

The genetic constitution of the retained human chromosome can also be verified by enzyme assays, electrophoretic analysis of the proteins, or immunological tests. For genetic analysis of higher mammals, particularly humans, where controlled mating is not feasible, the availability of somatic hybrids has opened a very productive approach. In the somatic hybrid cells, allelism and synteny or linkage can be shown. If two or more human genes are consistently expressed in a particular retained chromosome, it is a safe conclusion that they are in the same chromosome. The diagram on preceding page

best explains the principle of gene assignment. Genes can be mapped to particular chromosome bands by deletions (see diagram next page). By somatic cell fusion, hybrids can be obtained between taxonomically very distant species. All kinds of animal cells may be fused with each other; plant and animal cells can also be fused. Some of these exotic hybrids may have, however, difficulties of continuing cell divisions.

Somatic cell fusion may make possible the study of recombination between various mitochondria and chloroplast DNAs in cells, whereas in generative hybrids, because of uniparental (female) transmission of the organelles, such analyses are not feasible. Since plant cells generally retain totipotency in culture, hybrid cells may be regenerated into intact organisms and can be further studied in favorable cases by the methods of classical progeny analysis. *See* cell fusion; fusion of somatic cells; HAT medium; human chromosome maps; IFGT; in situ hybridization; mapping, genetic; microfusion; mitochondrial mutation; monoclonal antibody; radiation hybrids; somatic embryogenesis. (Vasil. I. K., ed. 1984. *Cell Culture and Somatic Cell Genetics of Plants.* Academic Press, San Diego, CA; Rédei, G. P. 1987. *McGraw-Hill Enc. Sci. Technol.* 6th ed., 16:628.)

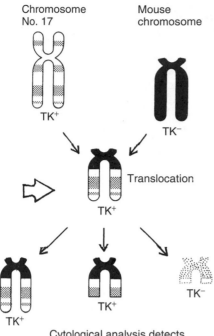

Chromosome No. 17

Mouse chromosome

TK⁺

TK⁻

Translocation

TK⁺

TK⁺

TK⁺

TK⁻

Cytological analysis detects missing segments

Regional mapping of the thymidine kinase (TK) gene to a segment of human chromosome 17. A translocation was obtained between human chromosome 17 and a mouse chromosome that was deficient for TK. The translocation chromosome was exposed to a chromosome-breaking agent that cleaved off segments of various lengths from the end of the translocation. When the segment containing the gene (TK) was removed, the cells carrying this broken chromosome no longer displayed TK activity and the gene's location was revealed at 17q23.2-q25.3. (*See* human chromosome map.) (Modified after F. H. Ruddle & R. S. Kucherlapati.)

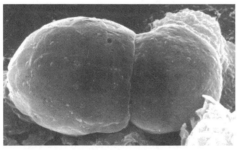

The process of fusion of plant protoplasts. (From Fowke, L. C., et al. 1977. *Planta* 135:257.)

somatic crossing over   *See* mitotic crossing over.

somatic embryogenesis   Formation of embryos either directly, or by first passing through a callus stage, from cultured adult plant cells or protoplasts. In about nine animal phyla, the germline is not separated from the somatic line; therefore, gametes develop from the somatic cell lineage. The situation in protists and fungi is similar. A strict germline does not exist in plants either. *See* apomixis; embryo culture; embryogenesis, somatic; germline. (Sata, S. J., et al. 2000. *Methods Cell Sci.* 22:299.)

somatic hybrids   *See* graft hybrids; microfusion; somatic cell fusion; somatic cell hybrids.

somatic hypermutations   Generally occur in a region of 1–2 kilobases around the rearranged V-J regions of the immunoglobulin genes and very rarely extend into the C (constant) sections. These mutations, usually transitions, and predominantly involving guanine, are most common in the complementarity-determining region (CDR), and the events usually take place in the germinal centers. The preferred hot spots are purine-G-pyrimidine-(A/T) sequences, although not all these sequences are hot spots for mutation. The serine codons AGC and AGT can represent a hot spot, but TCA, TCC, TCG, and TCT are not hot spots. Codon usage in the CDR region appears to be evolutionarily determined to secure the maximum complementarity to antigens. It is not entirely clear what determines the targeting of hypermutations to the special area, but transcriptional enhancers appear to be involved. Hypermutations are limited to the coding strand and occur in a downstream polarity. Deletions and insertions are rare. For somatic hypermutation, V(D)J rearrangement is a prerequisite. Somatic hypermutation, along with recombination, is the major source of antibody variation and increases the defense repertory in mammals. *See* CDR; enhancer; germinal center; hot spot; hypermutation; immunoglobulins; lymphocytes; mosaic; transition. (Harris, R. S., et al. 1999. *Mutation Res.* 436:157; Meffre, E., et al. 2001. *J. Exp. Med.* 194:375; Papavasiliou, F. N. & Schatz, D. G. 2002. *Cell* 109:S35; Di Nola, J. & Neuberger, M. S. 2002. *Nature* 419:43.)

somatic mutation   Occurs in the body cells. It is undetectable if recessive, unless the individual is heterozygous. The mutation is manifested by sector formation. This procedure for testing mutability is particularly effective if multiple-marker heterozygotes are treated. It is not inherited to generative progeny unless it also occurs or expands to the germline. It has been successfully used for studies of mutation in stamen hairs of the plant *Tradescantia*. In fungal cultures, mutation in mitotic cells may appear as sectorial colonies. Somatic mutation can be studied in in vitro cell cultures. Recessive mutations are detectable for X-linked genes in hemizygous cells such as in the male. The procedure is effective if it is directed toward loci with selectable products. Molecular methods (PCR) may permit the identification of mechanisms involved in such point mutations, chromosomal deletions, and rearrangement. The frequency of mutation may be affected by age, and various environmental factors can be analyzed. *See* bioassays in genetic toxicology; cancer; Knudson's two-mutation theory; LOH; mosaic; nucellus; paramutation; PCR; pseudodominance; SNIP; somatic hypermutations; *Tradescantia* stamen hairs; transposable elements. (Orive, M. E. 2001. *Theor. Popul. Biol.* 59[3]:235; Koike, H., et al. 2002. *EMBO Rep.* 3:433.)

somatic pairing   Generally only the meiotic chromosomes pair during prophase (possibly late interphase), but in some tissues, such as the salivary glands, the chromosomes are always tightly associated. Also, mitotic association of the chromosomes is a requisite for mitotic (somatic) crossing over in a few organisms where this phenomenon has been analyzed. (*Drosophila*, some fungi, *Arabidopsis*, etc.). *See* intimate pairing; mitotic recombination; pairing; parasexual mechanisms; polytenic chromosomes. (Burgess, S. M., et al. 1999. *Genes Dev.* 13:1627.)

somatic recombination *See* mitotic crossing over.

somatic reduction Reduction of chromosome numbers during mitosis in polyploids.

somatic segregation Unequal distribution of genetic elements during mitoses. *See* chloroplast genetics; mitotic crossing over; mosaic; somatic mutation; sorting out.

somatogamy Fusion of sexually undifferentiated fungal hypha tips. *See* fungal life cycles.

somatomedin Second messenger-type polypeptide. In association with other binding proteins, it is involved in the stimulation of several cellular functions. *See* insulin-like growth factor; pituitary dwarfness; second messenger. (Deng, G., et al. 2001. *J. Cell Physiol.* 189:23.)

somatoplastic sterility In certain plant hybrids, the nucellus may show excessive growth and chokes the embryo to death. The embryo can be rescued if excised early and transferred to in vitro culture media. *See* embryo culture.

somatostatin 14–20-amino-acid-long hypothalamic neuropeptide inhibits the release of several hormones (somatotropin, thyrotropin, corticotropin, glucagon, insulin, gastrin) in contrast to growth hormone-releasing factor, which stimulates production of growth hormone. The somatostatin gene has been mapped to human chromosome 3q28 and to mouse chromosome 16. Two somatostatin receptors, SSTR1 and SSTR2 (391 and 369 amino acids, respectively), with very substantial homology, have been identified. They are 7-pass-transmembrane proteins frequently bound to G-proteins and distributed all over the body, particularly at high levels in the stomach, brain, and kidney. The somatostatin and dopamine receptors may cooligomerize. *See* animal hormones; dopamine; GHRH; G-proteins; seven-membrane proteins; signal transduction. (Patel, Y. C. 1999. *Front. Neuroendocrinol.* 20:157.)

somatotropin Mammalian growth hormone (GH, $M_r \approx$ 21,500); it can correct some dwarfisms when its level is increased. It also stimulates milk production. Actually, there are three human growth hormones, all coded at chromosome 17q23-q24. Their mRNAs display about 90% homology and their amino acid sequences are shared as well. Placental lactogen protein is an even more effective growth hormone that is located nearby and it is highly homologous. The human growth hormone gene is transcribed only in the pituitary, whereas the homologues are expressed in the placental tissues. Human and other somatotropin genes have been cloned. Transformation of mice with rat growth hormone genes increased body size substantially. *See* dwarfism; pituitary dwarfism; stature in humans. (Yin, D., et al. 2001. *J. Anim. Sci.* 79:2336.)

somites Paired mesoderm blocks along the longitudinal axis (notochord) of an embryo, giving rise to the vertebral column

and other segmented structures. After migration, they may form the skeletal muscles. The development of the somites is controlled mainly by the Notch family and associated proteins. *See* Notch. (Pourquié, O. 2001. *Annu. Rev. Cell Dev. Biol.* 17:311.)

somitogenesis Differentiation of the somites. *See* somites.

sonicator Ultrasonic ($\approx$20 kHz) equipment used for disrupting cells to extract contents.

sonic hedgehog (*Shh*) Vertebrate gene that provides information in head-tail direction for development; it is a rather general signaling protein of animal differentiation. *Shh* contributes to the specification of the notochord, brain, lung, and foregut. It is homologous to the *Drosophila hedgehog* (*hh*) gene; its receptors are *patched* (*ptc*) and *smoothened* (*smo*) signaling factor genes. A freely diffusible Shh is modified by cholesterol, and a balance of Patched and Hedgehog-interacting proteins (Hip) regulates it. Defects in patched may cause various carcinomas, medulloblastoma, and rhabdomyosarcoma. These genes have regulatory functions for oncogenic development. *See* GLI; *hedgehog*; holoprosencephaly; medulloblastoma; notochord; rhabdomyosarcoma. (Villavicencio, E. H., et al. 2000. *Am. J. Hum. Genet.* 67:1047.)

son of sevenless *See* SOS.

sonography Method of ultrasonic prenatal analysis of possible structural and other defects in heart, kidney, bone, sex organ, umbilical chord, and body movement; verifies pregnancy, etc. Presumably it entails no appreciable risk. *See* fetoscopy; prenatal diagnosis; ultrasonic.

SopE Guanyl-nucleotide exchange factor for Rho and Rac GTPase proteins. It participates in the reorganization of the cytoskeleton and mediates bacterial entry into mammalian cells. After entry of the bacteria, SptP (GTPase-activating protein) restores the cytoskeleton. *See* cytoskeleton; GTPase; Rac; Rho.

Sordaria fimicola (*n* = 7) Ascomycete with linear spore octads. It has been extensively used for genetic recombination. A large number of mutants is available. In the 1940s, it was suspected that sexuality was relative in this fungus, but the poor maters were just weak mutants.

sorghums Arid, warm-region crops. *S. bicolor* (and kaoliang), $2n = 2x = 20$. *S. halepense* (Johnson grass) is tetraploid. Compared to other grain crops, their nutritional value is lower because of the tannin content and low lysine level of the grain, and the folding of its proteins (kafirin) lowers digestibility. Mutant varieties exist that correct these deficiencies. (Oria, M. P., et al. 2000. *Proc. Natl. Acad. Sci. USA* 97:5065; <http://www.tigr.org/tdb/tgi.shtml>.)

sori *See* sorus.

sorrel 1. Light chestnut fur color of horses determined by homozygosity for *d* gene; similar brownish color in other mammals. 2. The sorrel plants *Rumex acetosella and R. scutatus* are used as tart vegetables. 3. The sorrel tree is *Oxydendron*. *See Rumex*.

Sorsby syndrome   *See* night blindness; Stargardt disease.

sortase   Bacterial enzyme that anchors surface proteins to the bacterial cell wall. These surface proteins promote interaction between the pathogen and the animal cell. These proteins also mediate escape from the immune defense system of the animal cell. (Muzmanian, S. K., et al. 2001. *Mol. Microbiol.* 40:1049.)

sorting   Mechanism that ensures molecules (isozymes) are directed into the appropriate cellular compartments (cytosol, nucleus, mitochondria, chloroplasts). The sorted proteins are generally equipped with special NH_2-end signals generated by differential transcription and/or translation. This process is assisted by trans-acting transmembrane proteins and glycosylphosphatidylinositol-linked proteins. *See* chloroplasts; mitochondria; transit peptide.

sorting out   Genetically different organelles (plastids, mitochondria) segregate into homogeneous groups of cells, or during embryogenesis cells of common origin reaggregate in order to form certain cell types and/or structures. The segregation of two different types, A and B, of mitochondria in a heteroplasmic cell line may be characterized by the formula of Solignac, M., et al. (*Mol. Gen. Genet.* 197:183): $V_n = \rho_0 (1 - \rho_0)(1 - [1 - 1/N]^n)$, where V_n is the variance of ρ (the fraction of A within a cell) at the nth cell generation, ρ_0 is the fraction of A in the original cell line, and N is the number of sorting-out units. In *Schizosaccharomyces*, the distribution of mitochondria is mediated by microtubules. In humans, the frequency of mutation in mtDNA is 10 times higher than in nuclear DNA, yet heteroplasmy is very rare except in some diseases. Despite the fact that the number of mitochondria in mammalian cells runs to the thousands per cell, usually the replication switches to one type. The number of the founder mtDNAs has been estimated within a wide

range: 1–6 and 20–200; in cattle and in *Drosophila*, 370–740. These founders than undergo a restriction/amplification type of replication, i.e., they pass through a bottleneck and therefore heteroplasmy is very limited. *See* ctDNA; heteroplasmy; mitochondrial genetics; mtDNA; plasmone mutation; plastid number; Romanovs. (Kowald, A. & Kirkwood, T. B. L. 1993. *Mutation Res.* 295:93.)

sorus   Group of sporangia, such as those found on the lower surface of fern leaves.

SOS (*son of sevenless*)   *Drosophila* gene that functions downstream from *sevenless*, encoding a receptor tyrosine kinase (RTK) in the light signal transduction pathway. SOS is a guanine nucleotide-releasing protein (GNRP; it is also called guanine-nucleotide exchange factor, GEF), and it interacts with RTK through the protein Drk receptor kinase, a homologue of the vertebrate Grb2, and SEM-5 in *Caenorhabditis*. These proteins function in a variety of signal transduction pathways involving EGF. SOS is frequently called a mediator protein. *See* BOSS; daughter of sevenless; DRK; GNRP; Grb2; receptor tyrosine kinase (RTK); rhodopsin; signal transduction. (Hall, B. E., et al. 2001. *J. Biol. Chem.* 276:27629.)

sos recruitment system (SRS)   Is a tool in proteomics. A temperature-sensitive cdc25-2 allele of yeast permits grows at 25°C but not at 36°. Normally this protein when localized to the plasma membrane facilitates Ras guanyl nucleotide exchange and via signal transduction events promotes cell growth. The human homolog (hSOS) is complementary for the mutant and secures growth at the otherwise nonpermissive regime, The hSos function requires protein–protein interaction that is secured by fusing a bait to the C-end of the truncated protein. The co-expressed bait and prey are targeted to the membrane. The prey is either an integral membrane protein or a soluble protein, which after myristoylation signaling, attached to the membrane and then growth is restored. *See* two-hybrid system; CDC25; proteomics. (Auerbach, D., et al. 2002. *Proteomics* 2:611.)

SOS REPAIR   Error-prone repair. *See* DNA repair.

Sotos syndrome   *See* cerebral gigantism.

source habitat   Some individuals contribute more to the future generations than the average individual. *See* habitat; sink habitat.

Southern blotting   DNA fragments cut by restriction endonucleases are separated on agarose gel by electrophoresis, then transferred to membrane filters by blotting to hybridize the pieces with radioactively (or fluorescent) labeled DNA or RNA. Physical sites of restriction fragments and genes are identified. The transfer to membranes may be achieved by capillary action of wicks and sucked through layers of filter papers on top or by vacuum-driven devices. *See* autoradiography; biotinylation; fluorochromes; membrane filters; nucleic acid hybridization; restriction enzyme; RFLP.

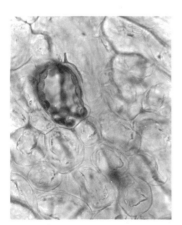 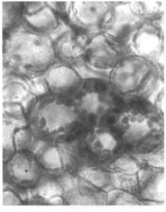

When variegation is caused by a mutation in a nuclear gene, all the plastids within a cell are either green or colorless (left). Mutations in the chloroplast genetic material display sorting out at the boundary of sectors, i.e., cells with both green and colorless plastids occur (mixed cells) and in a nonstochastic process cells with all colorless and all green plastids appear in later sectors (right).

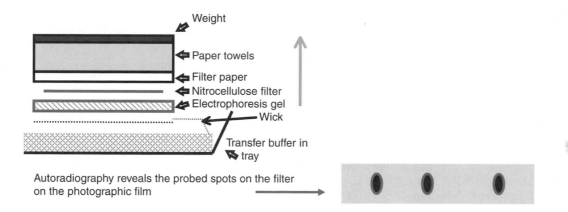

Autoradiography reveals the probed spots on the filter
on the photographic film

Southern hybridization   *See* Southern blotting.

southwestern method   Simultaneously labeling cDNA and binding proteins (transcription factors). The screening is carried out by hybridizing labeled probes of DNA to bind to polypeptides immobilized on nitrocellulose filters. *See* Southern blot; Western blot.

Soviet genetics   Not meant to define a special genetics because science does not have ideological, political, or ethnic attributes and it transcends all boundaries. A sad exception is Soviet genetics, a misnomer because it was not genetics at all and it collapsed before the implosion of the political system that nurtured and enforced it. Genetics in the Soviet Union had a very remarkable and successful beginning. In 1944, L. C. Dunn, professor of zoology at Columbia University, noted: "There are today literally hundreds of trained genetical investigators in the U.S.S.R., certainly more than in any other country outside the U.S.A" (*Science* 99:2563). This outstanding research and teaching establishment was destroyed, however, in 1948 and geneticists suffered humiliation, persecution, and almost total physical annihilation for a period of over 20 years by lysenkoism. *See* lysenkoism; Mitchurin.

SOX   SRY-type HMG box. Mammalian genes (∼30) encoding proteins with over 60% similarity to the HMG box of SRY. The SOX9 gene is critical for the differentiation of Sertoli cells and chondrocytes. Sox genes apparently encode transcription factors and are capable of transactivation of genes involved in gonadal differentiation. SOX is involved in cartilage formation and induces testis development in chromosomally XX mice (Vidal, V. P. I., et al. 2001. *Nature Genet.* 28:216). The bacterial *Sox* genes are involved in superoxide responses. The mammalian SOXs bind to an 5′-(A/T)(A/T)CAA(A/T)G site in the DNA and regulate the expression of LINE-1 retrotransposons in human cells. The different SOX genes use different short DNA sequences for binding specific transcription factors. Several human syndromes have mutations in Sox genes. *See* anti-Müllerian hormone; campomelic dysplasia; Hirschsprung disease; HMG; LINE; Sertoli cells; Shah-Waardenburg syndrome; SRY; Wolffian duct. (Kamachi, Y., et al. 2000. *Trends Genet.* 16:182; Takash, W., et al. 2001. *Nucleic Acids Res.* 29:4274; Wilson, M. & Koopman, P. 2002. *Current Op. Genet. Dev.* 12:441; Laumonnier, F., et al. 2002. *Am. J. Hum. Genet.* 71:1450.)

soybean   *Glycine max*, $2n = 40$; the basic chromosome number may be $x = 10$. Some related species are tetraploid. This crop is of great economic significance because of the high oil (20–23%) and high protein contents (39–40%) of the seed. (gene index: <http://www.tigr.org/tdb/tgi.shtml>.)

sp   As a prefix, indicates *Schizosaccharomyces pombe* (fission yeast) DNA, RNA, or protein.

Sp1   Mammalian protein; general transcription factor for many genes recognizing the DNA sequence $\frac{GGGCGG}{CCCGCC}$. Sp1 elements also protect CpG islands of housekeeping genes from methylation. Sp1 requires the general transcription factor TFIID and the complex CRSP. Sp3 and Sp4 are very similar; the latter is expressed in the brain. *See* CRSP; DNA methylation; LCR; MDM2; transcription factors; transcription factors, inducible. (Nicolás, M., et al. 2001. *J. Biol. Chem.* 276:22126.)

spaced dyads   Many regulatory sites include a pair of conserved trinucleotides spaced by a nonconserved tract of fixed width. Genes with functions coordinated by a common regulatory element are expected to share upstream binding sequences. These elements contain fixed sites for a linker domain and the dimerization domain of transcription factors. Their analysis facilitates the identification of coregulated genes. (van Helden, J., et al. 2000. *Nucleic Acids Res.* 28:1808.)

space flight   Genetic effects are not entirely clear. The microgravity in cooperation with space radiation may have adverse stress consequences. (*See* White, R. J. & Averner, M. 2001. *Nature* 409:1115.)

spacer DNA   Nontranscribed nucleotide sequences between genes (IGS) in a cell. In the rDNA, there are spacers within (internal transcribed spacers, ITS) and between gene clusters (external transcribed spacers, ETS). These were thought of earlier as untranscribed tracts, but actually they represented very short transcripts. Animal mtDNA genes generally have very short (few bp) spacers. In fungi, mtDNA spacers are common and variable in length because of recombination and slippage during replication; they may also be mobile. Plastid genes of higher plants are organized into operons without interruptions. Insertions, deletions, and unequal recombinations determine the length of the spacers. *See* ribosomal RNA; rrn.

spasmoneme   Rod-like thin bundle of filaments (2 nm each) forming a cytoplasmic organelle of 2–3 mm in extended form. When exposed to calcium, it contracts and serves as an engine for moving different structures within the cell of cicliated protozoa. *See* centrin. (Maciejewski, J. J., et al. 1999. *J. Eukaryot. Microbiol.* 45[2]:163.)

spastic paraplegia (Strumpell disease, SPG)   Collection of paralytic diseases encoded by at leat 14 loci under a variety of genetic controls recessive, dominant autosomal, and Xq22-linked. Genes responsible for the disease have been mapped to 14q (SPG3), 2p21-p22 (SPG4), 15q (SPG6), 12q13, and 8q (SPG8). The most prevalent form (40–50%), SPG4, encodes an AAA protein called spastin. SPG13 is due to mutation in mitochondrial Hsp60. *See* AAA proteins; heatshock proteins; Pelizaeus-Merzbacher disease; Silver syndrome. (Vazza, G., et al. 2000. *Am. J. Hum. Genet.* 67:504; Svenson, I. K., et al. 2001. *Am. J. Hum. Genet.* 68:1077.)

spaying   Neutering of a female. *See* castration.

SPB   *See* spindle pole body.

Spearman rank-correlation test (SRC)   Determines the relations between two variables without elaborate calculations:

Pairs	Trait X	Rank	Trait Y	Rank	Difference of Ranks (d)	d^2
1	8	3	14	2	1	1
2	10	5	20	7	−2	4
3	12	7	17	5	2	4
4	9	4	15	3	1	1
5	6	1	13	1	0	0
6	11	6	18	6	0	0
7	7	2	16	4	−2	4

Sum $d = 0$ Sum $d^2 = 14$

The SR correlation coefficient,

$$r = 1 - \frac{6 \text{ Sum } d^2}{n^3 - n},$$

in this example,

$$1- = \frac{6 \times 14}{7^3 - 7} = 0.75$$

In case of ties, correction can be used,

$$T = \frac{m^3 - m}{12},$$

where m stands for the number of measurements or classifications of identical values. If, for example, there are three measurements of 5, two measurements of 7, and two amounting to 8, the correction for ties among the X trait variables becomes

$$\text{Sum } Tx = \frac{3^3 - 3}{12} + \frac{2^3 - 2}{12} + \frac{2^3 - 2}{12},$$

and in a similar way the correction for ties can be determined in the Y series. Now we can obtain the terms

$$\text{Sum } X^2 = \frac{n^3 - n}{12} - \text{Sum } Tx,$$

and

$$\text{Sum } Y^2 = \frac{n^3 - n}{12} - \text{Sum } Ty,$$

and

$$r = \frac{\text{Sum } X^2 + \text{Sum } Y^2 - \text{Sum } d^2}{2\sqrt{\text{Sum } X^2 \text{Sum } Y^2}}$$

It is a nonparametric test and can be used for comparison of traits that cannot be measured but can be classified subjectively (on the basis of their appearance). If the traits are quantified by measures, they are ranked according to their relative magnitude, i.e., assigning the highest rank to the largest. If an exact measurement is impractical (e.g., degree of susceptibility), they are simply ranked. In case of ties, an equal rank is assigned to both. The differences of the rank scores are squared and summed. Tables constructed for various degrees of freedom can be used to determine the probabilities. *See* correlation; nonparametric tests; statistics.

specialized transduction   The transducer temperate phage picks up a piece of the host DNA at the immediate vicinity of its established prophage site (and generally leaves behind a comparable length of its own). When this modified phage infects another bacterium, it may integrate the gene that it picked up from the previous host into the new host genome. Lambda-phage (λdgal; meaning a defective lambda that carries *gal*) is a typical specialized transducer with a preferred site near 17 minutes on the map next to the *gal* locus, and it is a specialized transducer of this gene. Another well-known specialized transducer phage is φ80trp, which can carry the tryptophan operon (at about 27 1/2 min). Specialized transducer phages must be temperate phages, and they transduce (only) the special gene that is at their integration site. Gene transfer in eukaryotes by transposable elements, genetic vectors, is similar to prokaryotic transduction. *See* bacterial recombination; double lysogenic; helper virus; high-frequency lysate; transformation, genetic; transposable elements; diagram next page. (Morse, M. L. 1954. *Genetics* 39:984.)

speciation   Process by which new species diverge from an ancestral species. Evolutionists may distinguish *conventional* methods of speciation when geographic isolation and accumulated mutations eventually cause reproductive isolation. According to the *quantum model* of speciation, the divergence begins with spatial isolation followed by the survival of a few new types of individuals that give rise to reproductively isolated new forms as mutations accumulate. *Saltational* speciation comes about by sudden major mutations. *Parapatric* (or stasipatric) speciation occurs without geographic separation and it is initiated from a relatively small number of individuals that are producing divergence under continued natural selection. *Sympatric* speciation occurs within the original area of dispersal, due to the emergence of genetic isolation mechanisms or sexual selection. The mechanism of speciation in plants and animals differs because in animals behavioral traits may also lead to speciation. In plants,

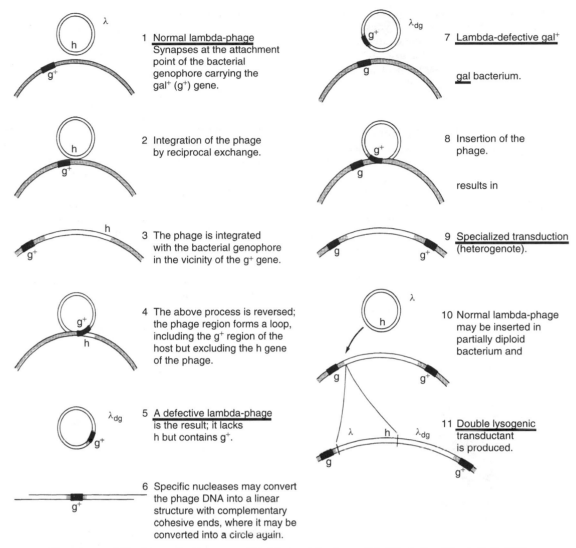

1 <u>Normal lambda-phage</u> Synapses at the attachment point of the bacterial genophore carrying the gal$^+$ (g$^+$) gene.

2 Integration of the phage by reciprocal exchange.

3 The phage is integrated with the bacterial genophore in the vicinity of the g$^+$ gene.

4 The above process is reversed; the phage region forms a loop, including the g$^+$ region of the host but excluding the h gene of the phage.

5 <u>A defective lambda-phage</u> is the result; it lacks h but contains g$^+$.

6 Specific nucleases may convert the phage DNA into a linear structure with complementary cohesive ends, where it may be converted into a circle again.

7 <u>Lambda-defective gal$^+$</u>

<u>gal</u> bacterium.

8 Insertion of the phage.

results in

9 <u>Specialized transduction</u> (heterogenote).

10 Normal lambda-phage may be inserted in partially diploid bacterium and

11 <u>Double lysogenic</u> transductant is produced.

Steps in specialized transduction exemplified by lambda-phage. (Modified after Stent, G. S. 1971. *Molecular Genetics*. Freeman, San Francisco, CA.)

sterility may not necessarily hinder propagation because some species reproduce primarily by vegetative means, and they can practice both autogamy and cross-pollination. Chromosomal rearrangements and the action of transposable elements may lead to hybrid sterility, sexual isolation, and eventually speciation. *See* cospeciation; evolution; fertility; neo-Darwinism; neo-Lamarckism; phylogeny; saltation; sibling species; species; theory of evolution.

species Potentially interbreeding population that shows reproductive isolation from other species. Transgenic technology somewhat confuses the classical definition because of the possibility of exchanging genes among totally different taxonomic categories. The number of eukaryotic species is not known, but about 10^6 have been described. The expectations are that from 3- to 10- fold more have not been discovered yet. *See* fertility; identification by genome; OTU; sexual isolation; sibling species; speciation; species, extant; transgene. profile: Watanabe, T., et al. 2002. *Genome Biol.* 3[2]:res.0010.1.

species, extant With the increase of agricultural, industrial, and changing land use, the currently alive (A), %

extinct (E), and % threatened (T) species indicate an alarming reduction of biodiversity: *molluscs* (A) 10^5, (E) 0.2, (T) 0.4; *crustaceans* (A) $4-10^3$, (E) 0.01, (T) 3; *insects* (A) 10^6, (E) 0.005, (T) 0.07; *vertebrates total* (A) 4.7×10^4, (E) 0.5, (T) 5; *mammals* (A) 4.5×10^3, (E) 1, (T) 11; *gymnosperms* (A) 758, (E) 0.3, (T) 32; *dicots* (A) 1.9×10^5, (E) 0.2, (T) 9; *monocots* (A) 5.2×10^4, (E) 0.2, (T) 9 (data from Smith, F. D. M., et al. 1993. *Nature* 364:494). Within these larger categories, some species are much more endangered than these figures indicate. These numbers are not exactly valid because new species are being discovered. Between 1980 and 1990, more than 100 new mammals were identified. The extinction percentages date from the year 1600. According to Dobson, A. P., et al. ([1997], *Science* 275:550, the number of endangered species in the United States is 503 plants, 84 molluscs, 57 arthropods, 107 fish, 43 herptiles, 72 birds, and 58 mammals. Important factors affecting the survival of the feral species is the annual increase of the human population (1.7%), urbanization, industrial and livestock production, pollution, and natural disasters. According to the World Conservation Union 1996 biennial report (http://www.iucn.org/themes/ssc/index.html), 5,205 species are endangered and risk extinction. Actually, the number of species is still unknown and the latest estimates vary from 3.5 to 10.5 million, about a third lower than some earlier estimates. *See* conservation; genetics; species. (*Nature* 405:207ff; *Proc. Natl. Acad. Sci. USA* [2001], 98:5389ff; Alroy, J. 2002. *Proc. Natl. Acad. Sci. USA* 99:3706; Pitman, N. C. A. & Jørgensen, P. M. 2002. *Science* 298:989.)

species, synthetic   Produced in the laboratory by crossing putative ancestors of amphidiploid forms that are partially isolated genetically. Doubling the chromosome number of the hybrids prevents sterility of the offspring. The synthetic species permit verification of the putative evolutionary path of existing polyploid species by studying the pairing behavior of chromosomes, the frequency of chiasmata, the number of univalents and multivalents formed, etc. These classical cytogenetic methods may be supplemented with nucleic acid hybridization, in situ hybridization, the study of proteins, nucleic acid sequences, etc. Some of the hybrid species, if they ever occurred in nature, were not maintained by natural selection. *See* allopolyploids; evolution; *Raphanobrassica* speciation; *Triticale*.

Middle: *Triticale* ($2n = 56$), Left: Wheat ($2n = 42$), Right: Rye ($2n = 14$). (Courtesy of Dr. Árpád Kiss.)

specific activity   Number of molecules of a substrate (μmol) acted on by enzymes (mg protein) per time (minutes) at standard temperature (25°C). Relative amount of radioactive molecules in a chemical preparation.

specific combining ability   *See* combining ability.

specific heat   Joules or calories required to raise the temperature of 1 g substance by 1°C.

specificity   Measure of discrimination among compounds. In DNA sequencing: the correctly predicted bases divided by the sum of the correctly and incorrectly predicted bases.

specificity, mutagen assay   Percentage of correct identification by mutagenesis of presumably 'noncarcinogenic' 'nonmutagenic' agents. *See* accuracy; bioassays in genetic toxicology; predictivity; sensitivity.

specific locus mutations assays (SLT)   Have been used to detect X-chromosomal mutations in hemizygous males (mouse); autosomal heterozygotes for special fur color genes such as the Oak Ridge stock heterozygous for a, b, p, c^{ch}, se, d, s or the Harwell tester stock (HT) heterozygous for six loci (a, bp, fz, ln, pa, pe); heterozygotes for thymidine kinase; or hypoxanthine-guanine phosphoribosyltransferase genes in different mammalian cell cultures (mouse, Chinese hamster ovary cells). Mutation of the wild-type allele is then immediately revealed in the first generation. Similar procedures are applicable in plants and other diploid systems. Animal assays are not as simple as microbial assays, e.g., the Ames test, but they are considered to be more relevant to human studies. *See* bioassays in genetic toxicology; doubling dose; mutations in human populations. (Cattanach, B. M. 1971. p. 535 in Hollaender, A., ed. *Chemical Mutagens*. Plenum, NY.)

specific rotation (α)   Degrees of the plane of polarized light by an optically active compound of specific concentration at 25°C: $(\alpha) = \dfrac{rv}{nl}$, where r = rotation in degrees, v = volume of the solution in cubic centimeters l = length in centimeters of the light path.

speckles, intranuclear (IGC)   Storage areas for primary RNA transcript processing factors; the speckles may not be the compartments of eukaryotic splicing, which instead takes place in association with transcription. *See* introns; paraspeckles; splicing. (Eilbracht, J. & Schmidt-Zachmann, M. S. 2001. *Proc. Natl. Acad. Sci. USA* 98:3849.)

SPECT   *See* tomography.

spectinomycin   Inhibits translocation of charged tRNA from the A site to the P site of the prokaryotic ribosome by interfering with elongation factor G.

Spectinomycin.

spectral genotyping   Uses molecular beacons labeled with different fluorophores. When the probes for the wild type (green) and for the mutant with one base substitution (red) are fluorophore labeled, the homozygous wild type (green) can be distinguished from the heterozygote's DNA (displaying both green and red) and the homozygous mutant showing only the other color (red). The potential color of the fluorophores can be chosen arbitrarily. This procedure can distinguish genotypes in clinical diagnostic analyses without DNA sequencing. *See* beacons, molecular; fluorophore. (Kostrikis, L. G., et al. 1998. *Science* 279:1228.)

spectral karyotyping   Chromosomes or chromosomal segments are hybridized with probes of fluorescent dyes in a combinatorial manner. Five dyes (Cy2, Spectrum Green, Cy3, Texas Red, and Cy5) provide enough combinatorial possibilities to "paint" each chromosome with a different color or shade of it. Although the human eyes cannot distinguish these different hues, by the use of optical filters, Sagnac interferometer, a CCD camera, and Fourier transformation, they were able to discriminate the special differences in standard classification colors using a computer program. The approach permitted classification of translocations that were not identifiable by other staining techniques. The lowest limit of differentiation by this technology is 500 to 1,500 kbp. The procedure is applicable to clinical laboratory testing and evolutionary analyses. *See* chromosome painting; FISH; GISH. (Schröck, E., et al. 1996. *Science* 273:494.)

spectrins Filamentous tetrameric proteins (220–240-kDa) present in the red blood cell membrane and may constitute 30% of it. Spectrins may mediate muscle and neuron organization by facilitating the binding of other proteins. They are also involved in the formation of a network on the cytoplasmic surface in cooperation with actin, dynein, ankyrin, and *band III* protein. βII-spectrin is the same as fodrin. Several human chromosomes encode spectrins. Spectrin defects may cause auditory and motor neuropathies. *See* ankyrin; cytoskeleton; dynein; elliptocytosis; fodrin; glycophorin; poikilocytosis; spherocytosis. (Parkinson, N. J., et al. 2001. *Nature Genet.* 29:61.)

spectrophotometry Estimating the quality and quantity of a substance in solution on the basis of absorption of monochromatic light passing through it.

spectrosome Cytoplasmic organelle anchoring the mitotic spindle, thus defining the spatial direction of cell division. It contains spectrins, cyclin A, and other regulatory proteins.

speech and grammar disorder (SPCH1) Human chromosome 7q31 dominant lack of coordination of face and mouth muscles, some cognitive impairment (low IQ), and articulation and expressive language. *See* dyspraxia; human intelligence; language impairment, specific; stuttering. (Lai, C. S. L., et al. 2000. *Am. J. Hum. Genet.* 67:357; Enard, W., et al. 2002. *Nature* 418:569.)

Spemann's organizer Embryonic tissue site of signals that mediate the organization of the body. Its signaling center is at the blastopore of the gastrula and releases a variety of polypeptides that control neural and dorsal or ventral mesodermal differentiation. *See* blastopore; gastrula; morphogenesis; organizer. (Niehrs, C. 2001. *EMBO J.* 20:631; De Robertis, E. M., et al. 2000. *Nature Rev. Genet.* 1:171.)

sperm Animal seminal fluid; geneticists generally understand it as the spermatozoon or plant male generative cells. The *Drosophila* spermatozoon far exceeds that of any other animal in length. (up to 58 mm in *D. bifurcata*, although that of *D. melanogaster* is about 1.91 mm; the human spermatozoon is about 0.0045 mm; Karr, T. L. & Pitnick, S. 1996. *Nature* 379:405). In a single normal human ejaculate, the number of spermatozoa is within the range of 20–40 million

in the seminal fluid volume of >2 mL. The human spermatozoon count has shown a decreasing tendency during the last decades. The cause of this appears to be the increase of environmental pollutants with hormone-like effects, such as PCB (polychlorinated biphenyl), an industrial carcinogen and pesticide, dioxin (a solvent), phthalates (may be present in cosmetics), bisphenols (used in manufacturing resins and fungicides). During fertilization the animal spermatozoon is attracted to the egg by chemotactic peptides, causing changes in the voltage of the membrane, the concentration of cAMP, cGMP, Ca^{2+}, and the activity of K^+ ion channels. A sperm chemoattractant protein, a member of the cysteine-rich secretory proptein (CRISP) family, allurin, has been isolated (Olson, J. H., et al. 2001. *Proc. Natl. Acad. Sci. USA* 98:11205). Cytoskeletal proteins of the fertilizing sperm trigger oocyte maturation and ovulation of animals. The number of pollen grains (containing two sperms) released by a plant may vary from a few hundred to tens of millions. *See* acrosomal process; DDT; environmental mutagens; fertilization; gametogenesis; infertility; oocyte, primary; polyspermic fertilization; RPTK; seleno-protein; semen. (Wassarman, P. M., et al. 2001. *Nature Cell Biol.* 3:E59; Ren, D., et al. 2001. *Nature* 413:603; Swallow, J. G. & Wilkinson, G. S. 2002. *Biol. Rev.* 77:153, <http://www.med.unipi.it/agp/siti/siti.htm>.)

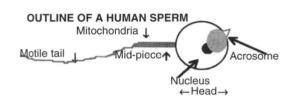

OUTLINE OF A HUMAN SPERM

sperm activation Conversion of the spermatids into moving spermatozoa. In *Caenorhabditis*, four gene loci (*spe-8*, −12, −27, and *spe-29*) control the process.

spermatheca Storage facility of insect females for sperm to be used at a later fertilization following an initial mating. *See* sperm competition; sperm displacement; sperm precedence; sperm storage.

spermatia Male sperm produced at the tip of hyphae within the spermatogonia of rust fungi; they are comparable to the microconidia. *See* conodia.

spermatid Cell formed by the secondary spermatocyte; it differentiates into a spermatozoon. *See* gametogenesis; sperm; spermatozoon.

spermatocyte (primary) Diploid cell that gives rise to the haploid secondary spermatocytes by the first division of meiosis, forming the spermatids, which differentiate into spermatozoa in animals. In plants, the spermatocytes function in a similar manner, thus producing the microspores, which develop into pollen grains, and within them the two sperm are formed either before or after the pollen tubes begin to elongate. *See* gametogenesis; gonads.

spermatogenesis *See* gametogenesis; spermiogenesis.

spermatogonia Primordials of the sperm cells; the secondary spermatogonia produce the primary spermatocytes. *See* gametogenesis; spermatocyte. in vitro culture: (Feng, L.-Xin., et al. 2002. *Science* 297:392, Zhao, G.-Q. & Garbers, D. L. 2002. *Developmental Cell* 2:537.)

spermatophyte Seed-bearing plant. *See* cryptogamic plants.

spermatozoon Fully developed (differentiated) male germ cell; a sperm. *See* fertilization; sperm; spermatid; oocyte, primary.

Spermatozoon.

sperm bank Human sperm depository for the purpose of artificial insemination in case of sterility of the husband or other conditions that may warrant their use. Thousands of sperm banks (gene banks) exist in the world; most of the deposited samples are anonymous and the genetic constitution of the donors is not known completely. Since the 1930s, H. J. Muller advocated the use of sperm banks for positive eugenics purposes. Accordingly, the donors should be selected on the basis of superior talents, mental ability, and physical constitution. This idea has not been widely accepted, however, because of moral objections and biological shortcomings. Most of the "superior" phenotypes cannot be evaluated by generally accepted criteria, and the phenotype may not fully represent the heritability of particular traits. Since artificial insemination of domestic animals has become routine, sperm banks have been exploited for animal breeding programs in order to produce the maximal number of progeny of high-performance males. Even for animals, this technology should be used with thorough consideration of population genetics principles in order to avoid narrowing the gene pool and inbreeding. *See* ART; bioethics; in vitro fertilization. (Critser, J. K. 1998. *Hum. Reprod. Suppl.* 2:55; Deech, R. 1998. *Hum. Reprod.* 13 Suppl. 2:80.)

sperm competition In animals in which the female may mate repeatedly during the period of receptivity, the spermatozoa of different genotype may have a selective advantage or disadvantage in fertilization. The success of particular types of spermatozoa is influenced by a large number of seminal proteins. The accessory gland proteins appear to be more polymorphic than other proteins, yet this may not distinguish between the interpretations of rapid evolution and limited selection. *See* accessory gland; certation; multipaternate litter; sperm displacement. (Fu, P., et al. 2001. *Proc. R. Soc. Lond. B Biol. Sci.* 268:1105; Simmons, L. W. 2001. *Sperm Competition and Its Evolutionary Consequences in the Insects.* Princeton Univ. Press.)

sperm displacement The female animals practice sperm storage. Depending on the genetic constitution of the sperm, one may interfere with the fertilization of a competing sperm already present in the spermatheca of the female. In this process, the genetic constitution of the female also has a selective role. *See* spermatheca; sperm competition; sperm precedence; sperm storage. (Gilchrist, A. S. & Partridge, L. 2000. *Evolution Int. J. Org. Evolution* 54:534.)

spermidine ($C_7H_{19}N_3$,N-[3-aminopropyl]-1,4-butanediamine) Polyamine regulating (+ or −) the binding of proteins to DNA, condensation of DNA, controlling gene expression, etc. Spermidine synthase is encoded by human chromosomes 1p36-p22 and 3p14-q21. *See* spermine.

spermine ($C_{10}H_{26}N_4$,N,N′-bis[3-aminopropyl]tetramethylenediamine) Its oxidative cleavage product is spermidine. Spermine synthase was assigned to human chromosome Xp22.1. Its mouse homologue *Gyro* [*Gy*]) is suspected in hypophosphatemia, resulting in rickets, hearing disorders, etc. *See* hypophosphatasia; hypophosphatemia; rickets; S-adenosylmethionine decarboxylase.

spermiogenesis Postmeiotic process of differentiation of mature spermatozoa. It is supposed that CREM is involved in the control of genes required for the process because CREM-deficient mice (obtained by recombination) cannot complete the first step of spermiogenesis and late spermatids are not observed. The defective spermatids are apparently eliminated by apoptosis. *See* apoptosis; CREM; gametogenesis; protamines; sperm; spermatozoon; transition protein. (Macho, B., et al. 2002. *Science* 298:2388.)

spermist *See* preformation.

sperm morphology assays in genetic toxicology Based on the expectation that mutagens and carcinogens may interfere with normal spermiogenesis, resulting in abnormal head shape, motility and viability of the treated or exposed sperm. This expectation is met with some agents but not all. Thus, sperm alterations may indicate mutagenic and/or carcinogenic properties, but not all mutagens/carcinogens seem to affect these sperm parameters within nonlethal doses. *See* bioassays in genetic toxicology. (Baccetti, B., et al. 2001. *Hum. Reprod.* 16:1365.)

sperm precedence In multiple copulation of a female with different males within a period of receptivity, one type of sperm (usually the last mating male's) may have a selective advantage in producing the offspring. *See* certation; last-male sperm precedence; sperm competition. (Price, C. S., et al. 2000. *Evolution Int. J. Org. Evolution* 54:2028.)

sperm receptor *See* fertilization.

sperm selection *See* sex selection.

sperm storage Insects commonly store sperm in the spermatheca for 2 weeks and fertilization may follow any time within this period. This is why geneticists tend to use virgin females for controlled matings. Sperm can be stored at the temperature of liquid nitrogen (−195.8°C) and retain its fertilization ability. Storage efficiency depends on seminal fluid proteins that the male transmits along with the sperm. Mouse spematozoa can be stored by freeze-drying and function is maintained. *See* sperm bank; sperm competition; sperm displacement; sperm precedence. (Yin, H. Z. & Seibel, M. M.

1999. *J. Reprod. Med.* 44[2]:87; Mortimer, D. 2000. *J. Androl.* 21:357; Corley-Smith, G. E. & Brandhorst, B. P. 1999. *Mol. Reprod. Dev.* 53:363; Kusakaba, H., et al. 2001. *Proc. Natl. Acad. Sci. USA* 98:13501.)

sperm typing   Permits analysis of recombination in diploids using the recombinant gametes. The method of analysis requires the analysis of PCR-amplified gamete DNA sequences of known paternal types. This type analysis can resolve recombination even between single base pairs. The results must be scrupulously studied because the PCR method may have inherent errors. Single sperm can be separated with the aid of fluorescence-activated cell sorters. If the sperm is subjected to primer extension preamplification (PEP) before analysis, enough material can be obtained to carry out multipoint tests. This procedure is not practical with egg cells. *See* cell sorter; crossing over; genetic screening; maximum likelihood applied to recombination; polymerase chain reaction; prenatal diagnosis; recombination frequency. (Hubert, R., et al. 1994. *Nature Genet.* 7:420; Shi, Q., et al. 2001. *Am. J. Med. Genet.* 99:34.)

SPF   Cell cycle S-phase promoting factor, a kinase. *See* cell cycle.

SPF condition   Specific pathogen-free condition of organisms maintained in a quarantine. *See* quarantine. (Yanabe, M., et al. 2001. *Exp. Anim.* 50[4]:293.)

Spfi-1   ETS family transcription factor. *See* ETS oncogenes.

SPH   Protein-binding DNA elements. *See* Simian virus 40.

S phase   Of the cell cycle, when regular DNA synthesis takes place. *See* cell cycle.

spherocytosis, hereditary (HS)   Most common hemolytic anemia in northern Europe. The basic defect involves both recessive and dominant mutations affecting ankyrin-1 (8p11.2) and spectrin. The β-spectrin gene was assigned to human chromosome 14q22-q23. The spectrin α-chain is responsible for elliptocytosis II (1q21). An autosomal-dominant form was assigned to human chromosome 15q15. There is also an autosomal-recessive type. *See* anemia; ankyrin; elliptocytosis; poikilocytosis; spectrin.

spheroplast   Spherical bacterial cell after (partial) removal of the cell wall. *See* protoplast.

sphingolipid activator protein (SAP, Saposin)   Cofactor for the physiological degradation of sphingolipids. The GM2 activators are encoded in different chromosomes and their deficiencies cause metachromatic leukodystrophy-like and Gaucher disease–like symptoms. *See* Gaucher disease; metachromatic leukodystrophy; sphingolipids. (Matsuda, J., et al. 2001. *Hum. Mol. Genet.* 10:1191.)

sphingolipidoses   Hereditary diseases involving the metabolism of sphingolipids with the enzyme defects indicated in parentheses:

Farber's disease (ceramidase)	Lactosyl ceramidosis (β-galactosyl hydrolase)
Fabry's disease (α-galactosidase)	Metachromatic leukodystrophy (sulfatase)
Gaucher's disease (β-glucosidase)	Niemann-Pick disease (sphingomyelinase)
Generalized gangliosidosis (β-galactosidase)	Sandhoff's disease (hexosaminidases A and B)
Krabbe's leukodystrophy (galactocerebrosidase)	Tay-Sachs disease (hexosaminidase A)

See These diseases under separate entries.

sphingolipids   Sphingosine-containing lipids with the following general structure, depending on the particular substitutions at X.

sphingosine — fatty acid (in neural cells most
commonly stearic acid)

|
X

hydrogen (ceramide)
glucose (glucosylcerebroside), [neutral glycolipid]
glucose and galactose (lactosylceramide), [neutral glycolipid]
complex of sialic acid, glucose, galactose, galactose amine (ganglioside G_{M2})
phosphocholine (sphingomyelin)

If sialic acid (acetyl neuraminic acid, glycolylneuraminic acid) is removed, asiogangliosides result. Sphingolipids mediate signal transduction, calcium homeostasis and signaling, traffic of secretory vesicles, cell cycle, etc. They are the lipid moiety of glycosylphosphatidylinositol-anchoring proteins. *See* ceramide; cerebrosides; gangliosides; Sandhoff disease; sphingolipidoses; sphingosine. (Dickson, R. C. & Lester, R. L. 1999. *Biochim. Biophys. Acta* 1438:305; Leipelt, M., et al. 2001. *J. Biol. Chem.* 276:33621.)

Neuraminic acid.

sphingomyelin   Phospholipid with the sphingosine amino group linked to fatty acids and the terminal OH group of sphingosine esterified to phosphorylcholine. *See* fatty acids; Niemann-Pick disease; sphingosine.

sphingosine Solid fatty acid–like component of membranes. The principal naturally occurring sphingosine is D(+) erythro-1,3-dihydroxy-2-amino-4-transoctadecene, $CH_3(CH_2)_{12}CH=CH(OH)-CH(NH_2)-CH_2OH$. In addition to the C_{17} sphingosines shown, the molecules may have 14, 16, 18, 19, and 20 carbons. The molecules may also be branched or may contain an additional OH group. Sphingosine-1 phosphate is a rather universal signaling molecule for cell proliferation, chemotaxis, differentiation, senescence, and apoptosis. The action of sphingosine kinase is reversed by sphingosine phosphatase. *See* sphingomyelin. (Brownlee, C. 2001. *Curr. Biol.* 11:R535; Merrill, A. H. 2002. *J. Biol. Chem.* 277:25843.)

Spi⁺ Wild-type λ-phage is sensitive to phage P2 inhibition. Phage-λ lacking the functions of *red* and *gam* can grow in P2 carrying–lysogens if it has *chi* (recombination sites for the *RecBC* system). *See* lambda phage.

SPI1 oncogene Spleen focus forming retrovirus homologue. It is in human chromosome 11p11.22. Spi1 is an ETS family transcription factor. *See* ETS; oncogenes; PU.1.

Spielmeyer-Sjögren-Vogt disease *See* ceroid lipofuscinosis.

spike (1) Inflorescence alternatively called head or ear; a typical example is the spike of wheat or some other grasses. The spikelets or flowers are sessile on opposite sides of the axis, which is called rachis. (2) Short-duration electrical variations along the nerve axon; a peak in electric potential. (3) Claw-like structure on the base plate of bacteriophages. *See* development.

spiked oligos Phosphoramidates are incorporated into deoxyribonucleotides and when these are used as primers for PCR-based mutagenesis random mutations may be selected that slow down the activity of the mutant gene products (proteins) and open a chance to screen for extragenic suppressors that restore better or full function. See phosphoramidate: PCR-based mutagenesis: suppressor gene. (Hermes, J. D., et al. 1989. *Gene* 84:143.)

spikelet Group of florets sitting on a common base in a spike. *See* spike.

Wheat spikelet.

spiking Information transfer by neurons.

spina bifida Developmental disability in various mammals, including humans, determined by autosomal-dominant inheritance and reduced penetrance. The spinal column is incompletely closed and in some instances this involves no serious problem and is detectable only by X-ray examination (spina bifida occulta). The more serious case is spina bifida aperta, in which the spinal cord, membranes (meninges), and nerve ends protrude (*myelomeningocele*). Hydrocephalus, incontinence, etc., frequently accompany this condition. The *meningocele* form involves only membrane extrusion and consequently it is a less severe defect. Defects in cadherins may be responsible for the anomalous differentiation. The overall frequency of these anomalies may be 0.2–0.3% in the general population. *See* cadherins; genetic screening; hydrocephalus; mental retardation; MSAPF; neural tube defects; prenatal diagnosis.

spinach (*Spinacia oleracia*) Dioecious plant, $2n = 12$ (XX or XY, included). Many of the trisomics can be identified without cytological examinations. All six pairs of the chromosomes have distinct morphology.

spinal and bulbar muscular atrophy (SBMA) *See* Kennedy disease.

spinal muscular atrophy (SMA) Degeneration of the spinal muscles; occurs in different forms and under different genetic controls. The adult type, proximal, is autosomal dominant and may be associated with different chromosomes. The juvenile type (Kugelberg-Welander syndrome) primarily affects the proximal limb muscles and frequently involves twitching. This form was assigned to human chromosome 5q11.2-q13.3. The prevalence of this type is about 1/6,000 in newborns. The literature also distinguishes Werdnig-Hoffmann disease; the various juvenile forms map to the same chromosomal segment, although the expressions may differ. Deletions of this area are frequently the basis of the disease. The combined estimated gene frequency is about 0.014. Spinal muscular dystrophy with microcephaly and mental retardation, spinal muscular dystrophy distal, SMA proximal adult type, and other variations appear to be autosomal recessive. SMA expression is positively associated with the activity of the *survival motor neurons* controlled by the centromeric SMN2 gene (human chromosome 5q12.2-q13.3.) The highly homologous SMN1 gene has a telomeric location. Mutation in SMA may affect one of the eight subunits of the Sm protein, which is a component of the U snRNA nuclear import system. Spinal muscular atrophy with respiratory distress (SMARD) is caused by mutation if an immunoglobulin (IgG)-binding protein (IGHMBP2) is encoded at 11q13.2-q13.4. *See* atrophy; dystrophy; gemini of coiled bodies; Kennedy disease; Kugelberg-Welander syndrome; muscular atrophy; muscular dystrophy; NAIP; neuromuscular diseases; SMAD; U RNA; Werdnig-Hoffmann disease. (Jablonka, S., et al. 2001. *Hum. Mol. Genet.* 10:497; Grohmann, K., et al. 2001. *Nature Genet.* 29:75; Feldkötter, M., et al. 2002. *Am. J. Hum. Genet.* 70:358; Frugier, T., et al. 2002. *Current Opin. Genet. Dev.* 12:294; Paushkin, S., et al. 2002. *Current Opin. Cell Biol.* 14:305.)

spindle System of microtubules (≈20–30 nm) emanating from the poles (centrioles of the centrosomes in animals) during mitosis and meiosis and attaching to the kinetochores within the centromeres, pulling the chromosomes toward the poles. In plants and in some animal oocytes, there are no centrosomes and the chromosomes assume some of the centrosome functions. In yeast, the spindle forms within the nucleus. Some of the microtubules reach from one pole to the other without attaching to the kinetochore. In *Drosophila* meiosis, the spindle originates from each of the chromosomes; as the prophase progresses, a bipolar spindle emerges. This stage uses a kinesin-like protein (NCD). The arrangement of the microtubules also requires the motor protein dynein. The meiotic pole is different from the mitotic centrosomes (DMAP60, DNAP190, and γ-tubulin are apparently absent).

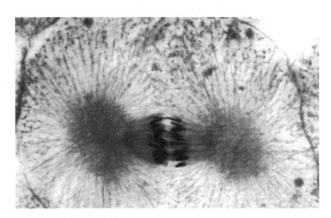

Spindle (Illustration by courtesy of Dr. P. C. Koller.)

The orientation of the spindle may be controlled in yeast by myosin V. In case of univalents, only a monopolar spindle is formed. In some species, the spindle origination from the chromosome is apparently suppressed by the centrosome. In other species, the chromosomes and the centrosomes cooperate in developing the meiotic spindle. The mitotic spindle may also need both chromosomes and centrosomes (echinoderms). Microtubule assembly is not an exclusive property of the kinetochores as holocentric chromosomes indicate. Anaphase and cytokinesis, however, can take place in cells after the chromosomes have been removed. On the oocyte chromosomes of several species, surface proteins (NOD, Xklp1) are found that stabilize premetaphase chromosomes, and in achiasmate meiosis they substitute for the chiasmata. According to current views, the information for meiotic disjunction resides within the chromosomes and not in the spindle apparatus. The kinetochore determines the transition from metaphase to anaphase. Tension of the kinetochore generates a checkpoint signal (Cdc20, fizzy, Cdc55), and it is supposed that a phosphorylated kinetochore protein attracts protein. At least six known genes seem to be involved in kinetochore functions. The X-chromosomes in XO cases of sex determination do not involve such a pause. The cytoskeleton is also involved in the correct organization of the spindle. *See* acenaphthene; achiasmate; APC; Bim; Cdc2; Cdc20; Cdc55; CENP; centromere; centromere proteins; centrosome spindle fibers; chromatid; cytoskeleton; dynactin; holocentric chromosome; kinesin; kinetochore; Mad2; meiosis; microtubule; mitosis; multipolar spindle; myosin; nucleus; NuMA; RAN; RMSA-1; securin; Stathmin; Swi6p; tubulins; univalent. (Sharp, D. J., et al. 2000. *Nature* 407:41; Shah, J. V., et al. 2000. *Cell* 103:997; Compton, D. A. 2000. *Annu. Rev. Biochem.* 69:95; Karsenti, E. & Vernos, I. 2001.

Science 294:543; Musachio, A. & Hardwick, K. G. 2002. *Nature Rev. Mol. Cell Biol.* 3:731.)

spindle fibers   Microtubules that are clearly visible (by appropriate techniques) during mitotic and meiotic nuclear divisions. The microtubules originate at the spindle poles, the aster in animals. Three classes of fibers emanate from the poles, the astral microtubules that radiate from the centrioles. The polar microtubules that meet at the divisional plane and appear to stabilize the spindle and the kinetochore microtubules that are anchored at the centromere of the chromosomes and at anaphase pull them toward the opposite poles. *See* aster; centrioles; centromere; kinetochore; meiosis; mitosis; spindle; tubulins.

spindle poison   Blocks the formation of spindle fibers and as a consequence polyploidy may result. *See* acenaphthene; colchicine; polyploidy; Stathmin.

spindle pole body (SPB)   Fungal equivalent of the centrosome. Cyclins and cyclin-dependent kinases promote and regulate their duplication. SPB assists in the assembly of membrane proteins in the meiotic prospores of yeast. Its defect results in genetic instability, and aneuploidy. SPB also affects mRNA metabolism. *See* centrosome. (Haase, S. B., et al. 2001. *Nature Cell Biol.* 3:38; Bajgier, B. K., et al. 2001. *Mol. Biol. Cell* 12:1611; Lang. B. D., et al. 2001. *Nucleic Acids Res.* 29:2567.)

spin infection   Adherent cells are exposed to the transformation vector mix by centrifuging.

spinobulbar muscular atrophy   *See* Kennedy disease.

spinocerebellar ataxia (SCA)   Autosomal-dominant defects involving CAG repeats (polyglutamine) in the coding region of ataxin genes (SCA1, SCA2, SCA3, SCA7). SCA causes nerve degeneration and loss of Purkinje cells and neurons in the brain. As a consequence, motor functions deteriorate. SCA6 affects a calcium ion channel. In a normal state, the number of repeats is 6−44 in disease 40−93. The ataxin gene (SCA1) encoded at 6p23 is transcribed into 792−869 amino acids, depending on the number of CAG repeats. Ataxin-1 binds to RNA and controls the expansion of polyglutamine sequences. SCA8 involves CTG repeats. SCA10 (22q13-qter) displays ATTCT repeats in variable numbers in intron 9. SCA12 (5q31-q33) is caused by CAG repeats in the regulatory subunit of protein phosphatase PP2A. SCA13 childhood ataxia with mental retardation with delayed motor functions is encoded at 19q13.3-q13.4. *See* ataxia; CACNA1A; Huntington chorea; Kennedy disease; Machado-Joseph disease; migraine; myotonic dystrophy; trinucleotide repeats. (Stevanin, G., et al. 2000. *Eur. J. Hum. Genet.* 8:4; Yue, S., et al. 2001. *Hum. Mol. Genet.* 10:25; Libly, R. T., et al. 2003. *Hum. Mol. Genet.* 12:41.)

spiracle   Breathing hole on the insect body.

spiralization   *a* pattern of winding of molecules or chromosomes. *See* coiling.

spirochetes (Spirochaeta)   Filiform bacteria (5−6 μm) cause sex ratio distortion in *Drosophila* by the killing effect of their toxin on the male flies. Some lower animals (e.g., *Schistosoma*) are also called spirochetes. *See Schistosoma.*)

spironucleoside (hydantocydin) Herbicidal growth regulator of *Streptomyces hygroscopicus*. *See* herbicide.

Spiroplasma *See* sex ratio; spirochetes; symbionts, hereditary.

spitz (*Drosophila* chromosome 2-54) Locus encodes the Spitz protein, a ligand of DER. *See* DER.

SPK-1 Protein kinase of the MAPK family. *See* signal transduction.

spleen Upper left abdominal, oblong (ca. 125 mm) ductless gland. At the embryonal stage, it participates in erythrocyte formation; in adults it also makes lymphocytes. By decomposing erythrocytes, it provides hemoglobin to the liver to form bile. The red pulp contains red blood cells and macrophages; the white pulp area carries lymphocytes. It has a role in the defense mechanism of the body. It is generally much enlarged in some lysosomal storage diseases, e.g., Gaucher disease. *See* asplenia; lysosomal storage diseases.

splice Resolution of a recombination intermediate (Holliday junction) resulting in an exchange of the flanking markers. *See* Holliday model; introns; spliceosome.

spliceosomal intron Utilizes spliceosomes for the removal of introns, in contrast to type II introns, which do it themselves. *See* introns; spliceosome; splicing.

spliceosome Protein-U snRNA (there are five main uridine-rich oligonucleotides) complex required for the folding of pre-mRNA into the proper conformation for removal of introns and splicing the transcripts of exons. The *majority* of eukaryotic primary transcripts contain introns with 5′ GU and AG 3′ splice sites, and the mammalian consensus is (A/C)**AG↓GUA**/GAGU; in yeast (**A/G**) ↓**GUAUGU**. Less than 1% of the mammalian splice junctures are GC-AG. In the excision, the spliceosome complex U1, U2, U4–U6, and U5 snRNA and non-snRNAs work together. Initially the U1 and U2 snRNA attaches by base pairing to the splice and to the branch sites, respectively. The U4–U6•U5 tri-snRNP complex joins in presplicing complexes. This is followed by snRNP-snRNP and mRNA-snRNP interactions. U6 base-pairs with U4. During spliceosome assembly, U1 and U4 are displaced; U6 pairs with the 5′ splice site and with U2 snRNA. Through the coordination of a divalent metal ion (Mg^{2+}), U6 snRNA contributes to splicing of the RNA transcript. The 5′ splice site and the branch nucleotide then move to each other and the 2′-OH group of the latter serves as an electron donor for the first step of splicing. The excised 5′-exon and the lariat intron-3′-exon are the reaction intermediates. This first step is followed by a reaction between the electron donor nucleophile and the electron-deficient electrophile at the 3′-splice junction

by the 3′-OH of the freed 5′-exon, resulting in the ligation of the exons and the removal of the intron.

A *minority* (0.1%) of the introns, the AT-AC introns, occur in some animal genes such as encoding PCNA. The 5′-splice site of such introns has the consensus **AT↓ACCTT**, and their branch site is TCCTTAAC. Their splicing complex includes U11 and U12 snRNP and one or more U5 snRNP variants. The PCNA branch site pairs with U12, and a loop of U5 aligns the exons of PCNA for ligation. U4 and U6 snRNAs are not used, but the highly divergent U4atac and U6atac take over their role. The reaction at the 5′-splice site is mediated by a metalloenzyme but not at the 3′ site. The spliceosome is a huge and variable complex of 40–50 or more proteins. For the recognition of the 3′-AG splice site, U2AF[35] (the 35 K subunit of the heterodimeric [M_r 65 K] U2AF[65] protein) is required in vivo; it is needed for viability and it is present in different organisms. U2AF was considered to be only an auxiliary factor (because in vitro it was not indispensable) to recognize the 5′-polypyrimidine sequences. For splicing, the U2AF protein must be in close vicinity of the polypyrimidine tract that is recognized by its 35 K subunits. The *Drosophila* protein Sex-lethal (SXL) controls dosage compensation by inhibiting splicing of the *male-specific-lethal-2* transcripts. When the large subunit of U2AF is displaced from the polypyrimidine tract–3′AG interval by SXL, the 35 K subunit can mediate the removal of the intron. *See* dosage compensation; exons; introns; PCNA; snRNA; splicing. (Hastings, M. L. & Krainer, A. R. 2001. *Curr. Opin. Cell Biol.* 13:302; Nagai, K., et al. 2001. *Biochem. Soc. Trans.* 29:15; Valadkhan, S. & Manley, J. L. 2001. *Nature* 413:701; Sträßer, K. & Hurt, E. 2001. *Nature* 413:648; Nilsen, T. W. 1998. *RNA Structure and Function*, Simons, R. W. & Grunberg-Manago, eds., p. 279. Cold Spring Harbor Lab. Press, Villa, T., et al. 2002. *Cell* 109:149; Brow, D. A. 2002. *Annu. Rev. Genet.* 36:333; <http://www.cse.usc.edu/research/compbio/>.)

splicing Joining of RNA with RNA or DNA with DNA at the sites of previous cuts. Constitutive splicing indicates that the exons are spliced in the same order as they occur in the primary RNA transcript, in contrast to alternative splicing, in which the exons may be joined in alternative manners, thus providing mRNAs for different proteins transcribed from the same gene. General scheme of pre-mRNA splicing:

The splicing factor family SR (named because they are rich in serine [S] and arginine [R]) contains an RNA recognition motif (RRM/RNP) and Ψ-RRM domain Ser-Trp-Gln-Asp-Leu-Lys-Asp separated by a Gly-rich tract. The Ser-Arg domains are well phosphorylated by the serine kinase SRPK1. The 3′-splice site is recognized by the U2AF[65]/U2AF[35] heterodimer. The former binds to the polypyrimidine sequence in the RNA, whereas the latter associates with the RS domain of other RS-containing factors or the U4/U6.U5 small ribonuclear complexes at the 5′-splice site. These proteins select the splice sites and are active parts of the spliceosome complex (*see* spliceosome). The ASF/SF2 family recruits U1 snRNP to the RNA transcript. SR proteins also communicate between introns across exons and affect alternative splicing. SR proteins are subject to regulation. In between the branch site and AG, there are polypyrimidine sequences that aid various proteins in defining the 3′-splice site. Some of the many relevant proteins are PTB (polypyrimidine tract–binding protein) and PSF (PTB-associated splicing factor). The splicing reaction is driven by various nucleotide triphosphatases. Plants also have splicing factors, but they are more variable than those of animals. Splicing precedes the export of RNA to the cytoplasm and requires the association of mRNA with the splicing factor Aly. *See* alternative splicing; introns; lariat; restriction enzyme; snRNA; speckles, intranuclear; spliceosome; U1 RNA; vectors. (Madhani, H. D. & Guthrie, C. 1994. *Annu. Rev. Genet.* 28:1; Kramer, A. 1996. *Annu. Rev. Biochem.* 65:367; Kim, N., et al. 2001. *EMBO J.* 20:2062; Thanaraj, T. A. & Clark, F. 2001. *Nucleic Acids Res.* 29:2581; Luo, M.-J., et al. 2001. *Nature* 413:644; Clark, T. A., et al. 2002. *Science* 296:907; <http://isis.bit. uq.edu.au>.)

splicing enhancer Repetitive GAA sequences associated with proteins and facilitate the joining of exons.

splicing inhibition May be brought about by (phosphorothioate) 2′-O-methyl-oligoribonucleotides or morpholino oligonucleotides that are resistant to RNase H. However, sometimes these analogs activate cryptic splice sites within exons.

Morpholine.

splicing juncture (splice junction) Sequences at the exon-intron boundaries. They are needed for selective export of

mature mRNAs from the nucleus and they assist in RNA surveillance, which leads to degradation of mRNAs with premature transcription termination. *See* exons; introns, nonsense-mediated decay; RNA surveillance; spliceosome. (Lykke-Andersen, J., et al. 2001. *Science* 293:1836.)

split gene Discontinuous because introns are intercalated between exons; the majority of eukaryotic genes contain introns. *See* exons; introns; spliceosome; splicing.

split-hybrid system Provides means for positive selection for molecules that disrupt protein-protein interactions. The genetically engineered construct is shown below.

If the binding of X and Y is prevented, the VP16 transactivator cannot turn on TetR, thus permitting *HIS3* gene expression, and it can be seen by the growth of yeast cells on a histidine-free medium. *See* one-hybrid binding assays; rtTA; three-hybrid system; tTA; two-hybrid method; VP16. (Goldman, P. S., et al. 2001. *Methods Mol. Biol.* 177:261.)

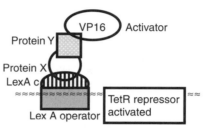

Prevents the expression of *HIS3* gene; no growth in a medium without histidine. *HIS* is not shown.

SPM Scanning probe microscope. Images of biological macromolecules under a thin layer of aqueous solution. *See* scanning electron microscopy.

Spm (*suppressor mutator*) Transposable element system of maize of an autonomous *Spm* and a nonautonomous (originally unnamed) element. The nonautonomous element cannot insert or excise by its own power because it is defective in the transposase enzyme. This system is the same as the *En* (*Enhancer*)-*I* (*Inhibitor*) system. The nonautonomous

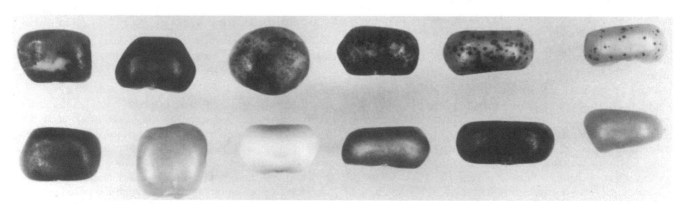

The dominant alleles of maize normally form a dark color, but the a^{m-1} allele possessing the *Spm* element may display alternative states. In the absence of an inactive *Spm*, any shade of solid color may appear (bottom row), but in the presence of an active *Spm*, a variety of sectors are displayed (top row). (Courtesy of Barbara McClintock.)

component has been called *dSpm* (*defective Spm*). The original name represents the fact that insertional mutations caused by the nonautonomous *dSpm* (*I*) element revert at high frequency only when *Spm* (*En*) is introduced into the genome because the latter has a functional transposase gene. The insertion does not always eliminate the function of the target gene and both a recessive mutation and the *Spm* element may be expressed. Such a case at the *A* (anthocyan) locus of maize (chromosome 3L-149) was designated as *a − m2* (*a* mutable) because of its frequent reversion to the dominant allele and displaying sectors in the presence of *Spm*. *Spm-dependent* alleles harbor a nonautonomous (*dSpm* or *I*) insertion element that may jump out (and cause reversion) only when functional *Spm* is introduced. *Spm-suppressible* alleles indicate the presence of a nonautonomous element that may or may not permit the expression of the gene, but the presence of active *Spm* allows the insert's removal. The terms *Spm-w* and *Spm-s* indicate weak and strong *Spm* transposase elements, respectively. *Inactivated Spm* has been called *Spm-i*, and elements that have alternating active and inactive phases are *cycling Spm* (*Spm-c*), whereas very stable inactive forms are *cryptic Spm* (*Spm-cr*). In addition, a *Modifier* factor has been named that enhances the activity of *Spm*.

The various forms of *Spm* alleles were found to alter their activity in time by developmental stage and extraneous factors, which cause chromosome breakage (radiation, tissue culture, etc.). The term *presetting* was applied to the phenomenon that *Spm* determined the expression of a gene even after its removal from the genome, but this effect later may fade away. Presetting was supposed to happen during meiosis and it was attributed to methylation. Indeed, the first exon of *Spm* is rich in cytidylic residues, the most commonly methylated base in DNA. Methylation may be instrumental in the variations and inactivation of the various *Spm* elements named above. A *Regulator* element is credited with the control of the extent of methylation. Although the various *Spm* alleles display some heritable qualities, they appear to represent labile "changes in states" of the element, except *dSpm* or other deletional forms. Transposition of *Spm* elements is not tied to DNA replication. The transposition favors new insertion sites within the same chromosome, although it may move to any other part of the genome. Integration of *Spm* reresults in a 3 bp duplication at the target site. The frequency of insertion and excision is also influenced by the base sequences flanking the target site. *Spm* appears to be modified by the process of excision and involves various lengths of terminal deletions, but subterminal repetitious sequences may also be involved. Deletion of the terminal repeats abolishes transposition of the element. The abilities of *Spm* to transpose and to affect gene expression are inseparable.

The *Spm* element is transcribed into a 2.4 kb RNA with 11 exons; it includes two close open-reading frames. The transcript may be alternatively spliced into four different open-reading frames, called *tnpA*, *tnpB*, *tnpC*, and *tnpD*, the latter being the longest. Transcripts *tnpA* and *tnpD* are necessary for transposition. The sequences downstream from the transcription initiation site are rich in GC base pairs and are susceptible to methylation. The active *Spm* elements at about 0.6 kb around the transcription start site are not methylated. The methylated elements are, however, inactive. The entire element (8.4 kb) is flanked by 13 bp inverted repeats (CACTACAAGAAAA and TTTTCTTGTAGTG). In a region

180 bp from the 5'-end and 299 bp from its 3' end, several copies of a receptive consensus CCGACATCTTA occur. The defective elements, *dSpm*, have various lengths of internal deletions covering the entire length or part of the two open-reading frames. Some partially deleted elements may still function as a weak *Spm-w* element. The function of an Spm element is regulated by sequences within the element and the location of the transposable element within the target genes. The function of the target genes depends on when, where, and how the *Spm* element and the target gene's transcripts are spliced. *See Ac-Ds*; hybrid dysgenesis; insertional mutation; transposable elements. (Fedoroff, N. 1989. *Mobile DNA*, Berg, D. E. & Howe, M. M., eds., p. 375. Am. Soc. Microbiol., Washington, DC.)

SPN Single-nucleotide polymorphism indicates single-nucleotide difference between two nucleic acids. *See* SNIP.

SPO Group of sporulation-mediating proteins in yeast. (Kee, K. & Keeney, S. 2002. *Genetics* 160:111.)

spondylocostal dysostosis (SD) Nonsyndromal short stature, vertebral and rib defects encoded at 19q13.1-q13.3. The defect in the homologue of *Drosophila* gene *Delta* (*Dll3*) is involved. *See* Alagille syndrome; Simpson-Golabi-Behmel syndrome.

spondyloepimetaphyseal dysplasia (SEMD) Heterogeneous group of hereditary skeletal bone and cartilage diseases involving defects in protein sulfation. A recessive type was assigned to chromosome 10q23-q24.

spondyloepiphyseal dysplasia (SED) Autosomal-dominant phenotype includes flattened vertebrae, short limbs and trunk, barrel-shaped chest, cleft palate, myopia (nearsightedness), muscle weakness, hernia, and mental retardation. Collagen defects are incriminated in many cases. *Autosomal-recessive* forms mimic arthritis-like symptoms (arthropathy) besides the short stature. Autosomal-recessive forms may not involve flat vertebrae. The defect was attributed to a deficiency of phosphoadenosine-5'-phosphosulfate and thus to undersulfated chondroitin. An *X-linked* SED (Xp22.2-p22.1) was also described. SED-type diseases were located to human chromosomes 5q13-q14.1, 12q13.11-13.2, and 19p13.1. The prevalence of the X-linked form is about 2×10^{-6}. *See* achondroplasia; arthropathy-campylodactyly; chondroitin sulfate; collagen; dwarfness; Schimke immunoosseus dysplasia. (Gedeon, A. K., et al. 2001. *Am. J. Hum. Genet.* 68:1386.)

spontaneous generation, current Before Louis Pasteur (1859–61), it was assumed by many scientists that microorganisms were formed from abiotic material even during the present geological period. The organisms found in broth and other rich nutrients grew — as he demonstrated — only when the solutions were not heated to a sufficiently high temperature, for a sufficient duration of time, and exposed to unfiltered air. His discovery has been fundamental to modern microbiology and medicine and proved that spontaneous generation is not responsible for the current variations in microbial cultures. *See* biogenesis; lysenkoism; pleomorphism; spontaneous generation, unique or repeated. (Farley, J. 972 *J. Hist. Biol.* 5:285; ibid., 95.)

spontaneous generation, unique or repeated

Explanation for the abiotic origin of life. It is assumed that first, after the earth was formed, water, carbon dioxide, ammonia, and methane—all simple molecules—were formed. Subsequently, in a reducing atmosphere or at deep oceanic vents containing reducing minerals (iron and nickel sulfides) and high temperature (300–800°C), molecular nitrogen could have been reduced to ammonia. When energy sources became available at the surface (ultraviolet light), simple organic acids (acetic acid, formic acid) arose. In the presence of ammonia, methane, hydrogen, hydrogen cyanide, and lightning energy amino acids could be formed. In the following steps, nucleotides could arise. Actually, under simulated early earth conditions, chemists could synthesize amino acids, polypeptides, carbohydrates, and nucleic acids. These simple organic molecules might have aggregated into some sorts of micellae (bubbles), and after self-replicating mechanisms came about the possibility opened up for the generation of an ancestral cell with a primitive RNA as the genetic material. It is not entirely clear when, where, how, and how many times these events took place. Life is estimated to begin 3–4 billion years ago. It is not known whether on other planets, under similar conditions to those of the earth, living cells evolved. *See* abiogenesis; biopoesis; evolution, prebiotic; exobiology; origin of life; spontaneous generation, current. (Lennox, J. 1981. *J. Hist. Philos.* 19:19; Harris H. 2002. Things Come to Life: Spontaneous Generation Revisited. Oxford Univ. Press, New York.

spontaneous mutation

Occurs at a relatively low frequency when no known mutagenic agent is or was present in the environment of the cell or organism; the cause of the mutation is thus unknown. The spontaneous frequency of mutation in mice for seven standard loci is 6.6×10^{-6} per locus. The frequency in humans is in the range of 10^{-5} to 10^{-6}, in *Drosophila* 10^{-4} to 10^{-5}, in yeast and *Neurospora* 10^{-5} to 10^{-9}, in bacteria 10^{-4} to 10^{-9}, and in bacteriophages 10^{-4} to 10^{-11}. In maize, the frequency is comparable to that in other eukaryotes. At some other loci and in other organisms, it may be substantially higher or lower. In higher eukaryotes, the frequencies appear lower than in microorganisms, but this does not seem to be due to intrinsic biological differences; rather it reflects the limitations of the size of the populations, which were amenable to screening. The extent of DNA repair may have a profound influence on recovered mutations. *See* diversity; mutation, spontaneous; mutation rate. (Wloch, D. M., et al. 2001. *Genetics* 159:441.)

spooling

Assumes that the triple- or quadruple-stranded naked DNA molecules are wound into the RecA protein filament for pairing and exchange. After the formation of heteroduplexes, the DNAs are released. *See* recombination, molecular mechanism.

sporadic

The rare occurrence of an off-type does not show a clear familial pattern, and the etiology (cause or origin) is unknown. *See* epidemiology.

sporangiophore

Sporangium-bearing branch. *See* sporangium.

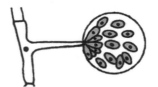

Sporangiophore and sporangium.

sporangium

(plural sporangia) Spore-producing and -containing structure in lower organisms (fungi, protozoa).

spore

Reproductive cell; generally the product of eukaryotic *meiosis*, or it may arise *mitotically* as fungal conidiospores (conidia). *Bacterial spores* are metabolically dormant cells surrounded by a heavy wall for protection under very unfavorable conditions.

spore mother cell

See sporocyte.

sporidium

"Sexual" spore of basidiomycetes fungi. *See* basidium.

sporocyte

Diploid cell that produces haploid spores as a result of meiosis. *See* gametogenesis.

sporogenesis

Mechanism or process of spore formation. *See* conidia; meiosis.

sporophore

Fruiting body capable of producing spores in fungi. *See* spore.

sporophyte

Generation of the plant life cycle that produces the ($1n$) gametophytes by meiosis. The common form of plants (displaying leaves and flowers, etc.) is the ($2n$) sporophytic generation. *See* gametophyte.

sporopollenin

Mainly polymerized carotenoids forming the exine of the pollen grains.

sporozoite

Infective stage of the protozoan life cycle; in malaria, they are formed within the mosquito. *See Anopheles*; malaria; *Plasmodium*.

sporulation

(1) In bacteria, the process of formation of morphologically altered cells that can survive adverse conditions and assure the survival of the sporulating bacteria. *Bacillus subtilis* is a typical spore-forming bacterium. Within its cell a new cell is pinched off to create the spore, which can eventually develop into a new regular cell. (2) Ascospore formation through meiosis in fungi. (<http://cmgm.stanford.edu/pbrown/sporulation/.)

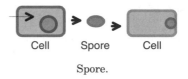

Cell   Spore   Cell

Spore.

spot test Variation of mutagen/carcinogen bioassays when the compound to be tested by bacterial reversions is added to the surface as a crystal or a drop after the petri plates have been seeded by the bacterial suspension and the S9 microsomal fraction has been added. If the substance to be tested is mutagenic, a ring of revertant cells should appear around the spot where it was added. This type of mutagenic assay is rarely used today. *See* Ames test; plate incorporation test; reversion assays of *Salmonella* photo below. (Ames, B. N. 1971. *Chemical Mutagens I*, Hollaender, A., ed., p. 267. Plenum, New York.)

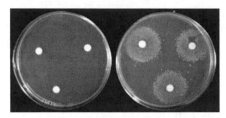

Reversion assay of *Salmonella* bacterium by the spot test. (Courtesy of Dr. G. Ficsor.)

spotting *See* KIT oncogene; piebaldism; variegation.

spp Abbreviated plural of species.

spreader Simple instrument for distributing microorganisms on the surface of agar plates.

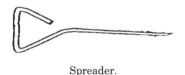

Spreader.

Sprekelia formosissima (Amaryllidecea) Subtropical plant, $n = $ ca.60; genome size 1.8×10^{11} bp.

spretus *Mus spretus*, a species of mouse commonly used for genetic analyses.

springer *see copia*.

S protein *See* vitronectin.

spruce (*Picea* spp) Timber tree species with $2n = 2x = 24$. *See* douglas fir; pine.

sprue Tropical intestinal disease, apparently due to infection(s).

SP score *See* MP score.

SPT Suppressor of transposition (e.g., of a Ty element) and regulator in SAGA function in yeast. The *Drosophila* homologues are *Dspt4* and *Dspt6*. This family of genes encodes histones H2A and H2B and TATA-binding proteins. There is evidence for their regulation of transcription,

replication, recombination, and some developmental processes. *See* histones; SAGA; TBP. (Yamaguchi, Y., et al. 2001. *J. Biochem.* 129:185.)

SptP *See* SopE.

Spt proteins Involved in the elongation of RNA transcripts on DNA. (Winston, F. 2001. *Genome Biol.* 2 [reviews]:1006.)

squalene *See* prenylation.

squamous Scaly, e.g., squamous cell carcinoma, a form of epithelial cancer.

squash *Cucurbita maxima* (winter squash), $2n = 24, 40$; *C. pepo* (summer squash), $2n = 40$; *C. moschata* (pumkin), $2n = 24, 40, 48$.

squash preparation For microscopic analysis of chromosomes in soft (softened) tissues there may be no need for sectioning, but the fixed and stained material can be examined after smearing it directly on the microscope slide. In some cases, the fixation is done by gentle heating on the slide followed by adding a small drop of aceto carmine or aceto orceine stain or even just placing the specimen into a drop of acetic stain and heating. These rapid procedures may permit an estimation of the stage of meiosis or mitosis. *See* microscopy; stains. (Belling, J. 1921. *Am. Nat.* 55:573.)

squelching Suppression or silencing adverse effects.

squirrel *Tamiasciurus hudsonicus streatori*, $2n = 46$; *Callospermophilus lateralis*, $2n = 42$; *Ammospermophilus*, $2n = 32$; *Citellus citellus*, $2n = 40$.

SR1 Cytoplasmic mutant of tobacco resistant to streptomycin. *See Nicotiana*; streptomycin.

S1 ribosomal protein Binds to U-rich sequences upstream of the Shine-Dalgarno sequence and may promote translation. *See* Shine-Dalgarno sequence; translation. (Boni, I. V., et al. 2001. *EMBO J.* 20:4222.)

S6 ribosomal protein Phosphorylated at about five serine residues near its C terminus. The phosphorylated state is correlated with the activation of protein synthesis on the ribosomes; the phosphorylation is stimulated by mitogens and growth factors. *See* ribosome. (Recht, M. I. & Williamson, J. R. 2001. *J. Mol. Biol.* 313:35.)

SRB Protein stabilizing RNA polymerase II binding to general transcription factors. It phosphorylates pol II RNA polymerase. SRBs occur in eukaryotes from yeast to humans and have the forms SRB 2, 4, 5, 6, 7, 10, 11. The CDK-like *SRB* genes involve mutation control and promotion/suppression of transcription by phosphorylating the carboxyl-terminal of RNA pol II. *See* CDK; kinase; mediator complex; open promoter complex; regulation of gene activity; RNA polymerase holoenzyme; transcription complex; transcription factors; TUP.

(Hampsey, M. & Reinberg, D. 1999. *Curr. Opin. Genet. Dev.* 9:132; Carlson, M. 1997. *Annu. Rev. Cell Dev. Biol.* 13:1.)

SRBC Sheep red blood cell.

SRC Rous sarcoma virus oncogene of chicken. Its product is a protein tyrosine kinase, a cellular signal transducer. Its homology domains SH2 and SH3 are present in several cytoplasmic mediator and adaptor proteins in the signal transduction pathways in different organisms. These domains bind phosphotyrosine or proline-rich residues. Phosphorylation, dephosphorylation, and proteolysis regulate SRC. In humans, SRC is in chromosome 20q12-q13. SRC may be involved in both RAS-dependent and RAS-independent signaling pathways and may lead through either FOS or MYC to transcription factors. This family of nonreceptor kinases includes Src, Yes, Fgr, Fyn, Lck, Hck, Blk, Zap70, and Tec. Csk⁻ cells display increased Src, Fyn, and Lyn activity. Src proteins may also have an autoinhibitory function. The Cbl oncogene acting downstream of Src is responsible for bone resorption in osteoporosis. c-SRC has been implicated in various types of cancers. *See* Blk; Cbl; Csk; Fgr; Fyn; Hck; Lck; oncogenes; osteoporosis; SH2; SH3; signal transduction; TCR; Tec; Yes; Zap-70. (Schlessinger, J. 2000. *Cell* 100:293.)

SRC-1 Steroid receptor coactivator-1 enhances the stability of the transcription complex controlled by the progesterone receptor. It is actually a coactivator of histone acetyltransferase and mediates the access of the transcription complex within the nucleosome. *See* histone acetyltransferase; N-CoR; progesteron; TGF; transcription. (Liu, Z., et al. 2001. *Proc. Natl. Acad. Sci. USA* 98:12426; Auboef, D., et al. 2002. *Science* 292:616.)

SRE Cis-acting enhancer element responding to serum induction: CC(AT)₆GG (CarG box) is present in all serum-response factor regulated genes. *See* MADS box; serum-response element; serum-response factor; TCF.

SREBP Sterol regulatory element-binding proteins are in a hairpin shape in the membrane of the endoplasmic reticulum (ER). The N-terminal domain is in the cytosol and acts as a basic helix-loop-helix transcription factor. The C-terminus is also in the cytosol and is complexed with the cleavage-activating protein SREBP-SCAP, which has eight membrane-spanning regions. When the ER is low on sterols, the complex is transferred to the cleavage compartment. In case of sterol overload, the complex is sequestered into the ER and there is no cleavage. S1P and S2P process SREBP. *See* lipodystrophy; Rip; SCAP. (Shimano, H. 2001. *Progr. Lipid Res.* 40:539; Dobrosotskaya, I. Y., et al. 2002. *Science* 296:879.)

SRF Serum-response factor is a trans-acting regulatory protein binding to SRE and regulating serum-induced gene expression. *See* serum-response element; trans-acting element. (Kim, S. W., et al. 2001. *Oncogene* 20:6638.)

SRK Self-compatibility protein receptor kinase. *See* self-incompatibility. (Takayama, S., et al. 2000. *Nature* 413:534.)

SR motif Serine/arginine-rich domains in RNA-binding proteins involved in splicing pre-mRNA transcripts. They are required in the early steps of spliceosome assembly. One Sr protein (SC35) alone is sufficient to form a committed complex with human β-globin pre-mRNA. Different single SR proteins commit different pre-mRNAs to splicing, and different sets of SR proteins may determine the alternative and tissue-specific splicing within an organism. SR proteins are regulated by phosphorylation/dephosphorylation. *See* DEAD-box proteins; DEAH-box proteins; ESE; introns; primary transcript; processing; spliceosome; splicing; tissue specificity. (Tian, H. & Kole, R. 2001. *J. Biol. Chem.* 276:33833.)

5S RNA *See* ribosomal RNA; ribosomes. (Artavanis-Tsakonas, S., et al. 1977. *Cell* 12:1057.)

7S RNA *See* RNA 7S.

S-RNase Ribonuclease responsible for pollen rejection (with factor HT) in self-incompatible plants. *See* self-incompatibility. (Luu, D. T., et al. 2001. *Genetics* 159:329.)

SRP Signal recognition particle is an element of polypeptide transport systems through the membranes of the endoplasmic reticulum in eukaryotes and to the plasma membranes in prokaryotes. Some of the polypeptides are inserted in the endoplasmic reticulum; others are destined toward the cell membrane for secretion. The subunit (Ffh) that recognizes the signal sequence and the α-subunit (FtsY) of its receptor (SR) are GTPases. SRP has a variable-length RNA component (4.5S in prokaryotes and 7SL in eukaryotes). The Ffh subunit has an N domain where the signal peptide binds and a G domain of the GTPases. Adjacent to the signal-binding pocket, the methionine-rich M domain forms a small globular structure, which folds into a helix-turn-helix type of motif for binding the RNA component. *See* endoplasmic reticulum; Fts; signal peptide; signal sequence recognition particle; 7SL RNA. (Batey, R. T., et al. 2000. *Science* 287:1232; Oubridge, C., et al. 2002. *Mol. Cell* 9:1251.)

SRY Sex-determining region Y, previously called testis-determining factor (TDF). Mammalian gene in the short arm of the Y chromosome (Yp11.3) responsible for testis determination and for the development of pro-B lymphocytes. The protein (223 amino acids) is a member of the high-mobility group proteins. In mice, this HMG protein includes a large CAG trinucleotide repeat tract that functions as a transcriptional trans-activator and is required for male sex expression. The expression of SRY initiates the formation of the Müllerian-inhibiting substance (MIS) and the synthesis of testosterone. SRY contains the DNA minor groove-binding domain, the HMG box that is conserved among mammals. Mutations affecting human sex reversal are generally within the HMG box. A human chromosome 17q24.3-q225.1-located gene SRA-1 (sex-reversal autosomal) may also be controlled by SRY. Another sex-reversal gene was identified in human chromosome 9p24. In rodents, several *tda* (testis-determining autosomal) alleles exist. In the short arm of the human X-chromosome (Xp21), there is the DSS (dosage-sensitive sex reversal) locus, which also may be involved in sex reversal (male→female) in case of its duplication or in case the SRY alleles are weak.

Deletion in the 160 kb DSS region (DAX) does not affect male development but may cause adrenal hypoplasia. Apparently, the DAX gene encodes nuclear hormone receptors. DAX1 expression ceases early in testis development but persists through the development of the ovaries. There is/are the SRYIF inhibitory factor(s) involved in gonadal differentiation. The voles (rodents) *Ellobius lutescens* $2n = 17$, XO constitution in males, and females, and *E. tancrei*, $2n = 32-54$, XX in both males and females, as a normal condition do not have SRY, whereas SRY is present in other rodents as well as in other eutherian and marsupial species. *See* adrenal hypoplasia; animal hormones; campomelic dysplasia; eutherian; high mobility group proteins; LINE; Müllerian ducts; SOX; Swyer syndrome; TDF; trinucleotide repeat; Wolffian ducts; ZFY. (Murphy, E. C., et al. 2001. *J. Mol. Biol.* 312:481; Yuan, X., et al. 2001. *J. Biol. Chem.* 276:46647.)

SSA   Single-strand annealing is a repair mechanism apparently employed by prokaryotes and eukaryotes when one or more units of tandem repeats are eliminated by bacterial RecBCD-mediated degradation or by other nucleases. The process apparently does not require DNA polymerase or helicase action. *See* DNA repair. (Van Dyck, E., et al. 2001. *EMBO Rep.* 2:905; Paques, F. and Haber, J. E. 1999. *Microbiol. Mol. Biol. Rev.* 63:349.)

Ssb   *See* Hsp70.

SSB   Single-strand (DNA)-binding protein; in yeast it is encoded by gene *RPA1*. *See* DNA repair (SOS repair); recombination, molecular mechanism; replication protein A. (Reddy, M. S., et al. 2001. *J. Biol. Chem.* 276:45959.)

Ss blood group   *See* MN blood group.

SSC $1 \times$ SSC is a solution of 0.15 M NaCl + 0.015 M sodium citrate that is frequently used as a solvent for nucleic acids. *See* DNA extraction.

Ssc   *See* Hsp70.

SSCA   Single-strand conformation analysis. *See* single-strand conformation polymorphism.

SSCP   *See* single-strand conformation polymorphism.

SSE   HSP. *See* HSP.

Ssh   *See* Hsp70.

Ssi   *See* Hsp70.

SSI-1   STAT-induced STAT inhibitor. *See* STAT.

SSLP   *See* simple-sequence-length polymorphism.

SSM   Slipped-strand mispairing. *See* unequal crossing over.

SSN6   Yeast factor abolishing glucose repression of SUC2 invertase and regulator of nucleosome positioning in chromatin. Other *SSN* genes are components of the RNA polymerase II complex and are negative regulators of transcription. *See* catabolite repression; SNF; *SUC2*. (Li, B. & Reese, J. J. 2001. *J. Biol. Chem.* 276:33788.)

SSR   Small-segment repeat or simple-sequence repeat. These clusters of single to multiple nucleotides within the genome occurring per ~6 to ~30 to ~80 kb in plants and per ~6 kb in mammals have been used as chromosomal markers in various types of genetic studies. They may show a high degree of mutability. *See* microsatellites. (Qi, X., et al. 2001. *Biotechniques* 31:358; Bacon, A. L., et al. 2001. *Nucleic Acids Res.* 29:4405.)

SsrA (tmRNA, 10Sa RNA)   *See* protein repair.

SSRP1   Structure-specific recognition protein is a high-mobility group protein that probably targets FACT to the nucleosomes and facilitates gene transcription by RNA polymerase II. *See* FACT, high-mobility proteins, transcription factors, nucleosome remodeling. (Bruhn, S. L., et al. 1993. *Nucleic Acids Res.* 21:1643.)

SSV   Simple sequence variation in DNA.

ST-1   Single-stranded DNA phage related to ϕX174 and G4. *See* G4; map; ϕX174.

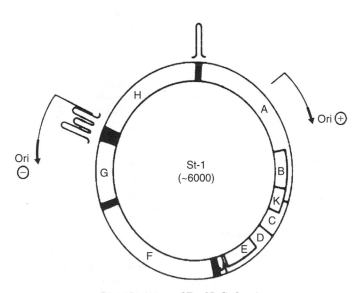

St-1 (Courtesy of Dr. N. Godson).

stab culture   The microbial inoculum is introduced into agar medium by a stabbing motion of the inoculation needle or loop for the purpose of propagation.

stabilizing selection   *See* selection types.

stable RNA   Ribosomal and tRNA that persists long in the cell in comparison to mRNA, which may be degraded in minutes. *See* mRNA; rRNA; tRNA.

STACK Sequence Tag Alignment and Consensus Knowledgebase. *See* gene indexing. (<http://ziggy.sanbi.ac.za/stack/stacksearch.htm.)

stacking gel Porous gel on top of SDS polyacrylamide electrophoresis running gel. It concentrates large volumes into a thin, sharp band at the beginning of the run, thus permitting sharper separation of proteins. *See* electrophoresis; SDS.

staggered cuts After the cut of a double-stranded DNA, the length of the two polynucleotides is unequal, such as

staggered extension process (StEP) Is a method for in vitro mutagenesis and recombination of polynucleotides. By its use mutant proteins can be generated in vitro. The template sequences are primed, then repeated cycles of denaturation and very short annealing and extension follows. In each cycle the extended fragments anneal to different templates, depending on complementarity, and the extension continues until full length is formed. The template switching generates recombined sequences from different parental sequences. *See* molecular evolution, directed mutation, RNA-peptide fusions. (Zhao, H., et al. 1998. *Nature Biotechnol.* 16:258; Xia, G., et al. 2002. *Proc. Natl. Acad. Sci. USA* 99:6597.)

STAINS For light microscopic examination of chromosomal specimens, aceto carmine, aceto orcein, or Feulgen stains are commonly used. Preparation: 0.5–1 g dry *carmine* powder is boiled for about 1/2 hour under reflux in 100 mL 45% acetic or propionic acid. Orcein 1.1 g is dissolved in 45 mL glacial acetic acid or propionic acid and filled up to 100 mL by H$_2$O. Filter and store stoppered at about 5°C. Feulgen: 1 g leuco—basic fuchsin is dissolved by pouring over 200 mL boiling H$_2$O, shake, cool to 50°C, filter, add 30 mL 1/N HCl, then 3 g K$_2$S$_2$O$_5$, allow to bleach in dark for 24 hr stoppered. Decolorize by 0.5 g carbon, shake 1 min, then filter and store stoppered in refrigerator. For carmine or orcein staining, fix specimens in Carnoy and stain. For Feulgen, fix in Farmer's solution for a day, rinse with water, hydrolyze at 60°C for 4–10 min (the duration of hydrolyzation is critical and may need adjustment for each species). Rinse with stain for 1–3 hr. Tease out tissue in 45% acetic acid, remove debris, flatten by coverslip, and examine. May need overstaining with carmine if Feulgen staining is poor. For histological staining, a variety of other stains may be used such as hematoxylin, methylene blue, ruthenium red, malachite green, sudan black, coomassie blue, fluorochromes, etc. *See* aequorin; C-banding; chromosome painting; FISH; fixatives; fluorochromes; G-banding; harlequin stain; light microscopy; Q-banding; sectioning.

stamen Male reproductive organs of plants composed of the anther, which contains the pollen and the filament. *See* anther; pollen.

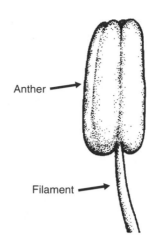

stamen hair assay *See Tradescantia.*

stamina Capacity to endure; vigor.

staminate flower Male flower of monoecious or dioecious plants.

stammering Most serious form of the speech defect *stuttering*. The sufferers usually cannot talk fluently and after involuntary stops repeat syllables or entire words. Stammering is very frequent among Japanese and very unusual among native Americans. Autosomal-dominant inheritance seems to be involved. *See* stuttering speech and grammar.

standard Accepted point of reference, e.g., standard wild type. *See* control.

standard deviation (parametric symbol σ) Measure of the variability of members of the populations ($s = \sqrt{\text{variance}}$). *See* normal deviate; standard error; variance.

standard error Measures the variation of the means of various samples of a population $s/\sqrt{m}$ (some authors use standard error and standard deviation as synonyms); standard error of proportions or fractions or frequencies is shown by the formula below where p = proportion or frequency and n is the population size (or number of measurements).

$$\sqrt{\frac{p(1-p)}{n}}$$

standardized fitness *See* selection coefficient.

standard type Generally the wild type used as a genetic reference. *See* wild type.

Stanford-Binet test Modified Binet intelligence test. *See* Binet test; human intelligence.

stanniocalcin Protein hormone inhibiting Ca^{2+} uptake and stimulating phosphate adsorption in fishes and mammals. (Varghese, R., et al. 1998. *Endocrinology* 139:4714.)

Staphylococcus aureus Bacterium responsible for toxic shock, scarlet fever, and hospital-acquired infections. *See* toxic shock syndrome. (Kuroda, M., et al. 2001. *Lancet* 357:1225.)

STAR **(1)** Signal transduction and activation of RNA is generally a 200-amino-acid protein domain associated with the cell and RNA splicing. The tripartite domain has a single RNA-binding site, a KH module of different length (2–14) flanked by QUA1 (80) and QUA2 (30) amino acid sequences. These complexes have been detected in a wide range of eukaryotic organisms and they appear to be regulators of translation. STAR family proteins may repress *tra-2* and cause masculinization in females. *See* Sam 68; sex determination. (Stoss, O., et al. 2001. *J. Biol. Chem.* 276:8665.) **(2)** Signature-tagged allele replacement may facilitate genetic analysis in cases in which mutational studies are impractical due to the essentiality of the gene for viability. *See* signature of a molecule; targeting genes. (Yu, Y., et al. 2001. *Microbiology* 147:431.) **(3)** Subtelomeric antisilencing region is an insulator-type sequence that protects a gene from the action of a silencer if it lays in between the silencer and the gene. *See* insulator; silencers.

StAR Steroidogenic acute regulatory protein (8p11.2) enhances mitochondrial conversion of cholesterol into pregnenolone, an intermediate in steroid biosynthesis. Mutation in the coding gene results in a deficiency of adrenal and gonadal steroidogenesis, and leads to congenital lipoid adrenal hyperplasia, an autosomal-recessive disorder. *See* adrenal hyperplasia; cholesterol; hormones; pregnenolone; steroid. (Petrescu, A. D., et al. 2001. *J. Biol. Chem.* 276:36970.)

starch *See* amylopectin.

starfish *Asterias forbesi*, $2n = 36$.

Starfish.

Stargardt disease Actually, a complex recessive degenerative disease of the retina evoked by environmental factors such as smoking and high cholesterol. In both the early-onset form (macular dystrophy with flecks) and age-related macular degeneration (AMD), mutations in the ATP-binding cassette transporter of the retina (ABCR) gene are involved. ABCR contains 51 exons in chromosome 1p13-p21. Physically the product is located in the outer segment of the retinal rods (also called rim protein). The disease is responsible for blindness in many cases. Sorsby fundus dystrophy, an autosomal-dominant disease, appears somewhat similar, but it is caused by a malfunction of the tissue inhibitor metalloproteinase-3 gene (TIMP3). *See* ABC transporters; eye diseases; macular degeneration; macular dystrophy; retinal dystrophy; retinitis pigmentosa.

START *See* cell cycle.

start codon AUG in RNA that specifies either formyl methionine (in prokaryotes) or methionine (in eukaryotic cells). Note that the Met codon in mitochondria varies in different organisms. In some organisms, other triplets may also initiate translation. *See* genetic code.

start point Position in the transcribed DNA where the first RNA nucleotide is incorporated. *See* transcription.

stasis Equilibrium state without change.

STAT Signal transducers and activators of transcription are cytoplasmic proteins that become activated by SRC (SH2, SH3)-mediated phosphorylation of tyrosine at around residue 700 and serine residues at the C-terminus through the action of Jak kinases, which are receptors for cytokine signals and enzymes at the cytosolic termini. PDGF, EGF, and CSF catalyze phosphorylation. The latter process requires the action of the 42 kDa MAPK or ERK2. Stat 1 mediates reactions to microbial and viral infections. Stat 4 protein is essential for interleukin-12-mediated functions such as the induction of interferon-γ, mitogenesis, T-lymphocyte killing and helper T-lymphocyte differentiation. Disruption of STAT 2 and 3 causes embryonic lethality. STAT 5A and B are required for breast development in mice and for the stimulation of T-cell proliferation. Stats 1, 3, and 4 recognize and activate different genes by binding to the TTCC(C/G) GGAA (TTN5AA) sequences. Rac1 GTPase may regulate STAT 3 activation through phosphorylation. Stat6 prefers TTN6AA. Cooperative binding of Stats makes the recognition of variations possible at different binding sites. Sequence-selective recognition resides in their amino-terminal domains. In humans, there are at least seven STAT genes, encoding proteins of 750 to 850 amino acids. The 130-amino-terminal residues bind to multiple sites in the DNA. Residues 600–700 are homologous to SH2 domains and mediate dimerization. Disruption of Stat activity leads to the loss of interferon-controlled immunity to pathogens. STAT genes can be found in mouse (m) and human (h) chromosomes: STAT1, STAT4: m 1, h 2q12-q33; STAT 3, STAT 5A & B:m 11, h 12q13-q14-1; STAT 2 & 6: m 10, h 17q11.1-q22. Stat genes also occur in *Drosophila* and *Dictyostelium*. The PIAS family of proteins includes negative regulators of STATs. *See* CSF; EGF; interferon; Jak-STAT pathway; leptin; lymphocytes; PDGF; RAC; SH2; SH3; signal transduction; SRC; SSI-1. (Davey, H. W., et al. 1999. *Am. J. Hum. Genet.* 65:959; Naka, T., et al. 1999. *Trends Biol. Sci.* 24:394; Levy, D. E. & Darnell, J. E. 2002. *Nature Rev. Mol. Cell Biol.* 3:651.)

stathmin (op18) Regulates microtubule polymerization by affecting tubulins. The activity of Stathmin is controlled through phosphorylation by chromatin. *See* spindle. (Gavet, O., et al. 1998. *J. Cell Sci.* 111:3333; Charbaut, E., et al. 2001. *J. Biol. Chem.* 276:16146.)

stathmokinesis Mitotic arrest. *See* spindle poison.

statins Inhibitors of 3-hydroxy-3-methylglutaryl-coenzyme A reductase, an important enzyme of cholesterol biosynthesis. Cholesterol-lowering drugs reduce the incidence of Alzheimer disease. *See* Alzheimer disease; cholesterol; lovastatin. (Fassbender, K., et al. 2001. *Proc. Natl. Acad. Sci. USA* 98:5856.)

stationary phase The population size is maintained without an increase or decrease. At this stage, the genetic material may not replicate and the mutations occurring in bacteria are expected to be due to recombination. Mutation (chromosome breakage) in the stationary-phase cells of higher eukaryotes may be one of the causes of cancer. *See* growth curve. (Bull, H. J., et al. 2001. *Proc. Natl. Acad. Sci. USA* 98:8334.)

stationary renewal process Based on a paper by R. A. Fisher (1947. *Phil. Trans. Roy. Soc. B* 233:55), assumption that crossing over is formed as a regular sequence starting from the centromere and the length between two adjacent crossovers always following the same distribution. The idea that crossing overs begin at the centromere and proceeds toward the telomere turned out to be incorrect, yet the stationary renewal process gained entry into several later models of recombination, interference and mapping functions. *See* mapping function. (Zhao, H. & Speed, T. P. 1996. *Genetics* 142:1369.)

statistic Estimate of a property of an observed set of data (e.g., mean) that bears the same relation to the data as the parameter does to the population. *See* statistics.

statistics Mathematical discipline that assists in collecting and analyzing data; guides in making conclusions and predictions on the basis of the analysis and reveals their trustworthiness by determining probability or likelihood. Classical genetic analyses usually require statistical methods. Statistics is used for collections of facts on demography, industrial productivity, trade, etc., such as in statistical yearbooks. *See* Bayes' theorem; likelihood; maximum likelihood; nonparametric tests; probability. (Robbins, L. G. 2000. *Genetics* 154:13; assistance for the use of the most frequently needed statistical procedures: <faculty.vassar.edu/~lowry/VassarStats.html>, <www.ruf.rice.edu/~lane/rvls.html>.)

statolith Granules that are believed to be sensing gravity in cells. *See* gravitropism.

stature, humans Influenced by environmental causes such as nutrition, disease, and injuries by the use of various medications and by simple or complex genetic factors. Heritability of human height commonly exceeds 80%. Prenatal anomalies of bone length may be determined by prenatal diagnosis. QTL analysis (Hirschorn, J. N., et al. 2001. *Am. J. Hum. Genet.* 69:106) of Scandinavian and Canadian populations involving 2,327 individuals revealed linkage of height to 6q24-q25 (lod score 3.85), 7q31.3-q36 (lod score 3.40), 12p11.2-q14 (lod score 3.35), 13q32-q33 (lod score 3.56), and 3p26 (lod score 3.17; Wiltshire S., et al. 2002. *Am. J. Hum. Genet.* 70:543). The most common types of

genetically determined human dwarfism and other defects involving reduced stature are achondroplasia, hypochondroplasia, achondrogenesis, osteochondromatosis, dyschondrosteosis, Russel-Silver syndrome, Smith-Lemli-Opitz syndrome, Opitz-Kaveggia syndrome, SHORT, Aarskog syndrome, Noonan syndrome, Hirschsprung disease, Turner syndrome, trisomy, Mulbrey nanism, hairy elbows, and Pygmy. *See* exostosis; growth hormone; growth retardation; limb defects; lod score QTL; regression.

STC Sequence-tagged connector. *See* genome project.

STCH *See* Hsp70.

STE Proteins are pheromone receptors and scaffolding proteins. They coordinate and organize the signal transduction paths in budding yeast. Ste3 and Ste2 are receptors of the α- and **a**-mating-type factors, respectively. Ste7 is MAPKK, Ste4 is the β-subunit, Ste18 is the γ-subunit of the G-trimeric proteins, Ste5 is a member of the MAPK cascade, Ste11 is MAPKKK, and Ste 20 is a PAK/MEKKK protein required for the MAPK activation of $G\beta\gamma$. A Ste20-like protein kinase, Msst, regulates chromatin structure and apoptosis. Ste12 is a transcription factor. In fission yeast, the homologues *Byr* are extracellular signal-regulated kinase homologues of MEK and ERK. (The *Ste* [*Stellate*, 1–45.7] locus of *Drosophila* encodes protein crystals in the primary oocytes). *See* FUS3; G proteins; KSS1; MAPK; mating type determination in yeast; MEK; MEKK; MP1; PAK; pheromones; signal transduction; (Graves, J. D., et al. 2001. *J. Biol. Chem.* 276:14909; Ura, S., et al. 2001. *Proc. Natl. Acad. Sci. USA* 98:10148; Ge, B., et al. 2002. *Science* 295:1291.)

steady state Reaction enzyme-substrate concentration and other intermediates appear constant over time, but input and output are in a flow.

steel factor (stem cell factor) 40–50 kDa dimeric protein produced by the bone marrow and other cells. It migrates to the hematopoietic stem cells. Its receptor is a transmembrane protein tyrosine kinase. *See* cell migration; hematopoiesis; Kit oncogene; microphthalmos; stem cell factor; tyrosine kinase.

steer Emasculated bovine male animal. *See* emasculation.

stele Core cylinder of vascular tissues in plant stems and roots. *See* root.

stem cell factor (SCF/M-CSF, 5q33.2-q33.3) Required for the normal development of B lymphocytes. The Kit oncogene product is its receptor. *See* B cells; Kit oncogene; lymphocytes; M-SCF. (Broudy, V. C. 1997. *Blood* 90:1345; Smith, M. A., et al. 2001. *Acta Haematol.* 105:143.)

stem cells Of animals, are not terminally differentiated, can divide without limit; when they, divide, the daughter cells can remain stem cells or can terminally differentiate in one or more ways, or they may have a restricted potential for differentiation (*transit-amplifying cell*). They occur in various tissues. Adult neural stem cells may differentiate into cells of diverse germ layers with broad developmental potentials. The balance between self-renewal of stem cells and differentiation seems to be determined by the signal receptors or their phosphorylation in the neighboring cells. In the ovaries of mammals, the initial stem cells are arrested after a finite number of divisions and then enter into meiotic prophase. Functionally they resemble the meristem in plants. Embryonic *germ cells* (EG) have been used for transfer into mouse blastocysts and are capable of differentiating into most types of fetal tissues, even including the germline. The feasibility of culturing pluripotent (endoderm, mesoderm, ectoderm) human embryonic stem cells (ES) opens new potential for the generation of tissues (Smith, A. G. 2001. *Annu. Rev. Cell Dev. Biol.* 17:435). They may be used for transplantation to treat neurodegenerative diseases, spinal cord injuries, liver diseases, diabetes, immunological diseases, cancer, and finding a means for the repopulating of hematopoietic cells, etc. (Krause, D. S., et al. 2001. *Cell* 105:369). Stems cells derived from another individual may lead to an adverse immune reaction. Allogeneic, hematopoietic bone marrow cells transplanted into host cells treated by ionizing radiation may, however, become permanently tolerant of the foreign cells. The intolerance to a third-party donor, however, remains. Adult bone marrow co-purifying with mesenchymal stem cells when injected into early blastocysts can develop into most types of cells (Jiang, Y., et al. 2002. *Nature* 418:41). It is conceivable that cotransplantation of hematopoietic stem cells and stem cells for other tissues using the same donor may become feasible. In order to maintain stem cells in culture in the pluripotent state, transcription factor Oct4 and the leukemia inhibitory factor (Lif) must both be expressed. When Lif is withdrawn, various types of differentiation may begin. TCF and LEF may also be important. Actually, this may be regarded as a teratocarcinoma type of growth. In order to force the cells into a specific type of differentiation, special culture conditions must be established, e.g., Retinoic acid → Insulin (triiodothyronine) passage may lead to the differentiation of adipocytes. Employing c-Kit (a transmembrane tyrosine kinase) + Erythropoietin may lead to the development of erythrocytes. Macrophage colony-stimulating factor + IL-3, IL-1 lead to the differentiation of macrophages. Fibroblast growth factor and epidermal growth factor combinations coax the ES culture to form astrocytes and oligodendrocytes. There is potential for genetically engineering embryonic stem cells (ES) for special medical purposes. Undifferentiated mouse ES cells may become tumorigenic, developing into teratomas or teratocarcinomas when introduced into an animal. A rich source of embryonic stem cells is the umbilical cord at birth. Epithelial cells, hair follicles, intestinal epithelium, multipotent brain cells, and hematopoietic cells may be employed as potential ES. The full clinical exploitation of stem cell technology requires technical improvements (Humpherys, D., et al. 2001. *Science* 293:95). On December 19, 2000, the British Parliament approved greater freedom in embryonic stem cell research (Ramsay, S. 2000. *Lancet* 356:2162). In the U.S., embryonic stem cell research can be funded by government agencies only on the existing ca. 60 cell lines (Aug. 9, 2001.). No embryos are supposed to be generated for the purpose of extracting stem cells; no fertilized and unused eggs generated by the process of in vitro artificial fertilization should be used for stem cell research. The objection to stem cell research originates from the belief that life begins at fertilization and extracting embryonic stem cells from 4–5-day-old blastomeres amounts to violation of the sanctity of life. The ethical problems may be dispelled if somatic stem cells occurring in placental, umbilical, or fat tissues are used. Another—albeit tenuous—possibility is to generate blastomeres by inserting diploid somatic nuclei into enucleated human eggs and cloning them. *See* adipocyte; embryo research; epidermal growth factor; erythropoietin; fibroblast growth factor; genetic engineering; hematopoiesis; hepatocyte; IL-1; IL-3; insulin; KIT; LEF; leukemia inhibitory factor; LIF; macrophage colony stimulating factor; meristem; mesenchyma; metaplasia; niche; nuclear transplantation; Oct; pluripotent; public opinion; regeneration in animals; retinoic acid; TCF; teratoma; therapeutic cloning; tissue engineering; transdetermination; transplantation of nuclei. (Fuchs, E. & Segre, J. A. 2000. *Cell* 100:143; Weissman, I. L. 2000. *Science* 287:1442; Mezey, É., et al. 2000. *Science* 290:1779; Edwards, B. E. B. A., et al. 2000. *Fertility & Sterility* 74:1; Lennard, A. L. & Jackson, G. H. 2001. *West. J. Med.* 175:42; Wakayama, T., et al. 2001. *Science* 292:740; Odorico, J. S., et al. 2001. *Stem Cells* 19:193; Blau, H. M., et al. 2001. *Cell* 105:829; Nichols, J. 2001. *Curr. Biol.* 11:R503; Toma, J. G., et al. 2001. *Nature Cell Biol.* 3:778; Edwards, R. G. 2001. *Nature* 413:349; *Nature* [2001], 414:87–131, Hochedlinger, K. & Jaenisch, R. 2002. *Nature* 415:1035; J. Cellular Biochem 85:S38 [2002.], Board on Life Sciences, National Research Council and Board on Neuroscience and Behavioral Health, Institute of Medicine 2002. Stem Cells and the Future of Regenerative Medicine. National Acad. Press, Washington, DC, stem cell registry: <escr.nih.gov>.)

stem-loop structure   Any DNA or RNA that may have nonpaired single-strand loops associated with a double-strand stem similar to a lollipop. In the stem, Watson-Crick pairing may not be perfect along its length. *See* palindrome; repeat, inverted.

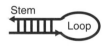

stem rust   Caused by infection of the basidiomycete fungus *Puccinia graminis* on cereal plants. The haploid spores produced on the wheat plant germinate on the leaves of barberry shrubs and form pycnia (pycnidium). The pycniospores of different mating types undergo plasmogamy and form dikaryotic aecia (aecidia) on the lower surface of the barberry leaves. The aeciospores infect the wheat leaves and form the dark brown rust postules, called uredia. The dikaryotic uredospores reproduce asexually and spread the disease. At the end of the growing season, karyogamy takes place and the diploid teliospores are formed. The teliospores overwinter and eventually undergo meiosis, liberating the haploid basidiospores that germinate on barberry and restarting the cycle. Stem rust may cause very substantial crop loss in wheat and other gramineae. The newer varieties that are genetically more or less resistant to the fungus. *P.* spp have chromosome numbers 3–6. *See* host-pathogen relationship; rust.

stenosis   Narrowing of a body canal or valve such as in the aorta, heart valve, pulmonary artery, vertebral canal, etc. *See* restenosis.

stenospermocarpy Genetically determined abortion of the embryo soon after fertilization, resulting in seedless normal-size berries, a desirable trait of table grapes. *See* seedless fruits.

STEP Single-target expression profile.

step allelomorphism Historically important concept that paved the way for allelic complementation and the study of gene structure. In the late 1920s, Russian geneticists discovered that partial complementation among allelic genes might occur in a pattern, which was inconsistent with the prevailing idea that the gene locus was the ultimate unit of function, mutation, and recombination and that alleles were stereochemical modifications of an indivisible molecule. *See* allelic complementation; Offermann hypothesis. (Carlson, E. A. 1966. *The Gene: A Critical History*. W. B. Saunders, Philadelphia, PA.)

step gradient centrifugation In the centrifuge tube, usually three different concentrations of CsCl or sugar are layered without allowing them to mix. The highest concentration is at the bottom of the tube. The different components of the mix, layered at the top, accumulate at the boundaries, which have a higher density than the separated component.

stepwise mutation model (SMM) Microsatellites may change by repeated gain or loss of a small number of nucleotide repeats. Electrophoretic variations of enzymes may also be fitted to such a model. *See* infinite allele mutation model; microsatellite; trinucleotide repeats. (Moran, P. A. P. 1975. *Theoret. Population Biol.* 8:318; Ohta, T. & Kimura, M. 1973. *Genet. Res.* 22:201.)

stereocilia Protoplasmic thin filaments like the ones in the inner ear.

stereoisomers Molecules of identical composition but with different spatial arrangements.

stereomicroscopy (dissecting microscopy) Used for visual analysis under relatively low magnification of natural specimens without sectioning. It has a special advantage for dissecting structural elements with binocular viewing and top or side illumination without fixation and/or staining. *See* confocal microscopy; microscopy; scanning electron microscopy.

stereotactic Action precisely positioned in space such as irradiation of a small spot in the body, surgical introduction of cells, a genetic vector at a defined location of the brain, etc.

steric-exclusion model Of DNA replication, states that Watson-Crick hydrogen pairing is not an absolute necessity for the faithful replication of DNA, but it is essential that the building block (not necessarily a purine or pyrimidine) will fit into the frame of the DNA double helix. A pyrene nucleoside triphosphate, with the size close to a nucleotide pair, has sufficient steric complementarity to fit into an abasic site and permits DNA replication. *See* abasic sites; DNA replication;

Pyrene nucleoside triphosphate.

hydrogen pairing; Watson and Crick model. (Matray, T. J. & Kool, E. T. 1999. *Nature* 399:704.)

sterigma Small stalk at the tip of a fungal basidium where spores come off. *See* basidium.

sterile insect technology (SIT) *See* genetic sterilization.

sterile RNA *See* germline transcript.

sterility Either the male or the female or both types of gametes (haplontic) or the zygotes (diplontic) have reduced or no viability caused by lethal or semilethal genes, chromosomal defects, differences in chromosome numbers, or incompatible cytoplasmic organelles. *See* azoospermia; cytoplasmic male sterility; deletion; incompatibility; infertility; inversion; self-incompatibility; semisterility; somatoplastic sterility; translocation.

sterilization *See* aseptic; autoclaving; axenic; birth control; *Cochliomya hominivorax*; ethanol; ethyleneoxide; filter sterilization; genetic sterilization; hypochlorate; pasteurization; radiation effects; sterilization humans.

sterilization, genetic *See* genetic sterilization.

sterilization, human Had been practiced by various societies for different reasons. The eunuchs of the Chinese imperial courts and of the Osmanic harems served as guardians of the privileges of tyrannical social structures. The castrati of the Italian opera stages were exploited for singing female roles in an era when women were banned from performing arts. In the 1880s, scientific justifications were sought by the publications of Sir Francis Galton for negative eugenics in order "to produce a highly gifted race of men by judicious marriages during consecutive generations." The 1890s initiated sporadic sterilization of institutionalized, mentally retarded persons. Starting in 1907, in about 14 U.S. states, laws were enacted for systematic sterilization of the mentally retarded, blind, deaf, crippled, those afflicted by tuberculosis, leprosy, syphilis, and chronic alcoholism. This "practical, merciful and inevitable solution" eventually degenerated into legal suggestions to eliminate criminal behavior, disease, insanity, weaklings, and other defectives and "ultimately to worthless race types." Strong moral objections gained noticeable ground in 1956 after nearly 60,000 human individuals were legally sterilized. Interestingly, the Oklahoma law exempted from mandatory sterilization offenses against prohibition, tax evasion, embezzlement, and political crimes. Several state laws advocated mandatory sterilization as a guard against illegitimacy, particularly by unwed recipients of the Aid to Families with Dependent Children (AFDC). Until 1965, judicial approval could be obtained for a "good cause" for forced sterilization of mentally retarded individuals whose family had hardship in supporting the offspring of illegitemate or promiscuous children. Although not all states rescinded the old laws, sterilization of humans is now practiced only voluntarily by ligation of the vas deferens (vasectomy), tubal constriction, ovariectomy, etc. Mandatory sterilization was practiced for eugenic and social reasons in several enlightened countries (e.g., Sweden) until the 1970s. Compulsory sterilization is objectionable on moral grounds because reproduction is a basic human right, although the society cannot support irresponsible reproductive behavior in cases of certain genetic defects. Reproductive rights must be balanced with the right of born and unborn children with a potentially severe genetic load (*see* wrongful birth).

Sterilization is particularly reprehensible when advocated as a selective measure against certain human races. The Third Reich annihilated millions and sterilized thousands for eugenic and other evil reasons. From a genetic perspective, it is controversial, since 83% of mentally retarded children are born to nonretarded parents. Also, selection against the majority of human defects is quite inefficient, since the vast majority of defective genes are in heterozygotes and many of the conditions are under polygenic control or are nonhereditary. Furthermore, there are no objective scientific or practical measures for the evaluation of most human traits. *See* eugenics; ovariectomy; polygenic; salpingectomy; selection; selective abortion; vasectomy. (Reilly, P. 1977. *Genetics, Law, Social Policy*. Harvard Univ. Press, Cambridge, MA.)

sternites Ventral epidermal structures of the abdomen. *See* *Drosophila*.

steroid 5-beta reductase (SRD5B1, 7q32-q33) Catalyzes the reduction of bile acid intermediates and steroid hormones.

steroid dehydrogenase-like protein (NSDHL, XDq28) Mutations affect cholesterol biosynthesis and may cause male lethality. Hydroxysteroid dehydrogenase (HSD3B1, 1p13.1) deficiency may involve adrenal hyperplasia, hypospadias, and gynecomastia. *See* cholesterol; hypospadias.

steroid doping Used by athletes to boost performance. Amphetamines increase alertness and may reduce onset of fatigue. Side effects are insomnia, exhaustion, violence, and potential heart disease. Health hazards are increased if anabolic steroids, insulin, insulin-like growth hormone, etc., are used simultaneously. Even nonsteroid anti-inflammatory drugs may be risky because they mask pain and may aggravate injuries. *See* anabolic steroids.

steroid hormones Derived by the pathway shown below. Number [3] is the principal hormone of the endocrine gland, the corpus luteum, in the ovarian follicle after the release of the ovum.

Steroid hormones regulate the expression of secondary sexual characters of females. Number [4] is the main male sex hormone produced in 6–10 mg quantities daily in men and ca. 0.4 mg in women. It is responsible for the production of facial hair and baldness and regulation of growth. Number [5] is formed by oxidative removal of C-129 from its precursor; it is primarily a female hormone occurring in the ovaries and placenta and is responsible for regulating—among other functions—bone growth, increased fat content, and smoother skin of females compared to men. This hormone is also present in the testes. In cooperation with progesterone, it regulates the menstrual cycle. Numbers [6] and [7] are synthesized in the kidney cortex and regulate—among others—mineral (Na^+ Cl^-, HCO_3^-) reabsorption. They are frequently called mineral corticoid hormones. Number [8] is a glucocorticoid that affects protein and carbohydrate metabolism and regulates the immune system, allergic reactions, inflammations, etc. Number [9] is an antiinflammatory glucocorticoid with a role in activating glucocorticoid receptors. The number of steroid hormones is about 50, and they are present in practically every cell of the body—besides those mentioned—and, along with thyroid hormones, have important roles in gene activation. Before 1966, the general assumption was that plants did not use steroid hormones. It has been demonstrated that brassinolids (related to cholesterol, ecdysone) mediate several developmental processes in plants, such as elongation, light responses, etc. The steroid receptor superfamily includes receptors for estrogen, progesterone, glucocorticoid, mineral corticoid, androgen, thyroid hormone, vitamin D, retinoic acid, 9-cis retinoic acid, and ecdyson. Steroid hormone receptors stimulate the formation, then stabilize the preinitiation complex of transcription. Most commonly, the condition of their binding to the hormone-response element is the binding to their appropriate ligands. Some, such as thyroid hormone receptor, can bind to DNA in the absence of a ligand. In the absence of the ligand, they function as silencers through interaction with the TFIIB transcription factor. *See* anabolic steroids; animal hormones; aromatase; brassinosteroids; coactivator; estradiol; hormone-response elements; PIC; plant hormones; prenylation; regulation of gene activity; silencer; SRC-1; steroids; transcription factors; transcriptional activator. (Lösel, R. & Wehling, M., 2003. *Nature Rev. Mol. Cell. Biol.* 4:46.)

steroidogenic factor-1 *See* SF-1.

steroid receptor *See* hormone receptors.

steroid sulfatase deficiency *See* ichthyosis.

steroids Contain a four-ring nucleus consisting of three six-membered rings and one five-membered ring. *See* brassinosteroids; steroid hormones.

$$| \rightarrow [4] \text{ Testosterone} \rightarrow [5] \text{ Estradiole}$$
$$[1] \text{ Cholesterol} \rightarrow [2] \text{ Prognenolone} \rightarrow [3] \text{ Progesterone} \rightarrow [6] \text{ Corticosterone} \rightarrow [7] \text{ Aldosterone}$$
$$\downarrow \qquad\qquad | \rightarrow [8] \text{ Cortisol}$$
$$[9] \text{ Dexamethasone}$$

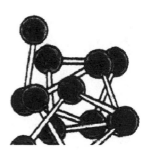

General structural formula of steroids.

sterols Lipids with a steroid nucleus. The concentration of free sterols determines the fluidity of eukaryotic cell membranes. Esterification of sterols prevents their participation in membrane assembly. The process is mediated by the ACAT complex (acy-CoA:cholesterol acyltransferase). An increase of ACAT activity may lead to hyperlipidemia and atherosclerosis. Sterol esterification may modify LDL receptors and potentiate atherogenic processes. It may limit intestinal sterol absorption. *See* atherosclerosis; cholesterol; hyperlipidemia; LDL; membranes. (Kelley, R. I. & Herman, G. E. 2001. *Annu. Rev. Genomics Hum. Genet.* 2:299.)

stg (*string*, map position 3-99) *Drosophila* gene locus controlling the first 10 embryonic divisions (similar to gene *Cdc28* in *Shizosaccharomyces pombe*); it is a cyclin gene. *See* cell cycle.

stick-and-ball model Representation of chemical structure as shown above.

stickiness, chromosome Observed as sorts of adhesion between any chromosome within a cell. *See* side-arm bridge.

Stickler Syndrome (arthroophthalmopathy, AOM) Early and strong progressive myopia (nearsightedness) and hearing deficit. Retinal detachment may result in blindness caused probably by a dominant mutation in the collagen (COL11A1) gene (human chromosome 1p21). Overlapping mutations are responsible for Marshall syndrome at 1p21. Stickler syndrome 3 is located at 6p21.3. The collagen type II (COL2A1, 12q13.11-q13.2) defects of Stickler syndrome involve achondroplasia, skeletal dysplasia, eye and hearing defects. *See* collagen; connective tissue disorders; eye disease; skin diseases. (Annunen, S., et al. 1999. *Am. J. Hum. Genet.* 65:974; Richards, A. J., et al. 2000. *Am. J. Hum. Genet.* 67:1083.)

sticky ends Double-stranded DNA with a single-stranded overhang to which complementary sequences are available so that they can stick by base pairing:

stigma Tip of the style that is normally receptive to the pollen of plants. In zoology, it means spot, such as a hemorrhagic small area on the body. *See* gametophyte, female; gametophyte, male; protandry; protogyny.

stigmasterol Plant lipid derivative formed by methylation of ergosterol. For guinea pigs, it is a vitamin necessary to avoid stiffness of the joints. *See* cholesterol; ergosterol.

stilbene *See* resveratrol.

stillbirth Birth of a dead offspring. It is caused by chromosomal defects in ca. 7% of stillborns or by other pathological conditions. *See* chromosomal breakage.

stipule Leaf-like bract at the base of a leaf.

Stipules.

Stk Macrophage-stimulating factor receptor. *See* macrophage.

STM *See* scanning tunneling microscope.

Stn1 Telomere length-determining protein factor of yeast working in concert with Cdc13. *See* Cdc13. (Grandin, N., et al. 1997. *Genes Dev.* 11:512.)

stochastic Corresponds to a random process; a process of joint distribution of random variables. In a population—in contrast to a deterministic model—random drift and other chance events may determine the gene frequencies. Generally, deterministic and stochastic processes run parallel and simultaneously. *See* deterministic model.

stochastic detriment of radiation Combined risk of cancer, genetic damage, and shortened life due to radiation exposure. The figures may vary according to tissues: for gonads, it may be 1.33, for bone marrow 1.04, for breast 0.24, for liver 0.16 (in 10^{-2} Sv^{-1}), etc. *See* radiation hazards.

stock (1) Genetically defined strain of organisms. (2) Root stock on which a scion is grafted.

stock, garden (*Matthiola incana*) *See Matthiola*.

stoke Unit of kinematic viscosity (the ratio of viscosity to density). *See* viscosity.

stolon Horizontal underground stem such as the tuber-bearing structures of potatoes.

stoma Small pore on the leaf surface surrounded by two guard cells that control opening and closing. The stoma permits gas exchange (CO_2 uptake) and release of water vapors (transpiration). The opening of the stoma requires an increase in the turgor of the guard cells. It was suggested that the process might be promoted by opening of K^+ and Cl^- channels and the subsequent influx of K^+ and Cl^-. A light-controlled proton pump activates the opening of the K^+ channel. The closure of the stoma is controlled by the hormone ABA, the influx of Ca^{2+}, and the efflux of K^+ and Cl^-. The calcium level is sensed by a cyclin-dependent protein kinase (CDPK). In the regulation, Ca^{2+}-dependent ATPases and GTPases have a major role. The processes involve changes in electric potentials (depolarization). In the control of the ABA response, syntaxin-like proteins play a role. The sphingosine-1-phosphate level signals to calcium mobilization. *See* ABA; aequorin; ATPase; calmodulin; cell cycle; cyclin; GTPase; ion channels; proton pump; sphingolipids; syntaxin. (Blatt, M. R., 2000. *Annu. Rev. Cell Dev. Biol.* 16:221; Schroeder, J. I., et al. 2001. *Nature* 410:327; Wang, X.-Q., et al. 2001. *Science* 292:2070; Schroeder, J. I., et al. 2001. *Annu. Rev. Plant Physiol. Plant Mol. Biol.* 52:627; Hetherington, A. M., 2001. *Cell* 107:711; Nadeau, J. A. & Sack, F. D. 2002. *Science* 296:1697.)

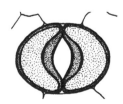

Open stoma.

stomatin Cation conductance protein in the cell membrane. *See* anesthetics.

stone *See* scaffolds in genome sequencing.

stop codon *See* genetic code; nonsense codon.

stoppers Mitochondrial mutations in *Neurospora* displaying stop-start growth. *See* poky.

stop signal *See* release factor (RF); stop codon; transcription termination in eukaryotes; transcription termination in prokaryotes.

STP *See* signal transfer particle.

STPβ Second-strand transfer protein. *See* recombination mechanisms eukaryotes; Sep 1.

STR Short tandem repeats, such as those found in micro- and minisatellites. Using the profiles of only 13 STRs can provide a rapid test for crime scenes. STRs are used for forensic analysis. They can also be used for population studies. *See* DNA fingerprinting; forensic genetics; microsatellite; minisatellite.

strabismus Anomalies of the eyes. The eyes may be either divergent or convergent, or one directed up and the other down because of the lack of muscle coordination. Some persons display strabismus only periodically. The pattern of inheritance is not entirely clear; most likely, dominant factors are involved. Recurrence among the offspring of convergent probands is higher than that among children of the divergent type. Incidence in the general population is $\sim$0.002. *See* Duane retraction syndrome; eye diseases.

strabismus.

strain Isolate of an organism with some identifiable difference from other similar groups. This term does not imply to any other stringent criteria.

strain distribution pattern (SDR) Distribution of two alleles of a diploid among the progeny where linkage is studied either by a backcross or by a recombinant inbred procedure. *See* backcross; linkage; recombinant inbred.

strand assimilation The *exo* gene of lambda phage codes for a 5′-exonuclease (M_r 24,000) that can convert a branched DNA structure to an unbranched nicked duplex by strand assimilation during recombination. Progressive incorporation of one DNA strand into another during recombination ($\rightarrow$). *See* lambda phage; recombination, diagram below.

strand displacement Type of viral replication involving the removal of the old strand before the new strand is completed. A similar mechanism is used by mtDNA, *See* D loop.

stratification (1) Layering. (2) In statistical analysis, studying the population by, e.g., age groups or other suitable

Double-stranded DNA

Heteroduplex at strand assimilation

attributes besides some other criteria of comparison, e.g., onset of a disease.

stratification artefact   The disease and control alleles are from different (ethnic) populations in case-control studies. Sib or parent or other family comparisons may correct the problems. Lower gene frequencies favor reliable results. *See* case-control method.

stratocladistics   Study of evolution on the basis of fossil records; minimizes the significance of homoplasy and lack of preservation of lineages that would preserve other lineages under examination. *See* cladistic.

Strauss family   Viennese composers and conductors of three generations. Johann Strauss the Elder (1804–1849) became celebrated for his light waltzes and other dance music. His son, Johann Strauss the Younger (1825–1899), is the author of the *Blue Danube* and many other waltzes. His brothers, Josef Strauss (1827–1870) and Eduard Strauss (1835–1916), were also famous conductors and composers. Son of Eduard, Johann (1866–1939), was a renowned conductor. *See* music talent.

strawberry (*Fragaria ananassa*)   About 46 *Fragaria* species with $x = 7$; the wild European *F. vesca* is diploid ($2n = 14$); *F. moschata* ($2n = 42$); some east Asian species are tetraploid; the American strawberries as well as the garden strawberries are $2n = 56$.

streak (primitive streak)   Sign on the early embryonal disc indicating the movement of cells and the beginning of the formation of the mesoderm and an embryonal axis. *See* organizer.

streaking   Spreading microbial cells on the surface on a nutrient agar medium to observe growth or lack of it.

Streaking of wild-type (left top) and mutant strains of yeast.

streptavidin   Conjugated with rhodamine; specifically binds to biotins (biotinylated nucleic acids, immunoglobulins) and permits their detection by fluorescence. The binding constant for biotin is $k_a = 10^{15}$ M^{-1}. *See* avidin.

streptavidin-peroxidase   Identifies biotinylated antibodies in ELISA, in immunochemistry in general, and in protein blots. *See* biotinylation; ELISA; genomic subtraction.

Streptococcus A (1.85 Mb)   Common pathogenic bacterium causing pharyngitis (sore throat). About 5–10% of the infections may involve necrotic lesions of various severities. In rare and extreme cases, it may cause death. Some strains secrete a substantial amount of pyrogenic exotoxin A, which stimulates the immune system as a superantigen. Excessive stimulation results in the overproduction of cytokines, which may damage the lining of the blood vessels, thus causing fluid leakage, reduced blood flow, and necrosis of the tissues because of the lack of oxygen. As a further consequence, fasciitis (inflammation of the fibrous tissues) and myositis (inflammation of the voluntary muscles) may follow. Destruction of the tissues may result in death within a very short period of time after infection by the extremely virulent strain of flesh-eating bacteria. *Streptococcus pneumoniae* (*Pneumococcus*) provided the first information on genetic transformation in 1928. Its sequenced genome (in 2001) of 2,160,837 bp contains 2,236 ORFs. Approximately 5% of its genome are insertion sequences. The completely sequenced genome of *S. pyogenes* M1 has 1,852,442 bp and encodes ~1,752 proteins. *See* necrosis; superantigen; transformation, genetic. (Tettelin, H., et al. 2001. *Science* 293:498; Ferretti, J. J., et al. 2001. *Proc. Natl. Acad. Sci. USA* 98:4658; Hoskins, J., et al. 2001. *J. Bacteriol.* 183:5709; Ajdic, D., et al. 2002. *Proc Natl. Acad. Sci. USA* 99:14434.)

streptokinase   Activator of plasminogen. *See* plasmin; plasminogen activator.

streptolygidin   Antibiotic that blocks the action of prokaryotic RNA polymerase.

streptolysin   Cholesterol-binding bacterial exotoxin that forms large holes through the mammalian plasma membrane. At low concentrations, it is suitable for introducing proteins through living cell membranes without irreversible damage to the cell. (Walev, I., et al. 2001. *Proc. Natl. Acad. Sci. USA* 98:3185.)

Streptomyces   Group of gram-negative bacteria of the actinomycete group characterized by mycelia-like septate colonies. Spore-bearing organs develop on these mycelial colonies. These bacteria somewhat simulate a multicellular type of development. Their genetic material, unlike the majority of prokaryotes, is a linear DNA. The sequenced genome of *S. coelicolor* A3(2) is 8,667,507 bp, containing an estimated 7,825 genes (Bentley, S. D., et al. 2002. *Nature* 417:141).

streptomycin   Antibiotic compound. It precipitates nucleic acids, inhibits protein synthesis, and interferes with proofreading, thus causing translational errors. Mutation in its S12 ribosomal protein-binding sites leads to improved translational precision. Some mutations may lead to streptomycin dependence. Streptomycin-resistant mutations in the ctDNA are maternally inherited; such mitochondrial DNA mutations may lead to hearing loss in humans. *See* antibiotics; mitochondrial diseases in humans; mtDNA.

Streptomycin.

streptozocin (a nitrosamide, 2-deoxy-2-[3-methyl-3-nitrosoureide]-D-glucopyranose, streptozotocin) Methylating, carcinogenic, antibiotic agent (effective even against fungi). Induces diabetes and poisons B lymphocytes.

stress Condition in which living beings must cope with difficult mental or physiological conditions. A major gene for panic and phobic disorders appears to be at human chromosome 15q24-q26. Stress or anxiety activates corticotropin-releasing factor (CRF/CRH) synthesis in the hypothalamus. CRF then stimulates CRF receptors (CRHR) in the pituitary, which turns on adrenocorticotropin hormone (ACTH) in the kidneys, leading to the production of glucocorticoids, which hinder the stress reaction by feedback to the brain. In case of failure to successfully respond with some type of a homeostatic mechanism, death or substantial harm may result. Stress activates sphingomyelinase to generate ceramide, and the latter initiates apoptosis. Disruption of the glucocorticoid receptor gene may lead to reduced anxiety. Stress also activates heat-shock proteins and glucose-regulated proteins (GRP). The GRP proteins are highly active during tumor progression. Their suppression may lead to apoptosis and rejection of the tumor cells. Stress signals may mediate the activation of genetic repair systems or in animals may proceed through one of three main pathways: the c-Abl, JNK, or p53 route. The first two are specific to different types of genotoxic agents; the p53 protein responds rather generally to various chemical stresses. Signal transducers eventually reach the DNA by the activation of transcription factors. Ionizing or excitatory (UV) radiations may directly cause chromosome breakage, resulting either in repair or apoptosis. In plants, stress stimulates the formation of elicitors and pathogenesis-related proteins. *See* adrenocorticotropic hormone; apoptosis; cAbl; CAP; ceramides; GADD153; glucocorticoid; heat-shock proteins; homeostasis; human brain; JNK; p38; p53; pathogenesis-related proteins; SAP; SAPK; sphingolipidoses; sphingolipids. (Smith, M. A., et al. 1995. *Proc. Natl. Acad. Sci. USA* 92:8788; Gratacòs, M., et al. 2001. *Cell* 106:L367; Dolan, R. J. 2002. *Science* 298:1191.)

stress proteins Heat-shock proteins. *See* heat-shock protein.

stretching chromosomes For more precise localization of FISH labels, the chromosome can be extended 5–20 times their highly coiled length using hypotonically treated, unfixed metaphase chromosomes and centrifugation. *See* FISH. (Bennink, M. L., et al. 2001. *Nature Struct. Biol.* 8:606.)

striated muscles The heart and skeletal muscles are made of sarcomeres and thus striated transversely. *See* smooth muscles.

string edit distance Determined upon addition, deletion, or replacement of one base symbol in order to transform a DNA sequence (a string) into another. *See* tree edit distance.

stringent Rigidly controlled.

stringent control In amino-acid-starved bacteria (auxotrophs), the product of the *relA*$^+$ gene shuts off ribosomal RNA synthesis as an economical device. In the presence of *relA*, amino acid synthesis is promoted (relaxed control) because ppGpp regulates the discriminator regions of the promoters. *See* discriminator region; fusidic acid; relaxed control; ribosomes. (Chatterji, D., et al. 1998. *Genes Cells* 3:279; Chatterji, D. & Ojha, A. K. 2001. *Curr. Opin. Microbiol.* 4:160; Barker, M. M., et al. 2001. *J. Mol. Biol.* 305:673.)

stringent genomes Nuclear, because each chromosome is normally replicated once during the cell cycle, and normally during mitosis each member of a diploid (or other euploid) chromosome set is partitioned equally between the daughter cells. *See* relaxed genomes.

stringent plasmid Its low copy number is genetically controlled.

stringent replication Limited replication of the low-copy-number plasmid DNA.

stringent response Under poor growth conditions, prokaryotic cells may shut down protein synthesis by limiting tRNA and ribosome formation. Synthesis of the *rrn* genes is generally mediated by binding ppGpp or pppGpp sequences to *rrn* promoters. *See* guanosine tetraphosphate; magic spot; Rel oncogene; ribosomes; *rrn*; stringent control; transfer RNA. (Chatterji, D. & Ojha, A. K. 2001. *Curr. Opin. Microbiol.* 4[2]:160; van Delden, C., et al. 2001. *J. Bacteriol.* 183:5376.)

STRL Coreceptor of HIV and SIV. *See* acquired immunodeficiency syndrome.

stRNA *See* RNAi.

stroke Causes 150,000 deaths/year and afflicts three times as many in the U.S. The major genetic factors involved are telangiectasia, Osler-Weber-Rendu syndrome, CADASIL, Ehlers-Danlos syndrome, polycystic kidney disease, Marfan syndrome, cardiovascular diseases, hypertension, and Melas syndrome. (Gretasdottir, S., et al. 2002. *Am. J. Hum. Genet.* 70:593 reqested a susceptibility locus at 5q22.)

stroma (1) Aqueous solutes within an organelle. (2) Pseudoparenchymatous association of fungal mycelia. (3) Supportive tissue of an organ; stroma cells in the bone marrow may produce collagen and extracellular matrix. *See* parenchyma.

stromelysin *See* transin.

strong-stop DNA When reverse transcriptase initiates transcription of first-strand DNA from RNA, a second strand is made on the first-strand DNA template. The transcriptase pauses after the atmospheric transcription of the R U5 segments (first strand). The U5 R, U3 (second-strand) "strongstop" DNA species accumulate (in the latter case, also including small portions of the RNA primer) before transcription continues to completion of the first, then the second strand, respectively. *See* retroviruses; reverse transcriptase. (Driscoll. M. D., et al. 2001. *J. Virol.* 75:672.)

strontium Earth metal with several isotopic forms, the ^{90}Sr, a β-emitter radioactive component of nuclear fallout, is readily substituted for calcium and thus may be concentrated in milk if cows grazed on contaminated pastures. Its half-life is 28 years. After the atmospheric bomb testings in the 1950s, it became especially threatening to children whose bones accumulated 2.6 $\mu\mu$Ci in contrast to adults (0.4 $\mu\mu$Ci). *See* Ci; isotopes; radiation hazards.

STRP Short tandem repeat polymorphism. *See* microsatellite.

structural classification, protein *See* ASTRAL; SCOP.

structural gene Primarily nonregulatory DNA sequence that codes for the amino acid sequence in a protein or for rRNA and tRNA.

structural genomics Develop high-resolution models of protein structure in order to understand catalytic and other functional mechanisms, ligands, and domains; reveal critical targets for site-directed mutagenesis; develop means for therapeutic interventions. The main tools are X-ray crystallography and nuclear magnetic resonance analysis. (Baker, D. & Sali, A. 2001. *Science* 294:93.)

structural heterozygosity Involves normal and rearranged homologous chromosomes within cells. *See* aberration, chromosomal; inversion; translocation.

structure-directed combinatorial mutagenesis PCR-based mutagenesis.

Strumpell disease *See* spastic paraplegia.

STS *See* sequenced-tagged sites.

STS-content mapping In physical mapping, the large sequences (YAC clones) contain STS tracts and their position can be mapped. *See* sequenced-tagged sites; YAC. (Chen, Y. Z., et al. 2001. *Genomics* 74:55.)

Stuart factor deficiency *See* antihemophilic factors; prothrombine deficiency.

student's t distribution Statistical test for assessing a hypothesis about the means of two populations. This distribution enables the statisticians to compute confidence limits for μ (the true mean of a population) when σ (the standard deviation of the true mean) is not known, and only the standard deviation of the sample s is available. The quantity of t is determined by the equation

$$t = \frac{\overline{x} - \mu}{s/n},$$

where $\overline{x}$ is the experimental mean and n is the population size. The critical t values are generally read from statistical tables after they have been quantified by the calculated t value,

$$t = \frac{d}{\sqrt{V}},$$

where d is the difference between means and V is the variance. Under practical conditions, the significance of the difference between two means is calculated by the formula

$$t = (\frac{\overline{x}_1 - \overline{x}_2)}{\sqrt{[s_1]^2 + [s_2]^2}},$$

where the x values stand for the two means and the s^2 values are the variances of the two populations. *See* arithmetic mean; paired t test; standard deviation; variance; table next page.

stuffer DNA Part of the phage-λ genome is not entirely essential for normal functions of the phage. Sequences between gene J and *att* representing about one-fourth of the genome can be removed and replaced (stuffed in) by genetic engineering without destroying viability of the phage. *See* lambda phage; vectors. (Parks, R. J., et al. 2001. *J. Virol.* 73:8027.)

sturt Unit of fate mapping. *See* fate maps.

stutter bands DNA slippage may occur during PCR replication (especially of long dinucleotide repeats) and may create shorter sequences than expected, making the identification of the heterozygotes difficult for microsatellite sequences. *See* microsatellite; PCR. (Miller, M. J. & Yuan, B. Z. 1997. *Anal. Biochem.* 251:50; Walsh, P. S., et al. 1996. *Nucleic Acids Res.* 24:2807.)

stuttering Transcription termination phenomenon; poly U may easily break U-A associations. *See* stammering.

stylopodium Bones of the humerus and femur.

stylus (style) Slender structure leading from the stigma to the ovary of plants and through which the pollen tube grows to the embryo sac. *See* gametophyte, female; gametophyte, male.

su*1 = *supD*, *su*2 = *supE*, *su*3 = *supF*, *su*4 = *supC*, *su*5 = *supG*, *su*7 = *supU *See su*$^-$.

t distribution The calculated value at the determined degrees of freedom (*df*) must be identical or greater than the closest value found on the pertinent *df* line in order to qualify for the probability shown at the top of the columns, e.g., for *df* = 10, and *t* = 3.169, *P* = 0.01, but if the *t* would be only 3.168, *P* would be only 0.05 according to the table and statistical conventions. The use of *t* charts or linear interpolation using the logarithms of the two-tailed probability values (Simaika, 1942. *Biometrika* 32:263) can obtain more precise *P* values. Remember that the *t test* indicates the probability of the null hypothesis that the two means would be identical.

d	OP → 0.900	0.500	0.400	0.300	0.200	0.100	0.050	0.010	0.001
1	0.158	1.000	1.376	1.963	3.078	6.314	12.706	63.654	636.620
2	0.142	0.816	1.061	1.386	1.886	2.920	4.303	9.925	31.599
3	0.137	0.765	0.978	1.250	1.638	2.353	3.182	5.841	12.924
4	0.134	0.741	0.941	1.190	1.533	2.132	2.776	4.604	8.610
5	0.132	0.727	0.920	1.156	1.476	2.015	2.571	4.032	6.869
6	0.131	0.718	0.906	1.134	1.440	1.943	2.447	3.707	5.959
7	0.130	0.711	0.896	1.119	1.415	1.895	2.365	3.500	5.408
8	0.130	0.706	0.889	1.108	1.397	1.860	2.306	3.355	5.041
9	0.129	0.703	0.883	1.100	1.383	1.833	2.262	3.250	4.781
10	0.129	0.700	0.879	1.093	1.372	1.812	2.228	→**3.169**	4.587
11	0.129	0.697	0.876	1.088	1.363	1.796	2.201	3.106	4.437
12	0.128	0.696	0.873	1.083	1.356	1.782	2.179	3.054	4.318
13	0.128	0.694	0.870	1.080	1.350	1.771	2.160	3.012	4.221
14	0.128	0.690	0.868	1.076	1.345	1.761	2.145	2.977	4.140
15	0.128	0.691	0.866	1.074	1.341	1.753	2.131	2.947	4.073
16	0.128	0.690	0.865	1.071	1.337	1.746	2.120	2.921	4.015
17	0.128	0.689	0.863	1.069	1.333	1.740	2.110	2.898	3.965
18	0.127	0.688	0.862	1.067	1.330	1.734	2.101	2.878	3.922
19	0.127	0.688	0.861	1.066	1.328	1.729	2.093	2.861	3.883
20	0.127	0.687	0.860	1.064	1.325	1.725	2.086	2.845	3.850
21	0.127	0.688	0.859	1.063	1.323	1.721	2.080	2.831	3.819
22	0.127	0.686	0.858	1.061	1.321	1.717	2.074	2.819	3.792
23	0.127	0.685	0.858	1.060	1.320	1.714	2.069	2.807	3.768
24	0.127	0.685	0.857	1.059	1.318	1.711	2.064	2.797	3.745
25	0.127	0.684	0.856	1.058	1.316	1.708	2.060	2.787	3.725
26	0.127	0.684	0.856	1.058	1.315	1.706	2.056	2.779	3.707
27	0.127	0.684	0.855	1.057	1.314	1.703	2.052	2.771	3.690
28	0.127	0.683	0.855	1.056	1.312	1.701	2.048	2.763	3.674
29	0.127	0.683	0.854	1.055	1.311	1.699	2.045	2.756	3.659
30	0.127	0.683	0.854	1.055	1.310	1.697	2.042	2.750	3.646
40	0.126	0.681	0.851	1.050	1.303	1.684	2.021	2.704	3.551
60	0.126	0.679	0.848	1.046	1.296	1.671	2.000	2.660	3.460
120	0.126	0.678	0.845	1.041	1.289	1.658	1.980	2.617	3.373
∞	0.126	0.674	0.842	1.036	1.282	1.645	1.960	2.576	3.290

su⁻ Wild-type allele of a suppressor mutation; the suppressor allele is *su⁺*.

su⁺ Suppressor allele at a locus in contrast to the wild type that is designated *su⁻*.

subcellular Organelle or other structure or site within a cell.

subcellular localization Expression levels of many genes are correlated with their subcellular site(s). Gene expression level is generally high in the cytoplasm, low in the nuclear membrane, and intermediate in the secretory pathways (endoplasmic reticulum and Golgi). Fluctuations occur in each group. *See* chloroplast genetics; FL-REX; mitochondria; mitochondrial diseases in humans. (Drawid, A., et al. 2000. *Trends Genet.* 16:426; Feng, Z.-P. & Zhang, C.-T. 2002. *Int. J. Biochem & Cell Biol.* 34:298.)

subcloning Recloning a piece of DNA. *See* cloning, molecular; cloning vectors.

subculturing Transferring of a culture into a fresh medium.

subcutaneous Beneath/under the skin.

suberin Corky complex polymeric material (of fatty acids but no glycerol associated with it) on the surface and within plant cells. In many plants, there is a subepidermal layer of suberin in air-filled cells and in various scar tissues. Suberin is frequently associated with cellulose, tannic acid, dark pigments (phlobaphenes), and inorganics. The commercial cork produced by the oak, *Quercus suber*, is suberin. *See* host-pathogen relation.

sublethal Only about 50% of the affected may live until sexual maturity.

sublimon Substoichiometric molecules of mtDNA that are supposed to be the products of recombination within short repeated sequences in this organelle. *See* mtDNA. (Kajander, O. A., et al. 2000. *Hum. Mol. Genet.* 9:2821.)

subline New colony of rodents set up in a new laboratory. *See* inbred; substrain.

Su blood type Occurs in pigs and resembles the Rh blood type in humans. *See* erythroblastosis fetalis; Rh blood type.

submetacentric Chromosome with two arms clearly unequal in length. *See* chromosome morphology.

Submetacentric chromosome.

submission signal In the majority of vertebrates, aggressive behavior generally ends when the weaker partner in the conflict displays the submission signal, e.g., dogs lay on their back. The human race does not employ such definite signals and thus the conflicts frequently end in violence. *See* aggression; behavior genetics; ethology; human behavior.

subspecies Group of organisms within a species distinguishable by gene frequencies, chromosomal morphology and/or rearrangement(s); may show some signs of reproductive isolation from the rest of the species. *See* species.

substance abuse Proclivity is genetically controlled. Morphine preference has at least three known QTLs; alcoholism also has several QTLs. Some of these conditions may be associated with variations in the serotonin transporter. A genetic factor for the latter is linked to the human ALPC2 locus, controlling vulnerability to alcohol. The conditions leading to depression may generally affect substance abuse; some of the quantitative trait loci mentioned do not appear, however, to act globally. *See* alcoholism; QTL; serotonin; steroid doping. (Uhl, G. R., et al. 2001. *Am. J. Hum. Genet.* 69:1290.)

substantia nigra Site in the middle part of the brain with gray pigment deposits.

substitution, disomic Two homologous chromosomes replaced by two others. *See* alien substitution; intervarietal substitution.

substitution, monosomic One entire chromosome is substituted for another. *See* substitution line.

substitution line One of its chromosomes (or pair) is derived from a donor variety or species. *See* alien substitution; intervarietal substitution.

substitution mutation One base pair is replaced by another. The evolutionary mutation rate in the human X chromosome is lower than in the Y chromosome, and the ratio is about 1.7 at the 95% confidence level according to one study based on molecular analysis. *See* base substitution; Li-Fraumeni syndrome; point mutation; SNIP; transition and transversion; photo below.

substrain Line separated from another after 8–19 cycles of inbreeding or when a single colony is different from the rest of the strain of rodents. *See* inbred; subline.

substrate (1) Compound on which an enzyme can act. (2) Culture medium for an organism or a particular surface. *See* pseudosubstrate.

substrate adhesion molecules (SAM) Bind as extracellular molecules to independent receptors on adhering cells.

substrate cycle *See* futile cycle.

substrate induction Enzyme synthesis is stimulated by the presence of the substrate of the enzyme. *See lac* operon; substrate.

subtiligase Enzyme capable of ligating esterified peptides in aqueous solutions.

subtilisin Protease enzyme that cuts at serine in the context Gly-Thr-**Ser**-Met-Ala-Ser; chymotrypsin also cuts at serine but within a different sequence. Subtilisin is translated as a pre-pro-polypeptide containing the IMC (intramolecular chaperone) sequence between the signal peptide and the mature enzyme. IMC is responsible for folding of the final enzyme without being present in the functional subtilisin,

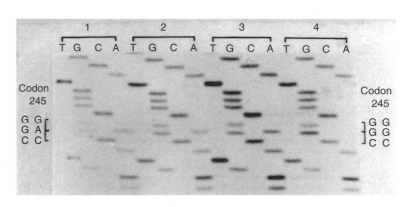

Base substitution mutations in the p53 gene in the noncancerous fibroblast cells in a family with Li-Fraumeni syndrome. (1) proband, (2) his brother, (3) their father, and (4) a normal control. In codon 245 GGC→GAC, mutations are evident in the two generations investigated. (Courtesy of Professor Esther H. Chang; see also *Nature* 348:747.)

which is a general scavanger molecule. *See* chymotrypsin; scavenger molecules; signal peptide.

subtraction, genomic   *See* genomic subtraction.

subtractive cloning (driver excess hybridization)   Provides information on genes selectively expressed in different tissues. A single-stranded RNA (or DNA, the so-called tracer) is hybridized to another nucleic acid (the driver) present in the reaction mixture at least 10-fold in excess. Usually, from the tester RNA a cDNA is generated by the use of poly(dT) primers. Then double-strand DNA is generated from the single-strand cDNA with the aid of a poly(dT) primer. The tester cDNA is mixed with an excess of driver cDNA, and the cDNAs are allowed to reanneal after denaturation. The double-strand DNA is removed by adsorption to a hydroxyapatite column. The hybrids and the unhybridized driver are selectively removed (subtraction), and the remaining single-stranded, unhybridized tracer is further enriched by hydroxyapatite, biotinylation and selection by streptavidin, chemical cross-linking, RNase H, and PCR. The process may be repeated until pure tracer-specific nucleic acid becomes available. This can then be used to screen a library for tracer specificity. The procedure bears much similarity to $c_0 t$ or $r_0 t$ analysis. Hyperchromicity or digestibility by S1 nuclease can monitor the single-stranded molecules. The technique may provide information on the number or kind of genes expressed in specific tissues. *See* cascade hybridization; $c_0 t$; genomic subtraction; hydroxyapatite; hyperchromicity; normalization; nucleic acid hybridization; positive selection; RFLP subtraction; $r_0 t$ analysis; S1 nuclease; streptavidin; subtractive hybridization; tissue-specificity. (Sagerström, C. G., et al. 1997. *Annu. Rev. Biochem.* 66:751.)

subtractive hybridization   *See* subtractive cloning.

subtractive suppression hybridization (SSH)   First, subtracted cDNA libraries are generated. Then selective/and or suppressive cycles of PCR are combined with normalization and subtraction in a single procedure. The normalization equalizes the abundance of cDNAs within the samples and the subtraction excludes the sequences common to the target and driver populations. The procedure is well suited for identification of disease, developmental or other differentially expressed genes. *See* PCR; subtractive cloning. (Diatchenko, L., et al. 1999. *Methods Enzymol.* 303:349; den Hollander, A. I., et al. 1999. *Genomics* 58:240; Nishizuka, S., et al. 2001. *Cancer Res.* 61:4536.)

subunits of enzymes (protomers)   Polypeptides that make up oligomeric proteins.

subunit vaccine   Contains only a fragment of the antigenic protein, which is sufficient to stimulate an immune response. *See* antigen; vaccines.

subvital   Has reduced viability, yet a 50% chance to survive up to the reproductive period of the species. *See* sublethal.

SUC2   Yeast invertase gene; it mediates the glucose effect and may be suppressed by *ssn* (suppressor of *snf*). *See* glucose effect; *SNF*; *SSN6*.

succinate dehydrogenase   Nonheme iron, inner mitochondrial dehydrogenase enzyme (subunits encoded at human chromosomes 11q23 and 1q21). It converts succinate to fumarate while flavin adenine dinucleotide serves as a H_2 acceptor. *See* mitochondrial diseases in humans; pheochromocytoma.

sucrose gradient centrifugation   *See* density gradient centrifugation.

sucrose intolerance   *See* disaccharide intolerance.

sucrose transporters   Deliver the photosynthetically produced sucrose within the plant body. *See* photosynthesis. (Truernit, E. 2001. *Curr. Biol.* 11:R169.)

Sudan black   Stains fatty tissues, wax, resins, cutins, etc., red in microscopic use.

sudden infant death syndrome (SIDS)   Unexpected death of healthy, normal infants within the first year of life during sleep. It appears to be associated with a deficiency in the binding of the muscarinic cholinergic receptors of the brain, resulting in accumulation of carbon dioxide or lack of oxygen in the blood. *See* muscarinic acetylcholine receptors.

SUG1   ATPase and activator of transcription; it can substitute in yeast for Trip1 and can interact with the transcriptional activation domain of GAL4 and herpes virus protein VP16. *See* ATPase; GAL4; signal transduction; transcriptional activator; Trip1; VP16.

sugar beet (*Beta vulgaris*) $2n = 18$. One of the greatest success stories of plant breeding. In the middle of the 18th century, the average sugar content of the plant was about 2%. This was increased by the 20th century to about 20%, and the sugar yield per hectare increased to about 4 metric tons. Some of the modern varieties have numerous agronomically important features (disease resistance, monogermy), and this is a rare plant where triploid varieties (besides bananas) are grown commercially. *See* banana; monogerm seed.

sugarcane (*Saccharum*, $x = 10$)   Its diploid forms are unknown and the cultivated varieties have a high and variable number of chromosomes (*S. spontaneum* $2n = 36-128$, *S. robustum* $2n = 60-170$, *S. officinarum* $2n = 70-140$), including polyploids and aneuploids. The modern cultivated forms are natural hybrids of *S. spontaneum* × *S. officinarum* (this hybrid is also called *S. barberi* in India and *S. sinense* in China). Genetic mapping requires DNA markers that — unlike the gene markers — can be studied despite the complex chromosomal situations. It is the most important source of saccharose or common sugar. *See* sugar beet.

suicidal antibody   The antibody is equipped with a targeting signal that can direct it to a degradative cellular compartment (e.g., lysosome, proteasome) where the bound antigen is destroyed. *See* lysosome; proteasome. (Larbig, D., et al. 1979. *Pharmacology* 18:1; Hsu, K. F., et al. 2001. *Gene Ther.* 8[5]:376.)

suicidal behavior   Appears to have a familial component. The suspected association between tryptophan hydroxylase and suicidal behavior is not supported by recent analysis (Lalovic, A. & Turecki, G. 2002. *Am. J. Med. Genet.* 114:533.)

suicide inhibitor   Molecule that inhibits enzyme action after the enzyme has acted upon it. In the original form, it is only a weak inhibitor, but after reacting with the enzyme, it binds to it irreversibly and becomes a very potent inhibitor. Allopurinol and fluorouracil are examples. *See* allopurinol; regulation of enzyme activity.

suicide mutagen   Uses a ^{32}P-labeled or other radioactive nucleotide that is incorporated into the genetic material and causes mutation by localized radiation. *See* magic bullet.

suicide vector   Delivers a transposon into the host cells in which the vector itself cannot replicate, but the transposon can be maintained and used for transposon mutagenesis. Additional use involves the delivery of the herpes simplex virus thymidine kinase gene (HSV-TK) and administering ganciclovir or acyclovir. HSV-TK is about three orders of magnitude more effective than cellular TK in phosphorylating these DNA base analogs, thus blocking DNA synthesis. Phosphorylated ganciclovir is also impaired in moving through cell membranes and thus has longer-lasting local effects. Eventually, some ganciclovir resistance may develop. The delivery of the gene for cytosine deaminase (CD) and the compound 5-fluorocytosine may produce a DNA inhibitor, 5-fluorouracil, within the target. Xanthine-guanine phophoribosyltransferase (XGPRT) may make the tumor cells more sensitive to 6-thioxanthene. P450-2B1 gene encoding a cytochrome converts cyclophosphamide into the toxic phosphoramide mustard. The *deoD* bacterial gene (purine-nucleoside phosphorylase) converts 6-methylpurine deoxyribonucleoside into the deleterious 6-methylpurine. Bacterial nitroreductases convert the relatively nontoxic monofunctional N-chloroethylamine derivative, C1954, into 10^4 times more active bifunctional alkylating agents. Employing HSV-TK, CD, and XGPRT to the same tumor, three different transgenic cell populations, a mosaic may result. Theoretically this could be used in a preventive program in putative cancer-prone cases. If the HSV-TK cells would become cancerous, ganciclovir would eliminate that cell subpopulation and the healthy tissues carrying the CD and XGPRT could take over as a prophylactic measure. Unfortunately, this mosaic approach has serious limitations. Targeting a foreign antigen (e.g., HLA-B7 protein) to the cancer cells may stimulate the immune system of the host and kill the tumor cells. The immune system may be stimulated by transformation with cytokine genes. *See* bystander effect; cancer gene therapy; cytokines; cytosine deaminase; ganciclovir; gene therapy; HLA; immunotherapy, adoptive; thymidine kinase; transposon mutagenesis. (Takamatsu, D., et al. 2001. *Plasmid* 46:140.)

sulfatase deficiency   *See* mucosulfatidosis.

sulfhydryl group   —SH, and when two are joined, the linkage is a disulfide bond. Sulfhydryl compounds, such as cysteine and cysteamine, are protectors against ionizing radiation. *See* cysteamine; cysteine; thiol.

sulfocysteinuria   Deficiency of sulfite oxidase. Restricted intake of sulfur amino acids in the diet greatly reduces sulfocysteine and thiosulfate in the urine and improves conditions.

sulfonylurea   Group of compounds that stimulate insulin production by regulating insulin secretion and lowering blood sugar level. It is used to treat patients with non-insulin-dependent diabetes. These drugs interact with the sulfonylurea receptor of pancreatic β-cells and inhibit the conductance of ATP-dependent K^+ ion channels. Reduction of potassium exit activates the inward rectifying Ca^{2+} channels and promotes exocytosis. Sulfonylureas are inhibitors of the acetolactate synthase enzyme; there are also sulfonylurea herbicides. *See* diabetes; herbicides; ion channels.

sulfur mustard   *See* formula below. Al is an alkyl group. If the two alkyl groups are chlorinated (Cl), the compound is called bifunctional. If only one alkyl group is chlorinated, the compound is monofunctional. *See* alkylating agents; nitrogen mustard. (Michaelson, S. 2000. *Chem. Biol. Interact.* 125:1.)

sulfurylase   Catalyzes the reaction $ATP^{3-} + SO_4^{-2} \rightarrow$ adenylyl sulfate (adenosine-5′-phosphosulfate, $C_{10}H_{14}N_5O_{10}PS$).

SUMO-1   Small ubiquitin-like modifier is a ubiquitin-like carrier protein (11 kDa). It is conjugated to a variety of proteins and seems to be involved in multiple functions such as cell cycle progression, DNA repair, etc. *See* IκB; IKK; monoubiquitin; PIC; sentrin; UBL. (Melchior, F. 2000. *Annu. Rev. Cell Dev. Biol.* 16:591; Müller, S., et al. 2001. *Nature Rev. Mol. Cell Biol.* 2:202.)

sum of squares   Sum of the squared deviations from their mean of the observations; it is used in statistical procedures to estimate differences. *See* analysis of variance; intraclass correlation.

sunburn   About 60% of sunburn (UV) cases cause keratoses. The p53 gene may suffer mutation(s), mainly C→T transitions. After clonal propagation, these p53 mutant cells, may develop into squamous cell skin cancer (SCC). In over 90% of the SCC cells, the p53 gene is mutant. Increased melanin production involved with ultraviolet light exposure may be correlated with the DNA repair system. *See* DNA repair; keratoses; melanoma; p53; pigmentation in animals; ultraviolet light.

SUNDS   *See* Brugada syndrome.

sunflower (*Helianthus*) Oil crop with about 70 xenogamous species, $2n = 2x = 34$. (Burke, J. M., et al. 2002. *Genetics* 161:1257.)

sun-red maize Develops anthocyanin pigment when the tissues are exposed to sunshine (e.g., through a stencil).

Sun-red maize ear.

SUP35 *See* prion.

supB Ochre/amber suppressor; it inserts glutamine.

supC Suppressor mutation for amber (5′-UAG-3′) and ochre (5′-U3′) chain-terminator codons. The mutation causes a base substitution in the anticodon of tyrosine tRNA (5′-GUA-3′) and changes to 5′-UUA-3′, which can recognize both the ochre and amber codons as if they were tyrosine codons. Note: Pairing is 5′-3′ and 3′-5′. For the recognition of the former, wobbling is required. *See* amber; anticodon; ochre; suppressor; wobbling.

supD Amber suppressor mutation that reads the amber (5′-UAG-3′) chain-termination codon as if it were a serine (5′-UCG-3′) codon because the normal serine tRNA anticodon (5′-CGA-3′) mutates to (5′-CUA-3′); therefore, instead of terminating translation, a serine is inserted into the amino acid chain. (Remember, the pairing is antiparallel.) *See* anticodon; suppressor.

supE Amber-suppressor mutation that reads the chain-termination codon 5′-UAG-3′ (amber codon) in the mRNA as if it were a glutamine codon (5′-CAG-3′) because the glutamine tRNA anticodon mutates to 5′-CUA-3′ from the 5′-CUG-3′. As a consequence, the translation proceeds and a glutamine is inserted at the site where the translation would have terminated in the presence of an amber codon in the mRNA. (Note: The pairing is antiparallel.)

superantigen Native bacterial and viral proteins that can bind directly (without breaking up into smaller peptides) to MHC class II molecules on antigen-presenting cells. The variable regions of T-cell receptor β-chains thus activate more T cells against, e.g., enterotoxins of *Staphylococci* — causing toxic shock syndrome or food poisoning — than normal antigens. Cellular superantigens are likely to be responsible for such diseases as diabetes mellitus. *See* antigen; diabetes mellitus; endogenous virus; enterotoxin; MHC; TCR; toxic shock syndrome. (Muller-Alouf, H., et al. 2001. *Toxicon* 39:1691; Alam, S. M. & Gascoigne, N. R. 2003. *Methods Mol. Biol.* 214:65.)

superchiasmatic nucleus Region near the optical center of the hypothalamus in the brain that controls circadian signals. *See* brain; circadian rhythm.

supercoiled DNA May assume the *positive supercoiled* structure by twists in the same direction as the original general (right-handed) coiling of the double helix, or it may be twisted in the opposite direction, i.e., *negative supercoiling*. Negative supercoiling (Z DNA) may be required for replication and transcription. The superhelical density expresses the superhelical turns per 10 bp, and it is about 0.06 in cells as well as in virions. Loss of positive supercoiling may be lethal. Localized negative supercoiling may be lethal. Localized negative supercoiling is essential for gene expression, replication, and many other functions of DNA. *See* DNA replication, prokaryotes; linking number; packing ratio; transcription; Z DNA. (Holmes, V. F. & Cozarelli, N. R. 2000. *Proc. Natl. Acad. Sci. USA* 97:1322.)

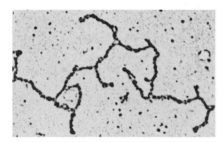

Supercoiled DNA plasmid of *Streptococcus lactis*. (Courtesy of Dr. Claude F. Garon.)

superdominance Overdominance or monogenic heterosis. *See* hybrid vigor; overdominance.

superfamily, gene Group of genes that are structurally related and may have descended from common ancestors, although their present function may be different.

superfemale (metafemale) *Drosophila*, trisomic for the X-chromosome (XXX) but disomic for the autosomes; she is sterile. *See* aneuploidy; supermale *Drosophila*.

superfetation Due to an apparently rare autosomal-dominant gene, ovulation may continue after implantation of the fertilized egg and an unusual type of twinning results. In animals, when a female may mate repeatedly with several males, the same litter may become multipaternal. Human dizygotic twins may also be of different paternity when during the receptive period the female had intercourse with two different males. *See* multipaternate litter; twinning. (Fontana, J. & Monif, G. R. 1970. *Obstet. Gynecol.* 35:585.)

supergenes Linked clusters of genes that are usually inherited as a block because inversions prevent the survival of recombinants for the clusters and thus have evolutionary and applied significance in plant and animal breeding.

superinfection A bacterium is infected by another phage. This is generally not possible in a lysogenic bacterium because of immunity, i.e., the superinfecting phage cannot enter a vegetative cycle within the lysogen. Infection by two T-even phages may fail in such an attempt by superinfection breakdown. Some higher organisms may be infected by parasites of different genotypes. The number of clones of the parasites may not increase with the progression of the disease, indicating competition among the superinfecting strains (e.g., *Plasmodium*). *See* immunity; phage; *Plasmodium*. (Hayes, W.

1965. *The Genetics of Bacteria and Their Viruses*. Wiley, New York, Vogt, B., et al. 2001. *Hum. Gene Ther.* 12[4]:359.)

supermale *Asparagus* Can be obtained by regenerating and diploidizing plants obtained from Y-chromosomal microspores or pollen by the techniques of cell culture. Thus, their chromosomal constitution is 18 autosomes + YY. These plants are commercially advantageous because of the higher yield of the edible spears. *See* YY *Asparagus*.

supermale *Drosophila* Has one X and three sets of autosomes, i.e., the fly is monosomic for X, but he is trisomic for all the autosomes; he is sterile. *See* sex determination; superfemale.

superman Homeotic mutation in *Arabidopsis* resulting in excessive development of the androecium at the expense of other flower parts. The DNA base sequences are the same as in the wild type, but the mutation evokes methylation. *See* androecium; flower differentiation. (Koshimoto, N., et al. 2001. *Plant Mol. Biol.* 46:171; Cao, X. & Jacobsen, S. E. 2002. *Current Biol.* 12:1138; Dathan, N., et al. 2002. *Nucleic Acids Res.* 30:4945.)

super-Mendelian inheritance Transmission of one of the alleles in a diploid is preferential because of the advantage of a gene at meiosis or postmeiotic steps of gametogenesis. This advantage may not be due to an intrinsic superiority of the allele. Homing endonuclease genes (HEG) may be preferentially transmitted horizontally but may or may not be propagated at a selective advantage. *See* homing endonuclease; meiotic drive; preferential segregation; segregation distorter. (Goddard, M. R. & Burt, A. 1999. *Proc. Natl. Acad. Sci. USA* 96:13880.)

supermutagen Efficient mutagen that primarily causes point mutations without inducing frequent chromosomal defects. *See* ethylmethane sulfonate; nitrosoguanidine; point mutation.

supermutation *See* antibody gene switching; hypermutation; somatic hypermutation.

supernatant Nonsedimented fraction after centrifugation of a suspension or in general any floating fraction derived from a mixture.

supernumerary chromosome *See* B chromosome; cat eye syndrome.

superoperon Regulatory complex tied together, e.g., photoreceptor pigment synthesis and photosynthesis. *See* operon.

superovulation By injection of gonadotropic hormones (into mice) the number of eggs produced may increase several-fold. Fertilization follows after about 13 hours and the eggs can surgically be collected for in vitro study of preimplantation development. Hormonal treatment may cause superovulation in usually monoparous animals. *See* twinning.

superoxide ($O_2^{-\bullet}$) Highly reactive species of oxygen. In the hypersensitivity reaction defense mechanism it appears to play an important role along with nitric oxide. *See* Fenton reaction; hydrogen peroxide; hypersensitivity reaction; nitric oxide; peroxides; peroxynitrite; ROS; SOD.

superoxide dismutase (SOD) Catalyzes the reaction $2O_2 + 2H^+ \rightleftarrows O_2 + H_2O_2$ and thus participates in the detoxification of the highly reactive (mutagenic) superoxide radical O_2^-. These enzymes are the main detoxificants of the free radicals. SOD overexpression confers resistance to ionizing radiation under aerobic conditions. There are three isozymes of SOD: cytosolic SOD-1 (human chromosome 21q22.1), mitochondrial SOD-2 (6q25.3), and extracellular SOD-3 (4pter-q21). In SOD deficiency, cardiomyopathy, brain damage, mitochondrial defects, Lou Gehrig's disease, and precocious aging may result. A nonpeptidyl manganese complex with bis(cyclohexylpyridine)—substituted macrocyclic ligand may mimic SOD activity. Inhibition of SOD enzymes may trigger suicide of leukemia cells, which have a higher rate of oxidative metabolism. Cell suicide may occur when the level of hydrogen peroxide is high or when it is converted to hydroxyl radicals in an Fe^{2+}-dependent process. Using the cytosolic SOD gene in an adenoviral vector may provide protection against ROS injury. *See* aging; amyotrophic lateral sclerosis; cardiomyopathies; granulomatous disease; host-pathogen relationship; oxidative stress; ROS. (Fridovich, I. 1975. *Annu. Rev. Biochem.* 44:147; Danel, C., et al. 1998. *Hum. Gene Ther.* 9:1487.)

super-repressed Bacterial operon cannot respond to inducer. *See* inducer; repression.

supershift *See* gel retardation assays.

supersuppressor Dominant suppressor acting on more than one allele or even on different gene loci. *See* suppressor. (Gerlach, W. L. 1976. *Mol. Gen. Genet.* 144:213.)

supervillin *See* androgen receptor.

supervised learning Finding shared patterns or motifs common to all positive sequences and absent from all negative ones. *See* motif; unsupervised learning. (Moler, E. J., et al. 2000. *Physiol. Genomics* 2[2]:109.)

supervital Its fitness exceeds that of the standard (wild) type.

supF Amber-suppressor mutation in tRNA^{Tyr} that recognizes the chain-termination codon (5′-UAG-3′) as if it were a tyrosine codon (5′-UAC-3′ or 5′-UAU-3′) because a mutation at the anticodon sequence in tyrosine tRNA changes the 5′-GUA-3′ into a 5′-CUA-3′; thus, tyrosine tRNA inserts a tyrosine into the growing peptide chain where translation would have been terminated. *See* πVX; suppressor tRNA.

supG Suppressor mutation for both amber (5′-UAG-3′) and ochre (5′-UAA-3′) chain-termination codons by a base substitution mutation in the anticodon of lysine tRNA from 5′-UUU-3′ to 5′-UUA-3′ that can recognize the amber and ochre codons in mRNA as if they were lysine (5′-AAA-3′ or 5′-AAG-3′) codons, thus inserting in the peptide chain a lysine rather than discontinuing translation. (Note: The pairing is antiparallel.)

support   As a statistical concept, it is synonymous with log likelihood. *See* likelihood.

Support Vector Machine (SVM) Computer program based on the principle of supervised learning techniques. It uses a training set to determine which data should be a priori clustered. The operations start with a set of genes that are already known to have a common function, e.g., genes that encode ribosomal proteins. Also, SVM picks another set of genes that are not members of the functional group specified. These two sets form the training examples, and the genes are marked positive if they fall in the specified functional class and negative if they do not. SVM thus learns from the two groups how to discriminate and classify any unknown on the basis of function. This procedure can classify genes on the basis of microarray hybridization data and identify functional clusters genome-wide. *See* cluster analysis; microarray hybridization; supervised learning. (Brown, M. P. S., et al. 2000. *Proc. Natl. Acad. Sci. USA* 97:262; Hua, S. & Sun, Z. 2001. *Bioinformatics* 17[8]:721; Cristianini, N. & Shawe-Taylor, J. 2000. *An Introduction to Support Vector Machines and Other Kernel-based Learning Methods.* Cambridge Univ. Press, New York.)

supportive counseling   Tends to alleviate the psychological problems involved with the discovery of hereditary disorders and birth defects in families. (See counseling genetic).

suppressor, bacterial   Protein product of the bacterial regulator gene (such as, e.g., *i* in the *Lac* operon) that prevents transcription when associated with the operator. *See ara* operon; *Lac* operon.

suppressor, extragenic   The suppressor mutation is outside the boundary of the suppressed gene. *See* second-site reversion; second-site suppressor; suppressor tRNA.

suppressor, informational   Interferes with the expression of another gene at the level of translation by affecting tRNAs, ribosomes, or peptide elongation factors.

suppressor, intragenic   The suppressor site is within the gene where suppression takes place. *See* frameshift suppressor.

suppressor, second site   Suppressor outside the boundary of the locus but usually within the same chromosome. *See* second-site reversion, suppressor gene.

suppressor gene   Restores function lost by a mutation without causing a mutation at the suppressed site. The suppressor can be intragenic (within the cistron but also at another site) or at any other locus, e.g., in the signal transduction pathway. The suppressor may act by reducing and overexpressing a gene product. *See* frameshift; mutation in organelle DNA;

suppressor, extragenic; suppressor, informational; suppressor, intragenic.

suppressor mutation   *See* frameshift suppressor; mitochondrial suppressor; *sup*; suppressor gene; suppressor RNA; suppressor tRNA.

suppressor RNA   Prevents translation of an mRNA by partial base pairing with a specific sequence of the target. Two 22- and 40-nucleotide-long tracts of the *lin-4* transcripts control the translation of gene *lin-14*. The latter is an early expressed nuclear protein in *Caenorhabditis*, but its expression is blocked during later stages by *lin-4* RNA, which itself does not operate through a protein. Several other genes are subject to RNA suppressors in other organisms. The mechanism of this suppression seems to be different from that of suppressor tRNA. *See* antisense DNA; antisense RNA; RNAi; suppressor gene; suppressor tRNA.

suppressor selection gene fusion vector   Carries nonsense codons in the structural gene and can be expressed only if the target genome carries nonsense suppressors. *See* gene fusion; transcriptional gene fusion vector; vectors.

suppressor T cell   Can suppress antigen-specific and allospecific T-cell proliferation by competing for the surface of antigen-presenting cells. *See* allospecific; T cell. (Maloy, K. J. & Powrie, F. 2001. *Nature Immunol.* 2:816.)

suppressor tRNA   Makes possible the translation of nonsense or missense codons in the original normal sense because a mutation in the anticodon of tRNA recognizes the complementary sequence in the codon, but its specificity resides in the tRNA molecule. Exceptions are possible, however. In the anticodon 5′-CCA-3′ of the tryptophan codon, 5′-UGG-3′ is not mutant, yet it may deliver a tryptophan to the opal position if a guanine is replaced by an adenine at position 24 of the D loop of tryptophan tRNA. Similarly, if in the GGG codon of glycine another G base is inserted by a frameshift mutagen, but subsequently an extra C is inserted in its anticodon (CCC), then tRNA may read the four bases normally as a glycine codon rather than a nonsense codon. *See* code, genetic; mutation in organelle DNA; phenotypic reversion; translation in vitro; tRNA. (Smith, J. D., et al. 1966. *Cold Spring Harbor Symp. Quant. Biol.* 31:479; Murgola, E. J., 1985. *Annu. Rev. Genet.* 19:57; Beier, H. & Grimm, M. 2001. *Nucleic Acids Res.* 29:4767.)

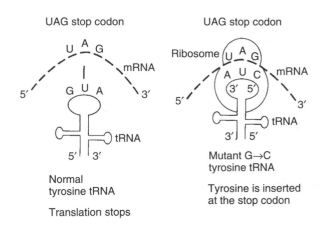

UAG stop codon

Normal tyrosine tRNA

Translation stops

UAG stop codon

Mutant G→C tyrosine tRNA

Tyrosine is inserted at the stop codon

supravalvar aortic stenosis (SVAS) Human chromosome 7q11 dominant mutation in an elastin gene with a prevalence of about 5×10^{-5} causing obstruction of the aortic blood vessels. The basic defect is due to a deficiency of elastin. It is pleiotropic and part of Williams syndrome. *See* cardiovascular diseases; coarctation of the aorta; elastin; Williams syndrome.

supU Suppressor of the opal (5'-UGA-3') chain-terminator codon with an anticodon sequence of 5'-UCA-3'. The mutation changes the anticodon of tryptophan tRNA from 5'-UCA-3' to 5'-CCA-3', thus permitting the insertion of a tryptophan residue into the polypeptide chain where it would have been terminated without the suppressor mutation (codon-anticodon recognition is antiparallel). This suppressor mutation is unusual because it recognizes both the tryptophan codon (5'-UGG-3') and the opal suppressor codon, i.e., its action is ambivalent.

SurA Periplasmic parvulin-type peptidy-prolyl isomerase involved in chaperoning outer membrane proteins. *See* parvulin; peptidyl-prolyl isomerase; periplasma.

surface antigen Generally, glycoprotein molecules on the cell surface that determine the identity of the cells for immunological recognition. The display of surface antigens is generally regulated at the level of transcription. *See* antigen; *Borrelia*; *Trypanosomas*; VSA.

surface plasmon resonance (SPR) Light coming in at a certain angle hits a hydrophilic dextran layer–covered gold surface. The interacting molecule is immobilized of this surface. As the injected molecule binds to this surface, the refractive index of the medium is increased and the change in the angle at which the intensity change occurs is measured in resonance units. The size of the macromolecule is positively correlated with the response. This analysis has numerous biological applications for the determination of protein-protein interactions such as antibody affinities, epitope, growth factor, signal-transducing molecule, receptor, ligand binding, etc. *See* biosensors; immunoprecipitation; microcalorimetry. (Heaton, R. J., et al. 2001. *Proc. Natl. Acad. Sci. USA* 98:3701.)

surfactant Common surfactants are soaps and detergents. In gene therapy, different surfactants such as perfluorochemical liquids and phospholipids may facilitate gene delivery to pulmonary tissues.

surrogate chains (ΨL, ΨH) Of immunoglobulins, are found in the progenitor (pro-) B lymphocytes. In the human and mouse fetal liver and adult bone marrow, an 18 kDa protein with ~45% homology to variable regions of the κ-, λ-, and heavy chains and a 22 kDa protein with ~70% homology to the constant region of the λ-chain have been found. The synthesis of these proteins (ΨL) ceases after the IgM chains appear on the lymphocyte surface. They participate in the formation of the B-cell receptor with other immunoglobulins and a short μ-chain containing only the N-terminal D(J)C sequences. The ΨL light chains are associated with a complex of 130 kDa/35–65 kDa glycoproteins (ΨH) and together make the surrogate receptors. *See* B lymphocytes;

immune system; immunoglobulins. (Meffre, E., et al. 2001. *F. Immunol.* 167:2151.)

surrogate genetics *See* reversed genetics.

surrogate mother Female who carries a baby to term for another couple. She may actually contribute the egg or she may just be a gestational carrier of a fertilized egg and have no genetic share in the offspring. In either case, moral, ethical, psychological, and legal problems must be pondered before this method of child bearing is chosen. Civil law generally keeps the maternal right of the gestational mother despite any contractual agreement contrary to natural parenthood. However, in case the gestational mother is not the donor of the ovum, the maternal right pertains to the biological ovum donor. Obviously, there are serious ethical problems beyond the principles of genetics. Society must protect the best interest of the child. *See* ART; oocyte donation; paternity testing.

surveillance, mRNA Mechanism of quality control for the elimination of defective proteins. *See* immunological surveillance. (Hilleren, P., et al. 1999. *Annu. Rev. Genet.* 33:229.)

survival The total survival probability per genetic damage is expressed by the equation

$$S(\rho, \partial) = \sum_{h=0}^{h=\partial} P(\rho, h, \partial!),$$

where ρ = hit probability, $\partial = VD$, V = total cell volume, D = density of the active events (dose), and P = likelihood of survival. *See* beneficial mutation; cost of evolution; fitness; genetic load; neutral mutation; target theory. (Alpen, E. L., 1998. *Radiation Biophysics. Acad. Press.*)

survival estimator *See* Kaplan-Meier estimator.

survival factors Interfere with apoptosis. The balance between survival and apoptosis is very complex, and a large number of proteins and ligands are involved through different pathways. *See* apoptosis; BAD; survivin; diagram next page.

survival of the fittest *See* fitness; neo-Darwinism; social Darwinism; social engineering.

survivin Inhibitory protein (16.5K) of caspases that are activated by cytochrome c. Survivin is expressed at the G2-mitosis phases of the cell cycle and is associated with the microtubules of the mitotic spindle. Apoptosis may be caused by the disruption of this association. Overexpression of survivin may lead to cancer. Disruption of survivin action may reduce melanoma growth. *See* apoptosis; cell cycle; spindle; survival factor. (Wheatley, S. P., et al. 2001. *Curr. Biol.* 11:886; Hoffman, W. H., et al. 2002. *J. Biol. Chem.* 277:3247; Altieri, D. 2003. *Nature. Rev. Cancer* 3:46; figure next page.)

suspension culture The cells are grown in a liquid nutrient medium.

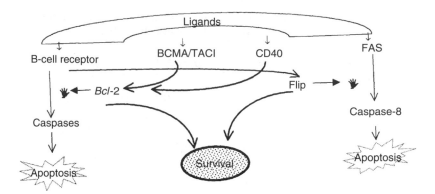

suspensor Line of cells through which the plant embryo is nourished by the maternal tissues. In general, anatomy ligaments may be called suspensors.

suture Junction of various solid animal and plant tissues.

SUV39H1 *See* histone methyltransferases.

Sv *See* Sievert.

SV2s Glycosylated transmembrane proteins homologous to prokaryotic and eukaryotic transporters. *See* transporters. (Jakobsen, A. M., et al. 2002. *J. Pathol.* 196:44.)

SV40 tag Simian virus 40 large T antigen. *See* simian virus 40.

SV40 vectors SV40 plasmids (vectors) can be packaged only if their DNA is within the range of 3,900 to 5,300 bp. Since these small genomes do not have much dispensable DNA, it is almost impossible to construct a functional vector with any added genes. Fortunately, functions provided by helper DNA molecules might help to overcome these problems. Simian virus 40 cannot replicate autonomously if the replicational origin (*ori*) is defective, yet it can integrate into chromosomal locations of green monkey cells and can then be replicated along the chromosomal DNA (such a cell is COS [cell origin simian virus]). Also, since the early genic region is normal (*see* simian virus 40 for structure), it may produce T antigens within the cell. If such a cell is transformed by another SV40 vector in which the viral early gene region was replaced by a foreign piece of DNA, the COS cell may act as a helper and replicate multiple copies of the second, the engineered, SV40 DNA. Since the late gene region of this plasmid is normal, the viral coat proteins can be synthesized within the cell. Availability of the coat proteins permits packaging of the engineered SV40 DNA into capsids. The virions so obtained can be used to infect other mammalian cells where passenger DNA can be transcribed and translated and foreign protein can be processed. The transformed cell can thus acquire a new function. An SV40 plasmid can be constructed by inserting a foreign gene with a desired function into the late gene cluster. This plasmid can then be used along with another SV40 plasmid with inactivated (deleted) early genes but with a good *ori* site. Upon coinfection, a mammalian cell with these SV40 plasmids can replicate to multiple copies and the inserted foreign gene can be expressed.

It is feasible to insert into a prokaryotic pBR322 plasmid the *ori* region of SV40 and another piece of DNA including all the necessary parts of a foreign gene. When this plasmid is transfected into a COS cell, the passenger gene can be transcribed, translated, and processed thanks to the replicated copies. With the assistance of SV40-based constructs, mammalian and other genes can be shuttled between mammalian and bacterial cells. A thymidine kinase–deficient (*tk⁻*) rodent cell is transformed by pBR322 bacterial plasmid carrying the *ori* of SV40 and the *TK⁺* (functional thymidine kinase), as well as an ampicillin-resistance gene (*amp*ᴿ) for bacterial selectability. The bacterial plasmid is integrated into a chromosome of the rodent cell, and this cell is fused with a COS cell. The integrated pBR322 plasmid is replicated into many copies, thanks to the presence of the COS nucleus. The hybrid plasmid, carrying a bacterial replicon, SV40 *ori*, *TK⁺*, and *amp*ᴿ, can infect *E. coli* cells and can be selectively propagated there. Thus, the shuttle function is achieved. Another SV40- and pBR322-based vector is the pSV plasmid. In a Pvu II and HinD III restriction enzyme–generated fragment, this contains the promoter signals and the mRNA initiation site. When any open-reading frame is attached to it, transcription can proceed. In addition, an intron of the early region provides splicing sites for other genes. The region contains transcription termination and polyadenylation signals. Some other pBR322 parts may be equipped with additional specific selectable markers. *See* shuttle vector; Simian virus 40; vectors; viral vectors. (Jayan, G. C., et al. 2001. *Gene Ther.* 8:1033; diagram next page.)

SVC *See* carbon dioxide (CO₂) sensitivity.

Svedberg units *See* sedimentation.

SVF-2 Member of the tumor necrosis factor receptor family. *See* Fas; TNF.

swapping genes Exchanging genes by horizontal transfer via plasmids. *See* site-specific recombination; targeting genes; transmission. (Nebert, D. W., et al. 2000. *Ann. NY Acad. Sci.* 919:148.)

swede (*Brassica napus*) (1) Leafy fodder crop in Northern climates. (2) Edible human vegetable; 2*n* = 38, AABB genomes. *See* rape.

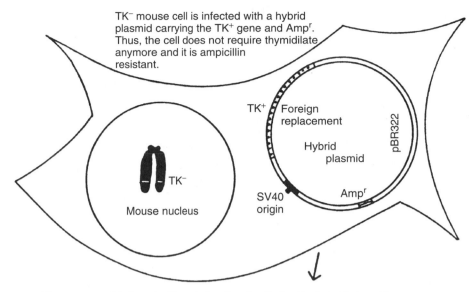

TK⁻ mouse cell is infected with a hybrid plasmid carrying the TK⁺ gene and Amp^r. Thus, the cell does not require thymidilate anymore and it is ampicillin resistant.

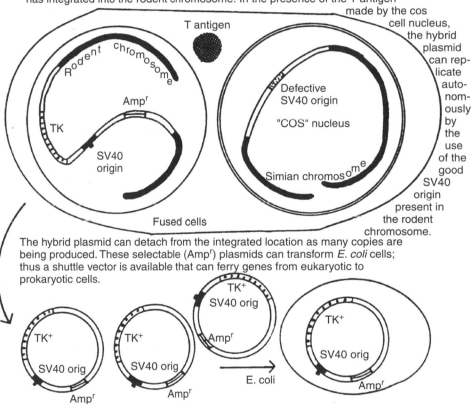

The mouse cell is fused with a simian "cos" cell after the hybrid plasmid has integrated into the rodent chromosome. In the presence of the T antigen made by the cos cell nucleus, the hybrid plasmid can replicate autonomously by the use of the good SV40 origin present in the rodent chromosome.

The hybrid plasmid can detach from the integrated location as many copies are being produced. These selectable (Amp^r) plasmids can transform *E. coli* cells; thus a shuttle vector is available that can ferry genes from eukaryotic to prokaryotic cells.

SV40 shuttle vector can ferry genes between mammalian cells and prokaryotes.

sweet clover (*Melilotus officinalis*) Fragrant leguminous plant because of its coumarin content. The coumarin-free forms (*M. albus*) are used as hay, $2n = 16$. *See* coumarin.

sweet pea (*Lathyrus odoratus*, $2n = 14$) Ornamental that had been exploited for studies on the genetic determination of flower pigments. Another peculiarity is that the "long"/"disc" pollen shape (L/ℓ) is determined by the genotype of the sporophytic (anther) tissue rather than by the gametophyte. Therefore, delayed inheritance is observed. *See* delayed inheritance.

Sweet pea.

sweet potato (*Ipomoea batatas*) Primarily a warm-climate vegetable with about 25 species with basic chromosome number $x = 15$ and a single genome; the most common cultivated form is hexaploid, although related species may be diploid or tetraploid.

swept radius *See* linkage.

SWI, SWI2/SNF2, SWI3 Yeast genes encoding transcriptional activators by chromatin remodeling. The homologous protein in *Drosophila* is BRAHMA. SWI2/SNF2 is a DNA-dependent ATPase. A heterodimer of Swi4 and Swi5 is SBF. A heterodimer of Swi6 and Mbp1 is MBF. *See* ACF; activator genes; CHRAC; chromatin remodeling; coactivator; Mbp1; nuclease-sensitive sites; nucleosome; NURF; *Polycomb*; SBF; transcriptional activator. (Muchardt, C. & Yaniv, M. 1999. *J. Mol. Biol.* 293:187; Sengupta, S. M., et al. 2001. *J. Biol. Chem.* 276:12636.)

Swi6 Centromeric repressor chromodomain protein in *Schizosaccharomyces pombe* that is essential for centromere function. *See* centromere.

swimming in bacteria Movement by counterclockwise rotation of the flagella.

swine Most commonly used name for the female animals of the *Sus* species (hogs). *See* pig.

SWI/SNF *See* nucleosome.

SwissProt database (protein information) <http://www.expasy.org/>; <http://www.ebi.ac.uk/swissprot/>.

switch, genetic An individual cell may initially express immunoglobulin gene $C\mu$, but in its clonal progeny it may change to the expression of $C\alpha$ as a result of somatic DNA rearrangement. A similar DNA switch may occur in the variable region of light-chain genes. Switching occurs during the mating-type determination of yeast and phase variation in prokaryotes. *See* antibody gene switching; *Borrelia*; epigenesis; Gene-Switch; immunoglobulins; mating type determination in *Saccharomyces*; phase variation; site-specific recombination; *Trypanosomas*. (Ptashne, P. M., 1992. *Genetic Switch*. Blackwell, Cambridge, MA.)

switching, phenotypic May occur without any change in the genetic material and involves only altered regulation of transcription, resulting in different phenotypes.

swivelase (topoisomerase type I) After a single nick in a supercoiled DNA, it permits the cut strand to make a turn around the intact one to relieve tension. *See* DNA replication; prokaryotes; topoisomerase. (Zhu, Q., et al. 2001. *Proc. Natl. Acad. Sci. USA* 98:9766.)

Swyer syndrome (gonadal dysgenesis XY type, Xp22.1-p21.2) Apparently a mutation or loss in the SRY gene is responsible for the anomaly. The afflicted individuals appear to be normal females until puberty but display deficient ("streak") gonads and fail to menstruate. *See* Denys-Drash syndrome; gonadal dysgenesis; sex determination; SRY; testicular feminization.

Sxl (*sex lethal*) Chromosomal location 1-19.2; controls sexual dimorphism in *Drosophila*; it is required for female development. *See* sex determination.

SXR Steroid xenobiotic receptor in humans. Its mouse homologue is Pxr. This protein is an inducer of the cytochrome P450 detoxifying enzyme CYP3A4 (7p22.1). Mice transgenic for SXR displayed enhanced protection against environmental and drug carcinogens. This system has an important role in the oxidative activation of procarcinogens such as aflatoxin B and many different drugs. Apparently SXR/Pxr is activated by the herb St. John's wort, which may lower the effectiveness of the antiseizure phenobarbital, the breast cancer drug tamoxifen, the contraceptive ethinyl estradiol, etc. *See* cytochromes. (Synold, T. W., et al. 2001. *Nature Med.* 7:536.)

SYBASE™ Computer program that links various databases for macromolecules such as GeneBank, EMBL. *See* databases; data management system. (<http://mysupport.sybase.com>.)

sycamore (*Platanus* spp) Large, attractive monoecious tree, $2n = 42$.

SYK (M_r 72 K) Signal protein for interleukin-2, granulocyte colony-stimulating factor, and other agonists. It is a protein tyrosine kinase of the SRC family and is indispensable for B-lymphocyte development. Syk is downregulated by the Cbl protein. Syk seems to modulate epithelial cell growth and may suppress human breast carcinomas. *See* agonist; B cell; B-lymphocyte receptor; BTK; Cbl; ZAP-70. (Sada, K., et al. 2001. *J. Biochem.* 130:177; Sedlick, C., et al. 2003. *J. Immunol.* 170:846.)

syllogism Form of deductive reasoning using a *major premise*, e.g., mice show graft rejection; the rejection of transplants is based on the presence of the MHC system (*minor premise*); therefore, mice must have a major histocompatibility system (*conclusion*). *See* logic.

symbionts Mutually interdependent cohabiting organisms, such as the *Rhizobium* bacteria within the root nodules of leguminous plants or algae within green hydra animals. Symbiosis may lead to loss or inactivation of genes that are no longer necessary in a cohabiting situation. On the other hand, new functions may be acquired or preexisting gene functions may be enhanced that are mutually beneficial. Symbiosis may be interpreted either as mere cohabitation or a mutually meaningful association (advantageous or parasitic). (Bermudes, D. & Margulis, L. 1987. *Symbiosis* 4:185; Currie, C. R. 2001. *Annu. Rev. Microbiol.* 55:357.)

symbionts, hereditary Occur in a wide range of eukaryotic organisms and their maternal transmission simulates extranuclear inheritance. They may be more widespread than recognized. The temperate viruses of prokaryotes and the retroviruses may also be classified along these groups. In some strains of the unicellular protozoa, *Paramecium aurelia*, carrying the *K* gene, bacteria (e.g., *Caedobacter taenospiralis*) live in the cytoplasm and are transmitted to the progeny. The first such infectious particles were named kappa (κ)-particles before their bacterial nature was recognized and they were supposed to be normal extranuclear hereditary elements. Many of these kappa particles contain R (refractive) bodies that are bacteriophages. The κ-particles with R bodies appear "bright" under the phase-contrast microscope. The nonbright cells may give rise to bright, indicating that the phages are in the free and infectious stage and the brights are in the integrated, proviral stage. The virus directs the synthesis of toxic protein ribbons that are responsible for killing kappa-free paramecia ($\kappa\kappa$). The strains carrying the dominant K gene are immune to the toxin. Other strains have been discovered carrying different infectious particles—lambda, sigma, mu—that also make toxins. The mu particles do not liberate free toxin and kill only the cells with which they mate (mate killer). Other symbionts delta—nu, and alpha—are not killer symbionts.

In *Drosophila* strains, Rhabdovirus σ may be in the cytoplasm and may be responsible for CO_2 sensitivity. Normal flies can be anesthetized with the gas for shorter periods without any harm. Those that carry the virus may be paralyzed and killed by the same gas treatment. This virus is similar to the vesicular stomatitis virus of horses that causes fever and eruptions and inflammation in the mouth and to the rhabdovirus of fish. In the nonstabilized strains only the females transmit the virus to part of the progeny (depending on whether a particular egg contains the virus). Some of the nonstabilized may become stabilized. Stabilized strains transmit it through nearly 100% of the eggs and even some of the males transmit σ with some of the sperm, yet the offspring of the male will not become stabilized. The *ref* mutants in chromosomes 1, 2, and 3 are refractory to infection. In some strains there are mutants of the virus that are either temperature-sensitive or constitutively unable to cause CO_2 sensitivity, although they are transmissible. Different ribosomal picornaviruses can be harbored in *Drosophila* that may reduce the life and fertility of the infected females. Females with the sex ratio (SR) condition produce no viable sons and the transmission is only maternal. In their hemolymph (internal nutrient fluids) the females carry spiroplasmas, bacteria without a cell wall. If the infection is limited to the XX sector of gynanders, they may survive, but not if the infection is in the XO sector. Triploid intersexes or females, phenotypically sterile males because of the genes *tra* (*transformer*, chromosome 3–45), *ix* (*intersex* gene located at 2–60.5), or *dsx* (*double-sex*, intersexes, chromosome 3–48.1), are not killed. Their special viruses may destroy the spiroplasmas. In plants (petunia, sugar beet), cytoplasmically inherited male sterility can be transmitted also by grafting. Some of the variegated tulips (broken tulips) are infected by viruses and have special ornamental value. During the 17th and 18th centuries, some rich Europeans, paid for the bulbs of the most attractive varieties equal to weights in gold to the mainly Turkish and Persian merchants. *See* broken tulips; cytoplasmic male sterility; *Drosophila*; extranuclear inheritance; lysogeny; meiotic drive; *Paramecia*, segregation distorter; sex determination; *Wolbachia*. (Preer, J. R. 1975. *Symp. Soc. Exp. Biol.* 29:125; Ehrman, L. & Daniels, S. 1975. *Aust. J. Biol. Sci.* 28:133.)

symbiosome   In legume root, nodules 2 to 5 μm structures enclosing (by peribacteroid membrane) 2 to 10 bacteroids. The fixed nitrogen is released through this membrane to the plant and reduced carbon is received by the bacteroids from the plant. *See* nitrogen fixation.

symbols   *See Drosophila*; gene symbols; pedigree analysis.

symmetric heteroduplex DNA   *See* Meselson-Radding model of recombination.

symmetry   *See* asymmetric cell division; axis of asymmetry.

sympathetic nervous system   Communicates with the central nervous system (CNS, brain) through the thoracic and lumbar parts of the spinal cord and controls blood vessels in various organs and involuntary movements (reflexes) of the body.

sympatric   Populations have overlapping habitats; it may be a beginning of speciation. *See* speciation.

sympatric speciation   Species that live in the same shared area for some reason become sexually isolated. *See* allopatric; parapatric.

sympetaly   Fused petals in a flower, see, e.g., snapdragon.

symplast   Multinucleate giant cell.

symplastic domain   Regulatory (supracellular) unit of several cells within the body.

symplesiomorphic   Two or more species sharing a primitive evolutionary trait. *See* apomorphic; plesiomorphic; synapomorphic.

symport   Cotransportation of different molecules through membranes in the same direction.

syn   (1) Prefix indicating union of tissues named after, e.g. syncytium (2) *See* ANTI.

synapomorphic   Species sharing an apomorphic trait. *See* apomorphic; plesiomorphic; symplesiomorphic.

synapomorphy   Shared derived characters that can be used to advance phylogenetic hypotheses. *See* symplesiomorphic. (Venkatesh, B., et al. 2001. *Proc. Natl. Acad. Sci. USA* 98:11382.)

synaps (synapse)   Site of connection between neural termini at which either a chemical or an electric signal is transmitted from one neuron to another (or to another type of cell). The neurotransmitter may diffuse across the synapse or the electric signal may be relayed from one cytoplasm to the other through a gap junction. The neurons must interpret the postsynaptic potentials (PSP) and integrate the excitatory (EPSP) and inhibitory (IPSP) paired pulse potentials. The leukocyte common antigen–related (LAR) protein, liprin (encoded by gene *syd-2* in *Caenorhabditis*), regulates the differentiation of the presynaptic vessels. This protein has tyrosine phosphatase activity. *See* gap junction; heregulin; ionotropic receptor; memory; metabotropic receptor; neuregulin; neurotransmitters; NMDA; tyrosine phosphatase. (Ziv, N. E. 2001. *Neuroscientist* 7[5]:365; *Science* [298:770–791 [2002].)

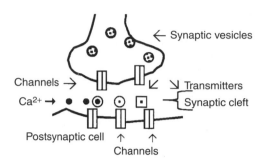

synapse, immunological Adhesion (gap) between the T-lymphocyte receptor and the antigen-presenting cell carrying the MHC-antigen complex or the killer cell inhibitory molecules and the special T-cell receptor. *See* killer cell; MHC; T cell. (Khan, A. A., et al. 2001. *Science* 292:1681; Bromley, S. F., et al. 2001. *Annu. Rev. Immunol.* 19:375.)

synapse, informational Specialized cell-cell junction mediating chemical or electric communication and displaying a supramolecular structure to mediate information transfer between cells. (Dustin, M. L., et al. 2001. *Annu. Rev. Cell Dev. Biol.* 17:133.)

synapsins Bind to actin filaments, microtubules, annexing, SH3 domains, calmodulins, and important regulators of synaptic vesicles. *See* actin; annexin; calmodulin; SH3; synapse; synaptic vesicles.

Synapsed homologous chromosomes (two pairs).

synapsis Intimate chromosome pairing during meiosis between homologous chromosomes that may lead to crossing over and recombination. In some instances, nonhomologous chromosomes or chromosomal regions may also associate. Protein-mediated synapsis involves recombination facilitated by integrases. *See* crossing over; illegitimate pairing; integrase; meiosis; pairing; *res*; synaptonemal complex; topological filter; tracking. (McClintock, B. 1930. *Proc. Natl. Acad. Sci. USA* 16:791; Romanienko, P. J. & Camerini-Otero, R. D. 2000. *Mol. Cell* 6:975; Baudat, F., et al. 2000. *Mol. Cell* 6:989.)

synapsis, bimolecular Occurs between complementary ends of different/separate transposable elements. *See* transposable elements.

synaptic adjustment The degree of synapsis may change during meiosis in certain chromosomal regions. *See* synapsis.

synaptic cleft Electric neuronal signals are transmitted from the presynaptic cell to the postsynaptic cell by the gap called the synaptic cleft. *See* neuron: synaptosome; diagram.

synaptic vesicles Originate from endosomes and store and release the neurotransmitters and other molecules required for signal transmission between nerve and target cells. Ca^{2+} regulates their function. *See* neurogenesis; neuromodulin; neurotransmitters; NSF; RAB; synaptophysin; synaptotagmins; syntaxin. (Cousin, M. A. & Robinson, P. J. 2001. *Trends Neurosci.* 24:659.)

synaptinemal complex *See* synaptonemal complex.

synaptobrevin (VAMP) *See* SNARE; syntaxin.

synaptogamins Integral membrane proteins that bind calcium and interact with other membrane proteins in the synaptic vessels of the nerves. Synaptogamins may interact with the different types of PtdIns and Ca^{2+}. *See* clathrin; Ipk1; neurexin; phosphoinositides; SNARE; synaptic vesicles. (Li, C., et al. 1995. *Nature* 375:594; Südhof, T. C. 2002. *J. Biol. Chem.* 277:7629.)

synaptogyrin Membrane protein that may be phosphorylated on tyrosine. (Zhao, H. & Nonet, M. L. 2001. *Mol. Biol. Cell* 12:2275.)

synaptojanin Neuron-specific phosphatase (M_r 145,000) working on phosphatidylinositol and inositols. Its putative role is in the recycling of synaptic vesicles. Its defect is responsible for Lowe's oculocerebrorenal syndrome. It binds to the SH3 domain of Grb2. *See* dynamin; Grb2; Lowe's oculocerebrorenal syndrome; synaptotagmin; syntaxin. (Khvotchev, M. & Südhof, T. C. 1998. *J. Biol. Chem.* 273:2306; Ha, S. A., et al. 2001. *Mol. Biol. Cell* 12:3175.)

synapton Synaptonemal complex.

synaptonemal complex Proteinacious element between paired chromosomes in meiosis. They consist of a central and two lateral elements. Denser spots within it are called recombination nodules and were supposed to have a role in genetic recombination. It is supposed that the complex holds in place the recombination intermediates rather than actively promoting the process. Actually, in some fungi (*Saccharomyces, Aspergillus*) no synaptonemal complex is observed and concomitantly chromosome interference is absent. These observations led to the assumption that the complex is responsible for interference rather than recombination. Actually, recombination may precede the formation of the synaptonemal complex. In *Drosophila*, the synaptonemal complex may be formed in the absence of meiotic recombination (crossing over or gene conversion); however, in the male flies, both the synaptonemal complex and meiotic recombination are absent, yet mitotic recombination may be observed. *See* association site; crossing over; interference; male

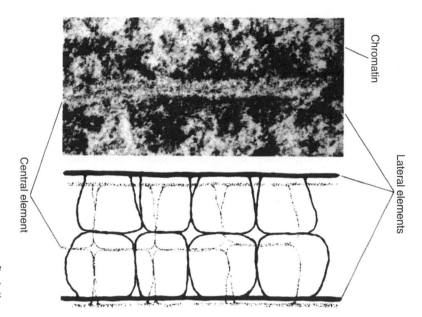

The tripartite structure may be visible from interphase through diplotene. (Electron micrograph courtesy of Dr. H. A. McQuade. Interpretative drawing after Comings, D. E. & Okada, T. A. 1970. *Nature* 227:451.)

recombination; meiosis; mitotic crossing over; recombination mechanism, eukaryotes; recombination nodule; single-end invasion; synapsis. (Westergaard, M. & von Wettstein, D. 1972. *Annu. Rev. Genet.* 6:71; Schmekel, K., et al. 1993. *Chromosoma* 102:669; Solari, A. J. 1998. *Methods Cell Biol.* 53:235; Heng, H. H., et al. 2001. *Genome* 44:293; Page, S. L. & Hawley, R. S. 2001. *Genes. Dev.* 15:3130; Lynn, A., et al. 2002. *Science* 296:2222.)

synaptophysin Membrane-spanning protein in the synaptic vessel involved in neurotransmitter release. *See* ceroid lipofuscinosis; neurotransmitter; synaptic vessel.

synaptosome Protein complex mediating interactions among neurotransmitters and receptors across the synaptic cleft. *See* neurotransmitter. (Husi, H. & Grant, S. G. 2001. *Trends Neurosci.* 24:259.)

synaptotagmin Proteins in the synaptic vesicles that have a role in Ca^{2+} -involved release of neurotransmitters, and in general in exo- and endocytosis. *See* synapse; synaptojanin; syntaxin. (Fernandez-Chacon, R., et al. 2001. *Nature* 410:41.)

synchronous divisions The cells are at the same stage of the cell cycle.

synchrotron Radiation emitted by high-energy, high-speed electrons accelerated in magnetic fields. The range varies from infrared to hard X-rays. Their high energy permits the study of their effects in monochromatic forms. *See* ionizing radiation.

synclinal *See* anticlinal.

syncytial blastoderm Early stage of embryogenesis in which the single layer of nucleated cytoplasmic aggregates does not yet have a cell membrane. *See* blastoderm.

syncytium Collection of nuclei surrounded by cytoplasm without the formation of separate membranes around each,

such as in early embryogenesis, among the progeny of a single spermatogonium, abnormal multinucleate cells, or the plasmodia of slime molds. *See* blastoderm; *Dictyostelium*; imaginal discs.

syndactyly Webbing or fusion between fingers and toes. In polysyndactyly, mutation in the polyalanine extension of the amino terminal of the human homeotic gene HOX13 is the responsible factor. Syndactyly 1 was located to 2q34-q36. *See* GLI3 oncogene; Greig's cephalopolysyndactyly; homeotic genes; limb defects; Pallister-Hall syndrome; Poland syndrome; polysyndactyly; Rubinstein-Taybi syndrome. (Bosse, K., et al. 2000. *Am. J. Hum. Genet.* 67:492.)

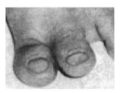

syndecans Membrane-spanning cell adhesion molecules and cofactor receptors bearing heparan sulfate proteoglycans distal from the plasma membrane. Syndecan promotes overeating and obesity in mice. Syndecan-1 opposes the effect of melanocyte-stimulating hormone. Syndecan-1, -2, -3, and -4 are encoded in human chromosomes 2p23-p24, 8q23, 1p32-p36, and 20q12-q13, respectively. *See* CAM; glypican; heparan sulfate; melanocyte-stimulating hormone; obesity; selectin. (Reizes O., et al. 2001. *Cell* 106:105.)

syndrome Collection of symptoms; traits caused by a particular genetic constitution. The individual symptoms of different syndromes may, however, overlap among a large number of genetic and nongenetic disorders. Therefore, the precise identification is often an extremely difficult task. More accurate identification will probably be possible when the genome projects can provide structural and topological evidence for all the loci concerned. The *unknown genesis syndromes* usually occur sporadically yet they may have genetic bases. *See* association; epistasis; genome projects; microarray; nonsyndromic; physical mapping.

synergids Two haploid cells in the embryo sac of plants flanking the egg. The pollen tube first penetrates one of the synergids, and after the rupture of that synergid, the vegetative and sperm nuclei fertilize the polar nuclei and egg nucleus, respectively. *See* gametogenesis; gametophyte; pollen tube. (Higashiyama, T., et al. 2001. *Science* 293:1480.)

synergistic action The participating elements enhance the reaction more than the separate strengths of their separate actions. *See* epistasis; interaction variance.

synexpression groups Genes that are expressed together spatially, temporally, or developmentally. Their identification is facilitated by microarray hybridization. These groups do not need to be genetically linked and therefore are different from gene clusters or operons of prokaryotes that are linked. *See* clustering of genes; macroarray analysis; microarray hybridization; operon. (Nirehrs, C. & Pollet, N. 1999. *Nature* 402:483.)

syngamy Union of two gametes in fertilization potentially leading to the fusion of the two nuclei in the cell. *See* karyogamy; plasmogamy; semigamy; synkaryon.

syngen Reproductively isolated group of ciliates.

syngeneic Antigenically similar type of cells (in a chimera). *See* antigen; immunoglobulins.

synkaryon Cell (zygote, fused conidia or spores) with a nucleus originated by the union of two nuclei. *See* dikaryon; heterokaryon; karyogamy.

synonymous codons Have different bases in the triplet, yet they specify the same amino acids. The 61 sense codons stand for 20 common amino acids. Some amino acids have up to 6 codons. Thus, mutation may not have any genetic consequence, except when the exon splice site is involved. *See* exon; genetic code; Grantham's rule; intron; nonsynonymous mutation; radical amino acid substitution; splicing.

synostosis Bone fusion. *See* noggin.

synovial sarcoma *See* SYT.

synpolydactyly *See* polydactyly; polysyndactyly; syndactyly.

syntaxin Synaptic membrane protein forming part of the nerve synaptic core complex along with the synaptosome-associated protein and synaptobrevin (VAMP), a vesicle-associated membrane protein. Syntaxins are involved in vesicular transport between the endoplasmic reticulum and the Golgi apparatus as target membrane receptors. Syntaxin-5 is an integral part of the endoplasmic reticulum–derived transport vesicles. A synaptobrevin-like gene (SYBL1) was located to the pseudoautosomal region of the human X chromosome. It recombines with the Y-chromosomal homologue and displays lyonization in the X chromosome; it is inactivated in the Y chromosome. A score of syntaxins have been identified in humans encoded at chromosomes 7q11.2, 17p12, 16p11.2, etc. *See* cystic fibrosis; endoplasmic reticulum; Golgi apparatus; lyonization; Munc; pseudoautosomal region; SNARE; stoma; synaps; synaptotagmin. (Bennett, M. K., et al. 1992. *Science* 257:255; Mullock, B. M., et al. 2000. *Mol. Biol. Cell* 11:3137.)

syntenic genes Within the same chromosome; they may, however, freely recombine if they are 50 or more map units apart. Gene blocks may display synteny among related taxonomic entities even when their chromosome number varies. Syntenic gene sets may provide information on phylogeny of the species. *See* crossing over; linkage.

syntenin Adaptor protein with PDZ domain. *See* IL-5; PDZ. (Zimmermann, P., et al. 2001. *Mol. Biol. Cell* 12:339.)

synthases Mediate condensation reactions of molecules without ATP. *See* synthetases.

synthetases Mediate condensation reactions that require nucleoside triphosphates as an energy source. *See* synthases.

synthetic DNA probes If the amino acid sequence in the protein is known but the gene has not been isolated, a family of synthetic probes may be generated to tag the desired gene. This probe is generally no longer than 20 bases because of the difficulties involved in synthesis. The genetic code dictionary reveals which triplets spell the amino acids. An amino acid sequence is selected that uses few synonymous codons. A computer match generally chooses the possible combinations, e.g., a probe for the His-Thr-Met peptide sequence would require eight polynucleotide sequences to consider all possible sequences for a probe (see below). The inclusion of methionine (having a single codon) simplifies the task. Histidine is relatively advantageous because it has only two synonymous codons. Threonine, with four codons, makes the work more difficult; leucine, serine, and arginine, containing parts of the proteins, should be avoided (because they have six codons), but tryptophan, also with a single codon, is highly desired. Insertion of ambiguous deoxyinosine nucleotides at some positions may facilitate the design of probes. *See* functional cloning; gene isolation; probe. (Ohtsuka, E., et al. 1985. *J. Biol. Chem.* 260:2605; Lichtenstein, A. V., et al. 2001. *Nucleic Acids Res.* 29[17]:E90.)

```
       His Thr Met
5′ CACACUAUG 3′
   CACACCAUG
   CACACAAUG
   CACACGAUG
   CAUACUAUG
   CAUACCAUG
   CAUACAAUG
   CAUACGAUG
```

synthetic enhancement   Basically an epistatic process by increasing or reducing interaction between gene products by using crossing, knockouts, transformation, etc.

synthetic genes   Produced in vitro by the methods of organic chemistry, by systematic ligation of synthetic oligonucleotides into functional units, including upstream and dowstream essential elements. The first successful synthesis involved the relatively short tyrosine suppressor tRNA gene. *See* suppressor tRNA *See* diagram below.

synthetic genetic array   Determines the functional relation between two genetic sites. A particular gene is crossed to a large number of different deletions. If the double mutants are inviable, the functional relationship of the two is revealed. On this basis, network maps can be constructed. (Tong, A. H. Y., et al., 2001. *Science* 294:2364.)

synthetic genetic networks   Constructs to simplify the understanding of regulated gene complexes. They are based on tools of nonlinear dynamics, statistical physics, and molecular biology. *See* genetic networks. (McMillen, D., et al. 2002. *Proc. Natl. Acad. Sci. USA* 99:679.)

synthetic lethal   Inviable only in certain genetic constitutions. Synthetic lethals may be involved in inbreeding depression and may be the cause of some hybrid inviabilities and sterility. A synthetic lethal test may be used in anticancer drug design by combining two different mutations to seriously impair cells in order to gain information on how to kill cancer cells. *See* inbreeding. (Jacobson, M. D., et al. 2001. *Genetics* 159:17.)

synthetic polynucleotides   Nucleic acid oligomers or polymers generated in the laboratory by enzymatic or other synthetic methods. Nucleic acid synthesizer machines produce some of them. (Benner, S. A., et al. 1998. *Pure Appl. Chem.* 70:263.)

synthetic seed   Somatic embryos encapsulated into a protective capsule (e.g., calcium alginate) and used for propagation in cases when regular seed is not available or homozygotes are difficult to obtain. *See* artificial seed.

synthetic species   Amphidiploids of presumed progenitors of existing species obtained by crossing and diploidization. Some synthetic species have never existed in nature before such as the *Raphanobrassica*, $2n = 36$, an amphidiploid of radish (*Raphanus sativus*, $n = 9$), and cabbage (*Brassica oleracia*, $n = 9$). *Triticales* are similarly new amphidiploids: $2n = 48$ or $2n = 56$, obtained by crossing tetraploid ($2n = 28$) or hexaploid ($2n = 42$) wheat (*Triticum*) with diploid rye (*Secale cereale*, $2n = 14$). Some synthetic species are only reconstructions of the evolutionary form, e.g., *Nicotiana tabacum*, *Hylandra suecica*, *Primula kewensis*, etc.

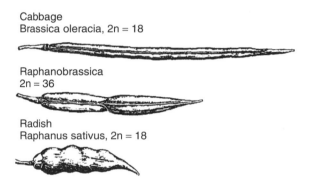

Cabbage
Brassica oleracia, 2n = 18

Raphanobrassica
2n = 36

Radish
Raphanus sativus, 2n = 18

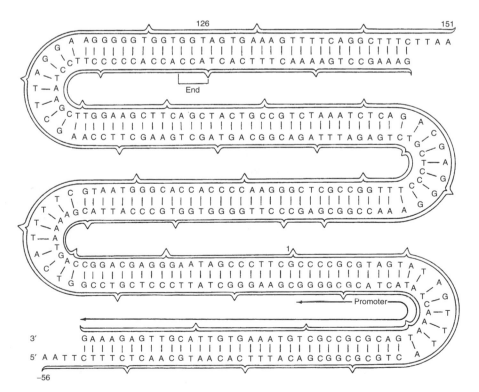

The complete (including promoter and terminator) synthetic structural gene of the tyrosine suppressor tRNA of *E. coli*. The projections along the sequence bracket the size of the fragments ligated together to form the complete gene. The gene when transformed into bacteria actually worked. (Redrawn after Khorana, H. G. 1974. *Proc. Int. Symp. Macromol.* p. 371, Mano, E. B., ed. Elsevier, Amsterdam; NL and Macaya, G., ed. 1976. *Recherche (Paris)* 7:1080.)

synthetic variety   Composed of several selected lines that may reproduce by outcrossing within the group. *See* polycross.

synthon   Synthetically produced molecule.

syntrophic   Can be maintained only by cross-feeding. On the basis of cross-feeding, metabolic pathways of microorganisms could be identified on culture media. In a metabolic pathway A′ B′ C′ D mutation blocked before D may cross-feed mutants blocked before C and B. Mutation blocked before C may facilitate the growth of mutants inactive in step B, and so on. A syntrophic hypothesis is that the mitochondria of eukaryotes evolved by the fusion of an achaeon, a δ-proteobacterium, and an α-proteobacterium. *See* channeling; cross-feeding.

synuclein (α-synuclein, 4q21)   140-amino-acid protein in the presynaptic neurons, a major constituent of the Lewy bodies. It interacts with synphilin-1 for normal function. In the oxidized or nitrated forms, synuclein aggregates may cause synucleinopathies such as Parkinson disease, Alzheimer disease, amyotrophic lateral sclerosis, and Huntington disease. Beta-synuclein gene in the brain in Alzheimer disease was assigned to 5q35; gamma synuclein in persyn (breast cancer) to 10q23.2-q23.3. *See* Lewy body; neurodegenerative diseases; Parkinson disease. (Touchman, J. W., et al. 2001. *Genome Res.* 11:78; Shimura, H., et al. 2001. *Science* 293:263.)

SYP   Tyrosine phosphatase.

syphilis   *see Treponema pallidum.*

syringa   *See* lilac.

syringe filter   Syringe equipped with a commercially available sterilizing filter block (0.45 or 0.20 μm pores); it removes microbial contaminations instantly without heating.

Syringe filter.

syringomelia   Rare autosomal-dominant or autosomal-recessive cavitations (formation of cavities) in the spinal cord. It may also be due to nonhereditary causes.

systematics   Method of classification such as taxonomy. Carl Linné (Linnaeus), Swedish botanist (1707–1778),initiated the classification and binomial nomenclature of organisms. (http://darwin.eeb.uconn.edu/molecular-ev>.)

systemic   Affects the entire cell or the entire body of an organism.

systemic acquired resistance   *See* SAR.

systemic amyloidosis, inherited   Extracellular deposition of fibrous proteins in the connective tissues under autosomal-dominant control. *See* amyloidosis.

systemic genes   Cell autonomous versus genes regulated by intercellular communication.

systemin   18-amino-acid signaling peptide for plant defense mechanisms. *See* host-pathogen relation; plant defense.

systemoid   Similar to a system; alternatively it is used to denote tumors that include different types of tissues. *See* teratoma.

systems biology   Study of the biological mechanisms by monitoring gene, protein, and informational pathways after systematically disturbing the biological systems using all possible, suitable means. The information is integrated into mathematical models. (Ideker, T., et al. 2001. *Annu. Rev. Genomics Hum. Genet.* 2:343; Shogren-Knaak, M. A., et al. 2001. *Annu. Rev. Cell Dev. Biol.* 17:405.)

systems of breeding   Sexual reproduction may be allogamous, autogamous, inbreeding, assortative mating, hermaphroditic, monoecious, or dioecious, but reproduction may also be asexual. *See* breeding system.

systole   Contraction of the heart that forces the blood into the arteries. *See* hypertension.

SYT (synovial sarcoma)   Oncogene in human chromosome 18q11.2. Translocations t(X;18) (p11;q11) are common. Gains of 8q and 12q as well as losses of 13q and 3p are frequent.

SZI   *See* micromanipulation of the oocyte.

Szostak model of recombination   Double break and repair model (Szostak, J. W., et al. 1983. *Cell* 33:25). It is applicable to transformational insertion of DNA molecules and can account for gene conversion and/or conventional recombination by outside marker exchange. The version of the model shown in the diagram does not explain why in

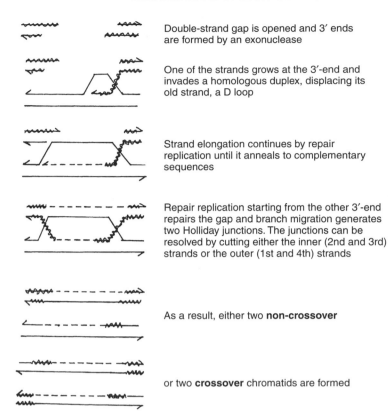

Double-strand gap is opened and 3′ ends are formed by an exonuclease

One of the strands grows at the 3′-end and invades a homologous duplex, displacing its old strand, a D loop

Strand elongation continues by repair replication until it anneals to complementary sequences

Repair replication starting from the other 3′-end repairs the gap and branch migration generates two Holliday junctions. The junctions can be resolved by cutting either the inner (2nd and 3rd) strands or the outer (1st and 4th) strands

As a result, either two **non-crossover**

or two **crossover** chromatids are formed

The Szostak, et al. model of recombination is based on double-strand breaks in contrast to the holliday or the meselson-radding models, which suggest single-strand breaks in the DNA.

yeast 5:3 gene conversion does not occur. (*Saccharomyces cerevisiae* has four spores per ascus, but *Schizosaccharomyces pombe* has eight unordered ascospores.) Other modified models of Szostak can account for these types of experimental data. Genetic recombination in eukaryotes generally occurs by double-strand break. Radiation and various chemicals can increase the frequency of double-strand breaks. Double-strand breaks can be increased in plant cells by transformation of a restriction endonuclease into the cell. *See* recombination, molecular models. (Szostak, J. W., et al. 1983. *Cell* 33:25; Smith, G. R. 2001. *Annu. Rev. Genet.* 34:243; diagram.)

E. B. Wilson 1896 The Cell in Development and Heredity. Macmillan, New York.

Now, chromatin is known to be closely similar to, if not identical with, a substance known as nuclein—which analysis shows to be a tolerably definite compound composed of nucleic acid (a complex organic acid rich in phosphorus) and albumin [protein]. And thus we reach the remarkable conclusion that inheritance may, perhaps, be effected by the physical transmission of a particular chemical compound from parent to offspring.

A historical vignette.

T

T *See* thymine.

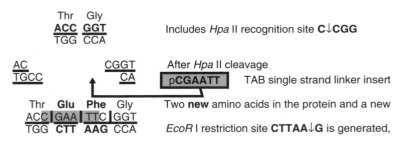

Thymine.

t Time.

T2 Virulent bacteriophage. *See* bacteriophages; T4.

T4 Virulent (lytic) bacteriophage of *E. coli*. It has double-stranded DNA genetic material of 1.08×10^6 Da (about 166 kbp) with a total length of about 55 μm. Its cytosine exists in hydroxymethylated and glycosylated form. Linear DNA is terminally redundant and cyclically permuted. The redundancy occupies more than 1% of the total DNA. It was an unexpected discovery that the thymidylate synthetase gene of T4 contains an intron. Introns are common in eukaryotes but exceptional in prokaryotes. The phage has over 80 genes involved in metabolism, but only about a fourth of them are indispensable. Metabolic genes control replication, transcription, and lysis. Other metabolic genes have functions overlapping those of the host. After infection, the phage turns off or modifies bacterial genes, degrades host macromolecules, dictates the transcription of its own genes, and utilizes the host machinery for its own benefit. For the synthesis of its own DNA, it relies on the nucleotides coming from the degradation of the host DNA. Bacterial RNA polymerase transcribes the phage genes. The viral protein Alc interacts with the β-subunit of the host and terminates transcription on templates with cytosine residues. Viral DNA contains hydroxymethyl cytosine that can serve as a template. The cytidylic acid residues of the host are prevented from incorporation into phage DNA by phosphatases, and a deaminase converts them to thymidylic acid or through a few steps to hydroxymethyl cytosine by a methylase enzyme. Hydroxymethylcytidylate is then glucosylated by a glucosyl transferase enzyme. The molar proportion of thymidylate is higher in T4 than in *E. coli*, presumably because of the conversion of cytidylate into thymidylate. A series of more than 50 genes are involved in morphogenesis. At least 40% of the genome is required for the synthesis and assembly of the viral particle. Head assembly is mediated by 24 genes; the base plate and tail require at least 31 genes. *See* bacteriophages; development; one-step growth; sliding clamp. (Calendar, R., ed. 1988. *The Bacteriophages*. Plenum, New York; Knipe, D. M. & Howley, P. M., eds. 2001. *Fundamental Virology*. Lippincott Williams & Wilkins, Philadelphia, PA.)

T7 Virulent bacteriophage with DNA genome size 39,937 bp, which encodes 59 proteins. *See* bacteriophages.

T18 Transcriptional activator protein.

2,4,5-T *See* agent orange.

Ta Copia-like element (5.2 kbp) in *Arabidopsis* flanked by 5 bp repeats without transposase function. *See* copia; retroposon; Tat1.

TAB1 Human TAK1 kinase-binding protein. Overproduction of TAB1 enhances the activity of the promoter of the inhibitor of the plasminogen activator gene, which regulates TGF-β, and increases the activity of TAK1 human kinase. *See* plasmin (plasminogen); TAK1; TGF-β.

Tabatznik syndrome Heart and hand disease II. *See* Holt-Oram syndrome. (Silengo, M. C., et al. 1990. *Clin. Genet.* 38:105.)

TAB mutagenesis (two-amino-acid Barany) DNA segment containing two sense codons introduced at a certain position into a gene in vitro, and probing the effect of the two-amino-acid modification regarding the function of the protein product. A brief outline of the essence of the procedure is shown below. *See* cassette mutagenesis; directed mutagenesis; localized site-specific mutagenesis. (Barany, F. 1985. *Proc. Natl. Acad. Sci. USA* 82:4202.)

TAC Transcriptionally active complex. *See* open-promoter complex; transcription complex; transcription factors; transcription factors, inducible.

TACC Centrosomal protein that regulates spindle formation in combination with other factors. *See* centrosome; spindle. (Gergely, F., et al. 2000. *Proc. Natl. Acad. Sci. USA* 97:14352.)

TAB mutagenesis.

TACE Tumor necrosis factor α-converting enzyme. Member of the ADAM metalloproteinase family of proteins involved in inflammatory responses. It removes ectodomains from cell-surface TNF-α receptors and ligands, selectins, etc., and cleaves amyloid precursor proteins. It proteolytically makes available soluble growth factors synthesized on the cell surface. *See* ADAM; β-amyloid; FAS; metalloproteinases; secretase; selectins; TNF. (Skovronsky, D. M., et al. 2001. *J. Neurobiol.* 49:40; Black, R. A. 2002. *Int. J. Biochem. & Cell Biol.* 34:1.)

tachykinin Peptide that mediates secretion, muscle contraction, and dilation of veins. *See* angiotensin. (Labrou, N. E., et al. 2001. *J. Biol. Chem. Oct. 12 Online.*)

tachytelic evolution *See* bradytelic evolution.

TACI Transmembrane activator and calcium modulator and cyclophilin ligand (CAML) is one of the TNF receptors that regulates the expression of transcription factors NF-AT, NF-κB, and AP-1. *See* TNF; TNFR. (Wang, H., et al. 2001. *Nature Immunol.* 2:632.)

TACTAAC box Highly conserved consensus in mRNA introns of *Saccharomyces* yeast.

tadpole Amphibian (frog/toad) larva at an early developmental stage. As it matures, it loses its gills, a more pronounced tail appears, then legs develop. Subsequently, the tail is lost from the free-swimming larva and the frog/toad emerges.

Tadpole.

TAE Tris-acetate-EDTA. *See* electrophoresis buffers.

TAF TATA box-associated factors TAF250, TAF150, TAF110, TAF60, TAF40, TAF30α, and TAF30β. In yeast, there are promoters that require TATA-box-binding protein (TBP) and TAFs (i.e., they are TAF-dependent). TAFs may require TIC for efficient function. Alternatively, other promoters are independent of TAFs and require only TBP. *See* core promoter; TBP; TIC. (Wassarman, D. A. & Sauer, F. 2001. *J. Cell Sci.* 114:2895.)

TAF$_{II}$ Transcription activating factors serve as coactivators of enhancer-binding proteins. TAF$_{II}$ appears to have sufficient homology to transcription factor TFIID, which in the absence of TAF$_{II}$ transcription can still proceed in yeast. TFIID is actually a complex of TBP and TAFs. TAF causes a conformational change in transcription factor TFIIB. The TAF$_{II}$250 subunit apparently modifies the H1 histone protein to facilitate the access of RNA polymerase to DNA in the chromatin. The acidic activator disrupts amino- and carboxy-terminal interactions within this molecule, resulting in exposure of the binding sites for the general transcription factors to enter into a preinitiation complex with TFIIB. TFIIB initiates the formation of an open-promoter complex. *See* coactivator; histones; open-promoter complex; regulation of gene expression; SAGA; TF; transactivator; transcription factors. (Frontini, M., et al. 2002. *J. Biol. Chem.* 277:5841.)

TAF$_{II}$ 230/250 Has histone H3, H4 acetyltransferase and protein kinase domains. The two bromodomains of the protein apparently accommodate acetyl lysines and support the activity of TFIID. The TFIID transcription complex regulates gene expression and several critical developmental processes. *See* bromodomain; histone acetyltransferase.

TAFE Transverse alternating-field electrophoresis. Used for pulsed-field gel electrophoresis when current is pulsed across the thickness of the gel. *See* pulsed-field gel electrophoresis.

taffazzin Fibroelastin group of proteins in the muscles. *See* Barth syndrome.

Tag Large T antigen, an early transcribed gene of SV40. It functions as an ATP-dependent helicase in the replication of DNA. The two Tag-binding sites each include two consensus sequences 5′-GAGCC-3′, separated by six or seven A = T base pairs. In order to start replication, Tag must bind to the replicational origin, *ori*, and to neighboring sequences of the virus. *See* simian virus 40.

tag Small t antigen of SV40, an early transcribed gene product.

Tag1 Autonomous transposable element (3.3 kb with 22 bp inverted repeats) of (Le-) *Arabidopsis*. It is a member of the Ac (maize), Tam3 (*Antirrhinum*), and hobo (*Drosophila*) family of elements. *See* Ac-Ds; *Arabidopsis*; hybrid dysgenesis; retrotransposons; Tam.

tagging Identifying a gene by the insertion of a transposon, an insertion element, a transformation vector, or by annealing with a DNA probe. These tags have known DNA sequences and can be detected on the basis of homology. When they are inserted within a structural gene or a promoter or other regulatory element of a gene, the expression of the gene may be modified or abolished. Therefore, their location may be detected by alteration in a specific function and may assist in the isolation and cloning of the target DNA sequence. *See* chromosome painting; FISH; insertional mutation; probe; transformation, genetic; transposon tagging. (Koncz, C., et al. 1990. *EMBO J.* 9:1337.)

tail bud Gives rise to the tail of the animal from epithelial cells of the mesenchymal tail bud. Its differentiation is usually completed after the basic organization pattern of the embryo has been realized.

tailing HOMO-A Eukaryotic mRNA generally contains a posttranscriptionally added polyA tail. PolyA or other homopolynucleotide sequences may be added to DNAs by terminal transferase. *See* mRNA; terminal transferase.

tailless *See* Brachyury, Manx.

tail-PCR Thermal asymmetric interlaced PCR is a polymerase chain reaction resembling inverse PCR. A nonspecific primer is paired with specific primers to obviate the need for circular genomic fragments. *See* inverse PCR; polymerase chain reaction. (Liu, Y. G. & Whittier, R. F. 1995. *Genomics* 25:674.)

tailpiece, secretory Immunoglobulins IgM and IgA possess an 18-amino-acid heavy-chain C-terminal extension, and 11 of these amino residues are identical. This tailpiece interacts with the J chain. *See* immunoglobulins; membrane segment. (Olafsen, T., et al. 1998. *Immunotechnology* 4[2]:141.)

Tajima's method Statistically tests the neutral mutation theory on the basis of DNA polymorphism. (Tajima, F. 1989. *Genetics* 123:585.)

TAK1 Human homologue of the kinase MAPKKK; an activator of the TGF-β signal. *See* TAB; TGF-β; MAP. (Wang, C., et al. 2001. *Nature* 412:285.)

talin Cytoskeletal protein binding integrin, vinculin and phospholipids. *See* adhesion. (Xing, B., et al. 2001. *J. Biol. Chem.* 276:44373.)

talipes *See* clubfoot.

***T* alleles** *See Brachyury.*

Tα Nonviral retrotransposable element. *See* retroposon; retrotransposon.

TAM Transcription-associated mutation. Rapid rate of transcription may involve an increase of mutation, apparently because translesion, nucleotide excision, or recombination may not repair the DNA damage. *See* Cockayne syndrome; DNA repair; translesion.

TAM Transposable element *Antirrhinum majus*. Responsible for the high mutability of genes controlling the synthesis of flower pigments, known in this plant since pre-Mendelian times. There are several *TAM* elements. The termini of *TAM1* and *TAM2* are homologous and their insertion results in 3 bp target-site duplication. Their termini are almost identical to those of the *Spm/En* transposons of maize and somewhat homologous to the termini of the *Tgm1* transposon of soybean. The *TAM3* transposon is different from *TAM1* and *TAM2* and 7/11 bp of its terminal repeats are homologous to the *Ac* element of maize. Both *TAM3* and *Ac* may generate 8 bp target-site duplications, although *TAM3* may also be flanked by 5 bp repeats. *TAM* elements, similarly to maize transposons, seem to move by excision and relocation. Excision is usually imprecise. Insertion within genes results in mutation, and the resulting mutant phenotype depends on the site of insertion. Excision results in more or less faithful restoration of the nonmutant phenotype depending on the extent of alteration left behind at the insertion site. *See* transposable elements, plants; transposons. (Coen, E. S., et al. 1989. p. 413 in *Mobile DNA*, Berg, D. E. & Howe, M. M., eds., *Am. Soc. Microbiol.*, Washington, DC.) (Photograph courtesy of B. J. Harrison and Rosemary Carpenter.)

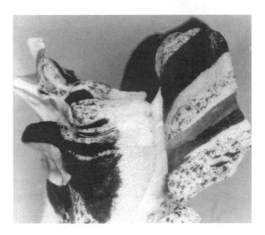

tamarins New world monkeys. *See* Callithricidae.

tamere Trans-allelic meiotic recombination is mediated by a targeted mechanism involving *loxP* sites in between two markers situated in trans, i.e., in homologus pairs of chromosomes. Recombinations may result in deletions and duplications. *See Cre/LoxP*. (Herault, Y., et al. 1998. *Nature Genet.* 20:381.)

Tamoxifen.

tamoxifen (2-[4-{1,2-diphenyl-1-butenyl}phenoxy]-N,N dimethyl-ethanamine) Selective estrogen modulator drug used for treatment of breast cancer. It may cause endometrial cancer, although the benefits may outweigh the risk. Tamoxifen binds to AP1 and to other estrogen-response elements in the uterus. *See* antiestrogens; AP1; breast cancer; estradiol; estrogen receptor; estrogen-response element; raloxifene. (Brewster, A. & Helzlsouer, K. 2001. *Curr. Opin. Oncol.* 13:420.)

TAN Translocation-7-9-associated *Notch* homologue. Located in human chromosome 9q34. It is involved in T-cell acute leukemia. *See* leukemia; morphogenesis in *Drosophila* (*Notch*); *Notch*; T cell. (Suzuki, T., et al. 2000. *Int. J. Oncol.* 17:1131.)

tandem duplications *See* tandem repeats.

tandem fusion Elements associated head-to-tail, following each other in the same direction.

tandem mass spectrometry Procedure selects one kind of peptide in a mixture by collision with argon or nitrogen gas. Then the fragments are processed in the tandem mass spectrometer, thus obtaining the MS/MS spectrum. *See* electrospray; MALDI; MS/MS; proteomics. (Kinter, M.

& Sherman, N. E. 2000. *Protein Sequencing and Identification Using Tandem Mass Spectrometry*. Wiley-Interscience, New York; Chace, D. H., et al. 2002. *Annu. Rev. Genomics Hum. Genet.* 3:17.)

tandem repeat Adjacent direct repeats (such as ATG ATG ATG) of any size and number. In *Caenorhabditis elegans*, 2.7% of the genome involves tandem repeats, and they occur on the average once per 3.6 kb. *See* repeat, inverted.

Tangier disease (HDL deficiency) Has a human chromosome 9q22-q31 recessive phenotype caused by a deficiency of the α-I component of apolipoproteins. The afflicted individuals have enlarged orange tonsils, liver, spleen, and lymph nodes and deficiency of the beneficial high-density lipoproteins. They accumulate cholesterol in their cells because of a defect in the ABC transporter (cholesterol-efflux regulatory protein [CERP]) that normally pumps out excessive amounts of cholesterol into the low-density lipoprotein fraction. Therefore, these patients are prone to develop coronary heart disease. *See* ABC transporters; apolipoprotein; cardiovascular disease; cholesterol; HDL; high-density lipoprotein; SNIPS. (Brooks-Wilson, A., et al. 1999. *Nature Genet.* 22:336; McManus, D. C., et al. 2001. *J. Biol. Chem.* 276:21292.)

tankyrase TRF1-interacting ankyrin-related ADP-ribose polymerase, (8q13). Telomere-associated protein binding a negative regulator of telomere length (TRF1) through the 24 ankyrin repeats of 33 amino acids involved in binding to TRF1, and at its C-terminal domain with homology to poly(ADP-ribose) polymerase. TRF2 is at 10q23.2. *See* ankyrin; PARP; telomeres; TRF. (Lyons, R. J., et al. 2001. *J. Biol. Chem.* 276:17172.)

tanning Ability to produce darker skin color depends largely on the activity of melanocyte-stimulating hormone (MSH) and its receptor (MC1R). Red-hair individuals — low in these activities — do not tan easily and are susceptible to UV damage (skin cancer). *See* albinism; hair color; melanin; melanoma; pigmentation in animals; UV.

T antigen Of SV40 virus, is a large multifunctional protein and an effector of DNA polymerase-α-function. It assists separating the DNA strands for replication and generates the replication bubble as the polymerase moves on. It has a special role in SV-induced tumors after tumor-suppressor proteins are eliminated or weakened. The T antigen and sequentially similar proteins are also nuclear localization factors. *See* nuclear localization sequences; simian virus 40; SV40.

t antigen Shares the same N-terminal sequences with the T antigen, but the carboxyl end is different. The t antigen has homology to the G_t protein, an α-subunit of the trimeric G-proteins involved in the activation of cGMP phosphodiesterase in photoreception and other processes. *See* G-proteins; simian virus 40.

TAP (1) Transporter-associated with antigen processing, encoded by loci TAP1 (6p21.3) and TAP2 (6p21.3) within the HLA complex. Delivers major histocompatibility class I molecule-bound peptides with the cooperation of β_2 microglobulin to the endoplasmic reticulum. When these peptides exit from the endoplasmic reticulum, they are carried to the T-cell receptors (TCR). Mutation in TAP may prevent antigen presentation. One member of TAP proteins is the retroviral constitutive transport element (CTE), which may mediate RNA (hnRNP, tRNA, mRNA) export from the nucleus. Efficient export, however, requires that RNA transcript be processed within the nucleus. Apparently, unprocessed RNA is better qualified for recruiting transport factors. *See* DriP; endoplasmic reticulum; HLA; hnRNP; immune system; major histocompatibility antigen; microglobulin; proteasomes; REF; RNA export; TCR. (Karttunen, J. T., et al. 2001. *Proc. Natl. Acad. Sci. USA* 98:7431.) **(2)** Tandem affinity purification is a method for rapid purification of protein complexes under native conditions. A tag is affixed to either the N or C end of one of the target proteins to facilitate the process. (proteomics, Puig, O., et al. 2001. *Methods* 24:218.)

Tap Bacterial transducer protein responding to dipeptides.

TAPA-1 Membrane-associated protein that mediates early immune reaction by B lymphocytes in association with CD19 protein and complement receptor 2 (CR2). *See* B lymphocyte; CD19; CD81; complement; immunity. (Dijkstra, S., et al. 2000. *J. Comp. Neurol.* 428:266.)

tapasin Transmembrane glycoprotein, encoded by an HLA-linked gene. Tapasin is involved with the endoplasmic reticulum chaperone calreticulin in processing the MHC class I–restricted antigens. *See* antigen processing and presentation; chaperone; HLA; major histocompatibility complex; MHC; TAP. (Turnquist, H. R., et al. 2001. *J. Immunol.* 167:4443; Vertigaal, A. C. O., et al. 2003. *J. Biol. Chem.* 278:139.)

tapetum Nutritive tissue lining of the anther, sporangia, or other plant or animal organs. During meiosis, they may become bi- or even multinucleate cells. The presence of B chromosomes may increase these irregularities.

tapir *Tapirus terrestris*, $2n = 80$.

TAQ DNA polymerase Single polypeptide chain, 94 kDa enzyme that extends DNA strands $5' \to 3'$; it also has $5' \to 3'$ exonuclease activity. The enzyme is obtained from the bacterium *Thermus aquaticus*. The commercially available, genetically engineered enzyme Ampli-Taq™ has temperature optima of 75 to 80°C. The enzyme is used for DNA sequencing by the Sanger method, for cloning and for PCR procedures of DNA amplification. For the latter applications, it is particularly useful because during the heat denaturation cycles it is not inactivated and it is not necessary to add new enzyme after each cycle. Phosphate buffers and EDTA are inhibitory to polymerization. *See* DNA sequencing; polymerase chain reaction. (Kainz, P. 2000. *Biochim Biophys. Acta* 1494:23.)

TaqMan (RT-PCR, real time PCR) Fluorogenic 5'-nuclease assay using a FRET probe generally consisting of a green fluorescent dye at the 5'-end and an orange quencher dye at the 3'-end of a DNA. In a PCR process when the probe anneals to the complementary strand, the Taq polymerase cleaves the probe and the dye molecules are separated. Thus, the quencher can no longer suppress the reporter (e.g., the green dye) and a fluorescence detector can quantitate the green emission; the green fluorescence directly correlates with the yield of the PCR product. In 7 minutes, sufficient quantities of DNA can be produced for the identification of pathogenic microbes. This technique may also be used for the molecular definition of deletions and SNIPs. *See* FRET; PCR; quenching; SNIPs; Taq DNA polymerase. (Medhurst, A. D., et al. 2001. *Brain Res. Mol. Brain Res.* 90[2]:125; Ranade, K., et al. 2001. *Genome Res.* 11:1262.)

TAR (1) Trans-activation responsive element. *See* acquired immunodeficiency; DNA-binding proteins; hormone-response elements; regulation of gene activity; transcription factors. (2) *See* transformation-activated recombination.

Tar Bacterial chemotaxis transducer protein with aspartate and maltose being attractants and cobalt and nickel repellents.

target Anything that is the place for an action, e.g., target cell, target organ, target for DNA insertion. The target of an X-ray machine is the surface hit by the electrons, then the electromagnetic radiation is emitted in the cathode tube. *See* insertion element; probe; transposons; X-rays.

targeted gene transfer Used for knockouts. *See* knockout; targeting genes.

targeted mutation recovery Mutations are induced by chemical or physical mutagens. Genetic alterations are then determined by polymerase chain reaction combined with denaturing high-performance liquid chromatography to distinguish between homo- and heteroduplex DNA sequences. If homozygotes are mutagenized, the presence of heteroduplexes indicates mutation at the selected locus. *See* HPLC; PCR. (Bentley, A., et al. 2000. *Genetics* 156:1169.)

targeted recombination *See* Cre / loxP; FLP / FRT; targeting genes.

target immunity (transposition immunity) Transposable elements usually do not insert within themselves or in the close vicinity of an existing element. *See* insertional mutation.

targeting Aiming at or transporting to a site of some molecules. The homing of free cancer cells recognizes endothelial surfaces by their peptide markers and permits organ selectivity. *See* lymphocytes; metastasis; mRNA targeting; site-specific mutagenesis; transit peptide; transit signal.

targeting, physiological Locates regulatory factors to the appropriate intracellular site. Protein domains may control these functions and may be hampered by chromosomal translocation and gene fusion.

targeting frequency Number of insertions formed at homologous or quasi-homologous sites in a genome by a transforming vector.

targeting genes Can be accomplished either by insertional mutagenesis or gene replacement. *Inducible gene targeting* can be carried out by introducing into an embryonic stem cell by homologous recombination the gene *loxP* to a flanking position of the desired target gene; *lox* facilitates the recognition of the sites for the *Cre* recombinase of phage P1. Then the mouse is crossed with a transgenic line expressing the *Cre* recombinase under the control of an interferon-responsive cell type–specific promoter. The tissue-specific recombinase (fused to a tissue-specific promoter) thus can remove the targeted gene from a particular type cell (floxing). The same procedure is applicable to other eukaryotic organisms using either the phage *Cre / loxP* or the yeast *FLP / FRT* system. Since the introduction of this site-specific alteration procedure, thousands of genes have been targeted, and in mice alone several thousands of targeted stocks have been generated. Gene targeting has become one of the most powerful tools in genetic analysis of eukaryotes. The general principles of the procedures are illustrated below.

By gene targeting through double crossover within the flanking chromosomal region, different copies of the gene can be inserted (replaced) or the gene can be placed under the control of a specific endogenous or foreign promoter.

Targeting mammalian genes is feasible, but the efficiency is fairly low (10^{-2} to 10^{-5}) compared to embryonic chicken stem cells (ES). In transfection of avian leukosis virus (ALV)-induced chicken pre-B cells, the efficiency of recombination between the exogenous DNA and target locus may be as high as 10 to 100%. When a single mammalian chromosome is

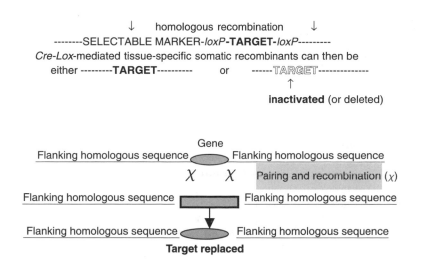

↓   homologous recombination   ↓
--------SELECTABLE MARKER-*loxP*-**TARGET**-*loxP*---------
Cre-Lox-mediated tissue-specific somatic recombinants can then be
either ---------**TARGET**---------   or   ------TARGET-------------
↑
inactivated (or deleted)

Gene
Flanking homologous sequence ⬭ Flanking homologous sequence
χ   χ   Pairing and recombination (χ)
Flanking homologous sequence ▭ Flanking homologous sequence
Flanking homologous sequence ⬭ Flanking homologous sequence
Target replaced

transferred to chicken cells by microcell fusion, in the somatic hybrid cell the recombination proficiency of the mammalian chromosome at the selected locus may increase up to 10–15%. The recombined chromosome can then be shuttled back to mammalian cells for analysis.

Another targeting procedure takes advantage of the bacterial tetracycline repressor gene that attaches to the promoter of some genes and keeps them silent unless tetracycline is applied, which binds to the repressor, and by inactivating it the genes are turned on. When this prokaryotic tetracycline repressor gene is inserted into a murine activator gene by transformation, the activator is incapacitated and the gene is silenced. Alternatively, inserting the tetracycline suppressor into a viral activator gene, all the genes of the transgenic mice that recognize the tetracycline suppressor-activator construct are turned on in the absence of tetracycline. Adding tetracycline to such a system, the antibiotic combines with the repressor-activator in the hybrid construct and the genes are now shut off because the suppressor-activator construct is removed from the activation position. Such a targeting construct can thus be used for on/off switching of particular genes. The success of targeting may be increased if the vector is an RNA-DNA hybrid molecule that pairs more efficiently with the target. By the PCR targeting procedure, a 20 bp DNA sequence tag may be generated using the photolithography procedure and DNA chips. The tag sequences are as different as possible yet possess hybridization properties to be identified simultaneously on high-density oligonucleotide arrays. Genomic DNA is isolated from a pool of deletions tagged and used as templates for amplification. For selectability, a resistance gene (aminoglycoside phosphotransferase) may be used. The targeting sequence is amplified by PCR with primers at the 3′-end homologous to the marker and at the 5′-end homologous to the target. This system is introduced into the cells by transformation. After homologous recombination at two flanks of the targeted open-reading frame, the target is replaced by a construct including the 20-base tag, the selectable marker, and the deletion mutation sequence.

The large number of tagged deletion strains can then be pooled and tested under a variety of conditions to test how the deletion affected the function of the gene. The molecular tags are amplified and hybridized to a high-density array of known oligonucleotides complementary to the tags. The relative intensity of hybridization reveals the relative proportion of the individual deletion strains in the pool and their fitness. Phenotypic methodological and other relevant information is contained in the TBASE (<http://www. jax.org/tbase>). Gene targeting should be extended to any animals in which embryonic stem cells can be successfully managed. Initially, this has not been realized with the majority of mammals with the exception of mice and sheep. An alternative approach by nuclear transplantation appears more practical (Kubota, C., et al. 2000. *Proc. Natl. Acad. Sci. USA* 97:990). *See* adeno-associated virus; chromosomal rearrangement; chromosome uptake; conditional targeting; *Cre/loxP*; DNA chips; *Flp/FRT*; gene replacement; gene therapy; GMO; homing endonucleases; homologous recombination; insertional mutagenesis; IRES; knockout; local mutagenesis; nuclear transplantation; photolithography; RID; RNA, double-stranded; site-specific recombination; TFO. (Thomas, K. R., et al. 1986. *Cell* 44:419; Sauer, B. 1998. *Methods* 14:381; Vasquez, K. M., et al. 2001. *Proc. Natl. Acad. Sci. USA* 98:8403; Rong, Y. S., et al. 2002. *Genes Dev.* 16:1568.)

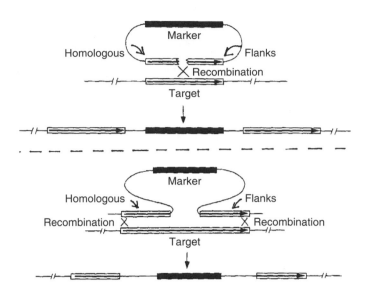

In *Drosophila*, gene targeting can be achieved by "ends-in" (upper part of the diagram) and "ends-out" (lower part of the diagram) procedures. The lower diagram indicates that parts of the 3′ and 5′ sequences of the target are deleted by the recombination. The requisites are (i) expressing a site-specific recombinase transgene, (ii) a transgene expressing a site-specific endonuclease, and (iii) a transgenic donor construct carrying recognition sites for both enzymes and the DNA of the locus targeted. This gene targeting mutates genes, which are not known by function but only by sequence. (Modified after Rong, Y. S. & Golic, K. G. 2000. *Science* 288:2013.)

targeting proteins *See* gel filtration; gel retardation; immunolabeling; immunoprobe; microcalorimetry; myristic acid; one-hybrid binding assay; prenylation; surface plasmon resonance; three-hybrid system; two-hybrid system. (Fischer, W., et al. 2001. *Infect. Immun.* 69:6769.)

targeting signal *See* signal peptide; signal sequence; signal sequence recognition particle.

targeting vector In a viral vector a section of the envelope protein gene is replaced by the coding sequences of, e.g., 150 amino acids of erythropoietin (EPO), thus improving its ability to recognize the EPO receptor. Other approaches involve pseudotyping or attaching special ligands to the envelope. Liposomal vehicles may be conjugated with special antibodies for target recognition. Bifunctional antibodies that recognize both viral epitopes and target cell antigens have been constructed. Another approach covalently linked biotin to recombinant adenovirus. Then the Kit receptor and the stem cell factor (SCF) were linked through an avidin bridge to the target to assure the proper tropism. *See* avidin; biotin; epitope; gene therapy; KIT oncogene; liposome; magic bullet; pseudotyping; stem cell factor. (Peng, K. W., et al. 2001. *Gene Ther.* 8:1456; Yu, D., et al. 2001. *Cancer Gene Ther.* 8:628.)

target theory Interprets the effect of radiations by direct *hits* on sensitive cellular targets. Physicists recognized that the amount of energy delivered to living cells and causing biological (genetic) effects is extremely low, and a comparable dose of heat energy would have no effect at all. Therefore, there must be some special sensitive targets in the cells that respond highly to ionizing radiations. Studies with irradiated sperm and cytoplasm of *Drosophila* indicated that the targets are the chromosomes and the genes. Between the 1920s and 1940s, these experiments paved the way for physical inquiries into the nature of genetic material. At this early period, it was hoped that the different radiation sensitivities among genes would permit the estimation of the size of these genes. It turned out, however, that radiation sensitivity of the same genes varied according to the physiological stage of the tissues (higher in imbibed seeds than in dry, dormant ones), and it was higher in spermatozoa than in spermatogonia. Furthermore, temperature, genetic background, and irradiation of only the culture media of microorganisms affected radiation sensitivity, indicating that radiation sensitivity is a more complex phenomenon and it does not precisely reveal the molecular nature of the gene. The direct action of radiation is proportional to the molecular weight of the target molecule: $(7.28 \times 10^{11})/D_{37}$, where D_{37} is the dose required to reduce the number of undamaged molecules to 37% of the initial total at a radiation dose of Gy^{-1}. DNA $(\times 10^8 \, mol \, s^{-1})$ interacts with radiolytic products of water: OH• 3, H• : 8×10^7, e_{aq}^- (hydrated electrons): 1.4×10^8. *See* physical mutagens; radiation, effects; radiation, indirect effects; survival. (Dessauer, F. 1954. *Quantenbiologie.* Springer, Heidelberg, Germany; Timoféeff-Ressovsky, N. W., et al. 1935. *Nachr. Ges. Wiss. Göttingen, Math. Phys. Kl. Biol.* 1:189; Lea, D. D. A. & Catcheside, D. G. 1945. *J. Genet.* 47:41.)

TAR syndrome *See* Robert syndrome.

TART Telomere-specific retroposon of *Drosophila* with an about 5.1 kb 3′-noncoding tract, which is homologous to HeT-A; it also encodes a reverse transcriptase. *See* Het-A; LINE; retroposon; reverse transcription; telomere. (Haoudi, A. & Mason, J. M. 2000. *Genome* 43:949.)

Tarui disease *See* glycogen storage diseases (type VII).

TAS (1) Termination-associated sequences are the signals for ending transcription. (2) Telomere-associated sequences involved in silencing of genetic functions. *See* silencing; telomere; transcription.

Tassel-seed Mutations (*ts*) in maize result in kernels on the normally male inflorescence (tassel) as a result of effemination. *See* sex determination.

taste Controlled by a signal-transducing G protein, gustducin. Both bitter and sweet tasting is mediated by gustducin. Ionic stimuli of salts and acids interact directly with ion channels and depolarize taste receptors. Sugars, amino acids, and most bitter stuff bind to specific receptors outside the cell membrane, and these are connected to G proteins. It is assumed that gustducin is involved with a phosphodiesterase. Phospholipase C appears to have a role in the taste circuits. Gustducin receptors are present in the tongue, stomach, and intestines. There are only a few thousand taste buds with 3 to 5×10^4 taste receptors. Chemosensory signals (ionic, caloric, toxic) often cooperate with olfactory signals. A locus in human chromosome 5p15 was associated with sensing the bitter taste of 6-n-propyl-2-thiouracil. Some individuals, however, fail to identify its bitter taste. Seven-transmembrane-domain taste-receptor

sequences were attributed to clusters in human chromosomes 7q31-q32 and 12p13. The T1R-1 taste receptor gene family is apparently specific for sweetness, whereas T1R2 is for bitter taste. The latter may include 50–80 genes. Mammals can taste sour, salt, sweet, bitter, and umami (monosodium glutamate). Genes for different taste receptors have been isolated. *See* degenerin; ion channels; olfactogenetics, signal transduction. (Dulac, C. 2000. *Cell* 100:607; Nelson, G., et al. 2001. *Cell* 106:381; Lindemann, B. 2001. *Nature* 413:219; Margolskee, R. F. 2002. *J. Biol. Chem.* 277:1; Bufe, B., et al. 2002. *Nature Genet.* 32:397.)

TAT Twin-arginine translocase is a ~600 kDa protein complex that moves proteins through thylakoid and prokaryotic membranes. *See* thylakoid. (Robinson, C. & Bolhuis, A. 2001. *Nature Rev. Mol. Cell. Biol.* 2:350.)

Tat 14 kDa primary regulator of the HIV virus. *See* acquired immunodeficiency.

Tat1 Transposon-like element (431 bp) in *Arabidopsis* flanked by 13 bp inverted repeats and 5 bp target-site duplications without any open-readingframe and thus incapable of movement by its own power. There is also a Tat1 human sulfate transporter and TAT1 and TAT2 yeast amino acid permeases. *See Arabidopsis*; open-reading frame; transposons. (Peleman, J., et al. 1991. *Proc. Natl. Acad. Sci. USA* 88:3618; Toure, A., et al. 2001. *J. Biol. Chem.* 276:20309; Schmidt, A., et al. 1994. *Mol. Cell Biol.* 14:697.)

TATA box Thymine (T)- and adenine (A)-containing binding sites for transcription factors and the RNA polymerase complex. In yeast, the TATA box is 40–120 bp, in the majority of other eukaryotes 25–30 bp ahead of the transcription initiation site. Many housekeeping genes and the RAS oncogene do not have this sequence. *See* asparagine synthetase; DPE; Hogness box; open-promoter complex; Pribnow box; promoter; PWM; transcription complex; transcription factors.

TATA box binding protein *See* TBP.

TATA factor (TF) *See* TATA box; transcription factors.

TATA Inr Core promoters may or may not contain these pyrimidine-rich transcription initiator elements. *See* core promoter; open-promoter complex; TATA box; transcription factors.

TAT-GARAT TAATGARAT enhancer motif of herpes simplex virus. *See* cigar.

Tatsumi factor Blood-clotting factor required for the activation of Christmas factor by activated PTA. It is controlled by an autosomal locus. *See* antihemophilic factors; blood-clotting pathways; hemostasis; PTA deficiency disease.

τ Yeast retroelement. *See* Ty.

TAU (MAPT, 17q21.1) ~352-amino-acid microtubule-associated protein. A pseudogene exists at 6q21. It seems to form tangles by virtue of the ^{306}Val-Gln-Ile-Val-Tyr-Lys^{311} motif in the six tau monomers in several types of nerve degenerative diseases (e.g., Pick disease, Alzheimer disease, progressive supranuclear palsy, corticobasal degeneration). Base substitution and splice-site mutations may lead to Pick disease, parkinsonism, and Alzheimer disease. In the tangle of the paired helical filaments, tau is hyperphosphorylated, causing the defect in microtubule assembly and mitotic arrest. The adverse effect of hyperphosphorylation may be prevented by trimethylamine N-oxide (TMAO) because this natural compound lowers the concentration of tubulin needed for assembly.

Hyperphosphorylated tau may self-assemble, causing the fibrillary tangle observed in the degenerated brain of Alzheimer patients. Cyclic-AMP-dependent protein kinase (PKA), glycogen synthase kinase 3β (GSK-3β), or Cdk5 may carry out phosphorylation. Phosphorylation of a Ser or Thr amino acid preceding Pro creates a binding site for the prolyl-isomerase Pin1. When Pin1 binds to this site in tau, it may deplete it in the brain, leading to some of the problems in Alzheimer disease. Neurofibrillary degeneration is increased when both Aβ and tau are expressed at the same time. In a *Drosophila* model of tauopathy, neurodegenerative symptoms appeared without the fibrillary tangle. *See* Alzheimer disease; amyloids; CDK; corticobasal degeneration; FTDP-17; microtubule; p35; palsy; parkinsonism; Pick disease; prion; secretase. (von Bergen, M., et al. 2000. *Proc. Natl. Acad. Sci. USA* 97:5129; Wittmann, C. W., et al. 2001. *Science* 293:711; Lewis, J., et al. 2001. *Science* 293:1487; Shahani, N. & Brandt, R. 2002. *Cell Mol. Life Sci.* 59:1668.)

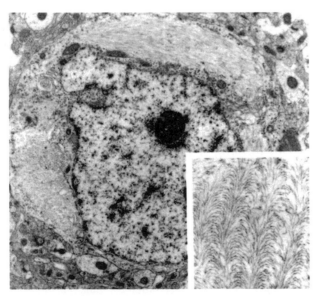

Transgenic mouse brain expressing neurofibrillary tangle (enlarged in inset) and Aβ plaques similar to those occurring in humans afflicted by Alzheimer disease. (Courtesy of Drs. Dennis W. Dickson and Wen-lang Lin, Mayo Clinic, Jacksonville, FL; I am indebted also to Mike Hampton, MD.)

tautomeric shift Reversible change in the position of a proton in a molecule, affecting its chemical properties; it may trigger base substitution and thus mutation in DNA. *See* enol form; hydrogen pairing; substitution mutation. (Watson, J. D. & Crick, F. H. C. 1953. *Cold Spring Harbor Symp. Quant. Biol.* 19:123.)

TAX1 Human T-cell leukemia virus (40 kDa) protein gene (1q32.1) with three 21 bp CRE-like sites, increasing DNA binding of transcription factors that contain a basic leucine-zipper domain. *See* CRE; CREB; HTLV; leucine zipper; leukemia. (Soda, Y., et al. 2000. *Leukemia* 14:1467.)

Taxol (paclitaxel) Spindle fiber blocking a natural substance is isolated from the yew *Taxus brevifolia*; it is a carcinostatic and radiosensitizing drug. It induces apoptosis. *See* apoptosis; carcinostasis; epothilone; microtubule; spindle. (Okano, J.-I. & Rustgti, A. K. 2001. *J. Biol. Chem.* 276:19555; Jennewin, S., et al. 2001. *Proc. Natl. Acad. Sci. USA* 98:13595.)

taxon (plural taxa) Collective name of taxonomic categories.

taxonomy Biological classification with a number of different systems. The bases of this classification are morphology, anatomy, genetics, biochemistry, physiology, cytology, and macromolecular structure (DNA, RNA, and proteins). Generally, five broad categories are recognized: prokaryotes and viruses, protists, fungi, plants, and animals. The taxonomic categories of eukaryotes include phylum, class, order, family, genus, species, and subspecies (such as varieties, cultivars, breeds). In the past, the classification was more rigid because the species was considered the mark of genetic isolation. Today, with somatic cell hybridization and transformation (transfection), there is no limit to genetic exchange between the various categories. There are over 300,000 plant and a million animal species named and classified by rules of nomenclature. For naming, binomial nomenclature is used. Capital first letter identifies the genus, then the species is designated in lowercase letters. This is sometimes followed by the name of the first taxonomists who classified the organisms, e.g., *Arabidopsis thaliana (L.) Heynh.*, indicating *Arabidopsis* as the genus, *thaliana* as the species, L. stands for Linnaeus, and Heynh. is the abbreviation for Heynhold, who suggested the current name. For taxonomic relations, see <http://www.ncbi.nlm.gov>; for sequence-based taxonomy:<http://www.ncbi.nlm.nih.gov/Taxonomy/taxonomy home.html>.

Tay-Sachs disease One of the most thoroughly studied biochemical diseases in human populations controlled by an autosomal-recessive gene. This defect occurs in all ethnic groups, but it is particularly common among Ashkenazi (eastern European) Jews, where the frequency of the gene is approximately 0.02 and the frequency of heterozygotes may be over 3%. The prevalence is about 1/2,500 to 1/5,000 births. Among the Sephardim Jews and other ethnic groups, the frequency is about 1/1 million. Since the afflicted individuals generally die by age 3 to 4, the high frequency indicates that some heterozygote advantage must have existed for this gene. The onset is at 6 months, when general weakness, extension of the arms in response to sounds with a scared look, muscular stiffness, and retardation appear. Then in rapid succession, paralysis, reduction of mental abilities, and vision problems leading to blindness become evident. One characteristic symptom is a cherry-red spot ($\downarrow$) on the macula (gray opaque part of the cornea [eye]) caused by cell lesions. All these symptoms are the result of a deficiency of β-hexosaminidase enzyme α-subunit that controls the conversion of ganglioside G_{M2} into G_{M3}. As a consequence, G_{M2} ganglioside accumulates, leading to a degeneration of myelin of the nervous system. Hexosaminidase A is composed of the α-subunits (human chromosome 15q23-q24) and hexosaminidase B is a multimer of the β-subunits (human chromosome 5q13). Sandhoff's disease involves both hexosaminidase A and B deficiencies or only hexosaminidase B deficiency, and it has somewhat similar symptoms with more rapid progression. Another (type 3) milder form of G_{M2} gangliosidosis (5q31.3-q33.1) with some hexosaminidase A activity may permit survival up to age ~15. In a mouse model, N-butyldeoxynojirimycin prevented the accumulation of G_{M2}. The older name of these gangliosidoses was amaurotic familial idiocy. *See* gangliosides; gangliosidoses; genetic screening; hexosaminidase; Jews and genetic disease; lysosomal storage disease; Sandhoff's disease; sphingolipids. (Mahuran, D. J. 1999. *Biochim. Biophys. Acta* 1455:105; Myerowitz, R., et al. 2002. *Hum. Mol. Genet.* 11:1343.)

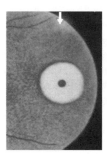

Cherry-red spot

Tay syndrome   *See* trichothiodystrophy.

T-BAM   CD40 ligand.

T-band   Telomeric regions of chromosomes with the highest concentrations of genes and $G + C$ in the genome. *See* band; chromosome banding; isochores.

T-bet   T-box expressed in T cells is a 530-amino-acid transcription factor that promotes the differentiation and activity of T_H1 cells and represses the formation of T_H2 lymphocytes. *See* T; T_H. (Mullen, A. C., et al. 2001. *Science* 292:1907.)

T box (Tbx)   Is a conserved ~14-nucleotide DNA domain upstream of the transcription terminator of Gram-positive bacteria in about 250 genes encoding aminoacyl-tRNA synthetases, amino acid biosynthetic and transport enzymes. Uncharged tRNAs appear to interact with the leader sequence of these genes by their amino acid-accepting termini at the middle of T-box (UGGN') to stabilize antitermination at the expense of the terminator. *See* antitermination; Brachyury; DeGeorge syndrome; Holt-Oram syndrome; MAR; T-bet. (Papaioannou, V. E. & Silver, L. M. 1998. *Bioessays* 20:9; Smith, J. 1999. *Trends Genet.* 15:154; Putzer, H., et al. 2002. *Nucleic Acids Res.* 30:3026.)

TBP (TFIIτ)   TATA-box-binding protein is a subunit of the general transcription factors, TFIID, SL1, and TFIIIB proteins, which bind to DNA like a saddle in two-fold symmetry. It has antiparallel β-sheets at the concave area where it forms a reaction with the special A- and T-rich sequence of DNA.

The convex surface provides opportunities for interaction with other proteins. TBP is highly conserved among different organisms from yeast to the plant *Arabidopsis* and to mammals. TBP may form a complex with several TAF proteins. It apparently nucleates the preinitiation complex of all three DNA-dependent RNA polymerases (pol I, II, and III). The TBF-like factors (TLFs) variously denoted as TRF2 or TRP are orthologs of TFL and appear functionally different from TBP. TLF appears to be required for differentiation and is needed for the transcription of some special set of genes. TBP may also bind the SL1 transcription factor of pol I, but this binding is exclusively either with TFIID or SL1. The pol III TBP complex is called TFIIIB. The Dr1 repressor binds to TBP and selectively inhibits pol II and pol III action, but pol I is not affected, because when pol II and III are repressed, the relative output of pol I appears higher. TFIID complex activity is restricted by the nucleosomal organization of chromatin. The C-terminus of TBP is higly conserved, whereas the N-region displays great variations. *See* pol; TAF; transcription factors. (Berk, A. J. 2000. *Cell* 103:5; Magill, C. P., et al. 2001. *J. Biol. Chem.* 276:46693; Zhao, X. & Herr, W. 2002. *Cell* 108:615; Martinov, I., et al. 2002. *Science* 298:1036.)

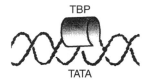

TBP-TATA box schematic representation. TBP OF *Arabidopsis* recognizes the minor groove of TATAAAAG.

TB-parse   Program for the identification of protein-coding sequences. (Krogh, A., et al. 1994. *Nucleic Acids Res.* 22:4768.)

TBR   Transforming growth factor receptors. *See* TGF.

TC4   Small nuclear G protein, with CD28 being its primary ligand. It is involved in the regulation of T lymphocytes. *See* G proteins; RAN. (Nieland, J. D. 1998. *Cancer Gene Ther.* 5:259.)

TCC   Terminal complement complex includes C5b-9, C5b-8, and C5b-7 complement components involved in complement-mediated killing of foreign cells. *See* complement; immune system. (2) Transitional cell carcinoma. (Dal Cin, P., et al. 1999. *Cancer Genet. Cytogenet.* 114[2]:117.)

TCCR   T-cell cytokine receptor mediates adaptive immune response of T_H1 lymphocytes. (Chen, Q., et al. 2000. *Nature* 407:916.)

T-cell receptor (TCR)   T-cell-surface glycoproteins recognize the antibody. The differentiation of the T cells begins with the differentiation of their receptors. At the beginning, the TCR is double negative, i.e., it is $CD4^-CD8^-$. The disulfide-linked heterodimers have α- and β- or γ- and δ-chains, containing variable and constant regions. They are homologous to the corresponding antibodies. TCRβ is rearranged, followed by the rearrangement of the α-chain. At this stage, allelic exclusion may be signaled by the β-chain, which means the end of rearrangements. The double positive stage follows, i.e., CD4$^+$CD8$^+$ TCR appears. Then a selection process results in an array of self-MHC-restricted and self-tolerant TCRs. The TCR-α chain is a transmembrane protein, and the cytoplasmic (carboxyl) end has two potential phosphorylation sites and an Src homology 3 (SH3) domain. The TCR-β chain regulates the development of the T cells in the absence of the α-chain. The $\alpha\beta$ TCR generally recognizes antigens bound to the major histocompatibility (MHC) molecules. The TCR complex also includes the CD3 protein required for signal transduction. It is made of the γ-, δ-, ε-, and ζ-chains. The ζ-chain plays a role in thymocyte development, but it is not indispensable for signal transduction. The chromosomal sites of the human α (14q11), β (6q35), γ (7p15-p14), δ (14q11.2), and ζ (1p22.1-q21.1) chains are shown in parentheses.

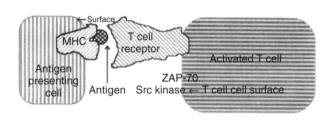

When the T-cell receptor binds the MHC-associated ligand on the antigen-presenting cell, activation is triggered. Costimulating signals may help. The CD3 component of the TCR may be altered, and protein tyrosine kinases (PTK) are activated. The PTKs may turn on the calcium-calcineurin, RAS-MAP, and protein kinase signaling pathways. These pathways may then activate the transcription factors NFAT, NF-κB, JUN, FOS (AP1), and ETS. They activate new genes; some of them are specific transcription factors that facilitate the release of cytokines, which in turn activate the clonal expansion of T cells. This is followed by the production of antibodies by B-cell or T-cell cytotoxicity. Immune memory, immune tolerance, anergy, and apoptosis are alternative functions to follow. Regulation of the development of TCR requires the cooperation of protein tyrosine kinases (Src family), phospholipase C (PLCγ), CD5, CD28, CD48, CD80, VCP, ezrin, VAV, SHC, PtdIns, PIP2, and PIP3. TCR may regulate apoptosis of T cells. For the TCRs, there are 42 variable (V) and 61 joining (J) segments at the α-chain immunoglobulin locus, and 47 V, 2 diversity (D), and 13 J segments for the β-chain genes. During rearrangements V$_\alpha$—J$_\alpha$ and V$_\beta$—D$_\beta$—J$_\beta$, deletions and additions of nucleotides, dimerization may increase the variations. In the blood there appears to be $\sim$10^6 different β-chains that may combine on the average with $\sim$25 different α-chains. In memory subsets, the diversity appears to be about one-third less. The estimated distinct TCR receptors may be 10^{12}. *See* α-CPM; antibody; Blk; B lymphocytes; caveolae; CD3; CTLA-4; Fgr; Fyn; $\gamma\delta$ T cells; Hck; ICAM; ICOS; immune system; immunoglobulins; immunological synapse; integrin; ITAM; Lck; LCK; LFA; lymphocytes; MHC; phosphoinositides; PIP; RAG; signal transduction; SLP-76; Src; T cells; TCR genes; Tec; thymus; Yes; Zap70. (Hennecke, J. & Wiley, D. C. 2001.

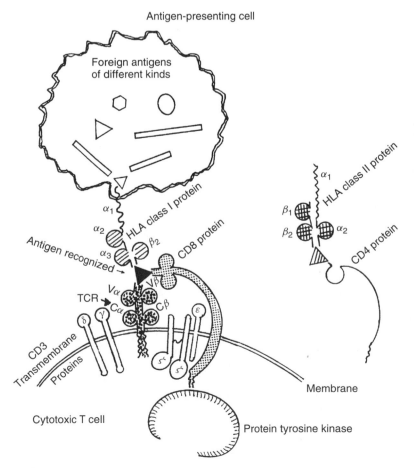

A general outline of the functions of the T-cell receptor (TCR) complex. Although diagrams always generalize beyond reality, they may be helpful to obtain a broad understanding. The foreign antigens are presented to the T lymphocytes either by the antigen-presenting cells or by macrophages. The macrophages are capable of partially degrading the large molecules or the invading cells. These cells associate with class II or class I MHC proteins, which are encoded by the HLA genes. The MHC molecules recognize the foreign antigens and bring them to the TCR and to the CD protein complex associated with the T-cell surface. Class I gene products use CD4 transmembrane proteins, whereas class II molecules rely on CD8. The COOH ends of the TCR chains are inside the T cell's double membrane. The NH$_2$ end is involved in the recognition of the antigen in association with the MHC and CD elements. The TCR complex also includes the CD3 transmembrane proteins. The ζ-subunit also serves as an effector in signal transduction. Protein tyrosine kinase is an important element in several signal transduction pathways. ICAM-1 on the antigen-presenting cell and LFA-1 on the CD8$^+$ cooperate in mediating the adhesion of the MHC complex. The right side of the incomplete diagram shows the association of class II and CD4 proteins with an antigen. The other elements of this system are very similar to those shown at the left main part of the outline.

Cell 104:1; Germain, R. N. 2001. *J. Biol. Chem.* 276:35223; Isakov, N. & Altman, A. 2002. *Annu. Rev. Immunol.* 20:761; Natarajan, K., et al. 2002. *Annu. Rev. Immunol.* 20:853.)

T cell, regulatory   May disarm non-tolerant naiv lymphocytes (*dominant tolerance*) and bring about *infectious tolerance* of tissue grafts. These T cells occur among CD4+ CD25+ and CD4+ CD25− cells. *See* immune tolerance; T cells. (Graca, L., et al. 2002. *J. Exp. Med.* 1951641.)

T-cell replacing factor   Lymphokine. *See* lymphokines.

T cells   Thymic lymphocytes control cell-mediated immune response (the foreign antigens are attached to them). T lymphocytes originate in the bone marrow, then differentiate in the thymus, and later migrate to the peripheral lymph nodes. While the early T cells are in the thymus, a *negative selection* eliminates those T cells that react with self-antigens. At about the same stage of T-cell development, a positive selection takes place under the influence of the MHC complex, securing the survival of those T cells that can interact with antigens associated with a type of MHC molecules presented to these cells. These encounters lead to the differentiation and activation of T cells. Cytotoxic T cells (CTL) are the major elements of the immune system; they cooperate with natural killer cells (NK) and degrade foreign antigens with immunoproteasomes. Heterodimeric PA28a and PA28b (proteasome-activating protein) activate immunoproteasomes.

Helper T_H (CD4+ T cells) and T-suppressor (T_S) cells mediate humoral (secreted) immune responses. On the surface of T cells are T-cell-surface receptors (TCR). T_H cells stimulate the proliferation of B (bursa) cells when they recognize their cognate antigens. The joint action of T_H-cell-surface receptors (TCR) and B-cell antigens brings about formation, growth, and differentiation of proteins in B cells. These lymphokines stimulate the propagation of B cells and the secretion of humoral antibody. T-cell-surface receptors recognize foreign antigens only if they become associated with major histocompatibility (MHC) molecules carried to the TCR by antigen-presenting cells (APC) or macrophages. TCR links to the various intracellular signaling pathways.

The activation of T cells requires phosphorylation by SRC tyrosine kinase of CD3 immunoglobulin chains. The activation involves costimulatory molecules CD28, ICAM-1, and LFA-1. For full activation, ZAP-70 protein-tyrosine kinase is also needed. The T cells are presented with a variety of antigens, including self-antigens carried by the APCs without discrimination. The T cells distinguish the self/foreign antigens. This ability begins to develop while the T cells are still in the thymus. The discrimination is a difficult task to achieve and complex phosphorylations are required for surveying the very large array of ligands of varying degrees of specificity (affinity). Alternatively, it is conceivable that first one peptide-MHC complex interacts with one TCR. This TCR then plays the role in a contact cap and detaches from the ligand, thus making it possible for the ligand to bind another TCR. The process is

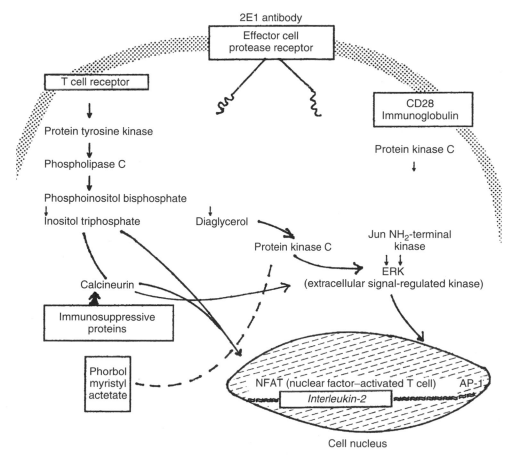

T-cell-signaling pathways. (Modified after Trucco, M. & Stassi, G. 1996. *Nature* 380:284.)

repeated in a serial manner and assembles a sufficient number of TCRs in the contact cap for productive signaling.

The activation of the T cell may be only partial in case there is a subtle change (e.g., one amino acid replacement) in the peptide-MHC. The subtle variants may inhibit the CD4[+] helper T cells to respond to the real antigen. Altered ligands or lack of costimulatory signals may cause anergy of the T cell and cannot be stimulated by the TCR but may proliferate under the influence of interleukin 2 (IL2). The ligands thus can be *agonists* that fully or partially activate the T cells, or altered ligands may be weakened agonists or even *antagonists* and reduce activation. The *null ligands* provoke no response. The weak agonists may not activate ZAP-70 and may have a different pattern of phosphorylation of the CD3 ζ-chain. For the fully active immune reaction, all the elements of the complex T-cell activation must be in place. Some viral infections (HIV-1, hepatitis B, etc.) may lead to the production of antagonist ligands, then the cytotoxic T cells (CTL) cannot protect the body against the invader. Some aspects of T-cell activation and regulation are outlined on page 1210.

The *T-cell receptor* associated with CD3 and CD4 immunoglobulins is on the cell membrane. This receptor TCR mediates phosphorylation, with the aid of a *protein tyrosine kinase* (PTK), *phospholipase* C-$\gamma 1$. The activated PLC-$\gamma 1$ cleaves phosphatidylinositol-4,5 bis-phosphate (PIP2) and generates *inositol trisphosphate* (IP3) and *diaglycerol* (DAG), which are second messengers. These second messengers activate *protein kinase C* and make cytoplasmic Ca^{2+} available for *calcineurin* (also called protein phosphatase IIB). As a consequence, transcription factor NFAT (*nuclear factor–activated T cell*) promotes the transcription of the *interleukin-2 gene*. Calcineurin also contributes to the activation of *ERK* (member of the mitogen-activated protein kinase family, MAPK). Costimulation is provided through the CD28 immunoglobulin system, which activates protein tyrosine kinase C, which—in collaboration with the RAS G protein—stimulates *Jun NH$_2$-terminal kinase* (JNK). It is also known that the *FOS* and *JUN* oncogenes contribute to the formation of the *AP* group of transcription factors probably by acting on protein kinase C (PKC). *Phorbol myristyl acetate* (PMA) is an adjuvant for the stimulation of the *IL-2* gene. These processes of the development of effector T cells may be downregulated when the *2E1 antibody* attaches to the *effector cell protease receptor-1*. Immunosuppressive agents such as FK506, cyclosporin, and OKT3 may cause further downregulation.

Another important player in the T-cell response is CTLA (cytotoxic T-lymphocyte antigen), a molecule with about 75% homology to CD28. While CD28 is a costimulator of T-cell activation, CTLA-4 is a negative regulator and protects against rampant lymphoproliferative disorders. The recognition of the role of CTLA offers an opportunity to neutralize its effect by specific antibodies and thereby accelerate the action of antitumor interleukin production without serious side effects. T-cell differentiation is accompanied by the appearance of surface markers in the order CD4, CD25, CD44, CD8, CD3. The commitment of T-cell differentiation requires tumor necrosis factor-α (TNF-α) and interleukin-1α (IL-1α). Understanding these circuits may lead to designing better drugs against infections and the suppression of tissue graft rejection and cancer. *See* AKAP79; antibody; AP; autoantigen; calcineurin; CD3; CD4; CD8; CD28; Cd117; CTLA-4; cyclosporin; EPR; ERK; Fas; FK506; $\gamma\delta$ T cells; GATA; HIV; HLA; ICAM; ICOS; immune system; immunoglobulins; immunophilins; interleukin; IP$_3$; killer cells; LCK; LFA; MAPK; MHC; NFAT; NF-κB; OKT; phorbol esters; phospholipase C; PIP; proteasomes; protein tyrosine kinase; RAD25; RANTES; RAS; signal transduction; SLAM TIL; SRC; TAP; TCR; T$_H$; thymus; vaccines; ZAP-70. (Dustin, M. L. & Chan, A. C. 2000. *Cell* 103:283; Staal, F. J. T., et al. 2001. *Stem Cells* 19:165; Barry, M. & Bleackley, R. C. 2002. *Nature Rev. Immunol.* 2:401; Sadelain, M., et al. 2003. *Nature Rev. Cancer* 3:35.)

TCF Ternary complex factors are coactivators of transcription such as Elk, Sap-1a, Sap-1b, ERP-1, and other members of the ETS family of transcription factors and oncoproteins. Some are members of the high-mobility group proteins. TCF usually binds to the AACAAAG sequences of the promoter. In case of a defect in the tumor suppressor APC gene (adenomatous polyposis) and/or in the β-catenin gene, Tcf-4 (responsible for crypt stem cells) is activated and malignancy may result. CBP seems to be a repressor of TCF. *See* catenins; CBP; ETS; Gardner syndrome; high-mobility group proteins; LEF; melanoma; mtTAF; *Wingless*. (Morin, S., et al. 2001. *Mol. Cell Biol.* 21:1036.)

TCGF *See* IL-2; interleukin-2.

TCID$_{50}$ Tissue culture infective dose causing (viral) infection to 50% of the cells.

Tcl *See* hybrid dysgenesis.

TCL1 Oncogene (14q32.1). It is normally expressed in fetal thymocytes, in pre-B and immature B cells, and weakly in CD19[+] peripheral blood lymphocytes. Chromosomal translocations or inversions near the enhancer element of TCR may cause leukemia or lymphoma. It enhances Akt kinase activity and promotes nuclear transport. *See* B lymphocyte; Akt; CD19; leukemia; lymphoma; TCR. (Pekarsky, Y., et al. 2001. *Oncogene* 20:5638.)

TCLo Toxic concentration low; the lowest concentration of a substance in air that produces toxic, neoplastic, and carcinogenic effects in mammals.

TC motif *See* AP1; AP2; transcription factors.

t-complex Of mice, consists of six complementation groups in chromosome 17, affecting tail development and viability. Homozygous mutants of the same complementation group are generally lethal. *See* brachyury.

T complex Products of the virD2 and virE2 agrobacterial virulence genes associated with the 5′-end of the transferred strand of agrobacteria. *See* agrobacterial virulence genes; T-DNA.

TCP-1 (CCT) Cytoplasmic α-subunit chaperonin coded for by the t-complex in mouse chromosome 17. A homologue of it occurs in the pea leaf cytosol. *See* brachyury; chaperone; chaperonins.

Tcp20 Synonymous with CCTζ and Cct6. *See* chaperonins. (Li, W. Z., et al. 1994. *J. Biol. Chem.* 269:18616.)

TCR genes T-cell receptors are glycoproteins and are similar to antibody molecules. The TCR α-chain has variable (V), diversity (D), junction (J), and constant (C) regions in the polypeptides generated with rearrangements of the gene clusters in human chromosome 14 in the proximity of the immunoglobulin heavy-chain genes (IgH). These recombinations take place in the switching regions, and the TCR genes have the same hepta- and nanomeric sequences as the Ig genes. TCRA (α) genes are in mouse chromosome 14, whereas the Ig heavy chains of the mouse are in chromosome 12. The TCR δ-chain locus is situated within the human α-gene between the V_α and J_α regions. When the δ-gene is excised, an excision circle is generated that includes excision signal joints (TREC). The human β-chain of TCR is encoded in chromosome 7q22-7qter. (In mice, its homologue is in chromosome 6.) The expression of β-genes also requires rearrangement of VDJ genes next to C genes. The γ1- and γ2-chain genes are in human chromosome 7p15-p14 area, whereas their homologues are in mouse chromosome 6. The early lymphocytes carry TCR built of γ- and δ-polypeptides, whereas later about 95% of the TCRs are built of α- and β-chains. The size of $\alpha\beta$-TCR is about 80 kDA built of four subunits. After the $\alpha\beta$-TCRs are formed, the $\gamma\delta$-TCRs are eliminated. $\alpha\beta$-TCR is part of the protein complex CD3 γ, δ, ϵ, and ζ. These chains also contain one ITAM motif except ζ, which has three. Phosphorylation of the ITAM motif facilitates signal transduction from TCR. Subunit ζ largely determines the specificity by specific series of phosphorylations.

In contrast to $\alpha\beta$-antigen receptors, which recognize only peptides (fragments) bound to MHC molecules, $\gamma\delta$-TCRs recognize polypeptides without MHC molecules and the MHC molecules without bound peptides. $\gamma\delta$-TCRs may use nonpeptide ligands such as phosphate-containing molecules. The TCR molecules are of a great variety, determined by the rearrangements just like in immunoglobulins. It appears that TCR does not mutate somatically, in contrast to Ig genes. Their association either with class I and class II HLA gene products (MHC) augments the specificity of the TCR. Class I HLA antigens are associated with cytotoxic (killer) cells, whereas helper lymphocytes attach to class II antigens. In the function of TCRs, an important role is played by CD4 (in class I associations) and CD8 (in class II associations). Both types of TCRs require the CD3 transmembrane protein complex also involved in signal transduction. CD peptides activate some protein tyrosine kinases, which are important elements of several signal transduction pathways. Many forms of cancer are associated with chromosome breakage in the regions where the TCR proteins are coded. The human TCRβ locus consisting of 685 kb has been sequenced. Besides the TCR elements, this large family includes other genes such as a dopamine-hydroxylase-like gene and eight trypsinogen genes. This large family includes besides the 46 functional genes, 19 pseudogenes, and 22 relics (genes with major lesions in one or more components), the large locus involves A portion of the locus is translocated from chromosome 7q22-7qter to 9. The V_β segments include promoters, the first exon as a signal peptide, with RNA splicing signals; the second exon is the V element and DNA rearrangement signal sequence. In some V_β families, a conserved decamer interacts with binding proteins. *See* antibody; antigen-presenting cell; CD4; CD8; HLA; immune response; immunoglobulins; ITAM; lymphocytes; MHC; SLAM; T-cell receptor. (Mak, T. W., et al. 1987. *J. Infect. Dis.* 155:418; Weiss, A. 1990. *J. Clin. Invest.* 86:1015; Moffatt, M. F., et al. 2000. *Hum. Mol. Genet.* 9:1011; Willemsen, R. A., et al. 2003. *Hum. Immunol.* 64:56.)

TCV *See* turnip crinkle virus.

T cytoplasm Texas male sterile cytoplasm of maize; almost 100% of the pollen is incapable of fertilization (sporophytic control of male sterility). *See* cytoplasmic male sterility.

TD$_{50}$ (toxic dose) Causes toxic (carcinogen) effects in 50% of experimental organisms. TD$_{50}$ is frequently identified as mmol/kg/day. *See* CASE.

TDF Testis-determining factor. *See* H-Y antigen; sex determination; SRY.

t distribution *See* student's *t* distribution, *t* value.

T-DNA Transferred DNA of the Ti plasmid is bordered by 24–25 bp incomplete direct repeats, TGGCAGGATATATT$_G^C$ X $_A^G$TTGTAAA for the left and TGGCAGGATATATT$_G^C$ X $_A^G$TTGTAAA for the right, in the octopine plasmids of *Agrobacteria*. The border sequences of nopaline plasmids are somewhat different. The left part of the sequences within the borders, T_L (14 kb), and right, T_R (7 kb), are distinguished. The left segment carries, among others, genes for plant oncogenicity (coding for the plant hormones indole acetic acid and the cytokinin, isopentenyl adenine) and either octopine or nopaline; the right segment contains genes for other opines and some others with unknown functions. The integration of T-DNA into plant chromosomes is mediated by virulence genes of the Ti plasmid and some chromosomal loci rather than by T-DNA sequences. T-DNA can integrate into the chromosome of plants by a process of illegitimate recombination at practically random locations. Most likely, the integration involves only one of the strands, the T-strand. Because of this transfer feature, T-DNA can be utilized as the most efficient plant transformation vector. The oncogenes or other sequences can be deleted and replaced by any desired DNA sequences (genes) and they are still inserted into the plant chromosome as long as the border sequences are retained. In *Agrobacteria*, besides T-DNA, multisubunit protein complexes are transmitted to the eukaryotic cell. *See* *Agrobacterium*; binary vectors; cointegrate vectors; opines; overdrive; Ti plasmid; transformation plants; virulence genes of *Agrobacterium*. (Koncz, C., et al. 1992. *Methods in Arabidopsis Research*, Koncz, C., et al. eds., p. 224. World Scientific, Singapore; Szabados, L., et al. 2002. *Plant J.* 32:233.)

TDT *See* transmission disequilibrium test.

TE *Drosophila* transposable elements cytologically localized in chromosomes 1, 2, and 3.

TE buffer Contains Tris-EDTA; the pH range is 7.2–9.1. *See* EDTA; Tris-HCl buffer.

TEC Family of nonreceptor tyrosine kinases required for signaling through the T-cell receptor. *See* MAPK; Src; T-cell receptor; TCR genes; Zap-70. (Mao, J., et al. 1998. *EMBO J.* 17:5638.)

technology transfer Converting basic or laboratory research results into industrial, agricultural, medical, pharmaceutical, or other applications.

tectorin Protein encoded at human chromosome 11q and mouse chromosome 9; defective in nonsyndromic deafness. *See* deafness.

TEFb Positive transcript elongation protein factor (P-TEFb). The *Drosophila* dimer has one ~43 and one ~120 kDA subunit. The small subunit is homologous to PITALRE and the large subunit is also called cyclin T. Protein TEFb phoshorylates the carboxyl-terminal domain of the largest subunit of DNA-dependent RNA polymerase II and thereby assures that transcript elongation proceeds with few or no pauses. *See* cyclins; DRB; DSIF; NELF; NusG; PITALRE; TFIIS; transcript elongation. (Lee, D. K., et al. 2001. *J. Biol. Chem.* 276:9978.)

tegument Protein layer between the viral capsid and the envelope.

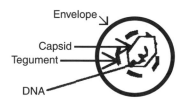

teichoic acid Constituent of the cell membrane and cells of some bacteria. The membrane techoic acid contains polyglycerol phosphate, linking glycerol units through phosphodiester, and there are glycosyl substitutions and alanine residues at some positions. The cell teichoic acids are more variable polymers of 6 to 20 units and include polyribitol phosphate chains. *See* gram-negative/positive.

TEL Phosphatidylinositol kinase of yeast. *See* PIK.

telangiectasia, hereditary hemorrhagic (Osler-Rendu-Weber syndrome) Generally nonlethal bleeding disease (except when cerebral or pulmonary complications arise) that is caused by lesions of the capillaries due to weakness of the connective tissues. The gene ORW1 in human chromosome 9q13 encodes a receptor (endoglin) for the transforming growth factor β expressed on vascular endothelium. ORW2 was mapped to chromosome 12, and it encodes an activin receptor-like kinase 1, a member of an endothelial serine/threonine kinase family. The prevalence of ORW syndrome in the U.S. is about 2×10^{-5}. *See* activin; ataxia telangiectasia; endoglin; hemostasis; Rothmund-Thompson syndrome; transforming growth factor β.

telangiectasis (telangiectasia) Defective veins causing red spots of various sizes. *See* glomerulonephritis; poikiloderma telangiectasia.

telemicroscopy Microscope linked to a computer, which then transmits JPEG-compressed images through the Internet network to distant viewers.

teleology Dogma attributing a special vital force and ultimate purpose to natural processes beyond material scientific evidence. (Lennox, J. 1981. *J. Hist. Philos.* 19:219.)

telethonin Z-disc protein in the sarcomeres. *See* sarcomere. (Faulkner, G., et al. 2000. *J. Biol. Chem.* 275:41234.)

teliospores Fungal spores protected by a thick wall. They are either dikaryotic or diploid. *See* stem rust; telium.

Dikaryotic and diploid teliospores.

telium Fruiting structure (sorus) of fungi that produces dikaryotic teliospores.

telocentric chromosome Has a terminal centromere (one arm). Telochromosomes can be used to determine which genes are located in that particular single chromosome arm if a telotrisomic female is used according to the scheme below. The segregation of *B:b* in both the 2*n* and 2*n*+ telo progeny is expected to be 1:1 (test cross).Among the 2*n* offspring, none or very few are expected to be *A* by phenotype because the telocentric egg cannot remain functional due to dosage effect of the essential genes in the missing chromosome arm. The *A* phenotype is usually due to recombination between the telo and the biarmed chromosomes. The 2*n* + telo offspring should be all *A* by *a* because of the dominant phenotype and none *a* because the dominant *A* allele is in the trisomic arm. In allopolyploids, the telochromosomes can be used to assign genes to the telochromosome and to determine recombination frequency between genes and centromeres. *See* centromere mapping in higher eukaryotes; misdivision of the centromere; Robertsonian translocation; tetrad analysis. (Photo is the courtesy of Dr. E. R. Sears.)

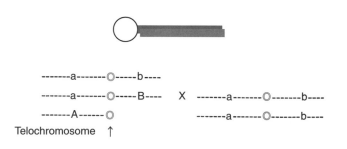

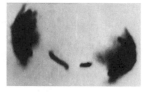

Misdivision generates telocentrics.

telochromosome Telocentric chromosome.

teloisodisomic In wheat, $20'' + ti''$, $2n = 42$ ($'' = $ disomic, t $=$ telosomic, i $=$ isosomic).

teloisotrisomic In wheat, $20'' + (ti)1'''$, $2n = 43$ ($'' = $ disomic, $''' = $ trisomic, t $=$ telosomic, $i = $ isosomic).

telomerase (TERT) Telomere reverse transcriptase enzyme synthesizes telomeric DNA. This enzyme is different from other replicases inasmuch as an RNA template (TERC) that is a part of telomerase (ribozyme) specifies telomeric DNA. RNA polymerase II enzyme transcribes this RNA and has a 5′-2,2,7-trimethyl guanosine cap. It has a binding site for Sm proteins that are characteristics for snRNPs. The transcription of human TERT (*hTERT*) is activated by the joint action of Sp1 and cMyc. It seems that the occupancy of the E box by either Myc or Mad1 determines the acetylation/deacetylation of chromatin in HeLa cells and thus the regulation of human hTERT. The RNA has A and C repeats and therefore T and G repeats characterize the telomeres. The human telomerase template has 11 nucleotides: 5′-CUAACCCUAAC. The telomere has (5′-TTAGGG-3′)$_n$ repeats.

In *Oxytricha*, the telomere consists of 36 nucleotides of which 16 form a single strand of 3′-G$_4$T$_4$G$_4$T$_4$ overhang that protrudes from the double-stranded remainder. The telomere-end-binding protein (TEBP) binds to the 3′-G$_4$T$_4$ tract and protects it from degradation. During replication, this TEBP is displaced for the replication to proceed. By rejoining the end, it displaces the telomerase and thus regulates telomere length. In budding yeast, the *TLC1* gene is responsible for telomerase activity and gene *EST1* is also needed for the maintenance of the telomeres. The *TLC1* (telomerase component) gene is also required for the preservation of the RNA template and normal telomerase function. The telomeres are made mainly of double-stranded DNA repeats; however, the far end has only single-strand G repeats. When the telomerase stays at the end of the chromosomes (capped state), even when the telomeres are short the cells remain viable. When the telomeres become uncapped, the cells exit from cycling and senesce. The replication of the telomere takes place near the end of the cell cycle. Telomerase elongates only the G-rich strand and the C-rich strand is filled in later. Protein RCF binds to telomeric DNA and is required for the replication of the leading strand by pol δ.

The Pif1 helicase is an inhibitor of yeast telomerase and may suppress healing of double-strand breaks by telomerase. Telomere-binding proteins (TEBP) bind either to the single-strand terminal repeats (*Oxytricha* proteins) or to the double-strand sequences (Rap1). The telomeres of the ciliate *Oxytricha fallax* terminate in duplex DNA loops. The TEBP heterodimer α-subunit (M$_r$ 56K) binds to the 3′-T$_4$G$_4$ single-stranded end, whereas the β-subunit (M$_r$ 41 K) attaches to other proteins. The heterogeneous nuclear ribonucleoprotein A1 (hnRNP A1) is also required for the maintenance of the normal length of the telomeres. The E6 protein of human papilloma virus-16 (HPV16) activates telomerase, but it does not immortalize the cells, although it expands their life span. The newly replicated telomeres are processed to size by proteins. Telomerase activity in tumor tissues is high, but it is low in somatic cells or in benign neoplasias (Artandi, S. E. & De Pinho, R. A. 2000. *Curr. Opin. Genet. Dev.* 10:39).

Targeting telomerase by antisense technologies may be an approach to anticancer therapy. The telomerase RNAs of *Tetrahymena*, *Euplotes*, and *Oxytricha* seem to be transcribed by polymerase III, and human telomerase RNA may be the product of pol II because it is sensitive to α-amanitin. The telomere end of *Tetrahymena*, where it was originally discovered, is somewhat different. In the RNA template there are 5′-CAACCC-3′ and in the telomere 5′-GGGGTG-3′ repeats occur. The Cdc13 protein of yeast mediates the access of the telomerase to the telomere. In the catalytic subunit of a telomerase of yeast (*EST2*, *ever-shorter telomeres*), reverse transcriptase-like motifs were identified. The telomeres are usually shortened as nuclear division proceeds, and aging has been attributed to reduced telomerase activity. Lack of telomerase does not prevent tumorigenesis; however, telomerase-deficient mice are resistant to skin tumorigenesis. Maintenance of normal telomere function is critical for the prevention of carcinogenesis. Usually, in normal cells telomerase activity is barely detectable, but it is well expressed in tumor cells. The catalytic subunits bear structural similarities to reverse transcriptases.

In mammals, reduction in telomerase activity is harmful to reproductive and blood-generating systems. Introducing an active telomerase into mammalian cells, may substantially extend the life of the cells. Telomerase defects increase damage susceptibility to ionizing radiation. Nevertheless, sustained telomerase activity does not result in malignancy. The ubiquitous c-MYC protein normally activates telomerase. The promoter of the TERT catalytic subunit has several c-MYC-binding sites. Sheep produced by nuclear transplantation apparently displayed shorter telomeres than normally generated animals, yet in small-scale experiments no premature aging was observed. Mutations in telomerase genes may result in senescence and late-onset sterility. When the telomeres are substantially lost, the chromosomes may fuse end-to-end. One class of *Caenorhabditis* telomerase mutants (*mrt*, *mortal germline*) is pleiotropic because it involves chromosomal loss and sensitivity to DNA-damaging agents, in addition to the symptoms displayed by other mutants in yeast and mice. Mutation *mrt-2* was identified to have a defect in checkpoint function that is apparently repaired when the telomerase is normal. Loss of the telomerase function leads to swift progressive deterioration in animals, but *Arabidopsis* plants may survive up to 10 generations without telomerase function. Due to chromosomal aberrations, eventually vegetative and reproductive fatal damage ensues. *See* aging; α-amanitin; antisense technologies; cap; E box; hTERT; Mad2/Mad1; malignant growth; MYC; Myc; nuclear transplantation; p16^{INK}; pol II; pol III; RAP; ribozyme; RNP; SP1; tankyrase; telomeres; *Tetrahymena*; tumorigenesis. (Collins, K. 1999. *Annu. Rev. Biochem.* 68:187; Herbert, B.-S., et al. 1999. *Proc. Natl. Acad. Sci. USA* 96:14276; Tzfati, Y., et al. 2000. *Science* 288:863; Cooper, J. P. 2000. *Curr. Opin. Genet. Dev.* 10:169; Betts, D. H., et al. 2001. *Proc. Natl. Acad. Sci. USA* 98:1077; Wenz, C., et al. 2001. *EMBO J.* 20:3526; Antal, M., et al. 2002. *Nucleic Acids Res.* 30:912; Arai, K., et al. 2002. *J. Biol. Chem.* 277:8538; Neidle, S. & Parkinson, G. 2002. *Nature Rev. Drug Discovery* 1:383.)

telomere mapping Eukaryotic chromosome has characteristic telomeric repeats and can be used for mapping RFLPs relative to the telomeres. *See* RFLP; telomeres.

telomere position effect *See* telomeric silencing.

telomeres   Special terminal structural elements of eukaryotic chromosomes rich in T and G bases. E-electron microscopic observations indicate that in mice and humans the ends of the telomeres bend backward and form a D loop/t loop as the telomeric single strands pair with double-stranded telomeric DNA. ==========⟹ In human chromosomes and all vertebrates, *Trypanosomas*, and fungi, the many times repeated telomeric box is CCCTAA/TTAGGG (ca. 300 bp). The telomeric region is highly conserved among diverse eukaryotes; in some species, however, variations exist. In *Drosophila*, the telomere is unusual inasmuch as it has a transposable element rather than having common repeats. This transposable element may assist the replication of telomeric DNA. In most organisms, middle-repetitive DNA is found (telomere-associated DNA, TA) proximal to the telomere that may display some similarity to *Drosophila* transposable sequences (Levis, R. W., et al. 1993. *Cell* 75:1083). The subtelomeric regions are repetitive yet highly variable. Ectopic recombination in this tract may lead to restoring defective ends, but it may be the cause of undesirable rearrangements (Heather, C., et al. 2002. *Nature Rev. Genet.* 3:91). The length of the telomeric sequences may vary among organisms, but variations exist during the life of the cells according to developmental stage. The telomeres usually become shorter in cultured cells as they age, although telomere length in mice may not be correlated with life span (Hemann, M. T. & Greider, C. W. 2000. *Nucleic Acids Res.* 28:4474).

Glowing telomeres probed by a fluorescent sequence.

Mutants of shorter telomeres may function reasonably well, although a transient pause may occur in cell division. The length of human telomeres may be different among homologous chromosomes (Londoño-Vallejo, J. A., et al. 2001. *Nucleic Acids Res.* 29:3164). The nonnucleosomal special chromatin, including the repeats of yeast, is called telosome. In mammals, the large telomeric DNA is nucleosomal, however. The major structural protein associated with yeast telomeres is Rap1. The same protein may be present at other locations of the chromosomes and acts either as a repressor or activator of transcription. Telomeres are required for the proper replication of linear eukaryotic chromosomes. This is probably why cytologists failed to find attachments of internal chromosomal pieces among the diverse types of chromosomal aberrations to the telomeres or telomere to telomere fusions, although chromosomes with broken ends may fuse into dicentric chromosomes after replication. Normally the telomeric repeat-binding factor, TRF2, or DNA-dependent protein kinase (DNA-PK) prevents telomeric fusions. In ciliates, a cap is built at the telomere by binding TEBPα and -β to 16-nucleotide single-strand DNA overhang. In fission yeast, the Taz1 protein protects the telomeric ends of the chromosomes from fusion, which otherwise might be mediated by the Ku proteins recognizing double-strand breaks (ends) of the DNA. In *taz⁻* cells, *rad22* may promote telomere fusion (Godinho Ferrira, M. & Promisel Cooper, J. 2001. *Molecular Cell* 7:55). In budding yeast, telomere capping is mediated by Cdc13 protein, which

recruits other proteins (Ten1/Stn1) to build a protective cap. DNA-dependent protein kinase may have a critical role in capping (Gilley, D., et al. 2001. *Proc. Natl. Acad. Sci. USA* 98:15084). In mammals, TRF2 protein has an end-protective role. It may remodel the chromosome ends by forming the *t loop*. The t loop of mammals is double-stranded DNA. In both fission yeast and mammals, the *POT1* (protection of telomere) wild-type allele encodes a telomere-capping protein, which binds to the G-rich telomeric single-strand tail and to the t loop. The mammalian single-strand DNA with a 3′-end may encompass 300 nucleotides and is located where the loop folds onto the main double-strand DNA (Baumann, P. & Cech, T. R. 2001. *Science* 292:1171).

The structure of the telomere may bear some similarity to the centromere. In some instances, (e.g., rye) it may function as a neocentromere. It may still be somewhat of a puzzle how the normally fragmented somatic chromosomes of *Ascaris* produce their telomeres and how the germline chromosomes control the apparently multiple intercalary telomeres. In the polyploid ciliate macronucleus, millions of centromeres may exist in some species. HIV-infected individuals appear to have shorter telomeres than healthy individuals of comparable age. Telomere loss or inactivation may play a role in senescence, and telomerase activation may be a mechanism of cellular immortalization. The replication of telomeric DNA is carried out by the ribozyme, telomerase. A specific telomere-binding human protein hTRF (60 kDa), recognizing the TTAGGG (and mammalian) sequences, has been isolated. TRF1 (8q13) is regulated by TIN2 (TRF1-interacting nuclear protein). The corresponding repeat in *Tetrahymena* is TTGGGG and in yeast it is TG_{1-3}. One of its domains is similar to the Myb oncogene product. Telomeric sequences may be shortened in several types of cancer cells, although in about 90% of malignant cell types the activity of the telomerase enzyme is higher than in normal cells. The enzyme poly(ADP)-ribose polymerase (PARP), which recognizes DNA interruptions, adds ADP-ribose units to the DNA ends. Defects in the encoding gene (ADPRP) and the protein PARP result in chromosomal instability and telomere shortening.

In *Caenorhabditis*, all chromosomes are capped by the same 4–9 kb tandem repeats of TTAGGC, but the sequences next to it are different among the chromosomes. Double-strand break repair proteins mediate chromosome capping. Intact telomeres cannot fuse, but a repair pathway may generate nonhomologous end-joining (NHEJ). When, however, NHEJ fuses two telomeres, unstable dicentric chromosomes are produced. Chromosomes lacking telomeres can perform most functions but lack stability and undergo fusion, degradation, loss at a high rate, and the chromatids may not be able to separate normally. If yeasts lose the *TRT-1* telomerase subunit gene, the cells may survive either by circularizing the chromosome or by restoring the function through recombination controlled by *RAD50* or *RAD51* genes. In the majority of tumors, centromeres are not shortened during consecutive divisions, whereas normal somatic human cells may lose 100 bp by each cell division, which may lead to discontinuation of cell divisions and cellular aging. In ciliates, telomeres may arise de novo, but in yeast this rarely occurs. In humans, telomerase may rarely replace lost telomeres. Genes that would normally be transcribed by pol I, pol II, or pol III are frequently repressed at the telomeric location. *See* aging; ALT; breakage-fusion-bridge cycle; chromosome diminution; dyskeratosis; HeT-A; immortalization; Ku; Myb; neocentromere; nonhomologous end-joining (NHEJ); plasmid telomere; pol; RAP; senescence; tankyrase; TART; telomerase; telomeric probe; TRF; UbcD1. (Griffith, J. D., et al. 1999. *Cell* 97:503; Pardue, M.-L. & DeBaryshe, P. G., 1999. *Chromosoma* 108:73; Varley, H., et al. 2000. *Am. J. Hum. Genet.* 67:610; Knight, S. J. L., et al. 2000. *Am. J. Hum. Genet.* 67:320; McEachern, M. J., et al. 2000. *Annu. Rev. Genet.* 34:331; Hodes, R. 2001. *Proc. Natl. Acad. Sci. USA* 98:7649; Shay, J. W., et al. 2001. *Hum. Mol. Genet.* 10:677; Blackburn, E. H. 2001. *Cell* 106:661; McEachern, M. J., et al. 2002. *Genetics* 160:63; Ren, J., et al. 2002. *Nucleic Acids Res.* 30:2307; Phan, A. T. & Mergny, J.-L. 2002. *Nucleic Acids Res.* 30:4618.)

telomere switching In *Trypanosoma*, the variable surface antigens (VSG) are encoded at the telomeres; frequent recombination of the chromosome ends can generate new variations and may affect the expression of VSG glycoproteins. *See* telomeres; *Trypanosoma*. (Rudenko, G., et al. 1998. *Trends Microbiol.* 6:113.)

telomere terminal transferase the ribozyme involved in replication telomeric DNA. (*See* telomerase.)

telomeric fusion End-to-end fusion of chromosomes does not take place between intact telomere-capped chromosomes. In double-strand repair-deficient cells, it may take place, however. *See* nonhomologous end-joining.

telomeric probe fluorochrome-labeled DNA sequences complementary to the telomeric repeats TTAGGG; can be used for cytological identification of short (cryptic) translocations. *See* chromosome painting; telomere.

telomeric silencing (telomeric position effect) Telomeres frequently reduce transcription of genes associated with it. This effect is similar to the silencing or position effect exercised by heterochromatin, although it may occur in this region even in the absence of heterochromatin. It may be due to higher-order chromatin arrangement. Rap1 (repressor and activator protein) recruits Sir3 and Sir4 (silent information regulators) and Rif1 (Rap1-interacting factor). Histones 3 and 4 and Ku proteins are among the best-known regulators. Mec1 (mitotic entry checkpoint) protein controls S-phase arrest of the cell cycle in reaction to DNA damage and regulates telomere length; some forms may also affect telomeric silencing. These genes/proteins are highly conserved but may have different names in various organisms. *See* checkpoint; heterochromatin; Ku; MEC1; position effect; RAP1A; silencers; telomeres. (de Bruin, D., et al. 2001. *Nature* 409:109; Lecoste, N., et al. 2002. *J. Biol. Chem.* 277:30421.)

telomutation (Dominant) premutation occurring in and transmitted through both sexes but expressed only in the offspring of the heterozygous female. *See* delayed inheritance; premutation. (Aleck, K. A. & Hadro, T. A. 1989. *Am. J. Med. Genet.* 33:155.)

telophase Final major step of nuclear divisions. The chromosomes have been pulled by the spindle fibers all the way to the poles. The microtubules attached to the kinetochores fade from view and the nuclear envelope appears again. The chromosomes relax and the nucleoli become visible again. *See* meiosis; mitosis.

telosome Telocentric chromosome; the nonnucleosomal chromatin at the end of the yeast and ciliate chromosomes is also called telosomal. *See* telochromosome; telomeres.

telotrisomic Trisomic having two biarmed and one telochromosome. *See* trisomy.

telson Most posterior part (opposite to the head) of the arthropod body. *See Drosophila*.

TEM (1) Transmission electron microscopy. *See* electron microscopy. (2) Triethylenemelamine. Alkylating clastogen. *See* alkylation; alkylating agent; clastogen.

TEM1 GTPase. *See* GTPase.

TEMED *See* acrylamide.

temperate phage Has both a lysogenic and a lytic lifestyle. The temperate phages may be of different types: (1) insert their DNA into one or few preferred sites like λ-phage; (2) insert their DNA into the host bacterial chromosome with the aid of a transposase at different sites like Mu-1; (3) or, e.g., P1, which does not insert its DNA into the host chromosome but is maintained as a plasmid; (4) e.g., P4, which can be either a plasmid or a prophage. *See* bacteriophage; lambda phage; lysogen; lysogeny; plasmid; prophage; specialized transduction.

temperature conversion $0.556°F - 17.8 \rightarrow °C$; $1.8°C + 32 \rightarrow °F$; $K - 273.15 \rightarrow C°$.

temperature-sensitive mutation (ts) The mutation causes such an alteration in the primary structure of the polypeptide chain that its conformation varies according to temperature; it is functional at either high or low temperature but not at nonpermissive temperature. Temperature-sensitive conditional lethal mutants are very useful for various analyses because the biochemical/molecular basis of the genetic defect can be analyzed at the permissive temperature range when the cells or organisms can grow (normally). Temperature-sensitive condition is correlated with the buried hydrophobic residues in protein. The majority of organisms (mesophiles) normally thrive best at temperatures <40°C; hyperthermophiles may exist at temperatures around 100°C. *See* cold hypersensitivity; hyperthermia; trichothiodystrophy.

template Determines the shape or structure of a molecule because it serves as a "mold" for it. The old DNA strands serve as templates for the new ones, or one of the DNA strands may be a template for mRNA and the other for another sense or nonsense RNA. Molecular biologists frequently call template (or antisense) DNA the strand of the DNA that serves for the synthesis of mRNA by complementary base pairing. T7 RNA polymerase can bypass up to 24-nucleotide gaps in the template strand by copying a faithful sequence of the deletion using the nontemplate strand. *See* coding strand; replication fork; semiconservative replication; sense strand.

template switch During replication, the polymerase enzyme may jump to another DNA sequence and copy elements that were not present in the original DNA tract. Such a switch may occur when plasmid DNA (T-DNA) is inserted into a chromosomal target site. As a consequence, deletions and rearrangements may follow. During primer extension, the displaced strand may reanneal onto the template and the extended strand may be partially dissociated. A single-stranded sequence, attached to the 5′ end of the displaced strand and complementary to the dissociated segment of the extending strand, can thus serve as an alternative template. *See* copy choice; primer extension; switch; T-DNA. (Negroni, M. & Buc, H. 2000. *Proc. Natl. Acad. Sci. USA* 97:6385.)

tenascin Large glycoprotein complex with disulfide-linked peptide chains. It either promotes or interferes with cell adhesion depending on the type of cell and the different protein domains. It also controls cell migration and axon guidance. *See* cell adhesion; cell migration. (Hicke, B. J., et al. 2001. *J. Biol. Chem.* 276:48644.)

Tendril.

tendril Plant organ that coils around objects and provides support.

tensin Cytoskeletal protein-binding vinculin and actin; it contains an SH2 domain. *See* actin; multiple hamartomas; PTEN; vinculin. (Chen, H., et al. 2002. *Proc. Natl. Acad. Sci. USA* 99:733.)

tension Force(s) generated by pulling mitotic/meiotic chromosomes to opposite poles; they are opposed by the attachment between homologues. *See* kinetochore; pole; spindle.

teosinte *See* maize.

TEP1 TGFβ-regulated and epithelial cell-enriched phosphatase is a tumor suppressor function of PTEN. *See* PTEN. (Sharrard, R. M. & Maitland, N. J. 2000. *Biochim. Biophys. Acta* 1494:282.)

tera- 10^{12}.

teratocarcinoma Malignant tumor containing cells of an embryonal nature; common in testes. Teratocarcinoma cells may differentiate into various types of tissues in vitro and have been used for studies of differentiation. *See* cancer; stem cells; teratoma. (Silver, L. M., et al., eds. 1983. *Teratocarcinoma Stem Cells*. Cold Spring Harbor Lab. Press, Cold Spring Harbor, NY.)

teratogen Agent that causes malformation during differentiation and development.

teratogenesis Malformation during differentiation and development. The inducing agents may be genetic defects, physical factors and injuries, infections, drugs, and chemicals. In frogs limb developmental anomalies have been traced to trematode (*Planorbella campanulata* and *Ribeiroira* sp.) infection stimulated by agricultural pesticides. *See* Kiesecker, J. M. 2002. *Proc. Natl. Acad. Sci. USA* 99:9900.

teratoma Mixed tissue group with cells of different potentials for development. It may be formed in various early or late animal tissues and may eventually become a malignant tumor. Teratomas are most common in germinal tissues. In plants, amorph, undifferentiated tumor tissue may be interspersed with differentiated elements, giving rise to shoots or roots. Teratomas may be formed in tumors induced by *Agrobacterium rhizogenes. See Agrobacterium.*

tergite Dorsal epidermis developed from histoblasts, not from imaginal discs. *See* histoblast.

terminal deoxynucleotidyl transferase (TdT) Enzyme that can elongate DNA strands at the 3-OH end with any base that is present in the reaction mixture. It is used in genetic vector construction to generate homopolymeric cohesive tails for the purpose of splicing a passenger DNA to a plasmid. The passenger and the plasmid are equipped with complementary bases such as polyA and polyT, respectively, and can readily anneal and be ligated. Terminal nucleotidyl transferase is also expressed in adult bone marrow and generates immunoglobulin diversity. The enzyme occurs primarily as a 58 kDa protein, but much smaller molecules were also found. Its N-terminus is homologous to the C-terminus of the breast cancer gene BRCA1. The N-terminus of TdT interacts with the heterodimeric Ku autoantigen, which is the DNA-binding component of a 460 kDa DNA-dependent protein kinase complex. This complex apparently has a role (besides RAGs) in the generation of immunoglobulin diversity and DNA double-strand break repair. *See* breast cancer; cloning vectors; double-strand break; immunoglobulins; Ku; RAG; T cell; vectors. (Delarue, M., et al. 2002. *EMBO J.* 21:427.)

terminal differentiation Usually irreversible; in plants, however, under culture of high levels of phytohormones, dedifferentiation may be possible. *See* dedifferentiation; differentiation.

terminalization, chromosome During meiosis, the bivalents may display one or more chiasmata which eventually (during anaphase) move toward the telomeric region in the process called terminalization. *See* meiosis; figure next page.

terminal nucleotidyl transferase Enzyme can add homopolymeric ends to DNA. It is useful in generating cohesive termini (in genetic plasmid vectors) if one of the reaction mixtures contains adenylic acids and the other (e.g., passenger DNA) thymidylic acid (or G and C, respectively). *See* cloning vectors; terminal deoxynucleotidyl transferase.

terminal protein (TP) Serine, threonine, or tyrosine residues of the terminal proteins provide OH groups for the initiation of replication instead of the 3′ OH group of a nucleotide in a linear double-stranded viral DNA. (Van Bzeukelen, B., et al. 2003. *J. Virol.* 77:915.)

terminal redundancy Repeated DNA sequences at the ends of phage DNA. *See* permuted and nonpermuted terminal redundancy.

Terminalization of chromosomes.

terminase Phage enzyme binds to specific nucleotide sequences and cuts in the vicinity of the binding at cohesive sites (*pac* or *cos*). *See* lambda phage. (de Beer, T., et al. 2002. *Mol. Cell* 9:981.)

termination codons *See* genetic code; nonsense mutation; terminator codons.

termination factors Proteins that release the polypeptide chains from the ribosomes.

termination of replication in bacteria Mediated by the homodimeric replication terminator protein (RTP). Two dimers bind to the two *Ter* inverted repeat sites in the DNA. The binding site has a strong core and an auxiliary site. One dimer binds to the core; binding of the second dimer follows this to the auxiliary site. When the replication fork encounters the RTP-*Ter* site, it is blocked, but not when it encounters it from the auxiliary site. The terminator prevents DNA unwinding by the helicase. There are several *Ter* sites. (Lemon, K., et al. 2001. *Proc. Natl. Acad. Sci. USA* 98:212.)

terminator Sequence at the end of a gene that signals for the termination of replication or transcription. The RNA polymerase recognizes this signal directly or indirectly by various proteins. The intrinsic terminator of *E. coli* (rho) uses two sequence motifs for the release of RNA from the DNA template: (i) a stem-loop hairpin and (ii) a tract of 8–10 nucleotides immediately downstream at the end of the released RNA. Before termination, there is generally a pause mediated by a U-rich transcribing DNA segment. *See* antitermination; pausing, transcriptional; protein synthesis; release factor; rho factor; RNA polymerase; transcription termination.

terminator codons UAA, UAG, and UGA in RNA (same as nonsense codons).

terminator technology Uses three different transgenes in plants to control their ability to bear seed as a means of protecting the interest of seed companies by preventing unauthorized use of their genetic stocks. One of the genes is a repressor of a recombinase; the other is a recombinase capable of deleting internal spacers; the third is a toxin gene. The seed company can grow seed-bearing plants because the toxin gene is inactivated by a spacer sequence between the promoter and the structural toxin gene as long as the recombinase

is suppressed. By treating the commercially sold seed with the antibiotic tetracycline, the synthesis of the suppressor is prevented, then the recombinase deletes the spacer and the toxin gene becomes activated, causing seed failure. This trick of biotechnology may become a safeguard for the property of the company but may hurt the potential consumer who cannot afford to buy the seeds annually. Similar, although simpler, technology has been practiced by the seed industry for decades in the commercially available double-flowered garden stocks. Although social activists condemn terminator technology, its principles may offer promises for shutting off undesirable genes or activating useful ones, and may serve beneficial roles in improving agricultural productivity. *See Matthiola*; promoter; recombinase; structural gene; suppressor; tetracycline; T-Gurt. (Kuvshinov, V. V., et al. 2001. *Plant Sci.* 160:517.)

termisome Nucleic acid terminal protein complexed with other proteins and DNA. *See* protein synthesis; terminator.

ternary Either made up of three elements or third in order.

ternary complex factors *See* TCF.

terpenes Hydrocarbons or derivatives (>30,000) with isoprene repeats; occur as animal pheromones and diverse types of plant fragrances. *See* fragrances; pheromones; prenylation. (Trapp, S. C. & Croteau, R. B. 2001. *Genetics* 158:811.)

terrestrial radiation Comes from the unstable isotopes in the soil, such as uranium-containing rocks. *See* cosmic radiation; ionizing radiation; isotopes; radiation effects; radiation hazard assessment.

terrific broth (TB) Bacterial nutrient medium containing bactotryptone 12 g, bactoyeast extract 24 g, glycerol 4 mL, H_2O 900 mL, and buffered by phosphate 100 mL.

ter **sites** *See* DNA replication, prokaryotes.

tertiary structure Three-dimensional arrangement of the secondary structure of the polypeptide chain into layers, fibers, or globular shape. A third order of complexity, folding or coiling the secondary structure once more. *See* protein structure.

testa (seed coat) Maternal tissue in hybrids; therefore, seed coat characters display delayed expression of recessive markers by one generation (e.g., to F_3 rather than F_2).

test cross Cross between a heterozygote and a homozygote for the recessive genes concerned, e.g., $(AB/ab) \times (ab/ab)$. *See* recombination frequency.

tester In a genetic cross, is intended to reveal the qualitative or quantitative gene content of the individual(s) tested. *See* combining ability; hybrid vigor; test cross.

testes Male gonads of animals that produce the male gametes. *See* gamete; gonad; sex determination. (Raymond, C. S., et al. 2000. *Genes Dev.* 14:2587.)

testicles Testes.

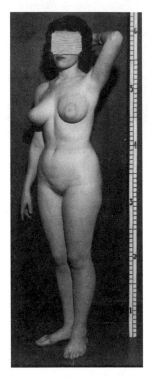

Testicular feminization. (Courtesy of Dr. McL. Morris.)

testicular feminization Developmental anomaly, that occurs in humans and other mammals. The chromosomally XY individuals display female phenotype including the formation of a blind vagina (no uterus), female breasts, and generally the absence of pubic hairs. Usually, the individuals affected by this recessive disorder (human chromosome Xq11.1-q12) appear very feminine but sterile. Generally, they develop abdominal or somewhat herniated small testes. About 1.5×10^{-5} of the chromosomally males have this disorder. The condition is the result of a complete or partial deficiency or instability of the androsterone receptor protein (917 amino acids). The function of this androgen receptor protein may be partially or totally missing (Reifenstein syndrome). This receptor appears to be highly conserved among mammals. The gene extends to about 90 kb DNA. The protein binds to DNA by two domains encoded by exons 2 and 3; five exons code for androgen binding while exon 1 has a regulatory function. *See* chromosomal sex determination; dihydrotestosterone; hermaphrodite; hormone receptors; pseudohermaphroditism; sex reversal; SF-1; Swyer syndrome. (Boehmer, A. L., et al. 2001. *J. Clin. Endocrinol. Metab.* 86:4151.)

testicular germ-cell tumor The TGCT gene at Xq27 conveys susceptibility to this cancer, affecting about 2×10^{-3} of men of Western European descent ages 15–40.

testosterone Most important androsterone. Testosterone increases the risk of coronary heart disease and atherosclerosis. If testosterone is converted to estrogen by aromatase, the risk is reduced. *See* animal hormones; hormonal effects on sex expression; progesterone.

Testosterone.

test tube baby *See* ART; intrafallopian transfer; intrauterine insemination; in vitro fertilization.

test weight Mass of 1,000 seeds or kernels randomly withdrawn from a sample. *See* absolute weight.

TET (1) Tetracycline antibiotic that inhibits protein synthesis; genetic (selectable) marker in pBR plasmids. *See* pBR322; tetracycline for formula. (2) Tubal embryo transfer is essentially the same as intrafallopian transfer of zygotes. *See* ART; intrafallopian transfer.

Tet A Tetracycline-resistance protein that controls the efflux of the antibiotic from the cells. *See* TetR; tetracycline.

tetany Highly stimulated condition of the nervous and muscular systems caused by low levels of calcium due to various diseases.

tethering Bringing together two distantly located nucleic acid sequences by DNA looping, catenanes, or RNA lariats. *See* introns; lariat RNA.

Tet-off *See* tetracycline.

Tet-on *See* targeting genes; tetracycline.

TetR Tetracycline repressor (homodimer) regulates its own expression as well as that of the antiporter (TetA, which exports the drug from the cells) at the level of transcription and it is activated by the $(Mg-Tc)^+$ complex. Tetracycline repressor binds to the dual operators of the TetR repressor and the TetA genes and blocks their transcription. Tetracycline alters the conformation of the repressor protein, which then is not capable of blocking transcription by RNA polymerase at the operators even in some eukaryotes. This system is thus suitable for exogenous regulation by tetracycline of a gene promoter fused by genetic engineering to the TetR system. The failure of the system in eukaryotes is determined by the toxicity of tetracycline to a particular organism. *See* antiport; tetracycline. (Scholz, O., et al. 2001. *J. Mol. Biol.* 310:979.)

tetracycline (Tet) Antibiotic that prevents the binding of amino acid–charged tRNAs to the A site of the ribosomes. The tetracycline repressor gene (TetR) interferes with the expression of tetracycline resistance. Tetracycline in the medium prevents the binding of TetR to DNA and relieves repression. Tetracycline is widely used as a tool in turning genes on and off, respectively. When the TetR gene is fused to the activation domain of herpes simplex virus protein VP16, it becomes a very effective Tet-responsive transactivator (tTA) of other genes. In the absence of Tet or Tet analogs, tTA binds the Tet operator (TetO) and initiates transcription. When Tet is present in the culture medium, the transactivator (TA) binds the antibiotic, DNA binding is disrupted, and transcription is prevented. A mutant form of TetR protein binds to TetO only in the presence of the antibiotic doxycycline, a Tet analog, and when fused to VP16 (rtTA), transcription is activated. *See* doxycycline; Gene-Switch; pBR322; protein synthesis; ribosomes; rtTA; targeting genes; transactivator; tTA. (Chopra, I. & Roberts, M. 2001. *Microbiol. Mol. Biol. Rev.* 65:232; Stebbins, M. J., et al. 2001. *Proc. Natl. Acad. Sci. USA* 98:10775.)

tetrad, aberrant Allelic proportions deviate from 2:2 because of polysomy, gene conversion, nondisjunction, suppression, etc. *See* gene conversion; nondisjunction; polysomy; suppression.

tetrad analysis The meiotic products of ascomycetes (occasionally some other organisms) stay together as a *tetrad* as the four products of single meiosis. In some organisms, the tetrad formation is followed by a postmeiotic mitosis within the ascus, resulting in spore *octads*. If the four spores are situated in the same linear order as produced by the two divisions of meiosis, they are an *ordered tetrad*.

In the ordered tetrad, considering two genes *A* and *B*, three arrangements of the spores (parental ditype [PD], tetratype [TT], nonparental ditype [NPD]) can be distinguished as seen in the figure next page. The parental ditype (PD) indicates no crossing over; tetratype (TT) reveals one recombination between the two genes; second-division segregation of the B/b alleles reveals recombination between the B/b gene and the centromere. The nonparental ditype (NPD) is an indication of double crossing over between the to gene loci. PD, TT, and NPD

may appear even if the genes are in separate chromosomes. An excess of PD over NPD is an indication of linkage. If the deviation from the 1:1 ratio between PD and NPD is small, a *chi-square test* may be used to test the probability of linkage by the formula:

$$\chi^2 = \frac{(\text{PD} - \text{NPD})^2}{(\text{PD} + \text{NPD})}.$$

By counting the number of tetrads of the above three types, *recombination frequency between the two loci* can be calculated as,

$$\frac{[1/2]\text{TT} + \text{all NPD}}{\text{all tetrads}}$$

and recombination frequency between the B/b *gene and the centromere can be calculated* as

$$\frac{\text{TT}[1/2]}{\text{all tetrads}}.$$

(Dr. Fred Sherman recommended using the formula of Dr. David Perkins for the gene-gene map distance [*Genetics* 34:607], i.e., $100\{[0.5(\text{TT}) + 6\text{NPD}]/[\text{PD} + \text{TT} + \text{NPD}]\}$.) The recombination frequencies (if they are under 0.15) multiplied by 100 provide the map distances in centimorgans. If the recombination frequencies are larger, mapping functions should be used. From the genetic constitution of the tetrads, a great deal of information can be revealed about recombination. When the four meiotic products are not in the order brought about by meiosis, the tetrad is *unordered*. For the estimation of gene-centromere distances from unordered tetrad data, one must rely on three markers; from those, no more than two are linked, and algebraic solutions are required (e.g., Whitehouse. 1950. *Nature* 165:893; *see* unordered tetrads). Tetrad analysis is most commonly used in ascomycetes (*Neurospora*, *Aspergillus*, *Ascobolus*, *Saccharomyces*, etc.), yet it can be applied to higher plants where the four products of male meiosis stick together (*Elodea*, *Salpiglossis*, orchids, *Arabidopsis* mutants). In *Drosophila* with attached X-chromosomes, half-tetrad analysis is feasible. Since several genomes of higher eukaryotes have been sequenced, molecular markers are available for tetrad analysis for the cases when

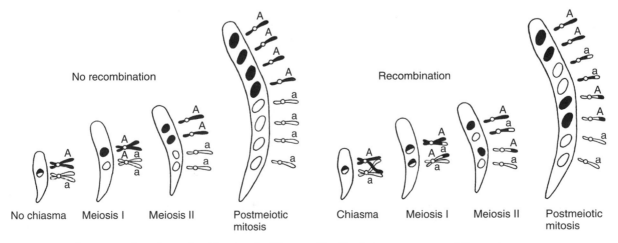

No recombination | No chiasma | Meiosis I | Meiosis II | Postmeiotic mitosis

Recombination | Chiasma | Meiosis I | Meiosis II | Postmeiotic mitosis

Spore tetrads and octads without and with recombination between a gene and the centromere.

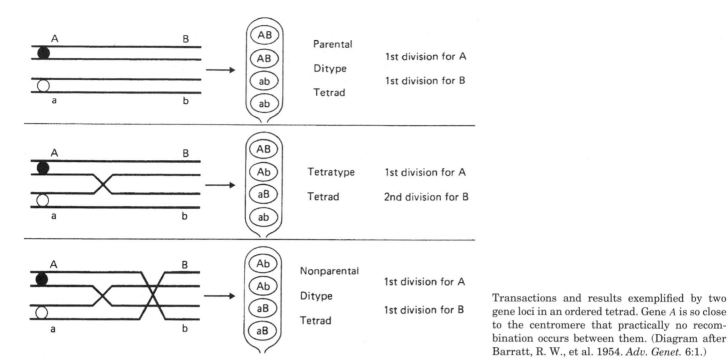

Transactions and results exemplified by two gene loci in an ordered tetrad. Gene *A* is so close to the centromere that practically no recombination occurs between them. (Diagram after Barratt, R. W., et al. 1954. *Adv. Genet.* 6:1.)

Octads of neurospora. (Courtesy of Dr. D. R. Stadler. 1956. *Genetics* 41:528.)

the products of individual meioses can be identified. *See* half-tetrad analysis; linkage; mapping; mapping function; meiosis; unordered tetrads; diagram next page.

tetrahydrofolate (THF)   Active (reduced) form of the vitamin folate, a carrier of one-carbon unit in oxidation reactions, and a pteridine derivative. *See* hyperhomocysteinemia)

Tetrahymena   pyriformis (2n = 10)   Ciliated protozoan with linear mtDNA. Its rRNA transcripts are self-spliced. *See* chromosome breakage, programmed; macronucleus; micronucleus; mtDNA; splicing; telomerase. (Turkewitz, A. P., et al. 2002. *Trends Genet.* 18:35.)

tetralogues   Paralogous gene groups originated by duplications (polyploidy), e.g., the families of multiple kinases. Some of these genes can be deleted without serious phenotypic consequence, yet their point mutations (dominant negative) may

be quite deleterious, especially if they have multiple interactive domains and a function in protein complexes. Some of the evolutionarily redundant genes can further evolve by selectable mutations. *See* dominant negative; paralogous loci; point mutation; polyploidy.

tetralogy of Fallot   *See* Fallot's tetralogy.

tetraloop   In structured RNAs, duplex runs are connected by loops of 5′GNRA (or UNCG or CUYG) tetranucleotides (N = any nucleotide, R = purine, Y = pyrimidine). They are involved in long-range molecular interactions such as in hammerhead ribozymes or introns. *See* introns; ribonuclease III; ribose zipper; ribozymes; stem-loop. (Baumrook, V., et al. 2001. *Nucleic Acids. Res.* 29:4089.)

tetranucleotide hypothesis   Historical assumption that nucleic acids are made of repeated units of equal numbers of the four bases (A, T/U, G, and C) and therefore monotonous structure does not qualify for being the genetic material. It was probably first proposed by the great German chemist Albrecht Kossel at around the turn of the 20th century, but through the brilliant yet erroneous work of Phoebus Levine it became a rather widely accepted view from 1909 to 1940. (Levine, P. A. 1909. *Biochem. Zeit.* 17:120; Levine, P. A. 1917. *J. Biol. Chem.* 31:591.)

tetraparental offspring   Results if in vivo fused blastulas of different matings are implanted together into the uterus. *See* allophenic; chimera; multiparental hybrids.

tetraplex   Consecutive guanine sequences may take four-stranded parallel or antiparallel conformations in the DNA or RNA. The tetraplex structure may have biological significance for the telomeres, specific recombination of the

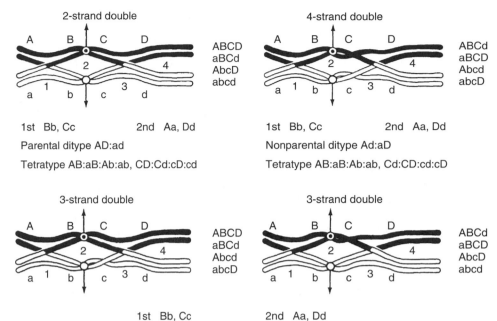

Four-point cross with genes in both arms of the chromosomes. it is a five-point cross if we consider the centromere as a genetic marker. From the spore order, we can recognize even if the chromatids rotated 180° after the exchange. (After Emerson, S. 1963. *Methodology in Basic Genetics*, Burdette, W. J., ed., p. 167. Holden-Day, San Francisco.)

immunoglobulin genes, dimerization of the HIV genome, etc. *See* HIV; immunoglobulins; telomere; tetrasomic. (Weisman-Shomer, P., et al. 2002. *Nucleic Acids Res.* 30:3672.)

tetraploid Has four sets of genomes per nucleus. *See* autopolyploidy; tetrasomic.

tetraploid embryo complementation Embryonic stem cells have a unique ability to complete embryonic development after nuclear transplantation or after injection into tetraploid host blastocysts. Tetraploid mouse blastocysts cannot autonomously develop into embryos normally, but they can do so when complemented by diploid embryonic stem cells. Success is substantially improved if the nuclei are not derived from inbred cells. *See* embryo culture; nuclear transplantation; stem cells. (Ewggan, K., et al. 2001. *Proc. Natl. Acad. Sci. USA* 98:6209.)

tetrasomic One chromosome is present in four doses in the nuclei. Tetrasomy exists in tetraploids for all chromosomes and may not involve any serious anomaly, although generally fertility is reduced. Tetrasomy for individual chromosomes may have more serious consequences because of genic imbalance. Tetrasomy is usually not tolerated by animals. Tetrasomic mosaicism for human chromosome 12p leads to developmental anomalies, mental retardation, defects in the central nervous system and speech, etc. *See* autopolyploid; cat-eye syndrome; polyploidy; sex-chromosomal anomalies in humans; trisomy.

tetrasomic-nullisomic compensation In allotetraploids the homoeologous chromosomes may compensate for each other; thus, if two chromosomes are missing (nullisomy) and one type of their homoeologues is present in four copies (tetrasomy), the individuals may function rather well, depending on the particular chromosomes involved. When, however, a nonhomoeologous chromosome is substituted in the presence of nullisomy, the condition is worsened. *See* chromosome substitution; nullisomic compensation. (Morris, R. & Sears, E. R. 1967. *Wheat and Wheat Improvement*, Quisenberry, K. S. & Reitz, L. P., eds., p. 19. Am. Soc. Agron., Madison, WI.)

tetraspanins Cell membrane molecules functioning in cell adhesion, motility, proliferation, differentiation signal transduction, and fertilization. The TM4SF2 protein encoded at Xp11.4 and involved in determining nonsyndromic mental retardation also belongs to the tetraspanin family of proteins. *See* CD9; CD63; CD82; mental retardation. (Cannon, K. S. & Cresswell, P. 2001. *EMBO J.* 20:2443.)

tetratrico sequences (TPR) Amphipathic α-helical amino acid tracts punctuated by proline-induced turns. CDC16, CDC23, and CDC27 of yeast and their homologues in other eukaryotes contain ~34 tandem-repeated residues. Such motifs exist in more than 100 proteins and may control mitosis and RNA synthesis in the cells. *See* APC; CDCs. (Wendt, K. S., et al. 2001. *Nature Struct. Biol.* 8:784.)

tetratype The meiotic products concerned with two genes show four types of combinations (e.g., *AB, Ab, aB,* and *ab*). *See* tetrad analysis.

tetrazolium blue Detects oxidation-reduction enzyme activity and thus identifies living cells and cancerous metabolism.

tetrodotoxin *See* toxins.

T-even phages Designation has even numbers such as T2, T4, etc. *See* bacteriophages.

Texas red Fluorochrome with excitation at 580 nm and an emission peak at 615 nm. *See* fluorochromes.

TF (1) transcription factors such as TF I, TF II, or TF III involved in the control of transcription by pol I, pol II, or pol III, respectively; TFs assist transcription by cooperation with other binding proteins. *See* open transcription complex; TIIFS; transcription factors. *See* transferrin.

Tfam *See* mtFAM.

TFD Transcription factor decoy is a double-stranded nucleotide sequence (e.g., $\frac{GGGACTTTCC}{CCCTGAAAGG}$) that may sequester a transcription factor, e.g., NF-κB, within the cytoplasm by virtue of the attachment of TF to a sequence homologous to its recognition site in the upstream region of the gene(s) in the nucleus. As a consequence, the nuclear gene may be (partially) silenced. Such an approach may be exploited for the control of oncogenes. *See* antisense technologies; NF-κB. (Mann, M. J. & Dzau, V. J. 2000. *J. Clin. Invest.* 106:1071.)

TFE Helix-loop-helix leucine-zipper transcription factor. *See* DNA-binding protein domains.

TFII *See* PIC; TBP; transcription factors.

TFIIS Eukaryotic transcript elongation stimulatory factor. It makes yeasts hypersensitive to 6-azauracil, which reduces the intracellular UTP and GTP pool. In *Drosophila*, the transcripts of two protein factors, N-TEF (negative transcript elongation) and P-TEF, regulate elongation. The positive effect is mediated by ATP-dependent phosphorylation of RNA polymerase. *See* azauracil; DRB; DSIF; elongation factors; NELF; PITSLRE; TEF; transcript elongation; transcription factors. (Lindstrom, D. l. & Hertzog, G. A. 2001. *Genetics* 159:487.)

TFIIτ TBP.

TFM *See* testicular feminization.

TFO Triple-helix-forming oligonucleotide may be used to interfere with the activity of genes. The effective TFO needs to recognize 10 to 17 bases in order to be gene specific. Some problems arise in vivo because genes in the chromosomes are organized as chromatin. TFO seems to mediate mutation, nucleotide excision repair, and transcription-coupled repair, consequently enhancing mutation rate induced by UV-A in the presence of psoralen dye sensitization at 365 nm. Psoralen is a bifunctional photoreagent. TFO may also stimulate targeted gene conversion using nucleotide excision repair and may be antiviral. Triplex-directed, site-specific mutagenesis is much improved when the phosphodiester backbone is replaced by cationic phosphoramidate linkages, particularly by employing N,N-diethylethylenediamine to target short polypurine tracts. *See* antisense DNA; antisense RNA; directed mutation; DNA repair; gene conversion; inhibition of transcription; peptide nucleic acid; pseudoknot; psoralen; targeting genes; triple helix formation; triplex; ultraviolet light; diagram above. (Vasquez, K. M., et al. 2000. *Science* 290:530; Vasquez, K. M., et al. 2001. *J. Biol. Chem.* 276:38536; Besch, R., et al. 2002. *J. Biol. Chem.* 177:32423.)

TFR Transferrin receptor. *See* transferrin.

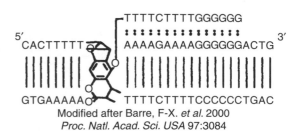

Modified after Barre, F-X. *et al.* 2000
Proc. Natl. Acad. Sci. USA 97:3084

TFTC TBP-free TAF$_{II}$-containing complex. *See* transcription factors. (Walker, A. K., et al. 2001. *EMBO J.* 20:5269.)

TGF Transforming growth factors are cell proliferation inhibitors and their loss may be involved in cancerous growth. In humans, the TGFA gene is in chromosome 2p13, TGFB1 in 19q13.1-q13.3, TGFB2 in 1q41, and TGFB3 in 14q24. A member of the family, Vg1, a factor localized to the vegetal pole of the animal embryo, may be involved in the induction of mesoderm formation in the embryo. The TGF-β family cytokines activate receptors of the heterodimeric serine-threonine kinases. TGF-β binds directly to TGF receptor II, a kinase. The TGF-β binds by TGF receptor I and becomes phosphorylated (**P**) by it. Type I receptor contains a repeated GS (Gly.Ser) domain and a binding site for FKBP12. TGFR I then transmits the cytokine signal along the signaling cascade and the target genes may be turned on. In humans, there are about 30 members of the TGFβ protein family, and various homologues, although in smaller numbers, occur in other species. SMAD3 protein affecting the pathway outlined in the diagram is also a coactivator of vitamin D receptor by forming a complex with steroid receptor coactivator-1 protein (SRC-1) in the cell nucleus. TGF-α is an epidermal growth factor–like molecule. Mutations leading to chondrodysplasia involve members of the TGF family (CDMP1). Beta-glycan and endoglin cell-surface proteins are also involved somehow with TGF. Endoglin and ALK1 (a type I receptor) mutations are found in hereditary hemorrhagic telangiectasia. TGFβ receptor I mutations have been found in different types of cancers, and TGFβ mutations may be responsible for heart disease, hypertension, osteoporosis, and fibrosis. The human genome includes 42 TGFβ genes

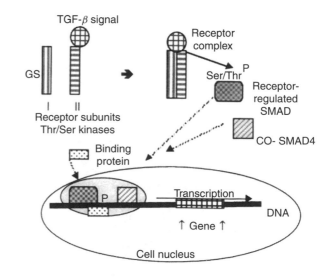

compared with 9 in *Drosophila* and 6 in *Caenorhabditis*. *See* activin; bone morphogenetic protein; cytokines; EGF; FK506; immunophilins; leukemia inhibitory factor; MDM2; Müllerian ducts; serine/threonine kinase; signal transduction; SMAD; SRC-1; telangiectasia, hereditary hemorrhagic. (Massagué, J. 1998. *Annu. Rev. Biochem.* 67:753; Massagué, J. & Chen, Y.-G. 2000. *Genes & Development* 14:627; Massagué, J., et al. 2000. *Cell* 103:295; Zwaagstra, J. C., et al. 2001. *J. Biol. Chem.* 276:27237, Wakefield, L. M. & Roberts, A. B. 2002. *Current Opin. Genet. Dev.* 12:22.)

TGN   *See* trans-Golgi network.

T-GURT   Trait-specific genetic-use restriction technology. Applied to products of plant breeding with the aid of biotechnology; makes the plants express special traits such as salt or drought tolerance on the condition that a special (chemical) treatment is applied. The plants will grow without the special treatment but do not express the agronomically advantageous special trait to the same extent. Such seed stocks protect the proprietary interests of the seed companies that developed them. *See* terminator technology.

T$_H$ (or Th)   T-helper cells. They can be T$_H$1-type-activating macrophages or T$_H$2, which primarily activate B cells. T-bet transcription factor regulates the activity of T$_H$1 cells. T$_H$1 cells produce IL-2, interferon-γ (IFN-γ), and tumor necrosis factor-β (TNF-β) and offer defense against intracellular pathogens by aiding cellular immunity. T$_H$2 cells make interleukin-4, -5, -6, -10, and -13 (IL-4, IL-5, IL-13) against extracellular pathogens and allergens, promote antibody formation, and have direct cytolytic activity. Interleukin-12 (IL-12) favors the development and maintenance of T$_H$1, whereas IL-4 promotes the differentiation of T$_H$2. IL-18 may both promote and hinder T$_H$2 activity by its effect on IL-13. T$_H$3 cells may express transforming growth factors. *See* B cell; IFN-γ; IL-4; IL-5; IL-6; IL-10; IL-12; IL-13; macrophage; T-bet; TCCR; T cells; TNF. (O'Garra, A. & Arai, N. 2000. *Trends Cell Biol.* 10:542; Mullen, A. C., et al. 2001. *Science* 292:1907; Helmby, H., et al. 2001. *J. Exp. Med.* 194:355; Ho, I-C. & Glimcher, L. H. 2002. *Cell* 109:S109.)

thalamus   Double-egg-shape area deep within the basal part of the brain involved in transmission of sensory impulses. *See* brain, human; fatal familial insomnia.

thalassemia   Hereditary defects in the regulation (or deletions) of hemoglobin genes, causing anemia. In the thalassemias, generally the relative amounts of various globins are affected because of deletions of the hemoglobin genes. *Thalassemia major*, the most severe form of the disease, is found in patients homozygous for a defect in the two β-chains with an excessive amount of F hemoglobin (HPFH: *hereditary persistence of fetal hemoglobin*). *Cooley's anemia* is also caused by β-chain defects. *Thalassemia minor* is a relatively milder form with some hemoglobin A$_2$ present, and usually a slight elevation of F hemoglobin is characteristic for the heterozygotes. In the β-thalassemias, different sections of the β-globin gene family from human chromosome 11p15.5 (5'-ε G$_\gamma$ $\psi\beta\delta\beta$-3') are missing. The deletions may involve only 600 bp from the 3'-end, as in β^0, or about 50 kbp, eliminating most of the family beginning from the 5'-end and retaining only the β-gene

at the 3'-end in human chromosome 11p. In the β-chain gene, about 100 point mutations and various deletions have been analyzed. In severe cases, the symptoms of the β-thalassemias are anemia, susceptibility to infections, bone deformations, enlargement of the liver and spleen, iron deposits, delayed sexual development, etc. They may appear within a few months after birth. In the α-thalassemias (in human chromosome 16p13), various members of the α-chain gene cluster (5'-$\zeta\psi\zeta\psi\alpha\psi\alpha\alpha2\alpha1\vartheta$-3') or even all four α-chains are defective, and the latter case results in hereditary hydrops fetalis (Bart's hydrops), entailing accumulation of fluids in the body of the fetus and severe anemia, resulting in prenatal death.

In α-thalassemia-1, both α1 and α2 genes are deleted; in thalassemia α-*thal-2*, only the left (α-*thal-2L*) or right end (α-*thal-2R*) of the α-gene(s) is lost. *Hb Lepore hemoglobin* is the consequence of unequal crossing over within the α-gene cluster, resulting in N-terminal-$\delta\beta$-C protein fusion. N-$\beta\delta$-C reciprocal protein fusion is called *Hb anti-Lepore* or *F thalassemia*. One of these types of hemoglobins was first reported in 1958 in the Lepore Italian family. Since then, a large number of Lepore-type hemoglobins were discovered in various parts of the world and named Greece, Washington, Hollandia, etc., hemoglobin. Thalassemias have a much higher incidence worldwide in areas where malaria is a common disease. Homozygotes for thalassemias rarely contribute to the gene pool; thus, the heterozygotes appear to have a selective advantage in the maintenance of this condition. Some recent studies, however, indicate that α-thalassemia homozygotes have a higher incidence of uncomplicated childhood thalassemia and splenomegaly (enlargement of the spleen, an indication of infection by *Plasmodium*). This increases susceptibility to the relatively benign *P. vivax*, which may provide some degree of immunization for later stages of life against the more severe disease caused by *P. falciparum* infection.

This relatively high incidence of thalassemias in tropical and subtropical areas and the well-known molecular genetics mechanism makes this disease a candidate for somatic gene therapy by either bone marrow transplantation or transformation. The prevalence of thalassemias varies from about $5-10 \times 10^{-4}$. In some geographically isolated areas in the Mediterranean region, the frequency of thalassemias may be much higher, however. Thalassemias are frequently associated with sickle cell anemias, which are β-globin defects and aggravate the conditions. Prenatal diagnosis is possible by the use of protein or DNA technologies. Fetal mutation may be identified after PCR amplification of chorionic samples (9–12 weeks) or in directly withdrawn cells from the amniotic fluid after the third month of pregnancy. Hydrops can be identified by ultrasonic techniques during the second trimester. Transformation by a functional β-globin gene may ameliorate the condition in a mouse model. *See* ATRX; genetic screening; hemoglobin; hemoglobin evolution; hydrops fetalis; Juberg-Marsidi syndrome; methemoglobin; PCR; plasma proteins; *Plasmodium*; sickle cell anemia; trimester; unequal crossing over. (Weatherall, D. J. 2001. *Nature Rev. Genet.* 2:245.)

thalidomide (N-phtaloyl glutamimide)   Sedative and hypnotic agent of several trade names. It had also been used experimentally as an immunosuppressive and antiinflammatory drug. Its medical use as a tranquilizer during pregnancy caused one of the most tragic disasters of severe malformations of primate embryos and newborns. It does not appear to be mutagenic and it is not teratogenic for other mammals. It is effective against multiple myeloma and possibly some other cancers. *See* amelia; phocomelia; teratogen. (Ashby, J., et al. 1996. *Mutation Res.* 396:45; Meierhofer, C., et al. 2001. *BioDrugs* 15[10]:681; Richardson, P., et al. 2002. *Annu. Rev. Med.* 53:629.)

thallophyte   Plant, fungus, or algal body, which is a thallus. *See* thallus.

thallus   Relatively undifferentiated colony of plant cells without true roots, stems, or leaves.

thanatophoric dysplasia   Dominant lethal human (4p16.3) dwarfism caused by deficiency of the fibroblast growth factor receptor 3, FGFR3. FGFR3 has a constitutive tyrosine kinase activity that activates STAT1 transcription and nuclear translocation. This system also controls the p21$^{\text{WAF1/CIP1}}$ protein involved in cell cycle suppression. Its prevalence is about 2×10^{-4}. *See* achondroplasia; dwarfism; fibroblast growth factor; p21; receptor tyrosine kinase; Schneckenbecken dysplasia; STAT.

t haplotype   *See* brachyury.

thea (*Camellia sinensis*) 82 species, $2n = 2x = 30$. The epigallocatechin-3 component of green tea appears to be an angiogenesis inhibitor and thus can have an anticancer effect. *See* angiogenesis; angiostatin; endostatin; polyphenols. (Kuroda, Y. & Hara, Y. 1999. *Mutation Res.* 436:69.)

theileriosis   African parasitic infection of the cattle's immune system, causing disease and death to infected animals. In the T lymphocytes, casein kinase II, a serine/threonine protein kinase, is markedly higher. This protein is not identical with the common casein kinase found in lactating animals. (Pipano, E. & Shkap, V. 2000. *Ann. NY Acad. Sci.* 916:484.)

thelytoky   Parthenogenesis from eggs, resulting in maternal females. *See* arrhenotoky; chromosomal sex determination; deuterotoky; sex determination. (Normark, B. B. 2003. *Annu. Rev. Entomol.* 48:397.)

theobromine   Principal alkaloid in cacao bean, containing 1.5 to 3% of the base. Present also in tea and cola nuts. The TDLo orally for humans is 125 mg/kg. It is a diuretic, smooth muscle relaxant, cardiac stimulant, and vasodilator (expands vein passages). *See* caffeine; TDLo.

therapeutic cloning   Generation of (human) stem cells; induces them to differentiate into a certain type of tissue that can be used to replace damaged tissues without the risk of rejection. Transfer stem cells are prepared by dislodging the nucleus of an oocyte and they are replaced by the nucleus of an adult cell of the individual that is supposed to be treated by the stem cells. Such a modified oocyte (any type, possibly even from a different species) is then grown in culture until the blastocyst stage. Then the inner cell mass is excised and cultivated in vitro until differentiated cells (muscle, neural, or hematopoietic cells) develop under the direction of the experimental conditions applied. Subsequently, the cells can be transplanted into the body of an individual. Nuclear transfer is possible between different species, but the few experiments carried out so far could not maintain steady normal development. An alternative to using oocytes is fusing an embryonic cytoplast with an appropriate karyoplast. The ES cells do not develop into an embryo even it is transplanted into a uterus because the required extraembryonic membranes cannot be formed. If such a stem cell is introduced into a blastocyst, chimeric embryos (host + ES) may form in the mouse. These cells may form different types of tissues (pluripotent) but are apparently not totipotent. Therapeutic cloning is thus similar in some respects to reproductive cloning, but current laws prohibit the latter in the majority of countries. Some of the cloned embryos suffered from (lethal) developmental anomalies. Although the majority of implanted stem cells develop rather normally, some may become teratogenic and carcinogenic when transferred into mice. *See* blastocyst; cloning animals and humans; cytoplast; grafting in medicine; graft rejection; karyoplast; nuclear transplantation; stem cell; totipotency. (Mitalipov, S. M. 2000. *Ann. Med.* 32:462; Illmensee, K. 2002. *Differentiation* 69:167; Rideout, W. M., III, et al. 2002. *Cell* 109:17.)

therapeutic index   Reveals the dose of a drug with optimal, threshold, maximal tolerated or lethal, etc., effect in a particular organism or tissue.

therapeutic radiation   *See* radiation hazard assessment.

therapeutic vaccine   Vaccination may decrease the level of an infective agent (e.g., HIV), then eventually the natural immune system may overpower the infection. *See* vaccines.

thermal asymmetric interlaced PCR   *See* tail-PCR.

thermal cycler   Automatic, programmable incubator used for polymerase chain reaction. *See* PCR.

thermal neutrons   *See* physical mutagens.

thermal tolerance   Genetically determined, and in plants it may be affected by fatty acid biosynthesis and abscisic acid deficiency, etc. Heat-shock proteins have also been credited with it. In *Tetrahymena thermophila*, a small cytoplasmic RNA (G8 RNA) is responsible specifically for thermotolerance independently from heat-shock response. *See* antifreeze protein; heat-shock proteins; RNA G8; temperature-sensitive mutation.

Thermatoga maritima   Thermophile (80°C optimum) bacterium; has a completely sequenced circular DNA genome of 1,860,725 bp with 46% G + C content; 24% of its genes appear homologous to archaea bacteria. (Nelson, K. E., et al. 1999. *Nature* 399:323.)

thermodynamics, laws of   (i) When a mechanical work is transformed into heat, or the reverse, the amount of work is equivalent to the quantity of heat. (ii) It is impossible to continuously transfer heat from a colder to a hotter body.

thermodynamic values Of chemicals, are measured in calories; 1 calorie = 4.1840 absolute joule. *See* joule.

thermoluminescent detectors For personnel monitoring a device in a small crystalline detector (lithium fluoride, lithium borate, calcium fluoride, or calcium sulfate, and metal ion traces as activators), radiation is absorbed in the crystalline body. Upon heating, it is released in the form of light. Useful range 0.003–10,000 rem. Radiophotoluminiscent (RPL) and thermally stimulated exoelectron emission (TSEE) detectors may be used instead of film badges in radiation areas. *See* radiation measurements.

Thermoplasma acidophilum Archaea bacterium with a completely sequenced genome of ~1.5 × 10^6 bp. Its temperature optimum is 59°C, pH optimum 2. Its cells do not have cell walls. *See* Archaea. (Ruepp, A., et al. 2000. *Nature* 407:508.)

thermosome *See* chaperonins.

thermotolerance Some mutants of plants display increased growth relatively to the wild type at higher temperature. Some of the temperature sensitivity is associated with differences in fatty acid biosynthesis. The tolerance of higher temperatures in plants is correlated with a reduced level in trienoic fatty acids. Trienoic acid synthesis is controlled by omega-3 fatty acid desaturase in the chloroplast membrane. If this desaturase is silenced, photosynthesis is improved at higher temperatures. *See* antifreeze protein; cold-regulated genes; fatty acids; temperature-sensitive mutation. (Mirkes, P. E. 1997. *Mutation Res.* 396:163.)

THF *See* tetrahydrofolate.

θ Symbol of recombination (and some other) fractions. *See* w.

theta replication θ, a stage of bidirectional form of replication in a circular DNA molecule. When the two replication forks reach about halfway toward the termination sites, the molecule resembles the Greek letter θ. *See* bidirectional replication; DNA replication; prokaryotes; replication; θ replication.

θ-type replication Occurs generally in small circular DNA molecules starting at a single origin and following a bidirectional course. *See* bidirectional replication; replication.

thiamin (vitamin B$_1$) Its absence from the diet causes beriberi, alcoholic neuritis, and Wernicke-Korsakoff syndrome. Its deficiency was the first auxotrophic mutation in *Neurospora*, and it is the most readily inducible auxotrophic mutation at several loci in lower and higher green plants such as algae and *Arabidopsis*. The vitamin is made of two moieties: 2-methyl-4-amino-5-aminomethyl pyrimidine and 4-methyl-5-β-hydroxyethyl thiazole. The pyrimidine requirement of the thiamin mutants cannot be met by the precursors of nucleic

acids. In *Arabidopsis*, more than 200 absolute auxotrophic mutations for thiamine have been isolated at five loci. This is remarkable because absolute auxotrophs in other metabolic pathways still need to be found in higher plants. *See* megaloblastic anemia; thiamin pyrophosphate; Wernicke-Korsakoff syndrome. (Rédei, G. P. & Koncz, C. 1992. *Methods in Arabidopsis Research*, Koncz, C., et al., eds., pp. 16–82. World Scientific; Miranda-Ríos, J., et al. 2001. *Proc. Natl. Acad. Sci. USA* 98:9736.)

thiamin pyrophosphate Coenzyme of vitamin B$_1$. It is an essential cofactor for pyruvate decarboxylase (alcoholic fermentation), pyruvate dehydrogenase (synthesis of acetyl Co-A), α-ketoglutarate dehydrogenase (citric acid cycle), transketolase (photosynthetic carbon fixation), and acetolactate synthetase (branched-chain amino acid biosynthesis). *See* thiamin.

thiazolidinediones Antidiabetic (type 2) drugs enhancing sensitivity to insulin. They are ligands to PPAR-γ receptors of adipocytes. *See* adipocytes; diabetes; insulin; PPAR.

Thiazolidinedione.

thin-layer chromatography (TLC) Separation of (organic) mixtures within a very thin layer of cellulose or silica gel layer applied uniformly onto the surface of a glass plate or firm plastic sheet and used in a manner similar to paper chromatography. The material is applied about 2 cm from the bottom, then the plate is dipped into an appropriate solvent (mixture). Generally, the substances are separated rapidly with excellent resolution. Identification is made generally on the basis of natural color or with the aid of special color reagents. *See* Rf.

thioester Acyl groups covalently linked to reactive thiol; they are high-energy acyl carriers in coenzyme A; also thioester may assist the formation of oligopeptide without any cooperation by ribosomes. There are suggestions that thioethers are the relics of prebiotic conditions when sulfurous (volcanic) environment existed and might have had a role in the origin of living cells as energy carriers. *See* disulfide; sulfhydryl.

thioguanine (2-amino-1,7-dihydro-6H-purine-6-thione) Antineoplastic/neoplastic agent. Its delayed cytotoxicity is attributed to postreplicative mismatch repair. After incorporation into DNA, it is methylated and pairs with either thymine or cytosine. The immunosuppressant azathioprine (6-[1-methyl-4-nitroimidazol-5-yl]-thiopurine) used in organ transplantation can be converted to thioguanine and may be the cause of the increased incidence of cancer after transplantation. *See* base analogs; DNA repair; hydrogen pairing; mismatch. (Nelson, J. A., et al. 1975. *Cancer Res.* 35:2872.)

thiol (SH⁻) Compounds may reduce DNA breakage by scavenging radiation-induced hydroxyl radicals and may repair the radicals chemically on DNA by hydrogen atom transfer. *See* sulfhydryl.

thioredoxins (TRX) ~12 kDa dithiol proteins that mediate the reduction of disulfide bonds in proteins. Also, light regulates photosynthesis through reduced thioredoxin linked to the electron transfer chain by ferredoxin. Light modulates translation in the chloroplasts by redox potential. In insects, it may substitute for glutathione reductase. *See* glutaredoxin; lysosomes; oxidative stress; photosynthesis; photosystems, pullulanase. (Yano, H., et al. 2001. *Proc. Natl. Acad. Sci. USA* 98:4794.)

thiostrepton ($C_{72}H_{85}N_{19}O_{18}S_5$) Antibiotic that binds to *E. coli* rRNA and to yeast ribosomes 10-fold less effectively. (Porse, D. T., et al. 1998. *J. Mol. Biol.* 276:391.)

thiotepa Alkylating agent formerly used as an antineoplastic drug. It induces a very high frequency of sister chromatid exchange. *See* alkylating agent; sister chromatid exchange.

thiouracil (4-hydroxy-2-mercaptopyrimidine) RNA base analog, carcinogen (?).

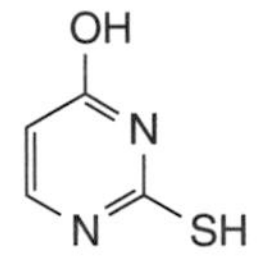

Thiouracil.

third messengers Propagate signals of the second messenger, thus activating or deactivating a series of genes. *See* second messenger.

Thomsen disease *See* myotonia.

thorax Chest of mammals, the segment behind (posterior) the head in *Drosophila*.

three-hybrid system Construct that detects the role of RNA-protein interactions (diagram below).

See one-hybrid binding assay; split-hybrid system; two-hybrid system. (Koloteva-Levine, N., et al. 2002. *FEBS Lett.* 523:73.)

three-point test cross It uses three genetic markers and thus permits the mapping of these genes in a linear order. *See* mapping, genetic; test cross.

three-way cross A single-cross (the F_1 hybrid of two inbred lines) is crossed with another inbred. The purpose is to test the performance of the single cross. *See* combining ability.

threonine This amino acid is derived from oxaloacetate via aspartate and immediately from

$$(HO)_2(PO)OCH_2CH(NH_2)COOH \quad \rightarrow \quad CH_3CH(OH)CH(NH_2)COOH$$
O-phosphorylhomoserine   *threonine*   threonine
synthase

Threonine dehydratase degrades threonine into α-ketobutyrate and NH_4^+, then five steps are required from α-ketobutyrate to produce isoleucine. The first of these steps is mediated by threonine deaminase. *See* isoleucine-valine biosynthetic steps; threoninemia.

threonyl tRNA synthetase (TARS) Charges tRNA^Thr by threonine. It is encoded in human chromosome 5 in the vicinity of the leucyl-tRNA synthetase gene. *See* aminoacyl-tRNA synthetase.

threshold traits Expressed conditionally when the liability reaches a certain level. These characters are frequently under polygenic control, yet they may fall into two recognizable classes, although with special techniques more classes may be revealed. Many pathological syndromes may fall into this category, making it very difficult to determine the genetic control mechanisms. *See* liability; syndrome.

thrombasthenia Autosomal-dominant and autosomal-recessive forms involve a platelet defect causing anemia and bleeding after injuries. The recessive form was assigned to human chromosome 17q21.32. *See* anemia; platelet; thrombophilia.

thrombin (fibrinogenase) Serine protease enzyme that converts fibrinogen to fibrin and causes hemostasis (blood clotting). Thrombin also has another role as an anticoagulant

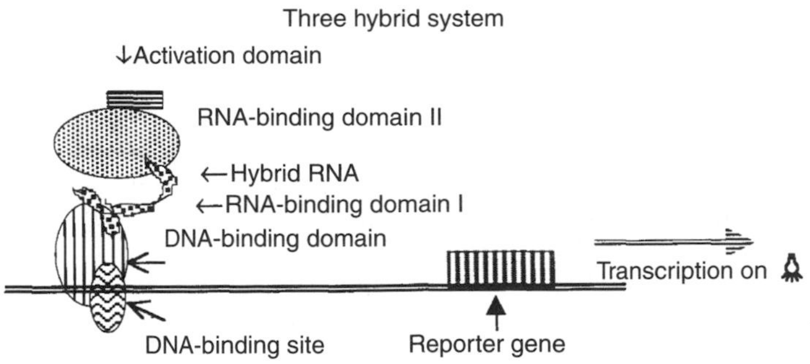

Three hybrid system

↓Activation domain

RNA-binding domain II
←Hybrid RNA
←RNA-binding domain I
DNA-binding domain

Transcription on

DNA-binding site   Reporter gene

The DNA-binding domain (e.g., LexA bacterial protein); the RNA-binding domain (e.g., MS2 phage coat protein); the top binding proteins contain an RNA-binding domain (e.g., IRP1[iron-response element]) and an activation domain (e.g., Gal4). The hybrid RNA (two MS2 phage RNAs) links the proteins together, and in case of function, the reporter gene (e.g., *LacZ*) is turned on. (Modified after D. J. S. Gupta, et al. 1996. *Proc. Natl. Acad. Sci. USA* 93:8496.)

by the proteolytic activation of protein C in the presence of Ca^{2+}. These seemingly conflicting functions are modulated by allosteric alteration brought about by Na^+ and the thrombomodulin protein, changing the substrate specificity of thrombin. The level of activated protein C is increased further by a compound named LY254603. The protease-activated G-protein-coupled receptor of thrombin (PAR1) is vitally important for embryonic development of mice. *See* anticoagulation; antihemophilic factors; fibrin; fibrin-stabilizing factor deficiency; hemostasis; PAR; plasmin; protein C. (Coughlin, S. R. 2000. *Nature* 407:258; Griffin, C. T., et al. 2001. *Science* 293:1666.)

thrombocytes   Platelets.

thrombocytopenia (TAR syndrome)   Autosomal-dominant bleeding disease (10p11.2-p12) caused by low platelet counts. Autosomal-recessive (11q23.3-qter deletion and 21q22.1-q22.2) forms are associated with heart and kidney anomalies and absence of the radius (the thumb-side arm bone), phocomelia, etc. Another thrombocytopenia is linked in human chromosome Xp11 and it is allelic to the Wiskott-Aldrich gene, WAS. Mutation in GATA1 may also cause the disease. The May-Hegglin anomaly, Fechtner syndrome, and Sebastian syndrome involve macrothrombocytopenia (giant platelets) and leukocyte inclusions besides some specific symptoms. All three of these diseases show mutations in nonmuscle myosin heavy chain 9. In some instances, Epstein syndrome and Alport syndrome display macrothrombocytopenia. *See* Alport disease; dyserithropoietic anemia; GATA; giant platelet syndrome; Holt-Oram syndrome; May-Hegglin anomaly; phocomelia; Roberts syndrome; thrombopathic purpura; Wiskott-Aldrich syndrome. (Heath, K. E., et al. 2001. *Am. J. Hum. Genet.* 69:1033.)

thrombomodulin   *See* anticoagulation; thrombin.

thrombopathia (essential athrombia)   Collection of blood-clotting anomalies caused by problems in aggregation of the platelets. *See* hemophilias; hemostasis; platelet abnormalities.

thrombopathic/thrombocytopenic purpura (TTP, also called von Willebrand-Jürgen's syndrome, 9q34)   Caused by an autosomal-recessive condition with symptoms resembling the bleeding and platelet abnormalities present in Glanzmann's disease. The primary defect may involve the platelet membrane. Other symptoms may vary from case to case. Blood transfusion may shorten the time of bleeding. The basic defect is in ADAMS13 zinc metalloproteinase genes. *See* Glanzmann's disease; hemophilias; hemostasis; platelet abnormalities; thrombocytopenia. (Levy, G. G., et al. 2001. *Nature* 413:488.)

thrombophilia   Complex blood-clotting disease that may occur as a consequence of mutation in any of the genes of antithrombin III, protein C, protein S, antihemophilic factor V, plasminogen, plasminogen activator inhibitor, fibrinogens, heparin cofactor, thrombomodulin, etc. Mutation at the 3′-untranslated region of prothrombin may result in increased mRNA level and protein synthesis due to this gain-of-function and to the disease. *See* giant platelet syndrome; prothrombin deficiency. (Gehring, N. H., et al. 2001. *Nature Genet.* 28:389.)

thromboplastin   *See* antihemophilic factors.

thrombopoietin   Regulates blood platelet formation and megakaryocytopoiesis. It is a member of a cytokine receptor superfamily and is similar to erythropoietin and granulocyte colony-stimulating factor receptors. *See* erythropoietin; G-CSF; megakaryocytes; platelet; thrombocytopenia. (Kato, T., et al. 1998. *Stem Cells* 16:322.)

thrombosis   Type of obstruction of the blood flow caused by aggregation of platelets, fibrin, and blood cells. May be caused by a mutation in serpin. Antihemophilic factors VII, VIII, IX, XI, XII, von Willebrand disease, tissue plasminogen activator, homocysteine, and the activated protein C ratio displayed genetic correlation with the incidence of thrombosis. *See* APC; cyclooxygenase; protein C; protein C deficiency; protein S; serpin; thrombocyclins. (Souto, J. C., et al. 2000. *Am. J. Hum. Genet.* 67:1452.)

thrombospondin (THBS)   Glycoprotein inhibitors of angiogenesis. THBS1 (15q15) is of 180 KMW platelet membrane protein, also occurring on the endothelium, fibroblasts, and smooth muscles. THBS2 (6q27) binds thrombin, fibrinogen, heparin, plasminogen, etc. THBS2 functions as a tumor inhibitor by curtailing angiogenesis. *See* angiogenesis; apoptosis; COMP. (Hawighorst, T., et al. 2001. *EMBO J.* 20:2631; Rodriguez-Manzaneque, C. C., et al. 2001. *Proc. Natl. Acad. Sci. USA* 98:12485; Adams, J. C. 2001. *Annu. Rev. Cell Dev. Biol.* 17:25.)

thrombotic disease   *See* protein C deficiency.

thromboxane   Induces the aggregation of platelets and acts as a vasoconstrictor. Thromboxanes are antagonists of prostacyclin G_2 and may mediate intrauterine growth retardation. *See* cyclooxigenase; prostaglandins.

Thy-1   19–25 kDa single chain glycoprotein expressed on mouse thymocytes but not on mature T cells of humans or rats.

thylakoid   Flat sac-like internal chloroplast membrane; when stacked, thylakoids appear as grana. Diacylglycerol galactolipids are common in these membranes but absent from others. An estimated 80 proteins constitute the intrathylakoid lumen. *See* chloroplast; girdle bands; grana. (Dalbey, R. E. & Kuhn, A. 2000. *Annu. Rev. Cell Dev. Biol.* 16:51; Schubert, M., et al. 2002. *J. Biol. Chem.* 277:8354; Kota, Z., et al. 2002. *Proc. Natl. Acad. Sci. USA* 99:12149.)

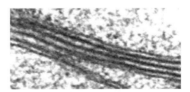

Electron micrograph of thylakoids.

thymidine   Nucleoside of thymine; thymine plus pentose (deoxyribose or ribose).

thymidine kinase (thymidine phosphorylase)   *See* TK.

thymidylate synthetase   Mediates the synthesis of thymidylic acid (dTMP) from deoxyuridine monophosphate (dUMP). Its gene has been assigned to human chromosome 18p11.32. Its inhibition may lead to apoptosis. A number of non-symbiotic archaea and bacteria lack this enzyme and use another protein, dependent on reduced flavin nucleotides rather than tetrahydrofolate as the source for reduction as it is the case in the majority of organisms. This is not a part of the salvage pathway. *See* apoptosis; salvage pathway. (Myllykallio, H., et al. 2002. *Science* 297:105.)

thymidylic acid (nucleotide of thymine)   Thymine + pentose + phosphate.

thymine   Pyrimidine base occurring almost exclusively in DNA; the exception is the T arm of tRNA. *See* pyrimidines; tRNA; formula at T.

thymine dimer   UV light–induced damage in DNA covalently cross-linking adjacent thymine residues through their 5 and 6 C atoms. This cyclobutane structure interferes with replication and other functions of DNA. Cytidine and uridine also can form similar dimers. The dimers can be eliminated by enzymatic excision and replacement replication (excision or dark repair). Alternatively, the dimers can be split by DNA photolyase, an enzyme that is activated by visible light maximally at 380 nm wavelength (light repair). *See* cyclobutane dimer; cys-syn dimer; DNA polymerases; DNA repair; pyrimidine-pyrimidinone photoproduct. (Medvedev, D. & Stuchebrukhov, A. A. 2001. *J. Theor. Biol.* 210:237.)

thymocyte   Precursor of T lymphocytes. Bone marrow stem cells migrate into the thymus and in response to antigens develop from naive CD4$^-$CD8$^-$ cells with the cooperation of T-cell receptors into double positive CD4$^+$CD8$^+$ cells that undergo clonal selection with the assistance of MHC molecules and eventually become specific T cells. *See* antigen-presenting cells; immune system; memory, immunological; MHC; T cell; T-cell receptor; thymus.

thymoma   Cancer of the thymus. *See* AKT oncogene.

thymosine β4   Binds monomeric actin and prevents its polymerization into filaments. Its expression increases when cancer cells metastasize. *See* actin; metastasis. (Hall, N. R. & O'Grady, M. P. 1989. *Bioessays* 11:141.)

thymus   Bilobal organ with three functionally important compartments: the subcapsular zone, the cortex, and the medulla. Immature lymphoid cells coming from the bone marrow invade it. The subcapsular space contains most of the immature CD4$^-$ and CD8$^-$ lymphoid stem cells. When they enter the cortex, they begin to express CD molecules and rearrange T-cell receptors, forming the different TCR αβ heterodimers. The epithelial cells in the cortex express MHC I and II molecules. The fate of the T cell is determined here in response to MHC molecules carried by antigen-presenting cells. CD4$^+$CD8$^-$ cells with TCR recognizing the MHC II complex or CD4$^-$ CD8$^+$ with TCR specific for MHC I will leave the thymus and populate the secondary lymphoid tissue. Cells that do not acquire self-peptide–MHC specificity will be eliminated and those that have strong avidity for self-peptide–MHC will die by apoptosis because they would be autoreactive (nonautoimmune). Aging (involution) reduces thymic tissues and it once was thought that their activity was lost too. Newer evidence indicates the thymus can generate new peripheral T cells after antiviral treatment of HIV patients. *See* autoimmune; HIV; MHC; spleen; T cell; T-cell receptor; TCR genes.

thyroglobulin   Iodine-containing protein in the thyroid gland that has hormone-like action upon the influence of the pituitary hormone. *See* animal hormones.

thyroid carcinoma   Transforming sequence was localized to human chromosome 10q11-q12, and a tumor suppressor gene in human chromosome 3p may be involved. Susceptibility loci to nonmedullary thyroid carcinoma have been revealed at 2q21, 19p13.2, and 1q21. (McKay, J. D., et al. 2001. *Am. J. Hum. Genet.* 69:440.)

thyroid hormone resistance   Dominant hyperthyroxinemia mutations (ERBA2) due to defects in the thyroid hormone receptor (3p24.3) are known. (Tsai, M. J. & O'Malley, B. W. 1994. *Annu. Rev. Biochem.* 63:451.)

thyroid hormone responsive element *See* goiter; hormone-response elements; hormones; regulation of gene activity; TRE.

thyroid hormone unresponsiveness   Autosomal recessive in humans. *See* hyperthyroidism.

thyroid peroxidase deficiency (TPO)   Group of recessive human chromosome 2p13 defects involving the incorporation of iodine into organic molecules.

thyroid-stimulating hormone (TSHB, 1p13)   The β-chain and its deficiency lead to hypothyroidism, goiter, and cretinism. *See* goiter.

thyroid transcription factor   TTF-2 defect is responsible for thyroid agenesis and cleft palate in mice. The human homologue FKHL1 (9q22) controls the same functions and cochanal atresia (closure of the nasal passageways).

thyronine (3p-[p(p-hydroxyphenoxy)-phenyl]-L-alanine)   Component of thyroglobulin of the thyroid hormone. It generally occurs as 3,5,3′-triiodothyrosine. *See* VDR.

thyrotropic   Affecting (targeting) the thyroid gland.

thyrotropin   Deficiency is apparently due to a defect in the thyroid-stimulating hormone β-chain defect at 1p13. Deficiency of thyrotropin-releasing hormone (TRH, 3q13.3-q21) results in hypothyroidism and various malformations including defects in the development of the central nervous system. Hyperthyroidism may involve fast pulsation of the heart, goiter, and adenoma. Hyperthyroidism may be transient during pregnancy. *See* adenoma; goiter.

thyroxin *See* goiter.

Thyroxin.

Thyroxin-binding globulin (TBG, Xq22.2) Its deficiency generally does not lead to disease, although in Graves' disease it may be more common. *See* goiter; Graves' disease.

TI antigen Stimulates antibody production independently of T cells and in the absence of MHC II molecules. TI type 2 (TI-2) has polysaccharide antigens. They are usually large molecules with repeating epitopes, activate the complement, and are rather stable. TI-1 type is mitogenic for mature and neonatal B cells. *See* antibody; antigen; MHC; T cell. (Vinuesa, C. G., et al. 2001. *Eur. J. Immunol.* 31:1340.)

TIBO (O-Tibo and Cl-Tibo) Nonnucleoside (benzothiadiazepin derivatives) inhibitors of HIV-1 reverse transcriptase. *See* acquired immunodeficiency syndrome; AZT; Neviorapine.

TIC Translation initiator-dependent cofactor. *See* TAF.

Tid50 56 kDa mitochondrial protein of *Drosophila* encoded at 2−104 in the nucleus by the *l(2)Tid* tumor suppressor gene. It is homologous to the DnaJ chaperone. *See* chaperones; DnaK; tumor suppressor factors.

Tie-1, Tie-2 Receptor tyrosine kinases expressed during endothelial cell growth and differentiation of the blood vessels. *See* Flk-1; Flt-1; signal transduction; transmembrane proteins; tyrosine kinase; vascular endothelial growth factor. (Lin, T. N., et al. 2001. *J. Cereb. Blood Flow Metab.* 21:690.)

tier Ordered arrangement by increasing stringency of a series of tests, e.g., the first tier provides an overview of potential mutagens, but subsequent tiers (using different techniques) may be necessary for clearance even when the first test might have been negative. Each tier may provide a different weight of evidence.

TIF1-α Interacts with the steroid hormone receptor. *See* hormone receptors; steroid hormones.

TIF-1B Transcription initiation factor of mouse binding to the core promoter of rRNA genes and controlling RNA polymerase I function. *See* transcription factors.

tiger (*Panthera tigris*) $2n = 38$. The tiger cat (*Leopardus tigrina*): $2n = 36$.

tight junction (zonula occludens) Forms a seal between adjacent plasma membranes and provides a barrier to paracellular leakage of membrane lipids and proteins, thus guarding cellular polarity. Tight junctions are apical domains of polarized epithelial and endothelial cells. The claudin gene family (CLAUDIN 1 senescence-associated membrane protein; CLAUDIN 3 *Clostridium perfringens* receptor [7q11]; CLAUDIN 4 *C. perfringens* enterotoxin receptor; CLAUDIN 5 velocardiofacial syndrome protein [22q11.2]; CLAUDIN 11 mediates sperm and nerve functions [3q26.2-q26.3]; CLAUDIN 14 deficiency responsible for deafness [21q22.3]; CLAUDIN 16 encodes paracellin, a paracellular conductance protein associated with hypomagnesia [3q27], and others) is distributed over the genome. *See* deafness; infertility; velocardiofacial syndrome. (Tsukamoto, T. & Nigam, S. K. 1999. *Am. J. Physiol.* 276:F737; Tsukita, A., et al. 2001. *Nature Rev. Mol. Cell Biol.* 2:285; Gonzalez-Mariseal, L., et al. 2003. *Progr. Biophys. Mol. Biol.* 81:1.)

tiglic acid Methyl-2-butenoic acid. It is present with geranyl in the ornamental geranium and as an ester in the oil of chamomile (*Matricaria chamomilla*), an herbal medicine used as a tea or disinfectant.

tiglicacidemia Defect in the degradation of isoleucine to propionic acid; large amounts of tiglic acid accumulate in the urine.

TIGR The Institute for Genomic Research in Rockville, MD. *See* databases.

TIL *See* tumor-infiltrating lymphocytes.

tiling Generating a longer (minimally overlapping) DNA fragment map spanning the overall length of the genome or a chromosome. *See* contig; DNA crystals; genome project; physical map; scaffold in genome sequencing; sequence-tagged connectors. (Siegel, A. F., et al. 1999. *Genome Res.* 9:297.)

tiller Lateral shoot of grasses arising at the base of the plant.

TILLING Targeting-induced local lesions in genomes attempt to induce mutations by chemical mutagens (ethylmethane sulfonate). Denaturing high-performance liquid chromatography (DHPLC) is used to detect base alterations by heteroduplex analysis. (McCallum, C. M., et al. 2000. *Nature Biotechnol.* 18:455.)

TIM Transfer inner membrane is a protein complex regulating the import of proteins into mitochondria. *See* mitochondria; TOM.

time-lapse photography Records events continuously as they take place in time.

timeless (Tim) Protein of *Drosophila* is involved in the circadian clock. *See* circadian clock.

time of crossing over Appears to coincide with the meiotic prophase (late leptotene and early diplotene, probably at zygotene). Some experimental data indicate that treatments at the S phase have an effect on the outcome. It is difficult to assess, however, whether these effects are direct or indirect. Meiosis is under the control of a long series of genes acting sequentially and cooperatively, and any of these may affect crossing over. *See* crossing over; recombination; recombination mechanisms. (Allers, T. & Lichten, M. 2001. *Cell* 106:45.)

TIMP Tissue inhibitor metalloproteinase. *See* metalloproteinases; night blindness (Sorsby). (Brew, K., et al. 2000. *Biochim. Biophys. Acta* 477:267.)

tinkering in evolution Idea that evolution does not proceed on the basis of purposeful plans. Rather, it uses the means at hand at a particular stage. *See* evolution. (Jacob, F. 2001. *Ann NY Acad. Sci.* 929:71.)

TIP (1) Tumor-inducing principle. *See Agrobacterium*. (2) Tail-interacting protein. TIP47 (47 kDa) recognizes the cytoplasmic domains of mannose-6-phosphate and binds to Rab9. It facilitates the endosome to Golgi transport and recruits effector proteins for appropriate membrane targeting. *See* endosome; Golgi; mannose-6-phosphate receptor; RAB. (Caroll, K. S., et al. 2001. *Science* 292:1373.)

Ti plasmid Large (about 200 kbp) tumor-inducing plasmid of *Agrobacterium tumefaciens* is responsible for crown gall disease of dicotyledonous plants. There are a few particularly important regions in this plasmid. The T-DNA of the octopine plasmid is divided into the left (T_L) and right (T_R) segments flanked by the two border sequences (B_L and B_R) and several genes are in between. The nopaline-type Ti plasmid (pTiC58) has only a single 20 kb T-DNA. The virulence gene cascade and its function are described under virulence genes of *Agrobacterium*. The origin of vegetative replication (*oriV*) functions during proliferation of the cells; the nearby *Inc* (incompatibility) site determines host specificity. *oriT* is the origin of replication operated during conjugation. It is often called *bom* (base of mobilization) or CON (conjugation) because the synthesis of transferred DNA begins there. The actual transfer starts when the mob site-encoded protein (Mob, synonym Tra) attaches to a specific nick site, a single-strand cut, and one of the strands (the T strand) is transported to the recipient bacterial cell through the cooperation of some pore proteins. The oriT and Mob complex includes proteins TraI, TraJ, and TraH. TraJ attaches to a 19 bp sequence in the vicinity of the nick and also binds TraI, which is a topoisomerase. TraH promotes these bindings. The process is a rolling circle-type replication and transfer. The integrity of virulence genes *virD* and *virB* is absolutely required for productive conjugation. *See Agrobacterium*; conjugation; rolling circle; T-DNA; transformation; virulence genes of *Agrobacterium*. (Hellens, R., et al. 2000. *Trends Plant Sci.* 5:446; Suzuki, K., et al. 2000. *Gene* 242:331.)

TIR Terminal inverted repeats such as occur in transposons: ABCD---DCBA.

TIS1 *See* NGFI-B; nur77.

TIS-8 Mitogen-induced transcription factor. *See* egr-1; NGFI-A.

TIS11 Transcription factor inducible by various hormones (similar to Nup475). *See* Nup475.

TIS11b Murine homologue of cMG1. *See* cMG1.

TIS11d Transcription factor with 94% identity to 367 amino acids in TIS11b. *See* TIS11b.

Tiselius apparatus Early model of electrophoretic separation equipment.

tissue culture In vitro culture of isolated cells of animals and plants. *See* aseptic; axenic; cell culture; cell fusion; cell genetics; embryogenesis, somatic; organ culture; somatic cell fusion.

tissue engineering Purposeful culture of stem cells for producing tissues and organs. *See* progenitor; stem cells.

tissue factor (TF) Cell-surface glycoprotein mediating blood clotting after injuries by its interaction with clotting factor VIIa. *See* antihemophilic factors; blood-clotting pathways.

tissue microarray (TMA) Adaptation of microarray hybridization to tissue samples in order to reveal the cellular location of gene activity at the DNA, RNA, or protein level. Cylindrical core specimens (biopsies) are acquired from formalin-fixed paraffin-embedded tissues and arrayed into high-density TMA blocks. Archival specimens are also suited for this analysis. Then up to 300 5-μm sections are prepared for probing with DNA, RNA, or protein using in situ hybridization (or FISH) or immunostaining. A single TMA allows the simultaneous study of targets in thousands of specimens on microscopic slides. TMA can be used to study cancerous

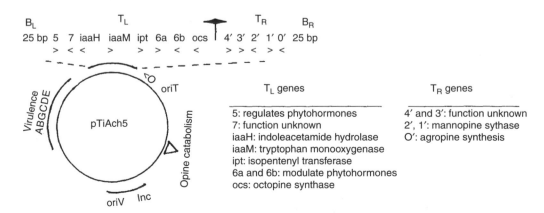

Major landmarks of the Ti octopine plasmid, pTiACH5. Arrowheads indicate the direction of transcription. ⊤ Marks the $T_L - T_R$ boundary.

tissue but it is not suitable for clinical diagnostics. *See* FISH, immunostaining, microarray hybridization. (Kallioniemi, O.-P., et al. 2001. *Hum. Mol. Genet.* 10:657.)

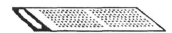

By TMA on a single slide the activity of one gene can be analyzed in 1000 tissue specimens. By microarray hybridization thousands of genes can be probed from a single tissue.

tissue plasminogen activator   Cleaves plasminogen to plasmin and enhances fibrinolysis; it controls blood coagulation. *See* blood-clotting pathways; PARs.

tissue remodeling   A coordinated series of events regulated by a balance between tissue proliferation and anti-proliferation factors such as cyclins a cyclin inhibitors and other metabolic regulators. Morphogenesis and various diseases involve alterations/anomalies in these processes. (Nabel, E. G. 2002. *Nature Rev. Drug Discovery* 1:587.)

tissue specificity   The expression of a gene is limited to certain tissues. Tissue specificity may be controlled at many levels (transcription initiation, differential recruitment of repressors or activators to the gene targets, promoter clearance, attenuator, transcription termination, etc.), e.g., the macrophage colony-stimulating factor receptor gene is constitutively expressed in many cell types, but elongation of the transcripts is permitted only in the macrophages. The evolutionarily determined size of a gene may have something to do with tissue specificity. In extremely fast dividing embryonic tissues, very large genes may not have enough time to be fully transcribed. The proportion of tissue-specific genes has been estimated by transformation of *Arabidopsis* with transcriptional and translational gene fusion vectors. Of 200 transgenic plants, about 10% displayed some degree of tissue-specific expression of the reporter gene (aph[3′]). *See* aph; Appendix II-10, cascade hybridization; constitutive mutation, housekeeping genes; isoenzymes; macrophage colony-stimulating factor; microarray hybridization; subtractive cloning; transcription, illegitimate. (Su, A. I., et al. 2002. *Proc. Natl. Acad. Sci. USA* 99:4465; Lunyak, V. V., et al. 2002. *Science* 298:1774.)

tissue-specific promoter   Facilitates gene expression limited to certain tissues or organs. *See* promoter.

tissue typing   Determining the genetic constitution of potential grafts before transplantation. Blood typing is used or DNA analysis (RFLP) or polymerase chain reaction. *See* blood typing; DNA fingerprinting; polymerase chain reaction.·

titer   Amount of a reagent in titration required for a certain reaction. Also, phage titer is the number of phage particles in a volume.

titin (connectin)   One of the largest protein molecules (3×10^6 M_r) along with nebulin; it forms a network of fibers around actin and myosin filaments in the skeletal muscles and may also be involved in chromosome condensation. The human titin gene contains 178/234 exons. Titin keeps myosin within the sarcomeres by being anchored to the Z discs (membrane bands in the striated muscles) and assures that the stretched muscles spring back. Titin is made up mainly of repeated modules, but at its C-end it has a threonine/serine kinase domain that specifically phosphorylates myosin. Calmodulin is required for its activation. The titin kinase domain itself is activated by phosphorylation at a tyrosine residue. Mutation in titin at 2q31 may cause dilated cardiomyopathy. *See* calmodulin; CaMK; cardiomyopathy; dystrophin; exon; glutenin; nebulin; sarcomere. (Labeit, S. & Kolmerer, B. 1995. *Science* 270:293; Machado, C. & Andrew, D. J. 2000. *J. Cell Biol.* 151:639; Gerull, B., et al. 2002. *Nature Genet.* 30:201; Hackman, P., et al. 2002. *Am. J. Hum. Genet.* 71:492.)

titration   Adding a measured amount of a solution of known concentration to a sample of another solution for the purpose of determining the concentration of the target solution on the basis of the appearance of a color or agglutination, etc. Also, the number of cells or phage particles in a series of dilutions can be determined by titration. *See* gene titration; titer.

TK   Thymidine kinase, which phosphorylates thymidine. Its gene in humans encodes the cytosolic enzyme that contains seven exons in the short arm of human chromosome 17q25-q25.3, but the mitochondrial TK gene is in chromosome 22q13.32-qter. The latter enzyme defect may be responsible for myoneurogastrointestinal encephalopathy. TK sequences are apparently highly conserved in different species. The herpes virus TK gene activates acyclovir and ganciclovir drugs and is used in gene therapy. *See* ganciclovir; gene therapy; HAT medium; myoneurogastrointestinal encephalopathy.

T$_L$ Left border of the T-DNA. *See* T-DNA; Ti plasmid.

TLC   *See* telomerase; thin layer chromatography.

TLE   *See* groucho; Tup.

TLF   *See* TBP.

t loop   *See* telomeres.

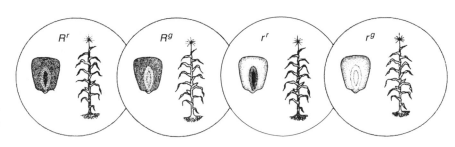

Tissue specificity of expression. Anthocyanin formation of four different *R* alleles of maize in the aleurone, embryo, stalk, tassel, and leaf tips.

TLV Threshold limit value. The upper limit or time-weighted average concentration (TWA) of a substance to which people can be exposed without adverse consequences.

T lymphocytes *See* T cells.

T$_m$ *See* melting temperature.

TM1, TM2 Transmembrane amino acid domains of *E. coli* transducers spanning the membrane layer inner space. TM1 is well conserved among the various transducers. TM2 is variable. *See* transducer proteins.

TMF TATA box modulatory factor. *See* promoter; TATA box; transcription.

TMP Thymidine monophosphate.

tmRNA (10Sa RNA) Basically a tRNA and an mRNA hybrid molecule. It is used in bacteria to protect against the detrimental effects of mRNAs that lack stop codons, which stall translation. tmRNA is first charged with alanine, then associates with EF-Tu. It binds to the A site of the ribosome and alanine is incorporated into the growing peptide chain. Simultaneously, the mRNA-like domain of tmRNA substitutes for the defective mRNA on the ribosome. tmRNA facilitates the inclusion of 10 additional amino acids into the growing chain, then the polypeptide is released when the translation is stopped. Proteins so formed are recognized by proteases because of the tmRNA tag and destroyed. This protects against defective proteins in the cell. This RNA is generated by RNase E cleavage. *See* aminoacylation; eEF; EF; EF-Tu.GTP; proteasomes; protein repair; regulation of gene activity; ribonuclease E. (Zwieb, C., et al. 1999. *Nucleic Acids Res.* 27:2063; Lee, S., et al. 2001. *RNA* 7:999; <http://www.indiana.edu/~tmrna>; <http://www.ag.auburn.edu/mirror/tmRDB>.)

TMS Tandem mass spectrometry is a method to detect defects in fatty acid metabolism, organic acidemias, and other human anomalies. *See* MALDI-TOF; mass spectrum.

TMV Tobacco mosaic virus is a single-stranded RNA virus of about 63,900 bases. Its cylindrical envelope contains 2,130 molecules of a 158-amino-acid protein. Its rod-shape particles are about 3,000 Å long and 180 Å in diameter. Historical reconstitution experiments from coat protein and RNA demonstrated that RNA can be genetic material. Its mutagenesis by nitrous acid contributed substantially to the genetic confirmation of RNA codons. Its genome encodes four

open-reading frames. Foreign genes attached to the regulatory tract of the coat protein or to the coat protein itself can be expressed in tobacco plants. Thus, tobacco plants may eventually be used to manufacture proteins, e.g., α-glycosidase that is deficient in Fabry disease patients or the antigen of Hodgkin lymphoma. *See* Fabry disease; genetic code; Hodgkin disease; plant vaccines, diagram. (Goelet, P., et al. 1982. *Proc. Natl. Acad. Sci. USA* 79:5818; Culver, J. N. 2002. *Annu. Rev. Phytopathol.* 40:287.)

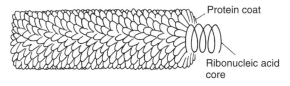

TMV.

Tn*3* family of transposons Genetic elements that can move within a DNA molecule and from one DNA molecule to another, carrying genes besides those required for transposition. The best-known representative is the Tn*3* element carrying ampicillin resistance (Ap^r). Tn*3* is 4,957 bp long with 38 bp inverted terminal repeats, leaving behind—after moving at the transposition target site—5 bp direct duplication. Although the terminal repeats of various Tn elements vary, some sequences are well conserved; the GGGG sequence is generally present outside the repeats and ACGPyTAAG is common inside one or both terminal repeats. In the Tn*3* group, an internal ACGAAAA is common. Normally transposition requires the presence of both terminal repeats; the presence of only one of them may still allow a lower-frequency transposition. The sites of integration may vary, yet AT-rich sequences are preferred and some homology between the terminal repeats and the insertional target may be needed. Some proteins of the host cell (IHF = integration host factor and FIS = factor for inversion stimulation) may facilitate transposase expression. Other members of the family are Tn*1*, Tn*2*, Tn*401*, Tn*801*, Tn*802*, Tn*901*, Tn*902*, Tn*1701*, Tn*2601*, Tn*2602*, and Tn*2660*, all about 5 kb length with 39 bp terminal repeats found in various gram-negative bacteria. Related is the $\gamma\delta$ (Tn*1000*) element in the F plasmid of 5.8 kb with 36/37 bp terminal repeats and IS*101* (insertion element *101*), a cryptic element. Tn*501* (8.2 kb) was the source of mercury resistance (Hg^r); Tn*1721*, Tn*1771* (11.4 kb) for tetracycline resistance (Tc^r); Tn*2603* (22 kb) for resistance to oxacillin (Ox^r), hygromycin (Hg^r), streptomycin (Sm^r), and sulfonamide (Su^r). Tn*21* (19.6 kb) carried resistance to Su^r, Hg^r, Ap^r, Sm^r, and Su^r. Tn*4* (23.5 kb) was endowed with genes for Ap^r, Sm^r, and Su^r. Tn*2501* (6.3 kb) was cryptic (i.e., expressed no genes besides the transposase). Tn*551* and Tn*917* (both 5.3 kb) from *Staphylococcus aureus* and *Streptococcus fecalis*, respectively, carried the erythromycin resistance (Ery^r) gene. The cryptic Tn*4430* (4.1 kb) was isolated from *Bacillus thüringiensis*, R46 from enterobacteria, and pIP404 from *Clostridium perfringens* plasmids coded for the resolvase protein. The terminal repeats are generally within the range of 35–48 bp. The transposition is usually replicative and its frequency is about 10^{-5} to 10^{-7} per generation. Integration of the element requires the presence of a specific target site, called the *res* site or IRS (internal resolution site), and genes *tpnA* (a transposase of about 110 MDa) and *tpnR*, encoding a resolvase protein (ca. 185 amino acids). The *res* site (about 120 bp) is where resolvase binds and mediates site-specific recombination, protecting DNA against DNase I. Within the *res* site are the promoters of *tnpR* and *tpnA* genes, functioning either in the same or in opposite directions, depending on the nature of the *Tn* element. Recombination between two DNA molecules requires the presence of at least two *res* sites in a negatively supercoiled DNA. Resolvase apparently has type I DNA topoisomerase function. After synapses, mediated by resolvase and multiple *res* sites, strand exchange and integration may result. The recipient molecule thus acquires the donor transposon. The transposition event requires replication of transposon DNA, then a fusion of the donor and the recipient replicons. This must be followed by a resolution of the cointegrate into a transposition product. *See* antibiotics; cointegration; gram-positive bacteria; resolvases; topoisomerase; transposable elements; transposon. (Sherratt, D. 1989. *Mobile DNA*, Berg, D. E. & Howe, M. M., eds., p. 163. Am. Soc. Microbiol., Washington, DC.)

Tn5 Bacterial transposon of 5.8 kb of the following structure:

and the transposon may be present as a monomer or as a dimer. Transposition may not require special homologous target sequences, yet some targets represent "hot spots" (displaying G•C or C•G pairs next to the 9 bp target duplications) because they are preferred for insertion. Less frequently used targets generally show G•C and A•T pairs at the ends. Tn5 insertions in general are almost random, yet it appears that transcriptionally active promoters may present favorable targets. Insertion of the transposon into active genes usually results in inactivation because of the interruption of coding sequences. Excision of the transposon may revert the gene to the original active state. This may occur at frequencies of 10^{-8} to 10^{-4} per cell divisions. The transposase gene alone does not mediate excision. It is independent from the bacterial *recA* gene, but it depends on the structure of inverted terminal repeats.

Replicational errors involving slippage of pairing between the new and template DNA strands may occur. The presence of some sequences in the target may also promote Tn excision, which may also involve flanking DNA sequences; in this case, wild-type function of the target gene is not restored.

```
      O    IS50 L  I            2.8 kb          I   IS50 R   O

     < ~ ~ ~ ~ ~ ~ ~ ------------------------------- ~ ~ ~ ~ ~ ~ ~ ~ >

         1.5 kb    p→    kan    ble    str        1.5 kb
                                            ← Transposase (tnp)
                                            ← Inhibitor (inh)
```

The inverted termini ($\sim\sim$) represent the IS50 insertion element that includes the *tnp* (58 kDa protein) and *inh* genes (product 54 kDa). The left (L) and right (R) IS50 elements are almost identical except that the L sequences contain an *ochre* stop codon in the *tnp* gene, rendering it nonfunctional, save when the bacteria carry an *ochre suppressor*. At the I (inside ends) site, binds the IHF (integration host factor) protein. *O* is the outside end of the IS sequence. Within the repeat beginning at nucleotide 8 is the bacterial DnaA protein-binding site $TTAT_A^C CA_A^C A$. DnaA product controls the initiation of DNA synthesis. Within the 2,750 bp central region are the genes for resistance to kanamycin (*kan*) and G418, bleomycin [(phleomycin) (*ble*)], and streptomycin (*str*) antibiotics; they are transcribed from the *p* promoter located within the Is L element at about 100 bp from the *I* end. These antibiotic resistance genes may not convey resistance in some cells, e.g., *str* may be cryptic in *E. coli*. The activation of this antibiotic resistance operon is contingent on the ochre mutation in the *tnp* gene. The inhibitor protein (product of *inh*) apparently interacts with the terminal repeats rather than with the transcription or translation of the *tnp* gene.

Tn5 can insert one copy at many potential target sites within a genome. IS terminal repeats are capable of transposition themselves without the internal 2.8 kb element. In *direct transposition*, the complete Tn5 will occur in the same sequence as shown above. In *inverse transposition*, mediated by the *I* ends, the 2.8 kb central element is left behind, away from the termini. Inverse transposition occurs 2–3 orders of magnitude less frequently than direct transposition. In5 can form cointegrates with plasmids or bacteriophages, and IS elements and the entire Tn5 may occur in these cointegrates. The orientation of the termini may be either:

direct:O←I $\boxed{kan}$ I ← O   or   indirect: O → I $\boxed{kan}$ I ← O

Several mutations in *E. coli* (*recB, recC, dam, mutH, mutS, mutD, ssb*) may promote excision, and mutation to *drp* reduces excision. Unlike Tn3, Tn5 does not have a resolvase function. DNA gyrase, DNA polymerase I, DnaA protein, IHF, and Lon (a protease effectively cleaving the SulA protein, a cell division inhibitor) may affect transposition. Transposition of Tn5 is substantially increased in *dam* mutant strains that are deficient in methylating GATC sequences. The *I* ends contain GATC sequences and thus can be affected by methylation. The *O* ends do not have methylation substrates, yet they are also affected by methylation in the *I* sequences (19 bp). *See* cut-and-paste; resolvase; Tn3; Tn7; Tn10; transposable elements, bacteria; transposable elements; transposon. (Berg, D. E. 1989. *Mobile DNA*; Berg, D. E. & Howe, M. M. eds., p. 185. Am. Soc. Microbiol., Washington, DC; Naumann, T. A. & Reznikoff, W. S. 2002. *J. Biol. Chem.* 277:17623; Peterson, G. & Reznikoff, W. 2003. *J. Biol. Chem.* 278:1904.)

Tn7 14 kb bacterial transposon of the general structure shown next page.

The Tn7 element has a high capacity to insert into the specific *att* Tn7 site of *E. coli* (25 kb counterclockwise from the origin of replication at map position 83). When this site is not available, it may transpose — at about 2 orders of magnitude lower frequency — to a *pseudo-att* Tn7 or to some other unrelated sites. At *att* Tn7, the right end is situated proximal to the bacterial *o* gene. Genes *tnsABC* mediate all transpositions but through different pathways; for transposition to *att* Tn7 and *pseudo-att* Tn7, in addition, the function of gene *D* is required, whereas for transposition to all other sites the expression of *ABC + E* genes is needed. (The name *tns* abbreviates **tra**nsposon **s**even.) The insertion at *att* Tn7 is also in a consistent orientation and is within an intergenic region; thus it does not harm the host. Insertion of

Left ⎡30 bp repeat⎤ > *dhfr aadA*   *tnsE tnsD tnsC tnsB tnsA* < ⎡30 bp repeat⎤ Right

dhfr = dehydrofolate reductase gene with much reduced sensitivity to trimethoprim, an inhibitor of the enzyme involved in the biosynthesis of both purines and pyrimidines and thus nucleic acids. *aada* = adenylyl transferase gene, encoding an enzyme, which inactivates aminoglycoside antibiotics, streptomycin and spectinomycin, thus conveying resistance to the cells. The *tns* genes are responsible for transposition.

Tn7 element protects the cell from an additional Tn7 insertion (immunity), yet under some conditions the immunity may not work. Integration results in 5 bp target site duplications that are different at *att* Tn7 and other sites. Several bacterial species besides *E. coli* have specific *att* Tn7 sites in their genomes. For transposition, both *L* and *R* terminal repeats are required. Elements with two *L* repeats do not move, whereas two complete *R* termini can assure insertion to *att* Tn7 sites. Because of site specificity, Tn7, is not very useful for insertional mutagenesis, but it has an advantage for inserting genes at the standard map position. Tn7 inserts preferentially in the vicinity of triple-helical sites. The Tn7 family of transposons includes Tn73, Tn*1824*, and Tn*1527* with substantial similarity or even identity. Tn*1825* is a clearly distinct member. *See* transposable elements; transposons, bacterial; triplex. (Craig, N. 1989. *Mobile DNA*; Berg, D. E. & Howe, M. M., eds., p. 211. *Am. Soc. Microbiol.*, Washington, DC.)

Tn*10* Bacterial (*E. coli, Klebsiella, Proteus, Salmonella, Shigella, Pseudomonas*, etc.) transposable element of 9.3 kb. It may move within a bacterial genome, from the bacterial chromosome to temperate phage or plasmid, and among different bacterial species. Its overall structure can be represented as

L ⎡1.3 kbp ⇒⎤ ← *tetR* → *tetA* ← *tetC* → *tetD* → ⎡← 1.3 kbp⎤ R

The boxed left insertion element IS*10* (L) is a defective transposase. The boxed right IS*10* element (R) is a functional transposase. Thus, Tn*10* is a composite transposable element. The *tetR*, *A*, *C*, and *D* genes are involved in resistance against tetracycline. *tetR* is a negative regulator and *tetA* encodes a membrane protein. The arrows indicate orientation and direction of transcription. The promoter of the Is*10* element may serve as a promoter for adjacent outside genes (pOUT).

take place only in some Tn*10* derivatives but not in the wild-type element. The rate of transposition for IS*10* is 10^{-4} and for Tn*10* it is 10^{-7} per cell cycle.

Both the IS*10* and Tn*10* elements can cause chromosomal rearrangements (see diagram next page). Insertion and transposon sequences may show a portable region of homology and can undergo homologous recombination (see diagram). These recombinations may generate deletions between or inversions in the regions at rates 2 orders of magnitude less frequent than transposition or may lead to the formation of cointegrates.

All DNA segments flanked by IS*10* can become transposable and thus may represent new composite transposons. IS*10* and Tn*10* may assist the fusion of different replicons and transfer information between bacterial chromosomes, plasmids, and phages. The transpositions may also generate new units of regulated gene clusters by the movement of structural genes under the control of other regulatory sequences. The excision of the transposon may be "precise" if the nucleotide sequence is restored to its preinsertion condition. Precise excision (average frequency 10^{-9}) may remove one copy of the 9 bp target site inverted duplications.

The near-precise excision events involving removal of most of the internal sequences of Tn*10* occur at a frequency of about 10^{-6} and may later be followed by precise excision of the remaining sequences. These nontransposase-mediated events are also independent from host RecA recombination functions. Precise excision is mediated by RecBC and RecF pathways. Tn*10* and IS*10* may transpose either by a nonreplicative or a replicative mechanism. In the former case, the whole double-stranded element is lifted from its original position and transferred to another site. In the second case, only one of the old strands of the transposon is integrated into the new position and the other strand represents the newly replicated one.

At the target site, apparently two staggered cuts are made 9 bases apart. This causes the 9 bp target duplications when the gap is filled and the protruding ends are used as templates:

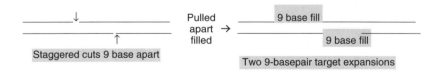

Insertion of Tn*10* within a gene, operon, or upstream regulatory element may abolish activity of the genes or may modify their transcription. In the so-called polar insertions, transcription is initiated within the Tn element, but it may be terminated when rho signals are encountered. In the nonpolar insertions, there are no rho sequences downstream to halt transcription. This type of *read-through* transcription may

The transposon (ca. 46 kDa) is coded within the IS*10* right terminus by about 1,313 nucleotides. The frequency of transposition may increase up to 5 orders of magnitude by increasing the expression of the transposase. The transposase action prefers being in the vicinity of the transposed sequences and its action is reduced by distance. Also, longer transposons are moved less efficiently than shorter ones. Each kb

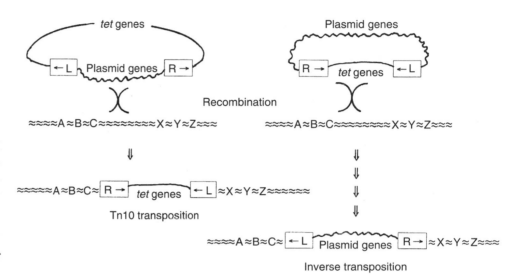

Generation of deletions and inversions by a portable region of homology represented by transposon Tn*10* (→ OR ←). Genes or sites are shown by capital letter in italic, χ indicates recombination.

Tn*10* transpositions. left: the normal event, right: the inverse transposition (or inside-out transposition). Some other events may lead also to deletions or deletions and inversions. The L and R boxes represent Tn*10* termini. The thin line stands for the transposon sequences, whereas the single jagged line indicates the sequences flanking the transposon in the original location of the plasmid. ≈≈≈≈ symbolizes the DNA sequences of the target.

increase in length involves about 40% reduction in movement within transposon sizes 3 to 9.5 kb. Proteins IHF, HU, and DNA gyrase also regulate transposition. Transposition frequency may be modified (within 3 orders of magnitude) by the chromosomal context, i.e., cis-acting sequences. The integration hot spots appear to have some consensus sequences within the 9 bp target site in three bases:

1st	2nd	3rd position			
G	C				
A	T	T	C	A	G
90%	98%	63%	23%	12%	2%, respectively

Transposition seems to avoid actively transcribed genes and thus the most essential ones of the recipient. The transposase gene is transcribed from a very-low-efficiency pIN promoter originating near the *O* end of the right *IS* element. The divergent pOUT promoter opposes its transcription. Apparently, on the average, each cell generation may not produce one molecule of transposase. Two *dam* (adenine) methylation sites (GATC) are located within pIN, and the activity of this promoter is facilitated in the absence of methylation by the host *dam* methylase (located in *E. coli* at 74 min). Actually, the transposase promoter is usually hemimethylated. An increase in the number of Tn*10* copies per cell reduces transposase activity because the pOUT promoter

generates an antisense RNA transcript that may pair to a 35-base complementary region of the pIN promoter. Transposase activity is also regulated by fold-back inhibition (FBI), hindering the attachment of the Shine-Dalgarno sequence of the transcript to the ribosome. The Tn*10* system is protected from transposase activation by inhibition of read-through.

Transposons inserted into a particular gene cause mutations because of the disruption of continuity of the coding sequences. Transposon mutagenesis has a considerable advantage over chemical or radiation mutagenesis because it induces mutation only at the site of insertion, whereas chemicals may simultaneously affect several genes. In addition, mutagenesis with Tn*10* labels the gene by tetracycline resistance, an easily selectable marker. Transposons may label foreign genes cloned in *E. coli* or *Salmonella*. The tetracycline-resistance gene has been used extensively for monitoring insertions that result in the loss of tetracycline resistance but retain other (selectable) markers in the cloning vector. *See* antisense RNA; FBI site; gyrase; HU; IHF; readthrough; transposable elements, bacteria. (Craig, N. 1989. *Mobile DNA*; Berg, D. E. & Howe, M. M. eds., p. 227. *Am. Soc. Microbiol.*, Washington, DC.)

TNA (L)-α-threo-furanosyl-('-2')-oligonucleotides. These molecules contain four-carbon (tetrose) sugars rather than pentose and are capable of pairing with RNA and DNA.

TNF Tumor necrosis factor. Proteins ($M_r \approx 17$ K) selectively cytotoxic or cytostatic to cancer cells of mammals. They are relatively harmless to normal cells by being lymphokines. They are defense molecules against intracellular pathogens. The TNF family includes diverse transmembrane proteins with high homology in receptor-binding regions. TNF was originally discovered in rodent cells infected by bovine *Mycobacterium* and then with endotoxin. The serum of these animals produced hemorrhagic necrosis and occasionally complete regression of transplanted tumors. Mature human TNFα consists of 157 amino acids after trimming off 73 amino acids from pre-TNF. It has receptor sites on the surface of tumors and its action is synergistic with γ-interferon. TNFα and TNBβ have similar functions and display $\approx 30\%$ homology. The two genes each have 3 introns in their 3 kb sequence. In mice, TNF is localized within *H-2* (histocompatibility cluster), and in humans in the homologous *MHC* region in chromosome 6p23 either between HLA-DR and HLA-A or proximal to the centromere. The genes were cloned and sequenced in the mid-1980s. TNFα and IL-1 are the major inflammatory cytokines, whereas IL-10 and TGFβ, IL-R, and TNF-R are antiinflammatory. TNF is produced mainly by macrophages. TNFβ deficiency may increase chromosomal instability and may lead to cancerous transformation. The tumor necrosis family of proteins includes FASL, CD40L, LTα and β, CD30L, CD27L, 4-1BBL, OX40L, TRAIL, OPGL, LIGHT, APRIL, and TALL. *See* arthritis; Crohn disease; cytokines; endotoxin; hemorrhage; histocompatibility; HLA; HVEM; interferons; LTα; lymphokines; lymphotoxin; macrophage; necrosis; NGF; TACE; TNFR; TRAF; TRAIL; transposase. (Kassiotis, G. & Kollias, G. 2001. *J. Clin. Invest.* 107:1507.)

TNFR Tumor necrosis factor receptor, p55, p75. Similar receptors in both animals and plants may be involved in processes of differentiation. TNFR-1 and -2 are distinguished; TNFR-1 mediates different effector functions through separate pathways. The intracellular portion of TNFR-1 contains a 70-amino-acid death domain that mediates signals for apoptosis and for the activation of NF-κB. TNF binds to the extracellular domain of TNFR-1, resulting in its trimerization. TNF1 attracts the adaptor TRADD (tumor necrosis factor receptor–associated death domain) and TRAF2 and TRAF1. Subsequently, TNFR1 recruits FADD (Fas-associated death domain). TRAF2 (tumor necrosis factor–associated protein 2) is contacted by TRAF1, and RIP (receptor-interacting protein). This complex then signals for apoptosis, JNK/SAPK and NF-κB activation. Protein SODD (457-amino-acid silencer of death domains) is expressed in all human tissues; interacts with the intracellular domain of TNFR-1 but not with TNFR-2, TRADD, FADD, or RIP; and interferes with all TNF signaling through TNFR-1. JNK does not participate in the apoptotic pathway, and the activated NF-κB works against apoptosis, but inflammation may result. *See* apoptosis; APRIL; BAFF; Blys; Fas; Jun; NF-κB; NGF; nitrogen fixation (ENOD); Paget disease; TNF; TRAF. (Locksley, R. M., et al. 2001. *Cell* 104:487.)

TNG Panel of 90 independent radiation hybrid clones constructed at Stanford University by irradiation of 50 Krad X-rays. This panel is used for the construction of the high-resolution STS map. *See* radiation hybrid panel; STS.

(Robic, A., et al. 2001. *Mamm. Genome* 12:380; Olivier, M., et al. 2001. *Science* 291:1298.)

Tnp Transposase enzyme. *See* Tn5.

Tnt Nonviral retrotransposable element. *See* transposable elements.

TNT Solution containing 10 mM Tris.HCl buffer (pH 8.0), 150 mM NaCl, 0.05% Tween 20.

Toad.

toad *Bufo vulgaris*, $2n = 36$; *Xenopus laevis*, $2n = 36$. *See* frogs.

TOAST Traced orthologous amplified sequence tags.

tobacco (*Nicotiana* spp) Smoking tobacco is an allotetraploid (*N. sylvestris x N. tomentosiformis*), $2n = 48$. The basic chromosome number is generally $x = 12$; however, diploid forms with $2n = 20$ (*N. plumbaginifolia*) and $2n = 18$ (*N. langsdorfii*), and in the *Suavolens* group, $2n = 36$ (*N. benthamina*), $2n = 46$ (*N. caviola*), $2n = 32$ (*N. maritima*), $2n = 36$ (*N. amplexicaulis*), $2n = 40$ (*N. simulans*), and $2n = 44$ (*N. rotundifolia*) are also found. For genetic studies, the diploid species are most useful (*N. plumbaginifolia*); for transformation, *N. tabacum* is used most commonly because of the easy regeneration of plants from single cells. Antibiotic-resistant chloroplast mutations are available. The S strain is a streptomycin-resistant mutant of the variety Petite Havana. Mutation, recombination, and transformation techniques are available for its plastid genome using primarily antibiotic-resistant ctDNA mutations. *See Nicotiana*; smoking.

tobacco mosaic virus *See* TMV.

tobacco necrosis virus (TNV) Icosahedral single-stranded RNA virus of ≈ 4 kb, a root plant pathogen. *See* icosahedral.

tobacco satellite necrosis virus (TSNV) 17 nm diameter; single-strand RNA (≈ 1.2 kb) virus that depends on TNV for its replication. *See* tobacco necrosis virus.

tocopherol-α (vitamin E) Its deficiency results in nutritional muscular dystrophy and sterility, although humans with regular diet normally do not show a need for it. Vitamin E may be beneficial as an antioxidant and for regulating nerve functions and atherosclerosis. Vitamin E is a lipid-soluble molecule and its major source is plant oils. *See* atherosclerosis; vitamin E.

T-odd virus   Bacteriophages with odd-number designations such as T3, T5, T7, etc. *See* bacteriophages.

toe printing   Procedure for mapping the translation initiation ternary complex (EI1A•GTP•tRNA) to the ribosome. In one approach, a ^{32}P-labeled oligonucleotide is annealed to the mRNA downstream from the presumed site of initiation and reverse transcriptase is used for the extension of the radioactive primer up to the position of the bound ribosome, where the chain growth stops. Various types of purification may be used for the isolation and identification of the initiation complex. *See* elongation factors; protein synthesis; ribosomes. (Ringquist, S. & Gold, L. 1998. *Methods Mol. Biol.* 77:283; Dinitriev, S. E., et al. 2003. *FEBS Lett.* 533/C:99.)

TOFMS   Time-of-flight mass spectrophotometry is a powerful technique for the analysis of the primary structure of proteins. *See* MALDI; mass spectrum; matrix-assisted laser desorption time of flight mass spectrometry. (Verentchikov, A. N. 1994. *Anal. Chem.* 66:126; She, Y.-M., et al. 2001. *J. Biol. Chem.* 276:20039.)

TOGA   **(1)** Total gene expression analysis is a completely automated technology for the simultaneous analysis of the expression of nearly all genes. Basically it selects a 4-base recognition endonuclease site and an adjacent 4-nucleotide parsing sequence (a syntactical determinant, e.g., for *Msp*I, CCGGN$_1$N$_2$N$_3$N$_4$) and their distance from the 3′-end of an mRNA (from the polyA tail). These generate a specific, single identity label for each mRNA. The parsing sequences serve as parts of the PCR primer-binding sites in 256 PCR-based assays, which determine the presence and concentration of that mRNA in a tissue. *See* microarray hybridization; RNA fingerprinting; SAGE. (Sutcliffe, J. G., et al. 2000. *Proc. Natl. Acad. Sci. USA* 97:1976.) **(2)** Tiger Orthologous Gene Alignment is a database generated by pair-wise comparison between tentative consensus sequences in different organisms. (<http://www.tigr.org/tdb/tgi.shtml>.)

togavirus   RNA plus-strand viruses of about 12 kb. The capsid proteins are synthesized only after completion of replication. This viral family includes rubella, yellow fever, and encephalitis viruses. *See* plus strand; RNA viruses.

toilet   Cleansing, clearing.

tolerance, antibiotic   The infectious organism does not die but stops reproducing.

tolerance, immunological   Nonreactivity to an antigen that under other conditions would evoke an immune response. Antigens provided to fetuses or neonates with immature immune systems can induce tolerance. In adults, a very high or very low dose of an antigen may cause tolerance. The tolerance is the result of either clonal elimination or inactivation of lymphocytes in the thymus. Liver transplantation may induce systemic immune tolerance for certain (kidney, heart) allografts. *See* allograft; antigen; immune response;

immunosuppressants; lymphocytes. (Gaunt, G. & Ramin, K. 2001. *Am. J. Perinatol.* 18[6]:299; Salih, H. R. & Nussler, V. 2001. *Eur. J. Med. Res.* 6[8]:323; Chang, C. C., et al. 2002. *Nature Immunol.* 3:237.)

TOLL   (*Drosophila*, 3–91, human homologue TRAF6)   Encodes a signaling protein operating through the NF-κB receptor in both the fly and humans and is mediated by the MyD88 adaptor protein. For the activation of NF-κB IRAK (interleukin-associated receptor kinase) and the TRAF6 (tumor necrosis factor associated receptor) protein, activated by IL-1 are required. This pathway is essential for the immune system. The human Toll-like receptors (hTLR) respond to bacterial lipoproteins (TLR2), lipopolysaccharides (TLR4), unmethylated CpG-DNAs (TLR9), etc., with the activation of apoptosis, bacterial killing, tissue injuries, and the induction of the innate immune response. The *Toll/spaetzle/cact*us gene cassette is involved in the control of antifungal peptide, drosomycin, production. *See* antimicrobial peptides; apoptosis; CD14; IL-1; IL-12; innate immunity; IRAK; morphogenesis {52} in *Drosophila*; MyD88; NF-κB; TRAF. (Kobayashi, K., et al. 2002. *Cell* 110:191; Takeda, K., et al. 2003. *Annu. Rev. Immunol.* 21:335.)

Tolloid   *See* bone morphogenetic protein.

TOM   Transfer outer membrane. Protein complex regulating transport through the outer layer of the mitochondrial membrane. *See* mitochondria; TIM.

tomato   (*Solanum lycopersicum*, 2*n* = 24)   Has about 8–10 related species. It is one of the cytologically and genetically best-known autogamous plants and is suitable for practically all modern genetic manipulations. (Genome size bp/n $\cong$ 6.6 × 10^8. (Gene index: <http://www.tigr.org/tdb/tgi.shtml>.)

tomato bushy stunt virus   Single-strand RNA virus of about 4,000 bases enveloped by an icosahedral shell consisting of 180 copies of a 40 kDa polypeptide. *See* icosahedral.

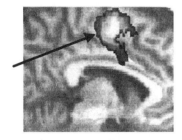

(Pet scan of the brain; note different activity area at the arrow.)

tomography   (body section radiography)   Conducted by a tomograph in which a source of X-radiation moves in the opposite direction to that of a film recording the image clearly only in one plane and blurring the rest of the images. In *computerized axial tomography* (CAT scan), the scintillations produced by the radiation are recorded on a computer disk and the cross section of the body is analyzed electronically. *Positron emission tomography* (PET) involves the use of positron-labeled metabolites (e.g., γ-ray-emitting glucose). Along the path of the radiation, positrons and electrons collide and the local concentration of the isotopes is recorded electronically. *Single-photon emission computed tomography* (SPECT) takes γ-ray photographs around the body and a computer reconstructs three-dimensional images, resulting in great resolution even of overlapping organs. *Ultrasonic tomography* uses ultrasound scanning. The radiation may not be without risk. *See* imaging; nuclear magnetic resonance spectroscopy; ultrasonic, sonography; X-rays. (Czernin, J. & Phelps, M. E. 2002. *Annu. Rev. Med.* 53:89; Cristofallini, M., et al. 2002. *Nature Rev.* Drug Discovery 1:415.)

tonoplast   Elastic membrane ($\approx$8 nm) that surrounds the vacuoles. The traffic through this membrane is controlled by several proteins. *See* vacuoles. (Maeshima, M. 2001. *Annu. Rev. Plant Physiol. Plant Mol. Biol.* 52:469.)

tooth agenesis   Caused by a dominant (chromosome 4p) mutation in the homeodomain of transcription factor MSX1, and it involves failure to form the second premolars and the third molars. Tooth development is under the control of many genes. Among them, fibroblast growth factor (FGF) and bone morphogenetic protein (BMP) play pivotal roles. FGF and BMP regulate MSX1 and PAX9 transcription factors. *See* amelogenesis imperfecta; BMP; FGF; homeodomain; hypodontia; MSX1; oligodontia; PAX; Rieger syndrome.

tooth-and-nail dysplasia (Witkop syndrome, TNS)   Most commonly, incisors, canine teeth, and some of the molars are poorly developed. In some cases, abnormal toenails in children accompany TNS. Mutations at an MSX1 (4p161) subunit seem to account for the phenotype. *See* dental no-eruption; denticle; dentin dysplasia; Hallermann-Streiff syndrome; hypodontia; MSX1. (Jumlogras, D., et al. 2001. *Am. J. Hum. Genet.* 69:67.)

tooth malposition   Apparently, an autosomal-dominant gene controls the various misplacements or underdevelopment of the incisors and the canine teeth. *See* amelogenesis imperfecta; cherubism; Hallermann-Streiff syndrome; Jackson-Lawler syndrome.

tooth size   Appears to be influenced by the human Y chromosome as indicated by observations on various sex-chromosomal dosages.

5'-TOP   5'-terminal oligopyrimidine tract. mRNAs are part of the protein synthetic machinery and they are translated under the control of S6 kinases that are targets of insulin signaling in mammals. Cell growth (not cell division) is controlled by this process. *See* insulin-like growth factors; S6. (Crosio, C., et al. 2000. *Nucleic Acids Res.* 28:2927.)

top agar   ~0.6% agar solution generally containing 0.5% NaCl and some organic supplements. About 2 mL of top agar with suspended bacterial cells is spread over the agar medium (30 mL/10 cm petri plates) to initiate selective bacterial growth, e.g., in Ames tests. *See* Ames test.

top-down analysis   Starting with a mutant phenotype, the physiological or molecular mechanism responsible for the alteration is investigated. In contrast, the bottom-up analysis first studies the molecules, then the analysis is extended to their relations to the phenotype. The major endeavors go beyond the role of individual molecules and major interest is being focused on the critical domains of the molecules. *See* reversed genetics.

top-down mapping   Uses either traditional genetic recombinational analysis or radiation hybrid maps. *See* bottom-up map; mapping, genetic; radiation hybrids.

topical reversion   The backmutation is the result of alterations within the gene rather than due to extragenic suppression. *See* reversion; suppressor, extragenic; suppressor, intragenic.

topoisomerase (TOP)   Enzymes that alter the tertiary structure of DNA without a change in the secondary or primary structure. Monomeric topoisomerase I (10q12-q13.1) nicks and closes single strands of DNA and changes the linking number in one strand; dimeric topoisomerase II can cut and reattach both strands of DNA, affecting linking number in both strands. Topoisomerase II (17q21-q22) disentangles DNA strands and has an important role in DNA replication, transcription and recombination, suppression of mitotic recombination, stabilization of the genome (chromosome breakage), regulation of supercoiling, eukaryotic chromosome condensation, control of segregation of the chromosomes, regulation of the cell cycle, and nuclear localization of imported molecules. Mitochondrially located Top1mt is encoded at 8q24.3. Topoisomerases are important objects in cancer therapy research. Prokaryotic topoisomerase I—in contrast to eukaryotic topoisomerase—requires Mg^{2+} and single-stranded DNA segments and relaxes only negatively supercoiled molecules. TOP I activity is required for the proper segregation of prokaryotic chromosomes (Zhu, Q., et al. 2001. *Proc. Natl. Acad. Sci. USA* 98:9766). Human topoisomerase I consists of 765 amino acids. Prokaryotic DNA topoisomerase III supports the movement of the DNA replication fork and

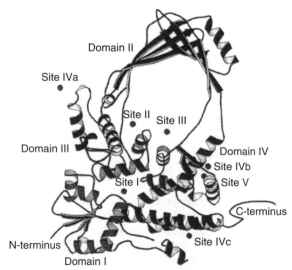

The 67 kDa fragment of DNA topoisomerase I of *E. coli* displaying the four major domains and the five nucleotide-binding sites. (From Feinberg, H., Changela, A. & Mondragón, A. 1999. *Nature Struct. Biol.* 6:961.)

can function as an RNA topoisomerase, interconverting RNA circles and knots. Eukaryotic topoisomerase III is homologous with prokaryotic topoisomerase I and has been located to human chromosome 17p12-p11.2. Mice with knocked-out DNA topoisomerase IIIβ are viable but senesce earlier and have about a 40% shorter life span. *See* camptothecin; DNA replication; gyrase; linking number; linking number paradox; mtDNA; p53. (Wang, J. C. 1996. *Annu. Rev. Biochem.* 65:635; Changela, A., et al. 2001. *Nature* 411:1077; Champoux, J. J. 2001. *Annu. Rev. Biochem.* 70:369; Wang, J. C. 2002. *Nature Rev. Mol. Cell Biol.* 3:430.)

topological filter Two-step synapsis. The assumption is that synapsis requires an interaction between DNA *res* site*s* and the three subunits of a resolvase enzyme (step 1). After this initial step, the II and III subsites of the resolvase-*res* dimers pair in an parallel manner and the subunits interwrap. Then the two I subsites of the resolvase dimer bind, resulting in a *productive synaptic complex* (step 2) capable of bringing about DNA change. *See res*; resolvase; synapsis. (Watson, M. A., et al. 1996. *J. Mol. Biol.* 257:317.)

topological isomers of DNA. *See* linking number.

TOR Target of rapamycin. Phosphatidylinositol kinases of yeast (TOR1, TOR2); also called RAFT1 and FRAP. TORs convey resistance to FKBP. The C-terminal domains of the TOR family are homologous to the catalytic domain of PI(3)K. TORs block turnover of nutrient transporters, autophagy, and the interphase of the cell cycle. They promote protein kinase C signaling, actin and cytoskeleton organization, tRNA and ribosome formation, transcription, and translation. *See* autophagy; cell cycle; cytoskeleton; FK506; PIK; PI[3]K; protein kinases; rapamycin. (Schmelzle, T. & Hall, M. N. 2000. *Cell* 103:253; Dennis, P. B., et al. 2001. *Science* 294:1102, Schalm, S. S. & Blenis, J. 2002. *Current Biol.* 12:632.)

tormogen Component cell of the bristle that secretes the bristle socket. The other bristle cells are trichogens, which secrete the bristle shaft, also a neuron; it contacts the shaft and connects with the nervous system through its axon.

toroid/toroidal Body or surface generated by the rotation of a plane curve or circle. It may assume the shape of a ring/doughnut or may resemble a barrel. *See* torus.

tortoiseshell fur color Develops in female cats heterozygous for the X-chromosome-linked genes *black* and *yellow* colors because of selective, alternate inactivation of the two mammalian X-chromosomes. Tortoiseshell animals have mixed patches of black and yellow fur. This fur pattern occurs in the XX (female) or exceptionally in XXY Klinefelter male cats. *See* calico cat; Lyon hypothesis; lyonization.

torus Ring-shaped emergence, swelling, or bordered pit. *See* toroid.

totipotency Characteristic of zygotic cells, which permits differentiation into any type of cell or structure, including the whole organism. Plant cells maintain their totipotency in diverse adult tissues and after dedifferentiation may initiate other types of differentiations; entire organisms may be regenerated from somatic cells in cultures. Totipotency of animal cells is much more limited, although lower animals such as hydra and earthworms may regenerate from differentiated tissues. Embryonic stem cells (ES) of mice come close to totipotency/pluripotency inasmuch as they can be transferred to mouse embryos and can contribute to the formation of various cell types, including the germline. After special treatments, differentiated animal cells may revert to stem cell status. *See* ES; morphogenesis; multipotent; nuclear transplantation; redifferentiation; somatic embryogenesis; stem cell.

touch-and-go pairing End-to-end synapsis of sex chromosomes in the heterogametic sex of some insects.

After Schrader, F. 1940. PNAS 26:634.

touch-down PCR *See* hot-start PCR.

touch sensitivity In mammals, is mediated by the brain sodium channel BNC1. *See* degenerin. (Welsh, M. J., et al. 2002. *J. Biol. Chem.* 277:2369.)

Toulouse-Lautrec, Henri (1864–1901) One of the most remarkable painters at the end of 19th-century Paris life, a colleague of van Gogh and an influential precursor of modern art. He was considered to be affected by pycnodysostosis. His parents were first cousins. This diagnosis of his genetic malady has been questioned (*Nature Genetics* 11:363), and it appears a deficiency of cathepsin K might have beeen involved. *See* cathepsins; pycnodysostosis.

Tourette's syndrome (Gilles de la Tourette disease, GTS) Human behavioral anomaly causing motor and vocal incoordination (tic, twitching), stuttering (echolalia), the use of foul language (coprolalia), obsessions, and hoarding. The onset is between 7 to 14 years of age and 75% of those affected are male. The genetic determination is apparently dominant with incomplete penetrance and expressivity. Male:female ratio of 4.3:1 has been reported. Genes in several chromosomes (4q, 5q, 8p, 18, 7, 9, 3, 11q23, 17q) have been implicated. The frequency of the defective gene(s) was estimated to be 0.4 to 0.9%, and a prevalence of 0.02 to 1% has been observed in different populations. Some of the mild cases are suspected to be responsible for male alcoholism and female obesity. *See* affective disorders. (Zhang, H., et al. 2002. *Am. J. Hum. Genet.* 70:896.)

Townes—Brocks syndrome (SALL1, 16q12.1) Highly variable dominant malformation with good penetrance. The symptoms include imperforate anus, polydactyly, syndactyly, abnormal ear lobes, hearing deficit, kidney and heart anomalies, and mental retardation. The majority of individuals display a mutation in the SALL gene encoding a 1,325 amino acid protein with double Zn-finger domains. The normal protein is a transcriptional repressor and is associated with heterochromatin. See heterochromatin; penetrance; zinc finger. (Netzer, C., et al. 2001. *Hum. Mol. Genet.* 10:3017.)

toxicogenomics Applies microarray hybridization techniques for the detection of genes that are turned on or off upon exposure of animals to toxic substances. See microarray hybridization. (Simmons, P. T. & Portier, C. J. 2002. *Carcinogenesis* 23:903.)

toxic shock syndrome Caused by an infection of *Staphylococcus aureus*/*Streptococcus pyogenes* gram-positive bacteria, affecting primarily menstruating women. It begins with sudden high fever, vomiting, diarrhea, and muscle pains (myalgia). Rash, hypotension, and death may follow. Although septic shock has been attributed mainly to the cell wall lipopolysaccharides (present in gram-negative bacteria) and peptidoglycan and lipoteichoic acid (present in gram-positive bacteria), actually unmethylated 5′-CpG-3′ prokaryotic DNA may be responsible for the immune reaction. These bacterial superantigens (SAG) bind directly to the class II histocompatibility complex on the antigen-presenting cells and to specific variable regions of the β-chain of the T-cell receptor, evoking a very severe response. Plasmids or phages carry the bacterial enterotoxin genes. See gram-negative/positive; peptidoglycan; *Staphylococcus*; *Streptococcus*; teichoic acid. (McCormick, J. K., et al. 2001. *Annu. Rev. Microbiol.* 55:77.)

toxin targeting See immunotoxin; magic bullet; toxins. (Goodsell, D. S. 2001. *Stem Cells* 19:161.)

toxins Organic poisons. A plasmid gene of *E. coli* produces colicins, and *Shigella* bacteria may affect sensitive bacteria in several ways. Cholera toxin produced by the bacterium *Vibrio cholerae* interferes with the active transport through membranes, thus causing the loss of excessive amounts of fluids and electrolytes in the gastrointestinal system. The *Bordatella pertussis* toxin, pertussin, contributes to a high level of adenylate cyclase and thereby to the symptoms of whooping cough. Bungarotoxin (from *Bungarus multicinctus*) and cobrotoxin (from *Formosan cobra*) in snake venoms block acetylcholine receptors or interfere with ion channel functions. Bungarotoxin has an LD$_{50}$ value in mice of 0.15–0.21 µg/g. Other bacterial toxins include tetanus toxin, diphtheria toxin, botulin, etc. Diphtheria toxin (of *Corynebacterium diphtheriae* if it carries a temperate phage with the *tox* gene) is one of the most dreaded human poisons (sensitivity is encoded in human chromosome 5q23) with a minimal lethal dose of 160 µg/kg in guinea pigs. Mice and rats are, however, insensitive to this toxin. It inactivates eukaryotic initiation factor eIF-2 in translation on ribosomes. The fungus *Amanita phalloides* toxin, amanitin (amatoxin), is a poison of RNA polymerase II. About two dozen cytochalasins are synthesized by different fungi and are composed of substituted hydrogenated isoindole rings fused with a macrocyclic ring (large organic compound). Colchicine is a plant alkaloid poison of the spindle fibers.

The seed toxins of the plants *Strophantus* and *Acocanthera* are blocking agents for membrane transport and are used as selective agents in mammalian cell cultures. Abrin (from a legume) and ricin (from Castorbean seeds) are inhibitors of the attachment of aminoacylated t-RNAs to the ribosomes. The piscine toxin tetrodotoxin (from *Spheroides rubripes*; LD$_{50}$ in mice 10 µg/kg) and the dinoflagellate *Gonyalux* species saxitoxin (LD$_{50}$ in mice 3.4–10 µg/kg) lock the sodium ion channels and block neurotransmission. The curare toxins were obtained originally as an arrow poison from the bark of the trees *Strychnos* and *Chondodendron*. The sources of other curare toxins, bamboo curare, pot curare, gourd curare, etc., are members of the *Menispermaceae* family and are highly poisonous muscle relaxants. They block acetylcholine receptors and some ion channels. Some of the curare toxins were used for the treatment of tetanus shock and in surgery to alleviate muscle rigidity. See abrin; acetylcholine receptors; aflatoxins; amatoxin; anthrax; antibiotics; colchicine; colicins; cytochalasins; diphtheria toxin; ion channel; laboratory safety; LD$_{50}$; neurotransmitters; ricin. (Schiavo, G. & van der Goot, G. 2001. *Nature Rev. Mol. Cell. Biol.* 2:530.)

toxoid Inactivated bacterial toxin that may still incite the formation of antitoxins and retain antigenicity.

toxoplasmosis Opportunistic infection by *Toxoplasma* protozoa that reduces nerve connections. It is frequently lethal in AIDS and other immune-compromised people. Meiotic recombination among different strains may greatly enhance their virulence. See AIDS; apicoplast. (Grigg, M. E., et al. 2001. *Science* 294:161.)

TP53 Tumor protein. It is now called p53, a tumor suppressor. See p53.

TPA (1) Phorbol ester (12-o-tetradecanoyl phorbol 13-acetate) that promotes neoplastic growth after induction has taken place. See cancer; carcinogens; phorbol esters. (2) Inducer protein; mitogen-activating protein kinase C. See protein kinases.

T1 phage Double-stranded DNA (48.5 kbp) virulent phage, a general transducer. Only 0.2% of its cytosines and 1.7% of the adenines are methylated. The terminal redundancy is about 2.8 kb. It infects *E. coli* and *Shigella* strains. See bacteriophages.

T2, T4, and T6 phages T-even virulent phages; they are closely related. The 166 kbp linear genome of T4 contains glycosylated and hydroxymethylated cytosine, and 1–5% of its DNA is terminal redundancy; it encodes about 130 genes with a known function, and about 100 additional open-reading frames have been revealed. See bacteriophages; development.

T5 phage (relatives BF23, PB, BG3, 29-α) Their genetic material is linear double-stranded DNA ($\approx$121.3 kbp) with terminal repeats ($\approx$10.1 kbp) but without methylated bases. Three internal tracts can be deleted without loss of viability. They are virulent phages with long tails. See bacteriophages; development.

T7 phage (T3 is related)   Virulent phage; it does not tolerate superinfection (would be required for recombination). Its 39.9 kbp genetic material is enclosed in an icosahedral head with a very short tail. It codes for about 55 genes. The T7 promoter is used in genetic vectors for in vitro transcription. Its replication requires DNA and RNA polymerase, helicase-primase complex, and single-strand-binding protein and endo- and exonuclease activity. *See* bacteriophages.

TPK1, TPK2, TPK3   Catalytic subunits of A-kinase. *See* protein kinases.

TPM   *See* two-phase mutation model.

TPN   Triphosphopyridine nucleotide; TPNH is the reduced form. They are synonymous with the analogs α-NADP and α-NADPH, respectively. Many enzymatic reactions require β-NADP and β-NADPH (nicotinamide adenine dinucleotide phosphates). *See* NAD, NADP$^+$.

TPO   Modulates megakaryocyte differentiation along with EPO and various cytokines. *See* EPO.

TPR   *See* tetratrico sequences.

TPX2   Microtubule-associated protein required for spindle assembly. Binding to importin-α inactivates it, but RAN•GTP reverses the binding. *See* importin; microtubule; RAN; spindle. (Wittmann, T., et al. 2000. *J. Cell Biol.* 149:1405; Gruss, O. J., et al. 2001. *Cell* 104:83.)

TR3   *See* nur77.

tra (*transformer*)   Gene (chromosomal location 3–45) of *Drosophila* controls sterile male development in XX flies; XY *tra/tra* males are, however, normal males. *tra* acts in cooperation with *tra2* (2–70). Tra and Tra2 proteins mediate sex-specific processing pre-mRNAs of *dsx* (*doublesex*, 3–48.1) and *fru* (*fruitless*, 3–62.0). The transcripts occur in both sexes but are processed differently. In the male germline, Tra2 is required for normal spermatogenesis. Tra2 also mediates sex-specific processing of *exu* (*exuperentia*, 2–93) and *att* (*alternative-testes-transcript*, chromosome 3 at 92E3-92E4). Prokaryotic *tra* genes in conjugative plasmids control the conjugal transfer of DNA. *See* sex determination; *tra* genes.

trabant   Terminal chromosomal appendage. *See* satellited chromosome.

Trabant.

trace elements   Required only in minute amounts.

tracer   Radioactively labeled molecule that permits identification of the fate of the molecules into which it has been incorporated. *See* isotopes.

trachea   Duct leading from the throat (larynx) to the lungs of animals or the duct system of insects through which air is distributed into the tissues. *See* FGF.

tracheid   Long, lignified xylem cell specialized for transport and support in plants.

trachophyte   Vascular plant endowed with xylem, phloem, and (pro)cambium in between. *See* cambium; phloem; xylem.

tracking   Mechanism that ensures transposition between two appropriate *res* (recombination sites). Experimental proofs for successful tracking (reporter rings) are not unequivocal. *See* reporter ring.

tracking dyes   In electrophoresis loading buffers permit the visualization of front migration toward the anode. Bromophenol blue in 0.5 × TBE buffer moves at the same rate as double-stranded linear DNA of 300 bp length. Xylene cyanol FF moves along with 4 kb linear double-strand DNA. *See* electrophoresis; electrophoresis buffers.

Tradescantia **species**   Occur in the polyploidy range of $2x$ to $12x$. The plants develop ca. 100 stamen hairs in their inflorescence. Each hair represents a single cell line. When plants heterozygous for anthocyanin markers are exposed to a mutagen, somatic mutations can be assessed as colored and colorless cell lines in ca. 150 flowers that single plants may form. Some of the species are favorites of cytologists. *See* bioassays in genetic toxicology; somatic mutation. (Photo courtesy of Dr. A. Sparrow.)

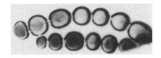

TRADD   Tumor necrosis factor receptor–associated death domain. *See* apoptosis; death domain; TRAF; tumor necrosis factor.

TRAF   Tumor necrosis factor–associated receptor and a signal transducer for some interleukins. TRAF5 is involved in CD40- and CD27-mediated signaling to lymphocytes. TRAF6 activates IκB kinase through a polyubiquitin chain. *See* ASK1; CD27; CD40; CRAF; IκB; interleukins; IRAK; Ire; MATH; NF-κB; TNF; TNFR; Toll; TRADD; ubiquitin. (Tada, K., et al. 2001. *J. Biol. Chem.* 276:36530; Li, X., et al. 2002. *Nature* 416:345.)

***tra* genes**   More than 17: mediate the conjugal transfer of F and other conjugative plasmids. *See* conjugation; ori$_T$; plasmids; relaxosome; *tra*.

TRAIL (Apo-2L, 3q26.1-q26.2)   Tumor necrosis factor–related apoptosis-inducing ligand that attaches to death receptor

DR4 and mediates apoptosis. DR4 does not respond to FADD like the Fas, TNFR-1, and DR3 systems. It has five receptors. TRAIL also activates NF-κB. The nontoxic TRAIL causes preferential killing of neoplastic cells in animal models, especially in combination with radiation therapy for cancer. It appears, however, that it does not discriminate sufficiently between normal human liver cells and liver cancer cells. *See* apoptosis; death domain; FADD; FAS; NF-κB; TNF; TNFR; TWEAK.

trailer sequence Follows the termination codon at the 3'-end of mRNA. It is not yet translated and may have a regulatory function. *See* polyA mRNA.

trait Distinguishable character of an organism that may be inherited.

TRAM (1) Translocating chain–associating membrane proteins are membrane-spanning glycoproteins associated with the nascent peptide chain while SRP mediates its transfer to the endoplasmic reticulum. *See* protein targeting; signal hypothesis; SRP; translocase; translocons. (2) Transverse rectus abdominis muscle.

TRAMP Apo-3-type TNF/NGF receptor. *See* Apo-3; NGF; TNF.

TRANCE (RANK/ODF/OPGL) Tumor necrosis factor–related activation-induced cytokine ligand (encoded at 13q14) that regulates T-cell-dependent immune reactions and bone differentiation, bone mass, and Ca^{2+} metabolism. *See* osteoclast; TNF. (Pearse, R. N., et al. 2001. *Proc. Natl. Acad. Sci. USA* 98:11581; Theill, L. E., et al. 2002. *Annu. Rev. Immunol.* 20:795.)

trans Position indicates that two genetic markers are not on the same molecule or not on syntenic parts of the chromatids. *See* chromatid; cis arrangement; synteny.

transacetylase Protein that transfers an acetyl group from acetyl coenzyme A (acetyl-CoA) to another molecule. The third structural gene of the lactose operon (*lacA*) encodes a 275-amino-acid polypeptide that forms a dimer of 60 kDa, which is a transacetylase. *See lac* operon.

trans-acting elements Proteins synthesized anywhere in the genome but regulating transcription by attachment to specific sites of a gene. *See* cis-acting element.

transactivation responsive element (TAR) In the HIV transcript, the Tat protein binds to the TAR sequence near the 5'-end. This binding then mediates an increased expression of the viral genes and the synthesis of more mRNA. The Rev protein binds to a specific RNA site, the rev-responsive element (RRE), and facilitates the export of the unspliced transcript to the cytoplasm where viral structural proteins and enzymes are made. *See* acquired immunodeficiency; transactivator; VP16; VDR.

transactivator Protein domain attached to a specific inhibitory protein may prevent its blocking of transcription

and may increase the transcription of the target gene(s) by several orders of magnitude. *See* p53; STAT; Switch-Gene; tetracycline; transactivation responsive element; two-hybrid method; VP16. (Devaux, F., et al. 2001. *EMBO Rep.* 2:493; Lottmann, H., et al. 2001. *J. Mol. Med.* 79:321.)

'transactive catastrophy' The similar base sequence in different organisms or even within the same genome may have a different functional meaning.

transaldolase deficiency (TALDO1, 11p15.5-p15.4, a pseudogene is at 1p34.1-p33) It cause cirrhosis of the liver, infantile hepatosplenomagaly (enlargement of the spleen and liver), and accumulation of metabolites of the pentose phosphate pathway (ribitol, D-arabitol, and erythrol) in the urine and blood plasma. *See* cirrhosis of the liver; pentose phosphate pathway. (Verhoeven, N. M., et al. 2001. *Am. J. Hum. Genet.* 68:1086.)

transaminases *See* aminotransferases.

trans arrangement of alleles Indicates that they are not in the same chromosome (DNA) strand (they are in repulsion), in contrast with the cis arrangement (coupling), when the two alleles are within the same strand. *See* cis; coupling; repulsion.

transcapsidation (heteroencapsidation) The viral coat protein and the enclosed genetic material are of different origin (wolf in a sheep's skin). *See* pseudovirion. (Quasba, P. K. & Aposhian, H. V. 1971. *Proc. Natl. Acad. Sci. USA* 68:2345.)

transchromosomal (transchromosomic) Cell that contains foreign chromosomal segments, e.g., a mouse cell with human chromosomal fragments. *See* chromosome uptake; somatic cell hybrid. (Tomizuka, K., et al. 2000. *Proc. Natl. Acad. Sci. USA* 97:722.)

transcobalamin (TCN1) Ligand and transporter of vitamin B12. It is encoded in human chromosome 11q11-q12.

transcobalamin deficiency (TC2) Recessive condition causing megaloblastic anemia located to human chromosome 22q11-qter. *See* anemia; megaloblastic anemia.

transconjugant The bacterial genetic material was (partly) derived by recombination during conjugation. *See* conjugation.

transcortin deficiency (CBG) Dominant human chromosome 14q31-q32; reduction in a corticosteroid-binding globulin. *See* corticosteroid.

transcribed spacer DNA element between genes that is transcribed but eliminated during the processing of the primary transcript. *See* primary transcript.

transcript RNA copied on DNA and complementary to the template. *See* transcription.

transcriptase DNA-dependent RNA polymerase, RNA-dependent DNA polymerase, or RNA-dependent RNA polymerase enzyme. *See* reverse transcription; RNA polymerase.

transcript cleavage factor   Induces cleavage, then release of the 3′-end of a 1- to 17-nucleotide RNA transcript while the 5′-end is still associated with the polymerase. Transcription may not be terminated by the process and it may be reinitiated without a deletion in the transcript. In prokaryotes, *GreA* and *B*, and in eukaryotes, transcription factor TFIIS, carry out such functions. The role of the process may be removal of wrong sequences, facilitating the transition from initiation to the elongation phase and the escape of the polymerase from the promoter complex. *See* promoter; RNA polymerases; transcription complex; transcription factors. (Conaway, J. W., et al. 1998. *Cold. Spring Harbor Symp. Quant. Biol.* 63:357.)

transcript elongation   Transcription has four basic phases: (1) promoter binding and activation of RNA polymerase; (2) RNA chain initiation followed by escape of polymerase from the TATA site; (3) transcript elongation; (4) termination of transcription and transcript release. The elongation phase in *E. coli* is marked by release of the σ-subunit of the polymerase, the exit of the polymerase from the promoter, and the tight association between the polymerase, DNA template, and nascent transcript ternary complex. Transcript elongation in eukaryotes is quite similar. The rate of elongation varies because of sites of pause, arrest, and termination. In eukaryotes, the rate of transcription in vivo is ∼1,200–2,000 nucleotides/minute, but in vitro it is only 100–300/minute. Some of the DNA-binding proteins (such as repressors, CCAAT-binding proteins, etc.), gaps in the DNA, and drugs may interfere with elongation, although few generalizations are possible at this time. In some instances, RNA elongation and DNA replication may conflict. A variety of controls regulate transcript elongation during development. Rapid early embryonal development involves relatively shorter transcripts. *See* attenuation; chromatin remodeling; Cockayne syndrome; DRB; ELL; elongator; elongin; error in transcription; inchworm model; leukemias (MLL); mediator complex; NELF; operon; promoter; RNA polymerase; σ; TEFb; TFIIS; transcription factors; transcription rate; transcription termination; TRAP; tryptophan operon; von Hippel–Lindau syndrome; walking. (Weliky Conaway, J. & Conaway, R. C. 1999. *Annu. Rev. Biochem.* 68:301; Wind, M. & Reines, D. 2000. *Bioassays* 22:327; Toulokhonov, I., et al. 2001. *Science* 292:730.)

transcripton   Synthesis of RNA complementary to a strand of a DNA molecule. In prokaryotes, the majority of transcriptionally active genes are located in the leading strand of replication and transcribed in the same direction as DNA synthesis. In the absence of a functional DNA helicase, genes involved in the replication of the lagging strand are hampered by the transcription complex fork and stalled for many minutes. If, however, the DNA helicase is present, the replication fork on the lagging strand can quickly pass the RNA polymerase complex. In prokaryotes, transcription and translation are coupled, unlike in eukaryotes, where mRNA must be released through the nuclear pore complex into the cytoplasm. Transcription is most commonly regulated by a variety of proteins (transcription factors). In the red clover necrotic mosaic virus (RCNMV) a subgenomic portion (sgRNA) of one of the two RNA genomes (RNA-1) and a 34-base portion of RNA-2 are required for transactivation of transcription of sgRNA. In eukaryotes, transcription takes place in higher-order transcriptional domains (16 in the mouse cell) that are usually independent from the replicational domains. *See* antitermination; chloroplast genetics; chloroplasts; chromatin remodeling; class II and class III genes of eukaryotes; inchworm model; mitochondria; mitochondrial genetics; open transcription complex; pause, transcriptional; pol; regulation of gene activity; replication fork; RNA polymerase; signal transduction; transcription complex; transcription factors; transcription rate; transcription termination. (*Cold Spring Harbor Symp. Quant Biol.*, Vol. 63, 1999; Lee T. I. & Young, R. A. 2000. *Annu. Rev. Genet.* 34:77; Johnson, K. M., et al. 2001. *Curr. Biol.* 11:R510; *Nature Rev. Mol. Cell Biol.* [2002], Vol. 3:11; Kapranov, P., et al. 2002. *Science* 296:916.)

transcription, ectopic   *See* transcription, illegitimate.

transcription, illegitimate (ectopic transcription)   Takes place when very low-level transcripts are detected in organs, tissues, or developmental stages where these special transcripts are not expected to occur. Usually, nested primers are employed for the amplification. *See* housekeeping genes; nested primer; tissue specificity. (Salbe, C., et al. 2000. *Int. J. Biol. Markers* 15:41.)

transcriptional activators   Proteins that facilitate the activity of DNA-dependent RNA polymerase(s) in prokaryotes and eukaryotes. Transcription activators may have two independent domains: DNA-binding domain and transcriptional activating domain. These domains may not have to be covalently associated, and dimeric association may carry out their function. In the abstract sketch below, the dimeric activators stimulate transcription by recruiting the components of the transcription complex to the TATA box, even

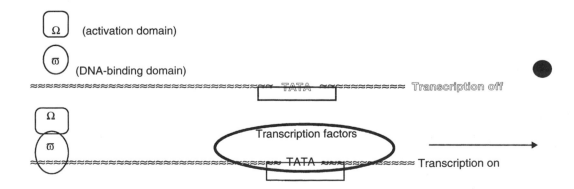

when in their absence or in the presence of the monomers transcription may be hindered. Employing several of the activators in a noncovalently bound bundle may enhance the potency of transcriptional activation. Adding six or eight additional amino acids to the yeast Gal4 activator generates artificial activators. The new activators may have higher or more specific activity. Many of the transcriptional activators (Myc, Jun, etc.) are unstable in the cell and they are destroyed by calpains, lysosomal proteases, and most commonly ubiquitin-mediated proteolysis. Actually, the activation and the ubiquitination domains functionally overlap. *See* calpain; catabolite activator; chromatin remodeling; coactivators; degron; destruction box; DNA-binding proteins; FKBP12; FK506; GCN5; lysosomes; N-degron; negative control; N-end rule; positive control; regulation of gene activity; suppression; SW; transcriptional modulation; transcription complex; transcription factors; transcription termination; ubiquitin; VDRI. (Ptashne, M. & Gann, A. 1997. *Nature* 386:569; Hermann, S., et al. 2001. *J. Biol. Chem.* 276:40127; Lu, Z., et al. 2002. *Proc. Natl. Acad. Sci. USA* 99:8591.)

transcriptional adaptor   Histone acetyltransferase, that acylates histone in the chromatin before activation of transcription. The yeast gene *GCN5* specifically affects histones 3 and 4. *See* histone; nucleosome; transcription; transcription initiation.

transcriptional coactivator   *See* coactivator; GCN; TAF$_{II}$.

transcriptional control   Regulation of protein synthesis at the level of transcription. *See* operon; regulation of gene activity; regulon; signal transduction; transcription factors.

transcriptional error   When cytosine is deaminated in the template DNA strand to uracil and the error is not repaired, adenine is incorporated in the RNA transcript in the place where guanine was supposed to be inserted by the RNA polymerase. This may be translated into a defective or aberrant protein. Such a mutation may take place in nondividing cells. *See* error in translation.

transcriptional gene fusion vector   Carries transcription-termination codons (stop codons) in front of the promoterless structural gene, so when the structural gene fuses to a host promoter and is thereby expressed (transcribed with the assistance of a host promoter), it will contain only the amino acid sequences specified by the inserted DNA. *See* gene fusion; read-through proteins; translational gene fusion; trapping promoters, see color plate in Color figures section in center of book.

transcriptional gene silencing (TGS)   Based on mechanisms of methylation of DNA and chromatin remodeling by proteins or RNA. *See* chromatin remodeling; dosage compensation; imprinting; methylation of DNA; silencer.

transcriptional modulation   In many transcription factors, proline- and glutamine-rich activation domains exist and modulating effects are attributed to them. *See* transcriptional activators; transcriptional suppressor; transcription factor; WD-40. (Rushlow, C., et al. 2001. *Genes Dev.* 15:340.)

transcriptional pause   *See* pause, transcriptional.

transcriptional regulation   *See* cosuppression, methylation of DNA; posttranscriptional gene silencing; quelling; regulation of gene activity; transcription factors. (Carlson, M. 1997. *Annu. Rev. Cell Dev. Biol.* 13:1; Vaucheret, H. & Fagard, M. 2001. *Trends. Genet.* 17:29; Myers, L. C. & Kornberg, R. D. 2000. *Annu. Rev. Biochem.* 69:729; <http://www.mgs.bionet.nsc.ru/mgs.dbases trrd4/>.)

transcriptional slippage   RNA polymerase transcribes longer or shorter RNA sequences than the actual template either during initiation or the elongation process. Slippage generally occurs when the template has homopolymeric sequences. (Larsen, B., et al. 2000. *Proc. Natl. Acad. Sci. USA* 97:1683.)

transcriptional suppressor   Can tightly bind to the operator or to other upstream elements of the DNA, thus preventing the initiation of transcription by elements (activators)

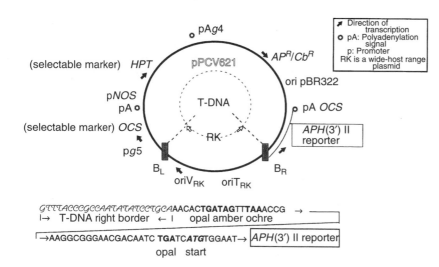

In the transcriptional gene fusion vector, the reporter (APH[3′]II, or luciferase or GUS) has no promoter and it is fused to the right border of the T-DNA. The reporter gene is expressed only if it integrates behind a plant promoter that can provide the promoter function. In front of the structural gene here, there are four nonsense codons to prevent the fusion of the protein with any plant peptides. Transcriptional fusion vectors are similar in other groups of organisms. For other symbols and abbreviations, *see* translational gene fusion vectors. (Based on oral communications by Dr. Csaba Koncz.)

of the transcription complex. Mot1 is an ATP-dependent inhibitor of the TATA box–binding protein and the members of the NOT complex inhibit the transcription machinery in various ways (TBP, TAF, etc.). Heterochromatin protein (HP1) and histone methyl transferase may also repress genes. *See* inhibition of transcription; MADS box; mating-type determination in yeast; nucleosome; silencer; transcriptional activator; transcriptional coactivator; transcriptional modulation. (Ma, Y., et al. 2001. *J. Biol. Chem. Online Sept. 27*; Hwang, K.-K., et al. 2001. *Proc. Natl. Acad. Sci. USA* 98:11423.)

transcription coactivator Activates RNA polymerase II but does not bind to DNA. (Oswald, F., et al. 2001. *Mol. Cell Biol.* 21:7761.)

transcription cofactor Links a transcription factor to the transcription complex without binding directly to DNA.

transcription complex The TATA box–associated complex has the components TFIIA, TFIIB, TFIID, TFIIE, TFIIF, TFIIH, TFIIJ, TFIIK, and RNA polymerase II. After the pol II enzyme moves downstream and away from the preinitiation complex and is phosphorylated by CTDK and TFIIH kinase action, it can continue transcription in the absence of other regulatory factors, although specific transcription factors (transactivators) may boost and regulate its activity. Usually, a low level of phosphorylation of the heptapeptide repeats of the C terminus (CTD) of the largest RNA polymerase subunit is conducive to attraction to the promoter sequences and to the initiation of transcription. The yeast protein FCP is phosphatase and is associated with the RNA polymerase II complex. It facilitates the association of the polymerase to the preinitiation complex. Some

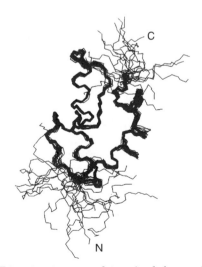

Human TFIIEβc structure as determined by nuclear magnetic resonance analysis. this transcription factor binds to DNA, where the promoter opens up when RNA polymerase II initiates transcription. (From Okuda, M., et al. 2000. *EMBO J.* 19:1346. Courtesy of Professor Y. Nishimura.)

transcription factors play a role in translation. *See* CTD; elongation factors; Hogness (-Goldberg) box; nucleic acid chain growth; open-promoter complex; Pribnow box; regulation of gene activity; RNA polymerase; snRNA; SRB; TBP; TCF; transcription factors; transcription shortening. (Wolfberger, C. 1999. *Annu. Rev. Biophys. Biomol. Struct.* 28:29.) See figure below.

transcription corepressor Represses RNA polymerase II without binding to DNA.

transcription-coupled repair *See* colorectal cancer; DNA repair; excision repair.

transcription factors Large number of different proteins that bind to short upstream elements, terminator sequences of the gene, or RNA polymerases and modulate transcription. *General transcription* factors have highly conserved sequences and are interchangeable even among such diverse organisms as mammals, *Drosophila*, yeast, and plants. TAF proteins by virtue of their different domains can modulate the transcription of different genes. By acetyltransferase activity, the large TAF_{II} protein associated with the TFIID complex can remodel chromatin structure and by its amino-terminal kinase activity it can transphosphorylate TFIIF, thus modulating transcription of protein genes in two different ways. The specific transcription factors may form a large combinatorial network with various promoter elements.

The transcription factors for RNA polymerase I, transcribing ribosomal RNA genes, show a relatively simple organization:

Transcription factors and RNA polymerase I.

The UBF (upstream-binding factor) binds upstream in the promoter and regulates transcription by acting as an assembly factor for the transcription complex. TIF-1 or its vertebrate homologue SL1 is required for the attachment of Pol I to the promoter; accessory proteins A and B assist the transcription.

The tRNAs, 5S rRNAs, and some other small RNAs are transcribed by RNA polymerase III (Pol III). This enzyme has a requirement for protein factor TFIIIB. In case of transcribing 5S rRNA, it also requires TFIIIA and TFIIIC for the assembly of the transcription complex. Proteins TFIIIA and C, however, are detached after TFIIIB binds, and only the latter stays on the DNA when Pol III lands and transcription begins. For the transcription of tRNAs, TFIIIA is not required. The Pol III transcription units also have internal control regions A (box 5′-TG GCNNAGTGG-3′) and B (box 5′-GGTCGANNC-3′) or similar sequences. Transcription factors required for RNA polymerase II (Pol II), which transcribes protein-coding genes, form the most elaborate complex:

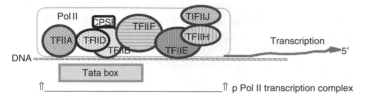

Transcription factors and RNA polymerase II.

The *general transcription complex* is initiated by binding the TBP (TATA box–binding protein) subunit of general transcription factor TFIID to the TATA box of eukaryotes (Hogness box). The TFIID complex exists in different forms (α and β) depending on its association with the TAF_{II}-30 protein (in β) or its absence (in α). The TATA box is present in upstream regions of genes, coding for protein and transcribed by RNA polymerase II. TFIID also attracts TFIIB. TFIID brings the cleavage-polyadenylation specificity factor (CPSF) to the preinitiation complex, and this assists also in the formation of the 3′-end of mRNA.

After this step TFIIF, TFIIE, and TFIIH proteins attach RNA polymerase II (Pol II) to the TATA box. TFIIH expresses a DNA helicase function (encoded by xeroderma pigmentosum genes XPB and XPD) and cyclin-dependent protein kinase activity (encoded by CDK7). TFIIF stimulates transcript elongation by stimulating the phosphorylation of polymerase II and it plays a role in Pol I transcription. TFIIF also stimulates a phosphatase specific for the largest subunit of RNA polymerase. TFIIS protein factor in eukaryotes and GreA and B stimulate transcript cleavage and read-through. It is a coactivator and regulator of transcription and it binds 3′ to the TATA box. Pol II is inactive at this stage until TFIIH phosphorylates the bound Pol II using ATP as a phosphate donor.

The targets of phosphorylation are several sites near the COOH-end of the largest subunit of pol II. TFIIH has a helicase, ATPase, and nucleotide excision repair activities. It also mediates promoter melting and promoter clearance. Some genes may be transcribed without the kinase activity of TFIIH, yet they need its helicase function. Eukaryotic RNA polymerase II generally contains 9 or 12 subunits. The largest subunit is usually about 200 kDa. There are 26 (yeast) to 52 (mammals) repeats (Tyr-Ser-Pro-Thr-Ser-Pro-Ser) close to the carboxyl end. The phosphorylated Pol II then moves out of the complex and can now initiate transcription. The specific transcription factors are operative in special genes and at special tissues and time frames. The transcription factors may bind a variety of other proteins before or during transcription and thus provide a great variety of fine-tuning of gene expression. The not-absolutely-essential TFIID TATA box–binding protein-associated factors $dTAF_{II}$ 42 and $dTAF_{II}$ 62 form a heterotetramer resembling the heterotetrameric core of the histone octamer in the nucleosome. The TBP protein subunit of TFIID may be dispensable in case $TAF_{II}30$ is present (named TFTC [TBP-free TAF_{II}-containing complex]).

The general transcription factors are the basic instruments of transcription initiation, but the modulation and regulation of transcription requires a large number of specific factors, activators, coactivators, suppressors, and their interactions. Transcription factors with altered specificities can be generated in the laboratory using structure-based design and molecular technology. Besides the general transcription factors that are involved with almost all RNA polymerase II transcribed genes, specific and inducible transcription factors, activators, and coactivators, as well as chromatin reorganization factors, have an important role. The genome of *Arabidopsis* has 1,533 (~5.9%) transcriptional regulators, whereas *Drosophila, Caenorhabditis*, and budding yeast have 635 (4.3%), 669 (3.5%), and 209 (3.5%), respectively. These molecules may be further explored for basic research and gene therapy. Viral repressors may prevent the association of RNA polymerase II with the transcriptional preinitiation complex. *See* ABC excinuclease; CAAT box; CDK; chromatin; chromatin remodeling; class II genes; coactivators; CTD; elongator; enhancer; FACT; GAGA; gene number; gene therapy; HMG; Hognes (-Goldberg) box; inchworm model; mediator; mtTF; mtTFA; NAT; nuclear receptors; nucleosome; open-promoter complex; pause, transcriptional; PITSLRE; pol II; Pribnow box; protein synthesis; regulation of gene activity; reinitiation of transcription; RNA polymerase; SAGA; SRB; SWI; TBP; TF; TIIFS; transactivator; transcript elongation; transcriptional activators; transcriptional modulation; transcription complex; transcription rate; xeroderma pigmentosum. (Myer, V. E. & Young, R. A. 1998. *J. Biol. Chem.* 273:27757; Riehmann, J. L., et al. 2000. *Science* 290:2105; Conaway, R. C. & Conaway, J. W. 1997. *Progr. Nucleic Acid Res. Mol. Biol.* 56:327; Hampsey, M. 1998. *Microbiol. Mol. Biol. Rev.* 62:465; Pauli, M. R. & White R. J. 2000. *Nucleic Acids Res.* 28:1283; Lemon, B. & Tjian, R. 2000. *Genes & Development* 14:2551; Misteli, T. 2001. *Science* 291:843; Svejstrup, J. Q. 2002. *Current Opin. Genet. Dev.* 12:156; Burley, S. K. & Kamada, K. 2002. *Current Opin. Struct. Biol.* 12:225; Warren, A. J. 2002. *Current Opin. Struct. Biol.* 12:107, <http://www.ifti.org>; <http://compel.bionet.nsc.ru/FunSite.html>; <http://transfac.gbf.de/TRANSFAC/>; <http://www.gene-regulation.com/pub/databases.html#transcompel>.)

transcription factors, designed Include DNA-binding zinc-finger domains fused to gene activation or suppression domains. Such constructs may up- or downregulate the expression of selected targets. *See* zinc finger. (Urnov, F. D., et al. 2002. *EMBO Rep.* 3:610; Rebar, E. J., et al. 2002. *Nature Med.* 8:1427.)

transcription factors, inducible Proteins that are synthesized within the cell in response to certain agents or metabolites. They bind to short upstream or downstream DNA sequences and affect transcription frequently by looping the bound DNA back to the promoter area and forming an association with the other protein factors and general transcription factors. Such transcription factors may be hormones, heat-shock proteins (DNA-binding site consensus [CNNGAANNTCCNNG]), phorbol esters (TGACTCA), serum-response elements (CCATATAGG), etc. These protein factors exert their specificity in gene regulation not only by discriminative ability for individual genes (since their numbers must be lower than those of the genes) but by their modular assembly. *See* genetic networks heat-shock proteins; hormone-response elements; HSTF; regulation of gene activity; transcription factors. (Mathew, A., et al. 2001. *Mol. Cell Biol.* 21:7163.)

transcription factors, intermediary Do not associate directly with the promoter but either affect the conformation of DNA or "adapt" other proteins to the transcription complex. *See* open-promoter complex; regulation of gene activity. (Steinmetz, A. C., et al. 2001. *Annu. Rev. Biophys. Biomol. Struct.* 30:329.)

transcription initiation The DNA forms a transcription bubble when the transcription begins. The phage T7 RNA polymerase undergoes a major conformational change at the amino-terminal 300 residues. This then entails the loss of the promoter-binding site and facilitates promoter clearance when the initiation is followed by transcript elongation. The RNA transcript peels off of a seven-base pair heteroduplex and an exit tunnel is created for the enhanced processivity of the elongation complex. *See* processivity; promoter clearance; replication bubble. (Yin, Y. W. & Steitz, T. A. 2002. *Science* 298:1387; Young, B. A., et al. 2002. *Cell* 109:417; Pokholok, D. K., et al. 2002. *Mol. Cell* 9:799, <http://elmo.ims.u-tokyo.ac.jp/dbtss>.)

transcription rate May vary from gene to gene and site to site. Based on fluorochrome labeling of β-actin RNA and serum induction of 1.1 to 1.4 kb per minute. Yeast RNA polymerase II can synthesize 1.2-kb-long sequences/min. Single-subunit bacteriophage RNA polymerase can incorporate in vitro 12−24 kb nucleotides/min. The bacterial enzyme can build in 3−6 kb/min in vivo and 0.6−2 kb/min in vitro. *See* error in transcription; RNA polymerase; transcript elongation; transcription factors. (Brem, R. B., et al. 2002. *Science* 296:752.)

transcription shortening RNA polymerase II (RNAP II) hydrolyzes the 3′-end of the transcript as part of the process of reading through pause signals and also secures fidelity of the transcription. It requires TFIIS protein.

transcription termination in eukaryotes The ribosomal gene cluster is generally terminated much beyond the 28S rRNA gene at about 200 bp upstream from the core promoter of the following pre-rRNA cluster. In the mouse, the pol I termination signal contains a Sal I box (5′-AGGTCGACCAG[T/A][A/T]NTCCG-3′) preceded by T-rich clusters. The actual termination is within the T-rich area assisted by the Sal I box and the T-rich sequences around it. In humans, the conserved repeats (5′-GACTTGACCA-3′) terminate pre-tRNA transcription. In *Xenopus* (5′-GACTTGC-3′), repeats and T-rich sequences in the spacer region bring about termination. Probably some proteins bind to the Sal I box. Polypeptide chain release factors (RF) have been identified in eukaryotes. The *Drosophila* NTEF protein releases the Pol II transcript in an ATP-dependent manner. The Reb-1 yeast protein stops the polymerase and mediates transcript release. The mouse TTF-1 protein, similarly to Reb-1, binds to the DNA and brings about termination. Thus, the termination mechanisms in different species vary substantially. In mtDNA to a 13-residue sequence embedded in tRNA$^{Leu(UUR)}$, a 34 kDa protein (mTERM) is bound and the complex is required for termination of transcription. The MAZ protein may regulate transcription from different promoters in closely spaced genes. The poly(A) signal is required for the termination of transcription, but before the actual termination, pretermination cleavage of the RNA transcript takes place. Therefore, before pol II is released from the DNA, it must transcribe the pretermination cleavage site and the poly(A) signal. *See* mRNA; polyadenylation signal; rho factor; transcription termination in prokaryotes. (Langst, G., et al. 1998. *EMBO J.* 17:3135; Dye, M. J. & Proudfoot, N. J. 2001. *Cell* 105:669.)

transcription termination in prokaryotes Can be rho-independent (intrinsic terminators exist in the RNA polymerase) and rho-dependent, i.e., RNA polymerase requires the cofactor rho for termination of transcription. The terminator regions in various systems have similar structures. They consist of palindromic sequences that can fold back into a hairpin. In the rho-independent terminator, there are one or more G≡C-rich sequences in the stem, and at the base of the stem there are about six consecutive U residues. This structure mediates a pause in the movement of the RNA polymerase, thus causing dissociation from the DNA template because the ribosyl-U of the transcript can make only weak hydrogen bonds with the deoxyribosyl-A in the DNA. In the rho-dependent termination, rho recognizes 50 to 90 bases before the hairpin facilitates termination. *E. coli* protein NusA promotes folding of the hairpin and the termination. The λN protein promotes antitermination. Polypeptide-release factors (RF) may be used in both prokaryotes and eukaryotes. On lambda-phage templates, the N-terminal of the 109-amino-acid Nun protein of phage HK022 blocks transcription by binding to BOXB on the nascent RNA transcript of the pL and pR operons and the C-terminal domain interacts with RNA polymerase. If the RNA polymerase ternary complex (at 3′-OH end) cleaves the transcript, transcription may be reinitiated, as long the upstream sequences remain firmly aligned with the DNA. The prokaryotic proteins GreA and GreB and the eukaryotic TFIIS may favor transcript cleavage. *See* antitermination; lambda phage. (Washio, T., et al. 1998. *Nucleic Acids Res.* 26:5456; Gusarov, I. & Nudler, E. 2001. *Cell* 107:437; Unniraman, S., et al. 2000. *Nucleic Acids Res.* 30:675; Kashlev, M. & Komissarova, N. 2002. *J. Biol. Chem.* 277:14501.)

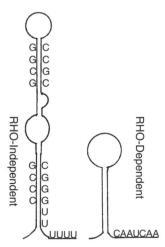

transcription unit DNA sequences between the initiation and termination of transcription of a single or multiple (cotranscribed, multicistronic) gene(s).

transcript mapping RNA transcripts are hybridized with specific DNA probes and subsequently the (not annealed) single strands are digested by S1 nuclease; thus, the resistant (annealed) fragments represent the homologous tracts and demarcate the transcript. For the process of mapping an entire genome, sequence-tagged sites (STS) of cDNAs are used. By the PCR method the STS sequences can be amplified and their position can be mapped either by YAC clones or by radiation hybrid panels. *See* nucleic acid hybridization; PCR; radiation hybrid; S1 nuclease; STS; YAC. (Barth, C., et al. 2001. *Curr. Genet.* 39[5–6]:355.)

transcriptome Collection of RNAs transcribed from the genome; transcriptomics is the generation and study of the mRNA profile of the cell. In a single human cell, ~43,500 genes may be expressed and a total of about 84,000 unique transcripts are expressed in a human body. The total number of transcripts in different tissue cells of these unique genes was found to be 134,135. Some genes were expressed only at 0.3 copies per cell; others had up to 9,417 transcripts. About 1,000 transcripts were expressed in 5 copies in all cells. In some cancer cells, some transcripts were present in about 10 copies but absent in normal cells; 40 genes were expressed in all types of cancer cells in about 3 copies, and this was twice the number for normal cells. *See* expression profile; RIDGE; SAGE; Atlas™ human cDNA. (Velculescu, V. E., et al. 1999. *Nature Genet.* 23:387; Caron, H., et al. 2001. *Science* 291:1289; Camargo, A. A., et al. 2001. *Proc. Natl. Acad. Sci. USA* 98:12103; Appendix II-10, Su, A. I., et al. 2002. *Proc. Natl. Acad. Sci. USA* 99:4465; Bono, H. & Okazaki, Y. 2002. *Current Opin. Struct. Biol.* 12:355; Wu, L. F., et al. 2002. *Nature Genet.* 31:255, <http://bioinfo. amc.uva.nl/HTM-bin/index.cgi>.) See color plate in Color figures section in center of book.

transcripton Unit of genetic transcription.

transcriptosome Complex of RNA processing proteins (capping enzymes, splicing factors, etc.) within the nucleus associated with the COOH-terminal domain of the large subunit of RNA polymerase II. *See* RNA polymerase, transcription factors, posttranscriptional processing, capping enzymes, splicing. (Halle, J. P. & Meisterernst, M. 1996. *Trends Genet.* 12:161.)

transcytosis Immunoglobulins (or other molecules) are transported within a vesicle from a secreting cell across the epithelial layer to another domain of the plasma membrane. (McIntosh, D. P., et al. 2002. *Proc. Natl. Acad. Sci. USA* 99:1996.)

transdetermination A particular pathway of differentiation is overruled by genetic regulation; thus, e.g., at the *Antp* (*Antennapedia* locus 3-47.5) "gain-of-function mutants" in *Drosophila*, the antenna is transformed into a mesothoracic leg or a wing may develop in the place of an eye, etc. These changes in the developmental pattern may be associated with chromosomal rearrangements. The breakpoints may be within the promoters. They may be altered as a cause of transcript heterogeneity and alternate splicing of the transcripts. Transdetermination occurs in plants (snapdragon, *Arabidopsis*, and others) by transforming anthers and pistil into petals and producing sterile full flowers, and in animals by changing *Drosophila* antennae into legs, etc. *See* determination; homeotic genes; imaginal disk; morphogenesis; stem cells. (Hadorn, E. 1978. *Sci. Am.* 219:110; Maves, L. &

Schubiger, G. 1998. *Development* 125:115; Wei, G., et al. 2000. *Stem Cells* 18:409; Glotzer, M., et al. 2001. *Annu. Rev. Cell Dev. Biol.* 17:351.)

The apical meristem developed into horn- or antler-shape structures rather than into a normal inflorescence (at left) in *Arabidopsis* mutants.

transdifferentiation Rare biological phenomenon of one type of differentiated cells is converted into another discrete type. *See* transdetermination.

transdominant molecules Used to selectively inhibit gene expression, e.g., antisense RNA, decoy RNA, ribozymes, suppressor proteins, single-chain antibodies. *See* antisense technologies; ribozymes; RNAi; suppressor gene; suppressor RNA; TNF; tumor-infiltrating lymphocytes; tumor suppressor factors. (Kamb, A. & Teng, D. H. 2000. *Curr. Opin. Mol. Ther.* 2:662.)

transducer proteins Respond to effectors in bacteria to relay information to cytoplasmic components of the excitation path to switch molecules. After the effector is diluted out, the adaptation pathway leads the restoration to the base condition. (Wang, C., et al. 2001. *Nature* 4121:285.)

transducianism *See* creationism.

transducin A G-protein, G_t, involved in transduction of light signals (RAS-related proteins) regulating cyclic GMP phosphodiesterase. It is activated by cholera toxin and inhibited by pertussis toxin. *See* cholera toxin; G_t protein; pertussis toxin; retinal dystrophy; transduction. (Norton, A. W., et al. 2000. *J. Biol. Chem.* 275:38611.)

transducing phage *See* transduction.

transductant Transduced cell. *See* transduction.

transduction, abortive Transduced DNA is not integrated into the bacterial chromosome and therefore fails to replicate among the bacterial progenies. It is diluted out during subsequent cell divisions. The nonreplicating abortively transduced genetic material is transmitted in a unilinear fashion. *See* transduction, generalized; transduction specialized. (Stocker, B. A. D. 1956. *J. Gen. Microbiol.* 15:575; diagram next page.)

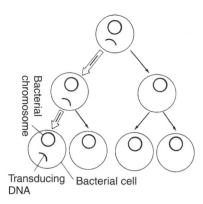

Transducing DNA / Bacterial cell

transduction, generalized Phage-mediated transfer of *unspecified* genes among bacteria (lysogenic or nonlysogenic). The virulent phage breaks down the host cell DNA into various-size fragments by the process called lysis. The transducing phage coat may then scoop up fragments of the DNA that fit into the capsule (head), which may not contain any phage genetic material. The DNA fragments enclosed in the phage head are picked up at random from the proper-size group of bacterial DNA without regard to the genes located in the fragments. For productive transduction, the transferred fragments must be stably integrated into the chromosome of the recipient bacteria. *See* generalized transduction; marker effect; pac site; specialized transduction; transduction, abortive; transduction mapping. (Lederberg, J., et al. 1952. *Cold Spring Harbor Symp. Quant Biol.* 16:413.)

transduction, specialized (or restricted) Temperate-phage-mediated transfer of *special* genes between bacteria. Transduction may be mediated in higher eukaryotes by horizontal transmission, transposable and retrotransposable elements. *See* specialized transduction; transposable elements. (Morse, M. L. 1954. *Genetics* 39:984.)

transduction mapping Determines the map position of very closely linked bacterial genes on the basis of cotransduction frequencies; e.g., the donor DNA is a^+b^+; the recipient bacterium is a^-b^-; then the recombination frequency is $[(a^+b^-) + (a^-b^+)]/[(a^+b^-) + (a^-b^+) + (a^+b^+)]$. *See* transduction, generalized.

transesterification An esterase enzyme catalyzes a replacement reaction. A nucleophile displaces an alcohol during the hydrolysis of an ester. A similar reaction occurs when phosphodiester exchanges take place at the splice junctions of exons and introns in nucleic acid processing. *See* introns; splicing.

transfection Originally, the term was coined for the introduction of viral RNA or DNA into bacterial cells and the subsequent recovery of virus particles. Today, it is used for introduction of foreign DNA into animal cells where the genes may be expressed. The delivery system may be microinjection, bombardment (gene gun), electroporation, DEAE dextran, calcium phosphate precipitation of the target cell membranes, liposomes, or polyamidoamine dendrimers (highly branched cationic polymers). After delivery, the DNA needs protection from elimination and degradation by nucleases, opsonins, and endocytosis. Nuclear targeting may be facilitated by polyethylene glycol. Nuclear localization signals combined with peptide nucleic acid may facilitate homing on the nucleus. There may be other hurdles to overcome such as cytotoxicity, DNA condensation, tissue targeting, etc. *See* biolistic transformation; DEAE dextran; electroporation; endocytosis; liposomes; microinjection; nuclear localization sequences; opsonin; peptide nucleic acid; polyethylene glycol; receptor-mediated gene transfer; transformation, genetic. (Földes, J. & Trautner, T. A. 1964. *Z. Vereb.-Lehre* 95:57; Lug, D. & Saltzman, W. M. 2000. *Nature Biotechn.* 18:33.)

transfectoma Hybridoma cell producing a specific mouse/human chimeric antibody. *See* antibody; hybridoma. (Sun, L. K., et al. 1991. *J. Immunol.* 146:199.)

transfer, clockwise/counterclockwise The bacterial F plasmid may be integrated in different orientations and at different locations in the bacterial chromosome; thus, Hfr strains may be formed that transfer the chromosome either clockwise or counterclockwise during conjugation. *See* conjugation; Hfr.

transfer, horizontal *See* transfer, lateral.

transfer, lateral Genetic information is transmitted by "infection" or by plasmids rather than by sexual means (vertical transfer). Many species of bacteria take up and maintain extracellular DNA, especially under conditions of starvation. *See* incongruence; lateral transmission. (Finkel, S. E. & Kolter, R. 2001. *J. Bacteriol.* 183:6288; Koonin, E. V., et al. 2001. *Annu. Rev. Microbiol.* 55:709.)

transfer, vertical Genetic information is transmitted by sexual means rather than by horizontal infection-type mechanisms.

transferase enzymes Move a chemical group(s) or molecule(s) from a donor to an acceptor.

transfer factors Bacterial plasmids capable of transferring information from one bacterial cell to another through conjugational mobilization. Some of the factors (e.g., ColE1) may not have genes for transfer, yet they may be transferred to other cells by the helper function of conjugative plasmids. These transfer factors may contain genes for resistance (transposable elements) and have great medical significance because of the transfer of antibiotic resistance, thus making the defense against pathogenic infection difficult. *See* antibiotics; colicins; resistance transfer factors; transposable elements, bacterial.

transfer line Polyploid species carrying a relatively short foreign chromosomal segment in its genome. The transfer is generally made by crossing over between homoeologous chromosomes in the absence of a gene or chromosome (chromosome 5B in wheat) that would normally prevent homoeologous pairing. It can also be obtained by (X-ray) induced translocation. Construction of such lines may have agronomic importance for introducing disease resistance or any other genes that are not available in the cultivated varieties or their close relatives. *See* alien addition; alien substitution; chromosome substitution; homoeologous chromosome. (Sears, E. R. 1972. *Stadler Symp.* 4:23.)

transferrin (TF) β-globulin   (M$_r$ ~ 75,000–76,000) that transports iron. The encoding gene is in human chromosome 3q21; the transferrin receptor (TfR) gene is nearby. Adenosine ribosylation factors affect the cellular redistribution of transferrin and endocytosis. *See* aceruloplasminemia; atransferrinemia; BLYM; endocytosis; hemochromatosis; RAC; receptor-mediated gene transfer. (Aisen, P., et al. 2001. *Int. J. Biochem. Cell Biol.* 33:940.)

transfer RNA (tRNA)   Genes coding for tRNAs are clustered in prokaryotes and eukaryotes. Some of the tRNA genes are located within the spacer regions of the ribosomal gene clusters. The majority of tRNA genes are clustered as a group in the DNA and frequently occur in 2–3 copies. In *Drosophila*, 284 tRNA genes have been identified. In humans, there are 497 (plus 324 pseudogenes); in *Caenorhabditis*, 584 tRNA genes. Some *E. coli* tRNA (86) gene clusters include genes for proteins. tRNA genes within the cluster are separated by intergenic sequences and transcribed as long pre-tRNA sequences. The primary transcript is processed at the 5′-end by RNase P and at the 3′-end by RNase D, BN, T, PH, RNase II, and polynucleotide phosphorylase. Before tRNAs are released to the cytoplasm, their integrity is ascertained and only the mature and structurally correct molecules are exported in a Ran-guanosine triphosphate–dependent manner. Aminoacylation

takes place before export from the nucleus. tRNAs are small molecules (70–90 nucleotides). They assume a cloverleaf secondary structure formed by single-stranded loops and double-stranded sequences.

The functioning tRNAs assume an L-shape configuration. After charged with amino acids (aminoacylation), they haul the amino acids to the ribosomes for translation of the genetic code (protein synthesis). The amino acids are attached to the protruding C–C–A–(OH) amino acid arm, and one of the C residues interacts directly at the P site with G2252 and G2253 of the 23S ribosomal subunit of prokaryotes. 3′-CAA is generally added to tRNA after transcription, but some bacterial tRNA genes encode this sequence. The anticodon loop contains a triplet complementary to the amino acid code word. This anticodon recognizes the code in mRNA on the surface of the ribosome. The D-arm (dihydrouracil loop) is the recognition site for the aminoacyl-tRNA synthetase enzyme, whereas the T-arm (a thymine-pseudouracil [ψ]-C consensus loop) recognizes the ribosomes. There is also a small variable loop (V arm).

The presence of modified nucleotides is characteristic for tRNAs and they modulate the anticodon domain structure for many tRNA species to accurately translate the genetic code (Yarian, C., et al. 2002. *J. Biol. Chem.* 277:16391). These modifications take place right after transcription or during processing. The number of tRNAs in prokaryotes is higher than the number of genetic code words (64); in prokaryotes, the number of different tRNA molecules may run into hundreds in the different species. In the nematode *Caenorhabditis elegans*, sequencing the entire genome identified 659 tRNA genes and 29 pseudogenes. Since there are only 20 amino acids, of the high number of tRNAs several deliver the

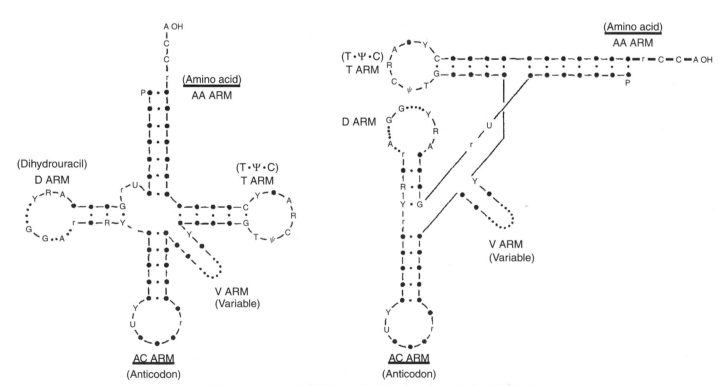

The general structural features of tRNAs. Left: The cloverleaf; Right: The L-shaped tertiary conformation. R stands for purine and Y for pyrimidine bases in all tRNAs; R and Y indicate the occurrence of these bases in many tRNAs; ψ is pseudouridine. (After Kim, S. H. 1976. *Progr. Nucleic Acid Res. Mol. Biol.* 17:181.)

same amino acid to the site of translation (isoaccepting tRNA). The amino acid accepting *identity* of transfer RNAs resides within the base sequences and structural features of the tRNAs, which are then recognized by aminoacyl tRNA synthetase enzymes. tRNAs possess dual functions: synthetase and editing, i.e., the removal of the wrong amino acid if attached. mRNA site recognition is the property of the anticodon.

The vast majority of animal and fungal mitochondria synthesize their own tRNAs (about 22). Their anticodon reads the codons by the first two bases and thus does not need isoaccepting tRNAs. Their structure usually lacks the pseudouridine loop, and there are other minor variations. The mitochondria of land plants, algae, *Paramecium*, *Tetrahymena*, and *Trypanosomes* partially rely on tRNA import. Organelle tRNAs, similarly to plants, posttranscriptionally add the 3′-CAA amino-acid-accepting terminus, in contrast to some cases of *E. coli*. Mitochondrial tRNAs in plants are generally quite variable and show similarities to chloroplast tRNAs. Chloroplast tRNAs of higher plants (about 30) may frequently be coded by more than one tract. The universal genetic code requires a minimum of 32 different tRNAs. Actually, however, 24–26 tRNAs (with special wobbles) may suffice for protein synthesis. The mammalian and some fungal mitochondria do not code for all the tRNAs required and import these nuclearly coded molecules from the cytosol. Many fungi code for 25–27 tRNA genes, however. In *Caenorhabditis*, the tRNA gene number and the number of tRNAs are highly correlated.

Prokaryotes and chloroplasts have a special tRNA^{F-Met} for the initiation of translation, and the same anticodon, CAU, may recognize not only the Met codon (AUG) but three other initiator codons. Nematodes initiate translation with UUG, which is a leucine codon. tRNAs are used as primers for reverse transcription. During amino acid starvation in bacteria, guanosine tetraphosphate (ppGpp) is made on the ribosomes with the aid of cognate, uncharged tRNA. Attenuation requires the cooperation of tRNA. In eukaryotes, a unique glutamyl-tRNA reductase mediates the formation of 5-aminolevulinic acid, which contributes to the porphyrin ring. *See* aminoacyl-tRNA synthase; attenuation; chloroplast genetics; chloroplasts; code, genetic; codon usage; fMet; isoaccepting tRNAs; mitochondrial genetics; modified bases; mtDNA; porphyrin; Ran; ribonuclease P; ribosome; *rrn*; stringent response; tRNA; wobble. (Giege, R., et al. 1993. *Progr. Nucleic Acid Res. Mol. Biol.* 45:129; Morl, M. & Marchfelder, A. 2001. *EMBO Rep.* 2:17; Grosshans, H., et al. 2000. *J. Struct. Biol.* 129:288; Beunning, P. J. & Musier-Forsyth, K. 1999. *Biopolymers* 52:1; Intine, R. V., et al. 2002. *Mol. Cell* 9:113.)

transformant   Cell or organism that has been genetically transformed by the integration of exogenous DNA into its genetic material. *See* transformation.

transformation, genetic   Information transfer by naked DNA fragments or plasmids, obviating traditional sexual or asexual processes in prokaryotes and eukaryotes. Transformation procedures may be transient when the introduced DNA is not being integrated into the genome of the cell. It may also be permanent when the exogenous DNA becomes an integral part of the recipient's genetic material.

BACTERIAL TRANSFORMATION: Genetic transformation was discovered in bacteria in the late 1920s. It became a widely used genetic method only in 1950. Originally, only naked bacterial DNA was used in fragments of 1/200 to 1/500 of the genome. This was provided to *competent* bacterial cells at a concentration of 5–10 µg/mL culture medium. Exogenous DNA can synapse with the bacterial genome and generally only one strand of transforming DNA is integrated into the recipient, although some bacteria (e.g., *Haemophilus influenzae*) preferentially take up double-stranded DNA from their own species but integrate only one of the strands. Recognition of homospecific DNA is mediated by uptake signal sequences (USS): 5′-AAGTGCGGT in the plus strand and 5′-ACCGCACTT in the minus strand. In the completely sequenced genome of 1,830,137 bp, 1,465 such USS were recognized. *Neisseria gonorrhoeae* also has USS elements (5′-GCCGTCTGAA).

Bacterial transformation generally may not involve an addition; rather it involves a replacement of part of the DNA of the recipient cell, except when plasmids are used. The nonintegrated parts of the donor DNA are degraded and the rest replicates along the genes as a permanent integral part of the bacterial chromosome. The frequency of bacterial transformation may be in the range of 1% or as low as 10^{-3} to 10^{-5}; however, using bacterial protoplasts (spheroplasts), up to 80% transformation is attainable. For bacterial transformation, most commonly various genetic vectors are used. Although different empirical procedures are employed in different laboratories, some general features of the methods are obvious. For transformation, either high-molecular-weight (DNase-free) DNA or plasmids (phage) dissolved in $1 \times$ SSC are used. Competence in recipient bacteria is induced by $CaCl_2$, $MnCl_2$, reducing agents and hexammine cobalt chloride, or competent cells are purchased in a frozen state from commercial sources. Highly competent cells may yield ca. 10^7 to 10^9 colonies per 1 µg plasmid DNA.

The success of transformation is improved by highly nutritious culture media and good aeration. The recognition of transformant cells is greatly facilitated by selectable markers. Transformation of bacteria by electroporation may be extremely efficient (10^{10} transformants/µg DNA). Cultures in mid-log phase are chilled and washed by centrifugation in low-salt buffer. The cells (3×10^{10}/mL) are suspended in 10% glycerol and can be stored on dry ice or at $-70°C$ for up to 6 months. Thawed aliquots of the cells are mixed with properly prepared donor DNA and exposed to high-voltage electric fields in small volumes (20–40 µL). Gram-positive bacteria such as *Bacillus subtilis* are more difficult to transform genetically than gram-negative bacteria, e.g., *E. coli*. *B. subtilis* attracted interest for cloning because it is not pathogenic for humans. In the presence of *B. subtilis recE*, transformation is facilitated if the cell contains a plasmid homologous to the vector. Also, spheroplasts in the presence of polyethylene glycol take up exogenous DNA much easier. Vectors derived from *Staphylococcus aureus* containing tetracycline (pT127) or chloramphenicol (pC194) resistance have been successfully used for the development of new vectors. *Staphylococcus aureus* is a serious pathogen. Shuttle vectors containing *E. coli* pBR322 and *S. aureus* plasmid elements have also been used. Some of the antibiotic-resistance genes, e.g., β-lactamase, have very different expression in different species of bacteria. *Streptomyces* are of substantial interest for transformation because of their efficient production of antibiotics. They can be transformed by a sex plasmid, liposomes, and phage vectors. See antibiotics; β-lactamase; cloning vectors; competence; DNA extraction; electroporation; liposome; sex plasmid; SSC; vectors.

FUNGAL TRANSFORMATION: It is not entirely different from that in prokaryotes. Transformation of *Neurospora* started during the early 1970s and caused genetic instabilities in the genome (see RIP). Transformation of budding yeast began in the late 1970s and became very useful for various types of studies (cloning, YACs, gene replacement, etc.). Yeast cells are grown to about 10^7 density/mL, then suspended in a stabilizing buffer containing 1 M sorbitol. Subsequently, the cell wall is removed by digestion with β-glucanase (an enzyme hydrolyzing glucan, the polysaccharide of the cell wall [yeast cellulose]). Donor DNA is added to the washed spheroplasts in sorbitol, in the presence of $CaCl_2$ and polyethylene glycol (PEG4000). After about 10 min incubation of the mixture, the cells are gently embedded in 3% agar and layered over a selective medium in a petri plate. The frequency of transformation depends a great deal on the type of vector used (from 1 to 10^6 colonies per μg DNA). Alternatively — although with lower yield — intact yeast cells have been treated with lithium salts before the DNA and polyethylene glycol are applied. This is followed by selection after spreading the cells onto the surface of selective media. This procedure does not require the production of spheroplasts and agar embedding. Both of these procedures may cause mutations. Similar methods of transformation have been used in other fungi (*Neurospora*, *Aspergillus*, *Podospora*) and in green algae. Shuttle vectors were advantageous for the transfer of genes between various fungi and between fungi and bacteria. *See* centromeric vector; episomal vector; gene replacement; integrating vector; replicating vector; shuttle vector; YAC.

TRANSFORMATION OF ANIMAL CELLS: The most commonly used procedures involve precipitation of the donor DNA with calcium phosphate or DEAE-dextran. The precipitated granules may enter animal cells by phagocytosis and up to about 20% of the cells may integrate the donor DNA into the chromosomes. By precipitation, physically unlinked DNA molecules can be transformed (cotransfected) into cultured animal cells. The polycation Polybrene (Abbott Laboratories trade name for hexadimethrine bromide) is also used to facilitate the transformation by relatively low-molecular-weight DNA (plasmid vectors) when some other procedures do not work. Electroporation also been successfully used for stable or transient introduction of DNA into the cells. Bacterial (or even plant) protoplasts can be used to bring about cell membrane fusion (in the presence of polyethylene glycol), which may be followed by transfer of plasmid DNA into animal cell nuclei. This procedure is less efficient than endocytosis mediated by calcium phosphate, and the plasmids are frequently integrated in tandem into the chromosome(s) of vertebrates. The exogenous DNA may be introduced by direct *microinjection* into (pro)nuclei or into embryonic stem cells (ES), thus generating chimeras (see gene transfer by microinjection). In the latter case, the transformed cells can be screened for insert copy number or the insert can be targeted to a specific site by homologous recombination. *Infection* of stem cells, bone marrow, zygotes, or early embryos by vectors or by isolated chromosomes is also feasible (see gene transfer by microinjection).

Gene replacement by homologous recombination (see targeting) is also an option. Sufficient information is required on the needs of critical cis-acting elements. The success of transformation of animal cells varies a great deal according to cell types used. Transformation of vertebrate cells became a very important tool of molecular biology and reversed genetics, but unfortunately the transformed cells cannot be regenerated into complete individuals, except when germline or ES cells are transformed. In *Drosophila*, an isolated gene can be inserted into the cloned *P* element with the aid of genetic engineering, and the element may be microinjected into a young embryo where the DNA can integrate into the chromosome, resulting in a stably transformed individual fly. The procedure is particularly effective if the P-element vector is equipped with a selectable marker. In mosquitoes, microinjection into the egg cells is feasible but not very effective. A more successful approach uses viral vectors with the vesicular stomatitis virus glycoprotein envelope, which binds to the cell membrane and delivers foreign DNA. Transgenic rhesus monkeys were produced by injecting pseudotyped replication-defective retroviral vector into the perivitelline space and later fertilized by intracytoplasmic injection of sperm (ICSI) into mature oocytes. See adenoma; ammunition; *Anopheles*; bovine papilloma virus vectors; DEAE-dextran; electroporation; ES; gene replacement; gene therapy; hybrid dysgenesis; ICSI; liposome; Polybrene; polyethylene glycol; pseudotyping; retroviral vectors; smart ammunition; SV40 vectors; targeting genes; transgenic; vesicular stomatitis virus. (Chan, A. W., et al. 2001. *Science* 291:309; for protocols: Ravid, K. & Freshney, R. I., eds. 1998. *DNA Transfer to Cultured Cells*. Wiley-Liss, New York.)

TRANSFORMATION OF PLANTS: Can be carried out by a variety of procedures. Most extensively used were the techniques of infecting leaf or root explants, protoplasts, or seeds by agrobacteria carrying genetically engineered plasmids. Practically all dicots can be readily transformed by agrobacteria; some monocots (*Dioscorea*, *Narcissus and Asparagus*) can also be transformed. The difficulty with monocot transformation by agrobacteria is apparently caused by the lack of secretion of substances needed for the activation of the virulence gene cascade or monocot cells fail to develop competence in response to infection by *Agrobacterium*. The DNA can be introduced, however, by biolistic methods.

The vector plasmids are either cointegrate or binary. Cointegrate plasmids contain the virulence genes of the Ti plasmids in cis, whereas in the binary vectors the virulence genes are carried by a separate small helper plasmid. A common feature of all these vectors is that the genes to be integrated into the plant chromosome are in between the two 25 bp inverted repeats of the T-DNA. The virulence genes and all other DNA sequences are not integrated into the host genetic material. The left and right border sequences are important for successful transformation, but only a few bp of the left border and either none or 1 to 3 bp of the right border are retained in the host (see *Agrobacterium*; Ti plasmid). The insertional target is not strictly specified in plants, yet a few base similarities are frequently found. It appears that the border repeats scout for appropriate target sites and they are appositioned there. The plasmid virulence genes direct this process and the bacterial chromosome has some genes that assist the transformation. The target suffers initial staggered nicks followed by degradation, and the DNA within the 25 bp borders of the T-DNA is integrated into the chromosome (see diagram below). In *Arabidopsis*, the histone-2A gene appears to be required for the integration of the T-DNA.

The genes within the T-DNA generally carry appropriate (plant-compatible) eukaryotic promoters and polyadenylation signals to be expressed in the plant cells. Some vectors may lack promoters and can be expressed only when fused (upon integration) in vivo with plant promoters. These may be translational or transcriptional fusion vectors. Translational fusion vectors lack the translation initiation methionine codon in order to facilitate the fusion of the structural gene in T-DNA with some amino acid sequences of the plant host. The purpose of these types of transformation is to study the strength and tissue specificity of different plant promoters and the function of fusion proteins. Transcriptional fusion vectors carry one or more translational stop codons (nonsense codons) in the nucleotide tract preceding the ATG (translation initiation codon). Therefore, the structural transgene is expressed if a genuine plant promoter drives it and no fusion protein is obtained. These vectors are also shuttle vectors; they can be propagated in agrobacteria and *E. coli*. Being shuttle vectors greatly facilitates various manipulations. The vectors can be replicated in *E. coli* because they have the origin of replication of the pBR322 plasmid and carry genes *oriV* (required for replication) and *oriT* (required for transfer) in *Agrobacterium* outside the boundaries of the two T-DNA sequences. The latter genes were derived from the promiscuous (wide host range) RK plasmid. Aseptically (axenically) grown plant tissues may start the transformation procedure. The vector cassettes generally carry selectable markers, most commonly for resistance against hygromycin B or kanamycin. The gene fusion reporter genes may be (bacterial or firefly) luciferase or GUS (β-glucuronidase) because of easy monitoring and the time and space of expression of the fused reporter gene. *See* transcriptional gene fusion vector; translational gene fusion vector.

Scalpels generally wound the plant explants as they are harvested. Then the tissues are dipped into a fresh bacterial suspension (grown to a density of about 10^6, washed and diluted to about half or less in plant nutrient solution). After the bacteria are blotted off, the plant material is incubated for 2 days on the surface of an agar medium in petri plates (*see* embryogenesis, somatic). Following incubation, the bacteria are stopped either by claforan (cefotaxime) or carbenicillin and the plant cells are grown further in media also containing hygromycin or kanamycin (G418) or another selective agent depending on the vector constructs. The regenerated plants

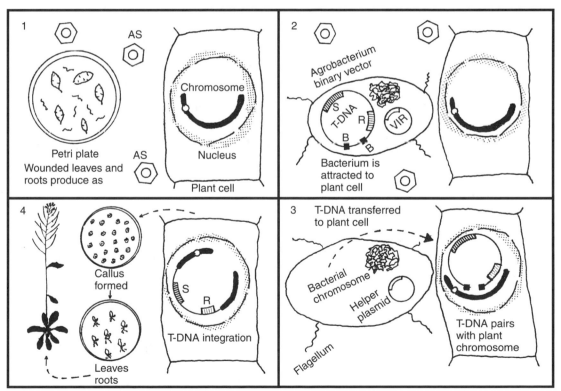

The major steps of transformation of plants by the use of agrobacterial binary vectors. Axenically grown leaves, roots, or stem segments are wounded. The wounding stimulates the production of phenolics such as acetosyringone (as) and attracts the bacteria carrying the engineered plasmids with selectable markers (S) and a reporter gene (R) placed between the two border sequences (B) of the T-DNA. A small helper plasmid carries the *vir* genes required for transfer and integration of the T-DNA. Two days after infection, the bacterial growth is stopped by antibiotics. The isolated plant organs develop callus, roots, and shoots, and eventually complete plants on the selective media only if the transformation was successful. The transformation is confirmed by the expression of the reporter gene. The transformants produce seed that develops into heterozygous progeny. The antibiotic markers are dominant in the plants.

may be grown axenically to maturity in test tubes (in case of *Arabidopsis*) or in soil (in case of larger plants). The transgenes usually segregate as dominant alleles if the plants do not have a corresponding native locus that might mask their expression.

Alternatively, agrobacterial infection can be applied to presoaked seeds of plants and eventually a small fraction of the embryos developing after meiosis will carry the transgene. Seedlings or plants can be infiltrated by agrobacterial suspensions and selection carried out on a large scale in soil cultures treated with an appropriate herbicide (Basta) against which the vector carries resistance (*in planta* transformation). Plants can be transformed by electroporation and by biolistic methods. Cauliflower mosaic virus (CaMV) and geminivirus vectors have been developed, but they are not widely used. *See* *Agrobacterium*; biolistic transformation; electroporation; floral dip; genetics of chloroplasts; genetics of mitochondria; gene transfer by microinjection; gene trap vectors; microinjection; transformation of organelles.

transformation, organelle

Mitochondria and plastids are amenable to transformation by exposing protoplasts to appropriate vectors in the presence of polyethylene glycol (PEG) or even more effectively by employing biolistic procedures. The efficiency of transformation is generally somewhat low. Transformation of organelles still struggles with methodological difficulties, yet it has great potential for improvement. The number of mitochondria (and plastids) within single cells may run into the hundreds or even thousands, which would make possible the amplification of economically important proteins. Transformation of organelles has the additional safety of containment of the transgenic organisms because mitochondria and plastids are usually not transmitted through the male. *See* Bellophage; biolistic transformation; chloroplast, genetic; gene therapy; genetic engineering; human gene transfer; metabolite engineering; mitochondrial genetics; protein engineering; transformation. (Bogorad, L. 2000. *Trends Biotechnol.* 18:257; Ruf, S., et al. 2001. *Nature Biotechnol.* 19:870.)

transformation, oncogenic

Change to malignant (cancerous) cell growth. Transformed animal cells are anchorage independent and typically free of contact inhibition. They may initiate tumors when implanted into immune-compromised animals. Oncogenic transformation results when tumor-suppressor genes are inactivated and oncogenes are activated. *See* cancer; carcinogen; gatekeeper; oncogenes; phorbol esters; tumor suppressor. (Hunter, T. 1997. *Cell* 88:333.)

transformation, stable

Produces cells that carry transforming DNA in an integrated form and thus the acquired information is consistently transmitted to the progeny. *See* transformation, genetic; transformation, transient.

transformation, transient

The introduced DNA may be expressed only for a limited time (1 to 3 days) in the recipient cell because it is not integrated into the host genetic material. Electroporation most commonly results in this kind of transformation. *See* electroporation; transformation, stable.

transformation-associated recombination (TAR)

Yeast cells are transformed simultaneously by a YAC vector (TAR)

with terminal human genomic repeats such as Alu and a long piece of human genomic DNA containing interspersed repeats (Alu). Recombination between the YAC and human genomic DNA within the homologous repeats yields large, stable, circular YACs that can be used for mapping or cloning. If the TAR vector contains an *E. coli* F-factor cassette, the vector can also be propagated in bacterial cells. *See* Alu; F factor; vector cassette; YAC. (Gonzáles-Barrera, S., et al. 2002. *Genetics* 162:603.)

transformation by protoplast fusion

See protoplast fusion.

transformation-competent artificial chromosome vector (TAC)

Can carry large (40–80 kb) inserts and can be maintained in *E. coli*, *Agrobacterium tumefaciens*, and expressed in plant cells. (Liu, Y.-G., et al. 1999. *Proc. Natl. Acad. Sci. USA* 96:6535.)

transformation mapping

Procedure in prokaryotes for determining gene order within the genetic region of integration of the homologous transforming DNA. It can be used as a three-point cross, but the additivity of recombination is generally imperfect in transformation. *See* bacterial recombination frequency.

transformation rescue

Recessive lethal phenotype is compensated for by introducing a viable allele or DNA sequences with the aid of transformation. By narrowing the transforming DNA to the minimal length that restores a viable phenotype (using restriction enzymes), the site of the damage can be estimated, cloned, and sequenced. *See* transformation, genetic. (de Vries, J. & Wackernagel. W. 1998. *Mol. Gen. Genet.* 257:606.)

transformation vectors

See cloning vectors; transformation, genetic; vectors.

transformed distance (TD)

UPGMA procedure for species i and j:

$$d_{ij'} = \frac{(d_{ij} - d_{ir} - d_{jr})}{2} + c$$

where r is a reference species within or outside the group and c is a constant to make $d_{ij'}$ positive. This TD formula is unsuitable for the estimation of branch length in evolutionary trees. *See* evolutionary distance; evolutionary tree; Fitch-Margoliash method for TD; UPGMA.

transforming growth factor

See TGF.

transforming growth factor β

Superfamily of proteins that induces the change of undifferentiated tissues into specific types of tissues. TGF-β1 peptide factor causes reversible arrest in the G1 phase of the cell cycle and thus is considered a tumor suppressor. TGF-β1 has been detected in cis-position to several genes. These proteins are serine/threonine kinases. *See* activin; bone morphogenetic protein; TGF; tumor suppressor.

transforming principle

Historical term used in the early bacterial transformation reports when it was not yet proven that DNA was the agent of transformation. (Alloway, J. L. 1931. *J. Exp. Med.* 55:91.)

transformylase   Enzyme that adds a formic acid residue to the methionine-charged fMet tRNA (tRNA^fMet) in prokaryotes. *See* protein synthesis.

transgene   Gene transferred to a cell or organism by isolated DNA in a vector rather than by sexual means. *See* transformation, genetic.

transgene mutation assay   Mouse transgenic for a prokaryotic reporter gene is exposed to mutagenic conditions (spontaneous or treated with an agent). The genomic DNA is isolated and rescued in phage-lambda vector or used in a plasmid rescue system. The cloning bacteria are then plated and the number of mutant reporter genes are compared with all the reporter genes analyzed to provide mutation frequency. Under experimental conditions, spontaneous mutation rates (*lacZ*) were observed within the range of about 6 to 80×10^{-6}, depending on the tissues from where the DNA was extracted. Some transgenic lines carry p53 tumor suppressor or RAS oncogene to test their effects on mutagenesis. Inserted genes of the P450 cytochromes may be helpful for the studies of the metabolism of promutagens and procarcinogens. *See* β-galactosidase; bioassays in genetic toxicology; host-mediated assays; *lac* operon; mutation detection; p53; P450; plasmid rescue; procarcinogen; promutagen; RAS; vectors. (Chroust, K., et al. 2001. *Mutation Res.* 498:169; McDiarmid, H. M., et al. 2001. *Mutation Res.* 497:39.)

transgenerational effect   Epimutations transmitted to the progeny. The agouti viable yellow allele A^{vy} of mice may be affected in a mosaic pattern by a retrotransposon in the female germline and the alteration is not cleared during subsequent meiosis. *See* epimutation.

transgenesis   Introducing a gene by genetic transformation. *Conditional transgenesis* introduces the desired gene by *Cre*-mediated recombinase under the control of a developmentally regulated promoter. Germline stem cell transplantation may also lead to transformation. *See* Cre; transformation, genetic. (Moon, A. M. & Capecchi, M. R. 2000. *Nature Genet.* 26:455; Brinster, R. L. 2002. *Science* 296:2174.)

transgenic   Carries genes introduced into a cell or organism by transformation. Transgenic animals can potentially produce therapeutically needed proteins such as human tissue plasminogen activator (tPA) or $α_1$-antitrypsin (ATT), human monoclonal antibodies, etc. Transgenic plants may have direct uses in agriculture by virtue of their resistance to herbicides, pathogens, or even by the production of biodegradable plastics, nutritionally safer fats, and carbohydrates; various antigens may even be substituted for standard vaccines by eating them. The availability of transgenic crops and farm animals has raised concerns by consumers and environmentalists about transfer of herbicide resistance to weeds, introducing antibiotic resistance genes into the food chains, and their potential hazards for fighting microbial infections, affecting the immune system of animals and humans, etc. Although the long-term consequences of the new technologies cannot be

precisely assessed yet, there appear to be more advantages than risks. The wide-scale application of antibiotics in medicine eventually was followed by the appearance of resistance to many antibiotics. It must not be forgotten, however, that antibiotics saved and are saving millions of lives since the introduction of penicillin after World War II. Also, pharmaceutical research has produced and is producing an ever-increasing variety of new antibiotics in order to compete with the evolutionary changes in the microbial world. While the danger of emergence of antibiotic-resistant pathogens must not be ignored, the reasonable medical use of these drugs remains a necessity. *See* ANDi; antibiotic resistance; bitransgenic regulation; nuclear transplantation; paratransgenic; transformation, genetic. (Transgenic mice: <http://tbase.jax.org/>.)

transgenome   Transformed eukaryotic cells by the isolated whole or parts of chromosomes containing transgenes in their nuclei. *See* transformation, genetic (animal cells). (Porteous, D. J. 1994. *Methods Mol. Biol.* 29:353.)

transgenosis   Alternative term for a controversial means of nonsexual transfer of genes. (Doy, C. H., et al. 1973. *Nature New Biol.* 244:90.)

trans-Golgi network (TGN)   Connection of the Golgi complex exit face to transport vesicles so that the molecules coming from the endoplasmic reticulum will be transported to their proper destination. *See* cis-Golgi; Golgi.

transgression   Some segregants exceed both parents and the F1 hybrids. (Burke, J. M. & Arnold, M. L. 2001. *Annu. Rev. Genet.* 35:31.)

transheterozygous noncomplementation   *See* nonallelic noncomplementation.

transient amplifying cells   The progeny of stem cells replicate but do not revert to stem cell status; rather, they generate differentiated cells. *See* stem cells.

transient expression DNA   *See* transformation, transient.

transilience, genetic   Rapid changes in fitness by a multilocus complex in response to changes in the genetic environment. (Templeton, A. R. 1980. *Genetics* 94:1011.)

transin   Cell-secreted metalloproteinase, a homologue of stromelysin. (*See* Luo, D., et al. 2002. *J. Biol. Cem.* 277:25527.)

transinactivation   *See* cosuppression.

transistor   *See* semiconductor.

transit amplifying cell   *See* stem cells.

transition mismatch   A purine pairs with a wrong pyrimidine. *See* mismatch; transition mutation; transversion mismatch.

transition mutation Either a pyrimidine is replaced by another pyrimidine, or a purine by another purine, in the genetic material leading to mutation. *See* base substitutions; transversion. (Freese, E. 1963. *Molecular Genetics*, Taylor, J. H., ed., p. 207. Academic Press, New York.)

transition proteins Basic proteins that replace (temporarily) histones during spermiogenesis along with protamines. *See* protamine; spermiogenesis.

transition state Unstable (lifetime ~10^{-13} second) intermediate between reactants and products of an enzymatic reaction:

REACTANT → TRANSITION STATE → PRODUCT(S).

Factors that stabilize the transition state relative to the reactant are expected to lower the activation energy. *See* ϕ value. (Komatsuzaki, T. & Berry, R. S. 2001. *Proc. Natl. Acad. Sci. USA* 98:7666.)

transit peptide A dozen to five-dozen amino acid residue leader sequences directing the import of proteins synthesized in the cytosol into mitochondria and chloroplasts. These peptides are generally rich in basic and almost free of acidic amino acids. Serine and threonine are usually very common. The transit peptide recognizes special membrane proteins but itself is not transferred into the target organelle and is cut off by a peptidase. The different transit peptides do not appear to have conserved sequences. The transit peptide is targeted posttranslationally. Some mitochondrial and plastid proteins do not have these cleavable N-terminal sequences. The routing within the target seems to be influenced by the carboxy terminus or inner sequences of the proteins. Mitochondria can import several plastid proteins, but the plastids do not import mitochondrial proteins. This mitochondrial import is not physiological because this organelle does not have essential specificity factors (Cleary, S. P., et al. 2002. *J. Biol. Chem.* 277:5562). The transit peptide engages several proteins localized in the organelle membranes. Subsequently, the protein inside the cell folds with the assistance of chaperonin 60. Some proteins targeted to the endoplasmic reticulum, Golgi membrane, peroxisome, etc., may carry the transit peptide at the C-end. *See* chaperones; signal peptide. (Jean-Benoît, P., et al. 2000. *Plant Cell* 12:319.)

translation Converting the information contained in mRNA nucleotide sequences into amino acid sequences of polypeptides on the ribosomes. mRNA is threaded through a channel wrapping around the 30S subunit of the prokaryotic ribosome, and translation initiation, polypeptide chain elongation, and other functions follow (Yosupova, G. Z., et al. 2001. *Cell* 106:233). The translational apparatus of eukaryotes consists of more than 200 macromolecules of varying importance. Some proteins, e.g., bacterial ribosomal proteins S10 and L4, participate in transcription and translation. *See* protein synthesis; rabbit reticulocyte in vitro translation; ribosomal proteins; translation, nuclear; wheat germ in vitro translation. (Sonnenberg, N., et al., eds. 2000. *Translational Control of Gene Expression*. Cold Spring Harbor Lab. Press, Cold Spring Harbor, NY.)

translation, in vitro Used to be accomplished by employing isolated mRNA and other factors required for translation (*see* rabbit reticulocyte; wheat germ). When the isolated gene is included in an appropriate expression vector, transcription and translation may be obtained in a single step from the plasmid construct. A much better defined system of in vitro translation employs highly purified and tagged protein factors that permits an efficient purification of the products by affinity chromatography. It may yield 160 μg protein per mL/hr. It can produce modified proteins by incorporating amino acid analogs with the aid of suppressor tRNA. *See* rabbit reticulocyte in vitro translation system; suppressor tRNA; translation; wheat germ in vitro translation system. (Shimizu, Y., et al. 2001. *Nature Biotechnol.* 19:751.)

translation, noncontiguous Generally, mRNA is translated in a collinear manner into amino acid sequences without skipping any parts. There are few exceptions to this continuity; 50 nucleotides in bacteriophage T4 *gene 60* are skipped during translation.

translation, nuclear Translation takes place on the ribosomes. In prokaryotes that do not have a membrane-enclosed nucleus, transcription and translation are coupled. In eukaryotes, intact ribosomes are limited to the cytoplasm. According to the traditional view, translation is limited to the cytoplasmic compartment. Evidence is accumulating in favor for the notion that even in mammals some translation may take place within the nucleus. The evidence for nuclear translation is based on the observation that in isolated, purified nuclei fluorescence-labeled proteins are not present outside the nuclei. Electron microscopy indicated the colocalization of nuclear translation sites with the eIF4E polypeptide elongation factor, the ribosomal subunit L7, and a β-subunit of proteasome. The presence of proteasomal activity was surprising inasmuch as it degrades proteins. It is conceivable that most of the nuclear-translated proteins are degraded normally. Increasing the concentration of ribonucleotides led to increased protein synthesis in the nucleus, indicating the possibility of coupled transcription and translation in the nucleus. Despite the critical evidence, the role and significance of nuclear translation is not entirely clear. *See* eIF4; proteasome; protein synthesis. (Iborra, F. J., et al. 2001. *Science* 293:1139.)

translational bypassing Processes two separate open-reading frames into one protein. (i) The charged peptidyl-tRNA and mRNA complex arrives at the P site of the ribosome and after dissociation the mRNA slides through the ribosome (takeoff). (ii) The peptidyl-tRNA searches the mRNA through the decoding center of the ribosome (scanning). (iii) The peptidyl-tRNA pairs with the appropriate codon after skipping some others (landing). (Herr, A. J., et al. 2000. *Annu. Rev. Biochem.* 69:343.)

translational control Protein synthesis is regulated during the process of translation on the ribosome; e.g., attenuation. *See* attenuation; closed-loop model of translation; masked RNA; regulation of protein synthesis; suppressor RNA; termination factors; terminator codons; translational termination. (Gale, M., Jr., et al. 2000. *Microbiol. Mol.*

Biol. Rev. 64:1092; Johnstone, O. & Lasko, P. 2001. Annu. Rev. Genet. 35:365; Szostak, J. W. 2002. Nature 419:890; <http://uther.otago.ac.nz/Transterm.html>.)

translational coupling When the secondary structure of the mRNA is such that the AUG site or the Shine-Dalgarno sequence is not readily amenable for translation at the first cistron, translation at the initiator codon of another cistron may open up for translation. Such a situation may occur in phages but rarely in eukaryotes. (Herr, A. J., et al. 2000. Annu. Rev. Biochem. 69:343.)

translational error See ambiguity in translation; error in aminoacylation.

translational gene fusion vectors Carry promoterless, 5′-truncated structural genes. When these are driven by the trapped host promoter, they direct the synthesis of fusion proteins containing amino acid residues coded for by both host and vector DNA sequences. See gene fusion; read-through proteins; transcriptional gene fusion; transformation, genetic; trapping promoters; diagram below.

translational hopping Occurs when a peptidyl-tRNA dissociates from its first codon and reassociates with another dowstream. See aminoacyl-tRNA synthetase; hopping; overlapping genes; protein synthesis; shunting; translational frameshift. (Herr, A. J. 2001. J. Mol. Biol. 309:1029.)

translational recoding (same as ribosomal frameshift) See overlapping genes.

translational research Applies something from basic research to a patient and determines the outcome of the treatment (Birmingham, K. 2002. Nature Med., 8:647.)

translational restart See reinitiation.

translational termination May take place by encountering stop codons, endonucleolytic cleavage, shortening of the poly(A) tail, premature decapping of the mRNA. See translational control.

translation error See ambiguity in translation; error in aminoacylation.

translation initiation Usually triggered by growth factors through signaling to the RAS/RAF G and MEK/MAPK proteins. Upon phosphorylation, 4E-BP1 releases the cap-binding protein eIF-4F and the mRNA cap associates with the 40S subunit of the eukaryotic ribosome. The phosphorylation of ribosomal protein S6 by protein kinase $p70^{s6k}$ is also required. Eukaryotic initiation factor eIF-2B is active when it is bound to GTP and ensures the supply of $tRNA^{Met}$. Insulin and other growth factors keep eIF2B attached to GTP, whereas glycogen synthase kinase (GSK) inactivates it because GSK inactivates insulin. In prokaryotes, the initiation begins when the ribosomal binding site of mRNA (including the Shine-Dalgarno sequence and AUG^{fMet} codon) binds to anti-Shine-Dalgarno sequence in 16S RNA of the 30S ribosomal subunit. The AUG codon is thus directly placed into the P pocket of the ribosome and can interact with the formyl-methionine-charged fMet-tRNA. Before the 30S subunit combines with

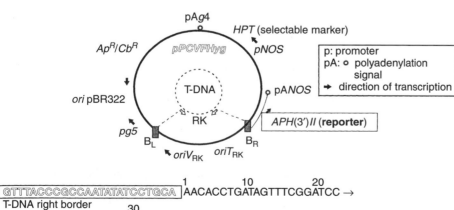

Agrobacterial translational in vivo gene fusion vector for plants. The critical feature is that the reporter gene is fused to the right border sequence of the T-DNA in such a manner that the translation initiator codon (AUG^{Met}) is deleted and the structural gene of the reporter begins with its second codon. It does not have a promoter either. The reporter gene can be expressed only when it is inserted and fused in the correct register into an active plant promoter. Since the AUG codon is missing, there is a good chance that the reporter protein will be fused with some plant (poly-) peptides. The use of such a vector permits an analysis of the expression of various fusion proteins on the reporter. Besides the structure, shown in detail at the lower part of the diagram, the transformation cassette contains selectable markers (e.g., HPT, permitting selective isolation of transformant on hygromycin media), ampicillin (Ap^R) and carbenicillin (Cb^R) resistance for selectability in bacteria and also the replicational origin of the E. coli plasmid pBR322. Outside the boundaries of the T-DNA, there are genes for both vegetative (oriV) and conjugational transfer (oriT) derived from the multiple-host range RK plasmid. Only the genes between the two border sequences (B_L and B_R) are inserted into the plant genome. The basic principle of this diagram has been exploited in designing vectors for other organisms. (After oral communication by Dr. Csaba Koncz.)

the 50S subunit, a number of other interactions take place. In eukaryotes, the first step of the initiation is the (1) dissociation of the 60S + 40S ribosomal subunit mediated by eIF6 and the attachment of eIF6 to the 60S subunit. Then (2) eIF3 attaches to the 40S subunit, (3) followed by the attachment of eIF1A. Next (4) eIF2 + GTP combines with tRNA^{Met}, and the complex joins the 40S subunit with eIF1A and eIF3 already in place, forming the **43**S subunit. (5) The capped mRNA plus elongation initiation factors eIF-4F, eIF-4A, and eIF-4B energized by ATP→ADP mediate the attachment of the capped mRNA to the small ribosomal subunit (now **48**S preinitiation complex). (6) While eIF5 and GTPase activating protein (GAP) mediate the release of eIF2● GDP● eIF3, (7) the small subunit scans the mRNA until the AUG^{Met} is located. (8) Capture of the free 60S ribosomal subunit, restoration of the 80S ribosome, and the start of translation follows this event. *See* cap; DEAD box; eIF; initiator tRNA; IRES; mRNA circularization; PABp; protein synthesis; ribosome; ribosome scanning; Shine-Dalgarno sequence; S6 kinase. (Gingras, A.-C., et al. 1999. *Annu. Rev. Biochem.* 68:913; Kimball, S. R. 2001. *Progr. Mol. Subcell. Biol.* 26:155; Pestova, T. V., et al. 2001. *Proc. Natl. Acad. Sci. USA* 98:7029; Walker, M., et al. 2002. *Nucleic Acids Res.* 30:3181.)

translation reinitiation   *See* reinitiation.

translation repressor proteins   May be attached to a site near the 5′-end of mRNA, preventing the function of the peptide chain initiation factors. *See* aconitase; eIF-2; protein synthesis; rabbit reticulocyte; trinucleotide repeats.

translation termination   Takes place in the decoding A pocket of the ribosome where the polypeptide release factors, RF1, recognizing prokaryotic stop codons UAG and UAA, and RF2 specific for UGA and UAA, or RF3 without selectivity (may cause misreading of all three stop signs), release the polypeptide chains. In eukaryotes, the eRF1 termination factor recognizes all three stop codons, and eRF1 and eRF3 are interactive. Several other proteins modulate the function of the RFs. RF3 has homology to elongation factors EF-G and EF-Tu. This fact seems to indicate that termination and chain elongation processes bear similarities; in one case, the stop codon is read; in the other, the sense codons. The RFs may have additional homology domains, e.g., with the acceptor stem, the anticodon helix, and the T stem of tRNAs, called tRNA mimicry. There is a conserved GGQ (Gly-Gly-Gln) group in eRF1s corresponding to the amino acyl group attached to the CCA-3′ end of tRNA. These homologies may assist their function. The yeast eRF3 is a prion-like element, psi⁺. Interestingly, the heat-shock protein 104, a molecular chaperone, can cure the cell from it. After protein synthesis is terminated, the termination complex and the ribosome are recycled. In the 16S rRNA of *E. coli*, mutation at nucleotide position C1054 causes translational suppression. Similarly, at the corresponding site in the 18S eukaryotic rRNA, substitutions of A or G resulted in dominant nonsense suppression while T substitution was a recessive antisuppressor and deletion of the site had a lethal effect. Although translation termination is mediated at the ribosomes, premature termination may result not just from nonsense codons but also by decay of the misspliced transcripts. Bacterial mRNA truncated at the 3′-OH and

without a termination codon may stall on the ribosomes. In such cases, the tRNA-like 10Sa RNA transcribed from *ssrA* gene of *E. coli* indirectly causes the degradation of the nascent peptide chain. The 10Sa binds to the ribosome and the ANDENYALAA amino acid sequence is added to the C-end of the peptide, making it a target for carboxyl-end-specific proteases. Translation termination may regulate gene expression with the aid of some weak internal termination codons that can be facultatively transpassed. Many human diseases are caused by improper translation termination, and modulating the process may have therapeutic potentials. *See* autogenous suppression; chaperone; EF-G; EF-Tu-GTP; PABp; phenotypic reversion; prion; protein synthesis; read-through; recoding; release factor; ribosome; sense codon; stop codon; translation initiation. (Song, H., et al. 2000. *Cell* 100:311.)

translesion pathway   SOS repair system of DNA; replication may lead to targeted mutation at the site of mismatches such as at a thymine dimer and at bases chemically modified or deleted by mutagens. The repair of the cis-syn cyclobutane dimers is mutational in about 6% of cases, whereas the pyrimidine 6-4 pyrimidinone adducts are repaired in a mutagenic manner in almost 100%. The Rev proteins of yeast and the UmuC (pol V) protein of *E. coli* are involved with translesion. The Rev polypeptides are subunits of DNA polymerase ζ involved in damage-induced mutagenesis. In humans, DNA polymerase η may protect cells against damage that may lead to skin cancer. In bacteria, DNA polymerases II, IV, and V are involved in translesion (Napolitano, R., et al. 2000. *EMBO J.* 19:6259). In bacteria, LexA represses translesion and RecA induces it. Which polymerase is selected depends on the damaging agent. The repair frequently involves mutation. In yeast, the Rad6 and Rad18 proteins are elevated upon UV irradiation, and the *REV3*-encoded polymerase ζ unit and polymerase η carry out the repair. The products of genes *REV7* and *REV1* are also required. In humans, pol ι and pol κ are additional repair polymerases. *See* cis-syn dimer; DNA polymerases; DNA repair; pyrimidine-pyrimidinone photoproduct; REV; ultraviolet photoproducts; UMU. (Livneh, Z. 2001. *J. Biol. Chem.* 276:25639; Pham, P., et al. 2001. *Proc. Natl. Acad. Sci. USA* 98:9350; Friedberg, E. C. 2001. *Cell* 107:9.)

translin   A protein binding to GCAGA[A/T]C and CCCA[C/G]GAC sequences at the translocation breakpoint junctions in lymphoid malignancies supposedly has a role in the rearrangement of the immunoglobulin−T-cell receptor. *See* immunoglobulins; lymphoma; T cell; T-cell receptor; TCR. (VanLock, M. S., et al. 2001. *J. Struct. Biol.* 135:58.)

transloading   Modification of cancer vaccines by including nonself peptides so that they may boost immunogenicity. *See* cancer gene therapy. (Buschle, M., et al. 1997. *Proc. Natl. Acad. Sci. USA* 94:3256.)

translocase   Protein complex mediating transport of proteins through cell membranes. *See* ABC transporters; ARF; protein targeting; SecA; SecB; SecY/E; signal hypothesis; SRP; translocon. (Mori, H. & Ito, K. 2001. *Trends Microbiol.* 9:494.)

translocation   Transfers codons of mRNA on the ribosomes as the peptide chain elongates. In general, any type of transfer of molecules from one location to another. *See* protein synthesis.

translocation, chromosomal
Segment interchange between two nonhomologous chromosomes.

Original chromosomes ∀ Reciprocal translocation

Broken chromosomes do not stick, however, to telomeres; the interchange must involve internal regions. Fragments may be inserted in between two ends of an internal breakpoint, and such an aberration is called *shift*. Translocations are detectable by the light microscope if the length of a chromosome arm is substantially altered. Translocations are usually reciprocal, but during subsequent nuclear divisions one of the participant chromosomes that does not carry essential genes may be lost. Heterozygotes for reciprocal translocations display a cross-shaped configuration in meiotic prophase.

(Photo courtesy of Brinkley, B. R. & Hittelman, W. N. 1975.)

Translocation heterozygotes generally display 50% pollen sterility in plants because alternate and adjacent-1 distributions occur at about equal frequency in the absence of crossing over, and adjacent-2 distributions, being nondisjunctional, are very rare. Note that inversion heterozygotes may also produce 50% male sterility, but it occurs only if recombination within the inverted segment occurs freely (*see* inversion). In animals, the gametes of translocation heterozygotes may succeed in fertilization, but the zygotes or early embryos resulting from such a mating are generally aborted.

Translocation homozygosity may not have any phenotypic consequence; however, it has been shown that many types of cancerous growth are associated with translocation breakpoints. Apparently, DNA rearrangements in the vicinity of the genes interfere with the normal regulation of their activity as a kind of position effect. Translocation breakpoints reduce the frequency of crossing over between the breakpoint and the centromere (*interstitial segment*). Recombination in translocation homozygotes may be normal. Because the reciprocal interchange physically alters the synteny of genes, linkage groups may be reshuffled as a consequence of the exchange. Because translocations partially join two linkage groups, they can be exploited for assigning genes to chromosomes. The number of required crosses to localize a gene to a chromosome may thus be lower. Also, the reduction of recombinations around the breakpoints may call attention to linkage over a somewhat larger chromosomal tract than a single marker. Furthermore, the association of certain genes with sterility may be used as a chromosomal marker for the breakpoints. The sterility marker may not always be useful because it can be recognized only late during development (after sexual maturity). (*See* diagram next page). (Tennyson, R. B., et al. 2002. *Genetics* 160:1363.)

translocation, heterozygote
At least two of the chromosomes of the genome are reciprocally exchanged (mutually translocated), whereas the corresponding homologous chromosomes are not involved in translocation within the same cell nucleus. *See* genome; homologous chromosomes; translocation, chromosomal.

translocation complex
Interchanged chromosomes of eukaryotes; members of the group are inherited as a complex that alone contributes viable gametes to the progeny.

translocation ring
Multiple reciprocally translocated chromosomes after terminalization are attached end-to-end, forming a ring of several chromosomes. *See* complex heterozygotes; ring, bivalent; terminalization.

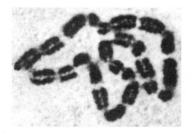

Six reciprocal translocations resulting in a ring of 12 in *Rheo discolor* (Sax, K. J. 1935. *Arnold Arboretum* 16:216.)

translocation test, heritable
See heritable translocation tests under bioassays in genetic toxicology. In some organisms, translocation testers have been developed to expedite linkage analysis. A clear and early marker is translocated to several (tester) chromosomes. Then the gene to be identified regarding its chromosomal position is crossed with all the translocation testers available. If according to previously obtained information with crosses involving nontranslocated chromosomes the marker shows independent segregation but crosses with the translocation testers it shows linkage, its chromosomal position is revealed. The frequency of translocation may vary a great deal according to species. In *Oenothera* plants, translocations are widespread, and in *Oenothera lamarckiana*, all the chromosomes are involved in translocations. According to some estimates, there is about 0.004 chance that a human baby will carry a translocation. Many types of tumors carry translocations and the pattern involved is not a haphazard one. Potential oncogenes (MYC, RAS, SRC) are frequently translocated into the 14q11 region, the location of the T-cell receptor (TCR) α and δ. MYC translocated to immunoglobulin genes

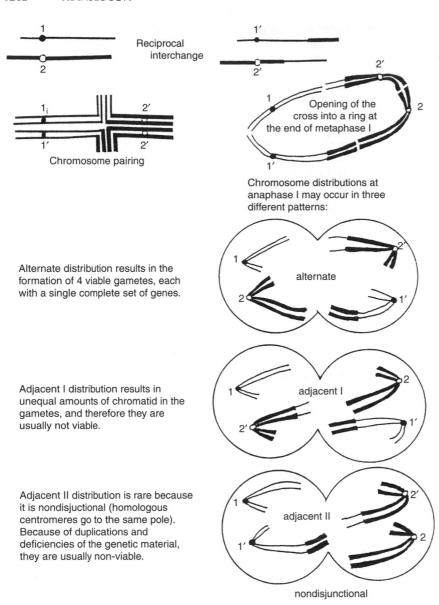

Reciprocal interchange

Chromosome pairing

Opening of the cross into a ring at the end of metaphase I

Chromosome distributions at anaphase I may occur in three different patterns:

Alternate distribution results in the formation of 4 viable gametes, each with a single complete set of genes.

alternate

Adjacent I distribution results in unequal amounts of chromatid in the gametes, and therefore they are usually not viable.

adjacent I

Adjacent II distribution is rare because it is nondisjuctional (homologous centromeres go to the same pole). Because of duplications and deficiencies of the genetic material, they are usually non-viable.

adjacent II

nondisjunctional

Translocation chromosomal.

is common in B-cell neoplasia and Burkitt's lymphoma. The chain formation question is intriguing of how these special translocations are controlled. One interpretation may be that the gene fusions may lead to the creation of highly selective combinations to stimulate proliferation. *See* adjacent disjunction; B chromosomes; Burkitt lymphoma; cancer; chromosomal rearrangement; multiple translocations; oncogenes; position effect; promoter swapping; synteny; telomere; trisomic, tertiary; unbalanced chromosomal constitution. (Generoso, W. M. 1984. *Mutation, Cancer and Malformation*, Chu, E. H. Y. & Generoso, W. M. eds., p. 369. Plenum, New York.)

translocon   Multiprotein complex (SecYp, SecGp, SecE in bacteria, Sec61p, Sbh1p, Ssh1p in yeast, sec61α, β, γ in mammals) involved in the transport of proteins through 40–60 Å diameter aqueous pores through membranes and the endoplasmic reticulum. In bacteria, DnaK chaperones keep the nascent polypeptide chains in shape until their synthesis is finished. *See* ABC transporter; ARF; DnaK; protein targeting; SRP; TRAM; translocase. (Hamman, B. D., et al.

1997. *Cell* 89:535; Heritage, D. & Wonderlin, W. F. 2001. *J. Biol. Chem.* 276:22655.)

transmembrane proteins   They generally have three main domains. The amino terminus reaches into the cytoplasm, where it usually associates with other cytosolic proteins. The hydrophobic domain generally makes seven turns within the cell membrane. The carboxylic end serves as a receptor for extracellular signals. *See* cell membrane; INT3 oncogene; KIT oncogene; MAS1 oncogene; membrane proteins; receptors; seven-membrane proteins; signal transduction. (Baldwin, S. A., ed. 2000. *Membrane Transport*. Oxford Univ. Press, New York.)

transmethylation   *See* methylation of DNA.

transmission   Indicates whether a particular gene or chromosome survives meiotic or postmeiotic selection and is recovered in the zygotes, embryos, or adults. Apoptotic mechanisms eliminate defective differentiating spermatogonia and elongating spermatids. During oogenesis, apoptosis is less active. This may be the source of the longstanding knowledge that many (chromosomal) defects are preferentially or to a large extent transmitted through the egg. Transmission is generally reduced if the chromosomes have deficiencies, duplications, or structural rearrangements. Monosomes and trisomes have impaired transmission as well as all defective genes. Segregation distorter genes and gametophyte factors may cause reduced transmission. Genetic factors located within cellular organelles and infectious heredity usually display uniparental (maternal) transmission. *Vertical transmission* indicates transfer by the gametes and *horizontal transmission* means spread of a condition through infectious agents without the involvement of the host genetic system. Horizontal transmission is indicated by sequenced genomes, e.g., in the archaebacterium *Holobacterium halobium*, the dihydrolipoamide dehydrogenase gene, displayed 50% homology to that of gram-positive eubacteria but only 25% to other archaebacteria. In the *Bacillus subtilis* genome, 10 nucleotide sequences were detected that are of infectious prophage origin.

Transformation by foreign DNA gives the best-documented case for horizontal transfer. Apparent homologies among genes carried by taxonomically different organisms may be interpreted as convergent evolution. The findings of transposable element (SINE) homologies in organisms widely separated taxonomically may be assumed to have origin in retroviral infections. Some of the homologous retrotransposons still carry the characteristic terminal repeats in, e.g., *Vipera ammodytes* and the bovine genome. The transfers might also have been mediated by parasites such as ticks (*Ixodes*) common to a very wide variety of vertebrates (from reptiles to humans). *See* apoptosis; certation; gene conversion; infectious heredity; intein; lateral transmission; megaspore competition; meiotic drive; preferential segregation; retroposon; retrotransposon; self-incompatibility; transformation; transposable elements; transposon.

transmission disequilibrium test (TDT)   Used to ascertain whether a tentative association between two traits is or is not transmitted from the heterozygous parents. This test is not applicable meaningfully to a population where one of the alleles has very high frequency, because in such a case the association will appear high, although no causal relationship may be present. The TDT test is not a genetic linkage test. The TDT test can be used to estimate quantitatively the distribution of k offspring carriers of a mutation among r affected progeny by truncated binomials:

$$\text{if } k = r \ t^k (1 - t^{s-r})/\{1 - [t^s + (1 - t)^s]\}$$

$$\text{if } 0 < k < r \ (\tbinom{r}{k}) t^k (1 - t)^{r-k}/\{1 - [t^s + (1 - t)^s]\}$$

$$\text{if } k = 0 \ (1 - t)^r [1 - (1 - t)^{s-r}]/\{1 - [t^s + (1 - t)^s]\}$$

where t = the segregation parameter, s = the size of the sibship. Segregation distortion can also be determined on the basis of m carriers of the mutation among s genotyped progeny: if parent is typed:

$$(\tbinom{s}{m}) t^m (1 - t)^{s-m};$$

if parent is inferred:

$$\frac{(\tbinom{s}{m}) t^m (1 - t)^{s-m}}{\{1 - [t^s + (1 - t)^s]\}}$$

(Formulas adopted from Hager, J., et al. 1995. *Nature Genet.* 9: 299.)

See association mapping; association test; binomial distribution; binomial probability; genomic control; SDT; segregation distortion; sib TDT. (McGinnis, R. 2000. *Am. J. Hum. Genet.* 67:1340.)

transmission genetics   It is actually a misnomer because genetics deals with inherited (transmitted) properties of organisms. In case there is no transmission, there is no genetics. The term has been used to identify those aspects of genetics that deal only with the transmission of genes and chromosomes from parents to offspring involving the study of segregation, recombination, mutation, and other genetic phenomena without the use of biochemical and molecular analyses. It is used in the same sense as classical or Mendelian genetics. *See* molecular genetics, reversed genetics.

transmitochondrial   Cells containing mitochondrial DNA introduced exogenously. The procedure is suitable to produce animal models for human mitochondrial diseases. *See* heteroplasmy; mitochondrial diseases in humans; mtDNA. (Hirano, M. 2001. *Proc. Natl. Acad. Sci. USA* 98:401.)

transmitter-gated ion channel   Converts chemical signals received through neural synaptic gates to electric signals. The channels in the postsynaptic cells receive the neurotransmitter. The process results in a temporary permeability change and a change in membrane potential, depending on the amount of the neurotransmitter. Subsequently, if the membrane potential is sufficient, voltage-gated cation channels may be opened. *See* ion channels.

transmogrification   Complete change of living creatures such as the mythological chimeras, satyrs, mermaids, etc. Genetic engineering and organ transplantation in medicine now bring into reality in some way the formerly imaginary beings; animals and plants expressing bacterial genes or vice versa. *See* allografts; chimera; gene fusion; homeotic genes.

transmutation   Changing one species into another (an unproven idea). Also, changing one isotope into another by radioactive decay or changing the atomic number by nuclear bombardment.

transomic   *See* transsomic.

transorientation hypothesis   Suggests a fourth site (D [decoding]) on the ribosome (besides A, P, and E) and assumes that the EF-G-GTP and tRNA ternary complex rotation (transorientation) moves the tRNA from the D site to the A site during protein synthesis. *See* ribosomes. (Simonson, A. B. & Lake, J. A. 2002. *Nature* 416:281.)

transpeptidation Transfer of an amino acid from the ribosomal A site to the P site. *See* aminoacyl-tRNA synthetase; protein synthesis.

transpiration Releasing water by evaporation through the stoma in plants; through exhalation, the skin, etc., in animals. *See* stoma.

transplacement Gene replacement with the aid of plasmid vectors. *See* gene replacement vector; localized mutagenesis; targeting genes.

transplantation, organelle Nuclei, isolated chromosomes, mitochondria, and plastids can be transferred into other cells by cellular (protoplast) fusion and by microinjection. In case of nuclear transplantation, the resident nucleus is either destroyed (by radiation) or evicted by the use of the fungal toxins, cytochalasins. The enucleated cell is called a *cytoplast* and the nucleus surrounded by a small amount of cytoplasm is a *karyoplast*. After introduction another nucleus into the cytoplast by fusion, a *reconstituted cell* is obtained. The individual components are labeled genetically, by radioactivity, by staining, or even mechanically (by 0.5 μm latex beads). The transferred organelles may express their genetic information and can be isolated efficiently and identified if selectable markers (e.g., antibiotic resistance) are used. Defective livers may be repopulated with normal liver cells expressing transgenic BCL-2 because of protection against apoptosis mediated by FAS. Without BCL-2, transplanted liver cells do not survive. Such a procedure may eventually become an alternative to liver transplantation. *See* apoptosis; cell fusion; nuclear transplantation; paternal leakage; rejection; transformation, genetic. (Kagawa, Y., et al. 2001. *Adv. Drug Deliv. Rev.* 49:107; Kuhholzer, B. & Prather, R. S. 2000. *Proc. Soc. Exp. Biol. Med.* 224:240.)

transplantation antigens Proteins on the cell surface encoded by the major histocompatibility (MHC) genes; they have a major role in graft (allograft) rejection in mammals. The rejection may depend on the perception of the foreign antigens (tissues) by the lymphoid organs. *See* microcytotoxicity assay; mixed lymphocyte reaction; HLA.

transplastome Plastid genome containing DNA introduced by transformation. *See* chloroplast genetics; chloroplasts; plastome.

transponder (microtransponder) A few hundred-micrometer-wide silicon chip–based device for memory storage. When prompted by laser light, it emits a radio signal that transmits its identification number. In a manner similar to DNA chips, it may assist identification of DNA sequences that are recognized by a probe. *See* DNA chips; probe.

transport elements, constitutive (CTE) Permit the transport of spliced and nonspliced RNAs (such as viral RNAs, U snRNAs, tRNAs), although unspliced mRNAs are not exported from the nucleus. *See* nuclear pore; RNA export; RNA transport; splicing.

transporters Permease proteins that assist the transport of various molecules and ions through membranes. *See* ABC transporters; ASCT1; CAT transporters; DNA transport; GLAST; GLYT; G-proteins; membranes, receptors; PROT; rbat/4F2hc; TAP. (Sprong, H., et al. 2001. *Nature Rev. Mol. Cell Biol.* 2:504; Alper, S. L. 2002. *Annu. Rev. Physiol.* 64:899, <plantst.sdsc.edu>.)

transportin 90 kDa protein distantly related to importin. It mediates nuclear transport with the M9, 38-amino-acid transport signal by a mechanism different from that of the importin complex. *See* export adaptor; importin; karyopherin; nuclear localization sequences; nuclear pore; RNA export. (Lai, M.-C., et al. 2001. *Proc. Natl. Acad. Sci. USA* 98:10154.)

transposable elements Occur in the majority of organisms. Their major characteristic is that they are capable of changing their position within a genome or may move from one genome to another. Transposable elements are classified into two major groups. Class I elements transpose with an RNA intermediate. Class II elements rely on a cut-and-paste mechanism. Elements that lack terminal repeats are unable to transfer horizontally. The estimated rate of transposition in eukaryotes is about 10^{-4} to 10^{-5} per element in *Drosophila*. The frequency of transposition may be regulated by the host genome in lower and higher organisms as well as by methylation of the transposase. The various types of elements may have intrinsic differences in mobility. Eukaryotic elements can be either retrotransposons (retrovirus-like) and have long direct terminal repeats (class I.1) or do not have long terminal repeats (class I.2) and are retroposons. Both types of class I elements have active or inactive reverse transcriptase. Class II elements have inverted terminal repeats that code for transposase. Transposition may take place through an RNA intermediate or directly by DNA. Transpositions may take place by homologous recombination between elements located at a different map position in the genome.

At least 35% of the human genome consists of transposons. Transposable elements may have an evolutionary role in the remodeling of genomes. The transposable elements are transmitted from generation to generation, but selection may act against them because insertions may damage the genes and equilibrium is generally reached. The I element of hybrid dysgenesis in *Drosophila* carries an internal sequence that regulates copy number. After about 10 generations of the inclusion of the first I element, transposition is tamed, i.e., this internal sequence slows down the movement of the transposon. The strength of selection for host alleles controlling transposition may be estimated according to Charlesworth and Longley (*Genetics* 112:359):

$$s \approx -\delta u = \left[\frac{\overline{n}(u - v)}{2H} + \frac{\overline{n}\pi}{2(1 - 2\pi)} \right]$$

where δu = change in the rate of transposition, n = copy number, u = rate of transposition, v = rate of excision per element, H = harmonic mean of the rate of transposition, and π = sterility or lethality caused by the transposition. The sliceosomes, telomerases, and the ability of immunoglobulin genes to transpose may have originated from transposons. Whole-genome sequencings revealed the existence of transposable elements that were not detectable by the methods of classical genetics. About 3% of the *Drosophila* genome are transposable. A survey of 13,799 human genes revealed that 533 (~4%) included some type of a transposable element (Nekrutenko, A. & Li, W.-H. 2001. *Trends Genet.* 17:619). Transposable elements frequently generate chromosomal aberrations such as deletions, duplications, inversions, translocations, etc. *See* cut-and-paste; genome defence model; Helitron; hybrid dysgenesis; immunoglobulins; isochores; second cycle mutation; selfish DNA; spliceosome; telomerase; transposable elements, animal; transposable elements, bacterial; transposable elements, fungal; transposable elements, plants; transposable elements, viral; transposase; transposon, conjugative; transposon footprint; transposon recombination; transposons. (Berg, D. E. & Howe, M. M., eds. 1989. *Mobile DNA.* Am. Soc. Microbiol., Washington, DC; Kidwell, M. G. & Lisch, D. R. 2000. *Trends Ecol. Evol.* 15:95; Lönning, W. E. & Saedler, H. 2002. *Annu. Rev. Genet.* 36:389.)

transposable elements, animal *See* copia; hybrid dysgenesis; immunoglobulins; LINE; P element; R2Bm; SINE.

transposable elements, bacterial May be classified according to the gram-negative host (Tn*3*, Tn*5*, Tn*7*, Tn*10*) or gram-positive host (Tn*554*, Tn*916*, Tn*1545*, Tn*55 1* and Tn*917*, Tn*4556*, Tn*4001*). The large transposons such as Tn*916* and Tn*1545* are capable of conjugative-like transfer to other cells. *See* insertion elements; nonplasmid conjugation.

transposable elements, fungal *See* transposable elements, yeast; Ty.

transposable elements, plant *See* Ac-Ds; controlling elements; Dt; Helitron; Mu; retroposons; somaclonal variation; Spm (En); Tam; transposons. (Wessler, S. R. 2001. *Plant Physiol.* 125:149; Jurka, J. & Kapitonov, V. V. 2001. *Proc. Natl. Acad. Sci. USA* 98:12315, Feschotte, C., et al. 2002. *Nature Rev. Genet.* 3:329.)

transposable elements, yeast *See* mating-type determination; Ω; *Schizosaccharomyces pombe*; Ty (including δ, σ, τ).

transposant Individual/line generated with the aid of a gene trap vector. *See* gene trap vector.

transposase Enzyme mediating the transfer of transposable genetic elements within the genome. The transposase function may be a part of the transposable element or it may be provided from trans-position for elements that are defective in the enzyme. *See* cut-and-paste; Tn*10*; transposable element; transposon.

transposition Transfer of a chromosomal segment to another position. The transposition may be *conservative* when the segment (transposon) is simply transferred to another location or it may be *replicative* when a newly synthesized copy is moved to another place while the original copy is still retained where it was. Transposition usually requires

that both terminal repeats of the transposon be intact. *One-sided transposition* — when one terminal repeat is lost — may still be feasible by replicative transposition. In *Drosophila*, ~80% of the mutations were attributed to transpositions. *Nonlinear transposition* takes place when the transposon ends are located in different molecules. The latter-type event may generate diverse chromosomal rearrangements. *See* hybrid dysgenesis; immunoglobulins; insertion element; mating-type determination in yeast; *Schizosaccharomyces pombe*; Tn; transposable elements; transposon recombination; transposons.

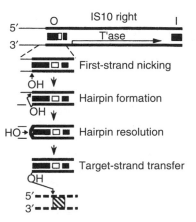

A model of transposition by IS10 of transposable element Tn*10*. (Courtesy of Mizuuchi, K. From Kennedy, A. K., et al. 2000. *Cell* 101:295.)

transposition immunity (target immunity) The transposable element does not move into a replicon, which already carries another transposon or the inverted terminal repeats of a transposon. Transposition immunity is overcome by high expression of the transposase or defects in the terminal repeats of the resident transposon. Transposition of Tn7 is inhibited by the presence of Tn7 sequences within the same replicon. *See* insertional mutation; Tn*3*; transposable elements; transposition. (Manna, D. & Higgins, N. P. 1999. *Mol. Microbiol.* 32:595.)

transposition induction The normal rate of transposition of the Ty1 yeast retrotransposon is about 10^{-5}–10^{-7} per element per cell division. Its rate can be substantially increased if in a multicopy plasmid pGTy1 the Ty element is under the control of an inducible *GAL1* promoter. Transposition of Ty is also controlled by *RAD25* and *RAD3*. *See* galactose utilization; *RAD3*; *RAD25*; transposase; Ty. (Staleva, L. & Venkov, P. 2001. *Mutation Res.* 474:93; Eichenbaum, Z. & Livneh, Z. 1998. *Genetics* 149: 1173.)

transposition site The target of insertion is generally not random. Bacterial Tn10 prefers 5′-NGCTNAGCN-3′. The mariner transposon selects the CAYA-**TA**-TRTG environment. Tn5 prefers a palindrome-like sequence flanked by A and T: **A**-GNTYWRANC-**T**. The insertion element IS231A likes a site within an S-shaped DNA. Retroviral elements show predilection for DNA around nucleosomes and cruciform DNA. Bacteriophage Mu is inserted nearly at random, yet some preferences for a pentamer within a 23–24 bp tract and avoidance of the *lacZ* control region have been noted. *See* cruciform DNA; mu bacteriophage; nucleosomes; transposons. (Haapa-Paananen, S., et al. 2002. *J. Biol. Chem.* 277:2843.)

transposome   Complex of a transposase, the transposon, and other proteins mediating the insertion of the transposon into a target DNA. *See* transposase; transposon. (Hoffman, L. M., et al. 2000. *Genetica* 108:19.)

transposon   *See* retroposon; retrotransposon; Tn; transposable elements; TRIPLES.

transposon, animal   *See* transposable elements, animal.

transposon, bacterial   DNA segments that can insert into several sites of the genome and contain genes besides those required for insertion; they are generally longer than 2 kilobases. It has been suggested that the introns of eukaryotic cells might have been introduced into the genes by broad host-range phages or transposons. *See* accessory proteins; insertion elements; Tn.

transposon, conjugative   Diverse group of broad host-range transposons varying in size from 18 to over 150 kb double-stranded DNA. They occur in different Bacteroides species. After excision, they form a circular intermediate molecule that can integrate into the DNA of another cell by a conjugation-like process. When they excise, they carry along a 6 bp adjacent sequence of the host. The excision may be followed by restitution without duplication at the original host site or it may leave a footprint at the original location. They may trigger the movement of other transposable elements. *See* conjugation; plasmid; transposable elements. (Hinerfeld, D. & Churchward, G. 2001. *Mol. Microbiol.* 41:1459.)

transposon, fungal   *See* Ty.

transposon-controlling elements, plant   The major transposable elements in maize are *Ac-Ds*, *Spm*, *Dt*, and *Mu*. There are much less well-defined controlling elements: *Bg* (Bergamo), *Fcu* (Factor Cuna), *Mr* (Mutator of R), *Mrh* (Mutator of a1-m-rh), *Mst* (Modifier of allele R-st), *Mut* (controlling element of *bz1-m-rh*), and *Cy* (regulatory element of *bz1-rcy*). *See* Ac-Ds; controlling elements; Dt; Mu; Spm; Ta; Tam.

transposon footprint   Short insertions left behind in the original target after the transposon exits from the sequences. These nucleotides may be the consequence of genetic repair after excision, e.g., sequence before the insertion CTGGTGGC after excision CTGGTGGC-TGGTGGC or CTGGTGG**gc**TGGTGGC. *See* second cycle mutation; transposable elements. (Plasterk, R. H. 1991. *EMBO J.* 10:1919; Hare, R. S., et al. 2001. *J. Bacteriol.* 183:1694.)

transposon mutagenesis   Transposable and insertion elements can move in the genome (mobile genetic elements) and may insert within the boundary of genes. Such an insertion, by virtue of interrupting the normal reading frame, may eliminate, reduce, or alter the expression of the gene; the event is recognized as a mutation. It has been shown that many of the insertions do not lead to observable change in the expression of the genes, or their effect is minimal and only sequencing of the target loci reveals their presence. Mutations so generated have great advantages for genetic analysis because the insertion serves as a tag on the gene, permitting its isolation and molecular study. Many of the insertions are retrotransposons. In plants, they are commonly located within introns. In animals, the comparable elements are frequently within intergenic regions. In *Caenorhabditis*, transposons—but not the retrotransposons—tend to be located within sequences of high recombination. In *Drosophila*, such differences were not confirmed. *See* gene isolation; gene tagging; insertional mutation; labeling; plasmid rescue; retroposon; retrotransposon; suicide vector; transformation. (Mills, D. A. 2001. *Curr. Opin. Biotechnol.* 12:503; Dupuy, A. J., et al. 2001. *Genesis* 30:82.)

transposon recombination   Transposons may induce various types of chromosomal rearrangements and deletions in prokaryotes and eukaryotes. The bacterial transposons (Tn), the *Drosophila* P elements, the budding yeast Ty elements, and under some conditions (but not under others) the plant transposons may enhance or even dramatically affect homologous recombination at the sites of their insertion. Recombination in plants may precede meiosis. *See* hybrid dysgenesis; Tn; transposon; Ty. (Xiao, Y.-L., et al. 2000. *Genetics* 156:2007.)

transposon tagging   Tagging a gene by the insertion of a transposon. Insertion disrupts the continuity of the gene, causing a mutation, and the success of the tagging is identified by the phenotype. Subsequently, using the labeled transposon as a probe can aid the isolation of the gene. *See* gene isolation; probe; transposon mutagenesis. (Long, D. & Coupland, G. 1998. *Methods Mol. Biol.* 82:315; Pereira, A. & Aarts, M. G. 1998. *Methods Mol. Biol.* 82:329; Kumar, A., et al. 2000. *Methods Enzymol.* 328:550.)

transposon vector   Can be used for introduction of genes into somatic cells of animals by microinjection into embryos of vertebrates or into invertebrates. These vectors must include transposase function, a selectable marker(s), and a chosen gene. The *Drosophila* Mariner-like, Tc1-like, or Sleeping Beauty vectors appeared successful and safer than viral vectors. *See* hybrid dysgenesis; mariner; P-element vector; piggyBAC; Sleeping Beauty; vectors. (Izsvak, Z., et al. 2000. *J. Mol. Biol.* 302:93; Grossman, G. L., et al. 2000. *Insect Biochem. Mol. Biol.* 30:909.)

transposon-based sequencing   Is used primarily for sequencing cDNA. Various transposons (Mu, Tn5) or repetitive DNAs are introduced at random into the cells and isolated on the basis of the selective markers (antibiotic resistance) within the transposon. Sequencing employs primers, which are specific for the ends of the transposon. *See* EST, DNA sequencing. (Yaron, S. N., et al. 2002. *Nucleic Acids Res.* 30:2460; Shevchenko, Y., et al. *Nucleic Acids Res.* 30:2469.)

transresponder   The ABL oncogene is activated (transresponds) by translocation to BCR in the Philadelphia chromosome and causes chronic myelogenous leukemia in more than 90% of cases. *See* ABL; BCR; leukemia; Philadelphia chromosome. (Gardner, D. P., et al. 1996. *Transgenic Res.* 5:37.)

transsensing Interaction between somatically paired homologous chromosomes affecting gene expression in diploids. *See* transvection. (Tartof, K. D. & Henikoff, S. 1991. *Cell* 65:201.)

transsexual Has an innate desire to change her/his anatomical sex to the other form. The volume of the central subdivision of the bed nucleus of brain strial terminals is larger in males than in females. In male-to-female transsexuals, this particular area of the brain is female-size. Thus, this anatomical condition may be a determining factor for transsexualism and sex hormone production. Estrogen family drug treatment may cosmetically help to improve breast size. *See* sex determination.

transsomic line Carries microinjected chromosomal fragments in the cell nucleus.

transsplicing Splicing together exons that are not adjacent within the boundary of the gene but are remotely positioned and may be in different chromosomes. *See* introns; regulation of gene activity; SL1; SL2. (Vandenberghe, A. E., et al. 2001. *Genes Dev.* 15:294; Denker, J. A., et al. 2002. *Nature* 417:667.)

transtranslation May occur if the stop codon and the preceding end of mRNA is lost and the translation is completed using another template RNA. *See* recoding; tmRNA. (Lee, S., et al. 2001. *RNA* 7:999.)

transvection Synapsis-dependent modification of activity in pseudoalleles. In paired chromosomes, genes in transposition may affect the expression of an allele. It has also been called trans-sensing. It has been interpreted as the result of interaction between DNA-binding proteins attached to the two synapsed promoters. *See* cis-trans effect; cis-vection; cosuppression; pseudoalleles; RIP; trans-acting element; transsensing. (Lewis, E. B. 1951. *Cold Spring Harbor Symp. Quant Biol.* 16:159; Matzke, M., et al. 2001. *Genetics* 158:451; Duncan, I. W. 2002. *Annu. Rev. Genet.* 36:521.)

transversion mismatch Mispairing involving either two purines or two pyrimidines. *See* mismatch; transition mismatch.

transversion mutation Substitution of a purine for a pyrimidine or a pyrimidine for a purine in the genetic material. *See* base substitution mutations; base substitutions. (Freese, E. 1959. *Brookhaven. Symp. Biol.* 12:63.)

TRAP (1) CD40 ligand. (2) Tryptophan RNA-binding attenuation protein. In *Bacillus subtilis*, when activated by L-tryptophan, this protein binds to the mRNA leader, causing a termination of transcription. This is in contrast to the situation in E. coli where the attenuation is brought about by an altered secondary structure of the nascent RNA transcript. Some sort of attenuation also takes place in eukaryotes, but the mechanism is not entirely clear yet. The *mtrB* gene in *B. subtilis* encodes the TRAP protein containing 11 identical subunits, and it binds single-stranded RNA. The β-sheet subunits form a wheel-like structure with a hole in the center and tryptophan is attached to the clefts between the β-sheets, resulting in circularization of the RNA target in which 11 U/GAG repeats are bound to the surface of this ondecamer (11-subunit) protein modified by tryptophan. TRAP may regulate both transcription and translation. Similar mechanisms occur in some other bacterial species. *See* attenuation region; tryptophan operon. (Antson, A. A., et al. 1999. *Nature* 401:235; Yakhnin, A. V. & Babitzke, P. 2002. *Proc. Natl. Acad. Sci. USA* 99:11067.)

TRAP/DRIP/ARC Part of a multiprotein complex of transcriptional regulators. (Crawford, S. E., et al. 2002. *J. Biol. Chem.* 277:3585.)

trapoxin Inhibitor of histone deacetylase. *See* histone deacetylase.

TRAPP Golgi-associated protein for docking vesicles. *See* Golgi; vesicles.

trapping promoters When a promoterless structural gene is inserted into a host genome with the assistance of a transformation vector, the inserted sequences may become "in-frame" located within the host chromosome and a host promoter may drive the transcription of the foreign gene that in the vector had no promoter. Since the promoter and upstream regulatory elements control transcription, directly or in association with transcription factors, the expression pattern (timing, tissue site) may be altered and the intensity of expression may be increased or decreased according to the nature of the promoter. *See* gene fusion; read-through proteins; transcriptional gene fusion vectors; translational gene fusion vectors. (Medico, E., et al. 2001. *Nature Biotechnol.* 19:579.)

Transgenic tobacco seedlings segregate for kanamycin sensitivity and resistance. A promoterless vector introduced the aminoglycoside gene into the cells. The structural gene was expressed only when it trapped a tobacco promoter. The strength of the promoters varied and consequently the degree of resistance too. Each petri plate was divided into two sections and in one section all the small, bleached, sensitive seedlings died. (From Y. Yao & G. P. Rédei, unpublished.)

traveler's diarrhea Caused generally by bacterial (*E. coli, Salmonella*) infection.

TRCF Transcription repair coupling factor is a eukaryotic repair helicase corresponding to *UvrA* in *E. coli*; it is encoded by yeast gene *MFD* (mutation frequency decline). (Li, B. H., et al. 1999. *J. Mol. Biol.* 294:35.)

TRD Transmission ratio distortion. *See* meiotic drive; segregation distorter.

TRE Thyroid hormone−responsive element in the rat growth hormone gene with a consensus of AGGTCA . . . TGACCT. *See* ERBA; hormone-response elements; regulation of gene activity. (Oofusa, K., et al. 2001. *Mol. Cell. Endocrinol.* 181:97.)

Treacher Collins syndrome Dominant (human chromosome 5q32-q33) complex defect of the face.

treadmill evolution *See* Red Queen hypothesis.

treadmilling Addition of microtubule subunits to the growing plus end and loss of subunits at the minus end. *See* dynamic instability; microtubules.

TREC TCR excision unit including recombination signals. *See* TCR genes.

tree edit distance Minimal weighted number of changes required to change one tree of descent into another. *See* evolutionary tree; string edit distance.

trehalose (α-D-glucopyranosyl-α-D-glucopyranoside) Nonreducing disaccharide, that accumulates in the yeast cell wall under conditions of stress. (Darg, A. K., et al. 2002. *Proc. Natl. Acad. Sci. USA* 99:15898.)

Trehalose.

TREMBL Computer annotated protein sequence extension of the SWISS-PROT database accessible by <http://srs.ebi.ac.uk:5000>. *See* SWISS-PROT in databases.

Treponema pallidum Spirochete bacterium with completely sequenced genome of 1,138,006 bp including 1,041 open-reading frames. It is responsible for the potentially deadly disease syphilis. (Fraser, C. M., et al. 1998. *Science* 281:375.)

trexon (transposed exon) Short duplicated modular units in the DNA with inverted terminal repeats. *See* exon; transposon.

TRF (1) *See* T-cell replacing factor. (2) Thyrotropic release factor. *See* corticotropin.

TRF1 (1) 60 kDA telomeric TTAGGG repeat-binding protein negatively regulates telomere extension and facilitates its interaction with the telomerase enzyme. *See* RAP; tankyrase; telomerase; telomeres. (Nakamura, M., et al. 2001. *Curr. Biol.* 11:1512.) (2) TATA box−binding protein related factor is a tissue- and gene-specific binding protein with preference for one of the two *Tudor* locus (2-[97]) promoters of *Drosophila*. The protein product of *tud* has a maternal effect and is expressed mainly in embryos and pupae. (Takada, S., et al. 2000. *Cell* 101:459.)

TRF2 Telomeric repeat-binding factor-2 is one of the proteins that protects telomeric ends of chromosomes. TRF2 inhibition may lead to apoptosis mediated by p53 and mutated ataxia telangiectasia genes. TRF2 is associated with RAD50, MRE11, and BS1 proteins. *See* apoptosis; ataxia telangiectasia; Mre; Nijmegen breakage syndrome; p53; RAD50; telomeres; TRF1. (Fairall, L., et al. 2001. *Mol. Cell* 8:351.)

Trg Bacterial transducer protein with attraction to ribose and galactose. (Beel, B. D. & Hazelbauer, G. L. 2001. *Mol. Microbiol.* 40:824.)

triabody Trimeric antibody built of three single-chain pairs of the variable heavy- and light-chain regions of antibody. *See* antibody, chimeric; diabody; recombinant antibody. (Le Gall, F., et al. 1999. *FEBS Lett.* 453[1−2]:164.)

triacylglycerols (triglycerides) Uncharged esters of glycerol and thus also called neutral fats. Triglycerides are energy storage compounds and contain four times as much energy in the human body as all the proteins combined. By lipase, they are hydrolyzed into glycerol and fatty acids. Lipolysis is controlled by cAMP in the adipose (fat) cells. Insulin inhibits lipolysis. Impaired long-chain fatty acid oxidation, triglyceride breakdowm defects, triglyceride transfer (MTP, 4q22-q24) in abetalipoproteinemia, hypertriglyceridemia (15q11.2-q13.1), a dominant hyperlipidemia, are diseases involved in triglyceride metabolism, and create a risk for heart disease. *See* adrenocorticotrophic hormone; epinephrine; fatty acids; glucagon; norepinephrine; triglycerol.

triage Assignment of priorities in medicine or in the regulation of cellular metabolism.

triallelic inheritance The manifestation of the recessive disease, e.g., Bardet-Biedl syndrome, may require the expression of three mutant alleles. *See* Bardet-Biedel syndrome; epistasis. (Katsanis, N., et al. 2001. *Science* 293:2256.)

triangulation number Represents the number of protein subunits (facets) in an icosahedral viral capsid. *See* capsid; icosahedral. (Paredes, A. M., et al. 1993. *Proc. Natl. Acad. Sci. USA* 90:9095.)

tribe Descendants of a female progenitor or a taxonomic group below a suborder or a group of primitive people with a common origin, culture, and social system.

Tribolium castaneum ($n = 10,200$ Mb) Flour beetle, object of cytological and population genetics studies. For a genetic map, see Beeman, R. W. & Brown, S. J. 1999. *Genetics* 153:333; <www.ksu.edu/tribolium/>.

TRiC Ring complex of eukaryotic chaperonin. TRiC-P5 is synonymous with CCTγ, Bin2p, and Cct3p. *See* chaperonins. (Dunn, A. Y., et al. 2001. *J. Struct. Biol.* 135:176.)

tricarboxylic acid cycle *See* Krebs-Szentgyörgyi cycle.

trichocyst Organ of protozoa that may extrude fibrous shafts and may serve as an anchor or defensive or offensive tool.

trichogen cell *See* tormogen.

trichogyne Hypha emanating from the protoperithecium to which conidia are attached prior to fertilization in some ascomycetes. *See* conidia; hypha.

trichome Hair or filament in plants, algae, and animals; some plant hairs may be single filaments or they may have tripartite termini. (Szymanski, D. B., et al. 2000. *Trends Plant Sci.* 5:214.)

Trichome.

trichorhinophalangeal syndrome (TRPS1) Dominant or recessive human chromosome 8q24 defect involving multiple exostoses (bone projections), mental retardation, protruding ears, sparse hair on the scalp, bulbous nose, and short stature. Mutation in a zinc-finger protein gene is responsible for TRPS1. TRPSIII is most severe. *See* Langer-Giedion syndrome.

Trichostatin A Antifungal antibiotic, an inhibitor of histone deacetylase of yeast. Trichostatin may reverse the effect of methylation and activate methylated genes. In some instances, low doses of 5-aza-2′-deoxycytidine along with trichostatin are required for substantial expression of originally methylated and silent cancer genes. *See* fragile X; histone deacetylase. (Marks, P. A., et al. 2001. *Curr. Opin. Oncol.* 13:477.)

trichothiodystrophy (TTD, 19q13.2-q13.3) Collective name for autosomal-recessive human diseases involving low-sulfur abnormalities of the hair. *Tay syndrome* also involves ichthyosiform erythroderma (scaly red skin), mental and growth retardation, etc. *Pollitt syndrome* (trichorrhexis nodosa or trichothiodystrophy neurocutaneous) displays low cystine content of the hair and the nails, and the head and the nervous

system are also defective. Xeroderma pigmentosum IV includes trichothiodystrophy and sun and UV sensitivity; also called PIBIDS. This type of mutation lacks helicase and excision repair activity because of the defect in the interaction between one of the xeroderma pigmentosa and the p44 protein subunit of the transcription factor TFIIH. Some TDD mutations are temperature-sensitive. *See* Cockayne syndrome; excision repair; hair-brain syndrome; ichthyosis; temperature-sensitive mutation; transcription factors; xeroderma pigmentosum. (Vermeulen, W., et al. 2001. *Nature Genet.* 27:299; de Boer, J., et al. 2002. *Science* 296:1276.)

Triclosan (trichlorinated diphenyl ether) Antibacterial and antifungal agent (blocking lipid biosynthesis) used in antiseptics, soaps, and other cosmetics.

tricyclo-DNA Tricyclo-DNA and -RNA can be used in antisense technologies to block selectively the expression of genes. *See* antisense technologies. (Renneberg, D., et al. 2002. *Nucleic Acids Res.* 30:2751.)

Tricyclo-DNA.

TRID TRAIL decoy receptor. *See* TRAIL.

tricuspid atresia Agenesis of the tricuspid valve, which connects the right atrium to the right ventricle of the heart. Some other heart defects may be associated with it. The condition is generally sporadic, but some cases are familial and involve defects of the Zfpm2/Fog2 zinc-finger protein.

Triethylene melamine.

triethylene melamine (TEM) Alkylating agent. *See* TEM.

trigger factor (TF, ~48 kDa) Prolyl isomerase enzyme (PPI) associated with the 50S ribosome unit of bacteria or with the GroL chaperone. It may aid translocation of molecules through the cytoplasmic outer membrane. In bacteria, TF and DnaK may cooperate in protein folding, but they are not indispensable at intermediate temperatures. Cyclosporin or FK506 does not inhibit TF and it is only moderately related to cyclophilins of FKBPs. *See* chaperone; Clp; cyclophilins; cyclosporin; DnaK; FK506; GroL; PPI. (Liu, Z., et al. 2001. *FEBS Lett.* 506:108; Patzelt, H., et al. 2001. *Proc. Natl. Acad. Sci. USA* 98:14244.)

triglyceride   Triaglycerol.

triglyceridemia   *See* familial hypertriglyceridemia.

trihybrid cross   The parental forms are homozygous altogether for three allelic pairs at (unlinked) loci, e.g., AABBdd x aabbDD, and therefore eight phenotypic classes may be distinguished in the F_2. *See* gametic array; Mendelian segregation.

tri-isosomic   In wheat, $20'' + i'''$, $2n = 43$ ($'''$ = disomic, $'''$ = trisomic, i = isosomic).

Trillium   Species have one of the largest normal chromosomes ($2n = 10$) in plants, about 50 times larger than *Arabidopsis*, a plant with one of the smallest chromosomes ($2n = 10$).

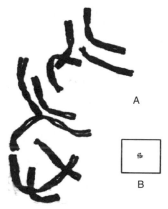

Trillium (A) is from Sparrow, A. H. & Evans, H. J. 1961. *Brookhaven Symp. Biol.* 14:76; *Arabidopsis* (B) karyotype is the courtesy of Lotti Sears.

trimester   Period of 3 months; the human pregnancy of 9 months includes 3 trimesters.

trimming   Processing of the primary RNA transcripts to functional mRNA or ribosomal and tRNA. The cleavage of pre-rRNA transcripts by RNase III into 16S, 23S, and 5S rRNA, as well as into the tRNAs, is contained within the spacer sequences of the cotranscripts. The cleavage takes place at the duplex sequences forming the stem of the rRNA loops. *See* introns; *rrn* genes; posttranscriptional processing.

trinomial distribution $(1 + 2 + 1)^n$ can be expanded

$$1(1 + 2)^n + \frac{n!}{1(n-1)!}(1 + 2)^{n-1} + \cdots + \frac{n!}{(n-1)!1}(1 + 2)^{n-(n-1)}$$
$$+ 1(1 + 2)^{n-n}$$

to predict the segregation of the genotypic classes (note that the quotients within parentheses must not be added!).

An example for three pairs of alleles:

$$1(1 + 2)^3 + \cdots + \frac{3!}{1(2!)}(1 + 2)^2 + 1$$
$$1 + (3 \times 2) + (3 \times 4) + 8 + 3 \times (1 + 2 + 2 + 4)$$
$$+ 3 \times (1 + 2) + 1$$

When rewritten in a symmetrical distribution:

$$1 : 2 : 1 : 2 : 4 : 2 : 1 : 2 : 1 : 2 : 4 : 2 : 4 :$$
$$8 : 4 : 2 : 4 : 2 : 1 : 2 : 1 : 2 : 4 : 2 : 1 : 2 : 1$$

the 27 terms indicate that triple heterozygotes are 8, double heterozygotes are 4, and single heterozygotes are 2 in a distribution in compliance with Mendel's law. *See* binomial, multinomials.

trinucleotide-directed mutagenesis (TRIM)   Introduction into the coding sequences of gene trinucleotide analogs such as 9-fluorenylmethoxycarbonyl (Fmoc) trinucleotide phosphoramidites. The synthetic analogs convey resistance to nucleases and are effective in induction of specific mutations. *See* phosphoramidates; trinucleotide repeats. (Sondek, J. & Shortle, D. 1992. *Proc. Natl. Acad. Sci. USA* 89:3581.)

trinucleotide repeats   Microsatellite sequences displaying some clustering in yeast genes. In more than 10 human neurodegenerative diseases, CAGs (glutamine codons) are repeated many times. The resulting polyglutamine (polyQ) sequences seem to be the cause of the diseases or the disease accumulates the polyglutamine tracts. The long polyglutamine sequences (most frequently beyond 35–40) interfere with CREB-dependent transcription by interacting with transcription factor $TAF_{II}130$. The polyglutamine proteins are more resistant to decay. Transglutaminase inhibitors (cystamine [← decarboxylated cysteine], monodansyl cadaverin

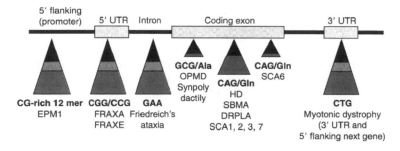

Location of expanded trinucleotide repeats in diseases. EPM: myoclonous epilepsy; FRAXA: fragile X syndrome; OPMD: muscular atrophy; HD: Huntington chorea; SBMA: Kennedy disease; DRPLA: dentatorubralpallidoluysian atrophy; SCA: spinocerebellar ataxia. The size of the triangles indicates the size of the repeat expansions. Top triangle: normal; gray in the middle: unstable premutational, bottom of the triangle: pathological condition. (Courtesy of Lunkes, A., Trottier, Y. & Mandel, J.-L. in Molecular Biology of the Brain, Higgins, S. J., ed. 1998. p. 149, © Biochemical Society, London, UK.)

[← decarboxylated lysine]) may alleviate apoptosis of the cells. RNAi may alleviate the problems normally associated with polyglutamine tracts (Caplen, N. J., et al. 2002. *Hum. Mol. Genet.* 11:175). A common feature of these diseases is that in successive generations the symptoms appear earlier and with greater severity (anticipation) as gain-of-function mutations. The repeats form a hairpin structure and interfere with DNA replication. The CpG sequences are likely to be methylated. The nature of the repeats may vary and may involve CGG, GCC, CAG, CTG sequences. These repeats may expand (from a few [5–50] in the normal to hundreds of copies) in an unstable manner in the 5′-untranslated region of the FMR1 gene and cause translational suppression by stalling on 40S ribosomal RNA. In *E. coli*, the larger expansions occur predominantly when the CTG trinucleotides are in the leading strands and deletions are mainly on the opposite lagging strands. The toxic effect of polyglutamine tracts can be genetically suppressed in *Drosophila* by proteins homologous to heat-shock protein 40 and a tetratricopeptide, both containing a chaperone-like domain. The instability caused by the repeats is more common in meiosis than in mitosis. In yeast, CAG/CTG repeat meiotic instability is based on double-strand DNA break repair. *See* anticipation; ataxia; chaperone; CREB; dentatorubral-pallidoluysian atrophy; dinucleotide repeats; epilepsy, ERDA1; FMR1 mutation; fragile sites; fragile X; FRAXA; FRAXE; Friedreich's ataxia; heat-shock proteins; human intelligence; Huntington's chorea; Jacobsen syndrome; Kennedy disease; Machado-Joseph disease; methylation of DNA; microsatellites; muscular dystrophy; myoclonic epilepsy; myotonic dystrophy; neurodegenerative diseases; polysyndactyly; premutation; RNAi; schizophrenia; slipped-structure DNA; spinocerebellar ataxias; SRY; TAF$_{II}$; tetratrico sequences; translation repressor proteins. (Claude, T., et al. 1995. *Annu. Rev. Genet.* 29:703; Orr, H. T. & Zoghbi, H. J. 2000. *Cell* 101:1; Cummings, C. J. & Zoghbi, H. Y. 2000. *Annu. Rev. Genomics Hum. Genet.* 1:281; Cleary, J. D., et al. 2002. *Nature Genet.* 31:37; *Hum. Mol. Genet.* [2002] 11:1909–1985.)

trioma   *See* heterohybridoma.

triose   Sugar with 3-carbon backbone.

triosephosphate isomerase deficiency (TPI1)   Encoded in human chromosome 12p13, but pseudogenes seem to be present at other locations. The level of activity of the enzyme varies. Null mutations are not expected to be viable, since this is a key enzyme in the glycolytic pathway. The symptoms may be general weakness, neurological impairment, anemia, recurrent infections, etc. *See* glycolysis.

Trip1   Thyroid hormone receptor. *See* Sug1.

tripeptidyl peptidase (TPPII)   Large protein complex outside the lysosomes with activity resembling the proteasome. *See* proteasomes.

triplasmy   Heteroplasmy for three different types of mtDNA. *See* heteroplasmy; mtDNA.

triple A syndrome   *See* achalasia-addisonianism-alacrima.

triple helix-forming oligonucleotides (TFO)   May bind to polypurine-polypyrimidine tracts in the major groove of the DNA helix by Hoogsteen or reverse Hoogsteen bonding and prevent the access of transcription factors. This may block transcription and cleave DNA but may enhance repair DNA synthesis. In the triplex sequences, mutation rate in SV40 increased more than an order of magnitude in the suppressor gene, supFG1, employed as reporter, with 30-nucleotide-long AG sequences (AG30). Shorter sequences or oligonucleotides of all four bases were either not effective or were much less effective. The triplex structure in xeroderma pigmentosum or in Cockayne syndrome cells was not effective for mutation enhancement, indicating the requirement of excision repair for the events. TFOs may be used for targeting specific genes and preventing their transcription or for inducing mutation. *See* antisense technologies; Cockayne syndrome; DNA kinking; DNA repair; Hoogsteen pairing; inhibition of transcription; pseudoknot; psoralen dyes; supF; SV40; targeting genes; TFO; TPO; triplex, xeroderma pigmentosum. (Grimm, G. N., et al. 2001. *Nucleosides Nucleotides Nucleic Acids* 20:909.)

TRIPLES   Transposon-insertion phenotypes is a database with information on phenotype, protein localization, and expression of transposon mutagenesis in yeast. *See* insertional mutation, transposon. (<http://ygac.med.yale.edu>.)

triple-stage quadrupole/ion-trap mass spectrometry   Proteome analytical procedure. *See* CID; electrospray; ESI; mass spectrometer.

triple test   Used for the identification of Down syndrome by assaying chorionic gonadotropin, unconjugated estriol, and α-fetoprotein levels. *See* Down syndrome; estriol; fetoprotein; gonadotropin.

triplet binding assay   Historically important method to determine the meaning of genetic triplet codons. A single type of radioactively labeled amino acid charged to cognate tRNA was allowed to recognize and bind to ribosomes with mRNA attached. Each type of charged tRNA then recognized only their code words and the ribosomes were then trapped on the surface of a filter. Synthetic polynucleotides (mRNA) of known base composition retained only the cognate aminoacylated tRNA, thus providing the base composition and sequence of the true coding triplets. *See* genetic code. (Nirenberg, M. W. & Leder, P. 1964. *Science* 145:1399; diagram next page.)

triplet code   *See* genetic code.

triplet expansion   *See* trinucleotide repeats.

triplex   (1) Three-stranded nucleic acid structure, e.g., an RNA oligonucleotide may bind within the strand or to double-stranded DNA, resulting in antisense effects. Triplex strands occur transiently in genetic recombination. Triplex-forming oligonucleotides may increase somatic mutation and recombination by the mechanism of nucleotide exchange repair. Some DNA polymerases (T7, Klenow fragment) can elongate a DNA strand from primers forming triple helices of 9–14 deoxyguanosine-rich residues. *See* antisense RNA; excision repair; H-DNA; Hoogsteen pairing; Klenow fragment;

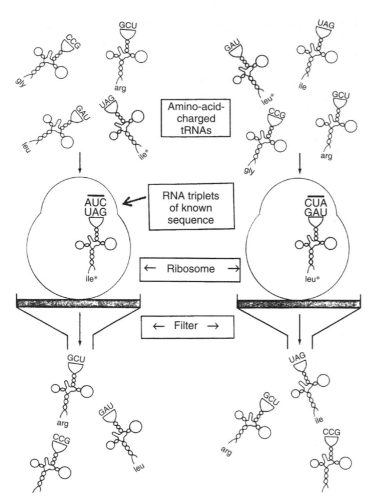

Triplet binding assay.

nodule-DNA; peptide nucleic acid; TFO. (Vasquez, K. M. & Wilson, J. H. 1998. *Trends Biochem. Sci.* 23:4; Luo, Z., et al. 2000. *Proc. Natl. Acad. Sci. USA* 97:9003; Datta, H. J., et al. 2001. *J. Biol. Chem.* 276:18018; Rocher, C., et al. 2001. *Nucleic Acids Res.* 29:3320; Knauert, M. P. & Glazer, P. M. 2001. *Hum. Mol. Genet.* 10:2243.) (2) Polyploid with three dominant alleles at a gene locus. *See* autopolyploid; trisomy.

H: hoogsteen pairing; **RH**: reverse hoogsteen; **R**: purine; **Y**: pyrimidine. Other configurations are also possible. (Hoyne, P. R., et al. 2000. *Nucleic Acids Res.* 28:770.)

triplicate genes Convey identical or very similar phenotype. In a diploid, when segregated independently, they display an F_2 phenotypic ratio of 63 dominant and 1 recessive.

triploblasts Animals with ecto-, meso-, and endodermal germ layers. *See* diploblasts.

triploid Cell or organism with three identical genomes. Triploids ($3x$) are obtained when a tetraploid ($4x$) is crossed with a diploid ($2x$). The majority of edible bananas, several cherry and apple varieties, and many sterile ornamentals (chrysanthemums, hyacinths) are triploid. Seedles watermelons produced by crossing tetraploids with diploids are triploids and have commercial value. Triploid sugarbeets are in large-scale agricultural production because of ≈10% or higher sugar yield per acre than the parental diploid varieties. Triploidy in humans may occur by fertilizing a diploid egg (digyny) by a monosomic (normal) sperm or by the union of a normal haploid egg and two spermatozoa (diandry). The green toad (*Bufo viridis*) is triploid but surprisingly reproduces bisexually. *See* trisomic. (Zaragoza, M. V., et al. 2000. *Am. J. Hum. Genet.* 66:1807; Stöck, M., et al. 2002. *Nature Genet.* 30:325.)

triplo-X Females (XXX) occur in about 0.0008 of human births. Their phenotype is close to normal and they can conceive. They are somewhat below average in physical and mental abilities and tend to be somewhat tall. With an increasing number of X-chromosomes beyond three, the adverse effects are further aggravated. *See* sex chromosomal anomalies in

Triradial chromosome (see next page).

humans; sex determination; trisomy; Turner syndrome. (Barr, M. L., et al. 1969. *Can. Med. Assoc. J.* 101:247.)

Tripsacum   *See* maize.

triradial chromosome   May be formed by fusion of broken translocated chromatids in a three-armed way. (Jenkins, E. C., et al. 1986. *Am. J. Med. Genet.* 23:531.) Figure preceding page.

Tris-HCl buffer   Contains tris-(hydroxy-methyl)aminomethane and hydrochloric acid; it is used in various dilutions within the pH range 7.2–9.1.

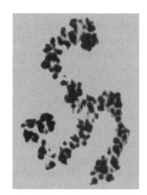

Triskelion (Redrawn after Ungewickell, E. & Branton, D. 1981. *Nature* 289:420.)

triskelion   Three-legged proteins on the surface of vesicles built of three clathrin and three smaller proteins. *See* clathrin; endocytosis. (Umgewickell, E. 1983. *EMBO J.* 2:1401.)

trisomic   Cell or organism with one or more chromosomes (but not all) represented three times. *See* triploid; trisomic analysis; trisomy.

Left: The four primary trisomics (1–4) and a telotrisomic (5) of *Arabidopsis*. at the lower right corner is a normal disomic individual (2*n*). All are in Columbia wild-type background, grown under short daily light periods. At right: The chromosome complement of a primary trisomic. Note the trivalent association at the upper right area. (The photomicrograph is the courtesy of Dr. Lotti Steinitz-Sears.)

trisomic, complementing   Can be of three types; one is an apparent trisomic only because there is a normal biarmed chromosome and the two other chromosomes are actually telosomes, representing the left and right arms of the intact, normal chromosome. The second complementing type also involves one biarmed normal chromosome but the two telosomes represent a pair of identical arms thus in essence this case is very similar to a secondary trisomic, having three identical arms and one single different arm. The third type is when there is a normal biarmed chromosome the second chromosome has an arm identical to the first but its second arm is a translocated segment from a non-homologous chromosome, the third chromosome has a different translocated arm linked to the centromere and to an arm of a normal chromosome. (See trisomic analysis, scheme above from G. P. Rédei's Lecture Notes 1980.)

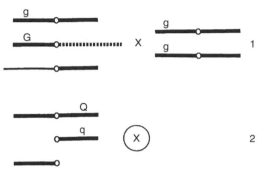

In the complementing trisomics there is only one normal representative of a particular chromosome. For the homologous arms it may be compensated by translocations containing jointly the entire chromosome (1) or by two complementary isochromosomes (2). The complementing trisomics can also be used to assign genes to chromosome arms. In case (1) the predominant fraction in the 2n progeny is expected to be recessive because only the non-translocation strand is transmitted by the monosomic gamete. The wild-type allele (G) can be transmitted only after recombination. In contrast, the 2n + 1 progeny should be almost all wild type (except when recombination occurs between gene and centromere). In selfing of genotype (2) the 2n and most of the 2n + 1 progeny are expected to be of dominant phenotype; however, when recombination occurs between (q) and centromere, the recessive allele is included in the normal chromosome. Such a setup as (2) can be exploited only if the two telo chromosomes are not distributed together in the majority of anaphase I's of meiosis.

trisomic, primary   Chromosome number is 2n + 1, and the triplicate homologous chromosomes are structurally normal. *See* trisomic analysis.

trisomic, secondary   The extra chromosome is an isochromosome, i.e., its two arms are identical. *See* isochromosome; trisomic analysis.

trisomic, tertiary   The extra chromosome is involved in a reciprocal translocation and is partly homologous with two standard chromosomes. *See* trisomic analysis.

trisomic analysis   Trisomics have one or more chromosomes in triplicate (*AAA*) and thus produce both disomic (*AA*) and monosomic (*A*) gametes for gene loci in those chromosomes. Male transmission of the disomic gametes is usually poor, rarely exceeding 1%. The transmission of disomic gametes through females varies according to the chromosome (genes) involved and depends on environmental conditions; most commonly, it is one-fourth or one-third of the normal (monosomic) gametes. The few viable human sex-chromosome trisomics (XXX, XXY, XYY) are either sterile or have very poor fertility because of the failure of normal development of the ovaries and testes (gonadal dysgenesis). From a single alternative allelic pair (A/a), disomics produce either A or a gametes. In contrast, trisomics potentially produce a maximum of five kinds of gametes: three disomic and two monosomic. The frequency of these five types of gametes also depends on the distance of the gene from the centromere, i.e., whether *chromosome segregation* (no crossing over between gene and centromere) or *maximal equational segregation* (random recombination between gene and centromere) occurs.

Gametic output of trisomics

Genotype ↓	Chromosome segregation					Maximal equational segregation				
	AA	Aa	aa	A	a	AA	Aa	aa	A	a
AAa (duplex)	1	2	0	2	1	5	6	1	8	4
Aaa (simplex)	0	2	1	1	2	1	6	5	4	8

In order to obtain two identical recessive alleles from a duplex, the chromosomes must form trivalents and recombination must take place between the gene and the centromere, then anaphase I distribution must move the two exchanged chromosomes toward the functional megaspore (equivalent to secondary oocyte in animals). At the most favorable coincidence of these events, in only half of the cases we can expect the two identical alleles to move into the same megaspore (equivalent to egg in animals). Thus, the maximal chance of having a tetrad with a double recessive gamete (aa) will be $1/3 \times 1/2 = 1/6$. These fractions are based on one reductional and two equational disjunctions at meiosis I (1/3), and two alternative disjunctions at anaphase II (1/2) are the determining factors of the outcome.

Phenotypic segregation can be derived from gametic output by random combinations. One factor, transmission difference between female and male gametes, generally seriously alters theoretically expected proportions.

Phenotypic proportions in the F_2 of trisomics (disomic and trisomic pooled)

Transmission[‡] & genotype	Chromosome segregation			Maximal equational segregation		
	Dom.	Rec.	aaa (%)	Dom.	Rec.	aaa (%)
DUPLEX (AAa)						
Male and female	35	1	0	22.04	1	1.39[¶]
Female only	17	1	0	13.40	1	1.39
None	8	1	0	8	1	0.00
Simplex (Aaa)						
Male and female	3	1	11[§]	2.41	1	13.9[*]
Female only	2	1	11	1.77	1	13.9
None	5	4	0	1.25	1	0.

[‡]refers to transmission of the disomic gametes
[¶]1/576 tetrasomics ($aaaa$) are not included
[§]1/36 tetrasomic ($aaaa$) is not included
[*]25/576 tetrasomics ($aaaa$) are not included

Phenotypic ratios in test-cross progenies (trisomics + disomics)

Genotypes of cross ↓	Only female transmission of disomic gametes	No transmission of disomic gametes
	Dom.: Rec.	**Dom.: Rec.**
AAa x aa	5:1	2:1
Aaa x aa	1:1	1:2

The genetic behavior of all chromosomes not present in triplicate in the trisomics is consistent with disomy. When the trisomic individuals are phenotypically distinguishable per se (do not require cytological identification of the chromosomal constitution), trisomy may be used very effectively to assign genes to chromosomes, irrespective of their linkage relationship. The segregation in a duplex will not be 3:1, but it will vary (most commonly) between 17:1 to 8:1 as predicted by the table at left. In this case, gene *a* is in the chromosome, which is triplicated. Genes in the disomic set of chromosomes are expected to display 3:1 segregation. In case telotrisomics are crossed with disomics (__o__a__, __o__a__, o__A__ × normal disomic (*a/a*), the dominant allele will be very rare among the diploid offspring, but it may be expressed in every individual of $2n$ + telochromosome. Mapping can also be done by properly constructed isotrisomics, tertiary trisomics, and compensating trisomics. Genetic mapping with the aid of multipoint trisomic data is feasible in humans. *See* chromosome segregation; disomic; Down syndrome; Edwards syndrome; gonadal dysgenesis; mapping by dosage effect; maximal equational segregation; monosomic; Pätau syndrome; trisomy. (Rédei, G. P. 1982. *Genetics*. Macmillan, New York; Li, J., et al. 2001. *Am. J. Hum. Genet.* 69:1255.)

trisomy Involves one or more but not all chromosomes of a genome in triplicate; trisomics are aneuploids. They may arise by selfing triploids when some of the extra chromosomes are lost. Nondisjunction during meiosis may also lead to trisomic progeny. Trisomy may exist in various forms. The phenotype of the trisomics varies depending on what genes are in the trisome. The phenotype may be very close, almost indistinguishable, from that of normal disomics, and in other cases it may be lethal. Only a few of the autosomal human trisomies permit growth and development beyond infancy. Trisomy 9 allows near-normal life expectancy, although it involves developmental and mental retardation. Individuals with trisomy 21 (Down syndrome) also reach adult age, but they remain mentally subnormal. Trisomy 22 may exist in mosaic form and the afflicted individuals are retarded in growth and mental abilities. About three-fourths of trisomy 8 cases are mosaics, mentally retarded, and affected by head abnormalities to a variable degree, according to the extent of mosaicism.

Trisomics for chromosome 11 have a variable extra long arm of chromosome 11 and accordingly brain and other internal organ damage of variable extent. Trisomics for the long arm of chromosome 3 have very short life and variable abnormalities in internal organs and body development and severe mental retardation. Trisomy for the short arm of chromosome 4 is characterized by serious malformations of the brain, head, extremities, genitalia, and these persons usually die early. About one-third of the abortuses, due to autosomal trisomy, involve chromosome 16; these are never carried to term. Trisomy 13 (Pätau syndrome) and trisomy 18 (Edwards syndrome) are live-born but die very early. Trisomy 16 is responsible for about 7.5% of spontaneous abortions. XYY trisomies are apparently never found in spontaneously aborted or stillborn fetuses. All other autosomal human trisomy leads to abortion at various stages after conception. Sex-chromosomal trisomy generally does not compromise viability of humans, yet they are usually sterile, displaying gonadal dysgenesis and a variety of physical and most commonly mental retardation as well. The vast majority of trisomics are the result of disomic female gametes. Paternal nondisjunction accounts for only a very small frequency of autosomal trisomy. However, 80% of $45 + X$ monosomics involve the loss of a paternal sex chromosome.

Trisomy is useful in plants for assigning genes to chromosomes. Microcell-mediated chromosome transfer permits the inclusion into mouse cells only fragments of the third chromosome. The fragments are produced by chromosome breakage with the aid of ionizing radiation. The size and contents of the fragments so generated are beyond precise experimental control; nevertheless, information can be obtained about the effects of segments smaller than an entire extra chromosome. Homologous recombination by gene targeting would appear to be a more desirable tool for generating segments of designed contents. Unfortunately, the frequency of this type of recombination is low. A chicken pre-B-cell line (DT40) is known, however, with 3–4 orders of magnitude higher homologous recombination. Technology is available for the transfer of human chromosomes to these chicken cell lines. The truncated human chromosomes from the chicken cell lines are reintroduced into human cells in culture. Another approach is the construction of human artificial chromosomes and minichromosomes. *See* aneuploidy; cat-eye syndrome; chromosome uptake; Down syndrome; Edwards syndrome; human artificial chromosome; Klinefelter syndrome; microcell; minichromosome; Pätau's syndrome; sex-chromosomal anomalies in humans; sex determination; targeting genes; trisomic analysis; trisomics. (Hassold, T. J. & Jacobs, P. A. 1984. *Ann. Rev. Genet.* 18:69; Robinson, W. P., et al. 2001. *Am. J. Hum. Genet.* 69:1245.)

trisomy, segmental   *See* segmental aneuploidy.

tristetrapolin   Zinc-finger protein inhibitory to TNF-α by destabilizing its mRNA, although TNF and agents that stimulate TNF also stimulate tristetrapolin production. *See* DNA-binding protein domains; TNF; zinc fingers.

tritanopia   *See* color blindness.

tritelosomic   In wheat, $20'' + t''$, $2n = 43$ ($'' = $ disomic, $''' = $ trisomic, t = telosomic).

Trithorax   *Drosophila* gene (3–54.2) regulating embryonic development (head, thorax, abdomen). Some of the alleles cause homeotic developments. The heterozygotes for the trx^D mutation display patchy variegation. *See* bithorax, polycomb; developmental-regulator effect variegation; homeotic genes; mll. (Petruk, S., et al. 2001. *Science* 294:1331.)

Triticale   Synthetic species produced by crossing wheat (*Triticum*) and rye (*Secale*) and doubling the chromosome number. The wheat parent may be tetraploid ($2n = 28$) or hexaploid ($2n = 42$) and the rye is generally diploid ($2n = 14$). Therefore, the amphidiploids may have the chromosome number of either $2n = 42$ (hexaploid) or $2n = 56$ (octaploid). As a crop variety, the former is used. On some soils suitable primarily for rye, *Triticale* may provide grains with a milling quality approaching that of wheat. The new hybrids generally have shrunken kernels because some rye genes cause poor endosperm development. Breeding efforts have largely solved these problems and commercial varieties have rather plump grains. *See* amphidiploid; chromosome doubling; *Secale*; species, synthetic; *Triticum*. (Gustafson, J. P. 1987. *Plant Breed. Rev.* 5:41.)

Triticolsecale   *See Triticale.*

Triticum (common wheat and relatives)   Form a series of allopolyploids (Huang, S., et al. 2002. *Proc. Natl. Acad. Sci. USA* 99:8133). The common bread wheat is *T. aestivum* (both winter and spring forms) and *T. durum* (*T. turgidum*) is used mainly for pastas (macaroni, spaghetti, etc). The wheat kernel contains ca. 60–80% carbohydrates (starch) and 8–15% protein, which is usually low in lysine, tryptophan, and methionine. Some of the wild relatives of cultivated wheat produce kernels with higher amounts of protein than commercial varieties, and these favorable traits can be transferred by chromosomal engineering. Wheats are the staple food for about one-third of the human population. *See* allopolyploids; photo and take next page; celiac disease; glutenin; monosomic analysis; nullisomic. (<http://www.tigr.org/tdb/tgi.shtml>, <www.ksu.edu/wgrc/>.)

tritium (H^3)   *See* isotopes.

TRITON X-100   Nonionic detergent (MW 537). It is used for the extraction of proteins and solubilization of biological materials. The reduced form is preferred when spectrophotometric measurement is required. Molar absorption is 1.46×10^3 at pH 8, 20°C, in 1% sodium dodecylsulfate (SDS) at 275 nm. Ammonium cobaltothiocyanate reacts with it as a blue precipitate.

Tritordeum   Is an allohexaploid hybrid of tetraploid wheat (A, B genomes) and barley (H) or allooctaploid (ABDH) constitution. *See* allopolyploid; triticum. (Hernandez, P., et al. 2002. *Genome* 45:198.)

trivalent   In a trisomic or polyploid individual, three chromosomes may associate during meiotic prophase. In trivalent association at any particular place, only two homologous chromosomes can pair, however, at any particular time and site.

Trivalent.

wheat species: From right to left: *Triticum monococcum, T. turgidum (durum), T. timopheevi, T. aestivum* (Chinese Spring [most widely used for cytogenetic studies]), *T. compactum. T. spelta.* (Courtesy of Dr. Carlos Alonso-Arnedo.)

The major species of triticum

Diploids (2n = 14)		genomic formula former designation* and species, respectively
T. monococcum	A	*T. boeoticum*, T. aegilopoides, T. thoudar. T. urartu*
T. speltoides	S (= G?)	*Aegilops speltoides*, Aegilops ligustica**
T. bicorne	Sᵇ	*Aegilops bicornis**
T. searsii	Sˢ	*Aegilops searsii**
T. longissimum	Sˡ	*Aegilops longissima** (Aegilops sharonensis ?)*
T. tauschii	D	*Aegilops squarrosa*, T. aegilops*

Tetraploids (2n = 28)		
T. turgidum	AB	*T. dicoccoides*, T. dicoccon, T. durum*, T. polonicum*, T. carthlicum*, T. persicum**
T. timopheevi	AG	*T. araraticum, T. dicoccoides var. nudiglumis T. armeniacum*

Hexaploids		
T. aestivum	ABD	*T. vulgare*, T. spelta*, T. macha*, T. sphaerococcum*, T. vavilovii**

trk **(TRK) oncogene**   Its product is a receptor in membranes (tyrosine kinase activity) for NGF (nerve growth factor). It is localized to human chromosome 1q23-q24. *See* neutrotrophin. (Nakagawara, A. 2001. *Cancer Lett.* 169:107.)

tRNA (transfer RNA)   Shortest RNA molecule in the cell (ca. 3.8S) consisting of about 76 to 86 nucleotides. They carry amino acids to the ribosomes during protein synthesis. The majority of cells have 40 to 60 types of tRNAs because most of the 61 sense codons have their own tRNA in the eukaryotic cytosol. The tRNAs that accept the same amino acid are *isoaccepting tRNAs*. In human mitochondria, there are only 22 different tRNAs and their number in the plant chloroplasts is about 30. tRNA is frequently called an adaptor molecule because it adapts the genetic code for the formation of the primary structure of protein. Rarely (ca. 1/3,000), a tRNA is charged with the wrong amino acid; in these cases, the complex is usually disrupted and the tRNA is recycled. tRNA genes may occur in between ribosomal RNA genes (promoter — 16S — tRNA — 23S — 5S — tRNA — tRNA...) or they frequently form independent clusters of different tRNA genes (promoter — tRNA^1 — tRNA^2 — tRNA^3 — •). The different cognate tRNA groups are generally identified by a superscript of the appropriate amino acid, e.g., tRNA^{Met}. Sometimes individual tRNA genes may be present in multiple copies. tRNA gene clusters are transcribed into large primary molecules that require successive cleavage and trimming to form mature tRNA. The endonuclease RNase P (a ribozyme) recognizes the primary transcript whether it is a single tRNA sequence or a cluster of rRNA or tRNA genes and cleaves at the 5′-terminus where the tRNA begins. For tRNA to become functional, RNase D cuts at the 3′-end and stops at a CCA sequence, if any, or the 3′-end receives one, two, or three bases through post-transcriptional synthesis (by tRNA nucleotidyl synthetase) and thus it is always finished by a 3′-CCA^{-OH} sequence. This 3′-OH end becomes the amino acid attachment site of tRNA (*see* transfer RNA). During the process of maturation, through modification of the original bases, thiouridine (S4U), pseudouridine (ψ), ribothymidine, dihydrouridine (DHU), inosine (I), 1-methylguanine (m^{1}G), 1-dimethyl-guanine (m^{1}dG), and N^6-isopentenyl adenosine (i^{6}A) may be formed within the tRNA sequence. *See* amino acylation; aminoacyl-tRNA synthetase; dihydrouridine; hypoxanthine; intron; isopentenyladenosine; mitochondrial diseases in humans; protein synthesis; pseudouridine; ribothymidine; ribozyme; transfer RNA; tRNA nucleotidyl transferase; trimming. (Söll, D. & Rajbahandary, U. L., eds. 1994. *tRNA Structure, Biosynthesis and Function.* AMS Press, Washington, DC.)

tRNA cleavage   T-even phages may cripple a bacterial host tRNA, then the viral-coded isoaccepting tRNA is substituted for that of the host. (Amitsur, M., et al. 1987. *EMBO J.* 6:2499; Kaufmann, G. 2000. *Trends Biochem. Sci.* 25[2]:70; Meidler, R., et al. 1999. *J. Mol. Biol.* 287:499.)

tRNA deacylase   Cuts off the tRNA D-amino acids from the polypeptide chain after the complex reaches the P site on the ribosome. *See* aminoacyl-tRNA synthetase; protein synthesis. (Ferri-Fioni, M. L., et al. 2001. *J. Biol. Chem.* 276:47285.)

tRNA mimicry Certain set of translation factor proteins resemble tRNA in shape and may even mimic tRNA in deciphering the genetic code. (Nakamura, Y. 2001. *J. Mol. Evol.* 53[4–5]:282.)

tRNA nucleotidyl transferase Attaches after transcription of CCA-3′-OH to the 3′-end of the amino-acid-accepting arm of tRNA without relying on a template. Before building this 3′-end of the tRNAs, a nuclease must remove the tail of the primary transcript at a discriminator position, which may be the 73rd base. Some prokaryotic tRNA transcripts already contain CCA ends and thus tRNA nucleotidyl transferase is not indispensable, but it is advantageous because it may repair the amino acid acceptor. The CCA transfer enzyme is not selective; it recognizes all tRNAs irrespective of their amino acid specificity. Most of the U2 snRNAs in humans also carry CCA ends. *See* transfer RNA; tRNA. (Vasil'eva, I. A., et al. 2000. *Biochemistry (Moscow)* 65:1157; Cho, H. D., et al. 2002. *J. Biol. Chem.* 277:3447.)

tRNA-SE Fast computer program capable of identifying 99–100% of tRNA genes in DNA sequences with extremely low false positives. It may be applied to the detection of unusual tRNA homologues such as selenocysteine tRNA, tRNA pseudogenes, etc.

trophectoderm Extraembryonic tissue at the blastocyst stage of mammalian development. *See* blastocyst.

trophoblast Surface cell layers of the blastocyst embryo connecting to the uterus.

trophozoite Growing and actively metabolizing cells of unicellular organisms vs. cysts.

tropic Indicates that something is aimed at it, e.g., T-tropic means that a virus targets T cells or is directed at a site in some way.

tropic hormone Stimulates the secretion of another hormone at another location.

tropism Growth of plants in the direction of some external factors.

tropomodulin Maintains actin filament growth by capping its pointed ends. (Littlefield, R., et al. 2001. *Nature Cell Biol.* 3:544.)

tropomyosin (TPM) Skeletal muscle fiber protein. TPM3 is encoded at 1q22-q23. Endostatin binds tropomyosin. *See* endostatin; troponin. (MacDonald, N. J., et al. 2001. *J. Biol. Chem.* 276:25190.)

troponin Ca^{2+}-binding regulatory polypeptide in the muscle tissue. Troponin C binds four molecules of calcium. Troponin I has an inhibitory effect on myosin and actin and binds tropomyosin, an accessory protein. In the relaxed muscles, troponin I binds to actin and moves tropomyosin to the position where actin and myosin would interact at muscle contraction. When the level of Ca^{2+} is high enough, troponin I action is blocked so that myosin can bind actin again, allowing the muscle to contract. Troponin C is related in function to calmodulin. In *nemaline* (thread-like) *myopathy* at 19q13.4, the sarcomeric thin-filament protein (TNNT1) is truncated. This recessive/dominant infant-lethal disease has an incidence of ~0.002 in Amish populations. *See* calmodulin; myopathy; myotonic dystrophy; receptor tyrosine kinase; signal transduction. (Johnston, J. J., et al. 2000. *Am. J. Hum. Genet.* 67:814; Hinkle, A. & Tabuman, L. S. 2003. *J. Biol. Chem.* 278:506.)

TRRP Transactivation/transformation domain–associated protein is a member of the ATM protein superfamily and a cofactor of cMYC-mediated transformation. The yeast homologue (Tra1) is a component of histone acetyltransferase (HAT), SAGA, PCAF, and NuA4. Trrap is essential for normal development. (Herceg, Z., et al. 2001. *Nature Genet.* 29:206.)

TRP Transient receptor potential. Plasma membrane ion channel components in control of active and passive Ca^{2+} stores are activated by 1,4,5-trisphosphate receptors. *See* phosphoinositides. (Montell, C., et al. 2002. *Cell* 108:595.)

Trp *See* tryptophan; tryptophan operon.

true breeding Absence of segregation among the offspring.

truncation Cutoff point; e.g., in artificial selection individuals before or beyond an arbitrarily determined point are discarded or maintained, respectively. A cloned eukaryotic gene without a polyadenylation signal or an incomplete upstream control element may be called truncated. *See* selection. (Crow, J. F. & Kimura, M. 1979. *Proc. Natl. Acad. Sci. USA* 76:306.)

Trypanosomas Protozoa spread by the tsetse fly (*Glossina*) cause sleeping sickness. Various developmental stages are distinguished on the basis of the relative position of the flagella (basal flagella: trypomastigote; median: epimastigote; apical: promastigote; no flagella: amastigote). *Trypanosomas* may reach a level of $10^9–10^{10}$ individuals per mL blood of mammals. The chromosomes are small and variable in number because minichromosomes are found in addition to stable chromosomes. The ca. 100 minichromosomes (50 to 150 kb) contain open-reading frames for variable-surface glycoproteins (VSG). These sequences are transcribed only when transposed to expression sites in the 0.2- to 6-megabase-long 20 maxichromosomes. Since some of the genes may be present in more than single copy in different chromosomes, these organisms may resemble allodiploids. The genes do not have introns and some of the tandem arrays of genes are transcribed as long polycistronic pre-mRNA. At the 5′-end, each primary transcript is capped by a 39-nucleotide spliced leader RNA (SLRNA). This cap itself is transcribed as a 139-base sequence, but the 100-base sequence is not used in this trans-splicing reaction. *Trypanosomas* have homologues of the mammalian U2, U4, and U5 small nuclear RNAs in the form of ribonucleoprotein particles (RNP). The mRNAs are polyadenylated.

In *T. brucei*, there is a family of about 1,000 genes involved in the recurring production of a great repertory of *variant antigen type* (VAT). At each flareup of division of the parasite, it switches on the production of a different type of antigen (serodeme). The more than million molecules of antigens on the surface of its cells are the phosphatidylinositol-anchored (about 60 kDa) *variable-surface glycoproteins* (VSG). All the different VSGs have at least one N-linked oligosaccharide and several cysteine residues near the N-end, and some similarities within the 50–100 amino acids at the C-terminus. Because of rapid switches in production to new antigens, the vertebrate cell's immunological surveillance system cannot adapt rapidly to contain the infection. In chronic infections, 50–100 different antigens may be produced. A particular *Trypanosoma* has only 1/million or less chance per cell division to switch to the production of a different antigen, yet because of their immense number, the parasite has a good chance to escape the immune system of the mammalian host. At any one time only one of the VSG genes is expressed in the protozoan. The switching is not a direct response to the host antibody but rather acts only as a selecting mechanism.

The switching of transcription may require a shutoff gene or a gene conversion-type process takes place. The expressed surface antigen gene, the so-called *expression-linked copy* (ELC), is always located in the telomeric regions of the chromosomes. Its promoter is located, however, 50 kbp upstream. In addition, non-VSG genes are located in the expression region; these are called *expression site–associated genes* (ESAG). The expression of these basic, silent genes requires a transposition into the activation region, in a manner similar to the mating-type switching in yeast. The switching apparently depends on expression-linked copies (ELC) of the mini-exon-dependent transcription into 140-base-long eukaryotic-type mRNA (also called medRNA). Because of this effective switching, there are serious problems in developing vaccines against *Trypanosomas*. Mutants can be produced that simultaneously turn on more than one type of surface glycoprotein. *Trypanosomas* take up transferrin (a β-globulin) from the host cells through about 20 homo- or heterodimeric transferrin receptors, which are activated alternatively just like the VSG genes. The different transferrin receptors make it possible for the parasite to adapt to different hosts. TbAT1 nucleoside transporter confers melaminophenyl arsenical susceptibility, whereas defects in TbAT1 render the cells resistant to the trypanocide. *See* Chagas disease; flagellar antigen; intron; kinetosome; *Leishmania*; mating type determination in yeast; transporters. (Coppens, I. & Courtoy, P. J. 2000. *Annu. Rev. Microbiol.* 54:129; Navarro, M. & Gull, K. 2001. *Nature* 414:759; Spadiliero, B., et al. 2002. *J. Cellular Biochem.* 85:798; Beverly, S. M. 2003. *Nature Rev. Genet.* 4:11. <http://www.tigr.org/tdb/tgi.shtml>.)

Trypanosoma.

tryphine (pollenkitt) Lipids and proteins filling the depressions of the pollen surface.

trypomastigote *See Trypanosoma.*

trypsin Proteolytic enzyme synthesized as an enzymatically inactive zymogen or trypsinogen that is activated by proteolytic cleavage. The 6,000 Mr pancreatic trypsin inhibitor inactivates it. Trypsin specifically cleaves polypetides at the carbonyl sides of Lys and Arg. Ser 195 and His 57 are at its active site.

trypsinogen deficiency (TRY1) 7q22-ter recessive hypoproteinemia and insufficient amino acid level. It is very similar to enterokinase deficiency. *See* enterokinase deficiency.

tryptases In the form of heparin-stabilized tetramers, they function similarly to trypsin-like serine proteases mainly in the mast cells. *See* heparin; mast cell; trypsin.

tryptic peptides Products of digestion of a protein by trypsin. *See* trypsin.

tryptophan Essential aromatic amino acid (MW 204.22), soluble in dilute alkali, insoluble in acids, and it is degraded when heated in acids. Its biosynthetic pathway (*with enzymes involved in parentheses*): Chorismate → (*anthranilatesynthase*) → Anthranilate → (*anthranilate–phosphoribosyltransferase*) → N-(5'-Phosphoribosyl) – anthranilate → (*N-(5'-phosphoribosyl)-anthranilateisomerase*) → Enol-1-*o*-carboxyphenylamino-1-deoxyribulosephosphate → (*indole–3–glycerolphosphatesynthase*) → Indole-3-glycerol phosphate → Indole → (*tryptophansynthase*) → Tryptophan. In *E. coli*, the first two enzymes constitute a single anthranilate synthase complex. Tryptophan is converted to formylkynurenine by *tryptophan dioxygenase* (tryptophan pyrrolase). Through the action of an *aminotransferase*, tryptophan gives rise to indole-3-pyruvate, which after *decarboxylation* forms indole-3-acetic acid, one of the most important plant hormones (auxin). Tryptophan and phenylalanine contribute to the formation of lignins, tannins, alkaloids (morphine), cinnamon oil, cloves, vanilla, nutmeg, etc., flavors. *See* attenuator; chorismate; fragrances; Hartnup disease; melanin; phenylalanine; pigments in animals; plant hormones; tyrosine; *tryptophan* operon.

tryptophanase (*tna*) Bacterial operon including major genes A, B, and a permease with a 319 bp leader encoded by tnaC preceding gene tnaA. It degrades L-tryptophan to indole, pyruvate, and ammonia. Tryptophan regulates it by induction and antitermination. *See* antitermination; induction. (Gong, F., et al. 2001. *Proc. Natl. Acad. Sci. USA* 98:8997.)

tryptophanyl tRNA synthase (WARS) Charges tRNA[Trp] by the amino acid tryptophan. The gene WARS is in human chromosome 14. *See* aminoacyl-tRNA synthetase.

Tryptophan operon Contains five structural genes. In *E. coli*, *Tryptophan* operons have been mapped (at 27 min) in exactly the same order as their sequence of action in the biosynthetic path (*see* tryptophan). This operon has a principal promoter and a secondary low-efficiency one. Between the operator and the proximal gene, there is a 162 bp leader sequence (*trpL*) including an attenuator site *a*. (*See* temperature-sensitive control sequences of transcription.) When there is a sufficient amount of tryptophan in the cell and all the tRNA^{Trp} are charged with the amino acid, transcription of the leader sequence stops at base 140. Thus, synthesis of the specific mRNA temporarily ceases when there is no immediate need for tryptophan.

The site of attenuation (*a*) is within the *trpL* (tryptophan leader) sequences. Before attenuation becomes effective, RNA polymerase pauses at the *tp* site in the *trpL* (*see* attenuation). Two main mechanisms, repression and attenuation, regulate expression of this operon. Repression prevents the initiation of transcription. The repressor is transcribed from the *trpR* gene (located at 100 min) and tRNA^{Trp} is coded by gene *trpT* (84 min), whereas aminoacylation of this tRNA is determined by gene *trpS* (74 min). The product of *trpR* is an aporepressor, i.e., it becomes active only when combined with tryptophan, its corepressor (*see* tryptophan repressor). The primary sequence of the leader sequence is as follows (the underlined sequences indicate ribosome binding, tryptophan codons and critical codons in outline):

The tryptophan repressor alone may reduce transcription by a factor of 70 and attenuation may decrease it 8- to 10-fold, but transcription may be reduced by 8×70 (=560) or 10×70 (=700) fold by the combination of these two controls. In *Rhodobacterium sphaeroides*, the tryptophan operon is shared between the two chromosomes of this bacterium.

Bacillus subtilis regulates seven genes of tryptophan biosynthesis. Six of them (*TrpEDCFBA*) are clustered within a 12-gene aromatic supraoperon. The seventh gene (*TrpG*) is unlinked and is in the folate operon. TRAP regulates by attenuation as it binds to a specific site in the leader sequence of the *trp* operon and facilitates a terminator formation of the RNA transcript. In addition, TRAP binds to the ribosome-binding sequence in *trpG* mRNA, thus inhibiting translation. An anti-TRAP (AT) protein signals to tRNA^{Trp} and then it is not charged with tryptophan. Uncharged tRNA induces the synthesis of AT. Linking AT to TRAP prevents TRAP binding to RNA. As a consequence, tryptophan biosynthesis is promoted (Valbuzzi, A. & Yanofsky, C. 2001. *Science* 293:2057.)

In *Neurospora*, there are also five distinct genetic loci controlling tryptophan biosynthesis. These genes are derepressed coordinately with histidine, arginine, and lysine loci and this phenomenon is called *cross-pathway regulation*. Humans do not have tryptophan synthetic genes and depend on food as their source of tryptophan (essential amino acid). *See* antitermination; attenuation; essential amino acids;

helix-turn-helix; *lac* operon; repressor; TRAP; tryptophan; tryptophan repressor. (Yanofsky, C. 1981. *Nature* 289:271.)

tryptophan repressor Consists of a repressor protein (using a helix-turn-helix motif) for binding to the operator site of the tryptophan operon and becomes active only when it is associated with tryptophan. Illustrated with geometric symbols: ∪ (aporepressor protein), ♦ (corepressor tryptophan), (active tryptophan repressor). Actually, the binding of tryptophan to the repressor protein facilitates the more intense binding of the complex due to a conformational change of the repressor protein. *See* arabinose operon; conformation; helix-turn-helix motif; *Lac* operon; negative control; positive control; *tryptophan* operon. (Khodursky, A. B., et al. 2000. *Proc. Natl. Acad. Sci. USA* 97:12170.)

tryptophan zipper Structural motif that can stabilize the hairpin structure of short peptide chains of 12 to 16 amino acids with the aid of cross-strand pairs of indole rings. It does not require metal or disulfide links.

ts Indicates temperature sensitivity of an allele. *See* temperature-sensitive mutation.

T$_S$ *See* T cell.

Tsarevitch Alexis Great-grandson of Queen Victoria who inherited her mutation causing classic hemophilia. *See* hemophilias; Queen Victoria.

tsBN2 Temperature-sensitive baby hamster kidney (BHK) cell line. *See* hamster.

TSC Reduced representation shotgun sequencing data set. *See* shotgun sequencing.

TSE Transmissible spongiform encephalopathy. *See* encephalopathy; prion.

tse-tse fly African fly of the genus *Glossina*, host of the parasitic *Trypanosomas*, causing sleeping sickness, a disease characterized by relapsing fever, enlargement of the lymph glands, anemia, severe emaciation, and eventually death in humans and domestic animals. *See* Trypanosomas. (Akman, L., et al. 2002. *Nature Genet.* 32:402.)

TSG Tumor susceptibility gene acts as a transcriptional cofactor and as a nuclear hormone receptor-mediated transactivator. There is ~94% homology between mice and human proteins. (Teh, B. T., et al. 1999. *Anticancer Res.* 19[6A]:4715.)

TSHB Thyroid-stimulating hormone; thyrotropin. *See* animal hormones.

Tsix *See* Xist.

```
                           Met                              Trp   Trp
       pppAAGUUCACGUAAAAGGGUAUCGACAAUGAAAGCAAUUUUCGUACUGAAAGGUUGGUGGCGCACUUC
Opal              end of pause↓         nucleotides beyond the box are not assembled if attenuation works

UGAAACGGGCAGUGUAUUCACCAUGCGUAAAGCAAUCAGAUACCCAGCCCGCCUAAUGAGCGGGUUUUUUU

                    Met starts trpE
    UGAACAAAAUUAGAGAAUGCAAACACAAAAACCGACUCUCGAA        _    → →
```

Tsr Bacterial transducer protein recognizing serine as an attractant and leucine as a repellent. *See* transducer proteins.

T-strand Single-stranded intermediate of T-DNA that is transferred from the Ti plasmid of *Agrobacterium* to the plant nucleus through the nuclear pores under the guidance of a virulence gene-encoded protein covalently attached to its 5′-end. *See* T-DNA; Ti plasmid; transformation.

tTA Tetracycline transactivator protein is a fusion protein of the *tet* (tetracycline) repressor of *E. coli* and the transcriptional activation domain VP16 of herpes simplex virus. This system is generally driven by the *tetP* promoter, which is actually a minimal immediate early cytomegalovirus (CMVIE) promoter preceded by seven copies of *tetO*, the tetracycline-resistance operator of transposon 10 (Tn*10*). In the presence of tetracycline, this system is expressed at a very low level, and by removal of tetracycline, the gene(s) under its control (e.g., luciferase, β-galactosidase) may be expressed at a three orders of magnitude higher level. The system can be used under the control of other promoters that are best suited for regulating the expression pattern of the gene of interest. *See* rtTA; split-hybrid system; tetracycline. (Gossen, M. & Bujard, H. 1995. *Science* 268:1766.)

t-test Used for the estimation of statistical significance of the difference(s) between means. The *t* value is the ratio of the observed difference to the corresponding standard error:

$$t = (\bar{x} - m)/(s/\sqrt{n})$$

where $\bar{x}$ and m are the two means, n = population size, and s = standard deviation. More commonly, the significance of the difference between two means is calculated as

$$t = \frac{m_1 - m_2}{\sqrt{[e_1]^2 - [e_2]^2}}$$

where m_1 and m_2 are the two means ($\bar{x}$) and e_1 and e_2 are the standard errors of the two means determined as

$$e = \frac{s}{\sqrt{n}}$$

and

$$s = \sqrt{V}$$

where

$$V = \text{variance} = \frac{\sum[(x - \bar{x})^2]}{n - 1}.$$

When the *t* value is available, the probability of the difference is determined with the aid of a *t* table. *See* student's *t* distribution.

TTK Tubulin-associated kinases. *See* tubulin.

TTP Tris-tetrapolin. cDNA shares 102 amino acid sequences in its product with TIS11 (insulin- and serum-responsive transcription factor). *See* insulin; serum-response element; TIS11; transcription factors.

T-tropic The T lymphocyte is targeted (e.g., by a virus).

TU Protein elongation factor (EF-TU) in prokaryotes binds aminoacyl-tRNA to the ribosomal A (acceptor) site. It is a guanine nucleotide-binding RAS-like protein. *See* aminoacyl-tRNA; EF-TU.GTP; protein synthesis; RAS; ribosome. (Zvereva, M. I., et al. 2001. *J. Biol. Chem.* 276:47702.)

tubal ligation Surgical fertility control by constriction of the fallopian tube, usually by placing a plastic ring on it. *See* salpingectomy; sterilization, human.

tube nucleus In the vegetative cell at the tip of the growing pollen tube. It has only a physiological role and no genetic role because it does not enter the embryo sac. *See* gametophyte.

tuber Underground enlarged stem specialized for food storage, e.g., in potato.

tuberculosis *See* mycobacteria.

tuberous sclerosis *See* epiloia.

tuboplasty Surgical repair of a defect on an internal tube such as the fallopian tube.

tubulins Globular polypeptides of α- and β-subunits (50 kDa, each with about 40% homology and very similar in structure [β-sheets surrounded by α-helices]) are G-proteins concerned with signal transduction of nerve cells and are components of microtubules such as the spindle fibers and the cytoskeleton. Both subunits can bind one guanine nucleotide, which is an exchange cable on the β- but not on the α-binding site. Folding of the tubulins requires cytosolic chaperonins, cofactors A, C, D, and E, ATP, and GTP. The bacterial plastid septation protein FtsZ filamenting temperature-sensitive septal peptidoglycan (Z ring) displays structures similar to tubulins and participates in the septation of the cell. FtsZ may have a GGGTGTG motif and has GTPase activity. With the GGGAGTG motif, FtsZ3 has no GTPase activity, and with the AGGTGTG sequence, FtsZ84 has reduced GTPase activity. FtsZ recruits additional cell division proteins. In the spore-forming bacteria (*B. subtilis*), initially two Z rings are formed, one at both poles, but only one is activated as the genetic material passes to the spore. In the chloroplasts, a nuclear-encoded ftsZ protein functions in fission. The mitochondria apparently do not need ftsZ. *See* chaperonins; chloroplasts; cytoskeleton; FtsZ; GTPase; mitochondria; peptidoglycan; septum; spindle. (Oakley, B. R. 2000. *Trends Cell Biol.* 10:537.)

Septation by the bacterial Z ring

tulip, broken Variegation caused by infection with the tulip-breaking virus. Floriculturists value plants displaying this sectoring. *See* tulip mania; tulips, broken.

tulip mania   *See* symbionts, hereditary; tulips, broken.

TULIP-PCR   Touch-up and loop-incorporated primers PCR. (Ailenberg, M. & Silverman, M. 1999. *Biotechniques* 29:1018.)

tumbling   Results when the bacterial flagella rotate clockwise.

TU/ml   Transforming unit per mL, i.e., the number of cells expressing a transgene. *See* transformation, genetic; transgene.

tumor   Abnormal clump of cells originated by benign or malignant growth. Malignant tumors may be invasive and show metastasis, which is common in many types of cancer. Mutant genes, chromosome breakage, and viral infections altering the normal regulation of cell proliferation may cause tumors. Tumorigenesis is generally a multiphase process involving activation of cell cycle–promoting genes and inactivation of tumor suppressors. Active MAPKK may stimulate tumorigenesis if transfected into mouse cells. The type of the tumor is different in different cancers and even within a single cancer it may vary a great deal because of frequent chromosomal breakage and other mutations as a consequence of cancerous growth. Their characterization using molecular markers may be facilitated by microarray analysis of the biopsies. Plant tumorous growth that is never metastatic, is often called callus although the crown gall tumors of plants may be spread by new foci of infection of the inciting bacterium, *Agrobacterium tumefaciens*. Plant viruses and higher concentrations of phytohormones may also cause tumors. *See* adenoviruses; *Agrobacterium*; cancer; cancer gene therapy; CATR1; genetic tumors; habituation; MAPKK; microarray hybridization; oncogenes; SV40.

tumor antigens   MHC-associated peptides recognized by CTLs as tumor antigens. The MHC class I tumor peptides are also called CTL epitopes. *See* cancer; CTL; dendritic cell vaccine; MHC; tumor vaccination.

tumor-associated antigen   Overexpressed normal self-proteins of cancer cells. They may incite immune reaction by breaking the self-tolerance limit. Examples: oncofetal, differentiation, or nuclear proteins such as carcioembryonic, melanoma-associated proteins, etc. *See* HER2; MAGE.

tumorigenesis   Formation of tumors. It is usually based on the activation of protooncogenes or the inactivation of tumor-suppressor genes. The former mechanism generally involves dominantly acting genes; the latter is usually based on recessive loss of function. Fourier transform infrared spectra (FT-IR) reveal structural modifications at many points in the DNA and marked differences between the primary and metastatic states. *See* apoptosis; cancer; chromosome breakage; DNA repair; environmental mutagens; FT-IR; neoplasia; protooncogenes; radiation effects; telomerase; tumor suppressor.

tumor-infiltrating lymphocytes (TIL)   Seek out tumors. They are isolated from solid tumors and cultured in single-cell suspension in a medium containing interleukin 2 (IL-2).

Through genetic engineering they may be equipped with the gene of tumor necrosis factor through transformation by retroviral or other vectors. The transformed cells may selectively kill cancerous tumor cells. *See* lymphocytes; retroviral vectors; tumor necrosis factor. (Smyrk, T. C., et al. 2001. *Cancer* 91:2417.)

tumor necrosis factor   *See* TNF.

tumor necrosis factor receptor   *See* TNFR.

tumorous hybrids   *See* genetic tumors.

tumor progression   *See* evolutionary clock.

tumor promoter   *See* phorbol ester.

tumor-specific antigen   Endogenous tumor cell-surface antigens that can be presented by major histocompatibility molecules to T cells. These antigens are absent from normal cells and are modified in response to viral transformation or genetic or somatic mutations in oncogenes. *See* antigen-presenting cell; MHC; T cell.

tumor suppressor factors   The development of tumors follows multiple routes, yet genes involved in the general control of differentiation, as revealed by studies of *Drosophila* morphogenesis, seem to be involved. Mammalian homologues of *Drosophila hedgehog (hh)*, sonic hedgehog (SHH), Indian hedgehog (IHH), and desert hedgehog (DHH) seem to be entailed in holoprosencephaly, a developmental anomaly connected to cancer. The receptor of *hedgehog* of *Drosophila* is *smoothened*, and its suppressor is *patched*, which may be related to the development of basal cell carcinoma and defects in the central nervous and skeletal systems. The *cubitus interruptus* (*ci*) *Drosophila* gene appears to function as an effector of *hh*; similarly, the GLI genes of humans are responsible for a type of brain tumor and for cephalopolysyndactyly syndrome. The decapentaplegic (*dpp*) *Drosophila* protein bears similarity to mammalian tumor growth factor (TGF) and to bone morphogenetic protein 4 (BMP4), causing defects in limb and gut formation. The effector of the mammalian TGF-β receptor (DPC4) involved in pancreatic tumors has its homologue in the *Drosophila mad* gene. The *Drosophila wingless* locus corresponds to the mammalian WNT loci controlling mammary tumors. The *Drosophila zeste-white-3* gene codes for a signal molecule similar to a mammalian glycogen synthase kinase, and the *Drosophila armadillo* gene, encoding β-catenin, is an oncoprotein controlling intestinal tumors. (Based mainly on Dean, M. 1996. *Nature Genetics* 14:245.)

tumor suppressor gene   Its loss, inactivation, or mutation permits neoplastic growth by deregulation. These genes most commonly have a role in the cell cycle or in the regulation of RNA polymerase II or III. Inhibition of peptide chain elongation may be a mechanism of tumor suppression. (Genes with cytostatic or cytotoxic effects are excluded from this category of tumor suppressors.) Actually, the majority of cancer cells display deletions that may indicate the loss of a tumor suppressor gene; for direct proof, further evidence is required for the loss of a tumor suppressor. Animal models are available for many human tumor suppressors (Hakem, R. & Mak, T. W. 2001. *Annu. Rev. Genet.* 35:149). Tumor suppressor genes are widely scattered in the human genome: p53/TP53 (17p13.1); retinoblastoma (RB1, 13q14.1-q14.2); adenomatous polyposis of colon (APC, 5q21-q22); deleted in colorectal cancer (DCC, 18q21.3); neurofibromatosis (NF1, 17q11.2); von Hippel–Lindau syndrome (VLH, 3p26-p25); Wilms' tumor (WT1, 11p13); breast cancer (BRCA1, 17q21); cyclin-dependent kinase inhibitor (CDKN2A, 9p21); patched homologue (PTCH, 9q22.3); tuberous sclerosis (TSC2, 16p13.3; TSC1, 9q34); etc. *See* apoptosis; breast cancer; cancer; caretaker genes; cell cycle; colorectal cancer; DPC4; ELL; elongin; gatekeeper genes; histone deacetylase; immunosuppression; LOH; malignant growth; oncogenes; p16; p16^{INK4}; p21; p53; p73; pol III; polycystic kidney disease; PPP2R1B; prostate cancer; PTEN; retinoblastoma (Rb); transformation, oncogenic; tumor suppressor factors. (Robertson, G. P., et al. 1999. *Mol. Cell Biol. Res. Comm.* 2:1; Varambally, S., et al. 2002. *Nature* 419:624.)

tumor susceptibility   May be determined by mutation of some major tumor suppressor (RAS, Cyclin D1, RB, pt3, p16, etc.) genes. Favorable combinations of minor genes may be responsible for "sporadic" cases of cancer. *See* oncogenes; tumor suppressor gene. (For a list of mouse susceptibility loci, see Balmain, A. & Nagase, H. 1998. *Trends Genet.* 14:139.)

tumor vaccination   Immunization by increasing the efficiency of tumor-specific antigen presentation or enhancing the activity of tumor-infiltrating T cells. The purpose is to generate antitumor immune reaction. The cancer cells of the body may be ex vivo genetically modified and reintroduced. Introduction into the cancer cells of a range of cytokines, e.g., IL-12, may enhance T-cell response. The T cells of the tumor can be genetically modified to secrete effector molecules to enhance immune response against the tumor or increase potentials for binding of tumor antigens. Introduction of the wild-type tumor-suppressor genes may also be an option. Transformation by GM-CSF may recruit monocytes/macrophages and APCs by *cross-priming*. Some tumors may mask the immunogenic potentials within the cells. In such cases, antisense technologies (against IGF, TGF, or IL-10), intrabodies, triple helix formation, and inactivation by the use of ribozymes may help in silencing the endogenous inhibitors. It is also possible to introduce activation enzymes into the tumor that can convert prodrugs into highly cytotoxic anticancer compounds or transfect the cells with toxin genes.

Because of pleiotropy and protein interactions, the genetic modifications may not always be favorable for the goals. Within the same individual, molecular heterogeneity may exist among the cancer cells. Therefore, only multivalent cancer vaccines may achieve the goals. Favorable results were obtained in animal models with the combinations of IL-4 + IL-12, costimulatory molecules + IL-12, interferon-γ + IL-2, etc. Many times, tumor cells have very weak or no antigen-presenting mechanism(s) and do not express the costimulatory proteins. Melanoma cells overproduce IL-10, a cytokine that downregulates the path of CTL formation. Fas ligand–expressing lymphocytes may be targets of upregulated Fas produced by tumor cells, especially when exposed to some types of chemotherapy. As a consequence, the apoptotic process may weaken immunotherapy against cancer. Cancer may involve tumor-specific immunodeficiency, but low doses of IL-2 administered over long periods may strengthen the immune system. Various treatments must be effective against the cancer cells without attacking the normal cells by balancing the antitumor and autoimmune responses. *See* antigen presentation; antisense RNA; apoptosis; autoimmune disease; cancer gene therapy; costimulator; cross-priming; CTL; dendritic cell vaccine; Fas; GM-CSF; IGF; IL-10; immune system; interferon; intrabody; LAK; lipids, cationic; lymphocytes; ribozyme; TGF; triple helix formation; tumor-infiltrating lymphocytes; vaccines. (Gunzer, M. & Grabbe, S. 2001. *Crit. Rev. Immunol.* 21[1–3]:133; Tada, Y., et al. 2003. *Cancer Gene Ther.* 10:134.)

tumor viruses   Induce or participate in the formation of tumors (cancer). Their genetic material can be DNA (such as SV40, adenoviruses, papilloma virus, hepatitis B virus, Epstein-Barr virus, adenoma virus, herpes virus, pox virus) or RNA (such as the retroviruses causing leukemia, lymphoma, AIDS, Kaposi's sarcoma, avian leukosis virus, mammary tumor viruses, etc). DNA viruses can integrate into the mammalian genetic material and activate cell replication by overwhelming the function of tumor-suppressor genes. DNA tumor virus genetic material has no counterpart in host genetic material. RNA tumor virus genetic material is replicated by reverse transcription and produces a double-stranded DNA counterpart of their genome. Viral RNA is then transcribed from the cellular DNA template. Most of the oncogenes in the cells (c-oncogenes) correspond to the v-oncogenes in the virus. Class I RNA tumor viruses themselves do not induce tumors unless they have picked up growth-regulating genes from cells. Class II RNA viruses do not contain oncogenes and induce cancer only when proviral DNA integrates in the vicinity of cellular oncogenes (c-oncogenes). *See* adenovirus; Epstein-Barr virus; oncogenes; papilloma virus; polyoma; retroviruses; Rous sarcoma; SV40. (Barbanti-Brodano, G., et al., eds. 1995. *DNA Tumor Viruses.* Plenum, New York; McCance, D. J., ed. 1998. *Human Tumor Viruses.* AMS Press, Washington, DC.)

TUNEL assay   Terminal deoxytransferase-mediated deoxy uridine nick end labeling usually uses biotin-labeled uridine and a streptavidin- or avidin-labeled enzyme, and at the reaction site color develops. *See* biotinylation; terminal nucleotidyl transferase. (Maciorowski, Z., et al. 2001. *Cytometry* 46[3]:159; Yamamoto-Fukud, T., et al. 2000. *Histochemistry J.* 32[11]:697.)

tunica   Cell layer in the plant apical meristem wrapping the inner corpus. See diagram below.

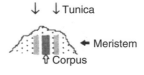

tunneling   Connecting through a structured path. *See* channeling.

TUP1 General suppressor of sugar (and other) metabolism of yeast; its product is a trimeric G-protein. It is similar to AAR1 and AAR2. TUP forms a repressor complex with yeast protein Cyc8, RNA polymerase II, and Srb10. *See* glucose effect; G proteins; Groucho; SRB; Tel. (Wu, J., et al. 2001. *Mol. Cell* 7:117.)

Tupaia Family of prosimii. *Tupaia,* $2n = 62$; *Tupaia glis,* $2n = 60$; *Tupaia montan,* $2n = 68$. *See* prosimii.

turbidity Commonly used for measuring total cell numbers in liquid cultures as long as the cell density is not excessive. It is measured as optical density $(OD) = x\ell$, where x = cell density and ℓ = light path in the cuvette of the spectrophotometer. Turbidity is expressed as $\log (I_0/I)$, where I_0 = incident light and I = transmitted light (at, e.g., 550 nm wavelength). The number of viable cells is determined by plating. *See* cell growth; plating efficiency.

turbid plaque *See* plaque.

Turcot syndrome (9p22, 54q21-q22, 3p21.3) Autosomal-recessive malignant tumor of the central nervous system (glioma) that is associated with polyposis. Defects in the mismatch repair system predispose for this tumor. *See* Gardner syndrome; mismatch repair; PMS1; polyposis, adenomatous.

turgid Expanded because of water uptake.

turgor Intracellular pressure caused by water absorption.

turkey *Meleagris gallopavo,* $2n \approx 80$.

Turner syndrome Based on an X0 chromosomal constitution in some female mammals. Its incidence in humans is ≈ 0.0003, but the frequency in abortuses may be 0.01 to 0.02. The missing short arm is critical for Turner symptoms. Turner females usually have short stature, webbed skin on a broad neck, underdeveloped genitalia and sterility, heart problems, proneness to kidney disease, diabetes, and hypertension, but generally they are of normal or near-normal intelligence. Retention of the paternal X usually results in better cognitive abilities (imprinting). In some instances, they are also fertile; most of these cases are probably X0/XX mosaics. Among 410 Danish women with Turner syndrome only the 45,X/46,XX or 46,XX with structural abnormality of one of the X-chromosomes gave birth to children after spontaneous pregnancy. After egg donation some Turner patients can deliver children although their may be a higher chance for chromosomal anomalies among the offspring (Birkabaek, N., et al. 2002. *Clinical Genet.* 61:35). The symptoms are similar even when only one of the short arms of one of the X-chromosomes is missing. The single X-chromosome is maternal in 75% of the cases. The underlying mechanism is puzzling because males normally have a single X, and even in the females one of the X-chromosomes is inactive, except during oogenesis and the first few weeks of embryogenesis. The majority of the studied female animals with single X (pigs, cattle, horses) are abnormal, but XO female mice appear rather normal and fertile. In *Drosophila,* the XO individuals are sterile males. In Turner syndrome, the most critical is the loss of the distal (pseudoautosomal) part of the short arm of the X-chromosome. Some human developmental difficulties may be normalized by estrogen treatment. Artificial insemination or intrauterine implantation may permit reproduction. About 15% of the X-chromosomes in Turner patients are dicentric. *See* ART; dicentric chromosome; human chromosome map; imprinting; MSAFP; Noonan syndrome; prenatal diagnosis; pseudoautosomal; sex-chromosomal anomalies in humans; sex determination; trisomy. (Zinn, A. R., et al. 1993. *Trends Genet.* 9:90.)

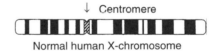

↓ Centromere

Normal human X-chromosome

turnip (*Brassica campestris*) Polymorphic species with chromosome number $2n = 2x = 20$; *Brassica napus* is $2n = 4x = 38$. *See* Brassica oleracea.

turnip crinkle virus (TCV) Plant RNA virus related to the tomato bushy stunt virus.

turnover Depletion — repletion cycle of molecules.

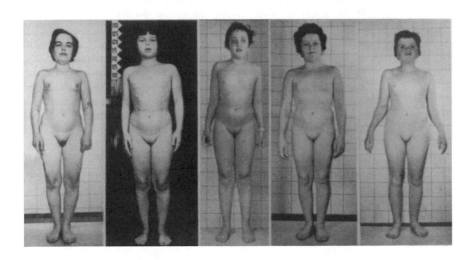

Turner syndrome females of different ages with three doses of the long arm but only one short arm of the X chromosome. Despite the variations in appearance, the similarities of the unrelated individuals are obvious. (Courtesy of Lindsten, J., et al. 1963. *Ann. Hum. Genet.* 26:383.)

turnover number Number of times an enzyme acts on a molecule in a unit of time at saturation.

turnover rate Pace of decay and replacement of a molecule.

TUSC Trait Utility System for Corn generates insertional mutations in maize with the aid of transposable elements like *Mutator*. *See* insertional mutation; *Mu*.

***Tus* gene product** Protein that senses *Ter* (transcription termination) sequence signals in DNA replication in, *E. coli. See* DNA replication; prokaryotes; replication, bidirectional; RTP. (Henderson, T. A., et al. 2001. *Mol. Genet. Genomics* 265:941.)

***t* value** (*t* test) Ratio of the observed deviation to its estimate of standard error: $t = \dfrac{d}{V}$. It is used as a statistical device to estimate the probability of difference between two means. In its most commonly used form, t is calculated as $t = \overline{x}_1 - \overline{x}_2 / \sqrt{[s_{\overline{x}_1}]^2 + [s_{\overline{x}_2}]^2}$, where $\overline{x}$ indicates the mean of the two sets of data and s stands for the standard deviation. *See* paired *t* test; standard deviation; Student's *t* distribution.

TV/RCAS vector The Rous sarcoma avian virus proviral vector (maximum carrying capacity 2.5 kb) first mediates in vitro the production of the avian leukosis virus (AVL) coat protein. The protein is required for the recognition of the avian retroviral receptor (tv-a), which has been transfected into mouse cells. The replication-competent virus thus can produce high titer in the infected mouse cells and can be injected into mice. *See* retroviruses; Rous sarcoma. (Fisher, G., et al. 1999. *Oncogene* 18:5253.)

tweak (Apo3-L, 17p38) TNF ligand with similarity to CD120 and CD95. The related motif of its mouse homologue displays greater similarity. *See* CD95; CD120; TNF. (Kaplan, M. J., et al. 2000. *J. Immunol.* 164:2997; Schneider, P., et al. 1999. *Eur. J. Immunol.* 29:1785.)

TWEEN 20 (polyoxyethylene sorbitan, monolaurate) Aninonic biological detergent with about 50% lauric acid; the rest is myristic, palmitic, and stearic acid.

TWIGDAM Technical Working Group on DNA Analysis Methods. It is concerned with forensic application of DNA analytical and statistical techniques *See* DNA fingerprinting.

twine *Drosophila* homologue of *Cdc25* meiosis-specific gene. *See* azoospermia; cdc25.

twin hybrids *See* complex heterozygotes.

twin meiosis May occur in diploid *Schizosaccharomyces* when after copulation of two nuclei undergo separate meioses.

twinning Phenomenon of developing two (or multiple) zygotes from a single impregnation in usually uniparous mammals. The frequency of twins is different in various ethnic groups. Among Nigerians it may be as high as 4.5% and in some South American and Far East populations it may be as low as 0.8%. In the U.S., its frequency among whites was about 0.89 to 0.94% and among blacks 1.37% before the widespread use of fertility drugs. After in vitro fertilization, about half of the births were multiple. Between 1980 and 1997, twinning increased by 52% and the frequency of triplets and higher-order gestations has quadrupled in the U.S.

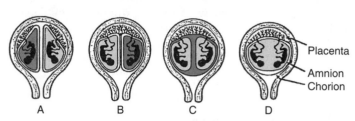

Human twins can be either Mono- or Dizygotic and the two conditions can be distinguished on the anatomical basis of the developing fetus. Monozygotic twins are surrounded by a common chorion (C and D) or even by a single amnion (D). (A) and (B) are most likely dizygotic. (modified after Stern, C. *Principles of Human Genetics*. Freeman. San Francisco.)

Multiple birth involves considerable medical risk to the babies and mother. The gestation time is frequently reduced by 4 weeks for twins, by 6 weeks for triplets, and by 10 weeks for quadruplets. Mortality rate (~16%) and developmental anomalies are higher. The lower birth weight involves a risk for physical and mental handicaps. Multifetal reduction during 9–12 weeks of gestation by injection of potassium chloride into one or more of the fetuses may alleviate the medical problems but may cause loss of the pregnancy entirely. The procedure involves serious ethical dilemmas, religious conflicts, psychological and other harm (Elster, N. J. D. 2000. *Fertility & Sterility* 74:617.)

Twins are either *monozygotic* (identical) or *dizygotic* (fraternal). The former ones are derived from a single fertilized egg and the latter ones develop from two separate eggs fertilized by different sperms. Monozygotic twins in the overwhelming majority of cases are genetically identical, whereas dizygotic twins are comparable to other siblings. It is possible that some dizygotic twins display higher similarity if one unfertilized egg gave rise to two blastomeres fertilized by separate sperms. Another possibility is that one of the polar bodies (identical to the egg) becomes an egg due to a developmental mishap. The frequency of twinning has a genetically determined component as the studies of various ethnic groups indicate. There are indications that the genetic component in dizygotic twin birth is higher than in monozygotic ones.

Monozygotic twins are expected to be of the same sex. Exceptionally, one of the XY male twin embryos may lose the Y chromosome and develop into an XO (possibly mosaic) female. The XO human cells develop into Turner syndrome females, unlike in *Drosophila* where XO individuals are sterile males.

Nonconcordance for sex in identical twins may be the result of mutation in autosomal or X-chromosomal sex-reversal genes. Sometimes it is difficult to distinguish between identical and nonidentical twins on the basis of phenotypic similarities. DNA fingerprinting may resolve the problem. Dermatoglyphics may not always be conclusive due to developmental differences in digital, palmar, or plantar (sole) ridge counts in monozygotic twins. Mono- and dizygotic twins provide useful tools for the study of inheritance of polygenically determined human traits. Among women of families with dizygotic twins, the rate of twinning is about double that of the general population, and it may be attributed to hereditary hormone levels. The inheritance of twinning through male parents is lower than through females. Monozygotic twinning may not have a genetic component. There are statistical methods for the discrimination between identical (MZ) and nonidentical (DZ) twins based on concordance of alleles (DNA or any other type). Maynard Smith and Penrose (*Ann. Human Genet.* 19:273) worked out the following formula for the probability of concordance for DZ twins:

$$P = \left(1 + 2\sum_{i=1}^{n} p_i^2 + 2\left[\sum_{i=1}^{n} p_i^2\right]^2 - \sum_{i=1}^{n} p_i^4\right)/4$$

where i stands for the phenotype of the markers, n = number of alleles, p_i = allelic frequencies calculated on the basis of the binomial distribution for the various types of matings. *See* concordance; cotwin; discordance; DNA fingerprinting; fingerprints; forensic genetics; freemartin; heritability estimation in humans; multiparous; multipaternate litter; ovulation; quadruplex; sex reversal; SRY; superfetation; zygosis. (Jones, H. & Schnorr, J. A. 2001. *Fertility & Sterility* 75:11; Boomsma, D., et al. 2002. *Nature Rev. Genet.* 3:972.)

twins   *See* twinning.

twin spots   Visible if (mitotic) somatic crossing over takes place between appropriately marked chromosomes or twin spots may be caused by nondisjunction. *See* mitotic crossing over.

Wheat leaf displaying white and dark green twin sectors on pale green background in a heterozygote for the hemizygous ineffective Neatby's virescent gene. *See* hemizygous, ineffective.

twintron   *See* intron group II.

TWIST (*Drosophila* gene *twi*, 2–100)   The lethal embryo is twisted in the egg case as the germ cell layers are defective. Mutation in the human homologue is responsible for Saethre-Chotzen syndrome. *See* Chotzen syndrome.

twisting number   Characterizes DNA supercoiling by indicating the number of contortions (writhing) and the number of twists, i.e., the number of nucleotides divided by the number of nucleotides per pitch. *See* DNA; supercoiling.

twitchin   Myosin-activated protein kinase.

two-component regulatory systems   In bacteria, pairs of proteins transduce environmental signals. In one of the proteins, ca. 250 amino acids at the C-terminus and ca. 120 amino acids at the N-terminus are conserved. They control chemotaxis, virulence, nitrogen assimilation, dicarboxylic acid transport, sporulation, etc. They function by autophosphorylation of a histidine residue by the γ-phosphate of ATP. The phosphate is then transferred to an aspartate in the *response regulator,* which modifies regulatory activity of the C-terminal output domain. The system is also called a phosphorelay. A specific phosphatase may reset the system. Frequently, more than two components are involved. *See* signal transduction. (Itou, H. & Tanaka, I. 2001. *J. Biochem.* 129:343.)

two-dimensional gel electrophoresis   The (protein) mixture is first separated by isoelectric focusing, then by size, using a slab of SDS polyacrylamide gel. Thus, all proteins, except those rare molecules that have identical charge and molecular size, are distinguished. For the detection of small quantities of the molecules, they are labeled radioactively or by some nonradioactive means. A single two-dimensional gel slab permits the separation of hundreds or thousands of proteins at a time. For the identification of proteins of very low abundance affinity, purification may be required. The technique is used extensively in proteomic analyses. The definitive identification of proteins is often hindered by the fact that the concentration of individual proteins may vary 5–7 orders of magnitude. *See* electrophoresis; FTICR; isoelectric focusing; proteome; microchannel plate. (Hoving, S., et al. 2002. *Proteomics* 2:127; Ros, A., et al. 2002. *Proteomics* 2:151; Gromov, P. S., et al. 2002. *Progr. Biophys. Mol. Biol.* 80:3; <http://www.expasy.ch/ch2d/>, <http://proteomics.cancer.dk/>.)

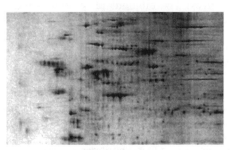

Two-dimensional gel electrophoresis of proteins. (Courtesy of ESA Inc. Chelmsford, MA 01824–4771.)

two-hit hypothesis Assumed that in order to develop cancer two genetic alterations must take place in succession. Some chromosomal aberrations have been called two-hit causes because two breaks are necessary to bring about an inversion or translocation. *See* kinetics; Knudson two-mutation theory.

two-hybrid method Genetic constructs of yeast facilitate the study of protein-protein interactions. The GAL4 protein is both an enzyme and an inducer. The native GAL4 protein contains an N-terminal UAS (upstream activator sequence), DNA-binding region, and carboxyl-terminal transcription-activating region. These regions—in close vicinity—are required for activation. Thus, fusing the N-terminal of a protein (bait) and the C-terminal of another protein (prey) enables study of the interaction between two proteins. If the two proteins interact, they reconstitute the link between the binding and activating domains, and transcription (expression) of the reporter gene (⚘) may proceed. Thus, to see expression, the DNA-binding domain (DBD) must bind to the UAS element, and contact is established with the other protein element, often called *prey*, which is attached to a transcriptional activator. The DBD + bait (hybrid I) and the prey + activator (hybrid II) separately are inactive.

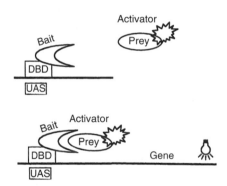

The expression of the downstream reporter gene requires the interaction between the *two-hybrid* proteins. The most commonly used binding component is derived from the Gal4 or LexA proteins, and as a reporter, bacterial LacZ or luciferase is employed. The receptor domain, usually called bait, may be on a plasmid that may also carry sequences to promote dimerization and thus the required protein interaction. The Gal bait (in contrast to the majority of LexAs) usually contains nuclear localization sequences. The efficient prey vectors may contain the VP16 activation domain or the Gal4 region II. The B42 bacterial activator is weaker than the other two, but it has affinity for a wider range of proteins and suppresses (squelches) the toxic effects that Gal4 and strong transcriptional activators may have on yeast. The two-hybrid method in yeast is very simple to use. Strains with the bait can be mixed and mated with strains with the prey, then plated on selective media where the interactions are readily detectable. The advantage of the two-hybrid method is that it can be used for testing protein interactions, determining the amino acid sequences critical for interactions, and screening gene libraries for binding proteins or activators.

The system is suitable for testing any molecule (including aptamers) that may affect the interactions, e.g., in the development of specific drugs. The system can be applied for studies of the cell cycle and transcription factors, tumorigenesis, tumor suppression, etc. In some instances, positive (i.e., nonrelevant for the purpose of the study) activation of the reporter gene may occur. This interference can be reduced or eliminated by selective systems. False negative interactions are observed when the protein-protein interactions are low, there are problems with intracellular folding of the proteins, or some other domains of larger proteins hinder the proper interactions. The cases of false negatives may be low yet may be very important in pharmacological studies. Some proteins may require a third (protein or nonprotein) element for stabilization, bridging, or modification, and then only a ternary complex is active. Systems have been constructed that prevent interactions between proteins. The URA3 reporter gene product converts fluoroorotic acid—with the assistance of other genes—into fluorodeoxyuridine monophosphate, an analog of thymidylate synthase that inhibits DNA synthesis and can be used for contraselection. The two-hybrid method applied sequentially or simultaneously to large gene pools may detect interacting systems of genes. It may also place functionally unknown open-reading frames into a biological context of a metabolic role. *Two-bait* systems have been developed for the possible detection of (allelic) variants of the same protein. Mammalian two-hybrid procedures may facilitate the detection of interacting proteins in mammalian systems that depend on posttranslational modifications not available in yeast cells. Protein-RNA interactions are studied by the *three-hybrid* systems. The *reverse two-hybrid* system detects mutations that are unable to bring the activation element to the DNA-binding domain and thus cannot convert a potentially toxic compound into a toxic one or into a suicide inhibitor. *See* aptamer; epistasis; four-hybrid system; galactose utilization; interaction trap; microarray hybridization; one-hybrid binding assay; pleiotropy; recruitment; reverse two-hybrid system; split-hybrid system; suicide inhibitor; suppressor second site; three-hybrid system; VP16. (Brent, R., et al. 1997. *Annu. Rev. Genet.* 31:663; Fields, S. & Song, O. 1989. *Nature* 340:245; Shioda, T., et al. 2000. *Proc. Natl. Acad. Sci. USA* 97:5220; Hirst, M., et al. 2001. *Proc. Natl. Acad. Sci. USA* 98:8726; Fang, Y., et al. 2002. *Mol. Genet. Genomics.* 267:142; <http://dip.doe-mbi.ucla.edu/>.)

two-phase mutatation model (TPM) The microsatellite instability may be gain or loss of a number of repeats. *See* IAM; microsatellite; SMM.

two-point cross Involves differences at two gene loci, e.g., *AB* × *ab* or *Ab* × *aB*.

two-step synapsis *See* topological filter.

Ty Nonviral retroelements (retrotransposons) in the nucleus of budding yeast (*Saccharomyces cerevisiae*). Ty elements occur in several related forms and are designated Ty1 (25–30 copies), Ty2 (10 copies), Ty3 (2–4 copies), etc. The elements are flanked by long terminal direct repeats (251–371 bp) that are designated as δ for Ty1 and Ty2, σ for Ty3. The open-reading frame between the repeats (LTR) in Ty1 and Ty2 is identified as ε. Elements that resemble these terminal repeats, δ (ca. 100 copies, σ (20–30 copies), and τ (15–25 copies), are also found. The terminal repeats contain sequences identified as U3 (unique for 3′), R (repeated), and U5 (unique for 5′-end) similarly to the designations in retroviruses. These LTRs contain the upstream gene activation sequence, TATA box, polyadenylation, transcription termination signals, and TG...CA bases involved in the integration of the reverse transcripts into the chromosomes. All the retroelements are generally flanked in the host by 5 bp target duplications. The open-reading frames, distinguished as TYA and TYB between the ends, are similar among them and resemble retroposons of other organisms. The TYA protein may be processed through proteolysis into several smaller proteins involved in the formation of the shell of the VLP (virus-like particle). TYB contains genes for protease (*pro*), integrase (*int*), reverse transcriptase (*rt*), and RNase H (*rnh*) but no *env* gene is present as would be expected for retroviruses. Although most of the Ty elements and the independently standing termini are intact, some contain deletions up to a few kb, and the LTR sequences may be truncated. There may be apparent insertions, duplications within the Ty elements, or inversion(s) involving parts of an LTR. The heterogeneity of the coding regions is due to base substitutions.

Ty retrotransposons transpose by synthesizing an RNA that is reverse-transcribed into DNA for integration. The Ty1 transcripts represent about 0.8% of the total RNA in the cell, but Ty cDNA is present in less than one copy per haploid cell before integration because of the inefficient processing of the transcript. The rate of Ty1 transposition is about 10^{-5} to 10^{-7} per element per cell division. Fus3 protein kinase, the transcription factor TFIIH, and nucleotide exchange repair modulate transposition of Ty. The reverse-transcribed DNAs can integrate into Ty elements with the aid of recombination factors of the host and can form tandem elements. RNA polymerase II transcribes them and the transcripts are polyadenylated. Transcription may be prevented by mutations symbolized as *spt* (suppressor of Ty). The transcription of Ty elements may be induced by sex pheromones synthesized by the *MATa* or *MATα* genes but not in the *MATa/Matα* diploids. *MAT* homozygotes do not affect transcription of Ty RNA. Ty RNA can be packaged into virus-like particles (VLP). Proteolysis is involved in the processing of TYA and TYB products required for the completion of VLPs. Reverse transcription appears to be primed by tRNA^{Met}. Degradation of the template RNA is due to RNase H. During reverse transcription, recombination, sequence modifications, and deletions may occur at high frequency. Transcription of Ty is regulated by several *PST* (suppressor of transposition), *ROC* (reducer of overproduction of transcripts), and *TYE* (Ty enhancer) genes. Transcription may be increased by exposure to UV light, ionizing radiation, chemical mutagens, and culture media and is elevated 20-fold at low temperature. The RAD52 group of recombinational repair genes inhibits transposition

of Ty. Insertion into and activation of these particular genes measures the frequency of transposition by genetically tagged Ty elements.

VLP is made in the cytosol, but it must enter the nucleus for transposition to take place. Insertion of Ty into structural genes eliminates their function, and reversion by excision is very rare because the target site is modified by the event. Insertion within noncoding sequences may activate or silence previously inserted elements. Ty1 and Ty2 elements appear to be distributed at random throughout the genome, but the family of Ty3 elements is much more restricted. The Ty3 element inserts at the transcription initiation sites of tRNA genes by pol III. The transcription factors TFIIIB and TFIIIC are required for this type of insertion. The Ty1 element has preference for integration into the 5′-sequences of pol II–transcribed genes. Ty, δ, σ, and τ are commonly targeted to the vicinity of tRNA genes (hot spots). The insertions are usually directed into the leader sequences (promoter sites) or near the 5′-end of the coding region. Among the abundant Ty elements, recombination may occur and cause chromosomal rearrangements and deletions. Recombination is most common between LTR (δ) repeats. Gene conversion may also occur between these elements.

Ty elements may incite deletions, inversions, and translocations with all the phenotypic consequences. The mutation rate within Ty elements was estimated to be 0.15 per Ty per replication cycle, or about 1/15 of the retrotranspositions result in some type of mutation. This rate of mutation is approximately 4 to 6 orders of magnitude higher than that of the rest in the nuclear genome but comparable to that in RNA viruses. This high mutation rate is attributable to the lack of proofreading ability of reverse transcriptase. The mutations in Ty (about 30% of all) occur within the seven bases of the primer-binding site (PBS) where the minus-strand replication is primed by a tRNA. Imprecise cutting by RNase H causes some of the mutations. Another error-prone event is the addition of nucleotides immediately adjacent to the tRNA primer-binding site. After the replication system reaches the 5′-end, nucleotide additions take place at the 3′-end. After the second-strand transfer, partial DNA·DNA duplexes are formed and the recessed 3′-ends prime the completion of double-strand synthesis. As a consequence of the addition of nucleotides, the 3′-end of the minus strand cannot anneal precisely with the plus-strand DNA template. In order to reconstitute the 5′-LTR of Ty, the mispaired 3′-primer ends are extended by reverse transcription. This process incorporates the mispairs into the Ty element that will integrate into the yeast chromosome. The mismatches must then be fixed by DNA repair.

Additional mutational mechanisms may occur. These mutational processes are not unique features of Ty elements, but other retroelements use them as well. Retroviruses are somewhat different, however, because they show frameshift mutations and complex rearrangements. Ty elements have been utilized as vectors (pGTy), as insertional mutagenic agents (transposon tagging), and for fusion of TyA proteins with certain proteins of interest to facilitate the purification of epitopes. Similar transposable elements occur in other fungi (*Candida*, *Pichia*, *Hansenula*). *See* DNA repair; epitope; fus3; integrase; mating type determination in yeast; pol

I; pol II; protease; retroposon; retrotransposon; reverse transcription; *Saccharomyces cerevisiae*; strong-stop DNA; TFIIH; transposition induction; VLP. (Boeke, J. D., et al. 1988. *Mol. Cell Biol.* 8:1432; Jordan, I. K. & McDonald, J. F. 1999. *Genetics* 151:1341; Wickner, R. B. 2001. *Fundamental Virology*, Knipe, D. M. & Howley, P. M., eds., p. 473; Lippincott Williams & Wilkins, Philadelphia, PA; Umezu, K., et al. 2002. *Genetics* 160:97.)

Tyk2   Nonreceptor tyrosine kinase. *See* Janus kinases.

tylosis   Formation of animal callus. *See* keratosis.

type III secretion system   *See* secretion systems.

typing   Determination of the blood group antigens and HLA. It is general classification by type. Also, DNA typing by fingerprinting using restriction enzyme–generated fragment pattern. *See* blood groups; DNA fingerprinting; HLA.

tyrosinase   *See* albinism.

tyrosine   Nonessential aromatic amino acid (MW 181.19) soluble in dilute alkali. Its biosynthetic pathway is *with enzymes involved in parentheses*): Chorismate → (*chorismate mutase*) → Prephenate → (*prephenate dehydrogenase*) → 4-Hydroxyphenyl pyruvate → (*amino transferase* with glutamate NH_3 donor) → Tyrosine. Tyrosine can also come from phenylalanine by dehydroxylation. Tyrosine is a precursor for norepinephrine, epinephrine, 3,4-dihydroxy-phenylalanine (dopa), dopamine, and catechol, which form the catecholamine family of animal hormones. *See* alkaptonuria; animal hormones; chorismate; goiter; phenylalanine; pigmentation of animals; tyrosine aminotransferase; tyrosine kinase; tyrosinemia.

$$HO—\bigcirc—CH_2---\underset{\underset{H}{|}}{\overset{\overset{NH_2}{|}}{C}}---COOH$$

tyrosine aminotransferase (TAT, Richner-Hanhart syndrome)   Converts tyrosine into p-hydroxyphenylpyruvate. Glucocorticoid hormones induce it. TAT deficiency occurs rarely in humans and it is controlled by a recessive gene (ca. 11 kb, 12 exons) in human chromosome 16q22.1-q22.3 (and mouse in chromosome 8). The condition involves elevated levels of tyrosine and in some cases an increased urinary excretion of p-hydroxyphenylpyruvate and hydroxyphenylacetate. The disease generally involves corneal ulcer, palm keratosis (callous skin), mental and physical retardation. The mitochondrial enzyme is under the control of another gene. A TAT regulator gene may be in the human X-chromosome and glutamic oxaloacetic transaminase (16q21) also regulates its activity. *See* tyrosine; tyrosinemia.

tyrosine hydroxylase (TYH)   Encoded in human chromosome 11p15 (mouse chromosome 7); controls the synthesis of dopamine from phenylalanine. Dopamine is a hormone involved with adrenergic neurons, the sympathetic nerve fibers that liberate norepinephrine when an impulse passes the nerve synapse. This enzyme may play a key role in fetal development and in manic depression. The microsatellite sequence in the first intron serves normally as a transcriptional enhancer of the gene. *See* DOPA; manic depression; tyrosine. (Albanèse, V., et al. 2001. *Hum. Mol. Genet.* 10:1785.)

tyrosine kinase (protein tyrosine kinase)   Activity is essential for many processes of signal transduction, tissue differentiation involving oncogenesis, and signaling to immunoreceptors. The most frequently activated proteins are phospholipase C-γ, phosphatidylinositol 3-kinase, and GTPase-activating kinase. *See* ABL; ARG; Blk; Btk; EGFR; ERBB1; FES; FGF; FGR; FHS; FLT; Fyn; hepatocyte growth factor; insulin (receptor β-chain); KIT; LCA; Lyn; MCSF; MET; morphogenesis in *Drosophila (tor)*; neutrotrophin; oncogenes; PDGF; RAF; receptor tyrosine kinase; RET; *sevenless*; signal transduction; *SOS*; SRC; Steel factor; Syk; T-cell receptor; trk; VEGF; YES; ZAP-70. (Latour, S. & Veillette, A. 2001. *Curr. Opin. Immunol.* 13:299.)

tyrosinemia (FAH)   The recessive gene responsible for the anomaly is in human chromosome 15q23-q25. Its prevalence is about 1/2,000 live births. FAH involves increased levels of tyrosine in the blood and a lack of p-hydroxyphenylpyruvate oxidase, but the primary enzyme defect appears to be fumarylacetoacetase deficiency leading to accumulation of succinylacetone and succinylacetoacetate. Defects in porphyrinogenesis appear secondary. Because of the deterioration of the liver, accumulation of methionine and other amino acids in the blood and urine is frequent. The particular odor of the body fluids may be due to α-keto-γ-methiolbutyric acid. Prenatal diagnosis of FAH relies on amniotic fluid analysis for succinylacetolactone or measurement of fumarylacetoacetase in cultured amniotic cells. Low tyrosine diet alleviates the symptoms, but to avoid liver cancer, liver replacement before age 2 is advisable. This type of tyrosinemia involves various kinds of chromosomal anomalies. Tyrosinemia II (16q22.1-q22.3) is a deficiency of tyrosine aminotransferase. Tyrosinemia type III (12q24-qter) is due to a deficiency of 4-hydroxyphenyl-pyruvate dioxygenase. This disease does not usually involve dysfunction of the liver. The condition can be detected prenatally, but carriers cannot be identified. Mice mutants homozygous for FAH but heterozygous for alkaptonuria were partially normalized in their hepatocytes. The tyrosine degradation pathway: tyrosine → (*tyrosinemia type II*) → 4-OH-phenylpyruvate → (*tyrosinemia type III*) → homogentisic acid → (*alkaptonuria*) → maleylacetolactate → fumarylacetolactate → (*tyrosinemia type I*) ⇒ fumarate and acetolactate. *See* alkaptonuria; amino acid metabolism; gene therapy; genetic screening; liver cancer; methionine adenosyltransferase deficiency; methionine biosynthesis; mosaic; tyrosine aminotransferase. (Jorquera, R. & Tanguay, R. M. 2001. *Hum. Mol. Genet.* 10:1741.)

tyrosine phosphatase (protein tyrosine phosphatase, PTP)   Family of enzymes with at least 40 known members. Some have properties of transmembrane receptor proteins (RPTP) and the cytosolic forms have a characteristic ca. 240-amino-acid catalytic domain. Each PTP also has a phosphate-binding, 11-residue motif containing catalytically active Cys and Arg. The proteins have a critical role in signal transduction relevant to growth, proliferation, and differentiation. Dimerization negatively regulates the activity of receptor tyrosine phosphatase-α. *See* PTK; signal transduction; synaps. (Stetak, A., et al. 2001. *Biochem. Biophys. Res. Commun.* 288:564.)

tyrosine protein kinase *See* phosphorylase; tyrosine kinase.

tyrosine transaminase *See* tyrosine aminotransferase.

tyrosinosis *See* tyrosine aminotransferase.

Ty δ Terminal repeat of the Ty insertion element. *See* Ty.

Aristotle reviews the ancient theories of sex determination, but finds them unsatisfactory:

"Some suppose that the difference [between sexes] exists in the germs from the beginning; for example, Anaxagoras and other naturalists say that the sperm comes from the male and that the female provides the place [for the embryo], and that the male comes from the right, the female from the left, since in the uterus the males are at the right and the females at the left. According to others, like Empedocles, the differentiation takes place in the mother, because, according to them, the germs penetrating a warm uterus become male, and a cold uterus female…" (*Generation of Animals*, Book IV, Part I, Para. 2.)

A historical vignette.

U

U (1) Uracil. *See* pyrimidines. (2) Uranium.

U-937 Human monocyte line. *See* monocytes.

UAA Ochre codon of translation termination. *See* code, genetic; translation termination.

U2AF Protein assisting U2 snRNA recognition. *See* U2 RNA. (Guth, S., et al. 2001. *Mol. Cell Biol.* 21:7673.)

UAG Amber codon of translation termination. *See* amber suppressor; code, genetic.

UAS Upstream activating sequences (regulate gene transcription). They behave similarly to enhancers. UAS encodes DNA-binding proteins, e.g., GAL4 UAS codes for a 100 kDa protein and protects its 17 bp palindromic sequence against DNase I digestion or methylation. These proteins bind to special DNA sequences and to other proteins in the transcriptional complex. Transcriptional activation and binding may rely on more than one tract of amino acids. *See* galactose utilization; promoter; two-hybrid method. (Blackwood, E. M. & Kadanoga, J. T. 1998. *Science* 281:61.)

uAUG In 3 to 10% of the RNA transcripts, translation initiation codons occur in the 5′-untranslated sequences (5′-UTR) of viruses, fungi, plants, and mammals. The open-reading frame due to uAUG may have an untranslated intercistronic region relative to the downstream ORF or may overlap with it or be in-frame. uAUGs may have regulatory functions. (Rose, J. K. & Iverson, L. 1979. *J. Virol.* 32:404.)

UBC2 (E2) Ubiquitin-conjugating enzymes. From UBC, ubiquitin is transferred to a lysine residue of the target protein. Ubc2/Rad6-Rad18 proteins have functions in both proteolyis and DNA repair. Ubc2 proteins occur in many isoforms within the same organism. *See* E2; isoform; Rad 18; ubiquitin. (Ptak, C., et al. 2001. *Mol. Cell Biol.* 21:6537.)

Ubc9 Ubiquitin interacting protein. It is involved in the control of the cell cycle from G2 → M and in the degradation of diverse proteins in a variety of organisms. *See* ubiquitin; UBL. (Kaul, S., et al. 2002. *J. Biol. Chem.* 277:12541.)

UBC3/Cdc34 Ubiquitination enzyme required for G1 → S phase transition in the cell cycle. *See* Cdc34; ubiquitin.

UbcD1 Ubiquitin protein of *Drosophila* degrades some telomere-associated proteins and thus controls the proper detachment and attachment of the telomeres during mitosis and meiosis. *See* telomere; ubiquitin. (Bocca, S. N., et al. 2001. *Biochem. Biophys. Res. Commun.* 286:357.)

UBE3A Ubiquitin ligase gene with a role in Angelman syndrome. *See* Angelman syndrome; ubiquitin. (Kishino, T., et al. 1997. *Nature Genet.* 15:70.)

Überoperon Includes gene neighborhoods where individual operons may show rearrangement in different species yet they remain in functional and regulatory context (Lathe, W. C., 3rd et al. 2000. *Trends Biochem. Sci.* 25:474). Rogozin, I. B., et al. 2002. (*Nucleic Acids Res.* 30:2212) called this phenomenon genome hitchhiking. The largest such neighborhood in prokaryotes included 79 genes. Most, albeit not all of these genes, share known functional role. *See* operon.

UBF Upstream binding factor. In association with protein SL-1, UBF controls the transcription of rRNA genes by RNA polymerase I. These protein factors differ even among closely related species; they are members of the high-mobility group proteins. *See* high-mobility group proteins; SL1; transcription factors. (Santoro, R. & Grummt, I. 2001. *Mol. Cell* 8:719; Chen, D. & Huang, S. 2001. *J. Cell Biol.* 153:169; Stefanovsky, V. Y., et al. 2001. *Mol. Cell* 8:1063.)

ubiquilin Protein, that stimulates the biosynthesis of neurofibrillary tangles in Alzheimer disease and Lewy bodies in Parkinson disease. *See* Alzheimer disease; Parkinson disease; tau. (Mah, A. L., et al. 2000. *J. Cell Biol.* 151:847.)

ubiquinones (coenzyme Q) Lipid-soluble benzoquinones mediating electron transport; their reduced form are potent lipid-soluble antioxidants capable of inhibition of lipid peroxidation. *See* cytochromes; lipids; peroxides. (Elias, M., et al. 2001. *J. Biol. Chem.* 276:48356.)

ubiquitin Acidic polypeptide (~76 amino acids) ubiquitously present in prokaryotic and eukaryotic cells. It may be associated with H2A histone, and the conjugate is called UH2A. Its function involves binding to other proteins and proteolytic enzymes, then degrading the ubiquitinylated proteins. In the majority of cases, the degradation takes place in the cavity of the (20S–26S) *proteasomes*, cylindrical complexes of proteases, or the ligand-bound molecule can be internalized into vacuoles and degraded by lysosomal enzymes. Ubiquitination is involved in many aspects of cellular regulation, DNA repair, stress response, cell cycle progression, the formation of the synaptonemal complex, signal transduction, and apoptosis. A cascade of conjugating, activating, and carrier enzymes recognizes ubiquitin. The E1 enzyme forms a thiol ester with ubiquitin using ATP energy. The E1-ubiquitin complex then transfers UB to the carrier E2. With the aid of E3 ligases, it mediates the association of E2 with the target proteins. E3 ligases use for catalysis either a HECT or RING-finger domain. Ubiquitin E3 ligase complexes may be SCF (Skp1-Cdc53/CUL1-F-box proteins), APC (anaphase-promoting complex), and VCB-like family (VHL-elongin C/elongin B) proteins. Several other molecules are also involved. Ubiquitin is degraded by deubiquitinating cysteine protease enzymes (DUBP). The signals for degradation are PEST sequence at the C-ends, N-end rule domains, and phosphorylation. The human ubiquitin genes are encoded at chromosome 17p11.1-p12 and polyubiquitin at 12q24.3. Some protein molecules may be degraded cotranslationally in case they are defective or if their folding is slow. Ubiquitin may have nondestructive functions by processing signaling molecules in the cell and facilitating the expression of genes (Finley, D. 2001. *Nature* 412:283). *See* antigen processing and presentation; APC; CDC34; Cullins; destruction box; E1; E2; E3; F-box; histones; IκB; lactacystin; lid; lysosomes; monoubiquitin; N-end rule; PEST; PIC; proteasomes; proteolytic; Rbx1; SCF; Skp1; sentrin; Socs-box; SUMO; UBP; Ubc; Ubl; VHL. (Hershko, A. & Ciechanover, A. 1998. *Annu. Rev. Biochem.* 67:425; Pickart, C. M. 2001. *Annu. Rev. Biochem.* 70:503; Weismann, A. M. 2001. *Nature Rev. Mol. Cell Biol.* 2:169; Conaway, R. C., et al. 2002. *Science* 296:1254.)

ubiquitous RNA   *See* URNA.

UBL   Ubiquitin (Ubc9)-interacting protein operating on many different proteins under various names such as sentrin, SUMO-1, etc. *See* E2; SUMO; Ubc; ubiquitin.

UBP   Group of (16 in budding yeast) ubiquitin-degrading enzymes or deubiquitinating enzymes or isopeptidases. UBPs may regulate gene silencing, differentiation and cell division (cyclins, Cdks, MPF, etc.) *See* CDK; cell cycle; cyclins; morphogenesis; MPF; silencer; ubiquitin. (Lin, H., et al. 2000. *Mol. Cell Biol.* 20:6568.)

UBR   An enzyme forms thioester with ubiquitin and UBC. *See* UBC. (Kwon, Y. T., et al. 1998. *Proc. Natl. Acad. Sci. USA* 95:7898.)

UCE   Upstream control elements regulate transcription. *See* regulation of gene activity.

UDG   Uracil-DNA-glycosylase is a repair enzyme capable of removing accidentally incorporated U or deaminated C, and the step is followed by removal of the apyrimidinic site with the aid of AP endonuclease. There are two nuclear-coded UDG enzymes in the mammalian cells; one of them is in the mitochondria. *See* AP endonuclease; DNA repair; excision repair.

UDP   Uridine diphosphate formed from uridine monophosphate and giving rise to uridine triphosphate by using ATP as the phosphate donor. *See* formula below.

UEP   (1) *See* unit evolutionary period. (2) Unique event polymorphism includes SNIPS, INDELs, and microsatellite sites that facilitate characterization of haplotypes.

UGA   Opal stop codon. *See* code, genetic.

UDP-5′-diphosphate Na.

UH2A   *See* ubiquitin.

UhpB   *E. coli* kinase involved in the regulation of sugar phosphate transport.

uidA   Gene for β-glucuronidase. *See* β-glucuronidase.

ulcer   Corruption of the surface or deeper cell layers of the body, e.g., gastric ulcer, an ulceration inside the stomach or diabetic ulcers of the legs, or genital ulcers in case of sexually transmitted disease, etc. *See* Crohn disease.

Ullrich-Turner syndrome   Turner syndrome.

ulna   Larger bone of the forearm opposite the thumb.

ultimate mutagen   Some chemical substances (promutagens) become mutagenic only after activation. In this process, proximal mutagens are formed that subsequently may be converted to a form (ultimate mutagen) that is genetically most reactive with the DNA. *See* activation of mutagens.

Ultrabar   *See Bar* mutation.

ultracentrifuge   Laboratory equipment suitable for the separation of cellular organelles and macromolecules by high (over 20,000 revolutions/minute) centrifugal force in an evacuated and refrigerated chamber. At maximal speed, it may exceed several hundred thousand times the gravitational force. The *analytical ultracentrifuge* monitors continuously or intermittently the boundary of movement of macromolecules in a solute (e.g., Cs_2SO_4). The more commonly used *preparative ultracentrifuge* fractionates organelles or macromolecules in sucrose or CsCl. The suspended material forms a band corresponding to its density and the density of the solute. Chloroplasts in sucrose gradients can be identified by color; DNA bands in CsCl can be identified by staining with the dye ethidium bromide. The bands, containing homogeneous material, can be removed by careful suction from the centrifuge tubes or by dropwise collection through a hole punctured at the bottom with a syringe. (Laue, T. M. & Stafford, W. F., III. 1999. *Annu. Rev. Biophys. Biomol. Struct.* 28:75.)

ultradian   Occurring periodically but less frequently than once a day. *See* circadian.

ultrasonic (ultrasound)   Radiation in excess of 2×10^4 hertz/second, generally 5×10^5 hertz/sec. It is used for breaking down cells, treatment of arthritis, and tomographic examination or sonography, etc. *See* arthritis; sonography; tomography.

ultrastructure   Fine structure beyond the resolution of the light microscope.

ultrathin-layer gel electrophoresis   Along with capillary microelectrophoresis, it is well suited for rapid automated separation of biopolymers, e.g., nucleic acids as needed for the genome sequencing projects. (Guttman, A. & Ronai, Z. 2000. *Electrophoresis* 18:3952.)

ultraviolet light (UV)   Emission below the wavelength of violet (400–424 nm). UV-A has emission 315–400 nm, UV-B 280–315 nm, and UV-C 200–290 nm wavelength. Nucleic acids have maximal absorption at about 260 nm. The absorption maximum may depend on base composition and pH. The maximal genetic effects of UV light coincide with the absorption maximum of nucleic acids. The major genetic effect of UV light is the generation of pyrimidine dimers and pyrimidinones and reactive peroxides. These compounds may damage the DNA and interfere with replication and transcription. They can cause mutation and induce cancer. UV-B has immunosuppressive potential, but UV-A is inert in this respect; it may suppress UV-B effects. This protective effect is due to the induction of skin heme oxygenase. Oxidation of the membrane proteins may affect the signal transduction pathways and activate genes of the cellular defense systems. In prokaryotes, the LexA repressor is inactivated, leading to the upregulation of genes that mediate mutation, recombination, and DNA repair. In yeast, cell cycle genes and various kinases are activated. In animal cells, through the RAS signal transduction pathways, AP and NF-κB transcription factors may be activated. In plant cells, UV exposure may increase the synthesis of UV-absorbing flavonoids and phenylpropanoids, activating the octadecanoid defense pathway of fatty acids and generating less than 0.2 cyclobutane dimers per gene. In plants, transposons like *Mu* may enhance the mutagenic effects of UV-B radiation. Ultraviolet light generally has low penetration into biological material and a few cell layers may completely trap it. *See* cyclobutane; DNA repair; electromagnetic radiation; fatty acids; flavonoids; genetic repair; JUN; light-response elements; light sensitivity diseases; *Mu*; phenylpropanoids; physical mutagens; pyrimidine dimer; RAS; signal transduction.

ultraviolet photoproducts *See* cis-syn photoproduct; cyclobutane ring; Dewar product; DNA repair; pyrimidine-pyrimidinone; translesion pathway.

ultraviolet sensitivity syndrome (UV^S S) RNA synthesis is inhibited in the cells by ultraviolet light. It appears different from xeroderma pigmentosum but similar to Cockayne syndrome. Ultraviolet sensitivity is caused by a defect in the RAD genes required for genetic repair. *See* Cockayne syndrome; xeroderma pigmentosum.

ultraviolet spectroscopy Measures the absorption of ($\approx$monochromatic) UV light and thus qualitatively or quantitatively identifies molecules such as DNA, RNA, and others.

UME Unique mutational event, a marker of evolutionary significance.

UMP Uridine monophosphate. *See* UDP.

UMU (UV mutagenesis) Genes *UmuC* and *UmuD* are involved in the repair of ultraviolet light damage to DNA. UmuD′ is a posttranslationally processed active form of the UmuD protein. *See* DNA repair (SOS repair); translesion pathway.

unbalanced chromosomal constitution Parents heterozygous for inversions or translocations are functionally normal but may transmit to their offspring (unbalanced) duplication-deficiency gametes, resulting in various physical and mental disabilities depending on the chromosomal region(s) involved. Unbalanced (duplication/deficiency and duplication and deficiency) gametes occur in more than 2% of the conceptuses. This is, however, the lowest estimate because very frequently pregnancy is terminated before the current methods of analysis can detect fertilization. Even more uncertain is the frequency of chromosomal anomalies in the human sperm and egg, which are extremely difficult to analyze cytologically. An unbalanced karyotype may be due to trisomy (in ~20–30% of spontaneous abortions). Monosomy has been observed in about 5 to 10%, triploidy 6 to 7%, and tetraploidy 2 to 4% of spontaneously aborted fetuses. There is no effective cure for unbalanced chromosomes. *See* inversion; translocation; photographs below. (Levy, B., et al. 1998. *Genet. Med.* 1:4.)

UNC-6 Protein of the netrin family in *Caenorhabditis elegans* guiding ventral migration of the axon growth cone. Similar is the function of the 407-amino-acid TGF-β-like protein encoded by the UNC-129 gene. *See* netrins; semaphorins; TGF. (Merz, D. C., et al. 2001. *Genetics* 158:1071.)

UNC-33 *Caenorhabditis* protein regulating axon extension. *See* axon; CRMP. (Ricard, D., et al. 2001. *J. Neurosci.* 21:7203.)

UNC-43 Encodes a calcium/calmodulin-dependent serine/threonine kinase type II. Its mutation is responsible for multiple behavioral defects in *Caenorhabditis*. (Sagasti, A., et al. 2001. *Cell* 105:221.)

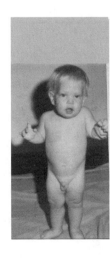 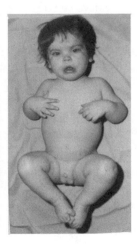 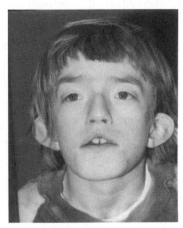

Unbalanced chromosomal constitution inherited from phenotypically normal carrier parents. Left: deficiency for the short arm of chromosome 18. Middle: Duplication of the tip of the short arm of chromosome 3. (Courtesy of Dr. Judith Miles.) Right: duplication of part of the long arm of chromosome 4, resulting from a translocation heterozygosity involving chromosomes 20 and 4. (Courtesy of Dr. D. L. Rimoin.)

unc-86 (uncoordinated) *Caenorhabditis* gene with a pivotal role in determining neural identities. (Burglin, T. R. & Ruvkun, G. 2001. *Development* 128:779.)

uncharged tRNA Has no amino acid attached to it. *See* tRNA.

uncoating During the initiation of infection, the viral genome dissociates from the other constituents of the virus particle.

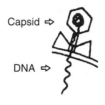

Uncoating.

uncompetitive inhibitor *See* regulation of enzyme activity.

uncoupling agent Uncouples electron transfer from phosphorylation of ADP, e.g., dinitrophenol. *See* cold-shock proteins.

uncoupling protein (mitochondrial UCP) Uncouples oxidative phosphorylation from ATP synthesis and generates heat to safeguard against cold. Noradrenaline and adrenaline regulate it. UCP-3 is a member of the mitochondrial transporter family in the skeletal muscles and controls metabolic rate and glucose homeostasis. UCP-1 is present in the brown adipose tissue and is involved in fat combustion and heat generation. UCPs have an apparent role in obesity, in macrophage-mediated immunity, and in the control of ROS. *See* ROS. (Jarmuszkiewicz, W. 2001. *Acta Biochim. Pol.* 48:145.)

underdominance Inferiority in performance or fertility or proportion of hybrids in feral populations. It may be caused by Robertsonian fusion, transposable element, meiotic drive, postimplantation selection, etc. *See* meiotic drive; overdominance; Robertsonian translocation. (Davis, S., et al. 2001. *J. Theor. Biol.* 212:83.)

underwinding Characterizes negative supercoiling. *See* supercoiling.

unequal crossing over Takes place between repeated sequences in the gene that may pair obliquely and therefore in the recombinant strands may have more or less copies of the sequences than in the parental ones:

The extra DNA material may not be needed for the species and thus can be used for evolutionary experimentation and

may contribute to the evolution of new function(s). The frequency of such events may vary; in *Arabidopsis*, $\sim 3 \times 10^{-6}$ has been observed. This estimate is within the range of mutation frequency and much lower than that found at the *Bar* locus of *Drosophila*, where unequal crossing over was first identified at a variable rate of 0.03 to 0.11. *See Bar* locus; duplication; intragenic recombination. (Bridges, C. B. 1936. *Science* 83:210.)

UniBLAST Provides information on UniGene clusters. *See* BLAST; UniGene.

unidentified reading frame *See* URF.

unidirectional replication The replication fork moves only in one direction, left or right from the origin. *See* bidirectional replication; replication fork. (Maisnier-Patin, S., et al. 2001. *J. Bacteriol.* 183:6065.)

unified genetic map Represents similarities in the distribution of nucleotide sequences across phylogenetic groups and is expected to provide tools for evolutionary studies; may assist in transferring economically advantageous genes to crops. The ancestral chromosomal pattern is shared between many species. Unique sequences characterize species that are more closely related. *See* CATS; comparative maps; evolution; integrated map; mapping, genetic; physical map; SCEUS. (O'Brien, S. J., et al. 1993. *Nature Genet.* 3:103.)

unfolded protein response (UPR) Accumulation of unfolded proteins in the endoplasmic reticulum activates the transcription of molecular chaperones such as BiP, GRP, calreticulin, and protein disulfide isomerase in the nucleus so that protein folding can take place as needed. UPR signaling is mediated by endoplasmic reticulum transmembrane protein, Ire1, which has both kinase and nuclease domains. Various adverse physiological conditions, toxins, inhibitors of the calcium pump, genetic defects altering protein structure, etc., evoke UPR. *See* BiP; calreticulin; endoplasmic reticulum; GRP; PDI; UPR. (Rüegsegger, U., et al. 2001. *Cell* 107:103; Ma, Y. & Hendershot, L. M. 2001. *Cell* 7:827; Calfon, M., et al. 2002. *Nature* 415:92; Kaufman, R. J., et al. 2002. *Nature Rev. Mol. Cell Biol.* 3:401; Harding, M. P., et al. 2002. *Annu. Rev. Cell Dev. Biol.* 18:575.)

UNG Uracil nucleotide DNA glycosylase is a DNA repair enzyme. It removes misincorporated uracil from DNA. *See* DNA repair; glycosylases. (Krokan, H. E., et al. 2001. *Progr. Nucleic Acid Res. Mol. Biol.* 68:365.)

uniformity principle *See* Mendelian laws.

UniGene Database of human ESTs and sequences of known genes to assist in identifying those which belong to the same cluster, and their functions are similar. *See* EST; <http://www.ncbi.nlm.nih.gov/UniGene/index.html>. (Schuler, G. D., et al. 1997. *J. Mol. Med.* 75:694; Zhuo, D., et al. 2001. *Genome Res.* 11:904.)

unilateral incongruity Special type of incompatibility when only the females of one of the related species are compatible with the males of the other species; it is a unidirectional compatibility. *See* self-incompatibility. (Bernacchi, D. & Tanksley, S. D. 1997. *Genetics* 147:861.)

Unimodal distribution.

unimodal distribution   Has only one major peak.

unineme   Single strand (of, e.g., DNA or postmitotic chromosome before the S phase).

uninformative mating   Does not shed light on, e.g., linkage relationship.

uniparental disomy (UPD)   Both homologous chromosomes are inherited from only one of the parents in a diploid. If these two chromosomes are identical, the case is called *isodisomy*; if they are different: *heterodisomy*. For example, in 10% of Russel-Silver syndrome patients, a 35 Mb segment at 7q31-ter is maternally inherited as an isodisome. Uniparental disomy may come about by the loss of one chromosome in a trisomic or duplication of a monosome. In human UPD, acrocentric isochromosomes and Robertsonian translocations between nonhomologous chromosomes predominate. *See* isochromosome; monosomic; nondisjunction; Robertsonian translocation; trisomy. (Kotzot, D. 2001. *J. Med. Genet.* 38:497.)

uniparental inheritance   In the absence of male transmission of plastids and mitochondria, the genetic material of these organelles is transmitted to the progeny only through the egg. Also, telochromosomes and large deletions can generally be transmitted only through the female if transmitted at all. *See* chloroplast genetics; ctDNA; doubly uniparental inheritance; mitochondrial genetics; mtDNA; paternal leakage.

unipolar depression   Periods of depression with generally debilitating consequences but usually for shorter duration. *See* affective disorders; bipolar mood changes.

unique DNA   Present in a single copy per genome. *See* singlet.

unit evolutionary period (UEP)   Time in million years (MY) required for the fixation of 1% divergence in two initially identical nucleotide sequences.

unitig   *See* scaffolds in genome sequencing; U-unitig.

univalent   Eukaryotic chromosome without a pair. In case the chromosomes in a hybrid are not sufficiently homologous, they may form univalents and their distribution to the poles in meiosis may be disorderly. The frequency of univalents may permit conclusions regarding the lack of relatedness of the parental forms. *See* Triticale.

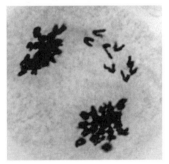

Hybrid of an octaploid ($2n = 56$) and hypoploid ($2n = 41$) hexaploid triticale. The female gamete contributed 28, the male only 20 chromosomes. Therefore, there are 20 bivalents and 8 univalents in meiosis. Seven out of the 8 univalents represent the *D* genome of wheat. Generally, the lagging univalents are not incorporated into the functional gametes because they fail to reach the poles. Usually, they divide belatedly. (Courtesy of Kiss, Á. 1966. *Z. Pflanzenzücht.* 55:309.)

universal bases   The natural nucleoside of hypoxanthine or the synthetic 6-hydroxy- and 6-amino-5-azacytosine nucleosides or 1-(2′deoxy-β-D-ribofuranosyl)-3-nitropyrrole

Pairing of the bases T (thymine), C (cytosine), A (adenine), and G (guanine) with N^8-(2′-deoxyribofuranoside) of 8-aza-7-deazaadenine, a universal base.

may serve for modified nucleotides in recognition of G, T, and U with equal efficiency for Watson-Crick pairing. The latter maximizes stacking while minimizing hydrogen pairing without sterically disrupting the double helix. Some of the analogs lower and others increase the melting temperature of the polynucleotides that contain them. These may be used to synthesize oligonucleotide probes and primers when the exact sequence needed cannot be inferred because of the redundancy of the genetic code. Oligonucleotides containing the 5-nitroindole base may be more easily detected by antibody. In SNIPs, the mismatched base can be identified more easily. Addition of the analogs to the 5′-end may enhance the stability of the analog oligonucleotide on the chips. The presence of these universal bases may alter nucleic acid-protein interactions. When 3-nitropyrrole replaced C, in an A:C mismatch, the fidelity of ligation increased substantially. Incorporation of nitroazole analogs may increase the stability of triplex sequences. Klenow fragment polymerase inefficiently incorporated nitropyrrole derivatives and the analog generally terminated further chain extension. *See* azacytidine; DNA chips; hydrogen pairing; PCR; primer; probe; SNIPs. (Loakes, D. 2001. *Nucleic Acids Res.* 29:2437; Seela, F. & Debelak, H. 2000. *Nucleic Acids Res.* 28:3224.)

universal code The majority of DNAs across the entire phylogenetic range use DNA codons in the same sense. Notable exceptions exist in the mitochondria and in a few species. *See* genetic code. (O'Sullivan, J. M., et al. 2001. *Trends Genet.* 17:20.)

universal donor *See* ABO blood group.

universal recipient *See* ABO blood group.

universal trees Display the evolution of orthologous proteins across taxonomic boundaries. (Brown, J. R., et al. 2001. *Nature Genet.* 28:281.)

unordered tetrads Do not contain the spores in a linear order as generated in the first and second meiotic divisions. In contrast to ordered tetrads, unordered tetrads require the presence of three genetic markers, two of them must be in different chromosomes to calculate gene-centromere distances. Again—like in ordered tetrads—we must determine the frequencies of tetratype tetrads for the at least three markers or more considered, and we designate p as the tetratype frequency of say a and b, and q = tetratype frequency of b and c; similarly r = tetratype frequency for a and c. The exchange frequency between a and its centromere = x, between b and its centromere = y, and between c and its centromere = z. Furthermore, we need the following three equations:

$$p = x + y - 3/2xy \quad q = y + z - 3/2yz \quad r = x + z - 3/xz$$

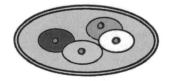

With the values p, q, and r being known, the unknown quantities, the recombination frequencies (x, y and z) between

the three genes and their centromeres, can be determined by resolving the three equations:

$$x = 2/3 \left(1 \pm \sqrt{\frac{4 - 6p - 6r + 9pr}{4 - 6q}} \right),$$

$$y = 2/3 \left(1 \pm \sqrt{\frac{4 - 6p - 6q + 9pq}{4 - 6r}} \right), \text{ and}$$

$$z = 2/3 \left(1 \pm \sqrt{\frac{4 - 6q - 6r + 9qr}{4 - 6p}} \right)$$

Once the recombination frequency between a gene and its centromere becomes available, the exchange frequencies between additional genes and their centromeres can be calculated by the formula

$$s = \frac{2(v - t)}{2 - 3t}$$

where s = the unknown recombination frequency between marker d and its centromere, t = the known recombination frequency between gene e and its centromere, and v = the tetratype frequency for the unmapped gene d and the mapped gene e. *See* tetrad analysis. (Emerson, S. 1963. *Methodology in Basic Genetics*; Burdette, W. J., ed., p. 167. Holden-Day, San Francisco, CA; Whitehouse, H. L. K. 1950. *Nature* 165:893.)

unrooted evolutionary trees Do not indicate the initial split of the branching. *See* evolutionary tree. (Steel, M. & McKenzie, A. 2001. *Math. Biosci.* 170:91.)

unsaturated fatty acid Contains one or more double bonds.

UNSCARE United Nations Committee on Effects of Atomic Radiation.

unscheduled DNA synthesis Replication of DNA is outside the normal S phase, indicating a repair replication. The tests generally carried out on cultured hepatocytes or fibroblasts exposed to certain treatment(s), then the incorporation of radioactive thymidine, are monitored either by autoradiography or by scintillation counting. The data are compared with concurrent controls that were not exposed to any mutagen. Although the procedure appears attractive, it is not very effective and practical for the identification of mutagens or carcinogens. *See* bioassay in genetic toxicology; DNA polymerases. (Zbinden, G. 1980. *Arch. Toxicol.* 46:139; Hoege, L., et al. 2002. *Nature* 419:135.)

Unstable genes.

unstable genes Have a higher than average mutation rate. Most commonly, the instability is caused by the movement of insertion or transposable elements. A higher mutation rate may also be due to deficiency of genetic repair or defects in DNA replication. *See* DNA repair; error in replication; fractional mutation; insertional mutation; transposable elements; variegation.

unsupervised learning Identifies a new, so far undetected, shared pattern(s) of sequences in macromolecules and determines which are the positive and negative representatives for the pattern(s). *See* supervised learning; support vector machine. (Wallis, G. & Baddeley, R. 1997. *Neural Comput.* 9:883; Hatzivassiloglou, V., et al. 2001. *Bioinformatics* 17 Suppl.1:S97.)

untoward pregnancy Terminates with congenital malformation or stillbirth or infant death within 14 days after delivery. *See* congenital.

untranslated regions (UTR) Leader sequences upstream from the first methionine codon and the downstream sequences beyond the stop codon of mRNA. The upstream and downstream regions include various control elements, and at the 3′-end in eukaryotes the polyadenylation signal is situated. The untranslated regions of mRNA do not code for any amino acid sequence. The 3′-UTRs in RNAs control translation, fate of germline cells, life cycle, anterior-posterior axis of development, and meiotic cycles by binding proteins similarly to DNA-binding proteins. *See* downstream; polyadenylation; stop codon; upstream. (<http://bighost.area.ba.cnr.it/Big/UTRHome/>.)

unusual bases Modified forms of the normal DNA or RNA bases; they may be common in tRNA. Their incorporation into DNA may lead to base substitution mutations. Methylation of C and A nucleotides may lead to imprinting, transient genetic variations in expression, and alteration of RFLP. *See* Bar; imprinting; methylation of DNA; RFLP; tRNA. (Barciszewski, J., et al. 1999. *Mol. Biol. Rep.* 26:231.)

Unverricht-Lundborg disease (EPM1) *See* myoclonic epilepsy.

unwindase Double-stranded RNA helicases that unwind in a single step the entire length of a molecule in an ATP-dependent process. *See* DexH; NPH. (Nishikura, K. 1992. *Ann. NY Acad. Sci.* 660:240.)

unwinding protein Facilitates unwinding of a DNA double helix and stabilizes single strands. The nucleotide triphosphate (NTP)-dependent RNA helicases are required for unwinding of double-stranded RNA and replication of many pathogenic viruses. *See* DNA replication; helicase. (Dillingham, M. S., et al. 2001. *Proc. Natl. Acad. Sci. USA* 98:8381.)

unzipping, DNA Separation of the double-stranded structure. Depending on the base composition, the threshold force required is about 12 pN (piconewton). One joule is the work done when a force of 1 N acts through a distance of 1 meter. *See* joule. (Cocco, S., et al. 2001. *Proc. Natl. Acad. Sci. USA* 98:8608.)

uORF Upstream open-reading frame; may have a regulatory role in protein synthesis.

up-and-down Structural arrangement of helical protein bundles comparable to the meander of β-sheets. *See* meander; protein structure.

UPE Upstream promoter element. *See* promoter.

UP elements Upstream elements. in the DNA (−40 to −150 sites from the initiation of transcription) are recognition sites for the α-subunit of prokaryotic RNA polymerase and boost (30- to 300-fold) the frequency of transcription initiation. *See* CRP; FIS; promoter; σ; transcription factors.

UPGMA Unweighted pair group method with arithmetic means are formulas for determining evolutionary distances. *See* evolutionary distance; transformed distance. (Kim, K. I., et al. 2002. *Animal Genet.* 33:19.)

UPR Unfolded protein response regulates gene expression when the endoplasmic reticulum does not function properly. It may control chaperones, phospholipid biosynthesis, secretory pathways, and degradation of proteins associated with the endoplasmic reticulum under stress or even without stress. *See* chaperones; endoplasmic reticulum; unfolded protein response. (Bertolotti, A. & Ron, D. 2001. *J. Cell Sci.* 114:3207.)

upregulation Increasing activity by regulation. *See* downregulation; regulation of gene activity.

U protein Unwinding protein. Unwinds the DNA strands at a distance from a nick. (Basak, S. & Nagaraja, V. 2001. *J. Biol. Chem.* 276:46941.)

upstream In the direction of the 5′-end of polynucleotides (DNA). *See* downstream.

upstream activation sequence *See* UAS.

upstream regulatory sequence (USR) Regulatory element in the promoter region.

uptag *See* bar code, genetic.

uptake Eukaryotic cells may incorporate nuclei, plastids and mitochondria, pseudovirions, plasmids, liposomes, and various other macromolecules besides smaller organic and inorganic molecules. Viruses move from cell to cell through the plasmodesmata and through the vascular system of plants. Plant viruses generally encode a movement protein that modifies the plasmodesmata, binds to single-stranded nucleic acids, and associates with the cytoskeleton and endoplasmic reticulum. (Tzfira, T., et al. 2000. *Annu. Rev. Microbiol.* 54:187.)

uracil Pyrimidine base in RNA; 2,4-dioxypyrimidine, MW 112.09, soluble in warm water but insoluble in ethanol. *See* pyrimidines.

URE Ureidosuccinate utilization. Cytoplasmic proteins involved in nitrogen metabolism in yeast are responsible for the production of a protein analogous to prion. *See* prion. (Baxa, U., et al. 2002. *Proc. Natl. Acad. Sci. USA* 99:5253.)

urea cycle Formation of urea ([NH$_2$]$_2$CO) from amino acids and CO$_2$; ornithine is converted to citrulline, which is converted to arginine. The hydrolytic cleavage of arginine produces urea and regenerates ornithine; thus, the cycle is completed. The urea cycle in the mitochondria secures homeostasis for ammonium with some independence of nitrogen intake. Mutation in the ornithine transporter (encoded at 13q14) involves the symptoms of hyperornithinemia-hyperammonemia-homocitrullinuria syndrome. *See* amino transferase; arginine; citrulline carbamoylphosphate synthetase deficiency; ornithine.

uredium Uredospore-producing sorus, a type of sporangium. *See* sorus; stem rust.

Uredium on wheat

Uredium.

uremia Urine in the blood caused by a variety of factors.

ureotelic Excretes urea.

urethan (CH$_3$H$_7$NO$_2$) Toxic (lethal dose 2 g/kg in rabbits) liquid (at 48–50°C). It causes chromosomal breakage and mutation; it is also antineoplastic.

URF Unidentified reading frame is capable of transcription (open), but the gene product has not been identified. *See* reading frame.

uric acid Degradation product of xanthine excreted in the urine in particularly high amounts in hyperuricemic individuals under dominant and polygenic control and in patients with glycogen storage diseases, in HPRT deficiency and gout. Birds and reptiles normally excrete high amounts. *See* glycogen storage diseases; gout; HPRT; Lesch-Nyhan syndrome; xanthinuria.

Uric acid.

uricotelic Excretes uric acid.

uridine Uracil + ribose, an RNA nucleoside. *See* pyrimidines.

uridine diphosphate glucuronosyl transferase Encoded in human chromosome 2; this group of enzymes includes glucuronate steroid hormones. *See* Crigler-Najjar syndrome.

uridine monophosphate synthetase deficiency (UMPS) *See* oroticaciduria.

uridylate Nucleotide of uracil; contains uracil + ribose + phosphate.

URL Uniform resource locator the generic name for an Internet resource, such as WWW page, Gopher menu, a file transfer protocol server, etc. Information about Internet use is obtainable at <http://home.netscape.com/home/about-the-internet.html>. For automatic monitoring of your interest on Internet use, see "The URL Minder": <http://www.netmind.com.URL-minder/URL-minder.html>. *See* bookmarks.

U RNA (**1**) Ubiquitous RNA. An snRNA; its transcript has a 2,2,7-trimethyl guanosine cap and may have modified U residues and no poly-A tail. *See* cap; polyA tail; snRNA; snurposome. (**2**) Uridine-rich nuclear RNA involved in transcript processing within the cell nucleus.

U1 RNA Has complementary sequences to the 5′-consensus sequences of the splice sites and presumably has a role in mRNA processing from the primary transcript. *See* hnRNA; RNA. (McNamara-Schroeder, K. J., et al. 2001. *J. Biol. Chem.* 276:31786.)

U2 RNA (snRNA) Apparently recognizes the 3′-end of introns at the lariat and seems to have a role in splicing. *See* introns; lariat; splicing.

U-RNP U-RNA associated with proteins involved in processing primary transcripts of genes within the nucleus.

urocortin Neuropeptide similar to urotensin and corticotropin-releasing factor. It evokes the synthesis of adrenocorticotropic hormone and thus is involved in stress-related endocrine, autonomic, and behavioral responses. *See* adrenocorticotropin; corticotropin-releasing factor; urotensin; (Parkes, D. G. & May, C. N. 2000. *News Physiol. Sci.* 15:264.)

urogenital Involved with the system of urine secretion and the reproductive organs.

urogenital adysplasia Autosomal-dominant failure to develop one or both kidneys, frequently coupled with wide-set eyes, low-set ears, and other anomalies. The incidence among newborns may be as high as 4.5% and among adults 0.3%. *See* kidney disease.

urokinase (uPA) Plasminogen activator; the urokinase receptor is a glycosylphosphatidyl-inositol-linked cell-surface protein that regulates cell adhesion. It has been suggested that it is a requirement for some cancerous growths. Some catechins (present in green teas) appear inhibitory to uPA. *See* CAM; intravasation; plasminogen. (Andreasen, P. A., et al. 1997. *Int. J. Cancer* 72:1; Plesner, T., et al. 1997. *Stem Cells* 15:398.)

uropathy   Diseases of the urogenital system.

urotensins   Short bioreactive peptides of 41 (urotensin I) and 12 (urotensin II) amino acids, respectively. *See* urocortin. (Lewis, K., et al. 2001. *Proc. Natl. Acad. Sci. USA* 98:7570.)

URS   Upstream regulatory/repressing sequences, binding sites for various transcription factors. *See* promoter; regulation of gene activity; transcription factors; UAS. (Hanna-Rose, W. & Hansen, U. 1996. *Trends Genet.* 12:229.)

urticaria, familial cold (Muckle and Wells syndrome, 1q44) Dominant, reoccurring, transient burning papules and macules on the skin, cold sensitivity, frequently associated with progressive deafness and renal amyloidosis (fibrillary protein deposits in the kidneys). *See* cold hypersensitivity.

use and disuse   *See* lamarckism.

Usher syndrome (USH)   Hereditary, yet the pattern of inheritance is quite variable, although in most cases it is probably autosomal recessive. Commonly it involves deafmutism, retinitis pigmentosa, mental disabilities, and ataxia. Three types of the syndrome are usually distinguished on the basis of the severity and onset of disease. The prevalence is about $4–5 \times 10^{-5}$. About 75% of afflicted individuals have a severe defect in the USH1B gene in human chromosome 11q13.5, encoding myosin VIIA. USH1A is in chromosome 14q32. Other types of the disease involving hearing deficit and some other anomalies were reported: USH1D and DFNB12 (10q21-q22), USH1F (chr. 10q21-q22, protocadherin), USH1E (21q21), USH2A (1q41). USH1D encodes a type of cadherin (CDH23) that causes stereocilia disorganization in waltzing mouse with hearing deficit (Bolz, H., et al. 2001. *Nature Genet.* 27:108). A milder form was assigned to 1q32. Myosin VIIA may be involved in the transport between the outer and inner layers of the eye photoreceptors. In type IIa, a 171.5 kDa protein is involved that has laminin epidermal growth factor and fibronectin type III motifs pointing to the involvement of defects in cell adhesion. USH1C has a defect in the PDZ domain (involving PSD-95, DLG, and ZO-1 proteins) of harmonin. PDZ modules interact with other proteins involved with signaling and the cytoskeleton. Deletions of the USH1C gene may involve infantile hyperinsulinism, enteropathy (intestinal disease), and deafness. USH3 (3q21-q25) encodes a 120-amino-acid protein concerned with recessive, progressive hearing loss and severe retinal degeneration. *See* ataxia; cadherin; CAM; deafmutism; deafness; EGF; fibronectin; hypoglycemia; laminin; myosin; retinitis pigmentosa; waltzing mouse. (Verpy, E., et al. 2000. *Nature Genet.* 26:51; Bolz, H., et al. 2001. *Nature Genetics* 27:108; Bork, J. M., et al. 2001. *Am. J. Hum. Genet.* 68:26; Ahmed, Z. M., et al. 2001. *Am. J. Hum. Genet.* 69:25; Joensuu, T., et al. 2001. *Am. J. Hum. Genet.* 69:673.)

USM   Ubiquitous somatic mutations are attributed to defects in DNA repair (mismatch repair) and replication. *See* DNA repair; unstable genes.

U-snRNP   Splicing factor of RNA transcript. *See* spliceosome; splicing. (Xue, D., et al. 2000. *EMBO J.* 19:1650.)

U3 snRNP   Most abundant U RNA ($\approx 10^6$ molecules/cell) processes near the 5′-end of the ribosomal RNA transcripts, stays on, and generates a 5′-knob characteristic for rRNA only. Binding U3 of the snRNP initiates processing, especially the transcripts of 18S rRNA. *See* ribosome; rRNA; snRNP. (Venema, J., et al. 2000. *RNA* 6:1660.)

U8 snRNP   Required for the upstream cleavage of 5.8S and for cutting off of 28S RNA ca. 500 nucleotides at the 3′-end. *See* ribosome; rRNA; snRNP. (Peculis, B. A. & Steitz, J. A. 1994. *Genes Dev.* 8:2241.)

U12 snRNP   Spliceosomal component along with other U snRNAs. *See* spliceosome. (Otake, L. R., et al. 2002. *Mol. Cell* 9:439.)

U14 snRNP   Maturase for 18S rRNA. *See* ribosome. (Newman, D. R., et al. 2000. *RNA* 6:861.)

U22 snRNP   Essential for processing both ends of 18S rRNA separated by about 2,000 nucleotides.

USP   Chromosome-specific unique sequence probes employ locus-specific fluorescent DNA sequences suitable for identification of small deletions and duplications. *See* chromosome painting; FISH; telomeric probes; WCPP.

Ustilago maydis ($n = 2$)   Basidiomycete that has been extensively used for meiotic and mitotic analysis of recombination, isolation of biochemical mutations, etc. This fungus causes the ear smut of maize. The haploid forms are nonpathogenic. Several other *Ustilago* species are pathogenic to other Gramineae. *See* fungal life cycles. (Banuett, F. 1995. *Annu. Rev. Genet* 29:179.)

UTase   Uridylyl transferase catalyzes the transfer of uridyl group to the P_{II} regulatory subunit of adenylyl transferase (ATase), an enzyme, which transfers an adenylyl group from ATP to a tyrosine-hydroxyl in glutamine synthetase. The complex ATase P_{II}-uridylyl transferase catalyzes phosphorolytic deadenylation of glutamine synthetase. Glutamine synthetase is an enzyme involved in many functions. *See* glutamate dehydrogenase; glutamine synthetase.

uterine cancer   Cancer of the uterus may be treated by antiestrogens similarly to breast cancer. The response to tamoxifen is very limited, however. Cervical cancer usually does not respond to such treatment, and surgery and radiation are used. *See* breast cancer.

uterus   Hollow female abdominal organ where the fertilized egg is embedded for the development of the embryo. The pear-shaped uterus ($\sim 5 \times 7.5$ cm in humans) is connected to the vagina through its inner gate, the cervix, permitting the entry of the spermatozoa and retention of the conceptus. The cervix opens to release the embryo at birth. The ovaries are connected to the uterus by the oviducts. The uterus is lined by the endometrium that feeds the early embryo. Its outer layer is shed during menstruation and regenerated afterward. *See* gonads; menstruation; vagina.

utility index for genetic counseling   If the mother is heterozygous for a recessive disease allele d/D and other linked alleles (M_1/M_2), i.e., she is DM_1/dM_2 and the frequency of recombination between the two loci $= r$, half of her sons will be afflicted by the disease and $1 - r$ frequency of M_2 sons are expected to express the disease (as long as the penetrance and expressivity are high). A small r helps predictability. If her husband's genotype is DM_1, among her M_1 M_2 daughters $1 - r$ will be carriers of the recessive disease allele d, but one must know for sure which of the codominant M alleles is syntenic with d. In case of *X linkage*, for prediction on the basis of M markers, the genetics counselor should know the genotype of the affected grandfather (dM_1 or dM_2). If the frequencies of the M_1 and M_2 markers is x_1 and x_2, respectively, and the grandfather is dM_2, then the probability that the mother being informative for genetic counseling is $= 2x_1x_2$. Roychoudhury and Nei call this the *utility index of a polymorphic locus* for genetic counseling. If the grandfather has the recessive allele m and the grandmother is either MM or Mm, their heterozygous (Mm) daughter must have inherited the M allele maternally and the expected frequency of informative mothers is $(1 - x)x$. In case both M and m grandfathers are considered, the utility index of the mother becomes $(1 - x^2)x$.

Genetic information regarding the mother can be obtained from her children. In case the X-linked disease gene is deemed to be in coupling with another marker, the Bayesian probability for coupling is $(1 - r)^2/[(1 - r)^2 + r^2]$ and the probability for repulsion is $r^2/[(1 - r)^2 + r^2]$ in case the mother already has two afflicted sons. In case the mother has both normal and afflicted sons, n_1 (DM_1), n_2 (DM_2), n_3 (dM_1) and n_4 (dM_2), the probability for coupling that she has $4(=n)$ sons with the genotypes above is $r^{n2+n3}(1 - r)^{n1+n4}/2^n$; in case of repulsion, the probability is $r^{n1+n4}(1 - r)^{n2+n3}/2^n$. The posterior probability that she is in coupling is $1/(1 + \rho^\alpha)$, where $\rho = r/(1 - r)$ and $\alpha = n_1 + n_4 - (n_2 + n_3)$; in case $\alpha = 0$, the linked markers will not help in the prognosis. For repulsion, the probability is $1/(1 + \rho^{-\alpha})$. The probability regarding the genotype of the next offspring depends on the tightness of linkage.

In case of *autosomal-dominant* disease in the presence of D, the disease is expected and both parents are informative. In case the mother is M_1M_2 and the father is M_1M_1, and the recombination frequency between the M and the D loci is r, the offspring of M_1M_1 genotype will have D with a probability of $1 - r$. DD homozygotes are expected very rarely because their occurrence depends on the product of the frequency of the D gene, which is usually in the 10^{-5} range. In case there are multiple alleles at a locus, the total frequency of informative parents is

$$1 - \sum x_i^2 - \left(\sum x_i^2\right)^2 + \sum x_i^4.$$

In a mating of $(Dm/dM) \times (dM/dm)$, the offspring homozygous for m is expected to carry the dominant disease gene D with

a probability of $1 - r$, but the one with the dominant M phenotype is expected to be D at a frequency of $(1 + r)/3$. If the affected parent D is heterozygous and the other parent is dd, the frequency of the informative families is $2(1 - x)x^4$. In case the mating is $(DdM_1M_2) \times (ddM_1M_1)$, the children are $n_1(DdM_1M_1)$, $n_2(DdM_1M_2)$, $n_3(ddM_1M_1)$, and $n_4(ddM_1M_2)$, respectively. The linkage phase can be estimated as shown above for X linkage [coupling: $1/(1 + \rho^\alpha)$, repulsion $1/(1 + \rho^{-\alpha})$]. All children of parent DdM_1M_2 will be informative, except when the spouse is ddM_1M_2 and all the progeny are heterozygous for the M locus, but the probability that all children would be of such genetic constitution is $(0.5)^n$. The proportion of informative families when $n > 1$ is $2x_1x_2(1 - 0.5^{n-1}x_1x_2)$. If there are multiple markers, the proportion of the informative families (with x_i frequency of the ith allele) is

$$2\sum_{i<j} x_ix_j(1 - 0.5^{n-1}x_ix_j).$$

In case of autosomal-recessive diseases, calculation of mathematical probabilities of informative families is more complex, and biochemical or molecular (DNA) analyses are preferred. *See* DNA fingerprinting; genetic counseling; paternity testing; risk. (Roychoudhury, A. K. & Nei, M. 1988. *Human Polymorphic Genes.* Oxford Univ. Press, New York.)

UTP   Uridine triphosphate. *See* UDP.

Uridine triphosphate.

UTR   *See* untranslated region.

utrophin   *See* dystrophin (DRP2).

U-unitig   *See* scaffolds in genome sequencing; unitig.

UV   *See* cyclobutane dimers; ultraviolet light.

uvomorulin (UM)   Transmembrane glycoprotein, also called E-cadherin. *See* cadherins.

UvrABC   Endonuclease complex of uvrA, uvrB, and uvrC. These enzymes are involved in excision of ultraviolet light–induced pyrimidine dimers. After the dimer is recognized, cuts are made on both sides, thus excising the damaged area of about 12 nucleotides. *See* ABC excinucleases; DNA repair; excision repair.

UV spectrophotometry of proteins Prepare a series of dilutions (20 to 3,000 µg/mL) from a pure 3 mg/mL standard bovine serum albumin (BSA).

$$\frac{O.D._{280}}{a_{280}xb}$$

Make a blank and a series of dilutions of the sample to be tested. Determine UV absorption at 280 nm. BSA, 3 mg/mL, is expected to have an absorption of 1.98. Calibrate the sample relative to the standard. In case the absorptivity of the protein to be tested is known, use the formula for the calculations. (a_{280} = the absorption in units of mg/mL per centimeter path b).

"The attainment ... of fundamental knowledge is usually of the utmost immeasurable practical importance in the end".

H. J. Muller 1916, quoted by Elof Carlson (30 years later [in 1946] Muller received the Nobel Prize)

A historical vignette.

V

V Maximal velocity of the reaction when the enzyme is saturated with substrate.

vaccines Suspension of killed or attenuated pathogens or recombinant protein or DNA for generating an immune defense system. The most successful vaccines (measles, mumps, rubella) are made of antigens generated against disease-causing microorganisms and injected into the bloodstream to stimulate the development of circulating or serum antibodies (immunoglobulin G). Efforts are being made to develop vaccines that activate mucosal immunity. Membranes of the body that cover the gastrointestinal tract, the air-intake organs, and the reproductive system are covered with mucosa. The purpose of this system is to trap infectious agents at the port of entry. The mucosa can develop sufficient quantities of immunoglobulin A. These new vaccines are expected to be delivered orally. Some vaccines may use live microorganisms that have been genetically engineered by removal of part of their genome so that they cannot cause the disease, yet they may promote the production of IgA and possibly IgG. Other approaches may include introduction into the cells of cytotoxic lymphocyte epitopes. The presence of CpG oligonucleotide motifs in the vector plasmid appears to enhance the immunogenicity of DNA vaccination in humans. Genetically engineered vaccines may be produced in transgenic plants and may eventually be edible. Mice after infection by the lymphocytic choriomeningitis virus may clear the LCMV and remain immune for life to reinfection because the DNA containing LCMV sequences is transcribed in the animal and represents a new type of DNA vaccination. Prophylactic vaccination against retroviral diseases is hampered by the apparent poor immune reaction against these agents. Bovine leukemia virus (BLV), which produces a relative paucity of variants that confound cytotoxic T cells (CTL), can, however, be vaccinated successfully by using a viral glycoprotein, gp51 epitope. Several poxviruses, modified vaccinia virus, replication-defective adenovirus, etc., appear to boost DNA vaccine effectiveness for primed CD^+ T cells. Some adjuvants (alum, monophosphoryl lipid A, oil/water emulsion, etc.) showed beneficial effects. A new approach to vaccination is the use of antibodies against cytokines that accumulate in inflamed or tumor tissues (Zagury, D., et al. 2001. *Proc. Natl. Acad. Sci. USA* 98:8024). *See* cancer prevention; CTL; cytokines; epitope; immune system; immunization; immunization, genetic; lymphocytes; memory, immunological; mucosal immunity; peptide vaccine; plant vaccines; subunit vaccine; therapeutic vaccine; tumor vaccination. (*Nature Medicine* 4(5) [1998], May supplement; Burton, D. R. 2002. *Nature Rev. Immunol.* 2:706.)

vaccinia virus Closely related to smallpox (variola) and cowpox viruses. Apparently, it does not occur in nature and can be found in the laboratory only where these two viruses are handled. Thus, it appears to have derived somehow from variola and cowpox. Vaccinia vectors have been used to express antigens of unrelated pathogens (AIDS virus, hepatitis B)

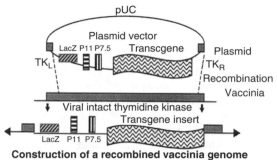

Construction of a recombined vaccinia genome
(LacZ bacterial reporter, P11 and P7.5 are promoters, TK: thymidine kinase fragments, pUC plasmid provides origin of replication and ampicillin resistance [β-lactamase])

and employ them for immunization. The virus contains about 190 kbp DNA genome with about 260 potential open-reading frames and about 200 bp telomeric sequences. These viruses direct replication and transcription in cell cytoplasm by viral-encoded enzymes. Therefore, insertion of the viral genome by recombination is not a great threat. Vaccinia virus appears to be relatively safe, yet periodic (10 years) vaccination of laboratory workers against it may be necessary, and handling should be under containment level 2. The vaccinia vectors appear effective against some tumors and metastasis when injected directly into the neoplasia. Vaccinia virus productively infects the majority of mammalian and avian cells, but Chinese hamster ovary cells are not infected and virus replication may not be completed in primary lymphocytes or macrophages. *See* β-galactosidase; biohazards; immunization; immunotherapy; Xgal.

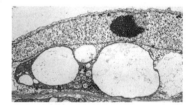

Large vacuoles in a plant cell (bottom) and an oblong nucleus with nucleolus (top).

vacuoles Vesicles within plant and fungal cells filled with various substances (nutrients, products of secondary metabolism, enzymes, crystals, solutes). Vacuoles may occupy minimal space in meristematic cells, whereas in older cells they take up to 90% of the cell inner volume. The elastic tonoplast membrane surrounds the vacuoles and permits the change of their size. They may regulate osmotic pressure of the cytosol by releasing smaller molecules or polymerizing them as needed to maintain a constant value in the cytoplasm. Vacuoles also regulate pH by a similar balancing action. The vacuoles supply the cells with storage nutrients, hydrolytic enzymes, anthocyanin pigments, and in some cases various toxic substances such as tannins, phenolics, alkaloids, etc. Vacuoles are inherited by cytoplasm partition, but if a particular cell becomes vacuole-free, its progeny may generate them de novo. *See* cells; lysosomes; membrane fusion. (Klionsky, D. J. & Ohsumi, Y. 1999. *Annu. Rev. Cell Dev. Biol.* 15:1; Rojo, E., et al. 2001. *Developmental Cell* 1:303.)

vagility   Ability of organisms to disperse in a natural habitat; it is thus a factor of speciation and survival.

vagina   Female organ of copulation beginning at the vulva and extending to the cervix of the uterus. *Vaginismus* is an involuntary painful contraction of the vaginal muscles and may cause severe pain by intercourse. An autosomal-recessive condition of vaginal atresia (absence of vagina) is known. In general, the term describes an anatomical sheath. *See* clitoris; egg; fallopian tube; migraine; ovary; uterus; vulvovaginitis.

vaginal plug   After successful copulation of mice, part of the male ejaculate forms a vaginal plug that closes the vagina for 16–24 hours and in some strains even for a few days. The presence of the plug reveals to the breeder the success of the mating. *See* mating plug.

valence (valency)   Of an antibody, indicates the number of antigen-binding sites. In chemistry it indicates the number of covalent bonds an atom can form; also called oxidation number, e.g., O^{2-} or Na^+ or Ca^{2+}.

validation   Confirmation of experimental results or working hypotheses by repeated tests.

valine biosynthesis   *See* isoleucine-valine biosynthetic pathway.

valinemia   *See* hypervalinemia.

valyl tRNA synthetase (VARS)   Encoded in human chromosome 9. VARS charging tRNA^{Val} by the amino acid valine. *See* aminoacyl-tRNA synthetase.

VAMOS   Variability modulation system is the expression of the *reactivity* of the females measured on the basis of the percentage of sterility in the daughters in hybrid dysgenesis. Reactivity is modulated by temperature, age of the females over the generations, and inhibitors of DNA synthesis and ionizing radiation, etc. *See* hybrid dysgenesis.

VAMP (synaptobrevin)   Synaptic protein. *See* SNARE; synaptobrevin; syntaxin.

Van der Waals force   Weak, short-range attraction between nonpolar (hydrophobic) molecules.

Van der Woude syndrome   Dominant human chromosome 1q32-q41-located cleft lip and palate syndrome. *See* cleft palate; harelip. (Schutte, B. C., et al. 2000. *Genome Res.* 10:81.)

Van Gogh, Vincent (1853–1890)   Famous Dutch painter might have been a sufferer of intermittent porphyria. The *Drosophila* gene named *vangogh* modulates the expression of *wingless*. *See* porphyria, *wingless*.

vancomycin ($C_{66}H_{75}Cl_2N_9O_{24}$)   Very potent glycopeptide antibiotic effective against gram-positive bacteria by blocking the cross-linking of adjacent peptidoglycan strands by peptide bonds in bacterial cell wall synthesis. It also inhibits transglycosylation, which connects existing glycan strands. *Streptococcus pneumoniae*, responsible for pneumonia, bacterial meningitis, and ear infection, as well as several *Enterococcus* species, have acquired tolerance or resistance to this antibiotic. Vancomycin tolerance is due to a mutation in the vancomycin signal transduction sensor kinase (VncS). Upon autophosphorylation in the presence of ATP, VncS-P is generated. As a consequence, the vancomycin sensor regulator (VncR) is phosphorylated to VncR-P and the defense genes (resistance to the infection) of the cells are turned off. In vancomycin (and some other antibiotic)-sensitive strains, VnsS dephosphorylates VncR. Vancomycin effectiveness as an antibiotic is due to its binding to the D-Ala–D-Ala moiety of bacterial peptidoglycan precursors, thereby hindering growth of the cell wall. In the resistant strains, the dipeptide is replaced by the depsipeptide D-Ala–D-Lac, diminishing the sensitivity to the antibiotic by three orders of magnitude. The levels of resistance depend on how many of the peptidoglycan precursor molecules carry this replacement. If D-Ala–D-Lac is selectively disrupted by small molecules (prolinol derivatives), sensitivity to vancomycin is restored. *See* antibiotics; tolerance to antibiotics. (Chiosis, G. & Boneca, I. G. 2001. *Science* 293:1484; Eggert, U. S., et al. 2001. *Science* 294:361.)

L-Prolinol [(s)-(+)2-pyrrolidinemethanol].

vanilla (*Vanilla planifolia*)   Tropical spice tree; $2n = 2x = 32$.

Váradi-Papp syndrome   *See* orofacial-digital syndrome V.

variability   Condition of being able or apt to vary.

variable number tandem repeats   *See* VNTR.

variable regions   Of the antibody, are situated at the amino end of both the light and heavy chains, and this region determines antibody specificity and antigen binding. *See* antibody; immunoglobulins.

variable surface glycoprotein   *See Trypanosoma*.

variance   Mean of the squared deviations of the variates from the mean of the variates:

$$V = \Sigma[(x - \bar{x})^2]/n - 1,$$

where x are the variates, $\bar{x}$ is the mean of the variates, and n is the number of variates (individuals). *See* analysis of variance; genetic variance; intraclass correlation; invariance; standard deviation; standard error; variate.

variant   Cell or individual different from the standard type.

variant detector array (VDA)   Uses oligonucleotide-labeled probes to locate/identify particular genes in microarrays. *See* microarray hybridization; oligo-labeling probes.

variate   Variable quantity measured in a sample of a population.

variation   May be *continuous*, and the individual measurements do not fall into discrete classes, e.g., the traits determined by polygenic systems. In case of *discontinuous* variation, the measurements can be classified into distinct classes such as the qualitative traits (black and white [and no gray]) in a segregating population. *See* continuous variation; discontinuous variation; genetic variance; variance.

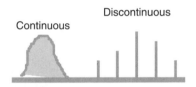

variation, mathematical   How can a cell express 3 different receptors of a pool of 10 receptors? The solution is

$$\frac{10 \times 9 \times 8}{1 \times 2 \times 3} = 120.$$

See combination.

variegation   Sector formation or mosaicism of somatic cells due to a number of different mechanisms such as nondisjunction, somatic mutation, segregation of organelles (chloroplasts), deletion, disease, etc. Heterochromatin may cause variegation of genes in its close vicinity by transient modulation of the exposure of DNA to transcription factors. *See* developmental regulator effect variegation; heterochromatin; lyonization; mitotic recombination; nondisjunction; piebaldism; position effect; transposable elements; tulips, broken; uniparental inheritance. (Ahmad, K. & Henikoff, S. 2001. *Cell* 104:839.)

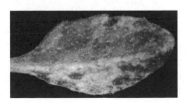

variety   Organism of a distinct form or function. *See* cultigen; cultivar.

variogram   Plot of genetic distance relative to geographic distance. *See* genetic distance.

Varkud plasmid   *See Neurospora* mitochondrial plasmids.

vascular cell adhesion molecule (VCAM)   *See* metastasis.

vascular diseases   *See* cardiovascular diseases.

vascular endothelial growth factors (VEGF)   Required for vasculogenesis. VEGFs are particularly active in some tumor tissues by providing the necessary blood supply for proliferation. VEGFs depend on their appropriate receptors (VEGFRs). The receptor VEGFRT-2 is a tyrosine kinase. In human primary lymphedema, VEGFR-3 tyrosine kinase activity is deficient. Mouse mutants with a defect in the receptor-3 of VEGF die of cardiovascular failure (9.5 d) before birth. VEGF has been considered as a potential benefit in case heart disease has damaged the blood vessels. Unfortunately, VEGF not only promotes tumor growth, but it may facilitate atherosclerosis. Endocrine gland–derived endothelial growth factor (EG-VEGF) is selectively expressed only in the ovary and other steroid-producing tissues (testis, adrenal and placental tissues). *See* angiogenesis; angiopoietin; atherosclerosis; Flk-1; Flt-1; hypoxia; KDR; lymphedema; *Peg-3*; polyposis hamartomatous; tyrosine kinase; VEGF. (Ferrara, N., 1999. *J. Mol. Med.* 77:527; Bellamy, W. T., et al. 1999. *Cancer Res.* 59:728; LeCouter, J., et al. 2001. *Nature* 412:877; Niethammer, A. E., et al. 2002. *Nature Med.* 8:1369.)

vascularization   Development of the veins and other vessels. *See* vasculogenesis.

vascular targeting   For the development and maintenance of tumors, ample blood supply is a requisite. Monoclonal antibody to bFGF and VEGF may block the required angiogenesis factors. Similarly, integrins (CD51/CD61) are required for angiogenesis and they can be interfered with by their cognate monoclonal antibodies. Antiendoglin antibody, especially with conjugated ricin, may lead to antitumor effects. *See* ADEPT; angiogenesis; angiostatin; endoglin; endostatin; FGF; integrin; VEGF.

vascular tissue   Of plants, includes the xylem, phloem, (pro)cambium, and surrounding fibrous parenchyma; in animals, the blood vessels are the primary vascular tissue.

vasculogenesis   Differentiation of mesodermal cells into hemangioblasts. *See* angiogenesis; blood formation; CXCR; hemangioblast.

vasculopathy   Vascular retinopathy, cerebroretinal vasculopathy, endotheliopathy with retinopathy, nephropathy, and stroke all map to 3p21.1-p21.3. Abnormalities of the vascular system cause Raynaud disease, migraine, retinal vein impairment, visual disease, renal disease, neurological problems, and possibly premature death. *See* Raynaud disease. (Ophoff, R. A., et al. 2001. *Am. J. Hum. Genet.* 69:447.)

vas deferens Excretory channel of the testis connected to the ejaculatory duct of the sperm. *See* CBAVD; P2X.

vasectomy Surgical removal of the vas deferens (ductus deferens), the excretory channel of the semen. It is a method of fertility control for males. *See* birth-control drugs. (Sandlow, J., et al. 2001. *Fertility & Sterility* 75:544; Weiske, W. H. 2001. *Andrologia* 33:125.)

vasodilator Causes expansion of (blood) vessels.

vasopressin *See* antidiuretic hormone; oxytocin.

$$\overbrace{S \underline{\qquad} S}$$
Cys-Tyr-Phe-Gln-Asn-Cys-Pro-Arg-GlyNH$_2$
Human vasopressin

VASP Profilin-binding protein. *See* profilin.

VAT Variant antigen type. *See Trypanosoma.*

V-ATPases Vascular ATPases in control of ionic homeostasis. *See* APPase; homeostasis.

vaults Large (42 × 75 nm, 12.9 MDa), predominantly cytoplasmic ribonuclein particles present across phylogenetic ranges. The mammalian vaults include the vault poly(A)DP-ribose polymerase, telomerase-associated protein 1, and one or more nontranslated RNAs. (Stephen, A. G., et al. 2001. *J. Biol. Chem.* 276:23217.)

VAV oncogene In human chromosome 19p13.2-p12, appears to be GDP→GTP exchange factor requiring tyrosine phosphorylation. It regulates lymphocyte development and activation. *See* B lymphocytes; oncogenes; T-cell receptor.

vBNS Very high-speed Backbone Network Service is a computer network linking five supercomputer centers to facilitate fast scientific communication and remote control by the use of special equipment.

vCJD Variant of Creutzfeldt-Jakob disease, a possible contagious form of mad cow disease, infectious also in humans, especially for those homozygous for codon 129 methionine of the prion. *See* Creutzfeldt-Jakob disease; encephalopathies; prion.

VCP Vasoline-containing protein is involved in lipid metabolism. *See* T-cell receptor.

V(D)J Variable (diversity) juncture. Sequences in immunoglobulins where antibody diversity is generated by recombination at the RSS sites. *See* immunoglobulins; RAG; RSS;

T-cell receptor. (Gellert, M. 1992. *Annu. Rev. Genet.* 26:425; Gellert, M. 1997. *Adv. Immunol.* 64:39; Bassing, C. H., et al. 2002. *Cell* 109:S45.)

VDR Vitamin D3 receptors control homeostasis, growth, and differentiation. VDRs preferentially bind well to response elements of direct repeats, palindromes, and inverted palindromes of hexameric core-binding domains when they are spaced by three nucleotides. They can dimerize with 3,5,3′-triiodothyronine, a thyroid hormone receptor that can direct sensitivity of ligands for transactivation. *See* hormone-response elements; transactivator; vitamin D.

vector, algebraic *See* matrix algebra.

vector cassette Transformation construct carrying all essential elements (including reporter genes, selectable marker, replicator, etc.). It can be used for insertion of different DNA sequences. *See* knockout; reporter gene; targeting genes; transformation; vectors.

vectorette Short DNA sequence serving as a specific linker-primer for PCR amplification. It generally contains an inner noncomplementary sequence (bubble) flanked by two short pieces of duplex DNA. The 5′-end may be either blunt or complementary to a restriction site, depending on the restriction enzyme used to digest DNA. An overhang may prevent ligation of the 3′-end. *See* amplification; blunt end; ligase; linker; overhang; polymerase chain reaction; primer; restriction enzyme. (Eggert, H., et al. 1998. *Genetics* 149:1427.)

vectors Molecular genetic constructs, generally circular plasmids that can introduce exogenous genetic material into prokaryotic and eukaryotic cells. They may be *cloning vectors* that only replicate DNA according to the plasmid replicon. *Expression vectors* carry genes complete with all the elements required for expression (promoter, structural gene, termination signals, etc.). *Gene fusion vectors* do not have promoters and the expression of the transferred gene is contingent on fusion with a host cell promoter. *Shuttle vectors* can carry DNA among different hosts (such as *E. coli* and COS cells, *Agrobacterium* and plants). All vectors must have as a minimum a replicator site, selectable marker(s), and mechanisms for introduction of parts of their sequences into the host genetic material. *See* BAC; BIBAC; cloning vectors; ColE1; cosmids; excision vectors; fosmids; HAC; NOMAD; PAC; pBR322; phagemids; plasmids; plasmovirus; pUC vectors; targeting vector; transcriptional gene fusion vector; translational gene fusion vector; transposable elements; transposon vector; viral vectors; YAC; yeast vectors. (<www.informaxinc.com>.)

vegetal pole Lower end of the animal egg where the yolk is concentrated. The opposite end of the egg is called the animal pole. After fertilization, the yolk moves to the central position and becomes the starting site of the differentiation of axes (anterior-posterior, dorsal-ventral, median-lateral) of the embryo. *See* morphogenesis *Drosophila*; pole cell.

vegetative cell Involved in metabolism but not in sexual reproduction. *See* gametogenesis in plants.

vegetative hybrids *See* graft hybrids.

vegetative incompatibility *See* fungal incompatibility.

vegetative nucleus (macronucleus) *See Paramecium*; tube nucleus.

vegetative petite *See* petite colony mutants.

vegetative reproduction Common practice in many species of plants (grafting, rooting) and lower organisms that uses fission for propagation. The advantage of this type of reproduction is that the progeny form a genetically homogeneous clone unless or until mutation takes place. *See* clone; grafting; regeneration; somatic embryogenesis; tissue culture.

vegetative state Asexual, unconscious, nonreplicating, noninfectious, etc., depending on context.

VEGF Smooth muscle cells synthesize vascular endothelial growth factor. It is somewhat related to PDGF. VEGF genes are divided among eight exons, and by alternative splicing three different proteins are produced. It may increase the growth of new blood vessels. *See* angiopoietin; neuropilin; PDGF; signal transduction; vascular endothelial growth factor; vascular targeting.

vehicle *See* vectors.

velans *See gaudens*.

velocardiofacial syndrome (VCFS) Heart, face, kidney, parathyroid, and thymus defect caused by deletion in human chromosome 22q11. The gene most commonly affected is UFD1-l (ubiquitin fusion degradation; 22q11.2). *See* DiGeorge syndrome; face/heart defects; tight junction.

vena Vein that carries blood toward the heart.

veneering, antibody Generation of humanized antibody where some of the surface regions of the mouse antibody framework regions are replaced by human sequences in order to reduce immunogenicity. *See* antibody; humanized antibody.

Venn diagrams Represent data by circles or ovals according to their common features, overlapping functions, or exclusion. The three circles generate seven $(2^3 - 1)$ areas that may be shaded or otherwise marked. Such diagrams may represent interactions of proteins or interactions among environmental effects.

venous malformation (VM) Caused by mutation at loci 1p21-p22 and 9p21. VM can occur in any tissue but is most common in the skin and muscle, causing pain, bleeding, and sometimes death.

vent DNA polymerase extracted from *Thermococcus litoralis*. *See* recursive PCR.

ventral Position at the side opposite the back.

ventricular Belonging to a ventriculus (cavity such as in the heart).

venture Variable number of repeating elements (40 to >150) of 14–15 nucleotides rich in guanine. The shorter VENTR alleles are associated with susceptibility to insulin-dependent diabetes mellitus (IDDM), but one of the long repeats ($14 \times 50 = 700$ nucleotides) appears to be protective against IDDM. *See* diabetes mellitus; imprinting; VNTR.

venules Small vessels that collect blood from the capillary veins.

venus mirror *See* pedigree, female.

Vermiculite™ Commercial silicate medium to grow plants under greenhouse conditions.

vermilion eye color of insects Controlled by the *v* locus of *Drosophila* encoding tryptophan pyrrolase, an enzyme that converts tryptophan to formylkynurenin. The recessive vermilion eye color is actually bright scarlet because the brown ommochrome is not formed. When transplanted into normal tissues, the *v* eye discs develop wild-type eye color. *See* animal pigments; ommochromes; tryptophan.

vernalization Some biannual or winter-annual species of plants have a low temperature requirement for the induction and completion of the bolting and flowering stage of development. This need can be satisfied in spring planting by exposing the germinating seeds to near-freezing temperature for a genetically determined and variable period. It is also called yarowization (in Russian, *yarowie kchleba* means spring cereal). In *Arabidopsis*, the *FLC* is one of the loci to control vernalization response. The other vernalization gene, *FRIGIDA* (*FRI*), has been cloned. VRN2 is a nuclear zinc-finger protein that mediates the level of FLC. Apparently, low temperature and other gene products modulate the level of its transcript supposedly by demethylation of specific DNA sites. *See* photomorphogenesis; photoperiodism. (Reeves, P. H. & Coupland, G. 2001. *Plant Physiol.* 126:1085; Gendall, A. R., et al. 2001. *Cell* 107:525; Macknight, R., et al. 2002. *Plant Cell* 14:877.)

versenes (EDTA) Widely used laboratory chelating agents that at higher concentrations may cause chromosome breakage. *See* EDTA.

$$\text{NaOOCCH}_2\diagdown\quad\diagup\text{CH}_2\text{COONa}$$
$$\qquad\qquad\text{NCH}_2\text{CH}_2\text{N}$$
$$\text{HOCH}_2\text{CH}_2\diagup\quad\diagdown\text{CH}_2\text{COONa}$$

vertical resistance The host plant is resistant to a specific race of the pathogenic microorganism.

vertical transmission *See* transmission.

very low-density lipoprotein *See* VLDL.

vesicles Membrane-surrounded sacs in the cell, generally with storage and transport functions.

vesicular stomatitis virus (VSV) Negative-strand RNA virus of 11,161 nucleotides enclosed by a nucleocapsid (N). N is a 35-turn helix within the membrane-surrounded oval particles. There is a transmembrane G protein on the surface of the virion for binding cell-surface receptors required for infection. VSV viruses can be engineered into useful genetic vectors. *See* CO_2 sensitivity; lentiviruses; viral vectors.

veto cell Recognizes T cells and inactivates them. *See* immune suppression. (Reich-Zeliger, S., et al. 2000. *Immunity* 13:507.)

Vg-1 Protein that sends signals in animals to develop head and other nearby organs.

V gene Codes for the variable region of the antibody molecule.

VHL *See* von Hippel–Lindau syndrome.

viability Ability to survive; this is a property of organisms depending on genetic, developmental, and environmental factors. A normal human fetus may become viable outside the womb after it has reached a weight of about 500 g at about 20 weeks after gestation. The viability of a mutant is often expressed as the survival rate relative to the wild type.

Vibrio cholerae Bacterium causing the disease cholera. The toxin is encoded in filamentous phage (CTXϕ), and the same extracellular protein secretion pathway that facilitates the horizontal spread of the phage, which releases it. It contains two chromosomes ($\sim$2.9 $\times$ 10^6 and $\sim$1.1 $\times$ 10^6 bp) encoding 2770 and 1115 ORF, respectively. The major virulence factor, CT, and the toxin-coregulated factor, TCP, are in the longer chromosome. The bacterium also contains a pathogenicity island (VPI) apparently not of viral origin.

The integron island involved in gene integration and dissemination is in the smaller chromosome. *See* cholera toxin. (Heidelberg, J., et al. 2000. *Nature* 406:477.)

Vibrio cholerae.

vicariance Occurrence of a species in a habitat other than expected or a function not expected by an organ.

Vicia faba Broad bean, $2x = 2n = 12$; its large chromosomes are well suited for cytological study. *See* favism; vicin.

vicine Alkaloid produced by *Vicia sativa* (vetch) may be the source of this cyanoalanine toxin, especially hazardous for people and animals on a low-sulfur diet. *See* favism.

VIGS Virus-induced gene silencing. Infecting viruses my silence host plant genes as well as transgenes in plants may silence viral genes by the mechanism of RNAi. *See* RNAi. (Fagard, M. & Vaucheret, H. 2000. *Plant Mol. Biol.* 43:285.)

villus Vesicular projections on a membrane. The amniotic villi near the end of the umbilical cord are sampled for genetic examination during prenatal amniocentesis. *See* amniocentesis.

vimentin Constituent of the filament network extending through the cytoplasm of eukaryotic cells. *See* intermediate filaments. (Perez-Martinez, C., et al. 2001. *J. Comp. Pathol.* 124:70; Mor-Vaknin, N., et al. 2003. *Nature Cell Biol.* 5:59.)

vinblastin and vincristine Antineoplastic alkaloids (interfere with microtubules and the cytoskeleton) from the shrub *Vinca rosea*. *See* cytoskeleton; microtubule.

Vinca rosea.

vinculin Protein-binding α-actinin, talin, paxillin, tensin, actin filaments, and phospholipids; mediates the assembly of the cytoskeleton. *See* adhesion. (Kálmán, M. & Szabó, A. 2001. *Exp. Brain. Res.* 139:426.)

violent behavior In humans, impulsive aggression has been attributed to reduced levels of 5-hydroxyindole-3-acetic acid in the cerebrospinal fluid, and a nonsense mutation in

MAOA enzyme resulted in aggressive behavior in a kindred. *See* behavior in humans; MAOA.

VirA   Agrobacterial kinase phosphorylating the product of virulence gene *VirG*.

viral cancer   *See* cancer; oncogene.

viral envelope   Protein-lipid coat of viruses. *See* viruses.

viral ghost   Empty viral capsids without their own genetic material, but they can be filled with DNA and become genetic vectors. *See* generalized transduction; transformation, genetic.

viral oncogene   *See* v-oncogene.

viral vectors   In vitro genetically modified viral DNA (e.g., adenovirus, SV40, bovine papilloma, Epstein-Barr, BK viruses, Baculovirus [Polyhedrosis] etc.) containing nonviral genes to be introduced into eukaryotic cells. *Autonomous stable viral vectors* have also been constructed that replicate in the cytoplasm. In order to prevent killing the cells, their copy number is limited by introducing copy number regulators. From the *bovine papilloma virus* (BPV), autonomous (episomal) and shuttle vectors have been constructed that maintain low (10–30) copies in the cytoplasm. *Shuttle vectors* can be rescued from the mammalian cells and can propagate various protein genes in another cell depending on a number of intrinsic and extrinsic factors. The *Epstein-Barr Virus* (EBV) vector can be propagated in the cytoplasm of various types of mammalian cells at low copy number (2–4) and is suitable for the study of gene expression, regulatory proteins, etc. It can also be maintained in the nucleus of, e.g., B lymphocytes. The vector has up to 35 kb carrying capacity and can be rescued. The *BK* (baby kidney) *virus* has been advantageously used for human cells. Truncated retroviral HIV vectors (may not be able to recombine and restitute pathogenic forms) may be useful because they can be introduced into nondividing cells. Lentivirus vectors can pass into nondividing cells such as hepatocytes, hematopoietic stem cells, and neurons, in contrast to the most widely used mouse leukemia retrovirus (MuLV) vectors, which require DNA replication for integration.

For human applications, *replication-deficient retroviral* vectors can accommodate 9 kb exogenous DNA and they are generally used in ex vivo studies. *Adenovirus vector* can carry 7.5 kb DNA and can be taken up by the cell by a specific virus receptor and $\alpha_V\beta_3$ or $\alpha_V\beta_5$ surface integrins. The adenoviral, herpes simplex, vaccinia, and autonomous parvovirus vectors are not integrated into the human genome and thus do not lead to permanent genetic change, and the treatment has to be reapplied periodically (in weeks or months). The *adeno-associated virus* can integrate into the chromosomes in dividing cells, but it is episomal in stationary-stage cells. Most of the current vectors may cause inflammation because of antivector cellular immunity. *Baculovirus (Polyhedrosis)* is used as a vector for insect cells. The targeting of viral vectors to specific tissues can be increased by genetically engineering into the envelope protein a special receptor for a target ligand e.g., into the Moloney murine leukemia retrovirus erythropoietin was inserted or into the avian retrovirus envelope an integrin sequence was added. The vesicular stomatitis virus (VSV) and the gibbon ape leukemia retrovirus (GALV) offer some target specificity, but they form only 10^7 to 10^9 cell-forming units (CFU/mL) and therefore all the target cells of the body cannot be reached. In order to target the vector to particular cell types, the envelope protein of the virus must be modified so that it will recognize the target cell membrane receptor. In order to pass the membrane and move into the cell, the envelope protein-receptor complex must undergo a conformational change. The modified envelope protein, however, may not form an effective complex, resulting in a very low level of passing of the vector into the target cell. New technology (ligands specific for fibronectin and collagen of the cell matrix) may facilitate enrichment of the vector in the extracellular matrix of the host cell and thus a more effective uptake. *See* adeno-associated virus; adenovirus; Bellophage; biohazards; biolistic transformation; episomal vector; erythropoietin; ex vivo; gene therapy; herpes; HIV; laboratory safety; lentiviruses; liposomes; microinjection; packaging cell line; parvoviruses; plasmid rescue; retroviral vectors; self-deleting vector; shuttle vector; transfection; transformation, genetic; vaccinia virus; vectors, genetic; virus. (Pfeifer, A. & Verma, I. M. 2001. *Fundamental Virology*, Knipe, D. M. & Howley, P. M., eds. Lippincott Williams & Wilkins, p. 353, Philadelphia, PA.)

viremia   Viruses in the blood.

VIR genes   *See* virulence genes, *Agrobacterium*.

virgin   Has not been mated or did not have prior sexual intercourse. In *Drosophila*, the females can store the sperm received by prior mating and therefore the paternal identity can be secured only if virgin females are used.

virgin T cell   *See* immune response.

virile   Has the characteristics of an adult male; masculine.

virion   Complete virus particle (coat and genetic material).

virion RNA   Cytomegalovirus/herpes virus-encoded transcripts packaged within the virion and delivered to the host cell to assure their immediate expression after infection.

viroceptors   Parts of the viral attack mechanisms directed against the host immune system. They mimic the cell receptors and tie up cytokines and chemokines destined to stimulate immune defense and weaken antibody production. *See* antibody; immune system; interferons. (Upton, C., et al. 1991. *Virology* 184:370.)

viroid   Nonencapsidated RNA (ca. 1.2×10^5 daltons) capable of autonomous replication and (plant) pathogenesis, such as the potato spindle tuber viroid. Viroids are single-stranded circular or linear RNA molecules with extensive intramolecular complementarity. These agents are localized in the nuclei of plants and do not seem to occur in animals, although for a while prions were supposed to be viroids. They are probably the smallest nucleic acid agents causing infectious diseases. (Elena, S. F., et al. 2001. *J. Mol. Evol.*

53[2]:155; Rezaian, M. A. 1999. *Curr. Issues Mol. Biol.* 1:13; <http://www.callisto.si.usherb.ca/~jperra>.)

virosomes Liposomes with associated viral proteins expected to be used as vehicles for gene therapy. *See* liposome. (Yomemitsu, Y., et al. 1997. *Gene Ther.* 4:631.)

virulence Determines or indicates the infectivity or pathogenicity of an organism. It was discovered that several bacterial species of different structures and functions acquired a shared mechanism for virulence. About 15–20 protein genes with relatively low G-C contents (below 40%) are assembled in a pathogenicity island of the bacterial chromosome or a plasmid. These genes encode the molecular machinery (type III virulence) to produce and transmit bacterial toxins to their target. In *Salmonella* species, a 65–100 kb plasmid carries the genes required for systemic infection. A *Yersinia* effector protein (Yopj), a cysteine protease, specifically blocks the host signal transduction system at MAPKK. Yopj-related proteins occur in other bacterial pathogens of animals and plants. *See* avirulence; host-pathogen relations; neurovirulence; SAR; signal transduction. (Mahan, M. J., et al. 2000. *Annu. Rev. Genet.* 34:139; Cotter, P. A. & DiRita, V. J. 2000. *Annu. Rev. Microbiol.* 54:519.)

virulence genes, *Agrobacterium* The Ti plasmid carries—in about 35 kb DNA—major (*A, B, C, D, E*) and minor (*F, H*) virulence genes that mediate the process of infection and T-DNA transfer. Gene *VirA* codes for a single protein that is a transmembrane receptor. Its N-terminal periplasmic region responds to sugars and pH, whereas the periplasmic loop between the two membrane layers responds to phenolic compounds (e.g., acetosyringone secreted by the wounded plant tissues). This substance plays an important role in the induction of the cascade of all *Vir* genes, although *VirA* itself is constitutive yet modulated by several factors. The C-terminus of VirA protein is autophosphorylated. *VirG* also codes for a single protein, a transcriptional regulator. For expression, it requires phosphorylation by VirA. *VirG* regulates phosphate metabolism by feedback mediated by *VirA*, and these two genes together are involved in conjugational transfer. The *VirB* operon encoding 11 proteins is also a conjugational mediator. VirB1 is a lysozyme-like protein. *VirC* determines host range and C1 protein binds to the *overdrive* repeats near the right border of some octopine plasmids. The *VirD* genes are responsible for four polypeptides. VirD1 is a topoisomerase, VirD2 is an endonuclease; in addition, this locus codes for a binding protein and a pilot protein guiding the T-DNA to the plant chromosome. *VirE2* codes for a binding protein, which coats the T-strand and mediates the transfer of single-strand DNA into the plant nucleus. VirE1 mediates VirE2 transport. *VirF* probably encodes an extracellular protein that regulates *Vir* functions. *VirH* product may metabolize plant phenolics. Nopaline plasmids contain the gene *tzs* (trans-zeatin secretion) and *pin* (plant-inducible) *F* loci. Chromosomal virulence loci (*ChvA, ChvB, Chv, psc*) are involved in the production of bacterial surface polysaccharides. Chromosomal genes *cbg* and *pgl*, when present, may enhance virulence. A single-stranded T-DNA complexed with proteins is transferred to plant cells. *See Agrobacterium*; crown gall; overdrive; T-DNA; Ti plasmid; transformation, genetic. (Winans, S. C., et al. 1999. *Pathogenicity Islands and Other Mobile Virulence Elements*;

Kaper, J. B. & Hacker, J., eds., p. 289. *Am. Soc. Microbiol.*, Washington, DC; Dumas, F., et al. 2001. *Proc. Natl. Acad. Sci. USA* 98:485.)

virulent In general, the poisonous form of prokaryotes; the virulent bacteriophages do not have the prophage lifestyle, and after reproduction they destroy the host bacteria by lysis. *See* lysis; prophage.

virus Small particle with double- or single-stranded DNA or RNA as the genetic material. Viruses are the ultimate parasites because they lack any element of the metabolic machinery and absolutely depend on the host for assistance to express their genes. They are generally so small in size (15–200 nm) that light microscopy cannot reveal them, except the pox viruses, which may be up to 450 nm in length. There are about 1,500 virus species. Bacterial, animal, and plant viruses exist. The majority of plant viruses have single-stranded RNA genetic material, but a few have double-stranded DNA, e.g., CaMO. *See* animal viruses; bacteriophages; cauliflower mosaic virus; oncogenic viruses; plant viruses; retroviruses; TMV. (Whittaker, G. R., et al. 2000. *Annu. Rev. Cell Dev. Biol.* 16:627, for a chronology of virology, Oldstone, M. B. A. & Levine, A. J. 2000. *Cell* 100:139; Sharp, P. M. 2002. *Cell* 108:305; <http://life.anu.edu.au/viruses/welcome.htm>; Virus Particle Explorer: <mmtsb.scripps.edu/viper/viper.Html>; <www.ncbi.nlm.nih.gov/PMGifs/Genomes/viruses.html>.)

virus, computer Deliberately generated destructive program that can be spread through borrowed computer disks as well as network services. The damage to the files can usually be prevented by the use of continuously updated virus monitoring and eliminating programs.

virus-free plants Plants regenerated under axenic conditions from apical meristems are generally free of virus disease until they are reinfected. Plants so obtained are commercially useful for the production of virus-free seed stocks. Antisense RNA constructs may render plants virus resistant. (Hammond, J. & Kamo, K. K. 1995. *Mol. Plant Microbe Interact.* 8:674.)

virus hybrid When the 2*b* gene of cucumber mosaic RNA virus is replaced by its homologue of tomato, virulence of the interspecific hybrid virus increases. *See* vaccinia virus.

virus morphology *See* development; lambda phage; retroviruses; T4.

virusoid 300–400-nucleotide-long RNA pathogenic to plants. virusoids accompany other plant viruses. *See* viroid.

virus receptors Cell-surface proteins that mediate viral entry into the host cell. The proteins can be cell adhesion molecules (CXCR4, CD4, dystroglycan, integrins, ICAM, major histocompatibility antigens, etc.), extracellular matrix proteins (heparan sulfate glycoaminoglycan, sialic acid derivatives), and complement-control proteins (CD46, CD55), aminopeptidase-N, lipoprotein receptors, coxsackie virus, and adenoviral receptors (CAR1), etc. Some receptors may serve several different viruses and some viruses can take advantage of more than one type of receptor. Single amino-acid replacements in the receptor either abolish or facilitate the uptake of the virus. (Baranowski, E., et al. 2001. *Science* 292:1102; Bomsel, M. & Alfsen, A. 2003. *Nature Rev. Mol. Cell Biol.* 4:57.)

virus reconstitution   *See* reconstituted virus.

virus resistance   Fv and Rfv genes of mice convey resistance against mouse leukemia (Friend virus) by encoding a retroviral envelope protein or viral gag (group-specific antigen). In the latter case, the Fv gene product blocks entry of the virus into the nucleus or these proteins may exercise a dominant negative effect on the virus that happened to be there. Introduction into the plant genome of tobacco mosaic virus coat protein genes restricts the virulence of the virus in the normally susceptible plants. (Lin, H. X., et al. 2003. *J. Gen. Virol.* 24:249.)

virus transport to the cell nucleus   Mediated by GAG matrix–associated protein (MA) and the VPR gene product in case of HIV. MA includes a nuclear localization sequence (NSL) and a signal targeting the cell membrane. When NLS is phosphorylated, MA becomes part of the preintegration complex. Further phosphorylation detaches MA from the membrane. *See* acquired immunodeficiency, nuclear localization sequence. (Bell, P., et al. 2001. *J. Virol.* 75:7683.)

viscosity   Internal friction of fluids expressed as dyne-seconds/cm^2 called poise unit. *See* dyne; stoke.

visible mutation   Can be identified by the phenotype seen.

vision   *See* rhodopsin.

VISTA   Visualization tool for alignment is a computer program.

vital genes   *See* lethal mutation.

vitalism   19th-century and earlier, postulates that living beings are controlled not only by physical and chemical mechanisms, but life is associated with a transcendental vital force. There still remain phenomena like the development of an embryo or regeneration from single cells that cannot be fully explained in physicochemical terms, although vitalism is no longer a viable idea since Friedrich Wöhler synthesized urea in 1828 from inorganic ingredients. (Hein, H. 1972. *J. Hist. Biol.* 5:159.)

vital stain   Colors living cells without serious damage to viability.

vital statistics   Involve birth marriage and death registrations. The information may assist in constructing human pedigrees and may be important for family histories, congenital and hereditary disease, longevity, etc.

vitamin A   Also called retinol. It is synthesized from carotenoids. Its deficiency results in visual and skin anomalies. Excessive amounts may be harmful. With the aid of three

Vitamin A (all-trans-retinol).

transformation vectors, the β-carotene pathway has been introduced into the carotenoid-free rice endosperm, enabling the production of provitamin A and alleviating deficiency in the diet. *See* retinal; retinol.

vitamin B$_1$ *See* thiamin.

Vitamin B$_1$.

vitamin B$_2$ *See* riboflavin; riboflavin retention deficiency.

vitamin B$_6$ *See* pyridoxine.

vitamin B complex   Includes thiamin, riboflavin, nicotinic acid (amide), panthotenic acid, pyridoxin, and vitamin B$_{12}$.

vitamin B$_{12}$ defects   Vitamin B$_{12}$ or its coenzyme form has a molecular weight of about 1,355. It is composed of a core ring with cobalt (Co^{3+}) at its center and a dimethyl benzimidazole ribonucleotide is joined to it through isopropanol. It also contains a 5′-deoxyadenosine. It is usually isolated as a cyanocobalamin because during the process of purification a

Vitamin B$_{12}$.

cyano group may be attached to the cobalt at the place where the 5'-deoxyadenosyl group is positioned in the coenzyme. B$_{12}$ is not synthesized in plants and animals; thus, it is not usually present in the diet. Intestinal microorganisms make it from the meat consumed and it is absorbed when the so-called intrinsic factor, a glycoprotein, is available in satisfactory amounts. If less than 3 μg/day is accessible, pernicious anemia develops in humans. Autosomal-recessive mutation may prevent the release of the lysosomally stored vitamin and B$_{12}$ deficiency may result. Cysteinurias are often called cobalamin F (cbl F) disease. In methylmalonicacidemia combined with homocystinuria, methylmalonyl-CoA mutase and homocysteine methyl-tetrahydrofolate methyltransferase (cbl C) deficiencies are involved. *See* amino acid metabolism; cysteinuria; methyl-malonicaciduria.

vitamin C   *See* ascorbic acid.

Vitamin C.

vitamin D   Antirachitic fat-soluble vitamin. Its deficiency leads to rickettsia and defects in bone development and maintenance. Vitamin D$_2$ (ergocalciferol) is formed upon irradiation of ergosterol and vitamin D$_3$ (cholecalciferol) from 7-dehydrocholesterol. Children need about 20 μg/day in their diet. Autosomal-recessive human defects in vitamin D receptors (12q12-q14) do not respond favorably to a vitamin D–fortified diet. Several cancer cells are inhibited by vitamin D$_3$ and it may induce apoptosis. *See* hormone-response elements; VDR; Williams syndrome. (McGuire, T. F., et al. 2001. *J. Biol. Chem.* 276:26365.)

Ergocalciferol.

vitamin E (tocopherols)   Vitamin E is an antioxidant (in several different forms) normally present in satisfactory amounts in a balanced diet. In an autosomal-recessive condition, vitamin E malabsorption was observed, causing intestinal and nervous anomalies and accumulation of cholesterol. The severe symptoms could be alleviated by 400–1,200 international units

(IU) of vitamin E. It appeared that the affected individuals lacked an α-tocopherol-binding protein required to build it into very low-density lipoproteins. In a transgenic mouse model, vitamin E reduced chromosomal damage and hepatic tumors. Familial vitamin E deficiency is under autosomal-recessive control in humans. *See* atherosclerosis; tocopherols; VLDL. (Azzi, A., et al. 2002. *FEBS Lett.* 519:8.)

Tocopherol.

vitamin K (phylloquinone)   Plant lipid cofactor of blood coagulation (vitamin K$_1$) and a related substance, menaquinone (vitamin K$_2$), is synthesized by intestinal bacteria of animals. Synthetic menadione (vitamin K$_3$) also has some vitamin K activity. *See* vitamin K–dependent blood-clotting factors.

vitamin K–dependent blood-clotting factors   Some autosomal-recessive bleeding diseases respond favorably to the administration of vitamin K. It is apparently required for posttranslational modification of at least six proteins involving the conversion of the NH$_2$-end of glutamic acid into γ-carboxyglutamic acid. Deficiency of this process may occur as a consequence of treatment by coumarin drugs (e.g., warfarin) used as anticoagulants. A vitamin K–dependent blood coagulation factor is encoded at 2p12. *See* antihemophilic factors; γ-glutamyl carboxylase; prothrombin deficiency; resistance to coumarin-like drugs; warfarin.

vitamins   Dietary supplements without measurable caloric value. They generally serve the role of coenzymes.

vitelline layer   Yolk (heavier) layer around the eggs; in mammals, it is (a thinner) zona pellucida. *See* egg.

vitellogenin   Yolk protein.

vitiligo   Autosomal-dominant (1p31.3-p32.2) halo skin spots (may be identical on opposite side of the body) present after birth that may spread or regress. It is an autoimmune disease with an incidence of ~1%. *See* nevus; piebaldism; skin diseases. (Alkhateeb, A., et al. 2002. *Hum. Mol. Genet.* 11:661.)

vitrification Procedure of protecting sensitive biological material (enzymes, seeds) from deterioration by coating with sugar mixtures (e.g., sucrose and raffinose) that, however, readily dissolve when needed. Vitrification literally means the formation of a glass-like structure. Human oocytes can be preserved in appropriate vitrification medium and frozen in liquid nitrogen. *See* cryopreservation. (Yokota, Y., et al. 2001. *Sterility & Fertility* 75:1027.)

vitronectin (S protein) Fluid-phase protein in the cell binding the C5b-9 complement component, which prevents its attachment to the membrane and thus interferes with lytic activity. *See* complement. (Tomasini, B. R. & Mosher, D. F. 1991. *Progr. Hemost. Thromb.* 10:269; Seger, D., et al. 2001. *J. Biol. Chem.* 276:16998.)

vivipary Giving birth to live offspring. In plants, the seed germinates before shedding from the fruits. This phenomenon is genetically controlled through abscisic acid metabolism. *See* ABA. (Paek, N. C., et al. 1998. *Mol. Cell* 8[3]:336; Jones, H. D., et al. 2000. *Plant J.* 21:133.)

V(J)D recombinase Assembles immunoglobulins and T-cell receptors. The activity of this enzyme is facilitated by histone acetylation. *See* accessibility; histone acetyltransferase; immunoglobulins; ligase; RAG; RSS; SCID; T-cell receptors. (Fugmann, S. D., et al. 2000. *Annu. Rev. Immunol.* 18:945; Gellert, M. 2002. *Annu. Rev. Biochem.* 71:101.)

VLDL Very low-density lipoprotein is the 55 nm precursor of LDL and triglycerides. Its core contains cholesteryl ester. In muscle capillaries and in the adipose tissues, VLDL triglycerides are removed and exchanges take place with other lipoproteins, resulting in the loss of proteins except apolipoprotein B-100 and a smaller (22 nm) particle size. It is now called LDL (low-density lipoprotein). The VLDLR receptor gene (9p24) is very similar to the LDLR, but it contains an additional exon. *See* Alzheimer disease; apolipoproteins; familial hypercholesterolemia; hypertension; low-density lipoproteins; sitosterolemia; sterol. (Merkel, M., et al. 2001. *Proc. Natl. Acad. Sci. USA* 98:13294.)

VLP Virus-like particle, e.g., some transposable elements (retroposons). *See* retroposons; Ty.

V_max *See* V.

v-mil *See* MIL; MYC.

vMIP *See* chemokines; herpes virus.

v-mos *See* Moloney mouse sarcoma.

VNC Viable but noncultivatable microbial cells. They may be transgenic (genetically modified) and may be protected by nucleic acids.

VNO *See* olfactogenetics; vomeronasal organ.

VNTR Variable number of tandem repeats. Loci are used in forensic DNA fingerprinting. These repeats may display hundreds of alleles per single loci and are extremely polymorphic in restriction fragments. They are useful as physical markers for mapping. Also, because matching patterns occur by a chance of 10^{-7} to 10^{-8} only, they are well suited for criminal or other personal legal identification on the basis of minute amounts of DNA extracted from drops of body fluids, blood, or semen. VNTR can be used for taxonomic and evolutionary studies in animals and plants. *See* diabetes mellitus; DNA fingerprinting; MLVA; MVR; SSM; trinucleotide repeats. (Le Stunff, C., et al. 2001. *Nature Genet.* 29:96.)

Vogt-Spielmeyer disease Batten disease. *See* ceroid lipofuscinosis.

vole *Microtus agrestis*, $2n = 50$; *Microtus arvalis*, $2n = 46$; *Microtus montanus*, $2n = 24$.

volicitin *See* biological control.

volt (V) Unit of electric potential. If the resistance is $1\ \Omega$ (ohm $= 1\ \text{V/A} = 1\ \text{m}^2\ \text{kg sec}^{-3}\ \text{A}^{-2}$) and the electric current is 1 ampere, the voltage is 1. *See* ampere; watt.

voltage-gated ion channels Opened/closed for the transport of ions in response to a change in voltage across cell membranes. *See* ion channels; signal transduction.

vomeronasal organ (VNO) Pair of chemosensory organs at the base of the nasal cavity in the majority of higher animals, except birds and some monkeys and apes. VNO is stimulated by pheromones. Signaling through GTP-binding proteins requires 1,4,5-trisphosphate rather than cAMP. In humans, pheromones play a subordinate role or no role and VNOs are apparently nonfunctional. *See* cAMP; olfactogenetics; pheromones; phosphoinositides. (Lane, R. P., et al. 2002. *Proc. Natl. Acad. Sci. USA* 99:291.)

v-oncogene Viral oncogene homologous to a c-oncogene but carried by oncogenic viruses that are capable of causing cancerous growth. *See* cancer; c-oncogene; oncogenes; retrovirus.

Von Gierke disease *See* glycogen storage disease.

Von Hippel–Lindau syndrome (VHL) Dominant phenotype (human chromosome 3p26-p25) involving tumorous growth primarily of the blood vessels of the eye (hemangioma) and the brain (hemangioblastoma). The central nervous system, kidneys (pheochromocytoma), and pancreas may also become tumorous. The incidence estimates vary between 0.00002 and 0.00003, and the mutation rate appears to be about $2{-}4 \times 10^{-6}$. The primary cause of the tumor is the inactivation of the VHL tumor-suppressor gene. VHL protein forms ternary complexes with elonginC, and elonginB, and CUL2. This complex marks HIF for degradation by a proteasome. RBX helps to recruit ubiquitinating proteins. With the assistance of normal pVHL, the α-subunit of HIF (hypoxia-inducible factor) is degraded and the level of oxygen increases. HIF, with the cooperation of other proteins, e.g., VEGF, promotes the formation of blood vessels that are required for cancerous growth. When pVHL protein formation is reduced or prevented, HIF remains active and creates the conditions for angiogenesis and cancerous growth. It is believed that VHL is a negative regulator of VEGF by association with complexes (SCF) that target proteins for degradation. The VHL gene falls to the same area as the RCC gene. VHL inactivation seems to affect TGF-α, which may stimulate the renal carcinogenic path. *See* angiogenesis; cullin; elongin; eye diseases; hypernephroma; kidney diseases; mutation rate; pheochromocytoma; proteasome; Rbx1; RCC; renal cell carcinoma; SCF; SKP; ubiquitin; VEGF. (Iwai, K., et al. 1999. *Proc. Natl. Acad. Sci. USA* 96:12436; de Paulsen, N., et al. 2001. *Proc. Natl. Acad. Sci. USA* 98:1387; Ivan, M., et al. 2001. *Science* 292:464; Friedrich, C. A. 2001. *Hum. Mol. Genet.* 10:763.)

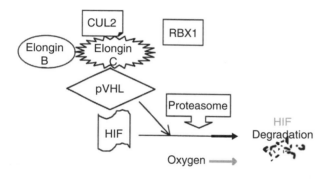

Von Neumann–Morgenstern gamble

Seriously or terminally ill patients may choose an experimental drug or treatment that may improve quality of life, but there is also a risk for further aggravating the illness and even shortening the life.

Von Recklinghausen disease (NF1)

Dominant human chromosome 17q11.2 neurofibromatosis, a skin tumor with characteristic café-au-lait spots. Some other possible features are pseudoarthritis, glioma, mental retardation, hypertension, hypoglycemia, etc. Mutation rate is ~10^{-4}. The basic defect appears to be in a cytoplasmic GAP protein. *See* café-au-lait spot; elephant man; GAP; neurofibromatosis.

Von Willebrand disease (VWD)

Complex hereditary bleeding condition based on the deficiency of a large antihemophilic cysteine-rich glycoprotein (2813 amino acids) in the blood plasma, platelets, and subendothelial connective tissue. It is different from hemophilias inasmuch as the bleeding from the gastrointestinal system, urinary system and uterus is prolonged. Several forms of this disease have been distinguished based on which component of the large gene is affected. The type III form causes very severe symptoms and it is the least common type of VWD. The most common VWD protein has a binding domain to antihemophilia factor VIII and its defect appears as a relatively rare recessive. There is also an X-linked form. The bleeding can be readily stopped upon supplying normal blood to patients. The most common dominant gene, which causes the reduction of this glycoprotein, was assigned to human chromosome 12pter-p12.3. In ~180 kb it contains 52 exons. Exon 28 is the largest, encoding domains A1 and A2. Domains D1-D2-D′-D3-A1-A2-A3-D4-B1-B2-B3-C-1-C2-CK are distinguished. Its highly homologous pseudogene is located in 22q11.2. At least 22 human genes carry homologies to the A domain of the chromosome-12 gene. The frequency of heterozygotes has been estimated from 1.4 to 5%, and thus VWD appears quite frequent, although some other estimates are much lower. Recurrence risk for the dominant form, in case one of the parents is affected, is about 50%. Both heterozygotes and homozygotes may express VWD. Percutaneous umbilical blood sampling (PUBS) may permit prenatal diagnosis. *See* antihemophilic factors; epiphyseal dysplasia; Glanzmann's disease; hemophilias; hemostasis; prenatal diagnosis. (Sadler, J. E. 2002. *Science* 297:1128.)

Von Willebrand–Jürgen syndrome

See thrombopathic purpura.

VP

Viral proteins such as the VP1, VP2, and VP3 of simian virus 40. *See* SV40.

VP16

(α-transinducing factor, α-TIF) Herpes simplex virus transcription activation domain that can boost the expression of other genes by a factor of 10^5. *See* transactivator. (Hirst, M., et al. 2001. *Proc. Natl. Acad. Sci. USA* 98:8726.)

Vpg

Genome-linked (polio) viral protein (22 amino acids) attaches to the 5′-end of RNA viruses and assists as a primer in the replication of the nucleic acid. *See* primer.

V-point

Progression of the cell cycle beyond the V-point (≈6 hr before S phase) requires no insulin but only IGF-1. *See* cell cycle; insulin; insulin-like growth factor.

VRE

Ventral-response element participates in the regulation of ventral development by preventing switching on of an activator that operates dorsal development.

Vrolik disease

See osteogenesis imperfecta type II.

VSG

Variable-surface glycoprotein. *See* Borrelia; telomere switching; *Trypanosoma*.

VSP

Very short patch repair. A prokaryotic repair involving T-G mismatches by restoring the original base pairs. *See* (Lieb, M., et al. 2001. *J. Bacteriol.* 183:6487.)

VSV *See* CO_2 sensitivity.

V-type ATPases Responsible for the acidification of cellular organelles (vacuoles, lysosomes, Golgi complex) by maintenance of the vacuolar type ATPase — proton pump in plant and animal cells. *See* ATPase; Golgi complex; lysosomes. (Gruber, G., et al. 2001. *J. Exp. Biol.* 204:2597.)

V-type position effect Variegated expression of genes transposed into the vicinity of heterochromatin. This phenomenon is common in *Drosophila*, but few cases have been demonstrated in plants. One of the most common causes of variegation is the movement of insertion or transposable elements or sorting out of plastid DNA-encoded mutations. *See* chloroplast genetics; heterochromatin; position effect; transposable elements; variegation.

vulva Outer region of the external female genital organ, the vaginal orifice, and organs associated with it.

vulvovaginitis Autosomal-dominant allergy to semen, resulting in vaginal inflammation after coitus, usually for several hours.

R.C. Punnett (Heredity 4:9) gives the following account: "I was asked why it was that, if brown eye were dominant to blue, the population was not becoming increasingly brown eyed: yet there was no reason for supposing such to be the case. I could only answer that the heterozygous browns also contributed their quota of blues and that somehow this lead to equilibrium. On my return to Cambridge I at once sought out G. H. Hardy with whom I was then very friendly. For we had acted as joint secretaries to the Committee for the retention of Greek in the Previous Examination and we used to play cricket together. Knowing that Hardy had not the slightest interest in genetics I put my problem to him as a mathematical one. He replied that it was quite simple and soon handed to me the now well-known formula $pr = -q^2$ (where p, $2q$ and r the proportions of AA, Aa, and aa individuals in the population varying for the A-a difference). Naturally pleased at getting so neat and prompt an answer I promised him that is should be known as 'Hardy's Law'—a promise fulfilled in the next edition of my Mendelism. Certain it is...that 'Hardy's Law' owed its genesis to a mutual interest in cricket."

A historical vignette.

w Symbol of map distance. *See* cM; map distance; mapping function; recombination; θ.

Waardenburg syndrome Autosomal-dominant forms may be distinguished on the basis of displacement (type I) and without displacement (type II) of the eyelids. Variegation in the color of the iris, white forelock, eyebrows, and eyelashes, syndactyly, heart problems, and hearing defects may occur as an autosomal-recessive anomaly. Dominant mutations in the SOX10 gene may affect neural crest–derived cell lineages. Waardenburg syndrome type 2 gene (MITF [microphthalmia-associated transcription factor]) converts fibroblasts into melanocyte-like cells by transactivation of a tyrosinase gene. If MITF is inactive, hypopigmentation occurs. Mutation in the PAX3 transcription factor is responsible for hearing and pigmentation defects of type I form caused by failure of transactivation of MITF. Human chromosomal locations: type I 2q35, type IIA 3p14.1-p12.3, type IIB 1p21-p13.3. Type III, called Klein-Waardenburg syndrome, is also at 2q35 location. The syndrome may be haploinsufficient. *See* DiGeorgesyndrome; eye defects; haploinsufficient; microphthalmos; PAX; polydactyly; Shah-Waardenburg syndrome. (Sanchez-Martin, M., et al. 2002. *Hum. Mol. Genet.* 11:3231.)

Waardenburg-Shah syndrome *See* Shah-Waardenburg syndrome.

WAF *See* p21.

WAGR *See* Wilms' tumor.

Wahlund's principle When two populations, each with different allelic frequencies and both in Hardy-Weinberg equilibrium, are mixed by migration, there will be an overall decrease of heterozygotes: $\overline{H} = 2\overline{p}\,\overline{q}[1 - (\sigma^2/\overline{p}\,\overline{q})]$. The decrease of overall heterozygosity indicates the degree of heterogeneity between the two populations and $(\sigma^2/\overline{p}\,\overline{q})$ is Wahlund's variance of gene frequencies. *See* allelic frequencies; migration. (Yasuda, N. 1968. *Am. J. Hum. Genet.* 20:1.)

Waldemar of Prussia One of the hemophiliac great-grandsons of Queen Victoria. *See* hemophilia A; Queen Victoria.

Waldenström syndrome *See* macroglobulinemia.

Walker boxes (P-loops) Nucleotide triphosphate-binding amino acid sequences in several proteins. Box A promotes branch migration in Holliday junctions during recombination mediated by the Ruv B protein. Walker A: GlyXX-GlyXGly**Lys**Thr; Walker B: AspGluXAsp. The Lys residue binds the γ-phosphate of nucleotides directly. *See* branch migration; Holliday junction; RuvABC. (Walker, J. E., et al. 1982. *EMBO J.* 1:945; Hishida, T., et al. 2001. *J. Biol. Chem.* 274:25335.)

Walker-Wagner syndrome Autosomal-recessive hydrocephalus (accumulation of fluid in the enlarged head) generally associated with retinal detachment, congenital muscular dystrophy, and lissencephaly. *See* head/face/brain defects; hydrocephalus; lissencephaly; Miller-Dieker syndrome; prenatal diagnosis.

Walker-Warburg syndrome (HARD, 9q31) Autosomal-recessive hydrocephalus agyria and retinal dysplasia that was originally described as lissencephaly. *See* Miller-Dieker syndrome. (Beltran-Valero de Barnabé, D., et al. 2002. *Am. J. Hum. Genet.* 71:1033.)

walking *See* chromosome walking.

walking of transcriptase Transiently halting the movement of the RNA polymerase ternary complex (polymerase, DNA, transcript) by sequentially providing subsets of the four ribonucleotides. *See* transcript elongation.

wallaby (1) *Wallabia bicolor*, $2n = 11$ in males and 10 in females; *Wallabia eugenii* $2n = 16$. (2) Nonviral retrotransposable element named after the jumping small Australian kangaroo. *See* retroposon.

walnut (*Juglans* spp) Both wild and cultivated forms; $2n = 2x = 32$.

walnut comb In poultry, determined by the genetic constitution $RrPp$. As a result of epistasis, it occurs in 9/16 frequency in F_2 after brother-sister mating of the same double heterozygotes. The other phenotypes in the segregating F_2 are rose $Rr/RR, pp$ (3), pea $rr, PP/Pp$ (3), and single pp, rr(1).

From left to right:l Walnut, Rose, Pea, Single Comb.

walrus *Odobenus rosmarus*, $2n = 32$

waltzing mouse Chromosome deletion causing involuntary movements.

wanda *See* fish orthologous genes.

wandering spots Older method of sequencing short oligonucleotides. (Le Gall, O., et al. 1988. *J. Gen. Virol.* 69:423.)

Ward-Romano syndrome (WRS, 11p15.5) Autosomal-dominant or -recessive LQT disease involving anomalous heart muscle fibrillations, fainting (syncope), and possibly sudden death. It is the same as long QT syndromes. *See* Andersen syndrome; electrocardiography; HERG; ion channels; Jarvell and Lange-Nielson syndrome; long QT syndrome; LQT.

warfarin (C$_{19}$H$_{16}$O$_4$) Slightly bitter water- and alcohol-soluble compound. Depresses the formation of prothrombin, which is necessary for blood clotting and may cause fragility of the capillary veins, leading to hemorrhages. It is used in certain surgeries and treatment of diseases that block arteries by blood clots. It is also a rodent poison. A single ingestion may not necessarily be very hazardous to humans, but rats or mice eating it repeatedly in baits suffer internal bleeding and die. The antidote of warfarin is vitamin K. Mutations in rodents may make them resistant to warfarin. *See* anticoagulation factors; chondrodysplasia; coumarin-like drug resistance; prothrombin deficiency; vitamin K-dependent clotting factors. (Van Aken, H., et al. 2001. *Clin. Appl. Thromb. Hemost.* 7:195; Loebstein, R., et al. 2001. *Clin. Pharmacol. Ther.* 70[2]:159.)

wasp *Habrobracon* spp. $2n = 20$ for female, $2n = 10$ for male. The males hatch from unfertilized eggs and are haploid. The females come from fertilized eggs and are diploid. They are heterozygous for any of the nine sex factors. As a result of inbreeding, occasionally sex factor homozygotes arise that are sterile biparental males. Gynandromorphs occur in wasps, but these are different from those in *Drosophila* because the haploid sectors are not necessarily male sectors as expected from the loss of one set of chromosomes. Although circulating sex hormones do not seem to exist in insects, some type of diffusible substance affects the chromosomally male sectors. Exceptional gynandromorphs may arise by fertilization of binucleate eggs. Gynandromorphic tendency is genetically determined. *See* honey bee; social insects. (Page, R. E., et al. 2002. *Genetics* 160:375.)

WASP (the right wings, legs, antenna were omitted from the picture).

WASP Wiskott-Aldrich syndrome proteins that regulate the assembly of actin monomers into filaments and regulate the cytoskeletal organization and motility of cells. WASF1 is encoded at 6q21-q22, WASF3 at 13q13. They are effectors for the signal transmission from tyrosine kinase receptors to the cytoskeleton. The function of WASF2 (1p36.11-p34.3) is also similar. The latter apparently has a pseudogene at Xp11.22. *See* actin; cytoskeleton; Wiskott-Aldrich syndrome.

watercress (*Rorippa nasturtium-aquaticum*) Northern European vegetable with $x = 16$, but the species may be diploid, sterile triploid, or tetraploid.

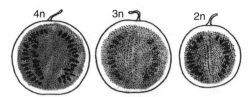

(Diagram modified after Eigsti, O. J. & Dustin, P. Jr. 1955. *Colchicine.* Iowa State College Press, Ames, Iowa.)

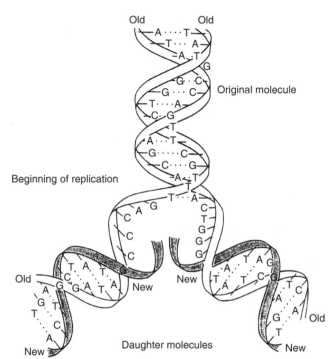

The double-stranded DNA molecule is joined through 2 and 3 hydrogen bonds between the A=T and G≡C nucleotides, respectively. The staircase-like ribbons represent the sugar-phosphate backbone of the double helix. During replication, the old plectonemic strands unwind and each old separated strand serves as a template for the formation of the new molecules that are composed from one old and one new single strand (*see* DNA replication; replication fork). It is therefore called the semiconservative mode of replication. The model is consistent with most genetic phenomena (mutation, recombination, gene expression, etc.). Since the model was originally proposed, the details of the mechanisms of DNA transactions have been worked out in greater detail, but basically none of the essential features had to be revised. This model served as a basis for the *central dogma of genetics*, indicating that the flow of information is from DNA to RNA and to protein. During the 1960s, it was discovered that through reverse transcriptase information can be directed by reverse transcription from RNA to DNA but not from protein to RNA and DNA. The discovery of prions makes the role of proteins in heredity somewhat ambiguous, however. All other proposals concerning hereditary molecules (besides DNA and RNA) have faded into oblivion. The Watson and Crick model is shown above.

watermelon (*Citrullus vulgaris*) Annual fruit, $2n = 22$. Triploids are grown commercially and the seeds are produced by crossing tetraploid plants with diploids. The fruits of the triploid plants are practically seedless and make eating more convenient. According to some reports, the triploids have a higher sugar content than either of the parental forms. *See* figure on top of this page.

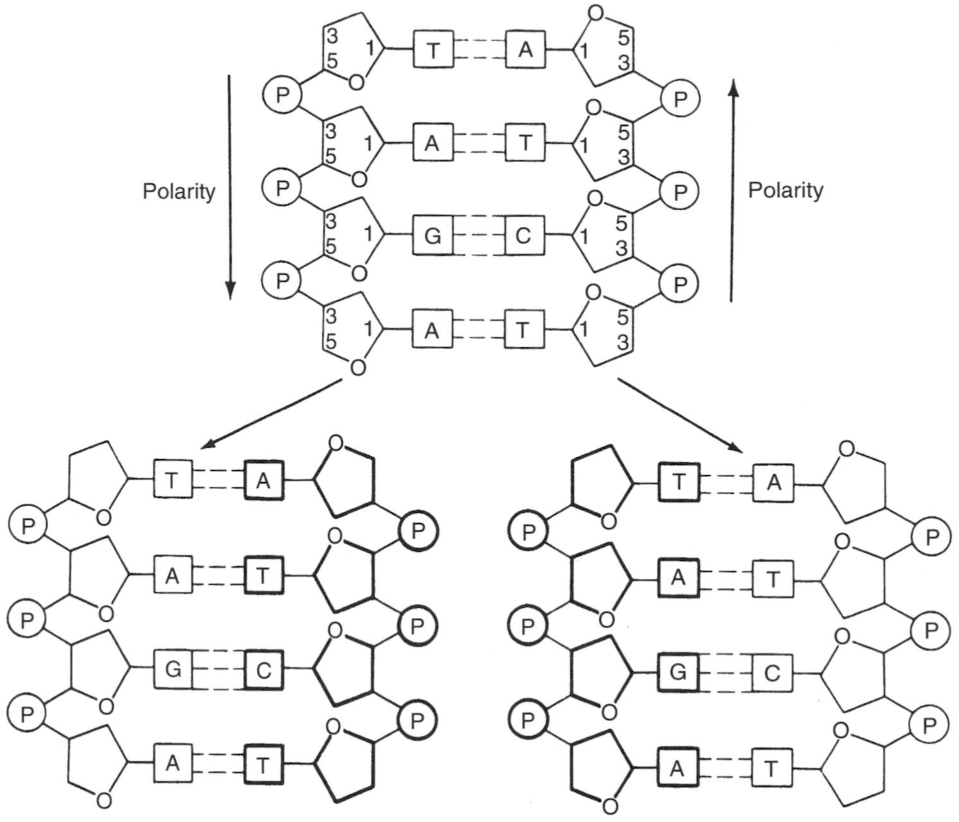

The Watson & Crick model as represented by Josse, J., Kaiser, A. D. & Kornberg, A. 1961. *J. Biol. Chem.* 236:864.

Watson and Crick model   As described by J. D. Watson and F. H. C. Crick, 1953, in *Nature* (171:964), it became perhaps the world's most famous biological model ever conceived. *See* diagrams on preceding and this page.

watt (W)   Product of volts and amperes in case of direct current. 1 W = 1 joule/sec = 0.293 calories/sec = 1/735 HP (horsepower). That is, 1 W power is generated by the electric potential between two points of 1 volt and 1 ampere current. *See* ampere; volt.

wax coat   *See* eceriferum.

W chromosome   Corresponds to the Y chromosome in the heterogametic females, WZ, in birds and butterflies. *See* chromosomal sex determination.

WCPP   Whole chromosome painting probe contains a mixture of many probes specific for a single chromosome and thus may label its entire length with color. The multicolor labeling probes may permit the differentiation of all chromosomes in a single karyotype. *See* chromosome painting; FISH; GISH; USP.

WD-40   Repeat (N) motif of tryptophan (W)–aspartic acid (D) in several eukaryotic regulatory proteins (absent in prokaryotes):

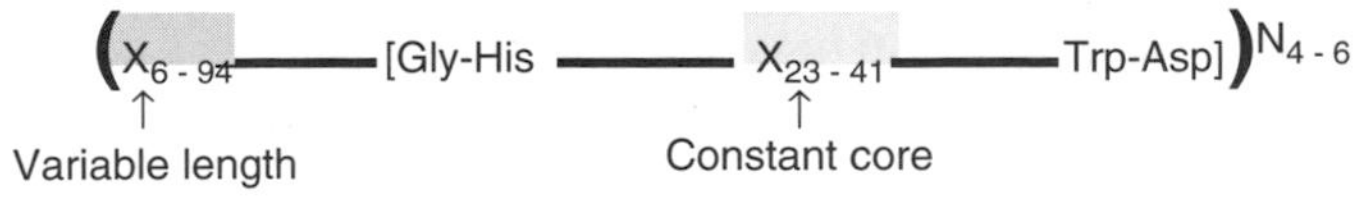

WD repeats are involved in signal transduction, RNA processing, developmental regulation, the cell cycle, vesicular traffic, etc. *See* signal transduction. (Neer, E. J., et al. 1994. *Nature* 371:2987; Smith, T. F., et al. 1999. *Trends Biochem. Sci.* 24:181.)

W-DNA   Left-handed zigzag duplex with the same directions as B-DNA, but other characteristics match that of Z-DNA. *See* DNA types.

weasel   *Mustela erminea*, 2n = 44; *Mustela frenata*, 2n = 44.

Webb (WB)   A very rare blood group involves an altered glycosylation of glycophorin. *See* blood groups; En; Gerbich; glycophorin; MN. (Reid, M. E., et al. 1985. *Biochem. J.* 232:289.)

WEE1   Protein kinase that inactivates *cdc2* gene product through phosphorylation of the tyrosine-15 residue. Wee1 is subject to proteolysis in a Cdc34-dependent way before S phase can be completed. *See* cdc; cell cycle; checkpoint; kinase; Mik1. (Tzivion, G., et al. 2001. *Oncogene* 20:6331; Batholemew, C. R., et al. 2001. *Mol. Cell Biol.* 21:4949.)

Wegener granulomatosis   Autosomal dominant, characterized by the presence of a ~29 kDa serine protease antigen identical with myeloblastin; a very serious ailment attacking primarily the upper and lower airways and kidneys. *See* myeloblast.

Weibel-Palade bodies   0.1–0.2 µm × 4 µm membrane-enclosed bodies found around the platelets or in the endothelium of all animals. They store the von Willebrand glycoprotein in their 150–200 Å tubules. *See* von Willebrand disease. (Denis, C. V., et al. 2002. *Proc. Natl. Acad. Sci. USA* 98:4072.)

weighted mean Calculated mean multiplied by the pertinent frequency, e.g., in a population the mean value of the homo- and heterozygous dominants (*AA and Aa*) is 250 and that of the homozygous recessives (*aa*) is 200; the respective frequencies are 0.8 and 0.2; the weighted mean of the population is $(250 \times 0.8) + (200 \times 0.2) = 240$. *See* mean.

Weigle mutagenesis Increase in mutation of phage by mutagenic treatment of the host. (Weigle, J. J. 1953. *Proc. Natl. Acad. Sci. USA* 39:628; Bhattacharyya, S. C., et al. 1991. *Can. J. Microbiol.* 37:265.)

Weigle reactivation Increase in phage survival when mixed with host cells exposed to low doses of UV light. *See* DNA repair; marker rescue; multiplicity reactivation. (Calsou, P. & Salles, B. 1991. *Mol. Gen. Genet.* 226:113.)

Weismannism Inheritance takes place by the transmission of the genetic determinants through the germline (Keimbahn); environmentally induced phenotypic variations are not inherited. *See* germline. (Weismann, A. 1985. *Die Continuität des Keimplasmas als Grundlage einer Theorie der Vererbung.* Fischer, Jena, Germany.)

Weissenbacher-Zweymuller syndrome (WZS, 6p21.3) Shows neonatal small jaws (micrognathia), hip and/or shoulder bone defects (rhizomelic chondrodysplasia), and general underdevelopment of bones that may improve in later years. Optic nerve defects, myopia, and deafness may also occur. Some the symptoms resemble those of Stickler syndrome. The basic defect is in collagen gene COL11A2. *See* collagen; Stickler syndrome.

Werdnig-Hoffmann disease Recessive (5q12.2-q13) infantile muscular dystrophy primarily affecting the spinal cord muscles. The prevalence is approximately 1×10^{-4}. The gene frequency was estimated to be about 0.014 and the frequency of heterozygotes about 0.02. It encodes a protein that shows homology to dystrophin. The survival rate varies depending on the severity of symptoms. A subunit of transcription factor TFIIH usually suffers deletions. *See* dystrophin; muscular dystrophy; neuromuscular diseases; spinal muscular atrophy; transcription factors.

Werner syndrome (WRN) Involves premature aging, hardening of the skin, cataracts, atherosclerosis, diabetes mellitus, etc. Cultured cells have higher chromosome breakage and mutability. Neoplasias develop frequently. Ulceration around the ankles and soft tissue calcification is symptomatic of this condition and is unrelated to aging. The recessive defect appeared to be controlled by human chromosome Xp12-p11.2, but current linkage information, including complete sequencing on the basis of positional cloning, confirms its location at 8p12. The inferred protein is required for focus-forming activity in DNA replication (FFA). It contains 1,432 amino acids and resembles RecQ-type ATP-dependent helicases (homologous to budding yeast *SGS1* and *SRS2* genes). The defects in the helicase (frameshift, nonsense mutation) explain most of the symptoms on the basis of a defect in DNA metabolism. WRN interacts with DNA polymerase δ. The amino-end domain of WRN shows, however, $3' \rightarrow 5'$ exonuclease activity, and it may be involved in genetic recombination and/or repair. Forced telomerase activity substantially extends the life span of cultured WRN cells. *See* aging; Bloom syndrome; cosuppression; exonuclease; FFA; helicase; progeria; RNAi. (Kamath-Loeb, A. S., et al. 2000. *Proc. Natl. Acad. Sci. USA* 97:4603; Shen, J.-C. & Loeb, L. A. 2000. *Trends Genet.* 16:213; Kawabe, Y., et al. 2001. *J. Biol. Chem.* 276:20364.)

Wernicke-Korsakoff syndrome Chromosome 3p14.3 recessive disorder involving transketolase deficiency caused by the reduced ability of the enzyme to bind thiamin pyrophosphate. Another transketolase deficiency was located to Xq28. This dysfunction of energy metabolism leads to neuropsychiatric anomalies, especially in alcoholics and when thiamin is low in the diet. *See* thiamin pyrophosphate.

West syndrome Involves an X-linked central nervous defect resulting in infantile spasms and mental retardation. *See* epilepsy.

Western blot (Western hybridization) Identification of polypeptides separated electrophoretically on SDS polyacrylamide gels, then transferred to nitrocellulose filter and labeled by immunoprobes such as radioactive or biotinylated antibodies. *See* far-Western hybridization; gel electrophoresis; immunoblotting. (Towbin, H., et al. 1979. *Proc. Natl. Acad. Sci. USA* 76:4350; Burnette, W. N. 1981. *Anal. Biochem.* 112:95.)

WEST NILE VIRUS (Flaviviridae) Single-stranded RNA virus. E-glycoprotein is its most important protein both structurally and for infection. Of the two strains (lineage 1) caused most of the health problems in USA. Outbreaks of epidemics have been described in Africa, the Mid-East and southern Europe. About 29 species of mosquitos transmit the virus to birds and from birds humans. The symptoms of infection include fever, headache, gastrointestinal and mental changes (memory loss, depression) but rash cough and others also may occur. The virus appears genetically quite stable. The fatality ranges from 4 to 14% but in older or immune-compromised persons the risks may be more than double. The most effective diagnosis reveals immunoglobulin M (IgM) in the cerebrospinal fluids. Effective medication is sill awaited. Mosquito control is key for prevention. (Petersen, L. R. & Marfin, A. A. 2002. *Annals Int. Med.* 137:E173; Brinton, M. A. 2002. *Annu. Rev. Microbiol.* 56:371.)

Weyers acrofacial/acrodental dysostosis Allelic variation in the gene locus at 4p16 encoding Ellis–van Creveld syndrome. *See* Ellis–van Creveld syndrome.

WGA Whole-genome assembly.

WGRH Whole-genome radiation hybrid is made when the donor is a diploid cell line for the generation of radiation hybrids. *See* radiation hybrid. (Vignaux, F., et al. 1999. *Mamm. Genome* 10:888.)

WGS whole-genome shotgun sequencing was developed by Craig Venter and associates at Celera Genomics, Rockville, MD. The genome is sheared into a few thousand base-pair long pieces and cloned into sequencing vectors. Each fragment end (~500 bp) is covered by the sequencing several times, then assembled into overlapping segments by end sequences. This way, oriented, contiguous sequences (contigs or scaffolds) are generated. Between the scaffolds, a physical gap may remain until the physical map is "finished" and the complete genome is reconstructed with the aid of high-power computers. The scaffolds are also mapped to eukaryotic chromosomes and chromosome arms with the aid of sequence-tagged sites. The heterochromatic regions (mainly around the centromeres and near the telomeres) are unstable when cloned and thus resist sequencing. These regions contain transposons and redundant ribosomal RNA genes with few interspersed ORFs. The function of the sequences is indicated by annotation using Genscan or Genie computer software. *See* annotation; gene ontology; genome projects; mate pair; ORF; ribosomal RNA; scaffolds in genome sequencing; sequence-tagged sites; shotgun sequencing; transposon. (Adams, M. D., et al. 2000. *Science* 287:2185.)

whale *Baleonoptera* species are $2n = 44$; *Kogia breviceps*, $2n = 42$.

wheat *See Triticum.*

wheat germ in vitro translation Cell-free wheat germ extracts can be used for the translation of viral prokaryotic and eukaryotic mRNAs into protein. The supernatant of the extract must be chromatographically purified from inhibitory endogenous amino acids and pigments before translation. Tritin is one of the endosperm proteins, which inactivates the ribosomes, thus reducing the efficiency of the system unless carefully removed. The extract contains tRNA, rRNA, and other factors required for protein synthesis. Phosphocreatine and phosphocreatine kinase additions are needed for supplying energy. Spermidine is added to stimulate translation efficiency and prevent premature termination of the polypeptide chain. Magnesium acetate and potassium acetate, mRNA (to be translated), and amino acids (including one in radioactively labeled form) are also necessary. Incubation is at 25°C for 1 to 2 hours. In general, the procedure is very similar to the rabbit reticulocyte system. *See* rabbit reticulocyte *in vitro* translation system. (Erickson, A. H. & Blobel, G. 1983. *Methods Enzymol.* 96:38.)

Whipple's disease Involves infection by *Tropheryma whippelii* bacteria.

white blood cell *See* leukocyte.

white forelock *See* forelock; white.

white leghorn (chicken) Has the constitution for genes controlling color of the plumage *CC, OO, II*, and the color is white. *C* and *O* both are needed for pigmentation, but *I* is an inhibitor of color. *See* white silkie; white wyandotte.

White leghorn rooster.

white matter *See* gray matter.

white silkie (chickens) Constitution *cc, OO, ii*; have white feathers because only one of the two dominant genes *O* required for pigmentation is present. *See* white leghorn; white wyandotte.

white wyandotte (chickens) Have the genes *CC, oo, ii* and are white because only one (*C*) of the two dominant genes is present but not the other (*O*). *See* white leghorn; white silkie.

whorl Circular or spiral arrangement of structures, such as various parts of flowers or the dermal ridges in a human fingerprint. *See* fingerprinting; flower differentiation.

widow's peak Autosomal-dominant pointed hairline in humans.

Wilcoxon's signed-rank test Nonparametric substitute for the *t*-test for paired samples. The desirable minimal number of paired samples is 10 and it is expected that the population would have a median, be continuous, and be symmetrical. The differences between the variates is tabulated and ranked; the largest receives the highest rank. In case of ties, each should be assigned to a shared rank. The smaller group of signed-rank values is then summed as the *T* value. This *T* is then compared with figures in a statistical table. If the figure obtained is smaller than that in the body of the table under probability and on the line corresponding to the number of pairs tested, then the null hypothesis is rejected and the conclusion is justified that the two samples are different. Example:

Pairs	Difference	Signed Ranks			Probability		
		+	−		n	0.05	0.01
1	+6	7			10	10	5
2	+5	6			11	13	7
3	+10	10			12	17	10
4	−3		4		14	25	16
5	+4	5			16	35	23
6	+7	8			18	47	33
7	−2		3		20	60	43
8	−1		0.5		22	75	55
9	+9	9			24	91	69
10	−1		0.5		26	110	84
			$T = \overline{8.0}$				

Since T is 8.0 according to the first line of the probability table, the difference between the two sets of data is significant at 0.05 probability but not at 0.01. A more general procedure for determining probabilities relies on determining the Z value for threshold probabilities or more precisely using a table of the cumulative normal variates (such as the *Biometrika Tables for Statisticians*, Vol. 1, Pearson, E. S. & Hartley, H. O., eds. Cambridge Univ. Press) Z values larger than 1.960, 2.326, and 3.291 correspond to $P = 0.05$, 0.01, and 0.001, respectively. These probabilities rule out the null hypothesis.

$$Z = \frac{\mu - T - 0.5}{\sigma}$$

and

$$\mu = \frac{n(n+1)}{4}$$

and

$$\sigma = \sqrt{\frac{[2n+1]\mu}{6}}$$

where n = the number of paired data. *See* Mann-Whitney test; nonparametric statistics; student's t distribution.

wild type   Standard genotype (that is most common in wild [feral] populations). *See* isoalleles.

Wildervanck syndrome   Appears to be an X-linked dominant deafness; frequently associated with other disorders. Approximately 1% of deaf females are affected by this syndrome. It does not occur in males, presumably because it is lethal when homo- or hemizygous. It has been suggested that the syndrome is polygenically determined with a still unknown mechanism of male exclusion. *See* deafness; imprinting; sex-limited.

wildfire disease of plants   Caused by the toxin (methionine analog) of the bacterium *Pseudomonas tabaci*, and it leads to necrotic spots on leaves. *See* Pseudomonas.

Williams factor (Flaujeac factor deficiency)   Autosomal-recessive mutation in human chromosome 3q26-qter causing a deficiency of a high-molecular-weight kininogen, a precursor of a blood-clotting factor. *See* antihemophilic factors; blood-clotting pathways; kininogen.

Williams syndrome (Williams-Beuren syndrome, WMS   Autosomal-dominant condition involving stenosis (narrowing) of the aorta, of the arteries, of the lung, elfin face (elfins are diminutive mythological creatures), malformation of teeth and stature, mental deficiency, and excessive amounts of calcium in the blood (hypercalcemia) and in some tissues. Cognitive abilities are impaired in an unusual fashion. During infancy, the patients perform poorly in language skills and relatively better in numerical intelligence tests. By adulthood, the trend is reversed, indicating that these abilities are under the control of two different developmental modules. Various types of deletions in different chromosomes (15, 4, 6) have been suspected. Chromosome-specific probes indicated a ~1.5 Mb deletion of 7q11.23. This region harbors Williams syndrome transcription factor (WSTF/WCRF/ACF) and bears structural similarity to a 180 kDa chromatin-remodeling factor. Lower calcium in the diet may alleviate some of the symptoms. Vitamin D2 anomaly is suspected. The function of Williams factor is unrelated. Findings indicate the involvement of LIMK protein kinases (carrying two LIM domains) that are serine/threonine/tyrosine kinases and regulate actin in the cytoskeleton. Under normal conditions, RAC-GTP activates LIM kinases, which in turn phosphorylate cofilin and inactivate it. Dephosphorylation permits the formation of active cofilin associated with actin. The actin-cofilin association is apparently mediated by phosphoinositides (PtdInsP$_2$). Defects (heterozygosity for a mutation) in LIMK also lead to abnormal neuronal connections. Elastin defects characterize the symptoms of Williams syndrome. *See* actin; cardiovascular diseases; cofilin, cutis laxa; dwarfism; face/heart defects; human intelligence; LIM domain, Marfan syndrome; module; phosphoinositides; supravalvular aortic stenosis, vitamin D; Williams factor. (Peoples, R., et al. 2000. *Am. J. Hum. Genet.* 66:47; Morris, C. A. & Mervis, C. B. 2000. *Annu. Rev. Genomics Hum. Genet.* 1:461; Sumi, T., et al. 2001. *J. Biol. Chem.* 276:23092; Urbán, Zs., et al. 2002. *Am. J. Hum. Genet.* 71:30.)

Wilms tumor (WT)   Usually associated with a deletion in the short arm of human chromosome 11 extending from 11p13 to 11p15.5 (WT1), apparently spanning several genes, and frequently designated as WAGR syndrome (Wilms' tumor−aniridia−genitourinary anomalies and RAS oncogene-like function). The cases usually involve symptoms of all or parts of the functions implied by WAGR. Wilms' tumor is caused by mutation in a cancer-suppressor transcription factor and splicing factor with 4 Zn-finger domains. Wilms' tumor is extremely complex and additional WT genes in chromosomes 16q; 1p, 4p, 8p, 14p, 17p; and q, 18q were also implicated. The transcript displays alternative splicings. The Wilms tumor-suppressor gene is expressed only in the maternally transmitted allele. Prevalence is about 10^{-4} during the first 5 years of life. The latest evidence indicates a WT gene in the 17q12-q21 region. The WT1 gene regulates muscle differentiation. It affects the SRY locus in the Y chromosome, which explains the symptoms concerned with genitourinary development when it is altered. Different functions are based on isoforms of the protein. *See* acatalasemia; aniridia; breast cancer; deletion; Denys-Drash syndrome; Frasier syndrome; hypertension; imprinting; isoform, kidney diseases; RAS; Rhabdosarcoma; zinc finger. (Hossain, A. & Saunders, G. F. 2001. *J. Biol. Chem.* 276:16817; Hammes, A., et al. 2001. *Cell* 106:319.)

Wilson disease (WD)   Recessive disease encoded in human chromosome 13q14.2-q14.3. It primarily affects persons age 30 and over, although juvenile forms have been described. The major symptom is cirrhosis of the liver and psychological ailments caused by a deficiency in ceruloplasmin, resulting in copper accumulation. In this and other diseases involving cirrhosis of the liver, at the upper and lower margins of the cornea a greenish narrow ring (↑) occurs. The basic defect is in a copper-transporting ATPase of the mitochondria. The prevalence of WD in the U.S. is about 3×10^{-5}. A similar anomaly has been observed in Long-Evans Cinnamon (LEC) rats that may serve as an animal model for study of the disease. Prenatal diagnosis of offspring of carrier parents may be possible by using linkage with DNA markers. *See* acrodermatitis; hemochromatosis; Menkes disease; mitochondrial disease in humans; neurodegenerative diseases.

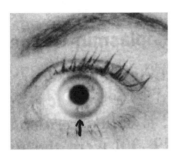

Wilson disease.

winged helix protein   Class of helix-turn-helix proteins that use a β-hairpin (a wing) to bind DNA. *See* DNA-binding protein domains. (Gajiwala, K. S. & Burley, S. K. 2000. *Curr. Opin. Struct. Biol.* 10:110.)

Wingless (*wg*/*Dint-1* 2–30)   *Drosophila* gene involved in morphogenetic signaling and homologues in all vertebrates and invertebrates. The mouse homologue is *Wnt*. Components of the *Wnt* cascade are altered in breast and colon cancers in mice and in human melanomas. The path of *Wg* function can be represented as shown below.

Fz (*frizzled trichomes*) and Dally (*division abnormally delayed*, controls heparan sulfate) are G protein-like receptors. Dsh (*dishevelled*) mitigates the block by Zw3 (*zeste white*). Armadillo is a β-catenin-like protein. TCF (*ternary complex factors*) coactivate transcription. Dsh blocks glycogen synthase kinase (GSK3) which would be a negative regulator of Armadillo. CK1 (a serine kinase, called casein kinase) phosphorylates Dsh~Zw. Actually, Wg is involved in more complex functions. Wg signals to the epidermal growth factor (EGF) receptor. WNT-4 duplication in humans and mice masculinizes XX individuals. In Sertoli and Leydig cells, DAX1, an antagonist of SRY, is upregulated. *See* adrenal hypoplasia congenita; catenins; EGF; Gardner syndrome; GSK; heparan sulfate; morphogenesis in *Drosphila*; organizer; sex reversal,

SRY; TCF; Wolffian duct; zeste, *Armadillo*. (Kühl, M., et al. 2000. *Trends Genet.* 16:279; Peifer, M. & Polakis, P. 2000. *Science* 287:1606; Wilkie, G. S. & Davis, I. 2001. *Cell* 105:209; van de Wetering, M., et al. 2002. *Cell* 109:S13; Peifer, M. & McEwen, D. G. 2002. *Cell* 109:271.)

Wiskott-Aldrich   syndrome (WAS, Xp11.23-p11.22)   X-chromosomal immunodeficiency disease causing eczema, reduced platelet size, bloody diarrhea, high susceptibility to infections, lymphocyte malignancies, and usually death before age 10. The prevalence is about 4×10^{-6}. Affected individuals are deficient in a 115 kDa lymphocyte membrane protein and the platelets are abnormally low in a glycoprotein (sialophorin). Carriers may be identified by linkage, lymphocyte analysis, and nonrandom inactivation of the X-chromosomes. WAS is allelic to the human Xp11.23 thrombocytopenia gene. This gene includes 12 exons in 9 kb genomic DNA and encodes 502 amino acids. The involvement of a CDC42 signaling defect is likely. WAS protein mediates cytoskeletal rearrangement and transcriptional activation of T cells. *See* cancer; CDC42; cytoskeleton; immunodeficiency; platelet; podosome; T cell; thrombocytopenia; thrombopathic purpura; WASP. (Silvin, C., et al. 2001. *J. Biol. Chem.* 276:21450; Devriendt, K., et al. 2001. *Nature Genet.* 27:313; Caron, E. 2002. *Current Opin. Cell Biol.* 14:82.)

Witkop syndrome   *See* tooth-and-nail dysplasia.

WL   Working level is used for the characterization of short-lived radon decay products in 1 liter of air resulting in the ultimate emission of 1.3×10^{-5} MeV of alpha-radiation energy. WL is also defined as 2.08×10^{-5} joule h m^{-3}. WLM (working level/month) = exposure of 170 h at 1 WL. *See* radon.

w **locus**   The first mutation discovered in *Drosophila* by Morgan is involved in the control, production, and distribution of brown (ommochrome) and red (pteridine) pigments of the eyes and ocelli and some other anatomical structures. The gene (at 1–1.5) apparently encodes an ATP-binding membrane transport protein for the precursors of the pigments. More than 200 alleles have been identified within the 0.03 centimorgan region, which has been mapped by intragenic recombination into seven domains. The wild-type allele is incompletely dominant over many mutant alleles. The alleles do not show partial complementation with the exception of the w^{sp} (white spotted) allele, which displays allelic complementation with the majority of other alleles in the presence of the z^a (zeste). The latter is a regulatory gene at 1–1.0 location, and

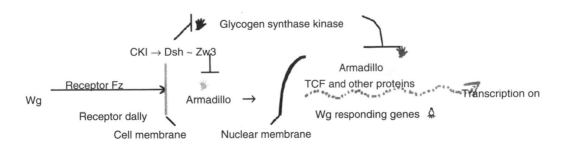

zeste encodes a specific protein binding to the promoters of *w*, *Ultrabithorax* (*Ubx*), and *decapentaplegic* (*dpp*). *See* eye color; map unit; morphogenesis in *Drosophila*; recombination; Tangier disease.

W mutagenesis Tendency of increased mutation after Weigle reactivation. *See* Weigle mutagenesis. (Yatagai, F., et al. 1983. *Adv. Space Res.* 3[8]:65.)

WNT1 (cysteine-rich glucoprotein ligand) The ~18 vertebrate genes have diverse roles in signaling to development, particularly along the anterior-posterior body axis. *See* Gardner syndrome; gonads; INT1 oncogene in mouse (*wingless* gene product in *Drosophila*); morphogenesis in *Drosophila* {63}; organizer; pattern formation. (Skromme, E. & Stern, C. D. 2001. *Development* 128:2915; <http://www.stanford.edu/~musse/wntwindow.html>.)

wobble The 5′-base of the anticodon can recognize more than one kind of base at the 3′-position of the codon, e.g., both U and C in the mRNA may pair with G, and both G and A may pair with U, or A, U, or C may recognize I (inosinic acid at the 5′-position in the anticodon). There is no AAA anticodon for Phe, but the GAA anticodon recognizes both UUU and UUC codons in mRNA. The GUU and GUC codons of Val are decoded by an anticodon AAC. Inosinic acid occurs in the anticodon of eight tRNAs of higher eukaryotes, in seven of yeast, and in the $tRNA_2^{Arg}$ of prokaryotes and plant chloroplasts. Inosine may be formed from adenosine by adenosine deaminases. In tRNAs, however, a different dimeric deaminase encoded in yeast by the *Tad2* and *Tad3* genes is acting. This deaminase is related to cytidine deaminase (CDA). According to the classical or universal genetic code of 61 sense and 3 missense codons, a minimum of 32 tRNAs would be required to recognize all the amino acids. Further simplifications permit protein synthesis, however, by 22–24 tRNAs. In the human genome, tRNA genes have been found all over, yet 140 tRNA genes are crowded in a 4 Mb region of chromosome 6. *See* anticodon; genetic code; hypoxanthine; isoacceptor tRNA; mtDNA; tRNA. (Crick, F. H. C. 1966. *J. Mol. Biol.* 19:548.)

Wolbachia Endocellularly infectious group of bacteria of arthropods and nematodes that are transmitted maternally and may cause feminization, cytoplasmic incompatibility, and thelytoky. Infected males cannot produce viable offspring with uninfected females because of cytoplasmic incompatibility. They are compatible, however, with infected females and produce offspring. Antibiotics can cure incompatibility in many instances. Removal of *Wolbachia* by antibiotics from parasitic wasps stops oogenesis. *Wolbachias* as endosymbionts of filarial nematodes are responsible for *river blindness* disease. *Wolbachia* infection of flies with some oogenesis-defect-causing *Drosophila Sex-lethal* alleles surprisingly restores fertility (Starr, D. J. & Cline, T. W. 2002. *Nature* 418:76). *See* cytoplasmic incompatibility; pronucleus; segregation distorter; symbionts hereditary; thelytoky. (Stoputhammer, R., et al.

1999. *Annu. Rev. Microbiol.* 53:71; Zimmer, C. 2001. *Science* 292:1093; Dedeine, F., et al. 2001. *Proc. Natl. Acad. Sci. USA* 98:6247; Saint André, V. A., et al. 2002. *Science* 295:1892, <www.wolbachia.sols.uq.au>.)

Wolcott-Rallison syndrome (WRS, 2p12) Involves a mutation in the translation initiation factor EIF2AK3. The recessive disorder affects neonatal or infantile insulin-dependent diabetes. Eventually, bone defects (epiphyseal dysplasia, osteoporosis), retarded growth, liver and kidney malfunction, mental retardation, and heart disease may complicate the condition. *See* EIF2.

wolf *Canis lupus*, $2n = 78$. It can form fertile hybrids with domesticated dogs (*Canis familiaris*, $2n = 78$) as well as with coyote (*C. latrans*, $2n = 78$) but not with foxes (*Vulpes vulpes*, $2n = 36$). Domesticated dogs appear to be much closer evolutionarily to wolves (on the basis of mtDNA) than to coyotes. *See* fox.

Wolffian ducts Develop as a precursor of the male gonads of vertebrates. This development is enhanced under the influence of testosterone. The male gonad (testes) is then formed by the Sertoli cells, which eventually surround the spermatogonia. Leydig cells (Yao, H. H.-C., et al. 2002. *Genes Dev.* 16:1433) secrete the steroid testosterone and Sertoli cells produce anti-Müllerian hormone (human chromosomal location 12q13), which causes regression of the uterus and fallopian tubes. *See* DSS; gonads; Müllerian ducts.

Wolf-Hirschhorn syndrome Involves a deletion (unequal crossing over, insertion) in one of the short arms of human chromosome 4p16.1 (usually the paternal), resulting in severe growth, mental, face, and genitalia defects, etc. The deletion generally eliminates HOX7 (homeobox 7), which is responsible for normal development in humans and mice. Hemizygosity for the gene is sufficient for the disease. *See* deletion; hemizygous; homeobox; homeotic genes; Huntington's chorea. (Näf, D., et al. 2001. *Am. J. Hum. Genet.* 10:91.)

Wolf-Parkinson-White syndrome Heart disease with a short P and long QRS phase of electrocardiography. It is also called preexcitation syndrome because the heart ventricles are excited prematurely. It may cause increased palpitations and sudden death. The causes are complex; dominant defects at chromosome 7q3 were implicated, and it may have a mitochondrial component. *See* electrocardiography; mitochondrial diseases in humans.

Wolfram syndrome (DIDMOAD) Mutation in the human chromosome 4p16.1 gene encoding a ~100 kDa transmembrane protein involves diabetes insipidus, diabetes mellitus, optic atrophy, and deafness. It has been proposed that a large mitochondrial deletion extending over several coding sequences (7.6 kb) is responsible for the diseases, but it could not be confirmed. *See* deafness; diabetes insipidus; mitochondrial diseases in humans; optic atrophy. (Strom, T. M., et al. 1998. *Hum. Mol. Genet.* 7:2021.)

Wolman disease (lysosomal acid lipase deficiency) Due to autosomal-recessive genes in the long arm of human chromosome 10q24-q25 and in mouse chromosome 19. Early-onset forms deficient in this enzyme (cholesteryl ester hydrolase) involve liver and spleen enlargement, failure to feed normally, and death by 2 to 4 months. The accumulation of cholesterol esters is caused by mutant alleles of the same locus. Some forms permit survival to the teens. *See* cholesterol; lipase; lysosomes. (Du, H., et al. 1998. *Hum. Mol. Genet.* 7:1347.)

woodchuck   *See Marmota monax.*

woodrats   *Several* species mainly with $2n = 52$.

Woods' light   Ultraviolet light source with nickel oxide filter and with a maximal transmission at about 365 nm while most other spectral regions are blocked. *See* ultraviolet light.

woolly hair   May be black (and autosomal dominant) or blond (and autosomal recessive). The hair in both cases is short and tightly curled.

working hypothesis   Experimentally testable assumption regarding a problem.

worm genetics   Informal reference to *Caenorhabditis elegans.* See *Caenorhabditis.*

Woronin bodies   Occur around fungal pores supposedly to protect against excessive leakage if the cells are damaged. (Tenney, K., et al. 2000. *Fungal Genet. Biol.* 31:205.)

wortmannin   Protein from *Penicillium fumiculosum*; it is a stimulator of neutrophils and an inhibitor of PIK and DNA-PK, thus inhibiting the repair of double breaks in DNA. *See* DNA-PK; neutrophil; PIK. (Wang, H., et al. 2001. *Nucleic Acid Res.* 29:1653.)

Woude syndrome   *See* van der Woude syndrome.

wound response   Upon wounding of one leaf, physiological changes take place in other parts of the plant body and the expression of proteinase inhibitors (Pin) is triggered by glycan, jasmonate, and peptide signals. The 18-amino-acid peptide signal, systemin, in tomato plants wounded by herbivorous insects stimulates proteinase expression. Salicylic acid may block the wound response. Wounding of plants may also induce a large number of genes of different functions. The plants may produce peroxides against microbes infecting the wounds. Phenolics may accumulate, photosynthesis may be reduced, and ethylene biosynthesis may be induced. *See* ethylene; glycan; host-pathogen relationship; insect resistance in plants; jasmonic acid; phenolics. (Zhou, L. & Thornburg, R. 1999. *Inducible Gene Expression in Plants*, Reynolds, P. H. S., ed., p. 127. CAD, New York.)

W point   Stage just before S phase when animal cells still have a serum growth factor requirement to enter the S phase. *See* cell cycle.

wrapping choice   Generalized transducing phage "chooses" to scoop up the host rather than viral DNA and "wraps" it into the phage capsid. *See* transduction.

W reactivation   Weigle reactivation.

Wright blood group   Very rare; the frequency of the Wr(a) antigen is about 3×10^{-4} in Europe. *See* blood groups.

Wright-Fisher model   *See* genetic drift.

wrinkled/smooth   Gene locus of pea immortalized by Mendel's discovery of monogenic inheritance. The recessive "wrinkled allele" turned out to be an insertional mutation. *See* pea.

writhing number   Indicates the contortion of a DNA double helix in a supercoiled state. It measures the helix axis in space. *See* linking number. (Kobayashi, S., et al. 2001. *Chem. Pharm. Bull. Tokyo* 49:1053.)

wrod score   Wrong lod score is obtained if linkage was estimated as lod scores on the assumption of an incorrect genetic model (ϕ). The genetic model in complex traits may easily be misspecified. *See* lod score; model, genetic; mod score. (Hodge, S. E. & Elston, R. C. 1994. *Genet. Epidemiol.* 11[4]:329.)

wrongful birth   Potential responsibility of physician or genetic counselor for negligence in informing or prenatal care of prospective parent(s) about risk involving childbirth. (Randall, K. C. 1979. *Hofstra Law Rev.* 8:257.)

wrongful life   Potential responsibility of parents, physicians, and genetic counselors for not preventing the birth of a child with serious hereditary disease or for illegitimacy, which may carry a social stigma. The affected offspring may sue. *See* confidentiality; counseling genetic; genetic privacy; paternity test. (Foutz, T. K. 1980. *Tulane Law Rev.* 54:480.)

WW domain Two-tyrosine motif (38–40 amino acids) of signaling proteins and a binding site for proline-rich peptides. The binding of a WW domain by Pin1 and Nedd4 proteins apparently does not require prolines for binding; rather the WW domains are binding sites for phosphoserine and phosphothreonine. Among the many diverse proteins, their ligands include Cdc25C phosphatase, microtubule-associated tau, carboxy-terminal of RNA polymerase II, etc. *See* Nedd; Pin1; pleckstrin; PTB; SH2; SH3; signal transduction. (Sudol, M. & Hunter, T. 2000. *Cell* 103:1001.)

WWW World wide web is the system available through the Internet with the aid of a browser program that makes possible the search through a computer linked to the system. *See* HTML; Internet.

wx gene (*waxy*) Occurs in various cereal plants and used as a chromosome marker (in the short arm of chromosome 9–56 of maize). In the presence of the dominant allele, starch is formed (stained blue by iodine stain). If it is replaced by the recessive allele, amylopectin is formed (stained red-brown) because of a defect in NDP-starch glucosyltransferase. *See* amylopectin; iodine stain.

Apparent mutations were recorded by the ancient literature. Aristotle says, "whoever does not resemble the parents is, in some respects, a monster, because in this case nature has deviated, to a certain degree, from the hereditary type." (*Generation of Animals*, Book IV, Part 3, Para. 1)

In 1844, W. H. Prescott retells the *History of the Conquest of Mexico* (Dutton, New York): "I must not omit to notice a strange collection of human monsters, dwarfs, and other unfortunate persons, in whose organization Nature had capriciously deviated from her regular laws. Such hideous anomalies were regarded by the Aztecs as a suitable appendage of state. It is even said, they were in some cases the result of artificial means, employed by unnatural parents desirous to secure a provision for their offspring by thus qualifying them for a place in the royal museum."

Thus, even induced mutation had been anticipated much before it had been experimentally demonstrated by H. J. Muller and L. J. Stadler in the mid-1920s.

A historical vignette.

X

x Basic chromosome number. *See* n; polyploids.

x̄ Arithmetic mean of the sample.

X^{1776} (bicentennial) Bacterial strain with an absolute requirement for diaminopimelic acid (a lysine precursor) needed for the growth of viable bacteria; thus, it cannot survive outside the laboratory. It was so named in the U.S. bicentennial year at the Asilomar Conference in 1976. *See* Asilomar Conference.

Xa Blood coagulation factor. Also a plasmid factor that specifically cleaves protein after Arg of the tetrapeptide Ile-Glu-Gly-Arg that connects the 31 amino terminal of phage-λ cII protein. *See* clotting; lambda phage. (Verner, E., et al. 2001. *J. Med. Chem.* 44:2753.)

xanthine Purine derived from either adenine through hypoxanthine by xanthine oxidase or from guanine by deamination. It is converted to uric acid by xanthine oxidase. *See* uric acid.

Xanthine.

xanthinuria (XDH, 2p23-p22) Recessive xanthine dehydrogenase/oxidase deficiency resulting in the excretion of excessive amounts of xanthine and xanthine stones in the kidneys. Uric acid content of the urine and serum is reduced. The 36-exon gene spans ~60 kb DNA. *See* kidney diseases; uric acid.

xanthomatosis *See* cerebral cholesterinosis.

xanthophyll Yellow carotenoid pigments that play an accessory role in light absorption. *See* photosystems. (Ruban A. V., et al. 2001. *J. Biol. Chem.* 276:24862.)

XBP Encodes the 89 kDa subunit of human transcription factor TFIIH, corresponding to yeast *SSL2* encoding a 105 kDa polypeptide. XBP-1 is required for the differentiation of plasma cells into B lymphocytes. *See* B lymphocytes; transcription factors.

X2BP MHC-II promoter-binding protein, along with NF-Y and RFX. *See* MHC; RFX. (Reimold, A. M., et al. 2001. *Nature* 412:300.)

XCAP-C and XCAP-E Proteins involved in condensation of chromosomes. *See* condensin complex. (Neuwald, A. F. & Hirano, T. 2000. *Genome Res.* 10:1445.)

X-chromosomal inactivation *See* Barr body; lyonization.

X-chromosome One of the sex chromosomes generally present in two doses in the females and in one in the males. In some species, the females are XY and the males are either XX or XO, or other chromosomal doses may be found. In species where the female is heterogametic, her sex-chromosomal constitution is often designated as WZ and the male as ZZ. The sex chromosomes apparently evolved from autosomes. The PAR region is a relic of the autosomal origin. There are 19 other genes shared between the X- and Y-chromosomes in mammals. The latter ones are located in three groups on both sex chromosomes, but they do not recombine and could be mapped only by radiation hybrids. The human X-chromosome has lower variability (~60%) than the autosomes. *See* autosome; Barr body; chromosomal sex determination; lyonization; Ohno's law; pseudoautosomal; radiation hybrid; sex chromosome; Y chromosome.

X-chromosome counting Mechanism of sex determination in case of XX versus XO sex determination. The *xol-1* gene of *Caenorhabditis* constitutes a switch mechanism that specifies the male developmental course if it is inactive. In XX nematodes, *xol-1* is repressed posttranscriptionally or the RNA-binding protein encoded by the fox-1 gene reduces its level of expression. The transcription of gene *xol-1* is suppressed primarily by SEX-1 hormone receptor protein, which binds to the promoter in XX nematodes. *See* chromosomal sex determination; dosage compensation; sex determination. (Maxfield Boumil, R. & Lee, J. T. 2001. *Hum. Mol. Genet.* 10:2225.)

XCID X-chromosome-linked severe combined immunodeficiency. *See* immunodeficiency; SCID.

xenia Expression of the gene(s) of the male in the endosperm (e.g., purple) following fertilization; the expression in the embryo may not yet be visible. *See* metaxenia.

Xenia.

xenobiotics Compounds that naturally do not occur in living cells.

xenogamy Fertilized by different neighboring plants.

xenogeneic Transplantation from another species (xenotransplantation). *See* allogeneic.

xenogenetics Study of the effect of environmental factors on conditions under polygenic control. *See* polygenic.

xenograft Transplantation of tissue from another species (e.g., animal → human). It poses problems of rejection and the possibilities of viral infection (e.g., porcine endogenous proviruses [PERV]), and possibly the emergence of new diseases. The complement cascade of the immune system mediates the rejection. When swine transgenic for the human complement were used as heart donors to baboons, the function of the heart was prolonged by several hours. The limited clinical evidence does not indicate substantial risk by pig → human grafts. *See* αGT; complement; epitope; gene therapy; immune system; PERV; transgenic; xenotransplantation. (Matsunami, K., et al. 2001. *Clin. Exp. Immunol.* 126:165.)

xenology Study of the mixture of original and foreign genetic sequences within an organism or group of organisms caused by horizontal transfer and transformation. *See* homology; infectious heredity; transformation; transmission. (Gogarten, J. P. 1994. *J. Mol. Evol.* 39:541.)

Xenopus *X. laevis*, South African clawed toad, $2n = 4x = 36$, DNA 3.1×10^9 bp. Species of frogs. Because of being tetraploid and a long generation time of several years, it is not a favorite object of genetic manipulations, although it is an excellent object of embryological studies. The size of the oocyte may reach 1.5 mm and it is thus visible by the naked eye, lending itself to various manipulations. Pioneering research was conducted with nuclear transplantation. Transformation of embryonic tissues was not as successful as expected because too few cells expressed the transgenes. Techniques have been developed for transformation of sperms, which will permit the study of dominant transgenes on various developmental and physiological processes. *Xenopus (Silurana) tropicalis* is the only diploid species (1.7×10^9 bp, generation time 4–6 months), but it is somewhat distantly related. *X. tropicalis* is very prolific (1,000–3,000 eggs per ovulation) and transformation is very efficient. *See* frog; transformation; transgene; toad; *Xenopus* oocyte culture. (<http://www.dkfz-heidelberg.de/abt0135/axeldb.htm>; <http://www.nih.gov/science/models>; <http://arrays.rockefeller.edu/xenopus>; <http://www.tigr.org/tdb/tgi.shtml>.)

***Xenopus* oocyte cultures** *Xenopus* frog oocytes are of about 1 μL in volume. They contain large amounts of DNA (12 pg in the nucleus, 25 pg in the nucleoli, 4 ng in mitochondria). They can synthesize 20 ng RNA and 400 ng protein daily. About 10^6 to 10^7 bp DNA can directly be injected into the *germinal vesicle* (the nucleus). The injected DNA is packaged into the nucleosomal structure, it is replicated and translated by cellular machinery, and the products can be ready for analysis within a few hours. Recombinant DNA can be introduced and studied the same way. Exogenous DNA is replicated according to the cell cycle of the oocytes and no reinitiation of the foreign DNA replication occurs. All foreign DNAs are replicated according to the oocyte cell cycles and all the foreign replicational signals are overridden. The replication in the oocytes is also extremely rapid because transcription is apparently halted during replication. *Xenopus* oocytes have been very successfully used for the structural

and morphological analysis of vertebrate embryogenesis. They have proven to be extremely useful for the molecular analysis of transcription, regulation, cell-to-cell communication, and early morphogenesis. *See* morphogenesis; oocyte; protein synthesis. (Romero, M. F., et al. 1998. *Methods Enzymol.* 296:17.)

xenotransplantation Transplantation of organs/tissues among different species may involve the potential danger of transferring new viruses. On the other hand, xenotransplantation may have potentials in curing disease, e.g., the use of resistant baboon livers in case of hepatitis B infections and baboon bone marrow for AIDS victims. Galactose α-1,3-transferase knockout lines may overcome the immune reaction of the recipient against the graft. More research is required in this field. *See* αGT; baboon; grafting in medicine; hyperacute reaction; immune tolerance; immunosuppression; microchimerism; OBA; PERV; rejection; xenogeneic; xenograft. (Sim, K. H., et al. 1999. *Can. J. Gastroenterol.* 13:11; Lai, L., et al. 2002. *Science* 295:1089; Cooper, D. K. C., et al. 2002. *Annu. Rev. Med.* 53:133.)

xenotropic retroviruses Replicated only in the cells of species other than the species from which the virus originated. *See* ecotropic and polytropic virus.

xeroderma pigmentosum (XP) The first symptoms may arise during the first year of life as very intense freckles induced by sunshine and become proliferative, appearing as skin cancer. In some cases, the central and peripheral nervous systems are also affected. Usually, A (9q22.2-q31), B (2q21), C (3p25), D (19q13.2-q13.3), E (11p12-p11, F (16p13.2-p13.1), and G (13q32-q33) types are distinguished; the other complementation groups (H and I) are less clear. Complementation group A encodes DNA damage binding protein B; D codes for helicases; C initiates global nucleotide excision repair and selectively repairs cyclobutane pyrimidine dimers rather than 6-4 photoproducts.

A mild form of xeroderma pigmentosum. The initial signs are usually very heavy freckles that gradually turn into different types of skin cancer. The progress of the disease is enhanced by sunshine. A variety of ailments may accompany the main symptoms and the afflicted persons rarely reach adulthood. (Courtesy of the March of Dimes — Birth Defects Foundation.)

XPC is homologous to yeast genes RAD23A and B and it is sometimes called XPC-HR23A and B. Types F and G are defective in a DNA repair endonuclease (homologous to yeast RAD1). XP-V patients, representing about 25% of the clinical cases of the disease, do not have a defect in nucleotide exchange repair and are not sensitive to UV radiation. In these individuals, the defect is caused by failure of replication of the leading strand through pyrimidine dimers. The replication (by polymerase ζ) of the lagging strand associated with the photoproduct can slowly proceed and results in an asymmetrical replication fork with an extended single-strand leading strand of the parental DNA molecule. Unfortunately, pol ζ is error-prone and the error-free pol η (encoded by *RAD30*) is defective in XP-V individuals. Humans have two homologues of RAD30. As a consequence, the patient's cells become susceptible to mutation and carcinogenesis. Also, another form with much milder symptoms is known with an apparently dominant inheritance. In some forms, the defect is not in the excision repair, but a postreplicational anomaly causes light sensitivity. Similar genes occur in mice and five *RAD* genes of yeast also have defects in excision repair. *See* ataxia telangiectasia; Bloom syndrome; Cockayne syndrome; complementation groups; cyclobutane dimer; DDB; DNA polymerases; DNA repair; excision repair; Fanconi syndrome; helicase; light sensitivity diseases; 5′,8-purine cyclodeoxynucleosides; pyrimidine-pyrimidinone photoproduct; *RAD*; thrichothiodystrophy; ultraviolet sensitivity syndrome. (Bootsma, D., et al. 1998. *The Genetic Basis of Human Cancer*, Vogelstein, B. & Kinzler, K. W. eds., p. 245. MacGraw-Hill, New York.)

xerophyte   Draught-tolerant plant; thrives at low-precipitation regions.

X-family of DNA polymerases   Are pol β-like nucleotidyl-
transferases. *See* DNA polymerases, nucleotidyl transferase. (Aravind, L. & Koonin, E. V. 1999. *Nucleic Acids Res.* 27:1609.)

XG (Xg[a])   Blood group antigen-determining dominant factor in the short arm of the human X-chromosome. It appears that this end of the X-chromosome can recombine with the Y-chromosome, and it does not undergo lyonization as do the majority of the X-chromosomal genes. *See* blood groups; ichthyosis; lyonization. (Fouchet, C., et al. 2000. *Immunogenetics* 51[8–9]:688.)

Xgal (5-bromo-4-chloro-3-indolyl-β-D-galactopyranoside)
When β-galactosidase enzyme hydrolyzes, this chromogenic substrate blue color is formed in a bacterial culture plate. *See* β-galactosidase.

XIAP   Caspase inhibitor, member of the human IAP family of proteins. *See* caspase; IAP. (Riedl, S. J., et al. 2001. *Cell* 104:791.)

Xic   X-chromosome inactivation center. Its effect is somewhat similar to *Xist*, but unlike *Xist*, which has a somewhat localized cis effect, *Xic* extends globally on the X-chromosome. *Xic* delays the replication of this X-chromosome and mediates the hypoacylation of histone 4, contributing to heterochromatinization of that chromosome. *Xic* controls chromosome counting in dosage compensation and selects one or more of the X-chromosomes for inactivation. *See* lyonization; *Xist*. (Prissette, M., et al. 2001. *Hum. Mol. Genet.* 10:31.)

XID   Sex-chromosome-linked immunodeficiency. *See* immunodeficiency; SCID.

Xist   X-chromosome inactive sequence transcript, human TSIX. This gene has no protein product, but its 15 kb RNA covers the genes of the inactive mammalian X-chromosome beginning from the *Xic* site (Xq13.2). Underacetylated histones, methylated CpG islands, and late replication of DNA cause inactivation of the X-chromosome. Xist-triggered inactivation is preceded by methylation of histone-3 at lysine 9. *Xist* acts in cis and may also be expressed in autosomes affecting neighboring genes. Since X-chromosomal inactivation is restricted to the copies of the X-chromosome present more than once, it has been suggested that *Xist* counts (senses) X-chromosomal dosage. Deletions downstream of *Xist* make it constitutive, i.e., it is expressed and causes inactivation of the X-chromosome concerned. From 15 kb downstream location, a 40 kb apparently antisense RNA *Tsix* is transcribed across the *Xist* locus in both X-chromosomes of diploids until the determination of inactivation. Prior to the beginning of the inactivation, *Tsix* is transcribed in antisense orientation only on the X-chromosome destined for inactivation and is shut down again after the inactivation begins. *Tsix* apparently determines the choosing of the X chromosome for inactivation by *Xce* element without affecting the silencing itself. Knockout of the paternal Tsix does not affect embryonal development. If the *Tsix* knockout is transmitted through the female, embryos of both sexes die because both X-chromosomes are silenced in the female offspring and also the single X in the males. *See* dosage compensation; histones; imprinting; lyonization; *Xic*. (Lee, J. T. 2000. *Cell* 103:17; Matsui, J., et al. 2001. *Hum. Mol. Genet.* 10:1393; Migeon, B. R., et al. 2001. *Am. J. Hum. Genet.* 69:951; Heard, E., et al. 2001. *Cell* 107:727; Wutz, A., et al. 2002. *Nature Genet.* 30:167; Plath, K., et al. 2002, *Annu. Rev. Genet.* 36:233; Shikata, S & Lee, J. T. 2003. *Hum. Mol. Genet.* 13:125.)

Xklp1   *See* NOD.

xl   Prefix for *Xenopus laevis* toad protein of DNA or RNA, e.g., xlRNA.

XLA   X-linked agammaglobulinemia. *See* agammaglobulinemia.

X-linked   The gene is within the X-chromosome. *See* Z linkage. *See* diagram next page.

XLNO 38   Protein with homologies to nucleoplasmin. Although it is not part of the mature ribosomes, it is associated with both small and large subunits of ribosomes. Its role appears to be chaperoning the assembly of basic proteins on ribosomal RNA precursors. *See* chaperone.

XML   Extensible markup language is a flexible communication system capable of sharing format and data through the www. *See* www.

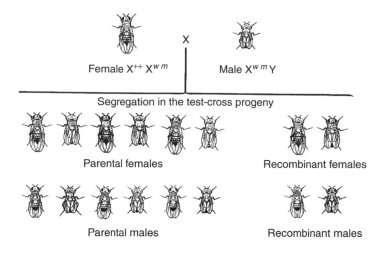

Female X^{++} X^{wm} X   Male X^{wm}Y

Segregation in the test-cross progeny

Parental females   Recombinant females

Parental males   Recombinant males

The female fly is heterozygous for the recessive white eye and the miniature body-size genes, the male X-chromosome carries the two recessive alleles. Since in the female fly recombination is normal (unlike in the *Drosophila* male), free recombination yields 16/4 = 0.25 recombinants in both sexes.

X-numerator Number of X chromosomes relative to autosomes (X:A) in the determination of sex. *See* chromosomal sex determination.

XO Having a single X chromosome in a diploid cell. *See* sex determination; Turner syndrome.

xolloid *See* bone morphogenetic protein.

XPA, XPD, XPF *See* ABC excinucleases.

XPB DNA helicase. Equivalent Ss12. *See* DNA repair; excision repair.

XPC-HR23A or B *See* xeroderma pigmentosum.

XPG Xeroderma pigmentosum endonuclease. *See* DNA repair; endonuclease; xeroderma pigmentosum.

Single- and double-strand breaks. (Courtesy of Brinkley, B. R. & Hittelman W. N.)

X-ray-caused chromosome breakage Major effect of ionizing radiations on living cells. Chromosome breakage has been extensively utilized for genetical analyses for knocking out genes, the production of radiation hybrids, deletion mapping, genomic subtraction, etc. The destructive effects of the radiation depend on its nature. Soft X-rays having higher density of linear energy transfer break the chromosome more effectively than the shorter wavelength and high-energy hard X-rays. The destructive effect also depends on the species

exposed, the type of tissues irradiated, the physiological conditions during the delivery of the radiation, the repair system, etc. In *Drosophila* and locusts, usually 1,000–5,000 R doses are employed to cause chromosome breakage in adults and embryos. In case of gonadal radiation, 100 to 500 R may be effective. In the testis cells of *Macaca mulatta* monkeys, an increase of radiation dose from 25 R to 400 R resulted in close to an exponential increase of chromosome breakage, reaching about 1.5% of the chromosomes at the highest dose. In the large nuclei and chromosomes of *Trillium* and *Vicia faba*, generally 100 to 500 R break the root-tip chromosomes, and lower doses may be sufficient if applied to pollen mother cells. For the smaller nuclei of maize, 800 to 1,500 R may be chosen. The very small somatic nuclei of *Arabidopsis* may tolerate 5–10-fold higher doses of irradiation than maize. *See* cathode rays; LET; radiation effects; radiation hazard assessment; radiation hybrid; radiation sensitivity. (Brinkley, B. R. & Hittelman, W. N. 1975. *Int. Rev. Cytol.* 42:49.)

X-ray crystallography *See* X-ray diffraction analysis.

X-ray diffraction analysis Used for the analysis of the structure of molecules by determining the angles of electron scattering upon exposure to X-rays. When a large number of a particular type of molecules in an array are irradiated, they will scatter the incident electrons. Where the scattered beams cancel each other, no bright image is formed. Where, however, the scattered electrons reinforce each other because they are diffracted by a certain common arrangement of crystals or molecules, a bright image is formed on the screen. Thus, the intensity of the spots on the screen provides a basis for calculating and determining the internal three-dimensional structure of the object. *See* nuclear magnetic resonance spectrography; X-rays. (<http://www.rcsb.org/pdb/>.)

X-ray hazard *See* radiation effects; radiation hazard assessment; radiation protection; X-ray-caused chromosome breakage; X-rays.

X-ray repair Involves excision repair mechanisms. A human X-ray repair gene locus was isolated through complementation of excision repair–deficient Chinese hamster ovary cells by human DNA. XRCC2 repairs DNA double-strand breaks by homologous recombination. The locus

XRCC1, encoding a ligase III polypeptide, was assigned to human chromosome 19q13.2-q13.3. XRCC4 locus (human chromosome 5) is involved in the determination of X-ray sensitivity in the G1 phase of the cell cycle, but its response in the S phase appears normal, indicating that the defective mutants (in Chinese hamster ovary cells) are deficient in the repair of DNA double strands. The XRCC9 allele is in the FANC-G complementation group of Fanconi anemia. *See* DNA repair; excision repair; Fanconi anemia; physical mutagens; radiation sensitivity; X-ray-caused chromosome breakage; X-rays; XRCC. (Tebbs, R. S., et al. 1999. *Dev. Biol.* 208:513.)

X-rays Ionizing electromagnetic radiation emitted by the cathode tubes of Röntgen machines within the range of 10^{-8} to 10^{-11} m wavelength. The shorter ones (hard rays) have greater penetration and lower ionization density while the longer wavelengths (soft rays) have reduced penetration and denser dissipation of the energy. Hard rays cause more discrete lesions to the genetic material; the soft rays are expected to cause more chromosomal breakage. The spectrum of the radiation may be controlled by filters. The effectiveness of the filters depends on the attenuation of the radiation by the nature of the filter. The fluence of the radiation can be defined as $I = (I_0)e - \mu x$, where I = fluence at a certain depth x; I_0 = fluence rate at the surface, μ = specific attenuation coefficient. And μ/p = mass attenuation coefficient, (p = density of absorbing material). At photon energy MeV = 0.1, the mass attenuation coefficients (cm^2/g) are for aluminum (0.171), iron (0.370), lead (5.400), water (0.171), and concrete (0.179). *See* cathode rays; Compton effect; ionizing radiation; Ku; radiation effects; radiation hazard assessment; radiation protection; X-ray repair.

Thickness of the protective shield needed in mm using lead, depending on voltage and amperage of the X-radiation.

Kilo Volt	milliAmpere		
	<5	5–10	30
50	0.5	0.6	0.7
125	1.5	3.0	3.5
250	6.0	7.0	8.0
400	16.0	18.0	21.0

X-ray sensitivity *See* nuclear size; X-ray-caused chromosome breakage; X-ray repair.

X-ray therapy *See* radiation therapy.

XRCC The XRCC2 (7q36.1), XRCC3 (14q32.3), XRCC4 (5q13-q14), and XRCC5 (Ku70, 2q35) genes in humans, along with RAD51 (15q15.1, 17q11-q12) and DMC, control recombination and repair. XRCC1 stimulates human polynucleotide kinase and promotes the repair of single-strand DNA breaks. XRCC3 protein, in concert with RAD51C, facilitates homologous pairing and recombinational repair. *See* DNA repair; Ku; RAD; RAG; X-ray repair. (Whitehouse, C. J., et al. 2001. *Cell* 104:107; Masson, J.-Y., et al. 2001. *Proc. Natl. Acad. Sci. USA* 98:8440; Brenneman, M. A., et al. 2002. *Mol. Cell* 10:387.)

XREF Cross-referencing model organism genes with human disease and other mammalian phenotypes. *See* databases.

Xrn1p Exonuclease degrading RNA beginning at the 5′-end.

X-stain Fluorochrome that stains the X-chromosome differently from other chromosomes. *See* chromosome painting; FISH.

XX males Occur as a normal condition in birds and some other species; it is a rare (0.000,05) condition of male mammalian (human) births. The recurrence risk is, however, about 25%. Actually, most of them (90%) have an X-chromosome–Y-chromosome short-arm translocation. Their phenotype and infertility resemble those of Klinefelter syndrome individuals, although the afflicted persons tend to be shorter in stature. Their distinction from hermaphrodites requires the use of an appropriate fluorochrome-labeled cytogenetic probe for the critical Y-chromosomal segment. *See* hermaphrodite; Klinefelter syndrome; sex chromosomal anomalies in humans; sex determination. (Vidal, V. P., et al. 2001. *Nature Genet.* 28:216.)

XXX *See* metafemale; sex-chromosomal anomalies in humans; triplo-X.

XXY *See* Klinefelter syndrome; sex-chromosomal anomalies in humans.

XY body *See* sex vesicle.

Xylella fastidiosa Bacterium causing citrus variegated chlorosis, affecting a broad spectrum of crop species. It has a 2,679,305 bp sequenced genome and two plasmids (51,158 bp and 1,285 bp). (*Nature* [2000] 406:151.)

xylem Transports tracheid vessels of plants carrying nutrients and water from the roots toward the leaves.

xylene (C$_6$H$_4$[CH$_3$]$_2$) Xylol; a highly flammable irritant and narcotic liquid used as a solvent in microtechniques. The permissible threshold of vapors in the air is 100 ppm.

xylene cyanole FF *See* tracking dyes.

xylose (wood sugar, C$_5$H$_{10}$O$_5$) Epimer of aldopentoses. It used in the tanning industry, as a diabetic carbohydrate food, and in clinical tests of intestinal absorption. *See* epimer.

xylulose (C$_5$H$_{10}$O$_5$) Intermediate in the pentose phosphate pathway; it accumulates in the urine of pentosuriac patients. *See* pentose phosphate pathway; pentosuria.

D-xylulose.

XYY *See* sex-chromosomal anomalies in humans.

Calvin B. Bridges (1889–1938) was one of the most talented geneticists of all time. Most everybody can identify him as the founder of cytogenetics through his studies on nondisjunction, or with his work on salivary maps. Fewer people remember that he initiated the currently used gene symbols in *Drosophila*. Bridges also discovered the basic types of chromosomal aberrations:

"The general term 'deficiency' is used to designate the loss or inactivation of an entire, definite, and measurable section of genes and framework of a chromosome. A case of deficiency in the X chromosome of *Drosophila ampelophila* [currently *melanogaster*] occurred in September 1914, and has given rise to a whole series of correlated phenomena. The first indication of this deficiency was the occurrence of a female which had failed to inherit from her father his sex-linked dominant mutant 'bar,' though she inherited in a normal manner his sex-linked recessive mutant 'white.' This female, when bred, gave only about half as many sons as daughters, the missing sons, as shown by the linkage relations, being those which had received that X which was deficient for bar" (*Genetics* 2:445, 1917).

Two years later, Bridges reported duplications and translocations in the *Abstracts of the Zoological Society* (published in *Anatomical Record*):

"... a section of the X-chromosome, including the loci for vermilion and sable, became detached from its normal location in the middle of the X-chromosome and became joined on to the 'zero' end (spindle fiber) of its mate. For certain loci this latter chromosome carries two sets of genes--those present in the normal location and also the duplicating set. If a male carries the recessive genes for vermilion and for sable in the normal loci and the wild-type allelomorphs in the duplicating loci, he is wild-type in appearance precisely as though he were an XX female heterozygous for vermilion and sable. ...

"A third case is the transposition of a piece of the second chromosome to the middle (spindle fiber) of the third chromosome. The genes of this duplication piece show linkage to both the second and the third chromosome at the same time" (*Anat. Record* 15:357–358, 1919).

The basis of a crossover reducer as an inversion of the third chromosome of *Drosophila* was identified by Alfred H. Sturtevant (1891–1970) in 1926 (*Biol. Zbl.* 46:697).

A historical vignette.

Y Symbol of pyrimidines in nucleic acid sequences.

YAC Yeast artificial chromosome vectors equipped with a yeast centromere and some (*Tetrahymena*) telomeres in a linear plasmid containing selectable markers, ARS (autonomously replicating sequence) for maintenance, and propagation of eukaryotic DNA inserts in cloning of even larger than 200-kb-size sequences. YACs have an important role for identifying contigs in physical mapping of larger genomes, in situ hybridization, map-based gene isolation, etc. The rate of instability of YAC was estimated to be about 2%. In mitotic yeast cells, YACs behave like other chromosomes. Meiosis can be analyzed by tetrads, although recombination appears to be reduced. YACs may be maintained through some cell divisions in the mouse cytoplasm and behave like *double minute* chromosomes or they may be integrated into the mouse chromosomes. *See* anchoring; chromosome walking; contigs; DM; in situ hybridization; pulsed-field gel electrophoresis; YAC library; diagram next page. (Peterson, K. R. 1999. *Methods Enzymol.* 306:186; Brown, W. R., et al. 2000. *Trands Biotechnol.* 18:218; Adam, G., et al. 1997. *Plant J.* 11:1349; Ragoussiz, J. & Monaco, A. P. 1996. *Methods Mol. Biol.* 54:157.)

YAC library Contains large restriction fragments of genomic DNA cloned in YAC vectors and separated by pulsed-field gel electrophoresis. A YAC library generally contains 100- to 250-kb (or larger)-size DNA fragments in multiple (at least five) copies if possible, covering the entire genome of an organism. Selecting the appropriate YAC clone for the purpose of finding the region of interest, the YAC library in yeast colony filter hybridization experiments is probed by RFLPs, PCR, inverse polymerase chain reaction probes, plasmid rescue, or other means. *See* colony hybridization; contig; genome project; inverse polymerase chain reaction; PCR; plasmid rescue; probe; pulsed-field gel electrophoresis; RFLP; YAC. (Larin, Z., et al. 1997. *Mol. Biotechnol.* 8:147.)

YAK (*Bos grunniens*) "Wild ox" but also an Asian draft animal, $2n = 60$.

YAM (*Dioscorea* spp) Tropical food crops. The Asian and African species are $x = 10$, but the Americans are $x = 9$. The actual chromosome numbers vary from diploid to decaploid.

YAMA Ced-3-like protease, also called CPP32β. *See* apoptosis.

Yang cycle Pathway of biosynthesis of ethylene from 2-keto-4-methylthiobutyrate (KMB) › S-adenosyl-L-methionine (AdoMet) → 5′-methylthioadenosine (MTA) → 5′-methylthioribose (MTR) → 5′-methylthioribose-1-phosphate (MTR-1-P) → KMB. AdoMet is then converted to 1-amino-cyclopropane-1-carboxylic acid (ACC) by ACC synthase, and ACC oxidase generates ethylene. *See* AdoMet; ethylene; plant hormones. (Pardee, A. B. 1987. *J. Cell Physiol. Suppl.* 5:107.)

yarowization *See* vernalization.

Y-box proteins Members of a family of transcription factors that bind to an inverted CCAAT box (Y box) that activates genes involved in cell proliferation and growth. Y-box proteins interact with other proteins and modulate transcription. (Ladomery, M. & Somerville, J. 1995. *Bioessays* 17:9.)

YBP Years before present.

Y-chromosomal linkage *See* holandric genes.

Y-chromosome One of the sex chromosomes generally present in males (XY). However, in some species (birds, insects, fishes), the females are XY and the males may be XX or XO. In organisms with homogametic males, their chromosomal constitution is commonly identified as WW. In the majority of species, the Y-chromosome has only genes for sex differentiation or sex determination. These genes in the nonrecombinant (NRY, ~35 Mb) part are involved with sex and fertility determination, whereas the other Y-chromosomal genes have homologies with a common segment of the X-chromosome as well as sequences scattered in autosomes. The major part of the Y-chromosome thus lacks homology to the X-chromosome. The TTY2 gene family includes at least 26 members arranged in tandem repeats. They are transcribed in the testis and in the adult kidney, but none of them are translated (Makrinou, E., et al. 2001. *Genome Res.* 11:935). Evolutionists suggested the general low gene number is the consequence of the absence of recombination. R. A. Fisher assumed that the absence of recombination and genes is due to the fact that recombination would mess up the sex-determination system and would lead to intersexes. The diminution of the gene content of the Y-chromosome has also been attributed to Muller's ratchet.

Molecular markers for the Y-chromosome (>250) are increasing and it is becoming possible to trace paternal lines of evolutionary descent on the basis of variations in these chromosomes. This analysis is analogous to the use of mtDNA for the development of evidence for (mitochondrial) Eve's origin (Su, B., et al. 1999. *Am. J. Hum. Genet.* 65:1718). Y-chromosome information must be carefully evaluated in pedigree analysis because the paternity may be equally likely for grandfather, brothers, cousins, etc., in the same family or even illegitimate male relatives. On the bases of Y-chromosome constitution, 10 lineages appear to account for more than 95% of current European human populations. Distribution of the Y haplotype shows more geographic than linguistic diversity because language influences were acquired more frequently than genes. The diversity of the human Y-chromosome is lower than that of any other chromosome. *See* azoospermia; Eve, foremother of molecular mtDNA; holandric genes; human

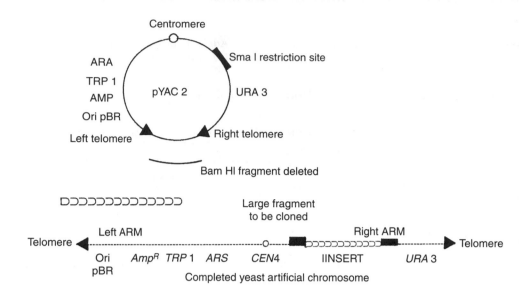

Top: A Circular yeast plasmid with *Tetrahymena* telomeres. A segment of the pBR322 prokaryotic plasmid contains the replicational origin and the *AMP^R* gene (selectable in *E. coli* for ampicillin resistance). The *ARS* (autonomous replication sequence), cloned centromere of yeast chromosome 4 (CEN 4), restriction enzyme recognition sites where the plasmid can be opened for insertion and later ligation are shown. The *TRP1* (tryptophan) and *URA 3* (uracil) genes of yeast serve to ascertain that both arms are present and the transformant can synthesize tryptophan and uracil. A piece of the plasmid between two bam sites and the yeast *HIS* (histidine) gene are deleted. This vector, because of the pBR322 sequences, replicates in *E. coli* as well as in eukaryotic cells. All details are not shown and the diagram is not on scale. Other YAC vectors are similar but not identical with this model. Bottom: completed linear YAC. (Modified after Burke, D. T., et al. 1987. *Science* 236:806.)

evolution; F_{ST}; Muller's ratchet; mutation rate; recombination, variations of; sex determination; SRY; UEP. (*Nature* 1998. 396:27; *Nature* 1999. 397:32; Owens, K. & King, M.-C. 1999. *Science* 286:451; Kayser, M., et al. 2001. *Am. J. Hum. Genet.* 68:173; Stumpf, M. P. H. & Goldstein, D. B. 2001. *Science* 291:1738; Underhill, P. A., et al. 2001. *Ann. Hum. Genet.* 65:43; Tilford, C. A., et al. 2001. *Nature* 409:943; Hurles, M. E., et al. 2002. *Genetics* 160:289; Bachtrog, D. & Charlesworth, B. 2002. *Nature* 416:323.)

Ycp   *See* yeast centromeric vectors.

Ydj1   Yeast homologue of DnaJ. *See* chaperones; DnaJ.

Yeast   (1) Budding yeast. *See Saccharomyces cerevisiae*. (2) Fission yeast. *See Schizosaccharomyces pombe*.

yeast artificial chromosomes   *See* YAC.

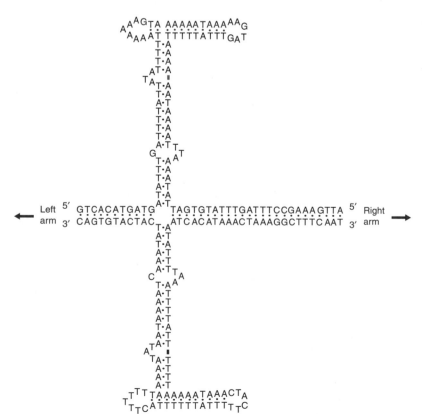

Yeast centromeric region 3. (From Clark, L., et al. 1981. *Stadler Symp.* 13:9.)

yeast cell density test   Cell suspension at 0.1 OD (optical density) at 600 nm corresponds to ~3×10^6 cells per mL.

yeast centromeric vectors (Ycp)   Carry centromeres, telomeres, and ARS; can be linear or circular; their stability improves with increased length. *See* vectors; YAC.

yeast episomal vector   Carries the origin of replication of the 2 μm yeast plasmid. *See Saccharomyces cerevisiae*; vectors; yeast plasmid 2 μm.

yeast integrating vector (YI)   Replicated only by the host; it behaves like a gene in the chromosome and can show homologous recombination, duplications, and substitutions like bacterial episomes. Although it can generate chromosomal rearrangements, its stability is extremely high; its ability to transform is extremely low. *See* vectors.

yeast plasmid 2 μm   Circular duplex DNA plasmid of 6,318 bp contains genes for replication and thus can maintain a copy number of about 50/cell, but it lacks any selectable marker in native form. By attaching it to the pBR322 *E. coli* plasmid, shuttle vectors (ca. 8.5 kbp) have been generated that carry the bacterial histidine operon (as a selectable marker), and it is expressed in bacteria as well as in budding yeast because a segment of this plasmid can serve as a promoter for the operon. *See* histidine operon; *Saccharomyces cerevisiae*; selectable marker; yeast vectors. (Velmurugan, S., et al. 1998. *Mol. Cell Biol.* 18:7466.)

yeast proteome   *See* YPD. (<www.proteome.com/>.)

yeast replicating vectors (Yrp)   Carry autonomously replicating sequences (ARS) and have moderate stability and relatively low copy number. They may be integrated into chromosomes. *See* vectors.

yeast transformation   May involve a number of different changes in the yeast chromosome at the site of transformation. *See* fungal transformation; transformation, genetic diagram below.

yeast transposable elements   *See* Ω; Ty.

yellow crescent   Cytoplasmic motions in the fertilized ascidian embryo may generate a yellowish area that is broken up and shared by the cleavage cells, thus indicating the origin of differentiating cell lineage.

Yep   *See* yeast episomal vector.

Yersinia   Group of gram-negative bacteria responsible for bubonic plague, gastroenteritis, etc. The sequenced genome of *Y. pestis* includes a 4.65 Mb chromosome and three plasmids of 96.2, 70.3, and 9.6 kb. It contains 150 pseudogenes. The genome shows frequent insertion sequences and intragenic recombinations. *See* biological weapons; virulence. (Parkhill, J., et al. 2001. *Nature* 413:323; Hinnebusch, B. J., et al. 2002. *Science* 296:733; Cornelis, G. R. 2002. *Nature Rev. Mol. Cell Biol.* 3:472.)

YES   Yeast–*E. coli* shuttle vector. *See* shuttle vector. (Elledge, S. J., et al. 1991. *Proc. Natl. Acad. Sci. USA* 88:1731.)

YES1 oncogene   In human chromosome 18q21. Its protein product is homologous to that of Roux sarcoma virus (SRC)

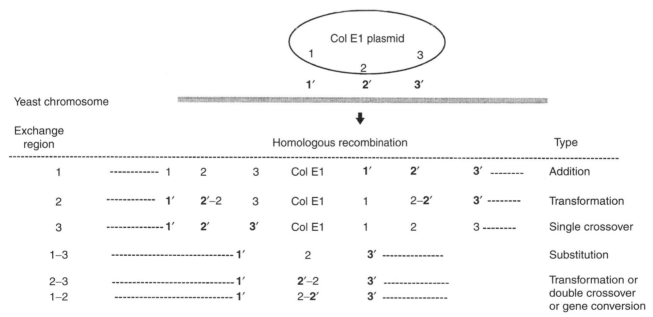

The various mechanisms of transformation and recombination between plasmid and chromosomal DNA in yeast. Gene 2′ is an auxotrophic marker; gene 2 is a prototrophic homologue; 1′ and 3′ are flanking chromosomal sequences homologous to 1 and 2 in the plasmid with a Cole1 or other replicator. (Modified after Hinnen, Botstein & Davis in *Molecular Biology of the Yeast Saccharomyces*, Strathern, et al., eds. 1981 Cold Spring Harbor Lab. Press, Cold Spring Harbor, NY.)

and it is also a protein tyrosine kinase. In association with other proteins (YAP), it is a transcriptional coactivator. *See* oncogenes; tyrosine kinases.

Y-family DNA polymerases Are error-prone enzymes such as the prokaryotic UmuD'$_2$C (pol V), Din B (pol IV) and the yeast proteins Rev1 and Rad30. They copy damaged DNA efficiently, make frequent incorporation error ($\sim 10^{-1}$ to 10^{-3}), rely on mispairing, use mismatches, misaligned templates, etc. *See* DNA polymerases; DNA repair; error in replication; error-prone repair; X-family of DNA polymerases. (Goodman, M. F., 2002. *Annu. Rev. Biochem.* 71:17.)

YI *See* yeast integrating vector.

yIF-2 Yeast eIF-5B; it is homologous to prokaryotic IF2. *See* eIF-5B.

Yin Yang (YY1, NF-E1) Multifunctional zinc-finger transcription factor of growth factors, hormones, and cytokines. (Santiago, F. S., et al. 2001. *J. Biol. Chem.* 276:41143.)

Y-linked *See* holandric gene.

Yohimbine Alkaloid produced by African plants of the families Rubiaceae and Apocinaceae. Their extracts have been used as a adrenergic-blocking medicine for arteriosclerosis and hypertension. It is supposedly an aphrodisiac. (Morales, A. 2001. *World J. Urol.* 19[4]:251.)

yolk Complex nutrients embedding the animal egg. *See* egg.

YPD (1) *See* yeast proteome. (2) Yeast nutrient medium containing g/L H_2O yeast extract 10, glucose 20, Bacto-Peptone 10 or 20.

YPGE Nutrient medium containing g/L H_2/O yeast extract 10, Bacto-Peptone 10, glycerol 20, and ethanol 10.

YPL.db Yeast Protein Localization database: <http://ypl.tugraz.at>.

YPT Homologue of Sec, and RAB. *See* RAB; Sec.

Y RNAs Have phylogenetically conserved secondary structure consisting of at least three stems and small internal loops. In humans, Y1, Y3, Y4, and Y5 are distinguished, but their function is unclear. They are suspected to regulate ribosomal protein synthesis.

Yrp *See* yeast replicating vectors.

Yt blood group Coded in human chromosome 7q.

YT bacterial medium H_2O 900 mL, bacto-tryptone 16 g, bacto-yeast extract 10 g, NaCl 5 g, pH 7.0 (adjusted by 5 N NaOH), filled up to 1 L and diluted to half before use.

Yohimbine.

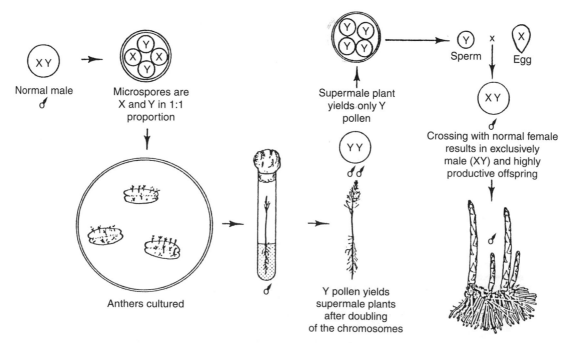

Production of (YY) all-male asparagus.

yttrium (Y) Very rare metal; ^{90}Y has a half-life of 64 hours and has been used for internally exposing cancerous tissues to β-radiation. *See* ionizing radiation; isotopes; magic bullet.

YUASA oncogene   Its human homologue is in chromosome 7.

YY asparagus   By regeneration of plants from microspores and chromosome doubling, plants can be obtained with this chromosomal constitution. The practical advantage of this reproducible vegetable is the approximately 30% higher yield of edible spears. *See* embryogenesis, somatic; diagram on preceding page.

Oscar Hertwig on genetics: "The hypothetical idioblasts... are according to their different composition, the bearers of different properties, and produce, by direct action, or by various methods of cooperation, the countless morphological and physiological phenomena, which we perceive in the organic world. Metaphorically they can be compared to the letters of the alphabet, which, though small in number, when combined form words, which in their turn, combine to form sentences or to sounds, which produce endless harmonies by their periodic sequence and simultaneous combination." (*The Cell*, Macmillan, New York, 1895, page 340.)

A historical vignette.

Z

Z (or z) Standard normal probability density function calculated by the formula

$$Z = \frac{1}{\sigma\sqrt{2\pi}} e^{-(Y-\mu)^2/2\sigma^2}$$

where Y is the normal variate, $\pi \cong 3.14159$, $e \cong 2.71828$, μ is the mean, and σ is the standard deviation. The z values are generally read from statistical tables. Z indicates the height of the ordinate of the curve and thereby the density of the items. *See* confidence intervals; F distribution; normal distribution; standard deviation.

ZAG (Zn-α_2-glycoprotein) Resembles MHC class I heavy-chain molecules, but it is different because it cannot bind β_2-microglobulin. It occurs in the majority of body fluids and appears to have a role in inducing fat loss in adipocytes. It accumulates in breast cancer cells and other cells in serious distress (cachexia). *See* HLA; MHC; obesity. (Kennedy, M. W., et al. 2001. *J. Biol. Chem.* 276:35008.)

ZAP-70 Zeta-associated protein 70 is a cytosolic protein tyrosine kinase expressed only in T cells and natural killer cells. By binding to phosphorylated ζ-chains of the CD3 T-cell antigen-receptor complex, it assists in the activation of T cells. Its defect may lead to severe combined immunodeficiency. ZAP-70 tyrosine kinase signals to the CXCR4 chemokine receptor and thus regulates the migration of T lymphocytes. *See* CD3 T cell; CXCR; immunodeficiency; killer cell; SHP-1; tyrosine kinases. (Ottoson, N. C., 2001. *J. Immunol.* 167:1857.)

Z-buffer $Na_2HPO_4.7H_2O$ 0.06 M, NaH_2PO_4 0.04 M, KCl 0.01 M, $MgSO_4.7H_2O$ 0.001 M, β-mercapto ethanol 0.05 M, pH 7. Do not autoclave it.

Z-chromosome Sex chromosome present in both sexes of species with heterogametic females (comparable thus to the X-chromosome). The males are ZZ and the females are WZ. *See* sex determination; W chromosome.

Z-disc *See* sarcomere.

Z-distribution Statistical device for testing the significance of the differences between correlation coefficients in case the null hypothesis is not $r = 0$. The relation of z to r was elaborated by R. A. Fisher as $z = (1/2)[ln(1 + r) - ln(1 - r)]$ and its standard error as

$$\sigma_z = \frac{1}{\sqrt{(n-3)}}.$$

Usually, tables are available in statistical textbooks for routine calculations. *See* covariance for correlation coefficient.

Z-DNA Relatively rare left-handed double helix that may be formed in short (8–62 bp) regions of alternating purines and pyrimidines (ATGTGTGT, GCATGCAT). The polyGC.polyGC sequences are most favorable for B → Z transition. In some species, CA/TG repeats are most conducive for Z-DNA formation. Alternating purine-pyrimidine tracts may also modulate transcription in a plus or minus way. The figure below shows space-filling models modified after Dickerson, R. E., et al. 1982. *Science* 216:475. *See* adar; DNA types; RNA editing. (Liu, R., et al. 2001. *Cell* 106:309; Rothenburg, S., et al. 2001. *Proc. Natl. Acad. Sci. USA* 98:8985.)

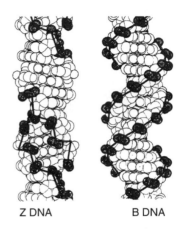

Z DNA   B DNA

The heavy black line accents the
sugar-phosphate backbone

Zea mays (L) *See* maize.

zeatin (6[4-hydroxy-3-methyl-cis-2-butenylamino]purine) Cytokinin plant hormone. In *Agrobacterium*, a T-DNA-located *ipt* gene encodes an isopentenyl transferase that mediates the synthesis of transzeatin and isopentenyl adenosine. In plant tissue culture media, it is a commonly used alternative to kinetin, benzylamino purine, and isopentenyl adenosine. *See* plant hormones.

zebra *Equus quagga*, $2n = 44$; it can form hybrids with both horses (*Equus caballus*, $2n = 64$) and donkeys ($2n = 62$). *Equus grevyi*, $2n = 46$; *Equus zebra hartmanniae*, $2n = 32$. The zebra duiker (*Cephalophus zebra*, $2n = 58$) is not a member of the Equidae family, but it is the male of a bovine species.

ZEBRA (Zta) Nonacidic activator protein of the lytic cycle of the Epstein-Barr virus; it promotes the assembly of the DA complex (TFIID-TFIIA) of transcription factors. *See* DA; open transcription complex; transcription factors.

zebrafish *Brachydanio rerio*, $2n = 50$. 3–4 cm tropical freshwater fish; the genome size is about 2×10^9 bp. It is easy to raise and sexually mature in 2–3 months (4 generations/year). The embryogenetic pattern is laid down in 12 hr, and it is well suited for the analysis of developmental pathways and cell lineages in this small vertebrate. Hundreds of eggs are laid externally. The embryos are transparent to permit viewing of gastrulation, development of the brain and heart, etc. Haploids survive for several hours and mutants are available. About 600 genes involved in its development have been identified, and a 3,350 cM map of 3.3 cM resolution became available in 1998. The zebrafish genome has extensive homology to that of humans. *See* EP; medaka. (Fishman, M. C. 1997. *Methods Cell Biol.* 52:67; Detrich, H. W., III, et al. eds. 1999. *The Zebrafish: Biology.* Academic Press; radiation hybrid map: Geisler, R., et al. 1999. *Nature Genet.* 23:86; Hukriede, N. A., et al. 1999. *Proc. Natl. Acad. Sci. USA* 96:9745; Woods, I. G., et al. 2000. *Genome Res.* 10:1903; Shin, J. T. & Fishman, M. C. 2002. *Annu. Rev. Genomics Hum. Genet.* 3:311. <http://zebra.sc.edu>; <http://zfish.uoregon.edu/ZFIN/>; <http://depts.washington. edu/~fishscop/>; <http://www.tigr.org/tdb/tgi.shtml>, <zfin. org>.)

zebu   *Bos indicus*, $2n = 60$. *See* Santa Gertrudis cattle.

zein   Prolamine protein in maize (homologous to gliadin in wheat) may make up to 50% of the grain proteins. It is a protein of low nutritional value (and thus undesirable) because of the low lysine and tryptophan content. It is deposited in zein bodies at the place of synthesis. In the high-lysine maize varieties (*opaque, floury*), prolamines are very low and nonprolamine proteins increase. *See* glutenin; high-lysine corn.

zeitgeber   Rhythmic external signal for circadian change. *See* circadian rhythm.

zeitnehmer   Internal signal for rhythmicity in response to the zeitgeber or even in its absence. *See* circadian rhythm; zeitgeber.

Zellweger syndrome (ZS)   Brain-liver-kidney (cerebrohepatorenal) recessive disease involving human chromosome 7q11.12-q11.13, but more than a dozen other loci may have similar effects. The basic defect is due to peroxisome anomalies. *See* cataract; chondrodysplasia punctata; microbodies; neuromuscular diseases; pseudo-Zellweger syndrome.

zero, absolute   Minimum lowest temperature; 0 Kelvin = −273.15°C.

zero time binding   Status of reassociation of two single-strand palindromic DNAs at the beginning of an annealing kinetics experiment. These are the fastest reassociating fractions because they are repeats and are close to each other. *See* c_0t curve.

zeste (*z*, chromosome 1.10 of *Drosophila*)   Protein kinase product apparently alters chromatin structure and affects the expression of *w*, *Ubx*, and *dpp* by attaching to their promoters. Its action bears similarity to *Polycomb*. The zeste-white 3 complex also regulates the spindle attachment to the cortical actin. *See Polycomb*; *w* locus. (McCartney, B. M., et al. 2001. *Nature Cell Biol.* 3:933.)

zeugopodium   Radius, ulna, tibia, and fibula bones.

ZFY   Zinc-finger Y is a sequence in the Y-chromosomes assumed to be involved in the maturation of testes or sperm. A 729 bp intron located immediately upstream of the zinc-finger exon shows no or possibly very little sequence variation in worldwide human samples. *See* zinc finger.

zidovudine   *See* AZT.

zif   *See* NGFI-A.

zig-zag inheritance   *See* criss-cross inheritance.

Z-inactivation   The Z-chromosome has the same role in sex determination in birds as the X chromosome in mammals, yet apparently its genes are not inactivated in homogametic individuals. *See* dosage compensation. (Kuroda, Y., et al. 2001. *Chromosome Res.* 9:457.)

zinc finger nuclease   Limited-specificity nucleases that can recognize certain DNA sequences and cause small deletions at selected targets. The Zn-fingers have preference for guanine-rich tracts, particularly for 5′-GNN-3′ triplets. Each nuclease units may have three fingers and each grabs one triplet thus a total of 9 bases. The dimeric structure of the nuclease cleaves both strands at a tract in between the G-rich sequences. Non-homologous end-joining may follow the double-strand breaks and thus somatic and germline mutations may occur. *See* non-homologous end-joining.

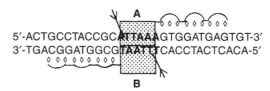

Diagram of a dimeric (A and B) zinc-finger endonuclease and its mode of cleavage (arrows) of DNA. The subunits recognize of the DNA sequence (bold) by the shaded square domain. The three zinc-fingers (diamonds within the semicircles) bind to the G-rich repeats. (Modified after Bibikova, M., et al. 2002. *Genetics* 161:1169.)

zinc fingers   Binding mechanisms of transcription factors and other regulatory proteins containing tandemly repeated cysteine and histidine molecules. These fold in a finger-like fashion cross-linked to Zn. Other highly conserved amino acids in zinc fingers are phenylalanine (F), leucine (L), and tyrosine (Y). Some zinc-finger proteins also bind RNA to DNA. About 0.7% of the proteins in budding yeast and the nematode *Caenorhabditis elegans* contain Zn-finger motifs; in yeast, most commonly 2/molecule, in *Caenorhabditis* up to 14/molecule. Using DNA microarray designs, putative Zn-finger transcription factor binding sites can be identified. By designing new zinc-finger motifs, regulation of gene expression may be achieved. Zinc is an essential element for life and it is a cofactor for several enzymes. *See* acrodermatitis enteropathica; binuclear zinc cluster; Cys_4 receptor; DNA-binding protein domains; GATA; hormone receptor; LIM; microarray hybridization; RING finger; ZFY. (Rubin, G. M., et al. 2000. *Science* 287:2204; Bulyk, M. L., et al. 2001. *Proc. Natl. Acad. Sci. USA* 98:7158; Pabo, C. O., et al. 2001. *Annu. Rev. Biochem.* 70:313.)

zinc-ribbon fold (CXXC(H)-15/17-CXXC) Present in diverse binding proteins including FIIB. *See* binding proteins; transcription factors.

zinc-ring finger *See* RING finger.

ZIP1, ZIP2 Protein factors that mediate phosphorylation of potassium ion channels by protein kinase C. *See* ion channel; protein kinase.

zipper domain Part of dimeric DNA-binding proteins where the two subunits are held together by repeating amino acid residues; in the other parts, the subunits are separated from each other. *See* leucine zipper.

Z-linkage Genes syntenic in the Z-chromosome. *See* X-linked.

zoidogamy Fertilization takes place by motile antherozoids. *See* siphonogamy.

Zollinger-Ellison syndrome (Wermer syndrome, MEN 1) Multiple endocrine adenomatosis (neoplasia) encoded in human chromosome 11q13. *See* adenomatosis multiple endocrine.

zona pellucida Yolky layer around the mammalian egg. *See* egg; fertilization (animals); vitelline layer. (Jovine, L., et al. 2002. *Nature Cell Biol.* 4:457.)

zonoskeleton Scapula, clavicle, and hip bones.

zonula occludens Tight cell junction mediated by ZO proteins associated with a 210−255 kDa membrane-tied guanylate kinase. *See* ELL; MAGUK. (Meyer, T. N., et al. 2002. *J. Biol. Chem.* 277:24855.)

zoo blot Southern hybridization experiments with probes derived from a variety of different species to test potential homologies. (Rijkers, T. & Ruther, U. 1996. *Biochim. Biphys. Acta* 1307:294.)

zoo fish Fluorescent in situ hybridization maps for several species to study evolutionary relations of their chromosomes. *See* FISH.

zoonosis Animal pathogens transferred to humans (by, e.g., grafts, food) may lead to the development of new human diseases under natural conditions. *See* acquired immunodeficiency; grafting in medicine; laboratory safety. (Weiss, R. A. 1998. *Nature Med.* 4:391.)

zoospore Motile (swimming) spore.

zootype hypothesis Claims that HOX-type homeobox genes are present in all metazoa. *See* homeotic genes.

ZP1, ZP2, ZP3 Zona pellucida glycoproteins. *See* fertilization; zona pellucida.

ZPA Zone of polarizing activity determines the anterior/posterior differentiation of the limbs and is located behind AER in the limb bud. *See* AER; limb bud.

ZPR Zinc-finger protein binding to epidermal growth factor receptor. *See* EGFR; zinc fingers.

Z-ring Fts Z ring. *See* tubulin.

Z-RNA May occur in double-stranded RNA molecules, and it is similar to Z-DNA. *See* Z-DNA. (Brown, B. A. 2nd, et al. 2000. *Proc. Natl. Acad. Sci. USA* 97:13532.)

Z-scheme Additional (zigzag) scheme to generate enough ATP by photosynthesis. By two steps (in photosystems I and II), an electron passes from water, but there is not enough energy in a single quantum of light to energize the electron directly and efficiently all the way from PS II to the top of PS I and make $NADP^+$. The leftover energy makes pumping H^+ possible across the membranes to capture some light energy for synthesizing ATP. The redox state of plastoquinone regulates the transcription of genes encoding the reaction center proteins of photosystems I and II. *See* photosystems diagram next page. (Prince, R. C. 1996. *Trends Biochem. Sci.* 21:121.)

Z-score *See* Z.

Zta *See* ZEBRA.

zuotin Z-DNA and tRNA-binding protein of yeast. It has structural homology with DnaJ and is functionally related to the mammalian chaperone MIDA1 required for cellular growth. *See* chaperones; DnaK; MIDA1. (Braun, E. L. & Grotewold, E. 2001. *Mol. Biol. Evol.* 18:1401.)

Z-value Natural logarithm of the ratio of two estimated standard deviations. *See* variance.

zwischenferment *See* G6PD.

zwitterion Dipolar ion with separated positive and negative poles.

zygDNA It has been reported that 0.1 to 0.2% of eukaryotic DNA may not be replicated until late leptotene−zygotene. This delayed replication involves dispersed 4−10 kb stretches. It was assumed that these segments (zygDNA) code for genes with products aiding chromosome pairing. A leptotene (L)

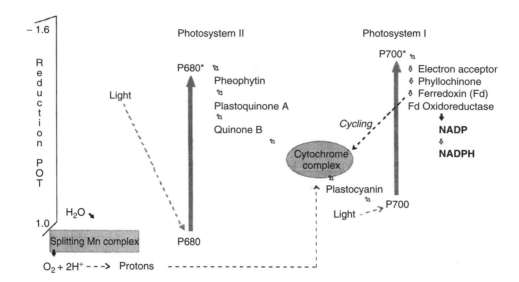

lipoprotein is suspected in the delayed replication. *See* (L) meiosis. (Hotta, Y., et al. 1985. *Cell* 40:785.)

zygomeres Hypothetical initiators of chromosome pairing in DNA.

zygomorphic Structure of bilateral symmetry, like a snapdragon flower.

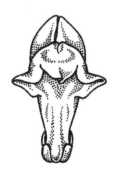

Zygomorphic.

zygonema Chromosome at the zygotene stage. *See* meiosis; zygotene stage.

zygosis Twins can be identical (monozygotic, MZ) or nonidentical (dizygotic, DZ). The distinction is not always simple because dizygotic twins (like any sibs) may show several identical features, depending on the genetic constitution (consanguinity) of the parents. If we designate the probability of monozygosis between twins with identity in a genetic marker as $P(A_1/B)$, where A_1 and B are different markers, and the probability of the twins being either dizygotic or erroneously assumed to be monozygotic is $1 - P(A_1/B)$, the calculation may be based on the formula

$$P(A_1/B) = \frac{1}{1 + [Q \times L]}$$

where Q is the dizygotic:monozygotic proportion in the population (DZ/MZ) and L is the likelihood ratio of the conditional probabilities for DZ and MZ twins that would be identical for a particular genetic condition. The conditional probabilities that if one of the twins is of a particular dominant type, the second would also be of the same type at dizygosis is 0.5–1.0, and at monozygosis 1.0. In case of recessive markers, these probabilities are 0.25 and 1.0, respectively. The probabilities also depend on the genetic constitution of the parents. Recessive markers may be expressed only if both parents carry that particular allele. If either of the parents is homozygous for a dominant marker, then both twins must carry that marker, irrespective of zygosity. L can be computed as

$$L = \frac{P[\text{DZ}]}{P[\text{MZ}]} L_1 \times L_2 \ldots L_n$$

where

$$\frac{P[\text{DZ}]}{P[\text{MZ}]}$$

is the empirical probability of the DZ:MZ proportions in the general population, and $L_1 \times L_2 \ldots L_n$ are the conditional probabilities for the genetic markers 1 to n used, either 0.25 or 1.0. Further complications may arise if either the penetrance or expressivity of the markers varies. The DZ:MZ proportions may vary generally from 0.65:0.35 to 0.70:0.30, but they are somewhat different in various ethnic groups, depending on the age of the mother and the use of fertility drugs and artificial insemination, etc. DNA markers, either by RFLP or the PCR method of typing, can better resolve the problem than using blood types or other genetic analyses. *See* concordance; conditional probability; discordance; dizygotic; DNA fingerprinting; likelihood; monozygotic; PCR; RFLP; twinning.

zygospore Formed by the fusion of two spores or two multinucleate gametangia. *See* gametangia; heterothallism; homothallism.

zygote Cell resulting in the union of two gametes of opposite sexes. *See* gamete.

zygotene stage Of meiosis, when the homologous chromosomes are supposed to begin to synapse. The pairing usually begins at the termini and proceeds toward the centromeric region. At this stage, the synaptonemal complex is detectable by electron microscopy. The intimate bivalent pairing appears to be a requisite for chiasma formation and presumably for genetic recombination. *See* association point; chiasma; meiosis; synaptonemal complex.

Zygotene stage (Illustration after K. Bělař).

zygotic combinations   *See* allelic combinations; Punnett square.

zygotic gene   During embryo development, it participates in the early control of differentiation in contrast to the maternal effect genes, which are transcribed from the maternal genome and their product is transfused to the embryo. *See* maternal effect genes; morphogenesis.

zygotic induction   Usually the integrated λ-prophage is inherited as an integral gene of the bacterial chromosome. When, however, a lysogenic Hfr cell carrying λ is crossed to a *nonlysogenic* F⁻ recipient, the prophage leaves the chromosomal position (induction) and becomes an infectious vegetative phage after replicating to about 100–200 particles. After zygotic induction, only the markers transmitted before the position of the prophage has changed are recovered in the recombinants. Zygotic induction does not occur if the *F⁻ cells are lysogenic*, irrespective of whether Hfr is lysogenic. The F⁻ recipient is immune to superinfection by free λ-phage, and it cannot support the vegetative development of the chromosomal λ-phage transmitted by the Hfr donor. This indicates that the F⁻ cytoplasm carries an immunity substance or a repressor. *See* F⁻; Hfr; lambda phage; lysogenic repressor; lysogeny. (Hayes, W. 1965. *The Genetics of Bacteria and Their Viruses.* Wiley, New York.)

zygotic lethal   Genetic factor that permits the function of the gametes but kills the zygote. The two systems shown at the right occur in the plant *Oenothera*, which carries translocation complexes. Letters printed in outline represent the inviable gametes and zygotes — indicates lack of zygote formation. *See* complex heterozygote; translocation.

zymogen   Inactive enzyme precursor. *See* regulation of enzyme activity.

zymogram   Electrophoretically separated isozymes are identified in the electrophoretic medium (starch, agarose) by supplying a chromogenic substrate in situ. This reveals the position of the functionally active enzyme bands. *See* electrophoresis; isozyme.

Gametic lethality

	A	**B**
A	--	--
B	(AB)	--

The gametic lethal factors **A** and **B** are unable to form selfed zygotes, and only the heterozygous embryos and viable. Only the female can contribute A viable *B* gamete and only the male produces viable *A* gametes

Zygotic lethality

	A	B
A	**AA**	(AB)
B	(AB)	**BB**

In zygotic lethality only the heterozygotes are viable and both types of homozygotes are lethal

zymolase   Hydrolyzes $1 \rightarrow 3$ glucose linkages such as those existing in yeast cell walls; it is not an exactly defined mixture of proteins extracted from *Athrobacter luteus*.

zymosan   Cell wall extract with a variety of components interfering with the C3 complement. *See* complement. (Ohki, K., et al. 2001. *Immunol. Cell Biol.* 79:462.)

zymotype   Electrophoretically determined pattern of enzymes (proteins) characteristic for individuals or a group of individuals. *See* electrophoresis; isozymes; zymogram.

"I have a rather strange feeling about our DNA structure. If it is correct, we should obviously follow it up at a rapid rate. On the other hand it will at the same time be difficult to avoid the desire to forget completely about nucleic acid and to concentrate on other aspects of life." (*J. D. Watson's letter to Max Delbrück on March 22, 1953. Quoted after H. F. Judson, 1979. The Eighth Day of Creation. Simon and Schuster, New York, p. 229.*)

Evelyn Witkin (2002) in reminiscing about the pre-Watson & Crick and Hershey & Chase era at the Cold Spring Harbor Laboratory writes (Annu. Rev. Microbiol. 56:1):

"Although Avery et al. (1) had demonstrated in 1944 that the genetic material is DNA, the prevailing attitude at Cold Spring Harbor had been respectful skepticism. Some suggested that the transforming DNA in their experiments had activated genetic information already present or somehow caused a directed mutation. Others believed that the minuscule trace of protein still contaminating the DNA was the active agent. Delbruck declared DNA to be a "stupid" molecule, incapable of carrying genetic information."

In 1962 Francis Harry Compton Crick, James Dewey Watson and Maurice Hugh Frederick Wilkins were jointly awarded by the Nobel Prize in Physiology or Medicine "for their discoveries concerning the molecular structure of nucleic acids and its significance for information transfer in living material".

In 1969 the Nobel Prize in Physiology or Medicine was shared by Max Delbrück, Alfred D. Hershey and Salvador Luria.

A historical vignettes.

GENERAL REFERENCES

Presented here is a list of books, published mainly during the last few years, selected on the basis of relevance to the topics covered in this volume. The groupings are is somewhat arbitrary because of the overlaps in scope and contents. The selection does not necessarily coincide with my views regarding the technical accuracy of the works listed. All books have positive and negative aspects and the evaluation may be subjective.

CONTENTS

- AGING
- BEHAVIOR
- BIOCHEMISTRY AND PHYSIOLOGY
- BIOLOGY
- BIOMETRY AND BIOINFORMATICS
- BIOTECHNOLOGY
- CANCER
- CONSERVATION
- CYTOLOGY CYTOGENETICS
- DEVELOPMENT

- DISEASES
- EVOLUTION
- GENETICS MONOGRAPHS
- GENETICS TEXTBOOKS
- HISTORY OF GENETICS
- IMMUNOLOGY
- LABORATORY TECHNOLOGY
- MAPPING
- MICROBIOLOGY
- MOLECULAR BIOLOGY
- MUTATION
- NEUROBIOLOGY

- ORGANELLES
- PHILOSOPHY
- PLANT AND ANIMAL BREEDING
- PLANT PATHOLOGY
- PLANT PHYSIOLOGY
- POPULATION GENETICS
- RADIATION
- RECOMBINATION
- SOCIAL AND ETHICAL ISSUES
- TRANSPOSABLE AND MOBILE ELEMENTS
- VIRUSES

AGING (*See also* Cancer, Diseases, Molecular Biology)

Arking, R. 1998. *Biology of Aging: Observations and Principles.* Sinauer, Sunderland, MA, USA.

Barnett, Y. A., and Barnett, C. R., eds. 2000. *Aging Methods and Protocols.* Humana Press, Totowa, NJ, USA.

Bellamy, D. 1995. *Aging: A Biomedical Perspective.* Wiley, New York.

Clark, W. R. 1999. *A Means to the End: The Biological Basis of Aging and Death.* Oxford Univ. Press, New York.

De Grey, A. D. N. J. 1999. *Molecular Biology Intelligence Unit: The Mitochondrial Free Radical Theory of Aging.* R. G. Landes, Austin, TX, USA.

Goate, A., and Ashall, F. 1995. *Pathobiology of Alzheimer's Disease.* Academic Press, San Diego, CA, USA.

Guarente, L. 2002. *Ageless Quest: One Scientist's Search for Genes That Prolong Youth.* Cold Spring Harbor Lab. Press, Cold Spring Harbor, NY, USA.

Hekimi, S. 2000. *The Molecular Genetics of Aging.* Springer, New York.

Holbrok, N. J., Martin, G. M., and Lockshin, R. A. 1995. *Cellular Aging and Cell Death.* Wiley, New York.

Holliday, R. 1995. *Understanding Aging.* Cambridge Univ. Press, New York.

Kanungo, M. S. 1994. *Genes and Aging.* Cambridge Univ. Press, New York.

Medina, J. J. 1996. *The Clock of Ages: Why We Age — How We Age — Winding Back the Clock.* Cambridge Univ. Press, New York.

Rattan, S. I. S., and Toussaint, O., eds. 1996. *Molecular Gerontology.* Plenum Press, New York.

Rose, M. R., and Finch, C. E. 1994. *Genetics and Evolution of Aging.* Kluwer, Norwell, MA, USA.

Smith, D. W. E. 1993. *Human Longevity.* Oxford Univ. Press, New York.

Tomel, D. L., and Cope, F. O., eds. 1994. *Apoptosis II. Molecular Basis of Apoptosis in Disease.* Cold Spring Harbor Laboratory Press, Cold Spring Harbor, NY, USA.

BEHAVIOR (*See also* Neurobiology)

Alcock, J. 2001. *The Triumph of Sociobiology.* Oxford Univ. Press, New York.

Avital, E., and Jablonka, E. 2001. *Animal Traditions: Behavioral Inheritance in Evolution.* Cambridge Univ. Press, New York.

Barondes, S. H. 1998. *Mood Genes: Hunting for the Origins of Mania and Depression.* Freeman, New York.

Begleiter, H., and Kissin, B. 1995. *The Genetics of Alcoholism.* Oxford Univ. Press, New York.

Boake, C. B., ed. 1994. *Quantitative Genetic Studies of Behavioral Evolution.* Univ. Chicago Press, Chicago, IL, USA.

Bock, G. R., and Goode, J. A., eds. 1996. *Genetics of Criminal and Antisocial Behavior.* Wiley, New York.

Bouchard, T. J., and Propping, P. 1993. *Twins as a Tool of Behavioral Genetics.* Wiley, New York.

Carew, T. J. 2000. *Behavioral Neurobiology: The Cellular Organization of Natural Behavior.* Sinauer, Sunderland, MA, USA.

Carson, R. A., and Rothstein, M. A., eds. 1999. *Behavioral Genetics: The Clash of Culture and Biology.* Johns Hopkins Univ. Press, Baltimore, MD, USA.

Cassidy, S. B., and Allanson, J. E., eds. 2000. *Management of Genetic Syndromes.* Wiley, New York.

Clark, W. R., and Grunstein, M. 2000. *Are We Hardwired? The Role of Genes in Human Behavior.* Oxford Univ. Press, New York.

Cloninger, C. R., and Begleiter, H., eds. 1990. *The Genetics and Biology of Alcoholism.* Cold Spring Harbor Laboratory Press, Cold Spring Harbor, NY, USA.

Crawley, J. N. 2000. *What is Wrong with My Mouse? Behavioral Phenotyping of Transgenic and Knockout Mice.* Wiley-Liss, New York.

de Waal, F. 1997. *Good Natured: The Origins of Right and Wrong in Humans and Other Animals.* Harvard Univ. Press, Boston, MA, USA.

Dixson, A. F. 1999. *Primate Sexuality: Comparative Studies of the Prosimians, Monkeys, Apes, and Human Beings.* Oxford Univ. Press, New York.

Dowling, J. E. 2001. *Neuron Networks: An Introduction to Behavioral Neuroscience.* Harvard Univ. Press, Cambridge, MA, USA.

Grandin, T., ed. 1997. *Genetics and the Behavior of Domestic Animals.* Academic Press, San Diego, CA, USA.

Jensen, A. R. 1998. *The g Factor: The Science of Mental Ability.* Praeger, Westport, CN, USA.

Levy, F., and Hay, D., eds. 2001. *Attention, Genes and ADHD.* Brunner and Routledge, Hove, East Sussex, UK.

Mackintosh, N. J. 1998. *IQ and Human Intelligence.* Oxford Univ. Press, New York.

Matsuzawa, T., ed. 2001. *Primate Origins of Human Cognition and Behavior.* Springer, Tokyo.

Plomin, R., DeFries, J. C., McClearn, G. E., and Rutter, M. 1997. *Behavioral Genetics.* Freeman, New York.

Quiatt, D., and Reynolds, V. 1995. *Primate Behavior.* Cambridge Univ. Press, New York.

Rodgers, J. L., et al., eds. 2000. *Genetic Influences on Human Fertility and Sexuality.* Kluwer, Dordrecht, The Netherlands.

Rutter, M., ed. 1995. *Genetics of Criminal and Antisocial Behavior.* Wiley, New York.

Scott, J. P., and Fuller, J. L. 1998. *Genetics and the Social Behavior of the Dog.* Univ. Chicago Press, Chicago, IL, USA.

Simmons, P., and Young, D. 1999. *Nerve Cells and Animal Behavior.* Cambridge Univ. Press, New York.

Sternberg, R. J., and Ruzgis, P., eds. 1994. *Personality and Intelligence.* Cambridge Univ. Press, New York.

Turner, J. R., Cardon, L. R., and Hewitt, J. K., eds. 1995. *Behavior Genetic Approaches in Behavioral Medicine.* Plenum, New York.

Wasserman, D., and Wachbroit, R., eds. 2001. *Genetics and Criminal Behavior.* Cambridge Univ. Press, Cambridge, UK.

Wilson, E. O. 1975. *Sociobiology: The New Synthesis.* Harvard Univ. Press, Cambridge, MA, USA.

BIOCHEMISTRY and PHYSIOLOGY (*See also* Laboratory Technology, Biotechnology, Molecular Biology)

Aidley, D. J., and Stanfield, P. R. 1996. *Ion Channels: Molecules in Action.* Cambridge Univ. Press, New York.

Annual Review of Biochemistry. Palo Alto, CA, USA.

Annual Review of Physiology. Palo Alto, CA, USA.

Banaszek, L. J. 2000. *Foundations of Structural Biology.* Academic Press, San Diego, CA., USA.

Barrett, A. J., et al., eds. 1998. *Handbook of Proteolytic Enzymes.* Academic Press, San Diego, CA, USA.

Beynon, R., and Bond, J. S., eds. 2001. *Proteolytic Enzymes.* Oxford Univ. Press, New York.

Blackburn, G. M., and Gait, M. J., eds. 1996. *Nucleic Acids in Chemistry and Biology.* Oxford Univ. Press, New York.

Bodansky, M. 1993. *Principles of Peptide Synthesis.* Springer-Verlag, New York.

Bowles, D. J., ed. 1997. *Essays in Biochemistry: Cell Signalling.* Portland Press, London, UK.

Branden, C., and Tooze, J. 1999. *Introduction to Protein Structure.* Garland, New York.

Burrell, M. M., ed. 1993. *Enzymes of Molecular Biology.* Humana Press, Totowa, NJ, USA.

Buslig, B. S. and Manthey, J. A., eds. 2002. *Flavonoids in Cell Function.* Kluwer/Plenum, New York.

Campbell, J. L., ed. 1995. *DNA Replication. Methods in Enzymology*, Vol. 262. Academic Press, San Diego, CA, USA.

Carafoli, E., and Klee, C., eds. 1999. *Calcium as a Cellular Regulator.* Oxford Univ. Press, New York.

Chapman, B. J., ed. 2002. *Advances in DNA Sequence Specific Agents.* Elsevier Science, St. Louis, MO, USA.

Chawnshang, C., ed. 2002. *Androgens and Androgen Receptors: Mechanisms, Functions, and Clinical Applications.* Kluwer, Boston, MA, USA.

Cohen, N. C., ed. 1966. *Guidebook on Molecular Modeling in Drug Design.* Academic Press, San Diego, CA, USA.

Conly, E. C. 1995. *Ion Channel FactsBook: Extracellular Ligand-Gated Ion Channels.* Academic Press, San Diego, CA, USA.

Couvreur, P., and Malvy, C., eds. 2000. *Pharmaceutical Aspects of Oligonucleotides.* Taylor & Francis, London, UK.

Dalbey, R., et al., eds. 2002. *Protein Targeting, Transport, and Translocation.* Academic Press, San Diego, CA, USA.

D'Alessio, G., and Riordan, J. F., eds. 1997. *Ribonucleases: Structures and Functions.* Academic. Press, San Diego, CA, USA.

De Maeyer, E., and De Maeyer-Guignard, J. 1988. *Interferons and Other Regulatory Cytokines.* Wiley, New York.

Devlin, T. M., ed. 2001. *Textbook of Biochemistry with Clinical Correlations.* Wiley, New York.

Dickinson, J. R., and Schweizer, M., eds. 1999. *The Metabolism and Molecular Physiology of Saccharomyces cerevisiae.* Taylor & Francis, Philadelphia, PA, USA.

Elliott, W. H., and Elliott, D. C. 1997. *Biochemistry and Molecular Biology.* Oxford Univ. Press, New York.

Ellis, R. J., ed. 1996. *The Chaperonins.* Academic Press, San Diego, CA, USA.

Fenniri, H., ed. 2000. *Combinatorial Chemistry. A Practical Approach.* Oxford Univ. Press, New York.

Findeis, M. A., ed. 2001. *Nonviral Vectors for Gene Therapy: Methods and Protocols.* Humana Press, Totowa, NJ, USA.

Fink, A. L., and Golo, Y. 1998. *Molecular Chaperones in the Life Cycle of Proteins.* Marcel Dekker, New York.

Fruton, J. S. 1999. *Proteins, Enzymes, Genes: The Interplay of Chemistry and Biology.* Yale Univ. Press, New Haven, CT, USA.

Hall, A., ed. 2000. *GTPases.* Oxford Univ. Press, New York.

Hammes, G. H. 2000. *Thermodynamics and Kinetics for Biological Sciences.* Wiley, New York.

Hardie, G. D., and Hanks, S. 1995. *The Protein Kinase FactsBooks.* Academic Press, San Diego, CA USA.

Harford, J. B., and Morris, D. R., eds. 1997. *mRNA Metabolism and Post-Transcriptional Gene Regulation.* Wiley, New York.

Hecht, S. M., ed. 1996. *Bioorganic Chemistry: Nucleic Acids.* Oxford Univ. Press, New York.

Hellwinkel, D. 2001. *Systematic Nomenclature of Organic Chemistry.* Springer, New York.

Helmreich, E. J. M. 2001. *The Biochemistry of Cell Signalling.* Oxford Univ. Press, New York.

Hunter, G. K. 2000. *Vital Forces: The Discovery of the Molecular Basis of Life*. Academic Press, San Diego, CA, USA.

Hurtley, S., ed. 1996. *Protein Targeting*. IRL Press, New York.

James, P., ed. 2001. *Proteome Research: Mass Spectrometry*. Springer, New York.

Juo, P. S., ed. 1996. *Concise Dictionary of Biomedicine and Molecular Biology*. CRC Press, Boca Raton, FL, USA.

Kornberg, A., and Baker, T. A. 1992. *DNA Replication*. Freeman, New York.

Kricka, L. J., ed. 1995. *Nonisotopic Probing, Blotting and Sequencing*. Academic Press, San Diego, CA, USA.

Kyte, J. 1995. *Structure in Protein Chemistry*. Garland, New York.

Lehninger, A. L., et al. 1999. *Principles of Biochemistry*. Freeman, New York.

Metzler, D. E. 1977. *Biochemistry: The Chemical Reactions of Living Cells*. Academic Press, New York.

Neidle, S., ed. 1999. *Oxford Handbook of Nucleic Acid Structure*. Oxford Univ. Press, New York.

Oettel, M., and Schillinger, E., eds. 1999. *Estrogens and Antiestrogens*. Springer, New York.

Pain, R. H., ed. 2001. *Mechanisms of Protein Folding*. Oxford Univ. Press, New York.

Papa, S., and Tager, J. M., eds. 1995. *Biochemistry of Cell Membranes: A Compendium of Selected Topics*. Birkhäuser, Cambridge, MA, USA.

Phoenix, D. A., ed. 1999. *Protein Targeting and Translocation*. Princeton Univ. Press, Princeton, NJ, USA.

Rabilloud, Th., ed. 2000. *Proteome Research: Two-Dimensional Gel Electrophoresis and Identification Methods*. Springer, New York.

Schomburg, D., and Schomburg, I. 2001. *Handbook of Enzymes*. Springer, Heidelberg, Germany.

Sen, C. K., et al., eds. 1999. *Antioxidant and Redox Regulation of Genes*. Academic Press, San Diego, CA, USA.

Simons, R. W., and Grunberg-Manago, M. 1998. *RNA Structure and Function*. Cold Spring Harbor Laboratory Press, Cold Spring Harbor, NY, USA.

Soyfer, V. N., and Potaman, V. N. 1996. *Triple-Helical Nucleic Acids*. Springer, New York.

Sperelakis, N., ed. 1998. *Cell Physiology Source Book*. Academic Press, San Diego, CA, USA.

Stryer, L. 2001. *Biochemistry*. Freeman, New York.

Tsai, C. S. 2002. *An Introduction to Computational Biochemistry*. Wiley, New York.

Van Eldik, L., and Watterson, D. M., eds. 1998. *Calamodulin and Signal Transduction*. Academic Press, San Diego, CA, USA.

Voet, D., and Voet, J. 1995. *Biochemistry*. Wiley, New York.

Wallsgrove, R. M., ed. 1995. *Amino Acids and Their Derivatives in Higher Plants*. Cambridge Univ. Press, New York.

Webb, E. C. 1992. *Enzyme Nomenclature*. Academic Press, San Diego, CA, USA.

White, D. 2000. *The Physiology and Biochemistry of Prokaryotes*. Oxford Univ. Press, New York.

White, J. S., and White, D. C. 2002. Proteins, *Peptides and Amino Acids Sourcebook*. Humana, Totowa, NJ, USA.

Woodgett, J. R., ed. 1995. *Protein Kinases*. IRL/Oxford Univ. Press, New York.

Zoref-Shani, E., and Sperling, O., eds. 2000. *Purine and Pyrimidine Metabolism in Man X*. Kluwer/Plenum, New York.

Zubay, G. L., Parson, W. W., and Vance, D. E. 1995. *Principles of Biochemistry*. Brown, Dubuque, IA, USA.

BIOLOGY AND CELL BIOLOGY (*See also* Microbiology, Molecular Biology)

Alberts, B., Bray, D., Lewis, J., Raff, M., Roberts, K., and Watson, J. D. 2002. *Molecular Biology of the Cell*. Garland, New York.

Asai, D. J., and Forney, J. D., eds. 2000. *Tetrahymena thermophila*. Academic Press, San Diego, USA.

Boal, D., ed. 2002. *Mechanics of the Cell*. Cambridge Univ. Press, New York.

Broach, J. R., Pringle, J. R., and Jones, E. W., eds. 1991. *Molecular and Cellular Biology of the Yeast Saccharomyces*. Cold Spring Harbor Laboratory Press, Cold Spring Harbor, NY, USA.

Calisher, C. H., and Fauquet, C. N., eds. 1992. *Stedman's / ICTV VirusWorlds*. Williams & Wilkins, Baltimore, MD. USA.

Cheresh, D. A., and Mecham, R. P., eds. 1994. *Integrins: Molecular and Biological Responses to the Extracellular Matrix*. Academic Press, San Diego, CA, USA.

Clemente, C. D., ed. 1985. *Gray's Anatomy*. Lea & Febiger, Philadelphia, PA, USA.

Cooper, G. M. 2003. *The Cell: A Molecular Approach*. Sinauer, Sunderland, MA, USA.

Cross, P. C., and Mercer, K. L. 1993. *Cell and Tissue Ultrastructure: A Functional Perspective*. Freeman, New York.

Desjardins, C., and Ewing, L. L., eds. 1993. *Cell and Molecular Biology of the Testis*. Oxford Univ. Press, New York.

Devor, E. J., ed. 1993. *Molecular Application in Biological Anthropology*. Cambridge Univ. Press, New York.

Dulbecco, R., ed. 1997. *Encyclopedia of Human Biology*. Academic Press, San Diego, CA, USA.

Epstein, H. F., and Shakes, D. C., eds. 1995. *Ceanorhabditis elegans: Modern Biological Analysis of an Organism*. Academic Press, San Diego, CA, USA.

Ferraris, J. D., and Palumbi, S. R., eds. 1966. *Molecular Zoology: Advances, Strategies and Protocols*. Wiley-Liss, New York.

Ferretti, P., and Geraudie, J. 1998. *Cellular and Molecular Basis of Regeneration*. Wiley, New York.

Graham, P. H., Sadowsky, M. J., and Vance, C. P. 1994. *Symbiotic Nitrogen Fixation*. Kluwer, Norwell, MA.

Groombridge, B., and Jenkins, M. D. 2002. *World Atlas of Biodiversity: Earth's Living Resources in the 21st Century*. Univ. California Press, Berkeley, CA, USA.

Grundzinskas, J. G., and Yovich, J. L., eds. 1994. *Gametes: The Oocyte*. Cambridge Univ. Press, New York.

Grundzinskas, J. G., and Yovich, J. L., eds. 1995. *Gametes: The Spermatozoon*. Cambridge Univ. Press, New York.

Hall, B. K., ed. 1994. *Homology: The Hierarchial Basis of Comparative Biology*. Academic Press, San Diego, CA, USA.

Holt, J. G., Bruns, M. A., Caldwell, B. J., and Pease, C. D., eds. 1993. *Bergey's Manual of Determinative Bacteriology*. Williams & Wilkins, Baltimore, MD, USA.

Hutchison, C., and Glover, D. M., eds. 1995. *Cell Cycle Control*. IRL, New York.

Jacobson, M. D., and McCarthy, N., eds. 2002. *Apoptosis: The Molecular Biology of Programmed Cell Death*. Oxford Univ. Press, Oxford, UK.

Jong, S.-C., Birmingham, J. M., and Ma, G., eds. 1993. *Stedman's ATCC Fungus Names*. Williams & Wilkins, Baltimore, MD, USA.

Kessin, R. H. 2001. *Dictyostelium: Evolution, Cell Biology, and the Development of Multicellularity*. Cambridge Univ. Press, New York.

Kitano, H., ed. 2001. *Foundations of Systems Biology*. MIT Press, Cambridge, MA, USA.

Kohen, E., Santus, R., and Hirschberg, J. G. 1995. *Photobiology*. Academic Press, San Diego, CA, USA.

Krstic, R. V. 1994. *Human Microscopic Anatomy: An Atlas for Students of Medicine and Biology*. Springer, New York.

Lackie, J. M., Dunn, G. A., and Jones, G. E., eds. 1999. *Cell Behaviour: Control and Mechanism of Motility*. Princeton Univ. Press, Princeton NJ, USA.

Larsen, W. J. 1993. *Human Embryology*. Churchill Livingstone, New York.

Lodish, H., Berk, A., Zipursky, S. L., Matsudaira, P., Baltimore, D., and Darnell, J. 2003. *Molecular Cell Biology*. Freeman, New York.

Mabberly, D. J. 1997. *The Plant Book: A Portable Dictionary of the Vascular Plants*. Cambridge Univ. Press, New York.

Mayr, E. 1997. *This is Biology: The Science of the Living World*. Harvard Univ. Press, Cambridge, MA, USA.

Merz, K. M., and Roux, B., eds. 1966. *Biological Membranes: A Molecular Perspective from Computation and Experiment*. Birkhäuser Verlag, Boston, MA, USA.

Moore, D. 1998. *Fungal Morphogenesis*. Cambridge Univ. Press, New York.

Peracchia, C., ed. 1994. *Handbook of Membrane Channels*. Academic Press, San Diego, CA, USA.

Pollard, T. D., and Earnshaw, W. C. 2002. *Cell Biology*. Saunders, Philadelphia, PA, USA.

Purvis, W. K., et al. 2000. *Life: The Science of Biology*. Sinauer/Freeman, New York.

Quick, M. M. 2002. *Transmembrane Transporters*. John Wiley & Sons, Hoboken, NJ, USA.

Raghavan, V. 1997. *Molecular Evolution of Flowering Plants*. Cambridge Univ. Press, New York.

Raven, P. H., and Johnson, G. B. 1992. *Biology*. Mosby, St. Louis, MO, USA.

Scott, T. A. (translator). 1966. *Concise Encyclopedia: Biology*. Walter de Gruyter, New York.

Stein, G. S., et al., eds. 1999. *The Molecular Basis of Cell Cycle and Growth Control*. Wiley, New York.

Thorpe, T. A., ed. 1995. *In Vitro Embryogenesis in Plants*. Kluwer, Norwell, MA, USA.

Wilson, J., and Hunt, T. 2002 *Molecular Biology of the Cell. A Problems Approach*. Garland Science, New York.

Wood, J. W., and de Gruyter, A. 1954. *Dynamics of Human Reproduction: Biology, Biometry, Demography*. Hawthorne, New York.

Yu, H.-S. 1994. *Human Reproductive Biology*. CRC Press, Boca Raton, FL, USA.

BIOMETRY and BIOINFORMATICS

Aitken, C. G. G. 1995. *Statistics and the Evaluation of Evidence for Forensic Scientists*. Wiley, New York.

Bailey, N. T. J. 1995. *Statistical Methods for Biologists*. Cambridge Univ. Press, New York.

Balding, D. J., Bishop, M., and Cannings, C., eds. 2001. *Handbook of Statistical Genetics*. Wiley, New York.

Barnes, M., and Gray, I. C. 2003. *Bioinformatics for Geneticists*. Wiley, Hoboken, NJ, USA.

Baxevanis, A. D., and Quellette, B. F. F., eds. 2001. *Bioinformatics: A Practical Guide to the Analysis of Genes and Proteins*. Wiley, New York.

Bishop, M., ed. 1998. *Guide to Human Genome Computing*. Academic Press, San Diego, CA, USA.

Bower, J. M., and Bolouri, H., eds. 2001. *Computational Modeling of Genetic and Biochemical Networks*. MIT Press, Cambridge, MA, USA.

Bürger, R. 2000. *The Mathematical Theory of Selection, Recombination, and Mutation*. Wiley, New York.

Dunn, O. J., and Clark, V. A. 2000. *Basic Statistics: A Primer for Biomedical Sciences*. Wiley, New York.

Durbin, R., et al. 1998. *Biological Sequence Analysis: Probabilistic Models of Proteins and Nucleic Acids*. Cambridge Univ. Press, New York.

Elands-Johnson, R. C. 1971. *Probability Models and Statistical Methods in Genetics*. Wiley, New York.

Elston, R. C., ed. 2002. *Biostatistical Genetics and Genetic Epidemiology*. Wiley, New York.

Emery, A. E. H. 1986. *Methodology in Medical Genetics: An Introduction to Statistical Methods*. Churchill Livingstone, Edinburgh, UK.

Evett, I. W., and Weir, B. S. 1998. *Interpreting DNA Evidence: Statistical Genetics for Forensic Scientists*. Sinauer, Sunderland, MA, USA.

Fall, C. P. et al. 2002. *Computational Cell Biology*. Springer, New York.

Forthofer, R. N., and Lee, E. S. 1995. *Introduction to Biostatistics: A Guide to Design, Analysis, and Discovery*. Academic Press, San Diego, CA, USA.

Gigerenzer, G. 2002. *Calculated Risk. How to Know When Numbers Deceive You*. Simon & Schuster, New York.

Hays, W. L., and Winkler, R. L. 1971. *Statistics: Probability, Inference and Decision*. Holt, Rinehart and Winston, New York.

Howson, C., and Urbach, P. 1993. *Scientific Reasoning: The Bayesian Approach*. Open Court, La Salle, IL, USA.

Kanehisa, M. 2000. *Post-Genome Informatics*. Oxford Univ. Press, New York.

Kempthorne, O. 1957. *An Introduction to Genetic Statistics*. Wiley, New York.

Lange, K. 2002. *Mathematical and Statistical Methods for Genetic Analysis*. Springer, New York.

Lengauer, T., ed. 2002. *Bioinformatics—From Genomes to Drugs*. Wiley, New York.

Lesk, M. 2002. *An Introduction to Bioinformatics*. Oxford Univ. Press, New York.

Letovsky, S., ed. 1999. *Bioinformatics: Databases and Systems 1999*. Kluwer, Boston, MA, USA.

Lindley, D. V. 1965. *Introduction to Probability and Statistics from a Bayesian Viewpoint*. Cambridge Univ. Press, New York.

Liu, B. H. 1998. *Statistical Genomics: Linkage, Mapping, and QTL Analysis*. CRC Press, Boca Raton, FL, USA.

Malécot, G. 1969. *The Mathematics of Heredity*. Freeman, San Francisco, CA, USA.

Mather, K. 1965. *Statistical Analysis in Biology*. Methuen, London, UK.

Mitchell, M. 1996. *An Introduction to Genetic Algorithms*. MIT Press, Cambridge, MA, USA.

Mount, D. 2001. *Bioinformatics: Sequence and Genome Analysis*. Cold Spring Harbor Laboratory Press, Plainview, NY, USA.

Norton, N. E., Dabeeru, C. R., and LaLuel, J.-M. 1983. *Methods in Genetic Epidemiology*. Karger, New York.

Paterson, A. H., ed. 1997. *Molecular Dissection of Complex Traits*. CRC Press, Boca Raton, FL, USA.

Pevzener, P. A. 2000. *Computational Molecular Biology: An Algorithmic Approach*. MIT Press, Cambridge, MA, USA.

Rao, D. C., and Province, M. A., eds. 2002. *Genetic Dissection of Complex Traits*. Academic Press, San Diego, CA, USA.

Riffenburgh, R. H. 1999. *Statistics in Medicine*. Academic Press, San Diego, CA, USA.

Salzberg, S., et al., eds. 1999. *Computational Methods in Molecular Biology*. Elsevier Science, New York.

Sham, P. 1997. *Statistics in Human Genetics*. Oxford Univ. Press, New York.

Shipley, B. 2001. *Cause and Correlation in Biology: A User's Guide to Path Analysis, Structural Equations and Causal Inference*. Cambridge Univ. Press, New York.

Shortliffe, E. H., et al., eds. 2000. *Medical Informatics*. Springer, New York.

Shoukri, M. M., and Edge, V. L. 1995. *Statistical Methods for Health Sciences*. CRC Press, Boca Raton, FL, USA.

Snedecor, G. W., and Cochran, W. G. 1967. *Statistical Methods*. Iowa State Univ. Press, Ames, IA, USA.

Sokal, R. R., and Rohlf, F. J. 1969. *Biometry*. Freeman, San Francisco, CA, USA.

Townend, J. 2002. *Practical Statistics for Environmental and Biological Scientists*. Wiley, New York.

Wang, J. T. L., Shapiro, B. A., and Shasha, D., eds. 1999. *Pattern Discovery in Biomolecular Data: Tools, Techniques, and Applications*. Oxford Univ. Press, New York.

Wastney, M. E., et al. 1998. *Investigating Biological Systems Using Modeling: Strategies and Software*. Academic Press, Orlando, FL, USA.

Waterman, M. S. 1995. *Introduction to Computational Biology: Maps, Sequences and Genomes*. Chapman & Hall, New York.

Woolson, R. F., and Clarke, W. R. 2002. *Statistical Methods for the Analysis of Biomedical Data*. Wiley, New York.

Yang, M. C. 2000. *Introduction to Statistical Methods in Modern Genetics*. Taylor & Francis, New York.

Young, I. D. 1991. *Introduction to Risk Calculation in Genetic Counseling*. Oxford Univ. Press, New York.

Zhang, W., and Shmulevich, I. 2002. *Computational and Statistical Approaches to Genomics*. Kluwer, Norwell, MA, USA.

Zwillinger, D. 1996. *CRC Standard Mathematical Tables and Formulae*. CRC Press, Boca Raton, FL, USA.

BIOTECHNOLOGY (*See also* Biochemistry, Laboratory Technology, Molecular Biology)

Atherton, K. T., ed. 2002. Genetically Modified Crops. Assessing Safety. Taylor & Francis, New York.

Bajaj, Y. P. S., ed. 1999. *Transgenic Medicinal Plants: Biotechnology in Agriculture and Forestry*, Vol. 45. Springer-Verlag, New York.

Bajaj, Y. P. S., ed. 1999. *Transgenic Crops, I*. Springer, New York.

Bajaj, Y. P. S., ed. 1999. *Transgenic Trees*. Springer, New York.

Bajaj, Y. P. S., ed. 2001. *Transgenic Crops, II*. Springer, New York.

Baxevanis, A. D., and Quellette, B. F. F. 2001. *Bioinformatics: A Practical Guide to the Analysis of Genes and Proteins*. Wiley, New York.

Borrebaeck, C. A. K., ed. 1995. *Antibody Engineering*. Oxford Univ. Press, New York.

Burden, D. W., and Whitney, D. B. 1995. *Biotechnology—Proteins to PCR: A Course in Strategies and LabTechniques*. Birkhäuser, Boston, MA, USA.

Craig, A., et al., eds. 1999. *Automation*. Academic Press, Orlando, FL, USA.

Crooke, S. T., and Lebleu, B., eds. 1993. *Antisense Research and Applications*. CRC Press, Boca Raton, FL, USA.

Eder, K., and Dale, B. 2000. *In Vitro Fertilization*. Cambridge Univ. Press, New York.

Endreß, R. 1994. *Plant Cell Biotechnology*. Springer, New York.

Factor, P. H., ed. 2001. *Gene Therapy for Acute and Acquired Diseases*. Kluwer, Boston, MA, USA.

Glazer, A. N., and Nikaido, H. 1995. *Microbial Biotechnology: Fundamentals of Applied Microbiology*. Freeman, New York.

Gresshoff, P. M., ed. 1997. *Technology Transfer of Plant Biotechnology*. CRC Press, Boca Raton, FL, USA.

Handler, A. M., and James, A. A. 2000. *Insect Transgenesis: Methods and Applications*. CRC Press, Boca Raton, FL, USA.

Heikki, M. T., and Lynch, J. M., eds. 1995. *Biological Control*. Cambridge Univ. Press, New York.

Howe, C. 1995. *Gene Cloning and Manipulation*. Cambridge Univ. Press, New York.

Kabanov, A., Seymour, L. W., and Feigner, P., eds. 1998. *Self-Assembling Complexes for Gene Delivery: From Laboratory to Clinical Trial*. Wiley, New York.

Kingsman, S. M., and Kingsman, A. J. 1988. *Genetic Engineering: An Introduction to Gene Analysis and Exploitation in Eukaryotes*. Blackwell Science, Oxford, UK.

Kreuzer, H., and Massey, A. 2000. *Recombinant DNA and Biotechnology: A Guide for Teachers*. ASM Press, Washington, DC. USA.

Lattime, E. C., and Gerson, S. L., eds. 2002. *Gene Therapy of Cancer*. Academic Press, San Diego, CA, USA.

Lauritzen, P., ed. 2001. *Cloning and the Future of Human Embryo Research*. Oxford Univ. Press, New York.

Letourneau, D. K., and Elpern Burrows, B., eds. 2002. *Genetically Engineered Organisms: Assessing Environmental and Human Health Effects*. CRC Press, Boca Raton, FL, USA.

Maclean, N., ed. 1995. *Animals with Novel Genes*. Cambridge Univ. Press, New York.

Machida, C. A., ed. 2002. *Viral Vectors for Gene Therapy. Methods and Protocols*. Humana, Totowa, NJ. USA.

Meager, A., ed. 1999. *Gene Therapy Technologies: Applications and Regulations—From Laboratory to Clinic*. Wiley, New York.

Meyers, R. A., ed. 1995. *Molecular Biology and Biotechnology: A Comprehensive Desk Reference*. VCH, New York.

Monsatersky, G. M., and Robl, J. M., eds. 1995. *Strategies in Transgenic Animal Science*. ASM Press, Washington, DC.

Murray, T. H., and Mehlman, M. J. 2000. *Encyclopedia of Ethical and Policy Issues in Biotechnology*. Wiley, New York.

Nielse, J., ed. 2001. *Metabolic Engineering*. Springer, New York.

Oksman-Caldentey, K.-M., and Barz, W. H., eds. 2002. *Plant Biotechnology and Transgenic Plants*. Marcel-Dekker, New York.

Old, R. W., and Primrose, S. B. 1994. *Principles of Gene Manipulation: An Introduction to Genetic Engineering*. Blackwell Science, Cambridge, MA, USA.

Percus, J. 2002. *Mathematics of Genome Analysis*. Cambridge Univ. Press, New York.

Pinkert, C. A., ed. 2002. *Transgenic Animal Technology*. Academic Press, San Diego, CA, USA.

Potrykus, I., and Spangenberg, G., eds. 1995. *Gene Transfer to Plants*. Springer, New York.

Quesenberry, P. J., et al., eds. 1998. *Stem Cell Biology and Gene Therapy*. Wiley, New York.

Ratledge, C., and Kristiansen, B., eds. 2001. *Basic Biotechnology*. Cambridge Univ. Press, New York.

Scanlon, K. J., ed. 1998. *Therapeutic Application of Ribozymes*. Humana Press, Totowa, NJ, USA.

Shuler, M. L., et al., eds. 1994. *Baculovirus Expression System and Biopesticides*. Wiley-Liss, New York.

Smith, J. E. 1996. *Biotechnology*. Cambridge Univ. Press, New York.

Stein, C. A., and Krieg, A. M., eds. 1998. *Applied Antisense Oligonucleotide Technology*. Wiley, New York.

Strauss, M., and Barranger, J. A., eds. 1997. *Concepts in Gene Therapy*. Walter de Gruyter, New York.

Suhai, S., ed. 1997. *Theoretical and Computational Methods in Genome Research*. Plenum, New York.

Suhai, S., ed. 2000. *Genomics and Proteomics*. Kluwer, Dordrecht, The Netherlands.

Sullivan, N. F. 1995. *Technology Transfer: Making the Most of Your Intellectual Property*. Cambridge Univ. Press, New York.

Swindell, S. R., Miller, R. R., and Myers, G. S. A., eds. 1996. *Internet for the Molecular Biologist*. Horizon Scientific, Portland, OR, USA.

Thomas, J. A., and Fuchs, R. L., eds. 2002. *Biotechnology and Safety Assessment*. Acad. Press, San Diego, CA, USA.

Tzotzos, G. T., ed. 1995. *Genetically Modified Organisms: A Guide to Biosafety*. CAB Internat, Oxford, UK.

Vining, L. C., and Stuttrad, C., eds. 1994. *Genetics and Biochemistry of Antibiotic Production*. Butterworth-Heinemann, Stoneham, MA, USA.

Wang, K., Herrera-Estrella, A., and van Montagu, M., eds. 1995. *Transformation of Plants and Soil Microorganisms*. Cambridge Univ. Press, New York.

CANCER (*See also* Aging, Cytogenetics, Cytology, Diseases, Molecular Biology, Mutation, and Viruses)

Bertino, J. R., ed. 2002. *Encyclopedia of Cancer*. Academic Press, San Diego, CA, USA.

Boultwood, J., and Fidler, C., eds. 2001. *Molecular Analysis of Cancer*. Humana Press, Totowa, NJ, USA.

Bronchud, M. H., et al., eds. 2000. *Principles of Molecular Oncology*. Humana Press, Totowa, NJ, USA.

Brooks, S. A., and Schumacher, U., eds. 2001. *Metastasis Research Protocols*. Humana Press, Totowa, NJ, USA.

Coleman, W. B., and Tsongalis, G. J., eds. 2001. *The Molecular Basis of Human Cancer*. Humana Press, Totowa, NJ, USA.

Clayson, D. B. 2001. *Toxicological Carcinogens*. Lewis Publishers, Boca Raton, FL, USA.

Eeles, R. Ponder, B., Easston, D., and Horwich, A. 1996. *Genetic Predisposition to Cancer*. Chapman & Hall, New York.

Ehrlich, M., ed. 2000. *DNA Alterations in Cancer: Genetic and Epigenetic Changes*. Eaton Publishing Co., Natick, MA, USA.

El-Deiry, W. S., ed. 2003. *Tumor Suppressor Genes*. Humana, Totowa, NJ, USA.

Fisher, D. E., ed. 2001. *Tumor Suppressor Genes in Human Cancer*. Humana Press, Totowa, NJ, USA.

Giordano, A., and Soprano, K. J., eds. 2002. *Cell Cycle Inhibitors in Cancer Therapy*. Humana, Totowa, NJ, USA.

Greaves, M. 2000. *Cancer: The Evolutionary Legacy*. Oxford Univ. Press, New York.

Habib, N. A., ed. 2000. *Cancer Gene Therapy*. Kluwer, Dordrecht, The Netherlands.

Hesketh, R. T. 1997. *The Oncogene and Tumor Suppressor Gene FactsBooks*. Academic Press, San Diego, CA, USA.

Hodgson, S., and Maher, E. 1999. *A Practical Guide to Human Cancer Genetics*. Cambridge Univ. Press, New York.

ICRP. 1999. *Genetic Susceptibility to Cancer*. Pergamon/Elsevier Science, New York.

Kastan, M. B., ed. 1997. *Checkpoint Controls and Cancer*. Cold Spring Harbor Laboratory Press, Cold Spring Harbor, NY, USA.

Kitchin, K. T., ed. 1999. *Carcinogenicity: Testing, Predicting, and Interpreting Chemical Effects*. Marcel Dekker, New York.

La Thangue, N. B., and Bandara, L. R. 2002. *Targets for Cancer Chemotherapy: Transcription Factors and Other Nuclear Proteins*. Humana Press, Totowa, NJ, USA.

Lattime, E. C., and Gerson, S. L. 1998. *Gene Therapy of Cancer: Translational Approaches from Preclinical to Clinical Implementation*. Academic Press, San Diego, CA, USA.

Lindahl, T., ed. 1996. *Genetic Instability in Cancer*. Cold Spring Harbor Laboratory Press, Cold Spring Harbor, NY, USA.

Marcus, A. I. 1994. *Cancer from Beef: DES, Federal Regulation, and Consumer Confidence*. Johns Hopkins Univ. Press, Baltimore, MD, USA.

Mendelsohn, J., Howley, P. M., Isreal, M. A., and Liotta, L. A., eds. 1994. *The Molecular Basis of Cancer*. Saunders, Philadelphia, USA.

Minson, A. C., Neil. J. C., and McCrae, M. A., eds. 1954. *Viruses and Cancer*. Cambridge Univ. Press, New York.

Mitelman, F. 1997. *Catalog of Chromosomal Aberrations in Cancer*. Wiley, New York.

Newton, R., et al. 1999. *Infections and Human Cancer*. Cold Spring Harbor Laboratory Press, Cold Spring Harbor, NY, USA.

Nickoloff, B. J., ed. 2001. *Melanoma Techniques and Protocols: Molecular Diagnosis, Treatment and Monitoring*. Humana Press, Totowa, NJ, USA.

Offit, K. 1998. *Clinical Cancer Genetics*. Wiley, New York.

Parsonnet, J., ed. 1999. *Microbes and Malignancy: Infection as a Cause of Human Cancer*. Oxford Univ. Press, New York.

Peters, G., and Vousden, K. H. 1997. *Oncogenes and Tumour Suppressor*. IRL/Oxford Univ. Press, New York.

Pettit, G. R., Person, F. H., and Herald, C. L. 1994. *Anticancer Drugs from Animals, Plants and Microorganisms*. Wiley, New York.

Ponder, B. A. J., Cavenee, W. K., and Solomon, E., eds. 1996. *Genetics and Cancer: A Second Look*. Cold Spring Harbor Laboratory Press, Cold Spring Harbor, NY, USA.

Pui, C.-H., ed. 1999. *Childhood Leukemias*. Cambridge Univ. Press, New York.

Rollins, B. J., ed. 1999. *Chemokines and Cancer*. Humana Press, Totowa, NJ, USA.

Schneider, K. 2001. *Counseling About Cancer: Strategies for Genetic Counseling*. Wiley, New York.

Schwab, M., ed. 2001. *Encyclopedic Reference to Cancer*. Springer, New York.

Sherbet, G. V., and Lakshmi, M. S. 1997. *Genetics of Cancer: Genes Associated with Cancer Invasion, Metastasis, and Cell Proliferation*. Academic Press, San Diego, CA, USA.

Steinberg, K., ed. 2001. *The Genetic Basis of Cancer*. Oxford Univ. Press, New York.

Tomatis. L. 1990. *Air Pollution and Human Cancer*. Springer Verlag, Berlin.

Tooze, J., et al., eds. 1999. *Infections and Human Cancer*. Cold Spring Harbor Laboratory Press, Cold Spring Harbor, NY, USA.

Utsunomiya, J., et al., eds. 1999. *Familial Cancer Prevention—Molecular Epidemiology: A New Strategy Toward Cancer Control*. Wiley, New York.

Vogt, P. K., and Verma, I. M., eds. 1995. *Oncogene Techniques*. Academic Press, San Diego, CA, USA.

Walther, W., and Stein, U., eds. 2000. *Gene Therapy of Cancer*. Humana Press, Totowa, NJ, USA.

Warshawsky, D., and Landolph, J. R., Jr. 2002. *Molecular Carcinogenesis*. CRC Press, Boca Raton, FL, USA.

Welschof, M., and Krauss, J., eds. 2002. *Recombinant Antibodies for Cancer Therapy. Methods and Protocols*. Humana, Totowa, NJ. USA.

Weinberg, R. A. 1996. *Racing to the Beginning of the Road: The Search for the Origin of Cancer*. Crown, New York.

Wickstrom, E., ed. 1998. *Clinical Trials of Genetic Therapy with Antisense DNA and DNA Vectors*. Marcel Dekker, New York.

Wolman, S. R., and Sell, S., eds. 1997. *Human Cytogenetic Cancer Markers*. Humana Press, Totowa, NJ., USA.

CONSERVATION (*See also* Evolution, Population Genetics)

Avise, J. C., and Hamrick, J. L., eds. 1996. *Conservation Genetics: Case Histories from Nature*. Chapman & Hall, New York.

Beattie, A., and Ehrlich, P. R. 2001. *Wild Solutions: How Biodiversity Is Money in the Bank*. Yale Univ. Press, New Haven, CT, USA.

Caughly, G., and Gunn, A. 1995. *Conservation Biology in Theory and Practice*. Blackwell Science, Cambridge, MA.

Clark, T. W. 1997. *Averting Extinction: Reconstructing Endangered Species Recovery*. Yale Univ. Press, New Haven, CT, USA.

Ehrenfeld, D., ed. 1995. *Plant Conservation*. Blackwell Science, Cambridge, MA.

Ehrlich, P. R. 2000. *Human Natures: Genes, Cultures and the Human Prospect*. Island Press, Washington, DC, USA.

Frankel, O. H., Brown, A. D., and Burdon, J. J. 1995. *The Conservation of Plant Biodiversity*. Cambridge Univ. Press, New York.

Frankham, R., Ballou, J. D., and Briscoe, D. A. 2002. *Introduction to Conservation Genetics*. Cambridge Univ. Press, New York.

Hawkes, J. G., Maxted, N., and Ford-Lloyd, B. V. 2000. *The Ex Situ Conservation of Plant Genetic Resources*. Kluwer, Dordrecht, The Netherlands.

Hunter, M. L., Jr. 1995. *Fundamentals of Conservation Biology*. Blackwell Science, Cambridge MA, USA.

Huston, M. A. 1994. *Biological Diversity: The Coexistence of Species in Changing Landscapes*. Cambridge Univ. Press, New York.

Juo, A. S. R., and Freed, R. D., eds. 1993. *Agriculture and Environment: Bridging Food Production and Environmental Protection in Developing Countries*. American Society of Agronomy, Madison, WI, USA.

Kammen, D. M., and Hassenzahl, D. O. 1999. *Should We Risk It? Exploring Environmental, Health, and Technological Problem Solving*. Princeton Univ. Press, Princeton, NJ, USA.

Kjellsson, G., and Amman, K. 1997–1999. *Methods for Risk Assessment of Transgenic Plants*. Birkhäuser, Basel, Switzerland.

Landweber, L. F., and Dobson, A. P., eds. 2000. *Genetics and the Extinction of Species*. Princeton Univ. Press, Priceton, NJ, USA.

Lawton, J. H., and May, R. M., eds. 1995. *Extinction Rates*. Oxford Univ. Press, New York.

Levin, D. A. 2000. *The Origin, Expansion, and Demise of Plant Species*. Oxford Univ. Press, New York.

Levin, S. A., ed. 2000. *Encyclopedia of Biodiversity*. Academic Press, San Diego, CA, USA.

Loreau, M., et al., eds. 2002. *Biodiversity and Ecosystem Functioning. Synthesis and Perspectives*. Oxford Univ. Press, Oxford, UK.

Meffe, G. K., and Carrol, C. R. 1995. *Principles of Conservation Biology*. Sinauer, Sunderland, MA, USA.

O'Riordan, T., and Stoll-Kleeman, S. 2002. *Biodiversity, Sustainability and Human Communities: Protecting Beyond the Protected*. Cambridge Univ. Press, New York.

Paehlke, R., ed. 1995. *Conservation and Environmentalism: An Encyclopedia*. Garland, New York.

Premack, R. B. 1998. *Essentials of Conservation Biology*. Sinauer, Sunderland, MA, USA.

Reaka-Kudle, M. L., Wilson, D. E., and Wilson, E. O., eds. 1996. *Biodiversity II: Understanding and Protecting our Biological Resources*. Joseph Henry Press (Natl. Academic Sci. USA), Washingon, DC, USA.

Samways, M. J. 1994. *Insect Conservation Biology*. Chapman & Hall, New York.

Shaw, I. C., and Chadwick, J. 1996. *Principles of Environmental Toxicology*. Taylor & Francis, London, UK.

Watson, P. F., and Holt, W. V., eds. 2000. *Cryobanking the Genetic Resource? Wildlife Conservation for the Future?* Taylor & Francis, New York.

Wilson, E. O. 2002. *The Future of Life*. Knopf, New York.

Winston, J. E. 2000. *Describing Species: Practical Taxonomic Procedure for Biologists*. Columbia Univ. Press, New York.

CYTOLOGY and CYTOGENETICS (*See also* Cancer, Evolution, Molecular Biology)

Allan, V. 2000. *Protein Localization by Fluorescence Microscopy: A Practical Approach*. Oxford Univ. Press, New York.

Appels, R., Morris, R., Gill, B. S., and May, C. E. 1998. *Chromosome Biology*. Kluwer Academic, Norwell, MA, USA.

Berezney, R., and Riordan, J. F., eds. 1997. *Nuclear Matrix: Structural and Functional Organization*. Academic Press, San Diego, CA, USA.

Berrios, M., ed. 1997. *Nuclear Structure and Function: Methods in Cell Biology*, Vol. 53. Academic Press, San Diego, CA, USA.

Bickmore, W. A., ed. 1999. *Chromosome Structural Analysis — A Practical Approach*. Oxford Univ. Press, Oxford, UK.

Blackburn, E. H., and Greider, C. W., eds. 1995. *Telomeres*. Cold Spring Harbor Laboratory Press, Cold Spring Harbor, NY, USA.

Borgaonkar, D. 1997. *Chromosomal Variation in Man: A Catalog of Chromosomal Variants and Anomalies*. Wiley-Liss, New York.

Chen, B. 2001. *Grauzone and Completion of Meiosis During Drosophila Oogenesis*. Kluwer, Dordrecht, The Netherlands.

Choo, K. H. 1997. *The Centromere*. Oxford Univ. Press, New York.

Cook, P. 2001. *Principles of Nuclear Structure and Function*. Wiley, New York.

Darlington, C. D., and Wylie, A. P. 1955. *Chromosome Atlas of Flowering Plants*. Allen & Unwin, London, UK.

Davis, K. E., and Warren, S. T., eds., 1993. *Genome Rearrangement and Stability*. Cold Spring Harbor Laboratory Press, Cold Spring Harbor, NY, USA.

DeGrouchy, J., and Turleau, C. 1984. *Clinical Atlas of Human Chromosomes*. Wiley, New York.

Double, J. A., and Thompson, M. J. 2002. *Telomeres and Telomerase: Methods and Protocols*. Humana Press, Totowa, NJ, USA.

Drlica, K., and Riley, M., eds. 1990. *The Bacterial Chromosome*. American Society of Microbiology, Washington, DC, USA.

Elgin, S. C. R., and Workman, J. L., eds. 2001. *Chromatin Structure and Gene Expression*. Oxford Univ Press, New York.

Epstein, C. J. 1986. *Consequences of Chromosome Imbalance: Principles, Mechanisms and Models*. Cambridge Univ. Press, New York.

Exbrayat, J.-M. 2001. *Genome Visualization by Classic Methods in Light Microscopy*. CRC Press, Boca Raton, FL, USA.

Fan, Y.-S. 2002. *Molecular Cytogenetics: Protocols and Applications*. Humana Press, Totowa, NJ, USA.

Fantes, P., and Beggs, J., eds. 2001. *The Yeast Nucleus*. Oxford Univ. Press, New York.

Fukui, K., and Nakyama, S., eds. 1996. *Plant Chromosomes: Laboratory Methods*. CRC Press, Boca Raton, FL, USA.

Gersen, S. L., and Keagle, M. B., eds. 1999. *The Principles of Clinical Cytogenetics*. Humana Press, Totowa, NJ, USA.

Glauert, A. N., and Lewis, P. R. 1999. *Biological Specimen Preparation for Transmission Electron-microscopy*. Princeton Univ. Press, Princeton, NJ, USA.

Goodhew, P., Keyse, R. J., and Lorimer, J. W. 1997. *Introduction to Scanning Transmission Electron Microscopy*. Springer, New York.

Gosden, J. R., ed. 1994. *Chromosome Analysis Protocols*. Humana Press, Totowa, NJ, USA.

Gustafson, J. P., and Flavell, R. B., eds. 1996. *Genomes of Plants and Animals. 21st Stadler Genetics Symposium*. Plenum, New York.

Hagerman, R., and Hagerman, P., eds. 2002. *Fragile X Syndrome*. Johns Hopkins Univ. Press, Baltimore, MD, USA.

Harris, H. 1999. *The Cell: The Development of an Idea*. Yale Univ. Press, New Haven, CT, USA.

Heim, S., and Mitelman, F. 1995. *Cancer Cytogenetics*. Wiley-Liss, New York.

Henriquez-Gil, N., Parker, J. U. S., and Puertas, M., eds. 1997. *Chromosomes Today*. Chapman and Hall, New York.

Herzberg, A. J., Raso, D. S., and Silverman, J. F. 1999. *Color Atlas of Normal Cytology*. Harcourt Brace & Co., New York.

Hsu, T. C., and Benirschke, K. 1967–1977. *An Atlas of Mammalian Chromosomes*. Springer, New York.

Hutchison, C., and Glover, D. M., eds. 1995. *Cell Cycle Control*. Oxford Univ. Press, New York.

Hyams, J. S., and Lloyd, C. W. 1994. *Microtubules*. Wiley-Liss, New York.

Jauhar, P. P., ed. 1996. *Methods of Genome Analysis of Plants*. CRC Press, Boca Raton. FL, USA.

Javois, L. C., ed. 1995. *Immunocytochemical Methods and Protocols*. Humana Press, Totowa, NJ, USA.

Jena, B. P. 2002. *Atomic Force Microscopy in Cell Biology*. Acad. Press, San Diego, CA, USA.

John, B., and Lewis, K. R. 1968. *The Chromosome Complement*. Springer, New York.

Kiernan, J. A. and Mason, I., eds. 2002. *Microscopy and Histology for Molecular Biologists. A User Guide*. Portland, London, UK.

Kipling, D. 1995. *The Telomere*. Oxford Univ. Press, New York.

Lacey, A. J., ed. 1999. *Light Microscopy in Biology. A Practical Approach*. Oxford Univ. Press, New York.

Macgregor, H. C. 1993. *An Introduction to Animal Cytogenetics*. Chapman & Hall, New York.

Mark, H. F. L., ed. 2000. *Medical Cytogenetics*. Marcel Dekker, New York.

Maunsbach, A. B., and Afzelius, B. A. 1998. *Biomedical Electron Microscopy*. Academic Press, San Diego, CA, USA.

Miller, O. J., and Thurman, E. 2000. *Human Chromosomes*. Springer, New York.

Moore, R. J. 1973. *Index to Plant Chromosome Numbers 1967–1971*. Oosthoek, Utrecht, The Netherlands.

Morel, G., and Raccurt, M. 2002. *PCR/RT-PCR in-situ Light and Electron Microscopy*. CRC Press, Boca Raton, FL, USA.

Obe, G., and Natarayan, A. T., eds. 1994. *Chromosomal Aberrations: Origin and Significance*. Springer, New York.

Olmo, E., and Redi, C. A., eds. 2000. *Chromosomes Today*. Birkhäuser, Basel, Switzerland.

Palasso, R. E., and Scahatten, G. P., eds. 2000. *The Centrosome in Cell Replication and Early Development*. Academic Press, San Diego, CA, USA.

Polak, J., and Van Noorden, S. 1997. *Introduction to Immunocytochemistry*. Springer, New York.

Popescu, P., et al., eds. 2000. *Techniques in Animal Cytogenetics*. Springer, New York.

Rautenstrauß, B. W., and Liehr, T. 2002. *FISH Technology*. Springer, Berlin, Germany.

Rieder, C. L., and Matsudaira, P., eds. 1999. *Mitosis and Meiosis*. Academic Press, San Diego, CA, USA.

Rooney, D. E., ed. 2001. *Human Cytogenetics*. Oxford Univ. Press, New York.

Rooney, D. E., and Czepulkowski, B. H. 1997. *Human Chromosome Preparations: Essential Techniques*. Wiley, New York.

Rooney, D. E., and Czepulkowski, B. H., eds. 2001. *Human Cytogenetics*. Oxford Univ. Press, New York.

Ruffolo, R. R., Jr., Poste, G., and Metcalf, B. W., eds. 1997. *Cell Cycle Regulation*. Harwood Academic Publishers, Langhorne, PA, USA.

Ruzin, S. E. 1999. *Plant Microtechnique and Microscopy*. Oxford Univ. Press, New York.

Shapiro, H. M. 1994. *Practical Flow Cytometry*. Wiley-Liss, New York.

Sharma, A. K., and Sharma, A. 1994. *Chromosome Techniques*. Harwood Academic Publishers, Langhorne, PA, USA.

Sharma, A. K., and Sharma, A., eds. 1999. *Plant Chromosomes: Analysis, Manipulation and Engineering*. Harwood Academic Publishers, Amsterdam, The Netherlands.

Sheppard, C., and Shotton, D. 1997. *Confocal Laser Scanning Microscopy*. Springer, New York.

Speel, E., et al., eds. 2003. *Chromosome Analysis Protocols*. Humana Press, Totowa, NJ. USA.

Stein, G., ed. 1998. *The Molecular Basis of Cell Cycle and Growth Control*. Wiley, New York.

Therman, E. 1993. *Human Chromosomes: Structure, Behavior, Effects*. Springer, New York.

Turner, B. M. 2001. *Chromatin and Gene Regulation: Mechanisms in Epigenetics*. Blackwell Science, Cambridge, MA, USA.

Van Driel, R., and Otte, A. P. 1997. *Nuclear Organization, Chromatin Structure and Gene Expression*. Oxford Univ. Press, New York.

Wagner, R. P., Maguire, M. P., and Stallings, R. L. 1993. *Chromosomes: A Synthesis*. Wiley-Liss, New York.

Wang, X. F., and Herman, B., eds. 1996. *Fluorescence Imaging Spectroscopy and Microscopy*. Wiley, New York.

Wegener, R-D., ed. 1999. *Diagnostic Cytogenetics*. Springer, New York.

Wolffe, A. 1998. *Chromatin. Structure and Function*. Academic Press, San Diego, CA, USA.

DEVELOPMENT (*See also* Molecular Biology, Neurobiology)

Annual Review of Cell and Developmental Biology. Palo Alto, CA, USA.

Arias, A. M., and Stewart, A. 2002. *Molecular Principles of Animal Development*. Oxford Univ. Press, New York.

Arthur, W. 2000. *The Origin of Animal Body Plans*. Cambridge Univ. Press, New York.

Bainbridge, D. 2001. *Making Babies: The Science of Pregnancy*. Harvard Univ. Press, Cambridge, MA, USA.

Barry, J. M. 2002. *Molecular Embryology: How Molecules Give Birth to Animals*. Taylor & Francis. Hamden, CT, USA.

Bate, M., and Martinez Arias, A., eds. 1994. *The Development of Drosophila melanogaster*. Cold Spring Harbor Laboratory Press, Cold Spring Harbor, NY, USA.

Beck, S. and Olek, A., eds. 2002. *The Epigenome. Molecular Hide and Seek*. Wiley, Hoboken, NJ. USA.

Birkhead, T. R., and Moller. A. P. 1998. *Sperm Competition and Sexual Selection*. Academic Press, Orlando, FL, USA.

Bowman, J., ed. 1994. *Arabidopsis: An Atlas of Morphology and Development*. Springer, New York.

Brookes, M., and Zietman, A. 1998. *Clinical Embryology: A Color Atlas and Text*. CRC Press, Boca Raton, FL, USA.

Campos-Ortega, J. A., and Hartenstein, V. 1997. *The Embryonic Development of Drosophila melanogaster*. Springer, New York.

Cardew, G., and Goode, J. A. 2001. *The Molecular Basis of Skeletogenesis*. Novartis Foundation Symposium 232. Wiley, New York.

Carroll, S. B., Grenier, J. K., and Weatherbee, S. D. 2001. *From DNA to Diversity: Molecular Genetics and the Evolution of Animal Design*. Blackwell Science, Malden, MA, USA.

Chadwick, D. J., and Cardew, G., eds. 1998. *Epigenetics*. Wiley, New York.

Coen, E. 1999. *The Art of Genes: How Organisms Make Themselves*. Oxford Univ. Press, New York.

Cronk, Q. C. B., et al., eds. 2002. *Developmental Genetics and Plant Development*. Taylor & Francis. London, UK.

Davidson, E. H. 2001. *Genomic Regulatory Systems: Development and Evolution*. Academic Press, San Diego, CA, USA.

Dickson, R. B., and Salomon, D. S., eds. 1998. *Hormones and Growth Factors in Development and Neoplasia*. Wiley-Liss, New York.

Duboule, D., ed. 1994. *Guidebook to the Homeobox Genes*. Oxford Univ. Press, New York.

Engel, E., and Antonarakis, S. E. 2001. *Genomic Imprinting and Uniparental Disomy in Medicine: Clinical and Molecular Aspects*. Wiley, New York.

Finch, C. E., and Kirkwood, T. B. L. 2000. *Chance, Development, and Aging*. Oxford Univ. Press, Oxford, UK.

Findley, J. K., ed. 1994. *Molecular Biology of the Female Reproductive System*. Academic Press, San Diego, CA, USA.

Fleming, T., and Muse, S., eds. 2002. *Cell-Cell Interactions*. Oxford Univ. Press, New York.

Fosket, D. E. 1994. *Plant Growth and Development: A Molecular Approach*. Academic Press, San Diego, CA, USA.

Gehring, W. J. 1998. *Master Control Genes in Development and Evolution: The Homeobox Story*. Yale Univ. Press, New Haven, CT, USA.

Gilbert, L. I., Tata, J. R., and Atkinson, B. G., eds. 1996. *Metamorphosis: Postembryonic Reprogramming of Gene Expression in Amphibian and Insect Cells*. Academic Press, San Diego, CA, USA.

Gilbert, S. F. 2000. *Developmental Biology*. Sinauer, Sunderland, MA, USA.

Gilbert, S. F., and Raunio, A. M., eds. 1997. *Developmental Diversity*. Sinauer, Sunderland, MA, USA.

Glover, T. D., and Barratt, C. L. R., eds. 1999. *Male Fertility and Infertility*. Cambridge Univ. Press, New York.

Gu, J. 1997. *Analytical Morphology: Theory, Applications and Protocols*. Springer, New York.

Hall, B. K. 1999. *The Neural Crest in Development and Evolution*. Springer, New York.

Hall, J. C., ed. 2003. *Genetics and Molecular Biology of Rhythms in Drosophila and other Insects. Adv. Genet.* 48: pp. 244.

Handel, M. A., ed. 1997. *Meiosis and Gametogenesis*. Academic Press, San Diego, CA., USA.

Harding, R., and Bocking, A. D. 2001. *Fetal Growth and Development*. Cambridge Univ. Press, New York.

Harrison, L. G. 1993. *Kinetic Theory of Living Patterns*. Cambridge Univ. Press, New York.

Heath, J. K. 2001. *Principles of Cell Proliferation*. Blackwell, Oxford, UK.

Held, L. I., Jr. 2002. *Imaginal Discs: The Genetic and Cellular Logic of Pattern Formation*. Cambridge Univ. Press, New York.

Howell, S. H. 1998. *Molecular Genetics of Plant Development*. Cambridge Univ. Press, New York.

Hunter, R. H. F. 1995. *Sex Determination, Differentiation and Intersexuality in Placental Mammals*. Cambridge Univ. Press, New York.

Jirasek, J. E. 2001. *An Atlas of the Human Embryo and Fetus: A Photographic Review of Human Prenatal Development*. Pathenon, New York.

Jowett, T. 1997. *Tissue In Situ Hybridization: Methods in Animal Development*. Wiley, New York.

Kalthoff, K. 1995. *Analysis of Biological Development*. McGraw-Hill, New York.

Kaufman, M. H. 1992. *The Atlas of Mouse Development*. Academic Press, San Diego, CA, USA.

Kaufman, M. H., and Bard, J. B. L. 1999. *The Anatomical Basis of Mouse Development*. Academic Press, San Diego, CA, USA.

Kolata, G. 1997. *Clone: The Road to Dolly and the Path Ahead*. Morrow, Allen Lane, New York.

Lash, J. 1999. *Interactive Embryology: The Human Embryo Program*. Sinauer, Sunderland, MA, USA.

Lawrence, P. A. 1992. *The Making of a Fly: The Genetics of Animal Design*. Blackwell Science, Cambridge, MA, USA.

Le Douarin, N. M., and Kalcheim, C. 1999. *The Neural Crest*. Cambridge Univ. Press, New York.

Martini Neri, M. E., Neri, G., and Opitz, J. M., eds. 1996. *Gene Regulation and Fetal Development*. Wiley-Liss, New York.

McElreavey, K., ed. 2000. *The Genetic Basis of Male Infertility*. Springer, New York.

Moody, S. A., ed. 1998. *Cell Lineage and Fate Determination*. Academic Press, San Diego, CA, USA.

Müller, W. A. 1997. *Developmental Biology*. Springer, New York.

Nüsslein-Volhard, C., and Dahm, R., eds. 2002. *Zebrafish. A Practical Approach*. Oxford Univ. Press, New York.

Ohlsson, R., ed. 1999. *Genomic Imprinting: An Interdisciplinary Approach*. Springer, New York.

Ohlsson, R., Hall, K., and Ritzen, M., eds. 1995. *Genomic Imprinting — Causes and Consequences*. Cambridge Univ. Press, New York.

Osiewacz, H. D. 2002. *Molecular Biology of Fungal Development*. Marcel Dekker, New York.

Pablo, de F., Ferrús, A., and Stern, C. D. 1997. *Cellular and Molecular Procedures in Developmental Biology*. Academic Press, Orlando, FL., USA.

Pigliucci, M. 2001. *Phenotypic Plasticity: Beyond Nature and Nurture*. John Hopkins Univ. Press, Baltimore, MD, USA.

Piontellin, A. 2002. *Twins — From Fetus to Child*. Taylor & Francis, New York.

Potten, C., ed., 1996. *Stem Cells*. Academic Press, San Diego, CA, USA.

Reik, W., and Surani, A. 1997. *Genomic Imprinting: Frontiers in Molecular Biology*. IRL/Oxford Univ. Press, New York.

Rossant, J., and Pedersen, R. A., eds. 1988. *Experimental Approaches to Mammalian Embryonic Development*. Cambridge Univ. Press, New York.

Rossant, J., and Tam, P. P. L., eds. 2002. *Mouse Development: Patterning, Morphogenesis and Organogenesis*. Academic Press, San Diego, CA, USA.

Russo, V. E. A., Brody, S., Cove, D., and Ottolenghi, S. 1992. *Development: The Molecular Genetic Approach*. Springer, New York.

Russo, V. E. A., Martienssen, R. A., and Riggs, A. D., eds. 1997. *Epigenetic Mechanisms of Gene Regulation*. Cold Spring Harbor Laboratory Press, Cold Spring Harbor, NY, USA.

Scherer, G., and Schmid, M. 2001. *Genes and Mechanisms in Vertebrate Sex Determination*. Birkhäuser, Basel, Switzerland.

Sharpe, P. T., and Mason, I., eds. 1999. *Molecular Embryology: Methods and Protocols*. Humana Press, Totowa, NJ, USA.

Slack, J. 2001. *Essential Developmental Biology*. Blackwell, Boston, MA, USA.

Sive, H. L., et al. 1999. *Early Development of Xenopus laevis: A Laboratory Manual*. Cold Spring Harbor Laboratory Press, Plainview, NY, USA.

Solari, A. J. 1994. *Sex Chromosomes and Sex Determination in Vertebrates*. CRC Press, Boca Raton, FL, USA.

Tomanek, R. J., and Runyan, R. B., eds. 2001. *Formation of the Heart and Its Regulation*. Birkhäuser, Boston, MA, USA.

Tuan, R. S., and Lo, C. W., eds. 1999–2000. *Developmental Biology Protocols*. Humana Press, Totowa, NJ, USA.

Veeck, L. L. 1999. *An Atlas of Human Gametes and Conceptuses*. Parthenon, New York.

Wachtel, S. S. 1993. *Molecular Genetics of Sex Determination*. Academic Press, San Diego, CA, USA.

Wassarman, P. M., ed. 1999. *Advances in Developmental Biochemistry*, Vol. 5. JAI Press, Stamford, CT, USA.

Wilkins, A. S. 1993. *Genetic Analysis of Animal Development*. Wiley-Liss, New York.

Wilkins, A. S. 2002. *The Evolution of Developmental Pathways*. Sinauer, Sunderland, MA, USA.

Williams, E. G., et al., eds. 1994. *Genetic Control of Self-Incompatibility and Reproductive Development in Flowering Plants*. Kluwer, Norwell, MA.

Wizemann, T., and Pardue, M.-L., eds. 2001. *Exploring the Biological Contributions to Human Health — Does Sex Matter?* National Acad. of Science Press, Washington, DC.

Wolffe, A., ed. 1998. *Epigenetics*. Wiley, New York.

Wolpert, L., Beddington, R., Jessell, T., and Lawrence, P. 2002. *Principles of Development.* Oxford. Univ. Press, New York.

Zon, L. I., ed. 2001. *Hematopoiesis: A Developmental Approach.* Oxford Univ. Press, New York.

DISEASES (*See also* Aging, Cancer, Immunology, Molecular Biology)

Ashcroft, F. M. 2000. *Ion Channels and Disease.* Academic Press, San Diego, CA, USA.

Bagasra, O. 1999. *HIV and Molecular Immunity: Prospects for the AIDS Vaccine.* Eaton, Natick, MA, USA.

Baker, H. F., ed. 2001. *Molecular Pathology of Prions.* Humana Press, Totowa, NJ, USA.

Baker, H. F., and Ridley, R. M., eds. 1996. *Prion Diseases.* Humana Press, Totowa, NJ., USA.

Barker, J., and McGrath, J., eds. 2001. *Cell Adhesion and Migration in Skin Diseases.* Gordon & Breach, Amsterdam, The Netherlands.

Barter, P., and Rye, K.-A., eds. 1999. *Plasma Lipids and Their Role in Disease.* Harwood Academic Publishers, Cooper Station, NY, USA.

Bazin, H. 2000. *The Eradication of Small Pox: Edward Jenner and the First and Only Eradication of a Human Infectious Disease.* Academic Press, San Diego, CA, USA.

Becker, R. 1996. *Alzheimer Disease: From Molecular Biology to Therapy.* Springer, New York.

Boué, A. 1995. *Fetal Medicine: Prenatal Diagnosis and Management.* Oxford Univ. Press, New York.

Bowcock, A. M., ed. 1999. *Breast Cancer: Molecular Genetics, Pathogenesis, and Therapeutics.* Humana Press, Totowa, NJ, USA.

Brioni, J. D., and Decker, M. W., eds. 1997. *Pharmacological Treatment of Alzheimer's Disease: Molecular and Neurobiological Foundations.* Wiley, New York.

Chadwick, R., et al., eds. 1999. *The Ethics of Genetic Screening.* Kluwer, Dordrecht, The Netherlands.

Clarke, J. T. R. 1996. *A Clinical Guide to Inherited Metabolic Diseases.* Cambridge Univ. Press, Cambridge, UK.

Collinge, J., and Palmer, M. S. 1997. *Prion Disease.* Oxford Univ. Press, New York.

Cossart, P., et al., eds. 2000. *Cellular Microbiology.* American Society of Microbiology, Washington, DC.

Crosby, A. W. 2002. *The Deadly Truth: A History of Diseases in America.* Harvard Univ. Press, Cambridge, MA, USA.

Dean, R. T., and Kelly, D. T., eds. 2001. *Atherosclerosis: Gene Expression, Cell Interactions and Oxidation.* Oxford Univ. Press, New York.

Doolan, D. L., ed. 2002. *Malaria Methods and Protocols.* Humana Press, Totowa, NJ, USA.

Dragani, T. A., ed. 1998. *Human Polygenic Disease: Animal Models.* Harwood Academic Publishers, Marston Book Services, Abingdon, Oxon, UK.

Edwards, R. G., ed. 1993. *Preconception and Preimplantation Diagnosis of Human Genetic Diseases.* Cambridge Univ. Press, New York.

Elles, R. 1996. *Molecular Diagnosis of Genetic Diseases.* Humana Press, Totowa, NJ, USA.

Emery, A. E. H., ed. 1998. *Neuromuscular Disorders: Clinical and Molecular Genetics.* Wiley, New York.

Emery, A. E. H., ed. 2001. *The Muscular Dystrophies.* Oxford Univ. Press, New York.

Evans, D. A. P. 1994. *Genetic Factors in Drug Therapy: Clinical and Molecular Pharmacogenetics.* Cambridge Univ. Press, New York.

Ewald, P. W. 1994. *Evolution of Infectious Diseases.* Oxford Univ. Press, New York.

Factor, P. H., ed. 2001. *Gene Therapy for Acute and Acquired Diseases.* Kluwer, Dordrecht, The Netherlands.

Gilbert-Barness, E., and Barness, L. A. 2000. *Metabolic Diseases: Foundations of Clinical Management, Genetics, and Pathology.* Eaton, Natick, MA, USA.

Giraldo, G., Bologesi, D. P., Salvatore, M., and Beth-Giraldo, E., eds. 1996. *Development and Applications of Vaccines and Gene Therapy in AIDS.* Karger, Basel, Switzerland.

Gorlin, R. J., et al., eds. 1995. *Hereditary Hearing Loss and Its Syndromes.* Oxford Univ. Press, New York.

Gupta, S., ed. 1996. *Immunology of HIV Infection.* Plenum, New York.

Haas, C., ed. 1999. *Molecular Biology of Alzheimer Disease: Genes and Mechanisms Involved in Amyloid Generation.* Gordon & Breach, Newark, NJ, USA.

Hall, L. L., ed. 1996. *Genetics and Mental Illness: Evolving Issues for Research and Society.* Plenum, New York.

Harper, J. C., et al. 2001. *Preimplantation Genetic Diagnosis.* Wiley, New York.

Harper, P. S., and Clarke, A. J. 1997. *Genetics, Society and Clinical Practice.* Bios Scientific, Cambridge, UK.

Harper, P. S., and Perutz, M. 2001. *Glutamine Repeats and Neurogenerative Diseases: Molecular Aspects.* Oxford Univ. Press, New York.

Hassold, T. J., and Patterson, D., eds. 1998. *Down Syndrome: A Promising Future, Together.* Wiley, New York.

Hebert, C. A. 1999. *Chemokines in Disease: Biology and Clinical Research.* Humana Press, Totowa, NJ, USA.

Hooper, N. M., ed. 2000. *Alzheimer's Disease.* Humana Press, Totowa, NJ.

Hunt, K. K., et al., eds. 2001. *Breast Cancer.* Springer, New York.

Iannaccone, P. M., and Scarpelli, D. G., eds. 1997. *Biological Aspects of Disease: Contributions from Animal Models.* Taylor & Francis, New York.

James, D. G., and Zumla, A., eds. 1999. *The Granulomatous Disorders.* Cambridge Univ. Press, New York.

Jeffery, S., Booth, J., and Butcher, P. 1997. *Nucleic Acid-Based Diagnosis.* Springer, New York.

Kalow, W. et al., eds. 2001. *Pharmacogenomics.* Marcel Dekker, New York.

Kaper, J. B., and Hacker, J., eds. 1999. *Pathogenicity Islands and Other Mobile Virulence Elements.* American Society of Microbiology Press, Washington, DC.

Karpati, G., et al. eds. 2001. *Disorders of the Voluntary Muscle.* Cambridge Univ. Press, Cambridge, UK.

Kase, H., et al., eds. 1999. *Adenosine Receptors and Parkinson's Disease.* Academic Press, San Diego, CA, USA.

Kashima, H., et al., eds. 2002. *Comprehensive Treatment of Schizophrenia. Linking Neuro-behavioral Findings to Psychosocial Approaches.* Springer, New York.

Khoury, M. J., Burke, W., and Thompson, E. J., eds. 2000. *Genetics and Public Health in the 21st Century.* Oxford Univ. Press, New York.

King, R. A., Rotter, J. I., and Motulsky, A., eds. 2002. *The Genetic Basis of Common Diseases*. Oxford Univ. Press, New York.

Kuschner, H. I. 1999. *A Cursing Brain? The Histories of Tourette Syndrome*. Harvard Univ. Press, Cambridge, MA, USA.

Licinio, J., and Wong, M. L., eds. 2002. *Pharmacogenomics: The Search for Individualized Therapeutics*. Wiley, New York.

Lowe, W. L., Jr., ed. 2001. *Genetics and Diabetes mellitus*. Kluwer, Dordrecht, The Netherlands.

McCance, K. L., and Huether, S. E. 1994. *Pathophysiology: The Biologic Basis of Disease in Adults and Children*. Mosby, St. Louis, MO, USA.

McCrae, M. A., et al., eds. 1997. *Molecular Aspects of Host-Pathogen Interactions*. Cambridge Univ. Press, New York.

Meniru, G. 2001. *Cambridge Guide to Infertility Management and Assisted Reproduction*. Cambridge Univ. Press, New York.

Miller, M. S., and Cronin, M., eds. 2000. *Genetic Polymorphisms and Susceptibility to Disease*. Taylor & Francis, New York.

Mooney, M. P., and Siegel, M. I., eds. 2002. *Understanding Craniofacial Anomalies. Etiopathogenesis of Craniosynostoses and Facial Clefting*. Wiley, New York Oxford Medical Databases.

Okazaki, I., ed. 2003. *Extracellular Matrix and the Liver*. Acad. Press, San Diego, CA.

Oxford Medical Databases. Oxford Univ. Press, New York.

Pawlowtzki, I.-H., Edwards, J. H., and Thompson, E. A. 1997. *Genetic Mapping of Disease Genes*. Academic Press, Orlando, FL, USA.

Pinsky, L., Erickson, R. P., and Schimke, R. N. 1999. *Genetic Disorders of Human Sexual Development*. Oxford Univ. Press, New York.

Prusiner, S. B. 1966. *Prions, Prions, Prions*. Springer, New York.

Pulst, S.-M., ed. 2002. *Genetics of Movement Disorders*, Acad. Press, San Diego, CA, USA.

Prusiner, S. B., ed. 1999. *Prion Biology and Diseases*. Cold Spring Harbor Laboratory Press, Cold Spring Harbor, NY, USA.

Quinn, S. 2001. *Human Trials: Scientists, Investors and Patients in the Quest for a Cure*. Perseus, Cambridge, MA, USA.

Robinson, A., and Linden, M. G. 1993. *Clinical Genetics Handbook*. Blackwell Science, Oxford, UK.

Roland, P. E. 1997. *Brain Activation*. Wiley, New York.

Ross, D. W. 1997. *Introduction to Molecular Medicine*. Springer, New York.

Royce, P. M., and Steinmann, B., eds. 2002. *Connective Tissue and Its Heritable Disorders: Molecular, Genetic and Medical Aspects*. Wiley, New York.

Scriver, C. R., et al., eds., 2001. *The Metabolic and Molecular Bases of Inherited Disease*. McGraw-Hill, New York.

Semenza, G. L. 1998. *Transcription Factors and Human Disease*. Oxford Univ. Press, New York.

Serjeant, G. R., and Serjeant, B. E. 2001. *Sickle Cell Disease*. Oxford Univ. Press, New York.

Shaw, K. J., ed. 2002. *Pathogen Genomics: Impact on Human Health*. Humana, Totowa, NJ, USA.

Stamatoyannopoulos, G., Nienhuis, A. W., Majerus, P. W., and Varmus, H. 1994. *The Molecular Basis of Blood Diseases*. Saunders, Philadelphia, PA, USA.

Steinberg, M. H., et al., eds. 2000. *Disorders of Hemoglobin: Genetics, Pathophysiology, and Clinical Management*. Cambridge Univ. Press, New York.

Stevenson, R. E., et al., eds. 1993. *Human Malformations and Related Anomalies*. Oxford Univ. Press, New York.

Stine, G. J. 1994. *Acquired Immunodeficiency Syndrome: Biological, Medical, Social, and Legal Issues*. Prentice Hall, Englewood Cliffs, NJ.

Tanzi, R. E., and Parson. A. B. 2000. *Decoding Darkness: The Search for the Genetic Causes of Alzheimer's Disease*. Perseus, New York.

Teebi, A. S., and Farag, T. I., eds. 1996. *Genetic Disorders Among Arab Populations*. Oxford Univ. Press, New York.

Terry, R. D., Katzman, R., and Bick, K. L., eds. 1994. *Alzheimer Disease*. Raven, New York.

Thakker, R. V., ed. 1997. *Molecular Genetics of Endocrine Disorders*. Arnold/Oxford Univ. Press, New York.

Theofilopoulos, A. N., and Bona, C. A., eds. 2002. *The Molecular Pathology of Autoimmune Diseases*. Taylor & Francis, New York.

Thoene, J. G., ed. 1996. *Physician's Guide to Rare Diseases*. Dowden, Montvale, NJ, USA.

Traboulsi, E. I., ed. 1999. *Genetic Diseases of the Eye*. Oxford Univ. Press, New York.

Vile, R. G. 1997. *Understanding Gene Therapy*. Springer, New York.

Vos, J.-M. H. 1995. *DNA Repair Mechanisms: Impact on Human Diseases and Cancer*. Springer, New York.

Wahlgren, W., and Perlmann, P., eds. 1999. *Malaria: Molecular and Clinical Aspects*. Gordon & Breach, Harwood Academic Publishers, Cooper Station, NY, USA.

Weiss, K. M. 1999. *Genetic Variation and Human Diseases: Principles and Evolutionary Approaches*. Cambridge Univ. Press, New York.

Winter, R. 1997. *Oxford Medical Databases*. Dysmorphology Photo Library 2.0. CD-ROM, Windows, Oxford Univ. Press, New York.

Young, D. B. 1998. *Genetics and Tuberculosis*. Wiley, New York.

Zaven, S. K., and Mesulam, M.-M., eds. 2000. *Alzheimer's Disease: A Compendium of Current Theories*. Annals, New York Acad. Sci. #924, New York Academy of Science, New York.

EVOLUTION (*See also* Cytology and Cytogenetics, Population Genetics)

Alters, B. J., and Alters, S. M. 2001. *Defending Evolution: A Guide to the Creation/Evolution Controversy*. Jones and Bartlett, Boston, MA, USA.

Anderson, E. 1949. *Introgressive Hybridization*. Wiley, New York.

Andersson, M. 1994. *Sexual Selection*. Priceton Univ. Press, Princeton, NJ, USA.

Arnold, M. L. 1997. *Natural Hybridization and Evolution*. Oxford Univ. Press, New York.

Avise, J. C. 1994. *Molecular Markers, Natural History and Evolution*. Chapman & Hall, New York.

Avise, J. 2000. *Phylogeography: The History and Formation of Species*. Harvard Univ. Press, Cambridge, MA, USA.

Bell, G. 1997. *Selection: The Mechanism of Evolution*. Chapman & Hall, New York.

Birkhead, T. 2000. *Promiscuity: An Evolutionary History of Sperm Competition and Sexual Conflict*. Faber & Faber, Winchester, MA, USA.

Bökönyi, S. 1974. *History of Domestic Mammals in Central and Eastern Europe*. Akadémiai Kiadó, Budapest, Hungary.

Brack, A., ed. 1999. *The Molecular Origins of Life: Assembling Pieces of the Puzzle*. Cambridge Univ. Press, New York.

Briggs, D., and Walters, S. M. 1996. *Plant Variation and Evolution*. Cambridge Univ. Press, New York.

Browne, J. 2002. *Charles Darwin: The Power of Place*. A. A. Knopf, New York.

Carroll, R. L. 1997. *Patterns and Processes of Vertebrate Evolution*. Cambridge Univ. Press, New York.

Cavalier-Smith, T. 1985. *The Evolution of Genome Size*. Wiley, New York.

Cavalli-Sforza, L. L. 2000. *Genes, People, and Languages*. Farrar, Straus & Giroux, New York.

Cavalli-Sforza, L. L., and Cavalli-Sforza, F. 1995. *The Great Human Diasporas: The History of Diversity and Evolution*. Addison-Wesley, Boston, MA, USA.

Cavalli-Sforza, L. L., Menozzi, P., and Piazza, A. 1994. *The History and Geography of Human Genes*. Princeton Univ. Press, Princeton, NJ, USA.

Cavalli-Sforza, L. L., et al., eds. 1999. *Human Evolution*. Cold Spring Harbor Laboratory Press, Cold Spring Harbor, NY, USA.

Clark, M., ed. 2000. *Comparative Genomics*. Kluwer, Dordrecht, The Netherlands.

Cockburn, A., Cockburn, E., and Reyman, T. A. 1998. *Mummies, Disease and Ancient Cultures*. Cambridge Univ. Press, New York.

Crandall, K. A., ed. 1999. *The Evolution of HIV*. Johns Hopkins Univ. Press, Baltimore, MD, USA.

Crow, T. J., ed. 2001. *The Speciation of Modern Homo Sapiens*. Oxford Univ. Press, New York.

Crozier, R. H., and Pamilo, P. 1996. *Evolution of Social Insect Colonies*. Oxford Univ. Press, New York.

Dayhoff, M. O., ed. 1972. *Atlas of Protein Sequences and Structure*, Vol. 5. National Biomed Research Fund, Washington, DC.

Delsemme, A. H. 2001. *Our Cosmic Origins*. Cambridge Univ. Press, New York.

DeSalole, R., Giribet, G., and Wheler, W., eds. 2001. *Techniques in Molecular Systematics and Evolution*. Birkhäuser, Basel, Switzerland.

Dobzhansky, Th., et al. 1977. *Evolution*. Freeman, San Francisco, CA, USA.

Dyson, F. J. 1999. *Origins of Life*. Cambridge Univ. Press, Cambridge, UK.

Eldredge, N. 2000. *The Triumph of Evolution and the Failure of Creationism*. Macmillan, New York.

Felsenstein, J. 2001. *Inferring Phylogenies*. Sinauer, Sunderland, MA, USA.

Fix, A. G. 1999. *Migration and Colonization in Human Microevolution*. Cambridge Univ. Press, Cambridge, UK.

Fleagle, J. G., and Kay, R. F., eds. 1994. *Anthropoid Origins*. Plenum, New York.

Ford, E. B. 1971. *Ecological Genetics*. Chapman & Hall, London, UK.

Futuyama, D. J. 1997. *Evolutionary Biology*. Sinauer, Sunderland, MA, USA.

Gerhart, J., and Kirschner, M. 1997. *Cells, Embryos and Evolution: Developmental Understanding of Phenotypic Variation and Evolutionary Adaptability*. Blackwell Science, Cambridge, MA, USA.

Gesteland, R. F., Cech, T. R., and Atkins, J. F., eds. 1999. *The RNA World*. Cold Spring Harbor Laboratory Press, Cold Spring Harbor, NY, USA.

Gibbs, A. J., Calisher, C. H., and Garcia-Arenal, F., eds. 1995. *Molecular Basis of Virus Evolution*. Cambridge Univ. Press, New York.

Gillespie, J. H. 1992. *The Causes of Molecular Evolution*. Oxford Univ. Press, New York.

Godfrey, A. 2000. *Life Without Genes*. HarperCollins, New York.

Gould, S. J. 2002. *The Structure of Evolutionary Theory*. Harvard Univ. Press, Cambridge, MA, USA.

Graur, D., and Li, W. 2000. *Fundamentals of Molecular Evolution*. Sinauer, Sunderland, MA, USA.

Hall, B. G. 2001. *Phylogenetic Trees Made Easy: A How Manual for Molecular Biologists*. Sinauer, Sunderland, MA, USA.

Hall, B. K., ed. 1994. *Homology: The Hierarchial Basis of Comparative Biology*. Academic Press, San Diego, CA.

Hall, B. K. 1999. *Evolutionary Developmental Biology*. Kluwer, Dordrecht, The Netherlands.

Harris, H. 2002. *Things Come to Life. Spontaneous Generation Revisited*, Oxford Univ. Press, New York.

Herrmann, B., and Hummel, S., eds. 1994. *Ancient DNA: Recovery and Analysis of Genetic Material from Paleontological, Archeological, Museum, Medical and Forensic Specimens*. Springer-Verlag, New York.

Hey, J. 2001. *Genes, Categories, and Species: The Evolutionary and Cognitive Causes of the Species Problem*. Oxford Univ. Press, New York.

Hillis, D. M., Moritz, C., and Mable, B. K., eds. 1996. *Molecular Systematics*. Sinauer, Sunderland, MA, USA.

Hochachka, P. W., and Somero, G. N. 2002. *Biochemical Adaptation: Mechanism and Process in Physiological Evolution*. Oxford Univ. Press, Oxford, UK.

Howard, D. J., and Berlocher, S. H., eds. 1998. *Endless Forms — Species and Speciation*. Oxford Univ. Press, New York.

Hughes, A. L. 2000. *Adaptive Evolution of Genes and Genomes*. Oxford Univ. Press, New York.

Jablonka, E., and Lamb, M. J. 1999. *Epigenetic Inheritance and Evolution: The Lamarckian Dimension*. Oxford Univ. Press, New York.

Jones, M. 2002. *The Molecule Hunt: Archeology and the Search for Ancient DNA*. Arcade Publishing Co., New York.

Jones, S., et al., eds. 1994. *The Cambridge Encyclopedia of Human Evolution*. Cambridge Univ. Press, New York.

Klein, J., and Takahata, N. 2001. *Where We Do Come From? The Molecular Evidence of Human Descent*. Springer, New York.

Lahav, N. 1999. *Biogenesis. Theories of Life's Origin*. Oxford Univ. Press, New York.

Lawton, J. H., and May, R. M., eds. 1995. *Extinction Rates*. Oxford Univ. Press, New York.

Lerner, I. M., and Libby, W. J. 1976. *Heredity, Evolution and Society*. Freeman, San Francisco, CA, USA.

Levinton, J. S. 2001. *Genetics, Paleontology, and Macroevolution*. Cambridge Univ. Press, New York.

Lewin, R. 1997. *Patterns in Evolution: The New Molecular View*. Freeman, New York.

Lewin, R. 1998. *The Origin of Modern Humans*. Freeman, New York.

Li, W.-H. 1997. *Molecular Evolution*. Sinauer, Sunderland, MA, USA.

Margulis, L. 1993. *Symbiosis in Cell Evolution*. Freeman, New York.

Margulis, L., and Sagan, D. 2002. *Acquiring Genomes: A Theory of the Origin of Species*. Basic Books, New York.

Maynard Smith, J. 1978. *The Evolution of Sex*. Cambridge Univ. Press, Cambridge, UK.

Maynard Smith, J. 1982. *Evolution and the Theory of Games*. Cambridge Univ. Press, Cambridge, UK.

Maynard Smith, J. 1998. *Evolutionary Genetics*. Oxford Univ. Press, New York.

Maynard Smith, J., and Szathmáry, E. 1995. *The Major Transitions in Evolution*. Freeman, New York.

Maynard Smith, J., and Szathmáry, E. 1999. *The Origins of Life: From the Birth of Life to the Origin of Language*. Oxford Univ. Press, New York.

Mayr, E. 1963. *Animal Species and Evolution*. Harvard Univ. Press, Cambridge, MA, USA.

Mayr, E. 2001. *What Evolution Is*. Basic Books, New York.

Mousseau, T. A. et al., eds. 2000. *Adaptive Genetic Variation in the Wild*. Oxford Univ. Press, New York.

Myamoto, M. E., and Cracraft, J., eds. 1991. *Phylogenetic Analysis of DNA Sequences*. Oxford Univ. Press, New York.

Nei, M. 1987. *Molecular Evolutionary Genetics*. Columbia Univ. Press, New York.

Nei, M., and Kumar, S. 2000. *Molecular Evolution and Phylogenetics*. Oxford Univ. Press, New York.

Niklas, K. J. 1997. *The Evolutionary Biology of Plants*. Univ. Chicago Press, Chicago, IL, USA.

Ohno, S. 1970. *Evolution by Gene Duplication*. Springer, New York.

Osawa, S. 1995. *Evolution of the Genetic Code*. Oxford Univ. Press, New York.

Page, R. D. M., and Holmes, E. C. 1998. *Molecular Evolution: A Phylogenetic Approach*. Blackwell Science, Oxford, UK.

Pagel, M., ed. 2002. *Encyclopedia of Evolution*. Oxford Univ. Press, New York.

Palumbi, S. R. 2001. *The Evolution Explosion: How Humans Cause Rapid Evolutionary Change*. Norton, New York.

Patthy, L. 1999. *Protein Evolution*. Blackwell, Oxford, UK.

Powell, J. R. 1997. *Progress and Prospects in Evolutionary Biology: The Drosophila Model*. Oxford Univ. Press, New York.

Raff, R. A. 1996. *The Shape of Life: Genes, Development, and the Evolution of Animal Form*. Univ. Chicago Press, Chicago, IL, USA.

Relethford, J. H. 2001. *Genetics and the Search for Modern Human Origins*. Wiley, New York.

Renfrew, C., and Boyle, K., eds. 2000. *Archaeogenetics: DNA and the Population Prehistory of Europe*. McDonald Institute for Archaeological Research, Cambridge, UK.

Ridley, M. 2000. *The Cooperative Gene: How Mendel's Demon Explains Evolution of Complex Beings*. Free Press, New York.

Robert, D., et al., eds. 1996. *Evolution of Microbial Life*. Cambridge Univ. Press, New York.

Roff, D. A. 1997. *Evolutionary Quantitative Genetics*. Chapman & Hall, New York.

Roff, D. A. 2001. *Life History Evolution*. Sinauer, Sunderland, MA, USA.

Rose, M. R., and Lauder, G. V., eds. 1996. *Adaptation*. Academic Press, San Diego, CA, USA.

Ruse, M. 1997. *Monad to Man: The Concept of Progress in Evolutionary Biology*. Harvard Univ. Press, Cambridge, MA, USA.

Sankoff, D., ed. 2000. *Comparative Genomics*. Kluwer, Dordrecht, The Netherlands.

Schlichting, C. D., and Pigliucci, M. 1998. *Adaptive Phenotypic Evolution*. Freeman, New York.

Shermer, M. 2002. *In Darwin's Shadow. The Life and Science of Alfred Russel Wallace*. Oxford Univ. Press, New York.

Simmonds, N. W., ed. 1976. *Evolution of Crop Plants*. Longman, London, UK.

Simmons, L. W. 2001. *Sperm Competition and its Evolutionary Consequences in the Insects*. Princeton Univ. Press, Princeton, NJ, USA.

Smith, J. M. 1998. *Evolutionary Genetics*. Oxford Univ. Press, New York.

Stebbins, L. G. 1950. *Variation and Evolution in Plants*. Columbia Univ. Press, New York.

Steele, A. J. 1981. *Somatic Selection and Adaptive Evolution: On the Inheritance of Acquired Characters*. Univ. Chicago Press, Chicago, IL, USA.

Strick, E. 2000. *Sparks of Life: Darwinism and the Victorian Debates over Spontaneous Generation*. Harvard Univ. Press, Cambridge, MA, USA.

Strickberger, M. W. 1995. *Evolution*. Jones and Bartlett, Boston, MA, USA.

Syvanen, M., and Kado, C. L., eds. 2002. *Horizontal Gene Transfer*. Academic Press, San Diego, CA, USA.

Wills, C., and Bada, J. 2000. *The Spark of Life: Darwin and the Primeval Soup*. Perseus, Cambridge, MA, USA.

Wolf, J. B., et al., eds. 2000. *Epistasis and the Evolutionary Process*. Oxford Univ. Press, New York.

Zeuner, F. E. 1963. *A History of Domesticated Animals*. Harper & Row, New York.

GENETICS MONOGRAPHS (See also Cytology and Cytogenetics, Evolution, Genetics Textbooks, Mapping, Molecular Biology, Mutation, Organelles, Radiation, Recombination, Transposable Elements)

Annual Review of Genetics. Palo Alto, CA, USA.

Annual Review of Genomics and Human Genetics. Palo Alto, CA, USA.

Progress in Forensic Genetics. Elsevier Sci., New York.

Baker, D. L., Schuette, J. L., and Uhlmann, W. R., eds. 1998. *A Guide to Genetic Counseling*. Wiley, New York.

Baraitser, M. 1997. *The Genetics of Neurological Disorders*. Oxford Univ. Press, New York.

Bartel, P. L., and Fields, S. 1997. *The Yeast Two-Hybrid System*. Oxford Univ. Press, New York.

Bennett, R. L. 1999. *The Practical Guide to the Genetic Family History*. Wiley, New York.

Bishop, J. 1999. *Transgenic Mammals*. Longman, Harlow, UK.

Bishop, M. J., ed. 1999. *Genetics Databases*. Academic Press, San Diego, CA, USA.

Bowling, A. T., and Ruvinsky, A., eds. 2000. *The Genetics of the Horse*. Oxford Univ. Press, New York.

Bradshaw, J. E., ed. 1994. *Potato Genetics*. Oxford Univ. Press, New York.

Brenner, S., and Miller, J. H., eds. 2001. *Encyclopedia of Genetics*. Academic Press, San Diego, CA, USA.

Bridge, P. J. 1994. *The Calculation of Genetic Risks: Worked Examples of DNA Diagnostics*. Johns Hopkins Univ. Press, Baltimore, MD, USA.

Brown, A. J. P., and Tuite, M. F., eds. 1998. *Yeast Gene Analysis*. Academic Press, San Diego, CA, USA.

Bulmer, M. G. 1980. *The Mathematical Theory of Quantitative Genetics*. Oxford Univ. Press, New York.

Camp, N. J., and Cox, A., eds. 2002. *Quantitative Trait Loci: Methods and Protocols*. Humana Press, Totowa, NJ, USA.

Charlebois, R. L., ed. 1999. *Organization of the Prokaryotic Genome*. American Society of Microbiology, Washington, DC.

Chaudharty, B. R., and Agarwal, S. B., eds., 1996. *Cytology, Genetics, and Molecular Biology of Algae*. SPB Academic, Amsterdam, The Netherlands.

Cooper, D. N. 1999. *Human Gene Evolution*. Academic Press, San Diego, CA, USA.

Danielli, G. A. 2002. *Genetics and Genomics for the Cardiologist*. Kluwer, Boston, MA, USA.

Dawkins, R. 1999. *The Extended Phenotype: The Long Reach of the Gene*. Oxford Univ. Press, New York.

Detrich, H. W., Zon, L. I., and Westerfield, M., eds. 1998. *The Zebrafish: Genetics and Genomics*. Academic Press, San Diego, CA, USA.

Edwards, A. W. F. 2000. *Foundations of Mathematical Genetics*. Cambridge Univ. Press, New York.

Epstein, H. F., and Shakes, D. C., eds. 1995. *Caenorhabditis elegans: Modern Biological Analysis of an Organism*. Academic Press, San Diego, CA, USA.

Esser, K., and Kuenen, R. 1967. *Genetics of Fungi*. Springer-Verlag, New York.

Falconer, D. S., and Mackay, T. E. C. 1996. *Introduction to Quantitative Genetics*. Longman/Addison Wesley, White Plains, NY, USA.

Fernandez, J., and Hoeffler, J. P., eds. 1998. *Gene Expression Systems: Using Nature for the Art of Expression*. Academic Press, San Diego, CA, USA.

Freeling, M., and Walbot, V., eds. 1994. *The Maize Handbook*. Springer, New York.

Freies, R., and Ruvinsky, I., ed. 1999. *The Genetics of Cattle*. Oxford Univ. Press, New York.

Goldstein, D. B., and Schlötterer, C., eds. 1999. *Microsatellites: Evolution and Applications*. Oxford Univ. Press, Oxford, UK.

Goldstein, L. S. B., and Fryberg, E. A., eds. 1994. *Drosophila melanogaster: Practical Uses in Cell and Molecular Biology*. AP Professional, Cambridge, MA, USA.

Guthrie, C., and Fink, G. R., eds. 2002. *Guide to Yeast Genetics and Molecular and Cell Biology. Methods in Enzymology 350*. Elsevier Science, St. Louis, MO, USA.

Harper, P. S. 1998. *Practical Genetic Counseling*. Butterworth-Heinemann/Oxford Univ. Press, New York.

Hawley, R. S., and Mori, C. A. 1998. *The Human Genome: A User Guide*. Academic Press, San Diego, CA, USA.

Hayes, J. D., and Wolf, C. R. 1997. *Molecular Genetics of Drug Resistance*. Harwood Academic Publishers, Cooper Station, NY, USA.

Jackson, J. E., and Linskens, H. F., eds. 2002. *Testing for Genetic Manipulations in Plants*. Springer, New York.

John, B., and Miklós, G. L. G. 1988. *The Eukaryotic Genome in Development and Evolution*. Allen & Unwin, London, UK.

Kang, M., ed. 2002. *Quantitative Genetics, Genomics and Plant Breeding*. CABI, Wallingford, Oxon, UK.

Kearsey, M. J., and Pooni, H. S. 1996. *The Genetical Analysis of Quantitative Traits*. Chapman & Hall, New York.

King, R. C., ed. 1974. *Handbook of Genetics*. Plenum, New York.

Klotzko, A. J. 2001. *The Cloning Sourcebook*. Oxford Univ. Press, New York.

Kohane, I. S., Kho, A. T., and Butte, E. J. 2003. *Microarrays for an Integrative Genomics*. MIT Press Cambridge, MA, USA.

Koncz, C., Chua, N.-H., and Schell, J., eds. 1993. *Methods in Arabidopsis Research*. World Scientific, Singapore.

Liebler, D. C. 2001. *Introduction to Proteomics: Tools for the New Biology*. Humana Press, Totowa, NJ, USA.

Lindsley, D. L., and Zimm, G. G. 1992. *The Genome of Drosophila melanogaster*. Academic Press, San Diego, CA, USA.

Lynch, M., and Walsh, B. 1998. *Genetics and Analysis of Quantitative Traits*. Sinauer, Sunderland, MA, USA.

Lyon, M. F., Rastan, S., and Brown, S. D. M. 1996. *Genetic Variants and Strains of the Laboratory Mouse*. Oxford Univ. Press, New York.

Malcolm, S., and Goodship, J., eds. 2001. *Genotype to Phenotype*. Academic Press, San Diego, CA, USA.

Martinelli, S. D., and Kinghorn, J. R., eds. 1994. *Aspergillus: 50 Years On*. Elsevier, Amsterdam, The Netherlands.

Mather, K., and Jinks, J. L. 1977. *Introduction to Biometrical Genetics*. Cornell Univ. Press, Ithaca, NY, USA.

McKusick, V. A., ed. 1978. *Medical Genetic Studies of the Amish*. Johns Hopkins Univ. Press, Baltimore, MD, USA.

McKusick, V. A. 1994. *Mendelian Inheritance in Man: A Catalog of Human Genes and Genetic Disorders*. Johns Hopkins Univ. Press, Baltimore, MD, USA.

Meyerowitz, E. M., and Somerville, C. R., eds. 1994. *Arabidopsis*. Cold Spring Harbor Laboratory Press, Cold Spring Harbor, NY, USA.

Miesfeld, R. L. 1999. *Applied Molecular Genetics*. Wiley, New York.

Neale, M. C., and Cardon, L. R. 1992. *Methodology for Genetic Studies of Twins and Families*. Kluwer, Dordrecht, The Netherlands.

Nicholas, F. W. 1987. *Veterinary Genetics*. Clarendon Press, Oxford, UK.

Nicholas, F. W. 1996. *Introduction to Veterinary Genetics*. Oxford Univ. Press, New York.

Nicholl, D. S. T. 2002. *An Introduction to Genetic Engineering*. Cambridge Univ. Press, New York.

Nicholoff, J. A., and Hoekstra, M. F., eds. 1998. *DNA Damage and Repair: DNA Repair in Prokaryotes and Lower Eukaryotes*. Humana Press, Totowa, NJ, USA.

Ohlsson, R., Hall, K., and Ritzen, M., eds. 1995. *Genomic Imprinting: Causes and Consequences*. Cambridge Univ. Press, New York.

Ostrer, H. 1998. *Non-Mendelian Genetics in Humans.* Oxford Univ. Press, New York.

Palzkill, T. 2002. *Proteomics.* Kluwer, Boston, MA, USA.

Phillips. M. I., ed. 2002 *Gene Therapy Methods.* Elsevier Science, St. Louis, MO, USA.

Piper, L., and Ruvinsky, A., eds. 1997. *The Genetics of the Sheep.* CAB/Oxford Univ. Press, New York.

Powell, K. A., Renwick, A., and Peberdy, J. F., eds. 1994. *The Genus Aspergillus: From Taxonomy and Genetics to Industrial Applications.* Plenum, New York.

Pringle, J. R, Broach, J. R., and Jones E. W., eds. 1997. *Cell Cycle and Cell Biology.* Cold Spring Harbor Laboratory Press, Cold Spring Harbor, NY, USA.

Reeve, E. C. R., ed. 2001. *Encyclopedia of Genetics.* Fitzroy Dearborn, London, UK.

Riddle, D. E., et al., eds. 1997. *C. elegans II.* Cold Spring Harbor Laboratory Press, Cold Spring Harbor, NY, USA.

Rimoin, D. L., Connor, J. M., and Peyeritz, R. E., eds. 1996. *Emery and Rimoin's Principles and Practice of Medical Genetics.* Churchill Livingstone, New York.

Rolland, A., ed. 1999. *Advanced Gene Delivery: From Concepts to Pharmaceutical Products.* Harwood Academic Publishers, Villiston, VT, USA.

Rotschild, M. F., and Ruvinsky, A., eds. 1998. *The Genetics of the Pig.* Oxford Univ. Press, New York.

Schindhelm, K., and Nordon, R., eds. 1999. *Ex Vivo Cell Therapy.* Academic Press, London, UK.

Silver, L. M. 1995. *Mouse Genetics: Concepts and Applications.* Oxford Univ. Press, New York.

Singer, M., and Berg, P., eds. 1997. *Exploring Genetic Mechanisms.* University Science Books, Sausalito, CA, USA.

Spector, T. D., Snieder, H., and Macgregor, A. J., eds. 2000. *Advances in Twin and Sib-Pair Analysis.* Greenwich Med. Media, London, UK.

Stevens, L. 1991. *Genetics and Evolution of the Domestic Fowl.* Cambridge Univ. Press, New York.

Strachan, T., and Read, A. P. 1999. *Human Molecular Genetics.* Wiley, New York.

Stubbe, H. 1966. *Genetik und Zytologie von Antirrhinum L. Sect. Antirrhinum.* Fischer, Jena, Germany.

Sykes, B., ed. 1999. *The Human Inheritance: Genes, Language, and Evolution.* Oxford Univ. Press, Oxford, UK.

Tsonis, P. A. 2002. *Anatomy of Gene Regulation: A Three-Dimensional Structural Analysis.* Cambridge Univ. Press, New York.

Weber, W. W. 1997. *Pharmacogenetics.* Oxford Univ. Press, New York.

Weil, J. 2000. *Psychosocial Genetic Counseling.* Oxford Univ. Press, New York.

Weir, B. S. 1996. *Genetic Data Analysis II.* Sinauer, Sunderland, MA, USA.

Wheals, A. E., Rose, A. H., and Harrison, S. J., eds. 1995. *The Yeasts, Vol. 6: Yeast Genetics.* Academic Press, San Diego, USA.

Young, I. D. 1999. *Introduction to Risk Calculation in Genetic Counseling.* Oxford Univ. Press, New York.

GENETICS TEXTBOOKS (*See also* Cytology and Cytogenetics, Genetics Monographs, History of Genetics, Molecular Biology)

Birge, E. A. 2000. *Bacterial and Bacteriophage Genetics.* Springer-Verlag, New York.

Brown, T. A. 2002. *Genomes.* Wiley, New York.

Brown, S. M., et al. 2003. *Essentials of Medical Genomics.* Wiley-Liss, Hoboken, NJ, USA.

Campbell, A. M. and Heyer, L. J. 2003 *Discovering Genomics, Proteomics, and Bioinformatics.* Cold Spring Harbor Lab. Press, Cold Spring Harbor, NY, USA.

Emery, A. E. H., and Malcolm, S. 1995. *An Introduction to Recombinant DNA in Medicine.* Wiley, New York.

Emery, A. E. H., and Mueller, R. F. 1992. *Elements of Medical Genetics.* Churchill Livingstone, New York.

Fincham, J. R. S., Day, P. R., and Radford, A. 1979. *Fungal Genetics.* Blackwell, Oxford, UK.

Gibson, G., and Muse, S. 2002. *A Primer of Genome Science.* Sinauer, Sunderland, MA, USA.

Griffiths, A. J. F., et al. 2000. *An Introduction to Genetic Analysis.* Freeman, New York.

Hartl, D. L., and Jones, E. W. 1996. *Essential Genetics.* Jones and Bartlett, Boston, MA, USA.

Hartwell, L., et al. 2004. *Genetics: From Genes to Genomes.* McGraw-Hill, Dubuque, IA, USA.

Hill, W. E. 2002. *Genetic Engineering. A Primer.* Taylor & Francis, New York.

Johanssen Mange, E., and Mange A. P. 1998. *Basic Human Genetics.* Sinauer, Sunderland, MA, USA.

Jorde, L. B., Carey, J. C., and White, R. L. 1997. *Medical Genetics.* Mosby, St. Louis, MO, USA.

Korf, B. R. 2000. *Human Genetics: A Problem-Based Approach.* Blackwell, Malen, MA, USA.

Kowles, R. 2001. *Solving Problems in Genetics.* Springer, New York.

Kresina, T. F., ed. 2000. *An Introduction to Molecular Medicine and Gene Therapy.* Wiley, New York.

Lewin, B. 1999. *Genes VII.* Oxford Univ. Press, New York.

Old, H. W., and Primrose, S. B. 1995. *Principles of Gene Manipulation.* Blackwell Science, Cambridge, MA, USA.

Pasternak, J. J. 1999. *An Introduction to Human Molecular Genetics: Mechanisms of Inherited Diseases.* Fitzgerald Science, Bethesda, MD, USA.

Primrose, S. B. Twyman, R., and Old, B. 2001. *An Introduction to Genetic Engineering.* Blackwell Science, Cambridge, MA, USA.

Rédei, G. P. 1982. *Genetics.* Macmillan, New York.

Rose, P. W., and Lucassen, A. 1999. *Practical Genetics for Primary Care.* Oxford Univ. Press, New York.

Russel, P. J. 1999. *Fundamentals of Genetics.* Addison-Wesley, Reading, MA, USA.

Schleif, R. 1993. *Genetics and Molecular Biology.* Johns Hopkins Univ. Press, Baltimore, MD, USA.

Serra, J. A. 1965. *Modern Genetics.* Academic Press, New York.

Singer, M., and Berg, P. 1991. *Genes & Genomes: A Changing Perspective.* University Science Books, Mill Valley, CA, USA.

Strickberger, M. W. 1985. *Genetics.* Macmillan, New York.

Thompson, J. N., et al. 1997. *Primer of Genetic Analysis.* Cambridge Univ. Press, New York.

Vogel, F., and Motulsky, A. G. 1996. *Human Genetics.* Springer, New York.

Whitehouse, H. L. K. 1972. *Towards an Understanding of the Mechanism of Heredity.* St. Martin's Press, New York.

Wilson, G. N. 2000. *Clinical Genetics: A Short Course.* Wiley, New York.

Winter, P. C., et al. 2002. *Genetics*. Bios Scientific, Oxford, UK.

HISTORY OF GENETICS (*See also* Genetics Monographs)

Allen, G. E. 1978. *Thomas Hunt Morgan: The Man and His Science*. Princeton Univ. Press, Princeton, NJ, USA.

Bearn, A. G. 1996. *Archibald Garrod and the Individuality of Man*. Oxford Univ. Press, New York.

Brenner, S. 2001. *My Life in Science*. BioMed Central, London, UK.

Brock, T. D. 1990. *The Emergence of Bacterial Genetics*. Cold Spring Harbor Laboratory Press, Cold Spring Harbor, NY, USA.

Cairns, J., Stent, G., and Watson, J. D. 1966. *Phage and the Origins of Molecular Biology*. Cold Spring Harbor Laboratory Press, Cold Spring Harbor, NY, USA.

Cambrosio, A., and Keating, P. 1996. *Exquisite Specificity: The Monoclonal Antibody Revolution*. Oxford Univ. Press, New York.

Carlson, E. A. 1966. *The Gene: A Critical History*. Saunders, Philadelphia, PA, USA.

Carlson, E. O. 1981. *Genes, Radiation, and Society: The Life and Work of H. J. Muller*. Cornell Univ. Press, Ithaca, NY, USA.

Carlson, E. O. 2001. *The Unfit: A History of a Bad Idea*. Cold Spring Harbor Laboratory Press, Cold Spring Harbor, NY, USA.

Comfort, N. C. 2001. *The Tangled Field: Barbara McClintock's Search for the Patterns of Genetic Control*. Harvard Univ. Press, Cambridge, MA, USA.

Crow, J. F., and Dove, W. F., eds. 2000. *Perspectives on Genetics: Anecdotal, Historical, and Critical Commentaries*. Univ. Wisconsin Press, Madison, WI, USA.

Davies, K. 2001. *Cracking the Genome: Inside the Race to Unlock Human DNA*. Free Press, New York.

Davis, R. H. 2000. *Neurospora: Contributions of a Model Organism*. Oxford Univ. Press, New York.

de Chaderevian, S. 2002. *Design for Life: Molecular Biology after World War II*. Cambridge Univ. Press, New York.

Dunn, L. C. 1965. *A Short History of Genetics — The Development of Some of the Main Lines of Thought: 1864–1939*. McGraw-Hill, New York.

Echols, H., and Gross, C. A. 2001. *Operators and Promoters: The Story of Molecular Biology and Its Creators*. Univ. California Press, Berkeley, CA, USA.

Emery, A. E. H., and Emery, M. L. H. 1995. *The History of a Genetic Disease: Duchenne Muscular Dystrophy or Meryon Disease*. Royal Society of Medicine Press, London, UK.

Fedoroff, N., and Botstein, D., eds. 1992. *Barbara McClintock's Ideas in the Century of Genetics*. Cold Spring Harbor Laboratory Press, Cold Spring Harbor, NY, USA.

Fisher, E. P., and Lipson, C. 1988. *Thinking About Science: Max Delbrück and the Origins of Molecular Biology*. W. W. Norton, New York.

Focke, W. O. 1881. *Die Pflanzenmischlinge*. Borntraeger, Berlin, Germany.

Fredrickson, D. S. 2001. *The Recombinant DNA Controversy: A Memoir, Science, Politics and the Public Interest*. American Society of Microbiology Press, Washington, DC.

Friedberg, E. C. 1997. *Correcting the Blueprint of Life: An Historical Account of the Discovery of DNA Repair Mechanisms*. Cold Spring Harbor Laboratory Press, Cold Spring Harbor, NY, USA.

Friedmann, T., ed. 1999. *The Development of Human Gene Therapy*. Cold Spring Harbor Laboratory Press, Cold Spring Harbor, NY, USA.

Gall, J. G., and McIntosh, J. R., eds. 2001. *Landmark Papers in Cell Biology*. American Society of Cell Biology Press, Bethesda, MD, USA.

Gillham, N. W. 2001. *Sir Francis Galton: From African Exploration to the Birth of Eugenics*. Oxford Univ. Press, New York.

Goldschmidt, R. B. 1956. *Portraits from Memory: Recollections of a Zoologist*. Univ. of Washington Press, Seattle, WA, USA.

Hall, M. N., and Linder, P., eds. 1993. *The Early Days of Yeast Genetics*. Cold Spring Harbor Laboratory Press, Cold Spring Harbor, NY, USA.

Harris, H. 1997. *The Cells of the Body: A History of Somatic Cell Genetics*. Cold Spring Harbor Laboratory Press, Cold Spring Harbor, NY, USA.

Harwood, J. 1993. *Styles of Scientific Thought — The German Genetics Community 1900–1933*. Univ. Chicago Press, Chicago, IL, USA.

Hayes, W. 1964. *The Genetics of Bacteria and Their Viruses*. Wiley, New York.

Holmes, F. L. 2002. *Meselson, Stahl, and the Replication of DNA: A History of the "Most Beautiful Experiment in Biology."* Yale Univ. Press, New Haven, CT, USA.

Jacob, F. 1995. *The Statue Within: An Autobiography*. Cold Spring Harbor Laboratory Press, Cold Spring Harbor, NY, USA.

Jacob, F. 1997. *La Souris, la Mouche et l'Homme*. Odile Jacob, Paris, France.

Johannsen, W. 1909. *Elemente der Exakten Erblichkeitslehre*. Fisher, Jena, Germany.

Judson, F. 1996. *The Eighth Day of Creation: Makers of the Revolution in Biology*. Cold Spring Harbor Lab. Press, Cold Spring Harbor, NY, USA.

Kay, L. E. 2000. *Who Wrote the Book of Life? A History of the Genetic Code*. Stanford Univ. Press, Stanford, CA, USA.

Lerner, I. M. 1950. *Genetics in the U.S.S.R: An Obituary*. Univ. British Columbia Press, Vancouver, Canada.

Lewis, R. A. 2001. *Discovery — Windows on the Life Sciences*. Blackwell Science, Malden, MA, USA.

Maas, W. 2001. *Gene Action: A Historical Account*. Oxford Univ. Press, New York.

Maddox, B. 2002. *Rosalind Franklin. Dark Lady of DNA*. HarperCollins, New York.

Magner, L. N. 1993. *A History of the Life Sciences*. Dekker, New York.

McCarty, M. 1985. *The Transforming Principle: Discovering That Genes Are Made of DNA*. Norton, New York.

McElheny, V. K. 2003. *Watson and DNA: Making a Scientific Revolution*. Perseus, Cambridge, MA, USA.

Medvedev, Zs. 1969. *The Fall and Rise of T.D. Lysenko*. Columbia Univ. Press, New York.

Morange, M. 1998. *A History of Molecular Biology*. Harvard Univ. Press, Cambridge, MA, USA.

Morgan, T. H. 1919. *The Physical Basis of Heredity*. Lippincott, Philadelphia, PA, USA.

Morgan, T. H., Sturtevant, A. H., Muller, H. J., and Bridges, C. B. 1915. *The Mechanism of Mendelian Heredity*. Holt, New York.

Müller-Hill, B. 1996. *The lac Operon: A Short History of a Genetic Paradigm*. Walter de Gruyter, New York.

Müller-Hill, B. 1998. *Murderous Science: Elimination by Scientific Selection of Jews, Gypsies and Others in Germany 1933–1945*. Cold Spring Harbor Laboratory Press, Plainview, NY, USA.

Neel, J. V. 1994. *Physician to the Gene Pool: Genetic Lessons and Other Stories*. Wiley, New York.

Olby, R. C. 1966. *Origins of Mendelism*. Schocken Books, New York.

Olby, R. 1974. *The Path to the Double Helix*. University of Washington Press, Seattle, WA, USA.

Orel, V. 1996. *Gregor Mendel: The First Geneticist*. Oxford Univ. Press, New York.

Perutz, M. F. 1998. *I Wish I'd Made You Angry Earlier: Essays on Science, Scientists, and Humanity*. Cold Spring Harbor Laboratory Press, Cold Spring Harbor, NY, USA.

Portugal, F. H., and Cohen, J. S. 1977. *A Century of DNA*. MIT Press, Cambridge, MA, USA.

Potts, D. M., and Potts, W. T. W. 1995. *Queen Victoria's Gene: Hemophilia and the Royal Family*. Alan Sutton, Dover, NH, USA.

Provine, W. B. 2001. *The Origins of Theoretical Population Genetics*. Univ. Chicago Press, Chicago, IL, USA.

Rabinow, P. 1996. *Making PCR: A Story of Biotechnology*. Univ. Chicago Press, Chicago, IL, USA.

Ridley, M. 1999. *Genome: The Autobiography of a Species in 23 Chapters*. HarperCollins, New York.

Roberts, H. F. 1965. *Plant Hybridization Before Mendel*. Hafner, New York.

Rushton, A. R. 1994. *Genetics and Medicine in the United States, 1800–1922*. Johns Hopkins Univ. Press, Baltimore, MD, USA.

Sarkar, S., ed. 1992. *The Founders of Evolutionary Genetics: A Centenary Reappraisal*. Kluwer Academic, Dordrecht, The Netherlands/Boston, MA.

Schrödinger, E. 1944. *What Is Life?* Cambridge Univ. Press, Cambridge, UK.

Shnoll, S. E. 1997. *Heroes, Martyrs and Villains in Russian Life Sciences*. Springer, New York.

Sinsheimer, R. L. 1994. *The Strands of Life: The Science of DNA and the Art of Education*. Univ. California Press, Berkeley, CA, USA.

Sirks, M. J., and Zirkle, C. 1964. *The Evolution of Biology*. Ronald Press, New York.

Stahl, F. W., ed. 2000. *We Can Sleep Later: Alfred D. Hershey and the Origins of Molecular Biology*. Cold Spring Harbor Laboratory Press, Cold Spring Harbor, NY, USA.

Stent, G. S., ed. 1981. *The Double Helix: Text, Commentary, Reviews, Original Papers*. Norton, New York.

Stubbe, H. 1972. *History of Genetics from Prehistoric Times to the Rediscovery of Mendel's Laws*. MIT Press, Cambridge, MA, USA.

Sturtevant, A. H. 1965. *A History of Genetics*. Harper and Row, New York.

Sulston, J., and Ferry, G. 2002. *The Common Thread: A Story of Science, Politics, Ethics and the Human Genome*. Joseph Henry Press, Washington, DC, USA.

Summers, W. C. 1999. *Félix d'Hérelle and the Origins of Molecular Biology*. Yale Univ. Press, New Haven, CT, USA.

Watson, J. D. 1997. *The Double Helix*. Weidenfeld & Nicholson, New York.

Watson, J. D. 2000. *A Passion for DNA: Genes, Genomes and Society*. Oxford Univ. Press, New York.

Watson, J. D. 2001. *Genes, Girls and Gamow*. Oxford Univ. Press, New York.

Watson, J. D., and Tooze, J. 1981. *The DNA Story: A Documentary History of Gene Cloning*. Freeman, San Francisco, CA, USA.

Weiner, J. 1999. *Time, Love, Memory: A Great Biologist [Benzer] and His Quest for the Origin of Behavior*. Random House, New York.

Weir, R. F., Lawrence, S. C., and Fales, E., eds. 1994. *Genes and Human Self-Knowledge: Historical and Philosophical Reflections on Modern Genetics*. Univ. Iowa Press, Iowa City, IA, USA.

Wilmut, I., Campbell, K., and Tudge, C. 2000. *The Second Creation: Dolly and the Age of Biological Control*. Farrar, Straus & Giroux, New York.

Wilson, E. B. 1925. *The Cell in Development and Heredity*. Macmillan, New York.

Witkowski, J., ed. 2000. *Illuminating Life*. Cold Spring Harbor Laboratory Press, Cold Spring Harbor, NY, USA.

Wood, R., and Orel, V. 2001. *Genetic Prehistory in Selective Breeding: A Prelude to Mendel*. Oxford Univ. Press, New York.

Zirkle, C. 1935. *The Beginnings of Plant Hybridization*. Univ. Pennsylvania Press, Philadelphia, PA, USA.

IMMUNOLOGY (*See also* Biology, Molecular Biology)

Annual Review of Immunology. Palo Alto, CA, USA.

Austyn, J. M., and Wood, K. J. 1993. *Principles of Cellular and Molecular Immunology*. Oxford Univ. Press, New York.

Beck, G., et al., eds. 2001. *Phylogenetic Perspectives on the Vertebrate Immune System*. Kluwer, Dordrecht, The Netherlands.

Bell, J. I., et al., eds. 1995. *T Cell Receptors*. Oxford Univ. Press, New York.

Benjamini, E., Sunshine, G., and Leskowitz, S. 1996. *Immunology: A Short Course*. Wiley, New York.

Birch, J. R., and Lennox, E. S., eds. 1995. *Monoclonal Antibodies*. Wiley-Liss, New York.

Bona, A., and Bonilla, F. A. 1996. *Textbook of Immunology*. Gordon and Breach & Harwood Academic Publishers, Toronto, Canada.

Bot, A., and Bona, A. A. 2000. *Genetic Immunization*. Kluwer, Dordrecht, The Netherlands.

Breitling, F., and Dübel, S. 1999. *Recombinant Antibodies*. Wiley, New York.

Cattaneo, A., and Biocca, S., eds. 1997. *Intracellular Antibodies*. Springer, New York.

Cochet, O., Teuillaud, J.-L., and Sautes, C., eds. 1998. *Immunological Techniques Made Easy*. Wiley, New York.

Coffman, R. L., and Romagnani, S., eds. 1999. *Redirection of Th1 and Th2 Responses*. Springer, New York.

Delves, P. J., and Roitt, I., eds. 1998. *The Encyclopedia of Immunology*. Academic Press, San Diego, CA, USA.

Durum, S. K., and Muegge, K., eds. 1998. *Cytokine Knockouts*. Humana Press, Totowa, NJ, USA.

Fernandez, N., and Butcher, G. 1997. *MHC1: A Practical Approach*. Oxford Univ. Press, New York.

Flint, S. J., et al. 2000. *Principles of Virology: Molecular Biology, Pathogenesis, and Control*. American Society of Microbiology Press, Washington, DC.

Frank, S. A. 2002. *Immunology and Evolution of Infectious Disease*. Princeton Univ. Press, Princeton, NJ, USA.

Gregory, C. D., ed. 1995. *Apoptosis and the Immune Response*. Wiley-Liss, New York.

Harlow, E., and Lane, D. 1999. *Using Antibodies: A Laboratory Manual*. Cold Spring Harbor Laboratory Press, Plainview, NY, USA.

Herbert, W. J., Wilkinson, P. C., and Sott, D. I., eds. 1995. *The Dictionary of Immunology*. Academic Press, San Diego, CA, USA.

Honjo, T., and Alt, F. W., eds. 1995. *Immunoglobulin Genes*. Academic Press, San Diego, CA, USA.

Janeway, C. A., Jr., and Travers, P. 1997. *Immunobiology: The Immune System in Health and Disease*. Garland, New York.

Kaufmann, S. H. E., Sher, A., and Ahmed, R. 2002. *Immunology of Infectious Diseases*. ASM Press, Washington, DC, USA.

King, D. J. 1998. *Application and Engineering of Monoclonal Antibodies*. Taylor & Francis, Philadelphia, PA, USA.

Kontermann, R., and Dübel, S., eds. 2001. *Antibody Engineering*. Springer, New York.

Krakauer, T., ed. 2002. *Superantigen Protocols*. Humana, Totowa, NJ, USA.

Kuby, J. 1997. *Immunology*. Freeman, New York.

Leffell, M. S., Donnenberg, A. D., and Rose, N. R. 1997. *Handbook of Human Immunology*. CRC Press, Boca Raton, FL, USA.

Lefranc, M.-P., and Lefranc, G. 2001. *The T Cell Receptor FactsBook*. Academic Press, San Diego, CA, USA.

Lefranc, M.-P., and Lefranc, G. 2001. *The Immunoglobin FactsBook*. Academic Press, San Diego, CA, USA.

Law, S. K. A., and Reid, K. I. M. 1995. *Complement*. IRL/Oxford Univ. Press, New York.

Liu, M. A., Hilleman, R. A., and Kurth, R., eds. 1995. *DNA Vaccines*, New York Academy of Science. Annals. No. 772.

Lowrie, D. B., and Whalen, R., eds. 1999. *DNA Vaccines*. Humana Press, Totowa, NJ, USA.

Male D., et al. 1996. *Advanced Immunology*. Mosby, St. Louis, MO, USA.

March, S. G. E., et al. 2000. *The HLA Facts Book*. Academic Press, San Diego, CA, USA.

McCafferty, J., Hoogenboom, H. R., and Chiswell, D. J. 1996. *Antibody Engineering: A Practical Approach*. IRL/Oxford Univ Press, New York.

Morgan, B. P., and Harris, C. L. 1999. *Complement Regulatory Proteins*. Academic Press, San Diego, CA, USA.

Morley, B. J., and Walport, M. J., eds. 2000. *The Complement Facts Book*. Academic Press, San Diego, CA, USA.

Nezlin, R. 1998. *The Immunoglobulins: Structure and Function*. Academic Press, San Diego, CA, USA.

Ochs, H. D., et al., eds. 1999. *Primary Immunodeficiency Diseases: A Molecular and Genetic Approach*. Oxford Univ. Press, New York.

Parmiani, G., and Lotze, M. T., eds. 2002. *Tumor Immunology: Molecularly Defined Antigens and Applications*. Taylor & Francis, London, UK.

Paul, W. E. 1993. *Fundamental Immunology*. Raven, New York.

Pillai, S. 2000. *Lymphocyte Development: Cell Selection Events and Signals During Immune Ontogeny*. Birkhäuser, Boston, MA, USA.

Powis, S. H., and Vaughan, R. W., eds. 2002. *MHC Protocols*. Humana, Totowa, NJ, USA.

Raz, E., ed. 1998. *Gene Vaccination: Theory and Practice*. Springer, New York.

Reid, M. E., and Lomas-Francis, C. 1997. *The Blood Group Antigen*. Academic Press, San Diego, CA, USA.

Ritter, M. A., and Ladyman, H. M., eds. 1995. *Monoclonal Antibodies: Production, Engineering and Clinical Application*. Cambridge Univ. Press, New York.

Roitt, I. 1991. *Essential Immunology*. Blackwell Science, Oxford, UK.

Rose, N. R., et al., eds. 2002. *Manual of Clinical Laboratory Immunology*, AMS Press, Washington, DC, USA.

Rowland-Jones, S. L., and McMichael, A. J., eds. 2000. *Lymphocytes: A Practical Approach*. Oxford Univ. Press, New York.

Rother, K. Till, G. O., and Hänsch, G. M., eds. 1998. *The Complement System*. Springer-Verlag, New York.

Sell, S. 2001. *Immunology, Immunopathology, and Immunity*. ASM Press, Washington, DC.

Shepherd, P. S., and Dean, C., eds. 2000. *Monoclonal Antibodies*. Oxford Univ. Press, New York.

Solheim, J. C., ed. 2000. *Antigen Processing and Presentation Protocols*. Humana Press, Totowa, NJ, USA.

Stites, D. P., Terr, A. I., and Parslow, T. G., eds. 1994. *Basic & Clinical Immunology*. Appleton & Lange, Norwalk, CT, USA.

Theofilopoulos, A. N., and Bona, C. D., eds. 2002. *The Molecular Pathology of Autoimmune Diseases*. Taylor & Francis, London, UK

Thèze, J., ed. 1999. *The Cytokine Network and Immune Functions*. Oxford Univ. Press, New York.

Tilney, N. L., Strom, T. B., and Paul L. C., eds. 1996. *Transplantation Biology: Cellular and Molecular Aspects*. Lippincott-Raven, Philadelphia, PA, USA.

Weir, D., ed. 1996. *The Handbook of Experimental Immunology*. Blackwell Science, Oxford, UK.

Westwood, O., and Hay, F., eds. 2001. *Epitope Mapping*. Oxford Univ. Press, New York.

Winkler, J. D., ed. 1999. *Apoptosis and Inflammation*. Birkhäuser, Basel, Switzerland.

Zanetti, M., and Capra, J. D., eds. 1995. *The Antibodies*. Harwood Academic Publishers, Langhorne, PA, USA.

LABORATORY TECHNOLOGY (*See also* Biotechnology, Cytology and Cytogenetics, Molecular Biology)

Alfa, C., et al., eds. 1993. *Experiments with Fission Yeast: A Laboratory Course Manual*. Cold Spring Harbor Laboratory Press, Cold Spring Harbor. NY, USA.

Amdur, M. O., Doull, J., and Klaassen, C. D., eds. 1991. *Casaretts and Doull's Toxicology: The Basic Sciences of Poisons*. Pergamon Press, Elmsford, NY, USA.

Andreeff, M., and Pinkel, D., eds. 2000. *Introduction to Fluorescence in Situ Hybridization: Principles and Clinical Applications*. Wiley, New York.

Ansorge, W., Voss, H., and Zimmermann, J., eds. 1996. *DNA Sequencing Strategies: Automated and Advanced Approaches*. Wiley, New York.

Aquino de Muro, M., and Rapley, R., eds. 2001. *Gene Probes: Principles and Protocols*. Humana Press, Totowa, NJ, USA.

Armour, M.-A. 1996. *Hazardous Laboratory Chemicals Disposal Guide*. CRC Press, Boca Raton, FL, USA.

Ausubel, F. M., et al., eds. 1987. *Current Protocols in Molecular Biology*. Wiley, New York.

Bagasra, O., and Hansen, J. 1997. *In situ PCR Techniques*. Wiley-Liss, New York.

Baldi, P., and Hatfield, G. W. 2002. *DNA Microarrays and Gene Expression*. Cambridge Univ. Press, New York.

Baldwin, S. A., ed. 2000. *Membrane Transport. A Practical Approach*. Oxford Univ. Press, New York.

Barbas, C. F., et al. 2001. *Phage Display: Laboratory Manual*. Cold Spring Harbor Laboratory Press, Cold Spring Harbor, NY, USA.

Barker, K. 2002. *At the Helm: A Laboratory Navigator*. Cold Spring Harbor Laboratory Press, Cold Spring Harbor, NY, USA.

Baxevanis, A. D. et al., eds. 2003. *Current Protocols in Bioinformatics*. Wiley, New York.

Bernstam, V. A. 1992. *Handbook of Gene Level Diagnostics in Clinical Practice*. CRC Press, Boca Raton, FL, USA.

Bhojwani, S. S., and Razdan, M. K. 1996. *Plant Tissue Culture: Theory and Practice*. Elsevier Science, New York.

Birren, B., and Lai, E. 1993. *Pulsed Field Gel Electrophoresis: A Practical Guide*. Academic Press, San Diego, CA, USA.

Birren, B., and Lai, E., eds. 1996. *Nonmammalian Genomic Analysis: A Practical Guide*. Academic Press, San Diego, CA, USA.

Birren, B., et al., eds. 1997. *Genome Analysis: A Laboratory Manual*. Cold Spring Harbor Laboratory Press, Cold Spring Harbor, NY, USA.

Birren, B., et al. 1998. *Detecting Genes: A Laboratory Manual*. Cold Spring Harbor Laboratory Press, Cold Spring Harbor, NY, USA.

Birren, B., et al., eds. 1999. *Cloning Systems: A Laboratory Manual*. Cold Spring Harbor Laboratory Press, Cold Spring Harbor, NY, USA.

Birren, B., et al. 1999. *Mapping Genomes: A Laboratory Manual*. Cold Spring Harbor Laboratory Press, Cold Spring Harbor, NY, USA.

Bishop, M., and Rawlings, C. 1997. *DNA and Protein Sequence Analysis: A Practical Approach*. Oxford Univ. Press, New York.

Bjornsti, M.-A., and Osheroff, N., eds. 1999. *DNA Topoisomerase Protocols*. Humana Press, Totowa, NJ, USA.

Bonifacino, J. S., et al., eds. 1999. *Current Protocols in Cell Biology*. Wiley, New York.

Bowtell, D., and Sambrook, J., eds. 2002. *DNA Microarrays: A Molecular Cloning Manual*. Cold Spring Harbor Laboratory Press, Woodbury, NY, USA.

Bozzola, J. J., and Russel, L. D. 1992. *Electron Microscopy: Principles and Techniques for Biologists*. Jones & Bartlett, Boston, MA, USA.

Bretherick, L. 1990. *Bretherick's Handbook of Reactive Chemical Hazards*. Butterworth, London, UK.

Brown, A. J. P., and Tuite, M. F., eds. 1998. *Yeast Gene Analysis*. Academic Press. San Diego, CA, USA.

Brown, T. 1998. *Molecular Biology Labfax: Recombinant DNA — Gene Analysis*. Academic Press, San Diego, CA, USA.

Brown, T., ed. 2001. *Essential Molecular Biology*. Oxford Univ. Press, New York.

Burke, D., et al. 2000. *Methods in Yeast Genetics: A CSHL Course Manual*. Cold Spring Harbor Laboratory Press, Cold Spring Harbor, NY, USA.

Butler, M. 1996. *Animal Cell Culture and Technology*. IRL/Oxford Univ. Press, New York.

Buxton, R. B. 2002. *An Introduction to Functional Resonance Imaging: Principles and Techniques*. Cambridge Univ. Press, New York.

Carraway, K. I., and Carothers Carraway, C. A., eds. 2000. *Cytoskeleton: Signalling and Cell Regulation. A Practical Approach*. Oxford Univ. Press, New York.

Celis, J., ed. 1997. *Cell Biology: A Laboratory Handbook*. Academic Press, San Diego, CA, USA.

Chalfie, M., and Kain, S., eds. 1998. *Green Fluorescent Protein: Properties, Applications and Protocols*. Wiley, New York.

Chen, B.-Y., and Janes, H. W. 2002. *PCR Cloning Protocols*. Humana Press, Totowa, NJ, USA.

Chrispeels, M. J., and Sadava, D. E. 1994. *Plants, Genes and Agriculture*. Jones and Bartlett, Boston, MA, USA.

Clark, M. 1997. *Plant Molecular Biology: A Laboratory Manual*. Springer, New York.

Clegg, R. A., ed. 1998. *Protein Targeting Protocols*. Humana Press, Totowa, NJ, USA.

Clynes, M., ed. 1998. *Animal Cell Culture Techniques*. Springer, New York.

Coligan, J. E., et al., eds. 1991. *Current Protocols in Immunology*. Wiley, New York.

Cotterill, S., ed. 1999. *Eukaryotic DNA Replication: A Practical Approach*. Oxford Univ. Press, New York.

Craig, A. G., and Hoheisel, J. D. 1999. *Automation: Genome and Functional Analyses*. Academic Press, San Diego, CA, USA.

Crowther, J. R. 1995. *ELISA*. Humana Press, Totowa, NJ, USA.

Dangler, C. A., ed. 1996. *Nucleic Acid Analysis: Principles and Bioapplications*. Wiley, New York.

Darby, I. A., ed. 2000. *In Situ Hybridization Protocols*. Humana Press, Totowa, NJ, USA.

Dass, C. 2000. *Principles and Practice of Biological Mass Spectrometry*. Wiley, New York.

Davis, L., ed. 2002. *Basic Cell Culture*. Oxford Univ. Press, New York.

Davis, L. G., Kuehl, W. M., and Battey, J. F. 1994. *Basic Methods in Molecular Biology*. Appleton and Lange, Norwalk, CT, USA.

deBoer, A. G., and Sutanto, W., eds. 1997. *Drug Transport Across the Blood-Brain Barrier: In Vitro and In Vivo Techniques*. Gordon & Breach, Amsterdam, The Netherlands.

Didenko, V. V., ed. 2002. *In Situ Detection of DNA Damage*. Humana, Totowa, NJ, USA

Dieffenbach, C. W., and Dveksler, G. S., eds. 1995. *PCR Primer: A Laboratory Manual*. Cold Spring Harbor Laboratory Press, Cold Spring Harbor, NY, USA.

Dodds, J. H., and Roberts, L. W. 1995. *Experiments in Plant Tissue Culture*. Cambridge Univ. Press, New York.

Doyle, A., and Griffith, J. B. 1998. *Cell and Tissue Culture: Laboratory Procedures in Biotechnology*. Wiley, New York.

Dracapoliu, N. C., et al., eds. 1994. *Current Protocols in Human Genetics*. Wiley, New York.

Draper, J., et al., eds. 1988. *Plant Genetic Transformation and Gene Expression: A Laboratory Manual*. Blackwell, Oxford, UK.

Echalier, G. 1997. *Drosophila Cells in Culture*. Academic Press, San Diego, CA, USA.

Edwards, R., ed. 2000. *Immunodiagnostics. A Practical Approach*. Oxford Univ. Press, New York.

Edwards, R., ed. 2000. *Immunodiagnostics. A Practical Approach*. Oxford Univ. Press, New York.

Edwards, S., and Collin, H. A. 1997. *Plant Cell Culture*. Springer, New York.

Evans, I. H., ed. 1966. *Yeast Protocols: Methods in Cell and Molecular Biology*. Humana Press, Totowa, NJ, USA.

Farrell, R. E., Jr. 1998. *RNA Methodologies: A Laboratory Guide for Isolation and Characterization*. Academic Press, San Diego, CA, USA.

Ferré, F. 1997. *Gene Quantification*. Springer, New York.

Fleming, D. O., and Hunt, D. L., eds. 2000. *Biological Safety: Principles and Practices*. AMS Press, Washington, DC.

Freshney, R. I., ed. 1992. *Animal Cell Culture*. IRL Press, New York.

Furr, A. K., ed. 1989. *CRC Handbook of Laboratory Safety*. CRC Press, Boca Ratoon, FL, USA.

Garman, A. 1997. *Non-Radioactive Labelling*. Academic Press, San Diego, CA, USA.

Glasel, J. A., and Deutscher, M. P., eds. 1995. *Introduction to Biophysical Methods for Protein and Nucleic Acid Research*. Academic Press, San Diego, CA, USA.

Gillespie, S. H., ed. 2000. *Antibiotic Resistance Methods and Protocols*. Humana Press, Totowa, NJ, USA.

Givan, A. L. 2001. *Flow Cytometry: First Principles*. Wiley, New York.

Glover, D. M., and Hames, B. D., eds. 1996. *DNA Cloning: A Practical Approach*. IRL/Oxford Univ. Press, New York.

Gold, L. S., and Zeigler, E., eds. 1997. *Handbook of Carcinogenic Potency and Genotoxicity Databases*. CRC Press, Boca Raton, FL, USA.

Goldstein, L. S. B., and Fryberg, E. A., eds. 1994. *Drosophila melanogaster: Practical Uses in Cell and Molecular Biology*. Academic Press, San Diego, CA, USA.

Golemis, E. A., ed. 2002. *Protein-Protein Interactions: A Molecular Cloning Manual*. Cold Spring Harbor Laboratory Press, Woodbury, NY, USA.

Gosden, J. R., ed. 1996. *Prins and In Situ PCR Protocols*. Humana Press, Totowa, NJ, USA.

Gosling, J. P., ed. 2000. *Immunoassays. A Practical Approach*. Oxford Univ. Press, New York.

Gottschall, W. C. 2001. *Laboratory Health and Safety Dictionary*. Wiley, New York.

Graham, C. A., and Alison, J. M. 2001. *DNA Sequencing Protocols*. Humana Press, Totowa, NJ, USA.

Grout, B. 1995. *Genetic Preservation of Plant Cells in Vitro*. Springer, New York.

Hames, B. D., and Higgins, S. J., eds. 1996. *Gene Probes 2*. Oxford Univ. Press, New York.

Hames, B. D., ed. 1998. *Gel Electrophoresis of Proteins: A Practical Approach*. Oxford Univ. Press, New York.

Hardie, D. G., ed. 2000. *Protein Phosphorylation. A Practical Approach*. Oxford Univ. Press, New York.

Hardin, C., et al. 2001. *Cloning, Gene Expression and Protein Purification. Experimental Procedures and Process Rationale*. Oxford Univ. Press, New York.

Harrison, M. A., and Rae, I. F. 1997. *General Techniques of Cell Culture*. Cambridge Univ. Press, New York.

Harwood, A. J., ed. 1994. *Protocols for Gene Analysis*. Humana Press, Totowa, NJ, USA.

Haynes, L. W., ed. 1999. *RNA — Protein Interactive Protocols*. Humana Press, Totowa, NJ, USA.

Henderson, D. S. 1999. *DNA Repair Protocols: Eukaryotic Systems*. Humana Press, Totowa, NJ, USA.

Higgins, D., and Taylor, W. 2000. *Bioinformatics: Sequence, Structure and Databanks*. Oxford Univ. Press, New York.

Higgins, S. J., and Hames, B. D. 1999. *Protein Expression. A Practical Approach*. Oxford Univ. Press, New York.

Hofker, M. H. and Deursen, van J., eds. 2003. *Transgenic Mouse. Methods and Protocols*. Humana Press, Totowa, NJ, USA.

Hogan B., et al. 1994. *Manipulating the Mouse Embryo: A Laboratory Manual*. Cold Spring Harbor Laboratory Press, Cold Spring Harbor, NY, USA.

Hope, I. A. 2000. *C. elegans — A Practical Approach*. Oxford Univ. Press, New York.

Hunt, S., and Livesey, F., eds. 2000. *Functional Genomics: A Practical Approach*. Oxford Univ. Press, New York.

Innis, M., Gelfand, D., and Sninsky, J., eds. 1999. *PCR Applications: Protocols for Functional Genomics*. Academic Press, Orlando, FL, USA.

Isaac, P. G., ed. 1994. *Protocols for Nucleic Acid Analysis by Nonradioactive Probes*. Humana Press, Totowa, NJ, USA.

Jackson, I. J., and Abbott, C. M., eds. 2000. *Mouse Genetics and Transgenics: A Practical Approach*. Oxford Univ. Press, New York.

Jenkins, N., ed. 1999. *Animal Cell Biotechnology: Methods and Protocols*. Humana Press, Totowa, NJ, USA.

Jezzard, P., et al., eds. 2001. *Functional Magnetic Resonance Imaging: An Introduction to Methods*. Oxford Univ. Press, New York.

Johnston, J. R. 1994. *Molecular Genetics of Yeast: A Practical Approach*. IRL Press, New York.

Jones, G. E., ed. 1996. *Human Cell Culture Protocols*. Humana Press, Totowa, NJ, USA.

Jones, P., ed. 1997. *Cloning Applications*. Wiley, New York.

Jordan, B. R., ed. 2001. *DNA Microarrays: Gene Expression Applications*. Springer, New York.

Joyner, A. L., ed., 2000. *Gene Targeting: A Practical Approach*. Oxford Univ. Press, New York.

Kannicht, C. 2002. *Posttranslational Modification of Proteins: Tools for Functional Proteomics*. Humana Press, Totowa, NJ, USA.

Kendall, D. A., and Hill, S. J., eds. 1995. *Signal Transduction Protocols*. Humana Press, Totowa, NJ., USA.

Kinter, M., and Sherman, N. E. 2000. *Protein Sequencing and Identification Using Tandem Mass Spectrometry*. Wiley, New York.

Kmiec, E. B. ed. 2000. *Gene Targeting Protocols*. Humana Press, Totowa, NJ, USA.

Kneale, G. G., ed. 1994. *DNA — Protein Interactions: Principles and Protocols*. Humana Press, Totowa, NJ, USA.

Knudsen, S. 2002. *A Biologist's Guide to Analysis of DNA Microarray Data*. Wiley, New York.

Kobayashi, T., Kitagawa, Y., and Okumura, S., eds. 1994. *Animal Cell Technology: Basic and Applied*. Kluwer, Norwell, MA, USA.

Kohler, J. M., and Mejewaia, T. 1999. *Microsystem Technology: A Powerful Tool for Biomolecular Studies*. Birkhäuser, Basel, Switzerland.

Körholz, D. and Kiess, W., eds. 2002. Cytokines and Colony Stimulating Factors. Methods and Protocols. Humana, Totowa, NJ, USA.

Krieg, P. A., ed. 1996. *A Laboratory Guide to RNA: Isolation, Analysis, and Synthesis*. Wiley-Liss, New York.

Kwok, P.-Y., ed. 2002 Single Nucleotide Polymorphisms. Methods and Protocols. Humana, Totowa, NJ, USA.

Landegren, U. 1996. *Laboratory Protocols for Mutation Detection*. Oxford Univ. Press, New York.

Larrick, J. W., and Siebert, P. D., eds. 1995. *Reverse Transcriptase PCR*. Ellis Horwood, London, UK.

Lasick, D. D. 1997. *Liposomes in Gene Delivery*. CRC Press, Boca Raton, FL, USA.

Laudet, V., and Gronemeyer, H. 2001. *The Nuclear Receptor FactsBook*. Academic Press, San Diego, CA, USA.

Levy, E. R., and Herrington, C. S. 1995. *Non-Isotopic Methods in Molecular Biology — A Practical Approach*. Oxford Univ. Press, New York.

Lincoln, P. J., and Thomson, J., eds. 1998. *Forensic DNA Profiling Protocols*. Humana Press, Totowa, NJ, USA.

Liu, Q., and Weiner, M. P., eds. 2001. *Cloning and Expression Vectors for Gene Function Analysis*. Eaton, Westborough, MA, USA.

Lo, Y. M. D. 1998. *Clinical Applications of PCR*. Humana Press, Totowa, NJ, USA.

Lowestein, P. R., and Enquist, L. W., eds. 1996. *Protocols for Gene Transfer in Neuroscience: Towards Gene Therapy of Neurological Disorders*. Wiley, New York.

Mak, T. W., ed. 1998. *The Gene Knockout FactsBook*. Academic Press, San Diego, CA, USA.

Maliga, P., et al., eds. 1995. *Methods in Plant Molecular Biology: A Laboratory Course Manual*. Cold Spring Harbor Laboratory Press, Cold Spring Harbor, NY, USA.

Malik, V. S., and Lillehoj, E. P. eds. 1994. *Antibody Techniques*. Academic Press, San Diego, CA, USA.

Markie, D., ed. 1995. *YAC Protocols*. Humana Press, Totowa, NJ, USA.

Mason, W. T. 1999. *Fluorescent and Luminescent Probes for Biological Activity*. Academic Press, San Diego, CA, USA.

Masters, J. R. W., ed. 2000. *Animal Cell Cultures*. Oxford Univ. Press, New York.

Matzuk, M. M., et al., eds. 2001. *Transgenics in Endocrinology*. Humana Press, Totowa, NJ, USA.

Maunsbach, A. B., and Afzelius, B. A. 1998. *Biomedical Electronmicroscopy: Illustrated Methods and Interpretations*. Academic Press, San Diego, CA.

McCarthy, D. A., and Macey, M. G., eds. 2001. *Cytometric Analysis of Cell Phenotype and Function*. Cambridge Univ. Press, New York.

McPherson, A. 1999. *Crystallization of Biological Macromolecules*. Cold Spring Harbor Laboratory Press, Cold Spring Harbor, NY, USA.

Meier, T., and Fahrenholz, F. 1997. *A Laboratory Guide to Biotin-Labeling in Biomolecule Analysis*. Springer, New York.

Methods in Cell Biology. Elsevier Science, St. Louis, MO, USA.

Methods in Enzymology (continuous series including wide areas of biology). Academic Press, San Diego, CA, USA.

Methods in Molecular Biology (continuous volumes). Humana Press, Totowa, NJ, USA.

Methods in Neurosciences (continuous volumes). Academic Press, San Diego, CA, USA.

Michels, C. A. 2002. *Genetic Techniques for Biological Research*. Wiley, New York.

Micklos, D., and Freyer, G. 2002. *DNA Science: A First Course in Recombinant DNA*. Cold Spring Harbor La. Press, Woodbury, NY, USA.

Milligan, G., ed. 1999. *Signal Transduction: A Practical Approach*. Oxford Univ. Press, New York.

Mills, K. I., and Ramsahoye, B. H. 2002. *DNA Methylation Protocols*. Humana Press, Totowa, NJ, USA.

Monaco, A. P., ed. 1995. *Pulsed Field Gel Electrophoresis*. Oxford Univ. Press, New York.

Montesano, R., Bartsch, H., Boyland, E., Della Porta, G., Fishbein, L., Griesemer, R. A., Swan, A. B., and Tomatis L., eds. 1982. Handling chemical carcinogens in the laboratory — Problems of safety. *Biol. Zbl.* 101:653–70.

Morgan, J. R., ed. 2001. *Gene Therapy Protocols*. Humana Press, Totowa, NJ, USA.

Murray, J. C., ed. 2001. *Angiogenesis Protocols*. Humana Press, Totowa, NJ, USA.

Mülhardt, C. 1999. *Der Experimentator: Molekularbiologie*. G. Fischer Verlag, Stuttgart, Germany.

Nagy, A., et al. 2002. *Manipulating the Mouse Embryo. A Laboratory Manual*. Cold Spring Harbor Laboratory Press, Woodbury, NY, USA.

Nielsen, P. E., and Egholm, M., eds. 1999. *Peptide Nucleic Acids: Protocols and Applications*. Horizon, Wymondham, UK.

New, R. R. C., ed. 1990. *Liposomes: A Practical Approach*. Oxford Univ. Press, New York.

Nuovo, G. J. 1994. *PCR in Situ Hybridization*. Raven, New York.

O'Brien, P. M., and Aitken, R., eds. 2002. *Antibody Phage Display*. Humana Press Totowa, NJ, USA.

O'Neill, L. A., and Bowie, A., eds. 2001. *Interleukin Protocols*. Humana Press, Totowa, NJ, USA.

Ormerod, M. G., ed. 2000. *Flow Cytometry: A Practical Approach*. Oxford Univ. Press, Oxford, UK.

Picard, D., ed. 1999. *Nuclear Receptors: A Practical Approach*. Oxford Univ. Press, Oxford, UK.

Poirier, J., ed. 1997. *Apoptosis Techniques and Protocols*. Humana Press, Totowa, NJ, USA.

Pollard, J. W., and Walker, J. M., eds. 1997. *Basic Cell Culture Protocols*. Humana Press, Totowa, NJ, USA.

Rampal, J. B., ed. 2001. *DNA Arrays: Methods and Protocols*. Humana Press, Totowa, NJ, USA.

Rapley, R., ed. 1996. *PCR Sequencing Protocols*. Humana Press, Totowa, NJ, USA.

Ravid, K., and Freshney, R. I., eds. 1998. *DNA Transfer to Cultured Cells*. Wiley-Liss, New York.

Ream, W., and Field, K. G. 1999. *Molecular Biology Techniques: An Intensive Laboratory Course*. Academic Press, San Diego, CA, USA.

Reinert, J., and Bajaj, Y. P. S. 1977. *Plant Cell, Tissue, and Organ Culture*. Springer, New York.

Rickwood, D., and Hames, B. D. 1990. *Gel Electrophoresis of Nucleic Acids: A Practical Approach*. IRL Press, New York.

Ridley, D. D. 1996. *Online Searching: A Scientist's Perspective—A Guide for the Chemical and Life Sciences*. Wiley, New York.

Roberts, D. B., ed. 1998. *Drosophila—A Practical Approach*. Oxford Univ. Press, New York.

Roe, S., ed. 2001. *Protein Purification Techniques: A Practical Approach*. Oxford Univ. Press, New York.

Ross, J., ed. 1997. *Nucleic Acid Hybridization*. Wiley, New York.

Rudin, N., and Inman, K. 2001. *Introduction to Forensic DNA Analysis*. CRC Press, Boca Raton, FL, USA.

Saccone, C., and Pesole, G. 2003. *Handbook of Comparative Genomics: Modern Methodology*. Wiley-Liss, Hoboken, NJ, USA.

Sambrook, J., and Russell, D. 2001. *Molecular Cloning: A Laboratory Manual*. Cold Spring Harbor Laboratory Press, Cold Spring Harbor, NY, USA.

Sansone, E. B., and Tewari, Y. B. 1978. The permeability of laboratory gloves to selected nitro-amines, pp. 517–43. In *Environmental Aspects of N-Nitroso Compounds*, Walker, E. A., Griciute, L., Castenegro, M., and Lyle R. E., eds., Int. Agency Res. Cancer, Sci. Publ. 19, Lyon, France.

Schaefer, B. C. 1997. *Gene Clonig and Analysis: Current Innovations*. Horizon, Wymondham, UK.

Schena, M. 2002. *Microarray Analysis*. Wiley-Liss, Hoboken, NJ, USA.

Schmidtke, J., and Krawczak, M. 1998. *DNA Fingerprinting*. Springer, New York.

Schumann, W., et al., eds. 2001. *Functional Analysis of Bacterial Genes: A Practical Approach*. Wiley, New York.

Seeley, H. W., VanDemark, P. J., and Li, J. J. 1991. *Microbes in Action: A Laboratory Manual of Microbiology*. Freeman, New York.

Settle, F., ed. 1997. *Handbook of Instrumental Techniques for Analytical Chemistry*. Prentice Hall, Upper Saddle River, NJ, USA.

Shepherd, P. S., and Dean, C., eds. 2000. *Monoclonal Antibodies: A Practical Approach*. Oxford Univ. Press, New York.

Simpson, R. 2002. *Proteins and Proteomics: A Laboratory Manual*. Cold Spring Harbor Lab. Press, Woodbury, NY, USA.

Slater, R., ed. 2002. *Radioisotopes in Biology*. Oxford Univ. Press, New York.

Smith, B. J., ed. 2002. *Protein Sequencing Protocols*. Humana Press, Totowa, NJ, USA.

Smith, M., and Sockett, E., eds. 1999. *Genetic Methods for Diverse Prokaryotes*. Academic Press, San Diego, CA, USA.

Spector, D. L., Goldman, R., and Leinwand, L., eds. 1998. *Cells: A Laboratory Manual*. Cold Spring Harbor Laboratory Press, Cold Spring Harbor, NY, USA.

Spector, D., and Goldman, R., eds. 2003. *Essentials from Cells: A Laboratory Manual*. Cold Spring Harbor Lab. Press, Woodbury, NY, USA.

Starkey, M. P., and Elswarapu, R., eds. 2001. *Genomics Protocols*. Humana Press, Totowa, NJ, USA.

Studzinski, G. P. 1999. *Apoptosis: A Practical Approach*. Oxford Univ. Press, New York.

Sullivan, W. Ashburner, M., and Hawley, R. S., eds. 2000. *Drosophila Protocols*. Cold Spring Harbor Laboratory Press, Cold Spring Harbor, NY, USA.

Trower, M. K. 1996. *In Vitro Mutagenesis Protocols*. Humana Press, Totowa, NJ, USA.

Tymms, M. J., ed. 1995. *In Vitro Transcription and Translation Protocols*. Humana Press, Totowa, NJ, USA.

Tymms, M. J., ed. 2000. *Transcription Factor Protocols*. Humana Press, Totowa, NJ, USA.

Tymms, M. J., and Kola, I., eds. 2001. *Gene Knockout Protocols*. Humana, Totowa, NJ, USA.

Vasil, I. K., ed. 1984. *Cell Culture and Somatic Cell Genetics of Plants*. Academic Press, San Diego, CA, USA.

Walker, J. M., ed. 1996. *The Protein Protocols Handbook*. Humana Press, Totowa, NJ, USA.

Ward, A. 2001. *Genomic Imprinting: Methods and Protocols*. Humana Press, Totowa, NJ, USA.

Wasserman, P. A., and DePamphilis, M. L., eds. 1993. *Guide to Techniques in Mouse Development*. Methods in Enzymology 225. Academic Press, San Diego, CA, USA.

Weigel, D., and Glazebrook, J. 2002. *Arabidopsis: A Laboratory Manual*. Cold Spring Harbor Laboratory Press, Woodbury, NY, USA.

Weising, K., et al. 1995. *DNA Fingerprinting in Plants and Fungi*. CRC Press, Boca Raton, FL, USA.

Westermeier, R., and Naven, T. 2002. *Proteomics in Practice: A Laboratory Manual of Proteome Analysis*. Wiley-VCH, New York.

White, B. A., ed. 1996. *PCR Cloning Protocols*. Humana Press, Totowa, NJ, USA.

Wilkinson, D. G., ed. 1998. *In Situ Hybridization: A Practical Guide*. Oxford Univ. Press, New York.

Wilson, Z. A., ed. 2000. *Arabidopsis. A Practical Approach*. Oxford U*niv. Press, New York.

Wu, W., et al. 1997. *Methods in Gene Biotechnology*. CRC Press, Boca Raton, FL, USA.

Yang, N.-S., and Christou, P. 1994. *Particle Bombardment Technology for Gene Transfer*. Oxford Univ. Press, New York.

MAPPING (*See also* Recombination)

Boultwood, J., ed. 1996. *Gene Isolation and Mapping Protocols*. Humana Press, Totowa, NJ, USA.

Cantor, C. R., and Smith, C. L. 1999. *Genomics: The Science and Technology Behind the Human Genome Project*. Wiley, New York.

Dear, P. H. 1997. *Genome Mapping: A Practical Approach*. Oxford Univ. Press, New York.

Haines, J. L., and Pericak-Vance, M. A., eds. 1998. *Approaches to Gene Mapping in Complex Human Diseases*. Wiley, New York.

Liu, B. H. 1997. *Statistical Genomics: Linkage, Mapping, and QTL Analysis*. CRC Press, Boca Raton, FL, USA.

Mather, K. 1957. *The Measurement of Linkage in Heredity*. Methuen, London, UK.

O'Brien, S. J., ed. 1993. *Genetic Maps: Locus Maps of Complex Genomes.* Cold Spring Harbor Laboratory Press, Cold Spring Harbor, NY, USA.

Ott, J. 1999. *Analysis of Human Genetic Linkage.* Johns Hopkins Univ. Press, Baltimore, MD, USA.

Paterson, A. H., ed. 1996. *Genome Mapping in Plants.* Academic Press, San Diego, CA, USA.

Pawlowitzki, I-H., Edwards, J. H., and Thompson, E. A., eds. 1997. *Genetic Mapping of Disease Genes.* Academic Press, San Diego, CA, USA.

Primrose, S. B. 1995. *Principles of Genome Analysis: A Guide to Mapping and Sequencing DNA from Different Organisms.* Blackwell Science, Cambridge, MA, USA.

Schook, L. B., Lewin, H. A., and McLaren. D. G., eds., 1991. *Gene-Mapping Techniques and Applications.* Marcel Dekker, New York.

Speed, T., and Waterman, M. S. 1996. *Genetic Mapping and DNA Sequencing.* Springer, New York.

MICROBIOLOGY (*See also* Laboratory Technology, Biology, Molecular Biology, Viruses)

Adolph, K. W., ed. 1995. *Microbial Gene Techniques.* Academic Press, San Diego, CA, USA.

Aktoris, K., and Just, I. 2000. *Bacterial Protein Toxins.* Springer, Berlin, Germany.

Amyes, S. G. B. 2001. *Magic Bullets, Lost Horizons: The Rise and Fall of Antibiotics.* Taylor & Francis, New York.

Annual Review of Microbiology, Palo Alto, CA, USA.

Atlas, R. M. 1993. *Handbook of Microbiological Media.* CRC Press, Boca Raton, FL, USA.

Baltz, R. H., Hageman, G. D., and Skatrud, P. L., eds. 1993. *Industrial Microorganisms: Basic and Applied Molecular Genetics.* American Society of Microbiology, Washington, DC.

Baron, E. J., et al., eds. 1994. *Medical Microbiology.* Wiley, New York.

Barnett, J. A., et al. 2000. *Yeasts: Characteristics and Identification.* Cambridge Univ. Press, New York.

Baumberg, S. 1999. *Prokaryotic Gene Expression.* Oxford Univ. Press, Oxford, UK.

Baumberg, S., Young, J. P. W., Wellington, E. M. H., and Saunders, J. R., eds. 1995. *Population Genetics of Bacteria.* Cambridge Univ. Press, New York.

Boyd, R. F. 1995. *Basic Medical Microbiology.* Little Brown, New York.

Bruijn de, F. J., Lupski, J. R., and Weinstock, G. M. 1998. *Bacterial Genomes: Physical Structure and Analysis.* Chapman & Hall, New York.

Brun, Y. V., and Shimkets, L. J., eds. 2000. *Prokaryotic Development.* American. Society of Microbiology Press, Washington, DC.

Bryant, D. A. 1994. *The Molecular Biology of Cyanobacteria.* Kluwer, Norwell, MA, USA.

Clayton, C. L., and Mobley, H. L. T. 1997. *Helicobacterium pylory Protocols.* Humana Press, Totowa, NJ, USA.

Clewell, D. B., ed. 1993. *Bacterial Conjugation.* Plenum, New York.

Collier, L., ed. 1998. *Topley and Wilson's Microbiology and Microbial Infections.* Arnold/Oxford Univ. Press, New York.

Cooper, G. M., Temin, R. G., and Sugden, R., eds. 1995. *The DNA Provirus.* American Society of Microbiology, Washington, DC, USA.

Cossart, P., et al., eds. 2000. *Cellular Microbiology.* American Society of Microbiology Press, Washington, DC. USA.

Dale, J. W. 1998. *Molecular Genetics of Bacteria.* Wiley, New York.

Dangl, J. L., ed. 1994. *Bacterial Pathogenesis of Plants and Animals: Molecular and Cellular Mechanisms.* Springer, New York.

Day, I. N. M. 2002. *Molecular Genetic Epidemiology — A Laboratory Perspective.* Springer, Berlin, Germany.

Donnenberg, M. S. 2002. *Escherichia coli: Virulence Mechanisms of a Versatile Pathogen*, Acad. Press, San Diego, CA, USA.

Dorman, C. J. 1994. *Genetics of Bacterial Virulence.* Blackwell Science, Cambridge, MA, USA.

Dunn, B. M., ed. 1999. *Proteases of Infectious Agents.* Academic Press, San Diego, CA, USA.

Elliott, C. G. 1993. *Reproduction of Fungi: Genetical and Physiological Aspects.* Chapman & Hall, New York.

England, R., et al., eds. 1999. *Microbial Signalling and Communication.* Cambridge Univ. Press, New York.

Funell, B. E. 1996. *The Role of Bacterial Membrane in Chromosome Replication and Partition.* Chapman & Hall, New York.

Gartland, K. M. A., and Davey, M. 1995. *Agrobacterium Protocols.* Humana Press, Totowa, NJ, USA.

Gilmore, M. S. 2002. *The Enterococci: Pathogenesis, Molecular Biology, and Antimicrobial Resistance.* ASM Press, Washington, DC.

Goset, F., and Guespin-Michel, J. 1994. *Prokaryotic Genetics: Genome Organization, Transfer and Plasticity.* Blackwell Science, Cambridge, MA, USA.

Gow, N. A. R., and Gadd, G. M., eds. 1994. *The Growing Fungus.* Chapman & Hall, New York.

Griffin, D. H. 1996. *Fungal Physiology.* Wiley, New York.

Hacker, J., and Heesemann, J., eds. 2002. *Molecular Infection Biology. Interactions Between Microorganisms and Cells.* Wiley-Liss, Hoboken, NJ, USA.

Hawkey, P. M., and Lewis, D. A., eds. 1989. *Medical Bacteriology: A Practical Approach.* IRL Press, New York.

Herbert, M. A., et al., eds. 2002. *Hemophilus influenzae Protocols.* Humana Press, Totowa, NJ, USA.

Heritage, J., Evans, E. G., and Killington, R. A. 1999. *Microbiology in Action.* Cambridge Univ. Press, New York.

Holt, J. G., ed. 1993. *Bergey's Manual of Determinative Bacteriology.* William & Wilkins, Baltimore, MD, USA.

Koehler, T. M., ed. 2002. *Anthrax.* Springer, New York.

Kwon-Chung, K. J., and Bennett, J. E. 1992. *Medical Mycology.* Lea & Febiger, Philadelphia, PA, USA.

Larone, D. 2002. *Medically Important Fungi.* ASM Press, Washington, DC, USA.

Lederberg, J., ed. 2000. *Encyclopedia of Microbiology.* Academic Press, San Diego, CA, USA.

Levine, J., et al., eds. 1994. *Bacterial Endotoxins.* Wiley-Liss, New York.

Lin, E. C. C., and Lynch, A. S., eds. 1996. *Regulation of Gene Expression in Escherichia coli.* Chapman & Hall, New York.

Maloy, S. R., and Taylor, R. K. 1996. *Genetic Analysis of Pathogenic Bacteria: A Laboratory Manual.* Cold Spring Harbor Laboratory Press, Cold Spring Harbor, NY, USA.

McCrae, M. A., et al., eds. 1997. *Molecular Aspects of Host-Pathogen Interactions*. Cambridge Univ. Press, New York.

McKane, L., and Kandel, J. 1995. *Microbiology: Essentials and Applications*. McGraw-Hill, New York.

Miller, V. L., Kaper, J. B, Portnoy, D. A., and Isberg, R. R., eds. 1994. *Molecular Genetics of Bacterial Pathogenesis*. American Society of Microbiology Press, Washington, DC, USA.

Moat, A. G., and Foster, J. W., and Spector, M. P. 2002. *Microbial Physiology*. Wiley, New York.

Mobley, H. T. L., et al., eds. 2001. *Helicobacter pylori: Physiology and Genetics*. ASM Press, Washington, DC, USA.

Murray, P. R., et al., eds., 1995. *Manual of Clinical Microbiology*. ASM Press, New York.

Neidhardt, F. C., et al., eds. 1966. *Escherichia coli and Salmonella*. American Society of Microbiology Press, Washington, DC, USA.

Nester, E. W. et al. 2004. *Microbiology: A Human Perspective*. McGraw-Hill, Boston, MA, USA.

Nickoloff, J. A., ed. 1995. *Electroporation Protocols for Microorganisms*. Humana Press, Totowa, NJ, USA.

Oliver, R., and Schweizer, M., eds. 1999. *Molecular Fungal Biology*. Cambridge Univ. Press, New York.

Parish, T., and Stoker, N. G., eds. 2001. *Mycobacterium tuberculosis Protocols*. Humana Press, Totowa, NJ, USA.

Sachse, K., and Frey, J., eds. 2002. *PCR Detection of Microbial Pathogens*. Humana, Totowa, NJ, USA.

Salyers, A. A., and Whitt, D. D. 2002. *Bacterial Pathogenesis: A Molecular Approach*. AMS Press, Washington, DC, USA.

Singleton, P. 1995. *Bacteria in Biology, Biotechnology and Medicine*. Wiley, New York.

Singleton, P., and Sainsbury, D. 2002. *Dictionary of Microbiology and Molecular Biology*. Wiley, New York.

Smith, I., Slepeczky, R. A., and Setlow, P., eds. 1989. *Regulation of Prokaryotic Development: Structural and Functional Analysis of Bacterial Sporulation and Germination*. American Society of Microbiology Press, Washington, DC.

Smith, M. C. M., and Sockett, R. E., eds. 1999. *Genetic Methods for Diverse Prokaryotes*. Academic Press, London, UK.

Snyder, L., and Champness, W. 2002. *Molecular Genetics of Bacteria*. ASM Press, Washington, DC, USA.

Somasegaran, P., and Hoben, H. J. 1994. *Handbook of Rhizobia: Methods in Legume-Rhizobia Technology*. Springer, New York.

Sonenshein, A. L., Hoch, J. A., and Losick, R., eds. 2001. *Bacillus subtilis and Its Closest Relatives: From Genes to Cells*. American Society of Microbiology Press, Washington, DC, USA.

Storz, G., and Hengge-Aronis, R., eds. 2000. *Bacterial Stress Responses*. American Society of Microbiology Press, Washington, DC, USA.

Streips, U. N., and Yasbin, R. E., eds. 2002. *Modern Microbial Genetics*. Wiley, New York.

Summers, D. K. 1996. *The Biology of Plasmids*. Blackwell Science, Cambridge, MA, USA.

Thomas, C. M., ed. 2000. *The Horizontal Gene Pool: Bacterial Plasmids and Gene Spread*. Taylor & Francis, New York.

Tortora, G. J., Funke, B. R., and Case, C. L. 1998. *Microbiology: An Introduction*. Addison Wesley Longman, Menlo Park, CA, USA.

Truant, A. L., ed. 2002. *Manual of Commercial Methods in Clinical Microbiology*. ASM, Washington, DEC, USA.

Vaillancourt, P. E., ed. 2002. *E. coli Gene Expression Protocols*. Humana, Totowa, NJ, USA.

Wackett, L. P., and Hershberger, C. D. 2001. *Biocatalysis and Biodegradation*. ASM Press, Washington, DC.

White, D. 1999. *The Physiology and Biochemistry of Prokaryotes*. Oxford Univ. Press, New York.

Wilson, M., ed. 2002. *Bacterial Adhesion to Host Tissues: Mechanisms and Consequences*. Cambridge Univ. Press, Cambridge, UK.

Wilson, M., McNab, R., and Henderson, B. 2002. *Bacterial Disease Mechanisms: An Introduction to Cellular Microbiology*. Cambridge Univ. Press, Cambridge, UK.

Woodford, N., and Johnson, A. 1998. *Molecular Bacteriology: Protocols and Clinical Applications*. Humana Press, Totowa, NJ, USA.

MOLECULAR BIOLOGY (*See also* Biotechnology, Biochemistry and Physiology, Laboratory Technology)

Abelson, J. N., ed. 1996. *Combinatorial Chemistry: Methods in Enzymology 267*. Academic Press, San Diego, CA, USA.

Agrawal, S. 1996. *Antisense Therapeutics*. Humana Press, Totowa, NJ, USA.

Alphey, L. 1997. *DNA Sequencing*. Springer, New York.

Archer, S. L., and Rusch, N. J., eds. 2001. *Potassium Channels in Cardiovascular Biology*. Kluwer, New York.

Arrigo, A. P., and Müler, W. E. G., eds. 2002. *Small Stress Proteins*. Springer, New York.

Ashcroft, F. M. 2000. *Ion Channels and Disease*. Academic Press, San Diego, CA, USA.

Ashley, R. H. 1966. *Ion Channels: A Practical Approach*. Oxford Univ. Press, New York.

Balkvill, F., ed. 2000. *The Cytokine Network*. Oxford Univ. Press, New York.

Bass, B. L., ed. 2001. *RNA Editing*. Oxford Univ. Press, New York.

Bauerle, P. A., ed. 1995. *Inducible Gene Expression: Hormonal Signals*. Birkhäuser, Cambridge, MA, USA.

Beckerle, M., ed. 2002. *Cell Adhesion*. Oxford Univ. Press, New York.

Beugelsdijk, T. J., ed. 1997. *Automation Technologies in Genome Characterization*. Wiley, New York.

Bishop, J. E., and Waldholz, M. 1990. *Genome*. Simon and Schuster, New York.

Blow, J., ed. 1996. *Eukaryotic DNA Replication*. IRL/Oxford Univ. Press, New York.

Bonas, C., and Revillard, J.-P. 2001. *Cytokines and Cytokine Receptors*. Taylor & Francis, New York.

Calladine, C. R., and Drew, H. R. 1997. *Understanding DNA: The Molecule and How It Works*. Academic Press, San Diego, CA, USA.

Callards, R., and Gearing, A. 1994. *The Cytokine FactsBook*. Academic Press, San Diego, CA, USA.

Carey, M., and Smale, S. T., eds. 2000. *Transcriptional Regulation in Eukaryotes: Concepts, Strategies, and Techniques*. Cold Spring Harbor Laboratory Press, Cold Spring Harbor, NY, USA.

Clapp, J. P., ed. 1995. *Species Diagnostics Protocols, PCR and Other Nucleic Acid Methods*. Humana Press, Totowa, NJ. USA.

Collado-Vides, J., and Hofestadt, R. 2002. *Gene Regulation and Metabolism*. MIT Press, Cambridge, MA, USA.

Conley, E. C 1995–1996. *The Ion Channel FactsBook*, Vols. 1–4 Academic Press, San Diego, CA, USA.

Cortese, R., ed. 1995. *Combinatorial Libraries: Synthesis, Screening and Application Potentials*. De Gruyter, New York.

Curiel, D. T., and Douglas, J., eds. 2002. *Vector Targeting for Therapeutic Gene Delivery*. Wiley-Liss, Hoboken, NJ, USA.

Dale, J., and Schantz, v. M. 2002. *From Genes to Genomes. Concepts and Applications of DNA Technology*. Wiley, Hoboken, NJ, USA.

De Murcia, G., and Shall, S. 2000. *From DNA Damage and Stress Signaling to Cell Death: Poly ADP-Ribosylation Reactions*. Oxford Univ. Press, Oxford, UK.

DePamphilis, M. L., ed. 1999. *Concepts in Eukaryotic DNA Replication*. Cold Spring Harbor Laboratory Press, Cold Spring Harbor, NY, USA.

Docherty, K., ed. 1996. *Gene Transcription: DNA Binding Proteins*. Wiley, New York.

Docherty, K., ed., 1996. *Gene Transcription: RNA Analysis*. Wiley, New York.

Doerfler, W. 2000. *Foreign DNA in Mammalian Systems*. Wiley-VCH, New York.

Doolittle, R. F., ed. 1996. *Computer Methods for Macromolecular Sequence Analysis: Methods in Enzymology 266* Academic Press, San Diego, CA, USA.

Grubin, D. G. 2000. *Cell Polarity*. Oxford Univ. Press, Oxford, UK.

Eckstein, F., and Lilley, D. M. J. 1997. *Catalytic RNA*. Springer, New York.

Eggleston, D. S., Prescott, C. D., and Pearson, N. D., eds. 1998. *The Many Faces of RNA*. Academic Press, San Diego CA, USA.

Ellis, R. J., ed. 1966. *The Chaperonins*. Academic Press, San Diego, CA, USA.

Eun, H-M. 1996. *Enzymology Primer for Recombinant DNA Technology*. Academic Press, San Diego, CA, USA.

Farrel, R. E., and Leppertt, G. 1997. *rDNA Manual*. Springer, New York.

Fitzgerald, K. A., et al. 2001. *The Cytokine FactsBook/Cytokine WebFacts*. Academic Press, San Diego, CA, USA.

Gelvin, S. B., and Schilperoort, R. A. 1994. *Plant Molecular Biology Manual*. Kluwer, Norwell, MA, USA.

Gesteland, R. F., et al., eds. 1999. *The RNA World*. Cold Spring Harbor Laboratory Press, Cold Spring Harbor, NY, USA.

Gomperts, B., et al. 2002. *Signal Transduction*. Academic Press, San Diego, CA USA.

Goodbourn, S., ed. 1996. *Eukaryotic Gene Transcription*. Oxford Univ. Press, New York.

Goodfellow, J. M., ed. 1995. *Computer Modeling in Molecular Biology*. VCH, New York.

Gribskov, M., and Devereux, J. 1991. *Sequence Analysis Primer*. Freeman, New York.

Griffin, A. M., and Griffin, H. G., eds. 1994. *Computer Analysis of Sequence Data*. Humana Press, Totowa, NJ, USA.

Gutkind, J. S., ed. 2000. *Signaling Networks and Cell Cycle Control: The Molecular Basis of Cancer and Other Diseases*. Humana Press, Totowa, NJ, USA.

Gutman, Y., and Lazarovici, P. 1997. *Toxins and Signal Transduction*. Harwood Academic Publishers, Langhorne, PA, USA.

Harris, D. A. 1997. *Molecular and Cellular Biology*. Horizon, Wymondham, UK.

Harris, D. A., ed. 1997. *Prions: Molecular and Cellular Biology*. Horizon, Wymondham, UK.

Hawkins, J. D. 1996. *Gene Structure and Expression*. Cambridge Univ. Press, New York.

Heilmeyer, L., and Friedrich, P., eds. 2001. *Protein Modules in Cellular Signaling*. IOS Press, Amsterdam, The Netherlands.

Hershey, J. W. B., Mathews, M. B., and Sonenberg, N., eds. 1996. *Translational Control*. Cold Spring Harbor Laboratory Press, Cold Spring Harbor, NY, USA.

Hertzberg, E. L., ed. 2000. *Gap Junctions*. JAI Press, Stamford, CT, USA.

Hideaki, H., et al., eds. 1996. *Intracellular Signal Transduction*. Academic Press, San Diego, CA, USA.

Hille, B. 2001. *Ion Channels of Excitable Membranes*. Sinauer, Sunderland, MA, USA.

Hills, D., and Moritz, C., eds. 1996. *Molecular Systematics*. Sinauer, Sunderland, MA, USA.

Horton, R. M., and Tait, R. C., ed. 1998. *Genetic Engineering with PCR*. Horizon, Wymondham, UK.

Houdebine, L. M., ed. 1997. *Transgenic Animals: Generation and Use*. Harwood, Newark, NJ, USA.

Howard, J. 2001. *Mechanics of Motor Proteins and the Cytoskeleton*. Sinauer, Sunderland, MA, USA.

Hoy, M. A. 1994. *Insect Molecular Genetics: An Introduction to Principles and Applications*. Academic Press, San Diego, CA, USA.

Huang, L., Hung, M-C., and Wagner, E. 1999. *Non-Viral Vectors in Gene Therapy*. Academic Press, San Diego, CA, USA.

Hughes, M. A. 1996. *Plant Molecular Genetics*. Addison Wesley, Boston, MA, USA.

Innis, M. A., Gelfand, D. H., and Sninsky, J. J., eds. 1995. *PCR Strategies*. Academic Press, San Diego, CA, USA.

Isenberg, G., ed. 2002. *Encyclopedic Reference to Cell Biology*. Springer, New York.

Jost, J.-P., and Saluz, H-P., eds., 1993. *DNA Methylation: Molecular Biology and Biological Significance*. Birkhäuser Verlag, New York.

Kannicht, C., ed. 2002. *Posttranslational Modification Reactions: Tools for Functional Proteomics*. Humana Press, Totowa, NJ, USA.

Kaplitt, M. G., and Loewy, A. D., eds. 1995. *Viral Vectors: Gene Therapy and Neuroscience Applications*. Academic Press, San Diego, CA, USA.

Kazazian, H. H., ed. 2002. *Wiley Encyclopedia of Molecular Medicine*. Wiley, Hoboken, NJ, USA.

Kirby, L. T. 1990. *DNA Fingerprinting: An Introduction*. Stockton Press, New York.

Kleanthous, C., ed. 2001. *Protein-Protein Recognition*. Oxford Univ. Press, Oxford, UK.

Kneale, G. G., ed. 1994. *DNA-Protein Interactions: Principles and Protocols*. Humana Press, Totowa, NJ, USA.

Korn, B., et al. 1998. *Positional Cloning by Exon Trapping and cDNA Selection*. Wiley, New York.

Krainer, A. R., ed. 1997. *Eukaryotic mRNA Processing*. Oxford Univ. Press, New York.

Latchman, D. S. 1999. *Eukaryotic Transcription Factors*. Academic Press, San Diego, CA, USA.

Lesk, A. M. 2000. *Introduction to Protein Architecture*. Oxford Univ. Press, New York.

Levy, S. B., ed. 1997. *Antibiotic Resistance*. Wiley, New York.

Lilley, D. M. J., ed. 1995. *DNA—Protein: Structural Interactions*. IRL Press, New York.

Litvak, S. 1996. *Retroviral Reverse Transcriptases*. Chapman & Hall, New York.

Locker, J., ed. 2001. *Transcription Factors*. Academic Press, San Diego, CA, USA.

Lund, P., ed. 2001. *Molecular Chaperones in the Cell*. Oxford Univ. Press, New York.

Mantovani, A., ed. 1999. *Chemokines*. Karger, New York.

Marsh, M., ed. 2001. *Endocytosis*. Oxford Univ. Press, New York.

Marshak, D. R., Gardner, R. L., and Gottlieb, D., eds. 2001. *Stem Cell Biology*. Cold Spring Harbor Laboratory Press, Cold Spring Harbor, NY, USA.

McDonald, C. J., ed. 1997. *Enzymes in Molecular Biology: Essential Data*. Wiley, New York.

Meager, T. 1998. *The Molecular Biology of Cytokines*. Wiley, New York.

Meijer, L., et al., eds. 2000. *Progress in Cell Cycle Research*. Kluwer, New York.

Meyers, R. A., ed. 1997. *Encyclopedia of Molecular Biology and Molecular Medicine*. Wiley, New York.

Mishra, N. C. 2002. *Nucleases. Molecular Biology and Applications*. Wiley-Interscience, Hoboken, NJ, USA.

Moley, J. F., and Kim, S. H. 1994. *Molecular Genetics of Surgical Oncology*. CRC Press, Boca Raton, FL, USA.

Morimoto, R. I., Tissières, A., and Georgopoulos, C., eds. 1994. *The Biology of Heat Shock Proteins and Molecular Chaperones*. Cold Spring Harbor Laboratory Press, Cold Spring Harbor, NY, USA.

Mullis, K. B., Ferré., F., and Gibbs, eds. 1994. *The Polymerase Chain Reaction*. Birkhäuser, Boston, MA, USA.

Nagai, K., and Mattaj, I. W., eds. 1995. *RNA–Protein Interactions*, IRL Press, New York.

Newton, C., and Graham, A. 1997. *PCR*. Springer, New York.

Nickoloff, J. A., and Hoekstra, M. F., eds. 2001. *DNA Damage and Repair*. Humana Press, Totowa, NJ, USA.

Nicola, N. A., ed. 1995. *Guidebook to Cytokines and Their Receptors*. Oxford Univ. Press, New York.

North, R. A., ed. 1994. *Ligand- and Voltage-Gated Ion Channels*. CRC Press, Boca Raton. FL, USA.

Parker, P., and Pawson, T., eds. 1996. *Cell Signalling*. Cold Spring Harbor Laboratory Press, Cold Spring Harbor, NY, USA.

Pena, S. D. J., ed. 1993. *DNA Fingerprinting: State of the Science*. Birkhäuser, Basel, Switzerland.

Pennington, S. R., and Dunn, M. J., eds. 2001. *Proteomics: From Protein Sequence to Function*. Springer, New York.

Privalsky, M. L., ed. 2001. *Transcriptional Corepressors: Mediators of Eukaryotic Gene Repression*. Springer, New York.

Ptashne, M. 1992. *Genetic Switch: Phage Lambda and Higher Organisms*. Blackwell, Cambridge, MA, USA.

Ptashne, M., and Gann, A. 2002. *Genes & Signals*. Cold Spring Harbor Laboratory Press, Cold Spring Harbor, NY, USA.

Ravid, K., and Licht, J. D. 2001. *Transcription Factors: Normal ans Malignant Development of Blood Cells*. Wiley, New York.

Richter, D., ed. 2001. *Cell Polarity and Subcellular Localization*. Springer, New York.

Richter, J. D., ed. 1997. *mRNA Formation and Function*. Academic Press, New York.

Rush, M., and D'Eustachio, P., eds. 2002. *The Small GTPase Ran*. Kluwer, Norwell, MA, USA.

Saluz, H. P., and Wiebauer, K. 1995. *DNA and Nucleoprotein Structure In Vivo*. CRC Press, Boca Raton, FL, USA.

Schenkel, J. 1997. *RNP Particles*. Springer, New York.

Schleef, M., ed. 2001. *Plasmids for Therapy and Vaccination*. Wiley-VCH, New York.

Sealfon, S. C. 1994. *Receptor Molecular Biology*. Academic Press, San Diego, CA.

Sedivy, J. M., and Joyner, A. L. 2000. *Gene Targeting*. Oxford Univ. Press, New York.

Setlow, J. K., and Hollaender, A., eds. 1979. *Genetic Engineering: Principles and Methods*. Plenum, New York.

Shinitzky, M., ed. 1995. *Biomembranes: Signal Transduction Across Membranes*. VCH, New York.

Sinden, R. R. 1994. *DNA Structure and Function*. Academic Press, San Diego, CA, USA.

Skalka, A. M., and Goff, S. P., eds. 1993. *Reverse Transcriptase*. Cold Spring Harbor Laboratory Press, Cold Spring Harbor, NY, USA.

Söll, D., et al., eds. 2001. *RNA*. Elsevier, Oxford, UK.

Söll, D., and Rajbahandary, U. L., eds. 1994. *tRNA: Structure, Biosynthesis and Function*. AMS Press, Washington, DC, USA.

Sonenberg, N., et al. 2000. *Translational Control of Gene Expression*. Cold Spring Harbor Laboratory Press, Cold Spring Harbor, NY, USA. (See also Hershey, J. W. B., et al., 1996.)

Tait, R. C. 1997. *An Introduction to Molecular Biology*. Horizon, Wymondham, UK.

Tomiuk, J., Wöhrmann, K., and Sentker, A. 1996. *Transgenic Organisms*. Springer, New York.

Vaddi, K., Keller, M., and Newton, R. C. 1997. *The Chemokine FactsBook*. Academic Press, San Diego, CA, USA.

Vega, M. A., ed. 1995. *Gene Targeting*. CRC Press, Boca Ratoon, FL, USA.

Wagner, R. 2000. *Transcription Regulation in Prokaryotes*. Oxford Univ. Press, New York.

Watson, J. D., et al. 2003. *Molecular Biology of the Gene*. Benjamin/Cummings, Menlo Park, CA, USA.

Watters, D., and Lavin, M., eds. 1999. *Signaling Pathways in Apoptosis*. OPA, Amsterdam, The Netherlands.

Weis, K., ed. 2002. *Nuclear Transport*. Springer, New York.

White, R. J. 2001. *Gene Transcription: Mechanisms and Control*. Blackwell, Oxford, UK.

Wilkins, M. R., et al., eds. 1997. *Proteome Research: New Frontiers in Functional Genomics*. Springer, New York.

Wilks, A. E., and Harpur, A. G. 1996. *Intracellular Signal Transduction: The Jak-STAT Pathway*. Hall, New York.

Wingender, E. 1993. *Gene Regulation in Eukaryotes*. VHC, New York.

Woodgett, J. R., ed. 1995. *Protein Kinases*. Oxford Univ. Press, New York.

MUTATION (*See also* Radiation)

Braman, J., ed. 2001. *In Vitro Mutagenesis Protocols*. Humana Press, Totowa, NJ, USA.

Brusick, D. 1987. *Principles of Genetic Toxicology*. Plenum, New York.

Choy, W. N., ed. 2001. *Genetic Toxicology and Cancer Risk Assessment*. Marcel Dekker, New York.

Cooper, D. N., and Krawczak, M. 1993. *Human Gene Mutation*. Bios Scientific, Oxford, UK.

Cotton, R. G. H., et al. 1998. *Mutation Detection: A Practical Approach*. Oxford Univ. Press/IRL, New York.

Friedberg, E. C., Walker, G. C., and Siede, W. 1995. *DNA Repair and Mutagenesis*. American Society of Microbiology Press, Washington, DC, USA.

Gold, L. S., and Zeiger, E., eds. 1996. *Handbook of Carcinogenic Potency and Genotoxicity Databases*. CRC Press, Boca Raton, FL, USA.

Halliwell, B., and Gutteridge, J. M. C. 1999. *Free Radicals in Biology and Medicine*. Oxford Univ. Press, New York.

Hollaender, A., ed. 1971 to date. *Chemical Mutagens*. Plenum, New York.

Klaassen, C. D., ed. 1996. *Casarett and Doull's Toxicology: The Basic Science of Poisons*. McGraw-Hill, New York.

Landegren, U., ed. 1996. *Laboratory Protocols for Mutation Detection*. Oxford Univ. Press, New York.

Li, A. P., and Heflich, R. H., eds. 1991. *Genetic Toxicology*. CRC Press, Boca Raton, FL, USA.

Loveless, A. 1966. *Genetic and Allied Effects of Alkylating Agents*. Pennsylvania State Univ. Press, University Park, PA, USA.

McPherson, M. J. 1991. *Directed Mutagenesis*. IRL Press, New York.

Pfeifer, G. P., ed. 1966. *Technologies for Detection of DNA Damage and Mutation*. Plenum, New York.

Ross, W. C. J. 1962. *Biological Alkylating Agents*. Butterworth, London, UK.

Sankaranarayanan, K., ed. 2000. *Protocols in Mutagenesis*. Elsevier Science, New York.

Schardein, J. L. 2000. *Chemically Induced Birth Defects*. Marcel Dekker, New York.

Singer, B., and Basrtsch, H. 2000. *Exocyclic DNA Adducts in Mutagenesis and Carcinogenesis*. Oxford Univ. Press, New York.

Taylor, G. R., ed. 1997. *Laboratory Methods for the Detection of Mutations and Polymorphisms in DNA*. CRC Press, Boca Raton, FL, USA.

Theophilus, B. D. M., and Rapley, R. 2002. *PCR Mutation Detection Protocols*. Humana Press, Totowa, NJ, USA.

Vogel, F., and Röhrborn, G., eds. 1970. *Chemical Mutagenesis in Mammals and Man*. Springer-Verlag, New York.

NEUROBIOLOGY (*See also* Behavior, Molecular Biology)

Ancill, R. J., Holliday, S., and Higgenbottam, J., eds. 1995. *Schizophrenia: Exploring the Spectrum of Psychosis*. Wiley, New York.

Andreasen, N. C. 2001. *Brave New Brain: Conquering Mental Illness in the Era of the Genome*. Oxford Univ. Press, New York.

Bellen, H., ed. 1999. *Neurotransmitter Release: Frontiers in Molecular Biology*. Oxford Univ. Press, London, UK.

Blum, K., and Noble, E. P., eds. 1997. *Handbook of Psychiatric Genetics*. CRC Press, Boca Raton, FL, USA.

Bohlen, v. O, and Dermitzel, H. O. 2002. *Neurotransmitters and Neuromodulators. Handbook of Receptors and Biological Effects*, Wiley, New York.

Bourtchouladze, R. 2002. *Memories Are Made of This*. Columbia Univ. Press, New York.

Brandt, T., et al., eds. 2002. *Neurological Disorders*. Acad. Press, San Diego, CA, USA.

Budnik, V., and Gramates, L. S., eds. 1999. *Neuromuscular Junctions in Drosophila*. Academic Press, San Diego, CA, USA.

Burrows, M. 1996. *The Neurobiology of the Insect Brain*. Oxford Univ. Press, New York.

Butler, A. B., and Hodos, W. 1996. *Comparative Vertebrate Neuroanatomy: Evolution and Adaptation*. Wiley, New York.

Cervós-Navarro, J., and Ulrich, H. 1995. *Metabolic and Degenerative Diseases of the Nervous System: Pathology, Biochemistry and Genetics*. Academic Press, San Diego, CA, USA.

Charney, D. S., Nestler, E. J., and Bunney, B. S. 1999. *Neurobiology and Mental Illness*. Oxford Univ. Press, New York.

Chiocca, E. A., and Breakfield, X. O., eds. 1998. *Gene Therapy for Neurological Disorders and Brain Tumors*. Humana Press, Totowa, NJ, USA.

Cowan, W. M., Südhof, T. C., and Stevens, C. F., eds. 2001. *Synapses*. Johns Hopkins Univ. Press, Baltimore, MD, USA.

Crick, F. 1994. *The Astonishing Hypothesis: The Scientific Search for the Soul*. Simon and Schuster, Old Tappan, NJ, USA.

Dominiczak, A. F., et al., eds. 1999. *Molecular Genetics of Hypertension*. Academic Press, San Diego, CA, USA.

Evers-Kiebomas, G., et al. eds. 2002. *Prenatal Testing for Late-Onset Neurogenetic Diseases*. Bios Scientific, Oxford, UK.

Finger, S. 2001. *Origins of Neuroscience: A History of Explorations into Brain Function*. Oxford Univ. Press, New York.

Friedman, J. M., et al., eds. 1999. *Neurofibromatosis: Phenotype, Natural History and Pathogenesis*. Johns Hopkins Press, Baltimore, MD, USA.

Galaburda, A. M., ed. 1993. *Dyslexia and Development: Neurobiological Aspects of Extra-Ordinary Brains*. Harvard Univ. Press, Cambridge, MA, USA.

Gershon, E. S., and Cloninger, C. R., eds. 1994. *Genetic Approaches to Mental Disorders*. American Psychiatric Press, Washington, DC.

Geschwind, D. H., and Gregg, J. P. 2002. *Microarrays for the Neurosciences: An Essential Guide*. MIT Press, Cambridge, MA, USA.

Harper, P. S., and Perutz, M., eds. 2001. *Glutamine Repeats and Neurodegenerative Diseases: Molecular Aspects*. Oxford Univ. Press, New York.

Higgins, S. J., ed. 1999. *Molecular Biology of the Brain*. Princeton Univ. Press, Princeton, NY, USA.

Honavar, V., and Uhr, L., eds. 1994. *Artificial Intelligence and Neural Networks: Steps Toward Principled Integration*. Academic Press, San Diego, CA, USA.

Hörnlimann, B., et al., eds. 2001. *Prionen und Prionkrankheiten*. Walter de Gruyter, Berlin, Germany.

Ingoglia, N. A., and Murray, M., eds. 2001. *Axonal Regeneration in the Central Nervous System*. Marcel Dekker, New York.

Johnston, D., and Wu, S. M.-S. 1994. *Foundations of Cellular Neurophysiology*. MIT Press, Cambridge, MA, USA.

Keynes, R. D., and Aidley, D. J. 2001. *Nerve and Muscle*. Cambridge Univ. Press, New York.

Latchman, D., ed. 1995. *Genetic Manipulation of the Nervous System*. Academic Press, San Diego, CA, USA.

Leboyer, M., and Bellevier, F., eds. 2002. *Psychiatric Genetics. Methods and Reviews*. Humana, Totowa, NJ, USA.

Lyon, G., Adams, R. D., and Kolodny, E. H. 1996. *Neurology of Hereditary Metabolic Diseases of Children*. McGraw-Hill, New York.

McGaugh, J. L., ed. 1995. *Brain and Memory: Modulation and Mediation of Neuroplasticity*. Oxford Univ. Press, New York.

Merchant, K. M. 1996. *Pharmacological Regulation of Gene Expression in the CNS: Towards an Understanding of Basal Ganglion Functions*. CRC Press, Boca Raton, FL, USA.

Methods in Neurosciences (continuous volumes). Academic Press, San Diego, CA, USA.

Micevych, P. E., and Hammer, R. P., Jr., eds. 1995. *Neurobiological Effects of Sex Steroid Hormones*. Cambridge Univ. Press, New York.

Nicholls, J. G., et al., eds. 2000. *From Neuron to Brain*. Sinauer, Sunderland, MA, USA.

Nieuwenhuys, R., et al. 1997. *The Central Nervous System of Vertebrates*. Springer-Verlag, Berlin, New York.

O'Rahilly, R., and Muller, F. 1999. *The Embryonic Human Brain: An Atlas of Developmental Stages*. Wiley-Liss, New York.

Pfaff, D. 1999. *Drive: Neurobiological and Molecular Mechanisms of Sexual Motivation*. MIT Press, Cambridge, MA, USA.

Pfaff, D. W., et al., eds. 1999. *Genetic Influences on Neural and Behavioral Functions*. CRC Press, Boca Raton, FL, USA.

Phillips, M. I., and Evans, D., eds. 1994. *Neuroimmunology*. Academic Press, San Diego, CA, USA.

Popko, B., ed. 1999. *Mouse Models in the Study of Genetic Neurological Disorders*. Kluwer Academic/Plenum, New York.

Potter, N. T. 2002. *Neurogenetics. Methods and Protocols*. Humana, Totowa, NJ. USA.

Pulst, S.-M., ed. 1999. *Neurogenetics*. Oxford Univ. Press, New York.

Purves. D., et al., eds. 2000. *Neuroscience*. Sinauer, Sunderland, MA, USA.

Rao, M. S. 2001. *Stem Cells and CNS Development*. Humana Press, Totowa, NJ, USA.

Revest. P., and Longstaff, A. 1997. *Molecular Neuroscience*. Springer, New York.

Sanes D. H., Reh, T. A., and Harris, W. A. 2000. *Development of the Nervous System*. Academic Press, San Diego, CA, USA.

Schachter, D. L. 1996. *Searching for Memory: The Brain, the Mind, and the Past*. Basic Books, New York.

Schmajuk, N. A. 1997. *Animal Learning and Cognition*. Cambridge Univ. Press, New York.

Smith, C. U. M. 1996. *Elements of Molecular Neurobiology*. Wiley, New York.

Snowling, M. J. 2000. *Dyslexia*. Blackwell, Oxford, UK.

Smythies, J. 2002. *The Dynamic Neuron: A Comparative Survey of the Neurochemical Basis of Synaptic Plasticity*. MIT Press, Cambridge, MA, USA.

Tolnay, M., and Probst, A., eds. 2001. *Neuropathology and Genetics of Dementia*. Kluwer, Dordrecht, The Netherlands.

Tuszynski, M. H., and Kordower, J., eds. 1998. *CNS Regeneration*. Academic Press, San Diego, CA, USA.

Vogel, F. 2000. *Genetics and the Electroencephalogram*. Springer, New York.

Wells, R. D., and Warren, S. T., eds. 1998. *Genetic Instabilities and Hereditary Neurological Diseases*. Academic Press, San Diego, CA, USA.

Wess, J. 1995. *Molecular Mechanism of Muscarinic Acetylcholine Receptor Function*. Springer, New York.

Wheal, H., and Thomson, A., eds. 1995. *Excitatory Amino Acids and Synaptic Transmission*. Academic Press, San Diego, CA, USA.

Yuste, R., et al., eds. 1999. *Imaging Neurons: A Laboratory Manual*. Cold Spring Harbor Laboratory Press, Cold Spring Harbor, NY, USA.

Zigmond, M. J., et al., eds. 1999. *Fundamental Neuroscience*. Academic Press, San Diego, CA, USA.

Zigova, T., et al., eds. 2002. *Neural Stem Cells for Brain and Spinal Cord Repair*. Humana, Totowa, NJ, USA.

ORGANELLES (*See also* Genetics Monographs)

Argyroudi-Akoyunoglu, J. H., and Senger, H., eds. 1999. *The Chloroplast: From Molecular Biology to Biotechnology*. Kluwer Academic, Dordrecht, The Netherlands.

Attardi, G. M., and Chomyn, A., eds. 1995. *Mitochondrial Biogenesis and Genetics: Methods in Enzymology*, Vol. 260. Academic Press, San Diego, CA, USA.

Beale, G., and Knowles, J. 1978. *Extranuclear Genetics*. Arnold, London, UK.

Brown, G. C., et al. 1999. *Mitochondria and Cell Death*. Princeton Univ. Press, Princeton, NJ, USA.

Copeland, W. C. 2002. *Mitochondrial DNA: Methods and Protocols*. Humana Press, Totowa, NJ, USA.

Darley-Usmar, V, and Schapira, A. H. V., eds. 1994. *Mitochondria, DNA, Proteins and Disease*. Portland Press, Chapel Hill, NC, USA.

Di Mauro, S., and Wallace, D. C., eds. 1992. *Mitochondrial DNA in Human Pathology*. Raven Press, New York.

Grun, P. 1976. *Cytoplasmic Genetics and Evolution*. Columbia Univ. Press, New York.

Gillham, N. W. 1994. *Organelle Genes and Genomes*. Oxford University Press, New York.

Herrmann, R. G., ed. 1992. *Cell Organelles*. Springer, New York.

Kirk, J. T. O., and Tilney Bassett, R. A. E. 1978. *The Plastids*. Elsevier, Amsterdam, The Netherlands.

Lestienne, P. 1999. *Mitochondrial Diseases*. Springer, New York.

Levings, C. S., III, and Vasil, I. K., eds. 1995. *Molecular Biology of Plant Mitochondria*. Kluwer Academic, Dordrecht, The Netherlands.

Margulis, L. 1970. *Origin of Eukaryotic Cells*. Yale Univ. Press, New Haven, CT, USA.

Möller, J. M., ed. 1998. *Plant Mitochondria from Gene to Function*. Backhuys, Leiden, The Netherlands.

Ostrer, H. 1998. *Non-Mendelian Genetics in Humans*. Oxford Univ. Press, New York.

Rochaix, J.-D., and Merchant, S., eds. 1998. *Molecular Biology of Chloroplasts and Mitochondria in Chlamydomonas*. Kluwer Academic, Dordrecht, The Netherlands.

Sager, R. 1972. *Cytoplasmic Genes and Organelles*. Academic Press, New York.

Scheffler, I. E. 1999. *Mitochondria*. Wiley-Liss, New York.

Singh, K. K., ed. 1998. *Mitochondrial DNA, Mutations in Aging, Disease and Cancer*. Springer, New York.

PHILOSOPHY (*See also* Biotechnology, Evolution, Social and Ethical Issues)

Adami, C. 1998. *Introduction to Artificial Life*. Telos (Springer), New York.

Avise, J. C. 1998. *The Genetic Gods: Evolution and Belief in Human Affairs*. Harvard Univ. Press, Cambridge, MA, USA.

Bayertz, K., ed. 1995. *GenEthics: Technological Intervention in Human Reproduction as a Philosophical Problem*. Cambridge Univ. Press, New York.

Fukuyama, F. 2002. *Our Posthuman Future. Consequences of the Biotechnology Revolution*. Farrar, Straus and Giroux, New York.

Gould, S. J. 1999. *Rocks of Ages: Science and Religion in the Fullness of Life*. Ballantine Press, New York.

Gratzer, W. 2000. *The Undergrowth of Science: Delusion, Self-Deception and Human Frailty*. Oxford Univ. Press, New York.

Holton, G. 1993. *Science and Antiscience*. Harvard Univ. Press, Cambridge, MA, USA.

Horgan, J. 1996. *The End of Science: Facing the Limits of Knowledge in the Twilight of the Scientific Age*. Helix (Addison-Wesley), New York.

Horrobin, D. 2001. *The Madness of Adam and Eve: How Schizophrenia Shaped Humanity*. Bantam, New York.

Jolly, A. 1999. *Lucy's Legacy: Sex and Intelligence in Human Evolution*. Harvard Univ. Press, Cambridge, MA, USA.

Kilner, J. F., Hook, C. C., and Uustal, D. B., eds. 2002. *Cutting-Edge Bioethics. A Christian Exploration of Technologies and Trends*. W. B. Eerdmans Publishing Co, Grand Rapids, MI, USA.

Kitcher, P. 1996. *The Lives to Come: The Genetic Revolution and Human Possibilities*. Penguin, New York.

Morange, M. 2001. *The Misunderstood Gene*. Harvard Univ. Press, Cambridge, MA, USA.

Moss, L. 2002. *What Genes Can't Do*. MIT Press, Cambridge, MA, USA.

Murphy, M. P., and O'Neill, L. A. J., eds. 1997. *What Is Life? The Next 50 Years—Speculations on the Future of Biology*. Cambridge Univ. Press, New York.

Park, R. L. 2000. *Voodoo Science: The Road from Foolishness to Fraud*. Oxford Univ. Press, New York.

Pennock, R. T., ed. 2001. *Intelligent Design Creationism and its Critics. Philosophical, Theological and Scientific Perspectives*. MIT Press, Cambridge, MA, USA.

Popper, K. R. 1959. *The Logic of Scientific Discovery*. Basic Books, New York.

Popper, K. R. 1983. *Realism and the Aim of Science*. Rowman and Littlefield, Totowa, NJ, USA.

Ruse, M. 2001. *Can a Darwinian be a Christian?: the Relationship between Science and Religion*. Cambridge Univ. Press, New York.

Stent, G. 1969. *The Coming of the Golden Age: A View of the End of Progress*. Natural History Press, Garden City, NY, USA.

Sterelny, K., and Griffith, P. E. 1999. *Sex and Death: An Introduction to Philosophy of Biology*. Univ. Chicago Press, Chicago. IL, USA.

Stock, G. 2002. *Redesigning Humans*. Houghton Miflin, Boston, MA, USA.

Turney, J. 1998. *Frankenstein's Footsteps: Science, Genetics and Popular Culture*. Yale Univ. Press, New Haven, CT, USA.

Wade, N. 2001. *Life Script. How the Human Genome Discoveries Will Transform Medicine and Enhance Your Health*. Simon & Schuster, New York.

Walters, L., and Palmer, J. G. 1997. *The Ethics of Human Gene Therapy*. Oxford Univ. Press, New York.

PLANT and ANIMAL BREEDING (*See also* Biometry and Bioinformatics, Biotechnology, Cytology and Cytogenetics, Plant pathology, Population genetics)

Axford, R. F. E., et al., eds. 2000. *Breeding for Disease Resistance in Farm Animals*. Oxford Univ. Press, New York.

Bahl, P. N. 1996. *Genetics, Cytogenetics and Breeding of Crop Plants*. Science Pubs. Bio-Oxford & IBH, Lebanon, NH, USA.

Clark, A. J., ed. 1998. *Animal Breeding: Technology for the 21st Century*. Gordon & Breach, Newark, NJ, USA.

Clement, S. L., and Quisenberry, S. S., eds, 1998. *Global Plant Genetic Resources for Insect-Resistant Crops*. CRC Press, Boca Raton, FL, USA.

Cooper, H. D., et al., eds. 2001. *Broadening the Genetic Base of Crop Production*. Wiley, New York.

Coors, J. G., and Pandey, S., eds. 1999. *The Genetics and Exploitation of Heterosis in Crops*. ASA-CSSA-SSSA, Madison, WI, USA.

Crabb, R. 1992. *The Hybrid Corn-Makers*. West Chicago Publishing Co., Chicago, IL, USA.

Elsevier's Dictionary of Plant Genetic Resources. 1995. Elsevier Science, New York.

Engels, J. M. M., et al., eds. 2002. *Managing Plant Genetic Diversity*. CABI, Wallingford, Oxon, UK.

Evans, L. T. 1996. *Crop Evolution, Adaptation and Yield*. Cambridge Univ. Press, New York.

Evenson, R. E., et al., eds. 1998. *Agricultural Value of Plant Genetic Resources*. CAB Internat., Wallingford, UK.

Fries, R., and Ruvinsky, A., eds. 1999. *The Genetics of Cattle*. CAB Internat., Wallingford, UK.

Hammond, J., et al., eds. 1999. *Plant Biotechnology: New Products and Applications*. Springer, New York.

Hayward, M. D., Bosemark, N. O., and Romagosa, eds. 1993. *Plant Breeding: Principles and Prospects*. Chapman & Hall, London, UK.

Kang, M. S., and Gauch, H. G., Jr. 1996. *Genotype-by-Environment Interaction*. CRC Press, Boca Raton, FL, USA.

Lamb, B. C. 2000. *The Applied Genetics of Plants, Animals, Humans and Fungi*. Imperial College Press, London, UK.

Le Roy, H. L. 1966. *Elemente der Tierzucht: Genetik, Mathematik, Populationsgenetik.* Bayerischer Landwirthschafts-Vlg., München, Germany.

Mason, I. L. 1996. *A World Dictionary of Livestock Breeds.* CAB Internat/Oxford Univ. Press, New York.

Mrode, R. A. 1996. *Linear Models for the Production of Animal Breeding Values.* CAB Internat./Oxford Univ. Press, New York.

Murray, J. D., et al., eds. 1999. *Transgenic Animals in Agriculture.* CAB Internat., Oxon, UK.

Nanda, J. S., ed. 2000. *Rice Breeding and Genetics.* Science Publishers, Enfeld, NH, USA.

Nevo, E., et al., 2002. *Evolution of Wild Emmer and Wheat Breeding. Population Genetics, Genetic Resources and Genome Organization of Wheat Progenitor, Triticum dicoccoides.* Springer, New York.

Poehlman, J. M., and Sleper, D. A. 1995. *Breeding Field Crops.* Iowa State Univ. Press, Ames, IA, USA.

Purdom, C. E. 1993. *Genetics and Fish Breeding.* Chapman & Hall, New York.

Renaville, R., and Burny, A. 2001. *Biotechnology in Animal Husbandry.* Kluwer, Dordrecht, The Netherlands.

Rothschild, M. F., and Ruvinsky, A., eds. 1998. *The Genetics of the Pig.* CAB Internat., Oxon, UK.

Simmonds, N. W. 1979. *Principles of Crop Improvement.* Longman, Harlow, Essex, UK.

Sobral, B. W. 1966. *The Impact of Plant Molecular Genetics.* Birkhäuser, Boston, MA, USA.

Stoskopf, N. C., Tomes, D. T., and Christie, B. R. 1993. *Plant Breeding: Theory and Practice.* Westview, Boulder, CO, USA.

Sybenga, J. 1992. *Cytogenetics in Plant Breeding.* Springer, Berlin, Germany.

Thomson, J. A. 2002. *Genes for Africa: Genetically Modified Crops in the Developing World.* Univ. Cape Town Press, Cape Town, South Africa.

Tsuchya, T., and Gupta, P. K., eds. 1991. *Chromosome Engineering in Plants: Genetics, Breeding, Evolution.* Elsevier Science, New York.

van Harten, A. M. 1998. *Mutation Breeding: Theory and Practical Applications.* Cambridge Univ. Press, New York.

Van Vleck, L. D., Pollak, E. J., and Oltenacu, E. A. B. 1987. *Genetics for Animal Sciences.* Freeman, New York.

Weller, J. I. 2001. *Quantitative Trait Loci Analysis in Animals.* Oxford Univ. Press, New York.

White, T. L., and Adams, W. T. 2001. *Forest Genetics.* Wiley, New York.

PLANT PATHOLOGY (*See also* Molecular Biology, Plant Physiology)

Agrios, G. N. 1997. *Plant Pathology.* Academic Press, San Diego, CA, USA.

Andersson, B., et al., eds. 1996. *Molecular Genetics of Photosynthesis.* Cambridge Univ. Press, New York.

Boller, T., and Meins, F., Jr., eds. 1992. *Plant Gene Research: Genes Involved in Plant Defense.* Springer-Verlag, New York.

Bradbury, J. F., and Saddler, G. S. 2002. *A Guide to Plant Pathogenic Bacteria.* Oxford Univ. Press, New York.

Brown, T. M., ed. 1996. *Molecular Genetics and Evolution of Pesticide Resistance.* Oxford Univ. Press.

Crute, I. R., Holub, E. B., and Burdon, J. J., eds. 1997. *The Gene-for-Gene Relationship in Plant–Parasite Interactions.* CAB Internat./Oxford Univ. Press, New York.

Daniel, M., and Purkayastha, R. P., eds. 1994. *Handbook of Phytoalexin Metabolism and Action.* Dekker, New York.

Datta, S. K., and Muthukrishnan, S. 1999. *Pathogenesis - Related Proteins in Plants.* CRC Press, Boca Raton, FL, USA.

Entwistle, P. F., et al., eds., 1994. *Bacillus thüringiensis: An Environmental Biopesticide—Theory and Practice.* Wiley, New York.

Gnanamanickam, S. S., ed. 2002. *Biological Control of Crop Diseases.* Marcel Dekker, New York.

Goodman, R. N., and Novacky, A. J. 1994. *The Hypersensitive Reaction in Plants to Pathogens.* The American Phytopathological Society Press, St. Paul, MN, USA.

Gurr, S. J., McPherson, M. J., and Bowles, D. J. 1992. *Molecular Plant Pathology.* IRL Press, New York.

Holliday, P. 2001. *A Dictionary of Plant Pathology.* Cambridge Univ. Press, New York.

Khetan, S. K. 2001. *Microbial Pest Control.* Marcel Dekker, New York.

Kranz, J. 2002. *Comparative Epidemiology of Plant Diseases.* Springer, New York.

Maloy, O. C., and Murray, T. D., eds. 2000. *Encyclopedia of Plant Pathology.* Wiley, New York.

Matthews, R., and Burnie, J. M. 1995. *Heatshock Proteins in Fungal Infections.* Springer, NewYork.

Narayanasamy, P. 2002. *Microbial Plant Pathogens and Crop Disease Management.* Science Publishers, Enfield, NH, USA.

Nobel, P. S. 1999. *Physicochemical and Environmental Plant Physiology.* Academic Press, San Diego, CA, USA.

Pedigo, L. 2002. *Entomology and Pest Management.* Prentice Hall, New York.

Prell, H. H., and Day, P. R. 2001. *Plant–Fungal Pathogen Interaction.* Springer, New York.

Scandalios, J. G., ed. 1997. *Oxidative Stress and Molecular Biology of Antioxidant Defenses.* Cold Spring Harbor Laboratory Press, Plainview, NY, USA.

Scheffer, R. P. 1997. *The Nature of Disease in Plants.* Cambridge Univ. Press, New York.

Scholthoff, K-B. G., et al., eds. 1999. *Tobacco Mosaic Virus: One Hundred Years of Contributions to Virology.* American Phytopathological Society Press, St. Paul, MN, USA.

Sigee, D. C. 1993. *Bacterial Plant Pathology: Cell and Molecular Aspects.* Cambridge Univ. Press, New York.

Singh, U. S., and Singh, R. P. 1995. *Molecular Methods in Plant Pathology.* CRC Press, Boca Raton, FL, USA.

Whitby, S. M. 2002. *Biological Warfare Against Crops.* Palgrave Macmillan, Houndmills, Basingstoke, Hamphire, UK.

PLANT PHYSIOLOGY (*See also* Biochemistry and Physiology, Biotechnology, Molecular Biology)

Basra, A. S., ed., 1994. *Stress-Induced Gene Expression in Plants.* Harwood Academic Publishers, Langhorne, PA, USA.

Böger, P., et al., eds. 2002. *Herbicide Classes in Development.* Springer, New York.

Buchanan, R., Gruissem, W., and Jones, R. L., eds. 2002. *Biochemistry and Molecular Biology of Plants*. Wiley, New York.

Clement, C., et al., eds. 1999. *Anther and Pollen*. Springer, New York.

Dashek, W. V., ed. 1997. *Methods in Plant Biochemistry and Molecular Biology*. CRC Press, Boca Raton, FL, USA.

Dey, P. M., and Harborne, J. B., eds. 1997. *Plant Biochemistry*. Academic Press, San Diego. CA, USA.

Fosket, D. E. 1994. *Plant Growth and Development: A Molecular Approach*. Academic Press, San Diego, CA, USA.

Gilmartin, P. M., and Bowler, C., eds. 2002. *Molecular Plant Biology*. Oxford Univ. Press, New York.

Hall, D. O., and Rao, K. K. 1994. *Photosynthesis*. Cambridge Univ. Press, New York.

Hooykaas, P. J. J., et al., eds. 1999. *Biochemistry and Molecular Biology of Plant Hormones*. Elsevier Science, Amsterdam, The Netherlands.

Howell, S. H. 1998. *Molecular Genetics of Plant Development*. Cambridge Univ. Press, New York.

Khripach, V. A., Zhabinskii, V. N., and de Groot, A. E. 1999. *Brassinosteroids: A New Class of Plant Hormones*. Academic Press, San Diego, CA, USA.

Kruger, N. J., et al., eds. 1999. *Regulation of Primary Metabolic Pathways in Plants*. Kluwer, Dordrecht, The Netherlands.

Lea, P. J., and Leegood, R. C., eds. 1999. *Plant Biochemistry and Molecular Biology*. Wiley, New York.

Li, P. H., and Palva, E. T, eds. 2002. *Plant Cold Hardiness. Gene Regulation and Genetic Engineering*. Kluwer/Plenum, New York.

Mohr, H., and Schopfer, P. 1995. *Plant Physiology*. Springer, New York.

Pirson, A., ed. 1993. *Encyclopedia of Plant Physiology*. Springer, New York.

Raghavendra, A. S., ed. 2000. *Photosynthesis: Comprehensive Treatise*. Cambridge Univ. Press, New York.

Reynolds, P. H. S., ed. 1999. *Inducible Gene Expression in Plants*. CABI Publishing, New York.

Romeo, J. T., and Dixon, R. A., eds. 2002. *Phytochemistry in the Genomics and Post-Genomics Eras*. Pergamon, New York.

Scheel, D., and Wasternack, C. 2002. *Plant Signal Transduction*. Oxford Univ. Press, New York.

Srivastava, L. M. 2002. *Plant Growth and Development*. Acad. Press, San Diego, CA, USA.

Taiz, L., and Zeiger, E., eds. 2002. *Plant Physiology*. Sinauer, Sunderland, MA, USA.

POPULATION GENETICS (*See also* Biometry, and Bioinformatics, Cytology and Cytogenetics, Evolution)

Baumberg, S., et al., eds. 1995. *Population Genetics of Bacteria*. Cambridge Univ. Press, New York.

Bonné-Tamir, B., and Adam, A., eds. 1992. *Genetic Diversity Among Jews: Diseases and Markers at the DNA Level*. Oxford Univ. Press, New York.

Cavalli-Sforza, L. L., and Bodmer, W. F. 1971. *The Genetics of Human Populations*. Freeman, San Francisco, CA, USA.

Cavalli-Sforza, L. L., Menozzi, P., and Piazzo, A. 1994. *The History and Geography of Human Genes*. Princeton Univ. Press, Princeton, NJ, USA.

Chadwick, D. J., and Cardew, G., eds. 1996. *Variation in the Human Genome*. Wiley, New York.

Christiansen, F. B. 2000. *Population Genetics of Multiple Loci*. Wiley, New York.

Crow, J. F., and Kimura, M. 1970. *An Introduction to Population Genetics Theory*. Harper and Row, New York.

Fisher, R. A. 1958. *The Genetical Theory of Natural Selection*. Dover Publications, New York.

Fisher, R. A. 1965. *The Theory of Inbreeding*. Academic Press, New York.

Goodman, R. M., and Motulsky, A. G., eds. 1979. *Genetic Diseases Among Ashkenazi Jews*. Raven, New York.

Haldane, J. B. S. 1935. *The Causes of Evolution*. Longmans Green, New York.

Hartl, D. L., and Clark, A. G. 1997. *Principles of Population Genetics*. Sinauer, Sunderland, MA. USA.

Henry, R. J., ed. 2001. *Plant Genotyping. The DNA Fingerprinting of Plants*. Oxford Univ. Press, New York.

Hoelzel, A. R., ed. 1998. *Molecular Genetic Analysis of Populations*. Oxford Univ. Press, New York

Kimura, M. 1983. *The Neutral Theory of Molecular Evolution*. Cambridge Univ. Press, Cambridge, UK.

Kimura, M., Takahata, N., and Crow, J. F., eds. 1994. *Population Genetics, Molecular Evolution and the Neutral Theory*. Univ. Chicago Press, Chicago, IL, USA.

Lerner, I. M. 1954. *Genetic Homeostasis*. Oliver & Boyd, Edinburgh, UK.

Levin, S. A., ed. 1994. *Frontiers of Mathematical Biology*. Springer, New York.

Li, C. C. 1976. *First Course in Population Genetics*. Boxwood Press, Pacific Grove, CA, USA.

Magurran, A. E., and May, R. M., eds. 1999. *Evolution of Biological Diversity*. Oxford Univ. Press, Oxford, UK.

Mettler, L. E., Gregg, T. G., and Schaffer, H. E. 1988. *Population Genetics and Evolution*. Prentice-Hall, Englewood Cliffs, NJ, USA.

Mitton, J. B. 2000. *Selection in Natural Population*. Oxford Univ. Press, New York.

Mourant, A. E., Kopec,, A. C., and Domaniewska-Sobczak, K. 1978. *Genetics of the Jews*. Clarendon Press, Oxford, UK.

Nei, M. 1975. *Molecular Population Genetics and Evolution*. American Elsevier, New York.

Papiha, S. S., et al., eds. 1999. *Genomic Diversity: Applications in Human Population Genetics*. Kluwer/Plenum, New York.

Real, L. A., ed. 1994. *Ecological Genetics*. Princeton Univ. Press, Princeton, NJ, USA.

Roychoudhury, A. K., and Nei, M. 1988. *Human Polymorphic Genes: World Distribution*. Oxford Univ. Press, New York.

Spiess, E. B. 1977. *Genes in Populations*. Wiley, New York.

Weir, B., ed. 1995. *Human Identification: The Use of DNA Markers*. Kluwer Academic, Boston, MA, USA.

Weiss, K., ed. 1996. *Variation in the Human Genome*. Wiley, New York.

Wilmsen Thornhill, N., ed. 1993. *The Natural History of Inbreeding and Outbreeding: Theoretical and Empirical Perspectives*. Univ. Chicago Press, Chicago, IL, USA.

Wright, S. 1968–1976. *Evolution and the Genetics of Populations*. Univ. Chicago Press, Chicago, IL, USA.

RADIATION (*See also* Mutation)

Alpen, E. L. 1998. *Radiation Biophysics*. Academic Press, San Diego, CA, USA.

Dunlap, R. R., Kraft, M. E., and Rosa, E. A., eds. 1993. *Public Reactions to Nuclear Waste*. Duke Univ. Press, Durham, NC, USA.

Fuciarelli, A. E., and Zimbrick, J. D., eds. 1995. *Radiation Damage in DNA*. Battelle, Columbus, OH, USA.

Heald, M. A., and Marion, J. B. 1995. *Classical Electromagnetic Radiation*. Saunders, Philadelphia, PA, USA.

Hendry, J. H., and Lord, B. L. 1995. *Radiation Toxicology: Bone Marrow and Leukemia*. Taylor & Francis, London, UK.

Lindee, S. 1995. *Suffering Made Real: American Science and the Survivors at Hiroshima*. Univ. Chicago Press, Chicago, IL, USA.

National Research Council, Committee on the Biological Effect of Ionizing Radiation. 1990. *Health Effects of Exposure to Low Levels of Ionizing Radiation*. Natl. Acad. Sci. and Natl. Res. Council, Washington, DC.

Peterson, L. E., and Abrahamson, S., eds. 1998. *Effects of Ionizing Radiation: Atomic Bomb Survivors and Their Children*. Joseph Henry Press (Natl. Acad. Sci. USA), Washington, DC.

Petryna, A. 2002. *Life Exposed. Biological Citizens of Chernobyl*. Princeton Univ. Press, NJ, USA.

Prasad, K. N. 1995. *Handbook of Radiobiology*. CRC Press, Boca Raton, FL, USA.

Shigematsu, I., et al., eds. 1995. *Effects of A-Bomb Radiation on the Human Body* (translation). Harwood Academic, Publisher, Langhorne, PA, USA.

Shull, W. J. 1995. *Effects of Atomic Radiation*. Wiley-Liss, New York.

Sources and Effects of Ionizing Radiation. United Nations Scientific Committee on the Effects of Atomic Radiation. UNSCEAR 2000 Report to the General Assembly, with Scientific Annexes. Vol. II: Effects. United Nations, New York.

Turner, J. E. 1995. *Atoms, Radiation, and Radiation Protection*. Wiley, New York.

Wilkening, G. M. 1991. Ionizing radiation, pp. 599–655, in *Patty's Industrial Hygiene and Toxicology*, Clayton, G. D., and Clayton, F. E., eds. Wiley, New York.

Wilson, M. A., ed. 1998. *Textbook of Nuclear Medicine*. Lippincott-Raven, Philadelphia, PA, USA.

RECOMBINATION (*See also* Mapping)

Bailey, N. T. J. 1961. *Introduction to the Mathematical Theory of Genetic Linkage*. Clarendon Press, Oxford, UK.

Bishop, M. 1998. *Guide to Human Genome Mapping*. Academic Press, San Diego, CA, USA.

Broach, J. R., Pringle, J. R., and Jones, E. W., eds. 1991. *Recombination in Yeast*. Cold Spring Harbor Laboratory Press, Cold Spring Harbor, NY, USA.

Leach, D. R. F. 1996. *Genetic Recombination*. Blackwell Science, Cambridge, MA, USA.

Mather, K. 1957. *The Measurement of Linkage in Heredity*. Methuen, London, UK.

Nelson, D. L., and Brownstein, B. H., eds. 1993. *YAC Libraries*. Oxford Univ. Press, New York.

Ott, J. 1999. *Analysis of Human Genetic Linkage*. Johns Hopkins Univ. Press, Baltimore, MD, USA.

Paszkowski, J., ed. 1994. *Homologous Recombination and Gene Silencing in Plants*. Kluwer Academic, Norwell, MA, USA.

Smith, P. J., and Jones, C. J., eds. 2000. *DNA Recombination and Repair*. Oxford Univ. Press, New York

Stahl, F. W. 1979. *Genetic Recombination: Thinking About It in Phage and Fungi*. Freeman, San Francisco, CA, USA.

Terwilliger, J. D., and Ott, J. 1994. *Handbook of Human Genetic Linkage*. Johns Hopkins Univ. Press, Baltimore, MD, USA.

Whitehouse, H. L. K. 1982. *Genetic Recombination: Understanding the Mechanisms*. Wiley, New York.

Wolff, S., ed. 1982. *Sister Chromatid Exchange*. Wiley, New York.

SOCIAL AND ETHICAL ISSUES (*See also* Philosophy)

Alper J. S. W., et al., eds. 2002. *The Double-Edged Helix: Social Implications of Genetics in a Diverse Society*. Johns Hopkins. Univ. Press, Baltimore, MD, USA

Andrews, L. B. 2001. *Future Perfect: Confronting Decisions about Genetics*. Columbia Univ. Press, New York.

Annas, G. J., and Elias, S., eds. 1992. *Gene Mapping: Using Law and Ethics as Guides*. OxfordUniv. Press, New York.

Ballantyne, J., Sensabaugh, G., and Witkowski, J., eds. 1989. *DNA Technology and Forensic Science*. Cold Spring Harbor Laboratory Press, Cold Spring Harbor, NY, USA.

Beckwith, J. 2002. *Making Genes, Making Waves: A Social Activist in Science*. Harvard Univ. Press, Cambridge, MA, USA.

Bonnicksen, A. L. 2002. *Crafting a Cloning Policy: From Dolly to Stem Cells*. Georgetown Univ. Press, Washington, DC, USA.

Bryant, J. A., et al., eds. 2002. *Bioethics for Scientists*. Wiley, New York.

Chadwick, R. F., et al., eds. 1999. *The Ethics of Genetic Screening*. Kluwer, Dordrecht, The Netherlands.

Chapman, A. R., and Frankel, M. S., eds. 2003. *Designing Our Descendants: The Promise and Perils of Genetic Modifications*. Johns Hopkins Univ. Press, Baltimore, MD, USA.

Claude, R. P. 2002. *Science in the Service of Human Rights*. Univ. Pennsylvania Press, Philadelphia, PA, USA.

Evans, J. H. 2002. *Playing God? Human Genetic Engineering and the Rationalization of Public Bioethical Debate*. Univ. Chicago Press, Chicago, IL, USA.

Frankel, M., and Chapman, A. R. 2000. *Human Inheritable Genetic Modifications: Assessing Scientific, Religious, and Politics Issues*. American Association for the Advancement of Science, Washington, DC.

Frankel, M. S., and Teich, A. H., eds. 1994. *The Genetic Frontier: Ethics, Law and Policy*. American Association for the Advancement of Science, Washington, DC.

Freeman, M., and Lewis, A. D. E. 2000. *Law and Medicine: Current Legal Issues 2000*. Oxford Univ. Press, New York.

Gaskell, G., and Bauer, M. W. 2001. *Biotechnology 1996–2000 The Years of Controversy*. National Museum of Science and Industry Ltd., London, UK.

Green, R. M. 2001. *The Human Embryo Research Debates: Bioethics in the Vortex of Controversy*. Oxford Univ. Press, New York.

Hindmarsh, R., and Lawrence, G., eds. 2001. *Altered Genes II. The Future?* Scribe Publications, Melbourne, Australia.

Holland, S., Lebacqz, K., and Zoloth, L., eds. 2001. *The Human Embryonic Stem Cell Debate: Science, Ethics, and Public Policy*. Cambridge Univ. Press, New York.

Human Cloning and Human Dignity. 2002. The Report of the President's Council on Bioethics. Perseus, New York.

Jackson Knight, H. 2001. *Patent Strategy for Researchers and Research Managers*. Wiley, New York.

Kass, L. R. 2002. *Life, Liberty and the Defense of Dignity. The Challenges of Bioethics*. Encounter, San Francisco, CA, USA.

Kitcher, P. 1996. *The Lives to Come*. Simon and Schuster, New York.

Little. P. 2002. *Genetic Destinies*. Oxford Univ. Press, New York.

Lynn, R. 2001. *Eugenics: A Reassessment*. Praeger, Westport, CT, USA.

Lurquin, P. F. 2002. *High Tech Harvest. Understanding Genetically Modified Food Plants*. Perseus, Boulder, CA, USA.

McSherry, C. 2001. *Who Owns Academic Work? Battling for Control of Intellectual Property*. Harvard Univ. Press, Cambridge, MA, USA.

Milunsky, A. 2001. *Your Genetic Destiny: Know Your Genes, Secure Your Health, Save Your Life*. Perseus, New York.

Quinn, S. 2001. *Human Trials: Scientists, Investors and Patients in the Quest for a Cure*. Perseus, Cambridge, MA, USA.

Reilly, P. 1977. *Genetics, Law and Social Policy*. Harvard Univ. Press, Cambridge, MA, USA.

Reilly, P. R. 2000. *Abraham Lincoln's DNA and Other Adventures in Genetics*. Cold Spring Harbor Laboratory Press, Cold Spring Harbor, NY, USA.

Reiss, M. J., and Straughan, R. 1996. *Improving Nature?* Cambridge Univ. Press, New York.

Resnik, D. B. 1998. *The Ethics of Science: An Introduction*. Routledge, New York.

Stansfield, W. D. 2000. *Death of a Rat: Understandings and Appreciations of Science*. Prometheus, Amherst, NY, USA.

Stephens, T., and Brynner, R. 2001. *Dark Remedy: The Impact of Thalidomide and Its Revival as a Vital Medicine*. Perseus, New York.

Stock, G., and Campbell, J. 2000. *Engineering the Human Germline: An Exploration of the Science and Ethics of Altering the Genes We Pass on to Our Children*. Oxford Univ. Press, New York.

Thompson, P., ed. 1994. *Issues in Evolutionary Ethics*. State Univ. New York Press, Alabany, NY, USA.

Walters, L., and Palmer, J. G. 1996. *The Ethics of Human Gene Therapy*. Oxford Univ. Press, New York.

Wilson, D. S. 2002. *Darwin's Cathedral. Evolution, Religion, and the Nature of Society*. Univ. Chicago Press, Chicago, IL, USA.

Watson, J. D. 2000. *A Passion for DNA: Genes, Genomes, and Society*. Cold Spring Harbor Laboratory Press, Cold Spring Harbor, NY, USA.

Winter, G. 1998. *Die Prüfung der Freisetzung von gentechnisch veränderten Organismen*. Erich Schmidt Vlg., Berlin, Germany.

TRANSPOSABLE AND MOBILE ELEMENTS (*See also* Genetics Monographs)

Arkhipova, I. R., Lyubomirskaya, N. V., and Ilyin, N. V. 1995. *Drosophila Retrotransposons*. CRC Press, Boca Raton, FL, USA.

Berg, D. E., and Howe, M. M., eds. 1989. *Mobile DNA*. American Society of Microbiology Press, Washington, DC.

Bushman, F. 2002. *Lateral DNA Transfer: Mechanisms and Consequences*. Cold Spring Harbor Laboratory Press, Cold Spring Harbor, NY, USA.

Capy, P. 1998. *Evolution and Impact of Transposable Elements*. Kluwer, Dordrecht, The Netherlands.

Craig, N. L., et al., eds. 2002. *Mobile DNA II*. American Society of Microbiology Press, Washington, DC.

Kaper, J. B., and Hacker, J., eds. 1999. *Pathogenicity Islands and Other Mobile Virulence Elements*. American Society of Microbiology Press, Washington, DC.

McClintock, B. 1988. *The Discovery and Characterization of Transposable Elements: The Collected Papers of Barbara McClintock*. Garland, New York.

McDonald, J. F., ed. 2000. *Transposable Elements and Genome Evolution*. Kluwer, Dordrecht, The Netherlands.

Saedler, H., and Gierl, A., eds. 1996. *Transposable Elements*. Springer, Berlin, Germany.

Sherratt, D. J., ed. 1995. *Mobile Genetic Elements*. IRL Press, New York.

VIRUSES (*See also* Microbiology, Molecular Genetics)

Adolph, K. W., ed. 1995. *Viral Gene Techniques*. Academic Press, San Diego, CA, USA.

Arvin, A., and Gershon, A. A., eds. 2000. *Varicella-Zooster Virus: Virology and Clinical Management*. Cambridge Univ. Press, New York.

Blumberg, B. S. 2002. *Hepatitis B: The Hunt for the Killer Virus*. Princeton Univ. Press, Princeton, NJ, USA.

Brown, S. M., ed. 1997. *Herpes Simplex Virus Protocols*. Humana Press, Totowa, NJ, USA.

Brunt, A., et al., eds. 1997. *Viruses of Plants. Descriptions and Lists from VIDE Databases*. CAB Internat., Willingford, UK.

Calendar, R., ed. 1988. *The Bacteriophages*. Plenum, New York.

Cann, A. J. 2001. *Principles of Molecular Virology*. Academic Press, San Diego, CA, USA.

Cann, A. J., ed. 2000. *RNA Viruses: A Practical Approach*. Oxford Univ. Press, Oxford, UK.

Cann, A. J., ed. 2001. *DNA Virus Replication*. Oxford Univ. Press, New York.

Coffin, J. M., Hughes, S. H., and Varmus, H. E., eds. 1997. *Retroviruses*. Cold Spring Harbor Laboratory Press, Cold Spring Harbor, NY, USA.

Crawford, D. H. 2000. *The Invisible Enemy: A Natural History of Viruses*. Oxford Univ. Press, New York.

Cullen, B., ed. 1993. *Human Retroviruses*. IRL/Oxford Univ. Press, New York.

Dalgleish, A. G., and Weiss, R. A., eds. 1999. *HIV and the New Viruses*. Academic Press, San Diego, CA, USA.

De Clercq, E. D. A., ed. 2001. *Antiretroviral Therapy*. ASM Press, Washington, DC.

Domingo, E., et al., eds. 1999. *Origin and Evolution of Viruses*. Academic Press, San Diego, CA, USA.

Flint, S. J., et al., eds. 1999. *Principles of Virology: Molecular Biology, Pathogenesis, and Control*. American Society of Microbiology Press, Washington, DC.